RSGB YEARBOOK 2025

Editor
Chris Danby, G0DWV

Co Editor
Ed Durrant, G8GLM/DD5LP/VK2JI

Advertising
Chris Danby, G0DWV, Danby Advertising

Front Cover
Kevin Williams, M6CYB

Production
Mark Allgar, M1MPA

Published by the
Radio Society of Great Britain,
3 Abbey Court, Fraser Road,
Priory Business Park, Bedford MK44 3WH
Website: *www.rsgb.org*
Tel: 01234 832700
Fax: 01234 831496

ISBN: 9781 9139 9560 7
ISSN: 1460-454X
Printed in Great Britain by CPI Antony Rowe

Contents

INFORMATION

OPERATING

UK Callbook 2024

Foreword

Last year I reported that the coming year promised to be a very interesting one for amateur radio in the UK as Ofcom had issued its consultation entitled "Updating the Amateur Radio Licensing Framework". This document made it clear that this was no mere tinkering around the edges but was a complete shift towards simplification of the licence and liberalisation of what we would be able to do in the future.

You will read elsewhere in this book much more about the detailed changes to the amateur licence but, suffice to say, the changes have been generally very well received and have opened up opportunities to showcase amateur radio in ways that were impossible in the past.

The RSGB Board is focusing on progressing the Society's revised strategy and the priorities that have been identified – namely, growth of amateur radio, membership of the RSGB, defence of the spectrum and the continued financial viability of the Society. Various workshops and meetings have been held and we have formed an Outreach Team to ensure that we re-engage with our members and start to reach new audiences. We must ensure that there is a flow of information across all levels of the Society: upwards, downwards and very importantly, outwards. Our detractors often accuse us of a lack of transparency but that is never intentional, and we are always willing to address any perceived shortcomings.

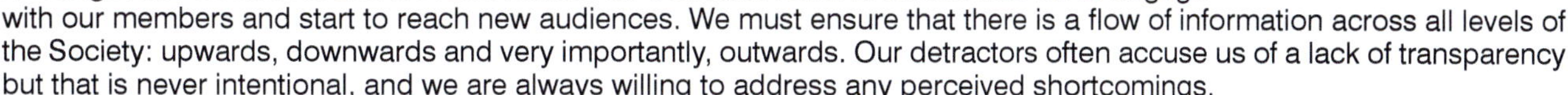

The world situation continues to be very unstable with the Ukrainian war continuing and increasing tensions in the Middle East. The war caused huge price increases resulting in high inflation which fortunately is now decreasing. This has, however, left a legacy of significant financial pressure which necessitates a close watch on our membership fees. We do appreciate that everyone is also under these pressures personally, and we will continue to monitor the situation closely to keep costs under control.

Our use of the spectrum is looked on enviously by others and consequently it is a very important part of the Strategic Priorities mentioned previously. The amateur spectrum must be rigorously defended and one prime example is in the case of 23 centimetres. Our use was likely to be wiped out by administrations who wished to impose impossible restraints on amateur users of the band, in order to protect the RNSS (Radio Navigation Satellite Service). After a huge amount of preparation and very long sessions at the WRC-23 conference, a successful compromise was reached which hopefully will permit our continued use of the band for years to come. RSGB volunteers were key to the negotiations, and we are grateful for the time and expertise they gave that helped to reach this solution.

Work is already well advanced in preparation for WRC-27 and you can be assured that our Spectrum Team will be fully represented at all levels. After all, no spectrum means no amateur radio!

One way that you can help to enhance and protect amateur radio is to remain a member of the RSGB, despite the increasingly tough calls on your finances. It would also be most helpful for you to encourage others to join the RSGB and keep it strong in these challenging times. Your membership helps the RSGB to fund services which enable you to develop your amateur radio skills and also supports the vital work described above with the IARU, ITU and other organisations in the international amateur radio community.

Volunteers are the lifeblood of the Society and an even wider range of people will be needed in the coming years to provide the skills, expertise and enthusiasm needed to share amateur radio with new audiences and to provide relevant and exciting activities for RSGB members. Please think how you could help and either respond to our call for specific volunteers or get in touch if you would like to get involved in a different way. You will help to protect the future of amateur radio in the UK and world-wide which can, at times, be frustrating, but is also incredibly rewarding.

One area that is supported in the most part by volunteers is the RSGB National Radio Centre (NRC) at Bletchley Park. The NRC typically has three or four volunteers working most days who, between them, spent more than 9,000 hours explaining and demonstrating amateur radio to visitors last year. From these impressive numbers you can see why the NRC plays such a key part in our growth of amateur radio strategic priority.

The role of President has opened a much wider picture to me and demonstrated the huge range of activities that we are involved with, and which need leadership to take forward. Please do consider putting your name forward for office when the time comes; as someone else once said, your country (or in this case your Society) needs you!

I look forward to meeting many of you at some of the events in the coming year and perhaps we will have a QSO along the way too.

I do hope that you enjoy this edition of the RSGB Yearbook and join me in thanking Chris Danby, G0DWV and the team of RSGB volunteers and HQ staff that compiled this edition.

John McCullagh MBE, GI4BWM

RSGB President

Radio Society of Great Britain

Benefits that keep you on the air

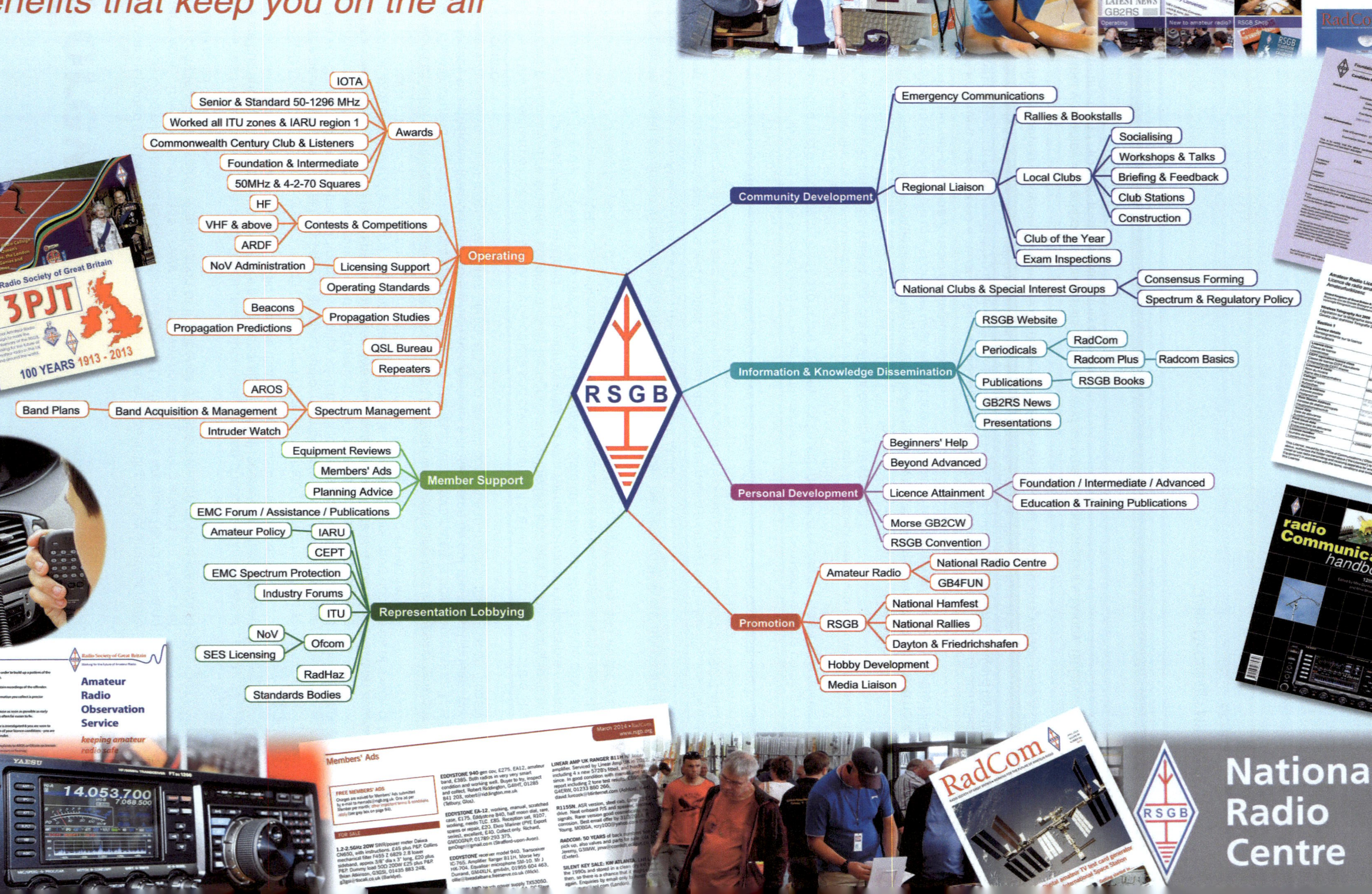

Introduction

Welcome to the 2025 edition of the RSGB Yearbook and UK Callbook, the information contained in this edition is accurate as far as possible but with any publication, changes will occur during the lifetime of an edition – so please check the pages of RadCom or the RSGB and club websites for relevant on-going information related to the sections of this yearbook, especially items such as GB2RS newsreaders, local QSL sub-managers, clubs or societies details and repeater / beacon listings etc.

Changes in 2024/5

We have an ever-changing hobby in an ever-changing world and several changes have come along in 2024 that affect us all as radio amateurs. The most major are those brought in through the year by Ofcom, starting in early 2024, with increased power limits for all licence class levels, changes to RSL usage (now optional except for "2" calls) but also a process to control the number of personal licences held by amateurs. New Intermediate licence holders to receive a call starting with M8 or M9 and the possibility for those holding a "2" call to choose an M8 or M9 to change to, clearly points to the wish from Ofcom to remove the "2" licences as soon as practical.

Microwave band plans have changed to work better when sharing of bands with other services have come out of the WRC and Ofcom actions.

Even local planning permissions rules have changed in the area of right-to-keep an antenna that does not have specific authorisation. Now it needs to be in place for at least 10 years, where the previous rule was 4 years.

Club Information pages

For administrative purposes the Society divides itself into regions, each comprising of four to six districts. The RSGB identifies these regions as, Regional Representatives (RR) and District Representatives (DR).

All the local information on all the clubs and societies can be found on pages 25-47 of this Yearbook.

Beacons

Listings are included of the UK VHF/UHF and Microwave Beacons as well as HF Beacons. You can keep up to date with any changes to the VHF/UHF/Microwave beacon listings at: *https://rsgb.org/main/operating/beacons-and-repeaters/vhf-uhf-and-microwave-beacons/*

Validating your Licence

Revalidate your licence to avoid revocation

Ofcom will soon implement their plans to revoke licences that have not been revalidated as required by the licence conditions at intervals of not more than 5 years. Therefore, this is an important action so that you remain legally allowed to operate on the amateur bands. To validate your licence or amend your licence, log in to: *https://ofcom.force.com/licensingcomlogin* or email: *amateur.validations@ofcom.org.uk* If you need assistance in the process, Ofcom staff are available to help, but please be patient during times of heavy workload.

If you have not yet registered to use the Ofcom Online Licensing Service you will need to do so in order to access your licence online. When registering for the first time you will need to have details of your lifetime licence number, which can be found on page 1 of your Licence document.

UK Callsign entries

The call signs listed in this Yearbook reflect the official records held by Ofcom. The call sign data was output mid-June and therefore any callsign changes made after this date will not appear in the call sign listing pages in this book.

Entries show correspondence addresses, not station addresses. Please note that some callbooks, CDs and websites rely entirely on updates being notified to them by individual licensees, so the information can be out of date.

To make this book readable and compact, we apply a few standard abbreviations but no changes are made to the substance of any entries. The society cannot accept direct input regarding an individual's entry. Please see the article on call sign formatting and how it is changing on page 186 of the call sign section of this yearbook for more information.

Details withheld

The words "Details Withheld" mean exactly what they say, remember entries are not within the control of the society. In some cases, there is a perfectly good reason for a licensee not to want his or her address published but in others it is possible that the 'call sign book consent' box was unchecked in error. Should you wish your details to be withheld from publication or released if they are blocked or there is an error in the substance of your entry, you can update the entry online or write to Spectrum Licensing to request them to take the necessary action. Their address is:

Spectrum Licensing, Riverside House,
2a Southwark Bridge Road,London SE1 9HA
Tel: 020 7981 3131

Email: *spectrum.licensingenquiries@ofcom.org.uk*
Web: *www.ofcom.org.uk/spectrum/radio-equipment/online-licensing-service/*

International listings

Amateurs sometimes ask about their entry in the International Listings of the Radio Amateur Call Book, sometimes known as the DX Listings. This is an entirely separate work published annually by an independent company. If you want to be listed in their publication, please write directly to them and not to the RSGB. Corrections are free, but they do make a small charge for special entries. Their address is:

Radio Amateur Callbook, mfcit GmbH
P.O.Box 1170, 34216 Baunatal, Germany
E-mail: *contact20@callbook.biz*
Web: *www.callbook.biz*

Acknowledgments

Finally my thanks go to all the contributors to the Yearbook, this includes a number of RSGB Staff, Committee Members, and Club Secretaries.

Chris Danby, G0DWV
Editor
Ed Durrant, G8GLM/DD5LP/VK2JI
Co Editor

At Your Service!

The Society provides a broad range of services for its members through its professional staff and management at Headquarters and a country-wide force of skilled, dedicated and knowledgeable volunteers. To get the best out of the RSGB it is important that you approach the correct part of the Society. On these pages you will find a practical guide to finding the right person for your enquiry.

YOUR RSGB

The Society's affairs are directed by the Board, supported by the Volunteer Leadership Team comprising the Regional Representatives, Committee Chairs and Honorary Officers. The Board Members and the Regional Representatives are elected by the membership in a postal and electronic ballot.

The day to day running of the Society is the responsibility of the General Manager, supported by the Board Members who liaise with each of the committees.

Each [Director/Board Member] has responsibilities for one or more Committees. The Board Liaison Member for each Committee ensures that communication between the Committees and the Board is maintained. In addition, each Director has a specific strategic focus on one of the 2022 goals.

Full details of Board and Committees' terms of reference can be found on our website.

For HQ staff, both email addresses and telephone details are provided, including the option to select when dialling through the RSGB switchboard (01234 832 700).

Committee Chairs & Honorary Officers

These are all volunteers and give their time freely to support the Society. Members should respect the fact that many also have full time day jobs, and so email is the appropriate method of communication.

General Manager

Steve Thomas, M1ACB
email: *steve.thomas@rsgb.org.uk*

Honorary Treasurer

Chris Wood FCCA, GD6TWF
email: *hon.treasurer@rsgb.org.uk*

Company Secretary

Stephen Purser, GW4SHF
email: *company.secretary@rsgb.org.uk*

WEBSITE

Main website: *www.rsgb.org*
Members Portal: To be able to log into the members only pages you must have completed your portal registration. To register for the first time you will need to use the callsign we know you by and the email address you already have associated with your membership account. If you have troubles with the registration process please contact:
membership@rsgb.org.uk

THE RSGB BOARD

President

John McCullagh, GI4BWM
email: *gi4bwm@rsgb.org.uk*

Chair

Stewart Bryant, G3YSX
email: *g3ysx@rsgb.org.uk*

Directors

Len Paget, GM0ONX,
email: *gm0onx@rsgb.org.uk*

Ben Lloyd, GW4BML
email: *gw4bml@rsgb.org.uk*

Mark Jones, G0MGX
email: *g0mgx@rsgb.org.uk*

Tony Miles, MM0TMZ
email: *mm0tmz@rsgb.org.uk*

Nathan Nuttall, 2M0OCC
email: *2m0occ@rsgb.org.uk*

Peter Ransom, M0SFZ
email: *m0sfz@rsgb.org.uk*

Peter Bowyer, G4MJS
email: *g4mjs@rsgb.org.uk*

Note: The General Manager, Company Secretary and Acting Honorary Treasurer are not Directors, but are in attendance at Board Meetings.

REGIONAL REPRESENTATIVES

Information on the Regional Representatives (RR) and District Representatives (DR), may be found elsewhere in the RSGB Yearbook and on the RSGB website, *www.rsgb.org*

Region 1 – Tony Miles, MM0TMZ , rr1@rsgb.org.uk

Region 2 – Jim Campbell, MM7BIW, rr2@rsgb.org.uk

Region 3 – Martyn Bell, M0TEB, rr3@rsgb.org.uk

Region 4 – Ian Bowman, G7ESY, rr4@rsgb.org.uk

Region 5 – Mark Savage, M0XIC, rr5@rsgb.org.uk

Region 6 – Simon Taylor, MW0NWM, rr6@rsgb.org.uk

Region 7 – Vacant

Region 8 – Peter Lowrie, MI5JYK, rr8@rsgb.org.uk

Region 9 – Ron White, G6LTT (Temp), rr9@rsgb.org.uk

Region 10 –Keith Bird, G4JED, rr10@rsgb.org.uk

Region 11 – Andrew Jenner, G7KNA, rr11@rsgb.org.uk

Region 12 – Brian Woolnough, M5ADQ, rr12@rsgb.org.uk

Region 13 – Bob Hambly, M0HAF, rr13@rsgb.org.uk

SPECIALIST AREAS

The many different activities of the Society are run by its committees, honorary officers and full-time staff. If you wish to take advantage of one of these services or have an administrative enquiry about any one of them, contact details are listed below.

Abuse and Poor Operating

Operating Advisory Service (OAS),
Ian Suart, GM4AUP, Coordinator
email: *oas@rsgb.org.uk, www.rsgb.org/oas/*

Amateur Radio Direction Finding

Bob Titterington, G3ORY,
Chair, ARDF Committee,
email: *ardf.chairman@rsgb.org.uk, www.rsgb.org/ardf/*

Awards

Lindsay Pennell, G8PMA, Awards Manager
email: *awards@rsgb.org.uk, www.rsgb.org/awards/*

Beyond Exams

Mark Burrows, 2E0SBM, BE Coordinator
email: *be.coordinator@rsgb.org.uk*
https://rsgb.org/main/beyond-exams-building-experience/

Contests

Ian Pawson, G0FCT, Chair,
Contest Support, email: *csc.chair@rsgb.org.uk, www.rsgb.org/radiosport/*

Nick Totterdell, G4FAL,
HF Contest Committee
email: *hfcc.chair@rsgb.org.uk*

Andy Cook, G4PIQ,
VHF Contest Committee
email: *vhfcc.chair@rsgb.org.uk*

EMC

John Rogers, M0JAV, EMC Committee,
email: *emc.chairman@rsgb.org.uk, www.rsgb.org/emc/*

Examination Standards Committee

Tony Kent, G8PBH, Chair
email: *esc.chair@rsgb.org.uk*

EQAM Exam Quality Assurance Manager

Dave Wilson, M0OBW EQA Manager
email: *eqam@rsgb.org.uk*

ESM Examination Standards Manager

Nigel Barker, M0HZR ESM Manager
email: *esm@rsgb.org.uk*

ESRG Exam & Syllabus Review Group

Andrew Lenton, G8UUG, ESRG Chair
email: *esrg.chair@rsgb.org.uk*

General Technical Matters
Andy Talbot, G4JNT, Chair, Technical Forum, email: *tech.chair@rsgb.org.uk, www.rsgb.org/technicalmatters/*

General Spectrum & Regulatory Matters
Murray Niman, G6JYB, Chair, Spectrum Forum, email: *spectrum.chairman@rsgb.org.uk www.rsgb.org/spectrumforum/*

GB2CW
Roger Cooke, G3LDI, GB2CW Coordinator email: *gb2cw@rsgb.org.uk*

GB2RS News Service Management
Steve Richards, G4HPE, GB2RS Manager email: *gb2rs.manager@rsgb.org.uk* (GB2RS news items should be sent to: *radcom@rsgb.org.uk)*

HF Matters
Ian Greenshields, G4FSU, HF Manager, email: *hf.manager@rsgb.org.uk*

Intruders to the Amateur Bands
Vaughan Ravenscroft, M0VRR, IW Coordinator email: *iw@rsgb.org.uk, www.rsgb.org/intruders/*

Legacy Committee
Richard Horton, G4AOJ, Committee Chair email: *legacy.chair@rsgb.org.uk www.rsgb.org/legacy/*

Microwave Matters
Barry Lewis, G4SJH, Microwave Manager, email: *microwave.manager@rsgb.org.uk*

Planning Advice
John Mattocks, G4TEQ, Chair, email: *pac.chairman@rsgb.org.uk, www.rsgb.org/planning/*

Propagation Studies
Steve Nichols, G0KYA, Chair, Propagation Studies Committee, email: *psc.chairman@rsgb.org.uk, www.rsgb.org/psc/*

Repeater and Data Communications
Andrew Barrett, G8DOR, Chair, ETCC, email: *etcc.chairman@rsgb.org.uk, www.ukrepeater.net*

Regional Forum
Keith Bird, G4JED, Chair email: *rr10@rsgb.org.uk*

Special Interest Group
Philip Hosey, MI0MSO, Manager email: *sig.manager@rsgb.org.uk www.rsgb.org/sig/*

Trophies
Mike Franklin, G3VYI, Trophy Manager email: *trophy.manager@rsgb.org.uk*

VHF Matters
John Regnault, G4SWX, VHF Manager email: *vhf.manager@rsgb.org.uk*

Youth Champion
Liam Robbins, G5LDR, Youth Champion email: *youth.champion@rsgb.org.uk www.rsgb.org/youth/*

Details of the Society's volunteer officers can be found in this Yearbook and on the RSGB website, *www.rsgb.org*

HEADQUARTERS STAFF

For HQ staff below, both email addresses and telephone details are provided, including the option to select when dialling through the RSGB switchboard (01234 832 700).

Subscription renewals
Telephone: 01234 832 700, Option 1

Sales department
(Membership, books and other products) email: *sales@rsgb.org.uk*
Telephone: 01234 832 700, Option 2

Amateur Radio Examinations
email: *exams@rsgb.org.uk*
Telephone: 01234 832 700, Option 3

Technical Amateur Radio Enquiries
email: *AR.dept@rsgb.org.uk*
Telephone: 01234 832 700, Option 4

Amateur Radio Licensing Enquiries
email: *AR.dept@rsgb.org.uk*
Telephone: 01234 832 700, Option 4

GB2RS and Club News
email: *radcom@rsgb.org.uk*
Telephone: 01234 832 700, Option 5

RadCom
(news items, feature submissions, etc) Edward O'Neil, M0TZX email: *radcom@rsgb.org.uk*
Telephone: 01234 832 700, Option 5

General Manager
email: *GM.dept@rsgb.org.uk*

HEADQUARTERS AND REGISTERED OFFICE

3 Abbey Court, Fraser Road,
Priory Business Park, Bedford MK44 3WH
Telephone: 01234 832 700
Main website: *www.rsgb.org*

Log in using your callsign as the user name and your membership number, without the leading zeros (see your *RadCom* address label) as the password.

QSL BUREAU ADDRESS

PO Box 5, Halifax HX1 9JR, England
Telephone: 01422 359 362
email: *qsl@rsgb.org.uk, www.rsgb.org/qsl*

PLAY YOUR PART IN YOUR RSGB

Have Your Say
Let us know how we're doing! Through 'Have Your Say' you can let us know your views and you will receive a reply from the General Manager or a Board Member.
email: *haveyoursay@rsgb.org.uk www.rsgb.org/haveyoursay*

Consultations
From time to time you will find we are consulting the Membership on aspects of Society policy. You can find current consultations at *www.rsgb.org/consultations/*

National Radio Centre
Don't forget to tell your friends about the National Radio Centre at Bletchley Park. Full details at *www.rsgb.org/nrc/*
RSGB Members can enter Bletchley Park for free by downloading the personalised voucher available from the *www.rsgb.org* home page

Licensing & Special Event Stations
Licensing and Notices of Variation (NoVs) for special event stations are handled by Ofcom, Tel: 0300 123 1000, *www.ofcom.org.uk,*
email: *Spectrum.Licensing@ofcom.org.uk*

FAQs
The RSGB has compiled the questions most frequently asked by Members at: *www.rsgb.org/faq/*

Band Plan
The latest version of the band plan is always available on the website at: *www.rsgb.org/band-plans/*

Good Operating Practice
The RSGB fully supports the code of conduct and encourages all amateurs to read the advice at *www.rsgb.org/op-guidelines*

RSGB Shop
All RSGB goods - books, filters, clothing etc - can be purchased online at: *www.rsgbshop.org/*

Club Finder
Use the website to find your nearest radio club and check out the facilities they have to offer. *www.rsgb.org/clubsandtraining/*

Yearbook
If you have moved home, if you would like your name and address to be withheld from future editions of the Yearbook (or released, for use in it), or if your callsign is not listed, you can **only** make the necessary changes to the database via Ofcom, direct by phone or via Ofcom's website below:
Tel: 0300 123 1000
www.ofcom.org.uk

RSGB National Radio Centre

The RSGB National Radio Centre (NRC) is a public showcase for radio communications technology. The Centre provides the opportunity for members of the public to get 'up close and personal' with the history and technology of radio communications.

This world-class radio communications centre is located within the grounds of the Bletchley Park museum in Buckinghamshire. From the first inventors in the late 19th century through to the latest radio developments, visitors will find interesting videos, interactive displays and hands-on experiments. Visitors learn about the basic principles of radio and discover the history of radio communication. They see how different parts of the radio spectrum have differing uses, can explore how radio works and experiment with the building blocks of a radio system. The NRC also allows visitors to find out about the role of radio amateurs, who push technology to the limits and have fun at the same time. You can watch some interesting videos about the NRC in the RSGB National Radio Centre playlist on the RSGB YouTube channel: youtube.com/theRSGB

Visitors

The NRC welcomed over 86,000 visitors in 2023 and we expect a slightly higher number during 2024. Whilst some of the visitors are radio amateurs specifically coming to the NRC, many are not, and it is a pleasure to introduce visitors to amateur radio, many for the first time. For those who have never made a radio QSO before, the new licensing laws allow a non-licensed person to 'have a go' - so come along and try it. Our volunteers are always on hand to explain things and are happy to chat with everyone, be it someone with a passing interest, a newly licensed amateur or to have an in-depth conversation with an experienced amateur. If you're a UK radio amateur, you are welcome to bring your licence (along with photo ID) and operate the GB3RS demonstration station. Foundation and Intermediate licence holders will need to be supervised by an NRC volunteer who has a Full licence, if they are free to do so.

International visitors

Over the years, the NRC has welcomed visiting radio amateurs from around the world. Many of the visitors come from Australia, Canada, Germany, Netherlands, New Zealand and the USA. If they bring a copy of their Full, Extra or Advanced Licence they can also enjoy operating GB3RS

The NRC experience

Starting in the Reception area, visitors learn first of the importance of wireless and then of the role that the Voluntary Interceptors and Military WI (Y) Stations played in intercepting the WWII Morse code messages which were brought to Bletchley Park for decrypting. In the main Theatre Zone, a short film describes the importance and many uses of wireless technology in society today – something easily overlooked in the modern world of internet and instant mobile phone communication.

Interactive displays

The RSGB National Radio Centre boasts a collection of interactive displays – both hardware and software – and experiments that show visitors the workings of a radio system. Interactive touch-screen presentations take visitors through key areas of radio technology, while hardware displays allow visitors to explore and discover the technologies that come together to make radio work.

New displays

Two of the newer displays have attracted a lot of visitor interest – one focuses on how to get started in amateur radio, and the other on receiving amateur TV signals via the QO-100 geostationary satellite. We are also delighted to have received by kind donation a Geochron display showing a huge amount of radio-related information, particularly about satellites.

GB3RS

Showcasing amateur radio in action, the NRC has a state-of-the-art demonstration radio station, GB3RS. Operational from 80m through to 70cm, GB3RS can communicate on the LF, HF, VHF & UHF bands using CW, SSB, FT8, JT65, FM, DMR and D-STAR. With supporting software, the NRC also demonstrates the application of many modern communication aids such as SolarHam, WSJT-X, DX Maps, KST Chat, webSDR and PSK Reporter. Additionally, the GB3RS station can also track and communicate using low earth polar-orbiting satellites, and the geostationary OO-100 satellite, as well as monitoring the International Space Station (ISS).

Special events

The NRC is proud to host a number of special events throughout the year including the international Enigma Reloaded event, the Arkwright Trust Foundation Licence training days, and several 'Build a Radio' workshops with support from the Radio Communications Foundation (RCF). Each year, the NRC also hosts the Special Operations Executive demo station, GB1SOE, and when the ISS astronauts link up with UK schools using amateur radio, this always proves popular with visitors.

Support and visitor entry

The RSGB is grateful to individuals, retailers and manufacturers who have donated amateur radio equipment to the NRC, making it a state-of-the-art centre. The RSGB National Radio Centre is located at Bletchley Park, Milton Keynes MK3 6DS. Entry to Bletchley Park museum and the NRC is free to RSGB Members on production of a printed downloaded voucher from the RSGB website: https://rsgb.org/bpvoucher For opening times and other details, see *www.nationalradiocentre.com*.

For non-members, entry price and travel details are available on the Bletchley Park website: *www.bletchleypark.org.uk*

Intruder Watch

RSGB Intruder Watch (IW) collects reports from licensed amateurs in the UK and Crown Dependencies about HF intruders. An intruder is a non-amateur transmission in an amateur band that is not entitled to be there, such as a military data link or over-the-horizon radar.

Fortunately most observed intruders disappear within minutes or hours of their arrival. However when an intruder in a primary amateur allocation is regular and persistent and has been observed by UK amateurs on three or more occasions (e.g. one amateur on three occasions or three amateurs once each), IW may report it to Ofcom's Spectrum Management Centre at Baldock, which has engineering staff on duty 24/365.

Following a report from IW, usually Ofcom will monitor the signal and attempt to get a 'fix' on the transmitter location with direction finders. In some cases an intrusion will last long enough for a formal complaint to be made by Ofcom to the administration concerned. From previous experience some of these transmissions will subsequently cease but others will persist.

Over-the-horizon radar continues to be a problem. Such transmissions are nearly always too short-lived for IW to ask Ofcom to investigate, but they are aware of the problem nevertheless. These radar intrusions come mainly from Russia, China, Iran and a UK base on Cyprus.

RSGB Intruder Watch forms part of Region 1 of the IARU Monitoring System. It has an online database which enables national coordinators and a limited number of collaborators to store observations of intruders in a searchable, shared log.

Information about intruders is also exchanged via a mailing list and a monthly newsletter. There is a comprehensive web site with links to these resources at: *https://www.iaru-r1.org/about-us/committees-and-working-groups/iarums/*

While input from other countries provides a useful stream of tip-offs about intruder activity, IW can only make reports to Ofcom on the basis of intruders heard by licensed amateurs in the UK and Crown Dependencies. Amateurs in other countries should report intruders via their own national monitoring systems.

Anyone can submit reports to IW on a one-off, occasional or regular basis as they wish.

The best way to submit reports of suspected intruders is by email to the RSGB Intruder Watch Coordinator at: *iw@rsgb.org.uk*

Abbey Court

The Radio Society of Great Britain continues to be one of a few radio societies in the world to maintain a full time staff. The Society is now administered from a modern, two-storey, open-plan office situated on the prestigious Priory Business Park in Bedford. No. 3 Abbey Court houses the General Manager's Department, Sales and Accounts, Examinations Department, Website Management, IT, *RadCom* and *RadCom Plus*.

Office hours are Monday to Friday, 8.30am to 4.30pm

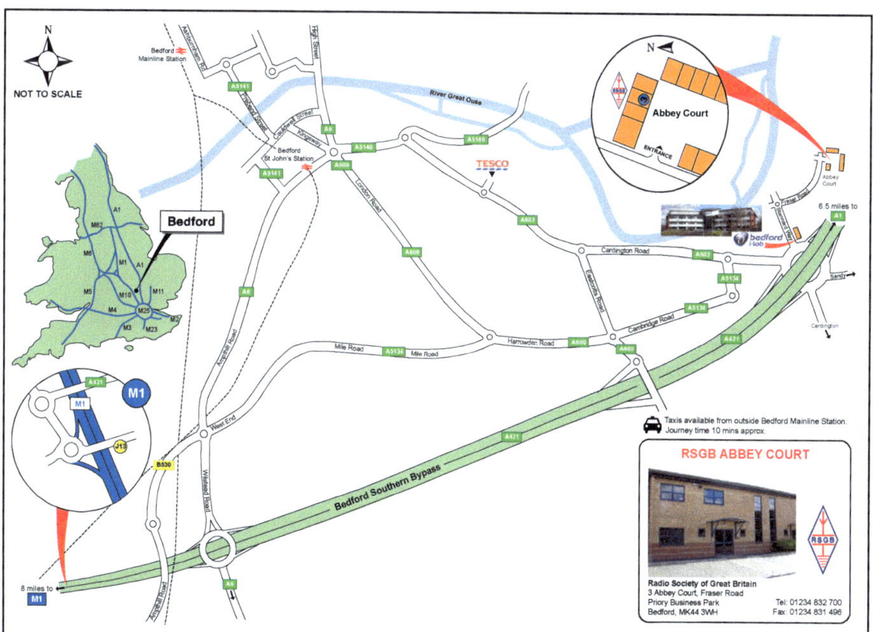

RSGB

3 Abbey Court, Fraser Road, Priory Business Park, Bedford MK44 3WH

Tel: 01234 832 700 **Web**: *www.rsgb.org*

RSGB online

In 2024, approximately 125,000 web pages were served to visitors to the RSGB website every month and we welcomed thousands of new users. There are hundreds of pages of information and links to resources from around the world, plus the very latest news from the world of amateur radio

LANDING PAGE

rsgb.org

Go straight to key areas of the website from our tablet and mobile-friendly front page. Access the latest news headlines, the main site index, guidance for newcomers and information on training and operating.

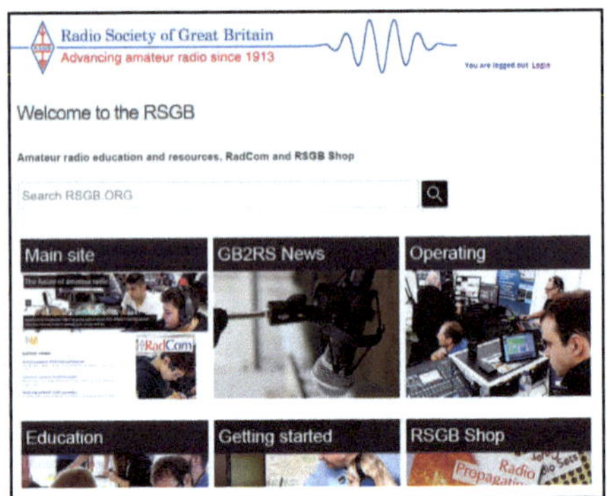

MAIN SITE

rsgb.org/main

The main site provides access to all the content in rsgb.org. The latest updates to the website are listed on this page along with our most important current events and activities.

NEWS

rsgb.org/news

The Society's flagship news bulletin GB2RS is published here every Friday and we add further news items throughout the week. Visit this section for local, regional, national and world news, plus upcoming special events and all the latest from the worlds of contesting and propagation.

LIVE NEWS

rsgb.org/live

The live news page brings together the RSGB newsfeed, YouTube channel, Facebook page and X (formerly Twitter) feed all in one easy-to-find place. If you don't have a social media account, this is where you can see the latest news, videos and discussions online and keep up to date.

ABOUT US

rsgb.org/main/about-us

A summary of who we are and what we do. You will find information about how we are organised, including details of our teams and regions.

RSGB BOARD ***rsgb.org/board*** You'll find a list of the current Board Directors and their areas of responsibility. If you have a question or concern about any of those areas, please contact the Board Director responsible via their email address shown on the page.

COMMITTEES The different committees can be found on the website by typing in rsgb.org/ and then adding the abbreviation of the committee's name e.g. rsgb.org/pac for the Planning Advisory Committee. These committees offer a wealth of knowledge and support on a wide variety of amateur radio topics.

HONORARY OFFICERS ***rsgb.org/main/about-us/honorary-officers/*** RSGB Honorary Officers, Managers, Coordinators and Advisors help to organise and run particular areas of interest within the hobby. You'll find a list on this page with information about their areas of expertise and how to contact them.

REGIONAL TEAM ***rsgb.org/regional-team*** The Regional Team provides support and guidance to current and prospective radio amateurs across 13 regions. In each region there are small teams of volunteers made up of District Representatives led by an elected Regional Representative.

Regions and Regional Team

The Society is organised on a Regional basis; in each Region there are small teams of volunteers made up of District Representatives led by an elected Regional Representative

Tap or click on a Region to meet your Regional Team

CLUBS

rsgb.org/main/clubs

This section includes information on how to affiliate your club with the RSGB, as well as details on the Society's insurance scheme for affiliated clubs. There is information on planning events, as well as details of special interest groups. You can use Club Finder to locate UK amateur radio clubs. Enter your location or postcode and select a travel distance to display clubs near you. Click the markers to display meeting and contact information that is provided and maintained by the clubs themselves.

TRAINING

rsgb.org/training

These pages bring you essential educational resources for all three levels of examinations (Foundation, Intermediate and Full) as well as the Direct to Full exam. The information and resources include all the latest syllabus updates following Ofcom licence changes. You'll find support material and advice for trainers as well as information about courses, exams and exam fees. Course and Exam Finder allows you to search for training in your area and this information is provided by clubs. We offer remote invigilation and paper exams and you can find all the information you need as well as the online booking process in this section.

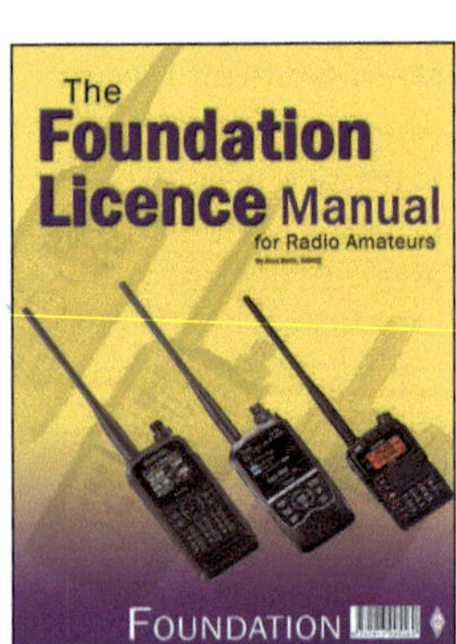

GET STARTED

rsgb.org/getting-started

Everything you need to know in one place if you are new to amateur radio, from getting licensed to setting up your first radio shack.

DISCOVERY

rsgb.org/Discovery

The Discovery Scheme has been updated and now aims to produce a fun, easy to follow, interesting programme that helps community members discover new aspects of the hobby. The Explorer level is available to download and enjoy now.

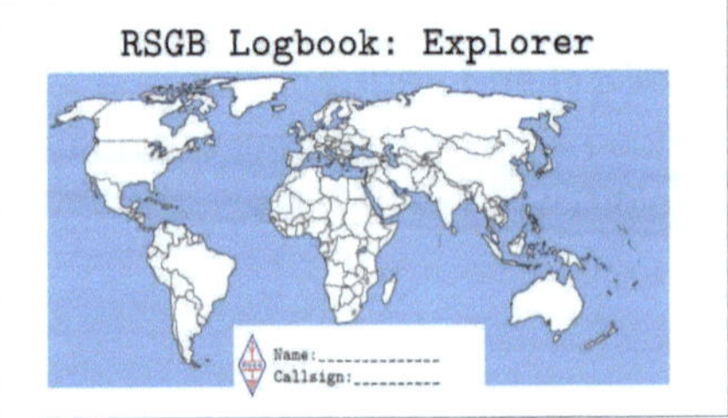

OPERATING

rsgb.org/operating

Information on band plans, awards, beacons and repeaters, emergency comms, the QSL bureau, planning, CW and NoVs, including online applications for selected NoVs.

TECHNICAL

rsgb.org/technical

Home to the EMC and propagation pages and a wide range of other specialist information, including space and satellites and microwave operation. There are also links to technical forums and useful apps.

YOUTH

rsgb.org/youth

We have dedicated web pages for young people. Find out about the radio clubs based at UK schools and universities, as well as Youngsters on the Air (YOTA) events and activity resources for other events such as British Science Week or National Coding Week. There are also details about who to contact if you'd like to get involved in youth activities.

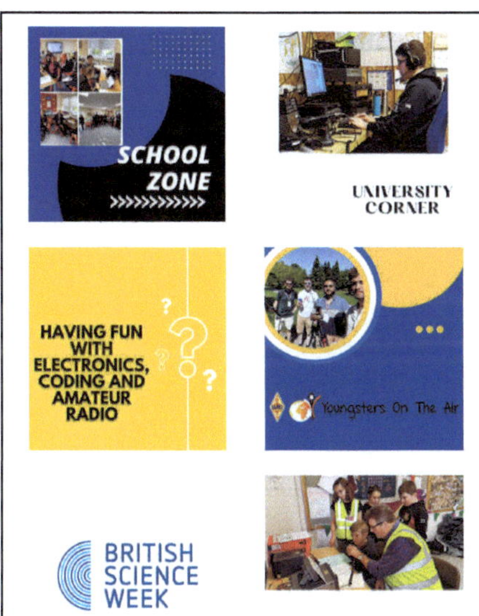

SOCIAL MEDIA

facebook.com/theRSGB

x.com/theRSGB

youtube.com/theRSGB

Our social media channels have a combined following of over 35,500 people. Join us on Facebook and X for breaking news, extra material and discussions. With multiple playlists on our YouTube channel you can enjoy a wide range of video content whatever your licence level or particular interests within amateur radio. You can also browse our *Tonight@8* live webinars, along with promotional videos for the RSGB National Radio Centre and a range of special events.

VIDEOS

www.rsgb.org/video

Our video library contains a wealth of videos from amateur radio promotional films to celebrations of special events. Our Foundation Practicals and Useful Practical Skills suites of videos, especially for new licensees, have been watched by thousands who have welcomed the clear and inspiring content. We have guidance and support videos to help with the Ofcom licence changes, EMF regulations, archive footage, video content aimed at young amateurs, as well as over 80 RSGB Convention lecture videos which clubs are welcome to use for their meetings.

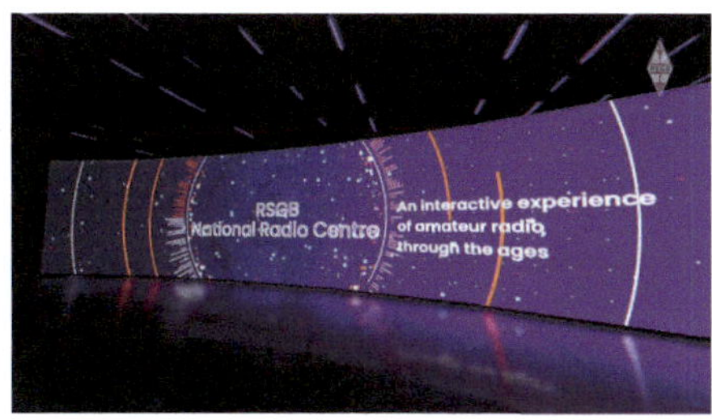

PUBLICATIONS

rsgb.org/main/publications-archives

Members can read the digital edition of RadCom and search the RadCom archive. Members can also read RadCom Basics and RadCom Plus, our two supplements that offer additional content for newer or more technical radio amateurs. You can also browse and purchase the full range of RSGB books.

ARCHIVES

rsgb.org/main/archive

The Events Archive includes preserved material from major RSGB events whilst the Publications Archive has back issues of our publications, including RadCom Basics and RadCom Plus. You'll find material relating to closed consultations in the Consultations Archive.

CONSULTATIONS

rsgb.org/main/rsgb-consultations

A list of current active consultations on matters of importance to the amateur radio community. You are invited to participate.

OFCOM LICENCE CHANGES

rsgb.org/licence-review

Information about the significant change to amateur radio licensing can be found on this page, including details of the 2023 Ofcom consultation and the RSGB's formal response. You can also find links to Ofcom's documents, RSGB guidance resources, as well as a number of video updates, two special *Tonight@8* live webinars and a summary of the key changes.

BE INSPIRED:

Looking for your next challenge in amateur radio? Or perhaps you'd like to see how other amateurs have progressed to the next licence level? Visit the following pages for your next inspiration: rsgb.org/your-stories, rsgb.org/student-stories and rsgb.org/volunteering

FAQs

rsgb.org/faq

If you have an amateur radio-related question, chances are it has been asked and answered before. In this section we answer your most common questions on amateur radio, DBS checking, exams and how to become a radio amateur

CONTACT

rsgb.org/main/contact

The RSGB's address, phone numbers, and departmental email addresses.

JOIN

rsgb.org/join

Sign up for RSGB Membership and become part of our circa 21,000 community working for the future of amateur radio.

YOUR MEMBERSHIP PREFERENCES

rsgb.org/members

Keep your details up to date via the Membership Services portal. Here you can subscribe to email updates, such as online events, RadCom alerts, GB2RS news and Membership service updates. This is also where you can update your RSGB account details, renew Membership, reset your login password, update your roles and preferences, read our digital publications and download a free admission voucher to Bletchley Park.

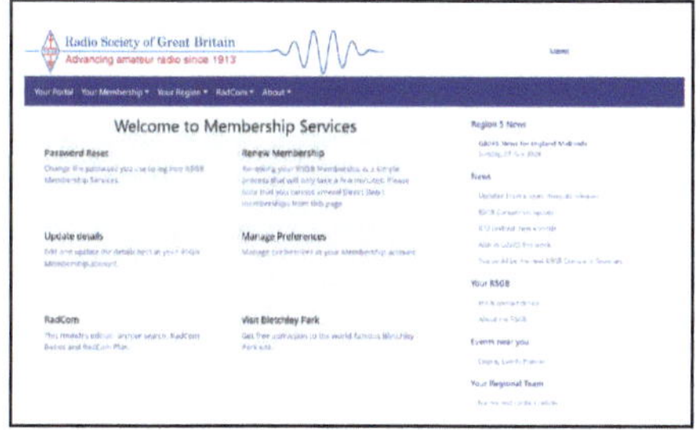

SHOP

rsgb.org/shop

Here you can join the Society, renew your Membership and buy RSGB and other publications with a members' discount. The shop also supplies EMC-related components, RSGB-themed polo shirts and baseball caps. All major cards acceptedd.

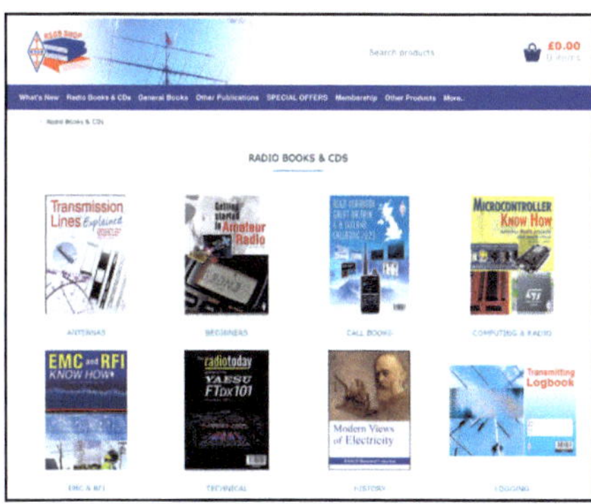

RSGB website: *www.rsgb.org*
RSGB Shop: *www.rsgbshop.org*

Radio Society of Great Britain

Advancing amateur radio since 1913

JOIN US TODAY

and get the best amateur radio magazine for FREE

RadCom

Icom IC-905 review

The RSGB is run by radio amateurs for radio amateurs (licensed or not), working for you, protecting your interests. We keep you informed of the latest amateur radio news, and amongst friends who understand the hobby.

Being part of the RSGB means that we post direct to your door each month the biggest and best UK amateur radio magazine, *RadCom*. Only available to RSGB members for less than the price of some other high street radio magazines, there is no better way to stay in touch with the world of radio.

Being a member of the RSGB is much more than a subscription to our magazine. You become part of a society which provides you with all the great benefits shown overleaf.

You also get online magazines

RadCom Basics: there are six of these popular magazines a year, with an array of material, which provide a host of getting started and understanding articles for the less experienced, or those who just want more.

RadCom Plus: This technical *RadCom* supplement is a members' only magazine that provides extended technical articles that have not appeared in *RadCom*.

If you want to get the most out of amateur radio, there is simply no better way to do that than by joining the Radio Society of Great Britain (RSGB).

That's Three Magazines Available Online

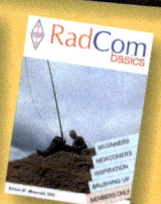

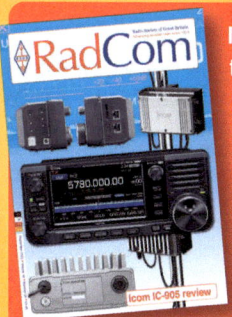

RSGB QSL Bureau

Whilst sending cards for a much-prized contact will always be quicker by direct mail, QSLing via the RSGB Bureau remains an extremely cost effective option, indeed the RSGB QSL Bureau enables members to exchange cards worldwide in the cheapest practical way.

How it works

QSL cards arriving at the central bureau are initially separated into UK and Foreign destinations. Overseas cards are sent in bulk to other member societies of the International Amateur Radio Union (IARU). Cards for stations within the UK are sorted into separate callsign groups and sent to the appropriate volunteer collection managers, on a quarterly schedule. They place cards in stamped addressed envelopes (SAEs) provided to them by the call holders.

Who can use The Bureau?

Unlike the RSGB, many other national societies make extra charges for using their QSL service. The RSGB QSL Bureau is an inclusive membership benefit as follows:

- **UK RSGB members,** including Channel Islands and Isle of Man only, can send and receive their personal cards without additional charges, subject to the conditions shown here.
- **UK non-RSGB members** can collect their personal cards only by using the '*Pay-to-Receive*' service but cannot send cards via the bureau. See RSGB website for details
- **Overseas RSGB members** can send their outgoing cards to the RSGB QSL bureau for distribution. Senders must include current membership information i.e. membership number and renewal date. UK call holders should collect in the normal way directly from their UK call sign. Non-UK call holders should arrange collection via a UK-based QSL manager, who must also be a member.
- **Overseas non-RSGB members** may send cards addressed to UK-based stations only.
- **Affiliated Societies and independent QSL Managers** can send their own cards and those for club members or stations for whom they act, but must include current membership or affiliated club confirmation for every station or group whose cards they wish to send. Cards included from overseas stations and intended for delivery outside the UK will not processed without proof of membership and will not be returned.

Available Destinations

A full list of IARU partner QSL bureaus can be found at: *www.iaru.org/iaruqsl.html* Keeping an up-to-date copy to hand is vital when deciding which route to send your card. For example, there are currently no bureaus in, Egypt, Kazakhstan, Morocco, and Mauritius, Sudan and several other African and Caribbean countries, plus many more smaller destinations.

Activity also relates to the frequency with which cards can be dispatched to a particular destination. This may range from monthly to annually, according to demand. Always check QRZ.com for QSL information for non-IARU destinations before sending cards via the bureau, to avoid disappointment.

Responsible QSLing

The Bureau handles approximately 1.5 million cards per year and is one of the busiest in the world. The Society has a policy of discouraging the sending of cards when they are not wanted or cannot be received.

Active Amateurs, GB and Special events, all Clubs and DXpeditions are strongly advised that 100% QSL outgoing is no longer desirable or cost effective.

Transporting large volumes of cards between bureaus, only to have them ultimately destroyed, returned or uncollected, is disappointing and not eco-friendly.

Tip: Ask yourself... Do I need to send a card for every contact before QSLing? Always ask the other station if they can receive a bureau card, before sending.

Log Book of the World (LOTW)

Receiving a nice card for a memorable contact is always a thrill, never matched by an electronic confirmation via the Internet. However, do consider the alternatives. Uploading your logs to Logbook of The World can automatically confirm some contacts, such as for contests and award purposes etc.

Confirmations via LOTW are easy and work well for everyone, if a few simple steps are followed. See: *www.arrl.org/logbook-of-the-world*

OQRS systems - the future of QSLing?

Many stations and most DXpeditions and rare calls are now using the worldwide OQRS network and only responding to requests for QSL cards. This online system means there is now no need to automatically send a card, to receive one via the bureau, or direct.

Using OQRS also speeds up the system so that it can now be only half the time it presently takes to send and receive a card, with the added benefit of not needing to send yours. Simply put cards are only sent in response to OQRS requests for a card. So if you are sending QSL cards you or your QSL manager will receive an email to generate a genuinely wanted card. This saves time and waste for both the user and the QSL system in general and is therefore recommended as good practice.

In the UK we are fortunate to have the free to use ClubLog, courtesy of Michael Wells, G7VJR and his team. Simply go to, *www.clublog.org* for more information or to register your call, club, GB station or event and start uploading your logs

Sending cards via the Bureau

Cards from RSGB members for both UK and worldwide should be sent, suitably packed, to the main UK bureau address: **RSGB QSL Bureau PO Box 5, Halifax HX1 9JR, England.** Members, clubs or DX groups wishing to send large or heavy packages to the Bureau via carriers other than Royal Mail should contact the Bureau for an alternative delivery address.

Fair Usage Policy

As part of their subscription, each Member can send up to 15kgs of cards through the Bureau each year (about 5000 standard cards).

Each Affiliated Club can send up to 20kgs through the Bureau each year. Additional cards will be charged at £6 per kilo or part thereof.

In the interests of fairness to others, members can only send to the Bureau a maximum of 1kg of cards (approx, 300) for any single country or DXCC entity per month.

Responsible QSLer...

Help us to speed up processing and cut waste for everyone,

- 10 simple things you can do.....
- Ask new contacts, "If I send you a QSL card, do you collect and how? – every time
- If you don't QSL, be polite but honest "Thanks but no thanks" is all it takes
- Please don't say you do when you don't, or ignore the other guy's kind offer
- QSLing 100% outgoing is costly for everyone. Half your cards could be wasted, please check before you send
- Create your own OQRS system or use ClubLog *www.clublog.org* to reply to incoming requests with a real card, save time and money for you or your club
- Make your QRZ.com QSL details clear, honest and visible
- Amend your online details if you change your QSL status - don't leave it
- QSL info– Direct or via...' is confusing. Please clarify what you really mean
- Clubs Calls – Collect only from the calls sub group. For, 'QSL-Direct,' show an address not a callsign as this can change, it's often confusing
- Always collect your cards. - even if you never send one, they will arrive

Be responsible, it only costs a stamp!

The bureau operates a 24-hour message line for members' QSL enquiries. Tel: 01422 359362.
When calling, please leave contact details, a brief message and if possible an email address.
email: qsl@rsgb.org.uk (please put your callsign series in the subject box)

More than 1kg of cards must be sent directly to the bureau in the relevant country (see IARU list on line) or direct to the call sign manager.
Heavy users such as, DXpeditions, affiliated clubs and groups should send large volumes direct to overseas bureaux. please see – 'Responsible QSLing' advice before QSLing 100% outgoing.

Every package of cards should contain ...

- Proof of current membership; that is an original RadCom address label, taken from the magazine wrapper or printed insert, showing: address, callsign and membership number, not more than 3 months older than the membership expiry date on the label. Photocopies must not be used as cards may not be processed or returned

  As Clubs receive the RSGB Yearbook each year in lieu of RadCom, they should include sufficient information for a check to be made against the Affiliated Societies' register, ideally in the form of a club letterhead, showing the membership number and renewal date. To speed status checking, clubs and groups are asked to ensure that they register club call and contact details at *My Account* directly in the group's name and not as secondary to a personal callsign, or qsl manager.
- Special event stations (GB) and single letter Abbreviated/Contest callsigns should include the membership number and call of the NoV license holder or affiliated club, for contact purposes.

Other important points

- Clubs and QSL Managers sending a bulk dispatch to the bureau should ensure that all callsign holders for whom they send cards are current members of RSGB and should enclose current membership details for every callsign with every batch of cards. All Clubs and Groups must include their Affiliated Society membership number.
- Members who operate from another station, typically a foreign club call or that of an individual overseas amateur, may send cards for contacts made from that station, provided they clearly identify themselves as the operator and state their UK callsign and membership number on each card.
- Listener report QSLs need sufficient information to be of genuine value to the transmitting amateurs. Reception reports relating to broadcasting stations cannot be accepted. *see* ' Card Issues..' for details of acceptable card details.

  The bureau system accepts standard cards only no letters, SAEs or money orders.
- All cards, whatever the quantity, should be pre-sorted into alphabetical and numerical country DXCC order (see the Prefix List pages and/or the *RSGB Prefix Guide* which also contains a complete cross reference and awards section).
- Countries with more than one prefix should be placed together. For example, JA and 7J cards (destined for Japan), F and TK-TM cards (destined for France) and SP, HF and 3Z (destined for Poland) may be grouped together.
- Cards for the USA need be sorted separately into call areas (numbers 0-9), regardless of the prefix letters.

NB: Exceptionally, cards in the number 4 series with either one or two letters before the number are handled by different bureaus and need to be separated, as are cards for Alaska (KL), Hawaii (KH6-7) and Puerto Rico (KP3-4).

- Cards for Russian Federation and former Soviet countries were traditionally grouped together. They now need to be separated into five individual groups; RA-RZ, UA-UI, UJ-UM, UN-UQ and UR-UZ, as they are no longer sent to a single destination in Moscow. See: *www.iaru.org/iaruqsl.html*
- **Cards for UK delivery G. M.2 etc must be banded separately from foreign destinations -see checklist above.** Our UK sorters currently have to split cards into 40 alpha-numeric categories. For this reason cards should be supplied to us pre-sorted as per the Sub Manager list on the following pages.
- Envelopes, paper or card dividers to separate countries or call groups are not required, as removing these can sometimes slow down the distribution process.
- Cards sent in date/time/logbook or random order are not acceptable, as they typically take up to five times longer to process. Similarly, those with small print or hand written callsigns can be very difficult to process, resulting in delays

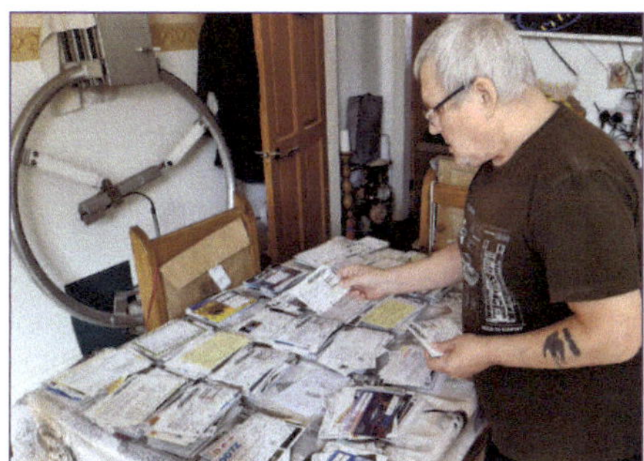
QSL card distribution relies on it's army of volunteer sub-managers for final sorting and distribution. Please support their efforts by collecting your cards. – see check list for details.

Checklist for sending cards

We need your help to sort more than a million cards each year and reduce delays.

A ***First, place your cards into three piles...***

1. UK destinations
Pre-sort G, M and 2 as per the Sub Managers list.

2. USA destinations
Sort by number only, 0-9, regardless of prefix. Separate cards for Alaska, Hawaii and Puerto Rica.

3. Rest of the World
In DXCC callsign prefix order.

4. Calls with numbers first
Sort in digit order, i.e. 3A, 4X, 8P, 9H etc.

5. Calls with one letter then one number
These come before two letter prefixes, i.e. S5 before SM, etc.

B ***Check ALL cards for possible ' Via' destinations***

Re-sort if necessary, and never rely on your computer print log, for example: F5/G3UGF isn't a French destination.

Africa, Caribbean and DX destinations are mostly QSL Direct only, or via a QSL Manager. Check *www.qrz.com*

C ***Pack your cards securely and don't forget***

A recent RadCom address wrapper as proof of membership.
Your callsign and return address on the package.
If you put more than ten cards in a C5 envelope, check the dimensions and weight at the Post Office, before sending - don't just post.

Whatever the quantity, never send unsorted cards!

Before you send, please check that we can deliver!

Check all Vias before you send, as your card may come back to you, or it may never arrive.

There are many world destinations, but only 190 IARU member and associate Bureaus worldwide.

The following IARU Bureaus are currently closed.

3B Mauritius	D4 Cape Verde
3DA Swaziland	HH 4V Haiti
4J-K Azerbaijan	HV Vatican City
7P Lesthoto	PZ Suriname
9L Sierra Leone	ST Sudan
A3 Tonga	SU, 6A-B Egypt
C2 Nauru	V3 Belize
C5 Nauru	V4 St Kitts and Nevis
C6 Bahamas	XY-XZ Myanmar
CN, 5C-G Morocco	Z2 Zimbabwe

Download your own Bureau list from: *www.iaru.org/iaruqsl.html*

Post outgoing cards to: RSGB QSL Bureau, PO Box 5, Halifax HX1 9JR.

for other users. The bureau reserves the right, at its discretion, to reject unsorted cards or those with callsign or routing information in small or difficult to read print. The minimum print size requirement is 12 point.

Tip: If you are unsure about your handwriting, why not ask someone else to check the cards to see if they can easily read the callsigns?

Card issues and some good advice

For economic shipment and to avoid possible transit damage, all cards need be single page, IARU recommended size and weight. That is, standard postcard size (140m x 90 mm) and no larger. Card weight and thickness is important and needs to be in the range, 230-350 grams.

Note: Per 10 cards, a single card should have an average weight not more than 3 grams.

The bureau reserves the right not to process or return, repeat batches of cards falling outside these guidelines and where a sender has been previously advised.

Large or unusual shaped cards are extremely difficult to process and most easily damaged when packed, or folded with others.

Thin, small, and paper cards are slow and extremely difficult to handle. They often stick to cards for other destinations, as do homemade cards using photo print or heat laminated paper. This type of card does not travel well, is difficult to write on and is very easily damaged when subjected to humidity or damp – not recommended!

Multi-page and cards not meeting the above IARU recommendations must not be sent via the bureau system as they increase workload and overheads. In fairness to others they should be sent direct only.

QSL Routing & QSL Direct

It sounds obvious, but the Bureau can only process outgoing cards if there is a destination to which they can be sent. Before sending cards, therefore, (particularly to rare stations or DXpeditions) please check the recipient's QSL policy. This is usually available on QRZ.com or via a websearch.

Many DXpeditions and rare callsigns only QSL direct, or respond to an OQRS request often via a QSL Manager who may be in another country. These stations are most often located where there is no bureau service and are operated by visiting non-resident Amateurs from another country. Some stations do not QSL at all, so it is vital to check before sending, whatever route you choose.

Please note that outgoing bureau cards where no destination bureau is available, or no clear 'Via' route is indicated, will be recycled.

Tip:

- Ask for the other station's QSL details at the time of the QSO, or by an Internet search before posting.
- Consider posting your most wanted cards direct or to an overseas bureau, if it's active. This helps to speed replies as most bureaus world-wide have backlogs. The IARU world-bureau list can be downloaded at: *www.iaru.org/qsl-bureas.html*. It's good practice to check the listings for changes at least twice a year.
- Always search the web and check *www.qrz.com* first before posting.
- Make sure that any "Via" information on your cards appears directly below, or next to, the station callsign, to avoid being missed. Using a different coloured ink for this purpose is a great help.

For guidance on what information to include on your card (and where), *see the example card on page 20.*

Using printed labels

Avoid cramming too much information on small printed labels. For health and safety reasons all callsigns should be a minimum of 12 point print size and in common, easy to read fonts such as Arial, Times New Roman, or clear, block capital hand written letters.

The bureau system is for the exchange of QSL cards only. Envelopes containing letters, photographs, money, stamped addressed envelopes (SAEs), awards, certificates and other items will not be processed and should be sent by other means.

Heavy users

Those sending more than a few thousand cards per year should send their largest volumes of cards directly to their top ten destination countries. The remaining balance can be sent via the normal bureau system.

The aim of this is to share some of the burden of cost, without penalising others who may only occasionally send a few more cards than normal. The bureau weighs and notes regular large consignments. Members or clubs may be contacted if their usage becomes excessive, with a request to follow the guidelines above. The IARU bureau list can be found at: *www.iaru.org/iaruqsl.html*

Packing and posting your cards

The bureau receives many damaged envelopes and packages from both UK and foreign amateurs. It also receives a significant number of requests each month from Royal Mail for payment of additional postage, which are always rejected.

Having first separated pre-sorted, UK destination cards from the rest of the world, please read on...

- Never post loose cards in lightweight or thin envelopes, as they will often cut through the edge of the envelope in transit.
- Always print return address details and callsign on your package, in case it arrives damaged.
- Secure batches of cards with a rubber band or - better still - a banknote style band of thin paper strip, folded around the cards.
- Never place two or more packs of cards side by side in a C5, A4 or larger envelope, as it will fold in transit and split down the middle, allowing the cards to spill out.
- Using lightweight 'Mail-Tuff' style plastic or 'Mail-Lite' style padded bags or Post-Pack envelopes usually avoids this problem.
- Always check the size and weight of your envelopes and packages, before posting as the Post Office now charge by volume as well as by weight.
- The current weight limit for a First Class stamp is 100g, but the package size is limited to 240mm x 165mm and the package should fit through a postal slot only 5mm in height.
- It is possible to send a large envelope A4, or a smaller envelope over 5mm in thickness. This type of envelope is considered to be a 'Large Letter.' Large Letter stamps costs more, but allows the letter thickness to be up to 25mm. 'Second Class Large' offers better value
- The Post Office can supply a paper/card copy of their pricing slot guide for a small charge. Frequent users are advised to obtain a plastic Helix HP5 'Pricing in Proportion' Ruler. It has postal slots built in, to check your packets.

Part of the overseas side of the bureau

Sending small numbers of cards in separate envelopes is not cost effective for the sender and means much more time spent opening, sorting and checking in the bureau. Sending not more than one pack per month, with your *RadCom* label, resolves many issues and can save you money .

Recorded delivery is not cost effective. We receive many packages and we are not always asked to sign for individual items, secure packing and a return address offers better value.

Receiving cards from the Bureau

RSGB is extremely fortunate to have around 30 dedicated volunteer Sub Managers who give freely of their time to support the work of the Bureau and in the service of their fellow radio amateurs. Members' cards are sent to the Sub Managers for onward distribution.

Sub Managers details are subject to change, so it a good idea to check the QSL section of the RSGB website from time to time for the latest information. From the RSGB Home Page click,'Operating' and follow the links.

Our system relies upon those wishing to receive cards depositing SAEs with Sub Managers, ready for each quarterly despatch. Members should use SAEs, as Sub Managers are not authorised or insured to accept

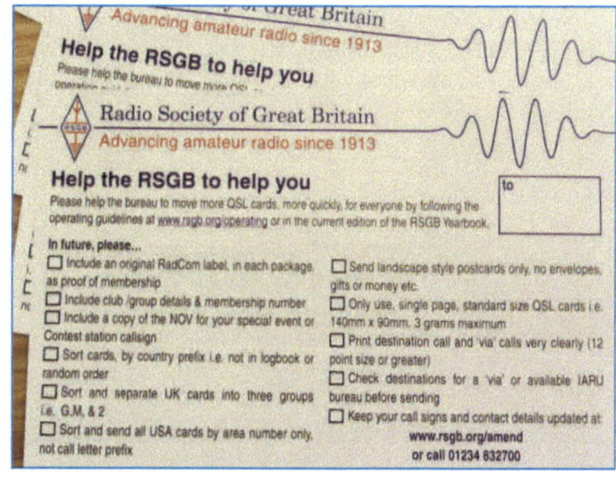

Radio Society of Great Britain
Advancing amateur radio since 1913

Help the RSGB to help you

Please help the bureau to move more QSL cards, more quickly, for everyone by following the operating guidelines at www.rsgb.org/operating or in the current edition of the RSGB Yearbook.

to

In future, please...

- ☐ Include an original RadCom label, in each package as proof of membership
- ☐ Include club /group details & membership number
- ☐ Include a copy of the NOV for your special event or Contest station callsign
- ☐ Sort cards, by country prefix i.e. not in logbook or random order
- ☐ Sort and separate UK cards into three groups i.e. G,M, & 2
- ☐ Sort and send all USA cards by area number only, not call letter prefix
- ☐ Send landscape style postcards only, no envelopes, gifts or money etc.
- ☐ Only use, single page, standard size QSL cards i.e. 140mm x 90mm, 3 grams maximum
- ☐ Print destination call and 'via' calls very clearly (12 point size or greater)
- ☐ Check destinations for a 'via' or available IARU bureau before sending
- ☐ Keep your call signs and contact details updated at: **www.rsgb.org/amend or call 01234 832700**

RSGB's new feedback card, is designed to help speed QSL throughput at one of the world's busiest bureaus. - If you receive one… "Please help us to help you."

money in lieu of postage stamps. RSGB is not liable in case of any loss or dispute.

The scheme is open to all RSGB members plus UK-based, pay-to-receive subscribers.

Collection Envelopes

- Envelopes need to be C5 size (160mm x 230mm) and of strong material (*see earlier*).
- Callsign or Listener number should be printed in the top left hand corner, followed by the a current membership number, immediately below.
- Print the name, delivery address and postcode clearly, as normal.
- Number each envelope sent to the Manager (eg '1 of 6', '2 of 6', '3 of 6', etc) always mark one of them 'Last', so that you will know when a fresh batch should be sent.
- Envelopes are normally despatched every quarter, subject to card availability.
- Always use stamps worded Second or First Class, rather than a numerical amount, as these will be honoured if the postal rate changes.

No delivery in that quarter means 'no cards waiting'.

Cards for amateurs who have not lodged envelopes are not returned to sender and will, at the Sub Manager's discretion, be recycled after a period of three months. Always keep your envelopes up to date. Many volunteer Sub Managers now operate their own websites, with links from the RSGB website, giving cards waiting, envelope status and next anticipated delivery details. RSGB requires these lists to be confidential. Members permission to display their callsign and details to others, is a condition of inclusion on any such listing, operated by a volunteer.

It is a good idea to note in your diary to check your Manager's list every quarter.

UK amateurs who do not wish to collect cards or those who use a separate QSL Manager are asked to notify the appropriate Sub Manager as a matter of courtesy and also make this clear at: *www.qrz.com*

More than one callsign?

Always collect from the callsign used.

Stations changing their callsign as a result of a licence upgrade, or other reason should inform RSGB of their change of status. Contact membership services direct or via the web site. Log in using current call and membership number and enter personal details at the, 'Update Details' page.

Amend the primary callsign and list all previous calls in the additional category, as soon as possible. Cards must not be diverted from one call to another. Envelopes should be maintained with both new and old QSL sub-Managers. Typically, envelopes for the old callsigns and membership number need to be available for up to five years after the old call is no longer the primary call.

Club stations should enter their callsign details in the club's name and not as a secondary call of a member or QSL manager, as this gives rise to confusion. Please avoid registering calls using optional club identifiers, such as X,S,C,N.

Stations operating from a different prefix, for example G9ABC as GW9ABC/P or GU9ABC/P, need to lodge envelopes with the appropriate Sub Managers for every area of operation, as cards may not be forwarded to the home call.

UK mainland stamps are not valid when sent from the Isle of Man or Channel Islands. Local stamps should be obtained during the period of operation, for use later. When operating outside the UK under CEPT rules, e.g. F/G9ABC, or more importantly with another callsign, it is vital to tell the QSO partner to 'QSL via G9ABC' and not simply state 'via home call'.

Registering any foreign calls separately, together with the QSL route and contact email address at: *www.qrz.com* is extremely helpful to others in these cases.

Requesting a 'Via' call route

In recent years there has been an explosion in the use of 'Via' requests from members, significantly those using multiple call signs, clubs and others using a QSL manager.

With so many cards passing through the bureau at any one time this massive growth in via activity contributes to card delays and is no longer sustainable.

- Cards can only be collected from the call sign used at the time of the contact.
- No need to give divert information during a contact -see exceptions below.
- No need to give divert information at QRZ.com.

The following are not subject to any diverts:

- Personal QSL cards ,including to a members other callsigns, a friend or family member.
- Club and Affiliated group calls,
- Contest calls
- Special Prefix calls (R.Q.O.V etc)

NB. Separate C5 collection envelopes, need to be lodged with the appropriate sub-manager for every call sign held. -see list

The following cards may be diverted:

- Calls used in another UK pre-fix location, back to the home call.
- Members personal overseas call signs sent to the UK home call.

Remember:

Even if you never send a QSL card, someone, somewhere, sometime will send a card to you.

It would be a shame not to receive it so, please send an SAE to your RSGB volunteer sub manager for each callsign you hold.

Card design

Whether you are designing and making your own QSL or having it made professionally, ***size, quality and design are the most important factors*** if you are hoping for a reply. Gone are the days when cards were printed in a single colour (black), with only a callsign and basic information and which took several weeks to produce. The advent of high resolution digital photography and computers has changed everything. High quality commercial QSL cards are now more interesting, colourful, easier, quicker and much cheaper to produce or change than ever before. What's more, professionally printed cards can and often do work out cheaper than making your own.

All the more reason to consider having a distinctive card that gives not just your station details, but perhaps reflects your radio and other interests, family, pets, location or some other part of your life. Cards can be simple, beautiful, artistic, funny, technical or even something completely unexpected. They make a statement about you - so what does your card say?

The range of choice has never been greater, so just use your imagination. Above all, make your QSL card something of quality that

Checklist for receiving cards

1. Register all your callsigns, past/present and always collect from the callsign as used, don't divert

Do this via the RSGB website at, 'Your membership ' page, or phone 01234 832700.

2. Send C5 stamped envelopes to each Sub Manager

In addition to your name and address...
Write your callsign and RSGB membership number at the top left.
Number each envelope at the bottom left.
Mark 'Last envelope' at the bottom left of the last envelope.
N.B. Sub Managers are not authorised to accept cash in lieu of SAEs.

3. Holidays and portable activations

We don't automatically divert cards to your home call. For temporary prefix operation (e.g. GM, MW, 2I etc), lodge separate envelopes with the relevant Sub Manager to collect your cards.
N.B. The Channel Islands and Isle of Man use different stamps.

4. Special event (GB) and abbreviated contest calls (G1A, etc)

No diversions apply, so please see the Sub Manager list.

5. Special Prefix NoV callsigns

For GR, MQ, 2O etc, no diversions apply. See the Sub Manager details.

Multiple callsigns can be listed on the same envelope.

G4EZT
TO: 4K9W
VIA: DL 6 KVA
CONFIRMING OUR QSO(s):

DATE (DAY : MONTH : YEAR)	TIME GMT	FREQ/MHz	MODE/2x	RST
29 12 2007	12:31	14·079	RTTY	599
16 10 2007	19:18	10·138	RTTY	599
18 11 2006	10:43	18·145	SSB	59
3 10 2004	10:03	24·945	SSB	57

Example of a QSL card that's well laid out, easy to read and easy to sort.

stands out; something that the other station will want to keep and display. If you are sending or receiving a 'gift', make it memorable. It's now possible to collect special interest cards showing planes, trains, ships, cars, families, pets, castles, churches, windmills, lighthouses, motorcycles and many other things, in addition to antenna farms, radios, vintage gear and shack interiors.

RSGB Bureau reserves the right not to accept, process, or return, cards from any source, containing images or content not directly relevant to Amateur Radio and which in its opinion may be likely to, or does cause offence to those handling or receiving them.

Tip: Remember to tidy up before you take a photo of your station!

The business side of the card is also very important. Here, simple clarity is the key to a good card and to receiving a reply. Use a clear type face that is easy to read. Don't put too much information or too many logos on the card, unless it's a special event when background information is always nice to see. Remember that English is not always a first language, so keep it simple, keep it relevant. Allow enough space to write or print the contact information clearly on the card, ensuring that the destination call is at the ***top right, with any via routing details immediately below.*** Many cards now have space to log more than one contact. This is a great eco-friendly idea. *See example card from G4EZT.*

Where to buy cards

The RSGB doesn't endorse any particular producer. Take a look at the cards you receive, as they will often include maker's details.

Apart from your local printer and checking with friends, there are now a whole range of specialist online makers offering superb, correctly sized card. We regularly see cards from UK stations being sent to us that have been designed online, some produced in other countries, and many are simply stunning. It is possible to download card making software from the Internet, but so much depends on the actual equipment used to make the card that the results are often disappointing or uneconomic, unless you have access to specialist print and cutting machinery. However, where practical, they do make possible one-off special, individual and personalised cards, for QSLing direct.

Remember: If you have invested time and energy on your station, isn't it right to do the same with your QSL card? Send something you would be pleased to receive.

Using www.QRZ.com

There's no doubt that QRZ.com is the go-to place for station information. Completing your outgoing cards either as a computer

The modern way to display your cards, using a digital photo frame. Scan the best of your collection, Create albums of interest, Antennas, Ships, Stations, Landscapes, people, places, DX, IOTA and more. Just use your imagination

run or handwritten, it's wise to check the QSL choices for the intended station, to avoid waste and disappointment.

- The information is only as good as the person who entered it so remember to keep your own up to date and easy to understand. n.b. every page also shows the last amended date.
- For, 'QSL bureau cards.' Avoid confusion, always state , 'QSL via RSGB Bureau.'
- For 'QSL Direct' – Always send to the station address, never to the page manager or to the bureau with a, via. It may be wrong or out of date
- If you wish to receive direct cards write "QSL direct to address above." To avoid confusion.
- Those giving no QSL details it must be assumed do not want or accept incoming cards.
- Always read the whole page as some put their QSL info at the end of the page, not at the top.

RSGB QSL Bureau Sub Managers

All details correct at time of press, but may be subject to change. For the latest information visit the QSL pages at the RSGB website

Abbreviated & Contest Calls

Mr. M. G. Coomber. G0NBI
3, Dolly Grove,
Blackdown Heights, Crimchard.
Somerset. TA20 1PF
grahamG0NBI@gmail.com

G0 Series

Mr. R. Rogerson G0OUC
8, Dearne Street,
Darton,
Barnsley
South Yorkshire S75 5 HL
Email: g0qsl@yahoo.com

G1 & G2 Series

Mr L. Pennell, GI3KME
21, Dundrum Road, Clough,
Downpatrick. Co. Down.
BT30 8SH
g8pma@pennell.eu

G3A-F

Mr P J Pasquet, G4RRA
Honey Blossom Cottage Spreyton
Devon EX17 5AL
gee4rra@gmail.com

G3G-L

Mr L. Pennell, GI3KME
21, Dundrum Road, Clough,
Downpatrick. Co. Down.
BT30 8SH
g8pma@pennell.eu

G3M-S

Mr G Coomber, G0NBI
3, Dolly Grove,
Blackdown Heights, Crimchard.
Somerset. TA20 1PF
grahamg0nbi@gmail.com

G3T -V

Mr. N. S. Cawthorne. G3TXF.
Dormers, Hinton Charterhouse,
Bath, Somerset BA2 7TJ
nigel@g3txf.com

G3W-Z

Mr J Peden, G3ZQQ
51A, Bewdley Road, Kidderminster.
Worcestershire DY11 6RL
g3zqq@yahoo.co.uk

G4A-F

Mr J J Pascoe, G4ELZ
3 Aller Brake Road, Newton Abbot,
Devon TQ12 4NJ
g4elz@blueyonder.co.uk

G4G-L

Mr I N Fugler, G4IIY
Lees Hill Farm, Lees Hill Brampton,
Cumbria CA8 2BB
ian.g4iiy@zen.co.uk

G4M-S

Mr C G Rowe, G4MAR
29 Lucknow Road, Willenhall,
West Midlands WV12 4QF
cliff1.g4mar@gmail.com

RSGB QSL Bureau Sub Managers

All details correct at time of press, but may be subject to change. For the latest information visit the QSL pages at the RSGB website

G4TAA-ZZZ Series

Mr P. Rivers, G4XEX
34 Coales Gardens,
Market Harborough,
Leicestershire.
LE16 7NY g4taa.g4zzz@gmail.com

G5 Series

Mr P J Pasquet, G4RRA
Honey Blossom Cottage, Spreyton, Devon
EX17 5AL gee4rra@gmail.com

G6 Series

Mr S Wellon, G6DMG
71 Toftdale Green, Lyppard Bourne,
Worcester WR4 0PE
g6dmg@hotmail.co.uk

G7 Series

Mr C. Flanagan, G7NRO
2 Wynyard House, Durham Road,
Wolviston. Billingham, TS22 5LP
g77nro@gmail.com

G8 Series

Mr D Helliwell, G6FSP
1 Beechfield Avenue, Torquay TQ2 8HU
dave@g6fsp.com

GBxAAA-ZZZ

Mr. M. G. Coomber. G0NBI
3, Dolly Grove,
Blackdown Heights, Crimchard.
Somerset. TA20 1PF
grahamG0NBI@gmail.com
www.gb-special-event-qsl-status.webs.com

GD, MD & 2D Series

Mr A Ames GD4SVD
20, Sunnybank Avenue, Onchan. Isle of Man
IM3 3BW gd.md.2dcards@outlook.com

GI, MI & 2I Series

Dr E H Squance, GI4JTF
11 Ballymenoch Road, Holywood,
Co Down, Northern Ireland BT18 0HH
gi4jtf@gmx.com

GJ, MJ & 2J Series

Mr M Roche, MJ0ASP
Flat 1 Stratscombe House, Le Quai Bisson,
St Brelade, Jersey JE3 8JT
mathieu.roche@hotmail.com

GM0 - GM3 Series

Mr F A Roe, GM0ALS
74 Willow Grove, Livingston,
West Lothian EH54 5NA
fred.roe190@googlemail.com

GM4-8 Series

Mr P Rose. GM3ZZA
4 Heatherfield Glade
Adambrae,
Livingston EH54 9JE
gm48.qsl@btinternet.com

GU, MU & 2U Series

Mr P F H Cooper, GU0SUP
1 Clos au Pre, Hougue du Pommier,
Castel, Guernsey GY5 7FQ
pcooper@guernsey.net

GW- Series

Mr. J. L. Lewis. GW0RAD
189,Heol y Gors, Cwmgors, Ammanford,
Carmarthenshire Wales SA18 1RF
gwmanager@sky.com

2E Series

Mr R Maltby, 2E1DFI
1 Briar Close Southfields, London Road,
Sleaford, Lincolnshire NG34 7NT
ray2e1dfi@aol.co.uk

2M Series

Mr. R. Roberts. MM0CPZ
16, Swanston Avenue,
Edinburgh
EH10 7BX
Email ; mmand2m@yahoo.com

2W Series

Mr. J. L. Lewis. GW0RAD
189 Heol Y Gors
Cwmgors
Ammanford, Carmarthenshire SA18 1RF
gwmanager@sky.com

M0A-L

Mr D E Mappin, G4EDR
13 Willow Close, Filey,
North Yorks YO14 9NY
g4edr@yahoo.com

M0M-Z

Mrs V Bates, G6MML
The Anvil.4 Eastgate,
North Newbald,
York YO43 4SD
g6mml@btinternet.com

M1 Series

Mrs. A. Eastwood M7ERT
3 Bowes Nook,
Buttershaw, Bradford. BD6 2BJ
West Yorkshire
M1-M7QSL@hotmail.com
groups.yahoo.com/group/M6_QSL/

M3 Series

Mrs. A. Eastwood M7ERT
3 Bowes Nook,
Buttershaw, Bradford. BD6 2BJ
West Yorkshire
M1-M7QSL@hotmail.com
groups.yahoo.com/group/M6_QSL/

M5 series

Mrs. A. Eastwood M7ERT
3 Bowes Nook,
Buttershaw, Bradford. BD6 2BJ
West Yorkshire
M1-M7QSL@hotmail.com
groups.yahoo.com/group/M6_QSL/

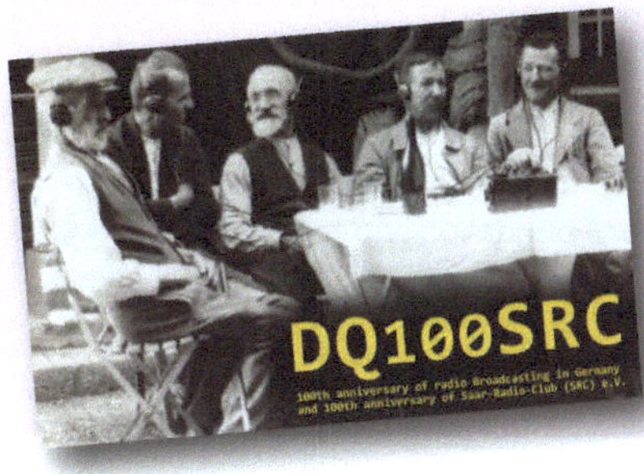

M6 - M7 Series

Mrs. A. Eastwood M7ERT
3 Bowes Nook,
Buttershaw, Bradford. BD6 2BJ
West Yorkshire
M1-M7QSL@hotmail.com
groups.yahoo.com/group/M6_QSL/

MM Series

Mr. R. Roberts. MM0CPZ
16, Swanston Avenue,
Edinburgh
EH10 7BX
Email ; mmand2m@yahoo.com

MW Series

Mr G Coomber, G0NBI
3, Dolly Grove, Blackdown Heights,
Crimchard. Somerset. TA20 1PF
grahamg0nbi@gmail.com

RS Receiving stations

Mr R Small, RS8841
13 Rydall Close, Stowmarket,
Suffolk IP14 1QX
rob@g3ali.co.uk

Special UK Prefixes - NoV Call Holders

R. Royal Weddings and Coronation
Q. Queen's Jubilee.
O. Olympic Games. **V**. RSGB Centenary.

Mr J. Peden, G3ZQQ
51A Bewdley Road, Kidderminster.
Worcestershire DY11 6RL
g3zqq@yahoo.co.uk

Note 1. The sub group for 2 letter suffix G Callsigns, has closed. All 2 letter calls are now sorted and distributed with 3 letter calls.

Ofcom licence review-

how the RSGB has supported the amateur radio community through the Consultation process and beyond

Early 2024 saw a landmark set of changes to the UK amateur radio licence. These changes offer new opportunities both for individual radio amateurs and for sharing amateur radio with those who don't yet have a licence. Are you aware of all the changes? Are you making the most of them?

The Ofcom Consultation

The Ofcom Plan of Work for 2023/24 included a major review of amateur licensing and related call sign policy. The Ofcom Consultation was launched on 23 June 2023, with replies required by 4 September 2023.

The RSGB's response

When the Consultation was announced, the Society acted immediately to ensure that all UK licensees were aware of the news. We undertook a number of major tasks, including analysing the Consultation document to see exactly what was being proposed. We also had to judge how it would affect our activities. Whilst the RSGB Spectrum Forum members did most of the work, we also had many discussions with other groups across the Society. The feedback we gathered was collated into the Society's full response to Ofcom.

Supporting the amateur radio community to respond

Ofcom had said it was keen to receive responses from individual radio amateurs as well as groups and, of course, the RSGB as the national society. To help individual radio amateurs understand the changes and to respond to the Consultation, the Society created a communications plan that included web resources, videos and regular updates.

We released a video in which RSGB Spectrum and licensing expert Murray Niman, G6JYB, explained more about the Consultation. Over 5,000 people have watched this, and it received great comments such as, *"A very clear and concise summary, with well-made points and a clear statement of what we all need to do. Very helpful."*

Ofcom Amateur Radio Consultation:
Updating the amateur radio licensing framework

RSGB Spectrum Chair
Murray Niman, G6JYB

Ofcom Consultation - video guidance by RSGB Spectrum Forum Chair Murray Niman, G6JYB

We also organised two special live *Tonight@8* events where, instead of the usual webinar-style, the events were forums with a panel of RSGB experts to answer questions shared on the YouTube live chat.

On Monday 31 July 2023, we looked at contests, operating and call sign policy, of particular interest to Full licence holders. The RSGB expert panel was: General Manager Steve Thomas, M1ACB; Spectrum Chair Murray Niman, G6JYB; Board Chair Stewart Bryant, G3YSX; and President John McCullagh, GI4BWM.

The focus of the second event in early August was on Foundation and Intermediate topics including call sign and exam changes, as well as the Consultation clauses that might lead to new opportunities for outreach activities for everyone. This event was also for people who aren't yet licensed but are thinking of taking an amateur radio exam in the next couple of years. This time the RSGB expert panel was: General Manager Steve Thomas, M1ACB; Board Chair Stewart Bryant, G3YSX; Spectrum Chair Murray Niman, G6JYB; and Examinations Standards Committee Chair Tony Kent, G8PBH.

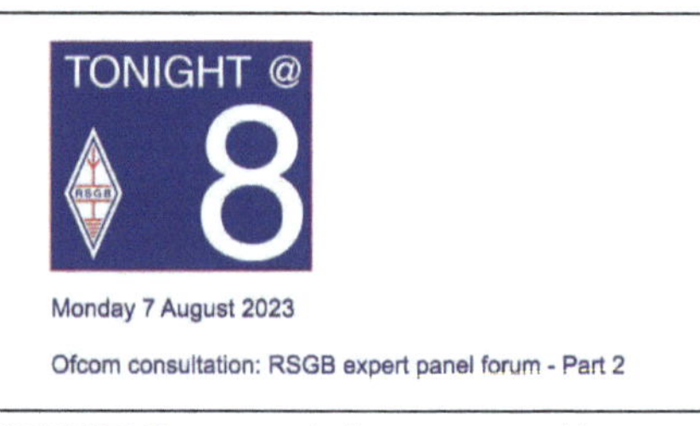

RSGB T@8 Ofcom consultation - expert panel forum

These special events have had over 5,000 views with a large number of very positive comments about how helpful they had been.

In October, Murray Niman also gave a presentation at the RSGB 2023 Convention on the Ofcom Consultation as well as looking ahead to other changes in prospect across HF to microwaves from IARU and the ITU World Radio Conference in the autumn. He outlined the RSGB's response to the Consultation and took questions about the next steps. We released this presentation onto our YouTube channel where it has had a further 1,000 views.

The December announcement

Ofcom issued a statement in December with the results of the Consultation, alongside a Summary of the Changes which gave a high-level overview of the policy changes Ofcom made to the amateur radio licensing framework. Also in December, it published its General Notice, giving notice of its proposal to vary all amateur radio licences, which had a deadline for representations of 22 January 2024. This was accompanied by the first draft of a completely new licence element – a Coordination Notice document. The RSGB submitted another formal response at this stage.

In response to Ofcom's statement, in January the RSGB released a video in which RSGB General Manager Steve Thomas, M1ACB chatted to RSGB Spectrum Forum Chair Murray Niman, G6JYB to recap some of the headline results of the Ofcom Consultation, including where changes had occurred to Ofcom's proposals. They also talked about what this could mean for radio amateurs in the future. This was a great example of how the RSGB is working not only to defend the spectrum, but also to support all radio amateurs to understand and make use of the privileges they enjoy through their amateur radio licence.

One comment on YouTube following the release of the video said,

"These critical negotiations and the supporting technical work and presentations are a great reason to be a member of the RSGB, supporting these long-term efforts and indirectly, those of the IARU."

Following receipt of all representations, Ofcom made a handful, but in some cases significant, changes as it prepared to finalise its proposals.

The licensing update

On Wednesday 21 February 2024 Ofcom released a document that set out its "decision to update the amateur radio licensing framework to ensure the policies and licences meet the needs of today and tomorrow's radio amateurs, while streamlining the licensing process". The new licence conditions took effect from that date. Ofcom also released an updated Coordinated document and new guidance to accompany the revised terms.

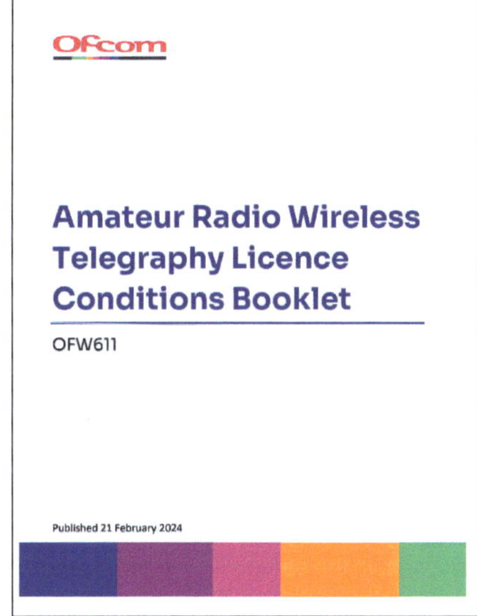

One of the key changes resulting from the Consultation is a new document, 'Notice of Coordination & Procedures', that includes additional licence conditions and replaces the former Schedule-2 for some bands (such as 70cm and 23cm), to ensure the protection of other radio users from Amateur Radio transmissions.

Next steps

Ofcom is introducing the changes in three stages with the revised terms already in effect. Later stages, mainly associated with call sign changes and Special Event Stations, are due later in 2024 and a further set in 2025.

You can see these stages summarised in the Ofcom document "Updating the Amateur Radio licensing framework: overview of key changes. What you need to know"and also outlined in Ofcom's video summary on its YouTube channel. The RSGB also has updated licensing advice on its website and YouTube.

Jack, aged 14, making a QSO under the supervision of a MADARS club member

Making the most of the licence changes

The day after the licence changes came into effect, the volunteers at the RSGB National Radio Centre were putting them into practice by encouraging some visitors, under supervision, to make QSOs on GB3RS. The NRC volunteers have continued to do this as part of their aim to demonstrate amateur radio in the most interesting and effective ways. Not only do they encourage visitors to make a QSO on the HF bands, but also on VHF, UHF, digital modes, and even satellites like the very popular QO-100.

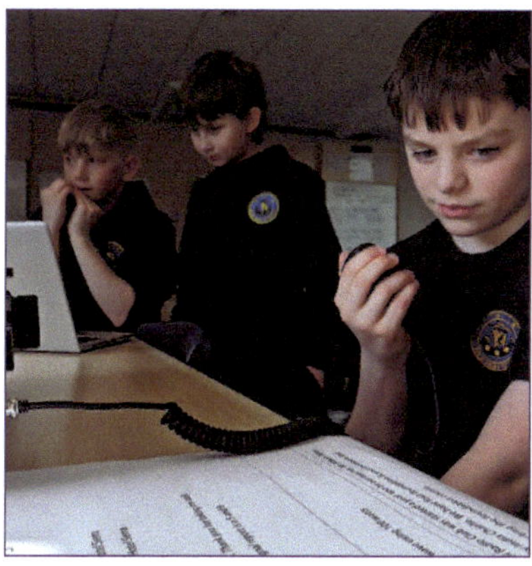

The photo shows some of the Wick High School Radio Club pupils making contacts.

It is great to see local clubs using the licence changes to showcase the hobby to new people. Mexborough and District Amateur Radio Society (MADARS) held an open event at its local library and many people enjoyed finding out more about amateur radio. The photo shows Jack, aged 14, making a QSO under the supervision of club member.

Schools are also starting to take advantage of the new opportunities. Wick High School Radio Club is currently preparing a new cohort of students for their Foundation Licence exam and, as part of their training, teacher Chris Aitken, MM0WIC got them on the air for the first time. They even had a pile up of 26 QSOs!

Need to refresh your memory?

We created a summary video of the licence changes which is an extended version of the one shown at the RSGB AGM in April. It covers the range of updates from power increases and call signs, to additional options for Foundation licensees and new opportunities for sharing amateur radio with people who don't have a licence. It also looks at the next phases of Ofcom's licence changes. If you don't feel you're making the most of the new licence conditions, or you're unsure about some of the changes, why not watch this short video and see what you could do!

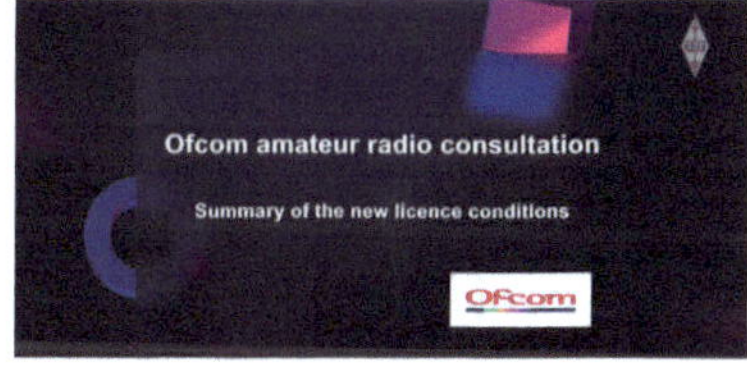

You can still access all the resources we've created including documents and videos on our special licence review web page at rsgb.org/licence-review or watch the videos and presentations on the RSGB YouTube channel in the "Ofcom licence changes" playlist.

Combined with the summary video we released in the summer of 2024, the video and webinar presentations we created to support the amateur radio community have received over 15,600 views.

Final word from the RSGB President

As RSGB President John McCullagh, GI4BWM said in his review at the AGM, "The new licence is a huge change, liberalising many aspects of the rules we work under and enabling us to connect with people in new ways. We should take every opportunity to do that and to grow amateur radio."

Heather Parsons
RSGB Communications Manager
comms@rsgb.org.uk

Morse Tuition, Practice and Assessment GB2CW

Whilst a Morse qualification is not needed by the present day licence, amateurs are realising that they are missing out on a lot of fun and DX by not using Morse and limiting themselves to part of the total amateur bandwidth available.

Introduction

Since the RSGB introduced the Certificate of Competency, there have been a number awarded. Some have been publicised in the RadCom, so the scheme is enjoying a considerable degree of success. Making the Certificate an 'award' rather than something mandatory in order to gain a licence has encouraged a lot of newly licensed amateurs to take up the challenge and achieve a skill that will enable them to work more DX, plus pass on their achieved skills to others. It can be compared to chasing DX certificates where effort is rewarded for the time spent in achieving the necessary qualifications.

The Morse certificate is no different - some will wish to obtain it and some won't. In the same vein, some clubs will embrace the idea and others will not. However, the skill will stay with you for the rest of your life.

International recognition is not an aspiration of the initial scheme, but if that develops later in a non-contentious manner it could be a clear additional benefit for some (provided that it is not abused to gain a higher class of licence overseas than is held here in the UK).

A Certificate of Proficiency for students when they pass a Morse test, this is not a legal licensing requirement, but does provide stimulus for further improvement and gives the student something to be proud of which can be displayed in theshack.

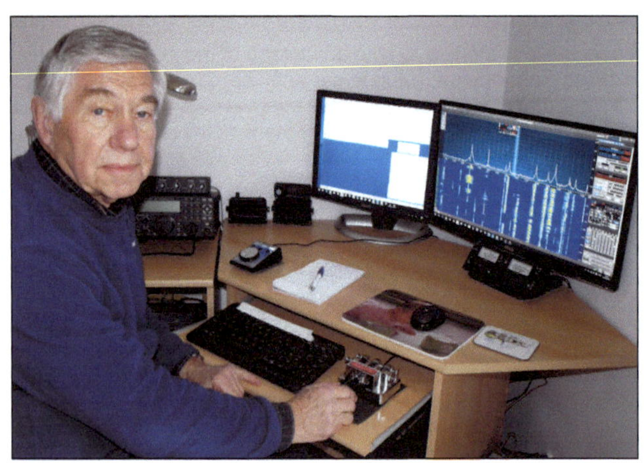

GB2CW volunteer, Malcolm Prestwood, G3PDH

The initial assessment will require the candidate to receive and send text, including some punctuation, for 3 minutes with no more than three uncorrected errors and will also include some figure groups (receiving and sending). The lowest speed for which the Certificate will be issued is 5WPM, although most people don't bother to start that low. Success in this will merit issuing the Certificate of Proficiency, after which endorsements (or a new certificate) may be obtained at 12, 15, 20, 25, 30WPM, etc. The test will be conducted with an Assessor only; an Adjudicator is no longer required. All assessments may be taken using equipment chosen by the candidate and appropriate for the speed being examined, including straight keys, paddles, bugs and semi automatic keys. Reception can be either on paper using a pen/pencil or using a computer with a suitable program such as Notepad.

To obtain an endorsement or a new certificate at higher speed, further assessments will also include receiving and sending proficiency in a basic rubber-stamp type QSO. This applies also to those wishing to take their initial assess ment at a speed higher than 5WPM.

Having passed the test, the Certificate will be mailed to the student as a PDF and he can print it out himself.

GB2CW Schedule

HF Transmissions

Monday	20.15	3.555	G4BSW	Kent (Margate)
Tuesday	20.00	3.555	GW0KZW	Prestatyn
Thursday	09.00	3.605	G3UKV	Telford
Friday	20.00	3.563	GW0KZW	Prestatyn
Sunday	19.00	3.555	GM5AUG	Glasgow

VHF Transmissions

Monday	10.00	145.250	Headcopy Class G3LDI	
	18.30	145.250	M0APY	Leeds
	20.00	144.600	GM5AUG	Glasgow
Tuesday	09.00	145.250	G4OOC	Pontefract
***	18.00	145.250	M0HAZ	Skegness
	18.30	145.250	M0APY	Leeds
	19.00	145.250	G5FM	Glastonbury
	19.00	145.250	MI0WWB	Newtownards
	19.00	145.250 Voice		
		145.250 CW	M0KZB	Shrewsbury
	19.00	145.250	M0PNN	Newport
	20.00	GB3NB	G3YLA	Norwich
Wednesday	09.00	145.250	G4OOC	Pontefract
	19:00	145.250	G0VCW	Lowestoft
Thursday	09.00	145.250	G4OOC	Pontefract
	11.30	145.250	MX0NCA	East Runton
	18.00	145.250	M0HAZ	Skegness
	18.30	145.250	M0APY	Leeds
	19.00	145.250	GM0EDJ	Johnstone
	19.00	145.250	G3XNE	Bude
	19.15	145.250	GM0UOU/ GM0EDJ	Elderslie, Renfrewshire****
19.30	145.250	G0TDJ	Crayford (Kent)	
	20.00	GB3NB	G3PDH	Norwich area (Advanced)
	20.00	144.600	GM5AUG	Glasgow
Friday	18.30	145.250	M0APY	Leeds
Saturday	08.00	145.250	G4OOC	Pontefract
Sunday	19:00	145.250	G0VCW	Lowestoft
	20.00	145.250 Voice		
		145.250 CW	G4PVB	St Albans

Modes: A1A/J3E 144.2508 and all HF transmissions
F2A/F3E All VHF transmissions

*** Excluding first Tuesday in each month

**** (The EARS club call sign MM0PYR will be used for the broadcast)

Regardless of speed, a requirement of every assessment taken will be that all sending will be pre-recorded, to guarantee the speed and ensure integrity of the assessment process. All assessments may be taken using equipment chosen by the candidate and appropriate for the speed being examined, including straight keys, paddles, bugs and semi automatic keys.

Tests over Skype

This will be via an online sound and video program such as Skype or similar

Training

The Society is not prescriptive about the method of training used to achieve the Certificate. There are numerous methods of learning Morse code and it is a personal choice as to the method used. Instructors and students will have preferences and individual teaching and learning styles. No written rules are made, but once the code has been learned it is absolutely necessary for candidates to practice regularly. This may be done in a group, such as at a club, by listening to Morse on the bands or by using one of the numerous computer programs available.

The student can supplement individual or group training at clubs; and (ideally) be further supported by the use of an active and well promoted GB2CW broadcast schedule. Regular attendance to a weekly tutorial on the air using GB2CW in an interactive way is extremely beneficial. 2m FM is preferred to achieve this activity and it is normally a lot of fun, especially with mutual competition with other students in the same class.

There is more comprehensive information which adds to and builds on this in the RSGB book Morse Code for Radio Amateurs, by Roger Cooke, G3LDI.

Assessments

When a candidate is ready to be tested they should complete the online application form on the RSGB Website. This can be found in the Operating Section by selecting “Morse”

The application form includes the option for a candidate to express a preference for a face-to-face or online test. It also enables them to choose their preferred Assessor and location. On receipt of the application, the Morse Test Coordinator will contact the applicant’s chosen Assessor and e-mail their details to the Assessor, The Assessor will then contact the applicant to make mutually agreeable arrangements for the Test to take place.

Following a successful test, the Assessor will contact the Morse Test Coordinator who will e-mail a Certificate of Proficiency to the applicant. Full details of the testing scheme can be found on the RSGB Website. The address for any enquiries is Eric Arkinstall M0KZB *morse.tests@rsgb.org.uk*

Consistency and integrity

Consistency, integrity and development of the scheme is monitored by a joint committee of the Regional Team and Amateur Radio Development Committee. The terms of reference for the ARDC encompass training and testing, and ensuring that the scheme is conducted in a thoroughly professional and competent manner. This is intended to guarantee that the Certificate is both desirable and that its reputation is respected both in the UK and internationally.

It may also be desirable that the ARDC will in due time extend its focus to further encourage the use of Morse

Learning Morse

Unlike the Foundation licence, where a course of a few hours learning will probably produce a pass for the candidate, learning Morse Code and becoming a proficient operator is akin to learning a musical instrument. Attending a class once a week will not produce results. Occasional listening on the air is also a waste of time. The student has to be motivated and disciplined. Learning a musical instrument requires constant practice, and that does not mean just ten minutes a day. If you aspire to become a top-notch CW operator, consider at least one or two hours per day practice, EVERY day, not just once a week. This must carry on for a few months to reach acceptable speeds. If you cannot meet those requirements, then Morse is not for you. This cannot be stressed enough. The results you obtain will be well worth that effort.

There is a new way of encouraging learning. In East Anglia, the Norfolk ARC started running two Bootcamps a year, and have around 15 students attended and it is normally a whole day event, running it in three classes and some on-air activity as well. Both sending and receiving are catered for with a range of keys, from straight to the latest in technology, the 9A5N paddle, and Begali paddles too. Essex CW Club copied this and was extremely successful such that they had to hire a village hall! Lots of fun can be had and XYLs usually provide a range of cakes and so on, with lots of tea and coffee. Try it in your Club if you can and it might catch on and provide incentive for more learning and practice, The latest one in Norfolk can be seen in the picture.

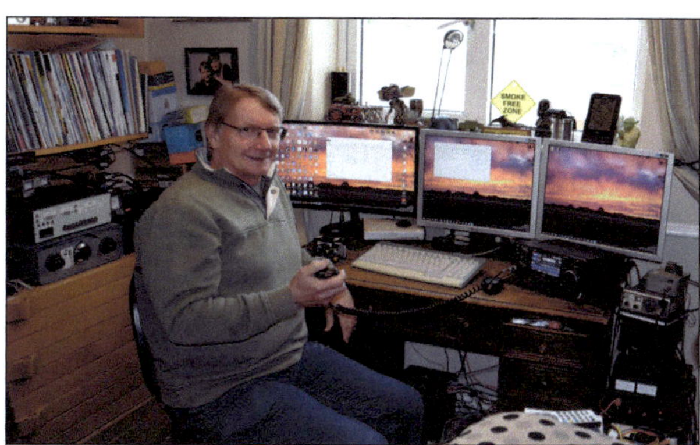

Phil G4LPP is shown at his station.

Volunteers for the GB2CW scheme

Norwich now has six classes running each week. Roger G3LDI runs a headcopy class on Monday mornings at 1000. This is a rapid fire decoding of a mixture of sending, including QSO format, cut numbers, EISH and 5 groups. The other classes cater for varying speeds, from raw beginners to those up to 30 wpm. One of my students has done so well that he is now a tutor running a beginner’s class. Phil G4LPP is shown at his station.

Two of the classes are on GB3NB and the others on 145.250MHz. Other tutors locally are Malcolm G3PDH and Jim G3YLA.

We were lucky to have a few new volunteers join in the past year including Roger MI0WWB, our first one from Northern Ireland. If you feel like becoming a volunteer yourself you would be most welcome. There can never be enough volunteers and using the GB2CW scheme it is so easy to do and also a lot of fun. It really isn’t that much to ask and the rewards come to both the student and tutor when progress is made. Additional volunteers are always needed to run GB2CW broadcasts, especially in some of the more remote parts of the UK. Broadcasts can take place on several bands, ranging from 3.5MHz to 50MHz. It may only take an hour of your time per week to ensure that amateur radio continues to have a flourishing pool of CW operators to ensure the future of the mode, so please consider helping.

As interest grows, more tutors are needed. It would be very nice to see volunteer instructors in every Club in the UK, and that is what I would like to see as Coordinator. There is a long way to go to achieve anything like that but in order to maintain this quintessential mode used in amateur radio we need a lot more tutors. Remember, some Elmer taught you, so now it’s your turn to be an Elmer!

Also, there are gaps in the coverage of Assessors. Assessors are needed in Regions 8 (Northern Ireland), and 11 (South West England and the Channel Islands). If you live in one of these areas, please consider joining this most worthwhile scheme. We have been fortunate in filling a few areas in the last year or so. However, more are always needed, not only for the vacant areas, but all areas, to act as backup. Full details, including an application form, can be found on the RSGB website at:

https://rsgb.org/main/operating/morse/certificate-of-competency/the-morse-test/

Volunteers are scarce and are perceived by some to be those with super human skills and speed in excess of 30WPM. This is far from the truth and if you have a good average skill level of around 15 to 20WPM, you could take on the role of instructor to that level anyway. Computer programs are used for instruc-

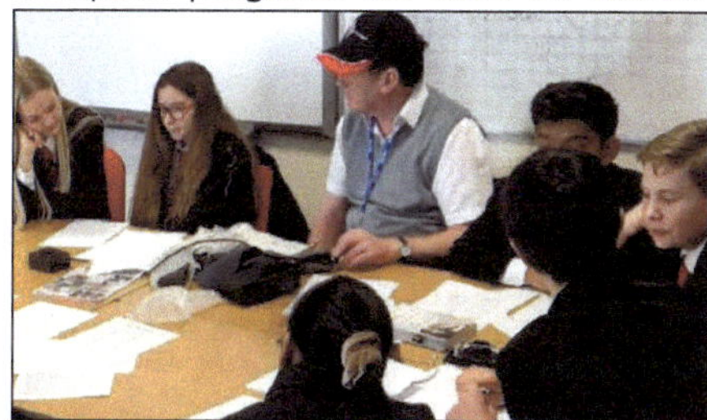
G4PVB teaching with one of the local Norfolk ARC Bootcamp, now a bi-annual event, spring and autumn. This is highly popular and normally has about 16 students attend an all-day session, all ranges from beginners to Full.

tion so therefore the Morse sent is perfectly formed so it is completely straightforward to implement on the air.

Two New Bootcamps In 2019

Bootcamps are catching on at last. Rich G4FAD held his first one in Hereford and it was a great success. This is what he had to say about it.

We held our first Hereford Morse Bootcamp with the kind permission of Geoff G8BPN at his QTH which used to be Geoff's electronic factory and it was an ideal building for our purposes and easily had room for the instructors and the 23 students. In the past every town or village had a person who used Morse code in their job and could help an aspiring CW operator. Today this is no longer true and a Bootcamp helps to fill that void.

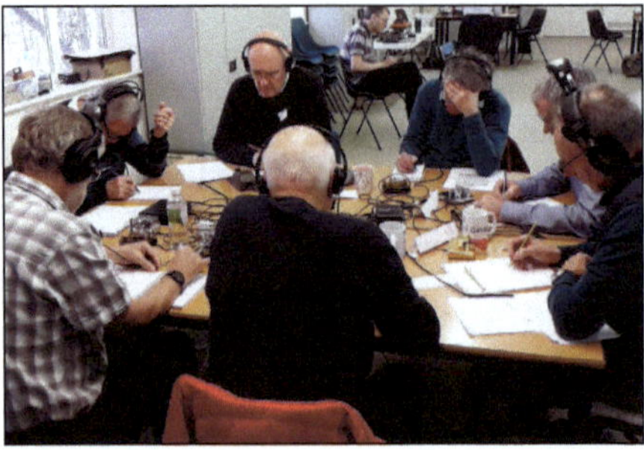

Judging by the number of people attending, CW is not a dying mode at all. This is one of the most enjoyable ways of learning Morse whilst having a good sociable day out. Try one in your club, you might be surprised.

There was also another new Bootcamp, this time in Scotland run by Gavin, GM0GAV. This too was successful with 12 students attending.

Since the Covid situation has now calmed down, it would be great to see new Bootcamps running. However, it needs some organising. It would be nice to see an increase each year. They provide an excellent training feature and also some stimulus to those attending to do even more to become proficient operators. The Al Slater G3FXB Memorial Award was presented to three Bootcamps for their contribution to CW. A silver salver was sent to me on behalf of the Norwich Club as we started Bootcamps nearly fifteen years ago now.

To offer your assistance, please contact the scheme's co-ordinator (details below).

The Morse Test

Taking the test is a simple three-stage process

Step 1 – When a candidate is ready to be tested, they should click the button below to complete the application form.

The application form includes the option for a candidate to express a preference for either a face-to-face or an online test.

Step 2 – The Morse Test Coordinator will contact an Approved Assessor and email contact details to the candidate who can then get in touch with the assessor and arrange the test.

Step 3 – Following a successful test, the assessor will notify the Morse Test Coordinator who will email a Certificate of Competency to the candidate.

Test structure

The lowest speed for which the Certificate is issued is 5wpm.

This has been chosen as the threshold to encourage learning the Code, to provide early confidence and to encourage moving forward to higher speeds.

It is specifically designed for those who are most comfortable with this rate of progress.

There is, however, no barrier to those who wish to enter the scheme at a higher speed.

Tests are available for 10, 12, 15, 20, 25 and 30 wpm.

The Certificate of Competency Morse test will:

- Require the receiving and sending of plain language*, for three minutes with no more than four uncorrected errors and;
- Require the receiving and sending of numbers, in five figure groups, for one minute with no more than three uncorrected errors.

*The plain language section of the test will include some numbers as well as the more commonly used punctuation marks such as oblique stroke and question mark.

Success in this test will merit issuing the Certificate of Competency, after which further certificates can be obtained for a successful test at a higher speed.

A requirement of every test taken—regardless of speed—will be that all sending by the assessor must be prerecorded, or generated by suitable computer software, application or keyer, at guaranteed speeds to ensure integrity of the testing process and thus prevent appeals in this regard.

Radio Society of Great Britain
Advancing amateur radio since 1913
Radio Society of Great Britain
Morse Proficiency Programme
This is to certify that
passed a Morse Code Proficiency Test on
sending and receiving at a speed of __wpm

Sending tests may be taken using equipment chosen by the candidate and appropriate for the speed being examined including straight keys, paddles, bugs and semi-automatic keys, etc.

Keyboard-generated Morse is not acceptable for sending tests.

The process of taking the test online is as follows:

- This will be via an online sound and video program such as Skype or similar
- The candidate and assessor must ensure that the video cameras are sited so that the assessor and candidate have a clear view of each other—in particular the assessor must be able to watch the candidate keying during the sending part of the test
- In the event of a failure of video during the test, it may continue using sound only, at the discretion of the assessor
- On completion of a successful test, the assessor will email confirmation to the Morse Test Coordinator who will email a Certificate of Competency to the candidate

It should be noted that The RSGB will not pay any expenses incurred by the candidate in connection with undertaking a Morse test.

If any of this causes any difficulties, please contact the Morse Test Coordinator via: *morse.tests@rsgb.org.uk*

Co-ordinator email: *morse.tests@rsgb.org.uk*
Latest GB2CW broadcast schedule:
www.rsgb.org/main/operating/morse/certificate-of-competency/gb2cw-broadcast-schedule/

Featured Clubs

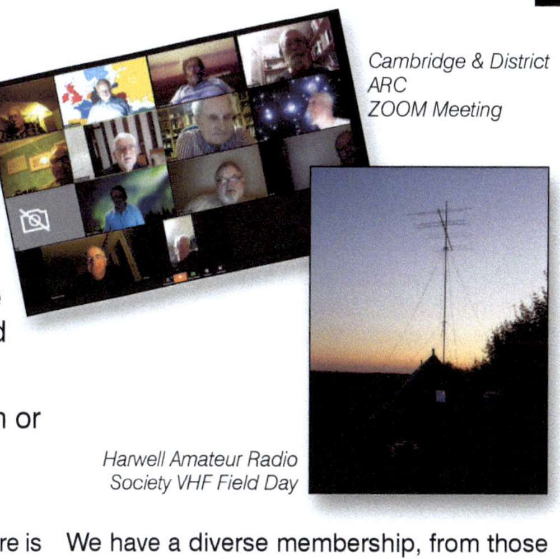

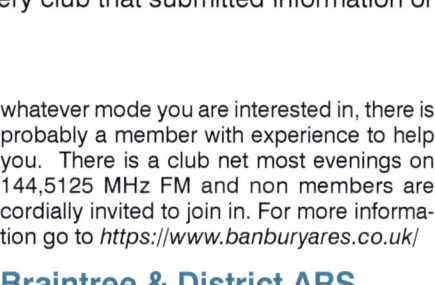

Cambridge & District ARC ZOOM Meeting

Harwell Amateur Radio Society VHF Field Day

Throughout the UK there are over 500 local clubs and societies affiliated to the RSGB. They vary from small local clubs, through repeater groups & online clubs to the large contest groups. Below are the details of some of these clubs, telling you a little about what they do for amateur radio and their members, and what they plan for 2023/24.

We are sorry that we cannot include every club that submitted information or every photo from those that did.

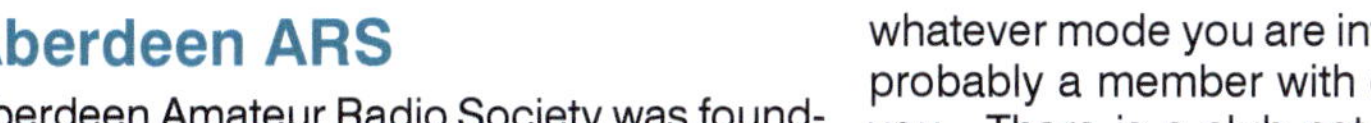

Aberdeen ARS

Aberdeen Amateur Radio Society was founded in 1946 and has brought together amateur radio enthusiasts from across the north-east of Scotland ever since. Meetings are held in Aberdeen on most Thursday evenings throughout the year and a warm welcome awaits any prospective new members who are interested in joining us. As the society approaches its 80th year thoughts are already looking towards how the club might celebrate this milestone.

During 2025 the International Tall Ships Race will be coming to Aberdeen, its only UK destination. Plans are underway for the AARS to celebrate this unique event by arranging a Special Event with an appropriate call sign. It is hoped this will attract some interesting QSO's both DX and nearer home. A special eQSL card is also being designed. It is intended that several bands will be used utilising SSB, CW and, hopefully, data modes.

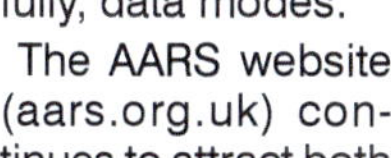

The AARS website (aars.org.uk) continues to attract both local and international interest and statistics indicate that our equipment sales and programme of events pages are of particular interest. The clubs archive pages and QSL card pages allow members past and present to recall events and members through our 'Mystery Photographs' pages.

Full details at; *www.aars.org.uk*

Banbury ARS

The Banbury Amateur Radio Society meet on the first four Wednesday evenings of each month, in the clubhouse located in the garden of 169 Bloxham Rd., Banbury. We welcome new members and encourage those curious about amateur radio as well as licensed amateurs, to have a look at what is on offer. The club is active in many areas including offering talks, demonstrations, operating in the clubhouse with up to date equipment and attending events such as Mills on the Air. We also offer training courses for all levels of licence and examinations are taken in the clubhouse. Members interests include data modes, morse, digital, repeaters, HF so whatever mode you are interested in, there is probably a member with experience to help you. There is a club net most evenings on 144,5125 MHz FM and non members are cordially invited to join in. For more information go to *https://www.banburyares.co.uk/*

Braintree & District ARS

BADARS is proud to celebrate its 50th anniversary in 2025. Located close to the boundary of North Essex and South Suffolk, as well as regular meetings, both over the air and in person, we participate in several special events throughout the year, including Mills on the Air and JOTA/JOTI.

We have an annual summer camp weekend in one of our members fields, which is very popular with members and visitors. Other regular events include a quiz night, barbecue evening, construction competition, rig testing evening, aerial testing evening and not forgetting the annual surplus equipment sale.

As part of the 50 years celebrations in 2025 we will be sharing club memories and achievements in our monthly club magazine, BARSCOM, which itself will be 45 years old in its current form. Our large scrapbook that has been maintained through the years will incorporate this BARSCOM edition as well as other historical material.

Building new foundations for the next 50 years will include a significant refresh of our website www.badars.co.uk expanding it to include more on our history and future plans. We will also organise club visits to radio related venues and potentially hold more on-air evenings.

All of our meetings, via radio and in person, are available via Zoom, which has allowed members who cannot attend in person or over the air, to still be a part of the evening - including one member who was mid-Atlantic on a cruise ship for one meeting!

We have a very broad range of experience and interests within our membership and whatever your focus is in amateur radio, we will have something to interest you and you will have like-minded members to talk to. We run a local dual mode repeater, GB3BZ which our members regularly monitor, so if you are travelling our way please call in - you will be made most welcome.

Bury Radio Society

Formed in 1938, Bury Radio Society is one of the oldest amateur radio clubs in the UK. We have a diverse membership, from those who are just starting out in amateur radio to those who have decades of practical experience. Our members include professional radio operators from the merchant navy. Other members have extensive experience of electronics, for example RF and electronics service engineers. This wealth of experience means that we are extremely well placed to answer technical questions, provide advice about antennas and equipment and mentor newly qualified licensees helping them get on the air for the first time.

We are always delighted to help new members who want to get started in our hobby, as well as those who wish to progress to the next licensing level. We provide guidance and support for the Foundation, Intermediate and Full licence exams. Bury Radio Society is an accredited member of the RSGB's Brickworks scheme, meaning that we arrange events and activities in a supportive and friendly environment throughout the year to help newly licensed members complete their Brickworks activities.

In addition to general operating, our members' interests include many specialist areas, including construction, ROTA (Railways on the air), CHOTA (Churches & Chapels on the air), RAYNET (emergency communications), data modes, DMR (digital voice) and many more facets of our hobby.

Bury Radio Society is an active club where members enjoy a full calendar of events throughout the year. These include talks and presentations, operating nights, exam and licensing support, social nights, day trips and operating 'portable' (when the weather permits!).

In 2025, our events will include Churches on the Air, Parks on the Air, Railways on the Air, Jamboree on the Air, a DF (direction finding) hunt and several field days from local high spots.

New members will always receive a warm welcome at our Society. If you would like to

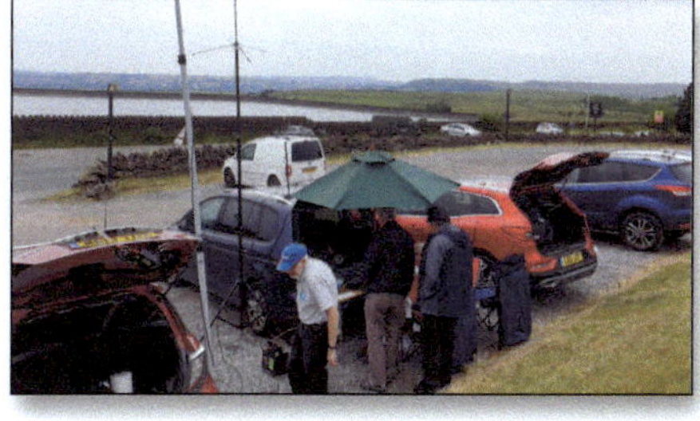

visit us, join us, or find out more about what we do please get in touch.

We look forward to meeting you at the Bury Radio Society.

Cambridge and District ARC

For over one hundred years Cambridge and CDARC have provided a focus for radio activities in the city. We meet both in-person and online every two weeks for talks and presentations, equipment demonstrations and social activities. Many club members also participate in weekly 2m, 80m and DMR on-air nets. Visitors are welcome at all our events.

We are very active in training and hold well-attended Foundation and Intermediate courses a couple of times each year. Mentoring is available for those studying for the advanced exam as is a buddy scheme for the newly licenced.

The club also provides internet connected remote transceiver/receivers at an electrically quiet location which can be used by our members who are troubled with noise issues or when away from their own shacks.

CDARC is active in contesting and field days, and we participate in various events such as Museums and Mills on the Air.

Our regular meetings are on the 2nd and 4th Friday evenings of each month at 7:30pm at Coleridge Community College, Radegund Road, Cambridge CB1 3RJ and online via Zoom. Anyone with an interest in amateur radio is welcome at our meetings, so please come along or contact us for further information.

Find us on Facebook or at our website *www.cdarc.org.uk.*

Email contacts are: *publicity@cdarc.org.uk* & *training@cdarc.org.uk*

Carmarthen ARS

Carmarthen ARS (CARS) is a medium sized club in Zone 7 (South Wales) which is run in accordance with its constitution. Its members go out of their way to provide a friendly and helpful atmosphere at their two club nights each month. We provide training at all three UK licence levels and partake in all sorts of events ranging from field days, to contests, to rallies, to BBQ's or even 10-pin bowling socials.

The club meet from 1830 (6:30pm) on the 1st and 3rd Tuesday of every month at the Cwmduad Community Centre, Cwmduad, Carmarthenshire, SA33 6XN (which is on the A484) and is located just north of Carmarthen at 51° 57' 20.96" N, 4° 21' 56.13" W for those with satnavs.

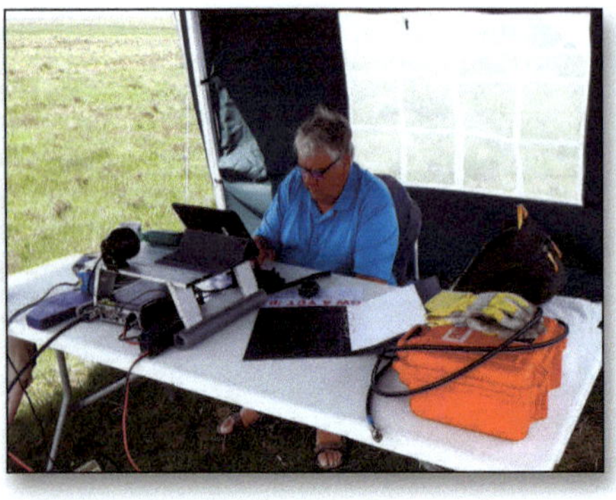

The 1st Tuesday is usually a social meeting and 'on air' radio night, whilst the 3rd Tuesday tends to be an activity night, either a talk, or another form of group activity. We have a 'show and tell' running every club night where members can bring along new items of kit they have acquired to show and discuss with other members.

CARS run training for all three UK licence levels, but not on club nights, preferring to run these on other days to avoid distractions for the students. We are an exam centre for all three licences.

The club call (GW4YCT), in use on club nights, by accessing the HF, VHF and UHF bands using our radio equipment and antennas.

Visitors and potential new members are made most welcome. Light refreshments are usually available.

Often prospective members visiting have questions which we try to answer as fully as possible by seeking out the club member with the most experience in the relevant aspect of the hobby. Answers are often accompanied by offers of practical help.

Anybody wishing to visit or join should contact Andy, GW0JLX on 07768 282880 or by email to carsmembershipsec@gmail.com

We hold a number of regular nets each week to allow members to maintain good contact with each other and experiment with different modes. Non-members are welcome to join in.

Our routine weekly nets comprise:

A multimodal digital net (currently D-STAR and YSF) on non-meeting Tuesdays using reflector XLX490D from 20:00 (8:00pm);

A 70cm FM net every Wednesday via GB3CM from 1930 (7:30pm);

A 2m FM net every Thursday via GB3FG from 20:00 (8:00pm);

An 80m SSB net from 14:30 (2:30pm) on Sunday afternoons.

The use of repeaters and NVIS techniques on HF are useful as our membership is spread over a wide geographic area that is challenging for 'normal' radio communications. Our member-maintained repeaters are moderately busy with traffic from a mix of both mobile, portable and fixed stations either ad-hoc, or on the scheduled nets as above.

We participate in contests such as the Practical Wireless 2 and 4 metre contests, 23cm contests, RSGB contests etc. We support some of the 'field day' events such as JOTA, CHOTA, SOTA and "Exercise Blue Ham".

The GB3RPE 10 GHz beacon for the Carmarthen area is operational on 10368.910 MHz.

Several of our members belong to RAYNET, supporting either the Pembrokeshire or the Carmarthenshire Raynet groups and their respective activities.

We all enjoy what we do for our hobby and invest a lot of time and effort into the club to make it work and meet members requirements. Come along and join in.

Anybody wishing to visit or join should check initially with Andy, GW0JLX on 07768 282880 or by email to *carsmembershipsec@gmail.com*

Cornish Radio Amateur Club

The club has been running successfully for over 80 years. The club is rich in history and well established in the county of Cornwall. Complemented often for the special event stations that we run and the training that we offer. We are very proud of our past, very few clubs have been running as long. The Cornish Radio Amateur Club meets on the first Thursday of the month and membership is open to anyone with an interest in radio communications, computing or electronics. We have members young and old all enjoying various aspects of the hobby. These range from traditional communications methods such as CW and voice to the more modern-day digital modes.

We try to put on as many special event stations as we can to promote the hobby and like to make our club meetings truly sociable for everyone. We have taken part in SOS week, Railways On The Air, Museums On The Air and JOTA. When possible, we try to attend local public events. In the past we have attended the West of England Steam Engine Rally, the Cornwall Volkswagen Owners Club Jamboree and the Cornwall Air Ambulance Trust event using this opportunity to promote the fact that Cornwall had the first Air Ambulance in the UK by using the call sign GB1AAT

Our biggest and proudest event that we run is International Marconi Day. In 1988 Norman Pascoe G4USB came up with the idea of running an event to celebrate the work and birthday of Guglielmo Marconi. He ran the idea past a close friend Monty G4ZKH, got a small team together and this was the beginning of International Marconi Day as we know it today. This is held on the weekend closest to Marconi's birthday, 25th April. Norman organised this worldwide event every year until he sadly became a silent key in 2016. Since then Steve G7VOH (G4CRC Chairman) has taken over the role and coordinates the event each year. In 2024 we had 69 stations registered with us from all over the world, which made this a truly major event in the radio amateur's calendar. As the organisers of the event we also run our own station GB4IMD.

We are proud to run the Cornish Radio Amateur Club's rally once a year at Penair School in Truro. Whilst rallies nowadays seem to be getting smaller we continue to have a good turnout. This is a great opportunity for all of the local clubs and groups to come together to create a day where we promote our hobby as well as catching up with friends old and new.

We offer training for Foundation, Intermediate and Full examinations. Most of our training is done via Zoom but we also offer in person training if required. We have a large selection of training videos for all levels on our YouTube channel. We have had fantastic feedback from people up and down the country who have successfully passed their exams using them. These are available for anyone to use free of charge and are updated as the syllabus changes.

To find out more about our club please visit our website www.gx4crc.com

Steve G7VOH Chairman Cornish Radio Amateur Club

Crawley ARC

Crawley ARC has about 70 members and benefits from having sole use of its own premises offering a meeting room, tea bar, modern lecture facilities and well-equipped contest standard radio shack with tower, beams and wire antennas.

Club members, some of whom are professional engineers, bring their combined technical expertise in RF engineering, electronics design & construction, antenna design, microwave engineering and electronic servicing to the membership. This experience is invaluable to members starting in the hobby.

The Ashdown Forest Repeater Group is an associated organisation well supported by our members who provide a wealth of experience with design updates and ongoing maintenance for both the D-Star/DMR/Fusion and FM repeaters. Call signs GB7MH and GB3MH respectively are co-located in the nearby village of Turners Hill and each have internet linking capability.

CARC hosts one of the four annual Microwave Round Tables bringing together microwave enthusiasts from a wide area with lectures and sale of components and general items of surplus equipment.

The Club's biggest outside event of every year is participation in the VHF National Field Day Contest in association with another local club, which involves the establishment of multiple stations to cover all bands and modes.

We also support Museums on the Air, Mills on the Air and Jamboree on the air.

In support of these activities, twice weekly meetings in our own premises cater for operating the club station, socialising, contesting, and monthly interesting talks by guest speakers.

We have a lot to offer, and we are a friendly helpful bunch! Please contact us at: *secretary@carc.org.uk* or come and visit on a Wednesday at 20:00hrs or on a Sunday at 11:00hrs local time.

Dartmoor Radio Club

Founded in 1983, the Dartmoor Radio Club welcomes anyone interested in amateur radio. Meetings are held monthly throughout the year at Yelverton on the western edge of Dartmoor. Members come from a wide area including some from Cornwall.

During the summer regular field days are held on Dartmoor using the Club's own radio equipment. During the winter members also operate indoors from Yelverton using the antenna permanently erected above the meeting room.

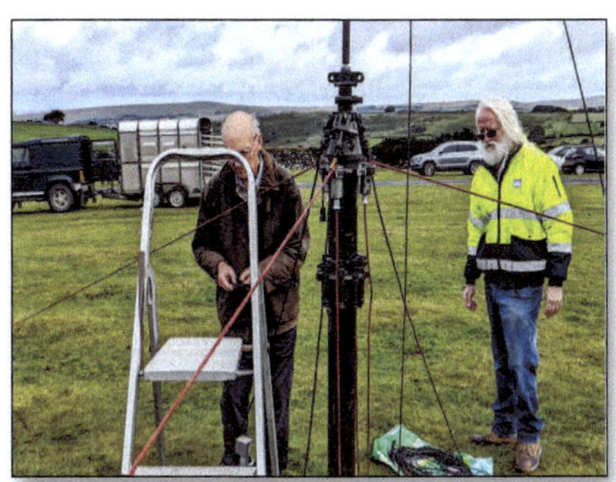

Contact details can be found on the Club's website: https://dartmoorradioclub.uk/

Dorking and District Radio Society

The Dorking and District Radio Society was formed in 1947 and offers a warm welcome to current and prospective radio amateurs in the Dorking, Leatherhead and Epsom areas of Surrey. We hold meetings monthly on the 4th Tuesday evening, normally in the Quaker's Hall, Dorking which are open to members and interested non-members. The topics vary between practical construction activities, visiting speakers and presentations by our own club members. We hold informal online social evenings over Zoom on the 3rd Wednesday of the month.

The club runs a daily net on 144.775 MHz at 10 am Monday-Saturday and around 3.772 MHz on Sundays at 8:15 am. This is a very popular way for local operators to have an open conversation on a wide variety of subjects.

We hold several in-person socials each year which attract member's families and give opportunities to go on-air as a club in rural locations such as a summer expedition to operate from close to the Devil's Dyke on the South Downs near Brighton. These trips also provide a social dining opportunity afterwards.

We have good relations with the National Trust and have a licensed arrangement to operate temporary stations from the "three peaks" of Leith Hill, Ranmore Common and Headley Heath around Dorking.

Visitors are always welcome at our meetings, which include an informal break to meet us and find out more about our activities.

Our website, *www.ddrs.org.uk* advertises upcoming events, reviews recent ones, includes articles about club activities and lists how to contact us for more information.

Durham & District Amateur Radio Society G4EUZ

Club Activities - G4EUZ, Durham and District Amateur Radio Society (DADARS) is located in the Bowburn Community Centre, Durham and welcomes new members interested in becoming licensed. The club runs weekly nets on 2m, 4m & 6m for members and where other radio amateurs can join in. Members work many DXpeditions and special event stations throughout the year and provide support to newly licenced members.

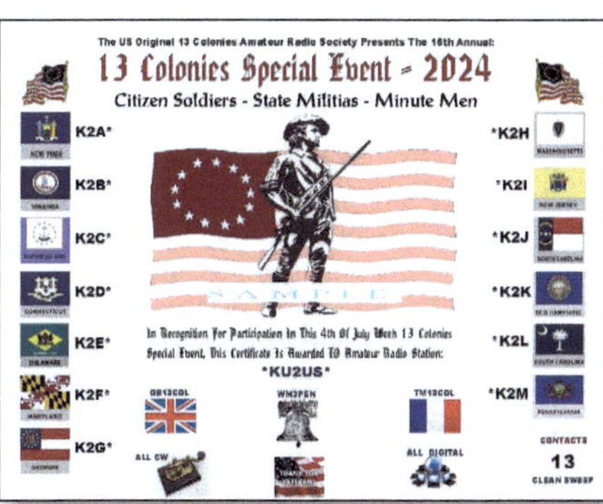

The club hosts an annual Amateur Radio Rally in May within the Community Centre, supported by a wide variety of vendors. There is excellent on-site food and beverages for visitors and access for those with disabilities is provided.

Special Event news - DADARS participates as one of the bonus stations in the hugely popular 13 Colonies Special Event. The special event call sign GB13COL runs during the 1st week of July 24/7. The primary focus of the event are the HF bands, however, VHF & UHF QSOs using SSB, CW, FM and various Digital Modes also take place.

Image 1. 2024 Certificate

The 13 Colonies event began in 2009 as a way of celebrating American independence from the original 13 colony states circa 1776. Since the UK was a major historical player in the Revolutionary War, GB13COL, from England presents an added historical significance and challenge for radio amateurs to make contact with. This event has rapidly become the premier Amateur Radio on-air activity in North America and beyond. In 2023, there were over 237,500 QSOs logged.

GB13COL 2024 QSL cards have been sponsored by Canny Components. A warm thank you to Amanda & Davey for this.

Image 2. GB13COL QSL Card-Front

Image 3. GB13COL QSL Card-Back

Ken Villone KU2US (operator of the K2A station) is the creator and manager of the event. Every year there is a different certificate theme for amateurs who make contact with either one or all of the participating stations, including the bonus stations.

Amateur radio operators taking part aim to get a "clean sweep" of all 13 stations: K2A (NY), K2B (VA), K2C (RI), K2D (CT), K2E (DE), K2F (MD), K2G (GA), K2H (MA), K2I (NJ), K2J (NC), K2K (NH), K2L (SC), K2M (PA), with WM3PEN, GB13COL & TM13COL operating as "bonus" stations. Contacts made are endorsed on the certificate along with the operators own call sign. The theme for 2024 was the "Citizen Soldiers, State Militias, Minute Men". There was also an opportunity to exchange QSL cards with all 13 colony state stations plus the bonus stations.

For further information regarding the 13 Colonies Special Event, please visit its website at, *www.13colonies.us*

If you would like to visit our club or want further information on GB13COL please contact Ray G0VLF on 07904196283 or Email *g0vlf@yahoo.co.uk*

Eryri DX

Who are we: The Eryri DX Contest Group (MW0IWU) is a mix of talented, skilled and time served licensed radio amateurs that wish to communicate on air in a more structured and competitive environment.

Situated in the Llandudno area of North Wales, we seek to build on the very best of the hobby and be active in operating together on all modes of communication for contest working, special event stations and for promoting amateur radio. Above all, to have fun and enjoy the hobby socially together.

We are informal in our approach, with no hierarchy or political aspirations but well organised. We support one another, any amateur or prospective radio amateur or club as much as we can. We do not compete with existing clubs in the area but rather, offer an alternative strategy.

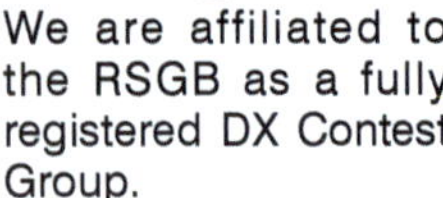

We are affiliated to the RSGB as a fully registered DX Contest Group.

Our mission statement:

To provide a friendly and informal social arena to enjoy the hobby of amateur radio.

To encourage and support education and technical proficiency from operating at contest level, as well as special event station activation using all modes of communication, especially CW.

To facilitate learning and development of CW to a contest operation standard by tuition and on-air practice in order to be able to participate globally.

To make regular professional standard operation of amateur radio together a priority.

To promote all aspects of amateur radio and electronics in a positive and meaningful way.

To enjoy ourselves in what we do.

Our Motto: "Having fun with RF without any drama.

Our website: Eryri DX's website is currently under construction. The URL is *http://eryridx.wales*

Essex CW ARC

Essex CW ARC's mission is to encourage the use and preservation of the CW mode on the amateur radio bands. Based and formed in Essex but open to all licensed radio amateurs worldwide who are interested in Morse code (CW). The club further aims to support those who strive to learn and improve their CW skills, as well as operators who are already competent at any level. Membership is free and on-air activities are publicised via the club's web site and via a regular electronic newsletter.

The club runs weekly CW training classes via Skype for beginners, intermediate and advanced operators. There are weekly on-the-air training nets on both 80m and 2m, as well as the weekly GB2CW QRS transmission on 80m. Each October we hold a one-day "boot camp" where CW enthusiasts of all abilities meet to practise and share their passion for CW. November has the annual activity week, which is a great opportunity for CW practice and to qualify for our operating award. For further information please visit the club's web site at *https://essexcw.uk/*

Greenock and District Scouts ARC

Every year we participate in Jamboree on We participate in Jamboree on the Air, Thinking Day on the Air and other Scout and Guide events throughout the year as MM0TSG.

We offer communications-based activities relating to the Cub, Scout, Brownie and Guide Activity Badges and parts of the Beaver, Cub and Scout Challenge Awards. We offer fun activities for Beavers and Rainbows. The club has developed a series of educational fun games relating to amateur radio which enables the young people to learn the requirements of their badge work whilst enjoying themselves. These games have been voted a great success by the young audience. The games are located in the badge work section of the club's website.

We prepare radio related activities for Explorer Scouts and Senior Section Guides.

In 2023 club members were invited to assist Stirling & District Amateur Radio Club operate the Special Event Station GB1CWC (Cycling World Championships). 4000+ contacts were made over the whole event.

At local campsite centenary celebrations, we were able to provide a drop-in session for anyone who wanted to try our activities. In addition, we supported our District Cub Camp by providing them radio related activities.

JOTA/JOTI (Jamboree on the air / Jamboree on the Internet) was a big success, as was TDOTA (Thinking Day on the air). The TDOTA event was extremely busy with young people from across the county dropping in as part of GirlGuiding Renfrewshire's 2024 Thinking Day celebrations. Girlguiding Renfrewshire's County Commissioner and Greenock Division Commissioner visited the shack and were very impressed with our amateur radio station and training programme.

We meet at 7.00 pm every Friday (term time only) at Greenock and District Scout Headquarters, 159A Finnart Street, Greenock, Inverclyde, PA16 8HZ.

You may contact us via Email at *mm0tsg@gmail*.com or via our Website *https://mm0tsg.org/*

Guildford & District RS

Founded in 1918 as the Guildford Wireless Alliance, club meetings are on the second, fourth and fifth Fridays of the month and are held at the Guildford Model Engineer's building, Burchett's Gate entrance, Stoke Park, London Road, Guildford GU1 1TU.

Guildford town is twinned with Freiburg. At Thursday lunchtimes GDRS members can be heard talking to friends in Freiburg. Other nets also operate and details can be found on our website *www.gdrs.net* together with other information, such as our programme of events.

GDRS can also be heard operating during various contests as "The Guildford Contest Group" using the call-sign G5RS.

We have the use of a recently modernized, extended clubhouse with access for the disabled and inside toilets, together with a patio area just outside and also a large area of grass, for BBQs etc. We have an Altron CM35 mast and a wire doublet for HF, a 2m beam and a 6m LFA Innov antenna. We have a club rig (Icom IC-7300) complete with antenna tuner, Daiwa keyer and Kent paddle key.

Members gather from about 7:30 pm (but not much before) and meetings start at 8:00 pm. We are a very friendly club and visitors are very welcome. We have plenty of free parking at the club house.

Coffee, tea and a selection of soft drinks and confectionary are available at most meetings.

Following a rule change GDRS can now offer Associate Membership for those who live a significant distance from the club house and attend meetings infrequently. For full details see our updated Constitution and Rules document, which can be found on our web site: *www.gdrs.net*

For enquiries please contact the secretary, Timothy Dabbs, G7JYQ Tel: 020 8241 9396 or via email: *secretary@gdrs.net*

Halifax and District ARS

Halifax and District ARS: A Triumph Over Adversity

Nestled within the warm confines of the Elim Church in Sowerby Bridge, Halifax and District ARS has recently transformed from a struggling entity to a flourishing community hub. Despite the initial hurdles during the COVID19 pandemic, the club has diligently worked to expand its membership, cultivating a sense of belonging and camaraderie among its diverse members.

Central to the club's revival was its engaging winter program, featuring a lineup of captivating talks that cater to various interests. These informative sessions not only imparted knowledge but also fostered unity and shared purpose among attendees.

An integral aspect of HADARS is our commitment to provide comprehensive training at all license levels. This commitment ensures that members have the necessary skills and knowledge to excel in the dynamic world of amateur radio.

Beyond its internal activities, HADARS plays a vital role in the local community by providing essential communication support for events such as the renowned Cragg Vale Challenge. This involvement underscores the club's dedication to serving the broader community and solidifies its reputation as a reliable resource in times of need.

Last years "Churches on the Air" event stands as a testament to HADARS renewed vitality, attracting a record number of stations and garnering widespread acclaim. This remarkable success not only showcases our club members' technical skills including antenna building but also highlights its ability to adapt and thrive in changing circumstances.

Weekly nets serve as a cornerstone of the club's activities, offering members a platform to connect, exchange ideas and foster friendships. These gatherings have become immensely popular, providing a sense of community and support in an increasingly digital age from FM voice to SSTV and FT8 digital modes.

For those interested in experiencing the warmth and camaraderie firsthand, Halifax and District ARS extends a warm invitation to visitors. Whether you're a seasoned enthusiast or a curious newcomer, you'll find a welcoming atmosphere at the club's weekly meetings.

We are based at Elim Pentecostal Church on Ryburn Street in Sowerby Bridge HX6 3AZ.

Our website is address is *https://hadars.org.uk/*

Our public Facebook group: *https://www.facebook.com/groups/1097194164482786*

Huntingdon Amateur Radio Society (HARS)

History: The Huntingdonshire Amateur Radio Society (HARS) was formed in the mid-1980s and amalgamated with the Local Repeater Group for GB3OV. Since then we have been based in several locations and now we meet in Buckden Village Hall, Buckden, Cambridgeshire.

HARS today: HARS has nearly 60 active Members with diverse and a wide variety of interests in amateur radio operating that include an active CW training and appreciation group, Digital Voice and Data Modes groups. Newcomers find a warm, friendly, relaxed and informal atmosphere with plenty of like-minded people willing to share their experiences and knowledge.

We hold club meetings on the 2nd and 4th Thursday of each month during the summer. These meetings include presentations, "Bring and Buy" sales and "Show and Tell" evenings.

The Club participated in four Special Event Stations in the last year:

GB2DWM - National Mills on the Air from Duloe Mill, Eaton Socon

GB2RRM - Museums on the Air from Ramsey Rural Museum

GB2RMR - Railways on the Air from the Riverside Miniature Railway, St Neots

GB0WYT - Airfields on the Air from Wyton Airfield

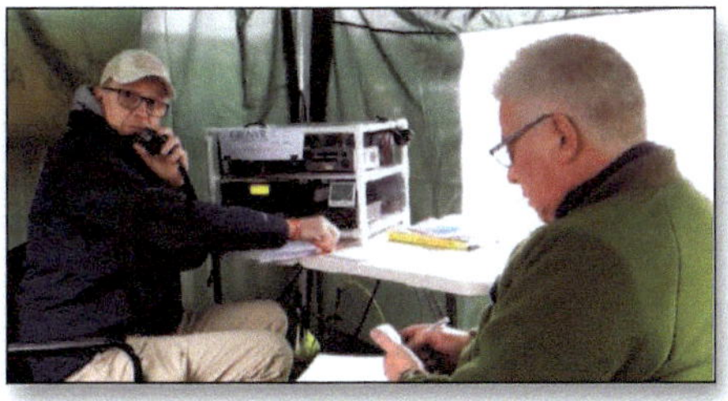

We are planning to increase this to six by the end of 2024.

GB0OV Repeater: The Society runs the Ouse Valley repeater GB3OV entirely from funds accrued at the annual rally on August Bank Holiday Monday currently held at Ernulf Academy, St Neots, Cambridgeshire.

HARS Contact: Malcolm Harrington (HARS Secretary)

c/o Tanglewood House, Station Road, Tilbrook, Cambridgeshire PE28 0JY, England

More information and contact available at *http://hunts-hams.weebly.com/*

Isle of Man ARS

The Society was affiliated to the RSGB on the 13 April 1948 which made 2023 the club's 75th anniversary.

The club has seen great changes in operating, from CW to AM to SSB and FM through to digital modes prevalent today and rigs from home-built valve equipment to the modern all singing and dancing radios from many manufacturers.

With currently about 30 active members, the society has had it's meetings at various venues over the years, and is currently at the Scout hut in Mill Road, Ballasalla. This venue has a large hall for meetings and demonstrations, accessible facilities, small mezzanine for classes and kitchen facilities and suits the Club very well. Meetings are held weekly on Wednesday evenings from 7pm.

We are registered for training and examinations.

The Club has another venue, more central on the Island, at the Sea Cadet hall in Tromode in Douglas and meets once a month on the second Tuesday from 7 pm. This smaller venue is ideal for talks and presentations.

To keep in touch with members on the Island and further afield, Nets are on the GB3IM repeater network after the 11:00 hours GB2RS news bulletin and talk-in Wednesday night at 18:40, Sunday on 3.715 Mhz and 7.115 at 09:30 and 22:00 hours. Visitors are always assured of a warm welcome at both venues.

Club news and events are on the website: *www.iomars.im*

Livingston and District ARS

Livingston and District Amateur Radio Society (LaDARS) is a group of radio amateurs living in West Lothian and the surrounding counties. The club was founded in the mid-1970s so we are looking forward to celebrating our golden anniversary in 2025 with a number of activities. Club meetings are held every Tuesday at 7 pm for training, operation and presentations.

The club has a comprehensive station with 2 HF transceivers and a VHF/UHF transceiver. We have a fan dipole for the 80, 40 and 30 metre bands and a 3-element Yagi for the 20, 17, 15, 12 and 10 metre bands as well as a white-stick antenna for the 2 metre and 70-centimetre bands. The society holds the callsign MS0LIV and all members may operate the station, under supervision if necessary. The club also hosts the GB7LV 2-metre digital voice repeater.

We run training sessions in small groups for all levels of license exam. This usually occurs in person during Tuesday meetings. Club members are willing to mentor new operators. One meeting a month is reserved for a presentation. Club members hold a weekly net on 2-metre FM at 7.30 pm on Thursdays and all local amateurs are welcome to call in. Details of meeting and net schedules are provided in each month's RadCom and on GB2RS. Anyone who has an interest in radio operation and related maker activities is welcome to visit on a club night.

LaDARS has been involved with the local Scout group providing a station for Jamboree on the Air and are always looking for other "on-the-air" events. At the time of writing we are working with Scottish Canals to run a station for Waterways on the Air 2024 at the Falkirk Wheel. We have been invited to work with the Guides for Thinking Day on the Air in 2025.

Find us on Facebook and through our website *https://www.ladars.org.uk* or contact us by e-mail at *secretary@ladars.org.uk*.

Lincoln Short Wave Club – LSWC

Lincoln Short Wave Club (LSWC) (g5fz.co.uk) meets every Wednesday evening and Saturday morning at their shack in the village of Aisthorpe, near RAF Scampton. A 2m net is also held on Thursday evenings at 20:00 using the local repeaters GB3LM & GB3LS. Wednesday meetings are general meetings with visiting speakers normally arranged once a month. Other Wednesday evening meetings are devoted to operating the shack, member feedback and social evenings.

Saturday mornings are filled with many aspects of the hobby from mentoring new licensees, repairing equipment and demonstrations of the Club's extensive range of radio equipment covering from 160m through to 23cm on CW, SSB, FM, Data & Digital modes and ATV. All are available for demonstration and use.

Through 2025 the LSWC will be continuing its drive to improve and expand the capability of the shack with improved antennas for HF operation and finalising its setup for working via satellites, including the ISS. The appearance of the club will continue to be improved by new and existing members modernising not just the tools and equipment but the shack and its facilities.

Over the last couple of years, the LSWC has been hosting its own radio rallies in Market Rasen during the winter and summer. These rallies have been well attended and will continue in 2025.

Training for Foundation and Intermediate examinations continues to be available in person and examinations for all three levels are arranged as needed. Support is provided for new licensees to enable them to fully enjoy the various aspects of amateur radio.

The club is active on most bands with the call signs: G5FZ, G6COL & GB2CWP. We run a number of special even stations throughout the year including the Dambuster Memorial Station (GB5DAM), Mills On The Air at Ellis Mill (GB5EM) and two country shows at Fir Park (GB5FPS & GB5WW).

We partake in the International Scout Jamboree "Poacher" event providing an amateur radio activity. The club's extensive array of portable equipment, including two trailer masts, is very useful at these events.

We maintain a radio shack at the Lincolnshire Heritage Aviation Museum (lincsaviation.co.uk) at East Kirkby (GB2CWP). The shack includes working equipment from WW-II including a complete R1155/T1154 station. The shack is normally open during summer weekends and when the museum holds their own special events as well as for museums on the air weekend. See our club website for details.

The club maintains two voice repeaters, GB3LM (2m), GB3LS (70cm) and two ATV repeaters, GB3VL (23cm ATV) and GB3LX (10GHz ATV) which are all located high above the city on Lincoln Cathedral.

LSWC is pleased to support the National Hamfest each year. Joining a number of volunteers from many of the local clubs without whose help the event would struggle.

Medway Amateur Receiving & Transmitting Society (MARTS)

The Medway Amateur Receiving & Transmitting Society is one of the oldest radio clubs in the UK. The club was founded in 1922, the same year as the BBC, by a group of local enthusiasts led by Bill Nutton G6NU as the Gillingham Wireless League and it became the Medway Amateur Transmitting Society in the late 1920's and finally adopting its current name in 1947 in recognition of the growing post war interest in short wave listening.

MARTS aims to promote and develop amateur radio in the Medway towns and surrounding areas. The club is committed to fostering programmes of providing information, technical reviews and discussion, guided construction activities, supported learning and the development of good operating techniques.

As a club, we respect the technical and operating skills of our members by promoting continuous development through a process of supportive encouragement. The club's traditions are based upon friendship, hospitality, equality, respect and upholding the values enshrined in the Amateur Code and the International Amateur Radio Union (IARU) Ethics and Operating Procedures for the Radio Amateur.

The club arranges a varied programme of talks, demonstrations, discussions, on-air activities, equipment sales, informal evenings as well as social events. We provide weekly Morse practice sessions for those wishing to further develop their skills. In addition to our meetings, we run daily club nets every weekday at 09:30 on 144.650 MHz FM using the club callsign G2FJA. Additional monthly nets are arranged using alternative amateur bands and modes, including HF and DMR where the flag ship call of G5MW can be heard.

Our membership is made up of individuals who have a wide range of backgrounds and expertise and specific technical knowledge areas and interests in the hobby. These include HF operation, VHF operation, propagation studies, working DX, satellite operation, CW, DMR, using digital data modes, antenna design, construction and contesting. Whatever the interest, our members are always keen to inform and offer developmental support to others and to promote the wide range of aspects that the hobby has to offer, including training for license examination.

The club meets every Friday evening from 19:30 at Tunbury Hall, Catkin Close, Chatham, Kent ME5 9HP. Members and visitors to the club have access to ample free parking spaces. Visitors will be sure to receive a friendly welcome including tea and often doughnuts as the MARTS is renowned for "marching on its stomach".

For further information on what's happening at MARTS, please visit our website *www.marts.org.uk* or follow us on Twitter(X) @ g8mwa. Alternatively, e-mail us at *secretary@marts.org.uk*

Norfolk ARC

Norfolk ARC has more than 200 members, a strong history and is thriving with a very active calendar of online talks, events, and special event station actions.

The club's calendar now includes in-person meetings at the CNS School in Norwich and virtual meetings and talks via Facebook and the BATC website – please check the schedule on the NARC website for dates.

Over the last year the club has had numerous online talks from experts around the world on subjects ranging from "Our Active Sun", by the chief scientist at the Rutherford Appleton Laboratory, to an "Update on Solar Cycle 25" by Carl, K9LA.

Other in-person club nights have included friendly and competitive direction finding "foxhunts", a Raspberry Pi "Jam", construction groups and a Super Science Saturday event at the CNS School.

The club runs special event stations, such as at Caister Lifeboat for "International Marconi Day" and Happisburgh Lighthouse for "Lighthouses on the Air".

The club continues with its Bright Sparks programme, aimed especially at helping youngsters learn to learn to solder, understand components and eventually gain their licences so that they can get on air.

The social side of the club is thriving with an annual "Radio by the Sea" day at West Runton, the annual Barford Radio Rally, friendly foxhunts and a Christmas dinner evening at a local pub.

The club has a very active contest group, taking part in the RSGB's 80m Club Championship (which it has won several times), plus CW and SSB VHF/HF Field Days and AFS events.

Other on-air activity includes a weekly 2m net, CW tuition and a monthly sked with the Koblenz Radio Club, one of Norwich's twinned cities. This has led to visits by members of both clubs.

The club has a strong heritage, dating back to the 1950s and its website has an interesting archive of photographs that go back through the decades. The club features many prominent amateurs, including RadCom authors and specialists.

Above all, the club has an ethos of its members helping and encouraging others.

If you are interested in radio and electronics and would like to find out more about NARC please visit our web site at *www.norfolkamateurradio.org* or e-mail *chairman@norfolkamateurradio.org*

The North Wales Amateur Radio Group

The North Wales Amateur Radio Group have been enjoying their many varied events and club nights, and our weekly nets run through GB3OR & GB3OG have played a major role in communicating to Amateur Radio enthusiasts within its exceptionally wide range, thanks to the repeaters position at the Summit Café Complex on the Great Orme Llandudno.

Maintenance has been carried out on both GB3OR & GB3GO over recent months and the Llandudno Repeater Group are now very pleased to report that maintenance work is now complete.

The NWARG hold regular on-air evenings with interesting talks by either members or invited guests and organise regular field trips or special event stations. We always welcome both licensed amateur radio operators and anyone who has an interest in radio, in particular people who are interested in obtaining an amateur radio license.

If you are interested in obtaining your amateur radio license we can help you with that. Amongst our membership we have a superb talent base and there will always be a member that can support you in almost every aspect of amateur radio.

The majority of our members live in the local area, can attend meetings and events although we do have some members that live overseas. New members need to have an interest in radio and above all need to agree to abide by our constitution, rules and group ethos of "radio and friendship". Why not pop along and see what North Wales Amateur Radio Group can offer you?

NWARG take access for the disabled seriously and our members provide a safe and friendly environment for all regardless of ability. One of our members is a trained nursing practitioner and is available on request.

The North Wales Amateur Radio Group (NWARG) meets weekly on Monday evenings

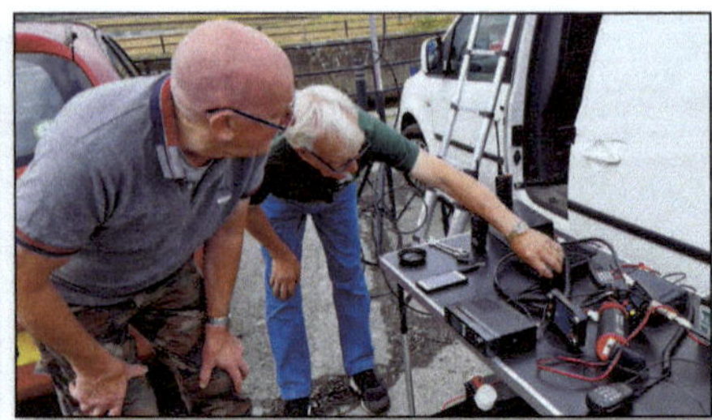

from 7 until 9 pm at the Bryn Cadno Community Centre located in Colwyn Heights, Conwy, North Wales LL29 6DW

The North Wales Amateur Radio Group would love to hear from any amateur who is in coverage area of the club. In person or via the club's UHF (GB3OR) or VHF(GB3OG) repeater where they can be assured of a friendly QSO with a NWARG & Llandudno Repeater Group member.

NWARG Weekly Nets are Sunday (UHF) 8–9pm on GB3OR, Thursday (VHF) 8–9pm on GB3OG

Club contact: *sandra.ted@zen.co.uk* or *https://www.facebook.com/groups/NWARG*

For the latest up to-date information regarding the North Wales Radio Group please visit: *www.nwarg.org*

Nuneaton & District ARC

This is our third year at our high location on Hartshill Ridge. At 140m ASL it overlooks North Warwickshire and South-West Leicestershire. We have a great mix of members from SWLs to professionals, young and old, all sociable and helpful. We have antennas installed at the premises including an 80m EFHW, 6 and 4m halos, a 2m/ 70cm colinear as well as a satellite dish for QO-100. All have permanent connections to the clubroom. Our club nights are the first Monday of each month, a social and operating night and third Fridays of the month is the full club night. Our offering covers practical antenna and operating/equipment workshops, visits and talks which are as varied as our members. We have an active VHF UKAC contest group and take part in RSGB and HF and VHF PW contests. For emergency communications we have just helped to form the expanded Warwickshire Raynet group.

New members and visitors at any level are welcome. Contact us via our email *(secretary@ndarc.co.uk)*, join our Facebook page or drop in and meet us on a club night from 7pm at the Windmill Sports and Social Club, Mancetter Road, Nuneaton, Warwickshire CV10 0HW.

Poole Radio Society

Poole Radio Society was formed over 45 years ago and continues to meet on a weekly basis throughout the year. We are made up of a wide variety of individuals with very different levels of expertise in the hobby. Many reside in the Poole area but also many much further afield. It's a strength of the club that many of our members remain with us, even after relocating elsewhere in the country – as a result, we also have a weekly club virtual meeting. We have a series of weekly operating nets – including 4m and 2m voice nets plus HF and 2m CW nets. The 2m CW and FM voice net is particularly aimed at beginners and for many, this is the first time they get on the air unaided after receiving their license. We have all been there – so the net is designed to cater for the nervous and the newcomer, who will receive a very supportive welcome.

The club activities are broadly split into two models – in the Winter months we concentrate on talks and demonstrations but in the summer months we arrange outdoor activi-

ties. These are a mix of social and training events and normally involve getting out into the countryside, using a whole range of equipment and antennas. It's a great way to learn, see and experience a wide variety of equipment, operational styles and different antennas.

We also focus on one or two large public events during the summer. We have regular events at Kingston Lacey and Corfe Castle which are held annually and give us the opportunity to show what Amateur Radio is to the general public. Occasionally we participate in commemorative events. In 2024 we were invited to participate in the Poole Lifeboat Festival to commemorate the 200th anniversary of the RNLI saving lives at sea. We co-operated with the British Amateur Television Club and Royal Signals Amateur Radio Club to offer the public something unusual. BATC operated through the QO-100 satellite using digital television and contacted a variety of very enthusiastic lifeboat and lifeguard related stations around the world. HF was used to participate in the SOS Radio week which was on at the same time.

We do not offer formal training for the license exams – there are many groups that do this better than we could, such as Bath-based distance learning – but we do offer mentoring and support to anyone preparing for the examinations. We offer practical training sessions – both indoors and outdoors – to supplement the academic training.

PRS is a very social and inclusive club and always aim to have an element of fun in everything we do in this remarkable hobby.

We meet at St.Aldhems Church Hall, Branksome, Poole, Thursdays at 7.30pm. For details of club membership, virtual meetings and our operating nets, please contact *secretary@g4PRS.org.uk*

Rhondda ARS

Rhondda Amateur Radio Society (GW2FOF) has been around for many years and over the years it has been located at a number of different venues. During 2023 the Society moved to the current venue, Rhondda Sports Centre, Ystrad CF41 7SY. The centre has plenty of parking spaces and is close to a bus route. There is ramp access at the building entrance and an internal lift is available to access the first floor. This is where our club meet on Tuesday evenings 7 to 9 pm. Friends old and new are always welcome to attend and to join us as members if they wish.

During the summer months we attend various public events in the locality and hold field events on the mountains around Rhondda. Whenever we can, throughout the year, we take part in 'On the Air' events. One that we always look forward to taking part in is the Scouts Jamboree on the Air (JOTA/JOTI), something the club has done for many years.

The club can be contacted via our Facebook page or via phone on 07394 994 555. If we are unable to answer your call, please leave a message and we will call you back.

Salop ARS

Salop ARS is a small club located in Shrewsbury which attracts members from a wide area of Shropshire and mid Wales.

We meet on Thursdays, normally from 8:00 pm until around 10:00 pm, with a mix of "Rag

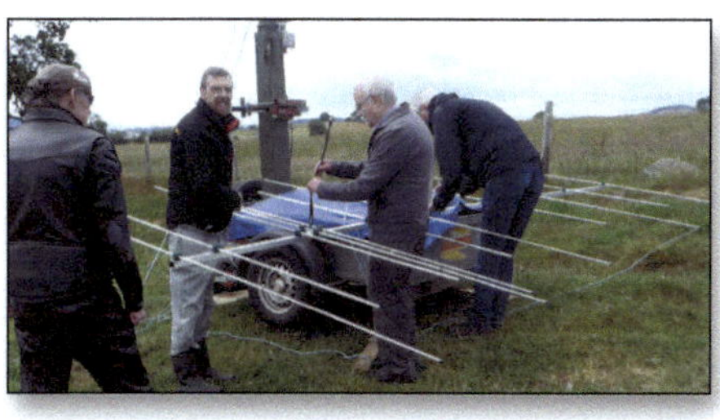

Chew" evenings and organised events. There is a bar at our meeting place serving hot and cold drinks, as well as an operational shack with HF and VHF equipment. Other activities include participation in Museums on The Air, VHF NFD and other contests.

Members reflect a wide range of interests within Amateur Radio and related fields and we welcome new members.

Please visit our website at *https://www.salopradiosociety.org/* for further information including location details and a current calendar.

Shirehampton ARC

Shirehampton club is now in its 56th year and still meets every Friday evening from 7pm at TS Enterprise Sea Cadets on the banks of the River Avon. We have members from across Bristol and a wide area stretching from Weston-Super-Mare to Berkeley. The club has a permanent shack with HF and VHF facilities and also a Kiwi SDR which can be remotely accessed.

The club provides a secure website (*https://shirehampton-arc.org.uk/*) for members, a private Groups.io group and WhatsApp chat facilities. There is a weekly club net each Sunday morning at 11am on 2m.

In recent years the club has designed and built projects to enable club members to improve their construction skills and take home a useful addition to their shack. These projects include a WSPR transmitter which has also been used for club WSPR contests, solar based DF equipment and most recently a shack barometer.

There are members of the club who actively restore old valve equipment and build their own new valve-based equipment. We have other members into software development with a number of programs which can be downloaded from the club website.

The club is not currently providing formal education but can assist anyone who comes along to the club requiring help to get through the examinations as we have a group of members who have assisted with education in the past.

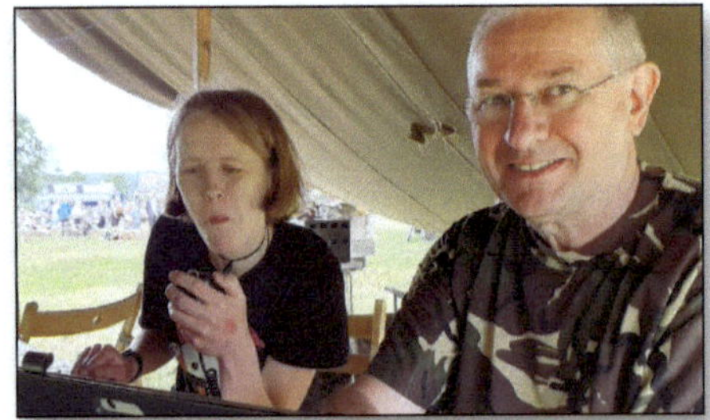

Outside of the club meetings we participate in Mills on the Air, Churches on the Air and Railways on the Air each year which are great social events for the club and a chance for members to operate in interference free locations in the countryside by setting up portable stations.

Other activities include participation and portable operation in the English Castles and English Lighthouse Award schemes. We have members who are part of the Bristol Contest Group and some who are specialists in the Microwave arena.

We plan at least 2 talks a month using club members and external speakers from around the Bristol area. Our website contains details of all future activities, so why not come along one Friday evening!

For further information please have a look at our website and contact *secretary@shirehampton*-arc.org.uk.

South Dorset RS

The club has enjoyed an active year but has been without its normal Clubhouse/G3SDS QTH, namely the Weymouth-West Scout Hut because it has been demolished and is being replaced with a brand-new building on the same site. The club has therefore been meeting at the United Reform Church Hall as a temporary measure.

Special events have included a RAFARS Airfields on The Air (AOTA) activation of the Battle of Britain airfield of RAF Warmwell (GB-0WWA), Marconi Day (GB0MPB) at Portland Bill and a very successful Scout Jamboree on the Air (JOTA) in October. In February 2024 we were invited (for the first time) to take part in the Scouts Winter Camp event which proved to be great fun. In September 2024 we embarked on a DXpedition to the Isle of Man.

Every Tuesday we host a club net (GX3SDS) from 11:00 to 12:00 local time on or near 3.784MHz LSB, preceded by a 15-minute AM net on 3.615MHz at 10:45 for club members who like to try out their vintage military sets.

During our monthly meetings, held on the second Friday of each month we maintain a stimulating program of presentations and demonstrations which has in the past included: building the QDX FT8 transceiver, the history of UK nuclear weapons, building a 1949-design valve CW transmitter, experiences with a doublet antenna and a personal journey with Hewlett Packard scientific calculators from 1968 to the present day.

The club publishes a detailed magazine called "Catswhisker" every month which contains a lot of relevant articles and personal experiences in Amateur Radio. This is sent to all members and prospective members via email.

All of our activities, presentations and news items are faithfully documented on our well managed website *www.SDRGS.org.uk* which also gives information on the South Dorset repeaters GB3DR (2m) GB3SD (70cm) and GB7SD (DMR 70cm), managed by the South Dorset Repeater Group. For this we thank our webmaster Mike (G0LQI).

Club dues remain at £15 per annum and are free for beginners (of any age!) until they qualify for their first licence.

Taunton & Somerset RAYNET

Taunton & Somerset RAYNET was established in 1978, currently has 32 members and holds the callsign G1SOM.

The User Services we work with include Avon & Somerset Police; NHS and approved health care providers and Exmoor Search & Rescue (Mountain Rescue), plus other utility organisations.

Following the guidance in the amateur licence, the club organises communication for local events as preparation and training for any emergency communication needs that may arise.

Regular activities include:

The Seaton Grizzly – a 20-mile running race attracting in excess of 1000 entrants from all over the world. Competitors run up and down the steeper bits of the south Devon coastline. Part of the course is along a pebbled beach. That's why it's called the Grizzly and is rated as one of the toughest such events in the UK.

A recent novel event was a marathon race; where the competitors were taken by steam train from Minehead to the starting point at the other end of the West Somerset Railway and then instructed to run back to the Bristol Channel resort.

On the more serious side, the club is engaged with local parish councils, assisting them with their Civil Contingency preparations. We helped them set up a licence-free communications network for use by their own members; in return they have installed an aerial system which can be used by Taunton & Somerset RAYNET as needed.

The club is working with Wessex Water utility company to cover their communications needs in case of failure of their existing system. "Civil Contingency" again.

We often work in a landscape of hills and deep valleys – challenging radio conditions. In these cases, repeater units are deployed at strategic spots – working on 70cm and linked on 4m.

The Club also sponsors GB3VS, GB7FI and GB7EQ 70cm repeaters.

For more information please visit *www.tauntonraynet.co.uk*

Wirral and District ARC

Wirral and District Amateur Radio Club meets face to face, twice monthly at Irby cricket club. In the weeks without club meetings, we have informal, social gatherings.

Despite being only a small club, we have twice been R.S.G.B. Regional Club of The Year.

Our youngest Foundation licence holder so far was eight years old when first licenced but we also have members more than ten times this age. Everyone is welcome.

We regularly run special event stations such as GB4LL for the International Lighthouses and Lightships Weekend, Youngsters on the Air and for the World Scouts Jamboree on the Air.

One activity that has become very popular with members in recent years is the club's participation in the UKAC weekly contests. Around a third of our members regularly take part in these challenges on different bands. In the last few years we have finished in the top 3 places nationally.

Throughout the year members hold regular direction-finding competitions across the Wirral and North Wales as well as an annual construction challenge, often featuring some of the projects that have been the subject of talks at our meetings.

Our members' interests range right across the amateur radio spectrum from the lowest frequencies right up to microwave and include all data modes, digital and television. For the more traditional, many are active using Morse code. Whatever you want to get involved with there will be a club member who can help and give advice.

We are affiliated to the Northern Amateur Radio Societies Association (NARSA) and our members are involved annually with the NARSA (Blackpool) Rally.

WADARC is a very active club and becoming more and more so as time goes on. We are always pleased to see newcomers. Most of our meetings are radio orientated evenings but there are also occasional, general interest talks with twice yearly equipment sales and joint events with our neighbours in the Chester Radio Society.

More details and useful links can be found on our club website and we can be contacted through it or via Facebook.

www.WADARC.Com

Worthing and District ARC

The group has an established range of interests, from contests to vintage radio restoration, construction projects to Morse code.

New members have brought new skills which have expanded the plans for the coming year. Contest participation across HF and VHF bands will increase, FT-8 guidance is available, along with beacon, repeater and network projects.

Our coffee mornings are a highlight.

Finally, members can count on assistance from all in the group, for example to move heavy equipment or for aerial erection/dismantling.

Contact: secretary@m0reg.co.uk

Bracknell Amateur Radio Club (BARC)

BARC has been an RSGB registered club for over 53 years and is glad to currently be seeing a considerable increase in new members, both those interested in the hobby for the first time and experienced operators, some rejoining the hobby after many years.

Our membership, of all ages, are friendly, social and welcoming and have a wide-range knowledge and experience of all aspects of Amateur Radio.

The club hosts a monthly meet-up, usually with a guest speaker, presentation or demo covering a wide range of Amateur Radio interests and disciplines such as CW, HF, VHF/UHF, Microwave, DMR and digital data modes.

Weekly radio 'Nets' are held at 8pm each Wednesday on 145.375MHz FM and at 5pm via the GB3BN Repeater. Our new monthly HF and 6m/4m Nets are also growing in popularity.

Many of our members are highly successful in regular RSGB and international contests, receiving several awards.

With our own Groups.io 'Reflector' and Members' WhatsApp group our online community are in regular touch with each other and always available for a friendly chat or to give advice. This is particularly useful for those studying for their amateur radio examinations.

Membership of BARC is open to anyone with an interest in Amateur Radio, whether already licensed, preparing for their exam or interested in radio and electronics in general.

BARC have recently been in touch with friends in the twinned town of Leverkusen, Germany and hope to establish a regular HF net between both towns.

For further information or to join BARC, please visit *https://g4bra.org.uk* or email secretary@g4bra.org.uk. We can also be found on Facebook *@bracknellarc*

Chelmsford Amateur Radio Society (CARS)

The Society holds monthly meetings at Danbury Village Hall on the first Tuesday of the month, usually taking the form of a technical talk by a Club member or a visiting speaker. CARS is a Brickworks accredited club, and our programme includes regular meetings with a focus on basic skills, often with a lively discussion. The meetings offer a "table top" sale area for equipment and the obligatory tea and biscuits. Full details of the meetings are published monthly on the CARS website and in the newsletter. Open club nets are regularly held on the other Tuesday evenings using VHF, UHF and HF - in turn.

With some 80 paid up members CARS is probably one of the largest and oldest clubs in Essex if not the Country. Formed in 1936 by engineers from the company then known as the Marconi's Wireless Telegraph Co., this is reflected in our well-known Club callsign G0MWT - all part of keeping the Marconi legacy alive in Chelmsford, the birthplace of radio. Louis Varney G5RV was one of our founder members.

CARS typically run a number of special event stations during the year. Recent activity has included operating GB9MWT to commemorate the 150th anniversary of Marconi's birth, GB5TLV for the International Lighthouse and Lightship weekend and GX0MWT from the former Marconi Research Centre in Great Baddow to commemorate International Marconi Day.

CARS have been actively involved with training since at least 1960. The society's reputation for quality training is well-deserved and our very experienced training team has achieved numerous successes. Courses for Foundation, Intermediate and Full licence exams continue to be available using training material designed and delivered by club members based on the current RSGB syllabus. The CARS website has a section dedicated to training and training materials are available for download.

CARS offers face-to-face training and exam sessions for all three exam levels based on our very successful "Fast Track" evening format for revision and pre-exam preparation. Online and written exam processes are supported. Our exam secretary and training coordinator John M0JOC is experienced and also a remote invigilator for the RSGB run, online exams

Interest in Morse code continues to thrive and classes are held weekly over Skype for the Beginners and for a social CW chat.

Our monthly newsletter gives excellent coverage of all our events and our website www.g0mwt.org.uk has full information regarding future activities, training, venues, dates etc.

On social media you can follow us on Twitter @ChelmsfordARS, @TrainWithCARS, and Facebook; *www.facebook.com/ChelmsfordARS*.

Paul G4PVM, Acting Secretary.

Chester & District Radio Society

The Society, as presently constituted, was founded in 1948 when wartime restrictions were lifted. It had existed, in previous incarnations, being active prior to the 1914 - 1918 war as Cheshire Radiographic Society and later as Chester Wireless Society with the callsign G5CH. This callsign is currently held by the society after recent reinstatement by Ofcom along with G3IZ and G8GV.

Despite its long history, the society is forward-looking and has an enthusiastic membership which meets weekly in the village of Waverton near Chester for talks by members and visitors. Subjects are many and varied and include technical topics both new and vintage. "Bring and Tell" evenings along with surplus equipment sales and video nights are popular. One series entitled "The other Man's Shack" consists of video recordings of the radio rooms of amateurs in the area. Once the lighter evenings arrive the society runs the club station at least once a month including Special Event Stations activations. An example of this is where the club has activated a local watermill with a portable HF station in recent years. The society has no official meetings during August but many members take this opportunity to erect and run portable stations to help promote and demonstrate amateur radio to the public often from a family friendly site on Frodsham Hill.

The society holds an annual construction competition which invariably produces imaginative and innovative projects. One example was the building, erection and testing of a compact multiband dipole designed by Dr Brian Austin which was later published in RadCom. We are looking to include software development in the contest in the future.

The club has many regular weekly nets on VHF and HF using the call GX5CH, an up to date schedule is kept on the website and members and non-members are very welcome to join the nets. We enjoy an active social life with events like our "Pie and Pint" night – which forms the last meeting before our August Break.

Visitors will always receive a warm welcome and anyone wanting further information can visit the Society's website at *www chesterdars.org.uk/*

Cockenzie & Port Seton Amateur Radio Club (CPSARC)

The Cockenzie and Port Seton ARC is structured in a non-traditional way in that since 1984 the club has operated without official positions, formal membership or the politics that sometimes invade formal structures. Its members like it that way and the results give evidence to that. Financial affairs are transparent and healthy. An extensive inventory of rigs, antennae, towers, and other equipment has been accumulated over the years to enable the club to mount a wide variety of activities.

The club holds regular monthly meetings and has successfully participated in contests, a Mini-Rally, training including CW training runs weekly via ZOOM. Exam training (Foundation, Intermediate & Full) takes place face to face.

Age Profile and Membership

The club has an informal membership with a good mix of ages participating in most events. Several members who started as teenagers, are now young adults and becoming the mainstay of activities. Retired members operating skills are much valued. Members come from a wide geographic area.

Training

A central activity of the club is training. Courses at all levels are conducted. CPSARC believes that the qualification stage is only the beginning and actively encourages new licensees to become active by giving strong encouragement in contesting, where some of younger members have become very competent. Having new, less experienced operators in contests can reduce winning chances but is seen as a price worth paying.

Spectrum Use

Members are active from LF through to UHF. HF, IOTA and CQWW are the contests we take part in the most.

Outreach and Promotion

The club is a social, technical and competitive group. We run our own annual Junk/Mini

Rally as well as several purely social activities annually during which we try to attract new people to the hobby. We support the British Heart Foundation, and our cumulative donations continue to climb, reaching almost £20,000 in 2023.

Our website at www.cpsarc.com contains news, discussion forums, and archives of our monthly, free to download, newsletter Elements. The website receives over 50,000 visits a year. On social media our Facebook page "CPSARC" displays up to the minute changes on events, etc

Contesting and Activities

As mentioned above, the club encourages less experienced members to join club contesting activity, with participation in VHF field day, IOTA, 6m Trophy, 2m QRP, and CQWW.

The activity nights which are run as regular monthly events, have become extremely popular encouraging all levels of licensee to participate.

ARDF (direction finding hunts) are held twice a year and construction nights are run throughout the year.

Summary

CPSARC is an active RSGB Affiliated club, with its non-traditional structure, is effective and healthy. Active training and examination of newcomers ensures new ideas in the club. Regular activities ensure the membership see value in the club and while the radio club is run in an informal way; there are no regular fees, no formal committee structure, just a group of like-minded people doing something they enjoy!

The club meets each month on the first Friday in the lounge of the Thorntree Inn from 7pm until late.

Colchester Radio Amateurs

CRA celebrated 60 years of amateur radio in 2023 with a return to a full training program, presentations and activities including an Island Activation, Field Days, contesting and club activities. 2024 will be no different with similar activities planned including another Island Activation and training.

2024 will also see an extension of our outside events including, but not limited to, GB6WLB to support Walton Lifeboat and GB6NT the Naze Tower activation.

Anyone wishing to join should have a good sense of humour and a wish to advance their knowledge.

We look forward to welcoming you, anyone wishing to visit or join can contact us by email at: *secretary@g3co.uk*

Cray Valley RS

The Cray Valley RS was founded in 1946 and one of the largest clubs in the UK with 150 members. Our membership comes from as far away as Devon, Gloucester, Stoke, Nuneaton and Gulfport USA. Conveniently located at Eltham, in the Royal Borough of Greenwich, it is 10 miles to the SE of Central London, about 35 mins travel by train to either Eltham or New Eltham railway stations.

Meetings are held on the first and third Thursdays of most months, from 7.00 – 9.30 pm at the 1st Royal Eltham Scout HQ, located between the rear of 61 - 71 Southend Crescent, Eltham, London SE9 2SD (what3words vital.vest.sounds).

For the meeting schedule and "how to find us, please check cvrs.org if new please introduce yourselves upon arrival. Our monthly "QUA" newsletter announces our lecture and activity programmes.

Radio amateurs visiting London and those new to the hobby are always welcome to join us. Advice is available to newcomers, to help them with rigs, antennas and getting on the air.

Training – CVRS is one of the few UK radio groups offering face to face training with examination at all licence levels. Foundation class lessons are offered in Spring; with Intermediate and Full alternating in Autumn. Formal CW training with the CW Academy is also available.

Special event stations – CVRS have run national special event stations including M2000A for the Millennium, 2012L for the London Olympics most recently, GB23C for the Coronation of HRH King Charles. Our focus in 2025 is to run smaller local special event stations e.g. the Crossness Museum as part of International Museums on the Air and Railways on the Air. Each year we run the Jamboree on the Air (JOTA) station GB2RE for the Greenwich District Scout group normally at Severndroog Castle on Shooters Hill. During the first weekend of January we run GB8KW, celebrating the UK's last major amateur radio manufacturer - KW Electronics of Dartford. More recently we have taken part in Exercise Blue Ham for the RAF Air Cadets.

Contesting – members are keen contesters and operate at the top level. The society has a quality inventory of equipment available for field and special event operations.

Technical Aspects – the club shack is extremely well equipped with contest level radios, amplifiers, computers and antennas. It is opened regularly for operation by members. We run an annual club construction project – to encourage all to "have a go". Within the membership, there are many highly qualified radio professionals, always ready to help newcomers.

DXpeditions – Both as a club and as individuals – CVRS members have operated from exotic locations including the Isles of Scilly, Botswana, The Minquiers, St Kilda & Rockall

Social Aspects – CVRS welcomes all wanting to be involved with amateur radio – lifelong friendships abound. Visits to places of interest, rallies and museums happen on an ad hoc basis. There is an annual family BBQ, a formal Christmas Dinner and a pub buffet. We are a welcoming society, dedicated to furthering the cause of amateur radio.

Dengie Hundred Amateur Radio Society (DHARS).

The last year has seen a change of Chairman and Secretary of the society.

The shack has undergone a rework allowing for more operating space and more room for club members to get involved. The antenna system has also recently had an overhaul.

The society meet weekly on a Thursday morning and have a regular net on Tuesday mornings on 145.4625 MHz at 10am UK time and then following-on on 7.125 MHz at approximately 10.30 am UK time. All are welcome to call in.

For more details please see *www.dhars.org.uk* or contact *secretary@dhars.org.uk*

Echelford Amateur Radio Society

The Society was formed in Ashford, Middlesex, UK in 1964, using the old English town name 'Echelford' to avoid confusion with other Ashfords in the UK.

In 2024 we celebrated our 60th Anniversary year with a membership of 54. We held a buffet tea at a popular local hostelry with kind sponsorship from our local amateur radio emporium and a very enjoyable time was had. We had a free draw for a donated M5CBR (SK) FT-990 transceiver in excellent condition and this was won by a very pleased John G1TFM. We trust that John and his daughter Emma M7ECB, also a member, will make good use of it.

On World Amateur Radio Day (18th April 2024) G0JOS and M1GWZ put club call signs G3UES and G2HS respectively on air to celebrate both the WARD and our anniversary year. Conditions were difficult but they made a good number of contacts across the UK and Europe and exchanged greetings.

We continue to do our best in the RSGB 80m club contests, entering with members contributing to SSB, DATA and CW modes and are pleased with our progress. We have just completed another CW NFD with a good score despite our limited (and diminishing) pool of CW operators (which we hope to enhance with our weekly slow CW training net). We have a number of keen DF'ers who enter and administer qualifying events for the national championships. Members have continued to give talks to a number of local clubs while we maintain Zoom meetings once a month and in-person gatherings twice monthly. We have a monthly newsletter ("All EARS"), a Twitter(X) account, a YouTube channel and a "groups.io" discussion forum, the latter open to members only. If you are looking for a friendly, active society in the North Surrey area, do not hesitate to contact us.

FDARS (Felixstowe and District Amateur Radio Society)

We are an active and friendly local radio club, primarily serving enthusiasts in the Ipswich and Felixstowe area but with members across East Suffolk. We are fortunate to have our own capably-equipped club shack. It is located at the Suffolk Aviation Heritage Museum in East Ipswich, formerly the RAF Martlesham Heath Wireless Telegraphy Communications site. This site has a rich heritage, and we do our best to support this in our Special Event activities – often aligned to events taking place at the museum.

We also make use of Broadway House – a

community venue in Felixstowe – which allows us to gather more comfortably in the winter months to enjoy a programme of talks, social discussion and practical construction activities.

FDARS has an established history of providing Amateur Radio Licence training and over the years has enabled many local amateurs to successfully study for and take their licence exams. By necessity, this training moved more online during the COVID years but based on demand we are starting to re-introduce more face-to-face sessions to ensure that candidates get to experience and practice the practical aspects of the hobby during their training.

Details of our activities and training can be found on our website, www.fdars.org.uk, and we very much look forward to welcoming you to one of our meetings soon, should you live within the vicinity – or if not, working you on the air with our club callsign GX4ZFR and special event call GB1FOX.

Sarah, 2E0ISJ (FDARS Chairman) Mark, M5BOP (FDARS Secretary)

Contact e-mail: *secretary@fdars.org.uk*

Guernsey ARS

Guernsey Amateur Radio Society was first formed sometime in the 1940's or 50's, and is currently located in an old WWII bunker at Beau Sejour Leisure Centre in St Peter Port, Guernsey, GY1 2DL.

The club meets every Friday at 7.30pm, and we have a lounge and a small, separate shack with an IC7300 and several VHF/UHF radios.

There are weekly nets on Tuesdays on 2m FM on 145.525MHz, and on UHF DMR on TG23515 every Thursday, both at 8pm.

There is a UHF FM repeater (GB3GU) on 435.525MHz.

We only have a small membership, but we try to be as active as possible. Every 5 years, we celebrate the Liberation of the Channel Islands on May 9th 1945, and put on a special call.

The club attends the Vintage Agricultural Show in the south of the island each August, where we put on a public demonstration of what amateur radio is all about to try and interest new members.

Our members have a wide range of skills, including computing, electronics and surface mount technology, antenna building, etc, and we also have a diverse range of interests, from QRP, CW, SSB, digital modes and satellite working, including QO-100.

Being a separate DXCC entity, means we often enjoy large pile-ups, and some visitors have been amazed to find that they can work DXCC easily in a weekend.

Visitors are most welcome to come along on a Friday and join in our discussions and have a go on the radio.

All contacts are logged using Cloudlog, and then uploaded to LoTW.

For more information, please contact Phil GU0SUP on: 07781 151747

You can also find us online here

gu3hfn@suremail.gg FaceBook - GsyARS

Twitter: @Guernsey_ARS

Website: gars.org.gg

Instagram: @guernseyradioclub

YouTube: search Guernsey Amateur Radio Society Website: gars.org.gg

Hilderstone RS

The Hilderstone Radio Society is made up of licensed amateur radio enthusiasts, students studying for their licence, short wave listeners, electronics & computer hobbyists.

We have a wide variety of interests including construction of electronic projects, computers, coding and contests. We are keen to advance our knowledge of radio communications as well as passing on that knowledge to local schools and organisations. Learning for members doesn't stop after passing licence exams nor is it limited by them alone.

We have been a RSGB regional & national club award winner.

In 2016 we helped Wellesley House school in Broadstairs with their amateur radio contact with Tim Peake while he was on the International Space Station. We hope to assist another school to again contact the ISS in 2023! As well as schools we have an established relationship with the local Scouting & Guiding groups supporting both JOTA/JOTI and Thinking Day on the Air. Members also support a local group of young girls who look at STEM subjects independent of school, the STEaMettes have regularly made presentations to the club on their projects.

Now fully established in our 'new' meeting place in a local Scout Hall we have invested in a club shack, equipped for HF, VHF & UHF, both for members to use & as a resource for training.

As training is an important thread within the club, we have recently acquired a second club call, G7HTI (Hilderstone Trainees & Instructors), to support that activity alongside the established club call of G0HRS. Face to face licence training as well as BRICKWORKS activity are promoted. A regular feature of club nights is to have a speaker both from within our own membership & guests on a wide variety of topics, not always amateur radio related but usually with a tangential link.

The club puts on several Special Event Stations throughout the year both to support local organisations & charities but also to demonstrate the hobby to a wider audience. The club is in the planning stage to get equipment to demonstrate the amateur satellite & amateur television service at these events, modes which particularly hold the attention of young people, almost more than operating a Morse key!

The club membership has been constantly growing over the past decade, amazingly including during the covid years.

We encourage new members to learn more about amateur radio, electronics or computing so come along and meet us. You will be sure of a warm welcome.

Contact: *secretary@g0hrs.org*

The Holsworthy Amateur Radio Club

The Holsworthy Amateur Radio Club is the cross-county club for North Devon and Cornwall with members from both counties.

The club meets at 19:30 on the first Wednesday of each month in the rural setting of the 'Old School Room' at the Methodist chapel, Milton Damerel North Devon. The club participates in a number of outdoor events throughout the year and is responsible for the two Holsworthy Radio Rallies. The local VHF Yaesu Fusion repeater GB3DN and the UHF DMR repeater GB3ND are owned and run by club members.

The club welcomes new members, licensed, unlicensed and short-wave listeners. Full club details are available on the club website *https://www.qsl.net/m0omc/* or Facebook page. Enquires to the club secretary *mØomc@mØomc.co.uk*

Medway Amateur Receiving & Transmitting Society (MARTS)

The Medway Amateur Receiving & Transmitting Society is one of the oldest radio clubs in the UK. The club was founded in 1922, the same year as the BBC, by a group of local enthusiasts led by Bill Nutton G6NU as the Gillingham Wireless League and it became the Medway Amateur Transmitting Society in the late 1920's and finally adopting its current name in 1947 in recognition of the growing post war interest in short wave listening.

MARTS aims to promote and develop amateur radio in the Medway towns and surrounding areas. The club is committed to fostering programmes of providing information, technical reviews and discussion, guided construction activities, supported learning and the development of good operating techniques.

As a club, we respect the technical and operating skills of our members by promoting continuous development through a process of supportive encouragement. The club's traditions are based upon friendship, hospitality, equality, respect and upholding the values enshrined in the Amateur Code and the International Amateur Radio Union (IARU) Ethics and Operating Procedures for the Radio Amateur.

The club arranges a varied programme of talks, demonstrations, discussions, on-air activities, equipment sales, informal evenings as well as social events. We provide weekly Morse practice sessions for those wishing to further develop their skills. In addition to our meetings, we run daily club nets every weekday at 09:30 on 144.650 MHz FM using the club callsign G2FJA. Additional monthly nets are arranged using alternative amateur bands and modes, including HF and DMR where the flag ship call of G5MW can be heard.

Our membership is made up of individuals

who have a wide range of backgrounds and expertise and specific technical knowledge areas and interests in the hobby. These include HF operation, VHF operation, propagation studies, working DX, satellite operation, CW, DMR, using digital data modes, antenna design, construction and contesting. Whatever the interest, our members are always keen to inform and offer developmental support to others and to promote the wide range of aspects that the hobby has to offer, including training for license examination.

The club meets every Friday evening from 19:30 at Tunbury Hall, Catkin Close, Chatham, Kent ME5 9HP. Members and visitors to the club have access to ample free parking spaces. Visitors will be sure to receive a friendly welcome including tea and often doughnuts as the MARTS is renowned for "marching on its stomach".

For further information on what's happening at MARTS, please visit our website *www.marts.org.uk* or follow us on Twitter(X) @ g8mwa. Alternatively, e-mail us at *secretary@marts.org.uk*

Norfolk Tank Museum Radio Group

Discover the world of military radio technology with the Norfolk Tank Museum's Radio Group. We are based at the Norfolk Tank Museum website: http://norfolktankmuseum.co.uk/ near Long Stratton in South Norfolk.

The Radio Group has a strong focus on military radio history and technology, and we're committed to preserving and showcasing this important aspect of military history. Our volunteers share a passion for military radio which they are keen to share with the public and fellow radio enthusiasts. Our membership is an eclectic mix of radio enthusiasts of diverse backgrounds including military and professional communications, academia and the electronic industry.

We aim to give visitors an insight into the importance of radio communication, from both the historical and modern perspective. The opportunity to try Morse Code operating is also available to younger (and older!) visitors.

The radio exhibits are housed in the front section of a Quonset Hut within the Norfolk Tank Museum's site and include examples of World War 2 interception receivers, famous radios such as the RAF R1155/T1154 and the Wireless set 18 and 19, together with NATO, Warsaw Pact, Indian and South East Asian equipment dating from WW2 to the 1980s.

We're always looking for new volunteers to join us. Whether you're a beginner seeking to progress or a licensed radio operator looking to share your knowledge you'll be warmly welcomed. The Radio Group is hands on with the radio collection's exhibits, transmitting by voice and CW modes using the group's permanent special event callsign GB2NTM whenever possible when the Museum is open to the public every Tuesday – Thursday and Sunday between Easter and the end of October, including additional external displays at the Museum's popular Armourfest weekend in August. We are usually on-site Wednesdays and Sundays if you are planning to visit specifically for the radio collection.

We are privileged to be an integral part of the Museum, enabling us to offer unique operating opportunities, including those from radio-equipped vehicles such as the Sultan Armoured Command Vehicle, using its original configuration radios.

In addition to a variety of aerial mountings on the Quonset Hut, the acres of open space around the museum provide tremendous potential to erect a range of both military and amateur radio configured aerials, always being aware of the somewhat unusual field day planning need to avoid snaring a passing tracked and armoured vehicle of course!

You can contact us through our QRZ.com page at *https://www.qrz.com/db/GB2NTM* or the museum to learn more about our military radio collection. Club/group visits are welcome, and it may be possible to arrange these outside normal opening hours; please do let us know in advance, we may even be able to host your club "away day" operating event! The Museum and Radio Group Volunteers look forward to meeting you; both as a visitor and as a new volunteer.

Paisley Amateur Radio Club

The Paisley Amateur Radio Club has a history dating back to late 1988. For six months during 2023, we had the opportunity to have a dedicated 24/7 premises which we named "The Alex Irvine Radio Centre" in honour of long-standing club member, Alex GM7OAW, who became SK during the run-up to the new premises project.

The centre is a fitting tribute to a highly respected radio amateur. Our time there was a great success.

We had to QSY from the Alex Irvine Radio Centre at the end of September 2023 to make way for the refurbishment of the building. Since then, we have had a club EGM to discuss our plans to properly establish the club at this site. We have lots of development ideas and are itching to get back into the building. Full re-decoration, installation of efficient antennas and a grand opening are all planned, as we move the club into a new chapter of its history.

The club is always looking for members keen to help and become involved with the running of the club. If this is of interest to you, just talk to a committee member - there's plenty to do. Our active membership covers many facets of the hobby. There are SOTA activators, IOTA activators, plenty of DXers, mobile operators, contesters, digital operators and military radio enthusiasts.

In 2024 we have members making DXpeditions to the Scottish islands, the US National Parks and several islands in the Caribbean. We had a group travel to Ham Radio 2024 in Friedrichshafen. We recently activated GB7PA, our new 70cm DMR repeater which has attracted much local interest.

The club were delighted to see three recent additions to the club manage further exam success. Chris Bryson and Daz Tilley passed their intermediate exam. Chris now operates as 2M0NZB and Daz as 2M0VSM. Eric 2M0VGT earned his full licence just a week after Chris' success. Eric now uses the callsign GM5RDX. As a result of his performance in contests as an Intermediate licence holder, he wasted no time in using a privilege of his new full licence, to apply for a special contest call sign - listen out for Eric as GM1R in contests.

We have our own channel in the Telegram messaging App to share lots of club information and chat including an active HF spotting page. We run a website, YouTube Channel and Facebook page.

Our 2m VHF FM club net is active Monday night on 144.550MHz from 20:30 across the Paisley area. The frequency is used often by club members during the week to keep in touch.

At time of writing we meet at the Methodist Central Halls in Smithhills Street, Paisley each Thursday night from 19:30-21:30.

Please check *www.parc.radio* for location updates.

Pontefract and District ARS

Pontefract and District Amateur Radio Society welcomes members from all over Yorkshire including Leeds and as far away as Hull.

For a small club, PDARS contributes much to amateur radio. Around 20 members make up the club some newer to the hobby than others and more would be welcomed.

The club meets in it's rented accommodation at the Grange Community Centre, Carleton, a short distance from Pontefract, on Tuesday and Thursday evenings from 6.30 onwards. As the rooms are rented we can have 24/7 access,and all the equipment is ready to use. This equipment includes an Icom 7300, a Yaesu 7800, a 160m delta loop antenna as well as the portable gear for our many events.

Every year we hold Special Event Stations, mainly in the community, for events such as Ackworth Vintage Steam Rally, Battle of Britain commemoration and Yorkshire Day. The club takes part in WAB contests but treats them as an enjoyable time on the air (including having rag chews) rather than serious contesting. Entering is fun, if we win it's a bonus. Thursday Club evenings consist of rag chews on many different subjects and aspects of Amateur Radio, while Tuesdays are for more formal education or construction projects. A few members have other hobbies apart from radio, including photography, motorbikes and computer science.

We also have a long history of construction. However mainly due to a lack of members we don't do much at the moment. The same applies to training for the radio licence exams. Although we don't run courses, we will help with topics on which the candidates are struggling, and the exams are held at the club.

We have recently re-arranged the main room creating a reconfigured space to give us more room so that we can accommodate more members.

Our aim is to encourage more members, including family members and more ladies. We have a new website promoting our cause and hope to continue introducing amateur radio to the local community. All are welcome

For more information visit our website: *www.pdars.co.uk*

Spalding and District Amateur Radio Society (SDARS)

Spalding and District Amateur Radio Society is a community of amateur radio enthusiasts based in Spalding, a market town in Lincolnshire. We are passionate about all aspects of amateur radio and provide a platform for individuals to explore, learn and share their interest in this fascinating hobby.

At SDARS, we offer a supportive and inclusive environment for both newcomers and experienced operators. Whether you're just starting out and looking to obtain your amateur radio license or an experienced operator seeking to expand your knowledge and connect with fellow enthusiasts, our society is here to help.

Our society conducts regular meetings, where members can come together to share their experiences, exchange ideas and participate in various activities. We often invite guest speakers to give presentations on topics related to amateur radio, technology and communications. These meetings provide an excellent opportunity to learn from experts in the field and stay up-to-date with the latest developments in amateur radio.

In addition to our weekly meetings, we also organise field days, where members can set up temporary stations at outdoor locations and engage in radio operations under different conditions. This allows the building of practical skills while experiencing the joy of amateur radio in the great outdoors.

Members are active in the POTA and UK Bunkers on the Air schemes and we enter many of the UKAC contests. This friendly competition has encouraged many to hone and improve not only their equipment but also operating style and confidence on air.

If you're interested in joining SDARS, whether you're a licensed operator or a beginner, we welcome you. This is a fantastic way to meet like-minded individuals, expand your knowledge and contribute to the amateur radio community.

G4DSP, G1DSP, G4OO *www.sdars.org.uk*

SEMARC

Surrey Electronic Makers and Radio Club (SEMARC) is a Surrey (UK) based club for those interested in making electronics projects related to radio communications.

The club was founded by M0REQ and G4NMD and the first meeting, held in September 2022, was attended by ten people with a further ten expressing interest by e-mail. After the founders shared their vision about a co-operative approach to making, learning, the pooling of knowledge, test equipment and books the group were asked what they would like out of such a club.

Out of that discussion came a variety of topics from construction techniques including kits and "from scratch" builds through understanding test equipment, microcontrollers in the shack and onto specialist modes and outside of amateur radio topics such as LoRa. The overriding factor was that the group was to be in-practice based.

The club now seeks to offer an active programme of workshop-based meetings led by the membership.

Workshop topics include or have included:

Antenna construction

Radio transceiver construction

Microprocessor (Arduino/ESP32)based projects.

Construction techniques (Fabrication, PCB design, 3D printing etc)

Home/shack automation

The club callsign is G8KVU which was originally the call of G4NMD but became the callsign of Madeley (father of G4NMD). Madeley was never afraid to try to make something work and never worried if it did not. He regarded it as a learning experience. In his late 80s and into his 90s he continued to grapple with modern radios and computer technology until becoming a SK in 2020. The call G8KVU is a fitting callsign for a club that learns by doing.

Meetings are held on the third Wednesday of month 19:30-21:30 at Grafham Room, Horsham Road, Grafham GU5 0LJ.

what3words address for the club is *https://w3w.co/hiring.hedgehog.prayers*

For further info on this practical club contact G4NMD or M0REQ

https://membermojo.co.uk/semarc

South Manchester Radio Club (SMRC)

South Manchester Radio Club is a member led club of over 30 members which meets every Thursday at 8:00pm at the Woodheys Club, 299 Washway Road, Sale M33 4EE.

SMRC runs a variety of monthly lectures interspersed with weekly discussions on a wide range of radio related topics. Interests and activities include but not limited to, communications on the LF bands right through to the Microwave spectrum and on a wide variety of modes from CW and SSB through to a large variety of digital modes. Many members have an interest in working at QRP power levels.

Weekly nets on the HF, VHF and Yaesu System Fusion (with many members working cross mode from DMR). Individual members also set up group schedules to supplement the fixed nets. This year the club will take part in RSGB on air competitions. Outside activities are planned for the summer months.

Club members often instigate club projects from aerial construction to radio related software, with the vast wealth of experience of the membership being utilised. The club website has a Wiki page that supports club projects information at https://smrcc42.groups.io/g/main/wiki

Every year the club runs a construction competition, projects include any radio related, self-built item. Entries come from a broad range of radio related projects, whether hardware or software based. The standard is high, being won this year by a home-brew 10m valve SSB transmitter.

The club regularly attends the NARSA rally at Blackpool and this year finished second in the club website category. More details of our club can be found at our website smrcc.org.uk. or via e-mail contact: - *contact@smrcc.org.uk*

SMRCC has the following call signs: - G2HW, G8SMR & G3UHF which are in regular use.

South Normanton, Alfreton & District ARC

The South Normanton, Alfreton & District Amateur Radio Club is more widely known as SNADARC in the local area. We are located on the border of Nottinghamshire and Derbyshire close to junction 28 of the M1 motorway, hence the name of our very successful radio rally "the Junction 28 Rally".

We meet at a local community hub called the Post Mill Centre and we have a mast for the sole use of the club at that location and use a meeting room as our club room, while the large bar area is used for our regular natter nights.

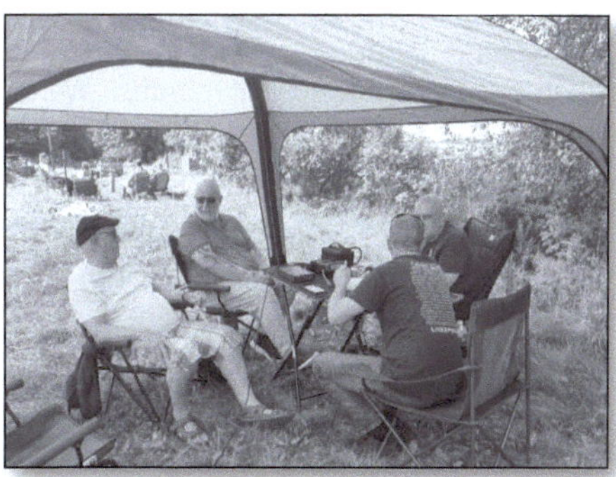

The club has around 60 members and interests cover all aspects of amateur radio, SWL and CB. We meet most Mondays and we have a variety of events from natter nights, radio operating, presentations and our very popular junk sales. The club owns a number of radios and it has been slowly updating them and now has base station and portable kit for members to use.

We are very active at weekends. Whenever the weather is fit you will find the club on a hilltop in Derbyshire operating across VHF and HF. We are proud members of the Royal Signals Amateur Radio Society with a number of our members being ex-forces and one of our outdoor operating locations being a memorial to the the Mercian Regiments (Crich Stand Memorial). The club will be running a number of special event stations throughout the year and we welcome everyone to come and join us.

Our website can be found at https://*www.snadarc.com*

email *secretary@snadarc.com*

Thames Amateur RG

Thames Amateur Radio Group is a South Essex based club, in regular discussion with other Essex clubs and RSGB Region 12 on making the most of all of our expertise and contacts. We have a lot of expertise and experience across our membership and a wish to help and develop others interested in the hobby. You are just as likely to find people who will delight in chatting to you about the hobby or life in general as you are to be steeped in radio activities alone, as we are a sociable group.

We live-stream all the face to face meetings we can, and have a good handful-plus of joiners through that route if they can't get out, or live at too great a distance. This development has been well received.

We meet in person on each month's 1st and 3rd Fridays, the 1st being presentation and Q&A based, the 3rd an activities and hands-on session where members bring in their gear, their questions on their radio projects, challenges or proud solutions, or their keen interest in our chosen themes. On Fridays in each month when we don't meet in person, we hold a net on GB3DA. Output 145.725 / Input 145.125 / CTCSS 110.90. Our back up if GB3DA isn't possible is GB3ER: 434.675MHz Input, 433.075MHz Output, CTCSS 110.9Hz.

Our Friday night nets are welcoming gatherings – anybody with an amateur radio licence who can reach us is welcome to join. A rota of members hosts them, so there is variety in styles, and nobody has to do the task more than once every couple of months. We get double figures of participants on most nets. We run them with the welcome use of the Essex Repeater Group's repeaters.

In the months since our AGM last November, to give you only a flavour of what we do, we have had a lively and very supportive CW evening, and talks on SSB, the use of and problems with radio in the emergency services, the joys of satellite tracking, microwave frequencies' set up, reach and versatility, tracking air traffic communications through your radio, the many uses of APRS, and what "going mobile" takes and why it's such fun – among many more! We will be part of the Military Machines event in mid-August 2023, and will take part in Gateways on the Air (GOTA.) We are found running radio activities at events around the Southend Airport based Vulcan bomber. We regularly support radio rallies, the Scouts' Jamborees on the Air (JOTA,) RAYNET deployments, and much more.

Contact: *Chair@thamesARG.org.uk* or *Secretary@thamesARG.org.uk*

Website: *www.thamesarg.org.uk*

The Lothians RS

The Lothians Radio Society GM3HAM, affiliated to the RSGB, Scottish Charity (SC045038), was formed in the early post war years of the 1940s. The current membership contains a pool of expertise not only covering amateur radio but extending into many other areas of science, commerce and engineering, who together continue to promote amateur radio as a vehicle for education and entertainment.

The society has successfully participated, in many contests since the 1940s, more recently specialising in VHF. Our formal meetings, which are open to the public, are held twice a month and consist of a lecture program drawing on speakers from a variety of backgrounds from amateur radio through to academics and professionals in the broadcast and communications community.

Each lecture considers different aspects from new technology, construction, test equipment, operating practice, application of radio in adjacent spaces through to historical topics.

Twice yearly the society holds a well-attended recycle and reuse surplus sale in a larger venue with a diverse selection of radio and test equipment content. Members and non-members can offer equipment for personal sale and the society also raises funds for itself and others as part of its silent key service of equipment disposal at a difficult time for families.

Within the society, there is a pool of specialist radio knowledge, with sage advice and support always available to other members. We run a top-band DF hunt each year which illustrates the breadth of experience and willingness to participate by the membership and has a high attendance.

Further interests of our membership include: the design, installation and maintenance of several beacons, HF, VHF, UHF and microwaves, SOTA, RAYNET, contesting, digital data modes (e.g. WSPR) and broadcasting GB2RS. As a group we strongly support others, being the key contributor to the annual Scottish Microwave Round Table, supporting the nearby Scottish Museum of Communications and having programme exchanges with other radio clubs and the local IET retired members group.

In addition to our Wednesday lecture programme we have informal social meetings in a local hostelry. To find out more about the meetings and venues visit our website *www.lothiansradiosociety.com* Our web site is always up to date and contains a diverse range of topics and historical items.

For a more personal approach email our syllabus secretary James Gentles GM4WZP using *secretary@lothiansradiosociety.com.*

Trowbridge & District ARC

We are a friendly Club that's been established for 40 years. Whether you already have an interest in Amateur Radio or a new member wishing to find out more please come and join us at one of our meetings or contact us if you prefer, everyone is welcome at our Club.

Our monthly talks and activities cover a wide range of topics, our members interests span from HF to microwaves with a range of operating modes. Our hobby is truly fun, exciting and diverse with an interest for everyone.

As a Club we aim to be proactive and diverse in various activities throughout the year. We take part in a number of activities, quite a number of members take place in the RSGB UKAC's on all bands from 6m to 3cm. In June we have an annual 2m direction finding competition. In November we hold our constructors cup competition for kit built and "built from scratch" projects and we award a trophy to the winner of each category. We must not forget another jewel in our Crown which is RSGB VHF NFD in July, this is a full weekend which can be enjoyed in many ways, this ranges from setting up and operating in the Contest, enjoying the beautiful views of the Mendip hills or simply attending and enjoying the company and banter amongst each other (there is even a nice hearty breakfast if nothing else tempts you!) We also provide CW training at every level of competence & experience

Naturally we have a daily morning net and Sunday evening net on 2m FM.

For any enquiries please don't hesitate to contact us, we look forward to hearing from you.

G0BKU Secretary, Trowbridge & District Amateur Radio Club

email: *secretary.tdarc@gmail.com*

Wales Digital Radio Group

The Wales Digital Radio Group was founded during COVID-19 as a place for people to meet virtually via video call. Not only did it prove popular but was a life-line for many who were isolated due to the restrictions in place at the time.

The WDRG is an online club with members not only in Wales. We are an all-inclusive group with members worldwide. Our website is *www.wdrg.uk* where details of online meetings, club nights, membership details and club events can be found. We offer help and advice on all radio related topics, DMR, D-STAR, C4FM as well as setting up hotspots and digital modes like FT8 etc. We also have members that have knowledge in working satellites.

The WDRG have some core events that we get involved with, SOS Radio Week, Railways on the Air as well as some new events planned for 2025 raising awareness to some organisations that various members are involved with.

We plan to be at NARSA, Northern Amateur Radio Societies Association Exhibition (Blackpool Rally) where we hope to operate from the RNLI station again and have members on hand to demonstrate some digital modes from the rally, Llanbedr (Shell Island, North Wales) for SOS Radio Week in May 2025 as well as Porthmadog for Railways on the Air in September 2025

If you need any technical help with any aspect of Digital modes we have a dedicated e-mail address which will put you in touch with one of our experienced members *tech@wdrg.uk*

For group membership please e-mail: *membership@wdrg.uk* or *info@wdrg.uk* for general enquiries.

Warrington ARC

Warrington Amateur Radio Club recently celebrated its 75th anniversary and a lot has changed.

We find ourselves now the other side of Covid but it has changed us. For the better. We now hold our main meetings as mixed format – face to face and zoom at the same time – it allows members across the country and beyond to participate.

It has had other effects too. In 2023 we expanded the HF and satellite stations and as a result have found ourselves lacking space in the club's shack so have moved the test bench out into the main room. Zoom has been good in this regard as it has freed up space in the main room – we are after all enjoying a communications-based hobby.

In 2024 we plan to continue regular talks interspersed with DF hunts, BBQs and trips – and are hoping to get back into contesting.

Having repurposed our QO-100 SSB ground station from the NARSA rally as a club facility, we can now look forward to more operations via the satellite as well as on HF

If you want to join us, call in on a Tuesday night for the main meeting, or Thursday morning for a coffee and a chat or Sunday afternoon for a technical session. We've also re-instigated club nets but with a twist – these are via YSF & P25 – more details on the club website: *warc.org.uk*

Welland Valley Amateur Radio Society

The Welland Valley Amateur Radio Society is a small club located in the town of Market Harborough in south Leicestershire. Despite being small we have a large presence in the town and support many local events.

We hold a net on 145.275 FM at 19:00 on the first Monday of every month and meet at Great Bowden Village Hall at 19:00 on the third Monday of every month.

We have a full Calendar of regular events every year starting in April when we activate the ROC bunker at Sutton Basset. In June we run a special event station from Symington's recreation ground in support of the Market Harborough Carnival. In August we operate a three-day special event station from the Market Harborough Showground in support of "Operation Market Harborough" which is a 1930's themed living history event. In October we again run a station from the ROC bunker in Sutton Bassett in support of the "UK Bunkers on the Air "event. We also try to fit a couple more field days in during the summer.

We have a very active construction group and every year at our October club meeting we build our "Project of the year". Details of previous construction projects can be found on our web site *http://www.wvars.com/index.html*

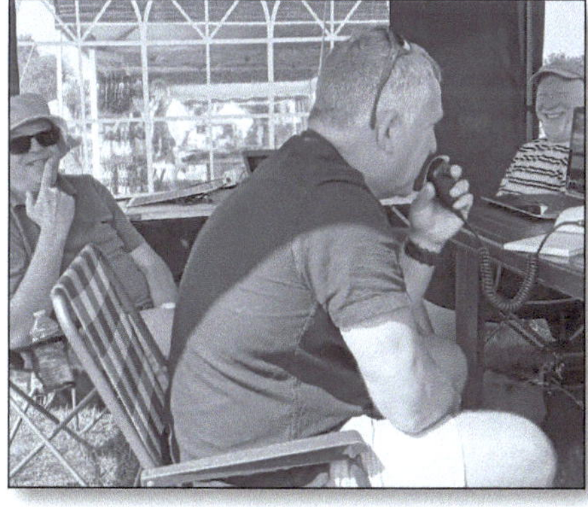

Members run several internet gateways for digital radio modes and a repeater in the town, again details can be found on our web site.

You can contact us via our club e-mail: interested*_in@wvars.com*

We have a public Facebook page which can be found at https:*//www.facebook.com/groups/3543791355677024* and a private Facebook page for members only.

Peter Rivers G4XEX Sec WVARS

West Glamorgan RAYNET-UK

West Glamorgan RAYNET was originally formed in 1976 and affiliated to the RSGB. Since 2018 we have also been members of RAYNET-UK. Our nominal area of operation includes Swansea, Neath, Port Talbot, the Gower & Llanelli Coast as well as the western end of the Brecon Beacons.

We have a wealth of experience supporting various community events, walks, marathons, cycle rides, motor sports etc. We provide callout services during periods of inclement weather. Much of the terrain we operate in is quite onerous, with heavily forested areas, deep and winding valleys and secluded beaches being a problem. This can be quite challenging requiring local knowledge and detailed planning.

Last years Duke of Edinburgh's Wales Gower Charity Walk proved to be quite a challenge. There had been significant route changes from previous events in the area, with sea mist reducing visibility to just 100m, alternating with strong sunshine and the organiser changing the cut-off time several times, gave rise to personnel safety issues. In contrast, the Admiral Swansea Bay 10k was a successful event with no issues. A Home Office exercise requiring communications from local authorities to the relevant Police HQ was disappointing as was an All-Wales communications exercise.

We make great use of Yaesu mobile transceivers in cross-band (2m/70cm) talk-through mode and used an FTM-500DE for the first time on the Gower Walk both in simplex and talk-through. The forward-facing speaker proving to be an excellent addition, along with just a 10A current draw on maximum power. We compliment these radios with a mobile "Repeater-in-a-Box", using a pair of Yaesu FT-8900R's on the RAYNET allocated in-band talk-through frequencies. Both transceivers link together with an Audio/PTT Interface and use a common antenna via a Duplexer.

Antennas vary from high gain whips on vehicle magnetic mounts to a 7.5m drive-on mast to support a dual-band collinear. Power varies from a portable/mobile generator at 230v with 10A-30A 13.8v DC power supply units as appropriate to 80Ah leisure batteries.

New members are always welcome, we usually meet on alternating second Sundays of the month in the first-floor conference room at Hazel Court, Sketty, Swansea SA2 8BP.

Find us on Facebook as West Glamorgan RAYNET or call 07967 043124 for further information.

West Kent ARS

A thriving club, with over 40 members, mainly from the Tonbridge, Tunbridge Wells & Crowborough areas, who enjoy a wide range of interests ranging from QRP HF CW with homebuilt gear to EME with a 3m dish! We hold monthly meetings involving amateur radio or social events at Bidborough Village Hall, where we now have HF & VHF antennas for live demonstrations. Members also meet on the first and third Friday evenings for practical sessions, building hardware & its firmware. Our nets are very popular and a sizeable portion of the Society support us from home in RSGB contests and at the HF & VHF Field Days & Trophy contests, for which the Society recently upgraded its HF contest gear with a FTdx10. By invitation, we support Nutley Windmill annually, running HF, VHF, QRP CW & FT8 stations for the Mills on the Air weekend as GB2NW. We are also seeking opportunities to extend our special event activities to Museums & Railways.

Formed on 7 April 1948, in 2023 the Society celebrated its 75th Anniversary with a gala dinner at Salomons Estate with its museum of the renowned inventor 2nd Baronet Sir David Salomons, famous for the introduction of domestic electrical lighting and, as a founder of the RAC, responsible for removing many of the legal restrictions on the use of the motor car in England. The Society call-signs GX-1WKS & GX3WKS, using a /75 suffix, enabled individual members to celebrate & promote the Society's 75th in 2023, with our Foundation & Intermediate licensees hosted by a Full Licence holder,with great success. A summer BBQ in conjunction with our VHF Field Day operation completed the celebrations.

Visitors are always welcome at our meetings, details of all the Society activities being on our website: *www.wkars.org.uk.*

West Manchester ARC

West Manchester Radio Club (WMRC) meet at Gin Pit Village (Astley) every Thursday evening from 7:30 pm until 10 pm. Our call sign is GX4MWC / G4MWC.

We are based at Astley and Tyldesley Miners Welfare Club which has a bar, meeting rooms and snooker (if you play). We are a friendly club and enjoy a beer and a chat when not operating in our shack. We have excellent facilities and operate on most bands using our tower (with multiple antennas). We built and operate a local UHF repeater (Yaesu System Fusion) call sign GB7WM. The repeater is connected to the North West Fusion Group and also accessible via the Wires-X fusion network.

Visitors are welcome to come down to the club three times as a guest before needing to join the club. We are one of the cheapest clubs in the UK with annual membership of just £20 (or £10 if you are over 60). This cost includes Membership to the Miners Welfare Club with its subsidised bar and other activities.

We have a good mix of abilities from beginners studying for the foundation license through to intermediate and full license holders. During the year we run a number of events including; The Red Rose Radio Rally, field trips, Museums on the Air (from Astley Mining Museum), annual bowl & BBQ and the mandatory Christmas Party!

Our web site (wmrc.co.uk) includes lots of information about the club's past and future events plus photos. We are always looking for new members so why not pop down for a drink and meet us one Thursday evening. We also have a Facebook page for members plus a public group called 'West Manchester Radio Group' if you want to connect.

Website: wmrc.co.uk. Contact: *secretary@wmrc.co.uk*

Club venue: Astley and Tyldesley Miners Welfare Club, Gin Pit Village, Astley M29 7DW

Winkleigh Repeater Group

Winkleigh Repeater Group was formed in Autumn 2023 to promote Amateur Radio in the Mid-Devon area. The group operates two repeaters at Winkleigh, (IO80AU) maintained by Simon G4MQQ, Phil G6DLJ, Cliff G4PZK and Greg M7GGZ. Both repeaters share a Diamond X30 antenna. The group is self-funded and rely on the generosity of local amateurs and the proceeds of its Winter rally.

GB3LZ (2m) is a Yaesu DR2-X running Yaesu System Fusion (C4FM and FM) and is normally connected to Wires-X/YSF room 418393. Set your radio to Rx 145.6625. Tx 145.0625. Analogue FM takes priority, FM users should use a 77Hz CTCSS tone to access and 77Hz tone to mute digital transmissions.

GB7LZ (70cm) is a Motorola DR3000 running Tier-2 DMR, Set your radio to Rx 430.9125. Tx 438.5125, Colour code 1, Slot 1 talk group 9 is linked to Southern Fusion Wires-X/YSF 41893. Slot 2 (also TG9) is linked to CQ-UK for local working and can also be changed using dial-on-demand to regional talk groups such as Phoenix TG810 (South West). The back-office network switching and routing is run by Dom M1AXM, sysop of GB7EQ.

Our Mid-Devon Amateur Radio and Electronics Fair is held in December at Winkleigh Sports Centre.

To contact us the group can be e-mailed at *wrg2024@hotmail.com* or using IP voicemail - Hams over IP ext. 20028

Online Clubs

Essex Ham (Region 12)

Essex Ham is more than a club, we're an online amateur radio community, and from our humble beginnings as a blog over ten years ago, the group now has over 3,400 members around the world.

It's fair to say that things have changed for clubs as a result of the pandemic, with amateurs and clubs now doing things a little differently. We've seen an increase in interest for video content, especially for newcomers, and at field events, we've met several new licence-holders who've studied online but now need some practical help and support.

We've recently formed a solid link with the East Essex Hackspace, which encourages people in Essex to gets hands-on with "making stuff" – the space has rooms for 3D printing, laser cutting, electronics, woodwork and metalwork, and as there's a natural overlap into our hobby, we're regularly on-air from the Hackspace to show off what we do, and interact with other "makers".

Our free online Foundation training courses continue to be a large part of what we do. We've now trained over 10,500 students. There's little doubt that the ability to train and sit exams online has made a massive difference to take-up of the hobby, due to the improved access to courses and exams, and we'd also like to acknowledge the hard-working volunteer RSGB exam invigilators and the exam team at RSGB HQ for helping to make it easy for students to take their exams.

We've taken part in some fun and interesting initiatives. Broadcasting started in Essex back in 1922, and we were on-air to commemorate the moment that the first broadcast took place, with live video and audio streams. We were also proud to support one of our members, Andrew M0ONH with a campaign to raise awareness for Prostate Cancer, a cause that affects 1 in 8 men. We ran several events for GB8PCA, and were joined by the GB1NHS team. We supported one of the largest JOTA events in the region, adding a "blindfold radio maze" challenge to the fun, using radio headsets from dBD Communications, providers of comms to the rail industry. Field events included our regular St George's Day get-together, a Jubilee event, and participation in the county-wide "Essex 2m Activity Days". Our weekly "Monday Night Nets" continue to be a popular on-air meeting place.

As well as being there for our members, we also actively support clubs around the UK. Our training slides, videos and handouts are available free to all tutors, and we have almost 200 tutors in our independent trainer's group who receive our regular Tutor's Newsletter. We also work hard to promote and champion the hobby.

Regardless of where you live, we'd love you to join our growing online community. Membership is free and open to all. Join us at *www.essexham.co.uk.* You'll also find us on Twitter, Facebook, YouTube and Discord.

Wales Digital Radio Group. (Region 7)

The WDRG is an online group with members in Wales, England, Scotland and Cyprus. We meet online via Google Meet every Thursday at 20:00. Where we have a general chat all things radio, as well as some technical help and advice.

The aim of the group is to encourage outdoor operating, either portable or from the back garden. In a group or on your own. The WDRG organise some field operations to take part in events like SOS Week, promoting the RNLI, and Railways On The Air celebrating the local rail heritage.

Contact : Daryll . MW0TTF
Darylldmellow@gmail.com

National Affiliated Clubs & Societies

AMSAT-UK *(G0AUK)*

This is the UK national society specialising in amateur radio satellite matters. It has approximately 700 members, produces a regular publication, Oscar News, for its members four times a year. It maintains a web site http://amsat-uk.org and holds weekly nets on 3,780 KHz +/- QRM on Sundays at 10.00 local UK time and also on 10,489,780 KHz on the QO-100 satellite, also on Sundays, normally at around 10.30 UK time. You do not have to be a member of AMSAT-UK to join either net - anyone with an interest in satellites is welcome.

In addition, AMSAT-UK organises an annual Colloquium. In recent years these have been held at the same dates and venue as the RSGB Convention. In 2020 and 2021, due to COVID restrictions, it held a very successful virtual webinar events. Both types of event provide an excellent chance to rub shoulders with experienced satellite operators and satellite designers. It is hoped in 2022 to hold a face to face event in October in Milton Keynes.

AMSAT-UK has been responsible for the development of a number of operational CubeSat spacecraft that follow the FUNcube format. These combine linear amateur transponders with STEM outreach telemetry Membership of AMSAT-UK is by donation, for which there is a suggested minimum. Any funds remaining after running expenses, are used to fund satellite design, building, and launching. Two forms of membership are available, either postal or electronic. Those with postal membership pay a little extra (to cover postage costs) and have a printed copy of Oscar News sent to them by Royal Mail.

Members with electronic membership down load Oscar News as a pdf file.

AMSAT-UK runs a small online shop, http://shop.amsat-uk.org, at which members, or new members, can make their donation. The shop also has a few items of interest to satellite operators. Enquiries and applications forms for membership can be sent with an sase to the Hon Sec, Jim Heck, G3WGM, Pickles Orchard, Great Hampden, Great Missenden, Bucks, HP16 9RE. He can also be contacted via email at: *g3wgm@amsat.org*

If you use amateur radio satellites be prepared to pay something to the organisation that builds and designs them – AMSAT!!

British Amateur Radio Teledata Group *(BARTG – G4ATG, GB2ATG)*

We are a web based datacoms group. Our aim is to encourage and promote the use of all types of amateur radio datacoms (such as RTTY, PSK, WSPR, Olivia and JT modes).

We run four Contests each year. In January we run our 24 hour RTTY Sprint Contest, in March we run our our 48 hour HF RTTY Contest, in April we run our 4 hour 75 Baud RTTY Sprint Contest and in September we run our 4 hour PSK63 Sprint Contest.

All Single Operator All Band (SOAB) results count towards the BARTG Championship. The leading station overall in the Championship will become the BARTG Diddler of the Year and will receive the prestigious BARTG FlexRadio Trophy kindly sponsored by FlexRadio.

Our Awards Scheme is open to both licensed amateurs and listeners. Its entry level will appeal to the occasional operator and its upgrades will appeal to the serious DX chaser. We have awards available for all data modes ranging from RTTY to the latest JT modes.

Our Quarter Century award is for making contact with 25 different ARRL DXCC countries. We have a series of Continent awards (African, Asian, European, North American, South American and Oceania) for contacting countries in those continents. Logs submitted to our contests can be used to apply for any of our awards. Award certificates are available for a modest fee through our website.

We have a small fund (dependent on income from fees for our award certificates and on donations to us) that we use to sponsor a few carefully chosen DXpeditions each year.

We are happy, via our For Sale/Wanted web page, to help people give away, sell or obtain datacoms equipment or seek technical advice.

Our website is central to everything we do so please visit it for more information about us, including our contests and awards.

www.bartg.org.uk

Secretary: Ian Brothwell G4EAN, 56 Arnot Hill Road, Arnold, Nottingham, NG5 6LQ.

bartg@bartg.org.uk

British Amateur Television Club *(BATC – RS38114)*

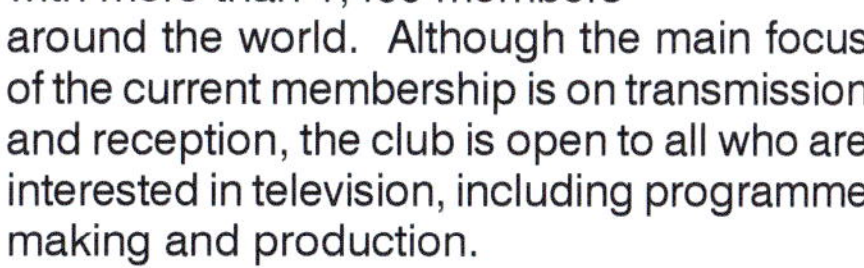

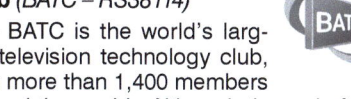

The BATC is the world's largest television technology club, with more than 1,400 members around the world. Although the main focus of the current membership is on transmission and reception, the club is open to all who are interested in television, including programme making and production.

Amateur TV can be transmitted on most bands above 50 MHz, and high definition TV pictures are routinely exchanged over distances of up to 50 km on 437 MHz, but DX of hundreds of km can be achieved under exceptional propagation conditions. TV pictures can also be transmitted and received through the QO-100 satellite to most of Europe, Africa and Asia using just a 1.2 metre dish.

The club sponsors a number of hardware and software projects (such as the Portsdown Transceiver and the Ryde receiver) to enable the construction of Amateur TV equipment by newcomers to this aspect of amateur radio. The club also grants bursaries to support capital expenditure for projects such as Amateur TV repeaters and open source software development.

In addition to producing a quarterly colour magazine (available to members as a hard copy or a download), the club runs a number of online services including a forum https://forum.batc.org.uk/, a Wiki *https://wiki.batc.org.uk/* and a video streaming site https://*batc.org.uk/live/* where ATV repeater outputs and members' streams can be viewed online. The club has a range of "hard to get" components and a selection of printed circuit boards for some of the club's projects available in our online shop: https://batc.org.uk/shop/

The BATC organises 2 "Conventions for Amateur Television" (CATs) each year. For 2025, an in-person Convention is planned for August, and an online convention, with technical lectures on ATV topics, is planned for October. The Conventions are open to both members and non-members with an emphasis on helping beginners to get started in this more challenging aspect of amateur radio.

BATC members run stands at major Radio Rallies throughout the year to demonstrate ATV and provide advice to those interested. In 2025 it is hoped that the rallies at Blackpool, Dunstable Downs, Newbury and Reading will have a BATC presence, as well as the

RSGB Convention and the Newark HamFest. The Club offers free membership (with access to online copies of CQ-TV) to students in full time education. For more information, or to join our club, see our website: *https://batc.org.uk/* or email the secretary:

memsec@batc.tv.

British Railways Amateur Radio Society

British Railways ARS (G4LMR) is a small and friendly society founded in 1966 by railway workers who were interested in amateur ra-

dio. Membership is open to anyone, licensed or listener, interested in railways of any gauge (including metros, light rail, trams and model railways) and amateur radio.

Our magazine, "Rails and Radio", is published four times a year and sent by post to our members. Articles and photos from members (and non-members) are always welcomed by our editor. We operate regular nets (see our website for frequencies and times) and all are welcome to join in.

Our members occasionally run special event stations during the year, usually with a railway theme and often at or near heritage railways.

BRARS is affiliated to the RSGB and also to FIRAC (Federation Internationale des Radio Amateurs Cheminots - International Federation of Railway Radio Amateurs Cheminots).

To find out more about us please go to our website or contact our Membership secretary: Richard Waterman G4KRW, 170 Station Road, Mickleover, DERBY, DE3 9FJ

membership@BRARS.info

www.BRARS.info

We are also on Facebook and X. Issued by Ian Brothwell G4EAN (BRARS secretary).

British Top Band DF Association (BT-BDFA)

The BTBDFA was formed in 2000 to centralise the organisation of amateur radio Direction Finding (DF) on the 160m band in the UK. The association organises the qualifying rounds and the final of the National DF championship. The winner receives the handsome RSGB Trophy, which dates back to 1951. The main aim of the association is to increase the popularity of Top Band DF, which it does by organising events, providing lectures to radio clubs, putting interested people in touch with each other, and by writing articles for magazines, etc.

website: *www.TopBandDF.org.uk*

Secretary, Bill Pechey, G4CUE.

Tel: 01491 680552.

email: *secretary@TopBandDF.org.uk*

British Young Ladies Amateur Radio Association *(BYLARA – M0BYL)*

BYLARA was formed on 29th April 1979 to further YL operation in Great Britain and to promote friendships with YLs and OMs worldwide. The association aims to encourage good operating, as well as persuade all YLs across the Great Britain and the World on to the air and to explore more of the hobby, renewing old friendships and making new ones, across the airwaves.

We would also like to encourage good operating techniques and courtesy to other operators at all times and also aim to encourage and introduce up and coming Guides, Brownies,

Scouts as well as other young people who would like to go into the Hobby. We now have Members worldwide and are growing all the time.

We have had our Callsign (M0BYL) since 1998 and more information can be found on our Web Page. Our Members take part in various Special Events, Dxpeditions, Contests and have our own Net Night. We are always open to new members, OMs and family groups.

For more information, contact:
The Chairmam – Carol, 2E1RBH on 01305 820400 or

email: *carolhodges1@btinternet.com*

www.bylara.org.uk

CDXC – The UK DX Foundation *(M0CDX)*

CDXC

CDXC is the UK's premier DX foundation and the biggest in Europe. We are dedicated to encouraging excellence in DXing and contest operating and have around 750 members – a quarter of whom are overseas. Formed in the 1980s, CDXC members share a common interest in HF/6m DXing and contesting.

There are FIVE reasons (at least!) why you would benefit from being a member of CDXC:

1. The CDXC community. Become part of the UK's active DX and HF contest commu¬nity. You will belong to a Club of like-minded people who are also passionate about your favourite hobby!

2. Help sponsor DXpeditions. CDXC sponsors DXpeditions to rare and hard to reach DXCC and IOTA entities. Part of your subscription goes towards making significant CDXC con¬tributions to Club Log, WRTC and YOTA. You will see us on the sponsors' page of all major DXpeditions.

3. The CDXC Digest magazine and Website. Receive, what we think is the best DX magazine every three months - the CDXC Digest is available printed and on-line. You can access back editions by following the link *www.cdxc.org.uk/CDXC_Digest_Ar¬chive*.

4. Learn from other CDXC members. Improve your DXing and contesting performance by learning from other members. CDXC's reflector provides advice and guidance and debate on the hot topics of the moment. We run a real-time Skype DX chat channel for members as well.

5. CDXC social events: great social events and get-togethers. CDXC annual convention and dinner, our stands at the RSGB convention and HAM RADIO Friedrichshafen are attended by well-known DXers and Contesters.

But most of all, CDXC is about being a member of a DX Community, of joining with like-minded people who want to see high standards of operating and want to work HF and 6m contests and make DX contacts on all modes and with anything from wires in trees (most of us!) to beams at 60ft.

You can sign up on line. Just go to our web¬site *www.cdxc.org.uk.* Follow us on X: @cdxcuk and find us on Facebook.

Chelsea Pensioners Radio Club

Club Call Sign *M0INU,*

Permanent Special Event Station GB4CP

e-mail: *royalhospitalradio@gmal.com*

Mike Hall 07970 724105

The club meets every Tuesday 14.00 to 16.00

Though can be on air by mutual agrement

The Duxford Radio Trust

The Duxford Radio Trust (DRT) is based in the East Anglia region of the UK and comprises an active group of radio specialists, plus supporters, many of whom are from the radio communications industry or Armed Forces. The group researches, collects, conserves, restores, displays, and demonstrates historic radio communica- tions, radar, and navigation equipment used in military conflict and civil emergencies, for the education of the public.

DRT also operates the permanent exhibition radio station G0PZJ/GB2IWM, using both modern and historic military equipment.

DRT was formerly the Duxford Radio Society (DRS), and previously provided the Radio Section at the Imperial War Museums, Duxford for 32 years. The group was founded in 1986 by Richard Pope G4HXH and Major John Brown G3EUR, renowned wartime designer of S.O.E. and clandestine radio equipment.

DRT is currently in the process of establishing new public exhibition and engineering facilities- at the Signals Museum RAF Henlow following its departure from IWM, prompted by a change of policy and direction at IWM Duxford and HMS Belfast, London.

DRT is entirely self-funded by its local and distant supporters and in addition to its public exhibition activities, normally publishes three issues of a historical/technical journal each year and a monthly newsletter.

The Trust welcomes new active members who share an interest in the history, technology, conservation, and restoration of military communication equipment, and who wish to explain and demonstrate this subject to the public and operate the radio transmitting station, using both vintage and modern equipment. DRT also welcomes distant supporters who are not able to act as active explainers, operators or restorers in the Cambridgeshire or Bedfordshire area.

Details can be found at *www.duxfordradio.org* and DRT can be contacted at: *duxfordradio@gmail.com*

Essex CW ARC

(ECWARC – G1FCW, G4C)

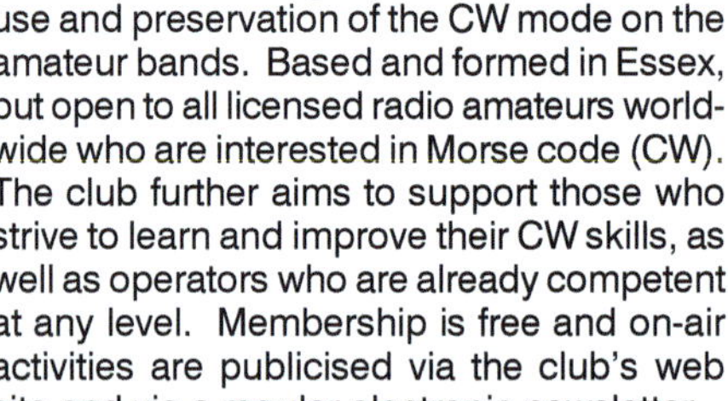

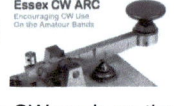

Essex CW ARC has as its mission to encourage the use and preservation of the CW mode on the amateur bands. Based and formed in Essex, but open to all licensed radio amateurs worldwide who are interested in Morse code (CW). The club further aims to support those who strive to learn and improve their CW skills, as well as operators who are already competent at any level. Membership is free and on-air activities are publicised via the club's web site and via a regular electronic newsletter.

The club runs weekly CW training classes via Skype for beginners, intermediate and advanced operators. There are weekly on-the-air training nets on both 80m and 2m, as well as a weekly GB2CW QRS transmission on 80m. Each October we hold a one-day "boot camp" where CW enthusiasts of all abilities meet to practise and share their passion for CW. In November is the annual activity week, which is a great opportunity for CW practice and to qualify for our operating award.

For further information please visit the club's website at: *http://www.essexcw.org.uk*

First Class CW Operators' Club
(FOC – G4FOC)

The First Class CW Operators' Club (FOC) was founded in 1938.

FOC is limited to approximately 500 members, who will typically meet both on air and in person. The Club holds a wide range of social gatherings throughout the year, in many different countries, with members also often meeting at major events such as Dayton, Friedrichshafen and the RSGB Convention. On-air activities include the FOC Marathon in February, and two FOC QSO Parties, which are normally held in March and September.

In 2023, FOC celebrated its 85th anniversary. The aim of the Club is to promote good CW (Morse code) operating, activity, friendship and socialising. The Club is UK-based, with many international members. The character of the club is best expressed in its motto, "A man should keep his friendship in constant repair" (Samuel Johnson, 1755).

Membership of FOC is achieved through a process of nomination by existing members, based on on-air activity and other criteria which are explained on the Club's website at *www.g4foc.org.*

The Secretary email: *secretary@g4foc.org*

FISTS CW Club
(GX0IPX, GX3ZQS & MX5IPX)

Since it was founded by Geo Longden, G3ZQS (SK) in 1987, FISTS CW Club has grown to be the largest open-membership CW group in the world. Based in the UK, the club has international chapters: FISTS North America, FISTS Down Under and FISTS East Asia.

The FISTS club promotes using Morse code on the air. The club encourages high standards of operating alongside accuracy in sending and receiving. Our membership is made up from beginners – who can benefit from tutoring and practice provided by club members – through established CW operators and those returning to the hobby after a break, to expert amateurs and those who were professional telegraphers.

The club now organises a wide range of activities under the direction of Activities Manager David, G4YVM. Alongside regular activities, such as the long-running ladder on two Sundays every month, there is a wide range of events – all friendly and some more competitive than others – along with awards schemes.

In the UK there is always a great deal of activity on 80m and 40m – 3.558 and 7.028MHz – especially on ladder Sundays. You'll also find FISTS members on VHF – 4m and 2m.

FISTS is an active supporter EuCW (*https://www.eucw.org*) and its on-air activities.

You can discover more about the FISTS CW Club, our activities and awards by visiting the comprehensive club website at *www.fists.co.uk.*

New members will always receive a warm welcome to FISTS. You will find encouragement and support wherever you may be on your Morse journey – whatever your experience or your code speed. There really is just one, single criterion for membership of FISTS: a love of the Morse code.

Brasspounder is the club's monthly newsletter and Key Note, its quarterly journal. Key Note is available on paper, delivered by post, or digitally. A sample copy may be requested from our web site – and you can also find out much more about FISTS at *www.fists.co.uk*

GMDX Group - Scotland's DX Association *(GM5A, GS8VL)*

The GMDX Group, based in Scotland, caters for all radio amateurs who have an interest in DXing, contesting and award chasing. The group has its own award scheme and produces a quarterly magazine, The GMDX Digest, which is sent out to members electronically. The Group has supported most of the major international DXpeditions and criteria for sponsorship is outlined on the group website.

There are usually two meetings per year, including a DX Convention held in April in the Stirling area. Membership is open to all radio amateurs or SWLs with an interest in Contesting and DXing.

Full details from Michael Topple GM5AUG Tel: 0141 286 0628. email: *gm5aug@topple.scot*

website: *www.gmdx.org.uk*

G-QRP Club

Formed in 1974, the GQRP Club is dedicated to low power communication, encompassing both operating and conducting technical investigations. The Club was founded by the Reverend George Dobbs, G3RJV (SK), and continues to thrive on a legacy of passion for QRP (low power) operations across a wide spectrum of frequencies and modes. 2024 marked 50 years of the club and its "Sprat" magazine.

The GQRP Club is renowned for its CW operations on the HF bands but its members are also active across LF, VHF, UHF, and SHF, utilising all modes including digital and offers many awards to members as a way of acknowledging their technical and operating achievements.

With a robust membership of around 4,000, including many active international members in addition to those from within the UK, the GQRP Club remains vibrant and engaged. Its full-colour, quarterly magazine, Sprat, focuses on QRP construction and operation including practical ideas, news, and updates from members. Additionally, club sales offer a variety of components, kits, books and other essential items for low power operation.

Members of the GQRP Club participate in a wide array of radio-related activities, including DXpeditions, Buildathons, contests and field day operations. Many are actively engaged in SOTA, IOTA and DXCC activating and chasing. The club promotes the sharing of knowledge and enjoyment of the hobby through training and construction initiatives.

The annual convention, a significant event in the club's calendar, takes place every September and features speakers, club sales and a Buildathon. The club's YouTube channel provides a valuable resource, archiving all talks from the previous year's convention to offer insights and information to its members.

For more information, visit the club's website at *www.gqrp.com* or contact the chairman at *g0fuw@gqrp.co.uk*

Ham Radio Network *(G5HAM)*

Ham Radio Network (HRN) is a virtual radio club formed during the very first Covid-19 lockdown in 2020. Its overarching purpose was to alleviate potential loneliness and a feeling of isolation for many like-minded radio enthusiasts who like to share, collaborate and experiment.

It was, and remains, intended to be a club that would offer a chance to meet via regular Zoom meetings, thereby attracting talks and demonstrations from all around the world, and to offer support and interesting topics via HRN's main interface on Facebook: *https://www.facebook.com/groups/517224945921950*

HRN continues to attract good discussions, talks, demonstrations, help and support, especially (but not restricted to) new licensees and returners to the hobby. still growing. HRN welcomes all amateurs from across the globe.

For more info: *www.hrn.world*

International Short Wave League *(ISWL – G4BJC, M1SWL)*

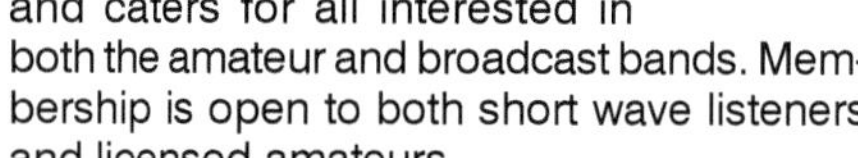

Known As the ISWL, the League was founded in October 1946 and caters for all interested in both the amateur and broadcast bands. Membership is open to both short wave listeners and licensed amateurs.

All members receive the monthly journal Monitor which includes columns of interest to both Short Wave Listeners and Licensed amateurs. The pages include Members Mailbag, reception reports for HF/VHF/UHF; SW BC Bands; Transmitting Topics; "In Days Gone By", a review of receivers from the past; articles of technical interest submitted by members; additional members contributions include "Meet the Member" and articles of interest; Stations to be looked for during the month together with QSL Managers and Addresses; information of general interest world-wide. A monthly report of the ISWL Nets activities during each month plus reservations of licensed members each month.

A full Contest programme and a comprehensive awards programme is open to all members, the latter free of charge to members, also available to non-members at a charge.

Weekly Nets are held on 80m and 40m SSB Sundays, Tuesdays and Saturdays; 160m SSB on Sundays; 80m or 40m CW Mondays.

Full details and a membership application can be found on the web: *www.iswl.org.uk* or available from the Membership Secretary Cliff Jobling, G-13557/G4YHP, "Joycliff", 20a Poplar Road, Healing, Grimsby, DN41 7RD, email *ch.jobling@ntlworld.com*, or, from The Hon Secretary, Peter Whiffing, G-13102/M0PGW, 38 Green Close, Stannington, Morpeth, Northumberland, NE61 6PE, email: *peter.whiffing@yahoo.co.uk*

Martello Tower Group

The Martello Tower Group is an amateur radio group which was originally formed in the late 1980s under the name ClackPak.

In recent times the group has set up an SSTV repeater (MB7TV), an FSTV beacon (GB3CZ) and the first licensed amateur radio to internet gateway in the UK (MB7UIV).

ClackPak was reborn in May 2009 as the Martello Tower Group with ambitious plans for other ground-breaking radio projects in the Tendring area.

In 2015 we installed a 70cms DMR repeater, GB7CL as well as a D-STAR repeater, GB7TE on 2m at the Martello Tower and in October 2017 it was moved to our Thorpe-le-Soken site.

Although affiliated to the RSGB, we are not a 'club' in the traditional sense. The Martello Tower where we meet is very limited in space hence we have to keep our membership numbers down.

Membership is strictly by invitation only.

www.martellotowergroup.com

NARSA (Northern Amateur Radio Societies Association)

NARSA is a non-profit making organisation run entirely by volunteers from our affiliated clubs. Our next annual rally will be held at the Norbreck Castle Hotel in Blackpool on Sunday 13th April 2025. All the usual traders will be there as well as over 40 clubs demonstrating what our hobby is all about. The RSGB bookstand is always very popular and a great place to arrange to meet somebody who you have only talked to on the air.

The NARSA rally is the largest single day rally in the country, regularly attracting well over 1,000 visitors. If you are a newcomer to the hobby we would urge you to come to the rally and join your local club. If your club is not affiliated to NARSA and you want to know more please contact us via our website.

We hold several competitions every year at the rally and award two prizes for construction entrants; one for experienced builders and one for novices. There is an award for the best club stand at the exhibition and to the club deemed to have had the best website throughout the year.

The RSGB president normally attends the rally and is usually persuaded to give a brief talk and present the prizes.

For further information visit our web site: http://narsa.org.uk/

National Hamfest (Lincoln) Ltd

The "National Hamfest" in association with The Radio Society Great Britain and supported by radio clubs in the East Midlands is held each Autumn at the Newark Showground. It is the UK's premiere radio rally and is unique in the support given by major distributors and retailers of amateur radio equipment and accessories in the UK.

The National Hamfest is renowned for its diverse array of exhibitors. Each year you'll find an impressive lineup of specialist traders and manufacturers showcasing the latest in radio technology, equipment, and accessories. From cutting-edge transceivers to vintage radio sets, there is no shortage of exciting products to explore.

In addition to the big-name brands, the National Hamfest features a vibrant selection of hobby traders offering unique, niche items. For those who love a bargain or the thrill of the hunt, the outdoor flea market is a must-visit. Here, you can discover hidden gems, rare components, and incredible deals that you won't find anywhere else.

The National Hamfest is not just about the equipment; it's about the people. This event is the perfect opportunity to connect with fellow radio enthusiasts from across the UK and beyond. Meet new friends, catch up with old ones and share your passion for radio in a welcoming and friendly environment.

The National Hamfest is the ultimate celebration of the radio community. With its rich mix of traders, manufacturers, hobbyists, demonstrations and social opportunities, it's an event that truly has it all.

https://www.nationalhamfest.org.uk

North West Fusion Group

North West Fusion Group operate a network of repeaters and gateways all connected to the NWFG room. This means we can offer superb coverage across the North West as well as enabling radio amateurs the ability to access our infrastructure from anywhere with internet access via Wires-X room 41755 or a hotspot connected to our reflector FCS00428

We are primarily based in the North West of England but we do have members all over the world.

If you are interested in adding a repeater or gateway to the network, or require any more information please email us at:

admin@nwfg.online and we will be happy to help.

Online Amateur Radio Community

We are a UK online based community, here to have fun and experiment with amateur radio – An amateur radio club providing Ham Support, Advice, and Community, without Committee.

For more information please checkout our website: *www.oarc.uk*

Radio Fraternity Lodge No. 8040

The Radio Fraternity Lodge (Lodge No. 8040) was formed in London in 1965 by one-time RSGB General Manager John Clarricoats, OBE, G6CL with other leading radio amateurs of the day who were also Freemasons. The Radio Fraternity Lodge is still a London based lodge and has always been principally composed of RSGB members. Since its foundation over fifty years ago, the Lodge has raised many thousands of pounds for charity including the Radio Communications Foundation.

The Radio Fraternity Lodge sees the importance of supporting amateur radio activity amongst the young and is looking for like-minded people to join them in raising money for such causes. If you are a radio amateur who is already a Freemason or interested in becoming one, then please email our lodge Secretary at *radiofraternityinfo@gmail.com* for more information about our activities.

Radio Amateur Old Timers' Association *(RAOTA – G2OT)*

The Radio Amateur Old Timers' Association (RAOTA) aims to maintain the traditions and spirit of amateur radio. Although we are passionately interested in the history and traditions of amateur radio and are fortunate that we can get a lot of this first-hand from our members, we are equally passionate about the future of our hobby. One of our magazine issues demonstrates this quite succinctly, with an article about trench communication in WW1 next to one about a DDS synthesiser kit.

One of the enduring myths about RAOTA is the need to have been licensed for 25 years to become a membe, but you don't. Anyone with an active interest in amateur radio is welcome to become a member. There is no need to hold (or be qualified to hold) an amateur radio licence.

We have two categories of membership - Associate and Full. Associate membership is open to anyone, regardless of how recently they have come into the hobby. Full membership is open to anyone who has been actively involved in Amateur Radio for at least 25 years. The annual subscription for each category is the same, the magazine received is the same and our warm welcome is the same. The only difference in terms of membership 'benefits' is that Full members have a vote in RAOTA's activities and Associate members do not.

Many say that the main benefit of being a member is our quarterly magazine "OTNews" which is produced to a high standard in terms of both content and print quality. We are fortunate that so many of our members are willing to contribute to "OTNews" by writing articles and letters for it. "OTNews" is printed in an A5 stapled booklet format and usually has at least 54 pages. It is also available on audio media for the benefit of members with limited eyesight.

RAOTA publishes a range of books, in A4 stapled format. In addition we produce a high quality log book with fifty double-sided pages and a 'lay flat' binding.

RAOTA has a number of nets operating on a variety of modes including SSB, CW and digital, on a range of HF bands. Our net controllers usually use our G2OT club callsign. If you hear one of our nets then please feel free to join in. Non-members are always welcome and there is no 'arm-twisting' to have you join.

Whether you meet us on the air, at a rally or via our web site, if the history, traditions and future of amateur radio interest you we hope you will become a member.

To find out more please go to our website at *www.raota.org* email, or write to: RAOTA Membership Secretary, 33 Swallow Drive, Louth LN11 0DN e-mail : *memsec@raota.org*

Radio Officers Association *(ROARS - M0ROA)*

The Association was formed in 1995 when the commercial use of Morse code for maritime communications

was being phased out. Some members of Portishead Radio and other ex-Radio Officers conceived the idea of forming a society to preserve the historical importance of this form of communication which had been in operation for nearly 100 years and had saved many lives.

It was discovered that a number of members were Radio Amateurs and the Radio Officers' Amateur Radio Section was formed to promote the use of Morse code on the bands and communicate with members using the form of transmission that we all knew so well. Any ex-R/O, whether from maritime, aeronautical or coast stations are welcome to get in touch. The Association has two get-togethers each year; the meeting in April incorporates the AGM and the September one is purely social. They are usually held in a hotel over three days in an appropriate city with a maritime connection. These meetings are very convivial occasions that allow members to recall their experiences when the UK had thousands of ships.

We have two CW nets each week, the main one being on Thursday evening at 19.00 local on 3565/7017kHz run by net controller Mike M6MPC and consists of weather reports, plain language and code groups, followed by a series of questions related to our time at sea. The Tuesday net is at 10.00 local on 3538/7017kHz and is just a general natter net.

The society has been involved in running special event stations and has a small but active contest group which takes part in the UKEICC, RSGB and other international contests. In 2024 the ROA has taken on the administration and web hosting for the Maritime Radio Day which takes place annually on the 14th and 15th of April. This event enables contact with ex-R/Os and coast station officers from around the world. Amateur stations are invited to join in as well as "Friends of MRD". A special certificate of participation (COP) is issued to every station which registers and submits results.

The ROA was delighted and honoured to be the recipient of the 2024 CW Ops award "in recognition of outstanding achievement in advancing the art of CW." A plaque was presented by the CW Ops President Stewart Rolfe GW0ETF to ROARS representatives at the ROA AGM in Liverpool.

In addition to the regular quarterly magazine "QSO", the ROA has also published two books titled "The Long Silence Falls". Volume one is back in print again and along with volume two contains many fascinating articles about the experiences of radio officers both in peace time and during war. These are still available and can be obtained from the Radio Officers Association.

The Association also has a web site *www.radioofficers.com* with a wealth of information and pictures showing the ships and radio rooms of members time at sea as well as details of nets in the ROARS section.

Membership Secretary - John Chalmers *johnnchalmers994@outlook.com*

Chairman ROARS - Peter Gavin M0URL *petertgavin@gmail.com*

Radio Security Services Memorial ARS

We are a virtual RSGB affiliated amateur radio club, callsign GI5TKA. Membership is open to any licensed radio amateur, or to anyone with an interest in amateur radio. As a virtual club we do not plan to introduce membership fees, but prefer to operate on a volunteer basis. Our activities will include three or four lecture talks per annum, which can be accessed live via Zoom. Occasionally on air activities will be arranged on the amateur radio frequencies and announced via this web site. We also organise an annual field day event to mark the role of the Radio Security Service. George Busby, author of 'Spies at Gilnahirk' is Patron of the Society.

For more information:

mi0wwb@btinternet.com or

https://rssmars.com

Royal Air Force Amateur Radio Society

ROYAL AIR FORCE amateur radio

(RAFARS – G8FC, G8RAF, G3RAF, G4RAF, G6RAF, G7RAF)

Formed in 1938, RAFARS, now with over 600 members worldwide, is an international society that aims to promote amateur radio activities within the Royal Air Force.

Through amateur radio it maintains and fosters the existing close bonds between radio amateurs still serving and those who have retired from, or have close associations with, the Royal Air Force, Commonwealth, NATO or Allied Air Forces. Additionally, all radio amateurs and short wave listeners are welcome to join as Associate members if they have an interest in military aviation and the RAF, and are willing to assist the Society in achieving its published aims.

The society runs its own QSL bureau and publishes its own call book, QRZ. An in-house Journal 'QRV', is published twice a year.

A monthly Newsletter will be found on our website. The Newsletter can also be emailed to subscribers.

The RAFARS website (*www.rafars.org*) contains information about the Society, including an abridged membership list, an application form to join, net schedules, monthly newsletters, details of awards and competitions Prospective members are welcome to join one of our daily or weekly nets.

Further information is available from:

The Secretary, G4DQP, HQ RAFARS,

RAF Cosford, Wolverhampton WV7 3EX.

Tel: 01902 372722 (answerphone).

email: *rafars.secretary@gmail.com*

Royal Naval Amateur Radio Society

(RNARS – GB3RN, GB0SUB, G3CRS, G1BZU, G3BZU, G7DOL)

The RNARS was formed in 1960 to promote amateur radio as an aid to technical education within the Royal Naval Service. In 1964 the society was considered to be sufficiently established to invite the Captain of HMS Mercury to become its first president. When HMS Mercury closed down the Signal School was transferred to HMS Collingwood. The last Commanding Officer of HMS Mercury, Cdre Paul Sutermeister DL RN is our current President. The society HQ is at HMS Collingwood, Fareham and the HQ Shack callsign is GB3RN. Membership is open to all Radio Amateurs with an interest in maritime affairs but particularly serving or veterans from:

Royal Navy, Royal Marines, Women's Royal Naval Service, Royal Naval Reserve, Royal Naval Auxiliary Service, Merchant Service, Sea Cadets, Sea Scouts, Nautical Training Corps, MOD in a civilian capacity, Commonwealth & other Navies.

A Free membership tier is available for those 25 and under.

We publish a newsletter four times a year, and radio nets are run on 2m, HF and Digital Modes. A number of awards are also available. Please see our Website for news and information:

www.rnars.org.uk

Contact the Secretary: Leading Seaman Martin Longbottom, M0EHL. *secretary@rnars.uk*

Membership Secretary: Joe Kirk, G3ZDF. *g3zdf@btinternet.com*

RNARS, Building 512, HMS COLLINGWOOD, Newgate Lane, Fareham, Hants, PO14 1AS.

01329 717627 (24hr answerphone)

RSGB Contest Club

The Contest Club is open to all RSGB Members and is for those interested in taking part in amateur radio contests and those who are keen to support contesting through the RSGB. Contest Club members who are not members of another participating Affiliated Society are welcome to enter any of the relevant contest series as representatives of the Contest Club – which is in the General Category of clubs. To join the Contest Club, you do not need any previous experience of contesting and you will not be obliged to enter any of the Affiliated Society events on behalf of the club.

The Contest Club is the custodian of five historic RSGB callsigns: G6XX, G6ZZ, G5WS, G5AT and G3DR. All five are held as club callsigns by the Contest Club. Members of the Contest Club who are full licensees may use these historic callsigns, or their regional variants, on behalf of RSGB, by prior arrangement; see "How to Join" below.

QSL details for all callsigns and their regional variants are available on QRZ.com. The logs are regularly uploaded to LoTW and Clublog. The Contest Club takes an active role in organising Radio Marathons on behalf of RSGB. These are International Events that encourage QSO activity, particularly on the HF bands. If you would like to be involved in the RSGB HQ station in future years please get in touch using the contact details below. This is a great opportunity to develop operating expertise in a team context. The club is affiliated to the RSGB as well as being organised from within the RSGB contest committees. The Club committee consists of the Chair of the RSGB HF Contest Committee, the Chair of the RSGB VHF Contest Committee and the Chair of the RSGB Contest Support Committee. Membership is open to any RSGB Member and there are no joining fees

or subscriptions. If you are an RSGB member and would like to join the RSGB Contest Club please send an email with your RSGB membership number to: *contestclub@rsgbcc.org*

Royal Signals Amateur Radio Society *(RSARS – G4RS)*

Formed in 1961 under the chairmanship of the late Major General Eric Cole CB CBE, G2EC, membership is open to: serving and past members of the British Regular and Territorial Army, Cadet Forces, civilian staff who have supported army telecommunications, Commonwealth Army signallers and licensed amateurs from other countries who have proven military connections, subject to status.

Members receive a high quality magazine, *Mercury*, three times a year, and the society runs its own contests and awards scheme. More information from: The Membership Secretary, (G3VBE), RSARS, 65 Montgomery Street, Hove, Sussex BN3 5BE.

Tel: **01273 703680**.
email: *RSARSMemSec@virginmedia.com*
website: *www.rsars.org.uk*

RAYNET-UK

RAYNET-UK is the UK's only national voluntary communications service provided for the community by licensed radio amateurs and other supporting volunteers.

RAYNET was formed in 1953 following the severe East coast flooding. It provides a way of organising the valuable resource that Amateur Radio is able to provide to the community.

Since then, it has grown into a very active organisation with around 2000 members, providing communication assistance to many hundreds of events each year.

RAYNET-UK is a company limited by guarantee in England (2771954). Charity registered in England and Wales (1047725) and in Scotland (SC046184).

Further details can be found on the Internet at: *www.raynet-uk.net*

Radio Security Service Memorial ARS

The patron of the Society is George Busby, author of 'Spies at Gilnahirk', the specific aims of the Society shall be to preserve and memory of Voluntary Interceptors (VIs) and staff employed in the RSS, to educate others in the amateur radio community of the role they played in defence of King and Country during the Second World War and to pass on their legacy to the next generation.

Given our very specific remit we are keen to research the background of radio amateurs who served as VIs and to endure that their contribution is preserved. Membership is open to all who share this specific interest.

The RSSM ARS is RSGB Affiliated and it operates as a virtual nation-wide club, with quarterly Zoom meetings at which there will be a special speaker. Annually each May the club activates a special event station GB0GLS at the site of a former Y Station at Gilnahirk. Inquiries about the Society can be made to: *mi0wwb@btinternet.com*. *https://rssmars.com*

Scarlett Point ARC, GT0SP

The Scarlett Point Amateur Radio Society seeks and promotes members from across the UK, currently having members in, GM, GD, GI, G and GW to not only operate radio in a portable settings throughout the UK but to keep the watch tower located at Scarlett Point, Isle of Man on air as part of the clubs operations. The club has a station there for visiting amateurs to operate (within CEPT regulations etc) including members ability to operate. The station is funded and maintained solely by the club and visitors and new members are welcome.

Scarlett Point Tower, Castletown, Isle of Man

Please contact the secretary Billy GM6DX at *gm6dx@outlook.com* or email the club at *gt0sp@outlook.com* for a member to reply.

The Duxford Radio Trust/Duxford Radio Ltd

The Duxford Radio Trust (DRT) and its wholly owned subsidiary Duxford Radio Limited (DRL) is based in the Cambridge region of the UK and comprises an active group of radio specialists plus supporters, many of whom are from the radio communications industry or the Armed Forces.

The group researches, collects, conserves, restores, displays, and demonstrates historic radio communications, radar, and navigation equipment used in military conflict and civil emergencies for the education of the public.

DRT/DRL operates the amateur radio station G0PZJ using both modern and historic military equipment and also holds the Notice of Variation (NOV) call sign GB0WMM for exhibition operation at the Waterbeach Military Heritage Museum near Cambridge, where it currently exhibits historic radio communications equipment.

DRT was formerly the Duxford Radio Society (DRS) and previously provided the volunteer radio section at the Imperial War Museum, Duxford for the 32 years up until 2020. The group was founded in 1986 jointly by Major John Brown G3EUR, renowned designer of World War 2 S.O.E. and clandestine radio equipment and by Richard Pope G4HXH.

DRT is entirely self-funded by its local and distant supporters and in addition to its public exhibition activities, normally publishes three issues of its historical/technical journal each year.

The Trust welcomes new active members who share an interest in the history, technology, conservation and restoration of communication equipment used in military conflict and civil emergencies and also wish to explain and demonstrate this subject to the public through operating the radio transmitting station.

DRT welcomes involvement from distant supporters who are not able to attend to be active explainers, operators or restorers in the Cambridgeshire/East Anglia area.

Details can be found at *www.duxfordradio.org* and DRT can be contacted at: *duxfordradio@gmail.com*

The UK Microwave Group

(UKuG) GX3EEZ

The UK Microwave Group acts as the focal point for radio amateurs and SWLs interested in frequencies above 1GHz. This includes optical communication - "nanowaves".

The primary aim of the Group is to develop and promote all forms of amateur interest in the microwave spectrum. This is not limited to traditional amateur radio activities, members are involved in a wide range of interests ranging from casual hilltopping, amateur television, data networking, propagation research, tropo and EME dxing, to monitoring deep space probes and radio astronomy.

All levels of UK licence allow operation on some, or all, of our microwave allocations. Microwaving, in all its forms, can be a fulfilling amateur radio activity for newcomers and experienced radio amateurs alike.

As a member of the RSGB Spectrum Forum, the Group works closely with the Society on band planning and licensing issues.

To focus activity, the Group organises Microwave Contests, and runs an Award Scheme which recognises all levels of achievement. It encourages microwave meetings (Roundtables) a mixture of flea market, social event, equipment test facility, and lecture stream.

A regionally based Technical Support Team is available, in which experienced members spread across the UK, with excellent test gear, provide facilities to those either beginning their exploration of the microwave spectrum, or wishing to improve existing equipment.

The Group's e-newsletter, 'Scatterpoint' has a worldwide reputation for providing excellent practical articles on microwave techniques, construction and operation, as well as news of recent activities. 'Scatterpoint' is accessible to all: contributors represent a cross section of the amateur microwave community, and the magazine reflects that.

The *www.beaconspot.uk* website is a unique facility supported and funded by UKuG. This aggregates 'spots' on the DXCluster, and inputs from other sources to provide a remarkably complete and up-to-date resource of information on beacons on the bands above 30MHz.

Beacons are an important feature of microwave operation, and the Group offers financial support on a 'payment by results' basis to individuals and groups building beacons.

The has also published designs for 2.4GHz and 10GHz equipment for using the QO-100 satellite.

Building equipment is aided by the Group's 'Chipbank' service which provides a means of obtaining some specialised components at zero cost. Membership of the Group is open to all those interested in amateur microwaves or a remarkably small subscription – currently £6.00 annually. For members under 21 years, there is no subscription.

For further information, please visit the Group's website: *www.microwavers.org* or contact our secretary – John Quarmby

G3XDY 12 Chestnut Close Rushmere St Andrew Ipswich IP5 1ED Tel 01473 717830 or email: *secretary@microwavers.org*

You can also follow us on Twitter @UKGHz and on the YouTube channel UKMicrowave- Group

UK Six Metre Group *(G5KW)*

The UKSMG was formed in 1982 with the primary aim of promoting 50MHz activity by amateurs worldwide. Today the group is the largest organisation in the world dedicated to 50MHz.

Through its quarterly journal Six News, the group provides comprehensive information on all aspects of the band, including DX and beacon news, propagation, awards, contests, technical articles, equipment reviews, QSL in- formation and DXpedition reports. For further information on "Six News" and to submit items for publication, contact John Rivers, G0GCQ (email: *editor@uksmg.org*).

A major objective of the group is to actively support 50MHz operation from new countries and from DX locations, as well as promoting the establishment of beacons in various parts of the world. Further information on sponsorship can be obtained from Trevor Day, EA5ISZ (email:*sponsorship@uksmg.org*).

The UKSMG website at *www.uksmg.org* carries a wealth of information about the 50MHz band and is guaranteed to provide interesting browsing for newcomers and old hands alike.

Further information about the group and membership details can be obtained from the website or from David Bondy, G4NRT. 19 Harriet Drive, Rochester, ME1 1DY (email: *secretary@uksmg.org*)

Vintage & Military Amateur Radio Society *(VMARS - M0VMW)*

The VMARS is an international society based in the UK with nearly 400 members. Its main aims are to restore and preserve historic communication and electronic equipment, from military, commercial and amateur sources, encourage and enable the use of historic equipment on the amateur bands, encourage the use of historic modes such as AM on the amateur bands and encourage research into radio history.

These aims are realised through the very wide range of VMARS members' interests. Whereas many are engaged in the operation of "vintage" military and amateur equipment on the air (19-sets, T1154/R1155, T1509, 52-sets, TCS TX/RX, LG300, Heathkit, KW equipment, RA17, AR88 etc) with many a rare item in evidence, and much of which is lovingly restored, there is a very strong interest in military man-packs, military wireless vehicles, spy-sets, radar, history and museums and not least in amateur radio construction of valved equipment. The Society runs a weekly AM net on 80m at 08.30 local time on Saturdays on 3615kHz ± QRM. All suitably licensed amateurs (whether VMARS members or not) are welcome. During the week, members may be found around 3615kHz at almost any time. There is also SSB activity, on Wednesdays, around 3615kHz and starting at 20.00 local time, primarily for those with ex-military USB-only equipment and on Fridays, a general SSB net, again around 3615kHz and starting at 19.30 local time.

The Society's website includes a virtual library of manuals for vintage equipment, many of which can be downloaded at no charge. This library is supplemented by many paper documents, for which copies are available to members. Further details may be obtained from the website or the Hon. Secretary, Peter Chadwick, G3RZP, Three Oaks, Braydon, Swindon, SN5 0AD. tel: +44 (0) 1666 860423 email: g8on@btinternet.com or honsec@vmars.org.uk Membership information may be obtained from the Membership Secretary, Ron Swinburne M0WSN, 32 Hollywell Road, Sheldon, Birmingham B26 3BX +44(0)1217 421808 *memsec@vmars.org.uk*

Hon Secretary is Peter Chadwick, G3RZP, Three Oaks, Braydon, Swindon, SN5 0AD.

email: *honsec@vmars.org.uk*

website: *www.vmars.org.uk.*

Worked All Britain Awards Group
(WAB – G3ABG, G4WAB, G7WAB)

The group was founded in 1969 by the late John Morris, G3ABG, to encourage greater amateur radio interest in Britain and to foster friendship. The group promotes an award programme and on-air activity including contests and nets. It makes regular donations to organisations such as the RAIBC who help the less fortunate members of the amateur radio community and other national charities. The group's motto is "To assist others".

The award scheme is open to licensed amateurs and SWLs and is based on the OS / OSNI grid squares of the UK and UK Crown Dependencies. It is in no way intended as, nor should it be taken as, a competition (except for WAB Contests). QSL cards are not required, only log entries and special tracker spreadsheets are available to assist in the claiming of awards.

Full details and checklists of all the squares and WAB awards are contained on the WAB website: *www.worked*-all-britain.org.uk and in the WAB Book/USB Key.

For further details please email the Membership Secretary at:

membership@worked-all-britain.org.uk

(WAB station call signs – G3ABG, G4WAB, G7WAB

World Association of Christian Radio Amateurs & Listeners
(WACRAL – M1CRA)

Dedicated to 'Friendship and Christian Fellowship among Radio Amateurs Worldwide', membership is open to committed Christians of all denominations and all nationalities.

Founded by the late Rev. Arthur Shepherd, G3NGF, members are encouraged to follow his example in spreading the 'Good News', fostering international relations, supporting disadvantaged amateurs and SWLs and providing a good role model for the new generations.

WACRAL publish a newsletter four times a year, featuring the news and activities of members around the globe. A members' QSL Bureau is provided and a variety of awards are available.

An annual activity period is open to all on HF and VHF, including CHOTA (Churches and Chapels On The Air) and conferences take place each year.

The popular WACRAL 'Good News' Nets meet at 07:30 UK local time every Sunday and Wednesday morning on 3.747kHz SSB and a schedule of national and international Christian nets is promoted on various bands on Saturday afternoons – all subject to propagation.

You are invited to go to their website at: *www.wacral.org* for full details of the Nets, and the latest news and details of membership, or contact the Membership Secretary:

Richard Paul G7KMZ

07519 508147

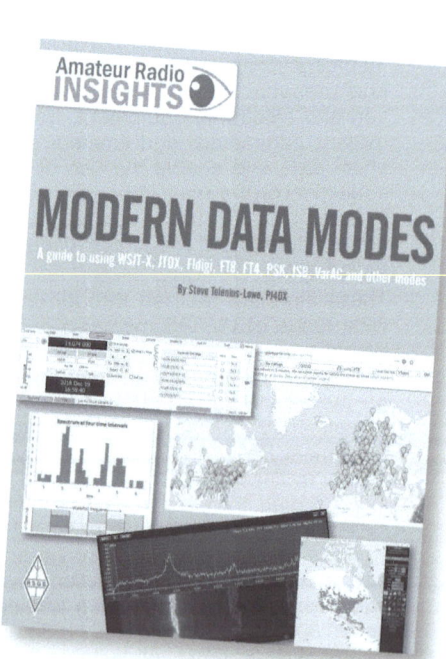

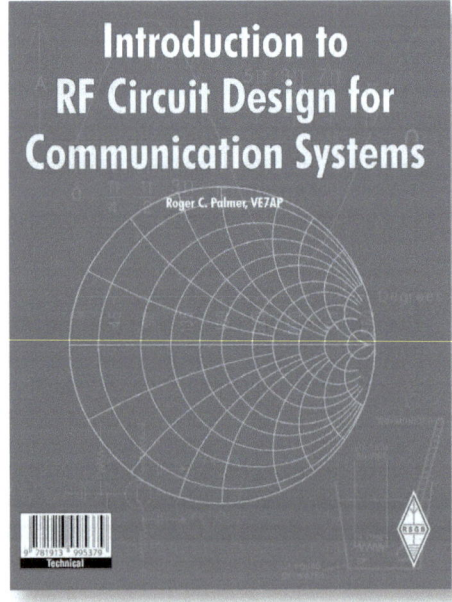

RSGB Club Affiliation

Many amateur radio societies and clubs choose to affiliate with the Radio Society of Great Britain because they see it as an effective way of demonstrating their support for the aims and aspirations of the National Society. The Society welcomes this support because it can only strengthen its claim to speak on behalf of amateurs and amateur radio.

The Society recognises that much of the vitality of amateur radio lies within clubs and wishes to encourage all clubs, societies and groups to join us to advance amateur radio. There are tangible benefits for the affiliated society. These include:

- Publicity for club activities through 'club news' in RadCom, via broadcasts on the Society's news service GB2RS, and on the RSGB website..
- Full facilities of the RSGB QSL Bureau for cards bearing the club station call.
- Purchase of publications at a discount with the RSGB.
- Receipt of the *RSGB Yearbook.*
- Freedom to participate in RSGB Affiliated Societies' contests.
- Third party liability insurance up to £10m whilst taking part in club events such as field days.
- Freedom to borrow RSGB DVDs, tapes and display materials. (This facility is also available to certain non-affiliated groups such as schools).

How to affiliate your club, group or society to the RSGB

Any UK club, group, society or emergency communication group may affiliate to the RSGB, provided it fulfills just a few requirements. The affiliation fee is currently £51.00 (with a £4 discount for those who pay by Direct Debit) and includes receipt of monthly email newsletters.

Procedure

(i) Please contact RSGB HQ or visit the website for an Affiliated Club membership application form. If your organisation has a callsign, please let us know on the application form. If it does not, we will issue a receiving station number for reference purposes.

Note that for UK clubs, groups and societies the RSGB region and RSGB district will be determined by the address given on the application form. Clubs, groups and societies near to, or spanning county boundaries should decide carefully with which county they wish to be associated and insert the appropriate choice in the address of the club they are entering on the application form.

Please also note that once your club address is on our files, we will regard it as information that can be freely given out to those seeking to contact clubs.

(ii) Please send to the appropriate Regional Manager the following:

- Your completed application form signed by the chairman or secretary
- A copy of the club's constitution or rules. There is no prescribed form of constitution but an example is provided for guidance if required.
- A list of current officers of the club
- A statement of the number of members, and the proportion who are RSGB members.

A list of Regional Managers is published on the RSGB website and in the RSGB Yearbook. The Regional Manager will vet your constitution/rules and if suitable will countersign your application form. He or she will then return the form and your constitution or rules to you.

Note that only the Regional Manager may countersign an application for a club, group or society. Overseas organisations should send their application form and constitution/rules direct to RSGB HQ addressed to 'The RSGB Secretary'.

(iii) Finally, please send your countersigned application form, constitution and remittance to:

Membership Secretary, 3 Abbey Court,
Fraser Road, Priory Business Park,
Bedford, MK44 3WH.

Downloadable application form: https://rsgb.org/main/clubs/club-affiliation/

Notes on Example Constitution (overleaf)

(1) It is recommended that the words 'amateur radio' appear in its title.

(2) It is useful to specify which groups have voting rights and whether reduced subscriptions apply, particularly for students in full time education.

(3) Alternatively, the subscription may be recommended by the Committee for ratification at the AGM.

(4) This period perhaps should not exceed one to three years to avoid placing an undue burden on future Committees

(5) There are great advantages in running the Club's finances on a strict basis, although a less formal arrangement may still be effective.

(6) There are two methods for electing the Committee: the more common is for the meeting to elect the Committee members and for the latter in turn to elect the officers from within the Committee; alternatively, the members may elect individuals to specific offices. The method adopted will need to be specified.

(7) The number of Ordinary Committee members should be related to the size of the Club. Remember that being a committee member is an essential part of the training of the future officers of the Club.

(8) These can replace elected Committee members who have left the Committee

(9) These can be people who need to be familiar within the work of the Committee such as the editor of the Club magazine or the press officer.

(10) This can be expressed either as a fixed number or, for example, as at least half or two-thirds of the full membership of the committee.

(11) This can be set either as a fixed number or a fixed percentage of the membership (state which members are to be included), or both "whichever is the smaller/greater". It is probably safer to make the numbers on the small side so as to ensure that the meeting can take place.

(12) Such as, among its members, to a charity, or to a club of similar interest.

Example Constitution for Affiliated Clubs

Guidance intended for those writing a constitution for their local club or society which will be acceptable for RSGB affiliation.

1. Name

The Club (1) shall be known as the ....

2. Aims

The aims of the Club shall be to further the interests of its members in aspects of amateur radio and directly associated activities.

3. Membership

Membership shall be open, subject to the discretion of the Committee, to all persons interested in the aims of the Club

(a) **Full members** Full members must be 16 years of age or over.

(b) **Honorary members** Honorary Life Membership may be granted to any person, who, in the opinion of the Committee, has rendered outstanding service to the Club, either directly or indirectly. Such membership shall carry the rights of full membership but shall be free from subscriptions.

(c) **Guests** Members may invite guests to meetings. No visitor may attend more than three meetings in each year.

All members shall abide by the constitution of the Club. The Committee shall have power to expel any member whose conduct, in the opinion of at least three-quarters of the full Committee, renders that person unfit to be a member of the Club. No Member shall be expelled without first having been given an opportunity to appear before the Committee.

4. Subscriptions

(a) The annual subscriptions for membership shall be set by the Committee (3).

(b) All subscriptions shall be due and payable at the beginning of the financial year. Members in arrears have no voting rights.

(c) The financial year shall be determined by the Committee

(d) A member shall be deemed to have resigned from the Club, if, by the end of the financial year, the subscription has not been paid.

(e) The Committee shall have the power to waive or reduce subscriptions in special circumstances for a period not exceeding...years at a time (4).

5. Finance

All money received by the club shall be promptly deposited in the Club's bank account. Withdrawals require the signature of the Club's Treasurer and one other nominated officer of the Club (5).

6. Membership of the Club's Committee

The Club's affairs shall be administered by a Committee elected at the Annual General Meeting (6). The Committee, in whom the Club's property shall be vested, shall consist of:

(a) A Chair who will preside at all meetings at which he is present.

(b) A Vice-Chair who will act as Chair in the absence of the Chair.

(c) A Club Secretary who will be responsible for:
- (i) keeping the minutes of all meetings of the Club.
- (ii) ensuring that all correspondence is correctly handled.
- (iii) maintaining the definitive register of members and honorary members.
- (iv) maintaining a register of Club equipment.

(d) A Treasurer, who will be responsible for:
- (i) keeping the Club's accounts.
- (ii) advising the Committee on all financial matters.
- (iii) preparing the accounts for audit and presenting them at the AGM.

(e) ......Ordinary Committee Members (8).

(f) Not more than......co-opted members who have full voting powers (8), and not more than......who are not permitted to vote (9).

7. Committee Standing Orders

(a) The quorum for the Committee shall be.....(10).

In the absence of a quorum, business may be dealt with but any decisions taken only become valid after ratification at the next meeting at which a quorum exists.

(b) Committee meetings may be called by the Chair, the Secretary or any vote.

8. Annual General Meeting

(a) The Annual General Meeting shall normally be held at the beginning of each financial year. At least 21 days notice shall be given to each member in writing.

(b) The quorum for the meeting shall be...... (11).

(c) The agenda for the meeting shall be:
- (i) Apologies for absence
- (ii) Minutes of the previous AGM
- (iii) Chair's report
- (iv) Club Secretary's report
- (v) Treasurer's report
- (vi) Election of the new Committee
- (vii) Election of auditors
- (viii) Other business

(d) Items (i) to (v) shall be chaired by the outgoing Chair, item (vi) by an acting Chair who is not standing for election to office, and the remaining business by the newly elected Chair.

(e) Nominations for Committee members will only be valid if confirmed by the nominee at the meeting or previously in writing.

(f) Items to be raised by members under other business must be notified to the Club Secretary not less than 21 days before the AGM.

9. Extraordinary General Meeting

(a) Extraordinary General Meetings may be called by the Committee or not less than......members of the Club, the date of the meeting being the earliest convenient as decided by the Committee. At least 28 days notice in writing must be given to the Secretary, who in turn shall give members at least 14 days notice in writing of the agenda. No other business may be transacted at the EGM.

(b) The quorum for the EGM shall be......(11).

10. Amendments to the Constitution

The constitution may be amended only at an EGM called for that purpose.

11. Winding-up of the Society

(a) The decision to wind up the Club may be taken only at an EGM.

(b) The funds of the Club shall, after the sale of all assets and the payment of all outstanding debts, be disposed of as directed by members at the final EGM (12).

* Club is used to denote any club or similar organisation wishing to apply for affiliation

Downloadable Model Constitution: www.rsgb.org/main/clubs/club-affiliation

Structure and Representation

For administrative purposes the Society divides itself into regions, each comprising four to six districts. Each region has an elected Regional Representatives (RR) and each district has an appointed District Representatives (DR). The map below identifies the RSGB Regions. On the following pages, listed by Region, you will find details of the relevant Regional Representatives (RR) and the District Representatives (DR) along with contact information on local clubs and societies, examination centres and Emergency Comms Groups.

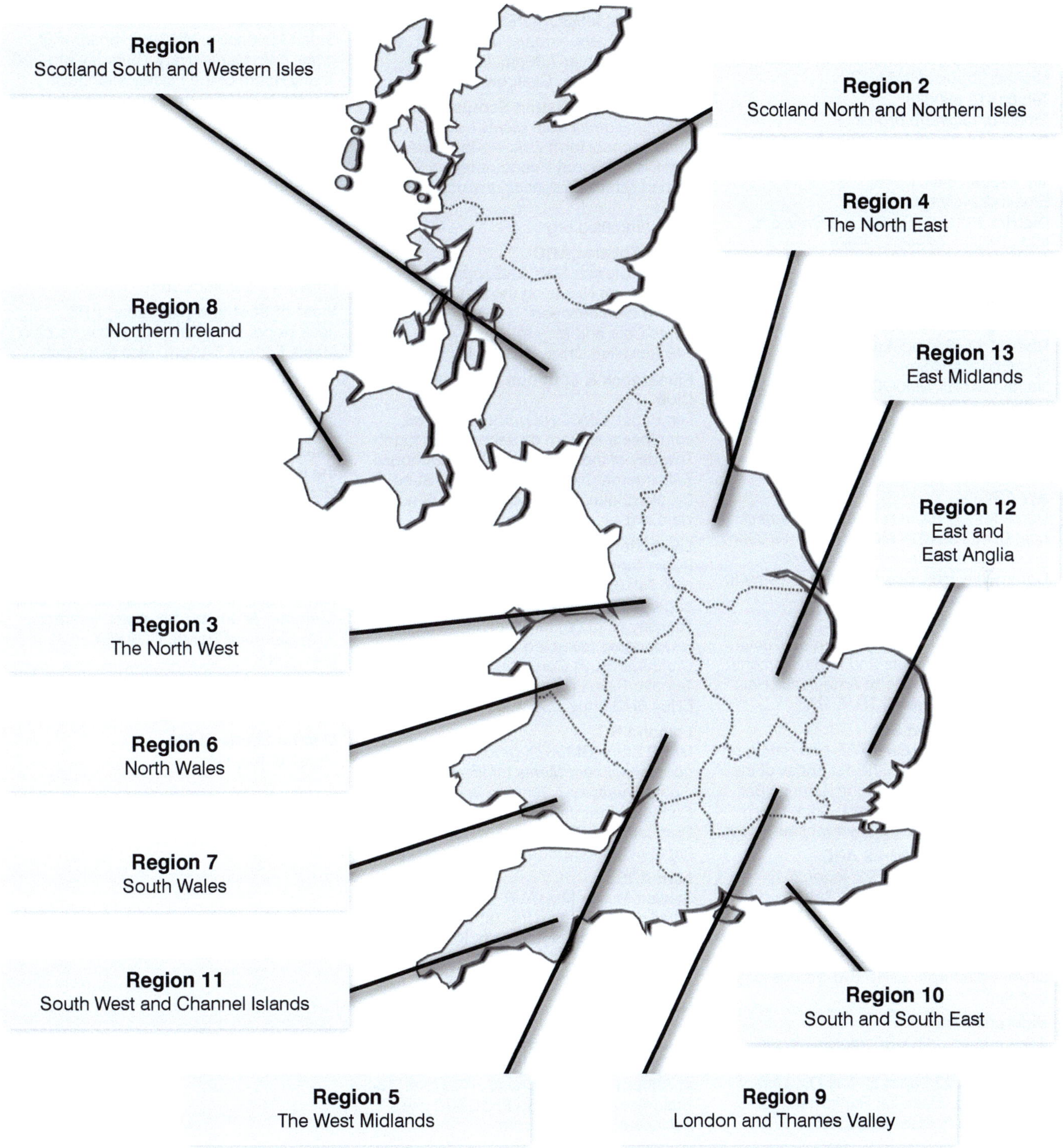

Region 1
Scotland South and Western Isles

Region 2
Scotland North and Northern Isles

Region 4
The North East

Region 8
Northern Ireland

Region 13
East Midlands

Region 12
East and East Anglia

Region 3
The North West

Region 6
North Wales

Region 7
South Wales

Region 11
South West and Channel Islands

Region 10
South and South East

Region 5
The West Midlands

Region 9
London and Thames Valley

Region 1 - Scotland South & Western Isles

(http://rsgb.org/region1)

RR

Tony Miles, MM0TMZ
Tel: 07702 134188
Email: *rr1@rsgb.org.uk*

DRs

District 11 (Renfrewshire and Inverclyde) William McCue, MM0ELF Tel: 01389 755758 Email: *dr11@rsgb.org.uk*
District 12 (Glasgow, Central and Lanarkshire) Michael Topple, GM5AUG Tel: 01412 860628 Email: *dr12@rsgb.org.uk*
District 13 (Ayrshire, Dumfries & Galloway) Michael Topple, GM5AUG **Tel: 01412** 860628 Email: *dr13@rsgb.org.uk*
District 14 (Dunbartonshire, Argyll & Bute, Western Isles) Barrie Spink, GM0KZX Tel: 01389 764401 Email: *dr14@rsgb.org.uk*
District 15 (Lothian) George Crawford, MM0JNL Tel: 07946 753985 Email: *dr16@rsgb.org.uk*
District 16 (Borders) George Crawford, MM0JNL Tel: 07946 753985 Email: *dr16@rsgb.org.uk*
District 17 (Stirling, Falkirk & Clackmannanshire) Nathan Nuttall, 2M0OCC Tel: 07952 819626 Email: *dr17@rsgb.org.uk*

Clubs & Societies

Ayr ARG
Derek MM0OVD, *derek.secaarg@gmail.com* 07482 994614 Meets 7.15pm-9:30pm on alternate Wednesdays Prestwick Community Hall, Briarhill Road, Prestwick, KA9 1HY *www.gm0ayr.org*

Borders ARS
Sandy Weddell GM1JFF *a.weddell764@btinternet.com* Meets 2nd Friday in month at 19.30 at the, St. John Ambulance Hall, Berwick-upon-Tweed, TD15 1NG

Cockenzie & Port Set
Mr Bob Glasgow, GM4UYZ. *bob.gm4uyz@talktalk.net* Meets on the 1st Friday of the month at the, Thorntree Inn, Lounge Bar, Old Cockenzie High Street, Cockenzie, East Lothian EH32 0DQ *www.cpsarc.com*

Edinburgh and District ARC
Secretary, 07740 946192; *jacobite@btinternet.com* Meets most Saturdays from 7pm. Corstorphine Club net on Mondays from 8pm. Contact the secretary for further details, Corstorphine, 160 Carrick Knowe Drive, Edinburgh, EH12 8AT groups.io/g/gs4ham

Elderslie ARS
John French, 2MOJSF *info@elderslie-*ars.org.uk Meets every thursday night at 19:00pm till 22:00pm at, The Old Library, Stoddart Sq, Elserslie, Renfrewshire PA5 9AS *www.elderslie*-ars.org.uk

Falkirk & DARS
P Howson, GM8GAX, *gm8gax@tiscali.co.uk* Meets every Friday 19:30. 62nd Forth Valley Scouts Hall, Denny Road, Larbert, Nr Falkirk, FK5 3AD

Galashiels & DARS
Mr Jim Keddie, GM7LUN, *mail@gm7lun.co.uk* Meets at 8.00pm on Wednesdays at the, Focus Centre, Livingston Place, Galashiels, Scottish Borders TD1 1DQ *www.galaradioclub.co.uk*

Glasgow & Clyde Raynet
Paul Lucas, MM3DDQ Tel: 01389 499972, *mm3ddq@yahoo.co.uk*, Meets 1st Tuesday of each month at 7.30pm, Braehead Shopping Centre, Glasgow, G51 4BN

Greenock and District Scouts ARC
mm0tsg@gmail.com Meets every Friday during school term time 7-10pm, Greenock and District Scout Headquarters, 159A Finnart Street, Greenock, Inverclyde PA16 8HZ
https://mm0tsg.org

Jaggy Thistles ARC
Charles Stewart MM0GNS, *cgnstewart@hotmail.com* Meets on the first Tuesday of each month between
September and May at
, 46 Rowallan Drive, KA3 1TU

Kilmarnock & Loudoun Amateur Radio Club
Len Paget, GM0ONX *gm0onx@gmail.com* Meets 7.30pm on second and fourth Tuesday of the month, KLARC Clubhouse, EAC Internal Transport, 34a Main Street, Crookedholm, Kilmarnock KA3 6JS *www.klarc.org*

Livingston & DARS
Robin Tannahill: robin_*tannahill@hotmail.com* Tel: 07880 992377 Also see group page on Facebook, Meets weekly on a Tuesday at 19.00, for Meetings, Training & Operating (disabled access available)., Crofthead Farm Community Centre, Templar Rise, Livingston, West Lothian EH54 6DG https:*//www.ladars.org.uk*

Lomond RC
Mr BP Spink,GM0KZX,*gm0kzx@googlemail.com* Meets for weekly club nights Thursday 7.30pm to 9pm, John Connolly Centre, 30 Main Street, Renton, Dumbarton G82 4LY
https:*//www.lomondradio.club*

Lorn ARS
Stewart McIver MM1AVR, *lornradioclub@gmail.com* Meets on the 1st and 3rd Wednesdays of each month, Kelvin Hotel, Oban, PA34 5AD
https://gm0lra.wordpress.com/

Lothians RS
Andy Sinclair *secretary@lothianradiosociety.com* meets normally on the 2nd and 4th Wednesday of the month at 7.30pm for 8pm., Braids Hills Hotel, 134 Braid Road, Edinburgh EH10 6JD *www.lothiansradiosociety.com*

Mid Lanark ARS
Kevin Mair, 2M0KVM Chair *mlarsclub@gmail.com* Meets every Friday 18.30 - 21.30, Newarthill Community Ed. Cent., High Street, Newarthill, Motherwell, Lanarkshire ML1 5JU *www.mlars.co.uk*

Na Fir Chlis ARC
Angus MacLeod, *mm0nfc@hotmail.com* Meets The club meets on an irregular basis., Heathbank Hotel, Northbay, Isle of Barra, Outer Hebrides HS9 5YQ
https:*//www.ms0nfc.yolasite.com*

Paisley ARC
Stuart McKinnon, MM0PAZ *mm0paz@gmail.com* Meets every Thursday at 19:30, Trinity Room of the Paisley Methodist Church Halls, 2 Gauze Street, Paisley, PA1 1EP
http://gm0pym.wix.com/paisleyarc

Stirling & DARS
secretary@gm6nx.com Meets Thursdays 7-9pm and Sundays 10am-1pm Times subject change, Stirling and District Amateur Radio Society Unit 6, Bandeath Industrial Estate, Throsk, FK7 7XY
https:*//www.gm6nx.com*

West of Scotland [Glasgow] ARS
Jack Hood, GM4COX, *info@wosars.club* Meets Fridays 8-10pm. Also technical support and licence guidance & support on Wednesday nights as SOLDER GROUP, located at the Electron Club, within The Centre for Contemporary Arts, 350 Sauchiehall Street, Glasgow, G2 3JD For the latest information see our Website, Garnethill Multicultural Centre, 21 Rose Street, Glasgow, Scotland G3 6RE
https:*//info@wosars.club*

Wigtownshire ARC
Bob Bower, GM4DLG: *gm4dlg@gmail.com* Meets 7.30pm on Thursdays at, Aird Unit, Stranraer Academy, Stranraer, (entrance from Cairnport Road), DG9 8BY
www.gm4riv.org

Repeater Groups

Central Scotland FM
Hazel McKay D3

Region 2 - Scotland North & Northern Isles

(http://rsgb.org/region2)

RR

Jim Campbell, MM7BIW
Tel: 07749 520320
Email: *rr2@rsgb.org.uk*

DRs

District 21 (Highlands) John Grieve, GM0OTI Tel: 07845 987304
Email: *dr21@rsgb.org.uk*
District 22 (Aberdeenshire, Moray) Peter Thomson, GM1XEA, 01224 740091
Email: *dr22@rsgb.org.uk*
District 23 (Angus, Fife. Perth & Kinross) Martin Krawczyk, 2M0KAU Tel: 07763 708933 Email: *dr23@rsgb.org.uk*
District 24 (Orkney) David Wishart, MM5DWW Tel: 01856 721422
Email: *dr24@rsgb.org.uk*
District 25 (Shetland) Peter Bruce, GM0CXQ Tel: 01595 880241
Email: *dr25@rsgb.org.uk*

Clubs & Societies

Aberdeen ARS
Fred Gordon GM3ALZ fred*_gordon@btinternet.com* Meets 19.30 - 21.30 on Thursdays at, 25th Scout Group, Oakhill Crescent Lane, Aberdeen, AB15 5HY *www.aars.org.uk*

Caithness ARS
Alistair Ross (2M0WTN). *ms0fnr@outlook.com* Meets 1st & 3rd Thursday of each month 19:30 to 21:30, Craigwell Farm, Skirza, Caithness, KW1 4XX *www.radioclubs.net/c.a.r.s./*

Dundee ARC
Jim Wilson *secretary@dundee*-amateur-radio.co.uk Meets every Tuesday throughout the year 19.00 to 21.00 during term time, Dundee and Angus College, Old Glamis Road, Dundee, DD3 8LE *www.dundee*-amateur-radio.co.uk

Glenrothes & DARC
Ron Murray 2M0DSU, *murrayphone@aol.com* Meets in New Football Pavilion Station Road, Thornton, Fife, KY1 4AX *www.gdarc.org.uk*

Inverness DARC
Adrian Hart MM0DHY *dhyclimber@yahoo.co.uk* Meets 2nd and 4th Wednesday at 7.30pm-9.30pm, DHM Blacksmiths, 7 Carsegate Road, Inverness, IV3 8EX

Montrose Air Station Heritage Centre
Ewan Cameron, MM0BIX, *rafmontrose@aol.com* Meets Sundays at 12.00pm at, Montrose Air Museum, Waldron Road, Broomfield, Montrose, Angus DD10 9BB *www.rafmontrose.org.uk*

Moray Firth ARS
Paul Furness *MFARS.secretary@gmail.com* Meets on first and second Tuesday in every month at 7.30pm, Reserve Forces Centre, Edgar Rd, Elgin, IV30 6YQ *www.mfars.co.uk*

Museum of Communication ARS
Ken Horne GM3YBQ Meets Wednesday and Saturday at 11am - 4am at, 131 High Street, Burntisland, Fife, KY3 9AA

Orkney ARC
Mr David Wishart MM5DWW, *oarc@live.co.uk* Meets on the first Wednesday of each month at the, Sea Cadets HQ, TS Vanguard, Scapa Beach, Kirkwall, Orkney Islands KW15 1SD
http://eu009.webplus.net/

Sutherland & District A.R.C
Frank Dinger GM0CSZ. *sadarc@sutherland*-arc.org.uk Meets evey Friday at 7:00pm at, Dunrobin Farm, Golspie, Highland, KW10 6RH

Repeater Groups

Grampian Repeater Group
gb3gn@btinternet.com The Grampian Repeater Group maintains the 2m Amateur Radio Repeaters GB3GN (Banchory) and GB3NG (Fraserburgh) located in NE of Scotland. We have no fixed meeting place or regular meetings. Our AGM is usually before the end of March each year in Aberdeen. Meeting Information: April, Once per yearInverurie, AB51 4FL https:*//www.grampianrepeatergroup.co.uk*

Region 3 - North West

(http://rsgb.org/region3)

RR

Martyn Bell, M0TEB Tel: 01229 833571Email: *rr3@rsgb.org.uk*

DRs

District 31 (Cumbria) Martyn Bell, M0TEB Tel: 01229 833571
Email: *dr31@rsgb.org.uk*
District 32 (Lancashire) Vacant
Email: *dr32@rsgb.org.uk*
District 33 (Greater Manchester) David Crowe, G4MVU Tel: 07518 913300
Email: *dr33@rsgb.org.uk*
District 34 (Cheshire) Mike Isherwood, G4VSS Tel: 07831 459299
Email: *dr34@rsgb.org.uk*
District 35 (Isle of Man) Chris Wood, GD6TWF Tel: 07624 483865
Email: *dr35@rsgb.org.uk*
District 36 (Merseyside) Richard Staples, G4HGI Tel: 07305 835107
Email: *dr36@rsgb.org.uk*

Clubs & Societies

Ashton In Makerfield ARC
Peter Williams M0RGN, *mx0htr@gmail.com* Meets fortnightly from 20:00 to 22:00, The Diamond Community Centre, Grey Road, Ashton-in-Makerfield, Wigan WN4 9QW https:*//www.aimarc.co.uk*

Bolton Wireless Club
Mark Bryant M0UFC, *boltonwireless@gmail.com* Meets at 7:00pm every 2nd and 4th, Monday (Bank Holidays excepted), Ladybridge Community Centre, Beaumont Drive, Bolton, BL3 4RZ https:*//www.boltonwireless.org.uk*

Bury RS
Mike G4GSY *mail@buryradiosociety.org.uk* Meets every Tuesday at 7.30pm, Hollins Social Club, brook lane, Bury, Gtr Manchester BL9 8BA https:*//www.buryradiosociety.org.uk*

Central Lancs ARC
Nonie Sinclair, G3TNN *secretary.clarc@gmail.com* Meets at the clubhouse at Ribble Steam railway, visits by arrangement only. Find us on Facebook. 'Central Lancs Amateur Radio Club And Friends'
, Ribble Steam Railway Musuem, Chain Caul Road, Ashton On Ribble, Preston PR2 2PD https:*//www.facebook.com/groups/443941206406534*

Chester & DRS
Philip Hughes G8IPT, *secretary@chesterdars.org.uk* Meets 1st, 3rd, 4th (&5th) Tues of each month except August, The Burley Memorial Hall, Common Lane, Waverton, Chester, or The Waverton Institute, Waverton, Chester CH3 7QN *www.chesterdars.org.uk*

East Lancs Pendle Radio Club
Neil Mooney M0NFI, *info@elrc.uk* Meets Monday evenings 19:00-21:00, Closed Bank Holidays, Higham Village Hall, Higham Hall Road, Higham, Burnley, Lancashire BB12 9EU *www.eastlancsradioclub.co.uk*

Furness ARS
Martyn Bell M0TEB. *info@fars.org.uk* Meets on the 2nd and 4th Wednesdays 7.30pm for practical evenings, talks and presentations. Social meetings every Monday at the Farmers arms, Newton-in-Furness LA13 0NB. FM Net around 145.325 Fridays at 8pm and some Wednesdays we chat on DMR TG23525. Regular breakfast meetings Wednesdays at the Furness Railway from 9.30am. During the summer we usually have outdoor activities., Hawcoat Park Sports Club, Hawcoat Lane, Barrow-in-Furness LA14 4HF *www.fars.org.uk*

Gilmore Radio Club
Heather Stanley: *info@m0juw.co.uk* Meets 1st Thursday of each month, St Marys church Hall, St Marys Drive, stockport, SK9 3PA https://*www.m0juw.co.uk*

Isle of Man ARS
info@iomars.im Meets every Wednesday from 7pm onwards at the Scout HQ, Mill Road, Ballasalla, IM92EG and on 2nd Tuesday at Sea Scout HQ Tromode, Scout HQ, Mill Road, Ballasalla, Isle of Man IM9 2EG https://iomars.im/

Leyland And District Amateur Radio
Phil 2E0DGP: *ladar@mail.com* Meets 2nd and 4th Wednesdays of the month 7-9pm , Chorley Sea Cadets, Heapy Road, Chorley, PR6 9BQ https://*www.ladar.club*

Macclesfield & DRS
Greg Acton M0TXX, *info@gx4mws.uk* Meets 19.30pm on Mondays, The Pack Horse Bowling Club, Abbey Road, Macclesfield, SK10 3AU *www.gx4mws.com*

Manchester Wireless Society
secretary@g5ms.com Meets on the first Monday at 8pm Cleveland Public House, Wilton Road, Crumpsall, M8 4WQ *www.g5ms.com*

Marine Radio Museum Society (Wallasey
W.H Cross 0151 207 1959 *g0elz@yahoo.com* Meets Tuesday and Friday each week 10am-4pm, Tug France Hayhurst Albert, Dock Harbourmasters Office, Liverpool L3 4AE

Merseyside Amateur Radio Society
Alan Birch G4NXG *g4nxg@btinternet.com* Meets 7.30pm on the 1st and 3rd Monday of the month, Mel Inn, 513 Hawthorne Road, Bootle, Merseyside L20 6JJ https://*www.marsradio.uk*

Mid Cheshire ARS
Peter Paul Fox, G8HAV,*midcars@woollysheep.org* Meets every Wednesday evening 19:30-22:30, Cotebrook Village Hall, Stable Lane, Cotebrook, nr Tarporley, Cheshire (NGR: SJ 571 655) CW6 0JJ *www.midcars.org*

Morecambe Bay ARS
Jon Allen M0IXK, *joninglemere@gmail.com* Meets 8:00pm on every Tuesday, Trimpell Sports & Social Club, Outmoss Lane, Morecambe, Lancs LA4 4UP https://*www.mbars.uk*

Newton-Le-Willows ARC
MX0NRC *www.nlwarc.co.uk* Lee Leland m0lgl@outlook.com Meets Thursdays at 13.00 - 20.00 Derbyshire Hill Family Centre, Derbyshire Hill Road,St Helens, Merseyside, WA9 2LU

Oldham ARC
president@oarc.org.uk Meets Thursdays between 19:30-22:00 hours. Training available on demand., 1855Squadron (Royton)ATC, Park Lane, Royton, Oldham OL2 6RE https://*www.oarc.org.uk*

Quantum Amateur Radio & Technology Society
Alison Hughes M6COV *info@quantumtech.club* Meets 1st & 3rd Thursday of each month at 19.30 for 20.00 to 22.00, Cottage Lane Mission, Ormskirk, Lancashire, L39 3NE *www.quantumtech.club*

Radio Millenium Lodge
Mr N Stackhouse, email: *g1scl@ntlworld.com* Meets the first Friday of Feb, Apr, Jun, Oct & Dec at 6:00pm, 1881 Building, 15 Westbourne Road, Urmston, Manchester, M41 0XQ https://*www.rml9709.org.uk*

Rochdale & DARS
Dave Carden *dave@cardens.me.uk* Meets 2nd and 4th Thursdays of the month from 19:30 until 21:30, Rochdale and District Amateur Radio Society, Crimble Croft Community Centre, Aspinall Street, Heywood OL10 4HW https://*www.g0roc.co.uk*

Sands Amateur Radio Communications Group
Brian Watson, G0RDH, *g0rdh@m0scg.org.uk* Meets every other Monday at 8pm at, The Owls Nest, Bare Lane, Morcambe, LA4 6DD *www.m0scg.org.uk*

South Cheshire Amateur Radio Society
Ronald Carter G7RC *info@g6tw.org.uk* Meets every 2nd & 4th Wednesday of the month at 7.30pm, Wilson House, Ford Lane, Crewe, CW1 3EH *www.g6tw.co.uk*

South Manchester Radio Club
David Crowe G4MVU *contact@smrcc.org.uk* Meets every Thursday. 20.00 to 22.00., The Woodheys Club., 299 Washway Road, Sale, Cheshire M33 4EE *www.smrcc.org.uk*

Southport & DARC
Matt Kerwin 2E0TXO Meets 8.00pm on 3rd Monday in the month at, St Marks Church Hall, Scarisbrick, Lancs, L40 9RE *www.sadarc.org.uk*

Stockport Radio Society
Tony Smithies M0SAV *info@g8srs.co.uk* Meets: 1st & 4th Tuesdays of the month(7pm) @ Walthew House, 3rd Tuesday of the month online (7.15pm), 2nd Sunday of the month (10am) @ Walthew House, Walthew House, 112 Shaw Heath, Stockport, Greater Manchester SK2 6QS https://*www.g8srs.co.uk*

Thornton Cleveleys Amateur Radio Society
Dave Ward G8KBH, *tcars.radio@gmail.com* Meets 8pm every Monday evening Except Bank Holidays, Cleveleys Community Centre and Church, Kensington Road, Cleveleys, Lancashire FY5 1ER https://*www.tcars.club*

Warrington & ARC
Dave Roberts G8KBB *dave.roberts@btinternet.com* Meets Tuesday 8 pm Thursday 10am and most Sundays 1:30pm , Grappenhall Community Centre, Bellhouse Lane, Grappenhall, Warrington, Cheshire WA4 2SG *www.warc.org.uk*

West Manchester Radio Cub
secretary@wmrc.co.uk Meets 8.00pm on Thursdays at the, WMRC (Astley and Tyldesley Miners Welfare Club), Gin Pit Village, Astley, Manchester M29 7DW *www.wmrc.org.uk*

Widnes & Runcorn ARC
Mark Keilty G8AA Meets every other Tuesday, Lostock Sports & Social Club, Works Lane, Northwich, Cheshire CW9 7NW *www.wararc.co.uk*

Wirral & DARC
Simon Richards G6XHF: *secretary@wadarc.com* Meets 8.00pm 2nd & 4th Wednesdays of month at, Irby Cricket Club, Mill Hill Road, Wirral, CH61 4XQ *www.wadarc.com*

Wirral ARS
William Davies *secretary.g3nwr@virginmedia.com* Meets on Tuesday & Wednesday evenings 19:00 - 21:00. Open seven days 10:00-14:00 for pre-arranged visitors, 24/7 for members, c/o William Davies, Building 27D Hooton Park, Airfield Way, Hooton, Cheshire CH65 1BQ *www.g3nwr.org.uk*

Wirral Peninsula ARC
wirralpeninsulararc@outlook.com Meets at 19.00 hrs every Wednesday, Unit 18 Peninsula Business Park, Reeds Lane, Moreton, CH46 1DW https://*www.wparc.co.uk*

Workington and District ARC
Peter Webster G8RZ, *pwebster220@gmail.com* Meets 1st & 3rd Wed of each month at 7pm Moorclose Community Centre, Needham Drive, Moorclose, Workington, Cumbria, CA14 3SE *www.mx0wrc.org*

Repeater Groups

UKFM Group Western
Julian Wolvern M0JPW *www.ukfmgw.org.uk*

Contest Groups

350 DX Club
Alan Birch G4NXG *g4nxg@btinternet.com*, Litherland Town Hall, Hontin Hill Road, L21

Duddon Contest Team
g3dct@outlook.com VHF, UHF, Microwave Contest club on the Duddon Valley area, LA16 7JG

Tall Trees Contest Group
Mr Bran Gale G3UJE *Bgale111s@googlemail.com*,

Travelling Wave Contest Group
Keith Haywood G8HXE, Skype

Region 4 - North East

(http://rsgb.org/region4)

RR

Ian Bowman, G7ESY Tel: 07710 992961
Email: *rr4@rsgb.org.uk*

DRs

District 41 (Northumberland, Tyne & Wear, Cleveland, Co Durham, Northallerton) Ian Bowman, G7ESY Tel: 07710 992961
Email: *dr41@rsgb.org.uk*
District 42 (East Yorkshire) John Baines, M0JBA Tel: 01482 842430
Email: *dr42@rsgb.org.uk*
District 43 (West Yorkshire) Vince Greatwood, M0VHG Tel: 07740 346777
Email: *dr43@rsgb.org.uk*
District 44 (South Yorkshire, NE Lincolnshire) Chris Drury, G1WSA Tel: 07821 761514 Email: *dr44@rsgb.org.uk*
District 45 (North Yorkshire) Mark Dutton, G0UOK Tel: 07900 503969
Email: *dr45@rsgb.org.uk*

Clubs & Societies

93 Contest Group
John Spurgeon, G4LKD Tel: 01405 704136 Meets last Saturday of each 14.00, Whitgift House, Whitgift, Goole, DN14 8HL

Angel of the North ARC
anarc.club@gmail.com Meets Mondays, fortnightly, 7-9pm, Whitehall Road Church, Bensham, Gateshead, NE8 4LH
https://*www.anarc.net*

Barnsley & DARC
David Gillot, G4TMZ, *g6aj@outlook.com* Meets every Monday 19.00 to 21.30, Higham Cricket club, Pog Well Lane, Barnsley, S Yorks S75 1PH

Bishop Auckland Radio Amateurs' Club
S.Bowman G7ESX *ian.bowman70@yahoo.co.uk* Meets 8.00pm Thursday evenings at the 1900 - 2200, Stanley Crook Village Hall, Rear High Road, Stanley, Crook, Co Durham DL15 9SN *www.barac.org.uk*

Brimham Contest Group
Neil Clark G8MC Meets 1st Tuesday of every Month at 7.30pm, Brimham Lodge, Brimham Rocks Road, Ripley, Harrogate, HG3 3HE *www.stevebb.co/brimham.htm*

Colburn & Richmondshire DARS
Chris Kirby, G4FZN *crdars@mailbox01.co.uk* Meets first and third Wednesday at 19.30, Hudswell Village Halll, Hudswell, Nr Richmond, N. Yorkshire DL11 6BL *www.crdars.org*

Durham & District Amateur Radio Society
Michael Wright, G7TWX *dadars@gmx.com* Meets 18.30 - 21.00 every Wednesday Bowburn Community Centre, Durham Road, Bowburn, Co. Durham, DH6 5AT
https://g4euz.com/

East Ardsley Radio Society
paul@m0xzt.radio Meets Wednesdays 6:30-8pm and Fridays 7-9pm, East Ardsley Cricket Club, Jeffrey Field, Royston Hill, East Ardsley WF3 2HB
http://www.ears.radio

Finningley ARS
M Hotchin M0HOM, *martin.m0hom@gmail.com* Meets Tuesdays 6-9pm, Sat 12-6pm at The Hurst Communications Centre, Belton Road, Sandtoft, North Lincolnshire, DN8 5SX https:*//www.g0ghk.com*

Goole R & ES
Ken G6YYN or Richard G0GLZ Meets every Wednesday. First Wednesday of the month at the Black Swan Inn, Main Street, Asselby DN17 7HE, White Horse Inn, Gilberdyke, HU15 2UP *www.gooleradioclub.btck.co.uk*

Grimsby ARS
Darren Hughes 2E0GUH *2e0guh@gmail.com* Meets Thursdays 7.30pm
, Cromwell Club, Cromwell Road, Grimsby, North East Lincolnshire DN31 2BA https://*www.gars.uk*

Guisborough District ARC
Mark Dutton (G0UOK) *rm.dutton@virginmedia.com* Meets Monday @ 7.30pm, Guisborough Cons Club, Chapel St, TS14 6QE

Halifax District Amateur Radio Society
Martin, M0GQB (Chairman) and Darren, M0WIT (Sec.) Meets every Tuesday between 19.00-21.00, at the Church of the Good Shepherd, New Road, Mytholmroyd, Halifax, West Yorkshire HX7 5EA
www.hadars.org.uk

Hambleton ARS
John Earland M6BHP *jearland@me.com* Meets at 7.30pm on alternative Wednesdays at, The Mencap Centre, Northallerton, N Yorks, DL6 1EG
hambletonars.wordpress.com

Hartlepool Amateur Radio Club
Tom Dyer / Trevor Sherwood 07469 710637 *hartlepoolclub@gmail.com*, Meets every Friday 7pm - 9pm, HQ Scout Centre, 236 Stockton road, Hartlepool TS25 1JW

Hornsea ARS
John Wresdell G3XYF *jhwresdell@gmail.com* Meets on Wednesday nights 7.30 for 8pm, Outdoor Bowling Club, Atwick Road, Hornsea, East Yorkshire HU18 1EL
www.hornseaarc.co.uk

Houghton-le-Spring
Mr George Thompson, M5GHT *m5ght@hotmail.co.uk* Meets Weekly on Tuesdays from 18.30 at the, Dubmire Royal British Legion, Dubmire, Fencehouses, Tyne & Wear DH4 6LJ

Hull & DARS
Richard Pike M0RKK *m0rrk@yahoo.com* Meets alternate Wednesdays at 19:30, Nellies, 173 -175 New Bridge Road, Hull HU9 2LR *www.hadars.co.uk*

Humber Fortress DX ARC
John G6LNV *club@hfdxarc.com* Meets Fridays of each week at 7pm, Mill House, Haven Road, Patrington, HU12 0PS
http://hfdxarc.co.uk/

Keighley ARS
Meets 8.00pm Thursdays at the, The Old Silent Inn, Hobb Lane, Stanbury, Keighley BD22 0HW *www.keighleyradio.co.uk*

Maltby & District ARS
Geoff Preston M7GMP *radio.dx@outlook.com* Meets at 18.00 every Friday, The Centenary Hall, Bateman Road, Hellaby, South Yorkshire, S66 8BH

Mexborough & DARS
Mrs S Saiger M0BOH *M0boh@aol.com* Meets Fridays 19.00-22.00 at, The Place, Castle Street, Conisborough, Doncaster DN12 3HH *www.madars.net*

Northumbria ARC
Mike Smith *mandmsmith5@talktalk.net*, 01670 861751 Meets Thursday evenings 7pm, Old Telephone Exchange, Cresswell Road, Ellington, Morpeth, Northumberland NE61 5HR
https://gx4aax.wordpress.com/

Otley ARS
M. Ross 2E0SNZ *michael.ross@ashfieldprimary.co.uk* Tel: 07768 996370, Meets 7pm-10pm Tuesdays at, The Clifton with Newall Village Hall, Newall Carr Road, Newall with Clifton, Otley, West Yorkshire LS21 2ES
www.otleyradio.org

Pontefract & District ARS
colin.g0nqe@btinternet.com or via website Meets Thursdays at the, Carleton Community Centre, Pontefract, West Yorkshire, WF8 3RJ *www.pdars.com*

Ripon & DARS
Mr David Cutter G3UNA. *d.cutter@ntlworld.com* Meets 7.30pm Thursdays at, The Bunker, rear of Ripon Town Hall, 21 Water Skellgate, Ripon, North Yorkshire HG4 1BH
www.ripon.org.uk

Scarborough Amateur Radio Society
Janet Porte, 2E0SCN Tel 01723 354502 Meets 7:30pm alternate Mondays, The Westover Club, 3 Westover Road, Scarborough, North Yorks, YO12 5AA
www.g4bp.org

Scunthorpe Steel ARC
Peter Jackson G3KNU, *peter.jackson17@ntlworld.com* Meets every other Tuesday at 8.00pm, App Frod Athletic Club, Off Ashby Road, Scunthorpe, North East Lincolnshire DN16 1AA *www.g4fuh.co.uk*

Sheffield And District Wireless Society
Pat Davies M7PAT, secretary Meets every Wednesday of each month, Heeley Green Community Centre, 304 Gleadless Road, Heeley, Sheffield S2 3AJ
https://sheffieldwireless.org

Sheffield ARC
Steve Webster M1ERS, *enquiries@sheffieldarc.org* Meets 7.00pm on Mondays at, Local repeater, Transport Sports Club, Greenhill Main Rd, Lowedges, Sheffield S8 7RH
https://www.sheffieldamateurradioclub.co.uk

Spen Valley ARS
Mr J R Wilde, G0FOI, *russell@wildegardens.co.uk* Meets first and third Thursday in the month at, Old Bank Club, Old Bank Road, MIRFIELD, West Yorkshire WF14 0HY *www.svars.org.uk*

SW Durham Raynet Grp
Mr Ian Bowman, G7ESY, *Ian.bowman@raynet*-uk.net *ian.bowman70@yahoo.co.uk* Meets 8.00pm 2nd Monday in the month at, Stanley Crook Village Hall, rear of High Road, Stanley Crook, Co. Durham DL15 9SN *http://g4ttf.uhp.me.uk/raynet.php*

Tynemouth ARC
Club Secretary. Email: *grahamerrington@sky.com* Meets 7pm to 9pm Fridays at, Tynemouth Scouts HQ, 31-35 Norfolk St, North Shields, Tyne & Wear NE30 1NQ https://*www.facebook.com/g0nwm/*

Wearside Electronics & Amateur Radio Society
Barry Young G0SCI *g0bnk@outlook.com* Meets on Mondays 6-10pm at, Herrington 2nd Scout Headquarters, Crow Lane, Herrington, Sunderland SR3 3TE https://facebook.com/groups/512400014309112/

York Amateur Radio Society
John Nowell G4FUO *radtronix@gmail.com* Meets 8pm Fourth Tuesday of each month , Guppy's Enterprise Club, 17 Nunnery Lane, York, YO23 1AB

White Rose ARS G3XEP
https://*www.whiteroseradio.uk*:5080 E Hannaby G0AEX, *g0aex@whiteroseradio.uk* Meets Wednesday evenings at the Moortown RUFC, Moss Valley, Kings Lane, Leeds, LS17 7NT

York Amateur Radio Society
John Nowell G4FUO *radtronix@gmail.com* Meets 8pm Fourth Tuesday of each month Guppy's Enterprise Club, 17 Nunnery Lane, York, YO23 1AB

York Radio Club
Mr Alan Hemenway G1VIZ *alan.hemenway@gmail.com* Meets 7.30pm Thursdays, Heworth and Bootham Conservative Club, 16-18 East Parade, York, YO31 7YK

Yorkshire Radio Friends
Steve Hall, M0HTH, *s.hall21@btinternet.com* Meets Tuesday 6.30 pm, Hatfield Woodhouse, Doncaster, DN7 6BP

Contest Groups

807 ARO
J Robert Brown M0JRB M0OTO,

Northern Fells Contest Group
Clive Davies, G4FVP, clive*_davies@ntlworld.com*,

Yorkshire Operator Brigade
Darran Chappell G0BWB *g0bwb@gobwb.comOnline* Virtual Club, *http://yobsradio.co.uk*

Region 5 - West Midlands

(http://rsgb.org/region5)

RR

Mark Savage, M0XIC Tel: 07812 212898
Email: *rr5@rsgb.org.uk*

DRs

District 51 (Staffordshire, Warwickshire)
Steve Turner, G0HCR Tel: 07804 596658
Email: *dr51@rsgb.org.uk*
District 52 (Central & East Birmingham)
Martin Hunt, M0JZT Tel: 07976 835351
Email: *dr52@rsgb.org.uk*
District 53 (Shropshire, North Worcestershire & West Birmingham)
John Alexander, M0XJA, Tel: 07384 369160
Email: *dr53@rsgb.org.uk*
District 54 (Gloucestershire, Hereford & South Worcestershire) Vacant
Email: *dr54@rsgb.org.uk*
District 55 (Gloucestershire, Hereford & South Worcestershire)
Leigh Preece, M5GWH Tel: 07534 640123
Email: *dr55@rsgb.org.uk*

Clubs & Societies

Birmingham, South Radio Society
gemmagordon.m6gkg@gmail.com Meets 8.00pm every Mon, Wed and Fri nights at c/o West Heath comm Association, West Heath, Birmingham, B31 3QY https://*www.radioclubs.net/southbirmgham*

Bromsgrove & DARC
Alan G4LVK. *alkel@btinternet.com* Meets Every Friday evening from 8pm local time, Avoncroft Arts Centre, Redditch Road, Stoke Heath, Worcestershire. B60 4JR *www.radioclubs.net/bdarc/*

Burton ARC
Bob McCracken *info@burton*-arc.co.uk Roger Smith Meets on each Wednesday of the month at 7.30pm at, Stapenhill Institute Club, 23 Main Street, Stapenhill, Burton Upon Trent DE15 9AP *www.burtonarc.co.uk*

Central Radio Amateur Circle
Martin Hallard, G1TYV, radio-*circle@live.co.uk* Meets Alt Saturdays 12:00 - 15:00, Bloxwich Memorial Club, 5 Harrison Street, Bloxwich, Walsall West Midlands WS3 3HP https://*www.theradioclub.co.uk*

Cheltenham ARA
Derek Thom, G3NKS *secretary@g5bk.uk* 01242 241099 Meets 7.30 for 8pm on the third Thursday of the month at, Robins Nest, Cheltenham Football Club, Whaddon Road, Cheltenham Gl52 5NA *www.caranet.org*

Coventry ARS
Mr John Beech, G8SEQ. *john@g8seq.com* Meets 1st, 2nd & 4th Friday each month at 20.00 3rd Fri are outdoor events or 2m net, St Bartholomews Church Hall, Brinklow Road, Coventry, CV3 2DT *www.coventryradio.org.uk*

Dudley and District ARS
Kevin Cartwright M0KCC *secretary@dadars.com* Meets 7.30pm every Tuesday at Ruiton Windmill, Vale Street, Dudley, West Midlands, DY3 3XF *www.dadars.co.uk*

Gloucester A R & E S
Dave Tunnicliffe *garesg4aym@aol.com* Meets twice a month on Monday evenings 7.30-9.30, Down Hatherley Village Hall, Down Hatherley Lane, Down Hatherley, Gloucester GL2 9QB https://*www.g4aym.org.uk*

Hereford Amateur Radio Society
enquiries@herefordradioclub.uk Meets 1st Friday of the month, 19:30 to 21:00, Hill House, Newton, nr Leominster, HR6 0PF https://*www.herefordradioclub.uk*

Malvern Hills RAC
Mike Allenson G3TGD *mike.g3tgd@gmail.com* Meets 8.00pm on 2nd Tuesday in the month at the, The Town Club, 30 Worcester Road, Great Malvern, Worcestershire WR14 4QW *www.mhrac.org*

Mid Warwickshire ARS
Don Darkes, G4CYG. *midwarwicks@gmail.com* Meets on 2nd and 4th Tuesday of the month 19.30 in Spring/Summer and 14.00 in Autumn/Winter, Warwick Ambulance Association HQ, 61 Emscote Road, Warwick, CV34 5QR

Midlands ARS
Ron Swinburne *M0WSN@aol.com* Meets every Wednesday evening from 7 - 9pm, Selly Park Baptist Church, 1041 Pershore Rd, Stirchley, Birmingham B29 7PS *www.radioclubs.net/mars*

Moorlands & DARS
Ian King 2E0IDK *m6idk@yahoo.com* Meets 8.30pm on Thursdays at the, Foxfield Railway, Caverswall Road Station, Caverswall Road, Blythe Bridge, Stoke-on-Trent, Staffs ST11 9BG

Nuneaton & District ARC
info@NDARC.co.uk Meets 1st Monday of the month as social and operating night, 3rd Friday of the month is club night,

Windmill Sports and Social Club, 145 Mancetter Rd, Nuneaton, Warwickshire CV10 0HP
www.ndarc.co.uk

Rugby ATS
Mr Stephen Tompsett G8LYB *stephen@tompsett.net* Meets every Tuesday from 20.00 to 22.30 and every Saturday from 14.00 to 18.00, 12th Rugby Scout Headquarters, Broughton-Leigh Community Junior School, Wetherell Way, Brownsover, Rugby CV21 1LT *www.rugbyats.co.uk*

Salop ARS
Les Griffiths M5LMG, *salopamateurradio@gmail.com* Meets Thursdays 8:00pm. Please check club website, as we occasionally have activities away our usual venue., The Telepost Club, Railway Lane, Abbey Foregate, Shrewsbury SY2 6BT
www.salopradiosociety.org

Solihull ARS
Stuart Hammonds G4KUR
solihullradioclub@gmail.com Meets every 3rd Thursday in the at 19.45 to 21.45, 1st Solihull Scouts HQ, Lode Lane, Solihull B91 2HZ
www.solihullradioclub.co.uk

Stafford & DARS
Nick Barnes G4KQK, *cnb@doctors.org.uk* Meets 7.30pm Wednesday at the, Wildwood Community Centre, Wildwood Gate, Stafford ST17 4RA *www.g3sbl.org.uk*

Staffordshire Portable ARC
Neville Briggs, M0VSP, 01922 449668 Meets Tuesdays from 7pm., Bolehall Manor Club, Amington Road, Tamworth, Lichfield, Staffs B77 3LH https://m0spa.co.uk

Stourbridge & DRS
Mr John Clarke Meets 8.00pm 1st & 3rd Mondays in month at the, Old Swinford Hospital/School, Stourbridge, West Midlands, DY8 1QX *www.g6oi.org.uk*

Stratford-upon-Avon
Clive Ousbey *sdrsinfo@talktalk.net* Meets second and fourth Mondays of most months 7.30pm for 8pm, Foundation House, Masons Road, Warwickshire CV37 9NF *www.stratfordradiosociety.freeserve.co.uk*

Sutton Coldfield ARS
Mr Robert Bird 2E0ZAP, *spirit.guide@hotmail.co.uk* Meets from 7 30pm till 10 30pm 2nd & 4th Mondays in the month (except bank hoildays), Sutton Coldfield Rugby Club, 160 Walmley Road, Nr Sutton Coldfield, Birmingham B76 2QA *www.g3rsc.co.uk*

Tamworth ARS
TG 'Robbie' Robertson *robbiebarb204@btinternet.com* Meets Drayton Village Club, Drayton Lane, Drayton Bassett, Tamworth, Staffordshire B78 3TX *www.tamworth*-ars.org.uk

Telford & DARS
John Humphreys, *M0JZH@yahoo.co.uk*, 01952 457234 Meets Wednesdays at 7.00pm. Events are at 8:00pm, Village Hall, Malthouse Bank, Little Wenlock, Telford TF6 5BG *www.tdars.org.uk*

Vulture Squadron C.G
Iain Kelly M0PCB *iain@m0pcb.co.uk* Meets 2nd Monday of each month at 7:30pm, Meeting venue varies, Old Chapel House, Malvern Road, Staunton, Glos GL19 3NZ

Wolverhampton ARS
A. Atkinson 2E0YEZ *secretary@wolverhmaptonars.com* Meets 7.30pm every Wednesday, The Royal British Legion Club, Vicarage Road, Wednesfield, Wolverhampton WV11 1SF
www.wolverhamptonars.co.uk

Wythall RC
Chris Skelcher, G3YHF *chris.bham@gmail.com* Meets weekly Tuesday and Friday 19.30 to 22.30, C/O Wythall Community Association, Wythall House, Silver Street, Wythall, Birmingham B47 6LZ
www.wythallradioclub.co.uk

Repeater Groups

Kidderminster Repeater Group
Ms P Dowie, G8PZT, *g8pzt@gb3kd.org.uk* 3rd Tuesday, monthly at 8pm at Queens Head Wolverley Village Nr Kidderminster Worcs DY11 5XB

Contest Groups

Bad Weather DX Group
Nicholas Pearce, G4WLC, chairmanMeets Tuesday 6pm, 4 Dunster Grove, Cheltenham, Glous.GL51 0PE
www.badwxdx.club

Radio GaGa Contest Group
John Warburton G4IRN
g4irn@dxdx.co.uk,

Redditch AR
Roy Jones G0RMG Meeting Information: On Air Only 144.700 most nights, B97 5NW
http://reddicams.co.uk/radio/about.htm

Region 6 - North Wales

(http://rsgb.org/region6)

RR

Simon Taylor, MW0NWM,
Tel: 07904 874652
Email: *rr6@rsgb.org.uk*

DRs

District 61 (Flintshire, Wrexham & Powys)
Mark Harper, MW1MDH, Tel: 07967 517892
Email: *dr61@rsgb.org.uk*

District 62 (Conwy, Denbigh)
Liz Cabban, GW0ETU Tel: 01690 710257
Email: *dr62@rsgb.org.uk*

District 63 (Gwynedd, Anglesey (Ynys Mon))
John Martin, MW0VTK, Tel: 07772 720099
Email: *dr63@rsgb.org.uk*

District 64 (Ceredigion & North Powys)
Ray Ricketts, GW7AGG, Tel: 01970 611853
Email: *dr64@rsgb.org.uk*

Clubs & Societies

Aberystwyth & DARS
Abby Pugh MW0ZXY *mw0zxy@abbypugh.uk* Meets 2nd Thursday of the month (except August) at 8.00pm-10pm we sometimes have visits or events elsewhere. Waunfawr Community Hall, Brynceinion, Waunfawr, Aberystwyth SY23 3PN
http://adars.org.uk

Dragon ARC
Simon Taylor MW0NWM, *darc.secretary@gmx.co.uk* Tel: 07904 874652, Meets 7.30 (for 8) 1st and 3rd Monday each month, at the Canolfan Esceifiog, Gaerwen, Llanfairpwll, Isle Of Anglesey LL6 0DE
http://www.radioclubs.net/dragonarc

Halkyn Radio Group
halkynradiogroup@yahoo.com Meets Tuesday evenings between 20.00 - 2200hrs, Britannia Inn X, Pentre Halkyn, Halkyn, Flintshire CH8 8BS
https://*www.gw0hrg.co.uk*

Holy Island Amateur Radio Club
Cath Thorley *mw0kbn@aol.com* Meets at 19.00 on 2nd and 4th Tuesdays each month, Boathouse Hotel, Newry Beach, Beach Road, Holyhead, Anglesey LL65 1YF

Meirion ARS
Mr R Smith GW0AYQ *monbob.bayq34@btinternet.com* Meets 1st Wednesday of each month (except August), Bryn Eiddion, Rhydymain, Dolgellau, LL40 2AS *www.meirion*-ars.info

Powys ARC
Michael.Lawton, *michael.lawton1@outlook.com* Meets 8.00pm 1st Thursday in the month, Berriew Community Centre, Welshpool, SY21 8AZ
www.parc.care4free.net

Rhyl And District Amateur Radio Club
secretary@radarc.uk
Meets for club net & activity nights - Sundays at 7pm 145.375 MHz FM, Rhyl Rugby Sports & Social Club, Tynewydd road, Rhyl, LL18 4AQ https://radarc.uk/

Wrexham and Marches ARS
Mr William Taylor 01244571425 Meets second and fourth Thursdays of each month at 19.30hrs at Black Park Community Centre, Halton, Chirk. LL14 5BB.Facebook page. *wrexhammarchesars@gmail.com*, Black Park Community Centre, Lon Graig, Halton, Chirk LL14 5BB
www.wrexham-ars.com/

Contest Groups

Mike-Whiskey DX Group
Online Team meetings 20.00 Monthly,

North Wales Amateur Radio Group
secretary@nwarg.org Every Tuesday 7-9pm at Mochdre Community Centre, Old Conway Rd, Colwyn Bay LL28 5HU
www.nwarg.org

Region 7 - South Wales

(http://rsgb.org/region7)

RR

Vacant
Email: *rr7@rsgb.org.uk*

DRs

District 71 (South Powys, Pembrokeshire, Camarthenshire, Swansea & West Glamorgan) Andy Digby, GW0JLX
Tel: 07768 282880
Email: *dr71@rsgb.org.uk*

District 72 (Mid Glamorgan, East Glamorgan, Cardiff)
Darren Jenkins, MW0MPD,
Tel: 07706 754474
Email: *dr72@rsgb.org.uk*

District 73 (Monmouthshire, Newport)
Nigel Paull, GW1CUQ Tel: 02920 892580
Email: *dr73@rsgb.org.uk*

Clubs & Societies

Aberdare and DARS
Andy Hall *gw7rkc@gmail.com* Meets every other Friday 18.30 to 21.00, Hirwaun YMCA, Manchester Place, Hirwaun, CF44 9RB

Aberkenfig And District ARC
Brian price GW0DVB Meets every Wednesday 19:00 to 22:00
, Tondu Railway Canteen, Station Approach, Tondu, Bridgend CF32 9DY

Barry ARS
Glyn Jones. *glyndxis@talktalk.net* Meets 7.30pm on Tuesdays at, Sully Sports & Leisure Club, South Road, Sully, S Glamorgan CF64 5SP

Blackwood & DARS
L.W.Wright GW8UAM *wynnwright7@aol.com* Meets on Fridays 1900 to 2100 School term time only 1900 to 2100, Islwyn High School, Waterloo, Oakdale, Blackwood Gwent NP12 0NU *www.gw6gw.co.uk*

Blaenau Gwent Radio Group ARS
Chris Taylor 2W0DOE *bgrg@gmx.co.uk* Meets 7pm every Tuesday, Bryn Farm Community House, 13 Heol Onen, Brynmawr, Gwent NP23 4TS

Brecon And Radnor ARS
Adam Tofarides Meets first Thursday of the month 7pm Llanddew village Hall
, Llanddew village Hall, Llanddew, Brecon, Powys LD3 9ST

Carmarthen ARS
Mr Andy Digby, GW0JLX
carsmembershipsec@gmail.com
Meets 1830 (6:30 pm) onwards 1st and 3rd Tuesdays each month, The Cwmduad Community Centre, Cwmduad, Carmarthenshire, which is on the A484. 51° 57' 20.96 N, 4° 21' 56 SA33 6XN

Chepstow & DARS
Dan Taylor GW0EGH *gw0egh@hotmail.com* Meets first and third Tuesday of the month at 19.30 - 22.00 pm, Chepstow Athletic Club, Chepstow, NP16 5JT
www.gw4lwz.co.uk

Cwmbran DARS
Ken Smith - *mc0yad.cadars@gmail.com* Meets 7-9pm every Wednesday Henllys Village Hall, Henllys Village Road, Cwmbran, NP44 6HX
www.mc0yad.co.uk

Highfields ARC
Darren Jenks - MW0MPD Meets 7.00pm Tuesdays, Meets 7.00pm Tuesdays, Rhiwbina Sports and Social Club, Lon-Y-Dail, Rhiwbina, Cardiff CF14 6EA
www.highfields-arc.co.uk

Newport ARS
Mr Paul Nicholls. Email: *nars@gw4ezw.org.uk* Meets every Thursday evening 19.00-21.00 at the chapel, Tabernacle Congregational Chapel, Rhiwderin, Nr Bassaleg, Newport NP10 8RH
www.gw4ezw.org.uk

No1 Welsh Wing ATC Amateur Radio Society
Mr Chris Stubbs, MW0LZZ,
onewingshotoff@hotmail.co Meets 1st Wednesday of each month at 7.00pm, HQ 1344 Sqn ATC, Ty Walter Cleall GC, Maindy Barracks, Cardiff CF14 3YE

Pencoed Amateur Radio Club
Ieuan Jones Meets Tuesdays at 7pm at Pencoed Rugby Club, Felindre Road, Pencoed, NR. BRIDGEND, CF35 5PB

Rhondda ARS
Mr John Howells, GW4BUZ *secretary@gw2fof.co.uk* Meets every other Tuesdays at 7.30 pm, Ystrad, Rhondda Cynon Taff CF41 7SY

St. Tybie ARS
Gareth Woods, GW4JPC, *gw4jpc@yahoo.co.nz* Meets fortnightly at Saron Hall at 7.30pm 01269 597524, Saron Hall, Saron Road, Saron, Ammanford, Carmarthenshire SA18 3LN *http://gc0vpr.clubbz.com*

Swansea & District Amateur Radio Club
Jeff Downer, GW6TYJ, *jeffdowner2017@gmail.com* Meets Tuesdays 7.00pm to 8.30pm, Llansamlet Community Centre, Church Rd, Llansamlet, Swansea SA7 9RH

Taff Vale ARC
Ashley Burns 01685 389434 *aburns02@btinternet.com* Meets every Monday 19.00-21.00, St Johns Ambulance Hall, Gwaun Farren, Merthyr Tydfil, CF47 8LX

Uskside Amateur Radio Club
Mr John Davies GW6RTV Meets Every Wednesday from 1800 - 2000 hrs, Newport Indoor Bowls Centre, Glebelands Stadium, Bank Street, Newport NP19 7HF
https://Coming Soon

Repeater Groups

Tenby Radio Repeater Group
John Rees, GW0JRF Meets 1st Saturday of the month at 1pmMan Shed, Pembroke Port Gate 4Fort Road, Pembroke Dock, SA72 6TH

Contest Groups

The Gower/Gwyr Contest Club
Paul Valerio, GW4KTT *paulvalerio@gmail.comEvery* VHF/UHF/SHF Contest evenings HF every major event, East Pilton Farm, Gower/GWYR, near Rhossili Swansea SA3 1PQ
www.gw4cc.wales

Region 8 - Northern Ireland

(http://rsgb.org/region8)

RR

Peter Lowrie, MI5JYK, Tel: 07929 204138
Email: *rr8@rsgb.org.uk*

DRs

District 81 (Co Antrim)
John Campbell, MI0WJC
Tel: 07712 115791
Email: *dr81@rsgb.org.uk*
District 82 (Co Down)
Roger Bradley, MI0WWB
Tel: 07788 207215
Email: *dr82@rsgb.org.uk*
District 83 (Co Londonderry, Co Tyrone)
Trevor Campbell, MI5TCC Tel: 07710 468835 Email: *dr83@rsgb.org.uk*
District 84 (Belfast)
Vacant Email: *dr84@rsgb.org.uk*
District 85 (Co Fermanagh, Co Armagh)
Vacant Email: *dr85@rsgb.org.uk*

Clubs & Societies

Antrim & District Amateur Radio Society
Darren Brown MI0YPT *drbniradio@gmail.com* Meets on the 2nd Friday of each month at 7.30pm at, Greystone Community Centre, 30 Ballycraigy Road, Antrim, BT41 1PW *www.adars.co.uk*

Ballymena ARC
Hugh Kernohan (Secretary) 07715539481 *hkernohan@aol.com*, Meets Thursdays night at 70 Nursery Road, Gracehill, Ballymena, Co. Antrim, BT42 2QA *http://gi3fff.synthasite.com/*

Bangor & DARS
Harry Squance GI4JTF, *gi4jtf@gmx.com* Meets 7.30pm to 9.30pm 1st Thursday of the month at the, Castle Park, 5 Abbey Street, Bangor, BT20 4JE *www.bdars.com*

Belfast RSGB Group
David Gillespie MI0FBI, *gwocni@hotmail.com* Meets 8.00pm on 3rd Wednesdays of each month September to June at the, Maple Leaf Club, Park Avenue, Standtown, Belfast BT4 1PU

Bushvalley Amateur Radio Club
Victor Mitchell Gi4ONL, *bushvalleyarc@gmail.com* Meets every 8pm 3rd Thursday of each month at the United Services Club, 8, Roe Mill Road, Limavady, Co. Londonderry, BT49 9DF *www.bushvalley*-arc.co.uk

Carrickfergus ARG
John Roberts GI0USX Meets every Tuesday at the, Downshire Community School, Downshire Road, Carrickfergus, BT38 7DA *www.radioclubs.net/carg/*

City of Belfast Radio Amateur Society
Frank Hunter GI4NKB *fthunter@virginmedia.com* Meets first Monday of each month at 8pm at, Shorts Social Club, Holywood Road, Belfast, BT4 1SL

Lagan Valley ARS
Mr Andrew Mulholland, *MI0BPB.gi4gty@hotmail.com* Meets every Wednesday 8pm at The Society Shack, Ballynahinch Road, Lisburn, Co Antrim, BT27 5LX *https://www.lvars.uk*

Lough Erne ARC
Alan Gault GI6PYP, *alan.r.gault@btinternet.com* Meets at the SHARE Centre., Smith's Strand, Lisnaskea, Co.Fermanagh, BT92 0EQ *https://www.lougherneradioclub.co.uk*

Mid Ulster ARC
Hazel McMullen *muarc.secretary@yahoo.co.uk* Meets Second Sunday of each month at 3pm, Ulster Aviation Society, Gate 3 – Maze Long Kesh, 94-b Halftown Road, Lisburn, BT27 5RF *www.muarc.com*

North West Amateur R.C
Mícheal, MI0HOZ *info@nwgarc.net* Meets every 6 weeks at 19.30pm, Lisneal College, 70 Crescent Link, Londonderry, BT47 5FQ

Rusheyhill Amateur Radio Society
Philip Holmes Meets the 4th Saturday of the month at 13:00 hours, 7 Rusheyhill Road, Lisburn, County Down, BT28 3TA

Strabane ARS
Terry White GI7THH *gi7thh@hotmail.com* Meets the second Thursday of each month at 20.00hrs at, 3a Park Road, Strabane, Co Tyrone, Northern Ireland BT82 8EL

Tri-County Amateur Radio Club
Stephen McFarland GI4RNP Meets Sunday weekly at 2pm, Mark Evans Plant Limited, 7 Sheepwalk Road, BT28 3RD

West Tyrone ARC
Philip Hosey MI0MSO *info@wtarc.org.uk* Meets 2nd Wednesday of each month at 20:00-21:30, Order of Malta Hall, Brook Street, Omagh BT78 5HE *www.wtarc.org.uk*

Contest Groups

Orchard County DX Club
Mrs Edith Simpson, MI0PRM *edithmioprm.es.45@gmail* All training and meetings are listed on *http://ocdclub.webs.com*, We are a web based club, BT62 2DD https:*//www.mn0ocg.co.uk*

Sperrin Amateur Radio Club
Mark Robinson MI0LNL *secretary@sarc.club* Meets at 19.30 last Wednesday of the month, The Hub, 14 Burn Road, Cookstown *www.sarc.club*

Region 9 - London & Thames Valley

(http://rsgb.org/region9)

RR

Ron White, G6LTT Tel: 07800 950175
Email: *rr9@rsgb.org.uk*

DRs

District 91 (North, East & West London)
Ron White, G6LTT
Tel: 07800 950175
Email: *dr91@rsgb.org.uk*
District 92 (Berkshire)
Alison Johnson, G8ROG Tel: 0118 9545368
Email: *dr92@rsgb.org.uk*
District 93 (Oxfordshire)
Malcolm Andrew, G8NRP
Tel: 01235 524844
Email: *dr93@rsgb.org.uk*
District 94 (Bedfordshire & Stevenage)
Terry Baldwin, G4UEM Tel: 07789 555514
Email: *dr94@rsgb.org.uk*
District 95 (South West London)
Garo Molozian, G0PZA
Tel: 07765 657542
Email: *dr95@rsgb.org.uk*
District 96 (Buckinghamshire)
Tom Cotton, M7BRW Tel: 07552 670486
Email: *dr96@rsgb.org.uk*

Clubs & Societies

Aylesbury Vale RG
Cathy Clark G1GQJ *cmc5146@gmail.com* Meets on Fridays from 8pm at, Hardwick Village Hall, Lower Road, Hardwick, Nr Aylesbury, HP22 4EA

Aylesbury Vale RS
Mr V Gerhardi, G6GDI, *avrs@rakewell.com* Meets every 2nd Wednesday of the month at 20.00 - 22.00, The Dog House Inn, Broughton Crossing, Broughton, Aylesbury HP22 5AR *www.avrs.org.uk*

B.A.R.S Banbury ARS
Stephen McGuigan *Stephen@mcguigan.com* Meets 1st & 3rd Wednesday of each month at 7pm - 9pm at, 169 Bloxham Rd, Banbury, Oxfordshire, OX16 9JU *www.banburyares.co.uk*

Bedford & DARC
Mr Glen Loake G0GBI *g0gbi.glen@gmail.com* Meets every Tuesday at 7.30 pm at, The Shack opposite the Plantation, Ravensden, Bedfordshire MK44 2RJ *www.badarc.net*

Bracknell ARC
Colin Ashley 2E0XDA *secretary@g4bra.org.uk* Meets on the 2nd Wednesday of each month at 8pm. We also hold weekly

Radio Nets at 8pm each Wednesday on 145.375 FM and each Sunday at 5pm via the GB3BN Repeater, Shepherds Lane, Bracknell, Berkshire RG42 2BU *https://g4bra.org.uk*

Burnham Beeches RC
Dave Chislett G4XDU, *bbradioclub@gmail.com*, Meets 8.00pm on 1st and 3rd Mondays in the month at the, Farnham Common Village Hall, Victoria Road, Farnham Common, Bucks SL2 3NL *www.bbrc.info*

Chesham & District ARS
Malcolm Appleby G3ZNU Meets second and fourth Wednesdays in the month, 8:00pm, Ashley Green Memorial hall, Two Dells Lane, Ashley Green, Bucks HP5 3PN *http://www.g3mdg.org.uk/*

Drowned Rats Radio Group
Carl Ratcliffe *info@drownedrats.uk* Meets second Wednesday of the month 20.00pm - 22.30pm, The Coy Carp, Copperhill Lane, Harefield, Middlesex UB9 6HZ *www.g3rat.com*

Dunstable Downs Radio Club
Mike Scarlett G4CAK *mikescarl@btinternet.com* Meets every friday at 8pm, St Fremunds Church Hall, 185 Westfield Road, Dunstable, LU6 1DR *www.dunstabledownsradioclub.org*

Edgware & DRS
Steve SlaterG0PQB *g0pqb25@gmail.com* Meets 8.00pm on 2nd and 4th Thursdays in the month at the, Watling Community Centre, 145 Orange Hill Road, Burnt Oak, Edgware, Middlesex HA8 0TR *www.g3asr.co.UK*

Farringdon Radio Group
secretary: *g0cag@hotmail.com* Meets 18.00 every Tuesday, c/o Hanwha Phasor, 4th Floor, Record Hall, 16 Baldwins Gardens, London EC1N 7RJ

Harwell Amateur Radio Society
Ann Stevens G8NVI *secretary@g3pia.net* Meets at 19.45 on the 2nd Thursday of each month 19.45 for 20.00hrs, Chilton Village Hall, Church Hill, Chilton, OX11 0SH *www.g3pia.net*

Maidenhead & DARC
Peter Hicks, G4KCX *G4KCX@talktalk.net* Meets 7.45 on 1st Thursday and 3rd Tuesday in the month at the, The Friends Meeting House, 14. West Street, Maidenhead, Berkshire SL6 1RL *https://www.madarc.org*

Mid Thames DFC G4MDF
www.topbanddf.org.uk John Mullins, *Chairman@TopBandDF.org.uk* Meets monthly in the winter at various pubs, currently, Dashwood Arms, Piddington,Bucks, HP14 3JG

Milton Keynes ARS (MKARS)
Francis Hennigan: *information@mkars.org.uk* Meets Mondays 19.30. Zoom Coffee Chats 10.30 on Fridays, Block H, Bletchley Park, Bletchley, Milton Keynes MK3 6EB https://*www.mkars.org.uk*

Newbury & DARS
Phill G6EES 07771 504738 *secretary@nadars.org.uk* Meets 4th Wednesday of every month (apart from December) at 19.30hrs, Acland Hall, Hermitage Rd, Cold Ash, Berkshire RG18 9JF https://*www.nadars.org.uk*

Oxford And District ARS
Graham Diacon G8EWT, *G8EWT@diacon.co.uk* Meets 7.30 for 8.00pm on the 1st and 3rd Tuesday of every month, at The Gladiator Club, 263 Iffley Road, Oxford, OX4 1SJ *www.odars.org.uk*

Reading and District ARC
https://radarc.org/contact-us/ Meets 2nd and 4th Tuesday of each month for social and practical sessions. Some events are held at Reading RFC, Sonning, 2nd Woodley Scout Group Hut Vauxhall Park, Vauxhall Drive, Woodley, Reading RG5 4EA https://*www.radarc.org*

RNARS London Group
D Goodison, G0LUH 0208 8470260 *info@gb2rn.org.uk*, Meets last Thursday of every alternate month, at 15.00hrs on HMS Belfast, Battleship Lane, Tooley Street, London SE1 2JH *www.gb2rn.org.uk*

RS of Harrow
Linda, G7RJL *info@G3EFX.org.uk* Meets 1st and 3rd Fridays of the month at 20.00 20.00 to 22.00, Blackwell Hall, Uxbridge Road, Stanmore, HA3 6DQ *www.g3efx.org.uk*

Shefford & District ARS
D Lloyd,G8UOD *davidg8uod@virginmedia.com* Meets at 8pm Thursdays (with summer break) at the, Community Hall, Ampthill Road, Shefford, Beds SG17 5AX *www.sadars.co.uk*

Silverthorn RC
Robin Bernard, M0HVC *m0hvc@protonmail.com* Meets every Friday evening at 7.30pm to 10.30pm on Fridays at, Friday Hall, 56 Friday Hill East, Chingford, London E4 6JT *www.silverthornradioclub.org.uk*

Southgate Amateur Radio Club
Keith Mendum G8RPA Tel: 02083603614 Meets 1st Monday of month at 19.30 hrs, 14th Southgate Scout Group Hut, 81 Green Road, Oakwood, London N14 4AP *http://southgatearc.uk*

Stevenage & DARS
Mr Rob McTait G2BKZ, rob*_g2bkz@talktalk.net* Meets 7.30pm on Tuesdays at the, Stevenage Resource Centre, Chells Way, Stevenage, Herts SG2 0LT *www.sadars.org*

Triple B Amateur Radio Contest Group
M Goodey, G0GJV - Tel: 07770938478 Meets Fridays 10pm, Jack O'Mewbury, Terrace Road South, Binfield, RG42 5PH

Verulam ARC
Greg Beacher: *secretary@verulam*-arc.org.uk Meets on the 3rd Monday of the month, The Boot (Inn / public house), Conservatory, High Street, 2 miles East of Harpenden, Hertfordshire SG4 8PT *http://www.verulam-arc.org.uk/*

Contest Groups

AMC Radio Club
David Millward M0PTP *David.m@pebbleltd.co.uk* Meets once every quarter and every day online, 69 Primrose Hill, Kings Langley, Hertfordshire WD4 8HX

Beds, Mid Contest A
Fred Handscombe *g4mbc@homeshack.freeserve.co.uk*, IP28 8LQ

Region 10 - South & South East

(http://rsgb.org/region10)

RR

Keith Bird, G4JED Tel: 01732 446331
Email: *rr10@rsgb.org.uk*

DRs

District 101 (Surrey, London South of the Thames) Alun Cross, G4WGE Tel: 07779 079503, Email: *dr101@rsgb.org.uk*
District 102 (Wiltshire)
Simon Harris, G4WQG Tel: 07498 213585
Email: *dr102@rsgb.org.uk*
District 103 (West Sussex)
Sean Pryer, 2E0XBT Tel: 07505 546810
Email: *dr103@rsgb.org.uk*
District 104 (Hampshire)
Alan Ball, G3UQW Tel: 07770 536975 Email: *dr104@rsgb.org.uk*
District 105 (Isle of Wight)
Fred Dawson, G1HCM Tel: 07885 634518
Email: *dr105@rsgb.org.uk*
District 106 (Kent)
Dave Lee, G8ZZK Tel: 07739 549822
Email: *dr106@rsgb.org.uk*
District 107 (East Sussex) Vacant
Email: *dr107@rsgb.org.uk*

Clubs & Societies

Andover RAC
Paul Philips, G4KZY, *arac@arac.org.uk* Meets first and third Tuesday of each month at 19:30 to 21:30, Tangley Parish Village Hall, Wildhern, Andover, Hants SP11 0JE https://*www.arac.org.uk*

Barbarian Drifters Radio Society
Philip Bourke M0IMA *barbariandrifters@gmail.com* Meets Saturday of each week at 12 noon, Meadow Bank, Rye Lane, Otford, Kent TN14 5JF

Basingstoke ARC
telephone 01256 883838 Meets 3rd

Monday of the month,7:30 meet in the bar for an 8pm start, May's Bounty Cricket Club, Fairfields Road, Basingstoke, Hampshire RG21 3DR *https://basingstokearc.wordpress.com/*

Brede Steam ARS
secretary@bsars.co.uk Meets 1st and 3rd Saturday of each month. Please check with secretary before visiting, as we might be participating in an event., Brede Steam Amateur Radio Society, The Scout Hut, Stubbs Lane, Brede, Nr Rye TN31 6EH *www.bsars.co.uk*

Bredhurst Rx & Tx Soc
secretary@brats-qth.org Meets Brats Shack, Brats Shack, Hurstwood Rd, Bredhurst, Gillingham ME7 3JZ *www.brats*-qth.org

Bromley & DARS
Andy Brooker *enquiries@bdars.co.uk* Meets 3rd Tuesday in the month 19:45 for 20:00, Victory Social Club, Kechill Gardens, Hayes, Bromley BR2 7NG *www.bdars.wordpress.com*

Chippenham and District ARC
Brian Tanner G6HUI *chairman@chippenhamradio.club* Meets every Tuesday from 8pm at the Tuesday 19.30 - 21.30, Kington Langley Village Hall, Church Road, Kington Langley, Chippenham SN15 5NJ *http://chippenhamradio.club*

Christchurch ARS
http://christchurchars.org.uk Richard M0RBF *vicechair@radioclub.uk* Meets 7.30pm on Thursdays at The Clubhouse adjacent to East Christchurch Sports and Social Club Grange Road, Christchurch, Dorset, BH23 4JE

Coulsdon Amateur Transmitting Society
Glenn Rankin G4FVL *secretary@catsradio.org* Meets 8.00pm on 2nd Mon in the month St Swithuns Church Hall, Grovelands Road, Purley, Surrey CR8 4LA *https://www.catsradio.org*

Crawley Amateur Radio Club
Phil Moore M0TZZ *secretary@carc.org.uk* Meets 8.00pm Wednesdays and, 11.00 am Sundays at the, Tilgate Forest Rec. Centre, Hut 18, Tilgate Forest, Crawley, West Sussex RH11 9BQ *www.carc.org.uk*

Cray Valley RS
Dave Lee, G8ZZK. *secretary@cvrs.org* Meets 1st & 3rd thursday of the month 19.30 - 21.30 at the, 1st Royal Eltham Scouts HQ, Rear of 61 - 71 Southend Crescent, Eltham, London SE9 2SD *www.cvrs.org*

Crystal Palace Radio And Electronics Club
Damien Nolan, 2E0EUI *cprec.g2lw@gmail.com* Meets 7.30pm on 1st Friday each month at the, All Saints Church, Beulah Hill, London SE19 3LG https://*www.cprec.org*

Darenth Valley RS
Phil Bourke M0IMA *info@darenthvalleyrs.org.uk* Meets 2nd & 4th Wednesday of the month 20.00 - 22.30, Crockenhill Village Hall, Stones Cross Road, Crockenhill, Swanley BR8 8LT https://*www.darenthvalleyrs.org.uk/joining*-or-listening-to-the-darenth-valley-club-net/

Dorking & District Radio Society
Sue Ellinor, *ddrs.secretary@gmail.com* Meets 7.45pm on 4th Tuesday of each Month at 19.45 pm, at The Friends Meeting House, Butter Hill, South St, Dorking, Surrey, RH4 2LE *www.ddrs.org.uk*

Dover Radio Club
Nathan Friend 2E0NFX, *secretary@darc.online* Meets on every Thursday 7-9pm. Talks and demos planned every 4 weeks or so, please visit the club calendar for more information, St Peter & St Pauls Church Hall, Minnis Lane, River, Kent CT17 0RG https://*www.darc.online*

DSTL Radio Club
Peter Allcock, *dstlradioclub@dstl.gov.uk* Meets 2nd Wednesday each month 12 noon, DSTL Porton Down, Crossley House, Porton Down, Wiltshire SP4 0JQ

East Kent Radio Soc.
Alan Perkins G7RBB *perkins.alan@gmail.com* Meets 8.00pm on the 2nd Wednesday (summer months only), The Herne Mill, Mill Lane, Herne Bay, Kent CT6 7DR *www.g0ekr.co.uk*

Echelford ARS
Philip Miller Tate, M1GWZ *m1gwz@icloud.com* Meets 7:30pm on 2nd & 4th Wednesdays of month, St Hildas Church Hall, Stanwell Road, Ashford, Surrey TW15 3QL *https://qsl.net/g3ues/*

Fareham & District Amateur Radio Club
C. Jenkins-Powell G7MFR, *chris@jenkins*-powell.com Meets 7.30pm on Wednesdays at the, Fareham Motorboat & Sailing Club, Lower Quay, Fareham, Hants PO16 0RA *www.fareham*-darc.co.uk

Farnborough & DRS
C. Andrews M7WJC, *treasurer@farnboroughradio.org.uk* Meets every 2nd and 4th Wed of the month at 7.30pm, The Aldershot Military Museum, Off Queens Avenue, Aldershot, Hants GU11 2LG *www.farnboroughradio.org.uk*

Fort Purbrook ARC
Chris Bryant G3WIE *g3wie@fparc.org.uk* Meets at 19.00 - 21.00 on the last Friday each month, Peter Ashley Lane, Portsdown Hill Road, Nr Cosham, Portsmouth, Hampshire PO6 1BJ *www.fparc.org.uk*

Guildford & District Radio Society
Timothy Dabbs, G7JYQ *secretary@gdrs.net* Meets on 2nd, 4th and 5th Fridays of each month. 1930 for 2000hrs, Guildford Model Engineers HQ, Stoke Park, London Road, Guildford, Surrey GU1 1TU *www.gdrs.net*

Hastings Electronics & Radio Club
John G0OZY, *herc.hastings@gmail.com* Meets every Wednesday in the local area for /P radio-related activities. The location and time changes from time to time, visit page 2 of our website on Wednesday mornings to find details for that day. TN38 0LQ *www.hastings-electronics-radio-club.com*

Hilderstone Radio Society
Ian Lowe, G0PDZ *secretary@g0hrs.org* Meets every Thursday (except August) 7pm for 7:30pm start. Visitors welcomed. Onsite parking & level access for the disabled, 1st Margate (St John's) Scout Group Hall, Durban Road, Margate, Kent CT9 2TE *https://g0hrs.org*

Hog's Back Amateur Radio Club
Simon Lambert M0XIE *hogsbackarc@gmail.com* Meets 2nd and 4th Mondays of the month (excluding Bank holidays), Scout Centre, Pankridge Street, Crondall, Hampshire, Surrey GU10 5RQ *https://www.hogsback-arc.org.uk*

Horndean & DARC
Mr Stuart Swain, *G0FYX@msn.com* Meets 1st & 3rd Friday in the month from 1830 to 2130, Deverell Hall, 84 London Road, Purbrook, Waterlooville, Hants PO7 5JU *www.hdarc.co.uk*

Horsham ARC
Mr Alister Watt, G3ZBU. *info@harc.org.uk* Meets 8pm 1st Thur in month at the, Guide Hall, Denne Road, Horsham, West Sussex RH12 1JF *www.harc.org.uk*

Isle of Wight RS
Steve Cownley, 2E0RQD. *iowradsoc@gmail.com* Meets 7.00pm Fridays at, Haylands Farm, Salters Road, Haylands, Ryde, Isle of Wight PO33 3HU *www.iowrs.org*

Itchen Valley ARC
Chris Ash, G7LWV. *chris.g7lwv@gmail.com* Meets 8.00pm on the 2nd and 4th Fridays of the month, Otterbourne Village Hall, Cranbourne Road, Otterbourne, Hampshire SO21 2ET *www.ivarc.org.uk*

Kent Active Radio Amateurs
Debbie Richardson G7EOZ *avckev@btinternet.com* Meeting will be regional and some regular meets the PS Medway Queen, Gillingham Pier, Pier Approach Rd, Gillingham, ME7 1RX

Kent Weald Radio Club
Mr P J Blunt, G0UXG *palybl@btinternet.com* MX0KWA For Club information *https://www.qrz.com/m0kwa*

Medway ARTS
secretary@marts.org.uk Meets weekly on Fridays from 19:30, Location Tunbury Hall, Catkin Close Chatham. ME59HP. Please see website for more detail and meeting schedule. Daily net at 09:30 on 144.650 FM, Tunbury Hall, Catkin Close, Walderslade, Chatham, Kent ME5 9HP *https://marts.org.uk*

Mid Sussex ARS
https://*www.midsussexars.org.uk* Alan Cragg *secretary@midsussexars.org.uk* Meets 19.30 - 21.30 at Cyprus Hall, Millfield Suite, Cyprus Road, Burgess Hill, West Sussex, RH15 8DX

Newhaven Fort Amateur Radio Group
Roger Parish, M1RPY, *newhavenarg@gmail.com* Meets daily at 10 am at, Newhaven Fort, Fort Road, Newhaven, East Sussex BN9 9DS

North Kent RS
Mr Stephen Osborn, G8JZT, *secretary@nkrs.org.uk* Meets 7.30pm on 1st and 3rd Tuesday of each month 19.30 to 22.00 hrs, Hurst Community Centre, Hurst Road, Bexley, Kent DA5 3LH *www.nkrs.org.uk*

Southampton Amateur Radio Club
Nigel Phillips, G7POC *sotonarc@gmail.*

com Meets at 7.30pm Third Wednesday of each month in 19.30-21.30hrs, Lounge Bar, The Peartree Inn, Peartree Road, Southampton SO19 7GZ https://*www.southampton*-arc.org.uk

Southampton University Wireless Soc
Andrew Barrett-Sprot *contact@suws.org.uk* Meets every Thurs at 18.00-20.00, University of Southampton, Buliding 40, University Road, Highfield, Southampton SO17 1BJ *www.suws.org.uk*

Southdown ARS
Tom Pitcher, *secretary@sars.club* Meets Every first Friday of the month, Old Town Community Centre, 1a Central Ave, Eastbourne, East Sussex BN20 8PL *http://sars.club*

Surbiton Heritage AR & Elec Society
Tony M0SHA email: *RSGB@m0sha.com* Meets 7.45 pm on 1st Wednesday of the month at the, The Coffee Bar, 1st Floor, Surbiton Hill Methodist Church, 39 Ewell Road, Surbiton, KT6 6AF *https://www.m0sha.com*

Surrey EARS
Caspar B Pierce, *radio@surreyears.co.uk* Meets Wednesdays 17:00 to 21:00. Open only to University students and staff, University of Surrey, 46-AB-05 EARS SHACK, 388 Stag Hill, Guildford, Surrey GU2 7XH https://activity.ussu.co.uk/ears

Surrey Electronics Makers & Radio Club
Revd Graham Smith G4NMD Meets 3rd Wednesday of the month at 1930-2130, Grafham Room, Horsham Road, Grafham, GU5 0LJ *https://www.facebook.com/hamradiobuilders*

Surrey Radio Contact Club
Quin Collier G3WRR *secretary@srcc.uk* Meets 7.45pm - 9.30pm on 1st and 3rd Mondays, Jubilee Room, St. Paul's United Reformed Church, Croham Park Avenue, South Croydon CR2 7HF *www.srcc.uk*

Sussex 4x4 Response
David Green G4OTV *dave.w.green@btinternet.com* Meets last Friday of each 2nd month 8pm, Hassocks Hotel, Station Approach East, Hassocks, West Sussex BN6 8HN

Sutton & Cheam RS
Chris Howard M0TCH. *secretary@scrs.org.uk* Meets Monday nights at 8pm, 7th Banstead Scout Hall, Woodgavil, Banstead, SM7 1DA *www.scrs.org.uk*

Swindon & DARC
Den, M0ACM *secretary@sdarc.net* Meets 7pm - 10pm every Thursday except July and August, Pinetrees Centre, Pinehurst Circle, Swindon, SN2 1RF *www.sdarc.net*

Trowbridge & DARC
Shaun Coles, G0BKU. *secretary.tdarc@gmail.com* Meets 1st & 3rd Wednesdays monthly at 8PM, Southwick Village Hall, Frome Road, Southwick, Wiltshire BA14 9QN *sites.google.com/prod/view/tdarc/home*

Waterside New Forest
Mr Tim Williams, G4YVY, 02380894278 Meets 8pm on 1st & 3rd Tues of month (not Aug) at, Applemore Scout HQ, near Hythe, Southampton, SO45 4RQ *www.watersidears.org.uk*

West Kent ARS
Phil Parkman G3MGQ, *secretary@wkars.org.uk* Meets 8pm on the 2nd Monday of the month, at Bidborough Village Hall, Bidborough, Kent, TN3 0XD *www.wkars.org.uk*

Wings Museum
B.Bloomfield, G4OKB, *bloomfieldbarrie@hotmail.com* Meets first Meets every Wed of each month at 11am, Wings Museum, Brantridge Lane, Balcombe, West Sussex RH17 6JT

Worthing & District ARC
secretary@wadarc.org.uk Meets every Tuesday doors open 7pm, TS Vanguard, 9a Broadwater Road, Worthing, W. Sussex BN14 8AD *www.wadarc.org.uk*

Worthing Radio Events Group
secretary@m0reg.co.uk Meets 1st Monday of the month at 8pm. Please contact the secretary for confirmation before visiting, Goring Conservative Club, 49 Mulberry Lane, Worthing, BN12 4RA *http://www.m0reg.co.uk*

Repeater Groups

Guildford UHF Repeater Group
Mr Alex Morris, G6ZPR *morris.alex@btconnect.comHolds* Yearly AGM Sanford ArmsEpsom Road, Guildford, Surrey *http://gb3gf.co.uk*

Kent Repeater Grp
chairman@krg.org.uk

Ridgeway Repeater Group
Robert Loss G4XUT *g4xut@rrg.org.uk* *https://www.rrg.org.uk*

Contest Groups

A1 Contest Group
Mr N Wilson, G4VVZ. *g4zap@aol.com*,

Addiscombe ARC
Mike Franklin G3VYI,

Invicta Contest Group
Ian Hope, *ian@dr.com* or Ian Lowe G0PDZ *g0pdz.ian@gmail.com* A Kent Based mainly VHF and above contest group. We normally enter contests such as the March 144/432, The May 144, 50mhz Trophy, VHF NFD, The Aug 144/432 Low power, the 144 trophy in Sep and the December 144 AFS. We currently do not enter HF contests. No Experience Required, Any Licence Level welcome. DA11 7EB *https://www.invictacg.co.uk*

Three A's Contest
Ian Pritchard, G3WVG, *g3wvg@btinternet.com*,

Vecta Contest Group ARS
Fred Dawson *fred.wp.dawson@googlemail.com* Church Crookham, Fleet Hants GU52 8LD

Region 11 - South West & Channel Islands

(http://rsgb.org/region11)

RR

Andrew Jenner, G7KNA
Email: *rr11@rsgb.org.uk*

DRs

District 111 (Cornwall and Scillies) Robert Searby, M0KLG Tel: 07972 144724 Email: *dr111@rsgb.org.uk*

District 112 (North Cornwall and North Devon) John Lovell, G3JKL, Tel: 01237 478410 Email: *dr112@rsgb.org.uk*

District 113 (South Gloucestershire, Bristol and South and East Somerset) Andrew Jenner, G7KNA Tel: 07838 695471 Email: *dr113@rsgb.org.uk*

District 114 (Dorset) Vacant Email: *dr114@rsgb.org.uk*

District 115 (Jersey) Peter Bertram, GJ8PVL Tel: 07829 722722 Email: *dr115@rsgb.org.uk*

District 116 (Guernsey) Jerry Bligh, MU0ZVV Tel: 01481 243322 Email: *dr116@rsgb.org.uk*

District 117 (East Devon and West Somerset) Vacant Email: *dr117@rsgb.org.uk*

District 118 (South Devon) Nigel Bennetts, M7NGL Tel: 07966 133537 Email: *dr118@rsgb.org.uk*

Clubs & Societies

Appledore & DARC
John Lovell, G3JKL (*john@g3jkl.co.uk*) Meets 7.30pm on 3rd Monday in the month at the, Appledore Football Club, EX39 1PA *www.adarc.co.uk*

Bath and District Amateur Radio Club
P D Carter, G4PDC *badarc@protonmail.com* Meets at 7.30 on last Wednesday of each month, Farmborough Memorial Hall, Bath BA2 0AH https://badarc.webs.com

Blackmore Vale ARS
Keith Chadwick M0TMO; *keith.m0tmo@btinternet.com* Meets 7.30pm every Tuesday Apr-Oct 2nd and 4th Tues Nov-Mar 19.30 to 21.30, New Remembrance Hall, Remembrance field, Charlton, Shaftsbury, SP7 0PL *http://www.bvars.org.uk/*

Burnham ARC - Mike Lang
Mike Lang G4DVK *mikelangphotography@gmail.com* Meets 1st & 3rd Wed of the month 7pm to 9pm, The Bay Centre Cassis Close, Highbridge, Burnham on Sea, Somerset TA8 1NN
http://burnhamradioclub.co.uk/

Callington ARS
secretary@callingtonradiosociety.org.uk Meets on the first Wednesday of each month, Callington Town Hall, PL17 7BD. Every Monday 8pm: Weekly net - GB3JL repeater. Every Wednesday: Lunch, The Engine House Café PL17 8EA from 1:30pm. Everyone welcome to all events, Council Chamber, Callington Town Hall, Callington, Cornwall PL17 7BD *https://www.callingtonradiosociety.org.uk/*

Cornish Radio Amateur Club
Steven Holland G7VOH, *g7voh@btinternet.com* Meets on the 1st & 3rd Thur of the month from 7.30pm at Gweal an Top School, School Lane, Redruth, TR15 2ER
http://gx4crc.com/

Dartmoor Radio Club
Please get in touch via https://dartmoorradioclub Meets at the Yelverton War Memorial Hall on the last Thursday of the month. On all other Thursdays we hold virtual meetings via Zoom, Yelverton War Memorial Village, Meavy Lane, Yelverton, Devon PL20 6AL
https://dartmoorradioclub.uk

Exeter ARS
Mr John Rooke, G4AP *rooke906@btinternet.com* Meets 1st and 3rd Tuesdays at 19:00, America Hall, De La Rue Way, Exeter, EX4 8PX *www.exeterars.co.uk*

Flight Refuelling
Sue Macdonald M0PSZ. *info@frars.co.uk* Meets twice a week Wednesdays and Sundays 19:00 to 22:00, Cobham Sports and Social Club, Merley Park Road, Merley, Wimborne, Dorset BH21 3DA
https://www.frars.co.uk

Gordano ARG
Malcolm Pitt *mal@g4kpm.co.uk* Meets fourth Wednesday of the month at 20:00, The Ship Inn, 310 Down Road, Portishead, Bristol BS20 8JT

Guernsey ARS
Phil Cooper GU0SUP 07781 151747 Meets 7.30pm the every Friday at The Bunker, Beau Sejour, Amherst, St Peter Port., GY1 2DL *www.gars.org.gg*

Holsworthy ARC
Ken Sharman G7VJA, *ken@g7vja.co.uk* Meets the 1st Wednesday of each month at 7.30pm at the, Milton Damerel Parish Hall, EX22 6PS *http://m0omc.co.uk/*

Jersey Amateur Radio Society
Peter Bertram, GJ8PVL (*gj8pvl@hotmail.com*) Meets 8.00pm every Friday, The German Signal Station, Le Chemin des Siganux, Rue Baal, La Moye, St. Brelade, Jersey JE3 8LQ *https://jerseyars.org.je/*

Kidderminster Repeater Group
www.gb3kd.org.uk Paula G8PZT *g8pzt@krg.club* Meets 3rd Tuesday of month at 8pm. Queens Head, Wolverley Village, Nr Kidderminster, Worcs, DY11 5XB

Mid Somerset ARC
Chris lavis G6HIQ *christopherlavis3@gmail.com* Meets Shepton Mallet Baptist Church: 2nd Monday of the month at 19:30 also meets Quarter Jack: last Monday of the month at 14:00, The Prayer Room, Shepton Mallet Baptist Church, Commercial Road, Shepton Mallet, BA4 5BU, Quarter Jack, 18 Priory Road, Wells BA5 1SY
www.midsarc.org.uk

Newquay & DARS
Terry 2E0XTM *newquayradioclub@gmail.com* Meets 2nd Wednesday of the month, Tel: 01841 540142, The Treviglas Community College, Bradley Road, Newquay TR7 3JA
https://newquayradioclub.co.uk

North Bristol Amateur Radio Club
Dave Bendrey *g7byn@blueyonder.co.uk* Meets every Friday at 19.00, SHE7 Building, Braemar Crescent, Northville, Filton, Bristol BS7 0TD *www.nbarc.org.uk*

Plymouth Radio Club
Martin Mills M0MLZ *martinmills8448@gmail.com* Meets on the 2nd Tuesday of each month 19.00 hrs, Weston Mill Oak Villa Social Club, Ferndale Road, Weston Mill, Plymouth PL2 2EL
www.radioclubs.net/g3prc

Poldhu ARC
secretary@gb2gm.org Tel: 01326 241 656 Meets on the 2nd Tuesday of the month, Marconi Centre, Poldhu Cove, Mullion, Cornwall TR12 7JB *http://gb2gm.org/*

Poole Radio Society
Charles Riley, G4JQX (*charles@fernpatch.com*) Meets 8pm every Thursday evening (except at Christmas and Easter) at, St Aldhelms Centre, Poole Road, Branksome, Poole BH13 6BT *www.g4prs.org.uk*

Riviera ARC
Ian Nelson M0IDP *rivieraARC@gmail.com* Meets first and third Thursdays of every month Precinct Centre 19.00 - 21.00, Church Road, St Marychurch, Torquay, Devon TQ1 4QY
www.rivieraarc.org.uk

Saltash District Arc
Mark Chanter M0WMB, *m0wmbsadarc@gmail.com* Meets second Tuesday of month at 7.30pm (Not Aug or Dec), Burraton Community Centre, Grenfell Avenue, Saltash, Cornwall PL12 4JB
www.sadarc.co.uk

Shirehampton ARC
Mr Chidgey *secretary@shirehampton*-arc.org.uk Meets 7.30pm on Fridays at, T.S Enterprise, Station Road, Shirehampton, Bristol, BS11 9XA
https://www.shirehampton-arc.org.uk

Sidmouth Amateur Radio Society
Dave Lee G6XUV Meets on Wednesdays at 19.00 throughout the year, our meetings vary during the year depending on seasonal variation., SARS, Sidford Community Hub, (rear of tennis courts), Byes Lane, Sidford Sidmouth EX10 9TB *https://www.facebook.com/groups/1207964666698144*

South Bristol ARC
secretariat@sbarc.co.uk - Meets 1st and 3rd Thursdays of the month from 19:30. Check our website (*www.sbarc.co.uk/calendar*) for event details, Novers Park Community Association, Rear of 124 Novers Park Road, Bristol, Avon BS4 1RN
https://www.sbarc.co.uk

South Dorset RS
Ray Coles M0XDL, *raycoles960@btinternet.com* Meets 2nd Friday of each month at 19:30, unless otherwise specified on the website, Weymouth West Air Scout Group Hall, Granby Close, Weymouth, Dorset DT4 0SW
www.sdrg.co.uk.

T.S.W.A.R.C
Tim Hugill G4FJK *tim@whitnole.com* Meets 8pm every Tuesday at, ROC Site, 7 Crosses, Tiverton, EX16 8JR
https://www.facebook.com/G4TSW/

Taunton & District ARC
Peter Robinson G0EYR, *g0eyr@hotmail.com* meets the first Tuesday of each month, Tangier Scout and Guide Centre, Tangier, Castle Street, Taunton TA1 4AS *https://tauntonarc.wixsite.com/taunton-radio-club*

Thornbury & Sth Glos
T J Humphreys, M0TJX, *secretary@tsgarc.uk* Meets Every Wed Evening 1930 - 2130, The Chantry, 52 Castle Street, Thornbury, BS35 1HB
https://www.tsgarc.uk/

Torbay ARS
John *membsec@tars.org.uk* Meets on Fridays from 19.30 at, Teignbridge District Scout Headquarters, Wolborough Street, Newton Abbot, Devon TQ12 1LJ
www.tars.org.uk

Weston Super Mare Radio Society
The Secretary (*westonradiosociety@gmail.com*) Meets Mondays at 7:30pm, Devonshire Road Social Club., Weston-super-Mare BS23 4LG
www.g4wsm.club

Yeovil ARC
Darren Mallinson 2E0EVU, *secretary@yeovil*-arc.com Meets 19.30 on Thursdays, Abbey Community Centre, The Forum, Abbey Manor Park, Yeovil, BA20 2BE
http://yeovil-arc.com

Contest Groups

Castel Contest Club
ichard Allisette GU4CHY
www.castelcontestclub.com

Isle of Avalon Amateur Radio Club
Matt Morse - 2E0FNT - *2e0fnt@gmail.com* Meets Fridays from 6pm. We monitor 145.475 FM, Ivor's Paddock and Yard, (contact Matt 2E0FNT for how to get there, this is a field not a building) Glastonbury BA6 9AF https://*www.avalonarc.org.uk*

Repeater Groups

Jersey AR Repeater Group
Peter Bertram, GJ8PVL (*gj8pvl@hotmail.com*)Meets 8.00pm every Friday at Old German Signal Station, Rue Baal, Moye, Jersey JE3 8LQ *www.radioclubs.net/gb3gi*

Kidderminster Repeater Group
Ms P Dowie, G8PZT, *g8pzt@gb3kd.org.uk* Meets 3rd Tuesday of month at 8pm. Queens HeadWolverley Village, Nr Kidderminster, Worcs DY11 5XB *www.gb3kd.org.uk*

Mid Cornwall Repeater Group
Paul Andrews G6MNJ *www.gb3nc.org.uk*

Region 12 - East & East Anglia

(http://rsgb.org/region12)

RR

Brian Woolnough, M5ADQM0MBD
Tel: 07740 663626
Email: *rr12@rsgb.org.uk*

DRs

District 121 (Cambridgeshire)
Mervyn Foster, G4KLE, Tel: 01480 878111
Email: *dr121@rsgb.org.uk*
District 122 (Norfolk)
Mark Taylor, G0LGJ, Tel: 07748 760535
Email: *dr122@rsgb.org.uk*
District 123 (Essex - North) Dave Cutts, M0TAZ (temp) Tel: 07506 035599
Email: *dr123@rsgb.org.uk*
District 124 (Essex - South)
Dave Cutts, M0TAZ, Tel: 07506 035599
Email: *dr124@rsgb.org.uk*
District 125 (Suffolk)
Iain Moffatt, G0OZS, Tel: 01449 766089
Email: *dr125@rsgb.org.uk*

Clubs & Societies

Bishops Stortford
Tony Judge, G0PQF, *g0pqf@hotmail.com* Meets 8.00pm on third Monday of the month (check on website for occassional changes), Farnham Village Hall, Rectory Lane, Farnham, Essex CM23 1HU *www.bsars.org*

Bittern DX Group
Linda Leavold G0AJJ *bdxg.secretary@gmail.com* Meets last Thursday in every month 19.30 to 21.30, Erpingham Arms, Eagle Road, Erpingham, Norfolk NR11 7QA *www.bittern*-dxers.org.uk

Braintree & DARS
Geoff Nurse G1GNQ *secretary@badars.co.uk* Meets 8.00pm on 2nd & 4th Tuesdays in the month at the, St Peters Church Hall, St Peters Road, Braintree, Essex CM7 9AR *www.badars.co.uk*

Bury St Edmunds ARS
Mr Melvin Green, *m0iid@bsears.co.uk* Meets on the third Wednesday each month at the, Rougham Tower Museum, Rougham, IP32 7QB *www.bsears.co.uk*

Cambridge & DARC
Bryan Davies, M0IPO *publicity@cdarc.org.uk* Meets on 2nd and 4th Fridays of the month 19.30 - 21.30, Coleridge College, Radegund Road, Cambridge, CB1 3RJ *www.cdarc.org.uk*

Cambridge Univ WS
Dan McGraw M0WUT, *chairman@g6uw.org* Meets Thursdays 8pm at the, Maypole Public House, Park Street, Cambridge, CB5 8AS *www.g6uw.org*

Chelmsford Amateur Radio Society
Paul Tittensor *secretary@g0mwt.org.uk* Meets first Tuesday of each month at 19.30 to 21.30, Danbury Village Hall, Main Road, Danbury, nr Chelmsford, Essex CM3 4NQ https://*www.g0mwt.org.uk*

Colchester Radio Amateurs
Graeme Chalklin G3CO *treasurer@g3co.uk* Meets 7.30pm on the third Thursday of each month at 19.30 - 21.30 facebook.com/groups/1928805160709742, Wilson Marriage Centre, Barrack Street, Colchester, Essex, CO1 2LR https://g3co.uk

Dengie Hundred ARS
Roger Jones G7RGR *rjonesa@aol.com* Meet Thursday's 09:30 - 13:00, Oak Tree Bungalow, The Endway, Althorne, Essex CM3 6DU *www.dhars.org.uk*

Felixstowe & DARS
Mark Riley M5BOP *m5bop@hisimage.co.uk* Meets 7:30pm on first, third (and if present, fifth) Mondays in month, Suffolk Aviation Heritage Museum, RAF Foxhall, Foxhall Road, IP16 4JU *www.fdars.org.uk*

Harlow & DARS
The Secretary, *secretary@g6ut.com* Meets 8.00pm every Friday until late, Mark Hall Barn, First (Mandela) Avenue, Harlow, Essex CM20 2LE https://dartmoorradioclub.uk

Harwich ARIG
Kevin Francis M0JVC *g0rgh@amsat.org* Meets at 8pm on the 2nd Wednesday of each month at, Park Pavillion, Barrack Lane, Dovercourt, Harwich, Essex CO12 3NS *www.harig.org.uk*

Havering & DARC
Alan Paul, *HaveringRadioClub@gmail.com* Meets every Wednesday 20.00 to 22.00, Fairkytes Arts Centre, 51 Billet Lane, Hornchurch, Essex RM11 1AX *www.haveringradioclub.co.uk*

Huntingdonshire ARS
Shaun Hagan, 2E0FFK: *hars.secretary@gmail.com* Meets on 2nd and 4th Thursdays in the month, 7.30pm - 9.30pm at, Buckden Village Hall, Burberry Road, Buckden, St Neots PE19 5UY *http://hunts-hams.co.uk*

Ipswich Radio Club
John Gee, G4BAV, *g4irc@icloud.com* Meets every Wednesday at 19.30 except 1st Wed of month at our contest site, Shrubbery Farm, Otley, IP6 9PD *www.qrz.com/db/G4IRC/*

Kings Lynn ARC
Mr Eric Allison G4JNQ 01485600587 Meets every Thursday 19:30 to 22:00, The Scout Hall, Chequers Lane, North Runcton, Kings Lynn, Norfolk PE33 0QN *www.klarc.org.uk*

Leiston ARC
Debbie Lucock 2E0ICX, *secretary@larc.org.uk* Meets 2nd Tuesday of the month at, Leiston Community Centre, Sizewell Road, Leiston, Suffolk IP16 4JU *www.larc.org.uk*

Loughton & Epping Forest (LEFARS)
David Priest, M0VID, *info@lefars.org.uk* Meets fortnightly on Fridays at 19.45 . Zoom Video Conference on alternate Fridays, All Saints House, Romford Road, Chigwell Row, Essex IG7 4QD *www.lefars.org.uk*

Lowestoft District & Pye ARC
Tim Ward,2E0TJW, *secretary@ldparc.co.uk* Meets Thursdays at 20.00 at, Clubhouse Victoria Road, Lowestoft, Suffolk, NR33 9LY *http://ldparc.co.uk/index.htm*

Martlesham RS
John Quarmby, G3XDY *g3xdy@btinternet.com* Meets last Saturday of every month at 9.30AM, IP3 9RZ, Also quarterly lecture programme at, BT Adastral Park, IP5 3RE *www.mrsap.org*

Norfolk ARC
David Palmer G7URP *radio@dcpmicro.com* Meets every Wednesday evening 19.00-21.30 at CNS School or via our online show NARC Live which starts at 19.30. Our website has details of our programme and meeting places. Watch live on BATC: https://batc.org.uk/live/NARC or via Facebook Live: https://*www.facebook.com/norfolkamateurradioclub*, City of Norwich School, Sixth form common room, Eaton Road, Norwich, Norfolk NR4 6PP *www.norfolkamateurradio.org*

Norfolk Coast ARS
Richard Leeds, G4RZN Tel: 07767 492456 Meets most Thursdays 10.00 to 14.00, Mill Lane, East Runton, Norfolk NR27 9PH *www.norfolkcoastamateurs.co.uk*

Norfolk Tank Museum
Tony Francis M0XTF *m0xtf@outlook.com* Meets most Wednesday mornings at 11:30am between April and October, Norfolk Tank Museum, Station Road, Forncett St Peter, Norwich, Norfolk NR16 1HZ *http://norfolktankmuseum.co.uk/radiocollection*

North Norfolk ARG
01263 824275, *g4nre.uk@gmail.com* Meets Thursdays at 10am, Mr Ward,

Cromer, 88 Central Road, Norfolk NR27 9BW https://*www.facebook.com/groups/134778936544632/*

Peterborough And District ARC
Alan Ralph *secretary@padarc.co.uk* Meets second & fourth Wednesday of the month Doors open at 19.00 to 21.30, Mace Road Church Hall, Mace Road, Stanground, Peterborough, Cambs PE2 8RQ *www.padarc.co.uk*

Saffron Walden Amateur Radio Society
https://*www.facebook.com/groups* Anthony Thompson G1RBO Meets at 19.00 1st Monday of every month, Lovecotes Farm, Chickney Road, Henham Village, Bishops Stortford, CM22 6BH

South Essex ARS
sears.enquiries@gmail.com Meets 2nd Tuesday of the month at 7.30pm, St Michaels Church, St Michaels Road, Daws Heath, Benfleet Essex SS7 2UW *https://g4rse.co.uk/*

Sudbury & DRA
Tony Harman G8LTY, *contact@sudburyradioamateurs.co* Meets 2nd Wednesday in the month at 8PM, Room 1, The Stevenson Centre, Stevenson Approach, Broom Street, Great Cornard, Sudbury CO10 0WD *https://sudburyradioamateurs.co.uk*

Thames Amateur Radio Group
Andy Atkinson M0IXY 07832 978681 Meets 1st Friday of the month at 20:00, Radioactive Night Meeting 3rd Friday of the month at 20:00. Club nets on GB3DA (or 145.300 if GB3DA not available) on other Fridays of the month at 20:00, Jubilee Hall, Waterside Farm Sports Centre, Somnes Avenue, Canvey Island SS8 9RA *www.thamesarg.org.uk*

Vange ARS
vars@live.co.uk Meets 1st Thursday of the month at 8:00pm, St Gabriels Youth And Community Centre, Rectory Road, Basildon, Essex SS13 2AA *https://www.vangeradio.org.uk*

Wisbech AR & Elec. C
Alan Bridgeland, M0DUQ *m0duq@talktalk.net* Meets Mondays 7.30 pm Elme Hall Hotel, 69 Elm High Rd, Wisbech, PE14 0DQ *www.warec.org.uk*

Repeater Groups

Cambridgeshire Repeater Gp
Phil Nice, G8IER Email: *secretary@cambridgerepeaters.net* Meets 7.30pm the first Wednesday of the month at The Carpenters Arms, Great Wilbraham, CB21 5JD. *www.cambridgerepeaters.net*

Essex Repeater Group
Mr Murray Niman, G6JYB *secretary@essexrepeatergroup.org.uk* Meets Danbury - Church Green c/o Danbury Village Hall, Chelmsford, CM3 4NQ *https://www.essexrepeatergroup.org.uk*

Contest Groups

Blackwater AR Contest Group
Clive Bennet M0BRT *cbenn10453@aol.com, www.barcg.org.uk*

Magnetic Fields Contest Group
Paul Marchant,

Secret Nuclear Bunker Contest Group
George Smart, *george@george*-smart.co.uk The group holds informal meetings at the J. J. Moons Wetherspoons Public House in Hornchurch, Essex., Kelvedon Hatch Secret Nuclear Bunker, Crown Buildings, Kelvedon Hall Lane,Brentwood Essex CM14 5TL *http://www.gb0snb.com/wordpress/snbcg*

Region 13 - East Midlands

(http://rsgb.org/region13)

RR

Bob Hambly, M0HAF Tel: 07447 945601
Email: *rr13@rsgb.org.uk*

DRs

District 131 (Leicestershire & Rutland) David Jacobs, 2E0DFJ, Tel: 07407 716445 Email: *dr131@rsgb.org.uk*
District 132 (South Derbyshire & South Nottinghamshire) Simon Strange, M0SYS Tel: 07856 809130 Email: *dr132@rsgb.org.uk*
District 133 (North Nottinghamshire & North Derbyshire) Dr John Rogers, M0JAV Tel: 07836 731544 Email: *dr133@rsgb.org.uk*
District 134 (Northamptonshire) Bob Hambly, M0HAF Tel: 07447 945601 Email: *dr134@rsgb.org.uk*
District 135 (Lincolnshire South) Graham Boor, G8NWC Tel: 07754 619701 Email: *dr135@rsgb.org.uk*
District 136 Vacant Email: *dr136@rsgb.org.uk*
District 137 (North of South Lincolnshire) Andrew Gilfillan, G0FVI Tel: 07909 680047 Email: *dr137@rsgb.org.uk*

Clubs & Societies

Bolsover ARS
Mr Alvey Street, G4KSY *streeta@suevin.com* Meets 1st & 3rd Wednesday of the month from 8pm, Bainbridge hall, Chapel street, Bolsover, Derbyshire S44 6PX https://*www.g4srb.co.uk*

Buxton RA
Mrs C Royle M0NHG *royle495@btinternet.com* Meets 8.00pm on 2nd and 4th Tuesday in the month at the, Leewood Hotel, Buxton, SK17 6TQ *https://buxtonradioamateurs.wixsite.com/buxton-radio-club*

Grantham ARC
www.garc.org.uk Kevin Burton G6SSN, *g6ssn@btinternet.com* Meets first Tuesday of the month Meet 19.30 for 20.00 start Grantham West Community Center, Trent Rd, Grantham, NG31 7QX

Heckington & District Radio Group
Carl Whitney *carl@heckingtonradiogroup.com* Meets on the last Wednesday of every month, The Pavilion, Heckington Howell Road, Heckington, Lincs, NG34 9RX

Hinckley ARES
Mark Burrows 2E0SBM *hinckleyradiosociety@gmail.com* Meets every 2nd and 4th Wednesday from 20.00, Britannia Scout Hut, Britannia Road, Burbage, Hinckley LE10 2HE *http://hinckleyares.co.uk*

Hucknall Rolls Royce ARC
Mark Attenborough, *Secretary@hrrarc.com* Meets at 8.30pm evey Friday but doors usually open for 8pm, Hucknall Rolls Royce Leisure Association, Gate 1, Watnall Road, Hucknall, Nottingham NG15 6BU *www.hrrarc.com*

Kettering & DARS
Les Moyles 2E0MNZ *les@moylehousehold.co.uk* Meets at 7:00pm on Tuesday evenings and from 10:00-12.00 Sundays, Harrington Aviation Museum, Sunnyvale Farm Nursery, Off Lamport Road, Harrington, Northants NN6 9PF *https://www.g5kn.org*

Leicester RS
Sandra Morley *nbcymar4@hotmail.co.uk* Meets every Monday at 19.00 - 22.00 pm, Gilroes Cottage, Groby Road, Leicester, LE3 9QJ *http://g3lrs.org.uk*

Lincoln Short Wave
Mrs Pam Rose, G4STO Meets 8.00pm Wednesdays (and from 9.00m to 12 noon on Saturdays in shack), LSWC C/o BSA Social Club, Village Hall Lane, Aisthorpe, Lincolnshire LN1 2SG *www.g5fz.co.uk*

Loughborough & DARC
Mr Chris Walker, G1ETZ, *g1etz@aol.com* Meets Tuesday evening from 19.30, Glenmore Community Centre, Thorpe Road, Shepshed, LE12 9LU *www.radioclub.org.uk*

Melton Mowbray ARS
Graham Mason G4PTK - 01664410733 - 07511416613 Meets 19.30 on 3rd Friday of the month, except July & August, The Edge Community Centre, Dalby Road (corner of Queesway), Melton Mowbray, Leics LE13 0BQ *https://www.facebook.com/groups/383420802558219*

Northampton Radio Club
Richard Kellow *M0RKJsecretary@northamptonradioclub* Meets on a Thursday from 20:00. If visiting please email *secretary@northamptonradioclub.co.uk* beforehand. Non-members are welcome to visit up to 3 times., The Grangewood Club, 50 Barn Owl Close, Northampton, NN4 0SL *www.northamptonradioclub.co.uk*

Northampton Scout
NSARG Team *nsarg@nsarg.co.uk* Meets 3rd Saturday every month, Overstone

Scout Activity Cntr., Northampton, NN6 0AF *http://nsarg.servehttp.com/*

Nunsfield House ARG
Mr Adrian Price, G1OXH. *sec@nharg.org.uk* Meets 7.45pm on Fridays at the, Nunsfield House, 33 Boulton Lane, Alvaston, Derby DE24 0FD *www.nharg.org.uk*

RAF Waddington ARC
Mr Bob Pickles, G3VCA Meets 7.30pm Thursdays at, The Pyewipe, Fossebank, Saxilby Road, Lincoln, LN1 2BG *www.g0raf.co.uk*

Sherwood Amateur Radio Club
c/o Edward Rippon *sherwoodarc@gmail.com* Meets at, 319 Beechdale Road, Nottingham, NG8 3FF *http://mx0gzd.org*

South Derby & Ashby Woulds ARG
sdawarg.chairman@gmail.com Meets 7.00pm on Wednesdays 19.00 to 21.00, Moira Replan Centre, 17 Ashby Road, Moira, Swadlincote, Derbyshire DE12 6DJ *www.sdawarg.org*

South Kesteven ARS
Stave Marsh Meets 1st and 3rd Saturdays of each month at 10am, Rookery Lane, Sudbrook, Grantham, Lincolnshire NG32 3RU *www.skars.co.uk*

South Normanton, Alfreton And District ARC
Alan Jones, M0OLT *secretary@snadarc.com* Meets Mondays 7:30pm, except Bank Holidays, Post Mill Centre, Market Street, South Normanton, Derbyshire DE55 2EJ *https://www.snadarc.com*

South Notts ARC
secretary@snarc.org.uk Contact Page on the website Meets 7.00pm Mondays see the website, Mapperley Plains Recreation and Social Club, Plains Rd, Mapperley, Nottingham NG3 5RH *www.snarc.org.uk*

Spalding & DARS
Graham Boor, G8NWC. *secretary@sdars.org.uk* Meets Every Friday at 19.30, West Pinchbeck Village Hall, Six House Bank, West Pinchbeck, Nr Spalding PE11 3GQ *www.sdars.org.uk*

The Gliding Centre Radio Amateur Society
email *gx1idr@outlook.com* Meet every 1st Saturday of the month 1400 onwards, The Gliding Centre, Husbands Bosworth Airfield Sibbertoft Road, Lutterworth, Leicestershire LE17 6JJ

Welland Valley ARS
Peter Rivers G4XEX, *info@wvars.com* Meets 7.30pm on third monday in the month at the, Great Bowden village hall, Great Bowden, Market Harborough, Leics LE16 7EU *www.wvars.com*

Worksop ARS
G.Lebond-Carroll M7WSP *gayle@lebond.co.uk* Meets on Tuesdays 6.30pm till 11.00pm, 59/61 West Street, Worksop, Notts, England S80 1JP *https://www.facebook.com/g3rcw*

Repeater Groups

Leicestershire Repeater Group
Phil Taylor M0VSE *phil@m0vse.uk* *www.leicestershirerepeatergroup.org.uk*

Contest Groups

Blacksheep Contest and DX ARS
Stephen Purser G4SHF *stephen@blacksheep.org*,

Five Bells Group
Mr B K Tatnall, G4ODA, Active in VHF/UHF contests

Leicestershire Foxes Contest Group
Adam Moss/Dabid Carter: *adyg6ad@gmail.comMeets* Thursdays at 20.00, Gilroes Cottages, The Chantry, Leicester LE3 9QT

Parallel Lines
g4lip@plcg.orgMeeting Information: Ad-hoc meeting times about once a month, The Hollybush, Main Street, Ashby ParvaLeicestershire LE17 5HS *https://www.plcg.org*

Weekend Contest Group (WCG)
Steve Hambleton G0EAK *g0eaksteve@gmail.com*
As this is a contest group we don't meet very often. The group was formed a few years ago mainly for VHF and up RSGB weekend contests. The members are mostly active UKAC contesters who submit their UKAC logs for their main club but that club doesn't do weekend contests. We mainly operate in the RSGB AFS Superleague and VHF Championship. Further information available from G0EAK or G1PPA, DN22 7DX

Training and Exams

As agreed by Ofcom, it's the RSGB which has the responsibility for making amateur radio exams available in the UK. We currently offer two routes to obtaining all the privileges associated with a Full level amateur radio licence: the three-tier route and direct to Full. The three-tier route is best suited to candidates who are interested in radio communications but have little or no prior experience. They can train and take examinations at Foundation level and then Intermediate level before moving on to Full. On passing each of the lower levels, the candidate can obtain from Ofcom an amateur transmitting licence, but with reduced privileges compared to Full, e.g. reduced maximum transmit power. Direct to Full is targeted at candidates who have had prior experience of the technical aspects of radio, e.g. through having a career in radio or electronics. Options for training include: self-training using the syllabuses as a guide and books and materials available from the RSGB and elsewhere; enrolling in face-to-face courses at a local amateur radio club which provides training; or enrolling on one of the online distance-learning courses, see below.

Exams

By far the majority (97%) of exams are taken online, in the comfort of the candidate's own home, generally on a day and time to suit them, using remote invigilation. Remote invigilators carry out those exact same functions as a face-to-face invigilator would do, the only difference being that they aren't in the same room as the candidate but are still able, using video conferencing facilities (e.g. Zoom), to keep a watchful eye on candidates taking an exam. The success of this scheme depended on being able to recruit a team of trusted volunteers to take on the actual remote invigilation and in this respect some 30 volunteers offered their services. Candidates who prefer taking their exam sitting in an open room whilst being monitored in-person by invigilators, can arrange their examination to take place in a club setting as previously when only paper exams were available. Such situations aren't limited to paper exams, subject to internet access at such a venue candidates can take the exam online.

The TestReach exam software used for online exams allows students to get used to the system ahead of their exam through an inbuilt tutorial. The software employs various techniques that prevent students from simply using their computer to look up answers online. The exams themselves are conducted under strict conditions, with the remote invigilator carrying out a virtual 'tour' of the student's exam area using the student's own webcam before the exam is allowed to start, and then continues to watch and listen to the student throughout their exam, again using the candidate's own webcam. The whole exam session is recorded for audit purposes, the recordings being destroyed 5 days after the exam.

On instruction from the invigilator, the exam starts, students are monitored throughout the process, and as soon as the student finishes their exam, they are presented with the outcome of their exam (pass or fail) their score, and as observed by many of the remote invigilators this often results in huge smiles, thumbs up and sighs of relief.

Confirmation of the exam pass is sent out from RSGB HQ to candidates via Royal Mail including a pass certificate, and at the same time the RSGB sends the candidate's data to Ofcom. Within a matter of just a few days, successful candidates can log into their Ofcom

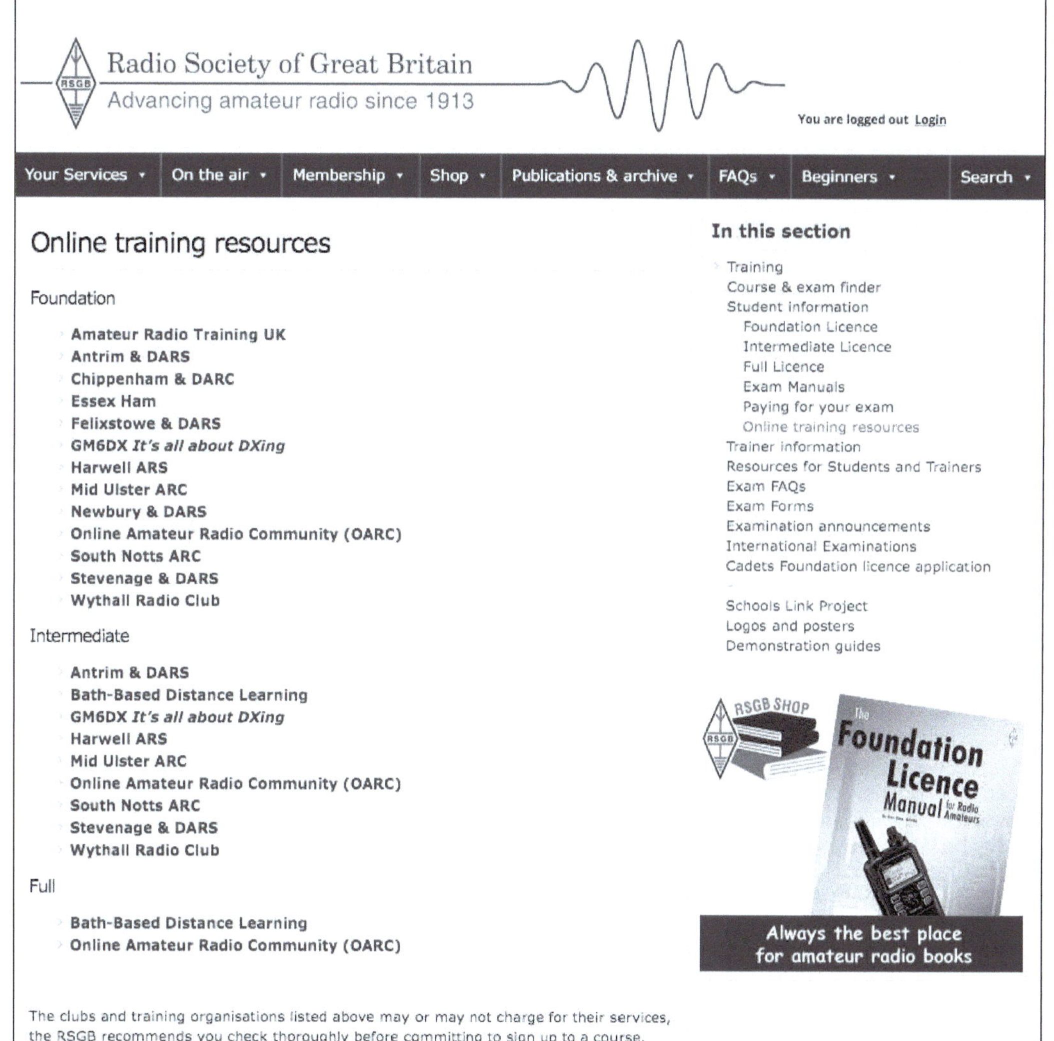

Online training resources

account, choose their new callsign, download a PDF of their licence, and be on the air.

To match the increased licence privileges, Intermediate and Full exams require the use of a second webcam, for example, using the video feature on a smartphone, to give an additional and different view of the candidate while they are taking their exam.

Distance Learning

Traditionally, the RSGB's exam syllabus has been taught to prospective new amateurs in village halls, scout huts and community centres across the UK by a network of volunteers at amateur radio clubs. A Foundation course could typically be taught over a weekend or spread across a few weekday evenings. A structured club course would normally include the hands-on once mandatory practical activities, mixed with "chalk-and-talk" presentations by club members.

With exams being possible to be taken at home, there was a sizeable demand for online training. Distance learning courses have been around for a while, and the Bath-based Distance Learning Course, pioneered by Steve Hartley G0FUW, had been successfully teaching the Full course nationally for some time and was responsible for a very significant fraction of all Full passes between 2011 and 2023. The online group Essex Ham had launched a remote Foundation course in 2015 to supplement the work of local clubs. The Online Amateur Radio Community (OARC) and GM6DX's "It's all about Dx-ing", as well as offering online Foundation courses also offer online Intermediate courses.

You can find a list of known online course providers on the RSGB site:

https://rsgb.org/main/clubs-training/for-students/online-training-resources-for-students/

With training being done via video conferencing or in online virtual classrooms, many of the online courses are free, as previous overheads such as venue hire costs don't apply online. With around a dozen clubs offering online training, many believe that getting started in amateur radio has never been easier, cheaper or more convenient.

The advantages of online study include the ability to study at a time of the students' choosing, the option to review course material again, and the opportunity to study "on the go" from a smartphone or tablet. Course videos have proven especially useful for those with dyslexia, who need extra help or find it hard to study from the written word alone.

During 2020 the RSGB launched a series of videos to help those that had trained online without being able to take part in hands-on practical learning – see link below. In 2021 a series of "Useful practical skills videos" was introduced, whilst many of the *Tonight@8* videos cover topics that newcomers should find useful – see links below

Online study isn't for everyone, and there's still very much a place for clubs in the new online world, as many students still prefer the personal touch, and potentially need extra help and support, e.g. hands-on practical training for some technical aspects of the syllabus. Those who've studied online will also no doubt still be keen to meet up with others, ask questions, get hands-on with radio equipment, and enjoy the social interaction that only a friendly local amateur radio club can offer.

Getting Results

Since the introduction of remotely invigilated online exams in April 2020, nearly 13,500 people have sat a remotely-invigilated exam, with over 8,000 newcomers passing their entry-level Foundation exam online. Over 2,600 have sat the Intermediate exam and over 1,700 candidates have sat the Full exam online all using remote invigilation.

What about candidates with special needs such as visual impairment, mobility issues etc? As has always been the case candidates with special requirements are encouraged to make contact with the exam office (01234 832 717 or *exams@rsgb.org.uk*) directly to discuss options that are open to them. Each case is looked at individually.

There's little doubt that online training and exams have been a significant success. The move to remotely invigilated online exams has brought many people into the hobby who've previously not been able to get involved, including those who don't have a club near to them, are restricted in terms of travel or mobility, have busy lives with little free time, or who appreciate the convenience of being able to study and pass at home at a time of their choosing.

For further information regarding training and examinations check out the Education pages on the RSGB web pages:

www.rsgb.org/main/clubs-training

Bath-Based Distance Learning

Background

The Bath-Based Distance Learning (BBDL) courses were developed over a number of years by the Bath Radio Classes Team for use in parallel with our classroom training. Since 2021 all training is carried out on-line. The BBDL Team includes around fifteen volunteer tutors spread throughout the UK, led by Steve Hartley, G0FUW.

The process is tried and tested and has been used by many Distance Learning students; it has helped over 1000 students to achieve exam passes at the Full Licence level since the first course started in Jan 2011. Between 2011 and 2023 some 29% of all UK exam passes at that level were BBDL students. Formal Intermediate courses using the same format only started in 2022, but they have proved equally popular.

Our successful students have come from a diverse cross-section of society and have included GCSE, A-level and university students, university professors, a street cleaner, a retired cat breeder, a welder, a member of a famous Brit-pop band, and a national TV presenter; very few have been electronics wizards or RF engineers!

Our track record shows that the vast majority of students who see the course through to the end pass the exam; we have consistently had pass rates higher than the national average at both Intermediate and Full level exams.

How It Works

Our training material is aligned with the RSGB syllabus and the Team guide students through the relevant RSGB textbooks. There are 16 weeks of study allocated to each of our Intermediate and Full level courses; this includes some revision weeks and the examination. We have no plans to offer any specific training for the Direct to Full examination.

Weekly work packages are posted using a Virtual Learning Environment called Moodle. We provide 'how to' guidance on how to use the Moodle system; no prior knowledge of the system is required.

We try to bring the theory to life with practical exercises where possible. These can involve making up some simple circuits and carrying out measurements, using your own radio equipment, or making use of on-line resources. These are not mandatory, as they are not required for the exam, but we include them because students have told us that they are extremely useful in understanding the theory that is set out in the textbooks.

BBDL does not charge any fees for the courses, we do it for the love

of amateur radio, and because we all had folk that helped us when we were learning.

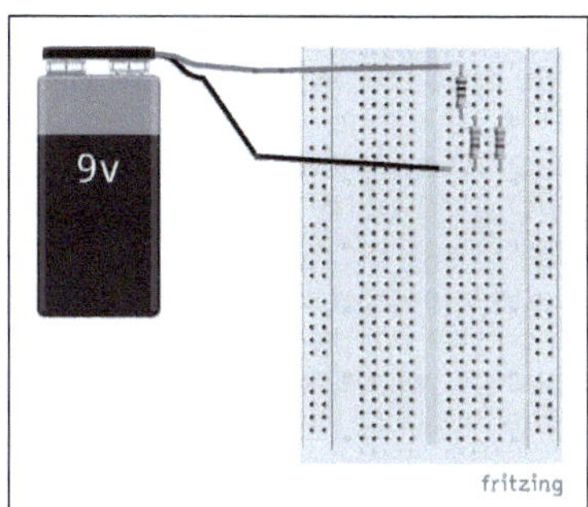

Volts and Amps in Series and Parallel

Pre-Course Work

Before each course, we have a pre-course classroom set up in Moodle for anyone wanting to join the courses. This allows us to check that potential students are able to access and use our systems and it allows potential students to decide if the format suits them.

The pre-course classroom is as near to the real thing as we can make it; some tutorial videos, some suggested reading, and a quiz; if potential students have watched the video and read the words, they should be able to answer the quiz questions. This work covers topics from the previous level that have been added since 2019, so anyone passing before then is able to get up to speed.

We have a limited number of places available and our courses have always been very popular. If we are over-subscribed, we now allocate places based on results in the pre-course classroom.

Student Commitments

Students must have internet access and are required to have their own copy of the RSGB textbook, a scientific calculator and a copy of the RSGB Exam Reference Data Booklet.

Students need to complete the weekly work packages within the allocated timescales. This normally involves:

Every fourth week there are additional revision questions looking back to the beginning of the course and, at the end of the course, we provide a number of Mock Exams for revision.

Potential students should not underestimate the time that is required to complete the weekly work packages; most students spend somewhere between three and four hours studying every week, some have said they spend up to eight hours a week studying and revising; doing a distance learning course is not a walk in the park and being able to dedicate time to studying is absolutely essential.

Tutor Commitments

Tutors commit to provide feedback and the 'worked answers' to quizzes within a few days of student submissions. Tutors answer specific questions about course material within a reasonable timescale, subject to the time they have available to do this voluntary work. The Tutors also monitor progress of all students and identify any weak areas for revision time.

Phil Miles, M7XCQ (now M1XCQ), accessing a BBDL tutorial. Phil was one of many who have completed both the Intermediate and Full Licence courses with BBDL.

Examinations

Students are responsible for making appropriate registration arrangements and paying for their exam. We advise against booking an exam until after Week 9; by then Tutors have a good idea if students are going to be ready, or not.

Timetable

Intermediate courses run January to May with the pre-course classroom open November and December. Full Licence courses run August to December with the pre-course classroom open June and July.

Up to date information can be obtained by sending an e-mail to *g0fuw@tiscali.co.uk* **Steve Hartley**, G0FUW

Scotland-Based Distance Learning - GM6DX

Based in Central Scotland GM6DX is run by Billy McFarland who's callsign this is. Training changed dramatically in 2020. Previously Billy had been used to conducting basic classes at Stirling and District Amateur Radio Society and spending some hours taking trainees through the practical assessment aspect of the training. With the lack of club attendance due to a national pandemic another platform was needed to provide my notes for people to access. GM6DX chose the thinkifc platform to access courses, mainly because its main purpose is online training and that it offers a lot of features at no cost.

After 6 weeks of continually editing and creating notes based on the Foundation examination syllabus this took the form of PDF documents and videos which helps explain the examined subjects. At the end of each section a small quiz allows the students to identify potential training gaps. As the student comes to the end of their course, they will find 10 or so mock examinations to give them a feel for what the actual exam may be like. What GM6DX didn't expect when creating this online course was the level of take up nationally.

From creation well over eight thousand people have signed up for the Foundation and Intermediate Courses with a very high success rate at both levels. Support to students who prefer a virtual classroom was created as a Facebook group, known as UK Intermediate Amateur Radio Licence. This group now numbers around 1000 members from various backgrounds. People post syllabus related questions which are explained in depth a true virtual classroom for all.

GM6DX now concentrates training mainly on Foundation and Intermediate although Billy is starting to offer Full licence for small group numbers when time permits. Overall, 2022 was a great success on the previous year and he hopes that many people make use at this free offering of amateur radio courses.

To enrol for GM6DX free licence courses then visit:
https://gm6dx.thinkific.com

Essex Ham - Foundation Online Courses

Essex Ham's Foundation Online

Within hours of the RSGB's announcement about remotely-invigilated exams, Essex Ham saw a massive spike in course enrolments. Before 2020, around 100 enquiries per month was common, but this quickly rose to 750 enquiries in the first month of online exams. As of 2024, over 12,000 students have enrolled onto one of the group's online courses.

The Foundation Online course originally launched in 2015, aimed at students with no prior experience of radio, electronics and RF theory. The course structure follows the same modular format as the RSGB's Foundation Licence Manual, for ease of cross-referencing. Three modules are released per week with a quick test at the end of each module, and mock tests along the way. With nine modules in total, the course runs for three weeks, and there is also a 'Fast Track' version available too.

The RSGB's popular 'Train the Trainers' sessions highlight that students study in different ways, which is why the course material is presented in different ways – in written format, in the form of bullet-point slides that can be printed, as a paperback and Kindle study guide and a series of videos covering the entire course content, as well as several topics outside of the exam syllabus that every new radio amateur needs to know about. The course uses a web-based virtual classroom that allows for group discussion of topics, plus private messaging if extra help is required.

The course is supplemented by frequent webinars where the training team answer common questions about the syllabus and exam, as well as demonstrating amateur radio with live contacts (simplex, and via a repeater), radio operation, data modes and operating 'nets.

The advantages of online study include the ability to study at a time of the students' choosing, the option to review course material again, and the opportunity to study "on the go" from a smartphone or tablet. The course videos have proven especially useful for those with dyslexia, who need extra help or find it hard to study from the written word alone.

Online study isn't for everyone, and in-person tuition is once again possible at local clubs, but the course coupled with the RSGB's online exams has allowed thousands were able to study for, and pass, their Foundation exam online in record time without the need to leave home.

For details of the course, and to book a place for a free Foundation Online course, go to: *www.hamtrain.co.uk*

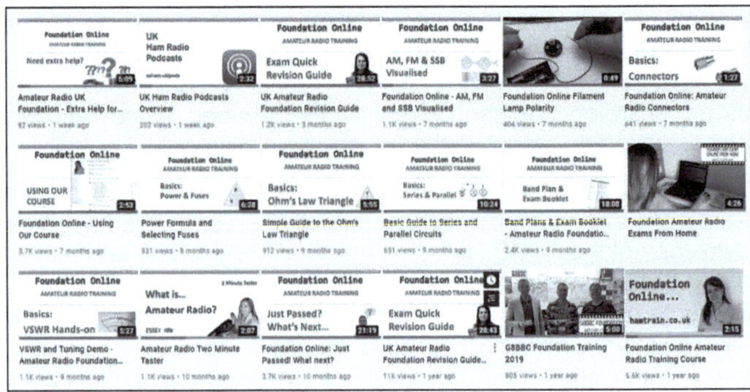

Essex Ham have a range of YouTube Training Videos for the Foundation Course

Online Amateur Radio Community (OARC)

Our courses are primarily delivered through weekly Zoom sessions, lasting up to two hours, supplemented by assigned readings from the RSGB licence manuals, interactive quizzes on Google Classroom, and vibrant discussions on dedicated Discord channels. Following the successful implementation of the Intermediate training, there arose a need for Foundation and Full courses. Despite having limited prior experience, our dedicated team of volunteer trainers diligently crafted these courses, incorporating engaging PowerPoint presentations and comprehensive mock exam questions. Presently, we provide continuous training for all levels of licences, with an impressive pass rate estimated to exceed 80%.

To date, we have successfully conducted 9 Foundation, 12 Intermediate, and 9 Full courses, with support to embrace the Direct to Full exam students who may elect to complete all or some of our course which run consecutively and offering additional pre-exam support on request. The current schedule is to run 2 courses at each level per year is planned. To find out more info please go to *www.OARC.UK*

The online amateur radio community thrives on various platforms, including Discord servers, where members come together to engage in a wide range of activities. These platforms serve as a hub for hams to connect, collaborate, and learn from one another. Within these Discord servers, hams actively participate in projects that push the boundaries of amateur radio. One notable project is the National Packet Radio Network, which utilize TNCs (Terminal Node Controllers) and other software modems for data communication. This project aims to establish a robust and reliable network for exchanging data among hams across the country. Additionally, hams contribute to an ADBS (Automatic Dependent Surveillance–Broadcast) flight tracking network, which involves monitoring and tracking aircraft positions using amateur radio technology. Over the past year, this network has grown significantly. See *https://nodes.ukpacketradio.network/packet-network-map.html*

To further enhance the social aspect of the community, monthly meet-ups on platforms like Zoom are arranged. These virtual gatherings bring hams together, allowing them to interact face-to-face, share stories, and forge deeper connections. It provides an opportunity to celebrate achievements, discuss ongoing projects, and plan for future endeavours.

Morse net, a weekly event, holds a special place in the online amateur radio community. It serves as a platform for hams to practice and improve their skills in Morse code communication.

To help spread the message about amateur radio and also to record other information, the community also has a YouTube channel which has many talks as well as community nights where we discuss whats happening within the community and through them you can get a flavour of what the community is all about: *https://www.youtube.com/@OnlineAmateurRadioCommunity*

Many members joining from Ireland meant that we have also set up an Ireland section of the community which has also got an Irish club callsign EI2OARC, this is as well as are UK club call of M0OUK.

Recently, the online amateur radio community, achieved a significant milestone by surpassing 2200 members.

As a prominent example of the online amateur radio community, offers a welcoming environment for hams from all over the world, of all skill levels and backgrounds, whether one is a seasoned expert or a beginner just starting their journey in amateur radio. OARC provides a wealth of resources and guidance to help individuals get started and actively participate in the community, this includes special channels within discord but also to the community WIKI where there is a number of articles on may topics: *https://wiki.oarc.uk/doku.php*

By visiting *oarc.uk* newcomers can access a plethora of information, including guides, tutorials, and resources tailored to different areas of interest.

So, if you are interested in amateur radio, don't hesitate to visit: *www.oarc.uk* and start your journey today.

For more information on Online training and resources and also FAQ please follow the links below:

https://rsgb.org/main/clubs-training/for-students/
https://rsgb.org/main/clubs-training/for-students/online-training-resources-for-students/
https://rsgb.org/main/clubs-training/exam-faq/
https://rsgb.org/main/clubs-training/course-exam-finder/

Operating Abroad

Assuming that the country allows Amateur operation, there are generally two routes that may apply to operating in it. For temporary operating a number of countries allow this under a European Conference of Postal and Telecommunication Administrations (CEPT) agreement. The second, that allows permanent operating, is to obtain a reciprocal licence.

CEPT agreement

CEPT is a group of countries from Europe near to Europe. Note that Russia and Belarus are currently suspended from CEPT.

Amongst their many actions, many but not all have agreed a common standard of amateur radio licence and rules (T/R 61-01) so as to facilitate temporary operation when visiting a fellow CEPT country, as well as a number of countries outside CEPT who have agreed and met the conditions to use T/R 61-01. Once each CEPT member's country has confirmed that their amateur radio licence conforms to the T/R 61-01 minimum standard, then its amateurs may operate temporarily in other countries which have also confirmed the recommendation.

CEPT T/R 61-01 operation does not replace reciprocal licensing - rather it supplements it. Only temporary operation (up to three months) is permitted under T/R 61-01, eg from holiday accommodation or mobile. Therefore, if you seek either longer-term (over three months) residence or additional facilities, you will still need to apply for a licence. Operating under CEPT regulations means that you are restricted by the regulations of the foreign country; thus the RSGB recommends that you get a copy of the licensing regulations for the countries in which you plan to operate.

Providing you hold a Full UK licence you can operate in countries which have implemented CEPT Recommendation T/R 61-01 in accordance with the terms of that recommendation.

To underline the point licences excluded from this arrangement are Full (Temporary Reciprocal) and Full (Club) along with the Intermediate and the Foundation licence holders.

If you satisfy the above then you must also ensure that at all times you comply with the requirements applicable to the use of amateur radio equipment at the location in the country concerned. Some vary from ours, so it is vital to check before your journey. Also, remember that if requested you must present your licence documentation plus any other documentation required to the relevant authorities in that country. Unless told otherwise you must use the call sign specified in section 1 of your licence after the appropriate host country callsign prefix.

To get the up-to-date list of countries (both CEPT and non-CEPT countries) that have implemented CEPT T/R 61-01, and to be sure you have the latest information check out the ERO website which has at the top of the page a downloadable version of CEPT Recommendation T/R 61-01. Another useful source of up-to-date information on licensing is OH2MCN's website.

Reciprocal licensing

A reciprocal licence is a licence issued by a foreign country to you because that country recognises the standards of the UK Full exam. A licence may be issued under the provisions of CEPT Recommendation T/R 61-02 or it may be issued under a bilateral reciprocal agreement between the UK and the other country.

Under T/R 61-02, the UK may issue a HAREC (Harmonised Amateur Radio Exam Certificate). This certificate confirms that you have passed the UK Full exam. It is recognised in the other countries that participate in T/R 61-02. The other country may issue its equivalent of the UK Full Licence against a UK HAREC. A UK HAREC is now appended to each Full Exam certificate that we issue. If you do not have a UK HAREC but have passed the UK Full exam, you may request a HAREC from Ofcom.

Countries that don't participate in T/R 61-02 but with which the UK has a bilateral agreement are listed in the UK amateur radio licence application form. Check with the licensing authorities in the overseas country what their licensing process is.

The callsign issued with a reciprocal licence may be your own callsign with the foreign country's suffix or prefix or may be allocated in the country's normal series of callsigns.

Overseas radio amateurs in the UK get a normal Full call sign if their licence is issued against a HAREC. Radio amateurs from other countries get a Temporary Reciprocal Licence.

Due to overseas post and administration delays, it is generally best to allow at least two to three months for your application to be processed - longer if it is a Third World country where amateur radio is not so sympathetically regarded. The use of air mail certainly helps in this regard.

Like operating under the CEPT T/R 61-01 agreement, HARECs are not available for Temporary Reciprocal, Foundation or Intermediate licence holders.

Other information

Most countries have a national society which looks after the wellbeing of that country's amateurs. A list may be found on the IARU website. Few are of the size of the RSGB - indeed many are staffed entirely by volunteers. Nevertheless, they will all give you as much assistance as they can.

There is usually little problem with customs. It certainly helps to be able to show that the equipment was purchased abroad and is not being exported. Unfortunately, neither a reciprocal licence nor operation under CEPT regulations is deemed an exemption from customs formalities. If in doubt, you should seek additional advice about importing/exporting equipment.

Ofcom's advice on operating abroad can also be found at:
https://www.ofcom.org.uk/__data/assets/pdf_file/0023/51296/fkm.pdf

Useful resources on operating in foreign countries can be found on the ARRL website at:

www.arrl.org/reciprocal-permit

and OH2MCN's website at:

www.qsl.net/oh2mcn/license.htm and also at: European Radiocommunications Office website:

https://www.cept.org/ecc/topics/radio-amateurs

Ofcom has published advice at:

https://www.ofcom.org.uk/__data/assets/pdf_file/0026/109547/guidance-become-radio-amateur.pdf

Section 2.20 et seq of the online Guidance:
https://www.ofcom.org.uk/__data/assets/pdf_file/0026/82637/amateur_radio_licence_guidance_for_licensees.pdf

see also the RSGB website

Operating Abroad:
https://rsgb.org/main/operating/licensing-novs-visitors/operating-abroad/

Operating for Visitors to the UK:
https://rsgb.org/main/operating/licensing-novs-visitors/operating-for-visitors/

Details of IARU Societies (on the IARU website): www.iaru.org/iaru-soc.html
European Radiocommunications Office website: https://www.cept.org/ecc/topics/radio-amateurs

Ofcom

Regulatory Principles

Ofcom was established as a public corporation by the Office of Communications Act 2002. Ofcom is the regulator for the UK communications industries, with responsibilities across television, radio, telecommunications and wireless communications services.

Ofcom's Statutory Duties Under the Communications Act 2003

"3(1) It shall be the principal duty of Ofcom, in carrying out their functions;

(a) to further the interests of citizens in relation to communications matters; and

(b) to further the interests of consumers in relevant markets, where appropriate by promoting competition"

Ofcom's specific duties fall into six areas:

1. Ensuring the optimal use of the electro magnetic spectrum
2. Ensuring that a wide range of electronic communications services - including high speed data services - is available through-out the UK
3. Ensuring a wide range of TV and radio services of high quality and wide appeal
4. Maintaining plurality in the provision of broadcasting
5. Applying adequate protection for audiences against offensive or harmful material
6. Applying adequate protection for audiences against unfairness or the infringement of privacy

Ofcom's Regulatory Principles

- Ofcom will regulate with a clearly articulated and publicly reviewed annual plan, with stated policy objectives.
- Ofcom will intervene where there is a specific statutory duty to work towards a public policy goal which markets alone cannot achieve.
- Ofcom will operate with a bias against intervention, but with a willingness to intervene firmly, promptly and effectively where required.
- Ofcom will strive to ensure its interventions will be evidence-based, proportionate, consistent, accountable and transparent in both deliberation and outcome.
- Ofcom will always seek the least intrusive regulatory mechanisms to achieve its policy objectives.
- Ofcom will research markets constantly and will aim to remain at the forefront of technological understanding.
- Ofcom will consult widely with all relevant stakeholders and assess the impact of regulatory action before imposing regulation upon a market.

Licences are issued on a lifetime basis. This is subject to the licence being validated with Ofcom at least once every five years. Amateur licensees are therefore reminded that their licence (now 'lifetime') must be validated at least once every 5 years in accordance with the terms, conditions and limitations of their licence. If your licence has been amended (e.g. by notifying Ofcom of a change of address) this will count as a validation. It only takes a few minutes to register your details and validate or make any necessary changes to your licence on-line (*https://ofcom.force.com/licensingcomlogin*) If you are experiencing difficulties or need assistance in processing your licence on-line, you can call: 0300 123 1000 or 020 7981 3131 or Textphone: 020 7981 3043 or
0300 123 2024 (please note that Textphone numbers only work with special equipment used by people who are deaf or hard of hearing).

Applying for an Amateur radio licence

You can apply for, vary, re-validate or surrender an amateur radio licence by either using the online system (subject to conditions), or by completing a paper based application form (new applications are subject to a small administrative fee, unless you are 75 years of age or over).

The free online system has eased administration and reduced costs for amateur users. This approach makes it easier for users to comply with legal obligations. More information on how to apply for an amateur radio licence is available on the Ofcom website (see below).

If you have any questions about your licence please contact:
Spectrum Licensing
PO Box 1285
Warrington WA1 9GL
Tel: 020 7981 3131
Email: *spectrum.licensing@ofcom.org.uk*
Website: https:*//www.ofcom.org.uk/manage-your-licence*

Varying the terms and conditions of a licence

Any licence changes affecting all amateurs are subject to Ofcom publishing a Notice of Variation (NoV) or a consultation (depending on the changes proposed). These are announced by Ofcom in the manner specified in the licence and will also appear on the Ofcom website.

Ofcom and the RSGB

Ofcom works closely with the RSGB and meets on a regular basis to ensure that where feasible, the amateur licence meets the needs of today's amateur. Ofcom continues to work closely with the Society to facilitate regulatory requirements on a number of ongoing developments in amateur radio. Spectrum allocations are listed in the Licence, which is published on the Ofcom website: *https://www.ofcom.org.uk/__data/assets/pdf_file/0015/214116/emf-amateur-licence-terms-and-conditions.pdf* Special Contests Call signs are available for individuals and clubs who regularly participate in Special Contests. These are administered for Ofcom by the RSGB and further information is available at: *http://rsgb.org/main/operating/licensing-novs-visitors/online-nov-application/application-for-a-special-contest-call-sign/* Ofcom also supports the Society in the international scene (see below).

International

There are three main areas of interest; Spectrum, Commonality of licensing conditions and Reciprocal agreements as detailed below.

Spectrum

Ofcom is responsible for administering the International Telecommunication Union (ITU) Radio Regulations in the UK, and ensures the effective management of spectrum taking into account international obligations.

Commonality of licensing conditions

Ofcom represents the UK within the European Conference of Postal and Telecommunications Administrations (CEPT) and continually works towards harmonisation of licence conditions and mutual recognition including countries outside Europe. The UK has not adopted any recognition agreements for the Foundation or Intermediate level licences.

Reciprocal agreements

A number of reciprocal agreements have already been negotiated which allow UK amateurs who have passed the 'Full' examination (or past equivalents) to operate abroad, although there are still a number of countries with which no agreement currently exists. In general, Ofcom prefers reciprocal operation under the CEPT arrangements. Non-participating countries (including those not in CEPT) may apply to CEPT to be included in the CEPT Reciprocal arrangement, governed by Recommendations T/R 61-01 and 61-02. This route is considered as being more effective and efficient than the process of numerous individual bi-lateral arrangements between two administrations. However, Ofcom will continue to review needs for new agreements as appropriate.

Due to the many differing arrangements in countries at levels below the full licence level, Ofcom does not recognise reciprocal agreements at Foundation and Intermediate level.

Spectrum Licensing website: https://www.ofcom.org.uk/manage-your-licence/radiocommunication-licences/online-licensing-service and also the online service portal at https://ofcom.force.com/licensingcomlogin

Repeater network

Frequency coordination and issuing of callsigns for repeaters is now undertaken by the RSGB on behalf of Ofcom. In most cases this no longer requires an onerous Notice of Variation (NoV) from Ofcom. However some bands such as 23cm require Frequency Clearance from Government departments. The Repeater Keeper remains responsible for compliance with the amateur licence.

Packet Data and Internet (voice) Linking

The latest licence terms enable licensees to operate Internet voice gateways and low power data systems up to 5Werp without an NoV. However they are subject to a general licence clause regarding coordination. The latter can be achieved via the RSGB-ETCC website, which can also facilitate specific callsigns for these systems.

Higher power Data Systems in the 5-25Werp range (such as AX25 packet or newer modes) require a fuller application and callsign from the RSGB, similar to the repeater process. Keepers remains responsible for compliance with the amateur licence.

Special Event Stations

Ofcom is responsible for the issuing of NoVs for Special Event Stations. Applications for a standard SES can be made and issued via the Ofcom online licensing system. Please see elsewhere in the Yearbook for full details. For further details regarding NoVs, please contact Spectrum Licensing at Ofcom.

Enforcement

The number of people who misuse amateur radio is fortunately small. However, there are some individuals who cause considerable problems to other amateurs or to other licensed radio users by transmitting in unauthorised frequency bands, using obscene language or generally using radio in an antisocial way. The majority of abuse is directed at the repeater network.

Repeater keeper responsibilities include taking steps to prevent and stop abuse. Details of repeater abuse should therefore be sent to the ETCC Chairman, c/o RSGB HQ. Other cases of abuse should also be taken up with Ofcom through the Operating Advisory Service (OAS). Subject to priorities, Ofcom may be able to take action, where there is a breach of licence terms and conditions. Having obtained evidence against an offender, Ofcom may issue formal warnings or Conformity Notices, initiate prosecution proceedings and/or revoke a licence, depending on the seriousness of the offence.

Provision of Information

Ofcom's policy is to provide its information in electronic form via its website, *www.ofcom.org.uk* Some of these information sheets are available free in paper form. The following relate to the amateur service.

Amateur application forms

OfW 346a Amateur Licence amendment, validation, surrender form

OfW 287 Application for a Notice of Variation for a Special Event Callsign

OfW 306 Application for an Amateur Radio Special Research Permit

Following an extensive amateur licence review in 2023, new licence terms were introduced in February 2024. The RSGB website has details of the review at rsgb.org/licencereview . The latest Ofcom terms and guidance are available at https://rsgb.org/main/operating/licensing-novs-visitors/ and on the Ofcom website at https://www.ofcom.org.uk/manage-your-licence/radiocommunication-licences/amateur-radio/amateur-radio-info

---

Licences

There are three levels of qualification for gaining an amateur radio licence; the three-tier structure consisting of Foundation, Intermediate and Full Amateur radio licence.

The licensing structure in the United Kingdom is now progressive, which means all new entrants have to enter the hobby at the Foundation level, progressing through the Intermediate licence and finally to the Full Amateur radio licence. There is also an option to take a Direct-to-Full (D2F) exam for those with a good technical background.

Foundation Amateur Radio Licence

Introduced in January 2002, the Foundation licence is designed as an entry point into amateur radio and forms the first part of the three-tier amateur licensing structure.

A Foundation licence can be obtained after a short course of study lasting some 12-15 hours. At the end of the course the candidate sits a short, multiple-choice exam, which lasts 60 minutes.

Available online, with no in-course practical assessment.

There are 26 straightforward questions to answer. The pass mark is 19 correctly answered questions. A 'pass' will allow operation of an existing Full licensee's station as a fully qualified Foundation Amateur immediately. You will be able to operate your own station when you receive your licence and M7 callsign.

Foundation Syllabus

1) Amateur Radio Licensing Conditions
2) Technical Aspects
3) Transmitters and Receivers
4) Feeder and Antenna
5) Propagation
6) EMC
7) Operating Practices and Procedures
8) Safety

Intermediate Amateur Radio Licence

The second level in the three-tier structure. Candidates must have completed the requirements of the Foundation licence syllabus and passed the associated examination as a pre- requisite to sitting the Intermediate Licence Examination.

To obtain the Intermediate licence it is advisable to take a training course. This is a little longer than the Foundation course, lasting some 20-25 hours. The aim is to teach many of the fundamentals of radio in a stimulating way, building on the experience gained as a Foundation licence holder.

After completing the practical assessments, a candidate will be ready to sit the Intermediate amateur radio licence examination. Once again this examination is multiple-choice, this time with 46 questions 90 mins and pass mark 28 correctly answered questions.

Intermediate Syllabus

1) Amateur Radio Licensing Conditions
2) Technical Aspects
3) Transmitters and Receivers
4) Feeder and Antenna
5) Propagation
6) EMC
7) Operating Practices and Procedures
8) Safety

Licence types and basic privileges

Type	Foundation	Intermediate	Full
Prefix(es)	**M3, M6, M7**	**2*0, 2*1, M8, M9**	**G, M0, M1, M5**
Amateur bands	136kHz-430MHz, excluding 472kHz & 5MHz, plus 2.4, 5 & 10GHz	(All bands excluding 472kHz & 5MHz)	All
Maximum output power	25W [1]	100W [1]	1000W [1]
TX modes	All permitted	All permitted	All permitted
Act as mailbox/BBS	No	No	Yes
Act as repeater?	No	No	Yes
Low power device/beacon?	Yes	Yes	Yes
CEPT reciprocal operation?	No	No	Yes (but check)
Other licensees allowed to operate?	Yes	Yes	Yes
Disaster comms permitted?	Yes	Yes	Yes
Control station remotely?	Yes	Yes	Yes

* Regional identifier letter. [1] Some bands have lower power limits.

Full Exam (formerly Advanced, or RAE C&G)

This is the highest level of licence that you can obtain. To gain a Full licence it is necessary to pass the Full radio communications examination, which contains 58 questions, pass mark 35 and lasts two hours. Once again the examination covers radio theory and licence conditions, but because holding an Full licence enables you to use 400 watts power output from your transmitter such subjects as Electro Magnetic Compatability (EMC), antenna design and safety issues are covered in some depth. The licence allows access to all the amateur allocations with full power.

When studying for the Full radio communications examination there is no requirement to take a formal training course because there is no practical element. It is possible to study at home on your own if you so wish, but you should recognise that a good understanding of the material is required.

Many local amateur radio clubs and societies and technical colleges run courses specifically for the Full radio communications examination. Alternatively, there are some correspondence and Internet courses available

Full Syllabus

1) Amateur Radio Licensing Conditions
2) Technical Aspects
3) Transmitters and Receivers
4) Feeder and Antenna
5) Propagation
6) EMC
7) Operating Practices and Procedures
8) Safety
9) Measurements

How and where to find a training course

The quickest ways to find a training course is to check on the RSGB website Clubs and Training web page or look at the Local Information section in this Yearbook.

Tutors may run courses using their own personal station. Local amateur radio clubs, Scouts, Guides or other organisations such as the Air Training Corps may also run courses. Youth organisations are likely to run courses solely for their own members, but it is always worth asking. If you are at school and your school does not run a course, suggest to someone that they do so. The RSGB will be happy to advise on how this might be done and point to sources of assistance and training.

In addition RSGB has also developed a 'Direct-to-Full' examination syllabus. This will satisfy the requirement that an applicant for a Full Licence must have passed all three levels of examinations. This new 'D2F' route was introduced in July 2022 for technically proficient candidates, who may for example have a professional background in electronics or telecomms.

Costs

There is likely to be a small charge for training course administration. It is best to ask before committing to attend, but the costs should not be high and certainly should not be a reason to miss out. Examinations and assessments cost £35.50 for the Foundation licence, £39.00 for the Intermediate licence and £45.00 for the Full licence, with £95 for Direct to Full.

Study material

The following titles are available from the RSGB Shop.

The Foundation Licence Manual for Radio Amateurs

The Intermediate Licence Manual for Radio Amateurs

The Full Licence Manual for Radio Amateurs

Exam Secrets for Radio Amateurs

Audio & Braille resources

UK Amateur radio has a charity dedicated to supporting the needs of people with disabilities – RAIBC :

RAIBC website: *www.raibc.org.uk*

Contact: Sandra Brown M6SDS, RAIBC Secretary, email: gensec@raibc.org.uk, Phone: 0800 028 8660

RAIBC Spoken word recordings

Exam Manuals: The RAIBC maintains an Audio Library which contains the Foundation, Intermediate and Full/ Advanced manuals. Please note that these are not available to download but are available on audio disk

https://rsgb.org/main/clubs-training/training-resources/audio-braille-resources/

Special needs candidates

Arrangements can be made for candidates with special needs who would otherwise have difficulties taking any of the amateur radio licence examinations. The key requirement is proper advice on what provisions should be made for the candidates concerned. This must be from an appropriate health, educational or other professional. The need is not for a statement of the candidate's circumstances, but what should be done in order to provide equality of access to the examinations.

Where a disability prevents a candidate from carrying out a practical task then advanced permission should be sought, again supported by proper advice, to waive that specific requirement. Not all requirements can be waived and this will be dealt with on a case by case basis.

It is important to start this process well in advance of requesting and exam, as it may take some weeks to obtain appropriate profes- sional advice. Advice should be sought from the RSGB Examination Department

RSGB Clubs and Training web page: *www.rsgb.org/clubsandtraining*

* Daisy players are available from the RNIB's Talking Book service and allow a visually impaired or blind person to listen to many hours of pre-recorded material. The player allows skipping between tracks, or in this case, pages, and bookmarks can be set to allow an easy return to a study point.

** Symphony players are available from the British Wireless for the Blind Fund (BWBF) and allow the listener to pause and return to the same point, provided the CD has not been changed.

Special Contest Callsigns

The holder of any UK amateur radio Full or Full (Club) licence may apply for a Special Contest Callsign (SCC), for use in contests.

Ofcom Policy and Procedure

The holder of a UK Amateur Radio Full Licence or Full (Club) Licence may apply for a SCC. This call sign may only be used in amateur radio contests of no more than 48 hours duration, run with the aim of contacting as many other stations as possible in a given period of time and run by an amateur radio club, national or international amateur radio association or another organisation (including amateur radio publications), generally accepted within the amateur radio hobby (locally, nationally or internationally) as being a bona fide contest organiser.

The call sign will be in the format 'G' or 'M', followed by a chosen digit and a chosen suffix letter. Examples are G8Z or M7R. We have made 519 call signs available, (G)(#)(&)(A-Z) and (M)(#)(&)(A-Z), where '#' is an applicable Regional Secondary Locator (if any) and '&' is a digit from 0 to 9.

Although variations are granted by Ofcom, the Radio Society of Great Britain (RSGB) administers, distributes SCCs and also advises Ofcom on applications. However, the RSGB's advice is only a recommendation, and the final decision remains with Ofcom.

Applicants should apply on the form at *http://rsgb.org/scc*, giving three choices of call sign in order of preference. Application forms must be emailed as attachments to *scc@rsgb.org.uk* for processing. Current availability of call signs may be checked at *https://www.rsgbcc.org/hf/information/scc.shtml*.

The qualifying criteria are divided into two parts, one for contest participation and the other for your achievement in the contest. **Both parts of the criteria must be met in order for a SCC to be granted**.

Part 1 - Individual licensees (or the licence-holder in the case of club licensees) must supply evidence of having entered at least five contests from the list in Table 1 within the last five years.

If you are requesting a variation to an individual licence (as opposed to a Full (Club) licence), you must have entered all of the contests as a single operator. If you are applying on behalf of a club, the contests may have been entered as a single operator or as a multi-operator team.

Part 2a (using a UK and Crown Dependency (UK&CD) call sign to enter a contest) - The applicant must have accrued five achievement points from the contests entered. One point is accrued if the licensee (or licence-holder) makes at least one fifth (20%) of the number of contacts of the UK&CD leader. Two points are accrued if the licensee (or licence-holder) makes at least one half (50%) of the number of contacts of the UK&CD leader.

Part 2b (using a non-UK&CD call sign to enter a contest) -The applicant must have accrued five achievement points from the contests entered. One point is accrued if the licensee (or licence-holder) makes at least one fifth (20%) of the number of contacts of the world leader. Two points are accrued if the licensee (or licence-holder) makes at least one half (50%) of the number of contacts of the world leader.

Any variation that Ofcom grants will be valid until 31st December 2029. Applications take up to 42 days to be processed.

When your variation expires, you may request a fresh variation for the following five-year period, keeping the same call sign, if you still qualify (see 'qualifying criteria', above). You must apply between 1st January and 31st December of the year of expiry. If a fresh application is not received, the call sign will be withdrawn for a period of 2 years before it is made available again.

If the Licence revalidation date falls due during the term of the SCC variation, the variation may be withdrawn if the licensee fails to revalidate the Licence.

Please email and questions and/or queries about any aspects of SCC to *scc@rsgb.org.uk*

ARRL DX CW	ARRL DX SSB	ARRL 1.8MHz CW only
ARRL 28MHz	ARRL RTTY Roundup	BARTG HF RTTY
BARTG SPRINT	BARTG SPRINT75	BARTG SPRINT PSK63
CQ WPX CW	CQ WPX RTTY	CQ WPX SSB
CW WW CW	CQ WW RTTY	CQ WW SSB
CQ WW 160m CW	CQ WW 160m SSB	IARU HF Championship
Russian DX	UK/EI CW DX Contest	UK/EI SSB DX Contest
WAE DX CW	WAE DX RTTY	WAE DX SSB
WW-DIGI	RSGB Commonwealth Contest	RSGB IOTA
RSGB DX	RSGB 1st 1.8MHz	RSGB 2nd 1.8MHz
144MHz Marconi Memorial	RSGB 50MHz Trophy	RSGB 144MHz Trophy
RSGB March 144MHz/432MHz (counts as one contest)	RSGB May 432MHz – 248GHz (counts as one contest)	RSGB 432MHz – 248GHz (Oct) (counts as one contest)

Table 1: *List of qualification contests*

Extract of Terms relating to the NoV for use of Special Contests Callsigns:

1. The Contest Call Sign may only be used to identify the station when the station is participating in an Amateur Radio Contest.
2. The Contest Call Sign may only be used for transmitting information directly related to participation in an Amateur Radio Contest and only on frequency bands set out in the rules of the Amateur Radio Contest.
3. Where the licence is an Amateur Radio (Full) (Club) licence, only members of the Club may use the Contest Call Sign, subject to all of the conditions of this variation and of the Licence.
4. Where a log is kept, the licensee must maintain a separate log in respect of the Contest Call Sign.
5. The Contest Call Sign may be used only within the United Kingdom, Guernsey, Isle of Man or Jersey.
6. A Regional Secondary Locator may also be used in conjunction with the Contest Call Sign.

For a list of current Special Contest Callsigns, see the Callsign Listings section of this Yearbook

Please send any questions about SCC to:
scc@rsgb.org.uk

The application form may be found and printed from the Internet at:
https://rsgb.services/public/nov/sccs/docs/application-for-a-special-contest-call-sign.docx

Special Event Callsigns

Do you really need a GB callsign?

All club stations are able to pass greetings messages sent by a non-licensed third party. This means that applying for a GB callsign no longer holds any advantages for a club!

Provided the club uses the prefix letters (see below), the club station is allowed to pass greetings messages and to operate simultaneously on more than one band. The club prefixes are very distinctive and create interest through their rarity. If a club regularly operates a special event station using the club callsign, this will help increase the club's identity. It will also benefit by being able to print its QSL cards in larger, more economic quantities.

Another advantage is that any suitably licensed and authorised club member may operate the club station. This gives greater flexibility over a GB callsign which is just a variation to an individual's licence.

Best of all, you don't have to fill in any forms or give 28 days notice of operation!

Old club prefix	New club prefix
G/M	GX/MX
GD/MD	GT/MT
GI/MI	GN/MN
GJ/MJ	GH/MH
GM/MM	GS/MS
GU/MU	GP/MP
GW/MW	GC/MC

GB callsigns

Ofcom issues and despatches Notices of Variation authorising special event (GB) callsigns. Consequently, all enquiries and correspondence should be addressed to Ofcom and not to the Society. Ofcom has stated that GB callsigns are issued for special event stations and that as such, they should normally be open to viewing by members of the public.

Applying for a GB callsign

You may apply for an NoV for a Special Event Stations online, via the licensing portal *https://ofcom.force.com/licensingcomlogin* These NoVs are available immediately. No charge is currently made by Ofcom for a special event callsign.

If you make a manual application, the form must be sent in at least 28 days prior to the start of the event.

Paper applications are normally processed shortly after receipt. If nothing has been received 14 days prior to the event, please contact Ofcom immediately. Please note that no authority exists until a Notice of Variation has been received.

A Notice of Variation will only be issued to licensees who hold a current Full, including a Club licence (ie not Foundation, Intermediate or Temporary Reciprocal). It will be valid for a maximum of 28 days. The station may only be established and operated at the location. This must be the address stated on the application form which must be detailed enough for anyone to find easily. Operation of a special event station from a licensee's home address is not normally permitted. This is because an SES is a showcase for the hobby and Ofcom plan to introduce new flexibility for SES in time for 2025..

Only the person responsible for the station need sign the form, as the authorisation is by Notice of Variation to that individual's licence. This person is required to be present to supervise the correct operation of the station. Additional operators need only sign and write their callsigns in the logbook.

If you have not used the callsign before, you can avoid last-minute disappointment by first contacting Ofcom, who can check that it is available and reserve it for you. A GB callsign may be reserved for up to six months in advance. When a GB callsign has been used it will not normally be re-issued to another amateur for use at a different event for a period of two years.

The holder of a Full Licence must apply for a GB callsign.

Subject to availability, special event callsigns are available in the following formats:

GB0 + 2 or 3 letters	GB1 + 2 or letters
GB2 + 2 or 3 letters	GB4 + 2 or 3 letters
GB5 + 2 or 3 letters	GB8 + 2 or 3 letters
GB6 + 2 or 3 letters	

Greetings messages

Ofcom licence conditions enable a Full or Full (Club) licence to facilitate a non-licensed person to pass a message on air. Such 'greetings messages' are on the basis that:

> a non-licensed person may speak into the microphone or even operate, but the licensed radio amateur must directly supervise and identify the station at all times.

Charitable events

It is recognised that some special event stations will be established at certain charitable events where a major concern will be the raising of funds.

Ofcom has agreed that the charity (if one is involved) or the reason for establishing the special event station may be mentioned 'on-air' provided that under no circumstances may a donation be requested during the con- tact, and sending of QSL cards must not be conditional upon the pledge of a donation. It is in the interests of everyone who holds a special event station licence that operators keep within the spirit of this by not asking for any money over the air.

The station may be sponsored per contact, ie the licensee may in advance of the event seek from his/her friends and relatives sponsorship assurances under the usual arrangements for sponsorship. You must not seek sponsors 'on-air' at any time.

QSL information

Special event stations generate many QSL cards, so it is important that you use the QSL Bureau correctly.

For instructions, see the 'QSL Bureau' pages in this edition of the *RSGB Yearbook*.

JOTA and Radio Scouting

Jamboree on the Air (JOTA) is an annual event designed to allow Scouts to send greetings messages to each other. Started in 1957, it now involves approximately 600,000 Scouts and Guides, with the help of over 23,000 radio amateurs in over 100 countries.

JOTA takes place on the third full weekend of October each year, officially between 00.00 Saturday and 24.00 Sunday, although most stations run for a period within these hours to suit their own requirements. The event is organised by the Scout movement, supported by radio amateurs or clubs. Their aim is to bring Scouts around the world closer together and to introduce them to the capabilities of amateur radio.

All amateur bands are used. Most stations use a special event or a club call, allowing the Scouts to pass greetings messages over the air. JOTA information packs are sent to all participating GB stations at the beginning of October. Any clubs taking part in JOTA wishing to receive this information pack should contact the SES Administrator at RSGB HQ.

The interest fostered by JOTA and World Jamboree has spread and many Scout camps and campsites boast amateur radio facilities. A number of proficiency badges in the radio, electronics and computer fields are available for Scouts. Several countries have permanent Scout Headquarters stations – for example the World Scout Bureau in Geneva has the callsign HB9S and Gilwell Park in the UK operates under the callsign GB2GP.

Many countries run periodic Scout nets. There are regular weekly UK and European nets aimed at Scouters who are also radio amateurs.

Usual Scout Net Frequencies

Band	SSB (Phone)	CW
80m	3.740 and 3.940*	3.590
40m	7.090	7.030
20m	14.290	14.070
17m	18.140	18.080
15m	21.360	21.140
12m	24.960	24.910
10m	28.990	28.190

The UK Scout net is on Saturdays, 3.740MHz at 0900 local time. The European Scout net is on Saturdays, 14.290MHz at 09:30GMT.

* USA only

Thinking Day On The Air

Thinking Day On The Air (TDOTA) is organised by The Guide Association on the third full weekend in February, to celebrate the birthdays of the founder of the movement, Lord Baden-Powell, and of his wife, Lady Baden-Powell, the World Chief Guide, on 22 February.

The aim of TDOTA is to encourage the girls to make Guiding friendships with members of other units and to introduce them to amateur radio. Station organisers are asked to keep these objectives in mind.

Guide amateur radio stations rely on the goodwill of radio amateurs in setting up stations, though the association has an increasing number of members of all ages holding callsigns.

Guiders interested in organising a TDOTA station can apply for a comprehensive information pack with suggestions for activities, logos for certificates and posters to report forms. Further information is published from time to time in the association's magazines.

Stations are requested to complete a brief report which is sent to Girlguiding UK HQ. All the information is collated into a National Report, which is sent to those who took part and contributed to the report. Copies of the current report are available on receipt of an A4 SASE or from the website (see below).

Amateur radio has a place in the programme for all age groups, encouraging girls to embrace the technical aspects and international perspectives of a world-wide movement. Girl- guiding UK supports the revised amateur radio licence structure and particularly welcomes the Foundation Licence.

While the main focus remains TDOTA, Guides can be heard on the air at other times of the year from camps, activity days and leader training courses.

Short-form call signs

Ofcom will assign a callsign with only two trailing letters or which starts with 'G2' only if the applicant previously held it.

For a list of permanent Special Event Callsigns, see the Callsign Listings section of this Yearbook

---

Prefixes and Common suffixes

Below is a list of Prefix characters for UK Callsigns for individual countries within the UK also Suffixes to be used. Approximate year of issue of UK callsigns is also listed below.

Common Prefixes

1st Character: Foundation: M3, M6, M7
Intermediate: 2, M8, M9
Full: G or M (except M3, M6 or M7)

2nd Character:

	Intermediate	Foundation/ Full	Club (Full)
England	E	(none)	X
Isle of Man	D	D	T
Northern Ireland	I	I	N
Jersey	J	J	H
Scotland	M	M	S
Guernsey	U	U	P
Wales	W	W	C

GB3 + 2 letters:	Repeaters
GB3 + 3 letters:	Beacons
GB7 + 2 letters:	Data repeaters
GB7 + 3 letters:	Data mailboxes
	Special event stations (class of call sign normally corresponds to the appropriate M format)
M/foreign call:	Reciprocal CEPT

Suffixes

-/A	Alternative address
-/M	Mobile (includes inland waterways and pedestrian)
-/P	Portable (temporary location)
-/MM	Maritime Mobile

Except in the case of the intermediate "2" call signs, the RSL is optional.

Approximate Issue Dates for UK Callsigns

Full Licence

Two letters

G2AA	1920-39
G3AA	1937-38
G4AA	1938-39
G5AA	1921-39
G6AA	1921-39
G8AA	1936-37

Three letters

G2AAA	Pre-war
G3AAA	1946
G3CAA	1947
G3EAA	1948
G3GAA	1949-50
G3HAA	1950-51
G3IAA	1951-52
G3JAA	1952-54
G3KAA	1954-56
G3LAA	1956-57
G3MAA	1957-58
G3NAA	1958-60
G3OAA	1960-61
G3PAA	1961-62
G3QAA	not issued
G3RAA	1962-63
G3SAA	1963-64
G3TAA	1964-65
G3UAA	1965-66
G3VAA	1966-67
G3WAA	1967
G3XAA	1967-68
G3YAA	1968-69
G3ZAA	1969-71
G4AAA	1971-72
G4BAA	1972-73
G4DAA	1974-75
G4EAA	1975-76
G4GAA	1977
G4IAA	1979
G4MAA	1981
G4QAA	not issued
G4RAA	1982
G4SAA	1983
G4WAA	1984
G0AAA	1985
G0EAA	1986
G0HAA	1987
G0JAA	1988
G0LAA	1989
G0MAA	1990
G0NAA	1991
G0SAA	1992
G0TAA	1993
G0VAA	1994
G0WAA	1995
M0AAA	1996
M0BAA	1997-98
M0CAA	1998-00
M5AAA	1999-00
G8AAA	1964-67
G8BAA	1967-68
G8CAA	1968-69
G8DAA	1969-70
G8EAA	1970-71
G8FAA	1971-72
G8HAA	1973
G8IAA	1973-74
G8JAA	1974-75
G8KAA	1975
G8MAA	1976-77
G8NAA	1977
G8OAA	1977-78
G8PAA	1978
G8QAA	not issued
G8TAA	1979
G8ZAA	1981
G6CAA	1981
G6QAA	not issued
G6RAA	1982
G1AAA	1983
G1DAA	1984
G1LAA	1985
G1QAA	not issued
G1SAA	1986
G1XAA	1987
G7AAA	1988
G7EAA	1989
G7FAA	1990
G7HAA	1991
G7MAA	1992
G7OAA	1993
G7SAA	1994
G7TAA	1995
G7WAA	1996
M1AAA	1996
M1CAA	1997
M1DAA	1998-99
M1EAA	1999-00

Intermediate

2E0AAA	1991
2E1AAA	1991
2E1BAA	1992
2E1CAA	1993
2E1DAA	1994
2E1EAA	1995-97
2E1GAA	1997-99
2E1HAA	1999-00

Foundation

M3AAA	2002
M6AAA	2008
M7AAA	2018

Note: *From April 2000, out-of-sequence callsigns could be requested, so calls in later series may be heard.*

GB2RS News

GB2RS is the weekly news service of the RSGB, delivering information to members and the wider radio community. It all began at 10am on Sunday 25th September 1955, when the Post Office authorised the first ever broadcast on 3600kHz from the home of Frank Hicks-Arnold, G6MB, in Walton-on-Thames, using the special callsign GB2RS. Today, the news is broadcast every Sunday by a team of over 100 volunteer readers, using a wide range of amateur bands and modes.

The latest schedule of GB2RS broadcasts can be downloaded from *https://rsgb.org/gb2rsschedule* National transmissions, on 160m, 60m and 40m, can be heard throughout the UK and in parts of western Europe. Many localised broadcasts take place on VHF and UHF, either simplex or via analogue and digital repeaters, and there is an international edition via the QO-100 satellite. Items of local news are read out as appropriate to the coverage area of the transmission. As well as traditional spoken readings, GB2RS is available via amateur television transmissions, and on the internet as both video and audio. There is also a digital voice broadcast on 80m.

About the broadcast

GB2RS carries items of interest to radio amateurs and listeners, together with current contest, special event and propagation information, provided by the RSGB's specialist teams or by direct submission. The news script is prepared each week by the RadCom editorial team at RSGB HQ and posted to the RSGB website on Friday afternoons. The majority of the GB2RS broadcasts take place on Sunday mornings, although a number extend throughout the rest of the day to suit listeners' free time and to make best use of propagation. The on-air broadcast frequencies are allocated to appointed newsreaders by the RSGB on behalf of Ofcom.

The frequencies are shown in the Schedule and all amateurs (including contest operators) are respectfully asked to avoid using these on Sundays so that the news transmissions can be heard as widely as possible. The readers are provided with a special GB2RS Notice of Variation which permits them to broadcast the news bulletin and, while using this callsign, two-way communication is not permitted. Therefore, readers will often convene nets before and after the news broadcast, using their own callsign. All stations are very welcome to call in. Listeners can also catch up with the news at their leisure. The GB2RS podcast from Jeremy G4NJH is available at *https://gb2rs.podbean.com* A catch-up visual version from Alison G8ROG can be seen at *http://gb2rs.apj1.co.uk/* and TX Factor provides a listen-again reading at *http://txfactor.co.uk/gb2rs-news.html*

Newsreaders

The organisation of the GB2RS News Service and the network of newsreaders is the responsibility of the GB2RS News Manager. Contact *gb2rs.manager@rsgb.org.uk* New readers are always needed. Interested amateurs need to be RSGB members and must hold a Full or Intermediate licence. The GB2RS News Manager will provide an application form and then discuss your news-reading proposal with you. Newsreaders must be able to deliver the script fluently and their station capability should match the proposed coverage area. The GB2RS News Manager will guide on time and frequencies of operation and arrange the required Notice of Variation.

Submitting news

News items for GB2RS, RadCom magazine and the RSGB website should be sent to the editorial team as far in advance as possible by email to: *radcom@rsgb.org.uk* This is also the destination for club and local news. The deadline is 10am on Thursdays.

GB2RS News Broadcast Schedule

The following broadcasts are made every Sunday

Time	Freq	Mode	Location	Reader(s)
NATIONAL BROADCASTS				
1000 UK	7.1270	LSB	HOY	GM8OFQ
			EXMOOR	G6ASK
			SUTTON COLDFIELD	G7LHK
1500 UTC	5.3985	USB	ROYSTON	G4HPE
			HOY	GM8OFQ
			MEPPERSHALL	G4JBD
		Note: Read in rotation		
2130 UK	1.9900	LSB	ROYSTON	G4HPE
			WEST BUTTERWICK	G8JET
			OSWESTRY	G4IOQ
			HOY	GM8OFQ
			MEPPERSHALL	G4JBD
		Note: Read in rotation		
NORTHERN ENGLAND				
0900 UK	50.8000	FM	VIA GB3WY WAKEFIELD	G0TKF
0900 UK	145.5250	FM	TYNE, WEAR, TEES	G4OLK G8YDC
0930 UK	145.5250	FM	BRADFORD	M0JPA
			CLECKHEATON	M0WIT
			HALIFAX	G6NTI G6YGV
1000 UK	145.5250	FM	HULL	G3GJA
1000 UK	1308.0000	DATV	VIA GB3EY HULL	G3GJA
		Note: 144.7750 ATV talkback frequency is also monitored		
		Also streamed live via BATC website https://batc.org.uk/live/gb3ey		
1030 UK	70.4250	FM	BURY	G4GSY
			DUKINFIELD	G0NAJ
			WIGAN	M0HDE M0OEG
1030 UK	145.5250	FM	BURY	G4GSY
			DUKINFIELD	G0NAJ
			WIGAN	M0HDE M0OEG
1200 UK	430.9125	FM	VIA GB3EG WIGAN	M0OEG
	430.0750		VIA GB3BY CHORLEY	
	430.8500		VIA GB3FC BLACKPOOL	
	430.9375		VIA GB3PG PRESTON	
2000 UK	430.9125	FM	VIA GB3EG WIGAN	M0OEG
	430.0750		VIA GB3BY CHORLEY	
	430.8500		VIA GB3FC BLACKPOOL	
	430.9375		VIA GB3PG PRESTON	
2100 UK	70.4250	FM	BURY	G4GSY
			DUKINFIELD	G0NAJ
			WIGAN	M0HDE M0OEG
2100 UK	145.5250	FM	BURY	G4GSY
			DUKINFIELD	G0NAJ
			WIGAN	M0HDE M0OEG
MIDLANDS				
0800 UK	3.6500	AM	MEPPERSHALL	G4JBD
			STOWMARKET	G0OZS
0930 UK	3.6500	LSB	PERSHORE	G8BGT
			WORCESTER	G4IDF
1200 UK	430.8750	DMR	GB7DC	G4SEB
	439.6500		GB7IN	
	439.6125		GB7RR	
		Note: Talk Group TG23590 - Time Slot 1		
1800 UK	145.5250	FM	STOKE ON TRENT	G0VVT
			NEWCASTLE-UNDER-LYME	M5GWH
1800 UK	70.4250	FM	STOKE ON TRENT	G0VVT
1830 UK	50.7900	FM	VIA GB3SX STOKE ON TRENT	G0VVT
1830 UK	439.5250	DMR	VIA GB7ST STOKE ON TRENT	G0VVT
		Note: Talk Group TG8 (local only) - Time Slot 2		
1830 UK	433.5250	FM	STOKE ON TRENT	G0VVT
1900 UK	145.6375	FM	VIA GB3IN HUTHWAITE	G4TSN G0LCG
2030 UK	145.5250	FM	NUNEATON	G8VHI
			BURTON-ON-TRENT	G8EKG
			NUNEATON	G4AEH
2030 UK	144.2500	USB	NUNEATON	G8VHI
		Note: 10 ele, beaming north		
2030 UK	433.5250	FM	NUNEATON	G8VHI
			BURTON ON TRENT	G8EKG
SOUTH EAST / EAST ANGLIA				
0800 UK	3.6500	AM	MEPPERSHALL	G4JBD
			STOWMARKET	G0OZS
0900 UK	3.6500	LSB	BRISTOL	G4TRN
			MANNINGTREE	G6UWK
			SANDFORD	G7NJX
			READING	G8ROG G4RDC
0900 UK	3.6400	DV	BURES	G6WPJ
		Note: FreeDV - mode 700E - set receiver to LSB		
0900 UK	51.5300	FM	READING	G8ROG G4RDC
0900 UK	70.4250	FM	READING	G8ROG G4RDC
0900 UK	433.0000	FM	VIA GB3BN BRACKNELL	G8ROG
			READING	G4RDC
0900 UK	3408.0000	DATV	VIA GB3HV FARNHAM	G8ROG
			READING	G4RDC
		Note: Also streamed live via BATC website https://batc.org.uk/live/gb2rs		
0900 UK	10355.0000	WFM	READING	G8ROG G4RDC
		Note: Four paths, each antenna at 60 deg beamwidth		
0930 UK	145.5250	FM	EAST PRESTON	M0KEL
			LANCING	G0TLU
			LITTLEHAMPTON	M0RDV
0930 UK	145.5250	FM	STOWMARKET	G0OZS
			WOODBRIDGE	G7CIY

**For GB2RS News Online, there are links from the RSGB main page.*

Time	Freq	Mode	Location	Reader(s)
			FELIXSTOWE	G4YQC
			IPSWICH	G0DVJ
		Note: Also covers parts of Norfolk, Essex and N Kent		
1000 UK	70.4250	FM	DUNTON	G4OXY
			LANGFORD	G1GSN
1000 UK	145.5250	FM	DUNTON	G4OXY
			LANGFORD	G1GSN
			HITCHIN	G4OXD
1015 UK	433.2250	FM	VIA GB3IW ISLE OF WIGHT	M0KEL M0RDV G0TLU
1100 UK	433.5250	FM	CLACTON ON SEA	G0NAD
1800 UK	433.5250	FM	FOLKSTONE	G4IMP
1900 UK	145.6250	FM	VIA GB3NB NORWICH	G4DYC G7URP G3LDI G3YLA
1900 UK	70.4250	FM	CLACTON ON SEA	G0NAD
1930 UK	145.5250	FM	EASTBOURNE	G0NQZ G1FBH M0LRE
SOUTH WEST				
0900 UK	3.6500	LSB	BRISTOL	G4TRN
			MANNINGTREE	G6UWK
			SANDFORD	G7NJX
			READING	G8ROG G4RDC
0900 UK	3.6400	DV	BURES	G6WPJ
		Note: FreeDV - mode 700E - set receiver to LSB		
0900 UK	51.5300	FM	READING	G8ROG
			READING	G4RDC
			SANDFORD	G7NJX
0900 UK	70.4250	FM	READING	G8ROG
			READING	G4RDC
0900 UK	145.5250	FM	EXMOOR	G6ASK
0900 UK	433.0000	FM	VIA GB3BN BRACKNELL	G8ROG
			READING	G4RDC
0900 UK	3408.0000	DATV	VIA GB3HV FARNHAM	G8ROG
			READING	G4RDC
		Note: Also streamed live via BATC website	https://batc.org.uk/live/gb2rs	
0900 UK	10355.0000	WFM	READING	G8ROG
			READING	G4RDC
		Note: Four paths, each antenna at 60 deg beamwidth		
0930 UK	145.7250	FM	VIA GB3NC ST. AUSTELL	G4BHD
				G4OCO
				G0PNM
0930 UK	430.8250	FM	VIA GB3ZB BRISTOL	G4TRN
	430.9250		and GB3FI CHEDDAR	G7NJX
1030 UK	145.5250	FM	CENTRAL BRISTOL	G4TRN
			SANDFORD	G7NJX
1100 UK	145.3375	FM	VIA MB7IPN WATERLOOVILLE	G7TEM
CHANNEL ISLANDS				
0900 UK	3.6500	LSB	BRISTOL	G4TRN
			MANNINGTREE	G6UWK
			SANDFORD	G7NJX
			READING	G8ROG G4RDC
0900 UK	3.6400	DV	BURES	G6WPJ
		Note: FreeDV - mode 700E - set receiver to LSB		
0900 UK	145.5250	FM	JERSEY	GJ0PDJ GJ8PVL
0930 UK	145.5250	FM	GUERNSEY	GU0SUP GU8ITE GU6EFB
WALES				
0900 UK	3.6500	LSB	BRISTOL	G4TRN
			MANNINGTREE	G6UWK
			SANDFORD	G7NJX
			READING	G8ROG
			READING	G4RDC
0900 UK	3.6400	DV	BURES	G6WPJ
		Note: FreeDV - mode 700E - set receiver to LSB		
0930 UK	3.6500	LSB	PERSHORE	G8BGT
			WORCESTER	G4IDF
1030 UK	3.6500	LSB	ST ASAPH	GW6VEI
			DENBIGH	GW0WWQ
1130 UK	145.5250	FM	BRIDGEND	2W0PQU
NORTHERN IRELAND				
0930 UK	145.5250	FM	CARRICKFERGUS	MI0AWL
1000 UK	3.6400	LSB	DUNGIVEN	GI0AZA GI0AZB
			STRABANE	MI0HOZ
1200 UK	439.5875	DMR	VIA GB7HZ STRABANE	MI0HOZ
	439.6000	DMR	VIA GB7WT OMAGH	
	430.8625	DMR	VIA GB7CX COLERAINE	
	439.7375	DMR	VIA GB7HI LISBURN	
	439.4875	DMR	VIA GB7MW CARRICKFERGUS	
		Note: Talk Group TG2354 - Time Slot 2		
1930 UK	430.9500	DMR+FM	VIA GB3OM OMAGH	MI0RWY
	439.6625	DMR	VIA GB7LY DERRY	2I0SJV
	439.5250	DMR	VIA GB7UL CARRICKFERGUS	MI0HOZ
	439.6250	DMR	VIA GB7HB TANDRAGEE	MI0RYL
	439.5875	DMR	VIA GB7HZ STRABANE	M1AIB
	439.6750	DMR	VIA GB7MF MAGHERAFELT	2E0SHZ
	439.5750	DMR	VIA GB7KP COMBER	2I0FVX
		Note: Talk Group TG880 - Time Slot 2		
ISLE OF MAN				
1600 UK	430.8250	FM	VIA GB3IM-C DOUGLAS	GD6ICR 2D0PEY MD0MAN
1600 UK	430.8750	FM	VIA GB3IM-P PEEL	GD6ICR 2D0PEY MD0MAN
1600 UK	430.8250	FM	VIA GB3IM-R RAMSEY	GD6ICR 2D0PEY MD0MAN
1600 UK	430.1250	FM	VIA GB3IM-S DOUGLAS	GD6ICR 2D0PEY MD0MAN
SCOTLAND				
0900 UK	145.5250	FM	PERTH	GM6MEN
0930 UK	70.4250	FM	EDINBURGH	GM4DTH
0930 UK	145.5250	FM	EDINBURGH	GM4DTH
			EAST WEMYSS	MM0TGB
0930 UK	433.5250	FM	EDINBURGH	GM4DTH
0930 UK	145.6500	FM	VIA GB3OC KIRKWALL	GM1BAN MM3YHA GM8OFQ
		Note: Repeater covers Caithness and Orkney		
1000 UK	145.5250	FM	ELGIN	GM4ILS
1000 UK	145.5250	FM	BEARSDEN	GM3VTB
			CARLUKE	GM4COX
			LENZIE	MM0TMZ
			WHITECRAIGS	MM0TSS
1000 UK	TG23559	DMR	CLYDEBANK	2M0PIJ
		Note: Talk Group TG23559 during the reading - Time Slot 2		
1930 UK	145.5250	FM	SOUTH WEST GLASGOW	GM5AUG
SATELLITES				
0800 UTC	10489.8550	USB	Es'hail-2 / QO-100	GU6EFB
		Note: Can also be monitored via WebSDR		G4BIP
			https://eshail.batc.org.uk/nb/	G0NAD
2100 UTC	10489.8550	USB	Es'hail-2 / QO-100	GU6EFB G4BIP G0NAD
		Note: Can also be monitored via WebSDR		
			https://eshail.batc.org.uk/nb/	
PODCASTS AND WEB				
Weekly			https://gb2rs.podbean.com	G4NJH DD5LP
0900 UK			https://batc.org.uk/live/gb2rs	G8ROG
		Note: Live TV news reading		
Weekly			http://gb2rs.apj1.co.uk/	G8ROG
		Note: Catch-up TV reading		
Weekly			http://txfactor.co.uk/gb2rs-news.html	G1IAR G0FGX 2E0FGQ
Weekly		CW	https://thersgb.org/go/gb2rsmorse	G4JBD
	Note: A Morse practice facility at various speeds with realistic HF noise backgrounds			

When submitting news items...

Do:

- Submit items by email, to: *radcom@rsgb.org.uk* This is the ONLY email address for all submissions.
- Always give a contact name, callsign (if any) and phone number.
- Say if a phone number is daytime only or evening only.
- Send GB2RS any last-minute details of your rally.
- Give the proper name of your club – there is a Wirral and District Amateur Radio Club and a Wirral Amateur Radio Society, so saying "the Wirral club" could lead to confusion.
- Always give the callsign of a speaker if he/she has one, not just 'Talk by John on aerials'.
- Provide full dates, not 'Last Friday in month'.
- Listen to GB2RS to hear for yourself the format used.

Don't:

- Forget to include the venue and opening time of a rally, details of talk-in (if any) and a contact name, callsign, and phone number.
- Send in 'to be confirmed' items. If they are not confirmed we assume that they are not taking place and they will therefore not be broadcast. Only send your item in when it has been confirmed.
- Give more than one contact person or telephone number. There is only time to broadcast one, so you decide which one to use.
- Mix regular club meetings with main news items, such as rallies and special event stations.
- Use GB2RS and *RadCom* as the only means of publicising club events to your own members – the main purpose of GB2RS should be to inform casual listeners and members of other clubs of the exciting things your club is doing.
- Use cryptic titles for talks, or 'in-jokes'. If we don't know what you mean, it is unlikely that anyone else will.

DO NOT SEND TO MORE THAN ONE EMAIL ADDRESS because, paradoxically, this increases the chances of your item getting lost. ONLY use *radcom@rsgb.org.uk*

Got a news item?
email: *radcom@rsgb.org.uk*

Got a network enquiry?
email: *gb2rs.manager@rsgb.org.uk*

Planning Advice

Many, if not most, radio amateurs never see the need to apply for planning permission for their aerials. After all the aerials work just as well without it and there is a school of thought that if you don't ask for planning permission the Planning Department can't be tempted to say no. This might seem an attractive argument if you use small visually unobtrusive wire aerials, but if you have aspirations of anything more substantial you are likely to fall foul of the local Planning Department.

Urban Myths

Unfortunately, holding an amateur radio licence in the United Kingdom does not convey any special 'rights' under planning legislation to have an aerial and there are a number of urban myths circulating regarding the need for planning permission.
Amateur radio aerials and masts are generally treated as householder development, exactly the same as a garage or conservatory, and will require planning permission unless they come under one of the following categories

Temporary

Unlike non-residential land which has a limit of 28 days, there are no specific time limits on how long a mast or aerial can be present and still be classed as temporary. It is the degree of permanence that is the deciding factor. The fact that the mast or aerial is installed in a ground socket and can be easily removed is not enough for it to be classed as temporary if it is in regular use.

De Minimis

This latin term means that something is of minor significance or is 'trifling' and should not need permission. A thin long wire would normally be regarded as such, but it is subject to the interpretation of the Planning Department. Also, in planning law, something which has 'no material effect on the external appearance of the building' is not classed as development and does not require planning permission. A wire dipole stretched between a house and a tree usually comes within that category which may also apply to roof or gable mounted verticals. If you have permission for a mast it is a 'building' and changing the aerials may not constitute a 'material effect' but it is better to seek advice before doing so.

Permitted Development

Minor development within the curtilage of a dwelling house, alterations and/or extensions to the dwelling are classed as 'permitted development' and does not require planning permission as long as certain conditions are met. The limits are set in what is called 'The General Permitted Development Order'. Different Orders apply in each of the four countries of the United Kingdom. Although no references are made in the Orders to amateur radio aerials and masts, some radio amateurs have successfully argued that an aerial, mast or pole to the rear of and attached to a dwelling house is an 'enlargement, improvement or alteration of a dwellinghouse' provided the aerial or mast does not protrude above the ridge of the roof. In England, it must also be of similar appearance to the existing dwelling. Any free-standing 'structure or erection' in the rear garden of a house is permitted up to 3m. high (2.5m. if within 2m of a boundary).
There is a class in the Order which refers to 'microwave antennas' which is intended to apply to satellite dishes and includes size restrictions. Should this be referred to by the Planning Department you may need to explain that there is no provision for HF or other aerials

A thing of beauty, but what do the neighbours think?

Mobile installation

The legal position regarding mobile, trailer mounted, masts is uncertain. It very much depends upon the circumstances of the case. Any installation must be truly mobile and not fixed to the ground or stayed in any way. It should be possible to demonstrate that it can be folded down and removed from site easily. If it cannot, and there is any degree of permanence it is likely that planning permission will be required. In such circumstances, especially if an enforcement notice is served, you should seek immediate advice from the Planning Advisory Committee.

The 4-year planning rule

The 4-year planning rule has offered a degree of protection for antennas and masts that have been erected without planning permission. Under this rule any installation which had been installed and unchanged for 4 years or more were protected against any planning enforcement action.
However, as of 25 April 2024, this rule ceased to exist in its current form in England under changes introduced under the *Levelling-up and Regeneration Act 2023 (LURA)*, being replaced by a more stringent ten-year period for the exemption from enforcement for residential dwellings. This change does not affect Wales, Scotland and Northern Ireland where the four-year rule will continue to apply.
The Levelling-up and Regeneration Act 2023 will apply immediately to all new installations not completed or substantially completed by 25 April 2024. After that date the installation will have to completed and unchanged for 10 years to avoid enforcement action for not having planning permission.
The new Act states that any installation that has been completed or substantially completed before the introduction of the Act will still be subject to the 4-year rule. For example if you completed the installation and can prove it to the satisfaction of the Local Planning Authority (usually the Council) by the 24 April 2024 and the Council chose to take enforcement action for you not having planning permission on the 23 April 2028, you could claim exemption under the previous 4 year rule, If it was not complete or substantially completed by the 26 April 2024, it would have to have been there and unchanged for 10 years or more before escaping enforcement action.
The wording of the Act effectively gives a grace period up to 25 April 2028 for completed or substantially completed installations completed by 24 April 2024.
There are a couple of points to note:-

- The installation has to remain present and unchanged for the 4 year/10 year rule to apply. Changing the antenna can restart the clock as can adding additional antennas.
- This Levelling-up and Regeneration Act 2023 only applies only in England. The existing 4 year rule will remain in force in the rest of the UK.
- The Act does not give any guidance on what 'substantially completed' means.

Applying for planning permission

Most planning applications are now made electronically using the Planning Portal at *www.planningportal.co.uk* but each local authority will also have their own planning permission application forms, but they generally follow a similar style. They will typically require you to complete a Householders Planning Application form, a site location plan(s) and a development plan(s) showing the dimensions of the proposed aerial and/or mast and the distances to your property and the boundary with neighbouring properties. The number of copies and scale for these plans will be specified by the Planning Department in their planning pack. The drawings need not be professionally prepared, as long as it is clear what your proposals are and they are to the scale specified by the Planning Department.
If you forget to show the aerial on your planning drawings you may receive planning permission for the mast only, without permission to attach any aerials.
You will also need to complete a neighbourhood notification form, detailing your 'notifiable neighbours'. A notifiable neighbour is someone who shares a boundary with your property or directly face any part your property from across the road. It is worth discussing your proposals with them before making your submission, so that when the official

notice comes through their door it will not be a surprise. If you have TVI issues get these resolved first, as although TVI is not part of the planning process experi- ence has shown neighbours will just object on other grounds, usually visual amenity.

Before formally submitting your planning application, ask if you can discuss the submission with your Case Officer. Minor changes at this stage may alleviate any concerns he/she may have, giving your application a better chance of success. You can also contact RSGB HQ to ask to be put in contact with a member of the Planning Advisory Committee, to discuss your proposals prior to submission. A letter of support from the RSGB for your proposed aerial or mast is also available on request.

Refusal to grant planning permission

Sadly, not all planning applications are successful and there is sometimes no apparent reason why one Planning Department will grant planning permission for an aerial and mast in one area and another in a neighbouring area will refuse planning permission for a near identical installation.

You will be told why your application was refused. Usually it's on the grounds of visual amenity. Consider if the Planning Department has a valid point. To a radio amateur a large beam is a thing of beauty and a joy to own, but what do your neighbours think? Does it overly dominate the area? The Planning Department has to weigh-up the rights of all involved, not simply take sides. You will usually be able to resubmit a revised application free of charge if it is less than 12 months from the original application. If appropriate, reconsider a less ambitious proposal.

If however you believe the Planning Department has treated your application unfairly you have the right of appeal to the Planning Inspectorate (England & Wales), the Planning Appeals Commission (Northern Ireland) or The Directorate for Planning and Environmental Appeals (DPEA), (Scotland).

The appeal must be made within 12 weeks from the planning decision and is usually made in the form of 'Written Submission'. No charge is made for the appeal. Most appeals are submitted electronically using the Planning Portal although it is still possible to fill in the appropriate form and to submit your evidence in writing.

To be successful you must state why you believe the original decision was unsound. Simply saying you disagree or that it will curtail your operations as a licensed radio amateur is not enough. You must establish that the Planning Department has failed to comply with planning law, policy or guidelines, or has sought to impose a different standard on your application than it has done for others.

The RSGB's Planning Advisory Committee can provide guidance to members in the preparation of a planning appeal if required. If you require assistance, contact RSGB HQ who will put you in contact with your nearest Committee member If your appeal is not upheld and you have not used up your free resubmission, you can submit a revised proposal free of change if it is still less than 12 months from the original application.

An 18m mast granted planning permission on appeal.

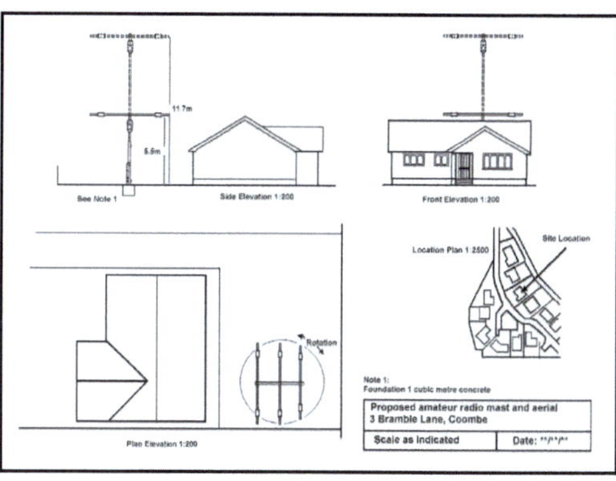

The kind of drawing that a council will want you to submit with your application.

Enforcement notices

If you have put up aerials without permission you might, at any time, receive a rather threatening letter from your local planning department along the lines of 'either apply for planning permission or take the aerials down within 28 days, or else'. Whatever you do, don't touch them, especially if they have been there unaltered for more than 4 years. Contact the Planning Advisory Committee for advice.

The Planning Department are likely to take enforcement action against you in two circumstances:

1. Where you have erected an aerial or mast which, in the Planning Department's opinion, requires permission and you have not obtained it.
2. Where the Planning Department alleges that you have breached a condition attached to the planning permission they have issued (for example, to keep a mast wound down when not in use).

The first is the most common. If you have not already submitted an application and had it refused the Planning Department will normally write to invite you to submit an application. It is usually worth doing so unless you want to argue that you have permitted development rights for the aerial or they are de minimalist. The Planning Department may serve on you a Planning Contravention Notice. This requires you to give certain information as to ownership or to attend the Planning Department's Offices at a specific date and time to give details of your installation and why you believe it does not need planning permission (for example, because it's permitted development or de minimis). You must comply with the Notice, because if you fail to do so you may be prosecuted.

If the Planning Department is not satisfied with your explanation they may elect to issue you an Enforcement Notice. Planning Departments can only do this if they can give reasons why they would not consider granting planning permission and may have to justify their decision to the Planning Inspectorate.

If an Enforcement Notice is issued it will set out what the Planning Department want you to do. Usually this will require you to remove the aerial and/or mast. Should you be served an Enforcement Notice you have two choices:

1. Comply by removing the offending aerial, mast, etc.
2. Appeal.

You must appeal within 28 days of receiving the Notice. Details on how to appeal are available from the Planning Inspectorate, Scottish Government and the Northern Ireland Planning Appeals Commission websites listed below. If the notice relates to a breach of conditions, the Planning Department may serve on you an ordinary enforcement notice, (against which you can appeal as above), or alternatively a Breach of Condition Notice, against which there is no appeal.

Failure to comply with an Enforcement Notice quickly can lead to legal action being taken against you, so don't ignore them. If the Planning Department considers that the aerial/mast has a severe environmental concern which requires immediate action they can apply to the Court for an injunction. If such an injunction is granted, you must comply or you will be prosecuted.

Planning Advisory Committee

The Planning Advisory Committee exists to assist RSGB members with planning applications, enforcement notices and planning appeals. Committee members will not actually prepare your planning application or submit an appeal on your behalf, but can check your application or provide you with a suggested appeal strategy.

The Committee also provides a guide to the planning process. This is available free of charge from RSGB HQ, or as a download from the member's only website.

Tenancy matters

For tenants, both planning permission and landlord consent are likely to be needed. Obtaining planning permission does not mean the landlord has to agree, so you could find yourself having incurred the expense of obtaining planning permission only to find that the landlord does not agree and so you cannot implement the permission.

Especially for private sector tenants, failure to obtain consent may give the landlord grounds to terminate the tenancy.

The Society cannot become involved in legal disputes between landlords and tenants, but will try to provide advice or to signpost members to other bodies who can help. You may therefore want to contact the PAC Chair before starting - contact details below.

	Online planning information	Appeals and enforcement notices
England/Wales	www.planningportal.gov.uk/	www.planning-inspectorate.gov.uk
Scotland	www.eplanning.scotland.gov.uk	www.dpea.scotland.gov.uk
Northern Ireland	www.planningni.gov.uk	www.pacni.gov.uk

RadCom Reviews

Items of amateur radio equipment are frequently reviewed in RadCom. The types of equipment range from antennas, through budget 'handies', software and ancilliary equipment, to top-of-the-line transceivers. Listed below are the reviews that have taken place in RadCom since January 1995.
Please refer to the relevant edition, if you need more information on any of the items.

A

AADE DFD (digital frequency display kit)	Apr 1998
AADE IIB (digital L/C meter kit)	Apr 2005
AAlog (logging software)	May 2006
ACE-HF (propagation prediction software)	Oct 2006
ACE-HF PRO propagation prediction software	May 2016
Acom 1000 (160m-6m linear)	Mar 2001
Acom 1010 (HF linear amplifier)	Aug 2005
Acom 1500 (160m-6m linear)	Dec 2012
Acom 2000A (HF auto-tuning linear)	Mar 2001
ADI AR-446 (70cm FM mobile)	Mar 1997
ADI AT-600D (2m+70cm handheld)	Apr 1997
Adonis AM-308 (desk microphone)	Jan 2006
Adonis AM-508 (desk microphone)	Jan 2006
Adonis AM-708 (desk microphone)	Jan 2006
AirNav RadarBox (virtual radar)	Mar 2009
Airspy SDR Dongle	Sep 2015
Aerial-51 ALT-512 QRP transceiver	Mar 2020
AKD HF3E (communications receiver)	Jun 1998
AKD Target HF3 (VLF-HF receiver)	Nov 1996
Ailunce HS-2 1.8MHz to 432MHz SDR transceiver	Dec 2021
Albrecht AE485S (10m multimode mobile)	Jan 2001
Alinco DM330MW PSU	Jun 2021
Alinco DJ-596 (2m+70cm FM handheld)	Dec 2004
Alinco DJ-C1 (2m FM micro handheld)	Feb 1998
Alinco DJ-C4 (70cm FM micro handheld)	Feb 1998
Alinco DJ-C5 (2m+70cm FM mini handheld)	Jul 1998
Alinco DJ-C7 (2m+70cm FM handheld)	Jan 2005
Alinco DJ-G5 (2m+70cm FM handheld)	Apr 1997
Alinco DJ-G7 (2m+70cm+23cm FM handheld)	Aug 2009
Alinco DJ-MD5	Feb 2019
Alinco DJ-V5E (2m+70cm FM handheld)	Sep 2000
Alinco DJ-V17 (2m FM handheld)	Jul 2008
Alinco DR-610E (2m+70cm FM mobile)	Feb 1999
Alinco DX-70TH (HF+6m multimode mobile)	Aug 1999
Alinco DX-SR8E(100W HF transceiver)	Jan 2014
Alinco DJ-VX50 dual band handheld	Apr 2020
Alinco switch mode power supplies	Jan 2016
Alpha 4510 (HF power/SWR meter)	Nov 2006
Alpha DX-Jr (lighweight portable 40m - 6m ant)	Feb 2013
Alpha SPID Rak (azimuth rotator)	Mar 2007
Alpin 100 (HF+6m linear amplifier)	Apr 2011
Ameritron ALS-500M (HF linear)	Dec 2005
Amlog 3 (logging software)	Sep 1996
Anan 8000DLE SDR 2000W HF Transceiver	Nov 2017
Antennas-Amplifiers 2m Band pass Filter	Jan 2023
AnyTone Ares II 10m transceiver	Aug 2023
Anytone AT-779 70MHz mobile transceiver	May 2022
Anytone AT-D878UV Plus handheld	Sep 2019
Anytone AT-D578UVPRO dual band mobile	Mar 2020
Anytone AT-779UV 2m/70cm transceiver	May 2021
AOR AR-DV1 Digital Voice Receiver	May 2016
AOR AR7030 (VLF-HF receiver)	Jul 1996
AOR ARD9000 Digital Voice (digital voice adapter)	Oct 2005
AOR ARD9800 Fast Data Modem (digital voice adapter)	Jul 2004
Arno Elettronica E-H Antennas (small HF antennas)	Sep 2003
Arno Elettronica Venus-80 (small HF antenna)	Aug 2005
Arno Elettronica Venus-160 (small HF antenna)	Aug 2005
Array Solutions Bandmaster (universal band decoder)	Jan 2011
Array Solutions PowerMaster (wattmeter)	Feb 2006
Ascel AE20401 frequency counter and power meter kit	May 2020
Astatic SWR Power Meter	Sept 2021
ATM Motion Picture (video grabber)	Nov 1996
Avair AV-20 (HF power/SWR meter)	Oct 2006
Avair AV-40 (2m-70cm SWR/power meter)	Jun 2006

B

Badger Boards Receiver (kit for Novice course)	Apr 1998
Basicomm CW Touch Paddle	Jan 2016
Begali CW Machine	Apr 2023
Begali HST (single lever paddle key)	Jul 2009
Begali Sculpture (iambic paddle key)	Feb 2008
Begali Simplex Mono (single lever paddle key)	Feb 2008
bhi Compact In-Line Noise Eliminating Module	Dec 2015
bhi 5W DSP noise cancelling inline module	Jan 2022
bhi DSPKR (noise reduction speaker)	Feb 2010
bhi groundbreaker	June 2019
bhi DSPKR 10W (noise reduction speaker)	Jan 2011
bhi ParaPro EQ20 DSP	Sep 2017
bhi NEDSP1061 (add-on DSP module for FT-817)	Dec 2003
bhi NEDSP1062 (add-on DSP module)	Jul 2005
bhi NES 10-2 (noise eliminating speaker)	Dec 2002
bhi NES 10-2 Mk2 (noise eliminating speaker)	Nov 2005
bhi NEIM1031 (noise eliminating inline module)	Mar 2004
bhi 1042 (switch box)	Mar 2004
bhi NES10-2 Mk 4 noise reduction speaker	May 2020
Bilal Isotron (small antennas for 80m and 40m)	Aug 2010
Buddipole (portable HF-VHF antenna)	Mar 2005
Butternut HF2V (HF multiband vertical antenna)	Mar 2005

C

Ceecom 10/11m Moxon antenna	May 2023
Chameleon CHA TD-Lite antenna	Dec 2023
Ciro-Mazzoni 'Stealth' antenna	Jult 2019
Ciro-Mazzoni 'Baby' Loop	Dec 2019
Cloud IQ SDR	June 2016
Comet CAT-300 ATU	Feb 2023
Comet CHA-250B (wideband vertical antenna)	Dec 2006
Comet CSW-201G (antenna switch)	Jul 2007
Comet CAT-300 ATU	Feb 2023
CommSlab μ-Modem (multimode radio modem)	Sep 1995
Crazy Daisy (Mag loop Antenna)	May 2014
CRKITS HT-1A Dual Band QRP Transceiver Kit	Aug 2019
Cross Country Wireless SDR (single band receivers)	Aug 2010
CRT SS6900 (10m multimode mobile)	Aug 2011
CT (contest logging software)	Nov 1999
Cushcraft 13B2 (2m beam)	Mar 2005
Cushcraft MA5B (compact HF beam)	Nov 1999
Cushcraft MA8040V (80m/40m vertical antenna)	Sep 2005
Cushcraft R8 (40m-6m vertical antenna)	Jun 2000
Cushcraft R7000 (HF vertical antenna)	Jan 1997
Cushcraft X7 ('Big Thunder' HF beam)	Nov 1998
CWMORSE 3D printers keys and paddles	May 2021

D

Daiwa CS-201A (antenna switch)	Jul 2007
DB6NT 13cm transverter (kit)	Jan 1998
Derek Stillwell Morse Key (handmade straight key)	Jun 1995
Diamond CA-35RS (lightning arrestor)	May 2006
Diamond CX-210A (antenna switch)	Jul 2007
Diamond SD300 (3-30MHz screwdriver antenna)	Nov 2011
Diamond SX20C (HF power/SWR meter)	Oct 2006
Diamond SX40C (2m-70cm SWR/power meter)	Jun 2006
Discovery TX-500 160-6m QRP transceiver	Nov 2021
DG8SAQ VNWA3 (vector network analyser)	Dec 2011
Digirig Mobile	Sept 2022
DK2DB 13cm PA (kit)	Jan 1998
DPRE4-6VL (cavity filters for 2m repeater)	Jun 2009
DSP-10 (DSP 2m multimode transceiver kit)	Feb 2000
DX4Win (station logkeeping software)	Aug 2000
DV Dongle (D-Star adapter)	Dec 2008
DV Access Point Dongle (2m 10mW D-Star node)	Mar 2011
DXAID 5.0 (propagation prediction software)	May 2004
DX Commander Rapide 7m antenna kit	Oct 2021
DX Engineering HEXX-5TAP-2 (5-band 2-ele hex beam)	Mar 2011

E

Elad FDM77 (software defined radio)	Nov 2005
Elecraft K1 (HF QRP CW transceiver kit)	Sep 2001
Elecraft K2 (HF QRP CW transceiver kit)	Mar 2003
Elecraft K3S	April 2016
Elecraft KX3 (HF/6m allmode transceiver)	Apr 2013

Elecraft KX2 (80-10m 10W allmode transceiver)	Jan 2017
Elecraft K4D HF and 50MHz transceiver	Jan 2022
Elecraft KPA500 (Solid State Amplifier)	Jan 2013
Elecraft KRC2 (band decoder kit)	Jan 2005
Elecraft N-gen (wideband noise generator)	Nov 2006
Elecraft P3 Panadapter & K3 Transceiver revisited	Oct 2012
Elecraft T1 (HF-6m QRP auto ATU)	Mar 2007
Elecraft XG2 (receiver test osc. / S-meter calibrator)	Nov 2006
ETO/Alpha 91B (HF linear)	Feb 1997
ETO/Alpha 87A (HF linear)	Feb 1997
EZMaster (SO2R interface)	Mar 2006
EZNEC 2 (antenna modelling software)	Sep 1998
F	
FA-VA5 Antenna Analyser	Sep 2018
FireSpot Digital Radio D-Star-DMR	Mar 2023
Flexradio SDR-1000 (software defined transceiver)	Jun 2006
Flexradio Flex-1500 (HF-6m software defined xcvr)	Apr 2011
Flexradio Flex-3000 (HF-6m software defined xcvr)	Aug 2009
FlexRadio Systems Maestro console	Aug 2016
Flexradio Flex-5000A (HF-6m software defined xcvr)	Jan 2008
Flexradio Flex-5000A upgrades (auto ATU and 2nd rx)	Mar 2009
Force 12 XR6/XR6C Yagi (11 ele six band ant)	May 2015
FoxRex 3500 ARDF Receiver	Sep 2017
FunCube Dongle Pro +	Feb 2013
G	
G1MFG ATV modules	May 2003
G3LIV Isoterm (data interfaces)	Sep 2009
G3WDG (23cm transverter kit)	Jun 2000
G4HUP L-C Meter Kit (test instrument)	Jul 2009
G4TPH Magnetic Loop Antennas (QRP loops)	Oct 2008
G4ZPY 3-in-1 Combo Keyer (electronic keyer)	Jun 1998
Garth 70cm and 23cm bandpass filters	Sep 2007
Gemini 23 (1296MHz linear amplifier)	Feb2016
GH Engineering PA1.6-16 (1.3GHz PA kit)	Jul 2006
Global CX201 (antenna switch)	Jul 2007
Goodwinch TDS (electric tower winches)	Jun 2011
Green Heron RT-21 (digital rotator controller)	Oct 2009
H	
HamGadgets MasterKeyer MK-1 (Morse keyer)	Oct 2011
HamGadgets PicoKeyer-Plus (Morse keyer kit)	Oct 2011
Ham Radio Deluxe (station control & logging software)	Apr 2005
Hamware AT-502 (remote auto ATU)	Mar 2009
Hamware AT-515 (remote auto ATU)	Mar 2009
Hands RTX/AMP (broadband HF linear amp kit)	Jan 1995
Hatley Crossed-Field Loop (antenna)	May 2002
Heil Proset Headphones (headphones with boom mics)	Aug 1996
Heil Proset 5 (headphones with boom mic)	Mar 2006
Heil Pro 7 communications headset	June 2015
Hexbeam Folding Antennas (20-10m multiband ant)	April 2015
HFx (propagation prediction software)	Oct 1999
HFx (propagation prediction software)	Jun 2000
High Sierra 1800/Pro (HF mobile antenna)	Aug 2004
Hilberling PT-8000A (HF-2m transceiver)	Nov 2013
Howes ASL5 (audio filter kit)	Jun 1995
Howes AT160, VF160 & MA4 (kits)	May 1997
Howes DC2000 (receiver kit)	Jul 1997
Howes DXR20 (receiver kit)	May 1995
Howes Tx2000 (transmitter kit)	Mar 1998
hupRF DG8 preamp and DCI-V bias injector	Oct 2016
I	
ICEPAK (propagation prediction software)	Jun 2000
Icom IC-207H (2m+70cm FM mobile)	Feb 1999
Icom IC-703 (HF-6m multimode portable)	Oct 2003
Icom IC-706 (HF-2m multimode mobile)	Nov 1995
Icom IC-706 Mk2 (differences from Mk1 version)	Jun 1997
Icom IC-705	Apr 2021
Icom IC-7100 (HF-70cm multimode)	Feb 2014
Icom IC-736 (HF+6m multimode base station)	May 1995
Icom IC-738 (HF multimode base station)	May 1995
Icom IC-746 (HF-2m multimode base station)	Mar 1998
Icom IC-756 (HF+6m multimode base station)	May 1997
Icom IC-756 Pro (HF-6m multimode base station)	Mar 2000
Icom IC-756 Pro II (HF-6m multimode base station)	Jun 2002
Icom IC-756 Pro III (HF-6m multimode base station)	Feb 2005
Icom IC-775DSP (HF multimode base station)	Jan 1996
Icom IC-821H (2m+70cm multimode base station)	Jan 1998
Icom IC-910 (2m-23cm multimode base station)	Jul 2001
Icom IC-7000 (HF-70cm multimode mobile)	Apr 2006
Icom IC-7200 (HF-6m multimode rugged /P/base)	Jan 2009
Icom IC-7300 (HF-6m multimode base station)	Aug 2016
Icom IC-7400 (HF-2m multimode base station)	Oct 2002
Icom IC-7410 (HF-6m multimode base station)	Jan 2012
Icom IC-7600 (HF-6m multimode base station)	Jun 2009
Icom IC-7700 (HF-6m multimode base station)	Jun 2008
Icom IC-7800 (HF-6m multimode base station)	May & Aug 2004
Icom IC-7851 HF-6m transceiver	Nov 2015
Icom IC-905 (2m to 10GHz transceiver	Sep 2023
Icom IC-9100 (HF-23cm base station)	Apr 2011
Icom IC-9100 (HF-23cm base station)	May 2012
Icom IC-E7 (2m+70cm handheld)	Jul 2006
Icom IC-E80 (2m+70cm FM + D-Star handheld)	Feb 2011
Icom IC-E90 (6m+2m+70cm FM handheld)	Dec 2004
Icom IC-E91 (2m+70cm FM handheld)	Jan 2007
Icom IC-E92 (2m+70cm FM + D-Star handheld)	Nov 2008
Icom IC-E2820 (2m+70cm FM + D-Star mobile)	Mar 2008
Icom IC-R3 (HF-microwave handheld receiver/TV)	Nov 2001
Icom IC-R20 (HF-microwave handheld receiver)	Sep 2004
Icom IC-T3H (2m handheld)	Jul 2002
Icom IC-T7E (2m+70cm handheld)	Apr 1997
Icom IC-T70E (2m+70cm handheld)	Nov 2010
Icom IC-T81E (6m-23cm FM handheld)	Sep 2000
Icom IC-V80E (2m handheld)	Nov 2010
Icom IC-V82 (2m FM/digital handheld)	Jan 2006
Icom IC-T10	Apr 2023
Icom ID-52E VHF/UHF D-Star transceiver,	Aug 2022
Icom ID-E880 (2m+70cm D-Star mobile)	Feb 2011
Icom PCR-1000 (remotable receiver)	Dec 1997
Idiom Press Rotor-EZ (rotator controller)	May 2001
IK Telecom DPRE4-6VL (cavity filters)	Jun 2009
Index Labs QRP+ (HF SSB/CW QRP transceiver)	Nov 1995
InnovAntennas 9-ele 2m LFA Yagi (antenna)	Mar 2012
InnovAnetnnas 5-element 15m OP-DES Yagi (antenna)	Aug 2012
Innovantennas 20/15/10 DESpole (antenna)	Dec 2013
INRAD W1 headset	Jan 2020
International Radio Roofing Filter (for FT-1000MP)	Jan 2005
I-PRO Home (multiband vertical dipole)	Jul 2011
I-PRO Traveller (portable vertical HF dipole)	Jun 2010
J	
JNC MC-750 portible antenna	Oct 2023
JPS ANC-4 (antenna noise canceller)	Aug 1996
JRC JST-245 (HF+6m multimode base station)	Oct 1997
JRC NRD-630 (professional MF/HF receiver)	Feb 2008
K	
Kanga Finningley (80m SDR kit)	Aug 2011
Kanga Foxx3 (QRP transceiver kit)	Sep 2010
Kanga QRP Pocket Transmatch 30	Jul 2023
Kenwood LF-30A (HF low pass filter)	May 2007
Kenwood TH-79E (2m+70cm handheld)	Apr 1997
Kenwood TH-D7E (2m+70cm FM handheld)	Sep 2000
Kenwood TH-F7E (2m+70cm FM handheld)	Dec 2004
Kenwood TH-D74E	Mar 2017
Kenwood TM-D710E (2m+70cm FM mobile)	Nov 2007
Kenwood TM-D710E + AvMap Geosat 5 Blu	Feb 2009
Kenwood TM-G707E (2m+70cm FM mobile)	Feb 199
Kenwood TS-480HX (HF+6m multimode mobile/base stn)	Mar 2004
Kenwood TS-570D (HF multimode base station)	Dec 1996
Kenwood TS-590S (HF+6m multimode base station)	Jan 2011
Kenwood TS-590SG (HF and 50MHz)	Mar 2015
Kenwood TS-870S (HF multimode base station)	Apr 1996
Kenwood TS-890S	Apr 2019
Kenwood TS-990S (HF+6m Flagship base station)	Jun 2013
Kenwood TS-2000 (HF-23cm multimode base station)	Apr 2001
Kinetics SBS-1eR (virtual radar)	Nov 2009
Kinetic SBS-3 (virtual radar)	Aug 2012
KK7P DSPx & KDSP-10 (DSP kit)	Jul 2005
Kuhne 23cm transverter	Dec 2007
Kuhne MKU23 G4 13cm band transverter	Sep 2016
Kuhne MKU LNA 131A HEMT (23cm preamp)	Jan 2008
Kuhne MKU10 G3 (10GHz transverter)	May 2008
Kuhne MKU 432 G2 (70cm transverter)	Apr 2011
Kuhne 2400MHz upconverter & 10GHz downconverter	Feb 2020
L	
Lake DTR7-5 (HF QRP transceiver)	Oct 1995
Lake Novice Receiver (kit)	Feb 2001
LAMCO DU1500L (HF ATU)	Feb 2012
LAMCO DU1500T (HF ATU)	Feb 2012
LDG AT-100 (automatic ATU)	Feb 2006
LDG Z11 (automatic ATU kit)	Aug 2002

LDG Z11 Pro (automatic ATU)	Jun 2009
Linear Amp UK Challenger (HF linear)	Feb 1997
Linear Amp UK Discovery 64 (6m/4m linear)	Apr 2010
Linear Amp UK Explorer 1200 (HF linear)	Feb 1997
Linear Amp UK Ranger 811 kit (HF linear)	Sep 2004
Little Tarheel II (80m-6m mobile antenna)	Aug 2008
LimeRFE front end and PA for LimeSDR	May 2022
Low cost digital oscilloscope kit	May 2019
Logger (station logkeeping software)	Aug 2000
Log-EQF (station logkeeping software)	Aug 2000
M	
M2 LEO pack 2m/70xm satellite antenna system	Nov 2022
Maha MH-C777 Plus-II (intelligent battery charger)	Jun 2007
Maldol HF, VHF & UHF mobiles (antennas)	Jan 2003
Maldol HVU-8 (80m-70cm vertical antenna)	Jan 2004
Maldol MFB-300 (broadband vertical antenna)	Apr 2007
Mastrant rope and antenna equipment	Sept 2020
MFJ-226 (graphical impedance analyser)	Nov 2015
MFJ-260C (1-650MHz 300W dummy load)	Feb 2008
MFJ-269 (HF-UHF SWR analyser)	May 2000
MFJ-270 (lightning arrestor)	May 2006
MFJ-393 (headphones with boom mic)	Mar 2006
MFJ-402 Nano keyer	Mar 2019
MFJ-461 (pocket Morse code reader)	May 2002
MFJ-704 (HF low pass filter)	May 2007
MFJ-781 (multimode DSP data filter)	Sep 1997
MFJ-784B (tunable DSP filter)	Feb 1996
MFJ-805 (RF current meter)	Jul 2006
MFJ-817C (2m-70cm SWR/power meter)	Jun 2006
MFJ-854 (RF current meter)	Jul 2006
MFJ-860 (HF power/SWR meter)	Oct 2006
MFJ-890 (DX beacon monitor)	Jul 2003
MFJ-935 (small loop tuner)	Apr 2005
MFJ-939	Mar 2016
MFJ-935B (small loop tuner)	Apr 2006
MFJ-936 (small loop tuner)	Apr 2005
MFJ-941 Tuner	Oct 2020
MFJ-989D (HF QRO manual ATU)	Jan 2007
MFJ-991B (automatic ATU)	Feb 2006
MFJ-993B (automatic ATU)	Jun 2009
MFJ-1234 RigPi Radio Station Server	Jan 2021
MFJ-1702C (antenna switch)	Jul 2007
MFJ-1786X Super Hi-Q Loop (HF loop antenna)	May 2007
MFJ-1897 (HF vertical antenna)	Sep 1995
MFJ 9020, 9420, 9140 & 9040 (HF QRP transceivers)	Oct 1995
MFJ 'Cub' (HF QRP transceiver kit)	Feb 2001
Microham Micro Keyer II (multimode digital interface)	Jan 2008
Microham Station Master & Six Switch (antenna switch)	Apr 2009
MicroKeyer (PIC-based electronic keyer)	Mar 1996
Microset CF-300 (DC-1GHz 300W dummy load)	Feb 2008
Microset PTS-124 (13.8V DC power supply)	Dec 2005
Microtelecom Perseus (software defined receiver)	Mar 2008
Microtelecom Perseus (software defined receiver)	May 2010
Miniprop Plus (propagation prediction software)	Jun 2000
miniVNA Tiny (network analyser)	Feb 2015
Miracle Ducker (antenna for FT-817 etc)	Nov 2004
Miracle Whip (antenna for FT-817 etc)	Feb 2002
Mizuho MX-14S (HF QRP transceiver)	Oct 1995
Moonraker HT-90E (2m FM handheld)	Sep 2011
Moonraker MT-270M (2m/70cm FM Mobile)	Dec 2015
Moonraker SPX-200 (HF-6m mobile antenna)	Jan 2007
Moonraker HT-500D (2m/70cm FM Mobile)	Oct 2017
Moonraker SHK-1 Straight hand key	Dec 2022
Morphy Richards 27024 (domestic DAB+DRM receiver)	May 2008
MW0JZE seven-band wide-spaced Hexbeam	July 2015
MyDEL AnyTone AT-5189 (4m FM mobile)	Apr 2011
MyDEL CG3000 (HF auto ATU)	Nov 2006
MyDEL HB-1A (compact 3-band QRP CW transceiver)	Oct 2010
MyDEL Multi-Trap Dipole (HF antenna)	Dec 1995
MyDEL SB-2000 (radio interface)	Sep 2010
MyDel SWR-001 & SWR-006 (SWR & Power Meters)	Aug 2017
MyDel JPC-7 and JPC-12 portable antenna kit	Oct 2022
MYDEL Windcamp Gipsy HF antenna	Feb 2023
M0CVO HW-20HP (off-centre fed dipole)	Jan 2012
N	
NA (contest logging software)	Nov 1999
nanoVNA-H series Vector Network Analyser	Feb 2021
O	
Optibeam OB9-5 (9-ele 5-band HF beam)	Aug 2003
Optibeam OB10-3W (10-element 20/17/15m beam)	Mar 2005
Optibeam OB10-5 (10-element 5-band HF beam)	Mar 2007
Outbacker Joey (QRP HF portable antenna)	May 2008
P	
Palstar AT1KM (HF ATU)	May 2005
Palstar AT1500CV (HF QRO manual ATU)	May 2005
Palstar AT1500CV (HF QRO manual ATU)	Jan 2007
Palstar AT-AUTO (HF auto ATU)	Sep 2006
Palstar KH-6 (6m FM handheld)	Oct 1997
Palstar ZM30 (digital antenna impedance bridge)	Sep 2005
Peak Electronics 'Atlas' Analysers (L, C & R meters)	Sep 2005
Peak Atlas DCA PRO (semiconductor analyser)	Mar 2013
Peak Atlas ZEN50 intelligent Zener diode tester	July 2015
Peet Bros Ultimeter 2100 (weather station)	Jun 2008
Piccolo 6m Transceiver Kit (synthesised FM)	Jun 1996
Pico Balun and Pico Tuner kits	July 2016
Pixie QRP transceiver kit	Jan 2016
PJ-80 (DF receiver)	Feb 2006
PJ-80 (DF receiver)	Apr 2007
PK-4 (Auto CW Pocket Keyer)	Nov 2015
PocketDigi (datamode software for PDA/smartphone)	Dec 2006
PolyPhaser IS-50UX-C0 (lightning arrestor)	May 2006
PolyPhaser IS-B50HN-C2 (lightning arrestor)	May 2006
PolyPhaser VHF50HN (lightning arrestor)	May 2006
Powerex MH-C9000 (charger for AA and AAA cells)	May 2010
Prepp Comm DMX-40 CW transceiver	Apr 2022
Procom DPF 2/33 & DPF 70/6 (2m & 70cm duplexers)	Sep 2009
Primetec Primesat Controller (satellite stn controller)	Dec 2005
Pro-Am HF Mobile Antennas (mobile whips)	Aug 1995
Pro-Am MM-3401 (mobile antenna mag-mount)	Aug 1996
Pro Antennas DMV Pro (portable antenna system)	May 2009
Pro Antennas Dual Beam Pro (5-band non-resonant beam)	May 2011
Pro Antennas I-pro Traveller (portable antenna system)	Jun 2010
Q	
Qpak Precision Tuner (mini ATU)	Oct 2004
QRP Labs QCX-mini QRP transceiver kit	Oct 2022
QDX Digital Transceiver	Dec 2022
R	
Red Pitaya STEMlab oscilloscope	Oct 2018
Ranger RCI-2950DX (10m/12m multimode mobile)	Feb 2012
RFSpace SDR-IQ (software defined receiver)	Mar 2008
RF Explorer spectrum analyser WSUB1	Oct 2021
RFinder B1 Dual Band DMR and analogue transceiver	Jan 2021
RigExpert AA-1000 antenna analyser (Antennas)	Aug 2012
RIGblaster Advantage (PC-radio interface)	Mar 2012
RigExpert TI-5000 Interface	Jun 2022
RIGrunner 4005 (12V distribution panel)	Mar 2012
Rigol DS2000 series (oscilloscope)	Aug 2013
Rigol DSA815-TG (spectrum analyser)	May 2013
Rigol DSA815 and RSA3030 spectrum analysers	Mar 2019
Rig Expert AA200 (antenna analyser)	May 2008
Rock-Mite QRP (40m Xtal controlled Kit)	May 2014
Rohde & Schwarz FSH3 (spectrum analyser)	Sep 2004
RT Systems AT-878 Programmer software	Sep 2023
S	
Samlex SEC-1223 (13.8V DC power supply)	Dec 2005
SD (contest logging software)	Nov 1999
SDR play	Mar 2016
SDRplay RSP2	Apr 2017
SDRplay RSPduo	July 2018
SDRPlay RSPdx	Jan 2020
SGC ADSP2 (noise eliminating speaker)	Nov 2003
SGC ADSP2 Mk2 (noise eliminating speaker)	Nov 2005
SGC SG-211 (internally powered HF-6m auto ATU)	Jun 2005
SGC SG-231 (HF-6m auto ATU)	Feb 2000
SGC SG-239 (budget HF auto ATU)	Jun 2005
SGC SG-500 (HF linear)	Dec 2005
SGC SG-2020 (HF 20W SSB/CW transceiver)	Mar 1999
SGC Stealth Kit (antenna)	May 2003
SG-Lab 2.3GHz transverter	Jan 2017
Shacklog (logging software)	Jan 1995
Shacklog (station logkeeping software)	Aug 2000
Sharman multiCOM AV-6075NF PSU	Jul 2022
Sharman AV-508 desktop microphone	Sep 2022
Sharman HLP-270 Halo antenna	Nov 2022
Shure 522 (desk microphone)	Jun 2007
Shure 550L (desk microphone)	Jun 2007
Shure 572B (fist microphone)	Jun 2007

Siglent SDG1062X waveform generator	Jan 2020
Siglent SDS 1202X-E oscilloscope	Nov 2019
Siglent SVA-1032X spectrum & network analyser	Jul 2020
Signal Hound SA44B (spectrum analyser)	Sep 2011
Signal Hound TG44 (tracking generator)	Sep 2011
SkySweeper (datamodes software)	Apr 2005
SkySweeper professional (datamodes software)	Feb 2008
Softrock V6.2 (HF SDR receiver)	Mar 2007
Sony ICF-SW100E (mini communications receiver)	Jul 1996
SOTA Beams 2m Portable Yagi (antenna)	Jul 2004
SOTAbeams LASERBEAM-DUAL CW filter	Nov 2016
SOTAbeams aerials, log book & battery monitor	Oct 2013
SOTAbeams WOLFWAVE	Oct 2019
SOTAbeams WSPRlite Antenna Tester	Jun 2017
SOTABEAMS 2m bandpass filter	Dec 2022
SPE Expert 1K-FA (linear amplifier)	Jun 2007
SPE Expert 1.3K-FA linear amplifier	July 2015
Spectran HF-6085 (hand-held spectrum analyser)	Jan 2010
Spiderbeam (5-band antenna kit)	Mar 2010
Spiderbeam 160-18-4WTH (160m vertical antenna)	Sep 2013
Spiderbeam Aerial-51 Model 404-UL (port ant 40m-6m)	May 2015
SRW CobbWebb (HF antenna)	Jun 1993
Standard C108 (2m FM handheld)	Jan 1997
Standard C408 (70cm FM handheld)	Jan 1997
Standard C568 (2m+70cm handheld)	Apr 1997
Standard C5900D (6m+2m+70cm FM mobile)	Jul 1997
StationMaster (station logkeeping software)	Aug 2000
SteppIR 3-element Yagi (14-52MHz beam)	Feb 2004
SteppIR 4-element Yagi (14-52MHz beam)	Oct 2007
SV4401A handheld vector network analyser,	Jan 2023
SunSDR2 PRO (HF to VHF transceiver)	Dec 2015
Super Antenna MP-1 (portable HF-VHF dipole)	Jun 2008
Super Keyer 3 (electronic keyer)	Jan 1997
Super Antenna MP1B Super-Stick	Jan 2013
T	
Talksafe (Bluetooth adapter)	Sep 2007
Telecom 23CM150 (23cm linear amplifier)	Oct 2009
Tennadyne T10 (13-30MHz log periodic HF beam)	Jan 2002
Teensy SWR/Power meter	Nov 2020
TenTec Argo (HF QRP transceiver)	Oct 1995
TenTec Eagle 599 (HF-6m compact multimode base stn)	Jul 2011
TenTec Jupiter 538 (HF multimode base station)	Jan 2004
TenTec Omvi VII (HF-6m multimode base station)	Sep 2007
TenTec Orion 565 (HF multimode base station)	Jun 2004
TenTec Orion 2 566 (HF multimode base station)	Aug 2006
TenTec 506 Rebel (CW QRP transceiver)	Dec 2015
TenTec RX340 (professional HF DSP receiver)	Mar 2002
TenTec 1320 (20m CW transceiver kit)	Aug 2000
The tinySA Spectrum analyser	Mar 2022
Thamway TX-2200A (136kHz transmitter)	Feb 2010
Timewave DSP-59+ (tunable DSP filter)	Feb 1996
Timewave TZ-900 (antenna analyser)	Nov 2009
Tokyo Hy-Power HL-1KFX (HF linear amplifier)	Oct 2005
Tokyo Hy-Power HL-2KFX (HF linear amplifier)	Oct 2005
Tokyo Hy-Power HL-50B (linear amplifier)	Sep 2002
Tokyo Hy-Power HL-100BDX (HF linear amplifier)	Oct 2005
Toyocom MS-5 (mobile hands-free kit)	Mar 2009
Trident 6M5L (6m long yagi antenna)	Jun 2003
TROPIC T-R (sequencer)	Aug 2012
TRlog (contest logging software)	Nov 1999
Turbolog III (software)	Nov 1997
Turbolog (station logkeeping software)	Aug 2000
TYT TH-UVF1 (2m+70cm handheld)	Dec 2010
Tytera MD380 DMR handheld	July 2016
TYT TH-UFV9 (dual band handheld)	Dec 2012
U	
uBITX V6 QRP transceiver	Dec 2020
V	
Vargarda 11EL2 (2m beam)	Mar 2005
Vectronics 1010K (10m FM receiver kit)	Feb 2001
Vectronics DL-300M (DC-150MHz 300W dummy load)	Feb 2008
Vectronics LP-30 (HF low pass filter)	May 2007
Vero VR-N7500 Dual band radio	Aug 2022
Videologic DRX-601E/ES (digital radio tuner)	Oct 2001
Vine Antennas LFA Yagis (loop fed VHF antennas)	Nov 2009
Vortex Whirlwind 6M4(6m delto loop)	Dec 2011
W	
Walford Berrow (QRP transceiver kit)	Apr 2014
Walford Brent (single band CW transceiver kit)	Feb 2005
Walford Compton (direct conversion receiver kit)	Jun 2002
Walford Langport (80m+20m CW+SSB transceiver kit)	Apr 2000
Walford Radio Today Chedzoy (receiver kit)	Feb 2001
Watson AT-715 (five-in-one power station)	Nov 2009
Watson CS-600 (antenna switch)	Jul 2007
Watson PBX-100 (portable HF antenna)	Nov 2002
Watson Power-Mite-NF (power supply)	Jun 2008 & Sept 2021
Watson SP-350V (lightning arrestor)	May 2006
Watson VAA-1 (antenna analyser)	Oct 2015
Watson W-25SM (13.8V DC power supply)	Dec 2005
Watson W-184 (headphones with boom mic)	Mar 2006
Watson W-8682 (radio controlled weather centre)	Sep 2008
Watson WM-S (mobile microphone system)	Nov 2005
Wavecom W-Code (datamodes decoding software)	Jul 2012
Wedmore 80m QRP Transceiver (kit)	Aug 1997
Wellbrook ALA1530 (active receiving loop)	Jan 2012
Whistler TRX-1 and TRX-2 scanners	Aug 2020
Wiber Mini 1300 digital antenna analyser	Feb 2022
WinCAP Wizard II (propagation prediction software)	Jun 2000
WinCAP Wizard III (propagation prediction software)	Dec 2002
WinRadio WR-G1DDC Excalibur (SDR receiver)	Oct 2010
WinRadio WR-G313i (PC-controlled receiver)	Mar 2005
Wimo Big-Wheel antenna	Jan 2016
Wonder Wand (antenna for FT-817 etc)	Jun 2004
Wouxun KG-699E (4m handheld)	Sep 2010
Wouxun KG-UV6D Pro Pack (2m/70cm handheld)	Jun 2012
Wouxun KG-UVD1P (2m/70cm handheld)	Sep 2010
Wouxan KG-UV8H 70/144 and 144/430MHz handhelds	Dec 2020
Wouxun tri-band VHF handhelds	Nov 2023
WriteLog (contest logging software)	Nov 1999
WSPR Desk Transmitter	Feb 2022
W2IYH (range of audio products)	Jan 2009
X	
Xiegu X5105 HF-6M QRP transceiver	Nov 2018
Xiegu G90 HF multi-mode transceiver	Feb 2020
Xiegu X6100 HF and 50MHz SDR transceiver	Jul 2022
Xiegu GNR1 digital noise reduction and filter	Oct 2022
Xiegu G106 HF transceiver	Nov 2022
Y	
Yaesu ATAS-100 (mobile antenna)	May 1999
Yaesu FT-5DE 144/430MHz C3FM/analogue handheld	Feb 2022
Yaesu FT-50R (2m+70cm handheld)	Apr 1997
Yaesu FT-60R (2m+70cm handheld)	Dec 2004
Yaesu FT-100 (HF-70cm multimode mobile)	Jun 1999
Yaesu FT-450 (HF-6m compact multimode base stn)	Oct 2007
Yaesu FT-450D (HF-6m compact multimode base stn)	Nov 2011
Yaesu FT-710 (HF, 50 and 70MHz transceiver)	Jun 2023
Yaesu FT-817 (160m-70cm multimode portable)	Jun 2001
Yaesu FT-847 (HF-70cm multimode base station)	Aug 1998
Yaesu FT-857 (HF-70cm mobile)	Jun 2003
Yaesu FT-891 (HF+6m multimode)	Mar 2017
Yaesu FT-897 (HF-70cm multimode /P/base station)	Apr 2003
Yaesu FT-920 (HF+6m multimode base station)	Aug 1997
Yaesu FT-950 (HF+6m multimode base station)	Dec 2007
Yaesu FT-991 (HF/VHF/UHF transceiver)	Feb 2016
Yaesu FT-1000MP (HF multimode base station)	Jan 1996
Yaesu FT-1000MP Mark-V (HF multimode base stn)	Oct 2000
Yaesu FT-1000MP Mark-V Field (HF multimode base)	Oct 2002
Yaesu FT-2000 (firmware upgrade)	May 2009
Yaesu FT-2000D (HF-6m multimode base station)	Mar 2008
Yaesu FT-8100R (2m+70cm FM mobile)	Feb 1999
Yaesu FTDX101D	Oct 2019
Yaesu FT-DX1200 (HF-6m multimode base station)	Mar 2014
Yaesu FT-DX3000 (HF-6m multimode base station)	Jan 2014
Yaesu FT-DX5000 (HF-6m multimode base station)	Jun 2010
Yaesu FT-DX9000D (HF-6m multimode base station)	Oct 2005
Yaesu FT-DX9000D (HF-6m multimode base station)	Dec 2006
Yaesu FTM-10R (2m+70cm FM mobile)	Feb 2008
Yaesu FTM-400DE dual band, dual mode mobile	Jan 2015
Yaesu FTM300 dual band mobile	Feb 2021
Yaesu VR-5000 (HF-microwave multimode receiver)	Aug 2001
Yaesu VX-5R (6m/2m/70cm FM handheld)	Sep 2000
Yaesu VX-7R (6m/2m/70cm FM handheld)	Oct 2003
YouKits FG-01 Antenna Analyser	Sep 2012
YP-3 (6-band 3-ele portable yagi)	May 2009
Z	
Zeus ZS-1 (SDR HF transceiver)	Dec 2013
ZUMspot RPi Digital Voice Hotspot	Mar 2021
ZUMspot-RPi Elite 3.5" LCD	May 2023

Radio Communications Foundation

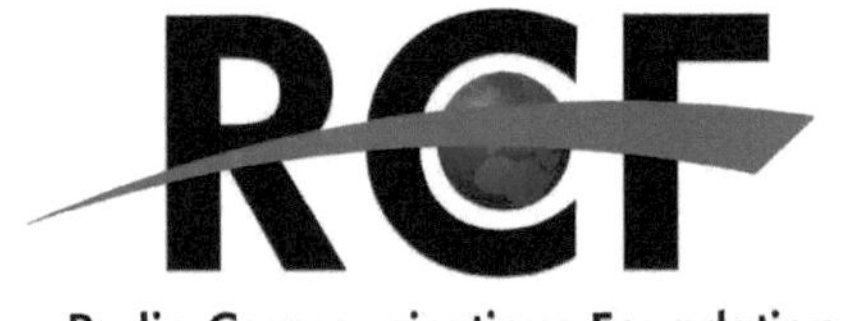

We were established in 2002 by the RSGB and formally incorporated as a Registered Charity in 2003. Although the RCF was established by the RSGB we are independent and we work to create a fund which can support efforts to heighten awareness of the importance of radio communications.

Mission

Our Trust Deed sets out our objective as "to advance the education of the public in the science and practice of radio communications and to promote the wider benefits to the public resulting from such education and training". In practical terms this means that we seek to increase the engagement of people, especially young people, in radio communications technology. We therefore encourage and assist students to pursue relevant higher education courses, leading to them being employed in the radio communications sector, and also work to raise public interest and involvement in radio communications and its associated science and technologies, including amateur radio.

The interested young person of today is the radio amateur of tomorrow and the engineer of the future.

Near and medium term objectives:

The objectives are to advance the RCF mission by engaging with four key groups:

- Those at school (and also those in uniformed groups such as the Scouts, Guides, Army Cadet Force, Royal Air Force Air Cadets, Air Training Corps and Ambulance Brigades) in order to develop an interest in radio communications
- Those planning or undertaking a university or higher education course to encourage the study of radio communications options
- Those planning or considering employment in radio communications
- The public more generally, including those with an interest in amateur radio

Radio communication is so widely practised that it is almost taken for granted, but a greater public understanding of it is vital for the UK economy. There is a serious shortage of radio communications scientists, engineers and technicians, all needed to exploit a myriad of commercial opportunities.

Radio communication is one of the vital technologies for the 21st century. It provides the backbone technology for the information economy. Every member of the public uses it. Many innovations were developed by scientists and engineers who had their interest aroused by a hands-on demonstration, perhaps at school, perhaps at an exhibition, perhaps through a demonstration by a radio amateur. Amateur radio is an underlining supporting strategy for us, and we see encouraging young people into amateur radio and helping them develop as not only supporting the longer term health of amateur radio itself, but directly underpinning achievement of our objectives.

Fund Raising

To the end of 2023, we had raised over £400k. That money came from:

- Members of the RSGB. Many members already make donations, some with their membership renewals, and donations can be increased when Gift Aid is applied to them
- Through bequests. Members can make bequests to the Foundation in their Wills and the Foundation rigorously respects any instructions made in a legacy
- Fund raising or club events
- Approaches to industry and public sector sources of grant aided money. We seek to work with industry by raising public awareness of the opportunities for jobs and careers in radio

Projects

The following are the sort of projects and activities the Foundation has supported in the past:

Grants to individual clubs or educational institutions

In recent years we have made grants to a wide range of people and groups. A school teacher highlighted that the exam fee was preventing some young people from taking

Two of the youngsters who passed their RCF funded exams.

their Foundation examination. The charity therefore launched an initiative to fund a number of exams each year for those in full time education. Applications need the support of a parent or guardian, or for older young people, a tutor can provide the supporting statement.

The science teacher at Wick Highschool secured a grant for some radio equipment and some exams for pupils at his school. The RCF funded a number of handhelds for pupils to borrow once they have passed their exams.

We have funded a number of workshops held at Bletchley Park with hands-on help from RSGB volunteers. These have proved very popular and the parents who attend are amazed at how their children become so engaged in soldering and building radios, or Morse tutors.

In 2024 we agreed to support a radio-astronomy project at Lancaster University. Undergraduates there are to study radio signals emitted by Jupiter as part of a NASA initiative.

The hope is that some of these young folk who would not otherwise have been exposed to radio communications and electronics will discover the fun and learning that can be had via amateur radio.

A happy young radio constructor at Bletchley Park

Details of how to apply for an RCF grant are set out on the website.

RCF Scholars

2023 Arkwright Scholars received their RCF sponsorships at a ceremony at the IET

Bursaries and scholarships

In partnership with the Smallpeice Trust, we continued to sponsor Arkwright Engineering Scholars. These students are approaching higher education and thinking about what universities they would like to attend. They have undergone a rigorous interview and selection process to become an Arkwright scholar and already boast impressive CVs. One of our Scholars told us about an induction powered projectile launcher project he was undertaking and several are involved in national robotics championships. Two of our past Scholars have secured engineering apprenticeships with James Dyson's company and a number have been selected to represent the UK at international Youngsters on the Air (YOTA) summer camps.

https://commsfoundation.org/arkwrightscholarships/rcf-arkwright-scholars/

We partner with the UK Engineering Skills Foundation to promote a competition at universities to recognise the best final year projects in the field of radio communications.

https://commsfoundation.org/ projects-2/ukesf-rcf-partnership/

The first prize winner in 2018 was a radio amateur who is now working in satellite communications. This illustrates the benefits of learning through amateur radio as a route to a rewarding career.

Furthermore, we can help fund students' university projects with grants towards test equipment or software. More details can be found on our website:

www.commsfoundation.org

How to donate

There are several ways:

- Through RSGB membership renewals
- Through the RSGB shop - *http://www.rsgbshop.org/acatalog/RCF_Donations.html*
- By the payroll giving scheme for Charities, which some employers offer as a route for regular donations
- By a bequest in a Will, so that an interest in amateur radio lives on for the benefit of others

Gift Aid

For every £1 you give to us, we get an extra 25p from the Inland Revenue. You must pay an amount of income tax and/or capital gains tax at least equal to the tax that the charity reclaims on your donations in the tax year for the RCF to receive this.

RCF Trustees

Registered charity number 1100694

http://commsfoundation.org/

Radio Communications Foundation,
3 Abbey Court, Fraser Road,
Bedford MK44 3WH

Website: *www.commsfoundation.org*

For more information,
email: *secretary@commsfoundation.org*

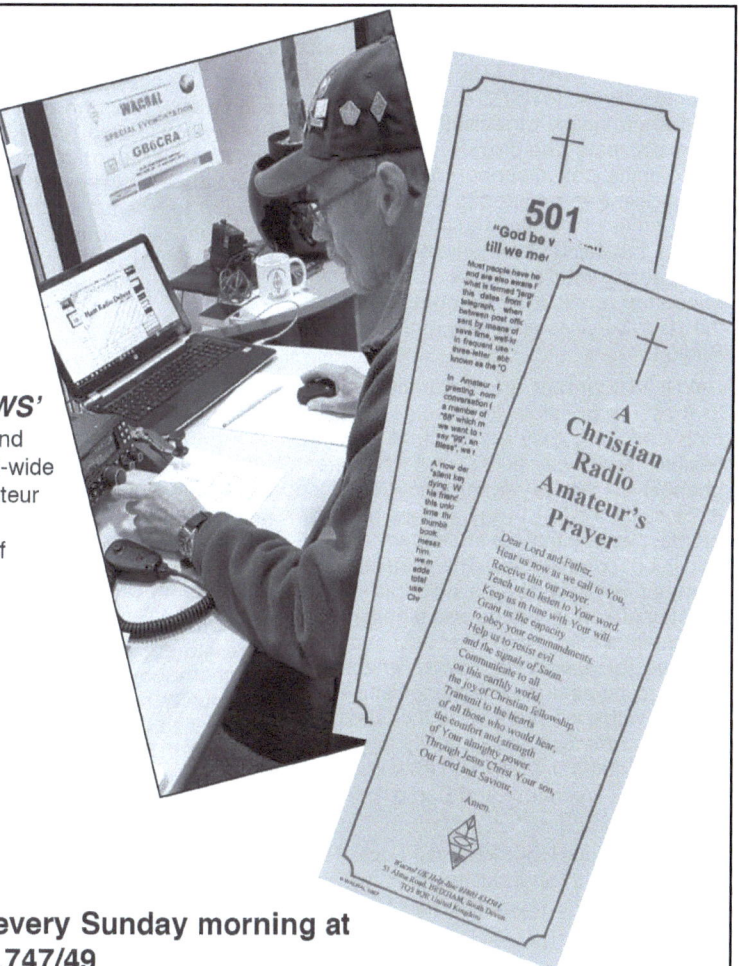

Public Service and Emergency Comms

RAYNET-UK

The Amateur Radio Service frequently provides services to the community in times of need, even when some think that technology has improved to the point where we are not needed, there are always situations where radio amateurs are in the right place at the right time to get a message through . Even the ITU continues to recommend that Amateur Radio should be involved in such events, because we are not dependent on any particular infrastructure. To provide the best services to the community it helps to be trained, recognised and organised and this is where RAYNET-UK comes in as an affiliated organisation to the RSGB.

Emergencies

Natural disasters of various scales occur all the time. Unfortunately, both natural and man-made events appear to be increasing in frequency with significant consequences on the communities affected but because media coverage has improved it feels like they are more 'business as usual' now. No matter whether the cause is severe weather or loss of power the quickest way for any crisis to turn into a disaster is to lose communications. The normal networks used by the public and the emergency services are vulnerable in varying degrees to loss of power, fallen pylons etc and it is here that the Radio Amateur with the flexibility and independence from other networks can provide valuable support to our services and communities with the many frequencies and tools available to us.

Amateur Radio emergency communications in the UK are usually provided by organised groups under the generic name of 'RAYNET'. RAYNET-UK is a Registered Charity, affiliated to RSGB and recognised as the UK's principal organisation comprising radio amateurs who provide voluntary radio communications in support of the activities of the User Services and to local communities in times of disaster and emergencies as well as providing communications support to local community events. It is the main conduit of amateur radio representation to the User Services specified in the Amateur Radio Licence.

RAYNET-UK groups are an attractive option to those responsible for emergency planning - because they are quick to respond, flexible, technically skilled, and with current changes to the way emergency resilience is being looked at in the UK we have the greatest benefit that we are already embedded in our communities. The vulnerability of communities to wide area incidents such as a National Power Outage was highlighted in a sequence of exercises last year and there is more focus now on how to enable the public to increase their resilience and ability to cope with such events. This includes guidance to communities to identify alternative means of communications in their area such as RAYNET operators and Radio Amateurs in general.

RAYNET groups work with their Local Resilience Forums throughout the country. These forums have the responsibility of planning 'resilience' or 'the ability to bounce back quickly' in the event of any type of disaster, co-ordinating the professional and volunteer response. They value dealing with organisations like RAYNET-UK who provide a single point of contact that can be tasked with dealing with a particular issue and whose members are a 'known quantity' providing a particular service.

This all combines with increasing collaboration between RAYNET-UK Groups and '4x4 Responders', Search and Rescue groups etc. to provide a service to the authorities and the communities when needed. This has added benefits for the hobby with some 4x4 group members or members of User Services taking out amateur licenses to increase their skill level and the service they can offer.

Community Events

Thankfully the number of disasters and emergencies in the UK is relatively low but this brings challenges in maintaining skills and training. Recognising this the Amateur Radio licence was relaxed many decades ago to allow Radio Amateurs to support our User Services by providing communications support to a range of local events such as fun runs, overnight hikes etc. These are used by RAYNET-UK groups as a means of practising procedures, testing equipment and honing skills, etc. They also provide a 'shop window' for Amateur Radio to demonstrate our abilities to Emergency Planning Officers, Police Officers etc. and also for RAYNET to learn from those services about how they approach situations. While any Radio Amateur or group can undertake these activities, being part of a larger organisation allows the most benefit to be obtained from the (hopefully) close relationships built over years with the Emergency Services as well as the Event Organisers.

The amateur radio licence changes which were introduced in 2024 increased the facilities the hobby can offer as a whole to the community, not just to send traffic on behalf of specified bodies, still referred to as 'User Services' in paragraph 10(hh) of the licence. The licence also liberalised the delivery of temporary repeaters, data stations etc. and these may be heard on the bands under an MB9xxx callsign when used on events by RAYNET groups.

However, there is a limit to what an individual amateur can achieve, so most will join local clubs or groups and pool their equipment and skills. All amateur radio operators are responsible for the safe operation of their station when in a public area.

RAYNET-UK groups enjoy the protection of public liability insurance, but such cover is not to be taken for granted and may be invalidated if carelessness is proven.

While this kind of activity was once intended to be a 'fill in' for RAYNET groups, it now forms the majority of some groups activities and is an excellent way of demonstrating the value and flexibility of our hobby to the general public.

No limits?

Emergencies know no boundaries and technically we should not limit ourselves in our response. While voice communications will always be the backbone of communication, our user services increasingly look for the ability to send email, pictures etc. just as they would do on a 'normal' day. With the launch of Es'Hail-2 providing geostationary satellite capabilities , the sky is now literally the limit in what amateur radio can provide. Groups are increasingly looking to build (or in many cases rebuild) independent data networks to meet the needs of local organisations . If you have an interest in data, video or microwaves, please get in touch with a local group and bring your specialist skills to emergency communications.

Since we may be asked to provide communications ranging from the very local to international, IARU and UK band plans allocate the following frequencies to emergency communications. VHF/UHF frequencies tend to be 'spot' frequencies while at HF they are defined as 'Centres of Activity' since no one, even emergency communications has an absolute right to a frequency. However given the potential urgency of the communications, and that our User Services are likely to be listening, the co-operation of all Radio Amateurs is requested to allow a clear channel for emergency communications. Even on what may sound like the quietest of public service events, if something goes wrong an emergency message will need to get through quickly and without error:

Most RAYNET activity takes place on those bands for which equipment is readily available, namely 2m and 70cm, although use is also made of 4m, 6m and 23cm. More use

6m: 51.65-51.75, 51.77 & 51.79 MHz

4m: 70.350, 70.375, 70.400 MHz

2m: 144.625-144.675MHz (12.5kHz steps),144.775, 145.200, 145.225, 144.800 (APRS), 144.260 (SSB)

70cm: 433.700-433.775MHz (12.5kHz steps)

In-Band 7.6MHz split temporary repeaters

430.800 (mobile)

438.400 (base)

In-Band 1.6MHz split temporary repeaters

434.375 (mobile)

432.775 (base)

of HF is also being made to supplement VHF & UHF, particularly for distance and hostile terrain. Increasingly, both 10m and 6m are being used for cross-band talk-through, in addition to the more usual 2m and 70cm. bands. These temporary repeaters are identified by the MB9xxx callsigns mentioned earlier, issued through collaboration between RAYNET-UK and the RSGB to RAYNET groups.

RAYNET is not restricted to UK operation and the lessons learned from emergency groups are shared internationally. An annual meeting of emergency communications groups is held at Friedrichshafen and the presentations from these and the preceding GAREC (Global Amateur Radio Emergency Communications) conferences can be found through the IARU website as www.iaru-r1.org.

International frequencies:

21.360 MHz – Global

18.160 MHz – Global

14.300 MHz – Global

7.110 MHz – IARU Region 1

3.760 MHz – IARU Region 1

3.663 MHz - UK Only.

The organisation

RAYNET-UK is affiliated to the RSGB and there is regular liaison between the two, promoting a coordinated approach to emergency communications. As the RSGB recognises RAYNET-UK's specialist nature, in return the RSGB is recognised as the national body representing all UK radio amateurs and the main route of amateur radio representation to the UK licensing authorities and bodies such as CEPT and the ITU. The RSGB is also the representative society of radio amateurs throughout the United Kingdom to the International Amateur Radio Union (IARU) who also co-ordinate emergency communications activities between countries.

The RSGB co-ordinate matters relating to changes in amateur radio licensing or band plans where such changes have an effect on the functioning of communication in community events, emergencies and disaster situations.

There are around 100 groups affiliated to RAYNET-UK around the country and at the time of writing ~1600 members who are committed to using their hobby to help others. The organisation is administered by a Committee of Management, comprising Zonal Co-ordinators who represent the nations and regions of the UK. Each group or county has its own organisational structure meeting their local needs.

To cover administration costs a small annual charge is levied nationally on each member, and in return offers combined liability and personal accident insurance, discounted supplies, as well as technical and training backup. Local groups may choose to cover this cost from donations or may choose to ask for more to help build out resilient networks and get specialist equipment.

RAYNET-UK provides a wide range of services to support Members, their Groups and liaisons with our User Services. This support is delivered at Member, Group and National level and is organised by the Committee of Management.

Becoming Involved

If you feel you would like to become involved on a regular basis, then you should consider joining a RAYNET-UK Group. First of all, find a Group near to you and contact them. Groups usually suggest that you come along to a few events and observe , and then perhaps apply to join. In fact you do not necessarily need to have an amateur radio licence, as there are often plenty of jobs that non-licensed persons can do and you may get radio experience by using a PMR446 licence free radio or even CB. ID cards are issued to members to identify them to User Services as a member of RAYNET-UK. Some groups adopt a readily identified dress code, which is often appreciated by User Services, giving them a quick means of identifying members on site at an incident. RAYNET-UK operates a Supplies service for all their members.

RAYNET-UK Groups in the UK

Contact information for RAYNET Groups can be found at: *www.raynet-uk.net*

EMC (Electromagnetic Compatibility)

The RSGB can offer help to members on EMC matters through its EMC Committee, which consists of volunteers who have professional as well as amateur radio experience in the field of EMC.

Introduction

Operating an amateur radio station in the 21st century in an urban or suburban environment presents particular challenges. Not only may there be limited space for antennas but the presence nearby of other electronic devices can result in emissions raising the noise level on the amateur bands, as well as breakthrough from amateur transmissions into other devices. EMC, or 'Electromagnetic Compatibility' is the term used to describe the ability of devices to co-exist without excessive interaction.

Fortunately, cases of breakthrough from amateur transmissions are becoming less frequent, and by following the "Good Radio Housekeeping" guidance can generally be managed. See Avoiding Interference below.

On the other hand emissions from the increasing number of electronic devices in every home have led to a marked increase of cases of interference to reception.

Particular threats

Almost any electronic device has the potential to cause emissions of some sort. Most are benign and conform to relevant Standards, but some have significant potential to cause problems:

- PLT/Powerline devices
- xDSL wired internet
- Plasma TVs
- Switch-mode power supplies (SMPSU)
- PV Solar Panels
- Wind Farms

– Plus a plethora of other electronic devices, such as:

- Remote controlled lamps
- LED low voltage lighting modules
- RF-excited lighting modules

Many of the sources of interference are familiar to members. The RSGB, through members of its EMC Committee, is represented on international standards bodies working to achieve standards which should allow coexistence of electronic devices with radio communications systems.

The Social Side

Many complaints of EMC problems can only be solved with the active cooperation of both parties. This requires diplomacy and tact. Complaints from neighbours of interference may be related to environmental impact of antennas, so see the Planning Advice pages in this Yearbook and discuss planning issues with the Planning Advisory Committee. Interference problems are not often understood by complainants or the owners of the offending apparatus, so help them to understand - be a good radio neighbour and be sensitive to their point of view.

Data Transmission Systems Using Telephone Lines & Electricity Cables

Technologies which use the telephone lines and the electricity cables to carry high speed data signals have been a source of concern to radio amateurs for many years. These notes give a brief outline of the radio interference (RFI) threat that may be expected from the various technologies.

Dial-up modems

These use audio signals on the phone line and have now been almost completely superseded by DSLs and fibre optic links. There are a whole family of DSLs, but the only ones of interest to us are ADSL and VDSL.

ADSL (Asymmetric Digital Subscriber Line)

Techniques and Frequencies used

Generally up to 1.1MHz, but could be up to 2.2MHz. ADSL is usually fed into the phone line at the local exchange, which could be up to 5km from the customer's premises.

Deployment

It was widely deployed in the UK with many millions of customers. It has now been largely superseded by VDSL.

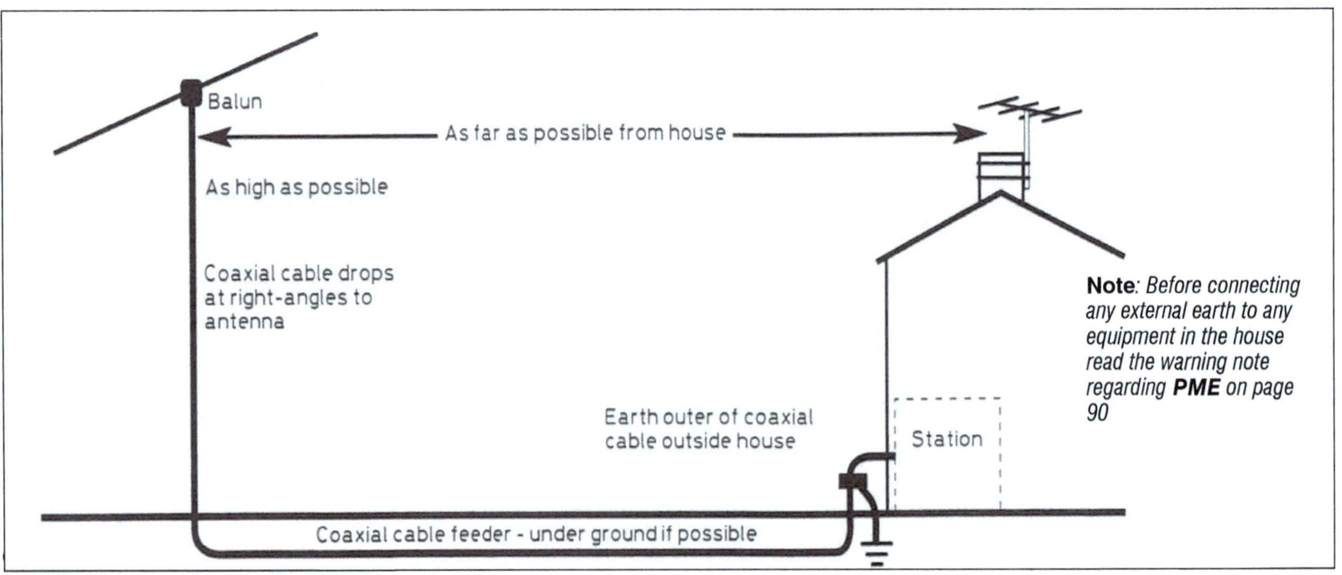

RSGB EMC Committee website: www.rsgb.org/main/technical/emc/

Interference Potential

In general, interference from ADSL is not a problem to amateur radio but there have been a number of reports of breakthrough to ADSL by amateur transmissions. More information can be found in the EMC Columns of RadCom.

VDSL (Very High Speed Digital Subscriber Line)

Techniques and Frequencies used

VDSL operates at up to about 17MHz and is launched into the telephone lines at the street cabinet (it is sometimes called FTTC Fibre-To-The-Cabinet). Since only a relatively short length of telephone cable from the street cabinet to the customer is involved (1km maximum), data speeds of up to 40Mb/s are possible.

Deployment

Deployment is well advanced in the UK and the service is now available in most urban and many rural locations.

Interference potential

Interference from VDSL is fairly common and usually takes the form of noise which sounds very much like white noise. This masks weak signals giving the effect of poor receiver sensitivity but with a higher than normal background noise reading on the "S" meter. Investigation and mitigation of VDSL interference has become a major issue for the EMC Committee. There is a large amount of information in the EMC column in RadCom and on the EMC web pages. Installations with underground connections seldom exhibit problems.

Identifying VDSL Interference

A key step is to identify whether the interference you may be experiencing, is from VDSL as opposed to other sources. The presence of VDSL interference can be proven using a software tool called 'Lelantos' developed by Dr Martin Sach, G8KDF—a member of the RSGB EMCC.

Lelantos is a stand-alone Microsoft Windows application. If you have an SDR receiver and a PC running Microsoft Windows, you can download it from the RSGB website. Further details of how Lelantos works can be found in the instruction manual and in a November 2018 RadCom article which is online along with the software at:

https://rsgb.org/main/technical/emc/vdsl-interference-reporting/

Systems using electricity cables

This is known as Power Line Telecommunications (PLT or PLC). In the USA it is usually known as BPL, Broadband over Power Lines. Low frequency signalling on the electricity mains has a long history, but so far as radio amateurs are concerned PLT refers to Internet access and computer networking.

In-House PLT

This makes use of modems which plug into a mains socket and communicate with one another via the electricity wiring in the house. The modems are called Power Line Adapters (PLAs).

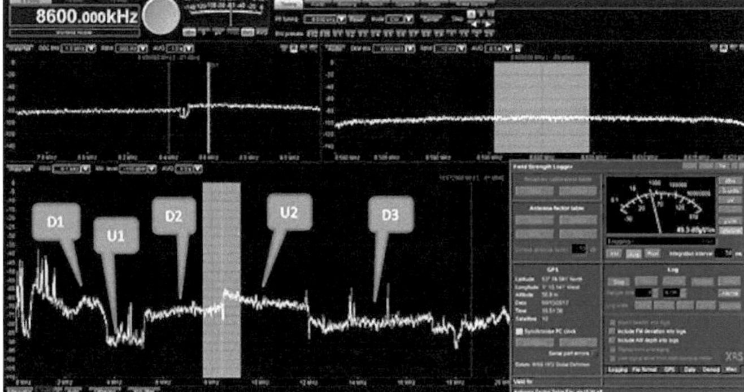

VDSL Interference Downlink and Uplink bands

Frequencies used

Systems vary, traditionally 4 to 28MHz. But newer devices go up to about 70MHz.

Deployment

Apart from computer networking, Power Line Adapters are widely used for video distribution in Internet TV systems (IPTV).

Interference potential

All PLAs reduce their launch power in the international amateur bands. This is known as 'notching'. This seems to be reasonably effective, though some filling due to intermodulation has been observed. Without the notching the interference on the amateur bands would be intolerable. Discussions on an EMC Standard specific to PLT have resulted in two new Standards EN50561-1 below 30MHz and EN50561-3 above 30MHz. Both seem to give a reasonable degree of protection to the amateur bands. This is too big a subject for these short notes and further information can be found on the RSGB website.

Other Potential Sources of Interference

The emphasis on preservation of the Environment has resulted in many schemes aimed at reducing the use of energy and harvesting of renewable sources. These have inevitably resulted in consequential environmental impact.

Solar panels

The Government incentives offered to house-holders and industrial users have encouraged many electrical power users to install Photo Voltaic (PV) panels on their roofs. These installations are a potential source of RFI. From the outset, it must be said that there are good RFI-free installations, and of course the converse is true.

An installation consists of the solar panels on the roof, and much more importantly an inverter, usually placed somewhere below the roof, which are connected by cabling. The inverter is the source of RFI and the cables are potentially the antenna that radiates the energy. The current UK Government and Ofcom view is that solar PV installations are comprised of separate items of apparatus rather than being an integrated fixed installation. The RSGB's view is that even so, an installer is responsible for ensuring the apparatus meets the EMC compliance requirements when the apparatus is first taken into service (see the section on the EMC Directive). In any case any member contemplating a Solar Energy Harvesting system should check that the installer understands the requirements of the EMC regulations. The industry has given some recognition to the potential RFI problems and lightweight inverters which can be installed within the roof space have been introduced. The interconnection between these is usually quite short and results in very low antenna efficiency, and low radiation. At the same time, much greater care has been taken to ensure that the leakage of RFI from the units is minimised. However, it is also true to say that the move towards so called 'transformer less' inverters has presented new challenges. These inverters, using solid state commutation to create the 50Hz AC signal, produce high frequency spikes which leak more readily from the unit housing.

Wind Farms

It is not necessary to travel very far in the UK to see a hilltop wind farm installation, and members have expressed concerns regarding how these will affect the amateur bands. An installation consists of the wind turbine itself and a complex control system at the base of the mast. There are a number of arrangements available for feeding 50Hz energy into the National Grid. Almost all of these involve complex electrical conversion of the voltages and current, with the inevitable switched mode power convertor

playing an important part.

The most probable cause of RFI from a wind farm is from the electrical control systems at the base of the tower, with once again the cables connected to the top acting as an antenna. Although these are usually screened within the metal structure, at ground level there may be feeds to the control systems that radiate.

The EMC Committee is gathering information from members and David Lauder will be making measurements on actual installations. These will be reported in his regular column.

Utility Services

A regular source of complaints to the EMC Committee comes from members who live in a rural, normally quiet location. An unexpected high noise level appears on the lower HF bands. In almost all cases the mains power feed is overhead.

As well as the possibility of arcing on the power line itself, frequently the cause has been found to be thyristor controlled motors installed in pumping stations operated by the water or sewage utility. Overhead power lines accentuate the radiation, acting as long wire antennas. Fortunately, the RFI is evident on MW broadcast stations, and is easily demonstrated with a portable radio as coming from an enclosure housing a pump. The advice from the EMCC has been to contact the Utility company who will usually be sympathetic to the problem.

Putting the RFI in context

Background noise on the HF bands

How much noise would one expect in a typical residential location?

Situations where there are continuous, high level broadband sources of interference are unusual in residential areas, though they are common in industrial/commercial premises. In residential locations broadband noise is usually relatively low, with occasional periods of high level noise. In addition there may be high levels of narrowband interference on specific frequencies. Where there is continuous broadband noise in a residential location it is likely to be something specific like an alarm system or some device such as a switch mode power supply.

There is an EMC Leaflet: EMC 16 Background Noise on the HF Bands. This can be found along with the other EMC Leaflets on the EMCC website at:

https://rsgb.org/main/technical/emc/emc-publications-and-leaflets/

Complaints Procedure

TV and Radio interference

The BBC has responsibility for investigating complaints of interference to domestic radio and television. All complaints should be made to the BBC. You can find the BBC's diagnostic guidance at the following address:

www.radioandtvhelp.co.uk/interference/rtis_tv/radcom_tools

This page also carries useful commentary for any of your neighbours who may be affected by your transmissions. There is also a facility to contact the Radio & Television Investigation Service (RTIS) where the basic diagnostic guidelines have not helped. If, following the investigation by the BBC, there is evidence of interference caused by something which is unlawful, the BBC may refer your case back to Ofcom for possible enforcement action.

Interference to amateur radio

Amateurs often mistakenly believe that the 'non-protected' status of the Amateur Radio Service means they are not entitled to any action in the case of interference caused to them. In fact, 'non-protected' is only in respect of interference from other authorised services operating in the same bands. Amateurs are as entitled to protection from external interference as any other radio user, although it must be accepted that Ofcom will have to give priority to safety of life and business radio users.

Avoiding Interference

Avoiding interference from the transmitter

Spurious Emissions

At one time, complaints of interference to TV from harmonics of amateur transmitters were a major concern in amateur radio. Nowadays complaints of this type are rare, mainly because TV is now transmitted Digitally via satellite or terrestrial signals, but also because transceivers, whether home brew or commercial, are designed with reduction of harmonics and other spurious emissions in mind.

Spurious emissions do still occasionally cause problems, for instance when harmonics of a 2m transmitter fall onto a UHF TV frequency or a harmonic of an HF or 50MHz transmitter might fall on a VHF radio frequency. Such cases are easy to identify by considering the frequency of the station being interfered with and the operating frequency of the amateur station. The solution is to check that the transmitter is working correctly and if necessary fit a low pass or band pass filter. Further information on spurious emissions can be found in the Radio Communication Handbook.

It is worth noting that interference to digital TV will not cause the typical picture and audio degradation which was associated with analogue TV, but will cause the picture to 'freeze', appear as blocks, or possibly disappear altogether until the receiver re-synchronises. These effects can also be caused by a number of signal degradation situations not related to amateur radio.

Breakthrough

When the fundamental signal from the transmitter gets into radio and electronic devices and causes interference it is usually called "breakthrough" to emphasise the fact that it not caused by a fault at the transmitter but a lack of immunity of the victim equipment. Breakthrough can be to either radio or non-radio equipment such as telephones or audio units. Most cases of HF interference to radio and TV are actually breakthrough, with the fundamental of the amateur signal getting in via the braid of the antenna coax or the mains lead, and causing overloading and inter-modulation effects.

There are two ways of tackling breakthrough problems.

By taking care to operate the amateur station so as to minimise RF energy getting into nearby radio and electronic equipment. This has been called good radio housekeeping and is covered in more detail in EMC leaflet **10**.

By increasing the immunity of the affected equipment.

Increasing Immunity

The simplest and, in most cases, the only way of increasing the immunity of radio or electronic equipment is by the use of ferrite chokes on the leads to the affected device. A choke is made by winding the lead onto a suitable ferrite ring. Where a lead comprises a pair of wires such as an audio lead the ferrite choke attenuates the common-mode currents picked up from the nearby transmitter while the wanted differential currents are not affected. Where the lead is a coaxial cable the same effect takes place and the wanted signals pass down the coax unaffected while the current on the braid is attenuated. This type of choke is often used on TV aerial leads. In this case they are called braid-breakers. The low-loss coax used for TV downleads is not suitable for winding on a ferrite ring so it is usual to use a short length of thinner 75 Ohm coax with connectors at each end. If possible 12 to 14 turns should be wound onto the core, though it is not necessary for the cable to be tight on the core. It is only necessary for the cable to pass through the ring to make a "turn". Ferrite rings available from the RSGB are about 12.7mm thick and one ring is sufficient. At one time thinner rings were popular and

two of these were stacked together to make a thicker core. More information of ferrite chokes can be found on the EMCC website.

In cases of breakthrough to neighbours' equipment it is particularly important to be diplomatic. Quite often complaints about breakthrough are exacerbated by other grievances such as unsightly antennas (from the neighbours point of view) or by unrelated causes of friction. Leaflets EMC 01, EMC 02, EMC 05 and EMC 08 are written with minimal technical jargon so that they can be given to neighbours if appropriate

Avoiding interference to amateur radio reception

There are three ways of dealing with interference to reception

1. Tackling the interference at source

This is the best option and should always be considered first. The object is to track down the source of interference and then persuade the owner to take action to suppress it or modify the use of the offending device so as to minimise the effect on your amateur operation. This will probably not be too much of a problem if the device is in your own home, but may be much more difficult if it is in a neighbouring property. Possible actions depend on whether the device is compliant with EMC regulations or not, but the golden rule is that any approach to neighbours should be diplomatic. It is not possible in these notes to do justice to this difficult subject. Further information can be found in EMC Leaflets **04** and EMC **09**. Contact the EMC helpline if you need specific help.

2. Reducing the coupling

The term good radio housekeeping was coined to cover breakthrough situations and especially to publicise the need to operate an amateur station with 'due care and attention' and to bear in mind the reasonable expectations of neighbours *see* EMC Leaflet **10**. For the purposes of these notes, good radio housekeeping has been expanded to include a discussion of the application of these principles to minimising received interference.

When the station is located in a residential area, siting the antenna in relation to surrounding properties is of major importance. Antennas should be as far from your own and neighbouring houses as possible, and as high as practical. This applies to both transmitting and receiving, since situations which cause breakthrough will also couple noise from the same wiring back to the antenna.

Some HF antennas can function near ground level, but this is not a good policy from the EMC point of view.

On HF there is, however, one big difference between transmission and reception. This is that, regardless of any local noise, there is an ambient noise level on the HF band, which greatly outweighs the thermal noise generated in the receiver front end. So, unless the antenna is very inefficient, the ambient noise dictates the received noise level. This means that it might be better for reception to mount a small, relatively inefficient, antenna in a place where local interference is least; high up and far from buildings. In special circumstances it might be worth considering an active antenna. Apart from this, good housekeeping rules for HF receiving and transmitting antennas are the same. They should be:

a – Horizontally Polarised. House wiring tends to look like an earthed vertical antenna and is more susceptible to vertical radiation. Likewise - but for rather more complex reasons - vertical receiving antennas tend to be noisier than horizontal ones.

b – Balanced. Out-of-balance currents on feeders generate vertically polarised radiation and likewise tend to pick-up vertically polarised noise.

c – Compact. So that one end is not much closer to the house than the other.

For most of us it is not possible to fulfil both conditions (b) and (c) at the lower HF frequencies, unless we have a very large or oddly shaped garden. However, they illustrate what to consider when making a compromise.

With VHF antennas there is a trade-off between antenna siting and feeder loss.

Where a high gain antenna is used, careful consideration must be given to the effective radiated power (ERP) and the proximity of nearby houses.

3. Actions to reduce the effects of interference at the receiver

It is usually better to tackle the interference at source, but if this is not possible the only option is to attempt to minimise the effect of the interference at the receiving end. First look at your radio housekeeping and at the same time check the whole antenna/earth and feeder installation for corroded joints. These can cause passive intermodulation products (PIPs), which, though not really interference, have the effect of increasing background noise. Interference can enter a receiver from the mains by unexpected common impedances and tests with a battery-operated receiver may give clues to what is happening. Don't forget that, in the absence of a signal, the receiver AGC will pull up the interference to a more or less constant level. This often leads to false conclusions.

If all else fails there are anti-interference measures which can be used at the receiver itself. Most amateurs are familiar with the function of the noise blanker and its much less effective grandfather the noise limiter. Modern transceivers include digital signal processing (DSP) which can be very effective with some types of interference.

In difficult cases it might be worth considering interference cancelling.This can be tricky to set up and operate but when functioning correctly is remarkably effective.

EMC leaflets: *www.rsgb.org/main/technical/emc/emc-publications-and-leaflets/*

Links to RSGB and Ofcom EMC problem reporting web pages:
www.rsgb.org/main/technical/emc/

EMC Helpdesk

If you are having problems with interference, the EMC help desk is there to help, and can be contacted by e-mail on:

helpdesk.emc@rsgb.org

The helpdesk, coordinated by Ken Underwood G3SDW, operates via e-mail, calling on many years of experience of dealing with interference problems. Please note that we are not able to carry out home visits except in very exceptional circumstances.

Note: Our helpdesk service is operated by volunteers in their own time via e-mail. Requests to correspond via the telephone, especially to mobile numbers are to be avoided.

The RSGB does not have any powers of enforcement, so if this is required, the regulatory body, Ofcom will have to be involved.

Additional Advice

Locating the source of interference, particularly if it is outside your own premises, will require a certain amount of "leg work". If you are unable to do this, you may be able to enlist the help of another amateur or local club.

The RSGB EMC Committee has produced a number of leaflets designed to help you, these can be found at *https://rsgb.org/main/technical/emc/emc-publications-and-leaflets*

Often, the problem is closer to home than you might think, so it is important to ensure that the source of the interference is not within your own premises. Resolving interference, particularly in urban environments, may at first seem a virtually impossible task. Nevertheless, by taking a logical approach, in most situations it should be possible to identify the source (or sources) of the interference. EMC Leaflet-04 explains how to do this.

You are likely to need the co-operation of your neighbours, so it is important to be on reasonably good terms with them.

Earthing and the Radio Amateur

EMC07 Basic Leaflet

1 **Disclaimer**

The leaflet EMC07 is intended for members of the RSGB who have passed the Radio Amateur's Examination or who are studying for it. It assumes a knowledge of electrical principles and safety practice.

RSGB Leaflets are made available on the understanding that any information is given in good faith and the Society cannot be held responsible for any misuse or misunderstanding. Where any doubt exists a suitably qualified electrical contractor shall be consulted.

UK Domestic Installations

With very few exceptions, electricity Installations in the UK will be one of three types. These are:

TN-S, TN-C-S and TT.

The letters stand for:

T = Terre, meaning earth in French coming from the Latin for earth terra, as in "terra firma".

N = Neutral

C = Combined, and

S= Separate

The three configurations are shown below

Prior to the late 1970s many properties in the UK were wired to the TN-S system where the earth and neutral are electrically separated all the way back to sub-station (Fig 1).

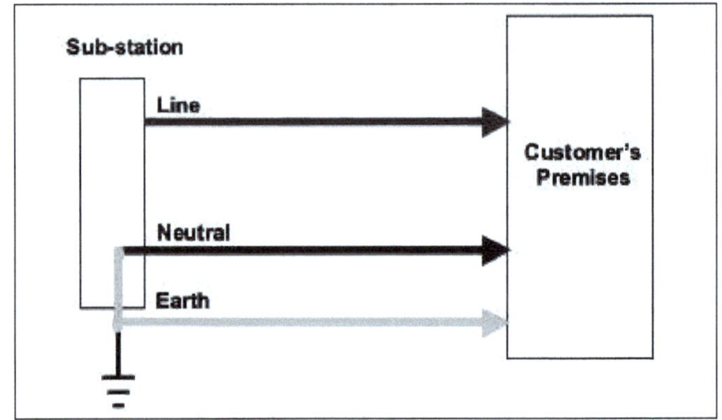

Fig 1: The TN-S Configuration

WARNING: Protective Multiple Earthing (PME)

Many houses in the UK are wired on what is known as the PME system. In this system the earth conductor of the consumer's installation is bonded to the neutral, close to where the supply enters the premises, and there is no separate earth conductor going back to the sub-station. Under certain rare supply fault conditions, a shock or fire risk could occur where external conductors such as antennas or earths are connected. For this reason, the supply regulations require additional bonding and similar precautions in PME systems.

Many houses in the UK were wired on the old TN-S system, where a separate earth goes back to the sub-station. In such systems there were no problems with connecting an external radio antenna or earth. It has recently become evident that changes to maintenance and installation practice mean that the inherent safety of the old TN-S systems cannot always be guaranteed. The situation is being reviewed. Until further information is available all installations should be treated as if they were PME.

Read EMC Leaflet 07 Earthing and the Radio Amateur before connecting any earth or antenna system to equipment inside the house.

If in doubt consult a qualified electrician

EMC Leaflet 07 is available on request from RSGB, or from the RSGB EMC Committee website.

The RSGB website has extensive guidance on EMC Matters: *www.rsgb.org/emc*

EMF – (Electromagnetic Field Exposure)

In May 2021, Ofcom issued new Amateur Radio Licence conditions which now require an assessment of **EMF compliance** for all station **equipment configurations** that you are currently using. Implementation has been rolled out over a period of 18 months, moving downward in frequency, so all bands down to 1.8MHz will be covered from November 2022 onwards.

What is EMF compliance?

Our radio transmitting stations communicate by generating electromagnetic fields (**EMF** for short) that propagate to other stations, often over considerable distances. This 'far field' becomes weaker with increasing distance, but closer to the antenna in the 'near field' there are potential hazards due to interactions of strong EM fields with the human body. Ofcom now requires us to comply with the recommendations of **ICNIRP**, the International Commission on Non-Ionizing Radiation Protection, and requires us to demonstrate compliance by making an EMF assessment for each **equipment configuration** that we currently use. You need to have these results available in case of any inspection.

What is an "equipment configuration"?

An **equipment configuration** is a list of the main factors that affect the EMFs around your station. These include:

- Frequency (typically, the middle of the amateur band in use)
- RF power delivered to the antenna (PEP output of transmitter, minus losses in feedlines, ATU etc)
- Mode of transmission and transmit/receive times (to calculate the RF power averaged over any rolling 6-minute period)
- Type and properties of the antenna (which radiates the EM field)
- How and where the antenna is installed, relative to nearby dwellings and other places where the general public may have access.

A change in any of the above will create a new equipment configuration and may well require a new assessment. However, if you can demonstrate compliance at the highest averaged RF power level that you currently use, that equipment configuration will also be compliant at lower power levels.

You do not need to update the EMF assessment or the equipment configuration if, for example, you replace your 100W transceiver with a different make and/or model. For these purposes, "100W is 100W" regardless of how that RF power is generated. However, you would need to review the assessment if you make a significant improvement to the feedline, because that increases the power delivered to the antenna.

What is an EMF assessment?

An EMF assessment is the combination of two steps:

1. A calculation of EM field strengths to define an **EMF Exclusion Zone** where no member of the general public should remain while you are transmitting.

This is combined with:

2. A practical way to ensure that no member of the general public can enter the Exclusion Zone while you are transmitting.

Most real-life situations are very straightforward. For example:

- "My antenna is on the chimney, inaccessible without a ladder."
- "Height of mast or tower makes the Exclusion Zone completely inaccessible."
- "My vertical antenna is ground-mounted, but the Exclusion Zone is entirely in my back garden and under my control. I can easily verify that no-one is present while I am transmitting."

EMF assessments introduce many new concepts that can only be briefly outlined here. For full details, downloadable calculators and practical advice, follow the RSGB EMF web pages at:

www.rsgb.org/emf

"**I have determined the extent of the Exclusion Zone around my vehicle [step 1 above]. I can thus make informed decisions on whether or not to transmit in any given circumstance.**"

RSGB's minimum separation guideline is 2.4m

For effective communication and to minimise interference, RSGB always recommends that antennas are installed as clear of obstructions as possible (see the EMC section above). To help meet the new requirements for EMF compliance – which begins with avoiding the risks of anyone touching the antenna while you are transmitting – RSGB now recommends that you should **aim for a minimum separation of 2.4m (8 feet) between any person and any part of the antenna**.

For example, if all parts of the antenna are more than 2.4m above ground, there is very little risk of anyone touching the antenna by accident. This minimum separation will also be a good start in ensuring compliance about the radiated EM fields. It obviously cannot guarantee compliance in every case, but a wide range of detailed assessments have shown 2.4m to be a useful guideline for time-averaged transmitter powers up to 100W.

Low-power compliance

The 2.4m separation guideline obviously cannot apply to hand-held or body-worn radios. For these and other low power situations, Ofcom allows compliance to be demonstrated a different way, by showing that the time-averaged EIRP is less than 10W (providing peak power is also less than 100W). If so, no further assessment is required – but you do have to calculate the EIRP and record that fact.

This same compliance route can often be used for low power operation on any band with simple low-gain antennas – but even at low power, operation on the VHF/UHF or microwave bands using high-gain beam antennas will very often exceed the 10W EIRP level, so more detailed assessment will be needed. This is where EMF calculators come in.

EMF Calculators

Ofcom and RSGB have both produced online EMF calculators to help you complete your assessments. The RSGB EMF calculator has been developed to offer extra help in numbers of areas:

- Help with entering the basic data that are pertinent to your equipment configuration;
- Help with navigating the ICNIRP guidelines, which vary considerably across the amateur bands from 1.8MHz to the high microwaves;
- A quick route to claim the low power exemption for typical VHF/UHF hand-held radios and some other kinds of low power operation;
- Formatted, downloadable copies of the completed calculations.

If the low power exemption does not apply, the RSGB calculator will then help you to choose the most appropriate calculator:

- A simplified calculator that gives compatible results to the Ofcom calculator;
- Guidance and links to more advanced methods, eg for higher-power stations, users of beam antennas, and stations in heavily built-up areas.

The "results" of any EMF compliance calculator are expressed as the size of the **EMF Exclusion Zone** (**EZ**) as noted earlier. Unfortunately, the use of any simplified EMF calculator (including either the Ofcom calculator or the compatible RSGB calculator) comes at a cost. For some equipment configurations, these calculators can over-estimate the size of the Exclusion Zone, giving a pessimistic impression about the possibilities for compliance.

If an overestimate of the size of the Exclusion Zone does not cause you any practical difficulty, then that's fine – save the results and you're done. But many UK amateurs operate in heavily built-up areas where an overestimate of even a few metres could cause practical difficulties… and that is why RSGB volunteers are developing advanced methods that can give more accurate results when needed.

Pre-Assessed Equipment Configurations (PAECs)

Advanced methods include the use of **Pre-Assessed Equipment Configurations** (**PAECs**), which are station configurations that have been assessed in great detail to identify the true size of the Exclusion Zone, using methods that are acceptable to Ofcom. When embodied in a calculator these methods are the simplest way to get an accurate prediction of your own local EZ.

PAECs are continually under development and the latest version of the RSGB EMF calculator will signpost you to the options that are currently available. Download the latest version from the RSGB EMF web pages at *www.rsgb.org/emf* and then follow the instructions.

RSGB EMF calculator

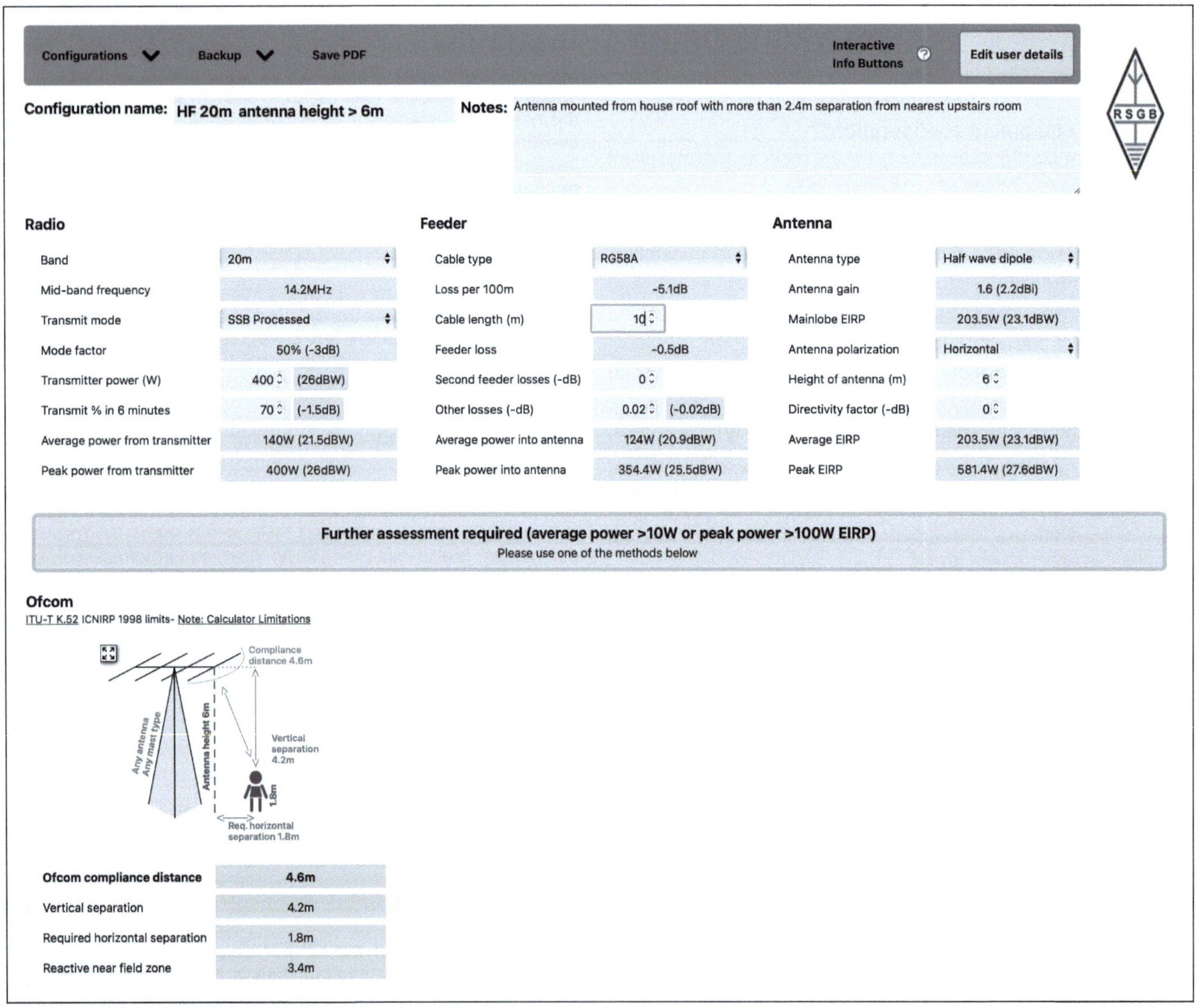

Configurations | Backup | Save PDF | Interactive Info Buttons | Edit user details

RSGB

Configuration name: HF 20m antenna height > 6m **Notes:** Antenna mounted from house roof with more than 2.4m separation from nearest upstairs room

Radio		Feeder		Antenna	
Band	20m	Cable type	RG58A	Antenna type	Half wave dipole
Mid-band frequency	14.2MHz	Loss per 100m	-5.1dB	Antenna gain	1.6 (2.2dBi)
Transmit mode	SSB Processed	Cable length (m)	10	Mainlobe EIRP	203.5W (23.1dBW)
Mode factor	50% (-3dB)	Feeder loss	-0.5dB	Antenna polarization	Horizontal
Transmitter power (W)	400 (26dBW)	Second feeder losses (-dB)	0	Height of antenna (m)	6
Transmit % in 6 minutes	70 (-1.5dB)	Other losses (-dB)	0.02 (-0.02dB)	Directivity factor (-dB)	0
Average power from transmitter	140W (21.5dBW)	Average power into antenna	124W (20.9dBW)	Average EIRP	203.5W (23.1dBW)
Peak power from transmitter	400W (26dBW)	Peak power into antenna	354.4W (25.5dBW)	Peak EIRP	581.4W (27.6dBW)

Further assessment required (average power >10W or peak power >100W EIRP)
Please use one of the methods below

Ofcom
ITU-T K.52 ICNIRP 1998 limits- Note: Calculator Limitations

Compliance distance 4.6m
Any antenna
Any mast type
Antenna height 6m
Vertical separation 4.2m
1.8m
Req. horizontal separation 1.8m

Ofcom compliance distance	4.6m
Vertical separation	4.2m
Required horizontal separation	1.8m
Reactive near field zone	3.4m

Input page of the RSGB EMF calculator including results (lower left) compatible with the Ofcom calculator.

Always use the latest version for the RSGB EMF calculator, from *www.rsgb.org/emf*

Upgraded versions are likely to appear during the lifetime of this Yearbook, and may be different from the example shown here. (Ofcom will accept calculations that have already been made using earlier versions of the Ofcom or RSGB calculators.)

Amateur Radio Direction Finding

Amateur direction finding in the UK goes back to the time between the wars when competitions were run using the 1.8MHz band. The arrival of the commercial VHF hand-held radio in the 1970s spawned a different kind of direction finding competition, generally using the 144MHz band. These events were organised by local clubs, many of which were affiliated to the RSGB. Usually a club member parked up in the countryside and made a series of transmissions using a mobile 144MHz radio. Other club members attempted to locate him and the evening often concluded comparing notes in a local hostelry.

Using cars in ARDF competitons

A lot has changed in the past fifteen years or so, both in terms of the Regulations attached to the Road Traffic Act and also the exclusions attached to the majority of motor insurance policies. As a responsible national society it is a role of the RSGB to draw the attention of its members and Affiliated Societies to factors that affect car based direction finding competitions, so that they can make their own decisions about the format of any event they wish to organise.

Section 12 of the Road Traffic Act says that both the competitor and the organiser of a race or trial of speed on the public highway is guilty of an offence.

The Regulations attached to the Road Traffic Act have changed in the last two decades and the Department of Transport has authorised the Royal Automobile Club Motor Sports Association Ltd to operate the Motor Vehicles (Competitions and Trials) Regulations 1969 on behalf of the Department. These Regulations cover many varieties of competitive activities on the public roads. A relatively new kind of event has been authorised and this is called a 'Navigational Scatter Event'.

The good news is that these are automatically authorised (so no permission from anyone is required) and participants are free to roam wherever they like on the public highways - exactly as one would need for a DF Competition.

Unfortunately there are three restrictions:

a. Timing is banned for any event which uses the public highway in part or in whole,

b. minimum distance covered is also not allowed in determining the winner,

c. participants must not all visit an identical set of control points - in motor sport typically 75% of the control points, as selected by individual participants, have to be visited.

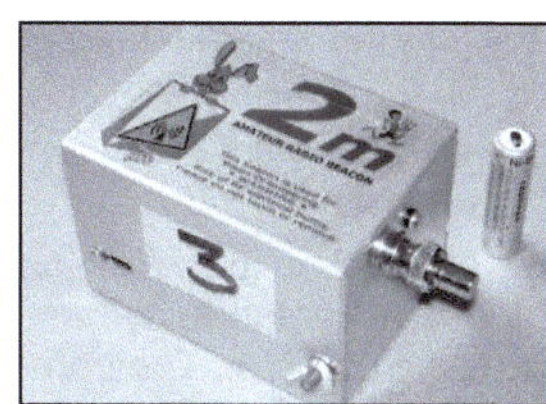

Fig 1: A typical 2m ARDF transmitter with an AA cell for size comparison.

Now it is possible to have a car based DF event and keep within these restrictions. For example, with one transmitter, five 'control points' could be specified. Participants would be required to visit three of these (of their choice) and take a bearing of the transmitter from each one. Competitors then have to give the grid reference of the transmitter location based on their bearings. The closest to the actual location can then be declared the winner.

TX	Minute 1	Minute 2	Minute 3	Minute 4	Minute 5	Minute 6
No.1	MOE (one dot)	Silent	Silent	Silent	Silent	MOE (one dot)
No.2	Silent	MOI (two dots)	Silent	Silent	Silent	Silent
No.3	Silent	Silent	MOS (three dots)	Silent	Silent	Silent
No.4	Silent	Silent	Silent	MOH (four dots)	Silent	Silent
No.5	Silent	Silent	Silent	Silent	MO5 (five dots)	Silent
	1 x 5-minute cycle (all five transmitters operate on the same freq)					

Table 1: The timing sequence of the five hidden transmitters.

There has been another development in the last fifteen years concerning third party motor insurance cover. The motor insurance market has become highly competitive. One way the insurers can offer cheaper cover is to reduce their risk. A way of doing this is to make participation in 'competitions' an exclusion on most policies (that is to say the policy is invalid when participating in any kind of competition). Of course it is possible to obtain third party cover without having this exclusion but that will come at the cost of an increased premium.

This is a requirement affecting the individual driver but only a small minority (if any) of the members of the average radio club are likely to have this cover. As a result it makes it very very difficult to run a fully legal car based DF competition. Readers may have noticed that there are very few traditional 'treasure hunts' taking place these days, for exactly the same reasons of third party insurance cover.

Club Competitions

It is still perfectly possible to have ARDF on the summer programme of your radio club. The RSGB has a set of 3.5MHz direction finding equipment available on loan. This comprises five transmitters suitable for the 'classic' format with each one transmitting for 1 minute in a 5 minute cycle and all transmitters on the same frequency. In addition ten 3.5MHz DF receivers are provided. The RSGB pays to have the kit shipped to you and your Club has to pay for the return. If you manage to break or lose anything your Club is expected to pay to have it repaired or replaced. The range of the transmitters is about 400m, so the kit is very suitable for use in an event on foot in a small country park and the limited distances make it attractive for a majority of Club members.

Contact *ardf.chairman@rsgb.org.uk* to arrange to borrow the kit.

International development of ARDF

After the end of WW2 things developed along a different track on the Continent. Back then, only in Great Britain, Eire, Gibralter and Czechoslovakia were amateurs allowed to operate on 1.8MHz, so countries wishing to introduce surface wave direction finding simply used the lowest frequency band available to them, which was 3.5MHz. Today, this choice is embodied in the set of rules supported by the IARU and also in thousands of DF receivers for this band across the world. Region 1 of the IARU has an ARDF Working Group responsible for the formulation of rules, since by far the greatest interest is in Europe. It is the custom for Regions 2 and 3 to adopt the rules originating in Region 1. The situation today is that ARDF is extremely vibrant in Europe. There have been nineteen bi-annual World Championships and twenty two Region 1 Championships, also a bi-annual event but in odd numbered years. Competitions are organised using the 3.5MHz band, where propagation is predictable and good bearings are generally obtained, and the 144MHz band, which exhibits significant multi-path propagation, sometimes leading to misleading bearings being obtained. The competitions take place entirely on foot and

no motor vehicles are involved. In Great Britain the RSGB made rather a late start to this international style of ARDF, with the first UK event being held in 2002. Many European countries have over 50 years of experience, especially in the old Soviet Bloc countries where there used to be significant state support for activities like ARDF that were also useful militarily. State support has ebbed away since 1989 but the Eastern European countries have inherited a long tradition of direction finding that makes them dominant at World level.

Introduction to the IARU Rules

Transmitters and timing

Five low power transmitters (3W output if 3.5MHz is being used and 800mW if 144MHz is chosen) are deployed in the area to be used. A typical transmitter is shown in Fig 1. All the transmitters operate on the same frequency, but not all at the same time. They transmit in sequence and send an identifier in Morse code for one minute each. Before the reader freaks out at the mention of Morse code, it should be pointed out that the identifier is simply a matter of dot counting.

The first transmitter sends the letters MOE in Morse. The first two letters are long ones in Morse and serve to keep the transmitter on the air for a while, to allow the competitor to swing the aerial carried and assess the direction of the transmitter. The last letter is a single dot, so one dot denotes transmitter 1. This transmitter sends for one minute before shutting down.

The second transmitter then radiates the morse sequence MOI. The last letter (i) is two dots, to denote transmitter number 2. Transmitters 3, 4 and 5 transmit MOS, MOH and MO5 respectively. The sequence is shown pictorially in Table 1.

In the UK it is a licence condition that the callsign of the supervising licensed amateur for the unattended transmitters, is radiated at the end of each transmission, so the one minute transmission terminates with a burst of higher speed Morse, which is this callsign.

Men	Women
M19	W19
M21	W21
M40	W35
M50	W45
M60	W55
M70	W65

Table 2: Age categories for competitors.

In addition to the five hidden transmitters there is a beacon transmitter operating on a different frequency, which radiates the letters MO repeatedly in Morse and is interrupted at intervals with the callsign of the supervising licensed amateur. This transmission is continuous and enables competitors who get hopelessly lost to simply DF the beacon to find their way to the finish.

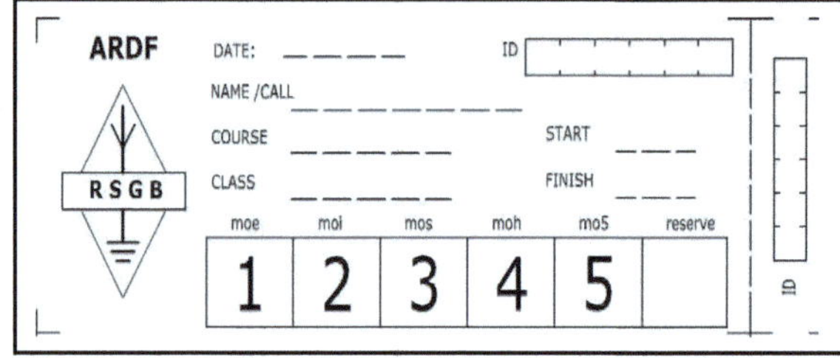
ARDF
DATE:
ID
NAME /CALL
COURSE
START
CLASS
FINISH
RSGB

moe	moi	mos	moh	mo5	reserve
1	2	3	4	5	

ID

Fig 2: Control Card for a five transmitter DF hunt.

All the transmitters use some form of omnidirectional antenna. For 3.5MHz an 8m vertical wire with an 8m counterpoise is frequently deployed, while on 144MHz a pair of crossed horizontal dipoles (aka a turnstile antenna) at a height of about 3m is commonplace.

Proof of finding the transmitters

Clearly it is necessary for the competitor to demonstrate that each assigned transmitter has been visited. This can be done in one of two ways:

1. The competitor carries a control card (see Fig 2) with a space for each of the five hidden transmitters plus one for the beacon if the latter is to be registered. At each transmitter there is a needle punch (see Fig 3) which is used to mark a unique pattern of needle holes in the card. In international competition the beacon will also be 'punched' but practice varies in domestic races.

2. Electronic timing equipment may be used. Each competitor carries a microchip (see Fig 4), which is inserted into a unit at each transmitter. The timing unit writes the identity of the transmitter plus the time of the visit to the microchip. On completion of the course, the competitor punches at the finish and then downloads all the data to a computer, which is able to print the time taken, the transmitters visited and all the split times.

Age categories

ARDF is organised into a series of age categories and the adult age categories in force are shown in Table 2. To explain how the system works, consider the age group M21. The M denotes a male age group. A man enters the M21 class on 1 January of the year in which he becomes 21 and leaves it on 1 January of the year in which he becomes 40 (M40 being the next age group).

There are a total of twelve adult age groups, with the older age groups hunting fewer transmitters over shorter distances than the younger age groups.

The result of this is that competition is against one's peers and this considerably broadens the appeal of this radio sport.

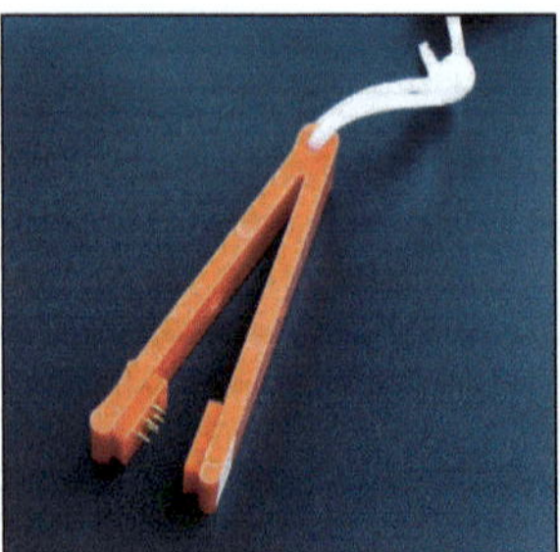

Fig 3: Pin punch found at the transmitter and used to mark the appropriate box on the control card. The punches at each transmitter carry a unique pattern of pins.

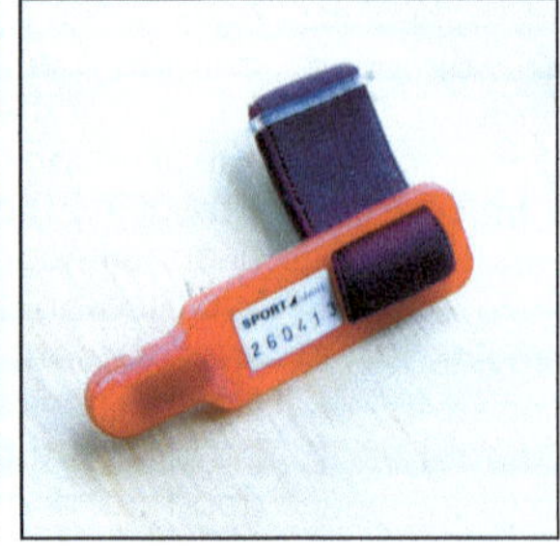

Fig 4: The electronic punching 'dibber' is carried on a finger of the competitor by an elastic strap. This is inserted into the unit at each transmitter to register that the competitor has been there.

Transmitter placement and the time limit

There are three further rules, which can be of great significance, depending on the shape and size of the area used for the competition. No transmitters can be placed within 750m of the start. In domestic competition this distance is frequently reduced to 400m, to avoid 'sterilising' a large part of a small wood as far as transmitter placement is concerned.

The second restriction is that there can be no transmitter within 400m of the finish. Finally, transmitters must be placed at least 400m apart.

A time limit is rigorously enforced for competitions, with the rule that any competitor over time is placed below a competitor who has found at least one transmitter and is within the time limit. It can be rather galling to find all five transmitters and finish one minute outside the time, to be beaten by someone who took nearly two hours to find just one transmitter. The time limit is decided by the course planner, but two hours is a frequent choice, with 90 minutes for easier areas. The object of this rule is to constrain those competitors who are determined to find all the transmitters at any cost, even if this sees them still hunting as darkness falls.

Equipment

It is obvious that competitors will require a receiver for the frequency band being used and a directional antenna for that band. These two items are normally combined into one unit and at UK events there is usually equipment available on loan, although it is sensible to confirm this with the organiser beforehand.

The receiver should be an AM receiver, since the direction to the hidden transmitters will be determined by swinging the antenna from side to side and noting changes in signal amplitude. The amplitude limiter in an FM receiver makes it less suitable for this task, although it can still be made to work in this application.

There is a need to plot bearings on the map provided at the start. Beginners usually plot more bearings than experienced competitors. To do this, the map should be taped to a lightweight board of some kind and either a spirit pen or a chinagraph (wax) pencil can then be used to mark the map. Lines can be drawn by both of these markers on any clear plastic covering used to protect the map. Neither of them will run if they later become wet, but only the chinagraph will make a satisfactory mark if the plastic is already wet.

A compass will be required to measure the bearings. The type with a rectangular base plate also doubles as a protractor. The compass is best looped round the wrist with the cord normally provided.

Proving the visit made to each transmitter involves carrying either a Control Card or an SI dibber. This 'dibber' is a plastic encased microchip used for electronic 'punching'. It is on a small elastic strap which allows it to be attached to the index finger. The control card is best pinned with safety pins to the front of the clothing. Many control cards are printed on tough, waterproof Tyvek paper and require no protection or strengthening. Control cards that are printed on thin card should be covered with Sellotape to both waterproof and strengthen them.

Moving to the desirable rather than the essential, a whistle should be carried. In some competitions is mandatory with a 'no whistle – no run' policy. The emergency signal is six blasts of the whistle at one minute intervals. Also in the desirable category is a circle stencil to mark the circles around the start and the finish, in which no controls can be placed. Obviously a different stencil is required for a map at 1:10,000 scale to one at 1:15,000 scale.

A check list is shown in Table 3.

Competition hints

Pre-start

There is normally five minutes after being given the map and before getting the signal to start, in which the competitor is able to:

a. waterproof and protect the map as deemed appropriate for the weather conditions,

b. on the map, draw a 750m circle around the start and a 400m circle around the finish (in domestic competition the 750m start circle is often reduced to 400m),

c. study the map to identify height features in particular.

Start + 5 minutes

After being given the start signal, the rules oblige the competitor to keep moving to the end of the start funnel. Once at the end the aim should be to listen to each transmitter in turn, assess the strength of the signal and plot the bearing – all within the 60 seconds that it is on the air. Prior practice at this procedure will pay big dividends for the beginner. If the 144MHz band is being used, bearings taken from high spots are more accurate than those taken from valleys. It may pay to sacrifice a complete transmitter cycle and climb to the top of a nearby hill or spur from which more accurate bearings can be obtained.

Decision time

Based on the information gained by listening just once to each of the transmitters, the most important decision of the day must be made. This is the choice of the first transmitter to be visited. In the case of a co-located start/finish, the penalties for getting it wrong are not as severe compared to a split start/finish. With a co-located start and finish, a poor choice can often be rectified on the route back from the furthest transmitters to the finish. When the start and finish are at separate locations, a bad decision may mean a lot of 'back tracking' and hence wasted time.

Bearing quality

The surface wave propagation on 3.5MHz during daytime leads to bearings which are generally pretty accurate. While it is wise to avoid wire fences and overhead power lines when taking bearings, the accuracy of the plotted bearing is determined by the equipment and skill of the competitor. This results in fast runners being able to get to the transmitters first, assuming that they also have reasonable direction finding skills.

On 144MHz there is a lot of multi-path propagation, with the signal being reflected or scattered from steep hillsides, rock outcrops and even the edges of wooded areas. The bearings obtained vary greatly in 'quality'. A sharp, clear peak in the signal as the antenna is swung from side to side is indicative of a single path signal and this is often the direct path from the transmitter. Multi-path propagation most often reveals itself as a rather diffuse bearing as the antenna is swung. Sometimes there may be more than one distinct peak to the signal and this is where an antenna with very low side and back responses comes into its own, to differentiate between the direct and the multi-path signals.

All this interpretation of bearings coupled with the need to view the bearings against the background of any high ground in the vicinity; leads to the winner needing to process all this information quickly and accurately. Hence, being able to run fast is no longer such a key quality to gain victory.

Item carried	Note(s)
Receiver	Usually fixed to the antenna.
Antenna	Usually fixed to the receiver.
Map	Normally issued at the start line, 5 or 10 minutes before starting.
Lightweight rigid board for the map	To deal with wet weather conditions, the competitor will need waterproofing (a plastic folder or sticky backed plastic film) to cover the map and possibly tape to fix the map to the board.
Compass	The type with a rectangular backplate doubles as a protractor.
Spirit pens and/or wax pencils	Will not run if it rains, but note that only wax pencils will write satisfactorily on plastic film that is already wet.
Circle stencil	750 and 400m circles at the map scale in use.
Control Card or SI 'dibber'	To register that the competitor has visited each assigned transmitter.
Whistle	Emergency signal is 6 blasts at 1 minute intervals.

Table 3: A checklist of the items you require for a competition.

Your first event

Event information

Finding out about competitions is clearly the first step and there are currently around 15-20 events held in the UK each year. There is usually a break around Christmas and January, with the 'season' generally commencing in February and the last event in November or December. The RSGB website is the gateway to information about ARDF events. The URL is given at the end of this article.

The events organised by the RSGB fall into two categories:

'National Events' in which there are usually two competitions in the day, with a Classic 144MHz event before lunch and then a less physically taxing event on 3.5MHz after lunch. There are about six of these in the year

In addition there are some introductory Regional events using the classic format on 3.5 MHz. In these events, newcomers are required to take bearings on all five transmitters before starting. One of the mentors present will check these bearings and give a 'good to go' message if the bearings plotted look OK. After this, the newcomers can cross the start line to commence the timed competition, secure in the knowledge that he or she has already taken and plotted five reasonable accurate bearings.

These introductory events are not available in all RSGB Regions. The ARDF Committee would welcome interest from any member willing to organise events in the Regions which currently have these events. There is a lot of support available.

Clothing and equipment

The issue of equipment has already been covered above. As far as clothing is concerned, for a first outing, stout shoes and outdoor attire is sufficient. If you become more committed, then studded orienteering shoes, gaiters to protect the lower leg against brambles and nettles and an orienteering suit are more appropriate. On occasions when the weather is particularly inclement (thankfully few) then a cagoule or other waterproof garment will be needed.

Registration

Events that are run in conjunction with an orienteering event will benefit from the direction signs to that event. ARDF events that are freestanding are not likely to be extensively signed, so the competitor should ensure

Andrew, G4KWQ 'flies the flag' at the World ARDF Championships.

that a copy of the Grid Reference and of the map extract that is frequently given with the event details, are carried on the journey to the venue.

Once there, it is necessary to register and pay the event fee (usually of the order of £5- £6). This provides for entry for all the competitions taking place on the day. There is usually a full scale event in the morning, followed by a more relaxed competition after lunch. All of this provides an excellent day of radio sport and makes it well worthwhile to travel a fair distance for a full day of radio in the open air.

At registration you may have to make a choice regarding the number of transmitters you wish to hunt. Details of the frequencies of the transmitters (hidden transmitters on one frequency and the homing beacon on a second frequency), the radius of the zone around the start in which transmitters may not be placed and your individual start time will be given to you. Finally, the time limit for the event should be noted.

If electronic timing is being used, it will probably be necessary for you to hire an electronic 'chip' (colloquially known as a dibber).

If pin punching is in use, then you will be given a control card. Fill this out with your details. If it is made of plain paper or thin card, protect and strengthen it with Sellotape and finally pin it to the front of your clothing with safety pins.

The map

An orienteering style map will be in use and the most common scale in domestic competition is 1:10,000. These maps show much more ground detail than an Ordnance Survey map. The first thing to note is that the white bits are trees – quite the opposite to an Ordnance Survey map. The white parts of the map denote runnable forest and various shades of green show less runnable areas, with dark green being really impenetrable and well worth avoiding. Fortunately you are very unlikely to have transmitters located in these latter areas.

Open and semi-open areas are shown in a yellow ochre colour.

Only the start (a triangle) and the finish (a double circle with a smaller circle inside a larger one) will be marked on the map.

After the start

In simple terms, listen to all of the transmitters to get a bearing and an idea of the signal strength of each one, decide which transmitter you wish to visit first and then head for the one selected.

Finding your very first hidden transmitter is a great moment and one to be remembered for a very long time. Keep an eye on the clock, so that you get back to the finish inside the time, and see if more transmitters can be located.

Newcomers are usually a bit erratic in their first events, as is to be expected. For some a brilliant performance at the first outing can be followed by poor and disappointing results at subsequent events. Experience tells us that it takes about six outings for the majority of competitors to settle in and be able to locate all the assigned transmitters inside the time on a reliable basis. In other words don't get discouraged by a few poor results; it will all come together for you with a bit of experience.

After finishing there is the opportunity to compare notes with other competitors and to get some tips on how to avoid any mistakes at future events.

Robert Vickers G3ORI shows that ARDF can be enjoyed by amateurs of any age.

Other Formats

There are two variants of the basic format. The first is Foxoring, which is a hybrid of Orienteering and Radio Direction Finding. A large number of very low power 3.5MHz transmitters are deployed and a circle is marked on the map within which each one will be audible. The competitor uses orienteering techniques to navigate to the area of the circle. Once the signals from the transmitter are picked up, direction finding enables the transmitter to be located.

The second variant is the sprint format. This provides two clusters of five transmitters operating on two different frequencies in the 3.5MHz band and keyed at different speeds. Each transmission lasts just 12 seconds and competitors have to return to a spectator beacon after finding all the transmitters in the first group and before they set out to find transmitters in the second group. The transmitters are keyed with the usual MOE, MOI etc. The format is very fast and furious and winning times of less than 15 minutes are not unknown.

UK coverage

There are many RSGB Regions where there is little or no ARDF taking place beyond the occasional Club based event. ARDF really lends itself to being a Regional activity and, for example in Region 13 (East Midlands) and Region 10 (Thames Valley), there are a series of five weekend events on 80m, one each month from May to September. There are never enough people in the average Radio Club who are interested in having more than an occasional evening event but across a whole Region this is no longer the case.

To promote this activity the RSGB Legacy Fund has financed the provision of two sets of transmitters and receivers available on long term loan (a year or so) to any Region wishing to run a similar series of events. Contact the ARDF Committee Chairman (*ardf.chairman@rsgb.org.uk*) if you have an interest in running such a series of events in your Region.

Further Information

A description of competitive direction finding cannot be exhaustive in the space available here. The RSGB book 'Radio Orienteering – The ARDF Handbook', goes into much fuller detail and is essential reading for the beginner. Event information is available on the RSGB website: www.rsgb.org>main>on the air>activity ardf. The ARDF pages also include the results of competitions, details of the big international events and information about sources of suitable equipment.

Among programmes that stimulate daily activity on the HF bands, two stand out head and shoulders above the others – DXCC for working countries, or 'entities' to use current terminology, and IOTA for contacting island groups. The programmes are similar in character – both are international in coverage, both have a strong rule structure and neither is open-ended. Moreover, in practical terms they complement and strengthen each other because activity to promote one often provides valid contacts for the other.

IOTA, or the Islands On The Air Programme to give it its full title, was created in 1964 by the late Geoff Watts, a leading British short wave listener and the only SWL in the CQDX Hall of Fame. When the programme was taken over, at Geoff's request, by the RSGB in 1985 it was already a favourite for many DXers. Its popularity has since grown each year, not only among ever-increasing numbers of island chasers but also among a rapidly expanding band of amateurs attracted by the possibilities for operating portable from islands. For both it is a fun pastime adding much enjoyment to on-the-air activity.

The basic building block for IOTA is the IOTA Group. The oceans' islands have been corralled into some 1200 IOTA Groups with, for reasons of geography, varying numbers of 'counters', i.e. qualifying islands, in each. Only in very few cases do the rules of IOTA allow single islands to count separately, DXCC island entities such as Barbados being one. The number of groups is now capped and further changes are expected to be minimal.

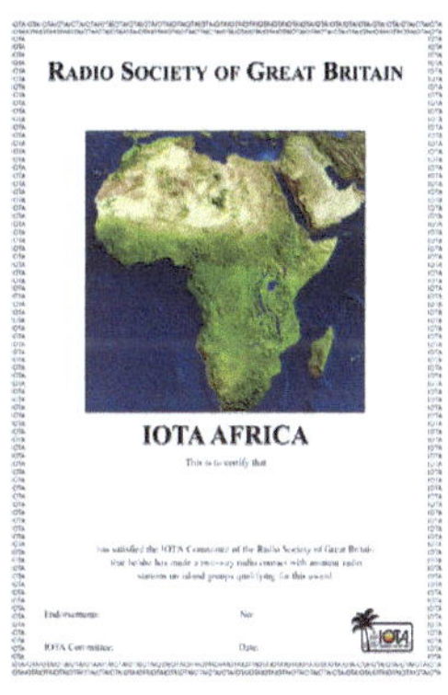

Each group activated has been issued with an IOTA reference number, for example EU-005 for Great Britain. Part of the fun of IOTA is that it is an evolving programme with new groups being activated for the first time. Currently some 1138 of the 1200 groups have been activated.

The objective, for the island chaser, is to make radio contact with at least one counter in as many of these groups as possible and, for the DXpeditioner, to provide island contacts. A wide range of separate certificates, graded in difficulty, is currently available for island chasers as well as two prestigious awards for high achievement (see the table overleaf). Applicants may be any licensed radio amateur (or SWL on a 'heard' basis) who has had confirmed contacts with the required number of IOTA Groups listed.

IOTA Directory

The IOTA Directory provides full programme rules as well as a comprehensive listing of IOTA groups together with the names of 15,000 qualifying islands. It can be downloaded from the IOTA website at *www.iota-world.org/iota-shop.html.*

Applying for an Award

IOTA has since 2017 run software which allows award credits to be given by electronic confirmation of contact by QSO matching with logs on Club Log and since May 2020 with logs on the ARRL's Logbook of the World (LoTW). The system however continues to accept confirmation by QSL cards. Award applicants should prepare and submit their applications electronically on the Internet. Full details of the application procedure and a list of checkpoints can be found on the IOTA website at *www.iota-world.org*. After signing off your application on-line for processing by your checkpoint, you should immediately send him any cards required for confirmation with return postage.

Island-chasing

1000 or more IOTA Groups may seem an enormous target. If you are a long time DXer who has worked it all and are looking for something new, you will already have amassed a very respectable IOTA score from among your DXCC contacts. If, however, you are new to the bands or one of the many amateurs who adopt a more relaxed approach to their operating, you can take full advantage of a very high level of IOTA activity, comprising easy and semi-rare groups, to launch you on your way. Well over 600 IOTA groups are usually activated over a three year period with, during a typical summer weekend, some 20/25 IOTA Groups being heard around the IOTA meeting frequencies. An enthusiast should be able to gain the IOTA Plaque of Excellence for working 750 groups in about six years, operating mainly at weekends. This must be a reasonable target to go for – after all, how long does it take to get to the top of the DXCC Honour Roll?

IOTA is one of the few award programmes that has an annual Honour Roll and other performance listings. These create a great deal of interest when they are published each spring. Many IOTA enthusiasts are more interested in participating in these listings than in collecting the certificates. All you need to enable you to participate is a registered score of at least 100 Island groups.

Operating from an island

Many amateurs are fortunate enough to live on an island and to be able to give out an IOTA every time they make a contact. Others are not so lucky. For both there is the lure of operating portable from a rare or semi-rare group – the fun of being at the other end of a pile-up for a few days. Many islands lie within a few hours' reach and, subject to the availability of suitable equipment, could be put on the air relatively easily.

Those amateurs lucky enough to be able to activate a rare or semi-rare IOTA Group can expect to generate huge pile-ups with thousands of contacts during even a short 2/3 day period. Rare groups are not all remote and difficult to access. Even in Europe and North America there are many that are needed by the chasers. For those interested, a list of most wanted IOTA Groups in each continent, ranked by rarity, can be viewed on the IOTA website.

Categories of application

IOTA began as an award for single operators working on the HF bands (1.8 to 30MHz). However, in response to demand, the IOTA Committee subsequently introduced categories specifically for club stations and for working on VHF/UHF (50MHz and above).

Address: Islands on the Air (IOTA) Ltd, La Quinta, Station Road, Chobham, Woking, Surrey GU24 8AR
Email: info@iota-world.org

IOTA meeting frequencies

Nobody and no group in amateur radio is entitled to reserved frequencies, but the IOTA community has adopted a number of 'meeting frequencies' which island stations are encouraged to use when they are free – and to operate close to, without causing interference, if they are occupied. The frequencies are 3755, 7055, 14260, 18128, 21260, 24950, 28460 and 28560kHz on SSB and 3530, 7030, 10115, 14040, 18098, 21040, 24920 and 28040kHz on CW.

IOTA contest

The IOTA Contest, first held in July 1993, has become enormously popular and now regularly attracts more than 2000 entries. It provides an opportunity annually, at the end of July, to work large numbers of rare and semi-rare IOTA Groups. Contest rules and results are available from the RSGB HF Contest Committee website.

IOTA Annual Listings

The following pages show the IOTA Annual Listings as of February 2024. The lists are divided as follows:

The Honour Roll is a list of the callsigns of stations with a checked score equalling or exceeding 50% of the total of activated IOTA groups, excluding those with provisional numbers, at the time of preparation.

The Annual Listing is a list of the callsigns of stations with a checked score of 100 or more IOTA groups but less than the qualifying threshold for entry into the Honour Roll.

The Club Listing is a list of the callsigns of club or multi-operator stations with a checked score of 100 or more IOTA groups.

The VHF/UHF Listing is a list of the callsigns of stations with a checked score of 100 or more IOTA groups on the VHF/UHF bands.

The SWL Listing is a list of SWLs with a checked score of 100 or more IOTA groups.

Listing in the 2024 tables was restricted to those participants who had updated their scores since February 2019. IOTA rules limit inclusion in the listings to those participants who have updated their scores at least once in the preceding five years and have opted to have their scores published. All participants should be reminded that the final decision on acceptance of credits is made at IOTA HQ and that this can mean downward adjustments to scores at any time to reflect corrections of one sort or another. Data-cleansing work is on-going and covers every participant's complete record, not just the latest credits added. Although efforts are made to alert participants to score changes, this cannot be guaranteed to happen in each case. Remember, the line of communication is via your checkpoint, so please do not route queries direct to IOTA HQ or the IOTA management. Always check first to see if the answer is in the IOTA Directory.

IOTA Awards

Award	Standard HF Category	VHF /UHF Category
IOTA 100 Islands of the World Certificate	100 Confirmed IOTA Groups including 1 from all 7 continents	100 Confirmed IOTA Groups including 5 continents
IOTA 200 Islands of the World Certificate	200 Confirmed IOTA Groups including 1 from all 7 continents	200 Confirmed IOTA Groups including 5 continents
IOTA 300 Islands of the World Certificate	300 Confirmed IOTA Groups including 1 from all 7 continents	
IOTA 400 Islands of the World Certificate	400 Confirmed IOTA Groups including 1 from all 7 continents	
IOTA 500 Islands of the World Certificate	500 Confirmed IOTA Groups including 1 from all 7 continents	
IOTA 600 Islands of the World Certificate	600 Confirmed IOTA Groups including 1 from all 7 continents	
IOTA 700 Islands of the World Certificate	700 Confirmed IOTA Groups including 1 from all 7 continents	
IOTA 800 Islands of the World Certificate	800 Confirmed IOTA Groups including 1 from all 7 continents	
IOTA 900 Islands of the World Certificate	900 Confirmed IOTA Groups including 1 from all 7 continents	
IOTA 1000 Islands of the World Certificate	1000 Confirmed IOTA Groups including 1 from all 7 continents	
IOTA 1100 Islands of the World Certificate	1100 Confirmed IOTA Groups including 1 from all 7 continents	
IOTA Plaque of Excellence (with shields for each additional 25 IOTA Groups)	750 Confirmed IOTA Groups including 1 from all 7 continents	300 Confirmed IOTA Groups including 5 continents
IOTA 1000 Islands Trophy (with shields for each additional 25 IOTA Groups)	1000 Confirmed IOTA Groups including 1 from all 7 continents	
IOTA Africa Certificate	75 African IOTA Groups	50 African IOTA Groups
IOTA Antarctica Certificate	75% of Antarctic IOTA Groups	50% of Antarctic IOTA Groups
IOTA Asia Certificate	75 Asian IOTA Groups	50 Asian IOTA Groups
IOTA Europe Certificate	75 European IOTA Groups	50 European IOTA Groups
IOTA North America Cert.	75 North American IOTA Groups	50 North American IOTA Groups
IOTA Oceania Certificate	75 Oceanian IOTA Groups	50 Oceanian IOTA Groups
IOTA South America Certificate	75 South American IOTA Groups	50 South American IOTA Groups
IOTA World Diploma	50% of the IOTA Groups in all 7 continents or 50 IOTA Groups for the continents where there are more than 100 IOTA Groups	
IOTA Arctic Islands Certificate	75 Arctic Island Groups	50 Arctic Island Groups
IOTA British Isles Certificate	75% of the British Isles Groups	50% of the British Isles Groups
IOTA West Indies Certificate	75% of the West Indies Groups	50% of the West Indies Groups

Further information and updates: *www.iota-world.org*

2024 Honour Roll

Pos.	Callsign	Total	Pos.	Callsign	Total	Pos.	Callsign	Total	Pos.	Callsign	Total	Pos.	Callsign	Total	Pos.	Callsign	Total
1	9A2AA	1133	90	VE7DP	1082	177	VE7SMP	1033	266	R7KM	978	357	UX2IQ	916	446	CU3EJ	832
2	I2YDX	1132	91	DL8FL	1081	180	AG9S	1032	269	I5YDO	977	358	S57L	915	447	UN7JX	830
3	I1JQJ	1131	91	ON5NT	1081	180	IK2QPR	1032	270	OE2VEL	976	359	UT1UY	914	447	UX1AA	830
4	G3KMA	1130	91	RU6K	1081	182	HA1RW	1031	270	SP6CIK	976	360	JM1XCW	913	449	DL2OE	828
4	K9PPY	1130	94	F6GCP	1080	182	HK3JJH	1031	272	JA3UCO	975	360	UA3DPM	913	449	I2AOX	828
6	W1NG	1129	94	HA5AGS	1080	184	I5HOR	1029	273	7N1GMK	972	362	EA7TV	912	449	JH1MXV	828
7	W5BOS	1128	96	R6AF	1079	185	DL6KVA	1028	273	IK4HPU	972	362	I2ZBX	912	452	CT1DKS	826
8	I1SNW	1127	96	UA1OIZ	1079	185	RU3EQ	1028	275	DL4FCS	969	362	R5AJ	912	452	IK8VRH	826
8	I4LCK	1127	98	DL8MLD	1078	185	W4ABW	1028	275	R7NB	969	365	I2PQW	908	452	OE3KKA	826
8	W9DC	1127	98	G3OCA	1078	188	K9MUF	1027	277	DL3APO	968	365	W7BEM	908	452	R0AZ	826
11	HB9AFI	1126	98	JE1DXC	1078	189	9A2NO	1026	277	OH2BCK	968	367	DL3JON	907	456	F6HQP	824
11	ON6HE	1126	101	EA4MY	1077	189	DL2RNS	1026	277	SM5AQD	968	368	HA3HP	906	457	JA2IHL	822
11	VE3XN	1126	101	JA8RJE	1077	191	AH6HY	1025	277	US4EX	968	368	IK2OVC	906	457	W9IXX	822
14	OM3JW	1125	101	OZ1BUR	1077	192	IZ7DOO	1024	281	DL2RU	967	368	ON7TK	906	459	DL5AN	821
15	EA8AKN	1124	101	PA3EXX	1077	193	DL2VPF	1022	281	RM0F	967	371	ON4CAS	905	460	LA8DW	820
15	HA0DU	1124	105	W4PKU	1076	193	IZ8DBJ	1022	281	SM5BFJ	967	372	UT7UW	904	461	N6AR	819
15	I2YBC	1124	105	WC6DX	1076	193	R7DX	1022	284	R7KC	966	373	JG3LGD	901	461	RZ3DX	819
15	OE3WWB	1124	107	HA5KG	1075	196	HA8IB	1021	284	VK3UY	966	373	OM7CA	901	463	JH1OCC	817
15	YT7DX	1124	107	JA1EY	1075	197	K9RR	1020	286	I5OYY	965	375	IK8BQE	897	464	GM7TUD	815
20	K9AJ	1123	107	KD1CT	1075	198	UA9LP	1019	286	K8CW	965	375	JA1NLX	897	464	VA7ZT	815
20	N8JV	1123	107	VE3JV	1075	199	DL8DSL	1018	288	JN3SAC	964	375	RU3SD	897	464	W2FB	815
22	I8XTX	1122	111	KD6WW	1074	199	JA1BPA	1018	289	G0WRE	963	378	R3AW	895	467	N8DX	814
22	IK8FIQ	1122	112	OZ4RT	1072	199	JA7BWT	1018	289	JA8BNP	963	378	UA0FO	895	467	PB1TT	814
24	AA5AT	1121	112	SM3NXS	1072	199	OH2BU	1018	289	RU3FM	963	380	DL1CL	893	469	CT1BOY	813
24	F6BFH	1121	114	UR3IFD	1071	203	IK6DLK	1017	292	EA3GP	962	380	JE3GUG	893	470	JA8ZO	812
24	W1DIG	1121	115	F5PAC	1069	203	JR2KDN	1017	292	OZ1ACB	962	380	SV1GYG	893	471	K8AJK	811
27	W4DKS	1120	116	DL7CM	1068	203	SP7GAQ	1017	294	I0SYQ	960	383	DL4FDM	892	471	RA9CMO	811
28	G3ZAY	1118	116	IT9EJW	1068	203	W5RQ	1017	295	YO7LCB	959	383	KJ3L	892	473	DL6MHG	809
28	N5JR	1118	116	JF4VZT	1068	207	HB9CEX	1016	296	S55SL	958	385	DK1BX	891	473	N9GKE	809
28	OE3SGA	1118	119	IK1AIG	1067	208	G3RTE	1015	297	5B4MF	957	386	AC0A	890	475	VE3ZZ	807
28	SM3EVR	1118	120	HA5WA	1066	208	UA9YJO	1015	297	UY5XE	957	387	W3TN	888	476	IZ5JMZ	806
28	W1JR	1118	120	UT7QF	1066	210	JR0DLU	1014	299	DK2BR	954	388	VK8NSB	886	476	JA9TWN	806
33	IK1ADH	1117	122	JA8MS	1065	211	DJ9HX	1013	299	G3XPO	954	389	HB9BGV	885	476	JE3AGN	806
33	K8NA	1117	123	IK4HLU	1064	212	UT5UGR	1012	299	RN3QN	954	390	I0MOM	878	479	G0JHC	805
35	DL8NU	1116	124	OE6IMD	1063	213	CT4NH	1011	302	I5ZGQ	953	390	RW3RN	878	479	HA0UZ	805
35	N5UR	1116	125	9A5CY	1062	213	DH5VK	1011	303	OZ8BZ	951	392	LX1NO	876	479	JH7DFZ	805
35	ON4AAC	1116	125	SM5FWW	1062	213	RA3DX	1011	303	RA1OD	951	392	N6PF	876	482	SM4AZQ	804
35	SM6CVX	1116	127	JA5IU	1061	213	VE3EXY	1011	303	UY0ZG	951	394	EA3WL	874	483	IZ1ANU	803
39	N6VR	1114	128	DK6NJ	1060	217	JA7DOT	1010	306	DF6QP	950	394	JH1IED	874	483	OZ4O	803
39	ON4IZ	1114	128	ON4ON	1060	217	SP8HXN	1010	306	DL1EJA	950	394	UY5BC	874	485	DK2LO	802
41	DL8USA	1113	128	R3OK	1060	219	HA9PP	1009	306	DL2DXA	950	397	IK2ILH	873	485	F4FEP	802
41	VE3LYC	1113	128	S51RU	1060	220	IT9YRE	1008	309	HB9ICC	949	397	IK2VUC	873	485	JG1OWV	802
43	IK4WMA	1112	132	K1HTV	1059	220	PA0ZH	1008	309	K0AP	949	399	UA4PT	872	485	OE2LCM	802
43	PY7ZZ	1112	133	RA9YN	1058	222	SM3NRY	1007	311	CT1CJJ	948	400	DL1AMQ	871	489	IW9HII	800
45	CT1EEB	1111	134	UA3AGW	1057	223	HA5DA	1006	312	JQ1ALQ	947	401	EA7TG	870	489	PE1NCP	800
45	F6FHO	1111	135	DL5CT	1056	223	SM6CMU	1006	313	EA3BT	945	401	SM6CUK	870	491	OM3DX	799
45	I4GAS	1111	136	DL1BKI	1055	225	DK1FW	1005	314	N4MM	944	403	HA0HW	869	492	F8AMV	798
45	UA9YE	1111	136	N7GR	1055	225	I1FY	1005	315	DL6KR	942	403	I2VGW	869	492	RZ3DJ	798
49	IK1JJB	1110	136	RZ3EC	1055	225	LZ1BJ	1005	316	HB9BIN	941	403	JA2CEJ	869	494	IK2RPE	795
50	IK8DDN	1109	136	VE7KDU	1055	228	VE6WQ	1004	316	VA3DXA	941	406	HB9AMO	868	495	R8DX	794
51	F6AJA	1108	140	DJ3XG	1054	229	SM6DHU	1003	318	JH1QVW	939	407	5B4AHJ	867	496	JA8COE	792
52	4Z4DX	1105	141	G3SJX	1053	230	HA0IH	1002	318	OE3EVA	938	407	G4NXG/M	867	497	I5KG	791
52	K5MT	1105	141	UR3HC	1053	231	IN3ASW	1001	320	JM1PXG	937	407	JF6WTY	867	498	UA4PF	790
54	IK5IWU	1104	143	EU7A	1052	231	UA0CW	1001	321	9A1DX	936	410	M0OXO	866	499	JA1GRM	789
54	K8SIX	1104	143	JH4IFF	1052	231	UA3TCJ	1001	322	JA9GPG	935	411	DK3GG	865	499	UA9JLL	789
56	K1OA	1102	143	UA3AKO	1052	231	UR5ZEL	1001	322	N6FX	935	411	G3KHZ	865	501	DL3KZA	786
56	SM0CXS	1102	143	W1CU	1052	231	VE3VHB	1001	324	R9OK	934	411	RW5C	865	501	JK1TCV	786
58	N7RO	1101	147	JA4UQY	1051	236	DL5MX	1000	324	SM6BZV	934	414	JH7VHZ	860	501	UX1UA	786
58	VE3LDT	1101	147	RY7G	1051	236	JA6LCJ	1000	326	KD3CQ	933	415	DL7VSN	859	504	SP5APW	785
60	GM3ITN	1100	149	AB6QM	1050	238	DK5WL	999	326	UW2ZM	933	416	CT1EGW	858	505	RV1CC	783
60	IT9DAA	1100	149	JA2KVB	1050	239	F5HNQ	996	328	JA1WPX	932	417	K8GI	857	506	SM5ELV	781
60	VE7QCR	1100	149	PT7WA	1050	239	JA1GHH	996	329	IK4DRR	931	418	OE3JHC	856	506	WX4G	781
63	AD5A	1098	149	UT5URW	1050	241	DL2CHN	995	329	IK8CVZ	931	419	DL3JPN	855	508	I5JRR	780
64	EA3JL	1096	153	DL5DSM	1049	242	I4GAD	993	331	AD6W	930	420	AI9Y	854	508	ON5JV	780
65	DL5ME	1095	153	N4WW	1049	242	IK5PWQ	993	332	JE8TGI	929	421	JI3MJK	853	510	F4EUG	779
65	JA1QXY	1095	155	F6DZU	1048	244	I8DVJ	992	332	JH2IEE	929	421	WB5JID	853	510	HB9DPZ	779
67	DF9ZN	1094	155	K5MK	1048	244	IZ4CZE	992	332	RA1QY	929	423	DF7GK	851	512	DL5KUD	777
67	N5ET	1094	157	DL6ZXG	1046	244	K6VVA	992	332	W2KKZ	929	424	RN1ON	850	512	EA3NW	777
67	UA4HBW	1094	157	JA1SKE	1046	244	RA6AR	992	336	IK2ZJN	928	424	UA9FAR	850	512	F8GB	777
70	G3OAG	1093	159	IK5ACO	1045	248	RA3RGQ	991	337	DJ4GJ	927	426	F5PAL	849	512	SM1TDE	777
71	I4MKN	1092	159	K2VV	1045	249	OM3XX	990	337	ON7DR	927	426	IK7NXM	849	516	ER1OO	776
71	WB2YQH	1092	161	JA3FGJ	1042	250	CT1BXX	989	339	JA1AML	926	428	RA0FF	847	516	IK1WGX	776
73	HB9BZA	1091	161	LZ1HA	1042	251	DL6JGN	988	339	UA0SFN	926	429	RL6M	846	518	JA5AQC	774
74	OK1ADM	1089	161	R4GM	1042	251	OH2BF	988	341	SM3GSK	925	430	BA4DW	845	518	N3NT	774
75	DK1RV	1087	161	SM4CTT	1042	253	EA1DFP	987	342	JJ0NCC	924	431	IZ2AMW	844	520	OZ8SW	773
75	G0ANH	1087	161	VK3QI	1042	254	RA3CQ	986	343	JA9BEK	923	432	K6FW	843	521	F4WBN	769
75	OH2BLD	1087	166	7K3EOP	1040	254	W5PF	986	343	RW0LT	923	433	OE3RPB	842	522	JH3CUL	768
75	S52KM	1087	166	G4BWP	1040	256	IZ8EFB	985	343	UA3EDQ	923	434	JH4DYP	841	522	JI3DST	768
75	SP6BOW	1087	166	IT9FXY	1040	256	JH8JYV	985	346	JE7JIS	921	434	MD0CCE	841	724	JA4LKB	767
80	DL1BKK	1085	166	N9BX	1040	256	RJ3AA	985	346	SM3DMP	921	434	NA5AR	841	724	OZ3SK	767
80	DL4MCF	1085	170	9A2EU	1039	259	K0DEQ	983	348	JE1LFX	920	434	ON4VT	841	526	IZ8FFA	766
80	IK8PGC	1085	171	IK8TWV	1038	260	RZ3FW	983	348	OM5FM	920	434	WT2O	841	526	KD7H	766
80	JF1SEK	1085	172	AB5EU	1037	260	SP5TZC	983	348	RK6AM	920	439	DL9UBF	840	528	HB9DKZ	765
80	UA0ZC	1085	172	DJ5AV	1037	262	G3UAS	981	348	SV1FJA	920	439	RA6ATZ	840	528	UR5EDX	765
85	DL6MST	1083	174	N6JV	1036	262	OZ1HPS	981	352	JR2UJT	919	439	WD8PKF	840	528	UR5LCZ	765
85	G0DQS	1083	174	WI8A	1036	264	EI7CC	980	353	EA3NT	918	442	JF1RDH	839	531	JA2FGL	762
85	G3HTA	1083	176	W5ZPA	1034	264	JA3FYC	980	353	RW4K	918	443	SM7DXQ	836	531	JH2RMU	762
85	N6AWD	1083	177	JA9IFF	1033	266	CT1EKY	978	355	I4KMN	917	444	IT9JNR	833	533	DF5JE	761
85	RZ1OA	1083	177	UA4CC	1033	266	JH4GJR	978	355	R7KW	917	444	JA6FIO	833	533	G3TXF	761

2024 Honour Roll

Pos.	Callsign	Total	Pos.	Callsign	Total	Pos.	Callsign	Total	Pos.	Callsign	Total	Pos.	Callsign	Total	Pos.	Callsign	Total
533	IN3NJB	761	577	F4BKV	727	620	DL9MRF	685	664	DL5ZL	641	707	UR4LTX	600			
533	K9RHY	761	579	DL9MKA	726	622	K7ACZ	683	664	G3KZR	641	708	JH3SIF	599			
537	IK8TMI	760	579	EC1AIJ	726	622	UT5ZY	683	664	I8LWL	641	709	DL5XL	598			
538	DL6ZFG	758	581	EI2HY	725	624	HB9IIO	681	667	WA1ZIC	640	710	DL3SUG	597			
539	DL6CNG	756	582	DH2PC	724	624	RJ9I	681	668	DL7VKD	638	711	IK2PZC	595			
539	E73Y	756	582	JA1FGB	724	626	A65CA	678	669	DL1ROJ	637	712	IZ8FQI	594			
539	I2BUH	756	582	K1ZN	724	627	LY3BG	677	669	SM4DDS	637	713	F6FYD	593			
542	CT1AHU	755	585	JN6RZM	722	627	SM5LNE	677	671	DL2VFR	636	713	I4YCE	593			
542	IV3RAV	755	586	OH3MKH	720	627	SP3CJS	677	672	UA6EX	636	713	KJ6P	593			
542	IZ1BII	755	587	W3UR	719	630	IZ2DVI	676	673	N3RW	635	716	IZ8XQC	592			
542	PA3C	755	588	DL5CW	717	631	4Z5AV	674	674	G3KYF	633	717	IZ3ETU	591			
546	IK8CNT	754	589	JE3GRQ	716	632	DL1NAX	672	675	G4KFT	632	718	RU6AI	590			
546	IK8YTA	754	590	N6UK	714	632	JF6XQJ	672	675	IK4DRY	632	719	JH9AUB	589			
546	JF2UPM	754	590	W5VFO	714	634	UR8IDX	671	675	JA2JRG	632	720	AA0FT	588			
549	JJ1CZR	753	592	JA3KZV	713	635	DJ9IN	669	675	VE2BR	632	720	DL4BBH	588			
549	SM0MPV	753	592	OH2FT	713	635	JA6WJL	669	679	HA5JP	630	722	RA6MQ	587			
549	UR7EZ	753	594	DK7MD	711	635	RT0F	669	679	IK0CNA	630	722	SV9AHZ	587			
552	R7LV	750	595	DL4FAY	710	638	DL4MN	667	679	UA3AB	630	724	DF7FC	585			
552	RV9CX	750	596	DL8UAT	708	638	DL8ZBA	667	682	UT3IW	629	724	UT4EK	585			
554	AA4V	749	596	SV2DGH	708	638	DS4DRE	667	683	7K3QPL	628	726	K4HB	584			
554	IK3OYU	749	598	RK9UN	707	641	E72U	664	683	EA3IM	628	727	G4JFS	582			
556	JA7BEW	745	599	YB5QZ	705	641	IZ8DFO	664	685	JG1UKW	626	727	KN7D	582			
557	PR7FB	743	600	DK3DUA	704	643	DK3DG	662	686	DL5DF	624	729	M0AID	580			
558	JH1OAI	742	600	WW8W	704	644	IK1RAE	661	686	JH7CFX	624	729	RM2A	580			
559	W4UM	741	602	HB9BXE	703	644	JA3DAY	661	686	UW5ZM	624	731	SP6MLX	578			
560	CT1JOP	740	602	JH2XQY	703	644	SP7BCA	661	689	DL4SZB	623	731	US0YA	578			
560	F5BOY	740	604	I2OGV	702	644	YO3APJ	661	689	IK2GPQ	623	733	IK2DOT	577			
560	HB9TKS	740	605	IW2FND	701	648	DB3LO	659	689	W8JRK	623	734	JE2VFX	575			
563	W1OW	738	606	OE1ZL	700	648	G4VMX	659	692	HB9AAA	621	735	GW4BKG	574			
564	DL1FU	737	607	F8NAN	699	650	JO1CRA	658	693	VK5MAV	619	736	JH6JMM	573			
565	DL6MIG	736	608	F6ACV	698	651	HB9DDZ	657	694	DL1TRK	618	736	K4PI	573			
565	W9ILY	736	608	HS0ZIV	698	651	RU5A	657	695	IW3SSA	615	736	RA6FG	573			
567	PA7RA	735	608	JA4GXS	698	653	PT7ZT	656	696	DL3TC	614	739	IZ4MJP	572			
567	UA9LBQ	735	611	F4HDR	692	654	OH2BN	655	697	N6VS	612	740	DF8HS	570			
569	DS5ACV	734	612	UN7ECA	691	655	IT9RTA	654	698	G4MFX	608	740	DK3BT	570			
570	JO3AXC	733	613	EV1R	690	656	HA5VZ	652	699	RU6B	607	740	K5WAF	570			
571	JA5CEX	732	614	HL4CBX	689	656	JJ1KZZ	652	699	SM5CBM	607	743	DG5LAC	569			
571	JF2OZH	732	615	W5GAI	688	658	UW5IM	650	701	JA3VPA	604	743	DM3ZF	569			
571	JP3AYQ	732	616	A65BR	687	658	I5HLK	649	702	EA6VQ	603	745	JF7RJM	568			
574	G4IUF	731	616	OE1WEU	687	660	OK2QA	648	702	ER5DX	603						
575	IK4MSV	729	618	IT9RZR	686	661	DJ1OJ	645	704	DL6JZ	602						
576	JH3GFA	728	618	W4HG	686	662	JA6CBG	643	704	K2AJY	602						
577	DL3BRE	727	620	DJ8VC	685	663	AA0MZ	642	706	JH3GCN	601						

Annual Update Deadline

IOTA enthusiasts are reminded that the last date for submitting applications or updates to checkpoints (and mailing cards and fees) for inclusion in the 2025 Honour Roll and other performance tables is 31 January 2025. If submitted/postmarked after that date, they will be processed in the normal way but the scores will be held over to the following year's listing. It is important that members who have not updated since the 2020 annual listings and wish to remain listed should make a submission on or before 31 January 2025.

Official sources of information

IOTA's website (*www.iota-world.org*) provides the following:

- A List of Frequently Asked Questions (FAQ) on IOTA
- IOTA Programme Rules
- A list of authorised IOTA checkpoints
- A detailed listing of IOTA groups by continent, region and country
- A listing of IOTA groups by short title only
- A listing of the most wanted IOTA groups by continent
- Information on how to obtain the IOTA directory

The IOTA website allows you to search the island listings by IOTA group number or island name – and returns the relevant listing together with details of rarity and past and future operations.

Radio Communication (RadCom), the monthly journal of the RSGB (see its website: *www.rsgb.org*), covers IOTA in the monthly HF column.

Sources of information on IOTA activity

DX-World.net

A very attractive website run by a dedicated team of DXers and IOTA enthusiasts, which is updated daily by Col McGowan, MM0NDX. It features a web-form for readers to submit their own DX or IOTA information.

DXnews.com

This is another great website with an IOTA section run by a team headed by Al Teimurazov 4L5A. Well worth bookmarking. As it says, "More than just DX News".

425 DX News

A weekly email DX bulletin issued by IOTA Adviser Mauro Pregliasco, *i1jqj.mauro@gmail.com*. This carries IOTA news – official news releases and listings, DXpedition activity reports, a calendar of forthcoming events – as well as a host of other useful material. For more information, check the 425 DX News website: *www.425dxn.org*.

The Daily DX

The first subscription-based email DX bulletin, published by Bernie McClenny, W3UR (email: *bernie@dailydx.com*). This provides a comprehensive daily commentary on the DX scene with a prominent IOTA section listing island activity. You may prefer to subscribe to The Weekly DX, which is also available. For full details, check Bernie's home page (*www.dailydx.com*).

GPDX IOTA News

Luis, CT4NH, maintains a weekly listing of upcoming IOTA operations under the Grupo Portugues DX at *gpdx.blogspot.com.*

Internet reflectors and forums

IOTA-chasers@groups.io

This provides a meeting-place for IOTA enthusiasts to exchange information and share views about different aspects of the IOTA Programme including island operations, QSL queries, etc.

Other information sources

The DX Summit web cluster *www.dxsummit.fi* is an invaluable way of keeping an eye on the bands when you are away from your QTH. It offers a listing of the last 50, 100, 500 and 1000 spots logged from clusters worldwide, as well as propagation reports. The search facility is a particularly valuable aid to IOTA DXing as you can check for spots for callsigns and IOTA numbers.

2024 Annual Listing

Pos.	Callsign	Total	Pos.	Callsign	Total	Pos.	Callsign	Total	Pos.	Callsign	Total	Pos.	Callsign	Total	Pos.	Callsign	Total
746	JR1WCT	566	827	K0JGH	485	910	N8OC	409	992	DK8MM	334	1074	AB1J	301	1155	EI3CTB	245
746	RA3TAR	566	829	AA1QD	484	911	UT5EL	407	992	IC8TEM	334	1074	DL2RVL	301	1155	IZ1QLT	245
748	IK1NLZ	565	830	K1HT	483	912	VK3GA	406	992	JH4JNG	334	1074	DL8JDX	301	1158	JH8RZJ	244
748	JM1GHT	565	831	DJ6OI	482	913	DL7UGO	405	992	JH7XRG	334	1074	DL9NEI	301	1158	JK1VXE	244
748	VK5CE	565	832	IU7QBB	480	913	JA3QOS	405	992	JI1HNC	334	1074	HL2KV	301	1158	RA4PQ	244
751	JP7EIP	563	832	UR7LY	480	913	LB2TB	405	997	W1GWN	333	1074	PG7M	301	1161	AC0CU	243
752	G3SJH	562	834	JK1KSB	479	916	9A1AA	404	998	DL4CF	332	1080	DH3RD	300	1161	DL2RZG	243
752	VE1AI	562	835	RX1AL	478	916	OH1TM	404	998	IU8FRF	332	1080	DK8IZ	300	1163	SM3OMO	242
754	JA1XZF	561	836	LZ2PEP	477	916	SM1NJC	404	1000	JE1VTZ	331	1080	HA3FMR	300	1164	DJ3XA	241
754	W7YAQ	561	836	MM0EAX	477	919	DS5DNO	403	1001	JK1FNN	330	1080	HB9EHJ	300	1164	OE1HHB	241
756	VE2ACP	560	838	DF1ZN	476	920	DL6FBR	402	1001	RX3DTN	330	1080	HK3W	300	1164	W6AER	241
757	JP1EWY	559	838	EA8AXT	476	920	G3ZSS	402	1003	JA8NSF	329	1080	IK5BSC	300	1167	DM2FX	240
758	AD1C	558	838	HB9JOE	476	922	DL3JXN	401	1004	JA9CHJ	328	1080	M0BUI	300	1167	IK5SRF	240
758	JH0JQS	558	841	G1VDP	475	922	EC7B	401	1004	JL3MCM	328	1080	W7MAE	300	1167	W6JZ	240
760	UN5J	557	842	DL2VM	474	922	JN1RFY	401	1004	K4WA	328	1080	YO3IPR	300	1170	IZ0FVD	239
761	DK7YY	555	842	I4KDJ	474	922	PT7YV	401	1007	DL1HBT	327	1089	F1BFD	297	1170	JA0RQV	239
761	UA9LAO	555	844	DL7UVO	473	922	PY6HD	401	1007	DS5TOS	327	1089	N7QT	297	1170	OE6CLD	239
763	DK6WA	554	845	UA0CID	470	922	UR5ZTH	401	1007	IW1ARB	327	1089	UR0IG	297	1170	W5VY	239
763	OE1PMU	554	846	H44MS	468	928	DF2LH	400	1007	JA0LFV	327	1092	G4FKA	295	1170	W6DE	239
763	R2RZ	554	847	DK7TX	467	928	DL2ASB	400	1011	I2KBD	326	1092	JA2MNB	295	1170	W7FN	239
766	M0URX	551	847	JI1FXS	467	928	RX3AEX	400	1012	DL9WO	325	1094	M0BJL	294	1176	IW0HQE	238
767	DL6MKA	550	849	JO7KMB	466	928	SV5DKL	400	1012	EI7JZ	325	1095	DL8UVG	293	1176	K0HB	238
768	DL9NC	549	849	KM6HB	466	928	VE7JH	400	1014	JI1BJB	324	1095	XE1EE	293	1176	LA9VFA	238
768	G3UHU	549	851	UT5IP	464	933	HL2IFR	398	1014	JR9LKE	324	1097	N7DED	291	1176	OZ0J	238
768	JG3SKK	549	852	WQ5C	463	934	JL3LSF	395	1016	DL2YBG	323	1097	XE1KK	291	1176	SQ8LUV	238
771	HB9BQB	545	853	DM3PKK	459	935	DL8BFV	394	1016	K9OT	323	1099	G4FCI	285	1181	G8AJM	237
772	N1RR	543	853	JJ6DGP	459	936	F5OHW	392	1018	IT9BLB	321	1099	I4FYV	285	1181	JJ0NSL	237
773	N7BT	542	855	DL2DWC	458	937	F5VHQ	390	1019	IZ1LBG	320	1101	4L6QL	284	1183	DL1HTW	236
774	I1YDT	540	855	JA1VSL	458	938	JH3KAI	388	1019	JA3AOP	320	1101	DL4FAP	284	1183	G4NAQ	236
775	JE3SSL	539	857	IW5AOT	457	939	IZ5FSA	387	1019	JG8IBY	320	1103	CT1EEQ	283	1183	K4KGG	236
776	W9OO	538	858	G3SBP	454	940	IZ2IPF	385	1022	HB9AIJ	318	1104	DJ2II	282	1186	IZ2USP	235
777	EA1N	534	859	HL2WA	453	940	R5DB	385	1022	KS4S	318	1104	F4FLF	282	1187	IK3DVY	234
777	PA0MBD	534	860	IZ8CCW	452	942	RV6ANI	384	1022	RD4A	318	1104	JA6CMQ	282	1187	IK6XEJ	234
777	XE1RBV	534	861	JF1MTV	446	943	JE6HCL	383	1025	IW7DOL	317	1107	IK1DFH	280	1189	DL2RUM	233
780	K1NU	531	862	F5LMJ	445	944	AA5JF	382	1025	JA2AYP	317	1108	9A5CB	278	1189	N2DPF	233
781	UX3IA	530	863	EA2WD	444	944	IK7EOT	382	1025	JH1FVE	317	1109	DL8ZAJ	277	1189	RX3X	233
782	CT1BLE	528	863	NL7V	444	946	DL1BSH	381	1028	JF0EBM	316	1109	F6HDH	277	1192	5B4AJT	232
783	DL2DQL	526	865	IV3VBM	443	946	G3XLF	381	1029	F1RAF	315	1111	G4DDL	276	1192	JE6HID	232
783	RN3CT	526	866	JR3CNQ	442	948	IZ2KPE	380	1029	PY7RP	315	1111	K9TF	276	1192	JH1CTV	232
785	CU3AC	525	867	G0TRB	440	948	R4FAN	380	1031	G4FFN	314	1111	LB5WB	276	1192	JK1UNZ	232
785	JM2LEI	525	867	IK6ZDF	440	950	G0PHY	379	1032	HB9DDO	313	1114	AL7KC	274	1192	N2ADE	232
785	VE6BMX	525	867	RN3RQ	440	950	JR3ADB	379	1032	I4JED	313	1115	IU8HEP	270	1197	DL1BSN	229
785	VK3HJ	525	867	RN3RY	440	952	JH1LPZ	377	1032	JA5IGX	313	1115	JA1NQU	270	1197	HA2MM	229
789	DL8JS	520	871	JG4OOU	438	953	G0THF	376	1032	KB1DMX	313	1117	DH1OK	268	1197	I8KRC	229
789	EA5HEU	520	872	IZ8QPA	437	954	F5AKL	374	1032	KO8SCA	313	1117	HB9IQB	268	1200	VA7DXX	228
791	NN7A	518	873	MW0RLJ	436	954	IT9VDQ	374	1037	DH0JAE	311	1117	JE1UMG	268	1200	WW7Q	228
792	JL1ELQ	517	874	EA7BUU	434	956	IK0HFO	370	1037	HL4CJG	311	1120	KL7TC	267	1202	F5AAR	227
793	W9RN	516	874	F5RAB	434	956	IK8GYS	370	1037	JH4HMG	311	1121	VE5MX	266	1202	JG0CQK	227
794	JR7FRW	515	876	JH1BSJ	433	958	DL9FCY	365	1037	JN1FRL	311	1122	WA3FRP	263	1202	LA6OP	227
794	M0YTT	515	877	G4PVM	431	958	IK4AUY	365	1037	N1LID	311	1123	KF6HI	262	1205	JO7BTV	226
796	DK1EI	514	877	IK4YCQ	431	958	JA3HZT	365	1037	WA4MIT	311	1124	DS1JFY	261	1205	WA3WZR	226
797	W3LL	513	879	DL1AY	430	958	SV1MO	365	1043	JE2LUN	310	1125	CT1APN	260	1207	IZ1MHY	225
798	JA5CPJ	511	879	SP7XK	430	962	EA1YO	363	1043	KB9LIE	310	1125	G4DBW	260	1207	K0TRL	225
799	DL8WEM	510	879	W1AL	430	963	DJ3CS	362	1043	OM8FR	310	1127	G4AYU	259	1207	VK3KTT	225
800	JH8GEU	509	882	JE1BJT	429	964	I4IKW	360	1046	G0DWV	309	1127	IZ1XEE	259	1207	W6ZL	225
800	K6UM	509	883	W6AFA	426	964	IK0PEA	360	1046	R6AW	309	1129	HB9TRR	258	1211	DL9ZWG	224
802	VK2DX	508	884	IK6HRB	423	964	JA3AVO	360	1046	VK5GR	309	1130	DG8HJ	257	1211	JA1TBX	224
803	CU3BL	507	885	JA5ALE	421	967	DL6UAA	359	1049	DL3FCG	308	1130	IZ2TBP	257	1213	DL1JPF	223
803	IK2RLS	507	885	N3RC	421	967	JA1MZL	359	1049	HB9ARF	308	1132	WX2CX	256	1213	IK3GIG	223
803	UT3IB	507	885	UA9CGL	421	969	NQ7R	358	1049	JA3UNA	308	1133	CT1DRB	255	1213	IZ2ESV	223
806	N0ODK	506	888	K3VAR	420	970	UT5ZC	357	1049	PY1NP	308	1133	IK2GAJ	255	1213	KM4VI	223
807	KM4HI	503	888	NI0C	420	971	KQ4EE	355	1049	SV2AEL	308	1135	AF2F	253	1213	KN7Y	223
807	US3LR	503	888	W4KVS	420	971	RL8C	355	1054	IZ5MKA	306	1135	JA6WIF	253	1218	KU5B	222
809	VK4CAG	502	891	F4GYM	419	971	W6KGP	355	1054	JA1VRY	306	1135	OE3CHC	253	1219	HB9RUZ	221
810	HB9DOT	501	892	RA9HM	418	974	I3VJW	353	1056	DL3MDJ	305	1135	SP2SGN	253	1219	IK2YJD	221
810	IV3ARJ	501	893	IK2CMN	417	975	K0TT	352	1057	JE2RBK	304	1135	VE1VOX	253	1221	7K4VPV	220
810	W1WBB	501	894	CT1AVR	416	976	DL7HKL	351	1057	N4UOZ	304	1140	DH2PG	251	1221	JA6MWW	220
810	YO9FNP	501	894	DL3LBM	416	976	RM2D	351	1057	VK3SU	304	1140	JJ1HHJ	251	1221	JS1IFK	220
814	DL5XAT	500	894	SP6DVP	416	978	KS1J	350	1060	G0AHC	303	1140	KM2O	251	1221	RZ0AF	220
815	IK5HHA	498	897	IK2YGZ	415	979	RD3AAD	349	1060	G4POF	303	1140	ND4V	251	1221	UA3SAQ	220
815	JG8FWH	498	898	DL2GAC	414	979	WA2BCK	349	1060	IZ8GCP	303	1144	9A2GA	250	1226	HL1VAU	219
817	JG3WCZ	497	898	JA6EXO	414	981	JA4CZM	342	1060	JE1QYI	303	1144	LA3MHA	250	1226	JR3XUH	219
818	K4KKL	495	898	M5BFL	414	982	JH1IFS	341	1060	JI1IXW	303	1144	NF1G	250	1228	DD5MA	218
818	NW7M	495	901	BA4TB	413	982	M5KJM	341	1060	K5KUA	303	1147	LA9PJA	248	1229	IV3IXN	216
820	CT1JOH	494	901	GW4TSG	413	984	IZ7ECL	339	1060	N6DHZ	303	1147	WA1NXC	248	1229	JA0CJK	216
820	UR5WCW	494	901	IZ0RVI	413	984	RM3DA	339	1060	YL2TD	303	1149	G4WGE	247	1229	N6VH	216
822	EI7BA	492	901	YO5BRZ	413	986	HB9DHG	338	1068	DL5NO	302	1149	W9ROG	247	1229	SM3CZS	216
823	G4GIR	490	905	M0KCM	412	987	JE2UFF	337	1068	JA9APS	302	1151	DF9VJ	246	1229	WB1ASL	216
824	JR6CSY	489	906	JF2AXT	411	988	N1EN	336	1068	JH1QKG	302	1151	EI4BZ	246	1234	IK8DNJ	215
825	W6OUL	487	906	W1RM	411	988	N9EAJ	336	1068	JL1QDO	302	1151	IK5AEQ	246	1234	JF1WLK	215
826	G3USR	486	908	EA3CCN	410	988	SV1VS	336	1068	KH2AR	302	1151	JQ1CIV	246	1234	PY8WW	215
827	JH3VWN	485	908	KJ6MBW	410	991	AA6RE	335	1068	RA4DAR	302	1155	EA5UJ	245	1237	IW3GJF	214

2024 Annual Listing

Pos.	Callsign	Total
1238	DL3MB	213
1238	IW2CAM	213
1238	IZ8FTW	213
1238	PA2TMS	213
1242	RN1CW	212
1242	VK3OHM	212
1244	IW2ENA	211
1245	IZ2MHT	210
1245	IZ4IRO	210
1245	JE8VZK	210
1245	JL1JVT	210
1245	N1AM	210
1245	SQ8GBG	210
1251	DL2DQN	209
1251	F5MZE	209
1253	DD0VU	207
1253	EI8IU	207
1253	IK7FPV	207
1253	NH6T/W4	207
1257	DL1USB	206
1257	DO7ES	206
1257	EA1NK	206
1257	IZ0DIB	206
1257	JA4GZK	206
1257	W7GSV	206
1263	F1VEV	205
1263	G8GHD	205
1263	PA3BFH	205
1266	IK1TTD	204
1266	JA1RRA	204
1268	IZ7KHR	203
1268	JA7NGE	203
1268	N2AE	203
1268	N2CJ	203
1272	HB9FKK	202
1272	IU8DKG	202
1272	JA3IWB	202
1272	JH6SCA	202
1272	R0CAF	202
1277	JA2NSH	201
1277	JS1ERB	201
1277	W1FNB	201
1277	YO8CRU	201
1281	DD0VE	200
1281	HL2DBP	200
1281	JR7COH	200
1281	K6EGF	200
1281	K6KZM	200
1281	K8WHA	200
1281	N7AME	200
1281	WL7CG	200
1281	WU4B	200
1290	K4PWS	199
1291	JL1DLQ	197
1292	K1ZE	196
1293	DL8MF	195
1293	K5FUV	195
1293	SM0LPO	195
1296	JE1WBA	194
1296	N4IJ	194
1296	W0PE	194
1299	RC2A	192
1300	DF2GH	190
1301	G4NBS	189
1301	G4VWI	189
1301	IT9IDR	189
1304	DL8UD	188
1305	K9BO	187
1306	G6BFP	186
1307	DL2RTU	185
1308	JP1KOA	184
1308	K6VXI	184
1308	N7AU	184
1308	W4PID	184
1312	SP4MPH	182
1313	IU8IYE	180
1314	N2SO	178
1314	NA9J	178
1316	EA3FZT	177
1316	K4OY	177
1318	JH3LIB	176
1318	N2IGW	176
1320	RN0F	175
1321	JG3KMT	174
1322	JF2VAX	172
1322	W8AKS	172
1324	KY6J	171
1325	K3FRK	170
1326	HB9DWR	169
1326	N4YHC	169
1328	DK3CG	168
1328	IZ5ILC	168
1328	K0BL	168
1331	G3WKL	167
1332	G4SJX	163
1332	IW3QRM	163
1332	IZ3AYS	163
1332	LA9RY	163
1332	OK2CSU	163
1332	VK4COZ	163
1338	NO2C	162
1339	EI2II	161
1339	G0BPK	161
1339	JA3AER	161
1342	DH7RG	160
1342	SV2CLJ	160
1344	G4HUN	159
1344	JI6BEN	159
1346	JA9EJG	158
1347	IZ2ABZ	156
1347	JR7ASO	156
1349	JA1CCJ	155
1350	IC8WIC	154
1350	VK3AWG	154
1350	W3MRL	154
1353	IZ3QFG	153
1354	CT4VB	152
1354	DL5JH	152
1354	JA1UBZ	152
1354	KC7CS	152
1354	SP5AMN	152
1359	JH1IHO	150
1359	VA3TTB	150
1361	KF7RO	149
1362	RD3TBQ	147
1363	CT2FEY	145
1364	DM4EZ	143
1364	G4BLI	143
1364	IZ1JIZ	143
1364	W4ER	143
1368	G5CL	142
1368	I3MDU	142
1370	EA4HKF	141
1370	G0VAX	141
1372	JJ0AEB	140
1373	AG5CN	138
1373	DL2DCX	138
1373	IZ0INX	138
1373	NI4Y	138
1377	AA6YQ	136
1377	EI3HA	136
1377	IK3MLF	136
1377	KC2IGE	136
1381	HB9GWJ	135
1381	JF7NXS	135
1381	N4FN	135
1381	SM6TKG	135
1385	DM4TJ	134
1386	KB3LAN	132
1387	G5HY	131
1388	G6AD	130
1388	JH1OZV	130
1390	6K5BXQ	129
1390	DK5LQ	129
1390	JA8XOD	129
1390	JG1JPE	129
1390	JJ3DKQ	129
1395	KJ2U	128
1396	JH1GBO	127
1396	YO8AAZ	127
1398	PV8AAS	126
1398	VA3CQG	126
1400	DD3KF	124
1400	G6GLP	124
1400	VK3YR	124
1400	W0RMS	124
1404	GI3SG	123
1404	I2OLV	123
1404	IT9FUN	123
1407	R6FFB	122
1408	KJ8O	121
1409	W7DDE	120
1410	F5PLR	119
1411	G0YCE	116
1411	W3DVY	116
1413	IZ8FGO	115
1413	JE6JZP	115
1415	7K1LUE	114
1415	R2IN	114
1417	DM1HR	113
1417	EA4EMC	113
1417	HL2DCM	113
1417	JL1CNY	113
1417	OH4UI	113
1417	PV8RR	113
1417	SV3QUP	113
1417	W7APM	113
1417	ZL1VAH	113
1426	K0UD	112
1427	DL3DUE	111
1428	AG2J	110
1428	DH0GHU	110
1428	DH0GME	110
1428	DK3ME	110
1428	EC1DD	110
1428	ES4AW	110
1428	F4FDA	110
1428	G0TSM	110
1428	G8HXE	110
1428	HA3OU	110
1428	HB9EFJ	110
1428	I1RJP	110
1428	I6COJ	110
1428	IK2UVR	110
1428	IK2WSO	110
1428	IK4MGP	110
1428	IU6PPX	110
1428	IW5BMS	110
1428	IZ8GEL	110
1428	JA1UAV	110
1428	JA1XEC	110
1428	JA2KAK	110
1428	JA4NIJ	110
1428	JA5OXV	110
1428	JA7DNO	110
1428	JE1NVD	110
1428	JE6AVT	110
1428	JF1CCH	110
1428	JG1RYQ	110
1428	JG5DHX	110
1428	JH3FEN	110
1428	JI3CJP	110
1428	JI4WHS	110
1428	JM8FEI	110
1428	K4DS	110
1428	K9OHI	110
1428	KA5WSS	110
1428	KF7ZN	110
1428	NB3R	110
1428	NS4C	110
1428	ON6AB	110
1428	OZ1AA	110
1428	R0JF	110
1428	RA1AOB	110
1428	RA3AV	110
1428	RC2SB	110
1428	RV3BV	110
1428	SM5FUG	110
1428	SP3EMA	110
1428	SQ2TOM	110
1428	UI4F	110
1428	UR4QFP	110
1428	US5CAO	110
1428	US8UA	110
1428	UY2ZZ	110
1428	VE3TG	110
1428	VE7UBA	110
1428	WX0Z	110
1428	WX3P	110
1428	XE3TT	110
1428	YO5LD	110
1489	JK1DDQ	109
1489	K4WSB	109
1489	UT3KW	109
1489	VA3VF	109
1493	2I0WAI	108
1493	G8GNI	108
1493	JA2QPD	108
1493	JJ1PFC	108
1493	N4LKB	108
1493	OH3OJ	108
1499	JI1NNE	107
1499	PE0JBE	107
1501	DJ8XB	106
1501	DL9YCS	106
1501	JA2FYO	106
1501	JL1KBS	106
1501	LU6XQB	106
1501	PV8ABC	106
1501	WV7S	106
1508	AC9GK	105
1508	DO9LX	105
1508	KN4DXT	105
1508	N7WEJ	105
1512	JS6TWW	104
1513	IC8AJU	103
1514	CE4WT	102
1514	JE6PJP	102
1514	WC6Y	102
1517	AA5H	101
1517	DF9TW	101
1517	DL2RPN	101
1517	EA3HRE	101
1517	HK5NLJ	101
1517	JE3WQU	101
1517	K3BAK	101
1517	PG4I	101
1517	UN7AW	101
1526	BI8CKU	100
1526	DC5IMM	100
1526	IU0LFQ	100
1526	IZ1XBB	100
1526	JA3MAT	100
1526	JI5USJ	100
1526	JS6SRY	100
1526	K5UHF	100
1526	KB6IGK	100
1526	KB7JJG	100
1526	OK1NYD	100
1526	R1TB	100
1526	UR5ECW	100
1526	UW7LL	100
1526	XE2SI	100

2024 Club Listing

Pos.	Callsign	Total
1	UT7WZA	1099
2	9A1CCY	1059
3	DK0EE	1040
4	DL0IOA	846
5	DK0PM	799
6	9A1HBC	701
7	IQ2VA	686
8	RO2E	629
9	RY9C	539
10	RN3D	524
11	HB9G	419
12	K6LY	383
13	HB9VC	333
14	UR4CWQ	309
15	LA4C	195
16	IQ2PB	144
17	DL0AVH	143
18	M0RSC	135
19	SK6LK	120

2024 VHF-UHF Listing

Pos.	Callsign	Total
1	IW9HII	236
2	JA6RJK	202
3	JH6BPG	184
4	SM6CMU	176
5	IK4WMA	174
6	G0JHC	171
7	IT9RZR	163
8	JA1UAV	161
9	JE3GRQ	158
10	JH4IFF	156
11	EA6VQ	154
12	K4PI	150
13	HB9RUZ	147
13	JA4LKB	147
15	JA6WJL	144
16	EA3GP	143
17	OZ1BUR	141
18	IK1UWL	135
19	F4BKV	134
19	JF4VZT	134
21	G3KMA	133
22	JI1IXW	130
23	G4BWP	129
24	JA1SKE	120
25	OK1RD	116
25	UT5URW	116
27	JA2FGL	115
27	JP7EIP	115
29	JR3REX	114
29	W1JR	114
31	JA4GXS	113
31	JA6LCJ	113
33	W1NG	109
34	DL5ME	108
34	W4UM	108
36	F4EUG	107
37	SP5APW	104
38	N4MM	103
39	MD0CCE	101
40	DL1EJA	100
40	I5KG	100

2024 SWL Listing

Pos.	SWL Number	Total
1	I1-21171	1132
2	BRS8841	1097
3	W1-7897	933
4	F-59706	867
5	R1A-644/MM	686
6	SM4-3434	645
7	JA4-4665	574
8	SM2-7734	196
9	JA1-22456	129

Trophy Winners

The Society is fortunate to have a large number of trophies

Many of these are awarded to winners of various contests, whilst others give public recognition to some particular aspect of Society work.

They are presented at a number of events - typically the RSGB Convention and the AGM.

Below you will find details of some of the many trophy winners, presented during 2022/2023

The Board

Calcutta Key
For work associated with international friendship through amateur radio

Barry Lewis G4SJH

Founders Trophy
For outstanding service to the society

Martyn Baker G0GMB

Special RSGB Award
For exceptional service to the RSGB and its members

(Not Awarded)

Training & Education Committee

Kenwood Trophy
For making a significant contribution to training and development in amateur radio within the UK

(Not Awarded)

ARDF Committee

3.5MHz Trophy

(Not Awarded)

144MHz Plate

(Not Awarded)

Sprint

(Not Awarded)

HF Contest Committee

1930 Committee Cup
International LP Contest Leading 10W Fixed Unassisted

Dave Cree G3TBK

AFS Club Data Challenge Trophy
AFS Data Contest winning team

Three As CG "A"

AFS Data Challenge Trophy
AFS Data Contest winner single-op

Andy Cook G4PIQ

Ariel Trophy
Club Calls Contest winning single-op

Andy Cook G4PIQ

Braaten Trophy
ARRL DX CW Contest Leading SO G

Nick Totterdell G6XX (G4FAL)

Bristol Trophy
HF NFD Winner LP Assisted Portable

Stockport RS

CDXC Geoff Watts Trophy
Leading Non-UK Fixed M1 or M2

DK3R - DL1ZKA & Friends

Col. Thomas Rose Bowl
Comonwealth Contest winer O, SO, Unassisted

Justin Snow G4A (G4TSH)

Commonwealth Medal
Commonwealth Contest most contribution

Phil Holliday ZL3P (ZL3PAH)

Cyril Leyden G4RYY - Memorial Trophy
Leading UK&CD DXpedition QRP

(Not Awarded)

David Hill G4IQM Memorial Trophy
Club Calls Contest winning team

Camb-Hams "A"

David King G3PFS Trophy
Leading UK&CD Non-DXpedition Mixed

Paul Tittensor G4PVM

Edgware Trophy
AFS CW Contest winning team

Three As CG "A"

Flight Refuelling ARS Trophy
AFS SSB Contest winning team

Three As CG "A"

G2QT Cup
HF Championship winner

Dave Cree G3TBK

G3MZV Memorial Cup
HF NFD winner LP Unassisted Portable SO

Allan Duncan GM4Z (GM4ZUK)

G3PSH Memorial Trophy
SSB FD winner QRP

Parrallel Lines CG

G3XTJ Memorial Trophy
RoLo CW Contest highest placed error-free entrant

Adam Moss G0ORY & Peter Lock M0RYB

G5MY Trophy
RoLo Contests highest combined score

Dave Cree G3TBK

G5RV Memorial Shield
Club Championship winning AFS local club

Norfolk ARC

G8KW Trophy
CQ WW DX CW Contest leading SO Unassisted

Mark Haynes G9W (M0DXR)

GM5VG Trophy
Leading UK&CD DXpedition M1 HP

Gordon Gray & Mark Jones MM1E (MM0GOR & M0UTD)

GMDX (high-power) Trophy
Leading UK&CD DXpedition HP

Pawel Zatylny M0DSL

GMDX (low-power) Trophy
Leading UK&CD DXpedition LP

Ian Pritchard MC5A (G3WVG)

GW4BLE Memorial Cup
ARRL DX SSB Contest leading SO

Keith Kerr GM5X (GM4YXI)

Henry Lewis G3GIQ Memorial Cup
Leading UK&CD Non-DXpedition SSB

Richard Brokenshaw MD7C (M5RIC)

Horace Freeman Trophy
Club Championship winning AFS General Club

Three As CG "A"

Houston Fergus Trophy
International LP Contest leading 10W Portable Unassisted

Norman Mackenzie GM3BSQ (GM3WIJ)

International 10w Trophy
International LP Contest leading 10W Non-UK Unassisted

Holger Wilhelm DL9EE

International QRP Trophy
International LP Contest leading 5W Portable Unassisted
Gediminas Lucinskas LY9A

IOTA Contest Manager Trophy
Leading Non-UK DXpedition M1/M2
Texel Contest Team

John Dunnington G3LZQ Trophy
Commonwealth Contest leading R, SO, Unassisted
John Cockrill G4CZB

Junior Rose Bowl
Commonwealth Contest winner R,SO,Unassisted
Dariusz Tatarski VE3BR

Lichfield Trophy
AFS SSBContest leading SO
Dave Lawley G4BUO

Marconi Cup
AFS CW Contest leading SO
Andy Cook G4PIQ

Milne Trophy
ARRL DX CW Contest leading SO Non-G
Allan Duncan GM4Z (GM4ZUK)

Newbury Trophy
Autumn Series winning AFS local club
Norfolk ARC

NFD Shield
HF NFD winning LP Unassisted, Portable
Sussex Downs CG

Northumbria Trophy
SSB NFD winner LP Unassisted Portable
Sussex Downs CG

Powditch Transmitting Trophy
DX Contest leading SO Mixed
Graham Bubloz G4FNL

Reading QRP Shield
HF NFD winner QRP Portable
Loxlot Club

Ross Cary G3DYY Memorial Trophy
Leading UK&CD Non-DXpedition CW
John Warburton G4IRN

Ross Cary Rose Bowl
Commonwealth Contest leading 12Hr, O, SO, Unassisted
Andy Cook G4PIQ

RSGB 80m Challenge Trophy
Club Championship highest scoring individual
Dave Cree G3TBK

RSGB CDXC Cup
CQ WW DX SSB Contest leading SO Unassisted
Eddy Howells MW0YVK

RSGB HF Contest Committee Trophy
International LP Contest leading 5W Fixed Unassisted
Graham Bubloz G4FNL

RSGB-IOTA Dxpedition Multi-2 Trophy
Leading UK&CD DXpedition M2 HP
Cockenzie & Port Seton ARC

RSGB-IOTA Multi-Op Fixed Trophy
Leading UK&CD Fixed M1/M2
Wisbech ARC

RSGB-IOTA Multi-Op LP Dxpedition Trophy
Leading UK&CD DXpedition LP M1/M2
Addiscombe ARC

RSGB-IOTA Non-UK Dxpedition HP Trophy
Leading Non-UK DXpedition HP
Petar Milicic 9A6A

RSGB-IOTA Non-UK Dxpedition M1 HP Trophy
Leading Non-UK DXpedition HP M1
Hjarno Contest Team

RSGB-IOTA Non-UK Dxpedition M2 HP Trophy
Leading Non-UK DXpedition HP M2
LZ Contest Team

RSGB-IOTA Non-UK Dxpedition QRP Trophy
Leading Non-UK DXpedition QRP
Oliver Droese SV9/DH8BQA

RSGB-IOTA Non-UK Non-Dxpedition CW Trophy
Leading Non-UK Non-DXpedition CW
SerhiyRebrov P3X (5B4AMM)

RSGB-IOTA Non-UK Non-Dxpedition Mixed Mode Trophy
Leading Non-UK Non-DXpedition Mixed
Ken Widelitz VY2TT (K6LA)

RSGB-IOTA Non-UK Non-Dxpedition SSB Trophy
Leading Non-UK NonDXpedition SSB
Alfio Scirto IT9VCE

RSGB-IOTA Single-Op HP Trophy
Leading Non-UK SO HP
Laszlo Vegh OM2VL

RSGB-IOTA Single-Op LP Trophy
Leading Non-UK SO LP
Ivo Novak 9A1AA

RSGB-IOTA Single-Op QRP Trophy
Leading Non-UK SO QRP
Martin Huml OL5Y

Senior Rose Bowl
Commonwealth Contest winner O,SO,Unassisted
Ron Vander Kraats CF2A (VE3AT)

Somerset Trophy
1st 1.8MHz Contest winner
Dave Cree G3TBK

Southgate Trophy
Low Power Contest leading 5W Portable Unassisted
Tim Raven G4ARI

Summer Isles Trophy
Leading Non-UK DXpedition LP
Norbert Moravansky 9A/OM6NM

T E Wilson G6VQ Trophy
DX Contest leading SO CW
Nick Totterdell M5DX (G4FAL)

The Lilliput Cup
Commonwealth Contest winner LP, SO, Unassisted
Richard Ferch VE3KI

The Rosebery Shield
Commonwealth Contest leading SO Assisted
John Sluymer VE3EJ

Victor Desmond Trophy
2nd 1.8MHz Contest winner
Andy Cook G4PIQ

VP8GQ Trophy
Commonwealth Contest leading Non-UK, 12Hr, O, SO Unassisted
Alan Ibbetson 9G5XA (G3XAQ)

Whitworth Trophy
DX Contest leading SO SSB
Andy Goldsmith G1A (M0NKR)

VHF Contest Committee

10GHz Trophy
10GHz Trophy Contest winner
John Quarmby G3XDY

144MHz BackPackers Trophy
144MHz Backpackers Contest winner 5B Section
Steve Clements G1YBB

1951 Council Cup
432MHz Trophy Contest winner
John Quarmby G3XDY

Arthur Watts Trophy
VHF NFD winner LP
Andover RAC

Bill Somerville G4WJS Trophy
Chris Gill G8EEM

Bolton Wireless Club Trophy
Local Club in UKAC leading AFS
Hereford ARS

Bryn Llewellyn G4DEZ Trophy
UKAC leading SO
Pete Lindsay G4CLA

Cockenzie Quaich
VHF NFD leading restricted resident GM
Lothians RS

Denis Jones G3UVR Trophy
SHF UKAC leading SO
Pete Lindsay G4CLA

Foundation Shield
144MHz UKAC leading foundation
Dave Jackson M7ALE

Four Metre Cup
70MHz Trophy Contest leading SO
Allan Duncan GM4ZUK

G0ODQ Trophy
432MHz AFS Contest leading AFS
John Quarmby G3XDY

G3JYP Memorial Award
70MHz CW Contest winner
Phil Guttridge G3TCU/P

G3MEH Trophy
144MHz AFS Contest leading AFS
Camb-Hams "A"

G5BY Trophy
VHF NFD winner M&M
Trowbridge & DARC

G6NB Trophy
144MHz UKAC leading local club
Colchester RA

G6ZR Memorial Microwave Trophy
2.3GHz Trophy Contest winner
John Quarmby G3XDY

Hadley Wood Contest Group Trophy
AFS Super League leading AFS local club
Camb-Hams "A"

Intermediate Shield
144MHz in UKAC leading intermediate
Pete Millard 2E0NEY

John Pilags Memorial Trophy
VHF Championship leading SO Fixed
Dave Butler G4ASR

Low Power VHF Championship Trophy
VHF Championship leading SO Fixed LP
Roger Rimmer MW0OMB

Martlesham Trophy
VHF NFD winner restricted
Windmill CG

Mitchell-Milling Trophy
144MHz Trophy Contest leading O
Reigate ATS & Crawley ARC

Racal Radio Cup
VHF Championship leading O
Colchester RA

Scottish Trophy
VHF NFD leading LP GM
Aberdeen ARS

SMC Six Metre Cup
50MHz Trophy Contest leading SO
Tim Hugill G8X (G4FJK)

Surrey Trophy
VHF NFD winner O
Colchester RA & A1 CG

Tartan Trophy
VHF NFD leading resident GM O
Cockenzie & Port Seton ARC

Telford Trophy
50MHz Trophy Contest winning O
Blacksheep CG

Thorogood Trophy
144MHz Trophy Contest leading SO
Allan Duncan GM4ZUK

UKAC 144MHz Club Trophy
144MHz UKAC leading AFS general club
Northern Fells CG

UKAC Club Challenge Trophy
UKAC leading AFS general club
Northern Fells CG

VHF Championship AFS Trophy
VHF Championship leading AFS club
Hereford ARS

VHF Contests Committee Cup
1.2GHz Trophy Contest winner
Colchester RA

VHF Manager's Trophy
70MHz Trophy Contest winner O
G4G (G4FZN & G8HQW)

Spectrum Forum

(HF Manager)

G5RP
For outstanding and consistent DX work
Stephen Armstrong 2E0SUG

ROTAB
For greatest progress in the field of HF DX
Andy Brown G4EZT

(VHF Manager)

1962 Committee Cup
For outstanding amateur development at VHF/UHF
Not Awarded

Don Cameron G4STT Award
For outstanding contribution to low power radio communications
"Tex" Swann G1TEX

Fraser Shepherd Award
For research into microwave applications for radio communications
Frank Schmaling DL2ALF

Harold Rose Trophy
For outstanding contribution to 50MHz
Chris Deacon G4IFX

Louis Varney Cup
For advances in space communication
Graham Shirville G3VZV

Technical Forum

Bennett Prize G8PF
Significant contribution furthering the art of radio communications
Brian Austin G0GSF

Courtenay Price Trophy
For outstanding technical contribution to amateur radio
Andy Talbot G4JNT & Martin Farrell G8ASG

Norman Keith Adams Prize
For the most original article published in RadCom
David Bowman G0MRF

Ostermeyer Trophy
For the description of a home constructed equipment in RadCom
Tim Forrester G4WIM

Wortley-Talbot Trophy
For most outstanding experimental work in amateur radio
Martin Peters G4EFE

Propagation Committee

Les Barclay Memorial Trophy
For contribution to propagation research
Gwyn Griffiths G3ZIL

UK Microwave Group

24GHz Cumulative Trophy
24GHz Trophy Contest leader
Not Awarded

47GHz Cumulative Trophy
47GHz Trophy Contest leader
Roger Ray G8CUB

Dain Evans, G3RPE Memorial Cup
10GHz Cumulatives Open winner
John Lemay G4ZTR

Dave Cox, G0RRJ Memorial Trophy
24GHz Cumulatives winner
Martyn Vincent G3UKV(SK)

G4EAT Trophy
UKuG Low Band Championships leading 1.3GHz
Combe Gibberlets M0HNA

Jack Brooker, G3JMB Trophy
10GHz Cumulatives Open winner
Adrian Dodd M0PAI

Tim Leighfield, G3KEU Trophy
5.7GHz Cumulatives leader
Telford & DARS G6ZME

G3EEZ Memorial Trophy
For Contributions to Microwave Communications
Neil Smith G4DBN

G3JVL Award
The newcomer who has made the greatest contribution to microwave communication
Dave Newman G4GLT

G3VVB Trophy
For the best home-constructed microwave equipment exhibited at a microwave roundtable or convention
Mark Hughes GM4ISM

Les Sharrock, G3BNL Trophy
For innovation or technical development of microwave equipment or techniques
Dave Austen G1EHF

Amateur Television

Radio Amateurs have been transmitting and receiving TV pictures for over 65 years. In most cases, simplified versions of the broadcast standards of the day have been used, perhaps tailored to a reduced bandwidth to fit within the amateur bands. All amateur bands above 50MHz are suitable for amateur TV (ATV), and advances in reduced bandwidth DVB-T transmission may allow that mode to be used for impressive DX contacts on 29 MHz and the lower VHF bands.

Licensing

No extra licence is required to transmit amateur TV, you just have to adhere to the existing rules of your licence. This does mean that Foundation licensees can transmit ATV in the 432MHz and 10GHz bands if their equipment meets the required standards. The regulations on transmitted content and station identification are the same as for voice, so most transmissions are pictures of the operators themselves, their equipment, their latest project or their surroundings. The station callsign needs to be transmitted every 15 minutes, either on-screen or in the digital station identification.

Transmission Modes

Although amplitude modulation (AM) was used for many years, it is now rarely employed because there is insufficient bandwidth to accommodate it within the 432MHz amateur band, and it is difficult to generate efficiently on the higher bands. As AM TV is no longer used commercially in the UK, receiving equipment is also harder to find.

Frequency modulation (FM), as used for early commercial satellite TV transmissions, can be used on the 1296MHz band and above, although restrictions to safeguard GNSS may limit the use of FM on 1296 MHz in some countries. Generally amateur FM ATV transmissions use half the deviation level of satellite transmissions (to reduce bandwidth occupancy and increase signal to noise ratio) with an FM audio subcarrier at 6MHz. The exception to this is at 5.6GHz, where repurposed hobby drone TV transmitters using the full deviation can be used without modification on 5665MHz.

Digital television modes, as used for current commercial satellite TV transmissions, have proved to be very robust and efficient for ATV transmissions. The DVB-S and DVB-S2 modes are both used, and can be generated from software defined radios (SDRs) driven by user-friendly Linux or Windows computers (including the Raspberry Pi). These transmissions can be received on a domestic satellite TV receiver (with an up-converter or downconverter for bands other than 1296MHz), or by using a specialised ATV receiver such as the PicoTuner or MiniTiouner.

A typical ATV Picture from MODTS

The published standard for Satellite TV does not include transmissions of less than 1.2MHz wide (generated using a digital symbol rate of less than 1 MSymbol/sec), however the amateur-developed tuners can decode signals of far lower bandwidths, and compatible transmissions are easily generated using computers and SDRs. The advantage of this is that the reduced bandwidth transmissions can be decoded from weaker signals than would be the case with wider bandwidth transmissions. These signals are often referred to as Reduced-Bandwidth TV (RB-TV) and typical symbol rates used are 500 kSymbol/sec (kS), 333 kS, 250 kS,125 kS and 66 kS.

Commercial terrestrial transmissions in the UK use the DVB-T and DVB-T2 transmission standards; these usually have a fixed 6 or 8MHz bandwidth, making it difficult to fit into some amateur bands and requiring higher power than the narrower RB-TV DVB-S and DVB-S2 modes. However, recent developments in commercial tuner technology have enabled amateurs to develop reduced bandwidth receivers for DVB-T and SDR software has been developed to enable compatible reduced-bandwidth transmissions. The use of reduced-bandwidth DVB-T should provide advantages in situations where multipath is a problem, such as for DX contacts on 29MHz and 50MHz or for mobile operation on 437MHz.

Receiving Digital Amateur TV

The most popular ATV receive system is based on a specific Satellite TV Receiver module with a USB interface to a computer running a receiver program. There are 2 variants; the recently-released PicoTuner and the older MiniTiouner. The PicoTuner uses a Raspberry Pi Pico to provide the USB interface and is capable of receiving 2 transmissions simultaneously (which is useful for monitoring QO-100 ATV contacts). The MiniTiouner is an older design that uses a more expensive FTDI module for the USB interface. There is free Windows and Linux software available, and most is compatible with both tuner versions. The exception is the aging MiniTioune Windows software which is only compatible with the MiniTiouner.

The recommended Windows software is "OpenTuner"; this has the dual reception capability when used with the PicoTuner, but is also compatible with the MiniTiouner. The Linux software is available in 2 pre-packaged variants for the Raspberry Pi 4: Ryde and Portsdown. The Ryde is a set-top box style receiver that uses a PicoTuner or MiniTiouner together with a Raspberry Pi 4 and an IR sensor to provide a remote-controlled receiver to drive any HDMI TV or monitor. The Portsdown is a DATV transceiver, again based on the Raspberry Pi 4, that uses a 7 inch touchscreen to display the pictures. Software for the Ryde and Portsdown is available from the BATC's GitHub site. OpenTuner software is available from the author's (ZR6TG) website.

The PicoTuner hardware needs to be self-built and has been designed as a beginner's construction project. The tuner and PCB are available from the British Amateur Television Club shop and the other parts are available from DigiKey or Mouser. Parts for the MiniTiouner are also available, although new builders are recommended to build the PicoTuner which can be built for under £100. Both designs are fully described on the BATC Wiki (*https://wiki.batc.org.uk/*).

The tuner input frequency range is 144MHz to 2650MHz so, with the additionof preamplifiers, it will receive on 4 amateur bands directly and it can be used withdown-converters (or up-converters) for the other bands. Being a satellite TV tuner, it is designed to operate with 50 dB of gain between it and the aerial, so it always benefits from a preamplifier for amateur use.

Transmitting Digital Amateur TV

The simplest option for transmitting Digital ATV is to use the BATC Portsdown software on a Raspberry Pi 4 with a touchscreen and

The PicoTuner Hardware

LimeSDR Mini or Pluto SDR. An alternative for Windows is the DATV Easy software with either of these SDRs. All of these solutions are broadband allowing the generation of signals anywhere in the frequency range 70MHz to 3450MHz. The signal from the SDR at about 0 dBm (1 mW) can either be amplified directly in a (very linear) power amplifier, or up-converted to the frequency required. For many microwave operators, existing transverters can be used with minimal modification. The PicoTuner or MiniTiouner can be tuned to the normal receive IF, and the transmit signal generated at the same IF..

Operating Standards

Commonly used operating frequencies and modes are listed below. Talkback is generally on 144.75 MHz FM, or the DXSpot.TV website *https://www.dxspot.tv/* can be used for chat at longer ranges.

Amateur TV By Satellite

The launch of Es'hail-2 with its QO-100 wideband transponder has enabled amateur TV contacts between stations in the UK and Europe, Africa and parts of Asia and South America. Reception of these transmissions is relatively easy with an 80 cm dish, a commercial LNB and a PicoTuner or MiniTiouner.

The wideband transponder downlinks between 10490.5MHz and 10499.5MHz using horizontal polarisation, and a normal "Universal" LNB converts this down to 740.5 – 749.5MHz when 18v is supplied to the LNB. This signal can then be directly demodulated by a PicoTuner or MiniTiouner. A good first signal to look for is the beacon on 10491.5MHz. The tuner parameters for this will be a frequency of 741.5MHz, symbol rate 1500, DVB-S2. Once the beacon has been received, other signals found on the BATC/AMSAT-UK online spectrum monitor *https://eshail.batc.org.uk/wb* can be tuned in.

The uplink for the wideband transponder is from 2401.0MHz to 2410.0MHz Right Hand Circular Polarised. Uplink transmissions should be DVB-S or DVB-S2 at less than 2 MS. Typically 30 W into a 1.2 meter dish is required to uplink a 333 kS digital ATV signal.

Further Information

The British Amateur Television Club (BATC) (*https://batc.org.uk/*) publish a quarterly magazine with all the latest ATV news and construction projects. They also maintain a Wiki (*https://wiki.batc.org.uk/*) with lots of useful information, and run a forum about amateur television (*https://forum.batc.org.uk/*). Membership of the BATC can be purchased online and is free to students in full-time education.

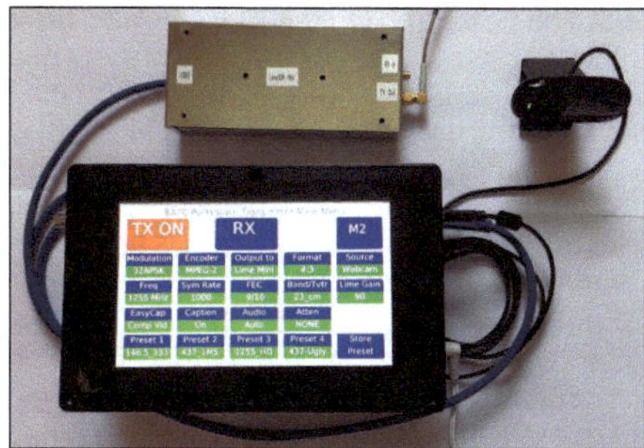

The BATC Portsdown Transmitter and LimeSDR

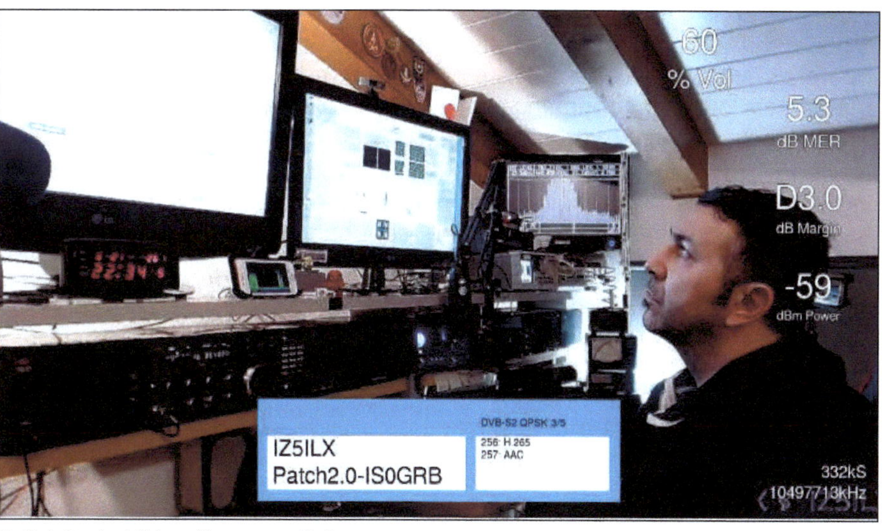

The Ryde Received Picture with on-screen-display

Frequency	Mode	Parameters	Notes
29.25 MHz	DVB-T	333 kHz QPSK	Max 500 kHz bandwidth
51.7 MHz	DVB-T	333 kHz QPSK	Max 500 kHz bandwidth
71.0 MHz	DVB-T	333 kHz QPSK	NoV required. 70.5 - 71.5 MHz
	DVB-S2	333 kS QPSK	
146.5 MHz	DVB-S2	333 kS QPSK	NoV required. 146.0 – 147.0 MHz
437.0 MHz	DVB-S2	333 kS QPSK	
		1 MS QPSK	Band Plan 436.0 – 438.0 MHz
1256.0 MHz	DVB-S2	Various SRs	GNSS-friendly frequency
2395.0 MHz	DVB-S2	Various SRs	Also used for ISS HamTV
3405.0 MHz	DVB-S2	333 kS	
5665.0 MHz	FM ATV	Wideband FM	Using FPV Drone equipment
5762.5 MHz	DVB-S2	333 kS	Using NB Transverters (from 146.5)
10370.5 MHz	DVB-S2	333 kS	Using NB Transverters (from 146.5)
24047.5 MHz	DVB-S2	333 kS	Using NB Transverters (from 143.5)
47090.5 MHz	DVB-S2	333 kS	Using NB Transverters (from 146.5)
75978.5 MHz	DVB-S2	333 kS	Using NB Transverters (from 146.5)

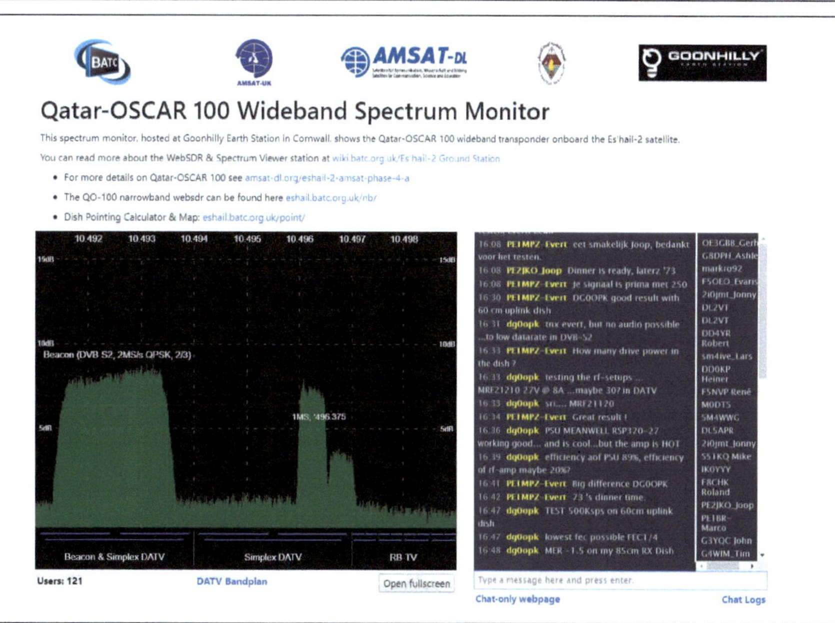

The Online Spectrum Monitor for the QO-100 Wideband Transponder

Amateur Satellites

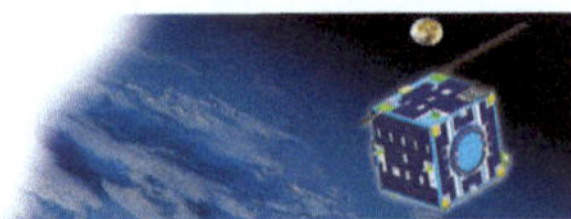

Since soon after the launch of the first artificial satellite, "Sputnik-1", radio amateurs have been constructing and operating amateur satellites. There have been more than 100 amateur satellites launched since then, of which more than 20 are currently operating and available to amateurs. In the early days, only a few well-equipped stations with operators well versed in orbital mechanics were able to make QSOs consistently.

Nowadays, modern technology takes the strain off the mathematics and physics, making listening to and communicating via amateur satellites far easier than it used to be. Indeed it's often the case that amateurs find that they already have all the equipment and knowledge required to start operating amateur satellites.

All radio amateurs licensed in the UK can communicate via satellites, Foundation licensees are actively encouraged to join in!

How far can I communicate via an amateur satellite?

How far a station can expect to communicate using an amateur satellite depends on a number of factors, but it is possible to operate almost globally using amateur satellites. One benefit of using amateur satellites as a communication medium is that it is possible to predict very accurately and consistently when a QSO can be made. Because satellite communication almost always demands line of sight between the ground station and the satellite, the vagaries of propagation present in many other DX modes can generally be ignored.

What's onboard an amateur satellite?

Different amateur satellites have differing capabilities in their payloads. Analogue satellites provide voice and CW transponders. Some analogue transponders simply receive a single channel FM signal on one band and retransmit the channel on another band. Others carry linear transponders designed for CW and SSB QSOs. Some satellites support an international version of the APRS network. Additionally, there are a rapidly increasing number of very small 'CubeSats' which generally provide simple CW telemetry. These are often only 10cm cubes and have a mass of only around 1kg. As technology progresses these small spacecraft are becoming more capable.

There are several amateur satellites providing digital capabilities. Some may simply provide telemetry information, while others provide two-way communications from simple single-channel digipeating to store-and-forward BBSs. Over the years, there have been a large number of different hardware modems required for digital satellite operation. Thanks to software DSP techniques, almost all of these modems are now emulated using a PC and a soundcard and, increasingly, Raspberry Pis and similar hardware.

Unlike a terrestrial repeater, almost all amateur satellites receive on one band and retransmit on another. This saves on payload weight on the satellite where potentially expensive, large and heavy filters would be needed. At the ground station end, it allows operators to listen easily to their own signals without de-sensing their receivers.

Operating full duplex is recommended practice on amateur satellites. If you can't hear your own signal on the satellite's downlink, it's difficult to know whether you are on frequency or even making it to the satellite at all.

ARISS

As well as autonomous spacecraft, amateur radio also has a permanent presence in space onboard the crewed International Space Station. ARISS (Amateur Radio on the International Space Station) is an international organisation comprised of national amateur radio societies, national AMSAT societies and five space agencies (NASA, ESA, CSA, JAXA and Roscosmos). It is designed to let students worldwide experience the excitement of talking directly with crew members of the International Space Station, inspiring them to pursue interests in careers in science, technology, engineering and mathematics (STEM), and engaging them with radio science technology through amateur radio. It is also possible for individual amateurs to hold random QSOs with astronauts - surely a highlight for any amateur's career!

2021 saw the installation of a new generation radio system called the Interoperable Radio System (IORS) for the US space station segment and the Russian space station segment has recently upgraded its radio to the IORS model. This has seen the return of activities that do not require the crew to be involved such as an enhanced cross band (U/V) repeater functionality as well as an upgraded digital packet (APRS) station and slow-scan television (SSTV) capabilities. It is envisaged that the cross band repeater mode and the packet/APRS mode will be active simultaneously in the near future.

Future upgrades and enhancements to the IORS are in various stages of design and development. These include an upgraded Ham Video system L-band (uplink) repeater, ground command operations capability, LimeSDR signal reception, a microwave 'Ham Communicator,' and a Lunar Gateway prototype experiment.

During 2016, Tim Peake, the first British ESA Astronaut, made contact with ten schools throughout the UK. Most of these also included live video from the HamTV installation. These were the first-ever ARISS contacts to have video as well as audio downlinks and this facility contributed greatly to the impact of the contacts. Great support was given by the ARISS UK Operations Team, the UK Space Agency and the STEM outreach teams of the RSGB itself.

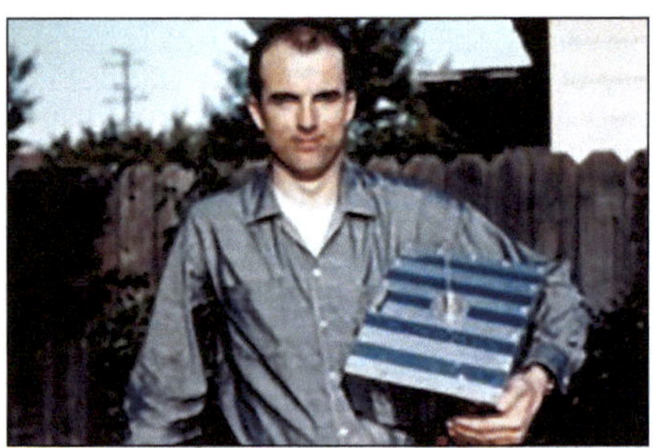

Lance Ginner, K6GSJ, poses with the flight model of Amateur Radio's first satellite, OSCAR I. The blue, stick-on label on the top of the spacecraft reads: "OSCAR I – AMATEUR RADIO BEACON SATELLITE"

ARISS school contacts continue to occur in the UK at a rate of on average once every year. Competition for inclusion is high, and schools that apply are expected to pursue a course of activities and learning opportunities for students in STEM/STEAM related subjects, including amateur radio. Applications to take part in the worldwide ARISS program occur in "windows" from January to March and September to November each year. Successful applicants can normally expect their contact to be carried out within approximately 12 months

Application forms and more information can be found at *http://www.ariss-eu.org*

Amateur satellite orbits

Although, these days, amateur satellite operators certainly don't need an understanding of orbital mechanics, knowing some basics of a satellite's orbit will help when operating.

Some satellites orbit higher in space than others. The Low Earth Orbit (LEO) satellites have coverage areas (or 'footprints') of between 3000 and 4500 miles range. Because of their low orbit, they complete an entire orbit in 90+ minutes, so they are only "visible" for a maximum of about fifteen minutes each orbit. Because of the short 'pass', QSOs and overs tend to be of short duration.

The 'Phase 3' High Earth Orbit (HEO) satellites such as AO-40 had the benefit of a much larger footprint because they were designed to operate from a much higher altitude. They also appeared to the observer on the ground

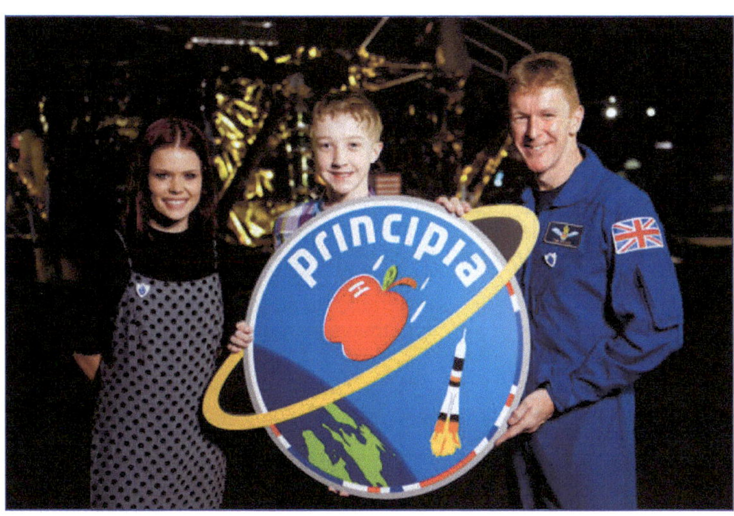

Blue Peter presenter Lindsey Russell, with competition winner Troy and UK ESA astronaut, Tim Peake with the patch.

to be hovering around for several hours at a time, perfect for a ragchew. These spacecraft were in a highly elliptical orbit but, at the time of writing, there are no operational Phase 3 orbiting satellites.

The recent launch of a "Phase 4" geostationary satellite, called Es'hail-2 or Oscar 100, represents a step-change for amateur satellite operation. It is in a geosynchronous orbit, located at 25.9° East and covers more than 1/3rd of the earth. It has been developed as a result of co-operation between the Es'HailSat Company, the Qatar Amateur Radio Society and AMSAT-DL. Its amateur payload comprises of two linear transponders.

These transponders operate with 2400MHz uplinks and 10450MHz downlinks. They comprise of a 500kHz bandwidth linear transponder intended for conventional analogue operations and an 8MHz bandwidth transponder for experimental digital modulation schemes and amateur television using DVB-S and DVB-S2 modulation at a variety of symbol rates.

This spacecraft appears stationary in the sky and is therefore available for use 24/7. It also does not require tracking antennas so the ground station implementation can be quite simple. More information about this unique amateur spacecraft, and how to use it, can be found at: *https://eshail.batc.org.uk* You can also listen to the downlink using the web SDR at the Goonhilly Earth Station *https://eshail.batc.org.uk/nb/*

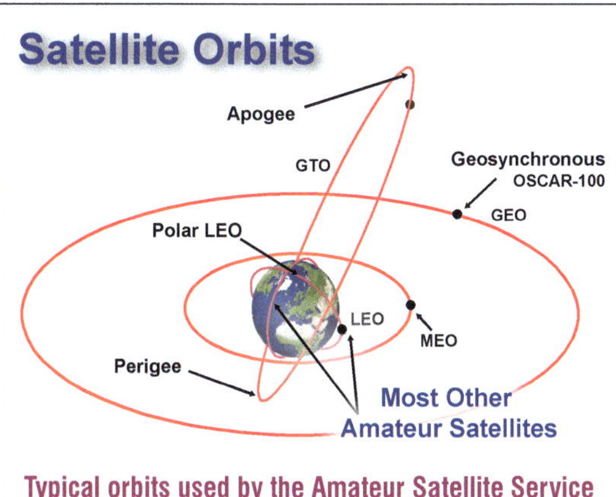

Typical orbits used by the Amateur Satellite Service

How do I find a satellite?

To figure out when a LEO satellite is passing over, and where it will be in the sky at a given moment in time, prediction software is used. If the station has directional antennas, most prediction software can also steer the antennas in real-time. It's important to ensure that prediction software has up-to-date orbital parameters, called Keplerian Elements, or often simply as 'Keps' or TLEs (Two Line Elements). These parameters are downloadable from the Internet and most software can be configured to obtain these updates automatically.

Because satellites are moving relatively rapidly compared to the ground station, there will be some degree of Doppler shift in the receiving and transmitting frequencies. Doppler shift is experienced in daily life when, for example, the tone of an emergency vehicle's siren drops as it passes. When operating amateur satellites, Doppler shift may be adjusted by manual tuning, or alternatively most prediction software can update the radio's frequency in real-time.

Ground Stations for operating amateur satellites

It is not necessary to have a large or expensive station to operate satellites. In its simplest form, a dual-band handheld FM transceiver can be used to make satellite contacts. When used with a handheld Yagi antenna, this simple configuration can make an effective station for portable contacts.

At the other end of the scale, a station may have a steerable antenna array under automatic computer control and a radio with satellite-specific functionality such as full-duplex operation, SSB capability and tracking uplink and downlink VFOs. The benefit of the larger station is that it allows the operator to use more satellites more consistently.

A well-equipped satellite station has antennas of similar size to a standard domestic TV and FM Band II antenna configuration, with perhaps eight elements on 70cm and four on 2m. This makes it surprisingly easy to make a neighbour-friendly and capable amateur satellite station. These antennas are usually crossed dipoles and can either be steerable in azimuth only or, for more consistent QSOs, in elevation too.

Rather than purchasing expensive Az/El rotators, antenna pointing is very often done manually with the antennas at ground level, especially in temporary or portable configurations. Because the satellite is above the horizon in the sky, locations often considered ineffective for terrestrial radio communication can be effective for amateur satellites.

As technology progresses, more satellites will be operating on the higher bands. For microwave use, a small dish is employed and because of their narrower beamwidths, it's essential to be able to point the antennas accurately both in azimuth and in elevation.

Perhaps the most important rule of thumb in satellite operation is to concentrate on the station's receiving equipment before investing time, money and effort in the transmitting side. The nature of satellite communications means that the old adage 'if you can't hear them, you can't work them' is especially true.

It is often tempting to improve your signal by increasing your station's ERP. It is likely that the transponder may already be limiting (eg, in linear transponders, the transponder's AGC has started to attenuate the passband so that its output can be maintained in the linear region), so more ERP is not going to be beneficial. Masthead preamps are always beneficial.

How are amateur satellites launched?

Historically, amateur satellites have been launched by generous space agencies with space available on their rockets. Often this space would otherwise have been dead weight ballast. Recently most of these free rides have dried up, but the launch costs for CubeSats can often be managed. The collaboration between radio amateurs and universities is also leading to having shared missions as described below.

Who makes amateur satellites?

Amateur satellites continue to be made by many organisations throughout the world, by individual national AMSAT societies, or a collection of national AMSAT societies. Many educational establishments have also launched amateur satellites.

AMSAT-UK is at the forefront of satellite build-

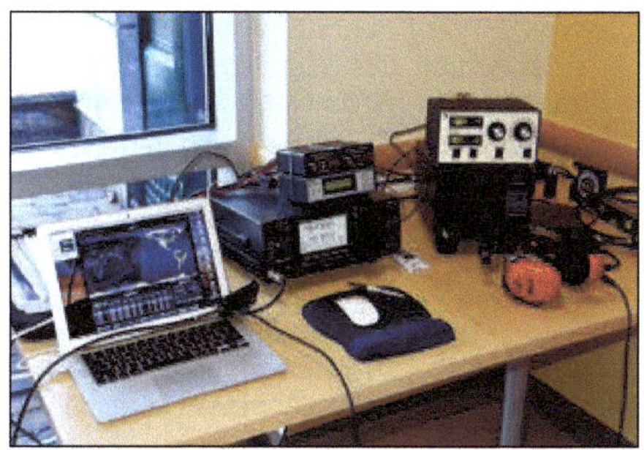
Picture from N1FD showing a typical, well equipped amateur satellite ground station.

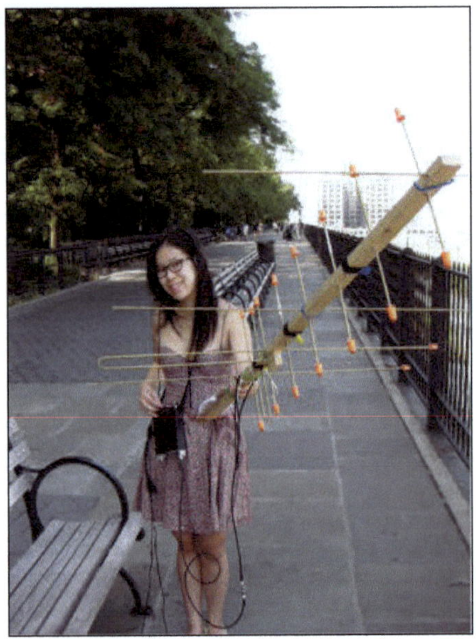

Picture from Makezine.com showing Diana Eng with her portable ground station and homemade yagi antenna.

ing and AMSAT-UK has already created the FUNcube-1 CubeSat in collaboration with AMSAT-NL which was successfully launched in late 2013. This also carries the name Oscar AO73 and acts as a linear 70cm to 2m transponder at night and during weekends and holidays. At other times it provides telemetry data for educational outreach for schools and colleges. The sharing of this resource is intended to encourage the uptake of STEM (Science Technology Engineering & Mathematics) subjects as well as increasing knowledge about and interest in amateur radio.

This spacecraft continues to operate nominally and has provided more than 1.6GB of data which is stored on a central Data Warehouse and which is available for research purposes.

AMSAT-UK also provided a similar FUNcube-2 payload for the UKube-1 spacecraft which was developed for the UK Space Agency. This spacecraft is also in orbit and although it has completed its science mission the transponder continued to function until mid-2018. Additionally, early 2017 saw the successful launch of Nayif-1 which provides a further addition to the FUNcube constellation.

Their latest launch, in late 2018, was JY1Sat. This is based on the FUNcube mission of having a transponder for radio amateurs and educational out-reach, with images in SSDV format, for schools and colleges. It was developed under the auspices of the Crown Prince Foundation in Jordan. At the time of writing this spacecraft is operating in transponder mode without any telemetry.

All artificial satellites have a limited lifetime. Solar panels gradually become less efficient as they are bombarded by the radiation in space, from which we are protected here on Earth. On other occasions, it might be that the rechargeable batteries fail first. Because of their inherent limited lifetime, there is a continual need for replacement satellites.

The longest surviving amateur satellite is AO-7, which was launched in 1974. AO-7 went silent for over two decades until Pat Gowen, G3IOR, heard it again in 2002. It is believed that AO-7 originally stopped functioning due to the battery going short circuit. After many years of the cumulative electrical and chemical stresses on the battery from the solar cells attempting to charge them, one of the cells in the battery went open circuit and now AO-7 is available again - although only when it is in sunlight. It is understood to be THE oldest earth orbiting spacecraft still in active use!

Where to find more information on amateur satellites

The AMSAT-UK and AMSAT-NA websites are very good sources of the most up-to-date information about satellites, and there is also a very active AMSAT reflector available on the internet. To join, send an email with the following text on the first line of your message: subscribe AMSAT- your request to *majordomo@amsat.org*

The RSGB has also published "HAMSATS & AMSATS", a book by Andre Barron, ZL3DW, which provides lots of useful information.

Who pays for amateur satellites?

Funding for amateur satellites comes from a number of sources, including the national radio societies and national AMSAT societies.

In the United Kingdom, AMSAT-UK represents the interests of amateur satellite operators, and, as discussed above, AMSAT-UK members have been instrumental in the development, manufacture and funding of several amateur satellites. With the generous donations and subscriptions of its members, AMSAT-UK helps to keep amateur satellites in space. So if you use amateur satellites, remember to contribute by joining AMSAT-UK! See *https://www.amsat-uk.org*

Which amateur satellites are currently operational?

The list can change very quickly with new launches taking place and some failing or burning up and de-orbiting... It is therefore not possible to publish a sensibly up to date list here. The *http://www.amsat.org/status/* page shows a large amount of detail for each spacecraft and this is updated by the hour. More details of each spacecraft can be found here *http://www.dk3wn.info/p/?page_id=29535* As will be seen, most satellites presently use frequencies in 2m or 70cm bands for which equipment is very readily available. A simple on-line satellite prediction service is provided here *http://www.amsat.org/track/* Be sure to enter your location first and then select the "amateur satellites" tab

Try amateur satellites for yourself!

Getting started on amateur radio satellites is straight forward, it is recommended that you read the AMSAT-UK an AMSAT 'beginners' web pages and the Oscar status page at: *http://www.amsat.org/status/*

Both operating schedules and modes of satellite operation regularly change. Over time, batteries, solar panels and other hardware can and do fail, and there's little opportunity to repair satellites once they're in space.

For first timers it is likely you will start on the FM satellites, the most common being AO-91 and AO-92. Both of these are equipped with single channel FM bent pipe transponders with a 70cm uplink and 2m downlink, however you will need a CTCSS tone of 67Hz to access, it's also worth noting that AO-92 also contains a 23cm uplink which is operated on a regular schedule.

To use either of these two satellites it's recommended that you operate in duplex mode, that is you can hear the satellite downlink at the same time as transmitting, this reduces interference to other users and increases your chances of a contact, Duplex can either be achieved with a handheld radio which supports it or two separate radios.

For an antenna, it's very beneficial to have a dual band antenna with some directional gain such as the Arrow or Elk antenna, if you prefer, you'll find plenty of online designs - a popular choice is the dual band satellite yagi design which is on M1GEO's website: *https://www.george-smart.co.uk/antennas/dual_band_satellite_yagi/*

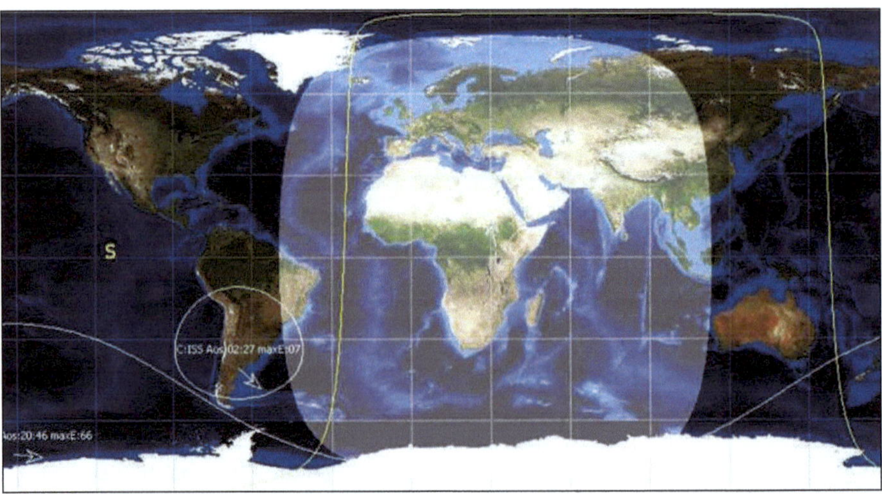

Earth Coverage of the Es'HailSat-2 also known as Oscar 100

The picture shows the well-engineered antenna system of Matthias DD1US on a sunny but, cold day.

Once you have those items, you will need a method to find when the satellite will pass over you, a simple method is the online prediction tool by AMSAT (*https://amsat.org/track/*) or you can use one of the many mobile applications for example Heavens Above or ISS Detector.

Before you transmit it's a good idea to follow some best practices that help all users, especially when a transponder is a single channel.

1. Share the Pass - FM satellites are just repeaters, only one person can work it at a time, so don't monopolise a pass, remember to let others have a go.
2. Let other QSOs Finish - Always let a QSO finish, transmitting over an on-going contact causes frustration and means there's less time for other QSOs.
3. Minimise Repeat QSOs - As FM satellites only have a single channel, it's best that on busy passes, once you have worked someone once that you do not call them on every pass. This gives everyone else a chance to work them, of course if the pass is quiet, a repeat contact is fine.
4. Don't call CQ - Please don't call "CQ Satellite" on an FM satellite. It's the same as calling CQ on a repeater; you just don't do it. Usually with FM satellites they are busy during the day time so you can just directly call someone, if not just announce your callsign.
5. Use Phonetics - Always use phonetics, ideally NATO phonetics that everyone recognises, just saying your callsign can cause delay or mis-understanding and incorrect logging.
6. Rare/Portable Stations take Priority - It is common for satellite operators to take their equipment with them to portable locations, to transmit from rare grid squares or other DX countries. Courtesy should be extended to these stations; they are providing a rare location to all satellite operators and will be at that location for a limited time. If you hear a station on from a rare grid or DXCC entity, use good judgement before calling stations in more common grids
7. Use only the Minimum Power Required - Generally 2.5-5W from a handheld with a directional yagi is more than enough to access a satellite, while it's tempting to run more it can damage the satellite longer term.
8. Work the New Stations - Remember to always listen and work the new stations, everyone remembers the excitement of their first satellite contact, remember to help others achieve theirs.

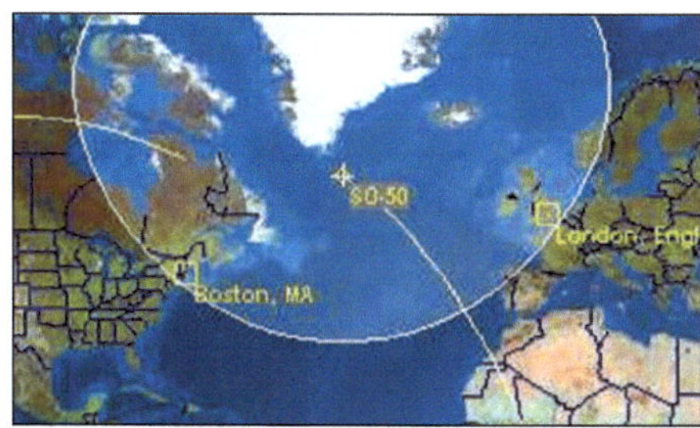

The footprint of Saudisat 1C (SO-50) showing co-visibility in both north east USA and the UK

AO-91 Frequency Chart

AOS 2	435.240 MHz	145.960 MHz
AOS 1	435.245 MHz	145.960 MHz
TCA	435.250 MHz	145.960 MHz
LOS 1	435.255 MHz	145.960 MHz
LOS 2	435.260 MHz	145.960 MHz

AO-92 Frequency Chart

AOS 2	435.340 MHz	145.880 MHz
AOS 1	435.345 MHz	145.880 MHz
TCA	435.350 MHz	145.880 MHz
LOS 1	435.355 MHz	145.880 MHz
LOS 2	435.360 MHz	145.880 MHz

Practice receiving the satellite for a few passes first. Remember to open the squelch of your receiver. Although they are not very weak, AO-91 & AO-92 will appear to suffer fading, and you will find that leaving the squelch open will aid reception. The AO-91 and AO-92 downlinks on 2m will mean you don't need to adjust for doppler although the uplink will need careful adjustment on 70cm and will need to be tuned in 5kHz steps for the correct Doppler. At the start of the pass it will be 10kHz lower and by the end of the pass 10kHz higher.

The biggest knack to learn is how to manoeuvre the antenna for the best signal. There is no hard or fast rule here, but keep in mind that you should know approximately where the satellite is in the sky, and that you are aiming to orient your antenna at the satellite. Also consider that because satellites spin like a gyroscope, their antenna polarisation changes. You'll find turning the antenna to try to match polarisations will prevent deep fades. It takes a couple of passes to get the hang of it. Keeping the pass listing, compass and your watch readily available is very useful here.

You may notice that both satellites have a background noise on their audio. This is their sub audible telemetry, which allows telemetry and voice at the same time. Also note that AO-91 & AO-92 require 67Hz CTCSS tone to be transmitted by the user to activate the transponder, if the satellite doesn't hear any CTCSS for 60 seconds the transponder will time out and you will hear just a voice announcement.

When it comes to conducting a QSO, 5W will easily be sufficient as long as the 'alligators' aren't about. This term is often used in the amateur satellite community to describe an operator who's all mouth and no ears. It is obvious from monitoring the downlink that these operators usually cannot hear the satellite at all! Both these satellites can be operated with less than 1W if no one is using high power.

Because you'll be operating full-duplex, take along some headphones. This will prevent feedback via the satellite when you're transmitting. If you have one, take a digital recording device with you for logging purposes. There's a lot to do, and recording the QSOs will allow you to keep your hands available for things other than logging. After the pass, you can write the QSOs into your logbook.

You will find that with satellite passes the more you practice the easier it becomes, a good way to look at a pass is like a contest, quick exchanges to get as many QSOs as possible in a 15 minute window.

Once you have honed your skills you will find that it's possible to work across Europe with ease and during periods of apogee (when the satellites furthest away from earth) you will be able to make QSOs into North America, Canada, Greenland and further into Europe.

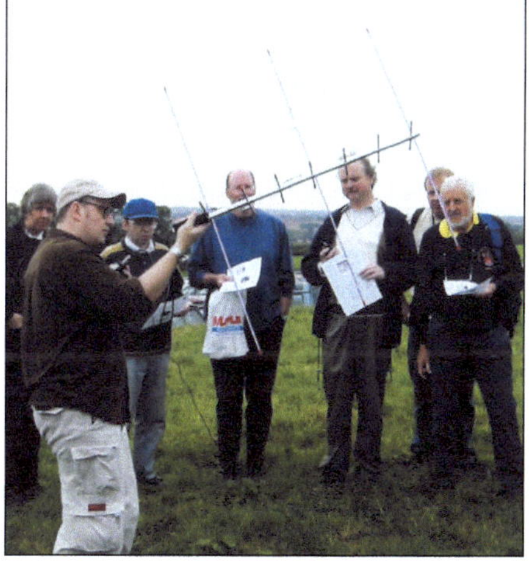

Howard Long, G6LVB, demonstrates amateur satellite operation with nothing more than a dual-band handheld transceiver and an 'Arrow' antenna, so-called because its elements are made from arrow shafts.

AMSAT-UK represents the amateur satellite community in the UK whose members not only operate amateur satellites, but also help to design, build and fund them

AMSAT-UK address: J D Heck, G3WGM, "Pickles Orchard", 30 Memorial Road, Great Hampden, Buckinghamshire HP16 9RE
email: *g3wgm@amsat-uk.org* (Jim Heck) website: *www.amsat-uk.org*

AMSAT-NA website: *www.amsat.org*

HF Awards

RSGB Awards

One of the most exciting and broad ranging facets of amateur radio operating is awards chasing. Awards are an area that everyone, whether individuals or clubs, participates in by simply being on air. Every contact you make has the potential to check off another requirement on many of the awards that are available through the RSGB. It is a major motivating force of so many QSOs that occur on the bands day after day. You should also realise that one QSO may qualify for more than that one RSGB award you are working towards. There are many awards worldwide that can enhance your enjoyment of your on-air interests.

Aside from the fun of operating itself, awards chasing is also a good way to get maximum performance from your station, become familiar with propagation, and learn about the geography, history or culture of places near and far.

Whilst the traditional method of proof of contact by QSL cards is still acceptable, LoTW confirmation is also available using the unique Record ID number. We now offer a QSL Card Checking service on request. If you require a card checker, the RSGB Awards Manager will contact you to arrange this.

When your claim has been approved by the RSGB Awards Manager you will be given a code number to be entered on the Order page to obtain your certificate.

Price and ordering information are available at the RSGB Shop at: *http://www.rsgbshop.org/acatalog/Awards.html*

NOTE 1: Information on how to apply for awards is on the RSGB Website *https://www.rsgb.org/awards/*

NOTE 2: Spreadsheets are very useful to track progress towards an award, and these have been made available for all of the RSGB's HF awards (as Check Sheets). Note that these have been updated in 2021 so it is wise to download the latest version from the awards website when you start working towards an award, and check for updates.

New style Award for all Class of Certificate

RSGB HF Awards

The HF Awards Programme aims to make the attainment of awards enjoyable, progressive, challenging and achievable in a reasonable time frame.

RSGB Youth Award

This RSGB Award is designed to encourage activity and exploration of the breadth of amateur radio, broadening the interest and experience of amateurs who are aged up to 26.

There are four classes of award:

Bronze: Use two modes with a minimum of 50 contacts

Silver: Use three modes with a minimum of 100 contacts

Gold: Use four modes with a minimum of 150 contacts

Platinum: Use five modes with a minimum of 300 contacts

Any Band or mode may be used

https://rsgb.org/main/operating/amateur-radio-awards/youth-award/

Foundation HF Award

This award is available to holders of UK The Foundation Awards, available to all Foundation licence holders, changed in 2023 and the new specification of Foundation Awards is as follows:

There are three levels of award :

Bronze: 40 contacts
Silver: 70 contacts
Gold: 100 contacts

Minimum of 10 contacts and maximum of 25 contacts on any four bands as specified below; and a minimum of 15 and maximum of 50 contacts in each mode used.
Modes: at least two of CW, SSB, FM, AM, FT8/FT4 (FT8 and FT4 count as one mode), JS8Call
Bands: any HF band below 50MHz permitted by the licence

Foundation All-Bands Award

Bronze – 50 contacts
Silver – 100 contacts
Gold – 150 contacts

Minimum of 10 contacts and maximum of 25 contacts on at least four bands; minimum of 15 and maximum of 30 contacts in each mode used.
Modes: at least four of CW, SSB, FM, AM, RTTY, any digital mode (but direct contacts only, not via a repeater or gateway); note that FT4 and FT8 count as one mode; similar applies to any fsk/psk/afsk/msk modes that are only differentiated by phase, tone or frequency shift rate.
Bands: any bands permitted by the licence

Intermediate 100 Award

This award is available to holders of UK Intermediate Licences and is designed to encourage activity by making 100 contacts across four designated HF bands during their first year of being licensed.

Contacts may be made using CW, SSB or Data modes, as appropriate, on 80m, 40m, 30m, 17m with a maximum of 25 contacts per band. Multiple contacts with the same amateur do not count.

IARU Region 1 Award

The purpose of this award is to generate activity on the HF bands whilst recognising the work of the International Amateur Radio Union Region 1. The award may be claimed by any licensed radio amateur who can produce evidence of having contacted amateur radio stations in the required number of countries whose national societies are members of the Region 1 Division of the International Amateur Radio Union (IARU).

There are 3 classes for contacts as follows:

Class 1 102 member societies
Class 2 60 member countries
Class 3 40 member countries

Confirmation of two-way contacts is required either by QSL cards or LoTW for contacts on 160, 80, 40, 30, 20, 17, 15, 12 and 10m.

IARU Region 1 28MHz Award

This special single band award is available under the same rules and 3 class structure as the IARU Region 1 Award. All contacts must be within the 10m band and have been made since 1st July 1983.

IARU Region 2 and IARU Region 3 also administer a regional award. Information can be found at their respective websites.

The Commonwealth Century Award (CCA)

This award may be claimed by any licensed radio amateur producing evidence of having contacted amateur radio stations in 100 (Century), 70 (Silver) and 40 (Bronze) Commonwealth call areas on the list current at the time of application.

The award can be claimed for contacts on the traditional 5 bands (80m, 40m, 20m, 15m and 10m) or over all 9 HF bands, (160m to 10m). *http://rsgb.org/main/files/2020/01/Commonwealth-Century-Award-2020.pdf*

Worked ITU Zones Award

The purpose of this award is to generate activity on the HF bands whilst recognising the work of the International Telecommunications Union.

The award may be claimed by any licensed radio amateur who can produce evidence of having contacted amateur radio stations in the required number of ITU zones.

There is a choice of 4 categories:

- Classic Worked ITU Zones
- 5 Band Worked ITU Zones
- WARC Band Award
- Top Band Award

Confirmation of two-way contacts is required either by QSL cards or LoTW for contacts on 160, 80, 40, 30, 20, 17, 15, 12 and 10m.

Worked All Continents Awards

This award, issued by IARU headquarters, may be claimed by any licensed radio amateur in the UK, Channel Isles or Isle of Man who can produce evidence of having contacted amateur radio stations in each of the 6 continents : North America, South America, Europe, Africa, Asia and Oceania. All contacts must be made from the same country or separate territory within the same continent. Various endorsements including "all 1.8MHz" are available. Applicants are advised to go to the ARRL website to download the WAC application form, enter details of the six confirmed contacts required, and then contact one of the UK ARRL-certified card-checkers to obtain their confirmation of your application. The RSGB is no longer able to certify applications on behalf of ARRL for this award.

IARU Region 1 Award (Classes 2/3) 2021-22

Callsign	Award
GM0SCA	IARU Region 1 Class 2
G0DDZ	IARU Region 1 Class 3
M0SNA	IARU Region 1 Class 3
M0TSV	IARU Region 1 Class 3
YC8GPH	IARU Region 1 Class 2
JL1JVT	IARU Region 1 Class 2
M0UPL	IARU Region 1 Class 2
2E0LMI	IARU Region 1 Class 2
NN6U	IARU Region 1 Class 3

IARU Region 1 28MHz Award 2020-21

Call	Class
2E0LMI	2
DJ3ZF	2
SV1QVA	3
JL1JVT	3

Other Awards

Callsign	Award
JL1JVT	Commonwealth Century Award
JL1JVT	Commonwealth Century Award
JL1JVT	Classic Worked ITU Zones
JL1JVT	Worked ITU Zones WARC Class 2
M7OFM	Youth Award
NN6U	Commonwealth Century Silver Award
M7JWD	Foundation Award Silver

Members of IARU Region 1 (from February 2020)

Prefix	Country	Nat. Soc.
3A	Monaco	ARM
3B8	Mauritius	MARS
3DA	Swaziland / Eswatini	RSE
3V	Tunisia	CAST
3X	Republic of Guinea	ARGUI
4J/K	Azerbaijan	FRS
4L	Georgia	NARG
4O	Montenegro	MARP
4X	Israel	IARC
5B	Cyprus inc UK Sov Bases	CARS
5H	Tanzania	TARC
5N	Nigeria	NARS
5X	Uganda	UARS
5Z	Kenya	ARSK
6W	Senegal	ARAS
7P	Lesotho	LARS
7X	Algeria	ARA
9A	Croatia	HRS
9G	Ghana	GARS
9H	Malta	MARL
9J	Zambia	RSZ
9K	Kuwait	KARS
9L	Sierra Leone	SLARS
9Q5	Congo DR	ARAC
9U	Burundi	ABART
9X	Rwanda	RARU
A2	Botswana	BARS
A4	Oman	ROARS
A6	United Arab Emirates	EARS
A7	Qatar	QARS
A9	Bahrain	
C3	Andorra	URA
C5	Gambia	RSTG
C9	Mozambique	LREM
CN	Morocco	ARRAM
CT	Portugal inc.CU CT3	REP
DL	Germany	DARC
E7	Bosnia & Herzegovina	ARABIH
EA	Spain	URE
EI	Ireland	IRTS
EK	Armenia	FRRA
EL	Liberia	LRAA
ER	Moldova	ARDM
ES	Estonia	EARU
ET	Ethiopia	EARS
EU	Belarus	BFRR
EX	Kyrgyz Republic	
EZ	Turkmenistan (ex UH)	LRT
EZ	Turkmenistan (ex UH)	LRT
F	France inc. TK	REF
G etc	United Kingdom	RSGB
HA	Hungary	MRASZ
HB	Switzerland	USKA
HB0	Liechtenstein	AFVL
HZ	Saudi Arabia	SARS
I	Italy	ARI
J2	Djibouti	ARAD
JT	Mongolia	MRSF
JY	Jordan	RJARS
LA	Norway	NRRL
LX	Luxembourg	RL
LY	Lithuania	LRMD
LZ	Bulgaria	BFRA
OD	Lebanon	RAL
OE	Austria	OEVSV
OH	Finland	SARL
OK	Czech Republic	CRC
OM	Slovakia	SARA
ON	Belgium	UBA
OY	Faeroe Islands	FRA
OZ	Denmark	EDR
PA	Netherlands	VERON
R	Russian Federation	SRR
S5	Slovenia	ZRS
S7	Seychelles	SARA
SM	Sweden	SSA
SP	Poland	PK
SU	Egypt	EARA
SV	Greece	RAAG
T7	San Marino	ARRSM
TA	Turkey	TRAC
TF	Iceland	IRA
TJ	Cameroon	ARTJ
TN	Congo	URAC
TR	Gabon	AGRA
TU	Ivory Coast	ARAI
TZ	Mali	CRAM
UN	Kazakstan	KFRR
UR	Ukraine	UARL
V5	Namibia	NARL
XT	Burkina Faso	ARBF
YI	Iraq	IARS
YK	Syria	SSTARS
YL	Latvia	LRAL
YO	Romania	FRR
YU	Serbia	SRS
Z2	Zimbabwe	ZARS
Z3	North Macedonia	RSM
Z6	Kosovo	SHRAK
ZA	Albania	AARA
ZB	Gibraltar	GARS
ZS	South Africa	SARL

VHF/UHF Awards

RSGB VHF/UHF Awards Programme

The RSGB VHF/UHF Awards Programme recognises successful operating achievements that depend on the special propagation modes which can be experienced at these frequencies.

http://rsgb.org/main/operating/amateur-radio-awards/vhfuhf-awards/

50MHz Squares, Continents and Countries Awards

The purpose of these awards is to encourage activity on the "magic" band and to recognise personal operating achievements. Stations are eligible for awards as: Fixed stations, Portable stations, and Mobile stations, but these categories cannot be mixed.

Squares awards start at 25 squares confirmed, and increase with incremental levels of 50, 75, 100, 200, 300, 400, 500, 600.

At the discretion of the Awards Manager, interim and higher levels can be added (a handful of stations in the past have achieved upto 800 squares and received commensurate awards).

Countries awards start at 10 countries confirmed and increment in 10s to 160. A new Continents & Countries Award, launched in 2021, additionally proceeds in increments of combined continents and countries from a start of 2/30, followed by incremental levels of 2/40, 2/50, 3/45, 3/50, 4/55, 4/65, 5/60, 5/65, 5/75, which can be extended at the discretion of the Awards Manager.

4-2-70 Squares and Countries

These are the traditional '4-2-70 Squares Awards' covering 4m, 2m and 70cms and are intended to mark successful vhf/uhf achievement. Initially, one certificate will be issued.

Further certificates will be issued as additional squares and countries are claimed. The title of each award gives the number of locator squares and countries needed to qualify for the award.

For example, to obtain the 144MHz 40/10 award you must have confirmed contact with 40 locator squares including 10 countries on 144MHz. Eligible countries are those shown in the countries list printed in the 'Year Book'. Stations are eligible for awards as:

Fixed stations, Portable stations, any location or Mobile stations, any location. Categories cannot be mixed.

4m band

70MHz	40/10
70MHz	50/15
70MHz	60/20
70MHz	70/25
70MHz	80/30
70MHz	90/35

2m band

144MHz	40/10
144MHz	100/20
144MHz	200/30
144MHz	300/40
144MHz	400/50
144MHz	475/50

70cm band

432MHz	40/10
432MHz	60/15
432MHz	100/15
432MHz	140/20
432MHz	160/20
432MHz	180/20

VHF/UHF Award Achievements Table

Callsign	Award
G4NBS	50MHz Countries
G4NBS	50MHz Continents & Countries
G4NBS	50MHz Squares
S57EN	V/UHF Activity Award
G8GNI	144MHz Squares and Countries
JL1JVT	50 MHz DX Countries
JL1JVT	50MHz Squares
GD0TEP	50MHz Squares

Radio Surfer Award

In 2021 a new award was introduced by the Youth team, the Radio Surfer Award, which is issued to enthusiasts, regardlless of whether they are licensed, who complete a selection of tasks from a list published on the Awards website pages.

Each task is worth a certain number of points, and the award is achieved when the applicant achieves a number of task-points equal to or greater than their age in years.

The tasks are to be supervised by a suitably qualified person (safety-vetted if the applicant is a child), and the achievement certified by them in the application.

VHF/UHF Activity Award

In 2022 a new award was created, the VHF/UHF Activity Award, which recognises operating activity and achievement for radio to radio (simplex) contacts on the vhf and uhf bands, particularly targeted at newcomers to the hobby but open to all classes of licensee:

Bronze: 30 Simplex contacts on 2m plus 20 Simplex contacts on 70cms using FM or SSB or a combination of these modes

Silver: 50 Simplex contacts on 2m plus 40 Simplex contacts on 70cms using any combination of FM, SSB, CW

Gold: 75 Simplex contacts on 2m, 60 Simplex contacts on 70cms, and at least 50 Simplex contacts on either of 6m or 4m, using any combination of FM, SSB, CW

Platinum: 100 Simplex contacts on three of the following bands: 6m, 4m, 2m, 70cm using at least 2 of the following modes on each band: FM, SSB, CW

Diamond: 100 Simplex contacts on all of the following bands: 6m, 4m, 2m, 70cms, plus 25 simplex contacts on 23cms using at least 2 of the following modes: FM, SSB, CW

Since introducing the award, Andy Callaghan, M0XTY and Ian Evans, GI0AZB, have received Bronze and Silver levels of award; more participation is strongly encouraged, to bolster activity on these bands and because they are a good fiirst step into VHF award chasing.

Foundation VHF Award

The Foundation Awards, available to all Foundation licence holders, changed in 2023 and the new specification of Foundation Awards is as follows:

Bronze	40 contacts
Silver	70 contacts
Gold	100 contacts

Minimum of 10 contacts and maximum of 30 contacts on at least three listed bands; minimum of 15 and maximum of 30 contacts in each mode used.

Modes: at least two of CW, SSB, FM, AM, RTTY, any digital mode (but direct contacts only, not via a repeater or gateway);); note that FT4 and FT8 count as one mode; similar applies to any fsk/psk/afsk/msk modes that are only differentiated by phase, tone or frequency shift rate.

Bands: any vhf/uhf/shf band 50MHZ and above permitted by the licence

A more detailed overview of all awards, including application forms and rules, are available on the

RSGB website at:

http://www.rsgb.org/awards

Microwave Awards

As of 2010, RSGB microwave awards were transferred to the UK Microwave Group (UKuG).

The following awards are intended to mark achievement on the microwave bands. Successful applicants will initially receive a certificate and one sticker. Further stickers will be issued as later claims are received.

Microwave Squares Awards

Existing RSGB records have been transferred so that claims for extra stickers to add to existing RSGB Squares Awards can be accepted. The existing sticker design has been retained.
Awards are available in 5 square increments on the following bands:

1.3GHz: 5 to 150 Locator squares
2.3GHz: 5 to 75 Locator squares
3.4GHz: 5 to 75 Locator squares
5.7GHz: 5 to 75 Locator squares
10GHz: 5 to 75 Locator squares
24GHz: 5 to 25 Locator squares

- Initial claims will require submission of QSL cards for all contacts claimed.
- Subsequent increments will need the additional cards and the countersigned check list from the previous claim.
- QSL cards must include the IARU QTH locator of the station worked.
- eQSL confirmations can be accepted from stations that have completed eQSL verification
- Contacts must have been made after 31 December 1978.
- All contacts must be two-way on the band in question, cross-band contacts are not eligible.
- Awards are available for fixed or portable/mobile stations but these categories cannot be mixed.
- Claims for portable operation must be for contacts made from one site, defined as anywhere within a 5km radius of the point operated from.
- Cards should be listed and sorted in IARU QTH Locator alphanumeric order.
- A self-addressed envelope must be included for return of the QSL cards, with sufficient postage value in UK stamps or IRCs.
- Certificates are free for members of the UK Microwave Group.
- Claims from non-members must include payment of £3 or $5 when an initial certificate is requested.
- Subsequent stickers will be enclosed with returned QSL cards and checklists.

Claims should be made using the application form and squares checklist available on the UK Microwave Group website: *www.microwavers.org* and sent to the address given below.

UKuG/SOTA Microwave Distance Awards

These new awards are jointly issued by UKuG and SOTA, and recognise achievement in working distance. They replace the previous RSGB/UKuG distance awards, and are available in 50km endorsements for all bands 1.3GHz and above.
For further information, please see: *http://www.sota-shop.co.uk/microwave.html*

Recording and recognising 'Firsts'

The UK Microwave Group has an award that recognises the achievements of British stations in making first contacts with other countries, and maintains a list of firsts as a historic record of the development of microwave operating techniques in Great Britain. Should you not wish to claim a certificate we would still be very grateful for details of your First to ensure the records are accurate.
Certificates will be awarded to stations that can demonstrate that they completed the first contact between one of the countries in Great Britain and Northern Ireland (Prefixes M, MM, MI, MW, MU, MJ, MD - and the G and 2E equivalents) and any other country on a particular band above 1GHz. Contacts within Great Britain are valid for this award (eg Scotland to Guernsey).
Only one 'First' will be recognised per band, irrespective of the propagation mode (eg tropo, rainscatter or EME).
Claims should be made using the application form available on the UK Microwave Group website.
A QSL will be required for award of a certificate (but are not required to provisionally enter a First in the database). QSL cards should not be sent until requested. eQSL confirmation will be accepted from stations that have completed eQSL verification.
All claims should be sent to John Quarmby, G3XDY. Details below.

Please note that data provided for the above awards will be held by the UK Microwave Group for the purposes of providing a published list on the UKuG website and in the group's newsletter *Scatterpoint*, of achievements by stations in Great Britain and Northern Ireland. Data will not be used for any other purposes.

You do not have to be a member of the UK Microwave Group to lodge a claim.
The decision of the UK Microwave Group committee is final on the validity of claims.
Prior to making a claim, please check the Firsts database on the UK Microwave Group website at: *www.microwavers.org.uk* for existing contacts.

Trophies & Awards

In addition to the certificates and distance/squares awards, and the trophies awarded for microwave contests by the RSGB Contest Committee, the UK Microwave Group also awards a number of cups and trophies, both for contest operating successes and for contribution to the facet of the hobby in both technical and supportive aspects. Most of these awards are presented annually at the UK Microwave Group Round Table event held at Martlesham, Suffolk, each April.
Photos of all the trophies can be found online at: *www.microwavers.org/trophies.htm*

Operating Trophies

5.7GHz - G3KEU

In memory of Tim Leighfield, (SK 2002), this cup is awarded annually to the winner of the UKuG 5.7GHz cumulative contest. See: *www.g3pho.free-online.co.uk/microwaves/keu.html*

10GHz - G3JMB

Awarded in memoriam to Jack Brooker MBE, (SK 2004) is awarded annually to the winner of the Restricted section of the 10GHz Cumulative contest. *www.r-type.org/g3jmb/*
This trophy is an attractive glass plaque, but due to the inadvisability of engraving it, each winner is awarded a small plaque which they retain in perpetuity.

10GHz - G3RPE

Awarded to the winner overall of the 10GHz Cumulative contest, this cup commemorates Dain S Evans, BSc, PhD, FIM and RSGB President in 1978. Dain was one of the prime movers of the expansion in 10GHz construction and activity in the UK and was the first, with G3ZGO, to break the 150km distance barrier, setting the bar for those who followed.

24GHz - G0RRJ

This cup, in memory of Dave Cox (SK 2009), is awarded to the winner of the 24GHz Cumulative contest.

24GHz and 47GHz - Trophy

These two cups are awarded to the winners of the 24 and 47GHz Trophy events (as distinct from the 24GHz Cumulative sessions).

Contribution Awards

There are four awards in this area:

G3BNL

In memory of Les Sharrock, this ornate decanter trophy is awarded by the UK Microwave Group committee for innovation or technical development of microwave equipment or techniques.

Send microwave award applications to the awards manager: awards@rsgb.org.uk

Fraser-Shepherd Award

This award, for research into microwave applications for radio communication, is presented by the RSGB on the nomination of the UK Microwave Group committee. Fraser Shepherd, GM3EGW was an early pioneer of UHF and microwave operation and techniques, and was a participant in the first GM 432MHz moonbounce activity.

G3EEZ

Awarded for contributions to Microwave Communications, this cup is in memory of Alan Wakeman.

G3VVB

Presented for the best piece of home constructed microwave equipment exhibited at a Round Table, this magnificent trophy is in memory of Cyril James, a keen home constructor who produced microwave artefacts of very high quality workmanship. The entries and judging of this event are a regular feature of Microwave Round Tables, with the trophy awarded at the RSGB Convention in October.

G3JVL

This award, introduced in 2020, is presented to the best newcomer each year who the committee considers has made the greatest contribution; be that in developing hardware/software or operating on the microwave bands.

Send microwave award applications to: John Quarmby, G3XDY, 12 Chestnut Close, Rushmere St Andrew, Ipswich IP5 1ED email: g3xdy@btinternet.com

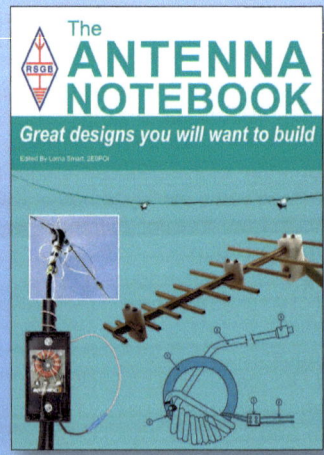

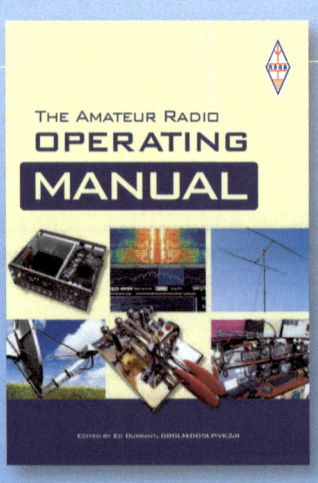

Microwaves

This an aspect of amateur radio where experimentation and construction are still alive and well. And it's never been easier to get started.

What are microwaves?

In amateur radio terms, frequencies above 1000MHz (1GHz).

Are microwaves only for line-of-sight communication?

Absolutely not! Try telling those operators who routinely work hundreds of kilometres with relatively modest antennas and power that they can't do it. They can, and so could you!

Microwaves is just another aspect of amateur radio which can be learned. You can take it in so many directions ...

All amateur radio equipment costs money, but a simple system of antenna and transverter could cost less than a new headset! Less if you are prepared to build your own from a kit.

So what can we do with microwaves?

Lots! That question is probably better answered by asking what we can't do! The microwave bands support a wide variety of propagation modes, some of which will be completely unfamiliar to people who haven't ventured above VHF.

On most bands troposcatter – scattering from the upper part of the lowest layer of the atmosphere, the troposphere, is important. This will reliably support contacts over several hundred kilometres with good modern equipment. With modest kit, perhaps a couple of watts to a discreet yagi antenna a couple of metres long on 1.3 or 2.3GHz, from a site with open horizons – troposcatter path losses are very dependent on the vertical angle of your horizon - will produce regular contacts up to around 200km without much effort. A similar power level to a repurposed satellite TV dish will provide contacts on 10GHz over the same range, or further.

The troposphere is involved in another way. At times the masses of air from different sources, some from the cold dry Arctic, some, perhaps, from the warm, wet Atlantic, move over each other. In regions sometimes thousands of kilometres across, these boundaries are able to 'duct' microwave signals over long distances. Ducting produces openings on the microwave bands just like those on VHF/UHF, and signals are often propagated over paths in excess of 1000km.

Other ways of propagating microwave signals include scatter from rainfall (and other forms of precipitation) and scattering from aircraft. Although rainscatter can occasionally be heard on 1.3 and 2.3GHz, it really starts to be useful at 3.4GHz and above. 'Aircraft scatter' can allow all year propagation up to about 700km.

On the higher frequency bands, at 24GHz and above absorption by water in the atmosphere becomes a challenge but it has not deterred UK microwavers from having a go. The UK DX record on the 24GHz band is almost 400km. Contacts have been made by UK radioamateurs on all of our allocated bands up to, and including 241GHz.

Microwave moonbounce (EME) currently produces worldwide DX from the UK on all of our bands up to 24GHz. Some operators apply for HF-style awards, such as WAC. Some UK stations are advancing towards DXCC on 1.3GHz. On the microwave bands, EME is practical with quite modest antennas. Significant numbers of contacts can be made on 23cms with single yagis or 2m diameter dishes. On 13cms and up even smaller antennas can be used. 10GHz has become a favourite for EME. There 50W and a 2.4m dish will allow a station to hear its own Moon echoes on SSB. 24GHz is starting to make headway as a moonbounce band. Bands up to 76GHz have been used experimentally.

Amateur microwavers can be found throughout the world, especially in the Americas, Western Europe, Australasia and Japan. Regular EME contacts are made between all of these areas, and there even DXpeditions bringing activity to other areas.

Amateur television has found its way into earth orbit with DATV transmissions being made from the ISS at 2.4GHz, while the geostationary Es'hailsat2 satellite is due to launch in 2017 with digital and linear transponders using 2.4GHz uplinks, and 10GHz downlinks.

A few UK radioamateurs have heard deep space satellites such as Voyager at the very edge of our Solar System. Others have discovered the joy of amateur Radio Astronomy.

Microwaving can be challenging, it can be fun, it can be educational, it can be social. Finally it's what you make it. It's a rapidly developing area of our hobby.

How did amateur microwaves come about?

As early as 1894 to 1896, Jagadish Chandra Bose, an Indian physicist, experimented on 60GHz over a one mile distance using primitive semiconductors! He was a truly remarkable man who is now seen as the 'Father of Microwaves'. It may seem strange, but microwaves pre-date HF radio!

In 1946 the first amateur microwave contacts were recorded in the USA, when W1LZV/2 worked W2JN over two miles on 10GHz and W1NVL/2 worked W9SAD/2 on 21.9GHz (800 feet!). The World 10GHz record was then set at 7.6 miles by W4HPJ/4 and W6IFE/3.

The first UK amateur microwave contact was made in 1949 by G3BAK and G3LZ over 27 miles, at the time a new world 10GHz record.

For the next twenty years operation on the bands above 3.4GHz centred mainly on the use of tubes, such as klystrons as local oscillators and transmitters with waveguide mounted diode mixers in wideband FM systems. Gunn diodes replaced the klystrons during the '70s, and the '80s saw a sharp spike in activity on 10GHz.

A modern 10 and 24GHz home system

Activity on the lower frequency microwave bands also grew, with many people using NBFM, or even AM, generated by solid-state varactor multipliers from 70cm.

At the beginning of the '80s, more advanced microwavers started to move to an approach more like that on the lower frequency bands, with CW and SSB becoming more common. The problem with low-power, wideband systems is that it is very difficult, particularly above 3.4GHz, to cover many non-line-of-sight paths.

More recently, the field has continued to develop, with narrowband standards in use on all bands. All DX operation is now on narrowband, often using weak-signal modulation schemes, such as K1JT's WSJT. Transmitters capable of several watts output are commonplace on 10GHz.

In addition commercial kits and ready-made modules have become available.

In the last decade, much focus has been on increasing frequency stability by locking oscillators to GPS stabilised standards. SDR technology operating directly at microwave frequencies is becoming more common.

The current State of Play: UK Microwave Terrestrial Records

Band (GHz)	Distance (km)
1.3	2617
2.3	1389
3.4	1137
5.7	1244
10	1429
24	408
47	203
76	129
122	35.9
134	35.6
145	1.29
241GHz	9.3
Light (red)	129.1

Although much has been made above about

the DX potential of the microwave bands, and many enthusiasts concentrate on this aspect of the hobby, there are other important user groups.

Conventional voice repeaters/beacons operate in the 23cm band. These have recently been joined by a number of digital voice repeaters using the 'Dstar' modulation scheme.

Amateur Television is very well established above 1GHz, with a network of repeaters. Many now employ Digital Video technology, although many analogue FM TV repeaters remain in service.

Digital Networking is an interest of an increasing number amateur radio operators, and several long haul networks exist using IEEE802 standards at both 2.4 and 5.6GHz. In many countries on the European mainland extensive amateur owned and operated high-speed data networks, linked to the Internet, operate on 1.3 and 2.3GHz.

More about the Microwave Bands

Many bands are shared with other users and are increasingly under threat from commercial interests. For example parts of the 2.3 and 3.4GHz bands have been subject to Public Sector Spectrum Release (PSSR) which has seen loss of access to some frequencies and other changes to the main UK amateur licence schedule. Users of parts of 2.3GHz (shown by ***) are now also specifically required to register with Ofcom. There are both limits to times of activity and to location, which make continued sharing with the main User possible. NoVs for new additional bands such as 2300-2302MHz and >275GHz are also available on request. (N)

Many of the bands shown in the following list contain amateur beacons and repeaters which are useful for frequency calibration and as indicators of propagation conditions. A more detailed spectrum allocation can be found in the Band plans Section of this Yearbook and on the RSGB website.

23cm	1240.0-1325.0MHz
13cm	2300.0-2302.0MHz(N)
13cm	2310.0-2350.0MHz***
13cm	2390.0-2450.0MHz
9cm	3400.0-3410.0MHz
6cm	5650.0-5680.0MHz
6cm	5755.0-5765.0MHz
6cm	5820.0-5850.0MHz
3cm	10.000-10.125GHz
3cm	10.225-10.500GHz
12mm	24.000-24.250GHz
6mm	47.00-47.2GHz
4mm	75.50-81.0GHz
2.5mm	122.25-123,
2.2mm	134-141
1.25mm	241-250GHz
also	>250GHz (N)

Ways into microwaves

The 1.3GHz (23cm) and the 10GHz (3cm) bands are the easiest ones on which to make a start. There is a lot of ready-made equipment, including antennas for these bands. The on-line auction and flea-market sites can be a great source of microwave surplus.

Simple 10GHz wideband gear

The days of simple Gunn oscillator transceivers are now past, if only because the intruder alarm modules which many people used in the past are no longer easily and cheaply available. The Gunn oscillator diodes used are obsolescent as low cost items. Modern intruder alarm modules are much less easy to modify. However, if old wideband gear is available, perhaps lurking in someone's attic, it can still be used, if you can find someone to cooperate with. In this case, it would be sensible to check that the gear is operating in a part of the band to which we still have access.

Satellite TV LNBs make very acceptable receive converters when combined with a cheap DATV "Dongle" or the more expensive Funcube Dongle. Many are good enough for narrowband use, but there is no obvious easily available companion transmitter. If you have a beacon or ATV repeater locally, a LNB, particularly if mounted in a dish, would make a great tool for initially exploring microwave propagation at 10GHz. You could also take part in UkuG contests, as 'one-way' contacts are allowed for 50% of the points of a two-way!

Another way of getting going, assuming that your licence allows it, is to consider using 5.6GHz and to exploit (cheap!) video-sender equipment. These often consist of a wideband, synthesised FM transmitter and a companion receiver, which can be tuned to a channel within the amateur band. Combine these with a suitable antenna, also available cheaply, and you have a simple transceiver which is capable of surprisingly good performance over line-of-sight paths.

There are groups around the UK using video sender technology, but if you can't locate one (try asking on 'ukmicrowaves') or they are too far away and you know someone close-by who is interested in experimenting, go ahead and try!

As with all worthwhile activities, there is a learning curve. You might not work very far initially, so it's worth finding someone with similar interests to experiment with. It's still an excellent, fun and cheap way to 'wet your feet' in microwaves. Don't forget you will also need something for talkback liaison, be it a mobile phone, computer with internet access, or even 2m/70cm!

Other bands for a beginner

1296MHz is the other band which can be recommended for people taking their first steps into the microwaving world. The technology is not too dissimilar to that used on VHF/UHF, and equipment needed to make a start is easily available. Recently, the Company owned by Bulgarian microwaver, Hristyiyan, LZ5HP, has been producing a simple 2W transverter – not a kit - of good performance, at a very attractive price *www.sg-lab.com/TR1300/tr1300.html*. This is small enough, and light enough to be mounted in a waterproof box close to the feed point of a 1296MHz antenna. From many locations this will allow regular QSOs up to around 250km – much further in a tropo opening.

M0EYT portable.

Microwave Antennas

On 1296MHz and 2.3GHz, yagi or sometimes dish antennas are the norm. High gain dishes and horns are the most commonly used antennas at 3.4GHz and above.

Microwave antennas give much higher gains than HF ones. 100 microwatts of 10GHz CW into a 60cm dish will be heard over any line-of-sight path in the UK by a receiver using a similar antenna.

On 10GHz, a 60cm ex-Satellite TV dish with a suitable feed horn will have a gain of around 35dB. This will give a potential range greater than 144MHz, for similar transmit power.

The extra gain comes from concentrating the transmitted energy into a narrow beam; an antenna with 35dB gain will have a -3dB beamwidth of about ±1.5°, so you have to point the dish more accurately than you would, say, a 2m yagi. The ability to go to a site, and to work out where to accurately point a dish is another skill which has to be learned as a microwaver!

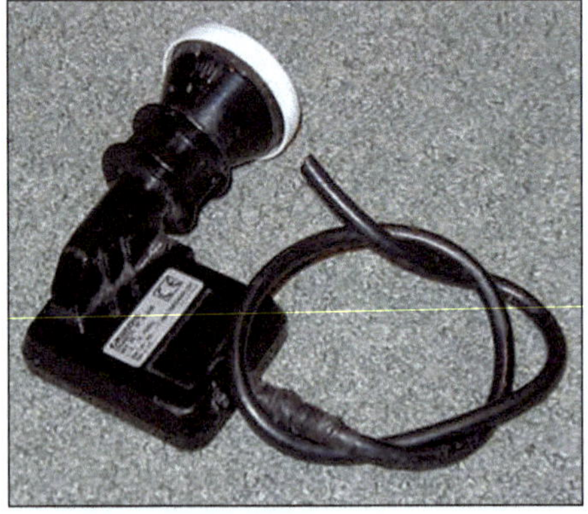

A more modern satellite TV LNB.

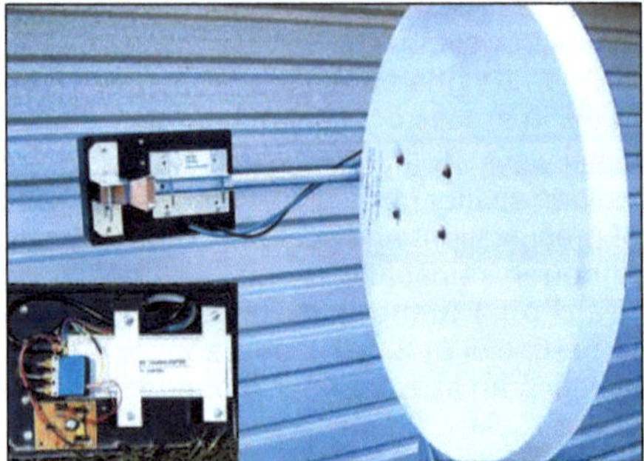

The 'Quickstart' microwave system

A small dish microwave EME system for 13cms

Equipment

Ready built transverters are available from a number of sources, such as Kuhne Electronik (DB6NT) in Germany, DEMI in the USA, and LZ5BP in Bulgaria. Kuhne and DEMI, along with Mini-Kits in Australia also market kits.

There is a market for second-user microwave equipment. A request on one of the online groups, such as:

https://groups.io/g/UKMicrowaves could well produce results.

Making QSOs, Is it like the lower frequency bands?

No! Perhaps on 1.3GHz, it can be, but at higher frequencies the beamwidth of antennas is so small that the chances of hearing another station randomly are not large. So, the use of talkback has developed. UK stations use 144.390 MHz SSB for on-air talkback but also use online tools. In most of the rest of Europe, 432MHz SSB was used. More recently, the ON4KST chat server *www.on4kst.com* has become very popular along with the use of Zello SHF Chat. While there are those who regret the way in which the Internet has replaced a 'pure' amateur radio talkback channel, it has its advantages.

The first is that as 2m propagation is often not the same as that at higher frequencies: the microwave bands are often 'open' when 2m is not, for well understood reasons. More prosaically, the use of a public talkback channel, rather than a semi-private VHF link, discourages people from trying to complete QSOs or even make QSOs over talkback.

On the microwave (and VHF/UHF) bands the completion of a QSO is defined rather differently to the usual practice on the lower frequencies. On the band in which you are operating you must exchange callsigns, some piece of unknown information, like a report or a QTH locator. Both stations must acknowledge the receipt of of the callsigns and unknown information from the other station. If you don't keep to that procedure, you haven't made a contact which would be seen as valid.

There are three sets of parameters you need to get right, if you are to work another station; the beam heading, the frequency, and who is to transmit and when.

The beam heading is easily obtained from a number of sources. Most conveniently this can be found by a couple of clicks in the ON4KST interface. But there are a number of apps available for all of the major operating systems which will convert an input in the shape of two QTH locators into a bearing and distance. The majority of antenna rotators – particularly those aimed at HF operators - do not have particularly accurate azimuth indications and some modification will be needed. Fortunately, some European manufacturers of rotators understand the need for more accurate calibration. For portable operation, a simple compass rose which can be calibrated by reference to a local landmark can produce an elegant solution.

Frequency is very important. Older transverters used either a free-running crystal oscillator or a crystal oscillator partially stabilised with a simple temperature control oven. That resulted in the frequent need to tune over perhaps 20kHz to find another station, even if both stations were fairly confident of their frequency. Many weak and transient signals were missed because of this. Modern practice is to phase-lock the transverter local oscillator to a GPS stabilised frequency reference, and even at 10GHz frequency accuracies of a very few Hz can be obtained.

If this all sounds very different to what you're used to, you can get help?'

The UK Microwave Group (UKuG) has a policy of supporting newcomers. Its runs Technical Support and free chip component services, covering a large part of the UK. This is a voluntary service to members – not a right of membership – and depends on the goodwill of the volunteers freely to provide (within sensible limits) their expertise, and often very extensive test equipment to support other members. Many of those providing this service are involved professionally with microwaves and have excellent laboratory facilities. In the commercial world the time they give would cost lots of money!

To find out more about UK Microwave Group you can find out more information by looking on the National Affiliated Clubs & Societies pages in this Yearbook or by searching their website:

www.microwavers.org

'Tell me more about UKuG'

The UK Microwave Group (GX3EEZ) was founded as a representative body for Amateur Radio Microwave activity in the late 1990s. It is a UK-wide national body affiliated to the RSGB, and works with that Society on issues related to the amateur radio allocations above 1GHz. It is a membership organisation, run by a Committee elected annually at the Group's AGM.

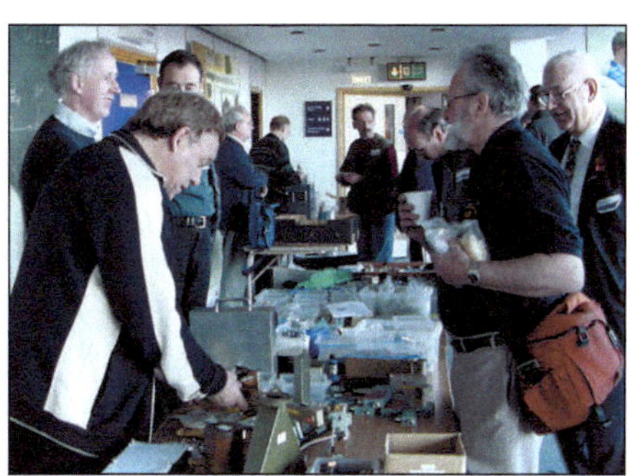
The Martlesham Round Table Fleamarket.

New-technology 10GHz PCB modules from GW4DGU

Apart from the important work with RSGB, and through that with Ofcom, the Group offers a number of services.

'Scatterpoint' is a monthly newsletter, usually of about 30pages, distributed electronically to UkuG members. It contains notices, operating and technical articles.

'Microwave Roundtables' are run at several locations around the UK each year. Here members can meet and socialise, exchange ideas, listen to lectures on various aspects of microwaving, and exchange and trade equipment and components.

'The Chip Bank' is a free-to-members benefit providing small quantities of components from an increasingly comprehensive stock donated by well-wishers in the industry. Even postage is free!

'UKuG Project Support'. Do you need some help with technical issues, or even finance relating to the development of a bona-fide microwave project, such as a beacon? UkuG has a considerable collective pool of knowledge and experience available. Financial support is considered in advance on a project-by-project basis on a 'cash by results' basis. However, the Group finances are not large enough to be able to provide ongoing support eg. for the running costs of a beacon.

'UkuG Technical Support' is a volunteer service provided by experienced microwavers who give their time and facilities to assist others.

It's a bargain at just £6.00 per year in the UK but membership is free for those under 21. If you want to join, please go to: *www.microwavers.org* and click on 'Membership'.

Have fun on microwaves!

Propagation

An explanation of the solar and geophysical information published by the RSGB.

Each week the GB2RS news bulletin includes a brief solar and propagation report and forecast prepared by members of the Society's Propagation Studies Committee. As carried on most GB2RS broadcasts and as posted on the Internet, these reports look back on the week up to the Thursday before transmission, and ahead to the week from the Sunday of transmission.

The usual format for the bulletins is a review of solar activity, geomagnetic activity and ionospheric data during the week. This is followed by the forecast.

Spacecraft data is used to compile the weekly propagation report, including imagery from the STEREO (Solar TErrestrial RElations Observatory) twin orbiters, which were launched in 2006. Unfortunately one of them has since developed a fault and imagery is currently only available from the STEREO Ahead spacecraft.

The Solar Dynamics Observatory (SDO), which was launched in February 2010, also helps us understand the Sun's influence on the Earth and near-Earth space.

Although a lot can happen in a few days (new regions appear, old ones decay), from now on more accurate forecasting will be possible.

Solar Activity

The general level of solar activity for a 24-hour period from midnight to midnight is described as:

Very low
Either no solar flares or only A- or B-class flares.

Low
C-class flares, which usually have little or no impact on propagation.

Moderate
Between one and four isolated M-class flares.

High
Five or more M-class flares or isolated (1-4) M5 or greater flares, including X-class solar flares.

Very high
Several flares of M5 or greater magnitude, including X-class flares.

A flare is a sudden eruption of energy on the solar disc emitting radiation and particles, which can last anything from a few minutes to some hours. Flares are classified as 'A', 'B', 'C', 'M' or 'X' according to their X-ray energy level. This is measured by satellites in terms of megaelectron volts (MeV).

There are four energy thresholds: 2, 10, 50 and 100MeV, and flares are classified by numbers, ie M3, X4 etc. Generally the broadcasts mention only M- or X-class events, as they are the most likely to have an impact on propagation.

However, flares are also classified by their optical importance, which gives a measure of their size and brilliance. Size is indicated by a number between 1 and 4, while brilliance is either 'F' = faint, 'N' = normal, 'B' = bright or 'S' meaning sub (so 'SF' indicates sub-faint). The energy and optical indicators combine to give the complete flare data, eg 'M3/2B'.

Major Flares

Flares above M9/3B and all X-class flares can be very disruptive to the ionosphere. They can lead to severely-degraded HF propagation and any associated Earth-facing coronal mass ejections (CMEs) can cause auroral events, usually after a lapse of between 30 and 50 hours, particularly if they are on the Sun's limbs or have just passed the central meridian. Flare effects may be reported as:

Sudden Ionospheric Disturbance (SID) or Short-Wave Fadeout (SWF)
HF propagation blacked out or degraded for between a few minutes and many hours, with the lower bands affected first and most severely, the higher bands affected less and recovering first, but LF (<500kHz) and VLF DX signals may be enhanced.

(Sudden) Storm Commencement
Increase or decrease in the northward component of the geomagnetic field, marking the beginning of a geomagnetic storm. The onset may be very sudden (SSC) or more gradual (SC).

Noise Bursts or Noise Storms
Enhanced emissions from the Sun at radio wavelengths, associated with major flare events or complex solar active regions. They may last only a few minutes or for many hours.

Proton Events
These may be mentioned if they have an energy level exceeding 10MeV. Proton events cause high absorption in the D region of the ionosphere, particularly affecting transpolar propagation due to polar cap absorption, which can be degraded for days or even weeks following such an event.

Coronal Holes

Coronal holes are holes in the Sun's outer corona through which material is ejected by various means. There are always holes at the Sun's polar regions but tongues sometimes extend to the equatorial regions, or small holes can form. The passage of these can cause a magnetic disturbance. This is particularly so if the interplanetary magnetic field is southerly, as this couples to the Earth's northward field. What have become known as 'Scottish' type auroras can generally be attributed to the passage of a coronal hole. If known about, coronal holes are always referred to in the text due to their importance. Coronal holes become more geoeffective when at low solar latitudes and are more numerous around the time of solar maximum and during the first few declining years after solar maximum.

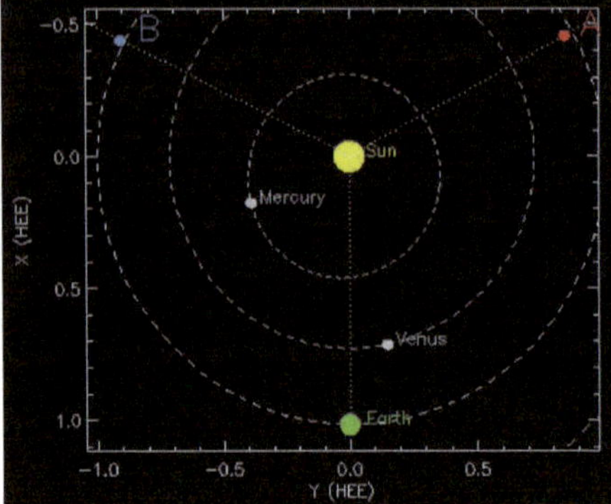

Fig 1: *Present locations of the STEREO spacecraft.*

Other Solar Events

Now and again reference is made to solar filaments. They appear as prominences on the solar limbs and as dark snaking strings of material against the limb as viewed in the light of hydrogen alpha. Occasionally the magnetic fields that hold filaments together break apart and fling the filamentary material into space. Filaments can last for several solar rotations before fading or erupting. These events can be sudden and are mostly unpredictable and can cause widespread auroras, ionospheric blackouts, and worldwide disruption to radio communication. Sometimes eruptive prominences are reported, but because they are located on the solar limbs they are not so geoeffective and therefore not so disruptive.

Satellite Data

Increasingly, new forms of data from satellites are supplementing or replacing traditional forms of ground-gathered data in explaining and predicting the 'propagation weather'. Some or all of the following may feature in a bulletin:

X-ray background flux
A more sensitive indicator than solar flux. It is reported on a rising scale of A1-9, B1-9 and C1-9. GB2RS reports the weekly average and any unusual levels.

>2MeV electron fluence
Referred to as 'high', 'normal' or 'low.' High levels adversely affect the HF bands.

Solar wind speed
The ACE satellite measures the speed of the flux of solar particles and magnetic fields moving outwards from the sun. Normal velocities are around 350-400km/s, though speeds exceeding 1000km/s have been recorded.

Particle densities
Under 10 per cm^3 are 'low'; 10-25 'moderate' and above 25 'high'.

Bz
The orientation of the interplanetary magnetic field measured in nanoteslas. A southerly (-ve) orientation, coupling with Earth's northern orientation, results in HF disturbances and auroras. A northerly (+ve) orientation has little or no effect.

K	0	1	2	3	4	5	6	7	8	9
A	0	3	7	15	27	48	80	140	240	400

Table 1: *Geomagnetic activity look-up table for a typical middle-latitude magnetic observatory.*

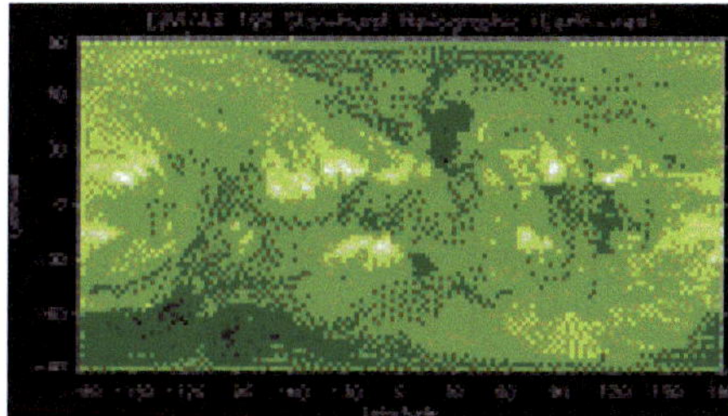

Fig 2: *A whole Sun image, created by combining images from Earth, STEREO-A and STEREO-B.*

Solar Flux

This is the 2800MHz radio noise output from the Sun at midday. This frequency is chosen because the radio Sun looks the same size as the visible Sun. The figure given is that obtained at Penticton (BC, Canada), which is the world standard. The level varies from (at the cycle's minimum) about 64 units to a maximum of around 300 units. The higher the level, the more intense is the Sun's ionising radiation and the higher the frequency that can be reflected from the ionosphere. Good HF band conditions require a high solar flux but the level of magnetic activity will also be a crucial factor. A 90-day average of flux levels is given, as this has been found to be best for home computer prediction programs.

Geomagnetic Activity

References to geomagnetic activity are made in terms of the worldwide or 'planetary' A index, expressed as 'Ap' units. During magnetic storms, the A-index may reach levels as high as 100. During severe storms, the A-index may exceed 200. Great 'rogue' storms may produce index values in excess of 300, although storms associated with indices this high are very rare indeed. Generally, an Ap index of 0-10 is considered 'quiet', 11-20 is 'unsettled', 21-50 as 'sub-storm', 51-80 'storm', 81 and above 'severe storm' or 'major storm'. High levels of geomagnetic activity - say roughly over 30 - are associated with poor HF conditions, especially on the higher bands. The greater the index over about 50, the greater the likelihood of aurora. And the higher the index, the further south auroral working may be possible. While auroral events, including auroral-E, are most likely to be found at 50, 70 and 144MHz, they can sometimes be found at 28MHz and, during major disturbances, contacts may be possible at 432MHz.

The Ap index is linear, unlike the alternative K index, used by WWV among others, which is quasi-logarithmic and open ended. Each observatory uses a look-up table created for that specific location, to convert an amplitude range into an associated K-index value. The table above shows the look-up table for a typical middle-latitude magnetic observatory.

Ionospheric Data

The critical frequency is the highest frequency reflected back from the ionosphere from a signal sent vertically upwards by an ionosonde. The maximum frequency that can be used for normal communication at equal latitudes is very roughly about three times the critical frequency (so a critical frequency of 7.0MHz indicates that it should be possible to work due east at single hop distance - 3000km - at 21.0MHz). For southerly working the multiplication factor would be higher but over northerly paths it would be lower. However, the actual level attainable also depends on the level of geomagnetic activity, the season and the time of day.

During the hours of darkness, the normally relatively smooth ionospheric layers can break up during magnetically disturbed periods. This is referred to as **Spread F**. The break-up can be vertical or horizontal, or both at once. The resulting holes give rise to deep fading. Northern circuits are more prone to this effect, which is more likely to occur during the early morning. References in the bulletins usually refer to the number of hours when Spread F has been present, or if it was very bad, on any particular day.

Blanketing E means that the E layer is so intensely ionised that the ionosonde cannot see through it. This effect is often associated with summertime sporadic-E or, for northern stations, with auroral conditions.

Absorption

Sometimes, for northern stations, complete absorption of the ionosonde signal occurs. This suggests that the D region is so heavily ionised that it is absorbing all but the strongest radio signals. These events can be associated with proton events, or high energy 'M' or 'X' flare events, or by electron precipitation from the Earth's radiation belts.

Seasonal Changes

The daily highs tend to be higher in winter and lower in summer. The darkness hour lows vary in the opposite way – high in summer and low in winter. The weekly average variations are balanced against these seasonal changes, and reference is made to any discrepancy when this applies. For the HF bands, the higher the daily 'highs', the better is the chance of DX on the higher HF bands.

The average times of the highest and lowest frequency recorded are given. The high times vary with the season, being around midday in the winter and early evening (about 2000UTC) in the summer. The low times do not vary much, being usually about 0400UTC.

Forecasts

Each week the bulletins include a forecast of expected events for the seven days following the Sunday of broadcast. This includes expected levels of solar flux, geomagnetic activity and the passage of any expected coronal holes. Maximum Usable Frequencies (MUFs) during daylight hours at equal latitudes are estimated for southern England. Scotland and Northern England will generally be down on these levels by around 3MHz at equal latitudes due to factors such as geomagnetic activity, which may affect northern areas more. In general, north-south paths will tend to be more readily workable than east-west, especially around the equinoxes. Bulletins usually include a forecast for one or more path.

The MUF given in these forecasts indicates the frequency up to which the path should be workable on 50% of days. On the better days, therefore, this value should be exceeded - at times by a considerable margin.

It also indicates a lower and more consistently reliable Optimum Working Frequency. This is the frequency on which it should be possible to operate on 90% of days. During months when sporadic-E propagation is prevalent - roughly May to August and around Christmas and the New Year - reminders are given of the likelihood of openings on 2, 4, 6 and 10m, and brief reports of major openings may be included.

Solar activity may be mentioned in the form "the quiet side of the Sun" or "the active side of the Sun"; which refers to the best chance of flare activity. For instance, the forecasts will attempt to predict the best chance of solar activity reaching moderate or high levels, therefore that being the active side of the Sun. This works in reverse for the quiet side.

The table published each month in *RadCom* shows path predictions from the UK to 31 locations around the world, 27 being short path and four being long path. All are F-layer predictions. The numbers indicate the expected reliability of the circuit; with a '1' representing between 1 and 19% of days, '2' between 20 and 29% of days, etc. The colours represent relative signal strength; a dot being no signal, black being a weak signal, blue being a fair signal, and red being a strong signal.

While every effort is made to ensure the table is as accurate as possible, it should always be remembered that propagation prediction is not an exact science. There will be times, for example during magnetically disturbed periods, when the table may be unduly optimistic. Alternatively, and especially during the peak of the solar cycle, conditions may for a time considerably exceed predicted values.

It should also be noted that these HF predictions are for F-layer propagation. They do not take into account either sporadic-E which, particularly during the summer, enlivens 28MHz, or auroral-E which occurs occasionally at HF (and VHF) during geomagnetic disturbances. They also do not include 'greyline' long-distance openings, which are a feature of the lower bands during the periods around dusk and dawn, as these are of too short a duration to appear in the two-hourly steps used in the tables. Neither can they take into the reckoning other short-lived phenomena like back-scatter and side-scatter. All these make the task of the forecaster more complex - but also make day-to-day working on the bands more varied and challenging. Fuller explanations of these and other solar and propagation phenomena, as well as near-real-time data, can be found on the Internet, notably at the Propagation Studies Committee (PSC) page.

RSGB PSC: www.rsgb.org/psc/

GB2RS propagation enquiries: Steve Nichols, G0KYA. Email: psc.chairman@rsgb.org.uk (or QTHR)

HF Propagation in 2024

2024 will be remembered for its unsettled geomagnetic conditions, a healthy display of sunspots and better upper HF band propagation than previous years.

The year was characterised by a large increase in the number of sunspots, with the SFI heading towards 200 a lot of the time, but also an increase in solar flares and coronal mass ejections. These often pushed the geomagnetic Kp index higher with a corresponding lowering of maximum useable frequencies at times.

We even had a night (Friday May 10) with widespread visible aurora from the UK as the Kp index hit nine, which will probably go down in history. I'm sorry if you missed it!

As always, my advice is to look for DX when the SFI is high and the Kp index is low, and pick the right time of year for your DXing. For example, autumn and winter are generally good seasons for HF DX, whereas the Summer can be lousy.

This seems to go against "common sense". After all, we need sunlight to ionise the ionosphere, so why isn't HF propagation better in the summer?

Basically, this is because of a chemical change in the summer ionosphere that means there are fewer monatomic species, which are easier to ionise. There are, however, more molecular, or diatomic components, which are more strongly bound and therefore harder to ionise.

Hence, HF propagation tends to better in the autumn and winter.

Current predictions are that solar maximum may occur some time around May-October 2025. But don't worry as we will have good sunspot numbers for quite a while after that.

The only worry is that geomagnetic disturbances can often increase on the downward part of the solar cycle, so it may not be all plain sailing.

Autumn tends to be the best time overall for HF propagation, with paths to North America and beyond opening up in late September.

The winter months, with lower levels of daylight, tend to be associated with good conditions on the lower bands too, such as 160, 80, and 40 metres.

What we can also say with some confidence is that the equinox months are best for North-South paths.

But you can prepare your HF activity with a range of tools available on the web. It is worth reminding you what they are.

Check out www.rsgb.org/voacap and see www.rsgb.org/proppy, which offer two alternative methods of predicting propagation to various locations around the world. These now allow you to prepare your own RadCom-type predictions, based on your preferred mode, power and antenna choice.

If you want access to the full-blown VOACAP prediction tool then go to www.voacap.com. Or if you want an alternative, which uses the newer ITURHFPROP engine to produce results, go to Proppy at https://soundbytes.asia/proppy/.

Now let's look at the whole year, season-by-season, band-by-band. From a propagation perspective, conditions are dependent upon the angle the Sun makes with the ionosphere, so the periods around both equinoxes are likely to be similar. There is a gradual change from one season type to another, so the periods listed below should not be taken too literally.

Winter period

(Jan-Feb / Nov-Dec)

These periods are when the low bands (160m, 80m and 40m) come into their own. While solar maximum is not the best time for low band propagation there will still be plenty of DX to be worked. Generally, winter is a good time for East-West paths on HF too..

160m (1.8MHz or Top Band)

Solar absorption will prevent skip during daylight hours. You should be able to work other UK stations out to about 50-80 miles via ground wave. The band will start to come alive around sunset and openings up to around 1,300 miles should be possible, with frequent openings up 2,300 miles. DX openings to the east from the UK should be possible around midnight and to the west before sunrise for well-equipped stations.

80m (3.5MHz)

Expect a similar pattern to Top Band, with DX openings at night with peaks at midnight and around sunrise (greyline openings). Openings around the UK and out to around 500 miles should be possible during the day and between 750-2,300 miles at night. A low, horizontal antenna will be useful for relatively local, NVIS (Near Vertical Incidence Skywave) signals, but lower angle radiation, such as obtained with a vertical, will be required for DX.

40m (7MHz)

Another great DX band at this time of year. 40m should open for DX in an easterly direction during the late afternoon and towards the south at sunset. Paths during the afternoon may also include W6 (west coast USA) in mid-winter. Openings to the west, including long path to VK/ZL, should

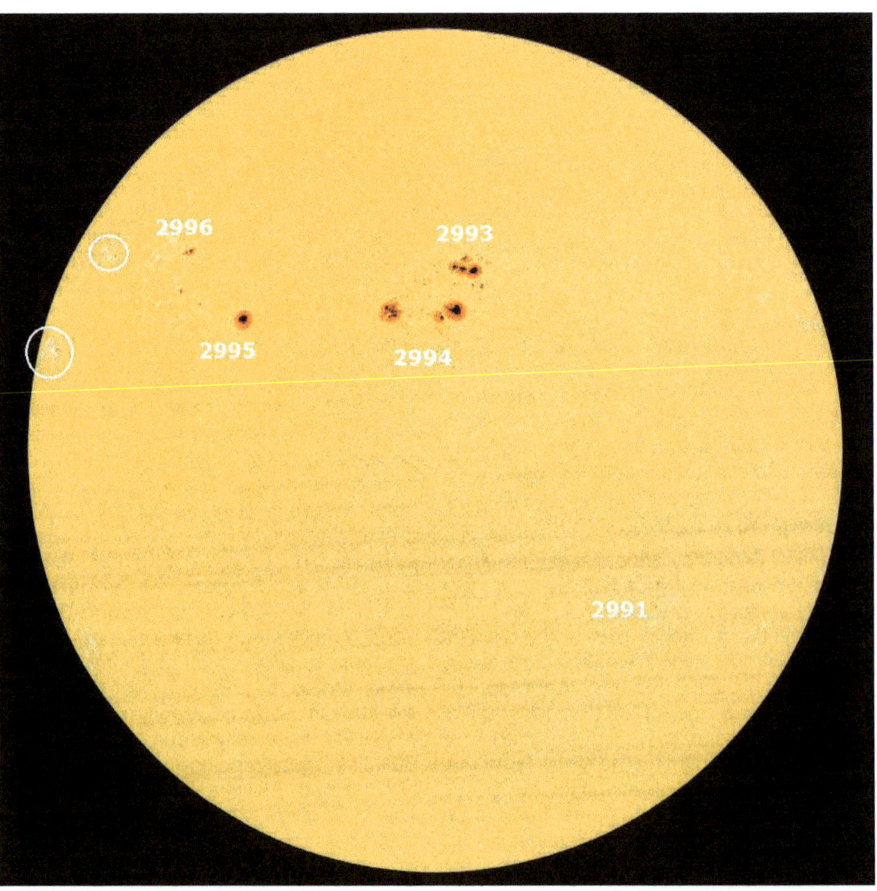

This image from the solar dynamics observatory (SDO) shows the sun in visible light and with a large number of sunspots

be possible after midnight and should peak just before sunrise. Relatively local contacts should be possible during the day as the daytime critical frequency is higher with the rising number of sunspots.

20m (14MHz)

This is likely to provide great DX openings during the hours of daylight. Peak conditions will be a couple of hours after sunrise for paths to the east and a couple of hours before sunset for paths to the west. Contacts up to 2,300 miles should be possible during daylight hours, and the band may remain open after sunset if the solar flux remains high. Occasional DX openings towards South America may be possible after nightfall. Watch out for higher levels of D-layer absorption around local noon, which may mean that you are better off heading for the higher bands at this time.

17m/15m (18MHz/21MHz)

Should provide good to excellent DX openings during daylight hours. The period from noon to late afternoon may be best, but both bands are likely to close soon after sunset and remain closed until some time after sunrise the following day, unless the solar flux is very high.

12m/10m (24MHz/28MHz)

If the solar flux index remains higher than 120 or so, which it should, good DX could be possible during daylight hours. A brief spell of Sporadic-E can sometimes occur in the New Year, resulting in very strong, but short-lived propagation on 10m out to around 1,300 miles.

Equinox periods

(Feb-May / Aug-Nov)

The equinox periods provide longer daytime periods than winter, but logically, shorter night-time periods too. These tend to be the best months for working North-South paths, such as UK to South Africa.

160m (1.8MHz or Top Band)

Look for short-skip and DX openings at night. Again, no daylight skip is possible due to absorption, but openings out to 1,300 miles and occasionally further afield can be expected at night, with conditions peaking around midnight and again at sunrise (greyline).

80m (3.5MHz)

Will generally follow the characteristics of Top Band at night, but will also provide good openings out to around 250 miles during the day. These will lengthen to around 500-2,300 miles at night, with fairly good DX opportunities at times.

40m (7MHz)

Should open to DX in an easterly direction at sunset. Openings to the west should be possible after midnight and should peak just before sunrise. Contacts should be possible during the day, and the higher critical frequencies caused by the high solar flux index will mean that it is possible to work other UK stations via NVIS – Near Vertical Incidence Skywave. Also, look for good paths near local midnight and at sunrise (greyline).

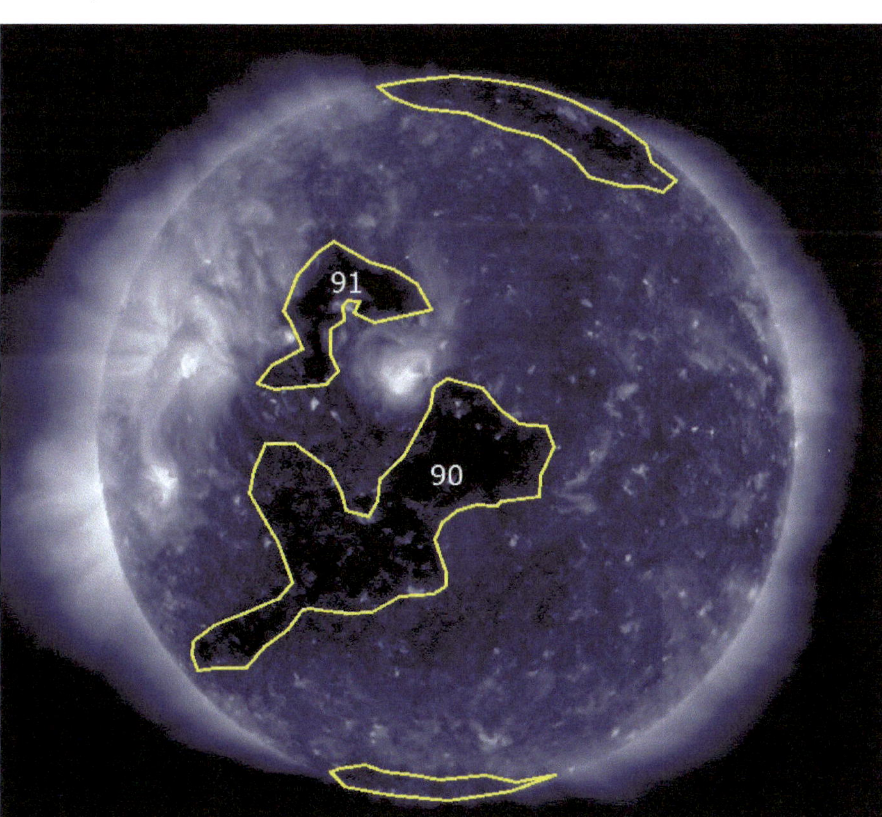

But in extreme ultra-violet this SDO image shows the extensive coronal holes, responsible for poor HF conditions.

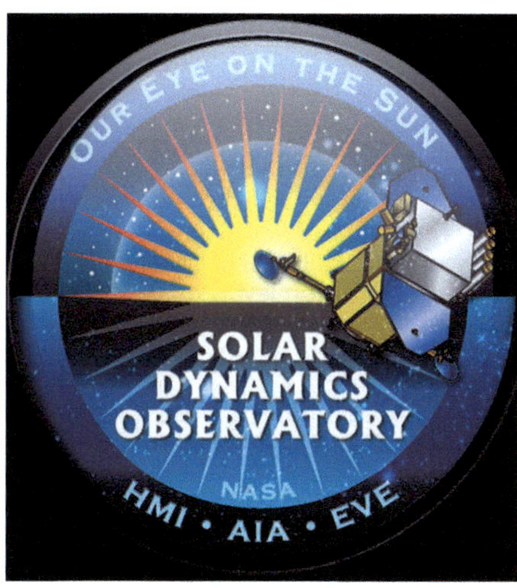

The Solar Dynamics Observatory mission logo.

20m (14MHz)

Likely to be the best DX band between sunrise and sunset. The band should remain open after dark if the SFI is high, perhaps giving openings to the southern hemisphere. Good openings will be possible during daylight hours out to around 2,300 miles.

17m/15m (18MHz/21MHz)

Should provide fairly good DX openings during daylight hours, especially to Africa and South America, with 17m being open more often than 15m. Once again, 15m could provide good openings if the SFI stays above about 100-120. Both bands are likely to close after sunset.

12m/10m (24MHz/28MHz)

If the solar flux heads above 120-130+ then openings will occur on both bands, although 24MHz will open first. If it keeps breaking the 130 mark then expect to see some excellent DX openings on 10m, especially in early spring/late autumn. Autumn should offer fine openings to the USA/Caribbean, even in the FM portion of the 10m band around 29.6MHz.

Summer Solstice period

(May-Aug)

Daytime MUFs are likely to be lower than those of winter. The so-called 'Seasonal Anomaly' is thought to be due to a large summer electron loss rate caused by an increase in the molecular/atomic composition of the ionosphere and the reaction rates being temperature sensitive.

It is not all bad news though. Night-time MUFs may be higher in summer than those in winter. Note that DX on the low bands, if possible, is unlikely to occur until around midnight or the early hours, due to the late sunset.

160m (1.8MHz or Top Band)

High levels of static and solar absorption mean that the band will not really support

sky-wave contacts during the day. During darkness, short-skip openings may occur, but DX may be a rarity. Occasional openings can occur during the hours of darkness, especially around local midnight/early hours. Not the best season for Top Band.

80m (3.5MHz)

Will generally follow the characteristics of Top Band with high levels of static. Absorption will grow to a maximum at midday for inter-G contacts, so you may be better going to 40m. DX capabilities will be poor to fair during the hours of darkness, compared with the winter.

40m (7MHz)

Will suffer from high static, caused by high numbers of thunderstorms. Nevertheless, night-time openings should be reliable from sunset to sunrise. Local daytime openings will be possible on the whole with 40m being the band of choice for contacts around the UK. Night-time skip distances are likely to be between 300 and 2,300 miles.

20m (14MHz)

Still likely to be a good DX band around the clock, although the band will be noisier than the winter period and perhaps not as reliable for long-haul contacts in the summer. The higher MUFs at night mean that 20m may remain open during the evening and night to DX. Short skip may also be possible due to summer Sporadic-E.

17m/15m (18MHz/21MHz)

Should provide a fair number of DX openings during daylight hours, especially to the southern hemisphere. Both bands are likely to close after sunset. Sporadic-E will provide good short-skip openings, predominantly in the May-June period.

12m/10m (24MHz/28MHz)

Sporadic-E openings will provide regular openings out to around 1,300 miles. Multi-hop Sporadic-E openings are possible, providing relatively good but short-lived paths to DX beyond this range. A typical multi-hop opening might provide brief contacts with the Middle East or USA, although they would be very hard to predict. Propagation via the F2 layer is likely to occur on north-south paths at times

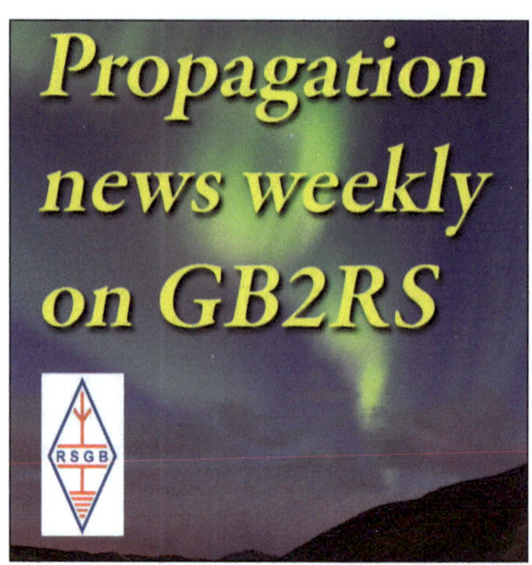

Steve Nichols, G0KYA
Chairman, RSGB Propagation Studies Committee

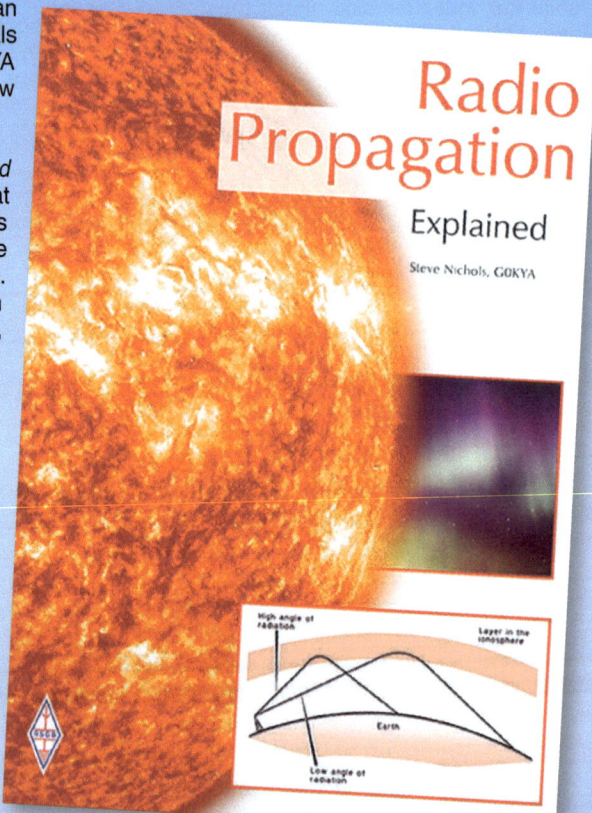

Data Communications

The Emerging Technology Co-ordination Committee (ETCC) exists to deal with all matters concerning amateur radio repeaters and data communications on behalf of the RSGB. It liaises with Ofcom in the processing of applications for Frequency Assignment Certificates (FAC) for repeaters, Internet Gateways and mailboxes. It is also responsible for the coordination of all requests for site and frequency clearances prior to submission to Ofcom were this is still required (Non-primary Amateur Bands).

The ETCC was formed in 2007 by the amalgamation of the Repeater Management Committee and the Data Communications Committee.

Because of the continuing and accelerating convergence of radio and computing and the emergence of amateur digital voice protocols it was considered that interests would be best served by having just one committee responsible for both analogue and digital proposals.

From 1999 to 2007 one of the ETCC's predecessors, the Data Communications Committee, acted as the body responsible for facilitating—by means of frequency co-ordination—simplex internet voice gateways.

At the time of writing there are operational gateways on 29 MHz, 50MHz, 70MHz, 145MHz, 430MHz, 431MHz, 434MHz and 1299MHz.

What is Packet Radio?

Packet radio is digital communications via radio. It began in 1978 in Canada and was introduced into the UK in the 1980s. Packet radio mailboxes (BBS) were first licensed in the UK in about 1988. The numbers grew until the late 1990's, when numbers then started to decline due to the widespread availability of broadband Internet to most of the UK population. However, some mailboxes do still operate in UK, so the following information should help any newcomers to the mode.

What can I do on Packet?

Mailboxes

Mailboxes allow amateurs to connect with their local mailbox and send and receive text messages. These messages can be sent as personal messages to another amateur anywhere in the world (or in space!). Alternatively, messages can be sent as a bulletin for any amateur to read.

Within the UK your messages will normally be relayed via other mailboxes using Internet gateways. These are used to forward mail to more distant continental mailboxes, so it is possible to exchange mail with amateurs on the other side of the world by simply logging into your local mailbox using a low power VHF transceiver and a simple aerial system, as well as a Terminal Node Controller (TNC) and computer.

File Transfer

Packet radio also allows you to be able to transfer files between amateur packet stations in both text and binary format.

DXCluster

There are a few DXCluster stations around the UK and, alongside, the 'automatic position reporting' component of the network. The provision of near real-time on a truly global network exchange of DX, band condition and stations heard/worked information is highly valued by the world's DX community.

APRS

Automatic Packet Reporting System (APRS) was developed by Bob Bruninga, WB4APR, to track mobile GPS stations with two-way radio. APRS can be used in a number of applications for data, communications and telemetry.

In the UK, the late Roger Barker, G4IDE, authored a protocol-compatible variant called 'UI-View'. This is extremely popular and is well supported with regular updates and third-party extensions that add increased functionality to the base product. Examples of add-ons include 'DXCluster spy', satellite telemetry decoding, rig control, rotator control the list is long! For full details see the website at www.ui-view.org

APRS remains one of the most active packet radio modes. ETCC issue NoV's for the operation of APRS digipeaters and internet gateways.

NoV's are now required for any unattended packet radio operation. Please see the ETCC website for details, as well as the on-line forms.

What will I need?

As well as a VHF or UHF FM radio transceiver, you will also need a TNC and a terminal or computer with some form of terminal software program or a specialist packet radio software program. And finally... a great deal of patience and willingness to learn.

Other data modes

RTTY or Radio Teletype

A frequency-shift-keying mode that has been in use longer than any other digital mode (except Morse). It uses a simple five-bit code to represent the alphabet, numbers, a few punctuation marks and a few control codes. As there is no error-correction, QRM and QRN can have seriously detrimental effects on copy. Despite all that, it is still the most popular digital mode. The bandwidth of a RTTY signal is 230Hz, and at 45.45 Baud it gives about 60 WPM throughput.

PSK31

This mode has the advantage of having a very narrow bandwidth and was designed for 'real-time' keyboard-to-keyboard QSOs. It can be used to send almost all of the characters shown on your keyboards and has even been used to send small pictures. If only lower case letters are used, you get about 53 WPM, but with all capitals, this is reduced to about 39 WPM. The bandwidth is as the name suggests 31Hz.

MFSK16

Uses 16 tones and has forward error correction, where it sends all data twice with an interleaving technique to reduce errors from such things as static crashes. It has a comparatively wide bandwidth of 316Hz, which allows faster baud rates, and has greater immunity to multipath phase shifts. This wider bandwidth gives you around 42 WPM.

Hellschrieber

Uses facsimile techniques to transmit and receive characters. It has been in use since the 1920s but has come on in leaps and bounds thanks to modern day DSP soundcard processing. Characters are painted on the screen in ticker-tape like fashion and are read directly from the screen, as opposed to being decoded and printed. Although this mode has a relatively small bandwidth of about 75Hz, it can handle about 35 WPM.

MT63

An excellent mode for sending text over propagation paths that suffer from fading and interference from other signals. It works by encoding text with a matrix of 64 tones over time and frequency. Although this is rather complicated, it does provide error correction at the receiving end, and gives about 100WPM. MT63 has a wide bandwidth of 1kHz.

Throb

Uses either 5 or 9 tones, depending on the version in use. The latest is the 9-tone version, and has speeds of 1, 2 and 4 Baud, enabling data rates of 10, 20 and 40WPM respectively. It appears to be quite good under poor propagation conditions and although it isn't commonly heard, it does seem to be gaining popularity.

23Hz RTTY

A relatively new thing, and hasn't proved as popular as hoped. It sounds similar to PSK31 but is a bit narrower in bandwidth. Although several programs now include this mode, it hasn't taken off well.

FT8 Digital Mode

FT8 is a weak signal mode for HF DXing. FT8 is simple to set up and works well even if the noise level is high.

FT4 Protocol for Digital Contesting

FT4 is an experimental digital mode designed specifically for radio contesting. Like FT8, it uses fixed-length transmissions, structured messages with formats optimized for minimal QSOs, and strong forward error correction. T/R sequences are 6 seconds long, so FT4 is 2.5 × faster than FT8 and about the same speed as RTTY for radio contesting. FT4 can work with signals 10 dB weaker than needed for RTTY, while using much less bandwidth.

Where to find the other data communication signals

In recent years there has been an explosion in the number of digital modes, with many different people writing software to decode

them. Some of the modes may be familiar to readers, others perhaps not. Many of you will know the sound of RTTY, and maybe that of PSK31, but do you know what Throb or MFSK sound like or where to find their signals in the amateur bands?

Within the amateur bands the data communications signals are to be found around the frequencies shown in the bandplans.

Throb, Hellschrieber and MFSK16 tend to congregate just below the RTTY segment.

These frequencies are approximate, but will give you an idea of where to start looking. A full set of datacommunications bandplans is available on the DCC website.

Getting started

The advent of soundcard software for the PC heralded a new era in the digital modes, especially since most of the software is free!

The advice given here is intended for people who are newcomers to the data modes and is therefore presented at a beginner's level. It does not explore the theory behind the data modes in any depth. Neither does it cover all aspects. Once you have understood the basic principles, you can start to learn the more advanced features at your own pace.

If you are not confident in wiring-up cables then you should either buy one of the commercial interfaces or get someone to do it for you.

These explanations will refer mainly to RTTY and PSK31, as they are the most popular of the digital modes in use at present. However, details are given of other modes present on the bands.

Background

Computer soundcards use Digital Signal Processing (DSP) to handle sound, and these techniques lend themselves very well to the processing and decoding of the audio data signals from the output of your radio.

Although RTTY, AmTOR and PacTOR have been around for quite some time, it is the general feeling that a renewed interest in the digital modes really came about when Peter Martinez, G3PLX, created the PSK31 program for the Windows operating system.

At first, it proved to be quite difficult to tune in a PSK signal, so much so that many gave up before they had even watched a QSO in progress. For those of us who persevered, however, it proved to be something of an enlightenment. At that point, tuning was aided only by turning the tuning knob, with the aid of the phase scope, and it took a great deal of patience and careful finger work to tune in one of these new sounds. You only needed to be off by as little as 5Hz to get garbage on the screen. After several months, Peter came up with a version that included a waterfall display, and that really made a big difference. It was probably a key turning point. Now you could see which way to turn the dial, and you had a fair chance of getting it right quite quickly. Later versions allowed you to point and click on a signal on the waterfall and it was tuned in instantly.

On the bands one was quite likely to meet up with G3PLX, and although he wouldn't hesitate to tell you that you were over-driving the soundcard, if indeed you were, he would offer suggestions, and between you, it was possible to adjust the levels during the QSO to an optimum state. Since then the number of operators has increased considerably. Unfortunately, many have not taken the advice offered by those with greater experience, especially with respect to the adjustments of sound levels and transmitter power. One common mistake is to leave the speech processor switched on, which will definitely cause you problems with the transmitted tones.

In the early days it was the norm to use between 5 and 10 watts of transmitted power and many used much less than that. This level of signal is perfectly adequate for world-wide communications, although these days many stations seem to use several hundred watts into large beams. Apart from the fact that it is not necessary, it also tends to reduce the bandwidth available to others.

Some new modes, such as MFSK16, have been developed in the past few years with the idea of replacing RTTY. Although they do have a following, RTTY has remained at the forefront of the digital modes. PSK31 has gained respect mainly because it is so good at low power levels, making it ideal for QRP work.

Operating the Data Modes

When you are ready to begin using the digital modes, if you are new to data communications it is suggested that initially you spend some time just listening and watching QSOs in progress. This will give you an idea of how they are conducted, what sort of phrases are commonly used and general etiquette.

When you are set up and as you begin operating you may find that RTTY and PSK give you greater scope for learning, mainly because they are more common, but also because they are easier to operate.

The world is waiting for you. It is quite feasible to achieve DXCC on digital modes and the good news is that there are many rare DX countries out there that regularly appear on one of the digital modes. A lot of the DXpeditions these days include RTTY and/or PSK31.

Remember that RTTY is 100% duty-cycle, so please watch your output power! PSK31 is about 80 per cent duty cycle, but as it is used at much lower power this is of less importance. Curiously, Hellschrieber has a duty cycle of only about 21 per cent, so is a lot more 'equipment friendly'.

If you change the output power when you change mode, you may need to adjust the mic gain, as the ALC may well have altered. Just hit the transmit key and adjust the gain so that the ALC is showing slightly, then back it off a touch. It shouldn't make any difference to the indicated output power, but your signal will be cleaner.

If you use the DX Cluster while you are using RTTY, you may well find that the spots are not where they are listed. This is because operators in Europe tend to use different standards from the rest of the world. In Europe, the norm is to use 'low tones' (1275 and 1445Hz) on USB whilst everyone else seems to use 'high tones' (2125 and 2295Hz) on LSB. Although the tone values won't make any difference to your receiving, the sideband will.

Many of the RTTY programs cater for the American market, which means that 'Normal' equates to LSB. If you intend using USB, you may need to find the 'Reverse' button to invert the tones. If you have a RTTY signal with the audio tuned in nicely but only get garbage on the screen, try hitting the 'Reverse' button and you will probably get clear text after that. It can be quite common to see someone operating 'Inverted', and although it will work equally well providing the other station is also inverted, it can reduce the chance of getting a reply to a CQ call.

If you want to work DX and increase your country count, a good way to do this is to enter one of the many RTTY contests. There are about 14 or 15 such contests per year, and many have sections for single operators with low power. No matter how many contacts you make, always try and submit your log, as this helps the contest organiser get an idea of popularity (it also allows them to verify other logs). If you are unsure about your log, then simply send it as a checklog.

One thing you will regularly see in PSK31 is the use of all capitals in the transmitted text. PSK uses an alphabet called 'Varicode', which has shorter codes for the more common letters (similar to Morse) and which also includes both upper and lower case letters. The lower case letters, being more common, have the shorter codes, so any given sentence takes longer to send in capitals. There really isn't any need to use capital letters at all in PSK, even for callsigns. RTTY uses ITC2 (5-element) code that doesn't include lower case, so all capitals is the only option.

Many operators use abbreviations much like in a CW QSO, and it really depends on the type of QSO you are having as to how you will operate. Don't worry about typing mistakes, as these are normal, and anyway who can tell if it wasn't a burst of static that caused his screen to mis-represent what you just typed?

If you are trying to contact a rare DX station, listen for a while and see if the operator is working in any kind of pattern. Listen to the callers and see how they are working. It may be that timing is the key to making the contact. Unless you have big antennas way up high, plus a big linear, and can drown everyone else out, don't try to be first in the pile; wait and let your call be the last one to be heard. That can often get you a response!

Don't forget that with most of the digital modes, if the transmissions of two or more stations overlap, all you get on screen is garbage. It is no good transmitting over the top of a QSO that is in progress, even though it may be nothing to do with the DX station, you will not be read and you will disrupt the QSO in progress.

When you do get through, the other station really doesn't need to know your working conditions or your life history, so keep your QSO short and leave space for others to have a go. It's a good idea to create a macro just for responding to a DX station, perhaps something like this: HISCALL DE MYCALL - TNX - UR ALSO 599 599, VY 73 ES GD DX DE MYCALL. That is all that is needed and anything more is only likely to get lost in the pile-up that will follow.

PSK31 is an excellent mode if you have a limited set-up and can operate only low power. 5 to 10 watts is quite practical with this mode and you can easily obtain DXCC at QRP levels.

To Sum Up:

1. Check and re-check your volume settings.
2. Listen BEFORE you transmit.
3. Watch your power levels.
4. Don't worry too much about spelling mistakes.

Operating Advisory Service

The Operating Advisory Service (OAS) is a service of the RSGB which is intended to assist radio amateurs or others who may be affected by problems which occur within the amateur bands or which develop on other frequencies as a result of amateur transmissions.

OAS works with a team of named volunteers, trained to offer support to the Regional Teams and to RSGB members and provide better education of amateurs who might be affected at some point, rather than focussing on those causing the problems. The aim is to get everyone better prepared and aware, thereby minimising the effect that these people have on the enjoyment of the amateur service by the overwhelming majority.

OAS is led by the OAS Coordinator, assisted by a team of OAS Advisers, located around the country and trained to help. The whole team communicates via an RSGB email list and so they can support each other in determining how best to resolve various problems. In this way, previous successful outcomes will assist others with similar problems to solve.

The team has developed good-practice guidelines for download by members and non-members alike. OAS additionally offers direct advice to RSGB members. They make themselves available to clubs in their region, to give presentations about what to do if problems occur. Requests from clubs for such a presentation are welcome, even if there is no immediate problem: prevention is always better than cure. Being prepared also means a swifter response by those directly affected, containing the problem early. Planning what to do if a fire breaks out is better than having a fire and then trying to find the buckets!

Getting support from OAS

The Regional Team is the first contact that most members will make to request any form of help, including a request for a club talk. This has the advantage that the Regional Team may be able to help directly but they can also help in a co-ordinated way. For example, they might want to bring EMC committee members to work alongside OAS members to help address your problem. They will also know who the ROA nearest to you is. You can also contact the OAS Coordinator, at oas@rsgb.org.uk. They will find a nearby Regional Operating Advisor and can additionally raise your problem with all the ROAs, to obtain the benefit of their collective experience. An acknowledgement of your request will be given swiftly, though it may take a little longer to consider how best to address the problem.

Information

A report to OAS should contain details of the issue and should if possible include dates, times, modes and details of what was heard or what happened to cause disruption.

It is important not to send sound or video files, nor to include a name, call sign or address of any alleged perpetrator of malpractice. These would be counter to RSGB's GDPR practice and will not be accepted.

Support offered

General guidance has been developed for downloading from the RSGB OAS web pages. Anyone is free to do this, whether or not they are RSGB members.

We strongly encourage those having problems to use OAS advice to formulate a plan and encourage local amateurs to cooperate in carrying out the plan. In most cases this will be key to obtaining a satisfactory result.

Given that most problems centre on repeater misuse, active OAS support will require the repeater keeper to be willing to work with OAS. The co-operation of regular users of that repeater will also greatly improve the outcome. OAS cannot solve problems for you but it can guide a committed local team to improve the chance of success.

Regional Operating Advisers are recruited and trained to help. They are expected to work courteously with you and will expect to be treated courteously in return. Like you, they are licensed users of the amateur service but they have no magic powers to solve any given issue. Please help them to help you.

OAS: 3 Abbey Court, Fraser Road, Bedford MK44 3WH. Email: *oas@rsgb.org.uk*

Your Time

We rely on the active support of a myriad of volunteers to enable the RSGB to provide the range of member services that we do. Indeed, we can only operate with the active support of dedicated and committed people who believe in the future of amateur radio and are prepared to give their time to making that future actually happen.

People like you

Putting something back into the hobby or passing on your experience, knowledge and specialised skill is a rewarding and fulfilling experience.

The RSGB has many areas in which your skills might be used to help others. Some of them require a level of specialised knowledge: EMC, Planning, Repeater Management and Data Communications. In other areas, enthusiasm and commitment is all that is required: GB2RS newsreading, Emergency Services or Deputy RSGB Regional Manager, to name but a few.

When it comes to training, why not consider becoming an RSGB Registered Instructor? There is a constant need for Instructors for the Foundation, Intermediate and Full amateur radio licence courses.

You decide when and where you can help (please note that all applications for Registered Instructors are, for security reasons, subject to a vetting procedure).

If you can't find time to volunteer to work for the Society, remember the other area where you can make a real difference - mentoring. Too often we hear of newly licensed amateurs who feel 'on their own' after completing their studies. You can pass on your experience and skill by 'taking under your wing' a new or prospective amateur and ensuring that he or she learns appropriate operating practice and behaviour. In that way we can ensure that our bands are properly used, to the greater enjoyment of everyone, and newcomers can make the most of amateur radio.

For further information on becoming a volunteer for the Society, contact the

RSGB General Manager,
Tel: **01234 832700** or
email: *GM.dept@rsgb.org.uk*

website: *www.rsgb.org/volunteer*

Locator System

The IARU Locator System, usually just called 'Locator', provides a means of pinpointing stations throughout the world. It is most often used by operators above 30MHz, as a means of calculating the distance between two stations. It is also used on the 136kHz band for the same reason. For use by operators on the upper microwave bands, it can have eight digits, though only the first six are dealt with here. The system is based upon latitude and longitude.

As the map and diagrams show, there are three sizes of 'rectangle'. The largest, known as a 'field', is 20° of longitude (east-west) by 10° latitude (north-south), and is designated by two letters. Most of Britain is in IO field. The next rectangle, known as a 'square' (though it is actually neither truly square nor rectangular!) is 2° of longitude by 1° of latitude. One hundred squares make up one field and, as the map shows, these are given numbers 00 in the south-west corner to 99 in the north-east. Dublin is in IO63. Finally, each square is divided into 576 'sub-squares', 5 minutes of longitude by 2.5 minutes of latitude, and given letters from AA to XX.

To find out your locator, first use a map of your area to determine your exact latitude and longitude, then use the map on this page and the squares diagram opposite to pinpoint your locator. Computer programs and online calculators are available to do this more easily, especially for those who operate from various locations.

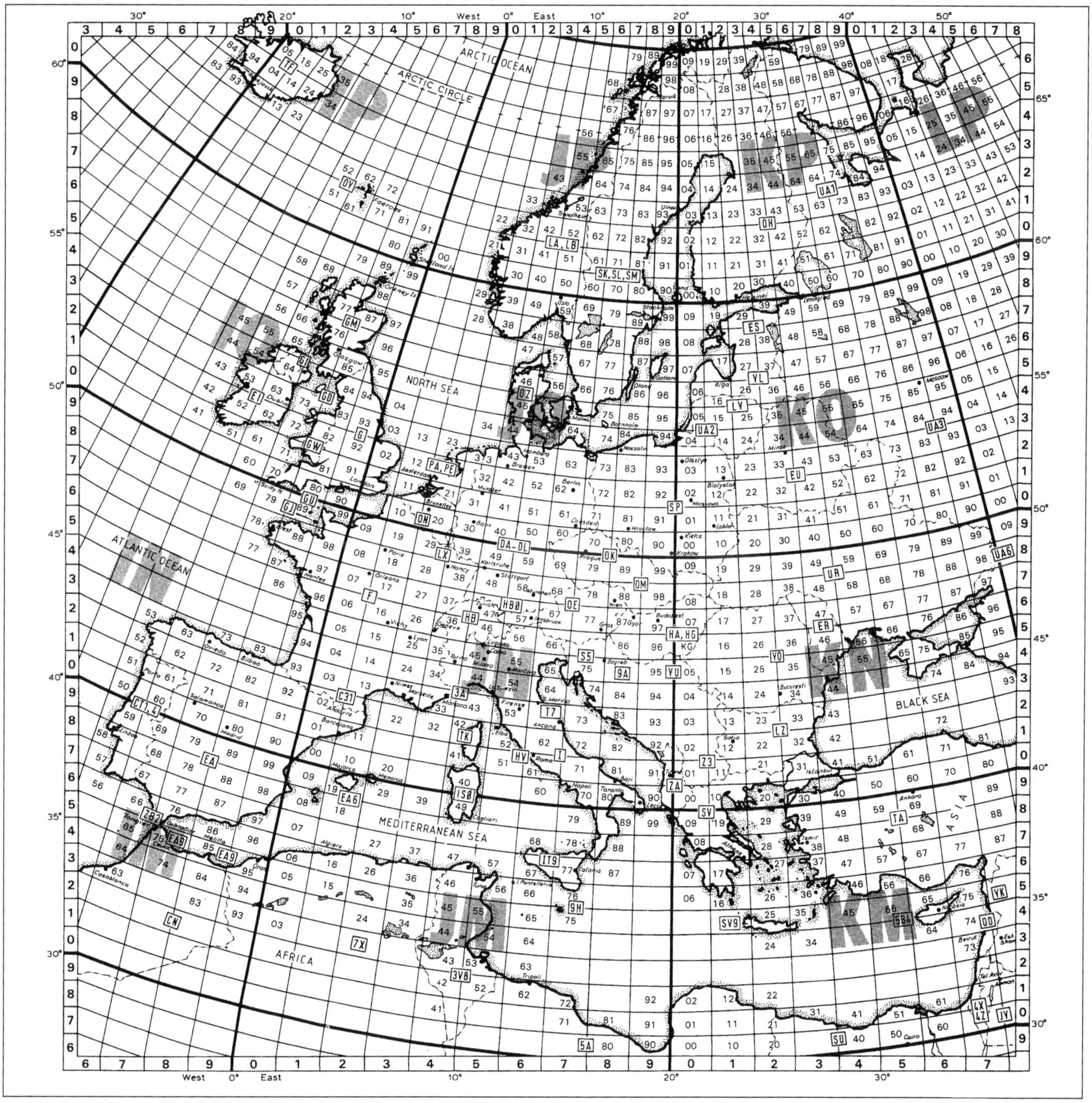

On-line Lat+Long to/from Locator calculators: www.arrl.org/locate/grid.html and www.amsat.org/cgi-bin/gridconv
On-line NGR to Locator calculator: www.ntay.com/contest/NGR2Loc.html

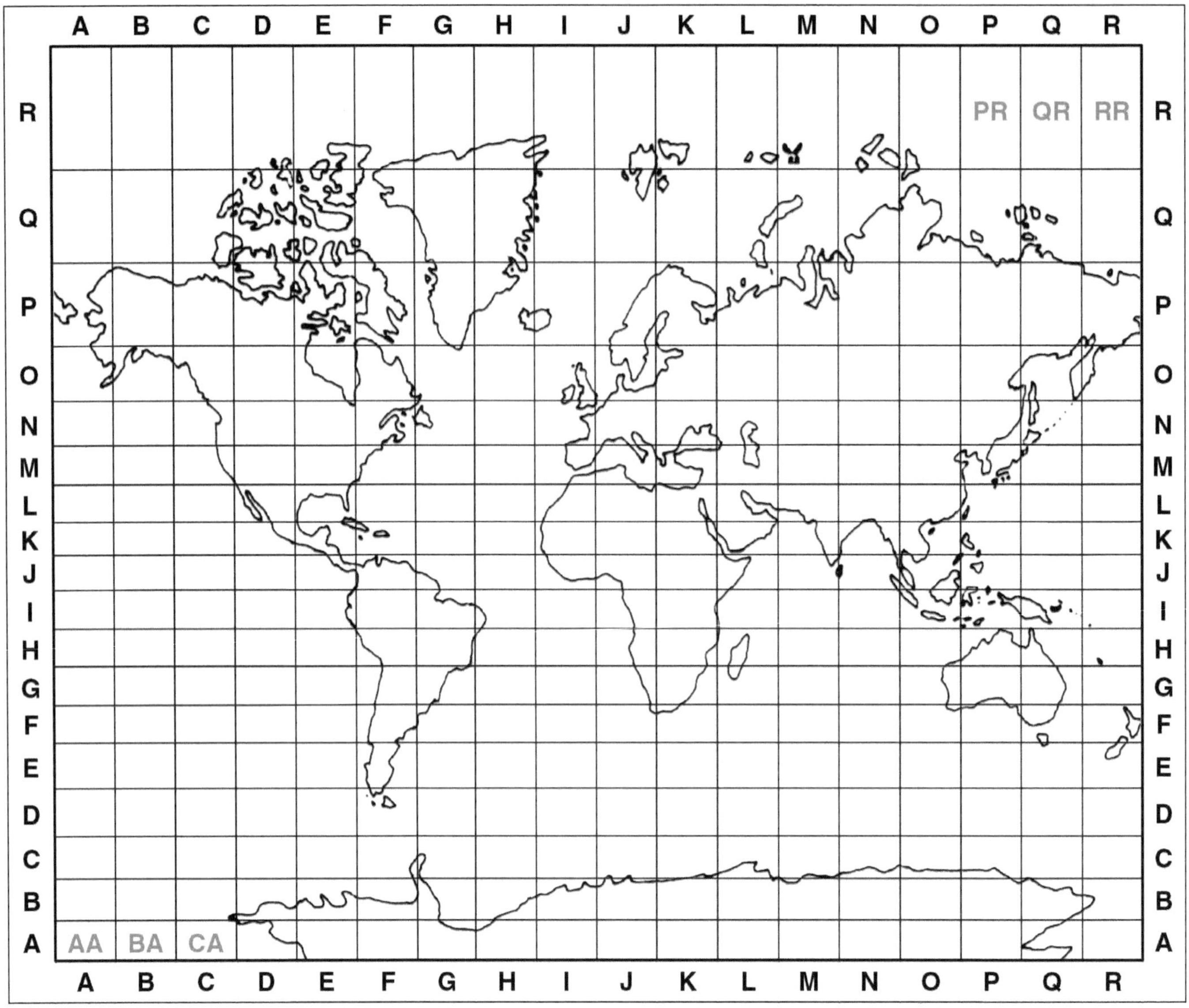

A B C D E F G H I J K L M N O P Q R

R Q P O N M L K J I H G F E D C B A

PR QR RR

AA BA CA

The IARU Locator system may be used throughout the world without repeats. The map above shows the fields that make up the first two letters of the Locator. Examples are shown at two of the corners. The map left shows numbering of squares within the fields.

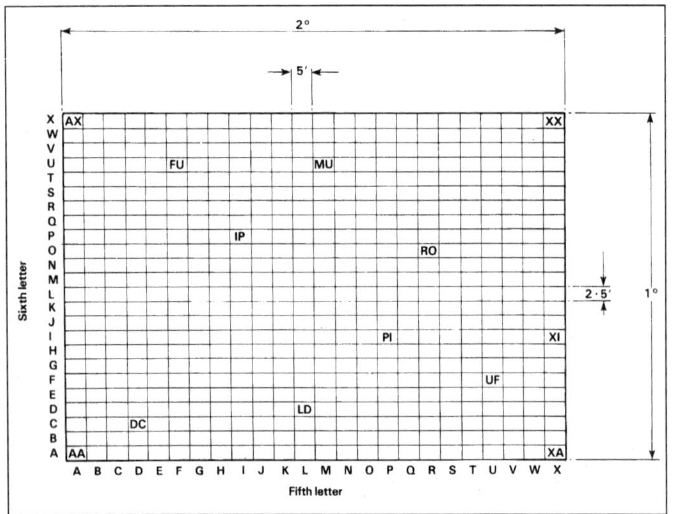

A square (the numbered part of the Locator) is divided into 576 sub-squares, designated AA to XX. Each sub-square is 5' W-E and 2.5' N-S.

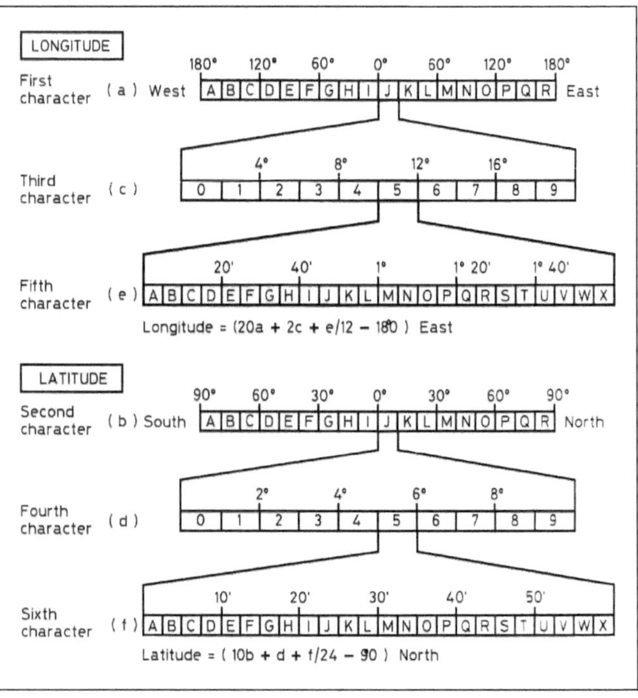

The final two letters may be calculated thus.

A large Locator map of Europe is available from the RSGB.
See www.rsgbshop.org or phone 01234 832700.

Contest Calendars

Contests are sporting events between amateur stations on specific bands and modes, conducted according to published rules. The activity appeals mainly to those with a competitive instinct, but construction and station optimisation are also important. For more information on the RSGB Contest Club *see* page 41.

2025 RSGB HF Contest Calendar

Date (2025)	Time (UTC)	Contest Name	Sections
Sat 4 Jan	1300-1700	**RSGB AFS 80m-40m Contests CW**	400W, 100W, 10W, Non-UK
Sun 12 Jan	1300-1700	**RSGB AFS 80m-40m Contests Datamodes**	400W, 100W, 10W, Non-UK
Sat 18 Jan	1300-1700	**RSGB AFS 80m-40m Contests Phone**	400W, 100W, 10W, Non-UK
Mon 3 Feb	2000-2130	80m CC SSB	100W Assisted, 10W Assisted
Sat 8 Feb	1900-2300	**1st 1.8MHz Contest**	UK Assisted, UK Unassisted, Non-UK Assisted, Non-UK Unassisted
Wed 12 Feb	2000-2130	80m CC DATA	100W Assisted, 10W Assisted, 100W Unassisted, 10W Unassisted
Thu 20 Feb	2000-2130	80m CC CW	100W Assisted, 10W Assisted, 100W Unassisted, 10W Unassisted
Mon 24 Feb	2000-2130	FT4 Series #1	100W UK, 100W Open UK, 10W UK, 100W Non-UK, 100W Open Non-UK, 10W Non-UK
Mon 3 Mar	2000-2130	80m CC DATA	100W Assisted, 10W Assisted, 100W Unassisted, 10W Unassisted
Sat 8-Sun 9 Mar	1000-1000	**Commonwealth Contest**	Open SO (12/24hr), Open SOA (12/24hr), Open Remote (12/24hr), Restr SO (12/24hr), Restr SOA (12/24hr), QRP SO, QRP SOA, Multi-Op, HQ
Wed 12 Mar	2000-2130	80m CC CW	100W Assisted, 10W Assisted, 100W Unassisted, 10W Unassisted
Mon 17 Mar	2000-2130	FT4 Series #2	100W UK, 100W Open UK, 10W UK, 100W Non-UK, 100W Open Non-UK, 10W Non-UK
Thu 27 Mar	2000-2130	80m CC SSB	100W Assisted, 10W Assisted
Sat 5 Apr - Sun 6 Apr	1200-1200	**FT4 International Activity Day**	100W UK, 100W Open UK, 10W UK, 100W Non-UK, 100W Open Non-UK, 10W Non-UK, 12hr or 24hr
Mon 7 Apr	1900-2030	80m CC CW	100W Assisted, 10W Assisted, 100W Unassisted, 10W Unassisted
Wed 16 Apr	1900-2030	80m CC SSB	100W Assisted, 10W Assisted
Thu 24 Apr	1900-2030	80m CC DATA	100W Assisted, 10W Assisted, 100W Unassisted, 10W Unassisted
Sat 26-Sun 27 Apr	1200-1200	UKEI DX CW	SO, SOA, 12hr, 24hr, High, Low, QRP
Mon 28 Apr	1900-2030	FT4 Series #3	100W UK, 100W Open UK, 10W UK, 100W Non-UK, 100W Open Non-UK, 10W Non-UK
Mon 12 May	1900-2030	80m CC SSB	100W Assisted, 10W Assisted
Mon 19 May	1900-2030	FT4 Series #4	100W UK, 100W Open UK, 10W UK, 100W Non-UK, 100W Open Non-UK, 10W Non-UK
Wed 21 May	1900-2030	80m CC DATA	100W Assisted, 10W Assisted, 100W Unassisted, 10W Unassisted
Thu 29 May	1900-2030	80m CC CW	100W Assisted, 10W Assisted, 100W Unassisted, 10W Unassisted
Sat 7-Sun 8 Jun	1500-1500	NFD	QRP Unassisted Portable, QRP Renewable Energy, Low Power Unassisted Portable, Low Power Assisted Portable, Fixed
Mon 2 Jun	1900-2030	80m CC DATA	100W Assisted, 10W Assisted, 100W Unassisted, 10W Unassisted
Wed 11 Jun	1900-2030	80m CC CW	100W Assisted, 10W Assisted, 100W Unassisted, 10W Unassisted
Mon 16 Jun	1900-2030	FT4 Series #5	100W UK, 100W Open UK, 10W UK, 100W Non-UK, 100W Open Non-UK, 10W Non-UK
Thu 26 Jun	1900-2030	80m CC SSB	100W Assisted, 10W Assisted
Mon 7 Jul	1900-2030	80m CC CW	100W Assisted, 10W Assisted, 100W Unassisted, 10W Unassisted
Sat 12-Sun 13 Jul	1200-1200	IARU HF Championships	GR2HQ Challenge for Individuals and Teams
Wed 16 Jul	1900-2030	80m CC SSB	100W Assisted, 10W Assisted
Sun 20 Jul	0900-1600	**International Low Power Contest**	UK Fixed 10W, UK Portable 10W, UK Fixed 5W, UK Portable 5W, Non-UK 10W, Non-UK 5W
Thu 24 Jul	1900-2030	80m CC DATA	100W Assisted, 10W Assisted, 100W Unassisted, 10W Unassisted
Sat 26-Sun 27 Jul	1200-1200	IOTA Contest	SO, SOA, Multi-1, Multi-2, SSB only, CW only, Mixed (SSB & CW), 12hr, 24hr, Island Fixed, Island DXpedition, World
Mon 28 Jul	1900-2030	FT4 Series #6	100W UK, 100W Open UK, 10W UK, 100W Non-UK, 100W Open Non-UK, 10W Non-UK
Mon 1 Sep	1900-2030	Autumn Series SSB	100W Assisted, 10W Assisted
Sat 6-Sun 7 Sep	1300-1300	SSB Field Day	QRP Unassisted Portable, QRP Renewable Energy, Low Power Unassisted Portable, High Power Assisted Portable, Fixed
Wed 10 Sep	1900-2030	Autumn Series CW	100W Assisted, 10W Assisted, 100W Unassisted, 10W Unassisted, 100W QRS, 10W QRS
Mon 15 Sep	1900-2030	FT4 Series #7	100W UK, 100W Open UK, 10W UK, 100W Non-UK, 100W Open Non-UK, 10W Non-UK
Thu 25 Sep	1900-2030	Autumn Series DATA	100W Assisted, 10W Assisted, 100W Unassisted, 10W Unassisted
Mon 6 Oct	1900-2030	Autumn Series CW	100W Assisted, 10W Assisted, 100W Unassisted, 10W Unassisted, 100W QRS, 10W QRS
Wed 15 Oct	1900-2030	Autumn Series DATA	100W Assisted, 10W Assisted, 100W Unassisted, 10W Unassisted
Thu 25 Oct	1900-2030	Autumn Series SSB	100W Assisted, 10W Assisted
Mon 27 Oct	2000-2130	FT4 Series #8	100W UK, 100W Open UK, 10W UK, 100W Non-UK, 100W Open Non-UK, 10W Non-UK
Sat 1 Nov - Sun 2 Nov	1200-1200	UKEI DX SSB	SO, SOA, 12hr, 24hr, High, Low, QRP
Mon 3 Nov	2000-2130	Autumn Series DATA	100W Assisted, 10W Assisted, 100W Unassisted, 10W Unassisted
Sat 8 Nov	2000-2300	**Club Calls (1.8MHz AFS)**	32W SSB only, 32W Mixed (SSB & CW)
Wed 12 Nov	2000-2130	Autumn Series SSB	100W Assisted, 10W Assisted
Sat 15 Nov	1900-2300	**2nd 1.8MHz Contest**	UK Assisted, UK Unassisted, Non-UK Assisted, Non-UK Unassisted
Mon 17 Nov	2000-2130	FT4 Series #9	100W UK, 100W Open UK, 10W UK, 100W Non-UK, 100W Open Non-UK, 10W Non-UK
Thu 27 Nov	2000-2130	Autumn Series CW	100W Assisted, 10W Assisted, 100W Unassisted, 10W Unassisted, 100W QRS, 10W QRS

HF Key to Special Rules

HF Championship	Special Rules for HF Championship
S1	Affiliated Societies contest
S2	Commonwealth Contest
AFS Super League 2020-21	Special Rules for AFS Super League 2020-21
AFS Super League 2022-23	Special Rules for AFS Super League 2022-23

HF Key to Multipliers

M1	DXCCs worked on each band
M2	UK Districts per band and mode
M3	DXCC & UK District Bonus
M4	IOTA Points

HF Key to Sections

A	10W Fixed
ALL	ALL
B	10W Portable
C	3W Fixed
D	3W Portable
HQ	HQ
LOW	100W maximum output power
LowPower	Low Power
Multi Operator	
Multi-Op	Multi Operator
Non-UK Open	Non-UK Open
Non-UK QRP	Non-UK QRP
Non-UK Restricted	Non-UK Restricted
Non-UK-Assisted	(b) Non-UK
Non-UK-Unassisted	(b) Non-UK
OPEN-SOA	Open - Single Operator Assisted
OPEN-SOU	Open - Single Operator Unassisted
Open	Open
Open (SSBFD)	Open
QRP	10W maximum output power
RESTRICTED-SOA	Restricted - Single Operator Assisted
RESTRICTED-SOU	Restricted - Single Operator Unassisted
Restricted	Restricted
Restricted (SSBFD)	Restricted
Single Operator Assisted	
Single Operator Unassisted	
UK Open	UK Open
UK QRP	UK QRP
UK Restricted	UK Restricted
UK-Assisted	(a) UK stations
UK-Unassisted	(a) UK stations

2025 RSGB VHF Contest Calendar

Date (2025)	Time (UTC)	Contest Name	Sections
Every 1st Tuesday	1900-1955 (L)	144MHz FMAC	FLL ITR FND
Every 1st Tuesday	2000-2230 (L)	144MHz UKAC	AO AR AL
Every 2nd Tuesday	1900-1955 (L)	432MHz FMAC	FLL ITR FND
Every 2nd Tuesday	2000-2230 (L)	432MHz UKAC	AO AR AL
Every 1st Wednesday	1700-2100	144MHz FT8 AC (4 hours)	AO AR AL
Every 1st Wednesday	1900-2100	144MHz FT8 AC (2 hours)	AO AR AL
Every 2nd Wednesday	1700-2100	432MHz FT8 AC (4 hours)	AO AR AL
Every 2nd Wednesday	1900-2100	432MHz FT8 AC (2 hours)	AO AR AL
Every 3rd Thursday	2000-2230 (L)	70MHz UKAC	AO AR AL
Every 2nd Thursday	2000-2230 (L)	50MHz UKAC	AO AR AL
Every 4th Tuesday (Jan-Nov)	1930-2230 (L)	SHF UKAC	SAO SAR
Every 3rd Tuesday	2000-2230 (L)	1.3GHz UKAC	AO AR AL
Sun 2 Feb.	0900-1300	432MHz AFS	SF O
Sat 1-Sun 2 Mar.	1400-1400	**March 144 432MHz** VHF Championship	O 6O SF SO 6S (HIGH/ LOW/QRP)
Sat 26-Sun 27 Apr.	1400-1400	MGM Contest	UKO EUO UKL EUL
Sat 3 May.	1400-2000	432MHz Trophy Contest VHF Championship	O SF
Sat 3-Sun 4 May.	1400-1400	**May 432MHz-245GHz Contest**	O SF
Sun 4 May.	0800-1400	10GHz Trophy Contest	SF O
Sun 11 May.	0900-1200	70MHz Contest CW	AO AR AL
Sat 17-Sun 18 May.	1400-1400	144MHz May Contest	O SF SO 6S 6O (HIGH/ LOW/QRP)
Sun 18 May.	1100-1500	1st 144MHz Backpackers	5B 25H
Sun 15 Jun.	0900-1300	2nd 144MHz Backpackers	5B 25H
Sat 21-Sun 22 Jun.	1400-1400	**50MHz Trophy Contest** UK (HIGH/LOW/QRP)	O SF SO 6O 6S Non
Sun 29 Jun.	0900-1200	50MHz Contest CW	AO AR AL
Sat 5-Sun 6 Jul.	1400-1400*	VHF NFD	Open R L M MS FSO FSR
Sun 6 Jul.	1100-1500	3rd 144MHz Backpackers	5B 25H
Sat 19 Jul.	1400-2000	**70MHz Trophy Contest**	O SO SF Non UK
Sat 2 Aug.	1400-1800	4th 144MHz Backpackers	5B 25H
Sat 2 Aug.	1400-1800	**144MHz Low Power Contest**	O SF SO
Sun 3 Aug.	0800-1200	**432MHz Low Power Contest**	O SF SO
Sat 6-Sun 7 Sep.	1400-1400	144MHz Trophy Contest	O SF SO 6O 6S (HIGH/ LOW/QRP)
Sun 7 Sep.	1100-1500	5th 144MHz Backpackers	5B 25H
Sun 21 Sep.	0900-1200	70MHz AFS Contest	SF O
Sat 4 Oct.	1400-2200	**2.3GHz Trophy**	O SF
Sat 4 Oct.	1400-2200	**1.2GHz Trophy**	O SF
Sat 4-Sun 5 Oct.	1400-1400	Oct 432MHz-245GHz Contest	O SF 6O(432) 6S(432)
Sun 19 Oct.	0900-1300	50MHz AFS Contest	SF O
Sat 1-Sun 2 Nov.	1400-1400	144MHz CW Marconi	SF O 6S 6O
Sun 7 Dec.	1000-1400	144MHz AFS	SF O
Sat 27 Dec	1500-1700	50MHz Christmas Contest	AR AO AL
Sun 28 Dec	1500-1700	70MHz Christmas Contest	AR AO AL
Mon 29 Dec	1500-1700	144MHz Christmas Contest	AR AO AL
Tue 30 Dec	1500-1700	432MHz Christmas Contest	AR AO AL

Events in **bold** qualify for the VHF Championship
L = Local

VHF/UHF Key to Sections

25H	25W Hill Toppers
5B	5W Backpacker
6O	6 hours Others
6S	6 hours Single Op Fixed
A	All
AL	UKAC Low Power
ALL	All
AO	UKAC Open
AR	UKAC Restricted
FLL	FMAC Full Licensees only
FND	FMAC Foundation Licensees only
FSO	VHF NFD Fixed Station Sweepers Open
FSR	VHF NFD Fixed Station Sweepers Restricted
ITR	FMAC Intermediate Licensees only
L	VHF NFD Low Power Section
LP	Single Operator, Fixed, 25W - Single Antenna
MM	VHF NFD Mix and Match Section
MS	VHF NFD Single Transmitter Section
Non-UK	Entrants outside of the UK&CD
O	Open
Open	VHF NFD Open Section
R	VHF NFD Restricted Section
SAO	SHF UKAC Open
SAR	SHF UKAC Restricted
SF	Single Op Fixed
SO	Single Op Others
Sweeper	VHF NFD Overall Sweeper Results

RSGB Contest Committee website: *www.rsgbcc.org*

Beacons

Beacons provide an essential service as propagation indicators, but also for development and testing – the later aspect being of particular importance in the microwave bands. There is an extensive network of coordinated propagation beacons in both the HF and VHF/Microwave bands. They feature in a wide range of band plans where segments are designated for beacons and where transmissions by any other users which may interfere with them are discouraged.

In the HF bands, systems such as the International Beacon Project (IBP) transmit CW at regular intervals across a range of bands. For HF there is also a strong IARU recommendation that with few exceptions, there should be no beacons below 14MHz

In the VHF/Microwave bands, beacon technology has advanced considerably and often incorporates GPS-locked frequency sources and various digital Machine Generated Modes (MGM) to facilitate automated monitoring.

The trend towards modern frequency sources at VHF and upwards has also resulted in beacons increasingly using Frequency Shift Keying, rather than Amplitude keying, for the CW ident along with a period of plain carrier. The latter can be useful in SDR waterfall displays where spreading or Doppler from aircraft scatter or other effects can be more readily identified. Alternate time periods may also feature transmission modes for the beacon callsign using MGM such JT65, JT4 or PI-4. These can be automatically detected at low signal levels by monitoring stations and fed into the DX cluster where sites such as Beaconspot.uk can automatically aggregate them and indicate openings.

Reception reports for VHF/Microwave illustrate that these bands are far from just simple line-of-sight. The GB3VHF 144MHz beacon can often be heard across the UK.

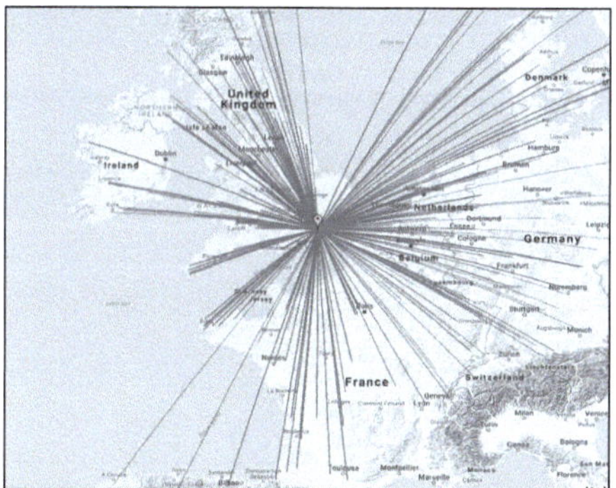

Figure 1 - GB3VHF on 144MHz Reception Reports aggregated and mapped by www.Beaconspot.uk

Timed / Synchronised beacons

HF- IBP

On HF there is a well-established network known as the International Beacon Programme (IBP). This was constructed by the NCDXF in cooperation with IARU

The network of 18 beacons (**Fig 2**) transmit on a repeating 3 minute cycle at staggered times on 14.100, 18.110, 21.150, 24.930, and 28.200MHz.

50MHz - SBP

In the 6m band the Synchronised Beacon Project (SBP) is gradually being deployed in the lower part of the 50MHz band, coordinated by the IARU Region-1 VHF Committee. This has been designed to accommodate a high number of sophisticated beacons that can accommodate the highly variable nature of propagation in the 50MHz band from flat tropo conditions to aurora and Sporadic-E. The SBP network is configured as a time shared multiplex, with ten frequencies available each of which has five 60s timeslots. Some beacons such as GB3MCB and GB3NGI when not transmitting on their assigned SBP time slot will additionally transmit further up the band in the 50.4-50.5MHz range. The mode most commonly used for SBP automated reception is PI4 - PharusIgnis4 - a digital modulation specifically developed for beacon purposes.

Setting up a beacon

The UK licence has a specific conditions and Coordination requirements associated with beacons. At powers below 5Werp there is a light touch licence category that can cater for unattended WSPR usage, DF competition beacons etc The more mainstream propagation beacons are in the 5-25W and higher-power licence categories and are assigned GB3xxx call signs. These are most prevalent in the microwave bands.

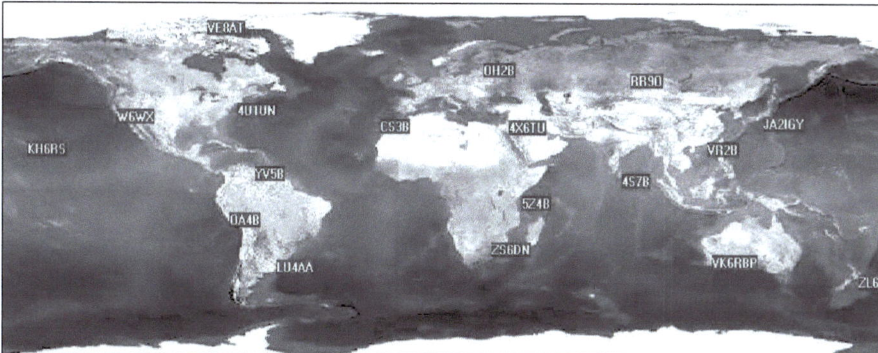

Figure 2 - The 18 sites used for the HF International Beacon Project (IBP)

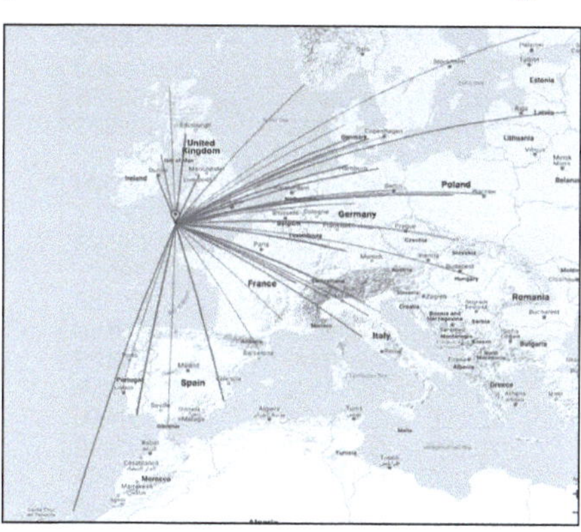

Figure 3 - GB3MCB on 50MHz is now one of a number of newer SBP systems and has reception distances enhanced by Sporadic-E and the use of PI-4 modulation

Ofcom conditions and coordination requirements may require frequency clearance and include some 50km radius exclusion zones around key locations for unattended usage. Establishing a permanent beacon can be a complex undertaking. Such an exercise is more suited to a group rather than an individual, but help is at hand. For example, the UK Microwave Group is very supportive of beacon equipment and keepers.

Another aspect when planning a beacon is that its site should be relatively remote from any other radio system or users as their 24/7 transmissions can be quite strong and de-sense local receivers. Other factors to be considered are that the high transmission duty cycle can be a significant consumer of electricity and IARU frequency coordination is typically required.

Further Information:-

There is a UK technical discussion group for supporting Beacon Keepers and developers at: *https://groups.io/g/UKBeacons*

NoV applications and renewals for UK beacons are handled via the RSGB-ETCC website, *https://www.ukrepeater.net/*. However liaison with RSGB Spectrum Managers is recommended in the first instance.

HF

- IARU List: *https://iaruhfbeacons.wordpress.com/*
- NCDXF/IBP: *http://www.ncdxf.org/pages/beacons.html*

VHF/UHF/Microwave

- BeaconSpot: *https://www.beaconspot.uk/* for UK/European reception reports/maps
- IARU-R1 list: *https://iaru-r1-c5-beacons.org/*

List of UK Beacons (by Frequency)

Freq, MHz	Call	Locator	Nearest Town	Height agl	Bearing	Polarisation	ERPw	Keeper
28.215	GB3MCB	IO70OJ	ROCHE	10	OMNI	V	14	G7KFQ
28.224	GB3IKL	IO81WL	KINGTON LANGLEY	12	OMNI	V	5	G6HUI
28.247	GB3NWB	JO02PP	NORWICH	3		H	3	M0ZAH
28.260	GB3MAT	IO82VO	BILBROOK	3	OMNI	H	7	M0IAW
28.268	GB3AAX	IO95FF	ASHINGTON	5	OMNI	H	1	G4NAB
28.287	GB3XMB	IO83SV	WADDINGTON	10	OMNI	V	10	G0SXC
50.000	GB3BUX	IO93BF	BUXTON	7	OMNI	H	14	G7EKY
50.002	GB3RMK	IO87AX	GOLSPIE	7	N/S	H	14	GM3WOJ
50.004	GB3NGI	IO65VB	LOGHGUILE	3	OMNI	V	15	GI6ATZ
50.005	GB3MCB	IO70OJ	ST AUSTELL	30	90/270	H	16	G7KFQ
50.006	GB3BAA	IO91PS	TRING HERTS	35	OMNI	V	14	M0ZPU
50.408	GB3MBA	IO93JC	MANSFIELD	5	0	H	26	G3PHO
50.416	GB3BAA	IO91PS	TRING HERTS	35	OMNI	V	14	M0ZPU
50.443	GB3MCB	IO70OJ	ST AUSTELL	30	90/270	H	16	G7KFQ
50.460	GB3RMK	IO87AX	GOLSPIE	7	N/S	H	14	GM3WOJ
50.462	GB3NGI	IO65VB	LOGHGUILE	3	OMNI	V	15	GI6ATZ
50.464	GB3LER	IP90JD	LERWICK	8	0/180	H	16	GM7AFE
70.000	GB3BUX	IO93BF	BUXTON	7	OMNI	H	14	G7EKY
70.007	GB3WSX	IO81VC	MERE WILTSHIRE	10	70 DEGREES	H	17	G3ZXX
70.016	GB3BAA	IO91PS	TRING HERTS	40	OMNI	H	14	M0ZPU
70.020	GB3ANG	IO86MN	DUNDEE	15	170	H	20	GM4ZUK
70.025	GB3MCB	IO70OJ	ST AUSTELL	27	45	H	16	G7KFQ
70.027	GB3CFG	IO74BS	CARRICKFERGUS	15	45/135	H	14	GI0GDP
70.050	GB3MBD	IO92RA	AMPTHILL BEDS	20	OMNI	H	9	G4FKI
144.430	GB3VHF	JO01EH	GRAVESEND	40	288 & 348	H	22	G0FDZ
144.453	GB3ANG	IO86MN	DUNDEE	15	170	H	20	GM4ZUK
144.462	GB3SEV	IO82UI	STOURPORT	6		H	20	M0VXX
144.469	GB3MCB	IO70OJ	ST AUSTELL	25	45 AND 180	H	16	G7KFQ
144.482	GB3NGI	IO65VB	LOGHGUILE	10	45 & 135	H	21	GI6ATZ
144.487	GB3WGI	IO64BL	DERRYGONNELLY NI	35	280	H	30	GI6ATZ
432.430	GB3UHF	JO01EH	GRAVESEND	40	288 & 348	H	22	G0FDZ
432.445	GB3FNY	IO93NN	DONCASTER	10	90 & 180	H	17	G3AAF
432.453	GB3ANG	IO86MN	DUNDEE	15	170	H	17	GM4ZUK
432.470	GB3MCB	IO70OJ	ST AUSTELL	20	45	H	13	G7KFQ
432.482	GB3NGI	IO65VB	LOGHGUILE	10	60 & 135	H	24	GI6ATZ
432.490	GB3LEU	IO92IQ	MARKFIELD	10	135 & 300	H	17	G3TQF
1296.830	GB3MHZ	JO02PB	MARTLESHAM HEATH	65	OMNI	H	27	G7OCD
1296.840	GB3IDT	IO92SA	AMPTHILL	8	90	H	8	G4FKI
1296.850	GB3FRS	IO91LC	FOUR MARKS	20	NNW	H	14	G3TCU
1296.860	GB3MCB	IO70OJ	ST AUSTELL	20	28	H	17	G7KFQ
1296.870	GB3USK	IO81VC	MERE WILTSHIRE	15	90	H	20	G3TCT
1296.890	GB3DUN	IO91SV	DUNSTABLE	6	OMNI	H	16	G3ZFP
1296.905	GB3CFG	IO74BS	CARRICKFERGUS	15	OMNI	H	15	GI0GDP
1296.905	GB3NGI	IO65VB	LOGHGUILE	10	OMNI	H	15	GI6ATZ
1296.965	GB3ANG	IO86MN	DUNDEE	15	170	H	16	GM4ZUK
1296.985	GB3CSB	IO75XX	KILSYTH	7	150	H	14	GM6BIG
1296.990	GB3EDN	IO85JV	EDINBURGH	10	OMNI	H	14	GM8BJF
2320.830	GB3MHZ	JO02PB	MARTLESHAM HEATH	65	OMNI	H	14	G7OCD
2320.870	GB3ZME	IO82RP	TELFORD	10	OMNI	H	19	G3UKV
2320.890	GB3ANT	JO02PP	NORWICH	52	OMNI	H	7	G8VLL
2320.905	GB3SCS	IO80UU	BLANDFORD	12	OMNI	H	8	G4JNT
2320.920	GB3FNM	IO91OF	FARNHAM	20	NNW/SSE	H	16	G4EPX
2320.925	GB3BSS	IO81SR	BRISTOL	65	OMNI	H	16	G4CJZ
2320.955	GB3LES	IO92IQ	MARKFIELD	12	OMNI	H	15	G3TQF
2320.985	GB3CSB	IO75XX	KILSYTH	7	150	H	14	GM6BIG
3400.830	GB3MHZ	JO02PB	MARTLESHAM HEATH	65	OMNI	H	14	G7OCD
3400.900	GB3OHM	IO92AJ	NORTHFIELD	12	OMNI	H	20	G8IFT
3400.905	GB3SCF	IO80UU	BLANDFORD	12	OMNI	H	11	G4JNT
3400.910	GB3ZME	IO82RP	TELFORD	10	OMNI	H	19	G3UKV
3400.935	GB3LPC	IO91FR	BAMPTON	30	OMNI	H	10	G3NNG
3400.955	GB3LEF	IO92IQ	MARKFIELD	12	OMNI	H	8	G3TQF
3400.985	GB3CSB	IO75XX	KILSYTH	7	150	H	14	GM6BIG
5760.830	GB3MHZ	JO02PB	MARTLESHAM HEATH	65	OMNI	H	14	G7OCD
5760.900	GB3OHM	IO92AJ	NORTHFIELD	12	OMNI	H	17	G8IFT
5760.905	GB3SCC	IO80UU	BLANDFORD	12	OMNI	H	8	G4JNT
5760.910	GB3ZME	IO82RP	TELFORD	10	OMNI	H	19	G3UKV
5760.920	GB3FNM	IO91OF	FARNHAM	20	NNW/SSE	H	14	G4EPX
5760.925	GB3KEU	IO93NN	DONCASTER	10	OMNI	H	14	G3AAF
5760.985	GB3CSB	IO75XX	KILSYTH	7	150	H	14	GM6BIG
10368.752	GB3FNY	IO93NN	DONCASTER	10	OMNI	H	10	G3AAF
10368.755	GB3CAM	JO02BI	CAMBRIDGE	30	OMNI	H	7	G4HJW
10368.770	GB3BED	IO92SD	BEDFORD	12	OMNI	H	10	G4FKI
10368.780	GB3OSW	IO82KV	OSWESTRY	8	OMNI	H	10	G3SMT
10368.790	GB3UVR	IO83KC	MOLD	11	NORTH EAST	H	13	GW0MDQ
10368.830	GB3MHZ	JO02PB	MARTLESHAM HEATH	65	OMNI	H	8	G7OCD
10368.850	GB3SEE	IO91VG	REIGATE	20	OMNI	H	5	G0OLX
10368.870	GB3KBQ	IO80LV97	TRULL	185	OMNI	H	10	G4UVZ
10368.895	GB3NGI	IO65VB	LOGHGUILE	10	OMNI	H	13	GI6ATZ
10368.905	GB3SCX	IO80UU	BLANDFORD	12	OMNI	H	7	G4JNT
10368.910	GB3RPE	IO71UU	CARMARTHEN	39	OMNI	H	8	GW8KCY
10368.935	GB3GCT	IO91IJ	NEWBURY	10	OMNI	H	10	G1EHF

Freq, MHz	Call	Locator	Nearest Town	Height agl	Bearing	Polarisation	ERPw	Keeper
10368.945	GB3PKT	JO01MT	ST OSYTH	12	OMNI	H	3	G0MBA
10368.955	GB3LEX	IO92IQ	MARKFIELD	12	OMNI	H	10	G3TQF
10368.960	GB3CMS	JO01GQ	CHELMSFORD	15	OMNI	H	7	G6JYB
10368.980	GB3MCB	IO70OJ	ST AUSTELL	20	67	H	10	G7KFQ
10368.985	GB3CSB	IO75XX	KILSYTH	7	150	H	14	GM6BIG
24048.830	GB3BED	IO92SD	BEDFORD	25	OMNI	H	14	G4FKI
24048.830	GB3MHZ	JO02PB	MARTLESHAM HEATH	65	OMNI	H	14	G7OCD
24048.850	GB3FNY	IO93NN	DONCASTER	10	OMNI	H	13	G3AAF
24048.870	GB3CAM	JO02BI	CAMBRIDGE	30	OMNI	H	14	G4HJW
24048.890	GB3DUN	IO91SV	DUNSTABLE	60	135 & 315	H	3	G3ZFP
24048.905	GB3SCK	IO80UU	BLANDFORD	12	OMNI	H	11	G4JNT
24048.910	GB3ZME	IO82RP	TELFORD	10	OMNI	H	19	G3UKV
24048.920	GB3FNM	IO91OF	FARNHAM	20	NNW/SSE	H	13	G4EPX
24048.940	GB3AMU	IO81JN	CARDIFF	10	135	H	-3	GW3TKH
24048.945	GB3PKT	JO01MT	ST OSYTH	10	OMNI	H	14	G0MBA
24048.960	GB3SEE	IO91VG	REIGATE	20	OMNI	H	14	G0OLX
24048.985	GB3CSB	IO75XX	KILSYTH	7	150	H	14	GM6BIG
47088.905	GB3SCQ	IO80UU	BLANDFORD	12	75DEG TRUE	H	6	G8BKE
47088.985	GB3CSB	IO75XX	KILSYTH	7	150	H	14	GM6BIG

List of UK Beacons (by Callsign)

Call	Freq, MHz	Locator	Nearest Town	Height agl	Bearing	Polarisation	ERPw	Keeper
GB3AAX	28.268	IO95FF	ASHINGTON	5	OMNI	H	1	G4NAB
GB3AMU	24048.940	IO81JN	CARDIFF	10	135	H	-3	GW3TKH
GB3ANG	70.020	IO86MN	DUNDEE	15	170	H	20	GM4ZUK
GB3ANG	144.453	IO86MN	DUNDEE	15	170	H	20	GM4ZUK
GB3ANG	432.453	IO86MN	DUNDEE	15	170	H	17	GM4ZUK
GB3ANG	1296.965	IO86MN	DUNDEE	15	170	H	16	GM4ZUK
GB3ANT	2320.890	JO02PP	NORWICH	52	OMNI	H	7	G8VLL
GB3BAA	50.006	IO91PS	TRING HERTS	35	OMNI	V	14	M0ZPU
GB3BAA	50.416	IO91PS	TRING HERTS	35	OMNI	V	14	M0ZPU
GB3BAA	70.016	IO91PS	TRING HERTS	40	OMNI	H	14	M0ZPU
GB3BED	10368.770	IO92SD	BEDFORD	12	OMNI	H	10	G4FKI
GB3BED	24048.830	IO92SD	BEDFORD	25	OMNI	H	14	G4FKI
GB3BSS	2320.925	IO81SR	BRISTOL	65	OMNI	H	16	G4CJZ
GB3BUX	50.000	IO93BF	BUXTON	7	OMNI	H	14	G7EKY
GB3BUX	70.000	IO93BF	BUXTON	7	OMNI	H	14	G7EKY
GB3CAM	10368.755	JO02BI	CAMBRIDGE	30	OMNI	H	7	G4HJW
GB3CAM	24048.870	JO02BI	CAMBRIDGE	30	OMNI	H	14	G4HJW
GB3CFG	70.027	IO74BS	CARRICKFERGUS	15	45/135	H	14	GI0GDP
GB3CFG	1296.905	IO74BS	CARRICKFERGUS	15	OMNI	H	15	GI0GDP
GB3CMS	10368.960	JO01GQ	CHELMSFORD	15	OMNI	H	7	G6JYB
GB3CSB	1296.985	IO75XX	KILSYTH	7	150	H	14	GM6BIG
GB3CSB	2320.985	IO75XX	KILSYTH	7	150	H	14	GM6BIG
GB3CSB	3400.985	IO75XX	KILSYTH	7	150	H	14	GM6BIG
GB3CSB	5760.985	IO75XX	KILSYTH	7	150	H	14	GM6BIG
GB3CSB	10368.985	IO75XX	KILSYTH	7	150	H	14	GM6BIG
GB3CSB	24048.985	IO75XX	KILSYTH	7	150	H	14	GM6BIG
GB3CSB	47088.985	IO75XX	KILSYTH	7	150	H	14	GM6BIG
GB3DUN	1296.890	IO91SV	DUNSTABLE	6	OMNI	H	16	G3ZFP
GB3DUN	24048.890	IO91SV	DUNSTABLE	60	135 & 315	H	3	G3ZFP
GB3EDN	1296.990	IO85JV	EDINBURGH	10	OMNI	H	14	GM8BJF
GB3FNM	2320.920	IO91OF	FARNHAM	20	NNW/SSE	H	16	G4EPX
GB3FNM	5760.920	IO91OF	FARNHAM	20	NNW/SSE	H	14	G4EPX
GB3FNM	24048.920	IO91OF	FARNHAM	20	NNW/SSE	H	13	G4EPX
GB3FNY	432.445	IO93NN	DONCASTER	10	90 & 180	H	17	G3AAF
GB3FNY	10368.752	IO93NN	DONCASTER	10	OMNI	H	10	G3AAF
GB3FNY	24048.850	IO93NN	DONCASTER	10	OMNI	H	13	G3AAF
GB3FRS	1296.850	IO91LC	FOUR MARKS	20	NNW	H	14	G3TCU
GB3GCT	10368.935	IO91IJ	NEWBURY	10	OMNI	H	10	G1EHF
GB3IDT	1296.840	IO92SA	AMPTHILL	8	90	H	8	G4FKI
GB3IKL	28.224	IO81WL	KINGTON LANGLEY	12	OMNI	V	5	G6HUI
GB3KBQ	10368.870	IO80LV97	TRULL	185	OMNI	H	10	G4UVZ
GB3KEU	5760.925	IO93NN	DONCASTER	10	OMNI	H	14	G3AAF
GB3LEF	3400.955	IO92IQ	MARKFIELD	12	OMNI	H	8	G3TQF
GB3LER	50.464	IP90JD	LERWICK	8	0/180	H	16	GM7AFE
GB3LES	2320.955	IO92IQ	MARKFIELD	12	OMNI	H	15	G3TQF
GB3LEU	432.490	IO92IQ	MARKFIELD	10	135 & 300	H	17	G3TQF
GB3LEX	10368.955	IO92IQ	MARKFIELD	12	OMNI	H	10	G3TQF
GB3LPC	3400.935	IO91FR	BAMPTON	30	OMNI	H	10	G3NNG
GB3MAT	28.260	IO82VO	BILBROOK	3	OMNI	H	7	M0IAW
GB3MBA	50.408	IO93JC	MANSFIELD	5	0	H	26	G3PHO
GB3MBD	70.050	IO92RA	AMPTHILL BEDS	20	OMNI	H	9	G4FKI
GB3MCB	28.215	IO70OJ	ROCHE	10	OMNI	V	14	G7KFQ
GB3MCB	50.005	IO70OJ	ST AUSTELL	30	90/270	H	16	G7KFQ
GB3MCB	50.443	IO70OJ	ST AUSTELL	30	90/270	H	16	G7KFQ
GB3MCB	70.025	IO70OJ	ST AUSTELL	27	45	H	16	G7KFQ
GB3MCB	144.469	IO70OJ	ST AUSTELL	25	45 AND 180	H	16	G7KFQ
GB3MCB	432.470	IO70OJ	ST AUSTELL	20	45	H	13	G7KFQ

Call	Freq	Locator	Nearest Town	Antenna	Height agl	Bearing	Polarisation	ERP	Mode	Keeper
GB3MCB	1296.860	IO70OJ	ST AUSTELL		20	28	H	17		G7KFQ
GB3MCB	10368.980	IO70OJ	ST AUSTELL		20	67	H	10		G7KFQ
GB3MHZ	1296.830	JO02PB	MARTLESHAM HEATH		65	OMNI	H	27		G7OCD
GB3MHZ	2320.830	JO02PB	MARTLESHAM HEATH		65	OMNI	H	14		G7OCD
GB3MHZ	3400.830	JO02PB	MARTLESHAM HEATH		65	OMNI	H	14		G7OCD
GB3MHZ	5760.830	JO02PB	MARTLESHAM HEATH		65	OMNI	H	14		G7OCD
GB3MHZ	10368.830	JO02PB	MARTLESHAM HEATH		65	OMNI	H	8		G7OCD
GB3MHZ	24048.830	JO02PB	MARTLESHAM HEATH		65	OMNI	H	14		G7OCD
GB3NGI	50.004	IO65VB	LOGHGUILE		3	OMNI	V	15		GI6ATZ
GB3NGI	50.462	IO65VB	LOGHGUILE		3	OMNI	V	15		GI6ATZ
GB3NGI	144.482	IO65VB	LOGHGUILE		10	45 & 135	H	21		GI6ATZ
GB3NGI	432.482	IO65VB	LOGHGUILE		10	60 & 135	H	24		GI6ATZ
GB3NGI	1296.905	IO65VB	LOGHGUILE		10	OMNI	H	15		GI6ATZ
GB3NGI	10368.895	IO65VB	LOGHGUILE		10	OMNI	H	13		GI6ATZ
GB3NWB	28.247	JO02PP	NORWICH		3		H	3		M0ZAH
GB3OHM	3400.900	IO92AJ	NORTHFIELD		12	OMNI	H	20		G8IFT
GB3OHM	5760.900	IO92AJ	NORTHFIELD		12	OMNI	H	17		G8IFT
GB3OSW	10368.780	IO82KV	OSWESTRY		8	OMNI	H	10		G3SMT
GB3PKT	10368.945	JO01MT	ST OSYTH		12	OMNI	H	3		G0MBA
GB3PKT	24048.945	JO01MT	ST OSYTH		10	OMNI	H	14		G0MBA
GB3RMK	50.002	IO87AX	GOLSPIE		7	N/S	H	14		GM3WOJ
GB3RMK	50.460	IO87AX	GOLSPIE		7	N/S	H	14		GM3WOJ
GB3RPE	10368.910	IO71UU	CARMARTHEN		39	OMNI	H	8		GW8KCY
GB3SCC	5760.905	IO80UU	BLANDFORD		12	OMNI	H	8		G4JNT
GB3SCF	3400.905	IO80UU	BLANDFORD		12	OMNI	H	11		G4JNT
GB3SCK	24048.905	IO80UU	BLANDFORD		12	OMNI	H	11		G4JNT
GB3SCQ	47088.905	IO80UU	BLANDFORD		12	75DEG TRUE	H	6		G8BKE
GB3SCS	2320.905	IO80UU	BLANDFORD		12	OMNI	H	8		G4JNT
GB3SCX	10368.905	IO80UU	BLANDFORD		12	OMNI	H	7		G4JNT
GB3SEE	10368.850	IO91VG	REIGATE		20	OMNI	H	5		G0OLX
GB3SEE	24048.960	IO91VG	REIGATE		20	OMNI	H	14		G0OLX
GB3SEV	144.462	IO82UI	STOURPORT		6		H	20		M0VXX
GB3UHF	432.430	JO01EH	GRAVESEND		40	288 & 348	H	22		G0FDZ
GB3USK	1296.870	IO81VC	MERE WILTSHIRE		15	90	H	20		G3TCT
GB3UVR	10368.790	IO83KC	MOLD		11	NORTH EAST	H	13		GW0MDQ
GB3VHF	144.430	JO01EH	GRAVESEND		40	288 & 348	H	22		G0FDZ
GB3WGI	144.487	IO64BL	DERRYGONNELLY NI		35	280	H	30		GI6ATZ
GB3WSX	70.007	IO81VC	MERE WILTSHIRE		10	70 DEGREES	H	17		G3ZXX
GB3XMB	28.287	IO83SV	WADDINGTON		10	OMNI	V	10		G0SXC
GB3ZME	2320.870	IO82RP	TELFORD		10	OMNI	H	19		G3UKV
GB3ZME	3400.910	IO82RP	TELFORD		10	OMNI	H	19		G3UKV
GB3ZME	5760.910	IO82RP	TELFORD		10	OMNI	H	19		G3UKV
GB3ZME	24048.910	IO82RP	TELFORD		10	OMNI	H	19		G3UKV

OK0EU

OK0EU is a cluster of five high-stability transmitters with 4 Hz separation, running for multipoint measurement. The call sign is sent 40 Hz up to identify ionospheric echoes. Output power 1 watt to dipole radiating N-S.

Transmitters are on:

- 3.594.492 Vackov (JO60EF),
- 3.594.496 Diouha Louka (JO60TP),
- 3.594.500 Panksa Ves (JO70GM),
- 3.594.504 Pruhonice (JN79GX),
- 3.594.508 Kasperske Hory (JN69SD).

DK0WCY/DRA5

DK0WCY 3579 operates 0720-0900UTC and 1600-1900UTC LOCAL time (ie UTC+1 winter and UTC+2 summer) DRA5 and DK0WCY on 10 144 are currently part-time. Transmissions on 3 579 are CW only, with beacon ID interrupted at ten minute intervals by a datagram giving the Kiel K, current fof2 and MUF at Juliusruh and indications of current solar-geophysical events. DRA5 and DK0WCY on 10 MHz carry the CW datagrams but, at 10 minutes after the hour, the datagram transmission is on RTTY and contains additional information on x-ray background and hf2, using RTTY. The H + 50 transmission is PSK31(BPSK) is used.

During auroral events the carrier at the end of the ID changes to a series of dots.

See also: *www.dk0wcy.de*

Schedule of IBP/NCDXF Beacon Transmissions

International Beacon Project beacons transmit for ten seconds on each frequency in turn in the sequence shown below. The whole cycle takes three minutes and then repeats. They send callsign at 22WPM and 100 watts, then four 1-second dashes at 100W, 10W, 1W and 0.1W. Equipment is a Kenwood TS-50, Cushcraft R-5 multiband vertical and a Trimble Navigation GPS receiver to ensure sychronization, with a control unit built by NCDXF.

LA2BCN

A2BCN operates 10-min cycle. Min 0-1t

Schedule of IBP/NCDXF Beacon Transmissions

		Frequency (kHz)				
Location	Call	14100	18110	21150	24930	28200
United Nations NY	4U1UN	00.00	00.10	00.20	00.30	00.40
Northern Canada	VE8AT	00.10	00.20	00.30	00.40	00.50
USA (CA)	W6WX	00:20	00.30	00:40	00.50	01:00
Hawaii	KH6WO	00.30	00.40	00.50	01.00	01.10
New Zealand	ZL6B	00.40	00.50	01.00	01.10	01.20
West Australia	VK6RBP	00.50	01.00	01.10	01.20	01.30
Japan	JA2IGY	01.00	01.10	01.20	01.30	01.40
Siberia	RR9O	01.10	01.20	01.30	01.40	01.50
China	VR2HK	01.20	01.30	01.40	01.50	02.00
Sri Lanka	4S7B	01.30	01.40	01.50	02.00	02.10
South Africa	ZS6DN	01:40	01.50	02:00	02:10	02:20
Kenya	5Z4B	01.50	02.00	02.10	02.20	02.30
Israel	4X6TU	02:00	02:10	02:20	02.30	02:40
Finland	OH2B	02:10	02:20	02:30	02:40	02:50
Madeira	CS3B	02.20	02.30	02.40	02.50	00.00 FN1
Argentina	LU4AA	02:30	02:40	02:50	00.00	00:10
Peru	OA4B	02.40	02.50	00.00	00.10	00.20
Venezuela	YV5B	02:50	00.00	00:10	00:20	00:30

Foot Note 1. Beacon destroyed https://www.iaru-r1.org/2023/hf-beacon-cs3b-madeira-destroyed-in-wildfire/

Foot Note 2. All the beacons are GPS locked and transmit a 10 frame continuously starting at 08, 18, 28, 38, 48, 58 minutes after the hour. The transmit frame is WSPR for 112 seconds; CW = call sign call sign Locator at 18 wpm; Slow Hell = call sign; FSKCW (QRSS) = call sign and 20 second calibration and start again.

Worldwide List of HF Beacons

(Updated June 2023)

The list is maintained by Dennis Green, ZS4BS, Region 1 HF Beacon Coordinator
Please notify errors/changes to *zs4bs@iaru-r1.org*

? = Activity uncertain; **INT** = Intermittent; **PT**=Part-time; **IRREG** = Irregular; **UC** = Under Construction; (T) **Non Op** = (Temporarily) NotOperational; **OP?** = Operational?; **V** = Varies; **EXP** = Experimental; **WkEnd** = Weekends during daylight; **LT** = Local Time; **SYNCH** = frequency sharing synchronised beacons: SYNCHn current sequence N4ESS (0,00 seconds), N4ES (0,20), WB4WOR (0,30), K7EK (0,40), N4ES (0,50). SYNCHx: K4UKB ID at 00 seconds and K4FUM at 30 seconds. ## K6FRC transmits on 28 245, 28 250, 28 300
+++ 10 minutes constant carrier followed by ID
zzz GB3WES at H+1, H+16 etc and GB3ORK at H+2, H+17 etc. Transmissions have a stepped power sequence.

IARU Region 1 discourages beacon operation on 1,8MHz
IARU Region 1 discourages beacon operation on 3,5MHz (DK0WCY, OK0EU OK!
IARU Region 1 discourages beacon operation on 7MHz (OK0EU, OK0EP OK!
IARU Region 1 discourages beacon operation on 10MHz (DK0WCY excepted)

Freq	Call	Nearest Town	Locator	ERPw	Antenna	Direction	Mode	Status
160 m								
1810.50	YR2TOP	Zlatita	KN94RU	100	Inv Vee		A1	24
1853.00	OK0EV	Near Prague	JN79EV	0.1	25m Vert	Omni	A1	PT
1875.00	DL3KR		JO63LV	5	Dipole		A1	PT
80 m								
3570.00	YO8RIX	Near Dorohoi	KN37FW	300mw	Lazi Loop	Omni	A1	24
3576.80	IZ3DVW	Monselice	JN55VF	0.5	Inv Vee		A1	IRREG
3579.00	DK0WCY	Scheggerott	JO44VQ	30	Dipole		A1	PT/zz??
3579.80	SM2IUF	Kalix	KP15NU	QRPP				15 - 07 UTC
3594.50	OK0EU	Panska Ves	JO70GM	100 mw	Mag Loop	N - S	A1	24
3600.00	OK0EN	Near Kladno	JO70AC	150mw	LW 41 m	SE - NW	A1	24
60 m								
5195.00	DRA5	Scheggerott	JO44VQ	30	Dipole		A1	06 - 24 LT
5195.00	DRA5	Scheggerott	JO44VQ	30	Dipole		PSK,RTTY	24
5205.25	LX0HF	Junglinster	JN39DR	5	Dipole		A1	24
5289.50	OV1BCN	10 km S Soroe	JO55SI	30	32 m F, Dip @ 1 m		USB/MT63	H + 4, 19, 34, 49
5290.00	ZS6SRL	Randburg	KG33WV	?			WSPR	
5290.50	OV1BCN	?						
5290.00	GB3ORK	Orkney	IO89JA	10<158 uw	Inv Vee		JT9A	24
5291.00	HB9AW	Sursee	JN43BA	10/5/1/,1/,001	1/2 Dip		A1 + PSK	5 m cycle
40 m								
7003.00	ES1VHF		KO29IK	?				
7034.80	PT9BCN		GG29RN	12			A1	OP?
7038.00	ZS1AGI	Kanaberg, Mossel Bay		KF06XB	500 mW	Inverted V		A1
7038.50	OK0EU	Panska Ves	JO70GM	1	Mag Loop	N - S	A1	OP?
7039.10	SK7CQ	TEST						
7039.20	IK1HGI	Cerano	JN61MM	0.1	QRSS			
7039.40	OK0EPB	Prague	JO70EC	10	Dipole@22m		A1	24
7039.80	IZ3DVW	Monselice Padua	JN55VF	0.5	Inv Vee		A1	24
7047.50	YD0MWK		OI33MQ	5	Inv Vee @ 15 m		A1	Planned
30 m								
10130.00	OK1IF	Liberec	JO70HG	0.5			A1	?
10133.00	SA6RR	Oxaback	JO67KI	0.5	GP	Omni	A1	INT
10133.50	HB4FV			0.5	H-pole		A1	?
10137.20	IK3NWX	Near Monselice	JN55VB	4.2	Rot Dipole	E - W	A1	24
10138.70	WSPR beacons around here							
10140.00	WSPR beacons around here							
10142.51	IK1HGI	Tricati	JN45IK	0.1	Dipole		QRSS3	OP?
10144.00	DK0WCY	Scheggerott	JO44VQ	30	Dipole		A1, PSK, RTTY	24 zz
10144.00	G0MBA	Essex	JO01NT	1	H 1/2 wave		A1	24 FN2
10144.60	G0PKT	Essex	JO01MT	1	1/2 w vertical		A1	24 FN2
10145.00	OK0EF			0.5				OP?
20 m								
14097.00	WSPR beacons around here							
14100.00	4U1UN	UN NY	FN30AS	100 - 0,1	Vertical	Omni	A1	IBP Cycle
14100.00	VE8AT	Eureka, Nunavut	EQ79AX	100 - 0,1	Vertical	Omni	A1	IBP Cycle
14100.00	W6WX	Mt Umunhum, CA	CM97BD	100 - 0,1	Vertical	Omni	A1	IBP Cycle
14100.00	KH6RS	Laie, Oahu	BL10TS	100 - 0,1	Vertical	Omni	A1	IBP Cycle
14100.00	ZL6B	near Masterton	RE78TW	100 - 0,1	Vertical	Omni	A1	IBP Cycle
14100.00	VK6RBP	Rolystone	OF87AV	100 - 0,1	Vertical	Omni	A1	IBP Cycle
14100.00	JA2IGY	Mt Asama	PM84JK	100 - 0,1	Vertical	Omni	A1	IBP Cycle
14100.00	RR9O	Novosibirsk	NO14KX	100 - 0,1	Vertical	Omni	A1	IBP Cycle
14100.00	VR2B	Hong Kong	OL72CQ	100 - 0,1	Vertical	Omni	A1	IBP Cycle
14100.00	4S7B	Colombo	NJ06CR	100 - 0,1	Vertical	Omni	A1	IBP Cycle
14100.00	ZS6DN	Vereniging	KG33XI	100 - 0,1	Vertical	Omni	A1	IBP Cycle
14100.00	5Z4B	Kiambu	KI88MX	100 - 0,1	Vertical	Omni	A1	IBP Cycle
14100.00	4X6TU	Tel Aviv	KM72JB	100 - 0,1	Vertical	Omni	A1	IBP Cycle
14100.00	OH2B	Lohja	KP20BM	100 - 0,1	Vertical	Omni	A1	IBP Cycle
14100.00	CS3B	Santo da Serra	IM12OR	100 - 0,1	Vertical	Omni	A1	T Nonop FN1
14100.00	LU4AA	Buenos Aires	GF05TJ	100 - 0,1	Vertical	Omni	A1	IBP Cycle
14100.00	OA4B	Lima	FH17MW	100 - 0,1	Vertical	Omni	A1	T Nonop
14100.00	YV5B	Caracas	FK06NK	100 - 0,1	Vertical	Omni	A1	T Nonop

Freq	Call	Nearest Town	Locator	ERPw	Antenna	Direction	Mode	Status
17 m								
18095.00	YO8RIX	near Dorohoi	KN37FW	150mw	Lazi Loop	Omni	A1	24
18095.50	HP1AVS	Cerro Jefe	FJ09HD	2	Inv Vee		A1	24
18100.00	IK6BAK	Montefelcino	JN63KR	0.1	Inv Vee	Omni	A1	?
18102.00	I1M	Bordighera	JN33UT	10	Vert	Omni	A1	24
18104.60	WSPR beacons around here							
18110.00	4U1UN	UN, NY	FN30AS	100 - 0,1	Vertical	Omni	A1	IBP Cycle
18110.00	VE8AT	Eureka, Nunavut	EQ79AX	100 - 0,1	Vertical	Omni	A1	IBP Cycle
18110.00	W6WX	Mt Umunhum, CA	CM97BD	100 - 0,1	Vertical	Omni	A1	IBP Cycle
18110.00	KH6RS	Laie, Oahu	BL10TS	100 - 0,1	Vertical	Omni	A1	IBP Cycle
18110.00	ZL6B	near Masterton	RE78TW	100 - 0,1	Vertical	Omni	A1	IBP Cycle
18110.00	VK6RBP	Rolystone	OF87AV	100 - 0,1	Vertical	Omni	A1	IBP Cycle
18110.00	JA2IGY	Mt Asama	PM84JK	100 - 0,1	Vertical	Omni	A1	IBP Cycle
18110.00	RR9O	Novosibirsk	NO14KX	100 - 0,1	Vertical	Omni	A1	IBP Cycle
18110.00	VR2B	Hong Kong	OL72CQ	100 - 0,1	Vertical	Omni	A1	IBP Cycle
18110.00	4S7B	Colombo	NJ06CC	100 - 0,1	Vertical	Omni	A1	IBP Cycle
18110.00	ZS6DN	Vereniging	KG33XI	100 - 0,1	Vertical	Omni	A1	IBP Cycle
18110.00	5Z4B	Kiambu	KI88MX	100 - 0,1	Vertical	Omni	A1	IBP Cycle
18110.00	4X6TU	Tel Aviv	KM72JB	100 - 0,1	Vertical	Omni	A1	IBP Cycle
18110.00	OH2B	Lohja	KP20BM	100 - 0,1	Vertical	Omni	A1	IBP Cycle
18110.00	CS3B	Santo da Serra	IM12OR	100 - 0,1	Vertical	Omni	A1	T Nonop FN1
18110.00	LU4AA	Buenos Aires	GF05TJ	100 - 0,1	Vertical	Omni	A1	IBP Cycle
18110.00	OA4B	Lima	FH17MW	100 - 0,1	Vertical	Omni	A1	T Nonop
18110.00	YV5B	Caracas	FK60NK	100 - 0,1	Vertical	Omni	A1	T Nonop
15 m								
21068.00	SK7OB				V,Dipole			IRREG
21094.60	WSPR beacons around here							
21145.70	IZ3DVW	near Monselice	JN55VF	2.6	Inv Vee, Dipole		A1	24
21149.00	F5ZHL		JO10SI					IRREG
21149.00	IT9ATQ/B	Catania	JM77MM	2	Bent Dipole		A1	24
21150.00	4U1UN	UN, NY	FN30AS	100 - 0,1	Vertical	Omni	A1	IBP Cycle
21150.00	VE8AT	Eureka, Nunavut	EQ79AX	100 - 0,1	Vertical	Omni	A1	IBP Cycle
21150.00	W6WX	Mt Umunhum, CA	CM97BD	100 - 0,1	Vertical	Omni	A1	IBP Cycle
21150.00	KH6RS	Laie, Oahu	BL10TS	100 - 0,1	Vertical	Omni	A1	IBP Cycle
21150.00	ZL6B	near Masterton	RE78TW	100 - 0,1	Vertical	Omni	A1	IBP Cycle
21150.00	VK6RBP	Rolystone	OF87AV	100 - 0,1	Vertical	Omni	A1	IBP Cycle
21150.00	JA2IGY	Mt Asama	PM84JK	100 - 0,1	Vertical	Omni	A1	IBP Cycle
21150.00	RR9O	Novosibirsk	NO14KX	100 - 0,1	Vertical	Omni	A1	IBP Cycle
21150.00	VR2B	Hong Kong	OL72CQ	100 - 0,1	Vertical	Omni	A1	IBP Cycle
21150.00	4S7B	Colombo	NJ06CC	100 - 0,1	Vertical	Omni	A1	IBP Cycle
21150.00	ZS6DN	Vereniging	KG33XI	100 - 0,1	Vertical	Omni	A1	IBP Cycle
21150.00	5Z4B	Kiambu	KI88MX	100 - 0,1	Vertical	Omni	A1	IBP Cycle
21150.00	4X6TU	Tel Aviv	KM72JB	100 - 0,1	Vertical	Omni	A1	IBP Cycle
21150.00	OH2B	Lohja	KP20BM	100 - 0,1	Vertical	Omni	A1	IBP Cycle
21150.00	CS3B	Santo da Serra	IM12OR	100 - 0,1	Vertical	Omni	A1	T Nonop FN1
21150.00	LU4AA	Buenos Aires	GF05TJ	100 - 0,1	Vertical	Omni	A1	IBP Cycle
21150.00	OA4B	Lima	FH17MW	100 - 0,1	Vertical	Omni	A1	T Nonop
21150.00	YV5B	Caracas	FK60NK	100 - 0,1	Vertical	Omni	A1	T Nonop
21151.00	I1M	Bordighera	JN33UT	10	2 5/8 Vert	Omni	A1	24
12 m								
24920.00	IY4M	Bologna	JN54QK				A1	H+30>H+60m
24930.00	4U1UN	UN, NY	FN30AS	100 - 0,1	Vertical	Omni	A1	IBP Cycle
24930.00	VE8AT	Eureka, Nunavut	EQ79AX	100 - 0,1	Vertical	Omni	A1	IBP Cycle
24930.00	W6WX	Mt Umunhum, CA	CM97BD	100 - 0,1	Vertical	Omni	A1	IBP Cycle
24930.00	KH6RS	Laie, Oahu	BL10TS	100 - 0,1	Vertical	Omni	A1	IBP Cycle
24930.00	ZL6B	near Masterton	RE78TW	100 - 0,1	Vertical	Omni	A1	IBP Cycle
24930.00	VK6RBP	Rolystone	OF87AV	100 - 0,1	Vertical	Omni	A1	IBP Cycle
24930.00	JA2IGY	Mt Asama	PM84JK	100 - 0,1	Vertical	Omni	A1	IBP Cycle
24930.00	RR9O	Novosibirsk	NO14KX	100 - 0,1	Vertical	Omni	A1	IBP Cycle
24930.00	VR2B	Hong Kong	OL72CQ	100 - 0,1	Vertical	Omni	A1	IBP Cycle
24930.00	4S7B	Colombo	NJ06CC	100 - 0,1	Vertical	Omni	A1	IBP Cycle
24930.00	ZS6DN	Vereniging	KG33XI	100 - 0,1	Vertical	Omni	A1	IBP Cycle
24930.00	5Z4B	Kiambu	KI88MX	100 - 0,1	Vertical	Omni	A1	IBP Cycle
24930.00	4X6TU	Tel Aviv	KM72JB	100 - 0,1	Vertical	Omni	A1	IBP Cycle
24930.00	OH2B	Lohja	KP20BM	100 - 0,1	Vertical	Omni	A1	IBP Cycle
24930.00	CS3B	Santo da Serra	IM12OR	100 - 0,1	Vertical	Omni	A1	T Nonop FN1
24930.00	LU4AA	Buenos Aires	GF05TJ	100 - 0,1	Vertical	Omni	A1	IBP Cycle
24930.00	OA4B	Lima	FH17MW	100 - 0,1	Vertical	Omni	A1	T Nonop
24930.00	YV5B	Caracas	FK60NK	100 - 0,1	Vertical	Omni	A1	T Nonop
24931.00	7Z1CQ	Jeddah	KL91ON	5	Vert, Dip	Omni	A1	IRREG
10 m								
28163.00	VK3XPT	Lang Lang	QF21SR		1/4 Vert			
28169.00	ZB2TEN	Gibraltar	IM76HD	4	1/4 Vert	Omni	A1	24?
28171.00	XE1FAS	Publa, PU	EK09UB	12	Dipole		A1	24
28173.00	IZ1EPM	27 km NE Turin	JN35WD	20	1/2 Vert	Omni	A1	24
28174.00	VE1VDM		FN85AA				A1	?
28175.00	VE3TEN	Ottawa, ON	FN25	10	GP	Omni	A1	24
28176.90	HP1RCP	Cerro Jefe	FJ09HD	5	Slope Dipole		A1	24
28177.00	IW1AVR	Cravanzana		5	Vertical		A1	?
28178.00	IQ0GV		JN61TR				A1	?
28180.30	I1M	Bordighera	JN33UT	5 W / 20 W	2 × 5/8 Vert	Omni	A1	30/60min
28182.40	SV3AQR	Amalias	KM07QS	4	GP	Omni	A1	24
28183.20	XE1RCS	Cerro Gordo	EK09OS	8	AR10	Omni	A1	24
28184.00	VE2REA	Quebec, QC	FN46IT				A1	24

For the latest HF beacon listing, see: *RSGB Website for link*

Freq	Call	Nearest Town	Locator	ERPw	Antenna	Direction	Mode	Status
28185.00	VA3SRC	Burlington, ON	FN03BH	5	Dipole		A1	PT
28187.60	VE7KC	Penticton, BC	DN09EL				A1	24
28188.00	OE3XAC	Kaiserkogel	JN78SB	20	7/8 GP @ 750 m	Omni	A1	24
28188.10	JE7YNQ	Fukushima	QM07					24
28188.90	SV5TEN	Raad	KM46CK	5	Vertical	Omni		24
28189.50	LU2DT	Mar del Plate	GF12	5	Vert, Dipole	Omni	A1/PSK	
28189.80	LU8XW	Ushuaia	FD55UE					?
28190.00	LU3HFA	Cordoba, CD	FF78UP	5	Vertical	Omni	A1	24?
28190.00	LU4VI	Villa Regina	FF60VI				A1	24
28191.50	A62ER	Sharjah, UAE	LL75QI				A1	OP?
28192.00	EP4HR	Shiraz	LL69GP	20/2/0,2	Dipole		A1	24
28193.00	EI7GR		IO53MG					?
28192.90	VE4ARM	Austin, MB	EM09HW	5	GP	Omni	A1	24
28193.50	A47RB	Oman	LL93FO	10	Vertical	Omni	A1	OP?
28193.50	LU2XPK		FF66DE				A1	?
28193.90	IW4EIR		JN54AS	1.5			A1	?
28194.00	IW4ERI		JN54AS					
28195.10	IY4M	Bologna	JN54QK	20	5/8 GP	Omni	A1	H>H+30m
28196.00	VA3ITA	Bramton, ON	FN03CW				A1	IRREG
28196.10	LU4JJ	Concordia, ER	GF08XO				A1	24
28196.70	LU5FB	Rosario, SF	FF97PB				A1	24
28197.00	VE7MTY	Vancouver, BC	CN89	5	Vertical	Omni	A1	24
28197.70	IK3OTW						A1	?
28193.00	LU2ERC	Ensenada	GF15AD	10	Vertical	Omni	A1	24
28198.00	LU9FE		FF98GR					?
28199.00	LW6DJD	La Plata	GF15AC	5			A1	?
28199.30	LU1FHH	El Trebol, SF					A1	24
28200.00	4U1UN	UN, NY	FN30AS	100 - 0,1	Vertical	Omni	A1	IBP Cycle
28200.00	VE8AT	Eureka, Nunavut	EQ79AX	100 - 0,1	Vertical	Omni	A1	IBP Cycle
28200.00	W6WX	Mt Umunhum, CA	CM97BD	100 - 0,1	Vertical	Omni	A1	IBP Cycle
28200.00	KH6RS	Laie, Oahu	BL10TS	100 - 0,1	Vertical	Omni	A1	IBP Cycle
28200.00	ZL6B	near Masterton	RE78TW	100 - 0,1	Vertical	Omni	A1	IBP Cycle
28200.00	VK6RBP	Rolystone	OF87AV	100 - 0,1	Vertical	Omni	A1	IBP Cycle
28200.00	JA2IGY	Mt Asama	PM84JK	100 - 0,1	Vertical	Omni	A1	IBP Cycle
28200.00	RR9O	Novosibirsk	NO14KX	100 - 0,1	Vertical	Omni	A1	IBP Cycle
28200.00	VR2B	Hong Kong	OL72CQ	100 - 0,1	Vertical	Omni	A1	IBP Cycle
28200.00	4S7B	Colombo	NJ06CC	100 - 0,1	Vertical	Omni	A1	IBP Cycle
28200.00	ZS6DN	Vereniging	KG33XI	100 - 0,1	Vertical	Omni	A1	IBP Cycle
28200.00	5Z4B	Kiambu	KI88MX	100 - 0,1	Vertical	Omni	A1	IBP Cycle
28200.00	4X6TU	Tel Aviv	KM72JB	100 - 0,1	Vertical	Omni	A1	IBP Cycle
28200.00	OH2B	Lohja	KP20BM	100 - 0,1	Vertical	Omni	A1	IBP Cycle
28200.00	CS3B	Santo da Serra	IM12OR	100 - 0,1	Vertical	Omni	A1	T Nonop FN1
28200.00	LU4AA	Buenos Aires	GF05TJ	100 - 0,1	Vertical	Omni	A1	IBP Cycle
28200.00	OA4B	Lima	FH17MW	100 - 0,1	Vertical	Omni	A1	T Nonop
28200.00	YV5B	Caracas	FK60NK	100 - 0,1	Vertical	Omni	A1	T Nonop
28200.50	VA3GMT	Toronto					A1	?
28200.80	AC7AV	Oak Harbour, WA					A1	?
28201.00	N7JUB						A1	?
28201.00	WB9OTX		EN55				A1	?
28201.30	PU2SUT	Sao Paulo	GG66TB	20	Dipole		A1	?
28202.00	WN2WNC	New Berlin, NY	FN22IO				A1	
28202.50	KA3BWP	Stafford, VA	FM18GK				A1	24
28203.00	K4MTP	Tannersville, PA					A1	
28203.00	PY2WFG	Ipisanga, SP	GG77FF				A1	24
28203.00	KB1QZY	Springfield, MA	FN32QC	2	Imax2000 V	Omni	A1	24
28203.00	KG8CO	Clinton, MI	EN82AB	5	Vertical	Omni	A1	24
28203.00	N6DXX	Sacramento, CA	CM98FM				A1	?
28203.00	K4MTP		FN21TA				A1	
28203.50	K6LLL	Mission Viejo, CA	DM13EO				A1	24
28202.00	SV2HNE		KN10LL	5	GP			?
28204.00	WA2NTK	Big Flats, NY	FN12NE			E-W	A1	24
28204.00	WL7N	Ward Cove, AK	CO45KK				A1	?
28204.00	KE4TWI	Watertown, TN	EM66VO	5			A1	24
28204.00	W0WF	St Charles, MO	EN02QB				A1	?
28204.00	W6CF	San Fancisco	CM87UU				A1	24
28204.00	KA1KNW	Windsor, CT	FN31RU	10			A1	?
28205.00	DL0IGI	Hohenpeissenberg	JN57MT	Varies	1/4 Vert	Omni	A1	24
28205.20	VN3NIA	Ridgway, PA	FN01PK	4	Dipole		A1	24
28205.90	HS0BBD	Bangkok	OK13					?
28206.10	VA3GRR	Brampton, ON	FN03	1.75	1/2 Vert	Omni	A1	24
28206.30	K9EJ	Toledo	EM59UG	2	Vert @ 4 m	Omni	A1	24
28206.50	HP1RIS	Panama City	FJ09GA	3	Vert Dipole	Omni	A1	24
28207.00	ON0RY	Binche	JO20CK	5	Vertical	Omni	A1	24
28207.30	KW7HR	Pasco, WA	DN06KG		A1	24?		
28207.80	W4CND	Jemison, AL	EM63QA	2	Vertical	Omni	A1	24
28208.00	KE6TE	Elk Grove, CA	CM98HK				A1	OP?
28208.00	AK2F	Randolph, NJ	FN20QT				A1	share
28208.00	WN2A	Budd Lake, NJ	FN20OU				A1	AK2F
28208.10	JR0YAN	near Toyama	PM86JW	25	Hor,Loop		A1	24+++
28208.20	IZ3LCJ	St Lucia di Piave	JN65DU	5	1/4GP@15m	Omni	A1	QRT?
28208.50	NB7A	Reno, NV	DM09BM				A1	24
28208.70	N8PVL	Livonia. MI	EN82GJ				A1	24?
28209.00	KH6AP	Kikei Maui, HI	BL10SS	20	3/8 Vert	Omni	A1	24
28209.50	K9CW	Thomasboro. IL	EN50WF	2	AR99@15□	Omni	A1	24
28209.80	KV6Q	San Diego	DM12JS	3			A1	?

Freq	Call	Nearest Town	Locator	ERPw	Antenna	Direction	Mode	Status
28210.00	PT2SSB	Brasilia	GH64CI				A1	?
28210.00	KB9UGA	Egg Harbour, WI					A1	24
28210.00	NJ4R						A1	?
28210.20	SB7W							?
28210.40	NT4F	Wilmington, NC	FM14AE	5			A1	24
28210.50	VE4TEN	Kelowna, BC		2	1/2 Vert	Omni	A1	24
28211.00	DB0FKS	near Frankfurt/M	JN40IT	0.2	DV27 Vert	Omni	A1	OP?
28211.00	K5ARC	Galvez, LA	EM40	20	Vertical	Omni	A1	24?
28211.00	CE1TUW	Antofagasta	FF46RQ				A1	?
28211.10	LA4TEN	near Hellvik	JO28WL	250	Vertical	Omni	A1	24
28211.50	VK4ADC	New Beith, QLD		10/5/2/1	1/4 Vert	Omni	A1	24
28211.50	VK4WSS	Mt Cotton, QLD						24
28211.80	AC7GZ	Chandler, AZ	DM43				A1	?
28212.00	W4AMA		EM63WA				A1	?
28212.00	7Z1AL	Dammam	LL56BK	10	Vertical		A1	OP?
28212.50	K0KP	Fredenberg, MN	EN36VW	0.5	GP	Omni	A1	OP?
28212.50	KJ4QYB	Rainbow City, AL	EM63WO				A1	?
28212.60	LU7DQP	Lanos Oest, BA	GF05TH				A1	24
28213.00	WA5SAT		EL09				A1	?
28213.30	KD8RKJ	Cleveland, OH	EN91CK	2	Vertical	Omni	A1	24
28213.50	KE4KAA	Big Stone Gap, VA	EM86OV	5			A1	24
28213.50	W3IK	Gray, TN	EM86				A1	24
28213.80	KF5KBZ	Austin, TX	EM10FB				A1	?
28213.80	WA5SAT		EL09					?
28214.00	N4PAL	Longwood, FL	EL98HQ	5	Vert@4,5m	Omni	A1	24?
28214.00	LA9TEN	Snertingdal	JP50EV	10	5/8 GP	Omni	A1	24
28214.50	FR1GZ	Reunion Island	LG79RC					IRREG
28215.00	LU5EGY	Buenos Aires	GF05QI				A1	24
28215.00	N8MIE		EM44				A1	?
28215.00	SR5TDM		KO01KX				A1	?
28215.00	KA9SZX	Paxton, IL	EN50VD	1	Antron99	Omni	A1	24
28215.00	W4JPL	Liberty. NC	FM05FV				A1	24
28215.00	GB3MCB	Cornwall	IO70OJ	25	Vertical	Omni	F1	24
28215.30	XE3D	Merida, YUC	EL50EG				A1	24
28215.50	KD5CKP	Olive Branch, MS	EM54BW	3	Vertical	Omni	A1	24
28215.80	K6WKX	Santa Cruz, CA	CM86XX	10	Horiz Dipole	Omni	A1	24
28216.00	K3FX	Neptune City, NJ	FN20XE	7	1/2 Vert	Omni	A1	24
28216.00	N7MA	Cataldo, ID	DN17SN	5	Vertical	Omni	A1	24
28217.00	LA2BCN	Telemark	JO48GX	8	5/8 Vert	Omni	A1	24$$$
28126.10	LA2BCN	Telemark	JO48GX	8	5/8 Vert	Omni	WSPR	24
28217.00	WB0FTL	Alden, MN	EN33FP	5	AR10@25□	Omni	A1	24
28217.00	W6GY	Star, ID	DN13RP				A1	24
28217.50	WA1LAD	West Warick, RI	FN41FQ	4.5	J-Pole	Omni	A1	24
28217.50	W8MI	Mackinaw MI	EN75	0.5	Vertical	Omni	A1	24
28218.00	IQ5MS	Marina da Massa	JN54AA		Vertical	Omni	A1	24
28218.30	AC0KC	Fort Lupton, CO	DN70OA	qrpp			A1	?
28218.30	KD8RKJ	Cleveland, OH		2	Vertical	Omni	A1	?
28218.30	KJ4LAA	Decatur, AL	EM64LN	3	Vertical	Omni	A1	?
28218.50	W5RDW	Murphy, TX	EM12QX	5	AR10@25□	Omni	A1	24
28218.60	KA4RRU	Catlett, VA	FM18EN				A1	24
28218.80	KN8DMK	Amanda, OH	EM89OO	3	Slope Dipole	NW/SE	A1	?
28219.60	PY2UEP		GG58GA				A1	?
28219.90	KB9DJA	Mooresville, IN	EM69RO	35	GP	Omni	A1	24
28220.00	5B4CY	Mandria	KM64KU	15	GP	Omni	F1	24
28220.00	W8VO	Sterling Hts, MI	EN82NU	5	Vert@50□	Omni	A1	OP?
28220.00	K4AQXI		FM03IL				A1	?
28220.40	WK4DS	Trenton, GA	EM74FU	2	Dipole	NE-SW	A1	24
28220.50	YM7KK	Giresun	KN90IV	4			A1	24
28220.50	N5FUN	Caroltton, TX	EM12				A1	?
28220.80	SV2MCG	Thessaloniki	KN10FC				A1	?
28221.00	WE4S	Rock Springs, CA					A1	24
28221.50	KC0TKS	Sedalia MO	EM38IQ	5	J-poleVert	Omni	A1	24
28221.50	GW7HDS		IO81IP	3			A1	IRREG
28221.90	W1DLO	Calais, ME	FN65JE	10	A99 Vert	Omni	A!	24
28222.00	TP2CE	Strasbourg	JN38VO	450mw	GP-7000	Omni	A1	24
28222.00	K6JCA	Carmel Valley CA	CM96DN	A1				QRT?
28222.00	IZ0KBA	Castel Madama	JN61KX	4	GP 350 m ASL	Omni	A1	24
28222.00	ZS1TEN	Cape Town	JF96FB	5	Vertical	Omni	A1	
28222.20	W4KLP						A1	?
28222.80	N4QDK	Sauratown Mt, NC	EM95	3	dipole		A1	24
28223.00	WY5B	Biloxi, MS	EM50NK				A1	24
28223.00	9H1LO	Malta	JM75	8	GP		A1	Irreg
28223.00	KP3FT	Ponce, PR	FK68	4	Vertical	Omni	A1	24
28223.30	XE3ACB	Hecelchakan	EL40				A1	24?
28223.50	KK6RE	Chico, CA	CM96VG				A1	?
28223.60	PB5A		JO21DE				A1	?
28224.00	WD0AKX	Albert Lea, MN	EN33	5	Vertical	Omni	A1	24
28224.00	WA3RNC	Lewistown, PA	FN10FO				A1	24
28224.20	YB9BWN	Denpasar, Bali	OJ13FK	2	Dipole	EU/VK	A1	24
28224.70	HA5BHA	near Budapest	JN97KO	5		Omni		24
28224.80	KW7Y	Marysville, WA	CN88SD	4	Vertical	Omni	A1	24
28224.80	IT9EJW	Sicily	JM77NN	3			A1	
28224.80	NT6T	Goleta, CA					A1	?
28225.00	YM7TEN		KN91RB	1	Vertical	Omni	A1	24
28225.00	K6FRC	Angels Common, CA	CM97GP		Vertical	Omni	A1	24??

Freq	Call	Nearest Town	Locator	ERPw	Antenna	Direction	Mode	Status
28225.00	K5GJR	Corpus Christi TX	EL17HR				A1	24
28225.40	KC0JCA	St Louis, MO					A1	?
28225.50	W2DLL	near Buffalo, NY	FN02PP	8	1/2 Vert	Omni	A1	24
28225.60	WB0LYV	Beatrice, NE	EN10		Delta Loop		A1	?
28226.00	ED1YCA		IN73AL	5	H Loop	Omni	A1	24
28226.00	PU4CBX	Baraco de Cocais	GH80DB				A1	Irreg
28226.20	WA6HXW	Westn Covina, CA	DM14AB				A1	irreg
28226.50	N7MSH	North Powder, OR	DN15				A1	24
28226.60	KC6WGN	Las Vegas, NV	DM26LD	10	Omni	A1	24	
28226.60	PY2RFF	Sco Pedro, SP	GG67AL	1.5	1/4 Vert	Omni	A1	24
28226.70	KU4A	Lexngton, KY	EM78SB				A1	?
28227.00	VE9AT	White Head Island		0.1	Dipole		A1	OP?
28227.00	KJ4HYV	Zellwood, FL					A1	24
28227.50	KC5MO	Austin, TX	EM10BF	2	Dipole		A1	24
28227.70	IW3FZQ	Monselice, PD	JN55VF	5	J-Pole@20m	Omni	A1	24
28228.00	ZL3TEN	Rolliston	RE66	10	1/2 Vert	Omni	A1	TNonOp
28228.00	OH5TEN	Kouvola	KP30HV	4	Horiz	Dip	A1	24
28228.30	TG9TEN		EK44				A1	OP?
28228.80	N3PV	Spring Valley CA	DM12LP				A1	24
28229.00	ZL2MHF	near Wellington	RE78NU	10	1/2 Vert	Omni	F1	24
28229.00	NG9Y	Vevay, IN	EM78LR				A1	?
28229.30	KA2LIM	Pine Valley, NY	FN12NP	NY			A1	?
28229.70	IQ8CZ	Catanzo	JM88HV	10	GP	Omni	A1	24
28230.00	WA4ZKO	Dry Ridge, KY	EM78PP	4	A99	Omni	A1	24
28230.50	AA0RQ	Pine, CO					A1	?
28230.00	KI4AED	Ocoee, FL	EL98FN	5	Antron 99	Omni	A1	24
28230.00	KQ4TG	Leland, NC	FM40XF				A1	24
28230.50	HP6RCP	Santiago	EJ98MB	3	AR99	Omni	A1	IRREG
28231.00	F5ZEH		IN88VA	,5/5/50	1/2 V ,3-el Yagi		A1	24
28231.80	WA4FC	Prince George, VA	FM17HD	5	Ringo@200□	Omni	A1	24
28232.00	N1FSX	Simla, CO	DM89AC	5	Vertical	Omni	A1	24
28232.60	N9BPE	Tuscaloosa, IL	EM59UT	2	1/2 Dip @20□	Omni	A1	24
28232.70	N2MH	West Orange, NJ	FN20UT				A1	24
28233.00	N2UHC	St Paul, KS	EM27JM	4	Vert Dipole	Omni	A1	24
28231.60	SV2AHT	Hortiatis	KN10NO				A1	24
28232.30	W7SWL	Tucson, AZ		5	Vertical	Omni	A1	?
28233.00	I0KNQ	Genzano di Roma	JN61FU	2	Turnstile	Omni	A1	24
28233.00	KB9GSY	Hammond, IN	EN61FP				A1	24
28234.00	K4DP	Covington, VA	FM07AR				A1	IRREG
28234.30	K4DXY	Birmingham, AL	EM63PP	2			A1	?
28234.80	VE1CBZ	Keswick Ridge, NB	FN65	2	A99 Vert	Omni	A1	24
28235.00	KI4AED	Ocoee, FL	EL98FN	5	Antron 99	Omni	A1	24
28235.00	KK6RE	Chico, CA					A1	?
28235.00	NP4LW	San Sebastian	FK68MI	15	Vertical		A1	?
28235.00	KQ4FM	Southlake, TX	EM12JW	5	GP		A1	24
28235.00	OY6BEC	Faroe Islands	IP62	20	Yagi		A1	24
28235.10	KI4HOZ	Pickens, SC	EM83XM				A1	?
28235.60	VE3GOP	Mississauga, ON	FN03GD	0.2	A99 Vert	Omni	A1	24
28236.00	W8YT	Martinsburg, WV	FM19AJ	5	Vertical	Omni	A1	24
28236.50	W0KIZ	near Denver, CO	DM79	5	Vertical	Omni	A1	24
28237.00	K7TIA	Houston, TX					A1	?
28237.50	WA2NEW	Beach Haven, NJ					A1	?
28237.60	LA5TEN	near Oslo	JO59JP	15	1/2 Vert	Omni	A1	24
28237.80	K7ZSA	Alger, WA	CN88UO	5	Vertical	Omni	A1	24?
28238.00	KB2SEO	Eton, GA	EM74OT	5	1/4 GP	Omni	A1	24
28238.20	G0PKT	Essex	JO01MT	1	1/2 w vertical		A1	24 FN2
28238.80	GB3PKT	Essex	JO01NT	1	1/2 w vertical		A1	24 FN2
28239.00	VA7PL	Crystal Mountain	DM09	5	GP	Omni	A1	24
28239.00	PP6AJM	Nosso Senhora da Socorro		HH19LD				?
28239.20	AL7FS	Anchorage, AK	BP51BD	3	1/2 Vert	Omni	A1	PT
28239.50	WA3HGT	Montoursville PA	FN11MG					
28239.80	N4LEM	Cocoa, FL	EL98	50	Vertical	Omni	A1	?
28239.80	IZ8RVA	Agropoli, SA	JN70LI				A1	24?
28240.00	I0KNQ	Rome	JN61FU				A1	24
28240.00	XE3OAX	Ocotlan, OAX	EK17PA	0,5?	+ sev,spurii		A1	24
28240.11	WE6Z	Granite Bay, CA	CM98JS	4	Vertical	Omni	A1	24
28240.50	N2DWS	Port Republic NJ	FM29SM				A1	24
28240.50	W4RKC	Winchester, VA	FM09VD	5	1/2 Vert	Omni	A1	24
28240.60	YO2X	Timisoara	KN05PS	2	GP	Omni	A1	09 - 15 UTC
28240.70	AJ8T	Sturgis, MI	EN71HS				A1	24
28241.50	F5ZUU	Malataverne	JN24IL	5	1/2 Vert	Omni	A1	24
28241.50	K5DZE	Erlanger, KY	EM78QS	5			A1	24
28242.00	IZ8DXB	Naples	JN70LN	6			A1	24?
28242.50	WD9CVP	Elgin, IL	EN52UA				A1	24
28242.70	F5ZWE	Foix	JN02TW	15	Vert Dipole	Omni	A1	24
28243.00	AA1SU	VT	FN34KL				A1	24
28243.50	G0MB	Essex	JO01OT	1	1/2 w vertical		A1	24 FN2
28244.00	WA6APQ	Long Beach, CA	DM13	30	Vertical	Omni	A1	24
28244.00	GB3TEN	Fleetwood	IO83LV	0.4	Dipole	Omni	F1A	24
28244.50	DV2FQN		KN10FC	5	GP			24?
28245.00	EB4YAK		IN80FK					24?
28245.00	DB0TEN	Bomlitz	JO42TW	2	1/2 GP	Omni	A1	24?
28245.60	SV2AHT	Hortiatis	KN1ONO	24				
28246.00	VE9BEA	Crabbe Mountain, NB	FN66	6	AR10@43□	Omni	A1	QRT
28246.00	KG2GL	Nutley, NJ	FN20UT	5	R5 Vert @4'	Omni	A1	INT

Freq	Call	Nearest Town	Locator	ERPw	Antenna	Direction	Mode	Status
28246.20	KI4LEV	Clarksville, TN	EM66	5			A1	?
28247.00	K6EMI		DM13AU				A1	?
28247.90	N1ME	Bangor, ME	FN54PS	5			A1	24
28248.00	K5DZE	Newman, GA					A1	?
28248.50	K5DDJ	San Antonio, TX	EL09	0.5	GP	Omni	A1	24
28249.00	N7LT	Bozeman. MT	DN45LQ	4/,4/,04	1/2 GP	Omni	A1	24
28249.00	KA3JOE	Bensalem, PA	FN29MB				A1	?
28249.10	ER1TEN	Chisinau	KN47IB	4	Vertical	Omni	A1	24
28249.50	PY3PSI	Porto Alegre	GF49KX	2.8	GP	Omni	A1	IRREG
26249.80	W4CJB	Santa Rosa Beach. FLEM60WR					A1	?
28249.90	W3ATV	Trevose, PA	FN20	1	Dipole		A1	24
28250.00	K1GND	Johnston, RI	FN41FT				A1	
28250.00	UB6LGR		JN54				A1	IRREG
28250.00	K8NDB	Somerton, AZ	DM22QQ	4	1/4 Vert	Omni	A1	24
28250.00	N4ES	Tampa, FL	EL88TA	20/2/,2/,02	Horiz		A1	SYNCH)n
28250.00	N4ESS	Zephyrhills, FL	EL88VG	A1			A1	SYNCH)n
28250.00	WB4WOR	Greensboro, NC	FM06BT	20/2/,2/,02	Horiz	Hor	A1	SYNCH)n
28250.00	K7EK	Graham, WA	CN87TB	25	1/2 Vert	Omni	A1	SYNCH)n
28250.00	N4ES	Clearwater, FL	EL88PA	20/2/,2/,02	Horiz	Horiz	A1	SYNCH)n
28250.00	K6FRC	Sutter Buttes, CA	CM97GP	10	Vertical	Omni	A1	24??
28250.00	KOHTF	Des Moines, IA	EN31DO	3	Inv,V@10□		A1	OP?
28251.00	ACOMO	Hutchinson, KS	EM18				A1	?
28251.15	WA4GEH	Clayton NC	FM05SN				A1	OP?
28251.10	ED4YAK		IN80FK	5	Vertical	Omni	A1	24
28251.50	KE5JXC	Pecan Island, LA	EL39SP	5	Vertical	Omni	A1	24
28252.00	WA2DVU	Cape May, NJ	FM29NC	NJ			A1	24
28252.50	K7OC	Fort Worth, TX	EL29	3.3	Vertical		A1	24
28252.00	WW9EE	Tremont, IL	EN50GK		A1	?		
28252.50	W6PC/4	Ocala, FL	EL89VD	10	Dipole	Omni	A1	24
28253.00	N3BSQ	Bethel Park, PA					A1	24?
28253.00	ED5YAU		IM98WN	5	Vertical	Omni	A1	24
28253.00	KG4YUV	Crandall, GA	EM74OW	7	A99 Vert	Omni	A1	24
28253.00	K8HWW	Warren, MI	EN82MN	3			A1	24?
28253.80	XE1USG	Puebla	EK09VB				A1	?
28254.00	W4CJB	Point Washington, Fl	EM60WR				A1	24?
28254.00	PI7BXM	Baron de la Marckstraat 7, 6095 AW BAexemJO21WF			1,5 Watt	1/2 wave vertical		Omni A1
28254.30	N1FCU	Windham, ME	FN43ST				A1	24?
28254.50	K4JEE	Louisville, KY	EM78				A1	24?
28254.50	K5AHH	Broken Bow, OK		OK			A1	?
28255.00	N0AR	St Paul, MN	EM73SW	0.5	I/2 Vert	Omni	A1	24
28255.50	K8HWW	Stirling Heights, MI	EN82MN	3	Vertical	Omni	A1	24
28256.00	C30P	Andorra	JN02SM	10	R5 Vert@2m	Omni	A1	24
28256.00	WI5V	Oklahoma City	EM15	0.5			A1	24
28256.50	VK3RMH	25 km NE Melbourne	QF22OH	20-Feb	1/2 Vert	Omni	A1	24
28256.50	K9JHQ	O'Fallon, IL	EM58AM	10	Vertical	Omni	A1	24?
28257.00	KB4UPI	Bessemer, AL	EM63MG				A1	24
28257.30	WA2DVU	Cape May, NJ		10	Mosley 57	45	A1	24
28257.50	N5WYN	Seven Points, TX	EM12VI				A1	
28257.50	DK0TEN	Sipplingerberg, near Überlingen			10			F1A
28257.80	WY5I	Port St Kucie, FL	EL97TF	5	7 dB Coll	Omni	A1	07 - 22:00 LT
28258.00	EA7JNC	La Linea de la Conception		IM76IE	8			24
28258.00	NM5TW	Albuquerque, NM	DM65RD	5	Vert Dipole	Omni	A1	24
28258.30	N1YPM	Corea, ME	FN64				A1	24
28259.00	K5TLL	Hattiesburg, MS	EM51GG				A1	?
28259.00	AB8CL	Arcanum, OH	EN79RA				A1	?
28259.00	AA4AN	Brentwood, TN	EM65NW	4	Vertical	Omni	A1	24
28259.00	F5ZVM	Valenciennes	JO10PA	5	3dbi Vert	Omni	A1	24
28259.30	VK5W1	Adelaide	PF95HG	10	GP	Omni	A1	24
28259.30	AF6PI	Indio, CA					A1	
28260.00	AD5KO	Mena, AR	EM24VS	20	Vert@20□	Omni	A1	24
28260.10	W7LFD/0	Shell Knob, MO		5	Vertical	Omni	A1	?
28260.80	NJ3T	Somerset, PA	FM09LX				A1	24?
28260.80	W5TXR	Schertz, TX	EL09VP				A1	?
28261.00	N7LF	Corbett, OR	CN85VI				A1	24
28261.50	N4VBV	Sumter, SC	EM93TW	5	Attic Dipole		A1	24
28261.60	RK3XWA	Kaluga	KO84DM	24				
28261.80	VK2RSY	Sydney	QF56MH	25	1/2 Vert	Omni	A1	24
28262.00	N4HFA	Ocala, FL	EL89VP	3	5/8 GP @ 25'	Omni	A1	24
28262.30	K8TK	Clarklake, MI	EN72TC	2	GP@15□	Omni	A1	24
28262.50	WF4HAM	Altamonte Springs, FLEL95HP		6	A99@40□	Omni	A1	INT
28263.00	VK3RRU	Mildura	QF15AT	20			A1	?
28263.00	ED4YBA	Cuneca	IN80WC	5	GP	Omni	A1	24
28263.50	N5YEY	Kilgore, TX	EM22OJ				A1	24
28263.50	W4JPL	Liberty, NC	FM05EW	4			A1	24
28264.00	AB8Z	Parma, OH	EN91DJ	5	5/8 Vert	Omni	A1	24
28264.00	VK6RWA	Carine, WA	OF78WB	20	5/8 Vert	Omni	A1	24
28264.50	K7NWS	Kent, WA	CN87TK	1	GP	Omni	A1	24
28264.50	W5ZA	Shreveport, LA	EM32DJ	3	Vert Dipole	Omni	A1	24
28265.00	DF0ANN	Moritzberg Hill	JN59PL	5	Dipole	E-W	A1	24
28265.00	VK4RRC	Woody Point QD	QG62NS	10	Vertical	Omni	A1	24?
28265.00	KJ3P	Schwenksville PA	FN20GG	5			A1	?
28265.00	PT9BCN	Campo Grande, MS	GG29RN	12	1/2 Vert		A1	24
28265.40	KR4HO	Lake City, FL	EM80QG	1	Vertical	Omni	A1	?
28265.00	NC4SW	Zebulon, NC	FM05				A1	24
28265.00	N7SCQ	Dixon, CA	CM98CK				A1	24

Freq	Call	Nearest Town	Locator	ERPw	Antenna	Direction	Mode	Status
28266.20	KB3ZI	Bloomsberg, PA					A1	24
28266.50	KA1EKS	Millinocket, ME	FN55OO	4	A99 GP	Omni	A1	24
28266.50	W5DJT	Pocola, OK	EM25SH				A1	24
28266.60	WN5KNY	Radium Springs NM	DM62LP	24			A!	24
28267.00	VK7RAE	TAS	QE38DT	10	Vertical	Omni	A1	24
28267.50	W5EFR	Houston, TX	EL29EW	2.75			A1	TQRT
28267.60	OH9TEN	Pirttikoski	KP36OI	20	1/2 GP	Omni	A1	24
28268.00	KB0QZ	Centralia, MO	EM39WE	5	Vertical	Omni	A1	24
28268.00	NM0R	St Genevieve, MO	EM47UV		A1		A1	
28268.30	VK8VF	Darwin, NT	PH57KP	25	1/4 Vert	Omni	A1	24
28268.50	KG4GXS	Coral Springs FL	EL96UG	3	Dip@23□	E-W	A1	24
28268.60	K7ZS	Hillsboro, OR	CN85MM				A1	24
28268.80	KD5ITM	Spring, TX		4	G5RV@50□		A1	
28269.00	WA2SFT	Cookville, TN	FN02OU				A1	
28268.90	AA1TT	Claremont, NH	FN33	5			A1	24
28268.90	SV6DBG	Ioannina	KM09KQ	2	5/8 Vertical	Omni	A1/ RTTY	24
28269.50	W3HH	near Ocala, FL	EL89VB	6	Hamstick	Omni	A1	24
28270.00	VK4RTL	Townsville	QH30JS	5	Vertical	Omni	F1	24
28270.50	PY4MAB	Pocos de Caldas	GG68RE	10	Vertical		A1	24
28271.00	OZ7IGY	Jystrup	JO55WM	10	Halo@90m	Omni	P14/CW	24
28271.70	W4TIY	Dallas, GA	EM73OW	4	¼ over 5/8	Omni	A1	24
28271.80	SV2HQL	Katakali-Grevens	KM09UV	5	5/8 GP	Omni	A1	24
28272.30	N1KON	Centerville, IN	EM79LT	5	Vertical	Omni	A1	24
28271.90	AC0RR	Springfield, MO	EM37IE				A1	?
28272.00	PY1RJ	San Goncalo, RJ	GG87ME	4			A1	24
28272.50	K5BTV	Cumming, GA	EM74	0.25	HF6V Vert	Omni	A1	24
28273.00	AC4DJ	Eustis, FL	EL98EU	20	Ringo	Omni	A1	24
28273.00	WF4HAM	Altamont Spr, FL		10	Ringo		A1	?
28273.00	DB0BER	Berlin	JO62QL	5	Dipole	Omni	A1	24
28274.00	LW1DZ	Escobar, BA	GF05OQ	10	Loop		A1	24
28274.70	N0UD	Halliday, ND	DN87SH				A1	24
28275.00	PY2EMG	Jacarel, SP	GG76AQ				A1	?
28275.00	NP2SH	St John, VI	FK78OI				A1	?
28275.00	KG4GVV	Summerville, SC	EM93		Vertical	Omni	A1	24
28276.00	K4UKB	Danville, KY	EM77NP	10	5/8 Vert	Omni	A1	SYNCHx
28276.00	K4FUM	Stone Mountain, GA	EM73WU				A1	SYNCHx
28276.50	XE2YBG	Victoria Tamaulipas	EL03				A1	?
28277.00	WB7RBN	Pasco, WA	DN06IG				A1	24
28277.00	WI4L	Dalton, GA	EM74MS				A1	24
28277.00	WD8AQS	Fremont, MI	EN73AL				A1	24
28277.30	KD4MZM	Sarasota, FL	EL87RG	3	Ringo@15□	Omni	A1	24
28277.60	DM0AAB	near Kiel	JO54GH	12	GP	Omni	F1	24
28278.00	WA4OTD	Carmel, IN	EM69	5	Indoor	Dip	A1	24
28278.20	KE4IFI	Lexington, SC					A1	24
28278.50	WA6MHZ	Crest, CA			Ringo@20□	Omni	A1	24
28279.00	DB0UM	Schwedt	JO73CE	2	SlopeDipV	Omni	A1	24
28280.00	KA3NXN	Charlottesville	FM08SA				A1	24
28280.00	K5AB	Goldthwaite, TX	EM01	20	5/8 GP @ 45'	Omni	A1	24
28280.00	PU5AAD	Nova Brasilia	GG51PS	10			A1	24
28280.00	N6SPP	Concord, CA	CM97	10	Vert Dipole	Omni	A1	24
28280.50	WB6FYR	Quartz Hill, CA	DM04VP	10	5/8 Vert	Omni	A1	24
28281.00	W8EH	Middletown, OH	EM79TL	7.5	Vert@40□	Omni	A1	
28281.20	IK6ZEW	Pescara	JN72OK				A1	?
28281.50	W4HEW	Millegeville GA	EM94LX				A1	24
28282.00	LA6TEN	Kirkenes	KP49XQ	10/1/,1	Omni		A1	OP?
28282.00	HP1ATM	Santiago	EJ98				A1	24
28282.00	XE2ES	Mexicali	DM22RP				A1	24
28282.00	N2IFC	Allamuchy					A1	?
28282.60	OK0EG	Hradec Kralove	JO70WE	10	GP	Omni	F1	24
28282.80	W0ERE	Fordlan, MO	EM36	5	Vertical	Omni	A1	24
28283.00	K7YSP	Gainsville, GA	EM84AH				A1	24
28283.60	KC9GNK	Madison, WI	EN53	4	Inv Vee		A1	24
28283.80	W5OM		DN28	98	3-el		A1	IRREG
28284.00	K2XG	Monticello, KY	EM76NU	5	V,Dip@25□	Omni	A1	24
28284.80	WD8AQS	Fremont, MI	EN73AL	5			A1	24
28284.80	WA3IIA	Bloomsberg, PA	FN11TA				A1	24
28284.50	KB9NK	Hudsonville, MI	EN72BU				A1	24?
28285.00	VP8ADE	Adelaide Is	FC52WK	10	1/4 Vert	Omni	A1	24
28285.00	W7IEW	Olympia, WA	CN87MC				A1	24?
28285.00	KD0GZJ	Loveland, CO	DN70KJ	1	Vertical		A1	24
28285.00	PT9BCN	Campo Grande, MS	GG29RN	1	Vertical	Omni	A1	24
28285.30	K5DRG	Lago Vista, TX	EM10CM			A1	?	
28285.80	WA4ROX	Largo, FL	EL87	0.75		A1	24	
28285.80	W0ILO	Fargo, ND	EN16					A1
28286.00	WI6J	Bakersfield, CA	DM05JJ	5	Vertical	Omni	A1	24
28286.00	N5AQM	Chandler, AZ	DM43AH	2	Vertical	Omni	A1	24
28286.00	N2PD	Middletown, NY	FN21	5			A1	?
28286.70	K3XR	Sinking Spring PA	FN10XH				A1	?
28287.00	GB3XMB	near Waddington	IO83SV	10	V,Dip@800□	Omni	F1A	24
28287.00	W6WTG	Bakersfield, CA	DM05MJ				A1	24
28287.30	W2SDX	Buffalo, NY	FM18LV				A1	
28287.50	N8PUM	Ishpeming, MI		0.5	Loop		A1	24
28288.00	WA7LNW	Harmony Mesa, UT	DM37	5	F Wave Loop	EU/PAC	A1	24
28288.00	K4LJP	W Palm Beach FL	EL96	5	AR99@30□	Omni	A1	24

Freq	Call	Nearest Town	Locator	ERPw	Antenna	Direction	Mode	Status
28288.00	RA3ATX		KO85NX					OP?
28288.50	ND3E	Newcastle, DE					A1	24
28289.00	KB9WA	Egg Harbour, WI					A1	
28289.00	WB5BXZ	Hattiesburg, MS					A1	?
28289.00	WJ5O	near Columbus, AL	EM72NE	2	Yagi	NE	A1	24
28289.50	N1KXR	Medway, MA	FN32	5			A1	24
28289.80	W0ERE	Highlandsville MO	EM36HX	Varies			A1	24
28289.00	PS8RF	Teresina, PI	GI84OW				A1	24
28290.00	N6UN	San Diego, CA	DM12	5			A1	24
28290.30	WB4WOR	Randleman, NC	FM06BT	3	Vertical	Omni	A1	24
28291.00	K5TLJ	Trumann, AR	EM45RQ	20			A1	?
28291.00	W6NIF	Fresno, CA			GP @ 25 Ft	Omni	A1	?
28291.80	N5MAV	Midland, TX					A1	
28292.00	VA3VA	Windsor, ON	EN82	5	Horiz Dipole		A1	24
28292.30	NH6HI	Kaleheo, HI		2			A1	24
28292.50	KM4GS	Kentucky Lake KY	EM56	0.5	Vertical	Omni	A1	24
28292.50	SK0CT	Sollentuna	JO89XK	10	GP	Omni	A1	24
28292.80	K7GFH	Damascus, OR	CN85SJ	3	Attic Dipole		A1	24
28293.40	ND4Z	Gilbert, SC	EM94JA	5	5/8@40▯	Omni	A1	24
28293.70	W4DJD	Woodbridge, VA	FM18HP				A1	24
28294.00	K7RON	Peoria, AZ	DM33				A1	24
28294.00	KE4IAP	Woodbridge, VA	FM18HP				A1	24
28295.00	KD1ZX	Central Falls RI	FN41HV	4	Vert@10▯	Omni	A1	24
28295.00	PU5ATX	Santa Catarina	GG51PR		Dipole		A1	06 to 23:50
28295.10	SK2TEN	Kristineberg	KP08FC	5	Vertical	Omni	A1	OP?
28295.40	K1SPD	La Vergne, TN	EM65RX			A1	24	
28295.50	K4IT	Flatwoods, KY	EM88PM	3.5		Dipole	A1	24
28295.50	IZ0CWW	Cervaro, FR	JN61VL	3				24
28295.70	W9MUP	WI	EN52				A1	?
28295.80	W3APL	Laurel, MD	FM19NE	10	Hor,Dipole	NE/SW	A1	24
28296.00	KA7BGR	Central Point OR	CN82				A1	24
28296.20	W5JDG	Washington, TX					A1	24
28297.00	NS9RC	Northfield, IL	EN62CC	5	1/2V@30▯	Omni	A1	24
28297.80	WA3BM	Valencia, PA	FN00SQ				A1	24
28298.00	V73TEN	Roi Namur I	RJ39RJ		Horiz OA50	Omni		24?
28298.00	K5TLL	Hattiesburg, MS	EM51GG	5	Vertical	Omni	A1	24
28298.15	WZ8D	Blachester OH	EM89OO				A1	24
28298.10	SK7GH	Bor	JO77BF	5			A1	24
28298.10	K7PO	Tenopah AZ	DM32	0.5	Vertical	Omni	A1	24
28298.50	K7FL	Battle Ground, WA	CN85SS	4	Horiz	Loop	A1	24
28298.60	K4JDR	Raleigh, NC	FM05	10	Vertical	Omni	A1	24
28299.00	N1SCA	Palm Bay, FL	EL97QX				A1	24
28300.00	K6FRC	Sutters Mountain, CA	CACM99				A1	24##
28300.00	HL1ZLL	Seoul	PM37MM	85	1/4 Vert	Omni	A1	24
28300.00	IK6ZEW	Pescara	JN72CK	0.4	Vertical		A1	?
28300.00	KF4MS	Tallahassee, FL	EM70VM	10	A99@25▯	Omni	A1	24
28300.30	PU5ZAA	Novegentes	GG53QE	3			A1	24
28301.00	PI7ETE	Amersfoort	JO22QD	0.5	Vertical	Omni	F1A	24
28320.97	IZ8HUJ	Pignola PZ	JN70VN	0.4	Windom			?
28321.00	I1YRB	Torre Bert (TO)	JN35UB	0.2	Vertical	Omni	QRSS3	24
28321.20	IS0GOV	Cagliari	JM49NF	0.3	Vertical	Omni	QRSS3	24
28321.20	IK3ERY	Vittorio, Veneto	JN65DX	0.1	Vertical	OMNI	QRSS3	24
28321.47	IZ1GJH	Casarza Lig, GE	JN44RG	0.1	Vertical	Omni	QRSS3	24
28321.65	IN3KLQ	near Trento	JN56RG	0.3	Vertical	Omni	QRSS3	24
28321.70	IW4EMG	Ferrara	JN54RU	0.1			QRSS3	24
28321.80	I3GNQ	Tencarola, PD	JN55VJ	0.4	GP	Omni	QRSS3	24
28321.85	IK0IXI	Aprilia LT	JN62VB	0.1	Dipole@30m	NE/SW	A1/F1	24
28321.86	IZ8JFA	Cosenza	JM89CH	0.3		N/NE	QRSS3	24
28321.94	IW9BAJ	Sicily	JM77NO	1			QRSS3	24
28321.95	IW0HK	Civitavecchia	JN52WD	1/,1	1/2 Vert	Omni	QRSS3	24
28321.70	IZ1TAA	near la Spezia	JN54AC	0.1	Dipole	NW	QRSS3	24
28322.00	IQ3QR		JN55XR	0.5			QRSS3	24
28322.00	IZ7AVU	Brindisi	JN80XP					?
28322.20	IZ5ILK		JN63BN	0.05	5/8 Vert	Omni	?	?
28322.00	IZ0ANE	Cassino		0.1				?
28322.01	IW1QIF	Davagna-Genova	JN44LL	0.1	Long Wire		QRSS3	24
28322.04	IT9YAF	Canicatti	JM67WI	0.1	Vertical	Omni	QRSS3	24
28322.05	IK1HGI	Trecate	JN45IK	0.1	Dipole	NNW/SSE	QRSS3	24
28322.08	IS0GSR	near Cagliari	JM49IN	0.1	Dipole	N-S	QRSS3	24
28322.11	IK8YTN	Salerno	JN70LG	0.1	Int Vee	E-W	QRSS3	24
28322.18	IZ0HCC	Roma	JN61FT	0.1	Vertical	Omni	QRSS3	?
28322.20	IK8SUT	Salerno	JN70JQ	0.1	Inv Vee	E - W	QRSS3	24
28322.20	IW7DEC	near Bari	JN81GF	0.1	1/2 Vert	Omni	QRSS3	24
28322.32	IW9FRA	Trapani, Sicily	JM68HA	0.1	Dipole	N-S	QRSS3	24
28322.36	IW3SGT	Trieste	JN65VP				QRSS3	24
28322.45	IW4EMG	near Ferrara	JN54RW		Vertical	Omni	QRSS3	?
28322.62	IK1BPL	Novara	JN45HK	0.1	GP	NNW/SSE	QRSS3	24
28322.63	IK0VVE	near Latina	JN61KN	0.1	Dipole	NNW/SSE	QRSS3	24
28322.70	G3ZJG	near Leicester					QRSS3	PT
29010.00	DU1EV	Metro, Manila		100	Vert GP	Omni		

Notes:

? = Activity uncertain; **INT** = Intermittent; **PT**=Part-time; **IRREG** = Irregular; **UC** = Under Construction; (T) **Non Op** = (Temporarily) NotOperational; **OP?** = Operational?; **V** = Varies; **EXP** = Experimental; **WkEnd** = Weekends during daylight; **LT** = Local Time;

SYNCH = frequency sharing synchronised beacons: SYNCHn current sequence N4ESS (0,00 seconds), N4ES (0,20), WB4WOR (0,30), K7EK (0,40), N4ES (0,50). SYNCHx: K4UKB ID at 00 seconds and K4FUM at 30 seconds. ## K6FRC transmits on 28 245, 28 250, 28 300

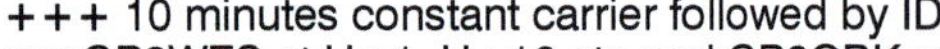
+++ 10 minutes constant carrier followed by ID

zzz GB3WES at H+1, H+16 etc and GB3ORK at H+2, H+17 etc. Transmissions have a stepped power sequence.

IARU Region 1 VHF/UHF and Microwave Beacons

This list of 50MHz Beacons is compiled for IARU Region 1 by G3USF. Thanks are also due to the VHF/UHF/Microwave Managers of radio societies across Region 1, beacon keepers, beacon co-ordinators and VHF/UHF DXers too numerous to mention.
Note that this list includes information regarding beacons notified to the R1 coordinator which operate (or are Planned to operate) outside of the IARU R1 band Planned beacon segments due to local licensing requirements.
Some member Societies may be unable to comply for local regulatory reasons. As a general principle, all beacons should move to band Planned frequencies as soon as possible.

Freq	Call	Nearest Town	Locator	ERPw	Antenna	Direction	H/V	Mode	Keeper
6M									
50000.000	YM1SIX	Tekirdag	KN30QT	16	Halo	omni	H	A1A	TA2NC
50000.000	YM2SIX	Ankara	KM69IV	16	Halo	omni	H	A1A	TA2NC
50000.000	YM3SIX	Izmir	KM38OJ	16	Halo	omni	H	A1A	TA2NC
50000.000	YM5SIX	Adana	KM76QX	16	Halo	omni	H	A1A	TA2NC
50000.000	YM6SIX	Merzifon	KN70QV	16	Halo	omni	H	A1A	TA2NC
50000.000	YM8SIX	Gaziantep	KM87QB	16	Halo	omni	H	A1A	TA2NC
50004.000	EI0SIX	Kilmolin, Enniskerry, Co. Wicklow	IO63VE	50	Ho Loop	omni	H	F1	EI7BIB
50005.000	PI7SIX	Rotterdam	JO21VF68	17	Crossed Dipole	Omni	H	F1A	PA2M
50006.000	IW9GDC	Messina	JM78SD	10	Big Weel	Omni	H	A1A	IW9GDC
50006.000	IW9GDC/B	Messina	JM78SD	10	Big Weel	Omni	H	A1A	IW9GDC
50012.000	OX3SIX	Tasiilaq city	HP15EO	25	Dipole		H	F1A	OZ1DJJ
50016.000	SV5SIX	Profitis Ilias Mt. / Rhodes	KM36XG	3	3el yagi	dir 330	H	A1A	SZ5RDS
50017.000	OH0SIX	Stålsby	JP90XI	3	Dipole	Omni	V	A1A	OH0JFB
50023.000	LX0SIX	Burscheid	JN39AV	10	Loop	omni	H	A1A	RL
50025.000	OH2SIX	Lohja	KP20DH	50	1/2 Vertical	Omni	V	A1A	OH2LNM
50039.000	FY7THF	Guyane	GJ35IG	10	Verticale	omni	V	A1A	FY1FL
50045.000	OX3VHF	Prince Christiansund	GP80KB	25	Dipole		H	F1A	OZ1DJJ
50046.000	SV9GPV	Rethimno / Crete	KM25EH	5	2el yagi	dir 330	H	F1A	SV9GPV
50047.000	OX6M	North East Greenland	HQ90AL	25	Dipole	N/S	H	F1A	OZ1DJJ
50052.000	SK2CP/B	Kiruna / Esrange Space Center	KP07MU	30		Omni	H	A1A	
50058.000	OE3XLB	Hohe Wand, Kleine Kanzel	JN78SB	10	Straight Dipole	N-S	H	A1A	OE3GWC
50060.000	SK6QW/B	Mariestad	JO68WR	8		Omni	H	A1A	
50066.000	OE3XAC	Kaiserkogel	JN78SB	10	M2 HoLoop	OMNI	H	A1A	OE3KLU
50067.000	OH9SIX	Rovaniemi Pirttikoski	KP36OI	35	2x X-Dipole	Omni	H	A1A	OH9LA
50069.000	FM1ZAC	Martinique	FK94NL	15	Turnstile	omni		A1A	FM1HM
50080.000	FK8SIX	Noumea	RG37GT	10	Verticale	omni	V	F1A	FK8HA
50400.500	IW3FZQ/B	Monselice PD	JN55VF	8	5/8 Vert.	Omni	V	A1A	IW3FZQ
50401.000	ON0VRN	DIKSMUIDE	JO11JA	5		omni	?	F1A	ON8SD
50401.000	IQ8KK/B	Cosenza	JM89DH	2	Halo	Omni	H	A1A	U
50402.000	OY6BEC	Sornfelli	IP62MB	15	Quad Loop	omni	H	F1A	OY9JD
50403.000	YM4SIX	Tepe, Alanya	KM66AO02	40	2 stacked triangle dipoles	omni	H	F1A	PE2M
50405.000	OE7XBI	Ranggerköpfl/Tirol	JN57OF	10	Groundplane	Omni	V	F1A	OE7NCI
50406.000	SV2JAO	Vermion Mt. / Imathia	KN10CL	10	3el yagi	dir 310	H	F1A	SV2JAO
50406.000	F5ZHQ	Le Pradet	JN33BC	10	Verticale	omni	V		F6FCE
50408.000	IK5ZUL/B	Poggio Spada GR	JN52JW	3	Dipole	90/270°	H	A1A	IK5ZUL
50412.000	OH1SIX	Ikaalinen	KP11QU	50	2x X-Dipole	Omni	H	A1A	
50414.000	OK0SIX	Ostrava Poruba	JN99CT	5	Groundplane	Omni	V	A1A	OK2WA
50415.000	SR5TDM	Tarczyn	KO01KX01	3.0 W PEP	cross dipole	OMNI	H	A1A	SP5XMU
50418.000	F1ZFE	Erching	JN39OC	8	Boucle	omni			?
50419.000	IZ1EPM/B	Saronsella TO	JN35WD	10	GH50 5/8	Omni	V	A1A	IZ1EPM
50420.000	IQ4FE/B	Fidenza PR	JN54AS	6	AR-6	Omni		A1A	IK4CIE
50422.000	S55ZRS	Kum	JN76MC	5	GP	omni	V	A1A	ZRS
50422.000	F5ZMT		IN88GN	10	Halo	omni	H	A1A	F5NLG
50425.000	PI7SIX	Rotterdam	JO21VF68	17	Crossed Dipole	Omni	H	F1A	PA2M
50427.000	IW0DTK/B	Latina	JN61TG	3	5/8 Vert.	Omni	V	A1A	IW0DTK
50430.000	IZ3OHR/B	Verona	JN55LJ	5	Dipole h.	Omni	H	A1A	IZ3OHR
50431.000	9A0BLH	Moslavacka gora	JN85JO	3.2	X dipole	omni	H	A1A	9A5R
50432.000	F5ZKY	Mt Alouettes	IN96LV	10	Verticale	omni	V	A1A	F6DBA
50433.000	OH7SIX		KP52			Omni	H	A1A	OH7GGX
50434.000	F8BHU	Nevers	JN17NA	1	GP	omni	V		F8BHU
50436.000	D4C	Monteverde, Mindelo	HK76MU79VB	30	5/8 Vertikal	OMNI	V	A1A	D4C (HB9DUR)
50437.000	ES0SIX	Vilsandi/Isl.Saaremaa	KO08WJ32MG	15.84	folded dipole	240/30	H	A1A	ES2PW
50438.000	OE5XHE	Sternstein/Mkr.	JN78DN	10	OA 50	OMNI	H	A1A	OE5ANL
50441.000	ON0SIX	VIEUX GENAPPE	JO20EP	50		omni	H	F1A	ON4KJV
50445.000	JW5SIX	Hopen	KQ26MM	10	Dipole	Omni	V	A1A	LA5RIA
50447.000	JW7SIX	Isfjord Radio	JQ68TB	30	3el Yagi	180	H	A1A	LA0BY
50448.000	FX4SIX	Neuville	JN06CQ	30	Turnstile	omni		A1A	?
50449.000	IQ0AM/B	M.te Serpeddi CA	JM49PI	5	GP	Omni	V	A1A	IS0BSR
50450.000	SV3BSF	Patras	KM08VA	5	2el yagi	dir 320	H	F1A	SV3BSF
50451.000	LA7SIX	Målselv	JP99EC	100	4el Yagi	190	H	A1A	LA5TFA

Freq	Call	Nearest Town	Locator	ERPw	Antenna	Direction	H/V	Mode	Keeper
50455.000	LA8SIX	Hvasser	JO59FB	50	Dipole	Dipole 0° / 180°	H	A1A	LA6LCA
50455.000	IQ0HV/B	Cecchina Roma	JN61HQ	8	Horned	NW-SE	H	A1A	IK0BDO
50456.000	YM0SIX	Bozcaada	KM39AT28TD	5	Halo	Omni	H	A1A	TA0G
50457.000	TF1VHF	Myrar, 80Km NNW Reykjavik	HP84WL	25	1/2 Dipole	Dipole NW / SO	H	F1A	TF1A
50458.000	IQ4AD/B	Parma	JN54DT	8	5/8 Vert.	Omni	V	A1A	I4YMB
50459.000	LA9SIX	Gjøvik	JP50EV	25	5/8 GP	omni	V	A1A	LB3RE
50468.000	SK3SIX	Östersund	JP73HC17RO	15	Crossed Dipole	Omni	H	F1A	SM3PXO
50471.000	OZ7IGY	Jystrup	JO55WM54UA	25	Big Wheel	omni	H	F1A	OZ7IS
50474.000	IQ5MS/B	Massa MS	JN54AA	3	Vert.	Omni	V	A1A	IW5ECP
50475.000	JW9SIX	Bjørnøya	JQ94LM	10	dipole	Omni	V	A1A	LA5RIA
50477.000	IZ3GWJ/B	Lendinara RO	JN55TC	5	J-pole ¼ wl	Omni	V	A1A	IZ3GWJ
50479.000	JX7SIX	Jan Mayen	IQ50RX	10	Dipole	150/330	H	A1A	LA7DFA
50481.000	F1ZFB	Ste Lheurine	IN95TM	10	dipole horiz.	Dipole 0° / 180°	H	A1A	F1MMR
50483.000	DB0DUB	nr Gangelt	JO31AA	1	Halo	omni	H	A1A	DL1KBL
50483.000	DB0ANN	Moritzberg	JN59PL	2	Ho-Loop	OMNI	H	A1A	DL8ZX
50485.000	SV1SIX	Athens	KM17UX	30	5/8 vertical	omni	V	F1A	SZ1SV
50485.000	OH5SHF	Kouvola	KP30HV	100	2 dBd	180	H	F1	OH5IY
50485.000	ED7YAN	Grenada	IM87GC	5	Loop	Omni	H	F1	EA7UU
50488.000	LA2SIX		JP53EG	25			V	A1A	
50488.000	YM7SIX	Ordu	KN90AW	16	Halo	omni	H	A1A	TA2NC
50490.000	F6IKY	Louhans	JN26OP	3	commut H/V	omni	H/V		F6IKY
50493.000	YM9SIX	Kars	LN10PM	16	Halo	omni	H	A1A	TA2NC
50495.000	SV9SIX	Vasiliko Mt. / Crete	KM25NH	25	Halo	omni	H	F1A	SZ1SV
50495.000	V53SIX	Swakopmund, Namibia	JG77II	H 20.0\V 10.0"	Ringo	omni		F1A	V51DM
60013.000	EI1KNH	Kilmolin, Enniskerry, Co. Wicklow	IO63VE	25	vertikal folded dipole	omni	V	F1A	EI4GNB
4M									
70008.000	OH1FOUR								
70012.000	OX4MB	Kangerlussuaq	GP47TA	25	Dipole		H	F1A	OZ1DJJ
70014.000	S55ZRS	Kum	JN76MC	5	ho-loop	omni	H	A1A	ZRS
70016.000	SV5FOUR	Profitis Ilias Mt. / Rhodes	KM36XG	5	3el yagi	dir 330	H	F1A	SZ5RDS
70018.000	OH2FOUR	Lohja	KP20DH	30	Dipole	Omni	V	A1A	OH2LNM
70021.000	OZ7IGY	Jystrup	JO55WM54UA	25	Big Wheel	omni	H	F1A	OZ7IS
70029.000	S55ZMB	Starše	JN76VK	5	4	omni	H	A1A	S51DI
70031.000	9A0BFH	Moslavacka gora	JN85JO	4	X dipole	omni	H	A1A	9A5R
70033.000	OH5RBG	Kouvola	KP30HW	15	7 dBi	232	H	F1	OH5IY
70035.000	OY6BEC	Sornfelli	IP62MB	20	Quad Loop	omni	H	F1A	OY9JD
70037.000	ES1VHF	Tallinn	KO39IK50RK	10	dipole	240°/30°	H	F1A	ES1NI
70040.000	SV1FOUR	Athens	KM17UX	15	5/8 vertical	omni	V	F1A	SZ1SV
70047.000	OX4M	North East Greenland	HQ90AL	25	Dipole	N/S	H	F1A	OZ1DJJ
70055.000	SV9FOUR	Rethimno / Crete	KM25EH	5	3el yagi	dir 330	H	F1A	SV9GPV
70057.000	TF1VHF	Myrar, 80Km NNW Reykjavik	HP84WL	25	1/2 Dipole	Dipole NW / SO	V	F1A	TF1A
70063.000	LA2VHF	Vassfjellet	JP53EG	25	GP	omni	V	A1A	LA1K
70065.000	LA5VHF	Kristiansand	JO48AD	15	GP	omni	V	A1A	LA4YGA
70067.000	LA9VHF	Gjøvik	JP50EV	25	5/8 GP	omni	V	A1A	LB3RE
70070.000	PI7RAZ	Zoetermeer	JO22FB	10			H		
70070.000	PI7RTD	Rotterdam	JO21FV68	10					PE1GHG
70075.000	LA4VHF	Egersund	JO28WL	50	Dipole	N/S	H	A1A	LA3EQ
70081.000	LA7VHF	Målselv	JP99EC	40	4-ele Yagi	190	H	A1A	LA5TFA
70087.000	SV2JAO	Vermion Mt. / Imathia	KN10CL	10	4el yagi	dir 310	H	F1A	SV2JAO
70114.000	ON0HVL	Heuvelland	JO10KS	10	Big Wheel 2 dBi	omni	H	F1A	ON7FLY
70115.000	SR5TDM	Tarczyn	KO01KX01	3.0 W PEP	cross dipole	OMNI	H	A1A	SP5XMU
70130.000	EI4RF	Kilmolin, Enniskerry, Co. Wicklow	IO63VE	30	2 Yagi System	Yagi E & S	H	F1	EI4GNB
70161.000	LX0FOUR	Burscheid	JN39AV	10	Loop	omni	H	A1A	RL
70242.000	ON0RCL	Leuven	JO20IV	5		omni	H	F1A	ON5CMB
70244.000	ON0BR	Montigny-Le-Tilleul	JO20EI				H	F1A	ON4UQ
2M									
144150.000	UR0UBA	Kyiv	KO50FK	5	Crossed Dipole	OmniH	H	F1A	UT4UWJ
144401.000	SV2HQL	Katakali / Grevena	KM09UV	10	Yagi	120° 315°	H	F1A	SV2HQL
144402.000	OY6BEC	Sornfelli	IP62MB	100	2el. HB9CV	SE	H	F1A	OY9JD
144403.000	YM4VHF	Tepe, Alanya	KM66AO02	400	2 x 4x3ele Yagi	30° 184° / 344°	H	F1A	PE2M
144404.000	DB0THE	Ohmgebirge - Birkenberg	JO51EL		4 x 2 Element	OMNI	H	"A1A	DF7AP
144404.000	IW1AU/B	Giaveno TO	JN35PA	5	Big wheel	Omni	H	A1A	IW1AU
144405.000	F5ZRB	Quistinic	IN87KW	40	Yagi 7 elts	Yagi 7 elts 135°	H	F1A	F6ETI
144406.000	SK6VHF	Tjörn Island	JO57TX	10	M2 loop	Omni	H	A1A	SM6CEN
144407.000	GB3SSS	Helston	IO70IA	200	8 OVER 8 SLOT YAGI	Yagi 300°	H		G7THT
144409.000	F5ZSF	Lannion	IN88GS	5	9 elts	9 elts 60°	H	A1A	F6DBI
144410.000	DB0MFI	Hainsfarth	JN58HW	2	Big Wheel	- OMNI	H	A1A	DK1MFI
144410.000	IK1FJI/B	Sant'Olcese GE	JN44LL	6	eggbeater	Omni	H	A1A	IK1FJI
144412.000	SK4MPI	Borlänge	JP70NJ	200	2 X 5 EL YAGIS	NV+NO	H	A1A	SM4HFI
144412.500	SZ1SV	Parnis Mt. / Athens	KM18UE	0.5	1/4 vertical-indoor	omni	V	F1A	SZ1SV
144415.000	DB0JW	Hellenthal	JO30EK	10	2 x Big Wheel	OMNI	H	A1A	DL3KAT
144415.000	SR5TDM	Tarczyn	KO01KX01	5.0 W PEP	cross dipole	OMNI	H	A1A	SP5MX
144415.000	IQ2MI/B	M.te Rosa VC	JN35WW	1	Dipole Kathrein	110°	H	A1A	IZ2JGB
144416.000	PI7CIS	Den Haag	JO22DC	17			H	FSK	PA0C

Freq	Call	Nearest Town	Locator	ERPw	Antenna	Direction	H/V	Mode	Keeper
144417.000	OH9VHF	Rovaniemi Pirttikoski	KP36OI	160	7 dBd	200	H	A1A	OH9LA
144417.000	F5ZXT		JN33	5					F6FCE
144418.000	ON0VHF	Louvain-La-Neuve	JO20HP	25		omni	H	F1A	ON4IV
144419.000	IQ2CY/B	Cremona	JN55AD	10	Big wheel	Omni	H	A1A	IK2THZ
144420.000	RB1CA	Priozersk/Russia	KP51BA	5	3 Quad Style Yagi 3 ele	135°	H	A1A	UA1CCB
144421.000	SK3VHF	Östersund	JP73HC17RO	500	2 X 3el Quad	60° 180°	H	F1A	SA3AZK
144422.000	OH0	Åland							
144425.000	SR9VHK	Siemianowice	JO90MH01	3.0 W PEP	Dipole	120° / 300°	H	F1A	SP9BGS
144425.000	F5ZAM	Blaringhem	JO10EQ	10	Trefle	omni	H	A1A	F6BPB
144425.000	I5WBE/B	Mt. Fiore PT	JN53LT28SV	3	2 x Halo	Omni	H	A1A	I5WBE
144428.000	DB0JT	Wildberg	JN67JT	30	4 Dipole	OMNI	H		DJ8QP
144429.000	IQ3MF/B	Cormons GO	JN65RW	3	2 Turnstile	Omni	H	A1A	IV3HWT
144430.000	GB3VHF	Fairseat,Kent	JO01EH08	30 / 100	2 X 3 Ele. Yagi	Yagi 288° / 348°	H	F1A	G0FDZ
144431.000	PI7BRG	Zevenbergen	JO21HP	16			H	FSK	PA3FSY
144431.000	9A0BVH	Moslavacka gora	JN85JO	3.8	V dipole	omni	H	F1A	9A5R
144432.000	GB3SEV	Stourport-on-Severn	IO82UI	10	Big Wheel	Omni	H	F1A	M0VXX
144433.000	OH7VHF	Tuupovaara / Joensuu	KP52HL	50	2x BigWheel	Omni	H	A1A /F1B	OH7DI
144433.000	IZ3DVW/B	Monselice PD	JN55VF	10	3 el. Yagi	30°	H	A1A	IZ3DVW
144434.000	DB0LBV	Leibzig	JO61EH	0.4	2 Dipole	OMNI	H		DC6WM
144435.000	IQ5MS/B	Massa	JN54AB	3	Big wheel	Omni	H	A1A	IW5ECP
144436.000	D4C	Monteverde, Mindelo	HK76MU79VB	50	3 X stacked dipoles	OMNI	H	A1A	D4C (HB9DUR)
144437.000	F1ZXK	Montigny	JN18AS	30/10/3/1.5	Trefle	omni	H	A1A	F4BUC
144438.000	SR8ZBW	Jasło	KN09RR67	5.0 W PEP	Crossed Dipole	OMNI	H	A1A	SP8WJW
144440.000	DB0OHZ	Osterholz/Scharmbeck	JO43JF	7.9	M² HOLOOP	OMNI	H	A1A	DL1BEB
144441.000	LA4VHF	Egersund	JO28WL	3	2 x 3el Yagi	40/220	H	A1A	LA1YCA
144442.000	IK4PNJ/B	Pianoro BO	JN54QK	10	2x Dipole	Omni	H	A1A	IK4PNJ
144443.000	OH2VHF	Inkoo	KP20BB	300/300	9/9dBi	230/20	H	A1A	OH2LNM
144444.000	DB0FGB	Schneeberg	JO50WB	15	4 x 3el. Yagi	OMNI	H	A1A	DG4NBI
144444.000	IQ5LU/B	Lucca	JN53GW	6	Big wheel	Omni	H	A1A	IK5AMB
144447.000	SK1VHF	Klintehamn	JO97CJ	10	BIG WHEEL	Omni	H	A1A	
144450.000	DM0HVL		JO62KI	7.8	Winkeldipol	OMNI	H		DL7AIG
144450.000	F5ZVJ	Le Pin	JN24GB	10	2 Trefles	omni	H	F1A	F5IHN
144451.000	LA7VHF	Tromsø	JP99KQ	1000	15el Yagi	185	H	A1A	LA5TFA
144451.000	LZ0STZ	YASTREBOVO	KN22UG	2	4 X HB9CH	omni	H	F1A	LZ1RT
144452.000	IQ7FG/B	San Severo FG	JN71TQ	10	Dipolo	Omni	H	A1A	IK7UXW
144453.000	GB3ANG	Dundee Scotland	IO86MN	20	4 ele Yagi	Yagi 160°	H	A1A	GM4ZUK
144455.000	DB0MMO	Breitsol	JN49RV	7.5	M² HoLoop	OMNI	H	F1A	DH4FAJ
144455.000	OE3XAC	Kaiserkogel	JN78SB	10	folded dipole	Dipole 270°	H	A1A	OE3KLU
144455.000	OH5ADB	Hamina	KP30NN	0,1	Dipole	NW/SE	H	A1A	
144455.000	F5ZXV	Nancy	JN38CO	2,5	Halo	omni	H	A1A	F5OOM
144456.000	OM0MVC	Skalka	JN98LR	5	MOXON	270	H	A1	OM7AC
144457.000	SK2VHF	Vindeln / Buberget	JP94TF		10 el Que-Deedir SSW	N+SV	H	A1A	
144457.000	IQ0HV/B	Cecchina Roma	JN61HQ	3	Big weel	Omni	H	A1A	IK0BDO
144458.000	F1ZAT	Brive	JN05VE	3	Trefle	omni	H	A1A	F1HSU
144461.000	SK7VHF	Sjöbo	JO65UQ	10	HALO	Omni	H	A1A	SM7DTT
144461.000	IW0FFK/B	Ostia RM	JN61DS	3	2 x Hentenna	Omni	H	A1A	IW0FFK
144463.000	LA2VHF	Vassfjellet	JP53EG	500	10el Yagi	15	H	A1A	LA1K
144464.000	F1ZDU	Pierre St Martin	IN92OX	1/0.25	Dipole	Dipole 0° / 180°	H		F6ENL
144465.000	DB0ANN	Moritzberg	JN59PL	0.3	Ho-Loop	OMNI	H	A1A	DL8ZX
144468.000	LA6VHF	Kirkenes	KP59AL	300	9el Yagi	210	H	A1A	LA4OO
144468.000	F1ZAW	Vercel	JN37EE	10	Trefle	omni	H	A1A	F4CQG
144469.000	GB3MCB	St. Austell, Cornwall	IO70OJ	8	3 ele Yagi	Yagi 45°	H	F1A	G7KFQ
144469.000	IQ9UI/B	M.te Lauro RG	JM77IA	10	2*4 el.dk7zb	355°	H	A1A	IT9CJC
144470.000	OH2VHH	Vantaa	KP20MH	1	Dipole	0/180	H	A1A	OH2FTJ
144470.000	ED7YAN	Grenada	IM87GC	5	Loop	Omni	H	A1A	EA7UU
144471.000	OZ7IGY	Jystrup	JO55WM54UA	50	2xBig Wheel	omni	H	F1A	OZ7IS
144471.000	IZ8EDE/B	Monte Pierfaone PZ	JN70VM	2	Halo	Omni	H	A1A	IZ8EDE
144473.000	YM0VHF	Bozcaada	KM39AT28TD	5	Halo	Omni	H	A1A	TA0G
144475.000	DB0HRF	Großer Feldberg Ts.	JO40FF	15	Turnstyle	OMNI	H	F1A	DK7FU
144475.000	OM0MVA	Bratislava	JN88NE	0,4	Dipole	225/45	H	A1	OM3ID
144475.000	IZ8GMP/B	Gambarie RC	JM78WE	10	2el.	310°	H	A1A	IZ8GMP
144476.000	F5ZAL	Pic Neulos	JN12LL	10	Halo	omni	H	A1A	F6HTJ
144477.000	IQ9GD/B	Monte Bonifato TP	JM67LX				H	A1A	IT9CIT
144478.000	LA3VHF	Mandal	JO38SA	120	9el Yagi	180	H	A1A	LA3PRA
144478.000	S55ZRS	Kum	JN76MC	1	X Dipol	omni	H	A1A	ZRS
144478.000	9A0BVS	Goli	JN75BA	4	Dipole	dipole N / S	H	A1A	9A4QV
144479.000	OE3XTR	Hohe Wand	JN87AT	10	2el. Quad	120° 90°	H	A1A	OE3KLU
144480.000	LA8VHF	Stavern	JO48XX	100	3 x 2el Yagi	150	H	A1A	LA6LCA
144480.000	IZ8IBC/B	Nocera Inferiore SA	JN70HR	2	Halo	Omni	H	A1A	IZ8IBC
144481.000	SR3VHX	Pozna□	JO82KL80	6.0 W PEP	4 X 2el Yagi	OMNI	H	A1A	SO3Z
144482.000	GB3NGI	Slieve Anorra	IO65VB27	20 / 125	2 x 4 ele. Yagi	Yagi 45° / 135°	H	A1A	GI6ATZ
144485.000	TK5ZMK	Coti Chiavari	JN41JS	5	Trefle	omni	H	A1A	TK5EP
144486.000	DM0PR		JO44JH	5	2 X 6el Yagi	N-S	H		DJ8ES
144486.000	SV3BSF	Patras	KM08VA	5	3el yagi	dir 320	H	F1A	SV3BSF

Freq	Call	Nearest Town	Locator	ERPw	Antenna	Direction	H/V	Mode	Keeper
144487.000	GB3WGI	Derrygonnelly NI	IO64BL	85 / 1000	2 x 6 ele. Yagi	Yagi 280°	H	F1A	GI6ATZ
144489.000	IT9GRR/B	Caltagirone CT	JM77GF	2	Halo	Omni	H	A1A	IT9GRR
144490.500	F6ABJ	P. de Beaurepaire	JN25NJ	50	2 X Trefle	omni	H	F1A	F6ABJ
144491.000	DB0XIT		JN39MI	3	2 x Winkeldipol	OMNI	H		DK4XI
144492.000	ON0VRN	DIKSMUIDE	JO11JA	5		omni	H	WSPR	ON8SD
144492.000	DB0LY	Retzow	JO63PF		2 times stacked BIG Wheel	Omni	H	WSPR	DL9SAU
144492.000	DM0ADA	Wöbbelin	JO53SJ04					WSPR	DM1AD
144847.000	SV5VHF	Profitis Ilias Mt. / Rhodes	KM36XG	1	5el yagi	dir 330	H	A1A	SZ5RDS
340888.000	OM0MSA	Bratislava	JN88NE	0,25	Slot	225/45	H	A1	OM3ID
70cm									
432400.000	OE3XAC	Kaiserkogel	JN78SB	10	7-El Yagi	53° 360°	H	A1A	OE3KLU
432401.000	SK2UHF	Buberget, Vindeln	JP94WG		2 x 5 el Que-Dee	S	H	A1A	
432401.000	F5ZBU		JN18KF	5	4 x 9 elts	omni	H		F2AI
432402.000	OY6BEC	Sornfelli	IP62MB	250	4xdipole	SE	H	F1A	OY9JD
432404.000	DB0THE	Ohmgebirge - Birkenberg	JO51EL	15	4 x 3 Element	OMNI	H	"A1A	DF7AP
432404.000	F5ZZI	Hyeres	JN33BD	5	Trefle	omni	H		F6FCE
432405.000	SK1UHF	Klintehamn	JO97CJ	30	ALFORD SLOT	Omni	H	A1A	
432405.000	ED3YBF	Sant Fost De Campsentelles	JN11CL	2.5	Big Wheel	omni	H	A1	EA3URC
432408.000	F5ZPH	Quistinic	IN87KW	15	4 elts	Yagi 135°	H	A1A	F6ETI
432410.000	DB0ZW		JN69AS	1	Schlitz	OMNI	H		DC9RK
432410.000	DB0JW	Hellenthal	JO30EK	10	4 x Doppelquad	OMNI	H	A1A	DL3KAT
432410.000	IW1AU/B	Giaveno TO	JN35PA	3	Big wheel	Omni	H	A1A	IW1AU
432412.000	DB0JG		JO31HS	1	Clover Leaf	OMNI	H		DL3QP
432412.000	SK6UHF	Varberg / Veddige	JO67EH	10	5 el yagi	210 deg	H	A1A	SM6ESG
432413.000	F5ZTX	Lacapelle	JN14EB	10/5/2.5	2 x 3 elts	Yagi 45° / 90°	H		F5AXP
432415.000	SR5TDM	Tarczyn	KO01KX01	5.0 W PEP	cross dipole	OMNI	H	A1A	SP5MX
432416.000	PI7CIS	Den Haag / Scheveningen	JO22DC	17.6			H	A1A	PA0C
432417.000	OH9UHF	Rovaniemi Pirttikoski	KP36OI	70	9 dBd	200	H	A1A	OH9LA
432418.000	F1ZQT	Moragne	IN95OX	1	Trefle	omni	H		F1MMR
432420.000	F5ZAS	Cerdagne	JN12BL	10	Trefle	omni	H	A1A	F6HTJ
432425.000	DB0MMO	Breitsol	JN49RV	10	M² HoLoop	OMNI	H	F1A	DH4FAJ
432430.000	DB0MFI	Hainsfarth	JN58HW	1	Big Wheel	- OMNI	H	A1A	DK1MFI
432430.000	DB0OHZ	Osterholz/Scharmbeck	JO43JF	4.5	Big Wheel	OMNI	H	A1A	DL1BEB
432432.000	D4C	Monteverde, Mindelo	HK76MU79VB	50	9 ele. Yagi	40° 40°	H	A1A	D4C (HB9DUR)
432432.000	OH6UHF	Uusikaarlepyy / Nykarleby	KP13GM	7	3x BigWheel	Omni	H	A1A	OH6NVQ
432432.000	PI7HVN	Heerenveen	JO22WW	0.5	Omni		H		
432435.000	IQ5MS/B	Massa	JN54AB	3	Big wheel	Omni	H	A1A	IW5ECP
432436.000	F5ZAA	Nerignac	JN06IH	20	Trefle	omni	H	F1A	F5EAN
432440.000	F1ZTV	Cloutons	JN24WX	2	Boucle	omni	H	A1A	F4AII
432440.000	IQ3ZB/B	M.te Cesen TV	JN65AW			Omni	H	A1A	IK3HHG
432441.000	LA5UHF	Flekkefjord	JO38HH	200	6el Yagi	180	H	A1A	LA3EQ
432443.000	OH2UHF	Inkoo	KP20BB	250/250	10/10 dBi	230/20	H	A1A	OH2LNM
432444.000	DB0FGB	Schneeberg	JO50WB	15	4 x Doppelquad	OMNI	H	A1A	DG4NBI
432447.000	DB0IH		JN39HJ	15	HoLoop	OMNI	H	CW	DF1VW
432447.000	S55ZRS	Kum	JN76MC	1	Omni	omni	H	A1A	ZRS
432449.000	OZ1UHF		JO57GH					F1A	
432449.000	I5WBE/B	Mt. Serra PI	JN53GS60KF	3	2x Ho-loop	Omni	H	A1A	I5WBE
432450.000	DM0HVL		JO62KI	6.5	2 x MalteserKreuz	OMNI	H		DL7AIG
432451.000	LZ0STZ	Yastrebovo	KN22UG	4	Big Wheel	omni	H	F1A	LZ1RT
432454.000	F5ZZY	Nancy	JN38CO	3.5	Halo	omni	H		F5OOM
432455.000	OM0MUC	Skalka	JN98LR	7	MOXON	270	H	A1	OM7AC
432455.000	SK3UHF	Nordingrå / Rävsön	JP92FW	50		Omni	H	A1A	
432458.000	IQ0HV/B	Cecchina Roma	JN61HQ	3	2x Ho-loop	Omni	H	A1A	IK0BDO
432459.000	F5ZHG	.	JO10UH	25	Halo	omni	H		F5HMS
432460.000	DB0LB		JN48OV	1	V- Dipol	OMNI	H		DK3PS
432460.000	SK4BX/B	Garphyttan / Storstenshöjden	JO79LH	50	4 X LOG PER	NESW	H	A1A	
432463.000	LA2UHF	Vassfjellet	JP53EG	150	10el Yagi	180	H	A1A	LA1K
432465.000	DB0ANN	Moritzberg	JN59PL	1	Ho-Loop	OMNI	H	A1A	DL8ZX
432467.000	IZ3DVW/B	Monselice PD	JN55VF	3,5	Dipole	45/225°	H	A1A	IZ3DVW
432468.000	LA6UHF	Kirkenes	KP59AL	40	15el Yagi	210	H	A1A	LA4OO
432470.000	ED7YAN	Grenada	IM87GC	5	Loop	Omni	H	A1A	EA7UU
432471.000	OZ7IGY	Jystrup	JO55WM54UA	75	3xBig Wheel	omni	H	F1A	OZ7IS
432473.000	ON0HVL	Heuvelland	JO10KS	10	Dipole	Dipole 180°	H	F1A	ON7FLY
432475.000	ES0VHF	Panga/Isl.Saaremaa	KO18ND46MQ	39.8	2xDouble Diamond	omni	H	A1A	ES0OK
432475.000	DB0XY	Boxberg/Harz	JO51EU25	15		omni	H		DF4OL
432475.000	IZ8GMP/B	Gambarie RC	JM78WE	10	3 EL. YAGI	320°	H	A1A	IZ8GMP
432478.000	LA3UHF	Mandal	JO38SA	50	13el Yagi	180	H	A1A	LA3PRA
432480.000	LA8UHF	Hvasser	JO59FB	50	8el Yagi	90/180	H	A1A	LA6LCA
432485.000	LA4UHF	Haugesund	JO29PJ	15	3el Yagi	250	H	A1A	LA9RY/NQ
432488.000	DB0AD	Salzburg/Westerwald	JO40AQ	1	2 X Yagi + LogPer	omni	H	A1A	DM8MM
432812.500	SZ1SV	Parnis Mt. / Athens	KM18UE	0.5	1/4 vertical-indoor	omni	V	F1A	SZ1SV
432838.000	9A0BVU	Moslavacka gora	JN85JO	1.0	V dipole	omni	H	A1A	9A5R
432875.000	OH7UHF	Kuopio	KP32TV	15	6 dBd	225	H	A1A	OH7VM
432886.000	OK0EP	Praded / Bruntal	JO80OB	2 X 5.0	2 X 4 Ele. Yagi	40° / 40° 280° / 150°	H	F1A	OK1VPZ

Freq	Call	Nearest Town	Locator	ERPw	Antenna	Direction	H/V	Mode	Keeper
432888.000	OM0MUA	Bratislava	JN88NE	0,08	Dipole	225/45	H	A1	OM3ID
502945.000	ON0ODR	Oudenaarde	JO10TT	25		omni	?	F1A	ON4PN
23cm									
1269875.000	ON0VHF	Louvain-La-Neuve	JO20HP	10		omni	H	F1A	ON4IV
1269985.000	DB0JW	Hellenthal	JO30EK	10	Wimo PA-23R	Beam 35°	H	A1A	DL3KAT
1296000.000	ON0EME	Lille	JO21JG	200			LHCP	F1A	ON7UN
1296050.000	S55ZSE	Kokoš	JN65WP	0.3	Slot	omni	H	A1A	S53MV
1296090.000	S55ZMS	Dolina	JN86CR	0.3	Slot	omni	H	A1A	S53M
1296380.000	S55ZRS	Kum	JN76MC	0.2	Slot	omni	H	A1A	ZRS
1296456.000	IZ1ERR/B	Bagnolo CN	JN34OS60UJ	0,2	Quad	45°	H	A1A	IK1YWB
1296800.000	DB0HEG		JN59JA	0.5	4*Schlitz	OMNI	H		DL2QQ
1296800.000	OE3XAC	Kaiserkogel	JN78SB	10	WG Slot	OMNI	H	A1A	OE3KLU
1296800.000	SZ1SV	Hymettus Mt. / Athens	KM17VW	10	2X 5/8 vertical	omni	V	F1A	SZ1SV
1296800.000	SK6MHI	Hönö	JO57TQ	30	Alford Slot	Omni	H	A1A	SM6EAN
1296805.000	DB0RIG		JN48WQ	50	4fach Kasten	OMNI	H		DG9SQ
1296805.000	SK6UHI	Tjörn Island	JO57TX	30	Alford Slot	Omni	H	A1A	SM6CEN
1296810.000	DB0ZW		JN69AS	1	Schlitz	OMNI	H		DC9RK
1296812.000	F1ZBI	Petit Ballon	JN37NX	0.8	Quad	180°	H	F1A	?
1296815.000	DB0VI		JN39NK	0.2		OMNI	H		DL3CM
1296815.000	SR5TDM	Tarczyn	"KO01KX01	4.0 W PEP	Alfrod Slot	OMNI	H	A1A	SP5MX
1296816.000	F1ZTF	Segonzac	IN95VO	10	Big Wheel	omni	H	F1A	F1MMR
1296820.000	LA5SHF	Karmøy	JO29OE	10	3-ele-yagi	180	H	A1A	LA9RY
1296820.000	IQ0RM/B	Roma	JN61FW	1	Alford	Omni	H	A1A	IW0CZC
1296825.000	F5ZRS	Chamrousse	JN25WD	0.1	Corner		H	A1A	F5LGJ
1296835.000	SKØEN/B	Väddö	JO99JX	4	Alford Slot	Omni	H	A1A	
1296840.000	DB0FGB	Schneeberg	JO50WB	1	3xDoppelquad	OMNI	H	A1A	DG4NBI
1296840.000	OH6SHF	Uusikaarlepyy / Nykarleby	KP13GM	30	4x EIA	180	H	A1A	OH6NVQ
1296840.000	IK5CON/B	Camaiore LU	JN53CV	10	Alford	Omni	H	A1A	IK5CON
1296845.000	DB0LBV	Leipzig	JO61EH	2	4*Schlitz	OMNI	H		DC6WM
1296846.000	I5MDE/B	Montignoso FI	JN53KM94HB	5	2x 11 el. Yagi	30°- 290	H	A1A	I5BLH
1296847.000	F5ZBM	Nangis	JN18MN	10	Slot	omni	H	F1A	F6ACA
1296850.000	DM0UB		JO62KK	4.5	Schlitzstrahler	OMNI	H		DL7AIG
1296850.000	OH3SHF	Tampere	KP11TM	50	6 dBi	Omni	H	A1A	OH3LWP
1296850.000	IQ3VO/B	Verona	JN55LL	2	Big weel	Omni	H	A1A	I3LDP
1296854.000	DB0JO		JO31SL	350	4*15 Yagi	W	H		DF1VB
1296854.000	F1ZBK	Nancy	JN38CS	4	Slot	omni	H		F6CXA
1296855.000	SK3UHG	Nordingrå / Rävsön	JP92FW	30	Slotted Waveguide	Omni	H	A1A	
1296857.000	ON0NR	NAMUR (Wépion)	JO20KJ	10		omni	H	F1A	ON6YH
1296859.000	F1ZAK	Istres	JN23MM	20	Slot	omni	H	F1A	F1AAM
1296860.000	DB0LB		JN48OV	1	Big Wheel	OMNI	H		DK3PS
1296860.000	LA8SHF	Hvasser	JO59FB	60	13dB Horn	180	H	A1A	LA6LCA
1296863.000	I7IWN/B	Martano LE	JN90DE	5	2x 26 el. Yagi	320°	H	A1A	I7IWN
1296865.000	DB0JK		JO30LW	40	4*8el Yagi	OMNI	H		DK2KA
1296869.000	IQ0AH/B	Monte Plebi OT	JN40RX	2	23 elem	30°	H	A1A	IS0YPW
1296870.000	LB2SHF	Mandal	JO38RB	90	Horn 8dBi	150	H	A1A	LB2S
1296872.000	F1ZMT	Le Mans	JN07CX	10	Panneau Trefle	180°	H		F1BJD
1296876.000	IZ8EDE/B	M.te Pierfaone PZ	JN70VM	10	Alford Slot	Omni	H	A1A	IZ8EDE
1296877.000	LA3SHF	Tromøy	JO48JK	100	2 x 15el Yagi	150	H	A1A	LA7LW
1296880.000	DB0MOT	Kleiner Feldberg im Taunus	JO40FF33	15	Big Wheel	Omni	H	F1A	DL1ZB
1296880.000	ON0SHF	Ellignies Sainte Anne	JO10UN	5		omni	H	F1A	ON5PX
1296883.000	DB0INN		JN68GI	1	Schlitz	OMNI	H		DL3MBG
1296885.000	DB0TUD		JO61UA		Quad	OMNI	H		DL4DTU
1296885.000	OY6BEC	Sornfelli	IP62MB	50	Hybrid quad	SE	H	F1A	OY9JD
1296886.000	F1ZBC	Adriers	JN06JG	15	A. Slot	omni	H		F1AFJ
1296888.000	OM0MLA	Bratislava	JN88NE	0,18	Dipole	225/45	H	A1	OM3ID
1296890.000	LA4SHF	Egersund	JO28XJ	45	collinear	180	H	A1A	LA3EQ
1296895.000	F5ZAN	Pic Neulos	JN12LL	10	Slot	omni	H	A1A	F6HTJ
1296900.000	OZ5SHF	Yding	JO45VX	10	Double Quad	omni	H	F1A	OZ9ZZ
1296900.000	OH1SHF	Salo	KP10NJ	2	6x 5/8	Omni	V	A1A	OH1BOI
1296900.000	IQ3ZB/B	Monte Cesen TV	JN65AW	10	Slot Alford 16	Omni	H	A1A	IK3HHG
1296903.000	OE3XTR	Hohe Wand	JN87AT	2	WG Slot	OMNI	H	A1A	OE3KLU
1296904.000	DB0THE	Ohmgebirge - Birkenberg	JO51EL		Ho-Loop	OMNI	H	A1A	DF7AP
1296905.000	DB0AD		JO40AQ						DM8MM
1296910.000	DB0UX	Kerlsruhe / Bergwald	JN48FX	1	4 fach Kastenantenne	OMNI	H	F1A	DK2DB
1296910.000	DB0XY	Boxberg/Harz	JO51EU25	15	4xDoppelquad	omni	H		DF4OL
1296910.000	LA1SHF	Drammen	JO59BR	3	Big Wheel	omni	H	A1A	LA3PNA
1296910.000	ES0VHF	Panga/Isl.Saaremaa	KO18ND46MQ	39.8	Alford Slot	omni	H	A1A	ES0OK
1296915.000	TK5ZMV	Coti Chiavari	JN41JS	10	Yagi	315°	H	A1A	TK5EP
1296918.000	PI7ALK	Alkmaar	JO22IP61HS	5		omni	H	F1A	PH0V
1296920.000	DB0VC	Schönwalde / Bungsberg	JO54IF60	15	2*Big Wheel	OMNI	H	A1A	DC6UW
1296920.000	IT9CIT/B	M.te Bonifato TP	JM67LX	14	2 x biquad	0/330°	H	A1A	IT9CIT
1296925.000	DB0AAT		JN67HU	1	Vertikal	OMNI	H		DL8MCG
1296928.000	OH2SHF	Inkoo	KP20BB	50	10/10 dBi	230/20	H	A1A	OH2LNM
1296930.000	OZ7IGY	Jystrup	JO55WM54UA	90	Slot WG	omni	H	F1A	OZ7IS

Freq	Call	Nearest Town	Locator	ERPw	Antenna	Direction	H/V	Mode	Keeper
1296933.000	F5ZBT	Pessac	IN94QT	10			H		F6CBC
1296935.000	DB0YI		JO42XB	3	Big Wheel	OMNI	H		DL4AS
1296935.000	LX0AO	Differdange	JN29WM	1.6	Yagi	36° NE	H	A1A	ADRAD
1296940.000	DB0MFI	Hainsfarth	JN58HW	5	4xBig Wheel	- OMNI	H	A1A	DK1MFI
1296945.000	DB0AJA		JN59AS	10		OMNI	H		DF6NA
1296945.000	LA7SHF	Bergen	JP20QJ	15		180		A1A	LA3QMA
1296945.000	OH9SHF	Rovaniemi Pirttikoski	KP36OI	30	10 dBd	200	H	A1A	OH9LA
1296950.000	SK1UHG	Klintehamn	JO97CJ	30	Alford Slot	Omni	H	A1A	
1296950.000	ON0TB	Weismes/Waimes (Robertville)	JO30BM	10		omni	H	?	ON6GPS
1296950.000	DB0GW	Duisburg / Universität	JO31JK	10	Slot 8	Omni	H	A1A	DL3YDP
1296955.000	OZ1UHF		JO57GH						
1296960.000	SK4BX/B	Garphyttan / Ånnaboda	JO79LI		Alford Slot			A1A	
1296963.000	LA2SHF	Vassfjellet	JP53EG	30	Dipole	bidir. 15/195	H	A1A	LA1K
1296965.000	DB0ANN	Moritzberg	JN59PL	0.5	4xDoppelquad	OMNI	H	A1A	DL8ZX
1296970.000	SR3LHY	Pozna□	JO82LJ83	10.0 PEP	Alfrod Slot	OMNI	H	F1A	SO3Z
1296975.000	ON0AZ	Antwerpen	JO21FE	10		omni	H	F1A	ON7BPS
1296980.000	LA9SHF	"Kolsås	JO59GW	10	Doppelquad	160°	H	A1A	LA8GKA
1296983.000	F5ZWX	Grand Cap	JN23XE	1	Slot	omni	H	F1A	F6FCE
1296985.000	SK2SHF	Vännäs / Granlundsberget	JP93VU	10	Alford Slot	Omni	H	A1A	
1296995.000	DB0WOS		JN68ST	5	4xDoppelquad	OMNI	H		DF8RU
1444905.000	ON0ODR	Oudenaarde	JO10TT	25		omni	H	WSPR	ON4PN
13cm									
2320050.000	S55ZSE	Kokoš	JN65WP	0.1	Slot	omni	H	A1A	S53MV
2320090.000	S55ZMS	Dolina nr. Murska Sobota	JN86CR	0.1	Slot	omni	H	A1A	S53M & S51ZO
2320800.000	SK6MHI	Göteborg	JO57XQ	150	Slotted Waveguide	Omni	H	A1A	SM6EAN
2320805.000	DB0ZEH		JO62PW	10		360?	H		DL2RUD
2320805.000	IW1GLM/B	Bagnolo CN	JN34OS60UJ	5	8 Slot	60°	H	A1A	IW1GLM
2320810.000	DB0ZW		JN69AS	1	6fach Schlitz	OMNI	H		DC9RK
2320810.000	DB0NCO		JN59JD	2	Dipol	45Grad	H		DL6NCO
2320815.000	DB0VI		JN39NK	0.2		OMNI	H		DL3CM
2320815.000	SR5TDM	Tarczyn	KO01KX01	3.0 W PEP	Waveguide 2X16	OMNI	H	A1A	SP5MX
2320816.000	F1ZQU	Segonzac	IN95VO	25	Slot	omni	H		F1MMR
2320820.000	IQ0RM/B	Roma	JN61FW	1	Alford	Omni	H	A1A	IW0CZC
2320825.000	DB0FKS		JN49IT	0.8	Doppelhelix 6dB	OMNI	H	A1A	DL2OCB
2320832.000	OH6SHF	Uusikaarlepyy / Nykarleby	KP13GM	40	7 dBd	180	H	A1A	OH6NVQ
2320833.000	DB0FGB	Schneeberg	JO50WB	15	8-fach Schlitz	OMNI	H	A1A	DG4NBI
2320835.000	F5ZAC	Cerdagne	JN12BL	5	Slot	omni	H	F1A	F6HTJ
2320840.000	DB0ODW		JN49LN	10		OMNI	H		DK2NO
2320840.000	F1ZYY	Mauroux	JN03KV	1	Panneaux	315° / 135°	H	F1A	F1MOZ
2320842.000	OH3SHF	Tampere	KP11VK	125	6 dBi	Omni	H	A1A	OH3LWP
2320845.000	DB0LBV	Leipzig	JO61EH	1.5	Doppelquad	SW-S	H		DC6WM
2320850.000	DM0UB		JO62KK	0.6	4fach Kastenant	OMNI	H		DL7AIG
2320850.000	DB0GW	Duisburg / Universität	JO31JK	10	Slot 8	Omni	H	A1A	DL3YDP
2320850.000	LA4SHF	Jæren	JO28UO	10	Log periodic	225	H	A1A	LA3EQ
2320850.000	I3EME/B	M.te Tomba TV	JN55WV	1	6 slot	180°	H	A1A	I3EME
2320852.000	DB0APD	Apolda	JO51SA25VI	0,1	WG Slot 20	Omni	H	A1A	DF4AE
2320855.000	DB0SHF		JN48WP	6	6er Feld	260?	H		DL1SBE
2320855.000	F1ZUM	Orleans	JN07WV	2		omni	H	A1A	F1JGP
2320857.000	PI7RTD	Rotterdam	JO21FV68	15.0	Slot	Omni	H		PE1GHG
2320860.000	LA8SHF	Hvasser	JO59FB	50	13dB Horn	180	H	A1A	LA6LCA
2320864.000	F5ZVY		IN93GJ	3.6	Slot	omni	H	F1A	F2CT
2320875.000	ON0VHF	Louvain-La-Neuve	JO20HP	10		omni	H	F1A	ON4IV
2320876.000	IZ8EDE/B	Monte Pierfaono PZ	JN70VM	3	4 Dipole H.	Omni	H	A1A	IZ8EDE
2320880.000	DB0YI		JO42XB	3	Big Wheel	OMNI	H		DL4AS
2320880.000	LA3SHF	Tromøy	JO48JK	1	6dB Horn	150	H	A1A	LA7LW
2320880.000	DB0MOT	Kleiner Feldberg im Taunus	JO40FF33	15	Big Wheel	Omni	H	F1A	DL1ZB
2320883.000	DB0INN		JN68GI	1	Schlitz	OMNI	H		DL3MBG
2320885.000	DM0TUD		JO61UC	10	Schlitz	OMNI	H		DG0DI
2320886.000	F5ZMF	Adriers	JN06JG	8	Slot	omni	H	F1A	F5BJL
2320895.000	PI7RMD	Melick	JO31AD	5	Doublequad	80° 270°	H	A1A	PE1KXH
2320900.000	DB0UX	Karlsruhe / Bergwald	JN48FX	1	4 fach Kastenantenne	OMNI	H	F1A 800Hz	DK2DB
2320900.000	DB0MJ		JO31UB	1	6-fach Schlitz	OMNI	H		DH1MJ
2320900.000	SK3UHH	Nordingrå / Rävsön	JP92FW			220 deg.	H	A1A	
2320900.000	F6DWG		JN19BQ	10	Slot	omni	H	F1A	F6DWG
2320900.000	F1ZCC	Maurepas	JN08XS	1	Slot	omni	H	F1A	F1PDX
2320903.000	IQ5FI/B	M.te Secchieta FI	JN53SR	5	Alford Slot	Omni	H	A1A	I5KRD
2320910.000	DB0UM		JO73CE	1	Schlitz	OMNI	H		DG1BHA
2320910.000	DB0XY	Boxberg/Harz	JO51EU25	15	4x Doppelquad		H		DF4OL
2320910.000	DB0WML	Reken	JO31MU						DK8QU
2320910.000	IW0REF/B	M.te San Pancrazio TR	JN62HK14PE	11	13dB	Omni	H	A1A	IW0REF
2320919.000	IQ9GD/B	Mte Bonifato TP	JM67LX	0,9	6 slot		H	A1A	IT9CIT
2320920.000	DB0VC	Schönwalde / Bungsberg	JO54IF60	15	Big Wheel	OMNI	H	A1A	DC6UW
2320920.000	PI7ALK	Alkmaar	JO22IP61HS	5		omni	H	F1A	PH0V

Freq	Call	Nearest Town	Locator	ERPw	Antenna	Direction	H/V	Mode	Keeper
2320928.000	OH2SHF	Inkoo	KP20BB	150/150	10/10 dBi	230/20	H	A1A	OH2LNM
2320933.000	F5ZEN	Pessac	IN94QT	5	cornets	?	H	F1A	F6CBC
2320940.000	DB0MFI	Hainsfarth	JN58HW	5	4xBig Wheel	- Omni	H	A1A	DK1MFI
2320960.000	DB0AJA		JN59AS	10		OMNI	H		DF6NA
2320960.000	SK4BX/B	Garphyttan / Ånnaboda	JO79LI		SLOTTED WAVEGUIDE		H	A1A	
2320965.000	DB0ANN		JN59PL	5	4xDoppelquad	OMNI	H		DL8ZX
2320975.000	DB0JL		JO31MC	2	Schlitz	OMNI	H		DB5KN
2320983.000	F5ZHX	Grand Cap	JN23XE	10	Slot	omni	H	F1A	F6FCE
2320985.000	SK2SHF	Vännäs / Granlundsberget	JP93VU	10	Parabol	190	H	A1A	
2400900.000	OZ5SHF	Yding	JO45VX	10	Double Quad	omni	H	F1A	OZ9ZZ
2400930.000	OZ7IGY	Jystrup	JO55WM54UA	30	Slot WG	omni	H	F1A	OZ7IS
9cm									
3400007.000	DM0HVL		JO62KI	7.5	12fach Schlitz	OMNI	H		DL7AIG
3400050.000	DB0JL		JO31MC	1	Helical	OMNI	H		DB5KN
3400050.000	S55ZSE	Kokoš	JN65WP	0.5	Slot	omni	H	A1A	S53MV
3400090.000	S55ZMS	Dolina nr. Murska Sobota	JN86CR	0.5	Slot	omni	H	A1A	S53M & S51ZO
3400800.000	DB0MOT	Kleiner Feldberg im Taunus	JO40FF33	15	Slot 12	Omni	H	F1A	DL1ZB
3400800.000	LA8SHF	Hvasser	JO59FB					A1A	LA6LCA
3400800.000	OH3SHF	Tampere	KP11VK	120	6 dBi	Omni	H	A1A	OH3LWP
3400810.000	DB0NCO		JN59JD	2	SBF	45?	H		DL6NCO
3400840.000	DB0ODW		JN49LN	10		OMNI	H		DK2NO
3400850.000	DB0GW	Duisburg / Universität	JO31JK	10	Slot 8	Omni	H	A1A	DL3YDP
3400850.000	LA4SHF	Jæren	JO28UO	2	Log periodic	225	H	A1A	LA3EQ
3400850.000	DB0KK	Berlin Lichtenberg	JO62RM76OP	13	Slot	Omni	H	A1A	DC7YS
3400883.000	DB0INN		JN68GI	1	Schlitz	OMNI	H		DL3MBG
3400885.000	DM0TUD		JO61UC	10	Schlitz	OMNI	H		DG0DI
3400900.000	OZ5SHF	Yding	JO45VX	2	Double Quad	omni	H	F1A	OZ9ZZ
3400910.000	DB0WML	Reken	JO31MU						DK8QU
3400912.000	DB0XY	Boxberg/Harz	JO51EU25	15		omni	H	A1AN	DF4OL
3400920.000	PI7RTD	Rotterdam	JO21FV68	14.5	Slot	Omni	H		PE1GHG
3400925.000	PI7ALK	Alkmaar	JO22IP61HS		Slotted Waveguide 6dBd	omni	H	F1A	PH0V
3400930.000	OZ7IGY	Jystrup	JO55WM54UA	50	Slot WG	omni	H	F1A	OZ7IS
3400940.000	DB0MFI	Hainsfarth	JN58HW	4	2X10 Slot	OMNI	H		DK1MFI
3400945.000	DB0AJA		JN59AS	10		OMNI	H		DF6NA
3400955.000	OZ1UHF		JO57GH						
3400975.000	DB0HRF	Großer Feldberg Ts.	JO40FF			OMNI	H		DF7KU
3456855.000	DB0SHF		JN48WP	0.5	Horn	260?	H		DL1SBE
3456965.000	DB0ANN		JN59PL				H		DL8ZX
4324875.000	F1ZBY	Roc Blanc	JN13TV	5	Trefle	omni	H	A1A	F4DVR
6cm									
5760010.000	S55ZSE	Kokoš	JN65WP	0.5	Slot	omni	H	A1A	S53MV
5760045.000	S55ZRS	Kum	JN76MC	0.2	Slot	omni	H	A1A	ZRS
5760060.000	F1ZAO	Plougonver	IN88HL	1	Fentes	omni	H	F1A	F1LHC
5760070.000	DB0JL		JO31MC	0.8	Schlitz	OMNI	H		DB5KN
5760090.000	S55ZMS	Dolina nr. Murska Sobota	JN86CR	0.5	Slot	omni	H	A1A	S53M & S51ZO
5760800.000	DB0MOT	Kleiner Feldberg im Taunus	JO40FF33	15	Slot 14	Omni	H	F1A	DI1ZB
5760800.000	SK4BX/B	Garphyttan / Ånnaboda	JO79LI				H	A1A	
5760800.000	SK6MHI	Göteborg	JO57XQ	10	SLOTTED WAVEGUIDE	Omni	H	A1A	SM6EAN
5760800.000	OH3SHF	Tampere	KP11VK	150	15 dBi	Omni	H	A1A	OH3LWP
5760800.000	IQ4AD/B	Monte Cassio PR	JN54AO	0,5	6 slot	35°	H	A1A	I4YMB
5760804.000	F1ZMF	Lure	JN24VC	10	Fentes	omni	H	F1A	F1OW
5760805.000	DB0RIG		JN48WQ	15		OMNI	H		DG9SQ
5760810.000	DB0NCO		JN59JD	2	Schlitz	OMNI	H		DL6NCO
5760816.000	IK1YWB/B	Bagnolo CN	JN34OS60UJ	0,6	Slot 16	55°	H	A1A	IK1YWB
5760820.000	F5ZBE		JN18HN	12	Fentes	omni	H	F1A	F5HRY
5760833.000	DB0FGB	Schneeberg	JO50WB	15	8-fach Schlitz	OMNI	H	A1A	DG4NBI
5760840.000	DB0ODW		JN49LN	10		OMNI	H		DK2NO
5760841.000	SK7MW/B	Gladsax	JO75DN	40		Omni	H	FSK	SM7DTE
5760845.000	F1ZBD	Orleans	JN07WV	2	Fentes	omni	H	F1A	F1JGP
5760850.000	DM0UB		JO62KK	9	12fach Schlitz	OMNI	H		DL7AIG
5760850.000	DB0GW	Duisburg / Universität	JO31JK	10	Slot 8	Omni	H	A1A	DL3YDP
5760850.000	LA4SHF	Jæren	JO28UO	2	Log periodic	225	H	A1A	LA3EQ
5760855.000	DB0SHF		JN48WP	0.5	Array	260?	H		DL1SBE
5760855.000	OE8XGQ	Gerlitzen, Carinthia	JN66XQ	3.2	Slot 10	Omni	H	A1A	OE8WOZ
5760860.000	DB0RDH		JN68KW	2.5		SW bis SO	H		DL2RDH
5760860.000	LA8SHF	Hvasser	JO59FB	25	13dB Horn	180	H	A1A	LA6LCA
5760860.000	F5ZUO	Pic Neulos	JN12LL	1	Fentes	omni	H	F1A	F6HTJ
5760880.000	LA3SHF	Tromøy	JO48JK	1		150	H	A1A	LA7LW
5760880.000	ON0VHF	Louvain-La-Neuve	JO20HP	10		omni	H	F1A	ON4IV
5760883.000	DB0INN		JN68GI	1	Schlitz	OMNI	H		DL3MBG
5760883.000	F5ZWY	Grand Cap	JN23XE	1	Fentes	omni	H	F1A	F6FCE
5760885.000	DM0TUD		JO61UC		Schlitz	OMNI	H		DL4DTU
5760888.000	I3EME/B	Monte Tomba TV	JN55WV	0,22	Slot 8	170°	H	A1A	I3EME

Freq	Call	Nearest Town	Locator	ERPw	Antenna	Direction	H/V	Mode	Keeper
5760889.000	F5ZIE		IN93GJ	0.1	Fentes	omni	H	F1A	F2CT
5760893.000	OE1XGA	Wien/Kahlenberg	JN88EG	7	Slot 10	Omni	H	A1A	OE4WOG
5760900.000	OZ5SHF	Yding	JO45VX	2	WG slot	omni	H	F1A	OZ9ZZ
5760900.000	ON0TB	Weismes/Waimes (Robertville)	JO30BM	0.2		omni	H		ON6GPS
5760900.000	IQ5FI/B	M.te Secchieta AR	JN53SR84XO	1	Slot,8 + 8		H	A1A	I5KRD
5760903.000	SKØCT/B	Väddö	JO99JX	80		Omni	H	A1A	
5760903.000	F6DWG		JN19BQ	10	Fentes	omni	H	F1A	F6DWG
5760910.000	DBOUM		JO73CE	1	Schlitz	OMNI	H		DG1BHA
5760910.000	DBOWML	Reken	JO31MU						DK8QU
5760910.000	IWOREF/B	M.te San Pancrazio TR	JN62HK14PE	1	10dB	Omni	H	A1A	IWOREF
5760915.000	PI7RTD	Rotterdam	JO21FV68	17	Slot	Omni	H		PE1GHG
5760917.000	F1ZMH	Berneuil	IN95QP	2.5	Fentes	omni	H	F1A	F1MMR
5760925.000	F5ZOI	Ste Fortunade	JN05VE	4	Fentes	omni	H	F1A	F6ETI
5760928.000	OH2SHF	Helsinki	KP20LE	65	10 dBi	Omni	H	A1A	OH2LNM
5760930.000	OZ7IGY	Jystrup	JO55WM54UA	50	Slot WG	omni	H	F1A	OZ7IS
5760930.000	F1ZWJ	Lacapelle	JN14EB	2	Fentes	omni	H	F1A	F1BOH
5760933.000	F5ZPR	Pessac	IN94QT	8	Cornet	130°	H	F1A	F6CBC
5760935.000	OH5TEN	Kouvola	KP30HV	32	12.9 dBd	240	H	A1A	OH5IY
5760935.000	PI7ALK	Alkmaar	JO22IP61HS	1	Slotted Waveguide 10dBd	omni	H	F1A	PH0V
5760940.000	DBOMFI	Hainsfarth	JN58HW	2	12fach Schlitz	OMNI	H		DK1MFI
5760945.000	DBOAJA		JN59AS	10		OMNI	H		DF6NA
5760965.000	DBOANN		JN59PL				H		DL8ZX
5760975.000	ON0GHZ	Tielt-Winge	JO20KV	5		omni	H	F1A	ON4IY
5769955.000	OZ1UHF		JO57GH						
10GHz									
10368033.000	I3CLZ/B	M Falcone VI	JN55OQ	0,6	Slot	Omni	H	A1A	I3CLZ
10368050.000	S55ZRS	Kum	JN76MC	0.8	Slot	omni	H	A1A	ZRS
10368057.000	I4BER/B	Piane di Mocogno MO	JN54IG	0,07	Slot 16		H	A1A	I4BER
10368070.000	S55ZMS	Dolina	JN86CR	0.8	Slot	omni	H	A1A	S53M
10368120.000	DB0JL		JO31MC	0.15	Schlitz	OMNI	H		DB5KN
10368282.000	I1TEX/B	Rossana CN	JN34QM58AU	1	Slot 8	65°	H	A1A	I1TEX
10368760.000	ON0NIV	Tielt-Winge	JO20DO	1		omni	H	F1A	
10368796.000	IQ0RM/B	Roma	JN61FW	0,2	Slot	0/180°	H	A1A	IW0CZC
10368800.000	SK6MHI	Göteborg	JO57XQ	1	SLOTTED WAVEGUIDE	Omni	H	A1A	SM6EAN
10368800.000	OH3SHF	Tampere	KP11VK	80	15 dBi	Omni	H	A1A	OH3LWP
10368800.000	IW3FZQ/B	Colli Euganei PD	JN55TG	0,5	Slot 14+14	60°-240°	H	A1A	IW3FZQ
10368805.000	DB0UL		JN48XK	4		OMNI	H		DL1SBM
10368805.000	ON0TNR	NAMUR (Temploux)	JO20KJ	0.2		omni	H	F1A	ON4TLW
10368805.000	DB0MVP	Schwerin	JO53QP		Slot 10dB	omni	H		DG3TP
10368809.000	IQ1GE/B	Monte Fasce GE	JN44MJ	1	Slot	340° - 160°	H	A1A	IZ1EVF
10368810.000	DB0GHZ	Helgoland	JO43UP	0.15	Slot	OMNI	H	A1A	DK7LJ
10368810.000	DB0ANU		JN59GG	10	Schlitz	OMNI	H		DL3NDX
10368815.000	DB0VI		JN39NK	0.3		OMNI	H		DL3CM
10368815.000	SR5TDM	Tarczyn	"KO01KX01	4.0 W PEP	Waveguide 2X8	OMNI	H	A1A	SP5MX
10368819.000	IZ1ERR/B	Bagnolo CN	JN34OS	1	Slot 20	55°	H	A1A	IZ1ERR
10368820.000	DB0MOT	Kleiner Feldberg im Taunus	JO40FF33	15	Slot 12	Omni	H	F1A	DL1ZB
10368820.000	LA5SHF	Stavanger	JO28UX	12	10dB Slot	240	H	A1A	LA3EQ
10368825.000	DB0HRO		JO64AD	0.2	Schlitz	OMNI	H		DL6KWN
10368826.000	IQ5FI/B	M.te Secchieta AR	JN53SR84XO	1,5	Slot 16	Omni	H	A1A	I5KRD
10368833.000	DB0FGB	Schneeberg	JO50WB	8	8-fach Schlitz	OMNI	H	A1A	DG4NBI
10368835.000	ON0VHF	LOUVAIN-LA-NEUVE	JO20HP	10		omni	H	F1A	ON4IV
10368840.000	DB0MMO	Breitsol	JN49RV	2	8fach-Schlitz	OMNI	H	F1A	DH4FAJ
10368841.000	SK7MW/B	Gladsax	JO75DN	40		Omni	H	FSK	SM7DTE
10368845.000	DO0CWC		JO60MT	10	8fach-Schlitz	OMNI	H	A1A	DO1CTL
10368847.000	SKØEN/B	Väddö	JO99JX	10000	33 dB dish	360 deg	H	A1A	
10368850.000	DB0GG		JN48NR	1	Schlitz	OMNI	H		DL5AAP
10368850.000	DM0UB		JO62KK	8.7	12fach-Schlitz	OMNI	H		DL7AIG
10368850.000	DB0GW	Duisburg / Universität	JO31JK	10	Slot 8	Omni	H	A1A	DL3YDP
10368850.000	LA4SHF	Jæren	JO28UO	25	2 x 13dB Horn	180/0	H	A1A	LA3EQ
10368850.000	SK1SHH	Klintehamn	JO97CJ	10000	SLOT WG	360 deg	H	A1A	
10368850.000	IQ6AN/B	Ancona	JN63QN	4	Horn	355°	H	A1A	IW6ATU
10368852.000	DB0APD	Apolda	JO51SA25VI	0,01	WG Slot 20	Omni	H	A1A	DF4AE
10368853.000	DL0WY		JN67AQ	1.5	12 Fach-Schlitz	O-N-	H		DJ8VY
10368855.000	DB0SHF		JN48WP	0.3	Horn	260?	H		DL1SBE
10368855.000	OE8XGQ	Gerlitzen, Carinthia	JN66XQ	0.6	Slot 8	Omni	H	A1A	OE8WOZ
10368855.000	IK7UXW/B	Martina Franca TA	JN80OQ	4	Slot 2x16	Omni	H	A1A	IK7UXW
10368860.000	DB0RDH		JN68KW	1.3		SW bis SO	H		DL2RDH
10368860.000	LA8SHF	Hvasser	JO59FB	10	13dB Horn	180	H	A1A	LA6LCA
10368865.000	DB0JK		JO30LW	200	Schlitz	OMNI	H		DK2KA
10368865.000	I3EME/B	Monte Tomba TV	JN55WV	2,2	Slot 12	170°	H	A1A	I3EME
10368868.000	IZ0CZW/B	Monte Calvo LT	JN61QI	0,25	Slot	0°- 180°	H	A1A	IZ0CZW
10368870.000	LB2SHF	Odderøya	JO48AD	25	Sector horn	unidir. 180	H	A1A	LB2S
10368870.000	IQ0AH/B	Monte Plebi OT	JN40RX	1,5	Slot 16	35°	H	A1A	IW0UAM

Freq	Call	Nearest Town	Locator	ERPw	Antenna	Direction	H/V	Mode	Keeper
10368875.000	DB0HW		JO51GT	0.04	Schlitz	WNO	H		DL3AAS
10368875.000	ON0EME	LILLE	JO21JG	10		omni	H	F1A	ON7UN
10368876.000	IZ8EDE/B	M.te Pierfaone PZ	JN70VM	1	Slot	Omni	H	A1A	IZ8EDE
10368878.000	IQ2CF/B	Bovezzo BS	JN55DO	0,25	Horn	180°	H	A1A	IK2BVP
10368880.000	DB0ODW		JN49LN	10	Schlitz	OMNI	H		DK2NO
10368880.000	OE1XGA	Wien/Kahlenberg	JN88EG	7	Slot 12	Omni	H	A1A	OE4WOG
10368880.000	IW9ARO/B	Caltagirone CT	JM77GF	6	Slot	Omni	H	A1A	IW9ARO
10368880.000	9A0BXP	Psunj	JN85QJ	0.2	Slot	omni	H	A1A	9A5R
10368883.000	DB0INN		JN68GI	1	Schlitz	OMNI	H		DL3MBG
10368885.000	DM0TUD		JO61UC	5	Schlitz	OMNI	H		DG0DI
10368885.000	OY6BEC		IP61PJ	750	Horn	SE	H	F1A	OY9JD
10368890.000	LA9SHF	Kolsås	JO59GW	10	10dB Horn	180	H	A1A	LA8GKA
10368890.000	OM0MXA	Bratislava	JN88NE	0,12	Slot	225/45	H	A1	OM3ID
10368890.000	9A0BXV	Velebitska pljesivica	JN74LT	0.2	Slot	omni	H	A1A	9A4QV
10368892.000	S55ZKP	Slavnik	JN65XM	0.4	Slot	omni	H	A1A	S51JN
10368895.000	LX0DB	Sandweiler	JN39CO	4	Slot	omni	H	F1A	RL
10368900.000	DB0ECA		JN58XD	1.5 / 8.0	8-Fach Schlitz	OMNI	H		DC8EC
10368900.000	DB0ELS		JO43FF	0.2	Schlitz 11dB	OMNI	H		DL9BL
10368900.000	DB0UX	Karlsruhe / Bergwald	JN48FX	1	Schlitz	OMNI	H	F1A 800Hz	DK2DB
10368900.000	OE3XTR	Hohe Wand	JN87AT	5	WG Slot	OMNI	H	A1A	OE3KLU
10368900.000	OZ5SHF	Yding	JO45VX	25	Slot WG	omni	H	F1A	OZ9ZZ
10368900.000	OH1SHF	Salo	KP10NJ	1	Corner reflector	100	H	A1A	OH1BOI
10368900.000	DB0UY	Karlsruhe							DL5IN
10368900.000	IW8PNY/B	Contessa Soprana CS	JM89CK	6	Slot 20+20	199°	H	A1A	IW8PNY
10368904.000	PI7RTD	Rotterdam	JO21FV68	10	Slot	Omni	H		PE1GHG
10368905.000	DB0SCS	Nürnberg							DG7NDV
10368910.000	DB0UM		JO73CE	1	Schlitz	OMNI	H		DG1BHA
10368910.000	DB0XY	Boxberg/Harz	JO51EU25	15	Schlitz		H		DF4OL
10368910.000	DB0WML	Reken	JO31MU	10	Slot 10dB	omni	H	A1A	DK8QU
10368910.000	IW0REF/B	M.te San Pancrazio TR	JN62HK14PE	0,7	Slot 2	Omni	H	A1A	IW0REF
10368915.000	ON0GBN	FLOBECQ/VLOESBERG	JO10US	1		omni	H	F1A	ON7FLY
10368920.000	DB0VC	Schönwalde / Bungsberg	JO54IF60	15	WG Slot 10	OMNI	H	A1A	DC6UW
10368920.000	PI7ALK	Alkmaar	JO22IP61HS	1		omni	H	F1A	PH0V
10368925.000	ON0GHZ	TIELT-WINGE	JO20KV	2.5		omni	H	F1A	ON4IY
10368928.000	OH2SHF	Helsinki	KP20LE	30	13.8 dBi	Omni	H	A1A	OH2LNM
10368930.000	OE3XAC	Kaiserkogel	JN78SB	10	WG Slot	OMNI	H	A1A	OE3KLU
10368930.000	OZ7IGY	Jystrup	JO55WM54UA	50	Slot WG	omni	H	F1A	OZ7IS
10368930.000	IK3TCH/B	San Mauro di Saline VR	JN55NO	1	Slot 16	Omni	H	A1A	IK3TCH
10368935.000	OH5TEN	Kouvola	KP30HV	23	15.9 dBd	240	H	A1A	OH5IY
10368935.000	DM0MAX	Holsen	JO42IG	15		omni	H	A1A	DL6XB
10368940.000	DB0DON		JN58JR	1	Schlitz	OMNI	H		DK1MFI
10368940.000	SR3XHY	Pozna□	JO82LJ83	6.0 PEP	Waveguide 2X8	OMNI	H	F1A	SO3Z
10368945.000	LA7SHF	Bergen	JP20QJ	10	17 dB Horn	180	H	A1A	LA3QMA
10368946.000	DB0AJA		JN59AS	10	Schlitz	OMNI	H		DF6NA
10368950.000	DB0OFG		JN48CO47	0.4	Schlitz	OMNI	H		DC5GF
10368952.000	IT9CIT/B	M.te Bonifato TP	JM67LX	1,16	Slot 12	0°	H	A1A	IT9CIT
10368955.000	OZ1UHF		JO57GH						
10368960.000	DB0MFI	Hainsfarth	JN58HW	1	Schlitzstrahler	- OMNI	H	A1A	DK1MFI
10368960.000	SK4BX/B	Garphyttan / Ånnaboda	JO79LI		Slotted Waveguide		H	A1A	
10368965.000	DB0ANN		JN59PL	0.2	12fach-Schlitz	OMNI	H		DL8ZX
10368965.000	ON0BRU	BRUSSEL	JO20EU			omni	H	F1A	ON4BHM
10368972.000	DB0MU	Nottuln	JO31QX	7	Slot 8	Omni	H	A1A	DJ2QZ
10368985.000	ON0HVL	Heuvelland	JO10KS	2	Dipole	Slot Antenna 10dBi		H	F1A ON7FLY
24GHz									
23209354.000	LX0AOK	Differdange	JN29WM	1	Yagi	36° NE	H	A1A	ADRAD
24048020.000	IQ5FI/B	M.te Secchieta FI	JN53SR84XO	0,8	Slot 8+8	Omni	H	A1A	I5KRD
24048033.000	I3CLZ/B	M Falcone VI	JN55OQ	0,6	Slot	135°	H	A1A	I3CLZ
24048044.000	S55ZMS	Dolina	JN86CR	0.17	Slot	omni	H	A1A	S53M & S51ZO
24048050.000	ON0GHZ	TIELT-WINGE	JO20KV	2.5		omni	H	F1A	ON4IY
24048065.000	IZ1OTT/B	Frassinetto TO	JN35TK24SI	0,6	20 dB Horn	105°	H	A1A	IZ1OTT
24048076.000	DB0QQ		JO31RI	0.13		OMNI	H		DJ5VW
24048078.000	I3EME/B	Monte Tomba TV	JN55WV	0,18	Slot 12	170°	H	A1A	I3EME
24048180.000	LX0CDF	Sandweiler	JN39CO	1	Slot	omni	H	F1A	RL
24048210.000	S55ZKP	Slavnik	JN65XM	0.4	Slot 10	omni	H	A1A	S56RGA
24048618.000	IK1YWB/B	Bagnolo CN	JN34OS60UJ	0,8	Horn	50°	H	A1A	IK1YWB
24048800.000	LA1SHF	Drammen	JO59BR					A1A	LA3PNA
24048800.000	SK6MHI	Göteborg	JO57XQ	10	Slotted Waveguide	Omni	H	A1A	SM6EAN
24048800.000	OH3SHF	Tampere	KP11VK	30	18 dBi	Omni	H	A1A	OH3LWP
24048805.000	IW3RMR/B	M.te Bernadia	JN66OF	0,1	Slot 10	180°	H	A1A	IW3RMR
24048810.000	DB0NCO		JN59JD	0.8	Schlitz	OMNI	H		DL6NCO
24048816.000	DB0VI		JN39NK	0.3		OMNI	H		DL3CM
24048820.000	DB0JL		JO31MC	0.01	Schlitz	OMNI	H		DB5KN
24048820.000	LA5SHF	Stavanger	JO28UX	0.2	10 dB Horn	180	H	A1A	LA3EQ

Freq	Call	Nearest Town	Locator	ERPw	Antenna	Direction	H/V	Mode	Keeper
24048820.000	PI7RTD	Rotterdam	JO21FV68	10	Slot	Omni	H		PE1GHG
24048825.000	DB0MMO	Breitsol	JN49RV	1	8-fach Schlitz	OMNI	H	F1A	DH4FAJ
24048832.000	DB0FGB	Schneeberg	JO50WB	5	8-fach Schlitz	OMNI	H	A1A	DG4NBI
24048833.000	IW3FZQ/B	Colli Euganei PD	JN55TG	0,16	Sector	210°	H	A1A	IW3FZQ
24048835.000	ON0VHF	Louvain-La-Neuve	JO20HP	1		omni	H	F1A	ON4IV
24048840.000	DB0ODW		JN49LN	10		OMNI	H		DK2NO
24048840.000	DB0WOL	Wollenberg / Barnim	JO62XR	12	Sector Horn	90° 240°	H	A1A	DL7VTX
24048843.000	SK7MW/B	Gladsax	JO75DN	70		Omni	H	FSK	SM7DTE
24048850.000	DB0GW	Duisburg / Universität	JO31JK	10	Slot 8	omni	H	A1A	DL3YDP
24048850.000	DB0KK	Berlin Lichtenberg	JO62RM76OP	6	Horn	270°	H	A1A	DC7YS
24048852.000	DB0MOT	Kleiner Feldberg im Taunus	JO40FF33	15	Slot 12	Omni	H	F1A	DL1ZB
24048852.000	DL0WY		JN67AQ	12	Sektorhorn	O-N-	H		DJ8VY
24048855.000	OE8XGQ	Gerlitzen, Carinthia	JN66XQ	0.9	Slot 15	Omni	H	A1A	OE8WOZ
24048860.000	LA8SHF	Hvasser	JO59FB	5	10dB Horn	120° +-90°	H	A1A	LA6LCA
24048864.000	DB0JK		JO30LW	1	2xH-Horn	OMNI	H		DK2KA
24048875.000	ON0EME	LILLE	JO21JG	10		omni	H	F1A	ON7UN
24048883.000	SK6SHG	Tjörn Island	JO57TX	2x1W	2xhorn	N / S	H	A1A	SM6CEN
24048884.000	DM0TUD		JO61UC	2	Schlitz	OMNI	H		DG0DI
24048900.000	DB0ECA		JN58XD	0.5 / 5.0	E-Horn 50□	E	H		DC8EC
24048900.000	OZ5SHF	Yding	JO45VX	7	Slot WG	omni	H	F1A	OZ9ZZ
24048902.000	9A5AA	Sljeme	JN75XV35	0.3	Slot	OMNI	H	F1A	9A5AA
24048910.000	DB0WML	Reken	JO31MU						DK8QU
24048910.000	IW0REF/B	M.te San Pancrazio TR	JN62HK14PE	0,15	Slot 2	Omni	H	A1A	IW0REF
24048912.000	DB0HHB		JN69LF	1	10 Slots	Omni	H		DL6NCO
24048920.000	DB0VC	Schönwalde / Bungsberg	JO54IF60	15	WG Slot 12	Omni	H	A1A	DC6UW
24048925.000	PI7ALK	Alkmaar	JO22IP61HS	1	slotted waveguide 12dBd	Omni	H	F1A	PH0V
24048928.000	OH2SHF	Helsinki	KP20LE	20	24-slot WG	Omni	H	A1A	OH2LNM
24048930.000	OZ7IGY	Jystrup	JO55WM54UA	20	Slot WG	Omni	H	F1A	OZ7IS
24048960.000	OE1XGA	Wien/Kahlenberg	JN88EG	10	Slot 12	Omni	H	A1A	OE4WOG
24048964.000	DB0ANN	Moritzberg	JN59PL			Omni	H	A1A	DL8ZX
24048985.000	ON0HVL	Heuvelland	JO10KS	1	Dipole/Slot Antenna 15dBi		H	F1A	ON7FLY
47GHz									
47088080.000	DB0AJA		JN59AS	10	Schlitz	OMNI	H		DF6NA
47088100.000	DB0QQ		JO31RI	10		3° 13°	H		DJ5VW
47088800.000	DB0FGB	Schneeberg	JO50WB	4	Sektorhorn	Beam 0°	H	A1A	DG4NBI
47088800.000	LA1SHF	Drammen	JO59BR					A1A	LA3PNA
47088800.000	OE1XGA	Wien/Kahlenberg	JN88EG	2.5	Sector Horn	"?° ÖW 100°"	H	A1A	OE4WOG
47088820.000	LA5SHF	Stavanger	JO28UX	0.2	10 dB Horn	180	H	A1A	LA3EQ
47088840.000	DB0WOL	Wollenberg / Barnim	JO62XR	12	Sector Horn	90° 240°	H	A1A	DL7VTX
47088850.000	DB0KK	Berlin Lichtenberg	JO62RM76OP	0.5	Slot	omni	H	A1A	DC7YS
47088850.000	DB0MOT	Kleiner Feldberg im Taunus	JO40FF33	10	Dish	2.6° 360° remote	H	F1A	DL1ZB
47088850.000	DB0GW	Duisburg / Universität	JO31JK	5	2 x Sectorhorn	0° / 180°	H	A1A	DL3YDP
47088853.000	DL0WY		JN67AQ	0.5	Sektorhorn	O-N	H		DJ8VY
47088865.000	DB0JK		JO30LW	0.1	2xH-Horn	OMNI	H		DK2KA
47088884.000	DM0TUD		JO61UC	1	Horn	SW	H		DG0DI
47088910.000	DB0WML	Reken	JO31MU						DK8QU
47088910.000	DB0OHL								
47088920.000	DB0VC	Schönwalde / Bungsberg	JO54IF60	10	WG Slot 15	Omni	H	A1A	DC6UW
47088925.000	PI7ALK	Alkmaar	JO22IP61HS	0.01	sector horn 17dBd	Horn 135°	H	F1A	PH0V
48088900.000	LX0CDE	Sandweiler	JN39CO	1	Slot	Omni	H	F1A	RL
76GHz									
76032050.000	DB0AJA		JN59AS	0,2	Slot 8	Omni	H	A1A	DF6NA
76032100.000	DB0QQ		JO31RI	10		3° 13°	H		DJ5VW
76032800.000	DB0FGB	Schneeberg	JO50WB	3	Sektorhorn	Beam 0°	H	A1A	DG4NBI
76032825.000	DB0MOT	Kleiner Feldberg im Taunus	JO40FF33	10	Dish	1.8° 360°: remote	H	F1A	DL1ZB
76032840.000	DB0WOL	Wollenberg / Barnim	JO62XR	12	Sector Horn	90° 240°	H	A1A	DL7VTX
76032850.000	DB0GW	Duisburg / Universität	JO31JK	2	2 x Sectorhorn	0° / 180°	H	A1A	DL3YDP
76032865.000	OE1XGA	Wien/Kahlenberg	JN88EG	"0.1	Sector Horn	?° ÖW 100°	H	A1A	OE4WOG
76032900.000	LX0CBC	SANDWEILER	JN39CO	0.1	Slot	Omni	H	F1A	RL
76032910.000	DB0WML	Reken	JO31MU						DK8QU
76032910.000	DB0OHL								

Abbreviations & Codes

Numerous abbreviations and codes are used by radio amateurs, especially when using Morse (CW) or Datamodes. The abbreviations listed below are internationally recognised and cross language barriers, enabling people without a common language to communicate. Listed are many of the more common ones in use.

For a list of acronyms, abbreviations and conventions as used in the Yearbook, RadCom and other RSGB publications please go to the RSGB website: *http://rsgb.org/main/publications-archives/radcom/supplementary-information/abbreviations-and-acronyms/*

Abbreviations

73	Best regards
88	Love and kisses
AA	All after (used after question mark to request a repetition)
AB	All before (similarly)
ARRL	American Radio Relay League
ABT	About
ADS	Address
AGN	Again
ANT	Antenna
BN	All between
BK	Break (to pause transmission of a message, say)
BUG	Semiautomatic mechanical key
C	Yes
CBA	Callbook address
CFM	Confirm
CLG	Calling
CQ	Calling any station
CS	Callsign
CUL	See you later
CUZ	Because
CW	Continuous wave
CX	Conditions
DE	From
DX	Distance (sometimes refers to long distance contact)
ES	And
FB	Fine business (Analogous to "OK")
FCC	Federal Communications Commission
FER	For
FM	From
FREQ	Frequency
GA	Good afternoon or Go ahead (depending on context)
GB	Good Bye
GE	Good evening
GM	Good morning
GND	Ground (ground potential)
GUD	Good
HI	Laughter
HR	Here
HV	Have
HW	How
LID	Poor operator
MILS	Milliamperes
N	No
NIL	Nothing
NR	Number
NW	Now
OB	Old boy
OC	Old chap
OM	Old man (any male amateur radio operator is an OM)
OO	Official observer
OP	Operator
OT	Old timer
OTC	Old timers club
OOTC	Old old timers club
PSE	Please
PWR	Power
QCWA	Quarter Century Wireless Association
R	I acknowledge or decimal point (depending on context. The origin of "Roger")
RCD	Received
RCVR	Receiver (radio)
REF	Refer to, Reference, Referring to.
RFI	Radio Frequency Interference
RIG	Radio apparatus
RPT	Repeat or report (depending on context)
RPRT	Report
RST	Signal report format (Readability-Signal Strength-Tone)
RTTY	Radioteletype
RX	Receive
SAE	Self-addressed envelope
SASE	Self-addressed, stamped envelope
SED	Said
SEZ	Says
SIG	Signal or signature
SIGS	Signals
SKED	Schedule
SN	Soon
SMS	Short message service
SRI	Sorry
STN	Station
TEMP	Temperature
TFC	Traffic
TMW	Tomorrow
TNX	Thanks
TT	That
TU	Thank you
TVI	Television Interference
TX	Transmit, transmitter
TXT	Text
U	You
UR	Your or You're (depending on context)
URS	Yours
VY	Very
WA	Word after
WB	Word before
WD	Word
WDS	Words
W	Watts
WDS	Words
WKD	Worked
WL	Will
WUD	Would
WX	Weather
XMTR	Transmitter
XYL	Wife
YL	Young lady (used for any female)

Q Codes

There are a huge number of three-letter Q codes. Radio amateurs use only a small percentage of them, as many are of relevance only to shipping, aircraft, the police etc. They fall into the following pattern.

QAA-QNZ Aeronautical, QOA-QQZ Maritime, QRA-QUZ All services, QZA-QZZ Other

All Q codes follow the form of a question and answer. The list below gives details of those in common use by radio amateurs.

Q-code	Question	Answer	Colloquial use (if different)/explanation
QRB	How far are you from my station?	The distance between our stations is ...	
QRG	What is my exact frequency?	Your frequency is ...	Frequency of operation
QRH	Does my frequency vary?	Your frequency varies	
QRI	How is the tone of my transmission?	The tone of your transmission is ...	
QRL	Are you (or is the frequency) busy?	I am (or the frequency is) busy	
QRM	Are you suffering interference?	I am suffereing interference	Man-made interference
QRN	Are you troubled by static?	I am troubled by static	Natural interference (atmospherics)
QRO	Shall I increase power?	Increase power	High power
QRP	Shall I decrease power?	Decrease power	Low power
QRQ	Shall I send faster?	Send faster	High speed
QRS	Shall I send more slowly?	Send slower	Low speed
QRT	Shall I stop transmitting?	Stop transmitting	To close down
QRU	Have you anything for me?	I have nothing for you	
QRV	Are you ready to transmit?	I am ready to transmit	
QRX	When will you call again?	I will call you again at ...	To stand-by
QRZ	Who is calling me?	You are being called by ...	
QSA	What is the strength of my signal?	The strength of your signal is ...	
QSB	Does the strength of my signals vary?	The strength of your signals varies	Fading
QSD	Is my keying defective?	Your keying is defective	
QSK	Can you hear me between your signals (and if so can I break-in)?	I can hear you between my signals (and it is OK to break-in on my transmission)	Break-in Morse operation
QSL	Can you acknowledge receipt?	I am acknowledging receipt	A card to confirm contact
QSO	Can you communicate with ... ?	I can communicate with ...	A contact
QSP	Will you relay to ...?	I will relay to ...	
QSX	Will you listen for ... (callsign) on ...?	I am listening for ... on ...	Split frequency operation
QSY	Shall I change frequency?	Change frequency to ...	
QTF	Will you give me the position of my station according to bearings taken?	The position of your station according to bearings taken is...	Beam heading (in degrees)
QTH	What is your position (location)?	My position (location) is ...	
QTR	What is the exact time?	The exact time is ...	

Phonetic Alphabet

The Phonetic Alphabet used by radio amateurs today was developed by NATO in the 1950s to be intelligible (and pronounceable) to all NATO allies. It replaced several other phonetic alphabets and is now widely used in business and telecommunications across Europe and North America.

A	Alpha	N	November
B	Bravo	O	Oscar
C	Charlie	P	Papa
D	Delta	Q	Quebec
E	Echo	R	Romeo
F	Foxtrot	S	Sierra
G	Golf	T	Tango
H	Hotel	U	Uniform
I	India	V	Victor
J	Juliet	W	Whiskey
K	Kilo	X	X-ray
L	Lima	Y	Yankee
M	Mike	Z	Zulu

ASCII Code

The American Standard Code for Information Interchange is the code used in computing and for packet radio. Standard ASCII consists of seven data bits, which gives 128 possible combinations. The first 32 characters are used for control and the remaining 96 are representable characters.

Dec	Hex	Chr	Ctrl	Dec	Hex	Chr	Dec	Hex	Chr	Dec	Hex	Chr
0	0	NUL	^@	32	20	SP	64	40	@	96	60	‘
1	1	SOH	^A	33	21	!	65	41	A	97	61	a
2	2	STX	^B	34	22	“	66	42	B	98	62	b
3	3	ETX	^C	35	23	#	67	43	C	99	63	c
4	4	EOT	^D	36	24	$	68	44	D	100	64	d
5	5	ENQ	^E	37	25	%	69	45	E	101	65	e
6	6	ACK	^F	38	26	&	70	46	F	102	66	f
7	7	BEL	^G	39	27	‘	71	47	G	103	67	g
8	8	BS	^H	40	28	(	72	48	H	104	68	h
9	9	HT	^I	41	29	)	73	49	I	105	69	i
10	0A	LF	^J	42	2A	*	74	4A	J	106	6A	j
11	0B	VT	^K	43	2B	+	75	4B	K	107	6B	k
12	0C	FF	^L	44	2C	,	76	4C	L	108	6C	l
13	0D	CR	^M	45	2D	-	77	4D	M	109	6D	m
14	0E	SO	^N	46	2E	.	78	4E	N	100	6E	n
15	0F	SI	^O	47	2F	/	79	4F	O	111	6F	o
16	10	DLE	^P	48	30	0	80	50	P	112	70	p
17	11	DC1	^Q	49	31	1	81	51	Q	113	71	q
18	12	DC2	^R	50	32	2	82	52	R	114	72	r
19	13	DC3	^S	51	33	3	83	53	S	115	73	s
20	14	DC4	^T	52	34	4	84	54	T	116	74	t
21	15	NAK	^U	53	35	5	85	55	U	117	75	u
22	16	SYN	^V	54	36	6	86	56	V	118	76	v
23	17	ETB	^W	55	37	7	87	57	W	119	77	w
24	18	CAN	^X	56	38	8	88	58	X	120	78	x
25	19	EM	^Y	57	39	9	89	59	Y	121	79	y
26	1A	SUB	^Z	58	3A	:	90	5A	Z	122	7A	z
27	1B	ESC		59	3B	;	91	5B	[	123	7B	{
28	1C	FS		60	3C	<	92	5C	\	124	7C	|
29	1D	GS		61	3D	=	93	5D	]	125	7D	}
30	1E	RS		62	3E	>	94	5E	^	126	7E	~
31	1F	US		63	3F	?	95	5F	_	127	7F	DEL

Morse Code

The standard alphabet and numbers, as required for the Foundation Licence Morse Assessment.

A	• ▬	N	▬ •
B	▬ • • •	O	▬ ▬ ▬
C	▬ • ▬ •	P	• ▬ ▬ •
D	▬ • •	Q	▬ ▬ • ▬
E	•	R	• ▬ •
F	• • ▬ •	S	• • •
G	▬ ▬ •	T	▬
H	• • • •	U	• • ▬
I	• •	V	• • • ▬
J	• ▬ ▬ ▬	W	• ▬ ▬
K	▬ • ▬	X	▬ • • ▬
L	• ▬ • •	Y	▬ • ▬ ▬
M	▬ ▬	Z	▬ ▬ • •

1	• ▬ ▬ ▬ ▬	6	▬ • • • •
2	• • ▬ ▬ ▬	7	▬ ▬ • • •
3	• • • ▬ ▬	8	▬ ▬ ▬ • •
4	• • • • ▬	9	▬ ▬ ▬ ▬ •
5	• • • • •	0	▬ ▬ ▬ ▬ ▬

Special Morse Characters

There are numerous procedural and punctuation characters. The following are those most commonly used by Morse operators.

$\overline{AR}$ (+)	[di-dah-di-dah-dit]	End of message (used before the final calls and is written as 'AR' or '+')
$\overline{CT}$	[dah-di-dah-di-dah]	Preliminary call
$\overline{BT}$ (=)	[dah-di-di-di-dah]	Separation signal (used in text and is written as 'BT' or '=')
$\overline{KN}$	[dah-di-dah-dah-dit]	Transmit only the station called (used after the final calls and is written as 'KN')
$\overline{VA}$	[di-di-di-dah-di-dah]	Transmission ends (written as 'VA' or 'SK')
?	[di-di-dah-dah-di-dit]	Question mark (written as 'IMI' or '?')
/	[dah-di-di-dah-dit]	Oblique stroke (can be used as part of a callsign and is written as '/')
.	[di-dah-di-dah-di-dah]	Full stop
Error	[di-di-di-di-di-di-di-dit]	Erases the word in which a mistake has been made
@	[di-dah-dah-di-dah-dit]	As used in Email addresses

Five-unit Code

The so-called Murray Code is used by RTTY operators and telex machines. It consists of one 'start' bit, five 'data' bits, then 1.5 'stop' bits. As there are only 32 possible combinations of code, a limited character set is available (e.g. no lower case). Traditionally, amateur RTTY is sent at 45.45 Bauds, which results in an element length of 22ms. It is known officially as the International Telegraphic Alphabet No.2.

Binary	Dec	Hex	Octal	Letter	Figure
00000	0	00	00	[Blank]	
00001	1	01	01	T	5
00010	2	02	02	[Carr. Return]	
00011	3	03	03	O	9
00100	4	04	04	[Space]	
00101	5	05	05	H	# (note)
00110	6	06	06	N	,
00111	7	07	07	M	.
01000	8	08	10	[Line Feed]	
01001	9	09	11	L	)
01010	10	0A	12	R	4
01011	11	0B	13	G	& (note)
01100	12	0C	14	I	8
01101	13	0D	15	P	0
01110	14	0E	16	C	:
01111	15	0F	17	V	;
10000	16	10	20	E	3
10001	17	11	21	Z	"
10010	18	12	22	D	$
10011	19	13	23	B	?
10100	20	14	24	S	Bell
10101	21	15	25	Y	6
10110	22	16	26	F	! (note)
10111	23	17	27	X	/
11000	24	18	30	A	-
11001	25	19	31	W	2
11010	26	1A	32	J	'
11011	27	1B	33	[Figure Shift]	
11100	28	1C	34	U	7
11101	29	1D	35	Q	1
11110	30	1E	36	K	(
11111	31	1F	37	[Letter Shift]	

Note:
The letters F, G and H in the Figures mode are not allocated internationally. Each country is free to use them as they see fit. The American usage is shown in the table above.

Repeater Listings

There are over 900 repeaters licensed in the United Kingdom which range in frequency from 28MHz. Many are traditional FM units, but there are also several for Amateur Television and increasingly Digital modes such as D-STAR, FUSION, DMR and other protocols. Many voice repeaters also offer internet access via computers or linking to other repeaters, using services such as Echolink and IRLP (Internet Radio Linking Project). These allow stations to communicate widely using radio into a local repeater or gateway, or by computer through a VoIP gateway. There is a network of APRS stations (Automatic Position Reporting System) which uses AX25 packet data to report the GPS location of users on a common frequency of 144.8000MHz. More information is available at: *https://ukrepeater.net*

Voice Repeaters (by output frequency)

Band	CH	Out (MHz)	In (MHz)	Callsign	CTCSS	Location	Keeper
10M	10M	29.63	29.53	GB3CQ	88.5	WORTHING	G4WTV
10M	10M	29.65	29.55	GB3TX	94.8	NORWICH	M0ZAH
10M	10M	29.66	29.56	GB3WP	82.5	LEEDS	M0YDG
6M	R50-01	50.72	51.22	GB3EF	110.9	SUFFOLK	G0OZS
6M	R50-02	50.73	51.23	GB3SL	103.5	KILSYTH	GM4COX
6M	R50-02	50.73	51.23	GB3XD	71.9	LOUTH	G7AJP
6M	R50-03	50.74	51.24	GB3UM	77	MARKFIELD	M1NAS
6M	R50-04	50.75	51.25	GB3LP	77	LIVERPOOL	M1SWB
6M	R50-05	50.76	51.26	MB5WP	82.5	WAKEFIELD	M0YDG
6M	R50-06	50.77	51.27	GB3DB	110.9	DANBURY	G6JYB
6M	R50-06	50.77	51.27	GB3FH	77	SOMERSET	G4RKY
6M		50.78	51.78	GB3PX	77	BARKWAY	M0ZPU
6M	R50-08	50.79	51.29	GB3SX	103.5	STOKE ON TRENT	G4SCY
6M	R50-09	50.8	51.3	GB3WY	82.5	WAKEFIELD	G1XCC
6M	R50-09	50.8	51.3	GB3ZY	77	BRISTOL	G4RKY
6M	R50-10	50.81	51.31	GB3FX	82.5	FARNHAM	G4EPX
6M	R50-10	50.81	51.31	GB3ZW	103.5	NEWTOWN POWYS	GW4IQP
6M	R50-11	50.82	51.32	GB3HM	71.9	BELPER	G8IQP
6M	R50-13	50.84	51.34	GB3AM	77	AMERSHAM	M0ZPU
6M	R50-14	50.85	51.35	GB3XS	77	NORTHAMPTON	G7SYT
6M	R50-15	50.86	51.36	GB3VI	67	DUDLEY	G8PYT
2M		145.575	144.975	2E0PEU	103.5	STOKE-ON-TRENT	2E0PEU
2M	RV47	145.5875	144.9875	GB3RT	118.8	RICHMOND Nth YORKSHIRE	G6IET
2M	RV47	145.5875	144.9875	MB5BW	82.5	WAKEFIELD	M0YDG
2M	RV48	145.6	145	GB3AS	77	LANGHOLM	GM6LJE
2M	RV48	145.6	145	GB3CF	77	MARKFIELD	M0VSE
2M	RV48	145.6	145	GB3HU	82.5	HUDDERSFIELD	G0ISX
2M	RV48	145.6	145	GB3LY	110.9	LIMAVADY	MI1VOX
2M	RV48	145.6	145	GB3RW	88.5	WORTHING	G4WTV
2M	RV48	145.6	145	GB3SS	67	ELGIN MORAY	GM4ILS
2M	RV48	145.6	145	GB3TX	94.8	WYMONDHAM	M0ZAH
2M	RV48	145.6	145	GB7PC	7	MULLION	G3XOU
2M	RV48	145.6	145	GB7RW	88.5	WHITBY	G4EQS
2M	RV48	145.6	145	M0WTV		WINTER HILL LANCS	M0WTV
2M	RV49	145.6125	145.0125	GB3CW	71.9	FAREHAM	M0LIH
2M	RV49	145.6125	145.0125	GB3EI	88.5	LOCHMADDY	GM8SAU
2M	RV49	145.6125	145.0125	GB3EW	77	EXETER	G6ATJ
2M	RV49	145.6125	145.0125	GB3NA	71.9	BARNSLEY	G4LUE
2M	RV49	145.6125	145.0125	GB3OA	82.5	SOUTHPORT	G4EID
2M	RV49	145.6125	145.0125	GB3SV	94.8	STAFFORD	G7PFT
2M	RV49	145.6125	145.0125	GB3TE	67	CLACTON ON SEA	G0MBA
2M	RV49	145.6125	145.0125	GB3UK	118.8	WASHINGTON	G1LBU
2M	RV49	145.6125	145.0125	GB3VM	103.5	TENBURY WELLS	G8XYJ
2M	RV49	145.6125	145.0125	GB7LV	1	LIVINGSTON	GM7HHB
2M	RV49	145.6125	145.0125	GB7VR	82.5	HARPENDEN	G7HMV
2M	RV50	145.625	145.025	GB3GD	110.9	SNAEFELL IOM	GD4HOZ
2M	RV50	145.625	145.025	GB3GM	67	GREAT MALVERN	G7KPR
2M	RV50	145.625	145.025	GB3HG	88.5	THIRSK	G8IMZ
2M	RV50	145.625	145.025	GB3KS	103.5	DOVER	M1CMN
2M	RV50	145.625	145.025	GB3MH	88.5	EAST GRINSTEAD	G7KBR
2M	RV50	145.625	145.025	GB3NB	94.8	NORWICH	G8VLL
2M	RV50	145.625	145.025	GB3NF	77	NOTTINGHAM	M0VUB
2M	RV50	145.625	145.025	GB3NG	67	FRASERBURGH	GM4ZUK
2M	RV50	145.625	145.025	GB3PA	103.5	PAISLEY	GM7GDE
2M	RV50	145.625	145.025	GB3PB	71.9	WIMBORNE	G7ICH
2M	RV50	145.625	145.025	GB3SF		PERTH	GM3NFO
2M	RV50	145.625	145.025	GB3SI	77	HELSTON	M1ERD
2M	RV50	145.625	145.025	GB7DW	103.5	DUNKERSWELL	G3ZXX
2M	RV51	145.6375	145.0375	GB3DN	77	STIBB CROSS	G1BHM
2M	RV51	145.6375	145.0375	GB3GJ	71.9	ST HELIER	GJ8PVL
2M	RV51	145.6375	145.0375	GB3IN	71.9	TANSLEY NR MATLOCK	G4TSN
2M	RV51	145.6375	145.0375	GB3MX	77	MARYPORT	M0LLC
2M	RV51	145.6375	145.0375	GB3VO	110.9	WREXHAM	GW0WZZ
2M	RV51	145.6375	145.0375	GB7CT	3	TRING	M0ZPU
2M	RV51	145.6375	145.0375	GB7DE	94.8	KENNOWAY FIFE	GM7RYR
2M	RV51	145.6375	145.0375	GB7RN		FAREHAM	G3ZDF
2M	RV51	145.6375	145.0375	GB7SW	2	CHIPPENHAM	G0RMA
2M	RV52	145.65	145.05	GB3AY	103.5	DALRY	MM0YET
2M	RV52	145.65	145.05	GB3HS	88.5	HULL	G3GJA
2M	RV52	145.65	145.05	GB3MN	82.5	DISLEY	G8LZO
2M	RV52	145.65	145.05	GB3OC	77	KIRKWALL	GM0HQG
2M	RV52	145.65	145.05	GB3PO	110.9	IPSWICH	G6PDE
2M	RV52	145.65	145.05	GB3SB	118.8	SELKIRK	GM7LUN
2M	RV52	145.65	145.05	GB3TR	94.8	TORQUAY	G8XST
2M	RV52	145.65	145.05	GB3WH	118.8	SWINDON	G0UWS
2M	RV52	145.65	145.05	GB7KB	3	HEATHFIELD	G1ERJ
2M	RV53	145.6625	145.0625	GB3AA	94.8	BRISTOL	G4CJZ
2M	RV53	145.6625	145.0625	GB3FE	103.5	STIRLING	GM0MZB
2M	RV53	145.6625	145.0625	GB3KI	103.5	HERNE BAY	G4TKR
2M	RV53	145.6625	145.0625	GB3LZ	77	WINKLEIGH	G4MQQ
2M	RV53	145.6625	145.0625	GB3PH	118.8	PWLLHELI	MW0XTK
2M	RV53	145.6625	145.0625	GB3PK	110.9	BALLYCASTLE	MI0CRR
2M	RV53	145.6625	145.0625	GB3SH	71.9	SOUTHAMPTON	G4MYS
2M	RV53	145.6625	145.0625	GB7AM		HIGHER POYNTON	G1DVA
2M	RV53	145.6625	145.0625	GB7HQ	71.9	HUTHWAITE	G4TSN
2M	RV53	145.6625	145.0625	GB7IC-C		HERNE BAY	G4TKR
2M	RV53	145.6625	145.0625	GB7PB		DURHAM	G7VIL
2M	RV54	145.675	145.075	GB3ES	103.5	HASTINGS	G6ZZX
2M	RV54	145.675	145.075	GB3LD	110.9	LANCASTER	G8TZJ
2M	RV54	145.675	145.075	GB3LU	77	LERWICK	GM0GFL
2M	RV54	145.675	145.075	GB3PE	94.8	PETERBOROUGH	M0ZPU
2M	RV54	145.675	145.075	GB3PR	94.8	PERTH	GM8KPH
2M	RV54	145.675	145.075	GB3RD	118.8	READING	G8DOR
2M	RV54	145.675	145.075	GB7IE	77	PLYMOUTH	G1LOE
2M	RV54	145.675	145.075	GB7NA-C		BARNSLEY	G4LUE
2M	RV54	145.675	145.075	GB7RB-C		COWBRIDGE	GW6CUR
2M	RV54	145.675	145.075	GB3BX	13	CLEOBURY NORTH	G4VZO
2M	RV55	145.6875	145.0875	GB3CD	118.8	CROOK	G0OCB
2M	RV55	145.6875	145.0875	GB3CN	94.8	NEWCASTLE EMLYN	GW6JSO
2M	RV55	145.6875	145.0875	GB3DC	71.9	DERBY	G7NPW
2M	RV55	145.6875	145.0875	GB3KE	103.5	GLASGOW	GM7SVK
2M	RV55	145.6875	145.0875	GB3SJ	103.5	NORTHWICH	M0WTX
2M	RV55	145.6875	145.0875	GB3WE	94.8	WESTON-S-MARE	G4SZM
2M	RV55	145.6875	145.0875	GB3XP	82.5	MORDEN	M0SGL
2M	RV55	145.6875	145.0875	GB7GD	1	ABERDEEN	MM0RDM
2M	RV55	145.6875	145.0875	GB7HD	3	TAVISTOCK	G3XOU
2M	RV56	145.7	145.1	GB3AR	110.9	CAERNARFON	GW4KAZ
2M	RV56	145.7	145.1	GB3BB	94.8	BRECON	MW0XDD
2M	RV56	145.7	145.1	GB3BT	118.8	BERWICK ON TWEED	MM0JNL
2M	RV56	145.7	145.1	GB3EV	77	DUFTON	G0TNF
2M	RV56	145.7	145.1	GB3HH	71.9	BUXTON	G7EKY
2M	RV56	145.7	145.1	GB3HI	103.5	ISLE OF MULL	MM0JRM
2M	RV56	145.7	145.1	GB3II	110.9	NR EYE	G0JSV
2M	RV56	145.7	145.1	GB3KN	103.5	MAIDSTONE	G3VFC
2M	RV56	145.7	145.1	GB3VA	118.8	BRILL	G8GQJ
2M	RV56	145.7	145.1	GB3WD	77	IVYBRIDGE	M0ZCP
2M	RV56	145.7	145.1	GB7CM	71.9	MILTON ABBAS	M0MRP
2M	RV57	145.7125	145.1125	GB3BM	67	DUDLEY	G8PYT
2M	RV57	145.7125	145.1125	GB3BM-2	67	BIRMINGHAM	G8PYT
2M	RV57	145.7125	145.1125	GB3FA	82.5	SHEFFIELD	M0PCF
2M	RV57	145.7125	145.1125	GB3FG	94.8	CARMARTHEN	GW8KCY
2M	RV57	145.7125	145.1125	GB3KY	94.8	KINGS LYNN	G1SCQ
2M	RV57	145.7125	145.1125	GB3LA	103.5	SANQUHAR	GM3SAN
2M	RV57	145.7125	145.1125	GB3MI	82.5	MANCHESTER	G0TOG
2M	RV57	145.7125	145.1125	GB3YA	94.8	CWMBRAN	MW0YAC
2M	RV57	145.7125	145.1125	GB7CG	77	WOBURN SANDS	G0WTZ
2M	RV57	145.7125	145.1125	GB7OK	3	BECKENHAM	G1HIG
2M	RV58	145.725	145.125	GB3AG	94.8	ANGUS	MM0XET
2M	RV58	145.725	145.125	GB3BI	67	MUIR OF ORD	MM0GKB
2M	RV58	145.725	145.125	GB3CG	118.8	GLOUCESTER	G3LVP
2M	RV58	145.725	145.125	GB3DA	110.9	DANBURY	G6JYB
2M	RV58	145.725	145.125	GB3LM	71.9	LINCOLN	M0TEF
2M	RV58	145.725	145.125	GB3NC	77	ROCHE	G4WVD
2M	RV58	145.725	145.125	GB3NI	110.9	BELFAST N.I.	MI1VOX
2M	RV58	145.725	145.125	GB3SN	71.9	ALTON HANTS	G4EPX
2M	RV58	145.725	145.125	GB3TP	82.5	SHIPLEY	G8ZMG
2M	RV58	145.725	145.125	GB3VT	103.5	STOKE ON TRENT	G8NSS
2M	RV59	145.7375	145.1375	GB3AL	77	AMERSHAM	M0ZPU
2M	RV59	145.7375	145.1375	GB3DR		WEYMOUTH	G3VPF
2M	RV59	145.7375	145.1375	GB3NX	77	NEAR HOLSWORTHY	G1BHM
2M	RV59	145.7375	145.1375	GB3RV	82.5	HASLINGDEN	M0LMN
2M	RV59	145.7375	145.1375	GB3ZA	118.8	HEREFORD	G0JWJ
2M	RV59	145.7375	145.1375	GB7BM-C	67	BIRMINGHAM	G8NDT
2M	RV59	145.7375	145.1375	GB7NM		ATTLEBOURGH	G0LGJ
2M	RV59	145.7375	145.1375	GB7SF-C		SHEFFIELD	M1ERS
2M	RV60	145.75	145.15	GB3BC	94.8	PONTYPRIDD	GW0UXJ
2M	RV60	145.75	145.15	GB3BG	118.8	CHULMLEIGH	G7SOJ
2M	RV60	145.75	145.15	GB3CS	103.5	MOTHERWELL	GM8HBY
2M	RV60	145.75	145.15	GB3FK	103.5	FOLKESTONE	M1CMN
2M	RV60	145.75	145.15	GB3HA	118.8	HEXHAM	M0TKF
2M	RV60	145.75	145.15	GB3MP	110.9	PRESTATYN	M0OBW
2M	RV60	145.75	145.15	GB3PI	77	ROYSTON	M0ZPU
2M	RV60	145.75	145.15	GB3WS	88.5	HORSHAM	G4LRP
2M	RV60	145.75	145.15	GB3FU	71.9	NOTTINGHAM	M5ADU
2M	RV60	145.75	145.15	GB3YC	88.5	FILEY	M0KXQ
2M	RV61	145.7625	145.1625	GB3IK	103.5	ROCHESTER	G6CKK
2M	RV61	145.7625	145.1625	GB3IR	88.5	RICHMOND YORKS	G4FZN
2M	RV61	145.7625	145.1625	GB3NE	118.8	NEWBURY	G6IBI
2M	RV61	145.7625	145.1625	GB3PL	77	PLYMPTON	M0WMB
2M	RV61	145.7625	145.1625	GB7SG	1	LIVERPOOL	M1SWB
2M	RV61	145.7625	145.1625	GB7YF	94.8	YEOVIL	M0VKR
2M	RV61	145.7625	145.1625	GB3EM	77	COLD ASHBY	M1FJB
2M	RV62	145.775	145.175	GB3DG	103.5	NEWTON STEWART	GM8CJG
2M	RV62	145.775	145.175	GB3GN	67	BANCHORY	GM4ZUK
2M	RV62	145.775	145.175	GB3IG	88.5	STORNOWAY	GM0LZE
2M	RV62	145.775	145.175	GB3NL	82.5	NORTH LONDON	G4DFB

Voice Repeaters (by output frequency)

Band	CH	Out (MHz)	In (MHz)	Callsign	CTCSS	Location	Keeper
2M	RV62	145.775	145.175	GB3RF	82.5	ACCRINGTON	M0NFI
2M	RV62	145.775	145.175	GB3WK	67	LEAMINGTON SPA	G6FEO
2M	RV62	145.775	145.175	GB3WT	110.9	OMAGH	GI3NVW
2M	RV62	145.775	145.175	GB7DA	1	AIRDRIE	GM4AUP
2M	RV62	145.775	145.175	GB7ER	77	EXETER	M0ZZT
2M	RV62	145.775	145.175	GB7IV		SOUTHAMPTON	G4MYS
2M	RV62	145.775	145.175	GB7MS	10	NEWCASTLE	M0DMP
2M	RV62	145.775	145.175	GB7RX	103.5	HASTINGS	M0JBR
2M	RV62	145.775	145.175	GB7TE		CLACTON ON SEA	G0MBA
2M	RV62	145.775	145.175	GB3WW	94.8	PORT TALBOT	MW0HAC
2M	RV63	145.7875	145.1875	GB3BF	77	BEDFORD	G8MGP
2M	RV63	145.7875	145.1875	GB3EB	88.5	NEWHAVEN	G0TJH
2M	RV63	145.7875	145.1875	GB3JB	103.5	MERE WILTSHIRE	G3ZXX
2M	RV63	145.7875	145.1875	GB3JL	77	LISKEARD	G4RKY
2M	RV63	145.7875	145.1875	GB3KD	118.8	KIDDERMINSTER	M0KRG
2M	RV63	145.7875	145.1875	GB3LB	118.8	LAUDER	GM7LUN
2M	RV63	145.7875	145.1875	GB3OG	110.9	LLANDUDNO	MW0JWP
2M	RV63	145.7875	145.1875	GB3YW	82.5	WAKEFIELD	G1XCC
2M	RV63	145.7875	145.1875	GB7YL	1	LOWESTOFT	G4TAD
70CM	DVU-R20	430.25	439.25	GB3XV		DAVENTRY	G8KHF
70CM	DVU-R20	430.25	439.25	GB7RU	123	EASTBOURNE	G6JME
70CM	DVU-R21	430.2625	439.2625	GB7RC		CHELTENHAM	M0URF
70CM	DVU-R21	430.2625	439.2625	GB7RO		HASLINGDEN	M0LMN
70CM	DVU-R22	430.275	439.275	GB7CC	2	CHELTENHAM	G1MAW
70CM	DVU-R22	430.275	439.275	GB7FO	1	BLACKPOOL	G0WDA
70CM	DVU-R22	430.275	439.275	GB7KE		CHATHAM	2E0SGG
70CM	DVU-R23	430.2875	439.2875	GB3PC	118.8	FARMBOROUGH	G4OTJ
70CM	DVU-R23	430.2875	439.2875	GB7HA	7	HARPENDEN	G7HMV
70CM	DVU-R23	430.2875	439.2875	GB7TH	5	MARGATE	M0LMK
70CM	DVU-R23	430.2875	439.2875	GB7YN	5	WAKEFIELD	M0YDG
70CM	DVU-R24	430.3	439.3	GB7BF	5	BLACKPOOL	M0DKR
70CM	DVU-R24	430.3	439.3	GB7KR	8	KIDDERMINSTER	M0KRG
70CM	DVU-R24	430.3	439.3	MB5KT	5	KEIGHLEY	M0DIT
70CM	DVU-R24	430.3	439.3	GB7RH	123	HEATHFIELD	G6JME
70CM	RU65	430.8125	438.4125	GB3BD	71.9	BANBURY	M0NCV
70CM	RU65	430.8125	438.4125	GB3BR	88.5	HOVE	G4WTV
70CM	RU65	430.8125	438.4125	GB3BW	82.5	WAKEFIELD	M0YDG
70CM	RU65	430.8125	438.4125	GB3FV	88.5	STIRLING	GM0WUR
70CM	RU65	430.8125	438.4125	GB3KR	118.8	STOURPORT ON SEVERN	G7SAI
70CM	RU65	430.8125	438.4125	GB3RX	77	PRESCOT	M0NFI
70CM	RU65	430.8125	438.4125	GB3TX	94.8	WYMONDHAM	M0ZAH
70CM	RU65	430.8125	438.4125	GB3VR	67	HARPENDEN	G7HMV
70CM	RU65	430.8125	438.4125	GB7AY	1	KILMARNOCK	GM0ONX
70CM	RU65	430.8125	438.4125	GB7GW	103.5	WASHINGTON	G1LBU
70CM	RU66	430.825	438.425	GB3CI	77	CORBY	G7HPE
70CM	RU66	430.825	438.425	GB3IM	110.9	DOUGLAS IOM	GD4HOZ
70CM	RU66	430.825	438.425	GB3IM	71.9	RAMSEY IOM	GD4HOZ
70CM	RU66	430.825	438.425	GB3KW	103.5	GLASGOW	GM7SVK
70CM	RU66	430.825	438.425	GB3RO	82.5	HASLINGDEN	M0LMN
70CM	RU66	430.825	438.425	GB3ZB	77	BRISTOL	G4RKY
70CM	RU66	430.825	438.425	GB3ZX	88.5	EASTBOURNE	G6ZZX
70CM	RU66	430.825	438.425	GB7CH	94.8	KIRRIEMUIR	MM0JNL
70CM	RU66	430.825	438.425	GB7CP	77	SLOUGH	M0WAQ
70CM	RU66	430.825	438.425	GB7NV	1	NOTTINGHAM	M0NGT
70CM	RU66	430.825	438.425	MB5CR	94.8	CROMER	G4LYB
70CM	RU67	430.8375	438.4375	GB3JS	94.8	GREAT YARMOUTH	G4TAD
70CM	RU67	430.8375	438.4375	GB3NY	67	HULL	M0DPH
70CM	RU67	430.8375	438.4375	GB3OR	103.5	LLANDUDNO	MW0JWP
70CM	RU67	430.8375	438.4375	GB3RL	88.5	RUSTINGTON	G4WTV
70CM	RU67	430.8375	438.4375	GB3VS	94.8	TRULL	G4UVZ
70CM	RU67	430.8375	438.4375	GB7EU	94.8	EDINBURGH	GM7RYR
70CM	RU68	430.85	438.45	GB3BS	118.8	BRISTOL	G4SDR
70CM	RU68	430.85	438.45	GB3BZ	110.9	BRAINTREE	G0DEC
70CM	RU68	430.85	438.45	GB3FC	82.5	BLACKPOOL	M0DKR
70CM	RU68	430.85	438.45	GB3GR	71.9	GRANTHAM	G0OJF
70CM	RU68	430.85	438.45	GB3KT	74.4	KEIGHLEY	M0DIT
70CM	RU68	430.85	438.45	GB3PD		WEYMOUTH	G1ZSE
70CM	RU68	430.85	438.45	GB3RR	67	DUDLEY	G8PYT
70CM	RU68	430.85	438.45	GB7EI	88.5	LOCHMADDY	GM8SAU
70CM	RU68	430.85	438.45	GB7GL	7	GLASGOW	GM7RYR
70CM	RU68	430.85	438.45	GB7JI	2	ST HELIER	GJ8PVL
70CM	RU69	430.8625	438.4625	GB3CM	94.8	CARMARTHEN	GW8KCY
70CM	RU69	430.8625	438.4625	GB3CY	88.5	YORK	G4FUO
70CM	RU69	430.8625	438.4625	GB3JH	77	TIVERTON	G6ASK
70CM	RU69	430.8625	438.4625	GB3LR	88.5	NEWHAVEN	G0TJH
70CM	RU69	430.8625	438.4625	GB3PT	94.8	PRESCOT	M0NFI
70CM	RU69	430.8625	438.4625	GB3RC	103.5	CHELTENHAM	M0URF
70CM	RU69	430.8625	438.4625	GB3SC	67	BURTON-UPON-TRENT	G8EKG
70CM	RU69	430.8625	438.4625	GB7CX	12	COLERAINE	MI0PKO
70CM	RU69	430.8625	438.4625	GB3BE	118.8	DUNS	GM7LUN
70CM	RU70	430.875	438.475	GB3BL	77	BEDFORD	G8MGP
70CM	RU70	430.875	438.475	GB3CH	77	LISKEARD	G4RKY
70CM	RU70	430.875	438.475	GB3IM	110.9	PEEL	GD4HOZ
70CM	RU70	430.875	438.475	GB3WA	103.5	DALRY	GM7GDE
70CM	RU70	430.875	438.475	GB3YL	94.8	LOWESTOFT	G4TAD
70CM	RU70	430.875	438.475	GB7HN	82.5	LEIGH	M0HOY
70CM	RU70	430.875	438.475	MB5HM	71.9	BELPER	G8IQP
70CM	RU71	430.8875	438.4875	GB3ED	94.8	EDINBURGH	GM4GZW
70CM	RU71	430.8875	438.4875	GB3EK	103.5	MARGATE	M0LMK
70CM	RU71	430.8875	438.4875	GB3HL	67	LESLIE	MM0TXO
70CM	RU71	430.8875	438.4875	GB3HO	88.5	HORSHAM	G4LRP
70CM	RU71	430.8875	438.4875	GB3IS	94.8	ISLE OF SOUTH UIST	GM8SAU
70CM	RU71	430.8875	438.4875	GB3KA	118.8	KIDDERMINSTER	M0KRG
70CM	RU71	430.8875	438.4875	GB3NP	77	TOWCESTER	G1IRG
70CM	RU71	430.8875	438.4875	GB3ZN	118.8	HEBBURN	M0OMT
70CM	RU71	430.8875	438.4875	GB7AF	71.9	HUCKNALL	G0LCG
70CM	RU71	430.8875	438.4875	GB7FI	3	AXBRIDGE	G4RKY
70CM	RU71	430.8875	438.4875	GB7FU		NOTTINGHAM	G0LCG
70CM	RU71	430.8875	438.4875	GB7JM	71.9	ALFORD	M0AQC
70CM	RU71	430.8875	438.4875	GB7TQ	94.8	TORQUAY	G8XST
70CM	RU71	430.8875	438.4875	MB5HR	110.9	MOLD	GW7VFJ
70CM	RU72	430.9	438.5	G4NAB	118.8	ASHINGTON	G4NAB
70CM	RU72	430.9	438.5	GB3BK	103.5	BROMLEY	G0WYG
70CM	RU72	430.9	438.5	GB3GL	103.5	GLASGOW	GM3SAN
70CM	RU72	430.9	438.5	GB3HY	88.5	HAYWARDS HEATH	G6DGK
70CM	RU72	430.9	438.5	GB3KC	67	STOURBRIDGE	G3PWJ
70CM	RU72	430.9	438.5	GB3MG		PONTYPRIDD	GW0UXJ
70CM	RU72	430.9	438.5	GB3PM	71.9	SHEFFIELD	M0HOY
70CM	RU72	430.9	438.5	GB3PY	77	CAMBRIDGE	M0ZPU
70CM	RU72	430.9	438.5	GB3PZ	82.5	DUKINFIELD	G4ZPZ
70CM	RU72	430.9	438.5	GB7AG	94.8	FORFAR	GM1TCN
70CM	RU72	430.9	438.5	GB7DG	7	PORTPATRICK	GM7RYR
70CM	RU72	430.9	438.5	GB7GJ	88.5	ST HELIER	GJ8PVL
70CM	RU73	430.9125	438.5125	G4IPE	77	SCUNTHORPE	G4IPE
70CM	RU73	430.9125	438.5125	GB3EG	82.5	WIGAN	M1EGH
70CM	RU73	430.9125	438.5125	GB3ML	103.5	STAINES UPON THAMES	M0TNY
70CM	RU73	430.9125	438.5125	GB3SG	118.8	STROUD	G1MAW
70CM	RU73	430.9125	438.5125	GB3SR	71.9	SKEGNESS	G7UVD
70CM	RU73	430.9125	438.5125	GB3UP	94.8	FISHGUARD	GW6TKK
70CM	RU73	430.9125	438.5125	GB3XX	77	DAVENTRY	G8KHF
70CM	RU74	430.925	438.525	GB3DM	103.5	DUMBARTON	GM7GDE
70CM	RU74	430.925	438.525	GB3FI	77	CHEDDAR	G4RKY
70CM	RU74	430.925	438.525	GB3HE	103.5	HASTINGS	G3TCG
70CM	RU74	430.925	438.525	GB3LN	118.8	LEICESTER	M0HEL
70CM	RU74	430.925	438.525	GB3PF	94.8	ACCRINGTON	M0NFI
70CM	RU74	430.925	438.525	GB3VN	103.5	LUDLOW	G4OYX
70CM	RU74	430.925	438.525	GB3XN	71.9	WORKSOP	G3XXN
70CM	RU74	430.925	438.525	GB7BR	3	RAMSEY IOM	GD4HOZ
70CM	RU74	430.925	438.525	GB7CA	2	DOUGLAS	GD4HOZ
70CM	RU74	430.925	438.525	GB7EC	11	ABERDEEN	GM0NRT
70CM	RU74	430.925	438.525	GB7EE	1	EDINBURGH	GM7RYR
70CM	RU74	430.925	438.525	GB7KK	1	BALLYCASTLE	MI0CRR
70CM	RU74	430.925	438.525	GB7NT	118.8	NEWBURY	G3WYW
70CM	RU74	430.925	438.525	GB7PY	3	CAMBRIDGE	M0ZPU
70CM	RU75	430.9375	438.5375	GB3LC	71.9	LOUTH	G7AJP
70CM	RU75	430.9375	438.5375	GB7DT	7	ROCHE	M1AEG
70CM	RU75	430.9375	438.5375	GB7IX	110.9	WESTCLIFF ON SEA	2E1HWE
70CM	RU76	430.95	438.55	GB3EX	77	SILVERTON	G7NBU
70CM	RU76	430.95	438.55	GB3NH	77	NORTHAMPTON	G1IRG
70CM	RU76	430.95	438.55	GB3OH	94.8	LINLITHGOW	GM0MZB
70CM	RU76	430.95	438.55	GB3OM	110.9	OMAGH	GI4SXV
70CM	RU76	430.95	438.55	GB3OY	82.5	BUCKHURST HILL	G7UZN
70CM	RU76	430.95	438.55	GB3PN	82.5	BLACKPOOL	M0DKR
70CM	RU76	430.95	438.55	GB3UO	110.9	WREXHAM	GW0WZZ
70CM	RU76	430.95	438.55	GB3WL	67	WYTHALL	G4VPD
70CM	RU76	430.95	438.55	GB7AC	103.5	LARGS	GM7RYR
70CM	RU76	430.95	438.55	GB7BD	3	BRISTOL	G4RKY
70CM	RU76	430.95	438.55	GB3WP-B	88.5	WAKEFIELD	M0YDG
70CM	RU77	430.9625	438.5625	GB3BU	77	BUDE	G1BHM
70CM	RU77	430.9625	438.5625	GB3CC	88.5	CHICHESTER	G4WTV
70CM	RU77	430.9625	438.5625	GB3EH	67	BANBURY	G4OHB
70CM	RU77	430.9625	438.5625	GB7SH	71.9	SHEFFIELD	M1ERS
70CM	RU78	430.975	438.575	G5BSD	82.5	SURREY	G5BSD
70CM	RU78	430.975	438.575	GB3AC	94.8	BRISTOL	G4CJZ
70CM	RU78	430.975	438.575	GB3BY	82.5	CHORLEY	M0DKR
70CM	RU78	430.975	438.575	GB3DQ	77	POLPERRO	G1YDQ
70CM	RU78	430.975	438.575	GB3EZ	110.9	BURY ST EDMUNDS	G1YFF
70CM	RU78	430.975	438.575	GB3JR	71.9	MANSFIELD	M5ADU
70CM	RU78	430.975	438.575	GB3KK	110.9	BALLYCASTLE	MI0CRQ
70CM	RU78	430.975	438.575	GB3KV	103.5	KILSYTH	GM3SAN
70CM	RU78	430.975	438.575	GB3WO	88.5	WORTHING	M0IAD
70CM	RU78	430.975	438.575	GB3ZI	103.5	STAFFORD	G6MDX
70CM	RU78	430.975	438.575	GB7GM	118.8	DUMFRIES	M0UKB
70CM	RU78	430.975	438.575	GB7TG	118.8	WEMBWORTHY	G7SOJ
70CM	RU78	430.975	438.575	GB7CN	67	CANNOCK	M0RKS
70CM	RU79	430.9875	438.5875	GB3BH	77	BEXHILL-ON-SEA	M0JDN
70CM	RU79	430.9875	438.5875	GB3DZ	71.9	DERBY	G7NPW
70CM	RU79	430.9875	438.5875	GB3WU	77	WHITEHAVEN	M0LLC
70CM	RU79	430.9875	438.5875	GB3YN	77	CLECKHEATON	M0YDG
70CM	RU79	430.9875	438.5875	GB7NI	77	NORTHAMPTON	2E0WSJ
70CM	RB0	433	434.6	GB3BN	118.8	BRACKNELL	G8DOR
70CM	RB0	433	434.6	GB3DT	71.9	BLANDFORD	G0ZEP
70CM	RB0	433	434.6	GB3LL	94.8	LLANGOLLEN	G7RPG
70CM	RB0	433	434.6	GB3MK	77	MILTON KEYNES	G6GEI
70CM	RB0	433	434.6	GB3NR	94.8	NORWICH	G8VLL
70CM	RB0	433	434.6	GB3NT	118.8	NEWCASTLE	M0MNE
70CM	RB0	433	434.6	GB3PU	94.8	PERTH	GM8KPH
70CM	RB0	433	434.6	GB3US	103.5	SHEFFIELD	M0RFT
70CM	RB01	433.025	434.625	GB3BV	82.5	HEMEL HEMPSTEAD	G7HMV
70CM	RB01	433.025	434.625	GB3HJ	118.8	HARROGATE	G4MEM
70CM	RB01	433.025	434.625	GB7HQ	71.9	HUTHWAITE	G4TSN
70CM	RB02	433.05	434.65	GB3AV	118.8	AYLESBURY	G8GQJ
70CM	RB02	433.05	434.65	GB3LS	71.9	LINCOLN	M0TEF
70CM	RB02	433.05	434.65	GB3LV	82.5	NORTH LONDON	G4DFB
70CM	RB02	433.05	434.65	GB3NN	94.8	WELLS NORFOLK	G0FVF
70CM	RB02	433.05	434.65	GB3UL	110.9	BELFAST N.I.	MI1VOX
70CM	RB02	433.05	434.65	GB3WM	67	LONGBRIDGE	M0JZT
70CM	RBW04	433.05	438.05	GB7DZ	103.5	CONGLETON	M0XDR
70CM	RB02	433.05	434.65	GB7ZE	103.5	HASTINGS	M0JBR
70CM	RB02	433.05	434.65	GB3HK	118.8	SELKIRK	GM7LUN
70CM	RB03	433.075	434.675	GB3ER	110.9	DANBURY	G6JYB
70CM	RB03	433.075	434.675	GB3KU	82.5	ASHTON-U-LYNE	M0NCZ
70CM	RB03	433.075	434.675	GB3TK	71.9	NOTTINGHAM	M5ADU
70CM	RB04	433.1	434.7	GB3IH	110.9	IPSWICH	G6PDE
70CM	RB04	433.1	434.7	GB3KL	94.8	KINGS LYNN	G0IJU
70CM	RB04	433.1	434.7	GB3LE	77	MARKFIELD	M1NAS
70CM	RB04	433.1	434.7	GB3NK	103.5	ERITH	G4EGU
70CM	RB04	433.1	434.7	GB3SP	94.8	PEMBROKE	GW3XJQ
70CM	RB04	433.1	434.7	GB3UB	118.8	BATH AVON	G4KVI
70CM	RB04	433.1	434.7	GB3VE	77	DUFTON	G0TNF
70CM	RB05	433.125	434.725	GB3GH	118.8	GLOUCESTER	G3LVP
70CM	RB05	433.125	434.725	GB3IC	67	WOLVERHAMPTON	G7CFC
70CM	RB05	433.125	434.725	GB3IM	110.9	DOUGLAS IOM	GD4HOZ
70CM	RB06	433.15	434.75	GB3CR	110.9	CAERGWRLE	M0OBW
70CM	RB06	433.15	434.75	GB3DI	118.8	DIDCOT	G8CUL
70CM	RB06	433.15	434.75	GB3ME	67	RUGBY	G7BQM
70CM	RB06	433.15	434.75	GB3SK	103.5	CANTERBURY	G6DIK
70CM	RB06	433.15	434.75	GB3SY	71.9	BARNSLEY	G4LUE

Voice Repeaters (by output frequency)

Band	CH	Out (MHz)	In (MHz)	Callsign	CTCSS	Location	Keeper
70CM	RB07	433.175	434.775	GB3AU	82.5	AMERSHAM	M0ZPU
70CM	RB07	433.175	434.775	GB3DE	110.9	IPSWICH	G1NRL
70CM	RB07	433.175	434.775	GB3NM	71.9	NOTTINGHAM	M0VUB
70CM	RB07	433.175	434.775	GB3OD	1	LLANDUDNO	MW0JWP
70CM	RB07	433.175	434.775	GB3TS	118.8	SUNDERLAND	G0SCI
70CM	RB07	433.175	434.775	GB7SJ	103.5	NORTHWICH	M0WTX
70CM	RB08	433.2	434.8	G1YFF	210.7	HAVERHILL	G1YFF
70CM	RB08	433.2	434.8	GB3AN	110.9	AMLWCH	MW0GCT
70CM	RB08	433.2	434.8	GB3RB	71.9	BOLSOVER	G1SLE
70CM	RB08	433.2	434.8	GB3SU	71.9	SOUTHAMPTON	G4MYS
70CM	RB08	433.2	434.8	GB3TF	103.5	TELFORD	M0JZH
70CM	RB09	433.225	434.825	GB3CL	67	CLACTON	G0MBA
70CM	RB09	433.225	434.825	GB3HD	82.5	HUDDERSFIELD	G0ISX
70CM	RB09	433.225	434.825	GB3IW	71.9	NEWPORT	M0RQD
70CM	RB09	433.225	434.825	GB3ST	103.5	STOKE ON TRENT	G8NSS
70CM	RB09	433.225	434.825	GB3TU	77	TRING	M0ZPU
70CM	RBW20	433.25	438.25	G4VZO	123	KINGSWINFORD	G4VZO
70CM	RB10	433.25	434.85	GB3AW	71.9	NEWBURY	G8DOR
70CM	RB10	433.25	434.85	GB3DD	94.8	DUNDEE	GM4FLP
70CM	RB10	433.25	434.85	GB3LI	82.5	LIVERPOOL	G3WIC
70CM	RB10	433.25	434.85	GB3LT	77	LUTON	G6OUA
70CM	RB10	433.25	434.85	GB3MW	67	LEAMINGTON SPA	G6FEO
70CM	RBW21	433.2625	438.2625	G3PWJ	67	STOURBRIDGE	G3PWJ
70CM	RB11	433.275	434.875	GB3AH	94.8	EAST DEREHAM	G0LGJ
70CM	RB11	433.275	434.875	GB3HT	77	HINCKLEY	M0RVD
70CM	RB11	433.275	434.875	GB3MY	82.5	ATHERTON	G4YYB
70CM	RB11	433.275	434.875	GB3RE	103.5	MAIDSTONE	G6RVS
70CM	RB11	433.275	434.875	GB3RH	94.8	AXMINSTER	G6WWY
70CM	RB11	433.275	434.875	GB3RU	118.8	READING	G8DOR
70CM	RB12	433.3	434.9	GB3DX	110.9	DERRY/LONDONDERRY	GI4YWT
70CM	RB12	433.3	434.9	GB3GF	88.5	GUILDFORD	G4EML
70CM	RB12	433.3	434.9	GB3WB	94.8	WESTON-S-MARE	G1VSX
70CM	RB13	433.325	434.925	GB3BA	71.9	BASINGSTOKE	G8GTZ
70CM	RB13	433.325	434.925	GB3CA	77	CARLISLE	G1XSZ
70CM	RB13	433.325	434.925	GB3DS	71.9	WORKSOP	G3XXN
70CM	RB13	433.325	434.925	GB3GU	71.9	GUERNSEY	GU6EFB
70CM	RB13	433.325	434.925	GB3HW	110.9	GIDEA PARK	G4GBW
70CM	RB13	433.325	434.925	GB3SM	103.5	STOKE ON TRENT	G4SCY
70CM	RB13	433.325	434.925	GB3VH	82.5	WELWYN GARDEN CITY	G4THF
70CM	RB14	433.35	434.95	GB3CB	67	BIRMINGHAM	G8NDT
70CM	RB14	433.35	434.95	GB3CE	67	COLCHESTER	G0MBA
70CM	RB14	433.35	434.95	GB3HR	82.5	HARROW	G3YXZ
70CM	RB14	433.35	434.95	GB3LF	110.9	KENDAL	G8TZJ
70CM	RB14	433.35	434.95	GB3MR	82.5	STOCKPORT	G8LZO
70CM	RB14	433.35	434.95	GB3SD	71.9	WEYMOUTH	G0ECX
70CM	RB14	433.35	434.95	GB3WF	82.5	OTLEY	M0RIQ
70CM	RB15	433.375	434.975	GB3AB	82.5	SHEFFIELD	M0GAV
70CM	RB15	433.375	434.975	GB3FN	82.5	FARNHAM	G4EPX
70CM	RB15	433.375	434.975	GB3HB	77	ROCHE	G4WVD
70CM	RBW30	433.375	438.375	GB3KH	103.5	CHATHAM	2E0SGG
70CM	RB15	433.375	434.975	GB3LH	103.5	SHREWSBURY	G8DIR
70CM	RB15	433.375	434.975	GB3MB	94.8	MERTHYR	GW0UXJ
70CM	RB15	433.375	434.975	GB3PP	82.5	PRESTON	M0NED
70CM	RB15	433.375	434.975	GB3TH	67	TAMWORTH	G8YUQ
70CM	RB15	433.375	434.975	GB3WI	94.8	WISBECH	M0DUQ
70CM		434	0	GB7TR		NEW ASH GREEN	G6VBJ
70CM	UR63	438.3875	430.7875	GB3MO	110.9	MORECAMBE	G0VGS
70CM	UR63	438.3875	430.7875	GB3RK	88.5	BELFAST	MI5DAW
70CM	UR63	438.3875	430.7875	GB3TD	118.8	SWINDON	G4XUT
70CM	UR63	438.3875	430.7875	GB3WN	67	WOLVERHAMPTON	G8NDT
70CM	UR63	438.3875	430.7875	GB7LC	71.9	LINCOLN	G7RUP
70CM	DVU12	439.15	430.15	GW7VFJ	110.9	HOLYWELL	GW7VFJ
70CM	DVU12	439.15	430.15	M0JXR	110.9	BRISTOL	M0JXR
70CM	DVU12	439.15	430.15	M0OFF	94.8	HUNTINGDON	M0OFF
70CM	DVU13	439.1625	430.1625	GB7BS	3	BRISTOL	G4SDR
70CM	DVU13	439.1625	430.1625	GB7NS	3	CATERHAM	G0OLX
70CM	DVU13	439.1625	430.1625	GB7TD	1	WAKEFIELD	G1XCC
70CM	DVU14A	439.2188	430.2188	GB7PU		PRESTON	M0VXT
70CM	DVU14A	439.2188	430.2188	GB7TB		BASTON	G6KSM
70CM	DMU27	439.3375	430.3375	GB3MC	110.9	LEASOWE	G4YWD
70CM	DMU27	439.3375	430.3375	GB7AI	103.5	HAWICK	GM8SJP
70CM	DMU27	439.3375	430.3375	GB7LB	7	ALEXANDRIA	MM0ELF
70CM	DMU27	439.3375	430.3375	GB3GB	67	SUTTON COLDFIELD	G8NDT
70CM	DMU28	439.35	430.35	GB7DC	71.9	DERBY	G7NPW
70CM	DMU29	439.3625	430.3625	GB3CV	67	COVENTRY	M1FJB
70CM	DMU29	439.3625	430.3625	GB7GX	103.5	PRESTONPANS	MM0DXE
70CM	DMU29	439.3625	430.3625	GB7LA		LANCASTER	G7CDA
70CM	DMU30	439.375	430.375	GB3OP	71.9	PORTSMOUTH	G7RPG
70CM	DMU30	439.375	430.375	GB3OV	94.8	GRAVELEY	M0OFF
70CM	DMU30	439.375	430.375	GB7WH	110.9	WREXHAM	M0XDR
70CM	DVU32	439.4	430.4	GB7AL	2	IPSWICH	M1NIZ
70CM	DVU32	439.4	430.4	GB7BM-B	8	BIRMINGHAM	G8NDT
70CM	DVU32	439.4	430.4	GB7EP	3	EPSOM	G0OXZ
70CM	DVU32	439.4	430.4	GB7LN	1	LINCOLN	G0RZR
70CM	DVU32	439.4	430.4	GB7LP	1	LIVERPOOL	M1SWB
70CM	DVU32	439.4	430.4	GB7LS		GALASHIELS	GM7LUN
70CM	DVU32	439.4	430.4	GB7MD		BLANDFORD FORUM	M0RHS
70CM	DVU32	439.4	430.4	GB7ZT	9	BALLANTRAE	GM4ZTO
70CM	DVU33	439.4125	430.4125	GB7BZ		BOZEAT	M5WOB
70CM	DVU33	439.4125	430.4125	GB7DN		DUNGIVEN	GI0AZB
70CM	DVU33	439.4125	430.4125	GB7GT	13	CLEE HILL	G3PWJ
70CM	DVU33	439.4125	430.4125	GB7JD	1	JEDBURGH	GM4UPX
70CM	DVU33	439.4125	430.4125	GB7MN	2	STOCKPORT	G8NSS
70CM	DVU33	439.4125	430.4125	GB7ND	1	GREAT ELLINGHAM	G0LGJ
70CM	DVU33	439.4125	430.4125	GB7RT	3	SUNNINGDALE	M0HBK
70CM	DVU33	439.4125	430.4125	GB7SD	1	WEYMOUTH	G3VPF
70CM	DVU34	439.425	430.425	GB3DP	88.5	BASTON	G6YCA
70CM	DVU34	439.425	430.425	GB7AE	3	MARLBOROUGH	M1CJE
70CM	DVU34	439.425	430.425	GB7AS	3	ASHFORD	M1CMN
70CM	DVU34	439.425	430.425	GB7DR	5	POOLE	G7ICH
70CM	DVU34	439.425	430.425	GB7FS		BILSTON	G0WEB
70CM	DVU34	439.425	430.425	GB7HS	2	HEPTONSTALL	G1XCC
70CM	DVU34	439.425	430.425	GB7WS	3	BILLINGSHURST	G4LRP
70CM	DVU34	439.425	430.425	GB7XY	10	NEWCASTLE	M0DMP
70CM	DVU34	439.425	430.425	GB7YY	4	DUNDRY	M0JXR
70CM	DVU35	439.4375	430.4375	2E0LNX	3	GOSPORT	2E0LNX
70CM	DVU35	439.4375	430.4375	GB7AV	3	AYLESBURY	G0RAS
70CM	DVU35	439.4375	430.4375	GB7BX	11	CLEOBURY NORTH	G4VZO
70CM	DVU35	439.4375	430.4375	GB7ES		EASTBOURNE	M0LRE
70CM	DVU35	439.4375	430.4375	GB7KN	1	EDINBURGH	GM7HHB
70CM	DVU35	439.4375	430.4375	GB7LD	15	LEYLAND	M0XRS
70CM	DVU35	439.4375	430.4375	GB7NR	1	NOTTINGHAM	M0VUB
70CM	DVU35	439.4375	430.4375	GB7RK	110.9	BELFAST	MI5DAW
70CM	DVU36	439.45	430.45	GB7BP		MILTON KEYNES	M1ACB
70CM	DVU36	439.45	430.45	GB7EF	103.5	LADYBANK	MM0DXE
70CM	DVU36	439.45	430.45	GB7EQ	10	ILMINSTER	M1AXM
70CM	DVU36	439.45	430.45	GB7GR	5	GRANTHAM	G6SSN
70CM	DVU36	439.45	430.45	GB7GS	3	DOWNHAM MARKET	G0FHM
70CM	DVU36	439.45	430.45	GB7HR	3	HEATHROW	G0OXZ
70CM	DVU36	439.45	430.45	GB7IC		HERNE BAY	G4TKR
70CM	DVU36	439.45	430.45	GB7II	3	INVERNESS	GM6GRE
70CM	DVU36	439.45	430.45	GB7NA-B	71.9	BARNSLEY	G4LUE
70CM	DVU36	439.45	430.45	GB7NE	3	ASHINGTON	G0UDZ
70CM	DVU36	439.45	430.45	GB7ST	1	STOKE ON TRENT	G8NSS
70CM	DVU36	439.45	430.45	GB7SU	8	SOUTHAMPTON	G4WFR
70CM	DVU37	439.4625	430.4625	GB7AN		ACCRINGTON	M0NFI
70CM	DVU37	439.4625	430.4625	GB7DI		IPSWICH	M0WRB
70CM	DVU37	439.4625	430.4625	GB7JB	1	WINCANTON	G3ZXX
70CM	DVU37	439.4625	430.4625	GB7NL		EXETER	M0ZZT
70CM	DVU37	439.4625	430.4625	GB7OR	77	KIRKWALL	GM4WMM
70CM	DVU37	439.4625	430.4625	GB7SK	1	LEICESTER	M1FJB
70CM	DVU37	439.4625	430.4625	GB7WL	3	AMERSHAM	M0ZPU
70CM	DVU37	439.4625	430.4625	GB7WR		GREAT MALVERN	M0XZS
70CM	DVU37	439.4625	430.4625	GB7XX	118.8	FELLING	G4MSF
70CM	DVU38	439.475	430.475	GB7FB	5	BIDEFORD	M0MFS
70CM	DVU38	439.475	430.475	GB7FE	1	STIRLING	GM0MZB
70CM	DVU38	439.475	430.475	GB7LR	1	LEICESTER	M1FJB
70CM	DVU38	439.475	430.475	GB7SE	3	THURROCK	M0PFX
70CM	DVU38	439.475	430.475	GB7WW	7	KINGTON LANGLEY	G6HUI
70CM	DVU38	439.475	430.475	GB7YZ		MOLD	M0WTX
70CM	DVU39	439.4875	430.4875	GB7CO	3	FAREHAM	G3ZDF
70CM	DVU39	439.4875	430.4875	GB7NF	1	NEWHAVEN	G0TJH
70CM	DVU39	439.4875	430.4875	GB7TF		TRING	G7HMV
70CM	DVU39	439.4875	430.4875	GB7WB	2	WESTON-S-MARE	G4SZM
70CM	DVU39	439.4875	430.4875	GB7WC		WARRINGTON	G4VSS
70CM	DVU39	439.4875	430.4875	GB7WF	13	BEWDLEY	G8OXG
70CM	DVU39	439.4875	430.4875	GB7ZP		CHELMSFORD	G0URK
70CM	DVU39	439.4875	430.4875	GB7HU		SOUTH CAVE	G0VRM
70CM	DVU40	439.5	430.5	GB7CZ	118.8	CROOK	G0OCB
70CM	DVU40	439.5	430.5	GB7FC		CLEVELEYS	G0GER
70CM	DVU40	439.5	430.5	GB7HF	5	NEWTON ABBOT	G4CPO
70CM	DVU40	439.5	430.5	GB7KT	1	ANDOVER	G8FHI
70CM	DVU40	439.5	430.5	GB7PE	3	PETERBOROUGH	M0ZPU
70CM	DVU41	439.5125	430.5125	GB7IT	1	WESTON-S-MARE	G4SZM
70CM	DVU41	439.5125	430.5125	GB7LO	3	BECKENHAM	G1HIG
70CM	DVU41	439.5125	430.5125	GB7WK	15	WOKING	M0MXC
70CM	DVU41	439.5125	430.5125	GB7WX	3	TARVIN	G7NEH
70CM	DVU41	439.5125	430.5125	GB7WY	82.5	WAKEFIELD	M0DUB
70CM	DVU42	439.525	430.525	GB7AR	15	NEW ASH GREEN	G6VBJ
70CM	DVU42	439.525	430.525	GB7EW	3	EXETER	G6ATJ
70CM	DVU42	439.525	430.525	GB7HC	5	HEREFORD	G3PWJ
70CM	DVU42	439.525	430.525	GB7KU	5	BOURNE	G6SSN
70CM	DVU42	439.525	430.525	GB7PA	1	PAISLEY	MM0PAZ
70CM	DVU42	439.525	430.525	GB7SO	1	GOSPORT	G7CHO
70CM	DVU42	439.525	430.525	GB7TC	2	SWINDON	G8VRI
70CM	DVU42	439.525	430.525	GB7UL	1	CARRICKFERGUS	GI6DKQ
70CM	DVU42	439.525	430.525	GB7YR	2	DONCASTER	M1DAH
70CM	DVU43	439.5375	430.5375	GB7IK	3	ROCHESTER	G6CKK
70CM	DVU43	439.5375	430.5375	GB7IS		WESTON-S-MARE	G4SZM
70CM	DVU43	439.5375	430.5375	GB7MC		ST AUSTELL	M1DNS
70CM	DVU43	439.5375	430.5375	GB7RJ	5	CHIRK	G7MHF
70CM	DVU45	439.5625	430.5625	GB7HI		LISBURN	GI0OKM
70CM	DVU45	439.5625	430.5625	GB7XN	1	WORKSOP	2E0KBZ
70CM	DVU46	439.575	430.575	GB8YDE	5	CHESHAM	G8YDE
70CM	DVU46	439.575	430.575	GB7BH	4	BRIGHTON	G7TXU
70CM	DVU46	439.575	430.575	GB7DY	11	DUDLEY	G4VZO
70CM	DVU46	439.575	430.575	GB7EB	2	BECCLES	M0JGX
70CM	DVU46	439.575	430.575	GB7HX	1	HUDDERSFIELD	G0ISX
70CM	DVU46	439.575	430.575	GB7HY	10	CHORLEY	M1CGI
70CM	DVU46	439.575	430.575	GB7KP	1	COMBER	GI0VKP
70CM	DVU46	439.575	430.575	GB7YI	10	HEXHAM	G1HZI
70CM	DVU47	439.5875	430.5875	GB7BN	3	BOGNOR REGIS	G0TJH
70CM	DVU47	439.5875	430.5875	GB7DH		NORTHWICH	M0XDR
70CM	DVU47	439.5875	430.5875	GB7SM	103.5	ST MONANS	MM0DXE
70CM	DVU47	439.5875	430.5875	GB7SV	82.5	HITCHIN	G6YIQ
70CM	DVU47	439.5875	430.5875	GB7YJ	3	LEAMINGTON SPA	G4USP
70CM	DVU48	439.6	430.6	GB3HN	103.5	LUDLOW	M0HOY
70CM	DVU48	439.6	430.6	GB7CW	5	WINCHESTER	G8IPG
70CM	DVU48	439.6	430.6	GB7MK	13	IPSWICH	M1NIZ
70CM	DVU48	439.6	430.6	GB7SC		SHEFFIELD	G4LUE
70CM	DVU48	439.6	430.6	GB7WT	1	OMAGH	GI3NVW
70CM	DVU49	439.6125	430.6125	GB7CE		GREAT CLACTON	G0DZZ
70CM	DVU49	439.6125	430.6125	GB7CQ	4	WORTHING	G4WTV
70CM	DVU49	439.6125	430.6125	GB7RR	1	KIMBERLEY	M0NGT
70CM	DVU49	439.6125	430.6125	GB7RS	4	MILTON KEYNES	G0GMB
70CM	DVU49	439.6125	430.6125	GB7TY	10	HEXHAM	G1HZI
70CM	DVU50	439.625	430.625	GB3DF	82.5	UPPER DENBY	M0ODM
70CM	DVU50	439.625	430.625	GB7CF	3	CLANFIELD	M0PPR
70CM	DVU50	439.625	430.625	GB7CY		CROOK	G0OCB
70CM	DVU50	439.625	430.625	GB7GG	1	AIRDRIE	GM4AUP
70CM	DVU50	439.625	430.625	GB7HB	1	TANDRAGEE	MI0IRZ
70CM	DVU50	439.625	430.625	GB7NH		ALDERSHOT	G1GRD
70CM	DVU50	439.625	430.625	GB7RV	2	BLACKBURN	M0NWI
70CM	DVU50	439.625	430.625	GB7SL	8	BASTON	G6YCA

Voice Repeaters (by output frequency)

Band	CH	Out (MHz)	In (MHz)	Callsign	CTCSS	Location	Keeper
70CM	DVU51	439.6375	430.6375	GB7BT	3	BRISTOL	G7FBD
70CM	DVU51	439.6375	430.6375	GB7CL	3	CLACTON ON SEA	G0MBA
70CM	DVU51	439.6375	430.6375	GB7HM	1	CAERGWRLE	G1SYG
70CM	DVU51	439.6375	430.6375	GB7JF		NORTHAMPTON	G1IRG
70CM	DVU51	439.6375	430.6375	GB7MH	2	EAST GRINSTEAD	G7KBR
70CM	DVU51	439.6375	430.6375	GB7MJ	5	ROMSEY	G3KYG
70CM	DVU51	439.6375	430.6375	GB7PF	3	PRINCETOWN	G3XOU
70CM	DVU51	439.6375	430.6375	GB7SR	2	SHEFFIELD	M0TWS
70CM	DVU52	439.65	430.65	GB7BC	3	BRIGHTON	G7TXU
70CM	DVU52	439.65	430.65	GB7DK		STRANRAER	GM0HPK
70CM	DVU52	439.65	430.65	GB7GP	5	HIGH WYCOMBE	M0GUY
70CM	DVU52	439.65	430.65	GB7IN	1	TANSLEY NR MATLOCK	M0NGT
70CM	DVU52	439.65	430.65	GB7KY	103.5	KIRKCALDY	GM0MMN
70CM	DVU52	439.65	430.65	GB7RA	5	NESTON	M0ORA
70CM	DVU53	439.6625	430.6625	GB7AB		ABERDEEN	GM7KBK
70CM	DVU53	439.6625	430.6625	GB7BE		BECCLES	M0JGX
70CM	DVU53	439.6625	430.6625	GB7BI	1	MUIR OF ORD	GM0UDL
70CM	DVU53	439.6625	430.6625	GB7DD	1	DUNDEE	MM0DUN
70CM	DVU53	439.6625	430.6625	GB7LE	2	LEEDS	G1XCC
70CM	DVU53	439.6625	430.6625	GB7MT		SOUTHAMPTON	G4MYS
70CM	DVU53	439.6625	430.6625	GB7SB	5	BIRMINGHAM	M0UEM
70CM	DVU54	439.675	430.675	GB3NS	82.5	CATERHAM	G0OLX
70CM	DVU54	439.675	430.675	GB7AA	1	BRISTOL	G4CJZ
70CM	DVU54	439.675	430.675	GB7BL	118.8	ALLENDALE	M0UKB
70CM	DVU54	439.675	430.675	GB7DL		SKEGNESS	G4HFG
70CM	DVU54	439.675	430.675	GB7ED	5	EXETER	M0ZZT
70CM	DVU54	439.675	430.675	GB7FK-B		FOLKESTONE	M1CMN
70CM	DVU54	439.675	430.675	GB7JT	4	LEIGHTON BUZZARD	M1CVZ
70CM	DVU54	439.675	430.675	GB7MF	3	MAGHERAFELT	MI0JPD
70CM	DVU54	439.675	430.675	GB7SN	1	SHEFFIELD	M1ERS
70CM	DVU54	439.675	430.675	GB7SQ	1	DALKEITH	MM1BJO
70CM	DVU55	439.6875	430.6875	GB7GF	3	GUILDFORD	G4EML
70CM	DVU55	439.6875	430.6875	GB7GY	5	ST. PETER PORT	GU7DAI
70CM	DVU55	439.6875	430.6875	GB7KD	3	HAILSHAM	G1ERJ
70CM	DVU55	439.6875	430.6875	GB7NB		NORWICH	G0LGJ
70CM	DVU55	439.6875	430.6875	GB7RD	3	YELVERTON	G4CPO
70CM	DVU55	439.6875	430.6875	GB7TP	1	SHIPLEY	M0IRK
70CM	DVU55	439.6875	430.6875	GB7WM		ASTLEY	G4WLI
70CM	DVU55	439.6875	430.6875	GB7WV	2	MARKET HARBOROUGH	G1IVG
70CM	DVU56	439.7	430.7	GB7AU		AMERSHAM	M0ZPU
70CM	DVU56	439.7	430.7	GB7CD		CARDIFF	GW6CUR
70CM	DVU56	439.7	430.7	GB7DO	2	DONCASTER	G1ILF
70CM	DVU56	439.7	430.7	GB7DX	3	NEW ROMNEY	G0GCQ
70CM	DVU56	439.7	430.7	GB7MB		MORECAMBE	G0VGS
70CM	DVU56	439.7	430.7	GB7VO	1	TENBURY WELLS	G8XYJ
70CM	DVU57	439.7125	430.7125	GB7BB	1	BANFF	MM0BUH
70CM	DVU57	439.7125	430.7125	GB7CB	4	BATHGATE	GM7RYR
70CM	DVU57	439.7125	430.7125	GB7EL	2	NELSON	G4BLH
70CM	DVU57	439.7125	430.7125	GB7EX	3	SOUTHEND ON SEA	G8YPK
70CM	DVU57	439.7125	430.7125	GB7FT	3	PORTSMOUTH	M0PPR
70CM	DVU57	439.7125	430.7125	GB7PT		ROYSTON	M0ZPU
70CM	DVU57	439.7125	430.7125	GB7RE	1	RETFORD	G0EAK
70CM	DVU57	439.7125	430.7125	GB7WD	11	SEDGLEY	G4VZO
70CM	DVU58	439.725	430.725	GB7BW	2	WAKEFIELD	M0YDG
70CM	DVU58	439.725	430.725	GB7ME	3	RUGBY	M0IJS
70CM	DVU58	439.725	430.725	GB7NW		SWINDON	G4XIB
70CM	DVU59	439.7375	430.7375	GB3ND	1	NEAR HOLSWORTHY	G1BHM
70CM	DVU59	439.7375	430.7375	GB7BJ	13	MALVERN	G3PWJ
70CM	DVU59	439.7375	430.7375	GB7BK	3	READING	G8DOR
70CM	DVU59	439.7375	430.7375	GB7CK	3	FOLKESTONE	M1CMN
70CM	DVU59	439.7375	430.7375	GB7HT	10	ASHINGTON	G4NAB
70CM	DVU59	439.7375	430.7375	GB7MR	2	BURY	G1XCC
70CM	DVU59	439.7375	430.7375	GB7PO		PORTSMOUTH	M0VMX
70CM	DVU59	439.7375	430.7375	GB7TM		EYE	G0JSV
70CM	DVU60	439.75	430.75	GB7DM	103.5	MONIFIETH	MM0DRA
70CM	DVU60	439.75	430.75	GB7HH	3	ROMFORD	M0UPM
70CM	DVU60	439.75	430.75	GB7MM	5	NAIRN	GM0RML
70CM	DVU60	439.75	430.75	GB7MP	3	HEYSHAM	G6CRV
70CM	DVU60	439.75	430.75	GB7NN		WORKSOP	M0GET
70CM	DVU60	439.75	430.75	GB7SP	3	SALISBURY	G8PCB
70CM	DVU60	439.75	430.75	GB7WN	8	WOLVERHAMPTON	G8NDT
70CM	DVU61	439.7625	430.7625	GB7AD		BRISTOL	G4CJZ
70CM	DVU61	439.7625	430.7625	GB7BA	13	BRADFORD	G1XCC
70CM	DVU61	439.7625	430.7625	GB7CU	1	MARYPORT	M0LLC
70CM	DVU61	439.7625	430.7625	GB7EG	2	EAST GRINSTEAD	G7KBR
70CM	DVU61	439.7625	430.7625	GB7FL		FRECKLETON	M0XDR
70CM	DVU61	439.7625	430.7625	GB7PI		ROYSTON	M0ZPU
70CM	DVU61	439.7625	430.7625	GB7SI	1	LEEK	G8NSS
70CM	DVU61	439.7625	430.7625	GB7GB	8	SUTTON COLDFIELD	G8NDT
70CM	DVU62	439.775	430.775	GB3JG	71.9	WEYMOUTH	G8IKP
70CM	DVU62	439.775	430.775	GB7EZ	14	CASTLE BROMWICH	M0SNR
70CM	DVU62	439.775	430.775	GB7PX	10	ROYSTON	M0ZPU
70CM	DVU62	439.775	430.775	GB7RL	10	LONGRIDGE	G7TRL
23CM		1290.65	1270.65	GB7IC-A		HERNE BAY	G4TKR
23CM	RM0	1297	1291	GB3NO	94.8	NORWICH	G8VLL
23CM	RM1	1297.025	1291.025	GB3XZ	77	DAVENTRY	G8KHF
23CM	RM2	1297.05	1291.05	GB3FM	100	FARNHAM	G4EPX
23CM	RM3	1297.075	1291.075	GB3PS	77	ROYSTON	M0ZPU
23CM	RM3	1297.075	1291.075	GB3SE	103.5	STOKE ON TRENT	G8NSS
23CM	RM12	1297.3	1291.3	GB7NA-A		BARNSLEY	G4LUE
23CM	RM14X	1297.35	1277.35	GB3AK	94.8	BRISTOL	G4CJZ
23CM	RM15	1297.375	1291.375	GB3WC	82.5	WAKEFIELD	G1XCC

All Repeaters (by Callsign)

MODES: A=ANALOGUE, M=DMR, D=D-STAR, F=FUSION, N=NXDN, T=TV, E=TETRA, X=PACKET, P=P25, 7=M17

Repeater	Band	Channel	TX	RX	Modes	QTHR	Where	CTCSS/CC	DMR CC
GB3AA	2M	RV53	145.6625	145.0625	A	IO81RO	BRISTOL	94.8	
GB3AB	70CM	RB15	433.3750	434.9750	A	IO93FK	SHEFFIELD	82.5	
GB3AC	70CM	RU78	430.9750	438.5750	A	IO81TS	BRISTOL	94.8	
GB3AG	2M	RV58	145.7250	145.1250	AF	IO86ON	FORFAR	94.8	
GB3AH	70CM	RB11	433.2750	434.8750	A	JO02KP	EAST DEREHAM	94.8	
GB3AK	23CM	RM14X	1297.3500	1277.3500	A	IO81RO	BRISTOL	94.8	
GB3AL	2M	RV59	145.7375	145.1375	AF	IO91QP	AMERSHAM	77.0	
GB3AM	6M	R50-13	50.8400	51.3400	A	IO91QP	AMERSHAM	77.0	
GB3AN	70CM	RB08	433.2000	434.8000	AF	IO73UJ	AMLWCH	110.9	
GB3AR	2M	RV56	145.7000	145.1000	AF	IO73VC	CAERNARFON	110.9	
GB3AS	2M	RV48	145.6000	145.0000	A	IO85MC	LANGHOLM	77.0	
GB3AU	70CM	RB07	433.1750	434.7750	A	IO91QP	AMERSHAM	82.5	
GB3AV	70CM	RB02	433.0500	434.6500	A	IO91OT	AYLESBURY	118.8	
GB3AW	70CM	RB10	433.2500	434.8500	A	IO91GH	NEWBURY	71.9	
GB3AY	2M	RV52	145.6500	145.0500	A	IO75OR	DALRY	103.5	
GB3BA	70CM	RB13	433.3250	434.9250	A	IO91	BASINGSTOKE	71.9	
GB3BB	2M	RV56	145.7000	145.1000	A	IO81	BRECON	94.8	
GB3BC	2M	RV60	145.7500	145.1500	A	IO81IO	PONTYPRIDD	94.8	
GB3BE	70CM	RU69	430.8625	438.4625	AF	IO85	DUNS	118.8	
GB3BF	2M	RV63	145.7875	145.1875	A	IO92	BEDFORD	77.0	
GB3BH	70CM	RU79	430.9875	438.5875	A	JO00FU	BEXHILL-ON-SEA	77.0	
GB3BI	2M	RV58	145.7250	145.1250	A	IO77RM	MUIR OF ORD	67.0	
GB3BK	70CM	RU72	430.9000	438.5000	A	JO01AK	BROMLEY	103.5	
GB3BL	70CM	RU70	430.8750	438.4750	A	IO92SD	BEDFORD	77.0	
GB3BM	2M	RV57	145.7125	145.1125	A	IO82	DUDLEY	67.0	
GB3BM-B	2M	RV57	145.7125	145.1125	A	IO92	BIRMINGHAM	67.0	
GB3BN	70CM	RB0	433.0000	434.6000	A	IO91OJ	BRACKNELL	118.8	
GB3BR	70CM	RU65	430.8125	438.4125	A	IO90	HOVE	88.5	
GB3BS	70CM	RU68	430.8500	438.4500	A	IO81TK	BRISTOL	118.8	
GB3BT	2M	RV56	145.7000	145.1000	A	IO85WT	BERWICK ON TWEED	118.8	
GB3BV	70CM	RB01	433.0250	434.6250	AF	IO91SR	HEMEL HEMPSTEAD	82.5	
GB3BW	70CM	RU65	430.8125	438.4125	A	IO93HO	WAKEFIELD	82.5	
GB3BX	2M	RV54	145.6750	145.0750	M	IO82	CLEOBURY NORTH		13
GB3BY	70CM	RU78	430.9750	438.5750	A	IO83QP	CHORLEY	82.5	
GB3BZ	70CM	RU68	430.8500	438.4500	AM	JO01GW	BRAINTREE		3
GB3CA	70CM	RB13	433.3250	434.9250	A	IO84OT	CARLISLE	77.0	
GB3CB	70CM	RB14	433.3500	434.9500	A	IO92BL	BIRMINGHAM	67.0	
GB3CC	70CM	RU77	430.9625	438.5625	A	IO90	CHICHESTER	88.5	
GB3CD	2M	RV55	145.6875	145.0875	AF	IO94DR	CROOK	118.8	
GB3CE	70CM	RB14	433.3500	434.9500	A	JO01KV	COLCHESTER	67.0	
GB3CF	2M	RV48	145.6000	145.0000	ADFN	IO92IQ	MARKFIELD	77.0	
GB3CG	2M	RV58	145.7250	145.1250	A	IO81VU	GLOUCESTER	118.8	
GB3CH	70CM	RU70	430.8750	438.4750	A	IO70SM	LISKEARD	77.0	

All Repeaters (by Callsign)

Repeater	Band	Channel	TX	RX	Modes	QTHR	Where	CTCSS/CC	DMR CC
GB3CI	70CM	RU66	430.8250	438.4250	A	IO92PM	CORBY	77.0	
GB3CJ	10M	10M	29.6400	29.5400	A	IO92QF	BOZEAT	77.0	
GB3CK	70CM	RB0	433.0000	434.6000	A	JO01JF	CHARING KENT	103.5	
GB3CL	70CM	RB09	433.2250	434.8250	A	JO01OT	CLACTON	67.0	
GB3CM	70CM	RU69	430.8625	438.4625	AF	IO71VW	CARMARTHEN	94.8	
GB3CN	2M	RV55	145.6875	145.0875	A	IO71TX	NEWCASTLE EMLYN	94.8	
GB3CO	2M	RV53	145.6625	145.0625	A	IO92QM	CORBY	77.0	
GB3CP	2M	RV59	145.7375	145.1375	A	IO64IG	FERMANAGH	110.9	
GB3CQ	10M	10M	29.6300	29.5300	A	IO90	WORTHING	88.5	
GB3CR	70CM	RB06	433.1500	434.7500	A	IO83	CAERGWRLE	110.9	
GB3CS	2M	RV60	145.7500	145.1500	A	IO85	MOTHERWELL	103.5	
GB3CV	70CM	DMU29	439.3625	430.3625	A	IO92FJ	COVENTRY	67.0	
GB3CW	70CM	RB04	433.1000	434.7000	A	IO82HL	NEWTOWN POWYS	103.5	
GB3CY	70CM	RU69	430.8625	438.4625	A	IO93KW	YORK	88.5	
GB3DA	2M	RV58	145.7250	145.1250	AF	JO01GR	DANBURY	110.9	
GB3DB	6M	R50-06	50.7700	51.2700	AF	JO01HR	DANBURY	110.9	
GB3DC	2M	RV55	145.6875	145.0875	A	IO92GW	DERBY	71.9	
GB3DD	70CM	RB10	433.2500	434.8500	AF	IO86MM	DUNDEE	94.8	
GB3DE	70CM	RB07	433.1750	434.7750	A	JO02NF	IPSWICH	110.9	
GB3DG	2M	RV62	145.7750	145.1750	AF	IO74UV	NEWTON STEWART	103.5	
GB3DI	70CM	RB06	433.1500	434.7500	A	IO91	DIDCOT	118.8	
GB3DM	70CM	RU74	430.9250	438.5250	AF	IO75QX	DUMBARTON	103.5	
GB3DN	2M	RV51	145.6375	145.0375	AF	IO70UW	STIBB CROSS	77.0	
GB3DP	70CM	RU69	430.8625	438.4625	A	IO92TR	BASTON	88.5	
GB3DQ	70CM	RU78	430.9750	438.5750	A	IO70RI	POLPERRO	77.0	
GB3DR	2M	RV59	145.7375	145.1375	AF	IO80SN	WEYMOUTH		
GB3DS	70CM	RB13	433.3250	434.9250	A	IO93KH	WORKSOP	71.9	
GB3DT	70CM	RB0	433.0000	434.6000	AF	IO80WU	BLANDFORD	71.9	
GB3DU	2M	RV61	145.7625	145.1625	ADMF	IO85	DUNS	1.0	
GB3DV	70CM	RB01	433.0250	434.6250	A	IO93JK	MALTBY	71.9	
GB3DW	2M	RV53	145.6625	145.0625	AF	IO72WT	HARLECH	110.9	
GB3DX	70CM	RB12	433.3000	434.9000	A	IO65HA	DERRY/LONDONDERRY	110.9	
GB3DY	70CM	RB10	433.2500	434.8500	AM	IO93FB	WIRKSWORTH	71.9	
GB3DZ	70CM	RU79	430.9875	438.5875	A	IO92	DERBY	71.9	
GB3EA	2M	RV55	145.6875	145.0875	A	JO02GE	WICKHAMBROOK	110.9	
GB3EB	2M	RV63	145.7875	145.1875	ADM	JO00AT	NEWHAVEN	88.5	
GB3ED	70CM	RB14	433.3500	434.9500	AD	IO85JW	EDINBURGH	94.8	
GB3EE	70CM	RB12	433.3000	434.9000	A	IO93	MANSFIELD	71.9	
GB3EF	6M	R50-01	50.7200	51.2200	A	JO02NF	STOWMARKET	110.9	
GB3EG	70CM	RU73	430.9125	438.5125	A	IO83QN	WIGAN	82.5	
GB3EH	70CM	RU77	430.9625	438.5625	A	IO92GC	BANBURY	67.0	
GB3EI	2M	RV49	145.6125	145.0125	ADMF	IO67IN	LOCHMADDY	88.5	
GB3EJ	23CM	RM04	1297.1000	1291.1000	A	JO02	CAMBRIDGE	77.0	
GB3EK	70CM	RU71	430.8875	438.4875	A	JO01QJ	MARGATE	103.5	
GB3EL	2M	RV48	145.6000	145.0000	A	JO01AM	LONDON	82.5	
GB3EM	2M	RV61	145.7625	145.1625	A	IO92LJ	COLD ASHBY	77.0	
GB3ER	70CM	RB03	433.0750	434.6750	AF	JO01GR	DANBURY	110.9	
GB3ES	2M	RV54	145.6750	145.0750	A	JO00	HASTINGS	103.5	
GB3EU	70CM	UR67	438.4375	430.8375	A	IO92LJ	COLD ASHBY	77.0	
GB3EV	2M	RV56	145.7000	145.1000	A	IO84SQ	DUFTON	77.0	
GB3EW	2M	RV49	145.6125	145.0125	A	IO80	EXETER	77.0	
GB3EX	70CM	RU76	430.9500	438.5500	A	IO80GT	SILVERTON	77.0	
GB3EZ	70CM	RU78	430.9750	438.5750	A	JO02GE	WICKHAMBROOK	110.9	
GB3FA	2M	RV57	145.7125	145.1125	A	IO93HI	SHEFFIELD	82.5	
GB3FC	70CM	RU68	430.8500	438.4500	A	IO83LU	BLACKPOOL	82.5	
GB3FE	2M	RV53	145.6625	145.0625	A	IO86BC	STIRLING	103.5	
GB3FF	2M	RV48	145.6000	145.0000	A	IO86	UPPER LARGO	103.5	
GB3FG	2M	RV57	145.7125	145.1125	AF	IO71	CARMARTHEN	94.8	
GB3FH	6M	R50-06	50.7700	51.2700	A	IO81OH	SOMERSET	77.0	
GB3FI	70CM	RU74	430.9250	438.5250	A	IO81OH	CHEDDAR	77.0	
GB3FJ	70CM	RU76	430.9500	438.5500	A	IO93XE	ASGARBY	71.9	
GB3FK	2M	RV60	145.7500	145.1500	AF	JO01	FOLKESTONE	103.5	
GB3FM	23CM	RM2	1297.0500	1291.0500	A	IO91OF	FARNHAM	100.0	
GB3FN	70CM	RB15	433.3750	434.9750	A	IO91OF	FARNHAM	82.5	
GB3FR	2M	RV62	145.7750	145.1750	A	JO03AE	SPILSBY LINCS.	71.9	
GB3FU	2M	RV60	145.7500	145.1500	A	IO92	NOTTINGHAM	71.9	
GB3FV	70CM	RU65	430.8125	438.4125	A	IO86	STIRLING	88.5	
GB3FX	6M	R50-10	50.8100	51.3100	A	IO91OF	FARNHAM	82.5	
GB3GB	70CM	DMU27	439.3375	430.3375	A	IO92BM	SUTTON COLDFIELD	67.0	
GB3GC	6M	R50-02	50.7300	51.2300	A	IO70TL	LISKEARD	77.0	
GB3GD	2M	RV50	145.6250	145.0250	A	IO74SG	SNAEFELL IOM	110.9	
GB3GF	70CM	RB12	433.3000	434.9000	A	IO91RF	GUILDFORD	88.5	
GB3GH	70CM	RB05	433.1250	434.7250	A	IO81VU	GLOUCESTER	118.8	
GB3GJ	2M	RV51	145.6375	145.0375	A	IN89WE	ST HELIER	71.9	
GB3GL	70CM	RU72	430.9000	438.5000	AF	IO75WU	GLASGOW	103.5	
GB3GM	2M	RV50	145.6250	145.0250	F	IO82UC	GREAT MALVERN	67.0	
GB3GN	2M	RV62	145.7750	145.1750	A	IO87TA	BANCHORY	67.0	
GB3GO	2M	RV63	145.7875	145.1875	AF	IO83BH	LLANDUDNO	110.9	
GB3GR	70CM	RU68	430.8500	438.4500	A	IO92QV	GRANTHAM	71.9	
GB3GU	70CM	RB13	433.3250	434.9250	A	IN89RK	GUERNSEY	71.9	
GB3HA	2M	RV60	145.7500	145.1500	AF	IO84	HEXHAM	118.8	
GB3HB	70CM	RB15	433.3750	434.9750	AF	IO70OJ	ROCHE	77.0	
GB3HC	70CM	RB06	433.1500	434.7500	AF	IO82PB	HEREFORD	118.8	
GB3HD	70CM	RB09	433.2250	434.8250	AF	IO93BP	HUDDERSFIELD	82.5	
GB3HE	70CM	RU74	430.9250	438.5250	AF	JO00HV	HASTINGS	103.5	
GB3HF	10M	10M	29.6900	29.5900	A	IO82	LLANGOLLEN	103.5	
GB3HG	2M	RV50	145.6250	145.0250	AF	IO94JF	THIRSK	88.5	
GB3HH	2M	RV56	145.7000	145.1000	A	IO93BF	BUXTON	71.9	
GB3HI	2M	RV56	145.7000	145.1000	A	IO76DK	ISLE OF MULL	103.5	
GB3HJ	70CM	RB01	433.0250	434.6250	A	IO94EB	HARROGATE	118.8	
GB3HK	70CM	RB02	433.0500	434.6500	A	IO85	SELKIRK	118.8	
GB3HM	6M	R50-11	50.8200	51.3200	A	IO93GA	BELPER	71.9	
GB3HN	70CM	RB01	433.0250	434.6250	A	IO82	LUDLOW	103.5	
GB3HO	70CM	RU71	430.8875	438.4875	A	IO91TB	HORSHAM	88.5	
GB3HR	70CM	RB14	433.3500	434.9500	A	IO91TO	HARROW	82.5	
GB3HS	2M	RV52	145.6500	145.0500	ADF	IO93RS	HULL	88.5	
GB3HT	70CM	RB11	433.2750	434.8750	A	IO92HN	HINCKLEY	77.0	
GB3HW	70CM	RB13	433.3250	434.9250	A	JO01CN	GIDEA PARK	110.9	
GB3HY	70CM	RU72	430.9000	438.5000	A	IO90WX	HAYWARDS HEATH	88.5	

All Repeaters (by Callsign)

Repeater	Band	Channel	TX	RX	Modes	QTHR	Where	CTCSS/CC	DMR CC
GB3IC	70CM	RB05	433.1250	434.7250	A	IO82	WOLVERHAMPTON	67.0	
GB3IE	70CM	RU68	430.8500	438.4500	D	IO70XJ	PLYMOUTH	77.0	
GB3IG	2M	RV62	145.7750	145.1750	A	IO68QE	STORNOWAY	88.5	
GB3IH	70CM	RB04	433.1000	434.7000	A	JO02OB	IPSWICH	110.9	
GB3IK	2M	RV61	145.7625	145.1625	A	JO01FJ	ROCHESTER	103.5	
GB3IM-C	70CM	RU66	430.8250	438.4250	A	IO74SD	DOUGLAS IOM	110.9	
GB3IM-P	70CM	RU70	430.8750	438.4750	A	IO74PF	PEEL	110.9	
GB3IM-R	70CM	RU66	430.8250	438.4250	A	IO74TI	RAMSEY IOM	71.9	
GB3IM-S	70CM	RB05	433.1250	434.7250	A	IO74SG	DOUGLAS IOM	110.9	
GB3IN	2M	RV51	145.6375	145.0375	ADMFPN	IO93GD	ALFRETON	71.9	
GB3IP	70CM	RU78	430.9750	438.5750	A	IO91WF	SOUTH NUTFIELD	82.5	
GB3IR	2M	RV61	145.7625	145.1625	A	IO94DJ	RICHMOND YORKS	88.5	
GB3IS	70CM	RU71	430.8875	438.4875	ADMF	IO67HC	ISLE OF SOUTH UIST	94.8	
GB3IW	70CM	RB09	433.2250	434.8250	A	IO90JO	EAST COWES	71.9	
GB3JB	2M	RV63	145.7875	145.1875	AM	IO81VC	MERE WILTSHIRE		5
GB3JL	2M	RV63	145.7875	145.1875	A	IO70SM	LISKEARD	77.0	
GB3JM	70CM	RU70	430.8750	438.4750	A	IO82	LEOMINISTER	118.8	
GB3JR	70CM	RU78	430.9750	438.5750	A	IO93HB	MANSFIELD	71.9	
GB3JS	70CM	RB01	433.0250	434.6250	A	JO02UO	GREAT YARMOUTH	94.8	
GB3KA	70CM	RU71	430.8875	438.4875	A	IO82	KIDDERMINSTER	118.8	
GB3KC	70CM	RU72	430.9000	438.5000	A	IO82WK	STOURBRIDGE	67.0	
GB3KE	2M	RV55	145.6875	145.0875	AF	IO75UV	GLASGOW	103.5	
GB3KI	2M	RV53	145.6625	145.0625	AD	JO01NI	HERNE BAY	103.5	
GB3KK	70CM	RU78	430.9750	438.5750	A	IO65VE	BALLYCASTLE	110.9	
GB3KL	70CM	RB04	433.1000	434.7000	AF	JO02FR	KINGS LYNN	94.8	
GB3KN	2M	RV56	145.7000	145.1000	A	JO01HH	MAIDSTONE	103.5	
GB3KR	70CM	RU65	430.8125	438.4125	A	IO82UI	STOURPORT ON SEVERN	118.8	
GB3KS	2M	RV50	145.6250	145.0250	AF	JO01PA	DOVER	103.5	
GB3KT	70CM	RU68	430.8500	438.4500	A	IO93AU	KEIGHLEY	82.5	
GB3KU	70CM	RB03	433.0750	434.6750	ADMFPN	IO83XM	ASHTON-UNDER-LYNE	82.5	
GB3KV	70CM	RU78	430.9750	438.5750	A	IO75XX	KILSYTH	103.5	
GB3KW	70CM	RU66	430.8250	438.4250	AF	IO75UV	GLASGOW	103.5	
GB3KY	2M	RV57	145.7125	145.1125	A	JO02FS	KINGS LYNN	94.8	
GB3LA	2M	RV57	145.7125	145.1125	A	IO85CJ	SANQUHAR	103.5	
GB3LB	2M	RV63	145.7875	145.1875	A	IO85	LAUDER	118.8	
GB3LC	70CM	RU75	430.9375	438.5375	A	IO93WH	LOUTH	71.9	
GB3LD	2M	RV54	145.6750	145.0750	A	IO84OA	LANCASTER	110.9	
GB3LE	70CM	RB04	433.1000	434.7000	AF	IO92IQ	MARKFIELD	77.0	
GB3LF	70CM	RB14	433.3500	434.9500	A	IO84PH	KENDAL	110.9	
GB3LG	6M	R50-12	50.8300	51.3300	A	IO82	LLANGOLLEN		
GB3LH	70CM	RB15	433.3750	434.9750	AF	IO82OP	SHREWSBURY	103.5	
GB3LI	70CM	RB10	433.2500	434.8500	A	IO83LL	LIVERPOOL	82.5	
GB3LL	70CM	RB0	433.0000	434.6000	A	IO82	LLANGOLLEN	94.8	
GB3LM	2M	RV58	145.7250	145.1250	AF	IO93RF	LINCOLN	71.9	
GB3LN	70CM	RU74	430.9250	438.5250	A	IO92JO	LEICESTER	118.8	
GB3LP	6M	R50-04	50.7500	51.2500	A	IO83MK	LIVERPOOL	77.0	
GB3LR	70CM	RU69	430.8625	438.4625	A	JO00AS	NEWHAVEN	88.5	
GB3LS	70CM	RB02	433.0500	434.6500	A	IO93RF	LINCOLN	71.9	
GB3LT	70CM	RB10	433.2500	434.8500	AF	IO91SV	LUTON	77.0	
GB3LU	2M	RV54	145.6750	145.0750	A	IP90JD	LERWICK	77.0	
GB3LV	70CM	RB02	433.0500	434.6500	A	IO91XP	NORTH LONDON	82.5	
GB3LY	2M	RV48	145.6000	145.0000	A	IO65NC	LIMAVADY	110.9	
GB3LZ	2M	RV53	145.6625	145.0625	AMF	IO80CV	LAPFORD	77.0	
GB3MA	6M	R50-01	50.7200	51.2200	A	IO83SM	ATHERTON	82.5	
GB3MB	70CM	RB15	433.3750	434.9750	AF	IO81HR	MERTHYR	94.8	
GB3MC	70CM	DMU27	439.3375	430.3375	AF	IO83KJ	LEASOWE	110.9	
GB3MD	70CM	RU77	430.9625	438.5625	AF	IO82GN	LLANFAIR CAEREINION	103.5	
GB3ME	70CM	RB06	433.1500	434.7500	A	IO92JJ	RUGBY	67.0	
GB3MG	70CM	RU72	430.9000	438.5000	F	IO81KP	PONTYPRIDD		
GB3MH	2M	RV50	145.6250	145.0250	A	IO91WC	EAST GRINSTEAD	88.5	
GB3MI	2M	RV57	145.7125	145.1125	AD	IO83	MANCHESTER	82.5	
GB3MK	70CM	RB0	433.0000	434.6000	AF	IO92OB	MILTON KEYNES	77.0	
GB3ML	70CM	RU73	430.9125	438.5125	ADMF	IO91RK	STAINES UPON THAMES		3
GB3MM	23CM	RM06	1297.1500	1291.1500	A	IO82XP	WOLVERHAMPTON	67.0	
GB3MN	2M	RV52	145.6500	145.0500	AF	IO83XH	DISLEY	82.5	
GB3MO	70CM	UR63	438.3875	430.7875	A	IO84	MORECAMBE	110.9	
GB3MP	2M	RV60	145.7500	145.1500	A	IO83	PRESTATYN	110.9	
GB3MR	70CM	RB14	433.3500	434.9500	A	IO83XH	STOCKPORT	82.5	
GB3MW	70CM	RB10	433.2500	434.8500	A	IO92FH	LEAMINGTON SPA	67.0	
GB3MY	70CM	RB11	433.2750	434.8750	AF	IO83SM	ATHERTON	82.5	
GB3NA	2M	RV49	145.6125	145.0125	ADF	IO93HN	BARNSLEY	71.9	
GB3NB	2M	RV50	145.6250	145.0250	A	JO02PN	NORWICH	94.8	
GB3NC	2M	RV58	145.7250	145.1250	AF	IO70OJ	ROCHE	77.0	
GB3ND	70CM	DVU59	439.7375	430.7375	M	IO70	HOLSWORTHY		1
GB3NE	2M	RV61	145.7625	145.1625	A	IO91HJ	NEWBURY	118.8	
GB3NF	2M	RV50	145.6250	145.0250	AMF	IO92	NOTTINGHAM	77.0	
GB3NG	2M	RV50	145.6250	145.0250	A	IO87XO	FRASERBURGH	67.0	
GB3NH	70CM	RU76	430.9500	438.5500	A	IO92NF	NORTHAMPTON	77.0	
GB3NI	2M	RV58	145.7250	145.1250	A	IO74CO	BELFAST N.I.	110.9	
GB3NK	70CM	RB04	433.1000	434.7000	A	JO01BL	ERITH	103.5	
GB3NL	2M	RV62	145.7750	145.1750	A	IO91XP	NORTH LONDON		
GB3NM	70CM	RB07	433.1750	434.7750	AMFN	IO92	NOTTINGHAM	71.9	
GB3NN	70CM	RB02	433.0500	434.6500	A	JO02JV	WELLS NORFOLK	94.8	
GB3NO	23CM	RM0	1297.0000	1291.0000	A	JO02PP	NORWICH	94.8	
GB3NP	70CM	RU71	430.8875	438.4875	AF	IO92LD	TOWCESTER	77.0	
GB3NR	70CM	RB0	433.0000	434.6000	A	JO02PP	NORWICH	94.8	
GB3NS	70CM	DVU54	439.6750	430.6750	A	IO91WG	CATERHAM	82.5	
GB3NT	70CM	RB0	433.0000	434.6000	AF	IO94FW	NEWCASTLE UPON TYNE	118.8	
GB3NU	70CM	DMU26	439.3250	430.3250	ADMFPN7	IO92	NOTTINGHAM	71.9	
GB3NY	70CM	RU67	430.8375	438.4375	AF	IO93RT	HULL	67.0	
GB3NZ	2M	RV48	145.6000	145.0000	A	JO02NN	WYMONDHAM	94.8	
GB3OA	2M	RV49	145.6125	145.0125	AF	IO83LP	SOUTHPORT	82.5	
GB3OC	2M	RV52	145.6500	145.0500	A	IO88LX	KIRKWALL	77.0	
GB3OH	70CM	RU76	430.9500	438.5500	ADMF	IO85EX	LINLITHGOW	94.8	
GB3OM	70CM	RU76	430.9500	438.5500	AM	IO64JQ	OMAGH		1
GB3OR	70CM	RU67	430.8375	438.4375	AF	IO83BH	LLANDUDNO	103.5	
GB3OV	70CM	RB05	433.1250	434.7250	A	IO92VG	ST NEOTS	94.8	
GB3OY	70CM	RU76	430.9500	438.5500	A	JO01AO	BUCKHURST HILL	82.5	
GB3PA	2M	RV50	145.6250	145.0250	A	IO75QV	PAISLEY	103.5	

All Repeaters (by Callsign)

Repeater	Band	Channel	TX	RX	Modes	QTHR	Where	CTCSS/CC	DMR CC
GB3PB	2M	RV50	145.6250	145.0250	A	IO90AT	WIMBORNE	71.9	
GB3PD	70CM	RU68	430.8500	438.4500	AF	IO80SN	WEYMOUTH		
GB3PE	2M	RV54	145.6750	145.0750	A	IO92TL	PETERBOROUGH	94.8	
GB3PF	70CM	RB12	433.3000	434.9000	A	IO83TR	ACCRINGTON	82.5	
GB3PH	2M	RV48	145.6000	145.0000	A	IO72SX	PWLLHELI	118.8	
GB3PI	2M	RV60	145.7500	145.1500	A	IO92XA	ROYSTON	77.0	
GB3PK	2M	RV53	145.6625	145.0625	A	IO65VE	BALLYCASTLE	110.9	
GB3PL	2M	RV61	145.7625	145.1625	A	IO70	PLYMPTON	77.0	
GB3PM	70CM	RU72	430.9000	438.5000	A	IO93	SHEFFIELD	71.9	
GB3PN	70CM	RU76	430.9500	438.5500	A	IO83PS	PRESTON	82.5	
GB3PO	2M	RV52	145.6500	145.0500	A	JO02OB	IPSWICH	110.9	
GB3PP	70CM	RB15	433.3750	434.9750	AMF	IO83	PRESTON	82.5	
GB3PR	2M	RV54	145.6750	145.0750	AF	IO86GI	PERTH	94.8	
GB3PS	23CM	RM03	1297.0750	1291.0750	A	IO92XA	ROYSTON	77.0	
GB3PT	70CM	RU69	430.8625	438.4625	A	IO83	PRESCOT	94.8	
GB3PU	70CM	RB0	433.0000	434.6000	AF	IO86GI	PERTH	94.8	
GB3PW	2M	RV62	145.7750	145.1750	A	IO82HL	NEWTOWN	103.5	
GB3PX	6M	R50-07	50.7800	51.2800	A	IO92XA	ROYSTON	77.0	
GB3PY	70CM	RU72	430.9000	438.5000	A	JO02AF	CAMBRIDGE	77.0	
GB3PZ	70CM	RU72	430.9000	438.5000	A	IO83XL	DUKINFIELD	82.5	
GB3RB	70CM	RB08	433.2000	434.8000	A	IO93IF	BOLSOVER	71.9	
GB3RC	70CM	RU69	430.8625	438.4625	A	IO81	CHELTENHAM	103.5	
GB3RD	2M	RV54	145.6750	145.0750	A	IO91JM	READING	118.8	
GB3RE	70CM	RB11	433.2750	434.8750	A	JO01HH	MAIDSTONE	103.5	
GB3RF	2M	RV62	145.7750	145.1750	ADMF	IO83TR	ACCRINGTON	82.5	
GB3RH	70CM	RB11	433.2750	434.8750	A	IO80MS	AXMINSTER	94.8	
GB3RO	70CM	RU66	430.8250	438.4250	A	IO83UQ	HASLINGDEN	82.5	
GB3RR	70CM	RU68	430.8500	438.4500	A	IO82	DUDLEY	67.0	
GB3RU	70CM	RB11	433.2750	434.8750	A	IO91LK	READING	118.8	
GB3RW	2M	RV48	145.6000	145.0000	A	IO90	WORTHING	88.5	
GB3RX	70CM	RU65	430.8125	438.4125	AF	IO83	PRESCOT	77.0	
GB3SA	70CM	RU77	430.9625	438.5625	A	IO92RP	STAMFORD	82.5	
GB3SB	2M	RV52	145.6500	145.0500	A	IO85	SELKIRK	118.8	
GB3SC	70CM	RU69	430.8625	438.4625	A	IO92ES	BURTON-UPON-TRENT	67.0	
GB3SD	70CM	RB14	433.3500	434.9500	A	IO80SQ	WEYMOUTH	71.9	
GB3SE	23CM	RM03	1297.0750	1291.0750	A	IO83	STOKE ON TRENT	103.5	
GB3SF	2M	RV50	145.6250	145.0250	F	IO86	PERTH		
GB3SG	70CM	RU73	430.9125	438.5125	A	IO81	STROUD	118.8	
GB3SH	2M	RV53	145.6625	145.0625	A	IO90IV	SOUTHAMPTON	71.9	
GB3SI	2M	RV50	145.6250	145.0250	A	IO70JB	HELSTON		
GB3SJ	2M	RV55	145.6875	145.0875	A	IO83RF	NORTHWICH	103.5	
GB3SK	70CM	RB06	433.1500	434.7500	A	JO01MH	CANTERBURY	103.5	
GB3SL	6M	R50-02	50.7300	51.2300	A	IO75XX	KILSYTH	103.5	
GB3SM	70CM	RB13	433.3250	434.9250	A	IO93BA	STOKE ON TRENT	103.5	
GB3SN	2M	RV58	145.7250	145.1250	A	IO91LC	ALTON HANTS	71.9	
GB3SO	70CM	RB0	433.0000	434.6000	ADMF	JO03AR	WITHERNSEA	71.9	
GB3SP	70CM	RB04	433.1000	434.7000	AF	IO71NQ	PEMBROKE	94.8	
GB3SR	70CM	RU73	430.9125	438.5125	A	JO03CB	SKEGNESS	71.9	
GB3SS	2M	RV48	145.6000	145.0000	AF	IO87	ELGIN MORAY	67.0	
GB3ST	70CM	RB09	433.2250	434.8250	A	IO83	STOKE ON TRENT	103.5	
GB3SU	70CM	RB08	433.2000	434.8000	A	IO90IV	SOUTHAMPTON	71.9	
GB3SV	2M	RV49	145.6125	145.0125	A	IO82WS	STAFFORD	94.8	
GB3SW	2M	RV57	145.7125	145.1125	A	IO80JQ	SIDMOUTH	77.0	
GB3SX	6M	R50-08	50.7900	51.2900	A	IO93BA	STOKE ON TRENT	103.5	
GB3SY	70CM	RB06	433.1500	434.7500	A	IO93	BARNSLEY	71.9	
GB3TA	2M	RV59	145.7375	145.1375	A	IO92DP	TAMWORTH	67.0	
GB3TD	70CM	UR63	438.3875	430.7875	A	IO91DL	SWINDON	118.8	
GB3TE	2M	RV49	145.6125	145.0125	A	JO01OU	FRINTON ON SEA	67.0	
GB3TF	70CM	RB08	433.2000	434.8000	AF	IO82RP	TELFORD	103.5	
GB3TH	70CM	RB15	433.3750	434.9750	AF	IO92	TAMWORTH	67.0	
GB3TJ	70CM	RB12	433.3000	434.9000	AF	IO85XA	CORBRIDGE	118.8	
GB3TP	2M	RV58	145.7250	145.1250	A	IO93CT	SHIPLEY	82.5	
GB3TR	2M	RV52	145.6500	145.0500	A	IO80FM	TORQUAY	94.8	
GB3TS	70CM	RB07	433.1750	434.7750	AF	IO94GX	SUNDERLAND	118.8	
GB3TU	70CM	RB09	433.2250	434.8250	A	IO91PS	TRING	77.0	
GB3TW	2M	RV58	145.7250	145.1250	AF	IO94DW	GATESHEAD	118.8	
GB3TX	10M	10M	29.6500	29.5500	A	JO02QP	NORWICH	94.8	
GB3UB	70CM	RB04	433.1000	434.7000	A	IO81UJ	BATH AVON	118.8	
GB3UK	2M	RV49	145.6125	145.0125	ADMF	IO94FV	WASHINGTON	10.0	
GB3UL	70CM	RB02	433.0500	434.6500	A	IO74CO	BELFAST N.I.	110.9	
GB3UM	6M	R50-03	50.7400	51.2400	A	IO92IQ	MARKFIELD	77.0	
GB3UO	70CM	RU76	430.9500	438.5500	AF	IO83LA	WREXHAM	110.9	
GB3UP	70CM	RU73	430.9125	438.5125	A	IO71MX	FISHGUARD	94.8	
GB3US	70CM	RB0	433.0000	434.6000	A	IO93	SHEFFIELD	103.5	
GB3VA	2M	RV56	145.7000	145.1000	A	IO91LT	BRILL	118.8	
GB3VE	70CM	RB04	433.1000	434.7000	A	IO84SQ	DUFTON	77.0	
GB3VH	70CM	RB13	433.3250	434.9250	A	IO91VT	WELWYN GARDEN CITY	82.5	
GB3VI	6M	R50-15	50.8600	51.3600	A	IO82	DUDLEY	67.0	
GB3VM	2M	RV49	145.6125	145.0125	AF	IO82QJ	TENBURY WELLS	103.5	
GB3VO	2M	RV51	145.6375	145.0375	AF	IO83LA	WREXHAM	110.9	
GB3VR	70CM	RU65	430.8125	438.4125	A	IO91TU	HARPENDEN	67.0	
GB3VS	70CM	RB03	433.0750	434.6750	A	IO80LV	TRULL	94.8	
GB3VT	2M	RV58	145.7250	145.1250	D	IO83	STOKE ON TRENT	103.5	
GB3VW	70CM	RU65	430.8125	438.4125	A	JO02NN	WYMONDHAM	82.5	
GB3WA	70CM	RU70	430.8750	438.4750	A	IO75OR	DALRY	103.5	
GB3WB	70CM	RB12	433.3000	434.9000	A	IO81MH	WESTON-SUPER-MARE	94.8	
GB3WC	23CM	RM15	1297.3750	1291.3750	A	IO93EP	WAKEFIELD	82.5	
GB3WD	2M	RV56	145.7000	145.1000	A	IO80	IVYBRIDGE	77.0	
GB3WE	2M	RV55	145.6875	145.0875	AD	IO81MH	WESTON-SUPER-MARE	94.8	
GB3WF	70CM	RB14	433.3500	434.9500	AF	IO93DV	OTLEY	82.5	
GB3WG	70CM	RU70	430.8750	438.4750	AF	IO81CP	PORT TALBOT	94.8	
GB3WH	2M	RV52	145.6500	145.0500	A	IO91EM	SWINDON	118.8	
GB3WI	70CM	RB15	433.3750	434.9750	AF	JO02	WISBECH	94.8	
GB3WK	2M	RV62	145.7750	145.1750	A	IO92FH	LEAMINGTON SPA	67.0	
GB3WL	70CM	RU76	430.9500	438.5500	ADF	IO92BJ	WYTHALL	67.0	
GB3WM	70CM	RB02	433.0500	434.6500	AF	IO92AJ	LONGBRIDGE	67.0	
GB3WN	70CM	UR63	438.3875	430.7875	A	IO82XP	WOLVERHAMPTON	67.0	
GB3WO	70CM	RU78	430.9750	438.5750	A	IO90TT	WORTHING	88.5	
GB3WP	10M	10M	29.6600	29.6500	A	IO93HP	WAKEFIELD	82.5	

All Repeaters (by Callsign)

Repeater	Band	Channel	TX	RX	Modes	QTHR	Where	CTCSS/CC	DMR CC
GB3WQ	70CM	RU71	430.8875	438.4875	A	IO84	WHITEHAVEN	77.0	
GB3WR	2M	RV48	145.6000	145.0000	A	IO81PH	CHEDDAR	94.8	
GB3WS	2M	RV60	145.7500	145.1500	A	IO91WB	HORSHAM	88.5	
GB3WT	2M	RV62	145.7750	145.1750	A	IO64JQ	OMAGH	110.9	
GB3WU	70CM	RU66	430.8250	438.4250	A	IO82VE	WORCESTER	118.8	
GB3WW	2M	RV62	145.7750	145.1750	AF	IO81CP	PORT TALBOT	94.8	
GB3WX	6M	R50-12	50.8300	51.3300	A	JO02NN	WYMONDHAM	94.8	
GB3WY	6M	R50-09	50.8000	51.3000	A	IO93EP	WAKEFIELD	82.5	
GB3XD	6M	R50-02	50.7300	51.2300	A	IO93WH	LOUTH	71.9	
GB3XL	70CM	RU71	430.8875	438.4875	AM	IO93CT	SHIPLEY	82.5	
GB3XN	70CM	RU74	430.9250	438.5250	A	IO93KJ	WORKSOP	71.9	
GB3XP	2M	RV55	145.6875	145.0875	AF	IO91VJ	MORDEN	82.5	
GB3XS	6M	R50-14	50.8500	51.3500	A	IO92MG	NORTHAMPTON	77.0	
GB3XV	70CM	DVU-R20	430.2500	439.2500	F	IO92KG	DAVENTRY		
GB3XX	70CM	RU73	430.9125	438.5125	AF	IO92KG	DAVENTRY	77.0	
GB3XZ	23CM	RM1	1297.0250	1291.0250	AD	IO92KG	DAVENTRY	77.0	
GB3YA	2M	RV57	145.7125	145.1125	A	IO81LP	CWMBRAN	94.8	
GB3YC	2M	RV60	145.7500	145.1500	A	IO94SE	FILEY	88.5	
GB3YL	70CM	RB14	433.3500	434.9500	A	JO02UL	LOWESTOFT	94.8	
GB3YW	2M	RV63	145.7875	145.1875	A	IO93EP	WAKEFIELD	82.5	
GB3ZA	2M	RV59	145.7375	145.1375	AF	IO82PB	HEREFORD	118.8	
GB3ZB	70CM	RU66	430.8250	438.4250	A	IO81QJ	BRISTOL	77.0	
GB3ZI	70CM	RU78	430.9750	438.5750	AF	IO82	STAFFORD	103.5	
GB3ZN	70CM	RU71	430.8875	438.4875	A	IO94GX	HEBBURN	118.8	
GB3ZW	6M	R50-10	50.8100	51.3100	A	IO82HL	NEWTOWN POWYS	103.5	
GB3ZX	70CM	RU66	430.8250	438.4250	A	JO00	EASTBOURNE	88.5	
GB3ZY	6M	R50-09	50.8000	51.3000	A	IO81QJ	BRISTOL	77.0	
GB7AA	70CM	DVU54	439.6750	430.6750	M	IO81RO	BRISTOL		1
GB7AB	70CM	DVU53	439.6625	430.6625	F	IO87VD	ABERDEEN		
GB7AC	70CM	RU76	430.9500	438.5500	AM	IO87	LARGS		3
GB7AD	70CM	DVU61	439.7625	430.7625	D	IO81RO	BRISTOL		
GB7AE	70CM	DVU34	439.4250	430.4250	D	IO91CJ	MARLBOROUGH		3
GB7AG	70CM	RU72	430.9000	438.5000	ADMF	IO86	FORFAR	94.8	
GB7AI	70CM	DMU27	439.3375	430.3375	AM	IO85	HAWICK		3
GB7AL	70CM	DVU32	439.4000	430.4000	M	JO02RD	IPSWICH		2
GB7AM	2M	RV53	145.6625	145.0625	DFP	IO83WI	HIGHER POYNTON		
GB7AN	70CM	DVU37	439.4625	430.4625	F	IO83TS	ACCRINGTON		
GB7AR	70CM	DVU42	439.5250	430.5250	DMFPN	JO01	NEW ASH GREEN		
GB7AS	70CM	DVU34	439.4250	430.4250	M	JO01KD	ASHFORD		3
GB7AU	70CM	DVU56	439.7000	430.7000	DF	IO91GP	AMERSHAM		
GB7AV	70CM	DVU35	439.4375	430.4375	M	IO91OT	AYLESBURY		3
GB7AY	70CM	DVU36	439.4500	430.4500	DMF	IO75TQ	KILMARNOCK		1
GB7BA	70CM	DVU61	439.7625	430.7625	M	IO93CS	BRADFORD		13
GB7BB	70CM	DVU57	439.7125	430.7125	DMF	IO87QP	BANFF		1
GB7BC	70CM	DVU52	439.6500	430.6500	DMFPN	IO90	BRIGHTON		3
GB7BD	70CM	RU76	430.9500	438.5500	M	IO81QJ	BRISTOL		3
GB7BE	70CM	DVU53	439.6625	430.6625	D	JO02TK	BECCLES		
GB7BF	70CM	DVU-R24	430.3000	439.3000	DMF	IO83LU	BLACKPOOL		5
GB7BH	70CM	DVU46	439.5750	430.5750	DMFP	IO90WU	BRIGHTON		4
GB7BI	70CM	DVU53	439.6625	430.6625	M	IO77RM	MUIR OF ORD		1
GB7BJ	70CM	DVU59	439.7375	430.7375	M	IO82	MALVERN		13
GB7BK	70CM	DVU59	439.7375	430.7375	M	IO91JM	READING		3
GB7BL	70CM	DVU54	439.6750	430.6750	ADMFPN	IO84VV	ALLENDALE		5
GB7BM	70CM	DVU32	439.4000	430.4000	MP	IO92BL	BIRMINGHAM		8
GB7BN	70CM	DVU47	439.5875	430.5875	DM	IO90PT	BOGNOR REGIS		3
GB7BP	70CM	DVU36	439.4500	430.4500	D	IO91PX	MILTON KEYNES		
GB7BR	70CM	RU74	430.9250	438.5250	M	IO74TI	RAMSEY IOM		3
GB7BS	70CM	DVU13	439.1625	430.1625	M	IO81TK	BRISTOL		3
GB7BT	70CM	DVU51	439.6375	430.6375	DMFP	IO81SK	BRISTOL		3
GB7BW	70CM	DVU58	439.7250	430.7250	DMF	IO93HP	WAKEFIELD		2
GB7BX	70CM	DVU35	439.4375	430.4375	M	IO82	CLEOBURY NORTH		11
GB7BZ	70CM	DVU33	439.4125	430.4125	D	IO92QF	BOZEAT		
GB7CA	70CM	RU74	430.9250	438.5250	M	IO74SD	DOUGLAS		2
GB7CB	70CM	DVU57	439.7125	430.7125	M	IO85	BATHGATE		4
GB7CC	70CM	DVU-R22	430.2750	439.2750	M	IO81	CHELTENHAM		2
GB7CD	70CM	DVU56	439.7000	430.7000	D	IO81JL	CARDIFF		
GB7CE	70CM	DVU49	439.6125	430.6125	F	JO01NT	GREAT CLACTON		
GB7CF	70CM	DVU50	439.6250	430.6250	M	IO90LW	CLANFIELD		3
GB7CG	2M	RV57	145.7125	145.1125	AF	IO92QA	WOBURN SANDS	77.0	
GB7CH	70CM	RU66	430.8250	438.4250	ADMF	IO85	DUNS		3
GB7CK	70CM	DVU59	439.7375	430.7375	M	JO01	FOLKESTONE		3
GB7CL	70CM	DVU51	439.6375	430.6375	M	JO01OU	FRINTON ON SEA		3
GB7CM	2M	RV56	145.7000	145.1000	AF	IO80UT	MILTON ABBAS	71.9	
GB7CO	70CM	DVU39	439.4875	430.4875	DMFPN	IO90JU	FAREHAM		3
GB7CP	70CM	RU66	430.8250	438.4250	F	IO91RO	SLOUGH	77.0	
GB7CQ	70CM	DVU49	439.6125	430.6125	DMF	IO90	WORTHING		4
GB7CS	70CM	DVU62	439.7750	430.7750	F	IO75VU	GLASGOW		
GB7CT	2M	RV51	145.6375	145.0375	M	IO91PS	TRING		3
GB7CU	70CM	DVU61	439.7625	430.7625	F	IO84GQ	MARYPORT		
GB7CV	70CM	DVU-R23	430.2875	439.2875	M	IO92GK	COVENTRY		5
GB7CX	70CM	RU69	430.8625	438.4625	DMF	IO65OD	COLERAINE		12
GB7CY	70CM	DVU50	439.6250	430.6250	F	IO94DR	CROOK		
GB7CZ	70CM	DVU40	439.5000	430.5000	M	IO94DR	CROOK		10
GB7DA	2M	RV62	145.7750	145.1750	DMF	IO85AV	AIRDRIE		1
GB7DB	70CM	DVU53	439.6625	430.6625	M	IO92QA	WOBURN SANDS		3
GB7DC	70CM	RU70	430.8750	438.4750	ADMFPN	IO92	DERBY	71.9	
GB7DD	70CM	DVU53	439.6625	430.6625	MN	IO86	DUNDEE		1
GB7DE	2M	RV51	145.6375	145.0375	AM	IO85	KENNOWAY FIFE	94.8	
GB7DH	70CM	DVU47	439.5875	430.5875	DF	IO83	NORTHWICH		
GB7DJ	70CM	DVU53	439.6625	430.6625	M	IO83RF	NORTHWICH		3
GB7DK	2M	RV55	145.6875	145.0875	AF	IO74LV	STRANRAER	103.5	
GB7DK-B	70CM	DVU52	439.6500	430.6500	D	IO74LV	STRANRAER		
GB7DL	70CM	DVU54	439.6750	430.6750	D	JO03DD	SKEGNESS		
GB7DM	70CM	DVU60	439.7500	430.7500	ADMF	IO86NL	MONIFIETH		1
GB7DN	70CM	DVU33	439.4125	430.4125	D	IO64MW	DUNGIVEN		
GB7DP	70CM	DVU47	439.5875	430.5875	M	IO71MV	TREFFGARNE		3
GB7DR	70CM	DVU34	439.4250	430.4250	M	IO90AR	POOLE		5
GB7DS	70CM	DVU34	439.4250	430.4250	ADMFP7N	JO02PP	NORWICH	94.8	
GB7DW	2M	RV50	145.6250	145.0250	AM	IO80JU	DUNKERSWELL		7
GB7DX	70CM	DVU56	439.7000	430.7000	M	JO00LW	NEW ROMNEY		3

All Repeaters (by Callsign)

Repeater	Band	Channel	TX	RX	Modes	QTHR	Where	CTCSS/CC	DMR CC
GB7DY	70CM	DVU46	439.5750	430.5750	M	IO82	DUDLEY		11
GB7DZ	70CM	RB02	433.0500	434.6500	AF	IO83	CONGLETON	103.5	
GB7EA	70CM	DMU30	439.3750	430.3750	AF	IO75SV	ERSKINE		1
GB7EB	70CM	DVU46	439.5750	430.5750	M	JO02TK	BECCLES		2
GB7EC	70CM	RU74	430.9250	438.5250	M	IO87	ABERDEEN		11
GB7ED	70CM	DVU54	439.6750	430.6750	M	IO80GR	EXETER		5
GB7EE	70CM	RU74	430.9250	438.5250	M	IO85JW	EDINBURGH		1
GB7EG	70CM	DVU61	439.7625	430.7625	M	IO91	EAST GRINSTEAD		2
GB7EH	70CM	RU75	430.9375	438.5375	M	IO92GC	BANBURY		8
GB7EI	70CM	RU68	430.8500	438.4500	ADMF	IO67IN	LOCHMADDY	88.5	
GB7EL	70CM	DVU57	439.7125	430.7125	M	IO83	NELSON		2
GB7EP	70CM	DVU32	439.4000	430.4000	M	IO91UI	EPSOM		3
GB7ER	2M	RV62	145.7750	145.1750	AF	IO80GR	EXETER	77.0	
GB7ES	70CM	DVU35	439.4375	430.4375	D	JO00DT	EASTBOURNE		
GB7EU	70CM	RU67	430.8375	438.4375	AN	IO85JW	EDINBURGH	94.8	
GB7EV	70CM	U94	431.1750	431.1750	M	IO85JW	EDINBURGH		11
GB7EW	70CM	DVU42	439.5250	430.5250	M	IO80	EXETER		3
GB7EX	70CM	DVU57	439.7125	430.7125	M	JO01GN	SOUTHEND ON SEA		3
GB7EY	70CM	DVU41	439.5125	430.5125	DMFPN	IO85	SELKIRK		1
GB7EZ	70CM	DVU62	439.7750	430.7750	M	IO92CM	ERDINGTON		14
GB7FB	70CM	DVU38	439.4750	430.4750	DMF	IO71VA	BIDEFORD		5
GB7FC	70CM	DVU40	439.5000	430.5000	D	IO83LU	CLEVELEYS		
GB7FD	70CM	DVU58	439.7250	430.7250	F	IO83LV	FLEETWOOD		
GB7FE	70CM	DVU38	439.4750	430.4750	DMF	IO86BC	STIRLING		1
GB7FG	70CM	RU76	430.9500	438.5500	AF	IO71WW	CARMARTHEN	94.8	
GB7FI	70CM	RU71	430.8875	438.4875	M	IO81OH	AXBRIDGE		3
GB7FK-B	70CM	DVU54	439.6750	430.6750	D	JO01OC	FOLKESTONE		
GB7FL	70CM	DVU61	439.7625	430.7625	DF	IO83	FRECKLETON		
GB7FO	70CM	DVU-R22	430.2750	439.2750	M	IO83LT	BLACKPOOL		1
GB7FR	70CM	DVU40	439.5000	430.5000	M	IO90	WORTHING		4
GB7FT	70CM	DVU57	439.7125	430.7125	M	IO90	PORTSMOUTH		3
GB7FU	70CM	RU71	430.8875	438.4875	AF	IO93	NOTTINGHAM		
GB7FW	70CM	DVU54	439.6750	430.6750	D	IO90LU	PORTSMOUTH		
GB7GA	70CM	RU72	430.9000	438.5000	ADMF	IO85	GALASHIELS		1
GB7GB	70CM	DVU61	439.7625	430.7625	M	IO92BM	SUTTON COLDFIELD		8
GB7GD	2M	RV55	145.6875	145.0875	DMF	IO87WE	ABERDEEN		1
GB7GF	70CM	DVU55	439.6875	430.6875	M	IO91RF	GUILDFORD		3
GB7GG	70CM	DVU50	439.6250	430.6250	DMF	IO85AV	AIRDRIE		1
GB7GJ	70CM	RU72	430.9000	438.5000	AM	IN89WE	ST HELIER	88.5	
GB7GL	70CM	RU68	430.8500	438.4500	M	IO75	GLASGOW		7
GB7GM	70CM	RU78	430.9750	438.5750	ADMF	IO85	DUMFRIES		3
GB7GP	70CM	DVU52	439.6500	430.6500	MF	IO91	HIGH WYCOMBE		5
GB7GR	70CM	DVU36	439.4500	430.4500	DMF	IO92QW	GRANTHAM		5
GB7GS	70CM	DVU36	439.4500	430.4500	MF	JO02EO	DOWNHAM MARKET		3
GB7GT	70CM	DVU33	439.4125	430.4125	M	IO82	CLEE HILL		13
GB7GW	70CM	RU65	430.8125	438.4125	ADMF	IO94FV	WASHINGTON		10
GB7GX	2M	RV59	145.7375	145.1375	ADMFPN	IO86	GLENROTHES		1
GB7GY	70CM	DVU55	439.6875	430.6875	M	IN89RL	ST. PETER PORT		5
GB7HA	70CM	DVU-R23	430.2875	439.2875	F	IO91TU	HARPENDEN		
GB7HB	70CM	DVU50	439.6250	430.6250	DMF	IO64	TANDRAGEE		1
GB7HC	70CM	DVU42	439.5250	430.5250	M	IO82	HEREFORD		5
GB7HD	2M	RV55	145.6875	145.0875	M	IO70	TAVISTOCK		3
GB7HH	70CM	DVU60	439.7500	430.7500	DMF	JO01CN	ROMFORD		3
GB7HI	70CM	DVU59	439.7375	430.7375	DMF	IO64	LISBURN		1
GB7HK	2M	RV50	145.6250	145.0250	AM	IO85	HAWICK		3
GB7HM	70CM	DVU51	439.6375	430.6375	M	IO83	CAERGWRLE		1
GB7HN	70CM	RU70	430.8750	438.4750	A	IO83RL	LEIGH	82.5	
GB7HR	70CM	DVU36	439.4500	430.4500	M	IO91SL	HEATHROW		3
GB7HS	70CM	DVU34	439.4250	430.4250	M	IO83XR	HEPTONSTALL		2
GB7HT	70CM	DVU59	439.7375	430.7375	M	IO95	ASHINGTON		10
GB7HU	70CM	DVU39	439.4875	430.4875	D	IO93RS	SOUTH CAVE		
GB7HX	70CM	DVU46	439.5750	430.5750	DM	IO93BP	HUDDERSFIELD		1
GB7IC-A	23CM	23CM	1290.6500	1270.6500	D	JO01NI	HERNE BAY		
GB7IC-B	70CM	DVU36	439.4500	430.4500	D	JO01NI	HERNE BAY		
GB7IC-C	2M	RV53	145.6625	145.0625	AD	JO01NI	HERNE BAY		
GB7IE	2M	RV54	145.6750	145.0750	AF	IO70WJ	PLYMOUTH	77.0	
GB7II	70CM	DVU36	439.4500	430.4500	M	IO77UL	INVERNESS		3
GB7IK	70CM	DVU43	439.5375	430.5375	M	JO01FJ	ROCHESTER		3
GB7IL	2M	RV53	145.6625	145.0625	M	IO77UL	INVERNESS		5
GB7IN	70CM	DVU52	439.6500	430.6500	M	IO93GD	ALFRETON		1
GB7IP	2M	RV52	145.6500	145.0500	M	IO92KP	LEICESTER		1
GB7IS	70CM	DVU43	439.5375	430.5375	F	IO81MH	WESTON-SUPER-MARE		
GB7IT	70CM	DVU41	439.5125	430.5125	M	IO81MH	WESTON-SUPER-MARE		1
GB7IV	2M	RV62	145.7750	145.1750	F	IO90IV	SOUTHAMPTON		
GB7JB	70CM	DVU37	439.4625	430.4625	M	IO81TB	WINCANTON		1
GB7JD	70CM	DVU33	439.4125	430.4125	DMN	IO85RL	JEDBURGH		1
GB7JF	70CM	DVU51	439.6375	430.6375	D	IO92NE	NORTHAMPTON		
GB7JI	70CM	RU68	430.8500	438.4500	D	IN89WE	ST HELIER		2
GB7JL	70CM	RU77	430.9625	438.5625	AM	IO83QL	WIGAN	82.5	
GB7JM	70CM	DVU47	439.5875	430.5875	DMF	JO03BI	ALFORD		3
GB7KB	2M	RV52	145.6500	145.0500	M	JO00CX	HEATHFIELD		3
GB7KD	70CM	DVU55	439.6875	430.6875	M	JO00DV	HAILSHAM		3
GB7KE	70CM	DVU47	439.5875	430.5875	MF	JO01GJ	GILLINGHAM		3
GB7KK	70CM	RU74	430.9250	438.5250	DMF	IO65VE	BALLYCASTLE		1
GB7KL	70CM	RU70	430.8750	438.4750	AF	IO84RD	KIRKBY LONSDALE	110.9	
GB7KN	70CM	DVU35	439.4375	430.4375	DMF	IO85	EDINBURGH		1
GB7KP	70CM	DVU46	439.5750	430.5750	DMF	IO74	COMBER		1
GB7KR	70CM	DVU-R24	430.3000	439.3000	M	IO82	KIDDERMINSTER		8
GB7KS	70CM	DVU-R20	430.2500	439.2500	M	IO81	STROUD		2
GB7KT	70CM	DVU40	439.5000	430.5000	M	IO91GE	ANDOVER		1
GB7KU	70CM	DVU42	439.5250	430.5250	DMF	IO92TT	BOURNE		5
GB7KY	70CM	DVU52	439.6500	430.6500	ADMFPN	IO86	KIRKCALDY		1
GB7LD	70CM	DVU35	439.4375	430.4375	M	IO83	LEYLAND		15
GB7LE	70CM	DVU53	439.6625	430.6625	M	IO93FU	LEEDS		2
GB7LF	70CM	DVU43	439.5375	430.5375	D	IO84OA	LANCASTER		
GB7LK	70CM	RU65	430.8125	438.4125	M	IO70SM	LISKEARD		5
GB7LL	70CM	DVU40	439.4969	430.4969	MN	IO82	LLANGOLLEN		12
GB7LN	70CM	DVU32	439.4000	430.4000	M	IO93RF	LINCOLN		1
GB7LO	70CM	DVU41	439.5125	430.5125	M	IO91	BECKENHAM		3
GB7LP	70CM	DVU32	439.4000	430.4000	M	IO83MK	LIVERPOOL		1

All Repeaters (by Callsign)

Repeater	Band	Channel	TX	RX	Modes	QTHR	Where	CTCSS/CC	DMR CC
GB7LR	70CM	DVU38	439.4750	430.4750	M	IO92IQ	LEICESTER		
GB7LS	70CM	DVU32	439.4000	430.4000	DMF	IO85	GALASHIELS		
GB7LT	70CM	U324	434.0500	434.0500	E	IO92	BOURNE		
GB7LV	2M	RV49	145.6125	145.0125	ADMF	IO85FV	LIVINGSTON		1
GB7LZ	70CM	RU73	430.9125	438.5125	AM	IO80CV	LAPFORD	77.0	
GB7MB	70CM	DVU56	439.7000	430.7000	F	IO84	MORECAMBE		
GB7MC	70CM	DVU43	439.5375	430.5375	DF	IO70	ST AUSTELL		
GB7MD	70CM	DVU32	439.4000	430.4000	F	IO80UT	BLANDFORD FORUM		
GB7ME	70CM	DVU58	439.7250	430.7250	D	IO92JJ	RUGBY		3
GB7MF	70CM	DVU54	439.6750	430.6750	DMF	IO64QR	MAGHERAFELT		3
GB7MH	70CM	DVU51	439.6375	430.6375	DMF	IO91WC	EAST GRINSTEAD		2
GB7MJ	70CM	DVU51	439.6375	430.6375	M	IO91GA	ROMSEY		5
GB7MK	70CM	DVU48	439.6000	430.6000	M	JO02OB	IPSWICH		13
GB7MM	70CM	DVU60	439.7500	430.7500	DMF	IO87BN	NAIRN		5
GB7MN	70CM	DVU33	439.4125	430.4125	M	IO83	STOCKPORT		2
GB7MP	70CM	DVU60	439.7500	430.7500	MP	IO84NB	HEYSHAM		3
GB7MR	70CM	DVU59	439.7375	430.7375	M	IO83	BURY		2
GB7MS	2M	RV62	145.7750	145.1750	M	IO94EX	NEWCASTLE UPON TYNE		10
GB7MT	70CM	DVU53	439.6625	430.6625	F	IO90IV	SOUTHAMPTON		
GB7NA-A	23CM	RM12	1297.3000	1291.3000	D	IO93HN	BARNSLEY		
GB7NA-B	70CM	DVU36	439.4500	430.4500	D	IO93HN	BARNSLEY		
GB7NA-C	2M	RV54	145.6750	145.0750	D	IO93HN	BARNSLEY		
GB7NB	70CM	DVU55	439.6875	430.6875	D	JO02PN	NORWICH		
GB7NC	70CM	DVU58	439.7250	430.7250	DMF	JO02OL	LONG STRATTON		3
GB7ND	70CM	DVU33	439.4125	430.4125	M	JO02LM	GREAT ELLINGHAM		1
GB7NE	70CM	DVU36	439.4500	430.4500	DMF	IO95FE	ASHINGTON		3
GB7NF	70CM	DVU39	439.4875	430.4875	M	JO00AS	NEWHAVEN		1
GB7NH	70CM	DVU50	439.6250	430.6250	F	IO91OF	ALDERSHOT		
GB7NL	70CM	DVU37	439.4625	430.4625	D	IO80FR	EXETER		
GB7NM	2M	RV59	145.7375	145.1375	DF	JO02LM	ATTLEBOURGH		
GB7NN	70CM	DVU60	439.7500	430.7500	F	IO93	WORKSOP		
GB7NR	70CM	DVU35	439.4375	430.4375	DM	IO92KX	NOTTINGHAM		1
GB7NS	70CM	DVU13	439.1625	430.1625	M	IO91WG	CATERHAM		3
GB7NT	70CM	RU74	430.9250	438.5250	AF	IO91IJ	NEWBURY	118.8	
GB7NV	70CM	RU66	430.8250	438.4250	ADM	IO93	HUCKNALL		1
GB7NW	70CM	DVU58	439.7250	430.7250	F	IO91DN	SWINDON		
GB7NY	70CM	DVU55	439.6875	430.6875	DMF	IO64	BESSBROOK		1
GB7OK	2M	RV57	145.7125	145.1125	DMF	IO91	BECKENHAM		3
GB7OR	70CM	DVU37	439.4625	430.4625	M	IO88LX	KIRKWALL		1
GB7PB	2M	RV53	145.6625	145.0625	D	IO94FR	DURHAM		
GB7PD	2M	RV59	145.7375	145.1375	ADF	IO71OV	MAENCLOCHOG	94.8	
GB7PE	70CM	DVU40	439.5000	430.5000	M	IO92TL	PETERBOROUGH		3
GB7PF	70CM	DVU51	439.6375	430.6375	M	IO80AN	PRINCETOWN		3
GB7PH	70CM	DVU37	439.4625	430.4625	DM	JO00	HAILSHAM		3
GB7PI	70CM	DVU61	439.7625	430.7625	D	IO92XA	ROYSTON		
GB7PM	70CM	DVU35	439.4375	430.4375	F	IO81PH	CHEDDAR		
GB7PN	70CM	DVU34	439.4250	430.4250	M	IO83HH	PRESTATYN		1
GB7PO	70CM	DVU59	439.7375	430.7375	F	IO90KT	PORTSMOUTH		
GB7PR	70CM	U324	434.0500	434.0500	E	IO83PS	PRESTON		
GB7PS	2M	RV55	145.6875	145.0875	M	IO90	PORTSMOUTH		3
GB7PT	70CM	DVU57	439.7125	430.7125	DMF	IO92XA	ROYSTON		
GB7PX	70CM	DVU62	439.7750	430.7750	DMP	IO92XA	ROYSTON		10
GB7PY	70CM	RU74	430.9250	438.5250	M	JO02AF	CAMBRIDGE		3
GB7RA	70CM	DVU52	439.6500	430.6500	DMF	IO83LH	NESTON		5
GB7RB-C	2M	RV54	145.6750	145.0750	D	IO81GK	COWBRIDGE		
GB7RC	70CM	DVU-R21	430.2625	439.2625	D	IO81	CHELTENHAM		
GB7RD	70CM	DVU55	439.6875	430.6875	M	IO70	YELVERTON		3
GB7RE	70CM	DVU57	439.7125	430.7125	M	IO93MH	RETFORD		1
GB7RJ	70CM	DVU43	439.5375	430.5375	DM	IO82LW	CHIRK		5
GB7RK	70CM	DVU35	439.4375	430.4375	DMFPN	IO74AN	BELFAST		3
GB7RL	70CM	DVU62	439.7750	430.7750	DMFPN	IO83RU	LONGRIDGE		10
GB7RM	70CM	U324	434.0500	434.0500	E	JO01CO	ROMFORD		
GB7RN	2M	RV51	145.6375	145.0375	D	IO90JT	FAREHAM		
GB7RO	70CM	DVU-R21	430.2625	439.2625	F	IO83UQ	HASLINGDEN		
GB7RR	70CM	DVU49	439.6125	430.6125	M	IO92	NOTTINGHAM		1
GB7RS	70CM	DVU49	439.6125	430.6125	M	IO91PX	MILTON KEYNES		4
GB7RT	70CM	DVU33	439.4125	430.4125	M	IO91	SUNNINGDALE		3
GB7RV	70CM	DVU50	439.6250	430.6250	M	IO83	BLACKBURN		2
GB7RW	2M	RV48	145.6000	145.0000	AD	IO94SO	WHITBY	88.5	
GB7RX	2M	RV62	145.7750	145.1750	AM	JO00GU	HASTINGS		3
GB7RY	70CM	RU76	430.9500	438.5500	AF	JO00HW	RYE	103.5	
GB7SA	70CM	DVU-R24	430.3000	439.3000	M	IO81	PORT TALBOT		5
GB7SB	70CM	DVU53	439.6625	430.6625	M	IO92	BIRMINGHAM		5
GB7SD	70CM	DVU33	439.4125	430.4125	M	IO80SQ	WEYMOUTH		1
GB7SE	70CM	DVU38	439.4750	430.4750	M	JO01DM	THURROCK		3
GB7SF-C	2M	RV59	145.7375	145.1375	D	IO93GK	SHEFFIELD		
GB7SG	2M	RV61	145.7625	145.1625	M	IO83MK	LIVERPOOL		1
GB7SH	70CM	RU77	430.9625	438.5625	AF	IO93GK	SHEFFIELD	71.9	
GB7SI	70CM	DVU61	439.7625	430.7625	M	IO83	LEEK		1
GB7SJ	70CM	RB07	433.1750	434.7750	A	IO83RF	NORTHWICH	103.5	
GB7SK	70CM	DVU37	439.4625	430.4625	M	IO92NO	LEICESTER		1
GB7SL	70CM	DVU-R21	430.2625	439.2625	DMFPN	IO92TR	BASTON		8
GB7SM	70CM	DVU47	439.5875	430.5875	ADMFPN	IO86	ST MONANS		1
GB7SN	70CM	DVU54	439.6750	430.6750	M	IO93GH	SHEFFIELD		1
GB7SO	70CM	DVU42	439.5250	430.5250	DM	IO90KT	GOSPORT		1
GB7SP	70CM	DVU60	439.7500	430.7500	M	IO91CB	SALISBURY		3
GB7SQ	70CM	DVU54	439.6750	430.6750	MN	IO85	DALKEITH		1
GB7SS	70CM	DVU47	439.5875	430.5875	M	IO87IP	ELGIN		1
GB7ST	70CM	DVU36	439.4500	430.4500	M	IO83	STOKE ON TRENT		1
GB7SU	70CM	DVU36	439.4500	430.4500	M	IO90IV	SOUTHAMPTON		8
GB7SV	70CM	RB11	433.2750	434.8750	AF	IO91UW	HITCHIN	82.5	
GB7SW	2M	RV51	145.6375	145.0375	D	IO81VL	CHIPPENHAM		2
GB7SZ	70CM	RU75	430.9375	438.5375	A	IO90AR	POOLE	71.9	
GB7TB	70CM		439.2188	430.2188	E	IO92	MORTON		
GB7TC	70CM	DVU42	439.5250	430.5250	M	IO91DL	SWINDON		2
GB7TD	70CM	DVU13	439.1625	430.1625	M	IO93EP	WAKEFIELD		1
GB7TE	2M	RV62	145.7750	145.1750	D	JO01OT	CLACTON ON SEA		
GB7TF	70CM	DVU39	439.4875	430.4875	F	IO91QT	TRING		
GB7TH	70CM	DVU-R23	430.2875	439.2875	M	JO01QJ	MARGATE		5
GB7TM	70CM	DVU59	439.7375	430.7375	F	JO02NH	EYE		

All Repeaters (by Callsign)

Repeater	Band	Channel	TX	RX	Modes	QTHR	Where	CTCSS/CC	DMR CC
GB7TP	70CM	DVU55	439.6875	430.6875	DMF	IO93CT	SHIPLEY		1
GB7TQ	70CM	RU71	430.8875	438.4875	AF	IO80FM	TORQUAY	94.8	
GB7TT	70CM	DVU40	439.5031	430.5031	N	IO82	LUDLOW		
GB7TX	70CM	U321A	434.0188	434.0188	E	JO02TK	BECCLES		
GB7TY	70CM	DVU49	439.6125	430.6125	DM	IO84XX	HEXHAM		10
GB7UL	70CM	DVU42	439.5250	430.5250	M	IO74	CARRICKFERGUS		1
GB7VO	70CM	DVU56	439.7000	430.7000	DMFN	IO82QJ	TENBURY WELLS		1
GB7VT	70CM	U94	431.1750	431.1750	M	IO83	STOKE-ON-TRENT		5
GB7WA	2M	RV55	145.6875	145.0875	DM	IO93WM	GRIMSBY		2
GB7WB	70CM	DVU39	439.4875	430.4875	D	IO81MH	WESTON-SUPER-MARE		2
GB7WC	70CM	DVU39	439.4875	430.4875	D	IO83QI	WARRINGTON		
GB7WD	70CM	DVU57	439.7125	430.7125	M	IO82	SEDGLEY		11
GB7WF	70CM	DVU39	439.4875	430.4875	DMFN	IO82	BEWDLEY		13
GB7WH	70CM	DMU30	439.3750	430.3750	AF	IO83	WREXHAM	110.9	
GB7WI	70CM	DVU33	439.4125	430.4125	M	IO93TR	HULL		1
GB7WK	70CM	DVU41	439.5125	430.5125	DMFPN	IO91	WOKING		15
GB7WL	70CM	DVU37	439.4625	430.4625	M	IO91QP	AMERSHAM		3
GB7WM	70CM	DVU55	439.6875	430.6875	F	IO83SM	ASTLEY		
GB7WN	70CM	DVU60	439.7500	430.7500	M	IO82XP	WOLVERHAMPTON		8
GB7WR	70CM	DVU37	439.4625	430.4625	D	IO82UB	GREAT MALVERN		
GB7WS	70CM	DVU34	439.4250	430.4250	M	IO91SA	BILLINGSHURST		3
GB7WT	70CM	DVU48	439.6000	430.6000	DMF	IO64HQ	OMAGH		1
GB7WV	70CM	DVU55	439.6875	430.6875	M	IO92NL	MARKET HARBOROUGH		2
GB7WW	70CM	DVU38	439.4750	430.4750	M	IO81	KINGTON LANGLEY		7
GB7WX	70CM	DVU41	439.5125	430.5125	DMF	IO83OE	TARVIN		3
GB7WY	70CM	DVU41	439.5125	430.5125	D	IO93FR	WAKEFIELD		
GB7XG	70CM	RU70	430.8750	438.4750	ADMFPN	IO86JF	GLENROTHES		1
GB7XX	70CM	DVU37	439.4625	430.4625	M	IO94	FELLING		10
GB7XY	70CM	DVU34	439.4250	430.4250	M	IO94EX	NEWCASTLE UPON TYNE		10
GB7YD	70CM	DVU55	439.6875	430.6875	M	IO80PW	YEOVIL		5
GB7YF	2M	RV61	145.7625	145.1625	AF	IO80PW	YEOVIL TOWN CENTER	94.8	
GB7YI	70CM	DVU46	439.5750	430.5750	M	IO84WX	HEXHAM		10
GB7YJ	70CM	DVU47	439.5875	430.5875	M	IO92FH	LEAMINGTON SPA		3
GB7YL	2M	RV63	145.7875	145.1875	M	JO02UL	LOWESTOFT		1
GB7YR	70CM	DVU42	439.5250	430.5250	M	IO93LM	DONCASTER		2
GB7YZ	70CM	DVU38	439.4750	430.4750	MF	IO83JE	MOLD		
GB7ZE	70CM	RB02	433.0500	434.6500	AM	JO00GU	HASTINGS		3
GB7ZP	70CM	DVU39	439.4875	430.4875	D	JO01FS	CHELMSFORD		

Echolink and IRLP Linking (by Callsign)

Call	Chan	Output MHz	In MHz	Mode	QTHR	Location	Keeper	ECHO	IRLP
GB3AG	RV58	145.725	145.125	AV	IO86ON	ANGUS	MM0XET	117931	
GB3AM	R50-13	50.84	51.34	AV	IO91QP	AMERSHAM	M0ZPU	4125	
GB3AR	RV56	145.7	145.1	AV	IO73VC	CAERNARFON	GW4KAZ	206003	
GB3AS	RV48	145.6	145	AV	IO85MC	LANGHOLM	GM6LJE	412685	
GB3BC	RV60	145.75	145.15	AV	IO81IO	PONTYPRIDD	GW0UXJ	39300	
GB3BN	RB0	433	434.6	AV	IO91OJ	BRACKNELL	G8DOR	1938	
GB3BW	RU65	430.8125	438.4125	AV	IO93HO	WAKEFIELD	M0YDG		5775
GB3CA	RB13	433.325	434.925	AV	IO84OT	CARLISLE	G1XSZ	412685	
GB3CA	RB13	433.325	434.925	AV	IO84OT	CARLISLE	G1XSZ		5280
GB3DC	RV55	145.6875	145.0875	AV	IO92	DERBY	G7NPW	92369	
GB3DQ	RU78	430.975	438.575	AV	IO70RI	POLPERRO	G1YDQ	418341	
GB3DQ	RU78	430.975	438.575	AV	IO70RI	POLPERRO	G1YDQ		5612
GB3DX	RB12	433.3	434.9	AV	IO65HA	DERRY/LONDONDERRY	GI4YWT	7125	
GB3EK	RU71	430.8875	438.4875	AV	JO01QJ	MARGATE	M0LMK	48360	
GB3FH	R50-06	50.77	51.27	AV	IO81OH	SOMERSET	G4RKY	228585	
GB3FH	R50-06	50.77	51.27	AV	IO81OH	SOMERSET	G4RKY		5361
GB3HH	RV56	145.7	145.1	AV	IO93BF	BUXTON	G7EKY	97616	
GB3IK	RV61	145.7625	145.1625	AV	JO01FJ	ROCHESTER	G6CKK	263025	
GB3IM	RU66	430.825	438.425	AV	IO74SD	DOUGLAS IOM	GD4HOZ	464453	
GB3IM	RU66	430.825	438.425	AV	IO74TI	RAMSEY IOM	GD4HOZ	464453	
GB3IM	RB05	433.125	434.725		IO74SG	DOUGLAS IOM	GD4HOZ	464453	
GB3IR	RV61	145.7625	145.1625	AV	IO94DJ	RICHMOND YORKS	G4FZN	1353	
GB3IR	RV61	145.7625	145.1625	AV	IO94DJ	RICHMOND YORKS	G4FZN		5562
GB3KC	RU72	430.9	438.5	AV	IO82WK	STOURBRIDGE	G3PWJ	430900	
GB3KD	RV63	145.7875	145.1875		IO82	KIDDERMINSTER	M0KRG	78750	
GB3KE	RV55	145.6875	145.0875	DM	IO75UV	GLASGOW	GM7SVK	5411	
GB3KE	RV55	145.6875	145.0875	DM	IO75UV	GLASGOW	GM7SVK		5410
GB3KL	RB04	433.1	434.7	DM	JO02FR	KINGS LYNN	G0IJU 77266		
GB3KS	RV50	145.625	145.025	DM	JO01PA	DOVER	M1CMN	346463	
GB3LR	RU69	430.8625	438.4625	AV	JO00AS	NEWHAVEN	G0TJH494669		
GB3LS	RB02	433.05	434.65	AV	IO93RF	LINCOLN	M0TEF	268511	
GB3LV	RB02	433.05	434.65	AV	IO91XP	NORTH LONDON	G4DFB	155403	
GB3LV	RB02	433.05	434.65	AV	IO91XP	NORTH LONDON	G4DFB		5600
GB3MH	RV50	145.625	145.025	AV	IO91WC	EAST GRINSTEAD	G7KBR	453929	
GB3MH	RV50	145.625	145.025	AV	IO91WC	EAST GRINSTEAD	G7KBR		5569
GB3MW	RB10	433.25	434.85	AV	IO92FH	LEAMINGTON SPA	G6FEO	749161	
GB3NK	RB04	433.1	434.7	AV	JO01BL	ERITH	G4EGU	54760	
GB3OA	RV49	145.6125	145.0125	AV	IO83LP	SOUTHPORT	G4EID 5302		
GB3OA	RV49	145.6125	145.0125	AV	IO83LP	SOUTHPORT	G4EID	5302	
GB3PA	RV50	145.625	145.025	AV	IO75QV	PAISLEY	GM7GDE	116678	
GB3PY	RU72	430.9	438.5	AV	JO02AF	CAMBRIDGE	M0ZPU	222303	
GB3PZ	RU72	430.9	438.5	AV	IO83XL	DUKINFIELD	G4ZPZ	2591	
GB3PZ	RU72	430.9	438.5	AV	IO83XL	DUKINFIELD	G4ZPZ		5400
GB3SB	RV52	145.65	145.05	AV	IO85	SELKIRK	GM7LUN	116678	
GB3SD	RB14	433.35	434.95	AV	IO80SQ	WEYMOUTH	G0ECX	112689	
GB3TR	RV52	145.65	145.05	AV	IO80FM	TORQUAY	G8XST		5582
GB3UB	RB04	433.1	434.7	AV	IO81UJ	BATH AVON	G4KVI 201135		
GB3VR	RU65	430.8125	438.4125		IO91TU	HARPENDEN	G7HMV	5939	
GB3WK	RV62	145.775	145.175	AV	IO92FH	LEAMINGTON SPA	G6FEO	749154	
GB3XN	RU74	430.925	438.525	AV	IO93KJ	WORKSOP	G3XXN	153126	
GB3XN	RU74	430.925	438.525	AV	IO93KJ	WORKSOP	G3XXN		5708
GB3ZB	RU66	430.825	438.425	AV	IO81QJ	BRISTOL	G4RKY		5429
GB7RW	RV48	145.6	145	DM	IO94SO	WHITBY	G4EQS	921135	
GB7SJ	RB07	433.175	434.775	DM	IO83RF	NORTHWICH	M0WTX	455339	
GB7SJ	RB07	433.175	434.775	DM	IO83RF	NORTHWICH	M0WTX		41360

Amateur TV Repeaters (by Callsign)

Call	Tx MHz	Rx MHz	QTHR	Location
GB3CT	3404	2435	IO93UM	Caistor
GB3CZ	2432	2346.5	JO01OT	Clacton
GB3DO	1310	1248	IO74DP	Bangor
GB3EN	1312	1249	IO91XP	North London
GB3ET	1322	1249	IO92GC	Edgehill nr Banbury
GB3EY	1308	1265	IO93RS	South Cave
GB3FT	1315	1249/1255/1265/1278	IO83LU	Blackpool
GB3FW	3406	2430	IO90LU	Portsmouth
GB3GG	1310	1280	IO93XN	Grimsby
GB3GV	1318.5	1249	IO92IQ	Markfield
GB3HV	1308	1248	IO91LD	Alton
GB3JT	1318	1249	JO00CS	Eastbourne
GB3JV	3404	2440	JO01AJ	Petts Wood
GB3KM	3406/10065/2440/1304	5665/2328/1280/10315	IO94EQ	Spennymoor
GB3LO	1316	1249	JO02UL	Lowestoft
GB3LX	10240	10425	IO93RF	Lincoln
GB3MV	1316	1249	IO92NF	Northampton
GB3NQ	1316	1249	IO70OJ	Roche
GB3NV	3406	2440	JO02PP	Norwich
GB3PV	1316	1249	JO02AF	Cambridge
GB3SQ	1304	1244	IO90BR	Bournemouth
GB3TB	1316	1249	IO80FM	Torquay
GB3TG	10240	10425	IO91PX	Milton Keynes
GB3TM	1316	1249	IO73UJ	Amlwch
GB3TN	1316	1249	JO02KS	Fakenham
GB3TT	1310	1278	IO93GK	Sheffield
GB3TV	1318.5	1249	IO91RU	Dunstable
GB3TZ	2440/2326	2397	IO91SV	Luton Beds
GB3UD	1318.5	1249	IO83VC	Stoke On Trent
GB3UT	1311.5	1249	IO81UJ	Bath
GB3VL	1310	1270	IO93RF	Lincoln
GB3VX	1310	1249	JO00	Eastbourne
GB3XY	10065	10315	IO93RS	Hull
GB3YT	1316	1276	IO93DP	Mirfield
GB3ZZ	1316	1249	IO81RM	Bristol City Centre

*Please check on: *https://ukrepeater.net/tvrepeaterlist.htm* for additional frequencies

As new applications are being made on a regular basis please visit the RSGB website and you will find more information on repeaters at: *www.ukrepeater.net*

RSGB Band Plan

EFFECTIVE FROM 1ST JANUARY 2024 UNLESS OTHERWISE SHOWN

In this edition of the Yearbook are the 2024 Band Plans. These are based on a combination of IARU Region-1 recommendations, as well as UK-specific usage and feedback by RSGB Spectrum Forum members.

They are also available on the RSGB website in various formats, where more detailed change notes can also be found to guide users in the Excel master.

The majority of changes have occurred in VHF and UHF as a result of the Ofcom changes. In summary the changes from last year are:

- HF: Editorial changes have been made in the 472kHz and 28MHz as a result of the licence changes. For now IARU has deferred a planned re-balance between CW/SSB versus digital usage, so there are no other HF updates at present.
- VHF: In order to accommodate licence-driven changes, the 144MHz band sees the introduction of a brand new category - two new 5W erp max ad-hoc repeater channels; as well as an extra gateway channel and a move to 12.5kHz channelling in the all-modes section.
- UHF: The 70cm band also sees a long overdue recognition of 12.5k channelling, which has been used by ETCC and IARU-R1 for some time. This also assists coordinated interleaving with the Primary User. The intention is that 70cm will be better able to accommodate demand.
- Microwave: Numerous bands have had their licence and coordination notes editorially updated. For the first time there are now full entries for the 122, 134 and 248GHz bands. It is also worth noting that the revised Foundation licence only gives access to 2.4 and not 2.3GHz.

In addition to the band plans, other operational and licensing aspects have changed, Therefore it is important that all amateurs familiarise themselves with the new licence conditions which include a new Ofcom Coordination condition and Notice.

Looking ahead we expect further changes in VHF/UHF including on gateway channels, a novel UHF repeater shift, and data experimentation. These will be reviewed by the RSGB Spectrum Forum, along with emerging proposals for HF.

Further Information

In addition to RadCom, the latest band plans can be found on the Operating section of the RSGB website at rsgb.org/bandplans - and, if you are unsure, by all means contact either the relevant spectrum manager:

hf.manager@rsgb.org.uk

vhf.manager@rsgb.org.uk

microwave.manager@rsgb.org.uk

Murray Niman, G6JYB,
RSGB Spectrum Chair

136kHz	NECESSARY BANDWIDTH	UK USAGE
135.7-137.8kHz	200Hz	CW, QRSS and Narrowband Digital Modes

Licence Notes: Amateur Service – Secondary User. 1 watt (0dBW) ERP.
R.R. 5.67B. The use of the band 135.7-137.8kHz in Algeria, Egypt, Iraq, Lebanon, Syrian Arab Republic Sudan, South Sudan and Tunisia is limited to fixed and maritime mobile services. The amateur service shall not be used in the above-mentioned countries in the band 135.7-137.8kHz, and this should be taken into account by the countries authorising such use. (WRC-19)..

472kHz (600m)	NECESSARY BANDWIDTH	UK USAGE
IARU Region 1 does not have a formal band plan for this allocation but has a usage recommendation (Note 1).		
472-479kHz	500Hz	CW, QRSS and Narrowband Digital Modes

Note 1: Usage recommendation – 472-475kHz CW only 200Hz maximum bandwidth, 475-479kHz CW and Digimodes.
Note 2: It should be emphasised that this band is available on a non-interference basis to existing services. UK amateurs should be aware that some overseas stations may be restricted in terms of transmit frequency in order to avoid interference to nearby radio navigation service Non-Directional Beacons.
Licence Notes: Amateur Service – Secondary User. Full Licensees only. **5 watts EIRP maximum**.
R.R. 5.80B. The use of the frequency band 472-479kHz in Algeria, Saudi Arabia, Azerbaijan, Bahrain, Belarus, China, Comoros, Djibouti, Egypt, United Arab Emirates, the Russian Federation, Iraq, Jordan, Kazakhstan, Kuwait, Lebanon, Libya, Mauritania, Oman, Uzbekistan, Qatar, Syrian Arab Republic, Kyrgyzstan, Somalia, Sudan, Tunisia and Yemen is limited to the maritime mobile and aeronautical radionavigation services. The Amateur Service shall not be used in the above-mentioned countries in this frequency band, and this should be taken into account by the countries authorising such use. (WRC 12).

1.8MHz (160m)	NECESSARY BANDWIDTH	UK USAGE
1,810-1,838kHz	200Hz	Telegraphy
1,838-1,840	500Hz	Narrowband Modes
1,840-1,843	2.7kHz	All Modes
1,843-2,000	2.7kHz	Telephony (Note 1), Telegraphy 1,836kHz – QRP (low power) Centre of Activity 1,960kHz – DF Contest Beacons (14dBW)

Note 1: Lowest LSB carrier frequency (dial setting) should be 1,843kHz. AX25 packet should not be used on the 1.8MHz band.
Licence Notes: 1,810-1,850kHz – Primary User: 1,810-1,830kHz on a non-interference basis to stations outside of the UK. 1,850-2,000kHz – Secondary User. 32W (15dBW) maximum..
Notes to the Band Plan: As on page 173.

3.5MHz (80m)	NECESSARY BANDWIDTH	UK USAGE
3,500-3,510kHz	200Hz	Telegraphy – Priority for Inter-Continental Operation
3,510-3,560	200Hz	Telegraphy – Contest Preferred. 3,555kHz – QRS (slow telegraphy) Centre of Activity
3,560-3,570	200Hz	Telegraphy 3,560kHz – QRP (low power) Centre of Activity
3,570-3,580	200Hz	Narrowband Modes
3,580-3,590	500Hz	Narrowband Modes
3,590-3,600	500Hz	Narrowband Modes – Automatically Controlled Data Stations (unattended)
3,600-3,620	2.7kHz	All Modes – Automatically Controlled Data Stations (unattended), (Note 1)
3,600-3,650	2.7kHz	All Modes – Phone Contest Preferred, (Note 1) 3,630kHz – Digital Voice Centre of Activity
3,650-3,700	2.7kHz	All Modes – Telephony, Telegraphy 3,663kHz May Be Used For UK Emergency Comms
Traffic		3,690kHz SSB QRP (low power) Centre of Activity
3,700-3,775	2.7kHz	All Modes – Phone Contest Preferred 3,735kHz – Image Mode Centre of Activity 3,760kHz – IARU Region 1 Emergency Centre of Activity
3,775-3,800	2.7kHz	All modes - Phone contest preferred Priority for Inter-Continental Telephony (SSB) Operation

Note 1: Lowest LSB carrier frequency (dial setting) should be 3,603kHz.
Notes to the Band Plan: As on page 173.

5MHz (60m)	AVAILABLE WIDTH	UK USAGE
5,258.5-5,264kHz	5.5kHz	5,262kHz – CW QRP Centre of Activity
5,276-5,284	8kHz	5,278.5kHz – May be used for UK Emergency Comms Traffic
5,288.5-5,292	3.5kHz	Beacons on 5290kHz (Note 2)
5,298-5,307	9kHz	
5,313-5,323	10kHz	5,317kHz – AM 6kHz maximum bandwidth
5,333-5,338	5kHz	
5,354-5,358	4kHz	Within WRC-15 Band
5,362-5,374.5	12.5kHz	Partly within WRC-15 band, WSPR
5,378-5,382	4kHz	
5,395-5,401.5	6.5kHz	
5,403.5-5,406.5	3kHz	

Unless indicated, usage is All Modes (necessary bandwidth to be within channel limits).
Note 1: Upper Sideband is recommended for SSB activity.
Note 2: Activity should avoid interference to the experimental beacons on 5290kHz.
Note 3: Amplitude Modulation is permitted with a maximum bandwidth of 6kHz, on frequencies with at least 6kHz available width.
Note 4: Contacts within the UK should avoid the WRC-15 band (5351.5 - 5366.5kHz) if possible
For the latest current guidance refer to the RSGB website
Licence Notes: **Secondary User, 100 watts maximum**. Note that conditions on transmission bandwidth, power and antennas are specified in the Licence. For the latest current guidance, refer to the RSGB website

Notes to the Band Plan: As on page 173.

7MHz (40m)	NECESSARY BANDWIDTH	UK USAGE
7,040-7,047	500Hz	Narrowband Modes (Note 2)
7,047-7,050	500Hz	Narrowband Modes, Automatically Controlled Data Stations (unattended)
7,050-7,053	2.7kHz	All Modes, Automatically Controlled Data Stations (unattended), (Note 1)
7,053-7,060	2.7kHz	All Modes, Digimodes
7,060-7,100	2.7kHz	All Modes, SSB Contest Preferred Segment Digital Voice 7,070kHz; SSB QRP Centre of Activity 7,090kHz
7,100-7,130	2.7kHz	All Modes, 7,110kHz – Region 1 Emergency Centre of Activity
7,130-7,200	2.7kHz	All Modes, SSB Contest Preferred Segment; 7,165kHz – Image Centre of Activity
7,175-7,200	2.7kHz	All Modes, Priority For Inter-Continental Operation

Note 1: Lowest LSB carrier frequency (dial setting) should be 7,053kHz.
Note 2: PSK31 activity starts from 7,040kHz. Since 2009, the narrowband modes segment starts at 7,040kHz.
Licence Notes: 7,000-7,100kHz Amateur and Amateur Satellite Service – Primary User.
7,100-7,200kHz Amateur Service – Primary User.

Notes to the Band Plan: As on page 173.

10MHz (30m)	NECESSARY BANDWIDTH	UK USAGE
10,100-10,130kHz	200Hz	Telegraphy (CW) 10,116kHz – QRP (low power) Centre of Activity
10,130-10,150	500Hz	Narrowband Modes Automatically Controlled Data Stations (unattended) should avoid the use of the 10MHz band

Licence Notes: Amateur Service – Secondary User.

The 10MHz band is allocated to the amateur service only on a secondary basis. The IARU has agreed that only CW and other narrow bandwidth modes are to be used on this band. Likewise the band is not to be used for contests and bulletins. SSB may be used on the 10MHz band during emergencies involving the immediate safety of life and property, and only by stations actually involved with the handling of emergency traffic. The band segment 10,120-10,140kHz may only be used for SSB transmissions in the area of Africa south of the equator during local daylight hours.

14MHz (20m)	NECESSARY BANDWIDTH	UK USAGE
14,000-14,060kHz	200Hz	Telegraphy – Contest Preferred 14,055kHz – QRS (slow telegraphy) Centre of Activity
14,060-14,070	200Hz	Telegraphy 14,060kHz – QRP (low power) Centre of Activity
14,070-14,089	500Hz	Narrowband Modes
14,089-14,099	500Hz	Narrowband Modes – Automatically Controlled Data Stations (unattended)
14,099-14,101		IBP – Reserved Exclusively for Beacons
14,101-14,112	2.7kHz	All Modes – Automatically Controlled Data Stations (unattended)
14,112-14,125	2.7kHz	All Modes (excluding digimodes)
14,125-14,300	2.7kHz	All Modes – SSB Contest Preferred Segment 14,130kHz – Digital Voice Centre of Activity 14,195 ±5kHz – Priority for DXpeditions 14,230kHz – Image Centre of Activity 14,285kHz – QRP Centre of Activity
14,300-14,350	2.7kHz	All Modes 14,300kHz – Global Emergency Centre of Activity

Licence Notes: Amateur Service – Primary User. 14,000-14,250kHz Amateur Satellite Service – Primary User.

Notes to the Band Plan: As on page 173.

18MHz (17m)	NECESSARY BANDWIDTH	UK USAGE
18,068-18,095kHz	200Hz	Telegraphy – 18,086kHz QRP (low power) Centre of Activity
18,095-18,105	500Hz	Narrowband Modes
18,105-18,109	500Hz	Narrowband Modes – Automatically Controlled Data Stations (unattended)
18,109-18,111		IBP – Reserved Exclusively for Beacons
18,111-18,120	2.7kHz	All Modes – Automatically Controlled Data Stations (unattended)
18,120-18,168	2.7kHz	All Modes, 18,130kHz – SSB QRP Centre of Activity 18,150kHz – Digital Voice Centre of Activity 18,160kHz – Global Emergency Centre of Activity

Licence Notes: Amateur and Amateur Satellite Service – Primary User. The band is not to be used for contests or bulletins.
Notes to the Band Plan: As on page 173.

21MHz (15m)	NECESSARY BANDWIDTH	UK USAGE
21,000-21,070kHz	200Hz	Telegraphy 21,055kHz – QRS (slow telegraphy) Centre of Activity 21,060kHz – QRP (low power) Centre of Activity
21,070-21,090	500Hz	Narrowband Modes
21,090-21,110	500Hz	Narrowband Modes – Automatically Controlled Data Stations (unattended)
21,110-21,120	2.7kHz	All Modes (excluding SSB) – Automatically Controlled Data Stations (unattended)
21,120-21,149	500Hz	Narrowband Modes
21,149-21,151		IBP – Reserved Exclusively For Beacons
21,151-21,450	2.7kHz	All Modes 21,180kHz – Digital Voice Centre of Activity 21,285kHz – QRP Centre of Activity 21,340kHz – Image Centre of Activity 21,360kHz – Global Emergency Centre of Activity

Note 1: 21,125-21,245 is also designated for use by amateur satellites
Licence Notes: Amateur and Amateur Satellite Service – Primary User.
Notes to the Band Plan: As on page 173.

24MHz (12m)	NECESSARY BANDWIDTH	UK USAGE
24,890-24,915kHz	200Hz	Telegraphy 24,906kHz – QRP (low power) Centre of Activity
24,915-24,925	500Hz	Narrowband Modes
24,925-24,929	500Hz	Narrowband Modes – Automatically Controlled Data Stations (unattended)
24,929-24,931		IBP – Reserved Exclusively For Beacons
24,931-24,940	2.7kHz	All Modes – Automatically Controlled Data Stations (unattended)
24,940-24,990	2.7kHz	All Modes, 24,950kHz – SSB QRP Centre of Activity 24,960kHz – Digital Voice Centre of Activity

Licence Notes: Amateur and Amateur Satellite Service – Primary User. The band is not to be used for contests or bulletins.
Notes to the Band Plan: As on page 173.

28MHz (10m)	NECESSARY BANDWIDTH	UK USAGE
28,000-28,070kHz	200Hz	Telegraphy 28,055kHz – QRS (slow telegraphy) Centre of Activity 28,060kHz – QRP (low power) Centre of Activity
28,070-28,120	500Hz	Narrowband Modes
28,120-28,150	500Hz	Narrowband Modes – Automatically Controlled Data Stations (unattended)
28,150-28,190	500Hz	Narrowband Modes
28,190-28,199		IBP – Regional Time Shared Beacons
28,199-28,201		IBP – World Wide Time Shared Beacons
28,201-28,225		IBP – Continuous-Duty Beacons
28,225-28,300	2.7kHz	All Modes – Beacons
28,300-28,320	2.7kHz	All Modes – Automatically Controlled Data Stations (unattended)
28,320-29,000	2.7kHz	All modes 28,330kHz – Digital Voice Centre of Activity 28,360kHz – QRP Centre of Activity 28,680kHz – Image Centre of Activity
29,000-29,100		All Modes – See Note 1 regarding 29,000-29,510kHz
29,100-29,200		All Modes – FM Simplex – 10kHz Channels
29,200-29,300		All Modes – Automatically Controlled Data Stations (unattended) 29,270kHz – Internet Gateways Channel 29,280kHz – UK Internet Voice Gateway (unattended) 29,290kHz – UK Internet Voice Gateway (unattended)
29,300-29,510		Satellite Links
29,510-29,520	Guard Channel	
29,520-29,590	6kHz	All Modes – FM Repeater Inputs (RH1-RH8)
29,600	6kHz	All Modes – FM Calling Channel
29,610	6kHz	All Modes – FM Simplex Repeater (parrot) – input and output
29,620-29,700	6kHz	All Modes – FM Repeater Outputs (RH1-RH8)

Note 1: Experimental wide bandwidth operation within 29,000 - 29,510 must be on a non-interference basis to other stations, including the amateur satellite service segment at 29300 - 29510 kHz.
Licence Notes: Amateur and Amateur Satellite Service – Primary User: Note specific conditions apply within 50km of NGR SK985640 (Waddington)within 50km of NGR SK985640 (Waddington).
Notes to the Band Plan: As on page 173.

50MHz (6m)	NECESSARY BANDWIDTH	UK USAGE
50.000-50.100MHz	500Hz	Telegraphy Only (except for Beacon Project) (Note 2) 50.000-50.030MHz reserved for Synchronised Beacon Project (Note 2) Region 1: 50.000-50.010; Region 2: 50.010-50.020; Region 3: 50.020-50.030 50.050MHz – Future International Centre of Activity 50.090MHz – Inter-Continental DX Centre of Activity (Note 1)
50.100-50.200	2.7kHz	SSB/Telegraphy – International Preferred 50.100-50.130MHz – Inter-Continental DX Telegraphy & SSB (Note 1) 50.110MHz – Inter-Continental DX Centre of Activity 50.130-50.200MHz – General International Telegraphy & SSB 50.150MHz – International Centre of Activity
50.200-50.300	2.7kHz	SSB/Telegraphy – General Usage 50.285MHz – Crossband Centre of Activity
50.300-50.400	2.7kHz	MGM/Narrowband/Telegraphy 50.305MHz – PSK Centre of Activity 50.310-50.320MHz – EME 50.320-50.380MHz – MS
50.400-50.500		Propagation Beacons only
50.500-50.700		All Modes 50.520MHz – FM/DV Internet Voice Gateway 50.530MHz – FM/DV Internet Voice Gateway 50.540MHz – FM/DV Internet Voice Gateway 50.600-50.700MHz – Digital communications 50.630MHz – Digital Voice (DV) calling
50.700-50.900	12kHz	50.710-50.890MHz – FM/DV Repeater Outputs (10kHz channel spacing)
50.900-51.200		All Modes
51.200-51.400	12kHz	51.210-51.390MHz – FM/DV Repeater Inputs (10kHz channel spacing) (Note 4)
51.400-52.000		All Modes 51.410-51.590MHz – FM/DV Simplex (Note 3) (Note 4) 51.510MHz – FM Calling Frequency 51.530MHz – GB2RS News Broadcast and Slow Morse 51.650 & 51.750MHz – See Note 5 (25kHz aligned) 51.970 & 51.990MHz – See Note 5

Note 1: Only to be used between stations in different continents (not for intra-European QSOs).
Note 2: 50.0-50.1MHz is currently shared with Propagation Beacons. These are due to be migrated to 50.4-50.5MHz, to create more space for Telegraphy and a new Synchronised Beacon Project.
Note 3: 20kHz channel spacing. Channel centre frequencies start at 51.430MHz.
Note 4: Embedded data traffic is allowed with digital voice (DV).
Note 5: May be used for Emergency Communications and Community Events.
Note 6: Digital experiments to support innovation may occur at 50.6, 51.0 or 51.7MHz with maximum bandwidths of 50, 200 and 400kHz respectively on a non-interference basis.
Licence Notes: Amateur Service 50.0-51.0MHz – Primary User. Amateur Service 51.0-52.0MHz – Secondary User. 100W (20dBW) maximum. Available on the basis of non-interference to other services (inside or outside the UK).
Notes to the Band Plan: As on page 173.

70MHz (4m)	NECESSARY BANDWIDTH	UK USAGE (NOTE 1)
70.000-70.090MHz	1kHz	Propagation Beacons Only
70.090-70.100	1kHz	Personal Beacons
70.100-70.250	2.7kHz	Narrowband Modes 70.185MHz – Cross-band Activity Centre 70.200MHz – CW/SSB Centre 70.250MHz – MS Centre
70.250-70.294	12kHz	All Modes 70.260MHz – AM/FM Calling 70.270MHz MGM Centre of Activity
70.294-70.500	12kHz	All Modes Channelised Operations Using 12.5kHz Spacing 70.3000MHz 70.3125MHz – Digital Modes 70.3250MHz – DX Cluster 70.3375MHz – Digital Modes 70.3500MHz – Internet Voice Gateway (Note 2) 70.3625MHz – Internet Voice Gateway 70.3750MHz – See Note 2 70.3875MHz – Internet Voice Gateway 70.4000MHz – See Note 2 70.4125MHz – Internet Voice Gateway 70.4250MHz – FM Simplex – used by GB2RS news broadcast 70.4375MHz – Digital Modes (special projects) 70.4500MHz – FM Calling 70.4625MHz – Digital Modes 70.4750MHz 70.4875MHz – Digital Modes

Note 1: Usage by operators in other countries may be influenced by restrictions in their national allocations.
Note 2: May be used for Emergency Communications and Community Events.
Licence Notes: Amateur Service 70.0-70.5MHz – Secondary User: 160W (22dBW) maximum. Available on the basis of non-interference to other services (inside or outside the UK).
Note that access to 70.5-71.5MHz by Full Licensees is also possible by NoV
Notes to the Band Plan: As on page 173.

144MHz (2m)	NECESSARY BANDWIDTH	UK USAGE
144.000-144.025MHz	2700Hz	All Modes – including Satellite Downlinks
144.025-144.100	500Hz	Telegraphy (including EME CW) 144.050MHz – Telegraphy Centre of Activity 144.100MHz – Random MS Telegraphy Calling, (Note 1)
144.110-144.150	500Hz	Telegraphy and MGM EME MGM Activity
144.150-144.400	2700Hz	Telegraphy, MGM and SSB 144.250MHz – GB2RS News Broadcast and Slow Morse 144.260MHz – See Note 10 144.300MHz – SSB Centre of Activity 144.370MHz – MGM MS Calling
144.400-144.490		Propagation Beacons only
144.490-144.500		Beacon guard band 144.491-144.493 Personal Weak Signal MGM Beacons (BW: 500Hz max)
144.500-144.794	12kHz	All Modes (Note 8) 144.500MHz – Image Modes Centre (SSTV, FAX, etc) 144.600MHz – Data Centre of Activity (MGM, RTTY, etc) 144.6125MHz – UK Digital Voice (DV) Calling (Note 9) 144.625-144.675MHz – See Note 10 144.750MHz – ATV Talkback 144.775-144.794MHz – See Note 10 144.794-144.794-
144.990	12kHz	MGM Digital Communications (Note 15) 144.800-144.9875MHz – MGM/Digital Communications 144.8000MHz – Unconnected Nets – APRS, UiView etc (Note 14) 144.8125MHz – DV Internet Voice Gateway 144.8250MHz – DV Internet Voice Gateway 144.8375MHz – DV Internet Voice Gateway 144.8500MHz – DV Internet Voice Gateway 144.8625MHz – DV Internet Voice Gateway 144.8750 - 144.9125 - Internet Gateways 144.9250MHz – Digital Usage 144.9375MHz – Digital Usage 144.9500MHz – Digital Usage 144.9625MHz – FM Internet Voice Gateway 144.9750, 144.9875 MHz Low power / Adhoc repeater inputs (Note 11)
144.990-145.1935	12kHz	FM/DV RV48-RV63 Repeater Input Exclusive (Note 2 & 5)
145.200	12kHz	FM/DV Space Communications (eg ISS) – Earth-to-Space 145.2000MHz – (Note 4 & 10)
145.200-145.5935	12kHz	FM/DV V16-V47 – FM/DV Simplex (Note 3, 5 & 6) 145.2250 MHz Internet gateways - and See Note 10 145.2375MHz – FM Internet Voice Gateway (IARU common channel) 145.2500MHz – Used for Slow Morse Transmissions 145.2875MHz – FM Internet Voice Gateway (IARU common channel) 145.3375MHz – FM Internet Voice Gateway (IARU common channel) 145.5000MHz – FM Calling (Note 12) 145.5250MHz – Used for GB2RS News Broadcast. 145.5500MHz – Used for Rally/exhibition Talk-in 145.5750, 145.5875MHz Low power / Adhoc repeater outputs (Note 11)
145.5935-145.7935	12kHz	FM/DV RV48-RV63 – Repeater Output (Note 2)
145.800	12kHz	FM/DV Space Communications (eg ISS) – Space-Earth
145.806-146.000	12kHz	All Modes – Satellite Exclusive

Note 1: Meteor scatter operation can take place up to 26kHz higher than the reference frequency.
Note 2: 12.5kHz channels numbered RV48-RV63. RV48 input = 145.000MHz, output = 145.600MHz.
Note 3: 12.5kHz simplex channels numbered V16-V47. V16 = 145.200MHz.
Note 4: Emergency Communications Groups utilising this frequency should take steps to avoid interference to ISS operations in non-emergency situations.
Note 5: Embedded data traffic is allowed with digital voice (DV).
Note 6: Simplex use only – no DV gateways.
Note 7: Not used.
Note 8: Amplitude Modulation (AM) is acceptable within the All Modes segment. AM usage is typically found on 144.550MHz. Users should consider adjacent channel activity when selecting operating frequencies.
Note 9: In other countries IARU Region 1 recommends 145.375MHz.
Note 10: May be used for Emergency Communications and Community Events.
Note 11: Ad-hoc Low power repeaters, 5W erp max
Note 12: DV users are asked not to use this channel, and use 144.6125MHz for calling.
Note 13: Not used.
Note 14: 144.800 use should be NBFM to avoid interference to 144.8125 DV Gateways.
Licence Notes: Amateur Service and Amateur Satellite Service – Primary User. Note specific conditions apply within 50 km of TA 012869 (Scarborough).
Notes to the Band Plan: As on page 173.

146MHz	NECESSARY BANDWIDTH	UK USAGE
Access to this band requires an appropriate NoV, which is available to Full Licensees only		
146.000-146.900MHz etc)	500kHz	Wideband Digital Modes (High speed data, DATV
modes		146.500MHz Centre frequency for wideband
		(Note 1)
146.900-147.000MHz	12kHz	Narrowband Digital Modes including Digital Voice 146.900 146.9125 146.925 146.9375 Not available in/near Scotland (see Licence Notes & NoV terms) 146.9500 146.9625 146.9750 146.9875

Note 1: Users of wideband modes must ensure their spectral emissions are contained with the band limits.
Licence Notes: Full Licensees only, with NoV, 50W ERP max – not available in the Isle of Man or Channel Isles. Note that additional restrictions on geographic location, antenna height and upper frequency limit are specified by the NoV terms.
It should be emphasised that this band is UK-specific and is available on a non-interference basis to existing services. Upper Band limit 147.000MHz (or 146.93750 where applicable) are absolute limits and not centre frequencies. The absolute band frequency limit in or within 40km of Scotland is 146.93750MHz – see NoV schedule
Notes to the Band Plan: As on page 173.

430MHz (70cm) IARU Recommendation	NECESSARY BANDWIDTH	UK USAGE
430.0000-431.9810MHz		430.0125-430.0750MHz – FM Internet Voice Gateways (Notes 7, 8)
All Modes		
		430.250-430.300 MHz UK DV 9 MHz reverse-split repeaters – Outputs 430.4000-430.7750 – UK DV 9MHz Split Repeaters – inputs
Digital Links 430.6000-430.9250		430.8000MHz – 7.6MHz Talk-through (Note 10)
Digital Repeaters		430.8250-430.9750MHz – RU66-RU78 7.6MHz Split Repeaters – outputs See Licence Exclusion Note; 431-432MHz 430.9900-431.9000MHz – Digital Communications 431.0750-431.1750MHz – DV Internet Voice Gateways (Note 8)
432.0000-432.1000	500Hz	432.0500MHz Telegraphy Centre of Activity Telegraphy, MGM
432.1000-432.4000	2700Hz	
SSB, Telegraphy MGM		432.2000MHz – SSB Centre of Activity 432.3500MHz – Microwave Talkback (Europe) 432.3700MHz – Meteor Scatter Calling
432.4000-432.4900	500Hz	Propagation Beacons only
432.4900-432.9940	12kHz	432.491-432.493MHz Personal Weak Signal MGM Beacons (BW: 500Hz max) 432.5000MHz – Narrowband SSTV Activity Centre
All Modes Non-channelised	(Note 11)	432.6250-432.6750MHz Digital Communications 432.7750MHz 1.6MHz Talk-through – Base TX (Note 10)
432.9940-433.3810	12kHz	433.0000-433.3750MHz (RB0-RB15) – RU240-RU270
FM repeater outputs in UK only (Note 1)	(Note 11)	FM/DV Repeater Outputs in UK Only
433.3940-433.5810	12kHz	433.4000MHz U272 – IARU Region 1 SSTV (FM/AFSK)
	(Note 11)	433.4250MHz U274
FM/DV (Notes 12, 13) Simplex Channels		433.450MHz U276 (Note 5) 433.4750MHz U278 433.5000MHz U280 – FM Calling Channel 433.5250MHz U282 433.5500MHz U284 – Used for Rally/Exhibition Talk-in 433.5750MHz U286
433.6000-434.0000 All Modes		433.6250-6750MHz – Digital Communications
433.8000MHz for APRS where 144.8000MHz cannot be used		433.7000MHz-433.7750MHz (Note 10) 433.8000-434.2500 MHz Digital communications & Experiments 434.0000 Low Power Non-NoV Personal Hot-Spot usage
434.0000-434.5940	12kHz	433.9500-434.0500MHz – Internet Voice Gateways (Note 8)
	(Note 11)	434.3750MHz 1.6MHz Talk-through – Mobile TX (Note 10) 434.4750-434.5250MHz DV Internet voice gateways (Note 8)
434.5940-434.9810	12kHz	434.6000-434.9750MHz (RB0-RB15) RU240-RU270
FM repeater inputs in UK only	(Note 11)	FM/DV Repeater Inputs in UK Only (Note 12)
435.0000-436.0000		Satellites only
436.0000-438.0000		Satellites and Experimental DATV/Data 437.0000 Experimental DATV Centre of Activity (Note 14)
438.0000-440.0000		438.8000 Low Power Non-NoV Personal Hot-Spot usage 438.0250-438.1750MHz – IARU Region 1 Digital Communications
All Modes		438.2000-439.4250MHz (Note 1) 438.4000MHz – 7.6MHz Talk-through (Note 10)

430MHz (70cm) IARU Recommendation	NECESSARY BANDWIDTH	UK USAGE Contd
		438.4250-438.5750MHz RU66-RU78 – 7.6MHz Split Repeaters – inputs 438.6125MHz – UK DV calling (Note 12) (Note 13) 438.8000 Low Power Non-NoV Personal Hot-Spot
usage		
		439.2500-439.3000MHz UK DV 9MHz reverse-split repeaters – Inputs 439.6000-440.0000MHz – Digital Communications 439.4000-439.7750MHz – UK DV 9MHz split repeaters – Outputs

Note 1: In Switzerland, Germany and Austria, repeater inputs are 431.0500-431.8250MHz with 25kHz spacing and outputs 438.6500-439.4250MHz. In Belgium, France and the Netherlands repeater outputs are 430.0250-430.3750MHz with 12.5kHz spacing and inputs at 431.6250-431.9750MHz. In other European countries repeater inputs are 433.0000-433.3750MHz with 25kHz spacing and outputs at 434.6000-434.9750MHz, ie the reverse of the UK allocation.
Note 2: Not used.
Note 3, Note 4: Not used.
Note 5: In other countries IARU Region 1 recommends 433.4500MHz for DV calling.
Note 7: Users must accept interference from repeater output channels in France and the Netherlands at 430.0250-430.5750MHz. Users with sites that allow propagation to other countries (notably France and the Netherlands) must survey the proposed frequency before use to ensure that they will not cause interference to users in those countries.
Note 8: All internet voice gateways: 12.5kHz channels, maximum deviation ±2.4kHz, maximum effective radiated power 5W (7dBW).
Note 10: May be used for Emergency Communications and Community Events.
Note 11: IARU Region-1 recommended maximum bandwidths are 12kHz to support 12.5kHz channel spacing.
Note 12: Embedded data traffic is allowed with digital voice (DV).
Note 13: Simplex use only – no DV gateways.
Note 14: QPSK 2 Mega-symbols/second maximum recommended.
Licence Notes: Amateur Service – Secondary User. Note specific conditions within 430-440MHz. Amateur Satellite Service: 435-438MHz – Secondary User. Exclusion: 431-432MHz not available within 100km radius of Charing Cross, London. Power Restriction 430-432MHz is 40 watts effective radiated power maximum.
Notes to the Band Plan: As on page 173.

1.3GHz (23cm)	NECESSARY BANDWIDTH	UK USAGE
1240.000-1240.500MHz –	2700Hz	Alternative Narrowband Segment – see Note 7 1240.00-1240.750MHz
1240.500-1240.750		Alternative Propagation Beacon Segment
1240.750-1241.000	20kHz	FM/DV Repeater Inputs
1241.000-1241.750	150kHz	DD High Speed Digital Data – 5 x 150kHz channels
All Modes		1241.075, 1241.225, 1241.375, 1241.525, 1241.675MHz (±75kHz)
1241.750-1242.000 All Modes	20kHz	25kHz Channels available for FM/DV use 1241.775-1241.975MHz
1242.000-1249.000 ATV		TV Repeaters (Note 9) New DATV Repeater Inputs Original ATV Repeater Inputs: 1248, 1249
1249.000-1249.250	20kHz	FM/DV Repeater Outputs, 25kHz Channels (Note 9) 1249.025-1249.225MHz
1250.00		In order to prevent interference to Primary Users, caution must be exercised prior to
using		
		1250-1290MHz in the UK 1260.000-1270.000
Amateur Satellite Service – Earth to Space		
		Uplinks Only
Satellites 1290.000		
1290.994-1291.481	20kHz	FM/DV Repeater Inputs (Note 5) 1291.000-1291.375MHz (RM0-RM15) 25kHz spacing
1291.494-1296.000 All Modes	All Modes	Preferred Narrowband segment
1296.000-1296.150 Telegraphy, MGM	500Hz	1296.000-1296.025MHz – Moonbounce
1296.150-1296.800 Telegraphy, SSB & MGM	2700Hz	1296.200MHz – Narrowband Centre of Activity 1296.400-1296.600MHz – Linear Transponder Input
(Note 1)		1296.500MHz – Image Mode Centre of Activity (SSTV, FAX etc) 1296.600MHz – Narrowband Data Centre of Activity (MGM, RTTY etc) 1296.600-1296.700MHz – Linear Transponder Output 1296.741-1296.743MHz Personal Weak Signal MGM Beacons
1296.800-1296.994		1296.750-1296.800MHz – Local Beacons, 10W ERP max 1296.800-1296.990MHz – Propagation Beacons only Beacons exclusive
1296.994-1297.481	20kHz	FM/DV Repeater Outputs (Note 5) 1297.000-1297.375MHz (RM0-RM15)
1297.494-1297.981 ing	20kHz	FM/DV Simplex (Notes 2, 5 & 6) 25kHz spac- 1297.500-1297.750MHz (SM20-SM30)
FM/DV simplex (Notes 2, 5, 6)		1297.725MHz – Digital Voice (DV) Calling (IARU recommended) 1297.900-1297.975MHz – FM Internet Voice Gateways (IARU common channels, 25kHz)
1298.000-1299.000 All Modes	20kHz	All Modes General mixed analogue or digital use in channels

1.3GHz (23cm) IARU Recommendation	NECESSARY BANDWIDTH	UK USAGE Contd
		1298.025-1298.975MHz (RS1-RS39)
1299.000-1299.750	150kHz	DD High Speed Digital Data – 5 x 150kHz channels
All Modes		1299.075, 1299.225, 1299.375, 1299.525, 1299.675MHz (±75kHz)
1299.750-1300.000	20kHz	25kHz Channels Available for FM/DV use
All Modes		1299.775-1299.975MHz
1300.000-1325.000		TV Repeaters (UK only) (Note 9)
ATV		New DATV Repeater Outputs
		Original ATV Repeater Outputs: 1308.0, 1310.0, 1311.5, 1312.0, 1316.0, 1318.5MHz

Note 1: Local traffic using narrowband modes should operate between 1296.500-1296.800MHz during contests and band openings.
Note 2: Stations in countries that do not have access to 1298-1300MHz may also use the FM simplex segment for digital communications.
Note 3, Note 4: Not used.
Note 5: Embedded data traffic is allowed with digital voice (DV).
Note 6: Simplex use only – no DV gateways.
Note 7: 1240.000-1240.750 has been designated by IARU as an alternative centre for narrowband activity and beacons. Operations in this range should be on a flexible basis to enable coordinated activation of this alternate usage.
Note 8: The band 1240-1300MHz is subject to major replanning. Contact the Microwave Manager for further information.
Note 9: Repeaters and Migration to DATV, inc option for new DATV simplex are subject to further development and coordination.
Note 10: QPSK 4 Mega-symbols/second maximum recommended.
Licence Notes: Amateur Service – Secondary User. Amateur Satellite Service: 1,260-1,270MHz – Secondary User Earth to Space only. Note specific conditions within 1240-1325MHz and within 50km of SS206127 (Bude), SE202577 (Harrogate), or in Northern Ireland.
Notes to the Band Plan: As on page 173.

2.3-2.302GHz IARU Recommendation	NECESSARY BANDWIDTH	UK USAGE

Access to this band requires an appropriate NoV, which is available to Full licensees only. Please note that the current NoVs last for up to three years prior to expiry.

2300.000-2300.400MHz	2.7kHz	Narrowband Modes (including CW, SSB, MGM) 2300.350-2300.400MHz Attended Beacons
2300.400-2301.800MHz	500kHz	Wideband Modes (NBFM, DV, Data, DATV, etc) Note 1
2301.800-2302.000MHz	2.7kHz	Narrowband modes (including CW, SSB, MGM) EME Usage

Note 1: Users of wideband modes must ensure their spectral emissions are contained within the band limits.
Note 2: Full licensees only with NoV, 400 watts maximum, not available in the Isle of Man. Note additional restrictions on usage are specified by the NoV terms. It should be emphasised that this is UK-specific and is available on a non interference basis to existing services.
Notes to the Band Plan: As on page 173.

2.3GHz (13cm) IARU Recommendation	NECESSARY BANDWIDTH	UK USAGE
2,310.000-2,320.000MHz		
	200kHz	2,310.000-2,310.500MHz – Repeater links
(National band plans)		
		2,311.000-2,315.000MHz – High speed data
		Preferred Narrowband Segment
2,320.000-2,320.800	2.7kHz	2,320.000-2,320.025MHz – Moonbounce
		2,320.200MHz – SSB Centre of Activity
		2,320.750-2,320.800MHz – Local Beacons, 10W ERP max
2,320.800-2,321.000		2,320.800-2,320.990MHz – Propagation Beacons Only
Beacons exclusive		
2321.000-2322.000	20kHz	FM/DV. See also Note 1
2,322.000-2,350.000		Wideband Modes including Data, ATV
2,390.000-2,400.000		All Modes
2,400.000-2,450.000MHz		2,435.000MHz ATV Repeater Outputs
Satellites		2,440.000MHz ATV Repeater Outputs

Note 1: Stations in countries which do not have access to the All Modes section 2,322-2,390MHz, use the simplex and repeater segment 2,320-2,322MHz for data transmission.
Note 2: Stations in countries that do not have access to the narrowband segment 2,320-2,322MHz, use the alternative narrowband segments 2,304-2,306MHz, 2,308-2,310MHz and 2400-2402MHz.
Note 3: The segment 2,433-2,443MHz may be used for ATV if no satellite is using the segment.
Licence Notes: Amateur Service – Secondary User. Users must accept interference from ISM users. Amateur Satellite Service: 2,400-2,450MHz – Secondary User. Users must accept interference from ISM users. Operation in 2310-2350 and 2390-2400 MHz are subject to specific conditions and guidance. Note specific conditions apply within 50km of SS206127 (Bude) or SE202577 (Harrogate). ISM = Industrial, scientific and medical.

Notes to the Band Plan: As on page 173.

3.4GHz (9cm) IARU Recommendation	NECESSARY BANDWIDTH	UK USAGE
3,400.000-3,400.800MHz	2.7kHz	Narrowband Modes (including CW, SSB, MGM, EME)
		3,400.100MHz – Centre of Activity (Note 1)
		3,400.750-3,400.800MHz – Local Beacons, 10W ERP max
3,400.800-3,400.995		3,400.800-3,400.995MHz – Propagation Beacons Only
3,400.000-3,401.000MHz	200kHz	3,401.000-3,402.000MHz Data, Remote Control
3,402.000-3,410.000		Wideband Modes including DATV Repeater
Outputs		
All Modes (Notes 2, 3)		

Note 1: EME has migrated from 3456MHz to 3400MHz to promote harmonised usage and activity.
Note 2: Stations in many European countries have access to 3400-3410MHz as permitted by the CEPT ECA Table.
Note 3: Amateur Satellite downlinks planned.
Licence Notes: Amateur Service – Secondary User. Subject to specific conditions and guidance.
Notes to the Band Plan: As on page 173.

5.7GHz (6cm) IARU Recommendation	NECESSARY BANDWIDTH	UK USAGE
5,650.000-5,668.000MHz		All Modes
Satellite Uplinks Only		Amateur Satellite Service – Earth to Space
5,668.000-5,670.000	2.7kHz	5,668.200MHz – Alternative Narrowband
Centre		
5,670.000-5,680.000		All Modes
5.755.000-5,760.000		All Modes
5,760.000-5,762.000	2.7kHz	Narrowband Modes (including CW, SSB, MGM,
EME)		
		5,760.100MHz – Preferred Centre of Activity
		5,760.750-5,760.800MHz – Local Beacons, 10W ERP max
5760.800-5760.995		5,760.800-5,760.995MHz – Propagation Beacons Only
Propagation Beacons		
5,762.000-5,765.000		All Modes
5,820.000-5,830.000		All Modes
5,830.000-5,850.000		All Modes
Satellite Downlinks	Amateur Satellite Service – Space to Earth Only	

Licence Notes: Amateur Service: 5,650-5,680MHz – Secondary User. 5,755-5,765 and 5,820-5,850MHz – Secondary User. Users must accept interference from ISM users. Amateur Satellite Service: 5,650-5,670MHz and 5,830-5,850MHz – Secondary User. Users must accept interference from ISM users. Note specific conditions apply within 50km of SS206127 (Bude) or SE202577 (Harrogate). ISM = Industrial, scientific and medical.
Notes to the Band Plan: As on page 173.

10GHz (3cm) IARU Recommendation	NECESSARY BANDWIDTH	UK USAGE
10,000.000-10,125.000MHz		Note 4
All Modes		10,065MHz ATV Repeater Outputs
10,225.000-10,250.000		
All Modes		10,240MHz ATV Repeaters
10,250.000-10,350.000		
Digital Modes		
10,350.000-10,368.000		10,352.5-10,368MHz Wideband Modes (Note 2)
All Modes		
10,368-10,370MHz	2.7kHz	10,368-10,370 Narrowband Modes (Note 3)
Narrowband Telegraphy		10,368.1MHz Centre of Activity
EME/SSB		
		10,368.750-10,368.800MHz – Local Beacons, 10W ERP max
10,368.800-10,368.995		10,368.800-10,368.995MHz – Propagation Beacons Only
Propagation Beacons		
10,370.000-10,450.000		10,371MHz Voice Repeaters Rx
All Modes		10,425 ATV Repeaters
10,450.000-10,475.000		10,400-10,475MHz Unattended Operation
All Modes & Satellites		10,450-10,452MHz Alternative Narrowband Segment (Note 3)
		10,471MHz Voice Repeaters Tx
10,475.000-10,500.000		
All Modes and satellites		Amateur Satellite Service ONLY

Note 1: Deleted.
Note 2: Wideband FM is preferred between 10,350-10,400MHz to encourage compatibility between narrowband systems.
Note 3: 10,450MHz is used as an alternative narrowband segment in countries where 10,368MHz is not available.
Note 4: 10,000-10,125MHz is subject to increased Primary user utilisation and restrictions.
Note 5: 10,475-10,500MHz is allocated ONLY to the Amateur Satellite Service and NOT to the Amateur Service.
Licence Notes: Amateur Service – Secondary User. Foundation licensees 1 watt maximum. Amateur Satellite Service: 10,450-10,500MHz – Secondary User. Note specific conditions apply within 50 km of SO916223 (Cheltenham), SS206127 (Bude), SK985640 (Waddington) and SE202577 (Harrogate).
Notes to the Band Plan: As on page 173.

24GHz (12mm) **UK USAGE**
IARU Recommendation

24,000.000-24,050.000MHz
Satellites — 24,025MHz Preferred Operating Frequency for Wideband Equipment

24,048.750-24,048.800MHz – Local Beacons, 10W ERP max

Propagation Beacons
24,050.000-24,250.000
All Modes

Licence Notes: Amateur Service: 24,000-24,050MHz – Primary User: Users must accept interference from ISM users. 24,050-24,150MHz – Secondary User. May only be used with the written permission of Ofcom. Users must accept interference from ISM users. 24,150-24,250MHz – Secondary User. Users must accept interference from ISM users. Amateur Satellite Service: 24,000-24,050MHz – Primary User: Users must accept interference from ISM users. Note specific conditions apply within 50 km of SK985640 (Waddington) and SE202577 (Harrogate).
ISM = Industrial, scientific and medical.
Notes to the Band Plan: As on page 173.

47GHz (6mm) **UK USAGE**
IARU Recommendation

47,000.000-47,200.000MHz — 47,088.2MHz – Centre of Narrowband Activity
47,088.000-47,090.000 — 47,088.8-47,089.0MHz – Propagation Beacons Only
Narrowband Segment

Licence Notes: Amateur Service and Amateur Satellite Service – Primary User. Note specific conditions apply within 50 km of SK985640 (Waddington) and SE202577 (Harrogate).

76GHz (4mm) **UK USAGE**
IARU Recommendation

75,500-76,000MHz
All Modes (preferred) — 75,976.200MHz – IARU Region 1 Preferred Centre of Activity
76,000.000-77,500.000 All Modes
77,500-78,000 — 77,500.200MHz – Alternative IARU Recommended Narrowband Segment
All Modes (preferred)
78,000-81,000
All Modes

Licence Notes:
75,500-75,875MHz Amateur Service and Amateur Satellite Service – Secondary User.
75,875-76,000MHz Amateur Service and Amateur Satellite Service – Primary User.
76,000-77,500MHz Amateur Service and Amateur Satellite Service – Secondary User.
77,500-78,000MHz Amateur Service and Amateur Satellite Service – Primary User.
78,000-81,000MHz Amateur service and Amateur Satellite Service – Secondary User.
Note specific conditions apply within 50 km of SK985640 (Waddington) and SE202577 (Harrogate).

122GHz (2.5mm) **UK USAGE**
IARU Recommendation

ARU Recommendation
122,250-122.251 MHz — IARU Region-1 preferred centre of activity
Narrowband modes

122,251-123,000 MHz — 122,256 / 122,400 MHz - UK centre of activity
All modes

Licence Notes:
122,250-123,000 MHz Amateur Service only **- Secondary User.**
Note specific conditions apply within 50 km of SK985640 (Waddington) and SE202577 (Harrogate).

134GHz (2mm) **UK USAGE**
IARU Recommendation

134,000-134,928 MHz
All modes — 134,256 / 134,400 MHz - UK centre of activity

134,928-134,930 — IARU Region-1 preferred centre of activity
Narrowband modes

134,930 -136,000
All modes

136,000 -141,000
All modes

Licence Notes:
134,000-136,000 MHz Amateur Service and Amateur Satellite Service - Primary User.
136,000-141,000 MHz Amateur Service and Amateur Satellite Service - Secondary User.
Note specific conditions apply within 50 km of SK985640 (Waddington) and SE202577 (Harrogate).

241GHz (1.2mm) **UK USAGE**
IARU Recommendation

241,000-248,000 MHz
All modes — 241,600 MHz +/-IF - UK centre of activity

248,000-248.001 MHz — IARU Region-1 preferred centre of activity
Narrowband modes

248,001-250,000 MHz — 248,800 MHz +/-IF - UK centre of activity
All modes

Licence Notes:
241,000-248,000 MHz Amateur Service and Amateur Satellite Service **- Secondary User.**
248,000-250,000 MHz Amateur Service and Amateur Satellite Service **- Primary User.**
Note specific conditions apply within 50 km of SK985640 (Waddington) and SE202577 (Harrogate).

Notes to the Band Plan
Note-1: Access to frequencies >275 GHz by Full Licensees is also possible by NoV

NOTES TO THE BAND PLAN

ITU-R radio regulation RR 1.152 and Recommendation SM.328 (extract):

Necessary bandwidth: For a given class of emission, the width of the frequency band which is just sufficient to ensure the transmission of information at the rate and with the quality required under specified conditions.

Foundation and Intermediate Licence holders are advised to check their Licences for the permitted power limits and conditions applicable to their class of Licence.

All Modes: CW, SSB and those modes listed as Centres of Activity, plus AM. Consideration should be given to adjacent channel users.

Image Modes: Any analogue or digital image modes within the appropriate bandwidth, for example SSTV and FAX.

Narrowband Modes: All modes using up to 500Hz bandwidth, including CW, RTTY, PSK, etc.

Digimodes: Any digital mode used within the appropriate bandwidth, for example RTTY, PSK, MT63, etc.

Sideband usage: Below 10MHz use lower sideband (LSB), above 10MHz use upper sideband (USB). Note the lowest dial settings for LSB Voice modes are 1843, 3603 and 7053kHz on 160, 80 and 40m. Note that on (5MHz) USB is used.

Amplitude Modulation (AM): AM with a bandwidth greater than 2.7kHz is acceptable in the All Modes segments provided users consider adjacent channel activity when selecting operating frequencies (Davos 2005).

Extended SSB (eSSB): Extended SSB (eSSB) is only acceptable in the All Modes segments provided users consider adjacent channel activity when selecting operating frequencies.

Digital Voice (DV): Users of Digital Voice (DV) should check that the channel is not in use by other modes (CT08_C5_Rec20).

FM Repeater & Gateway Access: CTCSS Access is recommended. Toneburst access is being withdrawn in line with IARU R1 recommendations.

Beacons Propagation Beacon Sub-bands are highlighted – please avoid transmitting in them!

MGM: Machine Generated Modes indicates those transmission modes relying fully on computer processing such as RTTY, AMTOR, PSK31, JTxx, FSK441 and the like. This does not include Digital Voice (DV) or Digital Data (DD).

WSPR: Above 30MHz, WSPR frequencies in the band plan are the centre of the transmitted frequency (not the suppressed carrier frequency or the VFO dial setting).

Transmitter setup and Linearity: Close attention should be given to power amplifier linearity to control the final transmitted bandwidth and avoid spectral regrowth affecting adjacent users. In particular this can be a major issue when operating digital modes. It is recommended that operators do not use more power than is necessary, and that care is taken to ensure sound cards, interfaces, and other equipment are properly set up so as to minimise the potential for interference.

CW QSOs are accepted across all bands, except within beacon segments (Recommendation DV05_C4_Rec_13).

Contest activity shall not take place on the 5, 10, 18 and 24MHz (60, 30, 17 and 12m) bands.

Non-contesting radio amateurs are recommended to use the contest-free HF bands (30, 17 and 12m) during the largest international contests (DV05_C4_Rev_07).

The term 'automatically controlled data stations' includes Store and Forward stations.

Transmitting Frequencies: The announced frequencies in the band plan are understood as 'transmitted frequencies' (not those of the suppressed carrier).

Centre of Activity (CoA): A guide to where users of a particular mode or activity tend to operate. The bandplan does not give such users precedence over other modes or activities.

Unmanned transmitting stations: IARU member societies are requested to limit this activity on the HF bands. It is recommended that any unmanned transmitting stations on HF shall only be activated under operator control except for beacons agreed with the IARU Region 1 Beacon Coordinator, or specially licensed experimental stations.

472-479kHz: Access is available to Full licensees only.

1.8MHz: Radio amateurs in countries that have a SSB allocation ONLY below 1840kHz, may continue to use it, but the National Societies in those countries are requested to take all necessary steps with their licence administrations to adjust phone allocations in accordance with the Region 1 Band Plan (UBA – Davos 2005).

3.5MHz: Inter-Continental operations should be given priority in the segments 3500-3510kHz and 3775-3800kHz. Where no DX traffic is involved, the contest segments should not include 3500-3510kHz or 3775-3800kHz. Member societies will be permitted to set other (lower) limits for national contests (within these limits). 3510-3600kHz may be used for unmanned ARDF beacons (CW, A1A) (Recommendation DV05_C4_Rec_12).

5MHz: Access is available to Full licensees only – see licence schedule for additional conditions.

7MHz: The band segment 7040-7060kHz may be used for automatic controlled data stations (unattended) traffic in the areas of Africa south from the equator during local daylight hours. Where no DX traffic is involved, the contest segment should not include 7,175-7,200kHz.

10MHz: SSB may be used during emergencies involving the immediate safety of life and property and only by stations actually involved in the handling of emergency traffic.

The band segment 10120kHz to 10140kHz may be used for SSB transmissions in the area of Africa south of the equator during local daylight hours.

News bulletins on any mode should not be transmitted on the 10MHz band.

28MHz: Operators should not transmit on frequencies between 29.3 and 29.51MHz to avoid interference to amateur satellite downlinks.

Experimentation with NBFM Packet Radio on 29MHz band: Preferred operating frequencies on each 10kHz from 29.210 to 29.290MHz inclusive should be used. A deviation of ±2.5kHz being used with 2.5kHz as maximum modulation frequency.

1.3GHz
The band is subject to re-planning. It is also shared with air traffic radar.

2.3GHz (2310-2350 & 2390-2400MHz)
Operation is subject to specific licence conditions and guidance – see also the Ofcom PSSR statement.

3.4GHz (3400-3410MHz)
Operation is subject to specific licence conditions and guidance – see also the Ofcom PSSR statement.

Innovation Bands: 70.5-71.5MHz, 146-147MHz, 2300-2302MHz and >275GHz
Access to these bands requires an appropriate NoV, which is available to Full licensees only.

The latest band plan information, including the master Excel files, can be found in the Operating section of the RSGB website.

Please ensure you only refer or link to the current Band Plans. Remove / delete any older versions you have locally or online.

Prefix List

Callsigns for the world's nations are determined by the International Telecommunications Union (ITU). This is the United Nations agency that co-ordinates radio activity for all spectrum users. The prefixes used by a country for both commercial and amateur radio purposes are determined from one or more ITU allocation blocks issued to that country. The amateur radio callsigns in use for a particular country might use one or a number of combinations derived from the authorised ITU allocation(s) for that country. The following list shows callsign prefixes currently in use. Most are derived from the callsign blocks allocated to administrations by the ITU for use within the countries, territories and dependencies for which a country is responsible. Also shown are some unauthorised prefixes which may be heard and which may or may not be recognised as a DXCC entity, eg 1A0 (SMOM). 1B (the Turkish area of North Cyprus) and 1Z (Karea State - Myanmar) are unofficial and are not recognised for DXCC purposes, so these are not shown.

Full information on prefixes is contained in the ***RSGB Prefix Guide***.

Prefix	Entity	Cont.	ITU	CQ	1.8	3.5	7.0	10.1	14	18	21	24	28	50	144	oth.
1A	Sov. Mil. Order of Malta	EU	28	15												
3A	Monaco	EU	27	14												
3B6, 7	Agalega & St. Brandon Is.	AF	53	39												
3B8	Mauritius	AF	53	39												
3B9	Rodriguez I.	AF	53	39												
3C	Equatorial Guinea	AF	47	36												
3C0	Annobon I.	AF	52	36												
3D2	Fiji	OC	56	32												
3D2	Conway Reef	OC	56	32												
3D2	Rotuma I.	OC	56	32												
3DA	Swaziland	AF	57	38												
3V	Tunisia	AF	37	33												
3W, XV	Vietnam	AS	49	26												
3X	Guinea	AF	46	35												
3Y	Bouvet	AF	67	38												
3Y	Peter 1 I.	AN	72	12												
4J, 4K	Azerbaijan	AS	29	21												
4L	Georgia	AS	29	21												
4O	Montenegro	EU	28	15												
4S	Sri Lanka	AS	41	22												
4U_ITU	ITU HQ	EU	28	14												
4U_UN	United Nations HQ	NA	08	05												
4W	Timor - Leste	OC	54	28												
4X, 4Z	Israel	AS	39	20												
5A	Libya	AF	38	34												
5B, C4, P3	Cyprus	AS	39	20												
5H-5I	Tanzania	AF	53	37												
5N	Nigeria	AF	46	35												
5R	Madagascar	AF	53	39												
5T	Mauritania	AF	46	35												
5U	Niger	AF	46	35												
5V	Togo	AF	46	35												
5W	Samoa	OC	62	32												
5X	Uganda	AF	48	37												
5Y-5Z	Kenya	AF	48	37												
6V-6W	Senegal	AF	46	35												
6Y	Jamaica	NA	11	08												
7O5	Yemen	AS	39	21												
7P	Lesotho	AF	57	38												
7Q	Malawi	AF	53	37												
7T-7Y	Algeria	AF	37	33												
8P	Barbados	NA	11	08												
8Q	Maldives	AS/AF	41	22												
8R	Guyana	SA	12	09												
9A	Croatia	EU	28	15												
9G	Ghana	AF	46	35												
9H	Malta	EU	28	15												
9I-9J	Zambia	AF	53	36												
9K	Kuwait	AS	39	21												
9L	Sierra Leone	AF	46	35												
9M2, 4	West Malaysia	AS	54	28												
9M6, 8	East Malaysia	OC	54	28												
9N	Nepal	AS	42	22												
9Q-9T	Dem. Rep. of Congo	AF	52	36												

Key to abbreviations of continents:
AF = Africa, **AN** = Antarctica, **AS** = Asia, **EU** = Europe,
NA = North America, **OC** = Oceania, **SA** = South America

Prefix	Entity	Cont.	ITU	CQ	1.8	3.5	7.0	10.1	14	18	21	24	28	50	144	oth.
9U	Burundi	AF	52	36												
9V	Singapore	AS	54	28												
9X	Rwanda	AF	52	36												
9Y-9Z	Trinidad & Tobago	SA	11	09												
A2	Botswana	AF	57	38												
A3	Tonga	OC	62	32												
A4	Oman	AS	39	21												
A5	Bhutan	AS	41	22												
A6	United Arab Emirates	AS	39	21												
A7	Qatar	AS	39	21												
A9	Bahrain	AS	39	21												
AP	Pakistan	AS	41	21												
B	China	AS	33, 42-44	23, 24												
BS7	Scarborough Reef	AS	50	27												
BU-BX	Taiwan	AS	44	24												
BV9P	Pratas I.	AS	44	24												
C2	Nauru	OC	65	31												
C3	Andorra	EU	27	14												
C5	The Gambia	AF	46	35												
C6	Bahamas	NA	11	08												
C8-9	Mozambique	AF	53	37												
CA-CE	Chile	SA	14,16	12												
CE0	Easter I.	SA	63	12												
CE0	Juan Fernandez Is.	SA	14	12												
CE0	San Felix & San Ambrosio	SA	14	12												
CE9/KC4	Antarctica	AN	67,69-74	S												
CM, CO	Cuba	NA	11	08												
CN	Morocco	AF	37	33												
CP	Bolivia	SA	12,14	10												
CT	Portugal	EU	37	14												
CT3	Madeira Is.	AF	36	33												
CU	Azores	EU	36	14												
CV-CX	Uruguay	SA	14	13												
CY0	Sable I.	NA	09	05												
CY9	St. Paul I.	NA	09	05												
D2-3	Angola	AF	52	36												
D4	Cape Verde	AF	46	35												
D6	Comoros	AF	53	39												
DA-DR	Fed. Rep. of Germany	EU	28	14												
DU-DZ	Philippines	OC	50	27												
E3	Eritrea	AF	48	37												
E4	Palestine	AS	39	20												
E5	N. Cook Is.	OC	62	32												
E5	S. Cook Is.	OC	62	32												
E6	Niue	OC	62	32												
E7	Bosnia-Herzegovina	EU	28	15												
EA-EH	Spain	EU	37	14												
EA6-EH6	Balearic Is.	EU	37	14												
EA8-EH8	Canary Is.	AF	36	33												
EA9-EH9	Ceuta & Melilla	AF	37	33												
EI-EJ	Ireland	EU	27	14												
EK	Armenia	AS	29	21												
EL	Liberia	AF	46	35												
EP-EQ	Iran	AS	40	21												
ER	Moldova	EU	29	16												
ES	Estonia	EU	29	15												
ET	Ethiopia	AF	48	37												
EU-EW	Belarus	EU	29	16												
EX	Kyrgyzstan	AS	30, 31	17												
EY	Tajikistan	AS	30	17												
EZ	Turkmenistan	AS	30	17												
F	France	EU	27	14												
FG, TO	Guadeloupe	NA	11	08												
FH, TO	Mayotte	AF	53	39												
FJ, TO	Saint Barthelemy	NA	11	08												
FK, TX	New Caledonia	OC	56	32												
FK, TX	Chesterfield Is.	OC	56	30												
FM, TO	Martinique	NA	11	08												
FO, TX	Austral I.	OC	63	32												
FO, TX	Clipperton I.	NA	10	07												
FO, TX	French Polynesia	OC	63	32												
FO, TX	Marquesas Is.	OC	63	31												

The letter ‘S’ against an ITU or CQ Zone indicates that the entity is split across several.

Prefix	Entity	Cont.	ITU	CQ	1.8	3.5	7.0	10.1	14	18	21	24	28	50	144	oth.
FP	St. Pierre & Miquelon	NA	09	05												
FR, TO	Reunion I.	AF	53	39												
FT/G, TO	Glorioso Is.	AF	53	39												
FT/J,E, TO	Juan de Nova, Europa	AF	53	39												
FT/T, TO	Tromelin I.	AF	53	39												
FS, TO	Saint Martin	NA	11	08												
FT/W	Crozet I.	AF	68	39												
FT/X	Kerguelen Is.	AF	68	39												
FT/Z	Amsterdam & St. Paul Is.	AF	68	39												
FW	Wallis & Futuna Is.	OC	62	32												
FY	French Guiana	SA	12	09												
G, GX, M, MX, 2E	England	EU	27	14												
GD, GT, MD, MT, 2D	Isle of Man	EU	27	14												
GI, GN, MI, MN, 2I	Northern Ireland	EU	27	14												
GJ, GH, MJ, MH, 2J	Jersey	EU	27	14												
GM, GS, MM, MS, 2M	Scotland	EU	27	14												
GU, GP, MU, MP, 2U	Guernsey	EU	27	14												
GW, GC, MW, MC, 2W	Wales	EU	27	14												
H4	Solomon Is.	OC	51	28												
H40	Temotu Province	OC	51	32												
HA, HG	Hungary	EU	28	15												
HB	Switzerland	EU	28	14												
HB0	Liechtenstein	EU	28	14												
HC-HD	Ecuador	SA	12	10												
HC8-HD8	Galapagos Is.	SA	12	10												
HH	Haiti	NA	11	08												
HI	Dominican Republic	NA	11	08												
HJ-HK, 5J-5K	Colombia	SA	12	09												
HK0	Malpelo I.	SA	12	09												
HK0	San Andres & Providencia	NA	11	07												
HL, 6K-6N	Republic of Korea	AS	44	25												
HO-HP	Panama	NA	11	07												
HQ-HR	Honduras	NA	11	07												
HS, E2	Thailand	AS	49	26												
HV	Vatican	EU	28	15												
HZ	Saudi Arabia	AS	39	21												
I	Italy	EU	28	15,33												
IS0, IM0	Sardinia	EU	28	15												
J2	Djibouti	AF	48	37												
J3	Grenada	NA	11	08												
J5	Guinea-Bissau	AF	46	35												
J6	St. Lucia	NA	11	08												
J7	Dominica	NA	11	08												
J8	St. Vincent	NA	11	08												
JA-JS, 7J-7N	Japan	AS	45	25												
JD	Minami Torishima	OC	90	27												
JD	Ogasawara	AS	45	27												
JT-JV	Mongolia	AS	32,33	23												
JW	Svalbard	EU	18	40												
JX	Jan Mayen	EU	18	40												
JY	Jordan	AS	39	20												
K, W, N, AA-AK	United States of America	NA	6,7,8	3,4,5												
KG4	Guantanamo Bay	NA	11	08												
KH0	Mariana Is.	OC	64	27												
KH1	Baker & Howland Is.	OC	61	31												
KH2	Guam	OC	64	27												
KH3	Johnston I.	OC	61	31												
KH4	Midway I.	OC	61	31												
KH5	Palmyra & Jarvis Is.	OC	61, 62	31												
KH5K	Kingman Reef	OC	61	31												
KH6,7	Hawaii	OC	61	31												
KH7K	Kure I.	OC	61	31												
KH8	American Samoa	OC	62	32												
KH8	Swains I.	OC	62	32												
KH9	Wake I.	OC	65	31												
KL, AL, NL, WL	Alaska	NA	1, 2	1												
KP1	Navassa I.	NA	11	08												
KP2	Virgin Is.	NA	11	08												
KP3, 4	Puerto Rico	NA	11	08												
KP5	Desecheo I.	NA	11	08												
LA-LN	Norway	EU	18	14												
LO-LW	Argentina	SA	14,16	13												

There are numerous instances of entities that do not count for DXCC before (or after) certain dates. Check the ARRL website for details: *www.arrl.org/country-lists-prefixes*

Prefix	Entity	Cont.	ITU	CQ	1.8	3.5	7.0	10.1	14	18	21	24	28	50	144	oth.
LX	Luxembourg	EU	27	14												
LY	Lithuania	EU	29	15												
LZ	Bulgaria	EU	28	20												
OA-OC	Peru	SA	12	10												
OD	Lebanon	AS	39	20												
OE	Austria	EU	28	15												
OF-OI	Finland	EU	18	15												
OH0	Aland Is.	EU	18	15												
OJ0	Market Reef	EU	18	15												
OK-OL	Czech Republic	EU	28	15												
OM	Slovak Republic	EU	28	15												
ON-OT	Belgium	EU	27	14												
OU-OW, OZ	Denmark	EU	18	14												
OX	Greenland	NA	5, 75	40												
OY	Faroe Is.	EU	18	14												
P2	Papua New Guinea	OC	51	28												
P4	Aruba	SA	11	09												
P5	DPR of Korea	AS	44	25												
PA-PI	Netherlands	EU	27	14												
PJ2	Curacao	SA	11	09												
PJ4	Bonaire	SA	11	09												
PJ5, 6	Saba & St. Eustatius	NA	11	08												
PJ7	St Maarten	NA	11	08												
PP-PY, ZV-ZZ	Brazil	SA	12,13,15	11												
PP0-PY0F	Fernando de Noronha	SA	13	11												
PP0-PY0S	St. Peter & St. Paul Rocks	SA	13	11												
PP0-PY0T	Trindade & Martim Vaz Is.	SA	15	11												
PZ	Suriname	SA	12	09												
R1/F	Franz Josef Land	EU	75	40												
S0	Western Sahara	AF	46	33												
S2	Bangladesh	AS	41	22												
S5	Slovenia	EU	28	15												
S7	Seychelles	AF	53	39												
S9	Sao Tome & Principe	AF	47	36												
SA-SM, 7S-8S	Sweden	EU	18	14												
SN-SR	Poland	EU	28	15												
ST	Sudan	AF	47, 48	34												
SU	Egypt	AF	38	34												
SV-SZ, J4	Greece	EU	28	20												
SV/A	Mount Athos	EU	28	20												
SV5, J45	Dodecanese	EU	28	20												
SV9, J49	Crete	EU	28	20												
T2	Tuvalu	OC	65	31												
T30	W. Kiribati (Gilbert Is.)	OC	65	31												
T31	C. Kiribati (British Phoenix Is)	OC	62	31												
T32	E. Kiribati (Line Is.)	OC	61, 63	31												
T33	Banaba I. (Ocean I.)	OC	65	31												
T5, 6O	Somalia	AF	48	37												
T7	San Marino	EU	28	15												
T8	Palau	OC	64	27												
TA-TC	Turkey	EU/AS	39	20												
TF	Iceland	EU	17	40												
TG, TD	Guatemala	NA	12	07												
TI, TE	Costa Rica	NA	11	07												
TI9	Cocos I.	NA	12	07												
TJ	Cameroon	AF	47	36												
TK	Corsica	EU	28	15												
TL	Central Africa	AF	47	36												
TN	Congo (Republic of the)	AF	52	36												
TR	Gabon	AF	52	36												
TT	Chad	AF	47	36												
TU	Cote d'Ivoire	AF	46	35												
TY	Benin	AF	46	35												
TZ	Mali	AF	46	35												
UA-UI1-7, RA-RZ	European Russia	EU	S	16												
UA2, RA2	Kaliningrad	EU	29	15												
UA-UI8, 9, 0, RA-RZ	Asiatic Russia	AS	S	S												
UJ-UM	Uzbekistan	AS	30	17												
UN-UQ	Kazakhstan	AS	29-31	17												
UR-UZ, EM-EO	Ukraine	EU	29	16												
V2	Antigua & Barbuda	NA	11	08												
V3	Belize	NA	11	07												

Prefix	Entity	Cont.	ITU	CQ	1.8	3.5	7.0	10.1	14	18	21	24	28	50	144	oth.
V4	St. Kitts & Nevis	NA	11	08												
V5	Namibia	AF	57	38												
V6	Micronesia	OC	65	27												
V7	Marshall Is.	OC	65	31												
V8	Brunei Darussalam	OC	54	28												
VA-VG, VO,VY	Canada	NA	2-4, 9, 75	1-5												
VK, AX	Australia	OC	55, 58-59	29,30												
VK0	Heard I.	AF	68	39												
VK0	Macquarie I.	OC	60	30												
VK9C	Cocos (Keeling) Is.	OC	54	29												
VK9L	Lord Howe I.	OC	60	30												
VK9M	Mellish Reef	OC	56	30												
VK9N	Norfolk I.	OC	60	32												
VK9W	Willis I.	OC	55	30												
VK9X	Christmas I.	OC	54	29												
VP2E	Anguilla	NA	11	08												
VP2M	Montserrat	NA	11	08												
VP2V	British Virgin Is.	NA	11	08												
VP5	Turks & Caicos Is.	NA	11	08												
VP6	Pitcairn I.	OC	63	32												
VP646	Ducie I.	OC	63	32												
VP8	Falkland Is.	SA	16	13												
VP8, LU	South Georgia I.	SA	73	13												
VP8, LU	South Orkney Is.	SA	73	13												
VP8, LU	South Sandwich Is.	SA	73	13												
VP8, LU, CE9, HF0, 4K1	South Shetland Is.	SA	73	13												
VP9	Bermuda	NA	11	05												
VQ9	Chagos Is.	AF	41	39												
VR	Hong Kong	AS	44	24												
VU	India	AS	41	22												
VU4	Andaman & Nicobar Is.	AS	49	26												
VU7	Lakshadweep Is.	AS	41	22												
XA-XI	Mexico	NA	10	06												
XA4-XI4	Revillagigedo	NA	10	06												
XT	Burkina Faso	AF	46	35												
XU	Cambodia	AS	49	26												
XW	Laos	AS	49	26												
XX9	Macao	AS	44	24												
XY-XZ	Myanmar	AS	49	26												
YA, T6	Afghanistan	AS	40	21												
YB-YH	Indonesia	OC	51,54	28												
YI	Iraq	AS	39	21												
YJ	Vanuatu	OC	56	32												
YK	Syria	AS	39	20												
YL	Latvia	EU	29	15												
YN,H6-7,HT	Nicaragua	NA	11	07												
YO-YR	Romania	EU	28	20												
YS, HU	El Salvador	NA	11	07												
YT-YU	Serbia	EU	28	15												
YV-YY, 4M	Venezuela	SA	12	09												
YV0	Aves I.	NA	11	08												
Z2	Zimbabwe	AF	53	38												
Z3	Macedonia	EU	28	15												
Z8	South Sudan (Rep of)	AF	48	34												
ZA	Albania	EU	28	15												
ZB2	Gibraltar	EU	37	14												
ZC4	UK Sov. Base Areas on Cyprus		AS	3 9												
20																
ZD7	St. Helena	AF	66	36												
ZD8	Ascension I.	AF	66	36												
ZD9	Tristan da Cunha & Gough I.	AF	66	38												
ZF	Cayman Is.	NA	11	08												
ZK3	Tokelau Is.	OC	62	31												
ZL-ZM	New Zealand	OC	60	32												
ZL7	Chatham Is.	OC	60	32												
ZL8	Kermadec Is.	OC	60	32												
ZL9	Auckland & Campbell Is.	OC	60	32												
ZP	Paraguay	SA	14	11												
ZR-ZU	South Africa	AF	57	38												
ZS8	Prince Edward & Marion Is.	AF	57	38												

For the latest DXCC list: www.arrl.org/country-lists-prefixes

Callsign Listings

Approximate Issue Dates for UK Callsigns

Full Licence

Two letters

G2AA	1920-39
G3AA	1937-38
G4AA	1938-39
G5AA	1921-39
G6AA	1921-39
G8AA	1936-37

Three letters

G2AAA	Pre-war
G3AAA	1946
G3CAA	1947
G3EAA	1948
G3GAA	1949-50
G3HAA	1950-51
G3IAA	1951-52
G3JAA	1952-54
G3KAA	1954-56
G3LAA	1956-57
G3MAA	1957-58
G3NAA	1958-60
G3OAA	1960-61
G3PAA	1961-62
G3QAA	not issued
G3RAA	1962-63
G3SAA	1963-64
G3TAA	1964-65
G3UAA	1965-66
G3VAA	1966-67
G3WAA	1967
G3XAA	1967-68
G3YAA	1968-69
G3ZAA	1969-71
G4AAA	1971-72
G4BAA	1972-73
G4DAA	1974-75
G4EAA	1975-76
G4GAA	1977
G4IAA	1979
G4MAA	1981
G4QAA	not issued
G4RAA	1982
G4SAA	1983
G4WAA	1984
G0AAA	1985
G0EAA	1986
G0HAA	1987
G0JAA	1988
G0LAA	1989
G0MAA	1990
G0NAA	1991
G0SAA	1992
G0TAA	1993
G0VAA	1994
G0WAA	1995
M0AAA	1996
M0BAA	1997-98
M0CAA	1998-00
M5AAA	1999-00
G8AAA	1964-67
G8BAA	1967-68
G8CAA	1968-69
G8DAA	1969-70
G8EAA	1970-71
G8FAA	1971-72
G8HAA	1973
G8IAA	1973-74
G8JAA	1974-75
G8KAA	1975
G8MAA	1976-77
G8NAA	1977
G8OAA	1977-78
G8PAA	1978
G8QAA	not issued
G8TAA	1979
G8ZAA	1981
G6CAA	1981
G6QAA	not issued
G6RAA	1982
G1AAA	1983
G1DAA	1984
G1LAA	1985
G1QAA	not issued
G1SAA	1986
G1XAA	1987
G7AAA	1988
G7EAA	1989
G7FAA	1990
G7HAA	1991
G7MAA	1992
G7OAA	1993
G7SAA	1994
G7TAA	1995
G7WAA	1996
M1AAA	1996
M1CAA	1997
M1DAA	1998-99
M1EAA	1999-00

Intermediate

2E0AAA	1991
2E1AAA	1991
2E1BAA	1992
2E1CAA	1993
2E1DAA	1994
2E1EAA	1995-97
2E1GAA	1997-99
2E1HAA	1999-00

Foundation

M3AAA	2002
M6AAA	2008
M7AAA	2018

Note: *From April 2000, out-of-sequence callsigns could be requested, so calls in later series may be heard.*

Please Read important Notices and Frequently Asked Questions on page 181

Callsign Formatting

Each year the RSGB requests callsign data from Ofcom for the production of this book. This data was output in the middle of June 2024 and therefore any callsign changes made after this date will not appear in the following pages. The RSGB makes no amendments to the data supplied by Ofcom other than formatting. This includes removing upper case from town names and text. First names and most titles (Mr, Mrs, Ms, etc.) some honorific titles (Sir, Lord, Dr, etc.) are though retained. This produces a consistent format presentation of the data.

Callsigns that are included in the 'Details withheld' section of the Yearbook are only formatted for appearance. In these instances the information about the holder is not passed to the RSGB so we are not able to release any further information. We are also not able to move individuals from one list to the other on request, as the data reproduced here is a reflection of the data held by Ofcom not the RSGB. Should you want to change this information this will need to be done via Ofcom and any change will then be reflected in future editions of the Yearbook.

It can be noted that we do add the (optional to use) Regional Secondary Locators (RSLs) which identify locations within the UK. This does not form part of the licence information so is no longer supplied by Ofcom to the RSGB. The RSGB uses UK postcodes to identify the station locations and the details of the changes are laid out in the table below. Where appropriate these appear as the suffix to the first character of the callsign. We only incorporate suffixes for individuals although we do recognise that in practice some clubs may use the appropriate suffix to identify themselves as such. We recognise that this process may contain a degree of inaccuracy and therefore apologise to anyone if their callsign is displayed incorrectly.

RSL Postcode Changes

Postcode	Suffix	Country	Notes
AB	M	Scotland	
BT	I	N. Ireland	
CF	W	Wales	
CH	W	Wales	Only CH5, CH6, CH7 & CH8
DD	M	Scotland	
DG	M	Scotland	
EH	M	Scotland	
FK	M	Scotland	
G	M	Scotland	
GY	U	Gurnsey	
HS	M	Scotland	
IM	D	Isle of Man	
IV	M	Scotland	
JE	J	Jersey	
KA	M	Scotland	
KW	M	Scotland	
KY	M	Scotland	
LD	W	Wales	
LL	W	Wales	
ML	M	Scotland	
NP	W	Wales	
PA	M	Scotland	
PH	M	Scotland	
SA	W	Wales	
SY	W	Wales	Only SY10, SY11, SY16, SY18, SY19, SY20, SY21, SY22 & SY23
TD	M	Scotland	Not: Berwick TD15
ZE	M	Scotland	

Frequently Asked Questions

Throughout the year the RSGB receives a number of standard queries about an individual's entry in the *Yearbook.* Here are the answers given:

Q: I have just gained my new licence, please include my new details in the next Yearbook.

A: *Your details will be passed on to us by Ofcom from the data you supply on your licence application form. There is no need to contact the Society directly.*

Q: I am an RSGB member. Thank you for sending *RadCom* to my new address, but why didn't you change my *Yearbook* entry?

A: *The Yearbook lists all UK amateurs, not just RSGB members, and the records are entirely separate. Your callbook address details come from Ofcom. Did you re-validate your licence with your up-to-date details or let them know you moved? (***read Important Note below***)*

Q: My entry shows me as being 'details withheld', but I want my full address to be listed.

A: *Please contact Ofcom Licensing and ask them to release your details.* **The Society cannot accept direct input from licensees**.

Q: You have published my address, but I would like it withheld.

A: *As above, but ask Ofcom Licensing to withhold it from* ***all*** *callbook publishers*.

Q: My callsign is not shown at all, please include it.

A: *If a callsign is not shown in the Yearbook it was not licensed at the time the data was supplied to the RSGB. Sometimes, through an administrative error or misunderstanding, an amateur believes that he or she is licensed but the licensing records show that the licence has lapsed. If this is the case, contact Ofcom Licensing and discuss the matter with them as it is important that you re-validate your licence if you intend using it.*

Important Note

We do not make amendments to the basic callsign data. Should the details that appear in the callsign section of this *Yearbook* need amending, you should contact the licensing authority direct. Your submission should be prior to the **1st June** to allow sufficient time for the amendment to be processed.

If you have not validated your licence, which you must do every 5 years, or if any of your details have changed, ie change of address, then your licence may be void and will not appear in the Callsign listings section of this *Yearbook*, and you may not be licensed to operate.

To validate your licence, log in to *https://services.ofcom.org.uk/* and amend any details as soon as possible, this will automatically validate your licence.

Ofcom Callsign Data Disclaimer

Please note that the use of this callsign data is entirely at your own risk. While every effort is made to ensure that the information provided to you is accurate, no guarantees for the currency or accuracy of information are made.

The callsign data is provided 'as is'. It is provided without any representation or endorsement made and without warranty of any kind, whether express or implied, including but not limited to the implied warranties of satisfactory quality, fitness for a particular purpose, non-infringement, compatibility, security and accuracy.

Ofcom does not accept any responsibility for any loss, disruption or damage to your data or your computer system which may occur whilst using the data provided by Ofcom.

In no event will Ofcom be liable for any loss or damage including, without limitation, indirect or consequential loss or damage, or any loss or damages whatsoever arising from use of loss of use of, data or profits arising out of or in connection with the use of information provided by Ofcom.

UK Callsigns

2#0
2E0 AAC G. Lockett 10 Lunesdale Street, Houghton le Spring DH5 0DB
2E0 AAF S. Rhenius Baythorne Cottage Baythorne End, Halstead CO9 4AB
2E0 AAI D. Simmons 23 Fairey Street, Cofton Hackett, Birmingham B45 8GU
2E0 AAJ R. Yarrow 27 Staplers Road, Newport PO30 2DB
2E0 AAK R. Berridge Bracklyn, St. Clare Road, Deal CT14 7QB
2E0 AAN M. Keilty 25 Lathom Avenue Wallasey, Wirral CH44 5UH
2E0 AAO S. Thirlwall 2 Crossfield Avenue Blythe Bridge, Stoke-on-Trent ST11 9PL
2M0 AAQ W. Ferguson 1D Macphail Drive, Kilmarnock KA3 7EJ
2E0 AAX E. Wills 1 Scott Close, High Street, Marlborough SN8 3AF
2E0 AAZ W. Elston 3 Ennerdale Close, Northampton NN3 6BL
2E0 ABD R. Mold 134 Kipling Avenue, Brighton BN2 6UE
2E0 ABL F. Hayes 88 Johns Road, Fareham PO16 0RX
2E0 ABT L. Simons Westwood, Faris Lane, Addlestone KT15 3DJ
2E0 ABU K. Williams 26 Charles Bassett Close, Helston TR13 8BG
2E0 ACA I. Craig Ferndown, Tilley Lane, Hailsham BN27 4UT
2E0 ACE A. Backhouse 6 Millfield, Brampton CA8 1TT
2E0 ACQ H. Barnes 24 Burleigh Place, Oakley, Bedford MK43 7SG
2E0 ACR H. Smallwood 14 Shaftesbury Avenue, Bedford MK40 3SA
2E0 ACV E. Barnes 3 Layton Road, Ashton-On-Ribble, Preston PR2 1PB
2E0 ADA R. Gaskell 18 Woodcroft Kennington, Oxford OX1 5NH
2E0 ADL J. Wresdell 4 Fairfield Close, Nafferton, Nafferton YO25 4JH
2E0 ADN M. Holmes 55 Fitzpain Road, West Parley, Ferndown BH22 8RZ
2E0 ADR S. Watson 8 Ventnor Road, Middlesbrough TS5 6DU
2E0 ADY A. Jones 67 Mandara Grove, Abbeydale., Gloucester GL4 5XT
2E0 AED A. Sumner 78 Woodlands Way, Southwater, West Sussex RH13 9HZ
2E0 AES B. Hyde 54 The Byway, Darlington DL1 1EQ
2E0 AFL M. Coxhead Barronsway, Parker Lane, Whitestake, Preston PR4 4JX
2E0 AFP R. Forrest-Webb 1 Trelasdee Farm Cottages, St. Weonards, Hereford HR2 8PU
2E0 AFT A. Rowe 3 The Badgers, St. Georges, Weston-Super-Mare BS22 7RE
2E0 AGB A. Boocock 25 Smallwood Road, Dewsbury WF12 7RU
2E0 AGI Dr E. Mclusky 11 Ripon Road, Killinghall, Harrogate HG3 2DG
2E0 AGQ B. Read 10 Shamblers Road, Cowes PO31 7HF
2M0 AGS G. Robbins 39 Locheil Gardens, Glenrothes KY7 6YL
2E0 AHB A. Beveridge Kernick Cottage, Sparry Bottom, Carharrack, Redruth TR16 5SH
2M0 AIE M. Hooks Braehead Mains, Main Street, Braehead Forth, Lanark ML11 8HA
2E0 AIK M. Footring 26 Ernest Road, Wivenhoe, Colchester CO7 9LG
2E0 AIL G. Harper 48 Norlands Lane, Widnes WA8 5AS
2E0 AIM J. Cowley 39 Alpine Way, Tow Law, Bishop Auckland DL13 4DS
2E0 AIT K. Lloyd 2 Bishopstone Drive, Beltinge, Herne Bay CT6 6RE
2E0 AJG A. Golding 44 Blendon Drive, Andover SP10 3NG
2E0 AJK D. Forsyth 20 Chapel View, Rowlands Gill NE39 2PN
2E0 AJM D. Martin Fawcett House 34 Forton Road, Chard TA20 2HL
2E0 AJP L. Humphrey 133 Hocombe Road, Chandler'S Ford, Eastleigh SO53 5QD
2E0 AJQ A. Nelson 29 Coxford Road, Southampton SO16 5FG
2E0 AJT W. Atkins 55 Park Grange Croft, Norfolk Park, Sheffield S2 3QJ
2E0 AJX C. Overson Studio Flat, 6 Grenville Street, Bideford EX39 2EA
2M0 AKI D. Hague 13 North Dell, Ness, Isle of Lewis HS2 0SW
2E0 AKJ S. Davin 40 Theynes Croft, Long Ashton, Bristol BS41 9NA
2E0 AKK E. Jones 43 Wesley Road, Wimborne BH21 2QB
2E0 AKL R. Mcdermott 2 Monument Close, Wellington TA21 9AL
2E0 AKN P. Newton 61 Ashbourne Crescent, Taunton TA1 2RA
2E0 AKO M. Newton 43 Hayfield Road, Minehead TA246AD
2E0 AKQ J. Paul East Park, Church Road, Cowes PO31 8HA
2M0 AKS Dr A. Curlis 94 Kirkhill Road, Aberdeen AB11 8FX
2E0 AKU R. Brisley 15 Elm Fields, Old Romney, Romney Marsh TN29 9SN
2W0 ALA N. Kind 32 Maesgarmon, Castle Caereinion, Welshpool SY21 9AN
2E0 ALB Dr M. Ali 4 The Crescent, Great Horkesley, Colchester CO6 4EH
2E0 ALD P. Girling The Gorse, Leiston Road, Aldeburgh IP15 5QE
2E0 ALF A. Taylor 113 Queensway, Grantham NG31 9RG
2E0 ALH J. Side Railway Crossing Cottage, Ash Road, Sandwich CT13 9JB
2E0 ALJ G. Brierley 35 Ochrewell Avenue, Deighton, Huddersfield HD2 1LL
2E0 ALL A. Sawyer 26 Dallas Brett Crescent, Folkestone CT19 6NE
2M0 ALS M. Gill Easter Templand, Fortrose IV10 8RA
2E0 ALW R. Hatcher 61 Holland Road, Oxted RH8 9AU
2W0 ALZ D. Gunning 19 Gordon Avenue, Prestatyn LL19 8RU
2W0 AMB J. Lloyd-Owen 34 Maes Y Castell, Llandudno LL30 1NG
2M0 AMD A. Dawson Colt House, Kinlochbervie, Lairg IV27 4RP
2W0 AMN C. Weaver 2 Ty Cerrig, Llanddoged, Llanrwst LL26 0TY
2E0 AMS A. Saville 1 Bellingham Close, Shaw, Oldham OL2 7UU
2E0 AMW G. Fowle 12 Lytham Road, Broadstone BH18 8JS
2E0 AND A. Miller East Lyn, Lazonby, Penrith CA10 1BX
2E0 ANM T. Headley 73 Tudor Green, Jaywick, Clacton-on-Sea CO15 2PB
2E0 AOK P. Godolphin Shepherds Cottage, Flakebridge, Appleby-in-Westmorland CA16 6JZ
2E0 AOL T. Money 119 Twyford Way, Canford Heath, Poole BH17 8SR
2E0 AOS E. Hedley 17 Chowdene Bank, Gateshead NE9 6JJ
2E0 AOU S. Tilly 24 Whinham Way, Morpeth NE61 2TF
2E0 AOZ I. Marshall 44 Cromwell Crescent, Pontefract WF8 2EJ
2M0 APB J. Macdonald 63 Perceval Road South, Stornoway HS12TL
2E0 APG G. Wilson 9 Newcomen Way, Telford TF7 5UB
2E0 APJ M. France 10 Grampian Avenue, Wakefield WF2 8JZ
2E0 APN T. Humble 43 Whiteley Crescent, Bletchley, Milton Keynes MK3 5DQ
2M0 APX S. Barbour 40 Mannerston Holdings, Linlithgow EH49 7ND
2E0 APY J. Rayson 18 Station Street, South Wigston, Wigston LE18 4TH
2E0 APZ M. Meyer 22 Orchard Grove, Newton Abbot TQ12 1FZ
2E0 AQE J. Brown 55 Barrington Road, Rubery, Birmingham B45 9EU
2E0 AQI D. Black 8 Cornwood Close, Finchley, London N2 0HP
2E0 AQU P. Webb 41 Lancaster Gardens, Wolverhampton WV4 4DN
2E0 ARA J. Griffin 35 Cottage Street, Kingswinford DY6 7QE
2E0 ARC R. Evans 32 Bracken Lane, Higher Bebington CH63 2LZ
2E0 ARJ R. Birch 15 Chester Street, Cirencester GL7 1HF
2I0 ARM A. Mccabe Scaddin House 15 Myra Road, Downpatrick BT30 7JX
2E0 ART P. Dudding 14 St. Levan Close, Marazion TR17 0BP
2E0 ARV I. Strachan 1 Gantley Avenue, Billinge, Wigan WN5 7AF
2E0 AST L. Compton 1 Southhaven, Beach Road, Bacton, Norwich NR12 0EP
2E0 ASU J. Bell 122 Howard Drive, Letchworth Garden City SG6 2DE
2E0 ASX R. Bannister 60 St. Johns Avenue, Bridlington YO16 4NL
2E0 ATT N. Watson 12 Gold Croft Gold Street, Barnsley S70 1TZ
2E0 ATY R. Browne 2 Martham Close, London SE28 8NF
2E0 ATZ R. Macdonald 4 Throwley Drive, Herne Bay CT6 8LP
2W0 AUC R. Henderson 150 Heol Maes Eglwys, Pantlasau, Swansea SA6 6NW
2E0 AUI D. Tyson 21 Providence Place, Filey YO14 9DU
2E0 AUM A. Peach 14 Northfield Avenue Rocester, Uttoxeter ST14 5LE
2E0 AUU P. Christopoulos Cosenda, Tye Green Village, Harlow CM18 6QY
2E0 AUV C. Brice 10 Swan Close, Weston-Super-Mare BS22 8XR
2E0 AVA T. Milne 3 Birch Road, New Ollerton, Newark NG22 9PU
2E0 AVB M. Orr 42 Mayfield Road, London W12 9LU
2E0 AVD W. Jukes 22 Hazelmere Road, Creswell, Worksop S80 4HS
2E0 AVK R. Harrall 47 Hawksmoor Road, Stafford ST17 9DS
2E0 AVQ J. Paul 67 Fleet Avenue, Upminster RM14 1PZ
2E0 AVS S. Panczel 11 Chauncy Road, Manchester M40 3GG
2E0 AVW R. SmithRenee . Shardlow Marina. London Road, Shardlow., Derby DE72 2GL
2W0 AVZ N. Heyne New House, Hope, Welshpool SY21 8JD
2E0 AWC K. Necchi 59 Winters Way, Waltham Abbey EN9 3HP
2E0 AWE J. Sanders 219 Station Road, Drayton, Portsmouth PO6 1PY
2E0 AWG S. Rope 12 Edrich Close, Norwich NR134JD
2E0 AWI R. Desbois 12 Dove Walk, Hornchurch RM12 5HH
2E0 AWK J. Jackson 35 Coppingford Close, Rochdale OL12 7PR
2E0 AWR J. Gibson 14 Douglas Bank Drive, Wigan WN6 7NH
2E0 AWS A. Storer 16 Eastfields, Braunston, Daventry NN11 7JN
2E0 AWT G. Griffiths 16 Back Lane, Winteringham, Scunthorpe DN15 9NW
2W0 AWW R. Evans 4 Green Terrace, Deiniolen, Caernarfon LL55 3LE
2M0 AWY J. Fulton 35 Heatherbank, Livingston EH54 6EE
2E0 AWZ R. Turton 31 Second Avenue, Woodlands, Doncaster DN6 7QQ
2E0 AXA K. Kerridge 31 Second Avenue, Woodlands, Doncaster DN6 7QQ
2E0 AXB M. Costello 25 Campden Green, Solihull B92 8HQ
2E0 AXN S. Jenner 4 Christie Close, Chatham ME5 7NG
2E0 AXT A. Smith 186 Longmead Drive, Nottingham NG5 6DJ
2E0 AXZ C. Cook 112 Waterford Road, Ipswich IP1 5NJ
2E0 AYI K. Williams 36 Castle St., Tiverton EX16 6RG
2E0 AYK M. Longbottom 32 Ann'S Hill Road, Gosport PO12 3JY
2E0 AYQ M. Walsh 25 Felstead Road, Orpington BR6 9AA
2E0 AYT R. Cole 2 Station Road, Sharpthorne RH19 4NY
2M0 AYU J. Bruce 33 Kintail Place, Dingwall IV15 9RL
2E0 AYW M. Pedley Comfort Cottage, Goadsbarrow, Ulverston LA12 0RE
2E0 AYX C. Penfold 14 Romney Road, Tetbury GL8 8JU
2E0 AYY P. Rainer 6 Highland Close, Folkestone CT20 3SA
2M0 AYZ C. Gajewski 7/2 Brunswick Road, Edinburgh EH7 5NG
2E0 AZA P. Livsey 33 Aldingham Walk, Morecambe LA4 4EW
2E0 AZB M. Livsey 26 Aldingham Walk, Morecambe LA4 4EW
2E0 AZD F. Russell 157 Durnford St., Plymouth PL1 3QR
2E0 AZF D. Whitworth 570 Dereham Road, Norwich NR5 8TE
2E0 AZJ V. Hocking 80 Barton Tors, Bideford EX39 4HA
2E0 AZK A. Cartwright 17 Greenheath Way, Wirral CH46 3RX
2E0 AZL L. Murphy 22 Shenley Fields Drive, Birmingham B31 1XH
2W0 AZM R. Harper 17 Conway Drive, Wrexham LL13 9HR
2W0 AZP D. Edwards 9 Ffos Y Cerridden, Nelson, Treharris CF46 6HQ
2E0 AZU A. Groat 23 Knightwake Road, New Mills, High Peak SK22 3DQ
2M0 AZW T. Galt 119 Cleeves Road, Glasgow G53 6NQ
2E0 AZZ A. Dalzell 9 Pyms Lane, Crewe CW1 3PJ
2E0 BAB G. Gragon 2 Greenacres Grove, Shelf, Halifax HX3 7RN
2I0 BAC C. Mcconnell 41 Moyra Road, Doagh, Ballyclare BT39 0SQ
2I0 BAD A. Hayes 35 Ballyhamage, Doagh BT39 0PZ
2E0 BAF D. Davies 35 Orchard Court Carlton, Nottingham NG4 1BD
2E0 BAH P. Mannering 101 Burford, Brookside, Telford TF3 1LJ
2E0 BAI R. Froud 48 Campbell Road, Maidstone ME15 6QB
2E0 BAJ F. Cappleman 8 The Woodlands, Lilleshall, Newport TF10 9EN
2E0 BAK M. Verrall Flat 4, 84 Briar Way, Skegness PE25 3PU
2W0 BAO T. Mulcuck 61 Hillside View, Graigwen, Pontypridd CF37 2LG
2E0 BAP S. Savage 450 Locking Road, Milton, Weston-Super-Mare BS22 8PS
2E0 BAQ D. Wills 37 Hawden Road, Bournemouth BH118RP
2E0 BAU C. Young 15 Shelton Avenue, East Ayton, Scarborough YO13 9HB
2E0 BAV A. Yorke 33 Avon Crescent, Stratford-upon-Avon CV37 7EX
2E0 BAX S. Amesbury 92 Range Court, Macclesfield SK10 2RR
2E0 BAY A. Harris 32 King Edward Road, Gillingham ME7 2RE
2E0 BBA I. Wengraf 45 School Lane, Higham, Rochester ME3 7JR
2E0 BBG D. Skinner 77 Rolleston Avenue, Petts Wood, Orpington BR5 1AL
2E0 BBI A. Hayward 23 Greenbank Close, Grampound Road, Truro TR2 4TD
2E0 BBL C. Lishman 6 Clarence Road, Accrington BB5 0NA
2E0 BBM S. Beales 49 Greenacres, Old Newton, Stowmarket IP14 4EJ
2E0 BBN S. Woodcock 2 Victory Drive, Farington, Leyland PR25 3DD
2W0 BBO L. Powell 2 Gelliderw Pontardawe, Swansea SA8 4NB
2E0 BBQ R. Gash 2 The Marsh, Benington, Boston PE22 0DH
2E0 BBS M. Janes 1 Cheswell Gardens, Church Crookham, Fleet GU51 5NJ
2W0 BBT S. Christie-Powell 2 Gelliderw Pontardawe, Swansea SA8 4NB
2E0 BBX J. Akinin 70 Valley Road, West Bridgford, Nottingham NG2 6HQ
2E0 BBY F. Coles 8 Moore Close, Church Crookham, Fleet GU52 6JD
2E0 BBZ M. Thyer 8 Roman Road, Taunton TA1 2BD
2E0 BCB D. Holland 13 Linley Drive, Boston PE21 7EJ
2E0 BCD B. Grice 17 Albion Field Drive, West Bromwich B71 4HN
2E0 BCE J. Ferris 39 Gladstone St., Beeston, Nottingham NG9 1EU
2E0 BCF Dr H. Donnelly 6 Famet Walk, Purley CR8 2DY

2E0 BCG Rvd. W. Walker 9 Malthouse Lane, Dorchester-On-Thames, Wallingford OX10 7LF
2E0 BCJ D. Neville 70 Alderwood Avenue, Liverpool L24 2UF
2E0 BCK Dr T. Gale 38 Lantree Crescent, Trumpington, Cambridge CB2 9NJ
2M0 BCL T. Mccall 119 Claremont, Alloa FK10 2ER
2E0 BCM B. Marston 38 Bulstrode Road, Ipswich IP2 8HA
2E0 BCO S. Norman 27 Ashburton Road, Ickburgh, Thetford IP26 5JA
2E0 BCQ R. Britt Thoroughfare House, South Burlingham Road, Norwich NR13 4FA
2D0 BCR R. Cunningham 3 Kellets Cottage, Lhergy Cripperty, Union Mills IM4 4NF Isle of Man
2E0 BCS Dr J. Schofield 6 Robin Royd Avenue, Mirfield WF14 0LF
2E0 BCW S. Bywater Birch Wood, Norwich Road, Cromer NR27 0HG
2E0 BCX J. Allen 20 Spa Hill, Kirton Lindsey, Gainsborough DN21 4BA
2E0 BDB A. Grosvenor 10 Neves Close, Lingwood, Norwich NR13 4AW
2E0 BDD A. Morrison 70 The Drive, Northampton NN1 4SP
2E0 BDI S. Moakes 46 Parsonage St., Stockport SK4 1HZ
2E0 BDJ D. Anthony 20 Parkfield Close, Leyland PR26 7XJ
2M0 BDN J. Walker 49 Great King St., Edinburgh EH3 6RP
2E0 BDO S. Grainger 15 Carr House Lane, Wirral CH46 6EN
2E0 BDP M. Saunders 70 Underwood Lane, Crewe CW1 3LE
2E0 BDQ W. Taylor 99 St. Marys Close, Littlehampton BN17 5QQ
2M0 BDR B. Keiller Da Cro, Branchiclate, Burra Isle ZE2 9LA
2E0 BDS B. Derbin-Sykes 1 , Lentons Lane, Friskney, Boston PE22 8RR
2M0 BDT A. Halcrow Da Cro, Branchiclate, Burra Isle ZE2 9LA
2E0 BDV J. Goulding 79 Dalston Drive, Manchester M20 5LQ
2E0 BEA J. Rushton 25 Garth Meadow, Catterick, Richmond DL10 7RT
2M0 BEC M. Mitchell Easter Kilwhiss Farm, Ladybank, Cupar KY15 7UR
2W0 BED C. Rosser 16 Thomas St., Penygraig, Tonypandy CF40 1EU
2E0 BEE A. Bell 2 Croft Foot, Sandwith, Whitehaven CA28 9UG
2E0 BEF M. Staton 52 School Road, Newborough, Peterborough PE6 7RG
2E0 BEG M. Davidson 19 Mason Street, Workington CA14 3EH
2E0 BEH C. Loughran 8 Douglas Road, Dover CT17 0BD
2E0 BEI C. Brett 35 Gossilin Street, Whitstable CT5 4LQ
2M0 BEL P. Mccluskey 119 Tower Drive, Gourock PA19 1SG
2E0 BEP A. Holmes 10 Doe Park, York YO30 4UQ
2E0 BEQ D. Whitehead 89, Cowpes Close, Sutton-in-Ashfield NG17 2BU
2M0 BET D. Boden 42 Kirkwynd, Maybole KA19 7AE
2E0 BEV I. Lowe 1 Hazelby Road, Creswell, Worksop S80 4BB
2E0 BEW D. Powis Fircroft, Pound Lane, Woodbridge IP13 0LN
2E0 BFA J. Mullarkey 41 Foyle Avenue, Chaddesden, Derby DE21 6TZ
2W0 BFC M. Williams 30 Elm Drive, Risca, Newport NP11 6HJ
2W0 BFD W. Corbett 27 Waunfawr Gardens. Crosskeys, Newport NP11 7AJ
2E0 BFF A. Bateman 154 Arnold Road Mangotsfield, Bristol BS16 9LB
2E0 BFG P. Davies 53 Lammas Road, Cheddington, Leighton Buzzard LU7 0RY
2E0 BFJ G. Swain 3 Flaxfield Drive, Crewkerne TA18 8DF
2E0 BFM A. Knights 81 Green Lane, Barnard Castle DL12 8LF
2E0 BFN W. Carty 49 Princess Gardens, Blackburn BB25EJ
2E0 BFS B. Townshend 9 Norfolk Place, Boston PE21 9JJ
2E0 BFT D. Doroba Flat 3, 305A London Road South, Lowestoft NR33 0DX
2W0 BFV W. Harries 18 Bro Teify, Alltyblacca, Llanybydder SA40 9SR
2E0 BFZ P. Hinde 19 Tadcaster Road, Sheffield S8 0RA
2E0 BGC N. Speight Flat 14, Cranbrook, London NW1 0LJ
2U0 BGE J. Bligh The Bounty, Salines Lane, St. Sampson GY2 4FL Guernsey
2W0 BGI M. Lewis 4 Coldwell Terrace, Pembroke SA71 4QL
2E0 BGJ M. Whitaker 5 Horns Drove, Rownhams, Southampton SO16 8AH
2E0 BGL B. Lloyd 104 Wootton St., Bedworth CV12 9DZ
2E0 BGM S. Russon 165 Billington Avenue, Newton-le-Willows WA12 0AU
2E0 BGO J. Woodruff 10 Bailey Close, Blackburn BB2 4FT
2E0 BGP E. Neil 5 Winsford Hill, Furzton, Milton Keynes MK4 1BJ
2W0 BGQ I. Canterbury Brynllethryd Bungalow, Senghenydd, Caerphilly CF83 4HJ
2W0 BGS J. Brydges 9 Twynygarreg, Treharris CF46 5RL
2E0 BGV R. Johnson 90 Regent Street Church Gresley, Swadlincote DE11 9PJ
2E0 BGX P. Sarll 81 Austendyke Road, Weston Hills, Spalding PE12 6BX
2E0 BGZ R. Dale 17 Spencer Gardens, Brackley NN13 6AQ
2E0 BHA P. Bennett 1 Queens Road, Carterton OX18 3YB
2E0 BHB R. Fallows Loughrigg, Cranmore Avenue, Yarmouth PO41 0XS
2E0 BHC L. Heenan 1 Howard Close, Daventry NN11 4TD
2E0 BHH R. Bullen 2 Redlands Cottages, East Coker, Yeovil BA22 9HF
2E0 BHJ J. List 41 Westbury Crescent, Dover CT17 9QQ
2E0 BHN J. Mccoll 6 Grenville Close, Bodmin PL31 2FB
2E0 BHP G. Boyd 26 Bluebell Drive, Littlehampton BN17 6UL
2E0 BHQ D. George Flat 9, The Old Court House, Waterloo, Frome BA11 3FE
2E0 BHS N. Tideswell 19 Wish Court, Ingram Crescent West, Hove BN3 5NY
2I0 BHT S. Quigg 100 Whispering Pines, Limavady BT49 0UF
2E0 BHU J. Hickman Ardoch, Harlestone Road, Northampton NN6 8AW
2E0 BHY S. Preston 14 Hulles Way, North Baddesley, Southampton SO52 9NS
2E0 BHZ T. Mildenhall 58 Montrose Avenue, Datchet, Slough SL3 9NJ
2E0 BIB R. Orton 38 Whitehill Avenue, Barnsley S70 6PP
2E0 BIC T. Humphries 10 Cropthorne Avenue, Leicester LE5 4QL
2I0 BID A. Jamison 11 Richmond Gardens, Newtownabbey BT36 5LA
2E0 BII R. Hathaway 38 Windsor Walk, Lindford, Bordon GU35 0SG
2M0 BIL W. Mccue 188 Redburn, Alexandria G83 9BU
2E0 BIM K. Hawke 111 Dorchester Avenue, Plymouth PL5 4AZ
2M0 BIN F. Coombes 44 Lochfield Road, Paisley PA2 7RL
2E0 BIO R. Aston 8 Parliament Road, Thame OX9 3TE
2I0 BIQ K. Blake Chestnut Lodge Care Home, 47 Carrickaness Road, Dungannon BT71 7NH
2I0 BIR J. Smyth 37 Ardfreelin, Newry BT34 1JG
2E0 BIS J. Cosson 25 Fox Brook, St. Neots PE19 6AL
2E0 BIU P. Rushby 16 Foxhill Lane, Selby YO8 9AR
2E0 BIY P. Lewis 16 Valley Road, St. Albans AL3 6LR
2E0 BJA B. Johnson 15 Oak Avenue, Willington, Crook DL15 0BJ
2E0 BJB A. Chaplin 33 The Crofts, Little Wakering, Southend-on-Sea SS3 0JS
2E0 BJD R. Rowe 16 Orchard Road, Plymouth PL2 2QY
2W0 BJE S. Merrifield 37 South View Drive Rumney, Cardiff CF3 3LX
2E0 BJK R. Robertson 2 Dunsley Close, Middlesbrough TS3 7DR
2E0 BJL M. Placidi 3 Eleanor Avenue, Epsom KT19 9HD
2E0 BJM B. Maggs 44 Coldharbour Road, Hungerford RG17 0AZ
2E0 BJP P. Norman 5 Stirling Close West Row, Bury St. Edmunds IP28 8QD
2E0 BJQ M. Roebuck 16 Dalton Street, Sowerby Bridge HX6 2HE
2W0 BJR R. Johnson 25 Lon Tyrhaul Llansamlet, Swansea SA7 9SF
2E0 BJS N. Roberts 40 Armour Road, Tilehurst, Reading RG31 6HN
2E0 BJT R. Carter 43 Sheldon Avenue, Standish, Wigan WN6 0LW
2M0 BJU G. Craig 1 Butt Avenue, Helensburgh G84 9DA
2E0 BJV T. Gabriel 57 West Down Road, Delabole PL33 9DT
2E0 BJW B. White 9 Springfield Close, Wirral CH49 7NJ
2E0 BJX G. Dyson 111 Chester Road, Ellesmere Port CH65 6SB
2E0 BJY S. Billingham Kewell House, Wombourne Road, Swindon, Dudley DY3 4NF
2W0 BKA M. Mainwaring 36 Oak Street, Gilfach Goch, Porth CF39 8UG
2E0 BKB B. Hall 6 Marshall Close, Parkgate, Rotherham S62 6DB
2E0 BKD G. Flack 20 The Pastures, Hardwick, Cambridge CB23 7XA
2E0 BKE A. Anderson 89A Malmesbury Park Road, Bournemouth BH8 8PS
2I0 BKI E. Kyle 2 Wattstown Crescent, Coleraine BT52 1SP
2E0 BKJ G. Douch 63 Greenaways Ebley, Stroud GL5 4UN
2M0 BKL S. Fradley 30 Polmont Park, Polmont, Falkirk FK2 0XT
2W0 BKM E. Thomas 29 Maes Y Wern, Carway, Kidwelly SA17 4HF
2W0 BKN M. Stokes 23 Goetre Fawr Road Killay, Swansea SA2 7QS
2E0 BKO A. Oxlade 27 Spenfield Court, Northampton NN3 8LZ
2E0 BKP S. Martin The Shieling, Bolton Low Houses, Wigton CA7 8PF
2E0 BKQ D. Griffin 101 Kingsway, Duxford CB22 4QN
2E0 BKT G. Milsom 31 Chichester Close, Bowerdean Road, High Wycombe HP13 6AU
2E0 BKU T. Coles 88C Dursley Road, Trowbridge Ba14 0Ns, Trowbridge BA14 0NS
2E0 BKV S. Bird 9 Almery Drive, Carlisle CA2 4EX
2E0 BKY F. Goodall 20 Halliday Court, Garforth, Leeds LS261ET
2E0 BKZ W. Owen 8 Sandhurst Avenue, Lytham St. Annes FY8 2DA
2W0 BLA J. Jones Bronydd, Blaenffos, Boncath SA37 0HZ
2E0 BLC J. Farina 9 Mallards Close, Alveley, Bridgnorth WV15 6JL
2E0 BLD V. Parton 51 Marston Grove, Stoke-on-Trent ST1 6EF
2E0 BLF A. Spaxman 70 Park View, Shafton, Barnsley S72 8PY
2W0 BLG R. Williams Plaen Cottage, Bodfari, Denbigh LL16 4BS
2E0 BLI Dr J. Skittrall 14 Tamarin Gardens, Cambridge CB1 9GH
2E0 BLJ A. Marks Grosvenor Hotel, 51 Grosvenor Road, Scarborough YO112LZ
2E0 BLK S. Heaton 13A Roche Avenue Bilton, Harrogate HG1 4ES
2E0 BLL M. Green 103 Preston Street, Kirkham, Preston PR4 2XA
2E0 BLN D. Blake 9 Malling Avenue, Eastfield, Scarborough YO11 3FA
2E0 BLQ J. Brewster 44 Beaulieu Close, Hounslow TW4 5EW
2E0 BLR R. Payne 74 Churchill Avenue, Newmarket CB8 0BY
2E0 BLS L. Turner 16 Woodland Place, Scarborough YO12 6EP
2E0 BLT J. Owen 90 Granville Drive, Kingswinford DY6 8LW
2E0 BLW R. Silcox 103 Oakdale Road, Downend, Bristol BS16 6EG
2E0 BLX G. Thorpe 81 Knoll Drive, Coventry CV3 5PJ
2E0 BLY A. Bailey 58 Billy Buns Lane, Wombourne, Wolverhampton WV5 9BP
2E0 BLZ D. Forster 23 Field Street, Padiham, Burnley BB12 7AU
2E0 BMB A. Bolla 11 Shelley Crescent, Blyth NE24 5RH
2E0 BMD J. Turner 17 Beechwood Road, Dronfield S18 1PW
2E0 BME C. King 21 Lowdham, Wilnecote, Tamworth B77 4LX
2M0 BMF F. Thomson 11 Carmichael Place, Irvine KA12 0XH
2E0 BMG W. Mcgill 49 Anthony Close, Colchester CO4 0LD
2E0 BMH B. Hawes 3 Orchard Close, Cassington, Witney OX29 4BU
2E0 BMJ R. Burrows 62 Fletcher Road, Burbage, Hinckley LE10 2PS
2M0 BMK J. Cosgrove The Cottages, Kirkinch, Blairgowrie PH12 8SL
2W0 BMM N. Hughes Mountain Farm Cottage, Clynderwen, Llandissilio SA66 7PX
2M0 BMN C. Lewis 9 Cessnock Road, Troon KA10 6NJ
2W0 BMO R. Rimmer Dwyfor, Heol Las, Llantrisant, Pontyclun CF72 8EG
2E0 BMP J. Redfern 19 Stanton Green, Shrewsbury SY1 4PL
2W0 BMR S. Williams Flat 28, Llys Celyn Cedar Crescent, Tonteg, Pontypridd CF38 1LF
2E0 BMT Rvd. B. Topham 2 Highgrove Gardens, Stamford PE9 2GR
2E0 BMU P. Wright 34 Coles Lane, West Bromwich B71 2QJ
2E0 BMW A. Rowe Southern Point, Grange View, Houghton le Spring DH4 4HU
2E0 BMY A. Farrar 8 Wensley Street, Thurnscoe, Rotherham S63 0PX
2E0 BNA N. Anderson 3 Phelipps Road, Corfe Mullen, Wimborne BH21 3NN
2E0 BNC C. Murray 7 Pilmoor Drive, Richmond DL10 5BJ
2E0 BND L. Woollard 46 Woodhill Lane, Morecambe LA4 4NN
2E0 BNE S. Sanderson 65 Holm Flatt Street, Parkgate, Rotherham S62 6HJ
2E0 BNF A. Webb 47 Granville Street, Gloucester GL1 5HL
2E0 BNH J. Hawkes 53 Mill Hill, Derby DE24 5AF
2E0 BNI G. Marshall 12 Arthur Avenue, Caister-On-Sea, Great Yarmouth NR30 5PQ
2E0 BNJ S. Scott Watercrook Bungalow, Natland, Kendal LA9 7QB
2E0 BNK J. Stoppard 15 South Lodge Court, Old Road, Chesterfield S40 3QG
2E0 BNT B. Trayhurn 15 Wight Drive, Caister-On-Sea, Great Yarmouth NR30 5UN
2E0 BNV K. Peel 123 Cunningham Road, Tamerton Foliot, Plymouth PL5 4PU
2E0 BNW P. Ashton 14 Poppy Close, Boston PE21 7TJ
2E0 BNZ W. Whitcher 17 Watermead, Stratton St. Margaret, Swindon SN3 4WE
2E0 BOB R. Hastings 2 Boleyn Way, Boreham, Chelmsford CM3 3JJ
2W0 BOC O. Williams 39 Camden Road, Maes-Y-Coed, Brecon LD3 7RT
2E0 BOD J. Tusler Kilima, Batts Corner, Farnham GU10 4EX
2E0 BOF P. Callaghan 41 Higher Ash Road, Talke, Stoke-on-Trent ST7 1JN
2E0 BOI R. Massimino 115 Trelowarren Street, Camborne TR14 8AW
2E0 BOJ M. Anthony Magpie Bungalow, Goongumpas, St. Day, , Redruth TR16 5JL
2W0 BOK A. Budding 54 Wern Isaf, Dowlais, Merthyr Tydfil CF48 3NY
2E0 BON J. Ball Manor House, Tolgus Hill, Redruth TR15 1AX
2E0 BOR R. Gee Flat 1D, Quarmby Road, Huddersfield HD3 4HQ
2M0 BOS S. Mccurdy 5 Kestrel Place, Greenock PA16 7BL
2E0 BOT A. Canning The Mount, Birmingham Road, Alcester B49 5EG
2E0 BOV T. Liu 52 Tenison Road, Cambridge CB1 2DW
2M0 BOY S. Kirkpatrick Sandyhill House Dunbae Farm, Stranraer DG9 8LX
2E0 BOZ D. Tidswell 1 Cherrytree Grove, Spalding PE11 2NA
2E0 BPE J. Freeman 38 City Road, Cambridge CB1 1DP
2E0 BPF F. Banner 14 Dale Road, Barnard Castle DL12 8LQ
2E0 BPG B. Dixon 21 Pankhurst Road, Hoo, Rochester ME3 9DF
2E0 BPI A. Schuler 6 Tatham Court, Taunton TA1 5QZ
2W0 BPJ R. Jones 5 Heol Llwyn Gollen, Merthyr Tydfil CF48 1LR
2E0 BPL D. Golding Windrush Cottage 84-85 Bradenstoke, Chippenham SN15 4EL
2M0 BPM M. Dickeson 44 Mossmill Park, Mosstodloch, Fochabers IV32 7JY
2E0 BPN J. Barker 26 Ardley Road, Fewcott, Bicester OX27 7PA
2I0 BPO J. Rice 42 The Crescent, Ballymoney BT53 6ES

2E0 BPP C. Bell 20 Kingfisher Close, Blackburn BB1 8NS
2E0 BPS J. Murray 2 The Cuttings Hampstead Norreys, Thatcham RG18 0RR
2E0 BPT L. Clark 16 Kibblewhite Crescent, Twyford, Reading RG10 9AX
2E0 BPU N. Phillips First Floor Flat, 116 Lodge Road, Croydon CR0 2PF
2M0 BPV N. Davidson 25 Hopetoun Court, Bucksburn, Aberdeen AB21 9QS
2E0 BPW J. Hall 1 Nash Close, Earley, Reading RG6 5SL
2E0 BPX D. Adshead 16 Moat Way, Swavesey, Cambridge CB24 4TR
2E0 BPY M. Rogers 41 Barton Hill Drive, Minster On Sea, Sheerness ME12 3NF
2E0 BQB T. Horsoo 5 Kelmarsh Court, Great Holm, Milton Keynes MK8 9EN
2E0 BQC K. Bindley 56 Iona Close, Beaumont Leys, Leicester LE4 0QY
2E0 BQE J. Callis 51 Pipistrelle Way, Oadby, Leicester LE2 4QA
2E0 BQF D. Gunn 40 The Pastures, Oadby, Leicester LE2 4QD
2E0 BQG S. Turner 12 Park Street, Morecambe LA4 6BN
2E0 BQH M. Bailey 17 Sparrowhawk Way, Hartford, Huntingdon PE29 1XE
2E0 BQJ D. Trudgian 18 Hart Close, Wootton Bassett, Swindon SN4 7FN
2E0 BQK M. Carr 25 Malvern Avenue, Fareham PO14 1QF
2E0 BQL B. Hall 64 Synehurst Crescent, Badsey, Evesham WR11 7XX
2E0 BQM N. Jewitt 10 Gorse Lane, Oadby, Leicester LE2 4RQ
2E0 BQN R. Suffling 9 Tamerton Square, Woking GU22 7SZ
2E0 BQO D. Green 12 Nostell Road, Ashton-In-Makerfield, Wigan WN4 9XD
2E0 BQQ J. Addy 12 Wortley Avenue, Swinton, Mexborough S64 8PT
2E0 BQR S. James 94 North Road, Hull HU19 2AY
2E0 BQU C. Langmaid Flat 4, Woodlawn High Street, Partridge Green West Sussex RH13 8HR
2E0 BQV J. Parrett 6 Shelley Road, East Grinstead RH19 1TA
2E0 BQW A. Chapman 24 Eaton Grange Drive, Long Eaton, Nottingham NG10 3QE
2E0 BQX N. Groat 138 Freedom Road, Sheffield S6 2XE
2E0 BQY A. Clinchant 14 Taylor Close Norton Fitzwarren, Taunton TA2 6TA
2E0 BQZ K. Puttock 12 Beechfields, School Lane, Petworth GU28 9DH
2E0 BRC E. Smith 13 Eagle Avenue, Waterlooville PO8 9UB
2M0 BRD B. Donnelly 19 Douglas Drive, Dunfermline KY12 9YG
2M0 BRE R. Docherty 207 Greengairs Road, Greengairs, Airdrie ML6 7SZ
2E0 BRF C. Dawson 9 Mulberry Close, Poringland, Norwich NR14 7WF
2E0 BRI B. Bott 15 Lansdowne Crescent, Darton, Barnsley S75 5PW
2E0 BRJ P. Shirlaw 32 West Street, Faversham ME13 7JG
2E0 BRK B. Kemp 4 Creek View, Basildon SS16 4RU
2E0 BRL R. Dewis 6 St Nicolas Close, Pevensey BN245LB
2E0 BRP G. Armitage Windmill Cottage, Greens Gardens, Nottingham NG2 4QD
2E0 BRQ K. Shaw 2 Montrose Avenue, Montrose Street, Hull HU8 7RY
2E0 BRS B. Simmonds 55 Pepys Road, St. Neots PE19 2EN
2E0 BRT P. Allen 21 Chase Vale, Burntwood WS7 3GD
2E0 BRU B. Pickering 7 Front Street Grindale, Bridlington YO16 4XU
2E0 BRY M. Edmond 12 Yeoman Close, Worksop S80 2RR
2I0 BSA D. Cooke 7 Killyclooney Road, Dunamanagh, Strabane BT82 0LZ
2E0 BSB T. Wooldridge 12 Redwood Avenue, Leyland PR25 1RN
2M0 BSE R. Ewing 7 Middlemas Drive, Kilmarnock KA1 3DZ
2E0 BSF B. Streeter Fairway, West Chiltington Road, Pulborough RH20 2EE
2E0 BSG B. Holland 11 Silverlands Park, Buxton SK17 6QX
2I0 BSH R. Vage 80 Chinauley Park, Banbridge BT32 4JL
2E0 BSI Dr C. Ferguson Royd Moor, Royd Moor Lane, Badsworth, Pontefract WF9 1AZ
2E0 BSJ M. Croxford Simmons 37 Queens Road, Askern, Doncaster DN6 0LU
2E0 BSK S. Everson 41 Westminster Lane, Newport PO30 5ZF
2E0 BSM T. Hall 18 Common Lane New Haw 18 Comman Lane New Haw, Addlestone KT153LH
2E0 BSN M. Blagg 17 Flint Avenue, Forest Town, Mansfield NG19 0DS
2E0 BSQ D. Gray 68 Endeavour Way, Hythe Marina Village, Southampton SO45 6LA
2E0 BSR R. Gay 47 Egerton Street, Flat C, Chester CH1 3ND
2E0 BSS Lady C. Windsor 44 Paragon Place, Norwich NR2 4BL
2E0 BST S. Clay Akers Lodge, 6 Penn Way, Rickmansworth WD3 5HQ
2E0 BSU J. Grosvenor 10 Neves Close, Lingwood, Norwich NR13 4AW
2E0 BSW C. Godwin 6, Red Earl Lane, Malvern WR142ST
2E0 BSX I. Colvin 80 Silvester Road, Cowplain, Waterlooville PO8 8TS
2E0 BTA C. Wynne 43 Lansdown Road, Broughton, Chester CH4 0NZ
2E0 BTB B. Bowen 2 Veronica Road, Manchester M20 6SU
2E0 BTD J. Wynne 43 Lansdown Road, Broughton, Chester CH4 0NZ
2W0 BTE J. Davies 122 Heol Frank, Penlan, Swansea SA5 7EG
2W0 BTF T. Banks 18 Leicester Road, Newport NP19 7ER
2E0 BTO A. King 6 Dunsfold Close, Crawley RH11 8EY
2W0 BTP B. Page 9 De Braose Close, Cardiff CF5 2DH
2W0 BTQ T. Rogers 59 Park Place, Newport NP11 6BN
2E0 BTR A. Passey 3 The Yard, Bayton, Kidderminster DY14 9LH
2E0 BTS G. Tyler Crofton, Stoney Ley, Worcester WR6 5NG
2I0 BTT R. Laverty 23 Hyacinth Avenue, Ballykelly, Limavady BT49 9HT
2E0 BTU J. Katz 8 Astor Drive, Birmingham B13 9QR
2E0 BTV J. Barbieri 20 Gilbard Court, Chineham, Basingstoke RG24 8RG
2E0 BTX P. Chamberlain 22 Stanedge Grove, Wigan WN3 5PL
2W0 BTZ D. Shipton Hillside, Tintern, Chepstow NP16 6TF
2E0 BUA J. Watson 20 St. Marys Gardens, Hilperton Marsh, Trowbridge BA14 7PG
2E0 BUD A. Hanson Pilgrim Cottage, South Road, Truro TR3 7AD
2E0 BUE S. Mather 35 Neargates Charnock Richard, Chorley PR7 5EY
2E0 BUF N. Ham 4 Heighes Drive, Alton GU34 2FJ
2E0 BUI Enfield CG c/o S. Chuter 13 Durham Lane, Eaglescliffe, Stockton-on-Tees TS16 0NB
2E0 BUJ C. Cherry 12 Scarisbrick New Road, Southport PR8 6PY
2E0 BUK M. Buckland 7 Heath Close, Newport PO30 1HN
2E0 BUN L. Knight Flat 4, Barrington House, Portsmouth PO2 7DD
2E0 BUO B. Jenson 10 Tintern Close, Paulsgrove, Portsmouth PO6 4LS
2E0 BUP G. Knight 131 Washington Road, Portsmouth PO2 7DF
2W0 BUQ S. Gibbon 39 Pen-Y-Groes, Penyrheol, Caerphilly CF83 2JL
2M0 BUT F. Pudsey 21/2 Bathfield, Edinburgh EH6 4DU
2E0 BUV A. Markettos 98 Foster Road, Trumpington, Cambridge CB2 9JR
2M0 BUX A. Young 4/4 Prestonfield Terrace, Edinburgh EH165EE
2M0 BUY Dr J. Henderson 7 Rowanhill Close, Port Seton, Prestonpans EH32 0SY
2E0 BUZ D. Burrows 19 Fleming Avenue, Bottesford, Nottingham NG13 0ED
2E0 BVB J. Roberts Worlds Wonder, Warehorne, Ashford TN26 2LU
2E0 BVD A. Burns 76, 76, Morprth NE65 0TF
2E0 BVE J. Mcbride 65 Dunston Road, Hull HU5 5ES
2E0 BVG J. Thomas 42 Brant Road, Lincoln LN5 8SJ
2E0 BVH M. Norris 35 Sudbrooke Road, London SW12 8TQ
2E0 BVJ W. Tunstall 89 Lever Street, Little Lever, Bolton BL3 1BA
2W0 BVK A. Bennett Erwenni, Llanbedrog, Pwllheli LL53 7PA
2M0 BVN I. Hepworth Bronte Cottage, Inverugie, Peterhead AB42 3DN
2E0 BVO A. Little Wisteria Cottage, 7 Shawfield Road, Havant PO9 2SY
2E0 BVQ G. Turner 28 Chapel Close, Needingworth, St. Ives PE27 4SH
2E0 BVR N. Breckons 5 Berrybut Way, Stamford PE9 1DS
2W0 BVS J. Smith 77 Nantgarw Road, Caerphilly CF83 1AL
2E0 BVU T. Loker 24 St. Albans Hill, Hemel Hempstead HP3 9NG
2W0 BVV T. Leaworthy 7 Maesderwen Rise, Stafford Road, Pontypool NP4 5SS
2E0 BVZ J. Harris 2 Elm Cottages, Boreham Lane, Wartling, Hailsham BN27 1RS
2E0 BWA L. Garnett 6 Tremaine Close, Norwich NR6 5EL
2E0 BWD E. Mehmet 8 Hailsham Road, Tooting, London SW17 9EN
2E0 BWH B. Harrison 24 Alderton Road, Nottingham NG5 6DX
2E0 BWI A. Tring 12 Ainsdale Close, Orpington BR6 8DJ
2E0 BWJ J. Swift 56 Leymoor Road, Huddersfield HD3 4SW
2E0 BWK K. Bushell 4 Birch Grove, Harrogate HG1 4HR
2E0 BWN C. Mason Apartment 23, Burnside House, Carleton Road, Skipton BD23 2BE
2E0 BWP D. Le Mare The Sycamore, Church Bank, Barnard Castle DL12 0AH
2E0 BWQ B. North 54 Parklands, Mablethorpe LN12 1BY
2E0 BWT A. Cockett 10 San Marcos Drive, Chafford Hundred, Grays RM16 6LT
2E0 BWU P. Moule 30 Hillview Road, Chelmsford CM1 7RX
2E0 BWV D. Jones 77 Brinkburn Grove, Banbury OX16 3WX
2E0 BWX T. Ward 20 Ollerton Road, Edwinstowe, Mansfield NG21 9QG
2E0 BWY A. Ferenc 2A Rosedene Avenue, London SW16 2LT
2E0 BXA J. Oakley 13 Worsley Street, Warrington WA5 0NA
2E0 BXC M. Wilkins 31 Stratton Audley Road, Fringford, Bicester OX27 8ED
2E0 BXD J. Bell 8 Firsleigh Park, Roche, St. Austell PL26 8JN
2E0 BXE R. Steele 175 Vale Road, Seaford BN25 3HH
2E0 BXF P. Freeman 24 Roe Green Close, Hatfield AL10 9PE
2E0 BXG H. Hughes 27 The Holt, Hailsham BN27 3ND
2I0 BXJ J. Steele 46 Circular Road, Newtownards BT23 4BN
2E0 BXK R. Hope 32 Winstanley Place, Rugeley WS15 2QB
2E0 BXM K. Foster 10 Bleaswood Road Oxenholme, Kendal LA9 7EY
2M0 BXN G. Moir 3 Three Wells Steading, Inverbervie DD10 0PH
2E0 BXQ L. Denham 92 Windermere Avenue, Southampton SO16 9GF
2E0 BXS B. Sims 4 New Cottages, Cranwich Road, Thetford IP26 5EQ
2E0 BXV M. Browne 60 Lindsay Avenue, Abington, Northampton NN3 2LP
2E0 BXW S. Johnston 67 Eversfield Road, Horsham RH13 5JS
2E0 BXX J. Searle 2 Tukes Avenue, Gosport PO13 0SE
2M0 BXY W. Taylor Garth Wood, Fishers Brae, Eyemouth TD14 5NJ
2E0 BXZ J. Reynolds 15 Chestnut Mead, Oxford Road, Redhill RH1 1DR
2E0 BYA D. Passey 5 The Croftings, Felton Close, Ludlow SY8 1DS
2E0 BYC M. Matthews 28 Kempson Drive, Great Cornard, Sudbury CO10 0YE
2E0 BYF J. Milne 9 Roman Road, Colchester CO1 1UR
2E0 BYG P. Bailey 103 Jarden, Letchworth Garden City SG6 2NZ
2E0 BYH S. Leadbetter 11 Cogos Park, Mylor Bridge, Falmouth TR11 5SF
2M0 BYI C. Williamson 31 Medrox Gardens, Cumbernauld, Glasgow G67 4AJ
2E0 BYJ E. Isaac 162B Hitchin Road, Stotfold, Hitchin SG5 4JE
2M0 BYK S. Troscheit 20 James Street, St. Andrews KY16 8YA
2I0 BYL B. Crozier 33 Cullentragh Road, Poyntzpass, Newry BT35 6SD
2E0 BYM M. Brough 12 Longfield Court, Barnoldswick BB18 5LP
2E0 BYN I. Patterson 63 Orchard Road, South Ockendon RM15 6HP
2E0 BYO J. Morris 96 Bradford Crescent, Durham DH1 1HW
2E0 BYQ J. Summerhill 43 Rangers Walk, Bristol BS15 3PW
2M0 BYT J. Cairney 5 James Street, Bannockburn, Stirling FK7 0NQ
2E0 BYW M. Bradley 55 Queensway, Penwortham, Preston PR1 0DT
2E0 BZA K. Moulder 51A Aston Cantlow Road, Wilmcote, Stratford-upon-Avon CV37 9XN
2M0 BZB S. Campbell 11 Convener Street , New Elgin, Elgin IV30 6BP
2E0 BZC P. Davies 68 Sidmouth Avenue, Stafford ST17 0HF
2E0 BZE T. Munro 71 Zig Zag Road, Liverpool L12 9EQ
2E0 BZG C. Warwick 104 Church Road, Formby, Liverpool L37 3NH
2E0 BZH R. Hopkins 17 Springside Court, Josephs Road, Guildford GU1 1BT
2E0 BZI J. Donnelly 23 Pitts Croft, Neston, Corsham SN13 9ST
2E0 BZJ B. Cooke 2 Harvey Place, Andover SP10 2BU
2M0 BZL C. Watkinson Hare House, Perth Road, Birnam, Dunkeld PH8 0AA
2E0 BZM M. Tew Willowell, Spring Valley Lane, Colchester CO7 7SD
2E0 BZO D. Chatzikos 53 Benbow Court, Shenley Church End, Milton Keynes MK5 6JE
2E0 BZS P. Burt 56 Winslade Road, Sidmouth EX10 9EX
2E0 BZT P. Burgess Tally Ho Cottage, High Street, Swindon SN4 0AE
2E0 BZU P. Fenney 8 Lynton Way, Bristolá BS16 1QP
2E0 BZV I. Govan 9 Willowbank, Sandwich CT13 9QA
2E0 BZX R. Crockford 17 Tadcroft Walk, Calcot, Reading RG31 7JR
2M0 BZZ W. Anderson 4 Brackendene, , Houston PA6 7DE
2E0 CAA C. Orange 79 Heath Avenue, Werrington, Stoke-on-Trent ST9 0HU
2E0 CAD C. Dodgson 16 Jefferson St., Goole DN14 6SH
2E0 CAE C. Frizzell 85 Gibbon Road, Newhaven BN9 9ER
2E0 CAJ C. Travis 4 Kingsdale, Worksop S81 0XJ
2E0 CAK A. Martyn 54 North Side, Hepthorne Lane North Wingfield, Chesterfield S42 5HY
2E0 CAL C. Bowley 2 Cottage, Middle Battenhall Farm, Worcester WR5 2JL
2E0 CAO M. Castanheira 7C Palace Road, London N8 8QH
2E0 CAP J. Williams 41 Overton Lane, Hammerwich, Burntwood WS7 0LQ
2E0 CAQ D. Arnold The Chase, Rectory Road, Penzance TR19 6BB
2E0 CAR C. Goulding 9 Lune Drive, Leyland PR25 5SX
2E0 CAS C. Spence 30 Chestnut Drive, Shirebrook, Mansfield NG20 8NH
2E0 CAT E. Taylor 39 Gill St., Newcastle upon Tyne NE4 8BH
2E0 CAU D. Holmes 5 The Cottages, Low Road, North Tuddenham, Dereham NR20 3DG
2E0 CAV C. Vernon 29 Alice St., Deane, Bolton BL3 5PJ
2E0 CAW S. Court 16 Worcester Road, Woodthorpe, Nottingham NG5 4HY
2E0 CBA P. Humphreys 30 The Chestnuts, Hinstock, Market Drayton TF9 2SX
2E0 CBB B. Clements 23 Croft Terrace, Egremont CA22 2AT
2M0 CBC C. Bryson 29 Roull Road, Edinburgh EH12 7JW
2M0 CBE C. Ellison 1 Newton Road, St. Fergus, Peterhead AB42 3DD
2E0 CBF A. Cunningham 18 Bradway, Whitwell, Hitchin SG4 8BE
2E0 CBG C. Gough 104 Canley Road, Coventry CV5 6AR

2E0 CBH J. Wooldridge 7 Heather Gardens, Belton, Great Yarmouth NR31 9PP
2W0 CBJ V. Holden 2 Anglesey Close, Tonteg, Pontypridd CF38 1LY
2E0 CBL I. Botham 12 Lairgill, Bentham, Lancaster LA2 7JZ
2E0 CBO P. Rogers 16 Begonia Close, Basingstoke RG22 5RA
2E0 CBQ R. Smith 5 Elizabeth Place, 13 Heath Road, Haywards Heath RH16 3AX
2E0 CBT M. Hersom Room 303, 95-98 Talbot Street, , Dublin DO1 WR94
2E0 CBU J. Horry 5 Donington Road, Bicker, Boston PE20 3EF
2I0 CBV A. Shilliday 26 Iskymeadow Road, Armagh BT60 3JS
2E0 CBX I. Connors 3 Wheatfield Way, Chelmsford CM1 2QZ
2W0 CBZ C. Nicholls 26 Maes Geraint, Pentraeth LL75 8UR
2E0 CCA C. Chambers 21 Sullington Way, Shoreham-by-Sea BN43 6PJ
2E0 CCB S. Connolly 82 Cheswood Drive, Minworth, Sutton Coldfield B76 1YE
2E0 CCC J. Restall 1 Johndory, Dosthill, Tamworth B77 1NY
2E0 CCD C. Raynerd 43 Astley Hall Drive, Ramsbottom BL0 9DF
2E0 CCF G. Wright 2 Hillcrest Drive, Castleford WF103QN
2W0 CCG G. Brierley 15 First Avenue, Flint CH6 5LP
2E0 CCJ C. Jones 6 Cleeve Park Cottages, Icknield Road, Reading RG8 0DJ
2W0 CCK M. Buxton 25 Pen Y Bryn, Sychdyn, Mold CH7 6EE
2E0 CCL S. Gillard 1 Chevening Close, Stoke Gifford, Bristol BS34 8NJ
2E0 CCQ T. Panesar 135A Hargate Way, Hampton Hargate, Peterborough PE7 8FL
2E0 CCR P. Alborough 7 Edrich Square, Andover SP10 5BS
2E0 CCW M. Shoyer 22 St. Andrews Road, Whitehill, Bordon GU35 9QN
2E0 CCY F. Gear 251 Abington Avenue, Northampton NN3 2BU
2E0 CDF B. Myler-Cook 11A York Street, Boston PE21 6JN
2E0 CDI S. Lee-Ray The Paddock, Sutton Road, Alford LN13 9RL
2E0 CDK D. Barnett 20 Middlemead Road, Great Bookham, Bookham, Leatherhead KT23 3DA
2E0 CDL C. Lote 8 Warren Place, Walsall WS8 6BY
2E0 CDM C. Mcnulty 91 Barn Hey Crescent, Meols CH47 9RW
2M0 CDO D. Fleming 4 Oxenrig Farm Cottage, Coldstream TD12 4EY
2E0 CDQ W. Haddock 5 Bradley Close, Middlewich CW10 0PF
2E0 CDR C. Russell 255 Leeds Road, Shipley BD18 1EH
2E0 CDS C. Small Riddings Barn, Hope Bagot, Ludlow SY8 3AE
2E0 CDT C. Taylor 1 Jasmine Gardens, Warrington WA5 1GU
2E0 CDU M. Hall 29 The Spinney, Wokingham RG40 4UN
2E0 CDV A. Drew 51 Hobart Road, Cambridge CB1 3PT
2E0 CDW C. Wade 31 Melton Green, Wath-Upon-Dearne, Rotherham S63 6AA
2W0 CDZ P. Gough 92 Pendwyallt Road, Whitchurch, Cardiff CF14 7EH
2E0 CEA M. Anderson 27 Laing Road, Colchester CO4 3UT
2E0 CEB B. Smithers 4 Bidmead Court Kent Way, Surbiton KT6 7SX
2W0 CED C. Davies 3 Penhydd Houses, Oakwood Avenue, Pontrhydyfen, Port Talbot SA12 9SE
2M0 CEE R. Renshaw Smithy House, Scotscalder, Halkirk KW12 6XJ
2W0 CEF F. Price 2 Bryniau Duon Estate, Llandegfan, Menai Bridge LL59 5PP
2E0 CEG I. Greathead 3 Helmington Terrace, Hunwick, Crook DL15 0LQ
2E0 CEH P. Blyth 20 Common Lane, Beccles NR34 9RH
2I0 CEI B. Craney 8A Drumhoy Drive, Carrickfergus BT38 8NN
2E0 CEJ R. Parrish 5 Kestrel Lane, Cheadle, Stoke-on-Trent ST10 1RU
2E0 CEK D. Mcauslan Casa Arco Iris, Via Variante Nascente, 8005-491 SANTA BARBARA DE NEX Portugal
2E0 CEM M. Miller Barn Cottage, Wingfield Hall, Manor Road, Alfreton DE55 7NH
2E0 CEN D. Mctaggart 59 Gainsborough Road, Richmond TW9 2DZ
2W0 CEO C. Gozzard Craig Dulas, Rhydyfoel Road, Llanddulas, Abergele LL22 8EG
2E0 CEP Dr J. Pelham 5 The Crescent, Shortstown MK42 0UJ
2E0 CER M. Everitt 10 Morris Close, Hatherleigh, Okehampton EX20 3NX
2E0 CEU Dr N. Hoare 5 Kelsey Head, Port Solent, Portsmouth PO6 4TA
2M0 CEX P. Rice 255 Eskhill, Penicuik EH26 8DF
2E0 CEY T. Edwards 17 The Green, Woodbastwick, Norwich NR13 6HH
2M0 CFA R. Mannifield 2 Plewlands Avenue, Edinburgh EH10 5JY
2M0 CFB I. Watson 10 Christie Place, Elgin IV30 4HX
2E0 CFD A. Mcneil 2 Palmerston Crescent, Liverpool L19 1RB
2E0 CFE T. Carpenter 11 Castle Road, Southwick, Fareham PO17 6EY
2E0 CFG Dr F. Fang 21 Loris Court, Cambridge CB1 9GF
2E0 CFH A. Andrew 80 Hamble Drive, Abingdon OX14 3TE
2E0 CFI M. Summers 21 Quantock Avenue, Caversham, Reading RG4 6PY
2E0 CFK P. Rimmington 28 Skipton Road, Swallownest, Sheffield S26 4NQ
2E0 CFL A. Norville 137 Foster Road, Trumpington, Cambridge CB2 9JW
2E0 CFM E. Vaughan 10 Evans Close, Manchester M20 2SQ
2E0 CFN A. Cornelius 16 Crown House, North Street, Nailsea, Bristol BS48 4SX
2E0 CFP J. Hitchens 57 Batchelor Way, Downton, Salisbury SP5 3FN
2E0 CFQ J. Hunt 14 Nevill Close, Hanslope, Milton Keynes MK19 7NY
2E0 CFT A. Wells 186 Manford Way, Chigwell IG7 4DG
2E0 CFV E. Beever 160 Granby Road, Buxton SK17 7TA
2I0 CFW D. Mcglone 10 O'Neill Terrace, Dromore, Omagh BT78 3AW
2E0 CFY A. Davis 101 Kenn Road, Clevedon BS21 6JE
2D0 CFZ C. Ingles 1 Hillberry View, Onchan IM3 3GB Isle of Man
2E0 CGB G. Grimshaw 1 Hardy Close, Pinner HA5 1NL
2M0 CGE A. Macdonald 1 Edinmore Cottage, Rothesay, Isle of Bute PA20 0QT
2E0 CGG S. Brown 4 Dorado Gardens, Orpington BR6 7TD
2E0 CGH W. Durrant 19 Rydal Rd, Gosport PO12 4ES
2E0 CGI J. Garrard 40 Wright Crescent, Bridlington YO164RG
2E0 CGJ D. Ramsell 36 West Street, Burton-on-Trent DE15 0BW
2E0 CGK N. Allen 59 Sherborne Road, Chichester PO19 3AN
2E0 CGL C. Lewis 3 Sovereign Way, Calcot, Reading RG31 4US
2W0 CGM N. Wells 52 Lowerdale Drive, Llantrisant, Pontyclun CF72 8DY
2E0 CGP I. Duffie Trebeighan Farm, Saltash PL12 5AE
2E0 CGR M. Cashman Flat 3, Linden Court Linden Road, Romsey SO51 8BR
2E0 CGS J. Goodyear 30 Ashburton Road, Alresford SO24 9HH
2E0 CGT J. Russell 81 Chapman Street, Loughborough LE11 1DD
2E0 CGU I. Albrighton 12 Clewley Road, Branston, Burton-on-Trent DE14 3JE
2E0 CGV D. Ingrey 1 Ponders Road, Fordham, Colchester CO6 3LX
2E0 CGW N. Bown 12 Sellwood Road, Abingdon OX14 1PE
2E0 CGX J. Shufflebotham 316 Stockport Road, Hyde SK14 5RU
2I0 CGZ D. Mudd 14 Bloomfield Road, Belfast BT5 5LT
2E0 CHA C. Nakajima 22 Royds Crescent, Rhodesia, Worksop S80 3HF
2W0 CHH C. Hill 9 Oliver Road, Newport NP19 0HU
2E0 CHK A. Wild 181 The Strand, Goring-By-Sea, Worthing BN12 6DY
2E0 CHL T. Chapman 1 East Dean Road, Lockerley, Romsey SO51 0JL
2E0 CHN P. Gale 37 Hazlebury Road, Poole BH17 7AX
2E0 CHQ S. Drake-Brockman 13 St. Johns Place, Bury St. Edmunds IP33 1SW
2E0 CHT Dr M. Cook 14 Speyside Close, Carterton OX18 1TT
2E0 CHU G. Hatt 4H Colman House, Earlham Road, Norwich NR4 7TJ
2W0 CHV S. Taylor 43 Toronnen, Bangor LL57 4TG
2E0 CHW C. West 1 Willetts Mews, Hoddesdon EN11 9DX
2E0 CHY G. Lyon 1 Eckersley Street, Wigan WN1 3PP
2E0 CHZ C. Jewell 3 Marsh Gate, Clee St. Margaret, Craven Arms SY7 9DU
2E0 CIA D. Campanario 3 Foxearth, Leek Road, Werrington, Stoke on Trent ST9 0DG
2E0 CID K. Percy 55 Buxton Avenue, Heanor DE75 7UN
2E0 CIG P. Holland 30 Knighton Park Road, London SE26 5RJ
2E0 CIH M. Verrechia 20 The Wyvern Grafham, Huntingdon PE28 0GG
2E0 CII T. Pearsall 16 Langdale Road, Leyland PR25 3AR
2E0 CIJ D. Simpson 50 Castle Hill, Berkhamsted HP4 1HF
2E0 CIK C. Storr 40 Weelsby Way, Hessle HU13 0JW
2E0 CIM S. Preece 14 Bettespol Meadows, Redbourn, St. Albans AL3 7EW
2E0 CIN A. Mawson 14 Pontop View, Delves Lane, Consett DH8 7JB
2E0 CIQ J. Chapman South View, Mill End Rushden, Buntingford SG9 0SU
2E0 CIR I. Sapstead 7 Shrubbery Grove, Royston SG8 9LJ
2E0 CIS D. Woodbine 29 Compass Tower, Munnings Road, Norwich NR7 9TW
2E0 CIT J. O'Brian 83 Bramdean Crescent, London SE12 0UJ
2W0 CIV M. Ireland Pen Y Gadlas, Ffordd Bryniau, Prestatyn LL19 8RD
2E0 CIX J. Rowsell 55 Scarlet Oaks, Camberley GU15 1RD
2E0 CJA L. Mcgaughey 1 Marsh Cottage, Keyingham Marsh, Keyingham, Hull HU12 9JZ
2E0 CJB C. Beresford 13 Chaseside Avenue, Twyford, Reading RG10 9BT
2E0 CJD C. Eyre 23 Nelson Street, Congleton CW12 4BS
2E0 CJF K. Lambert 38 Whittleford Road, Nuneaton CV10 9HU
2W0 CJG J. Loughlin 453 Heol-Y-Waun, Penrhys, Ferndale CF43 3NW
2W0 CJJ R. Squires Hillcrest, Pontfadog, Llangollen LL20 7AS
2E0 CJK P. Mullen 14 Anderson Road, Hemswell Cliff, Gainsborough DN21 5XP
2E0 CJM K. Nicholson 11 Lancaster Way, Skellingthorpe, Lincoln LN6 5UF
2E0 CJO M. Reynolds 24 Burton Close, Corringham, Stanford-le-Hope SS17 7SB
2E0 CJP C. Price 10 St. James Park, Lower Milkwall, Coleford GL16 7LG
2E0 CJQ T. Willis 8 Windsor Court, York YO31 7RY
2E0 CJS P. Hanman 7 Tremenheere Road, Penzance TR18 2AH
2E0 CJV A. Sharam 30 Heywood Avenue, Maidenhead SL6 3JA
2E0 CJW C. Wright Home Farm, Bretford CV230LB
2W0 CJZ J. Smith 38 Marshfield Street, Newport NP19 0GX
2I0 CKB C. Kelly 7A Bancran Road, Draperstown, Magherafelt BT45 7DT
2E0 CKC A. Bradshaw 130 Low Lane, Morecambe LA4 6PS
2E0 CKE A. Cooke Cooke Towers, 9 Nyetimber Crescent, Pagham PO213NN
2E0 CKI A. Pickles 87A Laburnum Road, Waterlooville PO7 7EW
2E0 CKJ P. Wright 16 Hainault Avenue, Giffard Park, Milton Keynes MK14 5PA
2E0 CKL M. Atherton 4 Bakers Park Saltney, Chester CH4 8FB
2E0 CKM A. Pawlak 8 Healey Close, Crewe CW1 4RS
2I0 CKN M. Allen 48 Kevlin Gardens, Omagh BT78 1QS
2E0 CKO Dr B. Le Page Apple Tree House, Reading Road North, Fleet GU51 4AG
2E0 CKP C. Prior 38 Windmill Road, Wombwell, Barnsley S73 8PP
2E0 CKQ J. Legrain 17 Route De La Cote, St. Laurent Sur Gorre 87310 France
2E0 CKR B. Cox 7 Wolsey Avenue, London E6 6HG
2E0 CKS A. Hickey 144 Gisburn Road, Barnoldswick BB18 5LQ
2E0 CKT B. Cunningham 33 Barry Street, Burnley BB12 6DT
2W0 CKV E. Price 1 Brynderi, Pontyates, Llanelli SA15 5SU
2E0 CKW R. Pringle 14 Marjorie Street, Cramlington NE23 6XQ
2E0 CKX Dr L. Schuy 32 Sudeley Street, Brighton BN2 1HE
2E0 CLD M. Shopland 128 Whitewood Park, Liverpool L9 7LG
2E0 CLE C. Edgson 59 Gilmour Crescent, Worcester WR3 7PJ
2M0 CLF C. Forsyth 14B Osborne Terrace, Edinburgh EH12 5HG
2E0 CLH S. Saunders 5 Park Court, Woking GU22 7NW
2E0 CLI M. Randle 2 Oak Gardens, White Farm Road, Sutton Coldfield B74 4Lq, Birmingham B74 4LQ
2W0 CLJ C. Jones 33 Graig Ebbw, Rassau, Ebbw Vale NP23 5SF
2E0 CLL C. Lee 46 Little Lane, Huthwaite, Sutton-in-Ashfield NG17 2RA
2M0 CLN C. Cosgrove 62 Cowgate, Tayport DD6 9DT
2E0 CLP R. Smith Five Elms, Lullington Road Edingale, Tamworth B79 9JA
2I0 CLS G. O'Neill 46 Ashgrove Road, Newtownabbey BT36 6LJ
2W0 CLT G. Williams 36 Park Street, Taibach, Port Talbot SA13 1TD
2M0 CLU R. Adamson 6 Camdean Crescent, Rosyth, Dunfermline KY11 2TJ
2E0 CLW D. Clewer 45 Ashfield Road, Andover SP10 3PE
2E0 CLY A. Page 207 Brooklyn Road, Cheltenham GL51 8DZ
2E0 CLZ I. Jones 21 Kennet Green, Worcester WR5 1JQ
2M0 CMA A. Campbell 1B Craig Road, Troon KA10 6DA
2E0 CMC A. Applegate 13 Deacons Close, Kings Stanley, Stonehouse GL10 3GA
2E0 CME R. Boardman 12 St. Margarets Road, Alderton, Tewkesbury GL20 8NN
2E0 CMF S. Mason 3 Grange Cottages, Low Road, Barrowby, Grantham NG32 1DL
2E0 CMH A. Brown 3 Alston Road, New Hartley, Whitley Bay NE25 0ST
2E0 CMK C. Norris 115 Sutton Road, Walpole Cross Keys PE344HE
2E0 CMN N. Curtis 102 Henley Meadows, Tenterden TN30 6EQ
2E0 CMO C. Overton 99 Hope Avenue, Goldthorpe, Rotherham S63 9DZ
2E0 CMP C. Pegrum 3 Bretland Road, Tunbridge Wells TN4 8PS
2E0 CMQ P. Chester 44 Kirtling Place, Haverhill CB9 0AU
2E0 CMT S. Metcalfe 55 Coventry Close, Corfe Mullen, Wimborne BH21 3UW
2E0 CMY F. Southgate 52 Jeffrey Lane, Belton, Doncaster DN9 1LT
2E0 CMZ C. Rawlin 5 Japonica Hill, Immingham DN40 1LT
2E0 CNA P. Davies 2 Lynfords Drive, Runwell, Wickford SS11 7PP
2E0 CNB Dr D. James Bramble Cottage, Tray Lane, Atherington, Umberleigh EX37 9HY
2E0 CNC T. Ward 1 Darrismere Villas, Edinburgh Street, Hull HU3 5AS
2E0 CND S. Bradley 6 Downing Street, South Normanton, Alfreton DE55 2HE
2E0 CNE M. Shepherd North Waver Cottage, Bells Road Belchamp Walter, Sudbury CO10 7AR
2E0 CNG C. Smith 44 Brooksfield, Bildeston, Ipswich IP7 7EJ
2E0 CNH N. Sinclair 8 Kareen Avenue, Scarborough YO12 4LD
2E0 CNL C. Lockyear 26 Wentworth Gardens, Exeter EX4 1NH
2E0 CNM C. Hunt 105 Worlds End Lane, Weston Turville, Aylesbury HP22 5RX
2E0 CNN D. Turnbull 63 Brecklands, Mundford, Thetford IP26 5EG
2E0 CNP F. Hatfull 16B Church Street, Easton On The Hill, Stamford PE9 3LL
2E0 CNQ C. Greenwood 21 Valley Drive, Thornhill Dewsbury WF120HE

2M0 CNS M. Hatfull Flat 2 1 Northumberland Place Lane, Edinburgh EH3 6LD
2E0 CNU T. Cocks 9 Mountfield Way, Westgate-on-Sea CT8 8HR
2E0 CNV T. Pelham 172A Gloucester Road, Patchway, Bristol BS34 5BG
2E0 CNW A. Palmer 18 Windsor Avenue, Great Yarmouth NR30 4EA
2E0 CNX I. Buckton 67 Tennyson Avenue, Middlesbrough TS6 7ND
2W0 CNY G. Fryer 9 Church Road, Chepstow NP16 5HP
2M0 CNZ J. Rayne 8 Bankton Grove, Livingston EH54 9DW
2E0 COA L. Layland 3 Thirlmere Road, Golborne, Warrington WA3 3HH
2E0 COB J. Cobbold 2 The Green, Blencogo, Wigton CA7 0DF
2E0 COD G. Glasgow 4 Beech Avenue, Culcheth, Warrington WA3 4JF
2E0 COF Dr J. Murray 69 Helsby Road, Lincoln LN5 8SN
2E0 COI P. Bailey 81A Kings Parade Holland-On-Sea, Clacton-on-Sea CO15 5JF
2E0 COJ K. Cartwright 53 Sedgley Road, Dudley DY1 4NE
2E0 COL Dr S. Britt-Hazard 9 Serbin Close, London E10 6JL
2E0 COM R. Hammond Bye Road Cottage Pot Kiln Chase Co93Bh, Gestingthorpe CO93BH
2E0 CON J. Middleton 16 Kyme Road, Boston PE21 8NQ
2J0 COQ L. Langlois Brookfield, La Rue D'Empierre, Trinity JE3 5QF Jersey
2E0 COX A. Cox 11 Windmill Drive, Audlem, Crewe CW3 0BE
2E0 COY A. Pointon 9 Parkwood Avenue, Stoke-on-Trent ST4 8PD
2E0 CPB C. Bond Tryfan, Vicarage Lane, Neston CH64 5TJ
2W0 CPD R. Briant Talarvor, Llanon SY23 5HG
2E0 CPE M. Phillips 59 Bradeley Road Haslington Crewe, Crewe CW1 5PX
2E0 CPF G. Evans 13 Lydgate Road, Sale M33 3LW
2E0 CPG C. Leviston 1 Great Carrs Close, Askam-in-Furness LA16 7FL
2E0 CPK K. Jackson 4 Milfoil Close, Marton-In-Cleveland, Middlesbrough TS7 8SE
2E0 CPL C. Luckett 257 Folkestone Road, Dover CT17 9LL
2M0 CPN A. Woodford Nordkette, Evnabrek, Levenwick, Shetland ZE2 9GY
2E0 CPP A. Collins 4 The Avenue, London W4 1HT
2E0 CPR R. Packman 62 Amsterdam Road, London E14 3JB
2E0 CPS W. Coburn 42 Hinton Wood Avenue, Christchurch BH23 5AH
2M0 CPV J. Hutchinson Hawthorn Cottage, Muirhall, West Calder EH55 8NL
2E0 CPX P. Strickland 100 Spitfire Road, Castle Donington, Derby DE74 2AU
2E0 CQB I. Mckean 14 Maltings Close, Cranfield, Bedford MK43 0BY
2E0 CQC D. Clavey 32 Apollo Close, Dunstable LU5 4AQ
2E0 CQD A. Kennington 63 Emmanuel Court, Scunthorpe DN16 2LR
2E0 CQG P. Jones Tallonhouse, Mill Lane, Pulham St. Mary, Diss IP21 4QY
2E0 CQH S. Baynton 50 Briton Way, Wymondham NR18 0TT
2M0 CQI J. Browne 33 Pilgrims Hill, Linlithgow EH49 7LN
2E0 CQJ N. Parry 125 Lawsons Road, Thornton-Cleveleys FY5 4PL
2E0 CQL K. Jeffery 9 Gordon Road, Tunbridge Wells TN4 9BL
2E0 CQN B. Mcglynn 22 Bracken Bank Way, Keighley BD22 7AB
2E0 CQO M. Mutkin 13 The Grove, Radlett WD7 7NF
2E0 CQQ I. Pryke 9 Charles Avenue, Grundisburgh, Woodbridge IP13 6TH
2E0 CQR P. Moye 13 Post Mill Gardens, Grundisburgh, Woodbridge IP13 6UP
2E0 CQS P. Elsey 62B Coleraine Road, London SE3 7PE
2E0 CQT Dr F. Agboma Flat 6, Mayfair Court, Edgware HA8 7UH
2E0 CQX A. Davis Old Malt Kiln House, Barden, Leyburn DL8 5JS
2E0 CQZ J. Street 22 Roman Acre, Wick, Littlehampton BN17 7HN
2W0 CRB K. Dutfield-Cooke Tan Yr Efail, Segurinside, Llandudno Junction LL31 9QE
2E0 CRC A. Bhakoo 4 Bryden Cottages, High Street, Uxbridge UB8 2NY
2E0 CRD C. Densham 27 Lloyds Crescent, Exeter EX1 3JQ
2E0 CRH P. Bull 87 Braemor Road, Calne SN11 9DU
2E0 CRI J. Parris 10, Wharfedale Grange, Ben Rhydding Drive, Ilkley, West Yorkshire LS29 8AR
2E0 CRM C. Murly C/O 1 Mount Pleasant, Middleton, Leeds LS10 3TB
2E0 CRN J. Cranston 7 Cowen Gardens, Gateshead NE9 7TY
2M0 CRQ C. Welsh 28 Peacock Wynd, Motherwell ML1 4ZL
2M0 CRR C. Rodger 23 Harrysmuir Road, Pumpherston, Livingston EH53 0NT
2E0 CRS C. Marsh 9 Westland Road Westwoodside, Doncaster DN9 2PE
2E0 CRU C. Redmond 6 Apsley Road, Southsea PO4 8RH
2E0 CRV S.Glam Rynt Grp c/o S. Hodder 19 Kingsclere, Huntington, York YO32 9SF
2E0 CRX S. Cross 31 Parkfields, Abram, Wigan WN2 5XR
2E0 CSD C. Davies 98 Central Avenue, Keighley BD22 7BD
2E0 CSE T. Chapman 12 Greenways Chilcompton, Radstock BA3 4HT
2E0 CSF C. Finnis 44 Disraeli Road, Christchurch BH23 3NB
2E0 CSG D. Pollard 8 Drammen Avenue, Burnley BB11 5EA
2E0 CSH G. Webster 15 Bridge Road, Chichester PO19 7NW
2E0 CSJ M. Catchpole Woodcote, Five Oaks Road, Horsham RH13 0RQ
2E0 CSK D. Rudling Rose Cottage, Ludwells Lane, Southampton SO32 2NP
2E0 CSL C. Lester 21 Barwell Way, Witham CM8 2TY
2E0 CSN G. Flinn 38 Fir Grove, Whitehill, Bordon GU35 9ED
2E0 CSO C. Opie 354 Beaumont Road, Plymouth PL4 9EN
2E0 CSQ N. Saunders 128 St. Michaels Gardens, South Petherton TA13 5BQ
2E0 CSU G. Wilman 8 Oldfield Drive, Mobberley, Knutsford WA16 7HB
2E0 CSV A. Parker 9 Milecastle Court West Denton, Newcastle-upon-Tyne NE5 2PA
2M0 CSX A. Thomson 5 Gib Grove, Dunfermline KY11 8DH
2E0 CSY S. Bradley 9 Crofton Road Crofton Road, Southsea PO4 8NX
2M0 CSZ S. Leighton 4 Earn Court, Alloa FK10 1PT
2E0 CTA A. Cross 12 Appleby Drive, Langdon Hills, Basildon SS16 6NU
2M0 CTB W. Scott 11 The Marches Armadale, Bathgate EH48 2PG
2E0 CTC C. Castle 32 Inglefield Road, Ilkeston DE7 5AP
2E0 CTD F. Davison 137 Hellis Wartha, Helston TR13 8WF
2E0 CTE C. Etchells 7 Woodlands Drive, Sandford, Wareham BH20 7QA
2E0 CTF P. Evans 67 Grenville Street, Stokport SK3 9ER
2M0 CTI M. Jamieson 5 Straid Bheag, Barremman, Helensburgh G84 0QX
2E0 CTJ C. Johnson 29 Linden Road Creswell, Worksop S80 4JT
2E0 CTK J. Schleswick 9 Wick House Close Saltford, Bristol BS31 3BZ
2E0 CTL K. Baker 27 St. Matthews Close, Cherry Willingham, Lincoln LN3 4LS
2E0 CTM C. Meakin 102 Ryknield Road, Kilburn, Belper DE56 0PF
2M0 CTN R. Tait 9 Fifth Avenueá Claythorná, Glasgow G12 0AS
2E0 CTO J. Killman 19 Moorland Avenue, Walkeringham, Doncaster DN10 4LG
2E0 CTQ T. Crew Lower Creedy, Upton Hellions, Crediton EX17 4AE
2E0 CTR T. Wright 11 Ash Close, Daventry NN11 0XH
2E0 CTT M. Ward 15 Northfield Crescent, Wells-next-the-Sea NR23 1LP
2E0 CTU D. Foley 1 Hill Rise Close, Harrogate HG2 0DQ
2E0 CTV I. Ftaiha 8 Parkside, London NW7 2LH
2E0 CTW J. Hobbs 82 Perry'S Lane, Wroughton, Swindon SN4 9AP
2E0 CTZ M. Carr 51 Langton Road, Holton-Le-Clay, Grimsby DN36 5BH
2E0 CUA J. Valle Espin 203 Broadway, Horsforth, Leeds LS18 4HL
2E0 CUB A. Clements 28 Durham Way Wyton, Huntingdon PE28 2EQ
2E0 CUC A. Hunter 22 Lindsay Court, Whitburn, Sunderland SR6 7LN
2E0 CUE R. Weaver 15 Sharps Field, Headcorn, Ashford TN27 9UF
2E0 CUH J. Farr 17 Bonython Close, Mylor Bridge, Falmouth TR11 5NF
2W0 CUJ J. Barry 7 Rockfield Rise, Undy, Caldicot NP26 3FG
2E0 CUL K. Richards 207 Beaulieu Gardens, Blackwater, Camberley GU17 0LG
2E0 CUO H. Hope 51 Margravine Gardens, London W6 8RN
2E0 CUS A. Cussen The Poplars, Mill End, Southminster CM0 7HJ
2E0 CUU M. Boon 45 Carlton Road, Wickford SS11 7ND
2E0 CUV J. Stewart 19 Salisbury Road, Dover CT16 1EX
2E0 CUW Dr R. Tofts Elmcroft, Redhill Road, Ross-on-Wye HR9 5AU
2E0 CUY S. Bache 62 Whittingham Road, Halesowen B63 3TP
2E0 CVC A. Chaplin 10 St. Leonards Road, Malinslee, Telford TF4 2EB
2E0 CVD C. Herd 10 Amethyst Close, Rainworth, Mansfield NG21 0GH
2W0 CVE C. Wilkinson High Breck Dolwyd, Colwyn Bay LL28 5HS
2E0 CVF A. Davies 4 Capella Path, Hailsham BN27 2JY
2E0 CVG C. Green 4 Lyme Grove, Knott End-On-Sea, Poulton-le-Fylde FY6 0AJ
2E0 CVJ R. Fidler 44 Windermere Avenue, Ramsgate CT11 0PF
2M0 CVK S. Muir Cairnside, Burnhead, Dundee DD3 0QN
2W0 CVL N. Edwards 17 Queensway, Garnlydan, Ebbw Vale NP23 5EE
2W0 CVM S. Broderick 179 Malpas Road, Newport NP20 5PP
2E0 CVN S. Treacher 9 Noelle Drive, Newton Abbot TQ12 1PS
2W0 CVO J. Pauline 54 Laurel Road, Bassaleg, Newport NP10 8NY
2E0 CVP J. Mason 77 Albutts Road, British WS8 7ND
2E0 CVQ S. Knott 15 Meadowlands Drive, Haslemere GU27 2FD
2I0 CVR J. Allen 192 Joanmount Gardens, Belfast BT14 6PA
2E0 CVU P. Swingewood 9 Goodall Grove, Great Barr, Birmingham B43 7PQ
2E0 CVV C. Sayles 11 Malton Close, Monkston, Milton Keynes MK10 9HR
2E0 CVW D. Levy Flat 36, Claydon House, London NW4 1LS
2E0 CVX G. Spicer 44 Cowley Lane, Chapeltown, Sheffield S35 1SY
2E0 CVY R. Smith 15 Hollybush Road, North Walsham NR28 9XT
2E0 CVZ P. Herron 102 Garden City Villas, Ashington NE63 0EU
2E0 CWB M. Fletcher Wray 70, Crook O'Lune Caravan Park, Caton Road, Lancaster LA29HP
2E0 CWC C. Lewis 9 Chatsworth Gardens, Sydenham, Leamington Spa CV31 1WA
2E0 CWI R. Last 30 Abbot Road, Bury St. Edmunds IP33 3UB
2E0 CWJ J. Taylor 90 Village Road, Gosport PO12 2LG
2E0 CWK J. Deery Flat 6, 33-34 Philbeach Gardens, London SW5 9EB
2W0 CWL K. Saltmarsh 15 Colbourne Road, Beddau, Pontypridd CF38 2LN
2E0 CWM G. Moore 40 Main Street, South Rauceby, Sleaford NG34 8QG
2E0 CWO B. Ledson 16 Caton Close, Southport PR9 9XF
2E0 CWQ J. Clarke 160 Hall Lane Estate, Willington, Crook DL15 0PP
2E0 CWR C. Ralphson 20 Monsal Grove, Buxton SK17 7TF
2E0 CWS K. Roberts 1 Tregerles Barns, Newquay TR8 4PW
2M0 CWV N. Ford 11 Kenmore Way, Coatbridge ML5 4FN
2E0 CWW A. Rowland-Stuart 86 Wiltshire House, Lavender Street, Brighton BN2 1LE
2E0 CWZ K. Legg Bennetts, High Street, Thorpe-Le-Soken, Clacton-on-Sea CO16 0EG
2E0 CXA D. Storer 13 The Square, Lower Burraton, Saltash PL12 4SH
2E0 CXD J. Johnson 49 Beach Priory Gardens, Southport PR8 2SA
2E0 CXE P. Naik 82 Misbourne Road, Uxbridge UB10 0HW
2M0 CXG A. Donald 10 Fraser Road, Burghead, Elgin IV30 5YN
2E0 CXH P. Arnold 20 Upper Seagry, Chippenham SN15 5EX
2M0 CXI D. Plummer 39 St. Nicholas Drive, Banchory AB31 5YG
2E0 CXJ C. Norman 15 Maple Close, Sedbergh LA10 5JE
2E0 CXK I. Nicholls 34 West Close, Bath BA2 1PY
2E0 CXL R. Chandler 12 Lyndhurst Drive, Hale, Altrincham WA15 8EA
2E0 CXM B. Cullen 25 Nea Close, Christchurch BH23 4QQ
2E0 CXN J. Bligh-Wall 3 George Street, Elworth, Sandbach CW11 3BL
2E0 CXO K. Ralph 11 Burrard Road, London E16 3QL
2E0 CXP R. Gill 45 Biggin Lane, Ramsey, Huntingdon PE26 1NB
2E0 CXQ T. Pettis 11 Curlew Drive, Hythe, Southampton SO45 3GB
2E0 CXU C. Ashworth 40 Fairholme, Sedbergh LA10 5AY
2W0 CXV Dr D. Morgan Ty Bettws, Kilgwrrwg, Chepstow NP16 6PN
2E0 CXW C. Wilson 87 Levensgarth Avenue, Fulwood, Preston PR2 9FP
2W0 CYE J. Jones Isfryn Bungalow, Glan-Y-Nant, Llanidloes SY18 6PQ
2E0 CYL R. Horvath 3 Back Knowl Road, Mirfield WF14 9SA
2W0 CYM A. Rowlands Lluest Wen, Penygarth, Caernarfon LL55 1EY
2E0 CYO P. Gregory 2 Lennox Close, Hunmanby, Filey YO14 0PY
2E0 CYP S. Challis 73 Rivenhall Way, Hoo, Rochester ME3 9GF
2E0 CYR M. Woodruff 5 Regency Heights, Caversham, Reading RG4 7RH
2E0 CYS P. Martin 108 Headlands Grove, Swindon SN2 7HP
2E0 CYT L. Starrett 50 Danes Road, Bicester OX26 2LP
2E0 CYU J. Powell 23 Park Road, Norton, Malton YO17 9DZ
2I0 CYW C. Wang 32 Broadlands, Carrickfergus BT38 7BL
2W0 CYX S. Rogers 30 Coed Celynen Drive, Abercarn, Newport NP11 5AU
2W0 CYY C. Smith 29 Heol Cwarrel Clark, Caerphilly CF83 2NE
2E0 CYZ M. Tarrant Wayside Cottage, Gabber Lane, Plymouth PL9 0AW
2E0 CZA G. Youll 4 Shaftsbury Court, Barnstaple Road, Scunthorpe DN17 1YB
2E0 CZC S. Ingledew 34 Sunningbrook Road, Tiverton EX16 6EB
2J0 CZD J. Crill Rocqueberg Farm, Rocqueberg Close, Jersey JE2 6LT
2E0 CZG S. Sissens 20 Fallow Drive, Eaton Socon, St. Neots PE19 8QL
2E0 CZI P. Patterson 3 Barnes Close, Southampton SO18 5FE
2E0 CZJ J. Chalmers 19 Brettenham Crescent, Ipswich IP4 2UB
2E0 CZK R. Silcock 18 Saxon Road, Southampton SO15 1JJ
2E0 CZN D. Endean 11 Forrester Drive, Brackley NN13 6NE
2W0 CZP A. Jones 31 Russell Terrace, Carmarthen SA31 1SZ
2E0 CZR D. Rogers 5 Semple Gardens, Chatham ME4 6QD
2E0 CZS P. Handley 97 Applegarth Avenue, Guildford GU2 8LX
2E0 CZT R. Morgan 14 Ash Road, Ashurst, Southampton SO40 7AT
2W0 CZU N. Adam Tan Ffordd, Mynydd Llandygai, Bangor LL57 4LX
2E0 CZW G. Street 105 Jeals Lane, Sandown PO36 9NS
2W0 CZZ P. Hampton Caretakers Flat, T.A. Centre, Newport NP20 5XE
2W0 DAA D. Wilson 94 Lon Hedydd, Llanfairpwllgwyngyll LL61 5JY
2E0 DAB D. Bambrook 18 Vervain Close, Bicester OX26 3SR
2M0 DAC D. Campbell 10 Balgate Mill, Kiltarlity, Beauly IV4 7GL
2E0 DAH D. Horner 21 Ainsworth Road, Little Lever, Bolton BL3 1RG

2E0 DAI D. Holman 20 Green Drive, Wolverhampton WV10 6DW
2E0 DAJ J. Bowley 2 Cottage, Middle Battenhall Farm, Worcester WR5 2JL
2E0 DAL D. Stringer 18 Townfield Close, Ravenglass CA18 1SL
2E0 DAO D. Oddie 5 The Bridleway, Forest Town, Mansfield NG19 0QJ
2W0 DAP T. Woodley 2 Parc Onen, Neath SA10 6AA
2E0 DAQ G. Clarke 11 Blackfordby Lane, Moira, Swadlincote DE12 6EX
2E0 DAR D. Robertson 53 Moor Lane, Weston-Super-Mare BS22 6RA
2E0 DAT D. Toyne 19 Poachers Rest, Welton, Lincoln LN2 3TR
2E0 DAW D. Williams 18 Lower Greave Road, Meltham, Holmfirth HD9 4DY
2E0 DAX D. Sobey Flat 2 73 Park Road, Blackpool FY1 4JQ
2E0 DAY S. Darby 4 Whately Mews, Whately Road, Lymington SO41 0XS
2E0 DBA D. Barnes 11 Yewside, Gosport PO13 0ZD
2E0 DBB D. Walker Flat 5, Seward Court, 380-396 Lymington Road, Christchurch BH23 5HD
2E0 DBH D. Hoare 47 High Street, Chalgrove, Oxford OX44 7SJ
2E0 DBI P. Marchant 16 Melrose Drive, Peterborough PE2 9DN
2I0 DBK D. O'Hale 6 Cochron Road, Newry BT35 6DD
2E0 DBL P. Kimberlee 24 Jacey Road, Shirley, Solihull B90 3LJ
2E0 DBM D. Mellor 28 Winster Road, Staveley, Chesterfield S43 3NJ
2E0 DBN D. Booth I3 Annandale Site Roman Bank, Skegness PE251HS
2E0 DBO J. Ledger 39Eascroft Drive, Sheffield S20 8JG
2E0 DBP D. Payne 31 Cockering Road, Canterbury CT1 3UP
2E0 DBQ D. Moger 23 Elmsleigh Road, Paignton TQ4 5AX
2E0 DBS D. Baines 21 Vera Road, Norwich NR6 5HU
2E0 DBW D. Whyatt 11 The Perrings, Nailsea, Bristol BS48 4YD
2E0 DBY C. Riley 2 Shottisham Hall Cottages, Alderton Road, Shottisham, Woodbridge IP12 3EP
2E0 DBZ D. Teasdale 43 Easington Road, Stockton-on-Tees TS19 8ES
2E0 DCA C. Price 9 Arlington Avenue, Aston, Sheffield S26 2AA
2E0 DCD B. Savage Rufford, Barnes Lane, Milford On Sea, Lymington SO41 0RR
2E0 DCF C. Lonie Jr 41 De La Hay Avenue, Plymouth PL3 4HS
2E0 DCH D. Hannington 3 Canadian Avenue, Gillingham ME7 2DN
2E0 DCJ D. Bishop 62 Brindley Crescent, Hednesford, Cannock WS12 4DS
2W0 DCK R. Powell Gerdd Y Don, Llansantffraed, Llanon SY23 5HS
2E0 DCL D. Clark 34 Magdalene Road, Owlsmoor, Sandhurst GU47 0UT
2E0 DCM D. Martin 14 Freeston Terrace, St. Georges, Telford TF2 9HD
2E0 DCN D. Filby 21 Knightlands Road, Irthlingborough, Wellingborough NN9 5SU
2E0 DCP D. Sproston 22 Oakland Avenue, Haslington, Crewe CW1 5PB
2E0 DCS D. Sharpen 52 Woodsend Road, Urmston, Manchester M41 8QT
2E0 DCV J. Hurlbutt 55 Prospect Avenue, Seaton Delaval, Whitley Bay NE25 0EL
2E0 DCX D. Cook 44 Statfold Lane, Fradley, Lichfield WS13 8NY
2E0 DCZ D. Mooney 41 Queenhill Road, South Croydon CR2 8DW
2E0 DDA D. Click 3 Brean Gardens, Bristol BS3 5ET
2E0 DDB C. Massey 23 Ladymead Lane, Langford, Bristol BS40 5EG
2E0 DDC D. Coe 199 Newark Road, North Hykeham, Lincoln LN6 8QS
2E0 DDD R. Hirst 106-108 Washerwall Lane, Werrington, Stoke-on-Trent ST9 0LR
2E0 DDF D. Pennison 69 Caneland Court, Waltham Abbey EN9 3DS
2E0 DDH D. De La Haye 4 Nicola Mews, Ilford IG6 2QE
2W0 DDJ D. Jones 8 Midsummer Road, Pontrhydyrun, Cwmbran NP44 1FR
2E0 DDL D. Curtis 7 Neale Close, Aylsham, Norwich NR11 6DJ
2E0 DDN B. Southern 25 Chilgrove Avenue, Blackrod, Bolton BL6 5TR
2E0 DDO D. Lyons 2 Goswick Farm Cottages, Berwick-upon-Tweed TD15 2RW
2W0 DDP R. Carpenter 2 Greenfield Terrace, Trinant, Newport NP11 3LJ
2E0 DDQ G. Mutch 94 Abbotswood Road, Brockworth, Gloucester GL3 4PF
2E0 DDR D. Randles 12 Wain Court. Rakeway, Saughall. Chester CH16BF
2M0 DDS D. Scott Farewell, Arnage, Auchnagatt, Ellon AB41 8UW
2E0 DDU S. Borrell Rose Cottage, Colchester Main Road, Colchester CO7 8DD
2M0 DDX D. Cunningham 3 Dallerie, Crieff PH7 4JH
2W0 DDZ D. May 12 Marl Crescent, Llandudno Junction LL31 9HS
2E0 DEC D. Chebsey 21 Shortlands Lane, Walsall WS3 4AG
2E0 DEE D. Nicholson 24 Barnmead, Haywards Heath RH16 1UZ
2E0 DEG A. Titmus The Old Police House, Arundel Road, Fontwell, Arundel BN18 0SX
2E0 DEH L. Mathlin 29 Wagtail Drive, Stowmarket IP14 5GH
2E0 DEI G. Watson 88 Avenue Road, Sandown PO36 8BE
2E0 DEK D. Gibson 25 Middleham Close, Ouston, Chester le Street DH2 1TA
2E0 DEM C. Wright 24 Charlemont Crescent, West Bromwich B71 3DA
2E0 DEO C. Phillips 14 Laburnum Way, Hatfield Peverel, Chelmsford CM3 2LP
2E0 DEP C. Garner 30 Pendula Road, Wisbech PE13 3RR
2E0 DEQ D. Pritchard Flat 23 Egan Court Price Street, Birkenhead CH416JA
2W0 DER D. Williams 10 Bronllys, Gaerwen LL60 6JN
2M0 DES G. Taylor 15 Ronaldsvoe, Kirkwall KW15 1XE
2E0 DEU M. Hanbuerger Arboris, New Road Hill, Reading RG7 5RY
2E0 DEV R. Barter 8 Orchard Close, Newton Abbot TQ12 3DF
2E0 DEX D. Rigby 32 Springs Road, Chorley PR6 7AN
2E0 DEZ D. Turner 29 Balmoral Road, Castle Bromwich, Birmingham B36 0JT
2E0 DFA B. Geall 129 Jewell Road, Bournemouth BH8 0JP
2E0 DFB D. Grundy 44 Heathend Road, Alsager, Stoke-on-Trent ST7 2SH
2E0 DFF M. Bookham 116 Clare Gardens, Petersfield GU31 4EU
2E0 DFG J. Tarrant 70 Sunnymead, Midsomer Norton, Radstock BA3 2SD
2E0 DFI M. Philpott Garmisch, Hazel Road, Aldershot GU12 6HP
2E0 DFJ D. Jacobs 7 Coppice Close Ravenstone, Coalville LE67 2NS
2E0 DFL D. Humm 15 Sherborne Road, Farnborough GU14 6JS
2W0 DFM R. Russell 1 Horeb Cottages, Rhiw Road, Colwyn Bay LL29 7TL
2W0 DFN G. Edwards 17 Glan Y Mor Road, Penrhyn Bay, Llandudno LL30 3NL
2E0 DFO P. Heiney 12 Church Lane, Walberswick IP186UZ
2E0 DFP D. Parker 53 Brisbane Way, Cannock WS12 2GR
2E0 DFQ P. Love 30 Salisbury Road, Canterbury CT2 7HH
2E0 DFS D. Smith 186 Weekes Drive, Slough SL1 2YR
2E0 DFT D. Fisher 34 Orange Croft, Tickhill, Doncaster DN11 9EW
2E0 DFV R. Rigby 11 Mallow Close, Thornbury, Bristol BS35 1UE
2E0 DGA P. Wilson 45 Newquay Close, Hartlepool TS26 0XG
2M0 DGB D. Baillie 126 Main St., Fauldhouse, Bathgate EH47 9BW
2E0 DGC D. Cowling 11 Shakespeare Avenue, Scunthorpe DN17 1SA
2E0 DGD D. Bailey 2B Queens Road, Enfield EN1 1NE
2E0 DGG M. Tinsell-Stanton 38 Comberton Road, Kidderminster DY10 3DT
2E0 DGH D. Humphrey 42 Ratcliffe Road Sileby, Loughborough LE12 7PZ
2M0 DGI P. Riddle Carngeal, Pitlochry PH16 5JL
2M0 DGJ Z. Bak 62/6 North Gyle Loan, Edinburgh EH12 8LD
2E0 DGL D. Lock1 Flat 33, Harrison Court, Harrison Close, Hitchin SG4 9SG
2E0 DGM S. Hopkins 3B Tolfa House Wellington Terrace, Truro TR1 3JA
2E0 DGQ P. Flanagan 71 Fellway, Pelton Fell, Chester le Street DH2 2BY
2E0 DGR A. Norton 44 Jessamine Road, Southampton SO16 6AL
2E0 DGS D. Smith 48 Shirley Gardens, Tunbridge Wells TN4 8TH
2E0 DGT R. Fripp 41 Sweyns Lease, East Boldre, Brockenhurst SO42 7WQ
2E0 DGU W. Alexander 81 Cherry Lane, Lymm WA13 0SY
2E0 DGV P. Hodson 21 Green Hill, London Road, Worcester WR5 2AA
2E0 DHA D. Atkins 20 Nappsbury Road, Luton LU4 9AL
2E0 DHB D. Bisson 48 Elmsfield Avenue, Rochdale OL11 5XN
2I0 DHC P. White 46 Pine Cross, Dunmurry, Belfast BT17 9QY
2E0 DHE R. Barnard 3 Heaths Close, Enfield EN1 3UP
2E0 DHF J. Jackson 49 Leafield Rise, Two Mile Ash, Milton Keynes MK8 8BX
2E0 DHG R. Holmes Coach House, Cadlington House Estate, , Waterlooville PO8 0AA
2M0 DHI R. Buchan 5 Fairview Terrace, Danestone, Aberdeen AB22 8ZH
2E0 DHJ T. Thompson 14 Queen Street, Northwich CW9 5JL
2E0 DHK B. Davies 12 Scalebor Gardens, Burley In Wharfedale, Ilkley LS29 7BX
2E0 DHO P. Chadwick 112 Sandy Lane, Warrington WA2 9JA
2E0 DHQ D. Rimmer 41 Ashburton Road, Wallasey CH44 5XB
2I0 DHR D. Richards 70 Cherryhill Avenue, Dundonald, Belfast BT16 1JD
2E0 DHS D. Sherwin 5 North Road, Buxton SK17 7EA
2E0 DHT J. Berry 245A Eaves Lane, Chorley PR6 0AG
2E0 DHV P. Walton 11 Parkfield Road, Northwich CW9 7AR
2E0 DHW A. Stabler 11 Lincolns Avenue, Gedney Hill, Spalding PE12 0PQ
2E0 DHX M. Watkins 108 Honiton Road, Exeter EX1 3EQ
2E0 DHY D. Ball 27 Bramble Close Aston, Birmingham B6 5HW
2E0 DHZ A. Hitchcott 121 Oakhurst Road, Acocks Green, Birmingham B27 7PB
2M0 DIB C. Morris 23 Sedgebank, Livingston EH54 6HE
2E0 DID D. Foyston 13 Hollinwood Road, Stoke on Trent ST71DQ
2M0 DIF I. Smith 32 Kaimes Avenue, Kirknewton EH27 8AU
2E0 DIG A. Taylor 130A Hazelwood Avenue, Eastbourne BN22 0UX
2E0 DIH E. Coles 41 Venn Court Brixton, Plymouth PL82AX
2E0 DII I. Phillips 124 Brookwood Drive, Stoke-on-Trent ST3 6LP
2E0 DIJ D. Yates 16 Sunnyfield Road, Prestwich, Manchester M25 2RD
2E0 DIL D. Yakub 42 Swift Close, Blackburn BB1 6LF
2E0 DIM D. Vainas 51 Magister Road, Bowerhill, Melksham SN12 6FD
2E0 DIP D. Webb 52 Simpkin Close, Eaton Socon, St. Neots PE19 8PD
2E0 DIQ A. Collins 14 Double Corner, Mendlesham Road, Cotton, Stowmarket IP14 4RF
2E0 DIS A. George 15 Ely Road, Croydon CR0 2LW
2E0 DIT N. Baulf 1 Lower Chart Cottages, Brasted Chart, Westerham TN16 1LS
2E0 DIU I. Jones 90 Preston, Cirencester GL7 5PR
2W0 DIV P. Williams 63 Trem Eryri, Llanfairpwllgwyngyll LL61 5JF
2E0 DIX B. Smith 7 Kestrel Avenue, Bransholme, Hull HU7 4ST
2W0 DJC D. Cole 14 Inner Loop Road, Beachley, Chepstow NP16 7HF
2E0 DJF D. Fagg 62 Hawkins Road, Folkestone CT19 4JA
2E0 DJH D. Harris 2 Horsecroft, , Ewyas Harold HR20EQ
2E0 DJI D. Oliver 20 Five Oaks Close, Malvern WR14 2SW
2E0 DJJ G. Mccaffery 7 Cliffe Court, Sunderland SR6 9NT
2I0 DJM J. Mcbride 22 Birchwood, Omagh BT79 7RA
2E0 DJQ M. Barnaby 8 Callowood Croft, Purleigh, Chelmsford CM3 6NZ
2E0 DJR J. Rudd 5 St Andrews Close Blofield, Norwich NR13 4JX
2W0 DJS D. James 10 Hafan Deg, Pencoed, Bridgend CF35 6YG
2M0 DJT D. Rodger 25 Wilson Road, Banchory AB31 5UY
2E0 DJU T. Bannister 6 Tanners Road, North Baddesley, Southampton SO52 9FD
2E0 DJV Uws RS c/o A. Wheeler 8 Elsworth Grove, Birmingham B25 8EJ
2E0 DJW D. Creech 29 Lake Road, Poole BH15 4LE
2E0 DJX S. Scott 13 Silver Close, Harrow HA36JT
2E0 DJY Norfolk ARC c/o D. Bennett 31 Park Road North, Urmston, Manchester M41 5AT
2E0 DJZ Dr B. Issac 9B Poplar Grove, Stockport SK2 7JD
2E0 DKA D. Carmichael 22 California Close Great Sankey, Warrington WA5 8WU
2E0 DKB D. Banks 41 East Road, Rotherham S65 2UX
2E0 DKD D. Saxon 53 Westmorland Road, South Shields NE34 7JJ
2E0 DKE C. Boyle 8 Westlees Close, North Holmwood, Dorking RH5 4TN
2E0 DKF A. Gow Flat 203, Viotti Heights Sandy Hill Road, London SE18 6PA
2E0 DKG E. Gomez Lozano 29 Wykeham Crescent, Oxford OX4 3SD
2E0 DKI D. Iveson 11 Newport Road, North Cave, Brough HU15 2NU
2E0 DKM W. Molloy 32 Millers Barn Road, Jaywick, Clacton-on-Sea CO15 2QB
2E0 DKO P. Taylor 32 Heliers Road, Liverpool L13 4DH
2E0 DKP A. Lutley Springfield, Rookery Hill, Ashtead KT21 1HY
2I0 DKQ G. Gardiner 60 Limestone Meadows, Moira, Craigavon BT67 0UT
2E0 DKR D. Reeves 67 The Cliff, Bryanston, Blandford Forum DT11 0PP
2E0 DKS C. Wilkes 2 Kings Crescent, Edlington, Doncaster DN12 1BD
2M0 DKU S. Boyd 269 Cloch Caravans Cloch Road, Gourock PA19 1AZ
2M0 DKV J. Flannigan 21 Kirkbean Avenue Rutherglen, Glasgow G73 4EA
2E0 DKW C. Nicholl 36 Eylewood Road, London SE27 9NA
2M0 DKX P. Bacon 12 The Greens, Maddiston, Falkirk FK2 0FN
2E0 DKY B. Cross 22 Park Avenue Washingborough, Lincolnshire LN 4 1 DB
2E0 DKZ D. Hyde 136 Station Road, Woodmancote, Cheltenham GL52 9HN
2E0 DLA P. Booth 7 Handley Crescent, East Rainton, Houghton le Spring DH5 9QX
2E0 DLC B. Tufnell 1 Moorlands Court Wath-Upon-Dearne, Rotherham S63 6DD
2E0 DLD D. Dewsbury 62 Yew Tree Drive, Leicester LE3 6PL
2W0 DLE T. Edwards Broadfield House, Vicarage Road, Tonypandy CF40 1HP
2E0 DLF A. Cook 84 Clent View Road, Birmingham B32 4LW
2E0 DLJ D. Johnstone 14 Carr Hey, Wirral CH46 6EL
2E0 DLK L. Sargent 18 Lyndhurst, Maghull L316DY
2E0 DLL M. Richardson 10 Cargill Close, Newton Aycliffe DL5 5ET
2E0 DLO S. Strange 94 Digby Avenue, Nottingham NG3 6DY
2E0 DLP D. Ion 78 Blackmore Street, Derby DE23 8AX
2E0 DLR D. Taylor Flat 27, Laurel Court, 24 Stanley Road, Folkestone CT19 4RL
2E0 DLT Willenhall & District ARS c/o M. Tucker 182 Salisbury Road, Amesbury, Salisbury SP4 7HW
2E0 DLV C. Lombao 82 Cirrus Drive, Shinfield, Reading RG2 9FL
2E0 DLZ D. Dunne 1 Burton Gardens, Brierfield, Nelson BB9 5DR
2E0 DMA D. Aldridge 5 Alpine Close, Paulton BS39 7SE
2E0 DMB D. Browne 27 Lewis Road, Emsworth PO10 7RP
2I0 DMC D. Mccloskey 1 Dernaflaw Cottages Dernaflaw Road, Dungiven, Londonderry BT47 4PP

2E0 DMD D. Mcarthur 7 Gore Avenue, Salford M5 5LF
2E0 DME D. Priestley 8 Cokefield Avenue, Nuthall, Nottingham NG16 1AU
2W0 DMG D. Griffiths 8 Heol Cynwyd, Llangynwyd, Maesteg CF34 9TB
2E0 DMI A. Yates 1 Roberts Court, Whitwell, , Hitchin, SG4 8AF
2E0 DMJ D. Gegg 84 Aberconway Crescent, New Rossington, Doncaster DN11 0JP
2E0 DMM D. Moran 37 Collingwood Road, Long Eaton, Nottingham NG10 1DR
2E0 DMP D. Powell 58, Lessingham Avenue, Wigan WN12HX
2E0 DMQ K. Henney 6 Peel House, Rusham Road, Egham TW20 9LP
2E0 DMS D. Stewart 79 Eastfield Road, Driffield YO25 5EZ
2E0 DMU D. Pearson 37 Elmridge, Leigh WN7 1HN
2E0 DMV G. Weston 131 Ringwood Road, Eastbourne BN22 8TQ
2E0 DMW R. Weir 130 Alexander Square, Eastleigh SO50 4BX
2E0 DMX D. Moffat 27 Cinque Ports Way, Seaford BN25 3UE
2E0 DMY D. Humphrey 11 Colborne Close, Poole BH15 1UR
2E0 DNB N. Brown 241 Bury Road, Tottington, Bury BL8 3DY
2E0 DNC D. Swift 31 Meadow Lane, Westbury BA13 3AE
2E0 DND D. Driver 34 Potter Avenue, Wakefield WF2 8HE
2E0 DNE B. Rajagopal 4 Balliol Road, Caversham, Reading RG4 7DT
2E0 DNF D. Featherby 14 Station Road, Sutton, Ely CB6 2RL
2E0 DNH R. Finch Garth Cottage North Cowton, Northallerton DL7 0HL
2W0 DNI W. Oliver Pwllmeyric, Chepstow NP16 6LE
2E0 DNJ D. Jarvice 15 Meden Avenue, Warsop, Mansfield NG20 0PS
2E0 DNL D. Neill 7 Ashbrow Road, Northampton NN4 8ST
2M0 DNM D. Mackenzie 4 Nabhar Laxay, Isle of Lewis HS2 9PJ
2E0 DNO D. Close 22 Station Road, Dodworth, Barnsley S75 3JE
2W0 DNR J. Hughes Brynawelon, Southgate, Aberystwyth SY23 1SE
2E0 DNS J. Cummins 55 Rowley Street, Walsall WS1 2AZ
2E0 DNU G. Cater 7 Seymour Street, Chelmsford CM2 0RX
2W0 DNV D. Hooper Nant Y Dryslwyn Cottage, Ty Mawr, Llanybydder SA40 9RD
2E0 DNW L. Stamper 22 Douglas Road, Workington CA14 2QY
2E0 DNX D. Smith 3 Cain Street, Bigrigg, Egremont CA22 2TP
2E0 DOD A. Dodd 14 Davies Street, Macclesfield SK10 1GE
2E0 DOF S. Black 7 Harwood Close, Gosport PO13 0TY
2E0 DOG R. Scholefield 4 Minnie Street, Haworth, Keighley BD22 8PR
2M0 DOI D. Speirs 45 Elmbank Crescent, Arbroath DD11 4EZ
2E0 DOJ D. Jenkins 1 Green End Road, Sawtry, Huntingdon PE28 5UX
2M0 DOL J. Marsh 8 Hazelton Way, Broughty Ferry, Dundee DD5 3BT
2E0 DOP A. Seelig 9 Warren House Court 17 St Peters Avenue Caversham, Reading RG4 7RG
2E0 DOQ M. Dinally 208 Ley Hill Farm Road, Birmingham B31 1UQ
2E0 DOW B. Lewin 68 Brackley Square, Woodford Green IG8 7LS
2E0 DOX D. Richards Flat 40, Leander Court, Teignmouth TQ14 8AQ
2E0 DOZ D. Logan Cedar House, Reading Road North, Fleet GU51 4AQ
2E0 DPD B. Scannell 60 Burnside Road, Dagenham RM8 1XD
2E0 DPH P. Hughes 111 Wisbech Road, Littleport, Ely CB6 1JJ
2W0 DPI A. Studdart 24 Wepre Park, Connah'S Quay, Deeside CH5 4HN
2E0 DPL C. Glass The Old Homestead, Havikil Lane, Knaresborough HG5 9HN
2E0 DPO D. Greenland 1 Hilltop, Tuesley Lane, Godalming GU7 1SB
2E0 DPR Dr D. Richardson 89A Bean Oak Road, Wokingham RG40 1RJ
2E0 DPS D. Seedhouse 8 Levett Road, Tamworth B77 4AB
2E0 DPT C. Appleton 140 St. James Road, Orrell, Wigan WN5 7AA
2E0 DPW D. Williams 6 Wellhouse Avenue, West Mersea, Colchester CO5 8GF
2E0 DPX W. Currie 24 Mill Lane, Walton on the Naze CO14 8PE
2E0 DPY Dr D. Pye 151 Smallbrook Lane, Leigh WN7 5PZ
2E0 DPZ R. Lyddall 102 Chapel Road, Brightlingsea, Colchester CO7 0HE
2E0 DQB A. Smith 116 Pilling Lane, Preesall, Poulton-le-Fylde FY6 0HG
2E0 DQH A. Robnett 38B Woodmere Avenue, Watford WD24 7LN
2E0 DQJ H. Woodfin 8 Bank Hall Close, Bury. BL8 2UL
2E0 DQL D. Coles 36 York Hill, Loughton IG10 1HT
2M0 DQN G. Mcleod 75 Grange Avenue, Wishaw ML2 0AH
2E0 DQO J. Marks Chantry End, Oak Hill, Epsom KT18 7BU
2E0 DQP Rvd. A. Lewis Four Winds Cottage, Main Street, Brough HU15 1RJ
2E0 DQQ S. Mcloughlin 7 Wilmots Way, Pill, Bristol BS20 0JT
2W0 DQT A. Williams 27 Hick Street, Llanelli SA15 1AR
2E0 DQU I. Lawton 11 Goosewell Terrace, Plymstock, Plymouth PL9 9HW
2E0 DQX J. Saunders 123 Medway Road, Ferndown BH22 8UR
2M0 DQY J. Dow 58 Beatty Crescent, Kirkcaldy KY1 2HS
2E0 DQZ T. Brooks 200 Kingsway, College Estate, Hereford HR1 1HE
2E0 DRA M. Draper 160 Chanctonbury Road, Burgess Hill RH15 9HA
2E0 DRB D. Barraclough Flat 18 89 Park Road, London SW19 2BD
2E0 DRD Dr P. Darlington 8 Uplands Road, Urmston, Manchester M41 6PU
2E0 DRE D. Dean 12 Abbeydale Road South, Sheffield S7 2QN
2D0 DRG C. Schofield Rockside, Dreemskerry Road, Maughold IM7 1BL Isle of Man
2E0 DRI K. Abel 7 Foldgate View, Ludlow SY8 1NB
2W0 DRK D. Machon 22 Albert Street, Caerau, Maesteg CF34 0UF
2E0 DRN D. Rayne 159A Arbury Road, Nuneaton CV107NH
2M0 DRO R. Drummond 11 Firwood Drive, Bo'ness EH51 0NX
2E0 DRQ D. Roberts 3 Heather Avenue, Melksham SN12 6FX
2E0 DRT C. Brock 30 Cromer Road, Norwich NR6 6LZ
2E0 DRW Dr W. Wightman 36 Holyoake Avenue, Woking GU21 4PW
2M0 DRY D. Drysder 37 Farburn Drive, Stonehaven AB39 2BZ
2E0 DRZ A. Potts 103 Etherstone Street, Leigh WN7 4HY
2E0 DSB D. Slade 22 Oaklands Road, Mangotsfield, Bristol BS16 9EY
2M0 DSG D. Gartshore 85 Springhill Street, Douglas, Lanark ML11 0NZ
2E0 DSH D. Hind 19 Ellington Road, Arnold, Nottingham NG5 8SJ
2E0 DSI D. Stocker 113 St. Marys Road, Bodmin PL31 1NH
2E0 DSK D. Manning 153 Pavilion Road, Worthing BN14 7EG
2M0 DSL D. Latto 8 Aspen Avenue, Glenrothes KY7 5TA
2W0 DSO C. Summerfield 11 Woodland Park, Penderyn, Aberdare CF44 9TX
2W0 DSP A. Elias 31 Banc Y Gors, Upper Tumble, Llanelli SA146BR
2E0 DSQ P. Taylor 104 Winstanley Drive, Leicester LE3 1PA
2E0 DSS W. Stewart 43 Newlands Drive, Halesowen B62 9DX
2M0 DSU R. Murray 18 Braids Road, Kirkcaldy KY2 6JE
2E0 DSV R. Nicholson 1 Isis Close, Aylesbury HP21 9LY
2E0 DSW D. Weight 15 Bourn Road, Caxton, Cambridge CB23 3PP
2M0 DSY S. Chacko 25 Wemyss Court, Rosyth, Dunfermline KY11 2LL
2E0 DTB D. Bradley 45 Fourth Avenue, Ketley Bank, Telford TF2 0AS
2E0 DTC D. Beck 94 Shaldon Crescent, Plymouth PL5 3RB
2E0 DTD T. Davies 7 Crescent Road, Warley, Brentwood CM14 5JR
2I0 DTE D. Best 13 Cranley Green, Bangor BT19 7FE
2E0 DTF M. Bostock 86 Beauvale Drive, Ilkest on DE7 8SJ
2E0 DTG D. Griffiths 44 Lowlands Road, Bolton Le Sands, Carnforth LA5 8HE
2E0 DTH D. Thompson 17 Sandpiper Close, Blyth NE24 3QN
2M0 DTJ G. Lewin Larch Cottage, Lein Road, Fochabers IV32 7NW
2E0 DTL C. Brink 138 Brookside, Burbage, Hinckley LE10 2TN
2W0 DTM N. Sugg 12 Caerleon Grove Castle Park Cf48 1Jh, Merthyr Tydfil CF48 1JH
2E0 DTN D. Niggemann 35 Holm Court, Twycross Road, Godalming GU7 2QT
2E0 DTO E. Bray 28, Henshall Avenue, Latchford, , Warrington WA4 1PY
2M0 DTP T. Burnett 45 The Murrays Brae, Edinburgh EH17 8UF
2E0 DTQ T. Kilfeather Flat 1, 57 Chalk Hill, Watford WD19 4DA
2W0 DTR D. Rosser 42 Cobden Street, Aberaman, Aberdare CF44 6EN
2E0 DTS B. Oldfield 60 Alexandra Road, Capel-Le-Ferne, Folkestone CT18 7LS
2E0 DTV M. King 6 Aylesbury Close, Hockley Heath, Solihull B94 6PA
2I0 DTW D. Williams 19 Ballymacruise Park, Millisle, Newtownards BT22 2NW
2E0 DTX L. Cook 43 Midge Hall Drive, Rochdale OL11 4AX
2E0 DTY M. Pesendorfer 13 Blake Road, London N11 2AD
2M0 DTZ J. Hogg 31 Woodlea Court, Crosshouse, Kilmarnock KA2 0ES
2E0 DUA K. Lynch Medindie, Woodside, Ryton NE40 4SY
2E0 DUB T. Price 40 East Street, Kidderminster DY10 1SE
2E0 DUD J. Storey 3 Woodside Road, Poole BH14 9JH
2E0 DUE I. Warnecke 12 Caxton Road, Margate CT9 5NP
2E0 DUF S. Lo 12 Kingston Wharf, Kingston Street, Hull HU1 2ES
2E0 DUH J. Shears 161 Park Road, Keynsham, Bristol BS31 1AS
2E0 DUI R. Sutton 80 Fishbourne Lane, Ryde PO33 4EU
2E0 DUJ S. Bluff 2 Astor Mews, High Street, Tidworth SP9 7TR
2W0 DUL D. Davies 2 Hendre Ddu, Manod, Blaenau Ffestiniog LL41 4BH
2E0 DUM I. Clarke 140 Cashmere Drive, Andover SP11 6SS
2W0 DUN T. Dungey 93 Boverton Road, Llantwit Major CF61 1YA
2E0 DUO A. Richards 16 Fruiterers Arms Caravan Park, Uphampton Lane, Ombersley, Droitwich WR9 0JW
2E0 DUP V. Downes 55 Ashfield Road Bromborough, Wirral CH62 7EE
2E0 DUQ A. Ali 42 Blease Close, Staverton, Trowbridge BA14 8WD
2I0 DUR A. Savage 469 Old Belfast Road, Bangor BT19 1RQ
2E0 DUS J. Fenton 4 Forest Hills, Newport PO30 5NG
2E0 DUU I. Holdford 46 Hildreth Road, Prestwood, Great Missenden HP16 0LY
2E0 DUW L. Jones 44 Althorpe Drive, Loughborough LE11 4QU
2E0 DUZ T. Hope 59 Chatsworth Crescent, Walsall WS4 1QU
2E0 DVD A. Swan 47 Warren Close, Whitehill, Bordon GU35 9EX
2E0 DVF L. Lemmon 4 Honington Close, Wickford SS11 8XB
2E0 DVH D. Hilton 6 Stow Villas The Street, Stow Maries, Chelmsford CM3 6RX
2E0 DVI G. Ridley 12 Garforth Avenue, Steeton, Keighley BD20 6SP
2E0 DVK R. Steele 244A Uttoxeter Road, Blythe Bridge, Stoke-on-Trent ST11 9LY
2E0 DVM D. Vincelli 90 Broadbottom Road, Mottram SK14 6JA
2E0 DVN J. Sanders 76 Fullerton Road, Plymouth PL2 3AX
2E0 DVO K. Wells 45 Laburnum Avenue, Yaxley, Peterborough PE7 3YQ
2W0 DVP D. Price 2 Heol Tyn-Y-Fron, Penparcau, Aberystwyth SY23 3RP
2E0 DVU M. Hughes 183 Station Road, Hednesford, Cannock WS12 4DP
2E0 DVV J. Allen 17 Inglemere Gardens Arnside, Carnforth LA5 0BX
2E0 DVW M. Roberts 5 Cliff Cottages, Cliff Road, Hessle HU13 0HB
2E0 DVX N. Malyon 50 Swanage Road, Southend-on-Sea SS2 5HY
2E0 DVY J. Poriyath 48 Howard Close, Cambridge CB5 8QU
2E0 DWA Dr A. Dickson 1 Roebuck Drive Baldwins Gate, Newcastle-under-Lyme ST5 5FE
2E0 DWB D. Belton 4 Sandown Road, Toton, Nottingham NG9 6GN
2M0 DWC D. Cowie 69 Broomfield Park, Portlethen, Aberdeen AB12 4XT
2U0 DWD A. Prosser Woodlands, La Vassalerie, , St. Andrew GY6 8XL Guernsey
2E0 DWE D. Toller Field Cottage, Ash Lane, Etwall, Derby DE65 6HT
2E0 DWF D. Leyland 20 Newton Heath, Middlewich CW10 9HL
2E0 DWG D. Gough 29 Belvedere Road, Biggin Hill, Westerham TN16 3HX
2E0 DWJ D. Johnston 34 Coxley Crescent, Netherton, Wakefield WF4 4LR
2E0 DWK C. Duckworth 15 St Johns Court, Burnley BB12 6QE
2E0 DWM D. Mardlin 13 Churchill Crescent Sonning Common, Reading RG4 9RU
2E0 DWP D. Prout 2 Pine Crest Way, Bream, Lydney GL15 6HG
2W0 DWR D. Rich 41 Ronald Road, Newport NP19 7GF
2E0 DWS D. Sewell 19 St. Leonards Way, Ashley Heath, Ringwood BH24 2HS
2E0 DWT D. Wells 40 Barnham Broom Road, Wymondham NR18 0DF
2E0 DWU A. Edmonds 20 Tomline Road, Ipswich IP3 8BZ
2E0 DWV E. Preda 34A Church Street, Willingham, Willingham CB24 5HT
2E0 DWW K. Gorringe 6 Buttercup Park, Pevensey Bay BN24 6BE
2E0 DWZ M. Davey Romea, Long Street, Great Ellingham, Attleborough NR17 1LW
2M0 DXC A. Brown 37 Sherwood Loan, Bonnyrigg EH19 3NF
2E0 DXJ D. Barrett 208 Doncaster Road, Rotherham S65 2UE
2E0 DXK J. Whitworth 31 Shirley Close, Chesterfield S40 4RJ
2E0 DXL I. Newton 16 Cross Close, Newquay TR7 3LB
2E0 DXO P. Higginson 9 St. Mildreds Way, Heysham, Morecambe LA3 2QJ
2E0 DXW M. Roper 26 Malpas Close, Bransholme, Hull HU7 4HH
2E0 DXZ L. Brearley Ash Tree Lodge, Snaith Road, Goole DN14 0AT
2E0 DYB D. Brough 38 Tynedale Avenue, Crewe CW2 7NY
2E0 DYD G. Lloyd 1 Holmside Terrace, Stanley DH9 6ET
2E0 DYG D. Young 35 Limber Hill, Cheltenham GL50 4RJ
2E0 DYJ M. Bentley 5 Stokewell Road, Wath-Upon-Dearne, Rotherham S63 6EL
2E0 DYM P. Cottam Eastleigh, Kings Nympton, Umberleigh EX37 9ST
2E0 DYN D. Jones Drove Farm, Sheepdrove, Lambourn, Hungerford RG17 7UN
2E0 DYP W. Hartley 30 Coltman Avenue, Beverley HU17 0EY
2E0 DYQ P. Sykes 16 Hill Fold, South Elmsall, Pontefract WF9 2BZ
2E0 DYU B. Healey 14 Orchard Close, Ferring, Worthing BN12 6QP
2E0 DYV P. Collier Flat 9, Henry House, Wyvil Road, London SW8 2TF
2E0 DYX T. Carpenter 1 Terriers Lane Hayling Island, Hayling Island PO11 0FF
2E0 DYY L. Hopgood 62 Briarwood Drive, Blackpool FY2 0EB
2E0 DZA H. Harclerode The Bungalow, Silverlace Green, Parham, Woodbridge IP13 9AD
2E0 DZC L. Davison 58 Priestley Court, South Shields NE34 9NQ
2E0 DZE M. White 9 Hawksworth Close, Rotherham S65 3JX
2E0 DZF P. Ridgers 231 The Greenway, Epsom KT18 7JE
2E0 DZJ G. Allen 14 The Parsonage, Sixpenny Handley, Salisbury SP5 5QJ

2E0 DZM R. Williams 10 Bramley Close, Twickenham TW2 7EU
2E0 DZN M. Gosi 49 Elms Drive, Marston, Oxford OX3 0NW
2E0 DZO P. Sargeant 6 Meldon Way, Winlaton., Blaydon-on-Tyne. NE21 6HJ
2E0 DZQ J. Ostapiuk 33 Kent Terrace, Haswell DH62EL
2E0 DZR P. Pain 12 Maple Drive, Bamber Bridge, Preston PR5 6RA
2E0 DZS Dr Z. Derzsi 217 Bensham Road, Bensham, Gateshead NE8 1US
2E0 DZT J. Emery Mulberry Cottage, Quarry Lane, Combe St. Nicholas, Chard TA20 3PH
2E0 DZV N. Harris 10 Pentland Drive, North Hykeham, Lincoln LN6 9TG
2M0 DZX A. Mccormick 79 Kennedy Crescent, Tranent EH33 1DN
2E0 DZY D. Atkins 32 Braybrook, Orton Goldhay, Peterborough PE2 5SH
2M0 DZZ D. Taylor 1 Mayfield Farm Cottages, Reston, Eyemouth TD14 5LG
2E0 EAA B. Matthews 30 Oaklands Drive, Brandon IP27 0NR
2M0 EAC J. Woods 12 Westbank Terrace, Macmerry, Tranent EH33 1QE
2W0 EAD D. Cook 19 Almond Avenue Risca, Newport NP11 6PF
2E0 EAF R. Bond 21 Coleridge Close, Bletchley, Milton Keynes MK3 5AF
2E0 EAI S. Bide 9 Greenway, Watchet TA23 0BP
2E0 EAL J. Oldman High Waters, Bentfield Green, Stansted CM24 8HX
2E0 EAN P. Fulbrook 167 Droitwich Road, Fernhill Heath WR3 7TZ
2I0 EAR E. Rainey 22 Cherry Gardens, Ballymoney BT53 7AS
2I0 EAS D. Elliott 15 Derrychara Park, Enniskillen BT74 6JP
2E0 EAU E. Aksamit 14 Popplewell Gardens, Gateshead NE9 6TU
2E0 EAV D. Blake Pound Farm, Swan Lane, Leigh, Swindon SN6 6RD
2E0 EAW E. Cross 12B Oakridge, Three Rivers Country Park, Clitheroe BB7 3JW
2E0 EAY A. Driver 12 Almond Tree Avenue, Malton YO17 7DF
2E0 EAZ E. Heath 63 Meadway, Dunstable LU6 3JT
2E0 EBA R. Aldridge 37 Vincent Road, Luton LU4 9AN
2I0 EBB A. Connolly 68 Willowbank Gardens, Belfast BT15 5AJ
2E0 EBD B. Clayton 26 Wood Walk, Mexborough S64 9SG
2E0 EBJ F. Taylor 55 Leicester Avenue, Horwich, Bolton BL6 5QX
2E0 EBK D. Levett 11 Love Lane, London SE25 4NG
2E0 EBL C. Haynes 25 Barnards Hill Lane, Seaton EX12 2EQ
2E0 EBP T. Stokes 1 Hunters Reach, Bradwell, Milton Keynes MK13 9BT
2E0 EBQ M. Priest 35 Albert Road, Chaddesden, Derby DE21 6SJ
2E0 EBR A. Buckland 21 Malton Close, Monkston, Milton Keynes MK10 9HR
2I0 EBS E. Mcknight 14 Marlacoo Beg Road, Portadown, Craigavon BT62 3TF
2M0 EBU R. Clow 25 Scott Street, Newcastleton TD9 0QQ
2M0 EBW D. Gemmell 36 Church St., Dumfries DG2 7AS
2E0 EBX K. Carter 50 Elliman Avenue Bottom Flat, Slough SL2 5BG
2E0 EBZ A. Chance 24 Doddsfield Road, Slough SL2 2AD
2E0 ECC E. Wright 2 Wulfrath Way, Ware SG12 0DN
2E0 ECD C. Dennis 1 West Villa, Crathorne, Yarm TS15 0BA
2E0 ECG T. Anthony 27 Evelyn Avenue, Doncaster DN2 6LN
2E0 ECI W. Canavan 9 The Ridings, Deanshanger, Milton Keynes MK19 6JD
2E0 ECJ H. Mcevoy 18 Brookfield Gardens, West Kirby CH48 4EL
2M0 ECK A. Falconer 61 Mountcastle Drive North, Edinburgh EH8 7SP
2E0 ECO A. Hawksworth 87 Bradeley Road, Haslington, Crewe CW1 5PX
2E0 ECP S. Andrews 24 Beesley Road, Banbury OX16 0HL
2E0 ECQ J. O'Donnell 24 Foxhill, Watford WD24 6SY
2E0 ECU P. Escott 84 Salisbury Avenue, Bootle L30 1PZ
2E0 ECV A. Jovanovic 33 Seward Road, London W7 2JS
2E0 ECW E. Williams 19 Raglans, Exeter EX2 8XN
2E0 ECY G. Bryant 11 Cadwallon Road, London SE9 3PX
2W0 ECZ M. Douglas 486 Malpas Road, Newport NP20 6NB
2E0 EDA J. Sim 11 Haven Close Istead Rise, Gravesend DA13 9JR
2E0 EDC D. Lavell 51 Kingfisher Close, Newport PO30 5XS
2E0 EDF T. Talbot-Humphries 22 Vicar Street, Wednesbury WS10 9HF
2E0 EDG N. Chamberlain 8 Southfields, Binbrook, Market Rasen LN8 6DX
2E0 EDI E. Stormes 1 Meadowbank, Belton, Doncaster DN9 1NW
2E0 EDL D. Simmons 8 Lower Grange, Huddersfield HD2 1RU
2E0 EDM E. Moore 44 Bridge Street, Oxford OX2 0BB
2E0 EDN V. Lucock 34 Wentworth Drive, Ipswich IP8 3RX
2E0 EDP E. Palo 13 Welwyn Close, St. Helens WA9 5HL
2D0 EDQ E. Quinney 69 Clagh Vane, Ballasalla IM9 2HF Isle of Man
2E0 EDR D. Lock 20 Jasmine Close Trimley St. Martin, Felixstowe IP11 0UY
2E0 EDS E. Kaye 119 St. Bernards Avenue, Louth LN11 8AS
2M0 EDV E. Stuart 6D Dundee Street, Letham DD8 2PQ
2E0 EDX I. Taylor 37 Wood Green Drive, Thornton-Cleveleys FY5 3DH
2M0 EDY E. Higgins 44A Mossvale St., Paisley PA3 2LR
2E0 EEB J. Cameron 29 Webster Road, Stanford-le-Hope SS17 0BE
2E0 EEF D. Finlay 23 Glen Way, Oadby, Leicester LE2 5YF
2E0 EEI M. Rose 149 Claremont Road, Blackpool FY1 2QJ
2E0 EEJ T. Scott Fiddlers Den, Ellingstring, Ripon HG4 4PW
2E0 EEM N. Bristow Flat 204, Gilbert House Barbican, London EC2Y 8BD
2E0 EEO Dr S. Leask 1 Collington Street, Beeston, Nottingham NG9 1FJ
2M0 EEQ A. Robson 100 Dawson Avenue, East Kilbride, Glasgow G75 8LH
2E0 EER J. Salter 20 Burrow Road, Chigwell IG7 4HQ
2E0 EES S. Rattley 2 Burnt Cottages Beanacre, Melksham SN12 7PT
2E0 EET P. Woodburn 21 The Row, Silverdale, Carnforth LA5 0UG
2E0 EEU A. Bullard 15 Rowan Drive, Lutterworth LE17 4SP
2M0 EEV W. Mcbain 56 Scotstoun Park, South Queensferry EH30 9PQ
2E0 EEW E. Thresher 18 Sandy Lane, Preesall, Poulton-le-Fylde FY6 0EH
2E0 EFA M. Elbourn 1 Downside, Gosport PO13 0JS
2E0 EFC D. Owings 11 Thingwall Road East, Thingwall, Wirral CH61 3UY
2M0 EFD A. Keogh 251 Main Street Plains, Airdrie ML6 7JH
2E0 EFG S. Ruddy 27 Grove Park Walk, Harrogate HG1 4BP
2E0 EFH T. Hull 12 Durley Road, , Gosport PO12 4RT
2M0 EFI F. Wenseth 2 Sunnybank Cottage, Logie Coldstone, Aboyne AB34 5PQ
2E0 EFK T. Bell 28 Inwood Drive, Coleford GL168EZ
2E0 EFN E. Almas 10, Rindal 6657 Norway
2E0 EFO M. Luper 49 Russell Road, Buckhurst Hill IG9 5QF
2E0 EFP S. Nelson 2 Boulsworth Road, North Shields NE29 9EN
2U0 EFR D. Robert Nos Treis 7 Liberation Drive, Route Des Clos Landais, St. Saviour, Guernsey GY7 9PH
2E0 EFS R. Hubbard Southbroom School House, Estcourt Street, Devizes SN10 1LW
2E0 EFU J. Leeson 2 Hawthorn Road, Radstock BA3 3NW
2E0 EFW H. Talbot Flat 11, Sedley Court Malta Road, Cambridge CB1 3LW
2E0 EFZ A. Haylor 43 Church Road, Old Windsor SL4 2PH
2M0 EGE R. Bisset 8/1 North Bughtlin Brae, Edinburgh EH12 8XH
2M0 EGI E. Ireland The Steading, Blairmains, Shotts ML7 5TJ
2W0 EGK K. Martin 12 Heol Fargoed, Bargoed CF81 8PP
2W0 EGL M. Lima Barbosa 65 Precelly Place, Milford Haven SA73 2BW
2E0 EGM D. Chapman 27 Cuff Crescent, London SE9 5RF
2I0 EGN R. Nelson 32 Rallagh Road, Dungiven, Londonderry BT47 4TT
2E0 EGO M. Read 3613 Clary Ave, Fort Worth 76111 United States
2E0 EGP J. Churchill West Winds, Brandheath Lane, New End, Astwood Bank, Redditch B96 6NG
2E0 EGQ D. Rowe 18 Burman Road, Wath-Upon-Dearne, Rotherham S63 7ND
2E0 EGR W. Smith 81 Hazelgrove Residential Park Milton Street, Saltburn-by-the-Sea TS12 1FE
2E0 EGS J. Booth 27 Moorlands Scholes, Holmfirth HD9 1SW
2E0 EHA H. Redington Galeholm, Whitecroft, Gosforth, Seascale CA20 1AY
2E0 EHB N. Livingstone 5 Drayton Court, High Street, Polesworth, Tamworth B78 1EX
2E0 EHD E. Delasalle 31 West Hill Road, Hoddesdon EN11 9DL
2E0 EHH N. Reeve 124 Greenhills Road, Eastwood, Nottingham NG16 3FR
2E0 EHJ S. Marsh 6 Sparsholt Road, Southampton SO19 9NJ
2E0 EHK J. Oldham 5 Amersham Rise, Nottingham NG8 5QG
2E0 EHN P. Norman 138 Meadowbank Road, Hull HU3 6XP
2E0 EHO D. Turford 51 Moorfield Avenue, Bolsover, Chesterfield S44 6EJ
2E0 EHP C. Hughes 41 Rotherham Road, Dinnington, Sheffield S25 3RG
2E0 EHQ N. Crudgington Appledore Blackness Lane, Keston BR2 6HL
2E0 EHR A. Lamont 89 Newlands, Whitfield, Dover CT16 3ND
2E0 EHS H. Butcher 12 Bath Road, Willesborough, Ashford TN24 0BJ
2E0 EHT G. Clark 65 Chyvelah Vale, Gloweth, Truro TR1 3YJ
2E0 EHV A. Macdonald Woodside Cottage, Horton Way, Verwood BH31 6JJ
2I0 EHW Dr P. Donaghy 41 Greenvale Manor Antrim, Antrim BT41 1SB
2E0 EHX C. Ogidih 89 West Road, Birmingham B43 5PG
2I0 EIB M. Rush 122 Cullaville Road Crossmaglen, Newry BT35 9AQ
2E0 EID D. Goodchild Gravel Lane, Ringwood BH24 1XY
2E0 EIE A. Goodchild Gravel Lane, Ringwood BH24 1LL
2I0 EIG J. Mcgoldrick 45 Stewarts Road, Dromara, Dromore BT25 2AN
2I0 EIU A. Dowling 74 Ashmount Gardens, Lisburn BT27 5DA
2E0 EIX R. Clark 67 Seymour Street, Chorley PR6 0RR
2E0 EIZ T. Baker 92 Conway Avenue, Derby DE723GR
2E0 EJA Blackwater Radio CG c/o E. Cole 24 Patrick'S Orchard Uffington, Faringdon SN7 7RL
2E0 EJC M. Dominguez 52 St. Leonards Road, Amersham HP6 6DR
2E0 EJJ E. Potter 3 Thomson Court Chadwick Close, Crawley RH11 9LH
2E0 EJK R. Keast 7 The Finches, Newport PO30 5GU
2U0 EJL J. Littlewood Wayland Les Martin, L'islet GY2 4XW
2E0 EJM E. Marsh 16 Laurel Close, North Warnborough, Hook RG29 1BH
2E0 EJO J. Wooldridge 6 Heskin Road, Lydiate, Liverpool L31 0BS
2W0 EJP Dr P. Messenger 9 South Parade, Maesteg CF34 0AB
2E0 EJQ M. Lisle 16 Collegiate Crescent, Sheffield S10 2BA
2E0 EJR R. Edwards 23 Queens Walk, Ruislip HA4 0LX
2I0 EJT J. Bingham 27 Carrickdale Gardens, Portadown, Craigavon BT62 3BN
2E0 EJV A. Jarvis 10 West Park, Wadebridge PL27 6AN
2E0 EJW E. Woodward 309 Hartfields Manor, Hartfields, Hartlepool TS26 0NW
2E0 EJZ J. Berrio 29 Scalborough Close, Countesthorpe, Leicester LE8 5XH
2E0 EKA J. Parton 6 Windmill Road, Atherstone CV9 1HP
2E0 EKB G. Patrick Athena, 121 Ringmer Road, Worthing BN13 1DX
2E0 EKC R. Fulcher 1 Edwards Close Hutton, Brentwood CM13 1BU
2E0 EKD T. Patatu 20 Compair Crescent, Ipswich IP2 0EH
2E0 EKI C. Abbott 38 Foxcover, Linton Colliery, Morpeth NE61 5SR
2E0 EKJ T. Smith Chy Crowshensy, Clifton Road, Redruth TR15 3UD
2E0 EKK G. Cliffe 5 Laurel Cottages, Ongar Hill Road, King's Lynn PE34 4JB
2E0 EKM P. Bray 24 Eldon Terrace, Bristol BS3 4NZ
2I0 EKN W. Martin 81 Camgart Road, Tempo, Tempo BT94 3EQ
2E0 EKP S. Crane 2 Wolsey Close, Ashton-In-Makerfield, Wigan WN4 8DL
2E0 EKR N. Harris 45 Sleigh Road, Sturry, Canterbury CT2 0HT
2E0 EKS D. Storton 50 Cromwell Road, Blackpool FY1 2RG
2E0 EKT M. Evans 9 Hollis Close, Long Ashton, Bristol BS41 9AZ
2E0 EKU J. Woods Invicta Cottage, Carbrooke Road, Thetford IP25 6SD
2E0 EKW R. Macdonald 31 Addison Drive, Stratford upon Avon CV377PL
2E0 EKX A. Trueman 10 Mountbatten Ave, Dukinfield SK165BU
2E0 ELA P. Milton 7 North Crescent, Steeple Bumpstead, Haverhill CB9 7DL
2E0 ELD E. Dalton 120 Goodway Road, Birmingham B44 8RG
2E0 ELE A. Jepson Edmonton Road, Mansfield NG21 9AH
2E0 ELI G. Hayers 87 Bradleigh Avenue, Grays RM17 5RH
2E0 ELK A. Hawkins 5 Ranworth Road, Great Sankey, Warrington WA5 3EH
2E0 ELO D. Smith 7 Kestrel Avenue Bransholme, Hull HU7 4ST
2J0 ELQ M. Thorpe Dolphin Cottage, Union Road, Grouville, Jersey JE3 9ER
2I0 ELR C. Adjey 1 Foyle Park, Portstewart BT55 7DL
2E0 ELT W. Foster 55 Drake Avenue Minster On Sea, Sheerness ME12 3SA
2E0 EMB L. Lee 8 William Avenue, Margate CT9 3XT
2E0 EME P. Smith 5 Olivers Hill, Cherhill, Calne SN11 8UR
2E0 EMF W. Johnson 10 Archdale Road, Nottingham NG5 6EB
2E0 EMG E. Macgurk 10 Elmore Road, Lee on Solent PO139DU
2E0 EMK M. Robino 61 Mill Hill Little Hulton, Manchester M38 9TN
2E0 EML D. Polley 6 Coneygear Road, Hartford, Huntingdon PE29 1QL
2M0 EMM E. Munro 55 Abergeldie Road, Aberdeen AB10 6ED
2E0 EMN C. Young 21A Union Crescent, Margate CT9 1NS
2E0 EMP C. Keszei 2 Blackmore Hill Farm Cottages, Calvert Road, Buckingham MK18 2HA
2E0 EMQ J. Arnold 17 Larch Road Roby, Liverpool L369TY
2E0 EMX Dr A. Holt 36 The Maltings, Malmesbury SN16 0RN
2E0 EMZ M. Brown 99 Apprentice Drive, Colchester CO45SE
2E0 ENB G. Wall Flat 2, Coniston House Holyoake Road, Worsley, Manchester M28 3DH
2E0 ENC E. Woolfenden 20 Belvedere Avenue Atherton, Manchester M46 9LQ
2W0 END M. Townsend 71 Elm Court, Newbridge, Newport NP11 5LU
2E0 ENE E. Smethurst 4 Lower New Row, Worsley, Manchester M28 1BE
2E0 ENF S. Klee 4 Abbots Road, Pershore WR10 1LL
2E0 ENG M. Winch 2 Cranleigh Gardens, Cowes PO31 8AS
2E0 ENI G. Howard 8 Paddock Road, Woodford, Kettering NN14 4FL
2W0 ENJ J. Turner 4 Imble Street, UK SA726QL

2E0 ENM N. Winfield 1 Southview, School Lane, Stoke Row, Henley-on-Thames RG9 5QX
2E0 ENN S. Burton 20 Flowerdown Avenue, Cranwell, Sleaford NG34 8HZ
2E0 ENP K. Matthews St. Helens Cottage, Flimby, Maryport CA15 8RX
2W0 ENQ M. Mckenna 25 Heaton Place Norton Road, Rhos On Sea, Colwyn Bay LL284TL
2W0 ENV C. Hurley 14 Magazine Street, Maesteg CF34 0TG
2E0 ENW E. Wilson 60 Seathorne, Withernsea HU19 2BB
2E0 ENZ P. Joynson 10 Rothesay Gardens, Prenton Hall Road, Prenton CH43 3DW
2E0 EOC L. Guy 34 Brindley Avenue, Wolverhampton WV11 2PB
2E0 EOD A. Billingham 6 Kemble Close, Lincoln LN6 0NR
2E0 EOF C. Westcott 50 Bentinck Street, Sutton-in-Ashfield NG17 4AZ
2E0 EOH C. Shipman 19 Westerkirk Drive, Madeley, Telford TF7 5RJ
2E0 EOI D. Rolfe 49 Hillrise Avenue, Sompting, Lancing BN15 0LU
2E0 EOL Rvd. D. Palmer 14 Walcot Parade, Bath BA1 5NF
2E0 EOM I. Garrard 33 Uplands Road, Hockley SS5 4DL
2E0 EON D. Hendy 12 Rumsam Gardens, Barnstaple EX32 9EY
2E0 EOP N. Haigh 10 Moor Park Gardens, Dewsbury WF12 7AS
2I0 EOS J. Kelso 32 Old Park Manor, Ballymena BT42 1RW
2E0 EOU J. Hinds 2C Telegraph Street, Stafford ST17 4AT
2E0 EOV A. Blackwell 27 Prince Charles Crescent, Farnborough GU14 8DJ
2E0 EOX R. Waterhouse Sunnymead, Beckley Road TN31 6JB
2E0 EOZ J. Green 26 Foxhill Road Burton Joyce, Nottingham NG14 5DB
2E0 EPA A. Bailey 52 Berkeley Road, Shirley, Solihull B90 2HT
2E0 EPB S. Duncan 5 Spring Vale Bilton, Hull HU11 4DN
2I0 EPC T. Crawford 21 Ardranny Drive, Newtownabbey BT36 6BD
2W0 EPE P. Plummer 26 Hill Road, Neath Abbey, Neath SA10 7NR
2E0 EPF A. Lane 26 Astral Gardens Sutton-On-Hull, Hull HU7 4YS
2I0 EPG I. Cairns 18 Molyneaux Avenue, Larne BT40 2TU
2E0 EPH C. Grant 27 Bulrush Close, Chatham ME5 9BN
2E0 EPJ Dr M. Heng 1 St. Giles Croft, Beverley HU17 8LA
2M0 EPK J. Wilson 16 Big Brigs Way, Newtongrange EH22 4DG
2E0 EPM J. Caswell 10 Beech Close, Scole, Diss IP21 4EH
2I0 EPN R. Hetherington 112 Screeby Road, Fivemiletown BT75 0LG
2E0 EPP P. Petersen 15 Kent Gardens, Birchington CT7 9RS
2E0 EPT R. Paris 14 Jarden, Letchworth Garden City SG6 2NP
2E0 EPU L. Pitman 14 Walton Village, Liverpool L4 6TJ
2E0 EPZ D. Taylor 8 Highfields, Mill Croft Close, Norwich NR5 0RU
2E0 EQB D. Baker 22 Cleveland Road, Plymouth PL4 9DF
2I0 EQC M. Mccourt 52A Moira Road, Crumlin BT29 4JL
2E0 EQD N. Peppe 5 Cherford Road, Bournemouth BH11 8SU
2E0 EQG M. Barrett 57 Marlborough Avenue, Hornsea HU18 1UA
2E0 EQH P. Catterall 117 Beech Hill Lane, Wigan WN6 8PJ
2E0 EQI P. Wright 25, Paynes Meadow, Whitminster, Gloucester U. K. GL2 7PS
2E0 EQJ Dr B. Minnis 16 Dene Tye, Crawley RH10 7TS
2E0 EQK Dr F. Harvey 14 Wilton Road, Hornsea HU18 1QU
2E0 EQL S. Daniels 48 Gason Hill Road, Tidworth SP9 7JX
2E0 EQN D. Ellens 150 Lumley Avenue, South Shields NE34 7DJ
2E0 EQP G. White 19 Abbots Hall Avenue Clock Face, St. Helens WA9 4UX
2W0 EQQ L. Pickering 14 Bryn Wyndham Terrace, Cardiff CF42 5NG
2I0 EQR D. Mcdonnell 52 Moira Road, Glenavy, Crumlin BT29 4JL
2I0 EQS T. Mcdonnell 52 Moira Road, Glenavy, Crumlin BT29 4JL
2E0 EQT T. Canning Whitegates, Mayfield Avenue New Haw, Addlestone KT15 3AG
2E0 EQU P. Hill 83 Gladeside, Croydon CR0 7RW
2E0 EQW J. Hope 4 Beaumont Way, Prudhoe NE42 6RA
2E0 EQZ B. Davies 16 Pearmains Close, Orwell, Royston SG8 5QY
2E0 ERB J. Neal 40 Channel View Road, Portland DT5 2AY
2E0 ERD R. Overy 62 Dykelands Road, Sunderland SR6 8ER
2E0 ERE M. Bruce 28 Pheasants Way, Rickmansworth WD3 7ES
2E0 ERF G. Sheppard 32 Bramble Drive, Hailsham BN27 3EG
2E0 ERG R. Harriman 9 Millers Close, Rushden NN10 9RP
2E0 ERJ D. Cassidy 30 Oakland Road, Botley, Southampton SO32 2SX
2E0 ERK E. Kissin 115 Aarons Hill, Godalming GU7 2LJ
2E0 ERL R. Carpenter Junior 218 Mansel Road West, Southampton SO16 9LR
2E0 ERM E. Milner 16 Spring Valley Court, Bramley, Leeds LS13 4TT
2E0 ERO A. Cattell 40 Collyweston Road, Northampton NN3 5ET
2E0 ERP A. Burton 12 Munden Grove, Watford WD24 7EE
2E0 ERQ M. Pittas 89 Seddon Road, Morden SM4 6ED
2E0 ERS R. Springall 18 Westbourne Park, Scarborough YO12 4AT
2E0 ERT R. Smith 21 Canal Road Crossflatts, Bingley BD16 2SR
2E0 ERU M. Curtis 9 Oaklands Crescent, Holt NR25 6UD
2E0 ERV S. Slack 10 Micheldever Road, Whitchurch RG28 7JG
2E0 ERX M. Hales 23 Raeburn Road, Sidcup DA158RD
2E0 ERZ E. Vrentzos 12 Floyer Close, Queens Road, Richmond TW10 6HS
2I0 ESA J. Mcgoldrick 45 Stewarts Road, Dromara, Dromore BT25 2AN
2E0 ESB A. Volkov Flat 8, Despard House, 43 Palace Road, London SW2 3EW
2E0 ESC M. Bennett-Blacklock 46 Friern Road, London SE22 0AX
2E0 ESI A. Ahuja 32 Riverview Gardens, London SW13 8QY
2E0 ESJ N. Evans 38 Cockster Road, Longton, Stoke-on-Trent ST3 2EG
2M0 ESL G. Robinson Asgard, 12 Upper Waston Road, Burray KW172TT
2E0 ESO M. Bull 14 Ermin Walk, Thatcham RG19 3SD
2E0 ESS E. Slevin Woodcock Hall, Cobbs Brow Lane, Newburgh WN8 7NB
2E0 ESU M. Eade Roemah-Kita, Strathcona Avenue, Leatherhead KT23 4HP
2E0 ESX J. Hawthorn 1 Tudor Close, Leigh-on-Sea SS9 5AR
2E0 ESY M. Clitheroe Green End, North Street, Castle Acre, King's Lynn PE32 2BA
2E0 ETA M. Shasby 19 Crawshaw Grange, Crawshawbooth, Rossendale BB4 8LY
2I0 ETB W. Curry 7 Ballyversal Road, Coleraine BT52 2ND
2E0 ETC L. Kay 20 Mytton Lane, Shawbury, Shrewsbury SY4 4JE
2E0 ETD R. Sindall 16 Chantrell Road, Wirral CH48 9XP
2E0 ETE A. Beardsley 10 Moreton Close, Church Crookham, Fleet GU52 8NS
2E0 ETG M. Landon 29 Portland Road, Hucknall, Nottingham NG15 7SL
2E0 ETH C. Marshall 51 Hedgerow Close, Redditch B98 7QF
2E0 ETI A. Thompson 25 Ardingly Road, Cuckfield, Haywards Heath RH17 5HD
2M0 ETJ G. Cartwright Redhouses Outertown, Annan DG12 5LN
2M0 ETL I. Ferguson 5 Grahamsfield Court Kirkpatrick Fleming, Lockerbie DG7 3BD
2E0 ETN R. Bradshaw 272 Councillor Lane, Cheadle Hulme, Cheadle SK8 5PN
2E0 ETP S. Cook 68, Barnsley S71 4RY
2E0 ETQ G. Hinson The Leys, Brierley Hill DY5 3UJ
2E0 ETT T. Horsten Kastelsvej 4, 2.Tv, Copenhagen E 2100 Denmark
2E0 ETU E. Dutson 19 Maythorn Drive, Cheltenham GL51 0QH
2E0 ETV K. Bianchini 10 St. Leonards Road, Headington, , Oxford OX3 8AA
2I0 ETW P. Moore 32 Kinnegar Rocks, Donaghadee BT21 0EZ
2E0 ETY C. Wrobel 33 Harsnett Road, Colchester CO1 2HS
2W0 ETZ W. Cooper 1 Bedw Street, Maesteg CF34 0TF
2E0 EUB C. Wooldridge 26 Grieg Close, Basingstoke RG22 4DU
2E0 EUD B. Clark 46 Fraser Close, Laindon, Basildon SS15 6SU
2E0 EUF F. Boyce 277 Manor View, Par PL24 2EP
2E0 EUH P. Davies 2 Lynfords Drive, Runwell, Wickford SS11 7PP
2E0 EUI D. Nolan Flat 7, Fonthill Court, Honor Oak Road, London SE23 3SJ
2E0 EUJ C. Cromie 140 Whalley Road Wilpshire, Blackburn BB1 9LJ
2E0 EUM T. Mann 6 Kenley Close, Wickford SS11 8XL
2E0 EUN E. Underhill 61 Goldthorne Avenue, Sheldon, Birmingham B26 3LA
2W0 EUO A. Jones 75 Hollybush Road, Cardiff CF23 6SZ
2E0 EUP J. Foxall 2 Millers Walk Pelsall, Walsall WS3 4QS
2E0 EUR S. Milner Pavilion House, School Lane, Ormskirk L40 3TG
2I0 EUV R. Skelton 16 Demiville Avenue, Lisburn BT27 5RE
2E0 EUW M. Augustus 3 Heathend Cottages Heathend, Wotton-under-Edge GL12 8AS
2W0 EUY P. Squire 18 Rayon Road Greenfield, Holywell CH8 7EQ
2E0 EUZ A. Fisher 63 Soloman Drive, Bideford EX39 5XY
2E0 EVA M. Knowles 12 Dalestorth Avenue, Mansfield NG19 6NT
2E0 EVC Dr C. Hall 6 Browning Road Church Crookham, Fleet GU52 0YJ
2E0 EVE D. Cattermole Blaxhall Hall Crossing, Little Glemham, Woodbridge IP13 0BP
2E0 EVF D. Redmayne 10 The Square, Kington HR5 3BA
2I0 EVH A. Porter 8 Ballyregan Avenue Dundonald, Belfast BT16 1JW
2E0 EVI G. Rodriguez 32 Mount Pleasant, Prestwich, Manchester M25 2SD
2E0 EVM E. Scott 64 Smallcombe Road, Paignton TQ3 3RX
2E0 EVP M. Brasher 48 Eldertree Road Thorpe Hesley, Rotherham S61 2TQ
2E0 EVQ G. Singleton 2 Rome Avenue, Burnley BB115LQ
2E0 EVR G. Round 128 Leicester Road, Shepshed, Loughborough LE12 9DH
2E0 EVU D. Mallinson 7 Abbots Way, Yeovil BA21 3HX
2E0 EVV T. Roberts 16 Malfort Road, London SE5 8DQ
2E0 EVW B. Skelton 1 Summer Hill Road, Bexhill-on-Sea TN39 4LN
2E0 EVX S. Hall 12 Lady Jane Grey Road, King's Lynn PE30 2NW
2E0 EWB J. Holley Lookers, Blenheim Road Littlestone, New Romney TN28 8PR
2E0 EWF S. Totham Cavaliers, 11 Dane Court Manor, School Road, Tilmanstone, Deal CT14 0JL
2E0 EWH M. Lovering 19 Chilton Avenue, Sittingbourne ME10 4TB
2E0 EWL J. Keefe 64 Heath Lane, Blackfordby, Swadlincote DE11 8AA
2E0 EWM E. Melman 177 Grantham Road, London E12 5NB
2E0 EWQ N. Booth 10 Games Walk, Wythenshawe, Manchester M22 1SN
2E0 EWS E. Sherwin 24 Bruce Avenue, Barnsley S70 4DZ
2E0 EWT C. Barker 11 Long Meadows, Chorley PR7 2YA
2E0 EWX J. Dowson 4 Thorntree Close, Goole DN14 6HJ
2M0 EWY A. Kinnersley 5A Regent Terrace, Dunshalt, Cupar KY14 7HB
2E0 EXA D. O Brien 5 Orson Leys, Rugby CV22 5RG
2E0 EXB R. Fearnley 31 Radburn Court, Dunstable LU6 1HW
2E0 EXC C. Wilson 21 New Road, Hythe, Southampton SO45 6BN
2E0 EXG R. Low 4A Bellevue Road, Romford RM5 3AY
2E0 EXH A. Cooke 17 St. Aldwyn Road, Seaham SR7 0AN
2E0 EXI C. Smith 37 Mace House, Union Lane, Isleworth TW7 6GP
2E0 EXJ S. Baldwin 143 Oxford Road, Swindon SN3 4JA
2E0 EXK M. Constantine 82 The Oval, Brough HU15 1DD
2E0 EXN T. Jones 2 Holly Road, Penketh, Warrington WA5 2AG
2E0 EXO T. Crocker 32 Godmanston Close, Poole BH17 8BU
2I0 EXP A. Boyd 27 The Meadows, Dungannon BT71 6PW
2E0 EXQ A. Parkin 22 Glorney Mead, Badshot Lea, Farnham GU9 9NL
2E0 EXU R. Coles Bay Cottage, St. Catherines Road, Ventnor PO38 2NE
2E0 EXV J. Kelly Flat 4, The Corner Place, 1 North Road, Harborne, Birmingham B17 9PA
2E0 EXW R. Whitehead 1 Smithy Site, Farnborough, Wantage OX12 8NS
2E0 EXX B. Smith 8 Mill Field Close, South Kilworth, Lutterworth LE17 6FE
2E0 EXY J. Sergeant 16 Green Park, Whalley, Clitheroe BB7 9TJ
2W0 EXZ D. Morgan 28 Harbour Village, Goodwick SA64 0DY
2E0 EYB M. Jenkins Flat 2, 66 Union Street, Ryde PO33 2LG
2E0 EYC C. Revell Westview, 1 Chapel Lane, Hull HU12 0US
2E0 EYE E. Ross Foundry Cottage, Crowders Lane, Battle TN33 9LP
2E0 EYF A. Kazmi 39 Herga Road, Harrow HA3 5AX
2E0 EYG J. Miller 9 St. Nicholas Road Tillingham, Chelmsford CM0 7SQ
2E0 EYH W. Rudge 33 Wyrley Rd, Wolverhampton WV11 3NY
2E0 EYI R. Wheeler 21 Abbey Drive Houghton Le Spring, Tyne&Wear DH45JZ
2D0 EYK N. Smith 4 Cooil Farrane, Douglas IM2 1NX
2E0 EYP J. Read 49 Bransdale Way, Macclesfield SK11 8QT
2E0 EYS R. Bewick 357 Franklands Village, Haywards Heath RH16 3RP
2E0 EYU S. Kembrey 101 Yew Tree Drive, Bristol BS15 4UF
2E0 EYY M. Murray 31 Feeny Street Sutton Manor, St. Helens WA9 4BJ
2M0 EZA R. Cantwell 78 New Street, Musselburgh EH21 6JQ
2E0 EZC A. De Mora 36 West Park, Minehead TA24 8AN
2E0 EZF G. Smith 92 Brighton Road, Banstead SM7 1BU
2E0 EZG M. Lewis 1 Kingsmead Stretton, Burton on Trent DE13 0FQ
2E0 EZL T. Newton 1 Brimley Park, Bovey Tracey, Newton Abbot TQ13 9DE
2E0 EZO M. O'Connor 5 Kesbrook Drive Ashwood Park, Overseal DE126NS
2W0 EZS P. Collier 42 Derwen Road Alltwen, Pontardawe, Swansea SA8 3AU
2E0 EZX W. Taylor 5 Council Bungalows, Churchtown, Belton, Doncaster DN9 1PD
2E0 EZY H. Foster 6 Lakeway, Blackpool FY3 8PF
2E0 EZZ G. Hutchings 28 Fortfield Road, Bristol BS14 9NT
2E0 FAA D. Mcglone 32 Shipley Mill Close, Kingsnorth, Ashford TN23 3NR
2E0 FAB J. Whalley Flat 9, 37 Church Street, Southport PR9 0QT
2E0 FAC P. Burke 38 Bosworth Square, Rochdale OL11 3QG
2E0 FAE A. Foote Flat One, Kimber'S Close Kennet Road, Newbury RG14 5JF
2E0 FAH P. Harris Flat 33, Buckingham Court Shrubbs Drive, Bognor Regis PO22 7SE
2E0 FAJ S. Kingstone 3 Roman Way, Tamworth B79 8NF
2E0 FAM C. Horridge 6 Back Sreet, East Stockwith, Gainsborough DN21 3DL
2E0 FAN P. Pearce 41 Tennyson Avenue, Boldon Colliery NE35 9EP
2W0 FAP A. Phillips 3 Pen Y Llys, Rhyl LL18 4EH
2E0 FAQ G. Austin 15K Hardie Avenue, Rugeley WS15 1NT
2W0 FAR A. Burgess 18 Fairmeadows, Maesteg CF34 9JL

2E0 FAS N. Duke 5 Hannay Close, Barrow-in-Furness LA14 1SZ
2E0 FAU B. Cairns 4 Spence Court, Great Ayton, Middlesborough TS9 6DW
2E0 FAV A. Barter 4 Blackberry Way Kingsteignton, Newton Abbot TQ12 3QX
2E0 FAY N. Fahey 5 Hillside, Felmingham, North Walsham NR28 0LE
2E0 FBA M. Mcphee 63 Mumford Close, West Bergholt, Colchester CO6 3HY
2E0 FBC F. Foy 4 The Square, East Rounton, Northallerton DL6 2LB
2E0 FBD D. Smith Heath Farm, Heath Road, Woolpit, Bury St. Edmunds IP30 9RL
2E0 FBE M. Goodman 80 Clay Street, Soham, Ely CB7 5HL
2E0 FBF R. Boyes 63 Larch Road, New Ollerton, Newark NG22 9SX
2E0 FBH Z. Yao Room 4C, Unit 12-2, Zhonghaibanshanxigu Garden, No.15 Zhongqing Rd., Yantian District, Shenzhen 518000 China
2E0 FBJ A. Tomczynski 66 Commercial Street, London E1 6LT
2E0 FBL F. Baker 275 Bye Pass Road, Beeston, Nottingham NG9 5HS
2E0 FBN T. Weston 4 The Pightle, Peasemore, Newbury RG20 7JS
2E0 FBO J. Raybould 33 Lincoln Road, Dorrington, Lincoln LN4 3PT
2E0 FBQ R. Copus 58 Colchester Rd, Holland on Sea CO15 5DG
2E0 FBR S. Davis 74 The Driveway, Canvey Island SS8 0AD
2M0 FBU K. Govan 30 Glenwell Avenue, Stranraer DG9 7BA
2I0 FBY I. Gillespie 32 Maghaberry Manor Moira, Craigavon BT67 0JZ
2E0 FCB Dr R. Hopkins 26 Seymour Road, Bordon GU35 8JX
2E0 FCE B. Clark 8 Langdale Close, Farnborough GU14 0LQ
2W0 FCF D. Clark 137 Llanedeyrn Road, Penylan, Cardiff CF23 9DW
2E0 FCG C. Staples 32 Browns Lane, Netherton, Bootle L30 5RW
2E0 FCH A. Finch 6 Clover Way, Thetford IP24 1LQ
2E0 FCJ C. Pendlebury 73 Pool Street, Wigan WN3 5BT
2W0 FCM L. Paschalis 45 Pencisely Road, Cardiff CF5 1DH
2E0 FCO M. Killoran 4 Victoria Road, Pudsey Leeds LS28 7SR
2M0 FCQ J. Degnan 6B Manse Road, Whitburn EH470QA
2E0 FCS C. Spencer 18 Coatsby Road Kimberley, Nottingham NG16 2TH
2M0 FCT S. Mulligan 17 Crawfurd Gardens, Rutherglen, Glasgow G73 4JP
2E0 FCZ C. Cooper 25 Waterside Close, Loughborough LE11 1LP
2E0 FDB R. Goldup 57 Partridge Way, Old Sarum, Salisbury SP4 6PX
2W0 FDC S. Quinn 1334 Carmarthen Road Fforestfach, Swansea SA5 4BR
2E0 FDD B. Evans 2 Hastings Road, Eccles, Manchester M30 8JR
2E0 FDE R. James 159 Orton Avenue, Birmingham B76 1JN
2E0 FDG B. Pettit 30A Station Road, Corton, Lowestoft NR32 5BE
2E0 FDH C. Ward 1 Granary Close, Codford, Warminster BA12 0PR
2M0 FDI D. Robertson Grindhus, , Gonfirth, Voe ZE2 9PY
2E0 FDS K. Bowyer 84 Old Chester Road Helsby, Frodsham WA6 9PG
2M0 FDT S. Doonan West Clanfin Farm Waterside, Kilmarnock KA3 6JQ
2E0 FDU D. Knowles 12 Forest Grove, Harrogate HG2-7JU
2E0 FDW F. Quinn Harley Cottage Beech Grove Gardens, Carlisle CA30LR
2M0 FDZ S. Mclaughlin 21 Shirrel Road, Motherwell ML1 4RD
2E0 FEB S. Scotching 26 Newton Way, Leighton Buzzard LU7 4YU
2E0 FEC G. Eycott 1 Ham Road, Wanborough, Swindon SN4 0DF
2E0 FEE W. Byers 37 Windermere Drive, Bletchley, Milton Keynes MK2 3NR
2E0 FEJ I. Davison 35 Newhouse Avenue, Esh Winning, Durham DH7 9JH
2E0 FEM S. Simpson 48 Weatherhill Road Smallfield, Horley RH6 9LY
2E0 FEO R. Hunter 3 Sandy Way, Croyde, Braunton EX33 1PP
2E0 FEP J. Harkins 22 Croomes Hill, Bristol BS16 5EH
2E0 FEQ K. Polston 10 Marsh Farm Road, South Woodham Ferrers, Chelmsford CM3 5WP
2E0 FEV D. Walton 106 Cumberworth Lane, Lower Cumberworth, , Huddersfield HD8 8PG
2I0 FEX D. Rantin 8 Buchanans Road, Newry BT35 6NS
2W0 FEY A. Fey 28 Bryn Rhedyn, Caerphilly CF83 3BT
2E0 FEZ A. Waller 155 Bridgemary Road, Gosport PO13 0UT
2E0 FFC D. Berry 27 Harcourt Terrace, Headington, Oxford OX3 7QF
2W0 FFI D. Evans 1 Heol Glyndwr, Fishguard SA65 9LN
2E0 FFJ G. Saunders Hastings Road, Battle TN330TA
2E0 FFK S. Hagan 5 High Street, Brampton, Huntingdonshire PE28 4TG
2W0 FFL Dr F. Labrosse 72 Ger Y Llan Penrhyncoch, Aberystwyth SY23 3HQ
2E0 FFM J. O'Reilly 116 Coleridge Way, Crewe CW1 5LF
2W0 FFN A. Prime 106 Snowden Road, Cardiff CF5 4PS
2E0 FFP G. Peters Curlew Court, Guys Head Road, Sutton Bridge, Spalding PE12 9QQ
2E0 FFS R. Baines 2 Lower Lune Street, Fleetwood FY7 6DA
2E0 FFV A. Hawkes 9 Ferneley Avenue, Hinckley LE100FE
2E0 FFW P. Drake Flat 1, Richmond Court, Eagle Close, Yeovil BA22 8JY
2M0 FFY C. Northcott 4/6 Castleview House, 2 Craigour Place, Edinburgh EH17 7RT
2E0 FGA P. Meanwell 20 Crow Park Avenue, Sutton-On-Trent, Newark NG23 6QG
2E0 FGH N. Speller 2 Hurst Rise Road, Oxford OX2 9HQ
2E0 FGI I. Kendrick Whybank 10 Bankhouse Drive, Congleton CW12 2BH
2E0 FGM S. Haigh 17 Glebe Street, Swadlincote DE11 9BW
2E0 FGO P. Offord 1 Adare Close, Dunmow CM6 2GR
2E0 FGQ N. Bennett 35 West Shepton, Shepton Mallet BA4 5UD
2E0 FGR M. Newman 274 Long Drive, Ruislip HA4 0HY
2E0 FGT M. Jones 93 Barrs Road, Cradley Heath B64 7HH
2E0 FGV G. Brindle 25 Chedworth Drive, Witney OX28 5FS
2E0 FGW D. Crouch 7 Tresco Road, Berkhamsted HP4 3JZ
2E0 FGY B. Bestwick 185 Ashbourne Road, Turnditch, Belper DE56 2LH
2E0 FHA M. Sprague 11 The Orchards, Witcham, Ely CB6 2LR
2E0 FHB N. Connolly 32 St. Oswald Road, Bridlington YO16 7SD
2E0 FHC R. Gaskell Cottonwood California, Baldock SG7 6NU
2E0 FHD L. Rodgers 27 Arden Houses Normanby-By-Spital, Market Rasen LN8 2HE
2E0 FHE I. Gould 2 Harps Avenue Minster, Sheerness ME12 3PF
2E0 FHH C. Wilkinson Flat 5, Hometide House Beach Road, Lee-on-the-Solent PO13 9BP
2E0 FHJ R. Lane 2 Dickiemoor Lane, Plymouth PL5 3NU
2E0 FHK I. Harding 7 Hawthorne Close, River, Dover CT17 0NG
2E0 FHM A. Papiewski 42 Balmoral Avenue, Spalding PE11 2RU
2E0 FHN J. Clarke-Stanley 6 Culpins Close, Spalding PE11 2JL
2E0 FHQ J. Gale 46 Ingshead Avenue Rawmarsh, Rotherham. S625BW
2E0 FHR K. Scott 86 Holcombe Drive, Burnley BB10 4BH
2E0 FHV A. Shaw 8 Champion Way, Mablethorpe LN12 2EJ
2E0 FHX M. Ross Whitewalls Off Golf Road, Mablethorpe LN12 1LP
2E0 FIA A. Smeed Flatt 8 9 Peir Terrace, Lowestoft NR330AB
2E0 FIB S. Watling 1 Chediston Green, Chediston, Halesworth IP19 0BB
2E0 FIF M. Hadfield 22 Mansfield Road, Clowne, Chesterfield S43 4DH
2E0 FIJ Dr D. De-Cogan 52 Gurney Road, New Costessey, Norwich NR5 0HL
2E0 FIK A. Holmes 23 Bowen Road, Darlington DL3 0TH
2E0 FIO G. Chalklin 18 Trinity Close, West Mersea, Colchester CO5 8RW
2I0 FIP J. Miskimmin 3 Orchard House, Mark Street, Newtownards BT23 4WS
2E0 FIR M. Firth 209 High Street, Wickham Market, Woodbridge IP13 0RQ
2E0 FIU J. Smith 54 Sneinton Hermitage, Sneinton, Nottingham NG2 4BS
2E0 FIV J. Rymell 9 Carter Avenue Ruddington, Nottingham NG11 6NP
2E0 FIW D. Peacock 41 Oxford Meadow, Sible Hedingham, Halstead CO9 3QW
2E0 FIY G. Walmsley 27 Camberwell Way, Hull HU8 0RU
2E0 FJA A. Ferriroli 142 Hillbury Road, Warlingham CR6 9TD
2E0 FJC A. Winton 10 Old Parsonage Court Guithavon Street, Withem CM8 1XP
2E0 FJD K. Dobson 1 Howarth Road, Ashton-On-Ribble, Preston PR2 2HH
2E0 FJE A. Pyatt 13 Summervale Road, Tunbridge Wells TN4 8JJ
2E0 FJG Z. Zhao Christs College, Cambridge CB2 3BU
2W0 FJG A. Heigh Janneen, Henry Street, Wrexham LL14 4DA
2E0 FJI A. Atkinson 21 Dennington Crescent, Basildon SS14 2FF
2E0 FJJ R. Norwood Flat 27, Cranfield Court, Galadriel Spring, South Woodham Ferrers Essex, Cm3 7Bd., Chelmsford CM3 7BD
2E0 FJK S. Chaney 54 Clementine Avenue, Seaford BN25 2XG
2E0 FJN D. Picker 12 Crown Close, Barnsley S70 4DB
2E0 FJP J. Park 18 Ladgate Grange, Middlesbrough TS3 7SL
2E0 FJQ P. Hall 11 Middleton Court, Mansfield NG18 3RN
2E0 FJT A. Fletcher 74 Devonshire Road, Maltby, Rotherham S66 7DQ
2E0 FJU J. Fletcher 74 Devonshire Road, Maltby, Rotherham S66 7DQ
2E0 FJV C. Walson 30 West Crescent, Duckmanton, Chesterfield S44 5HE
2E0 FJW J. Mcbride 78 Sherwood Drive, Redcar TS11 6DY
2E0 FJX D. Roake 9 Falcondale Walk, Westbury On Trym, Bristol BS93JG
2E0 FKB M. Edge 15 Littlemoor Avenue, Kiveton Park, Sheffield S26 5NZ
2E0 FKC R. Chappell 24 Woodend Drive, Shipley BD18 2BW
2E0 FKD J. Main 15 Byron Road, Lydiate, Liverpool L31 0DB
2E0 FKE D. Buchan 40 Bradstone Road, Winterbourne, Bristol BS36 1HQ
2M0 FKF D. Simpson 35 Westbourne Avenue Tillicoultry Clackmannanshire Fk136Pu, Tillicoultry FK136PU
2E0 FKH S. Caddy 133A Barnby Gate, Newark NG24 1QZ
2E0 FKJ R. Robinson 4 Limetree Court, Taverham, Norwich NR8 6QY
2I0 FKM G. Colgan 42 Loughview Village, Carrickfergus BT38 7PD
2E0 FKN K. Bramley 21 Charles Street, Sutton-in-Ashfield NG17 4LG
2E0 FKS B. Hoare 2 St. Peters Close, South Newington, Banbury OX15 4JL
2E0 FKT L. Allcock 7 Arundel Court, Chesterfield S43 3UY
2E0 FKU C. Gibson 11 Parkside Avenue, Queensbury, Bradford BD13 2HQ
2W0 FKW D. Delve 28 Cader Avenue Kinmel Bay, Rhyl LL18 5HY
2E0 FKY P. Robbins 2 Bramble Drive Claremont Park, Berrow, Burnham-on-Sea TA8 2NH
2E0 FLA D. Walker 290 Shannon Road, Hull HU8 9RY
2E0 FLF C. Wilson 12 Desmond Avenue, Hornsea HU18 1AF
2M0 FLG M. Bradshaw 32 Greycraigs, Cairneyhill, Dunfermline KY12 8XL
2E0 FLH M. Crees Flat 3 Fox House, Fox Lane North., Chertsey, KT16 9GY
2W0 FLI Dr E. Flikkema 7 St. James Mews, Great Darkgate Street, Aberystwyth SY23 1DW
2M0 FLJ J. Morris 10 Middlemas Road, Dunbar EH42 1GJ
2E0 FLK H. Bond Flat 4, Athrington Court, First Avenue, Felpham, Bognor Regis PO22 7LB
2E0 FLL B. Murtagh 4 Sandcroft Court, 76 Garlands Road, Redhill RH1 6GZ
2E0 FLM G. Powell 7 Donstan Road, Highbridge TA9 3LA
2E0 FLN J. Horn 8 Princess Close, Watton IP25 6XA
2I0 FLO I. Nicholl 7 Killyclooney Road, Dunamanagh, Strabane BT82 0LZ
2E0 FLQ O. Phillips 54 Marshall Road, Cambridge CB1 7TY
2E0 FLR S. Rhenius Baythorne Cottage, Baythorne End, Halstead CO9 4AB
2E0 FLU D. Chivers 12 Highburn Close, Burnham on Sea TA8 1LU
2E0 FLV M. Osband 22 Samian Crescent, Folkestone CT19 4JW
2W0 FLW D. Flewin Huntingdon Way, Swansea SA2 9HN
2E0 FMA A. West 33 Mundays Row, Waterlooville PO8 0HF
2E0 FMB T. Huntriss 1 Threefields, Ingol, Preston PR2 7BE
2E0 FME D. Bennett 29 Margraten Avenue, Canvey Island SS8 7JD
2E0 FMG P. Booker 17 Colton Copse, Chandler'S Ford, Eastleigh SO53 4HQ
2E0 FMH D. Gaskell 126 Higher Green Lane Astley, Tyldesley, Manchester M29 7JB
2E0 FMI W. Terry 11 Crescent Ave, Overhulton., Bolton BL51EN
2E0 FMK F. Hennigan 18 Friary Gardens, Newport Pagnell MK16 0ZJZ
2E0 FML A. Smith 6 Rawlinson Avenue, Caistor, Market Rasen LN7 6NQ
2I0 FMN J. Mcniece 14 Ballybracken Road, Doagh, Ballyclare BT39 0SE
2E0 FMO E. Harriott 21 Slindon Croft, Alvaston, Derby DE24SD
2E0 FMP T. Leatherbarrow 17 Egerton, Skelmersdale WN8 6AA
2W0 FMQ G. Goodbourn The Gables Cwmduad, Cwmduad, Carmarthen SA33 6XJ
2W0 FMR C. Worka 9 Bertha Street, Pontypridd CF37 1TS
2E0 FMS G. Bailey 61 Great Ranton, Pitsea, Basildon SS13 1JS
2E0 FMT P. Young 11 St. Andrews Avenue, Washington NE37 1AH
2E0 FMV V. Hayes 68 Billingsley Road, Birmingham B26 2EA
2E0 FMX N. Shajan 19 Sturgess Avenue, London NW4 3TR
2E0 FMY F. Masters 91 Mayfair Avenue, Worcester Park KT4 7SJ
2W0 FNA K. Byast 10 Chapel Street, Amlwch Port, Amlwch LL68 9HT
2E0 FNB F. Nuttall 4 Kingholm Gardens, Bolton BL1 3DJ
2E0 FNG M. Grice 48 St. Ives Road, Coventry CV2 5FZ
2E0 FNH D. Flood 506 Preston Old Road, Blackburn BB2 5LY
2M0 FNI Capt. R. Smith Buttview Cottage, South Bragar, Isle of Lewis HS2 9DH
2E0 FNJ C. Wilkinson 67 Middleton Park Grove, Leeds LS10 4BG
2E0 FNM M. Bamber 10 Sedgeley Mews, Freckleton, Preston PR4 1PT
2I0 FNN D. Wilson 81 Parknasilla Way, Aghagallon, Craigavon BT67 0AU
2E0 FNO R. Bishop 12A Goseley Avenue, Hartshorne, Swadlincote DE11 7EZ
2E0 FNQ J. Mason 11 Scriven Grove, Haxby, York YO32 3NW
2E0 FNT M. Morse 5 Northload Terrace, Glastonbury BA6 9JW
2E0 FNV A. Price 87 Derricke Road, Bristol BS14 8NH
2E0 FNY D. Chatterton 3 Hunt Close, South Wonston, Winchester SO21 3HY
2E0 FOD P. Dekkers 21 Nodens Way, Lydney GL15 5NP
2E0 FOE C. Pascoe Treleven, Primrose Hill, Goldsithney, Penzance TR20 9JR
2W0 FOG M. Haywood-Samuel 38 Tanygraig Road, Llanelli SA14 9LH
2E0 FOH C. Donachie 22 Glasgow Street, Hull HU3 3PR
2E0 FOI P. Pearce Criggion Mw Radio Station, Back Lane, Criggion, Shrewsbury

SY5 9BE
2E0 FOK D. Sutherland 22 Cherry Street, Wigston LE18 2BB
2E0 FOL W. Bartle 5 Crosskeys Row. Chapel Milton, Chapel En le Frith SK23 0QQ
2E0 FOP J. Bovey 17 North Bank Road, Bingley BD16 1UH
2E0 FOR I. Forester 35 Thackeray Street, Sinfin, Derby DE24 9GY
2E0 FOT S. Reason 49 Highfield Road, Pudsey LS28 7JW
2E0 FOX B. Hopkins 60 Hales Gardens, Birmingham B23 5DF
2E0 FPA R. Etchells 6 Woodbank Court, Canterbury Road, Manchester M41 7DY
2I0 FPB D. Neill 8 Castle Meadows Carrowdore, Newtownards BT22 2TZ
2E0 FPF M. Waite 108A Chantry Gardens, Southwick, Trowbridge BA14 9QS
2M0 FPI M. Drever 7 Abbotswell Crescent, Aberdeen AB12 5AQ
2E0 FPJ J. Phillips 1 Wath Cottages, Cundall, York YO61 2RL
2E0 FPO B. Bruce 1 Stone Cottages, Chilmington Green, Great Chart, Ashford TN23 3DW
2E0 FPQ E. Turner 2 Shepherds Close, Fen Ditton, Cambridge CB5 8XJ
2I0 FPT D. Hawthorne 58 Seagahan Road, Collone, Armagh BT60 2BH
2E0 FPW W. Garvey 254 Bury Road, Tottington, Bury BL8 3DT
2E0 FPX M. Howard Flat 10, Oakdale, 6 Westgate Road, Beckenham BR3 5DY
2E0 FPY A. Capitan 41 Cunningham Drive, Runcorn WA7 4DL
2E0 FQA L. Scully 17 Fenwick Close, Woking GU21 3BY
2E0 FQC J. Roberts 41 Mcneill Avenue, Crewe CW1 3NW
2E0 FQE A. Brook Inkerman House 113 Clovelly Road, Bideford EX39 3BY
2E0 FQG K. Lambert Flat 6, Vernons Court, Vernons Lane, Nuneaton CV10 8BB
2E0 FQI S. Wilson-King Maple House, Newton Reighny CA11 0AY
2E0 FQJ R. Goldston 30 Vincent Road, Liverpool L21 7NX
2W0 FQM B. Davies 10 Long Acre Court, Bishopston, Swansea SA3 3AY
2E0 FQP D. Daniel 38 Netherthorpe Lane, Killamarsh, Sheffield S21 1DA
2E0 FQS P. Dyke 10 Elmtree Close, Ashurst, Southampton SO40 7FD
2E0 FQT J. Mumby 68 East Common Lane, Scunthorpe DN16 1QH
2E0 FQU A. Thomas Penberthy Road, Portreath TR164LU
2I0 FQW M. Alsallal 9 Grove Street, Lisburn BT27 4YQ
2M0 FRA C. Fraser-Hopewell 2/1 70 Albert Road, Glasgow G42 8DW
2E0 FRB D. Munday 29 Coombe Park, Wroxall, Ventnor PO38 3PH
2E0 FRC F. Clements 40 Ellison Fold Terrace, Darwen BB3 3EB
2E0 FRD D. Collins-Cubitt 75 Anthony Drive, Norwich NR3 4EW
2E0 FRF A. Church The Willows, Warboys Road, Huntingdon PE28 3AH
2E0 FRG F. Noble 1045, 45Th Street Apartment A, California 94608 United States
2M0 FRJ R. Fair 6 Fairways, Stewarton, Kilmarnock KA3 5DA
2E0 FRK D. Church The Willows, Warboys Road, Huntingdon PE28 3AH
2E0 FRL Rvd. L. Williams 36 Royd Court, Mirfield WF14 9DJ
2E0 FRO P. Froggatt 11 Goldsmith Road, Walsall WS3 1DL
2E0 FRP K. Hiscock 45 Gloucester Road, Reading RG30 2TH
2M0 FRR A. Macintyre 2 Memorial Square, Main Street, Castletown, Thurso KW14 8TU
2E0 FRS R. Haughton 5 Fairfield Way, Wesham, Preston PR4 3EP
2E0 FRU C. Andrew 85 Priory Street, Corsham SN13 0BA
2E0 FRY C. Fryer 14 Perks Road, Wolverhampton WV11 2ND
2M0 FSB D. Kelly 21 Dhailling Road, Dunoon PA23 8EA
2E0 FSC C. Bugariu 17 Foxon Lane, Caterham CR3 5SG
2E0 FSD D. Park 29 Tresco Close, Blackburn BB2 4RT
2E0 FSE L. Beaney Brookside, Walkers Lane, Shorwell, Newport PO30 3JZ
2E0 FSG B. Jones 9 St. James Close, Hanslope, Milton Keynes MK19 7LF
2E0 FSH C. Winfield 11 Lowe Avenue, Smalley DE7 8PW
2E0 FSI D. Woolger 17 Wyvern Close, Tangmere, Chichester PO20 2GQ
2E0 FSK G. Hardill 107 Leicester Road Whitwick, Coalville LE67 5GN
2E0 FSM M. Bruce 27 Blaenant, Emmer Green, Reading RG4 8PH
2E0 FSN H. Hamilton Flat B, 9 Cambridge Drive, London SE12 8AG
2M0 FSP S. Paterson Springbank, 2 Mains Of Cuffurach, Clochan, Buckie AB56 5HP
2E0 FSQ D. Smith 7 Oakley Grove, Wolverhampton WV4 4LN
2E0 FSS D. Booth 19 Oak Avenue, Elloughton, Brough HU15 1LA
2M0 FSV K. Winkler 108 Gilmore Place, Edinburgh EH3 9PL
2E0 FSX F. Riches 4 Priory Close, Chelmsford CM1 2SY
2M0 FTA L. Pinkowski 69 Nelson Avenue, Livingston EH54 6BZ
2E0 FTB A. Price Barn Owl Roost, Astwith, Chesterfield S45 8AN
2E0 FTC J. Forbes Weald Barkfold Farm, Plaistow, Billingshurst RH14 0PJ
2E0 FTD T. Raymond 13 Riverside Wolsingham, Bishop Auckland DL13 3BP
2M0 FTH S. Macdonald 110 High Street Cuminestown, Turriff, AB53 5YH
2M0 FTI D. Yeaman 7 Brimmond Crescent, Westhill AB32 6RD
2E0 FTL M. Dorrington 19 Shaftesbury Drive, Wardle, Rochdale OL12 9LT
2E0 FTM D. Fletcher 97 Wallace Crescent, Carshalton SM5 3SU
2E0 FTO R. Parker 21 Yates Way, Ketley Bank, Telford TF2 0AZ
2E0 FTP N. Busley Busley, South Drove, Spalding PE11 3BD
2E0 FTQ A. Street 110 Magdalen Street, Colchester CO1 2LF
2E0 FTS P. Johnson 12A Beechroyd, Pudsey LS28 8BH
2E0 FTT M. Thompson 10 Kilchurn, Consett DH88TQ
2E0 FTU R. Dutton 3 Kilkenny Road, Guisborough TS14 7LE
2E0 FTV S. Barber 31 Aysgarth Avenue, Crewe CW1 4QE
2E0 FTW J. Northall 5 West Winds Road, Winterton, Scunthorpe DN15 9RU
2E0 FTX J. Krol 40 Hampton Gardens, Southend-on-Sea SS2 6RW
2E0 FTZ B. Louder Flat 1, 30 Mill Street, Bideford EX39 2JJ
2E0 FUA Dr R. Pediani Old School House Great Coxwell, Faringdon SN7 7NB
2E0 FUD Dr R. Blackwell Vikings Hall, Baylham, Ipswich IP6 8JS
2E0 FUE A. Taylor-Roberts 38 Old Coach Road, Bulford, Salisbury SP4 9DA
2E0 FUG R. Yarrow 7 Pitts Close, Binfield, Bracknell RG42 4ES
2E0 FUH Dr C. Pomfrett 17 Manifold Close, Sandbach CW11 1XP
2E0 FUJ M. Webster Summerfield Cottage, Walker Lane Wadsworth, Hebden Bridge HX7 8SJ
2E0 FUN S. Southern 37 Conway Road, Calcot, Reading RG31 4XP
2E0 FUO D. Shaw 81 Chesterfield Road Tibshelf, Alfreton DE55 5NJ
2E0 FUQ J. Matthews 2 Farm Close, Bungay NR35 1JG
2W0 FUR J. Pattinson 79 Waterloo Road, Penygroes, Llanelli SA14 7PN
2M0 FUS S. Mason 31 Clifton Road, Lossiemouth IV31 6DJ
2I0 FUT M. Crozier 33 Cullentragh Road, Poyntzpass, Newry BT35 6SD
2E0 FUZ D. King 215 Hartland Road, Reading RG2 8DN
2E0 FVB K. Smith 30 Azalea Drive, Swanley BR8 8HZ
2E0 FVC G. Gee 17 Portherras Villas, Pendeen, Penzance TR19 7TJ
2E0 FVF J. Dooley 29 The Drive, Alsagers Bank, Stoke-on-Trent ST7 8BB
2M0 FVJ M. Butterworth 97A Dunearn Drive, Kirkcaldy KY2 6AL
2E0 FVK J. Wright 149 Wroslyn Road, Freeland, Witney OX29 8HR
2E0 FVL P. Penycate 8 Campbell Road, Tangmere, Chichester PO20 2HX
2E0 FVO M. Burge 24 Oakdale Road, North Anston, Sheffield S25 4EY
2E0 FVR R. Farrington Sunny Brook, Broadway Rd, , Evesham WR11 7RN
2E0 FVS S. Horne Beaucroft, Keswick Road, Benfleet SS7 3HU
2E0 FVT G. Smith 148 Firhill Road, London SE6 3SQ
2E0 FVV M. Roberts Flat 6, 463 Brighton Road, Lancing BN15 8LF
2I0 FVX J. Robinson 15 Union Court, Lurgan, Craigavon BT66 8EE
2E0 FVZ A. Rawson 64 Dukes Mead, Fleet GU51 4HE
2E0 FWB R. Saddler 41 Clapham Close, Swindon SN2 2FL
2E0 FWC F. Clark 14 Warwick Road, Bude EX23 8EU
2E0 FWD C. Board Pinmoor, Moretonhampstead, Newton Abbot TQ13 8QA
2E0 FWK N. Woodruffe 139 North Home Road, Cirencester GL7 1DY
2E0 FWN E. Hunter 38 Blewitt Street, Hednesford, Cannock WS12 4BD
2E0 FWR F. Waller 19 Wortley Avenue S738Sb, Wombwell S738SB
2E0 FWS M. Le-Petit 3 Stanley Drive, Hatfield AL10 8XX
2E0 FWT T. Gore 22 Stoppard Road, Burnham-on-Sea TA8 1QB
2E0 FWW J. Haslam 25 Lulworth Road, Eccles, Manchester M30 8WP
2E0 FWY D. Smith 7 Oakley Grove, Wolverhampton WV4 4LN
2E0 FWZ A. Rowlands 35 Dublin Croft, Great Sutton, Ellesmere Port CH66 2TD
2E0 FXA P. Richardson 91 Park Road, Blackpool FY1 4JE
2E0 FXE C. Troughton 20 Oakleigh Road, Uxbridge UB10 9EL
2W0 FXF P. Brind Golwg Y Bryn, Tyn Y Morfa, Gwespyr, Holywell CH8 9JN
2E0 FXH A. Williamson 32 Beech Close Eastfield, Scarborough YO11 3QZ
2E0 FXJ D. Parkes 4 Round Saw Croft, Rubery, Birmingham B45 9TT
2M0 FXL F. Lane 23 Mayfield Avenue, Tillicoultry FK13 6HB
2E0 FXM Dr C. Fletcher 7 Highfield Crescent, Baildon, Shipley BD17 5NR
2E0 FXO D. Kerr 19 Park Lane, Sutton Bonington, Loughborough LE12 5NQ
2E0 FXP S. Westley 76 Rockingham Close, Birchwood, Warrington WA3 6UY
2E0 FXU A. Campbell 21 Sherbrook Gardens, London N21 2NX
2E0 FXV J. Keates Pickersleigh Court, North End Lane, Malvern WR14 2ET
2E0 FXW G. Davies 11, Liverpool L12 3HS
2M0 FXX M. Mcgrorty 59 Craighall Street, Stirling FK8 1TA
2E0 FXY Rvd. W. Hackman Kynance, Barden Road, Speldhurst, Tunbridge Wells TN3 0QB
2E0 FYA M. Champion 155 Walton Road, Walton on the Naze CO14 8NF
2E0 FYC D. Jarman 62 Combe Drive, Dunstable LU6 2AE
2E0 FYD M. Harper 67 Ludsden Grove, Thame OX9 3BY
2E0 FYE R. Fye 201 North Wing The Residence, Kershaw Drive, Lancaster LA1 3SY
2M0 FYF J. Fyfe 53A Ware Road, Glasgow G34 9AR
2M0 FYG J. Rodger 7 Heathryfold Circle, Aberdeen AB16 7DQ
2E0 FYL P. Catton 97 High Street, South Hiendley, Barnsley S72 9AN
2E0 FYO K. Hedges 28 Hill Park, Congresbury, Bristol BS49 5BT
2E0 FYP G. Stewart 35 Castle Crescent, Thornhill WF120EQ
2E0 FYR G. Hepworth 12 Fourlands Gardens, Bradford BD10 9SP
2E0 FYS S. Smith 133 Radcliffe New Road, Whitefield, Manchester M45 7RP
2E0 FYT A. Bent Lime Garth Sherburn Road, Durham DH1 2JR
2E0 FYU I. Lowbridge 4802 East Ray Road #23-513, Phoenix 85044 United States
2E0 FYW A. Shakesby 14 Dawnay Road, Bilton, Hull HU11 4HB
2M0 FYX J. Burton 6 Kilfinnan Lodges, Spean Bridge PH34 4EB
2E0 FYZ F. Islam 88 Ladies Grove, St. Albans AL3 5UB
2M0 FZA A. Chambers 35 Echline Grove, South Queensferry EH30 9RU
2E0 FZB C. Sjostedt 14 New Road, Marlow SL7 3NG
2W0 FZC J. Williams 12 Pantycelyn, Fishguard SA65 9EH
2E0 FZE G. Ashcroft 8 Launceston Close, Winsford CW7 1LY
2E0 FZF F. Cairns 20 St. Davids Close, Maidenhead SL6 3BB
2E0 FZG Dr J. Tromp St. Mary'S Street Practice Ltd, 63 St. Mary Street, Chippenham SN15 3JF
2E0 FZJ J. Foster Westfield House, 23 High Street, Cumnor OX2 9PE
2E0 FZK A. Mcbirnie 25 Ulverston Road, Swarthmoor, Ulverston LA12 0JB
2E0 FZM C. Ramsdale 87 Mill Lane, Kirk Ella, Hull HU10 7JN
2I0 FZO A. Mckenzie 32 Abbot Gardens, Newtownards BT23 8UL
2E0 FZS S. Carro 149 Ruston Rd, London SE18 5QY
2E0 FZT A. Craven 14 Deverel Road, Charlton Down, Dorchester DT2 9UD
2E0 FZU J. Davidson 78 Old Heath, Shrewsbury SY1 4SE
2M0 FZW M. Love 17 Lindsay Road, East Kilbride, Glasgow G74 4HZ
2E0 FZY S. Whiting 56 Station Road, Branston, Lincoln LN4 1LH
2E0 FZZ D. Sandland 54 Bishopdale Drive, Rainhill, Prescot L35 4QH
2E0 GAC N. Bexon 60 Whitwell Road, Nottingham NG8 6JT
2E0 GAF G. Sole 16 Beech Crescent, Hythe, Southampton SO45 3QG
2E0 GAH G. Hudson 26 Griffins Brook Lane, Birmingham B30 1PU
2E0 GAL R. Hughes 19 Pendine Crescent, North Hykeham, Lincoln LN6 8UW
2M0 GAN G. Prior 41 Beechwood, Linlithgow EH49 6SD
2E0 GAP A. Pilkington 26 Ryelands Close, Market Harborough LE16 7XE
2W0 GAQ I. Williams 5 Fron Goch, Llanberis, Caernarfon LL55 4LE
2E0 GAR G. Farrar 174 Houghton Road, Thurnscoe, Rotherham S63 0SA
2E0 GAU G. Cooper Holmfield, Chelmorton, Buxton SK17 9SG
2E0 GAV C. Holmes 8 Byron Way, Caister-On-Sea, Great Yarmouth NR30 5RW
2E0 GAW G. Webster 84 Sparrows Herne, Basildon SS16 5EN
2W0 GAY A. Ferguson The Mount Stables, Salem, Llandeilo SA19 7HD
2E0 GBA I. Stevenson 79 Lunedale Road, Darlington DL3 9AT
2E0 GBB G. Foster 22 Bradley Cottages, Consett DH8 6JZ
2E0 GBE G. Bell 83 Coopers Green, Bicester OX26 4XJ
2E0 GBF P. Dawes 49 Altofts Lodge Drive, Altofts, Normanton WF6 2LB
2E0 GBG D. Gillingham 5 Hillfield, St. Marks, Cheltenham GL51 7BQ
2E0 GBH M. Brinnen 82 Victoria Road, Mablethorpe LN12 2AJ
2E0 GBI L. Catterall 14 Dunham Drive, Whittle-Le-Woods, Chorley PR6 7DN
2E0 GBJ B. Cave 2 Beaufort Close, Newcastle upon Tyne NE5 3XL
2E0 GBK Dr J. Bell 6 Highfields, Fetcham, Leatherhead KT22 9XA
2D0 GBM P. Birchall 7 Richmond Close, Douglas IM2 6HR Isle of Man
2E0 GBN G. Barton 9 Tees Crescent, Stanley DH9 6HX
2E0 GBO M. King 7 Battismore Road, Morecambe LA4 4QG
2E0 GBP P. Lindley 17 Swallow Lane, Aston, Sheffield S26 2GR
2E0 GBT F. Woods 30 Hurst Close, Chandler'S Ford, Eastleigh SO53 3PA
2E0 GBU D. Torrance 30 St. Norbert Drive, Ilkeston DE7 4EH
2E0 GBV G. Brotherhood 17 Baldwin Close, Forest Town, Mansfield NG19 0LR
2E0 GCB G. Buxton 18 Savernake Close Rubery, Rednal, Birmingham B45 0DD
2I0 GCC G. O'Reilly 20 Lower Clonard Street, Belfast BT12 4NH
2E0 GCE G. Elsworthy 40 Moorfield Way, Wilberfoss, York YO41 5PL

2M0 GCF J. Brown 78 Egilsay St., Glasgow G22 7RG
2E0 GCG H. Vecenans 155 Upper Dale Road, Derby DE23 8BP
2E0 GCH S. Shreeves 20 Selly Oak Road, Jordanthorpe, Sheffield S8 8DU
2E0 GCI A. Clarke 57 Welland Avenue, Grimsby DN34 5JP
2E0 GCJ G. Jacks 2 Corve View, Fishmore, Ludlow SY8 2QD
2E0 GCL A. Finn 202 Northgate Road, Stockport SK3 9NJ
2E0 GCM N. Baker 56 Chalklands, Bourne End SL8 5TJ
2I0 GCN S. Morrow 769 Farranseer Park, Macosquin, Coleraine BT51 4NB
2E0 GCO S. Lake 85 Clarkson Road, Norwich NR5 8ED
2E0 GCP G. Piddington 45 Pleasant View Road, Crowborough TN6 2UU
2E0 GCW B. Barker 44 Falcon Crescent, Bilston WV14 9BE
2E0 GCY G. Cornish 78 Kerry Avenue, Ipswich IP1 5LD
2W0 GDA R. Davison 2 Marlow Terrace, Mold CH7 1HH
2W0 GDB G. Brookes Hafan Deg, Caergeiliog LL653YD
2E0 GDF M. Joynson-Ellis 20 Morland Court, Skaters Way, Peterborough PE4 6GW
2E0 GDG Dr G. Di Genova 7 Marina Way, Abingdon OX14 5TN
2E0 GDH S. Hunter 9 Gelt Burn, Didcot OX11 7TZ
2E0 GDL G. Ludlow 13 Laburnum Walk, Gilberdyke, Brough HU15 2TU
2E0 GDM G. Martin Flat 16, Sorrel House, Birmingham B24 0TQ
2E0 GDN K. Young 51 Haven Road, Barton-upon-Humber DN18 5BS
2E0 GDO G. Cochrane 133 Cotman Fields, Norwich NR1 4EP
2E0 GDT A. Shaw 9 Stamford Lane, Warmington, Peterborough PE8 6TW
2E0 GDV Dr J. Hunt 10 Couzens Close, Chippenham SN15 1US
2E0 GDY V. Dealey 11 Ashcombe Close, Witney OX28 6NL
2E0 GEB S. Marr 49 Gallows Hill, Ripon HG4 1RG
2E0 GEC C. Whatmough 11 Blackchapel Drive, Rochdale OL16 4QU
2E0 GEE N. Patel 78 Wesley Close, South Harrow, Harrow HA2 0QE
2E0 GEF G. Winterbottom 35 Abingdon View, Worksop S81 7RT
2E0 GEG G. Bramham 1 Watson Avenue, Dewsbury WF12 8PZ
2M0 GEJ G. Jamieson 6 Maryville Park, Aberdeen AB15 6DU
2M0 GEK J. Wright 43 Spey Court, Stirling FK7 7QZ
2E0 GEL J. Willetts 102 Welch Road, Cheltenham GL51 0EG
2W0 GEM P. Murray 45 Commercial Street, Risca, Newport NP11 6AW
2E0 GEN A. Askam 8 The Pastures, Weston-On-Trent, Derby DE72 2DQ
2W0 GER T. Doak 27 Hill St., Gilfach Goch, Porth CF39 8TW
2E0 GES L. Tarlow 59 Crouch Hall Road, London N8 8HD
2E0 GET D. Baker 25 Hitherspring, Corsham SN13 9UT
2E0 GEV A. Sherman 31 Peartree Avenue, Kingsbury, Tamworth B78 2LG
2E0 GEX G. Evans 19 Windsor Street, Thurnscoe S63 0HB
2W0 GEZ N. Shepherd Prospect, Newchapel, Llanidloes SY18 6JY
2E0 GFB A. Durrant 22 Supple Close, Norwich NR1 4PP
2M0 GFC P. Davis 4 Daisy Park, Baltasound, Unst, Shetland ZE2 9EA
2E0 GFE G. Teale Flat 3, 89-91 Barrack Road, Christchurch BH23 2AJ
2E0 GFF A. Banks 2 Holt Close, Farnborough GU14 8DG
2E0 GFJ G. Frost The Old Chestnut, 51 Knightcott Gardens, Banwell BS29 6HD
2E0 GFK I. Graham 49 Eagle Close, Leighton Buzzard LU7 4AT
2I0 GFO G. Craig 103 Moyle Parade, Larne BT40 1ET
2E0 GFQ R. Haynes 28 Ridgeway View, Montgomery SY15 6BF
2I0 GFR G. Ramsey 6 Creamery Park, Lisbellaw BT94 5BU
2M0 GFT Dr V. Vyshemirsky 2103 Great Western Road, Glasgow G13 2XX
2E0 GFU C. Bourdiec 135 Eglesfield Road, South Shields NE33 5PU
2E0 GFV J. Butcher 14 Park Road, Lowestoft NR32 1SW
2E0 GFW G. Watson The Dell, Nova Scotia Road, Great Yarmouth NR29 3QD
2W0 GFX J. Duffain 37 Mount Pleasant Street, Dowlais, Merthyr Tydfil CF48 3AF
2E0 GFY D. Horne Alias, Lowthorpe, Southrey, Lincoln LN3 5TD
2E0 GFZ B. Gentry 32 Abbey Meadow, Sible Hedingham, Halstead CO9 3QS
2E0 GGA G. Amos Willow Tree House, Deers Green, Clavering, Saffron Walden CB11 4PX
2W0 GGG M. Brady Ty Mawr Uchaf, Dulas LL70 9DQ
2E0 GGI R. Gleave 52 Cranborne Avenue, Warrington WA4 6DE
2E0 GGM A. Mansfield 20 High Street, Broughton, Kettering NN14 1NG
2E0 GGN N. Bentall 15 Maple Close, Oxford OX2 9DZ
2E0 GGO G. Jones 109 Montgomery Avenue, Lowestoft NR32 4DU
2E0 GGP G. Peters Flat 6, Park Court, 46 North Park Road, Harrogate HG1 5AD
2E0 GGQ G. Wilson 28 Stanley Grove, Redcar TS10 3LN
2E0 GGT G. Townsend 23 Lodgefield Park, Stafford ST17 0YE
2E0 GGU C. Braiden Flat 5, Lansdowne House, 12 Twickenham Close, Swindon SN3 3FQ
2E0 GGV C. Leek 46 The Hollies, Holbeach, Spalding PE12 7JQ
2E0 GGW G. Willard 4 Varrier Jones Place, Papworth Everard, Cambridge CB23 3XP
2M0 GGY A. Espie 70 Everard Rise, Livingston EH54 6JD
2E0 GGZ G. Talbot 14 Woodbury View, Exeter EX2 9JQ
2E0 GHA G. Hill-Adams 6 Broadleaze Way, Winscombe BS25 1JX
2E0 GHB G. Bourne 72 Cornish Way, Royton, Oldham OL2 6JY
2E0 GHC Dr T. Hoban Hillside Barn, London End, Priors Hardwick, Southam CV47 7SL
2W0 GHD A. Burleton 11 Orchard Close, Caldecott NP26 4BH
2E0 GHE J. Merrick 8 Maldowers Lane, Bristol BS5 7QT
2M0 GHF G. Fleming 169 Oldtown Road, Inverness, Scotland IV2 4QD
2E0 GHG Dr G. Turner 8 Scarborough Terrace, York YO30 7AW
2E0 GHH C. West 12 St. Georges Road, Wallington SM6 0AS
2E0 GHJ Dr D. Pegler September Cottage, East Bank, Winster DE42DT
2E0 GHK D. Hindle 18 Haig Street, Selby YO8 4BY
2E0 GHP J. Phillips The Manor, Blackwoods, York YO61 3ER
2E0 GHR R. Gill 84 Leypark Road, Exeter EX1 3NT
2E0 GHS S. Stanhope 61 Heathfield St., Manchester M40 1LF
2W0 GHV G. Thomas 4 Blenheim Court, Picton Road, Neyland, Milford Haven SA73 1QR
2E0 GHX S. Aucoin 296 Turkey Road, Bexhill-on-Sea TN39 5HY
2I0 GHY I. Gibb 1 Shankill Road, Garvary, Enniskillen BT94 3DB
2E0 GHZ J. Wakefield Oakhurst, Lower Common Road, Romsey SO51 6BT
2W0 GIA D. Owen Tanralt, Blaenpennal, Aberystwyth SY23 4TP
2E0 GIE R. Harper 19 Tennyson Avenue, King's Lynn PE30 2QG
2I0 GIF G. Clarke 12 Church Green, Dromore BT25 1LL
2E0 GIH E. Peck 11 Blake Road, Stapleford, Nottingham NG9 7HN
2W0 GIW G. Williams 99 Maes Llwyn, Amlwch LL68 9BG
2W0 GIX A. Hodgson Browns Holiday Park Towyn Road, Conwy LL22 9HD
2E0 GIY T. Orzechowski 25 Lutterworth Road, Northampton NN1 5JR
2E0 GJE G. Groves 5 Beech Road, Ashurst, Southampton SO40 7AY
2M0 GJG R. Hudson 128 Neilston Road 2/2, Paisley PA2 6EP
2M0 GJJ G. Johnson Speur Mor, Gifford Road, Longformacus, Duns TD11 3NZ
2E0 GJN M. Stokes 207 Sunderland Road, South Shields NE34 6AQ
2E0 GJP B. Coley 17 Livingstone Road Handsworth, Birmingham B20 3LS
2E0 GJQ N. Napiorkowski 74 Wigley Road, London TW13 5HE
2W0 GJR G. Reason 454 Cowbridge Road West, Cardiff CF5 5BZ
2E0 GJT N. Cooper Romer Cottage, Long Reach, Ockham, Woking GU23 6PF
2E0 GKA P. Houghton 6 Olivers Court, Calne SN11 0FL
2I0 GKB Colchester Radio Amateurs Club c/o G. Black 45 Meeting House Lane, Lisburn BT27 5BY
2M0 GKD A. Mccreadie 37 Beddie Crescent, Wigtown, Newton Stewart DG8 9HX
2E0 GKF M. Feakins 29 Whitesfield Road Nailsea, Bristol BS48 2DY
2E0 GKL K. Lord 21 Norfolk Road, Littlehampton BN17 5PW
2E0 GKM R. Ley 23 Heronbridge Close, Westlea, Swindon SN5 7DR
2E0 GKN J. Duck 24 Shearwater Grove, Innsworth, Gloucester GL3 1DB
2E0 GKQ S. Dean 154 Broad Lane, Walsall WS3 2TQ
2E0 GKR L. Bullen11 5 West View, Long Sutton TA10 9LT
2E0 GKT K. Taylor 56 Gibraltar Lane, Haughton Green Denton, , Manchester M34 7GG
2E0 GKU Dr A. Baker Victoria Cottage, Swan Lane, Aughton, Ormskirk L39 6SU
2E0 GLA M. Gladders 2 Albion Mansions, Saltburn-by-the-Sea TS12 1JP
2I0 GLC G. Crabbe 39 Arran Avenue, Ballymena BT42 4AP
2E0 GLD A. Goold 6 The Elms, Kempston, Bedford MK42 7JN
2M0 GLI G. Irvine 120A Shore Road, Innellan, Dunoon PA23 7SS
2E0 GLJ G. Jenkin 24 Coronation Avenue, Camborne TR14 7PE
2E0 GLL D. Firth 5 Mowhay Gardens, Hatherleigh EX20 3FE
2E0 GLM G. Parr 50 Broadmeadow Close, Birmingham B30 3NG
2E0 GLR E. Rimmer 26 Kenmore Road, Prenton CH43 3AS
2E0 GLS G. Stevens 17 Manston Close, Ernesettle, Plymouth PL5 2SN
2E0 GLT G. Cheeran 37 Farnol Road, Dartford DA1 5NG
2W0 GLV C. Tanner Pen Y Gogarth Llaneilian, Amlwch LL68 9NH
2E0 GLW G. Whittle 22 Warwick Street, Leigh WN7 2NH
2I0 GLY N. Sands 6 The Granary, Waringstown, Craigavon BT66 7TG
2E0 GMA G. Marsden 38 Sandhill Road, Rawmarsh, Rotherham S62 5NT
2M0 GMB I. Birse North Milton Of Corsindae, Midmar, Inverurie AB51 7QP
2E0 GMD M. Drury 19 Cuffley Avenue, Watford WD25 9RB
2E0 GMF S. Dearne 26 Dilly Lane Barton On Sea, New Milton BH25 7DQ
2E0 GMG Viscount A. Andover Bishoper Farmhouse, Brokenborough, Malmesbury SN16 9SR
2E0 GMM J. Cook 42 Pampas Close, Colchester CO4 9ST
2E0 GMN J. Rufes Flat 1 & 3-8 12 Smyrna Road, London NW6 4LY
2E0 GMS G. Brooks 14 Chalton Crescent, Havant PO9 4PT
2E0 GMU G. Murch 79 Alderson Crescent Formby, Liverpool L37 3LY
2E0 GMW S. Coombs 34 Mast Drive, Hull HU9 1ST
2W0 GMZ H. Hughes 21 Maes Geraint, Pentraeth LL75 8UR
2E0 GNC B. Chandler 1 Rambridge Farm Cottages, Weyhill, Andover SP11 0QF
2W0 GNG P. Smith 3 Islington Road, Bridgend CF31 4QY
2E0 GNH G. Spiers 2 Sponnes Road, Towcester NN12 6ED
2E0 GNI L. Mercer 5 Brinchcombe Mews, Plymouth PL9 7FB
2E0 GNJ W. Neill 21 Geelong Close, Weymouth DT3 6RE
2E0 GNL S. Bayntun 15 West Croft, Addingham, Ilkley LS29 0SP
2E0 GNN A. Glover 103A Latimer Street, Liverpool L5 2RF
2E0 GNO W. Walton 1 West Hall, Yeadon, Leeds LS19 7AJ
2E0 GNQ M. Hughes 16 Marlborough Road, Urmston, Manchester M41 5QG
2E0 GNS G. Sandell 20 Kirkby View, Sheffield S12 2NB
2E0 GNU E. Brook 30 Pitchstone Court, Farnley, Leeds LS12 5SZ
2E0 GNV M. Wanless 3 Bromlow Hall Barns, Bromlow, Minsterley, Shrewsbury SY5 0DX
2E0 GNW A. Land 12 Beverley Close Holton Le Clay, Grimsby DN36 5HG
2M0 GOE T. Macdonald Main Road Farm, Balephuil, Isle of Tiree PA77 6UE
2E0 GOG B. Garrison 121 Bromford Lane Erdington, Birmingham B24 8JP
2I0 GOK S. Lannie 81 Huntingdale Green, Ballyclare BT39 9FL
2E0 GOL A. Goldsmith 61 Fengate Drove, Weeting, Brandon IP27 0PW
2E0 GOM C. Blackburn 158 Dyas Road, Great Barr, Birmingham B44 8SW
2E0 GON P. Gonczarow 25 Ribchester Avenue, Burnley BB10 4PD
2E0 GOO J. Barker Pearl Bungalow, Killerby Cliff, Cayton Bay, Scarborough YO11 3NR
2E0 GOP S. Mansfield The Old Piggery, Ham Lane, Compton Dundon, Somerton TA11 6PQ
2E0 GOQ R. Gibbs 7 Thornhill, Eastfield, Scarborough YO11 3LY
2E0 GOS N. Gostling 49 Roundhouse Road, Dudley DY3 2AX
2E0 GOW C. Gowing 5 Curson Road, Tasburgh, Norwich NR15 1NH
2M0 GOY A. Mcdonald Kettlehills, Cupar KY15 7TW
2E0 GOZ R. Cichocki 12 Crossland Crescent, Wolverhampton WV6 9JY
2E0 GPA M. Phillips 36 Hyde Heath Court, Crawley RH10 3UQ
2E0 GPB G. Beacher 22 Trowbridge Gardens, Luton LU2 7JY
2E0 GPC G. Coleman 15 Redwood Drive, Ormskirk L39 3NS
2E0 GPD P. Dimes 5 Meadowbrook, Oxted RH8 9LT
2E0 GPE G. Pearson 41 Myrica Grove, Hoole, Chester CH2 3EW
2M0 GPF G. Fleming Tarbat View, Achavandra Muir, Dornoch IV25 3JB
2E0 GPG G. Bates 230 Brook Street, Erith DA8 1DZ
2E0 GPH G. Hart 11 Sadlers Ride, West Molesey KT8 1SU
2E0 GPI R. Mackay 7 Darwin Close, Lee-on-the-Solent PO13 8LS
2E0 GPJ A. Allan The Oxford Health Co Ltd, Unit 4, Longlands Road, Bicester OX26 5AH
2E0 GPK G. Kendall 6 Kershaw Road, Walsden, Todmorden OL14 7QF
2E0 GPL T. Arrow Crystalwood, Stonemans Hill, Newton Abbot TQ12 5PZ
2I0 GPQ D. Boyd 11 Abbey Gardens, Belfast BT5 7HL
2E0 GPS G. Perkins Gamekeepers Cottage, Snarehill, Thetford IP24 2QA
2E0 GPT R. Hampson 12 Oakhays, South Molton EX36 4DB
2E0 GPU A. Taylor 16 Bellmans Road Whittlesey, Peterborough PE7 1TY
2E0 GPV A. Alty 20 The Hawthorns Eccleston, Chorley PR75QW
2E0 GPX G. Matthews 81 Kipling Avenue, Goring-By-Sea, Worthing BN12 6LH
2E0 GPY N. Shepherd 12 Barnfield Close, Radcliffe, Manchester M26 3UA
2I0 GQA R. Bestek 111 Cloughwater Road, Ballymena BT43 6SZ
2E0 GQC D. Hughes 14 Holts Lane, Clayton, Bradford BD14 6BL
2E0 GQD J. Reynolds 3 Ardleigh, Basildon SS16 5RA
2E0 GQE L. Walsh 4 Musbury Crescent, Rossendale BB4 6AY
2M0 GQH Dr Z. Yang 48 Foxglove Road, Newton Mearns, Glasgow G77 6FP
2E0 GQI A. Hicks 15 West Road, Ruskington, Sleaford NG34 9AL

2E0 GQK C. Schroth Flat 3, 15B Cavendish Road, Bournemouth BH1 1QX
2E0 GQL R. Abbott The Bungalow, Wrangaton, South Brent TQ10 9HH
2E0 GQM R. Laidler Swallow Cottage, Wiverton, Plympton, Plymouth PL7 5AA
2E0 GQN D. Shingleton 6 Newsham Walk, Manchester M12 5QB
2E0 GQO S. Dudley 3 Lake Croft Drive Meir Heath, Stoke on Trent ST3 7SS
2E0 GQP I. Thompson 26 Countrymans Way, Shepshed, Loughborough LE12 9RB
2I0 GQR G. English 15 Murrays Hollows, Ballyroney, Banbridge BT32 5ES
2M0 GQS J. Rae 4 Hillside Crescent, Langholm DG13 0EE
2E0 GQT I. Alderman 107 Manton Drive, Luton LU2 7DL
2E0 GQU A. Rudgley The New House, Plymouth Road, Buckfastleigh TQ11 0DB
2E0 GQW E. Tart Sunnybank Farm, Wattlesborough Heath, Shrewsbury SY5 9EG
2I0 GQZ C. Kelly Bt457Dt, Magherafelt BT45 7DT
2E0 GRA G. Hayward 129 Nipsells Chase, Mayland, Chelmsford CM3 6EJ
2M0 GRE G. Lailvaux 4 Oxenfoord Avenue, Pathhead EH37 5QD
2E0 GRF G. Hewis 10 Albert Road, New Malden KT3 6BS
2E0 GRH S. Walters 87 Fairbourne Close, Bransholme, Hull HU7 5DH
2E0 GRI G. Reywer 1 Tiverton Close, Houghton le Spring DH4 4XR
2M0 GRK J. Black 13 Dunlop Street, Greenock PA16 9BG
2E0 GRL L. Spear 57 Station Road, Melbourne, Derby DE73 8EB
2E0 GRM N. Hallwood 32 Hawthorne Avenue, Ripley DE5 3PJ
2E0 GRN A. Young 48 Sussex Street, Cleethorpes DN35 7NP
2E0 GRP G. Priestley 53 Millfield Gardens, Crowland, Peterborough PE6 0HA
2E0 GRQ A. Waters 21 Kings Road, Lee-on-the-Solent PO13 9NU
2E0 GRR C. Hebden 1 Ringwood Avenue, Newbold, Chesterfield S41 8RA
2E0 GRS G. Street Flat 9, Weavers Cottages, Congleton CW12 1AG
2E0 GRW M. Harvey 129 Goldthorn Hill, Wolverhampton WV2 4PS
2E0 GRX G. Kennedy 4 Calder Crescent, Whitefield, Manchester M45 8LH
2E0 GRY G. Collis 16 Hill Grove, Barrow Hill, Chesterfield S43 2NW
2E0 GRZ G. Green 90 Princes Way, Fleetwood FY78DX
2E0 GSA G. Smith 2 Hawthorn Rise, Groby, Leicester LE6 0EX
2E0 GSB G. Bertola 17 Caraway Drive Branston, Burton-on-Trent DE14 3FQ
2E0 GSC G. Chaffey 63 Underwood Road, Eastleigh SO50 6FX
2W0 GSE R. Smith Hendafarn, Sarnau, Llanymynech SY22 6QJ
2E0 GSF T. Gadd 7 Dolina Road, Swindon SN25 1TL
2I0 GSG G. Gregg 30 Claremont Avenue, Moira, Craigavon BT67 0SS
2E0 GSH J. Haigh 1 Smithy Site, Farnborough, Wantage OX12 8NS
2E0 GSJ K. Andrews 80 Hollywell Road, Lincoln LN5 9DA
2E0 GSK M. Silver 52 Park Crescent, Elstree, Borehamwood WD6 3PU
2E0 GSL Rvd. L. Clark 226 Philip Lane Tottenham, London N15 4HH
2E0 GSN T. Savidis 32 Peace Close Rosedale, Cheshunt Waltham Cross EN7 5EQ
2E0 GSO T. Allen 11 Church View Highworth, Swindon SN6 7ER
2E0 GSP M. Carter 32 Victoria Avenue, Brighouse HD6 1QT
2E0 GSR G. Iredale Ship Cottage, Main Street, Maryport CA15 7DX
2E0 GST G. Starling 4 Three Corner Drive, Norwich NR6 7HA
2E0 GSW T. Rowlands 7 Northfield Crescent, Beeston, Nottingham NG9 5GR
2E0 GTA A. Holbrook 6 Birch Tree Way, Maidstone ME15 7RR
2E0 GTB P. Rigden 11 Railway Cottages, Station Road, Whitstable CT5 1JZ
2E0 GTE G. Cockburn 20 Hexham Avenue, Hebburn NE31 2HN
2E0 GTJ G. Mylonas 68 Lancaster Gate, Cambourne CB236AT
2E0 GTL G. Taylor 31 Ashfurlong Crescent, Sutton Coldfield B75 6EN
2E0 GTM G. Moon 1 Brankenwall, Muncaster, Ravenglass CA18 1RG
2E0 GTN G. Norbury 3 Sherard Croft, Birmingham B36 0LS
2I0 GTO G. Shaw 49 Cloughey Road, Portaferry, Newtownards BT22 1NQ
2M0 GTR S. Higgins 24 Centre Street, Kelty KY4 0EQ
2E0 GTU J. Haskell 60 Blenheim Drive, Witney OX28 5LJ
2E0 GTZ A. Blamire 21 The Laurels, Banstead SM7 2HG
2E0 GUA J. Hammond 8 Rowntree Way, Saffron Walden CB11 4DG
2E0 GUC M. Vowles Calle Mar Serena, 29, Torre de Banagalbon, Malaga 29738 Spain
2E0 GUF A. Bennett 112 Vicarage Crescent, Redditch B97 4RP
2E0 GUH D. Hughes Flat 2, 13 Thorgam Court, Grimsby DN31 2EU
2E0 GUJ S. Noller 3 Thor Road, Norwich NR7 0JS
2M0 GUL J. Hume 51 Vogrie Road, Gorebridge EH23 4HL
2E0 GUN A. Price 23 Greenway, Wingerworth, Chesterfield S42 6NP
2E0 GUR G. Urban 33 High Meadow, Hathern, Loughborough LE12 5HW
2E0 GUT G. Taljaard Flat 140, 105 London Street, Reading RG1 4QD
2E0 GUU P. Edwards 4 Stodart Road, London SE20 8ET
2E0 GUV P. Brown 1 Octavian Close, Hatch Warren, Basingstoke RG22 4TY
2E0 GUY S. Mellor 11 Bolton Meadow, Leyland PR26 7AJ
2E0 GVC G. Clayton The Forge , High Street, Moreton-in-Marsh GL56 0LL
2E0 GVI S. Dickson 11 Benfield, Grasmere, Ambleside LA22 9RD
2W0 GVK J. Doyle 40 Llanbeblig Road, Carrnarfon LL55 2LW
2W0 GVN J. Bradley Ty Newydd, Tregarth, Bangor LL57 4AF
2E0 GVO G. Owen 9 Stargate Close, St. Helens WA9 5XJ
2E0 GVP T. Hunt 36 Alverton Avenue, Poole BH15 2QG
2E0 GVR V. Hodge 37 Monkswood Crescent, Tadley RG26 3UE
2E0 GVW S. Gadd 62 Erica Way, Copthorne RH10 3XQ
2E0 GVZ J. Earye 28 Halls Drift, Kesgrave, Ipswich IP5 2DE
2I0 GWA A. Cummings 19 Bachelors Walk, Keady, Armagh BT60 2NA
2E0 GWB G. Bunting 31 Hardwick Avenue, Allestree, Derby DE22 2LN
2E0 GWC Dr A. Colman 5 Burn Heads Road, Hebburn NE31 2TB
2E0 GWD N. Reeves Flat 2, Delamore, Ivybridge PL21 9QT
2E0 GWE G. Watson 20 Windermere Drive, West Auckland, Bishop Auckland DL149LF
2E0 GWF B. Whitemore 24 Rectory Close, Wraxall, Bristol BS48 1LT
2E0 GWI N. Davies 27 Grafton Road, Ellesmere Port CH65 2BD
2M0 GWJ A. Stanley 22 Braid Mount, Edinburgh EH10 6JJ
2E0 GWK K. Jones 45 Liskeard Way, Freshbrook, Swindon SN5 8NL
2W0 GWM M. Martin 1 Y Gorlan, Bryn Street, Newtown SY16 2HN
2W0 GWO C. Marrs 49 Abraham Court, Lutton Close, Oswestry SY11 2TH
2E0 GWP G. Prescott 3 View Fields, Station Road, Doncaster DN9 3AE
2E0 GWR B. Bosson 80 White Horse Road, Marlborough SN8 2FE
2E0 GWS G. Salter 9 Spring Gardens, Malvern Link, Malvern WR14 1AP
2M0 GWU I. Colborn Gardeners Cottage, Craighouse, Isle of Jura PA60 7XG
2E0 GWV N. Parker 18 Sandown Road Bishops Cleeve, Cheltenham GL52 8BZ
2E0 GWZ C. Aldous 342 Feltham Hill Road, Ashford TW151LW
2E0 GXB G. Beaver 106 Dobede Way Soham, Ely CB7 5FN
2E0 GXC G. Hardman 12 Fernleigh Chorley New Road, Horwich, Bolton BL6 6HD
2E0 GXD J. Marchant 129 Highbury Grove, Clapham, Bedford MK41 6DU
2E0 GXE B. Thomson 40 Northgate Road, Stockport SK3 0LQ
2E0 GXF G. Fernando 1 Rosemary Avenue, West Molesey KT8 1QF
2E0 GXG G. Eaton 16 Holly Walk, Nuneaton CV11 6UU
2W0 GXI D. Burt 2 Cae Masarn, Pentre Halkyn, Holywell CH8 8JY
2E0 GXK P. Blagden 30 Charlecote Avenue, Tuffley, Gloucester GL4 0TH
2E0 GXM S. Elton 55 Dugdell Close, Ferndown BH22 8BQ
2E0 GXP P. Brennan 2 Ethel Road, Birmingham B17 0EL
2E0 GXQ S. Lovell 20 Courtenay Walk, Weston-Super-Mare BS22 7TQ
2E0 GXT D. Killingley 17 Colbert Drive, Leicester LE3 2JB
2E0 GXX I. Bardell 17 Stanton Avenue, Bradville, Milton Keynes MK13 7AR
2E0 GXY M. Ferenc 2A Rosedene Avenue, London SW16 2LT
2M0 GXZ G. Sinclair 33 Keptie Road, Arbroath DD11 3EF
2W0 GYB L. Brown 13 Station Road Loughor, Swansea SA46TR
2I0 GYC R. Mcknight Ardralla, Church Cross, Skibbereen P81 RK12 Ireland
2E0 GYF D. Hudson 30 Sarmatian Fold, Ribchester, Preston PR3 3YG
2E0 GYH P. Bates 34 Lamerton Road, Reading RG2 8AS
2E0 GYJ G. Booton 69 The Street, Deal CT14 0AJ
2E0 GYK J. Coates 2 Holstein Drive, Scunthorpe DN16 3TT
2I0 GYL G. Mccormick 24 Warren Park Drive, Lisburn BT28 1HF
2M0 GYM J. Branson 129 Thornhill Road, Elgin IV30 6DX
2M0 GYN A. Connell 39 Glebe Crescent, Maybole KA19 7HZ
2E0 GYP J. Wells 18 Roewood Road, Holbury, Southampton SO45 2JH
2W0 GYQ B. Williams Bryn Celyn, Hendre Road, Conwy LL32 8RJ
2E0 GYV G. Johnson 4 Delta Park Drive, Hesketh Bank, Preston PR4 6SE
2E0 GYW T. Clark 23 Shelley Grove, Bradford BD8 0JZ
2E0 GYX J. Margarson 26 David Street, Grimsby DN32 9NL
2E0 GYY L. Monshall 70 Oakley Road, Harwich CO12 4QU
2E0 GYZ W. Rittman 70 Market Street, Chapel-En-Le-Frith, High Peak SK23 0HY
2M0 GZA S. Hargreaves 4 Oxenfoord Avenue, Pathhead EH37 5QD
2E0 GZD J. Norris 55 Main Road, Great Leighs CM3 1ND
2E0 GZF D. Higton 24 Holly Avenue, Bradwell NR31 8NL
2E0 GZH Dr G. Howling Sysonby Knoll Hotel, Asfordby Road, Melton Mowbray LE13 0HP
2E0 GZJ C. Little 28 Cecil Avenue, Warmsworth, Doncaster DN4 9QW
2E0 GZM L. Wardle 35 Woodland Close, Barnstaple EX32 0EG
2E0 GZN D. Williams Lower House Farm Huntington, Kington HR53PU
2E0 GZP M. Stockdale 3 Manor Close, Sproatley, Hull HU11 4PY
2M0 GZQ T. Goodenough 83 Craufurdland Road, Kilmarnock KA3 2HU
2E0 GZT T. Marshall 63A Newport Road, Ventnor PO38 1BD
2M0 GZU M. Gallon 43/18 Viewcraig Gardens, Edinburgh EH8 9UW
2I0 GZX P. Brennan 23 Ardchrois, Donaghmore, Dungannon BT70 3LB
2E0 HAB H. Atifeh 57 Lincoln Drive, Rugby CV23 1BS
2W0 HAC C. Thomas 2 Ffordd Donaldson, Copper Quarter, Swansea SA1 7FJ
2E0 HAE A. Hodgeon 30 Rock Bank, Buxton SK17 9JF
2E0 HAF J. Raehse Felstead 10 Rubens Close, Aylesbury HP19 8SW
2E0 HAG C. Hall 28 Tidebrook Place, Stoke-on-Trent ST6 6XF
2E0 HAH M. Mckenna 54 Whickham Road, Hebburn NE31 1QU
2E0 HAJ S. Cordner 29 Buxton Road, Aylsham, Norwich NR11 6JD
2W0 HAK K. Vaughan 26 Mount Pleasant, Bargoed CF81 8UU
2E0 HAL G. Smith 7 Kestrel Avenue, Bransholme, Hull HU7 4ST
2E0 HAN H. Hopkins 3 Colegrave Road, Bloxham, Banbury OX15 4NT
2E0 HAP A. Craven 45 Benhams Drive, Horley RH6 8QT
2W0 HAS R. Williams 34 Maendu Terrace, Brecon LD3 9HH
2E0 HAT Clifton ARS c/o R. Hatton 18 Tangier Road, Guildford GU1 2DF
2E0 HAV S. James 35 Prospect Road, Dronfield S18 2EA
2E0 HAW C. Cross 131 Arnold Lane, Gedling, Nottingham NG4 4HF
2E0 HAZ G. Hazlewood 102 Throne Road, Rowley Regis B65 9JX
2E0 HBB H. Russell 4 Dearnsdale Close, Stafford ST16 1SD
2E0 HBD D. Hardy Flat 10, Bridport House, Hillwood Road, Birmingham B31 1DN
2E0 HBE D. Newman 78 Clapham Court, Gloucester GL1 3DE
2E0 HBJ D. Howard 31 White Mullein Drive, Redlodge IP288XP
2E0 HBK P. Jones 2 Whitley Place, Stoneley Park, Crewe CW1 4GH
2I0 HBO K. Mikicki 17 Glenhoy Drive, Belfast BT5 5LB
2E0 HBP M. Gorrill 30 Higher Dunscar Egerton, Bolton BL7 9TF
2E0 HBQ S. Benniman 5 Round Hill Lane, Shrewsbury SY1 2NE
2E0 HBT T. Richley 30 Chicheley Road, Harrow HA3 6QL
2E0 HBV J. Haynes 16 Mountsfield, Frome BA11 5AR
2W0 HBW J. Turobin-Harrington Michaelmas Barn, Sawmills, Kerry, Newtown SY16 4LL
2E0 HBX H. Gruen Allfrey House, Herstmonceux, Hailsham BN27 4RS
2E0 HBY A. Forster 62 Spa Road, Atherton, Manchester M46 9NQ
2E0 HCC G. Belgium 590 Wells Road, Bristol BS14 9BD
2M0 HCF Dr C. Moir 3 Farmfield Terrace, West Kilbride KA23 9ED
2E0 HCL G. Holland 6 Moorfield Road, Widnes WA8 3JE
2E0 HCM W. Rowland 5 Gwinear Downs, Leedstown, Hayle TR27 6DJ
2E0 HCO D. Shelsher Gables, Colchester Road, Ardleigh, Colchester CO7 7PQ
2E0 HCS C. Salisbury 9 Oakville Road, Heysham, Morecambe LA3 2TB
2E0 HCV M. Britton Butterwick Low, Hales Street Tivetshall St. Margaret, Norwich NR15 2EE
2E0 HCW C. Haynes 4 Thorn Close, Rugby CV21 1JN
2E0 HCY S. O'Neill 33 Norlands Park, Widnes WA8 5BH
2E0 HCZ P. Kilby 1 Home Farm Barns, South End, Milton Bryan, Milton Keynes MK17 9HS
2W0 HDB H. Bancroft Stop And Call, Goodwick SA64 0EX
2W0 HDC N. Orchard 152 Garden Suburbs, Trimsaran SA174AF
2E0 HDD M. Davies 40 Lake Crescent, Daventry NN11 9EB
2E0 HDE D. Edmondson 21 Hawthorne Close, Heathfield TN21 8HP
2E0 HDF A. Brain 20 South St., Spennymoor DL16 7TU
2E0 HDI J. Birkinshaw 57 Walnut Tree Avenue, Hereford HR2 7JU
2M0 HDL J. Howes 22 Criffel Drive, Lincluden, Dumfries DG2 0PE
2E0 HDM P. Allin 8 Kiln Close, Dove Holes SK178FQ
2W0 HDP M. Waldman 9 David Street, Cwmbwrla, Swansea SA5 8NX
2E0 HDQ J. Warburton 82 Hampton Drive, Newport TF10 7RF
2E0 HDU P. Loomes 107 Main Street, Sedgeberrow, Evesham WR11 7UE
2E0 HDW M. Langham 12 Cornflower Drive, Chelmsford CM1 6XY
2E0 HDX S. Hewick 56 Hemswell Avenue, Hull HU9 5JZ
2E0 HDY A. Hardy 54 Trueway Drive Shepshed, Loughborough LE12 9HG
2E0 HDZ I. Hammond 88 Great Innings North, Watton At Stone SG14 3TD
2E0 HEA J. Heagren 14 Pepperbox Rise, Whaddon, Salisbury SP5 3BF
2E0 HEC P. Mellers 34 Heasman Place, Southwater, Horsham RH13 9FT

2E0 HEF D. Robinson Height End Farm, Kirk Hill Road, Haslingden, Rossendale BB4 8TZ
2E0 HEG P. Sellick 1, The Gallery, Northwick Park, Blockley GL56 9RJ
2E0 HEI T. Bartholomew 6 Kilmaine Road, Harwich CO12 4UZ
2E0 HEM R. Milton 66 Beech Road, Hoo Marina Park, Hoo ME39TG
2M0 HEO L. Davis-Edmonds 6 Barlockhart Park, Glenluce, Newton Stewart DG8 0JQ
2E0 HEP J. Hobbs 2 Eccles Road, Wittering, Peterborough PE8 6AU
2E0 HEQ D. Barras 36 Carew Close, Yarm TS15 9TJ
2E0 HES R. Heslop 7 Fieldfare Close, Clanfield, Waterlooville PO8 0NQ
2E0 HET H. Taylor 25 Northolme Avenue, Nottingham NG6 9AP
2E0 HEU M. Gardiner 21 Bell Chase, Aldershot GU11 3GY
2E0 HEX J. Ash 47 Stein Road, Emsworth PO10 8LB
2I0 HEZ P. Robinson 119 Avenue Road, Lurgan, Craigavon BT66 7BD
2E0 HFA J. Wilson Flat 5, Blake House, London SE1 7DX
2E0 HFG A. Wale 17 Roundhouse Close, Welford, Northampton NN6 6NN
2E0 HFJ D. Sampson 116 South Mossley Hill Road, Liverpool L19 9BJ
2E0 HFK C. Steele 40 Landor Road, Whitnash, Leamington Spa CV31 2JX
2E0 HFN A. Toal 29 Highland Drive, Oakley, Basingstoke RG23 7LF
2E0 HFO S. Bracegirdle Flat 28, West Fryerne, Parkside Road, Reading RG30 2BY
2E0 HFS H. Starling 23 Lathom Road, Manchester M20 4NX
2E0 HFT Manx Sthrn DX G c/o D. Merridale The Granary, Falledge Lane, Upper Denby HD8 8YH
2E0 HGA R. Beardmore 28 Broadway, Ilkeston DE7 8TD
2E0 HGD J. Thirlwell 6A Brook Street, Warminster BA12 8DN
2E0 HGE A. Hitchens 16 Harrisons Place, Northwich CW8 1HX
2I0 HGF T. Smyth 139 Tullyreagh Road, Gorteen, Tempo, Enniskillen BT94 3PH
2E0 HGG D. Rennie 27 Orrell Road, Liverpool L21 8NQ
2I0 HGI A. Glasgow 17B Loy Street, Cookstown BT80 8PZ
2E0 HGK A. Dykes 84 Tennyson Road, Stafford ST17 9SR
2E0 HGM G. Heath 2 Lower Drake Fold, Westhoughton, Bolton BL5 2RE
2E0 HGO P. Niewiadomski Flat1 79A Dartmouth Road, London SE23 3HT
2E0 HGP G. Hawes 59 Hartlands, Bedlington NE22 6JG
2E0 HGQ R. Morris 45 St. Kildas Road, Bath BA2 3QL
2E0 HGR S. Rayner 109 Peckover Drive, Pudsey LS28 8EQ
2E0 HGT A. Todd 5 Dom Pedro Cottages, Normanton WF6 1RS
2E0 HGU P. Culmer 56 Main Road Washingborough, Lincoln LN4 1AU
2E0 HGW M. Spreafico 5A Bradfords Close, Bottisham, Cambridge CB25 9DW
2E0 HGX D. Clark 12 Wilson Crescent, Lostock Gralam, Northwich CW9 7QH
2E0 HHA P. Freeman Grove Orchard, Knapp Lane Coaley, Dursley GL11 5AR
2E0 HHB D. Matulewicz-Boyle 11 The Birches, Farnborough GU14 9RP
2E0 HHC R. Duff 55 Bigland Drive, Ulverston LA12 9PD
2E0 HHE J. Cryan 12 Stamford Avenue, Sunderland SR34AT
2E0 HHF A. Stevens Flat 19 Aylesbury House, London SE15 1RW
2E0 HHG D. Rudgley The New House, Plymouth Road, Buckfastleigh TQ11 0DB
2E0 HHJ A. Prestwich Highfield, Exminster, Exeter EX6 8AT
2E0 HHK S. Collins 5 Fernleigh Gardens, Stafford ST16 1HA
2W0 HHN D. Evans Bleddfa, Maes Meyrick, Heolgerrig, Merthyr Tydfil CF48 1RZ
2E0 HHP J. Cumming Farley Court, 100 Homer Close, Gosport PO13 9TL
2E0 HHT R. Flevill 62 Grosvenor Way, Horwich, Bolton BL6 6DJ
2E0 HHU J. Newham 6 Belsfield Gardens, Jarrow NE32 5QB
2E0 HHW C. Fox 209 Sulgrave Road, Washington NE37 3DE
2M0 HIB D. Hibberd 18 Whitestripes Path, Bridge Of Don, Aberdeen AB22 8WF
2M0 HIC Dr L. Campbell 21 Fordyce Way, Auchterarder PH3 1BE
2M0 HIE S. Scott 32 Dixon Terrace, Whitburn EH47 0LH
2E0 HIG B. Higgins 2 Bishops Yard, High Street, Huntingdon PE28 3JB
2E0 HIJ I. Halliday 14 Matterdale Gardens, Barming, Maidstone ME16 9HW
2E0 HIK J. Hawbrook 7 Birkdale, Norwich NR4 6AF
2E0 HIL D. Taylor 62 Floyds Lane, Walsall WS4 1LE
2E0 HIO Dr A. Ward 22 Whitecross Drive, Weymouth DT4 9PA
2E0 HIP D. Slater 13 Longford Close, Rainham, Gillingham ME8 8EW
2E0 HIQ M. Pope Drove Lodge 39 The Drove Barroway Drove, Downham Market PE38 0AJ
2E0 HIR M. Jenkin 13 Fore Street, St. Columb Major TR96RH
2E0 HIS G. Fenton 40 High Street, Easington Lane, Houghton le Spring DH5 0JN
2E0 HIT I. Petrie 88 Vicarage Road, Henley-on-Thames RG9 1JT
2E0 HIW C. Wilson Collingwood Avenue, Tolworth KT59PT
2E0 HIY A. Zahid Begum 20 Hill Top, Sutton SM3 9JH
2E0 HIZ J. Easdown 38 North Street, Barming, Maidstone ME16 9HF
2E0 HJB B. Halliwell 9 Tennyson Rd, Rothwell, , Kettering NN14 6JH
2E0 HJC J. Howarth 67A Crawford Ave, Manchester M298ET
2E0 HJE K. Earwicker Greenbank, Cripplestyle, Fordingbridge SP6 3DU
2M0 HJG J. Burns 6 Bonnyvale Place, Bonnybridge FK4 1DG
2E0 HJI E. Bray 56 The Links, Gosport PO13 0DX
2E0 HJJ H. Jackson Wingfield Farm, Cublington Road, Leighton Buzzard LU7 0LB
2E0 HJK A. Scott 19 Estuary Drive, Felixstowe IP11 9TL
2E0 HJN E. Buret 22 Russell Close Laindon, Basildon SS15 6AS
2M0 HJP H. Phillips Maplebank, Leithen Road, Innerleithen EH446NJ
2M0 HJS J. Smith 10 High Street, Portknockie, Buckie AB56 4LD
2E0 HJV J. Moore 65 Hamsey Green Gardens, Warlingham CR6 9RT
2E0 HJY P. Latham 135 Ashgate Road, Chesterfield S40 4AN
2E0 HJZ M. Pridmore 61 Gillards, Bishops Hull, Taunton TA1 5HH
2E0 HKC A. Chambers 5 Blandford Road, Shepton Mallet BA4 4FB
2E0 HKE P. Wightman 11 Sloway Lane, West Huntspill, Highbridge TA9 3RJ
2E0 HKG C. Cheung 23 Mosley Road, Timperley, Altrincham WA15 7TF
2U0 HKI R. Price Les Vaurioufs, St. Martins GY4 6TE Guernsey
2E0 HKK P. Buick Poundgate Farm, Uckfield Road, Crowborough TN6 3TA
2E0 HKL H. Klettke 13 Hastings Close, Wythall, Birmingham B47 6AW
2E0 HKP J. Coles 4 Carloggas Close, St Mawgan. Newquay, Truro TR8 4HJ
2E0 HKR A. Hill 11 Pelham Drive, Hull HU9 2AS
2I0 HKW R. Mcelholm 19 Summerhill, Prehen, , Derry BT47 2PL
2E0 HKZ J. Beer 92 Ashby Road, Hull HU4 7JT
2E0 HLC T. Winter 8 Thorpe Street, Hartlepool TS24 0DX
2E0 HLF J. Taylor 41 Waters View, Yarwell Mill, Yarwell, Peterborough PE8 6EU
2E0 HLG A. Allmark 44 Butterwick Fields, Horwich, Bolton BL6 5GZ
2E0 HLH H. Hall 7 Front Street Grindale, Bridlington YO16 4XU
2E0 HLM K. Schmidt Church, Corner, Mareham-le-Fen PE22 7RA
2E0 HLO A. Rowan 14 Craven Lea, Liverpool L12 0NF
2E0 HLP P. Hallas 37 Oakfield Road, Bromborough, Wirral CH62 7BA
2E0 HLR L. Richardson 35 Vidgeon Avenue, Hoo, Rochester ME3 9DE
2E0 HMA C. Cowley 103 Acre Road, Kingston upon Thames KT2 6ES
2E0 HMC Dr H. Coghlan 1Bell View Cross Houses, Shrewsbury SY5 6JJ
2E0 HMG H. Partridge 19 Dickens Drive, Melton Mowbray LE13 1HZ
2E0 HML P. Venables 9 Spring Gardens, Higham Ferrers NN10 8EP
2E0 HMM D. Lawson 2 The Blossoms, Fulwood, Preston PR2 9RF
2E0 HMP R. Scott 34 Moorfield Road, Birmingham B34 6QY
2E0 HMQ D. Allen 19 Brooklands Close, Uttoxeter ST148UH
2E0 HMR W. Andes 141 Sunningdale Avenue, Hanworth, Feltham TW13 5JS
2W0 HMS E. Barnes 2 Trem Y Garnedd, Bangor LL57 1NA
2M0 HMU G. Mcara 24 Balfour Street, Alloa FK10 1RU
2M0 HMV R. Rothon 112 Ravenswood Rise, Livingston EH54 6PG
2E0 HNC H. Clark 36 Market Oak Lane, Hemel Hempstead HP3 8JL
2E0 HNF D. Glover 24 Cadeby Road, Sprotbrough, Doncaster DN5 7SD
2E0 HNG M. Bell 11 Greaves Road, Sheffield S5 9DB
2E0 HNH A. Clark 62 New Road, London SE2 0QG
2E0 HNJ P. Harris 33 Kirklea Road, Houghton le Spring DH5 8DP
2E0 HNK H. Baird 35 St. Peters Road, Wolvercote, Oxford OX2 8AX
2M0 HNM M. Cowie Larachmohr, Hawkhill, Keiss, Wick KW1 4XF
2E0 HNP Dr D. Potts 90 Albemarle Road, Willesborough, Ashford TN24 0HN
2E0 HNR T. Bickerstaff 43 Riverway, Durrington, Salisbury SP48ES
2E0 HNX C. Shearan 59 Arnold Crescent, Mexborough S64 9JX
2E0 HOB J. Sparrow 12 St. Huberts Close, Gerrards Cross SL9 7EN
2E0 HOG M. Stroud 1 Sefton Court, Welwyn Garden City AL8 6WW
2W0 HOH P. Gostelow 6 Tree Field Caerau Farm, Llanidloes SY18 6LL
2E0 HOK K. Hough 15 Moorside Road Endmoor, Kendal LA8 0EN
2E0 HOL K. Craner 46 Meadowhill Crescent, Redditch B98 8HT
2E0 HOM A. Banwell Hazel Cottage, Chapel Lane, Hinton, Chippenham SN14 8HD
2E0 HOO H. Coram 16B Park Road, Dawlish EX7 9LQ
2E0 HOQ N. Lambert 17 Starcross Road, Weston-Super-Mare BS22 6NY
2M0 HOS J. Wallace 323 High Street, Dalbeattie DG5 4DX
2W0 HOT K. Jones 1 Alway Crescent, Newport NP19 9SX
2E0 HOU S. Houssart Flat 3, Virginia Court, London SE16 6PU
2E0 HOV K. Taylor 44 Main Street, Willoughby, Rugby CV23 8BH
2E0 HOX S. Hogg 38 The Gables, Widdrington, Morpeth NE61 5RA
2E0 HOY S. Hoyle 7 Horderns Road, Chapel-En-Le-Frith, High Peak SK23 9ST
2E0 HPB P. Brundrett 114 Ack Lane East, Bramhall, Stockport SK7 2AB
2E0 HPD R. Hodson 99 Alcester Road, Hollywood, Birmingham B47 5NR
2E0 HPE J. Blakemore 16 Brentwood Avenue, Newbiggin-by-the-Sea NE64 6JH
2E0 HPF J. Scannell 53 Morley Croft, Farington Moss, Leyland PR26 6QS
2E0 HPI C. Gorse Apartment 11 , 30 Stockton Road, Hartlepool TS25 1RY
2E0 HPJ J. Whiteside The Old Antique Shop, Bank Street Pulham Market, Diss IP21 4TG
2W0 HPK R. Hopkins 132 Laurel Road Bassaleg, Newport NP10 8PT
2W0 HPL P. Evans Flat 1 Red Cow Annex, Lloyds Terrace, Adpar, Newcastle Emlyn SA38 9EH
2W0 HPM M. Melody 7 Tegfan, Pontyclun CF72 9BP
2E0 HPO A. Riches 84 Elgar Drive, Shefford SG17 5RA
2E0 HPR I. Hooper 25 Honey Lane, Buntingford SG9 9BQ
2W0 HQA C. Cowling 14 Shelley Drive, Bridgend CF31 4QA
2E0 HQB D. Dobbie 24 Harrow Road, Leighton Buzzard LU7 4UQ
2W0 HQD R. Wilkes 9 Brynmawr Road, Ebbw Vale NP23 5FF
2E0 HQF R. Jones St. Fillans, The Warren, East Horsley, Leatherhead KT24 5RH
2E0 HQJ Rvd. C. Sherwood 1 Savile Road, Elland HX5 0LA
2E0 HQO R. Olive Lorien, The Ridge, Thatcham RG18 9HZ
2E0 HQQ J. Kelly 8 Greyhound Lane, Overton, Basingstoke RG25 3LE
2E0 HQU L. Russell 11 Hartbushes Station Town, Wingate TS28 5GA
2E0 HQV S. Ranger 137 Warren Avenue, Southampton SO16 6AF
2E0 HQY T. Jones 7 Sedum Close, Huntington, Chester CH3 6BL
2E0 HRB E. O'Neill 13 Goodwood Close, Market Harborough LE168JF
2E0 HRD V. Greatwood 11 The Green, Long Preston, Long Preston, Skipton BD23 4PQ
2E0 HRE S. Holroyd 31A Cross Street, Tenbury Wells WR15 8EF
2W0 HRG P. Sherwood High Croft Jeffreyston, Kilgetty SA68 0RG
2W0 HRH M. Bloore Halfway House, Hyfrydle Road, Talysarn, Caernarfon LL54 6HG
2E0 HRI R. Suchocki 8 High Street, Bluntisham, Huntingdon PE28 3LD
2E0 HRJ R. Huelin 15 Hill Chase, Walderslade, Chatham ME5 9HE
2W0 HRL H. Leonard 11 Newton Road Grangetown, Cardiff CF11 8AJ
2I0 HRM J. Mercer 32 Templemore Avenue, Belfast BT5 4FT
2M0 HRN R. Maccormack 90 Wardlaw Crescent East Kilbride, Glasgow G75 0PY
2M0 HRP I. Morrison Coach House, Olivers Brae, Stornoway HS12SX
2E0 HRR A. Hannen 9 Noble Road, Hednesford, Cannock WS12 4RW
2E0 HRS T. Dunne 23 Warstone Lane, Birmingham B18 6JQ
2I0 HRV T. Darrah 42 Pinewood Avenue, Carrickfergus BT38 8EW
2E0 HRY H. Roxbrough 17 Stanwell Close, Sheffield S9 1PZ
2E0 HSB D. Murphy 23 Lowndes Close, Stockport SK2 6DW
2E0 HSC R. Coombes 1 Green Lane, , Shipston on Stour CV36 4HG
2E0 HSE M. Ding 130 Windermere Avenue, Warrington WA2 0NE
2E0 HSG K. Dillow 101 Martins Lane Hardingstone, Northampton NN4 6DJ
2I0 HSL N. Davis 19 Toberhewny Hall, Lurgan BT66 8JZ
2E0 HSP N. Hawkins Deganwy Hardwick Road, King's Lynn PE30 5BB
2E0 HST C. Bolton 201 Lime Tree Avenue, Crewe CW1 4HZ
2W0 HSU C. Vugts Coed Coch, Llangammarch Wells LD4 4BS
2E0 HSV N. Silveston 28 Quernstone Lane, Northampton NN48UN
2E0 HSW H. Scott Whittle 7 Skyline House, Dickens Yard, Longfield Avenue, London W5 2BJ
2E0 HSY J. Housley Lesede Cottage, The Town, Carsington, Matlock DE4 4PX
2E0 HTB H. Banasiak 16 St Christophers Close, Bath BA2 6RG
2E0 HTC G. Carter 19 Brathay Crescent, Barrow-in-Furness LA14 2BG
2E0 HTG C. Gouveia 86A Spareacre Lane, Eynsham, Witney OX29 4NP
2E0 HTI A. Ariyapala 87 Lord Avenue, Ilford IG5 0HN
2E0 HTJ T. Hare 2 Primula Drive, Norwich NR4 7LZ
2E0 HTK A. Dowley 8 Paulet Place, Old Basing, Basingstoke RG24 7LA
2E0 HTM A. Taylor 14 The Lawns, Hinckley LE10 1DY
2E0 HTN D. Palmer 12 Grange Road, Southampton SO16 6UH
2E0 HTO A. Hine Tall Trees 13, Wilton Crescent, Alderley Edge SK9 7RE
2E0 HTR H. Moore 52 Limefield Street, Accrington BB52AF
2E0 HTS S. Davison 5 Denby Drive, Baildon, Shipley BD17 7PQ

2E0 HTV G. Radulescu 41 Sherard Road, London SE9 6EX
2E0 HTX A. Meredith 47 Northwick Road, South Oxhey, Watford WD19 6NE
2M0 HUD A. Allan 117 Bruce Gardens, Inverness IV3 5BD
2E0 HUH K. Pugh 4 Salt Boxes, Pinvin, Pinvin WR102LB
2M0 HUJ A. Hood 26 Annan Avenue, East Killbride G75 8XT
2E0 HUK M. Pratchett 119 Swindon Road, Wroughton, Swindon SN4 9AD
2W0 HUL K. Hulme 13 Lime Street, Gorseinon, Swansea SA4 4AD
2E0 HUM S. Crabb 22 Mary Warner Road Ardleigh, Colchester CO7 7RP
2E0 HUN A. Instone 63 Larch Road, New Ollerton, Newark NG22 9SX
2E0 HUR J. Jefferies Millfield Cottage, Wild Meadow Bolnhurst Road, Colmworth, Bedford MK44 2LF
2E0 HUT B. Hudson 12 Elmfield Road, Hebburn NE31 2DY
2W0 HUU P. Harper 11 Meas Stanley, Bodelwyddan LL18 5TL
2E0 HUY A. Moore 178 Park Road, Bedworth CV12 8LA
2E0 HUZ H. Hughes 24 Broadley Green, Windlesham GU20 6AL
2E0 HVB T. Evans 68 Wildbrook Road, Little Hulton, Manchester M38 0FU
2E0 HVD P. Dutton 10 River Lane, Partington, Manchester M31 4DB
2E0 HVE M. Simonsohn 5 Pitt Close, Blandford St. Mary, Blandford Forum DT11 9PS
2E0 HVF S. Pearson 3 Berkeley Road, Shirley, Solihull B90 2HS
2M0 HVG C. Coutts 1 Springfield Avenue, Duns TD11 3BF
2E0 HVH C. Sidaway 6 Rookery Chase Deepcar, , Sheffield S36 2NF
2E0 HVL H. Ly 11 Nodes Drive, Stevenage SG2 8AL
2E0 HVM P. Ashton 32 Sycamore Road, New Ollerton Nr Newark, Notts NG22 9PS
2E0 HVQ G. Tant 61 Western Way, Sandy SG19 1DU
2E0 HVT R. Lewis 46 Graydon Avenue, Chichester PO19 8RG
2E0 HVU P. Westall 1 Manor Gardens Morrison Road, Swanage BH19 1JT
2E0 HVV M. Dutton 10 Braemore Close Shaw Ol2 7Na, Oldham OL2 7NA
2E0 HVW S. Downe 9 Danesway, Exeter EX4 9ES
2E0 HVY A. Riddle 19 Cottey Crescent, Exeter EX4 9DT
2E0 HVZ J. Godfrey 24 Walton Way, Barnstaple EX32 8AE
2E0 HWC H. Cheesman 49 Front Street, Chirton, North Shields NE29 7QN
2E0 HWE J. Emery Flat 3, 12 Buxton Road, Ashbourne DE6 1EX
2M0 HWI R. Rae 1 Jura Drive, Tweedbank, Galashiels TD1 3ST
2E0 HWJ B. Kemble 3 Red House Close, Chudleigh Knighton, Chudleigh, Newton Abbot TQ13 0RH
2E0 HWN D. Wardman 45 The Grainger, North West Side, Gateshead NE8 2BG
2E0 HWO Dr H. Orridge Stonecross Cottage, Wadworth Hall Lane, Wadworth, Doncaster DN11 9BH
2W0 HWQ P. Rogers 109 North Road, Pontywaun, Cross Keys, Newport NP11 7FS
2E0 HWS I. Hemmens The Laurels, Barnstaple Road, South Molton EX36 3RD
2I0 HWW P. Ford 25 Carnhill, Londonderry BT48 8BA
2E0 HWY T. Kerswill 11A Vernon Drive, Prestwich, Manchester M25 9RA
2E0 HXC H. Castree 31 Fairview Thickwood, Colerne, Chippenham SN14 8BS
2E0 HXF C. Shepherdson 149 Scarborough Road, Norton, Malton YO17 8AD
2E0 HXG P. Stokes 26 Ashford Road, Hastings TN34 2HA
2E0 HXH R. Spooner 20 Hazledine Way, Bridgnorth WV16 5AE
2E0 HXJ S. Skelton 7 Sycamore Avenue, The Elms, Torksey, Lincoln LN1 2NJ
2D0 HXL D. Wilson 23 Snugborough Avenue, Union Mills, Braddan, Isle of Man IM4 4LT
2E0 HXP M. Savage 5 Mere Hall Barns, Mere Lane, Enville, Stourbridge DY7 5JL
2M0 HXQ A. Sloan 36 Paterson Avenue, Irvine KA12 9JJ
2E0 HXU D. Sutton 18 Knights Hill Severn Stoke, Worcester WR8 9JD
2E0 HXW S. Miller Foxglove Mead, North Common, Sherfield English, Romsey SO51 6JT
2E0 HXY G. Jones 158 Withington Lane, Aspull, Wigan WN2 1JE
2E0 HYC T. Rutt Granthorpe, Hull Road, Hull HU11 5RN
2E0 HYD D. Conner 20 Birch Avenue, Rochdale OL12 9QH
2E0 HYE T. Byers 1 Hazelwood Avenue, Sunderland SR5 5AH
2E0 HYF J. Mc Elhinney 231 Eastern Avenue, Sheffield S2 2GP
2E0 HYH D. Livingstone 4 Hatfield Road, Southport PR8 2PE
2E0 HYI M. Webb 10 Sausthorpe Street, Lincoln LN5 7XW
2E0 HYJ W. Lockley-Gardiner 193 Westbourne, Telford TF7 5QP
2E0 HYM A. Mears Waterways. Scotland Yard, Priors Leaze Lane, Hambrook, Chichester PO18 8RQ
2E0 HYO Prof. R. Houlston 51 Adelaide Road, Surbiton KT6 4SR
2E0 HYP S. Bailey 23 Maple Avenue, Tolladine, Worcester WR4 9RD
2E0 HYS Z. Gong Room 116, Walmsley Studio, 218 Saint John Street, London EC1V4AT
2E0 HYU A. Hitchcott 19 Corwen Road, Liverpool L4 7TL
2E0 HYY K. Mills 3 The Seasons, Summerway, Exeter EX4 8DQ
2M0 HYZ S. Magee 30 Burnfield Drive, Mansewood, Glasgow G43 1BW
2E0 HZA A. Fox 8 Giles Road, Swindon SN25 1QD
2W0 HZC R. Davis 19 Ael-Y-Bryn, Caerphilly CF83 2QX
2E0 HZE F. Stepien 11A Wirral Gardens, Wirral CH63 3BD
2E0 HZH S. O'Connor 4 Charlton Grove, Bradford BD20 0QG
2E0 HZK M. Ratcliffe 30 Newton Cross Road, Newton In Furness, Barrow-in-Furness LA13 0NB
2M0 HZL H. Mckay 3 Flemington Gardens, Whitburn, Bathgate EH47 0NS
2M0 HZO A. Moir 2/1 7 Stewartville Street, Glasgow G11 5PE
2E0 HZP D. Westwood 28 Weybridge Mead, Yateley GU46 7UY
2E0 HZQ Swansea & District ARC c/o B. Capewell 18 Westminster Road, Kidderminster DY11 6HG
2M0 HZR N. Mansfield Blencathra House, Township Road, Auckengill KW14XP
2E0 HZS H. Smith 11 Kettlebrook Road, Tamworth B77 1AB
2E0 HZU J. Howarth 19 Farnham Croft, Leeds LS14 2HR
2E0 HZW C. Norris Heather View, Forest Front, Hythe, Southampton SO45 3RJ
2E0 IAF S. Pryke 12 Seaward Avenue, Leiston IP16 4BB
2M0 IAH I. Macdonald 20 Newbigging Terrace, Auchtertool, Kirkcaldy KY2 5XL
2E0 IAJ I. Lockyer 11 Lorina Road, Ramsgate CT12 6DD
2E0 IAK H. Jeffery Albany, Cranbrook TN17 3JR
2E0 IAL I. Macdonald Broomhill Mill Lane, Worthing BN13 3DH
2E0 IAN I. Campbell 19B Elphinstone Road, Southsea PO5 3HP
2W0 IAO A. Oconnell 31 North Avenue, Tredegar NP22 3HF
2M0 IAQ J. Campbell 113 Brent Field Circle, Ellon AB41 9DB
2E0 IAS E. Alexander 25 Dunelm Road, Hetton-Le-Hole, Houghton le Spring DH5 9LB
2E0 IAZ D. Goldthorpe 6 West End Grove Haydock, St. Helens WA11 0AP
2E0 IBA S. Jones 11 Croft Close, Rowton, Chester CH3 7QQ
2E0 IBB B. Beard 8 Monks Close, Newcastle-under-Lyme ST5 3QU
2E0 IBI A. Douglas 3 Beech Avenue, Bilsborrow, Preston PR3 0RH
2W0 IBM A. Mcelroy 36 Bod Offa Drive, Buckley CH7 2PB
2E0 IBN L. Thacker Hunters Moon, Reigate Road, Horley RH6 0HU
2E0 IBP N. Whittaker 3 Westwood Dr Hellesdon, Norwich NR6 5DE
2E0 IBR M. Perescu 8 Marauder Road, Norwich NR66HD
2E0 IBT J. Ferrol The Hayloft, Brunstock CA64QG
2E0 IBU A. Buckley 25 Queensway, Pilsley, Chesterfield S45 8EJ
2M0 IBW I. Macdonald The Cottage, High Craigton, Glasgow G62 7HA
2W0 IBY L. Rodgers 16 Pembrey Gardens Pontllanfraith, Blackwood NP12 2LR
2E0 ICB I. Buchner 33 Blakewell Gardens Tweedmouth, Berwick-upon-Tweed TD15 2HJ
2E0 ICC I. Chilton 47 Mayfield Crescent, Eaglescliffe, Stockton-on-Tees TS16 0NH
2W0 ICD I. Davies 3 Keteringham Close, Sully, Penarth CF64 5JW
2E0 ICE P. Hallas 53 Thornleigh Avenue Eastham, Wirral CH62 9AZ
2E0 ICI I. Lippett 41 Springvale, Gillingham ME8 0JG
2E0 ICK A. Morris 71 Lurdin Lane, Standish, Wigan WN6 0AQ
2E0 ICN H. Lee Pantos Logistics, Attn50924, 776 Buckingham Avenue, Slough SL1 4NL
2E0 ICP I. Pass 69 Cotswold Road, Bath BA2 2DL
2E0 ICR D. Burton 16 Cage Lane, Great Staughton, St. Neots PE19 5DB
2E0 ICT M. Gascoyne 31 Dale View, Hemsworth, Pontefract WF9 4TA
2E0 ICU D. Marsh 16 Laurel Close, North Warnborough, Hook RG29 1BH
2E0 ICX D. Lucock 34 Wentworth Drive, Ipswich Borough Council IP8 3RX
2E0 ICY E. Moore 33 Avon Drive, Congleton CW12 3RQ
2E0 IDA I. Sharples 2 Moor Road Croston, Leyland PR26 9HQ
2E0 IDC I. Cosham 54 Hawkins Crescent, Shoreham-by-Sea BN43 6TP
2E0 IDE B. Smith Bees Corner, Ideford, Chudleigh, Newton Abbot TQ13 0AZ
2E0 IDF I. Firth 124 Viking Road, Bridlington YO16 6TB
2I0 IDJ K. Mclaverty 123A Castle Road, Antrim BT41 4NG
2E0 IDK I. King 7, Greenacres Avenue, Blythe Bridge, Stoke on Trent ST11 9HU
2E0 IDL D. Garner 1 Rowland Avenue, Field Street, Hull HU9 1HR
2E0 IDM D. Melbourne 160 Markfield, Court Wood Lane, Croydon CR09DW
2E0 IDN I. Norfolk Arwelfa, High Street, Uckfield TN22 3LP
2E0 IDO Y. Wang 50 Sleaford Street, Cambridge CB1 2PU
2E0 IDR I. Reeve 36 Stone Pippin Orchard, Badsey, Evesham WR11 7AA
2W0 IDT I. Booth 18 Clos Y Wiwer, Pentre Cwrt, Llantwit Major CF61 2SG
2E0 IEA E. Aspden 4 Lilac Close, Newcastle ST5 7DH
2E0 IEB I. Beales 6 Edge Well Rise, Sheffield S6 1FB
2M0 IEC E. Cohen 234 Allison Street, Glasgow G42 8RT
2E0 IED A. Wedge 30 Primrose Way Locks Heath, Southampton SO31 6WX
2E0 IEE S. Spice Ebor Lodge, Udimore Road, Rye TN31 6BX
2E0 IEI R. Adams 40 Lichfield Road, Gloucester GL4 3AL
2E0 IEL N. Raptopoulos 12 Wolversdene Road, Andover SP10 2AX
2E0 IEM J. Paine 11 Ferndale Park Fifield Road, Bray, Maidenhead SL6 2DZ
2E0 IEO M. Sanderson 2 East Crescent, Canvey Island SS8 9HL
2W0 IEP P. Roberts 18 Maes Mawr, Llanrwst LL26 0HW
2E0 IEQ J. Law 58 Westfields, Zeals BA126PW
2E0 IER B. Brooks 54 New Hills Road, Doncaster DN2 5TR
2E0 IET C. Blount 55 Silverthorne Drive, Caversham, Reading RG4 7NR
2E0 IEW J. Milner 30 Rowena Drive, Thurcroft, Rotherham S66 9HT
2E0 IEX T. Barris Vogelgartenstr. 41/1, Eislingen/Fils 73054 Germany
2E0 IFA C. Bolton 1A Salterns Terrace, Bideford EX39 4AG
2E0 IFB A. Wood Berrybrook Northall Road, , Eaton Bray LU6 2DQ
2E0 IFC N. Burnet 27 Mackenzie Way, Tiverton EX16 4AW
2E0 IFF P. Webster 15 Napier Street, Workington CA14 2PT
2E0 IFG A. Evans 7 Hillyer Grove, Combe Down, Bath BA2 5FQ
2E0 IFK D. Foxcroft 8 Cowlard Close, Launceston PL15 7EQ
2E0 IFN Dr D. Rodriguez 23 Thorpe Way, Cambridge CB5 8UJ
2E0 IFO N. Cranston 42 Curling Vale, Guildford GU27PH
2E0 IFP D. Greenwood 7 Royal Gardens, Ramsbottom, Bury BL0 9SB
2E0 IFS I. Stephens 14 Vardon Close Kingston Hill, Stafford ST16 3YW
2E0 IFT G. Saunders 140 Highbridge Road, Burnham-on-Sea TA8 1LW
2E0 IFV P. Kearney 22 Kingston Drive, Cheltenham GL51 0UB
2E0 IFW E. Evans 313 Parkgate Road, Chester CH1 4BE
2E0 IFZ I. Huggett 104 Hinchcliffe Orton Goldhay, Peterborough PE2 5SS
2W0 IGC I. Curnock 62 Heol Y Banc, Bancffosfelen, Llanelli SA15 5DL
2E0 IGE A. Insarov Broadway Chambers, 20 Hammersmith Broadway, London W6 7AF
2E0 IGH P. Frankowski Flat 7, Swift House, Chigwell Road, London E18 1TP
2E0 IGL P. Penn-Bixby 11 Iveston Road, Consett DH8 7HS
2E0 IGM G. Benford 34 Victoria Gardens, Colchester CO4 9YD
2W0 IGN R. Buchan-Terrey Godre'R Coed, Aberhosan, Machynlleth SY20 8RA
2E0 IGO R. Fountain 4 Den Hill, Eastbourne BN20 8SY
2E0 IGP Dr J. Harmer 1 Wynford Rise, Leeds LS16 6HX
2E0 IGQ I. Jameson 9 Pine Hey, Neston CH64 3TJ
2E0 IGT A. Williamson 4 Garden End, Melbourn, Royston SG8 6HD
2E0 IGW G. Taylor 39 Gill St., Newcastle upon Tyne NE4 8BH
2I0 IGX I. Kondrashenkov 44 Forge Manor, Magheralin, Craigavon BT67 0XP
2E0 IHB Dr R. Bleaney 40 Broadstone Road, Harpenden AL5 1RF
2E0 IHE D. Loveridge 2 Abbots Mead, Cholsey OX10 9RJ
2E0 IHH I. Hutchinson 41 Eastbourne Avenue, Gosport PO12 4NU
2E0 IHI C. Colless 128 Ditton Lane, Fen Ditton, Cambridge CB5 8SS
2E0 IHL A. Prime 68 Langley Drive, Crewe CW2 8LN
2E0 IHM S. Nelson-Smith Welbeck, Wickham Road, Fareham PO17 5BU
2E0 IHN N. Gooding 41 The Crescent, Wolverhampton WV6 8LA
2E0 IHO K. Brown 143 Princes Road, Ellesmere Port CH65 8EP
2E0 IHR O. Williams 6 Millfield Drive, Bristol BS30 5NR
2E0 IHS G. Sangwell 34 Gatcombe Close, Calcot, Reading RG31 4XQ
2E0 IHT C. Scouller 72 Brookmans Avenue Brookmans Park, Hatfield AL9 7QQ
2E0 IHU I. Leyland 61 Stacey Avenue, Milton Keynes MK12 5DN
2E0 IHV S. Jones 36 Woodbury, Lambourn, Hungerford RG17 7LT
2E0 IHW M. Rogers The Blue Mushroom, L E E, Ilfracombe EX34 8LR
2E0 IID I. Dodds 54 Philip Road, Newark NG24 4PD
2E0 IIF G. Gribbin 96 Old Church Lane, Stanmore HA7 2RR
2M0 IIH J. Jarvie Berryhill Farm, Tak-Ma-Doon Road, Kilsyth, Glasgow G65 0RY
2E0 IIJ C. Bassett 2 Culverden Square, Tunbridge Wells TN4 9NS
2E0 IIP E. Lewis 49 Ridgebourne Road, Shrewsbury SY3 9AB

2I0 IIQ A. Bandeja 16, Craigavon Crescent, Dungannon BT71 7BD
2E0 IIT G. Coleman 398 Chemin De Meynot, la Sauvetat Du Dropt 47800 France
2W0 IIU P. Harding 6 Ridgeway View, Newport NP205AW
2E0 IIV S. Lilley 1 Ormonde Road, Chester CH2 2AH
2E0 IIY A. Wade 6 Elliott Grove, Brixham TQ5 8RT
2E0 IIZ S. Martin 6 Cherry Tree Drive, Sedgefield, Stockton-on-Tees TS21 3DN
2W0 IJD T. Hughes 1 Craig Y Don, Amlwch LL68 9DN
2E0 IJG C. Nutu 37 Brading Road, London SW2 2AP
2E0 IJP C. Simon 55 Shirecroft Road, Weymouth DT4 0NH
2E0 IJQ R. Bedford 29 Kent Road, Brookenby, Market Rasen LN8 6EW
2E0 IJT R. Darby 39 Leonard Road Wollaston, Stourbridge DY8 3LU
2E0 IJV K. Thwaites 65 Chessington Park Hill, Chessington KT9 2BJ
2E0 IJW G. Williamson 26 Portland Mews, Bridlington YO16 4EH
2E0 IJX D. Wilderspin 14 Tannery Court, North Street, Crewkerne TA18 7AY
2E0 IJZ K. Cullen 29 Colman Road, London E16 3JY
2E0 IKB B. Wilkes 2 Kings Crescent, Edlington, Doncaster DN12 1BD
2E0 IKC E. Cooper 51 Lakeside Hightown, Ringwood BH24 3DX
2E0 IKG R. Butt Waterloo Cottage, Barrack Hill, Little Birch, Hereford HR2 8AX
2E0 IKH T. Anderson 45 Dudley Road, Clacton-on-Sea CO15 3DW
2E0 IKM M. Moffat 7 Lingfell Avenue, Cockermouth CA13 9BE
2E0 IKN C. Burls 86 All Saints Road, Kings Heath, Birmingham B14 7LN
2E0 IKO M. Tarrant Holly Cottage Nanstallon, Bodmin PL30 5JZ
2E0 IKS J. Mccosh Ch611Dl, Merseyside CH611DL
2E0 IKT C. Chisholm Hill Top Kellah, Haltwhistle NE49 0JL
2E0 IKU Dr S. Mitchell 56 Mill Green Road, Amesbury, Salisbury SP4 7RE
2E0 IKV S. Rush 88 Mountview, Borden, Sittingbourne ME9 8JZ
2E0 IKW G. Cannon 30 Main Street, Flixton, , Scarborough YO11 3UB
2E0 IKX P. Scarratt 4 Sandy Lonning, Maryport CA157LW
2E0 ILC C. Ng 45 Langdown Road, Southampton SO45 6EX
2E0 ILD R. Abeykoon 5 Challney Gardens, Luton LU4 8QQ
2E0 ILE M. Percival 10 Binder Close, Higham Ferrers NN108PH
2E0 ILF B. Campbell 9 Granams Croft, Liverpool L30 0PH
2E0 ILJ Rvd. P. Campbell 116 Kingsway Park, Urmston, Manchester M41 7FH
2E0 ILL J. Robinson 96 White Lodge Park, Shawbury, Shrewsbury SY4 4NU
2E0 ILN D. Bee 25 Blatcher Close, Minster On Sea, Sheerness ME12 3PG
2E0 ILO M. Noblet 1 Lingdale Road, Wirral CH48 5DG
2W0 ILQ W. Stevenson 310 Argosy Way, Newport NP19 0LP
2E0 ILX M. Bennett 33 Charles Street, Redditch B97 5AA
2M0 ILZ M. Greig 22 Rowanhill Close, Port Seton, Prestonpans EH32 0SY
2I0 IMB I. Barr 64 Owenreagh Drive, Strabane BT82 9DT
2E0 IMC I. Coggon 45, Ansten Crescent, Doncaster DN4 6EZ
2E0 IMD H. Ilie Sos. Iancului Nr. 33 Bloc 105A Scara B Apt 63, Bucharest 21717 Romania
2E0 IMG I. Gough Holythorn Cottage, Moreton Morrell CV359AR
2E0 IMH M. Williams 1 Montague Road Broughton Astley, Leicester LE9 6RL
2E0 IMJ P. Coppin 3 Firtree Close, Rough Common, Canterbury CT2 9DB
2W0 IMM C. Milne 18 Chapelfield, Deganwy LL31 9BF
2D0 IMN T. Hardwick 3 Poplar Terrace, Douglas, Isle of Man IM2 4AR
2I0 IMO T. Mcnaughter 36 Elms Park, Coleraine BT52 2QE
2M0 IMP S. Robertson 20 Knockard Place, Pitlochry PH16 5JF
2E0 IMS S. Edwards 68 Hampshire Road, Droylsden, Manchester M43 7PL
2E0 IMV D. Pelling 8 Fowler Close Earley, Reading RG6 7SS
2E0 IMW I. Walker 24 Hawthorn Road, Norwich NR5 0LP
2W0 IMX P. Lowe 9 Felin Uchaf, Dolgellau LL40 1NS
2E0 IMZ I. Ross 48 Henry Drive, Leigh-on-Sea SS9 3QF
2I0 INA A. Mcguigan 34 Ardymagh Road, Ballyclare BT39 9TJ
2E0 INC C. Jenkins 25 Longmeadow Grove West Heath, Birmingham B31 4SU
2E0 IND P. Ind 30 Thompson Road, Stroud GL5 1SY
2M0 INE M. Scott 28B Highfield Place, Birkhill, Dundee DD2 5PZ
2E0 INJ D. Blencowe 25 Harrington Road Kelmarsh, Northamptonshire NN6 9LX
2E0 INK J. Everatt 50 Barnsley Road, Thorpe Hesley, Rotherham S61 2RR
2E0 INP G. Inman 11 Sylvan Way, Gillingham SP8 4EQ
2M0 INS C. Ralph 37 Seaview Terrace, Edinburgh EH15 2HE
2E0 INT P. Kerton 15 Barncroft Way, Havant PO9 3AA
2E0 INX Prof. I. Neal 8 Rushey Gill, Brandon, Durham DH7 8BL
2W0 INY A. Noakes Westra Holt, Westra, Dinas Powys CF64 4HA
2E0 INZ A. Brooking 1 West Villas Cotford St Luke, Taunton TA4 1DF
2M0 IOB R. Kennedy 45 Rodney Road, Gourock PA19 1XG
2I0 IOE J. Allen 8F Marlborough Avenue, Belfast BT97HB
2E0 IOG Ts Gambia / Thorne Sea Cadets c/o D. Mathewson 89 Lower Mead Drive, Burnley BB12 0ED
2I0 IOI P. Gibson 118 Coleraine Road, Portstewart BT55 7HS
2E0 IOJ S. Bowen Flat 6, Wyndley House, 9 Welshmans Hill, Sutton Coldfield B73 6RY
2M0 IOK S. Webb Fawn View, Kennethmont, Huntly AB54 4PF
2E0 IOL R. Morley 7 Eliot Road, St. Austell PL25 4NL
2D0 IOM D. Allcote 18 Maynrys, Castletown IM9 1HR Isle of Man
2E0 ION A. Chlebikova St. Catharine'S College, Cambridge CB2 1RL
2E0 IOS R. Smith 62 Norwich Avenue, Plymouth PL5 4JQ
2E0 IOU D. Savage 72 Pitmore Road, Eastleigh SO50 4LW
2E0 IOX M. Cowsill 23 Laithes Croft, Earlsheaton, Dewsbury WF12 8BN
2E0 IOY S. Bone 20 Keld Avenue, Uckfield TN22 5BN
2E0 IOZ A. Nikitits 270A The Ridgeway, St. Albans AL4 9XQ
2E0 IPA D. Bradley 37 Leamoor Avenue, Somercotes, Alfreton DE55 1RL
2I0 IPB A. Kincaid 428 Cushendall Road, Ballymena BT43 6QE
2E0 IPC W. Catney 92 Second Avenue, Liversedge WF15 8JW
2M0 IPF R. Barlow Iphs, Westray KW17 2DW
2E0 IPG N. Head-Jenner 5 Ponders Road, Colchester CO63LX
2E0 IPK T. Bone 104 Rowdowns Road, Dagenham RM9 6NH
2E0 IPL C. Panaitescu 131 Stafford Road, Croydon CR0 4NN
2E0 IPW P. White 13 Field Court, Oxted RH8 0PD
2E0 IPX P. Ashby 45 Seaton Road, Felixstowe IP11 9BS
2E0 IPY M. Cooper Flat 6, Claydon House, 9 St. Quentin Close, Swindon SN3 4FA
2E0 IQE R. Jarrott 11 Stewkins, Stourbridge DY8 4YW
2W0 IQF T. Williams Sea View House, Calon Fawr, Lon Masarn, Tycoch, Sketty, , Swansea SA2 9EX
2E0 IQK J. Mcfarlane 16 Lake View Houghton Regis, Dunstable LU5 5GJ
2E0 IQL S. Austin Littlecroft, Luddington Avenue, Virginia Water GU25 4DF
2E0 IQM J. Watts 6 Hallfield Park Great Sutton, Ellesmere Port CH66 3TA
2E0 IQO M. Vardy 1 Sutton Middle Lane, Kirkby-In-Ashfield, Nottingham NG17 8FX
2E0 IQP M. Smith 30 Pitfold Road, London SE12 9HX
2E0 IQQ C. Camilleri Unit 123585, Po Box 6945, London W1A 6US
2M0 IQR H. Chen Iq Fountainbridge Room 118E 114 Dundee Street, Edinburgh EH11 1AD
2E0 IQT S. Magrys 13 Norfolk Road, Kidsgrove, Stoke-on-Trent ST7 1EZ
2M0 IQU W. Curry 36 Banklands Newburgh, Cupar KY146DN
2E0 IQW D. Parker 51 Parsonage Road, Henfield BN5 9HZ
2E0 IQX A. Emmerson 8 Weston Close, Heath Hayes, Cannock WS11 7YX
2W0 IQY N. Jones 11 Wood Green, Mold CH7 1UG
2M0 IRC J. Livingstone 17 Livingstone Drive, Bo'ness EH51 0BQ
2E0 IRD D. Korzeniewicz 6 Barley Down Drive, Winchester SO22 4LS
2E0 IRE M. O'Donovan Wyllsden House, Stroud GL52PA
2M0 IRG Dr M. Sutcliffe 11 Low Borland Way, Eaglesham, Glasgow G76 0BP
2E0 IRH S. Cooper 19 Old Chapel Road, Smethwick B67 6JA
2E0 IRJ I. Jones 33 Cobham Avenue, Liverpool L9 3BP
2E0 IRN J. Glicklich 86 Ainsdale Road, Bolton BL3 3ER
2I0 IRO G. Clarke 2 Gort Lomie, Clonlara V94 H96H Ireland
2E0 IRX G. Harman 58 Laurence Avenue, Witham CM8 1JB
2M0 ISA I. Watson 64 Anstruther Street Law, Carluke ML8 5JG
2E0 ISB S. Brown 18 Goring Ave. Gorton., Manchester M18 8WW
2E0 ISE I. Briley 12 Stirling Avenue, Waterlooville PO7 7NH
2E0 ISK K. Hyde 8 Pennant Close, Birchwood, Warrington WA3 6RR
2E0 ISN S. Narwan 14 Fermoy Road, Greenford, London UB6 9HX
2E0 ISO I. Butler 11 School Close, Basingstoke RG22 5FY
2E0 ISQ R. Lilley 3 Coultshead Avenue, Billinge, Wigan WN5 7HS
2E0 ISS P. Dann 4 Middlefields Court, Middlefields, Letchworth Garden City SG6 4NQ
2E0 ISU M. Dunstan 5 Polgooth Close, Redruth TR15 1QL
2M0 ISY I. Phillips 16 Seton Court, Port Seton, Prestonpans EH32 0TU
2W0 ISZ R. Priamo 58 Ffordd Glyn, Coed-Y-Glyn, Wrexham LL13 7QW
2E0 ITC A. Checketts 77 Shenstone Avenue, Stourbridge DY8 3EH
2E0 ITF C. Owen 3 The Flexton, Ottery St. Mary EX11 1DJ
2E0 ITG A. Woods 4 Sedgwick Close, Atherton M46 9EG
2W0 ITH M. Findlay 26 Swan Street, Llantrisant, Pontyclun CF72 8ED
2M0 ITI Dr J. Wills 2 Old Dalmore Gardens Auchendinny, Penicuik EH26 0RR
2E0 ITJ T. Johnson 15 Tennyson Road, Creswell, Worksop S80 4DW
2W0 ITM I. Mitchell 9 Rhiw Grange, Colwyn Bay LL29 7TT
2E0 ITN R. Goodall 76 Beaconfield Road, Plymouth PL2 3LF
2M0 ITO R. Bell 10 Lorraine Drive, Cupar KY155DY
2E0 ITQ J. Swales 90 Earlswood Road, Dorridge, Solihull B93 8RN
2E0 ITU P. James Flat 2, Marigold House, 2 Ironbridge Road, Twigworth, Gloucester GL2 9GS
2E0 ITV G. Moorhouse 1 Woodlands Avenue, Spilsby PE23 5EP
2E0 ITX D. Gibbeson 33 Treefield Walk, Barnstaple EX32 8PE
2I0 ITY D. Given 15 Middle Road, Lisburn BT27 6UU
2E0 IUA B. Chapman 62 Spring Grove, Loughton IG10 4QE
2E0 IUC N. Morris 31 Endeavour Place, Stourport-on-Severn DY13 9RL
2M0 IUE A. Young 2 Dunvegan Place, Ellon AB41 9TF
2E0 IUF C. Lundregan 45 Hildyard Close, Gloucester GL2 4PZ
2E0 IUH A. Morris 22 Dixon Avenue, Newton-le-Willows WA12 0NE
2E0 IUI D. Marshall 7 Lancaster Close, Newton-le-Willows WA12 9EY
2M0 IUJ A. Carmichael 18 Waterside Gardens Carmunnock, Clarkston, Glasgow G76 9AL
2E0 IUK D. Grayson 79 Errington Avenue, Sheffield S2 2EA
2E0 IUL I. Fleming 82 Malden Road, Sidmouth EX10 9NA
2E0 IUM J. Binks 12 Barnards Farm, Beer, Seaton EX12 3NF
2W0 IUN I. Jones 21 Albert Street, Maesteg CF34 0UF
2E0 IUO D. Stone 6 Beechwood Gardens, Bournemouth BH5 1NF
2E0 IUR R. Whiteley 3 Coreway Close, Sidford, Sidmouth EX10 9SX
2E0 IUW P. Jones 8 Normanton Close, Edwinstowe, Mansfield NG21 9PF
2E0 IVC P. Negros 14 Amersham Road, London SE14 6QE
2E0 IVD A. Beecroft 10 Marchant Close, Beverley HU17 9GE
2E0 IVE M. Williams 19 Cotebrook Drive, Upton, Chester CH2 1RA
2E0 IVF J. Gosling 11 Pinfold Place, Harby, Melton Mowbray LE144BX
2E0 IVK M. Kidd Sunny Orchard Farm, Kentmere Road, Staveley, Kendal LA8 9JF
2E0 IVO W. Gissing 2 Yeo Moor, Clevedon BS21 6UQ
2M0 IVP J. Pike Mosside Farmhouse, Culsalmond, Insch AB52 6TU
2E0 IVR I. Ross 15 Earlswood Close, Madeley, Telford TF7 5SF
2M0 IVS G. Brass 114 Torbrex Rd Cumbernauld, Glasgow G67 2JS
2E0 IVU K. Darker 19, The Horseshoe, Hemel Hempstead HP3 8QT
2E0 IVX D. Frost 169 West Auckland Road, Darlington DL3 0SP
2E0 IVY D. Slee Turnpike, Brearton, Harrogate HG3 3BX
2W0 IVZ R. Vials 46-48 Park Place, Gilfach, Bargoed CF8 8NA
2E0 IWB I. Bunting 14 Mill Pightle, Aylsham, Norwich NR11 6LX
2M0 IWD I. Davidson 43 Whitehall Avenue, Cardenden, Lochgelly KY5 0PH
2E0 IWF I. Francis 34 Furlong Road, Bourne End SL85AA
2E0 IWI D. Gilbertson 6 Lewens Lane, Wimborne BH21 1LE
2E0 IWJ F. Bila-Nicola 48 Shepherds Pool, Evesham WR11 4JG
2E0 IWL Dr T. Lachlan-Cope 38 Cambridge Road Waterbeach, Cambridge CB25 9NJ
2W0 IWM I. Miles 40 Seymour Street, Caegarw, Mountain Ash CF45 4BL
2E0 IWN S. Pitcher 32 Regent Court, South Shields NE33 5RX
2E0 IWR P. Perkins 47 Ranulf Road, Flitch Green, Dunmow CM6 3GR
2E0 IWS I. Smith 8 Tregarth Road, Chichester PO19 5QU
2E0 IWT M. Champness 10 Isaac Square, Great Baddow, Chelmsford CM2 7PP
2M0 IWU D. Spooner Glachbeg, Allanglach Wood, North Kessock, North Kessock IV1 3XD
2E0 IWY D. Reed Ivy Cottage, Foundry Yard, Ridsdale, Hexham NE48 2TG
2E0 IWZ A. Hanna 35 Orchard Drive, Mayland, Chelmsford CM3 6EP
2E0 IXA J. Masterton Nobes Avenue, Gosport PO130HX
2E0 IXC D. Watson 10 Gimson Close, Tuffley, Gloucester GL4 0YQ
2E0 IXH R. Maas 15 Pine Court, Attleborough NR17 2HU
2E0 IXI R. Forss Flat 1, 18 Park Lane, Bath BA1 2XH
2E0 IXJ M. Embleton 23 Holloway Gardens, Plymouth PL9 9TS
2E0 IXK D. Harris 12A Clevelands Park, Northam, Bideford EX39 3QH
2E0 IXM Dr D. Buckley 7 Clary Meadow, Northwich CW8 4XG
2E0 IXS I. Smith 50 Aylton Road, Liverpool L36 2LX
2M0 IXT S. Kelso 98 Highmains Avenue, Dumbarton G82 2QB

2E0 IXU W. Wei University Of Cambridge, Storey'S Way, Cambridge CB3 0DS
2E0 IXV Dr C. Roe-Bullion 96A Nunnery Street, Castle Hedingham, Halstead CO9 3DP
2E0 IXW M. Charnley 33 Cromdaleway Great Sankey, Warrington WA53NR
2E0 IXX D. Cattermole 39 Moor Lea, Braunton EX33 2PF
2E0 IXZ C. Talbot 24A Calder Road, Stourport-on-Severn DY13 8QD
2E0 IYA N. Olsson 60 Bebington Road, Birkenhead CH42 6PX
2E0 IYC N. Wache 11 Horselees Road, Boughton-under-Blean ME13 9TG
2E0 IYG S. Howell 22 Wootton Rivers, Marlborough SN8 4NQ
2I0 IYH A. Wilson 108A Salia Avenue, Carrickfergus BT38 8NE
2M0 IYI L. Thom Flat 804D, 110 St. James Road, Glasgow G4 0PS
2M0 IYK M. Binns Old Tileworks House, Ochiltree, Cumnock KA18 2NN
2E0 IYL S. I'Anson 5 Heather Gardens, Leeds LS14 3HU
2E0 IYM A. Murphy Linnets, Prospect Road, Upton OX11 9HT
2E0 IYN I. Melling 33 Brookfield Road, Market Harborough LE16 9DU
2E0 IYS C. Brown 17A Langley Drive, Derby DE74 2DN
2E0 IYU S. Swindells 95 Harrogate Road, Ripon HG4 1SX
2M0 IYV R. Pease The Treehouse, Inverkirkaig IV274LR
2E0 IYY H. Burnett 10 Galton Avenue, Christchurch BH23 1JU
2E0 IYZ M. Shutenko 48 Prenton Village Road, Prenton, Wirral CH43 0SZ
2E0 IZA J. Ross 5 Eastgate, Heckington NG349RB
2E0 IZB P. Lewis 37 Speedwell Close, Melksham SN12 7TE
2E0 IZF W. Taylor 10 Miller Close, Redbourn, St. Albans AL3 7BG
2E0 IZG G. Woods 4 Newton Gate, Sturminster Newton DT10 2EU
2E0 IZH P. Eachus 251 Forton Road, Gosport PO12 3HD
2I0 IZI A. Ismay 21 Hillsborough Drive, Belfast BT6 9DS
2E0 IZJ A. Dunn 185 Waterloo Road Yardley, Birmingham B25 8LH
2E0 IZK A. Lindsay 36 Hodgson Gardens, Guildford GU4 7YS
2E0 IZN C. Lavery 12 Jackson Close, Kesgrave, Ipswich IP5 2QL
2E0 IZP I. Pipe 9 Sherlock Hoy Close, Broseley, Telford TF125JB
2E0 IZR J. Bridson 10 Clegg Street, Astley, Tyldesley, Manchester M29 7DB
2E0 IZT M. Faben 15 Monks Close, Doncaster DN7 4QL
2E0 IZU S. Stray 9 Highfields Mews, Great Gonerby, Grantham NG31 8XA
2E0 IZW I. Whiteley 2 The Meade, Manchester M21 8FA
2W0 IZX C. De Winton Tymawr Farm, Llanfrynach, Brecon LD3 7BZ
2E0 IZZ P. Smith 14 Highfield Crescent, Kettering NN15 6JS
2E0 JAA J. Ahmed 59 Ramsgate, Lofthouse, Wakefield WF3 3PX
2E0 JAB J. Bradbury 11 Ravensbourne House, Arlington Road, Twickenham TW1 2AX
2E0 JAC J. Crank 38 Harley Avenue, Harwood, Bolton BL2 4NU
2E0 JAF J. King 22 Latchmere Gardens, Leeds LS16 5DN
2W0 JAI J. Griffiths 85 Tudor Estate, Maesteg CF34 0SW
2E0 JAJ J. Johnson 1 Stretton Lodge Gordon Road, London W13 8PR
2M0 JAL C. Maclean 16 Glamis Avenue, Elderslie, Johnstone PA5 9NR
2E0 JAM J. Dodds 8 York Road, Rowley Regis B65 0RR
2E0 JAN G. Martin 12 Poolside, Phase 1, St. Joseph Trinidad and Tobago
2E0 JAO J. Lockyear Berrow Bank, Bromsberrow, Ledbury HR8 1SG
2I0 JAP E. Rantin 8A Buchanans Road, Newry BT35 6NS
2E0 JAQ J. Bosworth 10 Aston Street, Leeds LS13 2BJ
2E0 JAR J. Rance 11 Orchard Lane, Longton, Preston PR4 5AX
2M0 JAT A. Murray 119 Carnarc Crescent, Inverness IV3 8SJ
2E0 JAW J. Phipps 42 Cavell Avenue, Peacehaven BN10 7NS
2E0 JAX J. Burdett 5 Winston Drive, Wainscott, Rochester ME2 4LJ
2E0 JAZ J. Sadler 10 Spindle Warren, Havant PO9 2PU
2E0 JBA J. Woods 3 Ingle Avenue Morley, Leeds LS27 9NP
2E0 JBC D. Croft 33 Roughaw Road, Skipton BD23 2PY
2E0 JBD J. Dukes 79 Jubilee Avenue, Boston PE21 9LE
2D0 JBE J. Mccartney 5 Riverside, Ramsey, Isle of Man IM8 3DA
2E0 JBF G. Brown 51 Arncliffe Drive, Knottingley WF11 8RH
2E0 JBI J. Bethell 19 Kiln Cottages, The Brickfields, Stowmarket IP14 1RY
2W0 JBJ J. Jenkins 13 Birch Hill, Newport NP20 6JD
2E0 JBK S. Glass 16 Norman Way, Colchester CO3 4PS
2E0 JBL J. Chatterton 6 Bayliss Road, Wargrave, Reading RG10 8DR
2E0 JBM J. Moppett 59 Piccadilly, Tamworth B78 2ER
2E0 JBQ D. Robinson 27 Abbeylea Drive Westhoughton, Bolton BL5 3ZD
2E0 JBS J. Byrne 45 Jenkins Drive, Sheffield S9 1AR
2E0 JBW S. Pascoe Trewyn, Carnmenellis, Redruth TR16 6PG
2E0 JBX J. Barratt 17 Main Road, Collyweston, Stamford PE9 3PF
2E0 JBY J. Bibby 12 James Street, Burton-on-Trent DE14 3SB
2M0 JBZ J. Coubrough 41 Bridge Court, Alexandria G83 0BZ
2E0 JCA J. Aubury 27 Gravel Walk, Tewkesbury GL20 5NH
2E0 JCB J. Waring 12 Mary Street, Farnhill, Keighley BD20 9AU
2E0 JCC J. Whitton 11 Dursley Road, Bristol BS11 9XB
2E0 JCD J. Dalgliesh 61 Clonners Field, Stapeley, Nantwich CW5 7GU
2E0 JCE J. Tompkins 3 Hartwell Road, Portsmouth PO3 5TN
2M0 JCF J. Forsyth 14B Osborne Terrace, Edinburgh EH12 5HG
2M0 JCG J. Rankin 17 Dippin Place, Saltcoats KA21 6AB
2I0 JCH Dr J. Henderson 76 Tirmacspird Road, Crimlin, Ederney, Enniskillen BT93 0FB
2E0 JCK E. Beechill Belleroyd Farm Blackshaw Head, Hebden Bridge HX7 7JP
2E0 JCL J. Swift Raf Holmpton Rysome Lane, Holmpton HU19 2RG
2E0 JCM J. Clark-Mcintyre 6 Belvedere Road, Blackburn BB1 9NS
2E0 JCO J. Crewe 22 Myrtle Tree Crescent Sandbay, Weston-Super-Mare BS22 9UL
2W0 JCP J. Percival Blue Cedars, Gresford, Wrexham LL12 8RN
2E0 JCQ J. Stevens 51 Cheddington Road, Pitstone, Leighton Buzzard LU7 9AQ
2E0 JCR J. Seear Old Gossington Chapel, Gossington, Slimbridge, Gloucester GL2 7DN
2E0 JCS J. Sanderson 54 Kelvedon Close, Chelmsford CM1 4DG
2E0 JCT J. Griffiths 83 Golborne Road Ashton-In-Makerfield, Wigan WN4 8XA
2E0 JCU J. Underwood 12 Forsythia Close, Bicester OX26 3GA
2W0 JCV J. Baldwin 94 King Street Abertridwr, Caerphilly CF83 4BG
2E0 JCW J. Wright 19 Halstead Close, Woodley, Reading RG5 4LD
2E0 JCX S. Jackson 5 Duchess Park Close, Shaw, Oldham OL2 7YN
2E0 JCY J. Connolly 2 Waring Avenue, St. Helens WA9 2QG
2E0 JCZ J. Robb 37 Wroxham Road, Woodley, Reading RG5 3AX
2E0 JDA J. Dale Corydon, Church Street, Sevenoaks TN14 7SW
2E0 JDC L. Short 10 Carville Crescent, Brentford TW8 9RD
2E0 JDD J. Delves 14 Stanthorne Avenue, Crewe CW2 8NH
2E0 JDE J. Hardingham 11 All Saints Close, Weybourne, Holt NR25 7HH
2E0 JDF A. Powell Crosstrees, Main Road, Theberton, Leiston IP16 4RX
2E0 JDH D. Harbron 6 West View, Penshaw, Houghton le Spring DH4 7HP
2E0 JDI J. Redhead 28 Sandfields, Frodsham WA6 6PT
2E0 JDK J. Hazeltine 21 Hassock Way, Wimblington, March PE15 0PJ
2W0 JDL G. Wilkins 7 Byron Road, Newport NP20 3HJ
2E0 JDM J. Matheson 24 Grange Road, Dacre Banks, Harrogate HG3 4HA
2E0 JDO J. Harriott Flat 10, 401 High Street, Cheltenham GL50 3FJ
2E0 JDP N. Pipkin 46 Charles Avenue, Albrighton, Wolverhampton WV7 3LF
2E0 JDY J. Boslem 14 Morrins Close, Great Wakering SS3 0DY
2D0 JEA J. Hill 54 Wybourn Drive, Onchan IM3 4AT Isle of Man
2E0 JED G. Maguire 39 Cooper St., Stretford, Manchester M32 8NA
2E0 JEE Dr J. Egleton 81 Pinchbeck Road, Spalding PE11 1QF
2E0 JEH J. Hewart 14 Kestrel Close, Marple, Stockport SK6 7JS
2E0 JEK J. Kay 36 Winnington Road. Marple, Stockport SK6 6PT
2E0 JEM J. Carvill The Lodge, Oldbury Road, Worcester WR2 6AA
2E0 JEN P. Halpin 50 Celtic Road, Deal CT14 9EF
2E0 JEQ S. Brunsden 43 Porthcawl Drive, Washington NE37 2LT
2E0 JES C. Ember 35 Mattock Lane, Ealing, London W5 5BH
2E0 JET R. Cunliffe 89 Colyers Lane, Erith DA8 3NG
2W0 JEV E. Jones 11 Alma Place Sebastopol, Pontypool NP45EA
2M0 JEY E. Black 20 Provosts Walk, Monifieth, Dundee DD5 4SH
2E0 JEZ J. Powell 46 Woodmancote, Yate, Bristol BS37 4LL
2M0 JFD R. Cosh Lyndhurst, Kingston Road, Neilston, Glasgow G78 3DY
2M0 JFI M. Gee Cruachan, Annfield Crescent, Kirkwall KW15 1NS
2E0 JFK J. Kirk 3 Knold Park, Margate CT9 3BH
2E0 JFL J. Lugsden 21 Overhill Way, Beckenham BR3 6SN
2I0 JFO W. Forde 35 Torr Gardens, Larne BT40 2JH
2E0 JFP J. Porter 45A Wealden Way, Bexhill-on-Sea TN39 4NZ
2E0 JFR S. Phillips 2 Watermill Drive, St. Leonards-on-Sea TN38 8WD
2E0 JFW J. Witchell 7 Watercombe Lane, Yeovil BA20 2ED
2E0 JFY Dr C. Holmes Old Vicarage Farmhouse, Course Lane, Wigan WN8 7LA
2E0 JGD J. Dinsbier Little Downs, Sandgate Lane, Pulborough RH20 3HJ
2E0 JGE J. Glover 12 Willow Street, London E4 7EG
2E0 JGG J. Gibbons 28 Deveron Gardens, South Ockendon RM15 5ET
2E0 JGH J. Hunt 15 Greenway, London SW20 9BQ
2E0 JGL J. Grant 129 Eastern Avenue, Lichfield WS13 6RL
2E0 JGM I. Mccorquodale 4 Wise Grove, Warwick CV34 5JW
2E0 JGP J. Paradi 168 Castle Road, Northolt UB5 4SG
2E0 JGS J. Seaton 88 Stanley Road, Cambridge CB5 8LB
2E0 JGW J. Whitehead 12 Polkerris Road, Carharrack, Redruth TR16 5RJ
2E0 JHC J. Clark 27 The Gabriels, Newbury RG14 6PZ
2E0 JHE L. Batten Y Felin Barn, Llawr-Y-Glyn, Caersws SY17 5RH
2E0 JHF P. Attwater 42 Danescourt Crescent, Sutton SM1 3EA
2E0 JHG J. Ginever 66 London Road, Maidstone ME16 8QU
2I0 JHH J. Hannon 75 Suffolk Road, Belfast BT11 9PU
2M0 JHN J. Brown 60 Laburnum Lea, Hamilton ML3 7LZ
2E0 JHO B. Hodgson 28 Grove Drive, Woodhall Spa LN10 6RT
2E0 JHP D. Porter 105 Shore Road, Littleborough OL15 9LJ
2E0 JHT J. Hill 45 Venus Street, Congresbury, Bristol BS49 5HA
2E0 JIA I. Boddy 5 Boverton Avenue, Brockworth, Gloucester GL3 4ER
2I0 JIE J. Evans 404 Foreglen Road, Dungiven, Londonderry BT47 4PN
2E0 JIG D. Harris 35 Itchenor Road, Hayling Island PO11 9SN
2I0 JIJ G. Travers 88 Palmerston Road, Belfast BT4 1QD
2E0 JIL G. Whitehead 29 Coulsons Road, Bristol BS14 0NN
2E0 JIP J. Pownall 75 Park Barn Drive, Guildford GU2 8ER
2E0 JIR S. Buckley 35 Marlborough Road, Irlam, Manchester M44 6HH
2E0 JIS S. Smart 26 Wycote Road, Gosport PO13 0TG
2E0 JIT D. Hook 6 Cliveden Close, Allington, Maidstone ME16 0PQ
2E0 JIW J. Biggin Galadean, Farriers Way, Newport PO30 3JP
2E0 JIX J. Packer 20 Shipman Road, Market Weighton, Market Weighton YO43 3RB
2E0 JIZ S. Phillimore 21 First Avenue, West Molesey KT8 2QJ
2E0 JJA J. Adler 1 Searles Meadow, Dry Drayton, Cambridge CB23 8BW
2E0 JJB J. Barton 93 Cardigan Road, Bridlington YO15 3JU
2E0 JJK J. King 18 Ross Road, Wallington SM6 8QB
2E0 JJN J. Nicholls 6 Marasca End, Holt Drive, Colchester CO2 0DL
2E0 JJP J. Walker Flat A 14 Elswick Road, London SE13 7SR
2E0 JJR R. Ellery 12 Sentry Close, St. Issey Wadebridge PL27 7QD
2E0 JJS J. Siddall 16 Alandale Avenue, Langwith Junction, Mansfield NG20 9RU
2W0 JJW R. Woodland Flat 14, Mays Court, Windsor Road, Neath SA11 1NG
2W0 JJY J. Young Rhos Owen, Llangristiolus, Bodorgan LL62 5RD
2E0 JKB S. Lucas 31 Lilian Close, Norwich NR6 6RZ
2E0 JKD K. Winward 123 Fulbeck Road, Middlesbrough TS3 0RL
2E0 JKG J. Gaskin Badgers Barn, Canterbury Road, Folkestone CT18 8DF
2E0 JKI Dr J. Kinrade 112 Ullswater Road, Lancaster LA1 3PX
2W0 JKN E. Jenkins 41 Park Street, Bridgend CF31 4AX
2E0 JKP J. Poole 61 Lower Vickers Street, Miles Platting, Manchester M40 7LX
2E0 JKR R. Whatmough 150 Whelley, Wigan WN1 3UE
2E0 JKS J. Slyne 3 Heaths Close, Enfield EN1 3UP
2E0 JKT J. Turner 34 Vaughan Road, Stotfold, Hitchin SG5 4EH
2D0 JKW Dr J. Wardle Cooyrt Vane, Ballamodha Straight, Ballamodha, Isle of Man IM9 3AY
2E0 JKY K. Young 48 Sussex Street, Cleethorpes DN35 7NP
2E0 JLB J. Bookham 116 Clare Gardens, Petersfield GU31 4EU
2E0 JLD J. Drinkell 11 Valley Walk, Kettering NN16 0LY
2E0 JLK K. Robinson 103 Recreation Street, Mansfield NG18 2HP
2E0 JLM J. Meek Rose Tree Cottage, Charlbury OX7 3RX
2E0 JLR B. Taylor 1 Whinchat Close, Stockport SK2 5UU
2M0 JLS J. Leitch 25 Lime Street, Grangemouth FK3 8LZ
2E0 JLW J. Williamson 7 Dale Close, Wrecclesham, Farnham GU10 4PQ
2E0 JLX J. Landless 2 Aspen Way, Banstead SM7 1LE
2E0 JMB J. Banham Timandra, Mill Road, Hardwick, Norwich NR15 2ST
2E0 JMC J. Mcinnes-Boylan 10 Faith Street, Leigh WN7 4TS
2E0 JMD J. Davies 66 Barnstaple Road, North Shields NE29 8QG
2E0 JME J. Smith 10 Wayside Road, Bridlington YO16 4BA
2E0 JMF M. Hardwick 34 The Pastures, Long Bennington, Newark NG23 5EG
2W0 JMH J. Hewitt 1 Highfield, Gloucester Road, Chepstow NP16 7DF
2M0 JMI J. Mclaren 33 Foulford Street, Cowdenbeath KY4 9NB

2E0 JMJ J. Jeffries 62 Rowans, Welwyn Garden City AL7 1PD
2W0 JMK M. Dean 4 Crud Yr Awel, Efailwen, Clynderwen SA66 7UX
2E0 JML J. Mould 30 Northfields, Dunstable LU5 5AL
2E0 JMR J. Rodley 268 Grovehill Road, Beverley HU17 0HP
2E0 JMW J. Walker 194A Mount Vale, Mount Vale, York YO24 1DL
2W0 JMX B. Preece 4 Waltham Close, Morriston, Swansea SA6 7PH
2E0 JMY J. Bickers 3 The Old Brickyard, West Haddon, Northampton NN6 7GP
2E0 JNE C. Arthur 25 Pump Hollow Lane, Mansfield NG18 3DU
2E0 JNG J. Green Belvoir Vale Cottage, Barrowby Stenwith, Grantham NG32 2HE
2E0 JNH J. Betteridge 57 Wood Road, Chaddesden, Derby DE21 4LY
2E0 JNM J. Morris Parsonage Farm, Croscombe, Wells BA5 3QN
2E0 JNR A. Davies 4 Erpingham Road, Poole BH12 1EX
2E0 JNU J. Nuttall 73 Severn Drive, Walton-Le-Dale, Preston PR5 4TD
2E0 JNY J. Warman 6 Down View Road, Denbury, Newton Abbot TQ12 6ER
2E0 JOA J. Adler 1 Searles Meadow Dry Drayton, Cambridge CB23 8BW
2E0 JOE J. Fletcher 32 Chapel Lane, Barwick In Elmet, Leeds LS15 4EJ
2E0 JOF J. Miles 11 Enborne Gate, Newbury RG14 6AZ
2E0 JOG G. Smith 36 Palma Park Homes, Shelly Street, , Loughborough LE11 5LB
2E0 JOH J. Hatton 49 Buxton Street, Morecambe LA4 5SR
2E0 JOI W. Pickles 31 Longfield Road, South Woodham Ferrers CM3 5JL
2M0 JOK J. Stewart 1 Barns Park, Dalgety Bay, Dunfermline KY11 9XX
2E0 JOP J. Priestman 198 Felmongers, Harlow CM20 3DW
2I0 JOS J. Millar 3, Ahoghill BT42 1JN
2E0 JOV A. Bogg 13, Emerald Grove, Hull HU35AE
2E0 JOX J. Jones 15 Kinnaird Road, Sheffield S5 0NN
2E0 JPA J. Wake 60 Cloverville Approach, Bradford BD6 1ET
2E0 JPC J. Clark 1 Brooklime Road, Liverpool L11 2YH
2E0 JPD R. Young 26 Silent Woman Park Coldharbour, Wareham BH20 7PE
2W0 JPE J. Elsmore 8 Clos Aberconway, Prestatyn LL19 9HU
2W0 JPF J. Freelove 12 Honeyborough Road, Neyland, Milford Haven SA73 1RE
2E0 JPH J. Hazell 90 Lichfield Road, St. Annes, Bristol BS4 4BN
2E0 JPM C. Monahan 48 Church Road, Earley, Reading RG6 1HS
2E0 JPO J. Owen 21 Marlborough Road, Luton LU3 1EF
2I0 JPP J. Page 259 Bridge Street, Portadown, Craigavon BT63 5AR
2E0 JPR P. Gaur 34 Queensberry Avenue, Copford, Colchester CO6 1YN
2E0 JPS J. Smith 38 New Waverley Road, Basildon SS15 4BH
2E0 JPT J. Taylor 6 Hawks Close, Walsall WS6 7LE
2E0 JPU J. Patient 4 Bucklebury Heath, South Woodham Ferrers, Chelmsford CM3 5ZU
2E0 JPW J. Websdale Blacksmith Cottage, Black Street, Lowestoft NR33 8EG
2E0 JPX J. Parkes 24 Kenilworth Road, Lichfield WS14 9DP
2E0 JQF S. Walrond 17 Madam Lane, Weston-Super-Mare BS22 6PW
2E0 JQG D. Start The Rise, Valley Lane Swaby, Alford LN13 0BH
2E0 JQI E. Lightbown 18 Formby Close, Blackburn BB2 3JZ
2E0 JQK P. Blackie 30 Queens Avenue, Ilfracombe EX34 9LS
2E0 JQW J. Wang Flat 31, 74 Arlington Avenue, London N1 7AY
2E0 JRA A. Jessop 4 Katherine Street, Thurcroft, Rotherham S66 9LG
2E0 JRB J. Barrow 45 Windermere Road, Seaham SR7 8HW
2E0 JRD J. Dowdeswell 18 Lechlade Gardens, Fareham PO15 6HF
2E0 JRE V. Vaznais 35 Lynwood Drive, London KT4 7AA
2W0 JRJ J. Jillings Cane Garden, Dolau, Llandrindod Wells LD1 5TE
2W0 JRK Dr J. Arkinstall Wrexham House, Domgay Road, Four Crosses, Llanymynech SY22 6SW
2E0 JRL J. Lambert 82 22Nd Avenue, Hull HU6 9LS
2E0 JRN J. Lynn 72 Badger Close, Guildford GU2 9WA
2E0 JRP J. Preston 5 Bodiam Crescent, Eastbourne BN22 9HQ
2E0 JRR J. Riley The Thatched House, 18 Bond Street, Norwich NR9 4HA
2E0 JRS J. Stallard 6 Richmond Crescent, Leominster HR6 8RX
2E0 JRT N. Wing 39 Whittington Road, Hutton, Brentwood CM13 1JX
2E0 JRU C. Denton 100 Lincoln Road Deeping Gate, Peterborough PE6 9BA
2E0 JRW M. Williams 48 Thackeray Drive, Tamworth B79 8HZ
2E0 JRZ B. Handley 68 Northfield Avenue, Rothwell, Leeds LS260SW
2M0 JSB J. Bence 5 Braeside Gardens, Hamilton ML3 7PN
2E0 JSE E. Riddle 37B Stubbs Lane, Braintree CM7 3NR
2M0 JSF J. French 19 Woodside Avenue, Bridge of Weir PA11 3PQ
2E0 JSG J. Godfrey 29 Ridgewood Drive, Harpenden AL5 3LJ
2E0 JSH J. Harris 23 Winchester Avenue, Chatham ME5 9AR
2E0 JSI B. Wolfe 60 Lindley Street, Rotherham S65 1RT
2E0 JSJ T. Peck 2 Primrose Lane, Miami Beach, Sutton on Sea LN12 2JZ
2E0 JSK S. Killian 1 Winsor Avenue, Saint Leonards Quarter, , Exeter EX2 4BL
2E0 JSM J. Preston 3 Essex Drive, Kidsgrove ST71HE
2E0 JSN J. Slattery 64 Rydal Avenue, Chadderton, Oldham OL9 0QX
2E0 JSO J. O'Shea 56 Crummock Gardens, London NW9 0DJ
2I0 JSQ J. Quinn 44 Gleanniseal, Dungannon BT70 3BE
2E0 JSR Dr Z. Rathore 7 Ashdale Avenue, Bolton BL3 4PH
2E0 JSS J. Symonds 14 Phillips Crescent Needham Market Ip6 8Tf, Ipswich IP6 8TF
2M0 JST J. Stuart 120 Wellbeck Crescent, Troon KA10 6AW
2E0 JSX J. Kelly Braeview, Buzzacott Lane, Combe Martin, Ilfracombe EX34 0NL
2E0 JSY B. Josyfon 25 Norrice Lea, London N2 0RD
2E0 JTB D. Baker 65 Madison Street, Tunstall ST6 5HS
2E0 JTG A. Gilbert 31 Elizabeth Road, Leamington Spa CV31 3LJ
2E0 JTH J. Thornhill 47 Hopton Lane, Mirfield WF14 8JP
2E0 JTM D. Bache 62 Whittingham Road, Halesowen B63 3TP
2E0 JTN M. Chivers 4 Hunters Lodge, Fareham PO15 5NF
2E0 JTO J. O'Connell 23 Halstead Road, Gosfield CO9 1PG
2E0 JTQ J. Cook 28 Wenny Estate, Chatteris PE16 6UX
2E0 JTS J. Walker-Wilson Rest Harrow Southside, Scorton, Richmond DL10 6DN
2E0 JTT J. Thompson 11 Foxbury Drive, Orpington BR66EJ
2E0 JTW J. Johnson 4 Wallace Close, Hullbridge, Hockley SS5 6NE
2E0 JTY J. Hewlett 21 Stedman Close Ickenham, Uxbridge UB10 8DY
2E0 JUD J. Boutros 15 Navigation House, Whiting Way, London SE16 7EG
2M0 JUH J. Heywood 2 Broaddykes Drive Kingswells, Aberdeen AB15 8UE
2E0 JUL J. Hardy Lambda House, Seanor Lane, Chesterfield S45 8DH
2M0 JUM C. Jump 15 The Lade, Bonhill, Alexandria G83 9JR
2E0 JUX A. Brown 17 Quail Ridge, Ford, Shrewsbury SY5 9LF
2E0 JVA A. Vasarhelyi 1 Eldon Close, Langley Park, Durham DH7 9FR
2E0 JVC N. Cochrane 3 Roman Bank Moulton Seas End, Spalding PE12 6LG
2E0 JVD J. Boyd 125 High Street, Great Wakering, Southend-on-Sea SS3 0EB
2E0 JVF K. Sutton 108 Blackley New Road, Manchester M9 8EG
2E0 JVM A. Mori 33 Valerian Road, Hedge End, Southampton SO30 0GR
2E0 JVP M. Mason 7 Langland Close, Malvern WR14 2UY
2M0 JVR J. Robertson 3 Richmond Place, Fochabers IV32 7HF
2E0 JVS J. Swanbrow 7 Manor Crescent, Rookley, Ventnor PO38 3NS
2E0 JVT S. Black 55 North Road, Hertford SG14 1NE
2E0 JVY P. Smythe 15 Falcon Drive, Trowbridge BA14 7GE
2E0 JWC J. Cater 5 Shady Grove, Hilton, Derby DE65 5FX
2E0 JWE J. Elsmore 142 St. Marks Road, Chester CH4 8DH
2E0 JWH J. Hope 40 Birch Lane, Rugeley WS15 1EJ
2E0 JWJ D. Shields 42 Studland Park, Westbury BA13 3HL
2E0 JWP J. Wishart 19 Chepstow Close, Chippenham SN14 0XP
2E0 JWS C. Shaw 15 Rosamund Avenue, Wimborne BH21 1TE
2E0 JWW J. Webster 72 Grosvenor Street, Derby DE248AT
2E0 JWY J. Willby 10 Sunbury Road, Birmingham B31 4LJ
2I0 JXA C. Matchett 28 Glendale Avenue East, Belfast BT8 6LF
2E0 JXB J. Bischoff 6 Harton Close, Bromley BR1 2UD
2M0 JXC J. Cockburn 5 Carribber Avenue, Whitecross, Linlithgow EH49 6JS
2E0 JXF D. Wells 96 Tennyson Avenue, Rugby CV22 6JF
2E0 JXG C. Millar 195 Worsley Road, Frimley, Camberley GU16 9BH
2M0 JXK J. Webster 6 Livingston Crescent, Winchburgh, Broxburn EH52 6FX
2E0 JXN P. Jackson Langsmead Barn, Eastbourne Road, Blindley Heath, Lingfield RH7 6JX
2I0 JXO D. Burke 7 Edinburgh Villas, Omagh BT79 0DW
2E0 JXT J. Townsend 16 Tower Gardens, Hythe CT21 6DG
2E0 JXU R. Sansom 72 Wannock Lane, Eastbourne BN20 9SQ
2E0 JYA J. Young 1 Pen Tye, Gwinear, Hayle TR27 5HL
2W0 JYB J. Byast Ty Arbennig Bull Bay Road, Amlwch LL68 9EA
2W0 JYC M. Coleman Felin Newydd Ciliau Aeron Lampeter Sa48 7Px, Lampeter SA48 7PX
2E0 JYE E. Brereton 21 Sydney Street, Colchester CO2 8UP
2W0 JYI Dr J. Blaxland 7 Maes Y Sarn, Pentyrch, Cardiff CF15 9QQ
2E0 JYM J. Downes 78 Astral Way, Kingston upon Hull HU74XZ
2W0 JYN S. Bobby 56 Ffordd Offa, Rhosllanerchrugog, Wrexham LL14 2EY
2E0 JYP J. Capstick 186 Forest Lane, Harrogate HG2 7EE
2E0 JZC J. Clark 21 Bilton Way Crewe Cheshire , Cheshire CW2 8SN
2E0 JZK J. Howlett 29 Little London, Heytesbury, Warminster BA12 0ES
2E0 JZU P. Holmes 82 Moore Avenue, Norwich NR6 7LG
2E0 KAB K. Bull 12 Hinksford Mobile Home Park, Kingswinford DY6 0BG
2E0 KAC D. Carr 19 Kingsmead Walk, Speedwell, Bristol BS5 7RL
2E0 KAF R. Dyson 4 Royston Lane, Royston, Barnsley S71 4NL
2E0 KAG M. Hughes 19 Pendine Crescent, North Hykeham, Lincoln LN6 8UW
2E0 KAK R. Hambly 144 Station Road, Irchester, Wellingborough NN29 7EW
2E0 KAL K. Lott 6 Centurion Close, Sandhurst GU470HH
2W0 KAP K. Gamwasam 01, Churton Drive, , Wrexham LL13 8RU
2E0 KAS K. Sharpe 18 Dudhill Road, Rowley Regis B65 8HT
2M0 KAU M. Krawczyk 19 Wishart Archway, Dundee DD1 2JA
2E0 KAX N. Griffiths 67 Warstones Drive, Wolverhampton WV4 4PF
2E0 KBA J. Fletcher 13 Lloyd George Grove, Cannock WS11 7GY
2I0 KBB J. Elliott 12 Drumbeg Drive, Lisburn BT28 1NY
2E0 KBD P. Smiths 12 Lambton Road, Stockton-on-Tees TS19 0ER
2E0 KBE J. Hatton 37 St. Cuthberts Way, Holystone, Newcastle upon Tyne NE27 0UZ
2E0 KBF B. Wiggins The Wigwam 13 Hastings Road, Bromsgrove B60 3NX
2E0 KBG K. Glaysher 66 Talbot Road, Farnham GU9 8RR
2E0 KBJ S. Inman 9 Colbert Avenue, Ilkley LS29 8LU
2E0 KBL S. Kneeshaw 39 Cherrytree Walk East Ardsley, Wakefield WF32HS
2E0 KBN K. Burness 4 Fenwick Street, Boldon Colliery NE35 9HU
2E0 KBP B. Al-Rawi Flat 17, Harrow Lodge, London NW8 8HR
2I0 KBS K. Boyle 764 Springfield Road, Belfast BT12 7JD
2E0 KBX K. Buxey 17 Gort Crescent, Southampton SO19 8LH
2E0 KBZ S. Stretton 9 Kilton Road, Worksop S80 2EG
2E0 KCB K. Smart 33 East Street, Littlehampton BN17 6AU
2E0 KCD N. Omer Flat 12, Mallard Court, 1 Piper Close, London N7 8TQ
2E0 KCF A. Hankins 16 Eastwick Barton, Nomansland, Tiverton EX16 8PP
2E0 KCN B. Somerville Roberts 21 Regency Way, Ponteland, Newcastle upon Tyne NE20 9AU
2E0 KCO K. Cornmell 19 Forest Road, Chandler'S Ford, Eastleigh SO53 1NA
2E0 KCP R. Hutton 22A Victoria Road, Maldon CM9 5HF
2E0 KCV K. Conlon 136 Chart Downs, Dorking RH5 4DG
2E0 KCW P. Walsh 181 Hermes Close, Hull HU9 4DR
2E0 KCZ R. Robinson 22 Riddings Court, Timperley, Altrincham WA15 6BG
2E0 KDB D. Barnett 12 Craig Walk Alsager, Stoke-on-Trent ST7 2RJ
2E0 KDE A. Lee 14 Bernice Avenue, Chadderton, Oldham OL9 8QJ
2E0 KDF A. Hamilton 56 Wyvern, Telford TF7 5QH
2E0 KDG C. Harman 58 Laurence Avenue, Witham CM8 1JB
2E0 KDH K. Hall 18 Brooklands Drive, Littleover, Derby DE23 1DN
2E0 KDI L. Azzopardi 11 Ilynton Avenue, Firsdown, Salisbury SP5 1SH
2W0 KDR M. Kidner 4 Tonypistyll Road, Newbridge, Newport NP11 4HJ
2E0 KDS K. Selvaratnam 15 Mark Close, Southall UB1 3QJ
2E0 KDT R. Simpson 48 Weatherhill Road, Horley RH6 9LY
2E0 KDV D. Craven 69 Markham Avenue, Rawdon, Leeds LS19 6NE
2E0 KEA T. Lupton 81 Home Farm Lane, Bury St. Edmunds IP33 2QL
2W0 KED S. Kedward 9 Lawrence Avenue, Aberdare CF44 9EW
2E0 KEG K. Stone 34 Daventry Grove, Birmingham B32 1JA
2E0 KEI K. Hastings 161 Cottingley Road, Allerton, Bradford BD15 9LD
2E0 KEK J. Russell Flat 1, Knapp Cottage, Roadwater, Watchet TA23 0QY
2E0 KEP T. Keep Coombe Cottage, Coombe Lane, Cradley, Malvern WR13 5JF
2W0 KEQ K. Mogford 49 Cefn Road, Rogerstone, Newport NP10 9AQ
2E0 KES R. Noakes 3 Firtree Road, Hastings TN34 3TR
2I0 KEW K. Mcdonald 37 Ardgarvan Cottages, Limavady BT49 0NF
2E0 KEY D. Ferguson 39 Fouracres, Maghull, Liverpool L317BP
2E0 KEZ D. Brough 57 Francis Road, Ashford TN23 7UP
2E0 KFA A. Walker 50 Parkstone Crescent, Hellaby, Rotherham S66 8HD
2E0 KFB Dr F. Kuttikkate 34 Shetland Crescent, Rochford SS4 3FJ
2I0 KFD K. Dorman 25 Blackthorn Road, Newtownabbey BT37 0GH
2E0 KFG C. Freebody 94 Fingringhoe Road, Colchester CO2 8EE
2E0 KFH J. Rowe 22 Treaty Road Glenfield, Leicester LE3 8LU
2E0 KFI S. Gilbert 34 Ramsey Road, St. Ives PE27 5RD

2E0 KFJ K. Jeffrey 26 Sandringham Drive, Paignton TQ3 1HY
2E0 KFK K. Kozlowski Woking Homes / Flat 2, Oriental Road, Woking GU22 7BE
2E0 KFL K. Ballard 34 Orange Street South Wigston, Leicester LE184QB
2E0 KFM A. Burfield 4 Eastern Crescent, Chelmsford CM1 4JQ
2E0 KFO S. Howard 11 Jolly Gardeners Court, Norwich NR3 3HD
2E0 KFR C. Pearcey 17 Peppercorn Close, Christchurch BH23 3BL
2E0 KFT J. Ferrol 29 Westlands, Haltwhistle NE49 9BS
2E0 KFV K. O'Connell 63 Hazelton Road, Colchester CO4 3DS
2E0 KFX M. Hill 4 Farmland Way, Hailsham BN27 1SP
2E0 KFY C. Day 14 Windsor Drive, Ramsey Forty Foot, Ramsey, Huntingdon PE26 2XX
2E0 KGA K. Gracey Gwendoline Close, Merseyside CH611DL
2E0 KGC K. Cossey 34 Pinewood Road, Hordle, Lymington SO41 0GP
2E0 KGD M. Tabberer 29 Chase Vale, Chasetown, Burntwood WS7 3GD
2E0 KGJ K. Green 12 Merton Place, Grays RM16 4HL
2E0 KGM A. Fielding 4 Poot Hall, Rochdale OL120AS
2E0 KGN J. Badham Comwillgur House Ross Road, Longhope GL17 0LP
2W0 KGP Dr P. Kelly Arosfa, Westminster Road, Moss, Wrexham LL11 6DN
2W0 KGQ J. Kelly Arosfa, Westminster Road, Wrexham LL11 6DN
2E0 KGT G. Tagg Tinkers Cottage, Nevendon Road, Wickford SS12 0QB
2E0 KGV C. Moulding 28 Queens Avenue, Highworth, Swindon SN6 7BA
2E0 KGY R. King 14 Roper Avenue, Heanor DE75 7BZ
2E0 KHA A. Hughes 8 Bullens Green Lane Colney Heath, St. Albans AL4 0QS
2E0 KHG K. Gibbs 30 George Road, Water Orton, Birmingham B46 1PE
2E0 KHI K. Hale 27 Heyes Street, Liverpool L5 6SE
2W0 KHK S. Rosser 25 Clos Tir-Y-Pwll, Newbridge, Pantside, Newport NP11 5GE
2E0 KHM K. Maddy 56 Coachwell Close, Telford TF3 2JB
2E0 KHO D. Pluright 21 Buscot Drive, Abingdon OX14 2BJ
2E0 KHW K. White 4 Top Birches, St. Neots PE19 6BD
2M0 KIA M. West Prieshach, Station Brae, Macduff AB44 1UL
2E0 KIL T. Lee 68 Wharton Drive, Springfield, Chelmsford CM1 6BF
2E0 KIM K. Matthews 4 George Croft, Gayton Le Marsh, Alford LN130NP
2W0 KIN A. Hawkins 5 Keats Road, Caldicot NP26 4LH
2E0 KIO D. Bloomfield 30 Wye Road, Clayton, Newcastle ST5 4AZ
2E0 KIS M. Mcnamara 50 Portsdown Road, Portsmouth PO6 4QH
2E0 KIT A. Gladman 19 Colchester Road, Wymering, Portsmouth PO6 3RH
2E0 KIV N. Allcorn 3 Larkswood Rise, St. Albans AL4 9JU
2E0 KJD K. Hale 44 Churchville, Micklefield Micklefield, Leeds LS254AP
2E0 KJG K. Gallyot 18, High Pines, St. Georges Close, Christchurch BH23 4LN
2E0 KJI M. Bidwell 3 Walsingham Place, London SW4 9RR
2E0 KJJ J. Kent Meadow Bank, Rye Lane, Sevenoaks TN14 5JF
2E0 KJK K. Kolesnik 15 Steer Road, Swanage BH19 2RU
2E0 KJL K. Lock 29 South Ave, Bideford EX394RB
2W0 KJO J. Orchard The Burrows, Spring Gardens, Whitland SA34 0HL
2E0 KJT K. Toohey 197 Broad Oak Road, St. Helens WA9 2AQ
2E0 KKC E. Brown Flat 8, Jacobs Court, High Street, Colchester CO7 0AQ
2E0 KKF F. Fitton 6 Leaford Close, Denton, Manchester M34 3QH
2E0 KKG A. Fraley 28 Riverside Park Colehouse Lane, Clevedon BS21 6TQ
2E0 KKH J. Fairbrother 6, The Dene, Bethersdon, , Ashford TN26 3AR
2E0 KKJ K. Juszczak 68 College Road, Sandy SG19 1RH
2E0 KKL Dr H. Chen Flat 37, Castel Mill, Rodger Dudman Way, Oxford OX1 1AD
2E0 KKM K. Minett Rosedene, Honey Hill, Wimbotsham, King's Lynn PE34 3QD
2E0 KKO T. Stack 31A Chester Road South, Kidderminster DY10 1XJ
2E0 KKQ W. Dunk 27 Trevor Drive, Maidstone ME16 0QW
2W0 KKR T. Rowan 12 Pencoed Avenue The Common, Pontypridd CF37 4AN
2E0 KKT P. Hodgson 32 Main Road, Seaton, Workington CA14 1HS
2E0 KKY R. Pas 12 Monument Way, Bodmin PL31 1NZ
2E0 KKZ D. Kirk 57 Sandringham Avenue, London SW20 8JY
2E0 KLB S. Arundale 5 Bowdon Street, Stockport SK3 9EA
2E0 KLD R. Delrosa 18 Rangers Avenue, Dursley GL11 4AS
2E0 KLF K. Francis 203 Colchester Road Lawford, Manningtree CO11 2BU
2E0 KLI J. Heng 1 St. Giles Croft, Beverley HU17 8LA
2M0 KLL R. Bertram 46 Main Street Main Street, Pathhead EH37 5QB
2E0 KLM D. Jackson 22 Trent Road, Walsall WS3 4DQ
2E0 KLN K. Nevins 4 Ubbanford, Norham, Berwick-upon-Tweed TD15 2LA
2E0 KLR A. Henshall 37 Marlow Drive, Branston, Burton-on-Trent DE14 3TX
2E0 KLS P. Elstub 4 Granville Lodge, Church Street, Telford TF2 9LX
2E0 KLT C. Taylor 9 Pendeen Crescent, Threemilestone, Truro TR3 6SP
2E0 KLV K. Brown 41 Church Street, Swinton, Mexborough S64 8EF
2E0 KLW L. Woods 193 Wimberley Street, Blackburn BB1 8HU
2E0 KLY K. Young 14 Beechwood Avenue, Chatham ME5 7HH
2E0 KMA K. Machen 18 Peveril Close, Whitefield, Manchester M45 6NR
2E0 KMB K. Bailey 58 Billy Buns Lane, Wombourne, Wolverhampton WV5 9BP
2E0 KMD K. Deans 31 Northcroft, Sandy SG19 1JJ
2E0 KMF K. Moysey 109 Langbrook Cottages, Langbrook, Ivybridge PL21 9JX
2E0 KMI K. Mills 6 West Coombe, Bristol BS9 2BA
2E0 KMN M. Wickens Haven Lea, Queens Drive, Windermere LA23 2EL
2E0 KMP A. Adams 45 Four Oaks Road Tedburn St. Mary, Exeter EX6 6AP
2M0 KMQ W. Donnelly 10 Sentry Knowe, Selkirk TD7 4BG
2E0 KMS K. Shenton 2 The Croft, Stramshall, Uttoxeter ST14 5AG
2E0 KMT K. Murphy 120 Elmway, Chester le Street DH2 2LQ
2W0 KMU G. Cattle 12 Claerwen Gelligaer, Hengoed CF828EW
2E0 KMY K. Murray 5 Princes Crescent, Basingstoke RG22 6DP
2E0 KMZ K. Missenden 47 Roseacre Drive, Elswick, Preston PR4 3UQ
2E0 KNA T. Clarke 63 Garratts Way, High Wycombe HP13 5YT
2E0 KNB M. Ronan 49 Dorset Street, Nottingham NG8 1PU
2E0 KNC S. Henson 297 Underwood Lane, Crewe CW1 3SG
2E0 KNE J. Dale 37 Bussey Road, Norwich NR6 6JF
2E0 KNF K. Sumner 18 Grange Road, Fleetwood FY7 8BH
2E0 KNH K. Holman 39 Trellech Court, Yeovil BA21 3TE
2E0 KNK K. Kariraman 17 Kipling Close, Galley Common, Nuneaton CV10 9SJ
2E0 KNL G. Knowles 29 Shepherds Cote Drive, Hepscott Park, Stannington, Morpeth NE61 6FN
2E0 KNM M. Hill 109 Kitchener Street, St. Helens WA10 4LU
2E0 KNT K. Royce 11 Church Lane, Stibbington, Peterborough PE8 6LP
2E0 KNU R. Weightman 2 Bannister Grove, Winsford CW7 1RJ
2E0 KOA P. Whitehead 29 Cleveland, Bradville, Milton Keynes MK13 7AZ
2E0 KOD A. Adams 27 George Street Stockton, Southam CV47 8JS
2E0 KOI H. Walker 9 Humphries Close, Leicester LE5 4LU
2E0 KOM L. Wilson 6 Marrick Road, Middlesbrough TS3 7RX
2W0 KOP A. Martin 117 Fforchaman Road, Cwmaman, Aberdare CF44 6NL
2E0 KOR L. Ross 133 Petersmith Drive, New Ollerton, Newark NG22 9SG
2E0 KOS A. Hollings 39 Rendham Road, Saxmundham IP17 1EA
2I0 KPA S. Frazer 2 Cavanballaghy Road, Killylea, Armagh BT60 4NZ
2E0 KPB P. Bridgwater 18 Angelica, Amington, Tamworth B77 3JZ
2E0 KPC P. Gagliardi 7 Saxon Way, Jarrow NE32 3QA
2E0 KPD P. Davies 60 Southerndown Road, Sedgley, Dudley DY3 3NA
2M0 KPE K. Page 43 Lothian Court, Glenrothes KY6 1LZ
2W0 KPF K. Foster Prince Of Wales House Short Bridge Street Sy18 6Ad, Llanidloes SY18 6AD
2W0 KPH K. House Liddington, Dehewydd Lane Llantwit Fardre, Pontypridd CF38 2EN
2E0 KPI N. Gregg 84 Aberconway Crescent, New Rossington, Doncaster DN11 0JP
2E0 KPL L. Kiddell 1 Sparham Hill, Sparham, Norwich NR9 5QT
2E0 KPM D. Roberts 19 East Avenue, Warrington WA2 8AD
2W0 KPN A. Thomas 10 Chapel Street, Gorseinon, Swansea SA4 4DT
2E0 KPO S. Warren 7 Crich Way, Newhall, Swadlincote DE11 0UU
2E0 KPP D. Fellows 147 Olive Lane, Halesowen B62 8LR
2E0 KPR G. Andrews 151 Great Gregorie, Basildon SS16 5QQ
2E0 KPS K. Parow-Souchon 87 Saxton Road, Abingdon OX14 5JD
2E0 KPT B. Harris 1 Newbridge Cottages, Midgehole, Hebden Bridge HX7 7AL
2E0 KPX D. Stephenson Flat 8, Thornwood Court 84-88 Hudson Road, Leigh-on-Sea SS9 5NF
2M0 KPZ N. Stewart 35 Newbattle Gardens, Dalkeith EH22 3DR
2E0 KQR A. Kinsella Apartment 28 241 Liverpool Road, Widnes WA8 7HL
2E0 KQV D. Harrison 64 Crosby Road, Grimsby DN33 1LU
2E0 KRB A. Cowan 217 South Park Road, Wimbledon, London SW19 8RY
2E0 KRD K. Dukes 127 Carlton Road, Boston PE21 8LL
2M0 KRG C. Mcintyre 70 Oldwood Place, Livingston EH54 6US
2E0 KRH C. Pierce 42 Staplehurst Road, Reigate RH2 7PY
2M0 KRI K. Waitz C/O Waitz-Rainey, 16 Inverkeithing Road, Aberdour, Burntisland KY3 0RS
2E0 KRK J. Birch 3 Partridge Way, High Wycombe HP13 5JX
2E0 KRN K. Pidwell 2 Welford Road, Chapel Brampton, Northampton NN6 8AF
2E0 KRR R. Rao 13 High Oaks, Enfield EN2 8JJ
2E0 KRS K. Tharanee 525 Burton Road, Littleover, Derby DE23 6FT
2E0 KRT K. Taylor 3 The Drive, Lichfield WS14 9QT
2E0 KRX K. Rosema Apartment 801, 25 Goswell Road, London EC1M 7AJ
2W0 KSA P. Terrell 82 Baglan Street, Treherbert, Treorchy CF42 5AR
2E0 KSC C. Moss 19 Tozer Close, Wallisdown, Bournemouth BH11 8RB
2E0 KSG K. Stevens 61A Main Road, Hoo, Rochester ME3 9AA
2E0 KSH K. Haywood 126 Derby Street, Sheffield S2 3NF
2E0 KSO A. Jackson 5 Woodside Lane, London N12 8RB
2E0 KSP D. Volkov 2 Robert Close, Potters Bar EN6 2DH
2E0 KSR R. Brimfield 6 Radways Close, Fairford GL7 4FZ
2E0 KSW K. Sewell 12 Haylands Square, South Shields NE34 0JB
2E0 KSX B. Elms-Lester Ferndale House Kerry'S Gate, Hereford HR2 0AH
2E0 KTD K. Davidson 5 Hanover Parc, Indian Queens, St. Columb TR9 6ER
2E0 KTG K. Gribben 33 Strawberry Close, Birchwood, Warrington WA3 7NT
2D0 KTH Dr J. Daniels 24 King Orry Road Glen Vine, Isle of Man IM4 4ES
2E0 KTK D. Van Dijk 76 High Street, Tetsworth, Thame OX9 7AE
2M0 KTL K. Lane 23 Mayfield Avenue, Tillicoultry FK13 6HB
2E0 KTQ K. Welch Jacobs Leys, Falcon Lane, Ledbury HR8 2JS
2E0 KTV B. Chauhan 45 Burnham Drive, Whetstone, Leicester LE8 6HY
2E0 KTW K. Wheeler 3 Praze Road, Porthleven, Helston TR13 9LR
2E0 KTX K. Browne 56 Moorhouse Avenue, Wakefield WF2 9QG
2E0 KUC D. Holman 38 Polyear Close, Polgooth, St. Austell PL26 7BH
2E0 KUE I. Vernon 7 Seaton Road Wick, Littlehampton BN17 7LG
2E0 KUF J. Inwood Flat 13, Kelvestone House, 47 Park Road, Cannock WS11 1NZ
2E0 KUH S. Steinhoefel 222 Stretford Road, Urmston, Manchester M41 9NT
2I0 KUJ T. Calka 71 Willowfield Street, Belfast BT6 9AW
2E0 KUK A. Mallinson 50 Cleveland Road, Huddersfield HD1 4PW
2I0 KUN A. Davidson 8 Rashee Court, Ballyclare BT39 9SE
2E0 KUR P. Fletcher 14 Orchard Avenue, Aylesford ME20 7LY
2E0 KVA A. Kossick 33 Verney Road, Winslow, Buckingham MK18 3BN
2E0 KVB M. Wall 227 Wayfield Road, Chatham ME5 0HJ
2E0 KVE K. Waterhouse 74 Clifford Road, West Bromwich B70 8JY
2E0 KVF K. Emery-Ford 48 Welham Grove, Retford DN22 6TS
2M0 KVI K. VielhaberNether Littlefold, Crieff PH7 3NY
2M0 KVM K. Mair 92 Graham Street, Wishaw ML2 8HR
2E0 KVR A. King 23 Tower Crescent, Lincoln LN2 5QF
2E0 KVS J. Fletcher 3 Moorend Glade Charlton Kings, Cheltenham GL53 9AT
2E0 KVV A. Taylor 60 Wood Ride, Petts Wood, Orpington BR5 1PY
2E0 KWA A. Miller Tile Barn, Straight Mile, Bourton, Rugby CV23 9QQ
2E0 KWB K. Wilkinson 9 Mersey Rd, Redcar TS101LS
2E0 KWC K. Comben 9 West Lane, North Baddesley, Southampton SO52 9GB
2E0 KWE S. Coppins 33 Honywood Road, Lenham, Maidstone ME17 2HH
2E0 KWG R. Knight Thrift, Ashwells Road, Brentwood CM15 9SG
2E0 KWL R. Barker 6 Trenoweth Cressent, Penzance TR184RY
2E0 KWM D. Wright 203 Winn Street, Lincoln LN2 5EY
2M0 KWP K. Rae 6 Dovecot Way, Pencaitland EH34 5HA
2E0 KWW K. Willson Ludpit Cottage, Ludpit Lane, Etchingham TN19 7DB
2E0 KXD C. Collins 2 Kew Crescent, Sheffield S12 3LP
2E0 KXH K. Heydon 34 Charles Road, Amble, Morpeth NE65 0SQ
2E0 KXI K. Dowling 8 Chathill Close, Morpeth NE61 2TH
2E0 KXL R. Bamber 28 Market Street, Cheltenham GL50 3NH
2I0 KXM K. Mitchell 39 Lissize Avenue, Rathfriland, Newry BT34 5DE
2W0 KXN K. Nicholls 35 Partridge Road, Cardiff CF24 3QW
2E0 KXR K. Golloghly 7 Jenner Close, Bungay NR35 1QR
2E0 KXV A. Jordan 19 Arrow Lane, Newhaven BN9 0FG
2E0 KXX B. Cook 11 Arbrook Lane, Esher KT10 9EG
2E0 KYB M. Tielemans 2 Parr Close, Grange Park, Swindon SN5 6JY
2E0 KYD R. Harwood 26 Barry Avenue, Bicester OX26 2DY
2W0 KYH D. Phillips 9 Baldwin Street, Newport NP20 2LT
2E0 KYI K. Armstrong 29 Thorntree Avenue, Crofton, Wakefield WF4 1NU
2E0 KYT S. Bishop 5 Mulberry Court, 266 Goring Road, Goring-By-Sea, Worthing BN12 4PF

2E0 KYX A. Tandler 6 Field View Cottages, Brimfield, Ludlow SY8 4LB
2W0 KYZ Dr A. Rykala 20 Mount Pleasant, Blaina, Abertillery NP13 3DD
2E0 KZB C. Kyle-Davidson Block C, Ingram Court, Garrowby Way, Heslington, York YO10 5DL
2E0 KZC M. Clack 42A Provost Street, Fordingbridge SP6 1AY
2E0 KZH D. Taylor 49 Boggart Hill Gardens, Seacroft, Leeds LS14 1LJ
2M0 KZI A. Broom 140 Green Road, Paisley PA2 9AJ
2E0 KZJ P. Lamb 13 Pool End, St. Helens WA9 3RE
2E0 KZK A. Whybrow 64 Church Road Sevington, Ashford TN24 0LF
2E0 KZL I. Hallatt 11 Cheshire St., Audlem, Crewe CW3 0AH
2E0 KZM M. Osborne 9 Sunningdale Court, Jupps Lane, Goring-By-Sea, Worthing BN12 4TU
2E0 KZN J. Spence Officers Mess, Royal Air Force, Brize Norton, Carterton OX18 3LX
2E0 KZQ M. Pott 29 Lilliebrooke Crescent, Maidenhead SL6 3XJ
2E0 KZS A. Jacobs 11 Office Road, Cinderford GL14 2HZ
2E0 KZU M. Macrae 91 Chosen Way Hucclecote, Gloucester GL3 3BX
2E0 KZV P. Mckie 65 White Cross, Hexham NE46 1JH
2E0 LAB L. Bain 45 Larpool Crescent, Whitby YO22 4JD
2W0 LAC S. Llewellyn 53 Cripps Avenue, Cefn Golau, Tredegar NP22 3PF
2E0 LAD M. Leech 11 Westlake Close, Torpoint PL11 2BZ
2E0 LAH B. Beckett 21 Horseshoes Lane, Langley, Maidstone ME17 1SR
2E0 LAI L. Potter 18 Elizabeth Gardens, Wakefield WF1 3SZ
2E0 LAL L. Larkins 34 Guycroft, Otley LS21 3DS
2M0 LAO S. Ramsay 6 Cross Road, Peebles EH45 8DH
2E0 LAR A. Davies 6 Gribble Road, Liverpool L10 7NF
2M0 LAS F. Davidson 27 Gordon Way, Livingston EH54 8JG
2E0 LAV A. Loyd Maple House, Pangbourne Road, Reading RG8 8LN
2M0 LAW A. Mccaig 46 Patterson Drive, Law, Carluke ML8 5LT
2E0 LAX A. Coombes 78 Wilton Crescent, Southampton SO15 7QE
2E0 LAY S. Lay 7 Hunt Street, Swindon SN1 3HW
2E0 LBA A. Polakovs 76 Sandringham Crescent, Leeds LS17 8DF
2E0 LBB L. Brown 20 Pinfold Lane, Mirfield WF14 9HZ
2W0 LBE M. Ryall 27 Clos Afon Twyi, Blackwood NP12 3FX
2E0 LBG J. Shatford 31 Pinner Park Avenue, Harrow HA2 6LG
2M0 LBH A. Hunsley 21 Dalkeith Street, Edinburgh EH15 2HP
2E0 LBI L. Bell 42 Ocean Road, Walney, Barrow-in-Furness LA14 3DX
2E0 LBJ L. Pinkney 18 Bridlington Road, Driffield YO25 5HZ
2E0 LBK L. Karthauser 17 Manor Close Abbotts Ann, Andover SP11 7BJ
2E0 LBL L. Lay 17 Herbert Road, Hornchurch RM11 3LD
2W0 LBN L. Au-Yeung 104 Fleet Street, Swansea SA1 3UX
2I0 LBS L. O'Sullivan 24 Swifts Quay, Carrickfergus BT38 8BQ
2E0 LBW R. Walker 159 Cuckfield Road, Hurstpierpoint, Hassocks BN6 9RT
2E0 LBY T. Littlebury 71 Hampden Drive, Kidlington OX5 2LT
2E0 LBZ D. Taylor 143 Sandhurst Road, London SE6 1UR
2E0 LCE I. Pilton Caleril Barn, Pool Foot Farm Haverthwaite, Ulverston LA12 8AA
2W0 LCJ L. Jones Ty'R Ysgol, Holland Street, Ebbw Vale NP23 6HT
2E0 LCK C. Cutting 5A Clifton Mansions, Clifton Road, Folkestone CT20 2EJ
2E0 LCN T. Cass April Cottage, South Eau Bank, Spalding PE12 0QR
2E0 LCP P. Rovensky 6 Linhay Close, Honiton EX14 2BJ
2E0 LCR L. Rimington 3 Amicombe, Wilnecote, Tamworth B77 4JJ
2E0 LCW T. Wright 2 Lilyville Road, London SW6 5DW
2I0 LDC P. Floyd 25 Glenside, Omagh BT79 7GL
2E0 LDE G. Matts 34 Barry Road, Leicester LE5 1FA
2E0 LDF R. Irving 2 Wasdale Close, Cockermouth CA13 9JD
2E0 LDJ L. Jepson 143 Walnut Avenue Weaverham, Northwich CW8 3DX
2E0 LDL D. Loftus 5 Knowle Mount, Burley, Leeds LS4 2PP
2E0 LDM L. Mason 9 Trenethick Avenue, Helston TR13 8LU
2E0 LDN K. Poulton 21 East View, London E4 9JA
2E0 LDQ L. Dobinson 20 Newholme Crescent, Evenwood, Bishop Auckland DL14 9RY
2E0 LDR L. Reynolds 12 Providence Crescent, Boundary Way, Hull HU4 6EF
2E0 LDV D. Butterfield 57 Holmes Road, Retford DN22 6QU
2W0 LDX S. Jones 14 Lower Cross Road, Llanelli SA15 1NQ
2E0 LDY D. Hayes 60 Shelby Road, Worthing BN13 2TT
2W0 LDZ T. Clapp Crunns Farm. Coxhill, Narberth SA67 8EH
2E0 LEA D. Anderson 35 Sycamore Road, East Leake, Loughborough LE12 6PP
2E0 LEE L. Tunstall 8 York Road, Rowley Regis B65 0RR
2E0 LEF N. Briggs 20 Broad Lane, Pelsall, Walsall WS4 1AP
2E0 LEG J. Landless 2 Aspen Way, Banstead SM7 1LE
2E0 LEL J. Jackson 7 Burford Crescent, Wilmslow SK9 6BL
2E0 LEM J. Francis 22 Acre Lane, Carshalton SM5 3AB
2W0 LEN L. Hayes 56 Snowden Road, Cardiff CF5 4PR
2E0 LEO L. Steer 51 Kings Chase, East Molesey KT8 9DG
2E0 LET R. Dunnaker 12 Dagger Lane, West Bromwich B71 4BA
2M0 LEW M. Strachan 62 Charleston Drive, Dundee DD2 2EZ
2M0 LEX A. Jenkins 19 Hogg Avenue, Johnstone PA5 0EZ
2E0 LEY L. Medley 9 Polyplatt Lane, Scampton, Lincoln LN1 2TL
2E0 LEZ L. Trend 140 Ardleigh, Basildon SS16 5RW
2W0 LFE L. Ansell 114 Bowleaze, Greenmeadow, Cwmbran NP44 4LG
2E0 LFF G. Phillips Labourn Fell Farm, Chopwell NE17 7AY
2E0 LFI A. Birch 22 Ullswater Road, Burnley BB10 4HX
2E0 LFK L. Brown 71 Chiltern Way, Nottingham NG5 5NP
2E0 LFM P. Dimmick Keir Hardie Court, Woodlea, Newbiggin-by-the-Sea NE64 6LH
2E0 LFR L. Rofix Birds Hill Cottage, Clopton, Woodbridge IP13 6SE
2M0 LFS A. Miles 9 Buchanan Drive, Lenzie, Kirkintilloch, Glasgow G66 5HS
2E0 LFT A. Norden 10 School Lane, Watton At Stone, Hertford SG14 3SF
2E0 LFW F. Simon 7 Yew Tree Court, Hemel Hempstead HP11NE
2E0 LFX J. Lamb 9A Matlock Road, Canvey Island SS8 0EW
2W0 LFY A. Jones 10 Dan Y Bryn, Caerau, Maesteg CF34 0UW
2E0 LGB G. Benson 2 Guisborough Road, Nunthorpe, Middlesbrough TS7 0LB
2W0 LGG L. Martin 2 Ael Y Glyn, Nant Road, Harlech LL46 2UJ
2E0 LGH S. Hallam 18A Market Street, Hoylake, Wirral CH47 2AE
2E0 LGL M. Rusu 1 Turnbull Road, March PE15 9RX
2E0 LGO P. Frost 26 Hollies Court, Britannia Road, Banbury OX16 5DR
2E0 LGR L. Rickman 42 Sycamore Close, Poole BH17 7YJ
2E0 LGS P. Schoenmaker 24 Greenheys Drive, London E18 2HB
2E0 LGT D. Page 17 Hedge End Walk, Havant PO9 5LS
2E0 LGV A. Meek 25 Somerset Close Oswaldtwistle, Accrington BB5 3AU
2E0 LGW A. O'Connell 5 Westend Terrace, Gloucester GL1 2RX
2E0 LGZ G. Cash 60 Vittoria Court, Birkenhead CH41 3LF
2E0 LHD L. Hoddinott 30 Deans Mead, Bristol BS11 0QX
2E0 LHE L. Horne 1268 Evesham Road Astwood Bank, Redditch B96 6AX
2E0 LHF K. Gibson 44 Churchville, Micklefield, Leeds LS25 4AP
2E0 LHI C. Smith 17 Essex Street Wash Common, Newbury RG14 6QJ
2E0 LHR S. Kapadia 7 Elms Lane, Wembley HA0 2NX
2E0 LHS G. Carver 4 Andrews Road, Farnborough GU14 9RY
2E0 LHY S. Raneses Grospe 8 Lincombe Road, Manchester M22 1GA
2E0 LHZ M. Johnson 2 Himley Close, Willenhall WV12 4LX
2E0 LIT C. Martin 20 Hall Green Road, West Bromwich B71 3LA
2E0 LIV A. Nicholson 24 Barnmead, Haywards Heath RH16 1UZ
2E0 LIW W. Sawyer 20 Park Terrace Willington, Crook DL15 0QL
2D0 LIX M. Behrman 60 Scarlett Road, Castletown IM9 1PN Isle of Man
2E0 LIZ E. Greatorex 22 Marlborough Way, Uttoxeter ST14 7HL
2E0 LJB L. Bedford 29 Kent Road, Brookenby, Market Rasen LN8 6EW
2W0 LJC C. Jones 19 Crud Y Castell, Denbigh LL16 4PQ
2W0 LJD J. Dyer Abergwawr House, Belmont Terrace, Aberdare CF44 6UW
2E0 LJG L. Goldsmith Hunters Cottage, 61 Fengate Drove, Weeting, Brandon IP27 0PW
2E0 LJH L. Halloway 82 Northwall Road, Deal CT14 6PP
2E0 LJJ L. Johnson 33 Yetholm Place, Newcastle upon Tyne NE5 4ED
2E0 LJK L. Marriott 94 Lyndhurst Road, Worthing BN11 2DW
2E0 LJL L. Lewis 12 Hall Drive, Middleton, Morecambe LA3 3LF
2I0 LJQ W. Phair 34 Sketrick Island Park, Newtownards BT23 7BN
2E0 LJR A. Siddall 12 Russell Gardens, Sipson, West Drayton UB7 0LS
2E0 LJS M. Fearon 70 George Street, Heywood OL10 4PW
2E0 LJT A. Tranter 122 Summerhill Road, Bristol BS5 8JU
2E0 LKC M. Balshaw 16 East Avenue, Heald Green, Cheadle SK8 3DL
2W0 LKD K. Doyle 81 St. Cenydd Road, Caerphilly CF83 2TA
2E0 LKE L. Kett 52 Northgate, Hornsea HU18 1EU
2E0 LKH K. Hull 3 Enas Crescent, Ena Street, Hull HU3 2TL
2E0 LKM L. Mcdonnell 108 Long Lane, Garston, Liverpool L19 6PQ
2E0 LKR R. King 12 South Road, Marden TN12 9EN
2W0 LKX W. Dabrowski 2 Underhill Crescent, Knighton LD7 1DG
2W0 LLA A. Jones 2 Erw Terrace, Bethel, Caernarfon. LL55 1YT
2E0 LLC L. Addison 25 Gladstone Street, Workington CA14 2XH
2E0 LLD E. Daniels 2 Garstons Close Titchfield, Fareham PO14 4EN
2E0 LLE R. Redmond 28 Common Lane, Polesworth, Tamworth B78 1LS
2W0 LLJ J. Henley Rhewin Glas, Llandysul SA44 6PR
2E0 LLK Dr D. Roberts Beggarwood House Ravensworth Park Estate, Gateshead NE11 0HQ
2W0 LLL C. Wadsworth Tyn Llwyn, Talybont LL43 2AN
2E0 LLM J. Proudman 61 Iffley Road, Oxford OX4 1EB
2E0 LLO J. Lovelock Sea Spray, Lighthouse Road The Lizard, Helston TR12 7NU
2W0 LLT L. Thomas 15 Blaenwern, Newcastle Emlyn SA38 9BE
2M0 LLU B. Gaudie Sunnyside, Harray, Orkney KW17 2JS
2E0 LLW P. Sampson 16 Rutland Place, Cirencester GL7 1PR
2E0 LLX J. Wright 2 Regent Road, Church, Accrington BB5 4AR
2W0 LLY P. Kyte 1 Dan Y Bryn, Caerau, Maesteg CF34 0UW
2E0 LLZ L. Andrews 72 Grange Road, Alresford SO24 9HF
2E0 LMD Wirral&Dist ARC c/o A. Bate 16 East Avenue, Heald Green, Cheadle SK8 3DL
2E0 LME G. Beardmore 9 Ashmore Drive, Gnosall, Stafford ST20 0RP
2E0 LMG C. Murray 6 Maple Grove, Stocksbridge S36 1ED
2E0 LMH L. Hudson 68 Eleanor Road, Harrogate HG2 7AJ
2E0 LMI N. Dorling 51 Haygarth Close, Cirencester GL7 1WY
2E0 LMK S. Wills 8 Amherst Road, Newcastle upon Tyne NE3 2QQ
2E0 LML A. Stevens 5 Ifield Mill Close, Stone Cross BN245PF
2W0 LMM P. Jones 115 Wordsworth Gardens, Pontypridd CF37 5HH
2E0 LMO M. Broyd 16 George Downing House, 4A Bickleigh Close, Plymouth PL6 5XJ
2E0 LMR G. Southall 28 Manor Road, Woodford Halse, Daventry NN11 3QP
2M0 LMY L. Mullaney 8 School Road, Wellbank, Broughty Ferry, Dundee DD5 3PL
2E0 LNC D. Lancaster Linkhill View, Frith Common, Eardiston, Tenbury Wells WR15 8JX
2M0 LNF I. Mcgurk 16 Smith Crescent, Alexandria G83 8BL
2E0 LNG R. Rider 25 Kimber Close, Lancing BN15 8QD
2E0 LNO T. Sidaway 6 Rookery Chase, Deepcar, Sheffield S36 2NF
2M0 LNR T. Couper 10 Sclandersburn Road, Denny FK6 5LP
2E0 LNU M. Durban 62 Westfield Way, Charlton, Wantage OX12 7EP
2E0 LNX N. Stone Flat, 97-99 Stoke Road, Gosport PO12 1LR
2I0 LNZ L. Craig 170 Donaghadee Road, Bangor BT20 4PP
2E0 LOA Capt. J. Pearce Clematis Cottage, 52 Cheselbourne, Dorchester DT2 7NP
2E0 LOE N. Chaplin 5 Maxwell Street, Bury BL9 7QA
2E0 LOG D. Whelan 431 Leeds Road, Huddersfield HD2 1XT
2E0 LOH P. Rasmussen 7 Portal Drive North, Upper Heyford, Bicester OX25 5TH
2E0 LOL L. Woolley 4 Robert Street, Warrington WA5 1TQ
2E0 LON I. Lonsdale 23 Hunts Field Clayton-Le-Woods, Chorley PR6 7TT
2I0 LOP Rvd. M. Donald Station Road, Garvagh BT51 5LA
2I0 LOR Dr A. Bell 4 Mount Pleasant View, Newtownabbey BT37 0ZY
2E0 LOW M. Duchar 13 Thirlmere Avenue, Chester le Street DH2 3ED
2E0 LOX S. Cilliers 61 Heathwood Gardens, London SE7 8ET
2E0 LPD D. Lester 171 Glenavon Road, Birmingham B14 5BT
2E0 LPE A. Monaghan Briar Patch 117 Elm High Rd, Wisbech PE14 0DN
2I0 LPG B. Bruce 19 Ashvale Heights, Hillsborough BT26 6DJ
2E0 LPJ A. Pomfrey-Jones 46 Hampton Road, Erdington, Birmingham B23 7JJ
2I0 LPO J. Mcerlean 24 Mccorley Road, Toomebridge, Antrim BT41 3NH
2E0 LPP P. Kenderes 43 Fairlight Road, London SW17 0JE
2E0 LPR M. Price 9 Herbarth Close, Liverpool L9 1JZ
2E0 LPW L. Walker 7 Stroudley Close, Ashford TN240TY
2M0 LPX M. Macfarlane 9 Dreghorn Park, Colinton, Edinburgh EH13 9PH
2E0 LPZ G. Harmath 101 Whitton Avenue East, Greenford UB6 0QE
2M0 LQF J. Bennet 24/11 Greenpark, Edinburgh EH177TA
2E0 LQR T. Longmore 3 Dairy Farm Cottages, Northlands Road, Gainsborough DN21 5DN
2E0 LQW R. Edwards 46 Lavers Oak, Martock TA12 6HG
2E0 LQY I. Gilmore 19 Green Sward Lane, Redditch B98 0EN
2E0 LQZ L. Woodward Oaklands, Ashford Lane, Hockley Heath, Solihull B94 6RH

2E0 LRA L. Ayre 30 Tithe Lane, Calverton, Nottingham NG14 6HY
2U0 LRB L. Bichard Brise De Mer, Les Rouvets De Bas, Guernsey GY7 9QF
2E0 LRD K. Bainbridge 29 Bluebell Grove, Calne SN11 9QH
2E0 LRG M. Parker Ridgeways, Mill Common, Westhall, Halesworth IP19 8RQ
2E0 LRJ A. Smith 101 Chaucer Drive, Lincoln LN2 4LT
2E0 LRK L. Kelsey 111-113 George Street, Mablethorpe LN12 2BS
2E0 LRL M. Russell 65 Court Farm Road, Bristol BS14 0EF
2I0 LRN K. Bell 3 Alexandra Crescent, Larne BT40 1NE
2M0 LRO J. Barton 58 Aberdour Road, Dunfermline KY11 4PE
2E0 LRP R. Hughes 7 Willow Place, Darlington DL1 5LX
2E0 LRR L. Rhodes 91 Bedonwell Road, Bexleyheath DA7 5PS
2W0 LRS L. Cole 14 Inner Loop Road, Beachley, Chepstow NP16 7HF
2E0 LRV L. Emanuel 20 Wychwood Drive, Redditch B97 5NW
2E0 LRZ P. Jones Sussex Cottage, The Limes, Felbridge, East Grinstead RH19 2QY
2E0 LSB A. Shepherd 39 Minehead Road, Dudley DY1 2NZ
2E0 LSC S. Deakin 20 Riccat Lane, Stevenage SG1 3XY
2E0 LSE G. Bystryakov 20 Elmhurst Gardens, Leeds LS17 8BG
2M0 LSG S. Gray 17 Pine Grove Cumbernauld, Glasgow G67 3AX
2E0 LSI N. Highfield 298 Mersea Road, Colchester CO2 8QY
2E0 LSL A. Wood 4 St. Andrews, The Common, Cranleigh GU6 8NX
2E0 LSO G. Coltman The Oaks Rushden, Buntingford SG9 0SN
2E0 LSR K. Titmarsh 8 Bainbridge Close, North Walsham NR28 9UP
2E0 LSS J. Lusted 17 St James Road, East Grinstead RH19 1DL
2E0 LST L. Timmins 83 Loxdale Sidings, Bilston WV14 0TN
2E0 LSV R. Derham Netherwood, Copse Lane, Long Sutton, Hook RG29 1SX
2E0 LSW L. Wellington 5 Pasture Road, London SE6 1JF
2E0 LSX A. Dale 37 Bussey Road, Norwich NR6 6JF
2E0 LTC C. Anderson 191 Waveney Road, Hull HU8 9NA
2E0 LTD S. Boyles 5 Penclose Road, Fleckney LE8 8TE
2E0 LTF L. Farrell 1 Meadway, Ince, Wigan WN2 2BZ
2E0 LTH L. Thornton 11 Polruan Road, Truro TR1 1QR
2E0 LTJ M. Brett Lindon, Bunkers Hill, Wisbech PE13 4SQ
2E0 LTL D. Cadden 24 Lindenbrook Vale, Stafford ST17 4QN
2E0 LTM P. Bone 11 Fox Hill Drive, Stalybridge SK15 2RP
2E0 LTP C. Brennan 19 The Furrow, Littleport, Ely CB6 1GL
2E0 LTR N. Rice 30 Oveton Way, Bookham, Leatherhead KT23 4ND
2E0 LTT M. Lovatt 3 Withington Close, Atherton, Manchester M46 0EZ
2E0 LTU A. Jakstas 36 Boxridge Avenue, Purley CR8 3AQ
2E0 LTX D. Akerman The Brick Barn Old Gore, Ross-on-Wye HR9 7QW
2E0 LUD P. Lloyd 8 Maydor Avenue, Saltney Ferry, Chester CH4 0AH
2E0 LUF A. Clack 3 Darwin Close, Swindon SN3 3NF
2M0 LUG L. Affleck 1 Fank Brae, Mallaig PH41 4RQ
2E0 LUJ S. Cottam 14 Barnard Close, Rednal, Birmingham B45 9SZ
2E0 LUK C. Cavalcante Pinheiro Filho Flat 7, Hollybush House, Hollybush Gardens, London E2 9QT
2E0 LUL L. Pearson 4 Brentwood Close, Thorpe Audlin, Pontefract WF8 3ES
2E0 LUN N. Evetts 35 Wood End Road, Kempston, Bedford MK43 9BB
2E0 LUT D. Murdoch 48 Bracklesham Gardens, Luton LU2 8QL
2E0 LUY L. Isaac Sneath 21 Garrick Close, Lincoln LN5 8TG
2E0 LVC D. Tofield 6 Church Grove, Barnstaple EX32 9DJ
2W0 LVE A. Moody Perthiteg, Cwmhiraeth, Llandysul SA44 5XJ
2E0 LVG M. Brooks 34 Roundmead, Stevenage SG2 9PH
2E0 LVL M. Bentley 5 Grendon Green, Stoke-on-Trent ST2 0EH
2E0 LVP I. Ahmed 122 South View Road, Sheffield S7 1DD
2E0 LVR Dr O. De Peyer Flat 5, Molasses House, Clove Hitch Quay, London SW11 3TN
2E0 LWA D. Mccorrie 145 Mildenhall Road, Fordham, Ely CB7 5NW
2M0 LWB L. Bradley Amon Sul, Kiltarlity, Beauly IV4 7HT
2M0 LWD G. Ledgerwood 28 Cannerton Park, Milton Of Campsie, Glasgow G66 8HR
2E0 LWL P. Gill 38 Claygate, Thorneywood, Nottingham NG36JX
2E0 LWQ G. Ford The Lodge, Home Farm Lane, Rimpton, Yeovil BA22 8AS
2E0 LWR A. Mundy 27 Yorke Road, Croxley Green, Rickmansworth WD3 3DW
2E0 LWT A. Teed 57 Lymington Road, Torquay TQ1 4BG
2E0 LXA A. Lowery 102 Brompton Park, Brompton On Swale, Richmond DL10 7JP
2E0 LXB R. Thompson Croft Michael Farm, Croft Mitchell, Troon, Camborne TR14 9JJ
2E0 LXC G. Goddard 13 Rosslyn Avenue, Coventry CV6 1GL
2E0 LXD D. Collins 30 Upham Road, Swindon SN3 1DN
2E0 LXF D. Shuttleworth 27 Union St., Egerton, Bolton BL7 9SP
2E0 LXI E. Pestano 38 Third Avenue, Bexhill-on-Sea TN40 2PA
2E0 LXQ J. Ratty 12 Challoners Close, East Molesey KT8 0DW
2E0 LXR L. Rowlands 6 St. Michaels Avenue, Clevedon BS21 6LL
2I0 LXS A. Logan 15 Park Lane, Saintfield, Ballynahinch BT24 7PR
2W0 LXT L. Thomas 64 Min Y Llan, Letterston, Haverfordwest SA62 5SP
2I0 LXW M. Lewis 7 Liester Park, Ballyrobert, Ballyclare BT39 9RZ
2M0 LXX C. Sturgeon 14 Jacks View, Redbrae, Maybole KA19 7BS
2E0 LXY D. Loxley 33 Longwood Road, Tingley, Wakefield WF3 1UG
2E0 LYB P. Lyba 6 Ambassadors Way, North Shields NE29 8ST
2E0 LYD B. Vile 24 Hudson Close, Dover CT16 2SG
2E0 LYF K. Fletcher Beverley Hotel, 55 Old Brumby Street, Scunthorpe DN16 2AJ
2E0 LYK A. Harrison 19 Marlborough Avenue Haxey, Doncaster DN9 2HL
2E0 LYM S. Owens 18 Harvester Way, Lymington SO41 8YD
2E0 LYR T. Martin 46 Hayes Crescent, Frodsham WA6 7PG
2E0 LYY I. Lloyd 30 Church View Gardens, Kinver, Stourbridge DY7 6EE
2E0 LYZ N. Goult 5 Colbourne Close, Bransgore, Christchurch BH23 8BW
2E0 LZB A. Highfield 38 Brunswick Gardens, Garforth, Leeds LS25 1HF
2E0 LZC D. Ross Flat A, 174 Estcourt Road, London SW6 7HD
2E0 LZD G. Kelley 31 Cherry Park, Brandon, Durham DH7 8TN
2E0 LZE E. Stone Oakley, Main Road, Salisbury SP4 6EE
2E0 LZG J. Kenny 29 Siskin Close, Bishops Waltham, Southampton SO32 1RP
2E0 LZI I. Duxbury 334 Linnet Drive, Chelmsford CM2 8AL
2E0 LZM J. Roberts 51 Bradfield Road, Broxtowe, Nottingham NG8 6GP
2E0 LZN J. Marshall 6 Foster Walk, Sherburn in Elmet LS25 6EU
2E0 LZQ L. Hollingworth 43 Wingfield Road, Hull HU9 4PR
2E0 LZT D. Young 20 Summerhouse, Tickenham, Clevedon BS21 6SN
2W0 LZU P. Matthews The Chateau, Wynnstay Hall Estate, Ruabon, Wrexham LL14 6LA
2E0 LZX J. Mutter 27 Snowdonia Way, Huntingdon PE29 6XP
2M0 LZY D. Clark 106 Braes Avenue, Clydebank G81 1DP
2E0 MAA M. Milne Flambards, Manor Road, Dunmow CM6 2JR
2E0 MAB M. Bridgeland 17 Oldfield Lane, Wisbech PE13 2RJ
2W0 MAC C. Mccarthy 7 Aneurin Avenue, Crumlin, Newport NP11 5HN
2E0 MAD M. Davidson 26 Hurford Drive, Thatcham RG19 4WA
2E0 MAF M. Beckett 59 Broadacre, Caton, Lancaster LA2 9NH
2E0 MAH M. Holbrook 9 Beechwood Mount, Hemsworth, Pontefract WF9 4ES
2E0 MAJ M. Jones 20 Chelsea Drive, Sutton Coldfield B74 4UG
2E0 MAL M. Frame 23 Greenside Court, Sunderland SR3 4HS
2E0 MAN R. Lomax 11 Sherbourne Drive, Heywood OL10 4ST
2W0 MAO M. Cowhey 11 Aspen Way, Newport NP20 6LB
2E0 MAP M. Woolley 84 Bowthorpe Road, Norwich NR2 3TP
2E0 MAQ M. Finn Atzenbach 38, Idar-Oberstein 54473 Germany
2M0 MAV J. Cattigan Lunan Home Farm Cottage, Lunan Bay, Arbroath DD11 5ST
2E0 MAX M. Mcsherry 5 Briery Croft, Stainburn, Workington CA14 1XJ
2E0 MAY D. Maydew 128 Thorne Road, Willenhall WV13 1AW
2E0 MAZ M. Stevenson 127 Walton Road, Chesterfield S40 3BX
2E0 MBA H. Anderton 69 Sycamore Grove, Lancaster LA1 5RS
2E0 MBB M. Bartley 19 South Avenue, Shadforth, Durham DH6 1LB
2E0 MBD N. Draper 107 Arkwrights, Harlow CM20 3LY
2M0 MBE C. Hebenton 43 East Avenue, Uddingston, Glasgow G71 6LG
2E0 MBG J. Girard 49 Beech Crescent, Hythe, Southampton SO45 3QF
2I0 MBI B. Mcdonald 20 Aughan Park, Poyntzpass, Newry BT35 6TW
2E0 MBK M. Lewis 73 Addenbrooke Street, Wednesbury WS108HJ
2E0 MBO M. Hughes 58 Grange Lane North, Scunthorpe DN16 1RW
2E0 MBQ A. Blamires 2 Foldings Grove, Scholes, Cleckheaton BD19 6DQ
2E0 MBS M. Strange 101 Southbroom Road, Devizes SN10 1LY
2E0 MBT M. Buchanan 36 Church Lane, Manby, Louth LN11 8HL
2E0 MBV M. Mchugh 51 Rutland Street, Hyde SK14 4SY
2E0 MBW M. Rickaby 57 Hylton Road, Jarrow NE32 5DN
2E0 MBX M. Bowell 28 Jubilee Close, Byfield, Daventry NN11 6UZ
2E0 MCA M. Addison 319 Long Lane, London N2 8JW
2W0 MCB M. Luxton 3 The Paddocks, Newgate Street, Brecon LD3 8DJ
2E0 MCC Dr H. Shekhdar Manora Lodge, Sea Bank Road, Skegness PE24 5QU
2M0 MCD M. Mcdonald 17 Ramsay Mews, Strathaven ML10 6GN
2E0 MCG J. Mcgill 7 Willow Crescent, Wistaston, Crewe CW2 8RL
2E0 MCH M. Bailey 34 Jephson Drive, Birmingham B26 2HW
2E0 MCJ M. Jeffrey 9 Stoney Lands, Plymouth PL125DF
2E0 MCK M. Bridgehouse 43 Age Croft, Oldham OL8 2HG
2E0 MCL M. Carney 2 Lilac Meadows, Lawley Village, Telford TF4 2NX
2E0 MCM B. Cameron 9 Finchale Road, Hebburn NE31 2HR
2E0 MCN M. Bridger 11, Beecham Close, Newcastle upon Tyne NE15 6LG
2W0 MCQ M. Ellis Heddwch Brithdir, Dolgellau LL40 2SF
2E0 MCS L. Allcock 26 Castleton Grove, Inkersall, Chesterfield S43 3HU
2W0 MCT A. Mctaggart Brick Hall, Hundleton, Pembroke SA71 5QX
2E0 MCW M. Carlin 44 Sileby Road, Barrow Upon Soar, Loughborough LE12 8LR
2E0 MD M. Davies 2 Ellins Terrace, Normanton WF6 1BL
2E0 MDC M. Crawley 16 The Meadows, Herne Bay CT6 7XF
2E0 MDE C. Cundall 43 High Street, Great Gonerby, Grantham NG31 8JR
2W0 MDG M. Griffiths Mandalay, Bromfield Street, Wrexham LL14 1NF
2E0 MDH M. Walters 65 Bannawell Street, Tavistock PL19 0DP
2E0 MDK A. Currie 31 Launceston Road, Bodmin SO50 6AY
2E0 MDN M. Bray 13 Rosebay Close, Hartlepool TS26 0ZL
2E0 MDR M. Bradley 13 Elizabeth Avenue, Bilston WV14 8EA
2E0 MDT B. Hiley 9 Pinfold Lane, Harby, Melton Mowbray LE14 4BU
2E0 MDU Rhyl District Amateur Radio c/o N. Alders 14 Forest Rise, Crowborough TN6 2ES
2E0 MDZ M. Smith 31 Atlantic Crescent, Sheffield S8 7FW
2E0 MED A. Medhurst 44 Battle Road, Hailsham BN27 1DS
2E0 MEG S. Ridley 123 Lanercost Drive, Newcastle upon Tyne NE5 2DL
2E0 MEH D. Howarth 6 Lyndhurst Close, Norton, Doncaster DN6 9PY
2E0 MEI M. Bennetts 2 Chywoone Terrace, Newlyn, Penzance TR18 5NR
2E0 MEK M. Prentice Flat 3, 13 Meadow Road, Salisbury SP2 7BN
2E0 MEL M. Mcgoldrick 7 Walnut Drive, Tiverton EX16 6HE
2E0 MEO A. Bond 6 Ainscow Avenue, Lostock, Bolton BL6 4LR
2E0 MEQ M. Rea 15 Wensleydale Close, Royton, Oldham OL2 5TQ
2E0 MES M. Skinner 5 Sycamore Avenue, Upminster RM14 2HR
2E0 MET G. Lewis 5 Framfield Road, Uckfield TN22 5AG
2E0 MEU A. Sharif 6 Buckle Rise, Seaford BN25 2QN
2E0 MEV M. Clarke 54 Stafford Grove, Shenley Church End, Milton Keynes MK5 6AZ
2W0 MEX R. Hicks 14 Carn Celyn Beddau, Pontypridd CF38 2TF
2E0 MEY M. Sadler 14 Woodlands Avenue, Water Orton, Birmingham B46 1SA
2E0 MEZ M. Marsh 25 Southdown Road, Seaford BN25 4PD
2E0 MFA F. Alfrey 16 Walls Road, Bembridge PO35 5RA
2E0 MFC L. Ross 2 Bedford Street, Blackburn BB2 4EU
2W0 MFD T. Heath 16 Beacons Park, Brecon LD3 9BR
2E0 MFF M. Clarke 4 Mill Lane, Brant Broughton, Lincoln LN5 0RP
2E0 MFG M. Mcgowan 48 Alderley Road, Thelwall, Warrington WA4 2JA
2E0 MFI M. Berrisford 5 Branwell Drive, Haworth, Keighley BD22 8HG
2I0 MFJ B. Traynor 94 Markville, Portadown, Armagh BT63 5SZ
2M0 MFK M. Cook 6 Kirkstead Drive, Dundee DD2 2FB
2E0 MFN P. Roberts 17 Cannon Hill, Prenton CH43 4XR
2E0 MFS M. Feast 30 Peveril Drive, Riddings, Alfreton DE55 4AP
2E0 MFT A. Hill 1 Rochester Close, Mountsorrel, Loughborough LE12 7UH
2E0 MFX Dr M. Fox 24 The Avenue, Sandy SG19 1ER
2E0 MFZ M. Fairchild Brooklyn Caravan Park 21 Almond Brow Gravel Lane Banks, Southport PR98BU
2E0 MGA M. Greensmith 14 Fountain Road, Draycott-In-The-Clay, Ashbourne DE6 5HP
2E0 MGC M. Chanter 7 Woodford Crescent, Plymouth PL7 4QY
2E0 MGI M. Isbell 20 Woodland Crescent, Wolverhampton WV3 8AS
2E0 MGL M. Talbot 26 Chevalier Grove, Crownhill, Milton Keynes MK8 0EJ
2M0 MGM M. Geldart 13B Greystone Place, Newtonhill, Stonehaven AB39 3UL
2E0 MGP G. Champion 34 Greenfields, Edenside, Kirby Cross., Frinton-on-Sea CO13 0SW
2E0 MGR M. Reeks 33 Madresfield Village, Madresfield, Malvern WR13 5AA
2E0 MGT J. Gardiner 31 Rodway Road Tilehurst, Reading RG30 6EH
2E0 MGU M. Guidolin Flat 3, 3 Grimston Gardens, Folkestone CT20 2PT
2E0 MGV L. Copeland 78 Penderyn Crescent Ingleby Barwick, Stockton-on-Tees TS17 5HQ

2E0 MGW M. Whitticombe 18 Fairclough Street, Burtonwood, Warrington WA5 4HJ
2E0 MGX M. Deeley Unit 4 Beechwood Business Park, Cannock WS11 7GB
2M0 MGY J. Boag 182 St. Fillans Road, Dundee DD3 9LH
2E0 MHC Dr G. Cardoso 11 Coppice Way, Aylesbury HP20 1XG
2E0 MHD G. Muirhead 13 Berry Street, Skelmersdale WN8 8QZ
2E0 MHE S. Snelson 212 Dickson Road, Blackpool FY1 2JS
2E0 MHG J. Gray 7 Ruskin Avenue, Melksham SN12 7NG
2E0 MHJ I. Jabegu 51 South Crescent, Blandford Camp, Blandford Forum DT11 8AJ
2E0 MHL M. Lacey 82 Bowerings Road, Bridgwater TA6 6HF
2E0 MHM M. Byard 1 Fieldside, Long Wittenham, Abingdon OX14 4QB
2M0 MHN N. Morris 23 Sedgebank, Sedgebank, Livingston EH54 6HE
2E0 MHT M. Thompson 4 Jubilee Court Ravenscroft, Holmes Chapel, Crewe CW4 7HA
2E0 MIB V. Ball 24 Carr Lane, Warsop, Mansfield NG20 0BN
2E0 MIC M. Steed 38 Rivelands Road Swindon Village, Cheltenham GL51 9RF
2E0 MID P. Staite Chestnut Farm, Eastville, Boston PE22 8LX
2M0 MIF D. Mifsud 25 Priory Road, Linlithgow EH49 6BP
2E0 MIH M. Humphries 5 Coppice Mead, Stotfold, Hitchin SG5 4JX
2E0 MIL I. Millman 3 Oyster Mews, 1-3 Forest Road, Poole BH13 6EN
2E0 MIS D. Smout Sunrays, Warbage Lane, Bromsgrove B61 9BH
2E0 MIU J. Marsh 14 Eyam Road, Hazel Grove, Stockport SK7 6HP
2E0 MIV B. Davies 60 Queensway, Blackburn BB2 4QT
2E0 MIX D. Edge 18 Sandringham Avenue, Whitehaven CA28 6XL
2E0 MIY P. Billingham 393 Landseer Road, Ipswich IP3 9LT
2E0 MIZ A. Bartlett 62 Kewstoke Road, Bath BA2 5PU
2W0 MJA A. Sneddon 3 Marigold Close, Gurnos, Merthyr Tydfil CF47 9DA
2W0 MJC M. Churcher 71 Twyn Road, Abercarn, Newport NP11 5JY
2E0 MJE M. Parker 1 Ham Road, Wanborough, Swindon SN4 0DF
2E0 MJG Rvd. M. Gillingham 14 Nethergreen Gardens, Killamarsh S21 1FX
2E0 MJH M. Holroyd 9 Coniston Green, Aylesbury HP20 2AJ
2E0 MJJ J. Jones 19 Southbank Street, Leek ST13 5LS
2E0 MJL M. Lee Up To Date House, Shore Road, Boston PE22 0NA
2E0 MJM M. Middleditch 56 Rowan Way, Yeovil BA20 2NR
2E0 MJO S. Marcomini 18 Chadwick Place, Long Ditton, Surbiton KT6 5RE
2E0 MJP M. Marsh 19 First Avenue, South Kirkby, Pontefract WF9 3EP
2E0 MJS S. Mcmurtrie 5 Hill Road, Carshalton SM5 3RA
2E0 MJT M. Troth 3 Laburnum Grove, Bromsgrove B61 8NB
2E0 MJX M. Cresswell 44 The Lea, Birmingham B33 8JP
2M0 MJY M. Yarrow Lomond Villa, Downies Village, Aberdeen AB12 4QX
2E0 MJZ D. Cook 15 Kendricks Fold, Rainhill, Prescot L35 9LX
2E0 MKB M. Ballard 41 Middlefield Ave, Halesowen B62 9QJ
2E0 MKE M. Gregory 65 Nursery Crescent North Anston, Sheffield S25 4BR
2E0 MKF M. Amphlett Highbanks, Charnes Road, Market Drayton TF9 4LQ
2W0 MKG M. Gray 15 The Circle, Two Locks NP44 7JP
2E0 MKH M. Heaton-Bentley 65 Brookfield Road, Thornton-Cleveleys FY5 4DR
2E0 MKI T. Palmer 70 Channel View Road, Eastbourne BN227LL
2E0 MKJ M. Johnson 7 Norfolk Wing, Tortington Manor, Ford Road, Tortington, Arundel BN18 0FD
2E0 MKK M. Kilkenny 23 Hazelhurst Road, Stalybridge SK15 1HD
2E0 MKT T. Walker 11 Banburies Close, Bletchley, Milton Keynes MK3 6JP
2E0 MKV M. Vardy 60 Hucklow Avenue, North Wingfield, Chesterfield S42 5PU
2E0 MKW M. Wiggins 2 Cherry Tree Close, Halstead CO9 2UA
2E0 MKX M. Keyte 3 Lower High St., Mow Cop, Stoke on Trent ST7 3PB
2M0 MKZ M. Devlin 4 County Place, Forgandenny, Perth PH2 9EP
2E0 MLA A. Highfield 29 Blewitt Street, Brierley Hill DY5 4AW
2E0 MLC S. Sedgwick Flat 2/A, St. Georges Court 44 Thorne Road, Doncaster DN1 2JA
2E0 MLD M. Dixon 37 Carlton Close, Parkgate, Neston CH64 6RB
2E0 MLE D. Pilkington 197 Saltings Road, Snodland ME6 5HP
2E0 MLF M. Raynor 21 Teversal Avenue, Pleasley, Mansfield NG19 7QQ
2W0 MLG S. Gordon 8 Maesteg, Cymau, Wrexham LL11 5EP
2E0 MLH M. Howse 12 Queens Road Moretonhampstead, Newton Abbot TQ13 8LP
2E0 MLJ M. Jones 2 Home Park Road, Saltash PL12 6BH
2E0 MLK M. Kipling 12 Jolly Brows, Harwood, Bolton BL2 4LZ
2E0 MLL A. Mccall 95 Newton Drive, Blackpool FY3 8LX
2E0 MLS M. Heywood 16 Edinburgh Drive, Hindley Green, Wigan WN2 4HL
2E0 MLV M. Grantham 7 Goodwin Close, Sandiacre, Nottingham NG10 5FF
2E0 MLX A. Boyes 12 Leyburn Grove, Stockton-on-Tees TS18 5NH
2M0 MMB M. Cleland 85 Carfin St., Motherwell ML1 4JL
2W0 MMD D. Holloway 14 Woodbrook Terrace, Burry Port SA16 0NF
2E0 MME R. Shulver 63 Hill Farm Way, Southwick, Brighton BN42 4YG
2M0 MMF R. Tripney 7 Sunnyside St., Camelon, Falkirk FK1 4BJ
2M0 MMG M. Gourlay 14 Holmes Holdings, Broxburn EH52 5NS
2E0 MMH M. Hall 67 Darlinghurst Grove, Leigh-on-Sea SS9 3LF
2E0 MMJ M. Majhail 3 Poynders Hill, Hemel Hempstead HP2 4PQ
2E0 MML M. Thompson 22 Churchfield Terrace, Barnsley S728JT
2M0 MMM M. Greig 7 St. Ronans Road, Forres IV36 1BQ
2M0 MMO M. Overthrow 63 Primrose Avenue, Larkhall ML9 1JX
2E0 MMP M. Porter Michael Porter C O 219 Burnley Road, Colne BB88JD
2I0 MMT M. Torley 4 Yew Tree Park, Newry BT34 2QP
2E0 MMU A. Moss Winstons, Mayfield Lane Durgates, Wadhurst TN5 6DG
2E0 MMX R. Hewson 10 Miriam Grove, Leigh WN7 3EX
2W0 MNA M. Mcdonald Falcondale, Pleasant Valley, Stepaside, Wisemans Bridge, Narberth SA67 8NT
2M0 MNB S. Wallace 28 Laurel Bank Terrace, Castle Douglas DG7 1BP
2E0 MNC M. Craner 46 Meadowhill Crescent, Redditch B98 8HT
2E0 MND A. Harrop 35 Langdale Crescent, Dalton-in-Furness LA15 8NR
2E0 MNG N. Giuliano 13 Walton Drive, Littleover, Derby DE23 1GN
2E0 MNH M. Harvey May Tree Cottage, Kelvedon Road, Tiptree, Colchester CO5 0LJ
2W0 MNJ M. Kenny 27 Brangwyn Crescent, Newport NP197QY
2E0 MNL S. Manley The Cottage, Crow Ash Road, Berry Hill, Coleford GL16 7RB
2E0 MNP M. Pomfret 5 Malvern Crescent, Ince, Wigan WN3 4QA
2E0 MNU P. Richardson 14 Portland Street, Worksop S80 1RZ
2E0 MNX C. Leece 101 Wellstead Way, Hedge End, Southampton SO30 2BH
2E0 MNY R. Parker 53 Tunstall Road, Canterbury CT2 7BX
2E0 MNZ L. Moyle 14 St. Lukes Close, Kettering NN15 5HD
2E0 MOB Radio Amateur Invalid And Blind Club c/o C. Larner 98 Allandale, Hemel Hempstead HP2 5AT
2M0 MOF T. Moffat 16 Drumellan Road, Ayr KA7 4XQ
2M0 MOK W. Fulton 15 Staffa Avenue, Port Glasgow PA14 6DT
2E0 MOL R. Moles 14 Dorsett Road, Stourport-on-Severn DY13 8EL
2W0 MON A. West Bryn Goleu, Gwalchmai, Holyhead LL65 4SW
2E0 MOR A. Willmore 31 Oaklands, Bugbrooke, Northampton NN7 3QU
2I0 MOU G. Moucka 5 Glebe Gardens, Moira, Craigavon BT67 0TU
2E0 MOW C. Naylor 16 Boston Avenue, Norbreck Blackpool FY29BZ
2E0 MOY R. Moys 12A Palmerston Avenue, Fareham PO16 7DP
2E0 MOZ M. Meadowcroft 8 Lamlash Road, Blackburn BB1 2AS
2E0 MPA M. Ashworth 123 Forest Road, Liss GU33 7BP
2E0 MPB R. Johnson 24 Fairfields, Upper Denby, Huddersfield HD8 8UB
2E0 MPC M. Carter 113 Old Road Tintwistle, Glossop SK13 1JZ
2E0 MPE R. Eaton 31 Pinfold Lane Ruskington, Sleaford NG34 9EU
2E0 MPF C. Le Marchant Bow House, Green Lane, Prestwood, Great Missenden HP16 0QE
2E0 MPG P. Mcgrath 24 Broadoak Drive, Lanchester, Durham DH7 0QA
2E0 MPJ J. Neal 75 Park Lane, Castle Donington, Derby DE74 2JG
2E0 MPN M. Nolan Bath Road Post Office, Post Restante. Bath Road, Devizes SN10 1QG
2E0 MPO K. O'Hara 41 Exeter Street, Blackburn BB2 4AU
2E0 MPR M. Rolls 3 Gleneagles Crescent, New Holland, Barrow-upon-Humber DN19 7TL
2E0 MPX P. Matthew 24 Jubilee Close, Pamber Heath, Tadley RG26 3HP
2W0 MPY P. Man 17A Cradock Street, Swansea SA1 3HE
2E0 MQA D. Rogers 9 Prospect Place, Stafford ST17 4HZ
2E0 MQB M. Boldero 1 Grasleigh Avenue, Allerton, Bradford BD15 9AR
2E0 MQC M. Le Moine 115 Rothesay Road, Blackburn BB1 2ER
2E0 MQT F. Walker 18 Cranmer Road, Manchester M20 6AW
2E0 MRA A. Armson 2 Windmill Gardens, St. Helens WA9 1EN
2E0 MRC D. Letton 21 Westfield, Bradninch, Exeter EX5 4QU
2E0 MRD D. Millard 112 Avenue Road, Sandown PO36 8DZ
2E0 MRF M. Keal 81 St. Catherines Grove, Lincoln LN5 8ND
2E0 MRG G. Whittle 22 St. Oswalds Close, Finningley, Doncaster DN9 3ED
2E0 MRJ M. Jarrett 17 Greenhill Gardens, Minster, Ramsgate CT12 4EP
2E0 MRM M. Tetley 27 Cunningham Hill Road, St. Albans AL1 5BX
2M0 MRO M. Reid 2 Pinkie Gardens, Newmachar, Aberdeen AB21 0QF
2E0 MRP M. Peters 25 Windsor Court, Falmouth TR11 3DZ
2E0 MRQ P. Gibbs 19 Lupin Close, Etherley Dene, Bishop Auckland DL14 0TP
2I0 MRY M. Ruddy C/O 6 Iveagh Park, Greysteel, Londonderry BT47 3DD
2E0 MRZ M. Roberts 13 Stanley Road, Portslade, Brighton BN41 1SW
2E0 MSA M. Statham Broad Oak Bungalow, Manston, Sturminster Newton DT10 1EZ
2M0 MSB S. Brown 21 Whiteford Avenue, Dumbarton G82 3JU
2E0 MSE M. Edmonds 60 Shenstone Road, Maypole, Birmingham B14 4TJ
2E0 MSI M. Sims 5 Sandy Leaze, Bradford-on-Avon BA15 1LX
2W0 MSL N. Berrall 41 Nantgarw Road, Caerphilly CF83 3FB
2E0 MSM M. Mohammed Shafi 575 Wood Lane, Dagenham, London RM8 1DR
2E0 MSN M. Boisriveau-Mitchell 16 Pochard Close, Quedgeley, Gloucester GL2 4LL
2I0 MSO P. Hosey 13 Glenelly Gardens, Omagh BT79 7XG
2E0 MSR M. Ramsay 28A Strathmore Drive, Charvil, Reading RG10 9QT
2E0 MSS M. Smith The Lawns Tylers Green, High Wycombe HP108BH
2M0 MSU R. Sutherland 25 Burns Wynd, Maybole KA19 8FG
2E0 MSZ M. Smith 78 Shiregate, Metheringham, Lincoln LN4 3DR
2E0 MTA T. Allman 46 Belmont Road, Rugby CV22 5NZ
2E0 MTB T. Beckett 95 Warrens Hall Road, Dudley DY2 8DH
2E0 MTC C. Mathewson 33 Thornton Road, Bootle L20 5AN
2W0 MTD M. Davies 11 High Street, Malltraeth, Bodorgan LL62 5AS
2W0 MTE E. Thomas 1 Cambrian Gardens Y Drenewydd, Newtown SY16 2AW
2E0 MTH M. Knowles 11 Thorneycroft Avenue, Birkenhead CH41 8HJ
2E0 MTL M. Leach 64 Grove Street, Wantage OX12 7BG
2E0 MTM T. Mloduchowski Flat 4, Gwynne House, London E1 2AG
2E0 MTN M. Newell 55 Station Road Brimington, Chesterfield S43 1JU
2M0 MTO R. Foulds 83 Croftfoot Road, Glasgow G44 5JU
2E0 MTR M. Reilly 26 Roman Way, Folkestone CT19 4JT
2E0 MTT Dr M. Nassau 4A London Road, Liphook GU30 7AN
2E0 MTX N. Challis 48 Brunsfield Close, Wirral CH46 6HE
2E0 MTY D. Raine 91 Lulworth Avenue, Jarrow NE32 3SB
2E0 MUA J. Anderson 121 Barton Road, Stretford, Manchester M32 9AF
2E0 MUD S. Sparks 25 Wilwick Lane, Macclesfield SK11 8RS
2E0 MUI M. Tsun Dyson'S Farm, Long Row, Tibenham NR16 1PD
2E0 MUN A. Munford 6 Column Mews, Alnwick NE66 1RZ
2M0 MUR G. Murray The Barn House, Springfield Farm, Carluke ML8 4QZ
2E0 MUS A. Sutton 3 Grotes Buildings, London SE3 0QG
2E0 MUT J. Merritt 41 Great Grove, Bushey WD23 3BQ
2E0 MUW J. Blaylock 23 Sunnyway, Blakelaw, Newcastle upon Tyne NE5 3QB
2E0 MUZ M. Colpman 20 Rochford Road, Basingstoke RG21 7TQ
2E0 MVD M. Denham1 209 Coppice Road, Arnold, Nottingham NG5 7HD
2E0 MVE D. Sharp Flat 22, Farnworth House, Manchester Road, London E14 3HY
2E0 MVH S. Smith 65 Furness Avenue, St. Helens WA10 6QF
2M0 MVI M. Vaci Davaar, Lochawe, Dalmally PA33 1AQ
2E0 MVN D. Mills 31 Gorseway, Hatfield AL10 9GS
2E0 MVT D. Harris 27 Ashley Road, Poole BH14 9BS
2M0 MVX D. Wilson Rivendell Lodge, Glenkindie, Alford AB33 8RN
2E0 MWA M. Austin 14 The Green, Brown Edge, Stoke-on-Trent ST6 8RN
2E0 MWB M. Bryant 284 Brantingham Road Chorlton Cum Hardy, Manchester M21 0QU
2E0 MWD M. Day 14 Windsor Drive, Ramsey Forty Foot, Ramsey, Huntingdon PE26 2XX
2E0 MWH M. Hetherington 18 Wesley Street, Low Fell, Gateshead NE9 5YN
2E0 MWJ M. Willis 51 Barnsdale Close, Loughborough LE11 5AN
2E0 MWN M. Singer 1 Bentley Road, Slough SL1 5BB
2E0 MWT R. Woolley 10 Hazelmoor Fold, Blackley, Elland HX5 0DR
2E0 MWW M. Wheal 7 Ryecroft Drive, Withernsea HU19 2LP
2E0 MXA Outer Hebrides Iota Group c/o M. Amos Willow Tree House, Deers Green, Clavering, Saffron Walden CB11 4PX
2E0 MXC M. Craven 78 Connaught Road, Brookwood, Woking GU24 0HF
2E0 MXG M. Glen 6 Willow View, Catterick, Richmond DL10 7PD
2E0 MXL M. Lee 2 Rupert Road Chaddesden, Derby DE214ND
2E0 MXM M. Meehan 14 Grosvenor Road, Walton, Liverpool L4 5RB

2E0 MXN R. Lovell Formby, Formby, Liverpool L37 4BP
2E0 MXP M. Palmer New Haven, Stoneraise, Durdar, Carlisle CA5 7AX
2E0 MXR A. Wilson 28 Langham Road, Bristol BS4 2LJ
2E0 MXW D. Platt 50 Poplars Road, Stalybridge SK15 3EN
2E0 MYA D. Mycroft 11 Paisley Walk, Church Gresley, Swadlincote DE11 9FF
2E0 MYB H. Ibbitson Tor View, Whitstone, Holsworthy EX22 6TB
2E0 MYD E. Chauvelaine 292 Mount Pleasant, Redditch B97 4JL
2E0 MYE D. Sykes 2 The Street, Claxton, Norwich NR14 7AS
2E0 MYG M. Young 72 Goddard Way, Saffron Walden CB10 2EB
2E0 MYH D. Morgan 87 Pool Hayes Lane, Willenhall WV12 4PX
2E0 MYK M. Knowles 86 West Shore Road, Walney, Barrow-in-Furness LA14 3UD
2E0 MYL J. Swann 5 Lanark Close, Hazel Grove, Stockport SK7 4RU
2E0 MYS M. Broad 29 Oakdale Close , Clay Cross, Chesterfield S45 9RY
2E0 MYT M. Corrigan 33 Westbourne Road, Knott End-On-Sea, Poulton-le-Fylde FY6 0BS
2E0 MYX S. Elliott 79 Somerton Road, Bolton BL2 6LN
2E0 MYZ J. Hughes 22 Marklay Drive, South Woodham Ferrers CM3 5NP
2E0 MZB B. Gutteridge 54 Malthouse Road, Southgate, Crawley RH10 6BG
2E0 MZC G. Fox 23 The Driveway, Canvey Island SS8 0AB
2E0 MZD B. Greenwood 13 Mayflower Street, Blackburn BB2 2RX
2E0 MZE M. Hartshorn 21 Bidford Road, Leicester LE3 3AE
2M0 MZK M. Keay Coomb Burn, Wamphray, Moffat DG10 9LZ
2E0 MZL G. Johnson 22 Beechwood Close, Blythe Bridge, Stoke-on-Trent ST11 9RH
2W0 MZS M. Simons 68 Harbour Village, Goodwick SA64 0DZ
2E0 MZU C. Smith 105 Netherton Road, Worksop S80 2SA
2W0 NAD O. Bross 8 Queens Drive, Buckley CH7 2LJ
2E0 NAE M. Ridgewell 75 Hillshaw Park Way, Ripon HG4 1JU
2E0 NAF N. Foster 18 Austen Ave, Sawley, Nottingham NG103GG
2E0 NAG T. Bown 16 Sandringham Court, Queen Elizabeth Road, Nuneaton CV10 9AR
2E0 NAI P. Turner 38 La Ferte Bernard Close, Louth LN11 0ZN
2E0 NAM N. Carey 28 Tremayne Road, St. Austell PL25 4NE
2M0 NAN S. Ram 28 Craigievar Gardens, Kirkcaldy KY2 5SD
2E0 NAP N. Deery 25 Ribblesdale Place, Preston PR1 3NA
2E0 NAQ N. Barnard 10 Whites Lane Kessingland, Lowestoft NR33 7TF
2E0 NAR R. Nagy 40 Oakhampton Road, London NW7 1NH
2E0 NAS N. Inglis 74 Runswick Avenue, Whitby YO21 3UE
2I0 NAT C. Mooney 12 Curragh Walk, Derry City BT48 8HX
2D0 NAU D. Smith Alwyn, Four Roads Port St. Mary, Isle of Man IM9 5LH
2M0 NAX A. Anderson 18 Selkirk Street, Wishaw ML2 8RA
2W0 NAY M. Williams 10 Clettwr Terrace Pontsian, Llandysul SA44 4TU
2E0 NAZ N. Azizoff 7 Spencer Close, London N3 3TX
2E0 NBC D. Waters 3 George Place Berry Hill Coleford, Lydney GL15 6NA
2E0 NBE N. Irvine 100 Cavendish Road, Sunbury-on-Thames TW16 7PL
2E0 NBG N. Newman 34 Winsford Road, Bury St. Edmunds IP32 7JJ
2E0 NBI N. Beresford 2 Meadow View, Great Addington, Kettering NN14 4BN
2E0 NBK N. Brooks 105B Upper Woodcote Road, Caversham, Reading RG4 7JZ
2E0 NBM N. Modi 20 Hereford Road, Basingstoke RG23 8QL
2E0 NBR S. Warren 7 Crich Way, Swadlincote DE11 0UU
2E0 NBX N. Cohen 8 Henry Gepp Close, Adderbury, Banbury OX17 3FE
2E0 NBZ N. Birnie 61 Pipers Croft, Dunstable LU6 3JZ
2W0 NCA N. Alward 22 Laugharne Court, Caldy Close, Barry CF62 9DW
2E0 NCB D. Lawson 30 Meadowcroft, St. Helens WA9 3XQ
2E0 NCC N. Croft 22 King Edward Crescent, Leeds LS18 4BE
2E0 NCD N. Chandler 4 Nursery Cottages, Woodgreen, Fordingbridge SP6 2AL
2E0 NCE D. Stanley 58 Wells Gardens, Basildon SS14 3QS
2E0 NCG M. Dumpleton 24 Barley Close, Newton St. Faith, Norwich NR10 3GY
2E0 NCI Sqdn. Ldr. B. Dowley 120 Capel Street, Capel-Le-Ferne, Folkestone CT18 7HB
2E0 NCJ C. Nicholson 97 Station Road, Burgess Hill RH15 9ED
2E0 NCK N. Taylor 212 Plantation Hill, Worksop S81 0HD
2M0 NCM N. Cunningham 11 Glendoune Street, Girvan KA26 0AA
2E0 NCN K. Clarke 112B John Street, Sheffield S2 4QU
2E0 NCO K. Tonge 25 Southcote Grove, Birmingham B38 8ED
2E0 NCR G. Paton 17 Blakeney Road, Stevenage SG1 2LH
2E0 NCS N. Sunley 1 East Lea View, Cayton, Scarborough YO11 3TN
2E0 NCT M. Steeples 44 Trunch Road, Mundesley, Norwich NR11 8JX
2E0 NCV G. Killpack 20 Fisher Close, Banbury OX16 3ZW
2E0 NCY J. Mooneapillay 354 Upper Elmers End Road, Beckenham BR3 3HG
2E0 NDA N. Ayre 58 Burford Avenue, Swindon SN3 1BN
2E0 NDG N. Graven 33 Sheldrake Road, Broadheath, Altrincham WA14 5LJ
2E0 NDH N. Hewitt 36 Kenilworth Road, Doncaster DN4 0UD
2I0 NDJ N. Jameson 15A Ednagee Road, Castlederg BT81 7QF
2M0 NDO D. Morrison Osbourne Cottage, Benderloch, Oban PA37 1QP
2E0 NDP N. Plunkett 11 Stoneleigh Gardens, Grappenhall, Warrington WA4 3LE
2E0 NDR N. Nash Roann, Bedmond Road, Pimlico, Hemel Hempstead HP3 8SH
2E0 NDT R. Thatcher 83 Westfield Drive, North Greetwell, Lincoln LN2 4RE
2E0 NDW P. Mackay 30 Main Road, Austrey, Atherstone CV9 3EH
2E0 NDY A. Williams 12 St. Wilfrids Crescent, Brayton, Selby YO8 9EU
2E0 NDZ A. Humphriss 44 Bishops Close, Stratford-upon-Avon CV37 9ED
2E0 NEC C. Humphries 44 Linksway, Folkestone CT19 5LS
2W0 NED K. Nedin 28 Llys-Y-Coed Birchgrove, Swansea SA7 9PR
2E0 NEI N. Yorke 30 Bramdene Avenue, Nuneaton CV10 0DH
2I0 NEJ D. Mulligan 10 Seaview, Ardglass, Downpatrick BT30 7SQ
2E0 NEL C. Nelson 14 Windy Harbour Road, Southport PR8 3DU
2E0 NEM P. Sanders 6 Primrose Hill, Warwick CV34 5HW
2E0 NEN B. Daniels Broomhill, Holmrook CA19 1UL
2M0 NEO N. Thomson Four Winds, Holland Bush Hightae, Lockerbie DG11 1JL
2E0 NEQ J. Clarke 24 Telford Court, East Howdon NE280JH
2E0 NER M. Straughan 21 Silcoates Avenue, Wrenthorpe, Wakefield WF2 0UP
2E0 NEV N. Griffin 54 Edinburgh Road, Newmarket CB8 0QD
2E0 NEY P. Millard1 Weavern House, Hartham Lane, Chippenham SN14 7EA
2E0 NFB N. Bisiker 31 Lansdowne Avenue, Waterlooville PO7 5BL
2E0 NFC A. Cockburn 52 Devon Road, Hebburn NE31 2DW
2E0 NFI N. Mooney 60 Rhyddings Street, Oswaldtwistle., Accrington BB5 3EY
2E0 NFS N. Stephens 3 Spinney House, College Road, Windermere LA23 1PX
2E0 NFX N. Friend 4 Bartholomew Street, Dover CT16 2LH
2E0 NFZ M. Levett 16 Firs Avenue, Waterlooville PO8 8RS
2E0 NGB N. Bland 63 Swindon Road, Wroughton, Swindon SN4 9AG
2E0 NGC L. Akred 25 Kitchener Street, Walney, Barrow-in-Furness LA14 3QW
2E0 NGG N. Clare 123 Cunningham Road, Tamerton Foliot, Plymouth PL5 4PU
2I0 NGK J. Allen 3 Malwood Close, Belfast BT9 6QX
2E0 NGL N. Green 44 Rushyford Drive, Chilton, Ferryhill DL17 0EQ
2I0 NGM A. Mckay 17 Thorn Hill Road, Banbridge BT32 3TL
2E0 NGN G. Nelson 4 Garnet Field, Yateley GU46 6FN
2M0 NGO J. Nattress 44 Broadlands, Carnoustie DD7 6JY
2E0 NGR N. Grey 131 Links Avenue, Hellesdon, Norwich NR6 5PQ
2E0 NGZ S. Lawrance 69 Athelstan Gardens, Wickford SS11 7EF
2E0 NHB N. Barker 17 Pippin Walk, Hardwick, Cambridge CB23 7QD
2E0 NHF N. Frost 5 George Street Elworth, Sandbach CW11 3BL
2E0 NHJ N. Heywood 38 Thurne Rise, Martham, Great Yarmouth NR29 4PU
2E0 NHM N. Meakin 60 Canberra Way, Warton, Preston PR4 1XY
2E0 NHR S. Sawyer 19 Malvern Close, Ashington NE630TD
2E0 NHS J. Kelly 12 Park Road, Milford On Sea, Lymington SO41 0QU
2M0 NIA N. Hague 11 Auchriny Circle, Bucksburn, Aberdeen AB21 9JJ
2E0 NIB N. Bennett 44 Glenmoor Road, Buxton SK17 7DD
2I0 NIE C. Morton 29 Lackaboy View, Enniskillen BT74 4DY
2E0 NIF G. Calder 41 Wood End Way, Chandler'S Ford, Eastleigh SO53 4LN
2I0 NIO J. Tipping 16 The Oaks, Portadown, Craigavon BT62 4HX
2M0 NIT D. De Freitas 14 York Street, Clydebank G81 2PH
2M0 NIX N. Robertson Ladyburn, Port William, Newton Stewart DG8 9QN
2E0 NJC N. Long 25 Blendworth Lane, Southampton SO18 5GY
2E0 NJE N. Gonzales 46 Whitton View, Rothbury, Morpeth NE65 7QN
2E0 NJI N. Isherwood 41 Livingstone Road, Blackburn BB2 6NE
2E0 NJJ D. Wharlley 15 Crampton Court, Grosvenor Road, Broadstairs CT10 2XU
2E0 NJK N. Kendall 19 Clowance Lane, Mount Wise, Plymouth PL1 4HU
2M0 NJM S. Spencer 55 Blackwell Court, Inverness IV2 7AR
2E0 NJO N. Jones 5 Montgomery Crescent, Quarry Bank, Brierley Hill DY5 2HB
2M0 NJS N. Sheridan Cemetery Lodge, Lochmaben, Lockerbie DG11 1RL
2E0 NJT N. Jones 63 Bernwell Road, London E4 6HX
2E0 NJV D. Villa 33 North Street, Tywardreath, Par PL24 2PW
2E0 NKC D. Ansell 30 Curzon Avenue, Horsham RH12 2LB
2E0 NKE N. Rotherham 4 Spenser Road, Cheltenham GL51 7EA
2E0 NKI N. Crabb 1 Council Houses, Hall Lane, Crostwick, Norwich NR12 7BB
2E0 NKK M. Remplakowski 11 De Montfort Road, Speen, Newbury RG14 1TA
2E0 NKM N. Morse 33 Tower Close, Bassingbourn, Royston SG8 5JX
2E0 NKN E. Coomer Torquay Road, Exeter TQ12 5EZ
2E0 NKP N. Palin 21 Ford Lane, Crewe CW1 3EQ
2E0 NKR M. Moss 27 Dunn Side, Chelmsford CM1 1DL
2W0 NKS M. Moyse 10 Clifton Rise, Abergele LL22 7DN
2E0 NKT N. Kent Flat 1, Manor House, Redruth TR15 1AX
2M0 NLA A. Cunningham 36 Station Brae Gardens, Dreghorn, Irvine KA11 4FB
2E0 NLB N. Brown 9 Devonshire Avenue, Wigston LE18 4LP
2E0 NLE M. Pack 59 West End Falls, Nafferton YO25 4QA
2E0 NLK N. Lake 64 Womersley Road, Norwich NR1 4QB
2E0 NLM P. Maybin 16 Appleby Road Canning Town, London E16 1LQ
2E0 NLP J. Watts 70 Castleway North, Leasowe, Wirral CH46 1RW
2E0 NLS N. Andre Churchill College, Storey'S Way, Cambridge CB3 0DS
2E0 NLT P. Austin Bush Farmhouse Clee St. Margaret, Craven Arms SY7 9DT
2E0 NLW S. Jones 30 Crown Fields Close, Newton-le-Willows WA12 0JW
2E0 NLX P. Cooke 26 Welby Way, Coxhoe, Durham DH6 4BT
2E0 NLY B. Plackett 36 Dartmouth Crescent, Brinnington, Stockport SK5 8BG
2E0 NMA M. Baynes 92 Belgrave Drive, Hull HU4 6DW
2M0 NMD T. Ormiston 22 St. Ronans Road, Innerleithen EH44 6LZ
2E0 NMK S. Bateson 2 Green Crescent, Coxhoe, Durham DH6 4BE
2E0 NNB A. Hopper 7 Holmesdale Villas, Swallow Lane, Dorking RH5 4EY
2E0 NNE D. Hanwell 28 Chipperfield Road, Norwich NR7 9RR
2E0 NNF N. Ferenc 2A Rosedene Ave, London SW162LT
2E0 NNG R. Turner 38 Cotman Road, Clacton-on-Sea CO16 8YB
2E0 NNH G. Bansil 15 Abington Close, Crewe CW13TL
2E0 NNI M. Bailey 54 Springfields Road, Stoke-on-Trent ST4 6RZ
2E0 NNP J. Pinto 7 Wright'S Way, Colchester CO6 4NS
2E0 NNQ J. Blamey 46 First Avenue, Canvey Island SS8 9LP
2E0 NNX D. Austin 31 Bark Street, Bolton BL1 2AE
2E0 NNZ J. Doughty 18 Kirk Place, Chelmsford CM2 6TN
2E0 NOC C. Arbon 8 Orchard Avenue, Ashford TW15 2JB
2E0 NOD N. Lightfoot 4 Prospect Close, Hatfield Peverel, Chelmsford CM3 2JE
2E0 NOK C. Smith 8 Pitts Street, Bradford BD4 9JJ
2E0 NOM R. Paterson 9 Hulver Court, Ipswich IP3 9LW
2E0 NON G. Fielding Chapel Court, Chapel Lane, Cradley, Malvern WR13 5HX
2M0 NOP N. Price 5 Haltree Cottage, Heriot EH38 5YD
2E0 NOS W. Warren 52 Harcourt Road, Swindon SN2 1DR
2D0 NOT D. Ali 25, Sunnydale Avenue, Port Erin IM9 6EU Isle of Man
2E0 NOU L. Mccarthy 34 Shawley Way, Epsom KT18 5PB
2E0 NOW M. Clarke 40 Fingringhoe Road, Langenhoe CO5 5AD
2E0 NOZ J. Norrington 32 Fulfen Way, Saffron Walden CB11 4DW
2E0 NPE J. Perfect 62 Warwick Close Holmwood, Dorking RH5 4NL
2E0 NPF R. Mcmanus 125 Whitaker Road, Derby DE23 6AQ
2E0 NPL A. Lyman 12 Chicheley Street, Newport Pagnell MK16 9AR
2E0 NPN N. Swift 59 Milton Avenue, Malton YO17 7LB
2E0 NPP P. Hayward 14 Micklewright Avenue, Crewe CW1 4DF
2E0 NPR N. Prater 100 Pitfold Road, London SE12 9HY
2E0 NPS C. Kenyon The Farmhouse, 10, Watermill Lane, Spilsby PE23 5AG
2U0 NPT N. Thomas 6 Tunstall Terrace, Gibauderie, St. Peter Port GY1 1XJ Guernsey
2E0 NPX N. Paxman 128 Coggeshall Road, Braintree CM7 9ES
2E0 NQA R. Clark 12 Ash Drive Haughton, Stafford ST18 9EU
2W0 NQE S. Walmsley 29 Shelley Court Machen, Caerphilly CF83 8TT
2E0 NQF J. Rookyard 3 Drapers House St. Johns Road, Banbury OX16 5BE
2M0 NQT M. Simon 14 Struan Place, Inverkeithing KY11 1NF
2E0 NQU E. Wagner 3 Sarre Road, London NW2 3SN
2W0 NRA M. Chell Mesen Fach, Llanybydder SA40 9TY
2E0 NRB M. Beckett 4 Sandcross Close, Orrell, Wigan WN5 7AH
2W0 NRE N. Holding Tir Nanog Crosswell, Crymych SA41 3UF
2E0 NRH N. Hickson 27 Cressing Road, Witham CM8 2NP
2E0 NRJ N. Johnson Belair, Western Road, Crediton EX17 3NB
2E0 NRL E. Tattersall 17 Malt Kiln Way, Sandbach CW11 1JL

2E0 NRN L. Lewis 653 Main Road, Dovercourt, Harwich CO12 4NF
2E0 NRS N. Roberts 34 Copeland Drive, Standish, Wigan WN6 0XR
2E0 NRW N. Waters 9 Shirley Road, Droitwich WR9 8NR
2E0 NRX S. Cook Deganwy, Hardwick Road, King's Lynn PE30 5BB
2M0 NSA A. Custura 14 Deansloch Terrace, Aberdeen AB16 5SN
2E0 NSC N. Smith 40 Fairdale Drive, Newthorpe, Nottingham NG16 2FG
2E0 NSG N. Gregson 4 Pollard House, Maldwyn Avenue, Bolton BL3 3RB
2E0 NSQ S. Beedham 27 Malpas Close Bransholme, Hull HU7 4HH
2E0 NSR E. Parrish 89 Delamere Drive, Macclesfield SK10 2PS
2E0 NSS M. Price 25 School Crescent, Lydney GL15 5TA
2M0 NSW N. White 2 Appleby Cottages, Whithorn, Newton Stewart DG8 8DQ
2E0 NSY S. O'Riordan 46 Grange Road, London HA20LW
2E0 NTC G. Bull 9 Kilburn Place, Dudley DY2 8HP
2I0 NTH C. Mccormick Flat 4, Legacorry House, Main Street, Armagh BT61 9RW
2E0 NTJ N. Jones 27 Gatcombe Gardens, West End, , Southampton SO183NA
2E0 NTM C. Hoult 11 Queen Street, Alnwick NE66 1RD
2I0 NTP N. Prentice 26 Claranagh Road, Claranagh, Enniskillen BT94 3FJ
2E0 NTR A. Butcher 31 Wittonwood Road, Frinton-on-Sea CO13 9JZ
2W0 NTS N. Hewelt 17 Saint Marys Road, Llandudno LL30 2UB
2E0 NTV Rvd. N. Wood 12 Spring Close, Verwood BH31 6LB
2E0 NTW C. Northwood Apartment 50, 2 Munday Street, Manchester M4 7BB
2M0 NTY C. Mcclymont 115 Glenavon Road, Flat13/1, Glasgow G20 0HT
2E0 NUD N. Cull 8 Eaton Road, Norwich NR4 6PY
2E0 NUG M. Wells 23 Eastmead, Bognor Regis PO21 4QT
2E0 NUL J. Unwin 39 Whinfell Drive, Normanby, Middlesbrough TS6 0BG
2M0 NUO Dr C. Brown 4 Damselfly View, Edinburgh EH17 8XH
2E0 NUQ M. Dickenson 6 The Pavilions, Blandford Forum DT11 7GF
2E0 NVB N. Betts 12 Sandy Lane, Worksop S80 1SW
2E0 NVK L. Bolton 9 Nab Crescent, Meltham, Holmfirth HD9 5LT
2E0 NVP M. Weir 153 Tyndale Crescent, Birmingham B43 7HX
2E0 NVT N. Tennant 22 The Lizard, Wymondham NR18 9BH
2E0 NWA N. Wong Montefiore House, Wessex Lane, Southampton SO18 2NU
2E0 NWB J. Benbow 44 Copthorne Park, Shrewsbury SY3 8TJ
2E0 NWE O. Price 5 Redicliff Way Sw, Redcliff T0J Canada
2W0 NWJ Dr N. Jones 54 Glanrhyd, Coed Eva, Cwmbran NP44 6TY
2E0 NWL A. May 12 Millfield Close, Ashby-de-la-Zouch LE65 2JS
2E0 NWN N. Nelson 40 Delph Mount, Leeds LS6 2HS
2I0 NWO D. Adams 65 Rose Park, Limavady BT49 0BF
2E0 NWR W. Westlake 2 Chegwin Court, Newquay TR7 2DE
2E0 NWT N. Topping 7 Beckstone Close, Harrington, Workington CA14 5QR
2E0 NWY S. Newhouse 28 Hillmorton Lane, Lilbourne, Rugby CV23 0SS
2E0 NXB N. Beck 2 Killerton View, Silverton, Exeter EX5 4JZ
2E0 NXE P. Symon 8 Glebe Close, St. Columb TR9 6TA
2E0 NXL P. Mcgee 2 Fishers Close, Norwich NR5 0QH
2E0 NXM N. Masters Wyngrove Vicarage Road, Wrawby, Brigg DN208RR
2E0 NXP Y. Weng 20 Kendal Grove, Leeds LS3 1NS
2W0 NXV B. Williams The Grange Llanddewi, Llandrindod Wells LD1 6SF
2E0 NYC S. Vzor 40 Henlow Road, Birmingham B14 5DS
2E0 NYE N. Whittaker 1 Edendale, Hull HU7 4BX
2E0 NYF D. Lamble 4 Laburnum Road, Chorley PR6 7BG
2E0 NYG N. Cox 182 North Tenth Street, Milton Keynes MK9 3AY
2M0 NYK I. Nicolson 1 Gullane Place, Dundee DD2 3BF
2I0 NYL L. Elliott 19 Gosford Road, Collone, Armagh BT60 1LQ
2E0 NYM C. Matthews 18 Tennyson Gardens, Darlington DL1 5BJ
2E0 NYP J. Dyson 77 Grantham Road, Southport PR8 4LT
2E0 NYT J. Woodcock 31 Moorway Lane Littleover, Derby DE23 2FR
2E0 NZA G. Lund 1 Thrush Close, Gloucester GL4 4WZ
2E0 NZD M. Phillips 44 Hilderic Crescent, Dudley DY1 2ET
2E0 NZT T. Reseigh 10 Higher Croft Parc, The Lizard, Helston TR12 7RL
2E0 OAA C. King 8A Barton Road, Bedford MK42 0NA
2M0 OAB S. Milne 5 Moriston Court, Grangemouth FK3 0JJ
2E0 OAC G. Dray 2 Mulberry Drive, Malvern WR14 4AT
2W0 OAG A. Graham 2 Heol Undeb, Beddau, Pontypridd CF38 2LB
2E0 OAH K. Johnson 32 Redmire Close, Bransholme, Hull HU7 5AQ
2E0 OAI D. Saunders 17 Sandy Lane, Prestwich, Manchester M25 9RU
2E0 OAK J. Burnett Hobart House, 16 Church Lane, Reepham, Lincoln LN3 4DQ
2E0 OAM A. Mullinex 84 Danesbury Crescent, Birmingham B44 0QS
2E0 OAO A. Al-Shakarchi 17 Fairfax Place, London NW6 4EJ
2E0 OAP M. Deary 7 Newbold Avenue, Sunderland SR5 1LG
2E0 OAR D. Edwards 212, Eastern Avenue North Kingsthorpe, Northampton NN2 7AT
2E0 OAS C. Sole 2 Shepley Street, Manchester M359DY
2I0 OAZ N. Armstrong 1 Diamond Cottages, Ardmore Road, Crumlin BT29 4QU
2E0 OBB O. Boar 19 Blyford Road, Lowestoft NR32 4PZ
2E0 OBC Dr C. Bridges 23 Bramley Vale, Cranleigh GU6 7FY
2E0 OBD M. Hopkins 27 Girtford Crescent, Sandy SG19 1HR
2E0 OBI P. Sherratt 39 Vimy Road, Leighton Buzzard LU7 1FQ
2E0 OBK D. Cull 6 Compass Way, Bromsgrove B60 3GP
2E0 OBL M. Orbell 21 Reedings Road, Barrowby, Grantham NG32 1AU
2E0 OBM D. Barwick 2 Sutton Close, Bury St. Edmunds IP327EP
2E0 OBO R. Blackman 32 Kingfisher Road, Sprowston NR7 8GX
2E0 OBP P. Percival 2 St. Catherine Street, Ventnor PO38 1HG
2E0 OBS B. Heath 108 Cow Lane, Bramcote, Nottingham NG9 3BB
2E0 OBZ D. Thomas 51 Sandringham Avenue, Vicars Cross, Chester CH3 5JF
2E0 OCB O. Carpenter-Beale 3 Linkden Cottages, Lomas Lane, Sandhurst TN18 5PU
2M0 OCC N. Nuttall 27 Bearside Road, Stirling FK7 9BY
2W0 OCF N. Smith 23 Pennyroyal Close, St. Mellons, Cardiff CF3 0NB
2E0 OCG O. Giles Holly Cottage, Main Road, Crewe CW4 8LL
2E0 OCH C. Howard 1 Beale Road, Cheltenham GL51 0JN
2E0 OCL L. Hendry 109 Grove Avenue, New Costessey, Norwich NR5 0HZ
2E0 OCM I. Johnson 9 Brook Road, Pontesbury, Shrewsbury SY5 0QZ
2E0 OCN H. Peberdy 53 Far Lane, Normanton On Soar, Loughborough LE12 5HA
2E0 OCP D. Drynski Flat 22, Nicholas Court, Corney Reach Way, Chiswick, London W4 2TS
2E0 OCS R. Wells 31 Bracklesham Road, Hayling Island PO11 9SJ
2E0 OCT A. Lawrence 5 Heighton Close, Bexhill-on-Sea TN39 3UP
2E0 OCW W. Joyce 2 Palmers Cottage, Main Street, Oakham LE15 8DH
2E0 OCY P. Gibbons 2A Gipsy Castle Estate, Hay-On-Wye, Hereford HR3 5EG
2E0 ODB D. Bambrough 7 Barnwell View, Herrington Burn, Houghton le Spring DH4 7FB
2E0 ODF J. Lashley 33 Goodes Avenue, Syston, Leicester LE7 2JH
2E0 ODJ D. Jones 20 Nightingale Road, Guisborough TS14 8HA
2E0 ODL K. Bedford 29 Kent Road Brookenby, Binbrook, Market Rasen LN8 6EW
2E0 ODN A. Swallow 16 Quarry Lane, Chesterfield S40 3AS
2E0 ODO E. White 3 Davy Drive, Maltby, Rotherham S66 7EN
2M0 ODP D. Pounder Birchbank, Lindean, Galashiels TD1 3PA
2W0 ODS D. Robins 25 Clos Mancheldowne, Barry CF62 5AB
2E0 ODT B. Hendry 109 Grove Avenue, New Costessey, Norwich NR5 0HZ
2E0 ODV D. Baddeley The Old Cottage, Sandy Lane, Baldwins Gate, Newcastle ST5 5DP
2E0 ODZ D. Ziaja 7 Earle Street, York YO31 8LJ
2E0 OEE J. Hendry 109 Grove Avenue, New Costessey, Norwich NR5 0HZ
2E0 OEL I. Robinson 12 Osprey Road, Flitwick MK45 1RU
2E0 OEM J. Summers 5 The Meadow, Bosvigo Lane, Truro TR1 3NG
2E0 OES R. Barnes 1 Moira Close, Chaddesden, Derby DE21 4RL
2E0 OET B. Mead 8 Wordsworth Road, Kettering NN16 9LB
2E0 OEV M. Cuff 1 Ivy Close, St. Leonards, Ringwood BH24 2QZ
2E0 OEZ I. Beresford 16A Holbeck Hill, Scarborough YO11 2XD
2E0 OFF G. Kenyon 2 Langdale Terrace, Stalybridge SK151EX
2E0 OFM P. Joyce 2 Harold Road, Cuxton, Rochester ME2 1EE
2E0 OFT J. Saunders 29 Mace Street, Cradley Heath B64 6HP
2W0 OFY B. O'Conor Drovers Tumble Ysceifiog, Holywell CH8 8NJ
2E0 OGA C. Ingram 17 Stackfield, Harlow CM202LA
2E0 OGB R. Evans 364 Aldermans Green Road, Coventry CV2 1NN
2E0 OGR M. Coombes 33 Woodside Road North Baddesley, Southampton SO52 9NB
2W0 OGT T. Cording 17 Maes Y Derwen, Llanrhaeadr Ym Mochnant, Oswestry SY10 0LE
2W0 OGY C. Hodgetts 16 Myrtle Drive, Rogerstone, Newport NP10 9EA
2E0 OGZ G. Wilkinson 50 Sherburn Road, Durham DH1 2JR
2E0 OHC M. Hill Buzon 169, Avd De La Condomina 53, Local 3, , Alicante 3540 Spain
2I0 OHE E. Paterson 1 Sycamore Grove, Belfast BT4 2RB
2E0 OHF K. Charlton 14 Rubens Close, Aylesbury HP19 8SW
2M0 OHL M. Cameron 27 Alwyn Avenue, Houston, Johnstone PA6 7LH
2E0 OHN N. Holden Plum Cottage, Avon Dassett, Southam CV47 2AP
2E0 OIK N. Titmus 68 Hobart Road, Cambridge CB1 3PT
2E0 OIL M. O'Connor 28 Cardigan Road, Southport PR8 4SF
2E0 OIN H. List 41 Westbury Crescent, Dover CT17 9QQ
2E0 OIR A. Birkett 21 Cedar Drive Wyke, Bradford BD12 9HL
2E0 OJB J. Bayliss 39 Elms Avenue, Littleover, Derby DE23 6FB
2E0 OJD K. Winton 130 George V Avenue, Worthing BN11 5RX
2E0 OJE A. Forbes Flat 10, Denby House, Paignton TQ4 6ES
2E0 OJH W. Waters 5 Pembroke Road, Newquay TR7 3HW
2E0 OJI J. Isaacs 20 Vine Street, Worcester WR3 7DY
2I0 OJK J. Kavanagh 11 Pinewood Crescent, Claudy BT47 4AB
2E0 OJN P. Field 63 Hartford Road, Davenham, Northwich CW9 8JE
2E0 OJS O. Squire 91 Victoria Road, London N22 7XG
2E0 OKB K. Barnett 22 Highclere Road, Southampton SO16 7AW
2E0 OKC A. Londors 112 Kingston Hill Avenue, Romford RM6 5QL
2E0 OKG K. George Wylye, Auberrow Wellington, Hereford HR4 8AN
2E0 OKH O. Hopkins Apartment 17, White Croft Works, Sheffield S3 7AH
2E0 OKK I. Day 137 Tuffley Lane, Tuffley, Gloucester GL4 0NZ
2E0 OKL A. Gates 46 Gloucester Place, Littlehampton BN17 7AL
2M0 OKO K. Biegun 81 Highfield, New Pitsligo AB43 6PZ
2E0 OKP S. Walsh 12 The Lawns Broadley Avenue Anlaby, Hull HU10 7HD
2E0 OKQ M. Broum 137 Culvers Avenue, Carshalton SM5 2BA
2E0 OKS M. Biadon 57 Fern Hill Road, Oxford OX4 2JW
2E0 OKY R. Eglington 33 Bradley Lane, Bilston WV14 8EW
2E0 OKZ A. Cammish 6 West Vale, Filey YO14 9AY
2E0 OLE O. Rofix 17 Potters Crescent, Great Moulton NR15 2HL
2E0 OLF M. Mills 17 Hornby Street, Plymouth PL2 1JD
2E0 OLG D. Rugen 19 Jacksons Close, Haskayne, Ormskirk L39 7LD
2M0 OLK O. Keast 6 Prospecthill Place, Greenock PA15 4DW
2E0 OLM G. Dewey West Riding Five Ash Down, Uckfield TN22 3AP
2W0 OLT O. Thomas Garth Celyn, St. Davids Road, Aberystwyth SY23 1EU
2I0 OMA J. Martin 23 Winters Gardens, Omagh BT79 0DZ
2E0 OME R. Hauk 6 Hall Road, Stowmarket IP14 1TN
2E0 OMG M. Robinson 10 Bramley Gardens, Poulton-le-Fylde FY6 7RD
2E0 OMI H. Kassier 26 Higher Port View, Saltash PL12 4BX
2D0 OMN R. Corrin 11 Cronk Y Berry, Douglas, Isle of Man IM2 6EY
2E0 OMR B. Withers 29 Yeoman Way, Trowbridge BA14 0QL
2M0 OMS M. Scullion 24 Langmuir Road, Kirkintilloch, Glasgow G66 2QE
2E0 OMT T. Baggley 16 Seaton Road, Seaton, Workington CA14 1DT
2E0 OMV J. Barton 37 Lytton Road, Sheffield S5 8AX
2E0 ONC G. Stephens 8 New Molinnis, Bugle, Saint Austell PL26 8QL
2E0 ONE P. Greenwood 1 The Garth, Whitby YO21 3PD
2E0 ONH A. Brown 51 Towncroft, Chelmsford CM1 4JX
2E0 ONI P. Cannings 30 Graham Gardens, Luton LU3 1NQ
2E0 ONM M. Kulakowski Flat 4 Clement Mellish House East Stockwell St., Colchester CO1 1GJ
2M0 ONS D. Anderson Dail Darach, Monydrain Road, Lochgilphead PA31 8LG
2E0 ONV J. Bonar Flat, 5A Friday Street, Minehead TA24 5UB
2M0 ONW K. Harper 3 Mosside Place, Kilmarnock KA3 2BA
2W0 ONY M. Knight 28 Ridgeway Hill, Newport NP20 5DG
2E0 OOC B. Cooper 71 High Street, Birstall, Batley WF17 9RG
2E0 OOH D. Meakin 27 Spencer Road, Long Buckby, Northampton NN6 7YP
2E0 OOI T. Hawksley Field Cottage, Stowey, Bishop Sutton, Bristol BS39 5TH
2E0 OOM M. Buist 23 St. Chads Drive, Gravesend DA12 4EL
2E0 OON C. May 16 Trelawn Road, London E10 5QD
2E0 OOO R. Clayton 9 Green Island, Irton, Scarborough YO12 4RN
2M0 OOR S. Boal 20 Fairinsfell, Broxburn EH526AL
2M0 OOT J. Ferrans 77 Knockinlaw Road, Kilmarnock KA3 2AS
2E0 OOU G. Cameron 33 Railway Terrace, York YO24 4BN
2E0 OOV S. Phillips 38 Corbet Ride, Leighton Buzzard LU7 2SJ
2E0 OOW P. Turner 8 The Retreat, Ramsgate CT126ET
2E0 OOX A. Tunley 5 Camborne Close, Lower Earley, Reading RG6 4EN
2E0 OOY P. Sullivan 25 Cavendish Drive, Birkenhead CH42 6RG
2E0 OPB O. Blackburn 128 High St., Crigglestone, Wakefield WF4 3EF
2W0 OPC O. Campbell 16 Whitcliffe Drive, Penarth CF64 5RY

2E0 OPM D. Whitehouse 6 Larch Close, Heathfield TN21 8YW
2E0 OPO O. Silva Flat 71, Long Acre House, Pettacre Close, London SE28 0PB
2E0 OPS D. Lapham-Crozier 109 Aylesbury Crescent, Plymouth PL5 4HX
2E0 OPT J. Nihill 14 Hereford Avenue, Clayton, Newcastle ST5 3ED
2E0 OPU K. Morris 80 Bridge Street, Chatteris PE16 6RN
2E0 OQD L. Sproule Dragonpits North Perrott, Crewkerne TA18 7TH
2E0 OQH D. Cooper 52 Meadow Lane, Birkenhead CH42 3YE
2M0 OQR P. Taylor 2 Laurel Grove, Aberdeen AB22 8YJ
2E0 OQV N. Connor 28 Church Street, Hungerford RG170JE
2E0 OQZ P. Hall 13 Sheard Avenue, Ashton-under-Lyne OL6 8DS
2E0 ORD R. Drage 4 Bruce'S Close Conington, Peterborough PE73QW
2E0 ORE C. Dodds 22 Cambridge Street, Wolverton, Milton Keynes MK12 5AJ
2W0 ORH O. Hopkin The Forge, Rock Road, , St. Athan CF62 4PG
2E0 ORI D. Dart 164A Salterton Road, Exmouth EX8 2PA
2M0 ORK M. Herridge The Hollies, Petticoat Lane, Orkney KW17 2RP
2E0 ORM T. Rowlands 7 Swan Delph, Aughton, Ormskirk L39 5QG
2E0 ORP M. Orpen Daymer, Tey Road, Colchester CO6 3RY
2E0 ORS P. Lewis 9 The Hill, Glapwell, Chesterfield S44 5LX
2W0 ORT B. Roberts 6 Trem Y Moelwyn, Tanygrisiau, Blaenau Ffestiniog LL41 3SS
2E0 ORX T. Hedger 1 Berry Terrace, Acton Square, Sudbury CO10 1HT
2M0 OSC J. Hanley 44 Waverley Crescent, Livingston EH54 8JN
2E0 OSE P. Pope 2 Hale Villas, Honiton EX14 9TQ
2W0 OSG J. Bellis 32 Broughton Road Lodge, Wrexham LL11 5NG
2W0 OSH S. Owen 8 Old Tanymanod Terrace, Blaenau Ffestiniog LL41 4BU
2M0 OSK A. Twort 17 Balallan, Isle of Lewis HS2 9PN
2E0 OSS S. Fraser Old Post Office, Mill Road Barton St. David, Somerton TA11 6DF
2E0 OST F. Limbert Knowlecroft Little Ribston, Wetherby LS22 4ET
2E0 OSX A. Logan 23 Cherry Tree Rise, Walkern, Stevenage SG2 7JL
2E0 OSY D. Coppenhall 55 Vicarage Lane, Elworth, Sandbach CW11 3BU
2E0 OTA S. Jenkinson 17A Britannia Road Burbage, Hinckley LE10 2HE
2E0 OTB P. Hateley 44 Painters Croft, Coseley, Bilston WV14 8AP
2I0 OTC B. Emerson1 67 Castlemore Avenue, Belfast BT6 9RH
2E0 OTG S. Ball 18 Lighthurst Avenue, Chorley PR7 3HY
2E0 OTI D. Scotcher 17 St. Dominics Square, Luton LU4 0UN
2E0 OTM O. Morris 1 Crawford Avenue, Peterlee SR8 5EG
2E0 OTP M. Anostalgia 136 Avenue Road Extension, Leicester LE2 3EH
2E0 OTR T. Rood The Beeches, Frodingham Road, Brandesburton, Driffield YO25 8QY
2E0 OTT J. Slobin 45 Dale Edge, Eastfield, Scarborough YO11 3EP
2I0 OTW P. Fallon 18 Church View, Killough, Downpatrick BT30 7RJ
2E0 OTY P. Lane 5 Swan Court, Middle Watch, Swavesey, Cambridge CB24 4AG
2E0 OTZ M. Wright 8 St. Wilfrids Road, Oundle, Peterborough PE8 4NX
2E0 OUD A. Marvel Dean House Farm, Nordan, Leominster HR6 0AW
2E0 OUH P. Loxton 32 Parkhill Crescent, Wakefield WF1 4EZ
2I0 OUI M. Doogan 54 Birchdale Manor, Lurgan, Craigavon BT66 7SY
2E0 OUK I. Westbrook 21 York Road, Cannock WS11 8ES
2E0 OUR D. Mills 91 Harp Road Hanwell, London W7 1JQ
2M0 OUU C. Mcconnochie 72 Duddingston Avenue, Kilwinning KA13 6RS
2E0 OVB R. Gowers 43 Tungstone Way, Market Harborough LE16 9GA
2M0 OVD D. Adamson 5 Central Quadrant, Ardrossan KA22 7DY
2E0 OVF S. Hedgecock 37 Tennyson Road, Maldon CM9 6BE
2E0 OVI O. Popa 5 Lanark Close, Horsham RH13 5RY
2E0 OVL J. Hirst 57 Newgate Street, Doddington PE15 0SR
2E0 OVR F. Farrer 16 High Street, Eagle, Lincoln LN6 9DH
2W0 OVT J. Jones 40 Ffordd Coed Marion, Caernarfon LL55 2EF
2M0 OVV S. Monaghan 13 Ballyhennan Crescent, Tarbet, Arrochar G83 7DB
2E0 OVW G. Sandell 1 St. Margaret Road, Ludlow SY8 1XN
2E0 OVX M. Walton 358 Old Heath Road, Colchester CO2 8DD
2E0 OWC S. Iles Bigbury Bay Holiday Park, Challaborough, Kingsbridge TQ7 4HS
2E0 OWE D. Fincham 2 Glebe Close, St. Columb TR9 6TA
2E0 OWG D. Burgfeld 20 Wilson Row, Crowthorne RG45 6WE
2E0 OWH T. Harrison 58 Ascot Drive, Cannock WS11 1PE
2E0 OWL S. Hurley 11 Beresford Avenue, Wirral CH63 7LR
2M0 OWT Z. Mckinnon 8 Rowanlea Avenue, Paisley PA2 0RP
2E0 OXF A. Comerford 21 New Cross Road, Headington, Oxford OX3 8LP
2E0 OXO D. Harden 59 Violet Avenue, Edlington, Doncaster DN12 1NW
2M0 OXQ S. Mckinnon 8 Rowanlea Avenue, Paisley PA2 0RP
2E0 OXT R. Ingram 17 Stackfield, Harlow CM20 2LA
2M0 OXX A. Berry 8 Hill Street Striling Fk7 0Dh, Striling FK7 0DH
2E0 OYG B. Nuttall 22 Countess Crescent, Bispham, Blackpool FY2 9LQ
2W0 OYL A. Owen Bro Dawel, Brynrefail, Caernarfon LL55 3NR
2E0 OYN R. Watson 60 Beresford Avenue, Surbiton KT5 9LJ
2E0 OYQ A. Gibbons 305 Monks Road, Lincoln LN2 5LB
2E0 OYR S. Roberts 77 Lambwath Road, Hull HU8 0HB
2M0 OYS R. Thomson Meadow Steading Tornaveen, Torphins, Banchory AB31 4PJ
2E0 OYY N. Weiner 3 Bluebell Court Lower Mardyke Avenue, Rainham RM13 8GF
2E0 OZE M. Crockford Centre Cottage Kelk, Kelk YO258HL
2E0 OZG K. Randle 36 All Saints Road, Sittingbourne ME10 3PB
2E0 OZI S. Carpenter 52 Mewstone Avenue, Wembury, Plymouth PL9 0JZ
2E0 OZK M. Kaszewski 87 Blenheim Road, Northolt UB5 4TS
2E0 OZM M. Collantine Little Mill Farm, Little Mill, Egremont CA22 2NN
2W0 OZO S. Hayward 22 Dewsland Street, Milford Haven SA73 2AU
2E0 OZQ A. Sherer 28 Baroness Road, Grimsby DN34 4DP
2E0 OZW K. Ozwell 109 Abbey Road, Grimsby DN32 0HN
2E0 OZX C. Russell 9 Sandfields Road, St. Neots PE19 1PF
2W0 OZY C. Osborne Thornleigh, Tremont Road, Llandrindod Wells LD1 5BH
2E0 PAA C. Jago 20 Glanville Road, Tavistock PL19 0EA
2E0 PAB P. Stone 60 Acorn Avenue, Braintree CM7 2LR
2I0 PAC P. Dallas 12 Glendun Crescent, Coleraine BT52 1UJ
2E0 PAD J. Stainton 40-41 Dyke End Golcar, Huddersfield HD7 4LA
2E0 PAE P. Illidge 55 East Park Road, Spofforth, Harrogate HG3 1BH
2E0 PAF P. Taylor 47 Pickhurst Park, Bromley BR2 0TN
2E0 PAH P. Hunt 33 Drakes Close, Bridgwater TA6 3TD
2E0 PAJ P. Kiernan 4 Bradley Street, Southport PR9 9HW
2E0 PAK P. Watson 10 Whitelands Crescent, Baildon, Shipley BD17 6NN
2W0 PAN A. Paffey 1 St. Vincent Road, Newport NP19 0AN
2E0 PAO D. Pike 46 Haymans Close, Cullompton EX15 1EH
2E0 PAP P. Woodyard 65 Raglan Street, Lowestoft NR32 2JS
2E0 PAS A. Spinks 10 Foxley Close, Norwich NR58DQ
2E0 PAT G. Dobson 4 Durley Gardens, Orpington BR6 9LL
2E0 PAU A1 CG c/o P. Dossett 92 Dale Valley Road, Southampton SO16 6QU
2E0 PAV S. Richards 18 Lowfields Staxton, Scarborough YO12 4SR
2E0 PAX E. Goodwin 55 Twickenham Road, Sunderland SR3 4JN
2E0 PBB W. Ramsell 8 Didcot Drive, Marchington, Uttoxeter ST14 8LT
2M0 PBC C. Montague 74 Holmbyre Road, Glasgow G45 9QD
2E0 PBE P. Edwards 791 Windmill Lane Denton, Manchester M34 2ER
2E0 PBH A. Billings 46 Thorley Drive, Cheadle, Stoke-on-Trent ST10 1SA
2E0 PBJ P. Mansfield 15 Earl Avenue, New Waltham, Grimsby DN36 4NE
2E0 PBL P. Ball 101 Chelwood Drive, Bath BA2 2PS
2I0 PBM P. Bingham 45 Gowanvale Drive, Banbridge BT323GD
2E0 PBN P. Dent 25 Clyde Avenue, Hebburn NE31 2JN
2E0 PBO P. Kay 30 Broadway, Grange Park, St. Helens WA10 3RX
2E0 PBP P. Jones 50 Clay Lane, Doncaster DN2 4RJ
2E0 PBR P. Alley 21 Orchard Court, Thorney PE6 0QW
2E0 PBS P. Sycamore 1 Clarence Road, Littlesea, , Weymouth DT4 9EE
2E0 PBT P. Burgess 61 Grosvenor Avenue, Torquay TQ2 7JX
2W0 PBU Dr W. Dickson The Rowans, Pwllmeyric, Chepstow NP16 6LA
2E0 PBV M. James 42 Doone Way, Ilfracombe EX34 8HS
2E0 PBW I. Yovchev 11 Beverley Drive, Edgware HA8 5NQ
2E0 PBY P. Barker 27 Sandwood Road, Sandwich CT13 0AQ
2I0 PBZ P. Bell 3 Alexandra Crescent, Larne BT40 1NE
2E0 PCC P. Garraway The Poplars, Crowell Road, Chinnor OX39 4HP
2W0 PCD P. Day 15-16 Troedrhiw-Trwyn, Pontypridd CF37 2SE
2W0 PCE P. Iles 150 Pen-Y-Bryn, Caerphilly CF83 2LA
2E0 PCF P. Faulkner 32 Manvers Road, Beighton, Sheffield S20 1AY
2E0 PCG P. Green 8 Grassthorpe Road, Sheffield S12 2JH
2E0 PCI P. Collins 16 Fern Grove, Haverhill CB9 9ND
2E0 PCL G. Owen 38 Trentham Drive, Bridlington YO16 6ES
2W0 PCN P. Nash 110 Aberporth Road, Cardiff CF14 2RY
2E0 PCO P. Coombes 2 Bissoe Cottages, Bissoe, Truro TR4 8SU
2E0 PCQ P. Morris 14 Marina Road, Darlington DL3 0AL
2E0 PCR C. Vincent 81 Trethannas Gardens, Praze, Camborne TR14 0LL
2W0 PCT S. Gau Disgwylfa, The Downs, Cardiff CF5 6SB
2E0 PCU P. Roberts 1 Ballard Road, Wirral CH48 9XU
2E0 PCV D. Whiting 133 Belfield Road, Accrington BB5 2JD
2M0 PCW Prof. S. Skerratt 51 Vogrie Road, Gorebridge EH23 4HL
2E0 PCX P. James 44 Narbonne Avenue Ellesmere Park Eccles, Manchester M30 9DL
2E0 PCZ P. Colyer 23 Florida Road, Torquay TQ1 1JY
2E0 PDG E. Aitken 20 Plover Drive, Bury BL9 6JH
2E0 PDH C. Macleod 21 Halsetown, St. Ives TR26 3LY
2E0 PDJ J. Sandon 461 Archer Road, Pin Green, Stevenage SG1 5QP
2E0 PDL M. Garry 34 Conway Road, Paignton TQ4 5LH
2E0 PDM Prof. P. March 46 Christchurch Road, Tilbury RM18 8XP
2E0 PDO A. Dingwall 48 Village Farm Caravan Site, Bilton Lane, Harrogate HG1 4DL
2E0 PDP J. Clarkson 56 Edward Bailey Close, Binley, Coventry CV3 2LZ
2E0 PDQ D. Carpenter 34B Carey Park, Killigarth, Looe PL13 2JP
2E0 PDU L. Fuller Rosemar Lodge Westford, Wellington TA21 0DX
2E0 PDX G. Swindells 15 Benedict Close, Salford M7 2GB
2E0 PDZ P. Harper 36 Barrow Close, Marlborough SN8 2BD
2E0 PEC B. Clayton 1 Maude Crescent Sowerby Bridge, Halifax HX6 1LB
2W0 PEE N. Tanner 3 Maes Y Tyra, Resolven, Neath SA11 4NN
2E0 PEF P. Freeman 57 Ruffa Lane, Pickering YO18 7HN
2W0 PEG J. Reason 158 Caerau Lane, Cardiff CF5 5JS
2W0 PEH B. Sellers 86 St. John Street, Ogmore Vale CF32 7BB
2E0 PEI I. Daraban 10 Holly Blue Close, Little Paxton, Saint Neots PE19 6TD
2E0 PEL S. Peel 21 Fairfield Avenue, Ormesby, Middlesbrough TS7 9BB
2E0 PEM P. Metters 137 Nevinson Avenue, South Shields NE34 8NE
2E0 PEP S. Hill 35 Longs Way, Wokingham RG40 1QW
2E0 PET P. Munson 8 Longley Lane, Spondon, Derby DE21 7AT
2E0 PEW P. Woolley 84 Bowthorpe Road, Norwich NR2 3TP
2E0 PEX D. Munn 36 Moor Lea, Braunton EX33 2PE
2D0 PEY G. Wilby Byways, Glenlough Circle, Glen Vine, Isle of Man IM4 4AX
2E0 PFA P. Fernie 39 North Parade, Falmouth TR112TE
2E0 PFB P. Browne 151 North Road, St. Andrews, Bristol BS6 5AH
2W0 PFD P. Devlin Brynteg, Fron Bache, Llangollen LL20 7BP
2E0 PFF S. Debenham 14 Aldergrove Close, Lincoln LN6 0SL
2E0 PFG P. Goddard 62 Woodlands Drive, Thetford IP24 1JJ
2E0 PFH P. Holmes 18 Raleigh Avenue, Whiston, Prescot L35 3PL
2E0 PFL P. Leng The Barn, Gildersleets, Settle BD24 0AH
2E0 PFO P. Noble 14 Park Street, Swallownest, Sheffield S26 4UP
2E0 PFR P. Ratcliffe 2 Newlands Avenue, Whitby YO21 3DX
2E0 PFT A. Perfect 3 Chelmarsh Close, Chellaston, Derby DE73 6PB
2E0 PFY R. Ball 7 Cliff Closes Road, Scunthorpe DN15 7HT
2E0 PGC P. Challans Flat 4, Sandringham Court, 2 Chandos Square, Broadstairs CT10 1QN
2E0 PGH P. Hill 14 Drovers Way, Woodlands, Ivybridge PL21 9XA
2E0 PGI G. Hartless 32 Long Acre, Mablethorpe LN12 1JF
2E0 PGJ P. Johnston 35 Staunton Ave, Hayling Island PO11 0EW
2E0 PGL P. Lewis 154 Meadow Head, Sheffield S8 7UF
2E0 PGM P. Mcfadden Maple Cottage, Leighton Buzzard LU7 9DZ
2M0 PGO P. Greig 22 Rowanhill Close, Port Seton, Prestonpans EH32 0SY
2E0 PGP W. Dover Windcrest, Fox Lane, Basingstoke RG23 7BB
2E0 PGR P. Grainger 36 Orchard Road, Wigton CA7 9JL
2E0 PGS P. Stevenson 6 Dighton Gate, Stoke Gifford, Bristol BS34 8XA
2E0 PGT P. Thompson 25 Pitclose Road, Birmingham B31 3HU
2I0 PHA S. Wright 3 Harryville, Portstewart BT55 7AU
2E0 PHB P. Haslam-Brunt 488 Klightwood Road, Lightwood, Stoke on Trent ST3 7EW
2E0 PHE D. Thomas 8 Cedar Avenue, Weston-Super-Mare BS22 8HL
2E0 PHK P. Bentley The Vauce Farm, Langley-On-Tyne, Hexham NE47 5NA
2E0 PHL P. Probst 37 Devonshire Street, Skipton BD23 2ET
2E0 PHM P. Meerman 24 Horseshoe Crescent, Burghfield Common, Reading RG7 3XW
2M0 PHO C. Hunter 1 North Gate Lodge, Erines, Tarbert PA29 6YL
2W0 PHP C. Maggs 15 Stuart Street, Treorchy CF42 6SN
2E0 PHS P. Sladen 25 Linden Grove, Beeston, Nottingham NG9 2AD
2E0 PHU P. Uttley 55 Dunce Park Close, Elland HX5 0PF

2E0 PHX P. Hunter 160 Pembroke Road, Northampton NN5 7ER
2E0 PIA M. Hill 56 Moorhouse Avenue, Wakefield WF2 9QG
2M0 PID T. Hamilton 57/6 North Street, Bo'ness EH51 0AE
2E0 PIH D. Kirk 26 Ingham Grove, Cramlington NE23 3LH
2M0 PIJ G. Wilson Flat 7 29 Second Avenue, Clydebank G813AB
2E0 PIK B. Pike 19 Cardigan Gardens, Reading RG1 5QP
2E0 PIO T. Nakagawa 7 Milton Street, Barrowford, Nelson BB9 6HE
2E0 PIP P. Marsh 16 Laurel Close, North Warnborough, Hook RG29 1BH
2E0 PIT C. Fox Millstone Cottage, Prior Wath Road, Scarborough YO13 0AZ
2E0 PIU P. Lounton 107 Browning Hill Coxhoe, Durham DH6 4SA
2E0 PIW R. Metcalfe 33 Midland Terrace, Hellifield, Skipton BD23 4HJ
2E0 PIX B. Wolff 7 Church Terrace, Reading RG1 6AS
2E0 PJC P. Carne 1 Curlew Close, Letchworth Garden City SG6 4TG
2E0 PJD P. Dawson 88 Urmson Road, Wallasey CH45 7LQ
2E0 PJE P. Elmore 8 Gray Street, Elsecar, Barnsley S74 8JR
2E0 PJH P. Holmes 1 Leonards Place, Bingley. BD16 1AD
2W0 PJJ P. Jones 23 Pinecroft Avenue, Aberdare CF44 0HY
2W0 PJM P. Mclaren 10 Haulfryn, Ruthin LL15 1HB
2E0 PJN P. Northover 181 Mullway, Letchworth Garden City SG6 4BD
2E0 PJR P. Radford 43 Bells Lane, Nottingham NG8 6EX
2E0 PJS P. Spilman 28 Staines Way, Louth LN11 0DF
2E0 PJT P. Tomlinson 11 Haynes Close, Clifton, Nottingham NG11 8JN
2E0 PJY P. Jay 6 Sportsmans Way, Princes Risborough HP27 9FZ
2M0 PKA P. Akula 19 Oldmill Crescent, Balmedie, Aberdeen AB23 8WB
2E0 PKB P. Beier 20 Markham Avenue, Armthorpe, Doncaster DN3 2AZ
2E0 PKF A. Goose 12 Brown Street, Rainham, Gillingham ME8 7JN
2E0 PKK P. Knight 73 Bramley Crescent, Southampton SO19 9LJ
2E0 PKL D. Cadet 2 Paddockside, Middleton, Ludlow SY8 3EB
2E0 PKM M. Brookes 73 Radwinter Road, Saffron Walden CB11 3HU
2E0 PKR S. Parker 57 Queen Street, Horncastle LN9 6BH
2E0 PKU D. Winstanley 43 Florence Street, St. Helens WA9 5NA
2E0 PLA P. Brown 11 Booth Crescent, Rossendale BB4 9BT
2E0 PLB P. Bromley Broadwood Treovis Upton Cross, Liskeard PL14 5BQ
2E0 PLC Dr T. Kyriacou 54 Sutton Avenue, Silverdale, Newcastle-under-Lyme ST5 6TB
2E0 PLE G. Dennis 21 Rydal Crescent, Scarborough YO12 4JJ
2E0 PLH P. Holmes 28 Ackworth Drive, Manchester M23 1LD
2E0 PLJ N. Bull Eldoret, Castle Street Bampton, Tiverton EX16 9NS
2E0 PLK A. Jedryka 71 West Royd Drive, Shipley BD18 1HL
2E0 PLR P. Tolcher 15 Langstone Close, Torquay TQ1 3TX
2E0 PLS J. Walczak 18 Heathfield, Chippenham SN15 1BQ
2E0 PLU T. Palmer Edison House, Bow Street, Great Ellingham NR17 1JB
2E0 PLV P. Le Vallois 14 London Row, Arlesey SG15 6RX
2E0 PLX P. Levy 43 Conroy Drive, Dawley, Telford TF4 2RW
2E0 PLY D. Hensman 130 Clittaford Road, Plymouth PL6 6DW
2E0 PMA N. Perkins 39 Ladychapel Road, Abbeymead, Gloucester GL4 5FQ
2E0 PMB P. Browne Ham Cottage, Hammingden Lane, Haywards Heath RH17 6SR
2E0 PMC C. Bowman 26 Albany Hill, Tunbridge Wells TN2 3RX
2E0 PMD P. Dowling 22 Chelkar Way, York YO30 5ZH
2E0 PME P. Martin 5 Shropshire Drive, Wilpshire, Blackburn. BB1 9NF
2E0 PMI P. Mccormick Fieldview, Crown East Lane, Lower Broadheath, Worcester WR2 6RH
2E0 PML C. Suddell Lynhurst, Littleworth Lane Partridge Green, Horsham RH13 8JX
2E0 PMM P. Mather 11 Odette Court, Bingley BD16 3QN
2E0 PMO P. Brindley 70 Atherfield Road, Reigate RH2 7PS
2E0 PMP P. Punjabi 62 Cleveland Road, London W13 8AJ
2M0 PMR A. Graham 27 Crichton Road, Pathhead EH37 5RA
2E0 PMU P. Muston 59 The Pastures Narborough, Leicester LE19 3DY
2E0 PMV P. Mansfield 27 Popplechurch Drive, Swindon SN3 5DE
2E0 PMX A. Marlow Yeomans Barn, Kingsbridge TQ7 3BH
2W0 PMZ P. Eckersley 14 Bro Nantfer, Gwaun Cae Gurwen, Ammanford SA18 1BR
2E0 PNA N. Hine 13 Wilton Crescent, Alderley Edge SK9 7RE
2E0 PNB P. Bozikis 336 Higham Hill Road, London E17 5RG
2E0 PNC G. Billington 47 Smithy Leisure Park, Cabus Nook Lane, Preston PR3 1AA
2I0 PND P. Doherty 69 Fivemile Straight, Draperstown BT457HT
2E0 PNG P. Green 34 Drydens Close, Titchmarsh, Kettering NN14 3DD
2E0 PNK C. Taylor 212 Plantation Hill, Worksop S81 0HD
2E0 PNN P. Bowen 12 Powell Place, Newport TF10 7BS
2E0 PNP R. Bartha 6 Chappell Close, Aylesbury HP19 9QA
2E0 PNR A. Ault 124 High Street, Aylesbury HP20 1RB
2E0 PNW P. Winkley 25 Grindley Avenue, Manchester M21 7NE
2E0 PNZ P. Norman Four Acres Bungalow Farm, Winwick Gated Road West Haddon, Northampton NN6 7BH
2E0 POB A. Carden Hazelgrove, South Allington, Kingsbridge TQ7 2NB
2I0 POD A. Hunter 38 Robinson Road, Bangor BT19 6NJ
2E0 POE Lord J. Williams 7 Southrop Road Kingsway, Quedgeley, Gloucester GL2 2HN
2E0 POQ R. Finch 19B Kiln Road, Newbury RG14 2LS
2E0 POU P. Daubaris 32 Chalcombe Road, Abbey Wood, London SE2 9QS
2E0 POZ P. Stilwell 3 Ridgeway Cottages, Foxhill, Swindon SN4 0DU
2E0 PPA P. Ridley 218 Lichfield Road, Rushall, Walsall WS4 1SA
2E0 PPB P. Perkins 52 Ashley Piece, Ramsbury, , Marlborough SN8 2QE
2E0 PPD P. Bengey 3 Millmead Road, Bath BA2 3JW
2E0 PPF S. Harding Acres Hill, Jacobs Lane, Alhampton BA4 6PZ
2E0 PPH P. Huband Flat 26, Tintern House, Selcroft Avenue, Birmingham B32 2BS
2E0 PPJ P. Joannou 5 Crowhurst Mead, Godstone RH9 8BF
2E0 PPK N. Green 11 Wythburn Way, Rugby CV21 1PZ
2W0 PPL A. Dighton 84 Trefelin, Aberdare CF44 8LF
2E0 PPM I. Barnes 35 Copley Road, Stanmore HA7 4PF
2E0 PPO J. Grint 9 Mountbatten Drive, Leverington PE13 5AF
2E0 PPR P. Rickwood 8 Bealeys Avenue, Wolverhampton WV11 1EG
2I0 PPW J. Macfarlane 5 Sunbeam Terrace, Lisnaskea, Lisnaskea BT92 0LL
2E0 PPY S. Evans 51 St. Georges Road, Dudley DY2 8EY
2E0 PPZ L. Westwood 28 Ash Crescent, Kingswinford DY6 8DJ
2E0 PQA P. Casado Arias 24 Oldridge Road, London SW12 8PJ
2E0 PQR P. Nathan 7 Pitt Drive, Seaford BN25 3JB
2W0 PQU E. Maher 25 Ffordd Cadfan, Bridgend CF31 2DP
2E0 PRA P. Randerson 7 Roman Crescent, Swindon SN1 4HH
2E0 PRC P. Craig 4 Poolside, Burston, Stafford ST18 0DR
2E0 PRD P. Denham Arlyn House, 10 Prince Alfred Avenue, Skegness PE25 2UH
2E0 PRG P. Garland Flat 4 Amante Court 190 Southwood Road, Hayling Island PO119QL
2I0 PRL P. Reid 1 Nettlehill Mews, Lisburn BT28 3HN
2I0 PRM E. Simpson 10 Woodview Park, Tandragee, Craigavon BT62 2DD
2E0 PRO P. Robinson 16 Bartlett Close, Liverpool L31 8BZ
2E0 PRS P. Shaw 32 Hardwick Road East, Worksop S80 2NT
2E0 PRV P. Bleasdale 12 Malvern Ave, Padiham BB127DT
2M0 PSA P. Smith 13 Newmills Grove, Balerno EH14 5SY
2E0 PSC P. Croxford 1 Meteor Close, Bicester OX26 4YA
2E0 PSD J. Eames 6 The Oaklands, Cold Meece, Stone ST15 0QH
2E0 PSH S. Storey 10 Amble Way, Trimdon Station TS29 6DZ
2E0 PSK S. Pierce 117 Victoria Avenue, Hastings TN35 5BS
2E0 PSM P. Smart 142 Finch Road, Chipping Sodbury, Bristol BS37 6JB
2E0 PSN G. James 28 Redcar Road, Romford RM3 9PT
2E0 PSO P. Sheffield 13 St. Winifred Road, Wallasey CH45 5EJ
2E0 PSP J. Anderson 57 Chapel Lane, Hadfield, Glossop SK13 1NX
2E0 PSR Dr P. Shaw 25 Headcorn Road, Platts Heath, Maidstone ME17 2NH
2E0 PSW J. Godfrey 4 Cherry Close, Houghton Conquest, Bedford MK45 3LQ
2E0 PSZ S. Macdonald Woodside Cottage, Horton Way, Verwood BH31 6JJ
2E0 PTA P. Chambers 257 Kings Acre Road, Hereford HR4 0SR
2U0 PTB P. Brucher No 2 Mon Desir, Les Monmains, Vale., Guernsey GY3 5TZ
2M0 PTE P. Pirie Willowbank, Kirkton Of Tough, Alford AB33 8ER
2E0 PTF A. Norton 9 The Common, West Tytherley, Salisbury SP5 1NS
2E0 PTI B. Parton 51 Marston Grove, Stoke-on-Trent ST1 6EF
2E0 PTM P. Millington 125 Telford Way, High Wycombe HP13 5SZ
2E0 PTS P. Boultwood 32 Makepiece Road, Bracknell RG42 2HJ
2E0 PTY S. Brace 2 Greenfields Cottages, Brockhampton Estate, Bringsty WR6 5TB
2E0 PTZ Prof. P. Curtis Cotswold, Salisbury Road, Abbotts Ann, Andover SP11 7NX
2E0 PUB A. Fulton 8 Priest Hill Gardens, Wetherby LS22 7UD
2E0 PUE A. Hannon 8 Circular Road West, Liverpool L11 1AZ
2E0 PUF K. Simkin The Flat, Cinque Ports, 49 High Street, Seaford BN25 1PP
2E0 PUG D. Cobbold 65 St. Olaves Road, Bury St. Edmunds IP32 6RR
2E0 PUH Dr M. Foster 58 New Terrace, Staverton, Trowbridge BA14 6NY
2E0 PUL D. Pullen 5 Weldon Close, Shotton Colliery, Durham DH6 2YJ
2E0 PUN J. Rideout 35 Colmead Court, Northampton NN38QE
2E0 PUS P. Ellis 40 Grasmere Road Royton, Oldham OL2 6SR
2E0 PUT S. Croston 12 Sefton Street, London SW15 1LZ
2E0 PUZ B. Titmus 68 Hobart Road, Cambridge CB1 3PT
2E0 PVK R. Banks 1 Holly Meadows, Ashford TN23 3QR
2E0 PVN P. Nicholls 23 Bishops Gate, Birmingham B31 4AJ
2E0 PVQ E. Rhodes The Old Forge, Stoke Gabriel, Totnes TQ9 6RL
2E0 PVU R. Coomer 30 Torquay Road, Kingskerswell, Newton Abbot TQ12 5EZ
2E0 PVW P. Armstrong 10 Shirdley Avenue, Liverpool L32 7QG
2E0 PWC P. Castle 3 Wye Road, Brockworth, Gloucester GL3 4PP
2E0 PWD J. Shaw 71 Rowntree Lodge New Earswick, York YO32 4AA
2E0 PWF C. Cousins 43 Avon Close, Little Dawley, Telford TF4 3HP
2E0 PWG P. Green 15 Dickenson Road, Chesterfield S41 0RX
2E0 PWI P. Warwick 24 Chiltern Close, Berinsfield, Wallingford OX10 7PZ
2E0 PWJ P. Williamson 22 Earls Road, Shavington CW2 5EZ
2E0 PWK S. Platts 59 Sea View Road, Drayton, Portsmouth PO6 1EW
2E0 PWL M. Mynn The Town House, Parsons Field, St. Mary'S, Hugh Town TR21 0JJ
2E0 PWM P. Mitchell 13 Ashorne Close, Matchborough, Redditch B98 0EY
2W0 PWO P. Oseland 6 Oaklands Close, Bridgend CF31 4SJ
2E0 PWP Dr P. Thompson 9 Thames Mews, Poole BH15 1JY
2W0 PWR D. Riley-Kydd 35 Pensyflog, Porthmadog LL49 9LB
2E0 PWU P. Wilson 90 Kenwood Crescent, Ingleby Barwick, Ingelby Barwick TS17 5BS
2E0 PXD P. Donaghy 67 Brockenhurst Way, Bicknacre, Chelmsford CM3 4XN
2M0 PXH Dr P. Holmes Maraval, Doune Road, Dunblane FK15 9AT
2E0 PXI L. Willis 12 Robartes Road, St. Dennis, St. Austell PL26 8DS
2E0 PXP J. Maudsley Knight Stainforth Hall, Little Stainforth, Settle BD24 0DP
2E0 PXW B. Smith 19 Alexandra Square, Winsford CW7 2YR
2E0 PXY C. Fox 45 Park Road, Wivenhoe, Colchester CO7 9LS
2E0 PXZ Dr C. Fox 45 Park Road, Wivenhoe, Colchester CO7 9LS
2E0 PYA W. Allen 109 Barston Road, Oldbury B68 0PU
2E0 PYC R. Watson 8 Bourne Close, Warminster BA12 9PT
2W0 PYL C. Thorley Helston, The Mountain, Holyhead LL65 1YR
2E0 PYM A. Nutt 77 Exeter Close, Stevenage SG1 4PW
2E0 PYN L. Scott 28 Cavendish Place, New Silksworth, Sunderland SR3 1JW
2E0 PYR P. Robinson 15 Cornelius Drive, Wirral CH61 9PY
2E0 PZK P. Kirby 11 Bembridge Court, Crowthorne RG45 6BN
2E0 PZM T. Menzies 4 Meadow Road, Muxton, Telford TF2 8JH
2E0 PZN D. Smith 22 Westmarsh Drive Cliftonville Margate, Margate CT9 3NX
2E0 RAA R. Keeley 17 Pembroke Avenue, Wirral CH46 0TP
2W0 RAD R. Miles 63 Phillip Street, Mountain Ash CF45 4BG
2E0 RAF A. Woodrup 440 New Hall Lane, Preston PR1 4TA
2E0 RAG A. Green 18 Harold Avenue, Ashton-In-Makerfield, Wigan WN4 9UZ
2E0 RAH R. Haynes 47 Alder Drive, Alderholt, Fordingbridge SP6 3EP
2E0 RAI R. Trim 23 Coleman Road, Bournemouth BH11 8EQ
2E0 RAJ P. Penfold 2 The Leas, Essenden Road, St. Leonards-on-Sea TN38 0PU
2E0 RAK P. Graham 28 Newburgh House, Highworth, Swindon SN6 7DW
2E0 RAL A. Clewes 20 Linden Drive, Crewe CW1 6HN
2E0 RAM R. Mason 73 Edinburgh Road, Chatham ME4 5BZ
2E0 RAS R. Shippey 43 Westbury Street, Bradford BD4 8PB
2E0 RBA Dr R. Bowman 48 Eliot Drive St. Germans, Saltash PL12 5NL
2E0 RBC D. Judge 18 Shepherd Street, Bacup OL13 8BH
2E0 RBG R. Rigden 36A Atherston, Bristol BS30 8YB
2E0 RBH M. Clifford 32 Tiverton Way, Chessington KT92QS
2E0 RBI R. Gilbert 61 Coltstead, New Ash Green, Longfield DA3 8LN
2E0 RBK B. Kelly 21 Hogarth Walk, Bristol BS7 9XS
2E0 RBN V. Steele 175 Vale Road, Seaford BN25 3HH
2E0 RBO R. Coleman 5 Meeting Lane, Burton Latimer, Kettering NN15 5LS
2E0 RBP R. Cobb 57 Adams Drive, Willesborough, Ashford TN24 0FX
2M0 RBQ R. Titmarsh Caberfeidh Clachan Na Luib, North Uist HS6 5HD
2E0 RBR C. Dunstan 67 Knights Way, Mount Ambrose, Redruth TR15 2BN
2E0 RBU R. Booth 142 Heath Lane, Earl Shilton LE9 7PD
2I0 RBV R. Montgomery 85 Rockfield Heights, Connor, Ballymena BT42 3GH
2W0 RBW R. Williams Bardsville, Porthdafarch Road, Holyhead LL65 2LL
2W0 RBX R. Jones 25 Whiteway Drive, Gresford, Wrexham LL12 8HW

2E0 RBZ R. Booker 6 Kipling Road, Dursley GL11 4QB
2E0 RCA B. Whiteley 2A Beechfield Close, Thorpe Willoughby, Selby YO8 9QJ
2E0 RCB R. Brown 9 Bayleaf Crescent, Oakwood, Derby DE21 2UG
2E0 RCC R. Chadwick 4 Gleneagles Drive, Haydock, St. Helens WA11 0YS
2M0 RCD S. Mckenzie 0/2 69 Glenkirk Drive, Glasgow G15 6AU
2E0 RCF R. Goody 113 Kenneth Road, Basildon SS13 2BH
2E0 RCI S. Lofthouse 30 Broughton Grove, Skipton BD23 1TL
2E0 RCL R. Buckland 34 Beechwood Drive, Meopham, Gravesend DA13 0TX
2E0 RCM R. Medland 5 Bay Tree Cottages, Hospital Road, Bude EX23 9BP
2E0 RCN R. Northway 8 Dean Close, Wick, Littlehampton BN17 7ND
2E0 RCO B. Walker Pilgrims Gore Road, Burnham-on-Sea TA8 2HL
2E0 RCR R. Rawson 30 Harty Road, Haydock, St. Helens WA11 0YY
2E0 RCT M. Russell 107 Cambridge Road, Hitchin SG4 0JH
2W0 RCU R. Gripp 23 Edmond Locard Court, Chepstow NP16 6FA
2E0 RCV R. Treacher 93 Elibank Road, London SE9 1QJ
2E0 RCW C. Wilson 31 Violet Road, South Woodford, London E18 1DG
2M0 RCZ A. Conlon Kilrae, Barrpath, Glasgow G65 0EX
2E0 RDA M. Salt 1 Chantry Close, Harrow HA3 9QZ
2W0 RDD R. Cotterell 49 Graham Court, Caerphilly CF83 1RF
2E0 RDE R. Seeley 2 Church Road, Folkestone CT20 3LH
2E0 RDF J. Bailey 22 Wilford Drive, Ely CB6 1TL
2M0 RDG R. Rogerson 93 Auchencrieff Road, Locharbriggs, Dumfries DG1 1UZ
2M0 RDH R. Hutton 2 Watson Place, Dunfermline KY12 0DR
2E0 RDI R. Topley 85 Stuart Road, Aylsham, Norwich NR11 6HW
2W0 RDJ R. Cole 14 Inner Loop Road, Beachley, Chepstow NP16 7HF
2E0 RDN R. Newton 38 Bedford Road, Denton, Northampton NN7 1DR
2E0 RDO R. Owen 23 Bevan Close, Stockton-on-Tees TS19 8RF
2E0 RDP M. Payne The Devonhurst, 13 Eastern Esplanade, Broadstairs CT10 1DR
2E0 RDQ R. Owen 4 Aldersleigh Drive, Stafford ST17 4RY
2I0 RDR Rvd. R. Rowe 31 Main Street, Brookeborough BT94 4EZ
2M0 RDT R. Tourish 8 Linnpark Gardens, Johnstone PA5 8LH
2E0 RDU J. Crosby 65 Bradwell Avenue, Stretford, Manchester M32 9RT
2E0 RDW R. Wyatt 297 Weston Road Weston Coyney, Stoke-on-Trent ST3 6HA
2E0 RDX B. Wilkes 9 Barnsley Avenue, Conisbrough, Doncaster DN12 3LB
2W0 RDZ R. Shipman 1 Lledfair Place, Heol Pentrerhedyn, Machynlleth SY20 8DL
2E0 REB R. Beardsley 10 Moreton Close, Church Crookham, Fleet GU52 8NS
2E0 REC M. Barker 18 Nickleby Road, Waterlooville PO8 0RH
2M0 REH R. Hay Roddach Cottage East, Cummingston, Burghead, Elgin IV30 5XY
2W0 REJ J. Richardson 15 Calland Street, Plasmarl, Swansea SA6 8LE
2E0 REL R. Allen 44 Overing Avenue, Great Waldingfield, Sudbury CO10 0RJ
2E0 REM R. Whalley 188 Astley Street, Astley, Tyldesley, Manchester M29 7AX
2E0 REN A. Pendle 17 Norrington Grove, Birmingham B31 5NY
2E0 RER R. Ridge Roskellan House, Maenlay, Helston TR12 7QR
2E0 RES R. Strong 2 Dean Avenue, Thornbury, Bristol BS35 1JJ
2E0 REU B. Robinson 11 Wimbledon Drive, Stockport SK3 9RZ
2W0 REX C. Moreton 20 Millbrook Court, Little Mill, Pontypool NP4 0HT
2E0 REY R. Milton 66 Hoo Marina Park, Vicarage Lane, Rochester ME3 9TG
2E0 REZ M. Ricketts 45 Jesmond Road, Grays RM16 2QS
2E0 RFD R. Dell 18 Greenacres, Fulwood, Preston PR2 7DA
2E0 RFE R. Parkinson 27 Hopton Avenue, Mirfield WF14 8JW
2E0 RFF M. Petchey 74 Avondale, Ellesmere Port CH65 6RW
2E0 RFG R. Gray Upper Bisterne Farmhouse, Bisterne, Ringwood BH24 3BP
2E0 RFH R. Henderson 9 Green Mead, South Woodham Ferrers, Chelmsford CM3 5NL
2E0 RFI A. Driscoll 82 Station Road, Langford, Biggleswade SG18 9PQ
2E0 RFK Rvd. R. Eardley Bridge Cottage, Martin, Fordingbridge SP6 3LD
2E0 RFL R. Tongs 9 Woodland Drive, Winterslow, Salisbury SP5 1SZ
2E0 RFM R. Corney Lavender Cottage, Worlds End, Hambledon, Waterlooville PO7 4QU
2E0 RFN P. Collins 7 Fitzmaurice Square, Calne SN11 8NL
2W0 RFT S. Beer 49 Central Street, Pwllypant, Caerphilly CF83 2NJ
2E0 RFU J. Byrne 316 Turncroft Lane, Stockport SK1 4BP
2W0 RGA R. Anderson 156 Cockett Road, Cockett, Swansea SA2 0FQ
2E0 RGB F. Birtwistle 27 Southwell Road, Wisbech PE13 3LF
2E0 RGC R. Charbonneau 22 Redgrave Drive, Crawley RH10 7WF
2I0 RGD B. Gilliland 28 Baird Avenue, Donaghcloney, Craigavon BT66 7LP
2E0 RGH R. Hillman 28 Alpha Street, Exeter EX1 2SP
2E0 RGK K. Harley Care Of: Mr K Harley 9 Amyas Way, Northam, Bideford EX39 1UT
2W0 RGL G. Lewis Bryn Cottage, Clydach, Abergavenny NP7 0LL
2I0 RGM R. Murphy 40 Stoneypath, Londonderry BT47 2AF
2E0 RGO M. White 7 Overthwart Crescent, Worcester WR4 0JW
2E0 RGR R. Hopwood 1 Shakespeare Grove, Cannock WS11 4BQ
2E0 RGS R. Simpson 22 Kenworthy Road, Stocksbridge, Sheffield S36 1BZ
2I0 RGT R. Todd 3 Granville Manor, Kells, Ballymena BT42 3JE
2M0 RGU D. Small 30 Caledonia Crescent, Ardrossan KA22 8LW
2D0 RGW D. Corkish 10 Vicarage Close, Ballabeg, Castletown IM9 4LQ Isle of Man
2E0 RGY P. Haydon 101 Blandford Street, Ashton-under-Lyne OL6 7HG
2E0 RHA P. Costall 3 Gaynsford Place, Little Canfield, Dunmow CM6 1WB
2E0 RHC R. Cook 16 First Square, Stainforth, Doncaster DN7 5RH
2E0 RHE R. East 6 Ashley Road, Worcester WR5 3AY
2E0 RHH R. Hawkins 3 Fairways Drive, Harrogate HG2 7ES
2W0 RHI A. Johnston 44 Cradoc Road, Brecon LD3 9LH
2E0 RHL R. Landragin 147 Taunton Road, Romford RM3 7PJ
2E0 RHM R. Murphy 23 Lowndes Close, Stockport SK2 6DW
2I0 RHN R. Dunwoody 16 Dernalea Road, Milford, Armagh BT60 4DZ
2E0 RHP R. Hanson 27 Almond Rise, Forest Town, Mansfield NG19 0NA
2I0 RHQ H. Selfridge 147 Culcrum Road, Dunloy, Ballymena BT44 9DT
2E0 RHS S. Hawkins Forest Edge, Deer Park, Milton Abbas, Blandford Forum DT11 0AY
2M0 RHT Dr R. Harkness Fernwood, 4 Cassalands, Dumfries DG2 7NS
2E0 RHW R. Haines-White 111 Glendower Crescent, Orpington BR6 0UP
2E0 RIA M. Landragin 101 Linden Gardens, Enfield EN1 4DY
2E0 RIH R. Hart Jays, South Street, Kington Magna, Gillingham SP8 5ET
2E0 RIL J. Riley 67 Moss Bank Road, St. Helens WA117DE
2E0 RIN R. Horner 21 Ainsworth Road Little Lever, Bolton BL3 1RG
2E0 RIO S. Rutt Granthorpe, Hull Road, Hull HU11 5RN
2E0 RIQ R. Back 79 Aspin Close Wellington Home, Somerset TA21 9EG
2I0 RIR E. Stevenson 25 Woodlands Manor, Portadown, Craigavon BT62 4JP
2E0 RIS J. Edwards 45 Bramshaw Gardens, Bournemouth BH8 0BT
2E0 RIT R. Royds 1 Castle Croft, Bolton BL2 3QT
2E0 RIW R. Cruse Howse 12 Queens Road Moretonhampstead, Newton Abbot TQ13 8LP
2E0 RIZ C. Chilton 5 Braithwaite Ave, Keighley BD22 6ET
2E0 RJA R. Ashman 61 Fairfield Road, Burgess Hill RH15 8NP
2E0 RJB B. Roaf 8 Weare Close, Portland DT5 1JP
2E0 RJD J. Dixon 23 Dee Way, Winsford CW7 3JB
2E0 RJE R. Earnshaw 53 Blue Waters Drive, Paignton TQ4 6JF
2E0 RJH R. Harrison 18-20 Hall Lane, Kirkburton, Huddersfield HD8 0QW
2M0 RJJ R. James The Garret, Alyth, Blairgowrie PH11 8HQ
2W0 RJL R. Lovesey 33 Ty Isaf Park Avenue, Risca, Newport NP11 6NB
2E0 RJM R. Millington Quaintways, The Avenue, Tarporley CW6 0BA
2E0 RJR R. Radley 20 Thorntondale Drive, Bridlington YO16 6GW
2E0 RJU A. Foster 7 Moira Dale, Castle Donington, Derby DE74 2PG
2E0 RJX R. Miller 23 Clarendon Road, Sevenoaks TN13 1EU
2E0 RJY R. Kumar 1 Stilton Close, Aylesbury HP19 8JH
2E0 RJZ R. Duthie 14 Kettles Close, Oakington, Cambridge CB24 3XA
2E0 RKB A. Turner 140 Thorndon Avenue, West Horndon, Brentwood CM13 3TR
2E0 RKD R. Dainton 1 The Woodlands, Stroud GL5 1QE
2W0 RKF W. Lewis 53 Leyshon Road, Gwaun Cae Gurwen, Ammanford SA18 1EN
2E0 RKK E. Whiten 17 Scott Close, Ashby-de-la-Zouch LE65 1HT
2W0 RKM M. Beasley Ffynnon Wen, Bontnewydd, Aberystwyth SY23 4JJ
2E0 RKO N. Jacobs 43 The Pines, Yapton, Arundel BN18 0EG
2E0 RKR G. Hirst 94 Upper Brighton Road Sompting, Lancing BN15 0LB
2E0 RKS R. Styles 4 Coningsby Close, Gainsborough DN21 1SS
2E0 RKV A. Gallop 75 Shearmans, Fullers Slade, Milton Keynes MK11 2BQ
2E0 RKX M. Hemming 11 Blackberry Way, Evesham WR11 2AH
2E0 RKZ F. Holmes 2 Station Lane, Hartlepool TS25 1AX
2W0 RLB M. Bannister 45 Queens Drive, Llantwit Fardre, Pontypridd CF38 2NT
2E0 RLE J. Earle 24 Brook Vale, Charlton Kings, Cheltenham GL52 6JD
2E0 RLG J. Berrisford 18 Trowels Ln, Derby DE22 3LS
2E0 RLJ R. Johnson 23 Friars Dene Road, Gateshead NE10 0DR
2E0 RLP L. Paston 138A Forest View Road, Manor Park, London E12 5HX
2E0 RLR R. Hanson 664 Leeds Road, Huddersfield HD2 1UB
2E0 RLT J. Taylor 4 North Villas, Dudley NE23 7QF
2E0 RLV R. Oliver 3 Histon Road Cottenham, Cambridge CB24 8UF
2E0 RLW R. Wood 7 Wishart Green, Old Farm Park, Milton Keynes MK7 8QB
2E0 RLX K. Lewis 15 Gilbert Scott Way, Kidderminster DY10 2EZ
2E0 RLY A. Rock 29 Marley Road, Kingswinford DY6 8RQ
2E0 RMC Dr R. Campbell 2 Hesketh Bank, York YO10 5HH
2M0 RMD R. Nelson 8 South Street Cambus Fk102Pa, Stirling FK102PA
2E0 RMF B. Hudson 19 The Ropery, Whitby YO224EY
2M0 RMH R. Hay 12 Mitchell Brae, Balmedie, Aberdeen AB23 8PW
2E0 RMI J. Salmon 25 Helston Road, Chelmsford CM1 6JF
2E0 RMJ J. Mitchell 11 Brookside Drive, Oadby, Leicester LE2 4PB
2I0 RMK A. Mackenzie 30 Dalriada Gardens, Ballycastle BT54 6DZ
2E0 RML R. Lunson 130 Pilgrims Way, Bedford MK42 9TZ
2E0 RMM J. Hodson 77 Waltham Road, Woodford Green IG8 8DW
2M0 RMN R. Nicoll 15 Redford Walk, Edinburgh EH13 0AF
2W0 RMO R. Orchard The Burrows, Spring Gardens, Whitland SA34 0HL
2M0 RMP J. Mclelland 24 Hyslop Street, Airdrie ML6 0ES
2W0 RMR M. Manley 25 Warwick Road, Brynmawr, Ebbw Vale NP23 4HW
2E0 RMS R. Stevenson 97 Queen Street, Crewe CW1 4AL
2E0 RMT M. Callow 4 The Firs, Canvey Island SS8 9TW
2M0 RMV J. Vine Seaview, Half Of 4 Kilvaxter, Kilmuir IV51 9YR
2E0 RMW R. Ambler 21 Whitley Spring Road, Ossett WF5 0QA
2W0 RMY A. Lewis 3 Aster View, Port Talbot SA12 7ED
2E0 RMZ R. Mansfield 8 Haysoms Drive, Greenham, Thatcham RG19 8EY
2E0 RNA P. Bannon 73 London Road, Worcester WR5 2DU
2E0 RNB R. Ballard 11 Thurston, Skelmersdale WN8 8QU
2M0 RND M. Gerrard 10 Whinhill Gardens, Aberdeen AB11 7WD
2E0 RNE R. Emery 67 Victoria Street, Gillingham ME7 1EW
2E0 RNF M. Faulkner 6 Stanley Avenue, Queenborough ME11 5DT
2E0 RNH I. Justice 4 Saxon Street, Droylsden, Manchester M43 7FR
2E0 RNI D. Taylor 10 Church Rd, , Northwich CW9 5NT
2E0 RNJ Dr J. Reynolds 38 Spring Lane, Hockley Heath, Solihull B94 6QY
2E0 RNM M. James 7 Pixey Place, Oxford OX2 8BB
2E0 RNO J. Franks 14 The Hamlet, Slades Hill, Templecombe BA8 0HJ
2E0 RNP R. Penaluna 113 Brocklesby Road, Scunthorpe DN17 2LW
2E0 RNR J. Shettler 504 Leeds Road, Huddersfield HD2 1YW
2E0 RNS P. Rainey 27 School Road, Silver End, Witham CM8 3RZ
2E0 RNT R. Taylor 22 Shakespeare Court, Chaucer Way, Hoddesdon EN11 9QS
2E0 RNU M. Nutt 110 Birkinstyle Lane, Shirland, Alfreton DE55 6BT
2E0 RNW N. Rapson 15 School Close, Bampton, Tiverton EX16 9NN
2I0 ROC A. Mccann 6 Bowens Meadow, Lurgan, Craigavon BT66 7UT
2E0 ROD R. Burton 23 Freston, Paston, Peterborough PE4 7EN
2E0 ROI W. Chorlton 25 Ash Grove, Orrell, Wigan WN5 8NG
2E0 ROJ A. Hodson 22 Walmley Ash Road, Sutton Coldfield B76 1HY
2E0 ROM R. Pasika 192 Longfield Lane, Cheshunt, Waltham Cross EN7 6AQ
2E0 ROO P. Purnell 89 St. Aubyns, Goldsithney, Penzance TR20 9LS
2E0 ROP P. Bolton 1 Acorn Rise, Hollesley, Woodbridge IP12 3JT
2E0 ROR A. Threlfall 6 Brookside Lane, High Lane, Stockport SK6 8HL
2E0 ROS C. Ross 27 The Meadows, Skegness PE25 2JA
2E0 ROV R. Van-Der-Wijst 6 Willow Street, Romford RM7 7LJ
2E0 ROX M. Johnson 27 Tyndall Walk, Birmingham B32 3UN
2E0 ROY R. Trzeciak-Hicks 24 Wolston Meadow, Middleton, Milton Keynes MK10 9AY
2E0 RPA R. Ashley 15 Wimbourne Drive, Gillingham ME8 9EN
2E0 RPC R. Cockayne 20 The Shrubbery, Rugeley WS15 1JJ
2E0 RPD I. Handley Rosedale, Chapman Street, Market Rasen LN8 3DS
2E0 RPE Dr A. Wallman Oakwood, 30 Elmsway, Bramhall, Stockport SK7 2AE
2E0 RPF R. Fullagar 6 Locke Way, Stafford ST16 3RE
2E0 RPH R. Hann Flat 4, Broadview Haye Down, Tavistock PL19 0NN
2E0 RPL R. Lyons 103A Oxney Road, Peterborough PE1 5NG
2I0 RPM R. Gault 7 Gardenmore Place, Larne BT40 1SE
2E0 RPN P. Whiterod 9 Brewhouse Lane, Soham, Ely CB7 5JD
2E0 RPO W. Donnelly 9 Old Laundry Mews, Laundry Lane, Ingleton, Carnforth LA6 3GH
2E0 RPR M. Roper 13 St. Cuthbert Street, Worksop S80 2HN
2E0 RPU C. Pallett 110 St. Leonards View, Polesworth, Tamworth B78 1JY
2E0 RPW R. Webb Norbury, Terrills Lane, Tenbury Wells WR15 8DD

2E0 RPY H. Kennedy 11 Green Road, High Wycombe HP13 5BD
2W0 RQC R. Chegwin 17 Cyncoed Crescent, Cardiff CF23 6SW
2E0 RQD S. Cownley 5 Pumphouse Lane, East Cowes PO32 6FJ
2E0 RQK B. Dare 1 St. Johns Villas, Sivell Place, Exeter EX2 5ES
2E0 RQN J. Foster 23 High Street, Cumnor, Oxford OX2 9PE
2E0 RQQ Dr R. Baldwin 20 Lynn Road, Shouldham, King's Lynn PE33 0BT
2E0 RRA M. Hall Ward 2 Berth 15, Royal Hospital, Chelsea Royal Hospital Road, London SW3 4SR
2E0 RRC R. Wilson 84 Sir Thomas Whites Road, Coventry CV5 8DR
2I0 RRE R. Rantin 8A Buchanans Road, Newry BT35 6NS
2E0 RRF D. White 2 Birchwood Avenue, Breaston, Derby DE72 3AQ
2E0 RRL K. Woodhams 83 Langdale Place, Newton Aycliffe DL5 7DY
2M0 RRO C. Duncan 131 Croftend Avenue, Glasgow G44 5PF
2W0 RRY P. Smith 19 Grandison Street, Neath SA11 2PG
2E0 RSB R. Blandford 60, Benomley Road, Almondbury, Huddersfield HD5 8LS
2E0 RSC J. Lawrence Flat 19, Badminton, 14 Manilla Crescent, Weston-Super-Mare BS23 2BP
2E0 RSD S. Ray 28 Stenbury View, Wroxall, Ventnor PO38 3DB
2E0 RSG S. Cowgill 17 Tooley Street, Boston PE21 6DP
2E0 RSH Dr R. Hodgkinson 39 Oxford Road, Carlton-In-Lindrick, Worksop S81 9BD
2E0 RSI R. Simms 3 The Byeway, London SW14 7NL
2E0 RSM M. Shortreed 13 Marshfield Road, Settle BD24 9DA
2W0 RSV G. Jones 31 Liverpool Road, Buckley CH7 3LH
2E0 RSX R. Mitchell 8 Prestwood Close, Benfleet SS7 3LD
2M0 RTA R. Turpie 11 Ashkirk Place, Dundee DD4 0TN
2E0 RTC C. Ring 29 Shelley Close, Newport Pagnell MK16 8JB
2M0 RTD D. Robertson 17 Keswick Drive, Hamilton ML3 7HN
2E0 RTE P. Ballington 7 Links Close, Sinfin, Derby DE24 9PF
2E0 RTG I. Sharpe 4 Low Dowfold, Crook DL15 9AE
2E0 RTH R. Hunter Poplars, March Road Guyhirn, Wisbech PE13 4DA
2M0 RTI I. Swanston 6 Roberts Grove, Galashiels TD1 2BJ
2W0 RTJ R. Johns 39 Tyla Coch, Llanharry, Pontyclun CF72 9LT
2E0 RTK Dr R. Tempo 35 Warminster Road, Bath BA2 6XG
2E0 RTM S. Key 159 Launcelot Road, Bromley BR1 5EA
2E0 RTN M. Symmonds 24 Woodville Grove, Stockport SK5 7HU
2M0 RTO M. Smith 2 Richmond Court, Dundee DD2 1BF
2E0 RTP R. Paster 8 Rachaels Lake View, Warfield, Bracknell RG42 3XU
2E0 RTU N. Sargent 21 St. Michaels Road, Claverdon, Warwick CV35 8NT
2E0 RTW R. Whitfield 47 Denchworth Road, Wantage OX12 9AY
2E0 RTY A. Loukes 14 Batchwood View, St. Albans AL3 5TD
2W0 RTZ T. Rowlands Caer Gog Farm, Bodfforddd LL77 7BX
2W0 RUA E. Jones 8 Ty Newydd Court, High Street, Ruabon, Wrexham LL14 6BF
2E0 RUD R. Rudd 11 Woodlands Way, Lepton, Huddersfield HD8 0JA
2E0 RUG J. Fry Deal Cottage, Ipswich Road, Long Stratton, Norwich NR15 2TF
2E0 RUI R. Lima Matos 37 Palace Road, London N8 8QL
2M0 RUP R. Hart Rosalis, Piperhill, Nairn IV12 5SD
2E0 RUS R. Garland 113 The Drive, Feltham TW14 0AH
2E0 RUT R. Rutland 53 Downs Avenue, Whitstable CT5 1RR
2E0 RUU N. Cooksley 10 Lion Close, Nailsea, Bristol BS48 2AL
2E0 RUX A. Rowland 5, Gwinear Downs, Leedstown, Hayle TR27 6DJ
2E0 RUZ R. Brierley 39 Hatfield Road, Alvaston, Derby DE24 0BU
2M0 RVF S. Wilson 8 Robertson Drive, Bellshill ML4 2EQ
2I0 RVH T. Nelson 25 Monaghan Road Annashanco Rosslea, Belfast BT927PT
2E0 RVI R. Miranda 33, Hill Rise, Sundon Park Leagrave, Luton LU3 3EA
2M0 RVM A. Johnstone 10 Earlspark Avenue, Bieldside, Aberdeen AB15 9BU
2I0 RVT R. Todd 58 Kilrea Road, Portglenone, Ballymena BT44 8JB
2E0 RVV D. Kemp 2 Darwin Close, Elston, Newark NG23 5PQ
2E0 RWB R. Whalley 65 Stanley Street, Nelson Lancashire BB9 7ET
2E0 RWC R. Clarke 84 Carol Crescent, Halesowen B63 3RP
2E0 RWE W. Eustace 34 Hertford Avenue, London SW14 8EQ
2W0 RWF T. Mitchell 9 Rhiw Grange, Colwyn Bay LL29 7TT
2I0 RWG R. Gilmore 11 Abbots Gardens, Newtownabbey, Belfast BT379QZ
2M0 RWH R. Humphrey 11 Pearce Grove, Edinburgh EH12 8SP
2E0 RWJ R. Jones 31 Mary Mead, Warfield, Bracknell RG42 3SZ
2E0 RWK R. Westrup 31 Robert Balding Road, Dersingham, King's Lynn PE31 6UR
2E0 RWN R. Nock 83 Coles Lane, West Bromwich B71 2QW
2E0 RWP R. Penny 93 Mirfield Grove, Hull HU9 4QR
2E0 RWR W. Rees 69 Pewley Way, Guildford GU13PZ
2E0 RWT A. White 3 Ipswich Place, Thornton-Cleveleys FY5 1SP
2E0 RWW M. White 130 Main Street Walton, Street BA16 9QX
2E0 RWX R. Williamson 23 Harrowdyke, Barton-upon-Humber DN18 5LN
2E0 RXC R. Codrai 46 The Dale, Waterlooville PO7 5DE
2E0 RXG M. Robson 270 Calder Road, Lincoln LN5 9TL
2E0 RXL R. Lester Finkle Street, Doncaster DN5 0RP
2E0 RXR R. Rich 54 Paddocks Way, Ferndown BH22 9FW
2E0 RXT M. Harrison 2 Rosemount Court, Holly Bank Road, York YO24 4EG
2E0 RXW R. Wells The Turkey House, Park Farm, Tudeley, Tonbridge TN11 0NL
2E0 RXX G. Acton 39 Craig Road, Macclesfield SK11 7YH
2E0 RYD P. Ryder Windrush Pudney Pie Lane Chalford, Stroud GL6 8FT
2M0 RYG R. Young 22 Station Road Armadale, Bathgate EH48 3LN
2M0 RYL R. Haynes 29 Invercauld Road, Aberdeen AB16 5RP
2E0 RYN D. Renshaw 25 Ashley Road, Worksop S81 7JS
2E0 RYP C. Barrett 1 Mead Road, Padgate, Warrington WA1 3TN
2E0 RYR R. Brown 2 Westgate, Leominster HR68SA
2E0 RYS R. Sayre 8 Lorne Road, Richmond TW10 6DS
2E0 RYX Dr I. Van Der Linde 77 Port Vale, Hertford SG14 3AF
2M0 RZE G. Smith 40 Pirleyhill Drive, Shieldhill, Falkirk FK1 2EA
2W0 RZL R. Higgins 44 Maeshyfryd Road, Holyhead LL65 2AL
2E0 RZM G. Kingstone 17 Ullswater Drive, Leighton Buzzard LU7 2QR
2E0 RZS R. Stone 63 Sands Lane, Lowestoft NR32 3ER
2I0 RZT J. Thompson 119 Rathkyle, Antrim BT41 1LN
2E0 RZX Leicester Amateur Radio Show c/o B. Forrest 32 Idonia Road, Wolverhampton WV6 7NQ
2E0 SAA S. Garrett 44 Wardle Crescent, Leek ST13 5PW
2E0 SAF S. Finch 25 Bluebell Avenue, Wigan WN6 8NS
2I0 SAI S. Barnes 191 Marlacoo Road, Portadown, Craigavon BT62 3TD
2W0 SAK S. Poyser Glandwr, Snowdon Street, Porthmadog LL49 9DF
2E0 SAN S. Neachell 59 Gilmour Crescent, Worcester WR3 7PJ
2E0 SAT M. Ward Flat 1, The Old Chapel, Chapel Street, Holsworthy EX22 6AY
2E0 SAW S. Williams Flat 35, Winterton House, London E1 2QR
2M0 SAX G. Sproul 25 Mulben Place, Glasgow G53 7UP
2E0 SAY M. Baker 20 Centurion Rise, Hastings TN34 2UL
2E0 SAZ S. Waller 17 Vere Road, Peterborough PE1 3DZ
2E0 SBB A. Southwell 56 Lambrook Road, Taunton TA1 2AF
2E0 SBC K. Smith 7 Rosebery Avenue, Morecambe LA4 5RU
2E0 SBD S. Shailes 9 Ingham Street, Padiham, Burnley BB12 8DR
2E0 SBH S. Ray 18 Crescent Way, Cholsey, Wallingford OX10 9NE
2E0 SBJ C. Didcott 9 Maid Marion Avenue, Selston, Nottingham NG16 6QH
2E0 SBK S. Harvey 40 Littlemoor Lane, Newton, Alfreton DE55 5TY
2E0 SBL S. York 1 The Cottage, Dogdyke Bank, Lincoln LN4 4JQ
2E0 SBM M. Burrows 4 Melton Street, Earl Shilton, Leicester LE9 7FP
2E0 SBN H. Buckley 10 Lower Hey Lane, Mossley, Ashton-under-Lyne OL5 9DE
2E0 SBO R. Raine 110 Stirling Avenue, Jarrow NE32 4HS
2M0 SBP K. Verrall 7 Roshven View, Arisaig PH39 4NX
2E0 SBS G. Hamilton 11 The Spinney, West Lavington, Devizes SN10 4HP
2E0 SBW S. Jeffery 79 Greenbank Road, Watford WD17 4FJ
2E0 SBX S. Elliott 74 Preston Avenue, Alfreton DE55 7JX
2E0 SBZ D. Smith 105 Princes Street, Dunstable LU6 3AS
2E0 SCA S. Waudby 31Bloomfield Avenue Hu5 5Nh, Hull HU5 5NH
2W0 SCB S. Elias 20 Attlee Way, Cefn Golau, Tredegar NP22 3TA
2E0 SCC D. Riman 22 Princess Road, Hinckley LE10 1EB
2E0 SCD J. Lord 5 Langworthy Avenue Little Hulton, Manchester M38 9GQ
2E0 SCF S. Faulkner Mount Pleasant, Elkstones, Buxton SK17 0LU
2M0 SCG S. Greenland Flat 12, Weymouth Court, 201 Weymouth Drive, Glasgow G12 0ER
2E0 SCH P. Hayes 4 London Road, Roade, Northampton NN7 2NL
2E0 SCJ C. Short 35 Whitley Willows Lepton, Huddersfield HD8 0GD
2E0 SCK M. Stillman 58 Highfield Road, Bognor Regis PO22 8PH
2W0 SCL A. Davies 25 Llanfair Road, Tonypandy CF40 1TA
2E0 SCM J. Stephenson 4 Carlow Drive West Sleekburn, Choppington NE62 5UT
2E0 SCN J. Porter Flat 2, 14 Trafalgar Square, Scarborough YO12 7PY
2E0 SCO J. Hepburn 32 Green Croft, Ashington NE63 8EF
2W0 SCP S. Peel 28 Dan Yr Allt, Llanelli SA14 8AT
2E0 SCQ S. Chandler 7 Clinton Road, Redruth TR15 2LL
2E0 SCR C. Petrie 14 Rotherfield Avenue, Eastbourne BN23 8JQ
2E0 SCS A. Mcallister-Bowditch 24 Morse Close, Chippenham SN15 3FY
2E0 SCV S. Shaw 3 Wellington Terrace, Littleborough OL15 9DA
2E0 SCW S. Whittaker 25 Cleveleys Road, Blackburn BB2 3JS
2E0 SCX S. Hassall 21 Bridgnorth Grove, Chesterton, , Newcastle under Lyme, ST5 7QP
2E0 SDA A. Sanderson 65 Holm Flatt Street, Parkgate, Rotherham S62 6HJ
2E0 SDC S. Cornwell 16 Chesterfield Way, Eynesbury, St. Neots PE19 2JY
2E0 SDD S. David 34 Hardwicke Walk, Kings Heath, Birmingham B14 5XX
2E0 SDK S. Allington 137 Marshall Lane, Northwich CW8 1LA
2E0 SDM C. Reid 28 Albion Road, London N16 9PH
2W0 SDO M. Johns 151 Somerset Street, Abertillery NP13 1DR
2E0 SDP R. De Vries Corner Cottage, Hillcrest Close, Sturminster Newton DT10 2DL
2E0 SDQ P. Weaver 1 Madeley Street, Newcastle ST5 6LS
2E0 SDR A. Lane 36 The Crescent, Eastbourne Estate, Coleford GL16 8BS
2E0 SDT S. Theaker 10 Grange Fields Mount, Leeds LS10 4QN
2E0 SDV J. Williams 41 Overton Lane, Hammerwich, Burntwood WS7 0LQ
2E0 SDW S. Whitehead 55 Crombie Road, Sidcup DA15 8AT
2E0 SDY J. Popple 11, Chapel Close, Waterbeach, Cambridge CB25 9JW
2E0 SDZ E. Ayre 1 Spring Gardens, Broadmayne, Dorchester DT2 8PP
2E0 SEA M. Bown 47 Ullswater Crescent, Weymouth DT3 5HF
2E0 SEB S. Gordon 6 Aspinall Grove, Hailsham BN27 3GP
2I0 SEC S. Carlin 9 Mullandra Park, Kilcoo, Newry BT34 5LS
2E0 SEE A. Seedig 21 Ambleside Close, Mytchett GU16 6DG
2M0 SEF S. Fleming 20 Park Road, Invergowrie, Dundee DD2 5AH
2E0 SEJ S. Johnson 2 North Square, Edlington, Doncaster DN12 1ED
2I0 SEK R. Thomson 1 Litchfield Park, Coleraine BT51 3TN
2E0 SEO S. Bates 6 Foxdell, Northwood HA6 2BU
2E0 SEP A. Howard 24 Ladybower Lane, Poulton-le-Fylde FY6 7FY
2E0 SER M. Edmonds 20 Tomline Road, Ipswich IP3 8BZ
2E0 SES Dr M. Cianni 121 Springfield Park Avenue, Chelmsford CM2 6EW
2E0 SET P. Setter 199 Southbourne Grove, Westcliff-on-Sea SS0 0AN
2E0 SEW G. Ferguson 31 Barton Court Road, New Milton BH25 6NW
2E0 SEY J. Averill 14 Shannon Close, Saltney, Chester CH4 8PJ
2W0 SEZ S. Ezard 59 Station Farm, Croesyceiliog, Cwmbran NP44 2JW
2I0 SFA N. Jenkinson 35 Scarvagh Heights, Scarva, Craigavon BT63 6LY
2W0 SFB P. Latham 20 Kenyon Avenue, Wrexham LL11 2ST
2E0 SFC R. Bates 61 Park View, Crowmarsh Gifford, Wallingford OX10 8BN
2E0 SFN A. Szot 90 Portland Avenue, New Malden KT3 6BA
2E0 SFP S. Finlayson 41 Low Catton Road, Stamford Bridge, York YO41 1DZ
2E0 SFQ Dr S. Pearce 15 Hillfield Court Road, Gloucester GL1 3QS
2E0 SFS M. Heenan 15 Woodacre Green Bardsey, Leeds LS17 9AB
2E0 SFT M. Shelton 92 Timberley Drive, Grimsby DN37 9QZ
2E0 SFV Dr S. Favell Lodge Farm, Moor Lane, Reepham, Lincoln LN3 4EE
2E0 SFZ S. Driscoll 21 Hall Drive, Middleton, Morecambe LA3 3LF
2E0 SGG S. Mallows 3 Village View, Chatham ME5 7TR
2E0 SGH S. Holt 14 Fir Street, Cadishead, Manchester M44 5AU
2E0 SGI A. Parker 43 Springfield Avenue, Shirebrook, Mansfield NG20 8LF
2I0 SGK J. Hunter 160 Ballynure Road, Ballyclare BT39 9AJ
2E0 SGL P. Hilton Shankly Cottage 161 Highgate, Jennings Yard, Kendal LA9 4EN
2I0 SGM S. Murray 117, Knockview Drive, Tandragee BT622BL
2M0 SGO S. O'Neill Flat 5-8 460 Sauchiehall Street, Glasgow G2 3JW
2M0 SGQ S. Gill 5 Ramornie Place, Kingskettle, Cupar KY15 7PT
2E0 SGS S. Smith 2 Burnsall Avenue, Blackpool FY3 7LQ
2E0 SGY S. Gregory 9 Croftlands Road, Wythenshawe M22 9YE
2E0 SHA J. Ridley 73 The Markhams, New Ollerton, Newark NG22 9QY
2E0 SHB S. Burrows 78A Coronation Road, Earl Shilton, Leicester LE9 7HJ
2E0 SHG D. Grindrod 6 Croft Way, Market Drayton TF9 3UB
2E0 SHH R. Devos 76 North Parade, Sleaford NG348AW
2E0 SHK S. Helm 10 St. Annes Avenue, Middlewich CW10 0AE
2I0 SHM S. Montgomery 2 Woodland Gardens, Lisburn BT27 4PL
2E0 SHN S. Pallister 32 Greensnook Lane, Bacup OL13 9DQ

UK Callsigns

2E0 SHO S. Hodgkiss Estren, Station Road, Wendling, Dereham NR19 2NE
2E0 SHP A. Sharp 30 Burbidge Close, Calcot, Reading RG31 7ZU
2E0 SHR R. Shepherd 19 Elford Avenue, Newcastle upon Tyne NE13 9AP
2E0 SHV A. Sidhu-Brar White Gates, Main Road, Northampton NN7 3NA
2E0 SHW A. Dransfield 34 Ernest Kirwood Close, Hull HU5 5XX
2W0 SHX J. Nicholas Reservoir House, St. Lythan'S, Wenvoe CF5 6BQ
2E0 SHY W. Shelley 91 Canterbury House, Stratfield Road, Borehamwood WD6 1NT
2E0 SHZ S. Lewis 58 Ocean Close, Fareham PO15 6QP
2E0 SIA S. Skirving 1 Hallington Close, Bolton BL3 6YH
2E0 SIB D. Sibley 85 Brick Crescent, Stewartby MK43 9GG
2E0 SID S. Frampton 20 Winslow Close, Boldon Colliery NE35 9LR
2E0 SIF S. Frost 4 Banister Way, Shipston-on-Stour CV36 4JU
2E0 SIH S. Hammond Ellsworth, Thrigby Road, Filby, Great Yarmouth NR29 3HJ
2E0 SIJ J. Simkins 37 St. Andrews Meadow, Harlow CM18 6BL
2E0 SIK N. Butler 75 Rutland Street, Derby DE23 8PR
2E0 SIM S. Matley 67 Alexandra Road, Chandlers Ford SO53 2BP
2E0 SIS A. Tanseli 157 Warwick Road, Rayleigh SS6 8SG
2E0 SIX S. Hannah 4 Station Road, Minsterley, Shrewsbury SY5 0BG
2E0 SIY S. Shaul 31 Chatterton Avenue Ermine West, Lincoln LN1 3SZ
2E0 SIZ S. Riley 51 Kenilworth Avenue, Reading RG30 3DL
2E0 SJE S. Etty 31 Mill Drive Leven, Beverley HU17 5NR
2E0 SJF S. Farnell 16 Lily Way, Lowestoft NR33 8NN
2E0 SJG S. Goodridge Trelane, Pelynt, Looe PL13 2LF
2E0 SJH S. Humphreys Flat 1, Winslow Court 100 Fordwych Road, London NW2 3NN
2E0 SJI S. Ibbotson 17 Marley Combe Road, Haslemere GU27 3SN
2E0 SJM S. Moore 33 Church Street, Heavitree, Exeter EX2 5EP
2E0 SJN S. Nash 3 Brightside, Waterlooville PO7 7BA
2E0 SJP S. Parker 36 Eton Close, Lincoln LN6 0YF
2E0 SJQ S. Carpenter Field View, Old Lyndhurst Road, Southampton SO40 2NL
2E0 SJS S. Spencer 55 Witton Lane, West Bromwich B71 2AA
2W0 SJT S. Tweddle 3 Bron Ffinan, Pentraeth LL75 8UT
2I0 SJV D. Parkinson 16 Beechwood Gardens, Moira, Moira BT67 0LB
2E0 SKA G. Brown 3 Willow Lane, Goostrey, Crewe CW4 8PP
2E0 SKE M. Swan 35 Colston Close, Plymouth PL6 6AY
2W0 SKG W. Lloyd 29 Duffryn Street, Mountain Ash CF45 3NU
2E0 SKI A. Skarzynski 1 River View Moorings, Bridge Road, Stoke Ferry, King's Lynn PE33 9TS
2E0 SKK A. Greig 17 Begbroke Lane, Kidlington OX5 1RN
2E0 SKL C. Hare 25 Southend Place, Sheffield S2 5FQ
2E0 SKP R. Ibbotson 33 St. Peters Avenue, Caversham, Reading RG4 7DH
2M0 SKR C. Aitken Windybraes, Upper Gills, Canisbay, Caithness, Canisbay KW1 4YB
2E0 SKZ J. Parfitt 5 Sheridan Road, Frimley, Camberley GU16 7DU
2W0 SLD D. Dash 36 Rockvilla Close, Varteg, Pontypool NP4 7QF
2E0 SLG M. Jordan-Reed Pump Corner, High Street Green, Sible Hedingham, Halstead CO9 3LG
2E0 SLH S. Hill 108 Hitchin Close, Romford RM3 7EQ
2E0 SLJ S. Legg 98 Shenstone Valley Road, Halesowen B62 9TF
2E0 SLK J. Blezard 10 North Row, Barrow-in-Furness LA13 0HE
2E0 SLM J. Stringer 31 Pipit Lane, Birchwood, Warrington WA3 6NY
2E0 SLN R. Nixon 2 Garmondsway Court, West Cornforth, Ferryhill DL17 9HE
2E0 SLO D. Rolph 17 Moorlands Park Ashby Road, Sinope LE67 3BD
2W0 SLP S. Williams 63 Trem Eryri, Llanfairpwllgwyngyll LL61 5JF
2E0 SLR B. Robins 99 Uplands Road, Dudley DY2 8BB
2E0 SLS K. Holloway 6 Britons Lane Close, Beeston Regis, Sheringham NR26 8SH
2E0 SLT S. Thompson 80 Mountain Road, Dewsbury WF12 0BP
2E0 SLV B. Gimbert 49 Ratcliffe Road, Loughborough LE11 1LF
2E0 SMA S. Harcourt 71 Ingleby Road, Long Eaton, Nottingham NG10 3DG
2I0 SMD Dr S. Davey 16 Maritime Drive, Carrickfergus BT38 8GQ
2E0 SMF C. Smith 21 Mill House Drive, Cheltenham GL50 4RG
2E0 SMK S. Kendrick 103A Latimer Street, Liverpool L5 2RF
2E0 SML S. Weightman 131 Leeds Road, Birstall, Batley WF17 0JZ
2M0 SMN S. Nicoll 15 Redford Walk, Edinburgh EH13 0AF
2E0 SMO S. O'Neill 1 Avenue Cottages, Winchester Road, Fareham PO17 5EX
2E0 SMS S. Stratford 23 The Fairway, Banbury OX16 0RR
2E0 SMX S. Hogg 57 The Grange, Burton-on-Trent DE14 2EX
2I0 SMY S. Dallas 101 Coagh Road Stewartstown, Dungannon BT71 5JL
2E0 SMZ S. Edwards 30 Morrison Road, Tipton DY4 7PU
2E0 SNA P. Harvey 35 Isaac Street, Liverpool L8 4TH
2E0 SNB S. Bodsworth 51 Berry Way, Andover SP10 3RZ
2E0 SNC M. Hemmings 62 Spencer Way, Stevenage SG2 8GD
2E0 SND A. Williamson 25 Manor Road, Rugby CV21 2SZ
2E0 SNE M. Bryan 13 Elmwood Avenue, Sunderland SR5 5AW
2I0 SNG S. Gilmour 14G Malcolm Road, Lurgan, Craigavon BT66 8DF
2E0 SNJ Nth West Irelan c/o S. Worger 6 Glendale Terrace, Mornington Road, Whitehill Bordon GU35 9AJ
2E0 SNM S. Hindmarsh 7 George Street Murton, Murton SR7 9BN
2E0 SNP S. Powell 3 Tadgedale Avenue, Loggerheads, Market Drayton TF9 4DD
2E0 SNS S. Button 6 Farfield, Retford DN22 7TL
2E0 SNU E. Collins 38 Meynell Road, Sheffield S5 8GN
2W0 SNW S. Williams Hillsboro Aberkenfig, Bridgend CF32 0EW
2E0 SNZ M. Ross 11 Queens Place, Otley LS21 3HY
2M0 SOE R. Baxter 7 Clegg Gardens, Troon KA10 7GZ
2E0 SOF W. Soffe 96 Urban Road, Doncaster DN4 0EP
2E0 SOJ J. Sykes 79 Owler Park Road, Ilkley LS29 0BG
2M0 SOP A. Gordon 26 East Millicent Avenue, Golspie KW10 6TL
2E0 SOR S. Orchard 30 Wilkes Court, Ipswich IP5 2EQ
2E0 SOT S. Gregory 11 Ribblesdale Avenue, Congleton CW12 2BS
2E0 SOX R. Cook 46 Wheatsheaf Road, Tividale, Oldbury B69 1SW
2E0 SOZ R. Mather 4 Vimy Road, Wednesbury WS10 9BQ
2E0 SPA S. Etheridge 16 Crown Meadow Way, Newton St. Faith, Norwich NR10 3GW
2E0 SPB D. Evans 21 Quilter Close, Bilston WV14 9AX
2E0 SPD S. Denman 12 Dyke Vale Road, Sheffield S12 4ER
2E0 SPE G. Bradley Greentree Cottage, Town End, Broadclyst, Exeter EX5 3HW
2E0 SPG M. Hennessey 3 Northgate Cottage, Falmer Road, Rottingdean, Brighton BN2 7DT
2E0 SPH D. Janowicz 20 Salisbury Road, St. Leonards-on-Sea TN37 6RX
2M0 SPL S. Ling Leadburnlea Leadburn, West Linton EH46 7BE
2E0 SPM S. Morris 23 De Courtenai Close, Bournemouth BH11 9PG
2E0 SPN J. Daniels 27 Hammerwater Drive, Warsop, Mansfield NG20 0DJ
2E0 SPR S. Randall 23 Onslow Road, Plymouth PL2 3QG
2E0 SPS J. Bogdaniec 3 Cavalry Chase, Okehampton EX20 1GR
2E0 SPT K. Tremain 26 Longbeech Park, Canterbury Road, Ashford TN27 0HA
2E0 SPU P. Saunders 62 Parkfield Avenue, Eastbourne BN22 9SF
2E0 SPW S. Woodmore 66 Imperial Way, Chislehurst BR7 6JR
2E0 SPZ R. Hinde 3 Baunhill Close, Northampton NN3 3EQ
2E0 SQB S. Stevens 40 Heath Road, Exeter EX2 5JX
2E0 SQK P. Henry 6 Greenwood, Bamber Bridge PR58JS
2M0 SQL P. Goodhall 12 Templand Road, Lhanbryde, Elgin IV30 8PP
2E0 SQN I. Franklin 23 Ingle Drive, Ashby-de-la-Zouch LE65 2LW
2E0 SRB S. Blaikie 22 Juno Close, Goring-By-Sea, Worthing BN12 4UB
2E0 SRC D. Barker 12 The Weavers, Denstone, Uttoxeter ST14 5DP
2W0 SRD D. Bray 24 Lon Gwesyn, Birchgrove, Swansea SA7 9LD
2M0 SRF R. Farrer 23 Upper Craigour, Edinburgh EH17 7SE
2E0 SRH S. Robinson 7 Higney Road, Hampton Vale, Peterborough PE7 8LZ
2E0 SRJ J. Statham Oakwoods, School Lane Upper Basildon, Reading RG8 8LT
2E0 SRO S. Rowland 6 Peach Hall, Tonbridge TN10 3HD
2E0 SRP P. Skidmore 36 Princes Drive, Harrow HA1 1XH
2E0 SRT C. Royle Eaton Bank House 3 Valerian Close, Buxton SK17 6PJ
2E0 SRV S. Vickers 22 Thistle Green, Birmingham B38 9TT
2M0 SRX S. Russell 3 Rankin Road, Wishaw ML2 8PG
2M0 SRY S. Young 6 Ramsey Cottages, Bonnyrigg EH19 3JG
2E0 SRZ S. Ahmed Flat 10 Jubilee Court 51 Eaton Road, London SM2 5AQ
2E0 SSA A. Dossa 24 Warwick Drive, Cheshunt, Waltham Cross EN8 0BW
2E0 SSC S. Charles 29 Woolford Close, Winchester SO22 4DN
2E0 SSD A. Gillard 4 Horton Avenue, Thame OX9 3NJ
2E0 SSE S. Evans 78 High Brooms Road, Tunbridge Wells TN4 9BN
2E0 SSG A. Butler 12 South Bank Cottages South Stoke, Reading RG8 0HX
2E0 SSI S. Irwin 18 Manse Way, Swanley BR8 8DD
2E0 SSJ S. Shortland 10 Highclere Court, Knaphill, Woking GU21 2QP
2E0 SSK S. Kamal 184 Aycliffe Road, Borehamwood WD6 4EG
2E0 SSL G. Sawyer 432 Rowood Drive, Solihull B92 9LQ
2E0 SSM S. Mcbain 13A St. Lukes Close, Cherry Willingham, Lincoln LN3 4LY
2E0 SSN B. Woods 28 Delph Drive, Burscough, Ormskirk L40 5BE
2E0 SST S. Slapper 1 Standards Keep, Standards Road, Bridgwater TA7 0EZ
2E0 SSX K. Baker 64 Pendle Drive, Basildon SS14 3LZ
2E0 SSY M. Hossell 80 Murray Road, Sheffield S11 7GG
2E0 SSZ S. Ziya 32 Latchford Road, Wirral CH60 3RW
2E0 STA S. Collis 21 Holme Hall Crescent, Chesterfield S40 4PQ
2M0 STB I. Learmonth 14 Deansloch Terrace, Aberdeen AB16 5SN
2E0 STH S. Holt 65 Brown Royd Avenue, Huddersfield HD5 9QA
2E0 STI S. Pearce 20 Barcote Walk, Plymouth PL6 5QE
2M0 STK T. Storkey 1 Hartington Gardens, Edinburgh EH10 4LD
2D0 STL S. Leslie 16 Little Meddow, Andreas IM7 4HY Isle of Man
2E0 STP B. Freeman 45 Pryor Road, Baldock SG7 6LH
2E0 STQ E. St Quinton Mill Cottage, The Thorofare, Woodbridge IP13 8BB
2E0 STT A. Burgess 1 Laurel Crescent, Long Eaton, Nottingham NG10 3NL
2M0 STV S. Mccormick 4 Poplar Avenue, Johnstone PA50EF
2E0 STX S. Jackson 64 Main Road, Moulton, Northwich CW9 8PB
2I0 SUB C. Williams 92 Whitehouse Park, Newtownabbey BT37 9SH
2E0 SUC C. Stokes 45 Chaucer Close, Basingstoke RG24 9DW
2E0 SUD A. Greenhalgh 61 Long Meadows, Chorley PR7 2YB
2E0 SUE S. Turford 1 Portland Crescent Bolsover, Chesterfield S446EG
2E0 SUF S. Batley 2 Boulge Road, Hasketon, Woodbridge IP13 6LA
2E0 SUH S. Halewood 12 Silver Street Riccall, York YO19 6PB
2E0 SUS S. Millard 112 Avenue Road, Sandown PO36 8DZ
2E0 SUT R. Sutton Yew Tree Farm, Paddol Green, Shrewsbury SY4 5QZ
2E0 SUX A. Suttle 61 Albert Street, Shildon DL4 2DN
2E0 SUZ S. Coombes 33 Clarence Park Road, Bournemouth BH7 6LF
2E0 SVB J. Ratcliffe 7 De Havilland Drive, Hazlemere HP15 7FP
2E0 SVG J. Savage 44 Hastings Road, Maidstone ME15 7SP
2E0 SVK L. Mikolka Flat, 1 Scotney Court, Romney Marsh TN29 9JP
2E0 SVP Splitters c/o V. Sterea 67 Lamberhead Road, Wigan WN5 9TU
2E0 SVT L. Tayler 22 Wheatley Road, Leicester LE4 2HN
2E0 SVV A. Kemp 2 Darwin Close, Elston, Newark NG23 5PQ
2W0 SVW D. Wilkinson Bryn Y Mor, North Road, Caernarfon LL55 1BE
2E0 SVZ A. Abson 117 Lysander Road, Rubery, Birmingham B45 0EN
2M0 SWA S. Anderson 33 Dryden Avenue, Loanhead EH20 9JT
2E0 SWB S. Bennett 155 Warstones Road Penn, Wolverhampton WV4 4LG
2E0 SWE O. Hall 2 Beverley Lodge, Paradise Road, Richmond TW9 1LL
2E0 SWF R. Jenkinson Esperance, West End Road, Doncaster DN9 1LB
2E0 SWK M. Sedgwick 44 Cundall Road, Hartlepool TS26 8LG
2M0 SWM P. Holmes 15A Ochlochy Park, Dunblane FK15 0DU
2E0 SWN S. Neale 43 Crompton Road Pleasley, Mansfield NG19 7RG
2W0 SWO S. Owen 19 Longtown Grove, Newport NP10 8HD
2E0 SWP S. Probert 10 The Green, Church Lawton ST73ED
2E0 SWS Rvd. S. Scotson 16 Merryfield Road, Dudley DY1 2PD
2E0 SWT T. Akay Caddebostan Mah. Plaj Yolu Sok. No25/15 Kad?K÷y, ?Stanbul (Asya) 34710 Turkey
2E0 SXC S. Ward 22 St. Margarets Close, Horstead, Norwich NR12 7ER
2E0 SXF S. Levsen 8 Craig Close Broughton, Brigg DN20 0SE
2E0 SXI S. Harriss 30 Chatsworth Place, Harrogate HG1 5HR
2M0 SXJ P. Duckles 26 Meadowpark Road, Bathgate EH48 2SJ
2I0 SXM G. Hutton 13 Meadowbank, Sepatrick, Banbridge BT32 4PZ
2E0 SXP S. Perring 16 Salford Road, Bolton BL5 1BL
2E0 SXS D. Truscott 37 Langley, Chulmleigh EX18 7BQ
2M0 SXT S. Thorogood 38 Forres Drive, Glenrothes KY6 2JU
2E0 SXY S. Lord 34 Alsop Street, Leek ST13 5NZ
2E0 SYA S. Yem 8 Beechwood Avenue, Wallasey CH45 8NX
2E0 SYB J. Redgrave 24 Burnham Close Trimley St. Mary, Felixstowe IP11 0XG
2E0 SYD K. Hunt 13 Beaumaris Court, Spondon, Derby DE21 7RG
2E0 SYE S. Greenheart 168 Beech Hill Lane, Wigan WN6 8PL
2M0 SYH S. Hancox 45 Park Road, Brechin DD9 7AE
2E0 SYI S. De Koster 4 Cambridge Way, Cullompton EX15 1GQ
2M0 SYL A. Haynes 29 Invercauld Road, Aberdeen AB16 5RP
2E0 SYM C. Symons 159 Middlecroft Road South Staveley, Chesterfield S43 3NF
2E0 SYN M. Hickey 100 Chester Road, Poynton, Stockport SK12 1HG

2E0 SYO B. Hender 11 Cambridge Road, Walton-on-Thames KT12 2DP
2E0 SYS M. Styles 46 Daffodil Drive, Lydney GL15 5RE
2W0 SYT M. Burnell Ty Talwyn Farm, Cefn Cribwr, Bridgend CF32 0BP
2E0 SYV S. Ambrosio 8 The Pines, Long Lane, Chester CH2 2QF
2E0 SYW S. Ward Wellbeck, Wheel Road, Alpington, Norwich NR14 7NH
2E0 SYX S. Young 26 Keel Close, Carlton Colville, Lowestoft NR33 8GT
2E0 SYY C. Gibson 3 Conway Drive, Billinge, Wigan WN5 7LH
2M0 SZC S. Kirkbride 18 North Roundall, Limekilns, Dunfermline KY11 3JY
2E0 SZG S. Gray The Pigsty, Cleverton, Chippenham SN15 5BT
2E0 SZH S. Hegarty 6 Wymbush Crescent, Bristol BS13 0BB
2J0 SZI M. Brown 77 Andium Court, Langtry Gardens, St. Saviours Hill, St. Saviour, Jersey JE2 7AH
2E0 SZL R. Williams 4 Bluebell Close, Barlborough, Barlborough S43 4WT
2E0 SZT Z. Szot 45 Ealing Park Gardens, London W5 4EX
2I0 TAA T. Boyd 40 Walnut Park, Larne BT40 2WF
2E0 TAB T. Balls Rostan, 99 Front Road, Murrow, Wisbech PE13 4JQ
2E0 TAC A. Paxton 20F Green End, Granborough, Buckingham MK18 3NT
2E0 TAD G. Aldridge Greenridge, Fore Street Bishopsteignton, Teignmouth TQ14 9QR
2E0 TAG J. Bingham 31 Wyre Close, Paignton TQ4 7RU
2W0 TAI T. Evans 27 Addison Road, Neath SA11 2AY
2E0 TAJ A. Turner 174 Preston Road, Standish, Wigan WN6 0NP
2E0 TAK D. Heathcote 154 High Street Harriseahead, Stoke-on-Trent ST7 4JX
2E0 TAL A. Walton 65 Broadway East, Rotherham S65 2XA
2E0 TAM A. Dodds 19 Westgate, Oldbury B69 1BA
2I0 TAN A. Kelly 19 Union Street Mews, Coleraine BT52 1EN
2E0 TAO G. Jones 57 Oxford Road, Banbury OX16 9AJ
2E0 TAQ R. Dicker 38 Inkerman Road, Southampton SO19 9DA
2W0 TAR K. Parry 80 Cripps Avenue, Cefn Golau, Tredegar NP22 3PB
2M0 TAS D. Branson Derelochy, Kingsteps, Nairn IV12 5LF
2E0 TAT A. Horton 11 Hilton Road, Tividale, Oldbury B69 1JU
2E0 TAU A. Smith 46 Mulberry Close, Goldthorpe, Rotherham S63 9LB
2E0 TAW T. Williamson 286 Glynswood, Chard TA20 1BX
2W0 TAX T. Mcintyre Coach House, Commercial Street, Griffithstown, Pontypool NP4 5JF
2E0 TAY A. Taylor 15 Woodford Glebe, Welford, Northampton NN6 6AF
2E0 TAZ A. Barron 80 Primrose Crescent, Norwich NR7 0SF
2E0 TBD A. Marshall 13 The Markhams, New Ollerton, Newark NG22 9QX
2E0 TBE E. Whitehouse 16 Rue Gaston De Caillavet, Paris 75015 France
2E0 TBF D. Bateson 19 Rothesay Road, Heysham, Morecambe LA3 2UR
2E0 TBH J. Smith 10, Seacroft Road Plymouth Pl51Ph, Plymouth PL51PH
2E0 TBI J. Lynch Beechway, Raddel Lane, Warrington WA4 4EE
2E0 TBL J. Wilson 35 Lawson Avenue, Jarrow NE32 5UF
2E0 TBQ A. Wong 43 Northern Road, Aylesbury HP19 9QT
2E0 TBR H. Hylton 214 School Road, Hall Green, Birmingham B28 8PF
2E0 TBS C. Burnham 93 Colliers Break, Emersons Green, Bristol BS16 7EB
2E0 TBV A. Hodgkinson 41 Allott Crescent, Jump, Barnsley S74 0LB
2E0 TBW B. Fitzgerald-O'Connor 14 Goldwing Close Custom House, London E16 3EQ
2E0 TBX C. Pulford 17 Canada Road, Cromer NR27 9AH
2E0 TBZ Dr C. Cowen Rosita, White Street Green, Sudbury CO10 5JN
2E0 TCB A. Bruton 29 Helyers Green, Wick, Littlehampton BN17 7HB
2E0 TCC A. Fleming 39 Urswick Green, Barrow-in-Furness LA13 0BH
2E0 TCF T. Guy 16 Cogdeane Road, Poole BH17 9AS
2E0 TCG A. Gilberts 22 Granby Road, Buxton SK17 7TW
2E0 TCI A. Ashton 46 Kingsland, Harlow CM18 6XL
2I0 TCJ T. Mckee 4 Earlford Heights, Newtownabbey BT36 5WZ
2E0 TCK P. Pritchard 7 Warbler Road, Leighton Buzzard LU7 4DA
2W0 TCM M. Digby 40 Waterloo Road, Ammanford SA18 3SF
2E0 TCN C. Lyne 4 Bridge Close, Catterick Garrison DL9 4PG
2E0 TCO T. Corcoran 191 Queensway, Rochdale OL11 2NA
2E0 TCQ M. Nicholas 2 Meadow Way, Harworth, Doncaster DN11 8PW
2E0 TCU B. Neal 6 Canterbury Street, Chaddesden, Derby DE21 4LG
2E0 TCV T. Curnow 5 Bosleake Row, Bosleake, Redruth TR15 3YG
2E0 TCX A. Wilson 38 Cotleigh Drive, Sheffield S12 4HU
2E0 TCY C. Maddex 154 Tintern Avenue, Westcliff, Westcliff-on-Sea SS0 9QF
2E0 TCZ A. Cudworth 30 Compass Tower, Munnings Road, Norwich NR7 9TW
2W0 TDD T. Dixon Troedyrhiw, Abercych, Boncath SA37 0EY
2E0 TDE A. Dilworth 808 Liverpool Road, Southport PR8 3QF
2W0 TDF W. Welch Kenilworth, School Lane, Gobowen, Oswestry SY11 3LD
2E0 TDH P. Thorley 57 Riverside Drive, Hambleton, Poulton-le-Fylde FY6 9EH
2E0 TDI D. Davies 32 Newlyn Crescent, Puriton, Bridgwater TA7 8BS
2E0 TDJ T. Kelly 32 The Foxgloves, Hedge End SO30 0UG
2I0 TDL T. Browne 7 Hawthorn Park Greysteel, Londonderry BT47 3YE
2E0 TDN S. Friend 1 Orford Road, Tunstall, Woodbridge IP12 2JH
2E0 TDO Dr T. Digman 74 Baddlesmere Road, Whitstable CT5 2LA
2E0 TDP A. Pickett 4 Trembel Road, Mullion, Helston TR12 7DY
2E0 TDR J. Mcmullen 281 The Broadway, Cullercoats, North Shields NE30 3LH
2E0 TDT A. Maclean 10 Elizabeth Close, West Hallam, Ilkeston DE7 6LW
2E0 TDV T. Millichamp 9 Wells Close, Bridgnorth WV16 5JQ
2E0 TED G. Cahill 81 Albemarle Road, Willesborough, Ashford TN24 0HJ
2E0 TEE A. Chapman 24 Oakford Park, Halnaker, Chichester PO18 0BF
2E0 TEF R. Ashman 85A Fakenham Road, Great Ryburgh, Fakenham NR21 7AQ
2E0 TEH T. Harris 12 Maple Close, Stourport-on-Severn DY13 8TA
2E0 TEI M. Bell 36 Schneider Road, Barrow-in-Furness LA14 5DW
2E0 TEJ A. Bell 36 Schneider Road, Barrow-in-Furness LA14 5DW
2E0 TEK G. Tecklenberg 33 Tilman Drive, Peterborough PE7 0LU
2E0 TEN A. Riddick 30 Britannia Road, Banbury OX16 5DW
2E0 TEU Dr S. Hill 36 The Woodlands, Market Harborough LE16 7BW
2E0 TEW J. Woodland 14 Kelham Green, Nottingham NG3 2LP
2E0 TEZ T. Kemp 30 Tawny Sedge, King's Lynn PE30 3PW
2E0 TFD T. Daniel 28 Polefield Circle Prestwich, Manchester M25 2WP
2E0 TFE J. King Plum Tree Cottage, Royston Place, Barton On Sea, New Milton BH25 7AJ
2M0 TFF T. Feltus 5/5 Moat House, Moat Drive, Edinburgh EH14 1NS
2E0 TFH T. Harrison 1 Tall Trees, Colchester CO4 5DU
2E0 TFI B. May 10 Farrier Place, Downs Barn, Milton Keynes MK14 7PJ
2E0 TFK J. Anthony 21 Belgrave Street, Denton, Manchester M34 3WP
2E0 TFM I. Whiteley 29 Harvey Avenue, Wirral CH49 1RT
2E0 TFN C. Tompkins 9 Billbrook Road, Hucclecote, Gloucester GL3 3QS
2E0 TFO R. Styles 52 Vernham Grove, Bath BA2 2TB
2E0 TFT S. Mayor 12 Yealand Avenue, Heysham, Morecambe LA3 2LT
2E0 TFX T. Fisk 2 Hall Farm Cottage, Caston Road, Caston, Attleborough NR17 1BW
2E0 TGA Dr M. Depardieu 4 Belvedere Fff, Bath BA1 5ED
2E0 TGB W. Millington 93 Feiashill Road, Trysull, Wolverhampton WV5 7HT
2E0 TGC G. Cooper 21 Thistle Bridge Road, Chivenor, Barnstaple EX31 4FL
2M0 TGD I. Currie 4 Greendyke Cottage, Falkirk FK2 8PP
2E0 TGF A. Mcgoff 55 Knights End Road, March PE15 9QA
2E0 TGG T. Sutton Yew Tree Farm, Paddol Green, Shrewsbury SY4 5QZ
2E0 TGI T. Griffiths 6 Witney Lane, Edge, Malpas SY148JJ
2E0 TGJ T. Jones 27 Chamberlain Grove, Fareham PO14 1HH
2E0 TGK R. Willis 12 Robartes Road St. Dennis, St. Austell PL26 8DS
2E0 TGL A. Wallace 17 Dennis Road, Liskeard PL14 3NS
2M0 TGM T. Mcconnell 2 Stewartgill Place, Ashgill, Larkhall ML9 3BB
2M0 TGN A. Barclay 21 Netherlea, Scone, Perth PH2 6QA
2E0 TGO T. Gentry 8 Bowldown Cottages, Bowldown Road, Westonbirt GL8 8UD
2E0 TGQ T. Garcia-Quismondo 11 Half Moon Lane, Worthing BN13 2EN
2E0 TGV D. Cook 35 Elmwood Drive, Breadsall, Derby DE21 4GA
2E0 THE S. Griffiths 37 Stourton Close, Knowle, Solihull B93 9NP
2E0 THF C. Collins 27 Arnold Close, Castle Gresley, Swadlincote DE11 9HF
2E0 THG L. Bennett 19 Campion Crescent, Cranbrook TN17 3QJ
2E0 THP A. Thorpe 8 Syke Avenue, Tingley, Wakefield WF3 1LU
2E0 THR J. Summers 263 Stroud Road, Gloucester GL1 5JZ
2E0 THS E. Graham 25 South End Road, Ottringham, Hull HU12 0DP
2E0 THT A. Hoyte 43 Orchard Drive Mayland, Chelmsford CM3 6EP
2E0 THZ M. Hopewell 4 Cotes Crescent Bicton Heath, Shrewsbury SY3 5AS
2W0 TID S. Cook 114 Caerphilly Road, Bassaleg, Newport NP10 8LJ
2E0 TIF C. Townsend 64 Burnbridge Road Old Whittington, Chesterfield S41 9LR
2E0 TIN D. Tinn 28 South Road, Kirkby Stephen CA174SN
2W0 TIR G. Williams 1 Pont Y Berllan, Talycafn Road, Llanrwst LL26 0EF
2E0 TIV S. Bryan 6 Reeds Place, Fore Street, Cullompton EX15 1LA
2E0 TIZ A. Tyrrell Flat 1, 61 Vicarage Road, Eastbourne BN20 8AH
2E0 TJC T. Catton 97 High St., South Hiendley, Barnsley S72 9AN
2E0 TJD T. Dix Willow Cottage 31 London Road, Woolmer Green SG3 6JE
2E0 TJE T. Ellis 10 Digby Crescent, Water Orton, Birmingham B46 1NP
2I0 TJF T. Ferguson 3 Wheatfield Park, Ballybogy, Ballymoney BT53 6NT
2E0 TJG J. Wong St Edmund'S College, Mount Pleasant, Cambridgeshire CB3 0BN
2E0 TJI S. Guest 19 Ellesmere Avenue, Ashton-under-Lyne OL6 8UT
2E0 TJJ T. Jinkerson 104 Foxcote, Finchampstead, Wokingham RG40 3PE
2I0 TJK J. Mackenzie 30 Dalriada Gardens, Ballycastle BT54 6DZ
2I0 TJM T. Mulholland 215 Finaghy Road North, Belfast BT11 9ED
2E0 TJN T. Newton 4 Manor Close, Bradford Abbas, Sherborne DT9 6RN
2M0 TJO T. Johnston The Old Schoolhouse, Luggate Burn, Haddington EH41 4QA
2E0 TJP P. Neal 14 Hilltop Close, Desborough, Kettering NN14 2LQ
2E0 TJQ D. Roberts 61 Teign Bank Road, Hinckley LE10 0ED
2I0 TJR T. Ruddell 30 Ballynacor Meadows, Portadown, Craigavon BT63 5UU
2E0 TJS A. Steer 51 Kings Chase, East Molesey KT8 9DG
2E0 TJT J. Thompson 32 Church Street, Warnham, Horsham RH12 3QR
2E0 TJU E. Duffield 92 Crosby Street, Stockport SK2 6SP
2E0 TJV T. Humphreys 93 Cornwall Crescent Yate, Bristol BS37 7RU
2E0 TJW T. Ward 25 Chislehurst Road, Carlton Colville, Lowestoft NR33 8BY
2E0 TJX T. Jones 15 Kinnaird Road, Sheffield S50NN
2E0 TJY J. Yeo 10, Bradgate Close, Leicester LE79NP
2M0 TKE R. Mckie 16 Silver Street, , Creetown, Newton Stewart DG8 7HU
2E0 TKF D. Long 25 St. Matthias Road, Deepcar, Sheffield S36 2SG
2W0 TKS S. Davies 5 Maldwyn Street, Cardiff CF11 9JR
2E0 TKV T. King 24 Royston Avenue, Basildon SS15 4EW
2E0 TKX A. Sood Parima, Sewardstone Road, London E4 7RA
2E0 TKY S. Potter 93 Church Lane, Bocking, Braintree CM7 5SD
2E0 TLB F. Smith 13 Heather Walk, Crowborough TN6 2HA
2E0 TLC C. Parry 27 Tynedale Close, Stockport SK5 7NA
2E0 TLD N. Sinclair 11 Primrose Close, Warton, Preston PR4 1EN
2M0 TLE K. Quillien 1/2 81 Laurel Street, Glasgow G11 7QX
2E0 TLF T. Lamb 11 Queensway, Carlisle CA2 6SQ
2W0 TLG D. James Coombe House, Coombe, Presteigne LD8 2HL
2E0 TLJ R. Mcwilliam 3 Fountains Close, Riccall, York YO19 6QN
2E0 TLM I. Williams 36 Telford Road, Tamworth B79 8EY
2I0 TLT W. Thompson 25 Darby Road, Carrickfergus BT38 7XU
2E0 TLW C. Conghos 11A Rowans Way, Leavenheath, Colchester CO6 4UU
2E0 TLX D. Burdsall 37 Fulmar Walk, Whitburn, Sunderland SR6 7BW
2E0 TLY I. Bain 45 Larpool Crescent, Whitby YO22 4JD
2E0 TMA T. Ahern 39 Essex Road, Romford RM7 8BE
2W0 TMB M. Woodington 2 Tir Founder Fields, Aberdare CF44 0DT
2E0 TMC S. Oram 7 West Bank, Main Street, Old Weston, Huntingdon PE28 5LJ
2E0 TMD P. Kirby 36 Durham Road, London E12 5AX
2I0 TME T. Evans 30 Seagoe Park, Portadown, Craigavon BT63 5HR
2E0 TMF A. Taylor 11 Fillingfir Drive, Leeds LS16 5EG
2E0 TMM J. Woodcock 1 Main Street Brandesburton, Driffield YO258RL
2E0 TMN A. Macnauton 27A Lincoln Road, Poole BH12 2HT
2E0 TMO T. Williams Moor Farm, Moor Lane, Lincoln LN3 4EG
2E0 TMS T. Mackenzie 2 Newcastle Street, Carlisle CA2 5UH
2E0 TMX A. Lee 21 Percheron Place, Westbury BA13 2GR
2E0 TMY T. Haley 3 Akeman Rise, Ramsden, Chipping Norton OX7 3BJ
2W0 TNB N. Williams 60 Denbigh Close, Wrexham LL12 7TW
2E0 TNC T. Humphries 32 Bonds Meadow, Oulton Broad NR32 3QL
2E0 TNE D. Camp Kidbrooke Lodge, Lewes Road, Forest Row RH18 5AF
2M0 TNM P. Rae Tigh-Na-Mara, East Kilbride, Isle of South Uist HS8 5TS
2E0 TNN M. Tinnion 3 Hillhead Road, Newcastle upon Tyne NE5 5AP
2E0 TNO R. Naden 10 Suffolk Close, Holland On Sea, Clacton-on-Sea CO15 5SQ
2E0 TNV M. Clough 8 Skeldyke Road, Kirton, Boston PE20 1LR
2E0 TNZ S. Swallow 16 Quarry Lane, Chesterfield S40 3AS
2E0 TOF J. Morgan Glas Y Dorlan, Pontrhydfendigaid, Ystrad Meurig SY25 6EJ
2E0 TOG B. Whelan 147 Lawsons Road, Thornton-Cleveleys FY5 4PL
2M0 TOK J. Mcgurk Pf2, 18 Warriston Road, Edinburgh EH7 4HN
2E0 TOL S. Tomlinson 8 Levett Road, Stanford-le-Hope SS17 0BB
2E0 TOM T. Wilcox 7 Jacob Court, Billinge, Wigan WN5 7GE
2E0 TOP R. Styles 16 Hatton Park, Bromyard HR7 4EY
2M0 TOR S. Gibson 41 Kane Place, Stonehouse ML9 3NR

2E0 TOT S. Dix 8 Beaumont Road, Longlevens, Gloucester GL2 0EJ
2E0 TOX P. Tivey 24 Short Street, Burton-on-Trent DE15 9LS
2E0 TOY L. Bodnar 47 Alchester Court, Towcester NN12 6RL
2E0 TPA A. Patrick 2 Beacon Grange, Malvern WR14 3EU
2I0 TPC T. Crozier 6 Garden Of Eden, Carrickfergus BT38 7LS
2E0 TPD F. Harwood 1, South Highall Cottage, Lincolnshire LN10 6UR
2M0 TPE B. Angus 6 Inshes Crescent, Inverness IV2 3SP
2E0 TPG A. Gravell 21 Wickridge Close, Stroud GL5 1ST
2E0 TPH T. Hazel 84 Rodwell Avenue, Weymouth DT4 8SQ
2E0 TPI T. Idiculla 3 Stanhope Road, Slough SL1 6JR
2E0 TPL A. Capon 1 Windermere Way, North Common BS305XN
2E0 TPN G. Grace 194 Lillechurch Road, Dagenham RM8 2EW
2E0 TPP G. Finney 4 Cressida Court, Braunstone Lane, Leicester LE3 3AP
2E0 TPR T. Prince 23 Redbrook Avenue, Stockton-on-Tees TS19 9HJ
2E0 TPY T. Pollett 10 Bridport Road, Poole BH12 4BS
2E0 TQB D. Holmes 4 Council House, Nidds Lane, Boston PE20 1LZ
2E0 TQF C. Bailey 55 Bridgend Park Brewery Road, Wooler NE71 6QG
2M0 TRA T. Mussell Dunelm, Thornhill Road, Cuminestown, Turriff AB53 5WH
2W0 TRD R. George 18 Bryndedwyddfa, Penygroes, Llanelli SA14 7PR
2E0 TRE T. Ellis 84 Revelstoke Road, London SW18 5PB
2E0 TRH A. Hargreaves 27 Meadow Head Close, Blackburn BB2 4TY
2E0 TRI N. Rostant 14 Gas Street, Leamington Spa CV31 3BY
2E0 TRJ R. Nicholson 8 Meadway, Maghull, Liverpool L31 8AX
2E0 TRK A. Cooper 23 Ash St., Manchester M9 5XY
2D0 TRL T. Leece Thie Sy Cheyll, Ballastrooan IM9 4NR Isle of Man
2I0 TRM S. Davison 60 Cornation Place, Craigavon BT66 7AN
2E0 TRO M. Tromans 10 Crofters View, Little Wenlock, Telford TF6 5AU
2E0 TRP T. Parsons 5 Blackmoor Road, Aubourn, Lincoln LN5 9SX
2W0 TRR T. Rees 4 Maes Y Glo, Llanelli SA149PZ
2W0 TRS Dr D. Humphreys 48 High Street, Abergwynfi SA13 3YW
2E0 TRU G. Truman 3 Mulberry Road, North Anston, Sheffield S25 4BH
2E0 TRW C. Jacobs Flat 33, The Lodge, Lavender Road, Waterlooville PO7 8BX
2E0 TRX R. Moldoveanu 17 Lynchford Road, Farnborough GU14 6AR
2E0 TRY H. Chawdhry Trinity College, Cambridge CB2 1TQ
2E0 TSA T. Stamp 187 Knapmill Road, London SE6 3TF
2E0 TSC M. Hemmings 212 High St., Pensnett, Brierley Hill DY5 4JF
2D0 TSE A. Elliott Round Table House, Ronague, Castletown, Isle of Man IM9 4HJ
2E0 TSI S. Gilham 32 Whitby Road, Lytham St. Annes FY8 3HA
2W0 TSJ S. Trott 6 Mounton Drive, Chepstow NP16 5EH
2E0 TSO I. Macfarlane 70 Ashby Drive, Rushden NN10 9HH
2E0 TSP R. Connolly 29 Gayer Street, Coventry CV6 7EU
2M0 TSR D. Murphy 38 Lothian Road, Stewarton, Kilmarnock KA3 3BT
2E0 TSU S. Watson 5 Birchwood Avenue, Whickham, Newcastle upon Tyne NE16 5QS
2E0 TSV A. Reynolds 5 Broadlands , Broadmeadows , South Normanton, Alfreton DE55 3NW
2E0 TSY I. Oxenham 16 Chester Long Court, Vaughan Road, Exeter EX1 3WU
2E0 TTA T. Arscott 29 Parsonsfield Road, Banstead SM7 1JW
2E0 TTB T. Bate 4 Yoxall Road, Newborough, Burton-on-Trent DE13 8SU
2E0 TTE C. Macrae 16 Handfield Road, Liverpool L22 ONX
2M0 TTF J. Mcmorland 382 Maryhill Road, Glasgow G20 7YQ
2E0 TTG T. Hoyle 60 Greenbank Crescent, Marple, Stockport SK6 7PB
2E0 TTH P. Troth Beuna Vista, Hawford Wood, Droitwich WR9 0EZ
2E0 TTM T. Moncaster 24 Minster Yard, Lincoln LN2 1PY
2E0 TTN M. Groves 15 Plains Lane, Littleport, Ely CB6 1RJ
2E0 TTP S. Bailey 39 Aintree Drive, Balby, Doncaster DN4 8TU
2E0 TTS D. Brook 140 Dearne Hall Road, Barugh Green, Barnsley S75 1LX
2E0 TTT A. Rosenschein 101 Christchurch Road, London SW14 7AT
2E0 TTW R. Glynn 106 Fairway Avenue, West Drayton UB7 7AP
2E0 TTY A. Smith 93 Sheriffs Highway, Gateshead NE9 6QN
2E0 TUB C. Austen Fenbank House, Roman Bank, Holbeach Clough, Spalding PE12 8DH
2E0 TUC B. Tucker 12 Alpha Place, Appledore, Bideford EX39 1QY
2E0 TUD R. Bolton 5 Beacon Hill Avenue, Harwich CO12 3NR
2E0 TUE N. Wadsworth Haygarth, Docker, Kendal LA8 0DF
2E0 TUF J. Caswell 3 Pavillion Court, Roydon, Diss IP22 5SP
2E0 TUG A. Barrett-Sprot 1 Malting End, Wickhambrook, Newmarket CB8 8YH
2E0 TUH R. Taylor Penhawger Park, Liskeard PL14 3LW
2I0 TUI C. Stockdale 3 Hightown Drive, Newtownabbey BT36 7TG
2E0 TUK T. Goddard 217 Speedwell Road, Bristol BS5 7SP
2E0 TUM T. Mcgoun 64 Buttfield Lane, Howden, Goole DN14 7DS
2E0 TUN A. Tunney 79 Scott Street, Burnley BB12 6NJ
2E0 TUO J. Tuohy 13 Ely Place, Woodford Green IG8 8AG
2E0 TUR T. Baines 10 Croydon Avenue, Leigh WN7 1TP
2E0 TUS R. Austen Fenbank House, Roman Bank, Holbeach Clough, Spalding PE12 8DH
2E0 TUT C. Howard 75 Gordon Road, Herne Bay CT6 5QX
2I0 TUV C. Bailie 26 Moatview Park, Dundonald, Belfast BT16 2BE
2E0 TUW R. Harris 7 Fosse Lane, Shepton Mallet BA4 4PS
2E0 TUX N. Trangmar 8 Maxstoke Close, Meriden, Coventry CV7 7NB
2E0 TVD S. Pettit 24 Bickington Lodge Estate, Bickington, Barnstaple EX31 2LH
2E0 TVG T. Gunnoo 76 Katherine Drive, Dunstable LU5 4NU
2E0 TVM T. Nott 87 Powney Road, Maidenhead SL6 6EG
2E0 TVR S. Martin 35 Hermitage Green, Hermitage, Thatcham RG18 9SL
2E0 TVS N. Simmonds 3 Noneley Hall Barns, Noneley, Wem, Shrewsbury SY4 5SL
2E0 TVW A. Wilson 11 Headland Way, Alton, Stoke-on-Trent ST10 4AN
2E0 TVZ D. Darby 116 Middle Road, Southampton SO19 8FS
2E0 TWA A. Weatherall The Old Telephone Exchange, The Street, Canterbury CT3 1ED
2E0 TWD P. Roche 8 Fold Mews Hazel Grove, Stockport SK7 4BU
2E0 TWI C. Wild 203 High Street, Saltney, Chester CH4 8SJ
2E0 TWK A. Champion 4 Oldcastle Croft, Tattenhoe, Milton Keynes MK4 3EN
2E0 TWO K. Monaghan The Bulstone Hotel, Branscombe, Seaton EX12 3BL
2E0 TWP P. Miller 46 Great Brooms Road, Tunbridge Wells TN4 9DH
2E0 TWQ J. Attwood 13 John Winter Court, Euston Road, Great Yarmouth NR30 1DU
2E0 TWS T. Wood 44 Wincobank Lane, Sheffield S4 8AA
2E0 TWT J. Colbourne 43, Westfield Road, Bilston WV14 6EW
2E0 TWW T. Larman 861B London Road, Westcliff on Sea SS0 9SZ
2E0 TWX S. Waterson 7 West Farm Road, Newcastle upon Tyne NE6 4JA
2E0 TWZ N. Houghton 100 Ellerburn Avenue, Hull HU6 9RW
2M0 TXA T. Dorricott 11 Dakota Way, Renfrew PA4 0NP
2I0 TXB J. Morrison 9 Coral Cottages, Kilkeel BT34 4FT
2E0 TXE D. Parker 51 Ruskin Road, Congleton CW12 4EA
2E0 TXG T. Ruddick Hazel Gill, Croglin, Carlisle CA4 9RR
2E0 TXH T. Holden 132 Sutherland Street, Barrow in Furness LA14 2BJ
2E0 TXI P. Deeprose 2 Denehurst Gardens, Hastings TN35 4PB
2E0 TXJ J. Hopkins 53 Sprules Road, London SE4 2NL
2M0 TXK Prof. T. Kerby 1 St. Mark'S Lane, Edinburgh EH15 2PX
2E0 TXL P. Dunnicliffe 19 Woodland Road, Chelmsford CM1 2AT
2I0 TXM A. Mcgarvey 66A Scaddy Road, Downpatrick BT30 9BS
2E0 TXP P. Cassells 49 Dodds Lane, Maghull, Liverpool L31 0BD
2E0 TXQ B. Roberts 20 Mitchell Road, Enderby LE19 4NX
2M0 TXR A. Mcneill 13 Spinkhill, Laurieston, Falkirk FK2 9JR
2E0 TXU M. Holmes Lower Farm, Stony Moor, Newton-On-Rawcliffe, Pickering YO18 8QJ
2M0 TXY A. Caldwell 180 Pappert, Alexandria G83 9LG
2E0 TYC T. Corcoran 50 Grange Road, Bracebridge Heath, Lincoln LN4 2PW
2W0 TYE M. Armstrong Tan Yr Efail, Segurinside, Llandudno Junction LL31 9QE
2W0 TYG C. Morris 17 Percy Road, Wrexham LL13 7EA
2E0 TYH T. Hill 14 Hunters Mead, Motcombe, Shaftesbury SP7 9QG
2E0 TYL M. Tyler 40 Bullards Lane, Woodbridge IP12 4HE
2E0 TYT C. Gibson 17 Clyde Court, Grantham NG31 7RB
2E0 TYY T. Bradshaw 12 Whitegate Hill, Caistor, Market Rasen LN7 6SW
2M0 TZB M. Lawson 23 Kirkfield View, Livingston Village, Livingston EH54 7BP
2E0 TZD A. Maiden 79 Green End Road, Manchester M19 1LE
2E0 TZE F. Simpson 36 Lyndhurst Ave, Blackpool FY4 3AX
2E0 TZG T. Grabiec 16 Jubilee Crescent, Clowne, Chesterfield S43 4ND
2E0 TZM A. Laister 23 Berry Street, Greenfield, Oldham OL3 7EF
2E0 TZR T. Tozer 9 Bainbridge Court, Plymouth PL7 4HH
2E0 TZY S. Crabb 1 Council Houses, Hall Lane, Crostwick, Norwich NR12 7BB
2E0 TZZ P. Moore 24 Plough Road, Dormansland, Lingfield RH7 6PS
2W0 UAA M. Uphill 1 Brynview Avenue, Ystrad Mynach, Hengoed CF82 7DB
2E0 UAB C. Kent 19 Coppice Rise, Harrogate HG1 2DP
2E0 UAC M. Timms 29 Sutherland Avenue, Coventry CV5 7ND
2I0 UAD A. Pritchard 16 Ballymaconnell Road South, Bangor BT19 6DQ
2E0 UAE J. Firth 36 Howley Grange Road, Halesowen B62 0HW
2E0 UAK A. Vaile 66 Grasmere Point, Old Kent Road, London SE15 1DU
2M0 UAL K. Mackenzie Alderwood, Braes, Ullapool IV26 2TB
2E0 UAM C. Byrne 31 Graham Drive, Castleford WF10 3EY
2E0 UAO M. Lewis 6 Remembrance Road, Newbury RG14 6BA
2E0 UAP P. Richards 22 Waterpump Court, Northampton NN3 8US
2W0 UAR P. Barnes 55 Erw Werdd, Birchgrove, Swansea SA7 0HF
2W0 UAS C. Walker Perthi, Llaneilian, Amlwch LL689LY
2E0 UAV T. Ward Flat 26, Bassett Court, Bassett Avenue, Southampton SO16 7DR
2E0 UAW A. Wragg 14 Grizedale Avenue Sothall, Sheffield S20 2DL
2E0 UAX A. Pursell 48 St. Stephens Avenue, Ashtead KT21 1PJ
2E0 UAY D. Levey Heriots Wood, The Common, Stanmore HA7 3HT
2I0 UBE L. Calderwood 43 Rathview Park, Mullybritt, Enniskillen BT94 5EW
2E0 UBH R. Tolman 10 Woodcote Way, Abingdon OX14 5NE
2E0 UBM B. Mcdowell 10 Vineyard Lane, Kingswood, Wotton-under-Edge GL12 8SB
2E0 UBN B. Shephard 13 Forest Street, Annesley Woodhouse, Kirkby In Ashfield, Nottingham NG17 9HE
2E0 UBT C. Bowes 11 Burghwallis Lane, Sutton, Doncaster DN69JU
2E0 UBU D. Hampton 6 St. Georges Lane South, Worcester WR1 1QZ
2E0 UBW J. Cardwell 35 Bush Lane, Freckleton, Preston PR4 1SB
2E0 UCB C. Barber 32 Ashcroft Road, Ipswich IP1 6AB
2E0 UCD N. Booth Laburnum, Landcross, Bideford EX39 5JA
2E0 UCE S. Roderick 47 Cuckmans Drive., St. Albans AL2 3AY
2E0 UCH W. Toher The Chapel, Station Road, Darlington DL2 1JG
2W0 UCL K. Ucele 54 Sebastopol Street, St. Thomas, Swansea SA1 8BL
2E0 UCR C. Rogers 1 Treguthen Close, Camborne TR14 7EB
2I0 UCS M. Edwards 15 Highgrove Road, Carrickfergus BT38 9AG
2E0 UCT C. Thorne 67 Devon Road, Cadishead, Manchester M44 5HB
2E0 UCV T. Kilroy 55 Summerfield Crescent, Brimington, Chesterfield S43 1HB
2I0 UCY L. Stock 15 Mahon Drive, Portadown, Craigavon BT62 3JB
2E0 UDA A. Cattell 2 St. James Close, Ruscombe, Reading RG10 9LJ
2E0 UDB D. Beard 1 Bond Close, Leonard Stanley, Stonehouse GL10 3GQ
2E0 UDE R. Brown 24 Malthouse Court, Wellington, Telford TF1 1QJ
2E0 UDM D. Meehan 47 Clinton Road, Shirley, Solihull B90 4RN
2I0 UDR J. Mcclarence 10 Cloyfin Park, Coleraine BT52 2BL
2M0 UEA E. Ewing Arisaig Priestland, Darvel KA17 0LP
2E0 UED E. Cook 10 Firwood Drive, Gloucester GL4 0AB
2E0 UEE R. Goldsack 5 Parc Dellen, Croft Farm Park, Luxulyan, Bodmin PL30 5EW
2E0 UEH Dr S. Smith 557, Riverside Island Marina, Isleham, , Ely, CB7 5SL.
2E0 UEL M. Tointon 13 Ridgeway, Broadstone BH18 8DY
2E0 UER S. Mauer Flat 18, Channing Court, Osborne Road, London W3 8SY
2E0 UET R. Taylor 68 Charter Road, Chippenham SN15 2RA
2E0 UFM A. Parkhouse 3 St. Margarets Avenue, Ashford TW15 1DR
2M0 UFO S. Mcdougall 6F Watchmeal Crescent, Clydebank G81 5EA
2E0 UFS J. Baines 34 Crompton Road, Stone ST15 8NL
2E0 UFT F. Tombling 48 Venables Road, Guisborough TS14 6LQ
2E0 UGE R. Evans 11 Swane Road Stockwood, Bristol BS14 8NQ
2E0 UGF M. Sims 133 Canterbury Road, Hawkinge, Folkestone CT18 7BS
2M0 UGL N. Rogers 108 Beechwood Road, Cumbernauld, Glasgow G67 2NP
2E0 UGO D. Hill 19 Farren Road, Birmingham B31 5HH
2E0 UGP G. Palin 104 Nelson Street, Crewe CW2 7LN
2E0 UHF N. Booth Greenfield, Westmancote, Tewkesbury GL20 7EP
2E0 UHJ P. Hopkinson 28 Stockdove Way, Thornton-Cleveleys FY5 2AR
2E0 UHL M. Simpson 32 Underhill Lane, Wolverhampton WV10 8NS
2E0 UHS I. Phillpott 14 Buttercup Close, Paddock Wood, Tonbridge TN12 6BG
2E0 UID U. Decruz Victoria Road, Lowestoft NR33 9LR
2E0 UIP D. Woodhouse 1 Low Metham Cottages, Metham DN14 7YA
2E0 UIT A. Rackett 151 Stoke Road, Gosport PO12 1SE
2E0 UJD K. Colman 10 South Rise, North Walsham NR28 0EE
2E0 UJM R. Banks 3 Parkhayes, Woodbury Salterton, Exeter EX5 1QS
2E0 UJR J. Risby 112 Stratton Heights, Cirencester GL7 2RL
2E0 UKA A. Reay 12 Victoria Avenue, South Hylton, Sunderland SR4 0QZ
2E0 UKB J. English 1 Niton Cottage Pound Lane, Meonstoke, Southampton SO32 3NP

2E0 UKD J. Parfrey 47 Ford Lane, Rainham RM13 7AS
2E0 UKK G. Radulescu 41 Sherard Road, London SE9 6EX
2E0 UKM M. Qassim Winchester Road, Kings Somborne, Stockbridge SO20 6NY
2E0 UKT N. Radulescu 41 Sherard Road, London SE9 6EX
2E0 UKU S. Crowther 4 Brigsteer Close, Clayton Le Moors, Accrington BB5 5GE
2E0 UKW D. Watts 3 Witney Road, Crawley RH10 6GJ
2E0 UKX A. Fletcher 2 Brow Crescent, Halfway, Sheffield S20 4GB
2E0 UKY G. Smith 47 Percy Road, Carlisle CA2 6ER
2E0 ULC P. Power 16 Mainstone Close, Redditch B98 0PP
2M0 ULD N. Chalmers 4 Hercules Place, Arbroath DD11 4HT
2E0 ULH D. Butchart Flat 52, Seldon House, Stewarts Road, London SW8 4DP
2E0 ULY A. Ulyatt Flat 3, 64-66 Queen Street, Seaton SY8 1
2E0 UMD A. Slee Foal Cottage Mare Hill Common, Pulborough RH20 2DX
2E0 UMG M. Ribbands Dyson'S Farm, Long Row, Tibenham NR16 1PD
2M0 UMH L. Mitchell Hynd Smithy House, Bruichladdich, Isle of Islay PA49 7UN
2E0 UMO K. Laing 16 Cherrywood Drive Gonerby Hill Footá, Grantham NG31 8QL
2E0 UMP I. Gilbert 84 Carlyle Road, London W5 4BJ
2E0 UMR U. Munir Flat 6, Horton House, Field Road, London W6 8HW
2E0 UNI G. Rigby Gas House Farm, Shavington Park, Market Drayton TF9 3SY
2W0 UNY L. Betts 12A Maesgwyn, Pontnewydd, Cwmbran NP44 1BQ
2E0 UOJ K. Barry 25 Delabole Road, Merstham, Redhill RH1 3PB
2E0 UOK A. Abraham Flat 2, 41 Francis Road, Birmingham B33 8SL
2E0 UOM S. Johnson 43 Terry Gardens Kesgrave, Ipswich IP5 2EP
2E0 UPA J. Van Der Elsen 6 Kent Close, Churchdown, Gloucester GL3 2HQ
2W0 UPH A. Williams 8 Old Tanymanod Terrace, Blaenau Ffestiniog LL41 4BU
2E0 UPK D. Harmer 98 King Georges Avenue, Coventry CV6 6FF
2E0 UPM P. Mynors 109 Walton Road, Frinton-on-Sea CO13 0AB
2W0 UPR S. Stupple 7 Glan Preseli, Llanddewi Velfrey, Narberth SA67 7PG
2E0 UPS A. Clarke Little Acre, Pound Lane, Hardwicke, Gloucester GL2 4RJ
2E0 UPT S. Thomas 103 Liverpool Road, Upton, Chester CH2 1BB
2E0 UPU A. Stirk 59 West Avenue, Lightcliffe, Halifax HX3 8TJ
2E0 URA R. Scott 157 Walton Back Lane, Walton, Chesterfield S42 7LT
2E0 URD J. Ratcliffe 63 Dickens Lane, Poynton, Stockport SK12 1NN
2E0 URF R. Freeman 9 Bramley Road, Wisbech PE13 3PA
2E0 URJ Capt. R. Jordan Pheasants Walk, Copyhold Lane, Haslemere GU27 3DZ
2M0 URP S. Gray-Jones Flat C, 7 Nelson Street, Aberdeen AB24 5EP
2M0 URT H. Knox Shalderha Holm, Orkney KW17 2SA
2E0 USA A. Jepson 24 Shortbrook Close, Westfield, Sheffield S20 8LE
2E0 USC S. Coleman 32 Southwell Road, Wisbech PE13 3LQ
2E0 USD S. Dean 39 Low Grange View, Leeds LS10 3DT
2E0 USG S. Green 11 Lavender Walk, Beverley HU17 8WE
2E0 USH A. Usher 3 Meadow View, Patrington HU12 0QG
2W0 USK T. Keenan 54 Burrium Gate, Usk NP15 1TN
2E0 USM J. Wright 32 Carlton Road, Nottingham NG10 3LF
2W0 USN J. Barry Flat 3, 31 Ely Road, Cardiff CF5 2JF
2E0 USS L. Sanduly 1 Archford Croft, Emerson Valley, Milton Keynes MK4 2EZ
2E0 USV D. Soames 40 Woodland Drive, North Anston, Sheffield S25 4EP
2E0 UTB M. Mcdonald 55 Bournemouth Avenue, Middlesbrough TS3 0NN
2E0 UTC J. Alincastre 14919 Hope Hills Lane, Cypress, Texas 77433 United States
2E0 UTD A. Mclean 47 Tarn Drive, Bury BL9 9QB
2M0 UTH P. Dower 1670 Maryhill Road, Glasgow G20 0HJ
2E0 UTL A. Salt 1 Chantry Close, Harrow HA3 9QZ
2W0 UTT B. Bull Swan Cottage, Swan Road, Welshpool SY21 0RH
2E0 UTX A. Emmerson 31 Culver Road, Stockport SK3 8PG
2M0 UTZ C. Mackie Hanover Court, 5/3 Slaeside, Balerno EH14 7HL
2W0 UUA I. Ward 37 Maes Gwydryn, Abersoch, Pwllheli LL53 7ED
2E0 UUW J. Douch 63 Greenaways, Ebley, Stroud GL5 4UN
2E0 UVK S. Roberts Errys, Stunts Green, Herstmonceux, Hailsham BN27 4PP
2E0 UVO A. Ruocco 16 Conyers Avenue, Grimsby DN33 2BY
2E0 UVW A. Randle 2B Truman Street, Kimberley, Nottingham NG16 2HA
2E0 UVZ K. Such 38 Hornby Grove, Hull HU9 4PG
2E0 UWI R. Wilmot 41 Milton Brow, Weston-Super-Mare BS22 8DD
2E0 UWK J. Hadley 75 Glendower Avenue, Coventry CV5 8BD
2E0 UWO J. Coleman 67 Church Meadows, Deal CT14 9QZ
2E0 UWT A. Chapman 10 Derwent Road, Seaton Sluice, Whitley Bay NE26 4JH
2E0 UXC J. Mckie 59 Leaholme Terrace, Blackhall Colliery, Hartlepool TS27 4AB
2E0 UXM K. Taylor 9 Whitcliffe Grange Richmond North Yorkshire, Richmond DL10 4ES
2E0 UXS C. Dutton Twillingate Farm, Tiptoe, Lymington SO41 6EJ
2E0 UXV I. Renton 6 Parham Road, Bournemouth BH10 4BB
2E0 UYB J. Smith 9 Trafalgar Road, Newport PO30 1QD
2E0 UYF E. Martin 28 Mountford Close, Wellesbourne, Warwick CV35 9QQ
2E0 UZK A. Sweet 3 Beechwood Grove, Blackpool FY2 0DZ
2W0 UZO D. White 222 St. Fagans Road, Cardiff CF5 3EW
2E0 VAA K. Britain Blenheim Cottage, Falkenham, Ipswich IP10 0QU
2W0 VAC C. Rayment Brambles, Alltami Road, Mold CH7 6RT
2E0 VAF J. Edmunds 17 Stephens Road, Liskeard PL143SX
2E0 VAG S. Smith 12 Stoneleigh Avenue, Hordle SO41 0GS
2E0 VAM M. Higham 6 Holbeck Avenue, Morecambe LA4 6NP
2E0 VAN D. Stinson 1 The Croft, Earls Colne, Colchester CO6 2NH
2E0 VAO S. Pankhurst 57 Barley Croft, Harlow CM18 7QZ
2E0 VAS V. Papanikolaou 104 West Drive Gardens, Soham, Ely CB7 5EX
2E0 VAT M. Raynor 68 Cambridge Street, South Elmsall, Pontefract WF9 2AR
2E0 VAV A. Burnett-Provan 6 The Park-Dodwell Park-Evesham Road, Stratford-upon-Avon CV37 9SR
2W0 VAW I. Kane 44 Hafod Arthen Estate, Brynithel, Abertillery NP13 2HY
2E0 VBB S. Coulson 17 Charlton St., York YO23 1JN
2I0 VBH V. Hazelton 12 The Elms Bush, Dungannon BT71 6UE
2E0 VBJ A. Barrett 2 Friars Close, Clacton-on-Sea CO15 4EU
2E0 VBK Dr B. Kalogerakis Inglewood, Madingley Road, Cambridge CB23 7PH
2E0 VBL D. Wood 3 Ripley Close, Wakefield WF3 2FG
2E0 VBM A. Bairstow 12 Danesfield Avenue, Waltham, Grimsby DN37 0QE
2E0 VBN K. Sloan Woodland Halt, Old Station Road, Winchester SO21 1BA
2E0 VBR J. Baughan Chestnut Farm, Eastville, Boston PE22 8LX
2E0 VBS K. Florence 30 Lancaster Gardens, Ealing, London W13 9JY
2W0 VBT M. Kveksas 4 Llys Coed Derw, Llantwit Fardre, Pontypridd CF38 2JB
2E0 VBW B. Whall 3 Farrow Close, Great Moulton, Norwich NR15 2HR
2E0 VBX P. Otterwell 50 Hythe Road, Staines-upon-Thames TW18 3EE
2E0 VBY D. Potter 30 Mersham Gardens, Goring-By-Sea, Worthing BN12 4TQ
2E0 VCA A. Hill 24 Keelham Drive, Leeds LS19 6SG
2E0 VCB B. Ashall 22 Bloomer Wood View, Sutton-in-Ashfield NG17 1HA
2E0 VCC D. Jacobs 5 Tor View, Tregadillett, Launceston PL15 7HB
2E0 VCD D. Jones 102 Bryce Road, Brierley Hill DY5 4ND
2E0 VCE B. Mcguirk 9 Almond Crescent Standish, Wigan WN6 0AZ
2E0 VCG P. Robins 20 Saffron Close, Chineham, Basingstoke RG24 8XQ
2E0 VCK K. Vickers 11 Kendal Drive, Rainhill, Prescot L35 9JQ
2E0 VCM C. Mallory 11 Baymead Meadow, North Petherton, North Petherton TA6 6QW
2E0 VCP C. Pinder 70 Highfield Road, Beverley HU17 9QR
2E0 VCU P. Pritchard 12 Easton Crescent, Billingshurst RH14 9TU
2E0 VCW C. Gray The Pigsty, Cleverton, Chippenham SN15 5BT
2E0 VCY G. Shakespeare 6 Waterworks Cottages, Clough Road, Hull HU6 7QB
2E0 VCZ Dr C. Jenkins 52 Warden Abbey, Bedford MK41 0SN
2E0 VDC N. Fairbairn 15 Hewitt Road, Dover CT16 1TH
2E0 VDD L. Divall 8 Holme Close, Weymouth DT3 5RW
2E0 VDE R. Ferguson 31 Barton Court Road, New Milton BH25 6NW
2J0 VDJ D. Mashev Flat 23, 1875 Wesley Street, St. Helier, Jersey JE2 4DA
2E0 VDL D. Lodwig 15 Frithwood Park, Brownshill, Stroud GL6 8AB
2E0 VDQ A. Bolster 45 Headlands Drive, Hessle HU13 0JP
2E0 VDV C. Lemin 44 Barton Road, Berrow, Burnham-on-Sea TA8 2LT
2W0 VDW N. Thomas 57 Brynhyfryd Street, Treorchy CF42 6DT
2E0 VEF P. Johnson 90 North Road, Southport PR9 8QR
2W0 VEH D. Thomas 57 Brynhyfryd Street, Treorchy CF42 6DT
2E0 VEK K. Lowcock 43 Larch Street, Nelson BB9 9RH
2D0 VES C. Glaister Balleigh Villa, Jurby Road, Isle of Man IM8 3NZ
2E0 VET S. Froggatt 140 Greenlea Court, Huddersfield HD5 8QB
2E0 VEW V. Wheatley 22 Woodlands Avenue, Shelton Lock, Derby DE24 9FQ
2E0 VEX A. Crawford 4 Trimpley Drive, Kidderminster DY11 5LB
2M0 VEY A. Dunlop 12 Kendal Avenue 2/1, Glasgow G12 0DL
2E0 VEZ P. Barrows 5A Magdalen Road, Willoughby, Rugby CV23 8BJ
2E0 VFA A. Colcombe 217 Church Drive, Quedgeley, Gloucester GL2 4US
2E0 VFD S. Harriott 188 Tean Road, Cheadle, Stoke-on-Trent ST10 1NQ
2E0 VFN B. Bewick 88 Bentley Way, Weston Road, Norwich NR6 6TS
2I0 VFO J. Sills 145 Ballycolman Estate, Strabane BT82 9AJ
2E0 VFT A. Middleton 89 Crediton Road, Okehampton EX20 1NU
2M0 VFV S. Young 103 Feorlin Way, Garelochhead, Helensburgh G84 0EB
2E0 VFX S. Bint 5 Chapel Corner, Hullavington, Chippenham SN14 6RT
2E0 VGA S. Unsworth Heath Terrace, Towcester NN128UP
2E0 VGB V. Greenway-Brown 207 Lowe Avenue, Wednesbury WS10 8NS
2E0 VGC D. Carter 18 Silent Woman Park, Coldharbour, Wareham BH20 7PE
2E0 VGF P. Williamson 17 Shaw Street Biddulph, Stoke-on-Trent ST8 6JE
2E0 VGK B. Lunn 204A Main Street, Horsley Woodhouse, Ilkeston DE7 6AX
2M0 VGT E. Smith 15 School Wynd, Quarrier's Village PA11 3NL
2E0 VGV G. Venugopalan 3 Southwater Close, London E14 7TE
2M0 VGY E. Blakeway 25 Allanton Grove, Wishaw ML2 7LL
2E0 VHA R. Parker 29 Hill Lea Gardens, Cheddar BS27 3JH
2E0 VHC M. Sands Room 3, 8 Upperton Gardens, Eastbourne BN21 2AH
2E0 VHF C. Craswell 49 Alexandria Walk, Cheltenham GL525LG
2E0 VHQ A. Lyke Nupton Farm Canon Pyon, Canon Pyon, Hereford HR4 8PW
2E0 VHV L. Kelly 9 Ham Lane, Farrington Gurney, Bristol BS39 6TW
2E0 VHZ J. Telfer 50 Agraria Road, Guildford GU2 4LF
2E0 VIA Dr L. Kirkcaldy 19 Itchen Avenue, Bishopstoke, Eastleigh SO50 8JW
2E0 VIB W. Pilling 12 Brooke Close, Accrington BB5 2QX
2E0 VID J. Brewer 1 Bentley Road, Forncett St. Peter, Norwich NR16 1LH
2E0 VIS A. Oliveira 13 A Lakefield Road, London N22 6RR
2E0 VIT A. Vitiello 8 Pegasus Road, Leighton Buzzard LU7 3NJ
2E0 VJB V. Bowkett 9 Gwealmayowe Park, Helston TR13 0PE
2E0 VJH R. Heming Milepost Cottage, Benenden Road, Ashford TN27 8BY
2D0 VJK R. Smith 3 Rheast Barrule, Castletown IM9 1HW
2W0 VJL V. Lea Rose Cottage, Y Ffor, Pwllheli LL53 6UR
2E0 VJO J. Sawyer 27 Croft Road, Wallingford OX10 0HN
2M0 VJS J. Scott 81 Kinellar Drive, Glasgow G14 0EU
2E0 VJX B. Walker 255 Packington Avenue, Birmingham B34 7RU
2W0 VKA A. Vincent 88 Lake Street, Ferndale CF43 4HE
2E0 VKB K. Davies 23 Egmanton Road, Meden Vale, Mansfield NG20 9QN
2E0 VKG A. Smith 305, 1414 5Th St Sw, Calgary T2R 0Y8 Canada
2E0 VKK R. Cresswell Meadow View, Hulver Road, Beccles NR34 7UW
2E0 VKM S. Latimer 40 Petersham Road, Long Eaton, Nottingham NG10 4DD
2E0 VKN I. Astley 6 Shay Court, Crofton, , Wakefield WF4 1SL
2M0 VKO S. Macdonald 366 Millcroft Road Cumbernauld, Glasgow G67 2QW
2E0 VKQ R. Vickerstaff 16 Sewell Wontner Close, Kesgrave, Ipswich IP5 2GB
2E0 VKR V. Rotaru 15 Leicester Drive, Glossop SK13 8SH
2E0 VKS D. Vickers 178 Bakewell Road, Matlock DE4 3BA
2E0 VKW V. Williams Moor Farm, Moor Lane, Lincoln LN3 4EG
2M0 VKY V. Fleming Tarbat View, Achavandra Muir, Dornoch IV25 3JB
2E0 VKZ M. Hillman Flat 5, 32 South Terrace, Littlehampton BN17 5NU
2E0 VLB Dr V. Boev Flat 4, Camborne House, Sutton SM2 6RL
2E0 VLD S. Sawyer 85 Beechwood Road, Sheffield S6 4LQ
2M0 VLF I. Mcglynn 25 Fairhill Avenue, Hamilton ML3 8JS
2E0 VLL L. Burbidge 33 Burcote Fields, Towcester NN12 6TH
2W0 VLO B. Henley Rhewin Glas, Llandysul SA44 6PR
2E0 VLR P. Moxhay 6 York Road, Torpoint PL11 2LG
2E0 VLT P. Honey 3 Peterswood, Harlow CM18 7RJ
2E0 VMA C. Skupski 57 Three Nooks, Bamber Bridge, Preston PR5 8EN
2W0 VMC J. Argent 7 Lloyds Hill, Buckley CH7 3ER
2E0 VMD H. Melhuish 22 Mayflower Close, Glossop SK13 8UD
2E0 VMG S. Brett 51 West End, Wirksworth DE44EG
2D0 VMN V. Matthewman Monte Rosa, 7 Ballaughton Close, Isle of Man IM2 1JE
2E0 VMV R. Vale 611 College Road, Birmingham B44 0AY
2E0 VNE S. Swinton 4 Fallow Road, Newton Aycliffe DL5 4SU
2E0 VNL M. Harris 5 Lynmore Close, Northampton NN4 9QU
2E0 VNN V. Nikolaidis 35-46 Ernst Chain Road, Manor Park, Guildford GU2 7YW
2E0 VNO D. Harwood 36 Seaview Drive, Great Wakering, Southend-on-Sea SS3 0BE
2E0 VNT V. Sheppard Flat 8, Riverside Court Cambridge Road, Harlow CM20 2AD
2E0 VNV D. Nicholls 12 Northview, Tufnell Park Road, London N7 0QB
2M0 VNW A. Sim 44 Hillmoss, Kilmaurs, Kilmarnock KA3 2RS
2E0 VNX A. Wright 32 Temple Grove, Leeds LS15 0HT
2E0 VNY V. Brindle 185 Brunshaw Road, Burnley BB10 4DL

2M0 VOB D. Johnson 8 Sandmartin Grove, Lenzie, Kirkintilloch, Glasgow G66 3WF
2W0 VOC A. Parsons 21 Rectory Drive, St. Athan, Barry CF62 4PD
2E0 VOD N. Mclean 21 Matlock Avenue, Wigston LE18 4NA
2I0 VOF P. Mcfadden 35 West Wind Terrace, Hillsborough BT26 6BS
2W0 VOG S. Cawsey 8 Hickman Road, Penarth CF64 2AJ
2I0 VOQ D. Sinton 34 West Link, Holywood BT18 9NX
2M0 VOZ C. Docherty 23 The Maltings, Haddington EH41 4EF
2E0 VPA P. Larcombe 55 Forest Drive, Weston-Super-Mare BS23 2UG
2E0 VPG J. Jackson 80A Clarence Road, Leighton Buzzard LU7 3EL
2E0 VPI V. Prystaj 69 Kingsdale Crescent, Bradford BD2 4DP
2E0 VPJ C. Rainbow 13 Audley Road, Talke Pits, Stoke-on-Trent ST7 1UG
2E0 VPL J. Ross-Monclus 2 Putland Place Harrietsham, Maidstone ME17 1SZ
2M0 VPM A. Cowan 32 Esk Valley Terrace, Dalkeith EH22 3FT
2E0 VPN D. Forbes Flat 3, Arundel Court 1 Cherrywood Drive, London SW15 6DS
2E0 VPO R. Dean 15 Gorge Road, Dudley DY3 1LF
2W0 VPT Keighley ARS c/o G. Taylorr 17, Llygad Yr Haul, Glynneath SA11 5RL
2M0 VPU J. Smith 82 Overton Road, Netherburn, Larkhall ML9 3BT
2E0 VRA K. Cope 9 Amber Heights, Ripley DE5 3SP
2E0 VRB A. Morehen 20 Castleton Grove, Inkersall, Chesterfield S43 3HU
2E0 VRC A. Newbould 9 Laburnum Road, Rudheath, Northwich CW9 7JT
2E0 VRD V. Bowen 4 Crossley Gardens, Halifax HX1 5PU
2E0 VRE D. Easden 20 Brunel Way, Calne SN11 9FN
2E0 VRO G. Bryant 15 The Clock Inn Park Lydeway., Devizes. SN10 3PP
2E0 VRP I. Bell 164 The Broadway, Herne Bay CT6 8HY
2E0 VRR R. Oxley 17 Hardhurst Road, Alvaston, Derby DE24 0LF
2E0 VRT J. Garwood 4 Ryedale, Carlton Colville, Lowestoft NR33 8TB
2E0 VRX C. Bradley 47 Long Meadow, Skipton BD23 1BP
2E0 VRZ A. Dean 17, The Grange, 51 Gwendolyn Drive, Coventry CV3 1QU
2E0 VSE Dr S. Vellaichamy 37 Briarswood, Chelmsford CM1 6UH
2M0 VSM D. Tilley Flat 2/1, 7 Porterfield Road, Renfrew PA4 8JB
2W0 VSW V. Wallace 10 Maes Llydan, Benllech, Tyn-Y-Gongl LL74 8RD
2E0 VSZ F. Zanchi Flat 2 4 Helios Road, London SM6 7BZ
2E0 VTC J. White 10 Meaux Road, Wawne, Hull HU7 5XD
2E0 VTF S. Campbell 50 Northwood Road, Whitstable CT5 2ES
2W0 VTK J. Martin 78 Llwyn Ynn, Talybont LL43 2AG
2W0 VTP M. Davies 26 Ty Nant, Caerphilly CF83 2RA
2E0 VTR N. Burton 11 Weldon Avenue, Stoke-on-Trent ST3 6PN
2E0 VTS P. Wilkes 8 Cloverdale, Stafford ST17 4QJ
2E0 VTT J. Owen 8 Highridge Crescent, Bristol BS13 8HN
2E0 VTV I. Roberts 15 Broadcroft, Hemel Hempstead HP2 5YX
2I0 VTZ S. Gore 5 Rosebrook, Dungiven, Londonderry BT47 4GA
2E0 VUB S. Daley 1 North Green, Calverton, Nottingham NG14 6NT
2E0 VUK B. Lewis 68 Irwin Avenue, Rednal, Birmingham B45 8QU
2E0 VUL K. Anand 19 Tempest Road, Upper Cambourne, Cambridge CB23 6HW
2E0 VUM D. Bailey 1 The Magpies, Maulden MK45 2EG
2M0 VUS C. Doherty 22 Castlelaw Street, Glasgow G32 0NF
2M0 VUV R. Fraser 72 Ferguson Drive, Denny FK6 5AG
2E0 VVA A. Amos Willow Tree House, Deers Green, Clavering, Saffron Walden CB11 4PX
2E0 VVB J. Hunter 53 Grove Road, Tiptree, Colchester CO5 0JJ
2E0 VVE K. Macmanus 23 Mount Pleasant Residential Park, Bloomhill Road, Moorends Doncaster DN8 4ST
2E0 VVF A. Blackburn Anvil House Strait Lane, Hurworth, Darlington DL2 2AH
2E0 VVJ T. Sandham 96 South Road, Morecambe LA4 6JS
2E0 VVK P. Dickson 23 Balmoral Avenue, Huddersfield HD4 5LL
2E0 VVM P. Hatfield 4 Fyfield Close, Blackwater, Camberley GU17 0HL
2W0 VVO S. Barry 1 Pearson Cottages, St. Brides, Haverfordwest SA62 3BN
2M0 VVS I. Lindsay 265 Stirling Street, Denny FK6 6QJ
2E0 VVX J. Platt 12 Tawny Grove Four Marks, Alton GU34 5DU
2E0 VWD G. Killeen 119 Sutton Road, Walsall WS5 3AG
2E0 VWF D. Smith 3 Skipton Close, Corby NN18 0NS
2E0 VWG M. Cross 11 Polyplatt Lane, Scampton, Lincoln LN1 2TL
2E0 VWJ J. Cooke The Old Cottage, Church Lane, Lower Moor, Pershore WR10 2PJ
2E0 VWK M. Poole 15 Roberts Place, Dorchester DT1 2JJ
2E0 VWL J. Campion 25 Suffield Road, Liverpool L4 1UL
2E0 VWT C. Poole 15 Devon Close, Macclesfield SK10 3HB
2E0 VWW S. Hegarty 31 Beaconsfield Road, Deal CT14 7DA
2E0 VWX T. Mcbride 53 Blackdown Grove, St. Helens WA9 2BD
2M0 VXB M. Al Saeed 9 Appin Place, Edinburgh EH14 1NJ
2E0 VXI D. Holden 24 Penny Gate Close, Hindley, Wigan WN2 3DP
2M0 VXL J. May 12 Clochbar Gardens, Milngavie, Glasgow G62 7JP
2E0 VXT A. Calvert 5 Pond Cottages Butts Pond, Sturminster Newton DT10 1BE
2E0 VXX T. Quiney 20 Britannia Gardens, Stourport-on-Severn DY13 9NZ
2E0 VYN M. Roberts 11 Oakleigh Road, Pinner HA5 4HB
2E0 VYW A. Willsher 1 Tolputt Court Gladstone Road, Folkestone CT19 5NE
2E0 VZL S. Haycock 51 South Crescent, Southend-on-Sea SS2 6TB
2M0 VZX D. Mcintosh Crunes Way, Greenock PA15 2WH
2E0 WAA O. Prin 19 The Colliers, Heybridge Basin, Maldon CM9 4SE
2E0 WAE A. Ward 29 Mainwaring Road, Wallasey CH44 9DN
2E0 WAF H. Burch 46 School Lane, Horton Kirby, Dartford DA4 9DQ
2E0 WAG D. Wagstaff 68 Braziers Quay, South Street, Bishop's Stortford CM23 3YW
2I0 WAH T. Quin 165 Marlacoo Road, Portadown, Craigavon BT62 3TD
2I0 WAI M. Mcerlean 38A Culbane Road Portglenone, Portglenone BT448NZ
2E0 WAJ W. Johnson 145 Netherton Road, Worksop S80 2SA
2E0 WAK P. Holton 66 Mill Road, Gillingham ME7 1JB
2E0 WAP A. Woodhouse 4 Grafton Close, St. Albans AL4 0EX
2E0 WAT A. Watmough 37 Heath Park Road, Buxton SK17 6NY
2E0 WAV A. Snelson 6 Rayleigh Close, Braintree CM7 9TX
2E0 WAY W. Davies 17 Oakdale Avenue, Harrogate HG1 2JN
2I0 WBD W. Mcdonald 14 Edenmore Park, Limavady BT49 0RG
2E0 WBE D. Buckley 99 Kings Drive, Bradwell, Great Yarmouth NR31 8TF
2I0 WBF W. Turkington 8A Drummullan Road, Moneymore, Magherafelt BT45 7XS
2E0 WBH W. Howie 152 Norwood Road, Birkby, Huddersfield HD2 2YD
2E0 WBI G. Birch 23 Hanson Street, Great Harwood, Blackburn BB6 7LP
2M0 WBJ W. Jackson 3 Annick Road, Dreghorn, Irvine KA11 4EY
2E0 WBL A. Riley 35 Ross Avenue, Wirral CH46 2SA
2E0 WBO W. Jones 50 Bridge Place, Croydon CR0 2BB
2E0 WBQ D. Hodgson 11 Harmony Place, Mountain, Queensbury, Bradford BD13 1LD
2E0 WBR W. Reeves 33 Pond Bank, Blisworth, Northampton NN7 3EL
2E0 WBS A. Whadcoat 38 Edwin Panks Road, Hadleigh, Ipswich IP7 5JL
2E0 WBT B. Weston 10 Clement Drive, Peterborough PE2 9RQ
2I0 WBU W. Bradley 16 Mullaghanagh Lane, Dungannon BT71 7NY
2E0 WBW R. Howard 13 Top Common, East Runton, Cromer NR27 9PW
2E0 WBX A. Coats 57 Mill Hill, Boulton Moor, Derby DE24 5AF
2E0 WBZ K. Hunt 11 De Marnham Close, West Bromwich B70 6RJ
2E0 WCB A. Williams 74 Broadfield Road, Bristol BS4 2UW
2E0 WCC C. Calvert 1 Moorsholme Avenue, Manchester M40 9BW
2E0 WCE M. Hall 259 Lambourn Drive, Allestree, Derby DE22 2UR
2E0 WCL O. Fallon 26 Central Avenue, Corfe Mullen, Wimborne BH21 3JD
2E0 WCM D. Neumunn 92 Miner Street, Walsall WS2 8QL
2E0 WCO P. Austen 20 Victor Close, Seaford BN25 2JQ
2W0 WCP Dr E. Harries Ty Traeth, Caerwedros, Llandysul SA44 6BS
2W0 WCQ E. Edwards 1 Brynhyfryd, Sarn, Bridgend CF32 9UR
2E0 WCX W. Dix 21 Pine Vale Crescent, Bournemouth BH10 6BG
2M0 WDA W. Shaw Shaws Farm, Selkirk TD7 4PR
2E0 WDB W. Bull 117 Walton Road, Wednesbury WS10 0EU
2I0 WDD D. Milligan 30 Belgrano Ahoghill, Ballymena BT42 2QQ
2M0 WDG W. Goodfellow 1 Yester Place., Haddington EH41 3BE
2E0 WDH Dr W. Henderson 14 Highfield Road, Newcastle upon Tyne NE5 5HS
2E0 WDI I. Woollen 33 The Oaks, Taunton TA1 2QX
2E0 WDM S. Thompson 64 Church Road, Fordham, Colchester CO6 3NJ
2E0 WDN S. Gibson Radio Licence, 122 Bridge Street, Whaddon, Royston SG8 5SN
2M0 WDP D. Park 54 Coblecrook Gardens, Alva FK12 5BL
2E0 WDR S. Lacey 2 Purbeck Cottages, Acton, Swanage BH19 3LU
2E0 WDS W. Sargeant 2 Church Mews, Judith Road, Kettering NN16 0QR
2E0 WDW W. Johnson Clock House, Main Road, Boreham, Chelmsford CM3 3JD
2E0 WDY C. Johnson 45 Gordon Road, Chelmsford CM2 9LN
2E0 WEC C. Newton 7 Moss Close, Bridgwater TA6 4NA
2E0 WED D. Richards 58 Holm Lane, Oxton, Prenton CH43 2HS
2E0 WEE N. Froggatt 6 Beech Grove, , New Malden KT3 3HR
2E0 WEF I. Chambers 2 Belford Road, Borehamwood WD6 4HY
2E0 WEG W. Gray 39 Guest Avenueá, Poole BH121JA
2E0 WEI B. Fitchett 1 Hilly Fields Mews, Parsonage Estate, Rogate, Petersfield GU31 5BF
2E0 WEJ W. Jefferies 26 Norcutt Road, Twickenham TW2 6SR
2E0 WEK K. Alabaster 16 Butlers Road, Horsham RH13 6AJ
2E0 WEL W. Easdown 11 Mulcaster Avenue, Kidlington OX5 2HG
2E0 WEO S. Kiel 32 Weavers Avenue, Frizington CA26 3AT
2E0 WES S. Weston 73 Priory Road, Ashton-In-Makerfield, Wigan WN4 9UP
2E0 WET A. Forrest 1 Errington Bungalows, Sacriston, Durham DH7 6NE
2M0 WEV G. Weir 95 White Street, Whitburn, Bathgate EH47 0BH
2E0 WEZ J. Weston 25 Cambridge Road, Orrell, Wigan WN5 8PL
2W0 WFB L. Bowman Chanrick, Penderyn Road, Aberdare CF44 9RU
2E0 WFC J. Paradas 1A Brocks Ghyll, Eastbourne BN20 9RQ
2E0 WFD P. Blundell 7 Benwood View, Crofton, Wakefield WF4 1NP
2E0 WFG W. Griffiths 68 Altcar Lane, Formby, Liverpool L37 6AY
2D0 WFH W. Hogg Medhamstead, Lhergydhoo, Isle of Man IM5 2AE
2D0 WFK R. Kissack 6 Falcon Cliff Court, Douglas, Isle of Man IM2 4AQ
2M0 WFN W. Noon 0/1 445 Royston Road, Glasgow G21 2DE
2E0 WFR G. Kirby East Road, West Mersea CO5 8EB
2E0 WFS R. Wormald 10 Erica Court, Woking GU22 0JB
2E0 WFW W. Fletcher-Wells 46 London Road, Buxton SK17 9NU
2E0 WFY K. Payne Eastern Esplanade, Broadstairs CT10 1DR
2E0 WGB G. Beale 34 Teville Road, Worthing BN11 1UG
2E0 WGC C. Watkins 25 Citadilla Close, Gatherley Road, Richmond DL10 7JE
2E0 WGD D. Warriner 1 St. Johns Avenue, North Hykeham, Lincoln LN6 8QR
2U0 WGE R. Batiste Asile De Paix, Clos Des Sablons, Sandy Lane, Guernsey GY2 4RN
2E0 WGI S. Sugihara Southfield, Park Lane, Wokingham RG40 4PY
2I0 WGL W. Leonard 57 Old Coach Road Mullanavehy, Enniskillen BT92 2EW
2I0 WGM G. Mccusker 10 Birchdale, Lurgan, Craigavon BT66 7TR
2W0 WGN C. Morris Hideaway Bettws Cedewain, Newtown, Powys SY16 3DS
2E0 WGO I. Paterson 11 Ocho Rios Mews, Eastbourne BN23 5UB
2E0 WGP W. Power 111 Woodlands Road, Ditton, Aylesford, Maidstone ME206EF
2E0 WGT B. Withington 20 Bond Way, Hednesford, Cannock WS12 4SN
2E0 WHA W. Armes 11 Rutland Road, Broadheath, Altrincham WA14 4HW
2E0 WHB B. Marks 167 Linnet Drive, Chelmsford CM2 8AH
2E0 WHD J. Gardner Silverdale, Vicarage Lane, Ormskirk L40 6HQ
2E0 WHF D. Cracknell 120 Woodhill, London SE18 5JL
2E0 WHH J. Cook 20 Huntingdon Close, Totton, Southampton SO40 3NX
2E0 WHO M. Wells 42 Eggesford Road, Stenson Fields, Derby DE24 3BH
2E0 WHP D. Cooper 58 Serpentine Road, Widley, Waterlooville PO7 5EF
2E0 WHT K. Snipe 5 Draycott Road, Chiseldon, Swindon SN4 0LT
2E0 WHU A. Hodgson 515 Ashingdon Road, Rochford SS4 3HE
2M0 WHX D. Strachan 30 Belhaven Park, Muirhead G69 9FB
2E0 WIE L. Shasby 19 Crawshaw Grange, Crawshawbooth, Rossendale BB4 8LY
2E0 WIG J. Shaw 54 Dicconson Street, Wigan WN1 2AT
2E0 WIL W. Whyatt 686 Whitchurch Lane, Whitchurch, Bristol BS14 0EJ
2E0 WIO D. Owen 8 Crag Bank Crescent, Carnforth LA5 9EQ
2E0 WIS D. Wiskow 15 Ferndale Close, Sandbach CW11 4HZ
2E0 WIV P. Sanders Alresford Road, Wivenhoe CO7 9JX
2E0 WIZ S. Keen 13 Ivy Road, Kettering NN16 9TG
2E0 WJB W. Bradley 4 Forest View Avenue, London E10 6DX
2E0 WJC W. Cromack 45 Southroyd Park, Pudsey LS28 8AX
2E0 WJE W. Ellis 16 Furlong Drive, Tean, Stoke-on-Trent ST10 4LD
2E0 WJF W. Furnell 4 The Gap, Canterbury CT13NN
2E0 WJH M. Hooper 2 Captains Parade, East Cowes PO32 6GT
2E0 WJI J. Dale 47 Mungo Park Road, Rainham RM13 7PD
2E0 WJL W. Lloyd 14 Lewis Grove, Wolverhampton WV11 3HR
2I0 WJM W. Mcclean 6 Alveston Drive Carryduff, Belfast BT8 8RL
2M0 WJP W. Paterson 1 Burnside Terrace, Stranraer DG9 8HH
2E0 WJR D. Howard 3 Ashdown Drive, Crawley RH10 5AB
2M0 WJS S. Wilson 2 Kinnear Court, Guardbridge, St. Andrews KY16 0UE
2E0 WJT W. Twemlow Flat 6, 27 Marmion Road, Liverpool L17 8TT
2E0 WJW J. Withington 20 Bond Way, Hednesford, Cannock WS12 4SN
2I0 WKE J. Wilkinson 11 Fairview Park, Dromore BT25 1PN
2E0 WKG A. Cole 104 Newport Road, Cowes PO31 7PS

2E0 WKM C. Wise 32 Commercial Street, Willington DL15 0AD
2E0 WKT G. Williams 18 Elmsleigh Road, Farnborough GU14 0ET
2E0 WKU G. Nightingale 21 Turners Close, Ongar CM5 9HH
2E0 WKV K. Dale 26 Warwick Place, Langdon Hills, Basildon SS16 6DU
2E0 WKW E. Stammers 40 Tillingbourne Road, Shalford, Guildford GU4 8EY
2I0 WKY P. Ffitch 1 Lisburn Road, Ballynahinch BT24 8BL
2E0 WKZ B. Drury 6 Ellen Grove, Harrogate HG1 4RH
2M0 WLA W. Lawson 60 Inglis Avenue, Port Seton EH32 0AQ
2E0 WLD W. Daley 27 Rosebery Street, Manchester M14 4UR
2E0 WLK R. Readman 1 Millside Close, Kilham, Driffield YO25 4SF
2E0 WLN J. Wilson 125 Langroyd Road, Colne BB8 9ED
2E0 WLQ J. Clark 4 Exeter Street, North Tawton EX20 2HB
2M0 WLX B. Ewart 94 Kirkness Street, Airdrie ML6 6ET
2E0 WLY M. Walters 39 Portland Place, Coseley, Bilston WV14 9TB
2W0 WMB A. Gibbs 105 Oak Place, Bargoed CF81 8NT
2I0 WMC W. Mccormick 6 Church Street, Rosslea, Enniskillen BT92 7DD
2E0 WMD M. Peters 9 Evelyn Close, Twickenham TW2 7BL
2E0 WMG K. Pugh Col Bern, Church Rd, Colchester CO7 8HS
2I0 WMH W. Hawkes 12 Meadow Court, Newtownards BT23 8YE
2M0 WMJ W. Mackenzie 7 Urquhart Grove, Elgin IV30 8TB
2M0 WML W. Little Burnside, Main Street, Lochans, Stranraer DG9 9AW
2E0 WMM J. Wilkinson 8 Hunters Point, Chinnor OX39 4TG
2I0 WMN W. Mcmullen 69 Lissize Avenue Rathfriland, Newry BT34 5DE
2E0 WMP M. Weaver 16 Avocet Drive, Kidderminster DY10 4JT
2E0 WMT R. Tew 61 Magna Road, Bournemouth BH11 9ND
2M0 WMU Dr M. Marino 35 Niddrie House Park, Edinburgh EH16 4UH
2E0 WMY C. Walmsley 6 Holly Close, Brighton BN1 6RZ
2M0 WNB W. Bennett 8 Laxay, Lochs, Isle of Lewis HS2 9PJ
2I0 WNC T. Marshall 13 Glenburn Park, Glynn, Larne BT40 3DH
2E0 WNI R. Karpinski 55 Cambridge Avenue, New Malden KT3 4LD
2E0 WNL S. Adams 1 Byford Way, Winslow, Buckingham MK18 3RJ
2E0 WNM W. Mccoo Ivy House, West Drove South, Walpole Highway, Wisbech PE14 7RA
2E0 WNO R. Springall 7 The Spinney Grange Park, Northampton NN4 5BT
2E0 WNP P. Newman Butlers Hill Farm Weights Lane, Redditch B97 6RQ
2E0 WNS C. Pryke 50 Raglan Gardens, Watford WD19 4LL
2E0 WNT T. Corker North Side, Wingerworth Hall Estate, Chesterfield S42 6PL
2E0 WNW A. Walker 4 Pretymen Crescent, New Waltham, Grimsby DN36 4NS
2D0 WNY C. Larkham Monte Rosa , 7 Ballaughton Close, , Douglas IM2 1JE
2E0 WOB R. Landragin 101 Linden Gardens Enfield En1 4Dy United Kingd101, England EN1 4DY
2W0 WOD W. Davies Foelallt, North Road, Aberystwyth SY23 2EL
2E0 WOF C. Elliott 51 Lower Street Quainton, Quainton HP22 4BL
2E0 WOH B. Pyatt 16 Colin Blythe Road, Tonbridge TN10 4LB
2E0 WOL W. Walther 57 A Lower Court Road, Epsom KT198SW
2E0 WOS W. Barnes Cushendall, Lyngate Road, North Walsham NR28 0DH
2E0 WOW D. Martin Kiln Close, Main Road, Lincoln LN4 4QH
2W0 WOY C. Wood 50 Heather Court, Cwmbran NP446JR
2E0 WOZ H. Greenhalgh 61 Long Meadows, Chorley PR7 2YB
2E0 WPB W. Bruen 25 Carlton Avenue Upholland, Skelmersdale WN8 0AE
2E0 WPD D. Woodhams 83 Langdale Place, Newton Aycliffe DL5 7DY
2E0 WPE M. Shaw 10 Beechwood Avenue, Shevington, Wigan WN6 8EH
2E0 WPH P. Holmquest 6 Rhyme Hall Mews, Fawley, Southampton SO45 1FX
2E0 WPI T. Hobson-Smith 15 Henconner Lane, Chapel Allerton, Leeds LS7 3NX
2E0 WPJ P. Joyner 3 Barton Road, Canterbury CT1 1YG
2E0 WPN A. Hunt 200 Chorley Old Road Whittle-Le-Woods, Chorley PR6 7NA
2E0 WPT M. Thompson 35 Princes Avenue, Desborough, Kettering NN14 2RQ
2E0 WPZ D. Chilvers Flint Cottage, Cherrytree Road, Plumstead, Norwich NR11 7LQ
2E0 WQK D. Blackie 8 Kingswood Road, Manchester M14 6SB
2E0 WRF W. Fuller 28 Birch Road, Normanton WF6 1LB
2E0 WRI P. Wright 4A Alma Street, Melbourne, Derby DE73 8GA
2E0 WRK J. Erinjeri 63 Butts Green, Westbrook, Warrington WA5 7XT
2E0 WRL B. Rickard Corminnow Lanjeth High Street, St. Austell PL26 7TE
2E0 WRO R. Woodford 7 Steer Road, Swanage BH19 2RU
2E0 WRQ S. Dawe Whiterocks House, St. Anns Chapel PL18 9HN
2E0 WRS S. Webber 59 Mincinglake Road, Stoke Hill, Exeter EX4 7DY
2E0 WRT C. Smith 2 Burley Gardens, Street BA16 0SN
2M0 WRX K. Glacken 14 Hailes Avenue, Edinburgh EH13 0NA
2E0 WRY L. Curtis 39 Mount Stewart Street, Seaham SR7 7NG
2E0 WSC W. Cuddeford 5 Rosevalley Threemilestone, Truro TR3 6BH
2I0 WSH W. Hamilton 9 Susan Street, Belfast BT5 4FE
2E0 WSJ J. Woodroof 37 Danefield Road, Northampton NN3 2LT
2M0 WSK J. Muchowski 71 The Braes, Tullibody, Alloa FK10 2TT
2E0 WSM J. Claydon 17 Canterbury Close, Weston-Super-Mare BS22 7TS
2E0 WSN D. Clempson 27 Spalding Road, Hartlepool TS25 2LD
2E0 WSP S. Peare 15 Clydesdale Gardens, Bognor Regis PO22 9BE
2E0 WSR A. Rigler 10 The Ball, Dunster, Minehead TA24 6SD
2E0 WSS L. Shand 52 Ten Acre Way, Rainham, Gillingham ME8 8TL
2E0 WST R. West 557 East Bank Road, Sheffield S2 2AG
2E0 WSW S. Thorne 2 Ellfield Close, Bristol BS13 8EF
2E0 WSX A. Holmes 49 Elm Grove South, Barnham, Bognor Regis PO22 0EJ
2E0 WSZ J. Hocking 26 Musket Road, Heathfield, Newton Abbot TQ12 6SB
2E0 WTA T. Wood 61 Berry Avenue, Watford WD24 6RU
2E0 WTD K. Metcalfe 33 Corsican Drive, Hednesford, Hednesford WS12 4SS
2M0 WTE P. Jackson 4 Wester Tarbat House, Kildary, Invergordon IV18 0GF
2E0 WTG D. Cooper Little Heath, Bradfield Common, North Walsham NR28 0QR
2E0 WTH P. Newth 3 Mulberry Court, Mulberry Street, Stratford-upon-Avon CV37 6RT
2M0 WTN A. Ross 29 East Banks, Wick KW1 5NL
2E0 WTQ T. Walsh 6 Brass Thill, Durham DH1 4DS
2E0 WTR B. Shackleton 7 Erringden Street, Todmorden OL14 6AW
2M0 WTT S. Paterson Free Church Manse, Church Street, Golspie KW10 6TT
2E0 WTU A. Kolesnyk 3 Hilary Close, Fulham Road, London SW6 1EA
2E0 WTY R. Clare Kimberley, Boston Road, Bicker, Boston PE20 3AP
2E0 WTZ D. Scott 198 Slade Green Road, Erith DA8 2JG
2E0 WUF A. Butkus 73A Hudson Road, Bexleyheath DA7 4PQ
2M0 WUI P. Bingham 129 Livingstone Terrace, Irvine KA12 9ER
2E0 WUK R. Mallinson 20 Moorside Court, Moorends, Doncaster DN8 4SL
2M0 WUL W. Murdoch 64 Cotton Street, Castle Douglas DG7 1AH
2E0 WUN R. Lester 17 Clarence Road, Capel-Le-Ferne, Folkestone CT18 7LW
2E0 WVD M. Bradshaw 118 Queens Road, Vicars Cross, Chester CH3 5HE
2E0 WVE M. Huggett 12 West View Cottages Lewes Road, Lindfield, Haywards Heath RH16 2LJ
2E0 WVG D. Millward 77A Meadowcroft, St. Albans AL1 1UG
2E0 WVL C. Lacey 2 Purbeck Cottages Acton, Langton Matravers, Swanage BH19 3LU
2E0 WVM M. Edwards Rouse Farm, Normans Lane, Warrington WA4 4PY
2E0 WVQ J. Rogers 288 Wareham Street, Middleborough 23462906 United States
2E0 WVS W. Symons Pammel-House, 4 Trevassack Court, Hayle TR27 4NA
2E0 WVW C. Ring Acorn Cottage Prospect Place, Helston TR13 8RU
2E0 WWA A. Kerr 9 Martindale Way, Sawston, Cambridge CB22 3BT
2I0 WWB W. Bradley 14 Ardmore Grange, Ballygowan, Newtownards BT23 5TZ
2M0 WWC W. Stevenson 28 Lightburn Road , Halfway, Cambuslang, Glasgow G72 8UE
2E0 WWD R. Dick 15 Havenwood, Arundel BN18 0AH
2E0 WWE Dr G. Geh 7 River Gardens, Carshalton SM5 2NH
2E0 WWF E. Field Sunnyside Farm, Shapwick Road, Westhay, Glastonbury BA6 9TT
2E0 WWG A. Higgins 7 Waterloo Terrace, Bideford EX39 3DJ
2E0 WWJ T. Evans 36A Swanmore Road, Ryde PO33 2TQ
2E0 WWK J. Sell 3 Powter Close, Elsenham CM226UT
2E0 WWL S. Wood 18 Eller Drive, West Winch, King's Lynn PE33 0NN
2M0 WWM R. Jowett Fearnoch Ardentallen, Oban PA34 4SF
2E0 WWN W. Northover 13 Dagenham Avenue, Dagenham RM9 6LD
2E0 WWP L. Rowe 4 Allandale View, Lincoln LN1 3RD
2W0 WWR P. Abram 2 Blackthorn Close Marford, Wrexham LL12 8LB
2E0 WWS D. Cox 9 Northbrook Copse, Bracknell RG12 0UA
2E0 WWT A. Walls 7 Waveney Grove, York YO30 6EQ
2M0 WWX A. Prentice 24 Victoria Road, Grangemouth FK3 9JN
2E0 WWY W. Hind 6 Whinfield Avenue, Shotton Colliery, Co Durham DH6 2HE
2E0 WWZ R. Alexander 118 Pepper Lane, Standish, Wigan WN6 0PW
2E0 WXD B. Wild 1 Sunnymount, Midsomer Norton, Radstock BA3 2AS
2E0 WXF M. Morton 40 Mulgrave Road, London SE18 5TY
2E0 WXK P. Marsh 30 Mount Pleasant, Aylesford ME20 7BE
2W0 WXM E. Lake 28 Hampsons Grove Ruabon, Wrexham LL14 6AN
2M0 WXS A. Mccall Shielhill, Greengairs, Airdrie ML6 7TJ
2E0 WXT T. Nimash 6 Wallingford Road, Bristol BS4 1SL
2E0 WXU S. Reed 101 Milton Road, Weston Super Mare BS23 2UX
2E0 WXY J. Freeman 46 Wargrove Drive, Sandhurst GU470DU
2E0 WXZ B. Harding 76 Grasmere Road, Bare, Morecambe LA4 6EN
2E0 WYG A. Wood 85 Love Lane, Rayleigh SS6 7DX
2E0 WYH Q. Wang 14, Marquis House, 45 Beadon Road, London W6 0BT
2E0 WYT T. Webster 1 Fen Close, Newton, Alfreton DE55 5TD
2E0 WYZ S. Melton 2A The Orchard, Bishopthorpe, York YO23 2RX
2E0 WZA N. Lee Whiteoak, Ashurst Drive, Tadworth KT20 7LN
2E0 WZR I. Bird 2 Church Street, Wiveliscombe, Taunton TA4 2LR
2U0 WZY S. Kirkpatrick Ste Helene Manor, St. Andrew GY6 8XN Guernsey
2W0 XAA D. Pollard 110 Rowan Way, Malpas, Newport NP20 6JN
2E0 XAE A. Oxborrow 24 Ickworth Crescent, Rushmere St. Andrew, Ipswich IP4 5PQ
2E0 XAG D. Robinson 1 Common Piece, Swinefleet, Goole DN14 8DE
2E0 XAH A. Harden 16 Shining Cliff Court, Bawtry, Doncaster DN10 6SW
2E0 XAI H. Parfitt 5 Sheridan Road, Frimley, Camberley GU16 7DU
2E0 XAL G. Davies 6 Bayleys Close, Empingham, Oakham LE15 8PJ
2E0 XAM A. Morgan 18 Keysworth Drive, Wareham BH20 7BD
2I0 XAN W. Nelson 44 Lacky Road, Tattynageeragh, Rosslea, Enniskillen BT92 7GA
2E0 XAO J. Cooke 9 Nyetimber Crescent , Pagham, Bognor Regis PO21 3NN
2E0 XAR S. Halliday 8 Newby Farm Road, Scarborough YO12 6UN
2E0 XAS J. Saxon 134 Sherwood Drive, Wigan WN5 9RS
2E0 XAV J. Martin 20 Hall Green Road, West Bromwich B71 3LA
2E0 XAW A. Winkley 77 Lechlade Road, Birmingham B43 5ND
2E0 XAY S. Greaves The Grange Farmhouse, Leicester Forest East LE3 3GA
2E0 XAZ J. Deacon 42A Fairfield, Christchurch BH23 1QX
2E0 XBA G. Harvey 55 Trelawney Road Hainault, Ilford IG6 2NJ
2E0 XBB R. Boland 10 Kenilworth Drive, Kidderminster DY10 1YD
2W0 XBC C. Powell 1 Llwyn-Onn, Penderyn, Aberdare CF44 9YJ
2M0 XBD S. Boyd 1 St. Marks Lane, Edinburgh EH15 2PX
2W0 XBE S. Best 38 Greensway Abertysswg, Rhymney, Tredegar NP22 5AR
2E0 XBG J. Walker 6 Wellington Terrace, Islip, Kettering NN14 3LJ
2E0 XBM C. Atkinson 7 Hamilton Road, Grantham NG31 9QG
2E0 XBN B. Johnson 6 Trevor Road, Swinton, Manchester M27 0YH
2E0 XBT S. Pryer 16 Wayside Avenue, Worthing BN13 3JU
2E0 XBW B. Woollett 23 Kinglake House, Southall UB24FZ
2E0 XBX M. Lee 46 Little Lane, Huthwaite, Sutton-in-Ashfield NG17 2RA
2E0 XBZ M. Stead 38 Park Road, Bracknell RG12 2LU
2E0 XCA C. Archer 31 Stoney Bank Drive Kiveton Park, Sheffield S26 6SJ
2E0 XCB D. Beech 23 Holding Crescent, Halmer End, Stoke-on-Trent ST7 8AS
2E0 XCC J. Blackwell 68 Scarborough Drive, Minster On Sea, Sheerness ME12 2NQ
2E0 XCD G. Coldham 27 Welsby Road, Leyland PR25 1JA
2E0 XCF I. Titchener 18 King Edgar Close, Ely CB6 1DP
2E0 XCH C. Hayes 7 Hadstock Close, Sandiacre, Nottingham NG10 5LQ
2E0 XCM R. Styles Padcroft, Weir OL13 8QL
2E0 XCN C. Norton 34 The Grove, Little Aston, Sutton Coldfield B74 3UD
2E0 XCO M. Macdonell 54 Cinque Foil, Peacehaven BN10 8DZ
2E0 XCP C. Parker The Grange, Watercombe Cornwood, Nr. Ivybridge PL21 9RB
2M0 XCS C. Sharp 12 Manse Place, Inverkeithing KY111AZ
2M0 XCT E. Whitaker Breal House, Drumindorsair, Beauly IV4 7AH
2E0 XCV M. Abberley 10 Cranesbill Close, Featherstone, Wolverhampton WV10 7TY
2E0 XCZ M. Hartley 24 Burnham Avenue, Bognor Regis PO21 2JU
2E0 XDC D. Copsey Fairview, Mill Lane, Hook End, Brentwood CM15 0PP
2E0 XDD D. Baseden Butt 29 Shearwater Way, Stowmarket IP14 5UG
2E0 XDF D. Ferrington 20 Innings Lane, Warfield, Bracknell RG42 3TR
2E0 XDG Dr G. Welch Amazonas, Sandy Lane, Hightown, Liverpool L38 3RP
2E0 XDH D. Hudson 34 Upton Gardens, Upton upon Severn WR8 0NU
2E0 XDI J. Peain 29 Wild Flower Way, Ditchingham, Bungay NR35 2SF
2E0 XDM D. Cole 39 Hillside Road, Southminster CM0 7AL
2I0 XDR R. Cross 15 Ballyfore Road, Larne BT40 3NF
2M0 XDS D. Suttie 37 Ullapool Crescent, Dundee DD2 4TT
2W0 XDT R. Snape Bodlondeb, North Road, Whitland SA34 0AX
2M0 XDX A. Lark 20 Lawfield, Coldingham, Eyemouth TD14 5PB
2E0 XDY C. Bass 51 Dane Park Road, Ramsgate CT11 7LP

2E0 XDZ G. Parsons 2 The Close, East Grinstead RH191DQ
2E0 XEA A. Welch 18 Monk Close, Tipton DY4 7TP
2E0 XEB R. Macgregor 125 Spring Lane, Birmingham B24 9BY
2E0 XEE A. King 25 Nash Road, Dibden Purlieu, Southampton SO45 4RS
2E0 XEM S. Christie 124 Bickershaw Lane, Abram, Wigan WN2 5PP
2E0 XEN J. Fautley 71 Pullman Lane, Godalming GU7 1YB
2E0 XES R. Vickery 17 Plain-An-Gwarry, Redruth TR15 1JB
2E0 XET J. Daws 1157 Evesham Road, Astwood Bank, Redditch B96 6DY
2E0 XEV N. Wright 10 Olden Mead, Letchworth Garden City SG6 2SP
2E0 XEW A. Diaz 29, Parkside Gardens, Widdrington, Morpeth NE61 5RP
2E0 XFD L. Austin 2 List Meadows, Littlebourne, Canterbury CT3 1XW
2E0 XFF M. Chamberlain 30 Roxton Rd, Great Barford MK44 3 LR
2W0 XFG S. Gair Forest View House, Woodcroft, Chepstow NP16 7PZ
2E0 XFH M. Ashton Lodge Farm Bungalow, Wattisham Road, Ipswich IP7 7LU
2M0 XFM B. Burrows 27 Bughtknowes Drive, Bathgate EH48 4DP
2E0 XFR J. Stamford 12 Springhead, Tunbridge Wells TN2 3NY
2E0 XFX J. Walters-Pennell Clopton Grange, Clopton, Woodbridge IP13 6QR
2E0 XGA G. White 89 Kings Drive, Thingwall, Wirral CH61 9QA
2E0 XGB M. Carpenter The Retreat, High Lane Manaccan, Helston TR12 6HT
2E0 XGO R. Mcgowan 67 Loop Road North, Whitehaven CA28 6LS
2E0 XGS G. Stanley 95 Old Vicarage, Westhoughton, Bolton BL5 2EG
2E0 XGW G. Whall 10 Hillcrest Court, Ipswich Road, Diss IP21 4YJ
2E0 XGX J. Parry 12 Kerrysdale Close Sutton, St. Helens WA9 3WA
2E0 XGY G. Young 6 The Maypole Thaxted, Dunmow CM6 2QZ
2E0 XHB B. Hobbs 2 Miller Court, Bedford MK42 9PB
2E0 XHG M. Bolton 11 Silvia Way, Fleetwood FY7 7JF
2E0 XHL P. Snook 7 Sandhurst Avenue, Kwazulu Natal 3610 South Africa
2E0 XIA N. Pilling 12 Brooke Close, Baxenden BB5 2QX
2E0 XIG Cwmbran Contest & DX Group c/o D. Cassidy 172 Lyde Road, Yeovil BA21 5PN
2E0 XII B. Wood 111 Beech Crescent, Castleford WF10 3RN
2E0 XIK J. Lambert Hurley 64 Henry Road, West Bridgford, Nottingham NG2 7ND
2E0 XIP I. Prior 81 Ladymeade, Ilminster TA19 0EA
2E0 XIS M. Morton 26 Elderberry Gardens, Witham CM8 2PT
2E0 XIT D. Fincham 2 Glebe Close, St. Columb Major TR9 6TA
2E0 XJJ J. Jones 2 Lavender Gardens, Warrington WA5 1BQ
2E0 XJL J. Welch 49 Walshs Manor, Stantonbury, Milton Keynes MK14 6BU
2E0 XJM J. Meredith Pearl House, 1 Quarry Close, Myddle SY4 3SB
2E0 XJN J. Neal 5 Shelley Close, Huntingdon PE29 1NF
2E0 XJP C. Dennis Hillsdene, Plex Lane, Ormskirk L39 7JY
2W0 XJQ D. Smith Tyddyn Bach, Bethel, Caernarfon LL55 1YD
2E0 XJR J. Raffill 34 Almond Road, Kettering NN16 9PF
2E0 XJT S. Ramsden 76 Brigg Lane, Camblesforth, Selby YO8 8HD
2E0 XJW J. Wood 3 Onslow Mews, Cranleigh GU6 8FD
2E0 XKC J. Cunningham Boleyn House, Erwarton, Ipswich IP9 1LL
2W0 XKL R. Jones Flat 2, Tan Y Geraint, 33 Princess Street, Llangollen LL20 8RD
2E0 XKM M. Koster Stalworthy Manor Farm, Suton Lane, Suton, Wymondham NR18 9JG
2E0 XKO P. Goodridge 22 Horefield, Porton, Salisbury SP4 0LE
2E0 XKS K. Spowage 3 Arcadia Avenue, Mansfield NG20 8JS
2E0 XKT K. Todman 12 Winscombe, Bracknell RG12 8UD
2E0 XKX S. Lyon 10 Sycamore Close, Preston, Hull HU12 8TZ
2E0 XLB L. Baker 1B The Parade, Moss Road, Askern, Doncaster DN6 0LF
2E0 XLG D. Moore 2 Queens Garth, Thornton In Craven, Skipton BD23 3TH
2E0 XLH A. Rutson-Edwards 89 Bemerton Gardens, Kirby Cross, Frinton-on-Sea CO13 0LQ
2E0 XLI R. Messen 45 Church Lane North Bradley, Trowbridge BA14 0TE
2E0 XLJ L. Justin Garth, Park View Road, Pinner HA5 3YF
2E0 XLM Dr S. Wing 107 Highlands Boulevard, Leigh-on-Sea SS9 3TH
2E0 XLO P. Ffitch 22 Beeching Close, Halwill Junction EX21 5XY
2E0 XLR L. Mcintyre 35 Coley Hill, Reading RG1 6AE
2E0 XLS O. Hutley 1 John Ray Street, Braintree CM7 9DZ
2W0 XLT W. Murphy 148 Caergynydd Road, Waunarlwydd, Swansea SA5 4RE
2E0 XLX J. Gascoigne 1 Mill Meadow, Aylesbury HP19 8GW
2E0 XLY M. Oxley 49 Dalton Crescent, Shildon DL4 2LE
2E0 XMC M. Callis 1 Webb Close, Letchworth Garden City SG6 2TY
2E0 XMF P. Foster 100 Howe Road, Norton, Malton YO17 9BL
2W0 XMG P. Provis Dingle Gardens, Croesbychan, Aberdare CF44 0EJ
2E0 XMH M. Hodgkinson 116 Bidwell Hill, Houghton Regis, Dunstable LU5 5EP
2E0 XMK M. Rose 19 Hawthorn Street, Peterlee SR8 3LY
2E0 XMO E. Martin 61 Uffington Avenue, Lincoln LN6 0AG
2E0 XMP M. Pearce 1 Hillside Close, Helsby, Frodsham WA6 9LB
2M0 XMQ L. Anderson 6 St. Keiran Crescent, Stonehaven AB39 2GQ
2W0 XMS M. Street The Lamb Inn, Hermon, Glogue SA36 0DS
2E0 XMT T. Wrenn Flat 3, York House, Nottingham Avenue, Maidstone ME15 7PX
2W0 XMW P. Hoath 8 Liverpool Terrace, Llithfaen, Pwllheli LL53 6NN
2E0 XNC R. Jeffery 126 Woodham Lane, New Haw, Addlestone KT15 3NQ
2E0 XNF N. Ferrington 20 Innings Lane, Warfield, Bracknell RG42 3TR
2E0 XNL G. Molendijk 47 Lodge Road, Scunthorpe DN15 7EN
2E0 XNT M. Galbraith Oaklands, Didcot Road Harwell, Didcot OX11 0DP
2E0 XOD I. Donnelly 17 Jessop Close, Horncastle LN9 6RR
2E0 XOE D. Hart 7 Penrose Road, Ferndown BH22 9JF
2E0 XOH V. Fox 48 Grenham Avenue, Manchester M15 4HD
2E0 XOJ M. Benson 11 Hield Grove, Aston By Budworth, Northwich CW9 6LN
2E0 XOK D. Mitchell 3 Ivy Cottage, Main Road, Theberton, Leiston IP16 4RX
2E0 XON H. Tranter Flat 6 Oak Apple Court 25 Acorn Road Catshill, Bromsgrove B61 0TR
2E0 XOR M. Hauser 23 Prince William Way Sawston, Cambridge CB22 3SZ
2W0 XOT J. Messenger 34 Goylands Close, Llandrindod Wells LD15RB
2E0 XPC P. Cole 28 Norfolk Gardens, Newcastle upon Tyne NE28 7HP
2E0 XPD P. Douglas 76 Woodside Avenue, Benfleet SS7 4NY
2E0 XPH P. Hennessey 11 Monmouth Drive, Eaglescliffe, Stockton-on-Tees TS16 9HU
2E0 XPJ J. Parfitt 12 Jodrell Place, Selsey, Chichester PO20 0FQ
2E0 XPK C. Park 197 Occupation Road, Albert Village, Swadlincote DE11 8HD
2E0 XPM P. Mullen 12 Poplar Grove, Conisbrough, Doncaster DN12 2JG
2E0 XPP P. Jarvis 24 St. Peters Gardens, Leeds LS13 3EH
2D0 XPS D. Heaton Rushen Vicarage Barracks Rd, Port St. Mary IM9 5LP Isle of Man
2E0 XPT T. Parfitt 5 Sheridan Road, Frimley, Camberley GU16 7DU
2E0 XPY P. Elliott 1 Speldhurst Gardens Cliftonville, Margate CT9 3HJ
2E0 XQA M. Ashford 11 Pond Close, Felixstowe IP112JW
2E0 XQK D. Goodfellow 60 Pickering Green, Gateshead NE9 7DX
2E0 XRD C. Darby Flat 4 146 Newark Road, Lincoln LN5 8QF
2E0 XRG G. Duffy 34 Twentyfifth Avenue, Blyth NE24 2QW
2E0 XRL S. Bennett 7 Holme Park, High Bentham LA2 7ND
2E0 XRM D. Bingham 33 Sheffield Road, Creswell, Worksop S80 4HN
2E0 XRO A. Dainty 6 St Nicholas Way Wygate Park Spalding, Spalding PE11 3GF
2E0 XRQ T. Jones 3 Langford Road, Liverpool L19 3RA
2E0 XRS R. Stevens 53 Keeble Way, Braintree CM7 3JX
2M0 XRV J. Caldwell 35, Clunie Drive, Larbert FK5 4UA
2M0 XRX A. Wright 17A James Court, 493 Lawnmarket, Edinburgh EH1 2PB
2M0 XRZ M. Nicholls Grahams Onsett Farm, Newcastleton TD9 0TT
2E0 XSB S. Bolton The Conifers, Methwold Road Methwold Hythe, Thetford IP26 4QW
2E0 XSD C. Catlin 27 Main Street, Frizington CA26 3SA
2E0 XSG C. Braisby 4 Langmans Way, Woking GU21 3QY
2E0 XSJ S. Jordan 74 Soane Gardens, South Shields NE34 8NN
2E0 XSL S. Looker 165 Mollison Drive, Wallington SM6 9GX
2E0 XSW S. Whall 17 Vicarage Road, Deopham, Deopham NR18 9DR
2E0 XSZ K. Edwards 24 Abbey Road, Halesowen B63 2HE
2E0 XTA J. Moore Flat 4, 219 Holland Road, Holland On Sea, Clacton-on-Sea CO15 6NL
2E0 XTB A. Blews 57 Highfield Grove, Stafford ST17 9RA
2E0 XTC K. Haworth 11 Petersfield Close, Bootle L30 1SG
2M0 XTH T. Henderson 139 Crewe Crescent, Edinburgh EH5 2JN
2W0 XTK A. Gill 2 Bank Road, Llangennech, Llanelli SA14 8UB
2E0 XTL M. Porter 20, Southfield Road, Much Wenlock TF13 6AX
2E0 XTM J. Maguire 14 Botha Road, St. Eval, Wadebridge PL27 7TS
2W0 XTP D. Willis 51 Fforchaman Road, Cwmaman, Aberdare CF44 6NG
2M0 XTS C. Robertson 5 Broomlands Place, Irvine KA12 0DU
2E0 XTV T. Benson 83 Glovers Road, Birmingham B10 0LE
2E0 XUA Dr M. Sinclair 40 Grotto Road, South Shields NE34 7AH
2E0 XUH Dr G. Thomas 3 The Croft, Wilton, Egremont CA22 2PW
2E0 XUI J. Ribbands Dyson'S Farm, Long Row, Tibenham NR16 1PD
2E0 XUK B. Chadwick 44 Glendale Drive, Mellor, Blackburn BB2 7HD
2W0 XUL M. Brooks Garth, Cemmaes, Machynlleth SY20 9PR
2E0 XUM W. Webb 84 Bruce Street, Swindon SN2 2EN
2E0 XUN A. Southern 89 Purlewent Drive Weston, Bath BA1 4BD
2E0 XUU R. Gopan 84 Hilmanton, Lower Earley, Reading RG6 4HN
2E0 XUZ N. Hanson-Collins 92 Howbury Lane, Slade Green, Erith DA8 2DR
2E0 XVD S. Jones 19 Runshaw Lane, Euxton, Chorley PR7 6AU
2E0 XVF J. Smith 8 Mayfields, Spennymoor DL16 6RN
2E0 XVJ I. Vernall 1 Owen Place, Bridge Street, Kington HR5 3DH
2E0 XVK D. Pounder 15 Eldon Grove, Hartlepool TS26 9LY
2W0 XVT C. Williams 1 South View, Freeholdland Road, Pontypool NP4 8LL
2E0 XVX M. Lawrence 16 Timson Close, Market Harborough LE16 7UU
2E0 XVZ A. Sibley 27 Sherwood Road, Tetbury GL8 8BU
2W0 XWD R. Hawkins Nook Cottage, Common-Y-Coed, Caldicot NP26 3AX
2E0 XWX D. Wilkins Malt Hill Cottage, Malt Hill, Warfield, Bracknell RG42 6JG
2E0 XXB S. Bunce 15 Downs View Road, Bembridge PO35 5QS
2E0 XXC A. Carter 36 Marriotts Close, Ramsey Mereside, Ramsey, Huntingdon PE26 2TX
2E0 XXI P. Wright 34 Tettenbury Road, Nottingham NG5 1LA
2E0 XXK K. Mitchell 1 Denstroude Cottages, Denstroude Lane, Canterbury CT2 9JX
2E0 XXM M. Jennings Springfield Farm, The Causeway, Stow Bridge, King's Lynn PE34 3PP
2E0 XXO S. Nesling 64 Ruskin Avenue, Lincoln LN2 4BT
2M0 XXP A. Pitkethley 99 Margaretvale Drive, Larkhall ML9 1EH
2E0 XXT T. Archer 1 Banks Road, Ashford TN234NR
2E0 XXX M. Davey 27 Earls Drive, Newcastle ST5 3QR
2E0 XYA P. Hodges 191 Broadstone Road, Stockport SK4 5HP
2E0 XYM M. Leonard 75 Skillings Lane, Brough HU15 1BA
2E0 XYT E. Wood 13 Rosedale, Welwyn Garden City AL7 1DW
2E0 XYX A. Austin 4 Cornwall Avenue, Oldbury B68 0SW
2E0 XYZ C. Edgar 9 Winchester Avenue, Morecambe LA4 6DX
2M0 XZA M. Saunders 8D Springhill Road, Port Glasgow PA14 5QP
2E0 XZI J. Ball Ponsharden Boatyard, Falmouth Road Ponsharden, Penryn TR10 8AB
2E0 XZM A. Ifrim 46 Northdown Road, Solihull B91 3NB
2E0 XZW C. Set Hughes Hall, Wollaston Road, Cambridge CB1 2EW
2M0 XZX G. Queen G/R 31 Provost Road, Dundee DD3 8AF
2E0 XZZ S. Rouse 7 Cranbrook Road, Thurnby, Leicester LE7 9UA
2W0 YAB S. Morgan 38 Ffordd Cadfan, Bridgend CF31 2DP
2W0 YAD R. Williams 92 Bowleaze, Greenmeadow, Cwmbran NP44 4LF
2W0 YAE G. Thatcher-Sharp 20 Dilys Street, Blaencwm, Treorchy CF42 5DT
2M0 YAF C. Tait 127 Bonnyton Road, Kilmarnock KA1 2NU
2E0 YAH S. Abdullah 24 The Grove, Walsall WS5 4BX
2E0 YAL D. Parker 16 Aldborough Road, Dagenham RM10 8AS
2E0 YAO L. Jex 26 Springdale Crescent, Brundall, Norwich NR135RA
2E0 YAP A. Kissin 115 Aarons Hill, Godalming GU7 2LJ
2E0 YAQ L. Iakovlev 20A Brownlow Road, London N3 1NA
2E0 YAR F. Lees 5 St. Winifred Road, Rainhill, Prescot L35 8PY
2E0 YAS J. Stacey 12 Kimbridge Road, East Wittering, Chichester PO20 8PE
2E0 YAV W. Jones 8 Oakbrook Close, Ewyas Harold, Hereford HR2 0NX
2E0 YAW M. Driscoll 14B Pretoria Road Hedge End, Southampton SO30 0BS
2E0 YAX A. Garn 5 Bassett Street, Walsall WS2 9PZ
2J0 YAY J. Bryant 5 Louiseberg Court Queen'S Road, St. Helier JE2 3GQ Jersey
2E0 YBB B. Brown 3 Swaledale, Worksop S810UY
2I0 YBH J. Mccourt 70A Sessiagh Scott Road Rock, Dungannon BT70 3JU
2E0 YBL L. Jones 8 Oakbrook Close, Ewyas Harold, Hereford HR2 0NX
2M0 YBR G. Fordyce 2 Church Street, East End, Earlston TD4 6HS
2E0 YBS B. Scroggs 3 Amroth Gardens, Berkeley Beverborne, Worcester WR4 0EP
2W0 YBZ P. Smith 29 Heol Cwarrel Clark, Caerphilly CF83 2NE
2E0 YCA E. Cooper 20 Manor Close, Baston, Peterborough PE6 9PH
2E0 YCD D. Ashton-Hilton 14 Weetwood Road, Congresbury, Bristol BS49 5BN
2M0 YCG C. Graham 64 Forgewood Road, Motherwell ML1 3TH
2M0 YCJ Prof. C. Jones 11B Ettrick Road, Edinburgh EH10 5BJ
2E0 YDA A. Bedford 1 Carder Crescent, Bilston WV14 0JT
2E0 YDB D. Bower 4 Winsford Road, Sheffield S6 1HT

2I0 YDF D. Foley 14 Chestnut Hall Court, Moira, Maghaberry BT67 0GJ
2W0 YDK D. Edwards 25 Bryn Coed, Gwersyllt, Wrexham LL11 4UE
2E0 YDM C. Preece 14 Dock Street, Widnes WA8 0QX
2E0 YDP D. Palmer 44 Milburn Street, Crook DL15 9DZ
2E0 YDT A. Carney 9 Hart Square, Sunderland SR4 8BS
2E0 YDX J. Leighton 27 The Pastures, Cayton, Scarborough YO11 3UU
2E0 YEA E. Raine Stable Cottage, St. Martins, Richmond DL10 4SJ
2E0 YEH D. Clough 44 Small Drove, Weston, Spalding PE12 6HS
2E0 YEK D. Noyek 34 The Spinney, Sidcup DA14 5NF
2E0 YEO R. Froggatt 39 Spitfire Way, Hamble, Southampton SO31 4RT
2E0 YEP S. Quinn 7 Poppleton Court, Tingley, Wakefield WF3 1UY
2M0 YEQ G. Pearce 2 Kirkriggs, Forfar DD8 2AT
2E0 YES M. Casey 7 Cobham Avenue, Manchester M40 5QW
2E0 YEW A. Mcewen 4 The Pantyles, Nightingale Lane, Sevenoaks TN14 6BX
2E0 YEY G. Matthews 255 Coach Road Estate, Washington NE37 2EU
2E0 YEZ A. Atkinson 2E Bagridge Road, Wolverhampton WV3 8HW
2W0 YFC J. Evans 11 Dew Crescent, Cardiff CF5 5PB
2E0 YFG F. Grande Restrup Engvej 26, Aalborg 9000 Denmark
2E0 YFR D. Cockburn 30 Queensberry Road, Burnley BB11 4LH
2E0 YFZ J. Blower 4 Lamorna Close, Luton LU3 2TH
2E0 YGB A. Birch 3 Partridge Way, High Wycombe HP13 5JX
2E0 YGC A. Cottell 3 Honeylight View, Swindon SN25 4XS
2E0 YGH F. Armstrong 38 Dovecote Drive, Haydock, St. Helens WA11 0SD
2E0 YGM M. Clayton Swallows 5 The Waldrons East Garston, Hungerford RG17 7JB
2E0 YGR G. Robertson 12 Chester Close Ince, Wigan WN3 4JP
2E0 YGS G. Moss 125 Lavender Avenue, Mitcham CR4 3RS
2E0 YGT S. Mucklow Blithbury House, Blithbury Road, Rugeley WS15 3HR
2E0 YHF J. Marvel Dean House Farm, Nordan, Leominster HR6 0AW
2E0 YHI R. Bragg 50 Moorclose Road Harrington, Workington CA14 5LB
2E0 YHN J. Penniston 1 Line Cottage The Causeway, Thorney, Peterborough PE6 0QJ
2E0 YHW R. Lloyd 18 Newbury Grove Blurton, Stoke-on-Trent ST3 3DD
2E0 YHZ K. Nelson 185 Headlands, Fenstanton, Huntingdon PE28 9LP
2W0 YIF D. Williams 371 Coed-Y-Gores, Llanedeyrn, Cardiff CF23 9NR
2E0 YIN D. Smith 5 Stocken Close, Hucclecote, Gloucester GL3 3UL
2M0 YIO B. Fullerton 55 Alexander Avenue, Stevenston KA20 4BG
2E0 YIP S. Chng Cambridge, CB3 9BB
2E0 YJB J. Blackall 2 Ryson Avenue, Blackpool FY4 4DN
2E0 YJF D. Moran 33 Pilgrim Drive, Bere Alston, Yelverton PL20 7DB
2E0 YJL J. Cairns 17 Alfred Avenue, Worsley, Manchester M28 2TX
2E0 YJW J. Williams 66 Oakfield Avenue, Hitchin SG4 9JD
2E0 YJY M. Kealey 24 Ben Nevis Road, Birkenhead CH42 6QY
2E0 YKC D. Forshaw 14 Hope Carr Road, Leigh WN7 3ET
2E0 YKD G. Lowe Stone Cottage, Esplanade Lane, Watchet TA23 0AH
2E0 YKX Dr J. Dunn 6 Webbs Way, Ashley Heath, Ringwood BH24 2DU
2E0 YLD L. Davis 9 High Street, Chapel-En-Le-Frith, High Peak SK23 0HD
2W0 YLE A. Doyle 54 Bro Syr Ifor, Tregarth, Bangor LL57 4AS
2E0 YLH J. Haystead 11 Lumley Close, Maltby, Rotherham S66 7SG
2W0 YLL R. Bowen 25 Maendu Terrace, Brecon LD3 9HH
2E0 YLP R. Campbell-Black 10 Wren Close, Towcester NN12 6RD
2E0 YLR J. Raynerd 3 Brooksbottoms Close, Ramsbottom, Bury BL0 9YP
2I0 YLT S. Mccormick 46 Lany Road, Moira, Craigavon BT67 0NZ
2D0 YLX D. Cain 7 Cronk Y Berry Mews, Douglas IM2 6HQ Isle of Man
2E0 YLZ N. Bacala Flat 3, 232A Seven Sisters Road, London N4 3NX
2M0 YMA Dr A. Brasier 10 Wellgrove Crescent, Westhill AB32 6TH
2I0 YME M. Smith Murreagh, Waterville V23 XY32 Ireland
2I0 YMF M. Foley 44 Gallows Street, Dromore BT25 1BD
2W0 YMG M. Graham 60 Heol Seward Beddau, Pontypridd CF38 2SR
2E0 YMM J. Griffiths 9 Cauldale Close, Middleton, Manchester M24 5SU
2W0 YMP S. Keeble Mynachlog, Tyn Y Gongl LL74 8SG
2E0 YMR L. Humphreys Flat 38, Wimborne House, 2 Stokewood Road, Bournemouth BH3 7NP
2W0 YMS M. Stevenson 64 Leslie Terrace, Porth CF39 9TE
2E0 YMT P. Horrox 39 Wilton Grove, Heywood OL10 1AS
2E0 YMV M. Vaites 19 Campion Drive, Sheffield S21 1TG
2E0 YMW M. Williamson 5 John F Kennedy Walk, Tipton DY4 0SF
2M0 YMZ J. Burwood 8 Barbours Park, , Stewarton KA3 5HS
2E0 YNC R. Dalumpines Flat 2, 27 Cotswold Way, Worcester Park KT4 8HD
2E0 YND B. Higgins 41 Lower Meadow, Harlow CM18 7RE
2E0 YNI S. Widdowson 19 Wayside, Brimington, Chesterfield S43 1BQ
2M0 YNK N. Kirtley 10 Millcroft Road, Auldearn, Nairn IV12 5TW
2E0 YNT A. Ashby 5 New Street, Osbournby, Sleaford NG34 0DL
2E0 YNY N. Oldrid 4 Bar Lane Mapplewell, Barnsley S75 6DQ
2E0 YOI I. Eliade 32 Meadow Sweet Road, Stratford-upon-Avon CV37 0TH
2E0 YOK C. Bietz 37 Alan Avenue, Newton Flotman, Norwich NR15 1PY
2E0 YOM J. Thresher Quarry Grange, Nuneaton Road Over Whitacre, Coleshill, Birmingham B46 2NH
2E0 YOP T. Court Eastgate Cottage, Perrys Lane, Norwich NR10 4HJ
2M0 YOY J. Moir 41 Brisbane Terrace, East Kilbride, Glasgow G75 8DL
2E0 YOZ J. Wilson 5 Queens Road Hoylake, Wirral CH47 2AG
2M0 YPE G. Burt 11 Keir Street, Cowdenbeath KY4 9DS
2E0 YPG R. Irwin 21 Penn St., Belper DE56 1GH
2M0 YPH P. Hickman 76 Ettrick Terrace, Selkirk TD7 4JR
2E0 YPJ P. Kirby 30 New Street, Eccleston, Chorley PR7 5TW
2E0 YPK E. Maddex 58 Bohemia Chase, Leigh-on-Sea SS9 4PP
2E0 YPU T. Galloway 1 Farley Close Shadoxhurst, Ashford TN26 1NB
2E0 YPW Dr P. Woodfin 14 Broomcroft Road, Bognor Regis PO22 7NJ
2E0 YQC G. Hope 27 Clearmount Drive Charing, Ashford TN27 0LH
2E0 YQT J. Best 24 Suggitts Lane, Cleethorpes DN35 7JJ
2E0 YRF Dr A. Saje 72 Bedworth Road, Bulkington, Bedworth CV12 9LL
2E0 YRM M. Radulov 14 Grove Road, Chatham ME4 5HS
2E0 YRT C. Williams 16 Mill Meadow, Tenbury Wells WR15 8HX
2E0 YRU K. Michael 4 Newcroft Gardens, Christchurch BH23 2AS
2E0 YRW J. Woods 21 Appleyard Crescent, Norwich NR3 2QN
2E0 YSB P. Bunting 29 Marion Avenue, Alverthorpe, Wakefield WF2 0BJ
2E0 YSF L. Metcalfe 40 St. Anns Court, Hartlepool TS24 7HY
2E0 YSK A. Bragg Buena Vista Low Moresby, Whitehaven CA28 6RR
2E0 YSO M. Mayson Hillview, Top Common, East Runton, Cromer NR27 9PR
2E0 YSP A. Gobey Nut Tree Cottage, Valley Rd, Ipswich IP8 4LR
2M0 YSR C. Phillips 8 The Square, Newtongrange, Dalkeith EH22 4QD
2E0 YSU G. Cummings 10 Perth Close, Skegness PE25 2HY
2M0 YTA J. Anderson The Grange, Leslie Road, Scotlandwell, Kinross, KY13 9JE
2E0 YTB R. Thomas 9 Spa View Terrace, Sheffield S12 4HG
2M0 YTD T. Davidson 141 Huron Avenue, Livingston EH54 6LQ
2E0 YTF G. Garman 11 Rye Close, Norwich NR3 2LF
2M0 YTN J. Keymer 7 Pladda Wynd, Broomlands, Irvine KA11 1DW
2E0 YTT R. Spooner 45 Shaftesbury Avenue, Southport PR8 4NH
2E0 YUD S. Nutt 77 Exeter Close, Stevenage SG1 4PW
2E0 YUN Y. Li Song 8 Birkin Court, Welwyn Garden City AL7 3FA
2M0 YUP S. Duff 5 Woodhall Farm Cottages, Innerwick, Dunbar EH42 1SH
2E0 YVR L. Palir 116 Carville Crescent, Brentford TW8 9RD
2E0 YVT P. Van Staveren 14 Fortune Green Road Flat 3, London NW6 1UE
2W0 YVY C. Ibarra-Rivadeneira 79 Heol Frank, Penlan, Swansea SA5 7AH
2E0 YWA I. Stanley 11 Reading Road, Burghfield Common RG7 3PY
2E0 YWN R. Cook Flat 2, 1 Duke Street, Salford M7 1PR
2E0 YWO M. Bruyneel 14 Riversmead, St. Neots PE19 1HA
2E0 YWP D. Spooner 30 Clover Road, Norwich NR7 8TF
2E0 YXB R. Barrett 18 Bullstake Close, Oxford OX2 0HN
2E0 YXO M. Bilverstone 12 Westlea Road, Sywell, Northampton NN6 0BY
2E0 YXX K. Ramminger 54 White Hedge Drive, St. Albans AL3 5TX
2E0 YXZ S. Christofi 19 Kingsland Avenue, Northampton NN2 7PP
2E0 YYA D. Crane 3 Middlemead Close West Hanningfield, Chelmsford CM2 8UR
2E0 YYB C. Norwood 94 Waverley, Woodside, Telford TF7 5LU
2E0 YYD P. Dumpleton 20 Cambridge Road North, Mablethorpe LN12 1QR
2E0 YYG D. Harwood 32 John Reid Road, South Shields NE34 9EB
2E0 YYK S. Hoyle 20 Brandsby Grove, Huntington, York YO31 9HL
2E0 YYM P. Boulding 16 Higher Croft Road, Lower Darwen, Darwen BB3 0QR
2W0 YYP P. Griffey 61 Cottesmore Way, Cross Inn, Pontyclun CF72 8BG
2E0 YYS S. Cole 13 Boscobel Road North, St. Leonards-on-Sea TN38 0NY
2E0 YYT G. Finney 121 School Lane, Caverswall, Stoke-on-Trent ST11 9EN
2M0 YYU A. Anderson 16 Walker Court, Glasgow G11 6QP
2E0 YYW A. Edge 79 Waterside Drive, Stoke-on-Trent ST3 3NU
2E0 YYY M. Hunter 126 Turner Street, Stoke-on-Trent ST1 2NE
2E0 YYZ S. Wall 26 Wallace Lane, Whelley, Wigan WN1 3XT
2E0 YZA B. Wilson 69 Rotherwood Road, Sheffield S21 2DT
2E0 YZC D. Matheson 21 Warren Hill Road, Woodbridge IP12 4DU
2E0 YZM G. Brown 134 Skipper Way, Lee-on-the-Solent PO13 8HD
2E0 YZQ A. Kent 4 Sellerdale Drive, Wyke, Bradford BD12 9DA
2M0 YZT P. Connon 4 Highfield Court, Stonehaven AB39 2PL
2E0 YZW M. Wandby 44 Windrush Road, Hollywood, Birmingham B47 5QA
2E0 YZX M. Galloway 2 Edendale Terrace, Horden, Peterlee SR8 4RD
2E0 YZY I. Hodgkiss 190 Ulverley Green Road, Solihull B92 8AD
2E0 YZZ P. Collingham 1 Wychwood Drive, Trowell, Nottingham NG9 3RB
2W0 ZAA S. Tozer 110 Glanffornwg, Wildmill, Bridgend CF31 1RL
2E0 ZAC M. Cotton 18 St. Oswalds Crescent, Brereton, Sandbach CW11 1RW
2W0 ZAE P. Mason 20 Coronation Road, Six Bells, Abertillery NP13 2PJ
2E0 ZAF R. Amos 6 Eccles Road, Wittering, Peterborough PE8 6AU
2E0 ZAH L. Almond 26 Ashbourne Drive, Desborough, Kettering NN14 2XG
2E0 ZAI B. Hardy 10 Spring Farm Road, Burton-on-Trent DE15 9BN
2E0 ZAJ S. Rafter 30 Monmouth Grove, St. Helens WA9 1QB
2E0 ZAL J. Moore Moorelake Lodge, Barholm Road, Stamford PE9 4RJ
2E0 ZAP R. Bird 78 Arden Road, Hockley, Tamworth B77 5JE
2E0 ZAU M. Southgate 107 Englands Lane, Loughton IG10 2QL
2E0 ZBB Dr M. Palmer 116 Claverham Road, Yatton, Bristol BS49 4LE
2W0 ZBC G. Jones 12 Wilson Place Ely, Cardiff CF54LN
2E0 ZBD J. Wells 15 Phillips Crescent, Needham Market, Ipswich IP6 8TF
2E0 ZBE J. Ellery 7 Midanbury Crescent, Southampton SO18 4FN
2M0 ZBF G. Irving 55 Gillbank Avenue, Carluke ML8 5UW
2M0 ZBH P. Mclaren 1 Morayvale, Aberdour, Burntisland KY3 0XE
2E0 ZBW R. Weaver 116 Carville Crescent, Brentford TW8 9RD
2E0 ZBZ M. Carvell 10 Burns Close, Stevenage SG2 0JN
2E0 ZCB C. Button 30 Southfall Close, Ranskill, Retford DN22 8NE
2E0 ZCG C. Gregory 81 Fiskerton Way, Oakwood, Derby DE21 2HY
2I0 ZCM A. Birkhead 21 Carson Villas, Upperlands Maghera BT46 5SH
2E0 ZCP C. Palawinna 3 Stirling Court Road, Burgess Hill RH15 0PS
2E0 ZDA D. Phillips 7 Broughton Road, Banbury OX16 9QB
2E0 ZDB D. Brownsea 47 Southill Road, Bournemouth BH9 1SH
2E0 ZDC Dr D. Cooke Apartment 9, 27 Sheldon Square, London W2 6DW
2E0 ZDE D. Kirkden 57 Crow Hill Road, Margate CT9 5PF
2E0 ZDH D. Hardwick 30 Halfcot Avenue, Stourbridge DY9 0YB
2E0 ZDJ L. Macrides 5 Apple Farm Lane, Weston-Super-Mare BS24 7TJ
2E0 ZDM D. Mason 94 Guessburn, Stocksfield NE43 7QR
2E0 ZDW D. Whitley 10 Kenmore Drive, Cleckheaton BD19 3EJ
2E0 ZDX P. Jones Stonehead Farm Over Wyresdale, Lancaster LA2 9DL
2M0 ZEB G. Barrie 24 Mauldeth Road, Broxburn EH52 6FB
2M0 ZEE P. Russell 21 St. Andrews Drive, Law, Carluke ML8 5GB
2E0 ZEH S. Hendy Flat 2, 33 Kingston Road, , Leatherhead KT22 7SL
2E0 ZEN H. Haydon 101 Blandford Street, Ashton-under-Lyne OL6 7HG
2M0 ZET H. Dally 3 Gremmasgaet, Lerwick, Shetland ZE1 0NE
2E0 ZEV B. Dexter 237 Wordsworth Avenue, Sheffield S5 8NE
2M0 ZFG S. Street Tangaroa, Fairfield Gardens, Kilcreggan, Kilcreggan G84 0HS
2E0 ZFV P. Whiteley 1 Newton Close, Fareham PO14 3LF
2I0 ZFZ R. Mckay 31 Squires Hill Crescent, Belfast BT14 8RE
2E0 ZGA G. Campbell 10 Welbeck Road, Rochdale OL16 4XP
2E0 ZGL A. Lunn 57 Greets Green Road, West Bromwich B70 9ES
2E0 ZGS Z. Sznober 9 Moor Road Dawley, Telford TF4 2AR
2E0 ZGX P. Beltrami 15 Woodroffe Square, Calne SN11 8PW
2E0 ZHG C. Barnes 23 South Street, Crewe CW2 6HN
2E0 ZHN E. Mady 130 Staveley Gardens, London W4 2SF
2E0 ZIP S. Kiley 178 Kingfisher Drive, Woodley, Reading RG5 3LQ
2E0 ZIV I. Vickers 3 Nesbit Road, St. Marys Bay, Romney Marsh TN29 0SF
2W0 ZJA D. Bowen 25 Maendu Terrace, Brecon LD3 9HH
2E0 ZJB M. Collier 32 London Road, Dereham NR19 1AW
2E0 ZJO J. Rawlinson Westfield Farm, Risden Lane, Cranbrook TN18 5DU
2E0 ZJQ A. Rawlinson Westfield Farm, Risden Lane, Hawkhurst, Sandhurst, Cranbrook TN18 5DU
2E0 ZKT P. Stone 2 Endeavour Close, Lower Stondon, Henlow SG166JR

2E0 ZLA A. Zeller Flat 1, 57 Chalk Hill, Watford WD19 4DA
2E0 ZLD Z. Dunne 1 Burton Gardens, Brierfield, Nelson BB9 5DR
2E0 ZLM L. Milburn 87 Caburn Court, Crawley RH11 8SX
2E0 ZLO M. Holmes 6 Wells Court, Saxilby, Lincoln LN1 2GY
2E0 ZMB M. Breslin 15 Acorn Gardens, East Cowes PO32 6TD
2E0 ZMI T. Kelly 2 Weaver House, Chester Road, Runcorn WA7 3EG
2E0 ZML J. Marr Touchstone, Heathfield Road, Bembridge PO35 5UW
2E0 ZMM D. Mainwaring 1 Buckingham Close, Didcot OX11 8TX
2E0 ZMO A. Maguire 132 Wigan Road, Ormskirk L39 2BA
2E0 ZMR I. Nicholson 2 Broom Close, Leyland PR25 5RQ
2E0 ZMS M. Strickland Ancoats, Piercy End, York YO62 6DQ
2E0 ZMT M. Thompson 133 Redford Avenue, Horsham RH12 2HH
2E0 ZMZ M. Coleman 3 Tummon Road, Sheffield S2 5FD
2M0 ZNQ W. Beaton 4 Moorfield Gardens, Springfield, Cupar KY15 5SH
2E0 ZNZ Dr G. Richardson Berwick Cottage, Bailes Lane, Guildford GU3 2AX
2E0 ZOM J. Skinner 36 Milton Road, Waterloo, Liverpool L22 4RF
2E0 ZOR A. Mccrystal 15 Amory'S Holt Way, Maltby, Rotherham S66 8RF
2E0 ZOT M. Toher The Chapel, Station Road, Darlington DL2 1JG
2E0 ZOZ A. Hunter 9 Gelt Burn, Didcot OX11 7TZ
2E0 ZPA P. Archer 31 Stoney Bank Drive, Kiveton Park, Sheffield S26 6SJ
2E0 ZPN L. Gibbs 37 Oxford Road, Fulwood, Preston PR2 3JL
2E0 ZPY R. Pyner 1 Avon Court, 63 Shakespeare Road, Bedford MK40 2DS
2E0 ZRB R. Brown 194 Wymersley Road, Hull HU5 5LN
2E0 ZRG R. Greaves 7 Eller Brook Close, Heath Charnock, Chorley PR6 9NQ
2E0 ZRL A. Briggs 3 Swallow Avenue, Leeds LS12 4RD
2E0 ZRM D. Morgan 2 Raymond Way, Plymouth PL7 4EG
2E0 ZRQ G. Crane 33A Carlisle Gardens, Horncastle LN9 5LP
2E0 ZRT T. Cooper 9 Websters Close, Shepshed, Loughborough LE12 9AT
2E0 ZRX C. Waterworth 4 Mossdale Road, Ashton-In-Makerfield, Wigan WN4 0EQ
2E0 ZSA S. Airs 6 The Willows, Culham, Abingdon OX14 4NN
2E0 ZSB S. Bannister 162 Dobcroft Road, Sheffield S11 9LH
2E0 ZSE P. Holmes 53 Bishops Hull Road, Bishops Hull, Taunton TA1 5EP
2E0 ZSH S. Hampson 12 Flying Fields Drive, Macclesfield SK11 7GE
2E0 ZSJ J. Gibson 22 Woodburn Drive Chapeltown, Sheffield S351YS
2E0 ZSK S. Kneller 12A Richard Street, Crewe CW1 3AF
2E0 ZSR S. Robottom-Scott 73 St. Bernards Road, Solihull B92 7DF
2E0 ZST S. Harris 61 Monks Park Avenue, Bristol BS7 0UA
2E0 ZSU Capt. P. Westwell Roden House, Dobsons Bridge, Whitchurch SY13 2QL
2E0 ZSY T. Symons Southgate, The Commons, Mullion TR12 7HZ
2E0 ZTD G. Berry 5 Oakholme Rise, Worksop S81 7LJ
2E0 ZTE S. Broom 128 Springhill Road, Wolverhampton WV11 3AQ
2E0 ZTG A. Hill 5 Park Road, Thurnscoe, Rotherham S63 0TG
2E0 ZTL C. Ingamells 41 Princess Anne Road, Boston PE21 9AP
2E0 ZTM T. Moore 16 Warwick Drive, Earby, Barnoldswick BB18 6LX
2E0 ZVG I. Browne 85 White Eagle Road, Haydon Leigh, Swindon SN25 1PY
2E0 ZVL V. Lynch 16 Okehampton Crescent, Sale M33 5HR
2E0 ZVR B. Bateman 27 Imperial Avenue, Kidderminster DY10 2RA
2E0 ZWA D. Tordoff 49 Dale Edge, Eastfield, Scarborough YO11 3EP
2W0 ZWR G. Spicer 6 Cromwell Road, Neath SA10 8DR
2E0 ZWW Dr W. Warwicker 28 Porters Wood, Petteridge Lane, Matfield, Tonbridge TN12 7LR
2E0 ZXG G. Carless Silver Cottage, Silver Street, South Petherton TA13 5BY
2E0 ZXJ J. Wildsmith 15 Harebell Close, Hartley Wintney RG27 8TW
2I0 ZXM M. Meagher 42 Mourne View Park, C.Down, Newry BT35 6BZ
2E0 ZXQ I. Talbot 41 Elmwood Close, Cannock WS11 6LX
2E0 ZXR M. Holbrook-Bull 66 Wayman Road, Corfe Mullen, Wimborne BH21 3PN
2E0 ZYG C. Richardson Heathercroft, Kirkby Mills, Kirkbymoorside, York YO62 6NN
2E0 ZYK M. Edge 19 Burton Av, Rushall, West Mids WS41NH
2E0 ZYL S. Allen Milverton, Mill Road, Pulborough RH20 2PZ
2E0 ZYX S. Entwisle 30 Arden Mhor, Pinner HA5 2HR
2E0 ZZC S. Alexander 13 Padgate, Thorpe End, Norwich NR13 5DG
2W0 ZZF D. Jones 1 Brig Y Nant, Llangefni LL77 7QD
2E0 ZZT N. Payne 19 Sid Park Road, Sidmouth EX10 9BW
2W0 ZZU E. Jones 39 Ger-Y-Llan, Velindre, Llandysul SA44 5YB
2E0 ZZZ J. Earnshaw Dunelm, Ayton Road, Irton, Scarborough YO12 4RQ

2E1

2E1 ABE H. Forder 4 Jackson Drive, Kennington, Oxford OX1 5LL
2E1 ABN J. Morrison 52 Kimberley Close, Dover CT16 2JW
2E1 ABQ N. Harman 7 Maple Avenue, Torpoint PL11 2NE
2E1 ABW S. Minnock 32 Sandwood Road, Sandwich CT13 0AQ
2E1 ABY M. O'Brien 14 Westdean Close, Dover CT17 0NP
2E1 ACG V. Hammonds 22 The Croft Meriden, Coventry CV7 7NQ
2E1 ACK D. Nye 5 Charles Road, Deal CT14 9AT
2W1 ACM D. Young 1 Hawthorn Road Llanharry, Pontyclun CF72 9JD
2E1 ACO K. Beech 44 Ashdene Close Willerby, Hull HU10 6LW
2E1 ACS N. Roberts 37A Rockley Avenue, Birdwell, Barnsley S70 5QY
2E1 ACT D. Thomas 4 Greenwood, Bamber Bridge, Preston PR5 8JS
2E1 ACW C. Pooler 18 Johnstone Close, Wrockwardine Wood, Telford TF2 7DA
2E1 ACZ R. Stanley 219 Fartown, Pudsey LS28 8NH
2E1 ADJ J. Bridgman 5 Drayton Avenue, Mackworth, Derby DE22 4JU
2M1 ADM A. Mair 8 Cockburn Crescent Whitecross, Linlithgow EH49 6JT
2W1 ADO K. Jenkins 79 Beaufort Road, Newport NP19 7PB
2E1 ADP T. Thompson 19 Park End, Summer Lane Caravan Park, Banwell BS29 6JD
2E1 ADQ C. Hammett 63 Treffry Road, Truro TR1 1WL
2E1 ADR D. Palmer 133 Victoria Road East, Thornton-Cleveleys FY5 5HH
2E1 ADT K. Barbery 17 Polbreen Avenue., St. Agnes TR50TR
2E1 AEC C. Vincent 134 Wolds Drive, Keyworth, Nottingham NG12 5DA
2E1 AEJ E. Jones 19 Foxhollow Bar Hill, Cambridge CB23 8EP
2E1 AEQ V. Parrish 89 Delamere Drive, Macclesfield SK10 2PS
2E1 AEU A. Prescott 43 Coombe Road, Southminster CM0 7AH
2E1 AFA J. Davis 5 James Way, Camberley GU15 2RQ
2E1 AFC J. Rossiter 7 Valley View, Bodmin PL31 1BE
2E1 AFI J. Charnley 30 Dunkirk Avenue, Fulwood, Preston PR2 3RY
2E1 AFN L. Swindale 17 Crofton Close, Bracknell RG12 0UR
2E1 AFR F. Batty 26 Kingsmead Park, Elstead, Godalming GU8 6DZ
2E1 AFS L. Jenkins 49 Harts Grove, Chiddingfold, Godalming GU8 4RG
2E1 AGE J. Prince Field House, 25 Chiltern Road, Slough SL1 7NF
2E1 AGQ J. Collins 61 Albemarle Road, Gorleston, Great Yarmouth NR31 7AS
2E1 AGV E. Muircroft 84 Longley Avenue West, Sheffield S5 8WF
2E1 AHK J. Robinson 35 Vegal Crescent, Halifax HX3 5PA
2E1 AHU T. Hassall 5 Ashworth Street, Bacup OL13 9LS
2W1 AID S. Williams 27 Barnardo Street, Maesteg CF34 0HT
2E1 AII D. Swann 9 A J Cooks Cottages, Rowlands Gill NE39 2DA
2E1 AIT J. Tonks 295 Quinton Road West, Quinton, Birmingham B32 1PG
2E1 AIY B. Whalley 46 Wayside, Woodside, Telford TF7 5NG
2E1 AJB A. Burgess 5 Wilkie Road, Birchington CT7 9HE
2E1 AKK T. Goodwin 37 Purdy Meadow, Long Eaton, Nottingham NG10 3DJ
2E1 AKW M. Bradley 7 Rippon Way, Bursledon, Bursledon, Southampton SO318PA
2E1 ALC A. Coward 11 Oaklands, Ardingly, Haywards Heath RH17 6UE
2I1 ALE D. Auld 37 Castlewellan Road, Rathfriland, Newry BT34 5EL
2E1 AMB A. Collins 141 Downside Avenue, Findon Valley, Worthing BN14 0EY
2E1 AMW D. Evans 9 Robin Close, Farnworth, Bolton BL4 0RG
2E1 ANG P. Jackson 55 Bomers Field, Rednal, Birmingham B45 8TQ
2E1 ANH C. Jackson 55 Bomers Field, Rednal, Birmingham B45 8TQ
2E1 ANN M. Kearney 18 Wayside Mews, Maidenhead SL6 7EJ
2E1 ANQ A. Bell 8 Silk Mill Green, Leeds LS16 6DU
2M1 ANY L. Waterall 3 Wavell Street, Grangemouth FK3 8TG
2E1 AOF G. Tutt 46 Heathcroft Avenue, Sunbury-on-Thames TW16 7TL
2E1 AOG J. Menday 3 Ash Grove, Guildford GU2 8UT
2E1 AOK R. Gill 45 Biggin Lane, Ramsey, Huntingdon PE26 1NB
2M1 AOL G. Sibbald 1 Ormiston Drive East Calder, Livingston EH53 0RN
2E1 APW D. Jenkinson 9 Dalton Street, Cockermouth CA13 0AR
2E1 APX D. Johnson 3 Plantation Avenue, Swalwell, Newcastle upon Tyne NE16 3JN
2E1 AQH S. Allgood 53 The Avenue, Leighton Bromswold, Huntingdon PE28 5AW
2E1 AQT R. Scott 39A Highgate Lane Goldthorpe, Rotherham S63 9BA
2E1 ARG M. Trigg 41 Veasey Road, Hartford, Huntingdon PE29 1TA
2E1 ARS W. Hornby Lindenstrasse 9, Allschwil 4123 Switzerland
2E1 ASF T. Stevens 20 The Butts, Crudwell, Malmesbury SN16 9HF
2E1 ATH J. Minnock 32 Sandwood Road, Sandwich CT13 0AQ
2E1 ATV R. Corden Konrad Cottage, Welburn, York YO60 7DX
2E1 AUN D. Austin 66 Homewood Avenue, Sittingbourne ME10 1XJ
2E1 AUQ E. Harding 17 Summerfield Close, Wokingham RG41 1PH
2E1 AVM S. Peacock 9 Lake View, Poolview Caravan Park, Buildwas, Telford TF8 7BS
2E1 AVT J. Hoggan 39 The Glade, Crawley RH10 6JL
2M1 AVZ N. Sinclair 16 Sycamore Glade, Livingston EH54 9JG
2E1 AWZ V. Hilton 232 Hurst Rise, Matlock DE4 3EW
2E1 AXD A. Richmond 57 The Fairway, Daventry NN11 4NW
2E1 AXE L. Richmond 57 The Fairway, Daventry NN11 4NW
2I1 AXH K. Bird 115 Halftown Road, Lisburn BT27 5RF
2E1 AXI F. Preece 8 Gregory Road, Hedgerley, Slough SL2 3XL
2E1 AXL W. Rose 7 Harby Close, Grantham NG31 7XA
2W1 AYO E. Phillips 2 Oak St., Newport NP9 7HW
2E1 AYS P. Cartwright 41 Sandgate Drive Kippax, Leeds LS257EX
2E1 AZA E. Williams 37 Danesby Crescent, Denby, Ripley DE5 8RF
2E1 AZK C. Morris 10 Hempits Grove, Acton Trussell, Stafford ST17 0SL
2E1 AZQ J. Perkins Highfield House, Newtown, Buxton SK17 0NF
2W1 AZU D. Williams 75 Queens Avenue, Maesgeirchen, Bangor LL57 1NH
2E1 AZW R. Roberts Rose Cottage, Castle Hill, Leyburn DL8 4QN
2E1 BAD B. Rowland Deacons Cottage, Bridleway, Croft LE9 6EE
2E1 BAE T. Ladley Marisdene, London Road, Faversham ME13 9LF
2M1 BBY J. Duncan 4 Lady Moss, Tweedbank, Galashiels TD1 3SB
2E1 BCC N. Kluger-Langer 23 Vernon Walk, Northampton NN1 5ST
2E1 BCF J. Brown 9 South Street High Spen, Rowlands Gill NE39 2HF
2E1 BDB P. Hudson 1 Dean Moore Close, St. Albans AL1 1DW
2E1 BDC P. Kennedy Spinney Bungalow, Thearne Lane, Woodmansey, Beverley HU17 0SA
2E1 BDV P. Mclusky 11 Ripon Road, Killinghall, Harrogate HG3 2DG
2E1 BEB J. Wright 4 Sweeters Field Road, Alfold GU6 8UD
2E1 BEV M. Norman 41 Avon Grove, Bletchley, Milton Keynes MK3 7BP
2E1 BFH M. Knight 359 Shelley Road, Wellingborough NN8 3EW
2E1 BFP J. Philpot 7 Providence Place, Ilkeston DE7 8AL
2E1 BFW P. Hyde 10 Highfield Crescent, Taunton TA1 5JH
2E1 BFX C. Mann 11 North River Road, Runham Vauxhall, Great Yarmouth NR30 1JY
2E1 BGN A. Rollitt-Smith 9 St Helens Road, Doncaster DN4 5EQ
2E1 BGQ M. Wynn 22 Matthews Drive, Wickersley, Rotherham S66 1NN
2E1 BHB A. Comis 178 Lordswood Road, Birmingham B17 8QH
2E1 BHC P. Comis 65 Montague Way Chellaston, Derby DE73 5AS
2E1 BHF J. Clifford 16 Park View Road, Birmingham B31 5AU
2E1 BHU P. Gosling 10 Prospect Road, Carlton, Nottingham NG4 1LY
2E1 BIL W. Harvey 76 Barlaston Road, Stoke-on-Trent ST3 3LF
2E1 BIM C. Faulkner 1 Westland, Martlesham Heath, Ipswich IP5 3SU
2E1 BIT C. Swain 4 Follows Way, Chesterfield S44 6ZJ
2E1 BJD R. Saunders 31 Greenwood Road, High Green, Sheffield S35 3GU
2E1 BJG S. Benson 45 Maple Way, Selston, Nottingham NG16 6FA
2E1 BKF G. Muircroft 62 New Road, Rotherham S61 2DU
2E1 BKK C. Berry Roseneath, Walcote Road, Lutterworth LE17 6EQ
2E1 BKP S. Martin The Shieling, Bolton Low Houses, Wigton CA7 8PF
2E1 BKT D. Jump 15 Swallowfield, Chorley New Road, Horwich, Bolton BL6 6HN
2E1 BLG M. Levinson-Withall The Bungalows, 20 Moor End Avenue, Salford M7 3NX
2E1 BLP B. Mulley 8 Drinkstone Road, Gedding, Bury St. Edmunds IP30 0QB
2E1 BMB B. Porthouse 10 Pecklewell Terrace, Maryport CA15 7QJ
2E1 BME Dr M. Riley 2 Keeble Drive, Washingborough, Lincoln LN4 1DZ
2E1 BMF D. Carskake 38 Loppets Road, Crawley RH10 5DW
2E1 BMJ D. Peters The Vicarage, Church Road, Cholsey, Wallingford OX10 9PP
2E1 BMV B. Mycock 69 Bentley Road, Uttoxeter ST14 7EN
2W1 BOG I. Skinner 4 St. Marys Crescent, Rogiet, Caldicot NP26 3TB
2E1 BOM M. Love 72A Hart Plain Avenue, Waterlooville PO8 8RX
2E1 BOO M. Constantine 18 Hillbeck, Halifax HX3 5LU
2E1 BOZ S. Razzaq 15 Wild Herons, Hook RG27 9SF
2E1 BPN A. Collins Flat 2, 49 Dukes Head Street, Lowestoft NR32 1JY
2E1 BPV D. Roberts 98 Pinkneys Road, Maidenhead SL6 5DN
2E1 BRA M. Scotton 15 Grove Road, Aston, Stone ST15 0DW
2E1 BRC G. Thornsby 25 Kipling Way, Stowmarket IP14 1TS
2E1 BRD M. Larcombe 52 Orchard Road, Burgess Hill RH15 9PL

2E1 BRG C. Sanderson 14 Hazelwood Avenue, York YO10 3PD
2E1 BSC C. Castle 26 Chestnut Walk, Pulborough RH20 1AW
2W1 BST S. Long 10 St. George Road, Bulwark, Chepstow NP16 5LA
2E1 BTG M. Joyner Brimar, Nelson Park Road, Dover CT15 6HL
2E1 BTK K. Creamer Ellipsis. 35 Salisbury Road, St Margaret'S Bay. Ct15 6DI, Dover CT15 6DL
2E1 BUJ P. Stott 12 Castle View, Ovingham, Prudhoe NE42 6AT
2E1 BUM F. Stone 7 Cherry Tree Close, Hilton, Derby DE65 5FD
2E1 BUR J. Izzard Sunnyside, West End, Hailsham BN27 4NH
2E1 BUV P. Fletcher 18 Woodside View, Holmesfield, Dronfield S18 7WX
2E1 BVJ E. Constantine 18 Hillbeck, Halifax HX3 5LU
2E1 BVQ N. Brown 55 Shakespeare Terrace, Chorley PR6 7AQ
2E1 BVY D. Fox 36A Northmere Road, Poole BH12 4DY
2E1 BWT A. Ross 42 New Heritage Way, North Chailey, Lewes BN8 4GD
2E1 BYH M. Hearn 118A Dover Road, Folkestone CT20 1NN
2E1 BYI S. Piccavey 611 Manchester Road, Linthwaite, Huddersfield HD7 5QX
2W1 BYK A. Sellors 12 Morfa View, Bodelwyddan, Rhyl LL18 5TT
2M1 BYW T. Conlan 12 Rowantree Road, Mayfield, Dalkeith EH22 5ER
2E1 BYY W. Kennedy 1 Lynton Road, Hindley, Wigan WN2 4EH
2E1 BZB D. Peter 41 Coleswood Road, Harpenden AL5 1EF
2E1 BZH G. Low 30 Laburnum Grove, Runcorn WA7 5EL
2E1 BZI D. Low 6 Pale Manor Close, Malvern WR141SZ
2E1 CAF S. Kirkpatrick 29 Barnfields, Gloucester GL4 6WE
2E1 CAH R. Richards 39 North Holme Court, Northampton NN3 8UX
2E1 CAJ J. Godding 70 Rodway Road, Tilehurst, Reading RG30 6DT
2E1 CAQ M. Porter 7 Dunkirks Mews, Hertford SG13 8BA
2E1 CAT L. O'Ryan 12 Minton Close, Congleton CW12 3TD
2E1 CAU A. Burt 20 Medina Breeze Walk, Binfield, Newport PO30 2GS
2E1 CAW R. Marshall 15 Whisby Court, Holton-Le-Clay, Grimsby DN36 5BG
2E1 CBH S. Stretch 5 Ledwych Road, Droitwich WR9 9LA
2E1 CBU C. Richards 11 Purvis Road, Rushden NN10 9QA
2E1 CCF P. Izzard 7 Yardley Drive, Northampton NN2 8PE
2E1 CCG R. Winship 32 Lytes Cary Road, Keynsham, Bristol BS31 1XD
2E1 CCI A. Murphy 34 Hawkenbury Way, Lewes BN7 1LT
2E1 CCN D. Hill 28 Pendarves Flats, St. Clare Street, Penzance TR18 2PL
2E1 CDK G. Langdon 43 Daniel St., Ryde PO33 2BH
2E1 CDS B. Allen 78 Bargates, Christchurch BH23 1QL
2E1 CDZ A. Woods 40 Windsor Way, Sandy SG19 1JL
2W1 CEE D. Whish 62 Marion Road, Prestatyn LL19 7DF
2E1 CEQ M. Sayers Flat 12, Quay Court, Loring Road, Christchurch BH23 2AU
2E1 CEU J. Columbine West Lodge, 166 Tollerton Lane, Nottingham NG12 4FW
2E1 CEZ J. Brown Kingsdown Cottage, Fron, Montgomery SY15 6SB
2E1 CFB D. Williams 100 Hills Lane Drive, Madeley, Telford TF7 4BX
2E1 CGS M. Smith 27 Oldway Lane, Slough SL1 5LA
2E1 CHX M. Reavill 11 Clarence Road, Beeston, Nottingham NG9 5HY
2E1 CIK D. Gillatt 1 City Mills, Skeldergate, York YO1 6DB
2E1 CIO C. Brooks 1 Gloucester Road, Guisborough TS14 7DZ
2W1 CIP R. Bufton 7 Laburnum Close, Rassau, Ebbw Vale NP23 5TS
2E1 CIT J. Watkins-Field Sharlions, 27 Bosvigo Road, Truro TR1 3DG
2E1 CIY B. Harratt 8 Aaron Wilkinson Court, South Kirkby, Pontefract WF9 3JT
2E1 CJB K. Jordan 11 Sandringham Place, Hucknall, Nottingham NG15 8EU
2E1 CJC R. Richardson 3 Cautley Drive, Killinghall, Harrogate HG3 2DJ
2E1 CJD P. Taylor 13 Mackenzie Crescent, Burncross, Sheffield S35 1UR
2E1 CJF S. Curtis 354 St. Helens Road, Leigh WN7 3PQ
2E1 CJJ S. Hull 1 Occupation Lane, New Bolingbroke, Boston PE22 7LW
2M1 CJL A. Jones 9 Firs Street, Falkirk FK2 7AY
2E1 CJN H. Hughes 46 The Boundary, Oldbrook, Milton Keynes MK6 2HT
2E1 CJZ Z. Hodges 12 Linwal Avenue, Houghton-On-The-Hill, Leicester LE7 9HD
2E1 CKH K. Riley 16 King St., Westhoughton, Bolton BL5 3AX
2E1 CKQ E. Swain 11 Blackdown, Fullers Slade, Milton Keynes MK11 2AA
2W1 CLC M. Holmes 77 Harlech Drive, Merthyr Tydfil CF48 1JU
2E1 CLG G. Harvey 3 Mulberry Way, Sittingbourne ME10 3TG
2E1 CLM E. Woolley 82 Pennycroft Road, Uttoxeter ST14 7ET
2E1 CML B. Norris 20 Laburnum Close, Guildford GU1 1NA
2E1 CNM A. Raxworthy 32 St. Marys Avenue, Alverstoke, Gosport PO12 2HX
2W1 CNN A. Gray 36 Heol Pentre Felen, Morriston, Swansea SA6 6BY
2E1 CNO L. Call 3 Southfield, Bramhope, Leeds LS16 9DR
2E1 COD B. Call 6 Harecroft Road. Otley, Leeds LS212BQ
2E1 COG D. Oakes 2 Hillcrest, Scotton, Catterick Garrison DL9 3NJ
2E1 COM M. Holt 20 Lingfield Mount, Leeds LS17 7EP
2E1 COV I. Cockshoot 72 Princess Margaret Avenue, Cliftonville, Margate CT9 3EF
2E1 CPB A. Cadey Wakeley House, High Street, Charing, Ashford TN27 0LS
2E1 CPC Worcester Radio Amateur Association c/o M. Sheppard Flat 3 Brayden Block 85 Burgess Road, Leicester LE2 8QL
2E1 CPF R. Kensall 16 Parkwood Close, Broadstairs CT102XN
2E1 CPI B. Rowley 7 Hall Farm Close, Castle Donington, Derby DE74 2NG
2E1 CPJ P. Fisk 38 Bedingfield Crescent, Halesworth IP19 8EE
2E1 CPP N. Newman 89 Sea Place, Goring-By-Sea, Worthing BN12 4BH
2E1 CPQ P. Goodwin 60 Dale Crescent, Congleton CW12 3EP
2W1 CPS D. Probert 32 Heol Penlan Longford, Neath SA10 7LB
2E1 CPV P. Porter 7 Long Road, Framingham Earl, Norwich NR14 7RY
2E1 CQM I. Hurst 14 Longhill Rise, , Kirkby in Ashfield NG17 9FL
2E1 CQP V. Brightwell 40 Streete Court Road, Westgate-on-Sea CT8 8BX
2E1 CQQ E. Whelan 54 Boroughbridge Road, Northallerton DL7 8BN
2E1 CRA M. Lewis 15 Highcliffe Avenue, Chester CH1 5DP
2E1 CRI J. Mosby 1 School Road, Golcar, Huddersfield HD7 4NU
2E1 CSD A. Smith 30 Lime Grove, Grantham NG31 9JD
2E1 CTU S. Crawshaw 1 Wardley Close, Burnley BB12 6ET
2M1 CUS R. Higgins 30 Wallfield Crescent, Aberdeen AB25 2JX
2E1 CVE R. Trow 5 Cranberry Drive, Stourport-on-Severn DY13 8TH
2E1 CWE M. Skewes 47 Pentrevah Road, Penwithick, St. Austell PL26 8UA
2J1 CWG J. Totty Flat 4, Beech Court, Woodlands Apartments, La Route Des Cotils, Grouville, Jersey JE3 9AY
2J1 CWH C. Totty Flat 4, Beech Court, Woodlands Apartments, La Rue Des Cotils, Grouville, Jersey JE3 9AY
2E1 CWJ D. Thatcher 6 Ivel View, Sandy SG19 1AU
2E1 CWN P. Kemble 88 Mayfield Road, Ipswich IP4 3NG
2E1 CWP J. Parker 19 Mayfair Close, Dukinfield SK16 5HR
2E1 CWQ P. Millward 28 Olive Grove, Burton Joyce, Nottingham NG14 5FG
2E1 CWX E. Parker 19 Mayfair Close, Dukinfield SK16 5HR
2E1 CXE J. Mortimer 14 Oakfield Road, Bourne End SL8 5QN
2E1 CXF A. Whittaker 62 Ingham Street, Padiham, Burnley BB12 8DR
2E1 CXI R. Maunder 12 Hamble Springs, Bishops Waltham, Southampton SO32 1SG
2E1 CXP E. Bradshaw 41 Sherwood Road, Woodley, Stockport SK6 1LH
2W1 CYC D. Tiltman 16 St. Georges Road, Heath, Cardiff CF14 4AQ
2E1 CYD G. Reilly 10 Gloucester Road, Huyton, Liverpool L36 1XX
2E1 CYE R. Stenhouse High Park, Common Road, Norwich NR16 1HH
2E1 CYI D. Pomery 3 Mayfair Park, Minorca Lane, St. Austell PL26 8QN
2E1 CYM N. Griffiths 85 Foljambe Road, Chesterfield S40 1NJ
2E1 CYP J. Bick 45 Gloucester Road, Almondsbury, Bristol BS32 4HH
2E1 CYS I. Limbert 9 Lyme Grove, Liverpool L36 8BN
2E1 CYT S. Gallagher-Willmer 14 Bellhurst Cottages, Bellhurst Cottages, Bellhurst Road, Robertsbridge TN32 5DN
2E1 CYU S. Barker 26 Rye Court Helmsley, York YO62 5DY
2E1 CYZ M. Baxter 5 Farnborough Street, Farnborough GU14 8AG
2E1 CZB B. Fido 7 Claires Walk, Parklands Mobile Hom, Scunthorpe DN17 1SW
2E1 CZF A. Pink 31 The Fairway, Daventry NN11 4NW
2E1 CZJ J. Bosworth 57 Livingstone Road, Derby DE23 6PS
2E1 CZO B. Coombs 10 Horseshoe Walk, Widcombe, Bath BA2 6DE
2E1 CZS S. Backhouse 49 Western Avenue, Felixstowe IP11 9SL
2E1 DAK C. Wilderspin 3 Ferndale, Eaglestone, Milton Keynes MK6 5AE
2W1 DAO J. Patrick 52 Huntsmans Corner, Wrexham LL12 7UH
2E1 DAP D. Pope 24 Wanstead Crescent, Blackpool FY4 4AR
2E1 DAR A. Cottle 66 Elmhurst Estate Batheaston, Bath BA1 7NX
2E1 DBP A. Gener 3E Dartmouth Terrace, Greenwich, London SE10 8AX
2E1 DBQ S. Walker 33 Parkside Somercotes, Alfreton DE55 4LA
2E1 DBS K. Budd 20 Marshal Road, Poole BH17 7HA
2E1 DBT Dr R. Verma 43 Farley Road, Derby DE23 6BW
2E1 DBZ S. Issatt 7 Birch Road, Doncaster DN4 6PD
2E1 DCV B. Shields 20 Gresley Court, Grantham NG31 7RH
2E1 DDJ J. Gallagher 71 Castle Hill, Beccles NR34 7BJ
2E1 DDZ S. Arter 17 Chester Road, Westgate-on-Sea CT8 8AR
2W1 DEA P. Smith 23 Gainsborough Close, Llantarnam, Cwmbran NP44 3BX
2E1 DEM N. Humphreys 4 Rose Cottages, Annscroft, Shrewsbury SY5 8AU
2E1 DEN P. Hart 18 Harewood Road, Rochdale OL11 5TG
2E1 DEP L. Darton 8 Foster Grove, Sandy SG19 1HP
2E1 DET A. Whyman 8 Staplers Close, Great Totham, Maldon CM9 8UN
2E1 DFB B. Williams Hillside, Wigmore Lane Eythorne, Dover CT15 4AW
2E1 DFE R. Scott 315 Ormskirk Road, Wigan WN5 9DL
2E1 DFS R. Bearcroft 45 Broad Marston Road, Pebworth, Stratford-upon-Avon CV37 8XT
2E1 DFZ M. Axon 48 Cowslip Road, Broadstone BH18 9QZ
2E1 DGL P. Lewis 166 Euston Road, Morecambe LA4 5LE
2W1 DGM J. Foster 56 Hillrise Park, Clydach, Swansea SA6 5DX
2E1 DHC M. Axworthy 3Shepherds Court, 1 Shepherds Street, St. Leonard's on Sea TN38 0ET
2M1 DHG G. Russell 76 Duffus Crescent, Elgin IV30 5PY
2E1 DHJ D. Jones 11 Kylemilne Way, Stourport-on-Severn DY13 9NA
2E1 DHX D. Walker 17 Pinehurst Avenue, Christchurch BH23 3NS
2E1 DIA J. Laffin 154 Blenheim Drive, Allestree, Derby DE22 2GN
2E1 DIH J. Bentley 4 Highway, Crowthorne RG45 6HE
2E1 DIJ T. Willard 4 Chapel Cottages, Cowfold Road, Bolney, Haywards Heath RH17 5QU
2E1 DIJ T. Willard 4 Chapel Cottages, Cowfold Road, Bolney, Haywards Heath RH17 5QU
2E1 DKU A. Whittle 9 Dale View, Littleborough OL15 0BP
2E1 DLA P. Craig Manor House, Ardens Grafton, Alcester B49 6DR
2E1 DLD E. Harrison 55 Hudson Close, Worcester WR2 4DP
2E1 DLM D. Wilson 76 Cheadle Road, Uttoxeter ST14 7BY
2E1 DLR R. Diaper 30 Holmcroft Road, Kidderminster DY10 3AG
2E1 DLR D. Carr 33 Livingstone St., Leek ST13 5JU
2E1 DLS J. Field 27 Lovelace Road, Barnet EN4 8EA
2E1 DLT P. Lilley 12 Trueman Gardens, Arnold, Nottingham NG5 6QT
2E1 DLX P. Thackray 20 Darfield St., Leeds LS8 5DB
2E1 DMH M. Rippin Gaverne, Welford Road, Stratford-upon-Avon CV37 8RA
2E1 DMI R. Dixon 14 Blackdown Close, Peterlee SR8 2JW
2E1 DMU S. Wilson 218 Bredhurst Road, Gillingham ME8 0RD
2E1 DMZ R. Laverick 12 Greenlands Road, Redcar TS10 2DG
2E1 DNB P. Lomas 42 Lane End, Pudsey LS28 9AD
2E1 DNC M. Harris Flat 22, Alexandra Court, The Royal Seabathing, Margate CT9 5NT
2E1 DNF J. Foy 2 Lark Rise, Northampton NN3 8QT
2W1 DNK A. Waller 4 Rose Court, Ty Canol, Cwmbran NP44 6JH
2E1 DNX M. Mackenzie 73 Newstead Road, Weymouth DT4 0AS
2E1 DOA P. Dalby Windfall, 11 Greensward Lane, Hockley SS5 5HD
2E1 DOG H. Godzisz 1 Jutland Place, Egham TW20 8ET
2E1 DOZ S. Brenchley 89 Thicket Mead, Midsomer Norton, Radstock BA3 2SL
2E1 DPK S. Hill 22B Strait Lane, Hurworth On Tees, Darlington DL2 2AL
2E1 DPQ R. Tattersall 70 Selwyn Street, Hillstown, Bolsover, Chesterfield S44 6LR
2E1 DQG D. Bennett 6 Chalton Place, London Road, Horndean, Waterlooville PO8 0JE
2E1 DQM N. Edwards 609 Upper Richmond Road West, Richmond TW10 5DU
2E1 DQQ D. Horsley1 1 Mead Close, Swanley BR8 8DQ
2E1 DQT A. Charles 6 Bridewell St., Wymondham NR18 0AR
2E1 DQZ C. Houlden 29 Court Barton, Portland DT5 2HJ
2W1 DRB P. Waller 4 Rose Court, Ty Canol, Cwmbran NP44 6JH
2E1 DRC P. Hatcher 32 Slough Road, Iver SL0 0DT
2E1 DRU K. Mccann Treverven, Back Lane, Selby YO8 6QP
2E1 DRV J. Wohlgemuth 37 Broadcoombe, South Croydon CR2 8HR
2E1 DRX R. Hauxwell 65 Harleston Way, Heworth, Gateshead NE10 9BQ
2E1 DRY G. Symonds Flat 13, Bradbury House, Norwich NR2 3PT
2E1 DSU J. Hewitt 49 Calder Drive, Sutton Coldfield B76 1YR
2E1 DSX C. Haddon 1 Victoria Place, Weston, Portland DT5 2AA
2E1 DTE Dr C. Folkerd 4 Beechwood, Shaw, Oldham OL2 8LP
2E1 DTF D. Folkerd The Old Manse, London W5 5QT
2E1 DTG V. Thomas 4 Greenwood, Bamber Bridge, Preston PR5 8JS
2E1 DTK R. Cleary 5 Gregson Road, Widnes WA8 0BX
2E1 DTN J. Bennett11 Rectory Cottage, Broad Street, Wrington, Bristol BS40 5LD

2E1 DUI G. Tunley 36 Hamilton Drive, Chippenham SN14 0XW
2E1 DUW M. Goldby Waylands Gate, St. Johns Road, New Milton BH25 5SD
2M1 DWK D. Keay Parkhill, Cromwell Park, Almondbank, Perth PH1 3LW
2E1 DWP B. Hanson 35 Market Lane, Wolverhampton WV4 4UL
2E1 DWW D. Chapman 38 Cranfleet Way, Long Eaton, Nottingham NG10 3RJ
2E1 DXB I. Humberstone 20 Kingswood Road, Colchester CO4 5JX
2W1 DXV A. Powers 9 Courtybella Gardens, Newport NP20 2GN
2E1 DYL M. Reid 24 Mitchel Way, Madeley, Telford TF7 5SN
2E1 DYT C. Block 13 Beatrice Road, Capel-Le-Ferne, Folkestone CT18 7LH
2E1 DZH J. Richards 1 Stocks Mead, Washington, Pulborough RH20 4AU
2E1 DZJ R. Morris 11 Trelispen Park Drive, Gorran Haven, St. Austell PL26 6HX
2E1 DZL R. Neville 4 Danson Gardens, Blackpool FY2 0XH
2E1 DZM Prof. S. Murdoch 12 Gwydir Street, Cambridge CB1 2LL
2E1 DZP A. Cannon 20 Gladwyn Street, Stoke-on-Trent ST2 8JZ
2M1 DZS C. Clark 3 Old Cottages, Seton Mains, Longniddry EH32 0PG
2E1 DZV J. Keegan The Cottage, 11 Condor Grove, Lytham St. Annes FY8 2HE
2M1 DZW B. Waugh 93 Denholm Road, Musselburgh EH21 6TU
2M1 DZX R. Stuart Tigh Na Coille, Bishop Kinkell, Conon Bridge, Dingwall IV7 8AW
2E1 EAG J. Oxley 97 Defoe Crescent, Colchester CO4 5LQ
2E1 EAK Y. Wood 173 Links Road, North Shields NE30 3TG
2W1 EAN G. Taylor 55 Haulfryn, Tregynwr, Carmarthen SA31 2DT
2E1 EAV I. Duggan 21 Chetwynd Road, Toton, Nottingham NG9 6FW
2M1 EAW M. Barnett 9 Cairncross Place, Coatbridge ML5 2FA
2E1 EAX R. Edwards Flat 10, Longford Court, 60 London Road, Sevenoaks TN13 2UG
2E1 EAZ D. Finch 29 Park Street, Chesterfield S40 2HD
2M1 EBJ D. Martin Inverleod, Avoch IV9 8PR
2E1 EBL K. Dignall 11 Mottershead Road, Widnes WA8 7LD
2E1 EBN S. Valvona 123 Binstead Lodge Road, Ryde PO33 3UB
2E1 EBR C. Smith 16 Meadowbrook, Ancaster, Grantham NG32 3RR
2E1 EBX L. Collinson 12 Victoria Avenue, Hunstanton PE36 6BX
2E1 EBZ M. Watkins 2 Griffin Close, Billingshurst RH14 9GS
2M1 ECF J. Mackenzie Rainbows End, Lochside, Lairg IV27 4EG
2E1 ECG A. Topping 30 St. Pauls Avenue, Nottingham NG7 5EB
2E1 ECL C. Gray 19 Marsh View, Newton, Preston PR4 3SX
2E1 ECM P. Fletcher 11 The Banks, Long Buckby, Northampton NN6 7QQ
2E1 ECN M. Chapman 15 Norwood Road, Somersham, Huntingdon PE28 3EY
2E1 ECV P. Elsey 129 Kingsway, Chandler'S Ford, Eastleigh SO53 5BX
2E1 EDA E. Kurtz 1 Leonard Medler Way, Hevingham, Norwich NR10 5LE
2E1 EDB M. Bradwell 6 Moorfoot Gardens, Gateshead NE11 9LA
2E1 EDD S. Lansdell 42 Marylebone Crescent, Derby DE22 4JX
2M1 EDM D. Martin 45 Tiree Place, Newton Mearns, Glasgow G77 6UJ
2J1 EDR A. Price Wheatlands, La Rue Des Longchamps, St. Brelade JE3 8BN Jersey
2M1 EDT J. Wilson 4F Langside Street, Clydebank G81 5HJ
2E1 EDV J. Blanche 11 Woodside, Southminster CM0 7RD
2E1 EDW J. Blanche 11 Woodside, Southminster CM0 7RD
2E1 EEK S. Paffett 15 Centaury Gardens Horton Heath, Eastleigh SO507NY
2W1 EEP J. Jones Glanrafon Garage, Pontfadog, Llangollen LL20 7AR
2E1 EET L. Edwards 39 Foskitt Court, Northampton NN3 9AX
2E1 EFG S. Philpot 93 Princess Drive, Grantham NG31 9QA
2E1 EFQ D. Bushby 66 Sandy Road, Everton, Sandy SG19 2JU
2E1 EFT L. Froggatt11 255 Rushton Road, Desborough, Kettering NN14 2QB
2E1 EGI G. Durrant 51 Raglan Avenue, Waltham Cross EN8 8DA
2E1 EGU W. Lupton Eaton Cottage, Shorts Green Lane, Shaftesbury SP7 9PA
2E1 EGV M. Townson 42 Osborne Grove, Shavington CW2 5BY
2E1 EHB I. Greenall 356 Warrington Road, Abram, Wigan WN2 5XA
2E1 EHF Rvd. J. Goodman 34 Hillside Close, Chalfont St. Peter, Gerrards Cross SL9 0HN
2E1 EHM D. Whittaker 68 Querns Road, Canterbury CT1 1PZ
2E1 EHP D. Moses 121 Badger Avenue, Crewe CW1 3JN
2E1 EHY N. Waters 23 Harold Road, Birchington CT7 9NA
2W1 EID R. Owens 11 Elm Grove, Newport NP20 6JF
2W1 EIN C. Bodley 38 Bryn Hawddgar, Clydach, Swansea SA6 5LA
2E1 EIO A. Godolphin Shepherds Cottage Flakebridge, Appleby-in-Westmorland CA16 6JZ
2E1 EIU D. Meddings 42A Argyle Road, Poulton-le-Fylde FY6 7EW
2E1 EIV A. Eyre St. Michael Mead, The Common, Norwich NR12 8BA
2E1 EIX A. Chruscinski 39 Sherwood Rise, Mansfield Woodhouse, Mansfield NG19 7NP
2E1 EJC T. Bain 23 Salisbury Crescent, Blandford Forum DT11 7LX
2E1 EJD W. Ling Valley Farm Equestrian Centre, Wickham Market, Woodbridge IP13 0ND
2U1 EJF A. Scheffer Foveat, Rue De Jardins, Les Prins, Guernsey GY6 8EZ
2E1 EJU D. Lumley 19 Bramley Avenue, Needingworth, St. Ives PE27 4UD
2E1 EJX M. Beynon-Tullett 9 Clarkes Croft, Dishforth, Thirsk YO7 3XB
2U1 EKE C. Ayres Rousay, Bailiffs Cross Road, St. Andrew GY6 8RY Guernsey
2U1 EKH K. Johnson 9 Clos Spurway, Victoria Avenue, Guernsey GY2 4AH
2M1 EKI K. Hughes 49 Marmion Drive, Kirkintilloch, Glasgow G66 2BH
2E1 EKM D. Baldwin 51 Queens Road, Broadstairs CT10 1PG
2E1 EKQ M. Eades 19 Egham Court, Horsa Street, Bolton BL2 2DE
2W1 EKR D. Pollard 19 The Drive, Bargoed CF81 8JX
2E1 ELE P. Williams 20 Elm Close, Great Haywood, Stafford ST18 0SP
2M1 ELU I. Sinclair Airdaniar, Kilchrenan, Taynuilt PA35 1HG
2E1 EMH V. Newton 60 The Lynch, Winscombe BS25 1AR
2E1 EMI G. Paterson 4 Rowallan Drive, Bedford MK41 8AW
2E1 EMK J. Roff 4 Needham House, Victoria Road, Devizes SN10 1FA
2E1 EMN J. Thompson 18 Winder Way, Micklefield, Leeds LS25 4FX
2M1 ENI D. Paterson Leuchlands Croft, Whitecairns, Aberdeen AB23 8UT
2M1 ENK D. Paterson 29 Arnothill Gardens, Falkirk FK1 5BQ
2E1 ENN G. Cattle 50 Oakland Avenue, York YO31 1DF
2E1 ENR S. Taylor 65 Cross Bank Road, Batley WF17 8PN
2E1 ENZ R. Baxter 20 Thorpe St., Thorpe Hesley, Rotherham S61 2RP
2E1 EOD N. Bridges Acomb, Station Road, Pershore WR10 3BB
2E1 EOI J. Allum 122 Long Chaulden, Hemel Hempstead HP1 2HY
2E1 EOK C. Blackman 7 Deanery Close, Ripley DE5 3TR
2E1 EOO D. Webb 11 Alfriston Road, Worthing BN14 7QU
2E1 EOQ D. Horton Glen View, New Road, Bude EX23 9LE
2E1 EOR K. Horton The Old School House, Kelly, Lifton PL16 0HJ
2M1 EOV M. Macleod 4 Portnaguran, Isle of Lewis HS2 0HD
2E1 EOZ R. Cannon 110 Great Gregorie, Basildon SS16 5QF
2E1 EPA P. Allaker 3 Eden Cottages, Watling Street, Consett DH8 6HZ
2E1 EPD M. Freedman Rivermeade, Irwell Vale, Bury BL0 0QA
2E1 EPE P. Freedman Rivermeade, Irwell Vale, Bury BL0 0QA
2W1 EPL G. Morris Flat 12, Windsor Court Crescent Road, Rhyl LL18 1TF
2W1 EPO F. Hodge Flat 7, Windsor Court, Rhyl LL18 1TF
2E1 EPQ T. Fanning 26 Mandeville Close, Tilehurst, Reading RG30 4JT
2M1 EPV D. Stewart 19 Arthur Street, Blairgowrie PH10 6PF
2E1 EQE D. Evans 5 Compton Drive, Streetly, Sutton Coldfield B74 2DA
2E1 EQI V. Collins 30 Upham Road, Swindon SN3 1DN
2E1 EQK P. Gibson 73 Marlborough Road Hadley, Telford TF1 5LN
2E1 EQQ K. Wood 52 Ashfield Avenue, Beeston, Nottingham NG9 1PY
2E1 EQR J. Jones 17 Dunkeld Drive, Shrewsbury SY2 5UZ
2E1 ERJ D. Lusty 104 Polstain Road Threemilestone, Truro TR3 6DB
2E1 ERU A. Ainslie 16 Ryeburn Close, Kessingland, Lowestoft NR33 7UH
2E1 ESK L. Hodge 11 Glebelands, Bampton OX18 2LH
2E1 ESM K. James 36 Lemon Hill, Mylor Bridge, Falmouth TR11 5NA
2E1 ESN C. Mclean 18 Chatfield Road, Gosport PO13 0TN
2E1 ESQ A. Mould 95 Stanton Road, Southampton SO15 4HU
2E1 ESW D. Wilkinson 139 Church Road, Jackfield, Telford TF8 7ND
2E1 ETB R. Moore Thorpefield Farm, 91 Thorpe Street, Rotherham S61 2RP
2E1 ETJ A. Willis Kilncroft, Broadlayings, Newbury RG20 9TS
2M1 ETM S. Mackie 25 Carlaverock Drive, Tranent EH33 2EE
2W1 ETN D. Jorgensen Unit 11, Llandow Trading Estate, Llandow, Cowbridge CF71 7PB
2W1 ETW R. Thomas 24 Heol Innes, Llanelli SA15 4LA
2E1 EUE E. Hunt 5 Freshfield Square, Southampton SO15 8QU
2W1 EUR A. Joynes 25 St. Annes Gardens, Maesycwmmer, Hengoed CF82 7QQ
2M1 EUV E. Clark 3 Old Cottages, Seton Mains, Longniddry EH32 0PG
2E1 EUY W. Partridge 15 Cranbourne Avenue, Wolverhampton WV4 6RJ
2E1 EVH F. Stevenson 56 Barrows Hill Lane, Westwood, Nottingham NG16 5HJ
2E1 EVJ R. Chipperfield 5 Lullingstone Close, Hempstead, Gillingham ME7 3TS
2E1 EVK R. Chipperfield 5 Clayton Avenue, Upminster RM14 2EZ
2E1 EVM J. Merrick The Birches, Dunn Street Bredhurst, Gillingham ME7 3NA
2E1 EWK B. Ashman 108 Eastwood Drive, Highwoods, Colchester CO4 9SL
2E1 EWN B. Storkey 9 Snatchup, Redbourn, St. Albans AL3 7HD
2E1 EXA P. Johnson 110 Rachel Clarke Close, Stanford le Hope SS17 7SX
2E1 EXI D. Taylor 188 Walstead Road, Walsall WS5 4DN
2E1 EXK J. Wilkes 47 Greenwood Park, Hednesford, Cannock WS12 4DQ
2E1 EXP C. Staerck 8 Cresswell Road, Worksop S80 1SU
2I1 EXU P. Robinson 20 Harwood Park, Carrickfergus BT38 7LZ
2E1 EYC M. Davis 15 Farmcroft Road Mansfield Woodhouse, Mansfield NG19 8QU
2E1 EYF E. Hedley 17 Chowdene Bank, Gateshead NE9 6JJ
2E1 EYL M. Harris 1 Brampton Court, Bowerhill, Melksham SN12 6TH
2E1 EYS B. Egglestone 4 Lancaster Close, Etherley Dene, Bishop Auckland DL14 0RP
2E1 EYY G. Allen 47 St. Botolphs Road, Sevenoaks TN13 3AG
2W1 EYZ S. Gray 36 Heol Pentre Felen, Morriston, Swansea SA6 6BY
2M1 EZA J. Boyle Flat 3/2, 33 St. Mungo Avenue, Glasgow G4 0PH
2E1 FAN K. Blanchard 17 Stephens Way, Sleaford NG34 7JN
2E1 FAT S. Hallsworth 27 Westfield Avenue, Heanor DE75 7BN
2E1 FBA R. Locke 12A Kennedy Court, Walesby, Newark NG22 9PQ
2E1 FBH C. Timms 47 Brittains Lane, Sevenoaks TN13 2JP
2E1 FBK C. Strange 12 Cricketts Lane, Chippenham SN15 3EF
2E1 FBS D. Seabridge 31 Charlestown Drive, Allestree, Derby DE22 2HA
2E1 FBY I. Roper 28 Bobbin Mill Close, Todmorden OL14 8PZ
2E1 FCC M. Williams Cornacres, Dodwell, Stratford upon Avon CV37 9ST
2E1 FCD M. Williams Cornacres, Dodwell, Stratford upon Avon CV37 9ST
2E1 FCE J. Lindsay 10 New Station Road, Swinton, Mexborough S64 8AH
2E1 FCO I. Brady 6 Bristow Close, Bletchley, Milton Keynes MK2 2XP
2E1 FDC L. Peck 17 Mill Lane, Barton Le Clay, Bedford MK45 4LN
2E1 FDD M. Warren 7 Ronksley Crescent, Sheffield S5 0HG
2E1 FDF G. Bilson 34 Bramlyn Close, Clowne, Chesterfield S43 4QP
2E1 FDJ K. Newnam 1 Wheatlands Close, Maulden, Bedford MK45 2AQ
2E1 FDK L. Rule 46 Meadowsweet Road, Poole BH17 7XT
2E1 FDM J. Mew 1 Council Houses, Norwich Road, Norwich NR9 4NY
2E1 FDP G. Westwood 6 Monkton Road, Borough Green, Sevenoaks TN15 8SD
2E1 FDT P. Brown 19 Huxley Close, Nottingham NG8 4PU
2E1 FDU M. Darton 8 Foster Grove, Sandy SG19 1HP
2E1 FDY K. Rankin 31 Gorsey Bank Road, Hockley, Tamworth B77 5JD
2E1 FEC F. Dunmore 36 Dove Rise, Oadby, Leicester LE2 4NY
2E1 FEF K. Rhodes 30 Priory Road, Louth LN11 9AL
2E1 FET N. Moore 84 Franklynn Road, Haywards Heath RH16 4DH
2E1 FFJ Five Towns ARC c/o R. Denniss Chapel House, Farlesthorpe, Alford LN13 9PH
2E1 FFL P. Harbinson 19 Pennine Road, Dewsbury WF12 7AW
2E1 FFX P. Fryer 19A Wintringham Way, Purley On Thames, Reading RG8 8BH
2E1 FFZ D. Fawcett 6 Wand Hill, Boosbeck, Saltburn-by-the-Sea TS12 3AW
2E1 FGB N. Ash 35 Fairford Road, Tilehurst, Reading RG31 6PY
2W1 FGR J. Clark 22 Heol Yr Wylan, Cwmrhydyceirw, Swansea SA6 6TB
2E1 FHO P. Sawyers 15 Park Terrace, Whitby YO21 1PN
2E1 FHQ C. Ward 18 Aspen Grove, Aldershot GU12 4EU
2E1 FHZ D. Hebb 45 Marklew Avenue, Grimsby DN34 4AD
2E1 FIE M. Robertson 98 Hawthorn Road, Bognor Regis PO21 2DG
2E1 FIQ H. Pook Beverley Friory, Friars Lane, Beverley HU17 0DF
2E1 FIV D. Layton 69 Elm Drive, Crewe CW1 4EL
2E1 FIX J. Shaw 27 St. Davids Avenue Romiley, Stockport SK6 3JT
2E1 FJL M. Gray 19 Marsh View Newton, Preston PR4 3SX
2W1 FJN I. Pearson Warren Cottage, Pontfadog, Llangollen LL20 7AT
2E1 FJP B. Melling 68 Westfield Drive, Ribbleton, Preston PR2 6TH
2E1 FJV I. Page 84 Beaulieu Close, Toothill, Swindon SN5 8AH
2W1 FJZ Prof. P. Edwards Cam O'R Afon, Dolywern, Llangollen LL20 7AD
2E1 FKD E. Merrington Cartref, Ball Lane, Frodsham WA6 8HP
2E1 FKJ J. Ewing 130 Uttoxeter Road, Hill Ridware, Rugeley WS15 3QX
2E1 FKM C. Andrews Flat 2, 1St Floor, 153 Thanet St., Chesterfield S45 9JT
2E1 FKT M. Boyes 19 Rowe Ashe Way, Locks Heath, Southampton SO31 7EY
2E1 FKZ A. Woodward 19 Hazel Grove, Winchester SO22 4PQ
2E1 FLD P. Whittaker 34 New Road, Newhall, Swadlincote DE11 0SP
2E1 FLN G. Mcmillan 10 Thornton Court, Girton, Cambridge CB3 0NS
2E1 FLW S. Burling Ongar Cottage, 28 Main Road, Macclesfield SK11 0BU

2E1 FMC R. Sanderson 92 Edge Lane, Dewsbury WF12 0HB
2E1 FMW A. Oughton 176 South Lodge Drive, Southgate, London N14 4XN
2E1 FNB F. Morgan 171 Town Road, London N9 0HJ
2E1 FNJ L. Carr 10 Bonds Road, Hemblington, Norwich NR13 4QF
2U1 FNQ J. Le Page Heathwick, Les Martins, St. Martin GY4 6QJ Guernsey
2E1 FNR D. Rowland 32 Treloweth Way, Pool, Redruth TR15 3TT
2E1 FNX J. Meredith 2 Hamilton Road, Dawley, Telford TF4 3NG
2E1 FNY C. Warren Clifden Farm, Quenchwell Road, Truro TR3 6LN
2E1 FON S. Simpson 20 Staveley Grove, Keighley BD22 7DH
2E1 FOU K. Perry 238 Sherwood Street, Warsop, Mansfield NG20 0HJ
2E1 FOW J. Thacker 15 Riverside Mews, Hall Yard, Stoke-on-Trent ST10 4FE
2E1 FOX J. Camp 1 Higher Tresillian Cottages, Tresillian, Newquay TR8 4PL
2W1 FPK A. Cartwright 7 Pen Parc, Malltraeth, Bodorgan LL62 5BG
2E1 FPM L. Noel 58 Easenhall Lane, Redditch B98 0BJ
2E1 FPP D. Paddon 21 Oak Park Drive, Havant PO9 2XE
2E1 FPU J. Overland 73 Butchers Lane, Walton on the Naze CO14 8UE
2E1 FPV J. Stringer The Cottage, High Street, Radstock BA3 5AL
2E1 FPW J. Green 2 Broadmeadow, Kingswinford DY6 7HG
2E1 FQB P. Elcombe 16 Blenheim Avenue, Martham, Great Yarmouth NR29 4TW
2E1 FQE C. Carter 331A Ordnance Road, Enfield EN3 6HE
2M1 FQI G. Robinson 12 Hannahston Avenue, Drongan, Ayr KA6 7AU
2E1 FQO M. Dunthorne 52 New Street, Oakengates, , Telford TF2 6ES
2E1 FQY O. James 24 Fryer Avenue, Leamington Spa CV32 6HY
2E1 FRC H. Doyle Hurst House, Stratford Road, Henley-in-Arden B95 6AB
2E1 FRE D. Franklin The Old Vicarage, Westville Road, Boston PE22 7HJ
2E1 FRI J. Wood 3 Harold Collins Place, Colchester CO1 2GQ
2E1 FRQ A. Storry 99 Swineshead Road, Wyberton Fen, Boston PE21 7JG
2E1 FRW T. Depledge 16 Pennington Walk, Retford DN22 6LR
2E1 FRY E. Young 7 The Quadrant, Fordingbridge SP6 1BW
2E1 FRZ T. Jennings 23 De Lacy Court, New Ollerton, Newark NG22 9RN
2E1 FSF J. Kinch 7 Fox Lane, Oakley, Basingstoke RG23 7BB
2E1 FSG R. Kinch 7 Fox Lane, Oakley, Basingstoke RG23 7BB
2E1 FSH C. Forber 32 Larch Avenue, Newton-le-Willows WA12 8JF
2E1 FSR G. Hammond 50 Fernhill Close, Woodbridge IP12 1LB
2E1 FSV D. Feetenby 32 Hawkins Close, Daventry NN11 4JQ
2E1 FSX D. Charlton 20 Bailey Crescent, South Elmsall, Pontefract WF9 2TL
2E1 FSY M. Fey 37 Winnards Close, West Parley, Ferndown BH22 8PA
2E1 FSZ T. Bashford 198 Uplands Road, West Moors, Ferndown BH22 0EY
2E1 FTA S. Goodall Red Roofs, 4 Chapel Street, Stapleton LE9 8JH
2E1 FTE C. Bingham 56 Newgate Lane, Mansfield NG18 2LQ
2E1 FTF E. Sheppard 4 Lindrick Avenue, Swinton, Mexborough S64 8TE
2E1 FTH R. Megone 16 Mercer Close, Basingstoke RG22 6NZ
2E1 FTI M. Burrows 40 Fairmile Road, Christchurch BH23 2LL
2E1 FTV J. Bailey 13 Newark Road, Mexborough S64 9EZ
2W1 FUD D. Green 17 Glyn-Y-Mel Pencoed, Bridgend CF35 6YA
2E1 FUH D. Hartley 2 Thirlmere Avenue, Burnley BB10 1HU
2E1 FUJ J. Hartley 23 Broomfield Road, Fleetwood FY7 7HA
2E1 FUQ J. Quartermaine 3 Markham Close, Duston, Northampton NN5 6TW
2E1 FUS A. Sherratt 28 Church Street Bollington, Macclesfield SK10 5PY
2W1 FVH R. Hallett Llamedos, 151 Trealaw Road, Tonypandy CF40 2NX
2E1 FVJ R. Halford 104 Gladstone Avenue, London N22 6LH
2E1 FVK T. Kiely 192 Morley Avenue, London N22 6NT
2E1 FVS M. Hotchin 122 Buckingham Avenue, Scunthorpe DN15 8NS
2E1 FVY D. Ripley 5 Rope Walk, Cranbrook TN17 3DZ
2E1 FWA C. Walker 12 Bradshaw Crescent, Honley, C/O Mr C Walker, Holmfirth HD9 6EG
2E1 FWD C. Booth 8 Heathfield Mews, Martlesham Heath, Ipswich IP5 3UF
2E1 FWM M. Foister 38 Nine Acres, Kennington, Ashford TN24 9JW
2E1 FWX G. Sessions 5 Luxton Court, Cullompton EX15 1FJ
2E1 FXN D. Boland 29 Broom House, Baneberry Road, Gloucester GL4 6UY
2E1 FYC D. Martyr 52 Parklawn Avenue, Epsom KT18 7SL
2E1 FYZ T. Wilkie Bramcote, Grange Road, Sutton on Sea LN12 2RE
2E1 FZC J. Charter 36 Northumberland Avenue, London E12 5HD
2E1 FZH M. Saunders 105 Raynham Road, Bury St. Edmunds IP32 6ED
2E1 FZU D. Peters 25 Corndon Crescent, Shrewsbury SY1 4LD
2E1 FZY J. Doyle Orchard Corner, Piddington Road, Ludgershall, Aylesbury HP189PJ
2W1 GAC W. Hicks 2 Second Avenue, Clase, Swansea SA6 7LN
2M1 GBG M. Thomson 194 East Main Street, Broxburn EH52 5HQ
2E1 GBM J. Lake 25 Erlensee Way, Biggleswade SG18 8GG
2E1 GBN A. Vincent 12 Broad Park Avenue, Ilfracombe EX34 8DZ
2E1 GCA P. Howden Goodyfield Farm, Tofts Lane, Stannington, Sheffield S6 5SL
2E1 GCB E. Muxlow 14 Maidwell Way, Grimsby DN34 5UP
2D1 GCC B. Smith 82 Royal Park, Ramsey, Isle of Man IM8 3UH
2E1 GCF C. Horn 12 Melbourne Road, Chichester PO19 7NE
2W1 GCU J. Bennett 28 Neyland Path Fairwater, Cwmbran NP44 4PX
2E1 GDA C. Norris 8 Sutton Lane, Middlewich CW10 9AU
2E1 GDB M. Jeffery 14 Holly Mount Shavington, Crewe CW2 5AZ
2E1 GDD D. Townson 4 Crawford Street, Bradford BD4 7JJ
2E1 GDF S. Okeefe 9 Ivernia Close, Derby DE23 1XF
2E1 GDG D. Baker 78 Station Road, Whittlesey, Peterborough PE7 1UE
2E1 GDK V. Potts 14 Topcliffe Mead, Morley, Leeds LS27 8UH
2E1 GDM M. Banner 23 Astral Grove, Hucknall, Nottingham NG15 6FY
2E1 GDO R. Hacker Flat 7, Dove Court Packers Lane, Ramsgate CT11 8QA
2W1 GDY A. Palmer 4A Clomendy Road, Cwmbran NP44 3LS
2E1 GES C. Knowlson 23 Hawthorne Avenue, Shipley BD18 2JB
2M1 GEZ R. Scott Kirklands, Craigend Road, Galashiels TD1 2RJ
2M1 GFG R. Dempster 42 Kirkhill Road, Edinburgh EH16 5DD
2E1 GFS S. Thomas 24 Jenwood Road, Dunkeswell, Honiton EX14 4UZ
2E1 GFW M. Jones 68 Hampton Park Road, Hereford HR1 1TJ
2E1 GGL E. Mills 70 Crescent Road, Rochdale OL11 3LG
2E1 GGT A. Walker 12 Bladen Close, Cheadle Hulme, Cheadle SK8 5RU
2E1 GHE A. Carter 50 Harberd Tye, Chelmsford CM2 9GJ
2E1 GHF A. Richardson 42 Gypsey Road, Bridlington YO16 4AZ
2E1 GHI E. Mills 80 Bransdale Road, Nottingham NG11 9JB
2E1 GHX C. Lewin The Hawthorns, Hawthorne Drive, Stafford ST19 9NQ
2E1 GHZ P. Mather 18 Watcombe Cottages, Richmond TW9 3BD
2I1 GIH L. Costford Aughakeerin, Derrygonnelly, Co. Fermanagh BT93 6FR
2E1 GIK A. Craven 4 Amanda Drive, Louth LN11 0AZ
2E1 GIZ E. Christieson September Cottage, Rushlake Green, Heathfield TN21 9PP
2E1 GJC A. Brown Hurst Farm, Ashworth Road, Rochdale OL11 5UP
2E1 GJD S. Burgess 14 Shrubcote, Tenterden TN30 7BA
2E1 GJE P. Lee 2 Dennis St., Worksop S80 2LL
2E1 GJG H. Kennedy 19 High Street, East Hoathly, Lewes BN8 6DR
2E1 GJJ A. Noon 12 Stoney Fold, Telford TF3 5GQ
2E1 GJL P. Lovesey 20 Lindsey Way, Louth LN11 8RP
2E1 GJP K. Turner 2 Bungalow, Dunston Fen, Lincoln LN4 3AP
2E1 GJT C. Wojcik 153 Netherton Road, Worksop S80 2SD
2E1 GJY S. Brazier 5 Crowbeare Meadow, Torrington EX38 7DX
2E1 GKB M. Brearley 136 Elmsfield Avenue, Rochdale OL11 5XA
2E1 GKE D. Shephard 107 Withywood Drive, Telford TF3 2HX
2E1 GKF N. Salter 10 Jarvis Place St. Michaels, Tenterden TN30 6DQ
2W1 GKJ M. Moore 56 Morfa Street, Bridgend CF31 1HD
2E1 GKP C. Dix 141A Jerningham Road, London SE14 5NJ
2E1 GKY A. Reed 32 Hollis Gardens, Cheltenham GL51 6JQ
2M1 GLD E. Clark 3 Old Cottages, Seton Mains, Longniddry EH32 0PG
2E1 GLQ J. Lutener 22 Heron Park, Basingstoke RG24 8UJ
2E1 GLR J. Halsall 83 Poole Road, Leeds LS15 7HD
2E1 GLS D. Brown 22 Hillcrest Road, Castleford WF10 3QX
2E1 GLT M. Simpson 20 Mount Pleasant Residential Park, Bloomhill Road, Moorends, Doncaster DN84ST
2W1 GLY C. Mills 26 Goossens Close, Ringland, Newport NP19 9JN
2E1 GMA S. Jordan 31 Rathmore Road, Birkenhead CH43 2 HE
2W1 GMM B. Richards 8 East Avenue, Griffithstown, Pontypool NP4 5AB
2E1 GMO M. Hastry 56 Kilsyth Close, Fearnhead, Warrington WA2 0SQ
2E1 GMQ J. Barrett 114 William Street, Long Eaton, Nottingham NG10 4GD
2E1 GMT J. Williams Hillsborough Lodge, Lower St. German'S Road, Exeter EX4 4PW
2E1 GMV S. Fletcher 5 Hayeswood Road, Stanley Common, Ilkeston DE7 6GB
2E1 GNE R. Wright 3 Ednall Lane, Bromsgrove B60 2DB
2E1 GNK G. Smith 106 Broadoak Road, Manchester M22 9PL
2E1 GNN P. Saben Tredinneck Moor, Newmill, Penzance TR20 8XT
2E1 GNR A. Taylor 16 Penny Lane, Collins Green, Warrington WA5 4DS
2E1 GNU A. Watson 7 Branksome Drive, Morecambe LA4 5UJ
2E1 GNZ D. Hoyle 31 Rochester Avenue, Bolton BL2 5ED
2E1 GOC S. Bedell 1 Pheasant Field Drive, Spondon, Derby DE21 7LR
2E1 GOE C. Hopkins 28 Pitcairn Road, Mitcham CR4 3LL
2E1 GOK K. Phillips 30 Allendale, Ilkeston DE7 4LE
2E1 GOM Dr J. Constable 1900 Olden Glen, de Pere 54115 United States
2E1 GOP B. Cartwright 14 Zealand Close, Hinckley LE10 1TJ
2E1 GOZ I. Press 19 Banwell Close, Keynsham, Bristol BS31 1JX
2E1 GPE M. Ruff Brambledown, Camberlot Road, Hailsham BN27 3QG
2E1 GPG D. Howells Flat 1-2, 130 Essex Road, London N1 8LX
2E1 GPO R. Dinnage 108 Chaddock Lane, Boothstown, Salford M28 1DF
2E1 GPV M. Allott 1 Charles Street, Lancaster LA1 4UU
2E1 GQB G. Meek 443 Springfield Road, Chelmsford CM2 6AP
2E1 GQD A. Smith 31 Haywards Place, Easterton, Devizes SN10 4PP
2E1 GQN T. Stevens 25 Avenue Road, New Milton BH25 5JP
2E1 GQU M. Forster 33 Deer Valley Road, Holsworthy EX22 6DA
2E1 GRA A. Merrill 66 Royal Oak Drive, Selston, Nottingham NG16 6RJ
2E1 GRG D. Brady 6 Bristow Close, Bletchley, Milton Keynes MK2 2XP
2E1 GRT M. Keeling 3 Waltham Hall Cottages, Norwich Road, Little Stonham IP14 5LX
2E1 GSC P. Sherratt 23 Paulina Avenue, Nottingham NG15 8JA
2E1 GSM D. Ridgway 145 Bodmin Road, Astley, Manchester M29 7PE
2E1 GSW J. Harratt 8 The Runcie Building, Ripon College, Oxford OX44 9EX
2E1 GSX H. Mustoe 5 Dartmouth Park Avenue, London NW5 1JL
2E1 GTB C. Barnes 533 Maidstone Road, Blue Bell Hill, Chatham ME5 9QP
2E1 GTD S. Keene 17 Chaffers Close Long Sutton, Hook RG29 1SY
2E1 GTE S. Hall 122 Norwich Road, New Costessey, Norwich NR5 0EH
2E1 GTF S. Silk 89 St. Barts Road, Sandwich CT13 0AS
2E1 GTI A. Cadier 28 Romney Avenue, Folkestone CT20 3QJ
2E1 GTQ A. Smith 5 Langton Avenue, Chelmsford CM1 2BW
2E1 GTT A. Davies Little Platt, Bodle Street Green, Hailsham BN27 4RA
2E1 GTW F. James 1 The Gorseway, Bexhill-on-Sea TN39 4PP
2E1 GUC A. Ibrahim 14 Dunvegan Road, London SE9 1SA
2E1 GUN S. Kirby 2 Kneeton Park, Middleton Tyas, Richmond DL10 6SB
2E1 GVB E. Artis 40 Nottingham Way, Great Yarmouth NR30 2SA
2E1 GVC G. Carlin 38 Balfour St., East Bowling, Bradford BD4 7JT
2E1 GVD G. Carlin 82 Kingsway, Drighlington, Bradford BD11 1ET
2E1 GVJ J. Millington 6 Fentonhouse Lane, Wheaton Aston, Stafford ST19 9NU
2E1 GVS G. Gallacher 112 Central Drive, Stoke-on-Trent ST3 2AJ
2E1 GW Rvd. G. Wellington 94 Arlington Road, London N14 5AT
2E1 GWX R. Shepherd 91 Saxon Road, Hastings TN35 5HH
2E1 GXE M. Rogers 45 Church Road, Westoning, Bedford MK45 5LP
2E1 GXH D. Webb 51 Garden Road, Walton on the Naze CO14 8RR
2E1 GXI J. Adams 35 Kimberley Road, Solihull B92 8PU
2E1 GXL P. Travis 42 Trafalgar Road, Wallasey CH44 0EB
2E1 GXQ P. Walker 78 Kirkby Road, Desford, Leicester LE9 9JG
2E1 GXS R. Tomlin 6 Gaviots Green, Gerrards Cross SL9 7EB
2E1 GXU R. Steven 23 Collingwood Road, Woodbridge IP12 1JL
2E1 GXV M. Rogers 2 Tudor Close, Framlingham, Woodbridge IP13 9SL
2M1 GXX G. Scott Kirklands, Craigend Road, Galashiels TD1 2RJ
2E1 GXY P. Batty 134 Plymouth Road, Scunthorpe DN17 1TS
2E1 GYB M. Peterson 29 Warwick Close, Saxilby LN1 2FT
2E1 GYC S. Jenkins 45 Dorchester Road, Solihull B91 1LN
2E1 GYD M. Hall 45 Dorchester Road, Solihull B91 1LN
2E1 GYG D. Abbott 5 Heathcote Gardens, Rudheath, Northwich CW9 7JB
2E1 GYO C. Stanley 9 Pyramid Caravan Site, Beeston, Nottingham NG9 1NS
2E1 GYR H. Webster 67 Greenways, Over Kellet, Carnforth LA6 1DE
2M1 GYX R. Macleod 13 Wyvern Park, Edinburgh EH9 2JY
2E1 GYZ L. Rowley 20 Long Leasow, Selly Oak, Birmingham B29 4LT
2E1 GZF J. Walker 1 Riverside Court, Louth LN11 7AG
2E1 GZV A. Ager 5 Matthews Close, Bedhampton, Havant PO9 3NJ
2E1 GZY A. Wilson 22 Ormesby Road, Raf Coltishall, Norwich NR10 5JY
2E1 GZZ M. Coles 29 Sydney Road, Exeter EX2 9AH
2E1 HAC M. Lucas 48 Sycamore Drive, Ash Vale, Aldershot GU12 5PR
2E1 HAL M. Farraway 35 Wingbourne Walk, Nottingham NG6 8DT
2E1 HAM D. Langmead 38 Milton Grove, London N11 1AX

2E1 HAQ K. Graffham 15 Hayes Road, Clacton-on-Sea CO15 1TX
2E1 HAS B. Heirene 9 Ryecroft Crescent, Barnet EN5 3BP
2E1 HAU M. Durrant 8 Drake Avenue, Great Yarmouth NR30 4BS
2E1 HAW J. Beith 18 Avenue Road, New Milton BH25 5JP
2E1 HBF L. Buggs 2 Archway Cottages, Valley Road, Leiston IP16 4AR
2E1 HBJ C. Masters 85 Petersham Road, Creekmoor, Poole BH17 7DW
2E1 HBS D. Treacher 6 Beech View, Whitwell, York YO60 7JW
2E1 HBT E. Marlow 17 Fellows Court, Weymouth Terrace, London E2 8LP
2E1 HCB D. Broom 114 Gammons Lane, Watford WD2 5HY
2E1 HCG A. Docherty The Flat, Winton House, Stoke on Trent ST4 2RQ
2M1 HCP A. Smith 4 The Terrace, Lhanbryde, Elgin IV30 8NY
2M1 HCQ A. Smith 4 The Terrace, Lhanbryde, Elgin IV30 8NY
2E1 HCT R. Cowlishaw 23 Aldrich Drive, Willen, Milton Keynes MK15 9HP
2E1 HDB M. Dimambro 26 Fetcham Court, Bank Top, Newcastle upon Tyne NE3 2UL
2E1 HDC P. Dimambro 26 Fetcham Court, Bank Top, Newcastle upon Tyne NE3 2UL
2E1 HDE A. Morris 71 Lurdin Lane, Standish, Wigan WN6 0AQ
2E1 HDF R. Harrison 36 Windermere Road, Farnworth, Bolton BL40QH
2E1 HDZ T. Lockett 14 Tildsley Crescent, Weston, Runcorn WA7 4RN
2E1 HEE H. Merrington Cartref, Ball Lane, Frodsham WA6 8HP
2E1 HEF J. Hastry 56 Kilsyth Close, Fearnhead, Warrington WA2 0SQ
2E1 HEG E. Mather 72 Cranleigh Road, Worthing BN14 7QW
2E1 HEK L. Taylor 127 Dundee Close, Fearnhead, Warrington WA2 0UJ
2E1 HEM C. Kulikovsky 42 Highlands Grove, Bradford BD7 4BG
2E1 HEO A. Austin Flat 4, De Cham Court, 33 De Cham Road, St. Leonards-on-Sea TN37 6JA
2E1 HER D. Webster 28 Longmeadow Lane Heysham, Morecambe LA32FH
2E1 HES M. Davenport 24 Willow Grove, Belper DE56 1LX
2E1 HEV S. Brandon 2 Moss Bank, Winsford CW7 2ED
2E1 HFA S. Emerton 18 Avenue Road, New Milton BH25 5JP
2M1 HFE C. Budas 20 Oak Avenue, Bearsden, Glasgow G61 3HD
2E1 HFH S. Overall Flat 74, Douglas Buildings Marshalsea Road, London SE1 1EL
2E1 HFN J. Newell 4 Honeyfield Drive, Ripley DE5 3JL
2E1 HFS P. Dunne 2 All Saints Road, Wyke Regis, Weymouth DT4 9EZ
2E1 HFV H. Shackleton Woodroyd, 66 Spring Ave, Keighley BD21 4TA
2E1 HFW M. Hemstock 4 Tavistock Avenue, Perivale, Greenford UB6 8AJ
2E1 HFX K. Taylor 46 Hunters Field, Stanford In The Vale, Faringdon SN7 8LX
2W1 HFZ L. Jones 52 George Street, Aberdare CF44 6SH
2E1 HGA T. Winwood 2 The Warren, Abingdon OX14 3XB
2E1 HGE A. Higgs 26 Avon Avenue, Ringwood BH24 2BH
2E1 HGF P. Crookes 11 Degens Way, Hugglescote, Coalville LE67 2XD
2E1 HGG P. Mead 9 Abraham Drive, Silver End, Witham CM8 3SP
2E1 HGJ J. Jordan 4 Tuckers Close, Loughborough LE11 2PG
2E1 HGM K. Cronin East Cottage Westville Rd, Thornton le Fen LN4 4YJ
2E1 HGR D. Edwards Whitehaven, High Street, Buxted, Uckfield TN22 4JU
2E1 HGT A. Ruaux 85 St. Catherines Road, Crawley RH10 3TB
2E1 HGU T. Lawrence 85 West Avenue, Clacton-on-Sea CO15 1HB
2E1 HGY M. Henson 35 Westbrook Drive Rainworth, Mansfield NG21 0FB
2E1 HHA A. Gittens Elyria, 22 Charles Lovell Way, Scunthorpe DN17 1YL
2E1 HHB D. Botterell 12 Selsey Avenue, Clacton-on-Sea CO15 1NQ
2E1 HHE Dr T. Carlson 15 White Hedge Drive, St. Albans AL3 5TU
2E1 HHG J. Bradbury 192 Greenwood Road, Bakersfield, Nottingham NG3 7FY
2E1 HHJ S. Brown 3 Marsden Drive, Scunthorpe DN15 8AD
2E1 HHK M. Tyldesley 118 Rutland Street, Grimsby DN32 7NF
2E1 HHL S. Williams Alwent Farm, Staindrop, Darlington DL2 3NS
2E1 HHY S. Day 10 Second Avenue, Wolverhampton WV10 9PP
2E1 HID D. Smith 23 Marlborough Road, Long Eaton, Nottingham NG10 2BS
2E1 HIL I. Finch 29 Sherwood Road, Grimsby DN34 5TG
2M1 HIN J. Mcphee 101 Birkenside, Gorebridge EH23 4JF
2E1 HIO W. Roberts 30 Park Boulevard, Clacton-on-Sea CO15 5RH
2E1 HIQ A. Mayes 126 Walesby Lane, New Ollerton, Newark NG22 9UU
2E1 HJA K. Mieske 10 Cowdrey Road, London SW19 8TU
2E1 HJE C. Clifton 35 Farrowdene Road, Reading RG2 8SD
2E1 HJL S. Young 33 Springbank Close, Barnsley S71 3HZ
2E1 HJO M. Hyde 10 Devonshire Drive, Barnsley S75 1EE
2E1 HJS R. Burns 130 Kingsway, Mapplewell, Barnsley S75 6EU
2E1 HJW L. Carey 5 Park Avenue, Harlow CM17 9NL
2M1 HKA S. Mcintyre-Stewart 4 Howie Crescent, Rosneath, Helensburgh G84 0RL
2E1 HKB K. Brunning 45 Dover Road, Ipswich IP3 8JQ
2E1 HKC K. Cullum 22 Orwell View Road, Shotley, Ipswich IP9 1NP
2E1 HKE R. Wilson Newstead Farm, Clay Lane, Norwich NR10 4PP
2E1 HKS S. Hall 1 George Place, Wellington, Telford TF1 2AJ
2E1 HKY D. Green 89 Upper Ratton Drive, Eastbourne BN20 9DJ
2E1 HLA J. Moran 8 Doffcocker Lane, Bolton. BL1 5RG
2M1 HLE G. Stephen 9 Carse Court Iv3 8Te, Inverness IV3 8TE
2E1 HLF E. Carder 45 Chalklands, Linton, Cambridge CB21 4JQ
2E1 HLH E. Gardner New House, Birdbush Avenue, Saffron Walden CB11 4DJ
2E1 HLL S. Batchelor 2 Belmont Avenue, Atherton, Manchester M46 9RR
2E1 HLO C. Mitchell-Watson 144 Shakespeare Crescent, Dronfield S18 1ND
2E1 HLP L. Lawrence 85 West Avenue, Clacton-on-Sea CO15 1HB
2E1 HLS J. Preen 12 Isaac Walk, Worcester WR2 5EQ
2E1 HLU T. Kimm 99 Midland Road, Bramhall, Stockport SK7 3DT
2E1 HLW M. Mcneany 29 Grenfell Road, Manchester M20 6TG
2E1 HMB T. Emmett Meadow Sweet, Ramsden Lane, , Offwell EX149RY
2E1 HMQ K. Vance 5 Riversdale Close, Birstall, Leicester LE4 4EH
2E1 HNB K. Hughes High Lane Cottage, Congleton Road, Macclesfield SK11 9RR
2E1 HNF J. Li 13 Tothill Street, Minster, Ramsgate CT12 4AG
2W1 HNH A. Chalk 42 Erskine Road, Colwyn Bay LL29 8EU
2E1 HNN T. Bennellick 18 Bailie Close, Abingdon OX14 5RF
2E1 HNS T. Middleton 31 Coltman Avenue, Long Crendon, Aylesbury HP18 9DP
2I1 HNZ M. Mccrum 2 New Road, Donaghadee BT21 0DR
2E1 HOF J. Johnson 2 Meadow Drive, Canon Pyon, Hereford HR4 8NT
2E1 HOK S. Langley 100 Pogmoor Road, Barnsley S75 2EF
2E1 HOO D. Anstie 20 Keyes Road, Norwich NR1 2JX
2E1 HOS T. Keating 65 Shorncliffe Avenue, Norwich NR3 2HT
2E1 HOT J. Tosh 65 Shorncliffe Avenue, Norwich NR3 2HT
2E1 HOU S. Clarke 14 Findley Drive, Wirral CH46 3SG
2E1 HPC S. Carr 10 Bonds Road, Hemblington, Norwich NR13 4QF
2E1 HPD D. Carr 10 Bonds Road, Hemblington, Norwich NR13 4QF
2E1 HPM L. Hall 9 Stone Court, South Hiendley, Barnsley S72 9DL
2E1 HPS J. Vandervord 13 Granville Avenue, Ramsgate CT12 6DX
2E1 HPT A. Spain 60 The Maples, Broadstairs CT10 2PE
2E1 HPZ W. Bellis Cliffe Bungalow, Barnsley Road, Barnsley S72 9JX
2E1 HQA F. Kennedy 19 High Street, East Hoathly, Lewes BN8 6DR
2E1 HQH J. Garner 30 Birds Avenue, Garlinge, Margate CT9 5NE
2E1 HQP M. Newth 13 Okement Close, West End, Southampton SO18 3PP
2E1 HQQ R. Woodruff 224A Spendmore Lane, 2Coppull, Chorley PR7 5BZ
2E1 HQV B. Cummings 8 Spey House, Criterion Street, Stockport SK5 6TD
2E1 HQW S. Hackney 8 Spey House, Criterion Street, Stockport SK5 6TD
2E1 HQY P. Procter 1 Brow Hey, Bamber Bridge, Preston PR5 8DS
2E1 HQZ A. Webb 52 Stanfield Road, Stoke-on-Trent ST6 1AT
2E1 HRB S. Dykes 15 Sunningdale Road, Chelmsford CM1 2NH
2E1 HRJ A. Hollis 89 Longfield Lane, Cheshunt, Waltham Cross EN7 6AN
2E1 HRM D. Morgan 171 Town Road, London N9 0HJ
2E1 HRN A. Fower 17 Oakwood Road, Leek ST13 8LW
2M1 HRS R. Duncan 90 Faulds Gate, Aberdeen AB12 5QT
2E1 HRY P. Odle 24 Longfellow Road, Gillingham ME7 5QG
2E1 HSA T. Newman Sometimes (The Workshop), South Pew, Dorchester DT2 9HZ
2E1 HSB C. Price 162 Stamshaw Road, Portsmouth PO2 8LX
2E1 HSD B. Manning Barton Farm, Barton Road, Wisbech PE13 4TL
2M1 HSG M. Douglas 195 Dumbuck Road, Dumbarton G82 3NU
2E1 HSJ M. Tulk 8 Cleves Close, Weymouth DT4 9JU
2E1 HSL T. Brodie 8 Meadow Way, Plymouth PL7 4JB
2E1 HSP A. Saville 4 Shannon Court, Downs Barn, Milton Keynes MK14 7PP
2E1 HSR M. Rogers 47 Tregarrian Road, Tolvaddon, Camborne TR14 0HD
2E1 HTE C. Hall 2 Brambling Lane, Wath-Upon-Dearne, Rotherham S63 7GT
2W1 HTK E. Bateman 32 Park Avenue, Bodelwyddan, Rhyl LL18 5TB
2E1 HTM J. Brice 10 Swan Close, Weston-Super-Mare BS22 8XR
2M1 HTR A. Pollard 12 Royfold Crescent, Aberdeen AB15 6BH
2E1 HTU R. Corbett 11 Old Office Close, Dawley Bank, Telford TF4 2QA
2E1 HTV C. Sargent 8 Gilwell Grove, Priorslee, Telford TF2 9SR
2E1 HTY A. Semple Linden Lea, Lydham, Bishops Castle SY9 5HB
2E1 HUB R. Semple The Inn On The Green, Wentnor, Bishops Castle SY9 5EF
2E1 HUC K. Jones 24 Brooke Avenue, Margate CT9 5NG
2E1 HUE D. Taylor 2 Delph Cottages, Barkisland, Halifax HX4 0BW
2E1 HUJ P. Adams 20 Grosvenor Road, London W4 4EH
2E1 HUQ K. Gerrard 8 Windsor Crescent, Little Houghton, Barnsley S72 0HG
2W1 HUX H. Huxley Hen Berllan, Nant Mawr Road, Buckley CH7 2BS
2E1 HVB A. Scothern 42 Griceson Close, Ollerton, Newark NG22 9BD
2E1 HVI G. Reilly B I Z Ltd, Millmarsh Lane, Enfield EN3 7QA
2E1 HVL G. Batsman 66A Merivale Road, Harrow HA1 4BH
2E1 HVM C. Jackson 76 Margards Lane, Verwood BH31 6JP
2E1 HVN P. Harris 17 Seymour Avenue, Great Yarmouth NR30 4BB
2M1 HVR C. Hextall 4 Hawthornbank, Cockenzie, Prestonpans EH32 0HZ
2E1 HVT R. Hicks 31 Arundel Road, Great Yarmouth NR30 4LD
2E1 HVU S. Hicks 50 Garfield Road, Great Yarmouth NR30 4JU
2E1 HVZ C. Stevens 82 Bembridge Drive, Alvaston, Derby DE24 0UQ
2E1 HWI C. Zdziech 200 Kensington Street, Rochdale OL11 1QS
2E1 HWJ T. Winch Whitehall Barn, Stowmarket Road, Stowmarket IP14 6BU
2E1 HWQ S. Cannon 22 Maid Marion Rise Warsop, , Mansfield NG20 0LD
2E1 HWU R. Noakes 14 Sackville Road, Immingham DN40 1EE
2E1 HWV M. Brett 30 Belmont Avenue, London N13 4HD
2E1 HXB B. Wheat 62 Havenwood Rise, Nottingham NG11 9HE
2E1 HXC A. Rodger 4 Burns Nurseries, Wootton Road, King's Lynn PE30 3BG
2E1 HXD A. Brett 1C Old Park Road, Palmers Green, London N13 4RG
2E1 HXM L. Tunstall 8 York Road, Rowley Regis B65 0RR
2E1 HXN J. Dodds 8 York Road, Rowley Regis B65 0RR
2E1 HXP M. Walsh 23 Moss Fold Road, Darwen BB3 0AQ
2E1 HXR E. Lacey 6 Weetshaw Close, Shafton, Barnsley S72 8PZ
2W1 HXT K. Jones 11 St. Davids Close, Gobowen, Oswestry SY11 3JF
2E1 HXY R. Ling 8 Spa Hill, Kirton Lindsey, Gainsborough DN21 4NE
2E1 HXZ P. Ling 8 Spa Hill, Kirton Lindsey, Gainsborough DN21 4NE
2E1 HYD A. Wilson 135 Britannia Road Morley, Leeds LS27 0DS
2E1 HYE D. Bruce 6 Princes Way, King's Lynn PE30 2QL
2E1 HYI P. Redfern 42 Newton Street, Retford DN22 7AD
2E1 HYX M. Hammond 10 Collingwood Close, Braintree CM7 9UG
2E1 HZI E. Sinclair 21 Longport Avenue, Manchester M20 1EN
2E1 HZM M. Fleming 54 Madison Avenue, Bradford BD4 0JJ
2E1 HZV E. Colley 14 Hawthorne Close, Tyldesley, Manchester M29 8PH
2E1 HZY P. Hartley 14 Medway Walk, Wigan WN5 9NQ
2E1 IAB J. Jackson 40 Lightbounds Road, Bolton BL1 5UN
2E1 IAC M. Dix 10 College Hill, Godalming GU7 1YA
2E1 IAE M. Hedges 14 Weavers Mill Way, Holmfirth HD9 7FB
2E1 IAI N. Revell York House, The Street, Great Saling, Braintree CM7 5FS
2E1 IAS S. Loane 30 St Wilfrids Road, Burgess Hill RH15 8BD
2E1 IAT B. Randell 32 Windsor Close, Rubery, Birmingham B45 0DA
2E1 IAX M. Frankland 90 Kensington Road, Coventry CV5 6GH
2E1 IAY J. Healey 28 Thatchers Place, Westlands, Droitwich WR9 9ED
2E1 IAZ P. Healey 28 Thatchers Place, Westlands, Droitwich WR9 9ED
2M1 IBE P. Woods 92 Preston Crescent, Prestonpans EH32 9RD
2M1 IBH G. Noonan 8 Johnston Terrace, Port Seton, Prestonpans EH32 0BB
2E1 IBJ D. Marshall 143 Middleton Road, Banbury OX16 3QS
2W1 IBN M. Tucker 21 Pen Y Fan Close, Pentwyn Crumlin, Newport NP11 3JQ
2E1 IBP J. Siddall 17 Dalebrook Road, Sale M33 3LD
2E1 IBS I. Grahame 24 Bushby Avenue, Broxbourne EN10 6QE
2E1 IBT J. Jordan 25 Morris Close, Loughborough LE11 1PU
2M1 IBX J. Millar 18A Dougall St., Tayport DD6 9JD
2E1 ICI C. Rengifo 61 Clarendon Road, Sale M33 2DY
2E1 ICM L. Kerswill 18C Jewell Road, Bournemouth BH8 0JQ
2E1 ICO A. Ryan 18 Belmount Avenue, Newcastle upon Tyne NE3 5QD
2E1 ICT G. Mears 4 Uplands Road, Woodford Green IG8 8JN
2E1 ICU R. Ropinski 38 The Leys, Little Eaton, Derby DE21 5AR
2E1 ICW A. Spiers 2 Adeline Cottages, Jacobs Well Road, Guildford GU4 7PD
2E1 IDC C. Hillier 15 Dabbs Hill Lane, Northolt UB5 4AQ
2E1 IDE G. Swain 33 Saville St., Blidworth, Mansfield NG21 0RW
2E1 IDG D. Greaves The White House, 25 London Road, Leicester LE8 9GF
2E1 IDH K. Reynolds 3 Lilac Close, Chelmsford CM2 9NY

2E1 IDI C. Welsh 56 Longacres, St. Albans AL4 0DR
2E1 IDM D. Cheetham 405 Jenkin Road, Sheffield S9 1AY
2E1 IEK E. Howie 102 Rushdean Road, Rochester ME2 2QB
2E1 IEM J. Cutler 11 Margaret Ashton Close, Manchester M9 4PZ
2E1 IFA F. Maddison 87 Godolphin Road, London W12 8JN
2E1 IFL C. Wynn 45 Hillcrest Road, Berry Hill, Coleford GL16 7RG
2E1 IFM A. Gow 11 Rodley Square, Lydney GL15 5AZ
2E1 IFN K. Gow 11 Rodley Square, Lydney GL15 5AZ
2E1 IFW D. Hyde The Grove, 7 Mill Lane, Kidderminster DY10 3ND
2E1 IGA E. Crane 161 Heath Way, Horsham RH12 5XX
2E1 IGG K. Horsley 6 Flatford Place, Kidlington OX5 1TH
2E1 IGI D. Willimot 5 Green Lane, Upton, Huntingdon PE28 5YE
2E1 IGJ Capt. H. Schnaar 18 Witham Lodge, Witham CM8 1HG
2E1 IGK C. Smith 21 Earl Spencer Court, Peterborough PE2 9PQ
2M1 IGO V. Gray 55 Prestongrange Road, Prestonpans EH32 9DD
2E1 IGP A. Wolstencroft 201 Walton Road, Sale M33 4ER
2M1 IGQ C. Bond Strathllan House, Doune Road, Dunblane FK15 9AR
2E1 IGU R. Ransom 97 Park Square West, Jaywick, Clacton-on-Sea CO15 2NU
2E1 IGY J. Kerr 14 Seafield Close, Barton On Sea, New Milton BH25 7HR
2E1 IHB W. Clarke 41 Upton Road, Atherton, Manchester M46 9RQ
2E1 IHE D. Siviter 72 Sandfield Road, West Bromwich B71 3NE
2E1 IHF L. Parrish 5 Kestrel Lane, Cheadle, Stoke-on-Trent ST10 1RU
2E1 IHJ D. Seeby 59 Dallamoor, Telford TF3 2EE
2E1 IHK J. Smith 54 Hillside Avenue, Bridgnorth WV15 6BU
2E1 IHO J. Willingham 3 Cherry Road, Nailsea, Bristol BS48 2EE
2E1 IHS A. Hardman 47 Oatlands Road, Manchester M22 1AH
2E1 IHT G. Large 11 Ranworth, King's Lynn PE30 4XD
2E1 IHW J. Eaton 184 Gore Road, New Milton BH25 5NQ
2E1 IHY C. Sneap 14 Calver Close, Nottingham NG8 1AT
2E1 IIA C. Lee 154 Grangeway, Runcorn WA7 5JA
2E1 IIC G. Foreman 41 Winnards Park, Sarisbury Green, Southampton SO31 7BX
2E1 IID G. Gudgeon 28 Park Close, Stevenage SG2 8PX
2E1 IIE D. Jones 31 Summerhill Drive, Liverpool L31 3DN
2E1 IIG H. Holmes De Wyvill Sinclair 27 Mount Crescent, Bridlington YO16 7HR
2E1 IIJ G. Colclough Little Hallands, Norton, Seaford BN25 2UN
2E1 IIL T. Rodgers 57 Knowles Hill, Rolleston-On-Dove, Burton-on-Trent DE13 9DY
2E1 IIM A. Bagg 1 Stone Road, Burnham-on-Sea TA8 1JU
2E1 IIP R. Mullen 58 Jasmin Avenue, Newcastle upon Tyne NE5 1TL
2E1 IIV G. Guinan 5A Temple Lane, Silver End, Witham CM8 3QY
2M1 IIW M. Smith 25 Charleston Crescent, Cove, Aberdeen AB12 3DZ
2E1 IJD C. Wilcockson Conybeare House, Willowbrook, Windsor SL4 6HL
2E1 IJE A. Wilcockson Conybeare House, Willowbrook, Windsor SL4 6HL
2E1 IJF J. Grubb Waterloo Farm, Foston-On-The-Wolds, Driffield YO25 8BH
2E1 IJK D. Butler Church Cottage, Church Road, Badminton GL9 1HT
2E1 IJL B. Gow 11 Rodley Square, Lydney GL15 5AZ
2E1 IJM E. Lawless 99 Ribbleton Avenue, Ribbleton, Preston PR2 6DA
2E1 IJN P. Nevard Millinder House, Westerdale, Whitby YO21 2DE
2E1 IJO B. Blackham 5 Reedham Drive, Bramley, Rotherham S66 2SW
2E1 IJQ J. Mallichan 17 Napier Road, Gillingham ME7 4HB
2E1 IJR B. Rendell 43 Springmead, Chard TA20 2EW
2E1 IJS A. Reed 139 Wigmore Road, Gillingham ME8 0TH
2E1 IJU E. Heagren 156 Eastbrooks, Pitsea, Basildon SS13 3QH
2E1 IJX J. Underwood 12 Forge Lane, Gillingham ME7 1UG
2E1 IJY A. Brown 24 Greenfields, Langley Mill, Nottingham NG16 4GJ
2E1 IKA M. Flanagan 33 Ullswater Road, Chorley PR7 2JB
2E1 IKB B. Randall 5 Percival Road, Eastbourne BN22 9JL
2I1 IKL D. Mcmichael 33 Shelton Road, Armoy, Ballymoney BT53 8YH
2E1 IKM K. Ralph 15 Hansell Road, Norwich NR7 0LY
2E1 ILH D. Ledger 10 Driftwood Gardens, Southsea PO4 9ND
2E1 ILM Dr A. Chandoo 59 Stanton Road, Birmingham B43 5HH
2E1 INC R. Davenport 10 Woodend Lane, Hyde SK14 1DT
2E1 INT D. Kenyon 88A Knutsford Road, Wilmslow SK9 6JD
2E1 INW I. Williamson 196 Bruntwood Lane, Heald Green, Cheadle SK8 3AS
2E1 IOS H. Handrick 13 Ormonde Way, New Rossington, Doncaster DN11 0SB
2E1 ITE G. Cahill 81 Albemarle Road, Willesborough, Ashford TN24 0HJ
2W1 ITI D. Quick Rosegarth, Woodbine Road, Blackwood NP12 1QH
2E1 IVT F. Handley 155 Heathcote Street, Longton, Stoke-on-Trent ST3 1AD
2E1 IWD D. Brooks 61 Carisbrooke High St., Newport PO30 1NR
2E1 IWG W. Goodwin 4 Southfields Close, Bishops Waltham, Southampton SO32 1EY
2E1 JAA J. Ahmed 59 Ramsgate, Lofthouse, Wakefield WF3 3PX
2E1 JAC J. Marter 4 Meadow Way, Seaford BN25 4QT
2E1 JBJ A. Berry 4 Newlands Park Way, Newick, Lewes BN8 4PG
2E1 JCM J. Matthews 9 Clive Green, Shrewsbury SY2 5QL
2E1 JEF G. Hensby Flat 12, Edel Quinn House, Wirral CH49 6PN
2E1 JEH J. Haimes 15 The Avenue, Hersden, Canterbury CT3 4HL
2E1 JGD J. Green 42 Longfield Ave, Newbarn DA3 7LA
2E1 JGM J. Moody 44 Frensham, Cheshunt, Waltham Cross EN7 6HB
2E1 JGW J. Wright 2 St. Leonards Park, East Grinstead RH19 1EE
2E1 JIM J. Deacon Spring Valley, Churt Road, Farnham GU10 2QU
2E1 JJM J. Martin 1 Collins Lane, West Harting, Petersfield GU31 5NZ
2E1 JJN J. Nicholson 6 Mill Gardens, West End, Southampton SO18 3AG
2M1 JKG J. Grieve 39 Kenmount Drive, Kennoway, Leven KY8 5HA
2E1 JKL J. Loader Furlong Farm, Henley, Langport TA10 9AX
2E1 JKP J. Peggram Cherry Trees, Broad Lane, Bracknell RG12 9BY
2E1 JLC J. Chaundy Ambleside, Barnes Lane, Lymington SO41 0RL
2I1 JMC J. Mccaw 62 High Street, Ballymena BT43 6DT
2E1 JME E. Marshall 60 Dudsbury Road, West Parley, Ferndown BH22 8RG
2E1 JMG J. Greaves The White House, 25 London Road, Leicester LE8 9GF
2E1 JMW J. Williams 73 Telford Road, Tamworth B79 8EY
2E1 JOD J. Antimano 23 Nupton Drive, Barnet EN5 2QU
2W1 JOL T. Philpott 6 Hillside, Fochriw, Bargoed CF81 9LQ
2E1 JON J. Sturgeon Windyridge, Linkside West, Hindhead GU26 6PA
2E1 JOY E. Nye 11 Barnhill Close, Marlow SL7 3HA
2E1 JPH J. Holden 47 Copse Hill, London SW20 0NJ
2E1 JRB J. Bancroft 7 Cordwainer Grove, Sedgefield, Stockton-on-Tees TS21 2JY
2E1 JRS J. Stew 19 Salisbury Close, Sittingbourne ME10 3BL
2E1 KAJ K. Johnson 22 Prior Road, Greatstone, New Romney TN28 8SB
2E1 KCC C. Coates 104 Orion Way, Willesborough, Ashford TN24 0DZ
2E1 KFS D. O'Hagan 44 Mulberry Close, Paignton TQ3 3GD
2E1 KID R. Gow 11 Rodley Square, Lydney GL15 5AZ
2E1 KIP J. Barker 5 Severn Avenue, Weston-Super-Mare BS23 4DH
2E1 KJB K. Blanch Sticelett Farm, Rolls Hill, Cowes PO31 8NE
2E1 KLT G. Clark 28 Manor Road, Woolton, Liverpool L25 8QG
2M1 KOJ C. Smith 68 Craigmore Street, Dundee DD3 0EA
2W1 KWK P. Stagg 8 Canal Terrace, Ystalyfera SA9 2LP
2E1 KYQ L. Fisher Annstuyvonne, Baghill Green, Wakefield WF3 1DL
2E1 LAM A. Lam Partridge Cottage, Redpale, Heathfield TN21 9NR
2W1 LCO M. Kinsey Hyfrydle, Cyffylliog, Ruthin LL15 2DW
2E1 LEC C. Cattel 21 School Hill, Chickerell, Weymouth DT3 4BA
2E1 LED F. Chorlton 25 Ash Grove, Orrell, Wigan WN5 8NG
2E1 LEN L. Kinley 100 Withington Lane, Aspull, Wigan WN2 1JE
2E1 LES L. Tomkins 46 The Boulevard, Great Sutton, Ellesmere Port CH65 7DZ
2E1 LEW S. Freestone 47 Salisbury Street, Gainsborough DN21 2RS
2E1 LEX A. Champkin 17 Blundell Place, Bedford MK42 9XB
2E1 LGA A. Edwards 23 Brittany Avenue, Ashby-de-la-Zouch LE65 2QY
2E1 LGE L. Edwards 8 St. Andrews Close, High Ham, Langport TA10 9DD
2E1 LGJ L. Gaston-Johnston 23A Nashleigh Hill, Chesham HP5 3JQ
2E1 LGV L. Verghese 19 Old Mansion Close, Eastbourne BN20 9DP
2E1 LIS E. Buckland 11 Veronica Close, Basingstoke RG22 5NW
2E1 LIZ E. Greatorex 22 Marlborough Way, Uttoxeter ST14 7HL
2E1 LJL L. Lewis 28 Brow Hey, Bamber Bridge, Preston PR5 8DS
2E1 LJW L. Walker 125 Devereux Road, West Bromwich B70 6RQ
2E1 LME M. Hulme 13 Cherry Tree Court, Diss IP22 4QW
2E1 LOZ L. Dodman 30 Cambridge St., Rugby CV21 3NQ
2M1 LPT C. Macnab 18 Lochport, North Uist, Western Isles HS6 5EU
2E1 LSI Prof. L. Soares Indrusiak 6 Trafalgar House, Piccadilly, York YO1 9QP
2E1 MAR M. Davies Newton Lodge, Newton, Ellesmere SY12 0PF
2E1 MAZ M. Adlington 21 Newstead Road, Stoke-on-Trent ST2 8HU
2E1 MDC C. Mackay 665A Edenfield Road, Rochdale OL11 5XE
2E1 MEL P. Cosens 34 Waterloo Road, Salisbury SP1 2JX
2E1 MEP M. Pearce 137 Westwood Road, Salisbury SP2 9HN
2E1 MFC M. Clews 16 Chestnut Street, Worcester WR1 1PA
2E1 MGB H. Knight 10 Welford Road, Barton, Alcester B50 4NP
2E1 MHB M. Balyuzi 48 Cleveland Gardens, London NW2 1DY
2E1 MIB D. Clegg 1 Green Croft, Hereford HR2 7NT
2M1 MIC M. Budas 20 Oak Avenue, Bearsden, Glasgow G61 3HD
2E1 MIN M. Dawson 140A Healey Road, Scunthorpe DN16 1HT
2E1 MJF M. Faulkner 49 Oakfield Road, Shrewsbury SY3 8AD
2E1 MJH M. Hickford 3 Ashen Road, Clare, Sudbury CO10 8LQ
2E1 MJM M. Marter 4 Meadow Way, Seaford BN25 4QT
2E1 MPB M. Bates Apartment 801, Imperial Point, Salford M50 3RB
2E1 MPN M. Nurse 81 Lexden Drive, Seaford BN25 3JF
2E1 MPQ Dr A. Gair-Harris Osterley, White Lackington, Piddletrenthide, Dorchester DT2 7QU
2W1 MRK M. Morgan 30 Hafan Werdd, Mornington Meadows, Caerphilly CF83 3BU
2W1 MSC M. Cook 9 Drenewydd, Park Hall, Oswestry SY11 4AH
2W1 MWS M. Southall 23 Ffordd Elias, Old Colwyn, Colwyn Bay LL29 9LA
2E1 MYK M. Nicholas 1 Buckland Barton Cottages, Newton Abbot TQ12 4SA
2E1 NAC N. Cosens 1 Bake Farm Cottages, Salisbury Road, Salisbury SP5 4JT
2E1 NEW A. Fleck 108 Hillsview Avenue, Newcastle upon Tyne NE3 3LA
2E1 NFD N. De-Thabrew 12 Balfour Road, Dover CT16 2NQ
2E1 NII C. Sims 7 Ainthorpe Lane, Ainthorpe, Whitby YO21 2JN
2E1 NJC N. Cook Idle Shores Springfield Road, Woolacombe EX34 7BX
2E1 NPH A. Arnold 2 Duck Lane, Haddenham, Ely CB6 3UE
2E1 NRL N. Loveridge 26 Haylands, Portland DT5 2JZ
2E1 NRQ J. Dean 9 School Hill, Chickerell, Weymouth DT3 4BA
2E1 OBI B. Atkins 30 Rishworth Rise, Shaw, Oldham OL2 7QA
2E1 ODG D. Goodall 19 Rossefield Avenue, Leeds LS13 3SG
2E1 OLI O. Bradley 19 Lincoln Close, Eastbourne BN20 7TZ
2E1 ORT M. Hickford Conifers, 3 Ashen Road, Sudbury CO10 8LQ
2E1 OUJ J. Walsh 155 Hunter Drive, Bletchley, Milton Keynes MK2 3NG
2E1 OZO D. Day Blakeney, Arbor Lane, Lowestoft NR33 7BQ
2E1 OZY O. Morris 44 Leamington Road, Weymouth DT4 0EZ
2E1 PAL A. Barnes 78 Greenhurst Drive, East Grinstead RH19 3NE
2E1 PAW P. Woodward 5 Lupin Way, Clacton CO167DX
2E1 PDH P. Hasney 15 Mansfield Court, West Boldon, East Boldon NE36 0PL
2E1 PDM P. Marney 5 Swanhill, Welwyn Garden City AL7 1PB
2E1 PDQ A. Pennington 16 Invicta Road, Margate CT9 3SL
2E1 PEC P. Cosens 1 Bake Farm Cottage, Salisbury Road, Salisbury SP5 4JT
2E1 PEW A. Pewsey Pineholm, High Close, Bovey Tracey, Newton Abbot TQ13 9EX
2E1 PGA P. Graham 556 Mather Avenue, Liverpool L19 4UG
2E1 PGB D. Brierley 69 Borough Road, Birkenhead CH42 9QA
2W1 PGL P. Lloyd Pentrip, Wynnstay Yard, Wrexham LL14 6DP
2E1 PHW P. Wade 94 Steyne Road, Bembridge PO35 5SL
2E1 PJJ J. Miller Flat 1 Block 2, St. Phillips Place, Eastbourne BN22 8LW
2E1 PMT J. Green 10 Holme Dene, Haxey, Doncaster DN9 2JX
2E1 PPK A. Boag 53 Castlewood Road, London N16 6DJ
2E1 PPM C. Martin 4 Chaloner Place, Aylesbury HP21 8NW
2E1 RAD J. Elliott 13 Ormonde Way, New Rossington, Doncaster DN11 0SB
2E1 RAF R. Walker 35 Romany Close, Letchworth Garden City SG6 4LA
2E1 RAO N. Brent 53 Middlewich Street, Crewe CW1 4DA
2E1 RBA A. Russell - Bishop 227 Ardleigh Green Road, Hornchurch RM11 2ST
2E1 RBH C. Hodges Marine Cottage, 1A Clements Lane, Portland DT5 1AS
2E1 RFS R. Starling 10 School Lane, Lawford, Manningtree CO11 2HZ
2E1 RIO R. Odle 24 Longfellow Road, Gillingham ME7 5QG
2E1 RJH Sir R. Heygate 34 Abercrombie Street, London SW11 2JD
2E1 RJS R. Spevack 87 Albany Road, Hersham, Walton-on-Thames KT12 5QG
2E1 RMD B. Debenham 80 Stewart Road, Chelmsford CM2 9BD
2E1 RMS R. Stevenson 97 Queen Street, Crewe CW1 4AL
2E1 RON C. Robinson 11 Poplar Way, Leeds LS13 4SU
2E1 ROO M. Ratcliffe 76 Churchill Road, Stone ST15 0DY
2E1 RSB R. Mather 76 Stavordale Road, Moreton, Wirral CH46 9PS
2E1 RSH R. Davies Lamb Cottage 3 Manor Barns, Snowshill, Broadway WR12 7JR
2W1 RSS R. Hark 5 Victoria Park, Bagillt CH6 6JS
2E1 RWC R. Cornwall 9 Bishop Close, Dunholme, Lincoln LN2 3US
2E1 RWN N. Rowan 27 Crieff Road, London SW18 2EB

2E1 SAZ S. Greatorex 54 Lilac Grove, Glapwell, Chesterfield S44 5NG
2E1 SBF C. Noon 24 Sunset Walk Bush Estate, Eccles-On-Sea, Norwich NR12 0SX
2M1 SCO R. Jeffrey 21E Harvie Street, Glasgow G51 1BW
2E1 SCR S. Ride 61 Pepper Street, Sutton-in-Ashfield NG17 5GD
2E1 SDI G. Baines 60 Parkdale Road, Thurmaston, Leicester LE4 8JP
2W1 SDR A. Rose 4 Ramsey Road, Barry CF62 9DF
2E1 SGK S. Knott 24 John Street, Leek ST13 8BL
2E1 SHA D. Mcmillan 1 Falcons Way Halewood Park, Runcorn WA7 2FF
2E1 SHE S. Brown 11 Wordsworth Avenue, Tamworth B79 8BZ
2E1 SIS L. Downes 6 Greenland Crescent, Beeston, Nottingham NG9 5LB
2M1 SJB S. Budas 20 Oak Avenue, Bearsden, Glasgow G61 3HD
2E1 SJG S. Grant 47 Coneyford Road, Shard End, Birmingham B34 7AY
2E1 SKA W. Pitt 237 Broadway, Dunscroft, Doncaster DN7 4HS
2E1 SKR A. Skrzypecki The Chestnuts, Birdcage Lane, Halifax HX3 0JQ
2E1 SKY P. Staerck 42 Plantation Hill, Worksop S81 0RJ
2E1 SOB A. Weatherall 1 Dean Place, Stoke-on-Trent ST1 3HS
2E1 SOX A. Hughes 4 Cobden Court, Birkenhead CH42 3YH
2E1 SPH S. Harris Lackington Drove, Dorchester DT2 7QU
2E1 SPY S. Palmer 21 Ibbett Close, Kempston, Bedford MK43 9BT
2W1 SRB S. Bowen 41 Bro Dawel, Merthyr Tydfil CF47 0YU
2E1 SRI S. Edwards 70 Summer Field Court Altona Close Stone, Staffs ST15 8AR
2E1 STK L. Meek 3 St. Johns Grove, Kirk Hammerton, York YO26 8DE
2E1 STO G. Stone 40 Friars Road, Stoke-on-Trent ST2 8DS
2E1 STU S. Green Meadow View, Ewleaze Farm, Tincleton, Dorchester DT2 8QR
2E1 SUE S. Macpherson 18 Mountbatten Avenue, Dukinfield SK16 5BU
2W1 SWB E. Jenkins 8 Ffordd Elias, Old Colwyn, Colwyn Bay LL29 9LA
2I1 SWD S. Mcauley Layde View, 19 Rathlin Avenue, Ballycastle BT54 6DQ
2E1 SWS S. Walters-Smith 83 Chesterfield Road, Tibshelf, Alfreton DE55 5NJ
2E1 TAB T. Brierley 6 Bridle Avenue, Wallasey CH44 7BJ
2E1 TAG D. Taggart 22 Worcester Drive, Sittingbourne ME10 2QP
2E1 TAP A. Weaver 14 Ferry View, Thorngumbald, Hull HU12 9GB
2W1 TBD T. Davies 72 Eversley Road, Sketty, Swansea SA2 9DF
2E1 TBW A. Walker 76 Greenway, Birmingham B20 1EQ
2E1 TCP T. Clayton 14 Medway Walk, Wigan WN5 9NQ
2W1 TDM T. Meredith 18 Hyde Place, Llanhilleth, Abertillery NP13 2RT
2E1 TIM T. Wightman Laithbutts Farm, Cowan Bridge, Carnforth LA6 2JL
2E1 TKD D. Hensby 28 Moorland Crescent, Whitworth, Rochdale OL12 8SU
2E1 TMB T. Bolderstone 20 Wellington Crescent, Sculthorpe, Fakenham NR21 7PU
2E1 TNE A. Millard 6 Connaught Road, Weymouth DT4 0SA
2E1 TOM T. Lake 77 Grafton Road, King's Lynn PE30 3EX
2E1 TON A. Steele 15 Roman Meadow, Downton, Salisbury SP5 3LB
2E1 TSO G. Smith 37 The Crescent, Bracebridge Heath, Lincoln LN4 2NP
2E1 TWB A. Brown 7 Brookfield Road, Wooburn Green, High Wycombe HP10 0PZ
2E1 UJE A. Dalzell 9 Pyms Lane, Crewe CW1 3PJ
2E1 UKT G. Eason Whitegates, Parsonage Road, Takeley, Bishop's Stortford CM22 6QX
2E1 UTD R. Smith 14 Oakfield Cottages, Brockton, Shrewsbury SY5 9JA
2E1 VAR R. Preece 5 Pavey Run, Ottery St. Mary EX11 1FQ
2M1 VFO M. Bartlett 93 Lumsden Crescent, Almondbank PH1 3UA
2W1 VMR M. Sides 17 Mayville Avenue, Llay, Wrexham LL12 0PW
2M1 VXB C. Andrew 23 Shore Street, Inverallochy, Fraserburgh AB43 8WA
2I1 WBC E. Paulikas 33 Clarefield, Dungannon BT71 6TQ
2E1 WCD W. Denny 86 Lloyds Avenue, Kessingland, Lowestoft NR33 7TR
2E1 WEB C. Webb 1 The Square, Eltisley Road, Sandy SG19 3BT
2M1 WEZ L. Cook Rockside, Locheport, Isle of North Uist HS6 5EU
2E1 WGB G. Barber 35 Lower Park Crescent, Bishop's Stortford CM23 3PU
2E1 WIN C. Wingfield 35 Causey Farm Road, Hayley Green, Halesowen B63 1EQ
2E1 WJB B. Walsh 20 Edge Fold Crescent, Worsley, Manchester M28 7EX
2E1 WNA I. Jones 151 Atherton Road, Hindley, Wigan WN2 3EE
2E1 WPW A. Newton 12 Hewett Street, Warsop Vale, Mansfield NG20 8XN
2E1 WRC W. Chorlton 25 Ash Grove, Orrell, Wigan WN5 8NG
2E1 WVF D. Fowler Millhouse, 8 Church Road, Stockport SK6 5PR
2E1 WWD E. Durkin 30 Douglas Road West, Stafford ST16 3NX
2E1 XDJ K. Senior 20A Union Street, Hemsworth, Pontefract WF9 4AP
2E1 XGX P. Hughes 17 Worcester Park, Bath BA1 6QU
2E1 XRM J. Thompson 3 Fern Avenue, Staveley, Chesterfield S43 3RH
2E1 XXX J. Day 49 Nelson Road, Hull HU5 5HN
2W1 YEG J. Thorne 11 Dowland Road, Penarth CF64 3QX
2E1 YES J. Glover 31 West Drive, Lancaster LA1 5BY
2E1 YHZ J. Hiltz 35 Highview Road, London W13 0HA
2E1 YRK C. Wright 55 Booth Street, Denton, Manchester M34 3HU
2E1 ZPR W. Toomer 10 Northfield, Tarrant Hinton, Blandford Forum DT11 8JD

G0

G0 AAA the Three A's CG c/o K. Pritchard 9 Golf Close, Pyrford., Woking GU22 8PE
G0 AAG W. Furnival 24 Moor Street, Hereford HR4 9LA
G0 AAM G. Willetts 2 Underlane, Boyton, Launceston PL15 9RR
G0 AAN Prof. L. Schnurr 42 Basin Road, Heybridge Basin, Maldon CM9 4RQ
G0 AAT P. Wheatley 44 Primrose Crescent, Worcester WR5 3HT
G0 AAU J. Blades 42 Ellesmere, Burnmoor, Houghton le Spring DH4 6EA
G0 AAW A. Whitehouse Forest Green, Cotchford Lane, Hartfield TN7 4DN
GM0 AAX G. Anthoney 10 Cedar Road, Kilmarnock KA1 2HP
G0 AAY I. Brooks The Lodge, Bilsborrow Lane, Bilsborrow, Preston PR3 0RP
G0 ABB M. Honeywell 23 Deverell Place, Waterlooville PO7 5ED
GW0 ABE P. Hughes 59 Jeffreys Road, Wrexham LL12 7PD
G0 ABI P. Green Camellia Cottage, The Challices, Chulmleigh EX18 7QX
G0 ABM T. Johnson 143 Queens Road, Tunbridge Wells TN4 9JY
G0 ABN A. Samuels 45 Mermaid Close, Chatham ME5 7PT
G0 ABP D. Orgill 32 Upland Avenue, Chesham HP5 2EB
GW0 ABT T. Thomas Tymawr Farm, Llanwern, Brecon LD3 7UW
G0 ABV D. Wood 18 Bankhouse Road, Nelson BB9 7RA
G0 ABW J. Harding Fen End Farm, High Street, Huntingdon PE19 4UE
G0 ACA J. Kliffen 8 West Park, Minehead TA24 8AW
G0 ACD M. Amos 41 Jocelyn Road, Richmond TW9 2TJ
GW0 ACH J. Cooper 157 Bryn Road, Brynmenyn, Bridgend CF32 9LU
G0 ACK D. Lamb 339 Victoria Road, Ruislip HA4 0DS
G0 ACQ M. Godden 20 Channel View Road, Portland DT5 2AY
G0 ACZ A. Hope Knab Hall Bungalow, Knab Hall Lane, Tansley Matlock DE4 5FS
G0 ADA N. Law 43 Canonsfield, Peterborough PE4 5AQ
G0 ADB C. Willis 5 Gower Drive, Biddenham, Bedford MK40 4PZ
GW0 ADC K. Shepherd 15 Gronant Road, Prestatyn LL19 9DT
GI0 ADD Chester And District RS c/o J. Ashe 49 Deans Walk, Richhill, Armagh BT61 9LD
GM0 ADF D. Mackinnon 60 Mount Stuart Drive, Wemyss Bay PA18 6DX
G0 ADG R. Bristeir 94 Burnthwaite Road, Fulham, London SW6 5BG
G0 ADH R. Razey 2 Park Farm Cottage, 26 St. Georges Road, Wallingford OX10 8HP
G0 ADJ D. Elsworth 34 Seal Road, Bramhall, Stockport SK7 2JR
G0 ADK J. Saueressig 8 The Ridgeway, River, Dover CT17 0NX
G0 ADL B. Barlow 134 Bury Road, Radcliffe, Manchester M26 2UX
G0 ADO A. Hodgson Arla Burn Farm, Middleton in Teesdale DL12 0QU
G0 ADP A. Wither 30 Mersey Road, Aigburth, Liverpool L17 6AD
GW0 ADS J. Jenkins Derwen Las, Llanwnnen, Lampeter SA48 7LG
G0 ADT I. Andrew 28 Beechnut Drive Blackwater, Camberley GU17 0DJ
G0 ADU B. Gibson 55 Ledward Street, Winsford CW7 3EN
G0 ADW P. Radford 42 Ashbury Drive, Weston-Super-Mare BS22 9QS
GM0 ADX Kilmarnock And Loudoun ARC c/o S. Mitchell 97 Barbieston Road, Auchinleck KA18 2ED
G0 ADZ D. Price Pippins, High St., Newmarket CB8 9DQ
G0 AED H. Gerard 18 Hunstanton Road, Dersingham, King's Lynn PE31 6HQ
GM0 AEG D. Greatorex 40 Robertson Road Lhanbryde, Elgin IV30 8PE
G0 AEL K. Horton 16 Linden Close West Parley, Ferndown BH22 8RS
G0 AEN S. Webb 7 Holbear Grange, Forton Road, Chard TA20 2ED
G0 AEP G. Roper 17 Slepe Crescent, Poole BH12 4DH
G0 AEU P. Tietz 5 Chevin Road, Belper DE56 2UW
G0 AEV Dr S. Reed Bridlands, Middle Common, Chippenham SN15 5NN
G0 AEW D. Arlette 12 Polmear, Par PL24 2AT
G0 AEX E. Hannaby 170A Weston Drive Otley, Leeds LS21 2DT
GM0 AEY D. Palmer 36 Kilsyth Road, Haggs, Bonnybridge FK4 1HE
GW0 AEZ J. Howarth 7A Liddell Drive, Llandudno LL30 1UH
G0 AFH I. Burns Little Delmar Farm, Leywood Road, Meopham DA13 0UD
G0 AFJ A. Brown 33 Marion Road, Haydock, St. Helens WA11 0PY
G0 AFN P. Howard 1 Avon Close, Bognor Regis PO22 6BX
G0 AFP D. Rayner 69 Lovelace Drive, Woking GU22 8QZ
G0 AFQ J. Cook 71 Richmond Avenue, Burscough, Ormskirk L40 7RB
G0 AFR C. Mears 11 Aberford Close, Reading North, Reading RG30 2NX
G0 AFT C. Bovey 12 The Mead, Beaconsfield HP9 1AW
G0 AFU M. Bousfield 5 South Waterside, Hexham NE48 1HF
G0 AFY R. Norman 98 Foxwell Drive, Headington, Oxford OX3 9QF
G0 AFZ C. Wilson 9-10 Daventry Street, Southam CV47 1PH
G0 AGB R. Ellis 22 Fern Road, Storrington, Pulborough RH20 4LW
G0 AGC A. Core 1 Partridge Ride, Loggerheads, Market Drayton TF9 2QX
G0 AGD P. Shortland 69 South Parade, Worksop S81 0BS
GW0 AGL E. Williams Caerbergam Cottage, Llanbedr LL45 2HT
GM0 AGN G. Speirs 43 Sheuchan View, Stranraer DG9 7TA
G0 AGO R. White 137 Fennells Harlow, Essex, Harlow CM19 4RR
G0 AGR North West Radio Group c/o R. Needs 13 Greenway, Great Horwood, Milton Keynes MK17 0QR
G0 AGU B. Humphries 22 Leander Close, Burntwood WS7 1PW
GM0 AGV A. Rollings 24 Millburn Court, Sheuchan St., Stranraer DG9 0DX
GW0 AGZ J. Squire Dyffryn, Llanymynech SY22 6EW
G0 AHA A. Pannell 3 Nethercourt Gardens, Ramsgate CT11 0RY
G0 AHB A. Bennett 2 Portland Place, Hertford Heath, Hertford SG13 7RR
G0 AHC B. Hewitt 32 Pinehurst Drive, Kings Norton, Birmingham B38 8TH
G0 AHD C. Richardson 122 Elmton Road, Creswell, Worksop S80 4DE
G0 AHE P. Moss Three Ways Wick Road, Langham CO4 5PE
G0 AHI T. Deacon 9 Mulberry Close, Woodley, Reading RG5 3LR
G0 AHJ J. Johnson 14 Park House Lane, Prestbury, Macclesfield SK10 4HZ
G0 AHK D. Layne 5 Howe Close, Christchurch BH23 3JA
G0 AHL R. Sears 19 Shepherds Grove Park, Stanton, Bury St. Edmunds IP31 2AY
G0 AHM P. Gregg 5 Rosevear Road Bugle, St. Austell PL26 8PH
G0 AHO C. Grant 20 Muriel Kenny Court, Hethersett, Norwich NR9 3EZ
G0 AHR P. Cuthbert 115 Tintern Avenue, Whitefield, Manchester M45 8WY
G0 AHU J. Tolson 1 Old Mill Court Station Road, Plympton, Plymouth PL7 2AJ
G0 AHV R. Gilling 24 Bellerby Road, Skellow, Doncaster DN6 8PD
G0 AIG M. Sutherland 28 Sycamore Way, Littlethorpe, Leicester LE19 2HT
G0 AIH R. Baker 38 The Front, Middleton One Row, Darlington DL2 1AU
GI0 AIJ I. Greenwood Deers Leap, 24 Tullyrusk Road, Crumlin BT29 4JQ
G0 AIL D. Penny 30 Belvedere Road, Yeovil BA21 5JB
G0 AIM S. Robinson 164 Leigh Road, Westhoughton, Bolton BL5 2LE
G0 AIN S. Withnell 85 Headroomgate Road, Lytham St. Annes FY8 3BG
G0 AIO A. Owens 69 Locomotion Way, North Shields NE29 6XE
GI0 AIQ F. Holland 413 Ballyoran Park, Portadown, Craigavon BT62 1JX
GM0 AIR Cambridge University Wireless Stn c/o D. Parker Devon Cottage, Main Street, Cupar KY15 7QX
G0 AIS J. Borg 94 Coldershaw Road, London W13 9DT
G0 AIX D. Westlake Chyvellin, Newmill, Penzance TR20 8XW
GW0 AIY R. Gibbons Cefnysgwyn, Capel Isaac, Llandeilo SA19 7UA
G0 AIZ W. Chesterton Homelea Farm, Fosse Way, Coventry CV7 9LR
G0 AJA A. Astley 16 Cedar Avenue, Ellesmere SY12 9PA
G0 AJB A. Botherway 4 Brodrick Drive, Ilkley LS29 9SN
G0 AJF S. Harrison 25 High Spring Road, Keighley BD21 4TF
G0 AJH J. Hornsby 15 Coronation Drive, Hornchurch RM12 5BL
G0 AJJ L. Leavold 8 Wilkinson Way, North Walsham NR28 9BB
G0 AJL B. Bromsgrove 34 Boundary Drive, Hunts Cross, Liverpool L25 0QD
GM0 AJT D. Rollings 2 Challoch Crescent, Leswalt, Stranraer DG9 0LN
GW0 AJU A. Underwood Rock Hill, Cefn Abbey, Llanarthne, Carmarthen, Carmarthen, UK SA32 8LJ
G0 AJW J. Welsby 47 Links Road, Knott End-On-Sea, Poulton-le-Fylde FY6 0DF
G0 AJX M. Coles 13 Dawnay Road, Bilton, Hull HU11 4HB
G0 AJZ B. Park 147 Castle Road, Ings Farm, Redcar TS10 2LT
G0 AKC Leicester Radio Group c/o T. Chamberlain 455 Norwich Road, Ipswich IP1 5DR
G0 AKF K. Farrance Clarewood, Tabley Road, Knutsford WA16 0NE
G0 AKH K. Hale 58 St. Stephens Road, Saltash PL12 4BJ
GM0 AKJ P. Seaton 51 Leachkin Avenue, Inverness IV3 8LH

G0 AKK J. Chivers 4 Laurel Drive, Bognor Regis PO21 3ND
G0 AKL R. King 4 Pinewood Drive, Horning, Norwich NR12 8LZ
G0 AKM K. Muchamore 3 Belfont Walk, London N7 0SN
G0 AKO R. Cleveland 2 Morse Close, Brundall, Norwich NR13 5LG
G0 AKR A. Poynter Hill Top Farm, Warren Road, Chatham ME5 9RD
G0 AKS D. Newell 7 Edward Road West, Clevedon BS21 7DY
G0 AKU R. Reanney 9 Stapleford Court, Ellesmere Port CH66 1RW
GW0 AKV J. Anderson 173 Gate Road, Penygroes, Llanelli SA14 7RW
G0 ALA D. Whitehouse 40 Fernleigh Crescent, Up Hatherley, Cheltenham GL51 3QL
G0 ALB C. Matcham 28 Buckingham Drive, Luton LU2 9RA
G0 ALC J. Richards 77 Poxon Road, Walsall Wood, Walsall WS9 9JR
G0 ALE Wey Valley ARG c/o P. Madagan 40 Lagham Park, South Godstone, Godstone RH9 8ER
G0 ALI A. Soars 8 Nomis Park Congresbury, Bristol BS49 5HB
G0 ALJ R. Prior 274 West View Lodge , Canterbury Road, Herne CT6 7HB
G0 ALQ J. Amer 1 Kingfisher Way, Watton, Thetford IP25 6SR
G0 ALR A. Wagenaar La Villette, Folles 87250 France
GM0 ALS F. Roe 74 Willow Grove, Livingston EH54 5NA
GM0 ALW G. Chalmers 38 Grove Hill, Kelso TD5 7AS
GM0 ALX M. Chalmers 38 Grovehill, Kelso TD5 7AS
GD0 AMD Hadley Wood CG c/o A. Dorman 1 Sprucewood Rise, Foxdale, Douglas IM4 3JP Isle of Man
G0 AMO M. Adams 9 Brancaster Avenue, Charlton, Andover SP10 4EN
G0 AMP R. Senft 11 Maltravers Drive, Littlehampton BN17 5EY
G0 AMS A. Sinclair 4 Blackbrook Park Avenue, Fareham PO15 5JJ
G0 AMU L. Parrott 3 Fox Gardens, Lymm WA13 9EY
G0 AMW C. Dunn 90 Mushroom Green, Dudley DY2 0EE
G0 AMX P. Appleyard 28 Romany Close, Letchworth Garden City SG6 4JZ
G0 AMY J. Sheppeck 25 Kingsleigh Road, Heaton Mersey, Stockport SK4 3QF
G0 AMZ K. Fay 2 Chapel Row, Hartley Wintney, Hook RG27 8NJ
GW0 ANA G. Jones Nirvana 2 Castle Precinct, Llandough, Cowbridge CF71 7LX
G0 ANE W. Burrows 35 Bamford Avenue, Barnsley S71 3SJ
GM0 ANG J. Fish Senekal, Alma Road, Fort William PH33 6HB
G0 ANH J. Wright 5 Abbeville Avenue, Whitby YO21 1JD
G0 ANK I. Smallwood 27 Cormorant Way, Herne Bay CT6 6HG
G0 ANL P. Ellis 8 Shropshire Close, Woolston, Warrington WA1 4DY
G0 ANM J. Mooney 2 Madford Lane, Launceston PL15 9EB
G0 ANN M. Viney 12 Palgrave House, Sherwell Road, Norwich NR6 6PU
G0 ANO D. Lawton Grenehurst, Pinewood Road, High Wycombe HP12 4DD
G0 ANP D. Guy 7 Park Avenue, Castle Cary BA7 7HE
G0 ANT Eden Valley RS c/o D. Shaw Cotehouse, Bleatarn, Appleby-in-Westmorland CA16 6PX
G0 ANW J. Hockley 6 King Edward Road, Birchington CT7 0EL
G0 AOA T. Marten 10 Chieveley Drive, Tunbridge Wells TN2 5HG
G0 AOB M. Beal 45 Castle Meadows, Launceston PL15 7DZ
G0 AOC S. Colledge 32 Wakeman St., Worcester WR3 8BQ
G0 AOD D. Heathcote 8 Ferrers Avenue, Tutbury, Burton-on-Trent DE13 9JR
G0 AOE N. Evans 16 Humbledon Park, Sunderland SR3 4AA
GM0 AOF Douglas Vall AR c/o R. Wallace 1 Holding West Kincardine, Crieff PH7 3RP
G0 AOH J. Abbruscato 22199 Pine Tree Ln, Hockley 77447 United States
G0 AOJ F. Fenwick 6 School Lane, Bonby, Brigg DN20 0PP
G0 AOK D. Gall 2 Norham Close, Wideopen, Newcastle upon Tyne NE13 7HS
G0 AOL G. Parsons 20 Fountayne Road, Hunmanby, Filey YO14 0LU
G0 AOM R. Day 3 Railway Cottages, Newby Bridge, Ulverston LA12 8AW
G0 AOO J. Butterwick 45 Fox Howe Coulby Newham, Middlesbrough TS8 0RU
G0 AOP P. Warriner 36 Eskdaleside, Sleights, Whitby YO22 5EP
G0 AOQ S. Boyd 25 Amos Drive, Pocklington, York YO42 2BS
G0 AOS M. Reynolds Willhay Cottage, Willhay Lane, Axminster EX13 5RW
G0 AOW S. Sands 33 High Street, Kinver, Stourbridge DY7 6HF
G0 AOX A. Sands Gabledown, Bridgnorth Road, Stourbridge DY7 6RW
G0 AOY T. Rudd Grasmere, Burgh, Woodbridge IP13 6SU
G0 AOZ R. Powell Town Pond Cottage, Town Pond Lane, Southmoor, Abingdon OX13 5HS
G0 APB P. Buckley 7 Callams Close, Rainham, Gillingham ME8 9ES
G0 API J. Fell 14 Rectory Avenue, Corfe Mullen, Wimborne BH21 3EZ
GM0 APN J. Loveday Crombiebrae, Inverurie AB51 0JT
G0 APP A. Pamment 5 New Captains Road, West Mersea, Colchester CO5 8QP
G0 APV V. Covell-London 15 St. Nicholas Way, Potter Heigham, Great Yarmouth NR29 5LG
G0 APY G. Flood 4 Campbell Crescent, Great Sankey, Warrington WA5 3DA
G0 AQA S. Baynes 1 Reeves Paddock, Townsend, Priddy, Wells BA5 3FG
G0 AQB J. Ireland The Grange, Grange Lane, Worcester WR2 6RW
GI0 AQD D. Burn 135 Main Road Portavogie, Newtownards BT22 1EL
G0 AQF D. Dolphin 16 Golden Cross Lane, Catshill, Bromsgrove B61 0LQ
G0 AQH G. Griffin 23 St. Giles Close, Shoreham-by-Sea BN43 6GR
G0 AQI E. Greenhalgh 19 Rooks Nest Lane, Therfield, Royston SG8 9QX
GW0 AQR J. Richards 21 Cae Gwyn, Caernarfon LL55 1LL
G0 AQS P. Goldthorpe 29 Broadoak Road, Ashton-under-Lyne OL6 8QN
G0 AQT V. Taylor 5 St. Matthews Drive, St. Leonards-on-Sea TN38 0TR
G0 AQU R. Mason 28 Vandyke, Great Hollands, Bracknell RG12 8UP
G0 AQZ D. Mcdonald 24 Wenning Street, Nelson BB9 0LE
GW0 ARA Epping Forest Raynet Group c/o R. Clews Maesygaer, Ciliau Aeron, Lampeter SA48 7SG
GM0 ARD J. Hoey 152 Muirhouse Avenue, Motherwell ML1 2LB
G0 ARF R. Canning Green Lane Cottage, Eardisland, Leominster HR6 9BN
GM0 ARH A. Haxton Lanimar, Dickson Avenue, Montrose DD10 9EJ
GW0 ARK K. Hudspeth 67 Bloomfield Road, Blackwood NP12 1LX
G0 ARL P. Shorland Jen-Lee, Tolvaddon, Camborne TR14 0EQ
G0 ARP A. Price Brook House, Drury Lane, Shrewsbury SY4 1DT
GM0 ART A. Gray 191 Greengairs Road, Greengairs, Airdrie ML6 7SZ
G0 ARU J. Lumb 2 Briarwood Avenue, Bury St. Edmunds IP33 3QF
G0 ARV A. Gates 278 Higher Road, Liverpool L26 9UF
GM0 ARY N. Hamilton Glenwood Cottage, Enzie, Buckie AB56 5BW
G0 ARZ A. Everard 3 St. Hild Close, Darlington DL3 8LD
G0 ASG P. Fautley The Old Reading Room Ashwater, Beaworthy EX21 5EF
G0 ASH J. Coupe 65 Irongate, Bamber Bridge, Preston PR5 6UY
G0 ASK S. Gray 29 Verity Walk, Stourbridge DY8 4XS
G0 ASL J. Gray 29 Verity Walk, Stourbridge DY8 4XS
G0 ASM N. Marston 14 Greystoke Avenue, Sunderland SR2 9DX
G0 ASN S. Hendry 5 Harvey Road, Great Totham, Maldon CM9 8QA
G0 ASP E. Mason 28 Pendil Close Wellington, Telford TF1 2PQ
G0 ASQ L. Williams 24 Alston Drive, Bare, Morecambe LA4 6QR
G0 ASX N. Ward 104 Kingsley Road, Bishops Tachbrook, Leamington Spa CV33 9RZ
G0 ASZ E. Gamble 87 Silverdale Drive, Waterlooville PO7 6DP
GM0 ATA Dr R. Mulholland 1 Larch Grove, Milton Of Campsie, Glasgow G66 8HG
G0 ATB V. Herbert 98 Blithdale Road, Abbey Wood, London SE2 9HL
G0 ATC Rafac ARC c/o D. Taylor 76 Heworth Village, York YO31 1AL
G0 ATD D. Welch 51 Verbena Way, Worle, Weston-Super-Mare BS22 6RL
G0 ATE A. Lovell Bogardesgatan 5, Goteborg 41654 Sweden
G0 ATG T. Goodyer 11 Upper Bere Wood, Waterlooville PO7 7HX
G0 ATK G. Atkins 97 South Street, Tillingham, Southminster CM0 7TH
GM0 ATL P. Smith 29 Rowan Drive, Bearsden, Glasgow G61 3HQ
G0 ATO Dr N. Gregorian 449 E Providencia Ave, Burbank, California 91501 2916 United States
G0 ATP C. Abela 14 Warren Road, Barkingside, Ilford IG6 1BJ
GM0 ATQ L. Morgan 12 Bayview Road, Gourock PA19 1XE
G0 ATS E. Green Chylean, Tintagel PL34 0HH
G0 ATW J. Ferrier 30 Grimsby Road, Laceby, Grimsby DN37 7DB
G0 ATZ C. Best 34 Julius Hill, Warfield, Bracknell RG42 3UN
G0 AUB F. Groves 14 Edenfield Road, Mobberley, Knutsford WA16 7HE
G0 AUE A. Knowles 3 The Avenue, Wighill Park, Tadcaster LS24 8BS
G0 AUF R. Griffiths 26 Hamilton Road, Morecambe LA4 6QG
G0 AUG T. Hopkinson Whitbarrow Hall, Caravan Park, Penrith CA11 0XB
G0 AUH M. Hopkinson Whitbarrow Hall, Caravan Park, Penrith CA11 0XB
G0 AUI C. Stiller 6 Barn Cottage Lane, Haywards Heath RH16 3QW
G0 AUJ M. O'Dell 19 Redwing Rise, Royston SG8 7XU
G0 AUK Amsat-UK c/o D. Johnson 96 Summerfields Avenue, Halesowen B62 9NR
GM0 AUL R. Mckenzie 26 Gladstone Place, Woodside, Aberdeen AB24 2RP
G0 AUN D. Taylor 24 Kingshill Drive, Hoo, Rochester ME3 9JP
G0 AUR A. Hogg 1 Champions Way, South Woodham Ferrers, Chelmsford CM3 5NJ
G0 AUT A. Utting 9 Sydney Road, Spixworth, Norwich NR10 3PG
G0 AUV A. Johnson 125 Charles Street Sileby, Loughborough LE12 7SH
G0 AUW R. Philpott 4 Dukeswood, Chestfield, Whitstable CT5 3PJ
G0 AUX K. Daly Granarogue, , Carrickmacross, A81 XD62 Ireland
GM0 AVB K. Graham 98 Dalswinton Avenue, Dumfries DG2 9NR
GW0 AVD P. Richards 11 Seabourne Road, Holyhead LL65 1AL
G0 AVE South Kesteven ARS c/o D. Brannon 10 Rochester Crescent, Crewe CW1 5YF
G0 AVH T. Mcguire 33 Sandy Lane, Hindley, Wigan WN2 4EJ
G0 AVJ M. Searley 49 Hollymount Close, Exmouth EX8 5PQ
G0 AVP P. Raxworthy 32 St. Marys Avenue, Alverstoke, Gosport PO12 2HX
G0 AVU G. Logan Fenton Hill Farm, Wooler NE71 6JL
G0 AWG W. Mcnamara 22 Cissbury Avenue, Peacehaven BN10 8TJ
G0 AWH N. Bergstrom-Allen Garden Cottage, Thicket Priory, Thorganby, York YO19 6DE
GI0 AWK D. Payne 10 Grovemount Court, Altnagelvin, Londonderry BT47 5JP
G0 AWM C. Warr 29 Barton Road, Lancaster LA1 4ER
G0 AWR D. Proud Willow Cottage, Tresevern, Truro TR3 7AT
GW0 AWT S. Richardson Glasfryn, Porthyrhyd, Llanwrda SA19 8DF
G0 AWW D. Elliott 188 Seaview Road, Wallasey CH45 5HB
G0 AWY R. Titmuss 70 Mallards Rise, Church Langley, Harlow CM17 9PL
G0 AWZ D. Richardson Beckside, Stockton Lane, York YO32 9UA
G0 AXB J. Moore 53 Thatchers Court, Westlands, Droitwich WR9 9EG
G0 AXC D. Lee 188 Manstone Ave, Sidmouth EX10 9TJ
G0 AXD Dr S. Waters 27 Mill Lane, Shepherdswell, Dover CT15 7LJ
G0 AXE M. Davenport 119 Gravel Lane, Wilmslow SK9 6EG
G0 AXI Dr N. Davies 15 Fyfield Road, Oxford OX2 6QE
G0 AXJ R. Mccluskey 29 Hotspur Avenue, Bedlington NE22 5TD
GM0 AXM Melton Mowbray ARS c/o J. Thomson 16 Ravelstone Terrace, Edinburgh EH4 3TP
G0 AXO R. Bainbridge 9 Hamilton Crescent, North Shields NE29 8DW
G0 AXQ P. Austin 28 Britannia Close, Sittingbourne ME10 2JF
G0 AXS S. Birch 29 Manners Road, Southsea PO4 0BA
GM0 AXX F. Gilhooly 26 Arnott Gardens, Edinburgh EH14 2LB
GM0 AXY E. Dons 37 Ashley Drive, Edinburgh EH11 1RP
G0 AXZ W. Johnson 29 Wentworth Park, Allendale, Hexham NE47 9DR
G0 AYA M. Chown 15 Hambleden Walk, Maidenhead SL6 7UH
GI0 AYB J. Throne Fascadail, 12 Mason Road, Londonderry BT47 2RY
G0 AYC C. Porter 10 Cotefield Drive, Leighton Buzzard LU7 3DS
G0 AYD D. Dixon 3 Towns End, Wylye, Warminster BA12 0RN
G0 AYF W. Collier 8 Douglas Street, Hindley, Wigan WN2 3HP
GI0 AYG M. Evans 12 Tullymore Park, Ballymena BT42 2AU
G0 AYI B. Spencer 161A West Lane, Hayling Island PO11 0JW
G0 AYM K. Luxton 2 Trinity Court, Westward Ho, Bideford EX39 1LT
GW0 AYP D. Graves 185 Rhyl Coast Road, Rhyl LL18 3US
GW0 AYQ R. Smith 4 Glan Ysgethin, Talybont LL43 2BB
GM0 AYT A. Mcdougall Ceol Na Mara, 3 Bowfield Road, West Kilbride KA23 9LB
G0 AYX P. Towell 25 Cedar Close, Grafham, Huntingdon PE28 0DZ
G0 AYY A. Perry Chala, Woodhouse Lane Hill, Lyme Regis DT7 3SX
GI0 AZA Dr E. Harper 404 Foreglen Road Dungiven, Londonderry BT47 4PN
GI0 AZB H. Evans Oville House 404 Foreglen Road, Dungiven, Londonderry BT47 4PN
GM0 AZC J. Sherry 26 Grahamshill Terrace, Fankerton, Denny FK6 5HX
G0 AZD G. Hayter 19 Austin Road, Cirencester GL7 1BT
G0 AZE L. Owen 68 Clevedon Road, Tickenham, Clevedon BS21 6RD
G0 AZG T. Wharton Onanole, Clitheroe Road, Clitheroe BB7 3DA
G0 AZH J. Wharton 66 Hayhurst Street, Clitheroe BB7 1ND
G0 AZM M. Walton 5 Home Farm Court, Home Farm Close, Leicester LE4 0SU
G0 AZP S. Tricker 1 Drewitt Court, 75 Godstow Road, Oxford OX2 8PE
G0 AZQ A. Pearce 21 Cherry Way, Nafferton, Driffield YO25 4PA
G0 AZR J. Norman The End Peg, The Street, Norwich NR11 7AQ
GM0 AZU J. Aiken 48 Kirkwall Avenue, Blantyre, Glasgow G72 9NX
GM0 AZV C. Norton 3 Nether Balfour Cottages, Durris, Banchory AB31 6BL
GW0 AZW R. Dooley 98 Gelli Aur, Treboeth, Swansea SA5 9DG
G0 AZX R. Pritchard 10 Dolphin Crescent, Paignton TQ3 1AE
G0 BAA East of Greenwich RAC c/o G. Gourley 6A Longsight Lane, Cheadle Hulme, Cheadle SK8 6PW
G0 BAF N. Quinn 15 Newham Lane, Steyning BN44 3LR

G0 BAG R. Cox 2 Yardlea Close, Rowland's Castle PO9 6DQ
GW0 BAH P. Bateman 48 Ogmore Drive, Nottage, Porthcawl CF36 3HR
G0 BAI P. Goodger 125 Mill Hill Wood Way, Ibstock LE67 6QD
G0 BAJ R. Edinborough Flat 5 Fairlawn Apartments, 10 Elmsleigh Park, Paignton TQ4 5AT
G0 BAK W. Schofield 24 Meltham Road, Honley, Holmfirth HD9 6HX
G0 BAM E. Smith 166 Tudor Way, Dines Green, Worcester WR2 5QY
G0 BAN J. Emerson 26 Cardwell Street, Roker, Sunderland SR6 0JP
G0 BAO M. Heath Brambles, 1 Bestwall Road, Wareham BH20 4HY
G0 BAP R. Harman 42 Newlyn Close, Bransholme, Hull HU7 4PQ
G0 BAQ I. Irving Fourwinds, Woodburn Drive, Leyburn DL8 5HU
G0 BAT W. Mcdonagh 35 Manor Way Deeping St James, Peterborough PE68PS
G0 BAU S. Craggs 79 Silverdale Road, Cramlington NE23 3LW
G0 BAW L. Haynes 9 Heather Gardens, Belton, Great Yarmouth NR31 9PP
G0 BAX Bomchil Castro Goodrich Claro Arosmena c/o A. Harrison 10 Knoll Road, Sidcup DA14 4QU
G0 BAY S. Parker 85 Highfield Road, Glossop SK13 8NZ
G0 BBB U. Grunewald Nuptown Orchard, Nuptown, Warfield, Bracknell RG42 6HU
GW0 BBC St Helens & District ARC c/o D. Thomas 88 Cefn Graig, Rhiwbina, Cardiff CF14 6JZ
G0 BBE L. Conlon 4 Hill Crest Drive, Slack Head, Milnthorpe LA7 7BB
G0 BBJ D. Fry 9 Brook Gardens, Emsworth PO10 7JY
G0 BBK P. Marshall 35 Rosewood Close, Burnham-on-Sea TA8 1HG
G0 BBN K. Hendry 23 Briscoe Way, Lakenheath, Brandon IP27 9SA
G0 BBO J. Stanbury 6 Waterside Apartments, Weech Road, Dawlish EX7 9FA
G0 BBR W. Kemp 4 Blacksmiths Field Crowhurst, Battle TN33 9AX
G0 BBT E. Snape 1 Stephen Crescent, Humberston, Grimsby DN36 4DS
G0 BBV H. Watts 44 Laurel Road, Norwich NR7 9LL
G0 BCF R. Pickering 14 Dalestorth Gardens, Skegby, Sutton-in-Ashfield NG17 3FT
G0 BCH P. Guppy 202 Exeter Road, Kingsteignton TQ12 3NJ
GD0 BCJ P. Mcgrath 69 Clagh Vane, Ballasalla IM9 2HF Isle of Man
GW0 BCL R. Johnson 55 Maes Yr Haf, Llanelli SA15 3NF
GD0 BCN H. Glaister 42 Barrule Drive, Onchan, Douglas IM3 4NR Isle of Man
GI0 BCP M. Jones 2 Pine Ridge, Donaghadee BT21 0QR
GW0 BCR J. Watts 39 Turnberry Drive, Abergele LL22 7UD
G0 BCS P. Rose 53 South St., Pennington, Lymington SO41 8DY
G0 BCT R. Sayer Vignouse, Paimpont 35380 France
G0 BCU D. Charnock 44 Bramshill Close, Birchwood, Warrington WA3 6TZ
G0 BCW D. Grevett 45 East Bridge Road, South Woodham Ferrers, Chelmsford CM3 5SB
G0 BCX G. Eden 83 Windle Hall Drive, St. Helens WA10 6QG
GM0 BCY A. Carslaw 51 Stonefield Drive, Paisley PA2 7QY
G0 BCZ R. Johnson 89 Arlington Gardens, Romford RM3 0EB
G0 BDB P. Stephens 259 Beaumont Road, Plymouth PL4 9EL
GU0 BDI D. Prosser Rustlings, Les Friquets, St. Andrew GY6 8SJ Guernsey
G0 BDJ K. Dower 16 Mills Drive, Wellington TA21 9ED
G0 BDK B. Steele 82 Margetts Road, Kempston, Bedford MK42 8DT
G0 BDM D. Briggs 72 Showell Grove, Droitwich WR9 8UD
G0 BDN A. Slater 1 Linglongs Avenue, Whaley Bridge, High Peak SK23 7DT
G0 BDP R. Wills 8 Owlswood, Ridingsmead, Salisbury SP2 8DN
G0 BDR M. Gee 100 Plantation Hill, Worksop S81 0QN
G0 BDS B. Scroggs 10 Lyon Close, Chelmsford CM2 8NY
GI0 BDU V. Fortune 10C Ards Drive, Newtownabbey BT37 0JN
GW0 BDW R. Kelsall 1+2 Tyn Rhos, Llanddona, Beaumaris LL58 8YG
GI0 BDZ D. Pavis 269 Lower Braniel Road, Belfast BT5 7NR
GI0 BEB T. Magee 10 Abernethy Park, Newtownabbey BT36 6QQ
G0 BEC S. Christmas Hitherto, Moats Tye Combs, Stowmarket IP14 2EY
G0 BEE S. Bloom 8 South Riding, Bricket Wood, St. Albans AL2 3ND
G0 BEJ P. Mallett Woodrow, Chalk Lane, Spalding PE12 9YF
GM0 BEL G. Lindsay 6 Netherhouse Avenue, Lenzie, Glasgow G66 5NG
G0 BEN D. Pykett 20 Rochester Drive, Lincoln LN6 0XQ
G0 BEP D. Boalch 90 Belle Vue Road, Wivenhoe, Colchester CO7 9EH
G0 BES S. Illsley 88 Arnold Road, Eastleigh SO50 5RR
G0 BEV Dr M. Hill Windrush, Jesmond Gardens, Newcastle upon Tyne NE2 2JN
G0 BEX A. Penfold 115 Applegarth Park, Seasalter Lane, Whitstable CT5 4BZ
GI0 BEY N. Turkington 16 Glenshesk Park, Bangor BT20 4US
GU0 BEZ V. Stamps Mill Cottage, Sark GY10 1SA Guernsey
GI0 BFA R. Mckersie 4 Burnside Park, Belfast BT8 6HU
G0 BFC C. Simkin Flat 33, Weymouth House Balfour, Tamworth B79 7BE
GI0 BFD A. Magee 70 Hillview Park, Enniskillen BT74 6EU
G0 BFJ J. Stocks 96, North Street, Lockwood, , Huddersfield HD1 3SL
G0 BFK K. James Yew Tree Cottage, Whiston, Stafford ST19 5QH
G0 BFM A. Anderson 23 Aldred Road, Sheffield S10 1PD
GD0 BFN J. Kneale 51 Maple Avenue, Onchan, Douglas IM3 3GA Isle of Man
GM0 BFW A. Mcclelland 4 Walkerston Avenue, Largs KA30 8ER
G0 BFZ A. Nock 57 Mushroom Green, Dudley DY2 0EE
G0 BGA G. Allan 24 Leadbetter Drive, Bromsgrove B61 7JG
G0 BGB M. Davis 1 Westwood View Crawcrook, Ryton NE40 4HR
G0 BGH I. Bamford 60 Coast Drive, Greatstone, New Romney TN28 8NX
G0 BGI S. Knight 14A Manor Road, Upton Lovel, Warminster BA12 0JW
G0 BGR R. Jones Glenroy, 5 School Road, Blackpool FY4 5DS
G0 BGV H. Eastwood 3 The Brambles, Thorpe Willoughby, Selby YO8 9LL
G0 BGX O. Pauley 235 Roughton Road, Cromer NR27 9LQ
G0 BGY J. Moult New Bungalow, Bar Bridge Lane Swineshead., Boston PE20 3PG
G0 BHA P. White 42 Abbey Road, Medstead, Medstead, Alton GU34 5PB
G0 BHH P. Gregory 20 Heyes Grove, Rainford, St. Helens WA11 8BW
G0 BHK E. Stiles 16 Henry Avenue, Rustington, Littlehampton BN16 2NY
G0 BHP P. Fambely 126 Ashton Lane, Sale M33 5QJ
G0 BHR D. Egan 44 Wrights Lane, Warley, Cradley Heath B64 6QX
G0 BHS D. Angell 12 Harrod Drive, Market Harborough LE16 7EH
G0 BHT W. Blake Tylers Farm, Grantham Road, Grantham NG33 5HG
G0 BHU M. Jamieson 3 Rowan Way, Bourne PE10 9SB
G0 BIA R. Armstrong 64 Churchill Drive, , Marske by the Sea TS11 6BE
G0 BIE D. Lucas 9 Newborough Road, Alvaston, Derby DE24 0LH
G0 BIN E. Ashley 2 School Road, Bulkington, Bedworth CV12 9JB
G0 BIQ R. Bellamy Emathu, No. 48 28Th April Street 1688, Xahhra XRA1033 Malta
G0 BIR A. Skinner Halfway Lock Cottage, Upper Gambolds Lane, Stoke Prior, Bromsgrove B60 3HB
G0 BIV J. Young 3 Kensington Close, Kings Sutton, Banbury OX17 3XB
G0 BIW M. Smith 394 Longbridge Road, Barking IG11 9EE
G0 BIX T. Dansey C/O 14 Stirling Park, Rochester ME1 3QR
G0 BJA G. Vincent-Squibb 7 Chudleigh Road, Henley Green, Coventry CV2 1AF
G0 BJD S. Wood 55 Megdale, Wolds, Matlock DE4 3TE
G0 BJI D. Peacock 15 Farmfield Road, Banbury OX16 9AP
G0 BJK D. Thomas 33 Chatsworth Road, Stretford, Manchester M32 9QF
G0 BJL D. Oxley 21 Ringwood Avenue, Chesterfield S41 8RA
G0 BJP T. Fox 206 Hollinsend Road, Sheffield S12 2EJ
G0 BJR G. Oliver 158 High Barn St., Royton, Oldham OL2 6RW
G0 BKA J. Leedham 27 St. Andrews Crescent, Stratford-upon-Avon CV37 9QL
G0 BKB V. Leedham 27 St. Andrews Crescent, Stratford-upon-Avon CV37 9QL
G0 BKC P. Glasper 2 Iris Close, Stockton TS18 1AX
G0 BKD B. Mawn 13 Micklethwaite Grove, Moorends, Doncaster DN8 4NU
G0 BKE F. Barker 17 Walders Avenue, Sheffield S6 4AY
G0 BKH B. Minton 8 Rosebank Walk, Barnton, Northwich CW8 4PU
GW0 BKJ M. Glover 4 New Hospital Villa, Hospital Road, Brecon LD3 0DU
G0 BKL E. Smith 40 Grays End Close, Grays RM17 5QR
G0 BKN I. Mciver 3 Asgard Drive, Bedford MK41 0UP
G0 BKP J. Dadswell Ivy House, Chivery, Tring HP23 6LE
G0 BKQ Utc Sheffield ARC c/o P. Rowe 45 Springhill Avenue, Wolverhampton WV4 4ST
G0 BKR Dr G. Clark 2 Roundheads End Forty Green, Beaconsfield HP9 1YB
GM0 BKS G. Christison 13/5 West Winnelstrae, Edinburgh EH5 2ET
G0 BKU S. Coles 88C Dursley Road, Trowbridge BA14 0NS
G0 BKW K. Weston 14 Auchinleck Court Burleigh Way, Crawley Down, Crawley RH10 4UP
GM0 BKX T. Stewart 104 Barrhill Road, Cumnock KA18 1PU
G0 BKZ D. Hyde 75 Bury Street, Stockport SK5 7RE
G0 BLB R. Baker Homelea, Upper Bristol Road, Clutton, Bristol BS39 5RJ
G0 BLM P. Mathews 25 Shore Mount, Littleborough OL15 8EN
G0 BLO B. Osborn 4 Highfield Close, Collier Row, Romford RM5 3RX
G0 BLQ H. Cooper 26 Badgers Way, Buckingham MK18 7EQ
G0 BLS A. Wilkie 7 Willow Drive, Droitwich WR9 7QE
G0 BLT G. Blackmoor 4 St. Godwalds Crescent, Aston Fields, Bromsgrove B60 2EB
G0 BLU E. Mustard 108 Allandale, Hemel Hempstead HP2 5AT
G0 BLV M. Puncer 17 St. Michaels Walk, Eye, Peterborough PE6 7XG
G0 BLW A. Crimlisk 14 Long Lane, Aughton, Ormskirk L39 5AT
G0 BMG X. Green 65 Rosamond Road, Bedford MK40 3UG
G0 BMH J. Mclean 24 Durham Drive, Oswaldtwistle BB5 3AT
G0 BML T. Garvey 162 Birchfields Road, Manchester M14 6PE
G0 BMM R. Howarth 95 Riverside Drive, Radcliffe, Manchester M26 1HY
G0 BMN K. Hutchins 23 Salisbury Road, Tunbridge Wells TN4 9DJ
G0 BMP L. Gyurgyak 7 Lakeside Close, New Park, Newton Abbot TQ13 9FE
G0 BMQ M. Armstrong 4 Medway Drive Preston, Weymouth DT3 6LF
G0 BMS I. Cooper Flat 1, Philip Howard Court, Glynne Street, Farnworth, Bolton BL4 7DQ
G0 BMT J. Walkley 10 Exton, Dunster Crescent, Weston-Super-Mare BS24 9EH
G0 BMU T. Drewitt 6 Copse View Cottages, Ascot Road, Maidenhead SL6 3JY
G0 BMZ B. Lindgren Berzeliigatan 26, Goteborg SE-41 53 Sweden
G0 BNE Dr A. Knell 13 Northumberland Road, Leamington Spa CV32 6HE
G0 BNF J. Holmes 63 Grange Avenue, Street BA16 9PF
G0 BNG L. Britton 36 Frampton Court, Trowbridge BA14 9HL
G0 BNJ B. Northway 3 Kingston Close, Kingskerswell, Newton Abbot TQ12 5EW
G0 BNK Wearside Electronics And ARS c/o B. Young 11 Gainsborough Avenue, Washington NE38 7EF
GW0 BNN J. Bulpin 12 Waungron Close, Treboeth, Swansea SA5 7DH
GW0 BNO D. Lindley 29 Belvedere Close, Kittle, Swansea SA3 3LA
GM0 BNQ D. Macdonald Greenbrae Cottage, Auchterless, Turriff AB53 8HD
G0 BNR N. Keightley Wavendon, Daintree Road, Ramsey St. Marys, Ramsey, Huntingdon PE26 2TF
G0 BNU D. Wright The Firs, Lutterworth Road, Wolvey, Wolvey LE10 3HW
G0 BNW W. Wheeler 201 Topsham Road, Exeter EX2 6AN
G0 BNY C. Lee 29 Meadow Dale, Chilton, Ferryhill DL17 0RW
G0 BNZ J. Hewitt 35 Birmingham Road, Alvechurch, Birmingham B48 7TB
G0 BOC E. Pitman 35 Brackley Way, Totton, Southampton SO40 3HP
GW0 BOE Dr W. Piotrowski Ty Nant, Abbeycwmhir LD1 6PH
G0 BOH D. Bowman 6 Linksfield, Denton, Manchester M34 3TE
G0 BOM A. Samouelle 4 Fox Road, Bourn, Cambridge CB23 2TU
G0 BON I. Rogers 33 Sandstone Road, Swindon SN25 2FE
G0 BOO B. Gilbert Silver Micha, Hunts Corner, Norwich NR16 2HL
G0 BOR J. Goodall 1 Purn Lane, Weston-Super-Mare BS24 9JG
G0 BOT D. Ashby 36 Mersey Grove, Birmingham B38 9LA
G0 BPA B. Lyford 48 Wilverley Place, Blackfield, Southampton SO45 1XW
GM0 BPF W. Johnstone Byre Cottage, Kilmichael Glassary, Lochgilphead PA31 8QL
G0 BPK N. Ferguson Royd Moor, Royd Moor Lane, Badsworth, Pontefract WF9 1AZ
G0 BPL D. Farnham 24 Downham Road, Watlington, King's Lynn PE33 0HS
G0 BPM D. Coffey 121 Worksop Road, Swallownest, Sheffield S26 4WB
G0 BPQ J. Jones 16 Laurel Avenue, Darwen BB3 3AG
G0 BPR S. Whitnear 55 Brier Crescent, Nelson BB9 0QD
G0 BPS R. Pascoe 12 Oak Rise, Terlingham Gardens, Hawkinge CT18 7FU
GM0 BPT A. Murray 1 Gordon Road, Edinburgh EH12 6NB
G0 BPU M. Johnson 65, Carlford Court, Parliament Road, Ipswich., Ipswich IP45EL
G0 BPX D. Norridge 125 Ferry Road, Marston, Oxford OX3 0EX
G0 BPZ P. Maple Beech House, Derby Road, Ashbourne DE6 1LZ
G0 BQB P. Smith 10 Denby Lane, Grange Moor, Wakefield WF4 4ED
G0 BQC W. Scrivener Folly Hall Farm, Kings Causeway, Nelson BB9 0EZ
G0 BQE J. Chown 15 Hambleden Walk, Maidenhead SL6 7UH
G0 BQG R. Jefferies 35 Lambrok Close, Trowbridge BA14 9HH
G0 BQI M. Chapman 6A Rees Street, Islington, London N1 7AR
G0 BQK M. Hall 31 Winchester Close, Stratton, Swindon SN3 4HB
G0 BQO J. Rawson 3 Cooks Close, Kesgrave, Ipswich IP5 2YT
G0 BQP Dr J. Simpson 18 Southdean Drive, Hemlington, Middlesbrough TS8 9HH
G0 BQV M. Ashdown 42 Alpine Avenue, Tolworth, Surbiton KT5 9RJ
G0 BQW S. Robinson 114 Hopefield Avenue, Sheffield S12 4XE
GI0 BQX D. Maguire 16 Kilmacormick Drive, Enniskillen BT74 6EP
G0 BQZ S. Smith 18 Stratford Drive, Eynsham, Witney OX29 4QJ
G0 BRA Banbury Amateur Radio c/o F. Humphris 169 Bloxham Road, Banbury OX16 9JU

G0 BRH C. Waters 468 Buckfield Road, Leominster HR6 8SD
GM0 BRJ D. Wilson Four Winds, High Barrwood Road, Kilsyth G65 0EE
G0 BRL M. Bevan 22 Spring Crescent, Brown Edge, Stoke-on-Trent ST6 8QH
G0 BRM N. Entwistle Sonunda House, Church Lane, Freckenham, Bury St. Edmunds IP28 8JF
GI0 BRO R. Wilson 68 Kensington Road, Belfast BT5 6NG
G0 BRQ A. Morris 108 Lytchett Drive, Broadstone BH18 9NR
GM0 BRS Border ARS c/o A. Scott 20 Treaty Park, Birgham, Coldstream TD12 4NG
G0 BRW B. Whatling 6 Rock Road, Dursley GL11 6LF
G0 BRX S. Shirras 72 Sefton Avenue, Poulton-le-Fylde FY6 8BL
G0 BRZ L. Rimmer 7 Dorfold Close, Sandbach CW11 1EB
G0 BSA M. Embling 23 Sinodun Road, Didcot OX11 8HP
G0 BSD D. Tringham 47 The Knoll Palacefields, Runcorn WA7 2UH
G0 BSF H. Rolfe 46 Great Gardens Road, Hornchurch RM11 2BA
G0 BSH R. Harman 22 Ridgebrook Road, Kidbrooke, London SE3 9QN
G0 BSJ E. New 6 Witchampton Close West Leigh, Havant PO9 5RY
G0 BSK K. Bryant 9 Cunningham Park, Mabe Burnthouse, Penryn TR10 9HB
G0 BSN R. Vincent-Squibb 1 Alexander Avenue, Earl Shilton, Leicester LE9 7AF
G0 BSP R. Snow 73 Boxtree Road, Harrow Weald, Harrow HA3 6TN
G0 BST J. I'Anson-Holton Lake View, Brookside Avenue, Telford TF3 1LA
G0 BSX Dr P. Meiring 18 Slayleigh Lane, Sheffield S10 3RF
G0 BSY J. Rennie 7 Aldermans Meadow, Quarry Moor Park, Ripon HG4 3FD
G0 BTA D. Round 21 Bunting Drive, Bradford BD6 3XE
GW0 BTB B. Botham The Cherries, Anglesey LL74 8SR
G0 BTD G. Owen 7 East View, Crossgates, Leeds LS15 8AY
G0 BTH P. Clark 180 Roselands Drive, Paignton TQ4 7RW
GM0 BTK Dr W. Rossmann Strathvale, Milton Of Ogilvie, Forfar DD8 1UN
G0 BTQ N. Burge 43 Bourn Rise, Pinhoe, Exeter EX4 8QD
G0 BTT E. Clay 23 New Street, Sleaford NG34 7HG
G0 BTU K. Tupman Magpies, 6 Larcombe Road, Petersfield GU32 3LS
G0 BTV S. Coleman 124 Greenacres, Kirkby-In-Ashfield, Nottingham NG17 7GF
G0 BUB Peterlee ARC c/o M. Sharpe New House, Highfield Farm, Grantham NG32 3SJ
G0 BUC K. Simpson 5 Cedar Close, The Elms, Lincoln LN1 2NH
G0 BUD J. Spearing 19 Elizabeth Square, London SE16 5XN
GM0 BUE A. Stark 30 Kelvin Way, Kilsyth, Glasgow G65 9UL
GM0 BUI G. Williamson 2 Laburnum Grove, Burntisland KY3 9EU
G0 BUJ R. Pelling The Orchard, Shirwell, Barnstaple EX31 4JR
G0 BUK I. Mitchell Cornerfield, Five Ash Down, Uckfield TN22 3AP
G0 BUV J. Howard 130 Coventry Road, Coleshill, Birmingham B46 3EH
G0 BUW P. Martin 39A The Grove, Bearsted, Maidstone ME14 4JB
G0 BUX R. Clinton Appletrees, Alexandra Road, Mayfield TN20 6UD
G0 BUZ E. Piper 44 Parsonage Estate, Rogate, Petersfield GU31 5HJ
G0 BVA S. Fawcett 34 Wantsume Lees, Sandwich CT13 9JF
G0 BVC G. Broadhurst Summerhayes, 10 Ottervale Close, Honiton EX14 9TA
G0 BVD P. Oakley Elmsleigh House, New Street, Great Torrington EX388BY
GM0 BVG J. Graham Clintpark, Lockerbie DG11 3JH
G0 BVK E. Evers 13 Grasmere Close, Penistone, Sheffield S36 8HP
G0 BVM J. Speers 187 Worsley Road, Eccles, Manchester M30 8BP
G0 BVO J. Teasdale 1 Newtown Bungalows, Newtown, Spennymoor DL16 7QS
G0 BVQ C. Hawkes 12 Summer Hall Ing, Wyke, Bradford BD12 8DN
G0 BVS M. Arthur 8 Hanbury Road, Bedworth CV12 9BX
G0 BVT A. Tonge 30 Cardigan Avenue, Morley, Leeds LS27 0DP
G0 BVU D. Davies 30 Bullpit Road, Balderton, Newark NG24 3LY
G0 BVV D. Clench 70 Shalbourne Crescent, Bracklesham Bay, Chichester PO20 8RG
G0 BVW Q. Curzon 154 Clophill Road, Maulden, Bedford MK45 2AE
G0 BWB J. Chappell 49 Midway, South Crosland., Huddersfield HD4 7DA
G0 BWC Bolton Wireless Club c/o R. Wilkinson 84 Park Road, Bolton BL1 4RQ
GW0 BWE D. Allen 48 Castle Road, Crickhowell NP8 1AP
G0 BWG J. Raynes 115 Deerlands Avenue, Sheffield S5 7WU
G0 BWJ K. Miller 8 Horsham Gardens, Sunderland SR3 1UJ
G0 BWK R. Lee 57 Hart Lane, Luton LU2 0JF
G0 BWO D. Wilkinson 139 Grosvenor Road, Dalton, Huddersfield HD5 9HX
G0 BWP B. Passmore 364 Franklin Road, Kings Norton, Birmingham B30 1NG
G0 BWQ K. Kitson 278 Cowcliffe Hill Road, Huddersfield HD2 2NE
G0 BWV J. Puttock Sutton & Cheam Rs, 53 Alexandra Avenue, Sutton SM1 2PA
G0 BWY G. Fearnside 16 Lee Court, Thwaites Brow, Keighley BD21 4TL
G0 BXC P. Hughes 123 Garth Road, Morden SM4 4LF
G0 BXD A. Wootton Dower House, Hilton, Bridgnorth WV15 5PB
G0 BXG R. Dring 22 Castle Street, Eastwood, Nottingham NG16 3GW
G0 BXH B. Hansell 7 Fry Road, Stevenage SG2 0QG
G0 BXJ A. Peachey 60 Upwell Road, March PE15 9EA
G0 BXL T. White 79 Elmbridge, Harlow CM17 0JY
G0 BXM M. Smyth Sunset Cottage, Cloonfree, Strokestown F42 YX92 Ireland
GM0 BXR T. Holland 7 Burnside, Flotta, Stromness KW16 3NP
G0 BXS R. Rennolds 6 Roman Way, Bourton-On-The-Water, , Cheltenham GL54 2EW
G0 BXU A. Collick 39 Beech Rise, Sleaford NG34 8BJ
G0 BXV D. Dixon Flat 19, Maple Court 3A Staunton Avenue, Hayling Island PO11 0EF
G0 BYA S. Houlding 90 Wordsworth Avenue, Stafford ST17 9UE
G0 BYF R. Aucote 20 Bridgewater Close, Billingshurst RH14 9EQ
G0 BYH G. Stokes 23 Maynard Close, Bradwell, Milton Keynes MK13 9HS
G0 BYK M. Jackson Stockswood, Stocks Lane, Southwold IP18 6UJ
G0 BYL J. Luxton 2 Trinity Court, Westward Ho, Bideford EX39 1LT
G0 BYQ S. Ratcliffe 173 Whinney Lane, New Ollerton, Newark NG22 9TJ
G0 BYU F. Cooper 23 York Close, Gillow Heath, Stoke-on-Trent ST8 6SE
G0 BYX C. Pavier 12A Friend Lane, Edwinstowe, Mansfield NG21 9QZ
GW0 BYZ R. Jones 9 Dennithorne Close, Merthyr Tydfil CF48 3HE
GW0 BZA K. Duckfield Ellesmere, 29 Esplanade Avenue, Porthcawl CF36 3YS
G0 BZB A. Volpe 7 Oakwell Terrace, Haltwhistle NE49 9LR
G0 BZC G. Smith 9 Blackett Avenue, Stockton-on-Tees TS20 2EX
G0 BZF D. Reid Nicolaas Beetsstraat 29, Hengelo Ov 7552HW Netherlands
G0 BZH G. Hodgson 16 Dockroyd, Oakworth, Keighley BD22 7RH
GW0 BZJ J. Newton 21 Village Court, Penrhiw Avenue, Blackwood NP12 0LU
GI0 BZM C. Colhoun 85 Whitehill Park, Limavady BT49 0QF
G0 BZN C. Johnson 23 Medlar Street, Weston Turville, Aylesbury HP22 5YQ
G0 BZP J. Bates 28 Westbourne Road, West Bromwich B70 8LD
GM0 BZS A. Ince Burnside, Braefield, Glen Urquhart, Inverness IV63 6TN
G0 BZT R. Targonski 4 Woodville Gardens, Dudley DY3 1LB
G0 BZU K. Barlow 105 Buller Street, Bury BL8 2BQ
G0 BZV E. Trett 11 Langton Avenue, Bierley, Bradford BD4 6BY
G0 BZW A. Porter 1 Bloomfield Terrace, Weston, Portland DT5 2AB
G0 BZX P. Smith 8 Avalon Close, Orpington BR6 9BS
GM0 CAD D. Hewitt Willowburn, Ardross, Alness IV17 0XN
G0 CAE K. Walsh 13 Weston Park Homes, Weston Road, Portland DT5 2DE
G0 CAG D. Talaber 54 Southfield Park, North Harrow, Harrow HA2 6HE
GI0 CAH P. Leonard 4 Clonurson Road, Enniskillen BT92 3BU
G0 CAJ T. Morgan 73 Winnards Park, Sarisbury Green, Southampton SO31 7BX
G0 CAK G. Russell 15 Warblington Close, Tadley RG26 3YW
G0 CAL R. Faulkner 10 Fell Wilson St., Warsop, Mansfield NG20 0PT
G0 CAM C. Chislett Woodview, Crofthandy, Redruth TR16 5PT
G0 CAS N. Clarke Lyme View, 3 East Cliff House, Dawlish EX7 9JB
G0 CAX I. Moore 25 Granby Close, Winyates, Redditch B98 0PJ
G0 CAY K. Arnold 14 Ford Close, St.Ive, Cornwall PL14 3FN
GM0 CBA B. Aitken 48 Kenilworth Rise, Livingston EH54 6JJ
G0 CBB M. Phillips 52 Rivington Drive, Burscough, Ormskirk L40 7RP
GM0 CBC J. Johnstone 12 Castle Acre, Ecclefechan, Lockerbie DG11 3DU
G0 CBD G. Roby 40 Lulworth Drive, Hindley Green, Wigan WN2 4QS
G0 CBI S. Mcarthur 36 Ingham Close, Bradshaw, Halifax HX2 9PQ
G0 CBJ A. Whittingham 61 Hillcroft Road, Altrincham WA14 4JE
G0 CBK N. Priestley 13 Orchard Close, Charfield, Wotton-under-Edge GL12 8TJ
GW0 CBL J. Follant 76 Wern Road, Skewen, Neath SA10 6DL
G0 CBM C. Wilkie Bramcote, Grange Road, Sandilands, Mablethorpe LN12 2RE
G0 CBN P. Forster 2 Rockingham Close, Birchwood, Warrington WA3 6UY
G0 CBO R. Hoffman 26 Penn Road Taverham, Norwich NR8 6NN
G0 CBP R. Mcmahon 34 Denmark Road, Poole BH15 2DB
G0 CBT S. Peat 64 Grange Road, Romford RM3 7DX
G0 CBU W. Drea 146 Slewins Lane, Hornchurch RM11 2BS
G0 CBW M. Bentley Nickers Hill Farm, Falhouse Lane, Dewsbury WF12 0NL
G0 CCA G. Coles Walnut Lodge, Staunton Lane, Bristol BS14 0QG
G0 CCB C. Gill 52 Southfield Road, Nailsea, Bristol BS48 1JD
G0 CCC Caversham CG c/o C. Young 18 Wincroft Road, Caversham, Reading RG4 7HH
G0 CCF J. Broome 35 Claygate Road, Cannock WS12 2RN
G0 CCG H. Toh 3 Priory Road, Dover CT17 9RQ
G0 CCJ J. Weinstock 24 Hamilton Road, Tiddington, Stratford-upon-Avon CV37 7DD
G0 CCL Cambridge Consultants ARC c/o L. Laprade 22 Langley Way, Hemingford Grey, Huntingdon PE28 9DB
G0 CCM M. Hurst 2 Poplar Road, Oughtibridge, Sheffield S35 0HR
G0 CCN P. Hurst 23 Cantilupe Crescent, Aston, Sheffield S26 2AS
GW0 CCO L. Holderness 8 Jubilee Gardens Templeton, Narberth SA67 8ST
G0 CCQ J. Smith Dobie Lodge, Rochford, Tenbury Wells WR15 8SR
G0 CCS J. Zissler 8 Norman Drive, Whittington, King's Lynn PE33 9TQ
G0 CCU L. Whitelegg 30 Chatsworth Road, Arnos Vale, Bristol BS4 3EY
G0 CCV J. Gaut 18 First Avenue, Clipstone Village, Mansfield NG21 9EA
G0 CCX A. Gilbert 6 Tarring Close, South Heighton, Newhaven BN9 0QU
G0 CDA M. Ryder 9 Lincoln Close, Woolston, Warrington WA1 4LU
G0 CDB J. May 6 Hodson Close, Paignton TQ3 3NU
GI0 CDM J. Neill 16 Rathlin Street, Belfast BT13 3DZ
G0 CDO J. Faulkner-Court Yew Tree Cottage, Avon Dassett, Southam CV47 2AT
G0 CDQ M. Murphy 10 Bayham Road, Sevenoaks TN13 3XA
G0 CDR J. Warwick 29 Hay Brow Crescent, Scalby, Scarborough YO13 0SG
G0 CDS C. Brind 8 Pezenas Drive, Market Drayton TF9 3UJ
GM0 CDV R. Evans 9 Courthill Farm Cottages, Kelso TD5 7RU
GM0 CDW A. Thomson 24 Craigmount Gardens, Edinburgh EH12 8EA
G0 CDY A. Hulme 71 Victoria Gardens, Ferndown BH22 9JQ
G0 CDZ E. Hicks-Arnold Ingleside, Junction Road, Salisbury SP5 3AZ
GM0 CEA R. Robertson St. Madoes Cottage, St. Madoes, Perth PH2 7NF
G0 CEB E. Hutchinson 58 Avon Rise, Retford DN22 6QH
G0 CEC P. Coenraats 54 Falstaff Avenue, Earley, Reading RG6 5TG
G0 CEF A. Sheard 8 Hazel Beck, Bingley BD16 1LZ
G0 CEG P. Worsdale 10 Manton Road, Lincoln LN2 2JL
G0 CEI P. Olliffe 4 Orpwood Paddock, School Road, Ardington, Wantage OX12 8RB
G0 CEJ C. Baguley 44 Royds Crescent, Rhodesia, Worksop S80 3HG
G0 CEL N. Evely 11 St.Margarets New Road, Teignmouth TQ14 8UE
G0 CEM T. Barry 26 Gatcombe Road, Hartcliffe, Bristol BS13 9RB
G0 CEN F. Field 19 The Maples, Nailsea, Bristol BS48 4RT
G0 CEO M. Ohta House No-0120 Marfori Ext. Puroku-01 Barangay Masiit, Calauan-City 4012 Philippines
G0 CEP D. Craker Brynamlwg, Llanddewi Brefi, Tregaron SY25 6PE
G0 CEQ D. Griffiths 297 Shurland Avenue, Barnet EN4 8DQ
G0 CER D. Harris 9 Garden City, Tern Hill, Market Drayton TF9 3QB
GW0 CES M. Smith 8 Ridgeway Avenue, Marford, Wrexham LL12 8ST
G0 CEU C. Hawkins 3 Offord Close, Tottenham, London N17 0TE
G0 CEV K. Towers 7 Copeland Road, Hucknall, Nottingham NG15 8EB
G0 CEW P. Allanson 16 Woodhouse Lane, Kirkhamgate, Wakefield WF2 0SE
G0 CEY G. Cadey 45 St. Mildreds Road, Westgate-on-Sea CT8 8RJ
G0 CFB R. Tyler The Firs, Laundry Lane Huntingfield, Halesworth IP19 0PY
GM0 CFC G. Percival 5 Murrell Terrace, Aberdour, Burntisland KY3 0XH
G0 CFD F. Dimmock The Cottage, Lutton Garnsgate, , Long Sutton PE12 9JP
G0 CFI R. Deans 23 Aldringham Park, Aldringham, Leiston IP16 4QZ
GM0 CFK C. Knight 8 Ednam Drive, Glenrothes KY6 1NA
G0 CFM B. Robbins 46 Purton Close, Kingswood, Bristol BS15 9ZE
G0 CFN T. Hastings 1 Pottle Close, Botley, Oxford OX2 9SN
G0 CFT T. Feaviour 57 Elizabeth Way, Felixstowe IP11 2PQ
GM0 CFW K. Moffat 11 Russell Court, Lochgelly KY5 9EU
G0 CGA W. Roberts 10 The Meadows, Station Road, Hodnet TF9 3QF
G0 CGD C. O'Neill 4 Bronte Walk, Backford, Chester CH1 6QJ
G0 CGE B. Griffin 75 Greenham Wood, Bracknell RG127WH
G0 CGH R. Hurley 57 Cannons Close, Bishop's Stortford CM23 2BQ
G0 CGI B. Redfern The Old Blacksmiths Forge, Old Newton, Craven Arms SY7 9PG
G0 CGM A. Cutcliffe 9 Dixon Close, Paignton TQ3 3NA
G0 CGO H. Jenman Royal Hospital Chelsea, Providence House, Royal Hospital Road, London SW3 4SR
G0 CGQ R. Hazlewood 9 The Brambles, Haslington, Crewe CW1 5RA
G0 CGS D. Morris 2 Willow Close, Brinsworth, Rotherham S60 5JU
G0 CGT B. Birch 59 Shepperson Road, Sheffield S6 4FG

G0 CGW G. Ashcroft 49 Tantallon, Birtley, Chester le Street DH3 2JG
G0 CGZ G. Allison 24 Southfield Road, Scartho, Grimsby DN33 2PL
G0 CHB B. Small 3 Katherine Crescent, Skegness PE25 3LF
G0 CHC M. Smith 12 High Street, West Wickham, Cambridge CB21 4RY
G0 CHE K. Piper Flat 5, Marine Court, 4 Marine Drive West, Bognor Regis PO21 2QA
G0 CHG R. Moate Garth House, Redbrook Street, Ashford TN26 3QS
G0 CHJ R. Poulter 19 Homestead Way, Winscombe BS25 1HL
G0 CHK R. Kerby Blackboy Lane, Chichester PO18 8BE
G0 CHL K. Dewhurst Rock Cottage, 82 New Street, Stoke-on-Trent ST8 7NW
GM0 CHM J. Stephen 26 Douglas Avenue, Elderslie, Johnstone PA5 9NE
G0 CHN J. Sutton 40 Dane Valley Road, Margate CT9 3RX
G0 CHO C. Ousbey 30 Hawthorn Way, Shipston-on-Stour CV36 4FD
G0 CHP R. Angus 25 Jellicoe House, Capstan Road, Hull HU6 7AS
G0 CHQ J. Pepper 7 East Towers, Pinner HA5 1TN
G0 CHR A. Steward 94 Whittington Avenue, Hayes UB4 0AE
G0 CHV T. Sherriff 5 Pembroke Avenue, Morecambe LA4 6EJ
G0 CHY P. Moulton 4 The Ridge, Withyham Road, Tunbridge Wells TN3 9QU
G0 CIG A. Aldridge 15 Doubletrees, St. Blazey, Par PL24 2LD
GM0 CII W. Rogers 170 Boswall Parkway, Edinburgh EH5 2JJ
G0 CIM S. Dodd 61 Church Road, Hove BN3 2BP
G0 CIR R. Jasper 84 Rose Green Road, Bognor Regis PO21 3EQ
G0 CIT G. Davies 18 Cheltenham Place, Brighton BN1 4AB
G0 CIX R. Wilton 30 Barrington Crescent, Birchington CT7 9DF
G0 CJA L. Harper 23 Beech Avenue, Hazel Grove, Stockport SK7 4QP
G0 CJD K. Spratley 92 Plantation Hill, Worksop S81 0QN
G0 CJG D. Slatter 13 Hill Burn, Henleaze, Bristol BS9 4RH
G0 CJO R. Nelson Flat 1, 17 Ashburnham Road, Hastings TN35 5JN
G0 CJQ R. Saw 33 Rectory Lane, Southoe, St. Neots PE19 5YA
G0 CJV C. Scholey 11A Guildford St., Grimsby DN32 7PL
G0 CJX J. Stott 12A Henley Close, Saxmundham IP17 1EY
G0 CJZ R. Waller 29 Valley Drive, Withdean, Brighton BN1 5FA
G0 CKA G. Waller The Pines, 18 Old London Road, Brighton BN1 8XQ
G0 CKD F. Wall 37 Newton Way, St. Osyth, Clacton-on-Sea CO16 8RQ
G0 CKE J. Centanni 5 Mickle Meadow, Water Orton, Birmingham B46 1SN
G0 CKF B. Woodward 10 Darley Close, Kilham YO25 4UA
G0 CKH A. Affolter 12 Newfound Drive, Norwich NR4 7RY
G0 CKI K. Tarbett 20 Leeholme, Houghton le Spring DH5 8HR
GW0 CKK D. Robinson Island View Caravan Park, Beach Road, Penarth CF64 5UG
GW0 CKL I. Gulyas 2 Eglwysilan Way, Abertridwr, Caerphilly CF83 4EQ
G0 CKM C. Gee 6 Canterbury Close, Dukinfield SK16 5RT
G0 CKV O. Lundberg Rowan House, Cavendish Road, Weybridge KT13 0JW
G0 CLC E. Rogers Room 21, Bodmeyrick Residential Home, Holsworthy EX22 6HB
G0 CLD H. Davey 6 Cambridge Grove, Otley LS21 1DH
G0 CLG S. Haynes 9 Heather Gardens, Belton, Great Yarmouth NR31 9PP
G0 CLH D. Lingard 17 Feltwell Road, Methwold Hythe, Thetford IP26 4QJ
G0 CLJ S. Banks 15 Hunters Way, Saffron Walden CB11 4DE
G0 CLM A. Greig 98 Appletree Lane, Redditch B97 6TS
G0 CLR R. Hunt 1 Avon View, Cotswold Grange Country Park, Twyning, Tewkesbury GL20 6DL
G0 CLT P. Hodgson 14 Catherine Howard Close, Thetford IP24 1TQ
G0 CLV S. Barr 11 Mallard Way, Wirral CH46 7SJ
G0 CLX G. Glotham 89 Mellish Road, Walsall WS4 2DF
G0 CMB R. Burling 28 Croydon Road, Arrington, Royston SG8 0DJ
GW0 CMI K. Voller Machlud Haul, Tanygroes, Cardigan SA43 2JF
G0 CMK N. Sherwood The Orchards, Barton Road, Brigg DN20 8SH
G0 CMM J. Bell The Coach House, Ellesmere Road, Weybridge KT13 0HY
GM0 CMO M. Murray 1 Gordon Road, Edinburgh EH12 6NB
G0 CMP V. Warren 129 Market Road, Thrapston, Kettering NN14 4JT
G0 CMR R. Melia 14 Friar Park Road, Wednesbury WS10 0TB
G0 CMT E. Gadeberg Hojmarksvej 35, Po Box 56, Horsens 8700 Denmark
G0 CMU C. Beecham Moorview, 2 Endor Crescent, Ilkley LS29 7QH
G0 CMW J. Groom Windyridge, High Road, Maidenhead SL6 9JF
G0 CNA M. Wyatt 32 Stafford Road, Bridgwater TA6 5PH
G0 CND I. Sharrott 31 The Fleet, Stoney Stanton, Leicester LE9 4DZ
G0 CNG C. Roberts 72 Nairn Road, Walsall WS3 3XB
GW0 CNJ S. Edwards 31 Eagleswell Road, Boverton, Llantwit Major CF61 2UG
GW0 CNK G. Orchard 8 Glyn Avenue, Prestatyn LL19 9NN
G0 CNL C. Rogers 34 Martin Court, Werrington, Peterborough PE4 6JS
GM0 CNP J. Mullen 46 Templars Crescent, Kinghorn, Burntisland KY3 9XS
G0 CNU J. Mckenzie 4 Hadrian Court, Humshaugh, Hexham NE46 4DE
G0 CNV T. Mcmanus 84 Beverley Road, Hessle HU13 9BP
GM0 CNW H. Taylor 20 Woodhaven Avenue, Wormit, Newport-on-Tay DD6 8LF
G0 COA G. Coates 1 Ash Brow, Flockton, Wakefield WF4 4TE
G0 COC R. Vallis Bartley Villa, Southampton Road, Southampton SO40 2NA
G0 COE E. Coe 20 Bitterne Way, Lymington SO41 3PB
G0 COG K. Stringer 33 Brookes Road, Flitwick, Bedford MK45 1BU
GW0 COH J. Rixon Bro Afallon, Rhoslefain, Tywyn LL36 9LY
G0 COI J. Atkinson 90 Priors Road, Cheltenham GL52 5AN
G0 COJ B. Ellery 384 Sutton Way, Great Sutton, Ellesmere Port CH66 3LL
G0 COL C. Shepherd 9 Wrea Head Close Scalby, Scarborough YO13 0RX
G0 COQ W. Kelsey-Stead 65/4 Moo 2, Rawai, Phuket 83130 Thailand
GW0 COU S. Jones 635 Clydach Road, Ynystawe, Swansea SA6 5AX
G0 COY T. Frearson 31 Paradise Street, Rugby CV21 3SZ
G0 COZ J. Wayman 1 Waterloo Close, Bredon, Tewkesbury GL20 7WL
G0 CPA A. Prichard 1 Poltondale, Swindon SN3 5BN
G0 CPD G. Forster 2 Rockingham Close, Birchwood, Warrington WA3 6UY
G0 CPF I. Whyte 36 Chestnut Road, Ashford TW15 1DG
G0 CPJ P. Walker 46 Ribble Avenue, Freckleton, Preston PR4 1RX
G0 CPN I. Gurton 28 Bloomfield Road, Harpenden AL5 4DB
G0 CPO S Norm Alfreton c/o G. Childe 20 Glenmore Drive, Stenson Fields, Derby DE24 3HE
G0 CPP J. Linfoot Flat, 10 Pembroke Court, Oxford OX4 1BY
G0 CPR J. Stewart 45 Dawn Crescent, Upper Beeding, Steyning BN44 3WH
G0 CPT A. Fielder 46 Route De Pontivy, 22570 Plelauff, Plelauff 22570 France
G0 CPU M. Cracknell 17 Windmill Fields, Harlow CM17 0LQ
G0 CPV S. Pocock 14572 West, 152Nd Place, Kansas 66062 United States
G0 CPZ B. Adams 85 Copperfields, Lydd, Romney Marsh TN29 9UU
G0 CQB International Shortwave League c/o G. Cant 4 The Mount, Docking, King's Lynn PE31 8LN
G0 CQC Prof. N. Wilding Lyncombe Ridge Lyncombe Vale Road, Bath BA2 4LP
G0 CQD A. Cooper 10 Buckthorne Court, East Ardsley, Wakefield WF3 2DD
G0 CQH T. Neal 34 Dene View, Ashington NE63 8JF
G0 CQI G. Baker 26 Gardeners Road, Halstead CO9 2TB
G0 CQJ J. Kilmister 9 Wheal Jane Meadows, Threemilestone, Truro TR3 6EN
G0 CQK J. Coombes 22 Chollerford Close, Gosforth, Newcastle upon Tyne NE3 4RN
GM0 CQL P. Rudd 41 Broadford Terrace, Broughty Ferry, Dundee DD5 3EF
G0 CQO K. Clark Flat 9, Middlewood Hall, Doncaster Road, Barnsley S73 9HQ
G0 CQP Dr R. Berrisford 19 Moorlands Drive, Mayfield, Ashbourne DE6 2LP
GM0 CQQ A. Herring Mountpleasant, 8 Linlithgow Road, Bo'ness EH51 0DD
G0 CQR P. Smith 93 Nottingham Road, Long Eaton, Nottingham NG10 2BY
G0 CQS S. Faulkner 96 Ashby Road East, Bretby, Burton-on-Trent DE15 0PT
G0 CQT J. Aulsebrook 38 South Road, Beeston, Nottingham NG9 1LY
G0 CQU M. Dean 3 Buxton Road West, Disley, Stockport SK12 2AE
GM0 CQV B. Hynes 1 Hillside Court, Hillside Place, Peterculter AB14 0TU
G0 CQY V. Tharp 14 Cumberland Road, Congleton CW12 4PH
G0 CQZ N. Gardner 9 Curbridge Road, Witney OX28 5JT
G0 CRB A. Paddock 15 Castle Close, Henley-in-Arden B95 5LR
G0 CRD M. Wallis Quernmore, Hammer Lane, Hailsham BN27 4JL
G0 CRE A. Green 59 London Road, Sleaford NG34 7LQ
G0 CRF T. Bailey 65 Edge Lane, Chorlton Cum Hardy, Manchester M21 9JU
G0 CRJ A. Reader 8 Daddyhole Road, Torquay TQ1 2ED
G0 CRK D. Bowles 9 The Meadows, Breachwood Green, Hitchin SG4 8PR
G0 CRL K. Moate 32 Bournewood, Hamstreet, Ashford TN26 2HL
G0 CRN B. Hopper 189 Western Road, Mickleover, Derby DE3 9GT
G0 CRO R. Crow 20 Victoria Grove, Wombourne, Wolverhampton WV5 9AJ
G0 CRT N. Jelley 64 Leicester Road, Broughton Astley, Leicester LE9 6QE
G0 CRU K. Tham Flat 4, Warwick Court, 4 Lansdowne Road, London SW20 8AP
G0 CRX K. Pearson Kalmia, Bishampton Road, Worcester WR7 4BT
G0 CRY T. Sparrey 16 Rosemary Road, Parkstone, Poole BH12 3HB
G0 CSK P. Senior 13 St. Michaels Avenue, Swinton, Mexborough S64 8NX
GM0 CSN R. Trussler 19 Royellen Avenue, Hamilton ML3 8QH
G0 CSS H. Shillitto 25 Commonside, Selston, Nottingham NG16 6FN
G0 CSU T. Chadwick 9 Ernest Street, Prestwich, Manchester M25 3HZ
G0 CSV J. Capindale 2 Rivan Grove, Grimsby DN33 3BL
G0 CSW W. Hudson 9 Nethergate, Dudley DY3 1XW
G0 CSY W. Dunstan 5 Twemlow Lane, Holmes Chapel, Crewe CW4 8DT
G0 CSZ F. Dinger 5A Knockroe, Delgany A63HK03 Ireland
G0 CTC E. Finnesey 17 Gilbert Close Formby, Liverpool L37 6FA
G0 CTD A. Shipperley 72 Hithercroft Road, Downley, High Wycombe HP13 5RH
G0 CTF W. Sargent Likoma, 32 Seaton Down Road, Seaton EX12 2SB
GW0 CTG G. Darrell 10 Clatter Brune Estate, Presteigne LD8 2LB
G0 CTH H. Churchman Westcroft, Church Road, Chester CH4 9NG
GI0 CTI M. Murphy 97 Longfield Road, Mullaghbawn, Newry BT35 9TX
G0 CTP J. Smart 4, Sycamore Close, Holmes Chapel CW4 7BT
G0 CTQ I. Whiffin 42 Canute Road, Birchington CT7 9QH
G0 CTR P. Martin 32 Warkworth Court, Ellesmere Port CH65 9EN
G0 CTS R. Gumb 17 Castle Lane Bolsover, Chesterfield S44 6PS
G0 CTZ E. Gray 18 Shepherds Court, Sheep House, Farnham GU9 8LF
G0 CUA M. Dissanayake 9 Sweyn Place, Blackheath, London SE3 0EZ
G0 CUB G. Smith 55 Countess Way, Euxton, Chorley PR7 6PT
G0 CUH S. Crane 70 Highertown, Truro TR1 3QD
G0 CUI L. Hobson 25 Bevan Close, Elsecar, Barnsley S74 8DR
G0 CUL B. Chilvers 99 Links Avenue, Hellesdon, Norwich NR6 5PQ
G0 CUN P. Connolly 21 Hartwood Green, Hartwood, Chorley PR6 7BJ
G0 CUO J. Hewitt Peel House, Sacriston Lane, Durham DH7 6TF
G0 CUX R. Bedford 26 Devon Avenue, Fleetwood FY7 7EA
G0 CUZ C. Morris 12 Turners Hill Road, Lower Gornal, Dudley DY32JU
G0 CVA C. Andrews 10 Hartford Road Hartley Wintney, Hook RG27 8QW
G0 CVB K. Albon 65 Belmont Road, Kirkby-In-Ashfield, Nottingham NG17 9DY
G0 CVC E. Buckley 34 Newstead Terrace, Halifax HX1 4TA
GM0 CVD F. Nicholl Trees, 7 Holmisdale, Isle of Skye IV55 8WS
G0 CVH V. Hughes 27 Billy Lane, Clifton, Manchester M27 8FS
G0 CVI S. Khandro 107 Castle Hill Gardens, Torrington EX38 8EX
G0 CVM W. Auty 22 Hurley Close, Great Sankey, Warrington WA5 1XG
G0 CVN I. Howsham West Street, North Kelsey, Lincoln LN7 6EL
G0 CVO Dr P. Jukes 30002 Aralia Drive, Fulshear TX 77423 United States
GM0 CVP O/B/O Rhosllannerchrugog Group c/o C. Phillips Lonnie Sanday, Orkney KW17 2BA
GW0 CVY D. Edwards 26 Russell Terrace, Carmarthen SA31 1SY
G0 CWA N. Strong 46 Malpas Drive, Great Sankey, Warrington WA5 1HN
G0 CWB M. Starkey 23 Arthur Street, Cannock WS11 5HD
G0 CWD I. Donachie 72 Gresham Road, Norwich NR3 2NQ
G0 CWF M. Warr 17 Moray Close, Peterlee SR8 1DQ
G0 CWH A. Carden Hazelgrove, South Allington, Kingsbridge TQ7 2NB
G0 CWL D. Chapaton 1 Rue De L'Angile, Lyon 69005 France
G0 CWO L. Gough 7 Congreve Road, Stoke-on-Trent ST3 2HA
G0 CWP K. Stather 5 St. Margarets Road, Bolton Le Sands, Carnforth LA5 8EN
G0 CWQ A. Nesbitt Amberley, Gabwell Cross, Stokeinteignhead, Newton Abbot. TQ12 4QS
GM0 CWR W. Pentland Cambir, Faskally, Pitlochry PH16 5LA
G0 CWS A. Hattersley The Bungalow, Top Lane, Buxton SK17 8LP
G0 CWU F. Humphreys 31 Springfield Way, Oakham LE15 6QA
G0 CWW R. Gooden 39 Heath Road, Ipswich IP4 5RZ
G0 CWX M. Shotter Peverley, Newport Road, Cowes PO31 8PE
GW0 CWZ T. Cattley Yew Tree Cottage, Ifton Heath, Oswestry SY11 3DH
G0 CXD T. Moore 13 Moss Lane, Hulland Ward, Ashbourne DE6 3FB
G0 CXJ A. Beasley 2 Ilmington Road, Blackwell, Shipston-on-Stour CV36 4PG
GW0 CXK F. Johns Manteg, Penslade, Fishguard SA65 9PB
G0 CXO S. Wellings 11 Matlock Road, Walsall WS3 3QD
G0 CXU R. Pitter 57 Greenhill Way, Farnham GU9 8TA
G0 CXV R. Clark Woodlands, Islet Road, Maidenhead SL6 8HT
G0 CXW T. Pearce 16 Beech Lodge, Rosewoodlane, Shoeburyness SS3 9FA
G0 CXX Wanlip Rad Club c/o J. Jopling 54 Redesdale Gardens, Gateshead NE11 9XH
GM0 CXY Castlerock ARS c/o C. Monteith 46 Lochryan Street, Stranraer DG97HR
G0 CYB P. Kelsall 200 Town Street, Middleton, Leeds LS10 3TJ
G0 CYC R. Curtis 125 Handside Lane, Welwyn Garden City AL8 6TA
G0 CYD W. Snow Willbeard Farm, Greenditch Street, Bristol BS35 4HJ

GW0 CYG D. Davies 40A Furzeland Drive, Neath SA10 7UG
G0 CYI J. Knight Urbanizacao Quinta Da Torre, Edificio Perola, Armacao de Pera 8365-184 Portugal
GW0 CYK S. Radford 11 South Parade, Maesteg CF34 0AB
G0 CYN R. Blackmoore 4 Haycocks Close Dothill Wellington, Telford TF13NN
G0 CYO G. Beddow 12 Wulfruna Gardens, Finchfield, Wolverhampton WV3 9HZ
G0 CYR D. Bramley 10 Thirlmere Close, Huncoat, Accrington BB5 6JQ
G0 CYX J. Faulkner 11 Valley View, South Elmsall, Pontefract WF9 2DD
G0 CZD M. Kinder 13 Oak Bank Close Willaston, Nantwich CW5 7JA
G0 CZR K. Laws 6 Crabbe'S Close Feltwell, Thetford IP26 4BD
G0 CZU H. Richardson 14 Melbreak Close, Whitehaven CA28 9TG
G0 CZY P. Heath The Cottage, Great Staughton Road, Bedford MK44 2BA
G0 DAB D. Buik 54A Buckshaft Road, Cinderford GL14 3DZ
G0 DAC D. Cowley 81 Ashtree Road, Walsall WS3 4LS
G0 DAE C. Tidwell 86 Powerscourt Road, North End, Portsmouth PO2 7JW
G0 DAF J. Randall 26 Marian Road, Boston PE21 9HA
G0 DAH B. Smith 1 High Street Over, Cambridge CB24 5NB
G0 DAI D. Isom 36 Deerfold, Astley Village, Chorley PR7 1UH
G0 DAS D. Spreadbury Little Reeds, Ford Lane, Trottiscliffe, West Malling ME19 5DP
G0 DAU M. Saunders 22 Humphreys Close, St. Cleer, Liskeard PL14 5DP
G0 DAV D. Paine Woodland View, St. Mellion, Saltash PL12 6RH
G0 DAX D. Burt 19B Midhurst Road, Eastbourne BN22 9HP
G0 DAY K. Banks 52 Hunter Avenue, Burntwood WS7 9AQ
G0 DAZ C. Mister Woodbine Cottage, Hanley Childe, Tenbury Wells WR15 8QY
G0 DBC J. Hudson 1 Linnet Way, Biddulph, Stoke-on-Trent ST8 7UF
G0 DBD J. West Stonecroft, 4 Trevella Road, Bude EX23 8NA
G0 DBE L. Marsland 154, Moss Lane, Litherland., Liverpool L21 7NN
G0 DBI K. Danks 28 Warnes Lane, Burley, Ringwood BH24 4EL
G0 DBJ G. Willetts Waterside, 48 Stourton Crescent, Stourbridge DY7 6RR
GM0 DBK D. Kerr 3 Glaisnock View, Cumnock KA18 3GA
G0 DBM S. Lovesey 10K Streamside, Slad Road, Stroud GL5 1UT
G0 DBP S. Hender 23 Broadacres, Honley, Holmfirth HD9 6ND
G0 DBS P. Le Feuvre 56 Greenfields Avenue, Alton GU34 2EE
GM0 DBW M. Bolton 11 Covenanters Drive, Corston, Aberdeen AB12 5AB
G0 DBX D. Beale 17 Rue Du Passolis, Montseret 11200 France
G0 DBY P. Dodge 425 Sutton Road, Maidstone ME15 8RA
G0 DCF R. Green Kenville, West Lane, Sheffield S26 3XS
G0 DCG M. Brown 31 Victoria Road, Littlestone, New Romney TN28 8NL
G0 DCI A. Merrylees 90 Grangehill Road, Eltham, London SE9 1SE
G0 DCJ J. Warren The Old Barn, Scotgate Close, Thetford IP24 1PF
G0 DCO M. Beirne 14 Swiss Cottage, Bollinbrook Road, Macclesfield SK10 3DJ
G0 DCP P. Holdaway 8 Beaufort Avenue, Market Deeping, Peterborough PE6 8JD
G0 DCR J. Frost 36 York Gardens, Braintree CM7 9NF
G0 DCS P. Ashton 27 Dunsby Road, Luton LU3 2UA
G0 DCU J. Faithfull 54 Cardiff Place, Bassingbourn, Royston SG8 5LR
G0 DCW A. Corallini 8 Britannia, Puckeridge, Ware SG11 1SY
G0 DCZ P. Richards 16 Charles Eaton Court, Bedworth CV12 0AX
G0 DDA P. White 11 Elms Road, Fareham PO16 0SQ
G0 DDE B. Dignum 16 Stirling Court Road, Burgess Hill RH15 0PT
G0 DDJ A. King 4 Tyne Close, Wellingborough NN8 5WT
GW0 DDK E. Down Silver Hill, Pen Y Cwm, Haverfordwest SA62 6JZ
GW0 DDL G. Jones 9 George St., New Quay SA45 9QR
G0 DDT J. Barnes 262 King Henrys Drive, New Addington, Croydon CR0 0AA
G0 DDU J. Norris 15 Liverpool Road North, Burscough, Ormskirk L40 5TN
G0 DDV N. Hewett 49 Harrow Way, Carpenders Park, Watford WD19 5EH
G0 DDW M. Hillier 28 Meadow Walk, Bridgemary, Gosport PO13 0YN
G0 DDY P. Piper 5 Goodwood Close, Midhurst GU29 9JG
G0 DDZ M. Eastman 23 Haughgate Close, Woodbridge IP12 1LQ
G0 DEB D. Bannister 60 St. Johns Avenue, Bridlington YO16 4NL
G0 DEC D. Willicombe 26 Falkland Court, Braintree CM7 9LL
G0 DEE R. Spilling 20 Saxonfields, Poringland, Norwich NR14 7JE
G0 DEF M. Mutton 39 Martin Road, Kettering NN15 6HF
G0 DEH P. Finbow 6 Down Road, Teddington TW11 9HA
G0 DEJ W. Rutt 24 Coopers Lane, Verwood BH31 7PG
G0 DEK F. Underwood Hobletts, Fen Lane, Grays RM16 3LT
GM0 DEM A. Jones 9 Firs Street, Falkirk FK2 7AY
G0 DEO W. Batey 13 Cassiobury Avenue, Feltham TW14 9JE
G0 DEP D. Jarrard 26 Lingmell Court, Tolladine, Worcester WR4 9YU
GM0 DEQ R. Alexander 9 Weston Place, Prestwick KA9 2ED
G0 DER D. Gillmore 4 Holly Ridge, Fenns Lane, Woking GU24 9QE
G0 DEU J. Tournant 47 High St., Linton, Cambridge CB1 6HS
GM0 DEX A. Goldie 87 Ardrossan Road, Seamill, West Kilbride KA23 9NF
G0 DEZ D. Watson 1 Castle House Drive, Stafford ST16 1DS
G0 DFA D. Allsopp 10 Chalfont Close, Middleton-On-Sea, Bognor Regis PO22 7SL
G0 DFC L. Cropley San Ferryann, Brundish Road, Wilby, Eye IP21 5LS
GI0 DFD R. Mcalister 78 Cairn Road, Carrickfergus BT38 9AP
G0 DFE J. Stone 12 Main Road, Hawkwell, Hockley SS5 4JN
G0 DFF J. Sidnell 17 Barlings Road, Harpenden AL5 2AL
G0 DFI D. Oakley 6 Staplehurst Gardens, Cliftonville, Margate CT9 3JB
G0 DFO J. Tomlinson 33 Belgrave St., Nelson BB9 9HR
G0 DFT J. Maw 10 Shamrock Close, Newcastle upon Tyne NE15 8TW
G0 DFV R. Cadd 27 Grove Road, Brafield On The Green, Northampton NN7 1BW
GW0 DFY R. Anthony 2 Barrfield Road, Rhuddlan, Rhyl LL18 2RY
G0 DGA D. Gullick Greenleas, Courthay Orchard, Langport TA10 9AE
G0 DGB L. Connell 24 Finchale Road, Framwellgate Moor, Durham DH1 5JN
G0 DGE J. Gullick Greenleas, Courthay Orchard, Langport TA10 9AE
G0 DGF M. Beakhust 63 Chadacre Road, Epsom KT17 2HD
G0 DGH G. Daniels 81 London Road, Clacton-on-Sea CO15 3SR
GW0 DGJ C. Riddle Flat 18, Windsor Mews, Adamsdown Square, Cardiff CF24 0HS
GM0 DGK A. Donaldson 30 Jeanfield Crescent, Forfar DD8 1JR
G0 DGQ G. Dudley 95 Alfreton Road, South Normanton, Alfreton DE55 2BJ
G0 DGU P. Brown 17 Freyden Way, Frettenham, Norwich NR12 7NB
G0 DGW D. Barham 64 Gorran Avenue, Rowner, Gosport PO13 0NF
GW0 DHA R. Waller 4 Rose Court, Ty Canol, Cwmbran NP44 6JH
G0 DHB R. Evans 20 Pulley Avenue, Eaton Bishop, Hereford HR2 9QN
GM0 DHD A. Lymer 16 Gerson Park, Greendykes Road, Broxburn EH52 6PL
GW0 DHG B. Ristic 168 Heather Road, Newport NP19 7QW
G0 DHI Wakefield And District RS c/o A. Rutherford 19 Briar Bank, Carlisle CA3 9SN
G0 DHJ J. Wraight 59 Sandy Lane, Walton, Liverpool L9 9AY
G0 DHL M. Mason 7 Clayhill Copse Peatmoor, Swindon SN5 5AL
G0 DHM D. Moore Stoke Hall Farm, Stoke On Tern, Market Drayton TF9 2DU
G0 DHR Prof. H. Rutt 3 Russell Place, Highfield, Southampton SO17 1NU
G0 DHS A. Kitching 1 Borrowdale, Albany, Washington NE37 1QD
G0 DHT K. Smith Lower Carniggey Farm, Greenbottom, Truro TR4 8QL
GI0 DHW C. Cleland 19 Sheskin Way, Belfast BT6 0ER
G0 DHZ H. Johnsen 47 Bondfields Cres, Havant PO9 5ER
G0 DIA A. Holland 156 Perry Rise, London SE23 2QP
G0 DIG P. Johnson 5 Brook Bank, Whitehaven CA28 8PZ
G0 DIH P. Strong 2 Jasper Cottages, Cornworthy, Totnes TQ9 7EY
G0 DIM A. Latham 49 Tithe Barn Road, Wootton, Bedford MK43 9EZ
G0 DIP J. Huggins 7 Coniston Drive, Jarrow NE32 4AE
GW0 DIQ M. Smith 7 Clos Gorsfawr, Grovesend, Swansea SA4 4GZ
G0 DIR S. Caslake Bishopwood Cottage, Wistow Common, Selby YO8 3RD
G0 DIS P. Pearce 42 Sinclair Garth, Wakefield WF2 6RE
G0 DIU T. Rumble 1 Victoria St., Brighouse HD6 1HH
GW0 DIV R. Griffiths 5 Heol-Y-Sarn, Llantrisant, Pontyclun CF72 8DA
GW0 DIX R. Rees 22 The Complex, Tan Y Bryn, Burry Port SA16 0HP
G0 DIY R. Allen 65, Fitzroy Drive, Leeds LS8 4AG
G0 DIZ R. Quaintance 18 Queens Avenue, Ilfracombe EX34 9LN
G0 DJA D. Ackrill 59 Moor Lane, Bolsover, Chesterfield S44 6EW
G0 DJC D. Collins 71 Trench Road, Tonbridge TN10 3HG
G0 DJF P. Griffin 63 Henley Crescent, Solihull B91 2JD
GM0 DJG J. Walker Mo Bhothan, Lochlibo Road, Uplawmoor, Glasgow G78 4AA
G0 DJK D. Keates 13 Willow Rise, Witham CM8 2LL
G0 DJL C. Blount 42 Penmere Drive, Newquay TR7 1QQ
G0 DJM M. Rawson 4 Fleet Lane, Tockwith YO26 7QD
G0 DJO A. Robson 19 Barnard Close, Bedlington NE22 6NE
G0 DJQ R. Salt Croft View, Little Cubley Cubley, Ashbourne DE6 2FB
G0 DJS H. Stemp 5 Depot Road, Horsham RH13 5HB
G0 DJT P. Odegaard Flat 4, 31 Birdhurst Road, South Croydon CR2 7EF
G0 DJV B. Joy 1 Riverbourne Road, Salisbury SP1 1NU
GW0 DJX A. Cullen 30 Llys Cyncoed, Oakdale, Blackwood NP12 0NQ
GW0 DKF R. Thomas 48 Maryport Street, Usk NP15 1AD
GW0 DKG R. Malpas 26 St. Davids Avenue, Whitland SA34 0AF
GM0 DKK A. Mackenzie 33 Castle Crescent, Denny FK6 6PN
G0 DKM S. Daniels 3 Warren Close, Hutton, , Weston-Super-Mare BS24 9QX
G0 DKN A. Mcclelland 12 Grove Heath North, Ripley, Woking GU23 6EN
G0 DKO G. Maskort 133 Borstal St., Rochester ME1 3JU
G0 DKR M. Cole 54 Ribble Road, Coventry CV3 1AU
G0 DKS R. Barnes Pentwyn, Graeme Road, Yarmouth PO41 0RX
G0 DKV E. Peberdy 18 Arden Road, Kenilworth CV8 2DU
G0 DKX G. Birk 30 Maple Drive, Alvaston, Derby DE24 0FT
G0 DKY J. Bass 8 Ann Close, Hassocks BN6 8NB
G0 DKZ B. Yates 9 Cloister Walk, Whittington, Lichfield WS14 9LN
GW0 DLA E. Stuckey 32 Stanley Road, Gelli, Pentre CF41 7NJ
G0 DLB R. Burdett 11 Fisher Avenue, Rugby CV22 5HN
G0 DLF A. Keon 72 Rochelle Way, Duston, Northampton NN5 6YW
G0 DLL W. Jackson 22 Cliff Gardens, Scunthorpe DN15 7PJ
G0 DLP P. Lee 3Holm Oaks Burndell Road, Arundel BN180GB
G0 DLS D. Sparey 21 Buxton Road, Ashbourne DE6 1EX
G0 DLT J. Aizlewood 7 Scott Avenue, Simonstone, Burnley BB12 7HY
GW0 DLW J. Goldsmith Woodlands, Candy, Oswestry SY10 9AZ
G0 DMA J. Slaney 50 Laburnum Road, Langold, Worksop S81 9RR
G0 DMB C. Cade 6 Court Close, Kirby Muxloe, Leicester LE9 2DD
G0 DME J. Hallam 34 Danethorpe Vale, Nottingham NG5 3DA
G0 DMH B. Cox 21 Shelthorpe Road, Loughborough LE11 2PB
G0 DMJ N. Beck 36 Grove Street, Great Hale, Sleaford NG34 9JZ
G0 DMK D. King 94 Western Road, Mickleover, Derby DE3 9GQ
G0 DMN S. Langham 43 Greenwood Drive, Kirkby-In-Ashfield, Nottingham NG17 8JT
G0 DMP D. Potter 102 Normandy Avenue, Beverley HU17 8PF
G0 DMS F. Spencer 35 Askew Grove, Repton, Derby DE65 6GR
G0 DMU I. Morison 4 Arley Close, Macclesfield SK11 8QP
G0 DMV R. Parrish 89 Delamere Drive, Macclesfield SK10 2PS
G0 DMW F. Cosgrove Denton Park Middle School, Linhope Road, Newcastle upon Tyne NE5 2NW
G0 DND N. Downs 10 Oak Street, Northwich CW9 5LJ
G0 DNF D. Fisher 69 Priors Orchard, Southbourne, Emsworth PO10 8GE
GM0 DNG G. Wallace 21 The Grange, Perceton, Irvine KA11 2EU
GM0 DNH P. Moore 105 Fintry Drive, Dundee DD4 9HQ
G0 DNI G. Wann Manor Barn, Shotatton, Shrewsbury SY4 1JH
G0 DNQ C. Anderton 19 Berrylands Close, Wirral CH46 7UT
G0 DNV R. Hammett 47 Sowden Park, Barnstaple EX32 8EJ
G0 DNY B. Whysker 21 Heyland Road, Manchester M23 1HF
G0 DOA C. Dale 11 Roman Close Newby, Scarborough YO12 5RG
G0 DOB R. Gray 3 Perth Way, Immingham DN40 1PW
G0 DOC H. Kiff Rock Cottage, Cloudside, Congleton CW12 3QG
G0 DOE T. Purcell 38A Moor Lane, Chessington KT9 1BW
G0 DOG C. Purcell 76 Wensleydale Road, Great Barr, Birmingham B42 1PL
G0 DOK R. Cornish 18 Rooksbury Croft, Havant PO9 5HU
G0 DOM D. Oskis 10 Moultrie Way, Cranham, Upminster RM14 1NB
G0 DOR P. Davidson 5 Derby Grove, Maghull, Maghull L31 5JJ
G0 DOU H. Grandfield 2 Bolshaw Road, Heald Green, Cheadle SK8 3PJ
G0 DOZ Dr J. Rozday 130 Walton Park Pannal, Harrogate, UK HG3 1RJ
G0 DPC J. Cook 88 South Avenue, Southend on Sea SS2 4HU
G0 DPE B. Aldersey 4 Salterbeck Terrace, Salterbeck, Workington CA14 5HP
G0 DPG D. Ganner 58 Kilngate, Lostock Hall, Preston PR5 5UW
G0 DPI D. Barkley 39 Fulbeck Avenue, Wigan WN3 5QN
G0 DPJ K. Ellis 7 Beech Close, Dudley DY3 1NG
G0 DPK P. Mccaldon 47 Merritt Road, Didcot OX11 7DF
G0 DPO K. Glazebrook 86 Deveraux Drive, Wallasey CH44 4DL
G0 DPQ A. Henning 24 Garfield Avenue, Draycott, Derby DE72 3NP
G0 DPS J. Fyrth 2 Merton Gardens, Farsley, Pudsey LS28 5DZ
G0 DPT M. Smith 32 Amesbury Drive, London E4 7PZ
GI0 DPV J. Mangan 141 Glen Road, Andersonstown, Belfast BT11 8BP
G0 DPW D. Welham 1 Torridge Road Keynsham, Bristol BS31 1QG
G0 DPX J. Brown 72 Whitcliffe Road, Cleckheaton BD19 3BY

G0 DPY A. Garner 7 Danes Court, Grimoldby, Louth LN11 8TA
G0 DQB R. White 27 Windsor Walk Scawsby, Doncaster DN5 8NQ
GM0 DQC H. Meikle 20 Muirsland Place, Lesmahagow, Lanark ML11 0FF
G0 DQH J. Colwill 18 Collingbourne Drive, Chandler'S Ford, Eastleigh SO53 4SW
G0 DQI D. Harding High Peak, Hillcrest Road, Deal CT14 8EB
GI0 DQJ D. Livingstone 16 Stronge Court, Portadown, Craigavon BT62 3QX
G0 DQM J. Williams Maes-Yr-Awel, Suttonfield Road, Doncaster DN6 9JX
G0 DQO W. Cartwright 50 Kings Road, Walsall WS4 1JB
G0 DQQ S. Power 10 Beach Road, Hartford, Northwich CW8 4BA
G0 DQS M. Glen 10 Field Lane, Dursley GL11 6JE
GW0 DQT A. Duck 15 Ambryn Road, New Inn, Pontypool NP4 0NJ
GM0 DQV G. George 13 Balmoral Terrace, Elgin IV30 4JH
GW0 DQW C. Hughes Cym Lane, Rogerstone, Newport NP10 9EN
GW0 DQY G. Hughes 39 Thornhill Close, Upper Cwmbran, Cwmbran NP44 5TQ
GM0 DQZ J. Butchart Woodville 49 Canaan Lane, Edinburgh EH10 4SG
G0 DRA D. Love 4 St. Chads Road, Lichfield WS13 7LZ
G0 DRD S. Carvin 43 Brackenstown Village, Dublin Ireland
G0 DRE J. Webster 3 Badby Road West, Daventry NN11 4HJ
G0 DRH B. Harris 9 Woodlands Close, Rayleigh SS6 7RG
GW0 DRI J. Smith 7 Clos Gorsfawr, Grovesend, Swansea SA4 4GZ
G0 DRJ J. Connell 24 Finchale Road, Framwellgate Moor, Durham DH1 5JN
G0 DRK K. Fox 54 Tuskar Street, London SE10 9UZ
G0 DRL G. Howarth 15C Shaftesbury Road, Southsea PO5 3JA
G0 DRM D. Cookson 70 Rope Lane, Wistaston, Crewe CW2 6RD
G0 DRN M. Jenkin 3A Westminster Street, Crewe CW2 7LQ
G0 DRO D. Roberts Flat 1, 129 Prestbury Road, Macclesfield SK10 3DA
G0 DRQ R. Hope 3 Farm Crescent, Sittingbourne ME10 4QD
G0 DRR O. Perry 9 Home Park Close, Bramley, Guildford GU5 0JP
GW0 DRS R. Ford 11 Lincoln Road, Ewloe, Deeside CH5 3RW
G0 DRT P. Quested Nethercroft, Southsea Avenue, Minster On Sea, Minster on Sea Sheerness ME12 2NH
G0 DRV W. Hoyle 10 Picton Gardens, Rayleigh SS6 7LB
G0 DRW T. Wright 73 West Street, Ryde PO33 2QQ
G0 DRX P. Hill 34 Church Road, Whitchurch, Bristol BS14 0PP
G0 DSB T. Winship 32 Lytes Cary Road, Keynsham, Bristol BS31 1XD
GI0 DSG W. Mckeever 17 The Hawthornes, Londonderry BT48 8TH
GW0 DSJ E. Shipton 30 Stephen Road, Prestatyn LL19 7EH
G0 DSN L. Nash Four Furlongs, Wells Road, Wells-next-the-Sea NR23 1QE
G0 DSO R. Calvert Shortley Close, Robin Hoods Bay, Nr Whitby YO22 4PB
GW0 DSP M. Lamb 4 Hadfield Close, Connah'S Quay, Deeside CH5 4JP
G0 DSQ Dr K. Myint 1 Belton Road, Camberley GU15 2DE
G0 DSR D. Jones 20 Marsh Green, Wigan WN5 0PU
G0 DSX N. Hanking 7 Clayside House, Kenton Court, South Shields NE33 4HP
G0 DTC Dr D. Coxon 13 Gate Farm Road, Shotley Gate, Ipswich IP9 1QH
G0 DTI K. Sumner 7 Largs Road, Shadsworth, Blackburn BB1 2JQ
G0 DTP L. Quantrill Innisfree, 1 Ironwell Lane, Hockley SS5 4JY
G0 DTQ W. Goldstraw 5 Council Houses, Wantage Road, Hungerford RG17 7DG
G0 DTT R. Mycock Beresford Crescent, Stockport SK56NU
G0 DTW S. Bolton 40 Claytonwood Road, Stoke-on-Trent ST4 6LD
G0 DUA S. Linden 4 Downing Drive, Great Barton, Bury St. Edmunds IP31 2RP
G0 DUB F. Mossop 4 Brookdale Way, Waverton, Chester CH3 7NT
G0 DUE A. Braybrook 4 Bodinar Road, Penryn TR10 8JD
G0 DUF R. Hoare 7 Springfield Close, Watlington OX49 5RF
G0 DUG D. Marsden 127 Morley Crescent, Kelloe, Durham DH6 4NP
G0 DUH P. Murrell 10 Irving Close, Braunton EX33 1DH
G0 DUI P. Smith 20 Deanscroft Way, Stoke-on-Trent ST3 5XW
G0 DUK K. Bennett 78 Rectory Road, Upper Deal, Deal CT14 9NB
G0 DUM D. Gibbons 17 Della Avenue, Barnsley S70 6LG
G0 DUN D. Wakeford 2 Rooley House Cottage, Rooley Lane, Sowerby Bridge HX6 1NS
GI0 DUP R. Miskimmin 15 Abbeydale Avenue, Newtownards BT23 8RT
G0 DUQ R. Wilkes 47 Greenwood Park, Hednesford, Cannock WS12 4DQ
G0 DUS M. Marlow 56 Harvest House, Cobbold Road, Felixstowe IP11 7SP
GM0 DUX R. Mcgowan 4 The Cottages, Ashfield, Dunblane FK15 0JS
G0 DVB Torbay ARS c/o J. Morris 18 Ellingdon Road, Wroughton, Swindon SN4 9HY
G0 DVC P. Robinson 256 Victoria Road, Ruislip HA4 0DW
G0 DVE S. Hutchings 5 Dales Close, Wimborne BH21 2JU
G0 DVG T. Callaghan 27 Thealby Lane, Thealby, Scunthorpe DN15 9AG
GM0 DVH N. Mcnulty 6 Main Road, Crookedholm, Kilmarnock KA3 6JT
G0 DVJ J. Mitchener Cabins, Wenham Road, Ipswich IP8 3EY
G0 DVL R. Nash 28 Squires Way, Wilmington, Dartford DA2 7NW
GM0 DVO A. Gemmell 23 Busby Road, Carmunnock, Glasgow G76 9BN
G0 DVP G. Gulliford 29 Windsor Road, Seaham SR7 8DG
G0 DVQ G. De-Wilton Homes 81 Oriel Avenue, Gorleston, Great Yarmouth NR31 7JJ
G0 DVS E. Holding 31 Popple Street, Sheffield S4 8JH
G0 DVT J. Brindle 1 Holywell Close, Bury St. Edmunds IP33 2LS
GI0 DVU J. Henry 3 Kirkwoods Park, Lisburn BT28 3RR
G0 DVY R. Ibbotson Fern Lea, Alford LN13 0JP
G0 DWB D. Wathen-Blower 61 Dykes End, Collingham, Newark NG23 7LD
G0 DWC S. Beadle 18 The Shrubberies, Cliffe, Selby YO8 6PW
G0 DWD J. Hawes Cherry Lodge, Woodwaye, Reading RG5 3HA
G0 DWE C. Brown 12 Forest Close, Newport PO30 5SF
G0 DWF D. Fouche 17 Burlington Gardens, Rainham, Gillingham ME8 8TA
GM0 DWH J. Cobley 15 Fintry Terrace, Bourtreehill South, Irvine KA11 1JD
G0 DWJ N. Hall 10 Newnham Road, Leamington Spa CV32 7SN
G0 DWM C. Pearsons 12 Abbey Way, Farnborough GU14 7DA
GI0 DWN M. Dougan 97 Redrock Road, Collone, Armagh BT60 2BN
G0 DWO P. Labron 22 Fourth Avenue, Morpeth NE61 2HJ
GW0 DWQ S. Outen 2 Heol Vaughan, Burry Port SA16 0HF
G0 DWR P. Bell 5 Sages Lane, Privett, Alton GU34 3NP
G0 DWS J. Tubbs 19 Greenhill Road, Northfleet, Gravesend DA11 7EZ
G0 DWV C. Danby Fir Trees, Hall Road, Norwich NR10 3LX
GM0 DWY R. Price 80 Eastern Avenue, Largs KA30 9EQ
G0 DWZ M. Sharp 50 Milton Drive, Southwick, Brighton BN424NE
G0 DX Burnham Beeches RC c/o U. Grunewald Nuptown Orchard, Nuptown, Warfield, Bracknell RG42 6HU
GM0 DXE H. Quin Flat/Edinbane Shop, Edinbane Shop, Portree IV51 9PW
GW0 DXG Prtsmth & DARS c/o D. Stoole Brookside Farm, Baltic Terrace, Cwmbran NP44 7AH
G0 DXH D X Hunter c/o C. Mortlock 27 Baldwin Road, Greatstone, New Romney TN28 8SY
GM0 DXI H. Lakhaney 5 Snowberry Fields, Thankerton, Biggar ML12 6RJ
G0 DXK M. Bedford 12 Winchester Drive, Mablethorpe LN12 2AY
GW0 DXO D. Oates 86 Queens Avenue, Maesgeirchen, Bangor LL57 1NG
G0 DXT T. Pearson 7 Lathkill Grove, Tibshelf, Alfreton DE55 5LU
GU0 DXX P. Guilbert Chinq, La Ferme Es Frases, Rue Des Issues GY7 9FS Guernsey
GW0 DXZ G. Stephens Ty Coch, Rhydwen Place, Clydach SA6 5RN
GM0 DYD D. Young 4 Primrose Avenue, Rosyth, Dunfermline KY11 2SS
G0 DYG D. Doyle 40 Howson St., Rock Ferry, Birkenhead L42 2BR
GW0 DYH S. Fergusson 37 Station Road, Old Colwyn, Colwyn Bay LL29 9EL
G0 DYL C. Jones 63 Hockenhull Avenue, Tarvin, Chester CH3 8LR
G0 DYM M. Rivers 5 Ann Carter Close, Hereford HR2 7LS
GM0 DYU M. Mcwhinnie 35 Morrison Place Cruden Bay, Peterhead AB423HZ
G0 DYW I. Dowse 57 Palmer Crescent, Leighton Buzzard LU7 4HY
G0 DZA P. Wentworth 46 Woodside Avenue, Cinderford GL14 2DW
G0 DZB P. Onion 18 The Maples, Bedford MK42 7JX
G0 DZC C. Cosgrif 53 Lower Manor Lane, Burnley BB12 0EF
G0 DZH D. Gough 20 Lawn Close, Ruislip HA4 6ED
G0 DZI C. Kuss 20 Windermere Road, Haydock, St. Helens WA11 0ES
GW0 DZL J. Davies 5 Talbot St., Llanelli SA15 1DG
G0 DZM P. Gainey Prencott, Harley Wood, Stroud GL6 0LD
G0 DZQ B. Stevens 24 Waverley Crescent, Wickford SS11 7LN
GW0 DZU P. Barker 2 The Uplands Pontrhydyfen Port Talbot, Port Talbot SA12 9TG
G0 DZV K. Rose 57 Cheriton Road, Winchester SO22 5AX
GM0 DZW R. Webster Tigh-Na-Darroch, Old Line Road, Aberdeenshire AB35 5UT
G0 DZX J. Huggins 12 Willow Chase, North Anston, Sheffield S25 4DQ
G0 DZY B. Sparrow 11B Croft Place, Mildenhall, Bury St. Edmunds IP28 7LN
G0 DZZ J. Hall 4 Dorking Crescent, Clacton-on-Sea CO16 8FQ
G0 EAE M. Taylor 26 St. Marys Road, Bozeat, Wellingborough NN29 7JU
G0 EAG A. Sammons 35 Fernlea Avenue, Herne Bay CT6 8UQ
GM0 EAH A. Mcdougall 16 Rotherwood Avenue, Glasgow G13 2RJ
G0 EAM A. Moore 69 Renfrew Avenue, St. Helens WA11 9RW
G0 EAN C. Bibb 54 Dorsett Road, Wednesbury WS10 0JF
G0 EAT S. Anderson 65 Sands Lane, Holme-On-Spalding-Moor, York YO43 4HJ
G0 EAU A. Surooprajally 26 Walton Avenue, North Cheam, Sutton SM3 9UB
GW0 EAW J. Humphries 19 Tai Newydd, Llanfaelog, Ty Croes LL63 5TW
G0 EBD M. Element 9 Longbridge Close, Shrewsbury SY2 5YD
G0 EBF L. Taylor 18 Manifold Road, Eastbourne BN22 8EH
G0 EBG R. Fuller 41 Burnham Road, Hullbridge, Hockley SS5 6BG
G0 EBI J. Speller 43 Castle Hill Park London Road, Clacton-on-Sea CO16 9QP
G0 EBL K. Mayes The Stone House, Goathland, Whitby. YO22 5AN
G0 EBP A. Bowmaker 1 Hestham Drive, Morecambe LA4 4QD
G0 EBQ N. Flatman 2 Deben Valley Drive, Kesgrave, Ipswich IP5 2FB
G0 EBS S. Artus 14 Bramley Road, East Peckham, Tonbridge TN12 5BW
G0 EBW D. Agar 122 Salisbury Road, Moseley, Birmingham B13 8JZ
G0 EBZ M. Whitfield Camp Farm, Elberton, Bristol BS35 4AQ
G0 ECB M. Hayhurst 3 Burton Gardens, Brierfield, Nelson BB9 5DR
G0 ECG M. Higgin 24 Tiverton Drive, Briercliffe, Burnley BB10 2JT
G0 ECI R. Matthews Sunrise, Back Bank, Whaplode Drove, Spalding PE12 0TT
G0 ECJ H. Hughes Asham, Walton Hill, Gloucester GL19 4BT
G0 ECK P. Jenkins 49 Ewell Park Way, Ewell, Epsom KT17 2NW
G0 ECL K. Clift 28 Redgate Road, Girton, Cambridge CB3 0PP
G0 ECM M. Bell 18 Linnet Close, Patchway, Bristol BS34 5RN
G0 ECN K. Watkinson Sunnyview, Beacon Way, Skegness PE25 1HL
G0 ECQ C. Quinnin 10 Willow Avenue, Blyth NE24 1PG
G0 ECS G. Westaby 2 Goodwood, Bottesford, Scunthorpe DN17 2TP
GM0 ECU R. Low 56 George Street, Whithorn, Newton Stewart DG8 8NZ
G0 ECW E. Wilson 20 Wivelsfield Road, Saltdean, Brighton BN2 8FQ
G0 ECX R. Mott 2 Dennis Road, Weymouth DT4 0NJ
G0 ECZ North Yorkshire YI Group c/o B. Clarke 4 Prospect Road, Langford, Biggleswade SG18 9NY
G0 EDC B. Hall 7 Ferndale Close, Penyffordd, Chester CH4 0NH
G0 EDE D. Proctor 7 Main Avenue, Westhill, Torquay TQ1 4HZ
G0 EDF G. Lunt 45 Malvern Road, Liverpool L6 6BN
GM0 EDJ P. Temple 23 Ramsay Place, Johnstone PA5 0EX
G0 EDK A. Lee 44 Lynn Road, North Shields NE29 8HS
G0 EDO R. Ormond 75 Desford Road, Newbold Verdon, Leicester LE9 9LG
GM0 EDQ J. Shaw 28 Drumcross Road, Bathgate EH48 4HG
GM0 EDR J. Bell 52 Turnberry Road, Glasgow G11 5AP
G0 EDS Men's Shed RC c/o A. Seabrook 63 St. Annes Road Willingdon, Eastbourne BN20 9NJ
G0 EDT J. Hopwood 53 St. Marys Road, Stratford-upon-Avon CV37 6XG
G0 EDU P. Martin 20 Eastbourne Terrace, Westward Ho, Bideford EX39 1HG
G0 EDY P. Gould 53 Green Road, Kidlington OX5 2EU
G0 EEA G. Reffell 26 Barnwood Road, Gloucester GL2 0RX
G0 EEF G. Legg 12 Churchill Road, Wimborne BH21 2AU
GM0 EEG D. Mctaggart 65 Oronsay Road, Airdrie ML6 8FX
GM0 EEH J. Hunter 2 Hagen Drive, Motherwell ML1 5RZ
G0 EEI A. Jones 18 Manor Road North, Nantwich CW5 5NW
G0 EEJ A. Mitchell 18 Burnards Court, Berrycombe Road, Bodmin PL31 2NU
GI0 EEO J. Ryan 11 Carnesure Heights, Comber, Newtownards BT23 5RN
G0 EET B. Henderson 34 Paget House Grove Place Upton Lane Nursling, Southampton SO16 0AQ
GM0 EEY D. Smith 23 Torness, Kirkwall KW15 1UU
G0 EEZ C. Wright 60 Grove Crescent, Hanworth, Feltham TW13 6LZ
G0 EFA C. Mansfield 1 The Courtyard, Deal CT14 9FB
GM0 EFC G. Perry 14B Meadowfoot Road, West Kilbride KA23 9BX
GM0 EFD C. Perry 14B Meadowfoot Road, West Kilbride KA23 9BX
G0 EFG B. Balmer 13 Chillingham Crescent, Ashington NE63 8BQ
GM0 EFH A. Buchan Flat 7, 1 Castlebank Court, Glasgow G13 2LA
G0 EFI V. Newman 14 Hilltop Close, Rayleigh SS6 7TD
G0 EFL L. Cohen 6 Branksome Walk Manor Branksomewood Road, Fleet GU51 4SW
G0 EFN H. Dunne 19 Bute Brae, Bletchley, Milton Keynes MK3 7TA
G0 EFO M. Shortland 4 Hillier Road, Guildford GU1 2JQ
G0 EFP N. Hughes 43A Wellhouse Road, Beech, Alton GU34 4AQ
GM0 EFQ H. Fisher 1 Millhill Lane, Musselburgh EH21 7RD
G0 EFR L. Wheeler 29 Belmont Park, Pensilva, Liskeard PL14 5QT

G0 EFS P. Hancock 7 Carlton Avenue, Hayes UB3 4AD
GM0 EFT R. Neilson 54 Macdonald Smith Drive, Carnoustie DD7 7TB
GI0 EFW J. O'Hara 284 Foreglen Road, Dungiven, Londonderry BT47 4PJ
G0 EFY B. Warman 177 Scotter Road, Scunthorpe DN15 8AU
G0 EFZ I. Gerrard 7 High Street, Wootton, Wootton, Ulceby DN39 6SG
G0 EGC W. Stormont 3 Bridge Cottages, Greenham, Crewkerne TA18 8QE
G0 EGE W. Trotter Bungalow, Stoupe Cross Farm, Whitby YO22 4JU
G0 EGG D. Jackson 41 Colman Avenue, Wolverhampton WV11 3RT
GW0 EGH D. Taylor 81 Goldfinch Close, Caldicot NP26 5BW
GM0 EGI B. Devlin 112 Benview, Bannockburn, Stirling FK7 0HJ
G0 EGP D. Williamson 25 Balfour Crescent, Wolverhampton WV6 0BJ
GW0 EGQ G. Peters 5 Roman Way, Buckley CH7 2EQ
G0 EGR C. Davis 49 Brackendale Road, Bournemouth BH8 9HY
G0 EGT G. Pitts 119 Rusper Road, Crawley RH11 0HW
G0 EGW V. Readhead White Lodge, Rendham Road, , Saxmundham IP17 2AA
GW0 EHA I. Pemberton 26 Stanley Grove, Ruabon, Wrexham LL14 6AH
G0 EHE B. Evans 27 Mulso Road, Finedon, Wellingborough NN9 5DP
G0 EHK G. Cheetham 172A Hesketh Lane, Tarleton, Preston PR4 6AT
GM0 EHL B. Armstrong 31 Old Abbey Road, North Berwick EH39 4BP
G0 EHO R. Mellor11 1 The Square, Lybury Lane, Redbourn, St. Albans AL3 7JB
G0 EHQ F. Skinner Halfway Lock Cottage Upper Gambolds Lane Stoke Prior, Bromsgrove B60 3HB
G0 EHR M. Clark 60A Clatterford Road, Newport PO30 1PA
GW0 EHS F. Fennah 7 Y Ddol, Llanbrynmair SY19 7DJ
G0 EHT J. Tommey The Birches, Upton Bishop, Ross-on-Wye HR9 7UF
G0 EHV E. Ashburner 8 Shellbark, Houghton le Spring DH4 7TD
G0 EHW B. Coles 32 Victoria Street, Lostock Hall, Preston PR5 5RA
G0 EHX T. Elliott 18 Hollinside Square, Sunderland SR4 8AU
G0 EIB L. Allen 28 Clarence Place, Maltby, Rotherham S66 7HA
G0 EIF P. Massheder 5 Hazel Close, Penwortham, Preston PR1 0YE
G0 EIG M. Holtham 37 Higher Efford Road, Plymouth PL3 6LD
G0 EIH T. Briley 9 Wheatfield Way, Cranbrook TN17 3LS
G0 EIM M. Heyes 11 Beech Close, Isleham, Ely CB7 5UU
G0 EIQ M. Richards 3 Derwent Crescent, Whetstone, London N20 0QN
G0 EIR G. Kemp 27 Shady Grove, Alsager, Stoke-on-Trent ST7 2NQ
GM0 EIT A. George 7 Mid St., Keith AB55 5AG
G0 EIY S. Pryce 40 Gains Avenue, Bicton Heath, Shrewsbury SY3 5AN
G0 EIZ W. Kenyon Flat 21 House 4, Copper Place, Manchester M14 7FZ
G0 EJD M. Moss 1 Orchard Rise, Beckingham DN10 4NG
GW0 EJE Saltash Dist Ar c/o E. Hollowell 54 Portfield, Haverfordwest SA61 1BW
G0 EJI L. Garden 9 Gateway Avenue, Smithville L0R 2A0 Canada
G0 EJO G. Nock 20 Chigwell Road, Bournemouth BH8 9HW
G0 EJR A. Love 15 Mountain Ash, Weston Park, Bath BA1 2UU
GI0 EJT S. Rafferty 81 Mullaghmore Drive, Omagh BT79 7PQ
GI0 EJU V. Hutchinson 10 Golan Road, Knockmoyle, Omagh BT79 7TJ
G0 EJV L. Hodges 34 Wiseholme Road, Skellingthorpe, Lincoln LN6 5TF
G0 EKH K. Harper 2 Vale Road, Decoy, Newton Abbot TQ12 1DZ
G0 EKK H. Wright 61D Clapgun Street, Castle Donington, Derby DE74 2LF
GM0 EKM C. Duncan Roadside Cottage, Hoswick, Shetland ZE2 9HL
G0 ELB B. Bray 34 Newlands Drive, Forest Town, Mansfield NG19 0HZ
G0 ELC P. Hall 28 Maria Drive, Stockton-on-Tees TS19 7JL
G0 ELJ D. Dawson Moonstone, Rottenstone Lane Scratby, Great Yarmouth NR29 3QT
G0 ELK E. Hausler 5 Balaton Place, Snailwell Road, Newmarket CB8 7YP
GM0 ELL N. Elliot Flat 1, Tarfside, Ascog, Isle of Bute PA20 9EU
G0 ELN S. Macdonald 61 Pavilion Road, Worthing BN14 7EE
G0 ELO J. Ellis 21 Coxway, Clevedon BS21 5AQ
GM0 ELP Warwickshire Avon Raynet Group c/o D. Maxwell 29 Ambleside Rise, Hamilton ML3 7HJ
G0 ELR East Lancashire Raynet c/o N. Isherwood 41 Livingstone Road, Blackburn BB2 6NE
G0 ELU K. Kyriacou 11 Mead Way, Bromley BR2 9EN
G0 ELX P. Wilson 39 Tintern Grove, Stockport SK1 4DS
GD0 ELY J. Brown Cleckheaton, Ballaragh, Laxey IM4 7PW Isle of Man
G0 ELZ W. Cross 31 Joshua Close, Liverpool L5 0TD
GW0 EMB H. Blore 17 Kendal Way, Wrexham LL12 8AF
GM0 EMC K. Mclaren 3 Bracany Gardens, Fogwatt, Dalriada, Elgin IV30 8SY
G0 EMF D. Brown 21 Scoular Drive, North Seaton, Ashington NE63 9SE
G0 EMJ G. Swann 22 Balmoral Court, Crawley RH11 9AX
G0 EMK M. Kendall 88 Coldnailhurst Avenue, Braintree CM7 5PY
G0 EML R. Bullock 40 Little Harlescott Lane, Shrewsbury SY13PY
G0 EMM K. Dockray 54 Kelsick Park, Seaton, Workington CA14 1PY
GM0 EMQ J. Vinton 2 Luncarty Place, Turriff AB53 4UD
G0 EMR P. Page 144 Cody Road, Farnborough GU14 0DD
G0 EMS H. Brown Priors Lea, Besford Road, Worcester WR8 9AN
G0 EMT M. Chapman 102 Fangrove Park, Lyne, Chertsey KT16 0BP
G0 EMU Hillbillies CG c/o N. Evely 11 St.Margarets New Road, Teignmouth TQ14 8UE
G0 EMV Franklin Radio Group c/o W. Van Aswegen 16 Spencer Road, Southampton SO19 6QX
G0 EMX E. Parr 74 Stanley Road, Coventry CV5 6FF
G0 ENA R. Procter 41 Bracken Road, Ferndown BH22 9PD
G0 ENB R. Felton 16 Quidenham Road, East Harling, Norwich NR16 2JD
G0 END R. Mcgarvie Croftlands, Thornthwaite, Keswick CA12 5SA
G0 ENF G. Crawshaw1 51 Templeway West, Lydney GL15 5JD
G0 ENJ D. Buckingham 208 Bannings Vale, Saltdean, Brighton BN2 8DJ
G0 ENM D. James 22 Gretton Road, Walsall WS9 0DT
G0 ENN B. Bowden 49 Springfield Drive, Westcliff-on-Sea SS0 0RA
G0 ENO K. Dempster 2 Hillstone House, 64 Graham Road, Malvern WR14 2HU
GM0 ENQ Over the Hill DX Group c/o W. Smith 10 Woodlands Place, Inverbervie, Montrose DD10 0SL
GW0 ENT J. Comerford Bod Elen, Bontnewydd, Caernarfon LL54 7YE
GW0 ENU D. Walters 132 Tan Y Bryn, Valley, Holyhead LL65 3ES
G0 ENV A. Wood 262 Egmanton Road, Meden Vale, Mansfield NG20 9PY
G0 ENW G. Fingerhut Wild Rose, Behind Hayes, Templecombe BA8 0BP
G0 ENY R. Ford 27 Albert Road, Millisons Wood, Coventry CV5 9AS
G0 ENZ T. Trudgeon 1 Bessy Beneath Cottages, Ruan High Lanes, Truro TR2 5JX
G0 EOF G. Potter 88 Highlands Close, Kidderminster DY11 6JU
G0 EOG J. Jennings 81 Newgate St., Burntwood WS7 8TX
G0 EOH D. Bristow Herniss Bungalow, Nr Penryn TR10 9DT
G0 EOI T. Froggatt Diestseweg 115, Geel 2440 Belgium
G0 EOJ Torbay ARS c/o D. Shore 9 Hawthorn Close, Clowne, Chesterfield S43 4SX
G0 EOK P. Gibson 44 Martindale Road, Hemel Hempstead HP1 2QR
G0 EOL W. Prater 44 Alundale Road, Winsford CW7 2QD
G0 EOM A. Deakin 36 The Ridgway, Romiley, Stockport SK6 3EY
G0 EON F. Cloke 9 Mill Close, East Coker, Yeovil BA22 9LF
G0 EOP L. Herf Old Chapel, Fore Street, South Molton EX36 3HL
G0 EOS P. Smith 35 Tanglewood Close, Birmingham B34 7QX
G0 EOX I. Hunton 123 Huddersfield Road, Diggle, Oldham OL3 5NU
G0 EOY S. Outterside 21 Coquet Grove, Throckley, Newcastle upon Tyne NE15 9JU
G0 EOZ P. Kerton Mooraless, 11 North Filham Cot, Ivybridge PL21 9DH
G0 EPA G. Whitham 55 Bisley Grove, Bransholme, Hull HU7 4PY
G0 EPC P. Eggleton 33 Elsham Way, Swindon SN25 4TJ
G0 EPE D. Cargill 41 Grosvenor Road, Skegness PE25 2DD
G0 EPL J. Love 191 High Street, Henley-in-Arden B95 5BA
GM0 EPO J. Shades 156 Cumbrae Drive, Doonfoot, Ayr KA7 4GA
G0 EPP A. Przybyla 18 Cherwell, Washington NE37 3LA
G0 EPR P. Wardale 104 Rectory Place, Woolwich, London SE18 5BY
G0 EPU East Cheam Wireless Society c/o C. Crosby 37 Malwood Way, Maltby, Rotherham S66 7HF
G0 EPV J. Collins 49 Alspath Road, Meriden, Coventry CV7 7LU
G0 EPY C. Hirst Sunnyside, Main Street Hellifield, Skipton BD23 4HX
G0 EQB K. Curson 20 Penelope Grove, Peterborough PE2 8XP
G0 EQC I. Williamson 4 Edgewell Road West, Prudhoe NE42 6JP
G0 EQD J. Archer 29 Easby Close, Bishop Auckland DL14 0RX
G0 EQE D. Cunningham 2 Fairmead Road Moreton, Wirral CH46 8TX
G0 EQH E. Shackleton 2 Culcheth Avenue, Marple, Stockport SK6 6NA
G0 EQI R. Mallinson 3 Captain Cooks Crescent, Whitby YO22 4HL
GM0 EQS Dr S. Palmer Fintry Schoolhouse, Turriff AB53 5RN
G0 EQV D. Buckley Cl/ Juan Ramon Jimenez 23, Formentera Del Segura, Alicante 1379 Spain
GM0 EQW T. Olsen Ard Chuan, Taynuilt PA35 1HY
GM0 ERB N. Calder 16 Camesky Road, Caol, Fort William PH33 7ER
G0 ERF K. Watkins 29 Saddlers Close, Billingshurst RH14 9GL
G0 ERL W. Marbus Elm Farm House, Debenham Road, Stowmarket IP14 5LP
G0 ERS R. Smith 79 Froxfield Road, Havant PO9 5PW
GM0 ERT R. Tannahill 62 Caroline Park, Mid Calder, Livingston EH53 0SJ
GM0 ERV S. Mclennan 6 Mull Terrace, Oban PA34 4YB
G0 ERW W. Spencer 46 Saxon Way, Bourne PE10 9QY
G0 ERY R. Saunders 24322 Augustin Street, Mission Viejo 92691 United States
G0 ESA W. Wilkinson 10 Chemin Des Mardeilles, , Chirac 16150 France
G0 ESD Royal Air Force Halton RS c/o T. Roberts Alcana 58.2 Bajo, La Romana, Alicante 3669 Spain
G0 ESF J. Southgate 36 Porch Way, Whetstone, London N20 0DS
G0 ESH G. Cooper The Bumbles, Shatterford, Bewdley DY12 1TR
G0 ESI D. Williams Miramare, Egremont Road, St. Bees CA27 0AS
GW0 ESK C. Williams 3 Lodge Orchard, Mona Street, Amlwch LL68 9RX
G0 ESL M. Musgrave 11 Hillside Drive, Yealmpton, Plymouth PL8 2NT
G0 ESO S. Llewellyn Eastfield Cottage, Mavis Enderby, Spilsby PE23 4EJ
GW0 ESU W. Lee 8 Bronheulog, Bodffordd LL77 7SU
G0 ESW R. Graham 454 Lobley Hill Road, Lobley Hill, Gateshead NE11 0BS
G0 ESY M. Perrett 2 Barn Park Ashwater, Beaworthy EX21 5EU
G0 ETA G. Luhman 31 Flexmore Way, Langford, Biggleswade SG18 9PT
GW0 ETF S. Rolfe Tynlon Minffordd, Bangor LL57 4DR
G0 ETI J. Bedford 41A Arden Road, Herne Bay CT6 7UW
G0 ETJ J. Bathurst 7 Princethorpe Road, Birmingham B29 5PU
G0 ETL G. Tatterson 2 Eden Road, Leeds LS4 2TT
GW0 ETM J. Davies 1 Mount View, Plas Road, Blackwood NP12 3RH
G0 ETP T. Howe 76 Birch Trees Road, Great Shelford, Cambridge CB22 5AW
G0 ETQ J. Curnow 4 Penmere Court, Falmouth TR11 2RN
GW0 ETU E. Cabban Garmonfa, Capel Garmon, Llanrwst LL26 0RG
G0 ETV G. Mcquire 43 Elmridge Crescent, Blackpool FY2 0NQ
G0 ETZ S. Gurney Crimond The Common, Exmouth EX8 5EE
G0 EUC Capt. R. Burnet 41 Douglas Crescent, Southampton SO19 5JP
G0 EUD C. Jackson Old Stable Cottage, Hall Road, Walpole Highway, Wisbech PE14 7QD
GI0 EUG E. Hagan 34 Coolshinney Road, Magherafelt BT45 5JF
G0 EUJ K. Neville 5 Coleville Avenue, Fawley, Southampton SO45 1DA
GM0 EUL J. Estibeiro The Joiners House Preston, Duns TD11 3TQ
GM0 EUM J. Mackinnon 60 Mount Stuart Drive, Wemyss Bay PA18 6DX
G0 EUN J. Nichol 58 Benson Crescent, Doddington Park, Lincoln LN6 3NU
G0 EUP M. Rigg 64 Hathaway, Blackpool FY4 4AB
G0 EUR B. Barber 1 Shore Place, Trowbridge BA14 9TB
G0 EUV R. Jones Greens Cottage, Luton Road, Offley, Hitchin SG5 3DR
G0 EUZ Pelican Radio Group Little Rissington c/o R. Jones 90 High Street, Cottenham, Cambridge CB24 8SD
G0 EVA D. Evans 113 Denby Dale Road, Wakefield WF2 8EB
G0 EVB J. Barnes 7 Parkfield Mews, Little Parkfield Road, Liverpool L17 8UD
G0 EVD J. Noble 22 Hadleigh, Letchworth Garden City SG6 2LU
GW0 EVE V. Berry 40 Yerburgh Avenue, Colwyn Bay LL29 7NB
G0 EVF J. Mowbray 44 Monkdale Avenue Cowpen Estate, Blyth NE24 4EB
GW0 EVG N. Berry 40 Yerburgh Avenue, Colwyn Bay LL29 7NB
G0 EVH A. Ferneyhough 30 Bedford Drive, Sutton Coldfield B75 6AU
G0 EVI S. Monk 310 Hinckley Road, Leicester LE3 0TN
G0 EVJ S. Evans 181 Curborough Road, Lichfield WS13 7PW
G0 EVM P. Stewardson Hyland, St. Kenelms Road, Halesowen B62 0NE
G0 EVN S. Parkin 13 Queens Drive, Nuthall, Nottingham NG16 1EG
G0 EVO F. Robinson 42 The Paddock, York YO26 6AW
G0 EVP K. Griffiths 44 Curzon Road, Poynton, Stockport SK12 1YE
G0 EVQ A. Sewell 53 Soame Close Aylsham, Norwich NR11 6JF
G0 EVR A. Kittrick 28 Jubilee Street, Hall Green, Wakefield WF4 3JZ
G0 EVS H. Angus 111 Great Elms Road, Hemel Hempstead HP3 9UQ
G0 EVT J. Hoban 3 Lake Lock Grove, Stanley, Wakefield WF3 4JJ
G0 EVU F. Samet 4 Pembroke Grove, Glinton, Peterborough PE6 7LG
G0 EVV D. Stansfield 22 Low Stobhill, Morpeth NE61 2SG
G0 EVW G. Watts 3 Maple Grove Knightsdale Road, Weymouth DT4 0FE
G0 EVX A. Cater 2 Turkdean Road, Cheltenham GL51 6AL
G0 EVY D. Strobel Dell Cottage, Copyholt Lane Stoke Pound, Bromsgrove B60 3AY

G0 EVZ S. Males 6 Lammas Path, Stevenage SG2 9RN
G0 EWD P. Kennedy 262 Green Road, Springvale, Sheffield S36 6BH
GI0 EWE P. Holland 35 Ashfield Road, Clogher BT76 0HJ
GM0 EWF D. Pettigrew 112 South Street, Armadale, Bathgate EH48 3JU
G0 EWH R. Newton 74 Walker Avenue, Stourbridge DY9 9EL
G0 EWI J. Daramy 6 Boulton Close, Chesterfield S40 4XJ
GI0 EWP J. Hartin 2 Berryhill Close, Dunamanagh, Strabane BT82 0GZ
G0 EWR D. Wale 33 Westground Way, Tintagel PL34 0BH
G0 EWT A. Coates 11 Canterbury Road, Brotton, Saltburn-by-the-Sea TS12 2XG
GM0 EWU C. Craig Knipoch Hotel, Knipoch, Oban PA34 4QT
G0 EWV T. Buckle 15 Gleaves Avenue, Harwood, Bolton BL2 4ET
GM0 EWW J. Moore 19 Mansefield Tyndrum, Crianlarich FK20 8RQ
GM0 EWX C. Macpherson 6 Borve, Skeabost Bridge, Portree IV51 9PE
GW0 EWY W. Woods 40 Ger-Y-Llan, Velindre, Llandysul SA44 5YB
G0 EWZ I. Mason 56 Evenlode Crescent, Coventry CV6 1BY
G0 EXA A. Bendall 3 St Michaels Gate, Brimfield, Ludlow SY8 4NE
G0 EXB P. Langdon Dahlia Cottage, Kidderminster DY14 9HP
GW0 EXD Dr C. Challinor Bryn Tirion, Lower Frankton, Oswestry SY11 4PA
G0 EXN J. Eden 4 Halescourt, Church Lane, Shifnal TF11 8RD
G0 EXU A. Davies-Jones 10 Ponsford Road, Knowle, Bristol BS4 2UP
G0 EYA J. Morris 31 Beldham Road, Farnham GU9 8TW
G0 EYE A. Lunn 45 St. Anthonys Avenue, Eastbourne BN23 6LN
G0 EYF B. Ford 15 Derby Road, Barnstaple EX32 7HW
G0 EYG C. Tite 13 Potter Way, Bedford MK42 9RG
GW0 EYH R. Dawson 74 Dylan Avenue Cefn Fforest, Blackwood NP12 3NG
G0 EYL W. Mcadam 2 Manor Orchards, Knaresborough HG5 0BW
G0 EYM R. Rowlett No 1 Bungalow, Main Road, Wisbech PE14 9JR
G0 EYO C. Pettitt 12 Hennals Avenue, Redditch B97 5RX
G0 EYP R. Francis 42 Carmarthen Road, Cheltenham GL51 3LA
G0 EYR P. Robinson 17 Highfield, Taunton TA1 5JE
G0 EYT M. Fox 49 Manor Drive, Esher KT10 0AZ
G0 EYU C. Talbot 59 Heywood Avenue, Austerlands, Oldham OL4 4AZ
G0 EYW D. Jones 4 Granville Crest, Kidderminster DY10 3QS
G0 EYX D. Southey 253 Sandon Road, Stafford ST16 3HQ
G0 EYZ P. Orchard 5 Vicarage Road, Bletchley, Milton Keynes MK2 2EZ
GW0 EZB S. Cowie 37 Rockfield Drive, Llandudno LL30 1PF
G0 EZI J. Pitfield 42 Spinney Green, Eccleston, St. Helens WA10 5AH
G0 EZJ D. Drake 60 Jessopp Avenue, Bridport DT6 4ES
G0 EZL D. Hodgkinson 16 Fortescue Avenue, Twickenham TW2 5LS
GW0 EZQ Llanelli ARC c/o P. Cavallucci 23 Pier Street, Swansea SA1 1RY
GM0 EZR P. Newton 115 Napier Road, Glenrothes KY6 1DU
G0 EZT E. Humphries 37 Grove Meadow, Cleobury Mortimer, Kidderminster DY14 8AG
G0 EZU A. Davies Flat 16, Beechcroft Salisbury Terrace, Teignmouth TQ14 8JA
G0 EZX C. Wood 34 Rosemary Lane, Stourbridge DY8 3EP
G0 EZY T. Jeacock 9 Parkwood Rise, Barnby Dun, Doncaster DN3 1LY
G0 FAB H. Kay 51 Colin Crescent, Colindale, London NW9 6EU
GW0 FAD J. Bowers 122 Tycroes Road, Tycroes, Ammanford SA18 3NS
G0 FAE R. Beer 65 Bridgefield Road, Whitstable CT5 2PH
G0 FAH W. Wright 46 Homestall Road, East Dulwich, London SE22 0SB
G0 FAJ L. Barnes 22 Holton Heath Park, Wareham Road, Poole BH16 6JS
G0 FAS G. West 33 Dorcis Avenue, Bexleyheath DA7 4RL
G0 FAU A. Doyle 2 Ferndown Way, Weston, Crewe CW2 5GS
G0 FAW H. Jenkinson Flat 2, 3 Kassima, Kissonerga/Paphos 8574 Cyprus
G0 FBB Spalding And District ARS c/o I. Burns Little Delmar Farm, Leywood Road, Meopham DA13 0UD
G0 FBC C. Hyatt 44 Barnes Lane, Sarisbury Green, Southampton SO31 7BZ
G0 FBG G. Hands 74 Berrington Road, Nuneaton CV10 0LB
G0 FBL B. Sharman 64 Collingwood Drive, Shiney Row, Houghton le Spring DH4 7LP
G0 FBM D. Lawson 52 Ryefield Road, Eastfield, Scarborough YO11 3DR
G0 FBO C. Roberts 86 Peake Road, Brownhills, Walsall WS8 7BZ
G0 FBQ W. Wootton 94 Dyas Avenue, Great Barr, Birmingham B42 1HF
G0 FBS G. Nicolson 34 Chalbury Close, Weymouth DT3 6LE
GW0 FBT C. Hughes Llanhennock Cheshire Home, Caerleon, Newport NP18 1LT
G0 FBW A. Armstrong 1 Montfalcon Close, Peterlee SR8 1DD
G0 FBX A. Leigh Roleystone, 7 Fieldfare, Gloucester GL4 4WH
G0 FCA I. Groom 12 Billington Avenue, Rossendale BB4 8UW
G0 FCB C. Tatlow Frenton Farm, Whitemoor, St. Austell PL26 7XQ
G0 FCG P. O'Neill 36 Grantley Gardens, Mannamead, Plymouth PL3 5BS
G0 FCH R. Last 39 Upham Road, Swindon SN3 1DJ
GM0 FCI P. Reid 129 Mckinlay Crescent, Irvine KA12 8DR
G0 FCJ M. Lawson 1A Hunters Close, Stroud GL5 4UW
G0 FCM I. Sircombe 4 Long Eights, Northway, Tewkesbury GL20 8QY
G0 FCO B. Cooke 49 Shepherds Croft, Slade, Stroud GL5 1US
G0 FCQ D. James 15 Kington Gardens, Birmingham B37 5HX
G0 FCT I. Pawson 3 Orion, Roman Hill, Bracknell RG12 7YX
G0 FCU S. Kennedy Laurel Cottage, Pond Lane, Peaslake, Guildford GU5 9RS
G0 FCV A. Woods 8 Wareham Road, Lytchett Matravers, Poole BH16 6DP
G0 FCX A. Traynor 2 Mansfield Road, Mossley, Ashton-under-Lyne OL5 9JN
G0 FCZ L. Standley 26 Ullswater Drive, Middleton, Manchester M24 5RL
G0 FDA R. Dresser 6 Acacia Avenue, Fencehouses, Houghton le Spring DH4 6JG
G0 FDD C. Shalley 4 Almond Walk, Lydney GL15 5LP
G0 FDE M. Jackson 2 Sunnybank, Watledge, Stroud GL6 0AP
G0 FDH R. Wishart 15 Plumer Avenue, Tang Hall, York YO31 0PX
G0 FDJ K. Castley Zone 2, (Opposite High School), Cagayan de Oro City 9000 Philippines
G0 FDP F. Parradine 87 Oakways Eltham, London SE9 2NZ
G0 FDS J. Johnson 66 Carlton Park, Manby, Louth LN11 8UQ
G0 FDT M. Flatman 8 The Pines, Cringleford, Norwich NR4 7LT
G0 FDV D. Jardine 11 Gorse Hill, Broad Oak, Heathfield TN21 8TW
G0 FDX Kent County Raynet c/o J. Lawson 14 Kentmere Avenue, Farington, Leyland PR25 3UH
G0 FDZ C. Whitmarsh 35 Dorchester Avenue, Bexley DA5 3AH
G0 FEH D. Thickett 23 Helmsdale Road, Leamington Spa CV32 7DN
G0 FEI V. Ward Romayne, St. Johns Road, Great Yarmouth NR31 9JT
G0 FEJ G. Marshall Birchlands, 3 Longbridge Close, Hook RG27 0DQ
G0 FEK R. Wilson 34 Belfairs Drive, Chadwell Heath, Romford RM6 4EB
GW0 FEM G. Felton 10 Penbodeistedd, Llanfechell, Gwynedd LL68 0RE
G0 FEO A. Quy 17 Fircroft, Kingsbury, Tamworth B78 2JU
G0 FEP Dr H. Maclean 15 Keystone Cres, Kew East 3102 Australia
G0 FEQ K. Howard 11 Station Road, Ulceby DN39 6UQ
GW0 FEU D. Davies Coedfryn, Halkyn, Holywell CH8 8ES
G0 FEV E. Kittrick 28 Jubilee Street, Hall Green, Wakefield WF4 3JZ
G0 FEY S. Goodwin 5 St. Wenefredes Green Bickley, Whitchurch SY13 4EB
G0 FEZ K. Wragg 11A Fall Road, Heanor DE75 7PQ
G0 FFB S. Maughan 17 Upper Dane, Desborough, Kettering NN14 2LB
G0 FFF R. Ford 15 Holloway, Pershore WR10 1HW
G0 FFK R. Mckenzie 40 Fairway Avenue, West Drayton UB7 7AN
G0 FFL R. O'Keeffe 40 Edinburgh Road, Maidenhead SL6 7SH
G0 FFN E. Fisher 19 Keats Way, West Drayton UB7 9DR
G0 FFQ R. Baldock 1A Thorneywood Road, Long Eaton, Nottingham NG10 2DZ
G0 FGA A. Walker 2 Chelwood Drive, Sandhurst GU47 8HT
G0 FGC J. Biggs 40 Packmore St., Warwick CV34 5BX
G0 FGE B. Bolt 106 Barley Farm Road, Exeter EX4 1NJ
G0 FGG J. Thompson 17 Fryer Crescent Haughton Le Skerne, Darlington DL1 2DX
G0 FGI H. Cromack 6 West Park, South Molton EX36 4HJ
G0 FGJ Royal Air Force ARS c/o D. Joyce 44 St. Marys Close, Marston Moretaine, Bedford MK43 0QZ
G0 FGK Clwud Portable Operating Group c/o B. Whitehouse Flat 53, Peel House, Tamworth B79 7BQ
GW0 FGO W. Waldron 100 Porthmawr Road, Cwmbran NP44 1NB
G0 FGP R. Bradwell Summer Fields School, Mayfield Road, Oxford OX2 7EN
G0 FGS Dr M. Bosley Crossroads Palmerston Road, Ross-on-Wye HR9 5PN
G0 FGW R. Clements 28 Willow Grove, Chippenham SN15 1AR
G0 FGX R. Mccreadie 45 Gwealhellis Warren, Helston TR13 8PQ
G0 FGZ Dr C. Newton Peartree Cottage, Little London, Longhope GL17 0PH
G0 FHC B. Trimmer Sydney Cottage, Salisbury Road, Romsey SO51 6EE
GM0 FHD E. Mottart 1 Muirake Cottages, Cornhill, Banff AB45 2BQ
GM0 FHF C. Ashdown Oliver Cottage 7 Fernilea, Carbost, Isle of Skye IV47 8SJ
G0 FHH P. Beatty 21 Avery Hill, Kingsteignton TQ123LB
GM0 FHJ Wirral Schools RC c/o R. Menzies 105 Yoker Mill Road, Glasgow G13 4HL
G0 FHK R. Peart 33 Fieldfare, Abbeydale, Gloucester GL4 4WH
GW0 FHL B. Thomas Plastirion, Padeswood Road, Buckley CH7 2JL
G0 FHO G. Taylor 93 Fengate Mobile Home Park, Peterborough PE1 5XE
G0 FHT G. Bate 7 Albany Court, Redruth TR15 2NY
G0 FHW T. Opie 3 Colborne Avenue Illogan, Redruth TR16 4EB
G0 FHX A. Hocking 79 Cornish Crescent, Truro TR1 3PE
G0 FHY J. Hocking 79 Cornish Crescent, Truro TR1 3PE
G0 FIC K. Tarry 38 Tresithney Road, Carharrack Tr165Qz, Reduth TR165QZ
G0 FIG A. Trusler 42 Mill Hill, Shoreham-by-Sea BN43 5TH
G0 FIJ C. Jones 46 Wilmington Close, Woodley, Reading RG5 4LR
G0 FIN A. Findlay 218 Lower Hillmorton Road, Rugby CV21 3TS
G0 FIP E. Tugwell 14 Martinique Way, Eastbourne BN23 5TH
GM0 FIQ H. Bohan 12 Loch Way, Kemnay, Inverurie AB51 5QZ
G0 FIT K. Menzel 8 Higher Bockhampton, Dorchester DT2 8QJ
G0 FIU R. Edwards Stanmore Cottage, Brockley Corner, Culford, Bury St. Edmunds IP28 6UA
G0 FIW I. Osborne Alacoo, Tan Lane, Clacton-on-Sea CO16 9PS
G0 FJA B. Samuels 63 Mill Road, Okehampton EX20 1PR
G0 FJB J. Bailey Powney Cottage, Powney Street, Ipswich IP7 7AL
G0 FJD D. Tyers 18 Gardeners Close, Kidderminster DY11 5DW
GW0 FJE R. Hill 13 Maesglas Grove, Newport NP20 3DJ
GW0 FJH C. Lonsdale 6 Oak Tree Close, New Inn, Pontypool NP4 0DG
G0 FJJ A. Winkler 10 Havenside, Shoreham-by-Sea BN43 5LN
GW0 FJP D. Stanley 9 Haywain Court, Bridgend CF31 2ED
GW0 FJQ A. Marshall 26 Avondale Road, Gelli, Pentre CF41 7TW
G0 FJR D. Paynter 6 Blacksmiths Close, Ramsey Forty Foot, Huntingdon PE26 2YW
G0 FJS P. Copeland 6 Waverley Road, Northampton NN2 7DA
G0 FJZ G. Bradley 59 Main Road, Watnall, Nottingham NG16 1HE
G0 FKF J. Smith Erw Las, The Cross Roads, Redruth TR16 5PN
G0 FKG C. Skillings 11 Curtis Road, Norwich NR6 6RB
G0 FKI A. Brown Panorama, Highway Lane, Redruth TR15 1SE
G0 FKJ C. Currey 1 Newport Close, Portishead, Bristol BS20 8DD
G0 FKK R. Parrish 11 Pitt Road, Maidstone ME16 8PA
GM0 FKP J. Tohill 71 Campsie Road, Kilmarnock KA1 3RY
G0 FKS K. Stancliffe 3 Upper Lambricks, Rayleigh SS6 8BP
G0 FKW A. Timms 63 High Street, Astcote, Towcester NN12 8NW
G0 FKX D. Warren 1 Ruby Terrace, Porkellis, Helston TR13 0LD
G0 FKY J. Merifield 84 Wareham Road Corfe Mullen, Wimborne BH21 3LG
G0 FLA W. Frewen Tantalus, Main Street, Rye TN31 6NB
G0 FLD K. Pitman Pump Cottage, High Street, Bridlington YO15 1JT
G0 FLG B. Parker 30 Caistor Road, Market Rasen LN8 3JA
G0 FLI N. Penistone 114 Long Lane, Worrall, Sheffield S35 0AF
G0 FLP J. Hammond Ashmond House, Queens Street, March PE15 8SN
G0 FLQ R. Cheetham 65 Avondale Avenue, Hazel Grove, Stockport SK7 4QE
G0 FLT A. Auker-Howlett 7 Caxton End, Eltisley, St. Neots PE19 6TJ
G0 FLU M. Crane 40 Dukes Way, Newquay TR7 2RW
G0 FLV R. Pennock 4 Millers Way, Heckington, Sleaford NG34 9JG
G0 FLW L. Williams 56 Meriden Avenue, Stourbridge DY8 4QS
G0 FLX D. Gray Jaig House, Mill Lane Hemingbrough, Selby YO8 6QX
G0 FMB B. Collins 28 Marlborough Road, South Woodford, London E18 1AP
G0 FMG J. Voss 4 Chaucer Avenue, Mablethorpe LN12 1DA
G0 FMI R. Friston Savmar, 72 Bradenham Road, Thetford IP25 7PJ
G0 FMJ R. Bennett 86 Westons Hill Drive, Emersons Green, Bristol BS16 7DN
G0 FMN A. Mcquarrie 348 The Manor, Billing Garden Village, The Causeway, Great Billing, Northampton NN39EX
G0 FMP I. Hodgkiss 41 Buckingham Rise, Worksop S81 7ED
G0 FMT D. Unwin 11 Carlton Rise, Melbourn, Royston SG8 6BZ
G0 FMU A. Turner 17 The Dell, Great Warley, Brentwood CM13 3AL
GM0 FMW D. Enderby 45 Oxgangs Park, Edinburgh EH13 9LF
G0 FMX A. Roberts Am Weinberg 5, Bergen 29303 Germany
G0 FNA C. Halliday 42 Denshaw, Upholland, Skelmersdale WN8 0AY
G0 FNB S. Mudd 91 Chalkwell Avenue, Westcliff-on-Sea SS0 8NL
G0 FND M. Cousins Wiccan Lodge, Mumbys Drove, Wisbech PE14 9JT
GM0 FNE T. Wilson 20 Mace Court, Stirling FK7 7XA
G0 FNF I. Gilbert 82 Abbotswood Road, Brockworth, Gloucester GL3 4PF
G0 FNH J. Toon 2 Home Farm Park, Burton-on-Trent DE13 9BJ

G0 FNJ D. Earnshaw 43 Bank Parade, Burnley BB11 1UG
G0 FNM E. Walker 216 Milnrow Road, Rochdale OL16 5BB
G0 FNP P. Radcliffe Hill View, Wilton, Pickering YO18 7LE
G0 FNS J. Brayshaw 26 Ashfield Avenue, Malton YO17 7LE
G0 FNV N. Moult 31 Stanhope Crescent, Arnold, Nottingham NG5 7AZ
G0 FOB N. Robinson 15 Hollins Bank, Sowerby Bridge HX6 2RU
G0 FOC M. Clowes 55 Landswood Close, Birmingham B44 0LF
G0 FOE P. Hume-Spry 23 Appledore Avenue, Wollaton, Nottingham NG8 2RE
G0 FOG C. Singleton 1 Abbey Close, Aslockton, Nottingham NG13 9AF
G0 FOH M. Holdsworth Merrill, Ringwood Road, Southampton SO40 7GY
G0 FOI J. Wilde 2 Bottoms Lane, Birkenshaw, Bradford BD11 2NN
G0 FOK R. Cox123 12 East Street, Thame OX9 3JS
GW0 FOL J. Brazier Flat 5, Llys Madryn Caernarvon Road, Pwllheli LL53 5LF
G0 FOT R. Gibbons 3 Fairfield, Gamlingay, Sandy SG19 3LG
G0 FOU G. Binns 21, Rydal Close, Winsford CW7 2SE
G0 FOY G. Grigg 12 Townfield Road, Mobberley, Knutsford WA16 7HF
G0 FPI J. Spence 60 Railey Road, Crawley RH10 8BZ
G0 FPM P. Fennell 45 Badby Road West, Daventry NN11 4HJ
G0 FPN D. Waller Apartment 23, 3 Woodbrooke Grove, Birmingham B31 2FG
G0 FPO W. Coulthard 1 Lambton St., Eccles, Manchester M30 8DD
G0 FPT A. Pettigrew 12 Pensford Close, Crowthorne RG45 6QR
G0 FPU M. Densham 69 Mortimer Way, Leicester LE3 1GR
GW0 FPY J. Bortowski 4 Bryn Deiniol, Valley Road, Llanfairfechan LL33 0SR
G0 FPZ R. Wileman 3 Primrose Way, Stamford PE9 4BU
G0 FQA A. Wallis 14 Varley Close, Wellingborough NN8 4UZ
G0 FQC P. Wood 26 Church Road, Bamber Bridge, Preston PR5 6EP
G0 FQD R. Harvey 42 Groomsland Drive, Billingshurst RH14 9HB
G0 FQF J. Rae Sunnybank House, Burnley Road East, Rossendale BB4 9PX
G0 FQI D. Breen 53 Swift Close, Grange Park, Northampton NN4 5AZ
G0 FQO D. Berry The Bungalow, Basil Road, Kings Lynn PE33 9RP
G0 FQP G. Owens 73 Edinburgh Road, Widnes WA8 8BG
GM0 FQQ B. Campbell 93 Treeswoodhead Road, Kilmarnock KA1 4PB
GM0 FQS A. Smith 60 Gordon Avenue, Bonnyrigg EH19 2PQ
G0 FQU A. Date 23 Gilpin Way, Olney MK46 4DN
GW0 FQZ C. Emblen 12 Gefnan, Mynydd Llandygai, Bangor LL57 4DJ
G0 FRB S. Gresty 4 Palace Road, Sale M33 6WU
GM0 FRC Falkirk & District ARS c/o P. Howson 1 Howetown Fishcross, Alloa FK10 3AW
G0 FRD S. Baldwin 18 Derwent Road, Leighton Buzzard LU7 2QW
G0 FRL R. Lowe 12 Cavenham Grove, Bolton BL1 4UA
G0 FRM H. White 51 New Close, Knebworth SG3 6NU
G0 FRN D. Oakes 56 Middle Way, Chinnor OX39 4TP
G0 FRO A. Medcalf 1 The Old School, School Lane, Didcot OX11 0ES
G0 FRR RC of Pabay c/o A. Baker Highleaze, Deans Drove, Poole BH16 6EQ
G0 FRS Farnborough CG c/o R. Konowicz 12 Ambleside Crescent, Farnham GU9 0RZ
GM0 FRT Funny Contest G c/o A. Duncan Barrhill House, Peterculter AB14 0LN
G0 FRU C. Osbourn Bourn Bungalow, Back Lane, Newmarket CB8 9NB
G0 FRV S. Adams 63 Turnbull Drive, Leicester LE3 2JU
G0 FRX G. Cowling Laissez Faire, Reedness, Goole DN14 8ET
G0 FRY J. Walker Wildersley, Wildersley Road, Belper DE56 1PD
G0 FRZ R. Swinney 27 Auckland Close, Houghton le Spring DH4 6GG
G0 FSA M. Hughes 2 Harrington Road, Desborough, Kettering NN14 2NH
G0 FSB S. Barton 154 The Hill Cromford, Matlock DE4 3QU
G0 FSD P. Robinson 198 Westfield, Plymouth PL7 2EJ
G0 FSF D. Cawser 26 Queen Street, Burton-on-Trent DE14 3LR
G0 FSG C. Easton Dallmore, Lockwood Beck Road, Saltburn TS12 3LE
GM0 FSH W. Mackenzie Nambrac, Cairnmount, Jedburgh TD8 6SA
G0 FSJ D. Hutchinson 6 Birdsall Avenue, Nottingham NG8 2EH
G0 FSL T. Fleet 51 The Crescent, Walsall WS1 2DA
G0 FSM J. Bent 32 Cross Waters Close, Wootton, Northampton NN4 6AL
G0 FSP J. Pears 19 Lichfield Close, Grantham NG31 8RS
GM0 FSV J. Mcgowan 26 Wallace Gardens, Stirling FK9 5LS
GM0 FSW N. Mcallister 36 Kinneff Crescent, Dundee DD3 9RG
GM0 FSY D. Macdonald 4A Brue, Isle of Lewis HS2 0QW
GM0 FSZ E. Sandilands Eric Sandilands, 12 Kerr Court, Girvan KA26 0BP
GM0 FTG R. Hill 9 Chambers Drive, Carron, Falkirk FK2 8DX
GM0 FTH H. Livingston Monthouse, Parkhead Road, Linlithgow EH49 7BS
G0 FTI R. Hughes 46 The Boundary, Oldbrook, Milton Keynes MK6 2HT
GM0 FTJ N. Mitchinson 85 Forest Road, Selkirk TD7 5DD
GM0 FTK W. Kirk 59 Silverbuthall Road, Hawick TD9 7BH
G0 FTN A. Lautman 3 Windsor Close, Leigh-on-Sea SS9 4EA
G0 FTO R. Morton 44 Cromer Drive, Atherton, Manchester M46 0QE
G0 FTR Brecon And Radnor ARS c/o D. Gill 80 Bramwell Street, Sheffield S3 7PB
G0 FTU C. Jones 2 Ladycross Cottages, Dormansland RH76PB
GM0 FTX I. Birkett 25 Darnhall Crescent Craigend, Perth PH2 0HH
G0 FUE D. Denton 7 Uplands Avenue, East Ayton, Scarborough YO13 9EU
G0 FUH R. Douglas 5 Portnalls Road, Coulsdon CR5 3DD
G0 FUI P. Abbott 170 Hangleton Valley Drive, Hove BN3 8FE
G0 FUN Kimbolton School ARC c/o W. Somerville Glendella, Wycombe Road Stokenchurch, High Wycombe HP14 3RP
G0 FUO D. Harrop 7 Haythorne Way Swinton, Mexborough S64 8SQ
G0 FUR D. Buckle 63 Ashley Drive South, Ashley Heath, Ringwood BH24 2JP
G0 FUS P. Fry Flat 2, National Westminster Bank, North Street, Wiveliscombe, Taunton TA4 2JY
G0 FUU P. Firmin 25 The Heights, Hastings TN35 5EP
G0 FUV J. Mills Smiths Hill Petrockstow, Okehampton EX20 3EZ
G0 FUW S. Hartley 5 Sydenham Buildings, Bath BA2 3BS
G0 FUY A. Firth 10 Holroyd Hill, Wibsey, Bradford BD6 1PQ
G0 FUZ C. Muten 20 Middlewich Road, Nantwich CW5 6HL
G0 FVB G. Stokes 87A Wimborne Road, Southend on Sea SS2 4JR
GW0 FVC J. Newman 57 Heritage Park, St. Mellons, Cardiff CF3 0DQ
G0 FVD R. Nichol 32 Greenwood Avenue, Harworth, Doncaster DN11 8HT
G0 FVF A. Howman 32 Dereham Road, Pudding Norton, Fakenham NR21 7NA
G0 FVH D. Dolling 41 Bullfinch Close, Poole BH17 7UP
G0 FVI Dr A. Gilfillan 57 Brant Road, Lincoln LN5 8RX
GM0 FVJ L. Martin 44 Glenearn Court, Pittenzie Street, Crieff PH7 3LE
G0 FVM P. Irons 15 Hanover Court, Kiln Row, North Cave, Brough HU15 2LQ
G0 FVN P. Johnston Woburn House, 38 Chatsworth Avenue, Pontefract WF8 2UP
G0 FVO J. Leach 19 Fyfe Crescent, Baildon, Shipley BD17 6DR
G0 FVS J. Jackson 165 Hall St., Briston, Melton Constable NR24 2LQ
G0 FVT D. Lisney 14 Clarendon Drive, Martham, Great Yarmouth NR29 4TD
G0 FVU M. Smith 31 Cromford Road, Crich, Matlock DE4 5DJ
G0 FWA P. Waddington 19 Olivers Drive, Witham CM8 1QJ
G0 FWD J. Adcock 102 Richmond Way, Newport Pagnell MK16 0LH
G0 FWF D. Stanton 53 Chester Road, London N17 6EH
G0 FWP J. Purvess 389 Otley Old Road, Leeds LS16 6BX
GM0 FWY P. Gray 21 Simpson Street, Glasgow G20 6XZ
GW0 FXC R. Rowland 60 Ombersley Road, Newport NP20 3EE
G0 FXD D. Harrison 41 North End Lane, Malvern WR14 2NG
G0 FXI R. Oram 4 Hardy Avenue, Bristol BS3 2BP
G0 FXK J. Paxton 27 Holborn View, Leeds LS6 2RD
G0 FXL D. Garratt 238 Hockley Road, Hockley, Tamworth B77 5EY
G0 FXM S. Vickery 17 Trenowah Road, St. Austell PL25 3EB
G0 FXQ I. Drury 263 Waxwing Lane, Strasburg VA 22657 USA
G0 FXR R. Clark 9 Kensington Avenue, Normanby, Middlesbrough TS6 0QQ
G0 FXS K. Rutter 6 Chetney Close, Stafford ST16 1XA
G0 FXT C. Richardson 47 Leighton Close, Crossgates, Scarborough YO12 4LA
G0 FXY M. Smith 19 Legion St., South Milford, Leeds LS25 5AY
GJ0 FYB C. Le Jehan Sundora, 4 St. Marys Village, St. Mary JE3 3BQ Jersey
G0 FYD I. Mccabe 99 Edgeway Road, Blackpool FY4 3NH
G0 FYE B. Moss 22 Battersby St., Ince, Wigan WN2 2NA
G0 FYH R. Butterworth 12 Strickland Drive, Morecambe LA4 6TB
G0 FYL J. Samuels 63 Mill Road, Okehampton EX20 1PR
GW0 FYO A. Williams 106 Garrod Avenue Dunvant, Swansea SA2 7XQ
G0 FYP C. Holland 44 Brightstowe Road, Burnham-on-Sea TA8 2HP
G0 FYU T. Liggins 55 Kirkland Street, Pocklington, York YO42 2BX
G0 FYW B. Morrin Flat 19 Elms Hall, Elms Rd, Morecambe LA4 6DD
G0 FYX S. Swain 40, Parkside, Havant PO9 3PL
G0 FZA N. Dickinson 14 Chelsea Mews, Lancaster LA1 2AS
G0 FZB J. Atherfold 42 Mansell Road, Shoreham-by-Sea BN43 6GP
G0 FZC G. Palmer 29 Lindsay Road, Leicester LE3 2EJ
G0 FZD C. Nichols 7 Royston Avenue, Owlthorpe, Sheffield S20 6SG
G0 FZE B. Gould 2 Parkdale, Ibstock LE67 6JW
G0 FZF C. Guinan 20 Putney Close, Oldham OL1 2JS
G0 FZG H. Eddleston 40 High Street Corby Glen, Grantham NG33 4LX
G0 FZH D. Moore 12 Newtown Park, Langport TA10 9TF
G0 FZM K. Blabey 19 Lud Lane, Tamworth B79 7EW
G0 FZO C. Lee 21 Sandholme, Market Weighton, York YO43 3ND
GI0 FZT P. Frend 41 Brunswick Manor Abbey Street, Bangor BT20 4JD
G0 FZU J. Tinsley 21 Peckforton View, Kidsgrove, Stoke-on-Trent ST7 4TA
GW0 FZY Dr J. Woolgar Glan-Yr-Afon, Cwmcerdinen, Felindre, Swansea SA5 7PU
G0 FZZ J. Foster 23 Shrewsbury Street, Hartlepool TS25 5RQ
G0 GAG M. Cowley 46 Mapletoft Avenue, Mansfield Woodhouse, Mansfield NG19 8HT
G0 GAJ F. Bunce 45 Hailes Road, Gloucester GL4 4RB
G0 GAL E. Howells 5 Bowland Close, Overdale, Telford TF3 5HE
G0 GAP W. Meecham 38 Douglas Road West, Stafford ST16 3NX
G0 GAQ B. Johnson 67 Nursery Lane, Northampton NN2 7PT
G0 GAR A. Robinson 28 Briarsleigh, Wildwood, Stafford ST17 4QP
GM0 GAT A. Thomasson 1 Eastside Green, Westhill AB32 6XY
GM0 GAV G. Taylor South Lodge, Fingask, Errol, Perth PH2 7SS
G0 GBC P. Dempster Flat 218A, Peachfield Road, Malvern WR14 4AP
G0 GBE C. Thompson 135 Stafford Road, Bloxwich, Walsall WS3 3PG
G0 GBG B. Wilkinson 3 Friarage Avenue, Northallerton DL6 1DZ
GM0 GBH P. Young 4 Primrose Avenue, Rosyth, Dunfermline KY11 2SS
G0 GBI G. Loake 81 Duchess Road, Bedford MK42 0SE
G0 GBL D. Taylor 38 Seward Road, Badsey, Evesham WR11 7HQ
G0 GBN J. Henshaw 7 Gorsefield Close, Wirral CH62 6BU
G0 GBP D. Porter 10 Broomholme, Shevington, Wigan WN6 8DT
G0 GBQ H. Whitfield 81 Tenter Balk Lane, Adwick-Le-Street, Doncaster DN6 7EE
G0 GBR S. Mattinson 76 Fairway Avenue, Tilehurst, Reading RG30 4QB
G0 GBU M. Mccallum 65 Whalley Road, Altham West, Accrington BB5 5DH
GM0 GBV D. Hackett 5 Craignair Park, Annan DG12 6ND
G0 GBW C. Oswald 3 Belvedere Drive, Bilton, Hull HU11 4AX
G0 GBY M. Francis 50 Edinburgh Avenue, Leigh-on-Sea SS9 3SG
G0 GCA M. Collins 22 Southfield Close, Woolavington, Bridgwater TA7 8HJ
G0 GCJ R. Mellor 2 Taxal View, Fernilee, High Peak SK23 7HD
G0 GCK M. Johnson 7 Ash Grove, Northallerton DL6 1RQ
GM0 GCO B. Carson 46 Tweed Drive, Bearsden, Glasgow G61 1EJ
G0 GCQ J. Rivers 16, Longshore Grove, New Romney TN28 8FP
GM0 GDD A. Mcmillan 10 Lyon Road, Erskine PA8 6HG
GI0 GDF E. Hooks 9 Curtis Walk, Lisburn BT28 1HE
GW0 GDI N. Erskine 302 Caerphilly Road, Cardiff CF14 4NS
G0 GDJ J. Brooks Bream, The Fitches, Saxmundham IP17 1UX
G0 GDL M. Rodgers 21 Dovedale Rise, Allestree, Derby DE22 2RE
G0 GDS J. Hyde 7 Wandells View, Brantingham, Brough HU15 1QL
G0 GDV J. Strickland 11 Chilworth Gardens, Waterlooville PO8 0LD
G0 GEB R. East Bramleys, 34 Boyd Avenue, Dereham NR19 1LU
GM0 GEE L. Stapleton 22 Ashie Road, Inverness IV2 4EN
G0 GEF P. Chipman 2 Cornforth Close, Trinity Road, Tamworth B78 2LA
G0 GEH L. Howarth 12 Thomas Street, Hemsworth, Pontefract WF9 4AY
GW0 GEI S. Jones Blaenpant, Dihewyd, Lampeter SA48 7PJ
G0 GEL J. Pruden 77 Browning Crescent, Ford Houses, Wolverhampton WV10 6BQ
G0 GEP G. Perry 123 Green Lanes, Wylde Green, Sutton Coldfield B73 5LT
G0 GEQ J. Rogerson 44 Romney Close, Clacton-on-Sea CO16 8YE
G0 GER R. Knighton 262 Victoria Road West, Thornton-Cleveleys FY5 3QB
G0 GEU S. Holt 22 Sulby Grove, Morecambe LA4 6HD
GW0 GEV T. Hurst Woodside, Parc Seymour, Caldicot NP26 3AB
G0 GEZ D. Creek The Coach House, Basketts Lane, Yarmouth PO41 0PY
G0 GFA C. Armitage 19 Park Road, Barlow, Selby YO8 8ES
G0 GFC R. Johnson 3 Lance Drive, Chase Terrace, Burntwood WS7 1FA
G0 GFD K. Venn 28 Streamleaze, Titchfield Common, Fareham PO14 4NP
G0 GFE B. Keechan 24 Tolley Road, Kidderminster DY11 7EW
G0 GFI C. Burrows 29 Hampden Road, Malvern Link, Malvern WR14 1NB
G0 GFK J. Denford 10 Churchill Road, Bideford EX39 4HG
GM0 GFL A. Marriott Parkview, Dunrossness ZE29JG
G0 GFM K. Sansoni 10 Charles Moor, York YO31 1BE
GW0 GFN D. Anderson Penrheol Farm, Meidrim, Carmarthen SA33 5NX

G0 GFP A. Fowler 43 Eastbourne Heights, Oak Tree Lane, Eastbourne BN23 8FB
G0 GFQ K. Martin 21 All Saints Close, Weybourne, Holt NR25 7HH
G0 GFR B. Holloway 28 Elmsdale Road, Ledbury HR8 2EG
GD0 GFV J. Angiolini Elm Lodge Patrick Road, St. Johns IM4 3BP Isle of Man
G0 GFY C. James 180 Mitcham Road, Croydon CR0 3JF
G0 GFZ P. Taylor 18 Redriff Road, Romford RM7 8HD
G0 GGA R. Parkins 5 Ferndale Grove, Hinckley LE100PH
G0 GGB A. Norman 13 Market Place, Great Yarmouth NR30 1LY
G0 GGE P. Burrow 9 Minsmere Road, Belton, Great Yarmouth NR31 9NX
G0 GGG N. Rogers 34 Broadway, Warminster BA12 8EB
G0 GGH F. Sell 17 Auriel Avenue, Dagenham RM10 8BS
G0 GGL D. Ellins 52 Littlewood, Stokenchurch, High Wycombe HP14 3TF
G0 GGM B. Fitzsimmons 64 Belle Vue Road Wivenhoe, Colchester CO7 9LD
G0 GGN R. Samways 7 St. Michaels Way, Steeple Claydon, Buckingham MK18 2QD
G0 GGQ G. Redding 50 Great Hill, Shefford SG17 5EA
GW0 GGW A. Patrick 1 Sunningdale Grove, Colwyn Bay LL29 6DG
GI0 GGY J. Porter 24 Cooleen Park, Londonderry BT48 8AQ
G0 GHB J. Graham Caxtonian, Brimbelow Road, Norwich NR12 8UJ
G0 GHD N. Houghton 1 Alma Street, Alfreton DE55 7HX
G0 GHE A. Goldspink Danehaven, Themelthorpe, Dereham NR20 5PS
GW0 GHF B. Williams 10 Pantycelyn Road, Llandough, Penarth CF64 2PG
GW0 GHG D. Roberts 16 Min Y Mor, Aberffraw, Ty Croes LL63 5PQ
G0 GHH P. Cross Balls Farm Cottage, Musbury Road, Axminster EX13 8TT
G0 GHK Finningley ARS c/o M. Hotchin 122 Buckingham Avenue, Scunthorpe DN15 8NS
G0 GHL T. Reeves Rowney Farm, Newcastle Road, Loggerheads, Market Drayton TF9 2QG
G0 GHM D. Coxon 7 Kingston Way, Nailsea, Bristol BS48 4RA
GM0 GHN T. Taylor 10 Woodside Drive, Forres IV36 2UF
G0 GHO W. Small 24 Barrowdale Close, Exmouth EX8 5PN
G0 GHT R. Pearce 52 Pearse Close, Hatherleigh, Okehampton EX20 3QW
G0 GHW G. Whitehead 27 Cheyney Walk, Westbury BA13 3UH
G0 GIA R. Keeley-Osgood 2A St. Anns Crescent, Gosport PO12 3JJ
GM0 GIB M. Gibb Am Fasgadh, Drumtian, Glasgow G63 0NP
G0 GIE D. Adams Penlon, Kingsley Gardens, Codsall, Wolverhampton WV8 2AJ
G0 GIF J. Catterson St. Sebastians Presbytery, Gerald Road, Salford M6 6DW
GW0 GIH J. Thomas 41 The Uplands, Brecon LD3 9HT
G0 GII B. Jackson 94 Nether Court, Halstead CO9 2HF
G0 GIL J. Carter 112 Landor Road, Whitnash, Leamington Spa CV31 2JZ
G0 GIN K. Mills 19 Thomas Bassett Drive, Colyford, Colyton EX24 6PN
G0 GIR P. Martin 47 Bryant Road, Rochester ME2 3EP
G0 GIT M. Jinks Caixa Postal 003, Campina Grande De Sul, Parana Brazil
G0 GJA D. Wright 110 Mancroft Road, Caddington, Luton LU1 4EN
G0 GJC M. Hole 110 North Boundary Road, Brixham TQ5 8JT
G0 GJD R. Radcliffe 23 Rose Vale, Heald Green, Cheadle SK8 3RN
G0 GJE R. Cleverley 22 The Tinings, Monkton Park, Chippenham SN15 3LX
G0 GJG B. Phillips 6 Greenways, Penkridge, Stafford ST19 5HD
G0 GJH G. Hughes 51 Kingsmead, Seaford BN25 2HA
G0 GJL S. Moyses 10 Jones Close, March PE15 9RZ
G0 GJM J. Rattigan 4 Grosvenor Street, Barrow-in-Furness LA14 4AH
G0 GJN T. Roberts 14 Glen Park Gardens, Bristol BS5 7NE
G0 GJR A. Beard1 9 Whitebeam Close, Basingstoke RG22 5FH
G0 GJV M. Goodey 62 Rose Hill, Binfield, Bracknell RG42 5LG
G0 GJW A. Sale 5 Kingswood Road, Gillingham ME7 1DZ
G0 GJX F. Norton 147 Wells Road, Glastonbury BA6 9AN
G0 GKH D. Birch 32 Union Street, Trowbridge BA14 8RY
G0 GKI C. Crawford 70 Westmoreland Avenue, Welling DA16 2QD
G0 GKK A. Pickering 5 West Farm Court, Medomsley, Consett DH8 6TL
G0 GKL M. Stevens 9 Highview Close, St. Leonards on Sea TN37 7HY
G0 GKN J. Dowse 46 Nantwich Road, Middlewich CW10 9HG
G0 GKO G. Lumsden Spencer Buildings, Front Street, Hartlepool TS27 4RT
GM0 GKR J. Stapleton Ravenscourt, 23 Strathkinness High Road, St. Andrews KY16 9UA
G0 GLA M. Hoppe 67 Belmont Road, Maidenhead SL6 6LG
G0 GLG H. Torunski 33 Brickhill Way, Calvert, Buckingham MK18 2FS
G0 GLH T. Mitchell 18 Park Avenue, Bedlington NE22 7EJ
GW0 GLI L. Edwards 23 Tan Y Bryn, Llanbedr Dc, Ruthin LL15 1AQ
G0 GLJ H. Robertson Flat 6, Hazelwood, Prince Of Wales Road, Cromer NR27 9HR
G0 GLQ P. Davenport 1 Boobery, Sampford Peverell, Tiverton EX16 7BS
G0 GLU M. Fry 1 Hebden Avenue, Warwick CV34 5XD
G0 GLW G. White 3 Kent Road, Gosport PO13 0SP
GW0 GLX M. Mcfarland Min Afon, Garreg Fawr Road, Caernarvon LL54 7ED
G0 GLZ Dr R. Sugden 7 Westbourne Grove, Goole DN14 6NA
G0 GMB M. Baker 25 Pentlands, Fullers Slade, Milton Keynes MK11 2AF
G0 GMC C. Cook 4 Woodlands Close Rustington, Littlehampton BN16 3ET
GM0 GMD T. Astbury 8 Auchinlay Holdings, Auchinlay, Dunblane FK15 9NA
G0 GME R. Turner 17 Wrose Brow Road, Shipley BD18 2NT
GM0 GMI J. Bavin Garvan, 10 Grampian Way, Glasgow G78 2DH
G0 GMJ J. Bowles 38 Rydal Grove, Liversedge WF15 7DN
G0 GML B. Ogden 14 Fermandy Lane, Crawley Down, Crawley RH10 4UB
GM0 GMN J. Bertram 20 Kyles View, Largs KA30 9ET
GM0 GMO G. Wylie 9 Friar Avenue, Bishopbriggs, Glasgow G64 2HP
G0 GMS A. Read 7 Grange Court, Hixon, Stafford ST18 0GQ
G0 GMY P. Whitelock 2 Shippards Road, Brighstone, Newport PO30 4BG
G0 GNA J. Weller Pytchley, Chichester Close, Dorking RH4 1LP
G0 GNE R. Maddison Tom Butt, Hope Street, Elstead, Godalming GU8 6DE
G0 GNF G. Frykman 8 Orchard Close, Bishops Itchington, Southam CV47 2QS
G0 GNI M. Anderson 17 Orchard Road, Seaview PO34 5JE
GM0 GNK Inverclyde ARG c/o A. Givens 5 Langhouse Place, Inverkip, Greenock PA16 0EW
G0 GNP B. Ackerman 31 Melton Mill Lane, High Melton, Doncaster DN5 7TE
G0 GNQ L. West 22 Lyndhurst Avenue, Margate CT9 2PS
G0 GNU J. Norton 7 Maudsley St., Bradford BD3 9JT
G0 GNV M. Mundy The Homestead, Homestead Lane, Burgess Hill RH15 0RQ
G0 GNW G. Goddard 113 Linden Walk, Louth LN11 9HT
GM0 GNY L. Graupner 5 Carden Close, Alves, Elgin IV30 8FE
G0 GOB R. Drage 13 Manor Ride, Brent Knoll, Highbridge TA9 4DY
G0 GOE P. Bozac La Vigne, Moternboeuf 16310 France
G0 GOH R. Cornell 81 Mercel Avenue, Armthorpe, Doncaster DN3 3HS
G0 GOI J. Macham 9 Bankfield Grove, Scot Hay, Newcastle ST5 6AR
GM0 GON A. Harrison 168 Ralston Road, Campbeltown PA28 6LQ
G0 GOO B. Evans 17 Clarence Place, Maltby, Rotherham S66 7HA
GM0 GOV R. Dinning South Brae Farm, Aiket Road, Dunlop, Kilmarnock KA3 4BP
G0 GOX F. Goodger 66 Selkirk Close, Wimborne BH21 1TP
G0 GOZ G. Reay 53 Tithe Barn Road, Stafford ST16 3PL
G0 GPB K. Love 16 Tenscore Avenue, Walsall WS6 7BX
G0 GPE D. Wells Browtop, Old Lane, Crowborough TN6 2AD
G0 GPF P. Crowley 45 Beeches Road Great Barr, Birmingham B42 2HJ
GI0 GPG D. Mckee 38 Nursery Road, Armagh BT60 4BL
G0 GPH Obo the Worcester Moonbounce Society c/o P. Butterworth 1 The Avenue, Bury BL9 5DQ
G0 GPK J. Burrows Applefields, Wilmingham Lane, Yarmouth PO41 0SL
GW0 GPQ R. Rees 5 Golwg Y Gaer, Salem, Llandeilo SA19 7PA
G0 GPR R. Wordsworth 59 Highgate Lane, Goldthorpe, Rotherham S63 9BA
G0 GPS I. Jones 107 Wolverley Road, Kidderminster DY11 5JN
G0 GPT S. Carter 8 St. Crispins Way, Ottershaw, Chertsey KT16 0RE
G0 GPV J. Barnes 24 Burleigh Place, Oakley, Bedford MK43 7SG
G0 GPX M. Wise 440 Tuahiwi Road Rd1, Kaiapoi 7691 New Zealand
GW0 GQC B. Morgan 5 Brynmawr, Bettws, Bridgend CF32 8SD
GI0 GQG J. Mills 60 Loughmacrory Road, Omagh BT79 0PH
G0 GQH A. Jenner 53 The Leys, Woburn Sands, Milton Keynes MK17 8QG
G0 GQI A. Orton 2 The Grove, Brampton Abbotts, Ross-on-Wye HR9 7JH
G0 GQJ T. Waters 42 Tregundy Road, Perranporth TR6 0EF
G0 GQK M. Evans St. Bega, Shay Lane, Newport TF10 8DA
G0 GQO S. Taylor 24 Catstree, Stirchley, Telford TF3 1XZ
G0 GQP D. Jackson 38 Chestnut Crescent, Bletchley, Milton Keynes MK2 2LA
G0 GQT M. Bishop 52 Lingley Drive, Wainscott, Rochester ME2 4NE
G0 GQV P. Leech Holly Tree Cottage, 32 Norwich Rd . Strumpshaw, Norwich NR13 4NF
G0 GQW H. Jones 21 Sandiways Road, Wallasey CH45 3HJ
G0 GQX M. Chappell 43 Aigburth Hall Avenue, Liverpool L19 9EA
G0 GQY D. Smith 62 Beresford Road, Dorking RH4 2DG
G0 GRB P. Attew 23 Kingsleigh Close, Trunch, North Walsham NR28 0QU
G0 GRC Grantham ARC c/o K. Burton 2 Council House, Stainfield Road, Kirkby Underwood, Bourne PE10 0SG
GM0 GRD R. Kelly Overmill John Allan Drive, Cumnock KA18 3AG
GM0 GRL D. Moore Parkview, Claredon Place, Dunblane FK15 9HB
G0 GRM G. Meanley Hemplands, 8 Bennett Drive, Warwick CV34 6QJ
G0 GRO B. Phillips 12 Fairview Avenue, Weston, Crewe CW2 5LX
GW0 GRQ J. Mitchell 44 Crossways St., Barry CF63 4PQ
G0 GRS G. Sawford 17 Church Road, Pytchley, Kettering NN14 1EL
G0 GRU R. Foster 30 Wimberley Way, South Witham, Grantham NG33 5PU
G0 GRV Y. Katoh 4-6-3 Minami-Aoyama, Minato-Ku, Tokyo 107-0062 Japan
GM0 GRW R. Young 8 Nursery Lane, Mauchline KA5 6EH
G0 GRX N.I DXer's Group c/o C. Ashlin 13 Brantfell Grove, Bolton BL2 5LY
G0 GRZ F. Pounder 29 Read Avenue, Beeston, Nottingham NG9 2FJ
G0 GSA R. Hall 47 Main Street, Stretton Under Fosse, Rugby CV23 0PE
G0 GSF Dr B. Austin 110 Frankby Road, West Kirby, Wirral CH48 9UX
GM0 GSG H. Cameron 36 Lynn Crescent, Kirkwall KW15 1FF
G0 GSH C. Summers 18 Hays Lane, Hinckley LE10 0LA
G0 GSJ D. Howie 22 Jason Street, Walney, Barrow-in-Furness LA14 3EJ
G0 GSK W. Blay 50 Fir St., Cadishead, Manchester M44 5AU
G0 GSL T. Ritchie 6 High Road East, Felixstowe IP11 9JT
G0 GSM T. Pritchard 21 Newlyn Drive, Bredbury, Stockport SK6 1EF
G0 GSN N. Pope 1 Knowsley Road West, Clayton Le Dale, Blackburn BB1 9PW
G0 GSR F. Johnson 9 Manor Close, Tavistock PL19 0PN
GW0 GST Northwich Repeater Group c/o A. Williams 34 Gwydyr Road, Llandudno LL30 1HQ
G0 GSU A. Dixon 33A Valley Road. Thornhill, Dewsbury WF12 0HY
G0 GSX C. Guerrero 183 Edinburgh House, Queensway Gibraltar
G0 GSY B. Thomsen 8 Richmond Road, Cleethorpes DN35 8PD
G0 GSZ P. Hunter 28 Hanover Court, Canterbury Way, Thetford IP24 1BZ
G0 GTI A. Dickinson 6 Church Lane, Bessacarr, Doncaster DN4 6QB
GM0 GTL T. Lorimer 443 Delgatie Court, Pitteuchar, Glenrothes KY7 4RW
G0 GTN J. Bumford 10 St. Alkmonds Square, Shrewsbury SY1 1UH
G0 GTV S. Moore 2 Sheppards Close, Heighington, Lincoln LN4 1TU
GW0 GTW G. Williams 1 Pont Y Berllan, Talycafn Road, Llanrwst LL26 0EF
GM0 GTY F. Stirling 42 Mcbain Place, Kinross KY13 8QZ
GW0 GUA J. Waddell 24 Gower View, Llanelli SA15 3SN
G0 GUC A. Russon 92 St. Georges Road, Dudley DY2 8ER
G0 GUD D. Law 10 Derwent, Tamworth B77 2LD
G0 GUE M. Tyou 19 Trinity Square, Margate CT9 1HU
G0 GUF J. Youde 4 Greenacre, Hixon, Stafford ST18 0QE
G0 GUG M. Grove 9 Foxcote Lane, Cradley Forge, Halesowen B63 2JJ
GM0 GUJ J. Cumming 24 Parkhill Wynd, Leven KY8 4LH
G0 GUN C. Critchley 3 Beaconsfield View, Robert Road, Slough SL2 3XT
G0 GUO J. Rosindale Treworder Farm, Ruan Minor, Helston TR12 7JL
G0 GUS S. Fellick Po Box 337, Kelmscott Wa 6995 Australia
G0 GUT N. Reed 30 Wrey Avenue, Liskeard PL14 3HX
GM0 GUU J. Stirling 42 Mcbain Place, Kinross KY13 8QZ
G0 GUV R. Murkin 59 Teagues Crescent, Trench, Telford TF2 6RF
G0 GUW B. Standen 43 Westover Gardens, St. Peters, Broadstairs CT10 3EY
GU0 GUX J. Scheffer Route De Carteret, Cobo, Castel GY5 7YS Guernsey
GW0 GUY E. Hollowell 54 Portfield, Haverfordwest SA61 1BW
G0 GVA J. Lawson 14 Kentmere Avenue, Farington, Leyland PR25 3UH
G0 GVB S. Balme 97 Backhold Drive, Halifax HX3 9DT
G0 GVE S. Thomas 14 Goodley, Oakworth, Keighley BD22 7PD
G0 GVF M. Davidson Cristina Bungalows, 26 Ave Del Pacific, Benelmadena Malaga Spain
G0 GVN C. Wackett 7 Newell Road, Hemel Hempstead HP3 9PD
G0 GVS N. Clayton 220 Milnrow Road, Rochdale OL16 5BB
G0 GVT J. Gibb 37 St. James Road, Melton, North Ferriby HU14 3HZ
G0 GVX B. Sherwood 363 Old Laira Road, Laira, Plymouth PL3 6DH
G0 GVZ A. Grimes 37 Cavendish Avenue, Cambridge CB1 7UR
G0 GWA Dr S. Browne 108 Whirley Road, Macclesfield SK10 3JL
G0 GWC D. Roberts 179 Southfield Avenue, Preston, Paignton TQ3 1JX
G0 GWD A. Brown 6 Laing Square, Wingate TS28 5JE
GW0 GWE R. Edwards 11 Trem Y Eglwys, Coed-Y-Glyn, Wrexham LL13 7QE

G0 GWG C. Partridge 44 Pine Close, South Wonston, Winchester SO21 3EB
G0 GWH A. Clampitt 139 Smiths Rd, Emerald Beach 2456 Australia
G0 GWI P. Bell 6 Dorchester Close, Hale, Altrincham WA15 8PW
G0 GWL D. Baker 2 Hall Lea Sedgefield, Stockton-on-Tees TS21 2AN
G0 GWM R. Sawkins 48 South Road, Saffron Walden CB11 3DN
G0 GWN J. Faulconbridge 32 Beridge Road, Halstead CO9 1LB
G0 GWP A. Horton 184 Mount Pleasant, Keyworth, Nottingham NG12 5ET
G0 GWS W. Duff Highfield, Maundown, Taunton TA4 2BU
G0 GWY G. Birch 33 Kenilworth Road, Scunthorpe DN16 1EY
G0 GXF P. Hirst 57 Etherington Drive, Hull HU6 7JT
G0 GXH P. Rogers 20 Clayfield Close, Nottingham NG6 8DG
G0 GXI S. Suter Fair View, Station Road, York YO62 4DG
G0 GXO H. Swaddle 12 Belmont Gardens, Haydon Bridge, Hexham NE47 6HG
GW0 GXQ J. Williams 91 Mold Road, Buckley CH7 2JA
G0 GXS R. Wareham 8 Meeting Lane Needingworth, St. Ives PE27 4SN
G0 GXT D. Mellor The Village Stores, Cleobury North, Bridgnorth WV166RP
G0 GXU M. Reeve 25 Chiltern Road, Hitchin SG4 9PJ
G0 GXX M. Smallwood 12 Rowan Walk, Hornsea HU18 1TT
G0 GXZ M. Butler 1A Springhead Avenue, Hull HU5 5HZ
G0 GYA C. Taylor 37 Manor Park Avenue, Allerton Bywater, Castleford WF10 2DN
G0 GYH D. Goodall 5 Coach Drive, Eastwood, Nottingham NG16 3DR
G0 GYI R. Griffith Wayside, Colchester Main Road, Colchester CO7 8DH
G0 GYJ M. Dawson 11 Eastholme Drive, York YO30 5SU
GM0 GYM T. Quinn 40 Drumry Road, Clydebank G81 2LL
GM0 GYN P. Gibson 7 Rogerhill Drive, Kirkmuirhill, Lanark ML11 9XS
G0 GYO M. Mcpherson 4 Highfield Place, Wideopen, Newcastle upon Tyne NE13 7HW
G0 GYP C. Fiedler 51 Fleet Lane, Tockwith, York YO26 7QD
GM0 GYQ Dr H. Garmany Woodside Cottage, Shielhill, Dundee DD4 0PW
GM0 GYT J. Ritchie 24 Kirkton Crescent, Knightswood, Glasgow G13 3AQ
G0 GYU P. Walters Stonecroft, Main Street, Knaresborough HG5 9LD
G0 GYY P. Markham 15 Victoria Road, Walton on the Naze CO14 8BU
G0 GZB J. Bartlett 44 Beverington Road, Eastbourne BN21 2SD
G0 GZE P. Wylie 15 Semley Road, Hassocks BN6 8PD
G0 GZF A. Pierce 17B Alderford Street, Sible Hedingham, Halstead CO9 3HX
G0 GZI R. Jeffery 7 Corfe Way, Winsford CW7 1LU
G0 GZL R. Daniels 8 Dun Cow Close, Brinklow, Rugby CV23 0NZ
G0 GZM M. Johns 3 Carew Close, Coulsdon CR5 1QS
G0 GZN L. Edmunds 27A Sea View Road, Parkstone, Poole BH12 3LP
G0 GZO K. Walters 18 Leabrooks Avenue, Sutton-in-Ashfield NG17 5HU
G0 GZP B. Lee 18 Boston Close, Eastbourne BN23 5RA
GW0 GZR Liverpool Raynet Group c/o M. Heel 27 Englefield Drive Oakenholt, Flint CH6 5SB
G0 GZU N. Harris 16 Gibbs Field, Bishop's Stortford CM23 4EY
G0 GZV K. Bailey 35 Edgehill Road, Chislehurst BR7 6LA
G0 HAE R. Isaac 12 Abbeyfields Close, Netley Abbey, Southampton SO31 5GR
G0 HAK P. Owens Flat 1, 45 The High Street, Enfield EN3 4EF
G0 HAL A. Paterson 1 Birch Grove, Timperley, Altrincham WA15 7YH
GM0 HAN K. Monaghan 10 Sauchiewood Cottages, Mintlaw, Peterhead AB42 5LR
G0 HAU E. Hazell 12 Fulford Close, Bideford EX39 4DX
G0 HAW C. Blackmoor 4 St. Godwalds Crescent, Aston Fields, Bromsgrove B60 2EB
G0 HBA R. Curzon 24 Edwards Drive, Wellingborough NN8 3JJ
G0 HBB D. Brown 15 Osborne Avenue, Tuffley, Gloucester GL4 0QN
GW0 HBD A. Kings 13 Fairfield Road, Bulwark, Chepstow NP16 5JP
GM0 HBF C. Fraser Rockside, Locheport, Isle of North Uist HS6 5EU
G0 HBJ R. Millett 8 Sidestrand Road, Newbury RG14 6HP
GM0 HBK C. Robertson 3 Sasaig, Teangue, Isle of Skye IV44 8RD
G0 HBL K. Alderman How Green Farm, Baldock Road, Buntingford SG9 9RH
G0 HBN W. Riley 18 Leyland Close, Trawden, Colne BB8 8TB
G0 HBO S. Holmes 17 Portland Gardens, Low Fell, Gateshead NE9 6UX
G0 HBS R. Darlington 7 Binbrook Place, Chorley PR7 2QU
G0 HBU A. Dwyer 10 Shaftway Close, Haydock, St. Helens WA11 0YQ
G0 HBV T. Sandilands 8 Cheviot View, Lowick, Berwick-upon-Tweed TD152TY
G0 HBW R. Fitzgerald 93 Finland Road, Brockley, London SE4 2JQ
G0 HBX J. Turbefield 126 Preston Road, Chorley PR6 7AU
GW0 HBZ D. Wright Drws-Y-Nant, St. Asaph Avenue, Kinmel Bay, Rhyl LL18 5EY
GW0 HCB P. Matheson 34 Pentwyn, Radyr, Cardiff CF15 8RE
G0 HCC Herts County C/ c/o T. Groves 31 Tunnel Wood Close, Watford WD17 4SW
G0 HCD T. Carroll 2 West Durham Cottages, Roddymoor, Crook DL15 9QX
G0 HCE M. Wilkinson 98 Whitestiles, High Seaton, Workington CA14 1LL
G0 HCI M. Bodle 10 Watts Road, Hedge End SO304EZ
GW0 HCK A. Morgan 1, Fryn Hir, Capel Dewi Road, Capel Dewi, Carmarthen SA32 8AY
GW0 HCN T. Hayden 7 Attlee Close, Garnlydan, Ebbw Vale NP23 5ES
G0 HCP Rvd. D. Goode 12 Hills Lane, Ely CB6 1AY
GM0 HCQ M. Gloistein 27 Stormont Way, Scone, Perth PH2 6SP
G0 HCR S. Turner 71 Valley Road Lillington, Leamington Spa CV32 7RX
G0 HCY Ci Bach DX Group c/o G. Vine 56 Colchester Road, St. Osyth, Clacton-on-Sea CO16 8HB
G0 HDC G. Evans 34 Coronation Drive, Donnington, Telford TF2 8HY
G0 HDD A. Adey 37 Cranmere Avenue, Wolverhampton WV6 8TR
G0 HDF R. Bartlam 34 Quarry Road, Selly Oak, Birmingham B29 5NX
G0 HDG A. Edwards 29 Larch Road, Maltby, Rotherham S66 8AZ
G0 HDH C. Worsfold 7 West View Road, Cowes PO31 8NR
G0 HDI B. Walker Lantilla, Elmfield Lane, Calshot, Southampton SO45 1BJ
G0 HDJ A. Douglas Threave House, Blind Lane, Barton St. David TA11 6BW
GI0 HDO M. Deehan 9 Farland Way, Londonderry BT48 0RS
G0 HDP B. Wainwright 19 Durkar Low Lane, Durkar, Wakefield WF4 3BL
G0 HDS R. Hurt 7 Atwood Close, Immingham DN40 2DQ
GW0 HDY R. Finnis 6 The Crwscent, Cwmbran NP44 7JG
G0 HDZ I. Rose 82 Little Brays, Harlow CM18 6ES
G0 HEA B. Griffin 26 Hamer Street, Radcliffe, Manchester M26 2RS
G0 HEE A. Salt New Meg, Greetwell Lane Nettleham, Lincoln LN2 2NQ
G0 HEF R. Mckeever 38 Brookfield Avenue, Runcorn WA7 5RF
G0 HEJ E. Lloyd 43 Queensway, Upton, Chester CH2 1PF
G0 HEL A. Nicholls 19B Dark Lane South, Steeple Ashton, Trowbridge BA14 6EZ
G0 HEM G. Gardner New House, Birdbush Avenue, Saffron Walden CB11 4DJ
G0 HEN G. Henstock 36 Cornwallis Road, Oxford OX4 3NW
G0 HER K. Kibblewhite 53 Woodcote, Bedford MK41 8EL
G0 HET P. Nutkins 31 Higher Spence Cottage, Bridport DT6 6DF
G0 HEU P. Stott 70 Wansbeck, Washington NE38 9EG
G0 HEV P. Brindley 6 Chaucer Close, Stowmarket IP14 1GH
G0 HEW J. Mchale 21 Bonython Road, Newquay TR7 3AW
G0 HEX D. Cheetham 4 Battersbay Grove, Hazel Grove, Stockport SK7 4QW
G0 HFA Dr J. Lansdowne 8 Nansloe Close, Helston TR13 8BP
G0 HFC F. Chadwick 59 Beech Avenue, Greenfield, Oldham OL3 7AW
G0 HFE T. Cadman 28 Denbigh Court, Ellesmere Port CH65 5DX
G0 HFK R. Fuller Flat 13, Samuel Lewis Trust Dwellings, London SW6 1BS
G0 HFL N. Major 10 Heather Close, Branston DE14 3FL
G0 HFO D. Hirst 10 The Rogers, Shanklin PO37 7HH
G0 HFX C. Parnell 29 Southfield, Southwick, Trowbridge BA14 9PW
GW0 HGC B. Roberts Ty Clyd, Ffordd Mela, Pwllheli LL53 5AP
G0 HGG S. Entwisle 30 Arden Mhor, Pinner HA5 2HR
G0 HGH J. Scott 3 Westminster Drive, Spalding PE11 2UW
G0 HGI B. Boden 71 Park Head Road, Sheffield S11 9RA
G0 HGM J. Jenkinson 4 Greenways, Ilfracombe EX34 8DT
GW0 HGN T. Jones 11 Lon Ogwen, Bangor LL57 2UD
G0 HGO D. Cady 45 The Hill, Wheathampstead, St. Albans AL4 8PR
GW0 HGP Dr L. Magfhogartai Talaharian, Abergwili, Abergwili, Carmarthen SA31 2JL
G0 HGV D. Burgess 243 Lichfield Avenue, Torquay TQ2 8AJ
G0 HHA P. Dixon 68 Chelsea Road, Sheffield S11 9BR
G0 HHC A. Gorton 7 Sterling Close, Colchester CO3 9DP
GW0 HHD G. Williams Gwastad Annas, Barmouth LL42 1DX
GI0 HHE J. Dowey 19 The Bridges, Newtownabbey BT37 0TD
G0 HHN P. Le-Brun The Granary, Henton, Chinnor OX39 4AE
GI0 HHV S. Johnstone 6 Gilbert Crescent, Bangor BT20 4PE
GW0 HHW B. Hayward Tyn Y Gerddi, Deiniolen, Caernarfon LL55 3ND
GI0 HHZ R. Fitzsimons 9 Ingledene Park, Newtownards BT23 8QT
G0 HIC J. East 35 Preachers Vale, Coleford, Radstock BA3 5PT
G0 HID D. Paine 35 Marfield Close, West Midlands B76 1YD
G0 HIF A. Edmonds 44 Blea Tarn Road, Kendal LA9 7NA
GM0 HIG C. Cameron 36 Lynn Crescent, Kirkwall KW15 1FF
G0 HIJ W. Roberts 13 Roseacre Road, Elswick, Preston PR4 3UD
G0 HIK N. Gregory Town End, Kirkby Road, Askam-in-Furness LA16 7EY
GM0 HIM J. Pert 56 Lochiel Drive, Milton Of Campsie, Glasgow G66 8ET
GW0 HIR A. Edwards Flat 6, Oaktree Court, Fields Road, Cwmbran NP44 3AZ
G0 HIU M. Valentine 34 Meadow Avenue, Preesall, Poulton-le-Fylde FY6 0HA
G0 HIW D. Ross 60 Kingsway, South Molton EX36 4AL
G0 HIZ D. Hughes 27 Thornfields, Crewe CW1 4TY
G0 HJB F. Little 26 Hoghton Road, Longridge, Preston PR3 3UA
G0 HJD J. Ford 42 The Grove, Walton-on-Thames KT12 2HS
G0 HJK J. Everett 92 Thackeray Road, Ipswich IP1 6JB
G0 HJL J. Levesley 96 Brookside Road, Bransgore, Christchurch BH23 8NA
G0 HJM R. Smith 100 Braemar Road, Billingham TS23 2AN
G0 HJR R. Filby 10 Malvern Avenue, Burton-on-Trent DE15 9EB
GM0 HJU J. Mackenzie 17 Maple Avenue, Milton Of Campsie, Glasgow G66 8BB
GM0 HJV C. Mcclure 27 St. Andrews Drive, Gourock PA19 1HY
G0 HJW P. Webber Palm Court Hotel, 1 Lansdowne Road, Falmouth TR11 4BE
G0 HJX P. Devine 41 Carodoc Road, Wingate TS28 5BT
G0 HJZ J. Durrell 8 Woodcote Cottages, Graffham, Petworth GU28 0NY
G0 HKB R. Conneely Harford House, Wells Road, Radstock BA3 4EX
G0 HKC E. Chambers 19, Courville Close, Alveston BS35 3RR
G0 HKE J. Boyton 3 Wenny Estate, Chatteris PE16 6UX
G0 HKF A. Roberts 149 Cannock Road, Burntwood WS7 0BB
G0 HKN R. Leigh 19 Richmond Court, Worthing BN11 4JB
GW0 HKQ D. Suddes Villa Dobochet, Holway Road, Holywell CH8 7DR
G0 HKW Oldham ARC c/o D. Spencer 48 Ingleborough Way, Leyland PR25 4ZR
G0 HKZ S. Liptrott 8 Fox Bank Close, Widnes WA8 9DP
G0 HLA T. Barker Peartree Cottage, Ash Hill Common, Romsey SO51 6FU
G0 HLB Dr R. Evans 191 St. Leonards Road East, Lytham St. Annes FY8 2HW
G0 HLI A. Dudley 7 St. Michaels Close, Willington, Derby DE65 6EB
G0 HLJ I. Barber 17 Copley Crescent, Scawsby, Doncaster DN5 8QW
GM0 HLK M. Borthwick 5/11 Dalgety Avenue, Edinburgh EH7 5UF
G0 HLL R. Steel 43 Westfield Avenue, Wigston LE18 1HY
G0 HLS S. Deacon 32 Westfield Way, Charlton, Wantage OX12 7EW
G0 HLU A. Mckay 48 Holdenby Close, Retford DN22 6UB
GM0 HLV D. Gill Slowbend Cottage, Loth, Helmsdale KW8 6HP
G0 HLW M. Neale 126 Brookvale Road, Solihull B92 7JB
G0 HMD M. Mills 44 East Acridge, Barton-upon-Humber DN18 5HH
G0 HME C. Statham 24 St. Johns Close, Heather, Coalville LE67 2QL
G0 HMF J. Tite 40 Dinglederry, Olney MK46 5ES
G0 HMK T. Angier 29 Sunnyhill Road, Herne Bay CT6 8LT
GM0 HMM F. Robertson-Mudie 18 Portnaguran, Isle of Lewis HS2 0HD
G0 HMO E. Cooke 190 Newark Crescent, Nottingham NG2 4NY
G0 HMX M. Wragg 27 Rosedale, Worksop S81 0TB
G0 HND A. Simmonds1 Fern Way, Purssells Meadow, Naphill, High Wycombe HP14 4SG
GW0 HNE K. Luke 19 Heol Y Gors, Cwmgors, Ammanford SA18 1PE
G0 HNG B. Horton 3 Cromer Road, Finedon, Wellingborough NN9 5LP
G0 HNI J. Buckler 181 Glen Road, Oadby, Leicester LE2 4RJ
GM0 HNJ J. Cowan 20A Harbour View, Invergordon IV18 0EY
G0 HNL T. Cannon 10 Badger Close, Guildford GU2 9PJ
G0 HNO T. Callanan 39 Greenlands Way, Henbury, Bristol BS10 7PH
GM0 HNP E. Bottomley 33 Duke Street, Coldstream TD12 4BS
G0 HNQ H. Brammeld School House, Rosley, Wigton CA7 8AU
GW0 HNS S. Yates 46 Y Berllan, Dunvant, Swansea SA2 7RW
GW0 HNT W. South Kimberlee, 45 Dan Y Bryn, Neath SA11 3PJ
GM0 HNV P. Mackenzie 16 Cedar Drive, Milton Of Campsie, Glasgow G66 8AY
G0 HNW P. Widger Notre Revie, Cinderhills Road, Holmfirth HD9 1EH
G0 HNZ R. Disney 25 Davos Way, Skegness PE25 1EL
G0 HOB J. Mc Glynn 1 Primrose Crescent, Leeds LS15 7QW
G0 HOC M. Harris Po Box 841, Hobart, Tasmania 7001 Australia
G0 HOD N. Colbourn 7 Brookwood Road, Farnborough GU14 7HH
G0 HOF K. Barnett 5 Morborne Road, Folksworth, Peterborough PE7 3SS
G0 HOJ J. Hoyland 67 Coronation Road, Wroughton, Swindon SN4 9AT
G0 HOP G. Evans 12 Marlowe Close, Stevenage SG2 0JJ
G0 HOQ M. Piper Applegarth Milford Road, Barton on Sea BH25 5PW

G0 HOS B. Tufnail 197 Portland Road, Wyke Regis, Weymouth DT4 9BH
G0 HOT M. Reeds 6 Leach Way, Riddlesden, Keighley BD20 5DB
G0 HOV A. Howes 11 Stretton Road, Wolston, Coventry CV8 3FR
G0 HOX I. Atkins 64 Twin Hill Lane, Stafford, Virginia United States
G0 HPA G. Smith 3 Park Lane, Featherstone, Pontefract WF7 6BL
GW0 HPC G. Evans 49 Pen Yr Ally Avenue, Neath SA10 6DS
G0 HPG L. Brookes 177 Charnwood Close, Rubery, Birmingham B45 0JY
G0 HPH P. Brookes 177 Charnwood Close, Rubery, Birmingham B45 0JY
GM0 HPK A. Gaston 9 Lochans Mill Avenue, Lochans, Stranraer DG9 9BZ
GM0 HPL T. Mccutcheon 25 Millburn Court, Sheuchan St., Stranraer DG9 0DX
G0 HPM R. Waddingham 59 Second Avenue, Frinton-on-Sea CO13 9LY
G0 HPN L. Denyer 7 Long Close, Chippenham SN15 3JY
G0 HPQ P. Delaney 27 Grasmere Road, Redcar TS10 1JA
G0 HPS G. Bukin Granary Cottage, Uffculme, Cullompton EX15 3DN
G0 HPV K. Hardy 12 Liquorpond Street, Boston PE21 8UF
GM0 HQF E. Kolonko West Manse, School Road, Kilbirnie KA25 7LB
GM0 HQG G. Flett Stenadale, Orphir, Orkney KW17 2RF
G0 HQH S. Mayer 7 Wright Avenue, Chesterton, Newcastle under Lyme ST5 7PB
G0 HQK J. Jones 28 Clares Lane Close, The Rock, Telford TF3 5DA
G0 HQN C. Wilson 34 Lyndhurst St., Salford M6 5YB
GM0 HQT B. Bell 4 Broadlee Bank, Tweedbank, Galashiels TD1 3RF
G0 HQU S. Broadhead 39 Haw Avenue, Yeadon, Leeds LS19 7XE
G0 HQX R. Rea 11 Wissage Lane, Lichfield WS13 6DQ
G0 HRD L. Lawrence 119 Ladywood Road, Lane End, Dartford DA2 7LP
G0 HRF B. Barwick 100 Westwood, Golcar, Huddersfield HD7 4JY
GW0 HRG Halkyn Radio Group c/o P. Clark 14 Lincoln Road, Ewloe, Deeside CH5 3RW
G0 HRH G. Vaughton Higher Woodhayne, Whitford, Axminster EX13 7PB
G0 HRJ Mid Lanark ARS c/o C. Mason 22 Eskdale Avenue, Halifax HX3 7NH
G0 HRK R. Eselgroth 15 Hedgerley Gardens, Greenford UB6 9NT
G0 HRL G. Ward 67 Sebright Road, Wolverley, Kidderminster DY11 5UA
G0 HRO J. Bland 5 King Street, Mansfield NG18 2PX
G0 HRR K. Mott 191 Joyners Field, Harlow CM18 7QD
G0 HRS Hilderstone Radio And Electronic Society c/o I. Lowe 54 College Road, Margate CT9 2SW
G0 HRT R. Harwood 4 Bartholomew Close, Walton Park, Milton Keynes MK7 7HH
G0 HRW Bromsgrove + District ARC c/o D. Barkley 39 Fulbeck Avenue, Wigan WN3 5QN
G0 HRX C. Deakin Birches, Lanreath, Looe PL13 2NX
G0 HSA A. Bennett 21 Hunstone Avenue, Norton, Sheffield S8 8GE
GI0 HSB W. Dickson 8 Drumglass Avenue, Bangor BT20 3HA
GM0 HSC H. Cumming 42 Hadrian Way, Bo'ness EH51 9QN
G0 HSD A. Shaw 15 Austerby, Bourne PE10 9JJ
G0 HSH D. Denford 20 Shaxton Crescent, New Addington, Croydon CR0 0NU
G0 HSK J. Hakes 86 Station Crescent, Rayleigh SS6 8AR
G0 HSN P. Broadley 8 Langley Gardens, Lowestoft NR33 9JE
G0 HSR Huntingdonshire ARS c/o M. Harrington Tanglewood House, Station Road, , Tilbrook PE28 0JY
G0 HSV S. Jacob 3 Broadfield Close, Gomeldon, Salisbury SP4 6LX
G0 HSW B. Statham 11 Old Woods Hill, Torquay TQ2 7NR
G0 HSX J. Harvey 38 Honey Close, Dagenham RM10 8TE
G0 HTD H. Todd 105 Brownhill Road, Blackburn BB1 9QY
G0 HTG P. Hague 14 Camellia Close, Driffield YO25 6QT
GM0 HTH J. Grieve Langamo Harray, Kirkwall KW17 2JU
G0 HTK T. St John-Murphy Sherbourne, Sherbourne Drive, Windsor SL4 4AE
G0 HTL B. Sargent 25 Jordans Way, Bricket Wood, St. Albans AL2 3SJ
G0 HTM Dr R. Bushell 121 Rickmansworth Road, Watford WD18 7JD
G0 HTO V. Morris 16 Wentworth Close, Longlevens, Gloucester GL2 9RB
G0 HTS S. Alder 56 Sparrowbill Way, Patchway, Bristol BS34 5AU
GM0 HTT A. Flett Shannon Dounby, Orkney KW17 2HR
G0 HTX A. Coyle 8 Farm Place, Kensington, London W8 7SX
G0 HUD D. Withers 141 Broadway, Walsall WS1 3HB
G0 HUF P. Peterson 13 Orchard Road, Smallfield, Horley RH6 9QP
G0 HUG P. Hemphill Chapel House, 19 Little Ditton, Woodditton, Newmarket CB8 9SA
G0 HUH J. O'Dowd 139A Town Lane, Denton M342DJ
G0 HUK G. Roberts 24 Cornwall Road, Bingley BD16 4RN
GW0 HUN M. Roberts 19 Bro Geirionydd, Trefriw LL27 0JE
G0 HUQ R. Monk 43 Lichfield Drive, Bury BL8 1BJ
G0 HUT T. Hutton 4 Victoria Gardens, Farnborough GU14 9UH
G0 HUW A. Dyson 24 Newborough Close, Austrey, Atherstone CV9 3EX
G0 HUX D. Griffin 62 Buckshaft Road, Cinderford GL14 3AX
G0 HVA J. Curtis Glebe Farm, West Knighton, Dorchester DT2 8PE
G0 HVB D. Clark 3 West Well Lane, Theale, Wedmore BS28 4SW
G0 HVC W. Hutchins 25 Manor Road, Herne Bay CT6 6RF
GI0 HVJ R. Cunliffe 82 Strabane Road, Newtownstewart, Omagh BT78 4JZ
G0 HVN D. Cottam 14 Barnard Close, Rednal, Birmingham B45 9SZ
G0 HVO R. Minter 15 Harold Close, Pevensey Bay, Pevensey BN24 6SL
G0 HVP D. Cooper 11 Downland Court, Magazine Road, Ashford TN24 8NF
GM0 HVS D. Kearns Mo Dhachaidh, Portnacroish, Appin PA38 4BL
G0 HVX R. Mcquillan 9 Sandpit Road, Welwyn Garden City AL7 3TW
GD0 HWA C. Howard 5 Ballure Grove, Ramsey IM8 1NF Isle of Man
GM0 HWB P. Lafferty 18 Chesters Crescent, Motherwell ML1 3QU
G0 HWC P. Young 14 Carisbrooke Avenue, Clacton-on-Sea CO15 4RZ
G0 HWI P. Mardle 55 Chelmsford Avenue, Southend-on-Sea SS2 6JG
G0 HWK M. Drew 10 Marina Drive, Dunstable LU6 2AH
GI0 HWO J. Crawford-Baker 131 Gobbins Road, Islandmagee, Larne BT40 3TX
G0 HWP A. Wilson 41 Bentley Drive, Walsall WS2 8RX
G0 HWQ P. Wood 2 Kingsbridge Close, Braintree CM7 5NB
G0 HWS P. Dowsett Furze Cottage, West Chiltington Road, Pulborough RH20 2PR
G0 HWT P. Littlechild 13 Castle Hill, Daventry NN11 4AQ
G0 HWU A. Edwards Higher Tregiddle Farm, Gunwalloe, Helston TR12 7QW
G0 HWV B. Lawton 9 Byron Court, Kidsgrove, Stoke-on-Trent ST7 4JF
G0 HWY P. Adams 229 Upper Selsdon Road, South Croydon CR2 0DZ
G0 HXC F. Pegg 20 Swaledale Avenue, Cowpen Estate, Blyth NE24 4DT
G0 HXD P. Halsall 22 Northway, Northwich CW8 4DF
G0 HXF A. Etheridge Wyngra, Haywards Heath Road, Haywards Heath RH17 6NJ
GI0 HXH A. Mccaldin 7 Mount Ida Road, Banbridge BT32 4HF
G0 HXL E. Calthorpe 49 Cross Coates Road, Grimsby DN34 4QH
G0 HXM A. Delves 11 Willoughby Road, Langley, Slough SL3 8JH
G0 HXR G. Martin 34 Parson Drove, West Pinchbeck, Spalding PE11 3QW
GW0 HXS T. Williams Bryndewi, Llanarth SA47 0QN
G0 HXU S. Dawson Hamlet Cottage, West End, Tadcaster LS24 9DL
G0 HYG R. Taylor 79 South Drive, Harwood, Bolton BL2 3NS
GW0 HYH V. Grayson Willow Lodge, Croeslan, Llandysul SA44 4SJ
GW0 HYL V. Underwood Rock Hill, Llanarthney, Carmarthen SA32 8LJ
GD0 HYM M. Dunning 55 Station Park, Colby IM9 4NL Isle of Man
G0 HYN D. Robertson 5 Chandlers, Orton Brimbles, Peterborough PE2 5YW
G0 HYP K. Ferguson 4 Honister Road, Whitehaven CA28 8HS
G0 HYR R. Deakin 5 Burton Road, Oakthorpe, Swadlincote DE12 7QU
G0 HYS S. Hutchinson 55 Richmond Avenue, Sheffield S13 8TH
G0 HYT P. Gray 2 Bryan Close, Sunbury Oppn Thames TW16 7UA
GW0 HYU K. Richards 7 Capel Edeyrn, Pontprennau, Cardiff CF23 8XJ
G0 HZA M. Rolph 60 Queen Street, Swaffham PE37 7BT
G0 HZB M. Roberts 4 Elba Close, Paignton TQ4 7LW
G0 HZC Dr G. Hutt 21 Bentley Crescent, Fareham PO16 7LU
G0 HZD W. Brown Longstone Cottage, Old Pound, St. Austell PL26 7XS
G0 HZE R. Howell 161 Coneygree Road, Stanground, Peterborough PE2 8LH
G0 HZG P. Sturgess 45 Queensmead Close, Groby, Leicester LE6 0YP
GM0 HZI N. Mclaren 10 Newton Avenue, Skinflats, Falkirk FK2 8NP
G0 HZK R. Muggleton 70 Front Street East, Wingate TS28 5AG
G0 HZL M. Hawkins Flat 8, Library Mews, Hawes Side Lane, Blackpool FY4 5RF
GM0 HZO T. Leckie 1 Dykehead, Port Of Menteith, Stirling FK8 3JY
G0 HZQ R. Eldrett 20 Hill Rise, Horspath, Oxford OX33 1TJ
G0 HZY H. Harding 29 Brighton Avenue, Elson, Gosport PO12 4BU
G0 IAA H. Singer Neptune Gap, Kent CT5 1EL
G0 IAC C. Caspell 28 Peel Terrace, Stafford ST16 3HD
G0 IAD A. Dobbyn Roadways, Selwick Drive, Bridlington YO15 1AP
G0 IAE D. Yeo 356 Radcliffe Road, Fleetwood FY7 7NH
G0 IAG A. King 2 Ebenezer Cottages, Thorney Road, Peterborough PE6 7UB
G0 IAH T. Preece White House, Bredwardine, Hereford HR3 6BY
G0 IAI P. Dean Down Farm, Lovaton, Yelverton PL20 6PT
G0 IAK M. Watts Hainault, Coronation Avenue, Bradford-on-Avon BA15 1AX
G0 IAL R. Scott 46 St. Albans Road, Hemel Hempstead HP2 4BA
G0 IAP P. Allen 25 Wayside Avenue, Hornchurch RM12 4LL
G0 IAS A. Hickman The Conifers, High Street, Retford DN22 8AJ
G0 IAX R. Bailey 3 Charlotte Close, Birstall, Batley WF17 9BX
G0 IAY L. Wiltshire 7 Burleaze, Chippenham SN15 2AY
GI0 IBC Lorn RAC c/o P. Hallam 95 Belfast Road, Carrickfergus BT38 8BY
G0 IBE R. Higgs 60 Lichfield Avenue, Evesham WR11 3EA
G0 IBG B. Armstrong 65 Asquith Road, Bentley, Doncaster DN5 0NT
G0 IBI R. Clarke 14 Cranwell Avenue, Cranwell, Sleaford NG34 8HG
G0 IBJ R. Bishop 25 Hennings Park Road, Poole BH15 3QU
G0 IBN A. Kersey 35 Sceptre Close Tollesbury, Maldon CM9 8XB
G0 IBR S. Dalley 5 Anstey Mill Close, Alton GU34 2QT
G0 IBS K. Brown Fernbank, Foreside Lane, Bradford BD13 4EY
G0 IBT H. Percy 13 Cherry Tree Walk, Astley Cross, Stourport-on-Severn DY13 0JT
G0 IBW D. Jones 5 Luccombe Close, Ingleby Barwick, Stockton-on-Tees TS17 0NL
G0 IBY F. Hubbard 187 Standhill Road, Carlton, Nottingham NG4 1LE
G0 IBZ M. Hackford Snuggles, Rockalls Road, Colchester CO6 5AR
G0 ICB J. Howell 21 Peaslands Road, Saffron Walden CB11 3ED
G0 ICC R. Shirvington 3 Main Street, Preston Bissett, Buckingham MK18 4LH
G0 ICD D. Matthews 32 Stewards Avenue, Widnes WA8 7BN
G0 ICE M. Knott 24 Walsingham Way, Ely CB6 3AL
G0 ICG Sir T. Lees Post Green, Lytchett Minster, Dorset BH16 6AP
G0 ICJ D. Dawkes 95 Houndsfield Lane, Wythall, Birmingham B47 6LX
G0 ICK Glasgow University Wireless Society c/o I. Kraven 55 Cranfield Crescent, Cuffley, Potters Bar EN6 4DZ
G0 ICP Dr J. Ker 1 The Willows, Ulcombe Road, Ashford TN27 9QR
G0 ICW M. Bagnall 6 Silver Fir Close, Hednesford, Cannock WS12 4SU
G0 IDB S. Jeffreys 26 Meadway, Esher KT10 9HF
G0 IDD D. Clarke 60 Dunedin Crescent, Burton-on-Trent DE15 0EJ
G0 IDE D. Dobson 79 Wood Green, Leyland PR25 2YL
G0 IDF D. Fletcher 12 Drayton Road, Dorchester-On-Thames, Wallingford OX10 7PJ
G0 IDH M. Frost Kernyk, Trevanion Terrace, Carn Brea Village, Redruth TR15 3BP
GM0 IDJ J. Low 56 George St., Whithorn, Newton Stewart DG8 8NZ
G0 IDL M. Bedell 1 Pheasant Field Drive, Spondon, Derby DE21 7LR
G0 IDP D. Ford 2 Bedelands Close, Burgess Hill RH15 8BL
G0 IDS S. Mcmaster 12 Gilpin Road, Newton Aycliffe DL5 5EQ
GD0 IDU N. Bayliss Teal, Tromode Road, Douglas IM2 5EH Isle of Man
GM0 IDV A. Price Upper Arsdale, Evie, Orkney KW17 2NN
G0 IDZ L. Bailey 22 Coventry Grove, Wheatley, Doncaster DN2 4QA
G0 IEB R. Neal 6 Wheatcroft Avenue, Scarborough YO11 3BN
G0 IEH T. Lister 6 Fordlands Crescent, Fulford, York YO19 4QQ
G0 IEN Dr R. Wharton 1407 Briar Bayou Dr, Houston 77077 United States
G0 IEO R. Woodberry 35 Whybridge Close, Rainham RM13 8BB
G0 IEQ G. Kennedy 1 Hornbeam Avenue, Great Sutton, Ellesmere Port CH66 2US
G0 IER B. Smith 73 Devon Street, Hull HU4 6PL
G0 IES D. Johnson 32 Middleton Road, North Reddish, Stockport SK5 6SH
GM0 IET M. Shell 15 Lundin View, Leven KY8 5TL
G0 IEW J. Rose 1 Nelson Place, Whiston, Prescot L35 3PP
G0 IEY S. Tribe 6 Privett Road, Waterlooville PO7 5HJ
G0 IFA K. Keitch 10 Sycamore Close, Willand, Cullompton EX15 2SH
GW0 IFB I. Griffiths 21 Queensway, Hope, Wrexham LL12 9PD
G0 IFC A. Burnett 2 Courtney Road, Tiverton EX16 6EE
G0 IFD T. Street Flat 18, 58 Mapledene Road, London E8 3LE
G0 IFF G. Spinney 7 Nightingale Gardens Nailsea, Bristol BS48 2BH
G0 IFL R. Finch 6 Clover Way, Thetford IP24 1LQ
G0 IFN Harig c/o R. Moriarty 46 Oak Avenue, Morecambe LA4 6HS
G0 IFQ P. Lait Glenside, 8 Kingston Lane, Shoreham-by-Sea BN43 6YB
G0 IFS N. Clayton Ringinglow, Fair Lawn, Whitstable CT5 3JZ
G0 IFT P. Nurse 67 Grasleigh Way, Allerton, Bradford BD15 9BD
GD0 IFU W. Corkish 23 Bollan Drive, Ballagarey, Glen Vine, Isle of Man IM4 4FE
G0 IFW M. Mcnamara 7 Moray Road, Chadderton, Oldham OL9 8AE
G0 IFX B. Galloway 2 Summers Close, Knutsford WA16 9AW
G0 IGA R. Clark 9 Windsor Close, Hove BN3 6WQ
G0 IGB V. Elliott 22 Kirkstead Road, Cheadle Hulme, Cheadle SK8 7PZ
G0 IGC J. Golightly 123 Littlefield Lane, Grimsby DN34 4PN

G0 IGH G. Golightly 123 Littlefield Lane, Grimsby DN34 4PN
GM0 IGJ J. Dickson Eilrig, Roberton, Hawick TD9 7PR
G0 IGK G. Knox 33 St. Lukes Road, Aller, Newton Abbot TQ12 4NE
G0 IGM M. Manning 31 Harcourt Crescent, Shrewsbury SY2 5LQ
G0 IGT H. Terry 7 Marlow Drive, Irlam, Manchester M44 6LR
G0 IGU G. Morton 23 Bembridge Road, Eastbourne BN23 8DX
G0 IHA G. Goodwin Thumpers, 16 St. Catherines Road, Winchester SO23 0PP
G0 IHC M. Nash 49 Oakfield Way, Sharpness, Berkeley GL13 9UT
G0 IHE Q. Reed 3 Carre Gardens, Worle, Weston-Super-Mare BS22 7YB
G0 IHF W. Hough 19 Farnham Close, Appleton, Warrington WA4 3BG
G0 IHI M. Eyers 190 Greenhill Road, Herne Bay CT6 7RS
G0 IHK I. Tough 15 Headlands Way Whittlesey, Peterborough PE7 1RL
G0 IHO R. Priestley Le Haut Courtigne, Parcay Les Pins, Maine Et Loire 49390 France
G0 IHU C. Smith 66 Bronte Farm Road, Shirley, Solihull B90 3DF
G0 IIA J. Stokes The Beehive, Debenham Road, Ipswich IP6 9TD
G0 IID Newbury Vintage Wireless Society c/o J. Batley 3 Folldon Avenue, Sunderland SR6 9HP
G0 IIE J. Colliton Po Box 17, Ramsgate 2217, Sydney 2219 Australia
G0 IIF J. Green 1 Huntley Grove, Sheffield S11 7LX
G0 IIG G. Fleming 168 Blythway, Welwyn Garden City AL7 1DU
G0 IIK N. Ackland 69 Great South West Road, Hounslow TW4 7NH
GM0 IIO G. Berrich Juno Lodge The Woods Diverswell Farm Fishcross, Alloa FK10 3AN
G0 IIP R. Chambers 12 Dorchester Close, Maidenhead SL6 6RX
G0 IIQ D. Pykett 35 Harneis Crescent, Laceby, Grimsby DN37 7BA
GI0 IJB J. Stevenson 22 Knockfergus Park, Greenisland, Carrickfergus BT38 8SN
GM0 IJD P. Taggerty 44 Ravenswood Drive, Glenrothes KY6 2PA
G0 IJI A. Muir Po Box 392363, Dubai 0 United Arab Emirates
G0 IJK B. Puncher Danbys Oast, Coldbridge Lane, Maidstone ME17 2AX
G0 IJN A. Slade Skerries, Summerhill Althorne, Chelmsford CM3 6BY
GM0 IJR I. Ross 14 Kilmundy Dr, Burntisland KY3 0JW
G0 IJU P. Elms 12 Suffield Way, King's Lynn PE30 3DE
GW0 IJY I. Roberts 7 Heol Bradwen, Four Mile Bridge, Holyhead LL65 2NF
G0 IJZ Dr M. Walden 181 Coleridge Road, Cambridge CB1 3PW
G0 IKB K. Brown 1 Deerwood Close, Macclesfield SK10 3RE
G0 IKC R. Hole 3 Holywell Park, Halwill, Beaworthy EX21 5UD
G0 IKD A. Galvin 27 Hill Top Lane, Tingley, Wakefield WF3 1HT
G0 IKE L. Dodson 1 Limmer Lane, Booker, High Wycombe HP12 4QR
G0 IKI M. Holbrough 21 Malyns Close, Chinnor OX39 4EW
G0 IKN S. Mcdonald 152 Stony Lane, Burton, Christchurch BH23 7LD
G0 IKP T. Clements 72 Stanbridge Road, Haddenham, Aylesbury HP17 8HN
G0 IKQ Daventry Amateur Radio Repeater Group c/o A. Cook 17 Lothersdale, Wilnecote, Tamworth B77 4HT
G0 IKR M. Best 21 Hubble Road, Corby NN17 1JD
GM0 IKY A. Diamond 51 Marchlands Avenue, Bo'ness EH51 9ER
G0 IKZ J. Pugh 1 The Lawns, Everton, Sandy SG19 2LB
G0 ILA N. Smith 165 Upper Deacon Road, Southampton SO19 5LN
GM0 ILB I. Brown Vadill, Brae, Shetland ZE2 9QN
G0 ILC J. Price 30 Pottery Close, Whiston, Prescot L35 3RW
G0 ILD B. Harrison 61 Foyle Road, Blackheath, London SE3 7RQ
G0 ILH J. Collins 14 Balmain Crescent, Wolverhampton WV11 1BG
G0 ILI G. Pomroy 17 Rock Close, Pengegon, Camborne TR14 7TT
G0 ILK B. Loram 12 The Finches, Castleham, St. Leonards-on-Sea TN38 9LQ
G0 ILN R. Putnam 95 Martyns Way, Bexhill-on-Sea TN40 2SH
G0 ILO Dr P. Taylor 16 Petrel Close, Herne Bay CT6 6NT
GM0 ILQ I. Ferguson C/O Mr Jb Ferguson, 8 Cleveden Crescent, Glasgow G12 0PB
G0 ILT T. Drummond 10 Delamere Road, Earley, Reading RG6 1AP
G0 ILV K. Betts 6, The Grove, , Sunderland SR5 3EG
G0 ILZ M. Hanraads 6 Oak Hill, Hollesley, Woodbridge IP12 3JY
G0 IMA P. Pearce 7 Otter Road, Clevedon BS21 6LQ
G0 IMB R. Battersby 4 Gorsey Brow, Urmston, Manchester M41 9QE
G0 IMD R. Clamp 276 The Parade, Greatstone, New Romney TN28 8UL
G0 IMG M. Gotch 44 Audley Road, Saffron Walden CB11 3HD
GM0 IMH C. Taylor The Grange Smithy, Errol, Perth PH2 7TB
G0 IMK N. Sparrey The Ashes, Mamble Road, Kidderminster DY14 9HX
G0 IMP N. Wheeldon 28 Constance Avenue, Lincoln LN6 8SN
G0 IMQ Rochdale & Disctrict ARS c/o N. Cummings 14 Cemetery Road, Gloucester GL4 6PB
G0 IMU D. Spearman 7 Farman Close, Salhouse, Norwich NR13 6QD
G0 IMV R. Hill Marclecote, Ledbury Road, Ross-on-Wye HR9 7BE
GM0 IMW A. Mcbride 4 Clova Place, Uddingston, Glasgow G71 7BQ
G0 IMX C. Lacey 69 Field Lane, Pelsall, Walsall WS4 1DQ
GM0 IMZ I. Mcewan Granary Cottage, Vogrie Grange, Gorebridge EH23 4NT
G0 INA E. Cairns 2 Stockhill Circus, Stockhill, Nottingham NG6 0LS
GM0 INC C. Connor 2 Clyde View, Girvan KA26 9DJ
G0 ING ACF/CCF Int RN c/o L. Parker 128 Northampton Road, Wellingborough NN8 3PJ
G0 INJ P. Shuttlewood 10 Church Lane, Costock, Loughborough LE12 6UZ
G0 INK S. Taylor 37 Crestfield Drive, Pye Nest, Halifax HX2 7HG
GW0 INN P. Garner 44 Tycoch Road, Sketty, Swansea SA2 9EQ
G0 INO D. Elvin 56 Cock Bank, Whittlesey, Peterborough PE7 2HN
G0 INQ C. Hannell Toat Lodge, Pulborough RH20 1BZ
G0 INT J. Lee 2 York Terrace, Birchington CT7 9AZ
G0 INV N. Wileman 3 Primrose Way, Stamford PE9 4BU
G0 INZ R. Ball 24 Healey Close, Abingdon OX14 5RL
GM0 IOA S. Aitken 52 Craigellachie Court, Glenrothes KY7 6XE
G0 IOE B. Haynes 6 Epping Walk, Furnace Green, Crawley RH10 6LX
G0 IOF G. Moore 64 Walmer Road, Seaford BN25 3TN
G0 IOH P. Thomas 1 Jervis Close, Daventry NN11 4LL
G0 IOI T. Maunder 19 St. Monica Road, Southampton SO19 8FF
G0 IOK G. Markeson 23 Chantry Lane, Tideswell, Buxton SK17 8NP
GD0 IOM R. Ferguson Moaney Moar House, Corlea Road, Ballasalla IM9 3BA Isle of Man
G0 IOO A. Willis 46 Maryland Road, Thornton Heath CR7 8DF
G0 IOP Dr D. Melville Little Buckden, Milberry Lane, Stoughton, Chichester PO18 9JJ
G0 IOQ Telford & District ARS c/o F. Cosgrove Denton Park Middle School, Linhope Road, Newcastle upon Tyne NE5 2NW
G0 IOR M. Robertson 12B Southfield Road, Grimsby DN33 2PL
GI0 IOT M. Rabbett 41 Richill Park, Londonderry BT47 5QY
G0 IOU B. Gleed 19 Silver Birch Caravan Site, Walters Ash, High Wycombe HP14 4UY
GM0 IOY E. Watt 43 Larkfield Road, Gourock PA19 1YA
G0 IOZ R. Southerington 4 Reculver Close, Sunnyhill, Derby DE23 1WN
G0 IPB L. Collins 32 Parkstone Road, Syston, Leicester LE7 1LY
G0 IPH D. Sutton 10 Normanhurst Road, Borough Green, Sevenoaks TN15 8HT
G0 IPJ C. Whitaker Cross Acres, Pottersheath Road, Welwyn AL6 9SZ
G0 IPK J. Marsh 44 Richmond Gardens, Harrow Weald, Harrow HA3 6AJ
G0 IPN A. Greenwood 4 Roach Place, Rochdale OL16 2DD
G0 IPO E. Landor Silverden, Silverden Lane, Cranbrook TN18 5LX
GM0 IPV B. Stephenson 11 Bishop Forbes Crescent, Blackburn, Aberdeen AB21 0TW
GM0 IPW Dr R. Dickie Taynish, 11 Churchill Drive, Stornoway HS1 2NP
G0 IPX Fists/International Morse Pre 'Soc' c/o J. Griffin 35 Cottage Street, Kingswinford DY6 7QE
GI0 IQA Dr S. Ruff 9 Cooleen Park, Newtownabbey BT37 0RR
GW0 IQC S. Steed 3 Claremont Road, Llandudno LL30 2UF
GM0 IQD D. Ross 24 Eriskay Road, Inverness IV2 3LX
G0 IQH F. Manning 180 Priestley Terrace, Wibsey, Bradford BD6 1QU
GM0 IQI P. James 6 Jenny Moore Road, St. Boswells Melrose TD6 0AL
G0 IQK W. Chewter 93 Eton Road, Ilford IG1 2UF
G0 IQM S. Atkinson 26 Skipton Road, Trawden, Colne BB8 8QS
G0 IQN C. Jones 189 Moor Lane, Cranham, Upminster RM14 1HN
G0 IQP P. Williams 28 Cross Likey, Church Stoke, Montgomery SY15 6AL
GW0 IQZ W. Williams 31 Stad Ty Croes, Llanfairpwllgwyngyll LL61 5JR
GW0 IRC D. Knibbs 24 Corbett Grove, Caerphilly CF83 1SZ
G0 IRH D. Pond 31 Quintilis, Bracknell RG12 7QQ
G0 IRI J. Wade 12 Kendal Road, Harlescott, Shrewsbury SY1 4ER
G0 IRJ A. Wilson 42 Bacheler St., Hull HU3 2TZ
G0 IRK N. Porter 23 Calder Court, 7 Britannia Road, Surbiton KT5 8TS
G0 IRM N. Chambers 78 Durley Avenue, Pinner HA5 1JH
GW0 IRP C. Evans 19 Bryn Terrace, Caerau, Maesteg CF34 0UR
G0 IRQ F. Wagner 4 Hamilton Close, South Walsham, Norwich NR13 6DP
GW0 IRT P. Williams 15 Rhymney Close, Rassau, Ebbw Vale NP23 5TF
G0 IRY W. Gallacher 143 Acre Street, Huddersfield HD3 3EJ
GM0 ISA D. Arcari 184 Fintry Drive, Fintry, Dundee DD4 9LP
G0 ISC R. Slyfield 359 Ringwood Road, Poole BH12 4LT
G0 ISE G. Carter 12 Somerford Road, Broughton, Chester CH4 0SZ
G0 ISG R. Hale 5 Land Oak Drive, Kidderminster DY10 2ST
G0 ISH R. Pownall Beechgrove, Field Road Whiteshill, Stroud GL6 6AG
G0 ISI R. Christopher 33 Grange Park, Albrighton, Wolverhampton WV7 3EN
G0 ISJ G. Parkin Mil-Rune, Marsh Gate, Cornwall PL32 9YN
G0 ISK M. Glover 50 Broadway, Brinsworth, Rotherham S60 5ES
G0 ISL Rvd. B. Shersby 4 Blenheim Gardens, Chichester PO19 7XE
G0 ISM R. Faversham 19 Pelwood Road, Camber, Rye TN31 7RU
G0 ISO M. Edmunds 27A Sea View Road, Parkstone, Poole BH12 3LP
G0 ISP M. Harrington 56 Harris Croft, Wem, Shrewsbury SY4 5DU
GI0 ISQ D. Christie 8 Ballytober Road Portballintrae, Bushmills BT57 8UX
GM0 IST J. Purtell 31 Daleally Crescent, Errol, Perth PH2 7QA
G0 ISX M. Williams 22 New Avenue, Huddersfield HD5 0JD
G0 ISY J. Davies Winter Cottage, Church Road, St. Ives TR26 3LE
GI0 ITJ Dr D. Linton 4 Elmwood, Cullybackey, Ballymena BT43 5PY
G0 ITL J. Dunkley 34 Sparrow Close, Ilkeston DE7 4PW
G0 ITM J. Dutton 16 Briarfield, Washington NE38 8RX
G0 ITO A. O'Shaughnessy Southby, Buckland, Faringdon SN7 8QR
G0 ITU M. Wood 78 Haycliffe Road, Bradford BD5 9HB
G0 ITZ E. Pieroni 395 Huddersfield Road, Millbrook, Stalybridge SK15 3HU
G0 IUD A. Wakeman 133 Stanshawe Crescent, Yate, Bristol BS37 4EG
G0 IUH P. Battershill 45 Winkworth Road, Banstead SM7 2QJ
G0 IUK E. Pickerill 6 Pitmore Walk, Moston, Manchester M40 0GB
G0 IUN C. Stain 6 Sutton Close, Sutton-in-Ashfield NG17 3DP
GI0 IUP G. Lyle 40 Enagh Crescent, Maydown, Londonderry BT47 6UG
G0 IUV K. Knowles 18 Croft Close, Pinxton, Nottingham NG16 6RF
G0 IUW R. Hill 7 Berkeley Close, Cashe'S Green, Stroud GL5 4SA
G0 IUY J. Tribe 6 Privett Road, Waterlooville PO7 5HJ
G0 IVB M. Llewellyn 28 North St., Clay Cross, Chesterfield S45 9PL
G0 IVD P. Mcgarvey 125 Esme Road, Sparkhill, Birmingham B11 4NJ
G0 IVI K. Edwards 39 King Edward Street, Sandiacre, Nottingham NG10 5BS
GI0 IVJ J. Mccausland 17 Dunavon Heights, Dungannon BT71 6TN
G0 IVO B. Singleton Langdales, Broadbury, Okehampton EX20 4LL
G0 IVP S. Parrish Ambleside, Carlton Avenue, Hornsea HU18 1JG
GM0 IVQ G. Mckinlay 68 Hillend Road, Clarkston, Glasgow G76 7XT
G0 IVR Itchen Valley ARC c/o K. Hastie 3 The Woodlands, Kings Worthy, Winchester SO23 7QQ
GW0 IVT T. Jones Heathbrook, Maesmawr Close, Brecon LD3 7JF
G0 IVV J. Sheldrake 66 Bibbys Way, Framlingham, Woodbridge IP13 9FD
G0 IVX H. Harrison 8 North Leigh, Tanfield Lea, Stanley DH9 9PA
G0 IVZ J. Fisher Farland, Rillaton, Callington PL17 7PA
G0 IWB J. Wells 12 Church Road, Grafham, Huntingdon PE28 0BB
GW0 IWD A. Aston Ty Newydd, Y Ffor, Pwllheli LL53 6UY
G0 IWF F. Oakton 180 Wragley Way, Stenson Fields, Derby DE24 3DZ
G0 IWI Dr D. Ranson 27 Kruses Road, North Warrandyte, Victoria 3113 Australia
G0 IWN D. Greenhalgh 8 Pleasant Road, Milton, Southsea PO4 8JU
GM0 IWX T. Lorimer 9 Orchard House, Orchard Grove, Leven KY8 5XA
G0 IWZ A. White 13 Woodcote Drive, Poole BH16 5RA
GW0 IXK R. Griffiths 71 Elder Grove, Llangunnor, Carmarthen SA31 2LH
GW0 IXM O. Williams 11 Hafod Road, Tycroes, Ammanford SA18 3QL
GM0 IXO G. Michie 2 Moncur St., Townhill, Dunfermline KY12 0HN
GW0 IXQ D. Graham 1 Maestir, Llanelli SA15 3NS
G0 IXS P. Turner-Hicks 64 Sharpley Avenue, Coalville LE67 4DT
G0 IXT G. Lambert Church House, Chapel Lane, Christchurch BH23 6BE
G0 IXV P. Biscombe Keverine, Victoria Road, Folkestone CT18 7JS
G0 IXZ D. Fleetwood Lynton House, Station Road Bolsover, Chesterfield S44 6BH
GM0 IYA W. Brown 29 Bankhead Crescent, Dennyloanhead, Bonnybridge FK4 1RY
G0 IYD P. Bates 29 Juler Close, North Walsham NR28 0SY
G0 IYE D. Chalmers 42 Thornbury Drive, Uphill, Weston-Super-Mare BS23 4YH
G0 IYK M. Harrison 12 Richmond Rise, Reepham, Norwich NR10 4LS
G0 IYM M. Linton 134 Eccles Old Road, Salford M6 8QQ
G0 IYO K. Stanmore 48 St. Michaels Avenue, Bishops Cleeve, Cheltenham GL52 8NX

GM0 IYP Sutherland And District ARC c/o C. O'Hennessy Savalbeg, Challenger Estate, Lairg IV27 4ED
G0 IYQ N. Mitchelson Trevena, Winskill, Penrith CA10 1PD
G0 IYS P. Willis 5 Binbrook Walk, Corby NN18 9HH
G0 IYT C. Savin Jenalri, Union Lane, Preston PR3 6SS
G0 IYU P. Kirk 38 Carleton Street, Morecambe LA4 4NY
G0 IYV T. Jones 159 Cobden View Road, Crookes, Sheffield S10 1HT
G0 IYW G. Farndon 28 Willow Close, Collycroft, Bedworth CV12 8BE
G0 IYX Capt. I. Roberts 2 Samuel Fold, Pendlebury Lane, Wigan WN2 1LT
G0 IYY M. Lindsay 5 Eatonhill, Norwich NR4 7PY
G0 IZC J. Carver 131 Rutland Avenue, High Wycombe HP12 3JQ
G0 IZE R. Bates 36 Maple Crescent, Alveley, Bridgnorth WV15 6LT
G0 IZG G. Bennett 21 Porlock Close Platt Bridge, Wigan WN2 5HY
G0 IZI A. Warwick 58 Longworth Avenue, Tilehurst, Reading RG31 5JY
G0 IZJ E. Whittaker 17 Packer St., Bolton BL1 3LD
G0 IZK M. Dennehy 45 Vine Road, Tiptree, Colchester CO5 0LR
GW0 IZL N. Procter 62 Oakfield Street, Pontarddulais, Swansea SA4 8LW
G0 IZN A. Halliday 12 Fowley Common Lane, Glazebury, Warrington WA3 5JJ
G0 IZP R. Mcdonald 4 The Paddocks, Baunton, Cirencester GL7 7DL
G0 IZQ J. Bradnock 5 Milverton Road, Knowle, Solihull B93 0HX
G0 IZR J. Bridson 10 Clegg Street, Astley, Tyldesley, Manchester M29 7DB
G0 IZV D. Gowers 32 Silver Fox Crescent, Woodley, Reading RG5 3JA
G0 IZY D. Hicks Woodlands, Crawley Ridge, Camberley GU15 2AJ
G0 JAA T. Watson 89 Addison Road, Wednesbury WS100LW
G0 JAC M. Hill 9 Longacre, Woodthorpe, Nottingham NG5 4JS
G0 JAF J. Foy 23 Lee Road, Nelson BB9 8SD
G0 JAG R. Glyn 171 Bull Close Road, Norwich NR3 1NY
GW0 JAI T. Jones 26 Treowain, Machynlleth SY20 8EJ
G0 JAJ G. West 39 Court Farm Road, Eltham, London SE9 4JL
G0 JAL W. Bell 244 Westbourne, Woodside, Telford TF7 5QR
G0 JAM J. Morrison 107 Crown Meadow, Colnbrook, Slough SL3 0LJ
G0 JAN J. Ilston 6 Dovedale, Canvey Island SS8 8HX
G0 JAO L. White 56 Grange Road, Leigh-on-Sea SS9 2HT
G0 JAP C. Collins 26 Nicholsons Wharf, Mather Road, Newark NG24 1FN
G0 JAQ G. Blackwood 63 Illingworth Avenue, Halifax HX2 9JH
G0 JAR R. Offord 116 Townsend Road, Snodland ME6 5RL
G0 JBA P. Boorman Gladstone House, Marshborough Road, Woodnesborough, Sandwich CT13 0PE
G0 JBC C. Collins 3 Hollin Hall Farm, Long Causeway, Denholme, Bradford BD13 4DX
G0 JBH Dr A. Bingham 5 Burns Rd, Southampton SO19 6QT
G0 JBJ A. Tungate 171 Marlborough Gardens, Faringdon SN7 7DG
G0 JBM L. Humphreys 19 Clinch Green Avenue, Bexhill-on-Sea TN39 5HN
G0 JBO B. Lees Preston Hall, Preston, Telford TF6 6DH
G0 JBP R. Lipscomb Redmoor, Bickley Road, Bromley BR1 2NF
G0 JBR E. Thorley 7 Drake Street, St. Helens WA10 4JG
G0 JBS M. Notman 3 Pilling Avenue, Lytham St. Annes FY8 3QF
G0 JBV A. Mcintosh 17 The Chase, Abbeydale, Gloucester GL4 4WP
G0 JBY J. Youd 25 Hanson Road, Andover SP10 3HL
G0 JBZ R. Stevens 51 Beacon Park Crescent, Poole BH16 5PB
G0 JCA J. Andress 46 Bridwell Road, Plymouth PL5 1AB
GW0 JCB C. Jones 8 West Walk, Barry CF62 8BY
G0 JCC A. Lancaster Little Edwards Barn, Newton St. Margarets HR2 0QG
G0 JCD Ansty Contest Club c/o P. Busby 7 Rimmer Green, Scarisbrick, Southport PR8 5LP
G0 JCF D. Smith 14 College Drive, Ruislip HA4 8SB
G0 JCG A. Butler 97 Hermitage Road, Saughall, Chester CH1 6AQ
G0 JCK W. Pattinson 2 The Green, Ticknall, Derby DE73 7GY
G0 JCN M. Lovatt 37 Hartland Avenue, Bilston WV14 9AN
G0 JCP D. Grey 7 Cemetery Lane, Tweedmouth, Berwick-upon-Tweed TD15 2BS
G0 JCQ B. Rimmer 8A Mallee Avenue, Southport PR9 8NL
GW0 JCT P. Jones Bronallt, Cenarth, Newcastle Emlyn SA38 9JS
G0 JCY P. Stuart Skylark Corner, Seaborough Hill, Crewkerne TA18 8PL
G0 JCZ M. Martin 2821 Bissonnet St, Houston 77005-4014 United States
GM0 JDB J. Ratter Foulawick Wethersta, Brae ZE2 9QS
G0 JDC J. Gwynn 117 Main Street, Goldthorpe, Rotherham S63 9JW
G0 JDD I. Douce 67 Glenbervie Drive, Leigh-on-Sea SS9 3JT
G0 JDE R. Doyle 61 St. Peters Road, West Mersea, Colchester CO5 8LN
G0 JDG A. Purseglove 122 Chesterfield Road, Huthwaite, Sutton-in-Ashfield NG17 2QF
G0 JDL J. Clarke 17-18 Sotherton Corner, Sotherton, Beccles NR34 8AP
G0 JDM T. Kewell The Old Skittle Alley Baker Street, Frome BA11 3BL
G0 JDO T. Thomas 26 Corfe Crescent, Torquay TQ2 7QX
G0 JDQ M. Cozens 1001 Starkey Rd, Lot 321, , Largo 33771 United States
GW0 JDS J. Stonehouse 54 Port Tennant Road, Port Tennant, Swansea SA1 8JF
GW0 JDW D. Willis 5 Dan Lan Rd, Llanelli, Dyfed SA16 0NF
GW0 JDY M. Lewis 12 Fern Rise, Neyland, Milford Haven SA73 1RA
G0 JEA R. Kaye 63 Coronation Drive, Birdwell, Barnsley S70 5RL
G0 JEC D. Naylor 6 Front St., Kirk Merrington, Spennymoor DL16 7HZ
G0 JEE B. Greer Willow Farm, Main Road, Burton-on-Trent DE13 9QD
GM0 JEF J. Fysh 7 Chestnut Place, Ellon AB41 9HF
G0 JEH S. Rosbottom 26 Wellington Street, Preston PR1 8TP
G0 JEK C. Kelland 11 The Meads, West Hanney, Wantage OX12 0LJ
GW0 JEQ R. Wicks Gooseberry Cottage, Llangunllo, Knighton LD7 1SW
G0 JEU S. Dunsmore 33 Church Street, Messingham, Scunthorpe DN17 3SB
GI0 JEV H. Kernohan 40 Lisnafillon Road, Gracehill, Ballymena BT42 1JA
G0 JEW P. Wells 12 Shelley Drive, Lutterworth LE17 4XF
G0 JEZ B. Wells 37 Elder Road, Denvilles, Havant PO9 2UW
G0 JFA A. Jones Highercombe West, Highercombe, Dulverton TA22 9PT
G0 JFC J. Aithison 14 Claymore Rise, Silsden, Keighley BD20 0QQ
G0 JFD G. Butcher 15 Leander Drive, Gosport PO12 4GG
G0 JFE P. Sixsmith 10 Wisbeck Road, Bolton BL2 2TA
GI0 JFF W. Mcbride 5 Aylesbury Road, Newtownabbey BT36 7YP
GM0 JFH J. Mair 43 Todhill Avenue, Onthank, Kilmarnock KA3 2EQ
GM0 JFK C. Harper Glencoul Cottage, Cullicudden, Dingwall IV7 8LL
GM0 JFL B. Harper 16 Brae Park, Munlochy IV8 8PJ
G0 JFM S. Nicholls Fieldway, The Street, Eyke, Woodbridge IP12 2QG
G0 JFP J. Scott 23 Botesworth Green, Milnrow, Rochdale OL16 3PJ
GW0 JFQ N. Williams Flat 4, 234-237 Chapmans High St., Swansea SA1 1NZ
G0 JGB St George's School c/o J. Barnett 43 Westsprink Crescent, Stoke-on-Trent ST3 5JD
G0 JGF S. Cox 60 Leawood Road, Trent Vale, Stoke-on-Trent ST4 6LA
G0 JGI C. Davison 28 Ashford Crescent, Hythe, Southampton SO45 6EU
G0 JGV K. Bewley 38 Great Innings South, Watton At Stone, Hertford SG14 3TF
GD0 JGX D. Ginsberg 26 Keeill Pharick Park, Glen Vine IM4 4EW Isle of Man
G0 JHD R. Jennings 54A Pensbury Street, Darlington DL1 5LH
GM0 JHE K. Hunter 2/2 206 Skirsa Street, Glasgow G23 5DJ
G0 JHG C. Holmes 17 Wenning Court, Morecambe LA3 3SH
GW0 JHH L. Ireland 109 Dan-Y-Cribyn, Ynysybwl, Pontypridd CF37 3EU
G0 JHJ W. Fowler 1 East Orchard, Sileby, Loughborough LE12 7SX
G0 JHK M. Hedges 24 Fletcher Avenue, St. Leonards-on-Sea TN37 7QX
G0 JHL R. Wilmot 43 A 2, PÒä□äSkylÒäNtie, JÒõmsÒõ 42100 Finland
G0 JHQ Dr A. Cotton Coburg Cottage, Barton Estate, East Cowes PO32 6NT
GI0 JHR H. Oxtoby 13 Castle Court, Cookstown BT80 8QJ
GI0 JHS N. Greer 11 Ardtrea Road, Stewartstown, Dungannon BT71 5LY
G0 JHT D. Talbot Southways, Tichborne Down, Alresford SO24 9PL
G0 JHU M. Stephenson 38 St. Helens Crescent, Low Fell, Gateshead NE9 6DH
G0 JHW J. Waterhouse 81 Barkham Road, Wokingham RG41 2RJ
G0 JIA A. Sharp 9 Higher Park, East Prawle, Kingsbridge TQ7 2DB
G0 JIB P. Moran 12 Sapphire Drive, Kirkby, Liverpool L33 1UW
G0 JIF A. Fennell 12 Vale Road, Ramsgate CT11 9LU
G0 JII C. Davis 10 Marnhull Road, Poole BH15 2EX
G0 JIL G. Hampson Flat 4, Priory Lodge, Stony Lane South, Christchurch BH23 1FA
G0 JIM J. King 4 Glenhurst Avenue, Ruislip HA4 7LZ
G0 JIR A. Potter 8 Oaklands, The Street, Ashford TN25 6NE
G0 JIS A. Myland 8 Burseldon Court, East Cliff Road, Devon EX7 0BP
G0 JIT R. Atherton Frensham, Grange Lane, Northwich CW8 2BQ
G0 JIW M. Edwards 98 Cornwallis Drive, Eaton Socon, St. Neots PE19 8TZ
G0 JJD G. Thorne 4 Barronwood Court, Tarleton, Preston PR4 6TR
G0 JJE H. Spratt 9 Kennedy Close, Halesworth IP19 8EG
GW0 JJF I. Mccormick The Old Railway Inn, Station Road, Gilwern NP7 0BY
G0 JJG J. Butt 29 Shearwater Way, Stowmarket IP14 5UG
G0 JJI P. Forshaw 73 Galloway Road, Hamworthy, Poole BH15 4JS
G0 JJK T. Atkins 46 Fallowfield, Ampthill, Bedford MK45 2TP
G0 JJM A. Eves 136 Thistle Grove, Welwyn Garden City AL7 4AQ
G0 JJO M. Wheatley 25 Sheringham Drive, Etchinghill, Rugeley WS15 2YG
G0 JJP R. Fowler 6 Salts Croft Hawksyard, Rugeley WS15 1SR
G0 JJQ W. Dillon 49 Goring Way, Greenford UB6 9NN
G0 JJR D. Briggs 130 Beresford Avenue, Skegness PE25 3JN
G0 JJS J. Smith 48 Oakleigh Gardens, Oldland Common, Bristol BS30 6RH
G0 JJV S. Hill 26 Harborough Way, Sheffield S2 1RG
G0 JJW J. Walsh Flints House, Coates, Peterborough PE7 2DD
G0 JJY I. Bowen 169 Clopton Road, Stratford-upon-Avon CV37 6TF
GD0 JKA A. Brook 9 Stonecrop Grove, Douglas IM2 7DX Isle of Man
G0 JKC K. Clarke 55 Compton Avenue, Aston-On-Trent, Derby DE72 2AU
G0 JKE S. Ward 27 Greenock Street, Sheffield S6 4NB
G0 JKG F. Shinn 23 Cygnet Close, Brampton Bierlow, Rotherham S63 6EY
G0 JKH D. Hinson 6 Nethergate, Stannington, Sheffield S6 6DJ
G0 JKI M. Edmunds 2 Jubilee Road, Bungay NR35 1RE
G0 JKJ S. Marshall 31 Postbridge Road, Coventry CV3 5AG
G0 JKL J. Hallendorff 8 Third Avenue, Kidsgrove, Stoke-on-Trent ST7 1BZ
G0 JKM P. Renvoize Flat 14 Block M, Peabody Buildings, Dufferin Street, London EC1Y 8NL
G0 JKP A. Whibley 8 Ticehurst Road, Brighton BN2 5PU
G0 JKU A. Bowering 137A Knole Lane, Brentry, Bristol BS10 6JN
G0 JKY B. Hadley 60 Chapel St., Pensnett, Brierley Hill DY5 4EF
G0 JLB P. Boorman 1 Stoney Lands, Landrake PL12 5DF
G0 JLE R. Adams 18 Dundridge Gardens, Bristol BS5 8SZ
G0 JLF J. Flowers 4 Ashleigh Crescent, Yatton, Bristol BS49 4DF
G0 JLI T. Davies 31 Burnbush Close, Bristol BS14 8LQ
GM0 JLJ E. Mottart 1 Muirake Cottages, Cornhill, Banff AB45 2BQ
G0 JLL N. Sheen 26 Springvale Rise, Hemsworth, Pontefract WF9 5HY
G0 JLP M. Bray 205 Woodlands Road, Gillingham ME7 2SW
G0 JLR I. Sabey 21 Althorpe Street, Radford, Nottingham NG7 3GN
G0 JLS M. Brown 4 River Gardens, Shawbury, Shrewsbury SY4 4LA
G0 JLU E. Wood 65 Walford Road, Rolleston-On-Dove, Burton-on-Trent DE13 9AR
G0 JLV D. Vaughan 23 Beckmeadow Way, Mundesley, Norwich NR11 8LP
GW0 JLX A. Digby 6 Melin Y Coed, Cilgerran SA43 2AQ
G0 JMD J. Davis 179 Bredon, Yate, Bristol BS37 8TG
G0 JME J. Pither 74 Bucklands Road, Teddington TW11 9QS
G0 JMI M. Parkin 17 Bolle Road, Alton GU34 1PW
GW0 JMJ W. Williams 16 Chapel Close, Elim Way, Blackwood NP12 2AD
G0 JMK M. Kemble 74 Teg Down Meads, Winchester SO22 5ND
G0 JML M. Pivac 72 Old Mill Road, Saffron Walden CB11 3ER
G0 JMN H. Marshall 23 Cranbourne Drive, Chorley PR6 0LJ
GM0 JMO J. Bell 5 Louisa Drive, Girvan KA26 9AH
G0 JMR D. Williams 27 Grindlestone Hirst, Colne BB8 8BF
G0 JMS M. Standen 11 Hazel Gardens, Sonning Common, Reading RG4 9TF
G0 JMW M. Williams 51 Crackley Hill, Coventry Road, Kenilworth CV8 2EE
G0 JMZ P. Farrar 2 Ancaster Avenue, Chapel St. Leonards, Skegness PE24 5SL
G0 JNA R. Janes 37 Valley View, Market Drayton TF9 1EA
G0 JNE T. Royle 35 Patrons Drive, Sandbach CW11 3AS
G0 JNG A. Stothard Lanshaw Farm, Otley Road, Harrogate HG3 1QX
G0 JNJ A. Denny 85 Delamere Drive, Macclesfield SK10 2PS
G0 JNK A. Powers 42 Newbridge Road Ambergate, Belper DE56 2GS
G0 JNQ S. Mitchell 78 Bellasize Park, Gilberdyke, Brough HU15 2XU
G0 JNR S. Doveton 2 Red Scar Drive, Scarborough YO12 5RQ
G0 JNT L. Keeton 66 Worlaby Road, Scartho Top, Grimsby DN33 3JP
G0 JNY H. Potter 720 Old Norwich Road, Ipswich IP1 6LB
G0 JNZ T. Bray 135 Fort Austin Avenue, Crownhill, Plymouth PL6 5NR
G0 JOC J. Lassemillante 25 Rissington Walk, Thornaby, Stockton-on-Tees TS17 9QJ
G0 JOD R. Degg 28 The Spinneys, Welton, Lincoln LN2 3TU
G0 JOG B. Wilson 41 Palmerston Close, Ramsbottom, Bury BL0 9YN
GM0 JOL Rvd. J. Lincoln 59 Obsdale Park, Alness IV17 0TR
G0 JOM J. Goddard 65 New Street, North Wingfield, Chesterfield S42 5JP
G0 JON J. Swain 1 Ganstead Way, Low Grange, Billingham TS23 3SY
G0 JOP J. Bryder 110 Georgelands, Ripley, Woking GU23 6DQ

G0 JOS E. Christmas 15 Norton Avenue, Surbiton KT5 9DX
GM0 JOV A. Farquhar 57 Woodcroft Avenue, Bridge Of Don, Aberdeen AB22 8WY
G0 JOX D. Sykes 449 Westdale Lane, Mapperley, Nottingham NG3 6DH
G0 JPE P. Roberts 61 Abbey Lane, Sheffield S8 0BN
G0 JPF S. Lee 15 Wilson Way, Earls Barton, Northampton NN6 0NZ
GM0 JPG D. Arnold 1 Knockenhair Road, Dunbar EH42 1BA
G0 JPH J. Holden 47 Copse Hill, London SW20 0NJ
G0 JPI S. Martin 34 Parson Drove, West Pinchbeck, Spalding PE11 3QW
G0 JPJ P. Smith S.V Kiwiroa, New Zealand Reg. Ship ON-876019 New Zealand
G0 JPL D. Smith 16 Leander Close, Nottingham NG11 7BE
G0 JPM J. Mitchell 8 Eldred Drive, Orpington BR5 4PF
G0 JPQ C. Barber Charity Farm House, Mill Lane, Skegness PE24 5NN
G0 JPT C. Hamlet 9 Covert Road, Manchester M22 4QS
G0 JPU Dr M. Ferguson 15 Squires Leaze, Thornbury, Bristol BS35 1TB
G0 JPY D. Polley 42 Lindfield Road, Eastbourne BN22 0AJ
G0 JPZ M. Kirk 5 The Paddock, Kirkby-In-Ashfield, Nottingham NG17 8BT
GM0 JQE I. Templeton 39 Cairngorm Court, Irvine KA11 1PN
G0 JQK J. Hughes 18 Monmouth Road, Wallasey CH44 3ED
G0 JQP G. Bradley 7 Copeland Row, Evenwood, Bishop Auckland DL14 9PY
GI0 JQQ E. Butler 59 Ballinlea Road, Maghernahar, Ballycastle BT54 6JL
G0 JQR D. Allison 12 Goosander Close, Kings Lynn PE31 7RF
G0 JQS A. Downing 1 Raglans, Alphington, Exeter EX2 8XN
GW0 JQT S. Lloyd 10 Park Crescent, Llanelli SA15 3AE
G0 JQX C. Ditchfield 8 Meerbrook Way, Quedgeley, Gloucester GL2 4QE
G0 JQZ S. Chamberlain 54 Henray Avenue, Glen Parva, Leicester LE2 9QJ
G0 JRB J. Reed 290 Messingham Road, Bottesford, Scunthorpe DN17 2QY
G0 JRC J. Clayton 49 Bramble Lane, Mansfield NG18 3NP
GI0 JRD B. Mcanespie 23 Ashton Park, Belfast BT10 0JQ
G0 JRE I. Perks 9 Atherton Close, Shalford, Guildford GU4 8HZ
GW0 JRF F. Rees Caerleon, Picton Road, Tenby SA70 7DP
G0 JRH P. Scott-Dickinson Ninicsu, 18 Pennington Drive, Weybridge KT13 9RU
GI0 JRI K. Murray 67 Sicily Park, Belfast BT10 0AN
G0 JRM C. Brown 8 The Elms, Horringer, Bury St. Edmunds IP29 5SE
G0 JRN A. Hansley 238 Milton Road, Cowplain, Waterlooville PO8 8SE
GM0 JRQ B. Baker The Ridge, Peat Inn, Cupar KY15 5LH
G0 JRR C. Curtis 1 Westover Drive, Burton-Upon-Stather, Scunthorpe DN15 9HH
G0 JRT T. Hanratty 12 Clarendon Street, Consett DH8 5LS
G0 JRV J. Broomfield 14 Woodfen Crescent, Leominster HR6 8SS
G0 JRX T. Cave 71 Cambo Drive, Cramlington NE23 6TW
G0 JRY C. Mulvany 25 Redwing Close, Bicester OX26 6SR
G0 JRZ T. Brookes Jemora, Littleham, Bideford EX39 5HN
G0 JSA R. Taylor 11 Yeadon Close, Accrington BB5 0FN
G0 JSC J. Wheatley 8 Winchester Close, Feniton, Honiton EX14 3EX
G0 JSE J. Edwards 49 The Fleet, Stoney Stanton, Leicester LE9 4DZ
G0 JSF B. Halmshaw 7 Gerrard St., Rochdale OL11 2EB
G0 JSG P. Holden 19 Briar Close, Lowestoft NR32 4SU
G0 JSJ R. Jackett Bourne House, 105 Moor Road, Leyland PR26 9HP
G0 JSK A. Wakefield Kuleana, Barnrigg, Carnforth LA6 2LJ
G0 JSL G. Brown 9 Western Drive, Leyland PR25 1YB
G0 JSM J. Brown 9 Western Drive, Leyland PR25 1YB
G0 JSO R. Wynne South Graceholme, High Lorton, Cockermouth CA13 9UQ
G0 JSP P. Fauchon 114 Petersfield Avenue, Staines-upon-Thames TW18 1DJ
G0 JSR S. Rickman 35 Cedar Way, Basingstoke RG23 8NG
G0 JST European Psk Club c/o J. Wheatley 8 Winchester Close, Feniton, Honiton EX14 3EX
G0 JSU A. Collett 5 Park View Drive, Lydiard Millicent, Swindon SN5 3LX
GW0 JSX R. Davies 8 Princes Park, Rhuddlan, Rhyl LL18 5RW
GJ0 JSY S. Smith-Gauvin 31 Le Jardin A Pommiers, La Rue De Patier, St. Saviour JE2 7LT Jersey
G0 JSZ J. Crellin 89 Wapshare Road, West Derby, Liverpool L11 8LR
G0 JTA D. Ellison Blackburn Hall, Grinton, Richmond DL11 6HH
G0 JTD R. Lyne 32 Davenwood Upper Stratton, Swindon SN2 7LL
GW0 JTE L. Horne 29 Station Terrace, Dowlais, Merthyr Tydfil CF48 3PU
GW0 JTF J. Hosking 14 School Terrace, Cwm, Ebbw Vale NP23 7QY
GW0 JTJ T. Watkins Ty Unig, Forest Road, Treharris CF46 5HG
G0 JTL J. Hinchliffe 19 The Terrace, Honley, Holmfirth HD9 6DS
G0 JTM P. Smith 999 Manchester Road, Linthwaite, Huddersfield HD7 5LS
G0 JTN C. Smith 35 Allendale Road, Earley, Reading RG6 7PD
G0 JTP A. Hill 159 Sandford Road, Bradford BD3 9NU
G0 JTR B. Thatcher 18 Harescombe, Yate, Bristol BS37 8UA
G0 JTT S. Hemsworth 4 Spoonhill Road, Stannington, Sheffield S6 5PA
GW0 JTU A. Lewis 33 Heol Helig, Brynmawr, Ebbw Vale NP23 4TY
G0 JUA J. Hardcastle 37 Caithness Road, Liverpool L18 9SJ
G0 JUE T. Cruse Watch Tower House, The Ridgeway, London NW7 1RS
G0 JUI E. Gibson 107 Church Avenue, Meanwood, Leeds LS6 4JT
G0 JUK N. Mayes Cliffe Cottage, Rotherham Road, Barnsley S71 5QX
G0 JUL G. Hogben Calle El Arado 7, Las Brenas 35570 Spain
G0 JUM S. Barker 11 Pennington Close, Copplestone, Crediton EX17 5NA
G0 JUN R. Softley 14 Topps Drive, Bedworth CV12 0DE
G0 JUQ M. Eborall 26 Bishopton Lane, Stratford-upon-Avon CV37 9JN
G0 JUR N. Barnett 60 Commercial Road, Spalding PE11 2HE
G0 JUT D. Horne 24 Ringwood Drive, Leeds LS14 1AP
G0 JUV A. Phillips 2 New Zealand Terrace, Bridport DT6 3PW
G0 JUY P. Barden 38 Silver Close, Tonbridge TN9 2UY
G0 JVB M. Jackson 114 Norman Street, Ilkeston DE7 8NL
GM0 JVC Dr M. Grant Monikie, Gryffe Road, Kilmacolm PA13 4BB
G0 JVF D. Cleaver 4 Wyvern Close, Devizes SN10 2UE
G0 JVH A. Puffett 142 Cheltenham Road East, Gloucester GL3 1AA
G0 JVI A. Mawson 38 Springbank Road, Gildersome, Leeds LS27 7DJ
G0 JVK R. Cook 15 Hucklow Court, Mansfield NG18 3QP
G0 JVL A. Sclater 6 Balmoral Close, Alton, Hampshire GU34 1QY
G0 JVN R. Horne 7 Alexander Road, Bentley, Walsall WS2 0HJ
G0 JVU N. Pattinson 4 Carlisle Road, Brampton CA8 1SR
GM0 JVV J. Stevenson 52 Fernbrae Avenue, Rutherglen, Glasgow G73 4AE
G0 JVW A. Thornton 1 Primula Close, Clifton, Nottingham NG11 8SL
GW0 JWC C. Jones 31 Tynewdd Nantybwch, Tredegar NP223SQ
G0 JWD J. Brambley Trail View, Biggin, Buxton SK17 0DH
GW0 JWF B. Matthews 25 Manor Park, Newbridge, Newport NP11 4RS
G0 JWG J. Gaunt Fenton House, Church Road, King's Lynn PE33 0HE
G0 JWJ T. Bridgland-Taylor 1 Overbury Court, Hereford HR1-1DG
G0 JWL K. Lindsay 11A Pyrford Close, Waterlooville PO7 6BT
G0 JWM J. Mcmullen 281 The Broadway, Cullercoats, North Shields NE30 3LH
G0 JWO T. Forbes 63 Wardle Drive, Annitsford, Cramlington NE23 7DE
GD0 JWR H. Richardson Pitcairn, Quarterbridge Road, Douglas IM2 3RQ Isle of Man
G0 JWV P. Bryant Crugsillick Cottage, Ruan High Lanes, Truro TR2 5JP
G0 JWY J. Curtis Conway, 27 Southgate, Hornsea HU18 1RE
G0 JXF A. Hancox 5 East Glade Close, Sheffield S12 4QL
GW0 JXG M. Price 4 Vale View, Woodfieldside, Blackwood NP12 0DB
G0 JXI M. Brown 11 Miles Close, Birchwood, Warrington WA3 6QD
G0 JXJ D. Copeland 2B Rose Road, Canvey Island SS8 0BP
G0 JXO R. Smith 24 Kirkstead Road, Carlisle CA2 7RD
G0 JXP K. Pile 14 Semper Close, Knaphill, Woking GU21 2NG
G0 JXQ D. Davis 17 Welbourne Close, Raunds, Wellingborough NN9 6HE
G0 JXR P. Keasley 55 Hillside, Hoddesdon EN11 8RW
G0 JXX M. Hoddy 52 Hayling Rise, High Salvington, Worthing BN13 3AG
G0 JXY T. Bartholomew 52 Lauderdale Avenue, Wallsend NE28 9HU
G0 JYC P. Hodgson Uhland Str 4, Neunkirchen, Seelscheid 53819 Germany
G0 JYE P. Foss 37 Ling Crescent, Ruddington, Nottingham NG11 6GG
G0 JYF S. Deakin Brook House, Ivy Lane, Lower South Wraxall BA152RZ
G0 JYH H. Ryan Fairview, Imperial Avenue, Sheerness ME12 2HG
G0 JYI A. Street 11 Leigh Gardens, Leigh-on-Sea SS9 2PX
G0 JYJ C. Hodgson 113 Roman Road, East Ham, London E6 3RY
G0 JYK J. Sharp 22 Boat Lane, Irlam, Manchester M44 6EN
G0 JYL J. Bartram 2 Reeves Piece, Bratton, Westbury BA13 4TH
G0 JYN S. Ashcroft 90 Kestrel Close, Chipping Sodbury, Bristol BS37 6XA
G0 JYQ M. Gregory 21 Jacaranda Close, Fareham PO15 5LG
G0 JYS A. Jepson 45 Cotefield Road, Manchester M22 1UR
G0 JYU D. Smith 104 Hanley Road, London N4 3DW
G0 JYV J. Rowlands 67 Woodside, Gosport PO13 0YX
G0 JYX T. Cloke 14 Bickley Close, Hanham Green, Bristol BS15 3TB
G0 JYZ I. Broomhall 49 Funtley Hill, Fareham PO16 7XA
G0 JZA N. Cox Flat 9, Fontigary, Harts Close, Teignmouth TQ14 9HG
G0 JZE A. Mcfadyen 26 Lewis Road, Chipping Norton OX7 5JS
G0 JZF N. Rogers 15 Templar Road Yate, Bristol BS375TF
G0 JZH R. Morris 96 Chandag Road, Keynsham, Bristol BS31 1QE
G0 JZJ F. Russell 37 Overpool Road, Ellesmere Port CH66 1JW
G0 JZL G. Galley 1 St. James Avenue, South Anston, Sheffield S25 5DR
G0 JZS G. Corbett 359 London Road, Stoke-on-Trent ST4 5AN
G0 JZT J. Chappell 2 Wayside, Knott End-On-Sea, Poulton-le-Fylde FY6 0DD
G0 JZU W. Etherington 15 East Bank, North End Road, Arundel BN18 0DJ
GM0 JZV J. Warden Westlea, Little Brechin, Brechin DD9 6RQ
G0 JZW A. Elford 10 Meadowlands, Lymington SO41 9LB
G0 KAB A. Pilkington Flat 17, Blackshaw House, Bolton BL3 5NU
G0 KAK I. Walker 6 Granary Court, Northampton NN4 0XX
GW0 KAM G. Jones 14 Plantation Drive, Croesyceiliog, Cwmbran NP44 2AN
G0 KAQ E. Walker 2 Newtown Road, Uppingham, Oakham LE15 9TS
G0 KAS M. Stevens 20 Melton Place, Epsom KT19 9EE
G0 KAT V. Chapman 20 St. Chad, Barrow-upon-Humber DN19 7AU
G0 KAU R. Crocker 10 Westhall Close Carlton-Le-Moorland, Lincoln LN5 9JD
GW0 KAX P. Owen 13 Highland Close, Sarn, Bridgend CF32 9SB
G0 KAY T. Davies Netaerial.Com, Europa House, Barcroft Street, Bury BL9 5BT
GM0 KAZ A. White 65 Orchard Street, Galston KA4 8EJ
G0 KBA I. Langtree 243 Devonshire Road, Atherton, Manchester M46 9QB
G0 KBJ E. Burndred 52 Everest Road, Kidsgrove, Stoke-on-Trent ST7 4DY
G0 KBK R. Sleigh 14 Brook House Flats, Chetwynd End, Newport TF10 7JD
G0 KBL S. Rudcenko 39 The Avenue, Sutton SM2 7QA
G0 KBM D. Manning 2 Sluice Farm Cottages, Kirton, Ipswich IP10 0QF
G0 KBN G. Beech 2 Whitely Avenue, Ilkeston DE7 8WB
G0 KBO V. Kravchenko Flat 16, Birchfield House, London E14 8EY
G0 KBP M. Dearing 1 Woodbine Villas, New Village Road, Cottingham HU16 4NF
GM0 KBR Obo Thames Amateur Radio c/o J. Mcfadyen 8 Ramsay Crescent, Bathgate EH48 1DD
G0 KBS L. Kay 2 Childwall Crescent Childwall, Liverpool L16 7PQ
G0 KBZ K. Harrison 6 Staveley Road, Alford LN13 0PN
G0 KCA I. Walder-Davis 93 Church St., St. Peters, Broadstairs CT10 2TX
G0 KCB D. Beckley Fen Hill, Hall Road, Great Yarmouth NR29 5NU
G0 KCC Dr W. Randolph 13 Links Road, Poole BH14 9QP
G0 KCD M. Leech 20 Walton Road, Frinton-on-Sea CO13 0AQ
G0 KCE V. Harding 17 St. Anns Road South, Heald Green, Cheadle SK8 3DZ
G0 KCF C. Fosbrook 4A Yew Tree Road, Hayling Island PO11 0QE
G0 KCG D. Hall 47 Sunningdale Road, Fareham PO16 9PA
G0 KCH M. Mccarthy 75 Taynton Drive, Merstham, Redhill RH1 3PX
G0 KCL Kings College ARS c/o J. Greenberg 12 Broadhurst Avenue, Edgware HA8 8TR
GM0 KCN White Rose ARS c/o D. Smith 12 Cannon Street, Selkirk TD7 5BP
GM0 KCY D. Michael 84 Bourtreehall, Girvan KA26 9EL
G0 KCZ W. Bowles Willow Grove, Little Common North Bradley, Trowbridge BA14 0TX
G0 KDA P. Cooper 6 Norwich Close, Scalby, Scarborough YO13 0PP
G0 KDB D. Greenhalgh Hillcroft, Colby, Appleby-in-Westmorland CA16 6BD
GM0 KDC R. Smith 21 Glen View Crescent, Gorebridge EH23 4BT
G0 KDD B. Woodward 58 Marine Drive, Bishopstone, Seaford BN25 2RU
GM0 KDF R. Thomson 25 Cheviot Road, Hamilton ML3 7HB
G0 KDG R. Simpson 29, Hampsfell Grange, Hampsfell Road, Grange-over-Sands LA11 6AZ
GI0 KDH A. Brown 3 Gargrim Road, Fintona, Omagh BT78 2EH
G0 KDI R. Steans 302 Walton Road, West Molesey KT8 2HY
G0 KDL W. Cooper 24 Ambleside Road, Lightwater GU18 5TA
GM0 KDO G. Kirkland 34 Langhouse Green, Crail, Anstruther KY10 3UD
GM0 KDP I. Dunbar Mabruk, 25 Kinord Drive, Aboyne AB34 5JZ
G0 KDQ M. Ward 2 Hollin Gate, Otley LS21 2DP
G0 KDR R. Lintott Upper Grove Farm, Rendham, Saxmundham IP17 2AS
G0 KDS S. Lindsay 27 Bagnell Road Stockwood, Bristol BS14 8PZ
G0 KDT P. Cracknell 54 Yannon Drive, Teignmouth TQ14 9JP
G0 KDV Darenth Valley RS c/o P. Bourke 26 Craylands Lane, , Swanscombe DA10 0LP
G0 KDW G. Burrett 10 Prospect Walk, Lower Burraton, Saltash PL12 4RG
G0 KDX B. Ashton Squirrel Wood, Anderton Mill, Chorley PR7 5PY

G0 KDY A. Spry Newlands Farm, Bradworthy, Holsworthy EX22 7RN
G0 KEB C. Frost 61 Selbourne Avenue, Surbiton KT6 7NR
G0 KEC H. Opitz 26 Holme Court, Lower Warberry Road, Torquay TQ1 1QR
G0 KED J. Bower Linwood, Stain Lane, Mablethorpe LN12 1QB
G0 KEE C. Simons 51 Moorville Drive South, Carlisle CA3 0AW
G0 KEI D. Kennard 8 Fraser Close, Southend on Sea SS3 9YS
G0 KEK B. Curtis Beggars Roost, Rea Barn Road, Brixham TQ5 9EE
GD0 KEO A. Birchenough 20 St. Stephens Meadow, Sulby, Sulby IM7 3DA Isle of Man
GM0 KEQ R. Crawford Glengarry, East Terrace, Kingussie PH21 1JS
G0 KEV K. Gallagher 8 Holme Grove, Burley In Wharfedale, Ilkley LS29 7QB
G0 KEX Sutton Area CG c/o A. O'Hara 26 Thompson Avenue, Ainsworth, Bolton BL2 5RJ
G0 KEY S. Cole 160 New Haw Road, Addlestone KT15 2DN
G0 KFD P. Bailey 273 Humberston Avenue, Humberston, Grimsby DN36 4JA
G0 KFF K. Field 50 Madrona, Amington, Tamworth B77 4EJ
GW0 KFL R. Rees 15 Taliesin Close, Pencoed, Bridgend CF35 6JR
G0 KFM J. Collins 19 Brookside Park Homes, Waterloo Road, Wimborne BH21 3SP
G0 KFQ Orchard County DX Club c/o B. Wilson 20 Peacock Way, Littleport, Ely CB6 1AB
G0 KFT C. Dickerson 1 Park Farm Lane, Nuthampstead, Royston SG8 8LT
G0 KFV M. Evans 40 Park Square East, Jaywick, Clacton-on-Sea CO15 2NN
G0 KFW W. Cole 5A Park Lane, Kemsing, Sevenoaks TN15 6NU
G0 KFY P. Elliot-West 135 Tunstall Road, Sunderland SR2 9BB
G0 KGA A. Danby Cornerstone, Foulbridge Lane, Snainton, Scarborough YO13 9AY
GW0 KGD Prof. V. Zakharov Ty-Brith, Cloddiau, Welshpool SY21 9JE
G0 KGE H. Johnson 2 Thirlmere, Kennington, Ashford TN24 9BD
G0 KGI J. Coleman 80 Ormston Avenue, Horwich, Bolton BL6 7ED
G0 KGL G. Lindsay 66 Jubilee Crescent, Mangotsfield, Bristol BS16 9AZ
G0 KGR R. Beadle 2 Edward Cottages, Great Munden, Ware SG11 1HT
G0 KGT B. Williams 8 Grimbald Road, Knaresborough HG5 8HD
G0 KHA K. Seddon 17 Dunmail Drive, Kendal LA9 7JG
G0 KHF P. Witley 18 Seagate Road, Hunstanton PE36 5BD
G0 KHH A. Rogers 3 Ripley Drive, Wigan WN3 6AJ
G0 KHJ J. Warburton 92 Worsley Road Farnworth, Bolton BL4 9LX
G0 KHK P. Shaw 15 Greenfield Avenue, Marlbrook, Bromsgrove B60 1HE
G0 KHQ P. Hughes 4 Millards Close, Hilperton Marsh, Trowbridge BA14 7UN
G0 KHR E. Forsyth 11 Brooklyn Road, Stockport SK2 6BX
G0 KHY J. Jenkins 3 Gosslan Close, St. Ives PE27 3YZ
G0 KHZ M. Crane Drewton House, Back Lane, Goole DN14 7HD
G0 KIA R. Harris 19 Old Bath Road, Sonning, Reading RG4 6SZ
G0 KIC B. Hayward 22 Waldron Street, Bishop Auckland DL14 7DS
GW0 KIG K. O'Reilly 14 Catherine Close, Abercanaid, Merthyr Tydfil CF48 1YY
G0 KIK S. Berry 85 Lake View Close, West Park, Plymouth PL5 4LT
G0 KIM J. West 242 Grane Road, Haslingden, Rossendale BB4 4PB
G0 KIN T. Harper 72 School Road, Salford Priors, Evesham WR11 8XN
GW0 KIR K. Jones 38 Pendre Close, Brecon LD3 9EW
G0 KIY N. Brook 2 Back Regent Place, Harrogate HG1 4QR
G0 KJC J. Clayton 49 Bramble Lane, Mansfield NG18 3NP
G0 KJF R. Warner Barley Hill Farm, Combe St. Nicholas, Chard TA20 3HJ
G0 KJG B. Bevington 12 Buckingham Road, Rowley Regis B65 9JN
G0 KJJ W. Ritchie 16 Avenue Mezidon Canon, Honiton EX14 2TT
G0 KJK K. Ranger Flat 12, Dulverton Hall Esplanade, Scarborough YO11 2AR
G0 KJM J. Richards 14 Southwood Drive East, Bristol BS9 2QP
G0 KJN J. Windebank 9 Townsend Place, St. Ippolyts, Hitchin SG4 7RQ
G0 KJP D. Scott 19 The Fillybrooks, Aston, Stone ST15 0DH
G0 KJR K. Robbertze 19 Shapton Close, Holbury, Southampton SO45 2QJ
GW0 KJT T. Lewis 2 Railway Terrace, Pontyberem, Llanelli SA15 5HN
G0 KJU J. Robertson 28 Frith Road, Bognor Regis PO21 5LL
GW0 KJZ J. Jones 64 Cleviston Park, Llangennech, Llanelli SA14 9UP
G0 KKC D. Browning 81 Bishop Road, Bishopston, Bristol BS7 8LU
G0 KKD A. Parker Old Rectory, 1 Church Lane, Matlock DE4 2GL
GM0 KKE I. Coulson 11 Redcliffs, Kingoodie, Dundee DD2 5DL
G0 KKF B. Hough 54 Woodbourne Road, Sale M33 3TN
G0 KKH J. Henderson Hodgkins Farm, Norton Heath Road, Willingale, Willingale CM5 0QG
G0 KKL P. Mayer Flat 7 Broomrigg, 5 Belle Vue Road, Poole BH14 8UE
G0 KKQ E. Birch 6 Totterton, Lydbury North SY7 8AN
G0 KKR B. Chapman Millbrooke Cottage, Covenham St. Bartholomew, Louth LN11 0PB
G0 KKS A. Nance 33 Oak Close, Copthorne, Crawley RH10 3QT
G0 KKT I. Osborne 19 Lumber Leys, Walton on the Naze CO14 8SS
G0 KKU J. Howard 111 Heath Road, Penketh, Warrington WA5 2DB
G0 KKV M. Lowe 11 Priory Walk Mancetter, Atherstone CV91QA
G0 KLA C. Thompson 425 Mortlake Ave, St. Lambert J4P 3C7 Canada
G0 KLD C. Wren 38 Green Street, Hyde SK14 1QX
G0 KLF N. Anderton11 29 Cliftonville Drive, Swinton, Manchester M27 5NA
G0 KLG R. Hodds 17 Oaklands Drive, Willerby, Hull HU10 6BJ
G0 KLH R. Griffin 59 Lakeland Gardens, Chorley PR7 2LS
G0 KLJ J. Leader 9 Southerwicks, Corsham SN13 9NH
G0 KLK A. James 14 Randle Drive, Sutton Coldfield B75 5LH
G0 KLN C. Ryalls 22 Carr Lanebd205Hn, Keighley BD205HN
GM0 KLO C. Grossart 11 Woodlands Drive, Brightons, Falkirk FK2 0TF
GM0 KLP J. Pentland 2 Glenniston Cottages, Auchtertool, Fife KY5 0AX
G0 KLQ D. Cross 15 Fernside Road, West Moors, Ferndown BH22 0EE
G0 KLT D. Rogers Jones 20 Birchwood, Leyland PR26 7QJ
G0 KLU P. Fairhurst 161 Daniells, Welwyn Garden City AL7 1QP
GW0 KLY R. Jones 20 Newfoundland Way, Blackwood NP12 1FS
GM0 KMA Dr M. Rainey 2 Shields Holdings, Lochwinnoch PA12 4HL
G0 KMB K. Bowdler 18 Cavendish St., Leigh WN7 1SG
G0 KMC A. Slaughter 42 Goss Avenue, Waddesdon, Aylesbury HP18 0LY
G0 KMF R. Holmshaw 142 Oakleigh Park Drive, Leigh-on-Sea SS9 1RU
GM0 KMJ P. Johnstone 26 Lomond Crescent, Stenhousemuir, Larbert FK5 4LT
G0 KMK M. Aslam 38 Grey St., Burnley BB10 1BA
G0 KML B. Hawes 201 Ridgeway, Plympton, Plymouth PL7 2HP
G0 KMN S. Hepworth 9 College View, Ackworth, Pontefract WF7 7LA
G0 KMP G. Aungiers 6 Woodlands Crescent, Barton, Preston PR3 5HB
G0 KMV H. King Que Lindo, Church Lane, Bristol BS39 5UP
G0 KMW H. Shepherd Whydown, 3 White House Close, Abingdon OX13 6LP
G0 KNJ R. Bygrave 69 Albert Gardens, Harlow CM17 9QG
G0 KNL W. Christlo 6 Nether Ley Gardens, Chapeltown, Sheffield S35 1AH
G0 KNN M. Gregg 22 Mayfields, Spennymoor DL16 6RN
GM0 KNT Dr A. Bates Caberfeidh, Balnageith, Forres IV36 2SG
G0 KNW C. Wiles Everest, Mile Road, Morpeth NE61 5QW
G0 KNX G. Allen 3 Ryton Close, Coventry CV4 8HF
G0 KNY K. Heaton 5 Perriams Place, Budleigh Salterton EX9 6LY
G0 KOC A. Kinson 6 Uplands Park, Broad Oak, Heathfield TN21 8SJ
G0 KOE T. Foxton Dunbar, Dam Lane, Malton YO17 9SJ
G0 KOF D. Henretty 13 Siskin Chase, Cullompton EX15 1UD
G0 KOI M. Cooper 15 Woodleigh Avenue, Harborne, Birmingham B17 0NW
G0 KOJ B. Thomas Harpley House, Police Road, Walpole St. Andrew, Wisbech PE14 7NN
G0 KOM A. Mcgonigle 5 Whitehouse Crescent, Chelmsford CM2 7LP
G0 KOO A. Mcdowell Fern Cottage, The Gride, Old Leake, Boston PE22 9LS
G0 KOU B. Arrowsmith 25 Watchouse Road, Chelmsford CM2 8PT
GI0 KOW R. Cummings 19 Bachelors Walk, Keady, Armagh BT60 2NA
G0 KOY B. Clues 8 Acland Avenue, Colchester CO3 3RS
GW0 KPD J. James 1 Pellau Road, Margam, Port Talbot SA13 2LF
G0 KPE Sparks At the Shed c/o C. Mcgowan Belvedere, Tylers Lane, Reading RG7 6TN
GI0 KPF B. Mccausland 5 Hollyfields, Dungannon BT71 7BH
G0 KPG R. Moore 34 Fishponds Road, Kenilworth CV8 1EZ
G0 KPH P. Keighley 15 Stuart Court, Warwick Terrace, Leamington Spa CV32 5NU
GD0 KPN J. Mcloughlin 8 Governors Hill, Douglas IM2 7AW Isle of Man
G0 KPQ J. Morgan 9 Consort Close, Plymouth PL3 5TX
GW0 KPU R. Harper 4 Gresford Road, Llay, Wrexham LL12 0NW
GW0 KPV G. Owen 2 Ffordd Beibio, Holyhead LL65 2EF
G0 KPY A. Baker-Munton 66 Stanway Road, Headington, Oxford OX3 8HX
G0 KPZ D. Portch 148 Brixham Road, Welling DA16 1EJ
G0 KQA P. Davies Silver Birches, Orchard Road, Basingstoke RG22 6NU
GM0 KQB R. Kemp 1 Grendon Court, Stirling FK8 2JX
GD0 KQE K. Jordan Engadine, Little Switzerland, Isle of Man IM2 6AG
G0 KQH B. O'Donoghue 30 Lake Drive, Hamworthy, Poole BH15 4LT
G0 KQI L. Painter 185 Albion St., St. Helens WA10 2HA
G0 KQK T. Chambers Autumn, Water Lane, Castle Bytham, Grantham NG33 4RT
G0 KQO G. Elliott 32 Chapel Street, Newport PO30 1PZ
G0 KQP J. Carroll 5 Montagu View, Leeds LS8 2RH
G0 KQR R. Bundell 24 Sylvan Avenue, East Cowes PO32 6PS
G0 KQS G. Griffiths Sherwood House, Buggen Lane, Neston CH64 6QB
G0 KQT A. Holdway 18 The Quantocks, Thatcham RG19 3SF
GW0 KQU G. Slatter 6 Glannant St., Penygraig, Tonypandy CF40 1JT
GW0 KQV D. Clark Martinique, Wolfscastle, Haverfordwest SA62 5DY
GW0 KQX W. Cook Fronoleu, Bryngwy, Rhayader LD6 5BN
G0 KQY D. Murrell 25 Waverley Road, , Hoylake CH47 3DD
G0 KRB G. Phillips 57 Hollytrees, Bar Hill, Cambridge CB23 8SF
G0 KRD D. Downes 7 Sandy Lane, Fakenham NR21 9ES
G0 KRG Special Communications (Ta) Association c/o L. Conlon 4 Hill Crest Drive, Slack Head, Milnthorpe LA7 7BB
G0 KRH C. Tarrant 91 Dunes Road, Greatstone, New Romney TN28 8SW
G0 KRK B. Cockfield 47 Aston Road, Willenhall WV13 3DG
GW0 KRL I. Capon Pentre Garreg Bach, Marianglas LL73 8PP
GW0 KRQ J. Cartwright 20 Castlefield Place, Cardiff CF14 3DU
G0 KRR J. Hough 1 Rock Lane, Linslade, Leighton Buzzard LU7 2QQ
G0 KRS Royal Air Force Cosford ARC c/o K. Conlon 4 Hill Crest Drive, Slack Head, Milnthorpe LA7 7BB
G0 KRT E. Masters 91 Mayfair Avenue, Worcester Park KT4 7SJ
G0 KRU A. Wright Cherry Tree Cottage, High Road, Beighton, Norwich NR13 3LA
G0 KRX P. Ruder 34 Chelmsford Road, South Woodford, London E18 2PL
G0 KRY D. Sanders 149 Sutton Road, Walsall WS5 3AW
G0 KSC Philips Tele Lt c/o J. Johnson 20 Sanders Road, Canvey Island SS8 9NY
G0 KSD R. Allgood 7 The Chase, Blofield, Norwich NR13 4LZ
G0 KSJ J. Graves 172 Hall Lane, Upminster RM14 1AT
G0 KSL R. Torr 68 East Towers, Pinner HA5 1TL
G0 KSN H. Dabhi 23 Shalgrove Field, Fulwood, Preston PR2 3SX
G0 KSS K. Nobbs 49 St. James Drive, Burton, Carnforth LA6 1HY
G0 KTC C. Ayres 219 Ashingdon Road, Rochford SS4 1RS
G0 KTD A. Bonney 6 Mitchell Road, St. Austell PL25 3AU
GM0 KTH D. Wakefield Millfield, , Burray KW17 2SU
GW0 KTL G. Edmunds 14 Avon Close, Bettws, Newport NP20 7BZ
G0 KTN T. Smithers 14 Georgian View, Bath BA2 2LZ
GM0 KTO J. Power 0/2 20 Eastercraigs, Glasgow G31 3LJ
G0 KTP R. Cockbill 45 Mills Road, Melksham SN12 7DT
G0 KTR R. Farnley 67 Barons Court Failsworth, Manchester M35 0LH
G0 KTS J. Wright 65 Groombridge Close, Welling DA16 2BP
G0 KTT P. Riley 11 Pinewood Court, South Downs Road, Altrincham WA14 3HY
G0 KTU A. Miller 11 Blackbrook Avenue, Paignton TQ4 7ND
G0 KTV J. Edgington 83 Woking Road, Guildford GU1 1QL
G0 KTW J. Moss 1 Millers View, Much Hadham SG10 6BN
G0 KTX A. Hornsby 328 Pelham Road, Immingham DN40 1PT
G0 KTY G. Salisbury Nythfa, 1 Stad-Y-Garnedd, Anglesey L60 6BB
G0 KUA V. Kathuria 2 Bevan Road, Lovedean, Waterlooville PO8 9QH
G0 KUC D. Bloomfield 14 Horsham Close, Luton LU2 8JH
G0 KUD P. Haith 17 Lime Tree Avenue, Grimsby DN33 2BB
G0 KUE P. Webb 119 Chipstead Valley Road, Coulsdon CR5 3BP
G0 KUF N. Buchanan Meadowside, Jacobs Well Road, Guildford GU4 7PD
GI0 KUH J. Mccabe 121 Garvaghy Road, Craigavon BT62 1EH
G0 KUI J. Wood 6 West Terrace, Stakeford, Choppington NE62 5UL
GM0 KUJ J. Mcgifford 52 Gartons Road, Glasgow G21 3HY
GM0 KUP F. Mann 12 Greenbank Court, Falkirk FK1 5DS
G0 KUQ F. Hills 112 Boxfield Green, Stevenage SG2 7DS
G0 KUU F. Gillham 260 Summerhouse Drive, Wilmington, Dartford DA2 7PB
G0 KUW C. Bennett 67 King St., Clowne, Chesterfield S43 4BS
G0 KUX P. Kay 97 Avenue Road, London N14 4DH
G0 KUY S. Crane 13 Kirkstone Drive, Royton, Oldham OL2 6TP
G0 KUZ B. Long 81 Easthorpe Street, Ruddington, Nottingham NG11 6LB
G0 KVA A. Sargent 25 Jordans Way, Bricket Wood, St. Albans AL2 3SJ
G0 KVC H. Crouch 21A Victoria Gardens, Horsforth, Leeds LS18 4PJ

GM0 KVD C. Mackay 5 Cromer Gardens, Glasgow G20 9JQ
GM0 KVE A. Dickson 17 Junction Road, Kinross KY13 8TA
G0 KVF R. Croucher 26 Edith Avenue, Peacehaven BN10 8JB
G0 KVG R. Neal 14 Saxton Close, Worksop S80 3DE
G0 KVJ Peterlee ARC c/o A. Pennell 99 Westheath Avenue, Sunderland SR2 9LQ
G0 KVK G. Cooper 33 Lawnswood Road, Wordsley, Stourbridge DY8 5PH
G0 KVM W. Gravenor 3 Foxhill Grove, Queensbury, Bradford BD13 2JN
G0 KVO D. Pallister 9 Curtis Hayward Drive, Quedgeley, Gloucester GL2 4WJ
GI0 KVQ G. Millar 1 Mullybrannon Road, Dungannon BT71 7ER
G0 KVR C. Mayo 118 Burden Road, Beverley HU17 9LH
G0 KVS Dr M. Gill 21 Priory Terrace London, London NW6 4DG
G0 KVU H. Sheratte Redcar Villa, 382 Buxton Road, Macclesfield SK11 7ES
GW0 KWA D. Clark 37 Rothersiade Road, Langland, Swansea SA3 4QW
G0 KWC R. Waite 95 Westlea Road, Leamington Spa CV31 3JE
G0 KWD K. Dyer 79 Station Road, Woolton, Liverpool L25 3PY
G0 KWE J. Knowles 10 Grove Hill, Hessle HU13 0RT
G0 KWF W. Taylor 14 Rossiters Lane, St. George, Bristol BS5 8TW
G0 KWG S. Lodge 7 Primrose Drive, Milkwall, Coleford GL16 7PU
GM0 KWL B. Mulleady 9 Elizabeth Crescent, Camelon, Falkirk FK1 4JF
GD0 KWM G. Brown 10 Albert Street, Ramsey IM8 1JF Isle of Man
GW0 KWO K. Williams 39 Lewis Drive, Caerphilly CF83 3FT
G0 KWQ P. Bates 46 Kingsley Avenue, Redditch B98 8PL
GM0 KWW J. Alexander Shore Cottage, Girvan KA26 9JH
G0 KXD H. Worden 3 Tower View, Darwen BB3 3GZ
G0 KXG J. Nicholls 93 Swan Road, Hanworth, Feltham TW13 6PE
G0 KXL S. Maton 117 Woodchurch Road, Birkenhead CH42 9LJ
G0 KXV G. Meredith Hedgerow, Watton Road, Larling, Norwich NR16 2AJ
G0 KXW A. Fitzmaurice 14 Welwyn Close, Thelwall, Warrington WA4 2HE
G0 KXY P. Wroe 44 Hillberry Crescent, Warrington WA4 6AF
G0 KXZ A. Sockett 35 Whernside Road, Woodthorpe, Nottingham NG5 4LB
G0 KYA S. Nichols 61B Norwich Common, Wymondham NR18 0SW
G0 KYB M. Kinder 58 Longridge Avenue, Stalybridge SK15 1HL
G0 KYD A. Hull 1 Occupation Lane, New Bolingbroke, Boston PE22 7LW
G0 KYE L. Landricombe 19 Crackston Close, Eggbuckland, Plymouth PL6 5SN
G0 KYG P. Willetts 197 Norwich Road, Fakenham NR21 8LR
G0 KYH J. Elliott Tregerrick, Martinstown, Dorchester DT2 9JN
G0 KYJ G. Liversidge 65 Rowelfield, Luton LU2 9HL
G0 KYK R. Beardsmore 2 Fitzmaurice Road, Wednesfield, Wolverhampton WV11 3EG
G0 KYL J. Lawrence 8 Murray Terrace, Dipton, Stanley DH9 9HB
G0 KYM W. Lowder 24 Plantation Lane, Bearsted, Maidstone ME14 4BH
G0 KYN R. Markham 25 Burndell Way, Hayes UB4 9YF
G0 KYR D. Cooper 7 Kendal Rise, Bedlington NE22 6PB
G0 KYS R. Edgar 45 Exeter Road, Dawlish EX7 0AB
GM0 KYU J. Robertson 143 Rankin Court, Greenock PA16 9AZ
G0 KYX Rvd. P. Morgan 4 Cromwell Mews, Burgess Hill RH15 8QF
GW0 KYY M. Warner 76 Rhodfar Eos, Cwmrhydyceirw, Swansea SA6 6SW
GJ0 KYZ Dr P. Mahrer 2 Oakley, La Rue Parcqthee, St. Lawrence JE3 1FR Jersey
G0 KZA E. Bishop 21 Mandalay St., Basford, Nottingham NG6 0BH
G0 KZD M. Withey 9 Marnhull Road, Longfleet, Poole BH15 2EX
GW0 KZE Dr C. Dublon Tyn-Y-Waun, Dare Road, Aberdare CF44 8UB
GW0 KZG A. Adams 1 Nant Y Ffynnon, Letterston, Haverfordwest SA62 5SX
G0 KZH E. Clayton 220 Milnrow Road, Rochdale OL16 5BB
G0 KZI J. Williams 18 St. Andrews Close, Holme Hale, Thetford IP25 7EH
G0 KZM D. Egan 56 Walker Avenue, Wollescote, Stourbridge DY9 9EL
G0 KZN A. Sargeant 27 Sandygate Crescent, Old Leake, Boston PE22 9RA
G0 KZO E. Lomas 2 Linney Road, Bramhall, Stockport SK7 3JW
G0 KZT A. Briers 33 Deans Walk, Coulsdon. CR5 1HR
GW0 KZW W. Jones Tanglewood, 2 Bryntirion Avenue, Prestatyn LL19 9PB
GM0 KZX B. Spink 9 St. Andrews Crescent, Dumbarton G82 3ER
G0 LAA E. Martin 90 Grand Drive, Herne Bay CT6 8LS
G0 LAD J. Parfett 65 Brompton Lane, Rochester ME2 3BA
G0 LAG J. Penney 2A St John St., Wainfleet, Skegness PE24 4DL
G0 LAK J. Rogers 186 Beavers Lane, Birleywood, Skelmersdale WN8 9BP
GI0 LAM G. Lamb 31 Dromara Road, Ballyward, Castlewellan BT31 9SJ
G0 LAN A. Taylor 24 Marlborough Drive, Mablethorpe LN12 2 BA
G0 LAP S. Jeffery 3 Cromwell Close Walcote, Lutterworth LE17 4JJ
G0 LAU J. Barber Jasmine Cottage, Spend Lane, Ashbourne DE6 2AS
G0 LAX A. Duggan 28 Higher Rads End, Eversholt, Milton Keynes MK17 9ED
G0 LAZ J. Pryer Apesford Crossing Cottage, Apesford, Leek ST13 7EX
GW0 LBA A. Hughes Derwen Las, Valley Road, Llanfairfechan LL33 0SS
G0 LBB R. Batty 31 Spring Lane, Balderton, Newark NG24 3NZ
G0 LBE S. Watkinson 40 Wharfedale, Westhoughton, Bolton BL5 3DP
GW0 LBI L. Smart Wordsley, Gwerthonor Road, Bargoed CF81 8JS
GM0 LBN Dr J. Clark 35 Jedburgh Avenue, Rutherglen, Glasgow G73 3EN
G0 LBO J. Ross The Gables, Jack Lane, Northwich CW9 8QA
G0 LBQ R. Perks 120 Cranes Park Road, Sheldon, Birmingham B26 3ST
GM0 LBR B. Gourlay Muirhead House, Chryston, Glasgow G69 9ND
G0 LBT K. Tromans 7 Heathfield, Heath Charnock, Chorley PR6 9LA
G0 LBZ P. Cahill 56 Dene Road, Headington, Oxford OX3 7EE
G0 LCB A. Cleaver 19 Newlands Drive, Grove, Wantage OX12 0NY
G0 LCC H. Grinter 16 Gladiolus Road, Langport TA109TA
G0 LCD P. Chinnock 30 Trelissick Road, Paignton TQ3 3GW
G0 LCE K. Robinson 33 Mirlaw Road, , Cramlington NE23 6UB
G0 LCG S. Sorockyj 8 Bowden Avenue, Bestwood Village, Nottingham NG6 8XN
G0 LCH M. Nash 22 Northleigh Close, Loose, Maidstone ME15 9RP
G0 LCJ B. Lucock 15 Mayfield Road, Newquay TR7 2DG
G0 LCN D. Prendiville 40 Caerleon Drive, Southampton SO19 5LF
G0 LCO R. Foster 18 Stokesay Way, Sutton Hill, Telford TF7 4QE
G0 LCS K. Rochester 22 Langford Road, Cockfosters, Barnet EN4 9DS
G0 LCT G. Moss 15 Coppice Avenue, Hatfield, Doncaster DN7 6AH
G0 LCU B. Walker 70 King George Road, Loughborough LE11 2PA
G0 LCV J. Fidoe 85 Sedgemoor Road, Bridgwater TA6 5NS
G0 LCX D. Weatherill North End Cottage, North End Road, Yatton, Bristol BS49 4AS
G0 LDB Obo Royal Air Force ARS E Riding Area c/o M. Mallinson 25 The Fairway, Banbury OX16 0RR
GI0 LDI D. Keys 71 Madison Avenue, Eglinton, Londonderry BT47 3PW
G0 LDJ D. Cansfield 1 Brook Walk, Calmore, Southampton SO40 2UY
G0 LDO R. Summerfield 64 Station Road, Broughton Astley, Leicester LE9 6PT
G0 LDP K. Starkey 13A Cardigan Road, Bedworth CV12 0LY
G0 LDR J. Marlow 21 Thames Rise, Kettering NN16 9JL
G0 LDU K. Allies 6 Alston Close, Hazel Grove, Stockport SK7 5LR
G0 LDY K. Jenkinson 2 Madeira Avenue, Codsall, Wolverhampton WV8 2DS
GW0 LDZ B. Garland 20 Bryn Avenue, Upper Brynamman, Ammanford SA18 1BD
GI0 LEC Poole RS c/o A. Duffy 81A Arney Road, Bellanaleck, Enniskillen BT92 2DL
G0 LEE R. Lee 7 Long Meadow, Little Hoole, Preston PR4 4RQ
G0 LEF T. Bell 16 North Seaton Road, Newbiggin-by-the-Sea NE64 6XT
G0 LEH G. Chatfield 1A Sheringham Way Orton Longueville, Peterborough PE2 7AH
G0 LEI E. Hughes 1 Leith Gardens Tanfield Lea, Stanley DH9 9LZ
G0 LEJ M. Huggett Rosslyn, Station Road, Brampton CA8 1EX
G0 LEL F. Sunley 39 Winton Road, Northallerton DL6 1QQ
G0 LEN Hucknall Rolls Royce ARC c/o P. Worsdale Emergency Planning Department, Fire Brigade Headquarters, Lincoln LN5 8EL
G0 LEP D. Stewart Buckskin, 16 Prescelly Close, Basingstoke RG22 5DN
G0 LES E. Simpson Laneside, Cliff Lane, Bridlington YO15 1JF
G0 LEU P. Johnson 5 The Hawthorns., Broadstairs. CT10 2NG
G0 LEV D. Painter Troutbeck, Mary Tavy, Tavistock PL19 9PR
G0 LEY R. Smith 63 Windsor Road, Wellingborough NN8 2ND
G0 LFA N. Swallow 178 Barcroft St., Cleethorpes DN35 7DX
G0 LFE Rvd. K. Gray 25 The Pastures, Blyth NE24 3HA
G0 LFF R. Hide Flat 22, Church Court Church Road, Haywards Heath RH16 3UE
G0 LFH P. Mustchin 6 Spinney North, Pulborough RH20 2AT
G0 LFI F. Cotton 49 Cornwall Road, Fratton, Portsmouth PO1 5AR
G0 LFM V. Sancto Meadowbank, 15A Spratling Street, Ramsgate CT12 5AW
G0 LFN S. Southwell Sullys, 12 Somerset Road, Southsea PO5 2NL
G0 LFP S. Courtney-Crowe 28 Brymore Close, Prestbury, Cheltenham GL52 3DY
G0 LFQ P. Mason Penrose House, Tavistock Road, Launceston PL15 9LE
G0 LFV P. Fisher Chevalier, Marks Corner, Newport PO30 5UH
G0 LFX M. Harrison 8 Browns Lane, Uckfield TN22 1RT
G0 LFY D. Recardo 1 Heronfield Close, Redditch B98 8QL
G0 LFZ A. Recardo 1 Heronfield Close, Redditch B98 8QL
G0 LGA R. Letts 28 Catlin Crescent, Shepperton TW17 8EU
G0 LGB J. Walker 22 Temperance Field, Wyke, Bradford BD12 9NR
G0 LGC L. Culshaw 15 Naunton Avenue, Leigh WN7 4SX
G0 LGE M. Brooman 25 Knockholt Road, Cliftonville, Margate CT9 3HL
G0 LGF T. Evennett The Homestead, Pound Green Lane Shipdham, Thetford IP25 7LS
G0 LGG N. Challacombe 17 Tanners Lane, Chalkhouse Green, Reading RG4 9AD
G0 LGJ M. Taylor 6 Welden Road, Scarning, Dereham NR19 2UB
G0 LGK E. Wall Shrubbery Cottage, Felderland Lane, Deal CT14 0BT
G0 LGO A. Turton 58 Highfield Lane, Quinton, Birmingham B32 1QT
GI0 LGV H. Magill 51 Ballybracken Road, Doagh, Ballyclare BT39 0TQ
G0 LGW R. Caine 148 Dumpton Park Drive, Broadstairs CT10 1RP
G0 LGZ B. Grimes Flat 12, Clyde House, Ventnor PO38 1QL
G0 LHB A. Okubo 1427-9-608, Yamazaki-Cho Machida City, Tokyo 195-0074 Japan
G0 LHD R. Caton 13 Goss Barton, Nailsea, Bristol BS48 2XD
G0 LHE R. Wilmot Elm Tres, Drayson Lane, Northampton NN6 7SR
G0 LHG A. Rayner 24 Syers Lane, Whittlesey, Peterborough PE7 1AT
G0 LHL M. Humphreys 25 Dalestorth Close, Sutton-in-Ashfield NG17 4EH
G0 LHM B. Tuffrey 53 Sheffield Road, Warmsworth, Doncaster DN4 9QR
G0 LHN J. Butterworth 38 Stuart Avenue, Moreton, Wirral CH46 9PF
G0 LHR L. Robinson 82 Grassholme, Wilnecote, Tamworth B77 4BZ
G0 LHU J. Lawton 37 Southway, Horsforth, Leeds LS18 5RN
G0 LHV R. Kay 24 Chapel Garth, Gilberdyke, Brough HU15 2UH
G0 LHX H. Passmore 26 Edgebury Woolavington, Bridgwater TA7 8ES
G0 LHZ J. Carter 22 Orchard Coombe, Whitchurch Hill, Reading RG8 7QL
G0 LIA R. James 77 Charlotte Close Mount Hawke, Nr Truro. TR4 8TT
G0 LIB R. Weston 38 Church Road, Peasedown St. John, Bath BA2 8AF
G0 LII S. Hodgson 4 Nikolaou Michael Street, Dasaki Achnas 5523 Cyprus
GW0 LIK C. Raymond 23 Castle Pill Crescent, Steynton, Milford Haven SA73 1HD
GM0 LIM J. Duffy 39 Kylerhea Road, Thornliebank, Glasgow G46 8AB
G0 LIN C. Smith 2 Ha'Penny Drive, Holbrook, Ipswich IP9 2TT
G0 LIQ J. Cunningham 219 Alfreton Road, Underwood, Nottingham NG16 5GX
GM0 LIR P. Woods 29 Yarrow Crescent, Wishaw ML2 7JX
GW0 LIS A. Wright 8 Bryn Mor Terrace, Holyhead LL65 1EU
G0 LIW P. Hopkins 2641 Suncoast Lakes Blvd, Port Charlotte 33980 United States
G0 LIY P. Smit 18 Owlwood Lane, Dunnington, York YO19 5PH
G0 LIZ E. Saunders 40 Walkley Street, Sheffield S6 3RG
G0 LJB P. Williams 44 Meadow Road, Mirehouse, Whitehaven CA28 8EP
G0 LJD B. Howard 8 St. Margarets Close, Seasalter, Whitstable CT5 4ST
G0 LJF M. Binks 24 Mill Lane, Stockport SK5 6UU
G0 LJG D. Green The Archways, St. Georges Road, Trowbridge BA14 6JQ
G0 LJH C. Holmes Manacor, 4 Dovefields, Uttoxeter ST14 5LT
G0 LJI G. Evans 241 St. Johns Road, Newbold Moor, Chesterfield S41 8PE
G0 LJJ D. Mackenny 21 Chilton Way, Hungerford RG17 0JR
G0 LJK W. Dancock 11 St. Davids Close, Stourport-on-Severn DY13 8RZ
G0 LJM D. Roebuck 8 Runnymede Court, Bradford BD10 9JW
G0 LJP P. Mcleod 4 Caple Avenue, Kings Caple, Hereford HR1 4UL
G0 LJU D. Koveos 235 Poynters Road, Dunstable LU5 4SH
G0 LJV S. Swinbourne 11 Stapleton Road, Warmsworth, Doncaster DN4 9LA
GW0 LJW D. Goodwin 25 Bevan Crescent, Blackwood NP12 1EW
G0 LKA C. Cruddas 81 Church Walk, Atherstone CV9 1PS
G0 LKI W. Cockerell 3 Churchford Road, Knowle, Braunton EX33 2LT
GW0 LKJ W. Halliwell 20 Llwynon Road, Oakdale, Blackwood NP2 0LX
GW0 LKN G. Weller Lluest Coch Tregynon, Newtown SY16 3PD
GM0 LKS E. Mcgreevy 47 Fairfield Drive, Renfrew PA4 0EG
GM0 LKT A. Ferris 60 Appin Crescent, Kirkcaldy KY2 6ES
G0 LKY G. Civil Whitehouse Farm, Magpie Lane, Brentwood CM13 3DZ
G0 LLB R. Smith 34 Churchill Rise, Chelmsford CM1 6FD
G0 LLC Reigate Amateur Transmitting Society c/o M. Bridges 7 Sun Road, Woodland, Bishop Auckland DL13 5NF
GW0 LLD H. Jones Dolau Bran, Cynghordy, Llandovery SA20 0LD
G0 LLE P. Ferris 116 Capel Road, London E7 0JS
G0 LLG D. Davies 4 Whitendale, Lancaster LA1 5JD
GM0 LLJ B. Borrows 27 Craigdimas Grove, Dalgety Bay, Dunfermline KY11 9XR
G0 LLL J. Roberts 69 Barnoldswick Road, Barrowford, Nelson BB9 6BQ
G0 LLP L. Proud 26 Drayton Court, The Green, Nuneaton CV10 0SL
G0 LLU A. Harrison 4 Hardwick Close, High Lane, Stockport SK6 8DG

G0 LLX A. Bassett 125 Stonyhill Avenue, South Shore, Blackpool FY4 1PW
G0 LLY S. Munro 10 Aykroft, Bourne PE10 0QX
G0 LMA S. Crooks 10 Mere Close, Mountsorrel, Loughborough LE12 7BP
G0 LMD M. Butler 44 East Stratton, Winchester SO21 3DU
G0 LMJ E. Garrott Lynden, Clappers Lane, Chichester PO20 7JJ
GI0 LMR W. Redmond 6 Hazelwood Crescent, Craigywarren, Ballymena BT43 6TA
G0 LMX Aberporth YMCA ARC c/o V. Denecker Kernanderry, Faringdon Road, Abingdon OX13 6QJ
G0 LNA R. Henderson 65 Rowarth Road, Manchester M23 2UL
G0 LNB G. Goodwin 16 Hucklow Avenue, Newall Green, Manchester M23 2YX
G0 LNE T. Stokes 22 Armada Close, Erdington, Birmingham B23 7PB
G0 LNI J. Stringer 2 West End, Marston Magna, Yeovil BA22 8BW
G0 LNK P. Bower 103 Henson Park, Chard TA20 1NJ
GW0 LNM P. Pentecost Brynhyfryd, Maes Y Bont Road, Gorslas, Llanelli SA14 7NA
G0 LNN D. Draycott 3 Sycamore Gardens, Dymchurch, Romney Marsh TN29 0LA
GW0 LNO Dr S. Feeney Tardd Y Dwr, Star Crossing Road Cilcain, Mold CH7 5NU
G0 LNS G. Robinson 9 Greenlands Court, Seaton Delaval, Whitley Bay NE25 0BU
G0 LNT M. Millward 50 Barnsley Road, Moorends, Doncaster DN8 4QT
G0 LNV Dr T. Appleyard 78 Chelsea Road, Sheffield S11 9BR
G0 LNW T. Horabin 69 Birchwood Avenue, North Gosforth, Newcastle upon Tyne NE13 6QB
G0 LNX I. Davison 20 Littlegreen Gardens Compton, Chichester PO18 9NP
G0 LOC T. Loraine Fieldgate, Coltstaple Lane, Horsham RH13 9BB
GM0 LOD G. Collier 64 Hadfast Road, Cousland, Dalkeith EH22 2NZ
G0 LOE S. Phillips 26 Belvedere Drive, Dukinfield SK16 5NW
G0 LOF F. James 6 Pinewood Close, East Preston, Littlehampton BN16 1HF
G0 LOH K. Dutson 3 The Barracks, Wynford Eagle, Dorchester DT2 0ER
G0 LOJ Dr C. Budd 18 Rossendale Close, Worle, Weston-Super-Mare BS22 9HA
GM0 LOK J. Leggat Ailach, St. Aethans Road, Elgin IV30 2YR
G0 LOL C. Kidger 25 Elmham Road Cantley, Doncaster DN4 6LF
GM0 LOO H. Hunter 25 Braehead Road, Kirkcaldy KY2 6XP
G0 LOP G. Tweedy 8 Greencliffe Drive, York YO30 6NA
GM0 LOT R. Clasper 32 Murieston Park, Livingston EH54 9DT
G0 LOU A. Hordle 152 Evering Avenue, Poole BH12 4JH
G0 LOW the Shortwave Shop c/o D. Kemp Zeacombe House Caravan Park East Anstey, Tiverton EX16 9JU
G0 LOZ I. Tomson 13 Valley View, Bewdley DY12 2JX
GM0 LPB J. Gault 25 Beech Brae, Bishopmill, Elgin IV30 4NS
G0 LPF R. Wilkins 20 Fairholme Drive, Yapton, Arundel BN18 0JH
G0 LPG B. Gaunt Po Box 50, Guildford GU1 2FJ
G0 LPN B. Alperowicz 20 Chemin Du Cabanis, Meynes 30840 France
G0 LPP C. Galea 36 Godwit Close, Gosport PO12 4JF
G0 LPQ J. Hagen 7 Oak Close, Whiston, Prescot L35 2YG
G0 LPT G. Wegg 23 Kerdane, Hull HU6 9EB
G0 LPU A. Newton 10 Rowan Court, Greasby, Wirral CH49 3QH
G0 LPV B. Chappell 49 Midway, South Crosland, Huddersfield. HD4 7DA
G0 LPX B. Garbutt 34 The Green, Tockwith, York YO26 7RA
G0 LQC D. Briggs 57 Charlton Drive, High Green, Sheffield S35 3PA
G0 LQD P. Valleley 9 Lavender Road, Basingstoke RG22 5NN
G0 LQI M. Murphy 133 Preston Road, Preston, Weymouth DT3 6BG
G0 LQK M. Hinchliffe 2 Ash Grove, New Longton, Preston PR4 4XJ
GD0 LQL S. Kewley 16 Hillcrest Grove, Onchan, Isle of Man IM3 3HY
G0 LQM J. Hipwell 5 Dolphin Crescent, Paignton TQ3 1AE
G0 LQN L. Fish 44 Maycroft Avenue, Poulton-le-Fylde FY6 7NE
G0 LQO R. Taylor 17 York Close, Clayton Le Moors, Accrington BB5 5RB
G0 LQT H. Smith 66 The Avenue, Clacton-on-Sea CO15 4ND
G0 LQU P. Fardell 90 Beechwood Avenue, St. Albans AL1 4XZ
G0 LQV M. Fordham 24A Main Street, Prickwillow, Ely CB7 4UN
G0 LQW P. Macolive 6 Pembroke Way, Hayes UB3 1PZ
G0 LQX T. Newstead 17 Aspland Road, Norwich NR1 1SH
G0 LQZ C. Walkup 1 Darley Hall, Luton LU2 8PP
GM0 LRA Lorn RAC c/o S. Mciver 9 Balvicar Road, Oban PA34 4RP
GI0 LRB P. Strawbridge 98 Moyola Drive, Londonderry BT48 8EF
G0 LRE J. Norman 9A St. Johns Grove, Heysham, Morecambe LA3 1ET
G0 LRI S. Kennedy Colmans Farm, Elmstone-Hardwicke, Cheltenham GL51 9TG
G0 LRJ P. Daymond 14 Philip Close, Plymouth PL9 8QZ
G0 LRK K. Wall 27 Broomfield Road, Fleetwood FY7 7HA
G0 LRM A. Littler 365 Westhorne Avenue, London SE12 9AB
G0 LRO D. Watmough 41 West Crayke, Bridlington YO16 6XR
G0 LRP P. Waters Unit3, Site2 Sandpitlane, Nr Beccles NR34 7TH
G0 LRR the East Coast Vhf Group c/o S. Greenwood Carter Place Farm, Hall Park, Rossendale BB4 5BQ
G0 LRU F. Alderson Old School House, Tattersett, King's Lynn PE31 8RS
G0 LRW M. Simmons 6 The Crescent, Bletchley, Milton Keynes MK2 2QD
GI0 LRZ Dr N. Mitchell 6 Brae Road, Newry BT34 1NZ
G0 LSA Fyle Coast R.G. c/o R. Knighton 262 Victoria Road West, Thornton-Cleveleys FY5 3QB
G0 LSE West Kent ARS c/o P. Harris 7 Rowan Avenue, Egham TW20 8AN
G0 LSI D. Peachey Thornbury Cottage, Ashmill, Beaworthy EX21 5HA
G0 LSJ C. Jones 179 Blandford Road, Efford, Plymouth PL3 6JZ
G0 LSK D. Taylor 22 Meon Road, Mickleton, Chipping Campden GL55 6TD
G0 LSP L. Pawlik 2 Woodcock Close, Bamford, Rochdale OL11 5QA
G0 LSQ D. Williams 41 Ravensgate Road, Charlton Kings, Cheltenham GL53 8NS
G0 LSU J. Hair 84 Oxford Street, Barrow-in-Furness LA14 5QQ
G0 LSV I. Limbert 9 Lyme Grove, Liverpool L36 8BN
G0 LSX D. Barber 179 Rye Hills, Bignall End, Stoke-on-Trent ST7 8LP
G0 LTB A. Veal 64 Hither Bath Bridge, Bristol BS4 5DJ
GW0 LTC G. Edwards 5 Lower Farm Court, Rhoose, Barry CF62 3HQ
G0 LTD B. Tugwell 14 Martinique Way, Eastbourne BN23 5TH
G0 LTE D. Prout 8 Ferenberge Close, Farmborough, Bath BA2 0DH
GI0 LTF H. Irwin 9 Edward Street, Armagh BT61 7QU
G0 LTO R. Summers 18B Rose Road, Canvey Island SS8 0BP
G0 LTP D. Freeman 88 Wharf Road, Wroughton, Swindon SN4 9LJ
G0 LTR Tamworth & Lichfield Raynet c/o R. Williams 76 Quince, Amington, Tamworth B77 4EU
GI0 LTT S. Beattie 28 Millers Lane, Newtownards BT23 7AR
G0 LTV R. Pearce Talara, Kent Street, Battle TN33 0SF
G0 LTX S. Painting Claytons, Inkpen, Newbury RG17 9QE
G0 LUB A. Nicholls 235 Thorpe Road, Melton Mowbray LE13 1SH
G0 LUC P. Dyke 5328 Malcolm Street, Oceanside 92056 United States
G0 LUD R. Spacey 18 Longdale Avenue, Ravenshead, Nottingham NG15 9EA
GM0 LUF T. Traill 30 Strathesk Road, Penicuik EH26 8EF
G0 LUH D. Goodison 33 Witham Road, Isleworth TW7 4AJ
G0 LUI P. Draper 265 Nottingham Road, Ilkeston DE7 5AT
G0 LUK D. Palmer Braidwood, Enborne Row, Newbury RG20 0LY
G0 LUL P. Clune 50 St. Marks Road, Mitcham CR4 2LF
G0 LUM W. Mitchell-Watson 144 Shakespeare Crescent, Dronfield S18 1ND
G0 LUN C. Stayt 100 Cromwell Way, Oddington, Kidlington OX5 2LL
G0 LUP K. Chambers 9 Village Farm Road, Preston, Hull HU12 8QH
G0 LUQ T. Vale Grange Farm Flat, Station Road, Bicester OX26 5DX
G0 LUU K. Blackburn 63 Robsons Drive, Huddersfield HD5 9JW
G0 LUY G. Oneill 16 Aldam Street, Darlington DL1 2HY
G0 LVF F. Talmage 54 Rodwell Avenue, Weymouth DT4 8SG
G0 LVG P. Nilan 15 Broomhall Road Pendlebury, Swinton, Manchester M27 8XP
GW0 LVH J. Wimpenny Gwili House, The Ropewalk, Milford Haven SA73 3LW
GM0 LVI D. Warburton Lawvista, High Street, Errol, Perth PH2 7QQ
G0 LVJ B. Bradshaw 18 Burley Avenue, Lowton, Warrington WA3 2ES
GM0 LVK L. Alexander 97 Land Street, Keith AB55 5AP
GM0 LVL C. Mcewan 42 Marionville Crescent, Edinburgh EH7 6AU
G0 LVR J. Russell 9 Pear Tree Lane, Rowledge, Farnham GU10 4DW
G0 LVT B. Thornber 78 Skipton Road, Silsden, Keighley BD20 9LL
G0 LVX T. Burns 52 Somerset Drive, Bury BL9 9DQ
G0 LVY J. Littler 39 Wigan Road, Golborne, Warrington WA3 3TZ
G0 LWC P. Timlett 10 Reynolds Gardens, Moulton, Spalding PE12 6PT
GM0 LWD L. Mcwilliams 38 Churchill Street, Alloa FK10 2JG
G0 LWE W. Short 6 Kensington Avenue, Normanby, Middlesbrough TS6 0QQ
G0 LWG N. Sheriden 10 Grimsby Road, Louth LN11 0DY
G0 LWI F. Butler 8 Bradwell Road, Buckhurst Hill IG9 6BY
G0 LWL D. Spooner 7 East Avenue, Althorne, Chelmsford CM3 6DD
G0 LWM A. Mothew 7 Ashfields, Loughton IG10 1SB
G0 LWN D. Parsons 107 Larkswood Road, Chingford, London E4 9DU
GI0 LWO Dr S. Magill 40 Gardners Road, Lisburn BT27 5PD
G0 LWU A. Scarr Kerrera, 33 Chapel Lane, Overton, Morecambe LA3 3JA
G0 LXC J. Searle 232 Park Lane, Frampton Cotterell, Bristol BS36 2EN
G0 LXF R. Turner 5 Darenth Court, Quilter Road, Orpington BR5 4NS
G0 LXG B. Clulee 25 Cloister Crofts, Leamington Spa CV32 6QG
G0 LXI C. Sidney 25 John Mcguire Crescent, Binley, Coventry CV3 2QG
G0 LXL G. Truckel 26 Elm Close, Chipping Sodbury, Bristol BS37 6HE
GI0 LXN W. Black 14 Killyliss Road, Fintona, Omagh BT78 2DL
G0 LXP R. Gant 25 Worcester Avenue, Garstang, Preston PR3 1FJ
G0 LXR B. Morgan 208 Main Street, Burley In Wharfedal, Ilkey LS29 7HS
G0 LXV M. Lee 23 Lyndale Road, Redhill RH1 2HA
G0 LXW A. Pollard 16 Bellenger Way, Kidlington OX5 1TR
G0 LXX J. Mayfield 9 Middlefell Way, Clifton, Nottingham NG11 9JN
G0 LXY J. Scarr Betula House, Barford Road Bloxham, Banbury OX15 4EZ
G0 LYC P. Hindle Three Ways, Oakwood, Hexham NE46 4LE
GW0 LYF D. Hobbs 9 Llwynypia Terrace, Tonypandy CF40 2JD
G0 LYG F. Aris 5 Horley Road, Mottingham, London SE9 4LF
GM0 LYH H. Cochrane 69 Balgray Avenue, Kilmarnock KA1 4QT
G0 LYI W. Stevens 97 Kelvin Grove, Portchester, Fareham PO16 8LF
G0 LYJ W. Hughes 26 Cambridge Cottages, Richmond TW9 3AY
GW0 LYK S. Parker Maesycoed, Blaenycoed, Carmarthen SA33 6ES
G0 LYN L. Roper 57 Burnt Hills, Cromer NR27 9LW
GM0 LYO J. Fletcher 1 Silverwood Farm Cottage, Kilmarnock KA3 6HJ
G0 LYQ L. Selman 156 Bradley Drive, Santa Cruz 95060 United States
G0 LYR P. Spencer 4 Jubilee Close, Duloe, Liskeard PL14 4PA
GM0 LYT A. Fegen Acharn, Losset Road, Blairgowrie PH11 8BU
G0 LYX D. Brown 67 Croft Road, Benfleet SS7 5RL
G0 LYZ C. Knaggs 29 Wansford Road, Driffield YO25 5NB
G0 LZD R. Wood 4 Burns Road, Royston SG8 5PT
GM0 LZE D. Morrison 27B Benside, Newmarket, Stornoway HS2 0DZ
G0 LZF J. Rivers 211 Upper Wickham Lane, Welling DA16 3AW
G0 LZG W. Miles-Williams 13 Cavendish Road, Chesham HP5 1RW
G0 LZI M. Tudor 32 Ringwood, Oxton, Prenton CH43 2LZ
G0 LZL D. Bates 92 Thirlmere Road, Partington, Manchester M31 4PT
G0 LZS T. Wright 8 Glentham Close, Lincoln LN6 8BX
G0 LZV C. Nicholas 19 Spring Crofts, Bushey WD23 3AR
G0 LZW D. Riddick 289 Hatfield Road, St. Albans AL4 0DH
G0 LZX R. Knowles 9 Malham Close, Southport PR8 6UP
G0 LZY R. Smith Pleasanton, Church Street, Halstead CO9 3AZ
G0 MAA S. Rawson The Cabin, Tully, Four Mile House, Roscommon Ireland
GM0 MAC A. Macfarlane Breadalbane, Ferrindonald, Isle of Skye IV44 8RF
G0 MAD Rosemary ARG c/o K. Moody 5 Moore Road, Mapperley, Nottingham NG3 6EF
G0 MAF M. Plaskitt 4 Church Lane, Immingham DN402EU
G0 MAH G. Humphrey 57 Haig Avenue, Leyland PR25 2DD
G0 MAL W. Bowden 43 Burlish Close, Stourport-on-Severn DY13 8XW
GD0 MAN Otley ARS c/o M. Dunning 55 Station Park, Colby IM9 4NL Isle of Man
G0 MAR N. Buchan Acorns, 37 Forge Rise, Uckfield TN22 5BU
G0 MAS A. Sedgbeer 7 Leofric Road, Pinnex Moor, Tiverton EX16 6JU
G0 MAT R. Johnson 30 Wheatlands, Titchfield Common, Fareham PO14 4SL
G0 MAY P. Holmes 11 Bingham Road, Cotgrave, Nottingham NG12 3JS
G0 MAZ M. Gurr Elan, Sandown Road, Sandwich CT13 9NY
G0 MBA A. Horsman 15 Hanwell Close, Clacton-on-Sea CO16 7HF
G0 MBB A. Cutter Hundred House, Pink Road Lacey Green, Princes Risborough HP27 0PG
G0 MBD J. Curzon 17 Bullfinch Way, Cottenham, Cambridge CB24 8AW
G0 MBG B. Drew 59 Coventry St., Kidderminster DY10 2BZ
G0 MBI A. Roxburgh Flat 1, Tudor Court, Sutton Coldfield B74 4AN
G0 MBK M. Kerr 83 Greenwood Crescent, Sheffield S9 4HE
G0 MBL A. Plant 99 Pegwell Road, Ramsgate CT110ND
GW0 MBN P. Salt Acorns, Llwyncelyn, Cardigan SA43 2PE
G0 MBP B. Vanson 25 St. Helens Road, Westcliff-on-Sea SS0 7LA
G0 MBQ S. Withers 14 Rushes Road, Petersfield GU32 3BW
G0 MBR Orkney ARC c/o I. Mciver 3 Asgard Drive, Bedford MK41 0UP
G0 MBS B. Sinclair 48 East Crescent, Duckmanton, Chesterfield S44 5ET

G0 MBU G. Buckwell 24 Ings Road, Redcar TS10 2DL
G0 MBV R. Buckwell 31 West Green Stokesley, Stokesley TS95BE
GW0 MBW M. Watkins Llwyn Onn, Wainfelin Road, Pontypool NP4 6DF
G0 MBY P. Eaton 3 Thirslet Drive, Heybridge, Maldon CM9 4YN
G0 MBZ M. Phillips 14 Kingsclere Drive, Bishops Cleeve, Cheltenham GL52 8TG
G0 MCE R. Daw Flat 11, Tong Court, Boscobel Crescent, Wolverhampton WV1 1QQ
GM0 MCJ H. Munro Flat 6, Charlie Devine Court Bridge Of Don, Aberdeen AB22 8WG
G0 MCM M. Sniezko-Blocki 18 Westwoods Hollow, Burntwood WS7 9AT
G0 MCO D. Belcher 7 Bower End, Chalgrove, Oxford OX44 7YN
G0 MCP K. Scott 16 Hawton Road, Newark NG24 4QB
G0 MCQ M. Quicke 53 Newfield Avenue, Farnborough GU14 9PJ
G0 MCT R. Craig 24 Avondale, Sunderland SR4 0LZ
G0 MCV S. Morley 9 Field Close, Thringstone, Coalville LE67 8PU
GM0 MDD J. Clough Obo Largs & District Ars, Redbank, Skelmorlie PA17 5DX
G0 MDJ K. Smithyes Roseville, Childs Ercall, Market Drayton TF9 2DG
G0 MDK C. Hobson 1 Martindale Avenue, Wimborne BH21 2LE
G0 MDM R. Robbins 3 North Approach, Watford WD25 0EH
G0 MDN B. Millward 5 Regency Close, Weddington, Nuneaton CV10 0DF
G0 MDO D. Ward 53 The Drive, Bingley BD16 2EY
GW0 MDQ P. Firmstone Bod Awen, Ffordd Top Y Rhos Treuddyn, Mold CH7 4NE
G0 MDR F. Lupton 51 Bullens Green Lane, Colney Heath, St. Albans AL4 0QR
G0 MDV M. Bellas 3 Elm Terrace, Penrith CA11 7JY
GM0 MDX W. Dempster 124 Chatelherault Crescent, Hamilton ML3 7PW
G0 MEA P. Smith 7 Prospect Drive, Keighley BD22 6DD
G0 MEC M. Spurgeon 11 Homestead Road, Bodicote Chase, Banbury OX16 9TW
G0 MEE B. Shelton 12 Meadowlands, Blundeston, Lowestoft NR32 5AS
G0 MEF M. Frear 18 Boulsworth Road, Preston Grange, North Shields NE29 9EN
G0 MEN A. Fitzgerald 39 Rue Marcel Miquel, Issy-Les-Moulineaux 92130 France
G0 MEO R. Davis 17 Welbourne Close, Raunds, Wellingborough NN9 6HE
G0 MEQ H. Rigby 33 Herne Rise, Ilminster TA19 0HH
G0 MET Bromsgrove And District ARC c/o D. Evans 2 Cottage Lane, Marlbrook, Bromsgrove B60 1DW
G0 MEV J. Thorndyke 23 Fordhams Close, Stanton, Bury St. Edmunds IP31 2EE
G0 MEW L. Whiteside 9 Nutfield Gardens, Ilford IG3 9TB
G0 MEX H. Horne 410 Bacup Road, Waterfoot, Rossendale BB4 7JA
G0 MEY M. Coulter 52 Pine Close, Brant Road, Lincoln LN5 9UT
G0 MEZ C. Thorndyke 23 Fordhams Close, Stanton, Bury St. Edmunds IP31 2EE
G0 MFH P. Lee 56 Cockshead Road, Liverpool L25 2RB
G0 MFQ Capt. G. Dunster 21 Brunel Quays, Great Western Village, Lostwithiel PL22 0JB
G0 MFR G. Ayre 1 Spring Gardens, Broadmayne, Dorchester DT2 8PP
G0 MFT G. Drake The Bungalow, Church Lane, Tydd St. Giles, Wisbech PE13 5LG
G0 MFY L. Choong 15 Hounsfield Lodge, 5 Chambers Park Hill, Wimbledon, London SW20 0QE
G0 MGC G. Clark Holly Cottage, New Road, Bristol BS35 4DX
G0 MGG S. Smith 60 Grange Road Tuffley, Gloucester GL4 0PG
G0 MGH A. Strevens 14 Larchfield Way, Waterlooville PO8 9HE
G0 MGI M. Goodall 2 Meadow Court, Littleport, Ely CB6 1JW
G0 MGJ K. Hancock 12 Westmorland Close, Stoke-on-Trent ST6 6UR
G0 MGL G. Lees 68 Green Lane, Oldham OL8 3BA
G0 MGM R. Dunne 8 Telston Close, Bourne End SL8 5TY
G0 MGN J. Brand 38 Canterbury Gardens, Hadleigh, Ipswich IP7 5BS
GM0 MGO D. Macdonald 22 Christie Place, Elgin IV30 4HX
GW0 MGQ G. Budge 47 Rhyd Y Defaid Drive, Sketty, , Swansea SA2 8AL
G0 MGT D. Carruthers 19 Creek View Avenue, Hullbridge, Hockley SS5 6LU
G0 MGU C. Brown The Cottage, Tylers Road, Harlow CM19 5LJ
G0 MGX M. Jones 8 Stanton Avenue, Belper DE56 1EE
G0 MGZ D. Chaloner 16 Farriers Road Middle Barton, Chipping Norton OX7 7EU
G0 MHA P. Bakrania 31 South Priors Court, Northampton NN3 8LD
GI0 MHB P. Mckee 168 Ballynamoney Road, Lurgan, Craigavon BT66 6LD
G0 MHC G. Ford Thornley Road, Trimdon Station TS29 6DA
GM0 MHD P. Overton Cluanie, Cairnballoch, Alford AB33 8HQ
GM0 MHE B. Hyde The Cottage, Killiechronan, Isle of Mull PA72 6JU
G0 MHF J. Bisson 14 Howbeck Drive, Oxton, Prenton CH43 6UY
GW0 MHK P. Lee 22 Bron Y Graig, Bodedern, Holyhead LL65 3SY
G0 MHN D. Tebay 19 St. Johns Road, Newport PO30 1LN
GM0 MHS D. Rendall Aranthrue, 17 Scapa Crescent, Kirkwall KW15 1RL
G0 MHY R. Jones 339 Glenashton Drive, Oakville L6H 4W2 Canada
G0 MHZ R. Pardoe 138 Fowler Road, Aylesbury HP19 7QJ
G0 MIA C. Murray Brookfield, Collaroy Road, Thatcham RG18 9PB
G0 MIB D. Hussey 15 The Ridings, Telscombe Cliffs, Peacehaven BN10 7EF
G0 MID R. Jeffery 3 New Road, Paddock Wood, Tonbridge TN12 6HP
G0 MIE R. Ebbs 25 Foxtail Close, Gloucester GL4 6DW
G0 MIF I. Buckle 28 Leybourne Road, Rochester ME2 3QG
G0 MIG N. May Spring Barn, Eastwood Park, Wotton-under-Edge GL12 8DA
G0 MIH P. Swift Minster View, Parish Road, Minster On Sea, Sheerness ME12 3NQ
G0 MIJ C. Nolan 95 Strodes Crescent, Staines TW18 1DG
G0 MIK M. Ritson 24 Chapel Road, Pawlett, Bridgwater TA6 4SH
GM0 MIS Dr S. Buchanan 7 Eilean Rise, Ellon AB41 9NF
G0 MIT G. Bromfield 63 Herondale Road, Mossley Hill, Liverpool L18 1JZ
GM0 MIW A. Mcnicol The Glebe House, Arbirlot, Arbroath DD11 2NX
G0 MIX M. Jones 15 Quadrant Close, Murdishaw, Runcorn WA7 6DW
G0 MIZ J. Munro Flat 3/4, 24 The Strand, Ryde PO33 1JD
G0 MJA G. Taylor 32 Main Road, Great Holland, Frinton-on-Sea CO13 0JL
G0 MJB R. Daynes 25 Redwood Close, Keighley BD21 4YG
G0 MJC A. Keeble 5 Thistledown Road, Horsford, Norwich NR10 3ST
G0 MJF M. Weaver 91 Mantle Street, Wellington TA21 8BB
G0 MJG S. Cartlidge 19 Thornfield Road Crosby, Liverpool L23 9XY
G0 MJJ L. Challis 30 London Road, Kirton, Boston PE20 1JA
G0 MJK D. Linnell 19 Beech Avenue, Northampton NN3 2HE
G0 MJO G. Lucas Flat 3, The Gateway, 2 Wilderton Road West, Poole BH13 6EF
G0 MJP R. Davies 59 Gaunts Way, Letchworth Garden City SG6 4PL
GM0 MJR E. Baviello 18 Glaskhill Terrace, Penicuik EH26 0EL
G0 MJT M. Tandy 10 Palace Close, Rowley Regis B65 9LG
G0 MJV A. Williams 16 Hoy Crescent, Seaham SR7 0JT
G0 MJX J. Harper 109 Baxter Avenue, Kidderminster DY10 2HB
G0 MJY D. Gourley 86 Upton Road, Kidderminster DY10 2YB
G0 MJZ J. Edwards 42 Tenterfield Road, Ossett WF5 0RU
G0 MKA T. Chapman 17 Trevor Road, Swinton, Manchester M27 0YH
G0 MKC W. Dunn 25 Magdalene Court, Seaham SR7 7DJ
G0 MKD K. Davenport 6 Malpas Road, Runcorn WA7 4AD
GI0 MKK M. Stockdale 18 Drumard Crescent, Lisburn BT28 2JP
G0 MKL R. Chell 3 Elderberry Close, Stourport-on-Severn DY13 8TF
G0 MKN K. Brady 17B Furzefield Road, Welwyn Garden City AL7 3RL
G0 MKP N. Grice 7 Brecon Avenue, Huddersfield HD3 3QF
G0 MKU G. Langford 11 Nearhill Road, Kings Norton, Birmingham B38 8LB
G0 MKW A. Jones 26 Clarendon Street, Bloxwich, Walsall WS3 2HT
G0 MKY Dr J. Herries Elmfold, Witney OX8 6PZ
G0 MKZ T. Pougher 8 Wensleydale, Hull HU7 6DE
G0 MLB K. Walters 14 Varo Terrace, Stockton on Tees TS18 1JY
G0 MLC W. Lowe 54 St. Lesmo Road, Edgeley, Stockport SK3 0TX
G0 MLE D. Sabin 1 West Nolands, Nolands Road, Calne SN11 8YD
G0 MLF P. Marshall 45 Haylings Road, Leiston IP16 4DJ
G0 MLJ M. Jones Mark Jones Eye Care, 21 Fisherton Street, Salisbury SP2 7SU
G0 MLL J. Lyons 40 Waddington Avenue, Burnley BB10 4LB
G0 MLM T. Leeman 5 Serlby Rise, Nottingham NG3 2LS
GW0 MLN E. Jones 55 Blackoak Road, Cyncoed, Cardiff CF23 6QU
G0 MLO K. Packard The Haven, Howe Green Road, Purleigh, Chelmsford CM3 6PZ
G0 MLQ D. Pearson Yewtree Cottage, Worcester Road, Chipping Norton OX7 5XX
G0 MMA K. Plumridge Flat 23, Barton Court, Tewkesbury GL20 5RL
GW0 MMB P. Evans 138 Pont Adam Crescent, Ruabon, Wrexham LL14 6EG
G0 MMC J. Cuthill 17 Elmwood Drive, Keighley BD22 7DN
G0 MMH P. Walker 11 Flixton Drive, Crewe. CW2 8AP
G0 MMI C. Underhill 5 Grove Way, Waddesdon, Aylesbury HP18 0LH
G0 MMJ D. Wilkins 18 Garendon Road, Loughborough LE11 4QD
G0 MMO N. Laud 3 Woodlands, Wirksworth, Matlock DE4 4PG
G0 MMQ H. Dadak 3 Cadogan Close, Holyport, Maidenhead SL6 2JS
G0 MMT L. Catherall New Haven, Peckforton Hall Lane, Tarporley CW6 9TF
G0 MMW L. Roberts 12 Deveron Close, Plymouth PL7 2YF
G0 MMX D. Hebden 7 Whitecroft Ave, Shaw, Oldham OL2 8HY
GW0 MMY W. Ellis Broad Oak Cottage, Llyndir Lane, Burton, Wrexham LL12 0AU
G0 MNA A. Munir 39 Gulberg V, Lahore Pakistan
G0 MNC J. Williams 34 Brassington Street, Clay Cross, Chesterfield S45 9NH
G0 MND T. Rogers 40 Rowell Way, Sawtry, Huntingdon PE28 5WB
G0 MNH M. Brown 15 Hamilton Row, Waterhouses, Durham DH7 9AU
G0 MNI A. Carlile Top Flat 13 B Mill Road, Cleethorpes DN35 8HZ
G0 MNN M. Franklin 7 Auburn Close, Bridlington YO16 7PN
GW0 MNO N. Bufton 7 Laburnum Close, Rassau, Ebbw Vale NP23 5TS
GW0 MNP M. Butler 1 Green Meadow, Cefn Cribwr, Bridgend CF32 0BJ
GM0 MNV Radio Fraternity Lodge (8040) c/o R. Gandy 102 The Henge, Glenrothes KY7 6XX
GM0 MNW K. Carmichael 8G Colonsay Terrace, Soroba, Oban PA34 4YL
G0 MNY A. Dagnall 10 Rosebury Avenue, Leigh WN7 1JZ
GW0 MOF G. Greenhalgh 6 Clifton Grove, Rhyl LL18 4AF
G0 MOH R. Greaves Paradise Meadow, Church Street, Okehampton EX20 1JF
G0 MOK R. Hamer 4 Maldon Road, Standish WN6 0EX
G0 MOM S. Kendall 220 Marsh St., Barrow in Furness LA14 1BQ
GW0 MOQ N. Brush 25 Heol Y Ffynnon, Efail Isaf, Pontypridd CF38 1AU
G0 MOR Dr S. Morrey 22 Wellpond Close, Sharnbrook, Bedford MK44 1PL
G0 MOU R. Clark 20 Oakcroft Gardens, Littlehampton BN17 6LT
GW0 MOW R. Harris 25 Twynyffald Road, Blackwood NP12 1HQ
G0 MOX D. Gleek 10 Castlereagh House, Lady Aylesford Avenue, Stanmore HA7 4FP
G0 MPA Dover Construction Club c/o B. Coram 18A Lake Green Road, Sandown PO36 9HW
G0 MPI S. Sutcliffe 142 Sandy Lane, Farnborough GU14 9JQ
G0 MPJ B. Osborne 12 Arminers Close, Gosport PO12 2HB
G0 MPK D. Knights 11 King Edward St., Kirton Lindsey, Gainsborough DN21 4NF
G0 MPM J. Claughton 14 Witch Close, East Stour, Gillingham SP8 5LB
G0 MPO A. Neenan 50 Middleton Road, Brownhills, Walsall WS8 6JF
G0 MPP J. Anderson Hazel Wood Lee Lane, Bingley BD16 1UF
G0 MPQ J. Wood Garthmere, 4 Hunters Lane, Lincoln LN4 4PB
G0 MPR F. Gibbons 26 Tenbury Close, Bentley, Walsall WS2 0NH
G0 MPT E. Roddy 1546 West St, Stoughton 2072 United States
G0 MPW J. Woods 26 Compton Road, Southport PR8 4HA
G0 MQC P. Capewell 191 Monyhull Hall Road, Birmingham B30 3QN
G0 MQD R. Field 10 Somerville Close, Waddington, Lincoln LN5 9QR
G0 MQE J. Peirce 600 Highland Avenue, Ottawa, Ontario K2A 2K3 Canada
G0 MQH D. Robinson 88 Cotton Lane Halton Lodge, Runcorn WA7 5JB
G0 MQI R. Ingle 232 North Park Avenue, Norwich NR4 7ED
G0 MQJ P. Robinson 92 Greasby Road, Greasby, Wirral CH49 3NG
G0 MQK V. Murton 4 Cross Park Road, Wembury, Plymouth PL9 0EU
G0 MQL D. Franklin 50 The Elms, Chatteris PE16 6JN
G0 MQM M. Hillier 5 Sinodun Road, Didcot OX11 8HP
GI0 MQN R. Browning 53 Caulside Park, Antrim BT41 2DR
G0 MQR I. Tuson 6 Buffs Lane, Heswall, Wirral CH60 2SG
GW0 MQU P. Smyth 19 North Avenue, Tredegar NP22 3HE
G0 MQV N. Cook 17 Moorside Road, Richmond DL10 5DJ
G0 MQW C. Mcwhinnie 32 The Horse Close, Emmer Green, Reading RG4 8TT
G0 MQX R. Bowers 54 Buxton Road, Dawley, Telford TF4 2EW
G0 MRA E. Southon 20 Edinburgh Crescent, Kirton, Boston PE20 1JT
G0 MRB R. Broughton 6 Lumley Place, Lincoln LN5 7UT
G0 MRD D. Gordon 152 Oldham Road, Ashton-under-Lyne OL7 9AN
G0 MRF D. Bowman 11 Crane Way, Twickenham TW2 7NH
GM0 MRJ M. Johnston 27 Denholm Court, Glenrothes KY6 1JP
G0 MRK J. Kelly 14 Arden Walk, Sale M33 5NY
G0 MRL L. Bradshaw 342 Manchester Road, Blackrod, Bolton BL6 5BG
G0 MRM E. Caligari 209 Ormskirk Road, Upholland, Skelmersdale WN8 0AA
G0 MRP D. Pidgeon 87 Suckling Green Lane, Codsall, Wolverhampton WV8 2BY
G0 MRR C. Denton-Powell 4 Korresia Walk, Bridgwater TA5 2GT
G0 MRY M. Hazzledine 52 Springfield Road, Repton, Derby DE65 6GP
G0 MRZ B. Rowell 73 Halsteads Road, Torquay TQ2 8HB
G0 MSA A. Hagland 11 Coppice View, Heathfield TN21 8YS
G0 MSF G. Obey 51 Chichester Close, Murdishaw, Runcorn WA7 6DQ
GI0 MSG T. Mc Geown 1 Drumcairn Road, Armagh BT61 7SA
GI0 MSH D. Mcelroy 81 Keady Road, Armagh BT60 3AA
GI0 MSI E. Nesbitt 47 Mossfield, Glenanne, Armagh BT60 2JF

GI0 MSK H. Rattray 20 Charlemont Gardens, Armagh BT61 9BB
G0 MSO A. Webb 12 Forthlin Road, Allerton, Liverpool L18 9TN
G0 MSR S. Rutt 3 Russell Place, Highfield, Southampton SO17 1NU
G0 MSS J. Taft 8 Dresden Close, Mickleover, Derby DE3 0RD
GM0 MST J. Scotter 3 George Street, Halkirk KW12 6YE
GW0 MSW E. Goodwin Tremayne, 11 Duchess Road, Monmouth NP25 3HT
GW0 MSY H. Duggan 41 Maesglas Road, Newport NP20 3DE
G0 MSZ J. Lycett 24 Milbank Court, Darlington DL3 9PF
G0 MTA Rvd. F. Bligh 60 Hoole Road, Chester CH2 3NL
G0 MTB P. Pearson 53 Station Road, Dersingham, King's Lynn PE31 6PR
G0 MTD S. Topping 14 Manor Park, Carleton, Penrith CA11 8AL
GI0 MTE P. Robinson 8 Annaboe Road, Kilmore, Armagh BT61 8NP
G0 MTF G. Sanders 18 Impey Close, Thorpe Astley, Leicester LE3 3SW
GW0 MTI M. White 52 James Street Trethomas, Caerphilly CF83 8FY
G0 MTJ J. Boothroyd Quince Cottage, Church Lane, Ashford TN26 1LS
G0 MTK I. Chapman 12 Guernsey Farm Lane, Bognor Regis PO22 6BU
G0 MTN L. Volante Richmond House, Icknield Street, Birmingham B38 0EP
G0 MTP A. Owen 26 Gresham Street, Coventry CV2 4EU
G0 MTQ J. Baker Moffat House, Church Road Broughton Moor, Maryport CA15 7SS
G0 MTT R. Williamson 47 Ochre Dike Walk, Rotherham S61 4DL
G0 MTV D. Wright Blakey Ridge, 2 Abbey Gardens, Wimborne BH21 2EA
G0 MTW F. Puffett 1 Far Sandfield Churchdown, Gloucester GL3 2JS
G0 MTY P. Stunden 81 Treloweth Way, Pool, Redruth TR15 3TS
G0 MUC S. Markwick Flat 64, Lakeside Court, 35 Dallington Road, Eastbourne BN22 9EJ
G0 MUD Darenth Valley RS c/o D. Layne 5 Howe Close, Christchurch BH23 3JA
G0 MUH S. Riley 7 Crow Wood Avenue, Burnley BB12 0JG
G0 MUJ East Lancashire ARC c/o S. Spink 12 Chaucer Court, Ewelme, Wallingford OX10 6HW
G0 MUK R. Sanders 1A Lychgate Drive, Waterlooville PO8 9QE
G0 MUN A. Collins Flat 7, Haydon Court, Newton Abbot TQ12 1GQ
G0 MUR R. Garrett 27 Victoria Park Road, Buxton SK17 7PU
G0 MUZ J. Lockyer Flat 1, 11 Birch Hill Court, Birchington CT7 9UQ
G0 MVC C. Neil 208 Stonelow Road, Dronfield S18 2ER
G0 MVD I. Nash 3 Thistledown, Wanborough, Swindon SN4 0BU
G0 MVE M. Storkey 9 Waterman Court, Acomb, York YO24 3FB
G0 MVM A. Frost 15 Church Street Lacock, Chippenham SN15 2LB
G0 MVP M. Isted 62 Chippers Road, Worthing BN13 1DG
G0 MVR M. Bentley 1 Cotswold Road, Lupset, Wakefield WF2 8EL
G0 MVT W. Brindley 41 Boon Hill Road Bignall End, Stoke-on-Trent ST7 8LA
G0 MVV C. Howes 8 Alder Way, Hazel Slade, Cannock WS12 0SX
G0 MVW W. Barker Fieldhead, School Lane, Thame OX9 2NE
G0 MVX J. Reeves 5 Arrows Crescent, Boroughbridge, York YO51 9LP
G0 MVY Dr J. Donnett 51 Beaumont Avenue, St. Albans AL1 4TT
G0 MWE R. Woodward 22 Maryport Road, Dearham, Maryport CA15 7EG
G0 MWH R. Atkins 24 Hill House Road, Norwich NR1 4BE
GM0 MWJ D. Robertson 73 Kettilstoun Mains, Linlithgow EH49 6SH
GD0 MWL A. Crowther 3 Lime Street, Port St. Mary, Isle of Man IM9 5ED
G0 MWM E. Bailey 8 Blackthorn Close, Thornton-Cleveleys FY5 2ZA
GW0 MWN Dr D. Harries Rhydiau, Pencader SA39 9BY
G0 MWS S. Mcphee 19 Lyttelton Road, Stourbridge DY8 3RP
G0 MWT Chelmsford ARS c/o P. Tittensor 47 St. Johns Road, Chelmsford CM2 0TY
G0 MWU B. Pebody 30 High St., Oakfield, Ryde PO33 1EL
G0 MWV D. Campbell Bychance, Higher Gardens, Corfe Castle, Wareham BH20 5ES
G0 MWW C. Murt 17 Drake Road, Padstow PL28 8ES
G0 MWZ R. Rutherford 26 St.Golder Road, Newlyn Coombe, Penzance TR18 5QW
G0 MXB D. Burnett Cloonmaghaura, Williamstown 907 Ireland
G0 MXD D. Wroe Tylers House, Coton, Whitchurch SY13 3LT
G0 MXE F. Jennings 2 Hickman Close, London E16 3TA
GW0 MXG P. Taylor 2 Pen-Y-Dre, Caerphilly CF83 2NZ
G0 MXH D. Kay 8 Meadowbrook Close, Lostock, Bolton BL6 4HX
GM0 MXP R. Park Flat 19 Allan Park House, Allan Park, Stirling FK8 2LT
G0 MXU S. Smith 99 Greenwood, Bamber Bridge, Preston PR5 8JX
G0 MXW D. Houghton 127 Melwood Drive, Liverpool L12 8RN
G0 MXX Prof. B. Clarke Linden Cottage, School Lane, Nottingham NG12 3FD
G0 MXY C. Erratt 60 Allen Court, Ridding Lane, Greenford UB6 0JZ
GM0 MXZ T. Goody Lambs' Park, Forgandenny, Scotland PH2 9HS
G0 MYA A. Gray 57 Dominie Cross Road, Retford DN22 6NH
G0 MYC R. Clifton Heathwood, Thrigby Road, Great Yarmouth NR29 3HJ
G0 MYD G. Hughes 292 Mount Pleasant, Southcrest, Redditch B97 4JL
G0 MYH J. Foster 5 Jacobs Close, Glastonbury BA6 8EJ
G0 MYJ D. Sutherland Devonia, Hamble Lane, Southampton SO31 8EL
G0 MYL Sir H. Pigott Brook Farm, Shobley, Ringwood BH24 3HT
G0 MYM P. Harrison 16 Bodiham Hill, Garforth, Leeds LS25 2LF
G0 MYN W. Hughes 60 Pineways, Appleton, Warrington WA4 5EJ
GM0 MYQ J. Frati 10 Benbecula Road, Aberdeen AB16 6FU
G0 MYR R. Worsley Omaru, Higher Pennance, Redruth TR16 5TQ
GM0 MZD A. Coutts 37 West High Street, Bishopmill, Elgin IV30 4DJ
G0 MZF Dr B. Nicholson 349 City Road, London EC1V 1LR
GM0 MZH Royal Air Force ARS (RAFARS) c/o R. Wallace 80 Tourhill Road, Kilmarnock KA3 2DA
G0 MZJ C. Earp 9 The City, Edington, Westbury BA13 4QQ
G0 MZK R. Little Mythe House, 129 Slad Road, Stroud GL5 1RD
G0 MZN J. Nunn 20 Somerton Gardens, Earley, Reading RG6 5XG
G0 MZP R. Kay 7 Alderson Road, Worksop S80 1UZ
G0 MZQ W. Greed 5 West View, Creech St. Michael, Taunton TA3 5QP
G0 MZY F. Woodhall 13 Whitegate Drive, Clifton, Manchester M27 8RE
G0 MZZ A. Benson 1 Oxford Close, Gomersal, Cleckheaton BD19 4RU
G0 NAA A. Leake Thorpe Garth, East Newton, Hull HU11 4SD
G0 NAD R. Naden 10 Suffolk Close, Holland On Sea, Clacton-on-Sea CO15 5SQ
GM0 NAE J. Carlin 24 Hillcrest Avenue, Paisley PA2 8QW
GM0 NAI J. Fisher High Birches, Culbokie, Dingwall IV7 8JS
G0 NAJ J. Neary 266 Yew Tree Lane, Dukinfield SK16 5DN
G0 NAP P. Howell 29 South View Park, Plymouth PL7 4JE
GM0 NAQ G. Furmage 25 Craigton Crescent, Alva FK12 5DS
G0 NAR C. Fenton Oakwell, Newtons Hill, Hartfield TN7 4DH
G0 NAS L. Aykroyd 3 Bank Cottages, Orton Road, Penrith CA10 3TW
G0 NAU M. Best 81 Maybury Road, Hull HU9 3LB
G0 NAX D. Edmonds 1 Ashtree Close, Chelmsford CM1 2RR
GM0 NAZ A. Heggie 75 Doon Walk, Craigshill, Livingston EH54 5AD
GM0 NBA T. Adam Burnbank, Cairnbaan, Lochgilphead PA31 8SQ
G0 NBB M. Watkins 7 Sand End, Whitstable CT5 4TH
G0 NBC L. Steenvoorden 1 Thornbury Road, Immingham DN40 1HH
G0 NBD A. Brown 20 Sheen Road, Wallasey CH45 1HA
G0 NBE R. Allen 5 Crompton Grove, Stoke-on-Trent ST4 8UZ
GM0 NBG J. Mcvittie 19 Beech Way, Girvan KA26 0BX
G0 NBH J. Goodwin Hankelow Court, Hall Lane, Crewe CW3 0JB
G0 NBI G. Coomber 3 Dolly Drove, Chard TA20 1PF
G0 NBJ N. Foster 20A Pear Tree Road, Ashford TW15 1PW
G0 NBP A. Stevens Gate Farm Barns, Earthcott Green, Bristol BS35 3TA
G0 NBW B. Nolan 4 Shetland Road, Blackpool FY1 6LP
GM0 NCA R. Pinkerton 20 St Andrew Drive, Castle Douglas DG7 1EW
G0 NCE O. Wheeler 56 Rochester Gate High Street, Rochester ME1 1JG
G0 NCH R. Magri 13 Roebuck Road, Chessington KT9 1JY
G0 NCL R. Harris 5 Campbell Close, High Wycombe HP13 5XY
G0 NCO D. Murray 11 Blandy Road, Henley-on-Thames RG9 1PH
G0 NCQ M. Hemmings 2 Holly Walk, Nuneaton CV11 6UU
G0 NCS C. Healey 22 Stirling Road, St. Budeaux, Plymouth PL5 1PD
G0 NCT F. Norman 101 Central Avenue, Canvey Island SS8 9QP
GW0 NCU S. David 142 Robert Street, Manselton, Swansea SA5 9NH
G0 NCW A. Judge 106 Bicknor Road, Park Wood, Maidstone ME15 9PD
G0 NCX R. Hughesdon 3 Lyndhurst Road, Gosport PO12 3QY
G0 NCY R. Hartley 23 Broomfield Road, Fleetwood FY7 7HA
GW0 NDA S. Frost Fron Hyfryd, Nebo, Caernarfon LL54 6EW
G0 NDB R. Evans 51532 Range Road 224, Sherwood Park T8C 1H5 Canada
G0 NDC R. Little Maranatha, Higher Moresk Road, Truro TR1 1BW
G0 NDD J. Jackson 26 Wadham Close, Peterlee SR8 2NN
G0 NDF C. Peters 347 Mile Oak Road, Portslade, Brighton BN41 2RD
G0 NDU J. Dykes 33 Mill House Drive, Cheltenham GL50 4RG
G0 NDV P. Timmins Flat 2 34 Lumley Avenue, Skegness PE25 2TH
G0 NDY R. Wayne Colkirk House, Manor House Street, Horncastle LN9 5HF
GW0 NDZ G. Davies 4 Crichton Street, Treorchy CF42 6DF
G0 NEB J. Dorning 51 O'Sullivan Crescent, St. Helens WA11 9RE
GW0 NEC V. Fletcher Brig Y Don, Pendre Road Penrhynside, Llandudno LL30 3BY
G0 NED E. Dudley 4 Lake Croft Drive, Stoke-on-Trent ST3 7SS
G0 NEE M. Stott Wellview, 12 Castle View, Prudhoe NE42 6AT
G0 NEF A. Chapman 13 Clayton Grove, Bracknell RG12 2PT
G0 NEM M. Purser 17 Firecrest Road, Chelmsford CM2 9SN
GW0 NEN G. Lewin 6 Trefeddian Terrace, Aberdyfi LL35 0SD
G0 NEO J. Boland 28 Vicarage Road, Orrell, Wigan WN5 7AX
G0 NEP P. Whitling 17 Balcomb Crescent, Margate CT9 3XJ
G0 NEQ Leicester Raynet Grp c/o L. Horton 4 Summer Lane, Walsall WS4 1DS
G0 NER S. Stones 38 Mill View, Ferrybridge, Knottingley WF11 8SR
G0 NES D. Bryant 35 Truemans Heath Lane, Hollywood, Birmingham B47 5QE
GM0 NET Surrey Raynet c/o T. Stewart 104 Barrhill Road, Cumnock KA18 1PU
G0 NEU A. Dawson 182 Ladysmith Road, Enfield EN1 3AE
G0 NEV M. Carter 1 Mill Lane Preston, Weymouth DT3 6DE
G0 NFA D. Gilbert 2 Greenfield Cottages, Bentley, Farnham GU10 5HZ
G0 NFB L. Raynor 19 West View, Doncaster Road, Worksop S81 9RA
G0 NFE R. Ransome 19 Victoria Hill, Eye IP23 7HH
G0 NFG D. Hopper 28 Western Avenue, Herne Bay CT6 8TU
G0 NFH J. Acton 63 Bevington Close, Patchway, Bristol BS34 5NP
GW0 NFI P. Edwards 6 Maes Burgedin, Arddleen, Llanymynech SY22 6FG
G0 NFL M. George 2 Jubilee Terrace, Isham, Kettering NN14 1HG
GD0 NFN J. Butler 15 Church Close, Lonan, Laxey IM4 7JY Isle of Man
G0 NFO R. Charteris 7 Kennedy Close, Kidderminster DY10 1LR
G0 NFR R. Glynn 118 Pelham Road, Birmingham B8 2PD
G0 NFV A. Hunt 1 Avon View Cotswold Grange Country Park, Downfield Lane, Twyning, Tewkesbury GL20 6DL
G0 NFY A. Cordwell 71 Tadcaster Road, Norton Woodseats, Sheffield S8 0RA
G0 NFZ R. Hodges 82 Frankholmes Drive, Shirley, Solihull B90 4YB
GW0 NGA P. Golder Havenlea Wesley Street Llanfair Caereinion, Welshpool SY21 0RX
G0 NGD C. Stenbacka 11 Mount View, Billericay CM11 1HB
G0 NGE W. Clarke 20 Langdale Road, Leyland PR25 3AR
G0 NGG R. Brown 20 King Edward Road, Stanford-le-Hope SS17 0EF
G0 NGI D. Johnson 16 Woodcote House. 188 Brookwood Farm Drive . Knaphill, Woking GU21 2FZ
GM0 NGJ A. Caldwell 7 Gladstone Terrace, New Deer, Turriff AB53 6TE
G0 NGK P. Walmsley Valley View, Longworth Ave, Chorley PR7 4PJ
G0 NGN C. Jordan Green Lane Cottage, Leintwardine, Craven Arms SY7 0NB
G0 NGP P. Wilson Jansil, Worthing Road, Littlehampton BN17 6JN
G0 NGQ J. Gibbs 3 Holts Green, Great Brickhill, Milton Keynes MK17 9AJ
G0 NGW R. Ramplin 17 Cross St., Langold, Worksop S81 9SL
G0 NHB Dr W. Stallard 28 Wheatfield Road, Stanway, Colchester CO3 0YJ
GU0 NHD K. Benton Keukenhof, Route De Carteret, Castel GY5 7YS Guernsey
GW0 NHE A. Smith 7 Clos Gorsfawr, Grovesend, Swansea SA4 4GZ
G0 NHG J. Godwin 22 Stonebeck Avenue, Harrogate HG1 2BW
G0 NHJ J. Robson 35 Melling Road, Cramlington NE23 6AS
G0 NHK K. Robson 35 Melling Road, Cramlington NE23 6AS
GM0 NHL C. Waldron 24/2 Vennel Street, Dalry KA24 4AF
G0 NHM N. Robertshaw Cherry Tree House, Church Street, York YO26 8DD
G0 NHO K. Crookes 64 Heron Drive, Audenshaw, Manchester M34 5QX
G0 NHP T. Maguire 1 Gosford St., Balsall Heath, Birmingham B12 9ER
G0 NHR Barry ARS c/o K. Clarke 55 Compton Avenue, Aston-On-Trent, Derby DE72 2AU
GM0 NHT H. Cherrie 6 Milking Hill, Tong, Isle of Lewis HS2 0HU
G0 NHZ A. Pollard 2 Forest Close, Cowplain, Waterlooville PO8 8JE
G0 NID E. Page Flat 29, Westfields, 212 Hall Lane, Manchester M23 1LP
G0 NIF S. Wilkins 15 Roundway, Egham TW20 8BS
G0 NIG N. Smith 45 The Gills, Otley LS21 2BY
G0 NIL D. Woods 30 Longridge Avenue, Stalybridge SK15 1HG
G0 NIN N. Beer 2 Marcelle Court, School Road, Hindhead GU26 6LR
G0 NIQ J. Naylor 6 Mallard Close, Christchurch BH23 4DD
G0 NIX W. Jones 13 Kilrush Terrace, Woking GU21 5EG
GW0 NIY R. Milton 49 Heol Y Deri, Rhiwbina, Cardiff CF4 6HD
G0 NJD R. Mallett 71 Olivet Road, Woodseats, Sheffield S8 8QR

G0 NJG K. Binns 31A Dellands, Overton, Basingstoke RG25 3LD

G0 NJJ N. Jones 19 Foxhollow, Bar Hill, Cambridge CB23 8EP

GM0 NJL R. Watson 24 Hillock Avenue, Redding, Falkirk FK2 9UT

G0 NJO J. Howard 37 Carisbrooke Road, St. Leonards-on-Sea TN38 0JN

G0 NJP M. Murakami 5-8 Takamidai, Takatsuki, Osaka 5691020 Japan

G0 NJQ P. Schlatter Churchgate House, Sutton Road, Maidenhead SL6 9SN

G0 NJS M. Davie 101 Upperfield Road, Welwyn Garden City AL7 3LR

G0 NJT J. Perkins Flat 2Block 122, Flat 2. Leachgreen Lane Rednal, Birmingham B45 8EH

G0 NJZ T. Dodds 33 Westgate, Warley, Oldbury B69 1BA

G0 NKC M. Connolly 3 Port Mer Close, Exmouth EX8 5RF

GW0 NKG M. York 9 Fox Hollows, Brackla, Bridgend CF31 2NE

GW0 NKH R. Hearne Mora, Rhydypandy Road, Morriston SA6 6NX

GW0 NKJ D. Davies 6 Dulais Fach Road, Tonna, Neath SA11 3JW

G0 NKK A. Dipper 31 Stratton Heights, Cirencester GL7 2RH

G0 NKM H. Monks 100 Crossefield Road, Cheadle Hulme, Cheadle SK8 5PF

G0 NKQ V. Nunns 8 Trevithick Road, Tregurra, Truro TR1 1RU

G0 NKU S. Fowler 58 Buxton Road, High Lane, Stockport SK6 8BH

G0 NKZ K. Everard Woodside, Staple Lane, Taunton TA4 4DE

G0 NLA R. Bryan 23 Quarry Lane, Halesowen B63 4PB

GW0 NLB W. Rees 51 Heol Capel Ifan, Pontyberem, Llanelli SA15 5HF

G0 NLG R. Chapman 49 Walden Way, Frinton-on-Sea CO13 0BH

G0 NLJ S. Oliver Chalk Lodge, Peters Lane, Princes Risborough HP27 0LG

G0 NLL J. Bowers 35 Peveril Gardens, Newtown, Stockport SK12 2RG

G0 NLM C. Ridley 20 Victoria Gardens, Ferndown, Dorset BH22 9JH

G0 NLN D. Ashworth 59A Normanby Road, London NW10 1BU

G0 NLQ P. Dunn 10 Endsleigh Close, Upton, Chester CH2 1LX

G0 NLT D. Wentworth 7 Gilbeys Close, Stourbridge DY8 4XU

GM0 NLU N. Harvey The Shieling, Tealing, Dundee DD4 0QU

G0 NLV F. Fairman 26 Marina Gardens, Cheshunt, Waltham Cross EN8 9QY

G0 NLX S. Pearson 18 New Road, Amersham HP6 6LD

G0 NMB A. Haberman 4 Allendale Drive, Copford, Colchester CO6 1BP

G0 NMC T. Neal Sunnymead, Ewyas Harold, Hereford HR2 0JA

G0 NMD Rvd. L. Austin 7 Kennedy Close, Chester CH2 2PL

G0 NMH B. Markey 164 Bourn View Road, Netherton, Huddersfield HD4 7JS

G0 NMJ J. Denniss 61 Checkstone Avenue, Bessacarr, Doncaster DN4 7JY

G0 NMP P. Chapman 1 Bader Close, Watton, Thetford IP25 6FF

G0 NMS J. Howes 39 Pound Hill Bacton, Stowmarket IP14 4LP

G0 NMY M. Longson 54 Beresford Street, Shelton, Stoke-on-Trent ST4 2EX

GW0 NNE R. Hart River View, Kilkewydd, Welshpool SY21 8RT

G0 NNG R. Westley 91 Lincoln Way, Daventry NN11 4SU

GI0 NNK J. Cairns Fridge Air, 10 Dunsilly Road, Antrim BT41 2JH

G0 NNN W. Forbes 29 High St., Maryport CA15 6BQ

G0 NNO M. Shore 12 Boscoppa Road, St. Austell PL25 3DR

G0 NNR B. Thomas Creekside, Greenbank Road, Truro TR3 6PQ

G0 NNS Ellesmre P&D Ar c/o C. Hendry 109 Grove Avenue, New Costessey, Norwich NR5 0HZ

G0 NNT Prof. V. Martinelli 23 St. Angelo Street, Sliema SLM1334 Malta

G0 NNU L. Payne 147 Upper Marehay Road, Marehay, Ripley DE5 8JG

G0 NNZ J. Belfield 17 Burtondale Road, Crossgates, Scarborough YO12 4JR

G0 NOB L. Leek 21 Riverside Drive, Solihull B91 3HH

G0 NOH J. Reed 23 Morehall Avenue, Folkestone CT19 4EQ

G0 NON T. Behan Maytree Cottage, Marley Lane, Haslemere GU27 3RG

GW0 NOO S. Coburn 54 Queensway, Hope, Wrexham LL12 9PE

GW0 NOP P. Coburn 54 Queensway, Hope, Wrexham LL12 9PE

G0 NOU Dr W. Ayers Peach Lodge, Foolow, Hope Valley S32 5QB

GI0 NOX S. Mcateer 33 Knocknamuckley Lane, Portadown BT63 5PF

G0 NPA R. Stephens 50 Windrush Way, Abingdon OX14 3SX

G0 NPC G. Hill 1 Gleneagles Court, Edwalton, Nottingham NG12 4DN

G0 NPE P. Hulme 92 London Road, Cowplain, Waterlooville PO8 8EW

G0 NPF D. Delacassa The Tree House, Easthams Road, Crewkerne TA18 7AQ

G0 NPG K. Heaviside 58 Arundel Drive, Ranskill, Retford DN22 8PQ

G0 NPI J. Podvoiskis 3 Barnview Drive, Irlam, Manchester M44 6WY

G0 NPJ L. Jackson 60 East Park Avenue, Darwen BB3 2SQ

G0 NPK D. Goulbourne Widowscroft Farm, Hollingworth, Hyde SK14 8LE

GW0 NPL S. Instone 61 Llanfach Road, Abercarn, Newport NP11 5LA

GW0 NPM H. Thomas 34 Upland Road, Pontllanfraith, Blackwood NP12 2ND

G0 NPN B. Harris 23 Pound Road, Highworth, Swindon SN6 7LA

G0 NPO I. Brown Egremont, Arterial Road, Basildon SS14 3JN

G0 NPP T. Watson 20 Ivanhoe View, Gateshead NE9 7TR

G0 NPQ H. Carruthers 55 Inskip Terrace, Gateshead NE8 4AJ

G0 NPV B. Ham 37 Lower Moor, Barnstaple EX32 8NW

G0 NPW M. Ambach Karwendelstrabe 7, Tyrol A-6130 Austria

G0 NPY P. Yates 31 Wallpark Close, Brixham TQ5 9UN

G0 NQA A. Gurbutt 16 Crabtree Lane, Sutton-On-Sea, Mablethorpe LN12 2RT

GI0 NQC A. Dornford-Smith 10 Carmorn Road, Toomebridge, Antrim BT41 3NX

G0 NQE C. Wilkinson 8 Westfield Avenue, Knottingley WF11 0JH

G0 NQG S. Latham 27 Rockside Gardens, Frampton Cotterell, Bristol BS36 2HL

G0 NQI J. Shepherd Yew Tree Cottage, The Stenders, Mitcheldean GL17 0JE

G0 NQJ D. Scaplehorn 9 Stockwell Avenue, Mangotsfield, Bristol BS16 9DR

G0 NQK R. Edwards Manor Cottage, Manor Road, Ipswich IP7 6PN

G0 NQN T. Fricker 8 Folly View Necton, Swaffham PE37 8LU

G0 NQV D. Litchfield 37 Graeme Road, Enfield EN1 3UU

G0 NQW D. Marshall 15 Whisby Court, Holton-Le-Clay, Grimsby DN36 5BG

G0 NQY S. Seggar 145 Mount View Road, Norton, Sheffield S8 8PJ

G0 NRA G. Lowe 25 Manor House Court, Kirkby-In-Ashfield, Nottingham NG17 8LH

G0 NRB R. Bellamy 4 Wimbourne Walk, Corby NN18 0BN

G0 NRF G. Stilgoe 47 Chesterton Close, Redditch B97 5XS

G0 NRI W. Hilton 5 North Street, Williton, Taunton TA4 4SL

G0 NRJ R. Croucher 17 Sundridge Road, Woking GU22 9AU

G0 NRK J. Butler 14 Fairfield Road, Barnard Castle DL12 8EB

G0 NRM R. Stout 7 Thornbridge Drive, Sheffield S12 4YF

G0 NRN G. Harrison 14 Hardy Avenue, South Ruislip, Ruislip HA4 6SX

GM0 NRT W. Cardno 52 Salisbury Terrace, Aberdeen AB10 6QH

GW0 NRW S. Chadwick 27 Refail Farm Estate, Four Mile Bridge, Holyhead LL65 2EX

G0 NRX S. Godbold 13 Dawn Crescent, Upper Beeding, Steyning BN44 3WH

G0 NRZ A. Pill 5 St. Leonards Close, Upton St. Leonards, Gloucester GL4 8AL

G0 NSA T. Brown 5 St. Valentines Close, Kettering NN15 5EG

G0 NSC W. Thornton 46 Lavender Court, Croft Road, Barnsley S70 3FG

G0 NSG P. Crespel Via Sandei 340, Lucca 55100 Italy

G0 NSH B. Piggott 33 Lawrence Close, Hertford SG14 2HH

G0 NSI J. Man 13 Cheriton Close, Barnet EN4 9TX

G0 NSK A. Pennell 99 Westheath Avenue, Sunderland SR2 9LQ

G0 NSL C. Russell 163 Halton Road, Runcorn WA7 5RJ

G0 NSO T. Barfield 91 Ollerton Road, New Southgate, London N11 2JY

G0 NSP B. Teasdale 18 Valley Forge, Washington NE38 7JN

GW0 NSZ J. Swinden 22 Heol Awel, Abergele LL22 7UQ

G0 NTA A. Jarvis Willowmead, Nugents Park, Pinner HA5 4RA

G0 NTB B. Jarvis Willowmead, Nugents Park, Middlesex HA54RA

GJ0 NTD G. Blake 29 Pied Du Cotil, St. Andrews Road, St. Helier JE2 3JF Jersey

G0 NTG D. Chawner 49 St. Anns Road, Middlewich CW10 9BY

G0 NTH A. Gardner Flat 6, Eastcliff Court, Shanklin PO37 6EJ

GM0 NTI L. Grieve 1 Orchard Way, Inchture, Perth PH14 9QB

G0 NTJ A. Williams 23 Lancaster Gardens, Aylsham, Norwich NR11 6LB

GM0 NTL R. Fraser Hopefield Cottage, Gladsmuir, Tranent EH33 2AL

GM0 NTR J. Harrison 17B High Street, Oban PA34 4BG

G0 NTT L. Lloyd 8 Coastal Rise Hest Bank, Lancaster LA2 6HJ

GM0 NTY D. Rankin 25 Tinto Avenue Bellfield, Kilmarnock KA1 3SJ

G0 NUA K. Franklin 7 Auburn Close, Bridlington YO16 7PN

G0 NUD B. Bell 74 Henderson Road, Carlisle CA2 4PZ

G0 NUH M. Darling 1 Roman Way, Highworth, Swindon SN6 7BU

GM0 NUI R. Honeyman 81 Glen Avenue, Largs KA30 8RH

G0 NUL D. Forster 33 Wigeon Lane, Walton Cardiff, Tewkesbury GL20 7RS

G0 NUN R. Barker 5 Wickridge Close, Uplands, Stroud GL5 1ST

G0 NUO G. Du Feu 17 Oak Road, Tavistock PL19 9LJ

G0 NUP K. Prince 59 Chantry Road, East Ayton, Scarborough YO13 9ER

GM0 NUQ R. Handyside 113 Stockiemuir Avenue, Bearsden, Glasgow G61 3LX

G0 NUR A. Bushell 121 Rickmansworth Road, Watford WD18 7JD

GW0 NUS G. Dyer 15 Park Road, Newbridge, Newport NP11 4RE

G0 NUT D. Mckay 43 Mordales Drive, Marske-By-The-Sea, Redcar TS11 7HT

G0 NUU R. Browne Fox Cottage, Bukehorn Road, Peterborough PE6 0QG

G0 NUZ L. Wildman 22 Berrys Wood, Newton Abbot TQ12 1UP

G0 NVA F. Stainsby 11 Stonehouse Park, Thursby, Carlisle CA5 6NS

G0 NVD J. Nothard Ashmount, Fockerby, Scunthorpe DN17 4RZ

G0 NVJ S. Winter 509 Stockwood Road, Brislington, Bristol BS4 5ES

G0 NVM J. Chandler 14 Highfield Road, Chelmsford CM1 2NQ

G0 NVO P. Oldham 59 Wellspring Dale, Stapleford, Nottingham NG9 7ET

G0 NVS M. Fletcher 31 Woodthorpe Court , Sherwood, Nottingham NG5 4DY

G0 NVT P. Boyle 99 Heath Road, Penketh, Warrington WA5 2BY

G0 NVV G. Price 58 Hollowfields Close, Redditch B98 7NR

G0 NVX C. Watts 41 Salter Street, Berkeley GL13 9BU

G0 NVY P. Hanson 10 Parkfield Road, Ruskington, Sleaford NG34 9HS

G0 NWC H. Cooper 24 Queens Road, Haydock, St. Helens WA11 0RH

G0 NWE G. Egan 11 Shepherds Row, Castlefields, Runcorn WA7 2LG

G0 NWF S. Webster 31 Park Estate, Shavington, Crewe CW2 5AW

GI0 NWG A. Williamson 23 Iskymeadow Road, Armagh BT60 3JS

GM0 NWI A. Cunningham 33 Broom Court, Stirling FK7 7UL

G0 NWJ G. Blomeley 13 Edale Grove, Sale M33 4RG

G0 NWL H. Jordan 33 Earlsbourne, Church Crookham, Fleet GU52 8XG

GI0 NWN M. Coyle 67 Glen Road, Londonderry BT48 0BY

GW0 NWR N.W.R.R.C. c/o E. Shipton 30 Stephen Road, Prestatyn LL19 7EH

G0 NWS A. Edwards 130 Bedowan Meadows, Tretherras, Newquay TR7 2TB

G0 NWT Stockport Rs c/o L. Nash Four Furlongs, Wells Road, Wells-next-the-Sea NR23 1QE

G0 NWV D. Brown 65 Warstones Drive, Penn, Wolverhampton WV4 4PF

G0 NWY I. Peters 62 Kingston Avenue, Seaham SR7 8NL

G0 NXA G. Herbert Savory Cottage, 1 Dingle Lane, Nr Tewkesbury GL20 6DW

G0 NXC R. Sillito 25 Naisbett Avenue, Peterlee SR8 4BW

G0 NXD P. Brazenall Flat 80, Clent Court, Dudley DY1 2AZ

G0 NXE F. Rogers 5 Station Gardens, Eckington, Pershore WR10 3EZ

G0 NXF D. Robinson 5 Hazel Grove, Welton, Lincoln LN2 3JX

G0 NXH P. Cunningham 2 The Park, Mistley, Manningtree CO11 2AL

G0 NXI L. Edgecumbe 51 Aller Park Road, Newton Abbot TQ12 4NH

G0 NXL B. Ewald Sto.Nino 2, Lower Casili Consolacion, Cebu 6001 Philippines

G0 NXM R. Najman 9 Bevin House, Alfred St., London E3 2BB

G0 NXN B. Mitchell 2 Mariners Court, Great Wakering, Southend-on-Sea SS3 0DR

GM0 NXO G. Fyall 105 St. Kilda Crescent, Kirkcaldy KY2 6DR

G0 NXQ W. Love 2 Longmead Cottages, Milborne St. Andrew, Blandford Forum DT11 0HU

G0 NXR A. Moreton 5 Gorse Avenue, Thornton-Cleveleys FY5 2PH

G0 NXS A. Ellis High Nentsberry, Alston CA9 3LZ

G0 NXT S. Platts 15 Holywell Avenue, Smisby, Ashby-de-la-Zouch LE65 2HL

GW0 NXW G. Scanlin 4 Eifion Close, Barry CF63 1RQ

G0 NXX J. Lynch 14 The Pastures, Cayton, Scarborough YO11 3UU

G0 NXY P. Martin 48 Mill Lane, Fazeley, Tamworth B78 3QD

GM0 NYD J. Burrow Bongate, Jedburgh TD8 6DU

G0 NYE J. Dixon 17 Marlowe Close, East Hunsbury, Northampton NN4 0QQ

G0 NYH J. Moseley 42 Burford Road, Chipping Norton OX7 5DZ

GI0 NYI J. Benson 18 Alexander Avenue, Armagh BT61 7JD

G0 NYJ S. Au Flat 2, 1St Floor, Block C, Greenland Garden, Tuen Mun, N.T. Hong Kong

G0 NYK J. Hoose 91 Brevere Road, Hedon, Hull HU12 8LX

G0 NYM M. Borer 37 Broadway, Ripley DE5 3LJ

GM0 NYP N. Purtell 31 Daleally Crescent, Errol, Perth PH2 7QA

G0 NYQ J. Pape 12A High Wiend, Appleby-in-Westmorland CA16 6RD

G0 NYR R. Cheetham 8 Fairway, Huyton, Liverpool L36 1UD

G0 NYS D. Cox 10 Calder Close, Bollington, Macclesfield SK10 5LJ

G0 NYY S. Nicholas 39 Cubitts Close, Welwyn AL6 0DZ

G0 NYZ S. Maloney 34 Keswick Road, Normanby, Middlesbrough TS6 0BN

G0 NZA M. Lowe 25 Manor House Court, Kirkby-In-Ashfield, Nottingham NG17 8LH

G0 NZE A. Benfield 12 St. Marys Court, Weald, Bampton OX18 2HX

G0 NZI C. Peake 3 Marigold Walk, Bermuda Park, Nuneaton CV10 7SW

G0 NZJ Rvd. P. Forbes 18 Francis Road, Hinxworth, Baldock SG7 5HL

GW0 NZN J. Smith 6 Cherry Grove, Croespenmaen, Newport NP11 3DF

G0 NZR D. Catterall 86 Broomfield Road, Swanscombe DA10 0LT

G0 NZT D. Nash 27 Sandling Avenue, Horfield, Bristol BS7 0HS

G0 NZU R. Blanning 38 Northville Road, Northville, Bristol BS7 0RG
GM0 OAA M. Wigg 1/1 2 Glencairn Drive, Glasgow G41 4QN
G0 OAB D. Griffith 21 City Bank View, Cirencester GL7 1LQ
GW0 OAJ J. Morrice 184 Rowan Way, Newport NP20 6JT
G0 OAS N. Goddard 15 Canada Road, Cobham KT11 2BB
G0 OAT R. Petri Tarnwood, Denesway, Gravesend DA13 0EA
G0 OAW W. Waldron Redstone Farm, Germans Week, Beaworthy EX21 5BQ
G0 OAZ J. Fox L'Auzisiere De St.Marsault, la Foret Sur Serve 79380 France
GW0 OBB W. Evans Brynawel, Cross Inn, Llanon SY23 5NB
G0 OBE J. Clarke 16 Silver Birch Avenue, Bedworth CV12 0AZ
G0 OBH H. Cox Windrush, Malthouse Lane, Great Yarmouth NR29 5QL
G0 OBJ G. Pratley 34 Druce Way, Thatcham RG19 3PF
G0 OBK K. Warnes11 3 Blue Bell Close Underwood, Nottingham NG16 5FN
G0 OBN C. Hodgson 25 Pembroke Court, Sunderland SR5 4DF
G0 OBO K. Weeks 11 Sandwich Road, Preston Grange, North Shields NE29 9HT
G0 OBP K. Hudson 2 Colwill Walk, Plymouth PL6 8XF
G0 OBQ G. Lang 63 Grosvenor Drive, Whitley Bay NE26 2JR
G0 OBT S. Fortt 59 Coombe Dale, Sea Mills, Bristol BS9 2JF
G0 OBV M. Roberts The Bourne, The Avenue, Reading RG7 6NN
G0 OCB Dr R. Dingle 29 Castle View, Witton Le Wear, Bishop Auckland DL14 0DH
G0 OCC G. Allen 2 Haworth Drive, Bootle L20 6EJ
G0 OCF R. Mccoye 26 Hansby Close, Oldham OL1 2UA
G0 OCK B. Pilkington 219 Brownhill Avenue, Burnley BB10 4QH
G0 OCL F. Pilkington 219 Brownhill Avenue, Burnley BB10 4QH
G0 OCR F. Batkin 24 Sandyfields Road, Sedgley, Dudley DY3 3LB
G0 OCS M. Stabbins Primrose Cottage, Carlidnack Lane, Falmouth TR11 5HE
G0 OCT L. Stabbins Primrose Cottage, Carlidnack Lane, Falmouth TR11 5HE
G0 OCW P. Brazier 1 Ravenshore Cottages, Holcombe Road, Rossendale BB4 4AN
G0 OCY P. Beeston 100 Suffield Road, High Wycombe HP11 2JL
G0 ODA G. Hill 163 Parsonage Rd, Castle Hill 2154 Australia
GM0 ODB J. Kane 21 Sersley Drive, Kilbirnie KA25 6EY
G0 ODE D. Williams 212 Birchanger Lane, Birchanger, Bishop's Stortford CM23 5QH
G0 ODG P. Duce Curlew Cottage, Weston, Hr6 9Je, Hereford HR6 9JE
G0 ODH D. Hibberd 180 King Street, Stoke-on-Trent ST4 3EN
G0 ODI R. Sutton 87 Downs Valley Road, Woodingdean, Brighton BN2 6RG
G0 ODK W. Everett 120 Wantage Road, Reading RG30 2SF
G0 ODM J. Chomer 14 Holly Park Gardens, Finchley, London N3 3NJ
G0 ODN M. Hall 31 Meendhurst Road, Cinderford GL14 2EF
G0 ODP P. Warman 107 Hillside Road, Corfe Mullen, Wimborne BH21 3SB
G0 ODQ J. Hall Two Chestnuts, Emmington, Chinnor OX39 4AA
G0 ODR Perth & Dist Ar c/o C. Hendry 109 Grove Avenue, New Costessey, Norwich NR5 0HZ
G0 ODS M. Treacher 6 Beech View, Whitwell, York YO60 7JW
G0 ODU K. Petherick 17 Castle Close, Totternhoe, Dunstable LU6 1QJ
G0 ODX D. Hughes Balshult, Eriksmala 361 94 Sweden
G0 ODY G. Fleming 27 Crawthorne Crescent, Huddersfield HD2 1LB
G0 OEA A. Moggridge Outer Bailey, Kingsland, Leominster HR6 9QN
G0 OEB C. Donald 8 Greenway, Walsall WS9 8XE
G0 OED A. Mardo 10 Meadow View, Uffculme, Cullompton EX15 3DS
GI0 OEH K. Patterson 8 Beechwood Gardens, Moira, Craigavon BT67 0LB
G0 OEI M. Hopkins 30 Commonside, Brownhills, Walsall WS8 7AY
G0 OEJ M. Garbutt 92 Owlet Road, Windhill, Shipley BD18 2LT
G0 OEK D. Spooner 60 St. Pauls Road, Staines TW18 3HH
G0 OEM G. Cunningham 7 Wykin Lane, Stoke Golding, Nuneaton CV13 6HN
G0 OEQ J. Blichfeldt 2 Duck Cottages, Rolvenden Road, Cranbrook TN17 4BT
G0 OER P. Roberts 1 Ardern Close, Bristol BS9 2QT
G0 OES D. Owen 5 Vicarage Walk, Rosliston, Swadlincote DE12 8LB
G0 OEW D. Rooke The Grange, 107 Wybunbury Road, Nantwich CW5 7ER
G0 OEY A. Kerrison 63 Stour Road, Harwich CO12 3HS
G0 OFA R. Dennis 8 Newton Hall, Coach Road, Newton Abbot TQ12 1ER
G0 OFB J. Jones 59 East Street, Long Buckby, Northampton NN6 7RB
G0 OFD J. Gilbert 6 Mill Hill, Brancaster, King's Lynn PE31 8AQ
G0 OFE J. Smith 38 Wilson Road, Bournemouth BH1 4PH
G0 OFF S. Hipkin 62 Woodberry Way, Walton on the Naze CO14 8EW
GW0 OFH Sunspots Rac c/o S. Williams 5 Brynmelyn Avenue, Llanelli SA15 3RU
GM0 OFL J. Wilkie Hope Cottage, 4 Main Street, Cupar KY15 4SS
GM0 OFM J. Park Rameldry Mill Cottage, Rameldry Mill Road, Rameldry, Kingskettle, Cupar KY15 7TY
G0 OFN I. Clabon 14 Melrose Avenue, Twickenham TW2 7JE
G0 OFR G. Borrowdale 30 Barton View, Penrith CA11 8AX
G0 OFT S. Duncan 10 Huntingdon Rise, Bradford-on-Avon BA15 1RJ
G0 OFW P. Lightfoot 18 Fields Close, Alsager, Stoke-on-Trent ST7 2ND
G0 OFX F. Rawlins 12 Arundel Road, Eastleigh SO50 4PQ
G0 OFY J. Wane 25 Holmdale Avenue, Crossens, Southport PR9 8PS
G0 OFZ A. Grace 7 Sandringham Heights, St. Leonards-on-Sea TN38 9UA
G0 OGB J. Wallis 10 Middlewood Road, Lanchester, Durham DH7 0HL
G0 OGE Wimbledon&Dis Rd c/o M. Free 2 St Pauls Court, Princess Street, Maidenhead SL6 1NX
GW0 OGI D. Keely 15 Ffordd Cerrig Mawr, Caergeiliog, Holyhead LL65 3LU
G0 OGJ K. Marshall 8 Porter Way, Northwich CW9 7JA
GW0 OGL S. Glanville Fron Haulog, Llanelian, Colwyn Bay LL29 8UY
G0 OGM S. Bowerman 24 Bingham Close, Emerson Valley, Milton Keynes MK4 2AU
GM0 OGN R. Hall 13 Cleat, Castlebay, Isle of Barra HS9 5XX
G0 OGP Y. Powell 18 Carrington Road, Stockport SK1 2QE
G0 OGS S. Malpass 21 Tollhouse Way, Wombourne, Wolverhampton WV5 8AF
G0 OGX J. Pennington 1 Chisel Close, Hereford HR4 9XF
GM0 OGZ R. Goodall 3 Croftcrunie Cottages, Tore, Muir of Ord IV6 7SB
G0 OHA A. White 11 Garden Close, Consett DH8 5PA
GM0 OHD M. Holding Craigvar Main Street, Lochfoot DG2 8NR
G0 OHF F. Wilson 3 Foundry Mews, Burgh Le Marsh, Skegness PE24 5HQ
GI0 OHG E. Bennett 53 Condiere Avenue, Connor, Ballymena BT42 3LD
GW0 OHJ D. Workman 4 Rhuddlan Road, Buckley CH7 3QA
G0 OHK N. King 7 Fountains Close, Biddick Village, Washington NE38 7TA
G0 OHQ R. Bunyan 11 Kenhill Close, Snettisham, King's Lynn PE31 7PA
G0 OHR J. Armitage 17 Worsbrough Road, Birdwell, Barnsley S70 5QR
GI0 OHT W. Stanley 95 Bangor Road, Newtownards BT23 7BZ
GI0 OHU R. King 28 Moss Road, Waringstown, Craigavon BT66 7QY
G0 OHW J. Vasek 20 West Hall Road, Richmond TW9 4EE
G0 OID F. Tett 2 Church Park, Bradenstoke, Chippenham SN15 4ER
G0 OIE M. Gathergood 54 Robin Lane, Bentham, Lancaster LA2 7AG
G0 OIF D. Read 17 Lumber Leys, Walton on the Naze CO14 8SS
G0 OII R. Pullen 1 Ridings Court, 5 Crown Crescent, Scarborough YO11 2BJ
G0 OIK P. King 96 Mancroft Road, Caddington, Luton LU1 4EN
G0 OIM B. Turnbull 56 Bryans Close Road, Calne SN11 9AB
G0 OIN A. Fairey 1 Imbert Close, New Romney TN28 8XP
G0 OIO J. Fuller 1 The Courtyard, Snape, Saxmundham IP17 1FB
G0 OIQ A. Welland 6 Chartwell Road, Stafford ST17 0AJ
G0 OIR G. Wicks 28 Old School Lane, Milton, Cambridge CB24 6BS
G0 OIS T. Smith 33 Kites Nest Lane, Lightpill, Stroud GL5 3PJ
G0 OIU A. Smith Jones 16 Armley Road, Liverpool L4 2UN
G0 OIV A. Sait 124 Dicksons Drive, Newton, Chester CH2 2BX
G0 OIW M. Palmer 28 Westfield Road, Caversham, Reading RG4 8HH
G0 OIX D. Wright 132 Longmoor Lane, Liverpool L9 9BZ
G0 OIY J. Smith 59 Charlecote Drive, Dudley DY1 2GG
G0 OJB L. Chadwick 1 Dovehouse Road, Haverhill CB9 0BZ
GW0 OJC M. Crowly 25A Brompton Avenue, Rhos On Sea, Colwyn Bay LL28 4TE
G0 OJF R. Shireby Norwood, Rookery Lane, Sudbrook, Grantham NG32 3RU
G0 OJG J. Omalley 8 Hawks Court, Hallwood Park, Runcorn WA7 2FR
G0 OJJ A. Green1 6 Goulds Close Palgrave, Diss IP22 1AR
GI0 OJL J. Leonard 10 Abbeyvale, Dungannon BT71 5BZ
G0 OJP R. Melton 4 Ashleigh Avenue, Maiden Newton, Dorchester DT2 0BP
G0 OJR B. Fox 6 Bury Gardens, Elmdon, Saffron Walden CB11 4LX
G0 OJS S. John 40 Elizabeth Avenue, Brixham TQ5 0AY
G0 OJT W. Jenkinson 7 Moortown Road, Watford WD19 6JH
G0 OJW J. Taylor C/O 16 Caroline Place, Plymouth PL1 3PS
G0 OJX P. Williams 3 Nutwell Cottages, Exmouth Road, Exmouth EX8 5AP
G0 OJY A. Hurt The Manse, High Oak Road, Ware SG12 7PD
G0 OKA D. Martin 67 Mill Street, Torrington EX38 8AL
G0 OKD R. Bradley 252 Alfreton Road, Blackwell, Alfreton DE55 5JN
G0 OKF S. Bolam 100 Bushfield Road, Scunthorpe DN16 1NA
G0 OKI R. Morris 4 Greenway Gardens, Kings Norton, Birmingham B38 9RY
GM0 OKJ J. Fraser 2 Barra Place, Stenhousemuir, Larbert FK5 4UF
G0 OKK B. Crowe-Haylett 13 Lynton Close, Ely CB61DJ
G0 OKL J. Collins 3 Burford Grove, Bristol BS11 9RT
GI0 OKM G. Nesbitt 205 Longstone Street, Lisburn BT28 1TY
G0 OKN R. Maloney Rosewell, Jacobstow, Bude EX23 0BN
G0 OKT C. Parker 1 Richmond Drive, Perton, Wolverhampton WV6 7RR
G0 OKV K. Cowell 4 St. Georges Close, Colne BB8 8DP
G0 OKX K. Gardner 13 Beanshaw, Eltham, London SE9 3HL
G0 OKZ J. Thorpe Four Jays, 46A High Street, Doncaster DN10 4BU
G0 OLD D. Mclaren 11 St. Matthew Close, Uxbridge UB8 3SR
G0 OLE Boothferry ARS c/o K. Mccann Treverven, Back Lane, Hemingbrough, Selby YO8 6QP
G0 OLL E. Platts 38 Swanbourne Road, Sheffield S5 7TL
G0 OLO D. Collinson 20 Carlisle Crescent, Penshaw, Houghton le Spring DH4 7RD
G0 OLR L. Roberts Rose Cottage, Castle Hill, Leyburn DL8 4QN
G0 OLS T. Humphries 23 Sycamore Drive, Lutterworth LE17 4TR
G0 OLT L. Tringale 19 Lysander Road, Kings Hill, West Malling ME19 4TT
G0 OLX D. Stanton 106 Scrapsgate Road, Minster On Sea, Sheerness ME12 2DJ
GW0 OLZ G. Smith 23 Gainsborough Close, Llantarnam, Cwmbran NP44 3BX
G0 OMB B. Walker 15 Infirmary Road, Workington CA14 2UG
GM0 OMC C. Cook Briarwood, 95 Old Edinburgh Road, Inverness IV2 3HT
G0 OMD A. Gilbert 19 Farrs Avenue, Andover SP10 2AH
G0 OMF D. Hupton 90 Warwick Road, Atherton, Manchester M46 9PQ
G0 OMH P. Burbeck 5 Wouldham Terrace, Saxville Road, St. Paul's Cray BR5 3AT
G0 OMM S. Adams 13 Bells Drove, Sutton St. James, Spalding PE12 0JG
G0 OMN G. Charman 4 Hornton Grove, Hatton Park, Warwick CV35 7UA
G0 OMZ R. Lomas 7 Chaunterell Way, Abingdon OX14 5PP
G0 ONA P. Nicholls 5 Dingle Road, Ashford TW15 1HF
G0 ONF V. Szendzielarz 5 Granville Road, Urmston, Manchester M41 0XY
G0 ONG J. Mobbs 5 Distaff Road, Poynton, Stockport SK12 1HN
G0 ONH B. Fellows 36 Balmoral Road, Stourbridge DY8 5HR
GM0 ONN I. Barnetson 14 Nasmith Crescent, Elgin IV30 4FG
G0 ONS J. Chinnery 31 Kingsway, Northampton NN2 8HD
G0 ONT R. Broom Staging Post Abbotskerswell, Newton Abbot TQ12 5NX
GW0 ONU D. Harris 2 Sheppard St., Pwllgwaun, Pontypridd CF37 1HT
GM0 ONX L. Paget 40 Davaar Drive, Kilmarnock KA3 2JG
GW0 ONY J. Edwards 3 High Street, Bryngwran, Holyhead LL65 3PL
G0 OOB D. Walpole 12 Damgate Lane, Acle, Norwich NR13 3DH
G0 OOD Elkstones ARS c/o T. Chapman 21, Links Close., Norwich.Norfolk. N6 5PJ
G0 OOF R. Williams Dyffryn Coed, Union Road, Coleford GL16 7QB
G0 OOI W. Humphries 76 Mortlake Road, Richmond TW9 4AS
G0 OON P. Healey 10 Wroxham Road, Great Sankey, Warrington WA5 3EE
G0 OOO Salop ARS c/o R. Clayton 9 Green Island, Irton, Scarborough YO12 4RN
G0 OOQ S. Whitehurst 28 Severn Way Cressage, Shrewsbury SY5 6DS
G0 OOR A. Jex 26 Springdale Crescent, Brundall, Norwich NR135RA
G0 OOS L. Marobin Flat 60, Tudor Court, King Henrys Walk, London N1 4NU
G0 OOU R. Field 34 Piltdown Close, Hastings TN34 1UU
G0 OPA J. Lee Holly Lodge, Carrhouse Road, Doncaster DN9 1PG
G0 OPC M. Marriott 188 Leverington Common, Leverington, Wisbech PE13 5BP
G0 OPG C. Knowlson 28 Hill Drive, Handforth, Wilmslow SK9 3AR
G0 OPI A. Bennett 32 Gainsborough Road, Bournemouth BH7 7BD
G0 OPL W. Cowell 72A The Malting, Ramsey, Huntingdon PE26 1LZ
G0 OPM G. Melia Sunnyside, Little Asby, Appleby-in-Westmorland CA16 6QE
GW0 OPP R. Owens 62 Ty Llwyd Parc Estate, Quakers Yard, Treharris CF46 5LB
G0 OPT P. Tennant 128 Devonshire Street, Keighley BD21 2QJ
G0 OPV R. Heatley 68 Jeckyll Road, Wymondham NR18 0WQ
GM0 OPX D. Mcferran Ardlair, Milltimber AB13 0ER
G0 OQE F. Porter 32 David Hobbs Rise, Market Harborough LE16 7YE
G0 OQI K. Zak 5 The Rookery, Sandy SG19 2UR
G0 OQK N. Garrod 121 Totteridge Lane, High Wycombe HP13 7PH
G0 OQP A. Caton 20 Lower Oxford Road, Newcastle ST5 0PB
G0 OQQ B. Wood 52 Ashfield Avenue, Beeston, Nottingham NG9 1PY
G0 OQR A. Glen 70 Moscow Road East, Stockport SK3 9QL
G0 OQT M. Jones 6 Eastleigh Close, Burnham-on-Sea TA8 2EW
G0 OQX J. East 30 Auckland Road, Scunthorpe DN15 7BT

G0 OQZ H. Dawson 6 Maer Top Way, Barnstaple EX31 1RZ
G0 ORC V. Shirley 160 Over Lane, Belper DE56 0HN
G0 ORD Dr E. Chantler Hilltop Gardens, High Beech Road, The Pludds, Ruardean GL17 9UD
G0 ORE RSGB Transatlantic Test Century Club No2 c/o N. Reddish 15 Drakes Close, Redditch B97 5NG
G0 ORG N. Robertson Clayhill Cottage, The Street, Aldham, Ipswich IP7 6NN
G0 ORJ J. Bamford 39 Skelldale View, Ripon HG4 1UJ
G0 ORK S. Humberstone 4 Rowcroft Road, Paignton TQ3 2RE
G0 ORL C. Rowley 31 Keepers Croft East Goscote, Leicester LE7 3ZJ
G0 ORM D. Birch 31 Grasmere Terrace, Maryport CA15 7QN
G0 ORO D. Martin The Shieling, Bolton Low Houses, Wigton CA7 8PF
G0 ORP M. Simpson 3 Front Street, Barnby, Newark NG24 2SA
G0 ORT D. Leonard Three Ashes Cottage, 442 Outwood Common Road, Billericay CM11 1ET
G0 ORV V. Wilton Fairthorn Trotts Ln, Pooks Green, Southampton SO4 4WQ
G0 ORX J. Melton 4 Charlwood Close, Copthorne, Crawley RH10 3TG
G0 ORY A. Moss 10 Shakespeare Drive, Leicester LE3 2SP
G0 OSA C. Wilkinson 10 Chemin Des Mardeilles, Chirac 16150 France
GW0 OSB I. Price 16 Carmarthen Court, Caerphilly CF83 2TX
G0 OSC G. Mason 18 Nithsdale Road, Liverpool L15 5AX
G0 OSD G. Alexander 15 Brackley Way, Totton, Southampton SO40 3HP
G0 OSG R. Brazier 9 Wheelers Walk, Blackfield, Southampton SO45 1WX
G0 OSI K. Pallant 7 Council Bungalows, Church Lane, Braintree CM7 5SH
GM0 OSJ W. Legge 57 High Street Archiestown, Aberlour AB38 7QZ
G0 OSK C. Saggers 49 Revels Road, Hertford SG14 3JU
G0 OSO P. Markham Moor Farm, Moor Lane, Lincoln LN3 4EG
G0 OSU J. Collier 27 Birdham Close, Bognor Regis PO21 5TD
G0 OSW R. Sainsbury Salem Park Farm, Southampton Road Landford, Salisbury SP5 2BE
G0 OSX N. Shackley 20A Pear Tree Road, Ashford TW15 1PW
GM0 OTB R. Pugh 28 Pladda Road, Saltcoats KA21 6AQ
GI0 OTC T. Doherty 37 Magheramenagh Drive, Portrush BT56 8SP
G0 OTE E. Bowell 7 Bede House Bank, Bourne PE10 9JX
G0 OTF G. George 211 Bromford Road, Birmingham B36 8HA
G0 OTH R. Topliss 12 Dorothy Avenue, Skegness PE25 2BP
GM0 OTI Dr J. Grieve 1 Orchard Way, Inchture, Perth PH14 9QB
G0 OTJ J. Cummins Tarr House, Lumb Lane, Matlock DE4 2HP
GM0 OTS W. Mcintosh 14 East Road, Hopeman, Elgin IV30 5SU
G0 OTT D. Mcdonald 118 Torrington Avenue, Tile Hill, Coventry CV4 9AA
GM0 OTU A. King 31 Pendreich Grove, Bonnyrigg EH19 2EH
GW0 OTY W. Cooper 50 Tennyson Road, Penarth CF64 2SA
G0 OUC R. Rodgerson 8 Dearne Street, Darton, Barnsley S75 5HL
GD0 OUD S. Hill 54 Wybourn Drive Onchan, Isle of Man IM3 4AT
G0 OUG I. Hole 50 Westcroft Drive, Westfield, Sheffield S20 8EF
GW0 OUH H. Griffiths 45 Jubilee Road, Godreaman, Aberdare CF44 6DD
G0 OUK J. Hinton Wayside, Beauchief Drive, Sheffield S17 4RJ
GI0 OUM R. Ferris 3 Kingsland Drive, Belfast BT5 7EY
G0 OUN P. Bingham 12 Sandpiper Road, Thorpe Hesley, Rotherham S61 2UN
G0 OUO S. Palk 10 Springfield Close, Andover SP10 2QT
G0 OUR Open University ARC c/o A. Rawlings 57 High Street, Nash, Milton Keynes MK17 0EP
GW0 OUV M. Williams 7 Heol Isaf, Nelson, Treharris CF46 6NS
GI0 OUZ B. Prunty 16 Old Portadown Road, Lurgan, Craigavon BT66 8RH
G0 OVA P. Crake 1 Ashdown Close, Bracknell RG12 2SE
G0 OVC B. Godfrey 291 Collier Row Lane, Romford RM5 3ND
GM0 OVD R. Darroch 36 Tweed Street, Dunfermline KY11 4NA
G0 OVE K. Mohammed 63 Shirley Gardens, Barking IG11 9XB
G0 OVK R. Mansell1 2 Ambrose Close, Willenhall WV13 3DQ
G0 OVQ A. Bannister 34 Morningside Drive, East Didsbury, Manchester M20 5PL
G0 OVT B. Navier 12 Brooklyn Avenue, Brooklyn Street, Hull HU5 1ND
G0 OVV M. Bolton 85 Oak Park Road, Wordsley, Stourbridge DY8 5YJ
G0 OVY P. Maggs 85 Helsby Road, Sale M33 2XF
G0 OWA J. Wright 10 Whalley Road, Heskin, Chorley PR7 5NY
G0 OWC P. Bush 8 Underwood Terrace, Farley Hill, Matlock, Matlock DE4 3LL
G0 OWE D. Matthews 54 The Wynding, Bedlington NE22 6HW
G0 OWH J. Dobbs 9 Highlands, Littleborough OL15 0DS
G0 OWI A. Hawkridge Thorntrees, 109 Allerton Road, Bradford BD8 0AA
G0 OWJ A. Cooper 28 Belmont Road, Pensnett, Brierley Hill DY5 4EX
G0 OWK S. Searle 14 Edison Gardens, Colchester CO4 0AJ
GM0 OWM Orkney Wireless Museum c/o E. Holt Ashwell, St. Ola, Kirkwall KW15 1SX
G0 OWP D. Edwards 9 Mark Road, Hightown, Liverpool L38 0BG
G0 OWR C. Howard 75 Westbury Park, Wootton Bassett, Swindon SN4 7DN
G0 OWU R. Wilkes 39 Hillside Road, Dudley DY1 3LE
G0 OWV J. Harbottle 42 Littlemede, Eltham, London SE9 3EB
G0 OXA G. Landen-Turner 59 Mill Road, Higher Bebington, Wirral CH63 5PA
G0 OXB P. Draycott 41 Ashleigh Avenue, Bridgwater TA6 6AX
GI0 OXK D. Taggart 106 Moorfield Road, Dromore, Omagh BT78 3LR
G0 OXL Dr C. Robinson 33 Windsor Rd, Wellesley, Ma 2481 United States
G0 OXP L. Matthews 6 Spotland Tops, Cutgate, Rochdale OL12 7NX
GM0 OXS M. Beith 30 Raith Road, Fenwick, Kilmarnock KA3 6DB
G0 OXT P. Hutchinson Rosebank Cottage, Marcombe Road, Torquay TQ2 6LL
G0 OXV K. Mahood 1A Heskin Lane, Ormskirk L39 1LR
G0 OXW V. Soutter 2 Hyde Barton, Churchill Way, Bideford EX39 1NX
G0 OXX Dr J. Berridge Bracklyn, St. Clare Road, Deal CT14 7QB
G0 OXY M. Gray 142 Harrowden Road, Bedford MK42 0SJ
G0 OXZ M. Stone 29 Chesterfield Road, Epsom KT19 9QR
G0 OYA M. Clapperton 99 Bath Road, Bridgwater TA6 4PN
G0 OYC K. Saunders 1 Chesham Way, Watford WD18 6NX
G0 OYF S. Harvey 68 Stuart Road, Rowley Regis B65 9HZ
G0 OYI G. Holden The House On The Green, Linstock, Carlisle CA6 4PZ
G0 OYJ T. Gonsalves 30 Cunnington Street, Chiswick, London W4 5EN
G0 OYL W. Waring 1 Innerhaugh Mews, Haydon Bridge, Hexham NE47 6DE
G0 OYM M. Trahearn 16 Grange Lane, Lichfield WS13 7ED
G0 OYN D. Hedley 42 Liphook Road, Lindford, Bordon GU35 0PP
G0 OYO D. James 7 Abbotts Road, Plymouth PL3 4PD
G0 OYP B. Barber 3 Catherine Avenue, Mansfield Woodhouse, Mansfield NG19 9AZ
G0 OYQ S. Lowe 14 Kensington Avenue, Kingswood, Hull HU7 3AF
G0 OYR N. Ashfield 167 Greville Road, Warwick CV34 5PU
G0 OYS D. Temple 4 Cameron Avenue, Abingdon OX14 3SR
GM0 OYU M. Chesters Blackhill, Blackhill Road, Kirkwall KW15 1FP
G0 OYX D. Medley 9 Northolme Crescent, Hessle HU13 9HU
G0 OYY G. Mantle 26 Graham Road, Wordsley, Stourbridge DY8 5PU
G0 OYZ R. Bray 10 Upwell Road, March PE15 9DT
GW0 OZB A. Gardner 28 Usk Court, Thornhill, Cwmbran NP44 5UN
G0 OZG D. Turner 27 Aylesbury Avenue, Langney Point, Eastbourne BN23 6AB
G0 OZJ G. Gourley 6A Longsight Lane, Cheadle Hulme, Cheadle SK8 6PW
G0 OZL B. Smith 19 Fieldstone Court, Howick, Northpark 1705 New Zealand
G0 OZM C. Rapson Kaloma, Northiam Road, Rye TN31 6EP
G0 OZO Walsall ARC c/o J. Harris 31 Grasby Road, Limber, Grimsby DN37 8LB
G0 OZP B. Salt 9 Ashville Gardens, Pellon, Halifax HX2 0PU
GI0 OZQ D. Gillespie 81 Lisfannon Park, Londonderry BT48 9DU
G0 OZR G. Markham 2 Edwin Avenue, Woodbridge IP12 1JS
G0 OZS I. Moffat The Hatchets, Brockford Street, Stowmarket IP14 5PE
G0 PAB P. Betts 14 Saltergate Road, Messingham, Scunthorpe DN17 3SZ
GM0 PAC G. Macaulay 22 Glencairn Drive, Glasgow G41 4PW
G0 PAD A. Jacobs 16 Clwyd Walk, Corby NN17 2LN
G0 PAE C. Hewitt 28 Amersham Avenue, Langdon Hills, Basildon SS16 6SJ
G0 PAG N. Page 54 Queensway, Old Dalby, Melton Mowbray LE14 3QH
G0 PAI I. Leitch 70 Hanover Road, Rowley Regis B65 9DZ
G0 PAN D. Elkington 45 Heathfield, Leeds LS16 7AB
G0 PAO C. Muddimer 7 Tots Gardens, Acton, Sudbury CO10 0DJ
G0 PAR D. How 25 Lovelace Road, London SE21 8JY
G0 PAS M. Lord 5 Wasdale Green, Cottingham HU16 4HN
G0 PAZ D. Utley 30 Station Road, Ackworth, Pontefract WF7 7NA
G0 PBB Richmond School Amateur c/o W. Bonser 24 Meend Garden Terrace, Cinderford GL14 2EB
G0 PBE D. Yates 101 Coach House Drive, Shevington, Wigan WN6 8AU
G0 PBF J. Brown 71 Piccadilly Road, Swinton, Mexborough S64 8LF
G0 PBH D. Mason 11 Bryony Close, Killamarsh, Sheffield S21 1TF
GW0 PBJ L. Wright Cedar House, Old Aston Hill Ewloe, Deeside CH5 3AL
G0 PBL P. Davies 85 Church Road, Byfleet, West Byfleet KT14 7NG
G0 PBM A. Razzell 96 Weston Road, Aston-On-Trent, Derby DE72 2BA
G0 PBN A. Moulder 10 Parsonage Road, Rainham RM13 9LW
G0 PBO A. Coleman 16 Cowley Close, Swineshead, Boston PE20 3ES
G0 PBP A. Evans 24 Oakleigh Avenue, Glen Parva, Leicester LE2 9TH
G0 PBQ D. Hodge 2 Leyland Close, Southport PR9 8AT
G0 PBR R. Clark 4 Haigh St., Cleethorpes DN35 8QN
G0 PBS D. Webber Lowenva, Shripple Lane, Winterslow, Salisbury SP5 1PW
G0 PBU Northeast ARS c/o D. Bradley 2A Mitchell Street, Kettering NN16 9HA
G0 PBV N. Plumb 35 Foamcourt Waye, Ferring, Worthing BN12 5RD
G0 PBW R. Brown 26 Lynnes Close, Blidworth, Mansfield NG21 0TU
G0 PBY R. Freer 15 Fosse Close Enderby, Leicester LE192AW
G0 PCA K. Godwin 11 St. Lukes Way, Allhallows, Rochester ME3 9PR
G0 PCB E. Godwin 11 St. Lukes Way, Allhallows, Rochester ME3 9PR
G0 PCD S. Farrow 7 Bakewell Close, Hull HU9 5LH
G0 PCE Dr R. Barnes Flat 113, Queens Quay, 58 Upper Thames Street, London EC4V 3EJ
G0 PCF B. Foxall 11 Cranley Gardens, Shoeburyness, Southend-on-Sea SS3 9JP
GW0 PCJ C. Watson 4 Brookland Close, Maesycwmmer, Hengoed CF82 7RH
G0 PCK Rvd. A. Lord 47 Nottingham Road, Trowell, Nottingham NG9 3PF
G0 PCM I. Calvert 16 Nab Wood Drive, Shipley BD18 4EJ
G0 PCP R. Baldock 19 Ferndale Close, Burntwood WS7 4US
G0 PCQ I. Yeo Chyventon, Smithams Hill, Bristol BS40 6BZ
G0 PCT D. Hambly Culver Park, Rattery, South Brent TQ10 9LL
GI0 PCU A. Stewart 1 Lislaynan, Ballycarry, Carrickfergus BT38 9GZ
G0 PCW J. Budden Fieldgate, Durnstown, Lymington SO41 6AL
G0 PCY J. Radford 93 Hook Road, Surbiton KT6 5AF
G0 PCZ B. Lody 41 Galsworthy Road, Chertsey KT16 8EP
G0 PDA W. Cole Y Marian, Bow Street SY24 5BE
GW0 PDB G. Griffiths Dolcoed, Llandysul SA44 4RJ
G0 PDE D. Livingstone 68 Brimley, Leonard Stanley, Stonehouse GL10 3NA
G0 PDH D. Hyde 9 Empress Avenue, Marple, Stockport SK6 7BG
GJ0 PDJ M. Turner 4 Le Clos Sara, St. Lawrence JE3 1GT Jersey
G0 PDK W. Marsden 8 Albert Road, Eston, Middlesbrough TS6 9QW
G0 PDM M. Glover 22 Fern Street, Sutton-in-Ashfield NG17 2DW
GD0 PDN D. Beedan Ashmawr, Mount Rule Road, Douglas IM4 4QZ Isle of Man
G0 PDP A. Farmer 76 Wood Lane, Kingsnorth, Ashford TN23 3AG
GM0 PDQ M. Kusin East Overhill Farm, Stewarton, Kilmarnock KA3 5JT
G0 PDS P. Sparling 4 Gilbert Mead, Hayling Island PO11 0RE
G0 PDV R. Netherway 2 Avon Court, Lawn Road, Bristol BS16 5BL
G0 PDZ I. Lowe 54 College Road, Margate CT9 2SW
G0 PEB R. Williams 10 Barton Close, Whippingham, East Cowes PO32 6LS
G0 PEC I. Tutt 1 Castle Road, Hadleigh, Ipswich IP7 6JH
G0 PEF I. Williams 6 Newport Road, Godshill, Ventnor PO38 3HR
G0 PEG J. Jenner 1 Bellflower Rise, Ashford TN24 0GS
G0 PEH A. Lifton 70 Scrapsgate Road, Minster On Sea, Sheerness ME12 2DJ
GM0 PEI A. Pollock 113 Gartmorn Road, Sauchie, Alloa FK10 3PD
G0 PEJ G. Ford 5 Rosslyn Close, Hockley SS5 5BP
G0 PEK K. Richardson 35 Vidgeon Avenue, Hoo, Rochester ME3 9DE
G0 PEP Lymington Community Association RC c/o P. Waters 9 Tudor Way, Hawkwell, Hockley SS5 4EY
G0 PEQ King Edward VII School c/o P. Cook 88 Sprowston Road, Norwich NR3 4QW
G0 PER K. Kreuchen 211 Creek Road, March PE15 8RY
G0 PEV R. Dawson 6 Oxton Lane, Tadcaster LS24 8AG
G0 PEW J. Lyne 157 Westwick Road, Sheffield S8 7BW
GM0 PEX P. Bendermacher 1 Cedar Drive, Milton Of Campsie, Glasgow G66 8AY
G0 PEY R. Pearson 26 Ammonite Drive, Needham Market, Ipswich IP6 8FJ
G0 PFA M. Sole 44 Chestnut Avenue, Ewell, Epsom KT19 0SZ
G0 PFD A. Davison 45 Cheyne Garth, Hornsea HU18 1BF
G0 PFE R. Lees Lyndric, 23 New Queen St., Scarborough YO12 7HL
GM0 PFH G. Spurr 6 The Granary, Glebe Street, Dumfries DG1 2LU
G0 PFI E. Ball 57 Cherry Tree Road, Sheffield S11 9AA
G0 PFJ F. Poynter 7 Howards Way, Cawston, Norwich NR10 4AZ
GI0 PFL S. Mcclean 22 Whiteways, Newtownards BT23 4UW
G0 PFM E. Ashworth 88 Hawthorn Avenue, Colchester CO4 3JP

G0 PFN D. Catchpole 43 Welsford Road, Norwich NR4 6QB
G0 PFO D. Butler 1901 Dean Avenue, Michigan 48842 United States
G0 PFQ S. Streluk 11 Ninefoot Lane, Belgrave, Tamworth B77 2NA
G0 PFT M. Farrell Hobberley House, Hobberley Lane, Leeds LS17 8LX
G0 PFU K. Wignall 4 Weavers Fold, Bretherton, Leyland PR26 9AP
G0 PFY R. Marshall 66 Oakwood Hill, Loughton IG10 3EP
GW0 PFZ A. Powell Rich Lyn, Carmel Road, Holywell CH8 7DF
G0 PGA C. Smith 5 Northfield Drive, Mansfield NG18 3DD
G0 PGB C. Hosking 32 Queen St., Penzance TR18 4BH
GI0 PGC J. Forsythe 1 Coulson Avenue, Lisburn BT28 1YJ
GM0 PGD A. Paterson 21 Kirkwood Avenue, Redding, Falkirk FK2 9UF
G0 PGI Dr D. Beckly Knighton, Buckland Monachorum, Yelverton PL20 7LH
G0 PGJ G. Smith 16 Weeth Lane, Camborne TR14 7JN
G0 PGK D. Lawrence 7 Richmond Road, Appledore, Bideford EX39 1PE
G0 PGL D. Blight 73 Stoke Road, Somerset TA1 3EL
G0 PGQ M. Molloy 20 The Lawn, Whittlesford, Cambridge CB22 4NG
G0 PGS P. Slater 1 Greyhound Road, Glemsford, Sudbury CO10 7SJ
G0 PGT J. Newman Sometimes (The Workshop), South Pew, Dorchester DT2 9HZ
G0 PGW G. Dunn 6 Rosewood Avenue, Haslingden, Rossendale BB4 5NG
G0 PGX S. Thomas Creekside, Greenbank Road, Truro TR3 6PQ
G0 PGY J. Underwood 56 Bassenhally Road, Whittlesey, Peterborough PE7 1RR
G0 PGZ B. Hill 48 Lackford Avenue, Totton, Southampton SO40 9BT
G0 PHC G. Woodhouse 12 Matthew Street, Alvaston, Derby DE24 0ER
G0 PHD C. Whitehead 27-28 St. Nicholas St., Scarborough YO11 2HF
G0 PHE P. Long 40D Curborough Road, Lichfield WS13 7NQ
GM0 PHG D. Mclaughlin 96 Craighlaw Avenue, Eaglesham, Glasgow G76 0HA
G0 PHI P. Hirst 4 Brook House, Brook House Lane, Huddersfield HD8 8LX
G0 PHO C. Wilson 448 Hythe Road Willesborough, Ashford TN24 0JH
G0 PHP K. Green 39 Fleetgate, Barton-upon-Humber DN18 5QA
G0 PHR M. Andrews 9 Irving Road, Solihull B92 9DQ
GM0 PHW M. Whitehead 185 Allanton Road, Allanton, Shotts ML7 5AX
G0 PHY O. Williams 30 Franklin Road, Biggleswade SG18 8DX
G0 PIA J. Brown 14 St. Georges Avenue, Hornchurch RM11 3PD
G0 PIB K. Simmonds 9 Packman Drive, Ruddington, Nottingham NG11 6GF
G0 PID B. Thomas 112 Pen Park Road, Bristol BS10 6BP
G0 PIK A. Clements 37 Sun St., Isleham, Ely CB7 5RU
G0 PIL T. O'Brien 45 Rossall Promenade, Thornton-Cleveleys FY5 1LP
G0 PIN A. Pinnock 1 Rutland Gardens, Ealing, London W13 0ED
G0 PIS J. Bird 12 Beresford Gardens, Romford RM6 6RX
G0 PIT A. Freeman 3 Greenleas Road, Wallasey CH45 8LR
G0 PIU G. Papadopoulos 1 Darenth Road, London N16 6EP
GM0 PIV M. Black Drumtochty, 37 Clepington Road, Dundee DD4 7EL
G0 PIY C. Pollock Flat 5, 93 Priory Grove, London SW8 2PD
GW0 PJA P. Baston 27 Higher Common Road, Buckley CH7 3NG
G0 PJC A. Jones 35 Orchard Way, Letchworth Garden City SG6 4RZ
GM0 PJD Dr P. Dobie Cairnview Cottage, Quothquan, Biggar ML12 6NB
G0 PJG J. Geraghty 61 Bridle Lane, Streetly, Sutton Coldfield B74 3QE
GI0 PJH W. Stewart 23 Sandy Grove, Magherafelt BT45 6PU
G0 PJI P. Wood 2 Central Crescent, Hethersett Nr93Ep, Norwich NR9 3EP
G0 PJM M. Hughes The Cottage, Astley Burf, Stourport-on-Severn DY13 0RX
G0 PJO M. Waller Olive Cottage, 6 Church Road, Ipswich IP9 1HS
G0 PJR P. Ruffle 55 Nailers Drive, Burntwood WS7 0ES
G0 PJS P. Spicer 86 Main St., Wilsford, Grantham NG32 3NR
G0 PJU J. Brown 20 Stamford Avenue, Seaton Delaval, Whitley Bay NE25 0PA
G0 PJW C. Wormald 22 Tulworth Road, Poynton, Stockport SK12 1BL
G0 PJY P. Graham 11 Raby Court, Ellesmere Port CH65 9DZ
G0 PJZ R. Dorling Aletheia, St. Marys Road, Colchester CO7 8NN
GM0 PKF P. French 7 Knockothie Hill, Ellon AB41 8BA
G0 PKJ D. Stallon 8 Hidcote Close, Eastcombe, Stroud GL6 7EF
G0 PKN T. Finneran 23 Longdales Road, Lincoln LN2 2JR
GM0 PKP W. Carroll 20 Pinewood Road, Mayfield, Dalkeith EH22 5HX
GM0 PKQ F. Grant Silverknowes, Arbeadie Road, Banchory AB31 5XA
G0 PKR K. Ritson 14 Dunsdale Road, Holywell, Whitley Bay NE25 0NG
G0 PKT C L P K c/o A. Horsman 15 Hanwell Close, Clacton-on-Sea CO16 7HF
G0 PKV E. Bennett 17 Bixhead Walk, Broadwell, Coleford GL16 7EB
GM0 PKW O. Fairgrieve 8 Aird, Point, Isle of Lewis H52 0EU
GM0 PKX E. Michael 8 Castlepark Grove, Kintore, Inverurie AB51 0SN
G0 PLA T. Reddish 72 Edgmond Close, Redditch B98 0JQ
G0 PLB K. Murray Viamory, Wistanswick, Market Drayton TF9 2BD
G0 PLC P. Gosnell 230 Rowley Gardens, London N4 1HN
G0 PLD T. Pogson 64 New North Road, Slaithwaite, Huddersfield HD7 5BW
GM0 PLH W. Chan 5 Lansdowne Drive, Cumbernauld, Glasgow G68 0JB
G0 PLK T. Kennedy Woodgreen, Williamstown Ireland
GD0 PLQ J. Mitchell 5 Westminster Drive, Douglas IM1 4EG Isle of Man
GD0 PLR W. Smith 1 High View Road, Douglas IM2 5BQ Isle of Man
G0 PLS I. Wallis 20 Gerard Avenue, Bishop's Stortford CM23 4DU
G0 PLX J. Parker 24 Egmont St., Salford M6 7LA
G0 PLZ D. Lindsay 33 Varna Road, Bordon GU35 0DG
G0 PMB G. Banks 10 Gregory Road, Glass Houghton, Castleford WF10 4PH
G0 PMF G. Dellbridge 19 Cleeve Close, Astley Cross, Stourport-on-Severn DY13 0NY
G0 PMG R. Dellbridge 4 Woodford Way, Wombourne WV5 8HB
G0 PMI R. Spencer 4 Barstow Avenue, York YO10 3HE
GI0 PML M. Mcpeake 7 Neillsbrook View, Randalstown, Antrim BT41 3FL
G0 PMM D. Carrott 5 Raeburn House, 42 Brighton Road, Sutton SM2 5JH
GM0 PMO A. Fawcett Glennairn, Stromness KW16 3EX
G0 PMP M. Overend 58 Church Road Roberttown, Liversedge WF157LP
G0 PMS R. Sweeney 33 Traherne Close, Lugwardine, Hereford HR1 4AF
G0 PMU R. Nolson 50 Shelf Hall Lane, Shelf, Halifax HX3 7NA
G0 PMX J. Garnham 20 Deans Walk, Durham DH1 1HA
G0 PMZ I. Brydon 12 Pearce Road, Maidenhead SL6 7LF
G0 PNA M. Cranwell 21 Cockhaven Road, Bishopsteignton, Teignmouth TQ14 9RF
G0 PNB R. Hope 7 Irwell Green, Taunton TA1 2TA
GW0 PNC H. Hartwell Heulwen, Llanfair Clydogau, Lampeter SA48 8LH
GW0 PNE D. Hutson Sandalwood, 60 Glyndwr Road, Colwyn Bay LL29 8TA
G0 PNF W. Warren 38 Stoneyhurst Drive, Curry Rivel, Langport TA10 0JH
G0 PNG G. Buckley 46 King Street, Portland DT5 1NH
GW0 PNI Dr D. Pitkin Highbury Aberporth, Cardigan SA43 2BZ
G0 PNM P. Sobye 2 Willowbank, Fraddon, St. Columb TR9 6TW
G0 PNO P. Studdart 656 Rayleigh Road, Hutton, Brentwood CM13 1SJ
GI0 PNP R. Pritchard 79 Harbour Road, Ballyhalbert, Newtownards BT22 1BW
G0 PNQ A. Varley 37 Forest Road, Cambridge CB1 9JA
G0 PNR G. Mcilroy 1 Belmont Walk, Worcester WR3 7HY
G0 PNS RC of Pabay c/o J. Harris 37 Long Orchard Way, Martock TA12 6FA
G0 PNT S. Poulter 119 Aragon Road, Morden SM4 4QG
GW0 POA M. Hale 5 Marchwood Close, Rumney, Cardiff CF3 3LZ
GI0 POB G. Eldridge 59 Beechwood Gardens, Bangor BT20 3JD
G0 POC Dr P. Elwood 55 Madan Road, Westerham TN16 1DX
GM0 POD W. Mccallum St Brides Way, Colyton, Ayr KA6 6QG
GW0 POG C. Gavin Hafod Wen, Bagillt Road, Bagillt CH6 6JE
G0 POK D. Quinnear 5 Heath Drive, Chelmsford CM2 9HA
G0 POM P. Harris 44 Boston Road, Heckington, Sleaford NG34 9JE
G0 POQ D. Kemp 7 St. Nicholas Avenue, Hull HU4 7AH
G0 POT M. Sansom 19 Baily Avenue, Thatcham RG18 3EG
G0 POU J. Crosby 9 Hermitage Close, North Mundham, Chichester PO20 1JZ
G0 POY A. Eskelson 90 Charlton Crescent, Barking IG11 0NL
GW0 POZ D. Morgan Coedybryn, Synod Inn, Dyfed SA44 6JE
G0 PPH W. Blythe 4 Beresford Road, Stubbington, Fareham PO14 2QX
G0 PPI D. Chenery 25 Aldreth Road, Haddenham, Ely CB6 3PW
G0 PPJ P. Johnson 20 Bearmore Road, Warley, Cradley Heath B64 6DU
G0 PPK W. Gill 21 Flockton Avenue, Standish Lower Ground, Wigan WN6 8LH
G0 PPL G. Lattka 9 The Row, Sutton, Ely CB6 2PD
G0 PPM K. Powell 86 Nortonwood, Forest Green, Nailsworth, Stroud GL6 0TB
G0 PPQ P. Jackson 8 Buttree Court, South Kirkby, Pontefract WF9 3NB
G0 PPR G. Whaling 3 Bell Close, Little Snoring, Fakenham NR21 0HX
G0 PPS Prudential ARS c/o D. Dyer 57 Garrison Lane, Felixstowe IP11 7RR
G0 PPU H. Bennett 32 Sculthorpe Road, Blakedown, Kidderminster DY10 3JR
G0 PPX J. Omara 18 Tarrant Grove, Quinton, Birmingham B32 2NW
G0 PPY N. Turner 31 Shamrock Avenue, Whitstable CT5 4EL
G0 PQB S. Slater 118 Danziger Way, Borehamwood WD6 5DG
G0 PQD K. Skuse 4 Barton Close, Berrow, Burnham-on-Sea TA8 2NN
G0 PQF A. Judge 44 Thorley Lane, Bishop's Stortford CM23 4AD
G0 PQG A. Harper 81 High Street, Great Houghton, Barnsley S72 0AU
G0 PQO Dr K. Martin 8 Taylors Close, Meppershall, Shefford SG17 5NH
G0 PQR C. Wardle P O Box N 3189, Nassau 0 Bahamas
GM0 PQV J. Maguire 64 High Street, Loanhead EH20 9RR
G0 PQW P. Bartholomew 29 Beatrice Avenue, East Cowes PO32 6HR
G0 PQX S. Shipley 102 Jackson Street, Goole DN14 6DH
G0 PQY A. Langford 53 Cambridge Avenue, Bottesford, Scunthorpe DN16 3PH
G0 PRF J. Goodwin 146 Grimescar Road, Ainley Top, Huddersfield HD2 2EB
GM0 PRG Perth Repeater Group c/o D. Morris Ash Cottage, Perth Road, Perth PH2 9LW
G0 PRH M. Grassi Little Ash, Sleight Lane, Wimborne BH21 3HL
G0 PRI L. Ward 20 The Green, Newby, Scarborough YO12 5JA
G0 PRK R. Weller 15 Richmond Avenue Highams Park, London E4 9RR
GW0 PRM B. Goodier 14 Meadowbank, Old Colwyn LL29 8EX
GM0 PRO P. Greenway 5 Java Place, Craignure, Isle of Mull PA65 6BG
GM0 PRQ M. Shield Castleshield, Fiscavaig, Isle of Skye IV47 8SN
G0 PRS Poole Radio Scouts (Prs) c/o C. Baverstock 43 Tatnam Road, Poole BH15 2DW
G0 PRU Prudential ARS c/o D. Dyer 57 Garrison Lane, Felixstowe IP11 7RR
G0 PRY D. Mcnab 10 Rainham Gardens, Alvaston, Derby DE24 0DJ
G0 PSD P. Hayward 6 Greenock Close, Westlands, Newcastle-under-Lyme ST5 2LG
G0 PSF P. Yeatman 73 Roundway, Waterlooville PO7 7QB
G0 PSG R. Carvell 26 Greenfield Avenue, Kettering NN15 7LL
G0 PSH A. Goldstraw 59 Lansbury Grove, Stoke-on-Trent ST3 6JY
G0 PSI J. Wood 18 Kennedy Avenue, Long Eaton, Nottingham NG10 3GF
G0 PSJ S. Jacques Torr Garth, 38 Cheyne Walk, Hornsea HU18 1BX
G0 PSK G. Hawkins 8 Broughton Road, West Ayton, Scarborough YO13 9JW
G0 PSO P. O'Nion 11 Capitol Close, Swindon SN3 4AB
GU0 PSP M. Dowding L'Ancrage, Les Marais, Guernsey GY7 9LD Guernsey
GW0 PSV G. Wardman 5 High Street, Trelewis, Treharris CF46 6AB
G0 PSY S. Brodie Waterloo Cottage, Tanners Green, Norwich NR9 4QS
G0 PSZ L. Banaszak 17 Stoney Piece Close Bozeat, Wellingborough NN29 7NS
G0 PTA R. Attwood 2 Elizabeth Road, Basingstoke RG22 6AX
G0 PTD A. Washington 22 Elm Tree Drive, Bignall End, Stoke-on-Trent ST7 8NG
G0 PTE Long Wave Club c/o P. Davidson 28 Daneswell Drive, Wirral CH46 1QH
G0 PTG J. Mattison 21 Maynard House, Dunmow Road, Dunmow CM6 2DL
G0 PTI H. Aigeldinger 14 Peregrine Avenue, Morley, Leeds LS27 8TD
G0 PTK D. Dunford 25 Northfields Lane, Brixham TQ5 8RS
G0 PTL D. Caley 5 Crosswood Close, Bransholme, Hull HU7 5BU
G0 PTM A. Baird 65 Waterpump Court, Thorplands, Northampton NN3 8UR
GI0 PTQ P. Keenan Drumbadreeuagh, Belleek, Enniskillen BT93 3FT
G0 PTR A. Ryland 19 Redwood Court, Northway, Tewkesbury GL20 8SN
G0 PTT K. Caunce 7 Trevanions Way, Totland Bay PO39 0JL
G0 PTU J. Davies 8668 Ne Orchard Loop Road, Leland 28451 United States
GM0 PTY A. Higgins 26 Waterton Road, Bucksburn, Aberdeen AB21 9HS
G0 PUD D. Shaw 27 St. Davids Avenue, Romiley, Stockport SK6 3JT
G0 PUK A. Johnson 3 Plantation Avenue, Swalwell, Newcastle upon Tyne NE16 3JN
GW0 PUM D. Jenkins Gwalia House, 143A Priory Street, Carmarthen SA31 1LR
GM0 PUN H. Heritage 6 Newton Place, Rosyth, Dunfermline KY11 2LX
GW0 PUP G. Brown 17 High Street, Senghenydd, Caerphilly CF83 4GG
G0 PUQ H. O'Hare 39 Crichton Road, Carshalton SM5 3LS
G0 PUW G. Taylor 33 Heol Aberwennol, Borth SY24 5NP
G0 PUY C. Duckworth 121 Mill Gate, Newark NG24 4UA
G0 PVB B. Sketcher 147 Moorside Road, Bradford BD2 3HD
G0 PVE K. Greaves 10 Chatsworth Drive, Syston, Leicester LE7 1HX
G0 PVF P. Benson 21 Farleigh Road, New Haw, Addlestone KT15 3HS
GI0 PVG T. Lyons 3 Clanbrassil Gardens, Portadown, Craigavon BT63 5YD
G0 PVJ E. Hewitt 8 Embleton Road, Headley Down, Bordon GU35 8AJ
G0 PVN C. Fleet 14 Fairwood Road, Penleigh, Westbury BA13 4EA
G0 PVO L. Hewitt Sunny Nook, Grains Road, Shaw, Oldham OL2 8JF
G0 PVP C. Duffy 590 Chorley Old Road, Bolton BL1 6AA
G0 PVQ P. Fuller 19 Greenwood Court, Webb Close, Crawley RH11 9JH
G0 PVR J. Davies 13 The Close, Stalybridge SK15 1HU
G0 PVT D. Henderson 7 Love Avenue, Dudley, Cramlington NE23 7BH
G0 PVU Rvd. J. Roberts 31 Seaton Way, Marshside, Southport PR9 9GJ
G0 PVW R. Hamer 2 Back Lane Pontesford, Pontesbury, Shrewsbury SY5 0UD

G0 PVY A. Heward 22 Ross Avenue, Leasowe, Wirral CH46 2SB
G0 PWA D. Williams 31 Piper Hill Avenue, Manchester M22 4DZ
G0 PWC B. Dawe 6 Ullswater Avenue, Stourport-on-Severn DY13 8QP
G0 PWH P. Hughes 21A Erua Road, Waiheke Island 1081 New Zealand
G0 PWK S. Alder 5 Inglesgarth Court, Spennymoor DL16 7UG
G0 PWL S. Wright 33 Virginia Avenue, Lydiate, Liverpool L31 2NN
G0 PWO A. Boyes 7 Thornwood Covert, Foxwood, York YO24 3LF
G0 PWQ W. Tonks 295 Quinton Road West, Quinton, Birmingham B32 1PG
GM0 PWS N. Doherty Cairdeas, Carrbridge PH23 3AA
G0 PWU G. Brown 21 Armada Drive, Teignmouth TQ14 9NF
G0 PWV H. Chorley 19 Cleeve Road, Priorswood, Taunton TA2 8DX
G0 PWW Rvd. T. Edwards Kenneggy Lodge, Polperro Road, Looe PL13 2JS
G0 PWX G. Richards 87 Woodlands Road, Ditton, Aylesford ME20 6EF
G0 PXA G. Petri Mount Holly, Castledon Road, Billericay CM11 1LH
G0 PXB C. Marsh White Rose Robin Hoods Walk, Boston PE21 9LW
G0 PXD A. Harrison 25 Lansbury Avenue, New Rossington, Doncaster DN11 0AA
G0 PXE M. Cook 37 Buttermere Crescent, Doncaster DN4 5QF
G0 PXF Scarborough ARC c/o K. Linsley 132 Rein Road, Tingley, Wakefield WF3 1JB
G0 PXG M. Hardman 47 Oatlands Road, Manchester M22 1AH
G0 PXH B. Wilkinson 22 Portree Crescent, Blackburn BB1 2HB
G0 PXI P. Rigby 41 St. Huberts Road, Great Harwood, Blackburn BB6 7AS
G0 PXK N. Pratt 23 Hall Lane, Whitwick, Coalville LE67 5FD
G0 PXL D. Martland 6 Omega Way, Trentham, Stoke-on-Trent ST4 8TF
G0 PXM G. Kirby Tralee, Main Road Brighstone, Newport PO30 4DJ
G0 PXO J. Morgan 5 Sealy Close, Wirral CH63 9LP
G0 PXP T. Cox 60 Seven Oaks Crescent, Bramcote, Nottingham NG9 3FP
G0 PXQ C. Bell 17 Jubilee Square, South Hetton, Durham DH6 2TR
GI0 PXS J. Madden The Cottage, 53 Clarendon Street, Londonderry BT48 7ER
G0 PXT E. Denman 15 Clare Way, Bexleyheath DA7 5JU
GM0 PXV P. Barclay 15 Craigmount Avenue North, Edinburgh EH12 8DH
G0 PXX E. Mason 36 Gattison Lane, New Rossington, Doncaster DN11 0NQ
G0 PXY D. Smith 27 Hanbury Close, Cheshunt, Waltham Cross EN8 9BZ
G0 PXZ G. Walker 54 Burnage Lane, Burnage, Manchester M19 2NL
G0 PYE Jersey ARS c/o A. Arnold 2 Duck Lane, Haddenham, Ely CB6 3UE
G0 PYF A. De Buriatte Tanglewood, East End, North Leigh OX8 6PZ
G0 PYI G. Bodaly 41 Robert Street, Northampton NN1 3BL
G0 PYJ J. Swatton 30 Squires Close, Crawley Down, Crawley RH10 4JQ
G0 PYL R. Harriss 38 Portland Drive Whittleford, Nuneaton CV10 9HY
GM0 PYM Paisley ARC c/o S. Mckinnon 8 Rowanlea Avenue, Paisley PA2 0RP
G0 PYS D. Rose 99 Blackfriars, Rushden NN10 9PF
GW0 PYU H. Clarke 3 Tanyrallt Avenue, Bridgend CF31 1PQ
G0 PYV M. Hainesborough 39 Princes Close, North Weald, North Weald CM16 6EW
G0 PYW A. Haworth Clayfoot, Collins Green, Knightwick, Worcester WR6 5PS
G0 PZB W. Hattrick 38 Nithsdale Road, Weston-Super-Mare BS23 4JR
G0 PZC D. Flitterman Flat 7, 1 Rutland Gate, London SW7 1BL
G0 PZD G. Holmes 6 Darleydale Drive, Eastham, Wirral CH62 8EX
G0 PZF J. O'Connell Apartment 5, Roxboro House, Bailick Road, Co Cork Ireland
G0 PZJ Duxford Radio Limited c/o D. Featherby 14 Station Road, Sutton, Ely CB6 2RL
G0 PZM N. Byron 2 St. Aidans View, Boosbeck, Saltburn-by-the-Sea TS12 3LS
G0 PZO C. Jordan 31 Rathmore Road, Birkenhead CH43 2HE
G0 PZP W. Rabbitt 21 Barnfield Road, Woolston, Warrington WA1 4NW
G0 PZR Caravan & Camping ARC c/o O. Prosser 2 Caroline Close, Ventonleague, Hayle TR27 4EX
GW0 PZS T. Edwards 6 Cottage Home, Newborough, Llanfairpwllgwyngyll LL61 6SY
GW0 PZT E. Allely Dwyfor, Rhiw, Pwllheli LL53 8AE
GW0 PZU A. Ward 158 Mold Road, Mynydd Isa, Mold CH7 6TF
G0 PZW Dr J. Birch 32 Poplar Grove, Scotter, Gainsborough DN21 3TZ
G0 PZX A. Dennis 44 Larksfield Road, Faversham ME13 7ES
GW0 PZZ M. Owen 90 Shakespeare Avenue, Penarth CF64 2RX
GW0 RAD J. Lewis 189 Heol Y Gors Cwmgors, Ammanford SA18 1RF
G0 RAE R. Walker 12 Hill Drive, Ackworth, Pontefract WF7 7LQ
G0 RAF RAF Waddington ARC c/o R. Pickles Bramcote Lorne, Rectory Lane Gamston, Retford DN22 0QQ
GU0 RAG K. De La Haye Flat 4, Forest Lodge Flats, Forest Lane, Guernsey GY1 1WJ
G0 RAL P. Vallis Gryphon, Dirtham Lane, Leatherhead KT24 5SD
G0 RAM M. King 65 Chepstow Close, Stevenage SG1 5TT
G0 RAN M. Jamil 29 Harrow Close, Bury BL9 9UD
GM0 RAO A. Williamson Cairn Cottage, Durris, Banchory AB31 6DT
G0 RAR A. Walton Flat 25 Albert Weedall Centre, 23 Gravelly Hill North, Birmingham B23 6BT
G0 RAS V. Maddex 3 The Vines, Shabbington, Aylesbury HP18 9HH
G0 RAT W. Barnes 17 Saxon Way, Bradley Stoke, Bristol BS32 9AR
G0 RAU D. Woodnutt 17 Hill Farm Road, Chalfont St. Peter, Gerrards Cross SL9 0DD
G0 RAV R. Ravenscroft 4 The Paddock, Lidlington, Bedford MK43 0RW
G0 RAX R. Preston 45 Long Meadow, Skipton BD23 1BP
G0 RBA E. Bannister 59 Home Farm Park Lee Green Lane, Church Minshull, Nantwich CW5 6ED
G0 RBB M. Batchelor 16 Clementi Avenue Holmer Green, High Wycombe HP15 6TN
GI0 RBC J. Thompson 3 Strandburn Park, Sydenham, Belfast BT4 1ND
G0 RBD D. Kiely 45 Redland, Chippenham SN14 0JB
G0 RBG R. Butt Waterloo Cottage, Barrack Hill, Little Birch, Hereford HR2 8AX
GW0 RBH Dr R. Hughes 17 Pentrosfa Road, Llandrindod Wells LD1 5NL
G0 RBI S. Ward Oaklands, Burtonwood Road, Warrington WA5 3AN
G0 RBJ P. Evans Flat 7, 150 Booker Avenue, Liverpool L18 9TB
G0 RBM C. Boland 13 Rushfield Crescent, Brookvale, Runcorn WA7 6BN
GI0 RBO J. Kernohan 17 Tullygrawley Road, Teeshan, Ballymena BT43 5NP
G0 RBQ R. Gibbs Lime Tree House, Top Road, Slindon, Arundel BN18 0RP
GI0 RBS Rvd. R. Rainey 4 Crossnadonnell Road, Limavady BT49 0BD
G0 RBV P. Brunton Flat 26, Absalom Court, Wright Close, Gillingham ME8 6XP
G0 RBW T. Jones 11 Coppice Close, Willaston, Nantwich CW5 6NL
G0 RCF E. Carrington 4 Lancaster Drive, East Grinstead RH19 3XF
G0 RCH T. Cullup 13 London Street, Whittlesey, Peterborough PE7 1BP
G0 RCI A. Gibson 1 Oakleigh Road, Grantham NG31 7NN
G0 RCJ J. Topham 23 St. Nicholas View, West Boldon, East Boldon NE36 0RF
G0 RCL O. Baldwin 23 Cherry Tree Walk, Tadcaster LS24 9HS
G0 RCN N. Allen 78 Bargates, Christchurch BH23 1QL
G0 RCP P. Mellors 64 Pinewood Way, North Colerne, Chippenham SN14 8QU
G0 RCU R. Thomas 164 Kings Head Lane, Bristol BS13 7BW
G0 RCW RSGB Contest Club c/o F. Mossop 4 Brookdale Way, Waverton, Chester CH3 7NT
G0 RCX C. Garbett 27 Burghley Drive, West Bromwich B71 3LX
G0 RCY M. Crimes 27 Dunmore Road, Little Sutton, Ellesmere Port CH66 4PD
GM0 RDA G. Adamson 10 Rossend Terrace, Burntisland KY3 0DQ
G0 RDB C. Fernie 2 Hopkins Close, Cambridge CB4 1FD
GM0 RDC C. Robinson Greenhouse Farm, Lilliesleaf, Melrose TD6 9EP
G0 RDD M. Prendergast 1 Olaman Walk, Peterlee SR8 2EA
G0 RDF L. Wolstenholme The Hollies, Avondale Road, Chesterfield S40 4TF
G0 RDG G. Kowalski 47 Graveney Place, Springfield, Milton Keynes MK6 3LU
G0 RDH B. Watson 7 Branksome Drive, Morecambe LA4 5UJ
GI0 RDJ I. Mcmullan 35 Howard Place, Lisburn BT28 1EX
G0 RDK C. Wiseman 42 Merlin Way, Kidsgrove, Stoke-on-Trent ST7 4YL
GI0 RDM K. Mcguckin 20 Lisnahull Park, Dungannon BT70 1UH
G0 RDN G. Johnston 11 Granville Street, Deal CT14 7EZ
G0 RDO J. Snell 5 Waverley Road, Newton Abbot TQ12 2ND
G0 RDP D. Peat 24 Brookland Ave., Mansfield NG18 5NB
G0 RDR B. Lowe 18 Lowther Drive, Swillington, Leeds LS26 8QG
G0 RDS A. Williams 30 Swan Close, Talke, Stoke-on-Trent ST7 1TA
G0 RDT D. Treen 13 Peveril Road, Duston, Northampton NN5 6JW
G0 RDU S. Emms 33 Whitworth Avenue, Stoke Aldermoor, Coventry CV3 1EQ
G0 RDV L. Davies 3 Rydalside, Kettering NN15 7DR
G0 RDX P. Walker Moze Cross Cottage, Beaumont Road, Harwich CO12 5BQ
G0 RDY G. Steel Long Close, 82 Whatton Road, Derby DE74 2DT
GM0 RDZ S. Smith 12 Home Avenue, Duns TD11 3HQ
G0 REA R. James Woodpeckers, Freshwater Lane, Truro TR2 5AR
G0 REB C. Salmon 14 Surrey Drive, Congleton CW12 1NU
GM0 RED Southport & District ARC (Sadarc) c/o J. Pert 56 Lochiel Drive, Milton Of Campsie, Glasgow G66 8ET
G0 REE D. Jones 120 Heathfield Road, Keston BR2 6BF
G0 REF Magherafelt ARS c/o J. Andrews 85 Little Cattins, Harlow CM19 5RN
G0 REL D. Gaskell 18 Woodcroft, Kennington, Oxford OX1 5NH
G0 REN C. Wienrich 94 Sandling Lane, Penenden Heath, Maidstone ME14 2EA
G0 REO P. Hill 16 Robins Way, Nuneaton CV10 8PA
G0 REP A. Blackburn 2 Blackthorn Road, Stratford-upon-Avon CV37 6TD
G0 REQ D. Hibberd 25 Manor Road, Rugby CV21 2SZ
G0 REU T. Lam 53 Beaufort Road, Upper Cambourne, Cambridge CB23 6FP
G0 REV A. Bowmaker Post Cottage, Ardley Road, Bicester OX25 6LP
GM0 REZ A. Dailey 82 Don Drive, Livingston EH54 5LP
G0 RFA S. Garczynski 19 Thornhill Croft, Leeds LS12 4JX
G0 RFE A. Moore 139 Argyle Street, Heywood OL10 3RS
G0 RFF C. Bourne Essams, 11 The Grove, Hailsham BN27 3HU
G0 RFG E. Hyde 63 Newlyn Drive, Sale M33 3LH
G0 RFI S. Brackley 12 Farnborough Grove, Liverpool L26 6LW
G0 RFL T. Pooley 133 Hardie Road, Dagenham RM10 7BT
G0 RFM J. Copplestone 25 Bruche Avenue, Paddington, Warrington WA1 3HX
G0 RFN Downside Scl AR c/o J. Taylor 121 Garesfield Gardens, Burnopfield, Newcastle upon Tyne NE16 6LQ
G0 RFQ G. Buck 3 James Street, Colne BB8 0HN
G0 RFS Dr C. Bradley 71A Bagley Wood Road, Kennington, Oxford OX1 5LY
G0 RFT R. Lagar 25 Neville Avenue, Warrington WA2 9BQ
G0 RFV K. Goodworth 12 Ewood Drive, Doncaster DN4 6AU
G0 RFX P. Walford Suite 184, 2 Old Brompton Road, London SW7 3DQ
G0 RFY D. Horton 21 St. James Street, Waterfoot, Rossendale BB4 7HN
G0 RGC J. Bridge Little House, Castle House Yard, Langport TA10 9PR
G0 RGE M. Jenkinson 25 Porchester Close, Hucknall, Nottingham NG15 7UB
G0 RGG J. Hubbard 4 Avondale, Ellesmere Port CH65 6RW
G0 RGH Harig c/o J. Mitchener Cabins, Wenham Road, Ipswich IP8 3EY
G0 RGJ R. Provins 42 Forest View Road, Tuffley, Gloucester GL4 0BX
G0 RGL D. Edmondson 64 Raleigh Avenue, Hayes UB4 0EF
G0 RGM J. Trice 71 Deerswood Road, Crawley RH11 7JP
G0 RGN B. Woodhead 16 Dow St., Hyde SK14 4BS
G0 RGO Rvd. J. Drummond 14 Bulls Head Cottages, Turton, Bolton BL7 0HS
G0 RGP A. Gibbs 17 Manor Bend, Galmpton, Brixham TQ5 0PB
G0 RGT J. Bellinger 108 Hulbert Road, Havant PO9 3TG
G0 RGU J. O'Gorman 141 Chesterfield Road, Huthwaite, Sutton-in-Ashfield NG17 2QF
G0 RGW R. Slatter Ashwell House, Stratford Road, Oversley Green, Alcester B49 6PG
G0 RGX J. Sandys 28 The Maultway, Camberley GU15 1PS
G0 RHB L. Mulford 55 Mill Farm Crescent, Hounslow TW4 5PF
GW0 RHC K. Dyer 34 Lundy Drive, West Cross, Swansea SA3 5QL
GW0 RHE S. Williams 5 Llys Yr Orsaf, Llanelli SA15 2LB
G0 RHF P. Ellwood Coire Cas, Marsh Lane, Poulton-le-Fylde FY6 9AW
G0 RHG D. Stewart 14 The Dell, East Grinstead RH19 3XP
G0 RHI B. Dooks 7 Manor Drive, Kirby Hill, Boroughbridge, York YO51 9DY
G0 RHJ R. Judson 27 Newcombe Drive, Arnold NG5 6RX
G0 RHK P. Ford 19 Swan Bank, Hay-On-Wye, Hereford HR3 5DW
G0 RHO J. Belling 77 Chantry Road, Marden, Tonbridge TN12 9JD
GM0 RHP D. Crooke 2 Main Street, Carnock, Dunfermline KY12 9JQ
G0 RHV J. Parish 83 Harold Road, Stubbington, Fareham PO14 2QS
G0 RIB A. Shaw 38 Longmead Gardens, Havant, PO9 1RR
G0 RIC R. Cannell 284 Archway Road, Highgate, London N6 5AU
G0 RIE D. Reilly 15 Shutewater Close, Bishops Hull, Taunton TA1 5EH
G0 RIF D. Barnes 11 Back Lane, Whittington, Lichfield WS14 9NH
G0 RII J. Spacey 43 Woodlands Road, Allestree, Derby DE22 2HG
G0 RIJ W. Sykes Summerfield, Second Avenue, Ross-on-Wye HR9 7HT
G0 RIK N. Stockwell 12 Weavers Mead, Great Cheverell, Devizes SN10 5TP
G0 RIP J. Austwick 22 Shurmer Street, Bolton BL3 4BW
G0 RIQ D. Wisbey 22 Rutland Drive, Hornchurch RM11 3EN
G0 RIR W. Lewis 88A Clifton Road, Grimsby DN34 4QN
G0 RIU P. Davis 21 Newton Way, St. Osyth, Clacton-on-Sea CO16 8RQ
G0 RIX B. Cook 7 Rosewood Gardens, New Milton BH25 5NA
G0 RIY G. Watson 52 Elmwood Park , Loddiswell, Kingsbridge TQ7 4SD
G0 RIZ B. Body 12A Elm Court Gardens, Truro TR1 1DS
G0 RJA K. Jones 10 Dale Terrace, Lingdale, Saltburn-by-the-Sea TS12 3EE
G0 RJC V. Fletcher 7 Highfield Crescent, Baildon, Shipley BD17 5NR
G0 RJE R. Enright 17 Ripston Road, Ashford TW15 1PQ
GM0 RJG E. Kelly Durness, Newbridge, Dumfries DG2 0QX
G0 RJI N. Rapson 21 Ashley Close, Penwithick, St. Austell PL26 8UB

G0 RJJ E. Foord 65 Dane Court Gardens, Broadstairs CT10 2SD
G0 RJL J. Hilton 177 Wilmot Road, Dartford DA1 3BP
G0 RJM J. Marchant 129 Highbury Grove, Clapham, Bedford MK41 6DU
G0 RJN H. Vicary The Brambles, Wrotham Road, Gravesend DA13 0QA
GI0 RJO L. Douglas 15 Bramhall Crescent, Londonderry BT47 5HE
G0 RJT Rvd. H. Leak 15 Sutherland Road, Tittensor, Stoke-on-Trent ST12 9JQ
G0 RJV G. Rogers Maes Gwersyll, Garthmyl, Montgomery SY15 6RS
G0 RJX E. Gaskell 18 Woodcroft Kennington, Oxford OX1 5NH
G0 RKB D. Roberts 20 Beech Grove, Trowbridge BA14 0HG
G0 RKC A. Alecio Flat 4, 19 The Beacon, , Exmouth EX8 2AF
GW0 RKD S. Gray Whispers, Front Street, Rosemarket, Milford Haven SA73 1JT
G0 RKE C. Burgess 12 Middleway, Grotton, Oldham OL4 5SH
G0 RKG R. Gaskell 18 Woodcroft Kennington, Oxford OX1 5NH
G0 RKN H. Burn 60 Uplands Croft, Stoke-on-Trent ST9 0LF
G0 RKP J. Aubin 46 Kenilworth Drive, Clitheroe BB7 2QN
G0 RKQ R. Plumtree 80 Dewsbury Avenue, Scunthorpe DN15 8BP
G0 RKS G. Goss Little Ashcroft, Parkgate Road, Dorking RH5 5DZ
G0 RKT D. Dukesell Mayfield, Ashbourne Road, Buxton SK17 9RY
GM0 RKU P. Craft 2 Luke Place, Broughty Ferry, Dundee DD5 3BN
G0 RKV V. Webley 2 Octavian Drive, Bancroft, Milton Keynes MK13 0PN
G0 RLA P. Harvey Rowlands Barn, Dunbridge Lane, Awbridge, Romsey SO51 0GQ
G0 RLB B. Stoneley 44 Ilthorpe, Hull HU6 9ER
G0 RLF G. Low 61 Fenwick Lane, Halton Lodge, Runcorn WA7 5YU
G0 RLH E. Miles 31 Winnipeg Road, Bentley, Doncaster DN5 0ED
G0 RLI J. Thomas 204 Watchouse Road, Galleywood, Chelmsford CM2 8NF
G0 RLJ P. Tyson 44 Windmill Avenue, Kilburn, Belper DE56 0PQ
G0 RLL T. Dyson 39 Newcastle Close, Stevenage SG1 4TL
G0 RLN K. Taylor 29 School Road, Pontefract WF8 2AJ
G0 RLO K. Conlon 4 Hill Crest Drive, Slack Head, Milnthorpe LA7 7BB
GW0 RLQ J. Ellwood 5 Smallwood Road, Baglan, Port Talbot SA12 8AP
G0 RLS P. Ashcroft 38A Wood End, Bluntisham, Huntingdon PE28 3LE
G0 RLT R. Taylor Flat 3, 3 St. Pauls Square, Southport PR8 1NQ
G0 RLV E. Jones 16 Fisher Avenue, Rugby CV22 5HN
G0 RLY J. Karkoszka 5 Wood Street Coldshaw, Haworth BD22 8BJ
GM0 RLZ C. Brown 41 Russell Avenue, Kingseat, Dunfermline KY12 0YX
GW0 RMB S. Ferris Temorfa, Y Ffor, Pwllheli LL53 6UB
G0 RMC M. Charlton 53 Dunstone View, Plymouth PL9 8TW
G0 RMD P. Calter 8 Exeter Road, Scunthorpe DN15 7AT
G0 RMG R. Jones 6 Wychwood Drive Hunt End, Redditch B97 5NW
G0 RMJ S. Hogg 38A High St., Ventnor, Isle of Wight PO38 1RZ
GM0 RML A. Smart 6 Alton Bank, Nairn IV12 5PJ
G0 RMN A. Younger 4 Esk Hause Close, West Bridgford, Nottingham NG2 6SG
G0 RMO M. Miller 8 Pilton Walk, Newcastle upon Tyne NE5 4PQ
G0 RMP R. Seal 2 Shaftesbury Road, Bridlington YO15 3NP
G0 RMR C. Rabey 23 Thorn Lane, Four Marks, Alton GU34 5BX
GM0 RMT G. Wilkie 25 Barn Rd., Stirling FK8 1EP
G0 RMU R. Clover Teffont, 42 Warren Road, Addlestone KT15 3UA
GM0 RMV M. Verity 19 Vivian Terrace, Edinburgh EH4 5AW
G0 RMX D. Esdale The Bell Inn, Central Lydbrook, Lydbrook GL17 9SB
G0 RNA T. Rawlinson 330 Blackpool Old Road, Poulton-le-Fylde FY6 7QY
G0 RNB N. Brooks 57 Mansel Crescent, Parson Cross, Sheffield S5 9QR
G0 RNC A. Pritchard 27 Walkley Crescent Road, Walkley, Sheffield S6 5BA
G0 RNF I. Hunnisett 69 Cornwall Road, Ruislip HA4 6AJ
G0 RNH M. Ahmed 75 Drove Road, Swindon SN1 3AE
G0 RNI Luton Rep Grp c/o D. Thorpe 70 Willow Way, , Ampthill MK45 2SP
GW0 RNK K. Williams 8 Trinity Place, Pontarddulais, Swansea SA4 8RD
G0 RNP D. Eves 64 Hillingdon Road, Gravesend DA11 7LG
G0 RNQ B. Willson 4 Caldew Grove, Sittingbourne ME10 4SL
G0 RNS J. White 24 Malines Avenue, Peacehaven BN10 7PS
G0 RNV B. Sherriff 27 Magellan Way, Spalding PE11 2FG
G0 RNX S. Onions 18 St. Cuthberts Crescent, Albrighton, Wolverhampton WV7 3HW
G0 RNY A. Attle 2 Watson Park, Spennymoor DL16 6NB
G0 ROA H. Seidner 7411 Morocca Lake Drive, Delray Beach, 33446 United States
G0 ROC Rochdale & Disctrict ARS c/o P. Hewitt 11 Thetford Close, Bury BL8 1XB
G0 ROD C. Reaney 81A Bargate Road, Belper DE56 1NE
G0 ROM R. Marcinkiewicz 8 Polefield Hall Road, Manchester M252WN
G0 RON R. Mcneil 3 Thorncliffe Gardens, Auckley, Doncaster DN9 3PE
G0 ROO Dover Construction Club c/o I. Keyser Rosemount, Church Whitfield Road, Whitfield, Dover CT16 3HZ
G0 ROS R. Kent 40 Waxes Close, Abingdon OX14 2NG
G0 ROT M. Davis 7 Walter Close, Chickerell, Weymouth DT3 4GU
GM0 ROU A. Butcher 224 Laird Street, Dundee DD3 9PL
G0 ROW A. Gurnhill 53 Millbrook Avenue, Denton, Manchester M34 2DQ
G0 ROX D. Lee 131 Abbotsbury Road, Weymouth DT4 0JX
G0 ROY R. Biddle 21 Kingsway West, Newton, Chester CH2 2LA
G0 ROZ the Terry Bain Memorial RC c/o C. Hardy 40 Beresford Road, Poole BH12 2HE
G0 RPA I. Mcavoy 74 Parkstone Heights, Poole BH14 0RZ
G0 RPD J. Barton 183 Windy Arbor Road, Whiston, Prescot L35 3SF
G0 RPF L. Smith 28 Chester Road, Stockton Heath, Warrington WA4 2RX
G0 RPG J. Riley 1 Chatsworth Avenue, Culcheth, Warrington WA3 4LD
G0 RPJ D. Wesil 8 Camber Way, Pevensey Bay, Pevensey BN24 6RW
G0 RPL N. Alison 9 South Drive, Burgess Hill RH15 9PY
G0 RPM N. Williams 11 Berkeley Gardens, London N21 2BE
G0 RPO R. Dowd Belgrano, 1 Watson Avenue, Warrington WA3 3QX
G0 RPU J. Symonds La Cumbre, 35 Byward Drive, Scarborough YO12 4JE
G0 RPV W. Till 97 Haslar Crescent, Waterlooville PO7 6DD
G0 RPW D. Wilson 39 The Wintles, Bishops Castle SY9 5ES
G0 RPY C. Button 8 Heywood Road, Diss IP22 4DJ
GW0 RQC R. Chegwin 17 Cyncoed Crescent, Cardiff CF23 6SW
G0 RQF K. Hales 3 New Barnfields, Hereford HR4 7AZ
G0 RQG J. Gill 24 Greenfields Court, Bridgnorth WV16 4JS
G0 RQH D. Hughes 31 Sussex Drive, Pagham, Bognor Regis PO21 4RN
G0 RQI S. Spragg 4 Valley Road Arleston, Telford TF1 2JP
G0 RQL D. Roomes View Field, Milton Damerel, Holsworthy EX22 7NY
G0 RQN P. Robertson 1 Yaffle Mews, Great Cambourne, Cambridge CB23 5HY
G0 RQO D. Hillyer 32A Belbroughton Road, Blakedown, Kidderminster DY10 3JG
GW0 RQP G. Ashford 26 Laura Street Treforest, Pontypridd CF37 1NW
G0 RQQ Hastings Electronics & RC c/o K. Ballinger 3 Cliff Court, Burton Road, Lincoln LN1 3NN
GW0 RQS L. Pritchard 86 Bryn Road, Markham, Blackwood NP12 0QE
G0 RQX D. Townend 38 Kingston Drive, Shrewsbury SY2 6SJ
G0 RQZ R. Lawrence 74 Principal Rise, Dringhouses, York YO24 1UF
G0 RRI I. Burden 2 Essex Road Flat 2, Lowestoft NR32 2HH
G0 RRL R. Leigh 33 Beddington Road, Orpington BR5 2TF
G0 RRM P. Brumby 69 Gilbert Walk, Nether Stowe, Lichfield WS13 6AU
G0 RRO J. Breingan 44 Farmstead Road, Corby NN18 0LG
G0 RRR J. Marsden 11 Firethorn Drive, Hyde SK14 3SN
G0 RRV A. Tomson 5 Fordham Close, Ashwell, Baldock SG7 5LJ
G0 RRZ R. Carrington 45 Crompton Road, Pleasley, Mansfield NG19 7RG
G0 RSA J. King 39 Nursery Gardens, St. Ives PE27 3NL
GM0 RSE Glenrothes & District ARC c/o T. Brown 11 Approach Row, East Wemyss, Kirkcaldy KY1 4LB
G0 RSG 1st Ringmer Scout Group c/o T. Mcconnell 51 Langney Road, Eastbourne BN21 3QD
GM0 RSI J. Ritchie 36 James Mitchell Place, Mintlaw, Peterhead AB42 5ES
G0 RSL K. White 25 Curson Rise, Kendal LA9 7PN
G0 RSR Reading Scouts Radio c/o S. French 22 Amity Street, Reading RG13LP
G0 RSS R. Simmonds 4 Corys Close, Kirby Road, Norwich NR14 7DP
G0 RSU G. Weston 2 Whitburn Road, Toton, Nottingham NG9 6HP
G0 RSV W. Webster 21 Quince Tree Way, Hook RG27 9SG
G0 RSW R. Waters 17 Wilson Road, Southend-on-Sea SS1 1HG
G0 RSY A. Gibbs Orchard Court, Woodside Road, Wootton Bridge, Ryde PO33 4JR
G0 RTA T. Arakawa 2-974-8-1502, Sayama, Osakasayama 589-0005 Japan
G0 RTC T. Chisholm 316 Birchfield Road East, Northampton NN3 2SY
G0 RTF I. Slaney 6 Little Shardeloes, High Street, Amersham HP7 0EF
G0 RTH A. Elcoate 9 Parsonage Lane, Laindon, Basildon SS15 5YN
G0 RTI S. Harriss 6 Redland Road, Leamington Spa CV31 2PB
G0 RTM P. Mcknight 39 Dunmail Drive, Kendal LA9 7JG
G0 RTN G. Lynch 52 Queens Road, Devizes SN10 5HW
GW0 RTP C. Llewellyn 16 Garth Street, Kenfig Hill, Bridgend CF33 6EU
G0 RTQ D. Lawrence 11 Pembroke Court St. Johns Road, Newbold, Chesterfield S41 8NX
GW0 RTR R. Rees 9 Langland Road, Mumbles, Swansea SA3 4ND
G0 RTU P. Kirkup 337 Wheatley Lane Road, Fence, Burnley BB12 9QA
GM0 RTY D. Inns 57 Craiglomond Gardens, Balloch., Alexandria. G83 8RP
G0 RTZ G. Hurst 35A Trewsbury Road, London SE26 5DP
GI0 RUC R. Kerr 194 Shore Road, Greenisland, Carrickfergus BT38 8TX
GW0 RUD Harlow & District ARS c/o P. Marriott 16 Heol Morlais, Llannon, Llanelli SA14 6BD
G0 RUF N. Taylor 30 Leonard St., Hull HU3 1SA
G0 RUH Dr M. Roberts 82 Glover Road, Scunthorpe DN17 1AS
G0 RUR P. Simpson Amber Lodge Nursing Home, 684-686 Osmaston Road, Derby DE24 8GT
G0 RUS R. Lenthall 182 Chelmsford Avenue, Grimsby DN34 5DB
G0 RUT Macclesfield & District ARS c/o R. Russell 4 Hinton Road, Newport PO30 5QZ
G0 RUV M. Gent 111 Portland Street Clowne, Chesterfield S43 4SA
GM0 RUW J. Coughtrie 61 Bells Burn Avenue, Linlithgow EH49 7LD
G0 RUX W. Taylor 21 Summerdale Road, Cudworth, Barnsley S72 8XG
G0 RUY A. Pritchard 41 Borough Close, Kings Stanley, Kings Stanley GL10 3LJ
G0 RUZ C. Farlow 4 Nether Road, Silkstone, Barnsley S75 4NN
G0 RVE A. Pierce 34 Church Close, Shawbury, Shrewsbury SY4 4JX
G0 RVH K. Dailey 55 Chesterton Avenue, Harpenden AL5 5SU
G0 RVI J. Davis 4 Stockbridge Close, Canford Heath, Poole BH17 8SU
G0 RVK M. Fogg 15 Elm Grove, Bisley, Woking GU24 9DG
G0 RVM A. Gawthrope 62 Meadow Way, Bradley Stoke, Bristol BS32 8BP
GW0 RVR R. Goodall 8 Heol Penderyn, Brackla, Bridgend CF31 2EA
G0 RVS B. Roff 1 Kennel Cottages, Arlington, Barnstaple EX31 4LP
G0 RWA B. Chorley 19 Cleeve Road, Priorswood, Taunton TA2 8DX
G0 RWI E. Johns 3 The Rowans, Portishead, Bristol BS20 6SR
G0 RWJ D. King 78 Andersey Way, Abingdon OX14 5NW
G0 RWL B. Mackenzie 73 Newstead Road, Weymouth Dorset DT4 0AS
G0 RWM R. Martin 82 Woodlands Avenue, West Byfleet KT14 6AP
GI0 RWO B. Madden 1 Skegoneill Drive, Belfast BT15 3FY
G0 RWQ N. Monument C/Isla Cabrera 14.1.10, Regia Roig Blq 2, Orihuela 3189 Spain
G0 RWS W. Scott Hunters Lodge, Broadmore Green, Worcester WR2 5TE
G0 RWT R. Pine Rhodanna, Tennis Court Road, Bristol BS39 7LU
GM0 RWU Dr J. Ponton Old Cottage Gardens, Legerwood, Earlston TD4 6AS
G0 RWW M. Barrass 11 Flintham Court, Mansfield NG18 4NB
G0 RWY Dr D. Ramsay 2 Old Church Road, Colwall, Malvern WR13 6ET
G0 RXA N. Roscoe 35 Kenilworth Road, Cheadle Heath, Stockport SK3 0QL
G0 RXQ F. Lockey The Dormers, Cirencester Road, Tetbury GL8 8HA
G0 RXU F. Nethercott 6 Laking Avenue, Broadstairs CT10 3NE
GM0 RYA G. Roberts 8 Parkview, Lhanbryde, Elgin IV30 8JZ
GM0 RYD J. Van Dyke 112 Alexander Avenue, Largs KA30 9EX
GI0 RYK R. White 1 Woodland Park, Lisburn BT28 1LD
G0 RYL Essex Raynet c/o R. Hodges 1A Clements Lane, Portland DT5 1AS
G0 RYM S. Goodwin 14 Greenhill, Alveston, Bristol BS35 2QX
G0 RYP C. Martin 7 St Lawrence Street, B'kara BKR1521 Malta
G0 RYQ P. Irwin 11 Cowdale Cottages, Cowdale, Buxton SK17 9SE
G0 RYR T. Ballinger 9 Somerville Court, Cirencester GL7 1TG
G0 RYS West Scotland Air Cadet RC c/o M. Vann Richmond School, Darlington Road, Richmond DL10 7BQ
GW0 RYT R. Pitman 7 Cleveland Drive, Risca, Newport NP11 6RD
GI0 RYU B. Millar 312 Churchill Park, Portadown, Craigavon BT62 1EY
G0 RYW S. Preston The Chapel, Robson St., Shildon DL4 1EB
G0 RZB D. Mcdonnell Glencoe, The Ridge, Salisbury SP5 2LN
G0 RZG R. Hayward Old School Farm, Wickham Market, Suffolk IP13 0HE
G0 RZI B. Easdon 20 Winder Gate, Frizington CA26 3QS
G0 RZM B. Judd 24 Haywood Way, Reading RG30 4QP
G0 SAC Sutton Area CG c/o A. Cross 31 Mountcombe Close, Surbiton KT6 6LJ
GW0 SAJ Dr H. Jones 6 Westfa Road, Uplands, Swansea SA2 0PR
G0 SAR South Anglia Raynet c/o D. Sparrow 23 Tranmere Grove, Ipswich IP1 6DU
G0 SAY C. Thorpe 78 Bowland Rd, Baguley, Manchester M23 1JX
G0 SBA D. Sant Marjar Marden, Hereford HR1 3EP
G0 SBB D. Barton Manuka, West Hill, Worthing BN13 3BZ

G0 SBC R. Harris 142 St. Nicolas Park Drive, Nuneaton CV11 6EE
G0 SBH T. Wernham 6 The Hill, Wangford, Beccles NR34 8AT
G0 SBK M. Jenkins 9 Tothill Road, Swaffham Prior, Cambridge CB25 0JX
G0 SBM South Devon Raynet Group c/o C. Coker 46 Clarendon Road, Ipplepen, , Newton Abbot TQ12 5QS
G0 SBO E. Hodgson 21 Royd Avenue, Mapplewell, Barnsley S75 6HH
G0 SBP F. Parkinson 28 Tillage Green, Darlington DL2 2GL
G0 SBU B. Wedgwood 40 Ford Street, Delves Lane, Consett DH8 7AE
G0 SBV R. Talbot 11 Whitefield Road, Holbury, Southampton SO45 2HP
G0 SBX E. Barclay 58 Stockton Road, Hartlepool TS25 1RW
G0 SBY J. Thompson 4 Ridgemont Fulwood, Preston PR23FQ
G0 SBZ W. Sandle 507B Harrogate Road, Leeds LS17 7DU
GM0 SCA S. Edwards The Old Police House, Broughton, Biggar ML12 6HQ
G0 SCG A. Leavey 14 Cherry Close, Ealing, London W5 4JW
G0 SCI B. Young 11 Gainsborough Avenue, Washington NE38 7EF
G0 SCK D. Britton 31 Clay Bottom, Bristol BS5 7EJ
G0 SCL S. Lawrence 4 Dale Park Rise, Leeds LS16 7PP
G0 SCM F. Binnington 7 Webbs Close, Combs, Stowmarket IP14 2NZ
G0 SCO Scottish Office ARC c/o J. Lefever 30 Holland Road, Melton Mowbray LE13 0LU
G0 SCQ D. Brusch 7 Tyrell Close, Stanford In The Vale, Faringdon SN7 8EY
G0 SCR Caterham Radio Group c/o P. Lewis 20 Annes Walk, Caterham CR3 5EL
G0 SCT R. Bricknell 82 Hills Road, Saham Hills, Thetford IP25 7EZ
G0 SCU F. Taylor 6 Shelley Close, Bolton Le Sands, Carnforth LA5 8HQ
G0 SCV G. Belt 3 Prospect Hill, Whitby YO21 1QE
GM0 SCW R. Anderson 10 Cyril Crescent, Paisley PA1 1GT
G0 SCX G. Hobbs 37 Winnards Park, Sarisbury Green, Southampton SO31 7BX
G0 SCY W. Best 61 Gainsborough, Hanworth, Bracknell RG12 7WL
G0 SDC Bath And District ARC c/o P. Robinson 11 The Avenue, Hambrook, Chichester PO18 8TZ
G0 SDD C. James 4 Hill Top, Bream, Lydney GL15 6JQ
G0 SDE B. Jupp 25 Briscoe Way, Lakenheath, Brandon IP27 9SA
G0 SDF J. Atkins 30 Bransby Road, Chessington KT9 2LA
G0 SDG Rvd. L. Wilkes Springfield, Sunnyfield Lane, Up Hatherley, Cheltenham GL51 6JE
G0 SDJ A. Mcmullon Carwood House, Hothersall Lane, Preston PR3 2XB
G0 SDL J. Wilson Appletrees, Combeinteignhead, Newton Abbot TQ12 4RE
G0 SDM P. Robinson 12 Maple Way, Donington PE11 4XL
G0 SDR S. Harriss 30 Chatsworth Place, Harrogate HG1 5HR
GM0 SDS B. Wills 24 Hopes Avenue, Dalmellington, Ayr KA6 7RN
G0 SDT J. Sparkes 18 Hermes Avenue, St. Erme, Truro TR4 9FW
G0 SDW M. Pattman 4 Branscombe Road, Bristol BS9 1SN
G0 SDX Willpower CG c/o J. Faulkner-Court Yew Tree Cottage, Avon Dassett, Southam CV47 2AT
G0 SEB J. Shepherd 25 Station Road, St. Helens, Ryde PO33 1YF
G0 SEC J. Curtis 24 Brisbane Road, Weymouth DT3 6RD
GM0 SEF D. Stolting 3 Eden Park, Cupar KY15 4HS
GM0 SEI R. Vennard 4 Braehead, Girdle Toll, Irvine KA11 1BD
GM0 SEP Waveney Wireless c/o R. Cowan 85 Eastwoodmains Road, Clarkston, Glasgow G76 7HG
G0 SET H. Pearson 110 The Gateway, Dover CT16 1LH
G0 SEU L. Payas C/O Jo-Anne Maclaren, 7801 Hibiscus Court, Gibraltar Gibraltar
G0 SEW K. Green 13 Knowle Road, Sheffield S5 9GA
G0 SEY E. Russell 60 Icknield Way, Tring HP23 4HZ
G0 SFA B. Hyde 108 St. Bedes Crescent, Cambridge CB1 3UB
G0 SFE K. Spring 18 Greenway, Woodmancote, Cheltenham GL52 9HU
G0 SFG T. Coneley 4 Beaconsfield Road Fareham, Fareham PO16 0QB
GD0 SFI B. Hull 75 Saddle Mews, Douglas, Isle of Man IM2 1HT
G0 SFJ A. Thomas 21 Great Bowden Road, Market Harborough LE16 7DE
GW0 SFP B. Rish 15 Bryn Marl, Deganwy, Llandudno Junction LL31 9BZ
GM0 SFQ J. Stirling 86 Obsdale Park, Alness IV17 0TR
GI0 SFT P. Mcdonald 13 Heathfield, Culmore, Londonderry BT48 8JD
G0 SFV D. Burton 3 Norwood, Carden Hill, Brighton BN1 8AH
G0 SGF J. Barrett 12 Trent View Gardens, Radcliffe-On-Trent, Nottingham NG12 1AY
GM0 SGH M. Brown 3 Arnott Road, Blackford, Auchterarder PH4 1QE
G0 SGI J. Sankey 56 Gorsey Lane, Mawdesley, Ormskirk L40 3TF
G0 SGP A. Danton 3 Cliffe Close, Ruskington, Sleaford NG34 9AT
G0 SGR S. Rice 56 St. Johns Avenue, Bridlington Y016 4NL
G0 SGT D. Huddleston 162 Manor Road, Newton St. Faith, Norwich NR10 3LG
G0 SGV J. Allen 57 Watford Road, Kings Langley WD4 8DY
G0 SGX F. Dingwall 20 Whitehills Road, Loughton IG10 1TS
G0 SHC M. Lane Cherry Tree House, Pipwell Gate Saracens Head, Holbeach, Spalding PE12 8BA
GM0 SHD G. Balfour 6 Kirkden Street, Friockheim, Arbroath DD11 4SX
G0 SHJ R. Harrison 22 East Anglian Way, Gorleston, Great Yarmouth NR31 6QY
G0 SHM B. Coates 74 Colescliffe Road, Scarborough YO12 6SB
G0 SHN G. Jacot Boucle De L'Observatoire, Le Grand Revard, Pugny-Chatenod 73100 France
G0 SHO B. Lawrence 70 Beacon Road, Rolleston-On-Dove, Burton-on-Trent DE13 9EG
G0 SHP A. Pratt 4 Chestnut Close, Braunton EX33 2EH
G0 SHT S. Rossi 21 Rattigan Gardens, Whiteley, Fareham PO15 7EA
G0 SHU G. Bennett 57 Princess Way, Euxton, Chorley PR7 6PL
G0 SHY M. Bamber 1 Penair Crescent, Truro TR1 1YS
GM0 SIA P. Brooks 3 Jamiesons Court, Kelso TD5 7EU
G0 SIE A. Swingler 9 Princess Drive, Wistaston Green, Crewe CW2 8HP
G0 SIG Signallers Interest Group c/o K. Prince 59 Chantry Road, East Ayton, Scarborough YO13 9ER
G0 SII T. Richards 142 Princes Mews, Royston SG8 9BN
GM0 SIM I. Simpson 1 West Abercromby Street, Helensburgh G84 9LL
G0 SIQ Rvd. R. Myerscough Hamer, The Street, Holt NR25 6NW
GW0 SIS K. Barrett Gorbio House, 47 West Road, Bridgend CF31 4HD
G0 SIU B. Durrant 3 Parklands, Shoreham-by-Sea BN43 6NN
G0 SIW B. Ellison 6 Eskdale Road, Ashton-In-Makerfield, Wigan WN4 8QT
G0 SIY A. Hopkinson 55 Nordale Park Norden, Rochdale OL12 7RT
G0 SJB S. Barraclough 67 Sude Hill, New Mill, Holmfirth HD9 7ER
G0 SJG S. Gunning 99 Mile Oak Road, Portslade, Brighton BN41 2PJ
G0 SJH S. Harris 19 Mundays Boro Road, Puttenham, Guildford GU3 1AZ
G0 SJP M. Windle 1A Prices Avenue, Margate CT9 2NS
G0 SJR R. Brand Foxgrove, 19A Mill End Close, Dunstable LU6 2FH
G0 SJU T. Bousfield 8 Harpington View, Mordon, Stockton-on-Tees TS21 2EZ
G0 SJV P. Gostick 25 Cashmere Lane, Cashmere, Queenslands 4500 Australia
G0 SKA C. Mitchell Nevada, Slad Lane, Lacey Green, Princes Risborough HP27 0PW
GW0 SKC B. Foote Red Roofs, 5 Woodland Avenue, Colwyn Bay LL29 9NL
G0 SKD T. Ward 4 Burrows Grove, Wombwell, Barnsley S73 8PS
G0 SKI M. Foy 335 South Avenue, Southend-on-Sea SS2 4HR
G0 SKJ K. Cockburn 11 Highlands Avenue, Barrow-in-Furness LA13 0AU
G0 SKK D. Chadwick 386 Tamworth Road, Amington, Tamworth B77 4AQ
G0 SKM M. Tunstall 24 Barbrook Avenue, Stoke-on-Trent ST3 5UG
G0 SKN P. Hartley 9 Weston Road, Wimborne BH21 2SF
G0 SKQ C. Haines 29 Woodlands Close, Aston, Stone ST15 0DX
G0 SKR J. Goodall Red Roofs, 4 Chapel Street, Stapleton LE9 8JH
G0 SKW K. Walker 20 Thornhill Close, Bramcote, Nottingham NG9 3FS
G0 SLB C.Mattison Mr. 14A Buckingham Drive, Colchester CO4 3YH
GW0 SLC R. Thomas 6 Grovers Close, Glyncoch, Pontypridd CF37 3DF
G0 SLD P. Westripp 2 Ridgeway, Horns Road, Cranbrook TN18 4RA
G0 SLH C. Shoesmith 2 Caravelle Gardens, Northolt UB5 6EU
G0 SLI T. Day 21 Mowbray Road, Ham, Richmond TW10 7NQ
G0 SLJ Dr D. Pepper 17 Cliffe House, Radnor Cliff, Folkestone CT20 2TY
G0 SLK E. Patterson 45 Sandhurst Road, Rainhill, Prescot L35 8NE
G0 SLL R. Petrie Royal Hospital Chelsea, Royal Hospital Road, London SW3 4SR
G0 SLN P. Grainger 11 Smith Grove, Ryhope., Sunderland SR2 0JU
G0 SLP M. Coultas 35 Monteigne Drive, Bowburn, Durham DH6 5QB
G0 SLQ S. Quinn 48 Aldsworth Close, Springwell Village, Gateshead NE9 7PG
G0 SLR R. Lisle 21 Porlock Close, Penketh, Warrington WA5 2QE
G0 SLU C. Barr 17 Knighton Road, Otford, Sevenoaks TN14 5LD
G0 SLW J. Waite 28 Overdown Rise, Portslade, Brighton BN41 2YG
G0 SLY C. Kratzer 9900 Dale Ridge Ct, Vienna VA 22181 United States
G0 SLZ A. Matthews 28 Sherwin Road, Stapleford, Nottingham NG9 8PQ
G0 SMH B. Marchant 20 Wrench Road, Norwich NR5 8AS
G0 SMJ M. Jackson Sunny Cot, 44 Dulwich Road, Holland-On-Sea, Clacton-on-Sea CO15 5NA
G0 SMM J. O'Nion 7 Ettington Close, Cheltenham GL51 0NY
G0 SMN A. Mckenzie 311 Weston Road, Weston Coyney, Stoke-on-Trent ST3 6HA
G0 SMO C. Cash 27 Robert Wynd, Bilston WV14 9SE
G0 SMP S. Pountain 21 Hayfield Road, Chapel-En-Le-Frith, High Peak SK23 0JF
G0 SMR J. Ballard 7 Chapelcroft Court, Liverpool L12 9GY
G0 SMS D. Barnham 35 Post Office Road, Frettenham, Norwich NR12 7AB
GI0 SMU A. Hanna 39 Dalton Crescent, Comber, Newtownards BT23 5HE
G0 SMZ R. Clews 99 Kilbury Drive, , Worcester WR5 2NG
G0 SNB W. Bonser 24 Meend Garden Terrace, Cinderford GL14 2EB
G0 SNF J. Culling 4 Ash Road, Princes Risborough HP27 0BQ
G0 SNG D. Hill 3 Morcar Road, Stamford Bridge, York YO41 1PR
G0 SNK A. Gill Bradgate, Kings Lane, Lymington SO41 6BQ
G0 SNM K. Killick 15 Popplechurch Drive, Swindon SN3 5DE
G0 SNO D. Lauder 20 Sutherland Close, Barnet EN5 2JL
G0 SNP J. Du Heaume 10 Water Lane, Pill, Bristol BS20 0EQ
G0 SNQ Surrey Raynet c/o H. Davis 44 Kenyon Street, Ashton under Lyne OL6 7DU
G0 SNS B. Harrison 8 Elm Park, Pontefract WF8 4LG
G0 SNU I. Gray 27 Meadow Close, Lavenham, Sudbury CO10 9RU
G0 SNV J. Worsnop 1217 Thornton Road, Thornton, Bradford BD13 3BE
G0 SNW P. Rogers 126 Bradford Road, Otley LS21 3LE
G0 SNX N. Johnson 12 Bleach Mill Lane, Menston, Ilkley LS29 6HE
G0 SNZ A. Flood 3 Ongar Walk, Blackley, Manchester M9 8JD
G0 SOA Statford-Upon-Avon & District RS c/o C. Ousbey 30 Hawthorn Way, Shipston-on-Stour CV36 4FD
G0 SOF G. Hedley 260E 100 Sts, Raymond, Alberta TOK 250 Canada
G0 SOG F. Mellings 4 Kiln Lane, Horley RH6 8JG
G0 SOK P. Mayer 248 Dimsdale Parade West, Newcastle ST5 8EA
G0 SON R. Peard 28 New Road, Shoreham-by-Sea BN43 6RA
G0 SOU J. Pendleton 17A Langley Drive, Kegworth, Derby DE74 2DN
G0 SOX P. Chapman 4 Churchill Close, Brightlingsea, Colchester CO7 0RS
G0 SOY A. Croft 34 Fourth Avenue, Wolverhampton WV10 9LZ
G0 SPA P. Benson 7 Crofton Close Attenborough, Nottingham NG9 5HX
G0 SPB G. Rusby 12 Park Meadow, Princes Risborough HP27 0EB
G0 SPC M. Chawner Timbertops, Churchfield Lane, Benson, Wallingford OX10 6SH
G0 SPF T. Hill 7 Broadlands Road, Paignton TQ4 5NY
G0 SPH K. Brooks 25 Lagos Grove, Winsford CW7 2BJ
G0 SPK D. Neal 490 Aureole Walk, Newmarket CB8 7BQ
G0 SPQ I. Wilson 3 Caring Lane, Bearsted, Maidstone ME14 4NJ
G0 SPS M. Forder 157 Kennington Road, Kennington, Oxford OX1 5PE
G0 SPX J. Sparks 34 Green Park Avenue, Skircoat Green, Halifax HX3 0SR
GW0 SPY J. Thorley 8 Bryn Gwyn, Abergele LL22 8JA
G0 SPZ M. Rickard-Worth 2 Harwood Road, Littlehampton BN17 7AT
G0 SQE E. Young 14 Hanover Court, Gateshead NE9 6TZ
G0 SQF J. Bubez 4 Southway, Burgess Hill RH15 9ST
G0 SQH D. Higbee 12 Shelley Close, Ashley Heath, Ringwood BH24 2JA
G0 SQI N. Blythe 14 The Green, South Creake, Fakenham NR21 9PD
G0 SQK M. Stuckey 212 Roughton Road, Cromer NR27 9LQ
G0 SQL R. Bishop 73 Broomgrove Gardens, Edgware HA8 5RJ
G0 SQP F. Ott 16 Hornbeam Close, Chelmsford CM2 9LW
G0 SQS M. Hewitt 1 Harpswell Hill Park, Hemswell, Gainsborough DN21 5UT
GW0 SQT A. Davis 8 Cook Road, Barry CF62 9HD
G0 SQX T. Donley 21 Elmridge, Leigh WN7 1HN
GW0 SQY S. Morgan Oakfield House, Barry Road, Pontypridd CF37 1HY
GW0 SRE D. Price 1 Rhas Cottages, Pontyates, Llanelli SA15 5SF
GW0 SRF D. Daniels 73 Bethania Road, Upper Tumble, Llanelli SA14 6DT
G0 SRG Sunderland Raynet Group c/o S. Green 133 Sevenoaks Drive, Sunderland SR4 9NQ
GI0 SRL A. Harbison 26 Ballymartin Road, Templepatrick, Ballyclare BT39 0BW
GM0 SRO I. Maclean 7 Thirlestane Crescent, Lauder TD2 6TT
GI0 SRP N. Averill 3 Edmund Court, Tobermore, Magherafelt BT45 5QA
GM0 SRQ D. Beacher 17 Cairn Grove, Crossford, Dunfermline KY12 8YD
G0 SRR M. Baugh 97 Wilson Avenue, Deal CT14 9NJ
G0 SRY J. Smart 60 Blaze Park, Wall Heath, Kingswinford DY6 0LN
G0 SRZ E. Hart 47 Northfield Crescent, Wells-next-the-Sea NR23 1LR
GI0 SSA J. Stevenson 27 Kinnegar Rocks, Donaghadee BT21 0EZ

G0 SSC G. Mills 49 Priestley Close, Doncaster DN4 9DQ
G0 SSE N. Morton Cami Terrapico 54, Roquetes 43520 Spain
G0 SSG R. Andre 54 Covertside, Wirral CH48 9UL
G0 SSJ A. Powell 61 Albert Road, Grappenhall, Warrington WA4 2PF
G0 SSK G. Johnson 503 Holden Road, Leigh WN7 2JJ
G0 SSL A. Mcevoy 12 Fountains Avenue, Haydock, St. Helens WA11 0RS
G0 SSN C. Ireland 14 Castlefields, Istead Rise, Gravesend DA13 9EJ
GM0 SSQ A. Winchester 23 Craigmount, Avenue North, Edinburgh EH12 8DL
G0 SSV G. Kendall 3 Brayton Avenue, Sale M33 5HF
G0 SSX P. Ellis 4 Rostwold Way, Norwich NR3 3NN
G0 SSY D. Webb 3 Cams Hill Lane, Hambledon, Waterlooville PO7 4SP
G0 SSZ Sandringham School ARC c/o R. Stanley 219 Fartown, Pudsey LS28 8NH
GM0 STB Scottish Tourist Board Radio c/o R. Aitkenhead 11 Elm Court, Quarter, Hamilton ML3 7FB
GI0 STC P. Dellett 14 Fox Park, Omagh BT79 0JX
G0 STF T. Clements 29 Bonchurch Drive, Wavertree, Liverpool L15 4PW
G0 STH Last Post Museum c/o R. Vaughan 6 Dellside Grove, St. Helens WA9 5AR
G0 STK A. Hardcastle 25 Aire Crescent, Cross Hills, Keighley BD20 7RW
GI0 STM K. Murray 17 Glebe Court, Dungannon BT70 3PU
G0 STR W. Shaw 161 Springwood Crescent, Edgware HA8 8SH
GI0 STS R. Todd 73 Lakeview Park, Drumgor, Craigavon BT65 4AL
G0 STW C. Kendrick 18 Ainger Road, Upper Dovercourt, Harwich CO12 4TS
G0 SUA A. Edwards 49 Griggs Meadow, Dunsfold, Godalming GU8 4ND
G0 SUB G. Thomas 83 Hollingthorpe Road, Hall Green, Wakefield WF4 3NW
G0 SUC B. Smith 89 Farndale Drive, Guisborough TS14 8JX
GM0 SUF H. Butcher 14 Newbattle Road, Dalkeith EH22 3DB
GM0 SUH J. Montgomery Woods 2, Tralee Bay Holidays, Benderloch, Oban PA37 1QR
G0 SUI G. Scarlett 10 Laythorpe Terrace, East Morton, Keighley BD20 5TL
G0 SUL J. Ward 8 Wilderness Road, Guildford GU2 7QN
GU0 SUP P. Cooper 1 Clos Au Pre, La Route De La Hougue Du Pommier, Castel, Guernsey GY5 7FQ
G0 SUQ I. Johnson 181 Broad Street, Bromsgrove B61 8NQ
G0 SUT Finningley ARS c/o J. Burdett 1 Main Street, Egginton, Derby DE65 6HL
G0 SUU J. Ogier 37 Hill Park Road, Torquay TQ1 4LD
GM0 SUY C. Auld 148 Echline Drive, South Queensferry EH30 9XG
G0 SVA D. Nock 431 Locking Road, Weston-Super-Mare BS22 8QN
G0 SVB P. Herrmann 5992 Royal Court, Lockport United States
G0 SVH E. Wright 26 Walmsley Close Church, Accrington BB5 4HL
G0 SVJ M. Creswick 5 Wheatlands Drive, Easington, Saltburn-by-the-Sea TS13 4PB
G0 SVK J. Hosfield 29 Whitecroft, Gosforth, Seascale CA20 1AY
G0 SVN N. Savin 138A Lillibrooke Crescent, Maidenhead SL6 3XH
G0 SVP G. Broadhurst 9 Sharples Street, Accrington BB5 0HQ
G0 SVQ A. Haydock 8 Corbridge Close, Blackpool FY4 5EZ
GM0 SVS Dr M. Whiteley 9 Pathfoot Avenue, Bridge Of Allan, Stirling FK9 4SA
G0 SVU R. Morris 57 Marten Drive, Huddersfield HD4 7JX
G0 SVX I. Roebuck 3 Lodge Hill Drive, Kiveton Park, Sheffield S26 5RU
G0 SVY D. Beardsley 1 Amber Villas, Sutton St. Nicholas, Hereford HR1 3DF
G0 SVZ M. Cooper 20 Bankfield Road, Widnes WA8 7UW
G0 SWB R. Atkinson 57 Jobling Avenue, Blaydon-on-Tyne NE21 4RR
G0 SWC R. Eeles 50 Nightingale Road, Guildford GU1 1EP
G0 SWE S. Whitbourn 50 Nightingale Road, Guildford GU1 1EP
G0 SWF J. Durrant 16 Bugdens Lane, Verwood BH31 6EY
G0 SWH B. Fletcher 58 Broomfield Avenue, Worthing BN14 7SB
G0 SWL C. Niles 23 Randsfield Avenue, Brough HU15 1BE
G0 SWN Gilwell Park Scout RC c/o S. Barber Homedale, St. Monicas Road, Tadworth KT20 6ET
G0 SWO B. Atkinson 165 Alliance Avenue, Hull HU3 6QY
G0 SWS T. Stow 38 The Strand, Mablethorpe LN12 1BQ
G0 SWU P. Broad 78 Lenham Road, Sutton SM1 4BG
G0 SWW M. Stevens Autumn Cottage, Silver Street, Horncastle LN9 5NH
G0 SWY M. Humphrey 5 Ventnor Court, Southampton SO16 3EB
G0 SXA W. Daly 85 Lordens Road, Huyton, Liverpool L14 9PA
GW0 SXE J. Williams 11 Courbet Drive, Connah'S Quay, Deeside CH5 4WP
G0 SXG K. Stammers 102 Eaton Road, Appleton, Abingdon OX13 5JJ
G0 SXK A. Dodd 20 Braemar Avenue, Chelmsford CM2 9PW
G0 SXM D. Smith River Meadow, Harlyn Bay, Padstow PL28 8SB
G0 SXN B. Watts 9 Weavers Walk, Cullompton EX15 1SS
GM0 SXO A. Segar 36 Kestrel Avenue, Dunfermline KY11 8JL
GM0 SXP Dr R. Cliff 32 Lochardil Road, Inverness IV2 4LD
GM0 SXQ M. Hepburn The Toll House, Corse, Lumphanan, Banchory AB31 4RY
GW0 SXS K. Wheeler The Glen, Dreenhill, Haverfordwest SA62 3XH
G0 SXU C. Bocock 20 Court Avenue, Stoke Gifford, Bristol BS34 8PJ
G0 SXW P. Cressey Skidby Hill Farm, Beverley Road, Cottingham HU16 5TF
G0 SXY P. Davies 46 Wagstaff Way, Olney MK46 5FB
GW0 SXZ J. Green 23 Litchard Park, Bridgend CF31 1PF
G0 SYF R. Pearce 14 Shepherds Leaze, Wotton-under-Edge GL12 7LQ
G0 SYI K. Treasure 30 Grace Park Road, Brislington, Bristol BS4 5JA
G0 SYP C. Steinhoefel 222 Stretford Road, Urmston, Manchester M41 9NT
G0 SYQ M. Wood 4 Gordon Road, Hastings TN34 3JN
G0 SYR B. Petifer 14 Wood Lane, Caterham CR3 5RT
G0 SYS D. Hall 72 Mansfield Road, Edwinstowe, Mansfield NG21 9NH
G0 SYT D. Firks Bryn Garth Cottage, Hereford HR2 8HJ
GM0 SYV J. Kelly 2 Ryeland Street, Strathaven ML10 6DL
GM0 SZA I. Stones 8 Cloverfield Place, Bucksburn, Aberdeen AB21 9RH
G0 SZE C. Andrews 33 Blackheath Road, Barnsley S71 3RH
G0 SZG J. Greene 308 Cedar Road, Nuneaton CV10 9DY
GI0 SZH I. Mcewen 234 Legahory Court, Legahory, Craigavon BT65 5DH
G0 SZI D. Green 81A Long Lane, Holbury, Southampton SO45 2ND
G0 SZJ S. Fletcher Apartado 39, Ourique 7670 Portugal
G0 SZK L. Joyce 106A Victoria Drive, Bognor Regis PO21 2EJ
GW0 SZN M. Lawrence 1 Greenwood Cottages, Gelligroes, Blackwood NP12 2JB
G0 SZO J. Everard Woodside, Staple Lane, Taunton TA4 4DE
G0 SZT D. Allibone Virginia, North Street, Langport TA10 9RH
GW0 SZU B. Osborne 163 Park Street, Bridgend CF31 4BB
G0 TAA W. Chandler 38 Falkland Road, Chandler'S Ford, Eastleigh SO53 3GD
GM0 TAE C. Porter 20 Baird Avenue, Kilwinning KA13 7AR
G0 TAG D. Beane 12 Strafford Court, Pondcroft Road, Knebworth SG3 6DF
G0 TAH K. Aldus 77 Springvale Road, Kings Worthy, Winchester SO23 7ND
G0 TAI I. Hawkins 1 Millhayes, Great Linford, Milton Keynes MK14 5EP
G0 TAK R. Walker 35 Romany Close, Letchworth Garden City SG6 4LA
G0 TAL S. Walsworth 4 The Homestead, Heckmondwike WF16 9JL
G0 TAM A. Farrow 18 The Green Trimingham, Norwich NR11 8ED
G0 TAN S. Venner 36 Broadwood Avenue, Ruislip HA4 7XR
G0 TAO R. Lokuge 11 Porchester Close, Southwater, Horsham RH13 9XR
G0 TAR B. Lucas 8 Gilbert Close, Hempstead, Gillingham ME7 3QQ
G0 TAS J. Taylor 19 Castle Close, Leconfield, Beverley HU17 7NX
G0 TAT M. Wilmot Southview, Roman Road, Weston-Super-Mare BS24 0AB
G0 TAX T. Welch 63 Vicarage Close, New Silksworth, Sunderland SR3 1JF
GM0 TAY O.B.O. Tayside Raynet c/o I. Strachan 238 Coupar Angus Road, Muirhead, Dundee DD2 5QN
G0 TAZ J. Burgess Combeside House, Symonsburrow, Cullompton EX15 3XA
G0 TBC S. Lawson 27 Broadlands Avenue, Eastleigh SO50 4PP
GM0 TBH J. Mcmaster 96 Cunningham Crescent, Ayr KA7 3JB
G0 TBI S. Mckinnon 145 Enville Road, Kinver, Stourbridge DY7 6BN
GW0 TBM J. Goulden Wenffrwd Cottage, Llangollen Road, Llangollen LL20 7UH
G0 TBO P. Clark 30 Alicia Avenue, Garlinge, Margate CT9 5JZ
G0 TBS W. Lucas 67 Tower Ride, Uckfield TN22 1NU
G0 TBU R. Brightwell 40 Streete Court Road, Westgate-on-Sea CT8 8BX
G0 TBW G. Davis 32 Medlock Close, Farnworth, Bolton BL4 9QW
G0 TCA R. Seaward 13 Blythe Close, Catford, London SE6 4UW
GM0 TCC K. Walker 174 South Seton Park, Port Seton, Prestonpans EH32 0BP
G0 TCE Dr P. Taylor 67 Rectory Park, South Croydon CR2 9JR
G0 TCF P. Loch 400 Loughborough Road, West Bridgford, Nottingham NG2 7FD
G0 TCH T. Noszkay 41 Brue Close, Weston-Super-Mare BS23 3BX
G0 TCI B. Hughes 12 Newcroft, Saffron Walden CB10 1LN
G0 TCJ T. Worrall 9 Barnstaple Close, Wigston LE18 2QX
GW0 TCL D. Parsons Aston Hall Res.Home, Lower Aston Hall Lane, Deeside CH5 3EX
G0 TCO M. Roper 1 The Cottages, Norwich Road, Norwich NR9 5BY
G0 TCP P. Maynard Seaton House, Lower Road, Westerfield, Ipswich IP6 9AR
G0 TCQ A. Cudlip The Oaks, 16 Bilberry Close, Southampton SO31 6XX
GM0 TCU C. Pirie 10 Annesley Park, Torphins, Banchory AB31 4HG
GW0 TCV A. Thomas 4 Gordon Terrace, King Street, Mold CH7 1LD
G0 TCW A. Furmston 46 Twydall Lane, Gillingham ME8 6JE
G0 TCY M. Templeman 44 Wisbeck Road, Bolton BL2 2TA
GW0 TDA P. Price 1 Brynderi, Pontyates, Llanelli SA15 5SU
G0 TDC S. Tinsley 48 Smithy Lane, Croft, Warrington WA3 7JG
G0 TDE R. Barrett Brookfield, Hobbacott Lane, Bude EX23 0ES
G0 TDG N. Hotson 1 Holly Court 24 Hastings Road, London E16 1GJ
G0 TDJ S. Smith Westgarth Flat 4, 145-146 Marina, St.Leonards-on-Sea TN38 0BT
G0 TDM J. Sutton 15 Lowther Street, Penrith CA11 7UW
G0 TDN S. Maclennan Shalam, Rising Sun, Callington PL17 8JE
GI0 TDP J. Driscoll 67 Whinney Hill, Holywood BT18 0HG
G0 TDQ P. Luscombe 33 Rea Barn Road, Brixham TQ5 9ED
G0 TDR T. Round 92 Church View Gardens, Kinver, Stourbridge DY7 6EE
G0 TDV R. King 10 Lansdown Road, Kingswood, Bristol BS15 1XB
G0 TDX Y. Kimoto 523 Yukinaga, Maizuru-City, Kyoto 625-0052 Japan
GM0 TEA A. Cherry 39 Clark Road, Edinburgh EH5 3AR
G0 TEB C. Sexton Lapswater, Marsh, Honiton EX14 9AL
G0 TED E. Thane 19 Churchill Road, Walton, Stone ST15 0EB
G0 TEE T. Speight 1 Lyndene Avenue, Worsley, Manchester M28 2RJ
G0 TEI R. Sparks Radio Licence Centre, Po Box 885, Bristol BS99 5LG
G0 TEL R. Bellenot 22 Roderick Avenue, Peacehaven BN10 8JT
G0 TEM A. Merrix 53 Pear Tree Road, Great Barr, Birmingham B43 6HX
G0 TEO L. Baker Flat 6, Francis Snary Lodge 12 Chesterton Close, London SW18 1SD
GD0 TEP A. Kissack 30 High View Road, Douglas IM2 5BH Isle of Man
G0 TES R. Dicks 10 Westfield Avenue, Raunds, Wellingborough NN9 6DQ
G0 TFB Widnes & Runcorn ARC c/o L. Wallen Lambda Works, 45A Whitehall Road, Ramsgate CT12 6DE
G0 TFC A. Stanley 145 Tribune Drive, Houghton, Carlisle CA3 0LF
G0 TFD D. Eyre 29 Old Acre Lane, Brocton, Stafford ST17 0TW
GM0 TFE C. Stewart 185 Newbattle Abbey Crescent, Dalkeith EH22 3LT
GM0 TFF A. Mcghie 16 Boyach Crescent, Isle Of Whithorn, Newton Stewart DG8 8LD
GD0 TFG J. Dowling 2 Ballabridson Park, Ballasalla IM9 2ES Isle of Man
G0 TFI D. Roberts 9 The Pound, Westoning, Bedford MK45 5JN
G0 TFK J. Sutcliffe 4 Lancaster Gate, Nelson BB9 0AP
G0 TFL J. Williams 107 Clay Lane, Rochdale OL11 5QW
GD0 TFO J. Bellis Jandakot, Old Castletown Road Port Soderick, Isle of Man IM4 1BB
G0 TFP J. Brett 11 Manor Road, Astley, Manchester M29 7PH
GM0 TFQ H. Wignall 7 Windyedge, Inverurie AB51 3WJ
G0 TFR A. Vining The Cedars, Thorney Road, Peterborough PE6 0LH
G0 TFT A. Gibson 28 Finchale Terrace, Jarrow NE32 3TX
G0 TFU M. Wilson 6 The Chase, Calcot, Reading RG31 7DN
G0 TFV T. Carvell 18 Park View, School Lane, Rye TN31 6UR
G0 TFX S. Roberts 20 Beech Grove, Trowbridge BA14 0HG
G0 TGB T. Blore Glendhoon, Laneham Street, Retford DN22 0JX
GM0 TGE I. Ross Idlewildeá, Inverurie AB51 0XA
GM0 TGG L. Mackenzie 90 Tay St., Newport on Tay DD6 8AP
G0 TGH S. Leak 12 Kentmere Approach, Leeds LS14 1JP
G0 TGM S. Fowler 38 Meadowcroft, Aylesbury HP19 9LN
G0 TGP W. Ford Flat 2, Hillyard Court, Wareham BH20 4QX
G0 TGQ O. Cubitt 97 Sutton Lane, Langley, Slough SL3 8AU
G0 TGR D. Greenacre 38 Toon Crescent, Bury BL8 1JB
G0 TGU W. Collier 20 Wainers Croft, Greenleys, Milton Keynes MK12 6AL
G0 TGX P. Ireland 2 The Ridings, Hull HU5 5HW
G0 THD P. Hart 39 Barley Drive, Burgess Hill RH15 9XG
G0 THF K. Greatorex 54 Lilac Grove, Glapwell, Chesterfield S44 5NG
G0 THH H. Hudders 7 Cedar Crescent, Thame OX9 2AX
G0 THI W. Davey 15 Park Avenue, Histon, Cambridge CB24 9JU
G0 THJ J. Stowell 6 Westage Lane, Great Budworth, Nr Northwich CW9 6HJ
GI0 THO W. Edmondson 7 Rathcavan Drive, Ballymena BT42 2QH
G0 THQ R. King 6 Mayon Green Crescent, Sennen, Penzance TR19 7BS
GI0 THR J. Mcdonald 54 Bettys Hill Road, Newry BT34 2ND
G0 THS S. Graham 25 South End Road, Ottringham, Hull HU12 0DP
G0 THV J. Coote 8 St. Francis Chase, Bexhill-on-Sea TN39 4HZ
G0 THW A. Brentnall Sandy Rise, Mill Lane, Woodbridge IP12 3LL
G0 THY M. Preston 15 Poplar Close, Kidlington OX5 1HH

GO TIA P. Lodge 79, Via Circonvallazione, 79, Milano 20090 Italy
GO TID C. Alexander 25 Diamedes Avenue, Stanwell, Staines-upon-Thames TW19 7JE
GO TIG H. Janes 91 Thorpe Bay Gardens, Southend-on-Sea SS1 3NW
GO TII S. Graham 4 Oakland Avenue, Ellenborough, Maryport CA15 7BU
GO TIJ D. Page 10 Davenport Road, Yarm TS15 9TN
GO TIL J. Parmenter 48 Honey Way, Royston SG8 7EU
GO TIP G. Thorne 19 Lapwing Lane, Brinnington, Stockport SK5 8JY
GO TIS P. Jones 61 North Road, Bourne PE10 9AU
GO TIW T. Parker The Bungalow, 178 Green End Lane, Hemel Hempstead HP1 2BQ
GO TIX P. Wilson 2 Staley Close, Stalybridge SK15 3HJ
GO TIZ E. Tometzki 11 Southey Close, Enderby, Leicester LE19 4QZ
GO TJC L. Taylor 35 Leafield Road, Darlington DL1 5DF
GO TJD A. Wedgwood 10 Milner Place, London N1 1TN
GO TJE S. Sullivan 7 Gosfield Road, Dagenham RM8 1JY
GO TJG M. Spafford Old School House, 11 Old School Lane, Chesterfield S44 5UE
GO TJH I. Foord 25 The Rose Walk, Newhaven BN9 9NH
GO TJI G. Woods 126 Luddenham Close, Ashford TN23 5SA
GO TJN T. Newland 80 Burnway, Hornchurch RM11 3SG
GO TJP A. Person 78 Green Street, Ston Easton, Radstock BA3 4BZ
GO TJQ C. Dowell 19 Field Top, Bailiff Bridge, Brighouse HD6 4EQ
GO TJR L. Ellams 131 Broadway, Dunscroft, Doncaster DN7 4HB
GO TJT R. Francis 661 Osmaston Road, Derby DE24 8NF
GI0 TJV A. Gibson 58 Cairnmore Park, Lisburn BT28 2DN
GO TJY Capt. T. Herd The Old Dairy, High Street, Ipswich IP8 3AP
GMO TKB L. Thomas Greengeo, Scarfskerry, Thurso KW14 8XN
GMO TKC T. Maxwell 28 Cedar Grove, Dunfermline KY11 8BH
GMO TKE J. Mcluckie 118 Lady Nairn Avenue, Kirkcaldy KY1 2AT
GO TKF W. Stuart 3 Rookery Vale, Deepcar, Sheffield S36 2NP
GO TKG D. Mawson 14 Windermere Close, Worksop S81 7QE
GO TKJ T. Clayton 40 Morrison Road, Darfield, Barnsley S73 9ED
GO TKL E. Roberts 43 Ashbourne Crescent, Sale M33 3LQ
GO TKR E. Martin 20 Easters Grove, Stoke-on-Trent ST2 7PF
GWO TKX A. Mason 101 Aneurin Bevan Avenue, Gelligaer, Hengoed CF82 8ET
GO TKZ R. Hamblin 5 Streatfield, Edenbridge TN8 5DF
GO TLA R. Lythall 6 Belmont Crescent, Little Houghton, Barnsley S72 0HT
GO TLI C. Vince 8 Kent Road, Swindon SN1 3NJ
GWO TLJ A. Mathias 75 Coombs Drive, Milford Haven SA73 2NU
GO TLN G. Sim 24 Fernmoor Drive, Irthlingborough, Wellingborough NN9 5TL
GO TLP A. Whitwam 9 Oak Apple Close, Stourport on Severn DY13 0JR
GO TLQ R. Parsons 10 Waterside, Isleham, Ely CB7 5SH
GWO TLS R. Upton 17 Bryn Gannock Deganwy, Conwy LL31 9UG
GO TLT D. Davies 14 Hammerwater Drive, Warsop, Mansfield NG20 0DJ
GO TLU P. Thompson Flat 4, 14 West End Way, Lancing BN15 8RL
GO TLZ J. Trefry 67 Axminster Close, Cramlington NE23 2UE
GO TMA R. Griffiths Flat 1, Buckhurst Court, 29 Buckhurst Road, Bexhill-on-Sea TN40 1QE
GO TME B. Park 69 Lea Villa Residential Park, Lea, Ross-on-Wye HR9 7GP
GO TMF M. Fleetwood 9 Reynolds Close, Swindon, Dudley DY3 4NQ
GO TMH C. Neary 3 Wordsworth Close, Torquay TQ2 6EA
GO TMJ E. Jones 1 Ivel View, Sandy SG19 1AU
GO TMK W. Howell 41 Chestnut Avenue, Shavington, Crewe CW2 5BJ
GO TML A. Rowley 32 Spring Lane, Flore, Northampton NN7 4LS
GI0 TMS M. Smyth 41 Coolaghy Road, Newtownstewart, Omagh BT78 4LG
GO TMT M. Tuttle 7 Mill Lane Horsford, Norwich NR10 3ES
GWO TMU D. Edwards 68 Heol Y Meinciau, Pontyates, Llanelli SA15 5RT
GWO TMV T. Vlismas Maes Yr Awel, Newport Rd, Crymych SA41 3RR
GO TMW G. Boswell 7 Chestnut Avenue, Wootton, Northampton NN4 6LA
GO TMZ N. Lilley 18 Beechwood Avenue, New Milton BH25 5NB
GO TNC G. Stephenson 54 Rock Road, Sittingbourne ME10 1JF
GO TNF M. Wills Chapel View, Cliburn, Penrith CA10 3AL
GO TNG P. Hayward 63 Devereaux Crescent, Ebley, Stroud GL5 4PX
GO TNH L. Crow 181 Foxlydiate Crescent, Batchley Estate, Redditch B97 6NS
GMO TNK D. Mackenzie 32 Miller Gardens, Inverness IV2 3DT
GO TNL A. Beattie Pine Croft, Blitterlees, Wigton CA7 4JJ
GO TNM R. Guppy 12 Highfield Road, Caterham CR3 6QX
GO TNO C. Billington 5 Lamers Road, Luton LU2 9BL
GO TNP P. Unstead 9 Buckingham Road, Swindon SN3 1HZ
GO TNQ A. Lewis 8 Newts Way, St. Leonards-on-Sea TN38 9TH
GO TNS H. Hauton 8 St. Catherines Crescent, Scunthorpe DN16 3LQ
GO TNY A. Rose Heronsgate, River Gardens, Maidenhead SL6 2BJ
GO TOB J. Horsfall 10 Derwent Close, Hebden Bridge HX7 7ED
GO TOC M. Litchman 26 Oak Tree Close, Loughton IG10 2RE
GO TOD T. Northover 13 Dagenham Avenue, Dagenham RM9 6LD
GO TOE A. Gallagher 1A Wynsome Street, Southwick, Trowbridge BA14 9RB
GMO TOF M. Farnworth 30E Roseangle, Dundee DD1 4LY
GWO TOI A. Lipian 10 Field Street Trelewis, Treharris, Treharris CF46 6AW
GO TOK B. Dixon 97 Sunny Blunts, Peterlee SR8 1LN
GWO TOM T. Beedle 2 Chestnut Grove, Maesteg CF34 0NT
GO TOO C. Richmond 11 Harewood Avenue, Morecambe LA3 1JH
GO TOQ S. Harrison 7 Harvey Street, Kingstone, Barnsley S70 6JT
GO TOS B. Harper 51 Cross Lane, Scarborough YO12 6DQ
GO TOT R. Claxton Purbeck View, 2 Hoburne Road, Swanage BH19 2SL
GO TOX C. Wheeler 190 Mount Pleasant, Redditch B97 4JL
GO TOY K. Li 16 Garthland Drive, Arkley, Barnet EN5 3BB
GO TPA A. Taylor The Grey House, Chipping Campden GL55 6XP
GO TPB B. Smith 34 Wheatlands, Fareham PO14 4SL
GO TPD J. Gayther 33 Greenways, Winchcombe, Cheltenham GL54 5LQ
GO TPE A. Davis 320 Preston Old Road, Blackburn BB2 2TX
GWO TPF N. Pugh 28 Hill View Road, Llanrhos, Llandudno LL30 1SL
GO TPG B. Taylor 3 Stonepits Lane, Hunt End, Redditch B97 5LX
GO TPH A. Horne 54 Manor Road, Desford, Leicester LE9 9JR
GMO TPI D. Harris 42 Shira Terrace, East Kilbride, Glasgow G74 2HU
GWO TPL D. Blundell 30 Heol Ffynnon Wen, Cardiff CF14 7TP
GO TPM P. Mercer 10 Holmcliffe Avenue, Huddersfield HD4 7RJ
GO TPN M. Padgett 97 Larks Hill, Pontefract WF8 4RP
GO TPO M. Cook 11 Atherton Close, Shurdington, Cheltenham GL51 4SB
GO TPP P. Pimblett 5 Edgeside, Great Harwood, Blackburn BB6 7JS
GWO TPR B. Carter Rhyd Y Mwyn, Cilgwyn Street, Llanerchymedd LL71 8ED
GO TPY J. Brunt 1 Dane Grove, Cheadle, Stoke on Trent ST10 1QS
GMO TQB M. De Vries Old Post Office House Knockbain, Munlochy IV8 8PG
GO TQC K. Sharman 7 Watkins Way, Paignton TQ3 3JJ
GI0 TQD J. Gough 50 Culmore Point, Londonderry BT48 8JW
GO TQJ C. Vernon 13 Greenlands Close, Tarporley CW6 0DA
GMO TQKá K. Traill 31 Sherwood Place, Bonnyrigg EH19 3JY
GO TQP J. Smith High Tree, Radford Lane, Wolverhampton WV3 8JT
GO TQR A. Baker 23 Trematon Drive, Ivybridge PL21 0HT
GO TQS C. Forsyth 8 Oriole Drive, Exeter EX4 4SJ
GO TQT J. Joll 16 Jephson Road, St. Judes, Plymouth PL4 9ET
GO TQV K. Taylor 1 Chapel Close, Reepham, Lincoln LN3 4EJ
GO TQZ M. Emm Highwood Cottage, Daggons Road, Fordingbridge SP6 3DJ
GO TRB R. Betts 15 Cleasby, Wilnecote, Tamworth B77 4JL
GO TRD T. Thorman Tir Na Nog, Coombe Ridings, Kingston upon Thames KT2 7JT
GO TRE N. Dimbleby 4 Rossetti Place, Holmer Green, High Wycombe HP15 6XA
GO TRG Thames ARG c/o A. Atkinson 21 Dennington Crescent, Basildon SS14 2FF
GO TRH D. Brooke 19 Albion Court, Anlaby Common, Hull HU4 7PL
GO TRI L. Pritchard The Granary, Greenways Farm, Ross-on-Wye HR9 6DH
GO TRJ R. Makepeace 8 Lethlean Close, Phillack, Hayle TR27 5AN
GO TRK J. Ogden 65 Elm St., Middleton, Manchester M24 2EQ
GO TRM C. Page 1 The Leeway Danbury, Chelmsford CM3 4PS
GO TRN J. De Frece 15 Kent Close, Highfield Road, Chesterfield S41 7HA
GO TRU A. Irvine 21 Rutland Road, Partington, Manchester M31 4NP
GO TRW R. Cross 7 The Island, Anthorn, Wigton CA7 5AN
GO TRY G. Lyon 33 Barlborough Road, Pemberton, Wigan WN5 9HZ
GI0 TSA D. Moore 3 Knightsbridge, Londonderry BT47 6FE
GO TSB B. Snell 30 Queens Crescent, Brixham TQ5 9PJ
GWO TSE L. Owen Cartref, Llangain, Carmarthen SA33 5AH
GO TSG P. Ryder 4 Edgeway, Nottingham NG8 6LY
GO TSH K. Hutt Fenwick Crossing House, Fenwick Lane, Fenwick, Doncaster DN6 0EZ
GO TSJ S. Ruud 5 Wood Street, Haworth, Keighley BD22 8BJ
GO TSK G. Wilkins 156 Buckingham Crescent, Bicester OX26 4HB
GWO TSL H. Chapman Flat 6, Archer Court, Phyllis Street, Barry CF62 5US
GO TSQ C. De Lacy Monday Cottage, Hammerwood, East Grinstead RH19 3QE
GO TSR Dr N. Depledge 29 Scargill Drive, Spennymoor DL16 6LY
GI0 TSS C. Tait 116 Aughnaskeagh Road, Dromara BT25 2NT
GO TST R. Hawtree 9 Stonechat Road, Billericay CM11 2NX
GO TSU D. Michael 5 Evelyn Close, Twickenham TW2 7BL
GO TTE S. Sutherland 11 Brentwood Road, Leicester LE2 6AD
GWO TTF W. Price Tramore, 67 High Street, Bridgend CF32 0HL
GO TTG M. Warriner 34 Cabrera Avenue, Virginia Water GU25 4EZ
GO TTI R. Rayment 145 Feeches Road, Southend-on-Sea SS2 6TF
GO TTL R. Booth The School House, Old School Lane, Wadworth, Doncaster DN11 9BW
GO TTM A. Radley 16 Kingsley Lane, Thundersley, Benfleet SS7 3TU
GWO TTN P. Brzenczek Flat 18, Ty Newydd House, Ty Newydd Court, Cwmbran NP44 1LH
GO TTO L. Chadwick 2 Auden Place, Longton, Stoke-on-Trent ST3 1SJ
GO TTQ B. Trivett 712 East Myrtle Ave, Foley 36535 United States
GO TTR Dr A. Bartle 10 Holme Dene, Haxey, Doncaster DN9 2JX
GO TTS P. Bulmer 61 Middleham Avenue, York YO31 9BD
GO TTW K. Yeates Newlands, Ashby Lane, Lutterworth LE17 4SQ
GMO TTY W. Mcburney 6 Hill Street, Tillicoultry FK13 6HF
GO TUC D. Wilkes 85 Moss Lane, Hesketh Bank, Preston PR4 6AA
GO TUE R. Gilchrist 53 High Cliff, Barrow-in-Furness LA14 4TH
GO TUI B. Hudson 5 Rylands Road, Southend-on-Sea SS2 4LW
GO TUJ W. Spencer 111 Rosmead Street, Hull HU9 2TE
GO TUL R. Woollard 68 Trunk Furlong, Aspley Guise, Milton Keynes MK17 8HX
GO TUM B. Cooper 13 Highbury Road, Leeds LS6 4EX
GO TUN A. Powney 16 Westbrook Way, Wombourne, Wolverhampton WV5 0EA
GO TUO M. Brough 24 St. Georges Road, Bletchley, Milton Keynes MK3 5EN
GO TUP N. Callow 3 Maunleigh, Forest Town, Mansfield NG19 0PP
GMO TUS R. Young 50 Berrywell Drive, Duns TD11 3HG
GO TUU R. Hudson Norton House, 27 Torne View, Doncaster DN9 3PQ
GO TUV L. Kennedy 28 Murton Garth, Murton, York YO19 5UL
GO TUW G. Woolfenden 11 Chesshire Close, Areley Kings, Stourport-on-Severn DY13 0EB
GO TUX J. Fossey 12 Hitchin Road, Arlesey SG15 6RP
GO TUZ W. Donovan 28 Nevill Road, Snodland ME6 5HX
GO TVB P. Rigg 1 Stones Hey Gate, Widdop Road, Hebden Bridge HX7 7HD
GO TVC A. Lanham 7 College Lane, Stratford-upon-Avon CV37 6DD
GO TVD M. Smith 9 Oak Green Way, Abbots Langley WD5 0PJ
GO TVL S. Birkenshaw 19A Vale Head Grove, Knottingley WF11 8JL
GO TVM A. Bashir 70 Smith Lane, Bradford BD9 6DQ
GO TVO B. Hewson 6 Sanworth St., Todmorden OL14 5BU
GO TVR C. Binnell 146 Hales Crescent, Warley, Smethwick B67 6QX
GO TVS C. Saunders 17 Bure Road, Friars Cliff, Christchurch BH23 4ED
GMO TVT R. Hemmings 2A Caversta, Isle of Lewis HS2 9QE
GO TVU M. Binns 49 Fairview, Pontefract WF8 3NT
GO TVV E. Graham Ibis Styles London, 543 Lea Bridge Road Reception Desk, London E10 7EB
GO TVW D. Rodman Flat 4, Heatherfield Court, 197 Baslow Road, Totley Rise, Sheffield S17 4DT
GWO TVX R. Elms 7 Manor Daf Gardens, St. Clears, Carmarthen SA33 4ES
GMO TWB I. Lindsay Fallady Cottage, Angus DD8 2SP
GO TWD Leeds Raynet Group c/o Rvd. J. Drummond 14 Bulls Head Cottages, Turton, Bolton BL7 0HS
GO TWE Rvd. T. Walker 1 Neville Turner Way, Waltham, Grimsby DN37 0YJ
GWO TWF M. Patterson 6 Devonshire Place, Port Talbot SA13 1SG
GO TWH M. Jewkes Rhuddlan Cottage, Coppenhall Lane, Crewe CW2 8UE
GWO TWI S. Foote Red Roofs, 5 Woodland Avenue, Colwyn Bay LL29 9NL
GWO TWL M. Lewis 75 Lon Maesycoed, Newtown SY16 1QQ
GWO TWO P. Murdoch 19 Penhelyg Road, Aberdovey LL35 0PT
GWO TWR C. Harrison 28 Brynau Wood, Cimla, Neath SA11 3YQ
GO TWT B. Lee Ridge Hill, Ledbury Road, Ledbury HR8 1ND
GO TWV K. Stewart 97 Chase Meadows, Blyth NE24 4LB
GI0 TWX Dr I. Chin 25 Meadowbrook, Islandmagee, Larne BT40 3UG

G0 TXA B. Carter 16 Hollow Street, Canterbury CT34DS
GM0 TXJ N. Service 13 Garden Terrace, Falkirk FK1 1RL
G0 TXL D X Hunter c/o P. Elliott 32 Crichton Avenue, Wallington SM6 8HL
G0 TXN M. Thoyts 17 Solent Avenue, Lymington SO41 3SD
G0 TXO D. Whitaker 19 Bradwell Fold, Glossop SK13 6HX
GW0 TXP A. Smith Courtlands, Llysonnen Road, Carmarthen SA33 5DR
G0 TXU G. Redmond 21 Grosvenor Gardens, Bognor Regis PO21 3EZ
G0 TXY A. Hicks 29 Oak Tree Close Strensall, York YO32 5TE
G0 TYM T. Allison12 2 Westlands, Stokesley, Middlesbrough TS9 5BU
G0 TYN M. Holmes 48 Woodpecker Close, Wirral CH49 4QP
G0 TYQ G. Clark 2 Keith Road, Swanton Morley, Dereham NR20 4NQ
G0 TYS S. Allanson 5 Kingsley Avenue Crofton, Wakefield WF4 1RN
G0 TYW P. Cocker 52 Heathlee Road, London SE3 9HP
G0 TYZ T. Turley 6 Rowan Grove, Oxford OX4 7FD
G0 TZC P. Sables 45 Carr Head Lane, Bolton-Upon-Dearne, Rotherham S63 8DA
G0 TZD R. Leah 64 Valley Road, Chatham N7L 5G3 Canada
G0 TZH M. Bushnell Rose Cottage, Street Ashton, Rugby CV23 0PH
G0 TZM K. Winfield Russettwalls, 58 Bretby Lane, Burton-on-Trent DE15 0QW
G0 TZO R. Nelson 61 Broken Cross, Charminster, Dorchester DT2 9QB
G0 TZP A. Seals 94A Snakes Lane, Southend-on-Sea SS2 6UA
G0 TZR J. Macknish 21A Knoll Rise, Orpington BR6 0EJ
G0 TZV G. Bryant 54 Drew Road, Stourbridge DY9 0UP
G0 TZY A. Davies 20 Peel Park Close, Accrington BB5 6PL
G0 TZZ C. Soames Stud Farm Bungalow, 3 The Street, Sporle PE32 2EA
G0 UAA I. Fallows 47 Melrose Avenue, Burnley BB11 4DN
G0 UAC C. Martin 17 Chambers Grove, Chapeltown, Sheffield S35 2TD
G0 UAD G. Rowe 8 Dove Close, Bolton-Upon-Dearne, Rotherham S63 8JL
GI0 UAG R. Anderson Derry Lodge, 2 Tullymally Road, Newtownards BT22 1JX
G0 UAI S. Marshall 68 Parkfield Avenue, Hampden Park, Eastbourne BN22 9SF
G0 UAK R. Thompson 51 Rydal Avenue, Ramsgate CT11 0PX
G0 UAO J. Smith 124 Parkside Avenue, Barnehurst, Bexleyheath DA7 6NL
G0 UAP P. Westbury 5 Smithfield Place, Bournemouth BH9 2QJ
G0 UAS A. Smith 3 Woodcourt Close, Sittingbourne ME10 1QT
G0 UAV P. Tristram 26 St. Andrews Road, Paignton TQ4 6HA
G0 UAY E. Fletcher 5 Butt Hill Court, Bury New Road, Manchester M25 9NT
G0 UBA R. Gibbs 357 Downham Way Bromley, Kent, , London BR1 5EW
G0 UBG D. Endean 6 Higher Westonfields, Totnes TQ9 5QY
G0 UBJ K. Beach Taylor Cove, Harwich Road, Clacton-on-Sea CO16 0AX
G0 UBK C. Carvell 6 Field Close, Whitby YO21 3LR
G0 UBL M. Stracey 9 Boundary Drive, Hutton, Brentwood CM13 1RH
G0 UBM J. Horsfield 8 St. Edmunds Road, Ipswich IP1 3QZ
G0 UBX A. Quince 9 Biscay Close, Irchester, Wellingborough NN29 7FD
G0 UBY L. Duffill 14 Leonard Street, Hull HU3 1SA
GW0 UCA Dr T. Ogden Tyn Llidiart, Brithdir, Dolgellau LL40 2RP
G0 UCC M. Sayegh11 1 Hoylake Road, East Acton, London W3 7NP
G0 UCD M. West 12 Jenny Gill Crescent, Skipton BD23 2RR
G0 UCE J. Swartz 15076 New Salem Bluff Road, Petersburg, Il 62675 United States
G0 UCF Dr G. Knox 117 Old Shoreham Road, Hove BN3 7AQ
G0 UCH C. Hawes 64 Whitmore Court, Whitmore Way, Basildon SS14 2TN
G0 UCI R. Jones 29 Avon Dale, Newport TF10 7LS
G0 UCK B. Linehan 28 Hurlstone Grove Furzton, Milton Keynes MK4 1EF
G0 UCN P. Turner 176 Crescent Road, Hadley, Telford TF1 5LF
G0 UCP Dr J. Seager 2 Waterford Road, Oxton, Prenton CH43 6UT
G0 UCS W. Dingley 63 Hurst Green, Mawdesley, Ormskirk L40 2QS
G0 UCT B. Obrien 59 Riddlesdown Avenue, Purley CR8 1JL
G0 UCX T. Taylor 1 Cooke Gardens, Branksome, Poole BH12 1QE
G0 UDB D. Birch 4 Godolphin Road, Helston TR13 8PY
G0 UDG K. Deegan 7 Oldcott Crescent, Kidsgrove, Stoke-on-Trent ST7 4HF
GW0 UDH R. Pritchard 1 Limetree Court, Station Road, Abergavenny NP7 5JA
G0 UDI J. Murphy 8 Spencer Avenue, Wribbenhall, Bewdley DY12 1DB
GW0 UDJ D. Jones Bryn Awelon, Brynsannan, Holywell CH8 8AX
GM0 UDL A. Cowan House Of Shannon, Wester Templands, Fortrose IV10 8RA
G0 UDO R. Dodd 33 Dogcroft Road, Stoke-on-Trent ST6 6PE
G0 UDP K. Fairbotham 32 Northolme Drive, York YO30 5RP
GM0 UDY A. Hyslop 9 Shalloch Square, Girvan KA26 0EA
G0 UDZ M. Baister 7B Hepple Road, Spital Estate, Newbiggin-by-the-Sea NE64 6ST
G0 UEA J. Heald 94 Haylings Road, Leiston IP16 4DT
G0 UEB R. Fisher 14 Colindeep Lane, Sprowston, Norwich NR7 8EG
G0 UEC J. Bird 56 Garsdale, Birtley, Chester le Street DH3 2EY
G0 UED I. Harkness 2 Trevor Drive, Maidstone ME16 0QP
GI0 UEG Leicester RS c/o R. Wilson 68 Kensington Road, Belfast BT5 6NG
G0 UEH R. Hislop 79 Norwood Avenue, Hasland, Chesterfield S41 0NJ
G0 UEK S. Payne 55 Binstead Lodge Road, Binstead, Ryde PO33 3TL
GW0 UEO J. Ewan 71 Kingston Drive, Connah'S Quay, Deeside CH5 4TN
GM0 UET Wey Valley ARG c/o R. Henderson 22 Bowmont Place, East Kilbride, Glasgow G75 8YG
G0 UEU A. Jackson 7 Rushfield, Sawbridgeworth CM21 9NF
G0 UEW Dunstable Downs RC c/o R. Jones Greens Cottage, Luton Road, Offley, Hitchin SG5 3DR
G0 UEY R. Muggeridge 46 Arun Road, Billingshurst RH14 9NB
G0 UFB P. Short 23 Barn Close, Hartford, Huntingdon PE29 1XF
G0 UFC Dr D. Meacock The Limes, Davids Lane, Boston PE22 0BZ
G0 UFD T. Reynolds 37 Clarendon Street, Rochdale OL16 4UB
G0 UFE S. Bird 15 Ludlow Drive, Stirchley, Telford TF3 1EG
G0 UFF R. Reynolds 5 Dymond Court, Bodmin PL31 2FP
G0 UFI G. Brady Thirn Grange, Thirn, Ripon HG4 4AU
G0 UFJ S. Glover 46 Merton Close, Oldbury B68 8NG
G0 UFL T. Reid Menwith Hill Station, Po Box 985, Harrogate HG3 2RF
G0 UFN P. Dean 80 Escallond Drive, Dalton-Le-Dale, Seaham SR7 8JZ
G0 UFP C. Beesley-Reynolds Kaos Roams, Palmerston Close, Kibworth Beauchamp LE8 0JJ
G0 UFU C. Jameson 35 Bilberry Grove, Taunton TA1 3XN
G0 UFV P. Shaw 15 Moorfield Road, St. Giles-On-The-Heath, Launceston PL15 9SY
G0 UFW J. Marvill 242 Hillmorton Road, Rugby CV22 5BG
G0 UFY J. Brown 3 Slipper Mill, Slipper Road, Emsworth PO10 8XD
G0 UFZ Dr J. Howard 32 Eastgate, North Newbald, York YO43 4SD
G0 UGA B. Moody 38 Bromwich Road, Willerby, Hull HU10 6SF
G0 UGD N. Boyd 16 Edensor Road, Eastbourne BN20 7XR
GM0 UGG A. Aird 3 Graystones, Kilwinning KA13 7DT
GM0 UGH M. Westland 142 Claremont, Alloa FK10 2EG
G0 UGI H. Argument 9 Oxley Close, Shepshed, Loughborough LE12 9LS
G0 UGJ D. Basford 91 Hollins Spring Avenue, Dronfield S18 1RP
G0 UGM E. Peasey Greenfield Cottage, Greenfield Road, Colne BB8 9PE
GW0 UGQ M. Webb 75 Bolingbroke Heights, Flint CH6 5AN
G0 UGR M. Chaloner Barnhay House, Newport, Berkeley GL13 9PY
G0 UGS F. Taberner 51 Canford View Drive, Colehill, Wimborne BH21 2UW
G0 UGW P. Grant 37 Glenmore Park, Dundalk, County Louth Ireland
G0 UGX J. Kinsella 85 Lowther St., Coventry CV2 4GL
G0 UGY S. Jackson 32 Sherwell Drive, Alcester B49 5HA
GM0 UHC I. Ropper 2 Deerhill, Dechmont, Broxburn EH52 6LY
G0 UHD K. Hore 15 Heriot Way, Great Totham, Maldon CM9 8BW
G0 UHF R. Darwent 139 The Oval, Sheffield S5 6SQ
G0 UHG R. Warne 32 Chance Street, Tewkesbury GL20 5RF
G0 UHI D. Norton 52 Letchworth Road, Leicester LE3 6FG
GW0 UHJ W. Griffiths 147 High Street, Tonyrefail, Porth CF39 8PL
G0 UHK Prof. M. Peiperl 45 High St., Harrow HA1 3HT
G0 UHM L. Ruddock 2 Cross Lane, Waterlooville PO8 9TJ
GW0 UHO P. Sage 4 Gladstone Terrace, Miskin, Mountain Ash CF45 3BS
G0 UHQ M. Minihane 60 Wolsey Drive, Walton-on-Thames KT12 3BA
G0 UHS R. Hatch 99-101 Hornby Road, Blackpool FY1 4QP
G0 UHU G. Bloyce 8 Olivers Court Olivers Close, Clacton-on-Sea CO15 3QX
GW0 UHX S. Jones 64 Springfield Gardens, Hirwaun, Aberdare CF44 9LY
G0 UID J. Parker 1 Church End, Syresham, Brackley NN13 5HU
G0 UIF G. Fairbrass 230 Kirkby Road, Barwell, Leicester LE9 8FS
GM0 UIG C. Cowan 85 Eastwoodmains Road, Clarkston, Glasgow G76 7HG
G0 UIH S. Lawman 44 Barnwell, Peterborough PE8 5PS
G0 UIL D. Brice 4 Bishop Fox Drive, Taunton TA1 3HQ
GW0 UIP N. Wallace Tan Y Bryn, Bryn Road, Flint CH6 5HU
G0 UIQ W. Furze 2 Lynewood Road, Cromer NR27 0EE
G0 UIW A. Jones 26 Grosvenor Road, , Harrogate HG1 4EG
G0 UIX A. Lambert 10 The Green, Cirencester GL7 1AU
GW0 UIZ B. Galsworthy 30 Pen Yr Yrfa, Morriston, Swansea SA6 6BA
G0 UJD M. Bartle 14 Litton Avenue, Skegby, Sutton-in-Ashfield NG17 3AB
GI0 UJG R. Stinson 51 Cloncarrish Road, Portadown, Craigavon BT62 1RN
G0 UJI D. Sparkes 31 Lockyers Drive, Ferndown BH22 8AL
G0 UJP J. Fleetwood 11 Chichester Road, Southampton SO18 6BB
G0 UJT J. Deal 19 Coniston Road, Gloucester GL2 0NA
G0 UJU D. Barlow 7 Prospect Terrace, Canal Road, Taunton TA1 1PH
G0 UKA J. Black 8 Cornwood Close, Finchley, London N2 0HP
GW0 UKC M. Price 50 Llangorse Road, Cwmbach, Aberdare CF44 0HR
GM0 UKD S. Munro 76 John Neilson Avenue, Paisley PA1 2SX
GW0 UKF J. Griffith Craig Artro, Llanbedr LL45 2LU
GW0 UKG A. Powell 80 St. Andrews Crescent, Abergavenny NP7 6HN
G0 UKK K. Stanyer 15 Wilbrahams Way, Alsager, Stoke-on-Trent ST7 2NR
G0 UKL P. Charlton 6 Maiden Road Shirebrook, Mansfield NG20 8GA
G0 UKM M. Russell 10 Baytree Grove, Ramsbottom, Bury BL0 9UF
G0 UKO R. Theakston 130 Greenshaw Drive Haxby, York YO32 2DG
G0 UKP B. Jopson 21 Richmond Street, Southend-on-Sea SS2 4NW
G0 UKS R. Towler 77 Glebe Road, Hull HU7 0DU
GW0 UKT Avon Valley Ara c/o W. Waldron Torfaen Scouts Arc, 23 Forest Close, Cwmbran NP44 4TE
GM0 UKZ C. Gibson 45 Tiree Place, Newton Mearns, Glasgow G77 6UJ
G0 ULA N. Foster 68 Brookfield Way, Lower Cambourne, Cambridge CB23 5ED
G0 ULG K. Dewing 124 Proctor Road, Norwich NR6 7PH
G0 ULH L. Harris 183A Painswick Road, Gloucester GL4 4AG
G0 ULI M. Kaliski 132 Woodland Road, Hellesdon, Norwich NR6 5RQ
G0 ULL E. Williams 50 Broad Lawn, New Eltham, London SE9 3XD
G0 ULM P. Wilson 56 Highfield Road, Blacon, Chester CH1 5AZ
G0 ULO P. Ravenscroft 23A Castleview Road, Bilston WV14 8LW
GW0 ULP L. Parsons 105 Victoria Road West, Prestatyn LL19 7DS
G0 ULQ W. Bucknell 119 Fossway, York YO31 8SQ
G0 ULS F. Mortimer 115 Dell Road, Lowestoft NR33 9NX
G0 ULY J. Stokes 76 Hakewill Way, Colchester CO4 5FY
G0 UMI E. Latter 76 Crossway, Plympton, Plymouth PL7 4HY
GM0 UMJ Dr S. Heerma Van Voss Blarachaorachan, Fort William PH33 6SZ
G0 UMK R. Adams 9 Chestnut Garth, Roos, Brooklands, Hull HU12 0LE
G0 UML N. Watling 36 All Saints Walk, Mattishall, Dereham NR20 3RF
G0 UMM N. Roberson 6 Long Lane, West Winch, King's Lynn PE33 0PG
G0 UMP M. Parker 102 Cavendish Road, Patchway, Bristol BS34 5HH
G0 UMS R. Petrie 11 St. James'S Close, Yeovil BA21 3AH
G0 UMV South Cheshire ARS c/o P. Johnson 52 Evesham Road, Cookhill, Alcester B49 5LJ
G0 UMY C. Lowe 37 Parsonage Brow, Upholland, Skelmersdale WN8 0JG
G0 UNB J. Jeffers 11 Polywell, Appledore, Bideford EX39 1SG
G0 UNC G. Hancock 3C Richmond Street, Hull HU5 3JY
G0 UND D. Scargill 10 Mendip Avenue, North Hykeham, Lincoln LN6 9SZ
G0 UNE J. Spencer 6 Redcar Road, Sunderland SR5 5QA
G0 UNF G. Taylor Flat 67A, Bramley Grange Hotel Flats, Guildford GU5 0BL
G0 UNG B. Massey 40 East St., Ashton In Makerfield, Wigan WN4 8ST
G0 UNK T. Wright 2 Regent Road Church, Accrington BB5 4AR
G0 UNW R. Martin 44 Threadneedle Cres, Willowdale, Ontario Canada
G0 UNY M. Lindley 23 Townend Lane, Deepcar, Sheffield S36 2TN
G0 UOB A. Chilinski 54 , Foxes Meadow, King's Lynn , Norfolk PE32 2AS
G0 UOD E. Sheather Clare Cottage, North End, Shaftesbury SP7 9HX
G0 UOI R. Fox 20 Levett Road, Polegate BN26 6NE
G0 UOK R. Dutton 3 Kilkenny Road, Guisborough TS14 7LE
G0 UOL D. Hards W7 B17 The Royal Hospital Chelsea Royal Hospital Road, London SW3 4SR
G0 UOM N. Taylor 16 Josephine Road, Rotherham S61 1BJ
G0 UOP Telford And District ARS c/o R. Linford Department Of Communication, And Electronic Engineering, Drake Circus Plymouth PL4 8AA
G0 UOQ P. Creissen 143 Hawthorn Bank, Spalding PE11 2UN
G0 UOS J. Butterworth 12 Ingswell Drive Notton, Wakefield WF4 2NF
GM0 UOU C. Muir 1/1 132 Falside Road, Paisley PA2 6JT
G0 UOV S. Porter 20 Newbridge Road, Ambergate, Belper DE56 2GR
G0 UOZ A. Morris 4 Pleasant Terrace, Lincoln LN5 8DA

G0 UPD R. Brinkley 70 Leopold Road, Felixstowe IP11 7NR
GM0 UPE Dr G. Sutherland 22 Montrose Drive., Bearsden G61 3LG
G0 UPG J. Dunne 40 Egmont Road, Poole BH16 5BZ
G0 UPK D. Harmer 98 King Georges Avenue, Coventry CV6 6FF
G0 UPL H. Summers 2 Hillview, Highgate Road, Forest Row RH18 5AZ
G0 UPO M. May Flat 4, Hyde Bank Court, Hyde Bank Road, New Mills, High Peak SK22 4NE
G0 UPP S. Bennett 102 Laflouder Fields, Mullion, Helston TR12 7EJ
G0 UPS P. Zimmermann 85 Wimborne Road West, Wimborne BH21 2DH
G0 UPV T. Berrisford 126 Star & Garter Road, Stoke-on-Trent ST3 7HN
G0 UPY M. Rogers 3 Wheelwright Gardens, Long Compton, Shipston-on-Stour CV36 5LN
G0 UQB D. Brittain 9 Highfield Road, Cookley, Kidderminster DY10 3UB
G0 UQC R. Birkett 14 Greta Street, Keswick CA12 4HS
G0 UQE R. Weaver 107 Patch Lane, Redditch B98 7XE
G0 UQF G. Merrils 2 East St., Darfield, Barnsley S73 9AE
GW0 UQH S. Provan 12 Trewarren Road, St. Ishmaels, Haverfordwest SA62 3SZ
G0 UQI L. Snowden 25 Brixham Road, Paignton TQ4 7HG
G0 UQJ N. Smith 36 Maple Road, Sutton Coldfield B72 1JP
GI0 UQK P. Sinclair 37 Willow Avenue, Banbridge BT32 4RE
G0 UQO G. Rekers 18 The Ridge, Purley CR8 3PE
G0 UQP F. Waters 96 Stockley Road, Barmston Village, Washington NE38 8DR
G0 UQQ N. Baskerville 10 Park Avenue, Sprotbrough, Doncaster DN5 7LW
G0 UQT S. Nash 26 Lyndhurst Crescent, Wembdon, Bridgwater TA6 7QG
G0 UQU J. Squires 44 St. Marys Road, Doncaster DN1 2NP
G0 UQV M. Parkin Chenet, 32 The Nooking, Doncaster DN9 2JQ
G0 UQY P. Cox 17 Hyde Lane, Upper Beeding, Steyning BN44 3WJ
G0 UQZ D. Eyre 41 Lindsay Road, Sheffield S5 7WE
G0 URB J. Morse 14 Strathmore Road, Bournemouth BH9 3NS
G0 URC P. Ball 12 Warren Way, Digswell, Welwyn AL6 0DH
GM0 URD I. Thomson 33 Aytoun Grove, Dunfermline KY12 9YA
G0 URF M. Dodsworth 359 Upper Town Street, Bramley, Leeds LS13 3JX
GI0 URI S. Robinson 19 Pattonville, Dungannon BT71 6DD
G0 URK J. Alderman 56 Edward Harvey Link, Beaulieu Park, Chelmsford CM1 6BU
GI0 URN the Royal Naval (Ulster) ARC c/o N. Mckee 54 Castlemore Park, Belfast BT6 9RP
G0 URO D. Fosh 8 The Pines, Horsham RH12 4UF
G0 URT S. Bradbury 39 Grosvenor Road, Hyde SK14 5AB
GM0 URU Dr J. Cartlidge 14 Davidson Street, Broughty Ferry, Dundee DD5 3AT
G0 URW A. Paone Viale Dei Quattro, Venti 128 C/O Int6, Roma 00 152 Italy
G0 USA L. Civita 53 Rockhurst Drive, Eastbourne BN20 8XD
GI0 USC J. Smith 5 Old Turn, Carrickfergus BT38 7EH
G0 USE J. Davy-Jones 1 Wensley Gardens, Emsworth PO10 7RA
G0 USF M. James 8 Swan Quay Bath Lane, Fareham PO16 0DX
GM0 USI A. Dimmick 02 120 Shakespeare Street, Glasgow G20 8LF
G0 USJ S. Johnson 36 Langar Woods, Langar, Nottingham NG13 9HZ
G0 USK P. Perera 13 Dalcross Road, Hounslow TW4 7RA
G0 USM B. Parker 24 Mayfield Road, Chorley PR6 0DG
GI0 USQ P. Fox-Roberts 8 Lynwood Park, Holywood BT18 9EU
GI0 USS K. Chambers 59 Ravenswood, Banbridge BT32 3RD
GI0 USW P. Mcdonald 13 Serpintine Road, Newtownabbey BT36 7HA
G0 UTA C. Gurney 9 Snowdrop Mews, Exeter EX4 2PN
G0 UTB Dr P. Cordrey Redline C.A Edif. Fannellis 5 Y 6, Av. Ricaurte, Cojedes 2201 Venezuela
GW0 UTC J. Lomas 5 Gwelfor, Rhos On Sea, Colwyn Bay LL28 4AJ
GM0 UTD H. Urquhart The Cless, Peebles EH45 8NU
GI0 UTE D. Auld 37 Castlewellan Road, Rathfriland, Newry BT34 5EL
G0 UTM B. Watson 6 Shakespeare Avenue, Scunthorpe DN17 1SA
G0 UTN S. Harrison 22 Dales Avenue, Sutton-in-Ashfield NG17 4BY
G0 UTP R. Walker 1 Farmcote Court, Hemlington, Middlesbrough TS8 9LJ
G0 UTR D. Coleman 1 Kerstin Close, Cheltenham GL50 4SA
G0 UTT Dengie Hundred ARS c/o A. Slade Skerries, Summerhill Althorne, Chelmsford CM3 6BY
G0 UTU P. Humphreys 47 Crescent Road, Locks Heath, Southampton SO31 6PE
GI0 UTV I. Ross 312 Castlereagh Road, Belfast BT5 6AD
G0 UTX P. Curtis 31 Dorking Close Ings Estate, Hull HU8 9DG
G0 UTZ R. Davies 36 Woodside Avenue, Brown Edge, Stoke-on-Trent ST6 8RX
G0 UUA Dr W. Hutchings Bitternsdale Farm, Bustomley Lane, Stoke-on-Trent ST10 4PE
GM0 UUB G. Matthews 88 Nevis Crescent, Alloa FK10 2BN
G0 UUC W. Slater 44 Hope St., Chesterfield S40 1DG
G0 UUF S. Errington 23 Pinewood Drive, Bletchley, Milton Keynes MK2 2HT
G0 UUI R. Newport 17 College Crescent, Oakley, Aylesbury HP18 9QZ
G0 UUM J. Lewis Burley Cottage, Shortwood, Stafford ST21 6RG
G0 UUN J. Lewis Burley Cottage, Shortwood, Stafford ST21 6RG
G0 UUP M. Stevens 24 Oakroyd Close, Burgess Hill RH15 0QN
G0 UUR C. Branch 10 Queens Road, Thame OX9 3NQ
G0 UUS G. Hanson 34 Old Garden Close, Locks Heath, Southampton SO31 6RN
G0 UUT I. Paim 8 Sparrow Close, Attleborough NR17 1GP
G0 UUU P. Earnshaw 7 Hampton Road, Scarborough YO12 5PU
G0 UUZ A. Breeze Drakelow Cottage, Drakelow Lane, Wolverley, Kidderminster DY11 5RU
GI0 UVD W. Bustard 66 Hertford Crescent, Lisburn BT28 1SQ
G0 UVE I. Reynolds Five Bars, Hereford Road Weobley, Hereford HR4 8SW
G0 UVG P. Ellis 9 Matilda Gardens, Shenley Church End, Milton Keynes MK5 6HT
GU0 UVH T. Bosher St Andrews 17 Le Banquage, Alderney GY9 3YP
G0 UVL L. Cartwright 14 Norley Hall Avenue, Wigan WN5 9TG
G0 UVN B. Goolding 10 Oakwell Close, Stevenage SG2 8UG
G0 UVR G. Akse Drive Cottage, Ebberston, Scarborough YO13 9PA
G0 UVT J. Goldie The Coachmans Cottage, Pulford Lane, Chester CH4 9NN
G0 UVX G. Valentine 19 Kingsway Stotfold, Hitchin SG5 4EL
G0 UWA E. Shanklin The Coachmans Cottage, Pulford Lane, Chester CH4 9NN
G0 UWB R. Moyle Aitchill House, Lower Brailes, Banbury OX15 5AP
GW0 UWD J. Matthews 42 Wexham Street, Beaumaris LL58 8HW
G0 UWF S. Bowles 37 Manor Road, Paignton TQ3 2HZ
G0 UWI H. Chipper 36 Newbridge Way, Truro TR1 3LX
G0 UWK I. Goodier 2 Chatterley Drive Kidsgrove, Stoke-on-Trent ST7 4HW
G0 UWO N. Winfield Oaklea, Chyvogue Meadow, Truro TR3 7JP
G0 UWS A. Sharman 3 Deben Crescent Haydonwick, Swindon SN25 3QB
G0 UWU M. Claridge 105 Barnwood Avenue, Gloucester GL4 3AG
GM0 UWV J. Reynolds The Paddock, Leswalt, Stranraer DG9 0LJ
G0 UWX D. Pollard 191 High Road, Halton, Lancaster LA2 6QB
GI0 UXD D. Burns 207 Rathfriland Road, Dromara, Dromore BT25 2EQ
G0 UXF C. Whittaker 3A Oak Avenue, Horwich, Bolton BL6 6JE
G0 UXG P. Blunt 17 Offens Drive, Staplehurst, Tonbridge TN12 0LR
G0 UXH A. Mawson 93 Glenridding Drive, Barrow-in-Furness LA14 4PA
G0 UXI M. Whitehead 7 Avalon Drive, Manchester M20 5WN
GW0 UXJ A. Burns 34 Lakeside Gardens, Merthyr Tydfil CF48 1EN
G0 UXO B. Hillman 2 Holmes Chapel Road, Congleton CW12 4NE
G0 UXR T. Gilmore 48 Ash Lane, Hale, Altrincham WA15 8PD
GW0 UXX D. Cumiskey 16 Delapoer Drive, Haverfordwest SA611HJ
G0 UXZ A. Walmsley Runaways Cottage, Ring Street Stalbridge, Sturminster Newton DT10 2LZ
G0 UYA S. Chamberlin 15 Bull Close, East Tuddenham, Dereham NR20 3LX
G0 UYC D. Rolph 3 Bell Close, Bawdeswell, Dereham NR20 4SL
G0 UYE A. Colton 9 Pineway, Bridgnorth WV15 5DS
G0 UYF P. Moss 34 Almond Tree Avenue, Goole DN14 9QR
G0 UYG A. Forster 9 Pinewood Walk, Stokesley, Middlesbrough TS9 5HU
G0 UYH A. Smart Nine Hamelin Street, Pye Green Road, Cannock WS11 2SE
G0 UYM Obo Thames Amateur Radio c/o J. Hall Pump Farm Cottage, Pump Lane South, Marlow SL7 3RB
G0 UYP R. Arkell 76, Wilnecote Lane, Tamworth B77 2JA
G0 UYQ M. Melbourne 42 Pasture Road, Stapleford, Nottingham NG9 8GL
GM0 UYS S. Jeffrey 33 The Rowans, Insch AB52 6ZD
G0 UYT J. Hemming 62 Beaumont Road Bournville, Birmingham B30 2DY
G0 UYV B. Smith 80A Bramcote Lane, Beeston, Nottingham NG9 4ES
GI0 UYY L. O'Flaherty 1 Ravensdale Villas, Newry BT34 2PG
GM0 UYZ J. Wheeler 54 Wittet Drive, Elgin IV30 1TB
G0 UZD A. Gisby 18 Westmead Gardens, West Avenue, Worthing BN11 5LP
G0 UZE K. Brookes Kohima, Spout Lane, Stoke-on-Trent ST2 7LR
G0 UZF I. Riley 102 Harrison Road, Chorley PR7 3HS
G0 UZK A. Rushton Rosemary Cottage, The Green, East End, North Leigh, Witney OX296PY
G0 UZL C. Clease 20 Medbourne Close, Blandford Forum DT11 7UA
GM0 UZV W. Cargill 23 Ceres Road, Craigrothie, Cupar KY15 5QB
G0 UZW N. Donald 20 Parkhill Road, Barnby Dun, Doncaster DN3 1DP
GW0 UZX F. Thompson 15 Aneurin Crescent, Twynyrodyn, Merthyr Tydfil CF47 0TB
G0 UZY J. Shotter Elthorne, Hawkswood Road, Hailsham BN27 1UN
GI0 VAB P. Moore 59 Belmont Avenue, Belfast BT4 3DE
G0 VAD J. Whitehall 29 Melrose Terrace, Newbiggin-by-the-Sea NE64 6XN
G0 VAE M. Clarke 21 Sycamore Road, Greenstead Estate, Colchester CO4 3NF
G0 VAG P. Mason 8 Westbourne Park, Scarborough YO12 4AT
G0 VAH A. Heyworth 3 The Knowl Churton, Chester CH3 6NE
G0 VAI R. Frow Valley Yard, Skeete Road, Folkestone CT18 8DS
G0 VAJ Warwickshire Wireless Society c/o S. Hobden 10 Turton Close, Brighton BN2 5DA
G0 VAL D. Wyatt 96 Woodlands Close, Clacton-on-Sea CO15 4RU
G0 VAM A. Witter 44 Regent St., Newton le Willows WA12 9LS
G0 VAR C. Wilson 15 Biddenden Close, Bearsted, Maidstone ME15 8JP
G0 VAS V. Ikonomou 18 Canhams Road, Great Cornard, Sudbury CO10 0EP
G0 VAU T. James 114 Broadway, Loughborough LE11 2JG
G0 VAV J. Farrington 6 The Gravel, Mere Brow, Preston PR4 6JX
G0 VAX B. Bowers 31 Gresford Avenue, Wirral CH48 6DA
G0 VAY G. Gower 14 Beck Garth, Hedon, Hull HU12 8LH
G0 VAZ R. Barker 56 Southend Road, Weston-Super-Mare BS23 4JZ
GM0 VBE B. Higton The Straith, Priestland, Darvel KA17 0LP
G0 VBG W. Chandler 6 St. Thomas Close, Hinton Waldrist, Faringdon SN7 8RP
G0 VBK M. Blackmore Flat 20 Hill View House, Bristol BS15 1TA
G0 VBN J. Cressey 32 Ballifield Road, Sheffield S13 9HX
G0 VBP F. Madely 138 Coningsby Drive, Kidderminster DY11 5LZ
G0 VBQ F. Webster 14 Redbank Avenue, Erdington, Birmingham B23 7JR
G0 VBR M. Dunstan 3 Coronation Road, Rawmarsh, Rotherham S62 5LW
G0 VBT G. Dawson 22 High St., Tean, Stoke on Trent ST10 4DZ
G0 VBX J. Collingwood 31 Corn Close, South Normanton, Alfreton DE55 2JD
G0 VBZ D. Stinton 43 The Meadows, Bidford-On-Avon, Alcester B50 4AP
G0 VCD L. Browne 7 Mornington Drive, Cheltenham GL53 0BH
G0 VCJ K. Emblen 17 Larch Close, Bognor Regis PO22 9LA
GM0 VCN W. Long 52 Stirling Road, Milnathort, Kinross KY13 9XG
G0 VCV J. Partridge 27 Leigh Road, Penhill, Swindon SN2 5DE
G0 VCW R. Evans 4 Cedar Drive, Carlton Manor Park, Chapel Road, Carlton Colville, Lowestoft NR33 8BL
G0 VDE Dr W. Rothwell 30 Wellbrook Way, Girton, Cambridge CB3 0GP
G0 VDJ E. Webster 33 Cherry Gardens, Bitton, Bristol BS30 6JA
G0 VDN Dr B. Logan 22 Chiltern House, Hillcrest Road, London W5 1HL
G0 VDO C. Pond 316 St. Faiths Road, Old Catton, Norwich NR6 7BL
G0 VDP K. Hutley Three Ways, 1 Walden House Road, Maldon CM9 8PJ
G0 VDQ A. Ramsden 7 Florence Road, Pakefield, Lowestoft NR33 7BX
G0 VDR L. Goffin The Hollies, Belaugh Green Lane, Norwich NR12 7AJ
G0 VDT P. Clark 21 Uplands, Welwyn Garden City AL8 7EN
G0 VDU J. Newman Shangrila, Treverbyn Road, Stenalees, St. Austell PL26 8TL
G0 VDV D. Steele Tanglewood, Beckingham Street, Maldon CM9 8LL
G0 VDZ N. Newby 167 Watersplash Road, Shepperton TW17 0EN
G0 VEC B. Hoare Kilclare, Carrick On Shannon, Drumaleague House, Co. Leitrim Ireland
G0 VEH Ariel RA GP LWS c/o J. Mulye 83 Forest Drive East, London E11 1JX
G0 VEI B. Davison Pond House, Moores Lane, Colchester CO7 6RF
GM0 VEK P. Davie 62 Monkland Avenue, Kirkintilloch, Glasgow G66 3BW
GW0 VEM A. Mccleverty 3 The Fold, Upper Thornton, Milford Haven SA73 3UE
G0 VEO R. Myatt 48 Adaston Avenue, Eastham, Wirral CH62 8BS
G0 VEP P. Steed 1 Falcon Green, Portsmouth PO6 1LW
G0 VEQ T. Arkadiusz Shio ^ I-Chq 5- 66, Yokosuka 238-0042 Japan
G0 VET P. Burnand 5 Northgate, Hornsea HU18 1ES
GW0 VEU G. Chantry Summerfield, The Avenue, Oswestry SY11 4LF
GW0 VEW D. Davies 27 Twyniago, Pontarddulais, Swansea SA4 8HX
G0 VEX P. Dickinson 34 Marshall Avenue, Bridlington YO15 2DS
G0 VFB D. Banks 6 Kirkstall Close, Walsall WS3 2SS
GM0 VFD A. Adam 10 Greenmount Road North, Burntisland KY3 9JQ

GW0 VFF S. Emanuel 98 Moorland Road, Cimla. Neath SA11 1JL
G0 VFL K. Gardner Str Matei Vasilescu 82, Drobeth Turnu Sevferin 1500, Mehedinti Romania
G0 VFS R. Bailey 13 Whiteland Rise, Westbury BA13 3HP
G0 VFU F. Cokayne 101 Neston Drive, Bulwell, Nottingham NG6 8QY
G0 VFV G. Marley 41 Scalby Road, Burniston, Scarborough YO13 0HN
G0 VFW T. Thirlwell 58 Chesham Road, Bovingdon, Hemel Hempstead HP3 0EA
GM0 VFY G. Stuart Easter Ardoe Cottage, Ardoe, Aberdeen AB12 5XT
G0 VFZ R. Barnes 18 Battle Road, Tewkesbury GL20 5TZ
G0 VGB D. London 113 Westbrooke Avenue, Hartlepool TS25 5HZ
G0 VGC B. Jenkins 27 Glendale, South Woodham Ferrers, Chelmsford CM3 5TS
G0 VGD J. Constance 4 Hopgarden Road, Tonbridge TN10 4QS
GM0 VGI G. Anderson 21 Bydand Gardens, Inverurie AB51 4FL
G0 VGJ Prof. D. Graham Parkburn, Colby, Appleby-in-Westmorland CA16 6BD
G0 VGK P. Jenkinson 5 Inglemere Drive, Stafford ST17 4QX
GI0 VGL G. Warnock 98 Skerriff Road, Altnamachin, Newry BT35 0PJ
G0 VGN A. Fawcett 26 Merlin Court, Oswaldtwistle, Accrington BB5 3TA
G0 VGP B. Davies 18 Wyresdale Gardens, Lancaster LA1 3FA
G0 VGR P. Tamplin 74 Asquith Road, Gillingham ME8 0JD
G0 VGT A. Trent 18 Castle Drive, Reigate RH2 8DQ
GI0 VGV A. Wright 29 Wynfort Lodge, Moira, Craigavon BT67 0QT
G0 VGY P. Keen 89 Raleigh Avenue, Hayes UB4 0EF
G0 VHF Milton Keynes ARS c/o J. Lemay Carlton House, White Hart Lane, Colchester CO6 3DB
GI0 VHG P. Hughes 17 Cardinal Dalton Park, Keady, Armagh BT60 3TS
G0 VHH G. Mann 10 Earsham Drive, King's Lynn PE30 3UZ
G0 VHK D. Godding 20 Southwood Gardens Burghfield Common, Reading RG7 3HY
G0 VHL P. Jones 28 Helena Road, Capel-Le-Ferne, Folkestone CT18 7LQ
G0 VHO Dr D. Ross Rosehill, Leyland Lane, Leyland PR26 8LB
G0 VHQ A. Harding Po Box 10620 Apo, Grand Cayman Cayman Islands
GM0 VHR C. Cook 2 Nortonhall Cottage Eildon, Melrose TD6 9HB
G0 VHT P. Morrison Primrose Bank, Holme Lacy HR26LJ
G0 VHY R. Wilson 7 Scarratt Close, Forsbrook, Stoke-on-Trent ST11 9AP
G0 VIA C. Cannon 1 Long Walk, Northstead, Scarborough YO12 6BQ
GI0 VIB A. Smith 42 Ballycullen Road, Moy, Dungannon BT71 7HT
G0 VID D. Morris 117 Lonsdale Avenue, Doncaster DN2 6HF
GI0 VIF D. Canning 10 Cotswold Close, Saintfield, Ballynahinch BT24 7FQ
G0 VIG L. Merrick 4 Berryfield Glade, Churchdown, Gloucester GL3 2BT
G0 VII G. Boundey 25 Ivy Place, Tantobie, Stanley DH9 9PT
G0 VIJ J. Johnson 1 Rosa Vella Drive, Dereham NR20 3SB
GD0 VIK D. Wood The Hawthorns, Droghadfayle Road, Isle of Man IM9 6EL
G0 VIM M. Rivers Snagsmount, Lambden Road, Ashford TN27 0RB
G0 VIQ E. Sully 10 The Paddock, Pound Hill, Crawley RH10 7RQ
GM0 VIT W. Henderson Strone View, Bridge Of Cally, Blairgowrie PH10 7JL
G0 VIX M. Rutland Roselea, 28 Eastfield Avenue, Fareham PO14 1EG
GM0 VIY Dr J. Oates 14 Craighlaw Avenue, Eaglesham, Glasgow G76 0EU
G0 VJB B. Vaughan 43 Bankfield Road, Shipley BD18 4AW
G0 VJC C. Smith 19 Wiscombe Avenue, Penkridge, Stafford ST19 5EH
GI0 VJE C. Hannigan 4 Silverhill Road, Strabane BT82 0AE
G0 VJH J. Herrington 84 Glenn Road, Poringland, Norwich NR14 7LU
G0 VJI R. Clay 38 Hubbards Road, Chorleywood, Rickmansworth WD3 5JJ
G0 VJJ S. Gould 87 Wentworth Drive, Bedford MK41 8QD
G0 VJK Dr M. Stanton 11 Eldean Road, Duston, Northampton NN5 6RF
G0 VJM A. Howell Beach Lawns Care Home, 67 Beach Road, Weston-Super-Mare BS23 4BG
G0 VJN P. Andres Flat 80, Northfield House, Bristol BS3 1XB
GJ0 VJP N. Collier Webb Chatelet, Les Marais Avenue, La Route De La Haule, Jersey JE3 1LE Jersey
G0 VJR R. Henshall 9 Murrayfield Drive, Willaston, Nantwich CW5 6QF
G0 VJY R. Welbourn 30 Bower Road, Swinton, Mexborough S648NU
G0 VKC K. Cunningham Claerwern Cottage, Three Ashes, Hereford HR2 8NA
G0 VKE A. Willis 5 Robin Hill, Shoppenhangers Road, Maidenhead SL6 2GZ
G0 VKF M. George 3 Oak Crescent, Cherry Willingham, Lincoln LN3 4AX
GM0 VKG Chippenham And District ARC c/o J. Clough Obo Largs & District Ars, Redbank, Skelmorlie PA17 5DX
G0 VKH H. Venus 45 St. Albans Road, Seven Kings, Ilford IG3 8NN
G0 VKI A. Hopkinson 34 Welby St., Fenton, Stoke on Trent ST4 4PL
G0 VKL R. Butler 34 Commissioners Road, Strood, Rochester ME2 4EB
G0 VKS H. Klein 30 Odenwaldstrasse, Frankfurt 60528 Germany
G0 VKX Dr R. Smith 16 Southbrook Road, Langstone, Havant PO9 1RN
G0 VKY D. Russell 1 Debden Close, Ernesettle, Plymouth PL5 2DB
G0 VLC Dr A. Betts 42 Goring Road, Steyning BN44 3GF
GI0 VLE W. Dalton 16 Junction Road, Randalstown, Antrim BT41 4NP
G0 VLF R. Mcaleer 44 Harvey Avenue, Durham DH1 5ZG
G0 VLI D. Taylor 118 Portsmouth Road, Lee-on-the-Solent PO13 9AF
G0 VLJ Prof. G. Heald 2 Holyrood Terrace, Weymouth DT4 0BE
G0 VLK A. Palfreeman 29 Boulby Road, Redcar TS10 5EB
G0 VLQ B. Close 74 Heston Avenue, Hounslow TW5 9EX
G0 VLR Moorlands&Dis A c/o A. Holmes 5 Launde Park, Market Harborough LE16 8BH
G0 VLV D. Hardman 15 Wordsworth Avenue, Bolton Le Sands, Carnforth LA5 8HJ
G0 VMA G. Skupski 57 Three Nooks, Bamber Bridge, Preston PR5 8EN
G0 VMC J. Williams 62 Hollows Close, Salisbury SP2 8JX
GW0 VMD K. Peacey 2 Robin Close, Cardiff CF23 7HN
G0 VME M. Macdonald 44 Hillesden Avenue, Elstow, Bedford MK42 9YX
G0 VMF P. Quirk 70 Sands Lane, Oulton, Lowestoft NR32 3HS
G0 VMK P. Nicholls 53 Fastolff Avenue Gorleston, Great Yarmouth NR31 7ND
G0 VMN D. Turton 68 Bartholomew Street, Wombwell, Barnsley S73 8LD
G0 VMP J. Roze 9 Ralfland View, Shap, Penrith CA10 3PF
G0 VMQ P. Ellis 104 Gravesend Road, Strood, Rochester ME2 3PN
GW0 VMR P. Smith Bron Awel, Brynisa Road, Wrexham LL11 6NS
GW0 VMS L. Beedle 2 Chestnut Grove, Maesteg CF34 0NT
G0 VMT Dr T. Moorhead 53 Childwall Priory Road, Liverpool L16 7PA
GM0 VMV E. Kennedy 1 Greenbank Gardens, Edinburgh EH10 5SL
GW0 VMW R. Price 8 Tanllwyfan, Old Colwyn, Colwyn Bay LL29 9LQ
GW0 VMZ J. Davies 34 Penlan View, Ynysfach, Merthyr Tydfil CF47 8NJ
G0 VNA G. Kent Plum Tree Cottage, Colesbrook Lane, Gillingham SP8 4HH
GW0 VND M. Goodridge 17 Charles St., Neyland, Milford Haven SA73 1SA
G0 VNE H. Stokes 9 Causeway Glade, Dore, Sheffield S17 3EZ
G0 VNH C. Langdon 652 Hotham Road South, Hull HU5 5LE
G0 VNI S. Williams The Croft, Ringwood Road, Southampton SO40 7LA
G0 VNJ M. Guy 38 Sandy Lane, Charlton Kings, Cheltenham GL53 9DQ
G0 VNK J. Harders Kalckreuthweg 17, Hamburg 22607 Germany
G0 VNO D. Johns 8 Hill Fold, Dawley Bank, Telford TF4 2QE
G0 VNQ W. Askam 10 Staunton Close, Castle Donington, Derby DE74 2XA
G0 VNW J. Carrington 74 Ogle Street, Hucknall, Nottingham NG15 7FR
G0 VNY J. Crawford 2C Papillon House, Balkerne Gardens, Colchester CO1 1PR
G0 VOB K. Fuller 28 Bradshaw Way, Irchester, Wellingborough NN29 7DP
G0 VOE C. Iles Moon Cottage, 2 Mountview Terrace, Pawlett, Bridgwater TA6 4SL
G0 VOF M. Walmsley 121 Roe Lee Park, Blackburn BB1 9SA
GW0 VOG D. Roberts 16 Pentre Isaf, Old Colwyn, Colwyn Bay LL29 8UT
G0 VOJ S. Williams 6 Oak Road, Clanfield, Waterlooville PO8 0LJ
G0 VOK N. Reilly 22 Lee Drive, Northwich CW8 1BW
GM0 VOL V. Nerurkar 26 Fothringham Drive, Monifieth, Dundee DD5 4SW
GM0 VOU P. Scott 30 Main St., Newmills, Dunfermline KY12 8SS
G0 VOV D. Hartwell 57 Wyatts Covert, Denham, Uxbridge UB9 5DJ
GU0 VPA R. Peeters 17 Grosse Hougue, Saltpans Road, St. Sampson., Guernsey GY2 4NS Guernsey
G0 VPC M. Davies 27 Bedford Road, Orpington BR6 0QJ
G0 VPE Reading And West Berkshire Raynet c/o D. Pibworth 20 Marathon Close, Woodley, Reading RG5 4UN
GM0 VPG N. Thackrey Rhumore, Kilmun, Dunoon PA23 8SB
G0 VPH M. Austin 107 Spicer Close, London SW9 7UE
G0 VPJ J. Stacey 16 Crane Drive, Verwood BH31 6QB
G0 VPO A. Robinson 2 Fort Street, Sandown PO36 8BA
G0 VPS D. Bennett 3 Sivilla Road, Kilnhurst, Mexborough S64 5TY
G0 VPT L. Smith Hillside, Kings Mill Lane, Stroud GL6 6SA
G0 VPU Dr M. Burbidge 3 Kirklands, Hest Bank, Lancaster LA2 6ER
G0 VPV G. Pesarini 53 Llanvanor Road, London NW2 2AR
G0 VPW M. Rhodes 102 Malvern Crescent, Little Dawley, Telford TF4 3JF
G0 VPX Halifax & District ARS c/o M. Worsfold 9 Montacute Road, Lewes BN7 1EN
G0 VPY E. Weston 33 William Street, Tunbridge Wells TN4 9RP
G0 VPZ J. Greenfield 36 Barttelot Road, Horsham RH12 1DQ
G0 VQA J. Groves 24 Gimble Way, Pembury, Tunbridge Wells TN2 4BX
G0 VQB M. Grainger 174 Woodlands Road, Gillingham ME7 2SX
G0 VQD P. Newberry 2 Gatewycke Terrace, Tanyard Lane, Steyning BN44 3RL
G0 VQG M. Thomas 10 Finchale Crescent, Darlington DL3 9SA
G0 VQH J. Bailey 9 Little Ditton Woodditton, Newmarket CB8 9SA
G0 VQJ J. Metcalfe 158 Barrowford Road, Colne BB8 9QR
G0 VQK S. Haigh 2 Locker Avenue, Warrington WA2 9PS
G0 VQL G. Guild 15 Canalside Cottages, Chester Road, Runcorn WA7 3AQ
G0 VQM C. Reid 54 Montacute Road, New Addington, Croydon CR0 0JE
G0 VQO N. Haynes 139 Hull Road Anlaby, Hull HU10 6ST
G0 VQR T. Cannon 35 Loddon Bridge Road, Woodley, Reading RG5 4AP
G0 VQS P. Beck 6 Holly Close, Little Bealings, Woodbridge IP13 6PL
G0 VQT F. Seabourne 11 Heyford Avenue, London SW20 9JT
G0 VQW A. Jack 1 Dockle Way, Upper Stratton, Swindon SN2 7LQ
G0 VQX D. Thomas 11 Fordwells Drive, Bracknell RG12 9YL
G0 VQY P. Wooding 31 Douglas Avenue, Brixham TQ5 9EL
GW0 VQZ N. Jones 59 Woodfield Terrace, Penrhiwceiber, Mountain Ash CF45 3YA
G0 VRE A. Challis 12 Moorland Crescent, Boultham Moor, Lincoln LN6 7NL
G0 VRF M. Waples 42 Butts Road, Wellingborough NN8 2PU
G0 VRH D. Bower 29 Worksop Road, Mastin Moor, Chesterfield S43 3DH
G0 VRK C. Seabridge 13 Hillside Avenue Forsbrook, Stoke-on-Trent ST11 9BH
GW0 VRL C. Saunders 14 Portway, Bishopston, Swansea SA3 3JR
G0 VRM A. Russell 3 St. Nicholas Close, North Newbald, York YO43 4TT
GM0 VRP R. Phillips 18 Broomridge Road St. Ninians, Stirling FK7 0DT
G0 VRQ R. Rendall 633 Moston Lane, Manchester M40 5QD
G0 VRS K. Gillen 33 Norwich Close, Ashington NE63 9RY
G0 VRT M. Ritson 14 Dunsdale Road, Holywell, Whitley Bay NE25 0NG
G0 VRU B. Gee 11 Spitfire Avenue, Grimoldby, Louth LN11 8UJ
G0 VRV S. Philipps 24 Acres End, Amersham HP7 9DZ
G0 VRW P. Wadhams Brickwood, Bigbury Road, Canterbury CT4 7ND
G0 VRX C. Jenkins 31 Ashbrook Crescent, Rochdale OL12 9AJ
G0 VRZ K. Tuer Broad Ing, Penrith CA10 2LL
G0 VSB A. Gibbs 18 Grange Close, Ludham, Great Yarmouth NR29 5PZ
G0 VSG I. Smyth Fuchsia Cottage Holywell Lake, Wellington TA21 0EL
G0 VSH M. Wright 3 St Anthony, Melleiha Malta
G0 VSJ P. Taylor 133 Beech Drive, Shifnal TF11 8HZ
G0 VSK A. Thomas 3 Barnfield Way, Cannock WS12 0PR
G0 VSL K. Hill 18 Baker Close, Chasetown, Burntwood WS7 4GU
G0 VSM T. Day Box 204, The Postroom, Calle San Jaime 5, Benijofar 3178 Spain
GW0 VSO P. Burchell 5 Brentwood Place, Ebbw Vale NP23 6JR
G0 VSS R. King 19 Greenhayes, Cheddar BS27 3HZ
GW0 VST P. Sandham 6 Skomer Close, Nottage, Porthcawl CF36 3QH
GW0 VSW J. Mason 2 Golwg-Y-Bryn, Off Woodland Road, Skewen, Neath SA10 6SP
G0 VSY G. Forster 3 Foxmoor, Bishops Cleeve, Cheltenham GL52 8SS
G0 VSZ D. Forster 3 Foxmoor, Bishops Cleeve, Cheltenham GL52 8SS
G0 VTA R. Collins 30 Upham Road, Swindon SN3 1DN
G0 VTC L. King 4 Glenhurst Avenue, Ruislip HA4 7LZ
G0 VTD S. Towler 77 Glebe Road, Hull HU7 0DU
G0 VTI T. Ibbitson 36 Knoll Park, East Ardsley, Wakefield WF3 2AX
G0 VTJ A. Jameson 9 White Laithe Green, Leeds LS14 2EP
G0 VTL H. Bradfield Glebe Farm, Wharf Street, Leicester LE4 8AY
G0 VTM J. Pearson 23 Glebe Avenue, Mitcham CR4 3DZ
GI0 VTS R. Boyle 53 Portaferry Road, Cloughey, Newtownards BT22 1HP
G0 VTV K. Groom 2 Ruins Barn Road, Tunstall, Sittingbourne ME10 4HS
G0 VUC C. Boughton 79 Hawfield Lane, Burton-on-Trent DE15 0BY
G0 VUH A. France 31 Broadway Swinton, Mexborough S64 8HD
G0 VUL A. Hardie Tana-Merah, Church Lane, Retford DN22 9NQ
G0 VUM P. Cooper 16 Mortomley Close, High Green, Sheffield S35 3HZ
G0 VUN T. Wooding 101 Park Farm Road, Ryarsh, West Malling ME19 5JX
G0 VUT C. Turner 28 Reading Street, Broadstairs CT10 3AZ
G0 VUX RSGB Transatlantic Test Centenary Club c/o C. Walker Bolton School Ltd, Chorley New Road, Bolton BL1 4PA
GM0 VUY R. Turnbull 6 Letham Gait, Dalgety Bay, Dunfermline KY11 9GT

UK Callsigns

G0	VVA	M. Newbold 10 Shaw St., Derby DE22 3AS
GI0	VVC	J. Serridge 21 Lassara Heights, Warrenpoint, Newry BT34 3PG
G0	VVF	D. Turner 2 Toulmin Drive, Swadlincote DE11 0BH
G0	VVG	A. Lomas 14 Ingledene Caravan Site, Lawsons Road, Thornton-Cleveleys FY5 4DL
G0	VVK	P. Neale 15 Kenmore Walk, Wibsey, Bradford BD6 3JQ
G0	VVP	R. Hatton Plot 2, Sutton Road, Hull HU7 5YY
G0	VVQ	J. Mahon 2 Rob Lane, Newton-le-Willows WA12 0DR
G0	VVR	C. Leigh 4 Corrick Close Draycott, Cheddar BS27 3UB
G0	VVT	E. Murphy 21 Standard Street, Stoke-on-Trent ST4 4NG
G0	VVX	Croham Callers c/o M. Samuel 71 Brighton Road, South Croydon CR2 6EE
G0	VVZ	Rvd. D. Matthiae 142 South Wing, Fairfield Hall, Kingsley Avenue, Fairfield, , Hitchin SG5 4FY
G0	VWB	M. Davies 23 Star Lane, Folkestone CT19 4QH
GW0	VWD	A. Ball 20 Hillcrest, Brynna, Pontyclun CF72 9SJ
G0	VWE	P. Whitfield 55 Greenways, Sutton, Woodbridge IP12 3TP
G0	VWF	C. Howell 24 Bellring Close, Belvedere DA17 6LP
G0	VWH	M. Wright 27 Ellesmere Close Hucclecote, Gloucester GL3 3DH
G0	VWP	T. Sayner 59 Horner St., York YO30 6DZ
G0	VWQ	D. Cockburn Spindleberry, Weare Giffard, Bideford EX39 4QR
GW0	VWS	G. Dickin 10 Pen Y Maes, Llanfechain SY22 6XL
G0	VWT	T. Holland 135 Newcastle Road, Stone ST15 8LF
GI0	VWU	A. Mccabe 40 Rydalmere St., Belfast BT12 6GF
G0	VWV	R. Giles 9 Bower Green, Lords Wood, Chatham ME5 8TN
G0	VWW	T. Robson 58 Burton Road, Lincoln LN1 3LB
G0	VWX	E. Collinson 40 Cock Robin Lane, Catterall, Preston PR3 1YL
GM0	VWZ	S. Lawrie 4 Glenavon Drive, Airdrie ML6 8QG
GM0	VXA	P. Marriott Craiglea Cottage, Omoa Road Cleland, Motherwell ML1 5LQ
G0	VXB	S. Cockburn 3 Inglebrook Heights, Westward Ho!, Bideford EX39 1GU
G0	VXC	M. Coles 133 Highthorn Road, Kilnhurst, Mexborough S64 5UU
G0	VXD	G. Clayton 27 Simpsons Lane, Knottingley WF11 0HG
G0	VXE	Eccentricity Radio Hams Club c/o D. Herbert 50 St. Leonards Crescent, Scarborough YO12 6SP
G0	VXG	R. Wilkinson 139 Church Road, Jackfield, Telford TF8 7ND
G0	VXJ	M. Finch 73 Cordingley Way, Donnington, Telford TF2 7LJ
G0	VXK	J. Davies 26 Beverley Close, Wylde Green, Sutton Coldfield B72 1YF
G0	VXM	G. Simmons 90 Pollards Fields, Knottingley WF11 8TD
GM0	VXQ	P. Mulheron 78 South Commonhead Avenue, Airdrie ML6 6PA
G0	VXS	C. Newman 89 Goose Cote Lane, Oakworth, Keighley BD22 7NQ
G0	VXV	P. Nicholls 21 Woodlea Close Yeadon, Leeds LS19 7NL
G0	VXW	M. Rendall 633 Moston Lane, Manchester M40 5QD
G0	VXX	A. Fry 96 Westcroft Gardens, Morden SM4 4DL
G0	VXY	G. Forde 22 Alloa Road, London SE8 5AJ
GM0	VXZ	I. Terris 1 Linden Avenue, Wishaw ML2 8SE
GM0	VYB	T. Cleghorn 9 Bridge Of Aldouran, Leswalt, Stranraer DG9 0LW
G0	VYC	M. Dawson 11 Owls Retreat, Colchester CO4 3FE
GW0	VYF	P. Simmons 153 St. Davids Road, Letterston, Haverfordwest SA62 5SS
GW0	VYG	Bury RS c/o T. Jones Penrhiw Bach, Bryngwran, Holyhead LL65 3RD
G0	VYK	R. Heesom 41 Ridgeway, Pembury, Tunbridge Wells TN2 4ER
GM0	VYL	P. Maver 69 Mayfield Crescent, Musselburgh EH21 6EX
G0	VYN	M. Guest 53 Ringwood Close, Furnace Green, Crawley RH10 6HQ
G0	VYP	W. Southworth 58 Moyse Avenue, Walshaw, Bury BL8 3BL
G0	VYQ	T. Mcinerney 41 Newton Way, Tongham, Farnham GU10 1BY.
G0	VYT	M. Garry 14 Adrians Close, Mansfield NG18 4HG
G0	VYU	R. Raynor-Smith 10 Marsh Road, Trowbridge BA14 7PR
G0	VYV	E. Smith Magnolia House, Main Road, Highbridge TA9 3QZ
G0	VYX	P. Rumsam 24 Hamer Avenue, Rossendale BB4 8QH
GM0	VYY	D. Macconnell Mine Cottage, Tashieburn Road, Lanark ML11 8ES
G0	VZA	B. Howlett 156 Lanarch Road, Waverley, Dunedin New Zealand
G0	VZB	Po Research Cnt c/o R. Ferris Polmarth House, Carnmenellis, Redruth TR16 6NT
G0	VZE	E. Cook 8 Calvert Grove, Newcastle ST5 8QA
G0	VZI	P. Robinson 5 Coppice Close, Haxby, York YO32 3RR
G0	VZK	C. Sampson 1 Warton Lane, Austrey, Atherstone CV9 3EJ
G0	VZL	E. Levring 3 Evelyn Croft, Wylde Green, Sutton Coldfield B73 5LF
G0	VZN	K. Voller 20 Browns Lane, Uckfield TN221RY
G0	VZO	J. Koops 64 Winchester Avenue, Nuneaton CV10 0DW
G0	VZT	S. Hopton 5 Wellington Close Marske-By-The-Sea, Redcar TS11 6NW
G0	VZV	D. A'Bear 7 Meadow Close, Bembridge PO35 5YJ
G0	VZX	P. Dart 208 Elburton Road, Plymouth PL9 8HU
G0	WAB	W. Neale 5 Gibbs Court, Dane Close, Wirral CH61 3XS
G0	WAC	K. Wells 42 Eggesford Road, Stenson Fields, Derby DE24 3BH
G0	WAD	C. Dyson 21 Highmoor, Kirkhill, Morpeth NE61 2AS
G0	WAE	D. Phillips 14 Seymour Road, Newton Abbot TQ12 2PU
GI0	WAH	W. Hutchman 35 Carlingford Park, Newry BT34 2NY
G0	WAL	W. Reed 10 Ashmeads Way, Wimborne BH21 2NZ
G0	WAM	S. Stevens 25 Busticle Lane, Sompting, Lancing BN15 0DJ
G0	WAN	M. Gill 23 Walmers Avenue, Higham, Rochester ME3 7EH
G0	WAS	A. Smith 9 Moor Lane Maulden, Bedford MK45 2DJ
G0	WAT	P. Brice-Stevens 31 Lodgefield, Welwyn Garden City AL7 1SD
G0	WAW	I. Oura The Quoins, Gloucester Road, Bath BA1 8AD
G0	WAX	L. Merrin 10 Hawkesbury Road, Canvey Island SS8 0EX
G0	WAY	R. Slimmon 22A Southbourne Place, Cannock WS11 4SA
G0	WBA	K. Fradgley 84 Church Road, Wordsley, Stourbridge DY8 5AU
G0	WBC	Dr C. Mortimer 6 Honeycomb Close, Narborough, Leicester LE19 3PS
G0	WBL	S. Hughes 43 The Cloisters, Rickmansworth WD3 1HL
G0	WBO	J. Jameson 2 Edgar Close, Swanley BR87JJ
G0	WBR	T. Johnson 7 Southover Way, Hunston, Chichester PO20 1NY
G0	WBS	Burnham Beeches c/o P. Wentworth 46 Woodside Avenue, Cinderford GL14 2DW
G0	WBT	J. Woodcock 54 Longworth Road, Horwich, Bolton BL6 7BE
G0	WBV	R. Burchell 1 Broad Forstal Farm Cottages, Tilden Lane, Tonbridge TN12 9AX
G0	WCB	A. Bathurst 81 Heatherstone Avenue, Dibden Purlieu, Southampton SO45 4LE
GI0	WCE	D. Cromie 11 Cherryvalley Park West, Belfast BT5 6PU
G0	WCH	B. Prestage 17 Moorgate Road, Hindringham, Fakenham NR21 0PT
G0	WCI	M. Hand 24 Ettingshall Road, Bilston WV14 9UJ
G0	WCJ	J. Slater 25 Croft Lane, Diss IP22 4NA
G0	WCK	Bewlay Bros ARC c/o D. Hornby 7 Milton Street, West Bromwich B71 1NJ
G0	WCO	P. Anness 18 Plaisir Place, Thurston Road, Lowestoft NR32 1RY
G0	WCR	Three Counties ARC c/o M. Knott 76 New Barns Avenue, Mitcham CR4 1LF
G0	WCS	S. Ormerod 6 New Wokingham Road, Crowthorne RG45 7NR
G0	WCU	R. Roberts 9 Stones Close, Hogsthorpe, Skegness PE24 5NZ
G0	WCZ	G. Sutherland 8 Dukes Avenue, London N3 2DD
G0	WDC	D. Clark Meadowcroft Bungalow, Ugthorpe, Whitby YO21 2BL
GM0	WDF	J. Dunlop West Dougliehill Farm, Dougliehill Road, Port Glasgow PA14 5XF
G0	WDG	L. Morgan Flat 30, Raleigh Court, Sherborne DT9 3EQ
G0	WDK	D. Hackett 131 Station Road South, Walpole St. Andrew, Wisbech PE14 7LZ
G0	WDQ	S. Collins 30 Upham Road, Swindon SN3 1DN
G0	WDT	R. Emery 10 Penarth Place, Newcastle ST5 2JL
G0	WDU	M. Round 2 Bell Mead, Studley B80 7SH
GW0	WDV	S. Steddy 1 Springfield Close, Wenvoe, Cardiff CF5 6DA
G0	WDW	W. Williamson Monfa Walford Heath, Shrewsbury SY4 2HT
G0	WDX	Usk Side RC c/o R. Morgan 56 The Meadows, Hull HU7 6EE
G0	WEA	R. Foster 60 Sandford Close Bransholme, Hull HU7 4HN
G0	WEB	Denby Dale And District ARS c/o R. Webb 25 Greenfield Croft, Bilston WV14 8XD
GM0	WED	E. Holt Ashwell, St. Ola, Kirkwall KW15 1SX
G0	WEF	N. Kenworthy 19 Parkside Close, Radcliffe, Manchester M26 2QS
G0	WEO	D. Waterfield 20 St. Andrew Road, Evesham WR11 2NR
GW0	WER	P. Moran 1 Plas Issa, Brook Street, Wrexham LL14 3EE
G0	WET	R. Wright Ul Dordoy 32/34 Kb 52, Bishkek 720082 Kyrgyzstan
G0	WEV	S. Mcmullen 70 Sylvan Avenue, Timperley, Altrincham WA15 6AB
G0	WEX	S. Elliott 1 The Gables, Oddfellows Road, Hope Valley S32 1DU
GW0	WEY	J. Doores 14 Parc Tyddyn, Red Wharf Bay, Pentraeth LL75 8NQ
GM0	WEZ	Dr P. Ewing Kildonan House, Caerlaverock Farm, Crieff PH5 2BD
GM0	WFA	T. Clark 23 Letham Place, St. Andrews KY16 8RB
GM0	WFB	J. Keenan 53 Clermiston Crescent, Edinburgh EH4 7DF
G0	WFD	M. Farrell 24 Lindsay Street, Stalybridge SK15 2LT
G0	WFE	S. Dean 42 North Street, Wareham BH20 4AQ
G0	WFF	C. Whelan 50 Garrick Close, Ings Road Estate, Hull HU8 0ST
G0	WFG	A. Vinters 106 Halifax Road, Ripponden, Sowerby Bridge HX6 4AG
G0	WFH	C. Gresswell 11 Dandy Dinmont Caravan Park Blackford, Carlisle CA6 4EA
G0	WFK	H. Bamford Upper Twynings Farm, Pumphouse Lane, Droitwich WR9 7EB
G0	WFL	M. Street 262 Carter Knowle Road, Sheffield S7 2EB
G0	WFM	P. Wallace 8 Wemmick Close, Rochester ME1 2DL
G0	WFO	D. Tarry 2 Kestrel Heights, Codnor Park, Nottingham NG16 5PW
G0	WFP	W. Painz 8 Warminger Court, Ber Street, Norwich NR1 3ED
G0	WFQ	M. Troy 22 Jackie Wigg Gardens Totton, Totton, Southampton SO409LZ
G0	WFT	D. Saunders 14 Shelton Avenue, Toddington, Dunstable LU5 6EL
G0	WFV	A. Corbett Lan Y Llyn, 10 Rutland Avenue, Waddington LN5 9FW
G0	WFX	S. Hayes 94 Kingshurst Way, Birmingham B37 6JQ
G0	WGA	M. Merrin 10 Hawkesbury Road, Canvey Island SS8 0EX
G0	WGB	J. Atkins 6 Carolina Gardens, Plymouth PL2 2ER
GW0	WGE	D. Thomas 85 Tan Y Bryn, Burry Port SA16 0LD
G0	WGH	J. Edwards 21 Ridgeway, Ottery St. Mary EX11 1DT
G0	WGJ	G. Jarvis 38 West Cliff Road, Dawlish EX7 9DY
G0	WGL	P. Furness 6 Westbrook Square, Manchester M12 5PU
GW0	WGM	T. Jones 26 Heol Maenofferen, Blaenau Ffestiniog LL41 3DL
GW0	WGN	J. Lyons 4 London Road, Pembroke Dock SA72 6DU
G0	WGP	R. Glover 5 The Vineries, Burgess Hill RH15 0ND
G0	WGV	I. French Penmellyn, Tarrandean Lane, Truro TR3 7NW
GW0	WGW	B. James 72 Park Place, Bargoed CF81 8NB
G0	WHC	W. Chesterton 61 Butt Lane, Blackfordby, Swadlincote DE11 8BG
G0	WHD	G. Hassall 3 Sunny Bank, Cark In Cartmel, Grange-over-Sands LA11 7PF
G0	WHL	E. Barnes 10 Cranbourne Road, Rochdale OL11 5JD
G0	WHN	J. Edwardes Pippins, 1 Horse Lane Orchard, Ledbury HR8 1PP
G0	WHO	R. Clutson 151 Stepney Road, Scarborough YO12 5NJ
G0	WHQ	K. Wilson 11 Harbour Close, Murdishaw, Runcorn WA7 6EH
G0	WHV	R. Bessell 6 Bayford Lodge, Wellington Road, Pinner HA5 4NJ
G0	WHY	Rvd. P. White 11A Dowse Road, Devizes SN10 3FN
G0	WHZ	T. Tipping Flat 24, The Moorings, Kingsbridge TQ7 1LP
GW0	WIB	M. Mcdermott Brynmeudwy, Llwyndrain, Llanfyrnach SA35 0AT
G0	WIC	E. Clark 8 Rose Cottages, Shotton Colliery, Durham DH6 2NF
G0	WIE	P. Swan The Old Rectory, Sampford Brett, Taunton TA4 4LA
G0	WIG	P. Ellison 28 Kingscote Road East, Cheltenham GL51 6JS
G0	WIL	M. Williams The Croft, Ringwood Road, Southampton SO40 7LA
G0	WIS	G. Woodbury 4 Henley Drive, Droitwich WR9 7RX
G0	WIT	M. Perry 52 Somerset Avenue, Chessington KT9 1PN
G0	WIW	N. Appleby Westholme, Asterby Lane Asterby, Louth LN11 9UE
G0	WIX	A. Beeching 174 Grove Road, Rayleigh SS6 8UA
G0	WIY	W. King 3 Grove Court, The Waterloo, Cirencester GL7 2PZ
GM0	WIZ	I. Waugh 2 Gilbeck Avenue, Dumfries DG1 4DN
G0	WJA	L. Coleman 1 Kerstin Close, Cheltenham GL50 4SA
G0	WJC	M. Briscoe 34 Winterton Drive, Low Moor, Bradford BD12 0UX
G0	WJD	D. Jackson 10 Wisdoms Green, Coggeshall, Colchester CO6 1SG
G0	WJH	J. Kemp 85 St. Andrews Way, Church Aston, Newport TF10 9JQ
GI0	WJI	R. Bicker 62 Spa Road, Ballynahinch BT24 8PT
G0	WJJ	D. Mullaney 62 Darby Road, Wednesbury WS10 0PN
G0	WJK	A. Cobb 37 Nordham, North Cave, Brough HU15 2LT
G0	WJN	P. Howland 73 Timberdine Avenue, Worcester WR5 2BG
G0	WJS	D. Mellings 36 Hillwood Drive, Glossop SK13 8RJ
G0	WJU	R. Crewe 11 Osbert Close, Norwich NR1 2NL
G0	WJV	C. Page 73 Two Saints Close, Hoveton, Norwich NR12 8QR
G0	WJX	R. Davies 84 Hob Hey Lane, Culcheth, Warrington WA3 4NW
G0	WJZ	I. Donachie 72 Gresham Road, Norwich NR3 2NQ
G0	WKA	M. Drever 66 Milton Road, Branton, Doncaster DN3 3PB
G0	WKH	M. Thomas 63 Corfe Way, Broadstone BH18 9ND
G0	WKI	H. Yearl 191 Tamworth Road, Kettlebrook, Tamworth B77 1BT
G0	WKJ	C. Towle 32 Charlwood Road, Luton LU4 0BU
G0	WKM	N. Purchon Mitchells Elm House, Wanstrow, Shepton Mallet BA4 4SN
G0	WKN	K. Whitmore Amosford, Lutton Gowts, Spalding PE12 9LQ
G0	WKQ	South Tyneside ARS c/o D. Harbron 48 Sheridan Road, South Shields NE34 9JJ
G0	WKT	M. Pugh 28 Marsh Road, Wilmcote, Stratford-upon-Avon CV37 9XR
G0	WKU	P. Yea 89 Laxton Road, Taunton TA1 2XF
G0	WKW	V. Pleshkevich Rya □ngavög 53, Rydeböck 25730 Sweden

GM0 WKZ B. Pybus 43 Kinacres Grove Carriden, Bo'ness EH51 9LT
G0 WLC B. Cole 499 Lightwood Road, Stoke-on-Trent ST3 7EN
G0 WLD M. Russell 23A St Ann'S Road, Barnes, London SW13 9LH
G0 WLF W. Storace-Rutter 46 Norbury Court Road, London SW16 4HT
G0 WLG P. Blizzard 12 Hilton Road, Malvern WR14 3NP
GW0 WLI J. Ruddle Flat 58, Thomas Court, Cardiff CF23 5EZ
GW0 WLN C. Purcell 31 Brookdale Court, Church Village, Pontypridd CF38 1RP
GW0 WLQ R. Evans 4 Llantrisant Road, Tonyrefail, Porth CF39 8PP
G0 WLR Dr M. Pettigrew 26 Victoria Court, Sheffield S11 9DR
G0 WLX A. Suckling 54 Lake Hill, Sandown PO36 9HF
G0 WMB A. Houghton 2 Beanhill Crescent, Alveston, Bristol BS35 3JG
G0 WMC P. Norman 53 Saddlers Park, Eynsford, Dartford DA4 0HA
G0 WMD M. De Silva 31 Rosemary Avenue, Hounslow TW4 7JQ
G0 WMG G. Studd 34 The Broadway, Lancing BN15 8NY
GM0 WMH A. Hughes 7 Eden Grove, Kirkpatrick Fleming, Lockerbie DG11 3AT
G0 WMJ J. Walker 29 Akenside Court, New Fort Way, Bootle L20 4UU
G0 WMN W. Mcnab 74 Parkhouse Road, Minehead TA24 8AF
G0 WMQ E. Williams 16 Birch Grove, Wallasey CH45 1JG
GW0 WMT 126 (City of Derby) Sqn ATC c/o S. Robinson 23 Thornhill Road, Cardiff CF14 6PE
G0 WMU R. Davis 49 Goodway Road, Great Barr, Birmingham B44 8RL
G0 WMW D. Roberts 17 Edgecombe Avenue, Weston-Super-Mare BS22 9AY
G0 WMX G. Giuliani 32 Davison Street, Newburn, Newcastle upon Tyne NE15 8NB
G0 WMY P. Hanmer 182 Tollemache Road, Prenton CH43 7SE
G0 WMZ R. Duckworth 39 George Lane, Bredbury, Stockport SK6 1AS
GW0 WNB S. Blumson 15 Bryn Colwyn, Colwyn Bay LL29 9LJ
G0 WND T. Sunouchi 200-20 Morooka-Cho, Kouhoku-Ku, Yokohama 222-0002 Japan
G0 WNF E. Clark Flat B, 145 Church Street, Whitby YO22 4DE
G0 WNJ M. Bottomley 21 Priory Park, Grosmont, Whitby YO22 5QQ
GM0 WNR A. Campbell 17 Arran Road, Motherwell ML1 3NA
GM0 WNS I. Calder Cut The Wind Cottage, Arbroath DD11 4RH
G0 WOA M. Lyon 128 Lydyett Lane, Barnton, Northwich CW8 4JU
G0 WOC J. Doxey 41 Shady Grove, Hilton, Derby DE65 5FX
G0 WOI Avon Scouts ARC c/o R. Laney 7 Downfield Close, Alveston, Bristol BS35 3NJ
G0 WOM T. Renshaw 1 Basford View, Cheddleton, Leek ST13 7HJ
G0 WON M. Kelly 9 Thetford Road, Brandon IP27 0BS
G0 WOP P. Dabell 21 Tatton Road North, Heaton Moor, Stockport SK4 4RL
G0 WOU D. Atkinson 20 Burnard Close, Plymouth PL6 6LG
GI0 WOW R. Cooper Rockmount, 10 Carmorn Road, Toomebridge, Antrim BT41 3NX
G0 WPC N. Crossley 55 Scholey Avenue Woodsetts, Worksop S81 8SF
G0 WPF J. Towle 36 Old Orchard, Haxby, York YO32 3DT
GM0 WPI A. Lister Heathfield Easter Hardmuir, Nairn IV12 5QG
G0 WPL J. Wozniak 40 Cockhill, Trowbridge BA14 9BQ
G0 WPM N. Gilboy 6 Talisman Close, Sherburn Village, Durham DH6 1RJ
G0 WPO N. Griffiths Orchard Marina (Nb T'Dreme) Robbo Farm School Road Rudheath, Northwich CW9 7RG
GM0 WPU K. Faloon Moss-Side Croft, 6 Rothiemay, Huntly AB54 5NY
GI0 WPV O. Price 18 Hill Crest Walk, Bangor BT20 4DF
GM0 WPW T. Halligan 4 Trainers Brae, North Berwick EH39 4NR
G0 WPX WPX CG c/o D. Beattie Hares Cottage, Woolston, Church Stretton SY6 6QD
G0 WQA J. Ward 49 Woodhall Drive, Lincoln LN2 2AE
G0 WQC J. Keeling 31 Tudor Drive, Otford, Sevenoaks TN14 5QP
GW0 WQP D. Taylor 18 Oakfield Avenue, Chepstow NP16 5NE
G0 WQQ D. Bennett Shrove Furlong Longwick Road, Princes Risborough HP27 9HE
G0 WQW J. Thompson 78 Lowestoft Road, Worlingham, Beccles NR34 7RD
G0 WQY L. Mansfield 25 Carlton Road, Derby DE23 6HB
G0 WRC Bay ARG c/o L. Volante Richmond House, Icknield Street, Birmingham B38 0EP
G0 WRE P. Scarratt 339 Utting Avenue East Norris Green, Liverpool L11 1DF
GM0 WRH E. Castle 38 Davieland Road, Giffnock, Glasgow G46 7LU
GW0 WRI Kelso ARS c/o R. Lewis 26 Bryn Road, Upper Brynamman, Ammanford SA18 1AU
G0 WRK P. Taylor Old Acres, Priory Road, Yeovil BA22 8NY
G0 WRL D. Webb 16 Burrowfield, Bruton BA10 0HR
G0 WRM Lichfield Raynet Group c/o M. White 3 High Street, Clay Cross, Chesterfield S45 9DX
G0 WRN J. Hodges 48 Beach Road, Severn Beach, Bristol BS35 4PF
G0 WRQ O. Baxter 5 Church Path, Bridgwater TA6 7AJ
GM0 WRR J. Scott 70 Montford Avenue, Glasgow G44 4PA
G0 WRS Warrington ARC c/o M. Isherwood 32 Franklin Close Old Hall, Warrington WA5 8QL
G0 WRT P. Winfield 150 Tinshill Road, Leeds LS16 7PN
GM0 WRU R. Holmes 3 Buchan Street, Wishaw ML2 7HG
GM0 WRV R. Spink 44 West Park, Inverbervie, Montrose DD10 0TT
G0 WSA J. Proctor Barfords Farm, Swineshead Road, Kirton Holme, Boston PE20 1SQ
G0 WSB M. Brimley 42 Grange Road, Netley Abbey, Southampton SO31 5FE
G0 WSC R. Connett 15 Channels Lane, Horton, Ilminster TA19 9QL
G0 WSD G. Swann 1 Beaver Close, Chichester PO19 3QU
G0 WSH R. Munt Box 166, Se-177 23 Jarfalla SWEDEN Sweden
G0 WSI Dr W. Warburton Marlins, Water Lane, Dibden Purlieu, Southampton SO45 4SB
G0 WSJ S. Jelly 10 Whitemarsh, Mere, Warminster BA12 6BP
GD0 WSK S. Kempner Sonnish-Ny-Marrey, St. Mary'S Avenue, Port St. Mary IM9 5ET Isle of Man
G0 WSP P. Croft 82 Granby Road, Buxton SK17 7TJ
GM0 WSR Strathclyde Regional Raynet Groups c/o D. Mackinnon 60 Mount Stuart Drive, Wemyss Bay PA18 6DX
G0 WSS Horsham Scout ARG c/o P. Head 36A Ashacre Lane, Worthing BN13 2DH
G0 WSY J. Adams 53 Princess Road, Rochdale OL16 4AY
G0 WTA N. Midworth 1 Highfields Drive, Loughborough LE11 3JS
G0 WTB C. Hicks 59 Elmsfield Avenue, Norden, Rochdale OL11 5XW
G0 WTC W. Clark 41 Brook Close, Jarvis Brook, Crowborough TN6 2ET
G0 WTD T. Stokes 33 Talbot Drive, Euxton, Chorley PR7 6PD
G0 WTF A. Wright 4 Wilshere Road, Welwyn AL6 9PX
G0 WTG D. Way 25 Kinson Avenue, Poole BH15 3PH
G0 WTI G. Lightfoot 1 Trewyn Road, Holsworthy EX22 6HX
G0 WTK R. Jenkins 11 Westfield Drive, Worksop S81 0JS
G0 WTL G. Fisher 6 Totternhoe Road, Dunstable LU6 2AG
G0 WTM D. Sutton 32 Queensway, Euxton, Chorley PR7 6PW
G0 WTO E. Birch 4 Kynnesworth Gardens, Higham Ferrers, Rushden NN10 8NH
GM0 WTP J. Whitecross 7 Northfield Crescent, Edinburgh EH8 7PU
G0 WTR M. Robertson Clayhill Cottage, The Street, Aldham, Ipswich IP7 6NN
G0 WTW G. Thompson Lyngrove, Seaton Lane, Seaham SR7 0LP
G0 WUA P. Harris 1 Newlands, Landkey, Barnstaple EX32 0NJ
G0 WUG G. Bishop 8 Bulstrode Place, Kegworth, Derby DE74 2DS
G0 WUH H. White 111 Beacon Glade, South Shields NE34 7QU
G0 WUI W. Cassidy 17 Catcheside Close, Whickham, Newcastle upon Tyne NE16 5RX
GW0 WUL C. Minard 3 Riverside Close, Aberfan, Merthyr Tydfil CF48 4RN
GW0 WUM E. Roobottom Puffin View, Abergwyngregyn, Llanfairfechan LL33 0LL
G0 WUO R. Berkeley 4 Raleigh Road, Leasowe, Wirral CH46 2QZ
GM0 WUP W. Steele 35 Devlin Court, Whins Of Milton, Stirling FK7 0NP
GM0 WUR G. Steele 1 James Street, Bannockburn, Whins Of Milton, Stirling FK7 0NQ
G0 WUS C. Rayns 18 Osprey Close, Leicestershire LE9 6RS
G0 WUU J. Purcell 2 Windsor Close, Cawood YO8 3WE
G0 WUV J. Dickinson 112 Stoneleigh Avenue, Longbenton, Newcastle upon Tyne NE12 8XQ
G0 WUW V. Claridge 105 Barnwood Avenue, Gloucester GL4 3AG
GM0 WUX J. Donnan 41 Annick Drive, Dreghorn, Irvine KA11 4ER
G0 WUY A. Williamson Millfield Lodge, 151 Hull Road, York YO10 3JX
G0 WVA D. Townsend Flat 32, Weymouth House, Eaton Place, Margate CT9 1FE
G0 WVD T. Lishman 13 Meadow Way, Sandown PO36 8QE
G0 WVE B. Richardson 12 Stoney Lane, Barrow, Bury St. Edmunds IP29 5DD
G0 WVM M. Brooker 4 May Close, Sidlesham, Chichester PO20 7RR
G0 WVT W. Carwood 57 Upton Road, Kidderminster DY10 2YB
G0 WVV D. Jones 4 Back Bower Lane, Gee Cross, Hyde SK14 5NS
G0 WVW D. Smith 16 Browning Drive, Great Sutton, Ellesmere Port CH65 7BW
G0 WVY F. Holt 22 First Avenue, Tottington, Bury BL8 3JA
G0 WWA D. Nicholas 41 Grayling Road, Stourbridge DY9 7AZ
G0 WWD D. Weston 25 Ley Lane, Kingsteignton, Newton Abbot TQ12 3JE
G0 WWE S. Fitton 7 Blackburn Way, West Wick, Weston-Super-Mare BS24 7GT
G0 WWF P. Baron 55 Church View, Brompton, Northallerton DL6 2RD
G0 WWH A. Houghton 3 Billinge Close, Bolton BL1 2JP
G0 WWL A. Babbage 10 Heather Close, Honiton EX14 2YP
G0 WWM Prof. T. Yukawa 5349 Route 12, Birch Hill, Richmond C0B 1Y0 Canada
G0 WWO C. Sawyer 4 Padley Close, Ripley DE5 3FG
G0 WWP S. Griffith 12 Beech Grove, Cliffsend, Ramsgate CT12 5LD
GW0 WWQ H. Burton Vale View, Porth Y Waen Bodfari, Denbigh LL16 4BU
G0 WWR R. Prior 20 Churchfield Road, Walton on the Naze CO14 8BL
G0 WWU R. Pratt 4 King John Avenue Gaywood, King's Lynn PE30 4QA
GM0 WWX R. Johnstone 8 Harris Court, Alloa FK10 1DD
G0 WWZ M. Charlesworth Po Box 841, 9506, Korondal City, South Cotabato Philippines
G0 WXA S. Everitt 125 Victoria Road, Warley, Oldbury B68 9UL
G0 WXC R. Needham 1 The Leas, Sedgefield, Stockton-on-Tees TS21 2DS
G0 WXD P. Atkinson 27 Ranworth Road, Bramley, Rotherham S66 2SP
G0 WXE A. Kelleher 144 Alibon Road, Dagenham RM10 8DE
G0 WXF P. Wright 60 Farnborough Road, Clifton, Nottingham NG11 8GF
G0 WXG J. Lamb 10 Lenhurst Avenue, Leeds LS12 2RE
G0 WXH S. Croot 58 Dixie Street, Jacksdale, Nottingham NG16 5JZ
G0 WXJ P. Badham 201 York Avenue, East Cowes PO32 6BH
G0 WXL E. Harper11 Po Box 5038, Orange 2800 Australia
GW0 WXO A. Davies 81 Brynifor, Mountain Ash CF45 3AB
G0 WXP S. Parkes 23 Curlew Close, Whittlesey, Peterborough PE7 1XQ
G0 WXZ D. Milne 22 Jessica Avenue, Verwood BH31 6LH
G0 WYA N. Taylor 61 Oldbury Road, Rowley Regis B65 0NP
GI0 WYB J. Simpson 66 Ballyportery Road, Dunloy, Ballymena BT44 9BN
G0 WYD R. Coleman 2 Chestnut Off Nepaul Road, Tidworth SP9 7EU
G0 WYF C. Rudge 1 Mill Lane, Alfington, Ottery St. Mary EX11 1PF
G0 WYG D. Biginton 67 Capstone Road, Bromley BR1 5NA
G0 WYI C. Sharpe 35 Fairmead Close, Nottingham NG3 3EQ
GI0 WYK W. Carress 12 Ashbourne Park, Newtownards BT23 7RE
G0 WYM A. Shields 8 Thames Drive, Melton Mowbray LE13 0DS
G0 WYN A. Godwin 27 Melbourne Avenue, Dronfield Woodhouse, Dronfield S18 8YW
GI0 WYO R. Kilgore 3 Summer Meadows Manor, Londonderry BT47 6SE
G0 WYP S. Barker C/O, 14 Cordova Avenue, Manchester M34 2WP
G0 WYQ J. Richardson 24 Brockhall Road, Kingsley, Northampton NN2 7RY
G0 WYR Dr C. Fox 36 Haig Drive, Slough SL1 9HB
G0 WYT I. Seabright 78 Lazy Hill, Birmingham B38 9PA
G0 WYU E. Martin 2 Bospowis, St. Martins Crescent, Camborne TR14 7HN
G0 WYV P. Little 24 Pickwick Crescent, Rochester ME1 2HZ
G0 WYY R. Bell 41 Ravenshill Road, West Denton, Newcastle upon Tyne NE5 5EA
G0 WYZ S. Gillen 33 Norwich Close, Ashington NE63 9RY
G0 WZA D. Wood 2 Buckingham Mews Flitwick, Bedford MK45 1TB
G0 WZB Cambridgeshire Repeater Group c/o B. Burdis 10 Johnston Avenue, Hebburn NE31 2LJ
G0 WZC S. Cooper 27 Polweath Road, Penzance TR18 3PW
G0 WZD Dr J. Paloschi 1 Lander Close, Milton, Cambridge CB24 6EB
G0 WZG K. Norris 15 East View, Choppington NE62 5UF
G0 WZH S. Frankum 39 Leighton Road, Wingrave, Aylesbury HP22 4PA
G0 WZJ J. Pickering 11 Smithson Court, Malton YO17 7BQ
G0 WZK N. Collis Bird 4 Manwell Drive, Swanage BH19 2RB
GW0 WZL J. Rushton 2 Fron Felen, Clogwyn Melyn, Penygroes, Caernarfon LL54 6PT
G0 WZM I. Kitchen 102 Riverview Road, Epsom KT19 0JP
GM0 WZO I. Finlayson 10B Flesherin, Isle of Lewis HS2 0HE
G0 WZV K. Aston 10 Browning Close, Lexden, Colchester CO3 4JJ
GI0 WZW R. Pollock 5 Brooke Grove, Banbridge BT32 3YA
G0 WZX B. Stevens 172 Gordon Avenue, Camberley GU15 2NT
G0 WZY M. Davies 45 Elm Grove, Swainswick, Bath BA1 7BA
GW0 WZZ M. Bobby Hafan, Church Street, Penycae LL14 2RL
G0 XAA Ansty Contest Club c/o K. Evans Littlefield House, Bolney Road, Haywards Heath RH17 5AW
G0 XAB R. Dawson 28 Calf Close, Haxby, York YO32 3NS
GI0 XAC Dr A. Chin 25 Meadowbrook, Islandmagee, Larne BT40 3UG
G0 XAD B. Wright 96 Ellenborough Close, Thorley, Bishop's Stortford CM23 4HU
G0 XAE M. Buckland The Homestead, Bollow, Westbury-on-Severn GL14 1QX
G0 XAF P. Coles 10 Springfield Close, Mangotsfield, Bristol BS16 9BZ
G0 XAG I. White-Sharman 62 Timberleys, Littlehampton BN17 6QB

G0 XAH D. Tecklenberg 1 Proby Close, Yaxley, Peterborough PE7 3ZF
G0 XAI S. Bennett Maggies Meadow, Hoe Lane, Flansham PO228NS
G0 XAK S. Curtis 389 Portway, Shirehampton, Bristol BS11 9UF
G0 XAN G. Aylward 53 Overdown Rise, Portslade, Brighton BN41 2YF
G0 XAO P. Cole 18 Mundys Field, Ruan Minor, Helston TR12 7LF
GW0 XAP B. Blake 39 Heol Sant Gattwg, Llanspyddid, Brecon LD3 8PD
G0 XAR S. Farthing 21 Cavell Close Swardeston, Norwich NR14 8DH
G0 XAS J. Gavin 22 Rotherwick Way, Cambridge CB1 8RX
G0 XAT R. Wallbank 32 Truro Place, Cannock WS12 3YJ
G0 XAU C. Martin 7 Mayfair Close, Dukinfield SK16 5HR
GM0 XAV A. Main 1 Border Avenue, Saltcoats KA21 5NH
G0 XAW A. Santillo 34 Wearde Road, Saltash PL12 4PP
G0 XAY R. Elford Prospects, Tormarton Road, Badminton GL9 1HP
G0 XAZ H. Simmons 96 Porlock Road, Southampton SO16 9JF
G0 XBA A. Hill 13 Sycamore Way, Winklebury, Basingstoke RG23 8AD
G0 XBC M. Soane 24 Nurseries Road, Wheathampstead, St. Albans AL4 8TP
G0 XBG A. Marinho 13 Pipkin Way, Oxford OX4 4AR
GI0 XBH K. Morgan 40 St. Marys Gardens, Belfast BT12 7LG
G0 XBL H. Watts 7 Hartwood Road, Liverpool L32 7QH
G0 XBO E. Smith 8 Nene Road, Hunstanton PE36 5BZ
G0 XBQ Sos Radio Group c/o C. Levingston 44 Lewis Road, South Australia, Glynde 5070 Australia
G0 XBV A. Nottage 99 Fermor Way, Crowborough TN6 3BH
G0 XCF C. Foley 12 Cross Street, Northam, Bideford EX39 1BS
G0 XDI M. Cabban 34 Sandycroft Road, Amersham HP6 6QP
G0 XDL G. Edwards 82 Regent Drive, Skipton BD23 1BB
G0 XDX G. Kendall 39 Foundry Gate, Wombwell, Barnsley S73 0LF
G0 XEG D. Riches 21 Brinkley Way, Felixstowe IP11 9TX
GM0 XFK J. Neary 17 Harkins Avenue, Blantyre, Glasgow G72 0RQ
G0 XGL G. Lawrence 20 Branewick Close, Fareham PO15 5RS
G0 XGM R. Burgess 22 Lee Close, Kidlington OX5 2XZ
G0 XIT B. Davis Westfield House, Wood End Road, Bedford MK43 9BB
G0 XJS J. Sliman 4 Coleridge Close, Exmouth EX8 5SP
G0 XKK K. Keenan 25 Harris Road, Harpur Hill, Buxton SK17 9JS
G0 XOX P. Thorndike 56 Durham Road, Southend-on-Sea SS2 4LU
G0 XPD A. Sutton Karena Gweek, Helston TR12 6UB
G0 XRC Exmouth ARC c/o A. Howell-Jones 11 Staffick Close, Kenton, , Exeter EX6 8NS
G0 XTA R. Skells 95 Sutton Road, Leverington, Wisbech PE13 5DR
G0 XTL T. Layphries 11 Crossfield, Fernhurst, Haslemere GU27 3JL
G0 XTM Dr N. Marshall 3 The Green, Dorking Road, Tadworth KT20 5SQ
G0 XVC R. Nixon 31 Ashfield, Shotley Bridge, Consett DH8 0RF
G0 XVL D. Veale 5 Heathfield Close, Dronfield S18 1RJ
G0 XXX Chester & District RS c/o D. Knowler Apartedo 1009, 8670 - 999, Aljezur Portugal
GW0 XYL J. Hockley 44 Brookfields, Crickhowell NP8 1DJ
GI0 XYZ Co-Antrim A.R.S DX Group c/o J. Hoey 66 Woodburn Road, Carrickfergus BT38 8PS
G0 XZT J. Gilbert 17 Phoenix Way, Portishead, Bristol BS20 7FG
G0 YBU Dr D. Cuff 7 Parsons Pool, Shaftesbury SP7 8AL
G0 YCE F. Ricci 4 Plymouth Street, Oldham OL8 1PP
G0 YDX R. Bates Apartment 608, 465 West Dominion Drive, Wood Dale 60191-2309 United States
G0 YKC G. Clampin Inverkeris, Tydd Road, Spalding PE12 0HP
G0 YKK K. Kelsall 56 Glenwood Avenue, Baildon, Shipley BD17 5RS
G0 YLO Wincanton Ladies CG c/o C. Monksummers 29 Cloverfields, Peacemarsh, Gillingham SP8 4UP
G0 YNC R. Dalumpines Flat 2, 27 Cotswold Way, Worcester Park KT4 8HD
G0 YOU C. Haye 15 Byron Close, Yateley GU46 6YW
G0 YRT B. Howarth 23 Yew Tree Road, Denton, Manchester M34 6JY
G0 YSS A. Jones 122 Slater St., Latchford, Warrington WA4 1DW
G0 YYH K. Yeung Flat 1, 2 Brunswick Hill, Reading RG1 7YT
G0 YYY R. Konowicz 12 Ambleside Crescent, Farnham GU9 0RZ
GI0 ZAK F. Tanner 4 Leopold Gardens, Belfast BT13 3XN
GM0 ZAM J. Glennon 68 Carronshore Road, Carron, Falkirk FK2 8EE
G0 ZAP R. Crozier 930-932 Burnley Road, Loveclough, Rossendale BB4 8QL
G0 ZAT W. Skipper 18 Central Avenue, South Shields NE34 6AZ
G0 ZDL D. Lister 7 The Fairways, Condover, Shrewsbury SY5 7BW
G0 ZEE C. Monksummers 29 Cloverfields, Peacemarsh, Gillingham SP8 4UP
GI0 ZER E. Robinson 32 Corrycroar Road, Pomeroy, Dungannon BT70 3DY
G0 ZGN M. Geernaert-Davies Bampton, Nettlestone Green, Seaview PO34 5DY
G0 ZHP Invicta CG c/o K. Jasinski 35 Friars Place Lane, London W3 7AQ
G0 ZIP F. Marston 1 Weaver Road, Leicester LE5 2RL
G0 ZMC M. Conlon 3 Selside, Brownsover, Rugby CV21 1PG
G0 ZMH Coulsdon Amateur Transmitting Society c/o M. Howell Orchard House, Blennerhasset, Wigton CA7 3QX
G0 ZPV J. Roberts Long Meadow, Kidderton Lane, Nantwich CW5 8JD

G1

G1 AAC P. Watterson 25 Church Lane, Mablethorpe LN12 2NU
G1 AAD N. Youd 8 Forest Road, Piddington, Northampton NN7 2DA
G1 AAG A. Withers 23 Fernie Road, Guisborough TS14 7LZ
G1 AAH P. Worledge 8 Forest Edge Road, Sandford, Wareham BH20 7BX
G1 AAK P. Webster 30 Belvoir Road, Widnes WA8 6HR
G1 AAL A. Woodward 40 Berwood Farm Road, Wylde Green, Sutton Coldfield B72 1AG
G1 AAP M. Wilson 18 Briars Close, Southwood, Farnborough GU14 0PB
G1 AAQ E. Writer 78 Henley Way, Ely CB7 4YJ
G1 AAR M. West The Lair, Lewes Road, Newhaven BN9 9AH
G1 AAV M. Young 325 Blair Avenue Unit A, Friday Harbor 98250 United States
G1 ABA D. Lambert 72 Johnson Drive, Barrs Court, Bristol BS30 7BS
G1 ABJ A. Norton 29 Long Lane, Chapel-En-Le-Frith, High Peak SK23 0TA
G1 ABM I. Macdonald 23 Cymberline Way, Warwick Gates, Warwick CV34 6FQ
G1 ABQ F. Macdonald Boothlands Farm, Newdigate, Dorking RH5 5BS
G1 ABW B. Webber 5 Sheldon Way, Berkhamsted HP4 1FG
GW1 ABX C. Passey 119 Ffordd Y Mileniwm, Barry CF62 5BD
G1 ACA J. Garner Cobwebs, Lewes Road, Haywards Heath RH17 7PG
G1 ACB G. Gifford 42 Green Park, Brinkley, Newmarket CB8 0SQ
G1 ACD P. Hughes 7 Dellfield Lane, Liverpool L31 6AS
GI1 ACN T. Hutton 23 Enniscrone Park, Portadown, Craigavon BT63 5DQ
GW1 ACV W. Harrison Top Flat, Craiglwyd Hall, Penmaenmawr LL34 6ER
G1 ACY M. Holley 95 Lyes Green, Corsley, Warminster BA12 7PA
G1 ADB M. Hunt 46 Cunningham Drive, Bury BL9 8PD
G1 ADE H. Kirk 34 Tilworth Road, Hull HU8 9BN
GM1 ADI R. Stevens 10 Tiel Path, Glenrothes KY7 5AX
GW1 ADY L. Rees 40 High St., Abergwili, Carmarthen SA31 2JB
G1 AEA A. Read 16 Western Close, Penton Park, Chertsey KT16 8QB
G1 AEB J. Savage 30 Green Meadows, Caravan Park, Cheltenham GL51 6SN
G1 AEF R. Smith 130 Winchester Road, Ford Houses, Wolverhampton WV10 6EZ
G1 AEI P. Stevenson 2 Lazonby Hall Cottages, Lazonby, Penrith CA10 1BA
G1 AEJ L. Smith 4 Penhale Road, Braunstone, Leicester LE3 2UU
G1 AEQ D. Lewis 10 Addington Road, Bolton BL3 4QZ
G1 AET F. Lawson 10 Avebury Close, Tuffley, Gloucester GL4 0TS
G1 AEU A. Lott 27 Queens Crescent, Brixham TQ5 9PJ
G1 AEX G. Muggeridge Gribble House, Wey Street, Ashford TN26 2QH
G1 AFI D. Mills 25 Lower Park Crescent, Poynton, Stockport SK12 1EF
G1 AFJ P. Keyte 11 Woodward Road, Pershore WR10 1LW
G1 AFK B. Kelsey 18 Parkside Avenue, Littlehampton BN17 6BG
G1 AFW R. Harris 15 Rodmer Close, Minster On Sea, Sheerness ME12 2BS
G1 AFX P. Gruet 12 Burgh Close, Crawley RH10 3TS
G1 AGA A. Gee 74 New Street, Milnsbridge, Huddersfield HD3 4LD
G1 AGB D. Gee 74 New Street, Milnsbridge, Huddersfield HD3 4LD
G1 AGK D. Woodruffe 7 Orchard Close, Melton, Woodbridge IP12 1LD
G1 AGM G. Williams 22 Moor Tarn Lane, Walney, Barrow-in-Furness LA14 3LP
G1 AGW D. Yeatman 302 Canterbury Road, Herne Bay CT6 7HD
GM1 AHF A. Mccormack 18 Harris Court, North Muirton, Perth PH1 3DD
GM1 AHG N. Mccormack 18 Harris Court, North Muirton, Perth PH1 3DD
G1 AHM A. Martland Knowleswood, Wrennals Lane, Chorley PR7 5PW
G1 AHQ D. Mobley 17 Butts Close, Aynho, Banbury OX17 3AE
G1 AHS A. Mitchell 7 Cross Park Street Horbury, Wakefield WF4 6AE
G1 AHT C. Wager-Bradley Swallowdale, Longhoughton Road, Alnwick NE66 3AT
G1 AIA E. Oliver 9 Taylor Terrace, West Allotment, Newcastle upon Tyne NE27 0EF
GW1 AIB E. Robinson 35 Bolgoed Road, Pontarddulais, Swansea SA4 8JF
G1 AIF R. Robinson 9 Prospect Terrace, New Kyo, Stanley DH9 7TR
G1 AIG S. Rimell 1 Francis Way, Bridgeyate BS30 5WJ
G1 AII J. Rushton 8 Keats Road, Stonebroom, Alfreton DE55 6JG
G1 AIO A. Peters 28 Drake Avenue, Didcot OX11 0AD
G1 AJC T. Howe 4 Willow Crescent, Broughton Gifford, Melksham SN12 8NB
G1 AJD M. Jacobsen 3 Green Lane, Tickton, Beverley HU17 9RH
G1 AJE M. Chambers 1 Green Lane, Tickton, Beverley HU17 9RH
G1 AJH A. Hardy 73 Lonnen Rd, Colehill BH21 7AU
G1 AJK D. Neale 161 Antrobus Road, Birmingham B21 9NU
G1 AJN G. Overy Flat2 37 Magdalene Rd, St. Leonards on Sea TN376ET
G1 AJQ D. Padfield 20 Gainsborough Crescent, Chelmsford CM2 6DJ
G1 AJS N. Parker 7 The Hollies, Clee Hill, Ludlow SY8 3NZ
G1 AJT C. Pickering 16 Ashworth Way, Newport TF10 7EG
G1 AJU J. Parsons 40 Tynings Close, Kidderminster DY11 5JP
G1 AJV E. Roberts 4 Willow Way, Redditch B97 6PH
G1 AJY A. Young 4 Woodlea, Leybourne, West Malling ME19 5QY
G1 AJZ P. Rayson 1 Grange Gardens, Taunton TA2 7EN
G1 AKA H. Rowley Apartment 27, The Bridges, Buxton Road, Macclesfield SK10 1FW
G1 AKB S. Ryder Corner Cottage, Crown Lane, Defford, Worcester WR8 9BE
G1 AKD A. Rideout 7 Beech Road, Martock TA12 6DT
G1 AKE P. Rowland 7 Maxwell Street, Bury BL9 7QA
G1 AKV T. Alexander 8 Greenway, Eastbourne BN20 8UG
G1 ALA J. Bird 17 Sherrards Way, Barnet EN5 2BW
G1 ALD A. Brown 40 Sutherland Road, Edmonton, London N9 7QG
G1 ALK R. Batchelor 20 Mallard Way, Lower Stoke, Rochester ME3 9ST
G1 ALL M. Bond 8 Alfred Street, Irchester, Wellingborough NN29 7DR
G1 ALR A. Stafford 24 Bourne Street, Croydon CR0 1XL
GM1 ALS J. Smith 63 Keith Street, Stornoway HS1 2JG
G1 ALU K. Spiers Robins Post, North Heath Lane, Horsham RH12 5PJ
GW1 ALV A. Shaw Derlwyn, Efailwen, Clynderwen SA66 7JP
G1 AML I. Tidey April Cottage, Cansiron Lane, East Grinstead RH19 3SE
G1 AMN N. Webb 3 Allens Lane, Norwich NR2 2JB
G1 AMS J. Winter Flat 23, Knightlow Lodge Knightlow Avenue, Coventry CV3 3HH
G1 ANA J. Coles 10 Westgate Hill Street, Bradford BD4 0SJ
G1 AND I. Dodds 54 Philip Road, Newark NG24 4PD
G1 ANF T. Crowe 15 Lambert Road, Kendray, Barnsley S70 3AA
GI1 ANG G. Carvill 18 Hospital Road, Newry BT35 8PW
G1 ANI M. Cooper 33 Park View, Royston, Barnsley S71 4AA
G1 ANK R. Bond 21 Dinglebank Close, Lymm WA13 0QR
G1 ANQ P. Fincher 11 Verney Mews, Reading RG30 2NT
G1 ANS K. Elvin 97 Jeans Way, Dunstable LU5 4PR
G1 ANV R. Edgar 19 Butt Hedge, Long Marston, York YO26 7LW
GW1 ANW D. Evans 6 Oakfield Terrace, Ammanford SA18 2NG
G1 ANZ P. Thirst1 The Haywain, Thirsts Farm, Happisburgh, Norwich NR12 0RU
G1 AOC R. Doughty 7 High Street, , Garlinge CT9 5LN
G1 AOE J. Darley 16 Ivydene, Knaphill, Woking GU21 2TA
G1 AOF R. Dean 10 Livingstone Road, Ellesmere Port CH65 2BE
G1 AOQ I. Gorsuch Elmstone Farm, Fosten Green, Ashford TN27 8ER
G1 AOR P. Grayshon 90 Park Lea Bradley, Huddersfield HD2 1QP
G1 AOZ J. Henderson The Bungalow, Appleby Grammar School, Battlebarrow, Appleby-in-Westmorland CA16 6XU
G1 APA C. Hibbert 19 Fern Road, Maia, Dunedin 9022 New Zealand
G1 APL R. Hopkins 15 Oak Meadow Shipdham, Thetford IP25 7FD
G1 APQ A. Howells 16 Oakley Wood Road, Bishops Tachbrook, Leamington Spa CV33 9RN
GW1 APU K. Pierson Cefn Glaniwrch Cottage, Llanrhaeadr Ym SY10 0DR
G1 AQF A. Matthews 44 Essex Close, Dines Green, Worcester WR2 5RW
G1 AQI J. Aldersey 36 Walls Road, Salterbeck, Workington CA14 5JA
G1 AQP A. Bell 1 Purbeck Drive, Lostock, Bolton BL6 4JF
G1 AQV J. Blackman 30 Parklands Way, Penrith CA11 8SD
G1 AQX K. Armstrong 30 Cobholm Place, Cambridge CB4 2UN
GD1 AQY S. Broad Ballaberragh, Farm, Ramsey IM73EB Isle of Man

G1 ARD G. Ashton 101 Wickenby Garth, Bransholme, Hull HU7 4RF
G1 ARF J. Makin 6 Cambridge House, Courtfield Gardens, London W13 0HP
G1 ARH R. Lupton 19 Avenue Close, Harrogate HG2 7LJ
G1 ARL C. Mandall 11A Hazel Road, Park Street, St. Albans AL2 2AH
G1 ARM R. Preece 5 Pavey Run, Ottery St. Mary EX11 1FQ
G1 ARU A. Judge 44 Thorley Lane, Bishop's Stortford CM23 4AD
GM1 ASA H. Kilpatrick 80 Livingstone Terrace, Irvine KA12 9DN
G1 ASD K. Knight 80 Winton Road, Reading RG2 8HJ
G1 ASG T. Stokes 21 Guildford View, Sheffield S2 2NZ
G1 ASN J. Stansfield Flat 12, Harty House Church Street, Eccles, Manchester M30 0LT
G1 ASR D. Thornton 8 Chestnut Court, Toft Hill DL14 0TG
G1 ASU A. Clews 47 Gaydon Road, Solihull B92 9BJ
GM1 ASY J. Christie 99 Meadow Crescent, Elgin IV30 6ER
G1 ATA D. Cotton 24 Stirling Rise, Stretton, Burton-on-Trent DE13 0JP
G1 ATC Air Cadet ARS c/o V. Tuff 8 Millcroft Court, Blyth NE24 3JG
G1 ATG C. Clarke 29 Huyton Hey Road, Huyton, Liverpool L36 5SF
G1 ATL K. Bishop 8 Sandbanks Grove, Hailsham BN27 3LS
G1 ATQ H. Rogers 74 Front Road, Murrow, Wisbech PE13 4HU
G1 ATU G. Krause 23 Daniel Crescent, Lincoln LN4 1QT
GM1 ATW E. King Marionville, Donibristle, Cowdenbeath KY4 8EU
G1 ATY B. Jarratt 4 Jazz Road, Aylesbury HP18 0EZ
GW1 ATZ G. Morris 18 Grosvenor Road, Shotton CH5 1NU
G1 AUH R. Holland 17 Eaton Square, Leeds LS10 4SN
G1 AUI C. Hayes 37 St. Radigunds Street, Canterbury CT1 2AA
G1 AUM G. Gumbrell 24 Tattershall Drive, Market Deeping, Peterborough PE6 8BS
G1 AUR H. Graham1 4 Eisenhower Road, Basildon SS15 6JR
G1 AUU S. Goan 15 Winchester Avenue, Chorley PR7 4AQ
G1 AUY R. Curzon 24 Edwards Drive, Wellingborough NN8 3JJ
GM1 AUZ G. Crockford 13 Forvie Street, Bridge Of Don, Aberdeen AB22 8TP
G1 AVA B. Carter 10 Opal Street, Keighley BD22 7BP
G1 AVB R. Davidson 3 Eastridge Drive Bishopsworth, Bristol BS13 8HQ
G1 AVC J. Dean 62 Melwood Drive, Liverpool L12 8RW
G1 AVF B. Eastick 30 Farmcote Road, Aldermans Green, Coventry CV2 1SA
G1 AVW H. Sterry 23 Eardisley Close, Matchborough, Redditch B98 0BX
G1 AVZ J. Toolan 12 Stillington Road, Huby, York YO61 1HW
G1 AW A. Wragg 14 Grizedale Avenue Sothall, Sheffield S20 2DL
G1 AWD T. Wells 2 Stephens Close, Mortimer Common, Reading RG7 3TL
G1 AWF A. Wyspianski 53 Alington Crescent, Kingsbury, London NW9 8JL
GW1 AWH I. Robinson 13 Clas Ty Wern, Cardiff CF14 4SB
G1 AWJ Dr J. Moyle Amberley, Cotswold Close, Shipston-on-Stour CV36 4NR
G1 AWK D. Moulson 15 Bramble Way, Leavenheath, Colchester CO6 4UN
G1 AWU A. Pickles 49 Hermitage Street, Crewkerne TA18 8ET
GM1 AXI T. Ball 3 Kersland Place, Glengarnock, Beith KA14 3BQ
GW1 AXU J. Cook 22 Northlands Park, Bishopston, Swansea SA3 3JW
G1 AXW C. Dann 113 Belle Vue Road, Cinderford GL14 3BL
G1 AYH M. Newell 15 The Grove, Luton LU1 5PE
G1 AYI T. Smith 19 Leaf Road, Houghton Regis, Dunstable LU5 5JG
G1 AYP K. Lawton Meadowbank, Sutton St. Nicholas, Hereford HR1 3BJ
G1 AYU C. Chambers 1 Saunders Close, Pound Hill, Crawley RH10 7AE
G1 AZA J. Cook 72 Valebridge Road, Burgess Hill RH15 0RP
G1 AZC M. Cannings 7 Whinlatter Place, Newton Aycliffe DL5 7DR.
G1 AZD A. Drage 51 Greenbank Avenue, Kettering NN15 7EF
G1 AZE B. Davies 22 Hillside Road, Four Oaks, Sutton Coldfield B74 4DQ
G1 AZZ G. Taylor 31 Ashfurlong Crescent, Sutton Coldfield B75 6EN
G1 BAA F. Whittaker 14 Mill Lane, Wombourne, Wolverhampton WV5 0LG
G1 BAB R. Wilson Barley Corners, Arundel Road, Seaford BN25 4LZ
GW1 BAI K. Lowe 18 Aberdovey Close, Dinas Powys CF64 4PS
G1 BAL J. Mahoney 27 Linby Drive, Bircotes, Harworth, Doncaster DN11 8PF
GM1 BAN D. Morrison 4 West Murkle, Murkle, Thurso KW14 8YT
G1 BAQ A. Miller 44 Spring Gardens, Newport Pagnell MK16 0EE
G1 BAR B. Norris 1 Earleswood, Benfleet SS7 1DN
G1 BAX J. Peters Ferndale Cottage, Brea, Camborne TR14 9AT
G1 BBA T. Tilley Mizpah, Carpalla Foxhole, St. Austell PL26 7TY
G1 BBC J. Duxbury 7 Osprey Close, Blackburn BB1 8LP
GW1 BBH J. Sharkey 33 Ffordd Morfa, Llandudno LL30 1ES
G1 BBI K. Ford 2 Ford Close, Ferndown BH22 8AA
G1 BBK B. Artingstall 19 Town Lane, Denton, Manchester M34 6AF
G1 BBT J. Dackham 4 Overbury Close, Weymouth DT4 9UE
G1 BBY W. Brown 53 Drummonds Close, Longhorsley, Morpeth NE65 8UR
G1 BCB A. Blake 58 Greenacres, Bath BA1 4NR
G1 BCE B. Brough 6 Higgs Road, Wednesfield, Wolverhampton WV11 2PD
G1 BCG P. Edwards 5 Playden Close, Brighton BN2 5GH
GW1 BCI A. Gray 69 Tyn Y Parc Road, Cardiff CF14 6BJ
G1 BCN A. Hopkinson 104 Everill Gate Lane, Wombwell, Barnsley S73 0YJ
G1 BCU R. Tagg 38 Salhouse Road, Rackheath, Norwich NR13 6QH
G1 BCW M. Williams 19 Cotebrook Drive, Upton, Chester CH2 1RA
GW1 BDF K. Jones 10 Trinity Road, Tonypandy CF40 1DQ
GW1 BDG O. Jones 10 Trinity Road, Tonypandy CF40 1DQ
GW1 BDH B. Jones 8 Walton Crescent, Llandudno Junction LL31 9ER
G1 BDI B. Jones 56 Mount Grace Road, Luton LU2 8EP
G1 BDP M. Broad 7 Steventon Road, Drayton, Abingdon OX14 4JX
G1 BDQ A. Bell Doddington Mill, Mill Lane, Nantwich CW5 7NN
G1 BDU A. Bradshaw Lyndale, 145 Alder Lane, Wigan WN2 4ET
G1 BDY M. Bragg 33 Mosley Street, Barnoldswick BB18 5BS
G1 BEB R. Cocking 22 Dunbeath Avenue, Rainhill, Prescot L35 0QH
G1 BEG J. Ellsmore 15 Greenbush Drive, Halesowen B63 3TJ
G1 BEJ I. Dixon 60A Woodlands Road, Allestree, Derby DE22 2HF
G1 BEK G. Death 105 Belvedere Road, Ipswich IP4 4AD
G1 BES J. Gibbard 2 Almond Court, Liverpool L19 2QZ
G1 BET A. Waters 12 Anvil Court, Whittonstall, Consett DH8 9JU
GI1 BEU L. Gough 76-78 Culmore Point, Londonderry BT48 8JW
GW1 BFB A. Frayne 20 Springfield Avenue, Upper Killay, Swansea SA2 7HW
G1 BFF T. Fishlock 62 Red Barn Road, Brightlingsea, Colchester CO7 0SJ
G1 BFG A. Harper 144 Ashfield Road, Bispham, Blackpool FY2 0EN
G1 BFK R. Else 134 Market Street, South Normanton, Alfreton DE55 2EJ
G1 BFS P. Sainsbury 103A Pilot Road, Hastings TN34 2AU
G1 BFV S. Harman Ivy Cottage, Grove Road, Friston, Saxmundham IP17 1PP
G1 BGC S. Javes 95 West Way, Lancing BN15 8LZ
G1 BGF D. Mclachlan 1 North Holme Court, Northampton NN3 8UX
G1 BGH P. Nicholls 40 Sedgefield Close, Worth, Crawley RH10 7XG
G1 BGJ E. Morgan 79 Mayland Avenue, Canvey Island SS8 0BU
G1 BGK G. Lang Belford Barn, Ashburton, Newton Abbot TQ13 7HT
G1 BGM P. Leslie 5 Maple Croft, Netherton, Huddersfield HD4 7HS
G1 BGO D. Lee 36 Westwick, Hedon, Hull HU12 8HQ
G1 BGQ P. Sampson 23 Westfield Road, Mirfield WF14 9PW
G1 BHB S. Matthews 222 Widney Lane, Solihull B91 3JY
G1 BHF S. Oldfield The Sycamores, Fulford Road, Stoke-on-Trent ST11 9QT
G1 BHG B. Osborne 12 Arminers Close, Gosport PO12 2HB
G1 BHO B. Palmer Flat 5, Brook Court Burcot Lane, Bromsgrove B60 1AD
G1 BHQ R. Pritchard 41 Greenland Avenue, Maltby, Rotherham S66 7EU
G1 BHR I. Rabbitt 66 Parkfield Avenue, Delapre, Northampton NN4 8QB
G1 BHV R. Green 20 Haygate Drive, Wellington, Telford TF1 2BY
G1 BHW C. Vaughan 11 Fremantle Road, Aylesbury HP21 8EH
G1 BIA H. Wilmshurst Langholm Lodge, Raydaleside, Darlington DL3 7SJ
G1 BIF T. Whelan 43 Martin Avenue, Little Lever, Bolton BL3 1NX
G1 BIM R. Ross 46 Arbour Close, Rugby CV22 6EH
G1 BIN B. Talbot 27 Shuttleworth Road, Clifton Upon Dunsmore, Rugby CV23 0DB
G1 BIU C. Utting 22 Yew Tree Drive, Bromsgrove B60 1AL
G1 BJE A. Stimpson 2 Church Avenue, Kings Sutton, Banbury OX17 3RJ
G1 BJK P. Lough 87 Finchley Road, Kingstanding, Birmingham B44 0LB
G1 BJN C. Stroud 6 Church Road Ideford, Chudleigh, Newton Abbot TQ13 0BB
G1 BJZ G. Garlick 1 Shannon Way, Burton Latimer, Kettering NN15 5SX
G1 BKB S. Stokes 52 Brantley Avenue, Wolverhampton WV3 9AR
G1 BKI M. York 38 Rannoch Way, Corby NN17 2LH
G1 BKJ J. Prior 6 Emfield Grove, Grimsby DN33 3BS
G1 BKL D. Ambler Corrig, 4 Old Main Road, Bridgwater TA6 4RY
GM1 BKR J. Rankin 3 Spalding Drive, Largs KA30 9BZ
G1 BKU R. Lawn 39 Wetherby Road, Knaresborough HG5 8LH
G1 BKZ W. Moore 10 Progress St., Darwen BB3 2DT
G1 BLB P. Thurman 11 Copperfield Drive, Langley, Maidstone ME17 1SX
G1 BLJ S. Lovell 12 The Holloway, Swindon, Dudley DY3 4NT
G1 BLK C. Ridley 14 Painswick Road, Hall Green, Birmingham B28 0HH
G1 BLO N. Swan 8 Tyrrells Court Bransgore, Bransgore BH238BU
G1 BLQ P. Lloyd 35 Westfield Road, Hertford SG14 3DL
G1 BLV S. Davies Swallow Cottage, The Chantry, Leyburn DL8 4NA
GM1 BLX I. Dewar 11 Abbotshall Road, Kirkcaldy KY2 5PH
G1 BMB G. Fallows 66 Ulverston Road, Swarthmoor, Ulverston LA12 0JF
G1 BMN N. Lamb 106 St. Davids Road, Leyland PR25 4XY
G1 BMP S. Norminton 63 Candish Drive, Plymouth PL9 8DB
G1 BMT P. Tuthill 12 Herbert Road, Salisbury SP2 9LF
G1 BMW P. Winterton 35 Paynesfield Road, Tatsfield, Westerham TN16 2AT
G1 BMZ S. Wilkins Garden House Westdown Farm, Exmouth EX8 5BU
GM1 BNA R. Main 14 Hunters Grove, East Kilbride G74 3HZ
G1 BNE A. Perkins 7 High Street, Wollaston, Wellingborough NN29 7QE
G1 BNG S. Marsh 28 Orcheston Road, Bournemouth BH8 8SR
G1 BNN S. Tilly 24 Whinham Way, Morpeth NE61 2TF
GM1 BNP R. Holt Whitlam Farmhouse, Newmachar, Aberdeen AB21 0RS
GM1 BNS W. Graham 6 Braemar View, Clydebank G81 3RR
G1 BNV S. Henderson 7 Havering, Castlehaven Road, London NW1 8TH
G1 BNX R. Huxley 83 Gleneagles Road, Wyken, Coventry CV2 3BH
G1 BOB C. Balsdon 4 Queens Hayes, Willey Lane, Okehampton EX20 2NG
G1 BOO F. Crompton 24 Alcester Road, Sale M33 3QP
G1 BOX M. Platten 48 Brier Road, Sittingbourne ME10 1YL
G1 BPD I. Attridge 12 Ascot Road, Orpington BR5 2JF
G1 BPE M. Barwick 32 St. Georges Road, Harrogate HG2 9BS
G1 BPS M. Rowell 1 Willow Street, Haslingden, Rossendale BB4 5NA
G1 BPU L. Staal 5 Hunt Court, 236 Chase Side, London N14 4PG
G1 BPV A. Stalker 1 Ganymede Lane, Brackley NN13 6RA
G1 BQG A. Bright 15 Cross Road, Maldon CM9 5EE
G1 BQH J. Murphy 34 Knights Hill, Walsall WS9 0TG
G1 BQI G. Smith 92 Lime Road, Accrington BB5 6BJ
GM1 BQP W. Macrobbie 44 Moray Park Terrace, Culloden, Inverness IV2 7RW
G1 BQQ J. Sowerbutts 22 Worsley St., Accrington BB5 2PA
G1 BQR M. Spinks 26 Church Hill, Royston, Barnsley S71 4NH
G1 BQV M. Sunderland 36 Moorlands Avenue, Leeds LS19 6AD
G1 BRB A. Patton 72 Sanctuary Way, Grimsby DN37 9RZ
G1 BRD J. Smith 127 Wolverhampton Road, Cannock WS11 1AR
G1 BRF P. Williams 292 Hagley Road, Hasbury, Halesowen B63 4QG
G1 BRP K. Crawley 27 West End, Northwold, Thetford IP26 5LE
G1 BRS Dorking And District RS c/o M. Stevens 16 Golf Links Road, Ferndown BH22 8BY
GM1 BSG J. Ross 16 Myreton Drive, Bannockburn, Stirling FK7 8PX
GI1 BSJ J. Cunningham 4 Garvaghy Road, Portglenone, Ballymena BT44 8EF
G1 BSY K. Morris 3 Moravian Close, Dukinfield SK16 4EW
G1 BSZ R. Nash Roann, Bedmond Road, Hemel Hempstead HP3 8SH
G1 BTF A. Hardy 14 Parsonage Road, Rainham RM13 9LW
G1 BTI K. Forrest 61 Woodbury Road, Bridgwater TA6 7LJ
GM1 BTL W. Erskine 30 Market Road, Kirkintilloch, Glasgow G66 3JL
G1 BTN I. Evans 37 Lyndale Avenue, Lostock Hall, Preston PR5 5UU
G1 BTV D. Holt 2 London Heights, Dudley DY1 2QZ
G1 BUJ A. Bates 17 Walkers Heath Road, Kings Norton, Birmingham B38 0AB
G1 BUQ J. Spink 38 Hemlingford Road, Sutton Coldfield B76 1JQ
G1 BUV M. Osborne 5 Wells Road, Riseley, Bedford MK44 1DY
GM1 BUY Bromley & District ARS c/o C. Rose 14 Spoutwells Drive, Scone, Perth PH2 6RR
G1 BUZ R. Scott 27 Woodford Av, Henderson, Auckland 610 New Zealand
GM1 BVA L. Nieto 23 Riverview Drive The Waterfront, Glasgow G5 8EU
GM1 BVT W. Pettett 15 Strude Howe, Alva FK12 5JU
G1 BVV G. Pemberton 8 Hotchin Road, Sutton-On-Sea, Mablethorpe LN12 2NP
G1 BWG R. Dodd 5 Halesworth Road, Wolverhampton WV9 5PH
G1 BWH P. Eaton 8 Chester Road, Barnwood, Gloucester GL4 3AX
G1 BWI J. Eastham 81 Park Lee Road, Blackburn BB2 3NZ
G1 BWP W. Webb The New Bungalow, Tram Road, Coleford GL16 8DN
GU1 BWW D. Ash Trigale House, La Trigale, St. Anne GY9 3TX Guernsey
G1 BWX P. Caton 39 Farmerie Road, Hundon, Sudbury CO10 8HA
G1 BWZ N. Fieldsend 47 Hollycroft, Barmston, Driffield YO25 8PP

GM1 BXI R. Clark 5 Abbotsfield Terrace, Auchterarder PH3 1DD
G1 BXQ J. Smith Noonhill Farm, Grove Road, Coventry CV7 9JE
G1 BXT I. Thomas Myrtle Cottage, Swan Lane, Sellindge, Ashford TN25 6EB
GW1 BXX A. Gillard 10 Parc Pendre, Brecon LD3 9ES
G1 BYI J. Moore 11 Sherborne Road, Wallasey CH44 2EY
G1 BYJ G. Noon 48 Sanderling Close, Letchworth Garden City SG6 4HY
G1 BYO S. Slaughter 96 Adys Road, London SE15 4DZ
G1 BYP P. Snitch 79 Albion Avenue, York YO26 5QZ
G1 BYQ D. Hatton 34 Avocet Way, Bicester OX26 6YP
G1 BYS A. Kempton 14 Lower Gravel Road, Bromley BR2 8LT
G1 BYT N. Kinselley 5 Helford Close, Bedford MK41 7TU
GM1 BZD P. Ward Burnside, Flotta Stromness KW163NP
G1 BZE C. Weeds Flat 1, Bank House, Campbell Street, Belper DE56 1AP
G1 BZM P. Endean 11 Forrester Drive, Brackley NN13 6NE
GM1 BZR D. Cameron 14 Queen St., Castle Douglas DG7 1HX
GI1 BZT Dr L. Gornall 14 Ballymoghan Lane, , Magherafelt BT45 6HW
G1 BZU Royal Naval ARS c/o J. Kirk 111 Stockbridge Road, Chichester PO19 8QR
G1 BZW R. Kimber 38 Greenmere, Brightwell-Cum-Sotwell, Wallingford OX10 0QG
GI1 CAI J. Mcbride Dunaree, 20 Oldcastle Road, Omagh BT78 4HX
G1 CAN P. Shonfield 242 Chickerell Road, Weymouth DT4 0QY
G1 CAR Mastcar ARC c/o J. Murray 20 East End, Cholsey, Wallingford OX10 9RT
G1 CAY D. Shea 38 Ranworth Avenue, Hoddesdon EN11 9NR
G1 CBB L. Walker The Biel, Furze Vale Road, Bordon GU35 8EP
G1 CBK Sugar Delta A.R.C c/o S. Mole 53 Parkfield Road, Rainham, Gillingham ME8 7TA
G1 CBL L. Mason Reflow, Unit 2 Spring Lane North, Malvern WR14 1BU
G1 CBS J. Hatt 77 Pentland Close, Basingstoke RG22 5BQ
G1 CBY S. Fountaine 142 Elvaston Road, North Wingfield, Chesterfield S42 5GA
G1 CCD M. Williamson 15 Nook Fields Harwood, Bolton BL2 4LN
GM1 CCI C. Watson 11 Ladybridge Houses, Banff AB45 2JR
G1 CCL D. Livsey 12 Aldingham Walk, Morecambe LA4 4EW
G1 CCM T. Mcmillan 47 Sandsend Road, Eston, Middlesbrough TS6 8AF
GM1 CCN C. Orr Easter Cowden Farm, Dalkeith EH22 2NS
G1 CCW F. Haselden 7 Chestnut Avenue, Gosfield, Halstead CO9 1TD
G1 CCX P. Kennedy 24 Leadhall Drive, Harrogate HG2 9NL
GW1 CDH D. Davies 10 Bryn Castell, Abergele LL22 8QA
G1 CDN D. Wormall 20 Greenfield Road, Hemsworth, Pontefract WF9 4RL
G1 CDO R. Wormall 17 Newstead Grove, Fitzwilliam, Pontefract WF9 5DS
G1 CDQ A. Sturman 3 Windward House, 73, Lytham St. Annes FY8 1LZ
GW1 CDY D. Forth 1 Upper Cwrt, Cwrt, Pennal, Machynlleth SY20 9LA
G1 CEI P. Hirons 27 Ashley Close, Crondall GU10 5RD
GM1 CEJ R. Stout 16 Ardoch Park, Balgeddie, Glenrothes KY6 3PJ
G1 CEO R. Day 17 Barry Avenue, Bicester OX26 2DZ
GI1 CET J. Barr 2 Willowvale Close, Islandmagee, Larne BT40 3SD
G1 CEU J. Clarke Timbers, Wayford Road, Norwich NR12 9LH
G1 CFA P. Middleton 36 Station Court, Railway Street, Hornsea HU18 1QD
G1 CFB K. Rhodes 34 Bannister Drive, Hull HU9 1EJ
G1 CFE A. Wyatt 32 Wensleydale Avenue, Blackpool FY3 7RS
G1 CFG C. Rigby 4 Humber Street, Longridge, Preston PR3 3WD
G1 CFJ G. Gardner 165 Brookhouse Road, Brookhouse, Lancaster LA2 9NY
G1 CFK D. Bowles Fiddlers Nook, Thurston Road, Bury St. Edmunds IP31 2PL
GW1 CFM Dr N. Bristow Bryn Gwydion, Pontllyfni, Caernarfon LL54 5EY
GW1 CGD S. Oliver 21 Hillside Court, Holywell CH8 7PJ
G1 CGH A. Rawlins 4 Low Park, West Woodburn, Hexham NE48 2SQ
G1 CGJ M. Davies Kuling, Bridgwater Road, Winscombe BS25 1NB
G1 CGP A. Gillard 28 Moor Tarn Lane, Walney, Barrow-in-Furness LA14 3LP
G1 CGU G. Fitzpatrick 12 Dunster Close, Minehead TA24 6BY
G1 CHE D. Ogarr 10 Ellastone Grove, Stoke-on-Trent ST4 5EE
G1 CHM C. Milburn Field House, Copper Hill, Hayle TR27 4LY
G1 CHN A. James The Red House, Gandish Road, Colchester CO7 6TP
G1 CHQ M. Wells 7 Vint Rise, Idle, Bradford BD10 8PU
GW1 CHS Washington ARC c/o J. Hughes Reservoir House, St. Lythan'S, Cardiff CF5 6BQ
GM1 CHT A. Hyde 19 Drum Brae Gardens, Edinburgh EH12 8SY
G1 CHV C. Compton 21 Vange Riverview Centre, Vange, Basildon SS16 4NE
G1 CHY P. Scott 12 Aire View Avenue, Bingley BD16 1NS
G1 CIA M. Ferentiuk 74 Fallowfield Drive, Rochdale OL12 6LZ
G1 CIM S. Hancock Monrad, Back Street East Stockwith, Gainsborough DN21 3DL
G1 CIT M. Whalley 6 Rookery Walk, Clifton, Shefford SG17 5HW
G1 CIV D. Owen 23 Munnings Drive, Hinckley LE10 0LG
GW1 CIY G. Evans Maes Yr Haf, Beulah, Newcastle Emlyn SA38 9QB
G1 CJC L. Gilbert Holmefield Cottage, Oker, Matlock DE4 2JJ
G1 CJI P. Arnold 36 Gopsall Road, Hinckley LE10 0DY
GW1 CJJ P. Williams 6 Parc Ffynnon, Llysfaen, Colwyn Bay LL29 8SA
G1 CJK R. Baiey 318 Plumstead Common Road, London SE18 2RT
G1 CJL B. Hallybone 38 Anvil House, Champion Way, Bedford MK42 9EH
G1 CKF J. Darby 97 Littlehaven Lane, Horsham RH12 4JE
G1 CKJ T. Martin 201 Gloucester Road, Kidsgrove, Stoke-on-Trent ST7 4DQ
G1 CKR T. Miller 31 Hampden Road, Malvern WR141NB
G1 CKT R. Nelson 11 Meadow Way, Plymouth PL7 4JB
GI1 CKU T. Gardiner 17 Grange Valley Gardens, Ballyclare BT39 9HE
G1 CKV N. Derbyshire 54 Windy Arbor Road, Whiston, Prescot L35 3SG
G1 CKY P. Turner 16 Pendragon Way, Leicester Forest East, Leicester LE3 3EY
G1 CLD S. Patterson Dunedin, Little Ness, Shrewsbury SY4 2LG
G1 CLJ P. Kennedy 12 Newbroke Road, Rowner, Gosport PO13 9UJ
G1 CLT R. Bokor 54 Granwood Road, Middlesbrough TS6 9HX
G1 CMC A. Lickley 19 Sandy Rise, Selby YO8 9DW
GM1 CMF P. Carnegie 29 Dalgetty Court, Muirhead, Dundee DD2 5QJ
G1 CMH J. Norwood Flat 28 The Manor, Church Road, Gloucester GL3 2HT
G1 CMZ S. Lewkowicz 7 The Mart, Locking Road, Weston-Super-Mare BS23 3DE
GM1 CNH N. Stewart 160 Carrick Knowe Drive, Edinburgh EH12 7EW
G1 CNI S. Dwyer Po Box 44, Tahmoor 2573 Australia
G1 CNN P. Beeson Flat 6, Oxford Court, London W3 0HH
G1 CNV T. Thornton Orchard Walk, 23 Crookham Road, Fleet GU51 5DP
G1 CNZ Pembrokeshire CG c/o R. Reid 34 Wellesley Street, Taunton TA2 7DT
G1 COD P. Stellings 10 Thornbrook Road Chapel-En-Le-Frith, High Peak SK23 0LX
G1 COE C. Howard 30 Torquay Road Kingskerswell, Newton Abbot TQ12 5EZ
GM1 COF P. Mcgowan 38 Mckenzie Crescent, Lochgelly KY5 9LT
G1 COV Coventry Raynet Group c/o D. Green 67 Coombe Park Road, Binley, Coventry CV3 2NW
G1 COW R. Penfold 1 Padworth Road, Burghfield Common, Reading RG7 3QE
G1 COX A. Berkeley 42 Ringley Drive, Whitefield, Manchester M45 7LR
G1 COY C. Robson 43 Longdyke Drive, Carlisle CA1 3HT
G1 CPA P. Nairne 137 Barden Road, Tonbridge TN9 1UX
G1 CPC J. Arthur St. Aubin, Plomer Green Lane, High Wycombe HP13 5XN
G1 CPD G. Ghetti 7 Rue De Provence, Paris 75009 France
G1 CPM E. Rose 26 Lavender Way, Bourne PE10 9TT
G1 CPO C. Haygarth 3 Rew Close, Ventnor PO38 1BH
G1 CPU G. Milligan 100 Churchfield Road, Gateacre, Liverpool L25 3SE
G1 CPX Granta CG c/o I. Clarke 19 Welbeck, Bracknell RG12 8UQ
G1 CQA R. Chaney 55 Bartlow Road, Linton, Cambridge CB21 4LY
GM1 CQC H. Smith 601 Ferry Road, Edinburgh EH4 2TT
G1 CQG D. Perry 5 Beech Hill, Wellington TA21 8ER
G1 CQK G. Knott 7 Nunney Close, Cheltenham GL51 0TU
G1 CQR D. Fuller 26 Longfields, Ely CB6 3DN
G1 CQT P. Turley 35 Alwinton Avenue, Stockport SK4 3PU
G1 CRH P. Everitt 10411 Tula Lane, Cupertino 95014 United States
G1 CRN W. Murray 91 Chaucer Avenue, Hounslow TW4 6NA
G1 CRT W. Cambridge 36 Selwyn Avenue, Richmond TW9 2HA
G1 CSA J. Walton 23 Keighley Avenue, Sunderland SR5 4BU
G1 CSN R. Beasley 26 Retford Close, Harold Hill, Romford RM3 9NA
G1 CSO J. Dent 18 New Street, St. Neots PE19 1AE
G1 CSR Silverthorn RC c/o N. Sanderson 54 Kelvedon Close, Chelmsford CM1 4DG
G1 CSS M. Wilson 210 London Road, Worcester WR5 2JT
G1 CSY E. Hartley 2 Lamberts Close, Weasenham, King's Lynn PE32 2TE
G1 CTF R. Fenton 37 Martingale Chase, Newbury RG14 2EN
GW1 CTO D. Powell 88 Church View, Chirk, Wrexham LL14 5PF
G1 CTQ N. Losardo 14 Arnside Close, Clayton Le Moors, Accrington BB5 5GG
GM1 CUC H. Mattinson 11 Riverside Park, Canonbie DG14 0UY
G1 CUG D. Laughton 2 Stamford Road, Careby, Stamford PE9 4EB
G1 CUM J. Rawlings Castle House, Barrow Haven, Barrow-upon-Humber DN19 7EY
GW1 CUQ N. Paull 6 Llys Caradog, Creigiau, Cardiff CF15 9JP
G1 CUZ S. Seal Crantock, Bellingdon, Chesham HP5 2XW
G1 CWI M. Kemp Casa Lucia, Vale Formosilho, S. Marcos Da Serra 8375-210 Portugal
G1 CWJ A. Burton 303 Heneage Road, Grimsby DN32 9NW
G1 CWP R. Hide Flat 22, Church Court Church Road, Haywards Heath RH16 3UE
G1 CWQ B. Wyatt 3 Shipley Close, Blackpool FY3 7UJ
G1 CWW G. Stone 37 Canterbury Drive, Ashby-de-la-Zouch LE65 2QQ
G1 CWZ D. Penrose 7 Two Ashes, Bayston Hill, Shrewsbury SY3 0QF
G1 CXQ C. Roberts 11 Adel Wood Drive, Leeds LS16 8JQ
GW1 CXT P. Lodge 1 Brookland Terrace, Nantymoel, Bridgend CF32 7SY
G1 CYQ B. Wheeldon 27 Lawrence Walk, Newport Pagnell MK16 8RF
G1 CYY T. Brien 54 Central Avenue, Fartown, Huddersfield HD2 1DA
GM1 CZE J. Jindra Rtrap 5 South Charlotte St Edinburgh, Edinburgh EH2 4AN
G1 CZH P. Harkins 27 Hillfoot Green, Liverpool L25 7UH
G1 CZN P. Burrows 7 Eton Terrace, Ince, Wigan WN3 4NS
G1 CZU M. Abram 28 Langport Drive.Vicarscross.Chester.Ch35Ly, Chester CH35LY
G1 CZW R. Silcocks 69 Kennaway Road, Clevedon BS21 6JJ
G1 DAE I. Rusby 12 Park Meadow, Princes Risborough HP27 0EB
G1 DAK S. Felton 8 Clancutt Lane, Coppull, Chorley PR7 4NS
G1 DAT P. Burnett 14 Hollywalk Drive, Middlesbrough TS6 0PL
G1 DAV D. Forsey 3 Northwood Drive, Newbury RG14 2HB
G1 DAX P. Costigan 10 The Paddock, Clevedon BS21 6JU
G1 DAZ S. Burchell 31 Thornton Road, Girton, Cambridge CB3 0NP
G1 DBH B. Cobb 28 Sandringham Road, Newton Abbot TQ12 4HA
G1 DBI G. Doig 78 Plane Tree Drive, Crewe CW1 4ES
G1 DBL L. Owen 27 Coniston Drive, Holmes Chapel, Crewe CW4 7LA
G1 DBR D. Ross 113 Nun House Drive, Winsford CW7 3LE
G1 DBZ R. Cooper 31 Erskine Crescent, Sheffield S2 3LQ
GM1 DCB M. Senior The Raw, Bridgend, Isle of Islay PA44 7PZ
GM1 DCI J. Griffith 19 Bayview, Machrihanish, Campbeltown PA28 6PX
G1 DCU P. Gardner 38Apley Rd Dy84Pa, Stourbridge DY84PA
G1 DCX M. Race 76 Lonsdale Road, Stamford PE9 2SG
G1 DCY S. Richmond 1042 Evesham Road, Astwood Bank, Redditch B96 6ED
G1 DCZ J. Sandall 7 St. Road, Compton Dundon, Somerton TA11 6PX
G1 DDA F. Wood 7 Yew Tree Park, The Rowe, Newcastle ST5 4EN
G1 DDF F. Griffin 77 Widmore Drive, Hemel Hempstead HP2 5JL
G1 DDK M. Abraham Skywave Marine Services, Unit 1, The Arcade, Falmouth TR11 2TD
G1 DDR R. Oakley 20 Halton Lane, Wendover, Aylesbury HP22 6AR
G1 DDS D. Seccombe 14 Millfield, Bedlington NE22 5DZ
G1 DEN P. Edinburgh 77 Westerley Lane, Shelley, Huddersfield HD8 8HP
G1 DEO B. Davies 12 Woodbine Close, Newport PO30 1AF
G1 DEP J. Dunhill 8 Brentwood Avenue, Thornton-Cleveleys FY5 3QR
G1 DEQ D. Gilbey 7 Victory Way, Cottenham, Cambridge CB24 8TG
G1 DER J. Hacker 4 Foxglove Close, Bamber Bridge, Preston PR5 6XR
G1 DES D. Smith 14 College Drive, Ruislip HA4 8SB
G1 DEU D. Hagger 11 Recreation Walk, Great Cornard, Sudbury CO10 0HH
G1 DEX H. Irvin 30 Bank View, Earlsheaton, Dewsbury WF12 8HH
G1 DEY C. Jacob 10 Wynchgate, Southgate, London N14 6RR
G1 DEZ P. Baxter 27 Manor Crescent, Brinsworth, Rotherham S60 5HG
G1 DFF M. Smith 22 Cedars Avenue, Wombourne, Wolverhampton WV5 0JX
G1 DFI J. Swift-Hook 12 Warwick Drive, Newbury RG14 7TT
G1 DFM A. Westlake 47 Quarry Road, Kingswood, Bristol BS15 8NZ
G1 DFN F. Wright 1 Old Engine Houses, Brusselton, Shildon DL4 1QA
G1 DFP G. Fielding 35 Amos Avenue, Litherland, Liverpool L21 7QH
G1 DFR A. Gemmill The Steps, Bliss Gate Road, Bliss Gate, Rock, Kidderminster DY14 9XT
G1 DFT Nthants Expd Gr c/o I. Hampson 293 Sandbrook Road, Southport PR8 3RP
G1 DFW D. Hoare 51 Hartington Road, Dronfield S18 2LE
G1 DFZ R. Jobbins 8 Newark Road, Hartlepool TS25 2LA
G1 DGL R. Simpson 51 Ramleaze Drive, Salisbury SP2 9PA
G1 DGW I. Johnson 24 York Road, Maghull, Liverpool L31 5NL
G1 DGY A. Koch 65 Collier Lane, Ockbrook, Derby DE72 3RP
G1 DHB R. Fagence 5 Balmoral Close, Billericay CM11 2LL
G1 DHM G. Miller 32 Belbroughton Close, Lodge Park, Redditch B98 7NH

G1 DHQ D. Palmer 60 Heathcote Drive, Sileby, Loughborough LE12 7ND
G1 DHY N. Roe 41 Highfield Lane, Chaddesden, Derby DE21 6PH
G1 DIA P. Rowe 5 Bramble Close, Great Boughton, Chester CH3 5XN
G1 DIF Devon Data Group c/o D. Roomes View Field, Milton Damerel, Holsworthy EX22 7NY
G1 DIG S. Cadman 71 Gayfield Avenue, Withymoor, Brierley Hill DY5 2BU
G1 DIK A. Smith Windycross, Newbourne Road, Woodbridge IP12 4PT
G1 DIL A. Witts Langdale 3 Forton Glade, Newport TF10 8BP
G1 DIM C. Smith 37 Ivory Close, Tuffley, Gloucester GL4 0QY
G1 DIO A. Heath-Anderson 12 The Medway, Daventry NN11 4QU
G1 DIR A. Wilkinson 15 St. Margarets Grove, Leeds LS8 1RZ
G1 DJI J. Short 7 Bushfields, Loughton IG10 3JT
G1 DJQ N. Lofthouse Cambridge Park, 8 Abbott Clough Avenue, Blackburn BB1 3LP
G1 DJU C. Whitby 7 Wentworth Way, Stoke Bruerne, Towcester NN12 7SA
G1 DKB K. Ball 74 North Street, Okehampton EX20 1BD
G1 DKE M. Spry 71 High Street, Topsham, Exeter EX3 0DY
G1 DKI M. Lindenbergh 26 Manston Drive, Perton, Wolverhampton WV6 7LX
G1 DKV G. Charlton 20 Bailey Crescent, South Elmsall, Pontefract WF9 2TL
G1 DKX A. Thomas 92 Singleton Crescent, Goring-By-Sea, Worthing BN12 5DJ
G1 DKY J. Miller 40 Central Avenue, Herne Bay CT6 8RX
G1 DLA R. Deacon 22 Islip Gardens, Northolt UB5 5BX
G1 DLB J. Desborough 106 Grand Avenue, Lancing BN15 9QD
G1 DLH M. Ogle 22 Warwick St., Daventry NN11 4AL
G1 DLJ G. Hope 3 Farm Crescent, Sittingbourne ME10 4QD
GW1 DLP W. Jones 160 Christchurch Road, Newport NP19 7SA
GM1 DLS C. Barry 32 Prospect Drive, Ashgill, Larkhall ML9 3AJ
G1 DMH L. Lees 3 Ockbrook Court, Muskham Avenue, Ilkeston DE7 8EY
G1 DMN P. Snow 14 Beechwood Avenue, Darlington DL3 7HP
G1 DMR R. Manser 53 Downs Barn Boulevard, Downs Barn, Milton Keynes MK14 7LL
G1 DMS D. Segal Flat 1, Masons House, London NW9 9NG
G1 DMW F. Latham Higher Lane, Parbold, Wigan WN8 7RA
G1 DNA J. Birkmyre Swarland, Morpeth NE65 9JW
G1 DNI W. Darling 2 Strathaird Avenue, Walney, Barrow-in-Furness LA14 3DE
G1 DNK B. Cunningham 14 Leeson Drive, Ferndown BH22 9QQ
G1 DNO C. Birtchnell Linnetts Roost End, Sturmer, Haverhill CB9 7XW
G1 DNP R. Collins 12 Bean Oak Road, Wokingham RG40 1RL
G1 DNT M. Cole 52 Lower Meadow, Quedgeley, Gloucester GL2 4YY
G1 DNY R. Clay 38 Hubbards Road, Chorleywood, Rickmansworth WD3 5JJ
G1 DNZ G. Clarke 150 Minver Crescent, Nottingham NG8 5PN
G1 DOA K. Chappell 17 Linton Close, Winyates, Redditch B98 0NA
G1 DOG I. Cheeseman 445 Uttoxeter Road, Blythe Bridge, Stoke-on-Trent ST11 9NT
G1 DOJ T. Brodrick 16 Wallenge Drive, Paulton, Bristol BS39 7PX
G1 DOL R. Breakspear 7 Woodside, North Leigh, Witney OX29 6SQ
G1 DON D. Macnamara 56 Macdonald Street, Orrell, Wigan WN5 0AJ
G1 DOT D. Barker 60 Rolvenden Road, Wainscott, Rochester ME2 4PG
G1 DOX J. Acton 63 Bevington Close, Patchway, Bristol BS34 5NP
G1 DPI A. Barratt 23 Wilberforce Road, South Anston, Sheffield S25 5EG
G1 DPJ C. Beasley 12 East Leys Court, Moulton, Northampton NN3 7TX
GW1 DPL M. Beer 67 Killan Road, Dunvant, Swansea SA2 7TH
G1 DPN D. Bettany 10 Redbrook Crescent, Melton Mowbray LE13 0EU
G1 DPT T. Cairney Dolgoch, Hall Lane, Lutterworth LE17 5RP
GW1 DPV M. Carter 14 Cerdin Avenue, Pontyclun CF72 9ER
G1 DPW S. Cmoch 25 Monro Place, Epsom KT19 7LD
G1 DPX R. Colley 12 Glenfield Road, Banstead SM7 2DG
G1 DQD C. Anderson 11 Swallowfield Drive, Hull HU4 6UG
G1 DQF P. Buckmaster 7 Yew Tree Close, New Ollerton, Newark NG22 9UP
G1 DQL R. Bradley 1 Audley Place, Sutton SM2 6RW
G1 DQQ D. Dwight 19 The Highway, Stanmore HA7 3PL
G1 DQU M. Elliott 52 Wellfield Road, Alrewas, Burton-on-Trent DE13 7EZ
GW1 DQV B. Emary 2 The Paddocks, Penarth CF64 5BW
G1 DRG University of Sussex ARS c/o G. Foster 19 Asquith Avenue, Burnholme, York YO31 0PZ
G1 DRI K. Gill 33 Hazel Croft, Werrington, Peterborough PE4 5BJ
G1 DRR N. Hamilton 8 Sudbury Court, Mansfield NG18 3RZ
G1 DRW D. Hart 71 Breinton Road, Hereford HR4 0JY
G1 DRY S. Cox 25 Church Close, Stoke St. Gregory, Taunton TA3 6HA
GW1 DRZ D. Rees 19 Waunfawr Road, Cross Keys, Newport NP11 7PG
G1 DSA J. Critchley 3 Beaconsfield View, Robert Road, Slough SL2 3XT
G1 DSB J. Darling 145 Hartlands, Bedlington NE22 6JJ
G1 DSF A. Daw 19 Rowan Close, Yarnfield, Stone ST15 0EP
G1 DSG M. Degerdon 25 Rosslyn Road, Billericay CM12 9JN
G1 DSJ P. Morgan 29 Brisbane Road, Reading RG30 2PE
GM1 DSK D. Keay Parkhill, Cromwell Park Almondbank, Perth PH1 3LW
G1 DSM A. Hicks 5 Restwell Avenue, Cranleigh GU6 8PQ
G1 DSP Spalding And District ARS c/o A. Hensman 24 Belchmire Lane, Gosberton, Spalding PE11 4HG
G1 DSZ J. Phillips 20 The Meadows, Broomfield, Herne Bay CT6 7XF
GW1 DTA M. Pilot 92 Llanllienwen Road, Cwmrhydyceirw, Swansea SA6 6LU
G1 DTE W. Merz 38 Lime Avenue, Colchester CO4 3NL
G1 DTF A. Middleton 2 Beccles Way, Bramley, Rotherham S66 2SJ
G1 DTG D. Gryta 5 Sefton House, Tenbury Fold, Bradford BD4 0BD
G1 DTS E. Kier 9 Newbridge Way, Truro TR1 3LX
G1 DUI P. Norman 3 Church View, Witchford, Ely CB6 2HH
G1 DUJ B. Oakley 6 Staplehurst Gardens, Cliftonville, Margate CT9 3JB
G1 DUO P. Richards 16 Fruiterers Arms Caravan Park, Uphampton Lane, Ombersley, Droitwich WR9 0JW
G1 DUS D. Roberts Westpark, 296 Westleigh Lane, Leigh WN7 5PW
G1 DUT J. Robertson 4 Pembroke Road, Macclesfield SK11 8RT
G1 DVD B. Marshall 1 Anglers Way, Chesterton, Cambridge CB4 1TZ
G1 DVH J. Knighton 90 Sherwood Crescent, Market Drayton TF9 1NP
GM1 DVO C. Hepworth 20 Station Avenue, Duns TD11 3HW
G1 DVU J. Green 788 The Ridge, St. Leonards-on-Sea TN37 7PS
G1 DWC Dr R. Everett 73 Fordwych Road, London NW2 3TL
G1 DWG R. Scicluna-Kerry 5 Laxton Drive, Sibsey PE22 0AF
GU1 DWO A. Smith La Cambrette, La Rue Des Reines, Forest GY8 0JB Guernsey
G1 DWT J. Dwight 59 Highfield Road, Bramley, Leeds LS13 2BX
G1 DWU A. Swales 90 Earlswood Road, Dorridge, Solihull B93 8RN
G1 DXD P. Darke 18 Colchester Close, Southend-on-Sea SS2 6HR
G1 DXH R. Crissell 1 Medlar Drive, South Ockendon RM15 6TS
G1 DXM D. Walling 37 Ulverston Road, Swarthmoor, Ulverston LA12 0JB
G1 DXN R. Walters 16 Lune Drive, Morecambe LA3 3RZ
G1 DXQ P. Postle 20 Courtenay Close, Norwich NR5 9LB
G1 DYC D. Winkley Southall Cottage, Hadley, Droitwich WR9 0AU
G1 DYL K. Tysoe Valeside, School Road, West Hanney, Wantage OX12 0LB
G1 DYN J. Snowling 5 Verbena Close, Beechwood, Runcorn WA7 3JA
G1 DYQ N. Prosser 35 Holmfirth Close, Belmont, Hereford HR2 7UG
G1 DYR R. Munday 12 Glisson Road, Hillingdon, Uxbridge UB10 0HH
G1 DZB N. Babbage 248 Molesey Avenue, West Molesey KT8 2ET
G1 DZC G. Baldwin 29 Ramshead Approach, Leeds LS14 1HH
G1 DZD L. Ball 14 St. Wilfrids Road, Burgess Hill RH15 8BD
G1 DZY K. Bricknall 21 Uplands Way, Springwell Village, Gateshead NE9 7NQ
G1 DZZ K. Bridle 8 Hardy Avenue, Dorchester DT1 1LL
G1 EAB A. Bolton 5 Willow Crescent, Gedling, Nottingham NG4 4BL
G1 EAE A. Broughton Cobwebs, The Fleet, Pulborough RH20 1HS
GM1 EAH W. Buchanan 38 Kenmount Place, Kennoway, Leven KY8 5LT
G1 EAJ M. Bunting 22 Ling Close Coltishall, Norwich NR12 7HZ
G1 EAM A. Bush 30 New Road, Smallfield, Horley RH6 9QN
G1 EAN A. Butler Ty Ni, Hall Lane, Leamington Spa CV33 9HG
GW1 EAV S. Davies Laburnum House Guilsfield, Welshpool SY21 9PX
G1 EAX R. Dawkins 17 Dacer Close, Stirchley, Birmingham B30 3BZ
G1 EBB A. Di Duca 8 Mount Pleasant Close, Kidderminster DY10 2HZ
G1 EBP C. Jermany 5 Lexington Close, Hemsby, Great Yarmouth NR29 4ES
G1 EBT A. Jones 34 Benbow Quay, Coton Hill, , Shrewsbury SY1 2DL
G1 EBV S. Dawswell 66 Priory Walk, Leicester Forest East, Leicester LE3 3PP
G1 EBW P. Challen 20 Drummond Road, Cawston, Rugby CV22 7TN
G1 EBX S. Challen The Vicarage, Chantry Lane, Towcester NN12 6YY
G1 EBZ R. Charlton Meer Booth Rd, Boston PE22 7AB
G1 ECC D. Chippendale 19 East Park Avenue, Darwen BB3 2SQ
G1 ECE B. Clark 9 Conigre, Chinnor OX39 4JY
G1 ECI J. Christy 1 Edinburgh Drive, Hindley Green, Wigan WN2 4HL
G1 ECK N. Preval 63 Dudley Avenue, Leicester LE5 2EF
G1 ECS C. Frettsome 16 Botany Avenue, Mansfield NG18 5NG
G1 ECV J. Gardener 32 Beckington Crescent, Chard TA20 2BU
G1 ECY G. Giles 74 The Larches, Uxbridge UB10 0DN
G1 EDA R. Goff 21 Findon Road, Elson, Gosport PO12 4EP
G1 EDH T. Hacker 179A Churchill Avenue, Chatham ME5 0DQ
G1 EDK P. Hammond 160 Westlands Caravan Park, Herne Bay CT6 7LE
G1 EDM J. Hargreaves 5 Nuttall Avenue, Little Lever, Bolton BL3 1PW
G1 EDP M. Hazell 15 Lords Hill, Coleford GL16 8BG
G1 EDT W. Hewitt 99 Derrydown Road, Perry Barr, Birmingham B42 1RY
G1 EDU A. Hicks 22 Manor Park, Mirfield WF14 0EW
G1 EDX M. Holtam 16 Cowley Close, Cheltenham GL51 6NP
G1 EEA S. Howcroft 23 Alderley Avenue, Blackpool FY4 1QG
G1 EEO M. Kirby Church Cottage, Burrington, Umberleigh EX37 9JG
G1 EET M. Davis 52 Belmont Avenue, New Malden KT3 6QD
G1 EEZ J. Lewis 516 Wellsway, Bath BA2 2UD
G1 EFF A. Marriott 75 St. Johns Road, Cudworth, Barnsley S72 8DE
G1 EFG C. Mcara 6 Winniford Close, Chideock, Bridport DT6 6SA
G1 EFK G. Means Ferry Farm, Witham Bank, Lincoln LN4 4QA
G1 EFL M. Medcalf 47 Paddock Drive, Chelmsford CM1 6UX
G1 EFO P. Hyde 24 Grassam Close, Preston, Hull HU12 8XF
G1 EFP A. Jarrett 4 Langstone Close, Horwich, Bolton BL65SZ
G1 EFS S. Newell 7 Edward Road West Walton Park, Clevedon BS21 7DY
G1 EFT P. Nicholson 20 Rowley Road, Torquay TQ1 4PX
G1 EFU A. Nixon 14 Carlton Road, Lowton, Warrington WA3 2EP
G1 EFX C. Nutkins Higher Spence, Bridport DT6 6DF
G1 EGB G. Page 23 Maskelyne Close, Battersea, London SW11 4AA
G1 EGE K. Pay 1 Swallow Walk Biddulph, Stoke-on-Trent ST86TY
G1 EGI D. Phillips Bethune, Rame Cross, Penryn TR10 9DZ
G1 EGK Dr J. Preece The Grange, Harewood Road, Wetherby LS22 5BL
G1 EGL R. Preston 45 Gaynor Close, Wymondham NR18 0EA
G1 EGR G. Roberts 32 Ancaster Court, Horncastle LN9 6HG
G1 EGZ A. Adams Radnor, Shorts Road, Carshalton SM5 2PB
G1 EHB P. Allcock 11 Hillbury Field Ticehurst, Wadhurst TN5 7AY
G1 EHE M. Appleton Flat 2, Black Swan Buildings, Winchester SO23 9DT
G1 EHF D. Austen Tudorlands, Silchester Road, Bramley, Tadley RG26 5DG
GW1 EHI R. Davies 1 Mount View, Plas Road, Blackwood NP12 3RH
G1 EHK R. Birch 17 White Wood Road, Eastry, Sandwich CT13 0JZ
G1 EHM Royal Air Force Henlow Radio & Electronics Club c/o P. Bird 4 Parkside Avenue, Tilbury RM18 8DT
G1 EHS B. Brodribb 1 Ponswood Road, St. Leonards-on-Sea TN38 9BU
G1 EHU M. Hostekens 1 Ponswood Road, St. Leonards-on-Sea TN38 9BU
G1 EHX C. Cameron Rose Cottageá Orchard Way Berry Hill, Coleford GL16 7AQ
G1 EIB N. Purkins 16 Nunburnholme Avenue, North Ferriby HU14 3AN
G1 EIG J. Ryan 71B Gunterstone Road, London W14 9BS
G1 EIH D. Samber 102 Midsummer Avenue, Hounslow TW4 5BB
G1 EIO B. Smith 43 Oak Avenue, Hindley Green, Wigan WN2 4LZ
G1 EIP G. Smith 52 Penhill Crescent, St Johns, Worcester WR2 5PX
G1 EIR R. Smith 29 Windmill Lane, Henbury, Bristol BS10 7XE
G1 EIV S. Stanley 11 Mandeen Grove, Mansfield NG18 4FA
G1 EIX H. Stephens 16 Addison Drive, Stratford-upon-Avon CV37 7PL
G1 EIZ M. Stewart 2 Patmore Link Road, Hemel Hempstead HP2 4PX
G1 EJA R. Stone 1 Poplar Close, Ashford TN23 3DY
G1 EJK G. Tomlinson 4 Werneth Close, Denton, Manchester M34 6LR
G1 EJQ A. Walker Gymru Fach, 51 The Crescent, Consett DH8 5JF
GW1 EKC M. Davis Minffordd, Oakeley Square, Blaenau Ffestiniog LL41 3PU
G1 EKM S. Evans 41 Clocktower Drive, Liverpool L9 1AG
G1 EKP G. Fowler 18 Lossie Drive, Iver SL0 0JS
G1 EKU N. Kernahan 15 Howgill Lane, Sedbergh LA10 5DE
G1 ELE R. Watt 88 Graham Crescent, Portslade, Brighton BN41 2YB
G1 ELF P. Wood 3 Elf Meadow, Poulton, Cirencester GL7 5HQ
G1 ELJ R. Williams 53 Springhill Park, Wolverhampton WV4 4TR
G1 ELK P. Wilson Barkstan Lodge, Quadring Bank, Spalding PE11 4RF
G1 ELP D. Allen 34 Meadow Way Rollesby, Great Yarmouth, Norfolk NR29 5HA
G1 ELQ B. Bentley Sandy Ridge, Church Street, Stoke-on-Trent ST7 4RS
G1 ELX Prof. A. Challinor 24 West End Rise, Horsforth, Leeds LS18 5JL

G1 ELZ M. Cook Well Cottage, Old Road, Alderbury, Salisbury SP5 3AR
G1 EME Obo the Worcester Moonbounce Society c/o W. Day 4 Queenswood Drive, Worcester WR5 3SZ
G1 EMF Emf Hams c/o J. Murray 20 East End, Cholsey, Wallingford OX10 9RT
G1 EML H. Hill 5 Wentworth Gardens, Alton GU34 2BJ
G1 EMM K. Hill 73 Yellow Birch Drive, Kitchener N2N 2M3 Canada
G1 EMW R. Dearsley Prince William Farm, Lynn Road, King's Lynn PE33 9BD
G1 ENA G. Edwards 22 Whalley Lane, Uplyme, Lyme Regis DT7 3UR
G1 END Dr M. Friedman Flat 28, Hertford Mews, Potters Bar EN6 1XW
GW1 ENG P. Gibson The Nook, Trimsaran Road, Kidwelly SA17 4EB
GW1 ENP P. Hewett Tarn Hows, 7 Lakeside Drive, Presteigne LD8 2EG
G1 ENR A. Deakin The Farmhouse, New House Farm, Tenbury Wells WR15 8TW
GM1 EOA J. Minaudo Meadowside, Newbridge, Dumfries DG2 0QX
G1 EOH S. Braybrooke 6 Tubbenden Lane, Orpington BR6 9PN
GW1 EOI G. John 31 Gellifawr Road Morriston, Swansea SA6 7PN
G1 EOJ M. Kay 59 Palmer Crescent, Leighton Buzzard LU7 4HY
G1 EOK E. Keeble 17 Moat Avenue, Green Lane, Coventry CV3 6BT
G1 EOM H. Kinghorn 29 Meadowview Road, Sompting, Lancing BN15 0HU
GI1 EOS P. Leitch 212 Belfast Road, Muckamore, Antrim BT41 2EY
G1 EPD D. Hathaway 46 Blackwell Avenue, Newcastle upon Tyne NE6 4DR
G1 EPF L. Marshall 75 Acacia Crescent, Wigan WN6 8NJ
G1 EPL R. O'Callaghan 47 Seabrook Drive, Thornton-Cleveleys FY5 3SE
G1 EPO J. Pragnell Sundale, Northampton Road, Brackley NN13 7TY
GW1 EPR R. Rees 16 Railway View, Caldicot NP26 5GB
G1 EPS North Yorkshire CG c/o M. Rhodes 155 High Park Road, Southport PR9 7BY
G1 EQF P. Uttridge Springers Rest, Beck Lane, Hull HU12 9RG
G1 EQJ M. Whittle Churchfield Cottage, West Road, Wareham BH20 5RY
G1 EQL P. Wootton 20 Oakhill Road, Dronfield S18 2EJ
G1 EQM R. Agacy 23 Highgate Lane, Bolton-Upon-Dearne, Rotherham S63 8HR
G1 EQU R. Percival 23 Plumtree Road, Thorngumbald, Hull HU12 9QG
GW1 ERA A. Price 2 Ger Y Coed, Brackla, Bridgend CF31 2LA
G1 ERF S. Rogers 31 Morgan Road, Southsea PO4 8JS
G1 ERM D. Salter 94 Clifton Street, Swindon SN1 3QA
G1 ERQ R. Stevens 172 Branksome Avenue, Stanford-le-Hope SS17 8DE
G1 ERS D. Strange 15 Truman Road, Bournemouth BH119BP
G1 ERU S. Mole 17A Marlborough, Seaham SR7 7SA
G1 ERY A. Moseley 15 Gillsway, Northampton NN2 8HT
G1 ERZ A. Moules 5 Hill Road, Borstal, Rochester ME1 3NJ
G1 ESC C. Mountain 16 Temples Court, Helpston, Peterborough PE6 7EU
G1 ESW K. Laughton 33 Tiverton Close, Radcliffe, Manchester M26 3UJ
G1 ESX K. Love 63 Buxton Road, Spixworth, Norwich NR10 3PP
G1 ETD Dr J. Newland 84 Waterman Way, London E1W 2QW
G1 ETQ B. Sweeney 51 Tristram Avenue, Hartlepool TS25 5PA
G1 ETZ C. Walker 1 Shepherds Close, Shepshed, Loughborough LE12 9SQ
G1 EUA B. Wall 3 Grenville Avenue, Teignmouth TQ14 9NJ
G1 EUD D. Wiles 62 Taylor St., Tunbridge Wells TN4 0DX
G1 EUF R. Wilson Street Farm, Henny Street, Sudbury CO10 7LS
G1 EUG D. Wolfe 48 Wilby Lane, Great Doddington, Wellingborough NN29 7TP
G1 EUH J. Woods 5 Sand Martin Avenue, Wesham, Preston PR4 3FE
G1 EUI M. Wright 71 Oakridge Road, High Wycombe HP11 2PL
G1 EUM S. Foote Harroway, South Hanningfield Road, Wickford SS11 7PF
G1 EUN A. Friend 43 Gildale, Peterborough PE4 6QY
G1 EUQ A. Gammon 5 Sommerville Close, Faversham ME13 8HP
G1 EUU M. Gibson 1 Oakleigh Road, Grantham NG31 7NN
G1 EVA M. Hattersley 190 Elmton Road, Creswell, Worksop S80 4DY
G1 EVI B. Green 49 Brockman Crescent, Dymchurch, Romney Marsh TN29 0UA
G1 EVR P. Lowe 155 Long Lane, Bolton BL2 6EU
G1 EVV C. Mylchreest 21 Bexhill Gardens, St. Helens WA9 5FQ
G1 EWC A. Webster 49 Uplands Croft, Werrington, Stoke-on-Trent ST9 0LF
G1 EWE T. Williams 20 Sandringham Drive, Dartford DA2 7WB
G1 EWH S. Bell The Haven, Low Street, Retford DN22 0LN
G1 EWM T. Conlin 5 Morland Drive, Rochester ME2 3LW
G1 EWV P. Dawson Honeysuckle Cottage, 16 Waterloo Street, Harrogate HG1 5JD
GW1 EWW K. Edwards 25 Woodland Road, Neath SA11 3AL
GW1 EWY B. Owen Llys Helen, Croesor, Penrhyndeudraeth LL48 6SR
G1 EXG J. Hare 1 The Copse, 50-52 Princes Road, Brighton BN2 3RH
G1 EXK S. Bussey 7 Ilderton Crescent, Seaton Delaval, Whitley Bay NE25 0FH
G1 EXM C. Bussey 6 Ray Court, Wimblington, March PE15 0FE
G1 EXR W. Cosgrove 62 Twyford Avenue, Great Wakering, Southend-on-Sea SS3 0EX
G1 EXU R. Cloke Woodlands, Roxton Lane, Keelby DN41 8JB
G1 EXV D. Cooper 75 Merevale Avenue, Nuneaton CV11 5LU
G1 EYD K. Rutter 12 Berwick Terrace, North Shields NE29 7AW
G1 EYG G. Shipperley 72 Hithercroft Road, Downley, High Wycombe HP13 5RH
G1 EYJ Rvd. I. Smith 21 Gorsehill Road, Wallasey CH45 9JA
G1 EYS Willpower CG c/o P. Jorquera 21 Highlands Road, Orpington BR5 4JP
G1 EYT P. Vickers 21 Blackwood Drive, Sutton Coldfield B74 3QP
G1 EYW S. Warren 43 Glebe Rise Kings Sutton, Banbury OX17 3PH
G1 EYY D. Whincup 172 Appleton Road, Hull HU5 4PF
G1 EYZ S. White 67 Wingfield Road, Lakenheath, Brandon IP27 9HR
G1 EZF University of Aston RS c/o M. Allmark 11 Potternewton Crescent, Leeds LS7 2DY
G1 EZI S. Armstrong 5 Dashpers, Brixham TQ5 9LJ
G1 EZJ C. Barker 52 Spode Street, Stoke-on-Trent ST4 4DY
G1 EZU D. Harpham 16 Scotts Way, Kirkby-In-Ashfield, Nottingham NG17 9DN
G1 FAA S. Jeffery 35 Lynton Avenue, Orpington BR5 2EH
G1 FAD T. Kenney 7 Hickin Close, Charlton, London SE7 8SH
GM1 FAF J. Marshall Drummorlie, Wallyford Toll, Musselburgh EH21 8JT
GM1 FAI A. Miller 21 Merker Terrace, Linlithgow EH49 6DD
G1 FBE D. Telford 9 Central Avenue, Carlisle CA1 3QB
G1 FBI J. Hughes 47 Stambourne Way, Upper Norwood, London SE19 2PY
GW1 FBL T. Boorman 43 Ffordd Taliesin, Killay, Swansea SA2 7DF
GM1 FBM B. Borland Beechwood Cottage, Muirhall Road, Perth PH2 7LL
G1 FBQ Dr S. Fraser Walnut Tree Cottage, Main Road, Fyfield, Abingdon OX13 5LN
GW1 FBU K. Goodchild 3 Mill Leat Lane, Gorseinon Swansea SA4 4QE
G1 FBW T. Howchen 1 Ash Road, Canvey Island SS8 7EA
G1 FBZ W. Hicks 7 Meadow Close, Thundersley, Benfleet SS7 3RJ
G1 FCN M. Nutton Rose Cottage, Henley Street, Luddesdown, Gravesend DA13 0XB
G1 FCU S. Reed 20 Mead Crescent, Bookham, Leatherhead KT23 3DU
G1 FCW Essex CW ARC c/o P. Tittensor 47 St. Johns Road, Chelmsford CM2 0TY
G1 FCX Dr N. Coleman The Old Croft Reeth, Richmond DL11 6TE
G1 FDD N. Groeber 113 Kings Road, Kings Heath, Birmingham B14 6TN
G1 FDL V. Kelk 7 Rowan Place, Garforth, Leeds LS25 2JR
G1 FDN W. Kenyon 22 Barons Way, Lower Darwen, Darwen BB3 0RG
G1 FDO T. King 32 Bagnall Avenue, Arnold, Nottingham NG5 6FT
G1 FEF C. Smith 27 Wye Road, Wooburn Green, High Wycombe HP10 0DU
GM1 FEM I. Smith 150 Eden Park, Clayton Caravan Park, St. Andrews KY16 9YB
G1 FEO G. Jones 8 Kenilworth Road Lighthorne Heath, Leamington Spa CV33 9TH
G1 FEP D. Twidale 18 Kinnaird Road, Wallasey CH45 5HN
G1 FET P. Taylor 29 Dunstall Road, Halesowen B63 1BB
G1 FEX J. Allsop 15 Woodland Grove, Mansfield Woodhouse, Mansfield NG19 8AZ
G1 FFH T. Firth 126 Tombridge Crescent, Kinsley, Pontefract WF9 5HE
G1 FFO P. Thomas 9 Awefield Crescent, Smethwick, W Midlands B67 6PR
G1 FFR S. Tucker 28 Peregrine Road, Stockport SK2 5UR
G1 FFU D. Woolmer 2 Muccleshell Close, Havant PO9 2HR
G1 FGC P. Chalkley 10 Preston Gardens, Luton LU2 7NL
G1 FGE N. Collier 104 Grosvenor Street, Kearsley, Bolton BL4 8DW
G1 FGI P. Escreet Wortley Cottage, Husthwaite, York YO61 4PY
G1 FGK W. Grint 15 Ivythorn Road, Street BA16 0TE
GM1 FGN R. Hussey 21 Maidenfield, Mossbank, Shetland ZE2 9TD
G1 FHH P. Johnson 30 Copplestone Grove, Longton, Stoke-on-Trent ST3 5UD
G1 FHI S. Murray 51 Huddersfield Road, Newhey, Rochdale OL16 3QZ
G1 FHK C. Peart 33 Fieldfare, Abbeydale, Gloucester GL4 4WH
G1 FHR R. Roots 14 Sussex Close, London Road, Sevenoaks TN15 6BB
G1 FHY S. Wise 23 Wordsworth Drive, Eastbourne BN23 7QP
G1 FIG J. Clark 58 Kelvin Grove, Fareham PO16 8LE
G1 FIM P. Mcdonnell 55 Lodge Hall, Harlow CM18 7SY
G1 FIP S. Rowlandson 48 Greville Road, Warwick CV34 5PB
G1 FJD A. Wood 2 Towning Close, Deeping St. James, Peterborough PE6 8HR
G1 FJF A. Bawden 67 Silo Drive, Farncombe, Godalming GU7 3NZ
G1 FJH P. Bruce 26 Queens Road, Wilbarston, Market Harborough LE16 8QJ
GW1 FJI R. Bullock 32 Tinmans Green, Redbrook, Monmouth NP25 4NB
G1 FJJ M. Cammish 20 Chantry Avenue, Hartley, Longfield DA3 8DD
G1 FJS K. Davis 16 West Field Close, Taunton TA1 5JU
G1 FKJ G. Portlock 1 Windmill Cottage, Oxford Street, Marlborough SN8 2DH
GW1 FKL G. Howells 70 Meadow Street, Treforest, Pontypridd CF37 1SS
G1 FKM J. Kendall 6 Wellington Street, Allerton, Bradford BD15 7QZ
G1 FKP D. Le Vine Anglecroft, Borough Road, Westerham TN16 2LA
G1 FKS C. Macliesh 33 River Way, Twickenham TW2 5JP
G1 FKT J. Mansley 2 Beech Tree Close, Cuerden Residential Park, Leyland PR25 5PA
GW1 FKY K. Eaton 21 Westminster Way, Bridgend CF31 4QX
G1 FLI C. Franklin 3 Park Road, Rugby CV21 2QU
G1 FLV M. Parr 5 Suffolk Grove, Leigh WN7 4TA
G1 FLW P. Partridge 18 Chaucers Drive, St Peters Field, Nuneaton CV10 9SD
G1 FLX S. Pickstone 48 Oak Tree Drive, London N20 8QH
GW1 FLY A. Rosier 2 Watkin Drive, Oswestry SY11 1SQ
G1 FMA A. Robinson 5 Alford Road, Heaton Chapel, Stockport SK4 5AW
G1 FMT C. Hardy 110 Jubilee Road, Waterlooville PO7 7RG
G1 FMU K. Harris 8 Trelawney Rise, Callington PL17 7PT
GM1 FMV G. Hind 135 Pilton Avenue, Edinburgh EH5 2HP
G1 FMW D. Lewis 4 Westwood Grove, Solihull B91 1QB
GM1 FMX W. Mccandlish Lingdowey, Stoneykirk Road, Stranraer DG9 7BX
G1 FNA L. Phillips 2 Stratton Green, Bedgrove, Aylesbury HP21 7EP
G1 FND N. Stephens 7 Quarry Road, Alveston, Bristol BS35 3JL
G1 FNF B. Walker 47 Coppice Avenue, Eastbourne BN20 9QJ
G1 FNN W. Ball 43 Fairfax Drive, Herne Bay CT6 6QZ
G1 FNP M. Bellas 3 Elm Terrace, Penrith CA11 7JY
G1 FNS B. Cutts 7 Lych Gate Close, Sandhurst GU47 8JH
G1 FNU J. Dodd 38 The Quadrant, North Shields NE29 7HP
GM1 FNX P. Ewing Arisaig Priestland, Darvel KA17 0LP
G1 FOA P. Franklin 14 St. Erth Hill, St. Erth, Hayle TR27 6EX
G1 FOE P. Holt 27 Sandown Close, Blackwater, Camberley GU17 0EN
G1 FOF M. James 9 Denham Crescent, Mitcham CR4 4LZ
G1 FOM P. Lyttle 3 Woodlands, East Ardsley, Wakefield WF3 2JG
G1 FON M. Mangan Schuetzenstr. 54, Kaiserslautern 67659 Germany
G1 FOW K. Worsley 102 Cabul Close, Warrington WA2 7SE
G1 FOZ M. Slattery 9 Barns Close, Walsall WS9 9BD
G1 FPC A. Sands 7 Kimberley Avenue, Seymour Street, Hull HU3 5PP
GM1 FPD D. King 18 Ford Spence Court, Benderloch, Oban PA37 1PY
G1 FPK R. Kerridge 80 Melton Road, Wymondham NR18 0DE
G1 FPP N. Sirkett Flat 2, Barfield House, Ryde PO33 2JP
G1 FPY A. Watson 9 Linthurst Newtown, Blackwell, Bromsgrove B60 1BP
G1 FPZ A. Wilcox 2 Dawkins Road, Poole BH15 4JD
G1 FQD S. Eldredge 2 Chelmsford Drive, Worcester WR5 1QX
G1 FQI D. Jackson 28A Wawne Road Sutton-On-Hull, Hull HU7 4YE
G1 FQX J. Lamb 205 Springfield Road, Sutton Coldfield B76 2SY
G1 FRD G. Bennett 14 Thessaly Road, Stratton, Cirencester GL7 2NG
GM1 FRG T. Watsonlolaire, Mannachie Grove, Forres IV36 2WE
G1 FRJ P. Bearne 59 Foxhole Road, Paignton TQ3 3TD
G1 FRL G. Mott 191 Joyners Field, Harlow CM18 7QD
G1 FRM R. Doughty 4 Trinity Road, Wisbech PE13 3UN
G1 FRS Farnborough CG c/o B. Dawson Isca, 12 Lestock Way, Fleet GU51 3EB
G1 FSE K. Stokes 33 The Crescent, Burntwood WS7 2PA
G1 FSF J. Williams 52B Pensford Drive, Eastbourne BN23 7NY
GI1 FSJ N. Colgan 11 St. Johns Park, Moira, Craigavon BT67 0NL
GM1 FSU I. Menzies 33 Lochside Drive, Bridge Of Don, Aberdeen AB23 8EH
G1 FSW R. Consolante 19 Chestnut Gardens, Stamford PE9 2JY
G1 FSX J. Nash 21 St. Marys Close, Peterborough PE1 4DR
GM1 FSZ K. Hall 12 Brockhill Rise, Inverurie AB51 5RH
G1 FTD J. Hobbs Fetchalls, The Green, Bury St. Edmunds IP30 9AF
GM1 FTG M. Lonnen Hunt Hall, Glendevon, Dollar FK14 7JZ
G1 FTH A. Marston 92 Sorrell Road, Nuneaton CV10 7AW
G1 FTJ S. Tonks Milestones Watling Street, Cannock WS11 1SH
G1 FTK N. Apps 71 De Cham Road, St. Leonards-on-Sea TN37 6HF
G1 FTU J. Pearson Largo, Hemming Green, Chesterfield S42 7JQ
G1 FTV A. Bryant 42 Kemerton Walk, Swindon SN3 2EA
G1 FTX E. Demeza 2 Adam Close, St. Leonards-on-Sea TN38 9QW

GM1 FTZ H. Simpson Clachan Farm Cottages, Rosneath, Helensburgh G84 0QR
G1 FUG A. Sagar 28 Rangoon Road, Solihull B92 9DB
G1 FUJ B. Jones 4 Caddick Close, Kingswood, Bristol BS15 4RT
G1 FVA K. Irons 14 Beech Grove, Houghton, Carlisle CA3 0NU
G1 FVC I. Batten 17 Cornfield Road, Birmingham B31 2EB
G1 FVE D. Ward 36 Croxby Avenue, Scartho, Grimsby DN33 2NW
G1 FVH M. Jordan 160 Beta Road, Farnborough GU14 8PH
G1 FVP R. Slone 55 Chilton Way, Hungerford RG17 0JR
G1 FVS P. Dean 10 Moor Terrace, Bradford BD2 4SG
G1 FVU K. Bloomfield 14 Manners Road, Fornham St. Martin, Bury St. Edmunds IP31 1TE
GW1 FWC W. Williams Llwynprenteg, Llanafan, Ceredigion SY23 4BQ
GW1 FWE J. Duggan-Keen Bodlondeb, Chapel Street Caerwys, Mold CH7 5AE
G1 FWF H. Morgan 2 Mayfield Park South, Fishponds, Bristol BS16 3NG
G1 FWR P. Harman 35 Point Clear Road, St. Osyth, Clacton-on-Sea CO16 8EP
G1 FWS F. Swaine 2 Norwich Close, Stevenage SG1 4NU
G1 FWU T. Norbury 19 Charles Cope Road, Orton Waterville, Peterborough PE2 5ER
G1 FWY C. Pitt 31 D'Arcy Way, Tolleshunt D'Arcy, Maldon CM9 8UD
G1 FWZ B. Lakey 3 Simons Close, Worle, Weston-Super-Mare BS22 6DJ
G1 FXB A. Turquand 63 Sundown Avenue, Dunstable LU5 4AL
G1 FXC R. Lambourne Rosedale, Townsend, Bicester OX27 0EY
G1 FXD O. Rogers The Barn, Millways Farm, Rosenannon, Bodmin PL30 5PJ
GW1 FXL S. Annetts Hengwm, Rhayader LD6 5LD
G1 FXM T. Young Elm Field Lodge, Tollesbury Road, Tolleshunt D'Arcy, Maldon CM9 8UA
G1 FXS P. Striplin 64 Ebrington Road, Malvern WR14 4NL
G1 FXT R. Robinson 5 Lilac Close, Newton Longville, Milton Keynes MK17 0DQ
G1 FXX R. Tunbridge 35 Coworth Close, Ascot SL5 0NR
G1 FYE C. Sterland 103 Main St., Distington, Workington CA14 5UJ
G1 FYF C. Bradley 7 Coltsfoot, Biggleswade SG18 8SR
G1 FYQ Hinckley Amateur Radio And Electronics Society c/o K. Taylor 29 School Road, Pontefract WF8 2AJ
G1 FYS K. Boothroyd 16, Kelvin Avenue, Dalton, Huddersfield HD5 9HG
G1 FYU M. Blockley 7 Stoke Lane, Stoke Bardolph, Burton Joyce, Nottingham NG14 5HR
G1 FZL P. Dyer 18 Christopher Close, Yeovil BA20 2EH
G1 FZR R. Delve 18 Thame Road, Piddington, Bicester OX25 1PX
G1 FZS R. Fleet 17 Crown Road, Portslade, Brighton BN41 1SJ
G1 FZV A. Ogden 5 Lower Bristol Road, Clutton, Bristol BS39 5PB
G1 GAD F. Mcloughlin 21 Darwin Crescent, Newcastle upon Tyne NE3 4TT
G1 GAN P. Cartwright 1 Railway Cottages, Sutton Bingham, Yeovil BA22 9QW
G1 GAR M. Kipping 46 Old Hardenwaye, Totteridge, High Wycombe HP13 6TJ
G1 GAS C. Kelley Dunkeld, Bridge Street Fenny Compton, Southam CV47 2XY
G1 GAT R. Fernihough 3 Sandpiper Close, Quedgeley, Gloucester GL2 4LZ
GW1 GAU M. Sullivan 14 Moorview Road, Gendros, Swansea SA5 8BU
G1 GAW R. Smith 18 Curlew, Wilnecote, Tamworth B77 5PL
G1 GBC W. Boucher 12 Highfield Terrace, Ilfracombe EX34 9LG
G1 GBF J. Delaney 31 Roose Road, Barrow-in-Furness LA13 9RG
GW1 GBH S. Parkins 100 Welsh Road, Garden City CH52HX
G1 GBI M. Pearson 34 Downside Road, Sutton SM2 5HP
G1 GBR A. Ziemacki 3 Wheatcroft Road, Rawmarsh, Rotherham S62 5JR
G1 GBV D. Evans 59 Watlington Road, Benfleet SS7 5DT
G1 GBX D. Vanbeck 101 Upper St., Islington, London N1 1QN
GM1 GCB Wacral (World Association of Christian Radio Amate c/o S. Raisey-Skeats 20 Gordon Street, Boddam, Peterhead AB42 3AY
G1 GCF F. Clough 2 Hudson Close, Tadcaster LS24 8JD
G1 GCJ G. Morris 21 Orchard Place, Deer Park, Ledbury HR8 2XD
G1 GCY G. Gowland 7 Canewdon Hall Close, Canewdon, Rochford SS4 3PY
G1 GDA M. Austin 10 Simon Place, Wideopen, (Division Of Tyneside Motor Sport, Newcastle upon Tyne NE13 7HT
G1 GDB D. Thwaytes 3 California Terrace, Bothel, Wigton CA7 2JF
G1 GDJ C. Godward 3 Court Close, Brighton BN1 8YG
G1 GDM J. Coombes 22 Chollerford Close, Gosforth, Newcastle upon Tyne NE3 4RN
GM1 GDO J. Morton 6 Deanpark Place, Balerno EH14 7ED
G1 GDR B. Smith 4 Planetree Close, Bromsgrove B60 1AW
G1 GDT C. Groom Woodstock Cottage, 2 Woodstock Terrace, Dursley GL11 5SW
GM1 GEQ T. Menzies 239 Eskhill, Penicuik EH26 8DF
G1 GER A. Goodings 2 Mulberry Grove, Bradwell, Great Yarmouth NR31 8QJ
GM1 GES W. Barbour 27 Drove Road, Langholm DG13 0JW
G1 GET F. Wood 40 Mill Lane, Spalding PE11 4TL
G1 GEV K. Argyle 62 Yew Tree Drive, Leicester LE3 6PL
G1 GEY D. Stoker Headlands, Front Street, Aldborough, Boroughbridge, York YO51 9ES
G1 GFA N. Pitt 37 Shelley Drive, Four Oaks, Sutton Coldfield B74 4YD
G1 GFC S. Bradley 75 New Road, Sawston, Cambridge CB22 3BN
G1 GFD A. Crook 54 Somerset Way, Paulton, Bristol BS39 7YX
G1 GFF V. Thomas Flat 38, Lazonby Court, St. Leonards-on-Sea TN38 0QP
GW1 GFO P. Carlisle 5 Pengry Road, Loughor, Swansea SA4 6PH
G1 GFW J. Jacobs 11 Delamere Close, Castle Bromwich, Birmingham B36 9TW
G1 GFZ R. Pelling Spring Cottage, Main Road, Hastings TN35 4SL
G1 GGB R. Phillips 58 Baranscraig Avenue, Patcham, Brighton BN1 8RE
G1 GGI S. Hargreaves 10 Reedham Crescent, Cliffe Woods, Rochester ME3 8HT
G1 GGK F. Coldham 5 Church Lane, Towersey, Thame OX9 3QL
G1 GGN R. Barkley 9 Eagle Close, Erpingham, Norwich NR11 7AW
G1 GGT Dr R. Sharp Arosa, The Park, Harwell OX11 0HB
G1 GHG K. Knibbs 8 Ferguson Way, Huntington, York YO32 9YG
GD1 GHK W. Corlett 9 Kerrocruin, Kirk Michael, Isle of Man IM6 1AF
G1 GHU T. Smith 58 Berkeley Road, Cleethorpes DN35 0NX
G1 GHY A. Laszkiewicz 38 Langley Lane, Ifield, Crawley RH11 0NA
GM1 GHZ Backpackers Radio Activity Group c/o P. Thompson 31 St. Marys Drive, Perth PH2 7BY
G1 GIA I. Sinclair 40 Holders Hill Gardens, London NW4 1NP
G1 GID G. Watt 48 Southdown Road, Portslade, Brighton BN41 2HN
G1 GIE R. Buckley 6280 Hawkes Bluff Avenue, Davie, Fort Lauderdale 33331-3419 United States
G1 GIJ D. Hadjidakis 19 Eastfield Road, Royston SG8 7ED
G1 GJD Mastcar ARC c/o S. Corson 9 St. Marys Way, Weedon, Northampton NN7 4QL
G1 GJT P. Jackson 10 Claremont Road, Nottingham NG5 1BH
G1 GKA R. Mason 32 Linden Drive, Evington, Leicester LE5 6AH
G1 GKF R. Mann Little Chysauster, Penzance TR20 8XA
G1 GKH G. Hinds 35 Lime Grove, Burntwood WS7 0HA
GI1 GKI T. Campbell 265 Ballynahinch Road, Lisburn BT27 5LS
G1 GKK S. Brooke 14 Saxton Avenue, Heanor DE75 7PZ
G1 GKN A. Tyler 41 Beadle Way, Great Leighs, Chelmsford CM3 1RT
G1 GKR A. Barlow 454 Shaw Road, Royton, Oldham OL2 6PG
GW1 GKV P. Jones Pen Y Galchen Farm, Pwlldu, Pontypool NP4 9SS
G1 GKW Furness ARS c/o E. Perryman Flat 4, Garth House, Bognor Regis PO21 1HQ
G1 GLG S. Padgham 4 Hollamby Park, Hailsham BN27 2LX
G1 GLN Iom ARS c/o D. Beasley 40 Susannah Street, London E14 6LS
G1 GLS R. Lees 28 Lyngarth House, Grosvenor Road, Altrincham WA14 1LH
G1 GLZ B. Start 9 Front Street, Corbridge NE455AP
GI1 GME A. Mcilwee 48 Carnduff Drive, Ballymena BT43 7AP
G1 GMF R. Sievert 5 Sandmoor Road, New Marske TS11 8BP
G1 GMG S. Gainswin 1 Buckfast Road, Buckfast, Buckfastleigh TQ11 0EA
G1 GMH D. Peat 23 Hill Bottom Close, Whitchurch Hill, Reading RG8 7PX
G1 GMM H. Barczynski 64 Kings Acre, Coggeshall, Colchester CO6 1NY
G1 GMQ A. Bolton 8 Turners Walk, Chesham HP5 3BT
G1 GMV A. Brewer 25 Ackerman Road, Dorchester DT1 1NZ
G1 GMX J. Mold Sycamore Lodge, Butt Lane, Goulceby, Louth LN11 9UP
G1 GNP D. Roe 9 The Orchard, Fairfield Road, Ilkeston DE7 6DD
G1 GNQ G. Nurse Orchard Bungalow, Priors Green, Stisted, Braintree CM77 8BP
G1 GNX G. Leonard 11 St. Leonards Drive, Timperley, Altrincham WA15 7RS
G1 GOP A. Abbott The Shieling, West Road, Weaverham, Northwich CW8 3HH
G1 GOQ S. Abbott 5 Heathcote Gardens, Rudheath, Northwich CW9 7JB
G1 GOY P. Miles 37 Central Avenue, Northampton NN2 8EA
G1 GPE Radio Delta Xray c/o D. Murray 8 Tweed Crescent, Rushden NN10 0GS
G1 GPL L. Pointon 10 Lovers Walk, Dunstable LU5 4BG
G1 GPM M. Stevenson 6 Charnock Crescent, Sheffield S12 3HB
G1 GPT A. Parsons 157 Deighton Road, Huddersfield HD2 1JT
G1 GQB J. Bagshaw 7 Queen Street, Gomersal, Cleckheaton BD19 4LG
G1 GQJ C. Clark 9 Conigre, Chinnor OX39 4JY
G1 GQP P. Rolfe 62 Wincanton Road, Reading RG2 8PB
G1 GQQ M. Rowbotham 37 Crawford Rise, Arnold, Nottingham NG5 8QF
G1 GQY A. Armstrong 18 Flaxfield Way, Kirkham, Preston PR4 2AY
G1 GQZ C. Armstrong 18 Flaxfield Way, Kirkham, Preston PR4 2AY
G1 GRB R. Arnold 26 Pinehurst Park, West Moors, Ferndown BH22 0BW
G1 GRM North London Six Metre Group c/o B. Sinclair 97 Lear Drive, Wistaston Green, Crewe CW2 8DS
G1 GRN L. Skorupinski 49 Pool Lane, Winterley, Sandbach CW11 4RZ
G1 GRT G. Thomas 83 Hollingthorpe Road, Hall Green, Wakefield WF4 3NW
G1 GRU W. Smith Flat 48, Winehala Court, 50A Sandbeds Road, Willenhall WV12 4GA
G1 GRZ W. Baker 26 Gardeners Road, Halstead CO9 2TB
G1 GSB P. Standley Bligh House, 1 Norwich Road, Norwich NR16 1DJ
G1 GSG M. Taylor 7 Marshall Road, Cropwell Bishop, Nottingham NG12 3DP
G1 GSJ W. Mclaren Ingleside, Waterloo, Whitchurch SY13 2PX
G1 GSK M. Baylis 45 Florence Avenue, Hove BN3 7GX
G1 GSN P. Bradfield 118 East Road, Langford, Biggleswade SG18 9QP
G1 GST J. Thomas 59 Cross Lane, Dudley DY3 1PD
G1 GSY G. Bridle 43 Cornflower Close Locks Heath, Southampton SO31 6SP
G1 GTA P. Butler 25 Harringdale Road, Workington CA14 4NU
G1 GTF E. Chilton 33 Kersall Court, Nottingham NG6 9DT
G1 GTH C. Clark 24 Daisy Royd, Huddersfield HD4 6RA
G1 GTK D. Towers 50 Westbeech Road, Pattingham, Wolverhampton WV6 7AQ
G1 GTM A. Turner Carnbrae, Woodhouse Hill, Lyme Regis DT7 3SL
G1 GTP B. Warnaby 69 Caledonian Road, Hartlepool TS25 5LB
G1 GTQ A. Clarke 18 Waterloo Road, Brighouse HD6 2AT
G1 GTR P. Clarke 13 Mitchell Street, Brighouse HD6 2AY
G1 GTS R. Clarke 47 Peartree Road, Enfield EN1 3DE
G1 GTX P. Cork 95 Barmston Way Barmston Village, Washington NE38 8DD
G1 GUI S. Watts 16 Northampton Close, Bracknell RG12 9EF
G1 GVJ P. Allen 17 Winfield Road, Sedbergh LA10 5AZ
GW1 GVM D. Gale 5 Gadlys Terrace, Glyncorrwg SA13 3BP
G1 GVP J. Gibbon 18 Eagle Street, Penn Fields, Wolverhampton WV3 7DN
G1 GWE A. Friel 10 Marvejols Park, Cockermouth CA13 0QR
G1 GWF B. Maxwell Hillcrest, Castle View, Egremont CA22 2NA
G1 GWJ A. Gillon 94 Pelham St., Ashton under Lyne OL7 0DU
G1 GWO J. Green 32 Elizabeth Close, Highwoods, Colchester CO4 9YU
G1 GWS J. Mossop 14 Websters Lane, Great Sutton, Ellesmere Port CH66 2LH
G1 GWX R. Patrick 9 Brant Avenue, Illingworth, Halifax HX2 8DL
G1 GXB K. Ray 4 Elm Road Bishops Waltham, Southampton SO32 1JR
G1 GXC S. Ray 75 The Meads, Edgware HA8 9HE
G1 GXF T. Scott 9 Walker Drive, Leigh-on-Sea SS9 3QS
GM1 GXH I. Sinclair Clan Sinclair House, Nosshead Lighthouse, Caithness KW1 4QT
GW1 GXQ R. Tulk Home Farm Lodge, Pen Y Lan, Wrexham LL14 6HS
G1 GXW B. Woodhouse 5 Filby Drive, Durham DH1 1LT
G1 GXX A. Mayes 31 Holsey Lane, Bletchley, Milton Keynes MK2 3FH
G1 GYC M. Hallsworth 87 Talbot Street, Hazel Grove, Stockport SK7 4BJ
G1 GYF D. Harvey 264 Rangefield Road, Bromley BR1 4QY
G1 GYH J. Hay 23 Manor Close, Wilmslow SK9 5PX
G1 GYJ F. Mallows 31 Booth Road, Hartford, Northwich CW8 1RD
G1 GYM P. Mcewen 7 Springfield, Longhoughton, Alnwick NE66 3NT
G1 GYQ A. Hayward 1 Cleveland Road, Basildon SS14 1NF
G1 GYT T. Down 1 Park View, East Tytherley Road, Romsey SO51 0LW
G1 GZG M. Newport 9 Highbury Park, Exmouth EX8 3EJ
G1 GZI A. Farmar Hawkes Place, Horslett Hill, Holsworthy EX22 6RS
G1 GZK K. Feay 19 Dorset Avenue, Diggle, Oldham OL3 5PL
G1 GZM I. Ford 97 Green Rock Lane, Walsall WS3 1NQ
G1 GZT N. O'Connor 39 Strickland Drive, Morecambe LA4 6TD
G1 HAB H. Birkmyre Swarland, Morpeth NE65 9JW
G1 HAC J. Hilton 32 Dowry St., Fitton Hill, Oldham OL8 2LP
G1 HAH B. Hodgson 40 Trentham Drive, Bridlington YO16 6ES
GW1 HAX N. Bevan Mountain View, Whip Lane, Oswestry SY10 8HU
G1 HBC T. Hopkins 58 Broom Grove, Knebworth SG3 6BQ
G1 HBD A. Hornby 2 Maple Close, Winnersh, Wokingham RG41 5PE
G1 HBE A. Howlett 43 Cheetham Hill Road, Dukinfield SK16 5JL

G1 HBF M. Hughes 2 Chaldon Road, Canford Heath, Poole BH17 8DB
G1 HBR K. Jackaman 14A Cloverdale Gardens, Sidcup DA15 8QL
G1 HBV E. Jones 37 Sluice Road, Denver, Downham Market PE38 0DY
G1 HCC E. Kent 100 Waskerley Road, Washington NE38 8DS
G1 HCI R. De Ste Croix 49 Oxford Street, Grimsby DN32 7JE
G1 HCJ G. De Ste Croix 49 Oxford Street, Grimsby DN32 7JE
G1 HCM F. Dawson 33 Oakwood Road, Ryde PO33 3JU
G1 HCU G. Gratton 5 Nursery Avenue, Ovenden, Halifax HX3 5SZ
G1 HDG P. Greed 12 Bailey Close, Windsor SL4 3RD
G1 HDK W. Akhurst 20 Newton Road, Faversham ME13 8DZ
G1 HDO A. Appleton Flat 9, 19-21 West Cliff Road, Bournemouth BH4 8AT
G1 HDR R. Stanford 1 South End, Bassingbourn, Royston SG8 5NG
G1 HDX M. Robertson 12 James Park Homes, Egremont CA22 2QQ
G1 HEA Dr A. Steele 18 Lace Crescent, Tiverton EX16 5FH
G1 HEJ J. Alexander 1 Locarno Road, Swanage BH19 1HY
G1 HEN D. Coates 2 Penfold Drive, Countesthorpe, Leicester LE8 5TP
G1 HEP C. Heptonstall Badger Cottage, 27 Bolster Moor Road, Huddersfield HD7 4JU
G1 HEQ K. Tucker 507 New North Road, Ilford IG6 3TF
G1 HER G. Dasilva-Hill 12 St. Stephens Crescent, Thornton Heath CR7 7NP
G1 HEU G. Tybora 37 Nunsfield Drive, Alvaston, Derby DE24 0GH
GW1 HEV D. Thomas 3 Oaklands Terrace, Wiston, Haverfordwest SA62 4PR
G1 HEW P. Travers 49 West Bank Drive, South Anston, Sheffield S25 5JG
G1 HEX J. Dunn 8 Ettrick Terrace North, Craghead, Stanley DH9 6BE
G1 HEY J. Todd Atlast, 7 Marine Avenue West, Mablethorpe LN12 2TX
G1 HFA R. Winder 176 Ambleside Road, Lancaster LA1 3ND
G1 HFE S. Wood 18 Rosemellin, Camborne TR14 8QF
G1 HFH D. Ward 10 Fulshaw Avenue, Wilmslow SK9 5JA
G1 HFK W. Willoughby 27 Foxwood Grove, Sheffield S12 2FN
G1 HFS K. Burgess 32 Hendon Street, Leigh WN7 1TS
G1 HFT E. Bennett 20 Cromford Road, Clay Cross, Chesterfield S45 9RE
G1 HFY B. Watson 20 St. Marys Gardens, Hilperton Marsh, Trowbridge BA14 7PG
G1 HGA K. Yates 3 Flaxland Crescent, Sileby, Loughborough LE12 7SB
G1 HGB J. Neville 44 Thorpe House Avenue, Sheffield S8 9NG
G1 HGC B. Newton 42 Heath Road, Widnes WA8 7NQ
G1 HGD M. Newell 189 Humber Road, Coventry CV3 1NZ
G1 HGF T. Standring 52 Beechcroft Road, Ipswich IP1 6BD
G1 HGT A. Berkerey 36 Erlesmere Gardens, London W13 9TY
G1 HGY Rvd. P. Parry Forge House, Church Road, Wellingborough NN9 6BQ
G1 HHB C. Brown 12 Forest Close, Newport PO30 5SF
G1 HHC D. Bolt C/O Old School Hse, Maristow Roborough, Plymouth PL6 7BY
G1 HHD E. Bolt Old School House, Roborough, Plymouth PL6 7BY
G1 HHG J. Pilfold-Bagwell 16 Anselm Close, Sittingbourne ME10 1EY
G1 HHH Hastings Electronics & RC c/o J. Tucker 17 Rye Industrial Park Harbour Road, Rye TN31 7TE
GW1 HHM K. Roberts Gwenallt, Lon Crecrist, Holyhead LL65 2AZ
G1 HHO G. Reeve 10 Badgers Copse, New Milton BH25 5PE
G1 HHQ Perth Repeater Group c/o J. Brown 14 The Green, Winscombe BS25 1AL
G1 HHS A. Burrows 1 Browns Avenue, Runwell, Wickford SS11 7PT
G1 HHT A. Benstock 10 Wike Ridge Avenue, Leeds LS17 9NL
G1 HHU N. Ball 140 Albert Avenue, Prestwich, Manchester M25 0HE
G1 HHW W. Curtis Rio Taibilla 11, San Miguel De Salinas, Alicante 3193 Spain
GD1 HIA P. Smith 98 Silverburn Crescent, Ballasalla IM9 2ED Isle of Man
G1 HIB M. Standing 7 Oxcliffe New Farm Caravan Park, Oxcliffe Road, Morecambe LA3 3EF
G1 HIG T. Ravelini 85 Glanfield Road, Beckenham BR3 3JT
G1 HIJ R. Dimmock 67 Meadway, Dunstable LU6 3JT
GW1 HIN R. Ellwood-Thompson 15 Skinner Street, Aberystwyth SY23 2JU
G1 HIO M. Horsfield 59 Queens Drive, Newton-le-Willows WA12 0LY
G1 HIP K. Horsfield 59 Queens Drive, Newton-le-Willows WA12 0LY
G1 HIU J. Clarke 3 Shelley Priory Cottages, Shelley, Ipswich IP7 5RQ
G1 HJD B. Taylor 111 High St., Warboys, Huntingdon PE17 2TB
GM1 HJL I. Copland 6 Dunadd View Kilmichael Glassary, Lochgilphead PA31 8QA
G1 HJO J. Cornall Fern Holme, Taylors Lane, Preston PR3 6AB
G1 HJP T. Carter 84 Colvile Road, Wisbech PE13 2EL
G1 HJS M. Todd 38 The Churchlands, New Romney TN28 8LB
GM1 HJX M. Williams 33/9 Marlborough Street, Edinburgh EH15 2BD
G1 HKF C. Maclennan 72 Sandsfield Lane, Gainsborough DN21 1DD
G1 HKM F. Woods 275 Scotter Road, Scunthorpe DN15 7EH
G1 HKR T. Whittam 27 Dimples Lane, Garstang, Preston PR3 1RD
G1 HKS D. Wilson 34 Belfairs Drive, Chadwell Heath, Romford RM6 4EB
G1 HKU C. Weatherley 4 The Orchard, Pine Tree Close, Cowes PO31 8DY
G1 HLP D. Plant 15 Heathcombe Road, Bridgwater TA6 7PD
G1 HLQ M. Edwards 11 Lightwood, Crown Wood, Bracknell RG12 0TR
G1 HLS W. Etherton-Scott 62 Spencer Road, Walthamstow, London E17 4BD
G1 HLT I. Fay 7 Oakridge Close, Forest Town, Mansfield NG19 0EY
G1 HLV J. Lee Deighton Manor, Deighton, Northallerton DL6 2SN
G1 HLY R. Powell 55 Lumley Road, Horley RH6 7JF
G1 HMI T. Rock 4 Hunters Gate, Much Wenlock TF13 6BW
G1 HML H. Willard M V Irma, Paglesham Boatyard, Paglesham Eastend SS4 2ER
G1 HMT G. Gray Home Farm, Furlong Drove, Ely CB6 2EQ
G1 HMW W. Gardner 25 Prospect Place, Wing, Leighton Buzzard LU7 0NT
G1 HMY P. Batty 14 Woodville Road, Penwortham, Preston PR1 9DR
G1 HMZ K. Breedon 17 Emmanuel Avenue, Arnold, Nottingham NG5 9QN
G1 HND M. Burling 28 Croydon Road, Arrington, Royston SG8 0DJ
GW1 HNF N. Bufton 7 Laburnum Close, Rassau, Ebbw Vale NP23 5TS
GW1 HNG P. Beesley 12 Bryngolwg, Aberdare CF44 0ER
G1 HNH P. Bannister 40 Regent St., Stowmarket IP14 1RJ
G1 HNN R. Cockman 31 Kensington Road, Southend-on-Sea SS1 2SX
G1 HNU M. Gray 28 The Close, Bradwell, Great Yarmouth NR31 8DR
GM1 HNZ A. Simmers Loanside, Crossroads, Keith AB55 6LP
G1 HOD A. Smith 14 Bridge Street, Shepshed, Loughborough LE12 9AD
G1 HOI N. Ballard 185 Nw Harwood, Spc #45, Prineville 97754 United States
G1 HOJ B. Baylis 118 Eastgate, Deeping St. James, Peterborough PE6 8RD
G1 HOL S. Cook 20 Gaymore Road, Kidderminster DY10 3TU
G1 HOP R. Chaston 157 Winston Avenue, Coventry CV2 1DL
G1 HOU A. Kirby 185 High Street, Dunsville, Doncaster DN7 4BU
G1 HPB R. Wearing 163 Birmingham Road Stratford Upon Avon Warwickshire, Stratford-upon-Avon CV37 0AP
G1 HPS T. Jones 25 Foxcotte Road, Charlton, Andover SP10 4AR
G1 HPU P. James 8 Pipers Wood Cottages, Little Missenden, Amersham HP7 0RQ
G1 HPV T. Jones 175 New Road, Great Wakering, Southend-on-Sea SS3 0AR
G1 HPZ I. Russell 10 Launceston Road, Bristol BS15 1EP
G1 HQE C. Close 3 Hay Green, Therfield, Royston SG8 9QL
G1 HQG A. Coley 5 Arundel Way, Highcliffe, Christchurch BH23 5DX
G1 HQH I. Church 8 Keats Drive Harwell, Didcot OX11 6FA
G1 HQJ D. Robinson 16 Green Lane, Platts Heath, Maidstone ME17 2NS
G1 HQK I. Richardson 1 Cedar Drive, Lowestoft NR33 9HA
G1 HQN J. Rattenbury Compton Lodge, High Ham, Langport TA10 9DH
G1 HQO Dr G. Spaven Spout House Farm, Macclesfield Road, High Peak SK23 7QU
G1 HQQ F. Jensen 79 The Drakes, Shoeburyness, Southend-on-Sea SS3 9NY
G1 HQW J. Kierman 26 Popes Lane, Gorefield Road, Wisbech PE13 5BD
G1 HRA D. Lloyd 35 Charles Close, Abbotts Barton, Winchester SO23 7HT
G1 HRD V. Allen 155, Reigate Avenue, Sutton SM13RJ
G1 HRH M. Gregory 45 Larksfield Avenue, Bournemouth BH9 3LW
G1 HRJ P. Deakes 108 Glaisdale Drive East, Nottingham NG8 4LZ
G1 HRL E. Dillow 18 Laburnum Grove, Warwick CV34 5TG
G1 HRM T. Davenport 36 Rydale Road, Nottingham NG5 3GS
G1 HRQ Ballymena ARC c/o B. Madore 66A West Street, Ryde PO33 2QF
G1 HRU S. Hill 26 Crescent Road, Dudley DY20NW
G1 HRV K. Higgins 22 Thatchers Lane, Cliffe, Rochester ME3 7TN
GM1 HRY A. Davis 3 High Shore, Banff AB45 1DB
G1 HSA S. Arnold 30 Pine Avenue, Newton-le-Willows WA12 8JE
G1 HSF M. Evans Fernlea, Pit Hill Lane, Bridgwater TA7 9BT
G1 HSG N. Evans 25 Chetwyn Avenue, Bromley Cross, Bolton BL7 9BN
G1 HSH M. Ellerby 3 Gilwern Court, Ingleby Barwick, Stockton-on-Tees TS17 5DJ
G1 HSI M. Glazier 19 West Place Brookland, Romney Marsh TN29 9RG
G1 HSJ L. Godden The Conifers, 14 Pirehill Lane, Stone ST15 0JN
G1 HSL J. Girt 22 Medway Road, Ipswich IP3 0QH
G1 HSM L. Heller 1 Princes Road, St. Leonards-on-Sea TN37 6EL
G1 HSO M. Hoey 14 Durham Rd, Blackpool Lands Fy1 3Qb, Lancs Fy1 3qb FY1 3QB
G1 HSP C. Hunt 700 Western Boulevard, Nottingham NG8 5FH
G1 HSX P. Kimber 16 Sycamore Close, Lydd, Romney Marsh TN29 9LF
G1 HTF H. Heron 43 Cheetham Hill Road, Dukinfield SK16 5JL
G1 HTL J. Foster 25 Hunter Road, Arnold, Nottingham NG5 6QZ
G1 HTM S. Froggatt 17 Queensway, Saxilby, Lincoln LN1 2QB
G1 HTN P. Farrow 10 St. Thomas Close, Chilworth, Guildford GU4 8LQ
G1 HTO R. Fortescue 7 Bodkin Lane, Weymouth DT3 6QL
G1 HTR A. Gee 4 Malkin Avenue, Radcliffe on Trent NG12 1DP
GU1 HTY B. Ayres Rousay, Bailiffs Cross Road, St. Andrew GY6 8RY Guernsey
G1 HUM R. Beech 6 Law Cliff Road, Birmingham B42 1LP
G1 HVL P. Howarth 7A Fox Court, Durkar, Wakefield WF4 3BH
G1 HVW J. Craft 3 Coltsfoot Drive, Royston SG8 9EU
G1 HWA Dr K. Harris 27 Middle Field Road, Rotherham S60 3JJ
G1 HWJ P. Milner 3 Larne Avenue, Cheadle Heath, Stockport SK3 0UJ
G1 HWK T. Mccarthy 25 Henley Avenue, North Cheam, Sutton SM3 9SG
G1 HWO T. Miller 27 Richmond Way, Oadby, Leicester LE2 5TR
G1 HWP R. Davies Room 7, Hedley Lodge, Monastery, Ruckhall Lane., Hereford HR2 9RZ
G1 HWR L. Mills 54 Petters Road, Ashtead KT21 1NE
G1 HWY M. Jupp 54 Shooting Field, Steyning BN44 3RQ
G1 HXN G. King 1 Sudan Cottage, Frogge Lane, Coltishall, Norwich NR12 7JU
G1 HXP J. Lennard 10 Orston Road East, West Bridgford, Nottingham NG2 5FU
G1 HXR R. Davis 6 Fairway Drive, Northmoor, Wareham BH20 4SG
G1 HXT G. Eden Heathend Cottage, Cromhall, Wotton under Edge GL12 8AS
G1 HXZ C. Cave 20 Meadow View, Banbury OX16 9SR
G1 HYA N. Cramp 3 Sowood Court, Ossett WF5 0TJ
G1 HYC D. Curson 25 Colbert Park, Swindon SN25 4YJ
G1 HYG B. Crowther 104 John St., Beamish, Stanley DH9 0QP
G1 HYM P. Bailey 10 Bell Drive Hednesford, Cannock WS12 4RA
GU1 HYN B. Bolderston 12 Hartlebury Estate, Steam Mill Lane, St. Martin GY4 6NH Guernsey
G1 HYO M. Green 26 Hunters Field, Stanford In The Vale, Faringdon SN7 8LR
G1 HYQ D. Chenoweth 20 Churchlands Road, Bedminster, Bristol BS3 3PW
G1 HYU K. Church 31 Riversway, King's Lynn PE30 2ED
G1 HYX A. Chance 18 Egdon Glen, Crossways, Dorchester DT2 8BQ
G1 HZD C. Legate 101 Butchers Lane, Walton on the Naze CO14 8UD
G1 HZI I. Dodd Gardeners Cottage, Sandhoe, Hexham NE46 4LU
G1 HZJ M. Devine 7 South Parade, Seascale CA20 1PZ
GM1 HZL M. Donaldson 43 Stair Street, Drummore DG9 9PT
G1 HZN C. Dadd 60 Rosaire Place, Scartho, Grimsby DN33 2JS
G1 HZR K. Farrar 8 Ascot Avenue Cantley, Doncaster DN4 6HE
G1 IAB S. Matthews 66 West End Road Epworth, Doncaster DN9 1LB
G1 IAD D. Morton 12 Pennygate Drive, Lowestoft NR33 9HL
G1 IAG P. Morris 18 Greenway Close, London NW9 5AZ
G1 IAL A. Plant 148 Chatsworth Road, Halesowen B62 8TH
G1 IAQ D. Memory 22 Marstown Avenue, South Wigston, Leicester LE18 4UH
G1 IAV P. Costello Newgrange, Poplar Road, New Milton BH25 5XP
GW1 IAW I. Woodward Corlander, Middle Road, Wrexham LL11 3TW
G1 IBF G. Tullock 16 Ward Lea Nafferton, Driffield YO25 4JZ
G1 IBJ C. Diaper 163 Edwin Road, Gillingham ME8 0AQ
G1 IBO D. Buss Rlc, Po Box 885, Bristol BS99 5LG
G1 IBP A. Heaysman 325 Broomfield Road, Chelmsford CM14DU
G1 IBS W. Chadwick 102 Feltham Road, Ashford TW15 1DP
G1 IBX C. Drayton The Lindens, Main Street, Kelfield, York YO19 6RG
G1 ICA D. Keable 90 King Edward Road, Rugby CV21 2TE
G1 ICH M. Adams 61 Monks Park Road, Northampton NN1 4LU
G1 ICI M. Thorpe 18 Sherrier Way, Lutterworth LE17 4NW
G1 ICK D. Winton 16 Lord Avenue, Clayhall, Ilford IG5 0HP
G1 ICQ A. Orgee 54 Riverview Close, Hallow, Worcester WR2 6DA
G1 ICX D. Palmer 18 Newfields, Sporle, King's Lynn PE32 2UA
G1 IDE S. Roy 28 Kingston Rise, New Haw, Addlestone KT15 3EY
G1 IDF A. Edwards 51 Redrock Road, Rotherham S60 3JN
G1 IDJ G. Perry 61 Ollands Road, Reepham, Norwich NR10 4EL
G1 IDQ G. Arnold 36 Market Street, Rugeley WS15 2JL
G1 IDR the Gliding Centre ARS c/o A. Gilmore Ashfields, Naseby Road, Clipston, Market Harborough LE169RZ

G1 IDV G. Bain 99 Longford Lane, Gloucester GL2 9HB
G1 IDZ D. Young 70 North Malvern Road, Malvern WR14 4LX
GW1 IEB L. Tatham Hebron Stores, Llangwnadl, Pwllheli LL53 8NW
G1 IEC P. Walton 2 Albert Road, Bromsgrove B61 7BE
GM1 IEL J. Bruce 24 South Green Drive, Airth, Falkirk FK2 8JP
G1 IEO J. Turner 2 Hilberry Road, Canvey Island SS8 7EL
G1 IEP G. Tomkins The Close, Broomfield Clayton, Bradford BD14 6PJ
G1 IEX D. Broughton 33 Queens Park Flats, Queens Park Close, Mablethorpe LN12 2AS
G1 IEY S. Reigate 9 Effingham Road, Croydon CR0 3NF
G1 IFF A. Rose 15 Elderwood Way Tuffley, Gloucester GL4 0RA
G1 IFH G. Reading 34A Harding Avenue, Rawmarsh, Rotherham S62 7ED
G1 IFV N. Ginger Barnlea, Fairwarp, Uckfield TN22 3DT
G1 IFW W. Gain 14 Clarence Road, St. Leonards-on-Sea TN37 6SD
G1 IFX D. Garratt 3 Fort Road, Mountsorrel, Loughborough LE12 7HB
G1 IGA C. Brooks 14 The Furlong, King Street, Tring HP23 6BX
G1 IGC S. Brookes 52 Larch Grove, Kendal LA9 6AU
G1 IGN G. Scroggs 52 Eastern Road, Burnham-on-Crouch CM0 8BT
G1 IGP G. Spinks 89 Uplands Road, Oadby, Leicester LE2 4NT
G1 IGW D. Cliff 23 Grey Towers Drive, Nunthorpe, Middlesbrough TS7 0LT
G1 IHA R. Stravens 75 Telford Road, London N11 2RL
G1 IHE J. Smith 65 Woods Avenue, Hatfield AL10 8QF
G1 IHI M. Godsave 35 Furlong Close, Midsomer Norton, Radstock BA3 2PR
G1 IHJ A. Homer 6 Ensall Drive, Wordsley, Stourbridge DY8 4XX
G1 IHL S. Hopkins 98 Court Road, Kingswood, Bristol BS15 9QP
G1 IHS C. Currie 33 Ashridge Drive, Bricket Wood, St. Albans AL2 3SR
G1 IHY J. Shilson 3 Hereford Close, Desborough, Kettering NN14 2XA
G1 III C. Smith 199A Richardshaw Lane, Stanningley, Pudsey LS28 6AA
G1 IIO B. Thornton 21 Valley Road, Banbury OX16 9BQ
GU1 IIW R. Loveridge Shamley, Route De Portinfer, Vale, Guernsey GY6 8LN
G1 IIX B. Lee 58 Shaw Avenue, Normanton WF6 2TT
G1 IIY D. Turner 154 Rowlett Road, Corby NN17 2BS
GW1 IIZ J. Underwood Rock Hill, Llanarthney, Carmarthen SA32 8LJ
G1 IJC M. Williamson 2 Lancaster Close, Fakenham NR21 8DW
G1 IJJ J. Lainchbury 17 Pearmain Avenue, Wellingborough NN8 4SF
G1 IJM M. Shoosmith 18 Pottery Close, Aylesbury HP19 7FY
G1 IJQ D. Martin 27 St. Andrews Road, Stratton, Bude EX23 9AG
G1 IJY R. Wise Flat 1, Capelia House, 18-21 West Parade, Worthing BN11 3RB
G1 IKF R. Blakemore 31 Millstone Rise, Liversedge WF15 7BW
G1 IKG P. Mcgahon 520 Chessington Road, West Ewell, Epsom KT19 9HH
G1 IKH I. Mclaughlin 28 Jarvis Avenue, Nottingham NG3 7BH
G1 IKL P. O'Sullivan Japonica Cottage, Ardens Grafton, Alcester B49 6DR
G1 IKT S. Elliott Manor House, Bewholme, Driffield YO25 8DX
G1 IKV B. Austin 16 Heathlands, Westfield, Hastings TN35 4QZ
G1 IKW G. Winn 16 Highbury Place, Bramley, Leeds LS13 4PW
G1 ILC S. Colley 8 Tennyson Road, Maltby, Rotherham S66 7LU
G1 ILF P. Ellis 48 Willington Road Skellow, Doncaster South Yorkshire DN6 8JE
G1 ILG B. Evans 12 The Mead Thaxted, Dunmow CM6 2PU
G1 ILH G. Farr 18 The Loont, Winsford CW7 1EU
G1 ILJ C. Wood White Cottage 2 Bainton Road, Tibthorpe , Driffield YO259LF
G1 ILO R. Bell 80 West Avenue, Lightcliffe, Halifax HX3 8TJ
G1 ILY C. Sims 226 Exeter Road, Exmouth EX8 3NB
G1 IMD M. Hall 6 Poplar Avenue, New Mills, High Peak SK22 4HR
G1 IME N. Hopkins 22 Cornfield Way, Ashton-Under-Hill, Evesham WR11 7TA
G1 IMI C. Foreman Thornham Farm, Wansford, Driffield YO25 8JJ
G1 IMM Covtry Tech ARC c/o A. Gee 24 Granhams Close, Great Shelford, Cambridge CB22 5LG
G1 IMS I. Stewart 34 Newgate Street Village, Hertford SG13 8RB
G1 IMY R. Laycock 24 Farmcroft Road, Mansfield Woodhouse, Mansfield NG19 8QT
G1 INA A. Lowe 47 Springfield Park Road, Chelmsford CM2 6EB
G1 INB P. Laker 207 Columbia Road, Ensbury Park, Bournemouth BH10 4EE
G1 IND V. Lowe 35 Elm Place, Armthorpe, Doncaster DN3 2DE
GI1 INH D. Munro 15 Main Street, Belcoo, , Enniskillen BT93 5FB
G1 INI B. Ginsburg 27 Park Crescent, Elstree, Borehamwood WD6 3PT
G1 INJ Dr R. Ginsburg 3 Basing Hill, London NW11 8TE
G1 INK S. Green 48 Queen Elizabeth Road Humberston, Grimsby DN36 4DE
GM1 INS B. Skakle 190 West Road, Fraserburgh AB43 9NL
G1 INU M. Sweet 67 Swinley Road, Wigan WN1 2DL
GD1 IOM Iom ARS(Iomars) c/o A. Morgan Thal'Loo Glass, Nassau Road, Mwyljyn Moddey IM7 4AQ Isle of Man
G1 IOO D. Camac 6 Wisbeck Road, Tonge Fold, Bolton BL2 2TA
G1 IOP A. Cheer 15 Stibbs Way, Bransgore, Christchurch BH23 8HG
GW1 IOT H. Harrison 2 Hendre, Newtown, Ebbw Vale NP23 5FE
G1 IOU G. Holland 16 Hancox Street, Oldbury B68 9LQ
G1 IPD D. Mobbs 64 Cranford Road, Kingsley, Northampton NN2 7QX
G1 IPE A. Medcalf 23 Allesborough Drive, Pershore WR10 1JH
G1 IPI R. Taylor Lower Manaton, South Hill Road, Callington PL17 7LW
GW1 IPJ I. Parry-Jones 21 Laurels Avenue, Bangor-On-Dee, Wrexham LL13 0BQ
G1 IPP M. Allen 23 Waterloo Crescent, Countesthorpe, Leicester LE8 5SU
G1 IPU G. Coote 10 Curlew Close, Clacton-on-Sea CO169EN
G1 IPY J. Rowlands 2 Wellfield, Longton, Preston PR4 5BX
G1 IQA G. Adkins 117 Connolly Drive, Rothwell, Kettering NN14 6TN
G1 IQE F. Angwin 171 Windsor Road, Wellingborough NN8 2LZ
G1 IQF M. Ames 7 Northgate, Leyland PR25 3NR
G1 IQG A. Bloodworth 79 Hands Road, Heanor DE75 7HB
G1 IQK B. Shaw 23 Lodge Drive, Culcheth, Warrington WA3 4ES
G1 IQN J. Spicer 5 Berries Mount, Bude EX23 8AP
GW1 IQS I. Jones 7 The Oaks, Quakers Yard, Treharris CF46 5HQ
G1 IQU D. Jolley 212 Eastern Esplanade, Southend-on-Sea SS1 3AD
G1 IRG S. Manning 11 Broomhill Crescent, Southfields, Northampton NN3 5BH
G1 IRQ N. Tansley 11 Juniper Close, Lutterworth LE17 4US
GM1 ISJ D. Barker 7 The Keys, Kildonan, Isle of Arran KA27 8SA
GW1 ISK F. Davies Hendref, Red Wharf Bay, Pentraeth LL75 8YG
G1 ISN T. Evans 22 Malthouse Lane, Ashover, Chesterfield S45 0AL
G1 ISP B. Etherington 24 Broomcroft Road, Ossett WF5 8LH
GW1 ISR Dr P. Kenington Trap Farm, Devauden, Chepstow NP16 6PE
G1 ISS B. Lyons 51 Wade Reach, Walton on the Naze CO14 8RE
G1 ISX C. Hall 9 Moneyhill Court, Dellwood, Rickmansworth WD3 7DY
G1 ITE P. Hayler 27 Birch Way, Heathfield TN21 8BB
G1 ITJ K. Edmett Solstice, Youngs Paddock, Salisbury SP5 1RS
G1 ITL D. Gilbey 34 Farnhurst Road, Barnham, Bognor Regis PO22 0JN
G1 ITS T. Williams 86 Hillcrest Road, Rochdale OL11 2QB
G1 ITV K. Ward 5 Clarence Court, Bolton BL1 2XP
G1 IUA N. Harris Sunnyside Lodge, Mongeham Road, Deal CT14 8JW
G1 IUD C. Sermons 17 Wellside, Marks Tey, Colchester CO6 1XG
G1 IUF M. Kilkenny 138 Stanbury Road, Hull HU6 7BW
G1 IUL B. Jackson 10 Wood End Croft, Coventry CV4 9RN
G1 IUT T. Christmas 3 Mount Road, Cosby, Leicester LE9 1SX
G1 IUW G. Diaper 89 East St., Sudbury CO10 2TP
G1 IUZ S. Groves 135 Ring Road, Crossgates, Leeds LS15 7QE
G1 IVF D. Lowe 21 Farndon Road, Market Harborough LE16 9NW
G1 IVG C. Lowe 22 Ryelands Close, Market Harborough LE16 7XE
G1 IVI M. Grey 7 Cemetery Lane, Tweedmouth, Berwick-upon-Tweed TD15 2BS
G1 IVK T. Garnham 45 Buscot Drive, Abingdon OX14 2BL
G1 IVL R. Hudson 80 Drake Avenue, Worcester WR2 5RR
G1 IVO L. Ladner 7 Polventon Close, Heamoor, Penzance TR18 3LD
G1 IVP J. Lamb 5 Honeycroft, Loughton IG10 3PR
G1 IVV G. Merrington Cartref, Ball Lane, Frodsham WA6 8HP
G1 IWE T. Coombs 114 Talbot Street, Whitwick, Coalville LE67 5AZ
G1 IWT R. Moore 9 Rowland St., Allenton, Derby DE24 9BT
G1 IXE V. Green 50 Alcove Road, Fishponds, Bristol BS16 3DR
G1 IXF I. Green 50 Alcove Road, Bristol BS16 3DR
G1 IXV C. Haver 18 Church Lane, Edenham, Bourne PE10 0LS
G1 IYA A. Greenwood 21 Ovenden Crescent, Halifax HX3 5PE
G1 IYB B. Haines 66 North Drive, Grove, Wantage OX12 7PN
G1 IYE I. Hawes 129 Manor Road, Ash, Aldershot GU12 6QB
G1 IYF C. Parfitt 12 Marigold Close, Basingstoke RG22 5RG
G1 IYO R. Reed 482 Baring Road, London SE12 0EG
G1 IZA D. Lamb 33 Cherston Road, Loughton IG10 3PL
G1 IZB F. Smith 6 Mill Close, Marshchapel, Grimsby DN36 5TP
G1 IZH A. Trueman 3 Higher Road. Woolavington, Bridgwater TA7 8EA
G1 IZN R. Mitchell 45 Kent Close, Mitcham CR4 1XN
G1 JAA R. Lees-Oakes 6 Tabley Street, Mossley, Ashton-under-Lyne OL5 9PD
G1 JAB J. Burke 48 Medina Road, Portsmouth PO6 3HD
G1 JAG K. Powell 89 Avenue Road, Leicester LE2 3EA
G1 JAH J. Hagues 7 Eastern Green Park Two, Eastern Green, Penzance TR18 3BA
G1 JAL P. Westbury 6 Bradford Road, Rode, Frome BA11 6PR
G1 JBB L. Richards 14 St. Julitta, Luxulyan, Bodmin PL30 5ED
G1 JBC N. Thompson 24 Braemor Road, Calne SN11 9DT
G1 JBE D. Blackburn 24 Reservoir St., Darwen BB3 1LQ
G1 JBG J. Beacon 14 Siskin Close, Bishops Waltham, Southampton SO32 1RQ
G1 JBJ S. Bartlett Lieu Dit La Gaule, Nantheuil 24800
G1 JBM M. Clark 28 Compton Crescent, West Moors, Ferndown BH22 0BZ
G1 JBW B. Ellison 931 Burnley Road, Todmorden OL14 7ET
G1 JBZ I. Halsey Cowtrott, Back Lane, Great Yarmouth NR29 5ED
G1 JCC I. Jefferson 19 Orchard Way, Flitwick, Bedford MK45 1LF
G1 JCL M. Munn 11 Foxley Road, Queenborough ME11 5AW
G1 JCP J. Pasfield Fairlands, White Lodge Crescent, Clacton-on-Sea CO16 0HT
G1 JCT S. Farrant The Bungalow, Brewery Yard, Stroud GL5 4JW
G1 JCW A. Duffy 14 Garden Street, Padiham, Burnley BB12 8NP
G1 JDE O. Graffham 106 Barford Road, Edgbaston, Birmingham B16 0EF
G1 JDF D. Gray 11 Field Close, Bollington, Macclesfield SK10 5JG
GM1 JDJ L. Mcleman 14 Flures Place, Erskine PA8 7DH
G1 JDO P. Oliver 3 Savile Walk, Brierley, Barnsley S72 9HJ
G1 JDQ P. Paterson Oak Lea, 11A Fletsand Road, Wilmslow SK9 2AD
G1 JDT G. Palmer 5 Dunstar Avenue, Audenshaw, Manchester M34 5LJ
G1 JDV J. Vasey 22 Rickleton Village Centre, Washington NE38 9ET
G1 JEA D. Hart 10 Marina Drive, March PE15 0AU
G1 JEH K. Schneider 65 Alpha Road, Birchington CT7 9ED
G1 JER Dr J. Johnson 5 Hunters Ride, Appleton Wiske, Northallerton DL6 2BD
G1 JEZ S. Taylor 72 Molyneux Drive, Wallasey CH45 1JT
GM1 JFF A. Weddell 10 High Street, Eyemouth TD14 5EU
G1 JFL M. Woolridge 23 Marina Drive, May Bank, Newcastle ST5 9NL
GW1 JFT R. Beaugie 32 Court Gardens, Rogerstone, Newport NP10 9FU
G1 JFU Wng. Cmdr. D. Bryant 22 Highfield Park, Heaton Mersey, Stockport SK4 3HD
G1 JGD N. Cullis 39 Gilbert Drive, Langdon Hills, Basildon SS16 6SP
G1 JGE M. Colley 118 Devon Crescent, Birtley, Chester le Street DH3 1HP
G1 JGF L. Cox 7 Timberdine Avenue, Worcester WR5 2BD
G1 JGM B. Easey 4 Ash Trees, East Brent, Highbridge TA9 4DQ
G1 JGR C. Fortnum 11 Ayr Close, Stamford PE9 2TS
G1 JGS M. Garland 4A St Andrews Way, Freshwater PO40 9NH
G1 JGT J. Giller 9 Alberta Crescent, Huntingdon PE29 1TL
G1 JGY H. Ketley 24 Farmcroft Road, Mansfield Woodhouse, Mansfield NG19 8QT
G1 JHB C. Lawrence Flat 22, St. Pauls Court, Salford M7 3NZ
G1 JHD M. Plant 7 Kendale, Hemel Hempstead HP3 8NN
G1 JHG T. Mccormick 33 Bryanston Road, Aigburth, Liverpool L17 7AL
GM1 JHM A. Harding Bandrum House, Roadside Of Catterline, Catterline, Stonehaven AB39 2UA
G1 JHN Brackwell Radio Amateur CG c/o F. Harris 4 Parc An Ithan, The Lizard, Helston TR12 7PA
G1 JHP H. Hamer 126 Mellor Brow, Mellor, Blackburn BB2 7PN
GI1 JHQ J. Selfridge 42 Cloonavin Green, Coleraine BT52 1RG
G1 JHS S. Barrett-Jolley 61 Eyston Drive, Weybridge KT13 0XE
GM1 JHU J. Adams 3A Glenpatrick Road, Elderslie, Johnstone PA5 9BH
G1 JHX A. Bennett 32 Park Road, Stretford, Manchester M32 8DQ
G1 JHY A. Potter 25 Robinsons Meadow, Ledbury HR8 1SU
G1 JHZ S. Potter 25 Robinsons Meadow, Ledbury HR8 1SU
GW1 JIE K. Robertson Tyn Y Pwll, Fachwen, Caernarfon LL55 3HD
G1 JIG S. Ridgard 5 Orchard Way, Luton LU4 9LT
G1 JIH E. Rowell 80 Kings Delph, Whittlesey, Peterborough PE7 2PD
G1 JIJ J. Passfield 2 Parker Road, Chelmsford CM2 0ES
G1 JIR D. Robinson 4 Mayorlowe Avenue, Stockport SK5 8DB
G1 JIW P. Spooner 62 Chester Crescent, Newcastle ST5 3RW
G1 JJA S. Bessent 347 Birchfield Road, Redditch B97 4NE
G1 JJE N. Banks 6 Bylands Place, Newcastle ST5 3PQ
G1 JJK P. Naylor 14 Wrockwardine Road, Wellington, Telford TF1 3DB

G1 JJQ J. Schulz 4 Collinge Close, East Malling, West Malling ME19 6QS
G1 JJR V. Smith 99 Dewhurst Road, Fartown, Huddersfield HD2 1BN
G1 JJT N. Stubbs 8 West Avenue, Hilton, Derby DE65 5FY
G1 JJZ S. Willey 34 Kernick Road, Penryn TR108NT
G1 JKE N. Knapton 4 Crabmill Lane, Easingwold, York YO61 3DE
G1 JKF N. Leaney 31 Saxon Way, Willingham, Cambridge CB24 5UR
GM1 JKJ A. Britton 15 Glenbrook, Balerno EH14 7JE
G1 JKL A. Crouch 107 Waterleat Avenue, Paignton TQ3 3UD
G1 JKN P. Cooper 5 Lower Leys Way, Leominster HR6 0SS
G1 JKO G. Cooper Hazeldene, Fleet Coy, Spalding PE12 0RU
GW1 JKP R. Coleman Rockwood, Clive Road, Fishguard SA65 9DB
G1 JKV T. Whittaker Flat 2, Paul Vanson Court, New Berry Lane, Walton-on-Thames KT12 4HQ
G1 JKX J. West 4 Coronation Terrace, Longhorsley, Morpeth NE65 8UN
G1 JLB M. Denison 9 Derwent Road, Harrogate HG1 4SG
G1 JLG B. Giddings 71 Tyrone Road, Southend-on-Sea SS1 3HD
G1 JLM S. Brosnan Arabella, Ballymacelligott, , Tralee V92V9H3 Ireland
GM1 JLP K. Robson 13 Woodstock Avenue, Galashiels TD1 2EE
G1 JLQ J. Yarnall 4 Parklands, Evesham WR11 2QJ
G1 JLX N. Povey 33 Church Lane, Fradley, Lichfield WS13 8NJ
G1 JMC B. Harris 55 Valiant Way, Melton Mowbray LE13 0GE
G1 JMD P. Hall 64 Synehurst Crescent, Badsey, Evesham WR11 7XX
G1 JMF A. Hooper 5 Nine Elms Road, Longlevens, Gloucester GL2 0HA
G1 JMH J. Hickey 36 Station Road, Alderholt, Fordingbridge SP6 3RB
G1 JMK M. Justice 6 Stanley Terrace, Devizes SN10 5AJ
G1 JMN R. Andrews Owls Rest, Park Lane, Worcester WR2 6PQ
G1 JMP R. Ainsworth 95 Heysham Close, Murdishaw, Runcorn WA7 6DT
G1 JMS J. Stoddart 24 Vicarage Close, Platt Bridge, Wigan WN2 5DW
G1 JMV P. Slark 11 Hillfield Walk, Bolton BL2 2UR
G1 JMW W. Smith 37 Peake Road, Brownhills, Walsall WS8 7BZ
G1 JMY F. Taylor Wold Lodge, Pocklington Road, York YO42 1YJ
GD1 JNB P. Clarke 5 Sumark Avenue Douglas, Isle of Man IM2 2AD
GM1 JNC A. Campbell 17 Moulin Circus, Cardonald, Glasgow G52 3JY
G1 JNG G. Eccles 1 Bridge Place, Amersham HP6 6JF
GW1 JNH W. Francis 8 George Street, Treherbert, Treorchy CF42 5AH
GW1 JNI A. Fennah 7 Y Ddol, Llanbrynmair SY19 7DJ
G1 JNQ P. Auld 80 Milestone Road, Stone, Dartford DA2 6DN
GW1 JNR P. Adcock Bleak House, Cefn Coch, Welshpool SY21 0AE
GM1 JNS G. Bartram 6 Craigewan Crescent, Peterhead AB42 1HL
G1 JNX S. Langston 4 Lagwood Close, Hassocks BN6 8HZ
G1 JNY A. Larkin 16 Thetford Close, Corby NN18 9PH
G1 JOA B. Marsh 96 Coopers Lane, Clacton-on-Sea CO15 2DA
G1 JOD R. Norton Middleton House, Bleathwood, Ludlow SY8 4LX
G1 JOJ B. Smith 146 Battram Road, Ellistown, Coalville LE67 1GB
G1 JOL B. Shane 7 Oakwood Glade, Holbeach, Spalding PE12 7JS
G1 JON J. Storey 34 Austin Rise, Longbridge, Birmingham B31 4QN
G1 JOO R. Seymour 5 Clifton Place, Easton, Bristol BS5 0SE
G1 JOR J. Ormsby-Rymer 109 Goldcrest Road, Chipping Sodbury, Bristol BS37 6XJ
GW1 JOV T. Moore Dan Bryn Coch, Llandyfan, Ammanford SA18 2TY
G1 JOW C. Oates-Miller 32 Burnlee Road, Holmfirth HD9 2PS
G1 JPC M. Ward Four Winds, 13 Westfield Avenue, Wellingborough NN9 6DQ
G1 JPI M. Taylor 26 Monks Close, Lancing BN15 9DD
GM1 JPJ R. Jamieson 6A Mary Street, Stonehaven AB39 2AD
G1 JPK T. Jefferies 98 St. Johns Road, Frome BA11 2BD
G1 JPP A. Hawes 25 Folly Close, Fleet GU52 7LN
G1 JPT B. Gleave 1 Fearnley Way, Newton-le-Willows WA12 8SQ
G1 JQK S. Gibbs 43 Reddish Vale Road, Stockport SK5 7EU
GI1 JQP J. Innes 22 Ashley Lodge, Dunmurry, Belfast BT17 0AF
G1 JQR D. Sell 17 Auriel Avenue, Dagenham RM10 8BS
G1 JRD J. Barber 7 Ash Green, Canewdon, Rochford SS4 3QN
G1 JRF D. Bruckshaw 18 Old Moat Drive, Northfield, Birmingham B31 2LY
G1 JRL H. Cook 31 Butley Road, Felixstowe IP11 2NY
G1 JRP C. Davis 5 Redwing Avenue, Chippenham SN14 6XJ
G1 JRR R. Chalker 182 Bridge Road, Chessington KT9 2EY
G1 JRU D. Evans 63 Malwood Road West, Hythe, Southampton SO45 5DL
G1 JRW D. Gilchrist 204 Great West Road, Heston, Hounslow TW5 9AW
G1 JRX Dr M. Girgis Rozel, Wilson Road, Kidderminster DY11 7XU
G1 JRZ G. Hobbs 3 Glebe Cottage, Bremhill, Calne SN11 9LD
G1 JSK P. Lees 2 Russet Close, Braintree CM7 1DR
G1 JSP J. Marston 29 Ise Road, Kettering NN15 7DU
G1 JST P. Johnson 156 Norby Estate, Norby, Thirsk YO7 1BQ
G1 JTC S. Baskerville Shalimar, Grove Lane, Leeds LS6 2AP
GM1 JTK A. Doig 18 Gotterstone Drive, Broughty Ferry, Dundee DD5 1QW
G1 JTM M. Ferris 22 Route De Matha, Aigre 16140 France
G1 JTX L. Sharman 14 Northlands Avenue, Orpington BR6 9LY
G1 JTZ B. Linn 35 New Street Carcroft, Doncaster DN6 8EH
G1 JUD R. Robinson 24 Affleck Avenue, Radcliffe, Manchester M26 1HN
G1 JUI M. Lister Beaconfield, Middle Road, Poole BH16 6HJ
G1 JUO I. Penney 11 Eclipse Drive, Sittingbourne ME10 2HR
G1 JUP M. Baker 9A, Manor Road, Alton GU34 2PF
G1 JUY M. Dyson 10 Smelter Wood Crescent, Sheffield S13 8RE
GW1 JVB G. Evans 2 Old Village Road, Barry CF62 6RA
G1 JVF G. Hannan 20 Arlington Drive, Stockport SK2 7EB
G1 JVG D. Jewsbury 68 Grainger Close, Basingstoke RG22 4EA
GW1 JVH C. Kirkman The Nant, Nantmawr, Oswestry SY10 9HN
G1 JVL J. Leggett 10 Home Park Road, Nuneaton CV11 5UB
G1 JVM D. Turton 8 Lightwoods Road, Warley, Smethwick B67 5AY
G1 JVN F. Ursell 110 Watt Lane, Sheffield S10 5RE
G1 JVO C. Ursell 110 Watt Lane, Sheffield S10 5RE
GM1 JVU R. Aitken 81 Rashgill, Locharbriggs, Dumfries DG1 1QN
G1 JVY A. Smith 87 Willian Way, Letchworth Garden City SG6 2HY
G1 JWD R. Osborne 24 Brockington Road, Bodenham, Hereford HR1 3LR
G1 JWG D. Mckay 15 Wellington Crescent, Baughurst, Tadley RG26 5PJ
GM1 JWJ R. Male 13 Briar Grove, Forfar DD8 1DQ
G1 JWL M. Warren Flat 15, Fulton Lodge, Harrogate HG3 2UT
G1 JWO A. Stone 32 Berrynarbor Park, Sterridge Valley, Ilfracombe EX34 9TA
G1 JWY T. Yorke 12 Shanklin Drive, Weddington, Nuneaton CV10 0BA
G1 JXA R. Whatley 7 Okefield Road, Crediton EX17 2DN
GI1 JXE G. Murray Ashgrove, 61 Monteith Road, Banbridge BT32 5RD
G1 JXG A. Mcmillan 183 Forest Road, Clipstone Village, Mansfield NG21 9DS
G1 JXL M. Phillips 71 Juniper Square, Havant PO9 1HZ
G1 JXP I. Wilson Meadow Lodge, Kilhallon, Par PL24 2RL
G1 JXS C. Bunkum 7 Goose Green Close, Wolvercote, Oxford OX2 8QT
G1 JXX H. Williams 24 Vaughan Close, Four Oaks, Sutton Coldfield B74 4XR
G1 JYB B. Cartledge Oysterber Farm, Burton Road, Lower Bentham, Lancaster LA2 7ET
G1 JYH M. Cherry 36 Meads Avenue, Hove BN3 8EE
G1 JYK S. Langdale Bramley House, Dishforth Road, Sharow, Ripon HG4 5BU
G1 JYR D. Fraley 1334 Warwick Road, Knowle, Solihull B93 9LQ
G1 JYZ A. Philpott 2 Ocean View Road, Ventnor PO38 1AA
G1 JZG M. Wilmshurst Chalklands, Main Street, Horncastle LN9 5PT
G1 JZL S. Beckett 15 Peaks Avenue, New Waltham, Grimsby DN36 4LJ
GM1 JZM D. Johnstone 7 Gleneagles Avenue, Glenrothes KY6 2QA
G1 JZN J. Jacklin 26 Rockmill End, Willingham, Cambridge CB24 5HY
G1 JZT R. Everitt 55 Risborough Road, Bedford MK41 9QR
G1 JZU D. Goulden 26 Derwent Walk, Greenacres, Oldham OL4 2DJ
G1 JZX J. Hesketh 735 Manchester Road, Over Hulton, Bolton BL5 1BA
G1 JZY T. Mitchell 22 Grundy Avenue, Prestwich, Manchester M25 9TG
G1 JZZ D. Porter 2 Flour Mill Close, Burscough, Ormskirk L40 5TL
G1 KAG A. Watson 60 Beresford Avenue, Surbiton KT5 9LJ
G1 KAK C. Buttery Yew Tree Cottage, Chapel Lane, Newport PO30 3DD
G1 KAO A. Hawxby 73 Anthea Drive, Huntington, York YO31 9DB
G1 KAR Southdown ARS c/o A. Seabrook 63 St. Annes Road Willingdon, Eastbourne BN20 9NJ
G1 KAS M. Hughes 9. Greenacres, Kirkby-in-Ashfield NG17 7GE
G1 KAT C. Lawrence 23 Brutus Drive, Coleshill, Birmingham B46 1UF
G1 KBC S. Barrington Fawley Cottage, Butt Lane, Loughborough LE12 5EE
G1 KBE T. Bradley 32 Laurel Gardens , Marlowe Road, Hartlepool TS25 4NZ
G1 KBF C. Halls 16 Stoats Close, South Molton EX36 4JU
G1 KBG D. Arthur Durnaford, Callington PL17 7HP
G1 KBH P. Bloy 29 Mannington Place, South Wootton, King's Lynn PE30 3UD
GM1 KBJ A. Wagstaff 2 Birnock Water, Moffat DG10 9DY
G1 KBL M. Rack 212 Willingham Street, Grimsby DN32 9PY
GM1 KBZ S. Crockford11 101 Cromwell Road, Aberdeen AB15 4UE
G1 KCA D. Thomas 2 Reynards Meadow, Sutton Hill, Telford TF7 4NQ
GM1 KCH W. Curran 10 The Laurels, Dundee DD4 0AD
G1 KCR J. Smith 125 De Montfort Way, Coventry CV4 7DU
G1 KCS A. Scrutton Ashleigh, Butt Hill, Southam CV47 8NE
G1 KCU J. Warrington 204 High St., Feltham TW13 4HX
G1 KCV C. Shepherd 18 Nichols Street Desborough, Kettering NN14 2QU
G1 KCW H. Elleray 34 Trent Close, Plymouth PL3 6PB
G1 KDO D. Cattell 22 Budmouth Avenue, Weymouth DT3 6JW
GI1 KDS K. Lewis 763 Antrim Road, Belfast BT15 4EP
G1 KEB R. Farmer 72 Bradleys Lane, Wallbrook, Bilston WV14 8YW
G1 KEI P. Smith 47 Bostock Road, Abingdon OX14 1DW
G1 KEV H. Denton 27 Melrose Gardens, Hersham, Walton-on-Thames KT12 5HF
G1 KEW Wythall RC c/o N. Coote 36 Summerfield Road, Manchester M22 1AF
G1 KFB T. Jesson The Hawthorns, The Outwoods, Burbage, Hinckley LE10 2UD
G1 KFG J. Feay 19 Dorset Avenue, Diggle, Oldham OL3 5PL
G1 KFH J. Richmond 11 Elm Avenue, Pennington, Lymington SO41 8BD
G1 KFQ P. King 10 Hockley Lane, Eastern Green, Coventry CV5 7FR
G1 KGA O. Himmo 227B Caterham Drive, Coulsdon CR5 1JS
G1 KGC P. Simpson 100 Pitchford Avenue, Maddington WA 6109 Australia
G1 KGE M. Phelps Windermere, Wyson, Ludlow SY8 4NQ
G1 KGL S. Dorrington Rosewood, 74 Hangleton Way, Hove BN3 8EQ
G1 KGO S. Coben 106 Fleming Mead, Mitcham CR4 3LW
G1 KGQ C. Buxton 3 Goodwood Way, Mansfield NG1 8 3BY
G1 KGU J. Amos Mite View, Ravenglass CA18 1SW
G1 KGV C. Ashcroft Wood End House, Wood End, Huntington PE28 3LE
GW1 KGW C. Walker Perthi, Llaneilian, Amlwch LL689LY
GI1 KHF R. Maternaghan 1 Pinegrove Crescent, Ballymena BT43 6TL
GW1 KHH Warrington ARC c/o H. Mosley 17 Cadwgan Road, Old Colwyn, Colwyn Bay LL29 9PY
G1 KHM K. Morgan 157 Headlands, Fenstanton, Huntingdon PE28 9LP
G1 KHS D. Tucker 5 Uplands Close, Hawkwell, Hockley SS5 4DN
GM1 KHU C. Wall 27 Golf Terrace, Insch AB52 6JY
G1 KHY R. Staszewski Green Rigg West Woodburn, Hexham NE48 2SG
G1 KIB J. Martin The Old Bakehouse, Shotteswell, Banbury OX17 1JA
G1 KII D. Beale 88 Long Innage, Halesowen B63 2UY
G1 KIJ D. Marsters 21 Cow Lane, Rampton, Cambridge CB24 8QG
G1 KIT J. Fisher 63 Rogers Avenue, Creswell, Worksop S80 4JR
G1 KIW J. Moss 42 Chantry Lane, Necton, Swaffham PE37 8ET
G1 KIZ T. Head 36A Ashacre Lane, Worthing BN13 2DH
G1 KJG C. Robinson Steinhalden 3, Muehlau 5642 Switzerland
G1 KJH J. Beach 8 Harvey Lane, Norwich NR7 0BQ
G1 KJQ M. Haymes 89 Everest Drive, Bispham, Blackpool FY2 9DJ
G1 KJX B. Hobbs 33 Hawthorn Drive, Heswall, Wirral CH61 6UP
G1 KKA P. Montgomery 7 Birchwood Close, Tavistock PL19 8DR
G1 KKD T. Elcock Little Grange, 33 Cromford Drive, Derby DE3 9JT
G1 KKE D. Rose 6 The Holt, Mollington, Banbury OX17 1BE
G1 KKF A. Lawes 15 Leybourne Road, Brighton BN2 4LT
G1 KKH P. Cunliffe 37 Rectory Road, Worthing BN14 7PE
GM1 KKI K. Johnston Innisfree, Gulberwick, Shetland ZE2 9JX
GW1 KKJ K. Taylor 23 Vardre Avenue, Deganwy, Conwy LL31 9UT
G1 KKS I. Gott Tayman House, The Street, Badminton GL9 1HH
G1 KLI M. Smith Sycamore Farm, 6 Station Road, Spalding PE12 0NP
G1 KLK A. Dutton 111 St. Michaels Road, Crosby, Liverpool L23 7UL
G1 KLP A. Jackson 81 Suffield Way, King's Lynn PE30 3DX
G1 KLW P. Golding 80 Birdbrook Road, London SE3 9QP
G1 KLZ D. Ellershaw 38 Lakeber Avenue, Bentham, Lancaster LA2 7JN
G1 KMJ J. Couzins 30 Camden Road, St. Peters, Broadstairs CT10 3DR
G1 KMN N. Thompson 10 Belmont Crescent, Swindon SN1 4EY
G1 KMS I. Millar The Grange, 105 High Street, Northampton NN3 3JX
G1 KNA J. Burdett Glencorse, 13 Fairfax Avenue, Selby YO8 4AZ
G1 KNI S. Martin 6 Prinsted Walk, Fareham PO14 3AD
G1 KNK G. Mellors 21 Church Close, Stoke St. Gregory, Taunton TA3 6HA

G1 KNQ G. Roberts 22 Grove St., New Balderton, Newark NG24 3AZ
G1 KNU P. Sharp Purbrook Cottage, Lyme Road, Axminster EX13 5BL
G1 KNX S. Jones 31 Church Street, Tewkesbury GL20 5PD
G1 KNZ J. Washby 2 Olivier Court, Council Avenue, Hull HU4 6RW
G1 KOD J. Rodgers 5 Bridge Avenue, Latchford, Warrington WA4 1RJ
G1 KOG E. Beir 17 Deansway, Hemel Hempstead HP3 9UE
G1 KOH G. Hall 22 Waterfield Close, Leicester LE5 4EN
G1 KON L. Mccoy 56 Curate Road, Anfield, Liverpool L6 0BZ
G1 KOP Liverpool Raynet Group c/o J. Gibbard 2 Almond Court, Liverpool L19 2QZ
G1 KOR A. Waddoups 20 Stevenson Way, Wickford SS12 9DY
G1 KOT M. Lynn 52 Vowler Road, Langdon Hills, Basildon SS16 6AQ
G1 KOX T. Williams 24 Elm Tree Close, Northolt UB5 6AR
G1 KPE R. Currinn 187 Coalpool Lane, Walsall WS3 1QJ
G1 KPI T. Houghton 24A Studley Road, Torquay TQ1 3JN
G1 KPU D. Skilton 137 Coast Drive, Lydd On Sea, Romney Marsh TN29 9NS
G1 KPV A. Rayner 147 Ramuz Drive, Westcliff-on-Sea SS0 9JN
G1 KPZ M. Gillott 2 Firthwood Avenue, Coal Aston, Dronfield S18 3BQ
G1 KQD S. Bennett Mumford Cottage, Wick Lane, Lower Apperley, Gloucester GL19 4DS
G1 KQE H. Sweet 5 Dence Close, Herne Bay CT6 6BH
G1 KQH S. Wigg 45 Cambrian Lane, Rugeley WS15 2XH
G1 KQN A. Bowyer 80 Holme Fen, Holme, Peterborough PE7 3PR
G1 KQP S. Price Whitton Paddocks, Pulley, Shrewsbury SY3 0AG
G1 KQU G. Gray 10A Albert Close, Rayleigh SS6 8HP
GW1 KQV C. Caudy 43 Graham Avenue, Pen-Y-Fai, Bridgend CF31 4NR
GW1 KQY B. Francis 4 Heol Tir Coch, Efail Isaf, Pontypridd CF38 1BW
G1 KQZ A. Quinn 581 Chorley Old Road, Bolton BL1 6BL
G1 KRU A. Graves 49 Robin Lane, Edgmond, Newport TF10 8JL
G1 KRX B. Piper 26 Hare Law Gardens, Stanley DH9 8DG
G1 KSC S. Harrison 44 Rosslyn Road, Whitwick, Coalville LE67 5PT
G1 KSE A. Robinson 2 Frome Close, Marchwood, Southampton SO40 4SL
G1 KSH P. Sherwood 43 Brighton Street, Penkhull, Stoke-on-Trent ST4 7HH
G1 KSI S. Bennington The Oaks, Lynn Road, Wisbech PE14 7DF
G1 KSK E. Mullin 26 Fearnhead Lane, Fearnhead, Warrington WA2 0BE
G1 KSN Derby & District ARS c/o V. Tankard 55 Filching Road, Eastbourne BN20 8SD
G1 KST J. Almond 49 The Promenade, Withernsea HU19 2DW
G1 KSW E. Giacani 21 Barton Terrace, Leeds LS11 8TP
G1 KTF D. Webb Fairway, Barkham Road, Wokingham RG41 4DH
G1 KTS G. Goodridge 110 Quarrendon Road, Amersham HP7 9EP
GW1 KTW C. Thomas 18 Acrefield Avenue, Guilsfield, Welshpool SY21 9PN
G1 KTY B. Rider Rose Cottage, Coley Road, Bristol BS40 6AP
G1 KTZ D. Robins Ayala, Higher Road, Liskeard PL14 5NQ
G1 KUG L. Preece Oakdean, Crow Ash Road, Berry Hill, Coleford GL16 7RB
GM1 KUI G. Smith 38 Crown Cottages, Stuartfield, Peterhead AB42 5HR
G1 KUN T. Sexton 41 St. Bedes Gardens, Cambridge CB1 3UF
G1 KUO G. Edwards 40 Dent Street, Hartlepool TS26 8AY
G1 KUQ P. Andrews 3 John Grace Street, Coventry CV3 5GZ
G1 KVC J. Norris 3 St. Pauls Close, Adlington, Chorley PR6 9RS
GW1 KVI D. Bedson Crud Yr Awel, Old Llanfair Road, Harlech LL46 2SS
G1 KVO M. Haydon 1 Glencrofts, Hockley SS5 4GN
G1 KVP B. Froggatt 59 Queens Road, Rushall, Walsall WS4 1HP
G1 KVQ Essex Raynet c/o A. Mallin 88 Highbridge Road, Burnham-on-Sea TA8 1LN
G1 KVR M. Wood Kviabol, Sheep Pen Lane, Seaford BN25 4QR
G1 KVW I. Wilson 45 Meadway, Halstead, Sevenoaks TN14 7EY
GM1 KWA B. Simpson 2 Cowden Way, Comrie, Crieff PH6 2NW
G1 KWF S. Watson 1 Owl Way, Hartford, Huntingdon PE29 1YZ
GM1 KWG J. Winterbourne Birkenbush, Clochan, Buckie AB56 5AL
G1 KWK M. Duerden 96 Chellow Grange Road, Bradford BD9 6NP
GM1 KWX C. Carter Annachmor House, Clynder, Helensburgh G84 0QD
G1 KXJ A. Poole 17 Adelaide Street, Stonehouse, Plymouth PL1 3JF
G1 KXP S. Lindsay-Smith 47 Shaftesbury Avenue, Timperley, Altrincham WA15 7NP
G1 KXQ M. Bloxham 34 Northcote Road, Farnborough GU14 9EA
G1 KXX A. Shaw 70 Field Gardens Steventon, Abingdon OX13 6TF
G1 KXZ A. Moore 103 Park Grove, Barnsley S70 1QE
G1 KYK M. Partridge Flat 2, Lymington Court Station Road, Sutton Coldfield B73 5JY
G1 KYN D. Jackson 37/39 Cloverhill, Sunniside NE16 5PT
G1 KYV S. Youngs Glenlovat, Oakley Road, Cheltenham GL52 6NZ
G1 KZA E. Kilner 3 Ruskin Close, Wath-Upon-Dearne, Rotherham S63 6NU
G1 KZD T. Asker 34 Post Office Road, Frettenham, Norwich NR12 7AB
GM1 KZG J. Southworth 2 School Street, New Pitsligo, Fraserburgh AB43 6NE
G1 LAN D. Roberts 192 Boothferry Road, Hull HU4 6EW
G1 LAO J. Flower 12 Balmoral Close, Alton GU34 1QY
G1 LAP J. Beecham Newholme, Whitemill Lane, Stone ST15 0EG
G1 LAR L. Rogers 37 Gilbey Road, Tooting, London SW17 0QQ
G1 LAT S. Kirkwood 1 Nether View Lodge Lane, Wennington, Lancaster LA2 8NP
G1 LAW E. Boyce 214 London Road, Benfleet SS7 5SJ
G1 LBH P. Tomkins 64 Glenwood Gardens, Bedworth CV12 8DA
GI1 LBI H. Budina 46 Dunderg Road, Macosquin, Coleraine BT51 4NE
G1 LBK M. Kirk Badgers Rise, Dunley Gardens, Stourport-on-Severn DY13 0LL
G1 LBM C. Taylor 28 Grace Court, Dial Lane, Bristol BS16 5UP
G1 LBU R. Parry 84 Sulgrave Road, Washington NE37 3BZ
G1 LCC J. Edwards 1 Herons Way, Sandymoor, Runcorn WA7 1UH
G1 LCE S. Turner Rainow Villa, Under Rainow Road, Congleton CW12 3PL
G1 LCN G. Clegg 16 The Pastures, Lower Westwood, Bradford-on-Avon BA15 2BH
G1 LCR Leicester Raynet Grp c/o D. Harrison 7 Shirley Close, Castle Donington, Derby DE74 2XB
G1 LCS J. Burrows 1 Browns Avenue, Runwell, Wickford SS11 7PT
G1 LCY R. Brown 3 Penmare Close, Hayle TR27 4PJ
G1 LCZ P. Milston 47 Wellington Road, Todmorden OL14 5EQ
G1 LDC P. Gibson 18 Vicarage Walk, Northwich CW9 5PS
G1 LDJ M. Ward 23, Tricketts Lane, Ferndown BH22 8AT
G1 LDN J. Burnet 41 Douglas Crescent, Southampton SO19 5JP
G1 LDY A. Smith 7 Gladstone Road, Broughton, Chester CH4 0RN
G1 LEC R. Gugi 47 Witton Crescent, Darlington DL3 0JQ
G1 LED K. Pinkard 2 Lonsdale Court, Lache Lane, Chester CH4 7LZ
G1 LEH M. Nicholson 7 Ellerbeck Close, Workington CA14 4HY
GW1 LEL A. Rowland 86 Dee Road, Connah'S Quay, Deeside CH5 4PA
G1 LEN L. Taylor 35 Leafield Road, Darlington DL1 5DF
G1 LEO J. Brittain 26 Saxby Close, Eastbourne BN23 7BH
G1 LES J. Buckle 14 Alder Close, Mapplewell, Barnsley S75 6JA
G1 LEX J. Blything Blythwood, 319 Great Brickkiln Street, Wolverhampton WV3 0PY
G1 LFD V. Middleton 5 Fieldhouse, Holmfirth HD9 1EN
G1 LFI G. Hepworth 3 College View, Ackworth, Pontefract WF7 7LA
G1 LFM D. Mawdsley 3 Chapel Lane, Cronton, Widnes WA8 4NT
GW1 LFN D. Rees 17 Cwm Mwyn, Gorslas, Llanelli SA14 7HY
G1 LFR D. Love 17 Longstaff Avenue, Rawnsley, Cannock WS12 0QE
GW1 LFX M. Lamb 4 Hadfield Close, Connah'S Quay, Deeside CH5 4JP
G1 LGB G. Gundry 181 Stoneleigh Avenue, Worcester Park KT4 8YA
G1 LGJ S. Stephens 41 Elliotts Lane, Codsall, Wolverhampton WV8 1PG
GI1 LGM C. Dowdall 24 Glasmullen Road, Glenariffe, Ballymena BT44 0QZ
G1 LGQ P. White 25 Witton Road, Ferryhill DL17 8QE
G1 LGY B. Bishop 26 Robin Gardens, Totton, Southampton SO40 8US
G1 LHD M. Goodes 17 Ashmead Close, Lords Wood, Chatham ME5 8NY
G1 LHE G. Courtney 24 Greenwood Close, Bognor Regis PO22 9DG
G1 LHL B. Mayson 19 Hudson Close, Sturry, Canterbury CT2 0HX
GW1 LHT G. Seldon 28 Monterey Street, Manselton, Swansea SA5 9PE
GW1 LHV J. O'Nions Pant-Glas, Gegin Lane, Wrexham LL11 3YT
G1 LIG Tatsfield Arts c/o M. Coker 5 Penling Close, Cookham, Maidenhead SL6 9NF
G1 LIK S. Church 24 Lovel End, Chalfont St. Peter, Gerrards Cross SL9 9PA
G1 LJL B. Duncan 13 Westwick Grove, Sheffield S8 7DP
GM1 LKD J. Craib 10 Cameron Road, Bridge Of Don, Aberdeen AB23 8QN
GW1 LKG Dr R. Cannon 43 Plas St. Pol De Leon, Portway Marina, Penarth CF64 1TR
G1 LKH R. Hilton 8 Hogshill Lane, Cobham KT11 2AQ
G1 LKJ P. Manning 1 Waverley Gardens, Ash Vale, Aldershot GU12 5JP
G1 LKK C. Wardle 16 Tedworth Avenue, Stenson Fields, Derby DE24 3BS
G1 LKL J. Wilkinson 160 Staines Road, Feltham TW14 9ED
G1 LLA C. Davis Fourwinds, Ringwood Road, Christchurch BH23 7BE
G1 LLI D. Hutchings 3 Willoway Grove, Braunton EX33 1AT
G1 LLQ P. Dollimore 96 Hayes Bridge Ct, Uxbridge Road, Hayes UB4 0JH
G1 LLU I. Olver 10 Celtic Road, Deal CT14 9EE
G1 LLW D. Rogers 36 Guessens Road, Welwyn Garden City AL8 6RH
G1 LLZ M. Jackson 146 Sea Road, Chapel St. Leonards, Skegness PE24 5RY
G1 LMC J. Trainer 86 Plessey Road, Blyth NE24 3HX
G1 LML J. Thornton 7 Queens Road, Vicars Cross, Chester CH3 5HA
G1 LMN S. Froggatt 255 Rushton Road, , Desborough NN14 2QB
G1 LMQ G. Donachie 36 Eastfields Narborough, King's Lynn PE32 1SS
G1 LMS J. Honeyball 3 Mill End Close, Warboys, Warboys PE28 2FP
G1 LMT S. Langson 7 John Street, Knutton, Newcastle ST5 6DT
G1 LMU S. Hodgetts 4 Stoke Park Court, Bishops Cleeve, Cheltenham GL52 8US
G1 LMW D. Partridge 44 Trumpet Terrace, Cleator CA23 3DY
G1 LMZ G. Clennell 69 Seventh Row, Ashington NE63 8HX
G1 LNA D. Wood 18 Rosemellin, Camborne TR14 8QF
G1 LNQ P. Bury 2 Manor Rise, Thornton In Craven, Skipton BD23 3TP
G1 LNR L. Button 37 Abbots Way, Preston Farm, North Shields NE29 8LU
G1 LOE K. Gosling 485B Blandford Road, Plymouth PL3 6JF
G1 LOK P. Blyth 12 Beulah Street, King's Lynn PE30 4DN
G1 LOL G. Davidson 18 Gotham Lane, Bunny, Nottingham NG11 6QJ
G1 LOU L. Vaisey Esperanza, 5 Hudson Close, Ringwood BH24 1XL
G1 LOV P. Foster 6 Croxteth Road, Bootle L20 5EA
G1 LOW P. Collick 39 Beech Rise, Sleaford NG34 8BJ
GM1 LOZ L. Hird 19 Alexandra Road, Keith AB55 5BX
G1 LPQ D. Thorpe 167 Southwell Road West, Mansfield NG18 4HD
G1 LPS T. Roxby 3 Coulton Terrace, Kirk Merrington, Spennymoor DL16 7HN
G1 LQB L. Clarke 10 Stonecliff Park Prebend Lane, Welton, Lincoln LN2 3JS
G1 LQC P. Chambers 26 Drummond Gardens, Christ Church Mount, Epsom KT19 8RP
G1 LQE A. Dickinson 5 Stone Croft, Barrowford, Nelson BB9 6BL
G1 LQH M. Lewis 8 Wetherdown, Herne Farm, Petersfield GU31 4PN
G1 LQM C. Costello The Coach House, The Green, Norwich NR12 9PZ
G1 LQP P. Gautrey 17 Heath Road, Market Bosworth, Nuneaton CV13 0NT
G1 LQT C. Matthews The Jasmine, School Hill, High Street, St. Austell PL26 7TP
G1 LQV G. Austin 10 St. Peters Close, Ruislip HA4 9JT
G1 LQX R. Saunders 3 Lancaster Road, Cressex Industrial Estat, High Wycombe HP12 3NN
G1 LRK C. Bassett 11 Redcastle Road, Thetford IP24 3NF
G1 LRM R. Cartmell 16 Churchfield Drive, Wigginton, York YO32 2FL
G1 LRU R. Jones 8 Mullen Avenue, Downs Barn, Milton Keynes MK14 7LU
G1 LRV Dr T. Edgar Gpo Box 37, Kelmscott 6991 Australia
G1 LSB P. Brockett 146 Winsover Road, Spalding PE11 1HQ
G1 LSK I. Wiseman Flat 4 Granville House, Glebe Avenue, Hunstanton PE36 6BS
G1 LSN M. Felton 1 Barnwell Close, Wistaston, Crewe CW2 6TG
G1 LSX J. Humphreys 44 Grosmont Grove, Hereford HR2 7EG
G1 LSZ P. Lambert 22 Cullingworth Avenue, Hull HU6 7DD
G1 LTC E. Rodd 32 Cranfield, Plympton, Plymouth PL7 4PF
G1 LTE G. Eades 117 Booths Farm Road, Birmingham B42 2NU
G1 LTG G. Perry 123 Green Lanes, Wylde Green, Sutton Coldfield B73 5LT
G1 LTH D. Williams 5 West Road, Ormesby, Great Yarmouth NR29 3RJ
G1 LTI P. Smith 12 Cleeves Court, Cleeves Way, Rustington, .Rustington BN16 3TS
G1 LTK Stoke On Trent ARS c/o A. Sturgess 11 Keats Close, Earl Shilton, Goonhilly, Leicester LE9 7DU
G1 LTL Dr D. Gardner New House, Birdbush Avenue, Saffron Walden CB11 4DJ
GM1 LTM U. Wallace No 1 Holding Wester, Kincardie, Tayside PH7 3RP
G1 LUC N. Spring Old Orchard House, Copyhold Lane, Dorchester DT2 9LT
G1 LUF J. Wright 298 Field Road, Bloxwich, Walsall WS3 3NB
G1 LUN P. Baker 651D Puketona Road, Paihia 204 New Zealand
G1 LUX C. Deacon 12 Russet Way, Burnham-on-Crouch CM0 8RB
GM1 LUZ C. Campbell 18 Parkview Avenue, Falkirk FK1 5JX
G1 LVH D. Barnett11 Steepholme, Front Street, South Clifton, Newark NG23 7AA
G1 LVR P. Cousins 38 Braunston Drive, Hayes UB4 9RB
G1 LVV A. Critchlow 51B West Road, Buxton SK17 6HQ
G1 LVW P. Parker 43 Meadow Close, Farmoor, Oxford OX2 9PA
GD1 LVY J. Dorman 1 Sprucewood Rise, Foxdale, Douglas IM4 3JP Isle of Man
G1 LVZ F. West 14 Ashley Drive, Twickenham TW2 6HW
G1 LWE J. Etchells 34 Link Avenue, Urmston, Manchester M41 9NJ
G1 LWF T. Finch 9 Halstead Road, Southampton SO18 2PQ
G1 LWH B. Whitehouse 105 Quarry Road, Birmingham B29 5LE
G1 LWL M. Patrick 39 Poplar Road, Healing, Grimsby DN41 7RE

G1 LWX M. Berry 133 Rectory Road, Ashton-In-Makerfield, Wigan WN4 0QF
G1 LWY K. Slater 56 Berners Road, Sheffield S2 2GB
GM1 LXA S. Sellick 1 Hatton Home Farm Cottages, Turriff AB53 8ED
G1 LXK P. Branston 19 Highfield Duddington, Stamford PE9 3QD
GM1 LXM S. Mcintier 2A Glenacre Drive, Largs KA30 9BH
G1 LYX N. Bristow 21 Dudley Street, Leighton Buzzard LU7 1SE
G1 LZF B. Dickinson 178 Wycliffe Gardens, Shipley BD18 3JB
G1 LZH P. Taylor 10 Pickenham Road, Birmingham B14 4TG
G1 LZL D. Atkinson 336 Moss Bay Road, Workington CA14 5AF
G1 LZS S. Stallworthy 6 Kenwood Close, Hastings TN34 2AT
G1 LZZ D. Dennett Redhill Cottage, 14 Main Road, Wareham BH20 5RN
G1 MAC M. Macbeth 58 The Common, Abberley, Worcester WR6 6AY
G1 MAD Moorlands And District ARS c/o C. Beesley 15 Byron Close, Cheadle, Stoke-on-Trent ST10 1XB
G1 MAL D. Simmons 47 Lower Street, Haslemere GU27 2NY
G1 MAR Fwe CG c/o N. Gutteridge 68 Max Road, Quinton, Birmingham B32 1LB
G1 MAS M. Crooks 19 The Nook Crookesmoor, Sheffield S10 1EJ
G1 MAV G. Seaman 22A Mount Road, Bexleyheath DA6 8JS
GW1 MAX R. Pullen 8 Carmarthen Court, Caerphilly CF83 2TX
G1 MBE C. Batty 32 The Warings, Heskin, Chorley PR7 5NZ
G1 MBM M. Kitson 54 Hollins Lane, Sowerby Bridge HX6 2RP
G1 MBN P. Doyle 11 Clifford Avenue, Longton, Preston PR4 5BH
GW1 MBV J. Follett 136 Westbourne Road, Penarth CF64 3HH
G1 MBW M. Smith 32 Amesbury Drive, London E4 7PZ
G1 MCG D. Minton 8 Rosebank Walk, Barnton, Northwich CW8 4PU
G1 MCI A. Rawson 43 County Road North, Hull HU5 4HN
GM1 MCN Canon C. Stanley Nazareth House, 34 Claremont Street, Aberdeen AB10 6RA
G1 MCT G. Williams 29 Coleridge Road, Barnby Dun, Doncaster DN3 1AN
G1 MCW W. Higgins 15 Redburn Close, Liverpool L8 4XR
G1 MCY C. Toogood 16 Penlea Avenue, Bridgwater TA6 6JU
G1 MDC D. Tommey 99 Fairfield Park Road, Bath BA1 6JR
G1 MDE D. Harrison 19 Oakwood Road East, Rotherham S60 3ER
G1 MDG Chesham & District ARS c/o M. Appleby 6 Mandeville Road, Prestwood, Great Missenden HP16 9DS
G1 MDJ J. Murch Downings, Prinsted Lane, Emsworth PO10 8HS
GM1 MDO J. Stewart 104 Barrhill Road, Cumnock KA18 1PU
G1 MDQ M. Mitchell 310 Parlaunt Road, Slough SL3 8AX
G1 MDS M. Lewis Westbank, 46 Weyside Road, Guildford GU1 1HX
G1 MET Dr R. Heywood 22 Catterall Close, Blackpool FY1 3RB
G1 MFK M. Morris 1 Fruitlands, Malvern WR14 4AH
G1 MGF A. Hall 172 Aldershot Road, Guildford GU2 8BL
GW1 MGI L. Long 18 Pentre Poeth Road, Bassaleg, Newport NP10 8LL
G1 MGN T. Jones 35 Manta Road, Dosthill, Tamworth B77 1PE
G1 MGU J. Studd 64 Moggs Mead, Herne Farm, Petersfield GU31 4NX
G1 MGZ M. Brophy 78 Foley Road West, Streetly, Sutton Coldfield B74 3NP
G1 MHA C. Corner 100 Monkseaton Drive, Whitley Bay NE26 3DJ
G1 MHB G. Patten 31 Sea View Road, Skegness PE25 1BN
G1 MHF G. Fleming 25 Waverton Avenue, Prenton CH43 0XB
G1 MHM W. Taylor Corsend Farmhouse, Corsend Road, Gloucester GL19 3BP
G1 MHN M. Dawson Steenbok, 9 Stockwell Gate West, Whaplode, Spalding PE12 6WG
G1 MHP D. Gilbert 87 Willow Close, St. Georges, Weston-Super-Mare BS22 7SF
G1 MIE K. Martin 21 All Saints Close, Weybourne, Holt NR25 7HH
GD1 MIP A. Morgan Thal'Loo Glass, Nassau Road, Mwyljyn Moddey IM7 4AQ Isle of Man
G1 MIY R. Lunnon 9 Hennerton Way, High Wycombe HP13 7UE
G1 MJA P. Griffin 147 Kingshayes Road, Aldridge, Walsall WS9 8SN
G1 MJI Magnetic Fields CG c/o D. Fry 23 South Lea, Braunton EX33 2HN
GI1 MJJ F. Gilliland 48 Malone Heights, Belfast BT9 5PG
G1 MJN M. Newbold Sawubona, Vicarage Lane, Skegness PE24 4JJ
G1 MJO A. Hemming 64 Haslucks Green Road, Shirley, Solihull B90 2EJ
G1 MJT R. Naylor 6 Alden Close, Morley, Leeds LS27 0SG
G1 MJV N. Liddiard Orchard End, Dunwich Lane, Peasenhall, Saxmundham IP17 2JP
GM1 MKC C. Strong Little Couchercairn St Katherines, Inverurie AB51 8TQ
G1 MKE R. Cox Carlingford, Brimley Drive, Teignmouth TQ14 8LE
G1 MKJ I. Benson Heathcliffe, Little Lonnen, Wimborne BH21 7BB
G1 MKP M. Pauley 235 Roughton Road, Cromer NR27 9LQ
G1 MKR Milton Keynes Raynet Group c/o J. Breen 68 Honeysuckle Way, Bedford MK41 0TF
G1 MKS P. Healy 43 Brockley View, London SE23 1SL
GW1 MKV A. Coe Woodbrook Cottage, Paddock Row, Clwyd LL14 6DD
G1 MKY J. Hargreaves 14 Newlands Close Blackfield, Southampton SO45 1WH
G1 MLC H. Taha 53 Barbers Hill, Werrington, Peterborough PE4 5ED
GM1 MLS A. Pearson 1 Bridgefield, Inverbervie, Montrose DD10 0SR
G1 MLV P. Kelsey 573 Stannington Road, Stannington, Sheffield S6 6AB
GM1 MLW C. Lindsay 51 Perrays Crescent, Dumbarton G82 5HP
GM1 MLY M. Bull 20 Grenitote, Lochmaddy, Isle of North Uist HS6 5BP
G1 MMA S. Butcher 155 Crow Lane East, Newton-le-Willows WA12 9UD
G1 MMD K. Morris 16 Goldcrest Road, Forest Town, Mansfield NG19 0GP
G1 MMI T. Ryder 117 Cotefield Drive, Leighton Buzzard LU7 3DN
GM1 MMK K. Cupples 16 Glebe Crescent, Airdrie ML6 7DH
G1 MMN C. Lovett 49 Tame St. East, Walsall WS1 3LB
G1 MMT Dr S. Quick C/O 9 Juniper Close, Ferndown BH22 9UB
G1 MMZ A. Roffey 32 Hertford Road, Digswell, Welwyn AL6 0DB
GW1 MNC A. Dykes 16 The Mercies, Porthcawl CF36 5HN
GW1 MNU I. Jukes 14 Beechwood Place, Narberth SA67 7EE
G1 MNX P. Lane 1 St. Davids Close, Lower Willingdon, Eastbourne BN22 0UZ
G1 MNY R. Smith 3 Florence Farm Mobile Home Park, London Road, Sevenoaks TN15 6BP
G1 MOB Dr N. Booth Greencotes, Warden, Hexham NE46 4SS
G1 MOK T. Dommett 9 Causeway Close, Woolavington TA7 8DW
GM1 MON J. Mcqueen Rowan Cottage Redcastle, Lunanbay, Arbroath DD11 5SS
G1 MOS H. Whitbread Foresters, Main Road, Martlesham IP12 4SL
G1 MOV M. Ballantyne 248 Calshot Road, Great Barr, Birmingham B42 2BX
G1 MOW D. Mallin 60 Arundel Road, Littlehampton BN17 7DF
G1 MOZ J. Nicholson 11 West View Rise, Huddersfield HD1 4UR
G1 MPC M. Human 28 Lincoln Drive, Croxley Green, Rickmansworth WD3 3NH
G1 MPD M. Champion 14 Meadow Court Road, Earl Shilton, Leicester LE9 7FF
G1 MPG C. Alefs 27 Millfields, Beckermet CA21 2YY
G1 MPI M. Hillman 28 Murndal Dr, Donvale 3111 Australia
G1 MPL C. Hitcham 27 Kirby Cane Walk, Lowestoft NR32 3EL
G1 MPP A. Baxter 94 Abbeyfield Drive, Fareham PO15 5PF
G1 MPT D. Bowden 4 Cornmill Close, Bardsey, Leeds LS17 9EG
G1 MPU P. Brolan 2 Mount Road, Barnet EN4 9RL
G1 MPW S. Cooke 21 Wealdon Close, Southwater, Horsham RH13 9HP
GM1 MQA R. Cooke Taigh Na Greine, Lower Bayble, Isle of Lewis HS2 0QB
G1 MQB P. Barker 7 Verbena Close, Nottingham NG3 4PZ
G1 MQC Devon County Council Emergency Plng c/o M. Newport 9 Highbury Park, Exmouth EX8 3EJ
G1 MQQ G. Stainton 168 Slades Road, Golcar, Huddersfield HD7 4JR
G1 MRC M. Johnson 24 Stjames'S Park, Wakefield WF1 4EU
G1 MRE P. Langford 33 Briscoe Road, Hoddesdon EN11 9DG
G1 MRI P. Gardiner 12 Weston Road, Aston Clinton, Aylesbury HP22 5EG
GM1 MRS L. Alexander 97 Land Street, Keith AB55 5AP
G1 MRX P. Young 30 Badminton Road, Maidenhead SL6 4QT
GM1 MRY W. Andrew 11 Eddington Gardens, Chryston, Glasgow G69 0JW
G1 MSA A. Taylor Apartment 125, Earlsdon Park Retirement Village, Albany Road, Coventry CV5 6LF
G1 MSB E. Turner 14 Lauderdale Gardens, Bushbury, Wolverhampton WV10 8AY
G1 MSD R. Newsome 6 Woodhouse Grove, Fartown, Huddersfield HD2 1AS
G1 MSG M. White 11 Beck Way, Loddon, Norwich NR14 6UZ
G1 MSK J. Riley Hillcrest, Norwich Road, Chedgrave, Norwich NR14 6BQ
GM1 MSN W. Clark 66 Winstanley Wynd, Kilwinning KA13 6EB
GM1 MSO D. Clark 50 Dalry Road, Kilwinning KA13 7HE
G1 MSR J. Cockroft 8 Harris Road, Standish, Wigan WN6 0QR
GM1 MSS Dr I. Coates 55 Whitehaugh Park, Peebles EH45 9DB
G1 MSY K. Ludgate 138 Halton Road, Runcorn WA7 5RW
G1 MTA R. Grey 23 Chapel Rise, Ringwood BH24 2BL
G1 MTB S. White 12 Morgan Close, Bexhill-on-Sea TN39 5EQ
GW1 MTH H. Williams 10, Moriah Street, Merthyr Tydfil CF47 8LU
G1 MTJ S. Tickle 14 Rothesay Drive, Crosby, Liverpool L23 0RF
G1 MTP N. Mcneil 3 Grosvenor Court, Water Lane, York YO30 6PX
G1 MTU D. Ward 27 Penzer St., Kingswinford DY6 7AA
G1 MUC C. Shingles 20 Spencer Close, Lingwood, Norwich NR13 4BB
G1 MUM H. Phillips 6 Peaks Down, Peatmoor, Swindon SN5 5BH
GU1 MUP P. Rudd Val Des Arquets, Les Arquets, St. Pierre Du Bois GY7 9HE Guernsey
G1 MUQ P. Hillier Bythan, Avenbury Lane, Bromyard HR7 4LB
G1 MUT P. Lawrence 2 Chapel Terrace, Station Street, Ashbourne DE6 1DF
GM1 MUY L. Coxon 40 Hamilton Street, Broughty Ferry, Dundee DD5 2RE
G1 MVE P. Tither 32 Manor Avenue, Marston, Northwich CW9 6DS
G1 MVF S. Rasmussen 10 The Pightle, Grafham, Huntingdon PE28 0UU
G1 MVG Dr F. Marshall Hartwell, Newgrounds, Fordingbridge SP6 2LJ
G1 MVI P. Bowe 197 Gloucester Avenue, Chelmsford CM2 9DX
G1 MVQ S. Hamilton-Cooper 4 Wren Close, Appleby Magna, Swadlincote DE12 7BD
G1 MVT D. Driscoll 25 Broom Close, Dawlish EX7 0RP
GW1 MVZ D. Petrie 48A Lower Quay Road, Hook, Haverfordwest SA62 4LR
GM1 MWK J. Grieve 10 Jubilee Court, Kirkwall KW15 1XR
G1 MWS R. Bell 92 Dean Drive, Wilmslow SK9 2EY
G1 MWT M. Clancy 34 High Meadows, Greetland, Halifax HX4 8QF
G1 MXC K. Sampson 40 Crisp Road, Lewes BN7 2TX
G1 MXD G. Waldron 55 Sheringham Road, Poole BH12 1NS
GM1 MXE G. Schafers Dahlsteven, Orkney KW17 2RD
G1 MXM I. Hunt Four Seasons, Westmarsh, Canterbury CT3 2LP
GM1 MYF R. Jones 46B Forest Road, Aberdeen AB15 4BP
G1 MYM E. Emons 18 Haig Road, Stanmore HA7 4EP
G1 MYO F. Ross 2 Mount Pleasant, Steeple Claydon, Buckingham MK18 2QS
G1 MYQ P. England Moonstones, Down Ampney, Cirencester GL7 5QS
GM1 MYR R. Cook 95 Old Edinburgh Road, Inverness IV2 3HT
G1 MZD D. Barlow 34 Mays Way, Potterspury, Towcester NN12 7PP
G1 MZG S. Banks 18 Sheerstock, Haddenham, Aylesbury HP17 8EU
G1 MZH E. Schamp 21 Beechwood Avenue, Melton Mowbray LE13 1RT
GD1 MZJ J. Sutherland Archallagan Park, Marown IO9 9SU Isle of Man
G1 MZM M. Bignell 53 Rosebay Avenue, Birmingham B38 9QT
G1 MZP A. Froggatt 16 Seagull Close Hull, Hull HU4 6SN
G1 MZT T. Tipper 114 Paddock Lane, Redditch B98 7XT
G1 MZW I. Gillson 13 Beech Green, Southcourt, Aylesbury HP21 8JG
GM1 MZZ J. Frearson 27 Miltonbank Crescent, Guardbridge, St. Andrews KY16 0XE
G1 NAA M. Crabtree 23 Ava Crescent, Richmond Hill, Ontario L4B 2X1 Canada
G1 NAB G. Rainy Brown Old Stores Cottage, Newbury RG20 8SE
G1 NAN A. Gateley 2 Langmere Road, Watton, Thetford IP25 6LG
G1 NAP J. Hopkins 3 De Havilland Road, Upper Rissington, Cheltenham GL54 2NZ
G1 NAQ A. Ashton 6 Lansdowne Crescent, Darton, Barnsley S75 5PW
G1 NAT D. Mcgowan 7 Eccles Close, Henley Green, Coventry CV2 1EF
G1 NAU J. Cawsey 134 Goddard Avenue, Swindon SN1 4HX
G1 NBK W. Winning Plump House, Terrington, York YO60 6QB
G1 NBO R. Mewis 52 Princess St., Burton on Trent DE14 2NP
G1 NBP Dr P. Allan 24 Farm Piece, Stanford In The Vale, Faringdon SN7 8FA
G1 NBT A. Bradley 59 Main Road, Watnall, Nottingham NG16 1HE
G1 NBU L. Wellbeloved 8 Orchard Close, South Wonston, Winchester SO21 3EY
GW1 NBW W. Morris 17 Fairway, Port Talbot SA12 7HG
G1 NBY Rvd. R. Roeschlaub 20 Pannatt Hill, Millom LA18 5DB
G1 NCG K. Powell 43 Mallard Close, Swindon SN3 5JG
G1 NCK D. Bates 71 Nicholas Crescent, Fareham PO15 5AJ
G1 NCL C. Holmes 16 Industrial St., Pelton, Chester le Street DH2 1NR
G1 NCM A. Urquhart 75 Springvale Road, Winchester SO23 7ND
G1 NCN H. Jones 8 Warren Close, Old Catton, Norwich NR6 7NL
G1 NCO P. Robinson 12 Mountain Ash Avenue, Leigh-on-Sea SS9 4SZ
G1 NCR Thurrock Acorns ARC c/o G. Gourley 6A Longsight Lane, Cheadle Hulme, Cheadle SK8 6PW
G1 NDK K. Dunn Marylands, Maidstone Road, Tonbridge TN12 0RH
G1 NDL K. Harrison 20 Springfield Avenue, Ashbourne DE6 1BJ
G1 NDQ M. Taylor 41 Hill View Gardens, Halifax HX3 7BT
G1 NDV R. Airey 30 White Horse Crescent, Grove, Wantage OX12 0PY
GW1 NED H. Anderson Penrheol Farm, Meidrim, Carmarthen SA33 5NX
G1 NEG P. Mcgarry 10 Douglas Avenue, Soothill, Batley WF17 6HG
G1 NEN A. Hefford 31 High Street, Rushton, Kettering NN14 1RQ

GM1 NET Strathclyde R G c/o R. Campbell 32 Harvie Avenue Newton Mearns, Glasgow G77 6LQ
G1 NEV D. Dawson 4 Hawksworth Lane, Guiseley, Leeds LS20 8HA
GM1 NEW J. Kerins 30 Beech Avenue, Newton Mearns, Glasgow G77 5PP
G1 NEZ B. Stiff 4 Timberlaine Road, Pevensey Bay, Pevensey BN24 6DE
G1 NFB D. Bagley 38 Ashchurch Road, Tewkesbury GL20 8BT
G1 NFE C. Duffy 25 Redcar Avenue, Thornton-Cleveleys FY5 2LG
G1 NFN R. Hammond 124 Maney Hill Road, Sutton Coldfield B72 1JU
G1 NFO B. Webb 63 Rother Road, Rotherham S60 2UZ
G1 NFQ R. Vivian Flat 1, 26 Beer Road, Seaton EX12 2PD
G1 NGE R. Nelson Woodlands, 35 Cromer Road, Hevingham, Norwich NR10 5QX
G1 NGI A. Jones Rotherhithe New Road Tissington Court Flat 5, London SE16 2AG
G1 NGL K. Roberts 4631 Chatham Se St, Salem 97302 United States
GW1 NGN P. Johansson 63 Grange Road, Rhyl LL18 4AD
G1 NGR R. Sharman Flat 1 11 Sherbourne Road, Blackpool FY12PW
G1 NHG Dr M. Hausler 5 Balaton Place, Snailwell Road, Newmarket CB8 7YP
G1 NHX P. Severn 310 Worlds End Lane, Birmingham B32 2SB
G1 NIC N. James America Lodge, Ridgeway Road, Torquay TQ1 2EU
G1 NIM R. Hill Cama De Mingot 3 La Alcoraia, Alicante 3699 Spain
G1 NIT M. Virtue 50 Borthwick Park, Orton Wistow, Peterborough PE2 6YY
G1 NIV J. Young The Estate Office, Granary Court, Chadwell Heath RM6 6PY
G1 NJG B. Asker 34 Post Office Road, Frettenham, Norwich NR12 7AB
G1 NJI R. Foster 35 Colin Road, Barnwood, Gloucester GL4 3JL
G1 NJV M. Wing 27 Hill Street, Hunstanton PE36 5BS
G1 NKF I. Spindler Birkett House, Weeton Village, Weeton, Preston PR4 3NB
G1 NKN Dr A. Mason 51 Benett Drive, Hove BN3 6UQ
G1 NKT P. Grant 37 Glenmore Park, Dundalk, County Louth Ireland
G1 NKV F. Smith 18-26 Hendon Rise, Thorneywood, Nottingham NG3 3AN
G1 NLQ D. Arter 18 Essex Road, Westgate-on-Sea CT8 8AP
G1 NLS G. Borrett 45 Yarwells Headland, Whittlesey, Peterborough Cambs PE7 1RF
G1 NLZ R. Brown 15 Johnson Road, Great Baddow, Chelmsford CM2 7JL
G1 NMI G Qrp Club c/o C. Whitehead Highfield Lodge, Fakenham Road, Brisley, Dereham NR20 4HU
G1 NML J. Hyde 71 Nunts Lane, Coventry CV6 4GZ
G1 NMN M. Durey 71 Orchard Road, Maldon CM9 6EW
G1 NMP N. Fenner 22 Gowers Field, Aylesbury HP20 2QT
G1 NMQ L. Gibbs 45 Woolavington Hill, Woolavington, Bridgwater TA7 8HQ
G1 NMW A. Stone 5 Bridge Street, Cheltenham GL51 9DQ
G1 NMY B. Hitchcock 111 Kingsdown Crescent, Dawlish EX7 0HB
G1 NNA B. Lloyd 17 Brooklands Road, Brantham, Manningtree CO11 1RN
G1 NNB G. Lloyd 9 Hornbeam Walk, Witham CM8 2SZ
G1 NNF R. Kenny 35 Broom Leys Road, Coalville LE67 4DD
G1 NNN S. Moore 104 Gloucester Avenue, Chelmsford CM2 9LF
G1 NNR D. Newman 78 Vale Road, Poole BH14 9AU
G1 NOO I. Allgood 53 The Avenue, Leighton Bromswold, Huntingdon PE28 5AW
G1 NOR G. Galbraith 44 Parker Road, Grays RM17 5YN
G1 NOS K. Brookes 20 School Avenue, Guide Post, Choppington NE62 5DN
G1 NPA P. Allan 214 Westwood Road, Sutton Coldfield B73 6UQ
G1 NPC G. Hammond 31 Earlsway, Macclesfield SK11 8RJ
G1 NPI C. Badcock 7 Heathfield Road, Chandler'S Ford, Eastleigh SO53 5RP
G1 NPJ K. Hyslop Smallgains Marina Prout Industrial Estate, Point Road SS8 7TJ
G1 NPN K. Mcdougal 51 Argyll Avenue, Wirral CH62 8EB
G1 NPX H. Nurse 46 Kelvedon Road Coggeshall, Colchester CO6 1RQ
G1 NQB C. Carpenter 10 D'Arcy Road, St. Osyth, Clacton-on-Sea. CO16 8QE
G1 NQH M. Cook Brooksdie Cottage, Brook Lane, Market Harborough LE16 8SJ
G1 NQN K. Benfold 56 Cornwall Avenue, Blackpool FY2 9QW
G1 NQO J. Jacques 65 Daggers Hall Lane, Marton, Blackpool FY4 4AX
G1 NQU M. Peacock 19 Ashfield Terrace, Haworth, Keighley BD22 8PL
G1 NRE South Devon Raynet Group c/o D. Paul 99 Wilkinson Road, Bedford MK42 7FR
G1 NRF M. Halloway 41 Trenoweth Estate North Country, North Country, Redruth TR164AQ
G1 NRG the Terry Bain Memorial RC c/o S. Manning 11 Broomhill Crescent, Southfields, Northampton NN3 5BH
G1 NRK P. Slater 12A Apsley Close, Bishop's Stortford CM23 3PX
G1 NRM A. Harrison 34 Marsh Lane, Mill Hill, London NW7 4QP
G1 NRN K. Johnson 98 Wroxham Road, Great Sankey, Warrington WA5 3NU
G1 NRX P. Hill 13 Onslow Road, Newent GL18 1TL
G1 NRY I. Leach 36 Harrowden, Bradville, Milton Keynes MK13 7DA
G1 NSB R. Winterburn Flat 15, Elms Farm, Mather Avenue, Manchester M45 8NT
G1 NSD D. Young 13 Crawshaw Park, Pudsey LS28 7EP
G1 NSG H. Goodwin 91 Grange Lane, Sutton Coldfield B75 5LD
G1 NSK F. Godfrey 50 Eskdale Drive, Worksop S81 7QB
G1 NSQ C. Joyce 70 Campbell Road, Twickenham TW2 5BY
G1 NST S. Dodd Chedburgh House, Hall Lane, North Walsham NR28 0RZ
G1 NSV D. Kemplen 2 Vicarage Close, Menheniot, Liskeard PL14 3QG
G1 NTI B. Longstaff 23 Chester Road Estate, Stanley DH9 0QD
G1 NTJ P. Worden 86 Buckwood Road Markyate, St. Albans AL3 8JB
G1 NTK P. Hull Hazelwood, 2 Cheats Road, Taunton TA3 5JW
G1 NTL J. Mcshane 12 Virginia Gardens, Middlesbrough TS5 8BT
G1 NTN A. Edwards 9 Lincoln Close, Woodley, Romsey SO51 7TJ
G1 NTP B. Haden 72 Charlton Road, Blackheath, London SE3 8TT
G1 NTR S. Mayer 41 Lowe Street, Macclesfield SK11 7NJ
G1 NTV D. Merry 18 Tremabe Park, Dobwalls, Liskeard PL14 6JS
G1 NTX S. Taylor 5 Collingwood, Farnborough GU14 6LX
G1 NUH R. Hardwick 5 Seaview, Oakmere Park, Little Neston CH64 0XP
G1 NUO E. Oram 31 Nathaniel Walk, Tring HP23 5DG
G1 NUS J. Thornley 270 Hurdsfield Road, Macclesfield SK10 2PN
G1 NVE V. Wood 175 Windleshaw Road Dentons Green, St. Helens WA10 6TP
G1 NVL G. Officer Flat 5, 8 Charlton Drive, Sale M33 2BJ
G1 NVN A. Parkin 1 Dunelm Walk, Leadgate, Consett DH8 7QT
G1 NVO B. Kimber 27 Court Road, Brockworth, Gloucester GL3 4ES
G1 NVS R. Pennington 5 Park Close, Northway, Tewkesbury GL20 8RB
G1 NVV M. Timlett Ashley Lodge, Church Lane, Strubby, Alford LN13 0LR
G1 NVY K. Peers 159 Reeve Court Retirement Village, Stratton Drive, Raiinhill, St. Helens WA9 5BP
G1 NWA C. Rickerby Brownsea, 113 Cliftonville Road, Woolston, Warrington WA1 4BJ
GW1 NWF R. Gray 36 Heol Pentre Felen, Morriston, Swansea SA6 6BY
G1 NWG C. Scates 17 Trecastle Way, Carleton Road, London N7 0EL
G1 NWH H. Walker 24 Castleton Avenue, Riddings, Alfreton DE55 4AG
G1 NWM J. Westwood 9 Landbeach Road, Milton, Cambridge CB24 6DA
G1 NWO J. Sharp Hunrosa, Crowlas, Penzance TR20 8DS
G1 NWT B. Worviell Cliddesden, The St., Shaftesbury SP7 9PF
G1 NWZ M. Spacey 4 Hickman Court, Copenhagen Close, Luton LU3 3TW
G1 NXB G. Fardoe 3 Park Avenue, Wallasey CH44 9DZ
G1 NXI C. Foster 203 Hempshill Lane, Nottingham NG6 8PF
G1 NXR V. Barrett 4 Alexandra Street, Heywood OL10 2AU
G1 NXS B. Martlew 15 Dunscar Close, Birchwood, Warrington WA3 7LS
G1 NXT K. Rushton 9 Laburnum Avenue, Woolston, Warrington WA1 4NY
G1 NYI T. Rozier 26 Watersmeet Way, London SE28 8PU
G1 NYJ D. Hopton 32 Braemar Avenue, Urmston, Manchester M41 6HP
G1 NYP M. Fleet 152 Bridge Road, Chessington KT9 2EY
GW1 NYS R. Parkhurst 13 Tasker Way, Haverfordwest SA611FB
G1 NYZ D. Robinson 4 Rushden Drive, Reading RG2 8LJ
G1 NZD P. Bates 132 Brownings Avenue, Chelmsford CM1 4HJ
GW1 NZF R. Stuckey 8 Gelli Crossing, Gelli, Pentre CF41 7UD
G1 NZH J. Edgecock 13 Holmsdale Close, Durgates, Wadhurst TN5 6UT
G1 NZK A. Davis 73A Milton Road, Taunton TA1 2JQ
G1 NZL C. Elliott 7 Elizabeth Diamond Gardens, South Shields NE33 5HX
G1 NZN J. Rolley 12 Ravenscroft Drive, Chaddesden, Derby DE21 6NX
G1 NZP E. Elliott 7 Red House Road, Hebburn NE31 2XS
G1 NZQ R. Finch 12 Simcox Street, Hednesford, Cannock WS12 1BG
G1 NZZ R. Nicol 37 Thicknall Drive, Stourbridge DY9 0YH
G1 OAE R. Steel11 7 Derwent Bank, Seaton, Workington CA14 1EE
G1 OAM T. Wilson 15 Whipperley Way, Luton LU1 5LB
G1 OAR P. Wallace 7 Trinity View, Ketley Bank, Telford TF2 0DX
G1 OAU V. English 29E High Street, Eye, Peterborough PE6 7UP
G1 OAW J. Smith 62 Elson Lane, Elson, Gosport PO12 4EU
G1 OAX L. Pugh 15 Didcott Way, Appleby Magna, Swadlincote DE12 7AS
G1 OAZ G. Sugden 247 Yorkland Avenue, Welling DA16 2LH
G1 OBA I. Bpophy 78 Foley Road West, Streetly, Sutton Coldfield B74 3NP
G1 OBC I. Evans 6 Park End, Lichfield WS14 9US
G1 OBM J. Miller 141 Discovery Road Mount Wise, Plymouth PL1 4PR
G1 OCH C. Hillman Mayjon, Crow, Ringwood BH24 3ER
G1 OCK R. Liepziger 21 Third Avenue, Woodside Park, Poulton-le-Fylde FY6 0PW
G1 OCL D. Potter 9 Beachcroft Place, Lancing BN15 8JN
GM1 OCR N. Law Kilbrannan, Benderloch, Oban PA37 1QU
G1 OCS A. Heap 56 Moorside Road, Eccleshill, Bradford BD2 3RB
G1 OCY K. Minihane 34 Wavell Road, Gosport PO13 0XR
G1 ODB D. Price 21 Orchard Road, Nailsea, Bristol BS48 2DZ
G1 ODD B. Westlake 47 Quarry Road, Kingswood, Bristol BS15 8NZ
G1 ODE G. Wheeler 14 Mina Road, Bristol BS2 9TB
G1 ODJ T. Boycott 17 Brook Street, Whitley Bay NE26 1AF
G1 ODK G. Cole 87 Chichester Road, Ramsgate CT12 6NZ
G1 ODN J. Thompson 14 Redwing Close, Horsham RH13 5PE
G1 ODQ C. Bond 6 Copse Close, Hugglescote, Coalville LE67 2GL
G1 ODT A. Pargeter Acres Cottage, Smallburn Road, Longhorsley, Morpeth NE65 8QH
G1 ODZ G. Fountain Tinkers Lane, Wigginton, Tring HP23 6JB
G1 OEB A. Orchard Flat No 3, 35 The High Street, Hemel Hempstead HP3 0HG
G1 OEF Dr E. Earland 7 Paxford House Square, Ottery St. Mary EX11 1BX
G1 OEM T. Webb 95 Devereaux Crescent, Ebley, Stroud GL5 4PX
G1 OEP J. Harris 109 Hook Rise South, Surbiton KT6 7NA
G1 OEQ D. Tribute 'Pathey', Lower Polstain Road, Truro TR3 6BQ
G1 OER A. Parkin 4 Waverley Road, Farnborough GU14 7EY
G1 OET Caversham Con G c/o K. Doswell 15A Queen Street, Desborough, Kettering NN14 2RE
G1 OFG P. Howard Cork Farm, Ruthern Bridge, Bodmin PL30 5LU
G1 OFL R. Hall-Osman 67 Livingstone Road, Gravesend DA12 5DN
G1 OFW S. Gadsby 30 Woodside Close, Knaphill, Woking GU21 2DD
G1 OFX W. Coates 3 Graysmead, Sible Hedingham, Halstead CO9 3NX
G1 OFY Capt. P. Hendy 4 The Pack, Burgh-By-Sands, Carlisle CA5 6BE
G1 OGB P. Campion 2 Woodside, Plymouth PL4 8QE
G1 OGC J. Kelly 18 Mount Pleasant, Riddings, Alfreton DE55 4BL
G1 OGE A. Wilson Moor Cottage, Ellastone Road, Stoke-on-Trent ST10 3ER
G1 OGH K. Atkinson 62 Fines Park, Stanley DH9 8QY
G1 OGR R. Welch 2 Broadlands Avenue, Waterlooville PO7 7JE
G1 OGV R. Bicknell-Thompson 4 Linden Court, Greenfrith Drive, Tonbridge TN10 3LW
G1 OGY D. Gilligan Two Worlds, Tan Lane, Clacton-on-Sea CO16 9PS
GM1 OGZ D. Madden Flat 2/2, 24 Collier Street, Johnstone PA5 8AR
G1 OHD B. Beswick 2 Ferndale Road, Peak Dale, Buxton SK17 8AY
G1 OHL Dr W. Morden Apartment 3, Fernside House 49 Hollington Park Road, St. Leonards-on-Sea TN38 0SE
G1 OHU J. Freeman 81 West Hill, Kimberworth, Rotherham S61 2EX
G1 OHV Royal Signals Swindon c/o R. De Havilland 11 Morvale Street, Stourbridge DY9 8DE
G1 OHX V. Pears 10 Fremantle Road, South Shields NE34 7RF
GW1 OIB R. Jones 10 Ferndale Crescent, Gobowen, Oswestry SY11 3PJ
GW1 OII S. Lloyd 10 Park Crescent, Llanelli SA15 3AE
GW1 OIK W. Jaggard 2 Aled Drive, Rhos On Sea, Colwyn Bay LL28 4UU
G1 OIO W. Benton 2 Regents Close, Seaford BN25 2EB
G1 OIS N. Ellis 1A Northcote Road, Croydon CR0 2HX
G1 OIZ R. Young 1 Croft Walk, Whitwell, Worksop S80 4UD
G1 OJB P. Critchley 4 Shandon Avenue, Northenden, Manchester M22 4DP
G1 OJD D. Facer 7 Lowry Close, Bedworth CV12 8DG
G1 OJL S. Evenden 11 Chapel Street, Tavistock PL19 8DX
G1 OJO D. Pearson 16 Hilldown Road, Hayes, Bromley BR2 7HX
G1 OJQ P. Brooks 7 Ashbourne Road, Underwood, Nottingham NG16 5EH
G1 OJS Dr A. Robinson 3 Clifton Mews, Fareham PO168TY
G1 OJT R. Barnish 64 Braithwell Road, Maltby, Rotherham S66 8JU
G1 OKB A. Ibbotson 62 Crag View Crescent, Oughtibridge, Sheffield S35 0GD
G1 OKF D. Cannon 44 Grange Bottom, Royston SG8 9UQ
G1 OKI D. Hyde 108 St. Bedes Crescent, Cambridge CB1 3UB
G1 OKP R. Lloyd 52 Roman Way, Ross-on-Wye HR9 5RL
G1 OKV G. Hendricks 105 Hillcrest Park, Wilbury Hills Road, Letchworth Garden City SG6 4LF
GW1 OKY J. Cottrell Bryn Dewi, Llanallgo, Moelfre LL72 8HB

G1 OLE D. Bates 10 Upton Gardens, Worthing BN13 1DA
G1 OLM J. Hesketh 87 Condor Grove, Blackpool FY1 5NA
G1 OLQ W. Dacey 69 Freshfield Gardens, Allerton, Bradford BD15 7PR
G1 OLY O. Brooks 34 Jubilee Road, Stokenchurch, High Wycombe HP14 3SJ
GI1 OMD ATC Scotland & N.Ireland Region ARC c/o T. Mcquaid 5 Edenamohill, Drumkeen, Enniskillen BT93 0FQ
G1 OMI J. Canning 130 Main Road, Duston, Northampton NN5 6RA
G1 OMX P. Cumiskey 1 York Terrace, Gateshead NE10 9NB
G1 OMY D. Ainscough 11 Tressel Drive, Sutton Manor, St. Helens WA9 4BS
G1 ONC P. Maitland 7 Spinners Court, Stalham, Norwich NR12 9EQ
G1 OND B. Hastry 56 Kilsyth Close, Fearnhead, Warrington WA2 0SQ
G1 ONE Bolton Wireless Club c/o D. Lewis 10 Addington Road, Bolton BL3 4QZ
G1 ONH F. Slater 32 Winthorpe Avenue, Morecambe LA4 4RE
G1 ONJ D. Tennant 128 Devonshire St., Keighley BD21 2QJ
G1 ONK L. Boston Lissa Park Sitesi No 16 (86/R), 965 Sokak Mustafa Kemal Bulvari, Calis FETHIYE 48300 Turkey
G1 ONQ F. Pearce 1B Council Street, Bozeat, Wellingborough NN29 7LS
G1 ONV R. Bonar 6 Harepark, Allerford, Minehead TA24 8HL
G1 OOB J. Clark The Bungalow, Sutton Lane, Walton, Street BA16 9RJ
G1 OOG A. Cooper Riverfield, Creek View Avenue, Hockley SS5 6LU
G1 OOJ M. Watson 15 Ellis Park, St. Georges, Weston-Super-Mare BS22 7FA
G1 OOM B. Woodcock 27 Main Street, Cosby, Leicester LE9 1UW
G1 OOS E. Rowthorn 4 Woburn Court, Rushden NN10 9HL
G1 OOU J. Pearce 19 Cawthorne Road, Kettlethorpe, Wakefield WF2 7HW
G1 OOW D. Streeter 78 Stockfield Road, Acocks Green, Birmingham B27 6BB
G1 OOZ D. Parsons 1 Carlyle Road, Rowley Regis B65 9BQ
G1 OPA D. Smith 15 Billington Close, Coventry CV2 5NQ
G1 OPD P. Elsom 7B Church Lane, Keelby, Grimsby DN41 8ED
G1 OPG K. Chappell 21 Victoria Street, Long Eaton, Nottingham NG10 3EW
G1 OPJ A. Golding Brown 17 Main Street, Withybrook, Coventry CV7 9LT
GM1 OPO G. Askew 49 Kittlegairy Road, Peebles EH45 9LX
G1 OPT P. Harvey 4 Linden Grove, Teddington TW11 8LT
G1 OPV P. Drew 20 Russell St., Accrington BB5 2NF
G1 OPW F. Cox 44 Mountain Wood, Bathford, Bath BA1 7SB
G1 OQB R. Tams 7 Hermitage Road, Abingdon OX14 5RN
G1 OQF C. Caines 6 Abel Smith Gardens, Branston, Lincoln LN4 1NN
G1 OQG D. Fryer 16 Elston Place, Aldershot GU12 4HY
G1 OQI C. Kill 169 Spring Road, Southampton SO19 2NU
G1 OQM M. Palmer 250 Kinson Road, East Howe, Bournemouth BH10 5EP
G1 OQO M. Williamson Greenfields Farm, Plumley Moor Road, Knutsford WA16 9SB
GM1 OQT J. Watson 64 Anstruther Street Law, Carluke ML8 5JG
G1 OQU A. Windsor 23 Pear Tree Way, Oakridge, Basingstoke RG21 5QE
G1 OQV J. Connor 28 Church Street, Hungerford RG17 0JE
G1 OQW A. Boot 63 Hunters Way, Stoke-on-Trent ST4 5EF
G1 ORB D. Tomsett 20 North Avenue, Bognor Regis PO22 6HG
G1 ORC Oldham Am Rad C c/o G. Oliver 158 High Barn St., Royton, Oldham OL2 6RW
G1 ORG D. Stanley 25 Kingsley Crescent, Bulkington, Bedworth CV12 9PS
G1 ORK D. Spicer 35 Strood Road, St. Leonards-on-Sea TN37 6PN
G1 ORL C. Barlow 16 Fosseway South, Midsomer Norton, Radstock BA3 4AN
G1 ORN F. Daniels 6 Middlemead, Stratton-On-The-Fosse, Radstock BA3 4QH
GW1 ORP J. Marlow West Bulthy, Bulthy, Welshpool SY21 8ER
G1 ORS B. Williams 3 Welton Close, Wilmslow SK9 6HD
G1 ORT Rvd. S. Smale The Old Vicarage, 68 Cardigan Road, Bridlington YO15 3JT
G1 OSA A. Sleigh 2 Rock Terrace, Buxton SK17 6HN
G1 OSE R. Howlett 37 Waveney Drive, Hoveton, Norwich NR12 8DP
G1 OSG W. Beilby 119 Beaconsfield, Withernsea, Nth Humberside HU19 2EW
G1 OSH G. Slater 12A Apsley Close, Bishop's Stortford CM23 3PX
G1 OSI J. Nicholson 117 Lower Meadow, Harlow CM18 7RF
G1 OSJ M. Howard 8 Abbotts Crescent, St. Ives, Huntingdon PE17 6YB
G1 OSL G. Hunter 2 Dilloway Street, St. Helens WA10 4LN
G1 OSO A. Durbridge 16 Nightingale Drive, Mytchett, Camberley GU16 6BZ
G1 OSP J. Woollons 28 Columbus Ravine, Scarborough YO12 7JT
GM1 OST T. Ferguson 40 Dallowie Road, Patna, Ayr KA6 7ND
G1 OTA D. Lee 25 Elm View, Steeton, Keighley BD20 6SZ
GW1 OTI K. Nickson 25 Burntwood Road, Buckley CH7 3EL
G1 OTN R. Weight 1 Crowland Road, Thornton Heath CR7 8RP
G1 OTZ J. Macdonald 42 Lion Lane, Haslemere GU27 1JD
G1 OUA H. Saunders 8 Norfolk Road, Luton LU2 0RE
G1 OUG K. Anderson 9 Bradford Park Drive, Bolton BL2 1PA
GW1 OUP D. George 24 Ty Fry Close, Brynmenyn, Bridgend CF32 8YB
G1 OUX P. Bolderson 113 Kirkdale Crescent, Leeds LS12 6AY
G1 OUY T. Willans 3 Highfield, Hatton Park, Warwick CV35 7TQ
G1 OVG T. Powell 11 Wymering Lane, Portsmouth PO6 3QT
G1 OVH N. Waud 32 Wellsfield, Huntingdon PE29 1LW
G1 OVK P. Beard 33 Sanctuary Close, Worcester WR2 5PY
G1 OVO R. Cox 60 Prospect Crescent, Whitton, Twickenham TW2 7EA
GM1 OVW R. Hetherington 37 Brockwood Avenue, Penicuik EH26 9AN
G1 OVY P. Mcclelland 30 Bowyer Road, Abingdon OX14 2EP
G1 OWD M. Forsyth 13 Hillside Close, Paulton, Bristol BS39 7PN
G1 OWI M. Branch 38 Kynaston Road, Didcot OX11 8HD
G1 OWJ A. Copsey 13 Monro Avenue, Crownhill, Milton Keynes MK8 0BB
G1 OWK G. Garner 8 Lansdowne Road, Swadlincote DE11 9DZ
G1 OWM G. Woodley 16 Albert St., St. Barnabas, Oxford OX2 6AY
G1 OWZ J. Scott 10 Beethoven Close Old Farm Park, Milton Keynes MK7 8PL
GM1 OXB D. Owen Ordiga, Wellheads, Clochan, , Buckie AB56 5HB
G1 OXF R. Lewis 17 Hollow Lane, Hayling Island PO11 9AA
G1 OXH A. Price 10 Low Meadow, Whaley Bridge, High Peak SK23 7AY
GW1 OXJ I. Jones 15 Victoria Road, Penygroes, Caernarfon LL54 6HD
G1 OXO J. Ilston 6 Dovedale, Canvey Island SS8 8HX
GM1 OXQ I. Mckune 16 Queensberry Court, Dumfries DG1 1BT
G1 OXT R. Faulkner 54 Clegg Hill Drive, Sutton-in-Ashfield NG17 2QA
G1 OYF V. Shirley 18 Crotch Crescent, Marston, Oxford OX3 0JJ
G1 OYG D. Crowe 18 Bengairn Avenue, Patcham, Brighton BN1 8RH
G1 OYH H. Grinter 16 Gladiolus Road, Langport TA109TA
G1 OYM A. Ingram 78 Kenwood Gardens, Gants Hill, Ilford IG2 6YG
G1 OYU B. Toon 2 Marstonlane Park, Rolleston On Dove, Staffs DE13 9BJ
G1 OYZ P. Smith 189 Rolleston Road, Burton-on-Trent DE13 0LD
G1 OZB A. Saunders 4, Severn Way, Bewdley DY12 2JQ
G1 OZD Dr J. Anderson 179 Rolleston Road, Burton-on-Trent DE13 0LD
G1 OZG D. Lewis 768 Rochdale Road Middleton, Manchester M24 2RF
G1 OZR I. Thomson 2 Casaubon Close, Dereham NR19 1EG
G1 OZV M. Jolly The Oaks, 6 Gwealhellis Warren, Helston TR13 8PQ
GW1 OZW B. Donovan Henysgol, Drope Road, Cardiff CF5 6EP
G1 PAF P. Foster 3 The Greenway, Ickenham, Uxbridge UB10 8LS
G1 PAK C. Parker 6 Chilham Close, Hemel Hempstead HP2 4UG
G1 PAT P. Chapman 24 Broad Lane, Moulton, Spalding PE12 6PN
GW1 PAV P. Grey 4 Lon Carreg Bica, Birchgrove, Swansea SA7 9QH
G1 PBB D. Smith 4 Field Rose Court Adlington, Chorley PR6 9SS
G1 PBF Marts c/o D. Pearce 32 Marshall Road, Willenhall WV13 3PB
G1 PBX E. Churchill 87 Bradley Crescent, Shirehampton, Bristol BS11 9SR
G1 PBY A. Parrott 54 Dockin Hill Road, Doncaster DN1 2QU
G1 PCA R. Blandford 16B Sherwood Road, Keynsham, Bristol BS31 1DB
GW1 PCD P. Dicken 8 Glan Llyn, Llanfairpwll LL61 5YX
G1 PCG J. Dunwell 8 Violet Grove, Thatcham RG18 4DQ
G1 PCN E. Benzie 6 Priors Park, Emerson Valley, Milton Keynes MK4 2BT
G1 PCQ East of England DX Group c/o A. Popplewell 38 Welbeck Street, Hull HU5 3SQ
G1 PCR C. Carter11 6 Turney Street, Aylesbury HP20 1AR
G1 PCU C. Mather 5 Knolles Road, Cowley, Oxford OX4 3HT
G1 PDA Dr E. Evans 4 Zig Zag Road, Wallasey CH45 7NZ
G1 PDS S. Cliffe 24 Dalehurst Road, Bexhill-on-Sea TN39 4BN
G1 PEE S. Jackson 71 Slyne Road, Bolton Le Sands, Carnforth LA5 8AQ
G1 PEI L. Berridge 33 Wesley Drive, Weston-Super-Mare BS22 7TJ
G1 PEK M. Sutton 17 Barton Close, Witchford, Ely CB6 2HS
GM1 PEL B. Taynton 32 Broomhall Road, Edinburgh EH12 7PD
G1 PER M. Payne 14 Linacres Drive, Chellaston, Derby DE73 6XH
G1 PEU G. Gibbons 43 Buckland Avenue, Basingstoke RG22 6JA
GW1 PFK A. Romano The Glen, Glen Road, Swansea SA3 5QJ
GM1 PFU C. Hewlett 6 Glenturret Terrace, Perth PH2 0AR
G1 PFY J. Gold 6 Woodland Avenue, Bournemouth BH5 2DJ
G1 PFZ W. Gill 2 Rufford Court, Rufford Avenue, Leeds LS19 7ED
G1 PGD K. Biggs 30 Elder Lane, Burntwood WS7 9BT
G1 PGH J. Baker 2882 Rikkard Dr, Thousand Oaks 91362 United States
G1 PGI C. Smith The Laurels, Maple Court, Rodmersham Green, Sittingbourne ME9 0LR
G1 PGJ D. Castle 8 Woodhall Court, Welwyn Garden City AL7 3TD
G1 PGN D. Brooks 10 Elmtree Road, Ruskington, Sleaford NG34 9BT
GM1 PGP R. Barbour 25/27 Drove Road, Langholm DG13 0JW
G1 PGQ R. Pennycook 28 Marine Court, Southsea PO4 9QU
G1 PGS D. Hands 45 Croft Avenue, West Wickham BR4 0QH
G1 PGV J. Davidson 5 Hanover Parc, Indian Queens, St. Columb TR9 6ER
G1 PGX M. Bingham 6 Bittern Close, Hull HU4 6SQ
G1 PHA G. Day 102 Meadlands Drive, Ham, Richmond TW10 7ED
GM1 PHD Dr N. Muir 25 Drylaw House Gardens, Edinburgh EH4 2UE
G1 PHJ H. Johnson 27 Ridgeway Avenue, Gravesend DA12 5BD
G1 PHK R. Baines 319 Pontefract Road, Featherstone, Pontefract WF7 5AB
G1 PHN Yeovil CG c/o M. Morley 8 The Becks, Alvechurch, Birmingham B48 7NE
G1 PHS P. Street 17 Roxby Gardens Thornton-Le-Dale, Pickering YO18 7SR
G1 PHU B. Burton Natson, Tedburn St. Mary, Exeter EX6 6ET
G1 PHV C. Marsh 85 Cromwell Crescent, Market Harborough LE169JW
G1 PIF D. Rogers 20 Chapel Close, Acomb, Hexham NE46 4RX
GW1 PIH B A E Systems Great Baddow ARC c/o H. Owen Llys Gwynedd, Bethel, Caernarfon LL55 1YB
G1 PII G. Cooper 39 Church Road, Harlington, Dunstable LU5 6LE
G1 PIN A. Clifford 397 Portland Road, Hove BN3 5SG
G1 PIX R. Bibby 40 Morval Crescent, Runcorn WA7 2QS
G1 PIY F. Cholerton 17 Stringer Crescent, Warrington WA4 1QN
G1 PJB T. Henderson 6 Sandford House, Sandford BH20 7DH
G1 PJC G. Siarey 23 Celsus Grove, Swindon SN1 4GE
G1 PJI S. Scott 11A Lodge Crescent, Orpington BR6 0QE
G1 PJJ S. Turvey 5 Ingestre Street, Harwich CO12 3JA
G1 PJK J. Foster 15 Parklands Way, Liverpool L22 3YX
GW1 PJL Dr D. Brookfield The Barns, Milwr, Holywell CH8 8HE
G1 PJM P. Mitchell 11 Wingle Tye Road, Burgess Hill RH15 9HR
G1 PJO K. Thompson 13 Kirby Walk, Bretton, Peterborough PE3 9UD
GW1 PJP P. Probert 7 Albany Road, Blackwood NP12 1DZ
G1 PJR J. Garrett 2 Wantsume Lees, Sandwich CT13 9JF
G1 PJT R. Wood Lynwood, Halley Road, Heathfield TN21 8TG
G1 PJV N. Saunders 24 Gateland Close, Haxby, York YO32 2ZZ
G1 PJZ J. Rogers 55 York Road, Driffield YO25 5AY
GM1 PKB W. Graham 98 Dalswinton Avenue, Dumfries DG2 9NR
G1 PKG A. Stockton 190 Sommerfield Road, Woodgate, Birmingham B32 3TA
GW1 PKM P. Miller Ddaugae Farm, Gwrhyd Road, Swansea SA9 2RY
GM1 PKN D. Gillies 10 Killeonan, Campbeltown PA28 6PL
G1 PKO C. Jones 52 The Drive, Bury BL9 5DL
G1 PKP M. Lloyd 243 Stand Lane, Radcliffe, Manchester M26 1JA
G1 PKR D. Bannister 11 Keats Drive, Swadlincote DE11 0DS
G1 PKV J. Sennitt 44 Pear Tree Avenue, Newhall, Swadlincote DE11 0NB
GW1 PKW P. Kingsley-Williams Banhadlen Uchaf, Back Road, Mold CH7 4QD
GW1 PLJ J. Morgan Holly Cottage, Old Racecourse, Oswestry SY10 7PQ
G1 PLU S. Goy 352 Chanterlands Avenue, Hull HU5 4ED
G1 PLV R. Cilia 18 London Fields House, Kensington Road, Crawley RH11 9NS
G1 PMA D. Harding 37 Junction Cottages, London Road, Hardham, Pulborough RH20 1LA
G1 PMF P. Felton 13 New St., Sudbury CO10 6JB
G1 PMJ D. Loon 18 Stourcliffe Road, Wallasey CH44 3AF
G1 PMK A. Mills 12 Sydney Street, Kimberley, Nottingham NG16 2LQ
G1 PML C. Roberts 21 Dryden Close, Grantham NG31 9QS
G1 PMU L. Ellis 4 Hunt Avenue Heybridge, Maldon CM9 4TY
G1 PNB S. Garwell 8 Shakespeare Road Prestwich, Manchester M25 9GW
G1 PNC T. Collins 1 Artillery Place, Hollyhedge Road, Manchester M22 4GG
GW1 PND K. Hassall 2 Hilton House, Steynton Road, Milford Haven, . Sa731Bd, Milford Haven SA73 1BD
G1 PNL D. Johnson 27 Ridgeway Avenue, Gravesend DA12 5BD
G1 PNX D. Staples 2 Bulcote Road, Clifton, Nottingham NG11 8FD
GM1 POA J. Jamieson 11 Binns Road, Glasgow G33 5HU
G1 POC C. Elsom 8 King Avenue, Maltby, Rotherham S66 7HX

G1 POD E. Mitchell 11 Wingle Tye Road, Burgess Hill RH15 9HR
G1 POJ K. Phillips Flat 18, Edith Ramsay House, 134 Duckett Street, London E1 4TD
G1 POM E. Baker 19 Ramsey Road, Thornton Heath CR7 6BX
G1 POR A. Porter 1125 Yardley Wood Road, Warstock, Birmingham B14 4LS
G1 PPB T. Roots 11 Windermere Avenue, Eastern Green, Coventry CV5 7GP
G1 PPD A. Shons 108 Southdown Road, Catherington, Waterlooville PO8 0NF
G1 PPG G. Joyner Valle De Los Nogales 609, Fraccionamiento Real Del Valle, Nuevo Leon ZP 66350 Mexico
G1 PPK J. Knight 183 Northumberland Avenue, Thornton-Cleveleys FY5 2JS
G1 PPO A. Wade 40 Throxenby Lane, Scarborough YO12 5HW
G1 PPQ S. Walker 22 Ward Close, Aylestone, Leicester LE2 8NJ
G1 PPU D. Walker 115 Kilby Road, Fleckney, Leicester LE8 8BP
G1 PPX G. Nicholls 2 Leybrook Croft, Hemsworth, Pontefract WF9 4JA
G1 PPZ A. Parton 17 Causey Farm Road, Halesowen B63 1EQ
G1 PQJ P. Wilson Orchard Cottage, Rectory Road, Norwich NR14 8HT
G1 PQK P. Harrison 20 Priory Road, Stanford-le-Hope SS17 7EW
G1 PQO S. Stevens 49 The Beeches, Upton-Upon-Severn, Worcester WR8 0QQ
G1 PQR J. Haynes 6 Gavin Close, Kettering NN16 9GA
G1 PQT J. Mayes 44 Foxwarren, Claygate, Esher KT10 0JZ
G1 PQX R. Moat 27 Pioneer Road, Dover CT16 2AR
G1 PQY G. Jones 25 Myvod Road, Wednesbury WS10 9BT
G1 PRF S. Haden 33 Poplar Avenue, Chelmsley Wood, Birmingham B37 7RD
G1 PRH M. Watkins 7 Sand End, Whitstable CT5 4TH
G1 PRL R. Williams 54 Windways, Little Sutton, Ellesmere Port CH66 1JF
G1 PRM A. Webber 37 Rasen Road, Tealby, Market Rasen LN8 3XL
G1 PRP A. Moore Wyndrush, Northend Lane, Southampton SO32 3QN
G1 PRS C. Ladley 25 Laburnum Crescent, Louth LN11 8SG
G1 PRW A. Swift 38 Knightsbridge Way, Stretton, Burton-on-Trent DE13 0WJ
G1 PRZ B. Saich 65 Orchard Rise West, Sidcup DA15 8TA
G1 PSH L. Walton 2 Church Road, Colmworth, Bedford MK44 2JX
G1 PSL C. Sparrow 112 Hill Cot Road, Bolton BL1 8RW
G1 PSS A. Laszkiewicz 13 Darwall Drive, Ascot SL5 8NB
GM1 PST P. Stanhope The Roundal, Alva FK12 5HU
GM1 PSU I. Manson 25 Etna Court, Armadale, Bathgate EH48 2TD
GW1 PSW A. Bunting 11 Lon Y Gaer, Deganwy, Conwy LL31 9RG
GM1 PSZ D. Liddle 9 Rullion Road, Penicuik EH26 9HS
G1 PUK Gm Flora And Fauna c/o D. Jones 100 Cop Lane, Penwortham, Preston PR1 0UR
G1 PUO D. West 30 Farm Avenue, Swanley BR8 7JA
G1 PUQ C. Tripp Kingshill House, Church Street, Keinton Mandeville TA11 6ER
GM1 PUR D. Wood 50 Riverside Road, Eaglesham, Glasgow G76 0DG
G1 PUU S. Pantall 27 Woodlands Drive, Foston, Derby DE65 5DL
G1 PUV C. Wiseman 42 Merlin Way, Kidsgrove, Stoke-on-Trent ST7 4YL
G1 PUY J. De Renzi Bankside, South Newington, Banbury OX15 4JE
G1 PUZ S. Box 103 Ilkeston Road, Bramcote, Nottingham NG9 3JT
G1 PVA G. Kemp 5 Gosselin Street, Whitstable CT5 4LA
G1 PVD H. Murray 93 Burridge Road, Burridge, Southampton SO31 1BY
GI1 PVE R. Law 147 Lone Moor Road, Londonderry BT48 9LA
GW1 PVN P. Jones 72 Lon Maesycoed, Newtown SY16 1QQ
G1 PVR R. Kelly Hogbrook Farm, Banbury Road, Leamington Spa CV33 9QL
G1 PVT D. Broad 14 Albion Road, Westcliff-on-Sea SS0 7DR
G1 PVZ J. Vincent Brookfield, North Street, Crewkerne TA18 7AX
G1 PWF D. King 79 Wootton Drive, Hemel Hempstead HP2 6LA
G1 PWH South Normanton, Alfreton & District ARC c/o A. Jones 16 Collumbell Avenue, Ockbrook, Derby DE72 3TF
GM1 PWL K. Robertson 7 Meadows Crescent, Lochgilphead PA31 8AG
G1 PWM J. Bulman 3 South View, Littlethorpe, Ripon HG4 3LL
G1 PWO P. Owens Flat 1, 45 The High Street, Enfield EN3 4EF
G1 PWS R. Manson Smavollen 11, Stavanger 4017 Norway
G1 PWU D. Dwyer 24 Alder Way, Melksham SN12 6UL
G1 PWY P. Gardner 2 South Road, Morecambe LA4 5RA
GW1 PXM Dr R. Blakeway Ty Nantglyn, Glascwm, Llandrindod Wells LD1 5SE
G1 PXQ A. Scivetti 10 Rippleside, Basildon SS14 1UA
G1 PXW P. Hollands 16 Hazel Crescent, Thornbury, Bristol BS35 2LX
GI1 PXX S. Argue 2 Lisnisky Walk, Portadown, Craigavon BT63 5FY
GW1 PYY A. Thomas 22 Sea Road, Abergele LL22 7BU
G1 PZA W. Grech-Cini Byram Garnge, Great North Road, Byram WF11 9PA
G1 PZD M. Pattison 8 Selsey Drive, Luton LU2 8HZ
G1 PZP C. Collett Yaffle, 26 Hertford Road, Hoddesdon EN11 9JR
G1 PZS J. Alltimes 9 Wrexham Road, Romford RM3 7YX
GM1 PZT G. Hardacre 242 Sutherland Way, Knightsridge, Livingston EH54 8JB
GI1 RAA Houghton ARC c/o T. Hourican 43 Burren Road, Warrenpoint, Newry BT34 3SA
G1 RAE M. Buckley-Brown 2 Brothertoft Road, Boston PE21 8HD
G1 RAF Royal Air Force Halton RS c/o A. Mockford 58 Wendover Heights, Wendover HP22 6PH
G1 RAG J. Jones 10 Huntington Close, Redditch B98 0NF
G1 RAO K. Hughes 20 Pickering Close, Bury BL8 1UE
G1 RAP R. Prosser 27 Dorset Gardens, Rochford SS4 3AH
G1 RAX D. Bendall Brambles, 17 Berryfield Road, Lymington SO41 0HQ
G1 RBA A. Wheatley 3 Woodsbank Terrace, Wednesbury WS10 7RQ
G1 RBH Buxton Radio Amateurs c/o K. Dunn 65 Lime Street, Sutton-in-Ashfield NG17 4GA
GI1 RBI W. Mckeown 15 Laragh Lee, Ballycassidy, Enniskillen BT94 2JA
G1 RBO A. Thompson 3 Jenkins Drive, Elsenham, Bishop's Stortford CM22 6JP
GM1 RBQ R. Johnson Wester Balquhandy House, Dunning, Perth PH2 0RB
G1 RBX R. Baker 3 Hazelton Close, Solihull B91 3GA
G1 RBY P. Hurp 55 Brooklyn Grove, Coseley, Bilston WV14 8YH
G1 RBZ G. Norris 26 Westwood Road, Leyland PR25 3NS
G1 RCE A. Hills 12 Heathway, Chaldon, Caterham CR3 5DL
G1 RCI P. Mason 34 Central Park Avenue, Wallasey CH44 0AQ
G1 RCN P. Wilson 146 Wilkinson Street, Nottingham NG8 5FJ
G1 RCV Cray Valley RS c/o A. Styles 6 Hill Brow, Crayford, Dartford DA1 3NX
G1 RCW J. Chetwynd 35 Cordelia Close, Dibden, Southampton SO45 5UD
G1 RCX P. Tweney 9 Dovehouse Close, Eynsham, Witney OX29 4EW
GM1 RDG J. Horsburgh 35 Watson Street, Aberdeen AB25 2QB
G1 RDJ C. Price 44 Poplar Road, Stourbridge DY8 3BD
G1 RDU P. Fanning 9 Fishermans Walk, Shoreham-by-Sea BN43 5LW
G1 RDX S. Newbold 7 Rookery Meadow Holmer Green, High Wycombe HP15 6XF
G1 REO J. Telford 85 Medway, Great Lumley, Chester le Street DH3 4HU
G1 RET R. Taylor 3 Solent View, Calshot, Southampton SO45 1BH
G1 RFB R. Palmer 55 Edinburgh Road, Freshwater PO40 9DL
G1 RFC R. Armitage 15 Northolmby St., Howden, Goole DN14 7JL
G1 RFI A. Venables 10 Tilehouse, Redditch B97 4PL
G1 RFQ D. Jenks C.A.T. Services Adva Andalucia 1, la Lafoquia 4661 SPAIN
G1 RFS T. Niner 281 Nightingale Road, Edmonton, London N9 8QL
G1 RFX K. Tonner Millstream Cottage, Golden Valley, Malvern WR13 6AA
G1 RGG P. King 124 Henley Grove Road, Rotherham S61 1RY
GM1 RGM D. Keddie 26 Daleally Crescent, Errol, Perth PH2 7QA
G1 RGT G. Williams 76 Eastern Avenue, Pinner HA5 1NJ
G1 RHB K. Sears 10 Canberra Crescent, Brooknby LN86ER
G1 RHE W. Berry The Bungalow, Basil Road, Kings Lynn PE33 9RP
G1 RHW T. Ibbitson 36 Knoll Park, East Ardsley, Wakefield WF3 2AX
GI1 RIB B. Rafferty 81 Mullaghmore Drive, Omagh BT79 7PQ
GM1 RIG I. Mcgowan Feddal Lodge, Braco, Dunblane FK15 9RA
G1 RIR L. Shears 7 Lower Furlongs, Brading, Sandown PO36 0DX
G1 RIV T. Dyson 39 Newcastle Close, Stevenage SG1 4TL
G1 RIX Dr I. Steele 10 Manor Road, Irby, Wirral CH61 4UA
G1 RIY M. Gray Brook House, Brandside, Buxton SK17 0SG
G1 RJA M. Johnson 28 Bittles Green, Motcombe, Shaftesbury SP7 9NX
G1 RJD Ilera c/o W. Blower 129 Kingsway, Kirkby-In-Ashfield, Nottingham NG17 7FH
G1 RJN S. Hall Little Dene, Eastwick Road, Leatherhead KT23 4BJ
GM1 RJS J. Hambrook 32 Blackdales Avenue, Largs KA30 8HU
G1 RJW R. Wicks 32 Shelley Close, Northcourt, Abingdon OX14 1PR
G1 RKD B. Allport 1 Percy Drive, Swarland, Morpeth NE65 9JN
GM1 RKI A. Morris 39 Old Town, Peebles EH45 8JE
G1 RKJ R. Wilson 107 Hamilton Avenue, Uttoxeter ST14 7FE
G1 RKR R. Luker Grantham Cottage, Haywards Heath Road, Lewes BN8 4DS
G1 RLA R. Hall Bliss Lodge, Worcester Road, Chipping Norton OX7 5XS
G1 RLB R. Lawson 28 Hallett Way, Bude EX23 8PG
G1 RLD D. Beeton 50 Hanson Avenue, Shipston-on-Stour CV36 4HS
G1 RLF R. Walter 10 Birch Meadow, Clehonger, Hereford HR2 9RH
G1 RLI P. Webb 41 Lancaster Gardens, Wolverhampton WV4 4DN
G1 RLK B. Ian 35 Stanley Road, Heysham, Morecambe LA3 1UR
G1 RLR P. Pedley 24 Appledore Road, Walsall WS5 3DT
G1 RLT P. Vipond The Old Forge, Nentsbury, Alston CA9 3LH
GM1 RLV D. Mackay Burnlea, Harrapool Broadford, Broadford IV49 9AQ
G1 RMC Farnborough & District c/o I. Jackson 5 Vivien Close, Chessington KT9 2DE
G1 RMN M. Richards 20 Tas Combe Way, Willingdon, Eastbourne BN20 9JA
G1 RNL M. Kinsella The Nook, Eaudyke Road, Boston PE22 8RU
G1 RNV B. Purse 28 Holford Road, Guildford GU1 2QF
G1 RNY P. Roe Kalmia, Sheffield Park, Uckfield TN22 3RA
G1 RNZ G. Saville 4 Shannon Court, Downs Barn, Milton Keynes MK14 7PP
GM1 ROB R. Rimmer Glenalty Cottage, Barrhill KA26 0QT
G1 ROD I. Lupton 19 Avenue Close, Harrogate HG2 7LJ
GW1 ROE I. Roe Tyddyn Berth, Chwilog, Pwllheli LL53 6RQ
G1 ROH D. Gower 68 Wood Common, Hatfield AL10 0UB
G1 ROK P. Court Hamara, Shortlands Grove, Bromley BR2 0LS
GM1 ROL S. O'Connor 2 Latch Farm Cottages, Off Leyden Road, Kirknewton EH27 8DQ
GM1 ROM D. O'Connor 2 Latch Farm Cottages Off Leyden Road, Kirknewton EH27 8DQ
G1 RON F. Donnachie 2 The Mall, Patrington Haven Leisure Park, Patrington HU12 0PT
GM1 ROX A. Donald South Sandlaw House, Alvah, Banff AB45 3UD
G1 RPE I. Smith 66 Aire Road, C/O Beech Croft, Wetherby LS22 7UE
G1 RPO C. Reed Colins, Throcking Road, Cottered, Buntingford SG9 9RA
G1 RPP C. Broadbent 7 Wharfe Park, Addingham, Ilkley LS29 0QZ
G1 RPT M. Tribe 11 Heathlands Close, Crossways, Dorchester DT2 8TS
G1 RPV R. Hardiman 27 Staithe Road, Martham, Great Yarmouth NR29 4PT
GM1 RQD D. Marwick 17 Laverock Road, Kirkwall KW15 1EE
G1 RQI P. Quirk 75 Harcourt Road, Folkestone CT19 4AF
G1 RQK S. Bland 6 Bryden Close, Northallerton DL6 1SF
GW1 RQM M. Aquilina 3 Aldergrove Close, Port Talbot SA12 8EY
G1 RRE J. Eckersley 88 New Heys Way, Bradshaw, Bolton BL2 4AQ
G1 RRG H. Johnstone 16 Riverside Crescent, Otley LS21 2RS
GM1 RRJ B. Elliott 195 Braehead Road, Cumbernauld, Glasgow G67 2BL
G1 RRR K. Bareham 19 Northfield Road, Ringwood BH24 1LS
G1 RRU P. Atkinson 19 Haggar Street, Wolverhampton WV2 3ET
G1 RRW A. Henderson Eastbury Farm House, Tarrant Gunville, Blandford Forum DT11 8JQ
G1 RSC S. Hall 17 Nevill Road, Rottingdean, Brighton BN2 7HH
G1 RSE J. Rod 42 Westwood Avenue, Ferndown BH22 9HN
G1 RSF A. Dean Les Monneries, Combieres, Charente France
G1 RSK C. Broughton 65 Manby Road, Immingham DN40 2SG
GI1 RSR C. Fogarty 46 Knockview Drive, Tandragee, Craigavon BT62 2BH
G1 RTW G. White 101 London Road, Hailsham BN27 3AH
G1 RTX I. Poole 8 Bates Close, Higham Ferrers, Rushden NN10 8HF
G1 RUE D. John 28 Cliff Road Winteringham, Scunthorpe DN15 9NQ
G1 RUG A. Kay Pear Tree Cottage, Hale House Lane, Farnham GU10 2JG
G1 RUL A. Cake 8 Carrick Close, Dorchester DT1 2SB
GW1 RVC R. Baker 24 Nant Road, Connah's Quay CH5 4AL
G1 RVF B. Dempster 5 Church Walk, Bozeat, Wellingborough NN29 7ND
G1 RVH C. Stancer 20 Overton Avenue, Willerby, Hull HU10 6AR
G1 RVK I. Brooks 10 Foxgloves, Deeping St. James, Peterborough PE6 8SH
GW1 RVP P. Scott 22 Powys Road, Llandudno LL30 1HZ
G1 RVT B. Kavanagh 73 Esh Wood View, Ushaw Moor, Durham DH7 7FD
G1 RWR S. Lee 7 Ridge Way Close, Rotherham S65 3NH
G1 RWT K. Whitton 11 Dursley Road, Shirehampton, Bristol BS11 9XB
G1 RWX K. Ikin 15 Broadway, Farnworth, Bolton BL4 0HQ
GI1 RXL C. O'Connell 15 Grange Road, Coleraine BT52 1NG
GI1 RXM A. Murphy 3 Church Lane, Crossgar, Downpatrick BT30 9PX
G1 RXV S. Mugele 19 Ambassador, Bracknell RG12 8XP
G1 RYF M. O'Callaghan 79 Kingsfield Avenue, Harrow HA2 6AQ
G1 RYM R. Knox 13 Grosvenor Road, Billingham TS22 5HA
G1 RYQ S. Marshall 15 Moulton Close, Belper DE56 0EA
G1 RYS I. Mcculloch11 5 Knighthead Point, The Quarterdeck, London E14 8SR
G1 RYY R. Howse 37 Great Eastern Road, Hockley SS5 4BX
GW1 RZE P. Morgan Flat 24, Ynysderw House, Swansea Road, Swansea SA8 4AA

G1 RZJ L. Ward 19 Spring Close View, Sheffield S14 1RJ
G1 RZZ R. Wood 40 Ashville Gardens, Pellon, Halifax HX2 0PL
G1 SAK M. Weatherley 95 Cambalt Road, Putney, London SW15 6EX
GW1 SAM A. Hodgkinson 64 Rhodfa Wen, Llysfaen, Colwyn Bay LL29 8LE
G1 SAT Open University ARC c/o R. Smith 32 Wolseley Gardens, London W4 3LR
GM1 SBD M. Angiolini Innis Chonain, Mill Road, Stirling FK7 9LP
G1 SBK J. Spink Highfields, Church Lane, Leeds LS16 8DE
G1 SBN J. Davison 29 Glenfield Avenue, Wetherby LS22 6RN
G1 SBW I. Case 4 Portside, Preston Brook, Runcorn WA7 3LE
G1 SBZ T. Lumley 32 Downland Road, Woodingdean, Brighton BN2 6DJ
G1 SCA S. Allen 28 Neville Road, Luton LU3 2JJ
G1 SCB K. Gray Donkleywood House, Donkleywood, Hexham NE48 1AQ
G1 SCL N. Stackhouse 16 Tintern Avenue, Urmston, Manchester M41 6FJ
G1 SCN Co-Antrim A.R.S DX Group c/o C. Taylor 4 Tunnel Road, Beaminster DT8 3BQ
G1 SCO R. Brown 52 Challenger Drive, Sprotbrough, Doncaster DN5 7RY
G1 SCQ K. Kent 5 Jubilee Road, Heacham, King's Lynn PE31 7AR
G1 SCR Shropshire Raynet c/o M. Jones 35 Pendle Way, Meole Brace, Shrewsbury SY3 9QS
G1 SCT M. Lane Cherry Tree House, Pipwell Gate Saracens Head, Holbeach, Spalding PE12 8BA
G1 SCV Leicester Raynet Group c/o A. Faulkner Northwood, Cranham, Gloucester GL4 8HB
G1 SCY F. West 9 Tregarland Close Coads Green, Launceston PL15 7NE
G1 SDJ C. Cooper Tapshays Cottage, Burton Street, Sturminster Newton DT10 1PS
G1 SDK S. Karpasitis Riverdene Blythe Road, Hoddesdon EN11 0BB
G1 SDN R. Mordue 29 Sycamore Close, Witham CM8 2PE
G1 SDX G. Taylor 3 Erica Drive, Torquay TQ2 8LP
G1 SEA D. Tucker 12 Chatsworth Way, Heanor DE75 7TJ
G1 SEF A. Fearnley 1 Dover Road, London E12 5DZ
G1 SEH South East Hampshire Raynet c/o J. Woonton 44 Jubilee Road, Southsea PO4 0JE
G1 SEO M. Lines 158 Nine Mile Ride, Finchampstead, Wokingham RG40 4JA
G1 SES M. Halden 19 Fenwick Lane Halton Lodge, Runcorn WA7 5YU
G1 SEW Lrgs ARC c/o A. Goatman 150 Merlin Park, Portishead, Bristol BS20 8RW
G1 SFU D. Coffey 14 Shawbridge, Harlow CM19 4NJ
G1 SGA R. Blunt 8 The Crescent, Wolverhampton WV6 8LA
GW1 SGE A. Morgan 1, Fryn Hir, Capel Dewi Road, Capel Dewi, Carmarthen SA32 8AY
GW1 SGG Lichfield A.R.S c/o M. Jones 48 Maes Alltwen, Dwygyfylchi, Penmaenmawr LL34 6UA
GW1 SGH R. Thorne 6 Cromwell Avenue, Rhyddings, Neath SA10 8DW
G1 SGM R. Parkin Craigside, 15 Holly Drive, Leeds LS16 6EF
G1 SGP N. Barnes 44 Cromford Road, Wirksworth, Matlock DE4 4FR
G1 SGR N. Marsh 16 Daytona Quay, Eastbourne BN23 5BN
G1 SGS B. Kimber 4 Nautilus Drive, Minster On Sea, Sheerness ME12 3NJ
G1 SGZ P. Gamble 9 Windmill Close, Ockbrook, Derby DE72 3TE
G1 SHH A. Compton Fairlight, 25 Framfield Road, Uckfield TN22 5AH
G1 SHI M. Kuik 196 Prestbury Road, Macclesfield SK10 3BS
G1 SHN G. Richardson 12 Northenhay Walk, Morden SM4 4BS
G1 SHT D. Wooster 34 New Road, Penn, High Wycombe HP10 8DL
G1 SHU N. Carr 7 Pear Tree Drive, Sedgeberrow, Evesham WR11 7GQ
G1 SID C. Siddons 423 London Road, Grays RM20 4AB
G1 SIG B. Scholte 266 Hednesford Road, Heath Hayes, Cannock WS12 3DS
G1 SIM Hounslow A.R. Instruction Centre c/o S. Hallam 46 Holte Road, Atherstone CV9 1HN
G1 SIO J. Robson Ealands, The Stanners, Corbridge NE45 5BA
G1 SIP R. Webb 54 Ashby Avenue, Chessington KT9 2BU
G1 SIU A. Bingley 51 Kirkby Folly Road, Sutton-in-Ashfield NG17 5HP
G1 SIX R. Foster 70 Mansfield Road, Edwinstowe, Mansfield NG21 9NH
G1 SJB C. Thompson 135 Stafford Road, Bloxwich, Walsall WS3 3PG
G1 SJD Lord N. Peirce 78 De Tracey Park, Bovey Tracey, Newton Abbot TQ13 9QT
G1 SJG E. Boydon 56 Oliver Leese Court, Ten Butts Crescent, Stafford ST17 9HP
G1 SJM Dr S. Mitchell 56 Mill Green Road, Amesbury, Salisbury SP4 7RE
G1 SJO A. Banthorpe 32 Long Close, Station Road, Henlow SG16 6JS
G1 SJT A. Long 23 Beech Road, Sutton Weaver, Runcorn WA7 3ER
G1 SJU D. Maciver 176 Burges Road, London E6 2BS
G1 SJZ P. Randall 7 Eastbourne Avenue, Featherstone, Pontefract WF7 6LQ
G1 SKE D. Turner 26 Carlton Way, Cleckheaton BD19 3DG
G1 SKI K. Weaver 26 Southdown Close, Haywards Heath RH16 4JR
G1 SKQ J. Forster 4 Rydal Road, Lemington, Newcastle upon Tyne NE15 7LR
G1 SKV B. Abell 5 Aldborough House Brook Street, York YO317QQ
G1 SKW R. Sidwell 81 Oakengates Road, Donnington, Telford TF2 7LQ
G1 SLA C. Baker 17 Dawlish Avenue, Chadderton, Oldham OL9 0RF
G1 SLE R. Drabble 37 Barton Street, Clowne, Chesterfield S43 4RS
G1 SLG M. Butcher 4 Sotheby Rise , Ecton Brook, Northampton NN3 5AD
G1 SLI B. Elliott 51 Allerhope, Hall Close Grange, Cramlington NE23 6SX
G1 SLO K. Turner 155 Bure Lane, Christchurch BH23 4HB
G1 SLP E. Methven 10 Woodbine Road, Durham DH1 5DR
G1 SLU K. Ford 123 Stockwood Lane, Bristol BS14 8SZ
G1 SMB M. Chitty Timbercroft, Faris Lane, Addlestone KT15 3DL
G1 SMC S. Mccloy 28 Ferndown Drive, Godmanchester, Huntingdon PE29 2LU
GW1 SMG A. Walker Blaenlluest Oakford, Llanath SA470RT
GW1 SMJ F. Beavan Uplands, Bronllys, Brecon LD3 0HN
G1 SMP P. Devlin Apartment 37, The Old Picture House, Tamworth Street, Lichfield WS13 6FL
GW1 SMT C. Manning Walker Road, Cardiff CF24 2EL
G1 SMY S. Fisher 19 Sandown Road, Sandown PO36 9JL
G1 SNI P. Smith 27 Briar Close, Gillingham SP8 4SS
G1 SNO D. Tubb 42 Hill Farm Road, Marlow SL7 3LU
G1 SNQ I. Boss 11 Penkridge Road, Church Gresley, Swadlincote DE11 9FH
G1 SNU D. Tunbridge 12 Burnham Road, Latchingdon, Chelmsford CM3 6EU
G1 SOB T. Yetton 7 Warwick Close, Canvey Island SS8 9YB
G1 SOG R. Stearn 18 Kings Avenue, Chippenham SN14 0UJ
G1 SOM Taunton & Somerset Raynet Group c/o D. Smith 47 Laburnum St., Taunton TA1 1LB
G1 SOX W. Stennett 26 Moorfield Road, St. Giles-On-The-Heath, Launceston PL15 9SY
G1 SOY J. Moseley 72 Wisden Road, Stevenage SG1 5JA
G1 SPA D. Harrison 7 Shirley Close, Castle Donington, Derby DE74 2XB
G1 SPJ A. Gibbs 86 Broadmark Road, Slough SL2 5PN
G1 SPM A. Hughes 8 Rigby Grove, Little Hulton, Manchester M38 0FQ
G1 SPT S. Tatem 55 Chelwood Road, Chellaston, Derby DE73 5SJ
G1 SPU A. Burnett 16 Shielding Way, Stafford ST16 3WG
GW1 SPW B. Saunders Bishops Mill, Llanwrda SA19 8AD
G1 SPX K. Shires 19 Prince Charles Avenue, Sittingbourne ME10 4NA
GM1 SQA J. Yates 67 Closeburn, Thornhill DG3 5HR
G1 SQC K. Boote 51 Sunnyfield Oval, Stoke-on-Trent ST2 7PA
G1 SQG P. Dennis Fuchsia House, 18 West View Road, Yelverton PL20 7DD
G1 SQI J. Bewley 21 Duloe Gardens, Pennycross, Plymouth PL2 3RS
G1 SQW R. Rawson 43 County Road North, Hull HU5 4HN
GM1 SQZ G. Pocock 1 Pitcairn Grove, East Kilbride, Glasgow G75 8TN
G1 SRA T. Binns Cross Farm, 1. Cross Lane Oxenhope, Keighley BD22 9LE
GW1 SRB K. Gough 2 Church Road, Abertridwr, Caerphilly CF8 4DL
G1 SRD P. Foulds 7 Bridge Road, Little Sutton, Spalding PE12 9EG
GM1 SRP J. Mcculloch Wester Curr Cottage, Dulnain Bridge, Grantown-on-Spey PH26 3LX
GM1 SRR M. Christmas Lindens, Smithy Loan, Dunblane FK15 0HQ
G1 SSL M. Belcher 52 Kynaston Road, Didcot OX11 8HD
G1 SSS Capt. J. Banfield 2 Laleham Close, Eastbourne, East Sussex BN21 2LQ
G1 SSZ F. Bowhill 78 East Gomeldon Road, Gomeldon, Salisbury SP4 6NB
G1 STK A. Goddard 65 Langley Hall Road, Solihull B92 7HE
G1 STP V. Trolan Hilbre, East Taphouse, Liskeard PL14 4NJ
G1 STQ J. Taylor C/O 22 Welland Grove, Newcastle ST5 4EP
G1 STW R. Gerrard 7 Wisteria Drive, Healing, Grimsby DN41 7JB
G1 SUH J. Speakman 130 Dicconson Street, Wigan WN1 2BA
GW1 SUK A. Dimmock Gwyndy, Llandegfan, Menai Bridge LL59 5PW
G1 SUM T. Cull 25 Queensway, Ponteland, Newcastle upon Tyne NE20 9RZ
GJ1 SUP T. Stone Les Quatres Saisons La Route De La Porte, St. John JE34DE Jersey
G1 SUS B. Cabay Suite 9559-488 Unit 9, Skyport Drive, West Drayton UB7 0LJ
G1 SVD R. Roper 57 Burnt Hills, Cromer NR27 9LW
G1 SVI T. Metcalfe 38 Station Road, Branston, Lincoln LN4 1LH
G1 SVJ C. Murphy 13 Northfield Road, Ringwood BH24 1LS
G1 SVL K. Stocker 22 Hadlow Down Close, Luton LU3 2PY
G1 SVN M. Churchman Westcroft, Church Road, Chester CH4 9NG
G1 SVP B. Howard 15 Four Acres, Bideford EX39 3RW
GM1 SVQ C. Pringle 12 Atkinson Road, Dumfries DG2 7DH
G1 SVR Durham And District ARS c/o E. Churchyard 11 Greenfields Drive, Bridgnorth WV16 4JW
GW1 SVV D. Osborne 22 Springfield Gardens, Hirwaun, Aberdare CF44 9LY
G1 SWE M. Foster 15 Parklands Way, Liverpool L22 3YX
G1 SWF S. Roberts 43 Lawn Close, Ruislip HA4 6ED
G1 SWH G. Schoof 6 Canal Row, Haigh, Wigan WN2 1NA
G1 SWI B. Gillett 18 Rookery Close, Fenny Drayton, Nuneaton CV13 6BB
G1 SWK T. King 10 Berkeley Close, Ipswich IP4 2TS
G1 SWP S. Pitt 105 Harebell Way, Lowestoft NR33 8EX
G1 SWR Warwickshire Avon Raynet Group c/o C. Ousbey 30 Hawthorn Way, Shipston-on-Stour CV36 4FD
G1 SWS S. Walters-Smith 83 Chesterfield Road, Tibshelf, Alfreton DE55 5NJ
G1 SWU G. Denham 36 Redstone Farm Road, Hall Green, Birmingham B28 9NT
G1 SWX C. Harrap Waverley Court, 22 Forth Avenue, Portishead, Bristol BS20 7NY
G1 SWZ R. Wright 61 Quarry Road, Hurtmore, Godalming GU7 2RW
G1 SXB G. Whetstone 60 Worple Road, Staines TW18 1EE
GM1 SXJ P. Duckles 26 Meadowpark Road, Bathgate EH48 2SJ
GW1 SXT M. Kerry 40 Oaklands Road, Sebastopol, Pontypool NP4 5BZ
GW1 SXU P. Janes 19 Fair View, Chepstow NP16 5BX
GM1 SXX Mid Warwickshire ARS c/o A. Copland 74 Whitehaugh Avenue, Paisley PA1 3SR
G1 SXY Y. Entwistle Sonunda House, Church Lane, Freckenham, Bury St. Edmunds IP28 8JF
GM1 SYC W. Graham 7 Brunt Place, Dunbar EH42 1RT
GI1 SYM G. Thompson 57 Rosepark, Donaghadee BT21 0BN
G1 SYP M. Thornton 58 The Green View, Shafton, Barnsley S72 8PW
G1 SYU M. Oubridge 54 Cantle Avenue, Downs Barn, Milton Keynes MK14 7QS
G1 SYV B. Goodier 42 Manchester Road, Clifton, Swinton, Manchester M27 6WY
G1 SYZ D. Setterfield 6 Murrayfield Close, Plymouth PL2 3FB
GI1 SZC Dr D. Mcmanus 38 Deanfield, Bangor BT19 6NX
G1 SZD T. Trengove 8 Kemp Close, Truro TR1 1EF
G1 SZK R. Frost 68 Wessex Road, Didcot OX11 8BP
GM1 SZM W. Robertson 28 Dewars Avenue, Kelty KY4 0BG
G1 SZT W. Dowkes Woodlea, Gillamoor Road, York YO62 6EL
G1 TAI P. Gabel 4 Blacksmiths Green, Shutlanger, Towcester NN12 7RS
G1 TAR C. Anderton 5 Leyland Avenue, Hindley, Wigan WN2 3SB
G1 TAU J. Drewry 10 Drayton Drive, Heald Green, Cheadle SK8 3LF
G1 TAY A. Shearer 101 Millside, Stalham, Norwich NR12 9PB
G1 TAZ F. Perkin 5 Highgrove, Trevadlock Hall Park, Launceston PL15 7PW
G1 TBE J. Kelday 20 Lowfield, Eastfield, Scarborough YO11 3LQ
G1 TBI W. Monk Brook House, River View, Buxton SK17 8SW
G1 TBK D. Watson 72 Dawes Avenue, West Bromwich B70 7LS
G1 TBN A. Malhi 517 Clifton Drive North, Lytham St. Annes FY8 2QX
G1 TBT D. Taylor Top Wath Laer, Top Wath Road, Harrogate HG3 5PG
GM1 TBW A. Napier Jehrada Cottage, Longhaven, Peterhead AB42 0NY
G1 TBX G. Williams 16 Coppice Road, Talke, Stoke-on-Trent ST7 1UB
G1 TCH C. Hinton 5 Vale Rise, Matlock DE4 3SN
G1 TCK D. Adams 28 Greenside, Stoke Prior, Bromsgrove B60 4EB
GM1 TCN T. Curran Miltonbank, East Cottage, Forfar DD8 3TU
GM1 TCP J. Campbell 16 Barony Road, Auchinleck, Cumnock KA18 2LL
G1 TDL M. Mundy The Homestead, Homestead Lane, Burgess Hill RH15 0RQ
G1 TDN T. Collinson 26 Westway Avenue, Hull HU6 9SA
G1 TDO F. Sanchez-Garci 74 Gorthorpe, Hull HU6 9EZ
G1 TDP J. Spry 21 Christchurch Gardens, Waterlooville PO7 5BT
GM1 TDT D. Robertson 131 Foxbar Road, Paisley PA2 0BD
GM1 TDU J. Rooney Rob Roy, Kinneff, Montrose DD10 0TG
GW1 TDV A. Potts 4 Bloomfield Close, Newport NP19 9ET
G1 TEX N. Swann 9 Alexandra Road, Parkstone, Poole BH14 9EL
GW1 TFB A. Hughes 38 Llys Dyffryn, St. Asaph LL17 0SX
GI1 TFC R. Mcmaster 43 Craigs Road, Carrickfergus BT38 9RL
GM1 TFF I. Hynd 48 Ben Ledi Crescent, Cumbernauld, Glasgow G68 9NG
GW1 TFL J. Robson 16 Dunraven Road, Sketty, Swansea SA2 9LG

G1 TFM J. Bishop 93 The Vale, Feltham TW14 0JY
G1 TFY D. Mackinnon 20 Saxon Grange, Sheep Street, Chipping Campden GL55 6BY
GM1 TFZ S. Bavin Garvan, 10 Grampian Way, Glasgow G78 2DH
GM1 TGY C. Christie Firlands, Spey Valley Drive, Aberlour AB38 9NU
G1 TGZ R. Mclintock 10 The Close, Riverhead, Sevenoaks TN13 2HE
G1 THA V. Collins Flat 2, Marsh Mead, Glebe Road, Petersfield GU31 5SB
G1 THD A. Simmons 22 Willow Way, Princes Risborough HP27 9AY
G1 THF the Ham Fellowship c/o A. Nixon 14 Carlton Road, Lowton, Warrington WA3 2EP
G1 THG D. Moore East View, The Common, Gillingham SP8 5NB
G1 THP M. Fudge 7 Shepherds Close, Winchester SO22 4HU
GM1 THR N. Harrison 127 Bruntsfield Place, Flat 1F3, Edinburgh EH10 4EQ
GM1 THS G. Slessor 27 Scurdie Ness, Aberdeen AB12 3NG
G1 THW A. Taylor 5 Brookside, Beare Green, Dorking RH5 4QH
G1 TIF J. Batey The Hemmel, Barrasford, Hexham NE48 4BD
G1 TIH F. Bell 143 Peter St., Blackpool FY1 3NN
G1 TIJ P. Seaman 18 Earlsford Road, Mellis, Eye IP23 8DY
G1 TIK D. Waters Gardeners Cottage, Sandhoe, Hexham NE46 4LU
G1 TIQ S. Crellin 89 Wapshare Road, West Derby, Liverpool L11 8LR
G1 TJH C. Barfoot11 6 Maldon Close Bishopstoke, Eastleigh SO50 6BD
GW1 TJK A. Evans Maes Yr Onnen, 134 Waterloo Road, Ammanford SA18 3RY
GJ1 TJP J. Poole Jardin Du Puits, La Longue Rue, St. Martin JE3 6ED Jersey
G1 TJR R. Bromley 33 Bromley Road, Lytham St. Annes FY8 1PQ
G1 TJT W. Adam 9 Maple Drive, South Ockendon RM15 6XE
G1 TJW B. Smith 25 The Ferns, Tetbury GL8 8JE
G1 TKE A. Gibson 9 Fishers Mead, Dulverton TA22 9EN
G1 TKQ I. Burdon 72 Greenway Road, Taunton TA2 6LE
G1 TKY P. Draper 4 Woodmans Croft, Hatton, Derby DE65 5QQ
G1 TLA B. Pattenden Inshallah, Abbeystrewery, Skibbereen Ireland
G1 TLC A. Myers 7 Hillside, Chelveston, Wellingborough NN9 6AQ
G1 TLE M. Rowe 11 Parkfield, Stillington, York YO61 1JR
G1 TLH D. Koopman 11 Tufts Field, Midhurst GU29 9BU
G1 TLW A. Lloyd 96 Fairdene Road, Coulsdon CR5 1RF
G1 TMF J. Firth 29 Curzon Street, Newcastle ST5 0PD
G1 TML P. Brooks 7 Beechcombe Close, Pershore WR10 1PW
G1 TMW A. Caspersz 25 Cheltenham Place, Harrow HA3 9NB
G1 TNK J. Flattley 53 The Drive, Bredbury, Stockport SK6 2ED
G1 TNP A. Karande Flat 1, 6 St. Dominics Close, Torquay TQ1 4UN
G1 TNR D. Jordan 21 Rosewood Park, Walsall WS6 7HD
GM1 TNT T. Glenny 115 Parkneuk Street, Motherwell ML1 1BY
G1 TOB T. Hall 3 Saville Road, Twickenham TW1 4BQ
G1 TOL S. Talbot 8 Thornford Drive, Swindon SN5 7BB
G1 TPC M. Bellamy 2 Nelson Drive, Rothwell, Kettering NN14 6DZ
G1 TPN R. Whateley 14 Eastfield Road, Delapre, Northampton NN4 8PE
G1 TPO R. Steel 15 Thornbury Avenue, Seghill, Cramlington NE23 7RT
G1 TPQ S. Codman 4 Farmland Road, New Costessey NR5 0HX
G1 TPV D. Juett 10 Leys Road, Cambridge CB4 2AU
G1 TQH K. Scroggins 44 Hillcroft Road, Herne, Herne Bay CT6 7EW
G1 TQN A. Hobbs The Sail Loft, 604 Blandford Road, Poole BH16 5EQ
G1 TQR J. Harris 48 Beech Close, Corby NN17 2AF
G1 TQT M. Costello Flat 24, Wisley House, London SW1V 2QS
G1 TQU C. Chambers 8 Dagtail Lane, Redditch B97 5QT
G1 TQY P. Bishop 2 Spruce Avenue, Whitehill, Bordon GU35 9TA
G1 TRI C. Curtis 89 Goldcroft Road, Weymouth DT4 0EA
G1 TRL J. Deacon 28 York Close, Bicester OX26 4XE
GI1 TRZ W. Hamilton-Sturdy 243 The Woods, Larne BT40 1BD
G1 TST P. Jackman 1 Palmer Road, Trowbridge BA14 8QP
G1 TSV S. Corrigan 163 Blackburn Road, Heapey, Chorley PR6 8EJ
G1 TTB G. Birkby 44 Lady Bay Road, West Bridgford, Nottingham NG2 5DS
G1 TTC K. Howard 73 Challacombe, Furzton, Milton Keynes MK4 1DP
G1 TTG J. Brickwood Datemachi House #301, 3-33-5, Tokyo 150 Japan
G1 TTH E. Roughton 18 Church Close, Braybrooke, Market Harborough LE16 8LD
G1 TTK G. Lewis 57 Edgcumbe Road, Roche, St. Austell PL26 8JH
G1 TTL F. Moss 64 Birch Avenue, Cuerden Residential Park, Leyland PR25 5PD
G1 TTX A. Docherty Sunnybrae, Wickhurst Road, Sevenoaks TN14 6LY
G1 TUI J. Tracy 18 Preston St., Kirkham, Preston PR4 2ZA
G1 TUL S. Neale 28 Needham Drive, Sutton St. James, Spalding PE12 0EG
G1 TUS R. Rodley Meadow Cottage, Cold Ashby Road, Northampton NN6 8QP
G1 TUU M. Dixon 118 Kings Ash Road, Paignton TQ3 3TU
G1 TUZ P. Nelson 42 York Avenue, East Cowes PO32 6RU
G1 TVW J. Halliday 24 Duncan Avenue, Otley LS21 3LN
G1 TWH S. Greenfield Byways, Brightlingsea Road, Colchester CO7 8JH
G1 TWS M. Dench 110 Eastwood Road, Rayleigh SS6 7JR
G1 TWT A. Scott 21 Hexham, Oxclose, Washington NE38 0NR
G1 TWW R. Reeves 40 Kennett Road, Romsey SO51 5PQ
G1 TWY B. Panton Lavers, Preston Road, Lavenham, Sudbury CO10 9QD
G1 TXO M. Riches 32 Wyncham Avenue, Sidcup DA15 8ER
G1 TYP E. Summers 262 Huddersfield Road, Stalybridge SK15 3DZ
G1 TYU R. Ward 1 Kirkcroft Close Thorpe Hesley, Rotherham S61 2UH
GI1 TYX S. O'Connor 27 Helenswood Court, Belfast BT17 0RZ
G1 TZS R. Haskell 76 Hull Grove, Harlow CM19 5RR
G1 TZZ M. Foster 7 Orion Way, Braintree CM7 9UR
G1 UAF A. Webster 7 Castlehythe, Ely CB7 4BU
G1 UAL C. Bagwell 1 Waldegrave Court, Movers Lane, Barking IG11 7UW
G1 UAY K. Varnals Regent Studio, Skidden Hill, St. Ives TR26 2DU
G1 UAZ C. Musson 14 Alfreton Road, South Normanton, Alfreton DE55 2AS
G1 UBC J. Pedley 24 Appledore Road, Walsall WS5 3DT
G1 UBH M. Howell 4 Chattisham Close, Stowmarket IP14 2RE
G1 UBL A. Camm 24 Croyde Avenue, Greenford UB6 9LS
G1 UBN S. Knight 46 Hollybank Road, Hythe, Southampton SO45 5FQ
G1 UBT E. Dunn 118 James Turner St., Birmingham B18 4NE
G1 UBV N. Brigden 176 Hulverston Close, Sutton SM2 6UA
G1 UCC D. Boot 13 Westland Street, Stoke-on-Trent ST4 7HE
G1 UCG W. Roberts 13 Roseacre Road, Elswick, Preston PR4 3UD
G1 UCI D. Roberts 56 Meadow Lane, Ainsdale, Southport PR8 3RS
G1 UCN M. Davis Flat 3, South Court, 29 Second Avenue, Bridlington YO15 2LW
G1 UCO D. Ralph 55 St. Marys Road, Warley, Smethwick B67 5DH
G1 UCR R. Irving 52 Holsworthy Square, London WC1X 0BG
G1 UCT D. Pye 95 Lansdowne Way, High Wycombe HP11 1UB
G1 UCZ D. Dewar 224 Seaside Road, Aldbrough, Hull HU11 4RY
G1 UDB G. Wardle 53 Braine Road, Wetherby LS22 6NP
G1 UDE D. Grayson 28 Chesterfield Road, Swallownest, Sheffield S26 4TL
G1 UDR J. Scott 47 Corinthian Road, Chandler'S Ford, Eastleigh SO53 2AY
G1 UDS P. Spence 17 Springvale Rise, Parkside, Stafford ST16 1TE
G1 UDT M. Boydon 56 Oliver Leese Court, Ten Butts Crescent, Stafford ST17 9HP
G1 UDW D. Upton 14 Ipley Way, Hythe, Southampton SO45 3LJ
G1 UDX M. Stasuik 30 Ramsey Drive, Arnold, Nottingham NG5 6QL
G1 UEA D. Mills 56 Canterbury Road, Birchington CT7 9AS
G1 UEO J. Aldred 1E North End Road, Steeple Claydon, Buckingham MK18 2PF
G1 UEQ J. Sims 345 Blandford Road, Hamworthy, Poole BH15 4HP
G1 UEV S. Brown 10 Walkers Lane South, Blackfield, Southampton SO45 1YN
G1 UFA B. Bailey 10 Milton Road, Waterlooville PO7 6AA
G1 UFH J. Coyne 20 Hawthorn Hill, Letchworth Garden City SG6 4HG
G1 UFJ R. Davis 10 Greenhill Gardens, Minster, Ramsgate CT12 4EW
G1 UFL K. Whistance 20 Solent Way, Milford On Sea, Lymington SO41 0TE
G1 UFM S. Pugh 18 Styal Avenue, Stretford, Manchester M32 9SJ
G1 UFS N. Chandler 7 Sherlock Avenue, Parklands, Chichester PO19 3AE
G1 UFT P. Zara 53 Castleton Road, Wigston LE18 1FQ
G1 UFX A. Hern Flat 9, Block P, Peabody Estate, London SE1 8DU
G1 UGB St George's Academy ARC (Sga-ARC) c/o J. Slater 57 Freshbrook Road, Lancing BN15 8DE
G1 UGG K. Blagg 2 Geldof Drive, Blackpool FY1 2AQ
G1 UGH T. Chaplin 21 Shillitoe Close, Bury St. Edmunds IP33 3DU
G1 UGJ Obo Largs & District ARS c/o B. Lowe 5 Kingfisher Drive, Necton, Swaffham PE37 8NN
G1 UGL P. Beardshaw 12 Haleswórth Close, Chesterfield S40 3LW
G1 UGO S. Lexton 70 Halsbury Road, Westbury Park, Bristol BS6 7SU
G1 UGV D. Bodman 34 Churchlands, North Bradley, Trowbridge BA14 0TD
G1 UGX G. Hodgkins 20 Broadstone Close, Barnwood, Gloucester GL4 3TX
G1 UHB S. Tomkins The Close, Broomfield, Bradford BD14 6PJ
GM1 UHF M. Perrett Otterburn House, Inverinate, Kyle IV40 8HE
G1 UHO H. Court 10 Dorset Road, West Kirby, Wirral CH48 6DJ
G1 UIB R. Moxon 16 Kielder Oval, Harrogate HG2 7HQ
G1 UID H. Kentfield 1 Torquay Avenue, Gosport PO12 4NS
G1 UIO M. Mcvittie 19 Alder Crescent, Poole BH12 4BD
GM1 UIR A. Perks The Lodge, Cemetery Drive, Dumbarton G82 5HD
G1 UJX J. Huddlestone 8 Wilmot Avenue, Chaddesden, Derby DE21 6PL
G1 UK A. Buck 10 Northfield Park, Mansfield Woodhouse, Mansfield NG19 8PA
G1 UKA M. Wilkie 1 Celandine Close, Billericay CM12 0SU
G1 UKH R. Rodgers 88 Norwich Road, Watton, Thetford IP25 6DW
G1 UKS D. Simmons 54 Rydal Court, Morecambe LA4 5LT
G1 UKW C. Crane 3 Hawkshead Drive, Royton, Oldham OL2 6TW
G1 UKZ P. Hall 32 Haslemere Road, Urmston, Manchester M41 6HA
G1 ULB G. Walker 20 Clough Drive, Prestwich, Manchester M25 3JL
G1 ULG D. Porter 44 Weir Road, London SW12 0NA
G1 ULP T. Garner 8 Brookside Park, Station Road, Hugglescote, Coalville LE67 2GB
G1 ULQ P. Rowsell Thomley Hall Farm, Worminghall, Bucks HP18 9JZ
G1 ULR D. Ross 2 Jemmetts Close, Dorchester-On-Thames, Wallingford OX10 7RA
G1 UMS J. Preece White House, Bredwardine, Hereford HR3 6BY
G1 UMY R. Wiltshire Danesboro, Stonehill Road, Chertsey KT16 0ER
G1 UNB P. Durrant Glenmore, 20 Linersh Wood Close, Bramley, Guildford GU50EG
G1 UNN P. Coldicott 45 Orrian Close, Stratford-upon-Avon CV37 0TT
G1 UNQ R. Davis Lowlands, Station Road, Evesham WR11 7QG
G1 UNU K. Etwell Hawthorn Cottage, Old Worcester Road, Kidderminster DY11 7XS
G1 UOD P. Farmer 22 Nortune Close, Birmingham B38 8AJ
G1 UOJ M. Lichtaowicz 219 Hamilton Drive West, York YO24 4PL
G1 UOR W. Rodgers 9 Hillcrest, Skelmersdale WN8 9JZ
GW1 UOV A. Ham Tynffordd, Blaencwrt, Lampeter SA39 9AZ
GW1 UOY W. Bagley Glan Severn, Trefeglwys Road, Llanidloes SY18 6HZ
G1 UPP M. Maude 6 Malin Parade, Portishead, Portishead BS20 7FW
G1 UPT J. Ravelini 15 Clarendon Green, Orpington BR5 2NY
G1 UPX T. Giles 108 Queensway, Didcot OX11 8SW
G1 UQC Capt. D. Rusbridge 1 Ray Bond Way, Aylsham, Norwich NR11 6UT
G1 UQF R. Davies 33 Melbourne House, Melbourne Road, Northampton NN5 5LW
G1 UQK N. Lewis 4 Wye Avenue, Malvern WR14 2SY
G1 UQT S. Alliott 32 Broad Gates, Silkstone, Barnsley S75 4HD
GW1 URD R. Ameson 9 Coed-Y-Fronallt Estate, Dolgellau LL40 2YG
GW1 URF Dr A. Jones Hafandeg, Southgate, Aberystwyth SY23 1RY
G1 URH S. Woodley Stables Edge, Lower Road, Chilton, Didcot OX11 0RR
G1 URJ N. Capon 13 West Croft, Berinsfield, Wallingford OX10 7NL
G1 URQ D. Traynor 5 Mount Street, Widnes WA8 6TL
G1 URR C. Gough 67 Pickmere Lane, Wincham, Northwich CW9 6EB
G1 URW J. Carman 5 Melbourne Road, Blacon, Chester CH1 5JQ
G1 URZ T. Bennett 16 Montgomery Avenue, Hemel Hempstead HP2 4HE
G1 USD R. Waugh 34 Wickham Street, Rochester ME1 2HH
G1 USF P. Napp 23 Harriot Drive, Newcastle upon Tyne NE12 7EU
GD1 USI A. Tawney Croym Dty Chione, The Howe, Isle of Man IM9 5PR
G1 USK R. Firth 40 Ashfield Road, Chippenham SN15 1QQ
G1 USN J. Challis 37 St. Ronans Drive, Seaton Sluice, Whitley Bay NE26 4HZ
G1 USV R. York 10 Severn View Road, Thornbury, Bristol BS35 1AY
G1 USW P. Daniels 29 Station Road, Wickwar, Wotton-under-Edge GL12 8NB
G1 USZ M. Trim 10 Oldends Lane, Stonehouse GL10 2DG
G1 UTC C. Thomas Hazel Mount, Lockhams Road, Curdridge, Southampton SO32 2BD
G1 UTF R. Beer 8 Littlestone Court, Grand Parade, New Romney TN28 8NF
G1 UTJ A. Lees Timbercroft, Elliotts Orchard, Warwick CV35 8ED
G1 UTM P. Yearsley 25 Dinmor Road, Manchester M22 1NN
G1 UTN C. King 7 Hillcrest, Hyde SK14 5LJ
G1 UTP S. Darlington 17 Eleanor Road, Royton, Oldham OL2 6BH
G1 UTS G. May 95 Moorfield Avenue, Denton, Manchester M34 7TX
G1 UTZ A. Peters 10 Hill View Close, Grantham NG31 7PH
G1 UUF I. Grounsell 23 Loughbrow Park, Hexham NE46 2QD
G1 UUJ D. Eyre 29 Old Acre Lane, Brocton, Stafford ST17 0TW
G1 UUK G. Perry 6 Morgan Close, Arley, Coventry CV7 8PR
G1 UUL S. Brough 9 Beech Tree Lane, Cannock WS11 1AZ
G1 UUO R. Hanson 1 Ashmore Road, Coventry CV6 1LH

G1 UUP A. Robins 38 Eastbourne Road, Willingdon, Eastbourne BN20 9NS
G1 UUS M. Carter 4 Asbury Road, Balsall Common, Coventry CV7 7QN
G1 UUT R. Bygate 91 Woodcote Avenue, Kenilworth CV8 1BE
G1 UUV R. Fairholm 63 Rugby Road, Clifton Upon Dunsmore, Rugby CV23 0DE
G1 UUZ M. Davis 20 Pavilion Avenue, Warley, Smethwick B67 6LA
G1 UVD R. Rothery 2 Highcroft, Mount Pleasant, Batley WF17 7NT
G1 UVE A. Shackleton 31 Ashton Street, Leeds LS8 5BY
G1 UVI M. Stanley 35 Moorgate Road, Kippax, Leeds LS25 7ET
G1 UVJ D. Rogers 11 Beech Crescent, Mexborough S64 9EH
G1 UVK A. Linfoot 19 Vicarage Close, Bubwith, Selby YO8 6LN
GW1 UVN J. Jones Silversprings, Llanelly Church Road, Abergavenny NP7 0EL
G1 UWD D. Peachey 28 Broad St., Truro TR1 1JD
GM1 UWE M. Peachey 8/7 Durar Drive, Edinburgh EH4 7HN
G1 UWQ J. Naughton 86 Brookford Avenue, Coventry CV6 2GQ
G1 UWV B. Dixon Terridene, Park Road, New Milton BH25 6QE
GW1 UXW M. Lewis Gorse Cottage, Graig Road, Cwmbran NP44 5AS
G1 UXZ Gravesend And Bean DX Group c/o J. Sage 8 Foxwood Road, Bean, Dartford DA2 8BH
G1 UYT S. Tomkinson 3 Heysham Close, Weston Coyney, Stoke-on-Trent ST3 6RG
GW1 UYW K. Wallis Cartref Newydd, Llanarmon D.C. LL20 X
G1 UYZ D. Payne 8 Gascoigne Drive, Spondon, Derby DE21 7GL
G1 UZC P. Jarrett 17 Wolmers Hey, Great Waltham, Chelmsford CM3 1DA
G1 UZD J. Yates 8 Holt Drive, Wickham Bishops, Witham CM8 3JR
G1 UZS E. Lees 11A Edale Avenue, Mickleover, Derby DE3 9FY
G1 UZW N. Boag 60 Harebell, Amington, Tamworth B77 4NA
G1 VAA N. Wills-Browne 2 Parkfield Road, Aigburth, Liverpool L17 8UH
G1 VAB D. Goodwill 94 Palmerston Street, Derby DE23 6PF
GM1 VAD S. Scanlain Crossraguel, Old Newton, Nairn IV12 5RA
G1 VAG A. Grant 26 Fountains Avenue, Boston Spa, Wetherby LS23 6PX
G1 VAJ A. Leach 8 Eskdale, Brownsover, Rugby CV21 1NJ
G1 VAL M. Pearson 5 Craven Court, Warwick Drive, Barnoldswick BB18 6WA
G1 VAN P. Van Falier 572 Stafford Road, Ford Houses, Wolverhampton WV10 6NN
G1 VAO A. Andrews 12 Kings Lea, Ossett WF5 8RY
GW1 VAW B. Williams 10 Tynybedw Terrace, Treorchy CF42 6RL
G1 VAY D. Hearn 90 Princes Drive, Valley Dip, Seaford BN25 2TX
GI1 VAZ G. Richardson 6 Cedarhurst Rise, Belfast BT8 7RJ
G1 VBA P. Buckle 63 Ashley Drive South, Ashley Heath, Ringwood BH24 2JP
G1 VBB R. Bedwell 24 Tiger Moth Drive, Southam CV47 1AS
GM1 VBD R. Scott Enzie Slackhead, Buckie, Moray AB5 2BJ
GM1 VBE Dr S. Clink Southsyde, Woodhead Avenue, Glasgow G71 8AR
G1 VBL R. Kelsall 11 Manor Road, Ducklington, Witney OX29 7YD
G1 VBO R. Piper Tatnam Farm, St. Mary'S Road, Romney Marsh TN29 0PW
G1 VBP P. Smith The Rufford Care Centre, Room 1, Gateford Road S81 7BH
G1 VBQ P. Wright 81 High St., Syston, Leicester LE7 1GQ
G1 VBY K. Moreton 29 Tiber Drive, Newcastle ST5 7QD
G1 VCU A. Smithson 118 High Street, Linton, Cambridge CB21 4JT
G1 VCZ A. Hewes 89 Fronks Road, Dovercourt, Harwich CO12 4EQ
G1 VDE D. Taylor 21 Munday Close, Bussage, Stroud GL6 8DG
G1 VDO S. Preston 50 Milton Avenue, Malton YO17 7LB
G1 VDP C. Colclough 53 St Marys Road, Nuneaton CV11 5AT
GW1 VDT P. Whatley Fairacre, Bishton Lane, Chepstow NP16 7LG
GW1 VDW L. Marquardt 2 Pembroke Terrace, Varteg, Pontypool NP4 7UJ
GM1 VDZ G. Neil 7 Dalfarson Avenue, Dalmellington, Ayr KA6 7TX
G1 VFB J. Bloy 25 Highfield Road, Minster On Sea, Sheerness ME12 3BA
G1 VFH M. Staley 128 Pinewood Crescent, Stoke-on-Trent ST3 6HZ
GM1 VFQ J. Murdoch 8 Primpton Avenue, Dalrymple, Ayr KA6 6EL
GM1 VFR H. Mcdonald 43 Southfield Road Cumbernauld, Glasgow G68 9DZ
G1 VFW I. Beeby 32 Edditch Grove, Bolton BL2 6BJ
G1 VGA K. Crane 92 Dimond Road, Southampton SO18 1JS
G1 VGI K. Waterson Bramleigh, 20 Cadogan Road, Bury St. Edmunds IP33 3QJ
G1 VGK M. Whittington 253 Kings Drive, Eastbourne BN21 2UR
G1 VGM S. Ball 16 Stonewood Gate, Field Lane, St. Helens, Ryde PO33 1FY
G1 VGO D. Holdsworth 28 Moorbank Close, Wombwell, Barnsley S73 8RX
G1 VGP G. Anderson 1 White Rose Mead, Garforth, Leeds LS25 2EG
G1 VHC P. Ward 6721 Pinetucky Road South, Mobile 36618 United States
G1 VHN D. Barnes 46 Lawnswood Avenue, Wordsley, Stourbridge DY8 5LR
G1 VHW W. Killeen 12 Meriden Grove, Lostock, Bolton BL6 4RQ
G1 VHY D. Symonds 79 Kingsway, Kirkby-In-Ashfield, Nottingham NG17 7EH
G1 VID T. Howe The Old School House, 3 Bairds Hill, Broadstairs CT10 3AA
G1 VIF C. Morphett 138 Healds Road, Dewsbury WF13 4HT
G1 VIG D. Carrick 5 Kings Close, Market Overton, Oakham LE15 7PS
G1 VII N. Pooley 64 Lynwood Grove, Orpington BR6 0BH
G1 VIN C. Kirkland 29 Shelley Road, Enderby, Leicester LE19 4QX
G1 VIO A. Eades 41 Woodhall Gate, Pinner HA5 4TX
G1 VIP M. Walker 43 Wimborne Road Cogdean, Corfe Mullen BH21 3DS
GW1 VIR I. Berry 1-2 Pottery Cottages, Trefonen, Oswestry SY10 9DF
G1 VIS S. Grimes 73 Ryston Road, Denver, Downham Market PE38 0DP
G1 VIT N. Cliff 12 New Church Road, Wellington, Telford TF1 1JH
G1 VIW R. Paterson 27 Copt Heath Drive, Knowle, Solihull B93 9PA
G1 VIY T. Depledge 13 Peel Drive, Astbury, Congleton CW12 4RF
G1 VIZ A. Hemenway 69 Sixth Avenue, Heworth, York YO31 0UR
GW1 VJB B. Flounders 76 Springfield Road, Sebastopol, Pontypool Torfaen NP4 5BX
GM1 VJD R. Terras 8 Haddington Gardens, Lawthorn, Irvine KA11 2EB
G1 VJE M. Sayers Sgs, Trrenchard Building, Jssu (Cy) BFPO59
G1 VJG C. Rhenius 8 Rutland Close, Cambridge CB4 2HT
G1 VJJ C. Silvey 106 Southchurch Avenue, Southend-on-Sea SS1 2RP
G1 VJN B. Jones 73 Tonge Road, Murston, Sittingbourne ME10 3NR
G1 VJQ T. Monaghan 15 Mulgrave Road, Worsley, Manchester M28 2RW
G1 VJY A. Birch Maximilien, Les Ecovets, Chesieres 1885 Switzerland
G1 VKB A. Morris 140 Astwood Road, Worcester WR3 8EZ
G1 VKC D. Close 60 Lead Lane, Ripon HG4 2LN
G1 VKG J. Howarth 10 Poplar Place, Penrith CA11 9HN
GI1 VKJ B. Duffy 45 Boulevard Green, Newcastle BT33 0FA
G1 VKN S. Matthews Matilda - Kings Marina, Mather Road, Newark NG24 1FW
G1 VKT E. Dale 29 Hulme Road, Leigh WN7 5BT
GM1 VLA A. Lee Sandana, Kirkpatrick Fleming, Lockerbie DG11 3BA
G1 VLD A. Burkitt 6 Chewells Close, Haddenham, Ely CB6 3XE
G1 VLS G. Turner 19 Hector Road Darwen, Blackburn With Darwen BB3 0AY
G1 VLU D. Green 5 New Mill Road, Holmfirth HD9 7SG
GW1 VMA M. Hills Ty Newydd, Rhos, Llandyssul SA44 5HE
GI1 VMF J. Oliver 29 Callan Bridge Park, Armagh BT60 4BU
G1 VMX E. Driver 39 Witham Road, Woodhall Spa LN10 6RW
G1 VNB D. Neeves 18 Beechwood Road, Bedworth CV12 9AG
G1 VNE S. Nocera Strada Provinciale, Mulazzano 88, Parma 43010 Italy
G1 VNH V. Palmer 57 Old Tiverton Road, Exeter EX4 6NG
G1 VNL A. Smith 112 Manor Lane, Charfield, Wotton-under-Edge GL12 8TN
G1 VNM S. Eyers 190 Greenhill Road, Herne Bay CT6 7RS
G1 VNS K. Baum 11 St. Ives Road, Wigston LE18 2JB
G1 VNU S. Exell 22 Woodside Road, Beare Green, Dorking RH5 4RH
G1 VNV M. Gold 14 Brewers Lane, Badsey, Evesham WR11 7EU
G1 VNZ A. Atkins 3 Stowe Drive, Bexhill-on-Sea TN39 4GL
G1 VOB M. Prior 36 Bassnage Road, Halesowen B63 4HQ
G1 VOC A. James 70 Martin Croft, Silkstone, Barnsley S75 4JS
G1 VOJ A. Duce 16 Gillmans Road, Orpington BR5 4LA
G1 VON S. Howard 95 Greenbarn Way, Blackrod, Bolton BL6 5TE
G1 VOP Dr D. Hettiarratchi 2 Carham Close, Gosforth, Newcastle upon Tyne NE3 5DX
G1 VOQ P. Tandy Old Channel Hill Farm, North End, Fordingbridge SP6 3HA
G1 VOR S. Twigg 16 Merlin Avenue, Nuneaton CV10 9JZ
G1 VOY R. Ward Overdale, Egton, Whitby YO21 1UE
GI1 VPA D. Smythe 57 Ballymacormick Avenue, Bangor BT19 6AY
G1 VPC M. Wigley 36 Wivelsfield Road, Haywards Heath RH16 4EW
G1 VPE R. Wilcockson Oaks Farm, Markham Road, Duckmanton, Chesterfield S44 5HP
G1 VPS M. Wright 27 Willow Road, Kettering NN15 7BA
G1 VQB G. Doughty 95 Buxton Road, Chaddesden, Derby DE21 4JL
G1 VQG G. Hannaford 8 Porthmellon Gardens, Callington PL17 7QL
G1 VQH G. King Mill Cottage, 48 Mill Street, Swadlincote DE12 8ES
G1 VQI C. Coates Ecalox Ltd, Hammonds Farm, Stapleford Road, Stapleford Tawney, Stapleford Abbotts, Romford RM4 1RR
G1 VQK E. Wright 94 Bachelor Gardens, Harrogate HG1 3EA
G1 VQV S. Gent 66 Apperley Way, Cradley, Halesowen B63 2PY
G1 VRA E. Jones Bramley Lodge, Back Lane, Royston SG8 6DD
G1 VRC T. Nicholson 8 East Street, High Spen, Rowlands Gill NE39 2HD
G1 VRJ J. Ager 20 Kirktonhill Road, Westlea, Swindon SN5 7AF
GW1 VRR J. Williams 31 Syr Davids Avenue, Cardiff CF5 1GH
GW1 VRW C. King 27 Gadlys Road West, Barry CF62 7HX
G1 VSD W. Bennett 90 Garwood Road, Yardley, Birmingham B26 2AW
G1 VSH G. Pettit 7 Dunster Crescent, Hornchurch RM11 3QD
G1 VSK A. Heyes 41 Coronation Drive, Penketh, Warrington WA5 2DD
G1 VSM D. Pratt 17 Worcester Gardens, Greenford UB6 0BH
G1 VSO C. Champ 31 Nobles Close, Oxford OX2 9DN
GM1 VSR A. Bain 4 Bawdley Head, Fraserburgh, Aberdeenshire AB4 5SE
G1 VSX G. Buck 1 Hobart Road, Weston-Super-Mare BS23 4QQ
G1 VTE M. Joynson 90 Fairhope Avenue, Bare, Morecambe LA4 6LA
G1 VTK R. Swan 13 Mendip Road, Torquay TQ2 6UQ
G1 VTN C. Peacock 1 Furnace Lane, Madeley, Crewe CW3 9EU
G1 VTO B. Kemp 193 Cavalry Park, March PE15 9DL
G1 VTP J. Bridgehouse 12 Castle Hall Close, Stalybridge SK15 2HR
G1 VTQ B. Gorman 40 Maudland Bank, Preston PR1 2YL
G1 VTS J. Smith 9 Birchway, Hayes UB3 3PA
G1 VUG M. Cooper Osborne House, Main Street, Louth LN11 0XF
G1 VUK S. Hunt 1 Lucknow Cottages, Northbridge Street, Robertsbridge TN32 5NP
G1 VUP A. Cheeseman Flat 5 Dubarry House, Hove Park Villas Hove Park Villas, Hove BN3 6HP
G1 VUY D. White 14 Waggoners Way, Bugbrooke, Northampton NN7 3QT
G1 VVB S. Patrick 9 Brant Avenue, Illingworth, Halifax HX2 8DL
G1 VVE G. Bindon 74 Cashford Gate, Taunton TA2 8QB
G1 VVH P. Fisher Flat 46 Morris House, Fairchild Close, London SW11 2SU
G1 VVL A. Proctor 448 Tuttle Hill, Nuneaton CV10 0HR
G1 VVM G. Brett 47 Windermere Avenue, Huncoat, Accrington BB5 6JG
G1 VVT H. Mawson 118 Byron Street, Loughborough LE11 5JW
G1 VVU J. Stephens 34 King Street, Seahouses NE68 7XR
G1 VVX I. Andronov 53 Broad Street, Ludlow SY8 1NH
G1 VVY J. Waters 2 Tea Caddy Cottages, Worthing Road, West Grinstead, Horsham RH13 8LG
GM1 VWA J. Gilruth 88 Fintry Crescent, Dundee DD4 9EX
G1 VWC G. Matthews 9 Sadlers Way, Hemingford Grey, Huntingdon PE28 9EW
G1 VWL R. Knowles 9 Malham Close, Southport PR8 6UP
G1 VWP S. Goodwin 75 Farleigh Hill, Tovil, Maidstone ME15 6AA
G1 VWU A. Driver Grace Barn, Pencarrow, Advent, Camelford PL32 9RZ
G1 VWZ R. Huntley 49 Main Street, Wetwang, Driffield YO25 9XL
G1 VXD R. Roche 98 Grange Road, Cheddleton, Nr Leek ST13 7NP
GM1 VXE A. Rennie 5 Barbieston Cottage, Drongan, Ayr KA6 7EF
G1 VXS S. Shirvington 22 Bassett Road, Northleach, Cheltenham GL54 3QJ
G1 VXX A. Powell 9A Shaftesbury Close, Bracknell RG12 9PX
G1 VXY Inmarsat ARC c/o T. Tomkins 40 Diksmuide Drive, Ellesmere SY12 9QA
G1 VYA H. Seddon 25 Thistledown Close, Wigan WN6 7PA
G1 VYB T. Smith 11 River Terrace, Wisbech PE13 1PZ
GM1 VYF P. Letters 23 East Lennox Drive, Helensburgh G84 9JD
GM1 VYG J. Mcdonald 4 Braeside Bowfield Road, Howwood, Johnstone PA9 1BP
G1 VYM D. Eveleigh 7 Malin Road, Littlehampton BN17 6NN
G1 VYS A. Walker 49 Selworthy Road, Stoke-on-Trent ST6 8PL
G1 VZB C. Lillis 6 Whitelake View, Urmston, Manchester M41 8UT
GM1 VZG T. Gilmour Fiold, Rope Walk, Kirkwall KW15 1XJ
G1 VZT B. Rayner 44 Foxhall Fields, East Bergholt, Colchester CO7 6QY
G1 VZW B. Hitchen 40 Methuen Avenue, Fulwood, Preston PR2 9QX
G1 WAB Derbys Workd All Britain G.ARC c/o J. Wainwright 8 Common Lane, Cutthorpe, Chesterfield S42 7AN
G1 WAC Wythall RC c/o M. Pugh 44 Simms Lane, Hollywood & Wythall, Birmingham B47 5HY
G1 WAE C. Rogers 63 Greenwell Road Haydock, St. Helens WA11 0SQ
G1 WAP B. Stott 35 Sheridan Road, Laneshawbridge, Colne BB8 7HW
G1 WAS P. Delaney 61 Lyndale Avenue, Eastham, Wirral CH62 8DG
G1 WAW Spen Valley ARS c/o P. Mitchell 3 High Howe Close, Bournemouth BH11 8NN
G1 WCY Ripon & District ARS c/o J. Reeves 5 Arrows Crescent, Boroughbridge, York YO51 9LP

G1 WDQ T. Hutton 23 Dines Close, Wilstead, Bedford MK45 3BU
G1 WEF R. Cornwell 13 Milford Road, Thurrock, Grays RM16 2QL
G1 WEV I. Henderson 1 Prestbury Road, Pennywell, Sunderland SR4 9DW
G1 WFA Dr C. Ward 7 Coates Close, Heybridge, Maldon CM9 4PB
G1 WFG S. Lake 42 Haling Park Road, South Croydon CR2 6NE
G1 WFJ A. Woodhouse 5 Dudley Road, Kingswinford DY6 8BT
G1 WFO P. Burden 110 Westbury Leigh, Westbury BA13 3SH
GI1 WFP N. Mcloughlin 44 Kilbroney Rd, Rostrevor BT34 3BL
G1 WFS Stockport Rs c/o S. Rogers 19 Stoke Street, Hull HU2 9BL
G1 WFU R. Dickson 49 Ashgrove, Peasedown St. John, Bath BA2 8EF
GI1 WGK W. Steele 19 William Street, Donaghadee BT21 0HL
G1 WGL S. Morris 9 Starling Close, Burgess Hill RH15 9XR
G1 WGM K. Perry 25 Hillary Drive, Crowthorne RG45 6QF
G1 WGO A. Smith 2 The Old Rectory, Church Lane, Kirkheaton, Huddersfield HD5 0BH
GW1 WGR Royal Naval ARS c/o M. Rowles 7 Gelli Deg, Bryncoch, Neath SA10 7PL
G1 WHT M. Watts 11 Bywood Place, Grimsby DN37 9RH
G1 WHU D. Baines-Jones 22-24 Grove Royd, Halifax HX3 5QU
G1 WHY J. Lowe 23 Hoylake Drive, Tividale, Oldbury B69 1QA
G1 WID J. Hartridge 15 Hundred Acres, Wickham, Fareham PO17 6JB
G1 WIS D. Mountain 45 Westway Gardens, Redhill RH1 2JB
G1 WIW R. Dowdeswell 5 Croft Close, Barwell, Leicester LE9 8EW
GU1 WJA W. Ayres Rousay, Bailiffs Cross Road, St. Andrew GY6 8RY Guernsey
G1 WJG J. Coates The Old Timbers, 23 Yoells Lane, Waterlooville PO8 9SG
G1 WJK G. Richards 3 Pleasant Close, Kingswinford DY6 9TQ
G1 WJO A. Blackwell 1 Gladstone Terrace, Hinckley LE10 1HE
G1 WJR W. Rollins 5 Little Clacton Road, Great Holland, Frinton-on-Sea CO13 0ET
GM1 WKH N. Graves The Lythe, 8 Tree Road, Tarves, Ellon AB41 7JY
G1 WKK J. Arnott 27 Main Road, Tadley RG26 3NJ
G1 WKO R. Reichmann 9 Rue Du Croteau, la Neuville Les Wasigny 8270 France
G1 WKS West Kent ARS c/o A. Lightbody 3 Elphicks Place, Tunbridge Wells TN2 5NB
G1 WKZ Dr G. Evans 4 The Mallards, Fareham PO16 7XR
G1 WLD S. Evans 4 The Mallards, Fareham PO16 7XR
GI1 WLJ G. Mccutcheon 73 Tullynagardy Road, Newtownards BT23 4TB
G1 WLN N. Roskruge 2 Lanner Green Terrace, Lanner, Redruth TR16 6DQ
G1 WLO M. Head 46 Ridgewood Gardens, Bexhill-on-Sea TN40 1TS
G1 WLU S. Brookes Ivy Cottage, Haselor, Alcester B49 6LX
G1 WLW A. Fordyce 41 Benscliffe Drive, Loughborough LE11 3JP
G1 WLX J. Goacher 41 Clay Hill, Two Mile Ash, Milton Keynes MK8 8AY
G1 WMK J. Mayo 10 Church Close, Fringford, Bicester OX27 8DR
G1 WMN K. Harvey 61 Westfield Road, Northchurch, Berkhamsted HP4 3PW
G1 WMS R. Cadwallader Rambla Grande, Los Reyes, Urcal 4691 Spain
GM1 WMU S. Webster 15 Forrest Place, Armadale, Bathgate EH48 2GZ
G1 WMV B. Catchpoole 8 Buckland Avenue, Basingstoke RG22 6JL
G1 WNL F. Thomas 38 Partridge Avenue, Yateley GU46 6PB
G1 WNZ G. Sollazzo 7 Miles Road, London N8 7SJ
G1 WOR Worthing & Dist ARC c/o P. Godbold 13 Dawn Crescent, Upper Beeding, Steyning BN44 3WH
G1 WOV A. Hart 15 Rendcomb Close, Milton, Weston-Super-Mare BS22 9QR
G1 WPG B. Groome The Old Smithy, High Street, Dorchester DT2 8JW
G1 WPH L. Eden 23 Elm Green Close, Worcester WR5 3HD
G1 WPL R. Balkwell 2 Franklyn Road, Droylsden, Manchester M43 6DS
G1 WPR T. Bromley 7 Brookside, Desborough, Kettering NN14 2UD
G1 WQC R. Pratt 11 Park Road, Ryde PO33 2BG
G1 WQH R. Booth 66 Fairburn Crescent, Pelsall, Walsall WS3 4PU
G1 WQL P. Kenyon-Brodie 17 Potterdale Drive, Little Weighton, Cottingham HU20 3UU
G1 WQN S. Mangan 48 Emblett Drive, Newton Abbot TQ12 1YJ
G1 WQU T. Gregg 27 Somerleaze Close, Wells BA5 1UD
G1 WQX A. Peek 7 De Havilland Road Upper Rissington, Cheltenham GL54 2NZ
G1 WQY K. Webber 2 Henniker Road, Ipswich IP1 5HD
G1 WRC Wisbech Amateur Radio & Electronics Club c/o J. Balls 70 Risegate Road, Gosberton, Spalding PE11 4EY
G1 WRD M. Simpson Orchard House, Todwick Grange, Sheffield S26 1JQ
G1 WRE R. Constantine 34 Lynden Close, Ripon HG4 1US
G1 WRF A. Jolly 27 Murrayfield Drive, Brandon, Durham DH7 8TG
G1 WRH G. Marshall 1 Portland Close, Braintree CM7 9NJ
G1 WRN University of Suffolk RS c/o T. Yorke 12 Shanklin Drive, Weddington, Nuneaton CV10 0BA
G1 WRO M. Smith 8 Milldale Road, Farnsfield, Newark NG22 8DQ
G1 WRS Wakefield And District RS c/o D. Burden 16 Milnthorpe Lane, Wakefield WF2 7DE
G1 WRU J. Jinks 27 Taryn Drive, Darlaston, Wednesbury WS10 8XY
GW1 WRV R. Marston 34 Kevin Ryan Court, Georgetown, Merthyr Tydfil CF48 1EE
G1 WRY Northwest Durham Raynet c/o A. White 34 Pain'S Way, Amesbury, Salisbury SP4 7RG
G1 WSA Dr C. Drury Elderberry House, Church Lane, Ulceby DN39 6TB
G1 WSC J. Bland 9 Earl Street, Grimsby DN31 2NB
G1 WSD D. Garratt 87 Garden Road, Eastwood, Nottingham NG16 3FY
G1 WSE J. Frizell 17 St. Johns Terrace, Lewes BN7 2DL
G1 WSF D. Pettican 52 Shepherds Way, Saffron Walden CB10 2AH
G1 WSN J. Spillett Mockbeggar Cottage, Mockbeggar, Ringwood BH24 3NQ
G1 WSW M. Flewitt 38 Laburnum Avenue, Newbold Verdon, Leicester LE9 9LQ
G1 WSZ V. Tosney 126 Norburn Park Witton Gilbert, Durham DH7 6SQ
G1 WTB E. Musson 110 Marples Avenue, Mansfield Woodhouse, Mansfield NG19 9DW
G1 WTH C. Piddock 118 Howley Grange Road, Halesowen B62 0HU
GW1 WTL J. Beachey 24 Trem-Y-Mynydd Court, Blaenavon, Pontypool NP4 9LX
G1 WTN R. Wroe 13 Silverdale Drive, Barnsley S71 2PP
G1 WTS M. Roper 19 Normay Rise, Newbury RG14 6RY
G1 WTW Prof. C. Underwood 4 Hawthorn Road, Godalming GU7 2NE
G1 WTX P. Egan 13 Beechcroft Drive, Guildford GU2 7SA
G1 WTY B. Parkes 11 Hampton Grove, Cheadle Hulme, Cheadle SK8 6DG
GW1 WTZ C. Green 11 Brookfield Close, Gorseinon, Swansea SA4 4GW
G1 WUC M. Garner 40 Studley Road, Harrogate HG1 5JU
G1 WUH K. Carr 41 Surrey Road, Dagenham RM10 8ES
G1 WUM R. Miles Haseley Lodge, Birmingham Road, Warwick CV35 7HF
G1 WUU J. Neate 23 Crossley Moor Road, Kingsteignton, Newton Abbot TQ12 3LE
G1 WUY J. Wilkins 21 Stocks Loke, Cawston, Norwich NR10 4BS
G1 WVD M. Shrago 12 Oakwood Road, Bricket Wood, St. Albans AL2 3PU
G1 WVK J. Power 45 Grace Gardens, Cheltenham GL51 6QE
G1 WVM R. Vowles 47 Tyndale Avenue, Yate, Bristol BS37 5EX
G1 WVP P. Marsh 182 Oldbrook Boulevard, Oldbrook, , Milton Keynes MK6 2HG
G1 WVR Cheltenham A.R.A c/o D. Lowe 21 Farndon Road, Market Harborough LE16 9NW
G1 WVS P. Gibson 17 Nene Side Close, Badby, Daventry NN11 3AD
G1 WVV R. Sutton 28 Shrubbery Gardens, Wem, Shrewsbury SY4 5BX
G1 WVW R. Jones 8 Downing Avenue, Newcastle ST5 0JY
G1 WVZ R. Mccutcheon 13 The Beeches, Rugeley WS15 2QY
G1 WWA R. Williams Coombe Farm Cottage, Stottesdon, Kidderminster DY14 8LS
G1 WWB R. Eeles 23 Elgin Avenue, Ashford TW15 1QE
GW1 WWE J. Peake Winley, 70 Higher Lane, Swansea SA3 4PD
G1 WWH A. Benn Burneston, Bedale DL8 2HT
G1 WWI M. Dronfield White Lodge Farm, High Bradfield, Sheffield S6 6LJ
G1 WWP J. Sharpe 10 Stocking Green Close, Hanslope, Milton Keynes MK19 7NH
GW1 WWW Lord W. Edmondson Glaslyn, Penysarn LL69 9YB
G1 WWY L. Donald 53 Andrews Way, Raunds, Wellingborough NN9 6RD
G1 WXC D. Blackman-Wells 15 Purbeck Place, Littlehampton BN17 5DP
G1 WXF J. Pearson 17 Hebden Avenue, Woodloes Park, Warwick CV34 5XD
G1 WXK M. Bell 151 Towngate, Ossett WF5 0PP
G1 WXS P. Springall 31 The Orchards, Epping CM16 7BB
G1 WXT M. Moorecroft 4 St. Davids Road, Locks Heath, Southampton SO31 6EP
G1 WXU G. Parsons 73 Worthing Avenue, Elson, Gosport PO12 4DB
G1 WXW P. Prescott 13 The Boltons, Waterlooville PO7 5QR
G1 WYA R. Webb 1, London Road, Oldham OL1 4BJ
G1 WYB Lord E. Coupe Killidina, 28 Wellington Road, Blackburn BB2 2NQ
G1 WYC S. Smith 82 Wignals Gate, Holbeach, Spalding PE12 7HR
G1 WYD A. Darlington 15 Kestrel Close, Carterton OX18 3LS
G1 WYG D. Biginton 67 Capstone Road, Bromley BR1 5NA
G1 WYM Dr M. Cheema Lower Ebford Barton, Ebford, Exeter EX3 0RA
G1 WYN F. Wilkinson Hollytree The Common Wellington Heath, Ledbury HR8 1LY
G1 WYP H. Milsom 1 Wyld Court, Blunsdon, Swindon SN25 2EE
GM1 WYV A. Henderson 30 Pentland Crescent, Larkhall ML9 1UR
GI1 WYZ R. Kennedy 3 St. Annes Crescent, Newtownabbey BT36 5JZ
G1 WZG J. Endicott 16 Packs Close, Harbertonford, Totnes TQ9 7TL
GW1 WZI J. Cartwright 20 Castlefield Place, Cardiff CF14 3DU
G1 WZK D. Heaton 39 Bridgwater Road, Romford RM3 7UB
G1 WZM C. Turner 2, Martins Mews, Haverhill CB9 7FU
G1 WZO L. Leach Leyland, The Street, Gloucester GL2 7ED
G1 WZQ A. Utting11 20 Davenport Road, Leicester LE5 6SA
G1 XAA H. Jacklin 26 Rockmill End, Willingham, Cambridge CB24 5HY
G1 XAJ J. Franklin 16 Mountbatten Drive, Colchester CO2 8BH
G1 XAL A. Dangerfield Brookside, High Street, Gloucester GL2 7LW
G1 XAM J. Bryant 12 Dale Tree Road, Barrow, Bury St. Edmunds IP29 5AD
G1 XAP P. Whittingham 28 Wedge Avenue, Haydock, St. Helens WA11 0DY
GW1 XAS J. Hume 2 Llain Wen, Pentrefelin, Amlwch LL68 9PD
G1 XBE T. Beecher 77 Grime Lane, Sharlston Common, Wakefield WF4 1EH
GW1 XBG P. Smith 35 Terrace Road, Swansea SA1 6HN
GM1 XBK K. Mcclure 9 Cumnock Road, Mauchline KA5 5AE
G1 XBL B. Darby Pippins, Green Street, Worcester WR5 3QB
G1 XBR S. Loney 4 Mendip Road, Southampton SO16 4BN
G1 XCB P. Davies 91 Station Road, Hadfield, Glossop SK13 1AR
G1 XCK W. Potter Flat, 1 Tabernacle Walk, Blandford Forum DT11 7DL
G1 XCY E. Knott 24 Walsingham Way, Ely CB6 3AL
G1 XDJ H. Opitz 26 Holme Court, Lower Warberry Road, Torquay TQ1 1QR
G1 XDS J. Fyson One Redlands Estate, Ibstock LE6 1HT
G1 XDV T. Gale 1A Aldridge Road, Streetly, Sutton Coldfield B74 3TU
GM1 XEA P. Thomson 13 Westwood Drive, Westhill AB32 6WW
GM1 XEB M. Mcculloch 6 Learmont Place, Milngavie, Glasgow G62 7DT
G1 XEH R. Burton 18 Churchfield, Harpenden AL5 1LL
G1 XEP L. Lambert 124 Frankland Road, Croxley Green, Rickmansworth WD3 3AU
G1 XES E. Turner 1104 Wimborne Road, Bournemouth BH10 7AA
G1 XET C. Burton 14 Fotherley Road, Mill End, Rickmansworth WD3 8QG
GW1 XFB D. Evans Bwthyn Bach, 2 Old Village Road, Barry CF62 6RA
G1 XFE J. King 40 Galway Avenue, Chaddesden, Derby DE21 6TR
G1 XFL K. Lanham 22 Ascot Close, Ladywood, Birmingham B16 9EY
G1 XFM J. Shaw 33 Park Farm Close, Horsham RH12 5EU
G1 XFO M. Price 67 Broadway, Oldbury B68 9DP
G1 XFR B. Bance 36 Leafy Oak Road, Grove Park, London SE12 9RS
G1 XGE D. Hutchinson 32 The Causeway, Kingswood, Hull HU7 3AL
G1 XGM M. De-Wynter 8 Eldon Place Cutler Heights, Bradford BD4 9JH
G1 XGN M. Brady 67 Boardman Fold Road Middleton, Middleton M24 1QD
G1 XGP S. Blinkhorn 12 Eloura Lane, New South Wales 2577 Australia
G1 XGW A. Gray 12 Peak Close, Oldham OL4 2TH
G1 XGZ D. Richards Flat 8, Thackeray Court, London SW3 3LB
G1 XHA J. De Bank 5 Horn Hill View, Beaminster DT8 3PJ
G1 XHO T. Williams 145 Bulwell Lane, Old Basford, Nottingham NG6 0BS
G1 XHR E. Goodwin Hankelow Court, Hall Lane, Crewe CW3 0JB
GM1 XHZ T. Valentine 4 Angus Cottages, Friockheim, Arbroath DD11 4SR
GI1 XIB J. Wilkinson 67 Glenwood, Ahoghill, Ballymena BT42 1GW
G1 XIC Callington ARS c/o P. Taylor Merrivale, Portuan Road, Looe PL13 2DW
G1 XIE R. Dyer 21 Allden Avenue, Aldershot GU12 4AG
G1 XIH L. Taylor 76 Sidney Road, Blackley, Manchester M9 8AT
G1 XII T. Lovatt 5 Acre Rise, Willenhall WV12 4SL
GM1 XIN W. Allan Corse Farm, Kininmonth, Peterhead AB42 4JU
G1 XIO A. Faram 4 Wellington Road, Gillingham ME7 4NN
G1 XIV G. Reynolds The Thatched Cottage, St. Thomas Drive, Bognor Regis PO21 4TN
G1 XIY Dr M. Nottingham 11 Taverners Drive, Ramsey, Huntingdon PE26 1SF
GM1 XJE Dr J. Hopkins 9 Pathfoot Avenue, Bridge Of Allan, Stirling FK9 4SA
GW1 XJJ H. Worgan 29 Mayfield Avenue, Laleston, Bridgend CF32 0LH
G1 XJK F. Tilley 37B Fant Lane, Maidstone ME16 8NP
G1 XJM R. Davidson 17 Willows Avenue, Alfreton DE55 7ER
G1 XJN D. Jones 12 Brockhill Close, Kettering NN15 7DS
G1 XJO N. Snowden 11 Marion Drive, Shipley BD18 2EY
G1 XJT A. Nichols Amherst Harnham Lane, Withington GL54 4DD
G1 XJZ D. Layton 7 Turton Street, Kidderminster DY10 2TH
G1 XKB N. Bowen 2 Thorncliffe Road Great Barr, Birmingham B44 9DB
G1 XKD G. Lawton 23 Fiske Court, Cavendish Road, Sutton SM2 5ER

G1 XKJ K. Higlett 3 Clover Way, Killinghall, Harrogate HG3 2WE
G1 XKL B. Smith 1 Hirsts Cottages, Spa Lane, Ormskirk L40 6JG
G1 XKN A. Chambers 34 Haunchwood Drive, Sutton Coldfield B76 1JR
G1 XKQ B. Neate 30 Berry Avenue, Paignton TQ3 3QN
G1 XKY E. Marsh 15 Beacon Close, Rubery, Birmingham B45 9DA
G1 XLE P. Bryan 2 Regency Court, Armitage Road, Brereton, Rugeley WS15 1PE
G1 XLG C. Proctor 24 Orchard Way, Southam CV47 1EX
GM1 XLH C. Cullingworth Lochmoss, Ythanwells, Huntly AB54 6HA
G1 XLL P. Greetham Flat 10, Hillman House, Coventry CV1 1FZ
G1 XLN J. Banks 2 Birchlands Road, Stoke-on-Trent ST1 6TW
G1 XLT R. Perrat 18 Petts Hill, Northolt UB5 4NL
G1 XLW A. Harrington 44 Fairburn Crescent, Pelsall, Walsall WS3 4PU
GD1 XMA M. Haley Yn Croit, Ballamanagh Road, Sulby, Isle of Man IM7 2HB
G1 XMH R. Fisher White House, Slough Road, Manningtree CO11 1NS
G1 XMI J. Brown 78 Park Way, St. Austell PL25 4HR
G1 XMP C. Brookes 58 Brookwood Drive, Stoke-on-Trent ST3 6HY
G1 XNC R. Gearing 6 Boughton Close, Gillingham ME8 6ND
G1 XNG C. Whitehead 6 Welbeck Street, Sutton-in-Ashfield NG17 4AY
G1 XNI N. Dingle 29 Castle View, Witton Le Wear, Bishop Auckland DL14 0DH
G1 XNK R. Rafter 8 Bishops Walk, Ilchester, Yeovil BA22 8NS
G1 XNN R. Harding Highview, High Road, Wallingford OX10 0QT
G1 XNX M. Cowell 105 Belgrave Road, Darwen BB3 2SF
G1 XOG I. Wilson 131 Weatherly Road, Torbay, Auckland 630 New Zealand
GM1 XOI Mid Lanark ARS c/o C. Welsh 28 Peacock Wynd, Motherwell ML1 4ZL
G1 XOT B. Blake Ty Capel Tynygraig, Ystrad Meurig SY25 6AE
G1 XOW S. Wragge Treetops, Priory Road, Thurgarton, Nottingham NG14 7GW
G1 XOZ C. Harding 24 Bryer Close, Bridgwater TA6 6UR
G1 XPD L. Wheatley 25 Hobbis House, Redditch Road, Birmingham B38 8LS
GM1 XPE J. Graham Lodge, Stronsay, Orkney KW17 2AN
G1 XPF A. Duell 3 Jail Lane, Biggin Hill, Westerham TN16 3SA
G1 XPI R. Wilson 10 Ringway Garforth, Leeds LS25 1BN
GI1 XPV Dr K. Nesbitt 2 Church Road Gracehill, Ballymena BT42 2NL
G1 XPW M. King 104 Green Lane, Vicars Cross, Chester CH3 5LE
G1 XQI K. Bates Flat 2, 26 Leinster Road West, Harolds Cross DUBLIN 6 Ireland
G1 XQP J. Jackson Greengable Upcott, Bishops Hull, Taunton TA4 1AQ
GW1 XQT F. Farmer 3, Old Post Office Gardens, Four Crosses, , Llanymynech SY22 6RG
G1 XRE S. Staton 6 Greenhowsyke Lane, Northallerton DL6 1HP
G1 XRF B. Rogers 5 Springfield Road, Ruskington, Sleaford NG34 9HG
G1 XRJ S. Hancock 8 Elanor Road, Sandbach CW11 3FZ
G1 XRM M. Braybrook 12 Bossington Close, Rownhams, Southampton SO16 8DW
G1 XRO C. Frost 110 Spring Hill, Weston-Super-Mare BS22 9BD
G1 XRQ J. Koenig 216 Bretch Hill, Banbury OX16 0LU
G1 XRT R. Taylor 24 Hoestock Road, Sawbridgeworth CM21 0DZ
G1 XSA S. Cattle 5 Highworth Drive, Newcastle upon Tyne NE7 7FB
G1 XSM P. Elwell Woodcroft, Vann Lake Road, Ockley, Dorking RH5 5JB
G1 XSQ C. Nightingale 3 Cedar Rise Crookham Common, Thatcham RG19 8DY
G1 XST Northants Raynet Group c/o S. Davies 64 Oakfields, Worth, Crawley RH10 7FL
G1 XSV M. Scarr 15 Biddesden Lane, Ludgershall, Andover SP11 9PG
G1 XTA S. Hodson Flagstones, 12 Duns Tew, Bicester OX6 4JR
G1 XTD I. Clark Flat, Redhill Farm, Penrith CA11 0DT
GI1 XTK B. Braniff 5 Cintons Park, Downpatrick BT30 6NS
GW1 XUD R. Andrews 270 Barry Road, Barry CF62 8BJ
G1 XUE G. Henne 57 Heaf Gardens, Bentley Close, Aylesford ME20 7SF
G1 XUH M. Thornton 46 Lavender Court, Croft Road, Barnsley S70 3FG
G1 XUU S. Bishop 22 John St., Brightlingsea, Colchester CO7 0NA
G1 XUW D. Austin 17 Patricia Avenue, Horstead, Norwich NR12 7EW
GW1 XVC S. Ward Beech Cottage, Saron Road, Penperlleni, Pontypool NP4 0BN
G1 XVD C. Snow 77 Oxford Drive, Hadleigh, Ipswich IP7 6AW
G1 XVF T. Pottage 18 Pennine Close, Huthwaite, Sutton-in-Ashfield NG17 2QD
G1 XVL B. Perry 152 Stanborough Avenue, Borehamwood WD6 5LR
GW1 XVM J. Duggan 112 Gaer Park Drive, Newport NP20 3NR
G1 XVR D. Briggs 17 The Lonnen, South Shields NE34 8EJ
G1 XVW A. Pogorzelski 28C Mosslea Road, Penge, , London SE20 7BW
G1 XVY R. Sacharewicz 15 Milford Close, Walkwood, Redditch B97 5PZ
G1 XWD A. Rhodes 2 Kent Avenue, Theddlethorpe, Mablethorpe LN12 1QE
G1 XWK A. Rich The Court House, Wadborough Road, Worcester WR7 4RF
G1 XWM M. Cox 17 Tybalt Close, Heathcote, Warwick CV34 6XB
G1 XWN G. Andrews 22 Arnhem Grove, Braintree CM7 5UQ
G1 XWO F. Williams 15 Hartsbourne Way, Stafford ST17 4NR
G1 XWS H. Seatory Ivydene, The Street, Woodbridge IP12 3QU
G1 XWZ F. Millbank Room 216 Kate House, Pitchill House Nursing Home, Evesham WR11 8SN
G1 XXE P. Yeates 9 Arlington Road, St. Annes, Bristol BS4 4AF
G1 XXF J. Ellison 68 Rocket Way, Forest Hall, Newcastle upon Tyne NE12 9RL
G1 XXH R. Tapp 26A Main Road, Grendon, Northampton NN7 1JW
G1 XXR S. Austin 5 Mercia Road, Baldock SG7 6RZ
G1 XXV T. Blackmore 56 Fraser Close, Shoeburyness, Southend-on-Sea SS3 9YS
G1 XXW P. King 2 Ebenezer Cottages, Thorney Road, Peterborough PE6 7UB
G1 XYD A. Bates 29 Juler Close, North Walsham NR28 0SY
G1 XYF C. Hudson 8 College Road, Bredon, Tewkesbury GL20 7EH
G1 XYG R. Butler 15 Bracknell Crescent, Nottingham NG8 5EU
G1 XYN I. Pritchard 8 Hoon Avenue, Newcastle ST5 9NY
G1 XYO S. Glazzard 109 Highfields Road, Chasetown, Burntwood WS7 4QS
G1 XYR D. Searle 33 Claypool Road, Kingswood, Bristol BS15 9QJ
G1 XYS A. Brown 5 Somersby Drive, Kenton, Newcastle upon Tyne NE3 3TN
G1 XYV M. Minshull 12 Dunnett Close, Attleborough NR17 2NG
G1 XYZ Kings Lynn ARC c/o E. Haskett 23 Gloucester Road, King's Lynn PE30 4AB
G1 XZB J. Rawlingson 1 Wadham St., Penkhull, Stoke on Trent ST4 7HF
G1 XZG D. Collins 5 Elmwood Close, Lincoln LN6 0LZ
GW1 XZI R. Magwood 13 Inverness Place, Cardiff CF24 4RU
G1 XZQ R. Carville 66 Ludlow Road, Paulsgrove, Portsmouth PO6 4AE
G1 XZV Air Training Corps c/o J. Heys 2 Oakenhill Walk, Bristol BS4 4LP
G1 XZW R. Hudson 7 Grange Avenue, Luton LU4 9AS
G1 XZX B. Strutt 9 Magdalen Road, Clacton-on-Sea CO15 3LZ
G1 YAB C. Rogers 221 Dales Road, Ipswich IP1 4JY
G1 YAE B. Thompson 1 Littlehoughton Farm Cottages, Littlehoughton, Alnwick NE66 3JZ
G1 YAF A. Tyler 16 Harridge Road, Leigh-on-Sea SS9 4HA
G1 YAH J. Mcsoley 88 Rodings Avenue, Stanford-le-Hope SS17 8DT
G1 YAS D. Elliott 48A Great Lane, Reach, Cambridge CB5 0JF
G1 YBA I. Hardaker 31 Shearing Hill, Gedling, Nottingham NG4 3GY
G1 YBB S. Clements 46 Brampton Road, Newton Farm, Hereford HR2 7DF
GW1 YBC S. Dewitt 21 Brownhills, Gorseinon, Swansea SA4 4AB
GW1 YBF L. Ward 11 Verlands Way, Pencoed, Bridgend CF35 6TY
G1 YBG Dr J. Ballance Orchid Bank, Woolhope, Hereford HR1 4RQ
G1 YBI A. Jones 43 Oakleigh Road, Droitwich WR9 0RP
G1 YBK J. Fyson 1 Redlands Estate, Ibstock, Leicester LE6 1HT
G1 YBM J. Pedley 92 Ashfield Drive, Moira, Swadlincote DE12 6HQ
G1 YBT J. Bagshaw 2 Boulton Court, Robin Hood Road, Skegness PE25 3QU
G1 YCK M. Travis 10 Victoria Road, Kearsley, Bolton BL4 8NR
G1 YCM A. Laughlan 33 Park House, Gorseyfields, Manchester M43 6DX
G1 YCN D. Lewis 81 Ashton Avenue, Rainhill, Prescot L35 0QR
G1 YCR R. Lawrence 82 Moseley St., Southend on Sea SS2 4NN
G1 YDA M. Davies 10 Rue Alphonse Delaveau, Pouzauges 85700 France
G1 YDD A. Forster 56 Tantobie Road, Denton Burn, Newcastle upon Tyne NE15 7DQ
G1 YDG A. Miles Yew Tree House, Main Street, Wantage OX12 0HT
G1 YDI C. Lambeth 36 Mill Lane, Oxford OX3 0QA
G1 YDJ T. Polley 9 Otter Road, Clevedon BS21 6LQ
G1 YDQ J. Carpenter 34B Carey Park, Killigarth, Looe PL13 2JP
GI1 YEA L. O'Flaherty 1 Ravensdale Villas, Newry BT34 2PG
G1 YED W. Ross-Fraser 47 Lichford Road, Sheffield S2 3LB
G1 YEH J. Davis 179 Bredon, Yate, Bristol BS37 8TG
G1 YEP F. Russell 7 Glenmore Avenue, Liverpool L18 4QE
G1 YES B. Underhay 24 Rutland Road, Southall UB1 2UP
G1 YEU M. Eales 32 Selston Drive, Nottingham NG8 1DE
G1 YEV Dr D. Martin 73 Summerfields Way, Ilkeston DE7 9HE
G1 YEW R. Hill Grange Barn, Funtington, Chichester PO18 9LN
G1 YEZ A. Lord 66 Salcombe Drive, Glenfield, Leicester LE3 8AF
G1 YFA J. Rymsza 24 Green Lane, Studley B80 7HD
G1 YFC P. Neades 57 Bullingham Lane, Hereford HR2 6RU
G1 YFD S. Lycett 27 Ropewalk, Alcester B49 5DD
G1 YFE J. Dent 90 Eastwood, Chatteris PE16 6RX
G1 YFG R. Szemeti The Stables, Hoarstone Court Trimpley Lane, Bewdley DY12 1RB
G1 YFI J. Simmonds 94 Gravel Hill, Tile Hill, Coventry CV4 9JH
GW1 YFP M. Hearne Mora, Rhydypandy Road, Morriston SA6 6NX
G1 YFQ D. Scothern 5 Wilkinson Drive Middle Rasen, Market Rasen LN8 3LD
G1 YFT R. Allsopp 271 Wigston Lane, Aylestone, Leicester LE2 8DL
G1 YGP S. Jarman 55 The Meadows, Todwick, Sheffield S26 1JG
GM1 YGV R. Johnstone 10 Lundy Road, Inverlochy, Fort William PH33 6NX
GM1 YGW Dr G. Craib 1C Cherry Bank, Dunfermline KY12 7RG
G1 YGY C. Weaver 11 Thirlmere, Swindon SN3 6LA
GW1 YHA P. George 24 Ty Fry Close, Brynmenyn, Bridgend CF32 8YB
G1 YHB J. Moggeridge 22 St. Michaels Court, Faircross Avenue, Weymouth DT4 0DS
G1 YHE D. Coate 74 Wimborne Road, Poole BH15 2BZ
G1 YHG M. Kennedy Milestones, Blandford Forum DT11 9DW
G1 YHI K. Davies 2 Orchard Close, Lytchett Minster, Poole BH16 6JH
G1 YHJ G. Williams 2 Cotton Close, Broadstone BH18 9AJ
GW1 YHL D. Crawshaw 12 Glanmor Crescent, Uplands, Swansea SA2 0PJ
G1 YHN S. Rhodes 221 Ormonds Close, Bradley Stoke, Bristol BS32 0DW
G1 YHP J. Hogg 1 Deepdale, Guisborough TS14 8JY
G1 YHV S. Briscoe 8B Corfe View Road, Corfe Mullen, Wimborne BH21 3LZ
G1 YIL A. Kaye 2 Church Place 135 Edward Road, Balsall Heath, Birmingham B12 9JQ
G1 YIQ J. Stafford 6 Gardners Drive, Hullavington, Chippenham SN14 6EL
G1 YIZ A. Ward 42 Felstead Crescent, Sunderland SR4 0AB
G1 YJB G. Evans 16 Kynaston Drive, Wem, Shrewsbury SY4 5DE
G1 YJF G. Moulds Mountain Ash, 2 Fremantle Close, Chelmsford CM3 5TY
G1 YJH M. Blackman Clough Head, Hollinsclough, Buxton SK17 0RG
G1 YJI P. Kay 97 Avenue Road, London N14 4DH
G1 YJJ R. Colman 197 Coppins Road, Clacton-on-Sea CO15 3LA
G1 YJL W. Pond The Wheatlands, Calais Street, Sudbury CO10 5JA
G1 YJQ J. Duffy Flat 14, The Sycamores, Newcastle upon Tyne NE4 7ER
G1 YJR J. Davies 70 Ash Road, Sandiway, Northwich CW8 2PB
G1 YJY P. Sengupta 48 Badger Close, Guildford GU2 9WA
GM1 YKE J. Campbell 16 Barony Road, Auchinleck, Cumnock KA18 2LL
G1 YKI J. Heathfield 82 Auriel Avenue, Dagenham RM10 8BT
G1 YKK S. O'Connor 32 Whitfield Cross, Glossop SK13 8NW
G1 YKL J. Brackenridge 21 St. Mark Road, Deepcar, Sheffield S36 2TF
GW1 YKT W. John 4 Heol Y Bryn, Rhiwbina, Cardiff CF4 6HY
G1 YKX M. Rouse 105 Great Spenders, Basildon SS14 2NS
GW1 YKY S. Jones 14 Plantation Drive, Croesyceiliog, Cwmbran NP44 2AN
G1 YKZ R. Burt 11 Long Common, Heybridge, Maldon CM9 4US
G1 YLB S. Doyle 2 The Greenways, Paddock Wood, Tonbridge TN12 6LS
G1 YLE P. Adams 25 Main Road, Kesgrave, Ipswich IP5 1AQ
G1 YLG A. Hodkin 18 Habershon Drive, Chapeltown, Sheffield S35 2ZT
G1 YLJ A. Hunt 14 Sandalwood Close, Willenhall WV12 5YJ
G1 YLM E. Bradshaw 38 Whiteford Drive, Kettering NN15 6HH
G1 YLN M. Swetman 11 Outer Circle, Taunton TA1 2BS
G1 YLV M. Bayliss 2 Plattens Court, Wroxham, Norwich NR12 8SQ
G1 YMA W. Scoles 26 The Close, Brancaster Staithe, King's Lynn PE31 8BS
G1 YMC S. Fitzpatrick 19 Claremont Falls, Killigarth, Looe PL13 2HT
GM1 YME J. Hein 78 Montgomery Street, Edinburgh EH7 5JA
G1 YMH M. Homer 86 Victoria Road, Brierley Hill DY5 1DB
G1 YMJ A. Smith 6 Norton Crescent, Towcester NN12 6DN
G1 YMP S. Ellis 6 Newark Road, Hindley, Wigan WN2 3HR
G1 YMR P. Webster 117 Warley Road, Blackpool FY1 2RW
G1 YMV H. Johnson 2 Greenbank Avenue, Storth, Milnthorpe LA7 7JP
G1 YMY A. Hussain 4 Riverside, Chadderton, Oldham OL1 2TX
G1 YNH C. Arundel 54 Broadmead, Castleford WF10 4SE
G1 YNJ M. Rocke Orchard House, 55 Tarvin Road, Chester CH3 5DY
G1 YNO B. Surtees 5 Haweswater Grove, West Auckland, Bishop Auckland DL14 9LQ
G1 YNQ J. Crow 71 Stockshill Road, Ashby, Scunthorpe DN16 2LQ
G1 YOA G. Hawkins 11523 Sun Ray Court, San Diego 92131 United States
G1 YOF A. Lockwood 9 Hartley Road, Exmouth EX8 2SG
G1 YOS M. Avenell Lime House, Worlds End, Newbury RG20 8SD

GJ1 YOT N. Paisnel 11 Bon Air Apartments, La Grande Route De La Cote, St. Clement JE2 6SE Jersey
G1 YOU S. Nicholls Fieldway, The Street, Eyke, Woodbridge IP12 2QG
G1 YOY J. Bowen 16 Cotham Lawn Road, Bristol BS6 6DU
G1 YPH A. Roberts 18 Surtees Grove, Stoke-on-Trent ST4 3HH
GM1 YPJ L. Davies 24 Ardgour Road, Caol, Fort William PH33 7PQ
G1 YPM R. Northcott 32 Lichfield Road, Exwick, Exeter EX4 2EU
G1 YPR G. Dearden 125 Campsall Field Road, Wath-Upon-Dearne, Rotherham S63 7ST
G1 YPT G. Hartshorn 11 Lime Avenue, Ripley DE5 3HD
G1 YPU E. Rowberry 69 Alpha Terrace, Trumpington, Cambridge CB2 9HS
G1 YPZ L. Pritchett Flat 75, Dalehead, London NW1 2JL
G1 YQI P. Bennett 7 Woburn Avenue, Firwood Ind Est, Bolton BL2 3AY
G1 YQL C. Stagg 559 Dividy Road, Stoke-on-Trent ST2 0BX
GW1 YQM R. Evans Maesyronnen, Sarnau, Llanymynech SY22 6QL
G1 YQN P. Mcshea Heathercot, Cross Drive, Maidstone ME17 3NP
G1 YQP M. West 27 Nidderdale Road, The Meadows, Wigston LE18 3XW
G1 YQU J. Stapleford 7 Garfield Road, Hugglescote, Coalville LE67 2HU
G1 YQY W. Oakes 2 Hillcrest Scotton, Catterick Garrison DL9 3NJ
G1 YRC York ARC c/o A. Palfrey 5 Ings View, York YO30 5XE
GM1 YRD J. Jones Kirkland Of Glencairn, Kirkland Moniaive, Thornhill DG3 4HD
G1 YRE S. Piper Willow End, The Street, Ipswich IP6 9HG
G1 YRF D. Buggs 2 Archway Cottages Valley Road, Leiston IP16 4AR
G1 YRJ M. Stott 8 Kingfisher Way, Stowmarket IP14 5BB
G1 YRM C. Mead 32 Sandy Road, Potton, Sandy SG19 2QQ
G1 YRQ R. Nock 43 Delph Drive, Brierley Hill DY5 2LQ
G1 YRR W. Fry 227 London Road North, Merstham, Redhill RH1 3BN
G1 YRY S. Roberts 23 Deal Court, Haldane Road, Southall UB1 3NT
G1 YSA M. Crick 85 Ashurst Road, London N12 9AU
GI1 YSG P. Kennedy 29A Barnfield Road, Lisburn BT28 3TQ
G1 YSX D. Taylor 103 Southend, Garsington, Oxford OX44 9DL
G1 YTG A. Thackray 39 Totnes Close, Corby NN18 8DB
G1 YTL H. Vyvyan 13 Shearwater Close, Peel Common, Gosport PO13 0RB
G1 YTO K. Wenman 2 Hythe Road, Sittingbourne ME10 2LR
G1 YTV S. Fisher Farlands, Lower Rillaton, Callington, Lower Rillaton PL17 7PF
G1 YTX T. Clayton 40 Morrison Road, Darfield, Barnsley S73 9ED
G1 YUB B. Harrison 15 Helmington Terrace, Hunwick, Crook DL15 0LQ
G1 YUL J. Robson 28 Eastfield Street, Sunderland SR4 7SA
G1 YUN K. Hetherington 29 Broomridge Avenue, Newcastle upon Tyne NE15 6QN
G1 YUS A. Lees 692 Walmersley Road, Bury BL9 6RN
G1 YUU A. Bagworth 127 Barnsley Road, Darfield, Barnsley S73 9PE
G1 YUX J. Garnett 21 Vicarage Close, Mossley Hill, Mossley Hill, Liverpool L18 7HU
G1 YVI K. Biddlecombe 29 Stone Close, Worthing BN13 2AU
G1 YVS C. Newby-Robson 1 Bramley Drive, Offord D'Arcy, St. Neots PE19 5SF
G1 YVV Elderslie ARS c/o I. Coleman 69 Glebelands, West Molesey KT8 2PY
G1 YVZ A. Bell 159 Hounslow Road, Hanworth, Feltham TW13 6PX
G1 YWI A. Williams 26 Matlock Road, Bloxwich, Walsall WS3 3QD
G1 YWN A. Whitworth 183 Logan St., Bulwell, Nottingham NG6 9FX
G1 YWY M. Jones 16 Cumnock Road, Castle Cary BA7 7FE
G1 YXA B. Dixon 16 Dyrham Parade, Patchway, Bristol BS34 6EF
G1 YXH R. Harrison 14 St. Leonards Avenue, Chatham ME4 6HL
G1 YXJ A. Palmer 14 Garibaldi Road, Redhill RH1 6PB
GW1 YXR N. Williams 31 Syr Davids Avenue, Cardiff CF5 1GH
G1 YXT C. Wise 28 Southlands, East Grinstead RH19 4BZ
G1 YXY L. White The Garden Flat, 47 Hamerton Road, Gravesend DA11 9DX
G1 YYC F. Karlinski 100 Lindsay Avenue, Wakefield WF2 8AS
G1 YYD E. Brown 25 Cork Road, Lancaster LA1 4BD
G1 YYH J. Heaton 85 Morris Green Lane, Bolton BL3 3JD
G1 YYL T. Mulloy 49 Barons Close Kirby Muxloe, Leicester LE9 2BW
G1 YYP M. Illston 4 The Sett, Oxhill, Warwick CV35 0RE
G1 YYU W. Fludgate Holly Lodge, Thorpe Bank, Little Steeping, Spilsby PE23 5BB
G1 YYY Braintree Raynet Group c/o D. Willicombe 26 Falkland Court, Braintree CM7 9LL
GW1 YZF D. Edwards 240 Berthin, Greenmeadow, Cwmbran NP44 4LB
G1 YZH R. Baxter 4 Kendal Gardens, Woodley, Stockport SK6 1BL
G1 YZJ R. Rennison Foxhall Cottage, Kelshall, Royston SG8 9SE
G1 YZT E. Fenlon 17 Hawes Avenue, Ramsgate CT11 0RN
G1 ZAA A. Tice 20 Manor Road, Middle Littleton, Evesham WR11 8LL
G1 ZAK A. Powney 16 Westbrook Way, Wombourne, Wolverhampton WV5 0EA
G1 ZAR S. Tyler 43 Wharf Road, Pinxton, Nottingham NG16 6LH
G1 ZAW M. Smoker 41 Queens Gardens, Dartford DA2 6HZ
G1 ZAY A. Robinson 31 Heathview Road, Socketts Heath, Grays RM16 2RS
G1 ZBB S. Brown 6 Pathfinder Way, Ramsey, Huntingdon PE26 1LX
G1 ZBG M. Bason 52 Wroslyn Road, Freeland, Witney OX29 8HH
G1 ZBH G. Brock 148 Lonsdale Drive, Rainham, Gillingham ME8 9HX
G1 ZBJ A. Manning 12 Clifford Drive, Heathfield, Newton Abbot TQ12 6GX
G1 ZBL N. Marsh 16 Laurel Close, North Warnborough, Hook RG29 1BH
G1 ZBO Dr E. Mclusky 11 Ripon Road, Killinghall, Harrogate HG3 2DG
G1 ZBP A. Whipp 114 Lower Manor Lane, Burnley BB12 0EF
G1 ZBU G. Newby 77 Darby Road, Garston, Liverpool L19 9AN
G1 ZBW W. Baker El-Granaro, Brampton, Appleby-in-Westmoreland CA16 6JS
G1 ZBY S. Parker 8 Greenbank Drive, Lincoln LN6 7LQ
G1 ZCC C. Feather 10 Thruffle Way, Bar Hill, Cambridge CB23 8TR
G1 ZCS E. Davis 10 Fairfield Drive, Lowestoft NR33 8QG
G1 ZDG P. Whittaker 156 Stoughton Road, Guildford GU2 9PG
G1 ZDR J. Angus 8 Gravel Road, Bromley BR2 8PF
G1 ZDT A. Gregory 9 Fordbridge Road, Ashford TW15 2TD
G1 ZDU B. Rowles 4 Milton Road, Aston Clinton, Aylesbury HP22 5LA
G1 ZDX M. Bodecott 1 The Park, St. Pegas Road, Peakirk, Peterborough PE6 7NG
G1 ZDY C. Tipp 27 Lakeland Avenue, Bognor Regis PO21 5FA
G1 ZEA P. Jones 21 Hill Top Rise, Harrogate HG1 3BW
G1 ZEC G. Stevens 25 Avenue Road, New Milton BH25 5JP
G1 ZED B. Haworth 139 Manchester Road, Accrington BB5 2NY
G1 ZEI J. Wyatt Ciampia, 10 St. Georges Hill, Perranporth TR6 0DZ
G1 ZEK D. Ault 68 Moira Dale Castle Donington, Derby DE74 2PJ
G1 ZEU S. Aspey 7 Foresters Path, School Aycliffe, Newton Aycliffe DL5 6TA
G1 ZEW D. Pentin 35 Seafore Close, Liverpool L31 2JS
G1 ZEX R. Davies 71 Higher Croft Road, Lower Darwen, Darwen BB3 0QT
G1 ZFB K. Barton 67, Grange Rd Flat3, Ramsgate CT11 9LP
G1 ZFD J. Davies 71 Higher Croft Road, Lower Darwen, Darwen BB3 0QT
G1 ZFF T. Voisey 26 Gorlands Road, Chipping Sodbury, Bristol BS37 6LA
G1 ZFG J. Stephenson 5 Hunstrete, Pensford, Bristol BS39 4NT
G1 ZFS N. Woolard 159 Medway Road, Worcester WR5 1LL
GW1 ZFX J. Milosevic 38 Thornhill Close, Upper Cwmbran, Cwmbran NP44 5TQ
G1 ZGF R. Jackson 37 Carisbrooke Road, Harpenden AL5 5QS
G1 ZGH J. Sharpe 204A Featherstone Lane, Featherstone, Pontefract WF7 6AH
G1 ZHD A. Gilmore Ashfields, Naseby Road, Clipston, Market Harborough LE169RZ
GW1 ZHI C. Owens 7 Frondeg, Southsea, Wrexham LL11 6RH
G1 ZHL Prof. M. Dharas 225 Redmile Walk, Peterborough PE1 4UR
G1 ZHN M. Griffiths 2 Muirway, Benfleet SS7 4LS
G1 ZHZ J. Hall 27 Quarry Hill Road, Ilkeston DE7 4DA
G1 ZIM S. Agnew Rose Mount, The Hill, Millom LA18 5HE
GM1 ZIV J. Large 9 Maitland Terrace, Kildrochat, Stranraer DG9 9EX
G1 ZJK D. Ellard 35 Edgehill Drive, Daventry NN11 0GR
G1 ZJP R. Offer Chapel Yard Cottage, Quadring Eaudyke, Spalding PE11 4QB
G1 ZJQ D. Smith 44 Yarmouth Drive, Cramlington NE23 1TS
GW1 ZKE M. Grindle 57 Islwyn Street, Cwmfelinfach, Newport NP11 7HY
GW1 ZKN R. Ogden Plas Yn Bonwm Farm, Holyhead Road, Corwen LL21 9EG
G1 ZKZ G. Kenealy 20 Penny Lane, Haydock, St. Helens WA11 0QS
G1 ZLA M. Roberts 18 Craster Drive, Nottingham NG6 7FJ
G1 ZLB J. Kynaston Smithy Cottage, Main Street, Nottingham NG12 5PY
G1 ZLC P. Ashby 12 Treeford Close, Solihull B91 3PW
G1 ZLD M. Bignell 57 Ramsey Road, Halstead CO9 1AS
GW1 ZLL D. Ball 38 Heol Sirhwi, Barry CF62 7TG
G1 ZLY N. Cooper 56 Kingfisher Road, Mansfield NG19 6EG
G1 ZME R. Coatman 3 Harold Avenue, Blackpool FY4 5HG
G1 ZMG R. Hoad Broad Lea, Amsbury Road, Maidstone ME17 4DN
G1 ZMJ Rvd. D. Roberts 31 Seaton Way, Marshside, Southport PR9 9GJ
G1 ZMW C. Tubey 2 Rowley Close, Swadlincote DE11 8LX
GW1 ZNC S. Elworthy 70 Maple Drive, Brackla, Bridgend CF31 2PF
G1 ZND A. Soble 6 The Glebe, Hildersley, Ross-on-Wye HR9 5BL
G1 ZNK A. Edwards 68 Middlemarch Road, Coventry CV6 3GF
GM1 ZNR V. Roberts 4 Ladieside, Brae, Shetland ZE2 9SX
G1 ZNT W. Barton 27 Hornby Crescent, Clock Face, St. Helens WA9 4RY
G1 ZNV R. Charteris 7 Kennedy Close, Kidderminster DY10 1LR
G1 ZNX R. Agnew 156 Goswell End Road, Harlington, Dunstable LU5 6NT
G1 ZNZ A. Steele 72 Park Lane, Knypersley, Stoke-on-Trent ST8 7AS
G1 ZOB R. Brown 28 Albertus Road, Hayle TR27 4JQ
G1 ZOS Weston Super Mare RS c/o C. Wood Wurzerstr. 180, Bonn 53175 Germany
GM1 ZOX N. Senior 36 Lathro Park, Kinross KY13 8RU
G1 ZOY M. Knowles 17 Stainmore Close, Birchwood, Warrington WA3 6TP
G1 ZPA N. Johanssen 10 Waverley Court, Verulam Place, St. Leonards-on-Sea TN37 6QR
G1 ZPC C. Rule 1 Park En Venton, Mullion, Helston TR12 7JH
G1 ZPJ P. Read 58 Godolphin Road, Helston TR13 8QJ
G1 ZPO A. Brookes 212 Pontefract Road, Featherstone, Pontefract WF7 5AG
G1 ZPQ J. Pitchford 7 Firecrest Drive, Leegomery, Telford TF1 6FZ
G1 ZPU R. Compton 18 Drove Road, Gamlingay SG193NY
G1 ZQE D. Marsden 94 Blackford Road, Shirley, Solihull B90 4BX
GM1 ZQF G. Milne 6 Alexandra Street, Alyth, Blairgowrie PH11 8AS
G1 ZQG P. Huntley 5 Beacon Avenue, Barton-upon-Humber DN18 5DP
G1 ZQN P. Gibson 60 Raglan Road, Bromley BR2 9NW
G1 ZQO C. Stokes 21 Deerswood Lane, Bexhill-on-Sea TN39 4LT
G1 ZQR R. Twyman Farmend, Halls Lane, Reading RG10 0JB
G1 ZQV M. Lowe 34 Woodbank Road, Groby, Leicester LE6 0BN
G1 ZRE R. Ellis 7 Bromley Close, Blackpool FY2 0SD
G1 ZRP M. Crook 21 Treyew Road, Truro TR1 2BY
G1 ZRQ D. Barrett 10 Trelawny Road, Menheniot, Liskeard PL14 3TS
G1 ZRR N. Youngman-Smith 12 Timber Way, Chinnor OX39 4EU
G1 ZRS J. Cantwell 10 Cathedral Drive, Fairfield, Stockton-on-Tees TS19 7JT
G1 ZRT C. Guymer 74 West Common Lane, Scunthorpe DN17 1DU
G1 ZSE K. Crocker 32 Godmanston Close, Poole BH17 8BU
G1 ZSF A. Lewis 76 Reading Road, Finchampstead, Wokingham RG40 4RA
G1 ZSG C. Bell 41A Handel Road, Canvey Island SS8 7HL
G1 ZSK T. Adams 26 Hillside Avenue, Plymouth PL4 6PR
G1 ZST L. Sherwood 50 Thornton Road, Manchester M14 7WT
G1 ZSV R. Mercer 23 Larne Road, Bilton Grange, Hull HU9 4UE
G1 ZSY A. Hughes 37 Brisbane Road, Reading RG30 2PE
G1 ZSZ H. Hossle Flat 91, Castlemeads Court, 143 Westgate Street, Gloucester GL1 2PB
GM1 ZTB W. Bell 77 Bongate, Jedburgh TD8 6DU
G1 ZTG P. Wilsdon 64 Chestnut Avenue, Euxton, Chorley PR7 6BS
G1 ZTJ G. Bailey 34 Newton Road, Bideford EX39 2LL
G1 ZTK W. Causer 47 Sandringham Road, Wombourne, Wolverhampton WV5 8EF
G1 ZTM A. Corp 158 Somerton Road, Street BA16 0SA
G1 ZTN P. Harvey 64 Privett Road, Gosport PO12 3SX
G1 ZUB P. Thompson Berry Brow, Wetherby Road, Leeds LS14 3AU
G1 ZUC Dr R. Johnson Mile House, Lansdown Road, Bath BA1 5SY
G1 ZUH M. Pomroy 21 Nook Farm Avenue, Syke, Rochdale OL12 0SH
G1 ZUS N. Watts The Vista, Churchill Way, Bideford EX39 1PA
G1 ZUU Hallam DX Group c/o S. Brookes Ivy Cottage, Haselor, Alcester B49 6LX
G1 ZUZ J. Rowling 11 Barncroft, Norton, Runcorn WA7 6RJ
G1 ZVC S. Beith 18 Avenue Road, New Milton BH25 5JP
G1 ZVE H. Barugh Westwinds, 40 Ruden Way, Epsom KT17 3LN
GM1 ZVJ J. Hilton 25 Alford Way, Dunfermline KY11 8BF
G1 ZVO W. Scott 16 Sweetbriar Lane, Holcombe, Dawlish EX7 0JZ
G1 ZVZ K. Fish 11 Little Meadow Way, Bideford EX39 3QZ
G1 ZWB B. Ward 6721 Pinetucky Road South, Mobile 36618 United States
G1 ZWH T. Sanders Sandalwood, 41 Tinney Drive, Truro TR1 1AT
G1 ZWQ J. Bowker 9 Scarthwood Close, Bolton BL2 4DU
G1 ZWY C. Roe 17 Hanbury Close, Chesterfield S40 4SQ
G1 ZXC J. Westgate 19 Granville Street, Gloucester GL1 5HL
G1 ZXD M. Williams 23 Sarum Avenue, West Moors, Ferndown BH22 0ND
G1 ZYJ A. Ainger 16 Hillside Road, Harpenden AL5 4BT
G1 ZYN N. Suffolk 2 Tamerton Road, Leicester LE2 9DD
G1 ZYS F. Woodland 12 Toll House Way, Chard TA20 1FH

G1 ZZA P. Harvey 10 Barnfield Close, Wirral CH47 7DA
G1 ZZC P. Golds 7 Selsey Close, Worthing BN13 1LQ
G1 ZZG B. Thomas 4 Gilbert Close, Torquay TQ2 6BS
G1 ZZL Essex Raynet (Braintree District) c/o M. Low 23 Larch Crescent, Tonbridge TN10 3NN

G2

GW2 ABJ G. Edwards 2 Heol Y Glo, Tonna, Neath SA11 3NJ
G2 ABR C. Mayman Greenacre, Stones Green Road, Harwich CO12 5BS
G2 AD East of England DX Group c/o A. Mcgonigle 5 Whitehouse Crescent, Chelmsford CM2 7LP
G2 AIW M. Lambeth 11 Ellerman Avenue, Twickenham TW2 6AA
GM2 AJW J. Jack Malindella, Main Road, Dumfries DG1 1RZ
G2 ALM R. Wilkins 36 Offington Gardens, Worthing BN14 9AU
G2 ALN L. Taylor 76 Sidney Road, Blackley, Manchester M9 8AT
G2 ALX A. Wilson 196 Spring Lane Lambley, Nottingham NG4 4PE
G2 AMG H. Mitchell Stone Cottage, Yeovil Road, Yeovil BA22 9RR
G2 ANC Peterborough Radio And Electronic Soc. c/o J. Bromiley 28 Clive Road, Westhoughton, Bolton BL5 2HR
G2 API H. Batty 64 North St., Scalby, Scarborough YO13 0RU
G2 AQJ R. Collins 17 Archers Court, Salisbury SP1 3WE
G2 ART F. Cawson 43 Trafalgar Road, Southport PR8 2HF
G2 ARU R. Loveland Apartment 8, Royal Bay Court, 86A Barrack Lane, Bognor Regis PO21 4DY
G2 ARV R. Bennett 16 Emily Street, St. Helens WA9 5LZ
G2 ARY G. Lee 16 Phoenix Chase, North Shields NE29 8SS
G2 AS Sheffield Hf DX Group c/o P. Day 38 Broomhill Road, Old Whittington, Chesterfield S41 9DA
G2 ASF Coventry ARS c/o J. Beech 124 Belgrave Road, Coventry CV2 5BH
G2 AXO W. Purser Bethel And Bethesda Residential Home, Equity Road East, Leicester LE9 7FY
G2 AZM E. Oakley 67 South Road, Northfield, Birmingham B31 2QZ
G2 BAR B. Hill 38 Westons Brake, Emersons Green, Bristol BS16 7BP
G2 BBC Ariel Rad Group c/o D. Pick 178 Alcester Road South, Kings Heath, Birmingham B14 6DE
G2 BBI L. Steel 1B Trinity Avenue, Westcliff-on-Sea SS0 7PU
G2 BGG J. Garner Barbon, Aigburth Hall Road, Liverpool L19 9DG
G2 BHG G. Harrison 13 High View Park, Cromer NR27 0HQ
G2 BHY A. Bonner 57 Downsview, Heathfield TN21 8PF
G2 BJK G. Brown 25 The Cloisters, South Street, Wells BA5 1SA
G2 BKZ R. Mctait 20 Rowland Road, Stevenage SG1 1TE
G2 BQP P. Gully 23 Lawrence Grove, Henleaze, Bristol BS9 4EL
G2 BQY Trowbridge And District ARC c/o D. Birch 32 Union Street, Trowbridge BA14 8RY
G2 BRS Sussex 4x4 Response c/o M. Stevens 16 Golf Links Road, Ferndown BH22 8BY
G2 BSJ R. Biltcliffe 3 Church View, Steeple Claydon, Buckingham MK18 2QR
G2 BSW R. Ward Serendipity, 17 Marlpit Lane, Seaton EX12 2HH
G2 BTZ E. Moreman 5 Sheridan Way, Longwell Green, Bristol BS30 9UE
G2 BUJ S. Greenwood 29 The Elms, Nine Elms, Swindon SN5 5XA
GM2 BWW A. Barrett Mains Of Glasclune Farm, Middleton Road, Blairgowrie PH10 6SF
GI2 BX City of Belfast Radio Amateur Society c/o F. Hunter Flat 9 50 Edenvale Crescent, Belfast BT4 2BH
G2 BXH A. Perkins 3 Greenway Close, Radcliffe-On-Trent, Nottingham NG12 2BU
G2 BZR R. Bassford 59 Watling St., Dordon, Tamworth B78 1SY
G2 CD R. Matthews 7 Coolgardie Avenue, Chigwell IG7 5AU
G2 CFC G. Fretwell 17 Cross Lane, Stocksbridge, Sheffield S36 1AY
GW2 CGF S. Griffiths 1 Nicholl Court, Mumbles, Swansea SA3 4LZ
G2 CHI W. Bailey 25 Lenham Road East, Saltdean, Brighton BN2 8AF
G2 CIW J. Moseley 33 Cathedral Court, London Road, Gloucester GL1 3QE
G2 CJK A. Clarkson 6 Mather Avenue, Accrington BB5 5AU
G2 CKR M. Garfitt 90 Wedderburn Road, Malvern WR14 2DQ
G2 CNN S. Ball 16 Stonewood Gate, Field Lane, St. Helens, Ryde PO33 1FY
G2 CO F. Cooknell 65 Coombe Valley Road, Preston, Weymouth DT3 6NL
G2 CP Scarborough ARC c/o D. Herbert 50 St. Leonards Crescent, Scarborough YO12 6SP
G2 CQX V. Pugh 8 Beech Close, Hanwood, Shrewsbury SY5 8RA
G2 CR Cambridgeshire Raynet c/o D. Burkin 26 Rampton Road, Cottenham, Cambridge CB24 8UL
G2 DAN S. Whiteley 142 Brisbane Road, Mickleover, Derby DE3 9JW
G2 DBH G. Dodd St Nicholas Cottage, 14 Bury Fields, Guildford GU2 4AZ
G2 DD L. James Pinecroft, Green Drive, Wokingham RG40 2HT
G2 DGB A. Short 12 Grosvenor Crescent, Dorchester DT1 2BA
GW2 DHM W. Andrews 69 Fairwater Grove West, Cardiff CF5 2JN
G2 DJ Derby & District ARS c/o R. Buckby 22 Woodstead, Embleton, Alnwick NE66 3XY
G2 DJM Dr N. Chilton 38 Kingswood Avenue, Newcastle upon Tyne NE2 3NS
G2 DLX D. Mitchell 1 Denstroude Cottages, Denstroude Lane, Canterbury CT2 9JX
G2 DML J. Crossfield Forest Lodge, Chopwell Wood, Rowlands Gill NE39 1LT
GW2 DNJ N. Brierley Minera, 6 Trinity Crescent, Llandudno LL30 2PQ
G2 DP Tingley Moonbounce Society c/o D. Parker 50 Rein Road, Tingley, Wakefield WF3 1HZ
G2 DPA M. Brashill 42 Bannister Street, Withernsea HU19 2DT
G2 DPY D. Silverson 63 Downside, Shoreham-by-Sea BN43 6HF
G2 DRM D. Mobey 21 Langdale Court 151 Windermere Drive, Wellingborough NN83XA
G2 DT Furness ARS c/o M. Bell 36 Schneider Road, Barrow-in-Furness LA14 5DW
G2 DWB N. Webster 1 Gratton Dale, Carlton Colville, Lowestoft NR33 8WP
G2 DWC Dubmire ARC c/o B. Wheeler 2 Rose Street, Houghton le Spring DH4 5BB
G2 DXU A. Wyatt Flat 3, Soper House, 9 Dart View Road, Galmpton, Brixham TQ5 0BQ
G2 DZH N. Talbot 105 Westwood Lane, Welling DA16 2HJ
G2 EC Blandford Garrison ARC c/o R. Carter 12 Glebe Close, Abbotsbury, Weymouth DT3 4LD
G2 FA North-Northants Raynet Group c/o Dr D. Pepper 17 Cliffe House, Radnor Cliff, Folkestone CT20 2TY
G2 FCP F. Varley 39 Nettleton Road, Mirfield WF14 9AW
G2 FFD D. Skipworth Melrose, West End Road, Boston PE22 0BU
G2 FGT R. Rogers 15 Tythe Barn Close, Westoning, Bedford MK45 5JT
G2 FHF J. Illsley Wootton Park Farm, Alcester Road, Wootton Wawen, Henley-in-Arden B95 6HJ
G2 FJA Marts c/o K. Earl 210 Churchill Avenue, Chatham ME5 0JS
G2 FKO Brimham CG c/o J. Lovell Kowloon, Slade, Bideford EX39 3LZ
G2 FKZ Radio Society of Great Britain c/o S. Thomas 2 Myrtle Cottages, Sandy Lane, Saxmundham IP17 1HR
G2 FLW M. Clarkson Causeway Cottage, Sawley Road, Clitheroe BB7 4RS
G2 FM Reading University ARC c/o C. Quarton Flaxton Gatehouse, Flaxton, York YO60 7QT
G2 FMW E. Baker 86 Osborne Gardens, Herne Bay CT6 6SE
GW2 FOF Rhondda ARS c/o J. Howells 13 Vicarage Road, Penygraig, Tonypandy CF40 1HR
G2 FQZ R. Day Resting Oak Cottage, Resting Oak Hill, Lewes BN8 4PS
G2 FSH B. Weeden 24 Berkeley Close, Rochester ME1 2UA
G2 FSJ K. Levitt 1 Charnwood Close, Andover SP10 2RB
G2 FSR J. Hunt 4 Warmdene Road, Brighton BN1 8NL
G2 FVL L. Carrick-Smith Highfields House, Sheffield Road, Clowne, Chesterfield S 43 4AP
G2 FWJ Stanford ARC c/o S. Simmons 48 Copland Road, Stanford-le-Hope SS17 0DF
G2 FXJ S. Moisy 15 Charles Street, Redditch B97 5AA
G2 FXQ S. Saddington South Ridding, Sibson Road, Atherstone CV9 3RE
G2 FXV M. Middleton Dolphin View Nursing And, Residential Home, Harbour Road, Morpeth NE65 0AP
G2 FXZ J. Hodgetts 59 Woodland Road, Halesowen B62 8JS
G2 FYO H. Terraneau 2653 Nutmeg Circle, Simi Valley 93065-1327 United States
GW2 HCA L. Sanders 2 Cae Neuadd, Penybontfawr, Oswestry SY10 0NS
G2 HCG B. Sykes Flat 7, Solent Pines Whitby Road, Milford On Sea, Lymington SO41 0UX
G2 HDF Midland CG c/o M. Waldron 32 Windmill Street, Upper Gornal, Dudley DY3 2DQ
G2 HFP S. Trudgill 267 Clifton Drive South, Lytham St. Annes FY8 1HW
GW2 HFR J. Kelly Bryn Llanerch, Pentre Celyn, Ruthin LL15 2HW
G2 HHH T. Bayliss 55 Foxlydiate Crescent, Redditch B97 6NJ
G2 HIX Dr D. Craig Flat 5, The Chapel, The Plains, Totnes TQ9 5DW
G2 HKQ A. Knight 17 Moorland Crescent, Upton, Poole BH16 5LA
G2 HKS R. Udall Longfield, 20 Upper Way, Rugeley WS15 1QA
G2 HKU E. Trowell 316 Minster Road, Minster On Sea, Sheerness ME12 3NR
G2 HLB C. Maltby The Willows Farm, Stallingborough Road, Grimsby DN40 1NR
G2 HLP D. Hearsum 1225 Duckview Court, Centerville 45458-2784 United States
G2 HMK T. Brown 99 Brinkburn Drive, Darlington DL3 0JY
G2 HNA J. Weaver 7 Cramer St., Stafford ST17 4BX
G2 HNI L. Hewitt 60 Shaftesbury Avenue, Southampton SO17 1SD
G2 HR Silverthorn RC c/o L. Butterfields 22 Horsley Road, London E4 7HX
G2 HS Echelford DX Group c/o Dr P. Miller Tate 19 Esher Avenue, Walton-on-Thames KT12 2SZ
G2 HX Gloucester Amateur Radio & Electronics Society c/o L. Harris 183A Painswick Road, Gloucester GL4 4AG
G2 IF W. Setterfield 54 Hallam Road, Nelson BB9 8AB
G2 JL T. Mortimer 10 Harold Road, Hayling Island PO11 9LT
G2 KF T. Harris Summerfield, Coombe Road, Lanjeth, High Street, St. Austell PL26 7TL
G2 KG C. Hill 47 Belswains Lane, Hemel Hempstead HP3 9PW
G2 KQ B. Hawes 3 Orchard Close, Cassington, Witney OX29 4BU
G2 KS K. Matthews St. Helens Cottage, Flimby, Maryport CA15 8RX
G2 LK L. Piggott 37 Moss Lane Worsley, Manchester M28 3WD
G2 LL Hastings Electronics & RC c/o J. Tucker 17 Rye Industrial Park Harbour Road, Rye TN31 7TE
G2 LW Crystal Palace Radio & Electronics Club c/o R. Burns 84 Portnalls Road, Coulsdon CR5 3DE
G2 MN J. Walker-Wilson Rest Harrow Southside, Scorton, Richmond DL10 6DN
GM2 MP North of Scotland CG c/o Prof. K. Kerr East Loanhead, Auchnagatt, Ellon AB41 8YH
G2 NF A. Canning 261 Loddon Bridge Road, Woodley, Reading RG5 4BL
G2 OA Southport & District ARC (Sadarc) c/o C. Staples 32 Browns Lane, Netherton, Bootle L30 5RW
GW2 OG J. Hogg Bwthyn Y Briallu, Ynys Ferw Bach, Gaerwen LL60 6NW
GW2 OP Pembrokeshire CG c/o M. Shelley Sunray, Pendine, Carmarthen SA33 4PD
G2 OU Highfields ARC c/o D. Tatlow Mulberry House, Bettys Grave, Cirencester GL7 5ST
G2 PA P. Dyke 5328 Malcolm Street, Oceanside 92056 United States
G2 PB P. Baron 55 Church View, Brompton, Northallerton DL6 2RD
G2 PK J. Ellison Jowsers, Northfield Lane, Wells-next-the-Sea NR23 1JZ
G2 RE R. Evered Ivy Cottage, Old Bristol Road, Wells BA5 3AL
GU2 RS R. Robilliard Moss Bank, La Mare, St. Andrew GY6 8XX Guernsey
G2 RSA P. King 32 Millstream Way, Leegomery, Telford TF1 6QR
G2 SH J. Shearme Chevin, Penn Street, Amersham HP7 0PY
G2 SR Surrey Raynet c/o T. Dabbs 4 Caverleigh, Cadogan Road, Surbiton KT6 4DH
G2 SU Royal Naval ARS c/o A. Robinson 9 Illingworth Close, Illingworth, Halifax HX2 9JQ
G2 SZ D. Goyder 8 Bloomsbury Walk, Southampton SO19 9GB
G2 TO Bury St Edmunds ARS c/o M. Green 4 Boundary Cottages, Great Finborough, Stowmarket IP14 3AG
G2 UG Halifax & District ARS c/o D. Baker 17 Woodroyd Gardens, Luddendenfoot, Halifax HX2 6BG
G2 UH D. Hayward Hope, Churchwell Street, Sherborne DT9 6RG
G2 UT K. Reid 4 Harles Acres, Hickling LE14 3AF
G2 VS R. Barrett 76 Westgate Park, Sleaford NG34 7QP
G2 XG E. Davie 7 Cranworth Crescent, Chingford, London E4 7HN
G2 XP Sutton & Cheam c/o J. Puttock Sutton & Cheam Rs, 53 Alexandra Avenue, Sutton SM1 2PA
G2 XV Cambridge & District ARC c/o B. Davies 16 Pearmains Close, Orwell, Royston SG8 5QY
G2 YC R. Mcknight Ardralla, Church Cross, Skibbereen P81 RK12 Ireland
G2 YL S. Quarton Flaxton Gatehouse, Flaxton, York YO60 7QT
G2 YT P. Fox Hillside House, Almshoebury, St. Ippolyts, Hitchin SG4 7NT

G3

G3 AAF K. Avery 4 Whiphill Close, Doncaster DN4 6DX
G3 AAS M. Glynn 39 Moor Allerton Drive, Leeds LS17 6RY
G3 AB A. Chadwick 5 Thorpe Chase, Ripon HG4 1UA
G3 ABG Wab Awards Group c/o J. Brooks 28 Avon Vale Road, Loughborough LE11 2AA
G3 ACQ H. Harmsworth 43 Cornelian Avenue, Scarborough YO11 3AN
GM3 AEI J. Rosselle 140 Main Street, Neilston G78 3JX
G3 AER G. Wright 70 Gunton Drive, Lowestoft NR32 4QB
G3 AFB D. Tait 34 Mount St., Dorking RH4 3HX
G3 AGC W. Curphey 8 Emily Davison Avenue, Morpeth NE61 2PL
G3 AGF R. Edginton 9 Churchill Road, Seaford BN25 2UL
G3 AH Riviera ARC c/o S. Crask 14 Southfield Road, Paignton TQ3 2SW
G3 AHE R. James 40 Barrack Road, Hounslow TW4 6AG
GM3 AHR A. Thomson Meadowrise, 4 Law View Gardens, Leven KY8 5SW
G3 AJD T. Moore 6 Old Parsonage Court, Otterbourne, Winchester SO21 2EP
G3 AJK R. Earland 7 Trews Weir Court, Exeter EX2 4JS
G3 AKF Reading & District ARC c/o V. Robinson 4 Hilltop Road, Caversham, Reading RG4 7HR
G3 AKI F. Knowles 1 Mayfield Close, Bishops Cleeve, Cheltenham GL52 8NA
G3 AKJ A. Wheele 4 Mannings Way, Barnstaple EX31 1QF
GM3 ALF A. Megson 2 Garden Cottage, Glendelvine, Murthly PH1 4JN
G3 ALG G. Starling 207 Shirley Road, Croydon CR0 8SB
G3 ALK E. Holmes 7 Castle Drive, Ilford IG4 5AE
GM3 ALZ F. Gordon Crofts Of Torrancroy, Strathdon AB36 8UJ
GJ3 AME P. Landor Lauge, Rue Des Raisies, St. Martin JE3 6AT Jersey
G3 AMH H. Green 9 Robert Avenue, Cundy Cross, Barnsley S71 5RB
G3 AMK B. Littleproud 25 Fern Avenue, Lowestoft NR32 3JF
G3 AMW Hull And District ARS c/o B. Atkinson 165 Alliance Avenue, Hull HU3 6QY
GI3 AMY J. Collett 10 Cronstown Road, Newtownards BT23 8QS
G3 APL J. Russon 59 Ridge Road, Kingswinford DY6 9RE
G3 APS L. Shergold 8 The Moors, Lydiard Millicent, Swindon SN5 3LE
G3 APU J. Andrews 44 Eastridge View, East The Water, Bideford EX39 4RS
G3 AQB W. Stephenson 20 Chapel Court, Chapel Row, Seahouses NE68 7TD
G3 AQF A. Kearns 8 Pennyfathers Lane, Welwyn AL6 0EN
G3 ARE F. Chubb 2 Brook Close, Plympton, Plymouth PL7 1JR
GW3 ARS J. Sagar 75 Hookland Road, Newton, Porthcawl CF36 5SG
G3 ASG R. Fautley 7 Kingfisher Road, Downham Market PE38 9RQ
G3 ASR Edgware & District RS c/o H. Haria 34 Larkfield Avenue, Harrow HA3 8NF
G3 AST J. Plowman 17 Orchardleigh, East Chinnock, Yeovil BA22 9EN
G3 ASV G. Pope 5 Penn Crescent, Haywards Heath RH16 3HW
GW3 ASW West Midland Rdf c/o B. Werrell 26 Glynhafod Street, Cwmaman, Aberdare CF44 6LD
G3 ASX D. Paine 43 Wilton Road, Muswell Hill, London N10 1LX
G3 ATC Royal Air-Force Air Cadets/Air Training Corps c/o S. Pankhurst 57 Barley Croft, Harlow CM18 7QZ
G3 ATI A. Williams 74 Broadfield Road, Bristol BS4 2UW
G3 ATX A. Perry The Cottage, The Green, Bristol BS48 3BG
GW3 ATZ G. Morris 18 Grosvenor Road, Shotton CH5 1NU
GM3 AUE A. Mcghie 1 Boyach Crescent, Isle Of Whithorn, Newton Stewart DG8 8LD
G3 AVE F. Flanner 1 Ludford Close, Sutton Coldfield B75 6DW
G3 AVL R. Reynolds 12 Eastham Rake, Wirral CH62 9AA
G3 AVN P. Parker Flat 15, The Rise Care Home, Dawlish EX7 0QL
G3 AWK N. Gough 3 Sycamore Close, Cherry Willingham, Lincoln LN3 4BJ
G3 AWP P. Gifford 21 Bengal Road, Bournemouth BH9 2ND
GM3 AXX A. Fraser 58 Rigghead, Stewarton, Kilmarnock KA3 3DQ
G3 AYL North Yorkshire Yl Group c/o R. Duffield 4 Crabmill Lane, Easingwold, York YO61 3DE
G3 AZI A. Mccann 105 Todd Lane North, Lostock Hall, Preston PR5 5UP
G3 AZW A. Bates 68 Hill St., Hilperton, Trowbridge BA14 7RS
G3 BAC R. Bastow 2A New Road, Meopham, Gravesend DA13 0LS
G3 BAR K. Brenchley 6 Windwards Close Lanreath, Looe PL132WP
G3 BAY Bay ARG c/o I. Maude 21 Colwyn Avenue, , Morecambe LA4 6EQ
G3 BBK J. Orrin Greenacres, Church Street, Heathfield TN21 9AL
G3 BBX D. Holloway 10 Spencers Orchard, Bradford-on-Avon BA15 1TJ
G3 BCE D. Nichols Marsh Farm, Camp Road, Templecombe BA8 0TH
G3 BDQ J. Heys White Friars, Friars Hill, Hastings TN35 4EP
G3 BDT A. Searle 30 Hawthorne Grove, Poulton-le-Fylde FY6 7PN
G3 BEX W. Short Highland Light, 26 Howard Crescent, Beaconsfield HP9 2XP
G3 BFL H. Siebert 3 Greenlands Road, Kingsclere, Newbury RG20 5RJ
G3 BGF R. Winkworth 1 Collingwood Drive, Mundesley, Norwich NR11 8JB
G3 BHA N. Taylor 8 Aragon Way, Bournemouth BH9 3SB
G3 BHM H. Kempson 8 Hounds Way, Hayes, Wimborne BH21 2LD
G3 BII A. Clark 19 Lakes Lane, Beaconsfield HP9 2LA
G3 BIK E. Chicken Ivy Thorn Cottage, Morpeth NE61 6LQ
G3 BJ D. Beattie Hares Cottage, Woolston, Church Stretton SY6 6QD
G3 BJD J. Maxwell 10 Castle View, Egremont CA22 2NA
G3 BKJ H. Alderson 31 Rumbold Road, Edgerton, Huddersfield HD3 3DB
G3 BLS D. Walker 32 South St., Osney, Oxford OX2 0BE
G3 BMO H. Speed 45 Willow Glade, Huntington, York YO32 9NJ
G3 BMQ G. Humphrey 56A Park Lane, Wallington SM6 0TN
G3 BNE G. Alderman 35 Eynswood Drive, Sidcup DA14 6JQ
G3 BNF A. Embleton 34 Riverdale Park, Bent Lane, Chesterfield S43 3UH
G3 BNW J. Bailey 13 Heywood Road, Alderley Edge SK9 7PN
G3 BOK S. Rutt Granthorpe, Hull Road, Hull HU11 5RN
G3 BPF A. Painter Cold Green, Rochford, Tenbury Wells WR15 8SP
G3 BPK Douglas Vall Ar c/o D. Snape 30 Culcross Avenue, Wigan WN3 6AA
G3 BPP R. Hampton 11 Greenlands, Hutton Rudby, Yarm TS15 0JQ
G3 BPQ E. Smith 23 The Ladysmith, Ashton-under-Lyne OL6 9AP
GM3 BQA North Berwick DX Club c/o B. Robertson Frian, Westerton, Dalcross, Inverness IV2 7JL
G3 BQE R. Fussey 9 Alicia Gardens, Harrow HA3 8JB
G3 BQT E. Hulme 21 Brookside Crescent, Greenmount, Bury BL8 4BG
G3 BRQ K. Tackley 1 Greenways, Fleet GU52 7UG
G3 BRS Bury RS c/o P. Stocks 12 Bredbury Drive Farnworth, Bolton BL4 7QD
G3 BSN P. Stanley 1 Thames View, Cliffe Woods, Rochester ME3 8LR
GM3 BSQ Andover Radio Amateurs Club c/o I. Munro 57 Craigiebuckler Avenue, Aberdeen AB15 8SF
GM3 BST J. Tuke 2/23 Hawthorn Gardens, Loanhead EH20 9EE
GW3 BV Q. Cruse Glas Y Dorlan, Llanfihangel-Y-Creuddyn, Aberystwyth SY23 4LA
G3 BVA E. Digman 75 Ramsden Road, Orpington BR5 4LU
G3 BVB D. Adair 3 Belmont Close, Shaftesbury SP7 8NF
G3 BWI W. Timms 22 Padway, Penwortham, Preston PR1 9EL
G3 BYG N. Williams Chapel Lake Halwill, Beaworthy EX21 5UF
G3 BZB R. Cunliffe 5 Silk Mill Lane, Tutbury, Burton-on-Trent DE13 9LE
G3 BZU Royal Naval ARS c/o J. Kirk 111 Stockbridge Road, Chichester PO19 8QR
G3 CAJ R. Prince 52 Mafeking Road, Southsea PO4 9BG
G3 CAZ J. Shaw 128 Perth Road, Ilford IG2 6AS
GW3 CBA H. Kellaway 34 Winston Road, Barry CF62 9SW
G3 CDM I. Gardner 30 Pierremont Crescent, Darlington DL3 9PB
G3 CEI Dr C. Brown Downlands, Off Hackwood Lane, Basingstoke RG25 2NH
GW3 CF Prestatyn ARC c/o K. Shepherd 15 Gronant Road, Prestatyn LL19 9DT
GI3 CFH Stockton & District ARG c/o D. Fulton 120 Dunnalong Road, Bready, Strabane BT82 0DP
G3 CFR J. Jowett Ashleigh, Kilmington, Axminster EX13 7ST
G3 CGD J. Yeend 30 St. Lukes Road, Cheltenham GL53 7JJ
G3 CGE R. Gardner 62 Rosewall Road, Southampton SO16 5DW
G3 CIK H. Romer 96 Mortlake Road, Richmond TW9 4AS
G3 CIL M. Holley 6586 196Th Street, Langley BC V2Y 1R3 Canada
G3 CIM S. Denney 52A Intwood Road, Cringleford, Norwich NR4 6AA
GM3 CIO Rsars - Royal Signals ARS c/o C. Hall 21 Peterculter Retirement Park Peterculter, Aberdeen AB14 0AB
G3 CKE S. Mason 46 Frankton Close, Redditch B98 0HJ
G3 CKR M. Ryder 9 Lincoln Close, Woolston, Warrington WA1 4LU
G3 CLW L. Hutton 46 Penwill Way, Paignton TQ4 5JQ
G3 CMH Yeovil ARC c/o D. Bowden 58 Southville, Yeovil BA21 4JF
G3 CMU H. Meyers Cornerways, 2 Old Mill Lane, Polegate BN26 5NS
G3 CNO Fort Purbrook ARC c/o M. Ponsford 83 Grant Road, Farlington, Portsmouth PO6 1DU
G3 CO Mapperley & District ARC c/o H. Yeldham 19 Wade Reach, Walton on the Naze CO14 8RG
GM3 COB J. Paterson 10 Hathaway Drive, Giffnock, Glasgow G46 7AE
G3 CON L. Crabbe 6 Node Hill, Studley B80 7RR
GM3 COQ D. Oswald 8 Redfield Road, Montrose DD10 8TW
G3 CPG L. Damon 18 Scafell Court, Dewsbury WF12 7PD
GW3 CPM C. Vugts Coed Coch, Llangammarch Wells LD4 4BS
G3 CPN M. Stevens 16 Golf Links Road, Ferndown BH22 8BY
G3 CPT D. Capp 46 Stoke Road, Bletchley, Milton Keynes MK2 3AD
G3 CQL M. Clarke 3 Shelley Priory Cottages, Shelley, Ipswich IP7 5RQ
G3 CQU K. Raffield 113 Waddington Avenue, Coulsdon CR5 1QP
G3 CRC Clacton RC c/o C. Day 35 Rochford Road St. Osyth, Clacton-on-Sea CO16 8PH
G3 CRH H. Sanders Little Orchard, 68A Park Road, Burton-on-Trent DE13 7AJ
G3 CRS Royal Naval ARS c/o J. Kirk 111 Stockbridge Road, Chichester PO19 8QR
G3 CSA Stromness Academy ARC c/o T. Saggerson 18 Ploughmans Way, Great Sutton, Ellesmere Port CH66 2YJ
G3 CSR Gxff Radio Group c/o N. Sanderson 54 Kelvedon Close, Chelmsford M1 4DG
G3 CSY K. Hill 30 Hestham Avenue, Morecambe LA4 4PZ
G3 CTP J. Swift 20 Leighlands, Crawley RH10 3DW
G3 CTQ H. Westwell 224 Dickson Road, Blackpool FY1 2JS
G3 CTZ A. Jones 17 Oaklea Way, Old Tupton, Chesterfield S42 6JD
G3 CUF H. Ashworth 97 Winchcombe Road, Sedgeberrow, Evesham WR11 7UZ
G3 CUR R. Collette 8A Woolwich Road, Belvedere DA17 5EW
G3 CUY E. Paul 91 Windmill Drive, Brighton BN1 5HH
G3 CVK P. Bolton 50 Meadow Road, West Malvern, Malvern WR14 2SD
G3 CWD J. Robinson 4 Phoenix Court The Mount, Taunton TA1 3NR
G3 CWH R. Rogers 107 Rotherham Road, Coventry CV6 4FH
G3 CWI R. Newstead 89 Victoria Road, Macclesfield SK10 3JA
G3 CXP R. Gill 45 Biggin Lane, Ramsey, Huntingdon PE26 1NB
G3 CYU J. Wilson 1 Beeches Farm Road, Crowborough TN6 2NY
G3 CYX P. Lambert 11 Marlborough Close, Musbury, Axminster EX13 8AP
G3 CZL R. Buckman Heathfield, Hang Hill Road, Lydney GL15 6LQ
G3 CZU Dorking & District RS c/o T. Ellinor 53 Hillside, Banstead SM7 1HG
G3 DAE C. Bland 84 Milton Road, Grimsby DN33 1DE
G3 DAQ R. Braithwaite 32 Rupert Crescent, Queniborough, Leicester LE7 3TU
G3 DAT D. Taylor 24 Quince, Tamworth B77 4EN
G3 DAV J. Waller 17 Spencer Close, Marske-By-The-Sea, Redcar TS11 6BD
G3 DBJ D. Buggs 2 Archway Cottages Valley Road, Leiston IP16 4AR
G3 DBV S. Hedges 25 Rudland Close, Thatcham RG19 3XW
G3 DCE F. Humphries Little Hayes, 1 Meadway, Sidmouth EX10 9JA
G3 DCO B. Coyne 58 Osborne Road, New Milton BH25 6AB
G3 DCT Duddon Contest Team c/o C. Leviston 1 Great Carrs Close, Askam-in-Furness LA16 7FL
G3 DCV A. Watson 93 St. Dunstans Drive, Gravesend DA12 4BJ
GM3 DDL J. Jackson 74 Cairngorm Crescent, Paisley PA2 8AW
G3 DEJ T. Wiseman 70 Dove House Lane, Solihull B91 2EG
G3 DEN R. Lea 90 Wroxham Gardens, Potters Bar EN6 3DL
G3 DEY E. Ford 177 Latters Orchard, Old Road, Maidstone ME18 5PR
G3 DFY N. Devine 46 Tytton Lane West, Wyberton, Boston PE21 7HL
G3 DID J. Doyle 16 Park Hall Crescent, Birmingham B36 9SN
GM3 DIE T. Dickson 91 Milton Road West, Edinburgh EH15 1RA
GM3 DIN A. Clark 11 Regent Park Square, Glasgow G41 2AF
G3 DIT Prtsmth & Dars c/o T. Mortimer 10 Harold Road, Hayling Island PO11 9LT
G3 DMO C. Earnshaw 35 Rogersfield, Langho, Blackburn BB6 8HB
G3 DNN G. Saville 2 Gaskell Close, Littleborough OL15 8EB
G3 DNS N. King 31 Great Norwood St., Cheltenham GL50 2AW
GM3 DOD A. Murray 50 Castlepark Drive, Fairlie, Largs KA29 0DG
G3 DOV D. Dove 3 Walnut Grove, Watton, Thetford IP25 6EY
G3 DPM D. Cooknell 23 The Hyde, Winchcombe, Cheltenham GL54 5QR
G3 DQQ D. Winterburn 47 Hilda Avenue, Tottington, Bury BL8 3JE
G3 DQT J. Ayres 8 Cornfield Road, Seaford BN25 1SW
G3 DQW Capel Battery Preservation Group c/o B. Vaughan 7 Oundle Road, Chesterton, Peterborough PE7 3UA
G3 DR RSGB Contest Club c/o N. Totterdell Moscar Cross House, Hollow Meadows, Sheffield S6 6GL
G3 DRN E. Allen 30 Bodnant Gardens, Wimbledon, London SW20 0UD

G3 DSZ A. Kent 23 Pagehall Close, Scartho, Grimsby DN33 2HF
G3 DT L. Boorman 2 Bull Lane Cottages, Bull Lane, Ashford TN26 3HA
G3 DTP A. Jackson Flat 6, St. Albans Court, Sparth Bottoms Road, Rochdale OL11 4HW
G3 DTU C. Prior 36 Bassnage Road, Halesowen B63 4HQ
G3 DTX I. Duck Chenies, Loudhams Wood Lane, Chalfont St. Giles HP8 4AR
G3 DUW R. Hodgson The Shealing, Forest Moor Drive, Knaresborough HG5 8JT
G3 DVF G. Cain 23 Wiltshire Avenue, Crowthorne RG45 6NR
G3 DWI G. Lusty Sundial House, High St., Chipping Campden GL55 6AG
G3 DXD D. Rolph 2 Victoria Court, Victoria Road, Marlow SL7 1DR
G3 DXZ C. Fletcher 12 Park Crescent, Retford DN22 6UF
G3 DYO N. Alder Greenwoods, Eastfield Road, Ross on Wye HR9 5JY
GW3 DZJ F. Pardy 5 Y Bryn, Glan Conwy, Colwyn Bay LL28 5NJ
G3 EAE G. Billington 75 Mount Vernon Road, Barnsley S70 4DW
G3 EBP M. Courcoux 116 Ameysford Road, Ferndown BH22 9QE
G3 EBV S. Squire Leafield, 4 Little Green Lane, Rickmansworth WD3 3JQ
GJ3 ECC R. Taylor 21 Samares Avenue La Grande Route De St. Clement, St. Clement JE2 6NY Jersey
G3 ECM P. Bowles 29 Coleman Avenue, Hove BN3 5ND
G3 ECP J. Brown Manor Cottage, 2 The Maltings, Huntingdon PE28 4DZ
GI3 ECQ G. Mcgarry 18 Marna Brae Park, Lisburn BT28 3PD
G3 EDM M. Marris 6 High Street, Wye, Ashford TN25 5AL
G3 EEH Dr J. Watkinson The Moorings, 63 Ruffa Lane, Pickering YO18 7HN
G3 EEO Nunsfield House - ARG c/o A. Price 10 Low Meadow, Whaley Bridge, High Peak SK23 7AY
GD3 EFD M. Thompson Whitehouse Cottage, St. Marks, Ballasalla IM9 3AH Isle of Man
G3 EFL W. Preston 8 Pencraig View, Greytree, Ross-on-Wye HR9 7JR
G3 EFS W. Borland Sloane Nursing Home, 28 Southend Road, Beckenham BR3 5AA
G3 EFX Rad Soc Harrow c/o C. Friel 102A Sharps Lane, Ruislip HA4 7JB
G3 EGF T. Kellett Braville, St. Ives Road, Consett DH8 7SJ
G3 EGV R. Staniforth 26 Winslow Road, Preston, Weymouth DT3 6NE
G3 EHQ H. Bone 2 Waterville Gardens, Orton Waterville, Peterborough PE2 5LG
G3 EHW J. Watkins 19 Barrow Grove, Sittingbourne ME10 1LB
GW3 EIZ C. Lyon Ardraeth, The Drive, Bodorgan LL62 5AW
G3 EJH W. Peatman 110 Cator Lane, Beeston, Nottingham NG9 4BB
GW3 EJR J. Armstrong Mirianog, 1 Bryn Bedw, Cardigan SA43 2NY
G3 EKE L. Stockley C/O Glebe Cottage, Baylham, Ipswich IP6 8JS
G3 EKJ H. Mattacks Fieldfare, Eastbourne Road, Lewes BN8 6PS
G3 EKL Colburn & Richmondshire District ARS c/o M. Vann 3 Mile Planting, Richmond DL10 5DB
G3 EKW Aylesbury Vale RS c/o S. Williams Haywood Community Centre, 46 Haywood Road, Nottingham NG3 6AD
G3 ELS B. Rudd Orchard Bungalow, 14 Walmer Close, Colchester CO7 0PE
G3 ELV Loughton & Epping Forest ARS c/o R. Walker 35 Romany Close, Letchworth Garden City SG6 4LA
G3 EME M. Wyszomirski 1 Chieftan Drive, Purfleet-on-Thames RM19 1PN
G3 ENO R. Green 8B The Beck, Elford, Tamworth B79 9BP
GM3 EOB C. Merrilees 6 Spoutwells Drive, Scone, Perth PH2 6RR
G3 EOO J. Hamlett 23 Riddings Road, Timperley, Altrincham WA15 6BW
G3 ERD Derby And District ARS c/o C. Gent 16 Coronation Avenue, Alvaston, Derby DE24 0LQ
G3 ESY P. Jones Fieldfarm House, Residential Home, Hampton Bishop, Hereford HR1 4JP
G3 ETP P. Woodyard 65 Raglan Street, Lowestoft NR32 2JS
G3 EUE E. Jones White Lodge, The Street, Steyning BN44 3WE
G3 EVA S. Ayers Rivendell 2 The Paddocks, Old Catton, Norwich NR6 7HF
G3 EVT R. Mutton Summer Hayes, Mill Lane, Alcester B49 6LF
G3 EWF A. Harris 5 Wickham Court, Stapleton, Bristol BS16 1DQ
G3 EWM P. Green 23 Tilton Road, Borough Green, Sevenoaks TN15 8RS
G3 EWT C. Tamkin 4 Stanmer Villas, Brighton BN1 7HP
G3 EXL D. Derham 3 Riverbank Cottages, Old Ferry Road, Saltash PL12 6BJ
G3 EZB J. Rackett Little Vectis, Folgate Lane, Norwich NR8 5DP
G3 FBT Eastbourne District Scouts ARC c/o S. Briggs 20 Bluebell Close, Newton Aycliffe DL5 7LN
G3 FBU W. Brown 79 Mill Hill, Deal CT14 9EW
G3 FCM A. Cowley 13 Steward Close, Stuntney, Ely CB7 5TW
G3 FDW A. Gibbings 16 Turnberry Avenue, Eaglescliffe, Stockton-on-Tees TS16 9EH
GW3 FDZ D. Whitehead Tyddyn Bach, Dyffryn Ardudwy LL44 2RQ
G3 FEW E. Rule 15 Norwich Road, Lenwade, Norwich NR9 5SH
GI3 FFF Ballymena ARC c/o J. Clarke 154 Galgorm Road, Ballymena BT42 1DE
G3 FGP R. Brooks 10 The Oval, Longfield DA3 7HD
G3 FHG M. Hopkins Hylton Cottage, Grafton, Tewkesbury GL20 7AT
G3 FHN E. Aldworth Glenaire, 15 Heather Way, Hastings TN35 4BL
G3 FHT A. Lewis 8 Lutyens Fold, Milton Abbot, Tavistock PL19 0NR
G3 FIA A. Lowden 3 Boscobel Road, Great Barr, Birmingham B43 6BB
G3 FIC J. Glover 53 Swanpool Lane, Aughton, Ormskirk L39 5AY
G3 FJE Lough Erne ARC c/o D. Ross 3 Little Lane, Clophill, Bedford MK45 4BG
GW3 FJI E. Jones 8 Merllyn Road, Rhyl LL18 4HH
G3 FJL J. Hall 250 Scraptoft Lane, Leicester LE5 1PA
G3 FJO A. Ellefsen 121 The Furlongs, Ingatestone CM4 0AL
GI3 FJX J. Davidson 7 Keel Point, Dundrum, Newcastle BT33 0NQ
G3 FKI E. Lambert 6 Abercorn Gardens, Kenton, Harrow HA3 0PB
GD3 FLH Iom ARS c/o A. Sinclair 1 Marathon Drive, Douglas, Isle of Man IM2 4BP
G3 FLV L. Keighley 24 St. Annes Road, Headingley, Leeds LS6 3NX
G3 FMO G. Elliott Oatlands, Southend Road, Chelmsford CM2 7TD
GW3 FMR C. Dwyer Ystrad, 29 The Oval, Llandudno LL30 2BU
G3 FMU D. Mcdiarmid 102 Shalloak Road, Broad Oak, Canterbury CT2 0QH
G3 FMW J. Stockley 22 Manor Gardens, Killinghall, Harrogate HG3 2DS
G3 FNL R. Grubb 7762 Brockway Drive, Boulder 80303 United States
G3 FNO G. Morgan 27 Kestrel Close, Downley, High Wycombe HP13 5JN
G3 FNZ J. Lambert 49 Rede Court Road, Strood, Rochester ME2 3SP
GW3 FPH J. Hayes 4 St. Marys Drive, Northop Hall, Mold CH7 6JF
G3 FPY J. Dew 62 Monks Park Avenue, Horfield, Bristol BS7 0UH
G3 FRE W. Frith 56 Ringleas, Cotgrave, Nottingham NG12 3NE
GM3 FRU D. Wark Flat 37A, Northwood House, Edinburgh EH9 2EL
G3 FRV R. Vaughan 1 Langstone Close, Maidenbower, West Sussex RH10 7JR
G3 FSA A. Davis Willow Cottage, Hedging, Bridgwater TA7 0DE
GW3 FSP L. Davies Glanmor, Brynna Road, Pencoed, Bridgend CF35 6PD
G3 FSX R. Ellis Laura House, 79 Sunte Avenue, Haywards Heath RH16 2AB
G3 FTK L. Gray 109 Foxholes Road, Poole BH15 3NE
GI3 FTT W. Brennan 10 Dunhugh Park, Londonderry BT47 2NL
G3 FUJ W. Scott 10 Pavilion Road, Littleover, Derby DE23 6XL
G3 FVA Sth Manchestr Rd c/o D. Armitage 12 Loughborough Close, Sale M33 5UF
G3 FVR R. Bannister 22 Manton Road, Hitchin SG4 9NW
G3 FWD B. Purchase 126 Renton Road, Wolverhampton WV10 6XH
G3 FWI W. Sutton Pendle, 6 Cuperham Close, Romsey SO51 7LH
G3 FWU L. Richardson Belmont Cottage, Christys Lane, Shaftesbury SP7 8NQ
G3 FXI P. Cardwell 3 Old Talbot, Llanwnog, Caersws SY17 5JG
GD3 FXN A. Radcliffe 3 Cronk Drine, Union Mills, Douglas IM4 4NG Isle of Man
G3 FYF P. Acke Kinghurst Farm, Holne, Newton Abbot TQ13 7RU
G3 FYQ Pontefract & District ARS c/o N. Ferguson Royd Moor, Royd Moor Lane, Badsworth, Pontefract WF9 1AZ
G3 FYX Rnars London (Hms Belfast) Group c/o R. Emery 30 Station Road Winterbourne Down, Bristol BS36 1EP
GW3 FZV R. Lewis 1 Victoria Avenue, Penarth CF64 3EN
G3 GAA W. Jeans 36 Pimms Grove, High Wycombe HP13 7EF
G3 GAF Dr C. Dollery 101 Corringham Road, London NW11 7DL
G3 GAH D. Johnson 31 Coniston Avenue, Penketh, Warrington WA5 2QY
G3 GAQ D. Bottomley 24 Midhope Road, Woking GU22 7UE
G3 GBD S. Hancock 53 Friary Grange Park, Winterbourne, Bristol BS36 1NA
GD3 GBG A. Moore 114 Ballabrooie Drive, Douglas IM1 4HQ Isle of Man
G3 GBN S. Feldman Flat 5 Maitland Joseph House, 35 Marlowes, Hemel Hempstead HP1 1LB
G3 GBS M. Sandoz Edelweiss, Broad Lane, Solihull B94 5DP
G3 GBU Stoke On Trent ARS c/o A. Allen 3 Wayfield Grove, Harpfields, Stoke -on- Trent ST46DB
GM3 GBZ Havering & District ARC c/o G. Balfour 6 Kirkden Street, Friockheim, Arbroath DD11 4SX
G3 GCU R. Adams 13 Birch Rise, Ashley Heath, Market Drayton TF9 4PZ
G3 GCW B. Jones 44 Winner Hill Road, Paignton TQ3 3BT
G3 GDB G. Bird 16 Simnel Road, London SE12 9BG
G3 GDH D. Silveston 192 Rosemary Avenue, Minster On Sea, Sheerness ME12 3HX
G3 GDZ R. Hooten 17 Clarence Road, Teddington TW11 0BQ
G3 GEF J. Andrews 45 Sandes Court, Sandes Avenue, Kendal LA9 4LN
G3 GEG E. Cooper Ciren, 19 Ventnor Road, Sandown PO36 0JT
G3 GEI Solihull ARS c/o Dr R. Hancock 80 Ulleries Road, Solihull B92 8EE
G3 GEJ L. Airey 32 Brookside Close, Bedale DL8 2DR
G3 GEX P. Burton 18 Tankerfield Place, Romeland Hill, St. Albans AL3 4HH
G3 GGG R. Bishop 31 Blenheim Close, Didcot OX11 7JQ
G3 GGH P. Horn Darfield, 50 Barrack Road, Bexhill-on-Sea TN40 2AZ
G3 GGI A. Laurence 70 Firs Avenue, London N11 3NQ
G3 GGL D. Wormald Long Acre, Mamble, Kidderminster DY14 9JY
G3 GGN D. Shute 100 Wick Street, Wick, Littlehampton BN17 7JS
G3 GGR J. Sykes 49 Chapel St., Pelsall, Walsall WS3 4LW
G3 GGS W. Waring 51 Church Road, Leyland PR25 3AA
G3 GGU G. Smith Greenacres, Top Road, Chesterfield S44 5AE
G3 GHN Hoover ARC c/o S. Fletcher 90 Westcombe Park Road, Blackheath, London SE3 7QS
G3 GIB A. Wake 42 Charles Avenue, Watton, Thetford IP25 6BZ
G3 GIH J. Bird The Old Stackyard, Daisy Green, Bury St. Edmunds IP31 3HX
G3 GIZ Chester & District RS c/o P. Holland Chatterton, Chapel Lane, Threapwood, Malpas SY14 7AX
G3 GJA C. Reynolds 49 Westborough Way, Anlaby Common, Hull HU4 7SW
G3 GJJ P. Watson 5 High Garth, Winston, Darlington DL2 3RY
G3 GJL Shetland CG c/o R. Moles 14 Dorsett Road, Stourport-on-Severn DY13 8EL
GW3 GJQ Verulam ARC c/o Sqdn. Ldr. R. Handley Flat 2, 11 Trinity Square, Llandudno LL30 2RA
G3 GJW T. Lundegard Saxby, Botsom Lane, Sevenoaks TN15 6BL
G3 GKS R. Christian 27 Howey Rise, Frodsham WA6 6DN
G3 GLA B. Mase 18 Norton Drive, Norwich NR4 6JD
G3 GLL T. Green 6 Woodrolfe Road, Tollesbury, Maldon CM9 8SB
G3 GLW P. Willis 26 Snellgrove Close, Calmore, Southampton SO40 2WD
G3 GLX J. Simmonds 99 Foljambe Avenue, Chesterfield S40 3EY
G3 GMC P. Mcvey 18 Worlebury Hill Road, Weston-Super-Mare BS22 9SP
G3 GML F. Murray 3 Rosemary Close, Tiptree, Colchester CO5 0QD
G3 GMM E. Mcfarland 60 Sutton Oaks, London Road, Crewe CW4 7AS
G3 GMS M. Thayne 14 Tynedale Avenue, Monkseaton, Whitley Bay NE26 3BA
G3 GMW L. Nichols 5 Middle Pasture, Peterborough PE4 5AU
G3 GMY F. Green 5 Silvercliffe Gardens, New Barnet, Barnet EN4 9QT
G3 GNA D. Macmillan Brook Farm, Broadwas, Worcester WR6 5NE
G3 GOS P. Peach The Firs, Goldsmith Lane, Axminster EX13 7LU
G3 GQC Obo Manfield Am RS c/o D. Riley 9 Century Avenue, Mansfield NG18 5EE
G3 GQK J. Wall P.O Box 631, Nambucca Heads 2448 Australia
GM3 GRG D. Rollo 25 Beaufort Drive, Kirkintilloch, Glasgow G66 1AX
G3 GRL S. Houlton New Tor Grange Street, Alfreton DE55 7HZ
G3 GRQ C. Hebden 129 Millers Way, Honiton EX14 1JB
G3 GRS Gravesend ARS c/o D. Lawley 5 The Limes, Buckland, Buntingford SG9 0PW
G3 GRV G. Halse 10 Charnock Close, Hordle, Lymington SO41 0GU
G3 GRY F. Wiseman 14 Parkway, Crowthorne RG45 6EN
G3 GTA J. Shute 32 Woodborough Drive, Winscombe BS25 1HB
G3 GTF B. Harris 6 The Priory, Lewes Road, Cross In Hand, Heathfield TN21 0FE
GM3 GTQ A. Mcphedran 3 Argyll Road, Bearsden, Glasgow G61 3JX
GI3 GTR R. Mckinty 3 Rhanbuoy Road, Craigavad, Holywood BT18 0DY
G3 GUE A. Dowling Church Cottage, Frittenden, Cranbrook TN17 2DD
G3 GUR J. Scully 1 Wyde Feld, Bognor Regis PO21 3DH
GW3 GUX J. Brimecombe Llwyn Onn, Llangoed, Beaumaris LL58 8PH
G3 GVM F. Robins 59 Titchfield Road, Stubbington, Fareham PO14 2JF
G3 GWB Northampton RC c/o J. Cockrill 28 Northampton Road, Harpole, Northampton NN7 4DD
G3 GWC E. Ramsdale 8 May Cottages, Monkswell Lane, Coulsdon CR5 3SX
G3 GWE A. Daum 100 Shawbridge, Harlow CM19 4NW
G3 GXG C. Lee 7 Lilac Avenue, Lower Quinton, Stratford-upon-Avon CV37 8US
G3 GXI Eccles And District ARS c/o Prof. C. Harrison 11 Ringley Park, Whitefield, Manchester M45 7NT
G3 GXQ W. Roberts 24 Leeds Road, Barwick In Elmet, Leeds LS15 4JD
G3 GYQ C. Spackman 10 Norton Drive, Warwick CV345FE

G3 GZT R. Moores 117 Horton Road, Brighton BN1 7EG
GW3 GZX A. Bladon 6 Quarry Bank, Mold Road, Denbigh LL16 4DT
G3 GZZ A. Bevan 14 Parsonage Road, Berrow, Burnham-on-Sea TA8 2NL
G3 HAA J. Morgan 10 Bamber Gardens, Southport PR9 7PQ
G3 HAL R. Parrott 3 Ash Grove, Chard TA20 1BZ
GM3 HAM Tintagel And District RAC c/o P. Bates 10 Swanston Avenue, Edinburgh EH10 7BU
G3 HAN M. Hitchman 12 Briar Walk, Oadby, Leicester LE2 5UE
G3 HCO G. Errock 307 Main Road, Emsworth PO10 8JG
G3 HCS H. Stratton 26 Marjorie Road, Chaddesden, Derby DE21 4HQ
G3 HCT J. Bazley C/O Mr P Chadwick, Three Oaks, Swindon SN5 0AD
G3 HCZ B. Edmondson 1 Harbour Lane, Turton, Bolton BL7 0PA
GW3 HDF K. Groves 6 Overleigh Drive, Buckley CH7 2PA
G3 HDM S. Campbell Carrer De Ses Sevines, 1 Bajo, Mallorca Spain
GM3 HDT J. Graham Lodge, , Stronsay KW17 2AN
G3 HEH E. Parker 39 Hellath Wen, Nantwich CW5 7BB
G3 HEJ D. Stanners Tanglewood Samarkand Close, Camberley GU15 1DG
GM3 HEN A. White Byeways Whiting Bay, Isle of Arran KA27 8QH
GW3 HEU D. Rickers 4 St. Marks Terrace, Wrexham LL13 0PQ
GD3 HFC F. Arrowsmith The Evergreens, South Cape, Laxey IM4 7JB Isle of Man
G3 HFM A. Vickers Foxcroft, 4 Woodlands End, Macclesfield SK11 9BF
GU3 HFN the Guernsey Ar c/o P. Cooper 1 Clos Au Pre, La Route De La Hougue Du Pommier, Castel, Guernsey GY5 7FQ
GM3 HGA J. Mccall 1 Pinewood Place, Aberdeen AB15 8LT
G3 HGD V. Best 3 Old Auction Mart, Kirkby Lonsdale, Carnforth LA6 2AF
G3 HGE T. Withers Woodpeckers, West Stow, Bury St. Edmunds IP28 6ER
G3 HHD T. Hayward Skirt Bank, Nether Silton, North Yorkshire YO7 2LL
G3 HHU Dr J. Ickringill 28 Deena Close, Queens Drive, London W3 0HR
G3 HIU Duxford Radio Limited c/o D. White 1 Whaddon Road, Shenley Brook End, Milton Keynes MK5 7AF
GI3 HJH R. Mcburney 8 Main Road, Ballymartin, Newry BT34 4NU
G3 HJP G. Cooper 25 Plantation Avenue, Shadwell, Leeds LS17 8TB
G3 HKA C. Booth 88 Green Drive, Thornton-Cleveleys FY5 1JD
G3 HKD D. Money 125 Wroxham Road, Norwich NR7 8AD
G3 HKF B. Ferris 5 Guildway, Todwick, Sheffield S26 1JN
G3 HKH M. Harrison 3 Stert Street, Abingdon OX14 3JF
G3 HKT A. Partner 10 The Tanners, Titchfield Common, Fareham PO14 4BH
G3 HLG D. Johnson Robins, 4 Station Road, Newark NG23 7RA
G3 HLI M. Bradford 101 Oxendon Way, Binley, Coventry CV3 2HA
G3 HLN P. Woods 145 Hollybush Lane, Welwyn Garden City AL7 4JT
G3 HMB I. Elliot Grange House, Manningtree Road, Ipswich IP9 2SW
G3 HMG A. Macgregor 14 Quantock Grove, Williton, Taunton TA4 4PD
G3 HMO J. Osborne 141 Chadwick Road, London SE15 4PY
G3 HMQ J. Robson 32 St. Stephens Road, Cold Norton, Chelmsford CM3 6JE
G3 HMR G. Moser 30 Blackhall Croft, Blackhall Road, Kendal LA9 4UU
G3 HMV N. Bolton 2 Selborne Villas, Clayton, Bradford BD14 6JZ
G3 HNC B. Dyer 30 Smithson Avenue, Castleford WF10 3HN
GM3 HNE G. Campbell 17 Roseburn Terrace, Edinburgh EH12 5NG
GI3 HNM C. Davies 121 Comber Road, Toye, Downpatrick BT30 9PD
GM3 HOM J. Reilly 30 Park Crescent, Bishopbriggs, Glasgow G64 2NS
G3 HPB F. Tooley 70 Langbury Lane, Ferring, Worthing BN12 6QA
G3 HPD F. Dews 341 Crossley Lane, Mirfield WF14 0NR
G3 HQG G. Atkins 20 Mansfield Road, Killamarsh, Sheffield S21 2BX
G3 HQS C. Baker Roffensis, 16 Boulderside Close, Norwich NR7 0JJ
G3 HQT P. Ball 68 Brook Lane, Warsash, Southampton SO31 9FG
G3 HQX J. Brodzky 3 Ropewalk House, Hyde Abbey Road, Winchester SO23 7XH
G3 HRE F. Watson 54 Tavistock Road, Cambridge CB4 3ND
G3 HRH R. Hills 2 The Dell, Otterbourne Road, Winchester SO21 2DE
G3 HRK D. Willies 17 Campion Way, Sheringham NR26 8UN
G3 HRX J. Hilling 24 Gloucester Road, Gaywood, King's Lynn PE30 4AB
GM3 HSO J. Yang Flat 6, 14 Bothwell Street, Edinburgh EH7 5PS
G3 HST G. Allen Moor Farm Cottage, Salcombe TQ8 8PW
G3 HSV D. Alesbury 23 Cullerne Road, Swindon SN3 4HU
G3 HTA J. Forward Sunrays, Barnstaple Cross, Crediton EX17 2EP
G3 HTB M. Squance Church Lane, Cubbington, Leamington Spa CV32 7JT
G3 HTC C. Storey 12 Vereker Drive, Sunbury-on-Thames TW16 6HF
G3 HTJ W. Walker 53 Wolfridge Ride, Alveston, Bristol BS35 3PR
G3 HTO R. Dolton 43 Jubilee Meadow, St. Austell PL25 3EX
G3 HTT W. Cheeseworth 10 Barton Mill Court, Station Road West, Canterbury CT2 7JZ
G3 HUB M. Harrison Rolling Hills, Brandy Lane, Lostwithiel PL22 0QH
G3 HUD M. Brown 10 Park House Mews, Congleton Road, Sandbach CW11 4SP
G3 HUK M. Morrissey 1 Hamilton Road, Church Crookham, Fleet GU52 6AS
G3 HUO K. Young 80 Darbys Lane, Oakdale, Poole BH15 3ET
G3 HUR D. Brough 18 Lark Hall Road, Macclesfield SK10 1QP
G3 HUX J. Matthews 4 Berrington Grove, Ashton-In-Makerfield, Wigan WN4 9LD
G3 HVA D. Pinnock 2 Oak Close, Oakley, Basingstoke RG23 7DD
G3 HVJ A. Chappell 22206 Del Valle St, Woodland Hills 91364-1515 United States
GM3 HVK J. Craig 147 Avon Road, Larkhall ML9 1RA
G3 HWF South And West Yorkshire Wing ATC c/o D. Taylor 76 Heworth Village, York YO31 1AL
G3 HWM J. Cowling 19 The Drive, Hullbridge, Hockley SS5 6LZ
G3 HWW York ARS c/o C. Rouse 86 Melton Avenue, Clifton, York YO30 5QG
G3 HXK P. Nethercot Ronhill, Stoodleigh, Tiverton EX16 9PJ
G3 HXN J. Crisp 371 Stroud Road, Tuffley, Gloucester GL4 0DA
G3 HYG D. Topping Bentley, Middle Street, Waltham Abbey EN9 2LB
G3 HYH S. Hay 27 Acres Road, Leicester Forest East, Leicester LE3 3HB
GM3 HYX C. Rattray 58 Aberdour Road, Dunfermline KY11 4PE
G3 HZP H. James 10 Playsted Lane, Cambourne, Cambridge CB3 6GA
G3 HZT P. Fraser 45 The Martlet, Hove BN3 6NT
G3 IAR M. Crowther-Watson The Snicket, 14 The Avenue, Sevenoaks TN15 8EA
G3 IAZ A. Wickham Apartment 6, Panama Reach, Eastbourne BN23 5PL
G3 IBI P. Scutt 62 Old Street, Fareham PO14 3HW
G3 IBQ K. Holt 61 Millford Avenue, Nepean, Ontario K2G-1C4 Canada
GM3 IBU 59 Degrees North ARG c/o E. Holt Ashwell, St. Ola, Kirkwall KW15 1SX
G3 IBY Dr T. Wilmshurst 4 Eastern Road, West End, Southampton SO30 3EQ
G3 ICA G. Adams Sue Marey, Selsley Hill, Stroud GL5 5JS
G3 ICB A. Bull 91 Lower Way, Thatcham RG19 3RS
G3 ICC I. Chilton 47 Mayfield Crescent, Eaglescliffe, Stockton-on-Tees TS16 0NH
G3 ICG K. Mcfarlane Clifton, 18 Needham Road, Harleston IP20 9JY
G3 ICN D. Collado Castells 37 Herbert Street, London NW5 4HB
G3 ICZ W. Clowes 144 Norton Lane, Norton-In-The-Moors, Stoke-on-Trent ST6 8BZ
G3 IDB A. Brooks 45 Northfield Road, Townhill Park, Southampton SO18 2QE
G3 IDW R. Reynolds 6 Church Way, Stratton, Swindon SN3 4NF
G3 IDY R. Robson 66 Tilstock Crescent, Shrewsbury SY2 6HQ
G3 IEJ S. Watson 6 Hope Street, Lytham St. Annes FY8 3SL
G3 IFX A. Cooke 9 Lee Crescent, Ilkeston DE7 5EF
G3 IGC A. Garforth 110 Foxdenton Lane, Chadderton, Oldham OL9 9QR
G3 IGQ University of Surrey Ears c/o N. Jones 61 Sutherland Close, Whitehill, Bordon GU35 9RE
G3 IGU K. Coates 76 Copley Crescent, Scawsby, Doncaster DN5 8QP
G3 IGV J. Birkbeck 4 Tregullan View, Bodmin PL31 1BH
G3 IGZ D. Bruce 22 Brownspring Drive, New Eltham, London SE9 3JX
G3 IHX N. Bond 333 Hillandale Drive, Charlotte 28270 United States
G3 IIN M. Griffin Michaelmas, Southdown Road, Freshwater PO40 9UA
G3 IIO D. Harriott 23 Hamsey Crescent, Lewes BN7 1NP
G3 IIV A. Davies Paarl, 129 Cotwall End Road, Dudley DY3 3YQ
G3 IIW M. Sands Beech Lea, St. Marks Road, Tunbridge Wells TN2 5LU
G3 IJA J. Allan 5 Terrington Court, Strensall, York YO32 5PA
G3 IJL A. Sephton 16 Bloemfontein Avenue, Shepherds Bush, London W12 7BL
G3 IJS J. Stratfull 55 Craigweil Lane, Aldwick, Bognor Regis PO21 4XN
G3 IJU E. Briggs 32 Lethbridge Road, Wells BA5 2FN
G3 IJV R. Harvey 16 Gatesgarth Close, Hartlepool TS24 8RB
G3 IKB D. Giddens 89 Pollards Oak Road, Oxted RH8 0JE
G3 IKL R. Craxton 103 Clifton Road, Rugby CV21 3QH
G3 IKQ R. Chilton 80 Plantation Road, Hextable, Swanley BR8 7SB
G3 ILE E. Marsh 63 Willows Lane, Accrington BB5 0SQ
G3 ILO S. Spencer 9 Vaisey Field, Whitminster GL2 7PT
G3 IMW S. Whitfield 7 Sir Alex Walk, Topsham, Exeter EX3 0LG
G3 IMX E. Jolliffe 96 Cowes Road, Newport PO30 5TP
G3 INP G. Stanway Bramble Edge Cottage, Bates Lane, Frodsham WA6 9LL
G3 INQ Capt. B. Podmore 6 Alfred Court, Furlong Road, Bourne End SL8 5AZ
G3 INR P. Buchan 79 Cavendish Avenue, Cambridge CB1 7UR
G3 INU R. Appleby 14 Truro Court, Canterbury Way, Stevenage SG1 4LF
G3 INY E. Tudor Mowhills House, 133 High Street, Bedford MK43 7ED
G3 INZ J. Tournier Avalon 13 Greenlands Flackwell Heath, High Wycombe HP10 9PL
G3 IOB P. Revell 54 Lytham Road, Perton, Wolverhampton WV6 7YY
G3 IOI N. Pascoe 36 Kilbirnie Road, Bristol BS14 0HS
G3 IOJ B. Rixon 1 Carde Close, Hertford SG14 2EU
G3 IOM R. Chidzey 8 Dormans Close, Dormansland, Lingfield RH7 6RL
G3 IOR P. Gowen 17 Heath Crescent, Norwich NR6 6XD
G3 IPD C. Oakley 4 Cross Keys Lane, Low Fell, Gateshead NE9 6DA
G3 IPG G. Phipps 12 Mill Close, Pulham Market, Diss IP21 4TQ
G3 IPL R. Winters 43 Manor Close, Harpole, Northampton NN7 4BX
G3 IPP M. Dance Golf Cottage, 8 St. Johns Road, Crawley RH11 7BD
G3 IQF R. Fowler 49 Westhorpe Park, Westhorpe, Marlow SL7 3RH
G3 IQX E. Popplewell 71 Thornbury Road, Southbourne, Bournemouth BH6 4HU
G3 IQY A. Rees 59 Hillside Gardens, Barnet EN5 2NQ
G3 IRA J. Wren 29 Carisbrook Terrace, Chiseldon, Swindon SN4 0LW
G3 IRQ P. Rackham Upyonda, Otley Bottom, Ipswich IP6 9NG
G3 ISB C. Brock 24 Glebelands Road, Knutsford WA16 9DZ
G3 ISD E. Hatch 147 Borden Lane, Sittingbourne ME10 1BY
G3 IST S. Turner 8001 Bayshore Drive, Seminole 34646 United States
G3 ISX C. Leal 61 Light Oaks Avenue, Light Oaks, Stoke-on-Trent ST2 7NF
GJ3 IT Richmond Raynet Group c/o M. Turner 4 Le Clos Sara, St. Lawrence JE3 1GT Jersey
G3 ITB T. Bartlett 19 Hardley Street Hardley, Norwich NR14 6BY
G3 ITF B. Freeman 47 Gorham Avenue, Rottingdean, Brighton BN2 7DP
G3 ITH R. Franklin 2 Berkeley Drive, Kingswinford DY6 9DX
G3 ITL J. Humpoletz 76 Marlborough Road, Braintree CM7 9LR
GM3 ITN Blandford Garrison ARC c/o L. Hamilton Halls Land, Cochno Road, Clydebank G81 6NR
G3 IUB Birmingham A R S c/o D. Cottam 14 Barnard Close, Rednal, Birmingham B45 9SZ
G3 IUC R. Mcmillan East Orchard, Almeley, Hereford HR3 6LF
G3 IUE M. Newell 35 Ingleside Crescent, Lancing BN15 8EN
G3 IUJ R. Rogerson 19 Martins Road, Shortlands, Bromley BR2 0EE
G3 IUO G. Allen 157 Lynton Road, Bedminster, Bristol BS3 5LN
G3 IUV G. Loveday 2 St. Aldwyns Close, Bristol BS7 0UQ
G3 IUW L. Pritchard Green Horizons, Send Hill, Woking GU23 7HR
G3 IUY J. Presland 6 Pippin Close, Sutton Ely CB6 2RX
G3 IUZ Rvd. H. Davis 6 St. Thomas Terrace, Wells BA5 2XG
G3 IVC A. Sycamore Fir Tree Cottage, Compton Valence, Dorchester DT2 9ES
GW3 IVK D. Evans 11 Hill View, Bryn-Y-Baal, Mold CH7 6SL
G3 IVP A. Page The Farmhouse, Budges Shop, Trerulefoot, Saltash PL12 5DA
G3 IW Worked All Britain Awards Group c/o W. Wilkie 14 Horseshoe Close, Northwood, Cowes PO31 8PZ
G3 IWE A. Wyse 29 Tregainlands Park, Washaway, Bodmin PL30 3AU
G3 IWH I. Hall 46 Bushmead Road, Luton LU2 7EU
GW3 IWM M. Holland 7 Willans Court, Willans Drive, Newtown SY16 4DB
G3 IWV J. Parker 472-474 Castle Lane West, Bournemouth BH8 9UD
G3 IWW R. Hopkins 34 Shelley Close, Abingdon OX14 1PR
GM3 IWX W. Ritchie 8 Cheviot Place, Grangemouth FK3 0DE
G3 IXI K. Landon 1 The Laurels, Leedons Park, Broadway WR12 7HB
G3 IXN M. Lovejoy 73 Stoneham Lane, Swaythling, Southampton SO16 2NZ
G3 IXZ R. Bowden 41 Brockington Road, Bodenham, Hereford HR1 3LP
G3 IYF D. Baker Long Haul, 3 Chapel Lane, Lincoln LN6 9EX
G3 IZA D. Allison 71 South Hill Road, Bromley BR2 0RW
G3 IZD I. Davies 13 Thurlow Way, Barrow-in-Furness LA14 5XP
G3 IZF D. Taylor 24 Woodville Avenue, Crosby, Liverpool L23 3BZ
G3 IZM J. Harper Bill 1 Shepherds Close, Staple Hill, Bristol BS16 5LE
G3 IZQ H. Hyman 19 Black Horse Drive, Acton 1720 United States
G3 JAL R. Taylor 304 Brigstock Road, Thornton Heath CR7 7JE
G3 JAU C. Davies 107 Talbot Road, Bournemouth BH9 2JE
G3 JBF L. Brown Ladygate, St. Michaels Road, Stafford ST19 5AH
GW3 JBJ F. Mathers 17 Penlon, Menai Bridge LL59 5LR
GW3 JBZ J. Brace 12 Heol Gwili, Gorseinon, Swansea SA4 4GE

G3 JCK F. Chilvers 5 Low Common Close, Foulsham, Dereham NR20 5TW
G3 JCM D. Bolwell 3 Mildmays, Danbury, Chelmsford CM3 4DP
G3 JCR K. Smith 20 Manor House Gardens, Abbots Langley WD5 0DH
G3 JDD R. Dobson 16 Howden Road, Fulham 5024 Australia
G3 JDO H. Martin 7 Nairn Street, Jarrow NE32 4HX
G3 JDT B. Read Glenside, 4 Hatton Lane, Warrington WA4 4BY
G3 JDY Obo Royal Air Force ARS E Riding Area c/o B. Atkinson 165 Alliance Avenue, Hull HU3 6QY
G3 JFD B. Brown 130 Ashland Road West, Sutton-in-Ashfield NG17 2HS
GM3 JFG Stornoway Repeater Group c/o N. Doherty Cairdeas, Carrbridge PH23 3AA
G3 JFR N. Cottrell 28 Colley Wood, Kennington, Oxford OX1 5NF
G3 JFS P. Cole 25 Wardlow Gardens, Plymouth PL6 5PU
G3 JFT B. Dare 128 Sancroft Road, Spondon, Derby DE21 7ES
G3 JFW P. Beevers Hill Farm Granary, Lower Somersham, Ipswich IP8 4PU
GW3 JGA J. Lawrence 40 Aberconway Road, Prestatyn LL19 9HL
GW3 JGE V. Owen 17 Knowles Avenue, Prestatyn LL19 8SG
G3 JGP E. Robinson 16 Shaw Green Storth, Milnthorpe LA7 7JB
G3 JHH S. Burgess 34 Redcliffe Road, London SW10 9NJ
G3 JHI R. Hathaway 30 Berkeley Drive, Hornchurch RM11 3PY
G3 JHP E. Allen 11 Newlands Close, Horley RH6 8JR
G3 JHU C. Pavey 3 Field Close, Chatham ME5 9TD
G3 JIE D. Youngs 12 Fox Grove, East Harling, Norwich NR16 2PS
GM3 JIG K. Hodge 66 Ardrossan Road, Seamill, West Kilbride KA23 9LX
GM3 JIJ Stornoway Repeater Group c/o N. Doherty Cairdeas, Carrbridge PH23 3AA
G3 JIP J. Hill Calle El Palomar 13, El Romeral, Malaga 29130 Spain
G3 JIR J. Hardcastle 8 Norwood Grove, Rainford, St. Helens WA11 8AT
G3 JIS R. Heaton 20 Tewkesbury Avenue, Urmston, Manchester M41 0RJ
GD3 JIU M. Thompson 3 Close Cam, Port Erin IM9 6NB Isle of Man
G3 JIX Dr K. Smith Staple Farm House, Durlock Road, Canterbury CT3 1JX
GM3 JJQ D. Millar 51 Tiree Crescent, Polmont, Falkirk FK2 0UX
G3 JJR J. Rickwood 44A The Bridle Path, Madeley, Crewe CW3 9EL
G3 JJT C. Kempson 8 Arle Gardens, Cheltenham GL51 8HR
G3 JKB D. Simmonds 73 Tor-O-Moor, Woodhall Spa LN10 6SD
GM3 JKC C. Cooper 28 Kippford St., Glasgow G32 9BW
G3 JKE G. Thomas 13 Essex Drive, Taunton TA1 4JX
G3 JKF K. Franklin 4 Princes Close, Seaford BN25 2EW
G3 JKL J. Lovell Kowloon, Slade, Bideford EX39 3LZ
G3 JKM D. Buckland 29, Longfields, Swaffham, , Swaffham PE37 7RH
GM3 JKS F. Claytonsmith 16 Templand, Crossmichael, Castle Douglas DG7 3BF
G3 JKX M. Street 12 Ullswater Close, Priorslee, Telford TF2 9RB
G3 JLK C. Jeffery 33 Thirlmere Road, Weston-Super-Mare BS23 3UY
G3 JLN F. Blain High Ridge, Howgate Lane, Bembridge PO35 5QW
G3 JLZ V. Ludlow 6 Raleigh Crescent, Stevenage SG2 0EQ
G3 JMJ D. Nunn Oak Lea, Crouch House Road, Edenbridge TN8 5EL
GM3 JMM J. Murdoch 4 Cedar Drive, Milton Of Campsie, Glasgow G66 8AY
G3 JMZ J. Hilton Windsor House, Preston Road, Charnock Richard, Chorley PR7 5HH
G3 JNB V. Brand 8 Greenway, Campton, Shefford SG17 5BN
G3 JNJ D. Platt Springview Residential Home 10 - 12, Crescent Road, Enfield EN2 7BL
G3 JNM T. Whittaker 16 Acresdale, Lostock, Bolton BL6 4PJ
GM3 JOB G. Bryce 3 West Bowhouse Way, Girdle Toll, Irvine KA11 1NJ
G3 JOE J. Brown 10 Park House Mews, Congleton Road, Sandbach CW11 4SP
G3 JOR V. Capell Endways, 15 Copse Road, Bexhill-on-Sea TN39 3UA
G3 JOT F. Whatley 1 Mill Close, Wroughton, Swindon SN4 9AR
G3 JOX A. Greaves Jacobs Well, Woodhill Road, Chelmsford CM2 7SF
G3 JPB C. Noden Brownhills Cottage Farm, Brownhills, Market Drayton TF9 4BE
G3 JPG R. Parker 6 Cambridge Road, Chingford, London E4 7BP
G3 JPJ J. Peerless 101 Greenside, Borehamwood WD6 4JD
G3 JPM B. Grainge 4 Maltings Close, Chevington, Bury St. Edmunds IP29 5RP
G3 JPO M. Fielding 68 Mitford Road, South Shields NE34 0EQ
G3 JPU D. Plant Briarfields, Raby Crescent, Shropshire SY3 7JN
G3 JPZ I. Denney 5 Howard Close, Harleston IP20 9HY
G3 JQ A. Webster 5 Brookside Court, 142 Prestbury Road, Macclesfield SK10 3BR
G3 JQC G. Hawksworth 16 Birkhead Street, Heckmondwike WF16 0BE
G3 JQK E. Jones Appleton Thorne, Lower Broad Lane, Redruth TR15 3HJ
G3 JQL J. Haggart 22 Alnwick Road, Newton Hall, Durham DH1 5NL
G3 JQS J. Guttridge Victoria Cottage, The Common, Cambridge CB21 5LR
G3 JRD R. Dancy 1 Ladds Corner, Eastcourt Lane, Gillingham ME7 2UW
G3 JRE F. Thomas 99 Eastfield Road, Wollaston, Wellingborough NN29 7RS
G3 JRH P. Horne The Annexe, Burntwood Farm, Winchester SO21 1AF
G3 JRK J. Knight 10 Lynton Drive, Burnage, Manchester M19 2LQ
G3 JRL F. Armstrong 4 Medway Drive Preston, Weymouth DT3 6LF
G3 JRM Lowestoft District And Pye ARC c/o R. Stone 63 Sands Lane, Lowestoft NR32 3ER
G3 JRS A. Kidd 35 Hollands Way, Kegworth, Derby DE74 2GQ
G3 JRY S. Auty 3 Rochford Crescent, Boston PE21 9AE
G3 JSA D. Wilcox 13 Richards Drive, Dartmouth NS B3A 2P1 Canada
GW3 JSG J. Gunn Flat 18, Bro Llewelyn, Penrhyndeudraeth LL48 6AL
G3 JSK D. Dean 8 Bradford Road, Corsham SN13 0QR
G3 JSU L. Sampson 107 South St., Lancing BN15 8AS
GW3 JSV D. Holmes Fair Oaks, Berriew, Welshpool SY21 8AU
G3 JTJ J. Jones Westerland, 1 Roborough Close, Plymouth PL6 6AH
G3 JTK G. Allen 119 Haymoor Road, Poole BH15 3NR
G3 JTQ R. Griffiths 7 Dever Way, Oakley, Basingstoke RG23 7AQ
G3 JTT P. Thompson 30 Farnol Road, Birmingham B26 2AF
G3 JUU D. Adams 23 Arlington Gardens, Attleborough NR17 2NH
G3 JUW C. Lovell 5 Montpelier Road, Ilfracombe EX34 9HP
G3 JUX J. Mcfarlane 141 Tyler Grove, Aston, Stone ST15 0JA
G3 JVC J. Cleeve 44 Ditton Hill Road, Long Ditton, Surbiton KT6 5JD
G3 JVL M. Walters 26 Fernhurst Close, Hayling Island PO11 0DT
G3 JVM R. Medcraft 134 Dulverton Road, Ruislip HA4 9AG
G3 JVN D. Keen 14 Penina Avenue, Newquay TR7 2LE
G3 JVP V. Purdy 99 Belmont Road, Uxbridge UB8 1QX
G3 JVR D. Nokes 16 Salisbury Grove, Giffard Park, Milton Keynes MK14 5QA
G3 JWB J. Boote 1 Shacklock Close, Arnold, Nottingham NG5 9QE
G3 JWI R. Page-Jones 34 Edwards Way, Hutton, Brentwood CM13 1BT
G3 JWN F. Walker 2 Croft Place, Brighouse HD6 4AP
G3 JWQ B. Maycock Hill House, Bullock Lane, Alfreton DE55 4BP
G3 JXC C. Gregory 51 Calgary Avenue, Blackburn BB2 7DS
G3 JYG J. Kirby 14 Grovelands Road, Hailsham BN27 3BZ
G3 JYS R. Finch 8 Chalfont Close, Allesley, Coventry CV5 9HL
G3 JZF J. Smith 17A, Sutton Coldfield B74 2QA
G3 JZL W. Montford 3 The Close Brandon, Coventry CV8 3JF
G3 JZT R. Cheetham 7 Parkway, Stockport SK3 0PX
G3 KAE J. Rowley 41 Main Street, East Ayton, Scarborough YO13 9HL
G3 KAF J. France 34 Ladythorn Road Bramhall, Stockport SK7 2ER
G3 KAG A. Parker Hillside Main Street, Roston, Ashbourne DE6 2EH
G3 KAN A. Shrewsbury 1 Dardis Close Kingsley Northampton, Northampton NN2 7DN
G3 KAP R. Taylor 2 Brenchley Mews, Charing, Ashford TN27 0JQ
G3 KAR D. Hammond Christen Mares, Willersey Hill, Broadway WR12 7PF
G3 KAU L. Laszkiewicz 38 Langley Lane, Ifield, Crawley RH11 0NA
GW3 KAX G. Mackrell Preseli Newchapel, Boncath SA37 0EH
G3 KBH Dr M. Hughes Northdean, Brimstone Lane, Gravesend DA13 0BW
G3 KBI T. Waller 12 Skelton Road, Brotton, Saltburn-by-the-Sea TS12 2TJ
GM3 KBP A. Kerr 47 Hillpark Avenue, Edinburgh EH4 7AH
GM3 KC Montrose Amateur Radio Station c/o B. Murray Sherwood Cottage, Farnell, Brechin DD9 6UH
G3 KCB B. Green 18 Kenilworth Road, Sale M33 5FB
G3 KCD P. Bedwell Narrowgate Rose Cottage, Court Road Rollesby, Great Yarmouth NR29 5HQ
G3 KCG D. Tyerman 20 Grace Gardens, Bishop's Stortford CM23 3EX
GW3 KCQ J. Williams Y-Fedw, Cwmann, Lampeter SA48 8DT
G3 KCT D. Blythe 6 Penn House, Mallory Street, London NW8 8SX
G3 KCV J. Saunders 7 Stone Lane, Yeovil BA21 4NN
GM3 KCY G. Buchanan 30 Gilmour Avenue, Clydebank G81 6AW
G3 KCZ W. Siertsema 21 Rowles Close, Kennington, Oxford OX1 5LX
GW3 KDB P. Miles Y Gorlan, Cross Inn, Llandysul SA44 6NP
G3 KDD V. Barrett 2 Carlisle Close, Sandy SG19 1TX
G3 KDP A. Bounds 32 Tregwary Road, St. Ives TR26 1BL
G3 KDU M. Crawford 95 Victoria Avenue, Princes Avenue, Hull HU5 3DW
G3 KDW H. Turnbull 13 Linden Court, Wessex Road, Southampton SO18 3RB
G3 KDY R. Folgate Stile Cottage, Wilkinson Drive, Market Rasen LN8 3LD
G3 KEG C. Rogers 100 Sparth Road, Clayton Le Moors, Accrington BB5 5QD
G3 KEK G. Carr 88 Woodrow Crescent, Knowle, Solihull B93 9EQ
G3 KEL R. Bray Croft House, Blencogo, Wigton CA7 0BZ
G3 KEP D. Pratt 11 Moorleigh Close, Kippax, Leeds LS25 7PB
G3 KEQ J. Mitchell Chellow Dene, Viewlands Avenue, Westerham TN16 2JE
G3 KEV M. Hamilton 2 Wordsworth Close, Scalby, Scarborough YO13 0SN
GM3 KEZ J. Little 33 Manor Court, Forfar DD8 1AD
G3 KFB N. Parkinson 16 Collinson Avenue, Scunthorpe DN15 8AB
G3 KFG H. Taylor 17 Rose Acre Road, Littlebourne, Canterbury CT3 1SY
G3 KFP A. Olds 43 Fourth Avenue, Teignmouth TQ14 9DT
G3 KFU P. Barry 21 Old Pasture Road, Frimley, Camberley GU16 8SA
GW3 KGI M. Bowen 24 Parklands View, Sketty, Swansea SA2 8LX
G3 KGP M. Palmer Fairways, 8 Gwealdues, Helston TR13 8JZ
G3 KGT J. Nicolson 24 Pottersfield Road, Woodmancote, Cheltenham GL52 9PY
GW3 KGV K. Bates 5 Ffordd Nant Goch, Llangadfan, Welshpool SY21 0PW
GM3 KHH W. Cecil Innes House Oran, Buckie AB56 5EP
G3 KHK D. Connolly Jonquil, 6 Sanderson Mews, Colchester CO5 7HF
G3 KHQ A. Langley 58 Dumbarton Road, Brixton Hill, London SW2 5LU
G3 KHR J. Fox 25 Langdale Crescent, Bexleyheath DA7 5DZ
G3 KHU R. Gabbitas 12 Thornyville Drive, Oreston, Plymouth PL9 7LF
G3 KHZ D. Cox 18 Station Road, Castle Bytham, Grantham NG33 4SB
G3 KII G. Lively 9 Wilson Road, Shurdington, Cheltenham GL51 4SN
G3 KIJ E. Lugmayer 17 Borough End, Beccles NR34 9YW
G3 KIL R. Messer The Shambles, Swinbrook Road, Oxford OX18 1DX
G3 KIP K. Grover 1 Powdermill Close, Tunbridge Wells TN4 9DR
G3 KIQ J. Elliot 2 Pennine Close, Blackley, Manchester M9 6HR
G3 KIW G. Jenner Pogles Wood Cottage, Paradise Lane, Reading RG7 6NU
G3 KJC R. Church Three Birches, Sandy Close, Thatcham RG18 9QP
GM3 KJE J. Scott 5 Garthdee Terrace, Aberdeen AB10 7JE
G3 KJK L. Wilkes Parc Crane, Penmenner Road, Helston TR12 7NN
GW3 KJN I. Winter 5 Uwch Y Nant, Mynydd Isa, Mold CH7 6YP
G3 KJO Huddersfield Technical College ARS c/o M. Lupton 44, Carr Road, Deepcar, , , Sheffield S36 2NR
G3 KJS W. Smith 32 Lumley Road, Chester CH2 2AQ
GW3 KJW P. Allely Dwyfor, Rhiw, Pwllheli LL53 8AE
G3 KJY J. York 48 Browhead Court, Shackleton Street, Burnley BB10 3DS
GM3 KJZ G. Paterson 3 Ferry Barns Court, North Queensferry, Inverkeithing KY11 1ET
G3 KKC A. Rumbelow 7 Hoof Close, Littleport, Ely CB6 1HU
G3 KKJ A. Shannon 23 Glebeland Close, West Stafford, Dorchester DT2 8AE
G3 KKP J. Burgess Moorend, Main St., Leeds LS20 8NX
G3 KKZ P. Champion 7 All Saints Drive, South Croydon CR2 9ES
G3 KLC J. Bennett Koivula, Station Road, Boston PE20 3QT
G3 KLD R. Russell 43 Ingestre Road, Hall Green, Birmingham B28 9EQ
G3 KLF I. Crowther 3 Glenelg, Fareham PO15 6JU
G3 KLH D. Alexander 20 Deans Court, Milford On Sea, Lymington SO41 0SG
G3 KLK B. Page 7 Marconi Way, Southall UB1 3JP
G3 KLN N. Whittaker 111 Burnley Road, Colne BB8 8DT
G3 KLP J. Young Woodglades, 34 The Demesne, Ashington NE63 9TP
G3 KLV G. Vine 4 Tollgate Close, Northampton NN2 6RP
G3 KLZ D. Enoch 7A Mount Road, Evesham WR11 3HE
G3 KMA R. Balister La Quinta, Mimbridge, Woking GU24 8AR
G3 KMD J. Bass 3 Tennyson Avenue, Grays RM17 5RG
GI3 KME L. Pennell 21 Dundrum Road, Clough, Downpatrick BT30 8SH
GM3 KMF G. Robbins 39 Locheil Gardens, Glenrothes KY7 6YL
G3 KMG D. Plumridge Rose Cottage, Castleside, Consett DH8 9AP
G3 KMI Southampton University Wireless Society c/o P. Crump 41 Vernon Way, Guildford GU2 8DE
G3 KML R. Whitfield 42 Greenwood, Tweedmouth, Berwick-upon-Tweed TD15 2EB
G3 KMM J. Crowther 15 Chemin De Bausses, Villelongue D'Aude, Limoux 11300 France
G3 KMO M. Birch 12 The Heath, Hevingham, Norwich NR10 5QW
G3 KMQ R. Heslop Fairways, Meadow Drive, Bude EX23 8HZ
G3 KMS D. Swain 3 Nevy Fold Avenue, Horwich, Bolton BL6 6QG
G3 KMV R. Birchall Willow Tree House, Poole, Nantwich CW5 6AL
G3 KND J. Hardy Vogelenzang, 1B Roberts Road, Aldershot GU12 4RD

G3 KNG A. Embrey 59 Oaken Lanes, Codsall, Wolverhampton WV8 2AW
G3 KNJ J. Otter 7 Longacre Road, Dronfield S18 1UQ
G3 KNU P. Jackson 7 Ferriby Road, Scunthorpe DN17 2EQ
GW3 KNZ A. Eccles 78 Uplands Avenue, Connah'S Quay, Deeside CH5 4LG
G3 KOA T. Robinson 32 Campbell Crescent, East Grinstead RH19 1JR
G3 KOB R. Goodman 8 Decouttere Close, Church Crookham, Fleet GU52 0UR
G3 KOD P. Kay 7 St. Regis Close, London N10 2DE
G3 KOJ R. Ezra 39 Buckland Close, Waterlooville PO7 6ED
G3 KOM F. Foulkes 27 Aspian Drive, Coxheath, Maidstone ME17 4JZ
G3 KOQ B. Parker 9 Yewdale Avenue, Heysham, Morecambe LA3 2LR
G3 KOS B. Faithfull 68 Lampton Road, Long Ashton, Bristol BS41 9AQ
G3 KOX N. Waite 7 Lanercost Close, Welwyn AL6 0RW
G3 KOZ W. Henderson 9 Chiselbury Grove, Salisbury SP2 8EP
G3 KPO Rosemary ARG c/o A. Thornton 78 Wellington Road, Ryde PO33 3QJ
G3 KPU E. Prince 9 Alwyn Road, Thorne, Doncaster DN8 5JG
G3 KPV J. Killeen 10 Den Brook Close, Torquay TQ1 3TP
G3 KQB RAF Digby ARC c/o P. Hanson 10 Parkfield Road, Ruskington, Sleaford NG34 9HS
G3 KQG Dr E. James The Meadows, Kennford, Exeter EX6 7TZ
G3 KQQ C. Mattacks 68 Middlesex Drive, Bletchley, Milton Keynes MK3 7EU
G3 KQV J. Ryley 30 St. Helens Drive, Leicester LE4 0GS
G3 KQY R. Disley 6 St. Margarets Road, Farington, Leyland PR25 4XT
G3 KRT G. Hodges 102 Torrington Road, Ruislip HA4 0AU
G3 KRW K. Whelan Killiney, Longsplatt, Corsham SN13 8DF
G3 KRX W. Addy 14 Cresttor Road, Liverpool L25 6DW
G3 KRZ J. Greenwood Lea Cottage, Meadow Close, Grimoldby, Louth LN11 8HY
G3 KSF R. Harper 21 Howard Oliver House, Harvey Gardens, Southampton SO45 3LS
G3 KSP P. Hooper 1 Victoria Mews, Morecambe LA4 5QD
G3 KTA P. Munt 130 Chipstead Way, Woodmansterne, Banstead SM7 3JR
G3 KTH M. Darkin 3 Adrian Close, Shell, Droitwich WR9 7AY
G3 KTI M. Rees Blue Pillars, 6 Grove Crescent, Coleford GL16 8AZ
G3 KTM R. Atthill Flat 23, Folland Court 70 Hamble Lane, Hamble, Southampton SO31 4JS
G3 KTP D. West 84 High Street, Castle Donington, Derby DE74 2PQ
G3 KTR A. Rock Licensee Address, Not Applicable, Resides in Usa . United States
G3 KTT M. Gallon 41 Dene Gardens, Newcastle upon Tyne NE15 8RL
G3 KTZ C. Lindsay 38 St. Vincents Close, Littlebourne, Canterbury CT3 1TZ
G3 KUD J. Duncan 9 Springhill Close, Westlea, Swindon SN5 7BG
G3 KUE Bracknell ARC c/o A. Mcphail 300 Fletcher Road, Preston PR1 5HJ
GI3 KVD D. Jones 5 Whitehill Park, Limavady BT49 0QF
G3 KVJ S. Tomlinson 31 The Quarry, Alwoodley, Leeds LS17 7NH
G3 KVP D. Kitchen Folkingham Place, Market Place, Sleaford NG34 0SE
G3 KVR S. Davis 3 Coronation Road, Banwell BS29 6AZ
G3 KVT A. Smith Winston House, Felthorpe Road, Norwich NR9 5TF
GW3 KWB R. Neville 35 Beechcroft Road, Newport NP19 8AG
G3 KWJ N. Valentine The White House, Dene Road, Ashtead KT21 1EB
G3 KWK R. Nolan 6 Plymouth Close, Redditch B97 4NP
G3 KWN Lt. Col. W. Nicoll Yonder, Milldown Road, Blandford Forum DT11 7DE
G3 KWO K. Dawson 44 Avondale Road, Darwen BB3 1NS
G3 KWT North Kent RS c/o I. Shaw The Hawthorns, Woodlands Drive, Leeds LS19 6JX
G3 KWW Dr R. Wilkinson 83 Palewell Park, London SW14 8JJ
G3 KWY A. Swain 5 Hilderstone Close Alvaston, Derby DE24 0SA
G3 KXB D. Pantony 71 South Street, Whitstable CT5 3EJ
G3 KXE E. Bettles 15 St. Francis Avenue, Southampton SO18 5QL
G3 KXF D. Wallis 17 Upper Belgrave Road, Seaford BN25 3AD
G3 KXI D. Keeler 16 Honeysuckle Way, Witham CM8 2XG
GM3 KXQ S. Floyd 3 Crarae Place, Newton Mearns, Glasgow G77 6XX
G3 KXS H. Perry 688 Durham Road, Madison 6443 United States
G3 KXV V. Johnston 9 Holbeck Avenue, Middlesbrough TS5 8DR
GW3 KXX R. Weaver 59 Broad St., Leckwith, Cardiff CF1 8BZ
G3 KYE J. Orr 102 Manor House Lane, Yardley, Birmingham B26 1PR
G3 KYF K. Sullivan 14 Wigston Road, Blaby, Leicester LE8 4FU
G3 KYM H. Stamper The Bungalow, School Hill, St. Austell PL26 7TP
G3 KYZ D. Clarke Primrose Mount, Old Neighbourhood, Stroud Glos GL6 8AA
G3 KZB M. Ward Flat 14, Meadrow Court, Godalming GU7 3HG
G3 KZC R. Harknett 28 Woodyleaze Drive, Hanham, Bristol BS15 3BY
G3 KZE J. Davies 45 Dahn Drive, Ludlow SY8 1XZ
G3 KZG A. Bills Brooklands 2 The Acre, Stourbridge DY7 6HW
GW3 KZO M. Dennis 11 Maes Yr Ysgol, Templeton, Narberth SA67 8TZ
G3 KZR I. Davies Lusty Hill Farm, Lusty Gardens, Bruton BA10 0BS
GW3 KZT A. James 143 Gaer Park Drive, Newport NP20 3NS
G3 KZU M. Dolan 15 Ringwood Road, Headington, Oxford OX3 8JB
G3 KZX L. Loveland 21 Roseland Close, Keyworth, Nottingham NG12 5LQ
G3 KZZ D. Forster 281 Mortimer Road, South Shields NE34 0DR
G3 LAA A. Sedman 69 Beechwood Avenue, Locking, Weston-Super-Mare BS24 8DS
G3 LAG H. Gow 43 Ringstead Crescent, Weymouth DT3 6PT
G3 LAI G. Livingston 24 Duncannon Drive, Falmouth TR11 4AQ
G3 LAS J. Butcher The Stables, Priory Farm, Thorpe Tilney, Lincoln LN4 3SL
G3 LAU F. Adkin 5 Cosway Mansions, Shroton St., London NW1 6UE
GM3 LAW W. Walker 45 Watts Gardens, Cupar KY15 4UG
G3 LAZ R. Gerrard Rnib Wavertree House, 211 Somerhill Road, Hove BN3 1RN
G3 LBM A. Mulcahy 8 Old Barn Close, Winkleigh, North Devon EX19 8JX
G3 LBS Dr G. Cleeton 24 Severn Drive, Newcastle ST5 4BH
G3 LCF P. Baldwin 49 King George Vi Mansions, Court Farm Road, Hove BN3 7QX
G3 LCH M. Pharaoh 1 Madeira Road, Mitcham CR4 4HD
G3 LCI H. Young 23 Willow Grove, Wirral CH46 0TU
G3 LCL Sqdn. Ldr. A. Baylis Queen Oak Inn, Bourton, Gillingham SP8 5AL
GW3 LCQ M. Williams Dwyros, 12 Penrhos Avenue, Llandudno Junction LL31 9EL
G3 LCY J. Tamlin 53 Hele Gardens, Plympton, Plymouth PL7 1JY
GW3 LDC J. Phillips 9 Trelawny Close, Usk NP15 1SP
G3 LDG B. Gee Daisy Bank, Carlton Road, Bedford MK43 7JL
G3 LDI R. Cooke The Old Nursery, The Drift, Swardeston, Norwich NR14 8LQ
G3 LDJ K. Day 45 Thick Hollins Drive, Meltham, Holmfirth HD9 4DR
GI3 LEG D. Wilson 189 Cregagh St., Belfast BT6 8NL
G3 LEK L. Kitching Woodsyde Lower Road, Harmer Hill, Shropshire SY4 3QX
G3 LEO G. Brigham Waterside House, West Tanfield, Ripon HG4 5LF
G3 LET P. Hobbs Honeysuckle Cottage, Stairbridge Lane Bolney, Haywards Heath RH17 5PA
GW3 LEW G. Weale Winfield, Templeton, Narberth SA67 8SP
G3 LFD R. Widders 82 Azalea Walk, Eastcote, Pinner HA5 2EH
GJ3 LFJ H. Mesny La Trigale, Route De L'Eglise, St. Lawrence JE3 1LA Jersey
G3 LFR M. Everett Thrupp Wharf, Cosgrove, Milton Keynes MK19 7BE
G3 LFV R. Manser 39 Long Meadow, Markyate, St. Albans AL3 8JN
G3 LFX Dr D. Pedder 37 Hersham Road, Walton-on-Thames KT12 1LE
G3 LGA M. Hayward Brindle, Romsey Road, Stockbridge SO20 8DB
G3 LGF G. Falding 10 Angel Court, Shaftesbury SP7 8HX
G3 LGK B. Sandall Amber Croft, Main Road Higham, Alfreton DE55 6EH
G3 LGQ P. Marsden 49 Southfield Park, North Harrow, Harrow HA2 6HF
G3 LGR M. Hooles 114 Cassiobury Drive, Watford WD17 3AQ
G3 LGT J. Tate Pine Holt, 34 Queens Road, Fleet GU52 7LE
GM3 LGU R. Pryde Room 5, Abbeyfield, , Paterson Court, Haddington EH41 3DU
G3 LGW D. Spencer Paladin, 89 Watling Street, Tamworth B78 3DE
G3 LHG E. Smith 3 Meadow Avenue, Wetley Rocks, Stoke-on-Trent ST9 0BD
G3 LHJ D. Webber 43 Lime Tree Walk, Milber, Newton Abbot TQ12 4LF
GW3 LHK G. Griffiths Glyndwr, Lampeter Road, Aberaeron SA46 0ED
G3 LHN R. Muir 19 Eastwick Drive, Bookham, Leatherhead KT23 3PY
G3 LHS L. Matthews 14 Mayforth Gardens, Ramsgate CT11 0LL
G3 LHU M. Dixon 27 Planets Lane, Cheltenham GL51 6GR
G3 LHZ Prof. M. Underhill Hatchgate, Tandridge Lane, Lingfield RH7 6LL
G3 LIK M. Puttick 21 Sandyfield Crescent Cowplain, Waterlooville PO8 8SQ
G3 LIO J. Gibbs 13 Bromley Road, Macclesfield SK10 3LN
G3 LIV J. Melvin 2 Salters Court, Newcastle upon Tyne NE3 5BH
GM3 LIW A. Wood 97A Fort St. Broughty Ferry, Dundee DD51DY
G3 LJD J. Davies 57 Madeira Court, Knightstone Road, Weston Super Mare BS23 2BH
GM3 LJR T. Saxton 38 William Street, Dalbeattie DG5 4EN
GW3 LJS Dr T. Bloxam 15 Cleveland Avenue, Mumbles, Swansea SA3 4JD
G3 LKV D. Locke 4 Glebe Close, Doveridge, Ashbourne DE6 5NY
G3 LKW D. Wiltshire 71 Ferndale, Waterlooville PO7 7PG
G3 LLD S. Collier 64 Slonk Hill Road, Shoreham-by-Sea BN43 6HY
G3 LLE K. Webster 25 Carlin Gate, Blackpool FY2 9QT
G3 LLG R. Loveday 42 Bridle Path, Woodcote, Reading RG8 0SE
G3 LLK J. Gale 66 Burys Bank Road, Crookham Common, Thatcham RG19 8DD
GM3 LLP B. Watson 4 Caldwell Road, West Kilbride KA23 9LE
G3 LLV J. Mcelvenney 10 Bignor Place, Sheffield S6 1JE
G3 LLZ D. Goacher 27 Glevum Road, Swindon SN3 4AA
G3 LME K. Taylor 7 Bowen Close, Cheltenham GL52 5EG
G3 LMH R. Wellbeloved 8 Orchard Close, South Wonston, Winchester SO21 3EY
G3 LMQ J. Hamer 7 Arundel Road, Coventry CV3 5JT
G3 LMR J. Eley 25 Peckleton View, Desford, Leicester LE9 9QF
G3 LMX T. Mitchell 27 Hanmer Road, Simpson, Milton Keynes MK6 3AY
G3 LNL P. Lovelady 14 Maunders Court, Liverpool L23 9YU
G3 LNM R. Scrivens 6 Highland Close, Cantley, Norwich NR13 3SW
G3 LNN J. Symes 19 Boundary Close, Kirkby-In-Ashfield, Nottingham NG17 8RS
G3 LNP A. Preedy 2 Hunters Gate, Much Wenlock TF13 6BW
GW3 LNR A. Gwynne 77 Edward St., Pant, Merthyr Tydfil CF48 2BB
G3 LNS G. Beasley Po Box 1344, Paphos 0 Cyprus
G3 LNW J. Mcguire 5 Primrose Way, Trevadlock Hall Park, Launceston PL15 7PW
G3 LOD D. Rowse 48 Oatlands Avenue, Bar Hill, Cambridge CB23 8EQ
G3 LOE W. Roberts 13 Brean Road, Stafford ST17 0PA
G3 LOF Dr G. Peskett 13 Warneford Road, Oxford OX4 1LT
G3 LOV M. Francis Cherry Tree Cottage, Atlantic Close, Tintagel PL34 0EL
G3 LPC Weald ARS c/o C. Desborough 22 Westland Road, Faringdon SN7 7EY
G3 LPL P. Sherdley 2 Stable Yard, Taylors Lane, Preston PR3 6AP
G3 LPN J. Hunt 28 Robins Bow, Camberley GU15 3NR
G3 LPT G. Woods Bamburgh House, Hunston, Bury St. Edmunds IP31 3EN
G3 LPU E. Burrell 20 The Avenue Richnond Wood Norton Eveham Road, Wood Norton Worcs. Wr11 4Ty, Evesham WR11 4TY
GU3 LPV T. Catts Po Box 1029, Alderney GY9 3JD
G3 LPY R. Nye Beech Cottage, Gorelands Lane, Chalfont St. Giles HP8 4HQ
G3 LQB K. Bishop Friedrich-Ebert Strasse 23, Osterode 37520 Germany
GW3 LQE A. Ernest 2 Osborne 7 Clive Crescent, Penarth CF64 1WW
G3 LQJ R. Cox 12A Kelling Close, Holt NR25 6RU
G3 LQO E. Harris 10 Girdle Road, Walsworth, Hitchin SG4 0AN
G3 LQP R. Brown 262 Fir Tree Road, Epsom KT17 3NL
G3 LQR S. Freeman West Farm, Cransford, Woodbridge IP13 9PQ
G3 LQS RAF Coningsby ARC c/o D. Bloomfield 14 Horsham Close, Luton LU2 8JH
G3 LQW K. Wallace 55 Lamborne Road, Leicester LE2 6HQ
G3 LQX M. Nicholls 6 Lyme Bay Road, Teignmouth TQ14 8RS
GI3 LQY J. Stronach 20 Monaville Drive, Lisburn BT28 2DR
G3 LRA C. Eley 1 Hilldale View, Gaisgill, Penrith CA10 3UE
GM3 LRG J. Gray 47 South Street, Greenock PA16 8QG
G3 LRH G. Frampton 1 Ludlow Road, Church Stretton SY6 6DD
G3 LRI J. Blakey 10 Wilson Terrace, Newcastle upon Tyne NE12 7JP
G3 LRL R. Bowell 16 Margarite Way, Wickford SS12 0ER
G3 LRQ M. Humphries 2 South View Close, Twyford, Reading RG10 9AY
G3 LRS Leicester RS c/o A. Moss 10 Shakespeare Drive, Leicester LE3 2SP
G3 LRU J. Miller 57 Clarendon Villas, Hove BN3 3RE
G3 LRX R. Durell Middleton Farm, Hubbards Hill, Lenham ME17 2EJ
G3 LSA D. Moore 5 Seahaven Springs Estate, Seaholme Road, Mablethorpe LN12 2QS
GD3 LSF E. Ellis Ballahams, 3 Glen Road, Laxey IM4 7AP Isle of Man
G3 LSJ C. Gerrard 6 Bridle Close, Sleaford NG34 7TD
G3 LSQ P. Aitchison Upper Weston House, Cot Lane, Chichester PO18 8SU
G3 LST P. Clarke Half Moon House, Church Street, Colchester CO6 4QH
G3 LSX G. Townsend 21 Grange Avenue, East Barnet, Barnet EN4 8NJ
G3 LTF P. Blair Woodleigh, Upper Wyke, Andover SP11 6EA
G3 LTM B. Moyler 1 Bay Walk, Aldwick, Bognor Regis PO21 4ET
GW3 LTX R. Savage Plas Gwyntog, Rhoslefain, Tywyn LL36 9ND
G3 LUA A. Knowles 73 Kingslea Road, Solihull B91 1TJ
G3 LUC E. Bate 5 Elm Road, Shildon DL4 1BH
G3 LUH K. Reader 21 Broadwater Avenue, Poole BH14 8QY
G3 LUK 59 Squadron Air c/o I. Haliwell 61 Cliffe Road, Shepley, Huddersfield HD8 8AG
G3 LUN Special Communications (Ta) Association c/o D. Smith The Old Forge, High Street, Brinkley, Newmarket CB8 0SE
G3 LUO C. Evans Peas Gill House, Gawthrop, Sedbergh LA10 5QB

G3 LUW B. Whittaker Woodlands, Newton Down, Lifton PL16 0AS
G3 LUZ F. Machin 70 Poors Lane, Hadleigh, Benfleet SS7 2LN
GM3 LVA D. Simpson Larchwood, Tomatin, Inverness IV13 7YR
G3 LVB G. Brooks 11 Prince Charles Way, Seaton EX12 2TU
G3 LVL Waters & Stanton Electronics c/o B. Ash 5 Church Close, Wickham Bishops, Witham CM8 3LN
G3 LVP K. Eastty 7 The Grange, The Reddings, , Cheltenham GL51 6RL
G3 LVW R. Smith 40 Highwoods Drive, Marlow Bottom, Marlow SL7 3PY
G3 LWD P. Stone Bramley, Stone Street, Hythe CT21 4JP
G3 LWF L. Franklin 1 Woodwell Cottages Woodwell Road, Bristol BS11 9UP
G3 LWJ C. Way 8 Stratford Place, Eaton Socon, St. Neots PE19 8HY
G3 LWM J. Harris 37 Long Orchard Way, Martock TA12 6FA
G3 LWR Clan Maclean ARS c/o J. Evans 18 Mandeville Road, Isleworth TW7 6AD
G3 LWT P. Buck Little Nook, Reed Close, Malvern WR14 2AG
GW3 LWU G. Brisbar 97 Chambers Lane, Mynydd Isa, Mold CH7 6UZ
G3 LXB S. Jones 43 New St., Chase Terrace, Walsall WS7 8BT
GW3 LXE J. Boden Plas Heulwen, Llanfair Road, Newtown SY16 3JY
G3 LXJ F. Fisher 7 Greenbank, Halesworth IP19 8RP
G3 LXQ D. Gallop 4 Volunteer Road, Theale, Reading RG7 5DN
GU3 LYC T. De Putron Shieling Cottage, La Rue Marquand, St. Andrew GY6 8RB Guernsey
G3 LYD Dr E. Henderson The Homestead, High Street, Ventnor PO38 3HZ
G3 LYG Dr A. Macgregor 10 Balroy Court Forest Hall, Newcastle upon Tyne NE12 9AW
G3 LYP Dr M. Scott The Magnolias, Marlow Road, , High Wycombe HP14 3JW
GW3 LYU Flight Refueling ARS c/o D. Price Llansilin, Oswestry SY10 7QB
G3 LYZ B. Currey 42 Westfield Avenue, Goole DN14 6JX
G3 LZC A. Stirland 98 Aldreds Lane, Heanor DE75 7HG
G3 LZI J. Oates Cherry Tree Cottage, Green Moor Wortley, Sheffield S35 7DQ
G3 LZM M. Bush 5 Quay Close, Hereford HR1 2RQ
G3 LZN G. Ellison Little Flushing, St. Peters Road, Falmouth TR11 5TJ
G3 LZO P. Thomas 20 Bleasdale Court, Longridge, Preston PR3 3TX
G3 LZR E. Speller 78 Chelmsford Road, Holland-On-Sea, Clacton-on-Sea CO15 5DJ
G3 LZZ A. Pomfret Flat 2, Ingwell House, Grange-over-Sands LA11 6DP
G3 MAE Dr A. Wilson 8 The Paddock Appleton Wiske, Northallerton DL6 2BE
G3 MAI R. Stevens 138 Grange Drive, Stratton, Swindon SN3 4LA
G3 MAJ E. Holden 10 Rowan Tree Close, Greasby, Wirral CH49 3AW
G3 MAR O/B/O Rhosllannerchrugog Group c/o N. Gutteridge 68 Max Road, Quinton, Birmingham B32 1LB
GM3 MAS A. Pringle 1 Falloch Road, Milngavie, Glasgow G62 7RR
G3 MAU J. Wardle 17 Frederick Neal Avenue, Coventry CV5 7EH
G3 MAV J. Bradley 17 Talboys Walk, Tetbury GL8 8YU
G3 MAZ H. Bell Downside, North Street Mere, Warminster BA12 6HH
GI3 MBB A. Mcmurtry 20 Towerview Crescent, Bangor BT19 6BA
GD3 MBC R. Wernham Fair Isle, Lhoobs Road, Douglas IM4 3JB Isle of Man
G3 MBD H. Dannatt-Brader 20 Shire Place, Northampton NN3 8DE
G3 MBK D. Underdown 26 Birch Road, Farncombe, Godalming GU7 3NT
G3 MBM J. Masters 8 Purbeck Terrace Road, Swanage BH19 2DE
G3 MBN B. Gibbs 15 Moor Barton, Neston, Corsham SN13 9SH
G3 MBO B. Aspinwall 33 Clipstone Crescent, Leighton Buzzard LU7 3LU
G3 MBU M. Standige 7 Hill Crest Avenue, Burnley BB10 4JA
G3 MCA D. Owen 1 Mosslea Road, Orpington BR6 8HP
G3 MCB A. Williams 1 Wyvern Road, Sutton Coldfield B74 2PS
G3 MCC K. Worrall 21 Northwood Avenue, Middlewich CW10 0HR
G3 MCD K. Holland Ravendale St. Lawrence, Bodmin PL30 5JL
G3 MCE L. Lee 34 Westby Way, Poulton-le-Fylde FY6 8AD
G3 MCK G. Stancey 22 Peterborough Avenue, Oakham LE15 6EB
G3 MCL C. Simpkins 6 Compton Way, Olivers Battery, Winchester SO22 4EY
G3 MCP P. Goadby 535 Welford Road, Leicester LE2 6FN
G3 MCV B. Vaughan 17 Richmond Close, West Town, Hayling Island PO11 0ER
G3 MCX J. Kennedy 22 Croham Park Avenue, South Croydon CR2 7HH
G3 MDD B. Mudge 9 Crossmead, Woolavington, Bridgwater TA7 8ER
G3 MDG Chesham & District ARS c/o M. Appleby 6 Mandeville Road, Prestwood, Great Missenden HP16 9DS
G3 MDI M. Plummer Kembali, 14 Turnberry Drive, Woodhall Spa LN10 6UE
G3 MDM G. Mcgee 2 Ilynton Avenue, Firsdown, Salisbury SP5 1SH
G3 MDR M. Hallet 33 Latimer Street, Romsey SO51 8DF
G3 MEC J. Pearce 86 Sopers Lane, Poole BH17 7EU
G3 MED F. Griffiths 105 Hillcroft Crescent, South Oxhey, Watford WD19 4PA
G3 MEH R. Piper 8 Osborne Way, Wigginton, Tring HP23 6EN
G3 MEV C. Cory Tekelex, Chapel Lane, Thatcham RG19 8BE
G3 MEY J. Lawrence 16 Waverley Court, Corsham SN13 9NN
G3 MFG D. Close 27 High St., Collyweston, Stamford PE9 3PW
G3 MFH G. Dale 20 Blythe Avenue, Stoke-on-Trent ST3 7JY
G3 MFJ G. Firth 13 Wynmore Drive, Bramhope, Leeds LS16 9DQ
G3 MFK M. Camp 82 Leicester Road, Hinckley LE10 1LT
G3 MFL A. Russell Pear Tree Cottage, Savernake Road, Marlborough SN8 3AS
G3 MFO P. Elliot 3 Shickle Place, Hopton, Diss IP22 2QR
G3 MFW H. Woodhouse 143 Bodmin Road, Truro TR1 1RA
G3 MGL A. Davis 22 Yarmouth Close, Crawley RH10 6TH
G3 MGQ P. Parkman 2A Frank Woolley Road, Tonbridge TN10 4LE
G3 MGS Prof. C. Stephens 12, Berkshire Road, Bristol BS7 8EX
G3 MGU A. Dodson 53 Simons Lane, Wokingham RG41 3HG
G3 MGW R. Wheeler 51 Seaview Road, Brightlingsea, Colchester CO7 0PR
G3 MGX J. Tomlinson 34 Bentley Road, Tacolneston, Norwich NR16 1DL
G3 MGY Dr M. Uotome 3-10-14 Kujayarna Sugimami-Ku, Tokyo 168-0082 Japan
G3 MHD A. Williams 9 Charlotte Cove Road Charlotte Cove, Tasmania 7112 Australia
G3 MHF M. Ockenden 16 Ripley Chase, 17 The Goffs, Eastbourne BN21 1HB
G3 MHT E. Landon 14 The Blackthorns, Broughton, Brigg DN20 0BB
G3 MHV Dr T. Langdon 58 Upper Marsh Road, Warminster BA12 9PN
G3 MHX M. Tate 48 Crossgates Bedwell Plash, Stevenage SG1 1LS
G3 MIP S. Heilbron 8 Beechwood Drive, Formby, Liverpool L37 2DG
G3 MJM A. Marshall 2 Westwood Way, Beverley HU17 8GE
G3 MJW C. Edmunds 23 Dorset Gardens, Northampton NN2 7PX
G3 MKE W. Smith 12 Benscliffe Drive, Loughborough LE11 3JP
GW3 MKT M. Hooks 1 Llwyn Castan, Pentwyn, Cardiff CF23 7DA
G3 MKU A. Bower 82 Anson Road, Shepshed, Loughborough LE12 9PU
G3 MKV C. Curtis 24 Rodney Road, Hartford, Huntingdon PE29 1RZ
G3 MLO P. Weatherall Woodside, Stone Street Stelling Minnis, Canterbury CT4 6DN
G3 MLQ S. Blundell 12 Brookfield St., Melton Mowbray LE13 0NB
G3 MLS D. Nappin New Edge Farm, Heptonstall, Hebden Bridge HX7 7PG
G3 MMA D. Mayes Flat 15, Hanover Court, Blackman Way, , Witham CM8 1JZ
G3 MME P. Whitford Three Pieces, Vernon Lane, Kelstedge, Chesterfield S45 0EA
GI3 MMF W. Mcaleer 90 Gortin Park, Belfast BT5 7EQ
GI3 MMG D. Noon 34 Rodney Park, Bangor BT19 6FN
G3 MMJ G. Browne 39A Cromwell Road, Canterbury CT1 3LD
G3 MMN B. Newman 101 Tally Ho Road, Shadoxhurst, Ashford TN26 1HW
G3 MMS G. Whiting 25 Obthorpe Lane Thurlby, Bourne PE10 0ES
G3 MMX E. Lawley 3 Barnicott Close, Newton Ferrers, Plymouth PL8 1BP
G3 MNB H. Benjamin 21 Sheephouse Green, Wotton, Dorking RH5 6QW
G3 MNS I. Swan Flat 6 Mason Court, Alford Road, Sutton-On-Sea, Mablethorpe LN122GY
G3 MNV P. Darragh 48 Goodwood Park Road, Northam, Bideford EX39 2RR
G3 MOA J. Ruff 17 Harts Close, Teignmouth TQ14 9HG
G3 MOL J. Lixenberg Orchard House, 77A Pembroke Crescent, Hove BN3 5DF
G3 MON D. Gent 12 Field Road, Billinghay, Lincoln LN4 4EA
GM3 MOR Dr R. Webster Meric, 7 Woodmuir Crescent, Newport-on-Tay DD6 8HL
G3 MOT J. Lambert Hurley 64 Henry Road, West Bridgford, Nottingham NG2 7ND
GW3 MOV Dr C. Smith 38 Plas Taliesin, Penarth CF64 1TN
G3 MPB A. Smith 10 Goodwood Road, Redhill RH1 2HH
G3 MPF C. Smith 29 Cloisters, Tarleton, Preston PR4 6UL
G3 MPN D. Johnson 54 Norwich Road, Wymondham NR18 0NT
GW3 MPP G. Price 17 Celtic Close, Undy, Caldicot NP26 3PB
G3 MPW A. Walker 14 St. Joans Drive, Scawby, Brigg DN20 9BE
G3 MQD P. Greed The Bungalow, Townsend, Devizes SN10 4RR
G3 MQI Essex Raynet c/o R. Gill 45 Biggin Lane, Ramsey, Huntingdon PE26 1NB
GM3 MQO G. Olesen 8 Rowallan Crescent, Prestwick KA9 2HE
G3 MQR D. Robinson 32 Bullock Wood Close, Colchester CO4 0HX
G3 MRQ D. Byne Storm Bay, Church Street Charwelton, Daventry NN11 3YT
G3 MRT R. Strafford Chy Lowarth, Sparnock, Truro TR3 6EB
GM3 MRV G. Carrick 4 Kingfisher Lane, Gretna DG16 5JS
G3 MRX Dr P. Robinson 9 Barton Close, Cambridge CB3 9LQ
G3 MRZ M. Crutchley 40 Ufton Crescent, Shirley, Solihull B90 3SA
G3 MSL Dorking & District RS c/o R. Ives 11 Coombe Drive, Fleet GU51 3DY
G3 MSO E. Tunstall 11 The Broadway, Charlton On Otmoor, Kidlington OX5 2UB
G3 MSW K. Ashcroft Fendley Corner, Common Lane, Harpenden AL5 5DW
G3 MTD B. Kissack 13 Church Street, The Old Saddlers, Braunton EX33 2EL
G3 MTG R. Prior 35 Hanson Drive, Fowey PL23 1ET
G3 MTJ R. Skoyles 2 Hay Close Great Oakley, Corby NN18 8HX
G3 MTL C. Knowles 18 Lancaster Close, Bolton BL1 1PJ
G3 MTP P. Gadsden Rose Cottage, Salwayash, Bridport DT6 5HX
G3 MTR B. Wolfe 24 Marchbank Drive, Cheadle SK8 1QY
GM3 MTW C. Wolstencroft 29 Fasach, Glendale, Isle of Skye IV55 8WP
GM3 MUA P. Lawlor Woodside, North Kessock, Inverness IV1 3XG
G3 MUO G. Gott 10 Churchill Crescent, Marple, Stockport SK6 6HJ
G3 MUX C. Benson Orchard Croft, 85 Runcorn Road, Warrington WA4 6UA
G3 MVE A. Bullimore 8 St. Georges Road, Felixstowe IP11 9PL
G3 MVM P. Pierson 7 Beehive Road, Goffs Oak, Waltham Cross EN7 5NL
G3 MVV N. Miller Avon, Gardiners Lane North, Billericay CM11 2XA
G3 MVX J. Burke 120 Seabourne Road, Bexhill-on-Sea TN40 2SD
G3 MVZ F. Garrett 18 Wolfe Close, Chichester PO19 6BY
G3 MWM Dr D. Murden Po Box 06, Curitiba, Parana 80011 970 Brazil
G3 MWO D. Beales 2 Wood Close, Tostock, Bury St. Edmunds IP30 9PX
GM3 MWX A. Winton 2 Castlehill Cottages, Brisbane Glen Road, Largs KA30 8SN
G3 MWZ J. Casling 19 Orchard Close, Tavistock PL19 8HA
G3 MXA B. Collins 2 Pilgrims Way, Ely CB6 3DL
G3 MXF P. Cutler 14 Verulam Road, Poole BH14 0PP
G3 MXH T. Downing 8 Auction Yard, Haughley, Stowmarket IP14 3GA
G3 MXJ D. Andrews Coupelle, Levignac de Guyenne 47120 France
GM3 MXN T. Sorbie 9 Lynn Court, Larkhall ML9 1QT
G3 MXP J. Palfrey Caprice, 4 Laverstock Park West, Salisbury SP1 1QL
G3 MXV H. Pierson 65 Station Road, Countesthorpe, Leicester LE8 5TB
G3 MYA A. Martindale 16 Charles Miller Court, Leiston IP16 4BY
G3 MYC C. Cheatle 56 Ashfurlong Crescent, Sutton Coldfield B75 6EN
G3 MYG R. Inman 60 Abercorn Road, Mill Hill, London NW7 1JL
G3 MYI J. Lewis 50 Robin Gardens, Waterlooville PO8 9XF
G3 MYM R. Micklewright 5 Sandringham Road, Yeovil BA21 5JE
G3 MYY S. Boston Beavers, Mill Lane, Ipswich IP8 4AU
G3 MYZ P. Nicholson 3 Welborn Court Main Street, Flixton, Scarborough YO11 3XA
G3 MZA E. Hamblen 64 Tollers Lane, Coulsdon CR5 1BB
G3 MZC C. Sutcliffe 1 Tollgate Road, Culham, Abingdon OX14 4NL
G3 MZI J. Hood 89 Freemens Way, Deal CT14 9DQ
G3 MZN R. Lightfoot 28 Wheal Gorland Road, St. Day, Redruth TR16 5LT
G3 MZO D. Rosen Barnham Manor Nursing & Residential Home 150 Barnham Road, Barnham Room 3, Bognor Regis PO22 0EH
GM3 MZX M. Pedreschi Clary Lodge, Carse Of Clary, Newton Stewart DG8 6BH
G3 MZZ A. Kightley 29 The Parkway, Gosport PO13 0PT
G3 NAE C. Richardson 10 Fielders Way, East Wellow, Romsey SO51 6EX
G3 NAI R. Norman 19 Hughes Croft, Bletchley, Milton Keynes MK3 5HA
G3 NAK G. Mallinson 145 Huddersfield Road, Meltham, Holmfirth HD9 4AJ
G3 NAN R. Henderson 48 Cartwright Crescent, Brackley NN13 6HA
G3 NAP B. Sowter 56 Alderminster Road, Coventry CV5 7JU
G3 NAQ Dr G. Grayer Bagatelle, 3 Southend, Newbury RG20 7BE
G3 NAT London Raynet G c/o A. Brooker 18 Honeybourne Way, Petts Wood, Orpington BR5 1EZ
G3 NAV E. Cook Edward R.Cook, 152 O St. Spc.51, Lincoln 95648 United States
G3 NAW J. Ryan 4 Ferry Lane, Bath BA2 4HS
G3 NAY S. Whithorn 53 Torbay Road, Allesley, Coventry CV5 9JY
GM3 NBE S. Hoar Windyridge, Lower Bayble, Isle of Lewis HS2 0QB
G3 NBL Dr J. Larson Nyhem, Whitton Village, Stockton-on-Tees TS21 1LQ
G3 NBN R. Weaving 7 Fairway Gardens Sparkwell, Plymouth PL7 5FE
G3 NBQ P. Burt 3335 Mountain Highway, North Vancouver V7K 2H4 Canada
G3 NBS A. Bairstow 27 Williams Way, Longwick, Princes Risborough HP27 9RP
G3 NBY H. Murray 36 Sterndale Road, Davenport, Stockport SK3 8QU
G3 NBZ K. Thorne 3 Cherry Gardens, Abstacle Hill, Tring HP23 4EA
G3 NCN J. Ellerton 7 Cotterell Close, Bracknell RG42 2HL

GM3 NCO A. Mustard Tigh Ard, Knockhouse Hill, Crossford, Dunfermline KY12 8PT
GW3 NCT R. Lord 8 Llys Steffan, Llantwit Major CF61 2UF
GW3 NDB G. Wyatt 3 Creidiol Road, Mayhill, Swansea SA1 6TZ
G3 NDC C. Deamer Gatehouse, Warren Lane, Stanmore HA7 4LD
G3 NDK R. Webb 142 Penrose Avenue, Carpenders Park, Watford WD19 5AA
G3 NDM B. Mahony 12 Orchard Green Marden, Hereford HR1 3ED
G3 NDN D. Newey 15 Clent View Road, Stourbridge DY8 3JE
G3 NDS R. Oliver Flat 21, Exeter Court, 52 Wharncliffe Road, Highcliffe, Christchurch BH23 5DF
GU3 NDX Castel Contest Club c/o R. Beebe San Grato, Les Houguette, Castel, Guernsey GY5 7DZ Guernsey
G3 NEG Key Performance Indicators c/o Dr N. Wilkinson 12 Woodlands Close, Grays RM16 2GB
G3 NEH J. Isles 3 Drovers Croft, Greenleys, Milton Keynes MK12 6AN
G3 NEO P. Bagshaw 48 Kiveton Lane, Todwick, Sheffield S26 1HL
G3 NEP C. Wager-Bradley Swallowdale, Longhoughton Road, Alnwick NE66 3AT
GM3 NEQ A. Finlay 19 Fraser Avenue, Newton Mearns, Glasgow G77 6HP
G3 NFB J. Leviston 9 Barnes Avenue, Fearnhead, Warrington WA2 0BL
G3 NFC Burton & District RS c/o G. Newstead 97 Hawthorn Crescent, Burton-on-Trent DE15 9QN
G3 NFJ M. Coward High Bank, 51 High Street, Warminster BA12 7AP
GI3 NFM K. Mcelhatton 2A Orpheus Drive, Dungannon BT71 6DR
G3 NFP L. Beckwith Westgate Burghill, Hereford HR4 7RW
G3 NFV R. Sykes 16 The Ridgeway, Fetcham, Leatherhead KT22 9AZ
G3 NFW J. Carroll White Lodge, Hunston, Chichester PO20 1PA
G3 NFY B. Twist 11 Church Street, Minehead TA24 5JU
G3 NGJ W. Epton 2 Eastcliff Road, Lincoln LN2 5RU
G3 NGK D. Chapman 6 Pickhurst Green, Hayes, Bromley BR2 7QT
GM3 NGW W. Webb 5 Thornlea Drive Giffnock, Glasgow G46 6DB
G3 NGX H. Hogg Crossways, Ferry Road, South Stoke, Reading RG8 0JL
G3 NGZ Pelican Radio Group Little Rissington c/o M. Grierson Gumstalls Barn Keynsham Lane Woolaston, Lydney GL15 6PR
G3 NHB Dr D. Bowyer 41A High Street, Trumpington, Cambridge CB2 9HR
G3 NHF J. Noble 27 Chestnut Avenue, Donington, Spalding PE11 4XH
G3 NHL C. Lewis The Anchorage, Quay Road, Devoran, Truro TR3 6PW
G3 NHP G. Peacock Hallowsgate House, Flat Lane, Tarporley CW6 0PU
GM3 NHQ T. Harrison 7 Cults Gardens, Broughty Ferry, Dundee DD5 1QT
G3 NHR H. Rogers Aughavore, Church Walk, Louth LN11 8LJ
G3 NHV D. Hare White Lodge, Mount Gabriel, Schull Ireland
G3 NHX G. Quarterman 2 Milton Avenue, Sutton SM1 3QB
G3 NIC K. Plant Rose Cottage, Lincoln Road, Lincoln LN2 2NE
G3 NID I. Douglas 6 Ansley Road, Houghton, Huntingdon PE28 2DQ
GM3 NIG D. Cram 61 Gailes Road, Troon KA10 6TB
G3 NII R. Porter 6 Clifton Road, Shefford SG17 5AA
G3 NIJ B. Barker 4 Glantlees, West Denton, Newcastle upon Tyne NE5 2PJ
G3 NIL G. Munden 126 Stanley Green Road, Poole BH15 3AQ
GW3 NIN J. Brogan 38 Graig Park Circle, Newport NP20 6HE
G3 NIQ R. Gorton 2 Clyde Court, Clyde Close, Redhill RH1 4AY
G3 NIW P. Ives Allt Na Crioch, Ockham Road South, Leatherhead KT24 6QJ
G3 NJA Yarmouth RC c/o D. Webber 43 Lime Tree Walk, Milber, Newton Abbot TQ12 4LF
G3 NJB Wacral (World Association of Christian Radio Amate c/o P. Jackson 8 Buttree Court, South Kirkby, Pontefract WF9 3NB
G3 NJG T. George 8 Lanehays Road, Hythe, Southampton SO45 5ER
G3 NJV P. Randall Myresyke, Ruan Minor, Helston TR12 7LU
G3 NJX R. Geeson The Grove, Main Road, Ripley DE5 3RE
G3 NJY Dr M. Bibby 47 Whitney Tavern Road, Weston 2493 United States
G3 NKC D. Sharred 4 Rufford Close, Wistaston, Crewe CW2 6XP
GM3 NKG A. Campbell 22 Saltire Crescent, Larkhall ML9 2LG
G3 NKH R. Dowling Orchard House, Oughtrington Lane, Lymm WA13 0RD
G3 NKJ Wpx CG c/o R. Gill 45 Biggin Lane, Ramsey, Huntingdon PE26 1NB
G3 NKL R. Jones 12 Crumpax Meadows Longridge, Preston PR3 3JG
GW3 NKM C. Jones 77 Margam Road, Port Talbot SA13 2LB
G3 NKQ C. Burchell 4 Bakers Way, Perry, Huntingdon PE28 0BS
G3 NKS D. Thom 78 Farmfield Road, Cheltenham GL51 3RA
G3 NKW H. White 16 Turnberry Close, Lymm WA13 9LY
GM3 NLB Dr F. Inglis 3 Fleming Road, Bishopton PA7 5HW
G3 NLY R. Smethers 46 Church Road, Burntwood WS7 9EA
G3 NMD Houghton ARC c/o I. Laidler 5 South Street, West Rainton, Houghton le Spring DH4 6PA
G3 NMH H. Perkins 31 Dorchester Road, Weybridge KT13 8PE
G3 NML M. Slater 46 Ladywood, Eastleigh SO50 4RW
GM3 NMN R. Dunlop 39 Braid Drive, Glenrothes KY7 4ES
G3 NMW T. Whateley 285 Harborne Road, Birmingham B15 3JB
G3 NMX D. Wills 4100 Jackson Ave, Apt 530, Austin 78731 United States
G3 NMZ G. Bath 11 Heron Way, Hickling, Norwich NR12 0YQ
G3 NN C. Bolt 147 Swan Avenue, Bingley BD16 3PL
G3 NNA M. Codd 1 Shaftesbury Place, Lancaster, LA1 4PZ
GW3 NNB R. Evans Cemlyn, Ffordd Dewi Sant, Pwllheli LL53 6EG
G3 NNG C. Desborough 22 Westland Road, Faringdon SN7 7EY
G3 NNN P. Mason 1 Morley Lane, Stanley, Ilkeston DE7 6EZ
G3 NNO M. George-Powell Old Church Lane Cottage, Pateley Bridge, Harrogate HG3 5LY
G3 NNT S. Pilkington The Quarries, Quarry Drive, Ormskirk L39 5BG
G3 NNV P. Swanson 11 Grassmoor Close, Wirral CH62 7JY
G3 NNW K. Taylor 34 Shore Road, Warsash, Southampton SO31 9FU
GM3 NNZ Dr B. East 26 Hyndford Road, Lanark ML11 9AE
G3 NOA P. Reynolds Brook Bushes, Bramshaw, Lyndhurst SO43 7JB
G3 NOC A. Waldie Gwyn Lyn, 85 Park Road, Coleford GL16 7AG
G3 NOI R. Cumming 21 Britannia Way, Woodmancote, Cheltenham GL52 9QP
G3 NOP D. Peacock Robin Hill, Cottingham HU16 5JG
G3 NOX J. Royle Keepers Cottage, Duddenhoe End, Saffron Walden CB11 4UU
G3 NPC Dr J. Swanson 23 Oatlands Road, Tadworth KT20 6BS
G3 NPM A. Macdonald 5 Arlington Close, Swindon SN3 3NB
GI3 NPP R. Gibson 109 Bush Road, Dungannon BT71 6QG
G3 NPS B. Harrad 32 Woodfield Avenue, Northfleet, Gravesend DA11 7QG
G3 NPT G. Bell 9 Humber View, Hessle HU13 0PY
G3 NPY J. Joslin 150 Roman Bank, Skegness PE25 1SE
G3 NPZ T. Griffiths 18 Lulworth Road, Lee-on-the-Solent PO13 9HU
G3 NQA S. Hall 76 Cheltenham Drive, Bromford, Birmingham B36 8QG
G3 NQF R. Fenton Harmins Green, France Lynch, Stroud GL6 8LZ
G3 NQK J. Beddows 17 Rue Francois Mitterand, Pleven 22130 France
G3 NQN M. Hartung 31 Ellenbrook Lane, Hatfield AL10 9RW
G3 NQT R. Levi 24 Stanmore Way, Loughton IG10 2SA
G3 NQX W. Brown 73 Church Avenue, Preston PR1 4UD
G3 NQZ Dr G. Lockhart 179 Poolbrook Road, Malvern WR14 3JZ
G3 NR A. Birt 36 Queens Road, Swanage BH19 2ET
G3 NRD J. Packer The Hayloft, Butts Bank Farm, Gulval, Penzance TR18 3BB
G3 NRH B. Perrin Apartment 37, Hardy Lodge, Coppice Street, Shaftesbury SP7 8GY
G3 NRM M. Moore 127 Adel Lane, Leeds LS16 8BL
G3 NRQ C. Higgins Billdoro, Mill Lane, Saltfleet, Louth LN11 7SA
G3 NRU D. Brook-Foster 246 St. Margarets Banks, High Street, Rochester ME1 1HY
G3 NRW I. Wade 7 Daubeney Close, Harlington, Dunstable LU5 6NF
G3 NRX R. Murphy 3 Lady Leasow, Shrewsbury SY3 6AB
G3 NRZ C. Hogg 7 Elm Grove, Erith DA8 3BL
G3 NSD Dr B. Styles York House, Bluntisham Road, Colne, Huntingdon PE28 3LY
G3 NSF T. Simpson 41 Benyon Grove, Orton Malborne, Peterborough PE2 5XS
G3 NSL I. Whitter The Old Hall, Hall Lane, Lincoln LN3 4HT
G3 NSO G. Brookes 27 Pineside Avenue, Cannock Wood, Rugeley WS15 4RG
G3 NSP Dr J. Lennox Kestrel, School Lane, Bicester OX25 4AW
G3 NSS T. Spain Manor View, Shotatton, Shrewsbury SY4 1JD
G3 NSW R. Kay 7 Lea Drive, Blackley, Manchester M9 7AR
G3 NTD A. Marsden 15 Northfield Way, Retford DN22 7LJ
G3 NTF I. Neary 65 Vicarage Road, Ashton-under-Lyne OL7 9QY
G3 NTI R. Blain 11 Mill Bank, Ness, Neston CH64 4BJ
G3 NTM W. Brown 18 Georgian Close, Staines TW18 4NR
G3 NUA J. Hogg 16 Moorston Close, Naisberry Park, Hartlepool TS26 0PJ
G3 NUB M. Bursnall Panorama, Church Lane, Bishops Castle SY9 5AF
G3 NUL V. Johnston 119 High St., Cheveley, Newmarket CB8 9DG
G3 NUN A. Langford-Brown 9 Orchard Lane, Corfe Mullen, Wimborne BH21 3SU
GW3 NUO Dr P. Williams Crud Y Gwynt, 27 Mynydd Garnllwyd Road, Swansea SA6 7PB
G3 NUQ I. Macarthur 2 Bramley Close, Bramhall, Stockport SK7 2DT
GM3 NUU Dr J. Reid Rochelle, Findon, Aberdeen AB12 3RL
G3 NVB A. Bryant 1 Downlands Road, Winchester SO22 4ET
G3 NVL R. Allen 692 Hitchin Road, Luton LU2 7UH
G3 NVM D. Arigho 81 Crookham Road Fleet Gu51 5Np Hampshire, Fleet GU51 5NP
G3 NVP B. Mapp 33 Cotswold Drive, Redcar TS10 4AG
GM3 NVQ G. Martin 39 St. Johns Drive, Dunfermline KY12 7TB
GI3 NVW W. Pollock J R Pollock & Co, 155 Doogary Road, Omagh BT79 0HF
G3 NVX R. Davison 76 Poplars Way, Beverley HU17 8PU
G3 NWG D. Stevens 8 Dane Road, Chelmsford CM1 2SS
G3 NWH A. Collis C/O 510 Lowther Road, Dunstable LU6 3LJ
G3 NWL A. Lock 7 Heather Close, St. Leonards, Ringwood BH24 2QJ
G3 NWR Wirral ARS c/o W. Davies Davies Electrical Services (N W) Ltd, 104 Bromborough Village Road, Bromborough, Wirral CH62 7EX
GW3 NWS F. Clare Glen View, Newport Road, Caldicot NP26 3BZ
G3 NWW M. Wakely Chyandour, 3 Ganges Close, Mylor Harbour, Falmouth TR11 5UG
G3 NWX K. Morgan 97 Elmwood, Sawbridgeworth CM21 9NN
G3 NWY D. Forster 79 Westbrooke Avenue, Hartlepool TS25 5HX
G3 NXC A. Plant 178 Clay Lane, Yardley, Birmingham B26 1DY
G3 NXK O. Diplock North Lodge, Messing Park, Colchester CO5 9TD
G3 NXL P. Lamming 25 Leconfield Garth, Follifoot, Harrogate HG3 1NF
G3 NXN F. Wickens 32 Kenilworth Avenue, Wimbledon, London SW19 7LW
G3 NXO A. Watt Little Owls, Singleton, Chichester PO18 0EX
GW3 NXR T. Miles West Uplands Lodge, Upland Arms, Dyfed SA32 8DX
G3 NXS F. Shaw 69 Finedon Road, Irthlingborough, Wellingborough NN9 5TY
G3 NXT W. Fletcher 3 Orchard Close, Metheringham, Lincoln LN4 3DT
G3 NXX I. Miller 11 Lynton Drive, High Lane, Stockport SK6 8JE
G3 NXZ J. Howe 18 Laburnum Grove, Conisbrough, Doncaster DN12 2JW
G3 NY North Yorkshire ARG c/o N. Knapton 4 Crabmill Lane, Easingwold, York YO61 3DE
G3 NYB W. Bingham 7 Bolton Hill Road, Doncaster DN4 6DQ
G3 NYD D. Coles 113 Berrow Road, Burnham-on-Sea TA8 2PH
G3 NYE A. Taylor 25 Burnside Road, Gatley, Cheadle SK8 4NA
GM3 NYG J. Fish 31 Oaklands Avenue, Irvine KA12 0SE
GI3 NYJ S. Currie 122 Belfast Road, Comber, Newtownards BT23 5QP
G3 NYK A. Melia 67A Deben Avenue, Martlesham Heath, Ipswich IP5 3QR
GI3 NYL L. Elliott 19 Gosford Road, Collone, Armagh BT60 1LQ
G3 NYM North Yorkshire CG c/o A. Duffield 4 Crabmill Lane, Easingwold, York YO61 3DE
G3 NYR D. Rayner 42 Canford Drive, Allerton, Bradford BD15 7AU
G3 NYS C. Whiteley 30 Lynch Hill Park, Whitchurch RG28 7NF
G3 NYX J. Heaviside 110A Cuckfield Road Hurstpierpoint, Hassocks BN6 9RZ
G3 NYZ A. Stafford Blakefield, Jawbone Lane, Derby DE73 1BW
G3 NZL H. Chapman 57 Athelstan Road, Southampton SO19 4DE
G3 NZP M. Harman 19 Hill House Close, Turners Hill, Crawley RH10 4YY
G3 NZR W. Young 5 Grasmere Grove, Frindsbury, Rochester ME2 4PN
G3 NZS H. Parkes 35 Dovey Road, Tividale, Oldbury B69 1NT
G3 NZV A. Park Waterside Cottage, Bowden Lane, High Peak SK23 0QF
G3 NZW S. James Beresford, Latchmoor Ave, Gerrards Cross SL9 8LJ
G3 NZY R. Shelley 4 Fairview Court, St. Martins Avenue, Scarborough YO11 2DA
G3 OAD T. Haydu Jones 1 Beggars Roost, Golf Course Road, Stroud GL6 6TJ
G3 OAF W. Jeffs Silver Jay, Colehill Lane, Wimborne BH21 7AN
G3 OAH Dr P. Whittlestone Turangi, 14A Croft Bank, Malvern WR14 4DW
GW3 OAJ C. Davies 11 Beaumaris Way Grove Park, Blackwood NP12 1DF
G3 OAL E. Lincoln Lynholme, Millbank, Newton Aycliffe DL5 6RF
G3 OAR G. Greenwood 1 Maltkiln Lane, Castleford WF10 4LF
G3 OAZ J. Randall 243 Paddock Road, Basingstoke RG22 6QP
GM3 OBC R. Thomson 1 Knowehead, Star, Glenrothes KY7 6LA
GM3 OBG P. Bridges 29 Kirkbank, Auchmithie, Arbroath DD11 5SY
G3 OBL J. Tyrrell 2 Briar Close, Yeovil BA21 5XA
GI3 OBO D. Waugh 16 Seaview Avenue, Millisle, Newtownards BT22 2BN
G3 OBV P. Harris 15 Ratliffe Road, Rugby CV22 6HB
G3 OBZ M. Birkett Hazelwood, Cromwell Ave, Woodhall Spa LN10 6TH
G3 OCA K. Frankcom 1 Chesterton Road, Spondon, Derby DE21 7EN
G3 OCB C. Bowden Tregwyn, Tregonning Road, Stithians, Truro TR3 7FG

G3 OCH J. Hulett 21 Exmoor Avenue, Leicester LE4 0BJ
G3 OCP D. Wallace 11 Station Road, Haddenham, Aylesbury HP17 8AN
G3 OCR S. Nutt 23A Hesketh Drive, Southport PR9 7JX
GW3 ODB A. Pritchard 41 Maes Cantaba, Ruthin LL15 1YP
G3 ODC D. Martin 7 Seaview Avenue, Eastham, Wirral CH62 0BD
G3 ODD E. Stables Manor Croft, Water Lane, Selby YO8 6QL
GM3 ODP Dr T. Salvesen Easter Catter, Croftamie, Glasgow G63 0EX
G3 ODX Fort Purbrook ARC c/o S. Clarke 18 Dunedin Drive, Caterham CR3 6BA
G3 OEB R. Downs 23 Old London Road, Benson, Wallingford OX10 6RR
G3 OEC Prof. C. Isham 2 Lime Grove, Ruislip HA4 8RY
G3 OEQ D. Bunn Heliophilia Gardens 1, Agias Annis St. 17, Kato Pafos 8036 Cyprus
G3 OFI B. Bisley 132-1919 St Andrews Place, Courtenay, British Columbia V9N 9J4 Canada
G3 OFP G. Cunnah 225 Springwell Lane, Balby, Doncaster DN4 9AJ
GM3 OFT P. Bower An Cluain, Ballplay Road Dg10 9Ju, Moffat DG10 9JU
G3 OFW H. Blake 19 Segsbury Grove, Harmans Water, Bracknell RG12 9JL
G3 OFX R. Welch 112 Copsewood Road, Bitterne, Southampton SO18 1QR
G3 OGE Southern Microwave Group c/o J. Rose 1 Westgate House, 22 Westgate, Hornsea HU18 1BP
G3 OGH A. Brooker-Carey 29 Byron St., Amble, Morpeth NE65 0ER
G3 OGK Dr G. Kennedy Thayers, Edwyn Ralph, Bromyard HR7 4LY
G3 OGP R. Powell Garlands Farm, The Haven, Billingshurst RH14 9BH
G3 OGX J. Allsop 17 Hambro Hill, Rayleigh SS6 8BN
G3 OGZ M. Beer 24 Byron Court, Beech Grove, Harrogate HG2 0LL
G3 OHC G. Badger 3 Hesketh Close, Cranleigh GU6 7JB
G3 OHH R. Hargreaves 46 Castle Road, Mow Cop, Stoke-on-Trent ST7 3PH
G3 OHL D. White Holme Fell Cottage, Hallbankgate, Brampton CA8 2NJ
G3 OHM South Birmingham RS c/o J. Storey 34 Austin Rise, Longbridge, Birmingham B31 4QN
G3 OHN K. Whitehouse 27A Howdles Lane, Brownhills, Walsall WS8 7PL
G3 OHP M. Winter 9 Higham Road, Cliffe, Rochester ME3 7SH
G3 OHS J. Perry 517 Longbridge Road, Barking IG11 9DD
G3 OHX I. Jackson Brattle House, Manor Road, Seer Green, Beaconsfield HP9 2QU
GM3 OIB K. Younger 183 Main St., Pathhead EH37 5SQ
G3 OIC I. Croxford 16 Chesterwood, Hollywood, Birmingham B47 5EN
G3 OIF P. Squires 191 Station Road, Knowle, Solihull B93 0PT
G3 OIH B. Shields 24 Churchfield, Fulwood, Preston PR2 8GT
G3 OIL M. Wills 23 Falcons Way, Salisbury SP2 8NR
GW3 OIN J. Nicholas 28 Hardy Avenue, Rhyl LL18 3BG
G3 OIP P. Holker 9 Limetree Grove, Braunton EX33 1HE
GM3 OIV W. Anderson 6 Winchburgh Road, Winchburgh, Broxburn EH52 6QB
G3 OJ J. Hobin 14 St. Martins Green, Trimley St. Martin, Felixstowe IP11 0UU
G3 OJG Dr P. Gale Garden Cottage, Sacombe Green, Ware SG12 0JQ
G3 OJI J. Sleight Orchard House, School Hill, Napton, Southam CV47 8NN
G3 OJK J. Bates 8 Spaxton Road, Durleigh, Bridgwater TA5 2AP
G3 OJL M. Plaster Combe House, Milton Lane, Wookey Hole, Wells BA5 1DG
G3 OJS H. Braham 10 Glebe Way, Frinton-on-Sea CO13 9HR
G3 OJV P. Waters 9 Tudor Way, Hawkwell, Hockley SS5 4EY
G3 OJZ B. Todd-White 3 Alexandra Road, Capel-Le-Ferne, Folkestone CT18 7LB
G3 OKA Dr J. Share 82 Birkenhead Road, Meols, Wirral CH47 0LB
G3 OKB M. Ireson 15 Digby Drive, North Luffenham, Oakham LE15 8JS
G3 OKD Z. Nilski The Poplars, Wistanswick, Market Drayton TF9 2BA
G3 OKH G. Hillman 504 Chester Road, Kingshurst, Birmingham B36 0LG
G3 OKS S. Smithies Moorcroft, Fernhill., Horley RH6 9SY
GW3 OKT Dr J. Thompson The Old Place, Old Racecourse, Oswestry SY10 7HL
G3 OKU M. Cross 39 Westfield, The Marld, Ashtead KT21 1RH
G3 OLB T. Boucher Hedgerows, Sheldon, Honiton EX14 4QS
G3 OLH A. Remsbury Nodali, 16 Little Green Lane, Chertsey KT16 9PH
G3 OLP B. Wadsworth 5 Birch Avenue, Todmorden OL14 5NX
G3 OLU J. Saunders Apartamento 6306, Forum Mare Nostrum, Camino De Pinxo 2, Alfaz Del Pi 3580 Spain
G3 OLW J. Burnett Wenrisc, Chapel Lane, Tewkesbury GL20 8HS
G3 OLX J. Parker Palfreys, Picquets Way, Banstead SM7 1AJ
G3 OMA Dr S. Kay 5 Chevalier Close, Swindon SN5 5TS
G3 OMB R. Spurgeon 57 Laburnum Crescent, Kirby Cross, Frinton-on-Sea CO13 0QH
G3 OMD A. Callegari Danebridge Nursery, Much Hadham SG10 6JG
G3 OMJ P. Judkins 18 St. Johns Square, Wakefield WF1 2RA
G3 OMK T. Kirk 54 Highfields Drive, Loughborough LE11 3JT
GW3 OMN M. Jenkins 25Stepneyroad, Swansea SA2 0FZ
G3 OMR M. Russoff Flat 3 Hartsbourne Court, Hartsbourne Road, Bushey Heath WD23 1PZ
G3 OMS Dr R. Simpson 23 Larkhill, Rushden NN10 6BG
G3 OMT A. Russell 5 Little Close, Swadlincote DE11 0EB
G3 OMY D. Hancock 17 Forestlake Avenue, Ringwood BH24 1QU
G3 OMZ D. Lee 19 Sarum Lodge, Three Swans Chequer, Salisbury SP1 1AL
G3 OND J. Denman 167 Minnis Road, Birchington CT7 9QD
GI3 ONF R. Sinton 35 The Rose Garden, Tandragee, Craigavon BT62 2NJ
G3 ONI D. Woods Flat 26, Chapel Court, Wilmslow SK9 5EN
GU3 ONJ A. Richmond The Cedars, 3 Holly Drive, Braye Road, St. Sampson GY2 4EF Guernsey
G3 ONL P. Brodribb 18 Ipswich Road, Debenham, Stowmarket IP14 6LB
G3 ONR B. Reynolds 17 Cresswells Mead, Holyport, Maidenhead SL6 2YP
G3 ONU D. Barry 2 Catherine Close, Shrivenham, Swindon SN6 8ER
G3 ONV J. Verity Tall Pine, Station Road Kirby Muxloe, Leicester LE9 2EN
G3 OOH G. Lander 132 Chemin De Saule, Bernex 1233 Switzerland
G3 OOK J. Plenderleith 3D Deluxe Court, Jalan Pahlawan Kepayan, Kota Kinabalu 88200 Malaysia
G3 OOL J. Hatch 628-707 Esquimalt Road, Victoria BC V9A 3L7 Canada
G3 OOP Dr B. Havenhand 15 Sandiway, Chesterfield S40 3HG
G3 OOU R. Burns 84 Portnalls Road, Coulsdon CR5 3DE
G3 OOW M. Docker Apartment 219, Clarence Park, 415 Worcester Road, Malvern WR14 1FU
G3 OPB M. Bues 7A Alice Parkins Close, Hadleigh, Ipswich IP7 6FE
GW3 OPC N. Ward 17 Heol Nant, Llanelli SA14 8EL
G3 OPG R. Tingay 18 Grove Road, Newbury RG14 1UH
G3 OPH R. Atkinson Lake Walk, Adderbury OX17 3PF
G3 OPJ C. Harrisson 129 Granville Way, Sherborne DT9 4AT
G3 OPW J. Cook Upwood Park, Black Moor Road, Keighley BD22 9SS
G3 OPX Sands Amateur Radio Communications Group c/o C. Green Gothic, Plymouth Road, Totnes TQ9 5LH
G3 OQC J. Woods 1 Dean Road, Cosham, Portsmouth PO6 3DG
G3 OQD M. Emmerson 6 Mounthurst Road, Hayes, Bromley BR2 7QN
G3 OQF R. Kay 7 Chemin Des Grands-Champs, Bogis-Bossey CH 1279 Switzerland
GM3 OQI J. Ramsay 150 City Road, Dundee DD2 2PW
GW3 OQK A. Fairgrieve 3 Pleasant Road, Gorseinon, Swansea SA4 9WH
G3 OQO D. Henley 36 Main Street, Newbold, Rugby CV21 1HW
GI3 OQR D. Gibson 93 Cavan Road, Dungannon BT71 6QN
G3 OQT R. Mclachlan Oak Trees, Park Lane, Rodsley, Ashbourne DE6 3AJ
G3 ORG I. Taylor 10 Westfield Road, Henlow SG16 6BN
G3 ORI South Manchester RC c/o J. Vickers 45 Willow Park Drive, Stourbridge DY8 2HL
G3 ORK R. Talbot 9 Bracebridge Drive, Southport PR8 6XH
GW3 ORL Lincoln Shortwave c/o D. Williams 14 Seymour Avenue, Parc Seymour, Caldicot NP26 3AG
G3 ORN W. Thomas 20 Vinnicombes Road, Stoke Canon, Exeter EX5 4BB
G3 ORP P. Pickering 21 Palmar Road, Maidstone ME16 0DL
G3 ORV M. Saunders 40 Archfield Road, Cotham, Bristol BS6 6BE
G3 ORY R. Titterington Wyclif House, St. Marys Road, Lutterworth LE17 4PS
G3 OS A. Boor 3 Croft Cottages, Beltoft, Doncaster DN9 1NA
G3 OSI D. Swanson 48 Moscow Drive, Liverpool L13 7DJ
G3 OSL Simon Langton Grammar School For Boys c/o Dr I. Stevenson 18 Sittingbourne Road, Wigan WN1 2RR
G3 OSP S. Plumtree Flat 18, Oliver Leese Court, Ten Butts Crescent, , Stafford ST17 9HP
G3 OSQ D. Beakhust Nonsuch Lodge, Morgans Vale Road, Redlynch, Salisbury SP5 2HU
G3 OST D. Wilson Chemin D'Arques, Ambrumesnil, Offranville 76550 France
G3 OTH C. Cook Swiss Cottage, Netherton Lane, Bedlington NE22 6DR
G3 OTK R. Harris 4 Alford Court, Hambridge, Langport TA10 0BS
G3 OTN P. Seaman 5 Berkeley Close, Maidenhead SL6 5JP
G3 OTR Royal Naval ARS c/o M. Beckley Mallards, Albury Road, Ware SG11 2DN
GI3 OTU A. Burge 38 Bayview Road, Bangor BT19 6AR
G3 OTV P. O'Kane 36 Coolkill, Sandyford, Dublin D18 P7F4 Ireland
G3 OTW W. Miller 418 Old Chester Road, Birkenhead CH42 4PD
G3 OTY Capt. R. Cogzell Flat 242, Clydesdale Tower Holloway Head, Birmingham B1 1UJ
G3 OUA D. Tarr 17 Allendale Avenue, Findon Valley, Worthing BN140AH
G3 OUC P. Painting 15 Turnpike Road, Shaw, Newbury RG14 2ND
G3 OUI I. Dickinson 64A Richmond St, College Park 5069 Australia
G3 OUT A. Walker High Beacon Farm, Fulletby, Horncastle LN9 6LB
GM3 OUU G. Rennie 60 Woodend Place, Aberdeen AB15 6AN
G3 OUV P. Perkins 47 Priory Avenue, High Wycombe HP13 6SN
G3 OVE Dr M. Brown 25 Carpenters Lane, West Kirby, Wirral CH48 7EX
G3 OVH A. Abbey 1 The Fairway, Kirby Muxloe, Leicester LE9 2EU
G3 OVL M. Hubbard 7 Creake Road Syderstone, King's Lynn PE31 8SF
G3 OVX H. Hammett 27 Courtman Road, Tottenham, London N17 7HT
G3 OWB J. Holland Carter 37 Highfield Avenue, Cambridge CB4 2AJ
G3 OWE D. Saunders 4A Ullswater Crescent, Radipole, Weymouth DT3 5HE
G3 OWJ P. Jarvis 44 Torrin Drive, Shrewsbury SY3 6AW
G3 OWQ J. Clarke 29 Long Brackland, Bury St. Edmunds IP33 1JH
GM3 OWU V. Stewart 9 Baberton Park, Juniper Green EH14 5DW
G3 OWX South Devon Raynet Group c/o J. Greany Flat 3 Crete Hill House, Cote House Lane, Bristol BS9 3UW
GM3 OXA A. Fosters 16 Reid Crescent Milnathort, Kinross KY13 9TB
G3 OXG D. Thompson 34 Sandy Road, Potton, Sandy SG19 2QQ
GM3 OXK Bangor And District ARS c/o J. Carson 23 Whinny Rig, Heathhall, Dumfries DG1 3RJ
G3 OXL D. Westbury Rose Cottage, Cruise Hill, Ham Green, Redditch B97 5UA
G3 OXN D. Swainson 4 Grasmere Avenue Spondon, Derby DE21 7JZ
G3 OXR P. Garthwaite 16 Newtown Avenue, Royston, Barnsley S71 4HF
GM3 OXX G. Burt Clunie Lodge, Netherdale, Turriff AB53 4GN
G3 OYB W. Waters 4 Calartha Road, Pendeen, Penzance TR19 7DZ
GI3 OYG J. Semple 5 Tullaghgore Road, Ballymoney BT53 6QF
G3 OYL D. Gilbert 348 Willington Road, Kirton End, Boston PE20 1NU
G3 OYN G. Saunders 17 Chester Street Caversham, Reading RG4 8JH
G3 OYT G. Clinton 2 Greenways, Abbots Langley WD5 0EU
G3 OYX M. Rignall Ashdown, Nupend, Stroud GL6 0PY
GM3 OZB A. Mckay 2 Osprey Drive Sparrow Plantation Kilmarnock Ka13 Lq, Kilmarnock KA1 3LQ
G3 OZC J. Holstead 72 Woodlands Avenue, Feniscowles, Blackburn BB2 5NN
G3 OZD P. Cross 5 Lings Lane, Hatfield, Doncaster DN7 6AB
G3 OZE J. Grainger 6 Fulford Cross, Fulford, York YO10 4PB
GM3 OZJ I. Morgan 43 Dalgety Gardens, Dalgety Bay, Dunfermline KY11 9LF
G3 OZK M. James 11 Shortborough Avenue, Princes Risborough HP27 9HU
G3 OZL Dr A. Jeavons Wadsley Grove, Worrall Road, Sheffield S6 4BE
G3 OZN E. Badger 20 Tennyson Drive, Worksop S81 0EE
G3 OZP P. Smith 39 Sherborne Avenue, North Shields NE29 8NT
G3 OZT R. German 10 Beverley Road, Dibden Purlieu, Southampton SO45 4HS
GI3 OZW P. Dynes 1 Rossin View, Donaghmore, Dungannon BT70 1SZ
G3 PAG J. Davies Cedar Croft, School Lane, West Malling ME19 5EH
G3 PAI J. Rabson 55 Severn Road, Ipswich IP3 0PU
GM3 PAK Dr M. Senior The Raw, Bridgend, Isle of Islay PA44 7PZ
G3 PAQ J. Davis 76 Allfarthing Lane, London SW18 2AJ
G3 PAX J. Barker 2 Barons Hall Lane, Fakenham NR21 8HB
G3 PBF J. Orford 63 Flowerhill Way, Istead Rise, Gravesend DA13 9DS
G3 PBI A. Davies 69 Sycamore Road, Chalfont St. Giles HP8 4LG
G3 PBR A. Green 6 Shipley Close, Woodley, Reading RG5 4RT
G3 PBT R. Hilsley 1 Chelmerton Avenue, Great Baddow, Chelmsford CM2 9RE
G3 PCG D. Askew Lapthorne, Adsborough, Taunton TA2 8RP
G3 PCJ T. Walford Vedal House, , Langport TA10 9FB
G3 PCL Avon Valley ARS c/o D. Shaw 3 Randolph Close, Cheltenham GL53 7RT
G3 PCT P. Hurst Anchorage House, Upper Wood Lane, Dartmouth TQ6 0DQ
G3 PCW M. Watling 8 Preetz Way, Blandford Forum DT11 7XG
G3 PCX B. Dodge 34 Downs Road, Penenden Heath, Maidstone ME14 2JN
G3 PCY J. Wilson 5 Huntsham Court Stables, Huntsham, Tiverton EX16 7NA
G3 PDC R. Curwen 53 Karslake Road, Liverpool L18 1EY

G3 PDD J. Dolby Oaklea, School Lane, Belper DE56 2AL
G3 PDH M. Prestwood Salatiga, Bell Lane, Salhouse, Norwich NR13 6RR
GI3 PDN R. Harbison 26 Ballymartin Road, Templepatrick, Ballyclare BT39 0BW
G3 PDP A. Ralls 12 Oakhill Close, Bursledon, Southampton SO31 1AP
GM3 PDX J. Barker 44 Priory Road, Linlithgow EH49 6BS
G3 PEJ P. Watson 37 Chestnut Bank, Scarborough YO12 5QJ
G3 PEK B. Simpson 20 Monterey Street, St. Ives NSW 2075 Australia
G3 PEM C. Thomson 109 Hillside Grove, Chelmsford CM2 9DD
G3 PET A. Widdowson 34 Highfields Road, Chasetown, Burntwood WS7 4QU
G3 PEW J. Hudson 68 Lower Street, Stansted CM24 8LR
G3 PEZ J. Gutteridge 66 Croft Drive, Moreton, Wirral CH46 0QT
G3 PFE G. Spriggs Brookbank Cottage, Newcastle Road, Nantwich CW5 7EJ
G3 PFH M. Blunden 24 Mill View Close, Woodbridge IP12 4HR
G3 PFM A. Baker Highleaze, Deans Drove, Poole BH16 6EQ
G3 PFO C. Barr Riders Way, Collum Green Road, Stoke Poges SL2 4AX
G3 PFT A. Heeley 108 Valley Lane, Lichfield WS13 6ST
GW3 PFV K. Robbins 1 Rhiw Parc Road, Abertillery NP13 1BS
G3 PFX C. Small Overlangs, Kingston, Kingsbridge TQ7 4PF
G3 PGA A. Hammond 23 St. Andrews Road, Fremington, Fremington EX313BS
G3 PGC R. Armstrong 6 Barnstaple Road, North Shields NE29 8QA
G3 PGJ R. Bashford Popplestone, 33 Penair View, Truro TR1 1XR
G3 PGK C. Pearless 26 Church Road, Preston, Weymouth DT3 6RP
G3 PGN H. Buckenham Tweed Cottage, Tilbury Road, Great Yeldham, Halstead CO9 4JG
G3 PGQ D. Yates 26 Lowestoft Road, Carlton Colville, Lowestoft NR33 8JD
GM3 PGY A. Mc Ewen 4 Reef Terrace, Crossapol, Isle of Tiree PA77 6UT
G3 PHD I. Gardiner 189 Brennan Road, Tilbury RM18 8BA
G3 PHG A. Gibbs 223 Crimea Street, Noranda WA 6062 Australia
G3 PHJ J. Johnston 9 Appleby Glade, Haxby, York YO32 3YW
G3 PHL B. Davies 17 Linksway, Leigh-on-Sea SS9 4QY
G3 PHO P. Day 38 Broomhill Road, Old Whittington, Chesterfield S41 9DA
G3 PIA Harwell ARS c/o C. Desborough 22 Westland Road, Faringdon SN7 7EY
G3 PID P. Chandler11 528 Goffs Lane, Goffs Oak, Waltham Cross EN7 5EW
G3 PIJ P. Mellett 16 Tutton Hill Colerne, Chippenham SN14 8DN
GM3 PIL R. Munro 20 County Cottages, Piperhill, Nairn IV12 5SE
G3 PIN J. Patten 8 Leacroft Road, Penkridge, Stafford ST19 5BX
G3 PIY C. Isaacs Holme View, Brick Lane, Christchurch BH23 8DU
G3 PIZ T. Watts 26 Woodger Close, Guildford GU4 7XR
G3 PJC C. Arnold 47 Peartree Lane, Danbury, Chelmsford CM3 4LS
G3 PJQ A. Aldridge 1 Mary Grove, Highnam, Gloucester GL2 8NH
G3 PJT Dr R. Whelan 36 Green End, Comberton, Cambridge CB23 7DY
G3 PJV P. Walsh 23 Moss Fold Road, Darwen BB3 0AQ
G3 PJW R. Unsworth 8 Coleridge Road, Billinge, Wigan WN5 7EB
G3 PJY R. Millman 103 The Crescent, Walsall WS1 2DA
G3 PKC J. Tinker 72 Jackson Avenue, Leeds LS8 1NS
G3 PKD R. Sharples 40 Greetham Road, Cottesmore, Oakham LE15 7DB
G3 PKL C. Fox 2 Mill Cottages, Wareham Road, Poole BH16 6ET
G3 PKQ J. Holmes 36 Hillside Gardens, Walthamstow, London E17 3RJ
G3 PKR K. Parker 263 High St., Hayes UB3 5ET
G3 PKY Rvd. P. Okelly The Ravel, School Lane, Drogheda Ireland
GW3 PLB R. Howe Brooklands, Caeffynnon, Kidwelly SA17 5EJ
G3 PLE D. Barlow Pine, Churchtown, Cury, Nr. Helston TR12 7BW
G3 PLJ P. Fairnington 30 Orchard St., Weston Super Mare BS23 1RQ
GI3 PLL R. Moore 818 Seacoast Road, Castlerock, Coleraine BT51 4SD
G3 PLN G. Smith 7 Coniston Avenue, Grimsby DN33 3EE
GM3 PLO J. Gray Norland, South End, Stromness KW16 3DJ
G3 PLP R. Cox 30 Brooks Road, Sutton Coldfield B72 1HP
G3 PLR D. Skye 16 Lulworth Avenue, Poole BH15 4DQ
G3 PLS Dr R. Fenby Apartment 9, Centennial Place, 28 Northwich Road, , Knutsford WA16 0AW
G3 PLT G. Lawes 7 Tormynton Road, Weston-Super-Mare BS22 9HU
G3 PLW J. Norton 32 Fismes Way, Wem, Shrewsbury SY4 5YD
G3 PLX J. Martinez High Blakebank Farm Underbarrow, Kendal LA8 8HP
G3 PLY G. Mcneil 168 Chobham Road, Ascot SL5 0HU
GM3 PMB W. Miller 15 Glenalla Crescent, Ayr KA7 4DA
G3 PMH Obo March & Dist ARS, British Legion Club, Rookswood Road c/o V. Cracknell 106 High St., Upwood, Huntingdon PE26 2QE
G3 PMJ S. Revell 11 Mere Fold, Worsley, Manchester M28 0SX
GD3 PML Dr D. Smith Cooilbane Cottage, Main Road, Sulby IM7 2HR Isle of Man
G3 PMR A. Jubb Psathi Village, Pafos 8749 Cyprus
G3 PMV A. Feist 1 Lowry Drive, Marple Bridge, Stockport SK6 5BR
G3 PMW K. Dews 14 Baddow Place Avenue, Great Baddow, Chelmsford CM2 7JN
G3 PND S. Appleyard Plumtree House, Mill Lane, Cromer NR27 9PH
G3 PNO I. Hawkins Victoria House, Victoria Street, Totnes TQ9 5EF
G3 PNP J. Ward 131 Monarch Road, Eaton Socon, St. Neots PE19 8GU
G3 PNQ A. Floyd 27 Beechfield Parbold, Wigan WN8 7AR
G3 PNT C. Durell 17 Ryders Avenue, Westgate-on-Sea CT8 8LW
G3 PNU E. Clark 1 Station Road, Drigg, Holmrook CA19 1XH
G3 POG St Christophers School ARC c/o D. Mawdsley 20 Cable Street, Formby, Liverpool L37 3LX
GM3 POI C. Penna North Windbreck, Deerness, Orkney KW17 2QL
G3 POM G. Morgan 7 Quantock Grove, Williton, Taunton TA4 4PD
G3 POQ P. Hayes 16 Melton Drive, Storrington, Pulborough RH20 4LU
GI3 POS G. Smyth 91A Gilford Road Lurgan, Craigavon BT66 7EB
GM3 POT J. Walford Chorcaill, Reay, Thurso KW14 7RG
G3 PPB P. Perkins 52 Ashley Piece, Ramsbury, , Marlborough SN8 2QE
G3 PPC D. Taylerson 18 The Grove, Teddington TW11 8AS
GM3 PPE Dr M. Eccles Newtonlees Bungalow, Kelso TD5 7SZ
G3 PPO L. Hook 79 Whiteley Crescent, Bletchley, Milton Keynes MK3 5DQ
G3 PPR Dr J. Beavon 24 Cromer Road Mundesley, Norwich NR11 8BE
G3 PPT L. Sear 4 Mount Pleasant Road, Threemilestone, Truro TR3 6BB
G3 PPU P. Smith 56 Alphington Avenue, Frimley, Camberley GU16 8LR
G3 PQA J. Rogers Dromore, Strande Lane, Cookham, Maidenhead SL6 9DN
G3 PQB S. Harbour 43 Warbon Avenue, Peterborough PE1 3DS
G3 PQC P. Turk 13 The Crescent, Farnborough GU14 7AR
G3 PQD D. St John 26 Henry Street Rainham, Gillingham ME8 8HE
G3 PQF D. Dell 7 Blunden Road, Cove, Farnborough GU14 8QJ
G3 PQJ B. Cole 17 Coburg Court, East Cowes PO32 6SS
G3 PQM M. Thorp Cecil, Thorrington Road, Clacton-on-Sea CO16 9ES
G3 PQP T. Foster 136 Sladepool Farm Road Kings Heath, Birmingham B14 5EF
G3 PQY J. Lawrence 2A Hall Road, Hull HU6 8SA
G3 PRC Merseyside ARS c/o P. Connor 20 Longfield, Lutton, Ivybridge PL21 9SN
G3 PRE W. Armstrong 24 Newbury St., South Shields NE33 4UE
G3 PRH M. Coward 51 Farleigh Road, Backwell, Bristol BS48 3PB
G3 PRI D. Quigley 1A Elizabeth Road, Bishop's Stortford CM23 3RJ
G3 PRK A. Yilmaz 7 Kerdistone Close, Potters Bar EN6 1LG
GW3 PRL D. Snow Rhwngyddwy Dre, Brynsiencyn, Llanfairpwllgwyngyll LL61 6TZ
G3 PRQ E. Wooden Mullins, Windsor Road, Alton GU34 5EF
G3 PRR Rvd. I. Partridge 4 Thames Street, Louth LN11 7AD
G3 PRU J. Nicholas-Letch 53 Hayden Road, Rushden NN10 0JH
GW3 PRW Loughborough And Disttrict ARC c/o J. Dolan 24 Pen Derwydd, Llangefni LL77 7QE
G3 PS A. Mccann 5 Arrowsmith Drive, Hoghton, Preston PR5 0DT
G3 PSB P. Bottomley 47 Stonelea, Barkisland, Halifax HX4 0HD
G3 PSC J. Holton 1204 Greenford Road, Greenford UB6 0HQ
G3 PSG North Riding Rafars c/o R. Armstrong 64 Churchill Drive, , Marske by the Sea TS11 6BE
G3 PSM C. Thomas 16 Fordlands, Thorpe Willoughby, Selby YO8 9PD
GM3 PSP Dr A. Masson 20 Frogston Avenue, Edinburgh EH10 7AQ
GI3 PSQ C. Bristow 58 Bristow Park, Belfast BT9 6TJ
G3 PSR M. Gibbs 62 Abinger Drive, Chatham ME5 8UL
G3 PSS M. Kent 99 London Road, Newington, Sittingbourne ME9 7RH
G3 PSU P. Martin 15 St. Lukes Close, Cannock WS11 1BB
G3 PSV D. Park 18 Widworthy Drive, Broadstone BH18 9BD
G3 PSZ K. Jones 24 Station Road, Okehampton EX20 1EA
G3 PTB A. Tomalin Chapel Street, Barford, Norwich NR9 4AB
G3 PTG R. Gealy 14 Wivelsfield, Eaton Bray, Dunstable LU6 2JQ
G3 PTI K. Atter 60 Hough Road, Barkston, Grantham NG32 2NS
G3 PTQ T. Chapman 5 Maple Close, Bottisham, Cambridge CB25 9BQ
G3 PTS Dr G. Holt 7 Beech Close, Olivers Battery, Winchester SO22 4JY
G3 PTX L. Buckley 188 Compstall Road, Romiley, Stockport SK6 4JF
G3 PTZ A. Bensley 13 Lime Grove, Cherry Willingham, Lincoln LN3 4BE
G3 PUO L. Rooks 17 The Close, Clayton Le Moors, Accrington BB5 5RX
G3 PUQ N. Semmens 4 South Park, Redruth TR15 3AW
G3 PUR R. Tarr 37 Warwick Avenue, Coventry CV5 6DJ
G3 PUX I. Champion Mill Bungalow, Billinghurst RH14 0DY
GM3 PUY I. Forsyth 68 Drumover Drive, Glasgow G31 5RP
G3 PUZ D. Hogan 17 Buckingham Mansions, Bath Road, Bournemouth BH1 2PG
G3 PVG J. Bennett 11 Enderby Road, Thurlaston, Leicester LE9 7TF
G3 PVH D. Sumner 64 Kelsey Avenue, Emsworth PO10 8NQ
G3 PVJ Dr H. Coltman 68 Cressex Road, High Wycombe HP12 4TY
G3 PVU J. Hunt 28 Harris Road, Lincoln LN6 7PN
G3 PWB I. Dufour 3 Western Close, Rushmere St. Andrew, Ipswich IP4 5UU
G3 PWJ R. Fisher 34 Doctors Hill, Stourbridge DY9 0YE
G3 PWN G. Grimshaw 1 Sandsacre Drive, Bridlington YO16 6UA
G3 PWP M. Bentley 5 Grendon Green, Stoke-on-Trent ST2 0EH
G3 PWS R. Dalton 23 Muswell Road, Mackworth, Derby DE22 4HN
G3 PWX A. Boyd 15 Northend, Batheaston, Bath BA1 7EE
G3 PWY D. Gresswell 10 Cherrywood Gardens, Flackwell Heath, High Wycombe HP10 9AX
G3 PXH M. Bartlett 3 Jessopp Avenue, Bridport DT6 4AN
G3 PXI A. Evans Apartado 286, Luz Lagos 8601-929 LUZ GS Portugal
GM3 PXK Mid Lanark ARS c/o M. Overthrow 63 Primrose Avenue, Larkhall ML9 1JX
G3 PXL A. Hickin 14 Churston Broadway, Dartmouth Road, Paignton TQ4 6LE
G3 PXU G. Grove 11 Croft Close, Warwick CV34 6QY
G3 PXV R. Wiseman 3 Springfield Road, Ruskington, Sleaford NG34 9HG
GW3 PYD D. Stephens 1 Awelfryn Terrace, Merthyr Tydfil CF47 9YP
G3 PYE Cambridgeshire Repeater Group c/o P. Nice 31 Elizabeth Drive, Chapel St. Leonards, Skegness PE24 5RS
G3 PYF J. Green 68 Magdalen Lane Wingfield, Trowbridge BA14 9LQ
G3 PYH A. Broadbent 52 Norman Street, Failsworth, Manchester M35 9EJ
G3 PYI D. Coy 26 Hardy Road, Bishops Cleeve, Cheltenham GL52 8BN
G3 PYL D. Justice 4 Birley Moor Avenue, Sheffield S12 3AQ
G3 PYO J. Dann 1 Ffinch Close, Ditton, Aylesford ME20 6ET
G3 PYW Rvd. A. Speight Glebe Cottage, Hollow Lane, Woodbridge IP13 8LZ
GW3 PYX J. Chetcuti 3 Beechwood Drive, Penarth CF64 3RB
G3 PYZ Rsars - Royal Signals ARS c/o M. Foster 7 Church Street, Fenstanton, Huntingdon PE28 9JL
G3 PZB A. Ash 34 Coronation Avenue, Northwood, Cowes PO31 8PN
G3 PZE C. Burkitt The Old Wheelwright, 17 Oxford Road, Breachwood Green SG4 8NP
G3 PZF G. Dale 16 Palfrey Close, St. Albans AL3 5RE
G3 PZL P. Brown Little Langford, Newlands Lane, Henley-on-Thames RG9 5PS
G3 PZN C. Wood 24 Talveneth, Pendeen, Penzance TR19 7UT
G3 PZU B. Brown 138 First Avenue, Sudbury CO10 1YU
G3 PZV P. Greed The Bungalow, Townsend, Devizes SN10 4RR
G3 PZX A. Ward 20 Tower Close, Costessey, Norwich NR8 5AU
G3 PZZ P. Smith 38 Leasway, Wickford SS12 0HE
G3 QI Worthing & Dist ARC c/o C. Quarton Flaxton Gatehouse, Flaxton, York YO60 7QT
G3 RAC Thales ARC c/o D. Waterworth 116 Reading Road, Woodley, Reading RG5 3AD
G3 RAF Royal Air Force ARS c/o M. Garrett 489 Dorchester Road, Weymouth DT3 5BP
G3 RAL Loughborough And Disttrict ARC c/o A. Harrison 44 Rosslyn Road, Whitwick, Coalville LE67 5PT
G3 RAM C. Langmaid Flat 4, Woodlawn High Street, Partridge Green West Sussex RH13 8HR
G3 RAU D. Moffatt Mill House, Middle Street, Glentworth, Gainsborough DN21 5BZ
G3 RBD F. Hanson 207 Grant Road, Liverpool L14 0LG
G3 RBJ A. Payne Laurel Bank, Sand Road, Wedmore BS28 4BZ
G3 RBP R. Parsons Netherhall Barn, Hallmoor Road, Darley Dale, Matlock DE4 2HF
G3 RBY A. Stagles 8 Goodwood Close, Cowplain, Waterlooville PO8 8BG
G3 RCB N. Kingsley 1 Wensleydale Gardens, Hampton TW12 2LU
G3 RCD C. Brockbank 31 Park Hill, Church Crookham, Fleet GU52 6PW
G3 RCE R. Allbright 50 Portsdown Road, Portsmouth PO6 4QH
G3 RCM Sheffield ARC c/o D. Littlewood 50 Industry Road, Sheffield S9 5FQ

G3 RCQ D. Cole Amber Lights, Market Lane, Walpole St. Andrew, Wisbech PE14 7LT
G3 RCV Cray Valley RS c/o A. Styles 6 Hill Brow, Crayford, Dartford DA1 3NX
G3 RCW Worksop Amateur RS c/o A. Bostock 26 Ingham Road, Bawtry, Doncaster DN10 6NW
G3 RCX L. Gibson 7 Heycroft Road, Eastwood, Leigh-on-Sea SS9 5SW
G3 RCZ G. Thompson 22 Warton Avenue, Heysham, Morecambe LA3 2LX
G3 RDA S. Whitehead 98 Oak Road, Fareham PO15 5HP
GW3 RDB Royal Engineers Swindon c/o T. George 80 Yew Street, Troedyrhiw, Merthyr Tydfil CF48 4EE
G3 RDC A. Wood 14 Little Paradise, Marden, Hereford. HR1 3DR
G3 RDF J. Jeffrey Old Church Cottage, Ipsden, Wallingford OX10 6AE
G3 RDG K. Michaelson 40 The Vale, Golders Green, London NW11 8SG
G3 RDH J. Barnes 4 Deepdene Drive, Dorking RH5 4AD
G3 RDP H. Cutts 50 Cropton Road, Hull HU5 4LP
G3 RDQ D. Griffiths Upcote Cottage, Chilbolton, Stockbridge SO20 6BA
G3 RDR P. Rudwick 29 Fuller Street, London NW4 4RR
G3 RDX Radio Delta Xray c/o A. Fulton 8 Priest Hill Gardens, Wetherby LS22 7UD
G3 RDZ J. Walker 94 Keys Park, Parnwell Way, Peterborough PE1 4SN
G3 RE Travelling Wave CG c/o S. Dixon 33 Medhurst Crescent, Gravesend DA12 4HJ
G3 REB R. Cole Lyndale, Brimscombe Lane, Brimscombe, Stroud GL5 2RF
G3 RED D. Sylvester 10 Ivy Grove Gunthorpe, Peterborough PE4 7TW
G3 REH H. Neale Thornlea Fishergate, Spalding PE12 0EZ
G3 REL B. Woodfield 49 Oakfield Road, Blackwater, Camberley GU17 9DZ
G3 REP R. Parkes 2 Saxon Road, Steyning BN44 3FP
G3 REU G. Hearn 70 Cranmer St., Long Eaton, Nottingham NG10 1NL
G3 REV R. Pulling 410 Leach Lane, Sutton Leach, St. Helens WA9 4NA
G3 REW D. Morris 66 Windmill Close, Brixham TQ5 9SQ
GM3 RFA D. Garrington 3 Sutherland Avenue, Fort William PH33 6JS
G3 RFH K. Randall 25 Kingsway, Thornton-Cleveleys FY5 1DL
GD3 RFK D. Dodd Ellan Geay, Ballayockey Lane, Regaby, Isle of Man IM7 3HP
G3 RFN G. Wild 17 New Church Close, Clayton Le Moors, Accrington BB5 5GH
G3 RGB A. Moon 6 Troon Close, Saltersgill, Middlesbrough TS4 3HX
G3 RGC T. Matthews 38 Foxhill, Grimsby DN37 9QL
G3 RGD R. Dobdinson 73 Watwood Road, Hall Green, Birmingham B28 0TW
G3 RGE K. King Blandford Garrison A.R.C., Cole Block, Blandford Forum DT11 8RH
G3 RGJ R. Weston 43 Pearce Avenue, Parkstone, Poole BH14 8EG
G3 RGM D. Mullins Flat 23, Kennington Palace Court, Sancroft Street, London SE11 5UL
G3 RGN L. Binns Leamar, 707 Halifax Road, Cleckheaton BD19 6LJ
G3 RGP R. Pratt 1 Colebrooke Avenue, Ealing, London W13 8JZ
G3 RGS D. Thomson Skippers Down, Old Coach Road, Wrotham, Sevenoaks TN15 7NR
GM3 RGU J. Connelly 9 Glenhead Crescent Hardgate, Clydebank G81 6LW
G3 RHP J. Garrett 5 The Courtyard, Sudbourne Park, Sudbourne, Woodbridge IP12 2AJ
G3 RHQ K. Vickers Hillview, Barton-upon-Humber DN18 5DZ
G3 RHR K. Drinkwater Brearton Lodge, Brearton, Harrogate HG3 3BX
G3 RHU M. Stanbridge 183 Charlton Park Midsomer Norton, Radstock BA3 4BR
G3 RHW C. Cushion 3 The Copse, Bridgwater TA6 4DW
G3 RHZ A. Wilkinson 41 Church Road, Nailstone CV13 0QH
GW3 RIB W. Huxley No2 Bungalow, Nant Mawr Rd, Clwyd CH7 2BS
G3 RID D. Nancarrow 6 Trythogga Road, Gulval, Penzance TR18 3NA
GW3 RIH W. Elton 15 Main Avenue, Peterston-Super-Ely, Cardiff CF5 6LQ
G3 RIK D. Carden 9 Wood Hey Grove, Rochdale OL12 9TY
G3 RIM T. Emeney 10 Kilnside, Claygate, Esher KT10 0HS
G3 RIR N. Ackerley 24 Macaulay Road, Lutterworth LE17 4XB
G3 RIX Aylesbury Ryn Group c/o M. Tetley 87 Main Street, Irton, Scarborough YO12 4RJ
GW3 RIY A. Chapman 14 Birch Walk, Porthcawl CF36 5AN
G3 RJE J. Hunt 33 Rainhill Road, Rainhill, Prescot L35 4PA
G3 RJF I. Walker 28 Norrington Road, Maidstone ME15 9RA
G3 RJH Dr R. Harding High Trees, Arrowsmith Road, Wimborne BH21 3BG
G3 RJI A. Paul 3 Brunswick Avenue, Upminster RM14 1NA
G3 RJM R. Cutts 60 Holmpton Road, Withernsea HU19 2QD
G3 RJS P. Barry 235 Manor Way, Aldwick, Bognor Regis PO21 4HT
G3 RJT C. Garland 48 Underbank End Road, Holmfirth HD9 1ES
G3 RJV Rvd. G. Dobbs 9 Highlands, Littleborough OL15 0DS
G3 RKF T. Roeves 33 York Crescent, Wilmslow SK9 2BB
G3 RKH Rvd. J. Marshall 166 Calton Road, Gloucester GL1 5ER
G3 RKJ N. Summers 126 Kestrel House, 1 Alma Road, Enfield EN3 4QE
G3 RKK Dr A. Shepherd 59 Lime Avenue, Camberley GU15 2BH
G3 RKL Dr A. Whitaker 160 Derbyshire Lane, Sheffield S8 8SE
G3 RKM J. Meaker 11 Woodend View, Mossley, Ashton-under-Lyne OL5 0SN
G3 RKQ A. Balmforth Leam Brink, Lutton Gowts Lutton, Long Sutton, Spalding PE12 9LQ
GW3 RKV R. Volck Maes-Y-Bryn, Rosebush, Clynderwen SA66 7QS
G3 RKZ B. Tibbert 99A Main Street Horsley Woodhouse, Ilkeston DE7 6AW
G3 RLA C. Phillips Bella Vista, The Moorings, Wirral CH60 9JT
G3 RLD R. Ramshaw 132 Main Road, Duston, Northampton NN5 6RA
G3 RLE B. Turner 56 Bamford Way, Rochdale OL11 5NB
G3 RLF D. Price 8 Newland Road, Droitwich WR9 7AF
G3 RLJ J. Harper East View, 35 John Street, Sutton in Ashfield NG17 4EN
G3 RLL D. Grindell 23 Park Hall Avenue, Walton, Chesterfield S42 7LR
G3 RLO D. Cadman 32 Breedon Hill Road, Derby DE23 6TG
G3 RLT W. Stewart 4 Denmark Road, Kingston upon Thames KT1 2RU
G3 RLV M. Vann 3 Mile Planting, Richmond DL10 5DB
G3 RLX N. Taylor 146 Morledge, Matlock DE4 3SD
G3 RMD F. Regan 7 Hilltop Road, Cheltenham GL504NW
G3 RMF B. Magill 14 Barry Street, Worcester WR1 1NR
GW3 RMJ P. Jennings Myrtle Cottage, Harpers Lane, Presteigne LD8 2AN
G3 RMK R. Ruaux Park View, Wallage Lane, Crawley RH10 4NG
G3 RMN M. Smith 121 Shirley Way, Croydon CR0 8PN
G3 RMQ J. Ingham High Croft, 1 Layton Crescent, Leeds LS19 6RJ
G3 RMX W. Hall 52 Barley Gate, Leven, Beverley HU17 5NU
G3 RMY J. Andrews 12 Gerald Close, Burgess Hill RH15 0NB
G3 RMZ A. Pink 37 Shute Park Road, Plymouth PL9 8RB
GI3 RNO P. Greenan 9 Ashville Park, Antrim BT41 1HH
G3 RNP J. Price 125 Oakfield Road, Malvern WR14 1DT
G3 RNX W. Walker 44 South Road, Weston-Super-Mare BS23 2HE
G3 ROC R. Collins Thorn Acacia, Rye Road Northiam, Rye TN31 6NJ
G3 ROD R. Davenport 7 Nether Close, Duffield, Belper DE56 4DR
G3 ROG G. Morgan 22 Monks Road, Winchester SO23 7EQ
G3 ROM B. Sweetman Cortijasa Los Perez 220, Benajarafe, Malaga 29790 Spain
G3 ROO I. Keyser Rosemount, Church Whitfield Road, Whitfield, Dover CT16 3HZ
G3 ROP M. Goodrick 18 Milford Street, Cambridge CB1 2LP
G3 ROQ East Kilbride ARS c/o R. Gill 45 Biggin Lane, Ramsey, Huntingdon PE26 1NB
G3 ROS H. Williams Roslyn, Whalley Road, Burnley BB12 7HT
G3 ROW S. Smith Po Box 2738, Silver City 88062 United States
G3 RPA J. Knowles Springhill, Gilpins Ride, Berkhamstead HP4 2PD
G3 RPB K. Spicer Grove Cottage, Dallinghoo, Woodbridge IP13 0LR
G3 RPD G. Clinch 2 Storrs Close, Bovey Tracey, Newton Abbot TQ13 9HR
G3 RPL T. Neyland 22 Pax Hill, Bedford MK41 8BT
GM3 RPM J. Mcavoy 120 Donaldswood Road, Paisley PA2 8EB
G3 RPO F. Seddon 23 Countessway, Euxton, Chorley PR7 6PT
G3 RPV T. Venn 22 Eaton Close Hartford, Huntingdon PE29 1SR
G3 RPZ H. Trunley Bijou House, 75 Belgrave Road, Leigh-on-Sea SS9 5EL
G3 RQF D. Keith 108 Lower Northam Road Hedge End, Southampton SO30 4FT
GM3 RQQ H. Robertson 102 Orchy Crescent, Bearsden, Glasgow G61 1RE
G3 RQR N. Kirtley 14 Byron Avenue, Winchester SO22 5AT
G3 RQS R. Rimmer 25 Haig Court, Chesterton, Cambridge CB4 1TT
GI3 RQU Dr S. Laverty 572 Antrim Road, Belfast BT15 5GL
G3 RQX P. Lewis 20 Osborne Road, Penn, Wolverhampton WV4 4AY
G3 RQZ P. Madagan 40 Lagham Park, South Godstone, Godstone RH9 8ER
G3 RR Hucknall Rolls Royce ARC c/o S. Sorockyj 8 Bowden Avenue, Bestwood Village, Nottingham NG6 8XN
G3 RRG P. Taylor 44 Leegate Road, Stockport SK4 4AX
G3 RRI R. Wilmot Swinside, Branthwaite Lane, Workington CA14 1HE
G3 RRM J. Hughes 41 Highfield Avenue, Great Sankey, Warrington WA5 2TW
G3 RRN Dr K. Jones Field House, Wragby Road, Lincoln LN2 2QU
G3 RRP R. Pine 21 Hatherden Avenue, Poole BH14 0PJ
G3 RRS Rutherford Appleton Laboratory ARC c/o J. Wright 2 Barnfield, Charney Bassett, , Wantage OX12 0HA
G3 RRW J. Francis 5 Central Park Avenue, Plymouth PL4 6NW
G3 RSB R. Scaife 7 Woodgates Close, North Ferriby HU14 3JS
G3 RSC Sutton Coldfield RS c/o B. Adkins 4 Orion Close, Ward End, Birmingham B8 2AU
G3 RSE C. Cheney 35 Metcalfe Road, Cambridge CB4 2DB
G3 RSF A. Notschild 8 Hillpark, Buckland Brewer, Bideford EX39 5HY
G3 RSI F. Mckeracher Wickets, 1 Marshal Close, Alton GU34 1RA
G3 RSM F. Burnett 4 Woodlands Drive Fulwood, Preston PR2 9SQ
G3 RST R. Southern 30 Barnfield, Crowborough TN6 2RY
G3 RSU D. Bindon Forth House, Water Street, Langport TA10 0HH
G3 RSV R. Dowsett 23 South Wootton Lane, King's Lynn PE30 3BS
G3 RSW W. Mullarkey The Barn, Skull House Lane, Wigan WN6 9DJ
G3 RTB R. Bell 14 Wacker Field Road, Rendlesham, Woodbridge IP12 2UT
G3 RTD J. Gailer Shelleys, King Stag, Sturminster Newton DT10 2BE
G3 RTE G. Kellaway 55 Ladbrooke Drive, Potters Bar EN6 1QW
G3 RTM A. Chaddock 4 Liddiard Close Kennington, Oxford OX1 5RY
G3 RTO N. Pratt 87A Dovecote Road, Newthorpe, Nottingham NG16 3QL
G3 RTP J. Pennington Brambling, Forest Road, Waterlooville PO7 6UE
G3 RTY Thorpe Camp Museum Radio Group c/o H. Meers 10 Lawnswood Avenue, Chasetown, Burntwood WS7 4YD
G3 RUD E. Workman Sunset, 2 Burnham Drive, Weston-Super-Mare BS24 9LW
GW3 RUE E. Edwards Ceris, Ruthin Road, Denbigh LL16 3EU
G3 RUG G. Twiss 9 Brae Head, Eaglescliffe, Stockton-on-Tees TS16 9HP
G3 RUH J. Miller 3 Bennys Way, Coton CB23 7PS
GM3 RUI R. Furness 43 Glebe Street, Leven KY8 4QN
G3 RUJ R. Powell 13 Bridges Drive, Bristol BS16 2UB
G3 RUO W. Williamson 84 Atfield Drive, Whetstone, Leicester LE8 3NE
GM3 RUP C. Morton 295 Byres Road, Hillhead, Glasgow G12 8TL
G3 RUV A. James 4 The Chestnuts, Aylesbeare, Exeter EX5 2BY
G3 RUZ D. Martin 4 Chilbolton Mews, 19 Chilbolton Avenue, Winchester SO22 5HU
G3 RVA R. Crowe 37 Huccaby Close, Brixham TQ5 0RJ
G3 RVC Prof. P. Cochrane Willow Barn, East Lane, Woodbridge IP13 6EB
GW3 RVG S. Sedgebeer 50 Minffrwd Road Pencoed, Bridgend CF35 6SD
G3 RVI J. Walch 52 Marsh House Road, Sheffield S11 9SP
GM3 RVL Dr H. Brash 5 Hillview Drive, Edinburgh EH12 8QW
G3 RVM C. Trusson 27A Roman Way, Thatcham RG18 3BP
G3 RVS G. Haynes Littleton House Cottage Blandford St. Mary, Blandford Forum DT11 9NB
G3 RVX J. Colegate 1 Oldmere Cottages High Street, Bathford, Bath BA1 7TJ
G3 RVY P. Colegate 65 Forest Road, Melksham SN12 7AB
G3 RWE T. Yates 3 Sycamore Crescent, Macclesfield SK11 8LL
G3 RWF P. Henwood Conifers, Church Road, Littlebourne, Canterbury CT3 1UA
G3 RWI Dr P. Cross Home Farm House, Icomb, Cheltenham GL54 1JD
G3 RWL R. Limebear 60 Willow Road, Enfield EN1 3NQ
G3 RWV M. Sanders 7 Netherby Close, Tring HP23 5PJ
G3 RWW G. Southern 27 Eldred Road, Liverpool L16 8NZ
GW3 RWX D. Thomas 88 Cefn Graig, Rhiwbina, Cardiff CF14 6JZ
G3 RXA J. Thomas Blair House, Market Place, Norwich NR16 2AN
G3 RXG R. Burgess 11 Beech Road, Shipham, Winscombe BS25 1SA
G3 RXI E. Blundell 7 Admiral Way, Alton GU34 1GH
G3 RXO R. Brown Lower Dicker, Hailsham BN27 4BG
G3 RXP D. Mason 5 Spa Top, Caistor, Market Rasen LN7 6RB
G3 RXS W. Scarlett 14 Warren Drive, Bingley BD16 3BX
GM3 RXU Prof. I. Macpherson 1 Broomie Dell, Earlston TD4 6BN
GI3 RXV N. Graham 3 Shilgrove Place, Castledawson, Magherafelt BT45 8AL
GM3 RXZ R. Marshall 52 Lumsdaine Drive, Dalgety Bay, Dunfermline KY11 9YU
GW3 RYE J. Harris Treweryn, Llwyndafydd Road, Llandysul SA44 6BT
G3 RYH J. Brodie Huntlands Farm, Gaines Road, Worcester WR6 5RD
G3 RYK I. Grayson 156 Little Brays, Harlow CM18 6EY
G3 RYP D. Craggs New House, Dacre Banks, Harrogate HG3 4EW
GW3 RYR Dr C. Morgan 33 West Grove, Merthyr Tydfil CF47 8HJ
G3 RYW D. Wardlaw 21 Tormey Street, Balwyn North VIC 3104 Australia
G3 RYZ M. Byrne 16 Downham Gardens Tamerton Foliot, Plymouth PL5 4QE
G3 RZC R. Pellett La Biochere, Aizenay 85190 France

G3 RZF D. Horton 26 The Crescent, Slough SL1 2LQ
G3 RZG M. Box 18 Stottingway Street, Weymouth DT3 5QA
G3 RZI M. Moss 1082 Evesham Road, Astwood Bank, Redditch B96 6ED
G3 RZJ G. Hall 185 Dialstone Lane, Stockport SK2 7LQ
G3 RZP P. Chadwick Three Oaks, Braydon, Swindon SN5 0AD
G3 RZV A. Lawrance 97 Dorchester Road, Oakdale, Poole BH15 3QZ
G3 RZY C. Abrey 31 Yew Tree Lane, Leeds LS15 9JD
G3 SAD Stevenage & District A.R.S c/o R. Mctait 20 Rowland Road, Stevenage SG1 1TE
GM3 SAE R. Mcmillan 54 Birchwood, Invergordon IV18 0BG
G3 SAH R. Matthews 2, Newton Close, Oakenshaw South, Redditch B98 7YR
GM3 SAN S. Weir 19 Ellismuir Road, Baillieston, Glasgow G69 7HW
G3 SAO Dr J. Midgley 3 Chipping Fold, Milnrow, Rochdale OL16 4YD
G3 SAR R. Warner Cubs Wood, Rycroft Lane, Sevenoaks TN14 6HT
GM3 SBC E. Murphy 65 Silverknowes Crescent, Edinburgh EH4 5JA
G3 SBF S. Eames 4 Dabey Close, Markfield LE67 9UJ
G3 SBL Stafford & Districts ARS c/o G. Reay 53 Tithe Barn Road, Stafford ST16 3PL
G3 SBM D. Turner 50 Hardings, Chalgrove, Oxford OX44 7TJ
G3 SBP R. Gynn Honeywood Belvidere Road, Exeter EX4 4RR
G3 SBT W. Turnbull 15 Marshallsay Road, Chickerell, Weymouth DT3 4BB
G3 SCB Rvd. R. Hinder 18 Mapledale Avenue, Croydon CR0 5TB
G3 SCD D. Dunn Mizpah Villa, 9 Low Toynton Road, Horncastle LN9 5LL
G3 SCJ D. Power 47 Marlborough Street, Gainsborough DN21 1BT
G3 SCL R. Houghton Hans-Miederer-Str. 10B, Schliersee 83727 Germany
GI3 SCM T. Mccullough 16 Mccormack Gardens, Lurgan, Craigavon BT66 8LE
G3 SCT Thurrock Sea Cadet Corp - Tilbury c/o Dr N. Wilkinson 12 Woodlands Close, Grays RM16 2GB
G3 SCV Rvd. G. Stanton 8 Kennett Close, Norwich NR4 7JA
G3 SCZ R. Brown 22 Lordswood Silchester, Reading RG7 2PZ
G3 SDC De Montfort University ARS c/o R. Titterington Wyclif House, St. Marys Road, Lutterworth LE17 4PS
G3 SDG J. Bottom 48 Chesterton Avenue, Harpenden AL5 5SU
G3 SDH P. Kelly Martyndale, The Street, Compton Martin, Bristol BS40 6JE
G3 SDL D. Court Connogue, River Lane, Shankill D18 W2R4 Ireland
G3 SDO K. Heathfield 2 Georgian Close, Broadway, Weymouth DT3 5PF
G3 SDS South Dorset Rs c/o G. Watts 3 Maple Grove Knightsdale Road, Weymouth DT4 0FE
G3 SDT D. Allen Chelsea Cottage, The Turnpike, Carleton Rode, Norwich NR16 1RS
G3 SDW K. Underwood Apartment 7, Imperial Court, Palermo Road, Babbacombe TQ1 3NW
G3 SDY G. Edinburgh 77 Westerley Lane, Shelley, Huddersfield HD8 8HP
G3 SEA P. Perretta 1511 Punahou St., Apt 208, Honolulu 96822 United States
G3 SED E. Devereux 191 Botley Road, Burridge, Southampton SO31 1BJ
G3 SEF R. Frew Sawley House, 82 Wormholt Road, London W12 0LP
G3 SEG W. Gordon 55 Trajan Avenue, South Shields NE33 2AN
G3 SEJ E. John Obo St. Dunstans Ars, 52 Broadway Avenue, Wallasey CH45 6TD
GM3 SEK Dr I. White 2 Appleby Cottages, Whithorn, Newton Stewart DG8 8DQ
G3 SEM P. Cort-Wright 11-13 Hardingham Street Hingham, Norwich NR9 4JB
G3 SEQ J. Crossfield Forest Lodge, Chopwell Wood, Rowlands Gill NE39 1LT
GM3 SER H. Bremner 2 Rowan Crescent, Lenzie, Glasgow G66 4RE
G3 SES P. Stevens 20 Abbots Park, Chester CH1 4AN
G3 SET G. Aram 5 Lancaster Green, Hemswell Cliff, Gainsborough DN21 5TQ
G3 SEY R. Mackey 44 South Street, Ossett WF5 8LF
G3 SEZ P. Luft Swan House, Livesey Road, Ludlow SY8 1EY
G3 SFB C. Hale 16 Windmill Court, East Wittering, Chichester PO20 8RJ
GW3 SFC A. Richards 30 Well Place, Aberdare CF44 0PB
G3 SFE P. Everett 58 Greenwood Avenue, Bognor Regis PO22 9EX
G3 SFG Southgate ARC c/o D. Berry 4 Holly Hill, Winchmore Hill, London N21 1NP
G3 SFK P. Kerry 251 Upper Rainham Road, Hornchurch RM12 4EY
GW3 SFQ R. Mugford 27 Highfield Close, Dinas Powys CF64 4LR
G3 SFU P. Woodfield 49 Oakfield Road, Blackwater, Camberley GU17 9DZ
G3 SFV E. Meachen 46 Rainsborough Gardens, Market Harborough LE16 9LW
GI3 SG J. Patty 3C Finwood Park, Belfast BT9 6QR
G3 SGA A. Jones Po Box 355, Crestholme 3652, Natal South Africa
G3 SGC G. Morris Norcrest, Beach Road, Norwich NR12 0AL
G3 SGF P. Casemore 9 Wellcroft Cottages, Church Lane, Albourne, Hassocks BN6 9BZ
GW3 SGK Dr B. King Ty Derwen Vinegar Hill, Undy, Caldicot NP26 3EJ
G3 SGL A. Isaacs Holme View, Brick Lane, Christchurch BH23 8DU
G3 SGR Dr J. Craig Ferndown, Tilley Lane, Hailsham BN27 4UT
G3 SGV J. Fallon 8 Tretower Close, Plymouth PL6 6BH
G3 SGX R. Bona 1 Maxwell Road, Broadstone BH18 9JG
G3 SGY A. Nesbitt 28 Fairfax Road, Middleton St. George, Darlington DL2 1HF
G3 SGZ T. Chapple 39 Maynards Park, Bere Alston, Yelverton PL20 7AR
G3 SHD L. Dray 1 Chalfont Close, Bradville, Milton Keynes MK13 7HS
G3 SHF B. Naylor 47 Chester Road Poynton, Stockport SK12 1HA
G3 SHK R. Pett 5 Kingford Close, Woodfalls, Salisbury SP5 2NQ
G3 SHL J. Harlow 3 The Fairways, Sherford, Taunton TA1 3PA
GM3 SHR J. Coster 17 Glamis Place, Dalgety Bay, Dunfermline KY11 9UA
G3 SHX R. West 15 St. Andrews Close, Margate CT9 4HA
G3 SHY University of York ARC c/o R. Cottrell 157 Ridge Lane, Watford WD17 4SU
G3 SHZ Dr J. Whittington Twyford Manor, Bicester Road, Buckingham MK18 4EL
G3 SIA B. Keyte 9 Swanns Meadow, Bookham, Leatherhead KT23 4JX
G3 SIG Rsars - Royal Signals ARS c/o A. Watt 5 Brambling Road, Horsham RH13 6AX
GW3 SIK K. Pugh Tanybanc, Blaenporth, Cardigan SA43 2BD
G3 SIR D. Durham 29 Waverley Road, Stratton St Margaret, Swindon SN3 4AY
G3 SIT R. Kressman 12 School Lane, Fenstanton, Huntingdon PE28 9JR
G3 SIU P. Hearson 14 Osgood Gardens, Orpington BR6 6JU
G3 SJH Dr C. Eyles 9 St. Peters Road, Harborne, Birmingham B17 0AT
G3 SJI M. Batt 9 Grange Park, Westbury-On-Trym, Bristol BS9 4BU
G3 SJJ C. Burbanks 16 Cotgrave Road, Plumtree, Nottingham NG12 5NX
G3 SJK S. Cherry 4 West Hill Road South, South Wonston, Winchester SO21 3HP
G3 SJR W. Tynan 22 Belchmire Lane, Gosberton, Spalding PE11 4HG
G3 SJW S. Haigh 29 West Street, Chichester PO19 1QS
G3 SJX P. Hart The Willows, Paice Lane, Medstead, Alton GU34 5PR
GM3 SJY C. Lawrenson Hollyburn, West Port, Cupar KY15 7BW
G3 SKI R. Bravery 19 Lindum Road, Worthing BN13 1LX
G3 SKN D. Naylor 52 Rue Du Port, Pontorson 50170 France
G3 SKR A. Gold 60 Wynnstay Gardens, London W8 6UU
G3 SKV S. Hobday 31 Sackville Crescent, Harold Wood, Romford RM3 0EJ
GD3 SKZ K. Manktelow Tramman House Main Road, Ballabeg IM94HA Isle of Man
G3 SLI A. Osborne 18A Cumnor Road, Boars Hill, Oxford OX1 5JP
G3 SLJ D. Parsons Am Dorfplatz 12, Winden Am Aign 85084 Germany
G3 SLK R. Pickering 147 Windermere Avenue, Nuneaton CV11 6HN
G3 SLL H. Tyreman U1 11Jenaya Place, Labrador 4215 Australia
G3 SLS A. Hancock 6 The Fairway, Mablethorpe LN12 1LL
G3 SLT D. Ormerod 21 Valletta Close, Chelmsford CM1 2PT
G3 SLX J. Smith 256 Stone Road, Stoke-on-Trent ST4 8NJ
G3 SMD R. Turner 7 Paddocks Lane, Cheltenham GL50 4NU
G3 SMF I. Hamill 74 Lampits Hill, Corringham, Stanford-le-Hope SS17 9AJ
G3 SMN R. Forster 28 Springbridge Road, Manchester M16 8PW
GW3 SMT P. Torry Pen-Y-Rhos Old Racecourse, Oswestry SY10 7HP
G3 SMV J. Smith 18 Hounslow Road, Mackworth, Derby DE22 4BW
G3 SMZ R. Hill 68 Chestnut Street, Chadderton, Oldham OL9 8HH
G3 SNA S. Andrew Berry Brow House, Berry Brow, Greenfield, Oldham OL3 7EJ
GJ3 SND B. Walster Le Ponterrin Cottage, La Rue Du Ponterrin, St. Saviour JE2 7HP Jersey
G3 SNG A. Ambler 12 Oakdene Road, Marple, Stockport SK6 6PJ
G3 SNH W. Harrison 44 Briar Road, Thornton-Cleveleys FY5 4NB
G3 SNN A. Woolford 39 Apple Orchard, Prestbury, Cheltenham GL52 3EH
G3 SNO G. Smith Stoneycroft, Godsons Lane, Napton, Southam CV47 8LX
G3 SNP M. Pitcher Sandycot, Cadsden Road, Princes Risborough HP27 0NB
G3 SNR G. Morgan Eaton House, Eaton Bank, Belper DE56 4BH
G3 SNT R. Dixon Copper Beeches, Witton Gilbert, Durham DH7 6TW
G3 SNU K. Selleck Westphalia, Dartington, Totnes TQ9 6DJ
G3 SOA W. Mccartney 158 Windingbrook Lane, Northampton NN4 0XF
G3 SOE R. Jennings 31 Copper Beech Drive, Wombourne, Wolverhampton WV5 0LH
GI3 SOO M. Foley 5 Woodland Drive, Cookstown BT80 8PL
G3 SOU Southampton ARC c/o M. Troy 22 Jackie Wigg Gardens Totton, Totton, Southampton SO409LZ
GW3 SPA R. Alban 73 Plymouth Road, Penarth CF64 3DD
G3 SPI I. Dawe 10 Selsden Close, Elburton, Plymouth PL9 8UR
G3 SPJ C. Wooff 55 Bostall Hill, Abbey Wood, London SE2 0QX
G3 SPL P. Lee 10 Antony Gardner Crescent, Whitnash, Leamington Spa CV31 2TQ
G3 SPN N. Collins Flat 1, Vista Mare West, 44 West Parade, Worthing BN11 5EF
G3 SPO P. Oneill Recreation Cottage, Slad Road, Stroud GL6 7QA
G3 SPP A. Minett 45 Patterdale Drive, Worcester WR4 9HS
GM3 SPT G. Mckay 152 Inveresk St., Greenfield, Glasgow G32 6TA
G3 SPV K. Richardson Brookfield Grove, The Dukes Drive, Bakewell DE45 1QQ
G3 SPY Echelford ARS c/o R. Harris Clevelands, Tamworth Road, Coventry CV7 8JJ
G3 SQN J. Grant 8 Thornhill Way, Mannamead, Plymouth PL3 5NP
G3 SQO D. Best Tanglewood, Showley Road, Blackburn BB1 9DP
G3 SQQ J. Franks 11 Thoresby Avenue, Kirkby-In-Ashfield, Nottingham NG17 7LY
G3 SQU C. Clarke 14 Woodlea Gardens, Newcastle upon Tyne NE3 5BY
G3 SQX E. Taylor 115 St. Albans Avenue, London W4 5JS
G3 SRC Surrey Radio Contact Club c/o Q. Collier 19 Grangecliffe Gardens, South Norwood, London SE25 6SY
GW3 SRF D. Woolen Rose Cottage, Newcastle Hill, Bridgend CF31 4EY
GW3 SRG A. Peake 70 Higher Lane, Langland, Swansea SA3 4PD
G3 SRJ G. Carlisle 3 Grimms Meadow, Walters Ash, High Wycombe HP14 4UH
GW3 SRM S. Hulme 64 Salem Street, Amlwch LL68 9BT
G3 SRN G. Bray 4 Ledway Drive, Wembley HA9 9TQ
G3 SRP G. Lewellen 11 High Street, Braunston, Daventry NN11 7HR
G3 SRQ R. Bisseker 24 Millgate, High Wycombe HP11 1GL
G3 SRR M. Rees 83 Salisbury Road, Farnborough GU14 7AE
G3 SRT Salop ARS c/o R. Golding 7 Belvidere Avenue, Shrewsbury SY2 5PF
GM3 SRV R. Tatton 17 Paties Road, Edinburgh EH14 1EF
G3 SRX N. Down 23 Christopher Close, Heckington, Sleaford NG34 9SA
GW3 SSK J. Williams 5A Derllwyn Close, Tondu, Bridgend CF32 9DH
G3 SSN J. Brand 133 Hatfield Avenue, Fleetwood FY7 7DU
G3 SSW Dr S. Erents 50 Blandy Avenue, Southmoor, Abingdon OX13 5DB
G3 SSZ L. Lavelle 49 Jones Road Goffs Oak, Waltham Cross EN7 5JT
G3 STF P. Sandiford 11 Calle Menendez Pelayo, San Javier 30730 Spain
G3 STJ Keighley Ray Gr c/o B. Riley 2 Watson Street, Swinton, Manchester M27 6AQ
GM3 STM S. Mitchell 97 Barbieston Road, Auchinleck KA18 2ED
G3 STP P. La Pierre 42 Berry Vale, South Woodham Ferrers, Chelmsford CM3 5GY
G3 STT W. Haynes 37 Hawthorn Grove, Southport PR9 7AA
G3 STZ C. Thorn 4 Riveredge, Framilode, Gloucester GL2 7LH
G3 SUA E. Winstanley 1 Drews Court Churchdown, Gloucester GL3 2LD
GW3 SUH K. Hughes 2 Graig Terrace, Ferndale CF43 4EU
G3 SUI J. Burrows 68 Grosvenor Road, Sale M33 6NW
G3 SUK M. Baker 8 Wynton Rise, Stowmarket IP14 2AB
G3 SUL D. Waller 66 Wallace Drive, Dunstable LU6 2DF
G3 SUN G. Hodgkinson Cfone Communications, 9 Adler Industrial Estate, Betam Road, Hayes UB3 1ST
G3 SUS S. Jacobs 16 Mayfield Park, Thorley, Bishop's Stortford CM23 4JL
G3 SUX D. Bradshaw 25 Meare Close, Tadworth KT20 5RZ
GM3 SUZ D. Mclean Whitecroft Farm, Barrs Brae, Port Glasgow PA14 5QG
G3 SVC Spen Valley ARS c/o J. Wilde 2 Bottoms Lane, Birkenshaw, Bradford BD11 2NN
G3 SVD A. Hewitt Mbe Redwood House, Adbury Holt, Newbury RG20 9BW
G3 SVI D. Davis 188 Eastwood Old Road, Leigh-on-Sea SS9 4RY
G3 SVJ Merseyside ARS c/o A. Barter 503 Northdown Road, Margate CT9 3HD
G3 SVQ A. Yallop Whitehill, 16 High Street, Bedford MK43 7JX
G3 SVR Royal Signals Museum c/o E. Churchyard 11 Greenfields Drive, Bridgnorth WV16 4JW
G3 SVW R. Smith 16 Coniston Avenue, Sale M33 3GT
G3 SVZ P. Laxton 52 Reddington Road, Plymouth PL3 6PT
G3 SWC B. Tinton 1 Bridge Road, Rudgwick, Horsham RH12 3HD
G3 SWH P. Whitchurch 21 Dickensons Grove, Congresbury, Bristol BS49 5HQ
GM3 SWK A. Shearer 12 Coolin Drive, Portree IV51 9DN
G3 SWU T. Heeley 34 Worlaby Road, Scartho Top, Grimsby DN33 3JT
G3 SWW H. Cooper 9 Fortyfoot, Bridlington YO16 7SA
G3 SXA J. Croft 14 Stanstead Road, Forest Hill, London SE23 1BW
G3 SXC A. Critchley 39 Westcliffe, Great Harwood, Blackburn BB6 7PH
G3 SXE L. Lethbridge 24 Furze Road, High Salvington, Worthing BN13 3BH

G3 SXH A. Henderson 50 Sylvan Road, Exeter EX4 6EY
G3 SXI D. Ashmore Flat 8, Carter Bench House, Clarence Road, Macclesfield SK10 5JZ
G3 SXP Capt. J. Redford Woodgates Harling Road, Gt Hockham IP24 1NP
G3 SXQ E. Rockett Devoran, The Causeway, Highbridge TA9 4QT
G3 SXR A. Read Readymoney Cove, Fowey PL23 1JH
G3 SXT Dr P. Sweeny 1A Market Street, Eckington S21 4EG
G3 SXV B. Vincent 18 Rowanhayes Close, Ipswich IP2 9SX
G3 SXW R. Western 7 Field Close, Chessington KT9 2QD
G3 SYA D. Ashworth 31 Belmont Avenue, Ribbleton, Preston PR2 6DH
G3 SYB H. Barker Azul Avion, Main Road, Alford LN13 0JP
G3 SYC Pontefract + District ARS c/o N. Ferguson Royd Moor, Royd Moor Lane, Badsworth, Pontefract WF9 1AZ
G3 SYD S. Beauchamp 1 Gosden Close, Furnace Green, Crawley RH10 6SE
G3 SYM D. Coltart The Sycamores St. Clether, Launceston PL15 8PP
G3 SYS Dr D. Emerson 4269 N.Soldier Trail, Tucson 85749 United States
G3 SYZ A. Rogers Draycott, Primrose Hill, Hastings TN35 4DN
G3 SZ King's Lynn ARC c/o E. Haskett 23 Gloucester Road, King's Lynn PE30 4AB
G3 SZE J. Evrall 38 Eastlang Road Fillongley, Coventry CV7 8ER
G3 SZF R. Frost 24 Mount Pleasant, Hertford Heath, Hertford SG13 7QU
G3 SZG J. Wright Flat 4 39A Sion Hill, Kidderminster DY10 2XT
G3 SZJ M. Shardlow 19 Portreath Drive, Allestree, Derby DE22 2BJ
G3 SZM J. Wuille 45 Keymer Crescent, Goring-By-Sea, Worthing BN12 4LD
G3 SZR C. Davis 148 Birkbeck Road, Beckenham BR3 4SS
G3 SZU K. Radford Birch Lodge, Mill Hill Lane, March PE15 9QB
G3 SZV B. Ward 138 County Road, Ormskirk L39 1NN
G3 SZY G. Douglas 169 High Street, Cheveley, Newmarket CB8 9DG
G3 TA C. Lambert Stonecroft Notch Rd Winstone, Cirencester GL7 7JU
G3 TAA K. Jessop 15 Courtenay Gardens, Newton Abbot TQ12 1HS
GI3 TAC D. Campbell 56 The Quay Killyleagh, Downpatrick BT30 9GB
G3 TAF D. Cassere 9 St. Marys Garth, East Keswick, Leeds LS17 9ER
G3 TAI C. Ward 50 Lakeside, Bracknell RG42 2LE
G3 TAJ R. Marchant Cascade, The Street, Canterbury CT3 1LN
GM3 TAL M. Hamilton 3 Charles Court, Limekilns, Dunfermline KY11 3LG
G3 TAO W. Eaton 8E St. Aubyns Road, London SE19 3AD
G3 TAQ N. Bullock 29 St. Marys Road, Stowmarket IP14 1LP
G3 TAW C. Wood 22 Habberley Road, Kidderminster DY11 6AA
G3 TAX J. Boydell 13 Lynch Road, Farnham GU9 8BZ
G3 TAY North Bristol ARC c/o A. Yarker 6 Moor Top Road, Halifax HX2 0NP
G3 TAZ R. Davies 69 Stopsley Way, Luton LU2 7UU
G3 TBF H. Wilkins 17 Bathleaze, Kings Stanley, Stonehouse GL10 3JN
G3 TBG G. Goulbourn 41 Rutland Road, Stamford PE9 1UP
G3 TBJ C. Webster Wayside, Rosenithon St. Keverne, Helston TR12 6QR
G3 TBK J. Cree 24 Old Lincoln Road, Caythorpe, Grantham NG32 3EJ
G3 TBL R. Ashman 85A Fakenham Road, Great Ryburgh, Fakenham NR21 7AQ
G3 TBW T. Westbury 6 Ellerdene Close, Redditch B98 7PW
G3 TCG M. Trundle 20 Denehurst Gardens, Hastings TN35 4PB
G3 TCI A. Bye 7 Larkfield Avenue, Gillingham ME7 2LN
G3 TCL M. Dawson 11 St. Georges Close, Brampton, Huntingdon PE28 4US
G3 TCO Dr A. Preece 12 South Dene, Bristol BS9 2BW
G3 TCT G. Kimbell Eastfield Farmhouse, Fair Place West Lydford, Somerton TA11 7DN
G3 TCU P. Guttridge 33 Franklyn Road, Godalming GU7 2LD
GW3 TCV J. Edwards Pen Y Maes, Trehelig, Welshpool SY21 8SG
GM3 TCW J. Kelly 144A Manse Road, Newmains, Wishaw ML2 9BL
G3 TCY J. Lewis 10 Sheringham Drive, Etchinghill, Rugeley WS15 2YG
G3 TCZ R. Freeman 3 Hoffmann Gardens, South Croydon CR2 7GE
G3 TDH R. Stevens 19 Canberra Road, Bramhall, Stockport SK7 1LG
G3 TDL R. Davis 105 Eldred Avenue, Brighton BN1 5EL
G3 TDM R. Mason 19 Lawrence Road, St. Agnes TR5 0XQ
G3 TDT I. Hollingsbee 89 Swift Road, Abbeydale, Gloucester GL4 4XJ
G3 TEB G. Addis 1 Smallridge, Newbury RG20 0LH
G3 TEC T. Rutherford 17 Rosedale Avenue, Sunderland SR6 8BD
G3 TEE F. Stork 20 Gay Meadows, Stockton On The Forest, York YO32 9UJ
G3 TEH A. Storey 23 Foster Street, Barnsley S70 3EW
G3 TEI Greater Manchester Raynet c/o J. Gaffney 77 South Street, Pennington, Lymington SO41 8DY
G3 TEL P. Mcpherson 2 Osborne Place, Lower Street, Merriott TA16 5NP
G3 TEP B. Atkinson 20 King Street, Seahouses NE68 7XP
G3 TEU A. Sherer 35 Beverley Road, Willerby, Hull HU10 6AW
G3 TEV M. Mills Shepton, 3 Tylers Way, Stroud GL6 8ND
G3 TEX P. Painter 1 Linden House, Barkleys Hill, Bristol BS16 1FB
G3 TFA G. Whenham Hogs Hollow, Welsh Road East, Southam CV47 1NF
G3 TFF G. Fuller 99 Stanbury, Keighley BD22 0HA
G3 TFL G. Rogers 19 Manor Road, Henley-on-Thames RG9 1LT
G3 TFO J. Auty 64 Ainley Road, Huddersfield HD3 3QX
G3 TFR J. Hardstone 17 Whitefield, Stockport SK4 2PE
G3 TFV E. Tokley 14 Maple Way, Earl Shilton LE9 7HW
G3 TFX R. Fusniak 35 High Street, Burwell CB25 0HD
GM3 TFY D. Guest 31 Newmills Crescent, Balerno EH14 5SX
G3 TGB B. Ely 375 Cressing Road, Braintree CM7 3PE
G3 TGD M. Allenson 4 The Orchard, Powick, Worcester WR2 4SE
G3 TGE D. Cahill Church Farm, Church Road, West Beckham, Holt NR25 6NX
G3 TGF C. Bonner 57 Downsview, Heathfield TN21 8PF
GM3 TGG T. Gratton 23 Culhorn Road, Stranraer DG9 8DB
G3 TGL A. Fantham 52 Calverley Road, Kings Norton, Birmingham B38 8PW
G3 TGN Dr P. Collar 5 Oak Tree Lane, Tavistock PL19 9DA
G3 TGO B. Vaughan 7 Oundle Road, Chesterton, Peterborough PE7 3UA
G3 TGS Worthing & District Scout ARC c/o N. Thrower 8 Upton Gardens, Worthing BN13 1DA
G3 THC Frst of Dean Ry c/o D. Stimson 94 Casterton Road, Stamford PE9 2UB
G3 THF B. Mchugh 283 Coppice Road, Poynton, Stockport SK12 1SP
G3 THG Wng. Cmdr. A. Kent The Coach House, Dipford, Taunton TA3 7NR
GM3 THI Dr R. Harkess Friarton Bank Rhynd Road, Perth PH2 8PT
G3 THQ B. Greenaway 5 Lansdowne Grove, Neasden, London NW10 1PL
G3 THS P. Last 66 Hare Close, Buckingham MK18 7EW
G3 THT E. Bennett 20 Cromford Road, Clay Cross, Chesterfield S45 9RE
G3 THV G. Swindells 4 Fitzhenry Mews, Norwich NR5 9BH
G3 THW P. Walters 22 Windmill Rise, Woodhouse Eaves, Loughborough LE12 8SG
G3 TIE A. Dutton 130 Wades Hill, Winchmore Hill, London N21 1EH
G3 TIG P. Turner 44 Bestwall Road, Wareham BH20 4JA
GI3 TIJ F. Eccles 31 Ballydawley Road, Moneymore, Magherafelt BT45 7NU
G3 TIK D. French 37 Warner Road, Ware SG12 9JN
G3 TIN B. Taylor Perry House, 188 Walstead Road, Walsall WS5 4DN
G3 TIR D. Stewart Apt 347, 8601-902, Luz-Lagos, Algarve Portugal
G3 TIX R. Hardy 47, Maple Mews Cressy Road, Alfreton DE55 7PL
G3 TJA R. Street 11 Royal Close, Rugeley WS15 2DD
G3 TJE P. Smith 7 Tower Walk, Weston-Super-Mare BS23 2JR
G3 TJH W. Bickham 22 Ash Crescent, Galmington, Taunton TA1 5PW
G3 TJI G. Roff 47 Penshurst Rise, Frimley, Camberley GU16 8XX
GI3 TJJ J. Boyce 19 Dunvale Park, Londonderry BT48 0AU
GI3 TJM R. Miller 47A Newtownards Road, Donaghadee BT21 0PY
G3 TJP D. Lankshear 28 Monmouth Place, Newcastle-under-Lyme ST5 3DF
G3 TJS P. Goodenough Llys Aderyn, 11 Guildford Road, Lightwater GU18 5RZ
G3 TJU L. Grant 4 Berry Close, Purdis Farm, Ipswich IP3 8SP
G3 TJX G. Tillson 95 Kelverlow Street, Oldham OL4 1LX
G3 TKA P. Duncan 18 Pickering Road, Hull HU4 6TL
G3 TKB J. Foster 4 Hookergate Lane, Rowlands Gill NE39 2AD
G3 TKF R. Thompson Walnut House, Chestnut Walk Saltford, Bristol BS31 3BG
GW3 TKG D. Locke 201 Tyn Y Tower, Baglan, Port Talbot SA12 8YE
GW3 TKH K. Winnard 208 Heol Hir, Thornhill, Cardiff CF14 9LA
G3 TKK Dr P. Doughty Mallows, Ballam Road, Preston PR4 3PN
G3 TKN V. Lear 53 Chaplains Avenue, Cowplain, Waterlooville PO8 8QH
G3 TKS J. Sanderson 28 Finmere, Hanworth, Bracknell RG12 7WF
GM3 TLA Dr D. Pearson 23 Binghill Road West, Milltimber AB13 0JB
G3 TLD M. Selwyn 50 Tufthorn Avenue, Coleford GL16 8PT
G3 TLH I. Brown 15 Juniper Close, Exeter EX4 9JT
G3 TLK A. Endacott Redacres Market Garden St. Marychurch Road Coffinswell, Newton Abbot TQ12 4SB
GW3 TLP I. Jones Tyddyn Brith, Star, Gaerwen LL60 6AL
G3 TLU J. Serlin 8/34 Ehud Manor, Netanya 4265935 Israel
G3 TLY S. Alexander Pinetrees Wilmslow Avenue, Woodbridge IP12 4HW
G3 TMA I. Buffham 62/70 Soi Sukhumuit 13, Sukhumvit Rd, Klongtoey Nua, Bangkok 10110 Thailand
G3 TMB J. Baker 29 Garstang Road, Southport PR9 9XW
G3 TMD E. Parsons 22 Colins Walk, Scotter Dn21 3Sr, Gainsborough DN21 3SR
GI3 TME R. Hargan 13 Drumlerry, Londonderry BT48 8GQ
GW3 TMJ A. Taylor 24 Emroch St., Goytre, Port Talbot SA13 2YE
GW3 TMP J. Jones Haulfryn Stryt-Cae-Rhedyn Leeswood, Mold CH7 4SS
G3 TMQ R. Harrison 57 Rue Des Bouviers, Mansle 16230 France
G3 TMR D. Emmett 22 Syljon, 114 Villiers Road, Walmer 6070 South Africa
GW3 TMS D. Smith 2 Glan Yr Afon Gardens, Sketty, Swansea SA2 9HY
G3 TMU C. Neale 63 Rosemary Gardens, Blackwater, Camberley GU17 0NJ
G3 TMX S. Bennett 12 Angel Lane, Bury St. Edmunds IP33 1RF
G3 TNE C. Wantling 28 Moss Green, Welwyn Garden City AL7 3TE
G3 TNI J. Clingan 41 Cranham Close, Headless Cross, Redditch B97 5AY
GI3 TNK S. Dornan 9 Clonallon Gardens, Belfast BT4 2BY
G3 TNN N. Sinclair 11 Primrose Close, Warton, Preston PR4 1EN
G3 TNQ C. Davis 963 Manchester Road, Bury BL9 8DN
GD3 TNS A. Sinclair 1 Marathon Drive, Douglas, Isle of Man IM2 4BP
G3 TNX V. Allison 24 Colston Gate, Cotgrave, Nottingham NG12 3JY
G3 TOA B. Otter Po Box 31191, Lusaka 3000 Zambia
GW3 TOB A. Coughlin 37 Parc Y Felin, Creigiau, Cardiff CF15 9PB
G3 TOJ G. Steel 10 Rossmere Avenue, Rochdale OL11 4BT
G3 TON A. Fentham 106 Elm Road, New Malden KT3 3HP
G3 TOP A. Peperell11 47 Glade Road, Marlow SL7 1DQ
G3 TOQ N. Taylor Ministro Raul Fernandes, 180 Apt 1805, Bothfogo, Rio de Janeiro 22260-040 Brazil
G3 TOV G. Miles 200 Ladybank Road, Mickleover, Derby DE3 0RR
GW3 TOW A. Hirst 17 Beech Hollows Lavister, Rossett, Wrexham LL12 0DA
G3 TOY R. Wright High View Cottage, Tatenhill Common, Burton-on-Trent DE13 9RT
G3 TPB Dr J. Knight 2120 North Pantops Drive, Charlottesville 22901 United States
G3 TPH R. Henville 67 Salisbury Road, Blandford Forum DT11 7LW
G3 TPI T. Wager 1 Sundown Close, New Mills, High Peak SK22 3DH
G3 TPJ O. Tillett 27 Cranbrook Drive , Gidea Park, Romford RM2 6AP
G3 TPO C. Ockendon 29 Garlies Road, Forest Hill, London SE23 2RU
G3 TPP B. Eyre 56A Ounsdale Road Wombourne, Wolverhampton WV5 8BH
G3 TPQ G. Harris 12 Highridge Close, Purton, Swindon SN5 4BS
G3 TPV F. Robinson 28 Homer Park, West Common, Soton SO45 1XP
G3 TPW S. Webb 1 The Green, Swinton, Malton YO17 6SY
G3 TQA A. Robinson 9 Illingworth Close, Illingworth, Halifax HX2 9JQ
G3 TQC J. Sunderland 7 Beavers Close, Guildford GU3 3BX
G3 TQF G. Findon 3 The Paddock, Newton, Rugby CV23 0EE
G3 TQL J. Jones 9 Stonehouse Avenue, Willenhall WV13 1AP
G3 TQQ J. Bottomley 27 Mill House , Potter Hill, Pickering YO18 8BJ
G3 TQX G. Grimshaw 50 Rembrandt Way, Bury St. Edmunds IP33 2LT
G3 TQY M. Knights Springside Farm, Tismans Common, Horsham RH12 3DU
G3 TQZ R. Allan Longfield, Upper Wick Lane, Rushwick, Worcester WR2 5SU
G3 TRB T. Barber 48 Newland Road, Droitwich WR9 7AZ
G3 TRC R. Collins 8 Sylvan Way, Redhill RH1 4DE
G3 TRD J. Bellamy 90 Colneis Road, Felixstowe IP11 9LG
G3 TRG R. Green 2 Ragley Walk, Rowley Regis B65 9NT
G3 TRH R. Farrance 63 Salisbury Close, Rayleigh SS6 9UH
G3 TRK D. Kitson 11 Deerstone Road, Nelson BB9 9LN
G3 TRL A. Green Rembrandt Stud, Clotton Common, Tarporley CW6 0HJ
G3 TRR A. Mills 207 Sutherland Drive, Wirral CH62 8EQ
G3 TRX C. Bailey 15 Seymour Avenue, Margate CT9 5HT
G3 TRY Mid-Thames Radio Direction-Finding Club c/o W. Pechey Jays Lodge, Crays Pond, Reading RG8 7QG
G3 TSA J. Denby 107 Station Road, Fenay Bridge, Huddersfield HD8 0DE
G3 TSC Trinity School RC c/o R. Evans 7 Westland Drive, Bromley BR2 7HE
G3 TSE D. Brealy 3 Springfield Road, Plymouth PL9 8EB
G3 TSF E. Glasscott 26 Columbus Circle, Bluffton 29909 United States
G3 TSM V. Mallows 13 Greatfield Way, Rowland's Castle PO9 6AG
G3 TSO M. Grierson Gumstalls Barn Keynsham Lane Woolaston, Lydney GL15 6PR
G3 TSR Col. P. Reader 42 Chiltley Way, Liphook GU30 7HG
G3 TSS C. Waters 1 Chantry Estate, Corbridge NE45 5JH

G3 TSV T. Clay 132 Underdale Road, Shrewsbury SY2 5EF
G3 TSZ A. Macwalter 142 Altrincham Road, Wilmslow SK9 5NQ
G3 TTB P. Clegg 6 Ricketts Drive, Billericay CM12 0HH
G3 TTC K. Orchard 32 Myton Crescent, Warwick CV34 6QA
G3 TTG York RC c/o V. Batchelor 31Bsupakarn Condo, 1057 Charoen Nakorn Road, Bangkok 10600 Thailand
G3 TTI L. Meikle 3 Hillcrest, West Woodburn, Hexham NE48 2RZ
G3 TTJ J. Barber 33 Midford Lane, Limpley Stoke, Bath BA2 7GR
G3 TTP B. Horsey Nethercotts, Gurney Street, Bridgwater TA5 2HW
G3 TTU R. Holt 50 Alverley Lane, Doncaster DN4 9AR
G3 TTY B. Field Greenleaves, 2 Duke Street, Southport PR8 1RS
G3 TUF F. Long 37 St. Catherines Road, Bitterne, Southampton SO18 1LS
G3 TUL J. Copson 16 Fothergill Way, Wem, Shrewsbury SY4 5NX
G3 TUU C. Keeble 86 Kirby Road, Walton on the Naze CO14 8RL
G3 TUW P. Moore 54 Herbert Ave, Palmerston North 4412 New Zealand
GU3 TUX C. Rees 2 Rue De La Saline, Alderney GY9 3XD Guernsey
G3 TUY M. Bruce 3 Redlands Place, Wokingham RG41 4ED
G3 TVC L. Rice Beechwood, 11 Barnoldby Road, Grimsby DN37 0JR
G3 TVD J. Shersby 29 Vale Square, Ramsgate CT11 9DE
G3 TVH J. Harknett 60 Windmill Drive, Croxley Green, Rickmansworth WD3 3FE
G3 TVI R. Stevens 64 Ferndale, Waterlooville PO7 7PB
G3 TVL P. Hunt 14 Walnut Close, Epsom KT18 5JL
G3 TVM H. Fletcher 20 Westfield Road, Great Shelford, Cambridge CB22 5JW
G3 TVN R. Williams 23A Acacia Avenue, Liverpool L36 5TN
G3 TVR E. Churchyard 11 Greenfields Drive, Bridgnorth WV16 4JW
G3 TVT I. Fraser 18 Savick Avenue, Bolton BL2 6JJ
G3 TVU I. Brown 63 Peak View Drive, Ashbourne DE6 1BR
G3 TVV A. Coates 35 Mogg St., St. Werburghs, Bristol BS2 9UB
G3 TVX D. Ashwood Apartment 43, Dane Court 21 Mill Green, Congleton CW12 1FS
G3 TVY J. Sutton 3 Sunrise Avenue, Nottingham NG5 1NH
G3 TWB R. Ballard 31 South Devon Avenue, Nottingham NG3 6FT
G3 TWJ M. Roach 35 Hartley Old Road, Purley CR8 4HH
GW3 TWN F. Mason Awel Mon, Bodffordd, Llangefni LL77 7LJ
G3 TWX D. Woodhouse 38 Jenny Road, Spixworth, Norwich NR10 3QW
G3 TWY G. Mills 11 Milton Street, Narborough LE19 3EZ
G3 TXC N. Harris April Cottage, Sheepcote Green, Saffron Walden CB11 4SJ
G3 TXE A. Parker 7 St. Peters Court, Claydon, Ipswich IP6 0HZ
G3 TXF N. Cawthorne Dormers, The Green, Hinton Charterhouse, Bath BA2 7TJ
G3 TXH B. Levett 18 Forge Road, Little Sutton, Ellesmere Port CH66 3SQ
G3 TXK C. Moss 2 Sutton Lane Adlington, Chorley PR6 9PA
G3 TXL A. Graham Woodtown, Sampford Spiney, Yelverton PL20 6LJ
G3 TXZ C. Tucker Flat 35, Martlets Court, Crowborough TN6 1JF
G3 TYA J. Grant Tanyanga, Wheal Leisure, Perranporth TR6 0EY
G3 TYG B. Winslow 10 Almond Walk, Hazlemere, High Wycombe HP15 7RE
GW3 TYI D. West 44 Glanmor Park Road, Sketty, Abertawe (Swansea) SA2 0QE
G3 TYO J. Stringer 1 Hazel Road, Tavistock PL19 9DN
G3 TYP I. Jackson 20 Daventry Road, Barby, Rugby CV23 8TR
G3 TZA J. Riley 38 West Broadway, Bristol BS9 4TB
GI3 TZB W. Mckinney 33 Heatherstone Road, Bangor BT19 6AE
G3 TZD R. Mansell 354 Allen Road, Salt Point 12578 United States
G3 TZE R. Armitage 3 Holst Mead, Stowmarket IP14 1TD
G3 TZG J. Glanville 3 Seneschal Road, Cheylesmore, Coventry CV3 5LF
G3 TZL P. Bowen White House, Durleigh Marsh, Petersfield GU31 5AX
G3 TZM W. Mahoney 61 Starbold Crescent, Knowle, Solihull B93 9LA
G3 TZO P. Holland Chatterton, Chapel Lane, Threapwood, Malpas SY14 7AX
G3 TZQ S. Ridgway 12 The Mead, Plymouth PL7 4HS
G3 TZT M. Mead 2 Market Court, Market Place, Wincanton BA9 9PB
G3 TZU J. Harding 5 Salisbury Road, Whitchurch SY13 1RQ
GI3 TZX W. Nesbitt 101 Belfast Road, Bangor BT20 3PP
GM3 UA A. Pairman Seabank, Largiebeg, Brodick KA27 8RL
G3 UAA D. Ramsey The Orchard, Carmen Grove, Leicester LE6 0BA
G3 UAE J. Gill 22 Maddever Crescent, Liskeard PL14 3PT
G3 UAF Dr M. Smith 138 Market St., Clay Cross, Chesterfield S45 9LY
GM3 UAG J. Davidson 3 Dunnottar Road, Ellon AB41 9JF
G3 UAP P. Parker Avenue Kersbeek 116, 1190 Brussels Belgium
G3 UAS T. Morgan 2 Park View, Hatch End, Pinner HA5 4LN
G3 UAX R. Stansfield 22 Reeds Avenue, Earley, Reading RG6 5SR
GW3 UAY C. Butters 39 Parc Y Ffynnon, Ferryside SA17 5TQ
GI3 UBA R. Reid 21 Ballymaconnell Road, Bangor BT20 5PN
G3 UBB D. Fill 2 Brook Close, Packington, Ashby-de-la-Zouch LE65 1WA
G3 UBD G. Higgins Lower Laithe Farm, Providence Lane, Keighley BD22 7QS
GW3 UBH J. Pugh 5 Pen Y Maes, Llanfechain SY22 6XL
G3 UBI M. Fisher Bank Top Farm, Cropton, Pickering YO18 8HH
GM3 UBJ Dr W. Hossack Kincrig, 39 Skene Street, Macduff AB44 1RP
G3 UBL C. Ledger Kinrara, Sandhills Road, Salcombe TQ8 8JP
G3 UBP C. Riches 28 Saxondale Avenue, Burnham-on-Sea TA8 2PS
G3 UBS B. Speakman Merrydown, Burley Lane, Derby DE6 4JS
G3 UBV D. Roberts 8 Churnet Close, Bedford MK41 7ST
G3 UBX P. Burden 68 Coalway Road, Wolverhampton WV3 7LZ
G3 UBY A. Clark Sans Souci, Fairmead Road, Saltash PL12 4JH
G3 UCA P. Sinclair 32 Barn Meadow, Bamber Bridge, Preston PR5 8DU
G3 UCD R. Pescod 7 Brian Close, Chelmsford CM2 9DZ
G3 UCF A. Passmore 16 Chaffinch Close, Basingstoke RG22 5QD
GM3 UCH W. Wright 460 Main Street, Stenhousemuir, Larbert FK5 3JU
GM3 UCI G. Mccallum 15 Quarry Road, Law, Carluke ML8 5HB
GW3 UCJ M. Evans 31 Cilmaengwyn Road, Pontardawe SA8 4QL
G3 UCK G. Downs 2 Dyehouse, Wilsden, Bradford BD15 0BE
G3 UCL University College London ARS c/o Dr G. Smart 30 Cornmills Road, Soham CB7 5AT
GM3 UCN F. Hetherington 4 Rosebery Place, Livingston EH54 6RP
G3 UCQ J. Farrar 2 Marsh Lane, Hayle TR27 4PS
G3 UCT M. Taylor Orchard House, Leigh, Sherborne DT9 6HL
G3 UCW M. Pettit 24 Pool Lane, Winterley, Sandbach CW11 4RY
G3 UD G. Bloor 26 Leveson Road, Hanford, Stoke-on-Trent ST4 4QP
G3 UDA K. Linney Sunnybank, Oak Lane, Bicton Heath, Shrewsbury SY3 5BW
G3 UDD Dr S. Chandler Malt House, Box, Stroud GL6 9HF
G3 UDH P. Butcher 55 Offington Lane, Worthing BN14 9RJ
G3 UDI Dr R. Butcher Temple Lodge, Six Mile Bottom Road, Cambridge CB21 5LD
GM3 UDK Dr C. Oliver 40 Charles Way, Limekilns, Dunfermline KY11 3LH
G3 UDN Mid Warwickshire ARS c/o Q. Wright 9 Browning Avenue, Warwick CV34 6JQ
G3 UDP M. Brown 4 Boyfields, Quadring, Spalding PE11 4QQ
G3 UDV P. Lindsley Oak Lodge, Cromer Road, Cromer NR27 9QT
G3 UED J. Jones 12 Francis Groves Close, Bedford MK41 7DH
G3 UEE D. Diamond 36 Darbys Lane, Oakdale, Poole BH15 3ET
G3 UEG D. Gould 2 Mayfield Close, Harlow CM17 0LH
G3 UEK J. Whitehouse 2884 Sequoyah Trl, Guntersville, Al 35976 United States
GW3 UEP R. Plimmer Fron Haul, Tregroes, Llandysul Sa44 4Ne, Llandysul SA44 4NE
G3 UEQ A. Hearn 53 Twyford Gardens, Salvington, Worthing BN13 2NT
G3 UES Echelford ARS c/o S. Roy 28 Kingston Rise, New Haw, Addlestone KT15 3EY
G3 UEU J. Holmes 23 School Lane Berry Brow Hd4 7Ra, Huddersfield HD4 7RA
GI3 UEX Bedford And District ARC c/o R. Thompson 94 Orangefield Crescent, Belfast BT6 9GJ
G3 UEY D. Browning11 13 Beechcombe Close, Pershore WR10 1PW
G3 UEZ R. Gilbert 18 Peckham Avenue, New Milton BH25 6SL
G3 UFB N. Brinkworth 11 Haycroft Road, Stevenage SG1 3JL
G3 UFF P. Rodway 37 Neville Avenue, Portchester, Fareham PO16 9NR
G3 UFI P. Conway 1 The Woodlands, The Ridge, Hastings TN34 2SF
G3 UFJ L. Symons 31 Springfield Way, Threemilestone, Truro TR3 6BJ
G3 UFQ D. Eckley 27 Apsley Grove, Dorridge, Solihull B93 8QP
G3 UFS C. Smith 50 Grand Avenue, Lancing BN15 9PZ
G3 UFV P. Crawshaw 35 Bishopton Avenue, Stockton-on-Tees TS19 0RA
G3 UFX H. Julian Brigantine, Lower Market Street, Penryn TR10 8BH
G3 UFY S. Knowles 77 Bensham Manor Road, Thornton Heath CR7 7AF
G3 UGC J. Smethurst 81 Springside Road, Bury BL9 5JG
G3 UGF R. Constantine 18 Hillbeck, Halifax HX3 5LU
G3 UGJ D. Smith 33 Rippington Drive, Marston, Oxford OX3 0RJ
G3 UGX R. Heaton Flat 5, 73 Belsize Park Gardens, London NW3 4JP
G3 UHF South Manchester RC c/o D. Thomas 33 Chatsworth Road, Stretford, Manchester M32 9QF
G3 UHJ R. Gordon 77 Alwyn Road, Darlington DL3 0AH
G3 UHK J. Baldwin 19 Lutyens Close, Stapleton, Bristol BS16 1WL
G3 UHN P. Neale 98 Meadway, Harpenden AL5 1JQ
G3 UHS C. Houltby 9 Bayard St., Gainsborough DN21 2JZ
GM3 UHT W. Garner Sarkshields Cottage, Eaglesfield, Lockerbie DG11 3AE
G3 UHU D. Hampton 9 Portwey Close, Weymouth DT4 8RF
G3 UHV C. Sutton Braehead, Old Lane, Stoke-on-Trent ST6 8TG
G3 UHW H. Tomlinson 32 Manor Road, Farnborough GU14 7EU
G3 UHX A. Thorpe 12 Newnham Lane, Ryde PO33 4ED
G3 UI L. Cobb 27 Moorlands Crescent, Halifax HX2 8AA
G3 UIB C. Hearn 8 The Poles, Upchurch, Sittingbourne ME9 7EX
G3 UID K. Baldock 284 Rocky Mountain High, Camano Island WA 98282 United States
G3 UIF G. Thorne Flagstaff House, Main Street, Hull HU12 0RY
GI3 UIH W. Aylward 37 Stewartstown Avenue, Belfast BT11 9GF
G3 UIK J. Young Shirley Lodge, 45 Graham Road, Malvern WR14 2HU
GM3 UIN M. Mcleman 6 Stenton Road, West Barns, Dunbar EH42 1UG
G3 UIS A. Stone West Lodge, The Downs, Poulton-le-Fylde FY6 7EG
G3 UIT B. Seedle 54 Normoss Road, Blackpool FY3 0AL
G3 UJA B. Mcclory 12 The Crescent, Mottram St. Andrew, Macclesfield SK10 4QW
G3 UJB B. Davis 2 Rawden Close, Harwich CO12 4BW
G3 UJE B. Gale Tall Trees Farm, Noah'S Ark Lane, Great Warford WA16 7AX
G3 UJI S. Turner 51 Hilton Road, Stoke-on-Trent ST4 6QZ
G3 UJM R. Banks 3 Parkhayes, Woodbury Salterton, Exeter EX5 1QS
G3 UJO P. Bradley 60 Weyland Road, Headington, Oxford OX3 8PD
G3 UJV R. Heath 26 Lancaster Avenue, Hadley Wood, Barnet EN4 0EX
G3 UJZ J. Mcnaught Ryton House, Glos GL7 3AR
G3 UK J. Whittaker Riverside, 48 Baunton, Cirencester GL7 7BB
G3 UKB R. Cowdery 80 Caxton End, Eltisley, St. Neots PE19 6TJ
G3 UKC Surbiton Heritage Amateur Radio c/o F. Barnes 4 Pound Close, Ducklington, Witney OX29 7TH
G3 UKD A. Golding 40 Unicorn Lane, Eastern Green, Coventry CV5 7LJ
G3 UKE P. Adams 34 Mount Pleasant Close, Lightwater GU18 5TP
GM3 UKG G. Grant 35 Inward Road, Buckie AB56 1DD
G3 UKH P. Hopwood 2 Spital Cottages Tyne Green, Hexham NE46 3RY
G3 UKI B. Curnow Flat 3A, Olivers Wharf, 64 Wapping High Street, London E1W 2PJ
G3 UKL M. Bennett Shireley, Munns Lane, Sittingbourne ME9 7SY
G3 UKM M. Leighton 85 Kemps Green Road, Balsall Common, Coventry CV7 7QF
G3 UKW M. Newton 11 Chestnut Close, Rushmere St. Andrew, Ipswich IP5 1ED
G3 ULD G. Cawkwell 50 Station Road, Patrington, Hull HU12 0NE
G3 ULN M. Hibbitt 123 Stanborough Road, Plymstock, Plymouth PL9 8PJ
G3 ULO I. Spencer Fichtenweg 10C, Much 53804 Germany
G3 ULT Reading & District ARC c/o J. Carter 22 Orchard Coombe, Whitchurch Hill, Reading RG8 7QL
GW3 UMD N. Maxwell 1 Nant Fawr Crescent, Cardiff CF2 6JN
G3 UMF Dr A. Simpson Forest Farm, Old Road, Shotover Hill, Oxford OX3 8TA
G3 UML L. Margolis 52 Park View Gardens, Hendon, London NW4 2PN
G3 UMM P. Hudson 105 Southlands, Weston, Bath BA1 4DZ
G3 UMT B. Turvey 90 Jenkinson Road, Towcester NN12 6AW
G3 UMV P. Johnson 52 Evesham Road, Cookhill, Alcester B49 5LJ
GU3 UMX D. Ozanne Eturs Lodge, Les Eturs, Castel GY5 7DT Guernsey
G3 UNA D. Cutter David Cutter Engineering, 34 Greengate Lane, Knaresborough HG5 9EL
G3 UNI T. Wood 4 Musgrave Road, Chinnor OX39 4PL
G3 UNK C. Hebden 1 Ringwood Avenue, Newbold, Chesterfield S41 8RA
G3 UNM A. Matthews Winsford, The Common, Stoke-on-Trent ST10 2PA
G3 UNS T. Mills 22 The Dingle, Crawley RH11 7JE
G3 UOA Southern DX Club c/o Dr P. Best 21 Greening Drive, Edgbaston, Birmingham B15 2XA
G3 UOC D. Brown Rexfield, Alcester Road, Henley-in-Arden B95 6BH
G3 UOD M. Spencer Cleeve House, Melton Road, Melton Mowbray LE14 3QG
G3 UOI J. Firby 19 Cliffe Avenue, Harden, Bingley BD16 1LN
G3 UOJ J. Hartwell Fulling Mill Oast, Caring Lane, Maidstone ME17 1TJ
G3 UOM D. Horsburgh 11 Delamare Way, Oxford OX2 9HZ
G3 UON D. Geere Tinos Premier Marinas Ltd, Western Concourse, Brighton BN2 5UP
GW3 UOO D. Rogers Green Tops, 69 Megs Lane, Buckley CH7 2AG
GU3 UOQ P. Le Boutillier Room 10 Maison L'Aumone, Castel GY5 7RT Guernsey

G3 UOS Dr A. Whitaker Univer Of Sheffield, Dept Of Elec Eng, Sheffield S1 3JD
G3 UPA M. Foden 10 Maud Road, Water Orton, Birmingham B46 1PD
G3 UPD H. Del Monte 5 Scotts Close, Colden Common, Winchester SO21 1US
GI3 UPG R. Mckimm 227 Millisle Road, Donaghadee BT21 0LN
G3 UPI T. Codling 21 Willow Close, Saxilby, Lincoln LN1 2QL
G3 UPJ D. Trainer 153 High Street, Cherry Hinton, Cambridge CB1 9LN
G3 UPM T. Burke 12 Worthing Road, Laindon, Basildon SS15 6AL
G3 UPN K. Snape Delamere, Ryston End, Norfolk PE38 9AX
G3 UPS R. Keyte 18 Mclean Drive, Kessingland, Lowestoft NR33 7TY
G3 UPW P. Smith 9 Ash Road, Shepperton TW17 0DN
G3 UPY D. Houghton 119 Welsby Road, Leyland PR25 1JD
G3 UPZ H. James 18 Clydesdale Road, Whiteley, Whiteley PO15 7BD
G3 UQD R. Whittington 65 King Edward Avenue, Worthing BN14 8DG
G3 UQL M. Baker 10 Catchpole Close, Greenleys, Milton Keynes MK12 6LR
G3 UQR D. Robinson 3 Marriott Close, Irthlingborough, Wellingborough NN9 5RB
G3 UQW A. Ball 3 Orchard Lea, Sherfield-On-Loddon, Sherfield on Loddon RG27 0ES
G3 URA R. Whittering Vrouhas, Crete GR - 72053
G3 URE J. Thexton 78 Greenfield Road, Newcastle upon Tyne NE3 5TQ
G3 URI Newbury Vintage Wireless Society c/o Dr M. Franks 13 Fifth Road, Newbury RG14 6DN
G3 URJ A. Moss 17 Surrey Drive, Finchfield, Wolverhampton WV3 9LW
G3 URK I. Campbell 27 Lewis Close, Adlington, Chorley PR7 4JU
G3 URL C. Adams 25 Avon Road, Cannock WS11 1LJ
G3 URN Dr M. Jolley 34469 N Circle Drive, Round Lake 60073 United States
G3 URQ J. Letts Bridgeways, Snows Lane, Leicester LE7 9JS
G3 URU Dr R. Edworthy 44 Middleton Avenue, Littleover, Derby DE23 6DL
G3 URV F. Stevens 38 Endhill Road, Birmingham B44 9RR
G3 URX J. Speake 211 Milton Road, Cambridge CB4 1XG
G3 URZ Dr B. Ewen-Smith 1 Kinnersley, Severn Stoke, Worcester WR8 9JR
G3 USA C. Taylor 39 School Road, Great Alne, Alcester B49 6HQ
G3 USC M. Hall Redthorn Bungalow, Upton, Langport TA10 9NL
G3 USD D. Mason 2A Devon Road, Bedford MK40 3DF
GI3 USK H. Kernaghan 1 Elizabeth Road, Holywood BT18 0PL
GM3 USL Cunninghame & District Amt RC c/o J. Walker Mo Bhothan, Lochlibo Road, Uplawmoor, Glasgow G78 4AA
G3 USO C. Walker St. Jude, Stoney Lane, Wilmslow SK9 6LG
G3 USR G. Rolland 3B Reeves Lane, Wing, Oakham LE15 8SD
G3 UST J. Turner Flat 4, Barnett Janner House, Leicester LE4 0UR
G3 USW W. Clough 32 Jackson Crescent, Rawmarsh, Rotherham S62 7EN
G3 USX M. Robertson 1 Lindvale Horsell Rise, Woking GU21 4BG
G3 UTA K. Smyth 154 Scrub Lane, Benfleet SS7 2JP
G3 UTC G. Farr 26 Burstead Drive, Billericay CM11 2QN
G3 UTE N. Wright-Williams Trinco, 9 Orpine Close, Bicester OX26 3ZJ
GW3 UTG A. Antley 12 Fairfield Avenue, Rhyl LL18 3EE
GW3 UTL R. Barker 51 Rockfield Drive, Llandudno LL30 1PF
G3 UTS T. Belshaw 20 Greencroft Road, Delves Lane Industrial Estate, Consett DH8 7DY
G3 UUB N. Bateman 10 Telford Crescent, Woodley, Reading RG5 4QT
G3 UUC J. Nurse 25 Dobson Road, Crawley RH11 7UH
G3 UUF J. Hansom 12 Torquay Avenue, Hartlepool TS25 3DP
G3 UUG E. Nightingale 61 The Cockpit, Marden, Tonbridge TN12 9TQ
G3 UUI M. Mapson 253 Central Avenue, Southend-on-Sea SS2 4ED
G3 UUL T. Jones 32 Oakwood Drive Hucclecote, Gloucester GL3 3JF
G3 UUM A. Page 22 Tower Road, Feniscowles, Blackburn BB2 5LE
G3 UUQ A. Clelland Rieschbogen 7, Hohenkirchen 85635 Germany
G3 UUR Dr D. Gordon-Smith The Chalet, Bell Road, Rockland St Peter, Attleborough NR17 1UL
G3 UUT J. Wilson 20B High Green, Great Shelford, Cambridge CB22 5EG
G3 UUU L. Newman Eastholme, Mill Street, Newton Abbot TQ13 8AR
G3 UUV R. Frost Chez Nous, Coldharbour Yard, Swindon Road , Kington Langley, Chippenham SN15 5LY
G3 UUY D. Wright St. Julians, 55 Old Road, Harlow CM17 0HD
GW3 UVA D. Knowles The Clappers, Spon Green, Buckley CH7 3BL
G3 UVB Tawney Hall Farm ARC c/o D. Barnes 27 Royal Court, Worksop S80 2DL
G3 UVM M. Simpson 36 Rectory Close, Newbury RG14 6DD
G3 UVQ N. Mercer 19 Sycamore Road, Brookhouse, Lancaster LA2 9PB
G3 UVR D. Jones 39 Pensby Road Heswall, Wirral CH60 7RA
G3 UVU J. Curry Clonlea, New Ridley, Stocksfield NE43 7RQ
G3 UVW Thornton Cleveleys ARS (Tcars) c/o R. Harris Clevelands, Tamworth Road, Coventry CV7 8JJ
G3 UVY L. Parkin 8 Smithfield Close, Ripon HG4 2PG
G3 UWE R. Simpson 30 Heath Lawns, Fareham PO15 5QB
G3 UWH Cmdr. J. Endicott The Mill House, Halse, Taunton TA4 3AQ
GW3 UWL Sir S. Grant The Court, 19 Marine Parade, Penarth CF64 3BE
G3 UWM P. Marchant 12 Laurel Way, Ickleford, Hitchin SG5 3UP
G3 UWP R. Pickering 41 Maiden Greve, Malton YO17 7BE
G3 UWR C. Bonsall Parkside, Lodge Road, Doncaster DN6 8EB
GW3 UWS De Havilland Heritage Radio Group c/o Dr T. Davies 44 Carnglas Road, Sketty, Swansea SA2 9BW
G3 UWT P. Myers 22 High Street Barnby Dun, Doncaster DN3 1DS
GM3 UWX J. Stirling 25 Maxwell Road, Bishopton PA7 5HE
G3 UWZ M. Newman 26 Highbank, Westdene, Brighton BN1 5GB
G3 UXH P. Carey 44 Monteney Gardens, Sheffield S5 9DY
G3 UXM J. Greaves 23 Woodhouse Road, Intake, Sheffield S12 2AY
G3 UXO Dr A. Eardley Flat E, 113 Sutherland Avenue, London W9 2QH
G3 UXR N. Goddard 1 Aston Mead, St. Catherine'S Hill, Christchurch BH23 2SP
G3 UXY A. Baker 1 Napier Road, Maidenhead SL6 5AR
G3 UYB M. Shaw Beech Farm Cottage Hawkhurst Road, Sedlescombe TN33 0QS
G3 UYC J. Peirson Ashfield Farm, Ulting, Maldon CM9 6QP
G3 UYD E. Clarke 65 Oakmount Road, Chandler'S Ford, Eastleigh SO53 2LJ
G3 UYE M. Richer 1 Station Road, Surfleet, Spalding PE11 4DA
G3 UYG J. Clegg 11 South Park Road, Gatley, Cheadle SK8 4AL
G3 UYK P. Kemble 74 Teg Down Meads, Winchester SO22 5ND
G3 UYL D. Knott 22 Linden Close, Prestbury, Cheltenham GL52 3DU
G3 UYN C. Malcolm Glen Mor, Trenance, St Keverne, Helston TR12 6QL
GM3 UYR Dr P. Gamble 21 St. Marys Drive, Perth PH2 7BY
G3 UYX J. Ball 8 Withybed Close, Alvechurch B48 7PL
G3 UYY E. Bradley 3 Windrush, Wargrave Road, Henley-on-Thames RG9 2LX
G3 UZB J. Shewan 42 Stirling Road, Redcar TS10 2JZ
GI3 UZJ D. Singleton 38A Cloughey Road, Portaferry, Newtownards BT22 1NQ
G3 UZK D. Bloomfield 22 Laurel Road Locks Heath, Southampton SO31 6QG
G3 UZM C. Haddock 26 Featherbed Lane, Exmouth EX8 3NE
GW3 UZS J. Diplock Cartref, 98 Pendwyallt Road, Cardiff CF14 7EH
G3 UZW R. Andrews 10 Hilltop Rise, Bookham, Leatherhead KT23 4DB
G3 UZX F. Mitchell 158 Cobham Road, Fetcham, Leatherhead KT22 9JR
GI3 VAF R. Best 6 Knightsbridge Court, Bangor BT19 6SD
G3 VAJ I. Gray 1 Greenside Avenue, Berwick-upon-Tweed TD15 1BZ
G3 VAK M. Sutcliffe 26 Weald Road, Burgess Hill RH15 9SP
GM3 VAL G. Talbot Whitethorn House, Milnathort KY13 9XU
G3 VAO M. Farmer Horton Brook Cottage, Horton Wem, Shrewsbury SY4 5NB
GM3 VAP C. Weston 18 Kirkbrae Mews, Cults, Aberdeen AB15 9QF
GI3 VAW R. Sherrard 39 Shanreagh Park, Limavady BT49 0SF
G3 VBA K. Hatton 38 Doric Avenue, Frodsham WA6 6QQ
G3 VBE F. Miles 65 Montgomery Street, Hove BN3 5BE
G3 VBG B. Morris 88 Newcastle Road, Leek ST13 7AA
G3 VBL C. Pedder Thorncliffe, 5 Royalty Lane, Preston PR4 4JD
G3 VBQ D. Wright 5 Padin Close, Chalford, Stroud GL6 8FB
GM3 VBT T. Logan 137 Buccleuch St., Garnethill, Glasgow G3 6QN
G3 VBU J. Lynch 11 Rosenthorpe Road, London SE15 3EG
G3 VBV S. Boyce 58 Woodbury Road, Halesowen B62 9AW
GM3 VBY F. Hindley The White House, 17 Main Road, Elgin IV30 8UR
G3 VC M. Bridge 5 Garrett'S Place, Donington, Spalding PE11 4YL
G3 VCA R. Pickles Bramcote Lorne, Rectory Lane Gamston, Retford DN22 0QQ
G3 VCG D. Wilks 36 Greenways, Chelmsford CM1 4EF
GI3 VCI M. Mcfadden 121 Greystown Avenue, Belfast BT9 6UH
G3 VCK J. Fenwick 78 Loveridge Road, London NW6 2DT
G3 VCL B. Clark 60 Somerset Avenue, Harefield, Southampton SO18 5FS
G3 VCM I. Anderson-Mochrie 10214 Hunt Club Lane, Palm Beach Gardens 33418 United States
G3 VCN P. Kalas 110A Underlane, Plympton, Plymouth PL7 1QZ
G3 VCP N. Kail 1 Siemons Street, One Mile, Queensland 4503 Australia
G3 VCQ C. Wilson 57 James Andrew Crescent, Sheffield S8 7RJ
G3 VCR C. Rooney 129 Drift Road Clanfield, Waterlooville PO8 0PD
G3 VCT R. Hemmings Wood View, Cryers Hill Road, Cryers Hill, High Wycombe HP15 6JR
G3 VCV D. Prout 7 Chemin Des Estimeurs Nord, Plan De La Dame, Valreas 84600 France
G3 VCX D. Bridgen 22 Maple Grove, Immingham DN40 2JH
G3 VCY Dr C. Clayton Wildfield, West Flexford Lane, Guildford GU3 2JW
G3 VDB J. Evans 7 Barncroft Close, Chelford, Macclesfield SK11 9SW
G3 VDE J. Sellers Blacktoft Grange, Blacktoft Grange Road, Sandholme, Brough HU15 2ZU
G3 VDF H. Gregory 44 Mowlands Close, Sutton-in-Ashfield NG17 5GH
G3 VDH R. Godwin Hopworthy Moor Cottage, Pyworthy, Holsworthy EX22 6XX
G3 VDK S. Bailey 6 Minnie Street, Keighley BD21 1HY
G3 VDL J. St Leger Warmbrook, Throwleigh, Okehampton EX20 2JF
G3 VDO I. Hacking 1 Pine Crescent, Poulton-le-Fylde FY6 8EB
G3 VDS R. Higham 34 Ashford Road, Wirral CH47 5AW
G3 VDU P. Bennett 56 Winchester Avenue, Weddington, Nuneaton CV10 0DW
G3 VDV N. Brinnen 134 Victoria Road, Mablethorpe LN12 2AJ
G3 VDZ T. Richardson Flat 11, The Moorings, 21 Albert Way, East Cowes PO32 6GA
G3 VEB R. Bridson 14 Zig Zag Road, Wallasey CH45 7NZ
G3 VEF Fareham And District ARC c/o C. Jenkins-Powell 43 Cambridge Road, Lee-on-the-Solent PO13 9DH
G3 VEH C. Morcom 15 Markson Road, South Wonston, Winchester SO21 3EZ
GM3 VEI I. Sheffield 37 Bellevue Court, Queens Road, Dunbar EH42 1YR
G3 VEK S. Holden Garden Flat 1 Emmanuel Road, Hastings TN34 3LB
G3 VER Verulam ARC c/o P. King 96 Mancroft Road, Caddington, Luton LU1 4EN
G3 VES H. Martin 1 Houghton Park Cottages, Ampthill MK45 2EY
G3 VET M. Langwade 19 South Wootton Lane, King's Lynn PE30 3BS
G3 VEV R. Butterfield 7 Kipling Close, Lincoln LN2 4EW
GM3 VEY F. Baxter 8 Northcote Park, Aberdeen AB15 7SX
G3 VFB A. Matthews 45 Kings Square, Taunton TA1 3FN
G3 VFC T. Chipperfield 5 Lullingstone Close, Hempstead, Gillingham ME7 3TS
G3 VFD C. Westwood Uplands, The Hillside, Orpington BR6 7SD
G3 VFF D. Hine Whirlwind, Chesboule Lane Gosberton Risegate, Spalding PE11 4EU
G3 VFH L. Moore 15 Elmete Drive, Roundhay, Leeds LS8 2LA
GW3 VFL A. Lightly 9 The Kymin, Monmouth NP25 3SD
G3 VFO T. Hart The Hawthorns, 163 Hastings Road, Battle TN33 0TP
G3 VFX D. Davison 28 Treve Avenue, Harrow HA1 4AJ
GW3 VFZ M. Hughes Cefn Dinas, Bangor LL57 4DP
G3 VG J. Wood 7 Sherring Close, Bracknell RG42 2LD
G3 VGD D. Jones 31 Meadow Road, Windermere LA23 2EU
G3 VGE M. Hickman 75 Carlton Road, Redhill RH1 2BZ
G3 VGH Dr B. Hutchinson 78 Strensall Road, Huntington, York YO32 9SH
G3 VGK K. Blackburn 57 Hope Street, Leigh WN7 1NB
G3 VGR D. Aldridge 62 Roding View, Buckhurst Hill IG9 6AQ
G3 VGW R. Buckby 22 Woodstead, Embleton, Alnwick NE66 3XY
G3 VGX Dr R. Orton 15 Middleton Close, Cambridge CB4 1DG
G3 VGY R. Ricketts 30 Water Lane, Tiverton EX16 6RB
G3 VGZ B. Duffell 7 Potto Close, Yarm TS15 9RZ
G3 VHE R. Evans 23 Hardwell Close, Grove, Wantage OX12 0BN
G3 VHH J. Delves 11 Willoughby Road, Langley, Slough SL3 8JH
G3 VHI G. Boultbee 6 Laxton Close, Heckington, Sleaford NG34 9TS
G3 VHK Dr J. Robinson 8 Lorraine Park, Harrow HA3 6BX
G3 VHL H. Buttress 132 Elan Avenue, Stourport-on-Severn DY13 8LR
GI3 VHM V. Addidle 23 Church Lodge, Moneyrea, Newtownards BT23 6ES
G3 VHN J. Burge 14 Robinson Place, Brant Broughton, Lincoln LN5 0SJ
G3 VHS J. Cobb Middle Cottage, Abingdon OX13 5LR
G3 VHU M. Herring Flat 6, 1 Royal Crescent, Bridlington YO15 2PG
G3 VHW N. Humphrey 10 Pembroke Close, Eastleigh SO50 4QY
G3 VHZ B. Neary 30 Laneham Close, Doncaster DN4 7HU
G3 VID T. Howe 33 Devon Gardens, Birchington CT7 9SR
G3 VIP G. Wood 47 Church Lane, Holton-Le-Clay, Grimsby DN36 5AQ
G3 VIR R. Brade 9 Magness Road, Deal CT14 9JF
G3 VIX T. Stevens 97 Broad Acres, Hatfield AL10 9LE

G3 VIY R. Vasper 31 Oakland Road, Forest Town, Mansfield NG19 0EJ
G3 VJE H. Cole 3 Canberra Crescent, Grantham NG31 9RD
G3 VJG M. Deutsch 1 Hodge Court, Kettering NN15 7EZ
G3 VJI J. Steel 4 Broom Close, Kendal LA9 6BN
G3 VJM A. Wood Danehill, Brookhill Road, Crawley RH10 3PS
G3 VJN A. Ryan 10 Skasmata, Koili 8543 Cyprus
G3 VJR J. Longstaff 23 Harlington Road, Adwick-Upon-Dearne, Mexborough S64 0NL
G3 VJV C. Hartley 16 Cyril Bell Close, Lymm WA13 0JS
G3 VJX M. Gill Upper Bean Hall, Church Road, Redditch B69 6RN
GM3 VJY J. Evans 64 Craigmount Avenue North, Edinburgh EH12 8DL
G3 VKB J. Orr 9 Chemin Des Postes, Villers Carbonnel 80200 France
G3 VKF K. Kelly 2 Longden Lane, Macclesfield SK11 7EN
G3 VKI F. Turner-Smith 26 Ash Church Road, Ash, Aldershot GU12 6LX
G3 VKK Mid Cheshire Ar c/o J. Otter 7 Longacre Road, Dronfield S18 1UQ
GW3 VKL Barry ARS c/o P. King 11 Lord Street, Penarth CF64 1DD
G3 VKM R. Basford Newgate, Thorpe Road Haddiscoe, Norwich NR14 6PP
GM3 VKN P. Mansell Broad Meadows, Fort Augustus PH32 4DW
G3 VKQ C. Mcewen 37 Malvern Way, Twyford, Reading RG10 9PY
G3 VKT R. Smith 32 Wolseley Gardens, London W4 3LR
G3 VKU D. Hollingsworth 4 Cairn View, Longframlington, Morpeth NE65 8JT
G3 VKV G. Jones 32 The Grove, Hales Road, Cheltenham GL52 6SX
G3 VKW K. Evans Littlefield House, Bolney Road, Haywards Heath RH17 5AW
G3 VLC C. Hawkins 2 Benett Drive, Hove BN3 6PL
G3 VLD T. Denney Spindrift-East-, Terrace, Walton on Naze CO14 8PX
G3 VLF Dr T. Beamond Park View, Middle Lane, Whatstandwell, Matlock DE4 5EG
G3 VLG Hinckley Amateur Radio And Electronics Society c/o C. Colclough 53 St Marys Road, Nuneaton CV11 5AT
G3 VLH J. Longhurst 13 Hophurst Drive, Crawley Down, Crawley RH10 4XA
G3 VLJ Dr A. Hansen 1829 Francisco St, Berkeley 94703 United States
G3 VLL G. Gauntlett 7 Riverside Drive, Sprotbrough, Doncaster DN5 7LH
G3 VLN J. Allin 57 Burleigh Road, West Bridgford, Nottingham NG2 6FQ
G3 VLO Dr J. Owen-Jones 16 Cotswold Close, Torquay TQ2 6UB
G3 VLR B. Rispin 37 Ferry Road South Cave, Brough HU15 2JG
G3 VLW P. Martin Orchard Rise, Littlemoor Road, Highbridge TA9 4NG
G3 VM D. Grace 107 Bush Avenue, Little Stoke, Bristol BS34 8NG
G3 VMI D. Pike 11 Cavalry Drive, March PE15 9EQ
G3 VMK N. Chadwick 1 St. Francis Meadow Mitchell, Newquay TR8 5DB
G3 VMP B. Mills Highlands Cottage, Crow Lane, Clacton on Sea CO16 9AN
G3 VMR R. Redding 53 Cadwell Drive, Maidenhead SL6 3YS
G3 VMU C. Davis 23 Vernon Walk, Northampton NN1 5ST
G3 VMV C. Whiting 5 Carlton, Elloughton, Brough HU15 1FF
G3 VMW S. Wilson 3 Crag Gardens, Bramham, Wetherby LS23 6RP
G3 VMY Dr E. Searle 203 Church Road, Earley, Reading RG6 1HW
G3 VMZ D. Nicholls 26 Highfield Close, Semington, Trowbridge BA14 6JZ
G3 VNB R. Thomas 7 Lane Gardens, Bushey Heath, Bushey WD23 1PE
G3 VNG D. Hind 4 Thornyville Villas, Plymouth PL9 7LA
G3 VNH P. Hardy Lambda House, Seanor Lane, Lower Pilsley, Chesterfield S45 8DH
G3 VNI S. Cammies 5 Sheringham Close, Allington, Maidstone ME16 0NF
G3 VNP P. Dowles 1A Queen Street, Maldon CM9 5DP
G3 VNQ M. Pritchard 9 Tamarack Drive, Cortlandt Manor 10567 United States
G3 VNT L. Pearson Hatherly, The Street Ashfield, Stowmarket IP14 6LX
G3 VNU J. Finch 286 Sea Front, Hayling Island PO11 0AZ
G3 VNY I. Walker 45 Terry Drive, Walmley, Sutton Coldfield B76 2PT
GW3 VNZ D. Jacklin 40 Westbourne Road, Penarth CF64 3HF
G3 VOB D. Vivian Belle View Cottage, Blandford Road North, Poole BH16 5PP
G3 VOF M. Foster 1 Clavering Court Lincombe Drive, Torquay TQ1 2HH
GW3 VOL J. Phillips 96 Maes Y Sarn, Pentyrch, Cardiff CF15 9QR
G3 VOM D. Lane 2 Eden Close, Wilmslow SK9 6BG
G3 VOO M. Barnett-Bone 7 Dorchester Hill, Milborne St. Andrew, Blandford Forum DT11 0JG
G3 VOS R. Cottrell Larkhill, 47 Bullsland Lane, Rickmansworth WD3 5BD
G3 VOT G. Webster Red House Farm, Ashford Lane, Bakewell DE45 1NJ
G3 VOU J. Barlow 68 Willow Avenue, Cheadle Hulme, Cheadle SK8 6AX
G3 VOV M. Lane 56 Main Street, Bushby, Leicester LE7 9PP
G3 VOW M. Fereday Spindlewood, Stoney Lane, Thatcham RG18 9HQ
G3 VPA M. Rose 59 Park Drive, Sittingbourne ME10 1RD
G3 VPE H. Pinchin Birchmere Retirement & Care Home 1270 Warwick Road, Knowle, Solihull B93 9LQ
G3 VPF E. Harland 5 Bramdon Lane, Portesham, Weymouth DT3 4HG
G3 VPG C. Jacob 18 Compton Way, Olivers Battery, Winchester SO22 4HS
G3 VPH J. Mayall 10 Manor Close, Droitwich WR9 8HG
G3 VPK W. Mcclintock 1 Small Horse Farm Close, Freshwater PO40 9FY
GM3 VPN J. Gardner Taringa, Edentown, Cupar KY15 7UH
G3 VPQ I. Westwood 14 Staplegrove Road, Taunton TA1 1DQ
G3 VPR R. Harrison 512 Broadgate, Weston Hills, Spalding PE12 6DA
G3 VPS P. Lennard 5 Parkside, East Grinstead RH19 1JG
G3 VPT P. Burgess 26 William Peck Road, Spixworth, Norwich NR10 3QB
GI3 VPV R. Aughey 30 Glen Road, Hillsborough BT26 6ES
G3 VPW J. Wright 2 Barnfield, Charney Bassett, , Wantage OX12 0HA
G3 VPX I. Sumner 132 Barrs Road, Cradley Heath B64 7EZ
G3 VQF J. Moorhouse 185 Aldermoor Road, Southampton SO16 5NQ
G3 VQG R. Beadle 5 Badgeney Road, March PE15 9AP
G3 VQO L. Allwood 9 Gorse End, Horsham RH12 5XW
G3 VQQ M. Hall 5 Kings Lea, Kingsway WF5 8RY
G3 VQR A. Henshaw 3 Lewens Close, Wimborne BH21 1JJ
G3 VQS R. Kirby 197 Longfield, Falmouth TR11 4SR
G3 VQW B. Fawkes 6 Oak Avenue, Worcester WR4 9UG
G3 VQY Cannock Chase ARS c/o J. Cumming Camelot, Cheltenham Road, Hockley SS5 5HJ
G3 VRB J. Nias 49 St. Margarets Road, Bishopstoke, Eastleigh SO50 6DG
G3 VRE Chippenham And District ARC c/o B. Tanner 2 Doveys Cottages Days Lane, Kington Langley SN15 5NT
G3 VRF J. Charlton 57 Victoria Road, Bidford-On-Avon, Alcester B50 4AR
G3 VRU P. Ford 15 Doles Lane, Whitwell, Worksop S80 4SN
G3 VRV M. Huish Becketts, Woodbury, Exeter EX5 1JD
G3 VRW P. Lamb 5 The Templars, Bridge End, Warwick CV34 6PF
G3 VRY J. Pitt 30 Hillcroft Road, Chesham HP5 3DJ
G3 VSB G. Jones Braemar, Alton Road, Uttoxeter ST14 5DH
G3 VSE K. Thompson 3 Parkside, Morecambe LA4 4TJ
G3 VSH D. Freedman 102 Collingwood Road, Sutton SM1 2RB
G3 VSI N. Prince 96 Foxglove Way, Springfield, Chelmsford CM1 6QR
G3 VSJ D. Chaloner 38 Barnfield Close, Hoddesdon EN11 9EP
G3 VSK T. Mccurry 148 Moorgate Road, Rotherham S603AZ
G3 VSL J. Arscott 122 Woodlands Road, Ashurst, Southampton SO40 7AL
G3 VSQ R. West 10 Hawkshill Drive, Hemel Hempstead HP3 0BS
G3 VSR T. Barraclough 27 Kestrel Park, Skelmersdale WN8 6TA
G3 VST F. Moore Causeway House, Risbury, Leominster HR6 0NG
G3 VSU A. Moore 48 Cransley Rise, Mawsley, Kettering NN14 1TB
G3 VSV D. Middleton 8 Fulmar Close, Bradwell, Great Yarmouth NR31 8JG
GM3 VTB V. Budas 20 Oak Avenue, Bearsden, Glasgow G61 3HD
G3 VTD R. Price 36, Hadleigh Rise, Pontefract WF8 4SJ
G3 VTE R. Swetmore 18 Tideswell Road, Stoke-on-Trent ST3 5EG
GM3 VTH D. Coutts 29 Barons Hill Avenue, Linlithgow EH49 7JU
G3 VTL J. Levett 56 St. Nicholas Avenue, Kenilworth CV8 1JW
G3 VTO M. Coombs 10 Horseshoe Walk, Widcombe, Bath BA2 6DE
G3 VTR A. Davis Fieldings, Bury Road, Bury St. Edmunds IP29 4PL
G3 VTS C. Walker 2 Georgian Close, Abbeydale, Gloucester GL4 5DG
G3 VTT C. Turner 84 Gravel Hill Way Dovercourt, Harwich CO12 4XN
G3 VUD P. Bentley 12 West Terrace Seaton Sluice, Whitley Bay NE26 4RE
G3 VUE T. Mowbray Elmhirst House, Lincoln Road, Horncastle LN9 5AW
G3 VUH M. Blackwell Room 2, Mulroy House Peaker Park, Market Harborough LE16 7FP
G3 VUI M. Harris 3 Ross Road East, Stanley Po Box 1064 FIQQ 1ZZ Falkland Islands (Malvinas)
G3 VUK R. Knight 8 Narromine Drive, Calcot, Reading RG31 7ZL
G3 VUL Dr J. Lotz 29 Burton Manor Road, Stafford ST17 9QJ
G3 VUN G. Ackerley The Paddock, Stoney Lane, Tarporley CW6 0SX
G3 VUO J. Mills 9 Sandpiper Walk, Chelmsford CM2 8XJ
G3 VUR K. Evans 68 Downs Road, Hastings TN34 2DZ
G3 VUS D. Latimer Braefoot, Lanercost Road, Brampton CA8 1EN
G3 VUY D. Bradley 4 Felthorpe Close, Upton, Wirral CH49 4GY
GW3 VVC J. Parry Ar-Allt, 19 Lon Hedydd, Llanfairpwllgwyngyll LL61 5JY
GM3 VVF A. Ross 17 Tarvit Green, Glenrothes KY7 4SJ
G3 VVL K. Lax 17 Malt Rise, Crew Green, Shrewsbury SY5 9EU
G3 VVR J. Grace Woodside, Easthorpe, Malton YO17 6QX
G3 VWA C. Marflow 13 Walthew Green, Roby Mill, Skelmersdale WN8 0QT
G3 VWC A. Marriott 28 Horseshoe Walk, Bath BA2 6DF
G3 VWD C. Bean 11 Nightingale Lane, Coventry CV5 6AY
G3 VWH B. Wilde 34 Grangefields Road, Shrewsbury SY3 9DB
G3 VWJ G. Westwood 133 Torrisholme Road, Lancaster LA1 2TZ
G3 VWK A. Hammett (Hammett), Ladock, Truro TR2 4PQ
G3 VWQ P. Forster 59 Woodland View, Stratton Strawless, Norwich NR10 5LT
G3 VWX E. Perks The Oaklands, Bromfield Road, Ludlow SY8 1DW
GM3 VWY I. Malcolm 2 Morton Crescent, St. Andrews KY16 8RA
G3 VXA M. Harrold 26 Leys Close, Harefield, Uxbridge UB9 6QB
G3 VXE G. Brindle 8 Peckover Drive, Pudsey LS28 8EF
G3 VXF B. Ellis 15A The Street , Stedham, Midhurst GU29 0NQ
G3 VXH R. Huffadine 19 Cumberland Street, Worcester WR1 1QE
G3 VXJ R. Rylatt 16 First Avenue, Worthing BN14 9NJ
G3 VXK R. Porter 16 Millcroft, Crosby, Liverpool L23 9XJ
G3 VXS D. Peach Flat 35, Homeshire House, 36 Sandbach Road South, Stoke-on-Trent ST7 2LP
G3 VXY Prof. B. Cotton 12 Tower Gardens, Bassett, Southampton SO16 7EL
G3 VYA B. Atkiss 47 Russell Road, Partington, Manchester M31 4DY
G3 VYD J. Bourne Tyndalls, 8 Kelvedon Road, Witham CM8 3LZ
G3 VYF M. Lee 11 Sturrocks, Vange, Basildon SS16 4PQ
G3 VYG R. Walpole 7 Springfield Road, Taverham, Norwich NR8 6QU
G3 VYI M. Franklin 6 Tor Road, Farnham GU9 7BX
GM3 VYJ T. Jameson 8 River View, Dalgety Bay, Dunfermline KY11 9YE
G3 VYK P. Frost 164 Newthorpe Common, Newthorpe, Nottingham NG16 2EN
G3 VYN M. Turner Plumtree Cottage, Spring Lane, Norwich NR15 2NT
G3 VYS A. Nell Teaselwood, Parkham, Bideford EX39 5PL
G3 VYU R. Chamberlain 1 Thornemead Werrington Meadows, Peterborough PE4 7ZD
G3 VYW D. Carter 13 Sturdy Close, Hythe CT21 6AG
G3 VYX C. Burr The Old Rectory House, Radipole Lane, Weymouth DT4 9RN
GI3 VYY B. Hamilton 4 Castleton Court, 16 Osborne Park, Belfast BT9 6HA
G3 VYZ L. Thompson 44 Tillmouth Avenue, Holywell, Whitley Bay NE25 0NP
G3 VZE D. Kennedy 79 High Street, Dunsville, Doncaster DN7 4BS
G3 VZF J. Adams Chilterns, Bellingdon, Chesham HP5 2XL
G3 VZG R. Golding 7 Belvidere Avenue, Shrewsbury SY2 5PF
G3 VZH Dr C. Doran 16 Wordsworth Road, Penge, London SE20 7JG
G3 VZJ A. Clemmetsen 6 Lord Louis Crescent, Plymouth PL9 9SH
G3 VZL R. Newman 20 Glapthorn Road Oundle, Peterborough PE8 4JQ
G3 VZM F. Houghton 14 Windfield Gardens Little Sutton, Ellesmere Port CH66 1JJ
G3 VZO V. Hartshorn 61 Fulmerton Crescent, Redcar TS10 4NJ
G3 VZR E. Thompson Meadowside, Bromsberrow Heath, Ledbury HR8 1NX
G3 VZU W. Mooney 538 Liverpool Road, Great Sankey, Warrington WA5 3LU
G3 VZV G. Shirville Birdwood, Heath Lane, Woburn Sands Heath Lane, Woburn Sands, Milton Keynes MK17 8TN
G3 WAB P. Harrison 8 Buxtons Lane, Guilden Morden, Royston SG8 0JU
G3 WAE I. Harris Orchard Cottage, The Street, Devizes SN10 2LD
G3 WAG D. Gillett 20 Redcar Avenue, Hereford HR4 9TJ
G3 WAH N. Hodgson 42 Tofts Grove, Rastrick, Brighouse HD6 3NP
G3 WAL J. Barker 76 Halebrose Court, Seafield Road, Bournemouth BH6 3DU
G3 WAM M. Taplin 146 Ashover Road, Old Tupton, Chesterfield S42 6HG
GM3 WAP A. Philp Philp House, High Street, Blairgowrie PH11 8DW
G3 WAS O.B.O. Mid-Cheshire CG c/o R. Smethers 46 Church Road, Burntwood WS7 9EA
G3 WBA I. Currell 47 Highdale Avenue, Clevedon BS21 7LU
G3 WBB E. Avery 2 Blythe Avenue, Thornton-Cleveleys FY5 2LL
G3 WBC R. Bryant 12 Laburnum Grove, Luton LU3 2DW
G3 WBG H. Hindle 6 Windsor Road, Conisbrough, Doncaster DN12 3DF
G3 WBI P. Lewis 15 Norwood Road, Lytham St. Annes FY8 2QN
G3 WBK Dr P. Tofts 48 Rugby Road, Brighton BN1 6EB
G3 WBL K. Weller Charbury House, Bayton, Kidderminster DY14 9LJ
G3 WBN A. Thurlow Chesnet House, Croydon CR0 5BA

G3 WBP J. Broadley 13 Portland Close, Bedford MK41 9NE
G3 WBQ T. Brook 22 Downside Road, Guildford GU4 8PH
GI3 WBR R. Mccrea 1 Killynoogan Terrace, Killynoogan, Enniskillen BT93 8DF
G3 WBS D. Thomson 2A The Landway, Kemsing, Sevenoaks TN15 6TG
GW3 WBU B. Vodden 22 Heath Avenue, Penarth CF64 2QZ
GW3 WCA P. Dunbar Pengwern Fach, Penrherber, Newcastle Emlyn SA38 9RL
G3 WCB D. John 27 Churchfields, Dartmouth TQ6 9HJ
G3 WCD C. Dillon 63 High Street, Toseland, St. Neots PE19 6RX
G3 WCE B. Edwards Elder Cottage Swanton Abbott Road Skeyton, Norwich NR10 5AU
G3 WCJ P. Hackett Po Box 1 Wooroloo Wa 6558, Wooroloo 6558 Australia
G3 WCL J. Croker 29 Alexandra Road, Bedminster Down, Bristol BS13 7DF
G3 WCM F. Chidlow 64 Mitchell Avenue, Northside, Workington CA14 1AA
G3 WCQ R. Bailey 43 Earlsdon Avenue South, Coventry CV5 6DR
G3 WCU J. Pealing 93 Fernside Road, Poole BH15 2JQ
GW3 WCV D. Howell 6 Douglas Close, Cardiff CF5 2QT
G3 WCY B. Smith 26 Sandhurst Lane, Blackwater, Camberley GU17 0DH
G3 WDD T. Horrobin 29 Ambleside Road Maghull, Liverpool L31 6BY
G3 WDE P. Ford 11 Brook Lane, Felixstowe IP11 7EG
G3 WDG Dr C. Suckling 314A Newton Road, Rushden NN10 0SY
G3 WDM C. Care 127 Brooklands Crescent, Fulwood, Sheffield S10 4GF
G3 WDN E. Fielding The Birches, 3 Sneath Road, Norwich NR15 2DS
G3 WDS D. Spooner 45 Otterburn Avenue, Whitley Bay NE25 9QR
G3 WDU I. Peterkin 243 Hampton Street, Hampton 3188 Australia
G3 WDX P. Hickey 11 Bembridge Place, Linden Lea, Leavesden, Watford WD25 7DN
G3 WEA A. Cross 34 Pinewood Drive, Potters Bar EN6 2BD
GM3 WED A. Rose Craiglea, Schoolcroft, Dingwall IV7 8LB
G3 WEF A. Beazley 24 Tealsbrook, Covingham, Swindon SN3 5AU
G3 WEG P. Webster 22 Whincroft Drive, Ferndown BH22 9LJ
G3 WEI D. Turner Birchwood, Heath Top, Market Drayton TF9 4QR
G3 WEJ S. Bradshaw 11 Meadow Park, Dawlish EX7 9BS
GI3 WEL R. Knox 91 Banbridge Road, Waringstown, Craigavon BT66 7RU
GI3 WEM V. Gracey 23 Cascum Road, Banbridge BT32 4LF
GW3 WEQ C. Collins 21 Bron Wern Llanddulas, Abergele LL22 8JD
G3 WEU K. Gregory 67 Clowne Road, Barlborough, Chesterfield S43 4EH
G3 WEW Dr R. Wood 8305 El Matador Drive, Gilroy 95020 United States
G3 WEY A. Nelson 4 Bell Close, Farmborough, Bath BA2 0AP
GW3 WEZ J. Lawrence 3 Siskin Crescent, Rogiet, Caldicot NP26 3UW
G3 WF D. Cockings Elettra, 207A Birchfield Road, Redditch B97 4LX
G3 WFF B. Tew 96 Mill Lane, Sawston CB223HZ
G3 WFH D. Morris 27 Albert Square, Bowdon, Altrincham WA14 2ND
GI3 WFP P. Mcalpine 20 Gransha Road South, Bangor BT19 7QB
G3 WFT D. Holland 32 Woodville Drive, Sale M33 6NF
G3 WFW K. Hampson 11 Gladstone Grove, Stockport SK4 4BX
G3 WGE Aberdare ARS c/o E. Law Brunanburgh, 1B Ponds Road, Galleywood., Chelmsford CM2 8QP
G3 WGH M. Reeve 18-20 Radford Road, Nottingham NG7 5FS
G3 WGK B. Wormwell 26 Windsor Avenue, Longridge, Preston PR3 3EL
G3 WGN D. Aslin Old Smithy, Cornworthy, Totnes TQ9 7HH
G3 WGV J. Linford Pennine View, Sleagill, Penrith CA10 3HD
G3 WGY H. Ashford 56 Guarlford Road, Malvern WR14 3QP
G3 WGZ G. Sowden Villa Clare, The Lizard, Helston TR12 7NU
GI3 WHA L. Hanna Igrangeville Park, Newtownards BT23 8TE
G3 WHB S. Christie 4 Dairy Court Holyport, Maidenhead SL6 2US
G3 WHG M. Key 12 Great Melton Road, Hethersett, Norwich NR9 3AB
G3 WHJ A. Johnson 49 Tennyson Drive, Malvern WR14 2UL
GM3 WHT M. Smith Vakterlee, Cumliewick, Shetland ZE2 9HH
G3 WI M. Hulme 44 Thirlmere Avenue, Ashton-under-Lyne OL7 9HN
G3 WIA R. Ottley 15 Orchard Way, Thrapston, Kettering NN14 4RE
G3 WII F. Clarke 8 Tristram Close, Chandler'S Ford, Eastleigh SO53 4TT
GM3 WIJ N. Mackenzie 57 Countesswells Terrace, Aberdeen AB15 8LQ
G3 WIK M. Shorland Baxhill Bungalow, Chase Road, Upper Colwall, Malvern WR13 6DL
G3 WIM Wimbledon & District ARS c/o S. Pelling 10 Narrow Way, Bromley BR2 8JB
G3 WIN Windscale ARS c/o R. Wood Abbots Croft, Abbey Road, St. Bees CA27 0EG
G3 WIO E. Obrien Tanglewood, Anthonys Way, Wirral CH60 0BP
G3 WIP Dr G. Bulger Flat C/21, Herbal Hill Gardens, 9 Herbal Hill, London EC1R 5XB
G3 WIS B. Day 54 South Avenue, Hope Carr, Leigh WN7 3BU
G3 WIU W. Bekenn 35 Blackdown Avenue, Rushmere St. Andrew, Ipswich IP5 1AY
G3 WIW A. Leach 199 Braemor Road, Calne SN11 9EA
GM3 WJE J. Thom 50 Dickson Avenue, Dundee DD2 4EG
G3 WJG G. Lean 54 Blacketts Wood Drive, Chorleywood, Rickmansworth WD3 5QH
G3 WJH W. Wilkinson Chiriqui, 15 Camerton Road, Workington CA14 1LP
G3 WJI P. White Linden House, Willisham, Ipswich IP8 4SP
G3 WJJ D. Finnemore 4 Purbeck Gardens, Felton Road, Poole BH14 0QS
G3 WJM B. Schoth 3 Solent Drive, Hythe, Southampton SO45 5FP
G3 WJN R. Hassell Bennett 30 Greenlands Avenue, Redditch B98 7QA
G3 WJS J. Starling 16 Queenscliffe Road, Ipswich IP2 9AS
G3 WKA N. Bardell 4 Church End, Arlesey SG15 6UY
GM3 WKB D. Topham Dairy Cottage, Kilmany, Cupar KY15 4PT
G3 WKE S. Braidwood Flat 229, Helen Gladstone House, Nelson Square, London SE1 0QB
G3 WKF M. Richards Wayside Cottage, Penwithick Road, Penwithick, St. Austell PL26 8UH
G3 WKI M. Hewins 37 Ringwood Close, Furnace Green, Crawley RH10 6HQ
G3 WKL Dr J. Gould 116 Wolverton Road, Newport Pagnell MK16 8JG
G3 WKP P. King Nirvana, Comprigney Hill, Truro TR1 3TX
G3 WKR M. Goodwin 6 Hobbs Hill, Rothwell, Kettering NN14 6YG
G3 WKS West Kent ARS c/o D. Green St. Annes Poundfield Road, Crowborough TN6 2BG
G3 WKW R. Thornton 26 Florence Road, Fleet GU52 6LQ
G3 WKX Maidenhead & District ARC c/o M. Palmer 28 Westfield Road, Caversham, Reading RG4 8HH
G3 WLA A. Macpherson 15A Monkstone Drive, Berrow, Burnham-on-Sea TA8 2NW
G3 WLD J. Hall 22 Haverhill Road, Stapleford, Cambridge CB22 5BX
G3 WLG Dr M. Griffiths The Oaklands, Hollybush Lane, Worcester WR6 6HQ
G3 WLH Dr C. Pell 1 Glenville Gardens, Hindhead GU26 6SX
G3 WLM R. Joyce 20 Barking Close, Luton LU4 9HG
GW3 WLN Club Hut c/o Dr J. Pritchard 13 Cefn Graig, Rhiwbina, Cardiff CF14 6SW
G3 WLO E. Denton 11 Highland Road, Amersham HP7 9AU
G3 WLT Dr D. Firth 3 School Lane, Shaldon, Teignmouth TQ14 0DG
G3 WLV J. Bushby 14 Clayton Drive, Thurnscoe, Rotherham S63 0RZ
G3 WLW R. Millar 1229 Leeds Road, Bradley, Huddersfield HD2 1UY
G3 WLY J. Harwood 12 Longwood Avenue, Cowplain, Waterlooville PO8 8HX
G3 WMA W. Shepperd 28 Tyne Road, Oakham LE15 6SJ
G3 WMD J. Whomes 44 Russell Close, Steeple Morden, Royston SG8 0NE
G3 WME M. Groom 409 Finchampstead Road, Finchampstead, Wokingham RG40 3RL
G3 WMJ G. Jillings 16 Greenwich Road, Diep River 7800 South Africa
GW3 WMP Dr J. Hopton 9 Bryneuraidd, Ammanford SA18 3TG
G3 WMQ M. Watson Chant House, Dark Lane, Nailsworth GL6 0DR
G3 WMS I. Vance Larkfield, Debden Road, Saffron Walden CB11 3RU
G3 WMT R. Dowling 80 Elmfield Way, Sanderstead, South Croydon CR2 0EF
G3 WMU Amberley Radio Group c/o E. Spicer 3 Golden Avenue Close, East Preston, Littlehampton BN16 1QS
G3 WMX C. Knott 7 Manor View, Crewkerne TA18 8JT
G3 WMY S. Smith Five Oaks, Sandy Lane, Henfield BN5 9UX
GM3 WNB J. Ohare 208 Gilmartin Road, Linwood, Paisley PA3 3ST
G3 WNC R. Todd 17 Tudor Road, West Bridgford, Nottingham NG2 6EB
G3 WND R. Aston 16 St. Johns Road, Mortimer Common, Reading RG7 3TR
G3 WNP T. Baker 54 Hamilton Road, Reading RG1 5RD
G3 WNQ E. Lingard Tedulf, Rotten Row, Theddlethorpe, Mablethorpe LN12 1NX
G3 WNR K. Grey 15 Woodbourne Avenue, Leeds LS17 5PQ
G3 WNS A. Willson Hilltop, Cryers Hill Road, Cryers Hill, High Wycombe HP15 6LJ
G3 WNV D. Field Rackhay, Prescott, Cullompton EX15 3BA
G3 WNW D. Bailey 31 Antrobus Street, Congleton CW12 1HE
G3 WOA North Lancashire Raynet Group c/o J. Goodman 15 Highway, East Taphouse, Liskeard PL14 4NW
G3 WOD J. Welford 303 Scalby Road, Scarborough YO12 6TF
G3 WOE M. White 76 Birch Row, Bromley BR2 8FG
G3 WOH E. Grossmith 4 Lincoln Way, Rainhill, Prescot L35 6PJ
GM3 WOJ C. Tran Achnacoille, Lamington, Invergordon IV18 0PE
G3 WOK D. Clifton 59 Grantham Road, Bracebridge Heath, Lincoln LN4 2LE
G3 WOM M. Muir 6 Broadstairs Court, Sunderland SR4 8NP
G3 WOO R. Brace 11 Cedar Close, Sawbridgeworth CM21 9NT
G3 WOR Worthing And District Amateur Club c/o A. Cheeseman Flat 5 Dubarry House, Hove Park Villas Hove Park Villas, Hove BN3 6HP
G3 WOS Neath & District Sea Cadet Unit c/o C. Gare Old White Lodge, 183 Sycamore Road, Farnborough GU14 6RF
G3 WOT M. Meads 12 Burlington Way, Hemingford Grey, Huntingdon PE28 9BS
G3 WOV G. Macnaught 30 West End Falls, Nafferton, Driffield YO25 4QA
GU3 WOW P. Hancock La Breloque, Les Grandes Rues, Les Buttes GY79EL Guernsey
GM3 WPA S. Hutchinson 4 Wiston Place, Dundee DD2 3JR
G3 WPB P. Smith 180 Victoria Road, Ferndown BH22 9JE
G3 WPD A. Smith 118 Bois Moor Road, Chesham HP5 1SS
G3 WPF R. Unsworth Spurs Lodge, Sagars Road, Styal, Wilmslow SK9 4HE
G3 WPG Livingstone District ARS c/o D. Dye 10 Headington Close, Bradwell, Great Yarmouth NR31 8DN
G3 WPH Dr M. Chamberlain 10 Clifton Rise, Wargrave, Reading RG10 8BN
G3 WPN V. Bennellick The Wolverns, Castle Frome, Ledbury HR8 1HG
G3 WPP D. Minett Melrose Cottage, South Road, Truro TR3 7AD
G3 WPQ M. Kaye Pucknell Lodge, Hollywell Lane Bayton Common, Kidderminster DY14 9NR
G3 WPR C. Richmond 1 Grangeway Gardens, Ilford IG4 5HN
G3 WPT R. Brown 65 Staining Rise, Staining, Blackpool FY3 0BU
G3 WPV D. Lamont 6B Route De Mailhac, Bize Minervois 11120 France
G3 WQG D. Chalmers 25 Willow Close, Flackwell Heath, High Wycombe HP10 9LH
G3 WQK Southdown ARS c/o A. Seabrook 63 St. Annes Road Willingdon, Eastbourne BN20 9NJ
G3 WQL A. Conway 17 Mountcastle Road Leicester, Leicester LE3 2BW
G3 WQU P. Mckay C/O Unifil, Po Box 5852, New York 10163-5852 United States
G3 WQY T. Codrai Sealand, Coast Road, Norwich NR12 0PD
G3 WRA S. Powell 9 Belgravia Gardens, Hereford HR1 1RB
G3 WRD R. Richardson Common Crest, Drapery Common, Sudbury CO10 7RW
GW3 WRE B. Jones 6 Pentyla, Maesteg CF34 0BB
G3 WRI Dr P. Brown 30 Applerigg, Kendal LA9 6EA
G3 WRJ R. Bacon Gyffen, Gosmore Road, Hitchin SG4 9AN
G3 WRK G. Oakes 13 Tidnock Avenue, Congleton CW12 2HN
G3 WRL E. Northwood 8 Derwood Grove, Werrington, Peterborough PE4 5DD
G3 WRO K. Haynes 34 Pear Tree Mead, Harlow CM18 7BY
G3 WRR Q. Collier 19 Grangecliffe Gardens, South Norwood, London SE25 6SY
G3 WRT Dr I. Dilworth Ashpound Cottage, Pound Lane, Ipswich IP9 2JB
G3 WSB Dr K. Band 11 Denewood Close, Watford WD17 4SZ
G3 WSC Crawley ARC c/o J. Pitty 12 St. Leonards Road, Horsham RH13 6EJ
G3 WSD A. Fisher 63 Spencer Close, Potton, Sandy SG19 2QR
G3 WSM B. Storry 508 Arleston Lane, Stenson Fields, Derby DE24 3AA
GM3 WSR V. Clark 6 Parkhill Circle, Dyce, Aberdeen AB21 7FN
GW3 WSU C. Beynon 16 Hardy Close, Barry CF62 9HJ
G3 WSV J. Lawson 3 Pearmains, Great Leighs, Chelmsford CM3 1QS
G3 WSW J. Holmes 37 Redwood Avenue, Leyland PR25 1RN
G3 WTB R. Baxter 10 Windsor Court, Oxford Road, Southport PR8 2JJ
G3 WTD J. Davis 71 Broughton Road, Croft, Leicester LE9 3EB
GI3 WTG G. Thompson 36 Crewcatt Road, Richhill, Armagh BT61 8QN
G3 WTN R. Limehouse 56 Lincoln Way, Daventry NN11 4SX
G3 WTO Dr J. Spencer 76 Durranhill Road, Carlisle CA1 2SZ
G3 WTP the Bedford And District ARC c/o H. Ehm 17 Stuart Road, Kempston, Bedford MK42 8HS
G3 WTQ P. Angold 10 Hartford Avenue, Wilmslow SK9 6LP
G3 WTR D. Wright 8 Calverley Park, Tunbridge Wells TN1 2SH
G3 WTS J. Smith Windycross, Newbourne Road, Woodbridge IP12 4PT
G3 WTT C. Goodwin 1 School Lane, Canwick, Lincoln LN4 2RP
G3 WTV K. Baker 33 Reading Road, Woodley, Reading RG5 3DA
G3 WTY P. Hodgkiss 28 Beaumont Rise, Worksop S80 1YA
GW3 WTZ M. Jones 55 Rowan Way, Malpas, Newport NP20 6JN
G3 WUA B. Lindop 56 Marina Court, 9-19 Mount Wise, Newquay TR7 2EJ
G3 WUB Dr P. Rice 23 Christchurch Square, Homerton, London E9 7HU
G3 WUG I. Elvins 203 Nuthatch Road, Evergreen CO 80439 United States
G3 WUH W. Dufton 22 Windsor Road, Bexhill-on-Sea TN39 3PB

G3 WUI G. Spink 60 Woodhouse Hill, Huddersfield HD2 1DH
G3 WUK J. Spencer Chapman Apartado De Correos 156, Mojacar 04638, Almeria, Spain ZZ2 3TP
G3 WUL R. Whillier 9 Tudor Drive, Yateley GU46 6BX
G3 WUN D. Holden 99 Sheerstock, Haddenham, Aylesbury HP17 8EY
GI3 WUO Dr L. Waring 16 Belfast Road, Holywood BT18 9EL
G3 WUW A. Papworth 3570 Corey Road, Malabar 32950 United States
GM3 WUX T. Robinson 82 Albert Road, Glasgow G42 8DR
G3 WVG K. Pritchard 9 Golf Close, Pyrford., Woking GU22 8PE
G3 WVM C. Loosemore 24 Myrtlebury Way, Exeter EX1 3GA
G3 WVQ J. Barratt 26 Johnstone Road, Newent GL18 1PZ
G3 WVR J. Green Honeysuckle Cottage, New Green, Braintree CM7 5EG
GW3 WVV R. Barker Henllys, Maenygroes, New Quay SA45 9RL
G3 WWG J. Ross 24 Raby Road, Stockton-on-Tees TS18 4JA
G3 WWH R. Taylor Shambles, Davids Lane, Benington, Boston PE22 0BZ
G3 WWI R. Oxley 1 Elm Grove, Maidstone ME15 7RT
G3 WWL B. Tipper 271 Blackberry Lane, Four Oaks, Sutton Coldfield B74 4JS
G3 WWS M. Southall 61 Grange Close, Horam, Heathfield TN210EF
G3 WWT J. Teed 47 West Cliff Road, Dawlish EX7 9DZ
GI3 WWY M. Anderson 8 Loughbrickland Road, Gilford, Craigavon BT63 6BH
GW3 WXA J. Gough Traleen, Rhydlewis, Llandysul SA44 5PN
G3 WXC P. Brooker 28 Uplands Road, Northwood, Cowes PO31 8AL
G3 WXD Dr C. Zammit 9 Sandbanks Drive, Hatch Warren, Basingstoke RG22 4UL
G3 WXG I. Habens 48 Carden Avenue, Brighton BN1 8NE
G3 WXH J. Arnold 6 The Spinney, Weston-Super-Mare BS24 9LH
G3 WXI A. Strong 3 Ellorslie Drive, Stocksbridge, Sheffield S36 2BB
G3 WXM M. Smith Linden, 86 Grove Road, Tring HP23 5PB
G3 WXN L. Mckown Flat D, 310 Oldham Road, Royton, Oldham OL2 5AS
G3 WXU S. Allbutt 8 Langton Close, Vinters Park, Maidstone ME14 5PG
G3 WXW C. Traveller 13 Cosy Corner, North Walsham NR28 0EN
G3 WYB A. Tring 1 Crownbourne Court, St. Nicholas Way, Sutton SM1 1JE
G3 WYD P. Patmore 141 Cannons Close, Bishop's Stortford CM23 2BL
G3 WYH R. Hutton 6 The Sidings, Ruskington, Sleaford NG34 9GA
G3 WYK P. Bysshe Orchard House, High Road, Maidenhead SL6 9JT
GM3 WYL A. Ritchie 83 Larkfield Road, Lenzie, Glasgow G66 3AS
G3 WYN J. Gibson Four Oaks, Tylers Green, Haywards Heath RH17 5DZ
G3 WYP D. Allan 283 Cliffe Lane, Gomersal, Cleckheaton BD19 4SB
G3 WYT M. Edwards 23 Burnside, Waterlooville PO7 7QQ
G3 WYW P. Bigwood 18 The Martins, Thatcham RG19 4FD
G3 WZE P. Cleary 531 Diamond Street, San Francisco 94114 United States
G3 WZG P. Murtha 9 Cross Street, Southport PR8 1HZ
G3 WZH N. Ghani 52B Stormore, Dilton Marsh, Westbury BA13 4BH
G3 WZI K. Reeves 9 Tibberton Close, Solihull B91 3UD
G3 WZJ A. Watt 30 The Hedgerows, Collingham NG237RL
G3 WZK S. Beal 16 Clovelly Avenue, Warlingham CR6 9HZ
G3 WZO J. Kyriakides 16 Wise Lane, London NW7 2RE
G3 WZP G. Budden 7 Ashburton Gardens, Bournemouth BH10 4HP
G3 WZR Dr R. Wright 2 Jackson Close, Devizes SN10 3AP
G3 WZS H. Williams 7 Munn Drive, Po Box 276, Tobermory N0H 2R0 Canada
G3 WZT J. Matthews 46 Park Lane West Grinstead, Horsham RH13 8LT
G3 WZW G. Laycock 1 Campsall Cottage, Churchfield Road, Doncaster DN6 9BY
G3 WZZ A. Huddleston Willow Bank Cottage, Willow Bank, Keighley BD20 5AN
G3 XAB D. Whittaker 2 Stone Edge, Halifax Road, Burnley BB10 3QH
G3 XAC C. Whitehead 10 Berkeley Drive, Read, Burnley BB12 7QG
G3 XAG J. Gibbon The Bungalow, Manless Terrace, Saltburn-by-the-Sea TS12 2DQ
G3 XAP A. Ashton 2 Wickham Road, Thwaite, Eye IP23 7EE
G3 XAQ A. Ibbetson Katallin, Town Lane Chartham Hatch, Canterbury CT4 7NN
G3 XAU T. Woodward 33 Common Road, Hemsby, Great Yarmouth NR29 4LT
G3 XAW M. Chouings 32 Nunney Close, Keynsham, Bristol BS31 1XG
G3 XAX A. Paley 19 Arbour Lane, Wickham Bishops, Witham CM8 3NS
G3 XBE H. Walton 12 Le Page Court, Nottingham NG8 3ES
G3 XBF Barking Radio And Electronics Society c/o S. Peat 64 Grange Road, Romford RM3 7DX
G3 XBH G. Thompson 25A Copleston Road, London SE15 4AN
G3 XBI P. Boast 19 Main Road, , West Huntspill TA9 3QU
G3 XBM R. Lapthorn 7 Mill Close, Burwell, Cambridge CB25 0HL
G3 XBN F. Chamberlain 43 Old Mill Close, Patcham, Brighton BN1 8WE
G3 XBQ A. Weseley Loves House, Goudhurst Road, Tonbridge TN12 9NB
G3 XBW M. Wells 15 Rivers Reach, Frome BA11 1AQ
G3 XBX D. Harris The White House, High Street, Newnham-on-Severn GL14 1BW
G3 XBY D. Harvey 38 School Road, Shirley, Solihull B90 2BB
G3 XBZ P. Ciotti 6 Bascott Road, Bournemouth BH11 8RH
G3 XCD H. Martin 17 Vyner Road, Wallasey CH45 6TE
G3 XCE E. Wells 23 Briarfield Road, Poulton-le-Fylde FY6 7PW
G3 XCJ W. Burden 44 Spekehill, Eltham, London SE9 3BW
G3 XCK J. Pegrum 14 The Leys, Langford, Biggleswade SG18 9RS
G3 XCO Dr D. Meldrum 34 Graham Road, Ipswich IP1 3QF
G3 XCS C. Squires 5 Frith Road, Saltash PL12 6EL
G3 XCT D. Dade 40 Compton Avenue, Brighton BN1 3PS
G3 XCW G. Winter 14 Drakes Lea, Evesham WR11 3BJ
G3 XCY K. Bristow 34 Stanier Road, Preston, Weymouth DT3 6PD
GI3 XCZ G. Martin 100 Drumconnelly Road, Gortaclare, Omagh BT79 0XS
GI3 XDD S. Crampton 135A Ballymena Road, Doagh, Ballyclare BT39 0TN
G3 XDK A. Maris 2 Court Ord Road, Rottingdean, Brighton BN2 7FD
G3 XDL A. Long 2A Hawthorndene Road, Bromley BR2 7DY
G3 XDM A. Benson 31 Oakhill Drive, Welwyn AL6 9NW
G3 XDP G. Wilkinson 509 Warrington Road, Culcheth, Warrington WA3 5QY
G3 XDS P. Wilde 5 Ruddington Court, Mansfield NG18 4QD
G3 XDU K. Whitbread 27 Duckmill Crescent, Duckmill Lane, Bedford MK42 0AF
GI3 XDX G. Mcdowell 13 Redford Road, Cullybackey, Ballymena BT43 5PR
G3 XDY J. Quarmby 12 Chestnut Close, Rushmere St. Andrew, Ipswich IP5 1ED
G3 XDZ Z. Skrobanski 1035 Pine Grove Pointe Drive, Roswell GA 30075-2704 United States
G3 XEC G. Grundy Route De L'Angouiniere, la Roche Sur Yon 85000 France
G3 XED C. Masters 79 Kings Head Lane, Bristol BS13 7DB
G3 XEF M. Fleetwood Hemmet, 235 Shingle Hill Way, Sidcup DA14 4ES
G3 XEI J. Hooper Long Barn House, Bolney Road, Horsham RH13 8AZ
G3 XEN P. Mullineaux 27 Ashfield Avenue, Lancaster LA1 5EB
G3 XEP White Rose ARS c/o E. Hannaby 170A Weston Drive Otley, Leeds LS21 2DT
GI3 XEQ J. Bailie 7 Houston Road, Belfast BT6 9SE
G3 XER D. Mannix 34 Ashby Road, Ticknall, Derby DE73 7JJ
G3 XEW G. Childs 115 Summerhouse Drive, Bexley DA5 2ER
G3 XEY A. Robinson 48 Colton Road, Shrivenham, Swindon SN6 8AZ
G3 XFD Dr R. Mannion Flat 1, 1 Spencer Road, Bournemouth BH1 3TE
G3 XFF E. Tuddenham 42 Garrison Lane, Felixstowe IP11 7RP
G3 XFU I. Hasman Fleetway, The Spinney, Newark NG24 2NT
G3 XG Braintree & District ARS c/o M. Kendall 88 Coldnailhurst Avenue, Braintree CM7 5PY
G3 XGC Dr G. Cottrell11 36 Davenant Road, Oxford OX2 8BY
G3 XGD G. Watson 6 The Avenue, Lyneal, Ellesmere SY12 0QJ
G3 XGE P. Greenhalgh 13 Primrose Avenue, Urmston M41 0TY
G3 XGH W. Jamison Horseshoe Cottage, Town Fold, Stockport SK6 5BT
G3 XGK C. Langley Clarence Cottage, Commodore Road, Lowestoft NR32 3NF
G3 XGU K. Hill 42 Greenleafe Avenue, Doncaster DN2 5RF
G3 XGV G. Fowles Ruby House, Broad Marston, Stratford-upon-Avon CV37 8XY
G3 XGW K. Yates Tibblestone Lodge, Ashton Road, Beckford, Tewkesbury GL20 7AU
G3 XGY B. Harris 36 Holland Way, Blandford Forum DT11 7RU
G3 XHB G. Bentley 2 Conan Drive, Richmond DL10 4PQ
GW3 XHD B. Walters 16 Broomhill, Port Talbot SA13 2US
GW3 XHG D. Griffiths 7 Canning Street Ton Pentre, Pentre CF41 7HF
G3 XHM A. Lewis Bradley Villa, 41 West Street, Ryde PO33 2UH
G3 XHW Dr J. Morris 2 The Corniche, Sandgate, Folkestone CT20 3TA
G3 XHY Dr C. Tinline 56 Appletree Lane, Redditch B97 6SE
G3 XHZ J. Farrer Woodside Cottage, Catmere End, Saffron Walden CB11 4XG
G3 XIA P. Bates 28 Salvington Hill, Worthing BN13 3AT
G3 XIB B. Johnson 30 Tamar Way, Gunnislake PL18 9DH
G3 XIG C. Graham 1 Arnhem Green, Dorchester DT1 2PS
G3 XIH W. Dixon 17 Chestnut Bank, Scarborough YO12 5QJ
G3 XII F. Harrison 78 Lancaster Lane, Leyland PR25 5SP
G3 XIP D. Aspinall 53 Springhill Road, Fen Drayton, Cambridge CB24 4SR
G3 XIQ K. Finch 116 Wisbech Road, Outwell, Wisbech PE14 8PF
G3 XIR I. Deane 26 Callin Court Grey Friars, Chester CH1 2NW
GW3 XIS Dr R. Belcher 8 Bishops Grove, Sketty, Swansea SA2 8BE
G3 XIV G. Bulleyment 30 Brackley Avenue, Fair Oak, Eastleigh SO50 8FL
G3 XIX J. Hobin 14 St. Martins Green, Trimley St. Martin, Felixstowe IP11 0UU
G3 XIY R. Hall 22 Cumbria Close, Thornbury, Bristol BS35 2YE
G3 XIZ C. Osborn 116 Holme Court Avenue, Biggleswade SG18 8PB
GW3 XJC B. Luke 33 Maiden Street, Cwmfelin, Maesteg CF34 9HP
GM3 XJE Dr P. Duffett-Smith 3 Simpson Place, Carnoustie DD7 7PJ
G3 XJI W. Wilkinson 1 Scafell Drive, Kendal LA9 7PE
G3 XJM J. Sawdy 41 Ashbarn Crescent, Winchester SO22 4QH
G3 XJN H. Duncombe La Rochelle, Thakeham Copse, Pulborough RH20 3JW
G3 XJP P. Rhodes Danvers House, Wigmore, Leominster HR6 9UF
GW3 XJQ M. Shelley Sunray, Pendine, Carmarthen SA33 4PD
G3 XJR A. Dickinson 1 Pearl Bank, Apartment #33-04, Singapore 169016 Singapore
G3 XJS P. Barville Felucca, Pinesfield Lane, West Malling ME19 5EN
G3 XJW L. Rix 63 Edendale Road, Melton Mowbray LE13 0EW
G3 XJZ C. Sykes 15 Morpeth Close, Wirral CH46 6HQ
GW3 XKB K. Bevan Renhold, 25 Bryn Gannock, Conwy LL31 9UG
G3 XKD M. King 15 Glebe Road, Prestbury, Cheltenham GL52 3DG
G3 XKE C. Evans 8 Blakelands Avenue, Sydenham, Leamington Spa CV31 1RJ
G3 XKG R. Stanton 16 Ashwood Park, Fetcham, Leatherhead KT22 9NT
G3 XKH B. Ward 12 Pagets Road, Bishops Cleeve, Cheltenham GL52 8AG
G3 XKL C. Gill Little Acre, Plough Lane Marston, Devizes SN10 5SR
G3 XKM Dr I. Mcmanus Orchard Cottage, Ecchinswell, Newbury RG20 4UQ
G3 XKS H. Grattan Grattan Grange, St. Breward, Bodmin PL30 3PN
G3 XKU M. Rose 71 Maryon Road, Ipswich IP3 9NJ
G3 XKV R. Stratton 60 Lateward Road, Brentford TW8 0PL
G3 XKX D. Wills 70 Hidcote Road, Oadby, Leicester LE2 5PF
G3 XKY G. Schrager Flat 4, 3 The Park, London N6 4EU
G3 XLE J. Vaughan Eastwood Lodge, Main Road, Boston PE22 7JU
G3 XLG R. Spreadbury Lings Farm, Blacksmiths Lane Forward Green, Stowmarket IP14 5ET
G3 XLI P. Holland 38 Marlin Square, Abbots Langley WD5 0EG
G3 XLL J. Lockwood 22 Egremont Road, Diss IP22 4NF
G3 XLN Dr D. Russell 29 Gold Street, Hanslope, Milton Keynes MK19 7LU
G3 XLP I. Richardson Brockwood, Grove Road, Ryde PO33 3LH
G3 XLR A. Bunyan 87 Seymer Road, Romford RM1 4LA
G3 XLW D. Powell Broomhill, Aveton Gifford, Kingsbridge TQ7 4NE
G3 XLX R. Littlewood Brewery Farm, Old Coach Road, Axbridge BS26 2EH
G3 XLZ J. Tozer 54 Ganges Road, Plymouth PL2 3AZ
G3 XMB R. Richardson 42 King Edwards Road, South Woodham Ferrers, Chelmsford CM3 5PQ
G3 XMC Thurrock Sea Cadet Corp - Tilbury c/o D. Brain Orchardleigh, Bristol Road, Bristol BS40 6HF
G3 XMG M. Graham 30 Moorlands Road, Thornton, Liverpool L23 1US
G3 XMH P. Brumfitt 85 Centenary Way, Copcut, Droitwich WR9 7TD
G3 XMK A. Flather 10 Oakleigh Court, Aston Lodge Park, Stone ST15 8LA
G3 XMM T. Morgan 32 Grasmere Road Longlevens, Gloucester GL2 0NQ
G3 XMP A. Brasier The Bears, Moor End Lane, Fakenham NR21 0EJ
G3 XMQ P. Eggleton 12 Newbold Road, Wellesbourne, Warwick CV35 9NZ
GM3 XMY D. Hobden 1 Larch Avenue, Glenrothes KY7 5TE
G3 XNE A. Smyth Flat 3, Tree View, Fore Street, Stratton, Stratton EX23 9DA
G3 XNN R. Jephcott 3 Chatsworth Park, Thornbury, Bristol BS35 1JF
G3 XNO Otley ARS c/o J. Castelow 7 Langford Close, Burley In Wharfedale, Ilkley LS29 7NP
GD3 XNU J. Craine Mwyllin Squeen, Station Road, Ballaugh IM7 5AH Isle of Man
G3 XNX D. Chivers 51 Alma Road, Brixham TQ5 8QR
G3 XOB D. Ellacott 39 Canford Lane, Bristol BS9 3DQ
G3 XOC M. Cooley 21 Castle Road, Newport PO30 1DT
G3 XOD R. Horsman 65 Pendennis Park, Brislington, Bristol BS4 4JL
G3 XOI A. Gordon 20 Hawkins Crescent, Shoreham-by-Sea BN43 6TP
GJ3 XOJ D. Gray La Brecque, La Grande Route De La Cote, St. Clement JE2 6FP Jersey
G3 XOK R. Kearney 32 Springfield Road, Lower Somersham, Ipswich IP8 4PQ
G3 XOP P. Featherstone 2 Firs Close, Whitchurch, Aylesbury HP22 4LH

GM3 XOQ P. Weller Mither Tap, Bridge Road, Kemnay, Inverurie AB51 5QT
G3 XOU D. Wright Cross Cottage, 208 Whitchurch Road, Tavistock PL19 9DQ
G3 XOV R. Johnson 29 Hungary Hill, Stourbridge DY9 7PS
G3 XPA R. Bevan Sitio Do Laranjeiro 616F, Moncarapacho, Olhao 8700 077 Portugal
G3 XPC R. Chapman 22 Windsor Ride, Finchampstead, Wokingham RG40 3LG
G3 XPD D. Smith 5 Peel Street, Stafford ST16 2DZ
G3 XPI B. Hallows 3 Southdown Close, Rochdale OL11 4PP
G3 XPJ K. George 34 Third Avenue, Northville, Bristol BS7 0RT
GW3 XPK J. Dore Henfaes Isaf, Llangurig, Llanidloes SY18 6SN
G3 XPM R. Tinson 1 Plompton Grove, Harrogate HG2 7DP
G3 XPQ G. Black 24 Mount Drive, Leyburn DL8 5JQ
G3 XPR I. Bassett-Smith Grey Gables, Southam Road, Cheltenham GL52 3BB
G3 XPT G. Symonds 45 Westfield Road, Dereham NR19 1JB
G3 XPU C. Woodley 170 Rugby Road, Burbage, , Hinckley, LE10 2ND
G3 XPW C. Moller 344 High Street, Cottenham, Cambridge CB24 8TX
G3 XPY A. Bagley 11 Glamis Road, Newquay TR7 2RY
G3 XPZ J. Appleton 66 Bolton Road West, Ramsbottom, Bury BL0 9ND
G3 XQJ G. Wren 2 Netheredge Close, Knaresborough HG5 9BZ
G3 XQM A. Finch 15 The Brooks, Burgess Hill RH15 8TR
GW3 XQO P. Salomon 28 Ansell Road, Wrexham LL13 9NQ
GM3 XQP Dr A. French 83/16 Hopetoun St., Edinburgh EH7 4NJ
G3 XQZ Sir P. Simpson 8 Church Road, Egleton, Oakham LE15 8AD
G3 XRC D. Carlsen 57 Chignal Road, Chelmsford CM1 2JA
G3 XRD G. Knight 57B Oliver Road, Kirk Hallam, Ilkeston DE7 4JY
G3 XRI P. Williams 2 Sycamore Avenue, Newton-le-Willows WA12 8LT
G3 XRJ S. Chappell Lyonesse, Trebehor, St. Levan, Penzance TR19 6LX
G3 XRK D. Griffin 12 Charles Road, Whittlesey, Peterborough PE7 2RG
GW3 XRM D. Dunn 9 Mill Bank Estate, Llandegfan, Menai Bridge LL59 5RD
GI3 XRQ Bangor And District ARS c/o R. White 28 Lord Warden'S Parade, Bangor BT19 1YU
G3 XSC Lowestoft District And Pye ARC c/o K. Southgate 10 Cott Road, Lostwithiel PL22 0ET
G3 XSD Prof. G. King 1 Spring Gardens, North Baddesley, Southampton SO52 9JG
G3 XSI T. Haslam 29 Backmoor Road, Norton, Sheffield S8 8LB
G3 XSJ K. Brooks Woodland House, 2 Beechfield Grove, Bristol BS9 2RZ
G3 XSN B. Donn 7 Thurne Way, Liverpool L25 4SQ
GW3 XSR A. Sutton 3 Pendre Walk, Tywyn LL36 0AY
G3 XSV A. Hydes Woodcroft Bath Road, Langford, Bristol BS40 5EB
G3 XSZ F. Mundy 5 The Paddocks, New Haw, Addlestone KT15 3LX
G3 XTC P. Borrett 21 Kenley Walk, Cheam, Sutton SM3 8ES
G3 XTH G. King 73 Grand Avenue, Hassocks BN6 8DD
G3 XTI J. Jarvie 11 Guild Road Aston Cantlow, Henley-in-Arden B95 6JA
G3 XTN R. Hough 1 Fiddler Hall, Newby Bridge, Ulverston LA12 8NQ
G3 XTP K. Lloyd 10 The Verneys, Cheltenham GL53 7DB
G3 XTQ M. Draycott 119A High Street North, Stewkley, Leighton Buzzard LU7 0EX
G3 XTR B. Dunn 12 Campbell Road, Westville 3629 South Africa
G3 XTT D. Field Daisy Cottage, Henton, Wells BA5 1PD
G3 XTZ G. Phillips 27 Stanley Road, Ashford TW15 2LP
G3 XUC A. Keohane 6 Birchwood Fields, Tuffley, Gloucester GL4 0AL
G3 XUD P. Kirby 7 Lillywhite Close, Burgess Hill RH15 8TF
G3 XUF A. Warner 79 Kelvin Grove, Portchester, Fareham PO16 8LF
G3 XUH R. Pearson 8 St. Benets Close, Walton-Le-Dale, Preston PR5 4UT
G3 XUM J. Moran 30 Elsie Street, Farnworth Bolton BL4 9HT
G3 XUP G. Everest 13 Noel Rise, Burgess Hill RH15 8BW
GM3 XUW R. Johnston 123 Craigmount Brae, Edinburgh EH12 8XW
G3 XUX E. Fitzgerald 4 Southwick Road, Wickham, Fareham PO17 6HS
G3 XVA D. Pickles 40 Jupiter House Hindhead Knoll Walnut Tree, Milton Keynes MK7 7FH
G3 XVB A. Vizoso Stones Farm House, Faringdon Road, Faringdon SN7 8NP
G3 XVC M. Collopy New Gardens, Burnt House Lane, Dartford DA2 7SP
G3 XVG T. Barraclough 52 Denshaw Avenue, Denton, Manchester M34 3NX
G3 XVH S. Franklin 337 Hendon Way, London NW4 3NB
G3 XVL C. Mccarthy 23 Hornbeam Road, Stowupland, Stowmarket IP14 4DL
G3 XVP P. Pimblott 40 Richmondfield Lane, Barwick In Elmet, Leeds LS15 4EZ
GW3 XVQ F. Jinks 28 St. Anns Road, Bonnie View, Blackwood NP12 3PG
G3 XVS R. Kitching Ashby Down, Streetway Road, Andover SP11 7EH
G3 XVV M. Salmon 54 Church Road, Rivenhall, Witham CM8 3PH
G3 XVW K. Ball 39 Spinney Close, Northfield, Birmingham B31 2JG
G3 XVX H. Dearing 18 Dowling Drive, Pershore WR10 3EF
G3 XVY P. Coull 40 Wear Bay Crescent, Folkestone CT19 6BA
G3 XWA J. Ennis 30 Hillcrest Avenue, Carlisle CA1 2QJ
G3 XWB C. Cadogan 8 Horncliffe Close, Rawtenstall, Rossendale BB4 6EE
G3 XWD D. Watts 40 Outlands Drive, Hinckley LE10 0TW
G3 XWG M. Shen 23 Back Lane, Whixley, York YO26 8BG
G3 XWH R. Horton 23 Back Lane, Whixley, York YO26 8BG
G3 XWL J. Cripps 3 Queens Court, Queens Road, Cranbrook TN18 4JE
G3 XWN G. Laycock 48 Marina Terrace, Golcar, Huddersfield HD7 4RA
G3 XWO T. Lowe 688 East J Street, Chula Vista 92010 United States
G3 XWU R. Pearson 10 Eastleigh Close, Boldon Colliery NE35 9NG
G3 XWV C. Hamilton The Stables Foinavon, Newsham, Richmond DL11 7RD
GW3 XXB A. Evans 74 Celyn Avenue, Cardiff CF23 6EQ
G3 XXC K. Rigelsford 14 Glebelands Avenue, South Woodford, London E18 2AB
G3 XXE P. Williams 4 Plantation Close, Aller, Newton Abbot TQ12 4NS
G3 XXF C. Vine 14 Hamilton Road, St. Albans AL1 4PZ
G3 XXG P. Sharpen 3 Western Road, Urmston, Manchester M41 6LE
G3 XXH Dr S. Watts 58 Cambridge Avenue, New Malden KT3 4LE
G3 XXM D. Richards 836 The Ridge, St. Leonards-on-Sea TN37 7PX
G3 XXN F. Pickersgill 3 Church St., Langold, Worksop S81 9NW
G3 XXO E. Birks 46 Curzon Drive, Worksop S81 0LP
G3 XXQ L. Dixon 24 Angerton Gardens, Newcastle upon Tyne NE5 2JB
G3 XXR P. Higton 13 Wilton Avenue Bradley Huddersfield, Huddersfield HD21RN
G3 XXX S. Bradnam 39 Pelham Way, Cottenham, Cambridge CB24 8TQ
G3 XYB G. Maitland 7 Battery Road, Cowes PO31 8DP
G3 XYC P. Crust 16 London Lane Wymeswold, Loughborough LE12 6UB
G3 XYD W. Gordon-Laycock 51 Overbrook, Swindon SN3 6AR
G3 XYE J. Clifton Romford Cottage, Romford, Verwood BH31 7LE
G3 XYF Invicta CG c/o J. Wresdell Bracey Bridge Farm, Harpham, Driffield YO25 4DE
G3 XYG Dr M. George 30 Northfield Park, Barnstaple EX31 1QA
G3 XYH J. Hill 35 Windmill Avenue, Marshalswick, St. Albans AL4 9SJ
G3 XYI D. Fearnley 14 Salforal Close, Rettendon Common, Chelmsford CM3 8EL
G3 XYJ R. Walker 27 Archer Close, Kings Langley WD4 9HF
G3 XYO S. Line Cottles House, Cottles Lane, Exeter EX5 1EE
G3 XYV I. Cooper 118 Stagsden Road, Bromham, Bedford MK43 8QJ
GW3 XYW D. Jones 22 Alltiago Road Pontarddulais, Swansea SA4 8HU
G3 XYZ Kings Lynn ARC c/o E. Haskett 23 Gloucester Road, King's Lynn PE30 4AB
G3 XZB N. Edwards 14 Churchill Close, Cowes PO31 8HQ
G3 XZG J. Browne 82 Cresswell Road, Chesham HP5 1TA
G3 XZK D. Gething 31 Lower Lodge Lane, Hazlemere, High Wycombe HP15 7AT
GI3 XZM D. Vance The Eaves, Reagh Island, Newtownards BT23 6EN
G3 XZO M. Rhodes 11 Morse Road, Norton Fitzwarren, Taunton TA2 6BU
G3 XZP D. Holburn 19 Whitwell Way, Coton, Cambridge CB23 7PW
G3 XZV J. Sonley Ravenscliffe, Lands Lane, Knaresborough HG5 9DE
G3 XZX J. Lowe 24 Candish Drive, Plymouth PL9 8DB
G3 XZY T. Garner 122 Wainfleet Road, Skegness PE25 3RX
GM3 YAC P. Howarth Kilbeg House, Teangue, Isle of Skye IV44 8RQ
G3 YAD M. Goodrich 2 Highworth Crescent, Yate, Bristol BS37 4EY
GW3 YAF T. Davies Rhyd Wen, Llangyndeyrn, Kidwelly SA17 5EN
G3 YAG W. Thompson 2 Fern Close, Frimley, Camberley GU16 9QU
G3 YAI T. Mills 16 Hunts Hill, Glemsford, Sudbury CO10 7RL
G3 YAJ D. Sellen Prospect House, Wignall Street, Manningtree CO11 2HX
GM3 YAO F. Offler Ar Dachaidh, 3 King David Drive, Montrose DD10 0SW
G3 YAR I. Gildersleve 7 Oak Park Road, Newton Abbot TQ12 1RQ
G3 YBA E. Cooper Flat, 23 Westminster Crescent, Sheffield S10 4EU
G3 YBE E. Gilbert 2 Church Field, Stanford, Ashford TN25 6UA
G3 YBG J. Rabjohns Quarries Bungalow, Barley Lane, Exeter EX4 1TA
G3 YBH P. Storey Po Box 47060, Denman Street Postal Outlet, Vancouver V6G 3E1 Canada
G3 YBK R. Donno 6 Mincinglake Road, Exeter EX4 7EA
G3 YBM R. Mitchell 98 Marlborough Drive, Burgess Hill RH15 0EU
GW3 YBN C. Davies 31 Park Prospect, Graigwen, Pontypridd CF37 2HF
G3 YBO R. Baines 10 Chartwell Ave, Wingerworth, Chesterfield S42 6SP
G3 YBP A. Young Hillcrest, Graynfylde Drive, Bideford EX39 4AP
GM3 YBQ K. Horne 10 Blair Place, Kirkcaldy KY2 5SQ
G3 YBR S. Cook 6 Essex Court, Marlton, New Jersey 8053 United States
G3 YBS R. Lindsay-Smith 58 Chalgrove Road, London N17 0JD
G3 YBU B. Whittle Holmlea, Main Road, Hull HU12 9NG
G3 YBY I. Mccarthy 76 High Street, Purton, Swindon SN5 4AD
GI3 YBZ J. Mccann 61 Glengawna Road, Glengawna, Omagh BT79 7WJ
GM3 YCB ACF/CCF RC c/o S. Riddell 16 Lewis Drive, Old Kilpatrick, Glasgow G60 5LE
G3 YCE J. Peck 7 Paddock Close, Radcliffe-On-Trent, Nottingham NG12 2BX
G3 YCH J. Sharman 11 Shapley Way, Liverton, Newton Abbot TQ12 6PN
G3 YCJ P. Sheard 52 Victoria Road, Elland HX5 0QA
G3 YCO R. Lewis 115 Chester Road, Whitby, Ellesmere Port CH65 6SB
G3 YCV J. Hibbert 5 Cliff View Road, Cliffsend, Ramsgate CT12 5ED
G3 YCX A. Cain 42 Wood Lane, Prescot L34 1LW
G3 YCY R. Barrett 47 Marshals Drive, St. Albans AL1 4RD
G3 YDD Hereford ARS c/o T. Bridgland-Taylor 1 Overbury Court, Hereford HR1-1DG
G3 YDE W. Bagwell 93 Broadley Drive, Torquay TQ2 6UT
GI3 YDH M. Mcintyre 36 Beechgrove Park, Belfast BT6 0NR
GI3 YDM J. Dunlop 34 Ballybentragh Road, Dunadry, Antrim BT41 2HJ
GM3 YDN D. Nutt Little Craigfin, Kilkerran, Maybole KA19 8LR
G3 YDO W. Mcnally 2607 The Highlands Dr., Sugar Land, Texas 77478 United States
G3 YDT W. Patterson 32 The Pagoda, Maidenhead SL6 8EU
G3 YDY P. Selwood 43 Keene Way, Galleywood, Chelmsford CM2 8NT
G3 YEC R. Edmondson 16 Orchard Close, Copford, Colchester CO6 1DB
G3 YED R. Nettleton 4 Sycamore Close, Stratford-upon-Avon CV37 0DZ
G3 YEG N. Sears 17 Walls Road, Bembridge PO35 5RA
G3 YEK R. Johnston Flat 906, Orchard Plaza, Poole BH15 1EH
GD3 YEO R. Rimmer 27 Manor Lane, Farmhill, Douglas IM2 2NP Isle of Man
G3 YEP R. Wakeley Fir Bank, Fell Lane, Penrith CA11 8BJ
G3 YEQ L. Miller 28 Arthur Road Cliftonville, Margate CT9 2EN
G3 YER D. Lowe Flat 14, Marklands, 37 Julian Road, , Bristol BS9 1NP
G3 YEU B. Short 83 Rowanfield Road, Cheltenham GL51 8AF
GM3 YEW D. Morris Ash Cottage, Perth Road, Perth PH2 9LW
G3 YFD W. Hewitt 22 Derby Road, Stockport SK4 4NE
G3 YFE J. Shaw 57 London Road, Amesbury, Salisbury SP4 7EE
G3 YFG S. Westell 2 Whiteacre Lane, Barrow, Clitheroe BB7 9BJ
G3 YFK P. Mcalister Lower Winnington Farm, Winnington, Shrewsbury SY5 9DJ
G3 YFL C. Wright Riverside Cottage, High Street, Meonstoke, Southampton SO32 3NH
G3 YFM M. Smyth 21 The Paddock, Longworth, Abingdon OX13 5BX
G3 YFO M. Bunce 36 Burlington Road, Burnham, Slough SL1 7BQ
G3 YFP J. Bottomley 42 Birkdale Court, Fornham St. Martin, Bury St. Edmunds I P28 6XF
G3 YFU E. Tomlin Indalo, Magna Mile, Market Rasen LN8 6AJ
G3 YFV P. I'Anson Flat 297, Latymer Court, London W6 7LD
G3 YFW P. Rosen 21 Hadley Heights, Hadley Road, Barnet EN5 5QH
G3 YGA E. Warwick-Oliver St. Madron, Throwleigh, Okehampton EX20 2HX
G3 YGB J. Coleman 19 Megdale, Matlock DE4 3JW
G3 YGC E. Elliott 18 Bear St., Lowerhouse, Burnley BB12 6NQ
G3 YGD D. Brown 5 Meadow Edge Barrowford, Nelson BB9 6BT
G3 YGE J. Okas Dipley Springs, Dipley Common, Hook RG27 8JS
G3 YGF Dr J. Gannaway Highview House, Winchester Road Fair Oak, Eastleigh SO50 7HB
G3 YGG J. Kelly 79 West Hill Avenue, Epsom KT19 8JX
GW3 YGH A. Hughes 40 Fontygary Road, Rhoose, Barry CF62 3DS
G3 YGJ D. Brierley 4 Waterloo Terrace, Bideford EX39 3DJ
G3 YGL F. Smith 34 Bridle Road Eastham, Wirral CH62 8BR
G3 YGM M. Osborne Cheriton, Alexandra Road, St. Ives TR26 1ER
G3 YGR C. Thomas Oakdene, School Lane, Reading RG7 3ES
G3 YGZ A. Walsh Green Royd, Saddleworth Road, Halifax HX4 8NU
G3 YHB J. Baker 109 Bermuda Road, Wirral CH46 6AX
G3 YHC W. Hermes 22 Mallinson Crescent, Harrogate HG2 9HP
G3 YHF C. Skelcher 51 Blenheim Road, Moseley, Birmingham B13 9TY
G3 YHG D. Harding 17 Summerfield Close, Wokingham RG41 1PH
G3 YHH J. Froud Summer Park, New Road, Teignmouth TQ14 8UF

G3 YHI R. Vale High House, Mounts Lane, Daventry NN11 3ES
G3 YHK J. Clemence Aroha, 67 Tomline Road, Felixstowe IP11 7NX
G3 YHM R. Harvey 26 Birkdale Road, Worthing BN13 2QY
G3 YHN C. Pedley 25 Fallowfield Road, Walsall WS5 3DH
G3 YHO R. Yaxley Fallow View, Swaffham Road, Dereham NR19 2LX
G3 YHQ D. Mercer 19 Kingsfield Drive, Didsbury, Manchester M20 6JA
GW3 YHR C. Briscoe Gadlas House, Ffordd Top Y Rhos, Treuddyn, Mold CH7 4NE
GJ3 YHU D. Robinson 16 Seagrove Court, La Rue De La Corbiere, St. Brelade JE3 8HN Jersey
G3 YHV C. Chidgey 46 Station Road, Shirehampton, Bristol BS11 9TX
G3 YIA M. Harris 100 Chapel Lane, Wymondham NR18 0DN
G3 YIB M. Knowler 2 Vulcan Street, Southport PR9 0TW
G3 YIC V. Sedgley 25 Avenue Road, Weymouth DT4 7JH
G3 YIE E. Lusty Stanley End Farm, Bell Lane, Stroud GL5 5JY
G3 YIF J. Weiner 1 Chippendayle Drive, Harrietsham, Maidstone ME17 1AD
GW3 YIH F. Cobb Mon Reve, Rhodfa Nant, Abergele LL22 9ND
G3 YII N. Smith Creekside, Nursery Lane, South Wootton, King's Lynn PE30 3NA
G3 YIK Dr J. Morgan Cedars, Springhill, Longworth, Abingdon OX13 5HL
G3 YIN E. Ellery 14 Four Acres, Bideford EX39 3RW
G3 YIQ R. Jones The Old Vicarage, Upper South Wraxall, Bradford-on-Avon BA15 2SB
G3 YIR the RC c/o A. Ward 20 Tower Close, Costessey, Norwich NR8 5AU
G3 YIW J. Gallop 55 Somervell Drive, Fareham PO16 7QW
G3 YIY R. Ingram 11 Bank Terrace, Mevagissey, St. Austell PL26 6QZ
G3 YJA B. Porter 49 Beverley Road, Leamington Spa CV32 6PW
G3 YJD J. Davies 25 Harkness Close, Bletchley, Milton Keynes MK2 3NB
G3 YJE P. Merriman The Old Croft, Brimpsfield, Gloucester GL4 8LD
G3 YJG G. Mason 8 Leighton Road, Sunderland SR2 9HQ
G3 YJN R. Hodge 36 Binswood End, Harbury, Leamington Spa CV33 9LN
G3 YJP D. Benham 19 Benham Road, Otis 1253 United States
G3 YJQ F. Bourne 78 Normandy Way, Plymouth PL5 1SR
G3 YJS M. Roche 8 Northdown Close, Penenden Heath, Maidstone ME14 2ER
G3 YJV Dr M. Hawthorne 49 Broom Close, Teddington TW11 9RL
G3 YJW R. Whitehouse White Court, Camp Hill, Tonbridge TN11 8LE
G3 YJZ A. Mitchell 89 Queen Annes Grove, Enfield EN1 2JU
GM3 YKA J. Wiewiorka 47 Albert Avenue, Grangemouth FK3 9AT
G3 YKB J. Hodgson 18 Nascot Place, Watford WD17 4QT
G3 YKC D. Fayers 1 Tismeads Crescent, Swindon SN1 4DP
G3 YKI K. Vickers Brick Kiln House, Ditton Priors, Bridgnorth WV16 6TW
G3 YKK C. Donne The Hideaway, Lease Lane, East Halton, Immingham DN40 3PT
G3 YKO D. Darwood Briarwood Cottage, Packhorse Lane, Birmingham B38 0DN
GM3 YKP W. Sutton Old Police Station, Dunbeath KW6 6EA
G3 YKS R. Butlin 48 Roman Way, Market Harborough LE16 7PQ
G3 YKW Capt. R. Walker 17 Ballantyne South, Montreal West QC H4X 2BI Canada
GW3 YKZ M. Biddiscombe 20 Arlington Close, Malpas, Newport NP20 6QF
G3 YLA J. Bacon 37 Burgh Lane, Mattishall, Dereham NR20 3QP
GM3 YLD J. Frew Queens Cottage, 87 Queen St., Dunoon PA23 8AX
GJ3 YLI A. Morrissey Flat 4, 1 Springfield Crescent, St. Helier JE2 4GL Jersey
G3 YLJ G. Whitehead 29 Coulsons Road, Bristol BS14 0NN
GJ3 YLN J. Speller Lindau, Gorey Village Main Road, Jersey JE3 9EP
GM3 YLU W. Taylor 4 Newbie Barns, Newbie, Annan DG12 5QL
G3 YLV P. Jones Oaklea, Cadney Lane, Whitchurch SY13 2LW
G3 YLW P. Lascelles 4 Meadowsweet Close, Snettisham, King's Lynn PE31 7UG
G3 YLY B. Hawes 15 Bridge Lane, Wimblington, March PE15 0RR
G3 YMC D. Sergeant 8 Toll Gardens, Bracknell RG12 9EX
G3 YMD Dover ARC c/o P. Love 2 Meadway, Dover CT17 0PS
G3 YMH R. Wainwright The Olives, High Street, Uckfield TN22 4LB
G3 YMM T. Campbell Davis 9 Cloister Road, London W3 0DE
G3 YMN J. Rhys 3 Oakbank Avenue, Walton-on-Thames KT12 3QY
GI3 YMT M. Higgins 1 Cairnshill Park, Belfast BT8 6RG
G3 YMU J. Hibberd Barn Cottage, School Lane, Crewe CW3 0BA
G3 YMV I. Machardie 20 Arley Close, Swindon SN25 4TP
G3 YMW D. Sapsworth 16 Laxton Avenue, Hardwick, Cambridge CB23 7XL
GI3 YMY N. Newell 18 Kilmaine Avenue, Bangor BT19 6DU
G3 YMZ Fife Raynet Group c/o A. Plummer 6 Shelley Place, Kallaroo, Perth 6025 Australia
G3 YNC C. Adams 18 Glenavon Road, Highcliffe, Christchurch BH23 5PN
GM3 YND I. Simpson 3 Ravenscraig Terrace, Steelend, Dunfermline KY12 9LU
G3 YNF B. Turner South Lodge, Stanford Road, Fineshade NN17 3BA
G3 YNJ C. Powell 38 Braeside Road, St. Leonards, Ringwood BH24 2PH
G3 YNK D. Evans Michigan Villa, Wall Road, Hayle TR27 5HA
G3 YNL O. Price 5 Redicliff Way Sw, Redcliff T0J Canada
G3 YNO M. Booth 28 Humber Road, North Ferriby HU14 3DW
G3 YNU Dr I. Stevenson 18 Sittingbourne Road, Wigan WN1 2RR
G3 YOA A. Adams The Gables, Chapel Road Trunch, North Walsham NR28 0QG
G3 YOC R. Moore 38 Sandygate, Wath-Upon-Dearne, Rotherham S63 7LR
GM3 YOI Dr K. Falconer Lumbo Farmhouse, St. Andrews KY16 8NS
G3 YOL S. Cole Halebrook, Bridgwater Road, Winscombe BS25 1NH
G3 YOM K. Beddoe 30 Tamella Road, Botley, Southampton SO30 2NY
G3 YON F. Webster 16 Pembroke Road, Dronfield S18 1WH
G3 YOO J. Webster 7 Hardwick Avenue, Allestree, Derby DE22 2LN
GM3 YOR A. Givens 5 Langhouse Place, Inverkip, Greenock PA16 0EW
G3 YOV T. Gammage 23 Artizan Road, Northampton NN1 4HU
G3 YOY H. Clark 2 Chestnut Close, Peakirk, Peterborough PE6 7NW
G3 YPD P. Chester 44 Richmond Drive, Lichfield WS14 9SZ
G3 YPE M. Greenwood 21 Dobb Top Road, Holmbridge, Holmfirth HD9 2PQ
G3 YPK S. Wallis Mas La Floride - Bp61068, 250 Chemin De Magarnaud, Sommieres 30250 France
G3 YPL D. Gray 19 Westbury Gardens, Higher Odcombe, Yeovil BA22 8UR
G3 YPM Dr R. Moore 20 Ebrington Road, Malvern WR14 4NL
G3 YPS S. Atkinson 13 Charles Street Dn21 2Ja, Gainsborough DN21 2JA
G3 YPT P. Tomes 86 Hurn Road, Christchurch BH23 2RP
G3 YPU P. Koker 1 Cedar Court, Bridlington YO16 6ZQ
G3 YPW P. Willingham 49 Creek Road, Hayling Island PO11 9RA
G3 YPY A. Head 32 Weald View Road, Tonbridge TN9 2NQ
G3 YPZ J. Petters 218 New House Farm Hospital Drove, Long Sutton PE12 9EN
G3 YQA F. Wilson 26 Humber Road, North Ferriby HU14 3DW
G3 YQB D. Rankin 105 Sparrowhawk Way, Hartford, Huntingdon PE29 1XY
G3 YQC J. Wood 14 Little Paradise, Marden, Hereford. HR1 3DR
G3 YQF R. Linford Department Of Communication, And Electronic Engineering, Drake Circus Plymouth PL4 8AA
GM3 YQG N. Waylett 3A Arabella, Tain IV19 1QH
GW3 YQH Firepower Museum c/o R. Evens 41 Heol Gwys, Upper Cwmtwrch, Swansea SA9 2XQ
G3 YQJ P. Burnet 166 Oundle Road, Thrapston, Kettering NN14 4PQ
GM3 YQK J. Dillon Steelbank Cottage, Dalgraven, Kilwinning KA13 6PL
G3 YQL P. Murfitt 53A Codnor Denby Lane, Codnor, Ripley DE5 9SP
GW3 YQM Dr D. Wynford-Thomas Coach House, The Hollies, Pentrepoeth Road, Newport NP10 8RT
G3 YQN R. Trott 120 Portland Road, Bromley BR1 5AZ
G3 YQO D. Kirkwood 205 Uplands Road, West Moors, Ferndown BH22 0EZ
GW3 YQP C. Hardie 3 Berth Glyd, Gyffin, Conwy LL32 8NP
G3 YQV C. Railton 14 Copford Lane, Long Ashton, Bristol BS41 9NF
G3 YQW M. Funnell 15 Mcindoe Road, East Grinstead RH19 2DD
G3 YQZ M. Johnson 36 Coventry Road, Bulkington, Bedworth CV12 9ND
G3 YRC Yarmouth RC c/o P. Nicholls 53 Fastolff Avenue Gorleston, Great Yarmouth NR31 7ND
G3 YRH B. Dodds 1 Croft View, Killingworth, Newcastle upon Tyne NE12 6BT
GI3 YRL J. Branagh 17 Rathmoyle Park West, Carrickfergus BT38 7NG
GW3 YRP I. Dudley Tynewydd, Llansantffraid SY22 6TW
G3 YRQ I. Parkinson 61 Cinnamon Lane, Fearnhead, Warrington WA2 0AG
G3 YRU P. Wilby 56 Main Street Thorner, Leeds LS14 3BU
G3 YRX I. Elston 11 Knowle Drive, Exwick, Exeter EX4 2DF
G3 YSD J. Murdoch 32 Scalegill Road, Moor Row CA24 3JL
G3 YSG M. Taylor 4 Yew Tree Court, Botley Road, Swanwick, Southampton SO31 1EA
GW3 YSI P. Tipping Pen Yr Enfys Ashdale Lane, Llangwm, Haverfordwest SA62 4NU
G3 YSK A. Button 13 Taplings Road, Weere, Winchester SO22 6HE
G3 YSM M. Davidson 14 Fuchsia Walk, Wirral CH49 3AG
G3 YSN H. Smith Carnview, Tregender Lane, Penzance TR20 8DJ
G3 YSQ A. Pratt 7 The Croft, West Hanney, Wantage OX12 0LD
G3 YSR C. Beattie Mayerin, Churchway, Aylesbury HP17 8RG
G3 YSW N. Thrower 8 Upton Gardens, Worthing BN13 1DA
G3 YSX Dr S. Bryant 154 London Road North, Merstham, Redhill RH1 3AA
GW3 YTC G. Rutherford 29 Sisial Y Mor, Rhosneigr LL64 5XB
GD3 YTE P. Gill Holly Bank, Sulby Bridge IM7 2AY
G3 YTG T. Blair 56 Rue Du Golfe De Barbareu, Etaules 17750 France
G3 YTI S. Cooper 24 Cambridge Street, Darwen BB3 3JH
GW3 YTL Bemerton Sc Grp c/o C. Lewis 26 Ffordd Gwenllian, Llay, Wrexham LL12 0UW
G3 YTN R. Hill 35 Coxwold View, Wetherby LS22 7PU
G3 YTQ C. Kidd 118 Segensworth Road, Fareham PO15 5EQ
G3 YTR M. Hatt 1 Larches Way, Crawley Down, Crawley RH10 4UJ
GM3 YTS R. Ferguson 19 Leighton Avenue, Dunblane FK15 0EB
G3 YTT W. Taylor 1 Milby Drive, Nuneaton CV11 6JR
G3 YTU C. Coward 27 Rothley Chase, Haywards Heath RH16 3PE
G3 YTW G. Clarke 117 Bermuda Village, Nuneaton CV10 7PW
G3 YTX G. Clamp 9 Furse Close, Camberley GU15 1BF
G3 YTY M. Edib 84 Connaught Gardens, London N13 5BT
G3 YTZ M. Readman Flat 1-8, 365 Wilmslow Road, Manchester M14 6AH
GW3 YUC D. Davies 30 Wern Isaf, Dowlais, Merthyr Tydfil CF48 3NY
G3 YUD P. Hawkins 37 Alexandra Road, Dorchester DT1 2LZ
G3 YUH R. Ayling 25 Nash Court Road, Margate CT9 4DH
GD3 YUM M. Parnell 1 Derwent Drive, Onchan, Douglas, Isle of Man IM3 2DF Isle of Man
G3 YUQ E. Elsley 25 Elmsdale Road, Wootton, Bedford MK43 9JW
G3 YUU Processed Audio Group c/o C. Lord Green Sleeves, Marley Road, Maidstone ME17 1BS
G3 YUX R. Moore 69 Ivatt, Tamworth B77 2HQ
G3 YUZ I. Wilson 23 Alyth Road, Bournemouth BH3 7DG
G3 YVA B. Edwards 774, Calle 21 S.O., Puerto Rico 921 Puerto Rico
G3 YVH A. Boyne 18 Crow Lane West, Newton-le-Willows WA12 9YG
G3 YVI R. Gilbert Mayville, New Copse, Alton GU34 5NP
G3 YVK H. Tabberer Apartment 228, The Hawthorns, Meadow Park, Tortoiseshell Way, Braintree CM7 1TD
GW3 YVN N. Little Brynhyfryd Llansteffan, Carmarthen SA33 5HA
GU3 YVV R. Outhwaite Le Courtillet, La Route De Sausmarez, Guernsey GY4 6SF
G3 YVW B. Blackwell 19 Tokely Road, Frating, Colchester CO7 7GA
GM3 YVX D. Coupar 32 Gillies Place, Broughty Ferry, Dundee DD5 3LE
G3 YVY D. Hanley 5 Hallcroft Close, Billingham TS23 1QN
G3 YVZ T. Gardner 64 Balmoral Terrace, Heaton, Newcastle upon Tyne NE6 5YA
G3 YWA E. Pepper 30 Westfield Drive, Harpenden AL5 4LP
G3 YWF P. Smith 18 David Avenue, Cliftonville, Margate CT9 3DU
G3 YWH F. Hill 24 Mount St. James, Guide, Blackburn BB1 2DR
G3 YWL Headcorn Aerodrome c/o C. Coverdale 2 Lillypilly Lane, Cooranbong 2265 Australia
G3 YWM P. Hubert 575 Bramford Lane, Ipswich IP1 5JX
G3 YWP Capt. P. White Po Box 86524 Al Jazeera Po, Ras Al Khaimah 86524 United Arab Emirates
G3 YWS J. Smith 16 Woodlands, Winthorpe, Newark NG24 2NL
G3 YWT P. Smith Beechwood, Clarendon Road, Salisbury SP5 3AT
G3 YWU S. Fisher Arkle, 31 Frith Avenue, Northwich CW8 2JB
G3 YWW A. Carpenter 17 Victoria Avenue Upwey, Weymouth DT3 5NG
G3 YWX I. Poole 17 Glebe Road, Dorking RH4 3DS
G3 YXH S. Marshall The Bunglaow, Llamedos Stables, Fieldhead Lane, Bradford BD11 1JL
G3 YXM D. Pick 178 Alcester Road South, Kings Heath, Birmingham B14 6DE
G3 YXN P. Whalley Southerly, The Common, Hanworth, Norwich NR11 7HP
G3 YXO Dr D. Watson Norton, Gote Lane, Ringmer, Lewes BN8 5HX
G3 YXQ Dr R. Ireland 31 St James Street, Sackville, New Brunswick E4L 4L7 Canada
G3 YXS D. Naylor 7 Ruthven Court, Litherland, Liverpool L21 2PE
G3 YXW P. Dunford 79 Summerdown Road, Eastbourne BN20 8DQ
GM3 YXY A. Thomson Roselea, Lanark ML11 7SE
G3 YYC Dr G. Sharples 3A Green Lane Park Homes, Breinton, Hereford HR4 7PN
G3 YYE H. Lawrence 14 Manor Lane, Dinnington, Sheffield S25 2SW
G3 YYG J. Bolton 3 Fyne Drive, Linslade, Leighton Buzzard LU7 2YG
G3 YYK R. North 13 Nicholson Way, Havant PO9 3AZ
G3 YYN P. Spurr Windmill Farm Barn, High Street, Milton Keynes MK14 5AX
G3 YYR Dr A. Parry 17 May Pole Knap, Somerton TA11 6HP

G3 YYW G. Wills 137 Aldermans Drive, Peterborough PE3 6BB
G3 YYZ J. Cuthbert Fulbeck, 48 Mayes Lane, Harwich CO12 5EJ
G3 YZK Dr G. West 6 Lammas Close, Cowes PO31 8DT
G3 YZN R. Awbery Dashwood, Beacon View Road, Godalming GU8 6DU
G3 YZO Dr R. Wilson Kellers, Duck Street, Wendens Ambo, Saffron Walden CB11 4JU
G3 YZQ P. Williams 41 Church Street, St. Georges, Telford TF2 9JZ
G3 YZR J. Porter Birklands, 16 The Oval, Scarborough YO11 3AP
G3 YZT A. Slaney 1 Marlborough Way, Goring-By-Sea, Worthing BN12 4HG
G3 YZV E. Woollard 24 Griffin Close, Twyford, Banbury OX17 3HR
G3 YZW A. Armstrong 423 Bideford Green, Linslade, Leighton Buzzard LU7 2TY
G3 YZZ K. Beverstock 16 Chaucer Close, Emmer Green, Reading RG4 8PA
G3 ZAB I. Pivac 427 Great North Road Glendene, Auckland 602 New Zealand
G3 ZAE S. Benstead 15 Les Congeries, Dournazac 87230 France
G3 ZAG B. Taylor 27 Ridgeway, Wellingborough NN8 4RU
G3 ZAJ D. Sutton Deer Wood, Canterbury Road, Ashford TN25 4DF
G3 ZAL L. Huggett 24 Hockers Lane, Detling, Maidstone ME14 3JN
G3 ZAU E. Lord 41 Daven Road, Congleton CW12 3RB
G3 ZAW K. Bird Leys House, Main Street, Northampton NN7 4SH
G3 ZAY M. Atherton 41 Enniskillen Road, Cambridge CB4 1SQ
G3 ZBB S. Jackson Old Fir Tree Inn Peacemarsh, Gillingham SP8 4EU
G3 ZBF M. Matthews 13 Bursill Crescent, Ramsgate CT12 6EZ
G3 ZBG G. Moorfield 43 Broadmark Lane, Rustington, Littlehampton BN16 2HH
G3 ZBI Nunsfield House c/o K. Frankcom 1 Chesterton Road, Spondon, Derby DE21 7EN
G3 ZBM W. Worthington 32 Princess Drive, Wistaston Green, Crewe CW2 8HS
G3 ZBP M. Baker 82 Folkestone Road, Copnor, Portsmouth PO3 6LR
GM3 ZBR K. Melvin 1 Charleston Village, Charleston, Forfar DD8 1UF
G3 ZBS J. Mccall 5 Sundew Close, Wokingham RG40 5YB
G3 ZBU A. Watt 5 Brambling Road, Horsham RH13 6AX
G3 ZBZ B. Cross 176 Outwood Road, Heald Green, Cheadle SK8 3LL
G3 ZCA D. Lake 9 Grafton Close, King's Lynn PE30 3EZ
G3 ZCD R. Fogg 24 Edinburgh Gardens, Windsor SL4 2AN
G3 ZCG K. Young Flat 36, Brookhurst Court, Leamington Spa CV32 6PB
G3 ZCH D. Hill 11 Chapeltown Road, Radcliffe, Manchester M26 1YF
G3 ZCI R. O'Brien 9 Holmwood Garth, Hightown, Ringwood BH24 3DT
G3 ZCJ M. Allerton Austin 13 Kilpin Green, North Crawley, Newport Pagnell MK16 9LZ
GI3 ZCK G. Ward 6B Stranmillis Road, Belfast BT9 5AA
G3 ZCL G. Hammersley 74 Cammel Road, West Parley, Ferndown BH22 8SB
G3 ZCT J. Beehlar 12 Dulverton Road, Leicester LE3 0SA
G3 ZCX P. Fort 1 Lowther Lane, Foulridge, Colne BB8 7JY
G3 ZCY R. Hogg 5 Bishop Way, Pateley Bridge, Harrogate HG3 5LH
G3 ZCZ Dr J. Kasser 39 Honeysuckle Drive, Hope Valley 5090 Australia
G3 ZDF J. Kirk 111 Stockbridge Road, Chichester PO19 8QR
G3 ZDG S. Cole 1 The Copse, Exmouth EX8 4EY
GM3 ZDH R. Dixon Flat 5, 10 Mavisbank Gardens, Glasgow G51 1HG
G3 ZDK P. Given 4 Elveden Drive, Ilkeston DE7 9JW
G3 ZDM R. Muriel 13 York Road, Sale M33 6EZ
G3 ZDQ I. Flemming Rudderhams Cottage, Blandford Hill, Winterborne Whitechurch, Blandford Forum DT11 0AA
G3 ZDT P. Morrison Saddlers, Upper Green Road, Tonbridge TN11 9PL
G3 ZDU G. Marshall Bassington, Hulne Park, Alnwick NE66 3JE
G3 ZDW R. Hyde 25 The Pastures, Cottesmore, Oakham LE15 7DZ
G3 ZDY D. Palmer Flat 3, 60 Millbank, London SW1P 4RW
G3 ZEB B. Robinson The Gables, The Street, Great Yarmouth NR29 4EA
G3 ZED A. Rothwell Brandon, Manor Brow, Keswick CA12 4AP
G3 ZEF R. Stephen 97 Hunters Field, Stanford In The Vale, Faringdon SN7 8ND
G3 ZEJ R. Smith 3 Willake Road, Kingskerswell, Newton Abbot TQ12 5AB
G3 ZEK Middlesex DX Grioup c/o M. Bailey 12 Bridgers Mill, Haywards Heath RH16 1TF
G3 ZEM R. Henderson Po Box 62155, Pafos 8061 Cyprus
G3 ZEN Dr A. Glaser 155 Little Breach, Chichester PO19 5UA
G3 ZEO S. Wilders Old Farm Barn, Silkstead Lane, Winchester SO21 2LG
G3 ZEQ M. Brenig-Jones Orchard House, Larters Lane, Stowmarket IP14 5HB
G3 ZER I. Mercer 28 West Way, Rickmansworth WD3 7EN
G3 ZES A. Downing Oaktree Bungalow, The Endway, Chelmsford CM3 6DU
GM3 ZET Melton Mowbray ARS c/o T. Goodlad 72 North Lochside, Lerwick, Shetland ZE1 0PJ
GM3 ZEU C. Clarkson 8 Moor Place, Portlethen, Aberdeen AB12 4TF
GD3 ZEX C. Douglas Sea Villa, The Promenade, Laxey IM4 7DF Isle of Man
G3 ZEZ G. Coleman 16 Kestrel Way, Clacton-on-Sea CO15 4JE
G3 ZFC C. Davis 3 Cross Road, Haslington, Crewe CW1 5SY
G3 ZFF R. Hornbuckle 54 Gladys Avenue, Cowplain, Waterlooville PO8 8HS
G3 ZFP R. Penberthy 10 Lancot Avenue, Dunstable LU6 2AW
G3 ZFR R. Harris Clevelands, Tamworth Road, Coventry CV7 8JJ
G3 ZFT A. Magnus Woodland Cottage, Linkside East, Hindhead GU26 6NY
G3 ZFV J. Watts Riverside, St. Georges Road, Barnstaple EX32 7AS
G3 ZFZ G. Gibson 174 Roose Road, Barrow-in-Furness LA13 0EE
G3 ZGA J. Hart Peter-Vischer-Str.9, Marktredwitz 95615 Germany
G3 ZGC R. Jolliffe 54 Glendale Avenue, Wash Common, Newbury RG14 6RU
GM3 ZGH R. Yeoman 162 Jamphlars Road, Cardenden, Lochgelly KY5 0ND
G3 ZGI T. O'Neill Braeside, Cookbury, Holsworthy EX22 7YG
G3 ZGN P. Swarbrick 1 Hill View, Charminster, Dorchester DT2 9QX
G3 ZGP R. Cridland 13 Clarendon Avenue, Redlands, Weymouth DT3 5BG
G3 ZGQ L. Mead 12 Ferniefields, High Wycombe HP12 4SP
G3 ZGT B. Druce 25 Boothgate Drive, Howden, Goole DN14 7EN
G3 ZGU K. Richens The Old Barn, Marsh Lane, Cheswardine, Market Drayton TF9 2SF
G3 ZGY G. Paddock 56 Clee View Road, Wombourne, Wolverhampton WV5 0BD
G3 ZGZ D. Woodhall 15 Cherrywood Avenue, Thornton-Cleveleys FY5 1SU
G3 ZHA G. Gillam 58 Downhall Road, Rayleigh SS6 9LY
G3 ZHB A. Stuart 207 Saunders Lane, Mayford, Woking GU22 0NT
G3 ZHC N. Willmot 2 Athlone Road, Walsall WS5 3QX
G3 ZHE A. Heyes 20 Walsingham Road, Penketh, Warrington WA5 2AQ
G3 ZHJ P. Moss 6 Windmill View, Houghton Conquest MK45 3GD
G3 ZHK T. Kellow Glenvale, St. Dominick, Saltash PL12 6TD
G3 ZHL Dr J. Morgan Cedars, Springhill, Abingdon OX13 5HL
G3 ZHO R. Wilton 23 Moorland View, Liskeard PL14 3TQ
G3 ZHP D. Marsden 14 Daisy Drive, Barnsley S70 4NY
G3 ZHS R. Ray 37 Doxey Fields, Stafford ST16 1HJ
G3 ZHT B. Lundean 13 Isis Close, Lympne, Hythe CT21 4JQ
G3 ZHU G. Clark Kenzie, Canterbury Road, Dover CT15 7HR
G3 ZHV R. Bowler 21 Pine Close, South Wonston, Winchester SO21 3EB
G3 ZHZ A. Macfadyen 19 Oldfield Road Lower Willingdon Bn20 9Qd, Eastbourne BN20 9QD
G3 ZIB D. Tye 21 Elmstone Drive, Tilehurst RG315NS
G3 ZIC J. Viney 20 South Drive, Upton CH49 6LA
G3 ZID A. Greathead 20 Westland, Martlesham Heath, Ipswich IP5 3SU
G3 ZIE E. Brown 21 Newbridge Way, Pennington, Lymington SO41 8BG
G3 ZIF H. Wilson 6 Risborrow Close, Etwall, Derby DE65 6HY
G3 ZIG Mid Beds R/Net c/o R. Reed Oak Cottage, Dereham Road, Dereham NR20 4AA
G3 ZII M. Rathbone 25 Halsall Road, Southport PR8 3DB
G3 ZIJ J. Stables 9 Milbanke Close, Ouston, Chester le Street DH2 1JJ
G3 ZIK A. Mather 15 Claughton Avenue, Bolton BL2 6US
G3 ZIL G. Griffiths 14 Bassett Close, Southampton SO16 7PE
G3 ZIM R. Wolten 12 Well Lane, Liverpool L16 5ET
G3 ZIN G. Spencer 16 West Lawn, Ipswich IP4 3LJ
G3 ZIO C. Harvey Ham Cottage, 1A Elstan Way, Croydon CR0 7PR
G3 ZIV K. Nolan West End Cottage, Woodhall, Selby YO8 6TG
G3 ZIY R. Drinkwater 15 Woodside Crescent, Smallfield, Horley RH6 9NA
G3 ZJF P. Broughton 7 Old Mill Way, Wells BA5 2JU
G3 ZJG J. Garner 50 Thorndale, Ibstock LE67 6JT
G3 ZJK C. Milner The Everglades, Sawbridge Road, Grandborough, Rugby CV23 8DN
G3 ZJO E. Bennett 44 Central Avenue, Northampton NN2 8DZ
G3 ZJP W. Fenton 50 Orion Road, Rochester ME1 2UH
G3 ZJQ Dr R. Walker 2 Chelwood Drive, Sandhurst GU47 8HT
GW3 ZJS J. Smith Llainfran, New Quay SA45 9RR
G3 ZJV M. Firth 63 Sycamore Road, East Leake, Loughborough LE12 6PP
G3 ZJW Dr B. Mccombe 208 Thorpe Road, Peterborough PE3 6LB
G3 ZJX B. Castle 10 Oakley Drive, Bromley BR2 8PP
G3 ZJY J. Greenwood 91 Keyhaven Road, Milford On Sea, Lymington SO41 0TF
G3 ZJZ J. Mason 35 Broad Way, Hockley SS5 5EL
G3 ZKD W. Ball 6 Coronation Drive, Penketh, Warrington WA5 2DD
G3 ZKG J. Riley 41 Church Avenue, West Sleekburn, Choppington NE62 5XF
G3 ZKH I. Bateman Jaysville, The Strand, Pershore WR10 3JZ
G3 ZKI A. Williams 38 Seneca Street, Bristol BS5 8DX
G3 ZKN D. Morgan 38 Ryder Close, Norman Hill, Dursley GL11 5SG
G3 ZKO P. Lee 20 Little Haseley, Oxford OX44 7LH
G3 ZKQ A. Walton 3 Fox Hill Close, Selly Oak, Birmingham B29 4AH
G3 ZKZ J. Shaw 2 Castle Close, Felixstowe IP11 9NN
G3 ZLD J. Barker 57 Broom Park, Teddington TW11 9RS
G3 ZLE D. Ward 11 Spruce Grove Avenue, Baden, Ontario N3A 3P7 Canada
G3 ZLF R. Nelson 225 Walton Road, Chesterfield S40 3BT
G3 ZLJ E. Dalton 29 Windmill Lane, Castlecroft, Wolverhampton WV3 8HJ
G3 ZLM R. Hook 35 Parkwood Crescent, Hucclecote, Gloucester GL3 3JH
G3 ZLP J. Campbell 4 Ladygrove Cottages Preston, Hitchin SG4 7SA
G3 ZLQ M. Adams 41 Primrose Lane, Yeovil BA21 5SH
G3 ZLR A. Ridley 28 Riverbank, Laleham Road, Staines TW18 2QE
G3 ZLS S. Craske Shallowford Holsworthy Road Hatherleigh, Okehampton EX20 3LE
G3 ZLX E. Jones 94 Westbrook End, Newton Longville, Milton Keynes MK17 0BX
GM3 ZMA J. Butler 11 Quartalehouse, Stuartfield, Peterhead AB42 5DE
G3 ZME Telford & District ARS c/o S. Bird 15 Ludlow Drive, Stirchley, Telford TF3 1EG
G3 ZMG J. Maughan 83 Oak Road, Peterlee SR8 3HU
G3 ZMH D. Mcauslan Golden Sedge, Street End, Bristol BS40 7TL
G3 ZMK J. Askew Po Box 4487, Linstead, St. Catherine Jamaica
G3 ZML M. Owen 3 Gordon Road, Mount Waverley, Victoria 3149 Australia
G3 ZMM R. Hodgkinson11 39 Oxford Road, Carlton-In-Lindrick, Worksop S81 9BD
G3 ZMO J. Callum Dales Barn Top, Town Head, Hawes DL8 3RH
G3 ZNB P. Hannam Bogg Hall, Oulston, York YO61 3RE
GM3 ZNC J. Mulheron 10 Devonview Place, Airdrie ML6 9DF
G3 ZND J. Bartlett Chickamauga, Tinnahinch, Clonaslee R32 T622 Ireland
G3 ZNE N. Ingle 81 Redmoor Close, Tavistock PL19 0ER
G3 ZNG S. Birt 7 Millbrook Court, Catterall PR3 1XR
G3 ZNH R. Coombes 7 Lower Grove, Whitsbury, , Fordingbridge SP6 3QA
G3 ZNK Ormiston Horizon Academy RC c/o D. Hainsworth 48 Greenacre Park, Rawdon, Leeds LS19 6AR
G3 ZNR D. Bailey 12 St. Philips Drive, Burley In Wharfedale, Ilkley LS29 7EN
G3 ZNT R. Brown 282 Luton Road, Dunstable LU5 4LF
G3 ZNU M. Appleby 6 Mandeville Road, Prestwood, Great Missenden HP16 9DS
G3 ZNW R. Blasdell 32 Fulham Close, Broadfield, Crawley RH11 9NY
G3 ZOG A. Elliott 15 Braemar Gardens, East Herrington, Sunderland SR3 3PX
G3 ZOH B. George 14 Pondfield Road, Orpington BR6 8HJ
G3 ZOI D. Deane 10 Stephens Road, Mortimer Common, Reading RG7 3TU
G3 ZOL J. Powell 156 Avon Way, Colchester CO4 3YP
GU3 ZOM D. Pearson Tequesta, York Avenue, Port Soif, Vale GY6 8HS Guernsey
G3 ZON K. Lacy 9 Rhodes Way, Tilgate, Crawley RH10 5DQ
G3 ZOT J. Hewitt 8 Charles Avenue, Scotter, Gainsborough DN21 3RR
G3 ZOW K. Clamp 12 Cowlishaw Close, Shardlow, Derby DE72 2GS
G3 ZOX P. Durham 18 Maldon Road Great Totham, Maldon CM9 8PR
G3 ZOY A. Alldrick 23 Coxs Close, Nuneaton CV10 7ET
G3 ZPA D. White 1 Whaddon Road, Shenley Brook End, Milton Keynes MK5 7AF
G3 ZPB P. Burton 202 Coulsdon Road, Coulsdon CR5 2LF
G3 ZPI G. Braund 184 Faversham Road, Kennington, Ashford TN24 9AE
G3 ZPJ M. Symons 11 Tudor Lodge Park, Truthwall, Penzance TR20 9BW
G3 ZPK H. Willis 12 Combe View, Hungerford RG17 0BZ
G3 ZPL Dr N. Richardson 501 Forest Avenue, Palo Alto CA 94301 USA
G3 ZPM A. Stormont The Hawthorns, Church Lane, Louth LN11 7JR
G3 ZPR D. Mason 26 Upton Road, Fleetsbridge, Poole BH17 7AH
G3 ZPS S. Shorey 47 Stanham Road, Dartford DA1 3AN
G3 ZPU A. Nightingale 42 Spilsby Road, Horncastle LN9 6AW
G3 ZPW M. Brown 6 Castle Court, Praa Sands, Penzance TR20 9SX
G3 ZQB A. Seabrook 63 St. Annes Road Willingdon, Eastbourne BN20 9NJ
G3 ZQC J. Smith 8 Upland Rise, Westbury BA13 3HW
G3 ZQF S. Carpenter 4 Mount Court, West Wickham BR4 9AH
G3 ZQH Dr D. Barrett Linden House, Clifton Lane, Nottingham NG11 6AA
G3 ZQI B. Downer 9 Crabapple Road, Dereham NR20 3GH
G3 ZQJ B. Stagg 1 Naunton Way, Leckhampton, Cheltenham GL53 7BQ

G3 ZQL S. Murray Old Court, Westleigh, Tiverton EX16 7HT
G3 ZQQ J. Peden 51A Bewdley Road, Kidderminster DY11 6RL
G3 ZQR B. Downton Lickwith Cottage, Monkokehampton, Winkleigh EX19 8SL
G3 ZQS Int. Morse Preservation Society c/o R. Walker Fists The International Morse Preservation Society, Po Box 6743, Tipton DY4 4AU
G3 ZQT J. Yu Hunterscombe, Dorking Road, Leatherhead KT22 8JT
G3 ZQU M. Goodrum Cedars, Church Lane, Little Stonham, Stowmarket IP14 5JL
G3 ZQW B. Barrington Pinlands Cottage, Bines Road, Horsham RH13 8EQ
G3 ZQY N. Clark Chelsworth, Heaton Grange Road, Romford RM2 5PP
G3 ZRA Dr R. Elliot 1321 East Bailey Road, Naperville 60565 United States
G3 ZRB D. Hill 872 Oldham Road, Rochdale OL11 2BN
G3 ZRE P. Ottewell 30 Cumberland Avenue, Leyland PR25 1BH
G3 ZRG I. Steward Keeper'S Cottage, Banville Lane, Cromer NR27 9RN
G3 ZRH A. Stokes 34 Shenfield Crescent, Brentwood CM15 8BW
G3 ZRJ A. Butler-Roskilly 2 North Road, Pennymoor, Tiverton EX16 8LQ
G3 ZRL A. Mcwatters 22 Church Close Swillington, Leeds LS26 8QJ
G3 ZRM M. Payne 3 Waterside Close, Bordon GU35 0HB
G3 ZRN D. Catherwood 14 Hatton Lane, Hatton, Warrington WA44BY
G3 ZRQ D. Maxfield 40 Fegg Hayes Road, Stoke-on-Trent ST6 6RA
G3 ZRR M. Samuel 71 Brighton Road, South Croydon CR2 6EE
G3 ZRS P. Rodmell 2 Meadow Way, Walkington, Beverley HU17 8SD
GM3 ZRT W. Strachan Inverlaroch Albert Road, Ballachulish PH49 4JR
G3 ZRX Basingstoke Makerspace c/o T. Lundegard Saxby, Botsom Lane, Sevenoaks TN15 6BL
G3 ZRY G. Stott 8 Willow Road, Chinnor OX39 4RA
G3 ZSB V. Poore 216 Powder Mill Lane, Twickenham TW2 6EJ
G3 ZSF A. Houltby 2 Sinderson Road, Humberston, Grimsby DN36 4UF
GM3 ZSH J. Donaldson 4 Wallsend Court, Dunfermline KY12 9BE
G3 ZSJ R. Troughton 4 Owletts, Worth, Crawley RH10 7SQ
G3 ZSQ R. Dunham 42 Marsdale Drive, Stockingford, Nuneaton CV10 7DE
G3 ZSS P. Bacon 3 The Grange, Woodmancote, Emsworth PO10 8UX
G3 ZST T. Surgey The Little Wood, Windmill Lane, Ladbroke, Southam CV47 2BN
G3 ZSU S. Scannell 20 Queens Road Wilbarston, Market Harborough LE16 8QJ
G3 ZSX Dr K. Craig 20 Alexander Close, Abingdon OX14 1XA
G3 ZSZ R. James 283 High St., New Whittington, Chesterfield S43 2AP
G3 ZTB R. Ranson Flat 1, The Steyne, Alexandra Road, Harrogate HG1 5JS
GW3 ZTH J. Ludlow 44 Fox Hollows, Brackla, Bridgend CF31 2NG
G3 ZTI K. Marshall 2 Keepers Mill, Woodmancote, Cheltenham GL52 9QS
G3 ZTJ C. Morgan The Villa, The Green, Wallsend NE28 7PH
GI3 ZTL F. Convery 2 Coolagh Road, Maghera BT46 5JR
GM3 ZTP S. Elwell-Sutton 17 Lintrathen Gardens, Dundee DD3 8EJ
G3 ZTR D. Lockwood 25 Thorntondale Drive, Bridlington YO16 6GW
G3 ZTT Mid Cheshire Ar c/o D. Bevan 46 Park Lane, Hartford, Northwich CW8 1PY
G3 ZTU J. Mace Westgate, Bignor, Pulborough RH20 1PQ
G3 ZTV P. Webster 3 Templemere, Norwich NR3 4EF
G3 ZTX P. Angell Star Hill House, Star Hill, Forest Green, Nailsworth, Stroud GL6 0NJ
G3 ZTY Flaxton Moor CG c/o J. Yale 15 Rectory Avenue, Corfe Mullen, Wimborne BH21 3EZ
G3 ZTZ P. Howell 1 Jasmine Close, Littlehampton BN17 6UP
G3 ZUB M. Davison 10 Springfield Close, Loughborough LE11 3PT
G3 ZUC D. Cardell 22 Millview Road, Heckington, Sleaford NG34 9JP
G3 ZUE A. Nicholas Verriotts Lane, Morecombelake, Bridport DT6 6DU
G3 ZUI M. Johnson Greentiles, Main Road, Maltby Le Marsh, Alford LN13 0JW
G3 ZUK Prof. R. Whitehead Church View, Church End, Cambridge CB21 5PE
G3 ZUL B. Kennedy 24 Mallards Close, Alveley WV15 6JL
G3 ZUM B. Lonnon 5 Mickle Meadow Water Orton, Birmingham B46 1SN
G3 ZUN D. Sharpe 12 Belgrave Crescent, Seaford BN25 3AX
G3 ZUO D. Ham Upton Manor Lodge, Upton Manor Road, Brixham TQ5 9QZ
G3 ZUS N. Ewer The Roost, Lyonshall, Kington HR5 3HZ
G3 ZUT M. Thorne Rossendale Main Road Easter Compton, Bristol BS35 5RE
G3 ZUZ M. Fox 41 Elgin Avenue, Ashton in Makerfield WN4 0RH
G3 ZVC B. Comer Corner House, High Street, Fairford GL7 4EQ
G3 ZVH D. Bedford 17 Tewkesbury Close Upton, Chester CH2 1NF
G3 ZVI P. Longhurst 18 Austen Close, Exeter EX4 8HB
G3 ZVK J. Simons 120 Bond Way, Hednesford, Cannock WS12 4SN
G3 ZVM A. Greenbank Grahamsley, Westburn, Ryton NE40 4EU
G3 ZVN G. Peck 4 Koonowla Close, Biggin Hill, Westerham TN16 3BJ
G3 ZVQ J. Bridge 8 Highfield Grove, Lostock Hall, Preston PR5 5YB
G3 ZVS E. Park Waterside Cottage, Bowden Lane, High Peak SK23 0QF
G3 ZVT G. Rabstaff Evesham, 26 Lichfield Drive, , Prestwich M25 0HX
G3 ZVV J. Gellatly 11 Archers Drive, Bilsthorpe, Notts NG22 8SD
G3 ZVW S. White Heatherleigh Crewkerne Road, Axminster EX13 5SX
GI3 ZVZ Dr D. Nicholls 2 Printshop Road, Templepatrick, Ballyclare BT39 0HZ
G3 ZWD P. Flicos 35 The Broadway, Northbourne, Bournemouth BH10 7EU
G3 ZWK D. Raimbach 12 Vineys Gardens, Tenterden TN30 7AZ
G3 ZWM I. Morrison The Vicarage, 1A Church Road, Sandy SG19 2JY
G3 ZWN A. Slingsby 10D Stanbury Road, Thruxton, Andover SP11 8NS
G3 ZWP R. Davies Flat 24, Hurley Court, Bracknell RG12 9QH
G3 ZWR N. Hay 19 Logan Road, Walkerville, Walkerville NE64SY
GM3 ZXB A. Robertson 77 Cobden St., Dundee DD3 6DD
G3 ZXD M. French Po Box 217, Leigh 947 New Zealand
G3 ZXF D. Corner 122 Stortford Hall Park, Bishop's Stortford CM23 5AP
GM3 ZXG J. Higgins 9 Waverley Street, Greenock PA16 9DH
GW3 ZXI S. Brennan 9 Swn Y Nant, Church Village, Pontypridd CF38 1UE
G3 ZXM M. Brown Curraghmore, Tullogher, Co. Kilkenny X91 R642 Ireland
G3 ZXO J. Burnie 1 Chapel Meadow, Buckland Monachorum, Yelverton PL20 7LR
G3 ZXV P. Veale 13 Lawford Gardens, Kenley CR8 5JJ
G3 ZXW J. Midmore 9 Whiteways, Wimborne BH21 2PQ
G3 ZXY P. Holtham 21 Sherborne Place, Chapel Hill QLD 4069 Australia
G3 ZXZ M. Stokes Chaos Cottage, Otley Road, Bingley BD16 3AY
G3 ZYC M. Sneap Ivy Farm Bungalow, Farm Close, Pentrich, Ripley DE5 3RR
G3 ZYD R. Alton 23 Cemetery Road, Belper DE56 1EJ
GM3 ZYE R. Bellerby Glenamour, Newton Stewart DG8 7AE
G3 ZYL Dr G. Bowhay Windwhispers, Lewannick, Launceston PL15 7QD
G3 ZYP A. Matheson 1 St. Edmunds Close, Bromeswell, Woodbridge IP12 2PL
G3 ZYQ A. Robinson 6 The Crescent, Minster On Sea, Sheerness ME12 3BQ
G3 ZYR A. Booer Lower Farm Barn, Duck End Lane, Witney OX29 5RH
G3 ZYX R. Offord 10 Barberry Way, Blackwater, Camberley GU17 9DX
G3 ZYY T. Day Box 204, The Postroom, Calle San Jaime 5, Benijofar 3178 Spain
G3 ZYZ Dr M. Joiner 22 Sanderstead Court, Addington Road, South Croydon CR2 8RA
GM3 ZZA P. Rose 4 Heatherfield Glade, Livingston EH54 9JE
G3 ZZD S. Ireland Po Box 55, Glen Forrest, Glen Forest 6071 Australia
G3 ZZF D. Woolley 21 Lulworth Avenue, Wembley HA9 8TP
G3 ZZH M. Battersby 25 Lowther Close, Emmbrook, Wokingham RG41 1JE
G3 ZZI G. Smith 34 Huria Lane, Woodend 7610 New Zealand
G3 ZZL Dr S. Keightley Riverrun, 4, Northern Burway Laleham Reach, Chertsey KT16 8RW
G3 ZZM M. Robinson 7400 Old Bunch Road, Wendell 27591 United States
GD3 ZZN M. Rickward Dunsandle 12 Fairway Close, Port Erin, Isle of Man IM9 6LS
G3 ZZP D. Matthews 7 Boulsworth Avenue, Hull HU6 7DZ
G3 ZZQ R. Ludwell Church Lodge, Thetford Road, Thetford IP24 2QX
G3 ZZS R. Wills 21 Woodford Road, Glenholt Park, Plymouth PL6 7HX
G3 ZZU C. Waldron 22 Windermere Road, Patchway, Bristol BS34 5PW
G3 ZZV G. Evans 20 Creekside View, Tresillian, Truro TR2 4BS
G3 ZZW P. Chimber 202 Wintersdale Road, Leicester LE5 2GP
G3 ZZX A. Evans Ashlea, Aston Munslow, Craven Arms SY7 9ER
G3 ZZZ J. Gibbs 32 Gresham Avenue, Margate CT9 5EH

G4

GM4 AAF Dundee ARC c/o J. Wilson 20 Ballumbie Gardens, Dundee DD4 0NR
G4 AAH K. Lawson 233 Southwell Road West, Mansfield NG18 4HF
G4 AAL J. Layton Meadow View, Martley, Worcester WR6 6QA
G4 AAQ P. Butterfield 29 Aire Street, Knottingley WF11 9AT
G4 AAR Ashford ARC c/o J. Wellard 19 South Motto, Kingsnorth, Ashford TN23 3NJ
G4 AAX Thornbury & S. Gloucestershire ARC c/o G. Emmerson 72 The Gables, Widdrington, Morpeth NE61 5RB
G4 ABC Thornbury & S. Gloucestershire ARC c/o P. Smart 142 Finch Road, Chipping Sodbury, Bristol BS37 6JB
G4 ABE J. Ellis 4 Hazelmount Crescent, Warton, Carnforth LA5 9HS
G4 ABL A. Howell 46 Acre Court, Andover SP10 1HH
G4 ABN T. Atkins 55 Havenbrook Blvd, Willowdale, Ontario Canada
G4 ABQ J. Hudson 46 High Street, Odell, Bedford MK43 7BB
G4 ABW L. Willey 7 Oaklands Road, Four Oaks, Sutton Coldfield B74 2TB
G4 ABX Dr B. Macaulay Old Chapel, Chapel Street, Swinford, Lutterworth LE17 6AZ
G4 ABY D. Green 2 Fossfield Winstone, Cirencester GL77JY
G4 ACF ACF/CCF Int R N c/o Capt. R. Cogzell Flat 242, Clydesdale Tower Holloway Head, Birmingham B1 1UJ
G4 ACI J. Blackburn 40 Carlton Avenue, Upholland, Skelmersdale WN8 0AE
G4 ACJ H. Reeve 11 Heather Drive, Ferndown BH22 9SD
G4 ACL D. Atkinson 38 Hornbeam Road, Theydon Bois, Epping CM16 7JX
GM4 ACM A. Miller 38 Randolph Road, Broomhill, Glasgow G11 7LG
G4 ACP J. Scherrer 26 Grange Way, Willington, Bedford MK44 3QW
G4 ACS Thornton Cleveleys ARS (Tcars) c/o G. Weale 11 Heather Drive, Kinver, Stourbridge DY7 6DR
G4 ACU M. Levy 34428 Yucaipa Blvd, E346, Yucaipa United States
G4 ACW N. Roe 10 Ramsdean Road, Stroud, Petersfield GU32 3PJ
G4 ACY R. Ratcliffe 173 Montague Road, Bilton Hill, Rugby CV22 6LG
G4 ACZ R. Mutton Summer Hayes, Mill Lane, Alcester B49 6LF
G4 ADD W. Ricalton 4 South Road, Longhorsley, Morpeth NE65 8UW
G4 ADE M. Woollin 14 St. Nicholas Drive, Hornsea HU18 1EW
G4 ADF P. Harrison Siedend÷rp 11, Fehmarn 23769 Germany
G4 ADG P. West 2 Quilp Drive, Chelmsford CM1 4YA
G4 ADJ P. Hampton 45 Mortlake Avenue, Redhill, Worcester WR5 1QB
G4 ADK R. Cutbush 60 Culver Way, Sandown PO36 8QL
GW4 ADL Dr T. Davies 44 Carnglas Road, Sketty, Swansea SA2 9BW
G4 ADM A. Maish 73 Edenfield Gardens, Worcester Park KT4 7DX
G4 ADP P. Mccurrie Lakefields, Drake Close, Southampton SO40 4XB
G4 ADR N. Ayres 17 Ludsden Grove, Thame OX9 3BY
G4 ADS J. Chisman 115 St. Lukes Avenue, Ramsgate CT11 7HT
G4 ADV Newquay & District ARS c/o K. Francks 63 Parc Godrevy, Newquay TR7 1TY
G4 AEB T. Baines 6 Brading Avenue, Clacton-on-Sea CO15 4PA
G4 AED B. Cator 9 Saham Road Watton, Thetford IP25 6EA
G4 AEE M. Bedford 4 Holme House Lane, Oakworth, Keighley BD22 0QY
G4 AEG I. Kemp 21 Rednal Road, Birmingham B38 8DT
G4 AEH J. Lee 44 Howard Road, Nuneaton CV10 7ES
G4 AEI G. Prater Heathfield, Wyndham Lane, Salisbury SP4 0BY
GM4 AEK P. Boswell Coralach, Dunvegan, Isle of Skye IV55 8WF
G4 AEL R. Cox 34 Ratcliffe Drive, Stoke Gifford, Bristol BS34 8UD
G4 AEM P. Ellis 96 Whitelands Avenue, Chorleywood, Rickmansworth WD3 5RG
G4 AEO P. Hunt 93 Park Road, Coalville LE67 3AF
G4 AEP W. Thomas High Field Claypit Lane, Froxfield GU32 1DD
G4 AER R. Sobey 5 Fairway Close, Liphook GU30 7XD
G4 AES K. Walker 3 Glen View, Stile, Sowerby Bridge HX6 1NL
G4 AEV R. Anderson Trinafour, Abingdon Road, Abingdon OX13 6NU
G4 AEY R. Twist 1 Birchwood Drive, Rushmere St. Andrew, Ipswich IP5 1EB
G4 AEZ B. Oughton 16 Thompson Way, Hertford SG13 8FX
G4 AFA Dr N. Porter 13 Charfield Close, Winchester SO22 4PZ
G4 AFE J. Tallentire 4 Alston Road, Middleton-In-Teesdale, Barnard Castle DL12 0UU
G4 AFF S. Cooper Fairfield, Southburgh Lane, Hingham, Norwich NR9 4PP
GI4 AFH G. Phillips 38 Sketrick Ind Park, Newtownards BT23 3BN
G4 AFI A. Cheetham 39 Burns Avenue, Church Crookham, Fleet GU52 6BN
G4 AFQ D. Warner Treeside, School Lane, Gimingham, Norwich NR11 8HJ
G4 AFR F. Nicholson 5 Friars Terrace, Barrow-in-Furness LA13 0BX
G4 AFS T. Bucknell 7 Alexander Court Sandbeds, Keighley BD20 5NW
G4 AFT D. Randles Long Reach, Westerns Lane, Harrogate HG3 3PB
G4 AFU P. Rollin Farthings, Burneston, Bedale DL8 2JE
G4 AFX A. Moore Garden Wing, Copinger Hall, Stowmarket IP14 3DJ
G4 AFY R. Perrin 8 Granville Crest, Kidderminster DY10 3QS
G4 AFZ V. Bott 25 Finkle Street, Hensall, Goole DN14 0QY
G4 AGC C. Wortham 57 Cranleigh Drive, Swanley BR8 8NZ
G4 AGE R. Evans Mansfield, 1 Horsehead Lane, Chesterfield S44 6HU
GM4 AGG W Scotland ARS c/o A. Stewart Three Acres, Cochno Road, Clydebank G81 6PX
G4 AGH S. Pearson 75 Gloucester Road, Thornbury, Bristol BS35 1JH

GM4 AGL W. Ferguson 72 High Parksail, Erskine PA8 7HX
G4 AGM R. Williams Flat 32, St. Johns Court, 59 Murray Road, Northwood HA6 2FY
G4 AGN J. Porter 109 Heacham Drive, Leicester LE4 0LL
G4 AGQ J. Billingham 14 St. Matthews Court, Sutherland Road, Brighton BN2 2EX
GM4 AGU Glasgow University Wireless Society c/o M. Topple Emmaus, 30 Robshill Court, , Newton Mearns G77 6UG
G4 AGY G. Rippengill 5 Bridge Farm Drive, Liverpool L31 9AL
GI4 AHD Dr F. Elder 44 Learmount Road, Claudy, Londonderry BT47 4AQ
G4 AHG Shirehampton ARC c/o C. Chidgey 46 Station Road, Shirehampton, Bristol BS11 9TX
G4 AHJ M. Downey 11 Woodlands Drive, Lepton, Huddersfield HD8 0JB
G4 AHK B. Palin 11 Ashgrove Close, Marlbrook, Bromsgrove B60 1HW
G4 AHM J. Stratton 10 Brownshill, Maulden, Bedford MK45 2BT
G4 AHN D. Lax 1 Gardeners Hill Road, Wrecclesham, Farnham GU10 4RL
G4 AHO K. Jones 13 Upland Grove, Bromsgrove B61 0EL
GI4 AHP T. Sloan 13 Mount Royal, Lisburn BT27 5BF
G4 AHT M. Niven 16 Treewall Gardens, Bromley BR1 5BT
G4 AHW A. Thomson 392 Glen Ross Road, Quinte West Ontario K0K 2C0 Canada
G4 AHZ J. Kynaston 19 Sharples Drive, Wrea Green, Preston PR4 2EL
G4 AIB P. Holt 20 Hedingham Close, Ilkeston DE7 5HR
G4 AIE W. Mackie 23 College Park, Horncastle LN9 6RE
G4 AIJ R. Jones Sycamores, The Sheet, Ludlow SY8 4JT
GI4 AIO R. Lindsay 67 Halfpenny Gate Road, Moira, Craigavon BT67 0HP
G4 AIR D. Bieber Tonkins Quay House, Lanteglos-by-Fowey PL23 1NB
GM4 AIS E. Bain 49 Paisley Avenue, Edinburgh EH8 7LQ
G4 AIU E. Morgan 12 Kitts, Wellington TA21 9AX
G4 AIW A. Scarsbrook 16 Greenbank Avenue, Uppermill, Oldham OL3 6EB
G4 AJA C. Hoare 16 Shrivenham Road Highworth Swindon Wilts Sn6 7Bz, Swindon SN6 7BZ
G4 AJB A. Bonwitt 60 Wellhouse Road, Beech, Alton GU34 4AG
G4 AJE P. Brown 33A March Road, Wimblington, March PE15 0RW
G4 AJG P. Perera 43 Hillside Avenue, Woodford Green IG8 7QU
G4 AJJ G. Smith 17 Marshall Drive, Ruddington NG11 6AJ
G4 AJO R. Finch 48 Allens Lane, Sprowston, Norwich NR7 8EJ
G4 AJQ Prof. N. Johnson 503-97 Lawton Blvd, Toronto, Toronto M4V 1Z6 Canada
G4 AJU I. Aldridge 28 Robert St., Williton, Taunton TA4 4PG
GM4 AJV M. Mackinnon 55 Fairbrae, Edinburgh EH11 3GZ
G4 AJW A. Wade 139 Gilbert Road, Cambridge CB4 3PA
G4 AJY D. Ellis 26 Drake Close, Benfleet SS7 3YL
G4 AKA M. Diprose 4A Russet Close, Staines TW19 6AX
G4 AKB M. Court Vereda Escorredor, 138, Dolores 3150 Spain
G4 AKC D. Starkie 5 Kidbrooke Avenue, Blackpool FY4 1QR
G4 AKD I. Alexander 46 Pettitts Lane, Dry Drayton, Cambridge CB23 8BT
G4 AKE C. Gent 16 Coronation Avenue, Alvaston, Derby DE24 0LQ
G4 AKG P. Fry 11 Park Road, Burgess Hill RH15 8EU
G4 AKR G. Slack 16 East Carr, Cayton, Scarborough YO11 3TS
G4 AKW G. Robinson 2 Hasketon Road, Woodbridge IP12 4JR
GW4 AKY D. Hayes Parth Y Barcud Blaenpennal, Aberystwyth SY23 4TR
G4 AL Secret Nuclear Bunker CG c/o J. Wood 18 Kennedy Avenue, Long Eaton, Nottingham NG10 3GF
G4 ALA J. Hardwick 455 Hatton Road, Feltham TW14 9QP
G4 ALB N. Castledine 1 Johns Close Burbage, Hinckley LE10 2LY
G4 ALC J. Balls 48 Collingwood Road, Great Yarmouth NR30 4LR
G4 ALD F. Donovan 4 Rembrandt Drive, Northfleet, Gravesend DA11 8NQ
G4 ALE Addiscombe ARC c/o M. Franklin 6 Tor Road, Farnham GU9 7BX
G4 ALF K. Law 93 Measham Drive, Stainforth, Doncaster DN7 5TQ
G4 ALR M. Down 5 Juniper Mead, Stotfold, Hitchin SG5 4RU
G4 ALT A. Taylor 21 Gould Avenue West, Kidderminster DY11 7HD
G4 ALY R. Bird 6 The Cross, St. Dominick, Saltash PL12 6SP
G4 ALZ R. Bridgland 20 Newling Way, Worthing BN13 3DG
G4 AMD C. Heavens 10815 N. Indian Wells Drive, Fountain Hills AZ 85268 United States
G4 AMF J. Cresswell 7 Glinton Avenue, Blackwell, Alfreton DE55 5HD
G4 AMI M. Hearn 63 Greswolde Road, Solihull B91 1DX
G4 AMJ Dr D. Evans 330 County Road 16 1/2, Longmont 805049467 United States
G4 AMN C. Wainwright 60 Main Street, , Hoby LE14 3DT
G4 AMP B. Flack Ave Des Hospitaliers De St Jean 7, Waterloo 1410 Belgium
G4 AMT T. GeorgeStoneways, Trevescan, Sennen, Penzance TR19 7AQ
GW4 AMX J. Barrett Flat 5, Rhos Abbey, Rhos Promenade, Colwyn Bay LL28 4QA
G4 AMY R. Briggs Nickey Nook View, Lancaster New Road, Preston PR3 1NL
GW4 AMZ P. Leach 41 Bryn Colwyn, Colwyn Bay LL29 9LJ
G4 ANB Dr J. Morris 4111 Eve Road, Simi Valley 93063 United States
G4 AND J. King Chetwynd, Henfield Road, Steyning BN44 3TF
G4 ANE H. Leach 30 Taywood Road, Thornton-Cleveleys FY5 2RT
GW4 ANK R. Davenport 14 Milward Road, Barry CF63 3QD
G4 ANN R. Hadfield 45 Erica Way, Copthorne, Crawley RH10 3XG
G4 ANP M. Valentine 10 Thellusson Avenue, Scawsby, Doncaster DN5 8QN
G4 ANT East Anglian CG c/o R. Reed Oak Cottage, Dereham Road, Dereham NR20 4AA
G4 ANU C. Columbine 5 Thornbury Drive, Mansfield NG19 6NB
G4 ANV P. Hudson 3 Rowan Drive, Kilburn, Belper DE56 0PG
G4 ANW T. Slack 16 Woodside Avenue, Alverstone Garden Village, Sandown PO36 0JD
G4 ANY D. Stephens Croeso Cottage, 31 Coton, Whitchurch SY13 2RA
G4 ANZ B. Warren 36 Hobbiton Road, Weston-Super-Mare BS22 7HP
G4 AOA H. Mason 9 Chatsworth Drive, Little Eaton, Derby DE21 5AP
G4 AOJ R. Horton 34 Rising Lane, Knowle, Solihull B93 0BZ
G4 AOK T. Winter 6 Cunliffe Drive, Brooklands, Sale M33 3WS
G4 AOL D. Harmer 4 Somerton Gardens, Earley, Reading RG6 5XG
G4 AOP D. Hibbin 95A Thorpe Acre Road, Loughborough LE11 4LF
GM4 AOR K. Henderson 97 Granton Road, Edinburgh EH5 3NH
G4 AOS Dr J. West Horsley House, Rochester, Newcastle upon Tyne NE19 1TA
G4 AP J. Rooke 12 Hellings Gardens, Broadclyst, Exeter EX5 3DX
G4 APB K. May 53 Shearwood Crescent, Dartford DA1 4SU
G4 APD Ofrac Baldock c/o P. Wells 12 Shelley Drive, Lutterworth LE17 4XF
GW4 APF M. Richards 9 Bank Road, Llangennech, Llanelli SA14 8UB
G4 APG M. Pellatt Old Thatch, Branscombe, Seaton EX12 3BL
GM4 API D. Hebenton Craigmill Cottage, 3 Craigmill Road, Dundee DD3 0PH
G4 APJ K. Punshon 24 Newcombe Road Ramsbottom, Bury BL0 9UT
G4 APL P. Lewis 20 Annes Walk, Caterham CR3 5EL
G4 APO Dr R. Hirst 21 Manor Farm Court, Thrybergh, Rotherham S65 4NZ
G4 APP W. Grogan 8 Fairway South Moor, Stanley DH9 7HP
G4 APS D. Fiander 2 Snowshill Close, Nuneaton CV11 4XQ
G4 AQA P. Hall 39 Mill Lane, Kirk Ella, Hull HU10 7JE
G4 AQB S. Macdonald 58A Tarbet Drive, Bolton BL2 6LT
G4 AQE P. Saunders Orchard Cottage, Vale Road, Broadstairs CT10 2JG
G4 AQG University of Sussex ARS c/o A. Maris 2 Court Ord Road, Rottingdean, Brighton BN2 7FD
G4 AQJ K. Gordon 96 Pear Tree Crescent, Shirley, Solihull B90 1LF
G4 AQK D. Davis 23 Matley Moor, Liden, Swindon SN3 6NL
G4 AQR I. Cordingley Orchard Cottage, Compton, Marldon, Paignton TQ3 1TA
G4 AQS M. Bliss 53 Rowallan Drive, Bedford MK41 8AS
G4 AQT J. Rowbotham 56 Longleat Crescent Beeston, Nottingham NG9 5EU
G4 AQZ G. Axford 24 Jack Branch Court, Wash Lane, Clacton-on-Sea CO15 1EJ
GW4 ARC Essex Raynet c/o A. Evans 4 Elm Grove, Rhyl LL18 3PE
G4 ARE Exeter ARS c/o J. Rooke 12 Hellings Gardens, Broadclyst, Exeter EX5 3DX
G4 ARF Furness ARS c/o M. Bell 36 Schneider Road, Barrow-in-Furness LA14 5DW
G4 ARI T. Raven 15 Preston Close Stanton Under Bardon, Markfield LE67 9TX
GM4 ARJ J. Ferguson 26 Cleuch Avenue, Tullibody, Alloa FK10 2RX
G4 ARN Blandford Garrison CW And DX Group c/o A. Hall 122 Norwich Road, New Costessey, Norwich NR5 0EH
G4 ARO T. Covey 68 Wellington Close, Walton-on-Thames KT12 1BB
G4 ARS Carlisle & Dis. ARS c/o C. Wolf 35A Moorhouse Road, Carlisle CA2 7LU
GM4 ARU J. Mcintyre 12 Johnstone Lane, Carluke ML8 4NR
G4 ARW C. Turner 2 Queens Way, Wotton-under-Edge GL12 7HA
G4 ARX B. Curley 22 Churchill Crescent, Sheringham NR26 8NQ
G4 ARY A. Langford 33 Briscoe Road, Hoddesdon EN11 9DG
G4 ASF R. Mccurrach Isa Coed, Bowden Lane, Bude EX23 9BJ
G4 ASG P. Bayley 9 Westbrook Green, Bromham, Half Acre, Chippenham SN15 2EF
G4 ASH I. Roberts 86 Federation Avenue, Desborough, Kettering NN14 2NX
G4 ASI F. Emery Room 10, Building 448, Westerham TN16 3BN
G4 ASK E. Rayland 40 Sycamore Close, Taunton TA1 2QJ
G4 ASL S. AylingKitnocks 89 Queen'S Road, Alton GU34 1JA
G4 ASM A. Murphy Apartment 28, Trinity Gardens, 1 Kingsmead Road South, Prenton CH43 6TA
GU4 ASO RSGB Transatlantic Test Centenary Club c/o R. Ayres Langaller, Rue Colin, Vale GY6 8LA Guernsey
G4 ASP J. Holding Old Pearmain, Eardisland, Leominster HR6 9DN
G4 ASQ Dr M. Jordan 4 Marchfont Close, Nuneaton CV11 6GA
G4 ASR D. Butler Yew Tree Cottage, Lower Maescoed, Hereford HR2 0HP
G4 ASW M. Yorke 8 St John Place, Port Washington NY 11050 United States
G4 ASX O. Perry 60 Malines Avenue, Peacehaven BN10 7RS
G4 ASY Strathclyde 4x4 Response c/o D. Yeaman Purbeck House, Trent, Sherborne DT9 4SW
G4 ASZ M. Hurst 21 Bankside, Dunton Green, Sevenoaks TN13 2UA
GM4 ATA J. Hotchin 2 Moorfield Place, Gatehead, Kilmarnock KA2 0AX
G4 ATB R. Shapland 14 Charney Court, Grange-over-Sands LA11 6DL
G4 ATG Bartg c/o Dr A. Thomas The Stone Barn, 1 Home Farm Close, Chesterton OX26 1TZ
G4 ATH Thornton Cleveleys ARS (Tcars) c/o J. Rodway 9 York Avenue, Thornton-Cleveleys FY5 2UG
G4 ATL D. Bloomfield 26 Preston Crowmarsh, Wallingford OX10 6SL
G4 ATQ G. Hawkins 18 Brook Street, Leighton Buzzard LU7 3LH
G4 ATU S. Brown Mullins View 1D Turnpike Road Aughton, Ormskirk L39 3LD
G4 AUB A. Smith 56 Longrood Road, Rugby CV22 7RE
G4 AUC S. Baugh 70 Madingley, Bracknell RG12 7TF
GW4 AUD A. Lacy Llanoris, Llanerfyl, Welshpool SY21 0EP
G4 AUE A. Rose 18 Highview Gardens, St. Albans AL4 9JX
G4 AUF C. Friel 102A Sharps Lane, Ruislip HA4 7JB
G4 AUG R. Mortimer 19 St. Monance Way, Colchester CO4 0PJ
G4 AUL East Dunbartonshire Raynet Group c/o G. Mitchell 10 Wealden Close, Hildenborough, Tonbridge TN11 9HB
G4 AUN R. Collett 70 Clifton Road, Darlington DL1 5DX
GM4 AUP I. Suart 37 Meldrum Mains, Glenmavis, Airdrie ML6 0QQ
G4 AUQ F. Barker 90 Hall Road, Hull HU6 8SB
G4 AUR J. Mcburney 4 Fownhope Road, Sale M33 4RF
G4 AUV G. Wing 105 Moore Avenue, Norwich NR6 7LG
G4 AUY P. Sherwood 43 Kingsland, Arleston, Telford TF1 2LE
GW4 AVC D. Bowers 31 Clarence Road, Wrexham LL11 2EU
G4 AVE L. Cates 45 Smoke Lane, Reigate RH2 7HJ
G4 AVF A. Fletcher 11 Little Oak Close, Lees, Oldham OL4 3LW
G4 AVJ G. Pople 3 Leighton Drive, Creech St. Michael, Taunton TA3 5DW
G4 AVK Raywell Park Scout ARS c/o S. Ripley 62 Palewell Park, London SW14 8JH
G4 AVL P. Newby 238A Wherstead Road, Ipswich IP2 8JZ
G4 AVS R. Wilson Aerial House, 1 The Fields, Woodbridge IP12 2HZ
G4 AVV G. Cluer 12 Bingham Road, Addiscombe, Croydon CR0 7EB
G4 AVX A. Newman 101 Washbrook Road, Portsmouth PO6 3SB
G4 AWA R. Payne 11 Beaconsfield Road, Christchurch BH23 1QT
GM4 AWB R. Macduff 7 Cairngorm Crescent, Bearsden, Glasgow G61 4EH
G4 AWF D. Wilson 4 Caradoc Meadow, Sellack, Ross-on-Wye HR9 6GJ
G4 AWG G. Higgs Firtree House, Perry Wood, Faversham ME13 9SE
G4 AWJ G. Thomas 9 Highcroft Crescent, Heathfield TN21 8HE
G4 AWK M. Roberts 2 Miners Garth, Liverton Mines, Saltburn-by-the-Sea TS13 4BU
G4 AWM D. Norfolk 13 Oakwood Crescent, Greenford UB6 0RF
G4 AWO R. Gray 10 Stone Park, Broadsands, Paignton TQ4 6HT
G4 AWU R. Lane 8 Town Street, Lound, Retford DN22 8RS
G4 AWW N. Shepherd Jo Kebi, Stonehall Road, Dover CT15 7JS
G4 AWY R. Mekka 57 St. Johns Road, Caversham, Reading RG4 5AL
G4 AWZ P. Matthews 22 Rydens Road, Walton-on-Thames KT12 3DA
G4 AXA N. Pope Silver Hill, Norwich Road, Great Yarmouth NR29 5PB
G4 AXC C. Burden Cedar Croft, Hengar Lane, Bodmin PL30 3PH
G4 AXD G. Edy 44 Roseholme, Maidstone ME16 8DR
G4 AXF J. Jacques 30 Centurian Way, Bedlington NE22 6LD
G4 AXL C. Gerrard 22 Kelso Drive, The Priorys, North Shields NE29 9NS
G4 AXO J. Wills 48 Fairfield Road, Winchester SO22 6SG
GM4 AXS P. Wilberforce 8 Ferryfield Road, Connel, Oban PA37 1SR

G4 AXU G. Parr Chesil Coppice, West Bexington, Dorchester DT2 9DD
GI4 AXV J. Doherty 172 Dunmore Road, Ballynahinch BT24 8QQ
G4 AXX M. Marsden Mill Cottage, Shrowle, East Harptree, Bristol BS40 6BJ
G4 AXY A. Mort 86 Longfield Road, Winnall, Winchester SO23 0NU
G4 AYB A. Kelle Urb.Sorries 10, la Massana AD.400 Andorra
G4 AYH G. Monks Weavers Cottage, 7 Town Street, Rawdon, Top Of Henley Close, Leeds LS19 6PU
G4 AYK Mid Severn Vall c/o P. Perrins 9 Merrick Close, Hayley Green, Halesowen B63 1JY
G4 AYL E. Lambert 41 Brand Hill Drive, Crofton, Wakefield WF4 1PF
G4 AYO M. Hewitt 10 Blacka Moor View, Sheffield S17 3GZ
GW4 AYQ J. Durrans 87 The Links Trevethin, Pontypool NP4 8DQ
G4 AYR T. Greenwood 30 Ringwood Road, Headington, Oxford OX3 8JA
G4 AYS A. Crook 153 Shortheath, Shortheath, Swadlincote DE12 6BL
G4 AYU N. Kenyon 74 Albert Road, Leyland PR25 4YJ
G4 AZA R. Winkworth 13 Bagley Close, Kennington, Oxford OX1 5LS
G4 AZC P. Martin Stoneovers, Wellow Top Road, Ningwood, Yarmouth PO41 0TL
G4 AZD A. Edgecock Sunnydene, Station Road, Colchester CO7 8JA
G4 AZG Dr G. Macdonald Pilgrims Cottage, Church Lane, Canterbury CT4 6HX
G4 AZH M. Bushnell Rose Cottage, Street Ashton, Rugby CV23 0PH
GW4 AZI D. Thomas Sunnydale, Scurlage, Swansea SA3 1BA
GD4 AZJ R. Troughton Flat 2, Waterfront Apartments, Mooragh Promenade, Ramsey IM8 3AN Isle of Man
G4 AZL P. Justin Garth, Park View Road, Pinner HA5 3YF
G4 AZM C. Wilson 17719 Phil C Peters Road, Winter Garden 34787 United States
G4 AZS A. Bayling 55 Shelton Road, Shrewsbury SY3 8SU
G4 AZT T. Barker 1 Links Road, Kennington, Oxford OX1 5RX
G4 AZU J. Tiller 21 Portal Road, Winchester SO23 0PX
G4 AZX J. Robinson 19 Sunnycroft Gardens, Cranham, Upminster RM14 1HP
G4 BAD Bath And District ARC c/o P. Carter 16 Alexandra Park Paulton, Bristol BS39 7QS
G4 BAN P. Godfrey 5 Parkway, Southgate, London N14 6QU
G4 BAO Dr J. Worsnop 20 Lode Avenue, Waterbeach, Cambridge CB25 9PX
GM4 BAP A. Beaton 46 Balmoral Place, Aberdeen AB10 6HP
G4 BAQ Gilwell Park Scout RC c/o R. Chambers 17 Exmoor Close, Worthing BN13 2PW
G4 BAS Stornoway Repeater Group c/o H. Ketley 24 Farmcroft Road, Mansfield Woodhouse, Mansfield NG19 8QT
G4 BAU R. Russell 228 Broomhill, Downham Market PE38 9QY
G4 BAV J. Gee Windmill Lodge, Mill Lane Witnesham, Ipswich IP6 9HR
G4 BBA P. Chilcott 321 Eastfield Road, Peterborough PE1 4RA
G4 BBD Dr M. Tooley 4 Shelley Road, Bath BA2 4RJ
GI4 BBE R. Bolton Ohmvilla, 69 Newcastle Street, Newry BT34 4AQ
G4 BBH R. Ferryman 25 Winant Way, Dover CT16 2AX
G4 BBI P. Nixon 8 White Edge Close, Chesterfield S40 4LE
G4 BBJ R. Ramsay 1 Sapho Park, Gravesend DA12 4NA
G4 BBL A. Thackery 19 Pyne Point, Clevedon BS21 7RL
G4 BBM A. Benn 10 Roman Way, Coventry CV3 6RD
G4 BBP Dr P. Howey Raybarrow Farm, Nettleton Shrub Nettleton, Chippenham SN14 7NN
G4 BBQ D. King 62 Ansley Road, Nuneaton CV10 8NU
G4 BBT Dr R. Hancock 80 Ulleries Road, Solihull B92 8EE
G4 BBU P. Whittle 20 Marlbrook Lane Marlbrook, Bromsgrove B60 1HN
G4 BBY R. Edwards 27 Provis Mead, Chippenham SN15 3UA
G4 BCA Dr D. Tunnicliffe 4 Chesford Drive Churchdown, Gloucester GL3 2BA
G4 BCB Dr K. Johnston 92C Mcdowalls Road, Yugar 4520 Australia
GW4 BCF R. Newman 32A Park Avenue, Porthcawl CF363EP
G4 BCG G. Wale 2 The Jordans, Coventry CV5 9JT
G4 BCH P. Burgess Tretawn, Kite Hill, Ryde PO33 4LG
G4 BCP L. Graves The Beach Hut, 6 Hauxley Links, Morpeth NE65 0JS
G4 BCT A. Gordon 4 Victoria Road West, Thornton-Cleveleys FY5 1BU
G4 BCV Essex Raynet c/o N. Smith Clare Cottage, White Ash Green, Halstead CO9 1PD
G4 BCX A. Helm 38 Blandford Road, Lower Compton, Plymouth PL3 5DU
GW4 BCZ J. White 13 Stokes Court, Ponthir, Newport NP18 1RY
G4 BDC K. Collerton 93 Nursery Lane, Leeds LS17 7EE
GM4 BDJ R. Mccartney Cairndhu Walter Street, Langholm DG13 0AX
GI4 BDL Milton Keynes Raynet Group c/o V. Simpson 25 Waringstown Road, Lurgan, Craigavon BT66 7HH
G4 BDQ P. Harris 76 Rozel Court, Southampton SO16 9QE
GI4 BDR Dr N. Evans 87A Oldtown Road, Castledawson, Magherafelt BT45 8BZ
G4 BDW J. Bagley 10 Dairy Close, Brixworth, Northampton NN6 9DR
G4 BDX M. Horoszko 1 Woodgarth Cottages, Reedness, Goole DN14 8EX
G4 BEB R. Browning Spracketts Orchard, Curry Rivel, Langport TA10 0PP
G4 BEH J. Eato 77 Rutland Avenue, Nuneaton CV10 8EG
G4 BEL R. Taylor 12 The Rampart, Haddenham, Ely CB6 3ST
G4 BEM S. Ford 3 Hill View, Stoke-on-Trent ST2 7AR
G4 BEO B. Hailstone 6 Larkswood Rise, St. Albans AL4 9JU
GW4 BEQ G. Hotchkiss 38 Arlington Road, Sully, Penarth CF64 5TQ
G4 BEU J. Small 20 Hastings Road, Birkdale, Southport PR8 2LW
G4 BEV R. Taylor 6 Churchill Crescent, Marple, Stockport SK6 6HJ
G4 BEZ J. Phillipson 3 Montrose Close, New Hartley, Whitley Bay NE25 0TA
G4 BFC A. Riddell 12 Sunrise, Malvern WR14 2NJ
G4 BFR D. Baldwin 112 Moorland View Road, Chesterfield S40 3DF
G4 BFS T. Sargent 15 Pound Lane, Blofield, Norwich NR13 4NB
G4 BFT C. Johnson 7 Field View, Braunston, Daventry NN11 7JS
G4 BFV D. Sinclair 46 Church Lane, Mablethorpe LN12 2NU
GM4 BFX Dr A. Milne 65 Lord Hays Grove, Aberdeen AB24 1WT
G4 BG A. Duckworth Ambergate, 2 Ashleigh Drive, Teignmouth TQ14 8QX
GI4 BGB P. Kelly 30 Cahore Road, Draperstown, Magherafelt BT45 7LY
G4 BGD Dr R. Williams Old Hall House, Old Hall Avenue, Littleover, Derby DE23 6EN
G4 BGH A. Ruddell 20 Davies Road, Longhedge, Salisbury SP4 6SF
G4 BGM C. Zeal 20, Hurst Park, Midhurst GU29 0BP
G4 BGP C. Barber 45 Cuerdale Lane Walton-Le-Dale, Preston PR5 4BP
GM4 BGS S. Liddell 49 Inchbrae Road, Cardonald, Glasgow G52 3HA
G4 BGT M. Staton 30 Shaftesbury Avenue, Chandler'S Ford, Eastleigh SO53 3BS
G4 BGV K. Graham 86 Broadway, Irlam, Manchester M44 6DQ
G4 BGW I. Wilson Whitethorn, Sandhurst Lane, Gloucester GL2 9NW
G4 BHC F. Stevens 11 Hen Wythva, Camborne TR14 7XN
G4 BHD T. Goldsworthy Trevarth, Atlantic Terrace, Camborne TR14 7AW
G4 BHE B. Macklin 4 Bramdown Heights, Basingstoke RG22 4UB
G4 BHJ M. Fochtmann 1 Chapmans Way, Over, Cambridge CB24 5PZ
G4 BHL J. Firth 10 Ridgway Avenue, Darfield, Barnsley S73 9DU
G4 BHP G. Benwell The Did, Tunley, Bath BA2 0DZ
G4 BHT M. Hulands 100 Avenue Road, Rushden NN10 0SJ
GM4 BHU D. Aitkenhead 37/3 Cavalry Park Drive, Edinburgh EH15 3QG
G4 BIA R. Hood 8 Fayre Meadow, Robertsbridge TN32 5AU
G4 BID W. Boyd 2 The Ramblers, Poringland NR14 7QN
G4 BII D. Williams 2 Main Street, Poundon, Bicester OX27 9AZ
G4 BIK P. Mellor 10 Greenfields, Earith, Huntingdon PE28 3QH
G4 BIM P. Bentley Blakes Hill, Limerstone Road, Brighstone, Newport PO30 4AE
G4 BIN N. Long Homedale, Bayford Hill, Wincanton BA9 9LS
GW4 BIS A. Davies 12 Church St., Troedyrhiw, Merthyr Tydfil CF48 4HD
GM4 BIT R. Wilson 5 Collins Drive, Loans, Troon KA10 7HA
G4 BIX Dr D. Price 34 Vanda Crescent, St. Albans AL1 5EX
G4 BIY M. Corbett 6 Windgap Lane, Haughley, Stowmarket IP14 3PA
G4 BIZ A. Paxton Cleveland House, Bartley Road, Southampton SO40 7GP
G4 BJB C. Hurst 28 Hengistbury Road, Barton-on-Sea BH25 7LU
G4 BJC International Shortwave League c/o A. Kinson 6 Uplands Park, Broad Oak, Heathfield TN21 8SJ
G4 BJD G. Overton 14 Aylestone Drive, Hereford HR1 1HT
G4 BJF B. Marshall 23 Sandgate Avenue, Birstall, Leicester LE4 3HQ
G4 BJG P. Smith 11 Chatsworth Avenue, Clowne, Chesterfield S43 4SR
G4 BJJ H. Tickell 26 Shear Brow, Blackburn BB1 7EX
GI4 BJK K. Patterson 1A Demesne Gate, Saintfield, Ballynahinch BT24 7BE
G4 BJN D. Harvey 23 Lapwing Close, Hemel Hempstead HP2 6DS
G4 BJO B. Greeves 65 Stowupland Road, Stowmarket IP14 5AN
G4 BJP S. Popek 42 Victoria Road, Polegate BN26 6DA
G4 BJS J. Loose Flat 30 Highbury Court, Howard Road East, Birmingham B13 0RQ
G4 BJT Dr M. Ware 20 Bath Road, Buxton SK17 6HH
G4 BJX W. Whatmore 51 The Fairways, Sherford, Taunton TA1 3PA
G4 BKA A. Neaves The Coach House, Cliff Hall Lane Cliff, Tamworth B78 2DR
G4 BKB G. Jessup 68 Danes Road, Bicester OX26 2LR
G4 BKE D. Wright 4 Wynne Close, Broadstone BH18 9HQ
G4 BKF T. Howarth 71 Ford Road, Wirral CH49 0TD
GW4 BKG S. Emlyn-Jones 26 Lime Tree Way, Porthcawl CF36 5AU
G4 BKH A. Chorley 354 Denton Lane, Chadderton, Oldham OL9 8QD
G4 BKI P. Evans 16 Cotgrave Road, Plumtree, Nottingham NG12 5NX
G4 BKO J. Francis 22 Earlswood Drive, Mickleover, Derby DE3 9LN
G4 BKQ R. Gubbins 29 Meadow End, Gotham, Nottingham NG11 0HP
G4 BKR W. Taggart Calle Zarauz 61, Urb. San Luis, Alicante 3180 Spain
G4 BKS P. Erkiert 129 Cannock Road, Aylesbury HP20 2AS
G4 BLD C. Croucher 13 Magnolia Way, Pilgrims Hatch, Brentwood CM15 9QS
G4 BLL P. Burnett 28 Crownest Lane, Bingley BD16 4HL
GM4 BLO G. Milne 65 Millburn Avenue, Clydebank G81 1ER
G4 BLS P. Appleby Flat 14, Maryan Court, Hailsham BN27 3DJ
G4 BLT R. Sterry 9 Finch Avenue, Wakefield WF2 6SE
G4 BM T. Searle 2 Woolfall Terrace, Seaforth, Liverpool L21 4PJ
G4 BMC D. Barrell 26 Yerville Gardens, Hordle, Lymington SO41 0UL
G4 BMD M. Hayes No 6 The Court, Dunboyne Castle, Dunboyne A86 FA02 Ireland
G4 BMK M. Kerry 2 Beacon Close, Seaford BN25 2JZ
GW4 BML B. Lloyd 1 Llys Melyn, Tregynon, Newtown SY16 3EE
G4 BMM P. Knight 75 Ashcroft Road, Luton LU2 9AX
G4 BMO D. Cloke Church Cottage, East Coker, Yeovil BA22 9LY
G4 BMP R. Sadler 9 Meade King Grove, Woodmancote, Cheltenham GL52 9UD
G4 BMQ D. Harrop 1 Edgecombe Crescent, Rowner, Hants PO13 9RD
GM4 BMS A. Redford 9 Broker, Isle of Lewis HS2 0EZ
G4 BMU S. East 2 Linscott House, 64D Russell Road, Buckhurst Hill IG9 5QE
G4 BMW C. Pescod 7 Brian Close, Chelmsford CM2 9DZ
G4 BNB R. Wynn 48 Darnley Road, Woodford Green IG8 9HY
GW4 BNC Dr G. Thomas Ash Barn Penylan Road Michaelston-Y-Fedw, Cardiff CF3 6XW
G4 BNE Dr R. Herring 96 St Fabians Drive, Chelmsford CM12PR
GW4 BNJ D. Williams 48 St. Hilary Drive, Killay, Swansea SA2 7EH
G4 BNK W. Wright 27 St. Johns Road, Farnborough GU14 9RL
G4 BNL R. Morley 63 Holt Park Crescent, Holt Park, Leeds LS16 7SL
G4 BNM S. Homans 3 Hilton Mews, Bramhope, Leeds LS16 9LF
G4 BNO M. Ayling 68 Littledown Avenue, Queens Park, Bournemouth BH7 7AS
G4 BNP J. Burgess 11 Winters Lane, Ottery St. Mary EX11 1AR
G4 BNS A. Collinson 30 Thornton Road, Pickering YO18 7HZ
G4 BNT G. Moore 50 Barley Rise, Strensall, York YO32 5AA
G4 BNW M. Knight 18 Friary Road, Abbeymead, Gloucester GL4 5FD
G4 BNX I. Middleton 9 Brentnall Court, Kirk Close, Nottingham NG9 5EZ
G4 BOB R. Ambler 21 Whitley Spring Road, Ossett WF5 0QA
G4 BOF P. Harry 5 St. Michaels Avenue, Kingsland, Leominster HR6 9QR
G4 BOH C. Cummings Castle View, Childs Lane, Congleton CW12 4TQ
G4 BOJ N. Greenstreet 223 Upperthorpe, Sheffield S6 3NG
G4 BOL R. Fineman 4 Sherbourne Avenue, Bradley Stoke, Bristol BS32 8BB
G4 BON J. Strutt 163 Scalby Road, Scarborough YO12 6TB
G4 BOO D. Rumens 3 Flecker Close, Thatcham RG18 3BA
G4 BOP Dr P. Berwick Beech Croft, West Hill, Ottery St. Mary EX11 1UY
G4 BOQ J. Hall 15 Main Street, Greetham, Greetham, Rutland LE15 7NJ
G4 BOU J. Chance-Read 15 Garrard Way, Wheathampstead, St. Albans AL4 8PE
G4 BOV A. Horton Martletts, 52 Lower Cookham Road, Maidenhead SL6 8JZ
G4 BOZ A. Brock 1 Carpenter Drive, St. Leonards-on-Sea TN38 9RX
G4 BP Scarborough ARS c/o M. Day 33 Ryndle Walk, Scarborough YO12 6JT
G4 BPE A. Evans Fairfield Main St, Claypole, Newark NG23 5BA
G4 BPJ B. Stone 12 Forbes Road, Newlyn, Penzance TR18 5DQ
G4 BPN N. Kerstein 40 Davidson Close Hythe, Southampton SO45 6JT
G4 BPO Po Research Cnt c/o C. Hoare 16 Shrivenham Road Highworth Swindon Wilts Sn6 7Bz, Swindon SN6 7BZ
G4 BPV P. Barker 2 Oriole Drive, Exeter EX4 4SJ
G4 BQA M. Morley 8 The Ridings, Seaford BN25 3HW
G4 BQB J. Crocker 4 Portland Terrace, Watchet TA23 0DD
G4 BQC B. Makeham 64 Benomley Road, Almondbury, Huddersfield HD5 8LS
GM4 BQD R. Muir 9 Craigs Court, Torphichen, Bathgate EH48 4NU
G4 BQF M. Duce 28 Thompson Avenue, Canvey Island SS8 7TS

G4 BQH D. Livsey 18 Tollards Road, Countess Wear, Exeter EX2 6JJ
GI4 BQI W. Mccullough 16 Ballylisk Lane, Portadown, Craigavon BT62 3RN
G4 BQJ A. Hill 3 Cambrai Avenue, Warrington WA4 6QU
G4 BQN N. Marsden 32 Chard Road, Drimpton, Beaminster DT8 3RF
GW4 BQQ L. Dean Tanraltt, Llangwyryfon, Aberystwyth SY23 4SP
G4 BQR W. Carmichael 47 Neath Drive, Ipswich IP2 9TA
G4 BQS B. Prichard The Gables, Wootton Lane, Canterbury CT4 6RT
G4 BQW W. Glover Swallows Meadow Court, 33 Swallows Meadow, Solihull B90 4PH
G4 BQY P. Aburrow 25 Hill Crescent, Worcester Park KT4 8NB
G4 BRA Bracknell ARC c/o M. Goodey 62 Rose Hill, Binfield, Bracknell RG42 5LG
GM4 BRB A. Stewart 121 William Street, Dalbeattie DG5 4EE
G4 BRC Ham Radio Network c/o T. Lundegard Saxby, Botsom Lane, Sevenoaks TN15 6BL
G4 BRF R. Mickleburgh 85 Carey Park, Killigarth, Looe PL13 2JP
GW4 BRH J. Sniadowski Bryn Bach Barn, Cwmdu, Crickhowell NP8 1RT
G4 BRK N. Whiting Forge End, Garford, Abingdon OX13 5PF
G4 BRL A. Moore 14 Heath Road, Ipswich IP4 5SA
GM4 BRM A. Long 34 Thornly Park Drive, Paisley PA2 7RP
GM4 BRN Kingdom ARS c/o P. Merckel 1 Mortimer Court, Dalgety Bay, Dunfermline KY11 9UQ
GW4 BRS Barry ARS c/o S. Trahearn 148 Gladstone Road, Barry CF62 8ND
G4 BRW M. Gordon 57 Taunton Road, Bridgwater TA6 3LP
G4 BSA M. Draper The Wallow, Mount Road, Bury St. Edmunds IP31 2QU
G4 BSC J. Wells Tredworth, Sunnyfield Lane, Up Hatherley, Cheltenham GL51 6JE
G4 BSD D. Hoose Leonard Cheshire He, Oaklands, Garstang PR3 1RD
G4 BSK M. Rhind-Tutt Oldfield, Moor Road, Bridgwater TA7 9AR
G4 BSM West Devon Raynet c/o S. Grove 31 Sheppard Way, Minchinhampton, Stroud GL6 9BZ
G4 BSS J. Spence 4 Langford Lane, Burley In Wharfedale, Ilkley LS29 7NR
G4 BSV A. Cox 175 Hillcrest, Weybridge KT13 8AS
G4 BSW N. Hadley 323 Canterbury Road, Margate CT9 5JA
G4 BTE M. Smith 24 Lea Bank, Wolverhampton WV3 9HN
GI4 BTG B. Davidson 106 Tudor Park, Newtownabbey BT36 4WL
G4 BTI D. Case 8 Fawley Road., Reading RG30 3EN
G4 BTK A. Whitehouse 690 Kingstanding Road, Kingstanding, Birmingham B44 9SS
G4 BTN C. Brion Passaford House, Hatherleigh, Okehampton EX20 3LU
GW4 BTW I. Jolly 1 Llewelyn Drive, Bryn-Y-Baal, Mold CH7 6SW
G4 BTX N. Monument 9 Tower Road, Felixstowe IP11 7PR
GM4 BUA Christchurch ARS c/o Dr T. Shepherd 1 Spruce Gardens, Cupar Muir, Cupar KY15 5WN
G4 BUB P. Cox 53 Boleyn Avenue, Enfield EN1 4HR
G4 BUE C. Page Cherry View 28 Ellerslie Lane, Bexhill-on-Sea TN39 4LJ
G4 BUF G. Jolley 70 Hempstead Road, Holt NR25 6DG
G4 BUH M. Banahan 2 The Paddock, Ely CB6 1TP
G4 BUI J. Simpson 19 Greenacres, Wetheral, Carlisle CA4 8LD
GI4 BUJ J. Sander 696 Doagh Road, Newtownabbey BT36 4TP
G4 BUK J. Turner 12 Purley Bury Avenue, Purley CR8 1JB
G4 BUO D. Lawley 5 The Limes, Buckland, Buntingford SG9 0PW
G4 BUP Dr P. Moss Amalrie, Franklin Road, Chelmsford CM3 6NF
G4 BUW K. Lamb 10 Malthouse Gardens, Marchwood SO40 4XY
G4 BUX Buxton Radio Amateurs c/o D. Carson 21 Harris Road, Harpur Hill, Buxton SK17 9JS
GW4 BUZ J. Howells 13 Vicarage Road, Penygraig, Tonypandy CF40 1HR
G4 BVB R. Pridham Treetops, Chilsworthy, Gunnislake PL18 9PB
GM4 BVD A. Sampson 47 Muirend Road, Perth PH1 1JD
GW4 BVE J. Clifford Dippers Barn Coppice Lane Pool Quay, Welshpool SY21 9JY
G4 BVF M. Sinclair 28 Roker Park Avenue, Ickenham, Uxbridge UB10 8ED
G4 BVG A. Young 90 Pine Ridge, Carshalton SM5 4QH
G4 BVI G. Chenery 44 Belstead Road, Ipswich IP2 8AZ
GW4 BVJ R. Mortimore 76 Cwmfferws Road, Tycroes, Ammanford SA18 3UA
G4 BVK K. Stevens 20 Coberley, Bristol BS15 8ES
G4 BVM C. Newman 19 Clare Road, Peterborough PE1 3DT
G4 BVP M. Noble Harbet, Shipley Road, Horsham RH13 9BG
G4 BVQ P. Kennedy 18 Rushmere Avenue, Levenshulme, Manchester M19 3EH
G4 BVS S. Overend Deepend Cottage, Lower Hone Lane, Bosham, Chichester PO18 8QN
GW4 BVT R. Osborne Plas-Y-Bryn, 1 Belle Vue Gardens, Brecon LD3 7NY
GM4 BVU N. Macdonald 3 Townhill Road, Hamilton ML3 9UX
G4 BVV P. Goben 1 Petal Close, Maltby, Rotherham S66 7HJ
G4 BVW A. Reilly 4 Moreton Drive, Poulton-le-Fylde FY6 8ED
G4 BVY I. Dixon 5 The Howsells, Lower Howsell, Malvern WR14 1AD
GM4 BVZ J. Davidson Rosemount, Whiting Bay, Isle of Arran KA27 8PR
G4 BWB R. Andrews Flat 19, 4 Salamanca Place, , London SE1 7HB
G4 BWC Scottish Office ARC c/o M. Bray 2 Camborne Drive, Fixby, Huddersfield HD2 2NF
G4 BWE S. Price 9 Spurcroft Road, Thatcham RG19 3XX
G4 BWF R. Johnson 29 Oakfield Avenue, Markfield LE67 9WH
G4 BWG S. Marsh 26 Station Road, Whyteleafe CR3 0EP
G4 BWL H. Morris 2 Brickwall Lane, Curry Rivel, Langport TA10 0NX
GI4 BWM J. Mccullagh 14 Parkgate Meadows, Parkgate, Ballyclare BT39 0FA
G4 BWN P. Funnell 6 Bolero Close, Wollaton, Nottingham NG8 2BZ
G4 BWO D. Tyler 5 Brentry Avenue, Bristol BS5 0DL
G4 BWP F. Handscombe Sandholm, Bridge End Road, Red Lodge, Bury St. Edmunds IP28 8LQ
G4 BWR M. Hildich 7 Claverham Park, Claverham, Bristol BS49 4LS
G4 BWV A. Burchmore 49 School Lane, Horton Kirby, Dartford DA4 9DQ
G4 BWX S. Egerton 15 Hyde Road, Torrisholme, Morecambe LA46NU
G4 BWY P. Willcocks 27 Manor Road, Barnet EN5 2LE
GI4 BXB R. Brown Apartment 10, Anchor Watch, Donaghadee BT21 0GA
G4 BXC H. Pearce 32 Marshall Road, Willenhall WV13 3PB
G4 BXD B. Nock 47 Oakfield Road, Kidderminster DY11 6PL
G4 BXH D. Hardy Box 52831, Dubai United Arab Emirates
G4 BXI C. Godden 84 Crescent Road, Ramsgate CT11 9QZ
G4 BXQ A. Pressley 22 Springbank Avenue, Farsley, Pudsey LS28 5LW
G4 BXS J. Morris 1 Maple Cottages, Martley, Worcester WR6 6QA
G4 BXY H. Barker 31 Briants Avenue, Caversham, Reading RG4 5AY
G4 BXZ Dr J. Howell 3 Gate Farm Road, Shotley Gate, Ipswich IP9 1QH
GW4 BYA P. Braham 23 Gilfach Y Gog, Penygroes, Llanelli SA14 7RJ
G4 BYB R. Penman 9 Southall Avenue, Worcester WR3 7LR
G4 BYD A. Atkinson 5 Ashfield Avenue, Skelmanthorpe, Huddersfield HD8 9BW
G4 BYE T. Miller 4 Jessop Road, Stevenage SG1 5NF
GM4 BYF P. Bates 10 Swanston Avenue, Edinburgh EH10 7BU
G4 BYG V. Lindgren 143 Hull Road, Anlaby., Hull HU106ST
G4 BYI A. Wilson 223 Waingaro Road, Rd 1, Ngaruawahia 3793 New Zealand
G4 BYL B. Smith 27 Thorneyholme Road, Accrington BB5 6BD
G4 BYM B. Buzzing 1 Westmead Close, Droitwich WR9 9LG
G4 BYO W. Tee 87 Higher Blandford Road, Broadstone BH18 9AE
G4 BYR I. Maslen 26 Millington Road, Wallingford OX10 8FE
G4 BYS Dr G. Warren 3 Nursery Lane, Eynsham OX29 4GL
GM4 BYT R. Cook 132 Clachtoll, Lochinver, Lairg IV27 4JD
G4 BYW J. Lekesys 4 Gleneagles Way, Fixby, Huddersfield HD2 2NH
G4 BYZ C. Mills North Lodge, Margery Wood Lane, Tadworth KT20 7BA
G4 BZB D. Parsons 27 St. Leodegars Way, Hunston, Chichester PO20 1PE
G4 BZE P. Bradley Woodlands, Longdown, Exeter EX6 7SR
G4 BZF M. Reed 1 The Cottages, Farm Lane, Plymouth PL6 5RJ
G4 BZG R. Smith11 17 Styrrup Road, Harworth, Doncaster DN11 8LL
G4 BZI R. Bracey 7 Park Estate, Shavington, Crewe CW2 5AW
G4 BZJ A. Mitchell 18 Malham Fell, Bracknell RG12 7DU
G4 BZL D. Simpson Ivy Cottage, Princess Street, Leeds LS19 6BS
G4 BZM M. Edwards 13 Lechmere Crescent, Malvern WR14 1TY
G4 BZP F. Partington 21 East Road, Wymeswold, Loughborough LE12 6ST
G4 BZR F. Jordan 16 Elterwater Crescent, Barrow-in-Furness LA14 4PH
G4 BZS M. Pasek 10 Prospect Place, Norwood Green, Halifax HX3 8QF
G4 BZU B. Beaven 7 Glamorgan Road, Up Hatherley, Cheltenham GL51 3JF
G4 CAA Nats & Caa RS c/o S. Rossi 21 Rattigan Gardens, Whiteley, Fareham PO15 7EA
GM4 CAB S. Reynolds 39 Panmure St., Broughty Ferry, Dundee DD5 2EU
G4 CAF D. Hogg Fairview, Dordale Road, Bromsgrove B61 9JT
G4 CAJ M. Farr 23 Waterfall Way, Barwell, Leicester LE9 8EH
G4 CAK M. Scarlett Maple House, Westford, Wellington TA21 0DT
GM4 CAM D. Hamilton 7 High Langside Holding, Craigie, Kilmarnock KA1 5ND
GM4 CAQ Milton Keynes ARS c/o R. Miles 15 Clark Avenue, Linlithgow EH49 7AP
GW4 CAT N. Schofield Maen Llwyd-Tan Yr Alt, Llanllyfni, Caernarfon LL54 6RT
GM4 CAU T. Wratten 89 Hilton Road, Aberdeen AB24 4HX
G4 CAX D. Borley 95 Meadow Lane, Moulton, Northwich CW9 8QQ
G4 CAY C. Parker 25 Meadow Dale, Chilton, Ferryhill DL17 0RW
G4 CAZ J. Lefever 30 Holland Road, Melton Mowbray LE13 0LU
G4 CBD J. Swanson 9 Park House Gardens, Twickenham TW1 2DF
GI4 CBG R. Smyth 58 Gilnahirk Road, Belfast BT5 7DH
G4 CBL P. Tomlinson 55 Reldene Drive, Hull HU5 5HS
G4 CBM G. Blakeley Stowe House, Preston Gubbals Road, Shrewsbury SY4 3LY
G4 CBO D. Aiken 16 Woodland Gardens, North Wootton, King's Lynn PE30 3PX
G4 CBQ P. Daniells Holly Villa, Foxholes, Wem, Shrewsbury SY4 5UJ
G4 CBT H. Wall 54 Little Harlescott Lane, Shrewsbury SY1 3PZ
G4 CBW A. Horsfall 60 Talke Road, Red Street, Newcastle ST5 7AH
G4 CBY T. Cooper Lincolnshire House, Brumby Wood Lane, South Humberside DN17 1AF
G4 CBZ A. Mepham 32 Brambletyne Avenue Saltdean, Brighton BN2 8EJ
GW4 CC the Gower/Gwyr Contesting Club c/o K. Dyer 34 Lundy Drive, West Cross, Swansea SA3 5QL
G4 CCA M. Fadil 25 North Parade, Horsham RH12 2DA
G4 CCC C. Young 18 Wincroft Road, Caversham, Reading RG4 7HH
G4 CCH H. Ling 8 Spa Hill, Kirton Lindsey, Gainsborough DN21 4NE
G4 CCI J. Chapman 7 Ravensthorpe Drive, Loughborough LE11 4PU
GM4 CCN T. Keats Tigh Na Luch, Skye Of Curr Road Dulnain Bridge, Grantown-on-Spey PH26 3PA
G4 CCQ M. Stanton 84 Forest Hill, Maidstone ME15 6TH
G4 CCT S. Hyman 49 Southover, Woodside Park, London N12 7JG
G4 CCY P. Fagg 113 Bute Road, Wallington SM6 8AE
G4 CCZ P. Simons Westwood, Faris Lane, Addlestone KT15 3DJ
G4 CDC E. Morton 6 Norfolk Avenue, Burton-Upon-Stather, Scunthorpe DN15 9EW
G4 CDD Denby Dale And District ARS c/o J. Chappell 49 Midway, South Crosland., Huddersfield HD4 7DA
G4 CDF M. Naylor 6 Holsworthy Close, Lower Earley, Reading RG6 3AH
G4 CDG A. Davidson Po Box Hm 150, Hamilton HM AX Bermuda
G4 CDH J. Brade 11 Old Farm Place, Ash Vale, Aldershot GU12 5SF
G4 CDI G. Boardman 9 Byron Road, Weston-Super-Mare BS23 3XQ
G4 CDJ P. Jarrett Roydan House 36 Ferndale Road, Teignmouth TQ14 8NH
G4 CDL F. Mepham Avenida Robleda 16/22, San Luis, Torrevieja 3180 Spain
G4 CDN Backpackers Radio Activity Group c/o R. Banester Fairfield, Church Road, Norwich NR12 9SA
G4 CDR C. Winstanley 3 Peter St., Blackburn BB1 5HQ
G4 CDU N. Huntley 26 Malin Parade, Portishead, Bristol BS20 7FW
G4 CDW G. Trickey 3 Fairleigh Rise, Kington Langley, Chippenham SN15 5QF
G4 CDX P. Wheeler 69 Waterside Road, Slyfield Green, Guildford GU1 1RQ
G4 CDY T. Giles Lanherne , Meaver Road, Mullion, Helston TR12 7DN
G4 CDZ J. Boden 2 The Coppice, Whaley Bridge, High Peak SK23 7LH
G4 CEC Dr P. Knight 26 Meadway, Harrold MK43 7DR
G4 CEI M. Baker 17 Whitehills Green, Goring, Reading RG8 0EB
G4 CEJ R. Moore 17 Somme Avenue, Flookburgh, Grange-over-Sands LA11 7LJ
G4 CEK J. Bird Panorama, Cobblers Lane, Swanage BH19 2PX
G4 CEL S. Hudson Frekes Cottage, Moorside, Sturminster Newton DT10 1HQ
G4 CEN D. Davies 35 Ruthellen Road, Chelmsford 1824 United States
G4 CEO M. Uotome 5-21-6-308, Kita Karasuama, Setagaya-Ku, Tokyo 157-0061 Japan
G4 CEP G. Morris 7 Manor Road, Sandy SG19 1DT
G4 CES Royal Air Force Cosford ARC c/o M. Farmer Horton Brook Cottage, Horton Wem, Shrewsbury SY4 5NB
G4 CEU D. Jarvis Flat 1, Gunnery House, 2 Chapel Road, Shoeburyness, Southend-on-Sea SS3 9SL
G4 CEX C. Durant 63 Ulleries Road, Solihull B92 8DX
G4 CEY J. Ball 68 Swallows Court, Pool Close, Spalding PE11 1GZ
G4 CFB K. Henry 80 Fernwood Rise, Westdene, Brighton BN1 5EP
GW4 CFC Gloucestershire County Raynet c/o L. Gruffydd 45 Maes Yr Hafod, Menai Bridge LL59 5NB
G4 CFD H. Garner 6 Blacksmiths Close, Barrow-upon-Humber DN19 7HG
G4 CFG P. Arnold 14 George Birch Close, Brinklow, Rugby CV23 0NN

G4 CFH J. Hill 10 Albert Clarke Drive, Willenhall WV12 5AU
G4 CFK L. Smith 36 Kinderton Park Cledford Lane, Middlewich CW10 0JS
G4 CFP W. Bones 22 Rotherhead Close, Horwich, Bolton BL6 5UG
GI4 CFQ J. Mcsweeney 109 Twaddell Avenue, Belfast BT13 3LG
G4 CFS G. Dodwell 9 Balfour Drive, Liss GU33 7BF
G4 CFV R. Hall Pinewood Lodge, 16 Tullyvarraga Hill, Co Clare V14 H292 Ireland
G4 CFW R. Raven 9 Southwood Close, Ferndown BH22 9HW
G4 CFY A. Nailer 12 Weatherbury Way, Dorchester DT1 2EF
G4 CFZ M. Stevens 3 Rip Croft, Portland DT5 2EE
G4 CGA D. Sellwood 47 Waterhall Avenue, London E4 6NA
G4 CGB D. Tromans 29 Cannon Road Wombourne, Wolverhampton WV5 9HR
G4 CGD A. Richardson 24 West House Close, Wimbledon, London SW19 6QU
G4 CGE W. Dore 87 Thame Road, Aylesbury HP21 8LY
G4 CGF W. Badz Bottom Flat, 36 Luckington Road, Bristol BS7 0US
G4 CGG R. I'Anson 87 Tranby Lane, Anlaby, Hull HU10 7DT
G4 CGH M. Davies 2 Manor Close, Berrow, Burnham-on-Sea TA8 2LN
G4 CGL J. Miller 29 Springhill Road, Wednesfield, Wolverhampton WV11 3AW
G4 CGM M. Duff Clittaford Club, Moses Close, Plymouth PL6 6JP
G4 CGO J. Pollock 71 Stevenson Street, Kew 3101 Australia
G4 CGP P. Wright 4 Avill Way, Wickersley, Rotherham S66 1DL
G4 CGR K. Davies High View, Alcester Road Wootton Wawen, Henley-in-Arden B95 6BH
G4 CGU R. Taylor 23 Ridgacre Lane, Quinton, Birmingham B32 1EL
G4 CGV C. Manklow 37 Brittons Crescent, Barrow, Bury St. Edmunds IP29 5AG
G4 CGW J. Dunglinson Blenheim, Willow Lane, Camberley GU17 9DL
GW4 CGZ D. Newman 138 Twyn Carmel, Merthyr Tydfil CF48 1PH
G4 CHD T. Adams 1 Francis Drive, Westward Ho!, Bideford EX39 1XE
GM4 CHF J. Magill 74 Garnqueen Crescent Glenboig, Coatbridge ML5 2SX
G4 CHG P. Ashton 7 Conway Grove, Cheadle, Stoke-on-Trent ST10 1QG
G4 CHI P. Robinson Longcroft House, Longcroft Lane, Burton-on-Trent DE13 8NT
G4 CHJ A. Williams 10 Olde Hall Road, Featherstone, Wolverhampton WV10 7BB
G4 CHL P. Howe 135 Rue De Pierrevert, Ste-Tulle 4220 France
G4 CHM R. Mcewan Fifth Acre, Carr Lane, Alfreton DE55 2DN
G4 CHS Cheltenham Hackspace ARS c/o D. Miller 50 Sandyleaze, Gloucester GL2 0PX
GM4 CHX J. Kyle 7 Fasaich, Strath, Gairloch IV21 2DH
GU4 CHY R. Allisette Lilyvale House, Rue Des Houmets, Castel GY5 7XZ Guernsey
G4 CIA W. Cooper 20 Planton Way, Brightlingsea, Colchester CO7 0LB
G4 CIB B. Woodcock 2 Poolhay Close, Gloucester GL19 4NY
G4 CIC S. Edmondson Fen, Rheda Park, Frizington CA26 3TA
GM4 CID R. Mcclements Eskdail, Newtown St. Boswells, Melrose TD6 0RY
G4 CIJ J. Chennells 10 Lower Cippenham Lane, Slough SL1 5DF
G4 CIO Dr M. Phillips Chapel House, The Cross, Stonehouse GL10 3TU
G4 CIZ A. Wallbank 1 Pollards Cottages, Clanville, Andover SP11 9JD
G4 CJJ M. Viner 15 St. Anthonys Drive, Hedon, Hull HU12 8NT
G4 CJK University of Surrey Ears c/o V. Roney 76 Hilton Lane, Great Wyrley, Walsall WS6 6DT
G4 CJM J. Alcock 1 Alma St., Fenton, Stoke on Trent ST4 4PH
G4 CJO A. Mountifield 6 Sawyers Close, Teg Down, Winchester SO22 5JX
G4 CJP V. Duffy 2 Moor View Close, High Harrington, Workington CA14 4NX
G4 CJR London Bbc Radio Group c/o C. Crick 19 The Drive, Coulsdon CR5 2BL
G4 CJT K. Hughes 4 Epsom Place, Cranleigh GU6 7ET
G4 CJV A. Kerton 8 Fabian Drive Stoke Gifford, Bristol BS34 8XN
G4 CJY B. Payne 78 Carver Hill Road, High Wycombe HP11 2UA
G4 CKB J. Banester Fairfield, Church Road, Norwich NR12 9SA
G4 CKH G. Jackson 86 Lloyds Avenue, Kessingland, Lowestoft NR33 7TR
G4 CKK P. Atkins 60 Wentworth Way, Harborne, Birmingham B32 2UX
G4 CKQ A. Horne 1 Upper Halliford Road, Shepperton TW17 8RX
G4 CKS D. Fitzgerald 36 Vardens Road, London SW11 1RH
G4 CKT R. Gwynne 17 Dorrington Close, Stoke-on-Trent ST2 7BZ
G4 CKX S. Taylor 5 Chiltern Avenue, Bishops Cleeve, Cheltenham GL52 8XP
G4 CLA P. Lindsay The Barn, Main Street Ashby Parva, Lutterworth LE17 5HY
G4 CLB C. Brown 31 Sapcote Road, Burbage, Hinckley LE10 2AS
G4 CLC D. Lewis Brandywine, Westbury BA13 4NY
G4 CLD G. Beaver The Gables, Reading Road, Reading RG7 3BU
G4 CLE T. Baker 18 Prescott Avenue, Rufford, Ormskirk L40 1TT
G4 CLF J. Bryant Hillhead Cottage, Calshot Road, Calshot, Southampton SO45 1BR
G4 CLG S. Whittingham 18 Northcroft, Shenley Lodge, Milton Keynes MK5 7AJ
G4 CLI Dr D. Sadler-Lockwood 14 Mountain Road, Dewsbury WF12 0BW
G4 CLJ P. Eccles 3 Ledger Lane, Outwood, Wakefield WF1 2PQ
G4 CLL R. Goodchild Grey Cliffe House, Owmby Cliff Road, Owmby-By-Spital, Market Rasen LN8 2HL
G4 CLM C. Le Marchant Bow House, Green Lane, Prestwood, Great Missenden HP16 0QE
G4 CLN Highfield ARC c/o P. Redfern 12, Wilbarn Road Wilbarn Road, Paignton TQ3 2BN
G4 CLO R. Armishaw Drift House, Cattons Drift, The Green, Felbrigg, Norwich NR11 8PN
G4 CLP J. Harrison 47 Mason Way, Padbury Wa 6025 Australia
G4 CLR I. Hewer 23 Thoresby Avenue, Tuffley, Gloucester GL4 0TD
G4 CLY N. Thompson 6 Miena Way, Ashtead KT21 2HU
G4 CMG T. Milne Lynwood, Clovelly Road, Hindhead GU26 6RP
G4 CMH D. Spendlove 22 Green Bank, Harwood, Bolton BL2 3NG
GM4 CMI R. Campbell 1 Gibraltar Terrace, Dalkeith EH22 1EE
G4 CMK R. Harker 140 Victoria Road, Beverley HU17 8PJ
G4 CML J. Livesley Rivendell 71B Hillfoot Road, Shillington SG5 3NS
G4 CMM C. Pope Silver Hill, Norwich Road, Great Yarmouth NR29 5PB
G4 CMP P. Lennon 53 Rycot Road, Speke, Liverpool L24 3TH
G4 CMT Rafac ARC c/o A. Russell 3 St. Nicholas Close, North Newbald, York YO43 4TT
G4 CMU G. Brind 9 Becket Wood, Newdigate, Dorking RH5 5AQ
G4 CMX P. Rossiter 36 Milton Drive Ravenshead, Nottinghamshire NG15 9BE
G4 CMY A. Mann 13 Rosedale Avenue, Stonehouse GL10 2QH
G4 CMZ K. Archer 24 Willson Road, Littleover, Derby DE23 1BZ
G4 CNH L. Carpenter 166 Abbey View Garsmouth Way, Watford WD25 9DZ
G4 CNI P. Geiger Lloyd Mount, Howard Drive, Altrincham WA15 0LT
GW4 CNL G. Goodfield 10 Lewis Street, Church Village, Pontypridd CF38 1BY
G4 CNZ D. Allen 344 Coventry Road, Hinckley LE10 0NH
G4 COE D. Smith 54 Warrington Road, Leigh WN7 3EB
GW4 COJ C. Roberts 8 Oaklands Park Drive, Rhiwderin, Newport NP10 8RB
GW4 COL I. Braithwaite Kew, Trefonen, Oswestry SY10 9DH
G4 COM J. Compton Aysgarth, Durley Brook Road, Southampton SO32 2AR
G4 COR I. Harvey 50 Callow Hill Way, Littleover, Derby DE23 3RL
G4 COS J. Hansell 87 Garratts Way, High Wycombe HP13 5XT
G4 COT S. Brett 8 Pinewood Grove, Hull HU5 5YY
G4 COV C. Cardwell 11 Manor Cottages, Heronsgate Road, Rickmansworth WD3 5BJ
GM4 COX J. Hood 4 Murray Road, Law, Carluke ML8 5HR
G4 CPA G. Hanson 11 Churchill Way, Cross Hills, Keighley BD20 7DN
G4 CPC S. Wright 20 Stillwell Grove, Wakefield WF2 6RN
G4 CPD G. Knox Glencairn, 6 Aldborough Road, York YO51 9EA
G4 CPE A. Turner 7 Slate Hall, Sundon, Luton LU3 3PY
G4 CPG M. Howkins 16 Beckett Court, Gedling, Nottingham NG4 4GS
G4 CPI J. Housden 5 Tothby Meadows, Alford LN13 0EH
G4 CPL C. Mcgee Zafra, The Knapp, Bromyard HR7 4BD
G4 CPM A. Fielding 95 Hillcrest, Weybridge KT13 8AS
G4 CPN J. Bird 166 Cowick Lane, Exeter EX2 9JF
G4 CPQ N. Scrogie 154 St. Albans Road, Derby DE22 3JP
G4 CPV R. Fisk 16 Sterry Drive, Thames Ditton KT7 0YN
G4 CPW P. Wilson 5 Pebble Close, Lowestoft NR32 4DR
G4 CPY N. Grassby 11 Eider Close, Whetstone, Leicester LE8 6YB
G4 CQA G. Angell 55 Golden Riddy, Leighton Buzzard LU7 2RH
G4 CQH J. Sperry 50 Lochinver, Hanworth, Bracknell RG12 7LD
G4 CQI A. Lanfear 120 Charlton Road, Kingswood, Bristol BS15 1HF
G4 CQM D. Hilleard Hazeldene, Bridgerule, Holsworthy EX22 7EW
G4 CQN M. Mawby 61 Carter Drive, Beverley HU17 9GL
G4 CQO S. Burgess Tretawn, Kite Hill, Ryde PO33 4LG
G4 CQQ R. Taylor 8 Park Avenue, Markfield LE67 9WA
G4 CQR D. Wood 49 Wolsey Crescent, Morden SM4 4TD
G4 CQS A. Rowsby 10 Echells Close, Bromsgrove B61 7EB
GW4 CQT D. Price Vine Cottage, Garth Road, Cwmbran NP44 7AB
G4 CQV P. Baldwin 26 Ashford Road, Fulshaw Park, Wilmslow SK9 1QE
G4 CQW A. Lane 21 Winterbourne Road, Poole BH15 2ES
G4 CQX L. Palfrey C/O Po Box 314, Cyprus XX99 1AA
GW4 CQZ M. Doig Helenfa, Ystrad Road, Denbigh LL16 3HE
G4 CRB W. Oxley Flat 147, Oceana Boulevard Orchard Place, Southampton SO14 3HW
G4 CRC Cornish RAC c/o S. Holland 49 Oxland Road, Illogan, Redruth TR16 4SH
G4 CRE D. Rush 8 Sheaf Place, Worksop S81 7LE
G4 CRG K. Burgin The Pike Lock House, Eastington, , Stonehouse, Glos Gl10 3rt GL10 3RT
GW4 CRH M. Worvill The Berwyns, Domgay Road, Four Crosses, Llanymynech SY22 6SL
G4 CRK R. Sellman 43 Mount Avenue, Stone ST15 8LW
G4 CRM J. Lennon 107 Andrew Crescent, Waterlooville PO7 6BG
G4 CRN Dr A. Hall Westhill, Bear Lane Longdon, Tewkesbury GL20 6BB
G4 CRP K. Tyler Pinfold House, 3 Pinfold Lane, Leeds LS25 1HE
G4 CRS E.C A.R.C. c/o A. Mackay 2 Highcliffe Grove, New Marske, Redcar TS11 8DU
G4 CRT L. Kirby 41 Woodville Road, Overseal, Swadlincote DE12 6LU
G4 CRW A. Holmes 4 Castle Avenue, Datchet, Slough SL3 9BA
G4 CSD P. Hyde 25 Merton Road, Basingstoke RG21 5UA
G4 CSE M. Lewis 10 Kenmore Drive, Bristol BS7 0TT
G4 CSM D. Chaplin 35 Lanes End, Totland Bay PO39 0AL
GI4 CSO J. Mccormack 12 Glengoland Crescent, Dunmurry, Belfast BT17 0JG
GI4 CSP G. Robinson 10 Ranfurly Avenue, Dungannon BT71 6PJ
G4 CST G. Hopkins Flat 9, The Cuttings, High Street, Hungerford RG17 0LU
G4 CSV J. Jackson 43 Ambleside, Boundary Court, Stockport SK8 1BA
GW4 CSY B. Vickery 6 Duffryn Close, St. Nicholas, Cardiff CF5 6SS
G4 CSZ M. Riley 5 Dunstarn Gardens, Leeds LS16 8EJ
G4 CTA A. Clewer 6 Frensham Close, Stanway, Colchester CO3 0HP
G4 CTC T. Cann Noahs Rough, Old Coach Road Wrotham, Sevenoaks TN15 7NR
G4 CTD C. Vernon 50 Copthall Road West, Ickenham, Uxbridge UB10 8HS
G4 CTE P. Bradshaw 43 Hill Top Road, Grenoside, Sheffield S35 8PE
G4 CTI P. Ashcroft 7 Kings Ripton Road, Sapley, Huntingdon PE28 2NU
G4 CTM P. Barrett 3 Bramshott Close, Hitchin SG4 9EP
G4 CTT Dr T. Thirst Thirsts Farm, Happisburgh Road, Happisburgh NR12 0RU
G4 CTU Obo West Kent Raynet c/o B. Hitchins 12 Parkland Avenue, Kidderminster DY11 6BX
GW4 CTV S. Mee Cysgod Y Gaer, Cwmsymlog, Aberystwyth SY23 3EZ
G4 CTY A. Lightbody 3 Elphicks Place, Tunbridge Wells TN2 5NB
G4 CTZ British Railways ARS c/o I. Cage 334 Stockton Lane, York YO31 1JW
G4 CUE W. Pechey Jays Lodge, Crays Pond, Reading RG8 7QG
G4 CUG R. Worsell 8 Waterworks Cottages, Old Willingdon Road, Eastbourne BN20 0AS
G4 CUI Dr G. Cook 1 St. Albans Road, Fulwood, Sheffield S10 4DN
G4 CUQ B. Hughes 30 Fuller Road, Dagenham RM8 2TU
GI4 CUV N. Atkins 38 Rosscoole Park, Belfast BT14 8JX
GM4 CUX North Wiltshire Raynet c/o G. Winchester 23 Craigmount, Avenue North, Edinburgh EH12 8DL
G4 CVA Rvd. J. Wardle 27 First Avenue, Bridlington YO15 2JW
G4 CVC J. Everist 11 Redding Close, Dartford DA2 6NB
G4 CVD P. Petty Bahnhof Strasse 17, Ralingen 54310 Germany
G4 CVF B. Sheppard 20 Lambourne Court, St. Johns Close, Uxbridge UB8 2UL
G4 CVG W. Bullock 14 Saxon Drive, Rillington, Malton YO17 8LZ
G4 CVM R. Watson 36 Abbots Close, Knowle, Solihull B93 9PP
G4 CVN D. Williams 2 Lassels Lane, Horsham RH12 1RF
G4 CVO W. Wyer 11 Nether Close, Wingerworth, Chesterfield S42 6UR
G4 CVS Dr B. Pearson 8 The Pastures, Edlesborough, Dunstable LU6 2HL
GW4 CVT C. Thorley Helston, The Mountain, Holyhead LL65 1YR
G4 CVU J. Swingewood 5 Blaze Park, Wall Heath, Kingswinford DY6 0LL
G4 CVX Dr R. Sims 345 Blandford Road, Hamworthy, Poole BH15 4HP
G4 CW Medway Amateur Receiving & Transmitting Society c/o F. Connor 134 Summerhouse Drive, Bexley DA5 2ES
G4 CWA Rvd. W. Burton 23 Purok 5, San Pedro Li, Pampanga Philippines
G4 CWB D. Andrews 100 Duchy Road, Harrogate HG1 2HA
G4 CWC R. Barrett Lumina, Bridegate Lane, Melton Mowbray LE14 3QA
G4 CWE South Staffordshire Ar Tutors Grp c/o A. Humm 32 Layton Road, Hounslow TW3 1YH

GW4 CWG G. Crossland 32 Long Bridge St., Llanidloes SY18 6AR
G4 CWH Dr C. Smithers 10 Grange Park, Bishops Stortford CM23 2HX
G4 CWM J. Pickles 111 Linden Avenue, Prestbury, Cheltenham GL52 3DT
G4 CWS S. Wood 16 Ramley Road, Lymington SO41 8GQ
GW4 CWU B. Heppenstall Gwelfor, Llanrhyddlad, Holyhead LL65 4BG
G4 CWV F. Parr 5 Benenden Road, Wainscott, Rochester ME2 4NU
G4 CXE P. Bolton 93 Westfields, Narborough, King's Lynn PE32 1SY
GM4 CXF J. Thomson 31 Teviot Place, Troon KA10 7EE
GW4 CXK R. Evans 74 Alexandra Street, Ebbw Vale NP23 6JF
G4 CXL R. Menday Huf House, Horseshoe Ridge, Weybridge KT13 0NR
GM4 CXP D. Dance 18 Masons Court, Kelso TD5 7NJ
G4 CXQ D. Dyer 26 Locking Road, Weston-Super-Mare BS23 3DF
G4 CXT M. Bell Quebec Cottage, Curlew Green Kelsale, Saxmundham IP17 2RA
G4 CXW G. Spencer 17 Rockland Road, Bristol BS16 2SW
G4 CXX M. Bullough 2 Colebrook Close, London SW15 3HZ
G4 CXZ A. Thompson 6 Ducks Walk, Twickenham TW1 2DD
G4 CYB F. Burnett Herons Siege, Blundies Lane, Enville, , Stourbridge DY7 5HU
G4 CYC K. Green Ao-Te-Aroa, 12 Hill Road, Fareham PO16 8LB
G4 CYF C. Tully Harmony, The Crescent, Clacton-on-Sea CO16 0EP
G4 CYG D. Darkes 70 Braemar Road, Lillington, Leamington Spa CV32 7EY
G4 CYI J. Palfrey Lower Trewince Farm, Newquay TR8 4AW
G4 CYO K. Robinson 3 The Woodhouses, Patshull Road, Wolverhampton WV6 7DU
G4 CYR S. Allen The Poplars, Wotton Underwood HP18 0RX
G4 CYY C. Lewis 54 Whelpley Hill Park, Whelpley Hill, Chesham HP5 3RJ
G4 CYZ L. Large Captains Farmhouse, Streat Lane, Hassocks BN6 8SB
G4 CZA K. Newman 2 Skys Wood Road, St. Albans AL4 9NZ
G4 CZB J. Cockrill 28 Northampton Road, Harpole, Northampton NN7 4DD
G4 CZH E. Brindley 150A Woods Lane, Derby DE22 3UE
GW4 CZK A. Mercer Flat 1, 3 Broad Street, Builth LD2 3DT
GI4 CZO G. Mccomb 1 Magheraboy Drive, Portrush BT56 8GP
G4 CZP R. Crossley 2 Jewel View, 31 Downside, Ventnor PO38 1AL
G4 CZR C. Redfern 6 Pont Croix, Mellionnec 22110 France
G4 CZU P. Hadler 30 Hillview Road, Whitstable CT5 4HX
GI4 CZW C. Corderoy 3 The Limes, Drumlyon, Enniskillen BT74 5NQ
G4 CZX I. Godden 163 Ringmer Road, Worthing BN13 1DZ
G4 CZZ R. Aggus 68 Conifer Walk, Stevenage SG2 7QS
G4 DAC D. Squires 91 Croham Valley Road, South Croydon CR2 7JJ
G4 DAF G. Walker 56 Goodwin Road, Croydon CR0 4EG
G4 DAM R. Dence 32 Hayeswood Road Stanley Common, Ilkeston DE7 6GB
G4 DAP C. Ison 19 Grays Close, Chalgrove, Oxford OX44 7TN
G4 DAQ W. Silvester 2 Tudor Close, Barton-Le-Clay, Bedford MK454NE
G4 DAT R. Davidson 5 St. Lucians Lane, Wallingford OX10 9ER
GI4 DAV D. Hart 31 Downshire Road, Carrickfergus BT38 7QD
G4 DAX D. Smith Red Roof, Goathland, Whitby YO22 5AN
G4 DAY D. Sawyer 2 Blunts Wood Road, Haywards Heath RH16 1NB
G4 DBD A. Borland 39 Green Lane, Willaston, Nantwich CW5 7HY
G4 DBE J. Clark 20 Sandy Lane, Irby, Wirral CH61 0HD
G4 DBF D. Freeston 20 Coningham Road, Whitley Wood, Reading RG2 8QP
G4 DBG F. Kneale 60 Summertrees Road, Great Sutton, Ellesmere Port CH66 2BJ
G4 DBM B. Mcgennity 46 St. Andrews Road, Boreham, Chelmsford CM3 3BY
G4 DBN N. Smith Birch Tree House Asselby, Goole DN14 7HE
G4 DBQ B. Roberts 7 North Square, London NW11 7AA
G4 DBR C. Ewing 130 Uttoxeter Road, Hill Ridware, Rugeley WS15 3QX
G4 DBS B. Middleton 10 Curtois Close Branston, Lincoln LN4 1LJ
G4 DBX L. Stubbs The Cottage, Middlewich Road, Crewe CW1 4RA
G4 DBY P. Walker 48 Whitefields Drive, Richmond DL10 7DL
G4 DBZ D. Martin 12 South Park, Redruth TR15 3AW
G4 DCB P. Mortimer 13A Elder Avenue, Wickford SS12 0LP
GI4 DCC W. Chesney 52 Taylorstown Road, Toomebridge, Antrim BT41 3RT
G4 DCD C. Stephenson 6 Livingstone Close, Rothwell, Kettering NN14 6HT
G4 DCE J. Sketchley 48 Coverdale, Whitwick, Coalville LE67 5BP
G4 DCF M. Booth 14 Blackstock Close, Sheffield S14 1AE
G4 DCH C. Tucker 29, Newton Abbot TQ12 1US
G4 DCI P. Hopewell 3 Hunts Orchard, Hathern, Loughborough LE12 5HQ
G4 DCJ D. Jarrett 15 Groveside, East Rudham, King's Lynn PE31 8RL
G4 DCK M. Holliday 1A Fairmile Gardens, Longford, , Gloucester GL29ED
GM4 DCL T. Main 15 Polton Road, Lasswade EH18 1AB
G4 DCM P. Rhodes Parcela 1352, Calle De Zurbaran 8, Alicante 3170 Spain
G4 DCN Dr J. Mason Keepers Cottage, Baldon Lane, Marsh Baldon, Oxford OX44 9LT
G4 DCP P. Hull Seymour Cottage, Forest Road, Waterlooville PO7 6UA
G4 DCS M. Grant Windmill House 27 Windmill Street, Whittlesey, Peterborough PE7 1QN
G4 DCW D. Walker 70 High Street, Cranfield, Bedford MK43 0DF
G4 DCX E. Trickey 53 Hollyguest Road, Hanham Green, Bristol BS15 9NN
G4 DCY W. Dransfield Flat 6, Heath Mount Hall, Ilkley LS29 9JN
G4 DDB E. Connor 8 Russell Street, Dover CT16 1PX
G4 DDC Dunstable Downs RC c/o P. Seaford 14 Nevis Close, Leighton Buzzard LU7 2XD
G4 DDD R. John 32 Hundred Acre Road, Streetly, Sutton Coldfield B74 2LA
G4 DDI C. Guy 7 Herrick Court, Clinton Park, Lincoln LN4 4QU
G4 DDK S. Jewell Blenheim Cottage, Falkenham, Ipswich IP10 0QU
G4 DDL M. Pemberton 37 Woodmancott Close, Forest Park, Bracknell RG12 0XU
G4 DDM R. Finch 1 Cherry Tree Cottage, Church Road, High Wycombe HP10 8LN
G4 DDN G. Leonard 65 Qualitas, Bracknell RG12 7QG
G4 DDP R. Clark 41 Avenue Road, Bexleyheath DA7 4
G4 DDT A. Ives 24 Johnson Crescent, Heacham, King's Lynn PE31 7LQ
G4 DDV R. Bowman 13 Wellington Road, St. Albans AL1 5NJ
G4 DDX R. Pratt 16 Thurlow Close, Stevenage SG1 4SD
G4 DDZ N. Turner Rose Cottage, Catshill Cross, Stafford ST21 6LT
G4 DEA P. Dunning Cold Harbour, Bishop Burton, Beverley HU17 8QA
G4 DEM D. Walker The Horseshoe Inn, 1 Horseshoe Court, Bristol BS36 2FD
G4 DEO A. Wallis 5 Nancevallon, Higher Brea, Camborne TR14 9DE
GW4 DEP D. Dabinett Pentre Isaf, Llangyniew, , Welshpool SY21 0JT
G4 DEQ A. Derrick 4 Hillside Cottages, Barrow Street, Bristol BS48 3RX
G4 DEU A. Fuge 6 Haythorne Court, Staple Hill, Bristol BS16 5QS
G4 DEV S. Newport 18 Chacewater Crescent, Worcester WR3 7AN
G4 DEW J. Males 49 Gunthorpe Road, Peterborough PE4 7TN
GM4 DEX J. Sharp 72 Broom Road, Rimbleton, Glenrothes KY6 2BQ
G4 DFA T. Ellinor 53 Hillside, Banstead SM7 1HG
G4 DFB D. Berry 4 Holly Hill, Winchmore Hill, London N21 1NP
G4 DFC C. Goldingay 71 Kingham Close, Redditch B98 0SB
G4 DFD K. Bailey 16 Chandos Drive, Martlesham, Woodbridge IP12 4ST
G4 DFE W. Raybould 33 Roberts Green Road, Dudley DY3 2BB
G4 DFG P. Gibbs Redhill Farm Stables, Redhill, Telford TF2 9NZ
G4 DFJ Dr J. Klein 5 Cranley Gardens, London N10 3AA
G4 DFN S. Widdett 21 Lower Howsell Road, Malvern WR14 1DX
G4 DFO S. Wainwright 39 Ascot Road, Birmingham B13 9EN
G4 DFP A. Morecroft 4 Arran Close, Bolton BL3 4PP
GW4 DFQ N. Dear Hollybush Cottage, Candy, Oswestry SY10 9BA
G4 DFS S. Booth 13 Milner Avenue, Penistone, Sheffield S36 9DB
G4 DFT R. Perrin 131 Acacia Avenue, Ottawa, Ontario K1M 0R2 Canada
G4 DFU F. Skillington 6 Okehampton Crescent, Nottingham NG3 5SE
G4 DFV D. Walters 11 King George V Avenue, Mansfield NG18 4ER
G4 DFX J. Taylor 26 Courthope Road, London NW3 2LD
G4 DFY R. Dedman 2 Forest Villas, Long Mill Lane, Sevenoaks TN15 8LQ
G4 DFZ K. Knight 61 Westbourne Road, Sutton-in-Ashfield NG17 2FB
G4 DGB T. Crute 26 Runcorn, Sunderland SR2 0BP
G4 DGF A. Matthews 14 Hardy Green, Wellington College, Crowthorne RG45 7QR
GI4 DGI Rvd. D. Coyle 16 Northland Avenue, Londonderry BT48 7JN
G4 DGL Dr E. Mills C/O Ac Clarke, 243 Barton Road, Cambridge CB23 7BU
G4 DGQ J. Dussart Seagarth, Cresswell, Northumberland NE61 5JU
GM4 DGT W. Stirling 16 Shire Way, Alloa FK10 1NQ
G4 DGW A. Gagnon 60 Woodruff Avenue, Hove BN3 6PJ
G4 DHF D. Johnson Deans Cottage, Dowsby Fen, , Fen Road, Dowsby, Bourne PE10 0TU
G4 DHK R. Stanleigh 15 Lansdown Terrace Golden Hill, Bristol BS6 7YW
G4 DHL C. Durnall 143 Green Lane, Wolverhampton WV6 9HB
GM4 DHN N. Macleod 54 Drum Brae South, Edinburgh EH12 8TB
G4 DHQ A. Becket 65 High Street, , Prestwood HP16 9EJ
G4 DHT R. Haverson Kerri, Sunton, Marlborough SN8 3DZ
G4 DHU D. Spender Rose Cottage, Huntsmans Lane, Sudbury CO10 7JX
G4 DHV C. Jones 49 Newport Pagnell Road, Hardingstone, Northampton NN4 6ER
G4 DHW P. Mcelroy 2 Donohue Lane, Manchester 6040 United States
G4 DIA B. Powell 1 The Heights, Market Harborough LE16 8BQ
G4 DIC R. Phipps 4 Mill Court, Wells-next-the-Sea NR23 1HF
G4 DIE I. Dredge Room 1, The Priory, Greenway Lane, , Chippenham SN15 1AA
G4 DIG D. Hine 6A Clifton Terrace, Southend-on-Sea SS1 1DT
G4 DIH R. Coates 57 Dalebrook Road, Burton-on-Trent DE15 0AB
G4 DII A. Excell 7 Kingslake Villas, Taunton Road, Bridgwater TA6 6BW
GM4 DIJ J. Howie 36 Clermiston Road, Edinburgh EH12 6XB
GM4 DIN N. Burns 24 Garioch Road, Inverurie AB51 4RQ
G4 DIP B. Chapman 83 Courtenay Road, Great Barr, Birmingham B44 8JB
G4 DIS K. Mills 7 Montgomery Close, Colchester CO2 8SJ
G4 DIT R. Siddall 79 The Knoll, Palacefields, Runcorn WA7 2UH
G4 DIU A. Walker 26 Sketchley Court, Nottingham NG6 7DL
G4 DIV Dr L. Day 86 Copperfield Road, Southampton SO16 3NY
G4 DIY R. Bennett 17 Truro Close, St. Helens WA11 9EL
GM4 DIZ H. Lydall 27 Calder Road, Edinburgh EH11 3PF
G4 DJB P. Roberts 10 Tintagel Drive, Frimley, Camberley GU16 8XQ
G4 DJC R. Baker 42 Rushleydale, Springfield, Chelmsford CM1 6JX
G4 DJD D. Carter 43 Sturton Road, Sheffield S4 7DE
G4 DJJ C. Callicott Clare House, Hepscott, Morpeth NE61 6LT
G4 DJK D. Corkill 1A Hardie Crescent Braunstone, Leicester LE3 3DQ
G4 DJP Dr J. Chivers 33 Hazelwood Road, Duffield, Belper DE56 4DP
G4 DJX A. Gray 5 Meadow Close, Marshalswick, St. Albans AL4 9TG
G4 DJY C. Steeden Parklands, Chapel Road, Blackpool FY4 5HT
G4 DJZ A. Petrie 3 Sharma Leas, Peterborough PE4 6ZH
G4 DKC T. Smith 57 St Andrew Street, Tiverton EX16 6PL
G4 DKD E. Pascoe 48 Bull Baulk, Middleton Cheney, Banbury OX17 2QQ
G4 DKH K. Hastie 3 The Woodlands, Kings Worthy, Winchester SO23 7QQ
G4 DKM R. Lacken 378 Wallisdown Road, Wallisdown, Bournemouth BH11 8PS
G4 DKP J. Etheridge 14A Areley Court, Stourport-on-Severn DY13 0AR
G4 DKQ J. Loughlin 24 Grassendale Rd, Liverpool L19 0NA
G4 DKV M. Pipes Thornville, 2 Eaton Court, Hulland Ward, Ashbourne DE6 3EF
G4 DKX N. Cartwright Little Bulmer Farm, Wiston Road, Colchester CO6 4LT
G4 DLA L. Turner 160 Sandbach Road, Church Lawton, Stoke-on-Trent ST7 3RB
GW4 DLC B. Bourne 15 Rhos Fawr, Morfa, Abergele LL22 9YH
G4 DLD M. Garwood 1 Orchard Terrace, Welford NN6 6AE
GM4 DLG R. Bower Tigh Na Bruaich, Port Logan, Stranraer DG9 9NE
G4 DLP R. Stoddon 34 Cromwell Road, Lancaster LA1 5BD
G4 DLT R. Hill Rose Lodge, 35 Colne Fields, Huntingdon PE28 3DL
GM4 DLU A. Mccudden 9 Dryburgh Lane, East Kilbride, Glasgow G74 1BQ
G4 DLY P. Collister Flat 11, Liffey Court, 165-173 London Road, Liverpool L3 8PZ
G4 DMB W. Green 38 Greenlands Way West, Sheringham NR26 8XP
G4 DMC R. Cleverley 13 The Close, Melksham SN12 6AG
G4 DMF J. Wright 10 Thorpes Road, Heanor DE75 7GQ
G4 DMG D. Griffiths 1 Shepherds Down, Alresford SO24 9PP
G4 DMH M. Horton 47 Checkstone Avenue, Doncaster DN4 7JY
G4 DMI C. Armistead 16 Rushfield, Sawbridgeworth CM21 9NF
G4 DML G. Moore Calvers Farm, Norwich Road, Thelveton IP21 4NG
G4 DMM I. Stinchcombe 2 Alexandra Drive, Bere Alston, Yelverton PL20 7DW
G4 DMP D. Pratt 11 Moorleigh Close, Kippax, Leeds LS25 7PB
GM4 DMQ J. Pritchard 36 Craigleith Hill Crescent, Edinburgh EH4 2JU
GW4 DMR D. Bevan 3 Trem Y Foryd, Kinmel Bay, Rhyl LL18 5JE
G4 DMT J. Southall 4 Tye Lane, Willisham, Ipswich IP8 4SR
G4 DND J. Kennedy Tor View Cottage, Postbridge, Yelverton PL20 6SY
G4 DNE G. Swaysland 35 Keyhaven Road, Milford On Sea, Lymington SO41 0QW
G4 DNG M. Whitaker 332 Milton Road, Cambridge CB4 1LW
G4 DNH J. Easteal The Chalkers, Ermin Street, Hungerford RG17 7TS
G4 DNJ A. Grisley 7 Arnhill Road, Gretton, Corby NN17 3DN
G4 DNK M. Lelliott Well Lane Corner, Lower Froyle, Alton GU34 4LJ
G4 DNP R. Travis 14 Elmstead Avenue, Wembley HA9 8NX
GI4 DNW Braunstone Troop Military Radio Group c/o M. Getty 34 Magheralave Park East, Lisburn BT28 3BT
G4 DNX D. Dyer 57 Garrison Lane, Felixstowe IP11 7RR
G4 DOA A. Mead 11 Yarnton Close, Nine Elms, Swindon SN5 5UQ

G4 DOC D. James 76 Grove Road, Harpenden AL5 1HD
G4 DOE J. Alford 26 Edmunds Avenue St. Pauls Cray, Orpington BR5 3LF
GM4 DOF R. Davidson 3 Hillcrest Avenue, Kirkcaldy KY2 5TU
GI4 DOH R. White 28 Lord Warden'S Parade, Bangor BT19 1YU
G4 DOJ N. Sanig 22 Bruntwood Avenue, Heald Green, Cheadle SK8 3RU
G4 DOL P. Atkins 28 Victoria Place, Easton, , Portland DT5 2AA
GI4 DOM D. Cafolla 87 Stockmans Lane, Belfast BT9 7JD
GM4 DON T. Donnelly 18 Birtwhistle Street, Gatehouse Of Fleet, Castle Douglas DG7 2JJ
GW4 DOO A. Kenyon 6 Abbey Road, Port Talbot SA13 1HA
G4 DOQ J. Willis Kilncroft, Broadlayings, Newbury RG20 9TS
GM4 DOZ T. Findlay 37 Adamton Road North, Prestwick KA9 2HY
G4 DPA G. Austin 16 Courtenay Road, Wantage OX12 7DN
G4 DPD Capt. C. Richardson Domaine De Calcat, Route De Cates, la Sauvetat Sur Lede 47150 France
G4 DPF I. Ross 37 County Rd, March PE158ND
G4 DPH G. Jones 7 The Avenue, Yatton, Bristol BS49 4DA
G4 DPJ D. Wear 84 Hulham Road, Exmouth EX8 3LA
GD4 DPK F. Quayle 1 Birch Hill Gardens, Onchan, Douglas IM3 4ET Isle of Man
G4 DPO A. Nixon 174 Davidson Road, Croydon CR0 6DE
G4 DPP P. Slade Derlee House, East Lane, Abbots Langley WD5 0QG
G4 DPT T. Upstone 10 Lythe Fell Avenue, Halton, Lancaster LA2 6NH
G4 DPU J. Pilling 223 Manchester Road, Accrington BB5 2PF
G4 DPV S. Ford 3 Hill View, Stoke-on-Trent ST2 7AR
G4 DPW P. Leslie-Reed 43 Milehouse Lane, Newcastle ST5 9JZ
G4 DPZ D. Johnson 96 Summerfields Avenue, Halesowen B62 9NR
G4 DQA D. Macken 17 Culvercroft, Binfield, Bracknell RG42 4DF
G4 DQB G. Wallis Ellerton Wood Farm, Little Soudley, Market Drayton TF9 2NB
G4 DQG A. Riley 378 Hungerford Road, Crewe CW1 6HD
GM4 DQJ R. Grant 31 Stormont Park, Scone, Perth PH2 6SD
G4 DQL N. Hall 5 Brooklyn Crescent, Cheadle SK8 1DX
G4 DQN G. Spenceley 168 Robin Hood Lane, Walderslade, Chatham ME5 9LA
G4 DQP Rvd. V. Lewis Four Winds Cottage, Main Street, Broomfleet, Brough HU15 1RJ
G4 DQQ W. Thomas 64 West End, Silverstone, Towcester NN12 8UY
GI4 DQT S. Taaffe 22 Skerriff Road, Cullyhanna, Newry BT35 0JG
G4 DQW J. Krzymuski 3079 Aberdeen Ct, Marietta 30062 United States
G4 DQZ W. Ellis Gillhams House, Gillhams Lane, Haslemere GU27 3ND
G4 DR D. Urquhart 7 Padwell Lane, Bushby, Leicester LE7 9PQ
G4 DRA S. Chester 63 Hawkshead Street, Southport PR9 9BT
G4 DRI I. Selby 2 Ashley Close, Welwyn Garden City AL8 7LH
G4 DRO T. Brosnan 168 Abbots Road, Edgware HA8 0SA
GW4 DRR G. Spencer Tyn Cae, Llanfwrog, Anglesey LL65 4YL
G4 DRS J. Wayman Oak Tree Lodge, Redbridge Road, Crossways, Dorchester DT2 8BG
G4 DRU B. Plastow 185 Allesley Old Road, Coventry CV5 8FL
G4 DRV J. Harris Flat 36, Colonel Stevens Court, 10A Granville Road, Eastbourne BN20 7HD
G4 DRX D. Mckone 12 Hawkshead Road, Knott End-On-Sea, Poulton-le-Fylde FY6 0QE
G4 DRZ G. Carney 94 Combe Avenue, Portishead, Bristol BS20 6JX
G4 DSA G. Kemp 4 Chapter Way, Monk Bretton, Barnsley S71 2HP
G4 DSC O. Boniface 11 Holmefield Road, Ripon HG4 1RZ
G4 DSD R. Woodman 89A Western Way, Ponteland, Newcastle upon Tyne NE20 9AW
G4 DSE P. Zollman 92 Well Lane, Curbridge, Witney OX29 7PA
G4 DSF S. Jones 11 Alba Close, Middleleaze, Swindon SN5 5TL
G4 DSI Dr I. Mcandrew South Winds, Outrigg, St. Bees CA27 0AN
G4 DSN J. Dryden 33 Old Station Road, Newmarket CB8 8DT
GM4 DSO T. Hughes 15 Boreland Road, Kirkcudbright DG6 4HL
G4 DSP Spalding And District ARS c/o A. Hensman 24 Belchmire Lane, Gosberton, Spalding PE11 4HG
G4 DSQ R. Coombe 150 Tean Road, Cheadle, Stoke-on-Trent ST10 1LW
G4 DSR B. Irwin 97 Offerton Lane, Stockport SK2 5BS
G4 DSY R. Miller 21 Woodstock Avenue, Sutton SM3 9EG
G4 DTB M. Bryan 58 Grandstand Road, Hereford HR4 9NF
G4 DTC R. Howgego 39 Harestone Valley Road, Caterham CR3 6HN
GM4 DTH P. Dick Napier House, 8 Colinton Road, Edinburgh EH10 5DS
GM4 DTJ R. Henderson 2 Burdiehouse Avenue, Edinburgh EH17 8AW
G4 DTL W. Young 56 Lincoln Road, Washingborough, Lincoln LN4 1EG
G4 DTM P. Marrable 6 Piccadilly Closeá, Northampton NN4 8RU
G4 DTP D. Pells 6 Clarence St., Stonebroom, Alfreton DE55 6JW
GW4 DTQ D. Gibbon 90 Grosvenor Road, Prestatyn LL19 7TS
G4 DTT W. Brooks 11 Lowther Grove, Garforth, Leeds LS25 1EN
GW4 DTU A. Roberts Brynlludw, Van, Llanidloes SY18 6NP
G4 DTW S. Parsons 54 Furze Cap, Kingsteignton, Newton Abbot TQ12 3TE
G4 DUA R. Bearne Gap House, Over Street Stapleford, Salisbury SP3 4LP
G4 DUB R. Harden 2 Diamond Ridge, Barlaston, Stoke-on-Trent ST12 9DT
G4 DUE A. Parker 5 Geddes Close, Hawkinge, Folkestone CT18 7QL
G4 DUF B. Phillips Woody Nook, Petworth Road, Godalming GU8 5TU
G4 DUI P. Wilson 6 Hereford Road, Colne BB8 8JX
G4 DUJ T. Morley 6 Bakers Close, South Woodham Ferrers, Chelmsford CM3 5JF
G4 DUL M. Coburn 16 Chapel Close Toddington, Toddington LU5 6AZ
G4 DUM V. Long 2 B Pinnacle Hill, Bexleyheath DA7 6AF
G4 DUO F. Taylor 7 Osterley Lodge, Church Road, Isleworth TW7 4PQ
G4 DUQ P. Keane 45 Bramblewood Road, Worle, Weston-Super-Mare BS22 9LW
G4 DUT D. Elliott 3 Oakland Walk, Dawlish EX7 9RS
G4 DUW J. Goldbey Waylands Gate, St. Johns Road, New Milton BH25 5SD
GM4 DUX K. Hampson 9 North Crescent, Garlieston, Newton Stewart DG8 8BA
G4 DVA T. Stanway 24 Fellbrook Lane, Bucknall, Stoke-on-Trent ST2 8AQ
GW4 DVB B. Price 156 Parc Bryn Derwen, Llanharan, Pontyclun CF72 9TX
G4 DVG J. Douglas 367 Wightman Road, London N8 0NA
G4 DVI M. Small 15 Cannock Drive, Stockport SK4 3JB
G4 DVJ R. Hall 12 Britannia Gardens, Westcliff-on-Sea SS0 8BN
G4 DVK M. Lang 52 Gloucester Road, Burnham-on-Sea TA8 1JA
G4 DVM M. Cartwright Seacue, 8 Adelaide Avenue, West Bromwich B70 0SL
G4 DVN S. Whalley 1 Radley Way, Werrington, Stoke-on-Trent ST9 0JN
G4 DVV J. Thomas 57 Bourton Avenue, Stoke Lodge, Bristol BS34 6EB
G4 DVX R. Farr 1 Lavender Road Up Hatherley, Cheltenham GL51 3BN
G4 DVZ T. Beaumont 39 Meadow Road, Garforth, Leeds LS25 2EN
G4 DWC D. Cannings 5 Rowan Close, Brackley NN13 6PB
G4 DWF Dr D. Faulkner 1 Westland, Martlesham Heath, Ipswich IP5 3SU
G4 DWM T. Hunt 45 Front Street Frosterley, Bishop Auckland DL13 2QR
GW4 DWN H. Richards 61, High Street, Abergwynfi, , Port Talbot SA13 3YN
G4 DWO W. Ingham Westfield Villa, Westfield Villas, Wakefield WF4 6EQ
G4 DWR M. Molloy 153 Palmdale Drive, Scarborough MIT 1P2 Canada
G4 DWU J. Blowers 28 Keld Close, Scarborough YO12 6UF
GW4 DWX M. Smith Tonn Marr, Bronybuckley, Welshpool SY21 7NQ
GI4 DWZ P. Tucker 1 The Courtyard, Dunadry Road, Dunadry, Antrim BT41 4QQ
G4 DXB B. Chester 147 Sanctuary Way, Grimsby DN37 9RX
GI4 DXK W. Gordon 17 Ballyheather Road, Ballymagorry, Strabane BT82 0BD
G4 DXN G. Williams 6 Nightingale Court, Leam Terrace, Leamington Spa CV31 1DQ
G4 DXO P. Jones 40 Furze Road, Worthing BN13 3BH
G4 DXP C. Howells 11 West Garth, Carlton, Stockton-on-Tees TS21 1DZ
G4 DXT T. Shaman 3 Padshall Park, Bideford EX39 3NE
G4 DXW Marine Radio Museum Society c/o R. Smith 29 George Street, Peterborough PE2 9PD
G4 DXY Duddon Contest Team c/o J. Spendlove 15 Grammer Street, Denby Village, Ripley DE5 8PQ
G4 DYC M. Cooke 4 Geddes Way, Mattishall, Dereham NR20 3RE
GI4 DYE E. Macintyre 115 Bell Doo, Strabane BT82 9QL
G4 DYG P. Rich 392 Doncaster Road, Stairfoot, Barnsley S70 3RH
G4 DYH G. Dunn Croft Cottage, Innocence Lane, Ipswich IP10 0PL
G4 DYI C. Titheridge 41 Church Walk, Worthing BN11 2LT
G4 DYJ J. Cope Brookfield, Willoughby Drive, Spilsby PE23 5EX
G4 DYM E. Auty Jesla, 5 Silverstone Way, Bristol BS49 5ES
G4 DYO B. Mccartney 123 Reading Road, Finchampstead, Wokingham RG40 4RD
G4 DYR R. Page 28291 Misty Morning Lane, Beloit, Ohio 44609 United States
G4 DYV B. Whiting Fourwinds Buttercake Lane Old Leake, Boston PE22 9QX
GW4 DYY R. Mander Meadowlands, Severn Lane, Welshpool SY21 7BB
G4 DZC Dr M. Bayes 25 Welby Gardens, Grantham NG31 8BN
G4 DZH H. Davies 33 Sandown Road, Ocean Heights, Paignton TQ4 7RL
G4 DZJ D. Paul The Caravan, Chy Leweth, Vorvas, St. Ives TR26 3HL
G4 DZK G. Stocker 8 Brook Drive Astley, Manchester M29 7HS
GM4 DZM I. Shewan Springbank, Distillery Road, Inverurie AB51 0ES
G4 DZS A. Watson 59 Merdon Avenue, Chandler'S Ford, Eastleigh SO53 1GD
G4 DZU D. Parker 50 Rein Road, Tingley, Wakefield WF3 1HZ
GM4 DZX R. Macleod 6 Swanson Drive, Wick KW1 5TF
G4 DZY R. De Tullio 34 Beech Cliffe, Warwick CV34 5HY
G4 EAB J. Blackburn 9 Pitchford Road, Albrighton, Wolverhampton WV7 3LS
GM4 EAF West Glamorgan Raynet c/o A. Hutton 4 Linn Road, Stanley, Perth PH1 4QS
G4 EAG S. Ruffle 39 Nightingale Avenue, Cambridge CB1 8SG
G4 EAK M. Betts 19 Maracas Cove, Western Australia 6028 Australia
G4 EAN I. Brothwell 56 Arnot Hill Road, Arnold, Nottingham NG5 6LQ
G4 EAQ Dr A. Churchley 46 Birchdale Road, Appleton, Warrington WA4 5AW
G4 EAS C. Ellery 17 Wessex Way, Dorchester DT1 2NR
GM4 EAU C. Murray 43 Malleny Avenue, Balerno EH14 7EJ
GM4 EAW J. Mathers 36 Alexander St., Dunoon PA23 7EW
G4 EAX J. Gell 21 Maylands Avenue, Breaston, Derby DE72 3EE
G4 EAZ P. Holliman 17 Arundel Road, Tewkesbury GL20 8AT
GD4 EBA D. Kinrade 8 Alfred Teare Grove, Douglas IM2 6EH Isle of Man
G4 EBE G. Harcourt Hard Farm, Little Marsh Lane, Field Dalling, Holt NR25 7LL
G4 EBF G. Reason 37 Park End, Croughton, Brackley NN13 5LX
G4 EBG B. Meredith 20 Kestrel Avenue, Thorpe Hesley, Rotherham S61 2TT
G4 EBI A. Hamm 166 Sylvan Road, London SE19 2SA
G4 EBK G. Smith 6 Fenby Close, Grimsby DN37 9QJ
G4 EBL R. Whitwell 14 Green Lane, Yarpole, Leominster HR6 0BG
G4 EBN M. Valente Glenville, Abbey Road, Durham DH1 5DQ
G4 EBO W. Gibbs 25 Belvedere Road, Exmouth EX8 1QN
G4 EBQ N. Talbot 59 Heywood Avenue, Austerlands, Oldham OL4 4AZ
GI4 EBS J. Mcnerlin 5 Rosendale Avenue, Limavady BT49 0AE
G4 EBT D. Taylor 3 Crofters Drive, Cottingham HU16 4SD
GM4 EBX P. Hopkinson The Coaches, Kingussie PH21 1NY
G4 EBY G. Head 7 Partridge Way Downley, High Wycombe HP13 5JX
G4 ECA M. Wrintmore 148 Westwick Road, Sheffield S8 7BX
G4 ECE Signallers Interest Group c/o J. Martin 38 Parklands, Mablethorpe LN12 1BY
G4 ECF G. Penney 8 Drake Park, Bognor Regis PO22 7QG
G4 ECO B. Palmer Small Pine, Hedgerow Lane, Leicester LE9 2BN
G4 ECS Yorkshire Radio Friends c/o A. Wisbey 12 Livingstone Road, Caterham CR3 5TG
G4 EDC R. Vane-Stobbs 2 Wood Cottages, Walford Heath, Shrewsbury SY4 3AZ
G4 EDD J. Fletcher 5 Hayeswood Road, Stanley Common, Ilkeston DE7 6GB
G4 EDG S. Taylor 80 Nadder Park Road, Exeter EX4 1NX
G4 EDH G. Rose 15 Pendennis Close, Winklebury, Basingstoke RG23 8JD
G4 EDK A. Ball Tinten House, 2 Tinten Lane, Dorchester DT1 3WP
G4 EDM W. Concannon 155 Walton Road, Sale M33 4FS
G4 EDN K. Currie 37 Golden Ridge, Freshwater PO40 9LF
G4 EDQ R. Gulliver 32 Lavender Close, Thornbury, Bristol BS35 1UL
G4 EDR D. Mappin 13 Willow Close, Filey YO14 9NY
G4 EDW P. Eaton Orchard House, Oxford Road, Sutton Scotney, Winchester SO21 3JG
G4 EDX J. Fletcher 69 Thackerays Lane, Woodthorpe, Nottingham NG5 4HU
G4 EDY M. Grindrod 20 Castle Mead, Kings Stanley, Stonehouse GL10 3LD
G4 EDZ W. Russell Blue Firs, Bower Road, Mersham, Ashford TN25 6NW
G4 EEE A. Wood Pezula, Brimpton, Reading RG7 4TR
G4 EEF S. Foster 6 Webster Close, Hornchurch RM12 6TF
G4 EEH D. Greer 5 Potto Close, Yarm TS15 9RZ
G4 EEJ Dr R. Arak 76 Halifax Road, Brighouse HD6 2EP
G4 EEL A. Cheshire 1 Westerby Lane, Smeeton Westerby, Leicester LE8 0RA
G4 EEQ Rvd. F. Robinson 26 Winstanley Road, Little Neston, Neston CH64 0UZ
G4 EES P. Smith Forge House And Stables Whistley Road, Potterne, Devizes SN10 5TD
G4 EET S. Greep 5 Berkswell Close, Solihull B91 2EH
G4 EEV D. Warwick Orchard Cottage, Colber Lane, Bishop Thornton, Harrogate HG3 3JR
G4 EEZ M. Bath 146 North Road, Hertford SG14 2BZ
G4 EFB C. Mccloud 34 St. Stephens Road, Portsmouth PO2 7PG
G4 EFD D. Stubbs 51 Shop Lane, Maghull, Liverpool L31 7BY
G4 EFE M. Peters 11 Filbert Drive, Reading RG31 5DZ
G4 EFG D. Watton 247 Bloxwich Road, Walsall WS2 7BB

G4 EFH A. Johnson Winchcombe, 3 Merrivale Lane, Ross-on-Wye HR9 5JL
GM4 EFR J. Moar Hansel, Stangergill Cres, Caithness KW14 8UT
G4 EFX A. Levitt The Coach House Strines Clough Farm, Blackshaw Head, Hebden Bridge HX7 7JA
G4 EFY J. Hurst 12 Dukes Mead, Fleet GU51 4HA
G4 EGB J. Fletcher 114 Scholes Park Road, Scarborough YO12 6RA
GM4 EGD I. Brownlie 16 Border Street, Greenock PA15 2EE
G4 EGG W. Higginson 7 Arundale, Westhoughton, Bolton BL5 3YB
G4 EGM R. Webster 230 Huyton Lane, Huyton, Liverpool L36 1TH
G4 EGN V. Coles 205 Farmers Close, Witney OX28 1NS
G4 EGQ P. Pennington 6 Highland Close Sandgate, Folkestone CT20 3SA
G4 EGR D. Barwood 41 Wingfield Road, Bristol BS3 5EG
GW4 EGS M. Price 19 Pencaerfenni Park, Crofty, Swansea SA4 3SE
G4 EGU P. Wolfe 90 Alderney Road, Erith DA8 2JD
GM4 EGX R. Howard 22 Kirkbrae Drive Cults, Aberdeen AB15 9RH
G4 EGY S. Liptrott 40 Mapperley Orchard, Arnold, Nottingham NG5 8AG
G4 EHD W. Tait 51 Broadley Crescent, Halifax HX2 0RL
G4 EHG C. Bryan 9 Brandy Hole Lane, Chichester PO19 5RL
G4 EHJ E. Wilby 5 Matuku Street, Heretaunga, Upper Hutt 5018 New Zealand
G4 EHK D. Goulbourne 8 Moor Avenue, Appley Bridge, Wigan WN6 9JS
G4 EHN Dr J. Axe 5 Hillgate Place, London W8 7SL
GM4 EHP I. Petrie Ugie Cottage, Victoria Road, Peterhead AB42 4NL
G4 EHQ M. Holley 76 Chatham Grove, Chatham ME4 6LY
G4 EHR D. Kirton 16 Silver Innage, Halesowen B63 2PP
G4 EHT W. Watson 12 Curlew Close, Uttoxeter ST14 8TR
G4 EHW Peterborough And District ARC c/o T. Ralph 15 Portchester Close, Stanground, Peterborough PE2 8UP
G4 EHX J. Fearn 37-39 Bourne Square Breaston, Derby DE72 3DZ
G4 EHY F. Greenough 7 Carnforth Avenue, Hindley Green, Wigan WN2 4LD
G4 EIA M. Wallis 34 St. Aidans Close, Bristol BS5 8RH
G4 EIC E. Calvert 163 Milner Road, Heswall, Wirral CH60 5RY
G4 EID M. Haworth 26 Willowhey Marshside, Southport PR9 9TW
GW4 EIE R. Francis 18 Iscoed, Beaumaris LL58 8HH
G4 EIG J. Vickerstaff 5 Luddington Road, Solihull B92 9QH
G4 EII A. Cunliffe 35 Coultshead Avenue, Billinge, Wigan WN5 7HT
G4 EIJ J. Rees 17 Finch Road, Chipping Sodbury, Bristol BS37 6JF
G4 EIK R. Currell Brookside, Treworga, Truro TR2 5NP
G4 EIL G. Oughtibridge 1 Lincoln Drive, Liversedge WF15 7NJ
G4 EIM J. Beaumont 132 Hull Road, Woodmansey, Beverley HU17 0TH
GW4 EIN D. Jones Vine Tree Cottage, Mill Lane, Abergavenny NP7 9SA
G4 EIO S. Harrison Manorlea, Main Street, Bickerton, Wetherby LS22 5ER
GD4 EIP Dr C. Baillie-Searle 2 Marguerite Place Foxdale, Douglas, Foxdale IM4 3HE
G4 EIV J. Sondhis 47 Emlyn Road, Horley RH6 8RX
GM4 EIW J. Dunnington 4 Woodburn Way, Cumbernauld, Glasgow G68 9BJ
G4 EIX D. Whalley 1 Lees Farm Drive, Madeley, Telford TF7 5SU
G4 EIY B. Thomas 8 Whitehill Road, Barton-Le-Clay, Bedford MK45 4PF
GI4 EIZ W. Stewart 56 Ballysillan Park, Belfast BT14 8HD
G4 EJD C. Bourne 5 Brempton Croft, Hilderstone, Stone ST15 8XL
G4 EJE J. Brown11 2 Coriander Gardens, Littleover, Derby DE23 2UB
G4 EJG I. Adams 2 Copper Hall Close, Rustington, Littlehampton BN16 3RZ
G4 EJH K. Middleton 92 South Road, Portishead, Bristol BS20 7DY
GM4 EJI G. Lucas 20 Myreside Gardens, Kennoway, Leven KY8 5TR
G4 EJK D. Reardon 65 Blenheim Road, Caversham, Reading RG4 7RP
G4 EJM M. West 19 Park Drive, Trentham, Stoke-on-Trent ST4 8AB
G4 EJP P. Sheppard 220 Beckfield Lane, York YO26 5QS
G4 EJU R. Hands 19 Orwell Road, Walsall WS1 2PJ
G4 EJW N. Perkins 231 Burnham Road, Burnham-on-Sea TA8 1LT
GM4 EJX A. Murray 67 Carronvale Road, Larbert FK5 3LH
G4 EKB D. Epton 61 Cartmel Drive, Dunstable LU6 3PT
GM4 EKC J. Mackinnon 185 Deeside Gardens, Aberdeen AB15 7QA
G4 EKD P. Spelman 68 Hardwick St., Tibshelf, Alfreton DE55 5QH
G4 EKF S. Sinclair Wayside, Alnwick Road Lesbury, Alnwick NE66 3PJ
G4 EKG M. Tittensor 16 Durcott Road, Evesham WR11 1EQ
GM4 EKI G. Marsh Riverview, Lewiston IV63 6UW
G4 EKJ C. Shaw 10 St. Helens Road, Harrogate HG2 8LB
G4 EKM S. Green 133 Sevenoaks Drive, Sunderland SR4 9NQ
G4 EKS R. Holtham 27 Peyton Close, Eastbourne BN23 6AF
G4 EKT Hornsea ARC c/o R. I'Anson 87 Tranby Lane, Anlaby, Hull HU10 7DT
G4 EKV M. Lobb 52 Ridge Park Avenue, Mutley, Plymouth PL4 6QA
G4 EKW M. Shaw 50 White Road, Nottingham NG5 1JR
G4 EKZ D. Saul 78 Ingleton Drive, Lancaster LA1 4QZ
G4 ELA Ashton In Makerfield ARC c/o R. Dawson 2 Bertram Drive, Wirral CH47 0LQ
G4 ELC G. Keay 9 Buchanan Avenue, Bournemouth BH7 7AA
G4 ELG D. Campbell 12 Newton Close, Newton Solney, Burton-on-Trent DE15 0SL
G4 ELI S. Brown Helford Lodge, The Fairway, Falmouth TR11 5LR
G4 ELJ Dr D. Clark 24B Heatherdale Road, Camberley GU15 2LT
G4 ELK A. Lewis 1 Springcroft, Parkgate, Neston CH64 6SF
G4 ELL R. James-Robertson 8 Whittington Road, Worcester WR5 2JU
G4 ELM E. Jewell 12 Patricks Copse Road, Liss GU33 7DL
G4 ELP D. Stockley 2 The Ridings Chestfield, Whitstable CT5 3PE
GI4 ELQ J. Cushnahan 34 Cornakinnegar Road, Lurgan, Craigavon BT67 9JN
G4 ELR East Lancashire ARC c/o N. Mooney 60 Rhyddings Street, Oswaldtwistle., Accrington BB5 3EY
GM4 ELV Dr D. Dhuglas 1 Micklehouse Road, Baillieston, Glasgow G69 6TG
G4 ELW I. Bontoft 5 Kings Drive Westonzoyland, Bridgwater TA7 0HJ
G4 ELY R. Panting 124 Loddon Bridge Road, Woodley, Reading RG5 4AW
G4 ELZ J. Pascoe 3 Aller Brake Road, Aller, Newton Abbot TQ12 4NJ
G4 EMA I. Welburn 33 Bowland Way, Clifton, York YO30 5PZ
G4 EMB S Norm Alfreton c/o N. Lockett 18 Seagers, Great Totham, Maldon CM9 8PB
G4 EMD R. Edge 4 Mortimer Hill Cleobury Mortimer, Kidderminster DY14 8QQ
G4 EMH G. Parsley 7 Rowan Road, Martham, Great Yarmouth NR29 4RY
G4 EMK G. Parker 15 Burton Road, Heckington, Sleaford NG34 9QR
G4 EML C. Durbridge 2 Send Villas, Sandy Lane, Send, Woking GU23 7AP
G4 EMQ J. Purchon 19 Warburton, Emley, Huddersfield HD8 9QP
G4 EMV P. Johnson 4 Chapel Lane, Blackwater, Camberley GU17 9ET
G4 EMW Prof. E. Warrington 3 Long Meadow, Wigston LE18 3TY
GM4 EMX C. Hall 21 Peterculter Retirement Park Peterculter, Aberdeen AB14 0AB
GM4 EMY J. Ferguson 26 Ashburnham Gardens, South Queensferry EH30 9LB
G4 ENA P. Asquith Well Cottage, The Green, Stroud GL5 5LN
G4 ENB C. Asquith 36 Sunningdale, Luton LU2 7TE
G4 ENC J. Fenton-Coopland 14 Chevril Court, Wickersley, Rotherham S66 2BN
GM4 ENF A. Fyffe 39 Watts Gardens, Cupar KY15 4UG
G4 ENH I. Hodgkiss 190 Ulverley Green Road, Solihull B92 8AD
G4 ENJ K. Hunter 1 Markers Park, Payhembury, Honiton EX14 3NL
G4 ENK P. Kelly 14 Manville Road, Wallasey CH45 5AY
G4 ENL P. Jewitt Colenco Power Engineering, Tafernstrasse 26, Baden CH5402 Switzerland
G4 ENM S. Camp 24 Cedar Avenue, Bournemouth BH10 7EF
GM4 ENN A. Rae 183 Campsie Street, Glasgow G21 4XY
GM4 ENP Dr J. Johnston 4 Lawhead Road West, St. Andrews KY16 9NE
G4 ENR K. Brook 154 Ridge Nether Moor, Swindon SN3 6NF
G4 ENS A. Morris 6 Barrowby Gate, Grantham NG31 7LT
G4 ENZ M. Church 2B Meadow Way, Churchdown, Gloucester GL3 2AU
G4 EOA Dr T. Strickland 22A Branksome Road, St. Leonards-on-Sea TN38 0UA
G4 EOB G. Lawrance 77 Bigland Drive, Ulverston LA12 9PD
G4 EOC Essex Raynet c/o R. Grunwald 20 Sunny Bank Road, Batley WF17 0LJ
G4 EOE R. Everest 2 Burley Road Parkstone, Poole BH12 3DA
G4 EOF S. Lawrence 34 Stanley Drive, Leicester LE5 1EA
G4 EOG A. Heritage 33 Peartree Lane Danbury, Chelmsford CM3 4LS
G4 EOJ T. Ilott 45 Parkside Snettisham, King's Lynn PE31 7QF
GU4 EON M. Allisette Les Amballes Lodge, Les Amballes, St. Peter Port, Guernsey GY1 1WU
G4 EOR P. Stevens 71 Tower Court, Westcliff Parade, Westcliff-on-Sea SS0 7QH
G4 EOT D. Bussell 26 Norbreck Crescent, Wigan WN6 7RF
GM4 EOU J. Smith 6 Rodger Street, Cellardyke, Anstruther KY10 3HU
GW4 EOW P. Baxter 103 Lon Conwy, , Benllech LL74 8RP
G4 EOX N. Davenport 25 Prairie Crescent, Burnley BB10 1EU
G4 EPA J. Pepper 52 King Style Close, Crick, Northampton NN6 7ST
G4 EPC J. Stevens 16 Brindles Field, Tonbridge TN9 2YS
G4 EPD R. Heeley 263 Barnsley Road, Cudworth, Barnsley S72 8JP
GW4 EPF J. Pile 5 Western Close, Mumbles, Swansea SA3 4HF
G4 EPH J. Splaine 765 Wells Road, Bristol BS14 0PB
GI4 EPK E. Coyle 14 Colby Avenue, Culmore, Londonderry BT48 8PF
G4 EPL L. Ward 49 Edgewood Drive, Hucknall, Nottingham NG15 6HY
G4 EPM N. Lewis 97 Orsett Road, Grays RM17 5HA
G4 EPN A. Wright 34 Webbs Way, Stoney Stanton, Leicester LE9 4BW
G4 EPU M. Gray 33 Claremont Drive, Pitsea, Basildon SS16 4TL
G4 EPW L. Goulding 24 Lancaster Drive, Lydney GL15 5SL
G4 EPX D. Chater-Lea Beech Rise, Church Road, Mortimer West End RG7 2HY
GI4 EQA E. Mooney 33 Piney Hill, Magherafelt BT45 6PY
G4 EQC B. Smith 11 Tean Close, Burntwood WS7 9JS
G4 EQD N. Smith 1 Park View, Messingham, Scunthorpe DN17 3TT
G4 EQE D. Smith 7 Demesne Gardens, Martlesham Heath, Ipswich IP5 3UA
G4 EQJ J. Lee 12 Gainsborough Close, Folkestone CT19 5NB
G4 EQK M. Hale 9 Cramer Gutter, Oreton, Kidderminster DY14 0UA
G4 EQL Dr M. Townsend 39 Main Street, Fleckney, Leicester LE8 8AP
G4 EQM W. Evans 9 Edwin Road, Didcot OX11 8LG
GI4 EQN C. Cupples 20 Westland Avenue, Ballywalter, Newtownards BT22 2TR
G4 EQP A. York 8 Granville Close, Hanham, Bristol BS15 3TJ
G4 EQS K. Dowson Fyling Hall Lodge, Fylingdales, Whitby YO22 4QN
G4 EQX J. Mcilroy 17 Brownsfield Road, Yardley Gobion, Towcester NN12 7TY
GM4 EQY J. Hately 10 Crags Road, Paisley PA2 6RA
G4 EQZ K. Faulkner 18 Milton Crescent, Talke, Stoke-on-Trent ST7 1PF
GW4 ERB B. Skidmore Milton Oak House, Oxland Lane, Milford Haven SA73 1LG
G4 ERD A. Hamilton 2905 Nancy Creek Road Nw, Atlanta 30327 United States
G4 ERF P. Jones Woodview, 10 Barrow Hill, Barrow, Bury St. Edmunds IP29 5DX
G4 ERH J. Perry C/O Wagons/Lits Apt168, San Antonio, Baleric Isles ZZ9 9CO Spain
G4 ERL E. Lawley 23 Briar Rigg, Keswick CA12 4NN
GI4 ERM K. Bones 54 Derryvolgie Park, Lisburn BT27 4DA
G4 ERO C. Leonard 24 Lower Road, Stuntney, Ely CB7 5TN
G4 ERP R. Marshall 40 Evesham Road, Bishops Cleeve, Cheltenham GL52 8SA
G4 ERQ T. Birchall 10 Avon Court, Alsager, Stoke-on-Trent ST7 2BA
G4 ERR J. Cummins 5 Blenheim Orchard, Shurdington, Cheltenham GL51 4TG
G4 ERS J. Gamblen Illfield, High Wych, Herts CM21 0HX
G4 ERT H. Marriott 108 Leicester Road, Quorn, Loughborough LE12 8BB
G4 ERV W. Coombes 33 Clarence Park Road, Boscombe East, Bournemouth BH7 6LF
G4 ERW D. Lurcook 2 Drury Road, Tenterden TN30 6QG
G4 ERX R. Elliott No 6 Aphrodities Rock Village, 19 Lykourisson Street, Paphos 8852 Cyprus
G4 ERY D. Tyson 12 Melbury Road, Woodthorpe, Nottingham NG5 4PG
G4 ERZ A. Wells 38 Sextant Road, Hull HU6 7BA
G4 ESG D. Neal 2 St. Margarets Avenue, Ashford TW15 1DR
GI4 ESI S. Mcclean 14 Bamber Park, Ballymena BT43 5HE
GW4 ESL P. Edwards 14 Northfield Close, Caerleon, Newport NP18 3EZ
G4 EST C. Cartmel 41 Lathom Drive, Rainford, St. Helens WA11 8JR
G4 ESU C. Rouse 86 Melton Avenue, Clifton, York YO30 5QG
G4 ESY D. Jackson 16 Melrose Park, Beverley HU17 8JL
G4 ETC P. Keeble 18 Shrubland Drive, Rushmere St. Andrew, Ipswich IP4 5SX
G4 ETD A. Firth 1 Wee Cottage Crook, Kendal LA9 8LH
G4 ETG D. Humphries 173 Herne Road, Ramsey St. Marys, Ramsey, Huntingdon PE26 2SY
G4 ETI J. Shaw 20 Castleton Grove, Jesmond, Newcastle upon Tyne NE2 2HD
G4 ETK C. Bourne 12 Sheepcoat Close, Shenley Church End, Milton Keynes MK5 6JL
G4 ETM J. Taylor 24 Marlborough Drive, Mablethorpe LN12 2 BA
G4 ETN B. Smith Cleeve Valley Farm, Chipstable, Taunton TA4 2QF
G4 ETO J. Roach 33 Pound Lane, Topsham, Exeter EX3 0NA
G4 ETP T. Pinch 1 Fernhill Close, Ivybridge PL21 9JE
G4 ETS J. Forsey 3 Orchard Leaze, Dursley GL11 6HY
G4 ETW Worthing Radio Events Group c/o M. Gibbons 117 Ettingshall Road, Bilston WV14 9XF
G4 ETX D. Ludlow 18 Springfield Court, Ravensbourne Place, Springfield, Milton Keynes MK6 3JJ
G4 ETZ F. Webb 166 Glastonbury Road, Yardley Wood, Birmingham B14 4DS
GW4 EUA G. Smith 13 Lapwing Close, Penarth CF64 5GA
G4 EUC G. Mendoza 32 The Circuit, Cheadle Hulme, Cheadle SK8 7LG
G4 EUF G. Mayo 28 Ring Fence, Shepshed, Loughborough LE12 9HY

G4 EUG G. Payne 28 Pollards Drive, Horsham RH13 5HH
G4 EUJ R. Whiteley Flat 30, Goldsmiths Court, 2 Briton Street, Southampton SO14 3ED
G4 EUK G. Adcock 2 Erringham Road, Shoreham-by-Sea BN43 5NQ
G4 EUL A. Clarke 5 Margil Way, Richmond 7020 New Zealand
G4 EUR M. Tout 25 Booth Rise, Northampton NN3 6HP
G4 EUW B. Keeling 43 Marennes Crescent Brightlingsea, Colchester CO7 0RU
G4 EUZ Durham And District ARS c/o R. Mcaleer 44 Harvey Avenue, Durham DH1 5ZG
G4 EVA C. Roberts Rosemary, York Road, West Byfleet KT14 7HX
G4 EVC K. Chadwick 1 Parklands, Southport PR9 7HX
G4 EVD E. Parry 60 Hunters Forstal Road, Herne Bay CT6 7DW
G4 EVE P. Webster The Old School, School Hill, Cirencester GL7 2LS
G4 EVI J. Howard 127 Goldcroft, Yeovil BA21 4DD
G4 EVK I. Shepherd 12 Watsons Lane, Harby, Melton Mowbray LE14 4DD
GW4 EVL T. Hopkins 39 Glen Road, West Cross, Swansea SA3 5PR
G4 EVN S. Garrett Church Farmhouse, The St., Stowmarket IP14 6LX
G4 EVP C. Mcpartland 55 Elliotts Lane, Codsall, Wolverhampton WV8 1PG
G4 EVR Dr A. Davies 10 New Street, Ludlow SY8 2NQ
GM4 EVS D. Johnstone Sycamore House, Kirk Loan, Perth PH2 6TD
GW4 EVX R. Price 19 New Brighton Road, Sychdyn, Mold CH7 6EF
G4 EVZ M. Powrie 31 The Grove, Billericay CM11 1AU
G4 EWB E. Bradfield St. Olafs, Beach Road, Bacton, Norwich NR12 0EP
G4 EWE D. Overton Toulouse Colwell Road, Totland Bay PO39 0AH
G4 EWI F. Warner 48 Brookfield Road, Walsall WS9 8JE
G4 EWJ B. Jordan 42 Ben Nevis Road, Birkenhead CH42 6QY
G4 EWK D. Mellor 18, Briar Close, Newhall DE110RX
GM4 EWL R. Macleod 9 Croftcroighn Gate, Glasgow G33 5JJ
GM4 EWM E. Mclean 21 Milnefield Avenue, Elgin IV30 6EJ
G4 EWT S. Mason 15 Northfield Close, Bishops Waltham, Southampton SO32 1EW
G4 EWV Mexborough ARS c/o I. Mcpherson 12 Victoria Crescent, Ashford TN23 7HL
G4 EWW T. James 2 The Green, Bottom Street, Southam CV47 2FJ
G4 EWZ J. Halford The Anchor Inn, Chesterfield Road, Alfreton DE55 7LP
G4 EXD I. Marsh 8 South Esk, Culgaith, Penrith CA10 1QR
GW4 EXE B. Hope Oriel, Moelfre LL72 8HN
G4 EXF A. Grindrod Mullions, Church Street, Stonehouse GL10 3HX
GI4 EXI G. Crothers 46 Culmore Point, Londonderry BT48 8JW
G4 EXK Prof. P. Bradbury Rosscairne, 42 Halfpenny Lane, Longridge, Preston PR3 2EA
G4 EXN L. Dolman 46 Norfolk St., Norwich NR2 2SN
G4 EXU D. Fisher 1 Francolin Close, Woodhaven, Natal, South Africa ZZ1 9FR
G4 EXZ R. Fidler 55 Sunnyvale Drive, Longwell Green, Bristol BS30 9YQ
G4 EYA C. Evans 64 Boyd Avenue, Toftwood, Dereham NR19 1ND
G4 EYB D. Fernie Shepherds Close, Reigate Road, Leatherhead KT22 8RD
G4 EYE A. Free Homeric, Harwich Road, Harwich CO12 5JF
G4 EYJ D. Davies 21 Russell Drive, Malvern WR14 2LE
G4 EYM J. Shardlow 19 Portreath Drive, Allestree, Derby DE22 2BJ
G4 EYN K. Wright 61 Albert Road, Chaddesden, Derby DE21 6SH
GW4 EYO C. Carver 8 Overlea Drive, Hawarden, Deeside CH5 3HS
G4 EYT C. Williams 35 Heath Crescent, Norwich NR6 6XF
G4 EYV Dr P. Skolar Apartment 12, Fircroft, Devenish Road, Ascot SL5 9GF
G4 EYX P. Davies 14 Saville Road, Blackpool FY1 6JP
G4 EZC H. Spencer 5 Carlyn Drive, Chandler'S Ford, Eastleigh SO53 2DJ
G4 EZE J. Hinton 7 The Glebelands, Crowborough TN6 1TF
G4 EZG Lord M. Mac Gregor Of Stirling 45 Woodpecker Way, Witney OX28 6NN
GM4 EZJ K. Glendinning 14 Craiglockhart Avenue, Edinburgh EH14 1HW
G4 EZM E. Green 6 Downham Place, Blackpool FY4 1QS
G4 EZN Prof. J. Keeler 67 Perne Avenue, Cambridge CB1 3RY
G4 EZP I. Melville 3 Crescent Road, Benfleet SS7 1JL
G4 EZQ C. Doman 6 Churnet Close, Bedford MK41 7ST
G4 EZU Dr W. Peterson 22 Weston Close, Potters Bar EN6 2BQ
G4 EZX D. Titheridge 2 The Oaks, Wilsden, Bradford BD15 0HH
G4 FAA L. Atkinson 56 The Spinney, Sidcup DA14 5NF
G4 FAB Burnham Beeches RC c/o S. Fox 16 The Teasels, Bingham, Nottingham NG13 8TY
G4 FAD R. Langford Foxholes, Parsonage Farm, Hereford HR4 8AJ
G4 FAE S. Hodgetts 79 Field Lane, Alvaston, Derby DE24 0GQ
G4 FAH D. Jones 41 Sorrel Walk, Brierley Hill DY5 2QG
G4 FAI A. Smith 13 Old Library Mews, Norwich NR1 1ET
G4 FAJ R. Sadler East Lynne, 202 Shire Oak, Walsall WS9 9PD
G4 FAL N. Totterdell Moscar Cross House, Hollow Meadows, Sheffield S6 6GL
G4 FAP R. Painting 15 Surrey Walk Aldridge, Walsall WS9 8JG
G4 FAQ D. Jones 7 Camrose Gardens, Pendeford, Wolverhampton WV9 5RN
G4 FAS G. Royle 56 Branksome Drive, Heald Green, Cheadle SK8 3AJ
G4 FAT N. Trollope 21 Glenmoor, Eckington, Pershore WR10 3BW
GM4 FAU Dr J. Walker West Lodge, Otterstone, Dunfermline KY11 7HZ
G4 FAV A. Bevan 330 Stourbridge Road, Halesowen B63 3QR
G4 FAW D. Cutts 2 Harvesters Way, Martlesham Heath, Ipswich IP5 3UR
G4 FAX R. Macfie 97 Chesford Road, Luton LU2 8DP
G4 FAZ G. Brownett Apartment 264, The Crescent, Hannover Quay, Bristol BS1 5JR
G4 FBB D. Ellis 17 Victoria Avenue, Yeadon, Leeds LS19 7AS
G4 FBC R. Heron 54 Basket Road, Kells, Whitehaven CA28 9AH
G4 FBG D. Shone 6 Windlehurst Road, High Lane, Stockport SK6 8AB
G4 FBI E. Creasy 16 Birchwood Close, Horley RH6 9TX
G4 FBK M. Kipp 55 Hollybrook Mews, Yate, Bristol BS37 4GB
G4 FBN B. Neale Badgers Sett, Harbertonford, Totnes TQ9 7PU
GM4 FBP J. Dean 22 Dalrymple Crescent, Edinburgh EH9 2NX
GW4 FBQ J. Hobley Lawley The Avenue West Felton, Oswestry SY11 4EQ
G4 FBS Horndean And District ARC c/o S. Tribe 6 Privett Road, Waterlooville PO7 5HJ
GM4 FBU H. Macdougall 17 Prospecthill St., Greenock PA15 4HH
G4 FBV D. Vaughan 6 Swallow Close, Felixstowe IP11 9LR
G4 FBY B. Sorger Courtlands, Monks Corner, Saffron Walden CB10 2RW
G4 FBZ W. Kitching 3 Prince Charles Crescent, Telford TF3 2JX
G4 FCA J. Haddon 8 Oaklands, Cradley, Malvern WR13 5LA
G4 FCB N. Edwards 40 Camden St., Walsall WS1 4HF
G4 FCC Stornoway Repeater Group c/o G. Freeman 12 The Haven, Beadnell, Chathill NE67 5AW
G4 FCD R. Girling Lee Gardens, Lower Road, Chalfont St. Peter, Gerrards Cross SL9 8LQ
G4 FCF W. Wade 11 St. Marys Road, Bluntisham, Huntingdon PE28 3XA
G4 FCI A. Cullup 201 Elm Low Road, Elm, Wisbech PE14 0DF
G4 FCL P. Lawson 1 Beehive Cottages, Wickham Road, Fareham PO16 7JF
G4 FCN C. Coker 46 Clarendon Road, Ipplepen, , Newton Abbot TQ12 5QS
G4 FCT J. Gunn 8 College Gardens, Hornsea HU18 1EF
G4 FCU D. Restall 7 Medway Close, Skelton-In-Cleveland, Saltburn-by-the-Sea TS12 2JZ
GW4 FCV Sheffield & District Wireless Society c/o R. Jones 2 Pen-Y-Cwarel Road, Wyllie, Blackwood NP12 2HP
G4 FCX B. Pearl 66 Benfleet Road, Benfleet SS7 1QB
G4 FCY I. Smith 17031 Los Cerritos, Los Gatos 95030 United States
G4 FCZ M. Thomas The Old School, Church Lane, Lowestoft NR32 5LL
G4 FDA W. Chapman 34 Saxon Close, Oake, Taunton TA4 1JA
G4 FDD J. Livingston 26 Dikelands Lane, Upper Poppleton, York YO26 6JB
G4 FDF V. Cunningham 9 Lacon Road, Bramford, Ipswich IP8 4HD
G4 FDG R. Taylor Trelawn, 26A Honiton Road Cullompton, Devon EX15 1PA
G4 FDI S. Giles Conifers, Kington Magna, Gillingham SP8 5EW
G4 FDK R. Palmer 26 Silverstone Way, Congresbury, Bristol BS49 5ES
GM4 FDM T. Wylie 3 Kings Crescent, Elderslie, , Johnstone PA59AD
G4 FDN P. Mcguinness 9 Farmdale Road, Carshalton Beeches SM5 3NG
G4 FDP R. Miller 65 West Road, Oakham LE15 6LT
G4 FDR Sqdn. Ldr. D. Roberts 236 Grantham Road, Sleaford NG34 7NX
G4 FDS J. Ingram 170 Churchill Road, Poole BH12 2JF
G4 FDU R. Mckinlay 54 Barn Meadow Lane, Bookham, Leatherhead KT23 3EY
G4 FDX I. Offer Brookside Cottage, Hare Lane, New Milton BH25 5AF
G4 FEA C. Beezley 19 Beech Avenue, Claverton Down, Bath BA2 7BA
G4 FEB D. Emery 424 Clement Avenue, Charlotte, North Carolina 8204 United States
GM4 FEI A. Marsden 63 Carlogie Road, Carnoustie DD7 6EX
G4 FEJ B. Fawcett 75 Ark Royal, Bilton, Hull HU11 4BN
G4 FEM P. Greatorex 2 Briar Briggs Road, Bolsover, Chesterfield S44 6SE
GM4 FEO J. Gaughan 12 Fernbank Avenue, Windygates, Leven KY8 5FA
G4 FEQ H. Stogdale 14 Main Street, Ledston, Castleford WF10 2AA
G4 FEU T. Southwell 12 Chequer Lane, Upholland, Skelmersdale WN8 0DE
G4 FEV D. Whitty 146 Avenue Road., Rushden. NN10 0SW
G4 FF Gxff Radio Group c/o A. Dodd 14 Davies Street, Macclesfield SK10 1GE
G4 FFA R. Harris 98 Evelyn Avenue, Ruislip HA4 8AJ
G4 FFC M. Packer Ricmaes Cottage, Chadwell End, , Pertenhall MK44 2AU
G4 FFE Bartg c/o L. Marriott 94 Lyndhurst Road, Worthing BN11 2DW
GM4 FFF Plymouth Radio Community c/o J. Phunkner 7 Plenshin Court, Glasgow G53 6QW
GI4 FFL J. Finnegan 15 Mossgreen, Richhill, Armagh BT61 9JX
G4 FFM D. Bailey 10 Manor Road, Stutton, Tadcaster LS24 9BR
G4 FFN Icars c/o C. Baker 78 Station Road, Whittlesey, Peterborough PE7 1UE
GM4 FFP I. Campbell 35 Radernie Place, St. Andrews KY16 8QR
G4 FFS D. Hodge 15 Buckland Close, Peterborough PE3 9UH
G4 FFW M. Betts 56 Kingswood Road, Fallowfield, Manchester M14 6RX
G4 FFX R. Clear 33 Cedars Road, Beddington, Croydon CR0 4PU
G4 FFY Maidenhead & District ARC c/o R. Howells 16 Handel Walk, Tonbridge TN10 4DG
G4 FGF J. Drakeley 31 Goldstar Way, Birmingham B33 0YP
GI4 FGH W. Tweedy 11 Beechgrove Rise, Belfast BT6 0NH
G4 FGJ G. Mcgowan 6 Caldecote Green, Upper Caldecote, Biggleswade SG18 9BX
GM4 FGL G. Williams 1 Fife St., Keith AB55 5EH
G4 FGM D. Lund P.O. Box 333, 23 Vander Avenue, Blenheim, on N0P 1A0 Canada
G4 FGO A. Oliver Beaver Lodge, Dale Road, Brough HU15 1HY
G4 FGR S. Porter 138 Broad Lane, Essington, Wolverhampton WV11 2RQ
GM4 FGS I. Douglas 47 Meadowpark, Ayr KA7 2LW
G4 FGW C. Hall 5 St. Edmunds Stamford Bridge, York YO41 1PW
G4 FGY J. Maltby Ingle Nook, Lenton Road, Grantham NG33 4HA
GI4 FHB Midland CG c/o W. Mcfaul 9 Durham Park, Londonderry BT47 5YD
G4 FHF J. Walker Roseberry, Owmby Road, Barnetby DN38 6BD
G4 FHK T. Knight 3 Eaton Close, Rainworth, Mansfield NG21 0AR
G4 FHN Inverclyde ARG c/o R. Lovell 16 North View, Staple Hill, Bristol BS16 5RU
G4 FHQ M. Hardy 6 Apple Tree Close, Bromyard HR7 4UL
G4 FI N. Seath 6 Harvester Way, Sibsey, Boston PE22 0YD
G4 FIA M. Mucklow 7 Burns Close, Newport Pagnell MK16 8PL
GW4 FIC D. Pearson Hope Cottage, Parkhouse, Monmouth NP25 4QD
G4 FIE P. Groom 2A The Chestnuts, Countesthorpe, Leicester LE8 5TL
G4 FIF D. Cherrington 4 Bloomfield Close, Wombourne, Wolverhampton WV5 8HQ
G4 FIG B. Callaway 44 Grover Avenue, Lancing BN15 9RQ
G4 FIH E. Fernandes 24 Keere Street, Lewes BN7 1TY
G4 FIN K. Mullaney 5 St. Peters Way, Cogenhoe, Northampton NN7 1NU
G4 FIQ G. Clegg 6 Vergette Court, Towngate West, Market Deeping, Peterborough PE6 8DJ
G4 FIT J. Chapman 83 High St., Sutton, Ely CB6 2NW
G4 FIV P. Morley Ash House, Germansweek, Beaworthy EX21 5BP
GM4 FIZ A. Murray Woodhouse, Mount High, Balblair IV7 8LH
G4 FJB J. Dodd 7 Hornbrook Grove, Solihull B92 7HH
G4 FJC R. De La Rue Linden Lea, Balls Chase, Halstead CO9 1NY
G4 FJF M. Thacker 164 Mongeham Road, Great Mongeham, Deal CT14 9LL
G4 FJH D. Powell 24 Beaconlea, Hanham, Bristol BS15 8NX
GD4 FJI R. Allison 20 Droghadfayle Park, Port Erin IM9 6EP Isle of Man
G4 FJJ D. Bayliss 20 Midhill Drive, Rowley Regis B65 9SD
G4 FJK T. Hugill West Whitnole House, Stoodleigh, Tiverton EX16 9QH
G4 FJP J. Perry 108 Elm Road, New Malden KT3 3HP
G4 FJT C. Cuthbert 44 Towse Close, Clacton-on-Sea CO16 8US
G4 FJV J. Barton 5 Buttermere Grove, Willenhall WV12 5FQ
G4 FJW C. Hook 11 Battlesmere Road, Cliffe Woods, Rochester ME3 8TR
G4 FJX I. Perera 1 Francis Road, Perivale, Greenford UB6 7AD
G4 FKA G. Plucknett 9 Oakwood Gardens, Coalpit Heath, Bristol BS36 2NB
G4 FKE C. White 111 Waterbeach Road, Slough SL1 3JU
G4 FKG G. Kirk 124 Star Road, Peterborough PE1 5HF
G4 FKH G. Williams 21 Borda Close, Chelmsford CM1 4JY
G4 FKI D. Thorpe 70 Willow Way, , Ampthill MK45 2SP
G4 FKP B. Tarry 6 Beech Gardens, Rainford, St. Helens WA11 8DJ
G4 FKQ M. Barnwell 77 Elmfield Road, Peterborough PE1 4HA
G4 FKR R. Hammond Southerly House, Littleton Lane, Sparsholt, Winchester SO21 2LS

G4 FKU K. Salter Alton, 12 Perinville Road, Torquay TQ1 3NZ
G4 FKX R. Abel 23 Edward Gardens, Wickford SS11 7EH
G4 FKY R. Sharpe 1 Park Copse, Horsforth, Leeds LS18 5UN
GI4 FLG K. Mayne 8 Grandmere Park, Bangor BT20 5RF
G4 FLM F. Crofts 43 Broadlands Drive, East Ayton, Scarborough YO13 9ET
GM4 FLP I. Strachan 238 Coupar Angus Road, Muirhead, Dundee DD2 5QN
G4 FLR D. Tanner 4 Duckpitts Cottages, Bramling, Canterbury CT3 1LY
G4 FLS A. Snow 1A Park Avenue, Longlevens, Gloucester GL2 0DZ
GM4 FLX A. Lovegreen 16 Grahams Avenue, Lochwinnoch PA12 4EG
G4 FLY G. Haynes 39 Zinzan Street, Reading RG1 7UG
GD4 FMB D. Taylor Burnt Mill House, Mount William, Summer Hill, Douglas, Isle of Man IM2 4PE
G4 FMC M. Constable 2 The Banks, Long Buckby, Northampton NN6 7QQ
G4 FMI F. Connor 29 Parkdale Road, Paddington, Warrington WA1 3EN
G4 FMJ L. Cooke 23 Widecombe Road, Stoke-on-Trent ST1 6SL
G4 FMM T. Walsh 106 Westgate, Elland HX5 0BB
G4 FMY D. Larsen C/O Salbu (Pty) Ltd, Private Bag X 2352, Wingate Park 153 South Africa
G4 FMZ E. Pearson 4 Forder Walk, Salisbury SP2 7FY
G4 FNC L. Harper Three Oaks, Braydon, Swindon SN5 0AD
G4 FND D. Yeates 61 Martins Hill Lane, Burton, Christchurch BH23 7NW
G4 FNG R. Walker South Moor Farm, Langdale End, Scarborough YO13 0LW
G4 FNI K. Nichols 11 Tregonwell Road, Bournemouth BH2 5NR
G4 FNJ P. Fuller 4 Whitworth Road, Minehead TA24 8EB
G4 FNK A. Jackson 6 Blandys Hill, Kintbury, Hungerford RG17 9UE
G4 FNL G. Bubloz 42 Hillcrest, Westdene, Brighton BN1 5FN
GW4 FNO G. Lloyd 15 Budden Crescent, Caldicot NP26 4PP
G4 FNP J. Guite 15 Marlborough Avenue, Falmouth TR11 2RW
G4 FNQ C. Wedgbury 32 Cloverdale, Stoke Prior, Bromsgrove B60 4NF
G4 FNR D. Rabone 6 Cranwell Grove, Kesgrave, Ipswich IP5 2YN
GI4 FNU M. Mcdowell 50 Dunraven Parade, Belfast BT5 6BT
G4 FNZ D. Bannister 7 Sudeley Close, Malvern WR14 1LP
G4 FOB J. Samuels 8 Holm Oaks, Butleigh, Glastonbury BA6 8UB
G4 FOH S. Foote 14 High Street, Chrishall, Royston SG8 8RP
GW4 FOI J. Doyle 54 Bryncatwg, Cadoxton, Neath SA10 8BG
G4 FOL J. Bell 60 Queens Close, West Moors, Ferndown BH22 0HN
GW4 FOM R. Rowles 37 Vincent Court, Vincent Road, Cardiff CF5 5AQ
G4 FON R. Goff 69 Frambury Lane, Newport CB11 3PU
G4 FOR D. Hawkes 19 Taj Court, Ottawa K1G 5K7 Canada
G4 FOS I. Gilmore 4 Borton Road, Blofield, Norwich NR13 4RU
G4 FOT H. Exley 16 Croft Street, Horncastle LN9 6BE
G4 FOW R. Strangeway 88 Old Manor Way, Portsmouth PO6 2NL
G4 FOX Melton Mowbray ARS c/o G. Mason 120 Scalford Road, Melton Mowbray LE13 1JZ
G4 FOY K. Scott 20 Tower Street, Alton GU34 1NU
GM4 FOZ D. Moodie 1 Lageonan Road, Grandtully, Aberfeldy PH15 2QY
G4 FPA J. Shorthouse 20 Boxgrove Road, Sale M33 6QW
G4 FPB C. Roper 52 Sandringham Avenue, Wirral CH47 3BZ
G4 FPE G. Butterfield 13 Windsor Walk, Batley WF17 0JL
G4 FPG R. Hawke Basque Close, Hastingleigh, Ashford TN25 5JB
G4 FPI B. Wood 193 Robin Way, Chipping Sodbury, Bristol BS37 6JU
G4 FPM E. Keeler 18 Clyde Road, Worthing BN13 3LG
G4 FPO K. Wilson 14 Stuart Grove, Eggborough, Goole DN14 0JX
G4 FPV S. Perkins 17 Lime Tree Avenue, Malvern WR14 4XE
G4 FPY K. Jones 58 Woodlands Road, Allestree, Derby DE22 2HF
GM4 FPZ M. Lisle 50 Lade Braes, St. Andrews KY16 9DA
GM4 FQE E. Thirkell 20 The Glebe, Crail, Anstruther KY10 3UJ
G4 FQF P. Herring 34 Woodlands Road, Romford RM1 4HD
GM4 FQG R. Mclaren Lethendry, North Road, Dunbar EH42 1AY
G4 FQH B. Nelmes Birchgrove, 17 Woodfield Road, Dursley GL11 6HB
G4 FQM S. Morris 23 Ellesmere Way, Carlisle CA2 6LZ
G4 FQN G. Kelly 7 Greenwood Road, Lymm WA13 0LA
G4 FQP C. Bamford 12 Lincoln Drive, Caistor, Market Rasen LN7 6PA
G4 FQT R. Gregory 24 Tilton Road, Borough Green, Sevenoaks TN15 8RS
GW4 FQU Dr I. Jones Rhandirmwyn, Llandygai, Bangor LL57 4LD
G4 FQV D. Gray 8 Foxglove Close, Wyke, Gillingham SP8 4TW
G4 FQW B. Dunn 17 Duke St., Clayton Le Moors, Accrington BB5 5NQ
G4 FQZ D. Simms 1 Old Barn Close, Little Eaton, Derby DE21 5AX
G4 FRA V. Lane 3 Lawnway, York YO31 1JD
G4 FRB Dr G. Morse Riversdale, High Street, Salisbury SP3 5JL
G4 FRD J. Walton 2 Billy Mill Avenue, North Shields NE29 0QX
G4 FRF K. Johnson 7 Bridge Croft, Clayton Le Moors, Accrington BB5 5XP
GW4 FRH R. Dawkins 22 Derwen Fawr, Crickhowell NP8 1DQ
G4 FRI G. Bird Holmwood, 101 Brookfield Road, Gloucester GL3 2PN
G4 FRK J. Rodway 9 York Avenue, Thornton-Cleveleys FY5 2UG
G4 FRL N. Ambridge 4 Staggs Road, Thame OX9 3AG
G4 FRM P. Hill 8 Davenport Park Road, Stockport SK2 6JS
G4 FRO G. Orford 29 Church Road, Stoke Bishop, Bristol BS9 1QP
G4 FRV R. Vincent Little Poulner, White Horse Lane, Bideford EX39 1NW
G4 FRW M. Nicholson 10 Beechfield Road, Cheadle Hulme, Cheadle SK8 7DS
G4 FRX J. Nelson11 Bank Cottage, Crew Green, Shrewsbury SY5 9AS
G4 FRZ A. Jarrett 73 Abbots Road, Abbots Langley WD5 0BJ
GM4 FSB G. Millar 30 Albert Crescent, Newport-on-Tay DD6 8DT
G4 FSD J. Creasey 144 Belthorn Road, Belthorn, Blackburn BB1 2NN
G4 FSE P. Biner 295 Daws Heath Road, Rayleigh SS6 7NS
GM4 FSF K. Horne The Coach House, Comely Park Lane, Dunfermline KY12 7HA
G4 FSG P. Murchie 42, Catherine Road, Woodbridge IP12 4JP
G4 FSH J. Bagnall Rainow, Under Rainow Road, Congleton CW12 3PL
G4 FSK Barking Radio And Electronics Society c/o P. Vallow 1 Carolbrook Road, Ipswich IP2 9JF
G4 FSN E. Walton 68 Mary Street West, Horwich, Bolton BL6 7JU
G4 FSQ 1st Pencoed Scout Group c/o J. Morley 65 Longfield Avenue, Golcar, Huddersfield HD7 4BT
G4 FSS D. Hocking 10 Garfit Road, Kirby Muxloe, Leicester LE9 2DE
G4 FSU I. Greenshields 3 Lovers Walk, Wells BA5 2QL
G4 FSX G. Down 20 Thackers Way, Deeping St. James, Peterborough PE6 8HP
G4 FTA R. Earle 10 Crosslands, Fringford, Bicester OX27 8DF
G4 FTG C. Norton 2 Heathlands Drive, Maidenhead SL6 4NF
G4 FTI A. Bowhill 9 West Park Drive East, Roundhay, Leeds LS8 2EE
G4 FTK N. Cridland Alfriston, 105 Elvetham Road, Fleet GU51 4HN
G4 FTL G. King 59 Rookery Lane, Northampton NN2 8BX
G4 FTN J. Grainger Kinlet Cottage 45 Stone Lane, Kinver DY7 6DU
G4 FTP E. Kraft 6 The Nook, Wivenhoe, Colchester CO7 9NH
G4 FTQ P. Clutterbuck 19 Warwick Close, Dorking RH5 4NN
G4 FTW Dr M. Rowland 16 Hayter Close, West Wratting, Cambridge CB21 5LY
G4 FTX G. Knock 31 Northmead, Ledbury HR8 1BE
G4 FTY R. Page Mercury House, 19 Green Lane, Birmingham B46 3NE
G4 FUA G. Cheater 34 Robbins Close, Bradley Stoke BS32 8AS
GI4 FUE C. Morrison 60 Windslow Drive, Carrickfergus BT38 9BB
G4 FUG P. Clark 42 Shooters Hill Road, Blackheath, London SE3 7BG
G4 FUI M. Rigby 16 Juniper Way, Penrith CA11 8UF
G4 FUJ G. Wright 35 Langdale Road, Cheltenham GL51 3LX
GI4 FUM Dr W. Hutchinson 40 Oldstone Hill, Muckamore, Antrim BT41 4SB
G4 FUO J. Nowell Crofters Cottage, Back Lane, Copmanthorpe, York YO23 3SH
G4 FUP N. Braeman 30 Oakley Road, Wimborne BH21 1QJ
G4 FUR Coulsdon Amateur Transmitting Society c/o A. Briers 33 Deans Walk, Coulsdon. CR5 1HR
G4 FUU M. Pothecary 61 Inglewood, Pixton Way, Croydon CR0 9LN
G4 FUY P. Bonson 26 Shreen Way, Gillingham SP8 4EL
G4 FUZ A. Mallows Kilmurry House, Kilmurry Estate, Kilmurry Fermoy 00 00 Ireland
G4 FVA P. Catling The Heights, 10 Adams Road, Cambridge CB25 0JU
G4 FVB I. Clabon 2 Farm View Lodge, Lodge Hill Lane, Rochester ME3 8NE
G4 FVI M. Willis 6 Walmsley House, Princess Street, Folkestone CT19 6QP
G4 FVK D. Sewell 11 Haddon Close, Stanground, Peterborough PE2 8LS
G4 FVL G. Rankin 25 The Chase, Coulsdon CR5 2EJ
GM4 FVM J. Edgar 7 Welltower Park, Ayton, Eyemouth TD14 5RR
GM4 FVO C. Evans East Cottage, Mount Melville, St. Andrews KY16 8NT
G4 FVP C. Davies 28 Neville Road, Darlington DL3 8HY
GM4 FVQ A. Dimmick 02 120 Shakespeare Street, Glasgow G20 8LF
GM4 FVS G. Cusiter 4 Elphin Hill, Ellon AB41 8BH
G4 FVU A. Sweetapple Bent Oak, Axminster Road, Axminster EX13 8AQ
G4 FVV B. Vincent 27 Naseby Walk, Leeds LS9 7SY
G4 FVW D. Hooper 8 Barn Close, Crewkerne TA18 8BL
G4 FVX R. Johns 42 Lansdown Road Redland, Bristol BS66NS
G4 FVZ Dr P. Gould 152 High Street North, Stewkley, Leighton Buzzard LU7 0EP
G4 FWA J. Beckett 9 Gleneagles Drive, Ipswich IP4 5SD
GM4 FWF A. Gurney 42 Fleming Way, Invergordon IV18 0LU
G4 FWK P. Mooney 57 Johnstown Road, Co Dublin, Dun Laoghaire Ireland
G4 FWM C. Webb 6 Chatsworth Avenue, Fleetwood FY7 8EG
G4 FWN N. May Sandock Nurseries, Middle Dimson, Gunnislake PL18 9NG
GD4 FWQ C. Matthewman 26 King Orry Road, Glen Vine, Isle of Man IM4 4ES Isle of Man
G4 FWR A. Johnson 86 Meadow Close, Thatcham RG19 3RL
G4 FWT S. O'Shanohun The Corner Stone, Treskinnick Cross, Bude EX23 0DT
G4 FXA V. Arnold 435 Manchester Road, Clifton, Manchester M27 6WH
G4 FXE R. Lott 83 Manor Road, Dover CT17 9LQ
GW4 FXF G. Swan Long Acre, New Road, Neath SA10 8HT
G4 FXI P. Overell 48 Bedgrove, Aylesbury HP21 7BD
GM4 FXL A. Docherty 10 Dumyat Road, Menstrie FK11 7DG
G4 FXM G. Farnie Barn End, Rughill, Wedmore BS28 4HL
G4 FXR W. Wunderlich 31 College Road, Bromley BR1 3PU
G4 FXT N. Burkitt 31 Loxwood, Earley, Reading RG6 5QZ
G4 FXU R. Napper 1 12 Brumell Drive, Lancaster Park, Morpeth NE61 3RB
G4 FXY P. Staton 52 School Road, Newborough, Peterborough PE6 7RG
G4 FYB S. Carter 48 Gorse Bank Road, Hale Barns, Altrincham WA15 0AS
G4 FYE G. Coggon 45 Ansten Crescent, Cantley, Doncaster DN4 6EZ
G4 FYG M. Newlands 72 Town Acres, Tonbridge TN10 4NG
GM4 FYH C. Waddington Wester Lathallan, Leven KY8 5QP
G4 FYI T. Fallick 44 Cypress Road, Newport PO30 1HA
G4 FYJ J. Lemon 30 Iveagh Court, Farm Hill, Exeter EX4 2LR
G4 FYM D. Wiggs 8 Bulbery Abbotts Ann, Andover SP11 7BN
G4 FYO T. Foley 16 Buckingham Road, Winslow, Buckingham MK18 3DY
G4 FYQ M. Robins 36 Wolverley Avenue, Wollaston, Stourbridge DY8 3PJ
G4 FYT D. Lawrence 23 Parkmead Road, Wyke Regis, Weymouth DT4 9AL
G4 FZA J. Bladen 4 St. James Close, Hanslope, Milton Keynes MK19 7LF
G4 FZC A. Chapman Majadilla Del Muerte 155, Malaga 29649 Spain
GI4 FZD P. Menown 34 Cairnburn Road, Belfast BT4 2HS
G4 FZF A. Grinling 32 Maybridge Square, Goring-By-Sea, Worthing BN12 6HR
G4 FZG B. Sirignano Eversholt, 22 Cleevelands Drive, Cheltenham GL50 4QB
GM4 FZH Dr C. Smith Renwickview, Throughgate, Dunscore, Dumfries DG2 0UG
G4 FZL L. Povoas 9 Masons Drive, Necton, Swaffham PE37 8EE
G4 FZM M. Davey 37 The Common West Wratting, Cambridge CB21 5LR
G4 FZN J. Kirby 2 Kneeton Park, Middleton Tyas, Richmond DL10 6SB
G4 FZP A. Drury 31 Brook Drive, Whitefield, Manchester M45 8FR
G4 FZR K. Dally Ealand Grange, Ealand, Scunthorpe DN17 4DG
G4 FZS H. Bulmer 10 Southfield Lodge, South End Villas, Crook DL15 8NN
G4 FZV Dr P. Redall 106 Stowey Road, Yatton, Bristol BS49 4EB
G4 FZZ D. Holmes 12 Chestnut Close, Rushmere St. Andrew, Ipswich IP5 1ED
G4 GA Radio Gaga CG c/o J. Warburton Peers Meadow, Plealey SY5 0UY
G4 GAB R. Padbury 8 Osbourne Drive, Holton-Le-Clay, Grimsby DN36 5DS
GW4 GAF A. Mccann Lower Fiddlers Green, Felindre, Knighton LD7 1YT
G4 GAI K. Taylor 31 Stonehill Drive, Rochdale OL12 7JN
G4 GAK M. Sykes 21 Croft Walk, Broxbourne EN10 6LD
G4 GAP H. Fitzherbert 36 Westover Road, Broadstairs CT10 3ES
G4 GAT B. Denton 2 Seacroft Road, Broadstairs CT10 1TL
G4 GBA C. Brookson Orchard View, The Street., Stonham Aspal, Stowmarket IP14 6AJ
G4 GBC F. Orchard 39B Breach Road Marlpool, Heanor DE75 7NJ
G4 GBE R. Blacker 20 Claremont Park, Lincoln Road, Sleaford NG34 8AE
G4 GBI A. Edwards 96 Bathurst Road, Winnersh, Wokingham RG41 5JF
G4 GBK C. Appleton 249 Devonshire Road, Atherton, Manchester M46 9QB
G4 GBP C. North Somerholme, Forest Road, Hale, Fordingbridge SP6 2NR
G4 GBW Dr J. Wilcox 533 Upper Brentwood Road, Gidea Park, Romford RM2 6LD
G4 GBX W. Greed 18 Nursteed Park, Devizes SN10 3AN
G4 GBY J. Robson 35 Hankin Avenue Dovercourt, Harwich CO12 5HE
G4 GCI N. Palmer 14 Cambria Drive, Dibden, Southampton SO45 5UW
G4 GCJ F. Fuller 7 Prestwick Close, Bletchley, Milton Keynes MK3 7RQ

G4 GCL J. Tyler 1 Mansefield Road, Tweedmouth, , Berwick-upon-Tweed TD15 2DX
GI4 GCN R. Booth 12 Priory Drive, Carrickfergus BT38 8HZ
G4 GCT North Bristol ARC c/o R. Elford Prospects, Tormarton Road, Badminton GL9 1HP
G4 GCU Z. Kowalczyk 6 St Georges Crescent, Redcar TS11 8BT
G4 GCX R. Reisch 18 St. Margarets Grove, Leeds LS8 1RZ
G4 GDB Dr A. Duncan 66 Ravenscroft Crescent, Sheffield S13 8PR
G4 GDC S. Wiles Conifers, Aisthorpe, Lincoln LN1 2SG
GM4 GDF J. Cain Carradale, Braehead, Kirkinner, Newton Stewart DG8 9AH
G4 GDG R. Smith 47 Windsor Road, Levenshulme, Manchester M19 2FA
G4 GDL M. Ellis 32 Pegholme Drive, Otley LS21 3NZ
GW4 GDM J. Owens Yr Hafan I Maes Gyn, An Llanarmon-Yn-Ial, Clwyd N Wales CH7 4PY
G4 GDO F. Lamb 13-336 Queen St. South, Mississauga, Ontario L5M 1M2 Canada
G4 GDP J. O'Shea 30 Sue Ryder Homes, Owning, Co Kilkenny BN2 7HA
G4 GDR Rvd. A. Heath 227 Windrush, Highworth, Swindon SN6 7EB
G4 GDS D. Jones 3 Kingfisher Drive, Benfleet SS7 5ES
G4 GDT D. Wood 30 Semley Road, Hassocks BN6 8PE
G4 GDU I. Hoskin 14 Trevingey Parc, Redruth TR15 3BZ
G4 GDX I. Smith 25 Windrush Avenue, Brickhill, Bedford MK41 7BS
G4 GDY M. Edwards 9 Earls Walk, Binley Woods, Coventry CV3 2AJ
G4 GED D. Richardson 68 Beech Tree Road, Holmer Green, High Wycombe HP15 6UT
G4 GEE Dr R. Nash 135 Farren Road, Coventry CV2 5EH
GI4 GEL R. Penn 9 Milltown Road, Donaghcloney, Craigavon BT66 7NE
G4 GEN A. Morriss Pipinford Park, Millbrook Hill Nutley, East Sussex TN22 3HX
G4 GEO C. Tomkinson Ridgeway, Towers Road, Poynton, Stockport SK12 1DD
G4 GEP V. Peake 24 Holyoke Grove, Leamington Spa CV31 2RB
G4 GET I. Jordan 70 Hungerhill Road, Kimberworth, Rotherham S61 3NP
G4 GEY J. Carter 30 Braemar Road, Hazel Grove, Stockport SK7 4QG
G4 GEZ R. Evans 2 Greyfriars Lane, West Common, Harpenden AL5 2QJ
G4 GFC S. Wright 163 Croham Valley Road, South Croydon CR2 7RE
G4 GFE D. Foulds 12 Royal Beach Court, North Promenade, Lytham St. Annes FY8 2LT
G4 GFI Wirral & District ARS c/o M. Broadway 91 Tattenham Grove, Epsom KT18 5QT
G4 GFJ L. Frankham 47 St. Marys Gardens, Hilperton Marsh, Trowbridge BA14 7PH
GW4 GFL Norfolk Scout Radio c/o A. Hopkins 30 Wavell Drive, Newport NP20 6QN
G4 GFM D. Hessom Flat 31, Town Mill, Overton, Basingstoke RG25 3JE
G4 GFN S. Dabbs 52 Hayling Rise, Worthing BN13 3AG
G4 GFV J. Simpson 19 Hollinside Close, Whickham, Newcastle upon Tyne NE16 5QZ
G4 GFY P. King 78 Gweal Wartha, Helston TR13 0SN
G4 GFZ S. Dunkerley Po Box Hm 2215, Hamilton ZZ9 9PO Bermuda
G4 GGC M. Marsh 21 Stour Gardens, Great Cornard, Sudbury CO10 0JN
G4 GGE D. Nicholson 41 Thurstons Barton, Bristol BS5 7BQ
GM4 GGF V. Mason 19 Sherwood Crescent, Bonnyrigg EH19 3LQ
G4 GGH P. Ledbury 12 Sandfield Close, Lichfield WS13 6BF
G4 GGI R. Williamson Burwood, Wych Hill Lane, Woking GU22 0AA
G4 GGL T. Grainger 34 Maple Avenue, Ripley DE5 3PY
G4 GGR F. Gemmell 89 Coach Road, Guiseley, Leeds LS20 8AY
G4 GGT M. Masterson 44 Highstone Avenue, London E11 2PP
G4 GGX S. Randall 66 Park Court, Harlow CM20 2PZ
G4 GGZ J. Birch 13 Alison Way, Aldershot GU11 3JX
G4 GHA J. Cleaton 1 Avon Drive, Northmoor, Wareham BH20 4EL
G4 GHB B. Kitchen 73 Birch Street, Ashton-under-Lyne OL7 0JD
G4 GHI Wessex DX Group c/o R. Crabb 29 Horsecastles Lane, Sherborne DT9 6BU
G4 GHK J. Donovan 6 Manor Place, Church, Accrington BB5 4DX
G4 GHL Prof. M. Ward 9 Woodshears Drive, Malvern WR14 3EA
G4 GHM J. Mills 2 Old Vicarage Close, Chilton Polden, Bridgwater TA7 9DY
G4 GHO S. Webb 10, Pilch Close, Norwich NR1 3FU
G4 GHQ P. Fisher 95 Slaithwaite Road, Thornhill Lees, Dewsbury WF12 9DN
G4 GHR D. Humphreys 64 Holne Chase, Plymouth PL6 7UB
G4 GHT M. Skyner 15 Dart Close, Alsager, Stoke-on-Trent ST7 2HY
G4 GHZ P. Collins 17 Tenterden Gardens, London NW4 1TG
GI4 GID J. Heasley 36 Collinbridge Gardens, Newtownabbey BT36 7SU
G4 GIG J. Mullany Flat 3 Michelle Close, Hollybank Road, Birmingham B13 0PR
G4 GIM B. Waters 60 Whitewood Way, Worcester WR5 2LN
GM4 GIO R. Marshall 9 Belford Terrace, Edinburgh EH4 3DQ
G4 GIR I. Frith 50 Rowallan Drive, Bedford MK41 8AS
G4 GIS J. Darbyshire 7 Sandle Road, Bishop's Stortford CM23 5HY
G4 GIX T. Kearns 7 Flitwick Grange, Milford, Godalming GU8 5DN
G4 GIY R. Harris The Willows, Nanny Lane, Church Fenton, Tadcaster LS24 9RL
GW4 GJA K. Austen 6 Caernarvon Grove, Merthyr Tydfil CF48 1JS
G4 GJE D. Davis 6 Regina Drive, Walsall WS4 2HB
GW4 GJI R. Whitley 22 Pen Y Bryn Road, Colwyn Bay LL29 6AF
G4 GJO D. Blampied 113 Green Street, Enfield EN3 7JF
G4 GJR T. Aldridge 180 Tickford Street Milton Keynes, Newport Pagnell MK16 9BG
G4 GJS W. Owens Dorfstrasse 49, Effeld D-41849 Germany
G4 GJU P. Moxham 233 Walsall Road, Aldridge, Walsall WS9 0QA
G4 GJV A. Horne 22 Hedingham Close, London N1 8UA
G4 GJY S. Simmonds 14 Lindsey Crescent, Kenilworth CV8 1FL
G4 GKC C. Willoughby 79 Liskeard Road, Walsall WS5 3ES
GM4 GKH G. Duke 3 Woodlands Grove, Westhill, Inverness IV2 5DU
G4 GKK A. Hawkins 101 Tobyfield Road, Bishops Cleeve, Cheltenham GL52 8NZ
G4 GKT Dr F. Delaney 6 Stour Road, Astley, Manchester M29 7HH
G4 GKU J. Cooper 44 Belvedere Road, Bridlington YO15 3NA
G4 GKX J. Trevett 12 Churchill Road, Blandford Forum DT11 7HH
G4 GKY C. Williams 12A Parc An Dix Lane, Phillack, Hayle TR27 5AB
G4 GKZ R. Revill 102 Hurst Drive, Stretton, Burton-on-Trent DE13 0EE
G4 GLC D. Hamilton Rome Lea, 4 Lane Ends, Settle BD24 0AG
GM4 GLG C. Edwards The Old Mill Ferry Road Sandbank, Dunoon PA23 8QH
G4 GLH D. Bennett Flat 3, Falcon Crag, Cowan Head, Kendal LA8 9HL
G4 GLI M. King 28 Topcliffe Way, Cambridge CB1 8SH
G4 GLM Dr G. Manning 63 The Drive, Edgware HA8 8PS
G4 GLN A. Bellfield 50 Highfield Road, Biggin Hill, Westerham TN16 3UU
G4 GLO A. Bridger 1, Fawn Rise, , Henfield BN5 9EZ
G4 GLP D. Dale-Green 31 Robins Bow, Camberley GU15 3NP
G4 GLQ J. Tysiorowski 52 Meadow Croft, Penrith CA11 8EH
GW4 GLU M. Norbury 16 Pont Aur, Ynyscedwyn Road, Flat, Swansea SA9 1BP
G4 GLW C. Redmayne 20 Kings Road, Accrington BB5 6BS
GM4 GM Clyde Valley DX Group c/o G. Mccallum 15 Quarry Road, Law, Carluke ML8 5HB
G4 GMB D. Hitchins 21 Colwell Court, Newton Aycliffe DL5 7PS
G4 GMI J. Seddon 8 Upper Elms Road, Aldershot GU11 3ET
G4 GMK M. North 10 Long Lane, Pott Shrigley, Macclesfield SK10 5SD
G4 GMN R. Caswell 15 Murtwell Drive, Chigwell IG7 5ED
G4 GMS L. Hicks 108 Northorpe, Thurlby, Bourne PE10 0HZ
G4 GMT A. Aedy 35 Ashlea Avenue, Brighouse HD6 3SR
G4 GMW M. Weaver 22 Greenhill Road, Alveston, Bristol BS35 3LZ
G4 GMZ Prof. J. Alder 104 Park Lane, Congleton CW12 3DE
G4 GNA D. Townend 442 Blackmoorfoot Road, Crosland Moor, Huddersfield HD4 5NS
G4 GND R. Culpan 23 Aldreth Road, Haddenham, Ely CB6 3PP
G4 GNG C. Pemberton 2 Henthorn St., Shaw, Oldham OL2 7AY
GD4 GNH R. Ferguson Moaney Moar House, Corlea Road, Ballasalla IM9 3BA Isle of Man
G4 GNO J. Callaghan Evergreen, Seale Lane, Farnham GU10 1LE
G4 GNP S. Mcgrory The Paddocks, High Street Hook, Goole DN14 5NY
G4 GNQ G. Sims 85 Surrey Street, Glossop SK13 7AJ
GM4 GNR W. Thow 11 St. Marys Place, Ellon AB41 8QW
G4 GNS Lowestoft And Gt Yarmouth Repeater Group c/o S. Henry 28 Marion Avenue, Shepperton TW17 8AY
GI4 GNT J. Taggart Windy Brae, 5 Glasvey Drive, Limavady BT49 9HQ
G4 GNU A. Cross 15 Louise Road, Rayleigh SS6 8LW
G4 GNV S. Jones 12 Yew Tree Close, Yeovil BA20 2PD
G4 GNW T. Hennigan 128 Dimsdale View West, Newcastle ST5 8EL
G4 GNX A. Baker 11 Fairfield Close, Shoreham-by-Sea BN43 6BH
GW4 GNY M. Davies Laburnum House Guilsfield, Welshpool, Powys SY21 9PX
G4 GOA J. Harris 4 Crispin Drive, Malvern WR14 1FB
G4 GOG T. Densham 37 Bovingdon Park, Roman Road, Hereford HR4 7SW
G4 GOJ G. Porter Ye Olde Homestede, High Street, Grimsby DN36 5PL
GI4 GOL G. Brennan 69 Kashmir Road, Belfast BT13 2SB
G4 GOM F. Smith 11 Reed Field, Bamber Bridge, Preston PR5 8HT
G4 GON Dr J. Guest 6 The Tyning, Bath BA2 6AL
G4 GOO M. Kimmitt Old Oaks, Tilston Road, Malpas SY14 7DB
G4 GOP D. Benn 36 Church Avenue, Horsforth, Leeds LS18 5LD
G4 GOR J. Cross 57F Grasmere Road, Blackpool FY1 5HP
GI4 GOS H. Sinclair 43 Edgcumbe Gardens, Belfast BT4 2EH
G4 GOT R. Bradbury-Harrison 11 Derwent Drive, Goring-By-Sea, Worthing BN12 6LA
G4 GOU M. Wilson Reedley Marina, Barden Lane, Burnley BB12 0DX
GI4 GOV Bristol Raynet c/o P. Barr 5 Rosewood Park, Belfast BT6 9RX
GM4 GOW R. Armstrong Lea Cottage, Charleston Village, Forfar DD8 1UF
G4 GOX R. Pearson 33 Livedge Hall Lane, Liversedge, West Yorkshire WF15 7DP
G4 GOZ E. Cockerill 6 Richmond Avenue, Barnoldswick BB18 5JB
GI4 GPA W. Otterson 34 Ashbourne Park, Coleraine BT51 3RE
G4 GPB R. Cooper 17 Cavendish Drive, Claygate, Esher KT10 0QE
GI4 GPC J. Ferguson 7 Lairds Road, Katesbridge, Banbridge BT32 5NN
G4 GPD Eastbourne & Wealden Raynet c/o W. Horn 9 Springwell View, Love Lane, Bodmin PL31 2QP
G4 GPF H. Winwood 16 Brook Lane Hackenthorpe, Sheffield S12 4LF
G4 GPJ N. Bailey 12 Carmarthen Close, Callands, Warrington WA5 9UU
G4 GPL A. Fish 32 Deacons Hill Road, Elstree, Borehamwood WD6 3LH
GM4 GPP C. Auty Valsgarth, Haroldswick, Shetland ZE2 9EF
G4 GPQ T. Stockill 26 Hunters Close, Chatteris PE16 6BD
G4 GPR A. Mills 116 Mays Lane, Barnet EN5 2LS
G4 GPV A. Brown 12 Winstone Gardens, Cirencester GL7 1GJ
G4 GPW B. Ainsworth 23 Cokeham Road, Sompting, Lancing BN15 0AE
G4 GPY S. Edwards 71 St. Leonards Road Molescroft, Beverley HU17 7HP
G4 GQA J. Chmielewski 2 Wolverton Avenue, Kingston upon Thames KT2 7QD
G4 GQE Barnsley And District ARC c/o N. Harris Mere Farmhouse, Matlaske Road, Norwich NR11 7BE
G4 GQL A. Schiffman Flat 48, Oakside Court, Ilford IG6 2PH
GM4 GQM G. Firmin 75 North Road, Lerwick, Shetland ZE1 0PQ
G4 GQP R. Foote 22 Hippings Vale, Oswaldtwistle, Accrington BB5 3LH
G4 GQR Brighton & Dist c/o P. Thompson Flat 4, 14 West End Way, Lancing BN15 8RL
G4 GQS B. Bentley 25 Edinburgh Drive, North Anston, Sheffield S25 4HB
G4 GQV J. Barrett 13 Church Bank, Church, Accrington BB5 4JQ
G4 GQY C. Lee 74 Ilkeston Road, Trowell, Nottingham NG9 3PX
G4 GQZ D. Tweedie 39 Frenchfield Way, Penrith CA11 8TW
GM4 GRC Glenrothes & District ARC c/o D. Francis 2 Morlich Crescent, Dalgety Bay, Dunfermline KY11 9UW
G4 GRG Grajon Radio Group c/o G. Badger 3 Hesketh Close, Cranleigh GU6 7JB
G4 GRJ D. Gower 2 Norview Road, Whitstable CT5 4DN
G4 GRK P. England 2A Firs Close, Cowes PO31 7NF
G4 GRM L. Horton 4 Summer Lane, Walsall WS4 1DS
G4 GRN T. Griffiths 75 Central Avenue, Waltham Cross EN8 7JJ
G4 GRP G. Gardiner 35 Westparkside, Goole DN14 6XN
G4 GRR Dr G. Searle Shalbourne, The Dell, Vernham Dean, Andover SP11 0LF
G4 GRS M. Williams Flat 2, High Point, Highgate N6 4BA
G4 GRT D. Mounter 36 Norwich Road, Watton, Thetford IP25 6DB
G4 GRU D. Jones 36 Moor Lane Woodford, Stockport SK7 1PP
G4 GRZ R. Marsh St2 0Qy Bentilee, Stoke-on-Trent ST2 0QY
G4 GSA P. Milsom 214 Ormonds Close, Bradley Stoke, Bristol BS32 0DZ
G4 GSB M. Hall 35 Bunns Lane, Dudley DY2 7RA
G4 GSC J. Osborne 3 Temple Gardens, Staines TW18 3NQ
G4 GSD A. Watkin 41 Brockwell Lane, Chesterfield S40 4EA
GW4 GSG E. Warner 99 St. Peters Park, Northop, Mold CH7 6YU
GW4 GSH M. Beynon 16 Hardy Close, Barry CF62 9HJ
G4 GSK P. Barnett Dunelm House, Barley Hill, Dunbridge, , Romsey, SO51 0LF
G4 GSL J. Foster 14 Braemar Grove, Heywood OL10 3RR
G4 GSO H. Elliott 40 Dene House Road, Seaham SR7 7BQ
GD4 GSR D. Roberts 7 Knock Rushen, Castletown IM9 1TQ
GW4 GSS R. Bennett Penrhiw Old Rd, Bwlchgwyn, Clwyd LL11 5UH
GI4 GST W. Johnston 3 Glenview, Comber, Newtownards BT23 5HR
G4 GSY M. Bainbridge 21 Cockey Moor Road, Bury BL8 2HD
G4 GSZ K. Court Vereda Escorredor 138, Alicante 3150 Spain
GW4 GTC Rsars - Royal Signals ARS c/o B. Davies Rhosyr, Llanfair Pg LL61 5JB
G4 GTD R. Ford 2 Jersey Avenue, St. Annes, Bristol BS4 4RA
GW4 GTE D. Evans Glendale, Mount Pleasant Road, Buckley CH7 3ET
G4 GTH M. Linda 16 Woodlinken Close, Verwood BH31 6BS
G4 GTN P. Reeve 2 Court Road, Tunbridge Wells TN4 8ED

G4 GTU S. Pocock 202-1969 Oak Bay Avenue, Victoria V8R 1E3 Canada
GM4 GTV N. Mackenzie 57 Countesswells Terrace, Aberdeen AB15 8LQ
G4 GTX W. Craigen 19 Nilverton Avenue, Sunderland SR2 7TS
G4 GTZ M. Phillips 12 Reydon Avenue, Wanstead, London E11 2JD
G4 GUA Dr J. Overton 1 Pigeon House Farm, Pigeon House Lane, Hampshire PO7 5SF
G4 GUC D. Bailey 12 Westbeck, Ruskington, Sleaford NG34 9GU
G4 GUE I. Pope P O Box 662, Durbanville 7551 South Africa
G4 GUG M. Meadows 8 Beeches Park Minchinhampton., Stroud GL6 9BA
GI4 GUH J. Clarke 1 Rathview, Banbridge BT32 4PY
G4 GUK K. Scott-Green 1 Pickwick, Corsham SN13 0JD
GM4 GUL S. Macdonald 5 Lower Glebe, Aberdour, Burntisland KY3 0XJ
G4 GUN G. Le Good 45 Kingsfield Crescent, Witney OX28 2JB
G4 GUO C. Brain 7 Elverlands Close, Ferring, Worthing BN12 5PL
G4 GUQ E. Crawford 28 Mccullogn Drive, Erin, Ontario N0B 1T0 Canada
G4 GUS J. Firmin Warren Cottage, Hill House Road, Norwich NR14 7EE
G4 GUV Dr J. Aindow 2 Cutlers Close, Sydling St. Nicholas, Dorchester DT2 9RG
G4 GUW G. Baggott 105 The Crescent, Walsall WS1 2DA
G4 GUX J. Kuipers 27 Shirley Street, Hove BN3 3WJ
G4 GUY T. Eaves 3 Barons Road, Dousland, Yelverton PL20 6NG
G4 GVE J. Hawkings Church Barn, Dingle Lane, Sandbach CW11 1FY
G4 GVG V. Gormley 24 Beech Road, Garstang, Preston PR3 1FS
GM4 GVJ G. Marshall Drummorlie, Wallyford Toll, Musselburgh EH21 8JT
GM4 GVK I. Munro 57 Craigiebuckler Avenue, Aberdeen AB15 8SF
G4 GVQ S. Eatough 48 Mount Marua Way, Upper Hutt 5018 New Zealand
G4 GVR R. Mason 3 Coronation Close, Hellesdon, Norwich, Nr65Hf, Norwich NR6 5HF
GI4 GVS P. Hallam 95 Belfast Road, Carrickfergus BT38 8BY
G4 GVV S. Fox Flat 3, Woodford House, Aldershot GU11 3EL
G4 GVW P. Gillen 8 Barton Hamlet, Great Barton, Bury St. Edmunds IP31 2PP
G4 GVZ D. Morris 40E Lansdown Crescent, Cheltenham GL50 2NG
G4 GWB I. Gibbs 9 The Square, Choppington NE62 5DA
G4 GWE J. Martin 57 Crescent Road East, Palm Beach, Auckland 1001 New Zealand
G4 GWF H. Haden 1 Bankside Close, Marple Bridge, Stockport SK6 5ET
G4 GWG D. Snape 30 Culcross Avenue, Wigan WN3 6AA
G4 GWH M. Steventon 22 The Beeches, Uppingham LE15 9PG
G4 GWI J. Sheehan 1 Osierground Cottages, Agester Lane, Canterbury CT4 6NP
G4 GWJ J. Butcher Mount Pleasant, Trampers Lane, North Boarhunt, Fareham PO17 6DG
G4 GWP B. Langford Dulce Verano, 29M San Jaime, Alicante 3720 Spain
GD4 GWQ A. Matthewman 26 King Orry Road, , Glen Vine IM44ES Isle of Man
G4 GWR A. Scott-Green 58B High Street, Sutton Benger, Chippenham SN15 4RL
G4 GWT A. Kittle 28 Clare Crescent, Towcester NN12 6QQ
G4 GWU T. Chapman 11 Ash Court, Brampton, Huntingdon PE28 4FH
G4 GWV R. Hookham 50 Billy Mill Avenue, North Shields NE29 0QN
G4 GWX Cotswold Raynet c/o J. Travis 4 Merrial Close, Bakewell DE45 1JB
G4 GWZ R. Whitehead 14 Southgate Crescent, Rodborough, Stroud GL5 3TS
G4 GXB P. Butcher 52 Chandos Road, Rodborough, Stroud GL5 3QZ
G4 GXD D. Travis 1 Hawthorn Close, Whixall, Whitchurch SY13 2ND
G4 GXI P. Pearson 58 Winchester Road, Grantham NG31 8AD
G4 GXK Saltash Dist Ar c/o K. Hale 58 St. Stephens Road, Saltash PL12 4BJ
G4 GXL S. Fletcher 31 Wesley Road, North Wootton, King's Lynn PE30 3XA
G4 GXM R. Corr 15 Waterdell Lane, St. Ippolyts, Hitchin SG4 7RA
G4 GXN Stanford ARC c/o M. Wright 5 Woodview Park, The Donahies, Dublin DUBLIN 13 Ireland
G4 GXO R. Taylor 16 Chestnut Close, Culgaith, Penrith CA10 1QX
G4 GXQ P. Swain 5 Cromley Road High Lane, Stockport SK6 8BP
G4 GXR J. Higginbotham Casas De Cabanes Y Las Fuentes, 117, Villena 3400 Spain
G4 GXW G. Cahill 21 Moresby Close, Westlea, Swindon SN5 7BX
G4 GXY E. Dowlman 4 Beald Way, Ely CB6 3DA
G4 GXZ A. Warrilow Gyse Lodge, Gussage All Saints, Wimborne BH21 5ET
G4 GYA R. Williscroft 91 Parkfield Crescent, Tamworth B77 1HB
G4 GYF G. Hiscoe 1 Greendale Close, Fleetwood FY7 8BQ
G4 GYI P. Ward 23 Ropewalk, Alcester B49 5DD
G4 GYJ R. Littlefield 7 Carron Mead, South Woodham Ferrers, Chelmsford CM3 5GH
G4 GYL M. Denby 13 Hunger Hills Avenue, Horsforth, Leeds LS18 5JS
G4 GYN Dr R. Colson 46 Westwood Drive, Amersham HP6 6RJ
G4 GYP L. Ratcliff 15 Spring Close, Biggleswade SG18 0HL
G4 GYS J. Plested 24 Farm Way, Bushey WD23 3SS
G4 GZA D. Ayris 16 Chapel Lane, Northorpe, Gainsborough DN21 4AF
G4 GZC P. Teanby 34 High Street, Belton, Doncaster DN9 1LR
GM4 GZD G. Smith Ardvourlie, Loaneckheim, Kiltarlity IV4 7JQ
G4 GZG L. Stringer 24 Seymour Drive, Torquay TQ2 8PY
G4 GZH D. Andrew Little Stone House, The Crescent, Steyning BN44 3GD
G4 GZK H. Dalton 24 Church Lane Coven, Wolverhampton WV9 5DE
G4 GZL D. Barker 79 South Parade, Boston PE21 7PN
G4 GZM A. Mcmillan 4 Aluric Rise, Newton Abbot TQ12 4FN
G4 GZN K. Andreang 62 Castleton Avenue, Barnehurst, Bexleyheath DA7 6QU
G4 GZO A. Thurbon 37 Lealand Road, Drayton, Portsmouth PO6 1LZ
GM4 GZQ J. Mcginty 77 Crawford Road, Houston, Johnstone PA6 7DA
G4 GZS K. Wallace 11 Orson Leys, Rugby CV22 5RG
G4 GZT P. Jensen 7 Union Street, Mosman, New South Wales 2088 Australia
G4 GZU R. Woodcock 143 Berry Hill Road, Mansfield NG18 4RT
GM4 GZW Dr E. Simon 12 St. Colme Road, Dalgety Bay, Dunfermline KY11 9LH
GW4 GZX J. Hunter 245 Heathwood Road, Heath, Cardiff CF14 4HS
G4 HAC C. Denscombe High Holme, 4 Kendricks Bank, Shrewsbury SY3 0EX
G4 HAG J. Long 9 Denbrook Avenue, Bradford BD4 0QH
G4 HAI P. Levitt 21 Station Road, Firsby, Spilsby PE23 5PX
G4 HAJ D. Magee 2 Holt Park Vale, Holt Park, Leeds LS16 7QX
G4 HAK P. Torrance 1 Clifton Lawn, Ramsgate CT11 9PB
GM4 HAM Edinburgh And District ARC c/o N. Stewart 160 Carrick Knowe Drive, Edinburgh EH12 7EW
GM4 HAO R. Mackean 10A Dick Place, Edinburgh EH9 2JL
G4 HAP H. Lavin 30 Greenslate Road, Billinge, Wigan WN5 7BG
G4 HAS D. Buck 4687 Bracknell Road, Burlington, Ontario L7M 0E5 Canada
GW4 HAT P. Jones 68 Pastoral Way Sketty, Swansea SA2 9LY
G4 HBA S. Horne Beaucroft, Keswick Road, Benfleet SS7 3HU
G4 HBD P. Trepess 3 Lawford Rise, Wimborne Road, Bournemouth BH9 2BZ
GM4 HBG I. Robertson 53 Carmuir Forth, Lanark ML11 8AP
G4 HBI F. Cassidy 55 High Bank Road, Droylsden, Manchester M43 6FS
GW4 HBK D. Lewis 23 Gelligroes Road, Pontllanfraith, Blackwood NP12 2JU
G4 HBL G. Hardy The Mill House, Thearne, Beverley HU17 0RU
GM4 HBQ A. Taylor 6 Bowling Green St., Methil, Leven KY8 3DH
G4 HBR J. Mcgee 3 Hedgelea Road, East Rainton, Houghton le Spring DH5 9RR
GW4 HBS S. Illidge 24 Maes Briallen, Llandudno LL30 1JJ
G4 HBT M. Foreman 83 Hawthorn Crescent, Yatton, Bristol BS49 4RG
G4 HBV A. Martin 21 Ashwood Way, Hucclecote, Gloucester GL3 3JE
G4 HBY M. Cotton Esterith, 113 Belvedere Road, Burton-on-Trent DE13 0RF
GW4 HBZ B. Clowes 7 Dukesfield Drive, Buckley CH7 3HN
GW4 HCA C. Prentice 1 Victoria Street, Maesteg CF34 0YP
G4 HCB Dr J. Harrison 36 Elmlea Avenue, Bristol BS9 3UU
G4 HCC M. Hodgkinson 34 Pennine Way, Brierfield, Nelson BB9 5DT
G4 HCD A. Reed 28 Russell Street, Sutton-in-Ashfield NG17 4BE
GM4 HCE K. Kirkland 11 Marchfield Park Lane, Edinburgh EH4 5BF
G4 HCG Dr R. Gordon Middle House, 9 Fotheringhay Road, Peterborough PE8 5HP
G4 HCI M. Foreman 39 Artists View Drive, Calgary, Alberta T3Z 3N4 Canada
G4 HCK N. Wilkinson 12 Woodlands Close, Grays RM16 2GB
GI4 HCN J. Clarke 154 Galgorm Road, Ballymena BT42 1DE
GM4 HCO V. Kusin East Overhill Farm, Stewarton, Kilmarnock KA3 5JT
GI4 HCX I. Magill 205 Whitechurch Road, Ballywalter, Newtownards BT22 2LA
GM4 HCY M. Stokes 22 Lothian Road, Jedburgh TD8 6LA
G4 HCZ L. Fellows 19 Grosvenor Road, Lower Gornal, Dudley DY3 2PS
GW4 HDB M. Greatrex 4 Lee Street, St. Thomas, Swansea SA1 8HQ
G4 HDD S. Rose 14 Highgate West Hill, London N6 6JR
G4 HDE S. Green 6 Poveys Mead, Kingsclere, Newbury RG20 5ER
GW4 HDF V. Hill 9 Cae Pant, Caerphilly CF83 2UW
GI4 HDJ B. Mcgarry 43 Umrycam Road, Feeny, Londonderry BT47 4TJ
G4 HDL N. Sedgwick Flat 3, Hartford Court, 33 Filey Road, Scarborough YO11 2TP
G4 HDO A. Kirkland 4 Laurelwood Road, Droitwich WR9 7SE
GW4 HDR A. Evans 4 Elm Grove, Rhyl LL18 3PE
G4 HDS P. Unwin Mycroft, Rochester, Newcastle upon Tyne NE19 1RH
G4 HDU Rvd. B. Keal 46 Eastway, Liverpool L31 6BS
G4 HDY Itchen Valley ARC c/o G. Burgess 44 Clifton Road, Winchester SO22 5BU
GW4 HDZ D. Birch 16 Llanharry Road, Brynsadler, Pontyclun CF72 9DB
G4 HEB P. Tuffs 48 Mackie Drive, Guisborough TS14 6DJ
G4 HEC P. Stracey 14 Portfield Road, Christchurch BH23 2AG
G4 HEE W. Dallas 21 Jubilee Avenue Asfordby, Melton Mowbray LE14 3RY
G4 HEH A. Smith 65 Rowelfield, Luton LU2 9HL
G4 HEJ W. Reid Comphurst, Comphurst Lane, Hailsham BN27 4TX
GW4 HER S. Rogers Green Tops, 69 Megs Lane, Buckley CH7 2AG
G4 HEV G. Cass 18 Rawcliffe Drive, York YO30 6PE
G4 HEW G. Hancock 12122-244Th Street, Maple Ridge BC V4R 1I1 Canada
G4 HFG G. Eckersall 40 Portland Drive, Skegness PE25 1HF
G4 HFI M. Roberts The Willows, Riverside, Hayle TR27 5JD
G4 HFO M. Blythe Trethullan Farmhouse, Sticker, Saint Austell PL26 7EH
G4 HFQ G. Freeth 9 South Avenue, New Milton BH25 6EY
G4 HFS M. Davies 4 Manor Close, Lewknor, Watlington OX49 5BL
G4 HFU P. Spooner The Birches, Wingrave Road, Aston Abbotts, Aylesbury HP22 4LT
G4 HFZ S. Mccann 6 Almond Grove, Scunthorpe DN16 2ES
G4 HGH A. Selmes 35 Windmill Rise, Hundon, Sudbury CO10 8EQ
GW4 HGJ G. Carruthers Henllys Farm, Cardigan SA43 2HR
G4 HGK J. Davis Hurstbourne, Westdown Road, Bexhill-on-Sea TN39 4DY
G4 HGL Dr J. Buckley Sandringham, Neston Road, Ness, Neston CH64 4AT
G4 HGM M. Gregory 1474 Profile Rd, Franconia NH03580 United States
G4 HGN D. Hoyle Pharmacy Cottage, Queen Street, Buxton SK17 8JT
G4 HGR M. Baker 39 The Cherry Orchard, Hadlow, Tonbridge TN11 0HU
GW4 HGS G. Passmore 18 Brickhurst Park, Johnston SA62 3PA
G4 HGT J. Wilkinson 7 Hilton Grange Bramhope, Leeds LS16 9LE
G4 HGV M. Leach 15 Beech Lea, Blunsdon, Swindon SN26 7DE
G4 HHA K. Stalley The Forge, Woodbridge Road, Woodbridge IP12 2JE
GW4 HHD J. Hutchinson 3 Erw Fawr, Henryd, Conwy LL32 8YY
G4 HHH P. Walker East Rigg, Fylingdales, Whitby YO22 4QG
G4 HHJ D. Thomas 64 Marconi Way, St. Albans AL4 0JG
G4 HHL V. Gorny 22 Park Road, Shirehampton, Bristol BS11 0EF
G4 HHM D. Ryder 96 Huttoft Road, Sutton-On-Sea, Mablethorpe LN12 2QZ
G4 HHO Rvd. C. Buckley Curraghmore, Model Farm Road, Co. Cork Ireland
G4 HHS L. May 20 Crescent Road, Marland, Rochdale OL11 3LF
G4 HHX R. Edmonds 14 Singledge Lane, Whitfield, Dover CT16 3EJ
G4 HHZ A. Harwood 55 Nichol Road, Chandler'S Ford, Eastleigh SO53 5AX
G4 HIA M. Nicholls 12 Bents Drive, Sheffield S11 9RP
G4 HIC M. Maddison 34 Maple Avenue, Sandiacre, Nottingham NG10 5EF
G4 HIE M. Hammond 53 Chiltern Road, Baldock SG7 6LT
G4 HIF D. Mallet 41 Kiln Close, Calvert, Buckingham MK18 2FD
G4 HIH R. Wilson 4 Dinmont Place, Hall Close Grange, Cramlington NE23 6DN
G4 HIJ R. Woolley 29 Belle Vue Road, Ashbourne DE6 1AT
G4 HIN R. Twiggs 31 Westlands Avenue, Slough SL1 6AH
G4 HIQ Dr A. Sturman 22 St. Crispins Avenue, Wellingborough NN8 2HT
G4 HIV B. Milne 11 Station Road Thorpe-On-The-Hill, Lincoln LN6 9BS
G4 HIW C. Vernon 2 Standing Butts Close, Walton-On-Trent, Swadlincote DE12 8NJ
G4 HIX P. Duncan 89 Felstead Crescent, Sunderland SR4 0AE
G4 HIZ J. Easdown 38 North Street, Barming, Maidstone ME16 9HF
G4 HJB C. Hall 10 Porlock Court, Cramlington NE23 3TT
G4 HJD A. Goy 352 Chanterlands Avenue, Hull HU5 4ED
G4 HJE S. Small Leydene House, 102 Crestway, Chatham ME5 0BH
G4 HJF W. Dredge 10 Lime Close, Locking, Weston-Super-Mare BS24 8BH
G4 HJH M. Hardaker 6 Caradon Place, Verwood BH31 7PW
G4 HJI J. Bright 18 B Rue De La Station, Aspach-le-Bas 68700 France
GM4 HJK R. Mitchell 9 Pine Way, Perth PH1 1DT
G4 HJL M. Zarattini The Pippins, Orchard Street, Derby DE3 0DF
GM4 HJO M. Mozolowski The Auld Manse, 8 Sandport, Kinross KY13 8DN
GM4 HJQ D. Mackenzie 58 High Street, East Linton EH40 3BH
G4 HJS P. Tempest 15 Charles Avenue, Leeds LS9 0AE
G4 HJT Dr D. Lloyd 39 High Street, 40 Bertrand Drive, Princeton 8540 United States
G4 HJV D. Miller 50 Sandyleaze, Gloucester GL2 0PX
G4 HJW B. Wright 39 High Street, Little Wilbraham, Cambridge CB21 5JY
G4 HJY M. Black 28 Cricketers Close, Chessington KT9 1NL
G4 HKB P. Turner 1 Longridge, Colchester CO4 3FD

UK Callsigns

G4 HKC I. Butson 60 Churnwood Road, Parsons Heath, Colchester CO4 3EY
G4 HKO Thurrock Acorns ARC c/o Dr N. Wilkinson 12 Woodlands Close, Grays RM16 2GB
G4 HKP C. Turner 37 Kabeljou Crescent, Randpark Ridge 2169 South Africa
G4 HKQ C. Marsh 33 Southview Road, Hockley SS5 5DY
G4 HKR A. Reed 85 Ringway, Garforth, Leeds LS25 1BZ
G4 HKS M. Lynch Wessex House, Drake Avenue, Staines-upon-Thames TW18 2AP
GM4 HKV J. Henderson 7 Lumsden Crescent, St. Andrews KY16 9NQ
GW4 HKX R. Rowlands 4 Glascoed, Hermon, Bodorgan LL62 5LF
G4 HKY L. Bower 1 Elmfield Drive, Skelmanthorpe, Huddersfield HD8 9BT
G4 HKZ Holsworthy Community College c/o J. Butcher Mount Pleasant, Trampers Lane, Fareham PO17 6DG
G4 HLA J. Sullivan 1 Godley Hill Road, Hye SK14 3BW
G4 HLB R. Hallam 16 Hall Road, Haconby, Bourne PE10 0UY
G4 HLF P. Westwell 11 Cheshire Park, Warfield, Bracknell RG42 3XA
G4 HLI J. Friend 62 St. Catherines Hill, Bramley, Leeds LS13 2LE
G4 HLL Southampton University Wireless Society c/o C. Willoughby 79 Liskeard Road, Walsall WS5 3ES
G4 HLN L. Bennett 26 Winchester Road, Burnham-on-Sea TA8 1HY
GW4 HLO W. Davies Erw Deg, 11 Madoc Street, Porthmadog LL49 9BU
G4 HLT M. Eckhoff 6 Ramsbury Drive, Earley, Reading RG6 7RT
G4 HLW K. Turnell 31 Greenbank Terrace, Ringstead, Kettering NN14 4DD
G4 HLX Dr N. Taylor 27 Alfredston Place, Wantage OX12 8DL
G4 HLZ M. Wood 48 High Street, Maryport CA15 6BQ
G4 HMA M. Smith 8A Duke Street, Cullompton EX15 1DW
G4 HMC J. Oliver Chalk Lodge, Peters Lane, Monks Risborough, Princes Risborough HP27 0LG
GM4 HML S. Mcluckie 12 Croft Place, Eliburn, Livingston EH54 6RJ
G4 HMM B. Dearing 44 Woodlands Way, Southwater, Horsham RH13 9HZ
GM4 HMN A. Cumming 18 South Covesea Terrace, Lossiemouth IV31 6NA
GW4 HMR D. Morris Hafodty Cottage, Tregarth, Bangor LL57 4NS
G4 HMS Rnars London (Hms Belfast) Group c/o C. Read 58 Somerset Road, Chiswick, London W4 5DN
G4 HMX J. Halliday 16 Ennerdale Drive, Congleton CW12 4FR
G4 HND A. Course 14A Wood Street, Geddington, Kettering NN14 1BG
G4 HNF D. Waterworth 116 Reading Road, Woodley, Reading RG5 3AD
G4 HNG G. Poulton Tresillian Morcombelake, Bridport DT6 6DY
GM4 HNK C. Ferguson Leckuary, Kilmichael Glassary, Lochgilphead PA31 8QL
G4 HNO S. Wilson 8 Hillcrest Avenue, Stockport SK4 3JS
G4 HNQ J. Bryden 32 Jerusalem Road, Skellingthorpe, Lincoln LN6 5TW
G4 HNU P. Vaughan 26 Canterbury Road, Worthing BN13 1AE
G4 HNW S. Walls 11 Copperfield Close, Malton YO17 7YN
G4 HNX E. Beal 49 Ambersham Crescent, East Preston, Littlehampton BN16 1AJ
G4 HNZ S. Bannister 14 Amery Close, Worcester WR5 2HL
G4 HOC M. Oliver 34 Manderley Close, Coventry CV5 7NR
G4 HOD M. Gunby 128 Heath Road, Runcorn WA7 4XL
G4 HOF P. Warrener 139 Louth Road, Holton Le Clay, Grimsby DN36 5AD
G4 HOI W. Skeels 141 Woodward Road, Dagenham RM9 4ST
G4 HOJ P. Hobson High Rising, 4 Dovecote Lane, Lincoln N5 0AD
G4 HOK J. Mckay 2 Bransghyll Terrace, Horton-In-Ribblesdale, Settle BD24 0HG
G4 HOL M. Holden Avda. Jardines Del Almanzora No 62, La Alfoquia De Zurgena, Zurgena 4661 Spain
G4 HOM F. Garratt 90 Brushfield Road, Birmingham B42 2QJ
G4 HON C. Ward 2 Arlington Drive, Stockport SK2 7EB
G4 HOP S. Fordham 61 Cemetery Road, Dronfield S18 1XX
G4 HOR N. Freer 2 Welham Croft, Shirley, Solihull B90 4UU
G4 HOU L. Anstead 21 Tickenor Drive, Finchampstead, Wokingham RG40 4UD
G4 HOW N. Cleaver 18 Old Cleeve, Minehead TA24 6HJ
G4 HOY J. Fennell Bajamar House, Belton Road, Doncaster DN9 1JL
GD4 HOZ D. Osborn Kionlough House, Kionlough Lane, Bride IM7 4AG Isle of Man
G4 HPA Hertfordshire Peak Assault (Scouts) c/o M. Wood 26 Parkfield Crescent Kimpton, Hitchin SG4 8EQ
G4 HPB R. Wilden 48A The Crescent, Cradley Heath B64 7JS
G4 HPD B. Constable Dukes Pleasure Long Headland, Ombersley, Droitwich WR9 0DX
G4 HPE S. Richards 6 Heathfield, Royston SG8 5BW
G4 HPH J. Littler 363 Atherton Road, Hindley, Wigan WN2 3XD
GM4 HPK D. Moore Rashfield Farm By Kilmun, Dunoon PA23 8QT
G4 HPN R. Baker Shadow Lodge, Upper Princes Road, Freshwater PO40 9EF
G4 HPS P. Barker 11 Dipton Gardens, Sunderland SR3 1AN
G4 HPT D. Oliver Ashdell, Newlands Lane, Birmingham B37 7EE
G4 HPX J. Trotter 29 Broad Park Road, Bere Alston, Yelverton PL20 7AH
G4 HPY R. Spragg 3 Truro Gardens, Luton LU3 2AP
G4 HQB P. Sandell 1 St. Margaret Road, Ludlow SY8 1XN
G4 HQC C. Wilcox 42 Kentmere Close, Cheltenham GL51 3PD
G4 HQD R. Bagley 8 Bishop Ruzar Furrugia Street, Xaghra, Xra 103 Malta
GM4 HQF D. Lindsay 39 Seamount Court, Aberdeen AB25 1DQ
G4 HQH S. Parker 20 Swaddale Avenue, Chesterfield S41 0SU
G4 HQM D. Waspe 28 Wilman Way, Salisbury SP2 8QS
GM4 HQU N. Gent 4 Eskview Villas, Eskbank, Dalkeith EH22 3BN
G4 HQX P. Morys 41 Salter Street, Berkeley GL13 9BU
GM4 HQZ A. Morrison Block 19, 2 Sandpiper Road., Edinburgh EH6 4TR
G4 HRB D. Taylor 8 Fambridge Close, Maldon CM9 6DJ
G4 HRC Havering & District ARC c/o D. Nuttall 92 Long Road, Lowestoft NR33 9DH
G4 HRE D. Hollow 8 Vermont Woods, Finchampstead, Wokingham RG40 4PF
G4 HRG R. Denley 50 Cranmere Avenue, Wergs, Wolverhampton WV68TS
G4 HRH A. Allen The Hollies Sedgeford, Whitchurch SY13 1EX
GM4 HRJ J. Mcniff East Cove Cottage, Main Road, Port Glasgow PA14 6XP
GM4 HRL A. Sergeant 24 Academy Road, Bo'ness EH51 9QD
G4 HRS Horsham ARC c/o J. Matthews 46 Park Lane West Grinstead, Horsham RH13 8LT
G4 HRU R. Profitt 10 Taunton Vale, Hunters Hill, Guisborough TS14 7NB
G4 HRY D. Farn 14 Corfe Close, Coventry CV2 2JG
G4 HS S. Hopper 16 Stanford Avenue, Hassocks BN6 8JL
G4 HSB P. Rovardi 8 Cambridge Road, Linthorpe, Middlesbrough TS5 5NQ
G4 HSC H. Hughes 16 Dalton Drive, Goose Green, Wigan WN3 6TQ
G4 HSD R. Smithers 16 Derby Road, Sutton SM1 2BL
GW4 HSH R. Williams 114 West Cross Lane, West Cross, Swansea SA3 5NQ
G4 HSK S. Glass 36 Pickwick Avenue, Chelmsford CM1 4UN
G4 HSM R. Hurrell 97 Dovercliffe Close Se, Calgary, Alberta T2B 1W4 Canada
G4 HSN A. Chorley Leycot, Cornells Lane, Saffron Walden CB11 3SP
G4 HSS P. Forshaw 54 The Park, Penketh, Warrington WA5 2SG
GJ4 HSW F. Le Quesne Brookhill House, Princes Tower Road, St. Saviour JE2 7UD Jersey
G4 HSX F. Cole 3 Wadsworth Avenue, Todmorden OL14 7NF
G4 HSZ P. Thacker 23 Lulworth Avenue, Leeds LS15 8LW
G4 HTB T. Rance 2 Glenavon Gardens, Slough SL3 7HN
G4 HTD L. Mason Forest Farm, Folly Drove, Stewley, Ashill, Ilminster TA19 9NW
G4 HTE E. Sergeant 13 Morven Close, Potters Bar EN6 5HE
G4 HTG A. Brunton 409 Outwood Common Road, Billericay CM11 1ET
G4 HTH R. Herringshaw 35 Oxley Close, Shepshed, Loughborough LE12 9LS
G4 HTL A. Mcculloch 14 Harbour Close, Blouberg Sands, Cape Town 7441 South Africa
G4 HTO I. Myford 33A Brick Kiln Lane, Mansfield NG18 5LA
GM4 HTU A. Langton 71 Gray Street, Aberdeen AB10 6JD
G4 HTV Itv West RC c/o R. Thompson Walnut House, Chestnut Walk Saltford, Bristol BS31 3BG
G4 HTW P. Mcveigh The Dale, Bowns Hill, Matlock DE4 5DG
G4 HTX R. Houghton Elmtrees, Church End, Bedford MK44 2RP
G4 HTY D. Stokes Flat 6, 35-37 Gratton Road, London W14 0JX
G4 HTZ S. Barrett 1 The Street, Ashen CO10 8JN
G4 HUA T. Ellam 3115 Carleton Street Sw, Calgary AB T2T3L5 Canada
G4 HUD J. Bramall 55 Wood Lane, Louth LN11 8RY
G4 HUE A. Nehan Danisway, Queens Road, Colmworth, Colmworth MK44 2LA
G4 HUF P. Baguley 16 Churchill Road, Broadheath, Altrincham WA14 5LT
G4 HUG W. Daniels 48 Mellanear Road, Hayle TR27 4QT
G4 HUH P. Chapman 1291 Los Amigos Avenue, California 93065 United States
GM4 HUL W. Savory 20 Broomfield, Carradale East, Campbeltown PA28 6RZ
G4 HUM D. Hazzard 34 Chessel Avenue, Bitterne, Southampton SO19 4DX
G4 HUN N. Whiteside 17 Shibleys Court, Fishers Lane, Norwich NR2 1EE
G4 HUO M. Bennett 9 Lavender Avenue, Blythe Bridge, Stoke-on-Trent ST11 9RN
G4 HUQ M. Crake 12 Bosburn Drive, Mellor Brook, Blackburn BB2 7PA
G4 HUT D. Consitt Saxtorpsvagen 210, Landskrona 26194 Sweden
G4 HUW S. Faulkner Vaarveien 8, Oslo 1182 Norway
GM4 HUX R. Lindsay 32A James Street, Alva FK12 5AL
GU4 HUY R. Sarre Le Clercs, Clos Du Murier, Rue De Bas, Guernsey GY2 4HJ
G4 HVC A. Kiddle 19 Old Lincoln Road, Caythorpe, Caythorpe, Grantham NG32 3DF
G4 HVF C. Bracewell Roseville Yoredale Avenue, Leyburn DL8 5BH
G4 HVG J. Phipps 5 Akeman Close, St. Albans AL3 4NJ
GI4 HVI A. Hamilton 11 Norwell Park, Castlerock, Coleraine BT51 4TS
GM4 HVM A. Douglas 24 Plane Grove, Dunfermline KY11 8RA
G4 HVN Powys ARC c/o R. Morby The Rectory, High Street, Edgmond, Newport TF10 8JR
G4 HVO J. Fitzwater The Olde Cottage, Babylon Lane, Tadworth KT20 6XE
G4 HVR G. Southwell 4A Neve Avenue, Wolverhampton WV10 9BU
GM4 HVS Dr R. Teperek 8 Forest Park, Stonehaven AB39 2GF
G4 HVT N. Wilkinson Breidablikkbakken 15, Porsgrunn 3911 Norway
G4 HVV Lswc Portable Group c/o C. Goadby Heligan, 12 School Road, Newmarket CB8 9RX
G4 HWA B. Morton Yew Tree House, 14, Baker Street, Gayton NN7 3EZ
G4 HWC E. King 2 Thornton Road, March PE15 8SH
G4 HWF R. Rudd 69 Stanford Avenue, Brighton BN1 6FB
G4 HWH A. Jandrell 21 Wildacres, Stourbridge DY8 3PH
G4 HWI M. Allin 50 Swallow Rise, Knaphill, Woking GU21 2LH
G4 HWJ M. Dawson Mulberry Cottage, The Hamlet, Ely CB6 1SB
G4 HWK F. Pilling 51 Lynton Road, Chingford E4 9EA
G4 HWM Dr D. Jeffery 14 Beechwood Crescent, Chandler'S Ford, Eastleigh SO53 5PA
G4 HWN R. Heath Flat 172, Hagley Road Retirement Village, 330 Hagley Road, Birmingham B17 8BN
GM4 HWO C. Wright 3 Stanedykehead, Liberton, Edinburgh EH16 6YE
G4 HWV T. Wiles Manor Farm, Manor Close, Middlesbrough TS9 5AG
G4 HWW R. Scott Flat 57 Tatton Cour, 35 Derby Rd, Stockport SK4 4NL
G4 HXC D. Edwards 179 Pallett Drive, Nuneaton CV11 6JA
G4 HXE A. Tilbee 9 Moorhead Court, Southampton SO14 3GQ
G4 HXH R. Pope 95 Northolt Avenue, Bishop's Stortford CM23 5DS
G4 HXK F. Rendell 64 Rivermead, Stalham, Norwich NR12 9PJ
G4 HXL Dr L. Manderson 16 Archery Avenue Foulridge, Colne BB8 7NH
G4 HXN D. Kelly 27 Keswick Road, Bookham, Leatherhead KT23 4BQ
GW4 HXO M. Probert 1 Ynys Dawel, Solva, Haverfordwest SA62 6UF
G4 HXQ G. Burlington Podgwell Cottage Seven Leaze Lane, Edge GL6 6NJ
G4 HXU D. Mcdermott 6 Chiltern Grove, Thame OX9 3NH
G4 HXX Belvoir Vale Ar c/o Dr C. Dollery 101 Corringham Road, London NW11 7DL
G4 HXY S. Simmons 48 Copland Road, Stanford-le-Hope SS17 0DF
G4 HYD Capt. A. Oakley 2 Manor Close, Beverley HU17 7BP
GM4 HYG C. Moulding Moorside Mosstowie, Elgin IV30 8TT
GI4 HYM C. Gill 11 Quay Street, Ardglass, Downpatrick BT30 7SA
GM4 HYR North Wiltshire Raynet Group c/o M. Bond 1 Saughtonhall Crescent, Edinburgh EH12 5RF
G4 HYT Dr P. Kurian 22A Lindisfarne Avenue, Blackburn BB2 3EH
G4 HYW A. Wilkes 48 Westfield Road, Lymington SO41 3QA
G4 HYY T. Jackson 33 Highcroft Road, Todmorden OL14 5LZ
GW4 HYZ B. Green 28 Sunnybank Road, Griffithstown, Pontypool NP4 5LT
G4 HZE E. Hill 14 Station Road, Saltash PL12 4DY
G4 HZF R. Scarlett 1 St. Martins Crescent, Grimsby DN33 1BG
G4 HZG M. White 3 High Street, Clay Cross, Chesterfield S45 9DX
GW4 HZH Dr D. Doherty 35 Ffordd Bryngwyn, Garden Village, Gorseinon, Swansea SA4 4EB
G4 HZI W. Backhouse 191 Wigmore Road, Gillingham ME8 0TL
G4 HZJ Iom ARS(Iomars) c/o L. Jackson 1 Belvedere Avenue, Atherton, Manchester M46 9LQ
GW4 HZM J. Styles 5 Heol-Y-Berth, Caerphilly CF83 1SP
G4 HZN T. Lockwood 8 St. Nicholas Road, Thorne, Doncaster DN8 5BS
G4 HZP A. Charlton11 5 Honeywood Lane, Carlisle CA2 6DD
G4 HZT T. Morton 3 Grandstand Road, Hereford HR4 9NE
G4 HZV R. Bagwell 30 Christmas Pie Avenue Normandy, Guildford GU3 2EN
G4 HZW A. Usher 14 Bucklow Avenue, Mobberley, Knutsford WA16 7ET
G4 HZX N. Squibb 127 Copers Cope Road, Beckenham BR3 1NY

G4 IAB A. Bell 10 Long Acre, Weaverham, Northwich CW8 3PT
G4 IAD D. Crompton The Beeches, 6 St. Johns Wood, Bolton BL6 4FA
G4 IAG T. Court Woodview Breach Oak Lane, Coventry CV7 8AU
G4 IAJ T. Jefferson Garden Flat, 5 Esplanade, Scarborough YO11 2AF
G4 IAO A. Robertson 7 Big Back Lane, Chedgrave, Norwich NR14 6BH
G4 IAQ J. Brooks 28 Avon Vale Road, Loughborough LE11 2AA
G4 IAR D. Brooks 28 Avon Vale Road, Loughborough LE11 2AA
G4 IAT B. Smith 69 Birch Hall Avenue, Darwen BB3 0JW
G4 IAU D. Lilley 65 Peel St., Horbury, Wakefield WF4 5AN
G4 IAY F. Whittaker 91 Oakdale, Worsbrough, Barnsley S70 5NR
G4 IBC Weymouth And District Short Wave Club c/o K. Marsh Highgrove, Creech Heathfield, Taunton TA3 5EW
G4 IBH D. Dockery 20 Saffron Way, Sittingbourne ME10 2EY
GM4 IBI Dr W. Mitchell Brownhill Of Ardo, Methlick, Ellon AB41 7HS
G4 IBM C. Murphy 15 Loders Close, Poole BH17 9BF
G4 IBN K. Pointon 65 Gypsy Lane, Castleford WF10 3PB
G4 IBS G. Baxendale Sarno, Granville Road, Darwen BB3 2SS
GI4 IBV S. Johnston 61 Ravenhill Park, Belfast BT6 0DG
G4 IBW R. Ropinski 38 The Leys, Little Eaton, Derby DE21 5AR
G4 ICB B. Clarke 59 Baden Powell Crescent, Pontefract WF8 3QD
G4 ICC M. Gater 17 Douglas Road, Northampton NN5 6XX
G4 ICE A. Mitchell 11 Poplar Lane, Cannock WS11 1NQ
G4 ICF A. Denison 40 Leysholme Drive, Leeds LS12 4HQ
G4 ICH C. Wickenden Chalfont, Little Whelnetham, Bury St. Edmunds IP30 0DG
G4 ICI R. Perks Drayton Lodge, Drayton Manor Drive, Tamworth B78 3TJ
G4 ICM Mexborough & District ARS c/o D. Stockley 2 The Ridings Chestfield, Whitstable CT5 3PE
G4 ICP R. Witney 36 Dapifer Drive, Braintree CM7 3LG
G4 ICU A. Jones 15 High Street, Sedgley, Dudley DY3 1RL
G4 ICZ B. Greatrix 12 Swainsfield Road, Yoxall, Burton-on-Trent DE13 8PT
GW4 IDC M. Rudge 8 Penrallt Estate, Llanystumdwy, Criccieth LL52 0SR
G4 IDD D. Dockar 49 Dixon Lane, Wortley, Leeds LS12 4RR
G4 IDF D. Hobro 60 Linksview Crescent, Worcester WR5 1JJ
G4 IDG G. Tonge 6 Bickford Close, Lapley, Stafford ST19 9JZ
G4 IDH I. Harris 47D Tower 2 Queens Terrace, 1 Queen Street, Sheung Wan 12345 Hong Kong
G4 IDJ D. Macgregor 29 Terrington Hill, Marlow SL7 2RE
G4 IDL T. Wade 47 Rig Drive, Swinton, Mexborough S64 8UL
G4 IDR D. Redman 13 Halifax Road, Golcar, Huddersfield HD7 4NS
G4 IDS I. Stanley 11 Reading Road, Burghfield Common RG7 3PY
G4 IDT F. Heywood 62 Southleigh Road, Leeds LS11 5SG
G4 IDU K. Kniveton 32 Minster Avenue, Bude EX23 8RY
GW4 IDV P. Brown 3 Lon Llewelyn, Abergele LL22 7DG
G4 IDW A. Compton Aysgarth, Durley Brook Road, Southampton SO32 2AR
G4 IDX D. Turner 36 Joyes Road, Folkestone CT19 6NX
G4 IEB C. Williamson 72 Granville Drive, Kingswinford DY6 8LL
G4 IEC A. Everard 2 Oak Wood Road, Wetherby LS22 7QY
GM4 IEF A. Hancock Pitlair House Nursing Home, Cupar KY15 5RF
G4 IEG C. Shearer 2 Perigrine Close, Basildon SS16 5HX
G4 IEH S. Lindell 60 Lakenheath, Oakwood, London N14 4RP
G4 IES W. Pitt 1 Windy Ridge, James Street, Stourbridge DY7 6ED
G4 IET J. French 10 Sunridge Avenue, Luton LU2 7JL
GW4 IEU W. Griffiths 3 Garreglwyd Park, Holyhead LL65 1NW
G4 IEV P. Gill 48 Meeting House Lane, Balsall Common, Coventry CV7 7FX
GW4 IEZ R. Senior 5 Cwm Arthur, Denbigh LL16 4BD
GW4 IFE Dr A. Strachan 1 Cornelius Close, South Cornelly, Bridgend CF33 4RQ
G4 IFI C. Loftus C/O, 15 Chappell Road, Manchester M43 7UQ
G4 IFJ M. Daniels 8 Hathersage Drive, Glossop SK13 8RG
G4 IFM Dr S. Petraitis 16 Brookbank Road, Dudley DY3 2RX
G4 IFQ A. Webb 15 Windsor Mead, Sidford, Sidmouth EX10 9SJ
G4 IFR P. Hanson 42 Oak Avenue, Newport TF10 7EF
G4 IFX Dr C. Deacon Spring Valley, Churt Road, Farnham GU10 2QU
G4 IGC L. Hall 57 Station Hill, Swannington, Coalville LE67 8RJ
GW4 IGF P. Higgs Oulton, Daisy Lane, Parkside, Rossett, , Wrexham LL12 0BP
G4 IGG N. Bennett 1 Burnham Avenue, Oxley, Wolverhampton WV10 6DX
G4 IGK M. Wickham 8 Verlands Close Niton, Ventnor PO38 2BG
G4 IGL R. Coombes 9 Beechwood Close, Evington, Leicester LE5 6SY
GM4 IGS R. Chapman 65 Lochgreen Avenue, Troon KA10 6UP
G4 IGT R. Roberts 37 Admirals Place, Gibraltar GX111AA Gibraltar
G4 IGU K. Blackett 46 Lansdown, Yate, Bristol BS37 4LR
GD4 IHC R. Furness Breryk, Windsor Road, Ramsey IM8 3EB Isle of Man
G4 IHI P. Ferrari Maggie, Back Road, Halesworth IP19 9DY
GW4 IHM I. Wingfield Keyhaven, 2 Belmont Close, Abergavenny NP7 5HW
G4 IHO D. Carson 21 Harris Road, Harpur Hill, Buxton SK17 9JS
G4 IHR N. Allen 8 Shoulbard, Fleckney, Leicester LE8 8TX
G4 IHS G. Donn Flat 31, Rex Cohen Court, Liverpool L17 1AB
G4 IHT R. Riddington Beech House, Tetbury GL8 8SN
GI4 IHY R. Clarkson 2 Massereene Gardens, Antrim BT41 4JQ
G4 IHZ M. Hyde 23 Northumberland Way, Ardsley, Barnsley S71 5DH
G4 IIA M. Stamford 1 Canon Drive, Norton Canon, Hereford HR4 7BJ
G4 IIB K. Marshall Alderbaran, Ruckcroft, Carlisle CA4 9QR
G4 IIC C. Clifford 11 Halfcot Avenue, Stourbridge DY9 0YB
G4 IID C. Eastland 40 Hillside Road, Bushey WD23 2HA
G4 IIH P. Henson 70 Mell Road, Tollesbury, Maldon CM9 8SR
G4 III P. Godwin Holgate, Selby Road, Goole DN14 0LN
G4 IIK C. Lodge 35 Beaumont Cottages, Kelsale, Saxmundham IP17 2NW
G4 IIN N. Evans 56 Homerton Road, Middlesbrough TS3 8LX
G4 IIO P. Howe 59 Days Road, Samford Valley 4520 Australia
GM4 IIR A. Nelson 5 Scarletmuir, Lanark ML11 7PS
G4 IIX C. Wherrett 14 Sails Drive, York YO10 3LR
G4 IIY I. Fugler (Fugler), Lees Hill Farm, Lees Hill, Brampton CA8 2BB
G4 IJA B. Barnes 28 Oaklands Park, Roughton Moor, Woodhall Spa LN10 6UU
G4 IJB R. Butterworth 3 Derriman Glen, Sheffield S11 9LQ
G4 IJD J. Seddon 38 Kemple View, Clitheroe BB7 2QD
GU4 IJF Dr N. Roberts Maison Du Cotil, Alderney GY9 3YZ Guernsey
G4 IJI M. Walker 19 Highbury Place, Headingley, Leeds LS6 4HD
G4 IJJ A. Spratt 8 Pheasant Rise, Copdock, Ipswich IP8 3LF
G4 IJM I. Arnold 44 Elwick Avenue, Acklam, Middlesbrough TS5 8NT
G4 IJO G. Gaunt 7 Marine Parade, Saltburn-by-the-Sea TS12 1DP
G4 IJR B. Moyse 1703 Twin Pond Circle, College Station, Texas 77845-3051 United States
G4 IJU J. Coles 84 Mansfield Lane, Calverton, Nottingham NG14 6HL
G4 IJV B. Dowling Box Cottage, Box, Stroud GL6 9HB
GI4 IKF T. Black 147 Old Westland Road, Belfast BT14 6TE
G4 IKI P. Gabriel 20 Barge Lane, Ryde PO334LB
G4 IKJ P. Edwards 34 Albion Road, Malvern Link, Malvern WR14 1PU
G4 IKL R. Hibbin 2 Phoenix Close, West Wickham BR4 0TA
G4 IKQ R. Kitchener 43 Haven Close, Swanley BR8 7JY
G4 IKX D. Thomas 18A Stockwell Lane, Aylburton, Lydney GL15 6DN
G4 IKY D. Sillars 34 Sandown Road, Stevenage SG1 5SF
G4 ILA Rvd. W. Mckae 3 Grantham Close, Wirral CH61 8SU
GM4 ILE J. Smy 2 Dungavel Gardens, Hamilton ML3 7PE
G4 ILF A. Hyde 68 Broxburn Road, Warminster BA12 8EZ
G4 ILH J. Acott 2 Park Hill Road, Sidcup DA15 7NL
G4 ILI G. Cratchley 2 The Maples, The Reddings, Cheltenham GL51 6RW
G4 ILM Dr M. Turnbull Southlea, Newbury, Gillingham SP8 4QJ
G4 ILN Phase Array DX Group c/o G. Fitt 15 Sidegate Avenue, Ipswich IP4 4JJ
G4 ILP C. Borkowski 25 Stroud Road, Wimbledon, London SW19 8DQ
G4 ILR C. Howett 38 Hillyfields, Dunstable LU6 3NS
GM4 ILS R. Adam 1 Woodlands Crescent, Bishopmill, Elgin IV30 4LY
G4 ILT G. Barnacle 58 Cotley Road, Leicester LE4 2LH
G4 ILW J. Dingwall 10 Loweswater Road, Gateshead NE9 6TN
G4 ILX S. Sliwinski 9 Oakhill Road, Sheffield S7 1SJ
GI4 ILZ W. Sharpe 22 Tweskard Park, Belfast BT4 2JZ
G4 IMB P. Gascoigne 108 Blandford Avenue, Castle Bromwich, Birmingham B36 9JD
GW4 IMC T. Waters 34 Woodlands Park, Betws, Ammanford SA18 2HF
G4 IMH V. Tatman 271 London Road, Bedford MK42 0PX
G4 IML M. Giles-Holmes 177B Babbacombe Road, Torquay TQ1 3SU
G4 IMM R. Steele 23 Peacocks Close, Cavendish, Sudbury CO10 8DA
G4 IMP A. Phillpott Southways, Stombers Lane Hawkinge, Folkestone CT18 7AP
G4 IMS J. Roe 5 Lawford Lane, Writtle, Chelmsford CM1 3EA
G4 IMU K. Holley 18 Sandford Avenue, Loughton IG10 2AJ
G4 IMV J. Mollart 8 Harrison St., Newcastle ST5 1NH
G4 INA P. Grice 48 Repington Road, Tamworth B77 4AA
G4 INB Dr B. Dupree 3 Hillary Road, Cheltenham GL53 9LB
G4 INF B. Walpole Bridge Farm, Stony Lane, Exeter EX5 1PP
G4 ING J. Hartley 50 Waverley Road, Hyde SK14 5AU
G4 INI J. Church Belle Vue, Gas Lane, Torrington EX38 7BE
G4 INU F. Haighton 2028 Cheviot Court, Burlington, on L7P 1W8 Canada
G4 INX A. Harada 3 Bazzleways Close, Milborne Port, Sherborne DT9 5FD
G4 IOA P. Hill 135 Village Road, Cheltenham GL51 0AE
GM4 IOB R. Smith Hestivald, Downies Lane, Stromness KW16 3EP
G4 IOD W. Marshall 92 High Street, Ossett WF5 9RQ
G4 IOE F. Stevenson Raakollveien 20A, Rolvsoy N-1663 Norway
G4 IOG North Norfolk Raynet c/o J. Blackett 70 Church Lane, Newington, Sittingbourne ME9 7JU
G4 IOJ M. Fielding 35 Windmill Grove, Fareham PO16 9HP
G4 IOK C. Marshall 100 Hailey Road, Witney OX28 1HQ
GD4 IOM Isle of Man ARS c/o M. Webb Coastguard House, 1 Mount Morrison, Peel IM5 1PN Isle of Man
G4 ION Horndean And District ARC c/o Dr E. Warrington Dept Of Engineering, University Of Leicester, Leicester LE1 7RH
GI4 IOO R. Chambers 32 Victoria Road, Sydenham, Belfast BT4 1QU
GW4 IOQ A. White Wddyn Cottage, Treflach, Oswestry SY10 9HQ
G4 IOV P. Emmerton 5 Portsmouth Wood Close, Lindfield, Haywards Heath RH16 2DQ
GM4 IPA International Police Association c/o J. Bertram 20 Kyles View, Largs KA30 9ET
G4 IPB P. Hodgkinson Woodedge, Snaisgill Road, Middleton-In-Teesdale, Barnard Castle DL12 0RP
G4 IPE R. Wilson Brookside, Stather Road, Burton-Upon-Stather, Scunthorpe DN15 9DH
G4 IPF L. Horseman 55 Sackville Avenue, Hayes, Bromley BR2 7JS
G4 IPH R. Bass 292 Thornhills Lane, Clifton, Brighouse HD6 4JQ
G4 IPI D. Foster 1 Thorn Court, Four Marks, Alton GU34 5BY
GM4 IPK A. Steven Pangdene, Virkie, Shetland ZE3 9JS
G4 IPL L. Winters 58 Larkhall Lane, Harpole, Northampton NN7 4DP
G4 IPM N. Terry 15 Baldwins Close, Bourn, Cambridge CB23 2TH
G4 IPN W. Flindall 3 Meadow Drive, Gressenhall, Dereham NR20 4LR
G4 IPR T. Jones 130 Turkey Street, Enfield EN1 4PS
G4 IPV G. Mayne 228 Tutbury Road, Burton-on-Trent DE13 0NY
G4 IPY Dr A. White 3 Guarlford Road, Malvern WR14 3QW
GW4 IQA R. Lloyd Llwyn Celyn, Pandy, Gwent NP7 8DN
GW4 IQB D. Fuller 9 Llwyn Onn, Croesyceiliog, Cwmbran NP44 2AL
G4 IQD N. Sivapragasam 1 Treve Avenue, Harrow HA1 4AL
G4 IQF S. Wilkinson 41, Church Road, Nailstone CV13 0QH
G4 IQJ P. Brannon 90 Jacksmere Lane, Scarisbrick, Ormskirk L40 9RS
G4 IQK G. Evans 14 Beach Priory Gardens, Southport PR8 1RT
G4 IQO C. Britton 271 Havant Road Farlington, Portsmouth PO6 1DB
G4 IQQ R. Phillips 2 The Close, Dartford DA2 7ES
G4 IQR N. Troop 8 Fox Green, Great Bradley, Newmarket CB8 9NR
G4 IQV G. Menzies 40 Epsom Lane North, Epsom KT18 5PY
G4 IQW N.W.R.R.C. c/o A. Langford 42 Amis Way, Stratford-upon-Avon CV37 7JF
G4 IQZ J. Long 51 Bratton Road, Westbury BA13 3ES
G4 IRC Ipswich Rad Club c/o J. Gee Windmill Lodge, Mill Lane Witnesham, Ipswich IP6 9HR
G4 IRD R. Richards 11 Purvis Road, Rushden NN10 9QA
G4 IRG E. Turner 9 Wallingford Road, Handforth, Wilmslow SK9 3JT
G4 IRH Wythall CG c/o T. Pendleton 17A Langley Drive, Kegworth, Derby DE74 2DN
G4 IRP F. Boocock 109 Northumberland Road, Harrow HA2 7RB
G4 IRS R. Ball 1 Mount Hindrance Close, Chard TA20 1DZ
G4 IRU N. Ashcroft Oaklands, 11 Greenway, Wilmslow SK9 1LU
G4 IRV J. Hastie 13 Thornlands, Easingwold, York YO61 3QQ
G4 IRY R. Gladden 145A Hampton Road, South Fremantle WA 6162 Australia
GI4 ISH M. Fearis 205 Dunluce Avenue, Belfast BT9 7AX
GW4 ISJ P. Martin 42 Leckwith Avenue, Cardiff CF11 8HQ
G4 ISK D. Brighton 39 Les Forets, Glenac 56200 France

GM4 ISM M. Hughes 6 Hawthorn Gardens, Larkhall ML9 2TD
G4 ISN A. Holmes 5 Launde Park, Market Harborough LE16 8BH
GI4 ISR C. Mcclurg 4 Gracefield Lodge, Dollingstown, Craigavon BT66 7UA
G4 ISS J. Proudfoot Laburnum Cottage, Corby Hill, Carlisle CA4 8PL
G4 ISU N. Whittingham The Lilacs, 4 Ridgedale Mount, Pontefract WF8 1SB
G4 ITB J. Stone 35 Landseer Avenue, Chapel St. Leonards, Skegness PE24 5QZ
G4 ITC C. Claydon 69 Abingdon Road, Dorchester-On-Thames, Wallingford OX10 7LB
G4 ITG B. Davey 31 Somervell Drive, Fareham PO16 7QL
GW4 ITJ C. Hard 3 Longbridge, Ponthir, Newport NP18 1GT
G4 ITP C. Owen 334 Beaumont Leys Lane, Leicester LE4 2BJ
GW4 ITQ B. Lindley 4 Stryd Y Brython, Ruthin LL15 1JA
G4 ITR K. Fisher 51 Edge Hill, Ponteland, Newcastle upon Tyne NE20 9RR
G4 ITV B. Dingle 74 Fenay Lane, Almondbury, Huddersfield HD5 8UJ
G4 ITX M. Payne 34 Thales Drive, Arnold, Nottingham NG5 7NF
G4 ITY D. Hardie 42 Lagoon Road, Pagham, Bognor Regis PO21 4TJ
G4 IUA J. Campbell 61 Telegraph Lane, Claygate, Esher KT10 0DT
G4 IUF M. Parker 23 Pannal Avenue Pannal, Harrogate HG3 1JR
G4 IUH R. Pye 7 Meadow View, Potterspury, Towcester NN12 7PH
G4 IUJ J. Wroe 25 Yew Tree Lane, Poynton, Stockport SK12 1PU
GW4 IUK H. Morley 63 Lewis Road, Neath SA11 1DJ
GW4 IUL D. Pullin 32 Clinton Road, Penarth CF64 3JD
G4 IUM G. Adams-Spink 55 Hawthorn Drive, Harrow HA2 7NU
GW4 IUN R. Janes 3 Greenway Avenue, Rumney, Cardiff CF3 3HQ
G4 IUP R. Limbert 21 Staincliffe Drive, Keighley BD22 6FF
GM4 IUS N. Bethune 9 Links Gardens, Leith, Edinburgh EH6 7JH
G4 IVC F. Wood 20A Lynwood Avenue, Felixstowe IP11 9HS
G4 IVD Rvd. A. James 6 Dovedale Close, Hardwicke, Gloucester GL2 4JH
GI4 IVI A. Kerr 29 The Rose Garden, Tandragee, Craigavon BT62 2NJ
G4 IVL Wythall CG c/o T. King Flat 1, 159 Cheriton Road, Folkestone CT19 5HG
G4 IVO R. Hargreaves 23 Bracken Road, Long Eaton, Nottingham NG10 4DA
G4 IVU A. Dixon 98 Seaview Chalet Park, Green Lane, Kessingland, Lowestoft NR33 7RG
G4 IVZ G. Harper 12 Bletchley Road, Stewkley, Leighton Buzzard LU7 0ER
G4 IWA J. Arrowsmith 16 Mancetter Road, Mancetter, Atherstone CV9 1NZ
G4 IWD G. Craig 83 Pearl Road, Walthamstow, London E17 4QY
G4 IWF G. Mason 51 Egerton Road, Streetly, Sutton Coldfield B74 3PG
G4 IWI J. Stocking Bildersbrook, Grove Road, Melton Constable NR24 2DE
G4 IWJ Capt. R. Towle Rye Hill Humshaugh, Hexham NE46 4BN
G4 IWN J. Andrews 5 Chapman Avenue, Maidstone ME15 8EG
G4 IWO N. Bradley 6 The Old School House, Shore, , Littleborough OL15 8EZ
GI4 IWP E. Maclaine 105 Bencran Road, Sixmilecross, Omagh BT79 9QA
G4 IWQ D. Cannon 57 Halswell Road, Clevedon BS21 6LE
G4 IWR S. Berry 40 Warrendale, Barton-upon-Humber DN18 5NH
G4 IWS C. Caine 10 Goodwood Close, Burghfield Common, Reading RG7 3EZ
G4 IWU J. Scrivens 130 Sea Lane, Rustington, Littlehampton BN16 2RZ
G4 IWV I. Parker 43 Longdown Road, Congleton CW12 4QH
G4 IXB C. Tuvey 1 Dorset Way, Heston, Hounslow TW5 0NF
G4 IXD I. Palgrave Brown The Abbey House, The Street, King's Lynn PE33 9HP
G4 IXE G. Walmsley Warwick Farm House, Cracknore Hard Lane, Southampton SO40 4UT
G4 IXF D. Toon 26 Reddish Avenue, Whaley Bridge, High Peak SK23 7DP
GM4 IXH Dr J. Finlayson 7 Abbotshall Road, Cults, Aberdeen AB15 9JX
G4 IXQ A. Constable Oakside, The Street, Bury St. Edmunds IP31 1NG
G4 IXT I. Jefferson 7 Bluebell Close, Rugby CV23 0UH
G4 IXY P. Beardsmore 2 Spencer Place, Sandridge, St. Albans AL4 9DW
G4 IYA M. Adams11 8 Boltons Close, Brackley NN13 6ND
G4 IYC B. Couchman 48 Eastfields, Blewbury, Didcot OX11 9NS
G4 IYE R. Smith 72 Worthing Road, Patchway, Bristol BS34 5HX
G4 IYK S. Dixon 33 Medhurst Crescent, Gravesend DA12 4HJ
GI4 IYO K. Burnside 4 Cuttles Road, Comber, Newtownards BT23 5YX
G4 IYP F. Dearden 22 Claremont Road, Chorley PR7 3NH
G4 IYS D. Burgess The Beeches, Half Moon Lane, Redgrave, Diss IP22 1RU
G4 IYT M. Melling 17 Heron Road, Leighton Buzzard LU7 4BY
GM4 IYZ J. Potts Eastwood Court, 1 Eastwoodmains Road, Glasgow G46 6QB
G4 IZA D. Howard 1700 3Rd Avenue West - Apt: 507, Bradenton 34205 United States
GI4 IZF M. Weller 58 Manse Road, Ballycarry, Carrickfergus BT38 9LF
G4 IZH P. Robinson 24 Haveroid Way, Crigglestone, Wakefield WF43PG
GW4 IZJ P. Rennick 41 Church Road, Pontnewydd, Cwmbran NP44 1AT
GD4 IZL G. Brookes 44 Magherchirrym, Port Erin, Isle of Man IM9 6DB
G4 IZQ A. Scarth 1 Beechwood Avenue, Whitley Bay NE25 8EP
G4 IZS Kent Raynet Group c/o R. Sexton 31 Fosters Lane, Woodley, Reading RG5 4HH
G4 IZU D. Byers 16 Tealby Court, Georges Road, London N7 8HY
G4 IZX P. Beards 3 Elm Drive Brightlingsea, Colchester CO7 0LA
G4 IZZ M. Eggleton 49 Gretton Road, Gotherington, Cheltenham GL52 9QU
G4 JA P. Stenning 9 Antoine Grove Richmond, Nelson 7020 New Zealand
G4 JAA P. Hawkins 38 Davidson Close, Great Cornard, Sudbury CO10 0YU
G4 JAC R. Emeny 28 Manor House Way, Brightlingsea, Colchester CO7 0QR
G4 JAJ B. Noble 19 Ayrton Avenue, Blackpool FY4 2BW
G4 JAQ M. Crofts 43 Broadlands Drive, East Ayton, Scarborough YO13 9ET
G4 JAR Dubmire ARC c/o I. Melville 3 Crescent Road, Benfleet SS7 1JL
G4 JAU S. Harris 92 Warwick Road, Wolston CV8 3GZ
G4 JAV W. Bird 42 Beaumont Way Norton Canes, Cannock WS11 9FQ
G4 JAX A. Lunn 11 Dibden Lodge Close, Hythe, Southampton SO45 6AY
G4 JBA Civil Service ARS c/o J. Alderman 38 Greenacres, Shoreham-by-Sea BN43 5WY
G4 JBD G. Laming 72 Fildyke Road, Meppershall, Shefford SG17 5LU
G4 JBE D. Lacey 78 New Road Sutton Bridge, Spalding PE12 9RQ
G4 JBF Dr G. Lester Lufflands, Yettington, Budleigh Salterton EX9 7BP
G4 JBG Yeovil CG c/o A. Dening 42 Grove Avenue, Yeovil BA20 2BD
G4 JBH A. Dening 42 Grove Avenue, Yeovil BA20 2BD
G4 JBK A. Maude 5 Darrowby Close, Thirsk YO7 1FJ
G4 JBL C. White Pegasus, Gotts Corner, Sturminster Newton DT10 1DD
GW4 JBQ J. Cleak Dantre, Newport Road, Cwmbran NP44 3AE
G4 JBR P. Dixon Hardwick House, New Road, South Molton EX36 4BH
G4 JBW D. Barber 3 Vestry Road, Street BA16 0HY
G4 JBY G. Bowden 78 Lynwood Avenue, Darwen BB3 0HZ
G4 JCA C. How 9 Chanctonbury Walk, Storrington, Pulborough RH20 4LT
G4 JCF G. Hoey Foehrer Strasse 8, Muenster 64839 Germany
G4 JCG P. Chapman 4 Chester Close, Garstang, Preston PR3 1LH
G4 JCH B. Hercombe 13 Dovecote, Shepshed, Loughborough LE12 9RW
G4 JCJ C. Newman 4 Winchilsea Drive, Gretton, Corby NN17 3BT
GW4 JCK N. Warnock 2 Sheepcourt Cottages, Bonvilston, Cardiff CF5 6TN
G4 JCL D. Bryan 3 New Lane, Skelmanthorpe, Huddersfield HD8 9EH
GM4 JCM Kingdom ARS c/o A. Glashan 35A Lochinver Crescent, Gourdie, Dundee DD2 4UA
G4 JCS J. Stevenson Highfields Farm, Saltburn by the Sea TS13 4UG
G4 JCX C. Gallacher 1 Brislands Lane Four Marks, Alton GU34 5AD
G4 JCY R. Thornton Binesfield, Bines Green, Horsham RH13 8EH
G4 JCZ A. Clifton 87 Aubrey Road, Quinton, Birmingham B32 2BA
G4 JDC L. Boddington Flat 33, Sorrel House, Birmingham B24 0TQ
GW4 JDE G. Evans 32 Radstock Court, Abergavenny NP7 5BQ
G4 JDF Dr P. Scovell 69 Nursery Road, Maidenhead SL6 0JR
G4 JDG C. Aitchison Upper Weston House, Cot Lane, Chichester PO18 8SU
G4 JDH R. Purbrick 2 Oyster Cottages, Tinnocks Lane, Southminster CM0 7NF
GM4 JDK M. Hopkinson The Coaches, Kingussie PH21 1NY
G4 JDO R. Tew 4 Chetwode Close, Allesley, Coventry CV5 9NA
G4 JDP S. Pallett 6 Lancaster Close, Coalville LE67 4TG
G4 JDS L. Radley 34 Queens Road, Chelmsford CM2 6HA
G4 JDT H. Lexton 11 Mulberry Close, Romford RM2 6DX
G4 JDW L. Nelson-Jones 15 Gainsborough Road, Bournemouth BH7 7BD
GW4 JDZ D. Samuel 87 Llewellyn Park Drive, Morriston, Swansea SA6 8PF
G4 JED K. Bird 25 Knowsley Way, Hildenborough, Tonbridge TN11 9LG
G4 JEF D. Wood Little Burgate Farm, Markwick Lane, Godalming GU8 4BD
G4 JEI N. Osborne 33 Dankton Gardens, Sompting, Lancing BN15 0DX
GM4 JEJ M. Thomson Ravenside, Mill Road, Carnoustie DD7 7SQ
GM4 JEM W. Redpath 69 Ulster Crescent, Edinburgh EH8 7JL
G4 JEO F. Kemp 42 Baker Road, Abingdon OX14 5LW
G4 JES M. Wells 16 Moorside Court, North Hykeham LN69XA
G4 JEY R. Furness Mermaid Lodge, 68/70 Brighton Lodge, West Sussex BN15 8LW
G4 JFC G. Hainsworth The Annexe, 16 Rowlandson Close, Northampton NN3 3PB
G4 JFD D. Featherstone 6 Claremont Gardens, Tunbridge Wells TN2 5DD
G4 JFF C. Webb 68 Higgs Field Crescent, Warley, Cradley Heath B64 6RB
G4 JFG J. May Midsummers Eve, Third Cliff Walk, Bridport DT6 4HX
GM4 JFH S. Draycott Kinmount, Whiting Bay, Isle of Arran KA27 8QH
G4 JFN R. Hudson 15 Fellows Road, Farnborough GU14 6NU
GI4 JFP D. Goodman 60 Castlewood Avenue, Coleraine BT52 1EW
G4 JFS J. Fitzsimons 27 Brese Avenue, Warwick CV34 5TS
G4 JFV R. Oldroyd Hambledon, 197 Inner Promenade, Lytham St. Annes FY8 1DW
G4 JFX B. Mount 4 Maplestone Road, Whitchurch, Bristol BS14 0HH
G4 JGF J. Fitzgerald 21 St. Aidans Avenue, Darwen BB3 2BS
G4 JGG J. Pether 7 Celina Close, Bletchley, Milton Keynes MK2 3LS
G4 JGH A. Allchin 9 Ashfield Road, Kings Heath, Birmingham B14 7AS
G4 JGQ J. Bevan 10 Streamdale, Abbey Wood, London SE2 0PD
G4 JGS S. Harding 9 Lightsfield, Oakley, Basingstoke RG23 7BL
GW4 JGU A. Green 9 Westbourne Grove, Sketty, Swansea SA2 9DT
G4 JGV S. Sharred Calle Pena De San Roque 23, Tinajo 35560, Lanzarote, Las Palmas, Birmingham 35560 Spain
GW4 JGW K. Simpson 59 Midland Place, Llansamlet, Swansea SA7 9QX
G4 JGX R. Calver Meadowbank, Lowertown, Helston TR13 0BY
G4 JHA R. Thomas 2 Woodlands Road, Astley, Manchester M29 7BH
G4 JHC P. Blunn 6 Cranbury Road, Eastleigh SO50 5HA
G4 JHE M. Green 1 Morley Hill, Stanford-le-Hope SS17 8HP
GU4 JHH R. Harvey Courtil Masse, Les Landes, Vale GY3 5JD Guernsey
G4 JHI D. Miller 10 Fair View, Horsham RH12 2PY
G4 JHN J. Unwin 28 Wallett Avenue, Beeston, Nottingham NG9 2QR
G4 JHP J. Hawes 13 Broadmead Road, Colchester CO4 3HB
G4 JHQ Dr C. Kear 60 Haywoods Lane, Somerset, Tasmania 7322 Australia
G4 JHS P. Hey 47 Hillcrest Road, Thornton, Bradford BD13 3PQ
G4 JHU N. Fineman Deansway, 2 The Drive, Rickmansworth WD3 4EB
G4 JHV S. Marsh Ravensknowle, Culgaith, Penrith CA10 1QF
G4 JHW D. Morrison Flat 1, 118 Anerley Park, London SE20 8NU
GI4 JIC P. Mcauley 68 Ballylenaghan Heights, Belfast BT8 6WL
G4 JIG E. White 12A Partridge Close, Great Oakley, Harwich CO12 5DH
G4 JIH K. Adams 12 Hawkewood Avenue, Waterlooville PO7 6EB
G4 JII R. Green Kingswood, Red House Lane, Doncaster DN6 7EA
G4 JIJ I. Kraven 55 Cranfield Crescent, Cuffley, Potters Bar EN6 4DZ
G4 JIK D. Bird 6 Wyebank, Bakewell DE45 1BH
G4 JIO K. Mason 5 Davenport Avenue, Hessle HU13 0RL
G4 JIQ W. Barker 69 Britten Road, Brighton Hill, Basingstoke RG22 4HN
G4 JIR A. Rixon 12 Vancouver St., Darlington DL3 6HN
G4 JIU I. Mcgarrigle 58 Langland Close, Corringham, Essex SS17 7LB
G4 JIV C. Davies 6 Valerie Avenue, Baulkham Hills 2153 Australia
GI4 JIW J. Ferrin 38 Dalewood, Newtownabbey BT36 5WR
G4 JIX J. Bentley 33 Lime Road, Ferryhill DL17 8DL
GI4 JJF K. McIlroy 69 Morston Park, Bangor BT20 3ER
G4 JJH J. Herbert 8 Falmouth Road, Springfield, Chelmsford CM1 6HY
G4 JJM M. Allison 19 Ash Grove, Kirklevington, Yarm TS15 9NQ
G4 JJP R. Thomas 28 Clarks Meadow, Shepton Mallet BA4 4FD
G4 JJQ J. Wheway 25 Mount View Avenue, Scarborough YO12 4EW
GW4 JJR L. James 65 Fflorens Road, Newbridge, Newport NP11 3DW
G4 JJS S. Harrison Seacroft Grange Care Village, The Green, Leeds LS14 6JL
GW4 JJV M. Bell 6 Owain Close, Cyncoed, Cardiff CF23 6HN
GW4 JJW A. Bell 6 Owain Close, Cyncoed, Cardiff CF23 6HN
G4 JJX M. Grange 6 Draysfield, Wormshill, Sittingbourne ME9 0TY
G4 JJY J. Carline 101 Cemetery Road, Scunthorpe DN16 1EB
G4 JKA J. Ewen-Smith 1 Kinnersley, Severn Stoke, Worcester WR8 9JR
GM4 JKB J. Barnes Capricorn, 13 Marchhill Drive, Dumfries DG1 1PP
G4 JKC P. Howard 72 Marlowe Way, Lexden, Colchester CO3 4JP
G4 JKE D. King Flat 21, Anchor Court, 2 Carey Place, London SW1V 2RT
G4 JKF B. Hodges Gramaur, Mucklestone Wood Lane, Market Drayton TF9 4ED
G4 JKH J. Phillips 57 New Sturton Lane, Garforth, Leeds LS25 2NW
GW4 JKK A. Bexley Pennar Fach Farm Plwmp, Llandysul SA44 6ES
G4 JKM D. Twigg 31 Parklands, Malmesbury SN16 0QH
G4 JKQ T. Bowen 40 Grange Road, Ibstock LE67 6LF

GW4 JKR D. Wilson 94 Lon Hedydd, Llanfairpwllgwyngyll LL61 5JY
G4 JKS M. Claytonsmith Hares Cottage, Woolston, Church Stretton SY6 6QD
GM4 JKT Dr O. Thores 5 Havens Edge, Limekilns, Dunfermline KY11 3LJ
GW4 JKV M. Rackham 31 Severn Road, Pontllanfraith, Blackwood NP12 2GA
G4 JKY E. Lennox Lowfield House, Low Street, York YO61 4QA
GM4 JKZ K. Leggett 3 The Steadings, Swinside Townhead Farm, Jedburgh TD8 6ND
GM4 JLD P. Woods 12 Dalriada Place, Kilmichael Glassary, Lochgilphead PA31 8QA
GI4 JLF R. Russell 1 Belmont Drive, Belfast BT4 2BL
G4 JLG Dr D. Yorke 40 Edge Fold Road, Worsley, Manchester M28 7QF
G4 JLJ L. Bailey 3 Eden Close, Hutton Rudby, Yarm TS15 0HT
G4 JLO H. Dyson 15 Swallow Grove, Netherton, Huddersfield HD4 7SR
G4 JLV J. Brower 37 High Street, Steventon, Abingdon OX13 6RZ
G4 JLX H. Braggs 47 Manor Road, Sandown PO36 9JA
GM4 JLZ E. Philip 12 Pinefield , Inchmarlo., Banchory AB31 4AF
G4 JMC J. Trickett 86 School Road, Thurcroft, Rotherham S66 9DL
G4 JMF D. Ollerhead 15 Kingsley Road, Chester CH3 5RR
G4 JMG J. Gorton 12 Apsley Close, Harrow HA2 6AP
G4 JMM J. Mcfadyen Flat 28, Dunstable Court, 12 St. Johns Park, London SE3 7TN
G4 JMO A. Oakley The Laithe, Coal Pit Lane, Colne BB8 8NR
G4 JMT M. Firth 6 Eastfield Drive, Woodlesford, Leeds LS26 8SQ
GM4 JMU K. Maxted 33 Woodpecker Grove Symington, Kilmarnock KA1 5SF
G4 JMY D. Liversidge 6 Yardley Way, Grimsby DN34 5UQ
GM4 JNB N. Baird 23 Scorguie Avenue, Inverness IV3 8SD
G4 JNE Dr C. Houghton 22 Rainow Road, Macclesfield SK10 2PF
G4 JNH R. Barker 171 Leicester Road, New Packington, Ashby-de-la-Zouch LE65 1TR
G4 JNK N. Kendall 13 Oaks Drive, Cannock WS11 1ET
G4 JNL P. Senior 9 Seely Close, Heighington, Lincoln LN4 1TT
G4 JNQ E. Allison 7 Abbey Road, Flitcham, King's Lynn PE31 6BT
GI4 JNS Dr D. Hughes 53 Cranley Grove, Bangor BT19 7EY
G4 JNT A. Talbot 15 Noble Road, Hedge End, Southampton SO30 0PH
G4 JNX N. Whyborn Kimberlin, Southwood Road, Norwich NR13 3AB
G4 JNZ C. Barron The Marling Pitts, Coughton, Ross-on-Wye HR9 5ST
G4 JOA K. Wood Flat 98, Harbour Tower, Gosport PO12 1HE
G4 JOB Obo West Glamorgan Cc c/o J. Barker 6 Larkswood Close, Tilehurst, Reading RG31 6NP
G4 JOD F. Rawlings 14 Haddon Way, Carlyon Bay, St. Austell PL25 3QG
GW4 JOG P. Truberg 106 Johnston Road, Llanishen, Cardiff CF14 5HJ
G4 JOI R. Tidnam 21 Manor Lane, Lewisham, London SE13 5QW
G4 JOO Cambridge & District ARC c/o C. Harman 46 Chandos Crescent, Edgware HA8 6HL
GI4 JOR J. Farrell 36 Cumber Park, Drumaness, Ballynahinch BT24 8GA
GW4 JOT S. Carfoot 24 Marble Church Grove, Bodelwyddan, Rhyl LL18 5UP
G4 JOU R. Bowden Flat 7, 39 Anstey Road, Alton GU34 2RD
G4 JOV J. Wedderburn 12 Victoria Avenue, Market Harborough LE16 7BQ
G4 JOW J. Butler Pickstock Manor, Pickstock, Newport TF10 8AH
G4 JPA R. Jarvis 2135 Oak Beach Blvd, 2135 Oak Beach Blvd, Sebring Fl 33875 United States
G4 JPB Canon J. Beaumont 9 Warren Bridge, Oundle, Peterborough PE8 4DQ
GW4 JPC G. Woods 178 Saron Road, Saron, Ammanford SA18 3LN
G4 JPE B. Hatley 9 Somerstown Court, Tilehurst Road, Reading RG1 7TY
GM4 JPG I. Wilson 11 Ellwyn Terrace, Galashiels TD1 2BA
GW4 JPJ H. Genon Dolau Cwerchyr, Penrhiwllan, Llandysul SA44 5NZ
G4 JPK J. Pymm Larkfield, Goxhill Road, Barrow-upon-Humber DN19 7EE
GW4 JPP E. Jones 1 Awel Y Mor, Cambrian Road, Tywyn LL36 0AG
G4 JPS Bristol Raynet c/o A. Williams 38 Seneca Street, Bristol BS5 8DX
G4 JPX I. Harrison 61 Charles Street, Golborne, Warrington WA3 3DF
GM4 JPZ C. Hall 42 Torridon Road, Broughty Ferry, Dundee DD5 3JG
G4 JQB R. Wickham 35 Ashley Road, Bathford, Bath BA1 7TT
G4 JQF M. Key 14 Ascot Road, Wigginton, York YO32 2QE
G4 JQJ R. Field 12 Granson Way, Washingborough, Lincoln LN4 1EY
G4 JQK S. Casey 14 Harrison Close, Emersons Green, Bristol BS16 7HB
G4 JQL S. Wayman Oak Tree Lodge, Redbridge Road, Crossways, Dorchester DT2 8BG
G4 JQN R. Ward 1 Dursley Road, Heywood, Westbury BA13 4LG
GW4 JQQ R. Henry 7 Gronow Close Neath Abbey, Neath SA10 7AD
G4 JQS C. Boulton Manor Cottage, Stratton, Dorchester DT2 9RY
G4 JQU Z. Pokusinski 362 Long Banks, Harlow CM18 7PG
G4 JQV C. Mee 26 De Lisle Court, Loughborough, Leicester LE11 4PP
G4 JQW F. Lobban 20 Evering Avenue, Poole BH12 4JQ
G4 JQX C. Riley 1 Coulston, Westbury BA13 4NX
GM4 JR A. Anderson 232 Annan Road, Dumfries DG1 3HE
GI4 JRA J. Harrigan 124 Drones Road, Pharis, Ballymoney BT53 8JT
G4 JRB M. Hahn 21 Stanley Road South, Rainham RM13 8AJ
G4 JRD R. De Muth 66 Perkins Road, Ilford IG2 7NQ
GM4 JRF H. Hamilton 8 Ardlui Gardens, Milngavie, Glasgow G62 7RL
G4 JRJ S. North 2 Robey Drive, Eastwood, Nottingham NG16 3DP
G4 JRW K. Burton 93 Truncliffe, Bradford BD5 8NX
G4 JRY T. Wislocki 30 Kingston Rd, Scunthorpe DN16 2BE
G4 JS British Aerospace ARS c/o W. Kenyon Flat 21 House 4, Copper Place, Manchester M14 7FZ
G4 JSD J. Hamilton 89 The Paddocks, Old Catton, Norwich NR6 7HE
G4 JSE R. Salaman 39 Arthur Street, Unley SA 5061 Australia
G4 JSK L. Welch 3 Sunnyfield Avenue, Cliviger, Burnley BB10 4TE
G4 JSM P. Hart 112 Shelton Avenue, Hucknall, Nottingham NG15 7QA
G4 JSP C. Perkins The Laurels, Higher Heath, Whitchurch SY13 2HZ
G4 JSQ D. Piper 102 Redhouse Lane, Walsall WS9 0DB
G4 JSS V. Waddington 1 Bridle Lane, Netherton, Wakefield WF4 4HN
G4 JST F. Ogden 11 Stocklands Close, Cuckfield, Haywards Heath RH17 5HH
G4 JSV N. Hingley 29 Mayfield Road, Hurst Green, Halesowen B62 9QW
G4 JSX M. Owen Thatched Cottage, Main Street, Rugby CV23 0JA
G4 JSZ D. Fry The Stocks, Lyth Bank, Shrewsbury SY3 0BE
G4 JTC J. Bautista 47 Valiant House, Varyl Begg Estate, Gibraltar GX11 1AA Gibraltar
G4 JTE Dr P. Djali 105 Pepper Lane Standish, Wigan WN6 0PW
GI4 JTF Dr E. Squance 11 Ballymenoch Road, Holywood BT18 0HH
G4 JTK J. Lee 203 Chester Road Whitby, Ellesmere Port CH65 6SE
G4 JTM J. Llewellyn Pier Road, Enniscrone, Co Sligo Ireland
G4 JTO H. Young 72 Perrinsfield, Lechlade GL7 3SD
G4 JTP G. Parker 14 Maplewood, Ashurst, Skelmersdale WN8 6RJ
G4 JTR V. Robinson 4 Hilltop Road, Caversham, Reading RG4 7HR
GI4 JTS R. Macrory 8 Manse Road, Newtownards BT23 4TP
G4 JTX P. Simon 19C High Street, Kilburn, Belper DE56 0NS
GW4 JUC H. Woodward 11 Pant-Yr-Odyn, Sketty, Swansea SA2 9GR
G4 JUD F. Loach 39 Park Road West, Wolverhampton WV1 4PL
GM4 JUE J. Cormack 16 Shore Lane, Wick KW1 4NT
G4 JUH R. Wilkinson 3 Anglesey Road, Dronfield S18 1UZ
GW4 JUI D. Draper Bryn Erin, Llangoed, Beaumaris LL58 8SU
G4 JUK M. Neville 103 Walsall Road, Great Wyrley, Walsall WS6 6LD
G4 JUM B. Buller 36 Grove Road, Ashtead KT21 1BE
GW4 JUN Leicester Raynet Group c/o V. Winton Ty Cerrig, Rhosesmor Road, Halkyn, Holywell CH8 8DL
G4 JUR R. Harrod 6 Carnforth Road, Barnsley S71 2RA
G4 JUV C. Bauers 21 Nethergate Street, Bungay NR35 1HE
G4 JUW W. Cole 5 Brook Furlong, Nesscliffe, Shrewsbury SY4 1BY
G4 JUZ N. Gabriel 156 Clarence Avenue, New Malden KT3 3DY
G4 JVA G. Butler 37 Turmore Dale, Welwyn Garden City AL8 6HT
G4 JVC I. Jones 4 Grove Crescent South, Boston Spa, Wetherby LS23 6AY
G4 JVD P. Hainsworth 74 Ravensbourne Drive, Woodley, Reading RG5 4LJ
G4 JVH G. Onions 3 Tower Rise, Tividale, Oldbury B69 1NP
G4 JVJ Dr R. Ashman 44 Conan Doyle Walk, Swindon SN3 6JB
G4 JVM F. Pearson Coach House, The Park, Manningtree CO11 2AL
GJ4 JVP J. Arthur 13 Les Quennevais Park, St. Brelade, Jersey JE3 8GB
G4 JVT G. Howell 25 Thornhill Road, Hednesford, Cannock WS12 4LR
G4 JVX D. Powell 2 Curlew Close, Winsford CW7 1SW
G4 JVZ M. Glennon 41 Moorway, Guiseley, Leeds LS20 8LD
G4 JWA D. Naylor 19 Bindbarrow, Burton Bradstock, Bridport DT6 4RG
G4 JWK L. Ball Tree Tops, Bodiam, Robertsbridge TN32 5UG
G4 JWL P. Woodward Le Rosey, Rolle 1180 Switzerland
G4 JWV N. Lyons 114 Spring Hill, Weston-Super-Mare BS22 9BD
GI4 JWW T. Martin 57 Oneill Road, Newtownabbey BT36 6UN
G4 JXC R. Butler 6 Woodland Avenue, Dursley GL11 4EW
G4 JXE P. King 21 Compton Way, Olivers Battery, Winchester SO22 4HS
G4 JXH P. Mcgivern 83 Birdhill Avenue, Reading RG2 7JU
G4 JXI H. Collier 12 Coronation Drive, Leigh WN7 2UU
G4 JXJ C. Blewitt 12 Salton Street, Secret Harbour 6173 Australia
G4 JXK D. Bonfield 14 Springdale Close, Brixham TQ5 9RL
GW4 JXN G. Roberts 4 Frondeg, Ffordd Penmynydd, Llanfairpwllgwyngyll LL61 5AX
GM4 JXP S. Green 48 Barclay Park, Aboyne AB34 5JF
G4 JXR G. Wilde 26 Fleetham Grove, Hartburn, Stockton-on-Tees TS18 5LH
G4 JXU K. Lee 34 Evergreen Way, Wokingham RG41 4BX
G4 JXZ I. Terrell 10 Red Lion Close, Cranfield, Bedford MK43 0JA
GM4 JYB B. Sparks Donlyn, Lyth, Wick KW1 4UD
G4 JYE D. Sargent 15 Wilton Road, Balsall Common, Coventry CV7 7QW
G4 JYF C. Golley 10 New Molinnis, Bugle, St. Austell PL26 8QL
G4 JYG the Three A's CG c/o T. Watson 59 Vincent Road, Norwich NR1 4HQ
G4 JYH A. Curtis 19 Donnelly Road, Bournemouth BH6 5NW
GI4 JYJ G. Mcmaw 26 Watch Hill Road, Ballyclare BT39 9QW
G4 JYK P. Leather 35 Somerset Close, Congleton CW12 1SE
G4 JYL G. Thomas 16 Fordlands, Thorpe Willoughby, Selby YO8 9PD
G4 JYN Waterside ARS c/o T. Williams 31 Manor Road, Holbury, Southampton SO45 2NQ
G4 JYP N. Shelley 25 Threeways, Cuddington, Northwich CW8 2XJ
G4 JYQ J. Tierney 39 Daneway, Southport PR8 2QW
G4 JYT R. Armstrong 38 Watson Avenue, Market Harborough LE16 9NA
G4 JYU N. Bourner 11 Richborough Road, Sandwich CT13 9JE
G4 JYW D. Proctor 36 Westlands, Pickering YO18 7HJ
G4 JZA S. Geary Bella Vista, The Square, Truro TR2 4DS
GM4 JZB D. Gardner 7 Croft Road, Auchterarder PH3 1EW
G4 JZF G. Taylor 1 Threshers Drive, Willenhall WV12 4AN
G4 JZL J. Adams 1 Powell Close, Creech St. Michael, Taunton TA3 5TE
G4 JZQ M. Noakes 333 St. Neots Road, Hardwick, Cambridge CB23 7QL
G4 JZR E. Williams 7 Laurel Drive, Willaston, Neston CH64 1TN
G4 JZV R. Bellamy 31 Shaftesbury Avenue, Lincoln LN6 0QN
GW4 JZY J. Price 18 Woodland Drive, Bassaleg, Newport NP10 8PA
G4 JZZ C. Gadd 40 Stanley Mount, Sale M33 4AE
G4 KAB L. Rose 2 Westglade Court, Woodgrange Close, Harrow HA3 0XQ
G4 KAE D. Wood 7 Mead Close, Cheddar BS27 3XN
G4 KAL B. Thompson 23 South Street, Keelby, Grimsby DN41 8HE
G4 KAM S. Greenwood Little Oaks, Green Lane, Axminster EX13 5TD
G4 KAR R. Jeffries 22 Ingrams Way, Hailsham BN27 3NP
G4 KAT M. Westwater 3 Burns Way, Harrogate HG1 3NA
G4 KAU Dr T. Mansfield 2 Stratford Crescent, Cringleford, Norwich NR4 7SF
GM4 KAV F. Bowles 40 Craigbarnet Road, Milngavie, Glasgow G62 7RA
G4 KAX M. Haswell 5 Westcombe Avenue, Leeds LS8 2BS
GW4 KAZ B. Davies 2 Glan Llyn Terrace Bethel, Caernarfon LL55 1YL
G4 KBA K. Boucher 22 Emery Close, Walsall WS1 3AL
G4 KBB B. Bristow 13 Princes Street, Piddington, High Wycombe HP14 3BN
G4 KBH R. Hodgson 29456 Trailway Lane, Agoura Hills 91301 United States
G4 KBI C. Wainman 9 Willson Drive, Riddings, Alfreton DE55 4AF
G4 KBK R. Fisher 80/72 Kangan Drive, Berwick 3806 Australia
GJ4 KBM B. Nelson 1 La Genetiere, La Route Orange, St. Brelade JE3 8GP Jersey
G4 KBP M. Ford Micalma, Anslow Lane, Burton-on-Trent DE13 9DS
G4 KBQ J. Haslam 20 Lightfoots Avenue, Scarborough YO12 5NS
GI4 KBW P. Henderson 7 Clonaslea, Newtownabbey BT37 0UL
G4 KBX C. Chapple Woodend, Hebron, Morpeth NE61 3LA
G4 KCC H. Holmden 29 Cambridge Road West, Farnborough GU14 6QA
G4 KCD B. Dean 3 Marchant Court, Gunthorpe Road, Marlow SL7 1UW
G4 KCF K. Sanderson 39 Kirkland Street Pocklington, York YO42 2BX
G4 KCM C. Sanders 13 Meadow Court, Whiteparish, Salisbury SP5 2SE
G4 KCN D. Salmon The Pines, 5A Westfield Avenue, Harpenden AL5 4HN
GI4 KCO Cheltenham Rynt c/o K. Wright 72 Elm Corner, Dunmurry, Belfast BT17 9PY
G4 KCP D. Appleton 28 Edgewood, Shevington, Wigan WN6 8HR
GW4 KCQ P. Evans 2 Cwmnantllwyd Road, Gellinudd, Swansea SA8 3DT
G4 KCR S. Dunn 4 St. Ronans Road, Harrogate HG2 8LE
G4 KCT B. Firth 8 Lyndale Avenue, Osbaldwick, York YO10 3QB
G4 KCU D. Greatbatch 1 Hilltop Way, Dronfield S18 1YL
GW4 KCV Dr R. Murray-Shelley 5 Dan Y Wern, Pwllgloyw, Brecon LD3 9PW

UK Callsigns

G4 KCX P. Hicks 7 North Croft, High Wycombe HP10 0BP
GW4 KCY P. Trimmer 15 Cypress Court, Landare, Aberdare CF44 8YB
G4 KCZ Dr C. Conduit 21 Shadybrook Lane, Weaverham, Northwich CW8 3PN
G4 KDE Dr A. Lamont 50 Hockley Road, Rayleigh SS6 8EB
G4 KDH K. Howe Woodlands, St. Peters Road, Hockley SS5 6AA
GW4 KDI R. Stanton 33 Brook Road, Shotton CH5 1HH
G4 KDK J. Riggs 28 Long Hill, Mere, Warminster BA12 6LR
G4 KDL A. Seago 50 Kimberley Road, Lowestoft NR33 0TZ
G4 KDM J. Pearson 110 Plover Mills, Lindley, Huddersfield HD3 3ZF
G4 KDN J. Phaff 28, Draycott Road, , Abingdon OX13 5BZ
G4 KDR I. Wassell 21 Speedwell Way, Horsham RH12 5WA
G4 KDS C. Lafferty Apw 0414427, Addresspal, The Beacon, Mosquito Way, Hatfield AL10 9WN
G4 KDU G. Baldwin 31 Kilnhurst Road, Todmorden OL14 6AX
G4 KDW I. Davidson 24 Queenswood Drive, Hitchin SG4 0LG
G4 KEB L. Bright 49 Fellows Avenue, Wall Heath, Kingswinford DY6 9ET
G4 KEC R. Cookson 4 Wellington Gardens Selsey, Chichester PO20 0EE
G4 KEE V. Tomkins 58 Chancellors Way, Beacon Hill, Exeter EX4 9DY
G4 KEG D. Fryer 28 Hudson Road, Eastwood, Leigh-on-Sea SS9 5NX
G4 KEI C. Gaston Seaward, Marshlands Lane, Heathfield TN21 8EY
G4 KEL S. Kell 11 Streatlam Road, Darlington DL1 4XG
G4 KEN K. Smith 32 St. Clements Road, Harrogate HG2 8LX
G4 KEP H. Haria 34 Larkfield Avenue, Harrow HA3 8NF
GI4 KEQ B. Mcmahon 26 Ballycraigy Road, Newtownabbey BT36 5ST
G4 KES B. Bloomer 2 Magor Hill Cottages, Magor Hill, Camborne TR14 0JF
G4 KEW R. Marshall 60 Drake Road, Harrow HA2 9EA
G4 KEX B. Hubbard 16 Shelf Moor, Halifax HX3 7PW
G4 KEY D. Turner 22 Westhawe, Bretton, Peterborough PE3 8BA
G4 KEZ B. Archer 86 York Road, Swindon SN1 2JU
G4 KFA T. Bearpark 19A Humber Lane, Patrington, Hull HU12 0PJ
G4 KFB M. Bird 84 Penwill Way, Paignton TQ4 5JQ
G4 KFC A. Scandrett 45 Merryhill, Northampton NN4 9YH
GW4 KFD B. Wilson 1A Treetops, Llanelli SA14 8DN
G4 KFF R. Hewson 2 Ribchester Way, Brierfield, Nelson BB9 0YH
G4 KFH E. Sinkinson 24 Old Hall Park, Langthorpe, York YO51 9BZ
GW4 KFI D. Bromfield 3 Warwick Road, Brynmawr, Ebbw Vale NP23 4AR
G4 KFJ C. Baker 5 Holly Bank Rise, Dukinfield SK16 5EG
G4 KFL R. Rowney 58 Wychdell, Stevenage SG2 8JD
G4 KFP J. Marshall 92 High Street, Ossett WF5 9RQ
G4 KFS T. Wood 47 Marsh View, Beccles NR34 9RT
G4 KFT M. Rothwell 3 Chiltern Road, Prestbury, Cheltenham GL52 5JQ
GW4 KFY J. Edwards 15 The Meadows, Llandudno Junction LL31 9LP
G4 KFZ R. Stanton 50 Plymstock Road, Plymstock, Plymouth PL9 7NU
G4 KGA M. Hattam 11 Dukes Wood Avenue, Gerrards Cross SL9 7LA
G4 KGC P. Suckling 314A Newton Road, Rushden NN10 0SY
G4 KGE J. Baldwin 30 Petters Road, Ashtead KT21 1NE
G4 KGF G. Brooks 1 Highfield Close, Pembury, Tunbridge Wells TN2 4HG
GM4 KGK N. Munro Windyridge, Lower Bayble, Isle of Lewis HS2 0QB
G4 KGL M. Lees 15 Blacklock, Chelmsford CM2 6QL
G4 KGN D. Mitchinson 11 St. Marys Avenue Hemingbrough, Selby YO8 6YY
G4 KGO R. Matthews 191 Valley Road, Ipswich IP4 3AH
G4 KGU B. Thomas 12 Link Road, Sale M33 4HP
G4 KGX W. Green 3 Amos Road, Leicester LE3 6NA
G4 KGY T. Lawford 20 Magdalen Court Ersham Road, Canterbury CT1 3DH
GM4 KGZ S. Low Gartwood, 14 Dundas Avenue, North Berwick EH39 4PS
GM4 KHE G. Phanco 28 Park Road, Clydebank G81 3LH
G4 KHG E. Scholes 19 Castle Hill, Newton-le-Willows WA12 0DU
GM4 KHI T. Ferrie 17 Bargarron Drive, Paisley PA3 4LL
G4 KHJ B. Geeson 24 Rydal Avenue, Poulton-le-Fylde FY6 7DJ
G4 KHK P. Martin 24 Heddington Close, Trowbridge BA14 0LH
G4 KHM J. Whitington 18 Somerset Road, Ferring, Worthing BN12 5QA
GW4 KHQ J. Woodland 7 Lighthouse Park, St. Brides Wentlooge, Newport NP10 8SL
G4 KHR R. North 21 St. Augustine Grove, Bridlington YO16 7DB
GM4 KHS Peterborough & District ARC c/o G. Chalmers 38 Grove Hill, Kelso TD5 7AS
G4 KHU P. Hawkins Temple View, High Street, Templecombe BA8 0JG
G4 KHX P. Winchester 27A Lower Road, Milton Malsor, Northampton NN7 3AW
G4 KIB J. Hambleton Monte Pascoal, Cci 264 St, Sao Teotonio 7630-583 Portugal
G4 KIF A. Sansom 1881-9 Avenue S E, Salmon Arm BC V1E 2J6 Canada
G4 KIH W. Bartlett 48 Barrymore Walk, Rayleigh SS6 8YF
G4 KII K. Harris St. Peters Church Vicarage, Haywood Road, Birmingham B33 0LH
G4 KIK D. Whyborn 33 Church Road, Trull, Taunton TA3 7LG
G4 KIL P. Williams 4 Church Court, 130 Nevill Avenue, Hove BN3 7NS
G4 KIM P. Newman 16 Oakwood Road, Westlea, Swindon SN5 7EF
G4 KIN P. Taylor 22 Windermere Drive, Rainford, St. Helens WA11 7LD
G4 KIP J. Ball Moss Nook Farm, Moss Nook Lane Road, St. Helens WA11 8AG
G4 KIQ A. Brooks 10 St. James Avenue East, Stanford-le-Hope SS17 7BQ
G4 KIR K. Chattenton 29 Wand Hill Gardens, Boosbeck, Saltburn-by-the-Sea TS12 3AP
G4 KIU N. Peacock 1 Highland Grange, Beacon Road, Crowborough TN6 1AT
GI4 KIX D. Gilmore The Overlook, 29 Ballymaconaghy Road, Belfast BT8 6SB
G4 KIZ D. Holmes Lancaster House, Magna Mile, Market Rasen LN8 6AD
G4 KJA B. Preston 24 Nursery Close, Hucknall, Nottingham NG15 6DQ
GI4 KJC N. Quinn 54 Moyle Road, Newtownstewart, Omagh BT78 4JT
G4 KJD I. Pitkin Clover Cottage, Kenny, Ashill, Ilminster TA19 9NH
G4 KJJ J. Smith 30 Rookery Close, St. Ives PE27 5FX
G4 KJK D. Oliver 15 Brixham Avenue, Cheadle Hulme, Cheadle SK8 6JG
G4 KJP L. Jordan Cami De Fuster No 1, Marxuquera Alta, Valencia 46700 Spain
G4 KJS A. Gregory 1-3 Nargate Street, Littlebourne, Canterbury CT3 1UH
G4 KJU R. Fisher 85 Larkway, Brickhill, Bedford MK41 7JP
G4 KKB K. Blamey 123 St. Edmunds Walk, Wootton Bridge, Ryde PO33 4JJ
G4 KKG J. Taylor 12 Glenthorne Avenue, Yeovil BA21 4PG
G4 KKJ H. Perryman 15 Queen Mary Crescent, Kirk Sandall, Doncaster DN3 1JU
G4 KKN P. Roberts 2 Samuals Fold, Pendlebury Lane, Wigan WN2 1LT
G4 KKO J. Walton 17 Wychperry Road, Haywards Heath RH16 1HJ
G4 KKR R. Page 4 Nursery Drive, March PE15 8EQ
G4 KKS A. Morris 4 Woodville Gardens, Wigston LE18 1JZ
G4 KKT J. Mahoney 18 Park Avenue, London N22 7EX
G4 KKU A. Imianowski 97 Bloomfield Road, Bristol BS4 3QP
GM4 KKV P. Rucklidge 8 Stanehead Park, Biggar ML12 6PU
G4 KKZ K. Robinson 13 Race Hill, Launceston PL15 9BB
G4 KLA Dr J. Nelson 67 Swarthmore Road, Birmingham B29 4NH
G4 KLB C. Watts 42 Truscott Avenue, Bournemouth BH9 1DB
G4 KLD C. Dewhurst 56 Collett Way, Priorslee, Telford TF2 9SL
G4 KLE M. Foster 7 Church Street, Fenstanton, Huntingdon PE28 9JL
G4 KLF A. Selmes 82 Beaufort Court, Beaufort Road, St. Leonards-on-Sea TN37 6PF
G4 KLJ D. Wellings 41 Wroxham Drive, Wollaton, Nottingham NG8 2QR
G4 KLM Aberystwyth & District RS c/o P. Raven Wedgewood, Green Lane West, Norwich NR13 6LT
GM4 KLN I. Moore 7 Greenside Avenue Rosemarkie, Fortrose IV10 8XA
GM4 KLO M. Mistofsky 18 Troon Place, Newton Mearns, Glasgow G77 5TQ
G4 KLT L. Jones 52 The Drive, Bury BL9 5DL
G4 KLX J. Naylor 35 Old River, Denmead PO7 6UX
G4 KMB A. Griggs 4 Raleigh Rise, Portishead, Bristol BS20 6LA
G4 KME J. Horley 50 Hillswood Drive, Endon, Stoke-on-Trent ST9 9BW
G4 KMF E. Colmer 31 Mosyer Drive, Orpington BR5 4PN
G4 KMH S. Cottis 61 Oaken Grove, Maidenhead SL6 6HN
G4 KMJ D. Edwards 72 Parkstone Road, Hastings TN34 2NT
G4 KMK RAF Holmpton Ara c/o R. Blower 133 Almondbury Bank, Huddersfield HD5 8EX
G4 KMM P. Northmore Flat 23, Margaret Hill House 77 Middle Lane, Hornsey N8 8NX
G4 KMP G. Ramsey 21 Goldsmith Road, Eastleigh SO50 5EN
G4 KMW R. Greenhough 8 Stella Gardens, Pontefract WF8 2SR
G4 KMX R. Cope 41 Hall Lane, Witherley, Atherstone CV9 3LT
G4 KNI D. Rickard 12 Dabryn Way, St. Stephen, St. Austell PL26 7PF
G4 KNN A. Leggett 3 Hayes Mead, Holbury, Southampton SO45 2JZ
G4 KNO A. Summers Broxwood, Bury Road, Bury St. Edmunds IP29 4PH
G4 KNQ H. Smith Grey Gables, Humphrey Gate, Buxton SK17 9TS
G4 KNR S. Mason 9 Bempton Close, Bridlington YO16 7HL
G4 KNS J. Wallett 46 Aldreth Road, Haddenham, Ely CB6 3PW
G4 KNT I. Morton 65 Manton Road, Hitchin SG4 9NP
GM4 KNU A. Torrance 306 Mearns Road, Newton Mearns, Glasgow G77 5LS
G4 KNV Dr D. Wilkinson Westview, Old Byland, York YO62 5LG
G4 KNX A. Bennett 4 Chelmarsh Close, Redditch B98 8SQ
G4 KNZ S. Davies 17 Haywood, Bracknell RG12 7WG
GW4 KOE R. Lines 19 Magnolia Close, Cardiff CF23 7HQ
GM4 KOI S. Milne 24 St. Ternans Road, Newtonhill, Stonehaven AB39 3PF
G4 KOJ J. Wilson 54 Devonshire Drive, Mickleover, Derby DE3 9HB
G4 KOK J. Stockley Clee View, Leys Lane, Leominster HR6 0AZ
G4 KON L. Butt 16A Kestrel Crescent, Oxford OX4 6DX
GM4 KOO S. Cawthorne 8 Captains Brae, Twynholm, Kirkcudbright DG6 4PE
G4 KOQ G. Birkhead 15 Crannog, Keshcarrigan, Carrick on Shannon N41K235 Ireland
G4 KOR A. Hughes 55 Welford Road, Shirley, Solihull B90 3HX
G4 KOT G. Lindsay 9 Pennine View, Sherburn Hill DH6 1QN
G4 KOU G. Martin Flat 1, Field House, Station Road, East Preston, Littlehampton BN16 3RU
G4 KOV H. Wright Sandpiper Cottage, Standard Road, Wells-next-the-Sea NR23 1JY
G4 KOW D. Mclachlan 48 Nursery Avenue, Bexleyheath DA7 4JZ
G4 KOY R. Gill 87 Penkett Road, Wallasey CH45 7QQ
GW4 KPD Dr A. Grant Chandlers, Welsh Street, Chepstow NP16 5LU
G4 KPE Southend&Dis Ar c/o P. Griggs 6 Nightingale Way, Sutton Bridge, Spalding PE12 9RG
G4 KPF T. Hart 15 Whitefriars Meadow, Sandwich CT13 9AS
G4 KPG W. Lam 2 Wistaria Road, Flat 3A, Kowloon Hong Kong
G4 KPH D. Lewis 4 Raymond Court, Pembroke Road, London N10 2HS
G4 KPI J. Lorton 14 Provis Mead, Chippenham SN15 3UA
G4 KPL M. Young 8 Tweed Close, Worcester WR5 1SD
G4 KPM M. Pitt 20 Little Halt, Portishead, Bristol BS20 8JQ
G4 KPP C. Kelly 115 Kingsdown Crescent, Dawlish EX7 0HB
G4 KPS Bishops Stortford Ar Society c/o C. Saunders 26 Henley Fields, St. Michaels, Tenterden TN30 6EL
G4 KPU G. Taylor 179 Bradway Road, Bradway, Sheffield S17 4PF
G4 KPV F. Dunn 12 Streete Court, Westgate-on-Sea CT8 8BT
G4 KPX R. Burton 28 Mulberry Way, Ely CB7 4TH
G4 KPZ V. Cracknell 106 High St., Upwood, Huntingdon PE26 2QE
GW4 KQ D. Phillips 37 Saint Margarets Park, Lower Ely, Cardiff CF5 4AP
GI4 KQA T. Moffitt 36 Greenview, Parkgate, Ballyclare BT39 0JP
G4 KQC G. Leatherbarrow 6 Queens Walk, Thornton-Cleveleys FY5 1JW
G4 KQD A. Down 5 Juniper Mead, Stotfold, Hitchin SG5 4RU
G4 KQE A. Mead 9 Abraham Drive, Silver End, Witham CM8 3SP
G4 KQH D. Howes 14 Manitoba Way, Eydon, Daventry NN11 3PR
G4 KQK Dr C. Barnes Glebe Farmhouse, Stafford ST18 9DQ
G4 KQL A. Daulman 2 Trentham Road, Hartshill, Nuneaton CV10 0SN
G4 KQO R. Ferguson 8 Rutland Gardens, Croydon CR0 5ST
G4 KQP S. Jones 114 Portland Road, Toton, Nottingham NG9 6EW
G4 KQQ R. Jones 2 Bubwith Walk, Wells BA5 2EN
GM4 KQS A. Smith 9 Woodmill, Kilwinning KA13 7PT
G4 KQV C. Hands 41 Coverdale Road, Solihull B92 7NU
G4 KQY M. Pearce 51 Grove Avenue, New Costessey, Norwich NR5 0JB
G4 KQZ T. Thorne 17 Pine St. South, Bury BL9 7BU
G4 KRD M. Khalaf 508 London Road, Thornton Heath CR7 7HQ
G4 KRF R. Moore Flat 15, Nelson Court, 130 Rowson Street, Wallasey CH45 2LZ
G4 KRG R. Bowden 38 Buxton Road, Furness Vale, High Peak SK23 7PF
G4 KRH R. Cane 24 South End, Longhoughton, Alnwick NE66 3AW
G4 KRJ E. Gaffney 54 Dockham Road, Cinderford GL14 2BH
G4 KRN A. Troy 1B Lidderdale Road, Liverpool L15 3JG
G4 KRT M. Davis 35 Mullion Croft, Kings Norton, Birmingham B38 8PH
G4 KRW R. Waterman 170 Station Road, Mickleover, Derby DE3 9FJ
G4 KSA D. Mountain 178 Wragby Road, Lincoln LN2 4PT
G4 KSG R. Ralph 62 Northdown Road, Solihull B91 3ND
GI4 KSH H. Morrow 2 Carnhill Grove, Newtownabbey BT36 6LS
G4 KSK R. Benyon C/ Grecia 17, Villalbilla, Madrid 28819 Spain
GI4 KSO D. Mawhinney 233 Ballynahinch Road, Annahilt, Hillsborough BT26 6BH
G4 KSQ B. Morris 22 Burdell Avenue, Headington, Oxford OX3 8ED
G4 KSR S. Norris 17 Montroy Close, Bristol BS9 4RS
G4 KST T. Hughes 42 Western Drive, Hanslope, Milton Keynes MK19 7LD
G4 KSU Dr K. Prettyjohns 315 High Street, Sheerness ME12 1UT

G4 KSY A. Street 43 Ridgedale Road, Bolsover, Chesterfield S44 6TX
G4 KTB T. Cottham 4 Talisman Close, Tiptree, Colchester CO5 0DT
G4 KTG H. Wilson 24 Clumber Avenue, Newark NG24 4DT
GW4 KTQ G. Davies 56 Ffordd Cynan, Bangor LL57 2NS
G4 KTR D. Burrell 67 Newfield Drive, Nelson BB9 9RR
GW4 KTT East Lancashire Raynet c/o P. Valerio Reynoldston, Gower SA3 1AE
G4 KTU K. White 22 Ridyard Street, Wigan WN5 9PA
G4 KTW E. Dale The Woodlands, Cotheridge, Worcester WR6 5LZ
G4 KTX J. Goldsmith The Maltings, Flacks Green, Chelmsford CM3 2QS
G4 KTZ P. Cullen 5 Swaledale Gardens, Fleet GU51 2TE
G4 KUC J. Goodier 20 Poleacre Lane, Woodley, Stockport SK6 1PG
G4 KUD B. Whittles 12 Locksley Gardens, Birdwell, Barnsley S70 5SU
G4 KUE C. Raspin 35 Allesley Hall Drive, Coventry CV5 9NS
G4 KUF A. Redman 42 Gallows Hill Lane, Abbots Langley WD5 0DA
G4 KUJ T. Groves 31 Tunnel Wood Close, Watford WD17 4SW
G4 KUL D. Hepplestone 19 Richlans Road, Hedge End, Southampton SO30 0HU
GI4 KUM W. Glenn 1 Meadowside, Antrim BT41 4HD
G4 KUQ P. Goodfellow 10 St. Agnes Walk, Knowle, Bristol BS4 2DL
G4 KUR S. Hammonds 22 The Croft, Meriden, Coventry CV7 7NQ
G4 KUX N. Peckett Fourwinds, Woodland, Bishop Auckland DL13 5RH
G4 KUY Loughborough & Distrct ARC c/o M. Hill Park Villa, Park Road, Tydd St. Giles, Wisbech PE13 5NH
GI4 KUZ W. Hamill 47 Gracefield, Gracehill, Ballymena BT42 2RP
G4 KVC R. Mitchell 6 Green St., Smethwick B67 7BX
G4 KVD J. Mcmahon 5 Victoria Walk, Wokingham RG40 5YL
G4 KVI C. Dunn 71 Redfield Road, Midsomer Norton BA3 2JH
G4 KVK P. Park 2 Leyburn Drive, High Heaton, Newcastle upon Tyne NE7 7AP
G4 KVL B. Tharme 4 Longcroft Avenue, Liverpool L19 4TB
G4 KVP A. Woodland 45 Walsingham Road, Wallasey CH44 9DX
G4 KVQ R. Scott 20 Forest Hill, Carlisle CA1 3HF
G4 KVR P. Bell 24 Onslow Gardens, Ongar CM5 9BG
G4 KVT J. Fairfax 382 Wells Road, Bristol BS4 2QP
G4 KVU M. Shearer Appleacre, Mill Road, Haverhill CB9 7NN
G4 KVX P. Bleiker Waterside, 31 North Shore Road, Hayling Island PO11 0HL
G4 KWE T. Peel Herongate, Derwent Lane, Hope Valley S32 1AS
G4 KWF E. Pickup 36 Werneth Road, Glossop SK13 6NF
G4 KWH C. Meadows 16 Dart Road, Bedford MK41 7BT
G4 KWJ J. Hakes Commonbank Cottage, Lancaster LA2 9AN
G4 KWK K. Hakes 2 Common Bank Cottages Dolphinholme, Lancaster LA2 9AN
G4 KWL T. Walter 11 Silver Street Congresbury, Bristol BS49 5EY
G4 KWM P. Deville Bexton Doncaster Road, Mexborough S64 0JD
G4 KWO G. Phillips 20 Eastfield Drive, Solihull B92 9ND
G4 KWQ A. Soltysik 24 Cottage Close, Hednesford, Cannock WS12 1BS
G4 KWT D. Pibworth 20 Marathon Close, Woodley, Reading RG5 4UN
G4 KWW J. Ilott Flat 3 Berkeley Court, Scotter Road, , Scunthrope DN15 7EG
G4 KWX B. Cox Sylvan House, Alton Road, , Farnham GU10 5EL
G4 KWY D. Gasser 49 Pennycress, Locks Heath, Southampton SO31 6SY
G4 KWZ G. Harris Windmill House, Ripon Road, Kirby Hill, York YO51 9DP
G4 KXG K. Jackson 45 The Crescent Bilsthorpe, Newark NG22 8QX
G4 KXK J. Ward 38 Stonechat Avenue, Abbeydale, Gloucester GL4 4XD
G4 KXL J. Redman 488 Blair Road, Georgia 30563 United States
G4 KXO M. Reynolds Ilex House, Redwick Road, Bristol BS12 3LQ
G4 KXP J. Brockett 17 Swan Drive, Droitwich WR9 8WA
G4 KXQ M. Wogden 28 Magnolia Close, Barnstaple EX32 8QH
G4 KXR A. Tipper 10 Tithebarn Copse, Exeter EX1 3XP
G4 KXU G. Robinson 25 Stable Way Kingswood, Hull HU7 3FA
G4 KXV D. Rigby1 145 Knightlow Road, Harborne, Birmingham B17 8PY
G4 KXW C L P K c/o G. Redhead 18 Paddock Way, Dronfield S18 2FF
G4 KYE T. Carhart C6 Tamar Park, Coxpark, Gunnislake PL18 9BD
G4 KYH A. Waddilove 2 Gwel Trencrom, Hayle TR27 6PJ
G4 KYI R. Shipton 3 Fiery Lane, Uley, Dursley GL11 5DA
GW4 KYK J. Jones 7 Frankwell St., Tywyn LL36 9EP
G4 KYO G. Barber 25 Queensway, Hayle TR27 4NJ
GW4 KYT D. Thomas 3 New Road, Trebanos, Swansea SA8 4DL
G4 KYU R. Ringrose Melford House, George Street, Ipswich IP8 3NH
G4 KYX D. Gee 13 Dart Road, Bedford MK41 7BT
G4 KYY P. Day Box 204 The Post Room, Calle San Jaime 5, Benijofar 3178 Spain
G4 KZB P. Hazelwood 12 Ryecroft, Stourbridge DY9 9EH
G4 KZD J. Young 30 Crofton Way, Enfield EN2 8HS
G4 KZI B. Clark 21 Church Road, Binstead, Ryde PO33 3TA
G4 KZK R. Smith 15 St. Anthonys Way, Brandon IP27 0DN
G4 KZO A. Keir Kingfisher House, Nantwich Road, Calveley, Tarporley CW6 9JT
G4 KZQ R. Bennett 16 Emily Street, St. Helens WA9 5LZ
G4 KZT B. Ashdown 1 The Warren, Little Snoring, , Fakenham NR21 0JU
G4 KZU N. Rathbone 7 Foreland Way, Keresley, Coventry CV6 2NN
G4 KZV Chorley & District A.R.S c/o J. Parkin 18 Bradnock Close, Birmingham B13 0DL
G4 KZW S. Haydock 60 Tong St., Bradford BD4 9LX
G4 KZX A. Still 17 Arundel Road, Newhaven BN9 0ND
G4 KZZ N. Roberts 13 Rosemoor Close, Hunmanby, Filey YO14 0NB
G4 LAB Leicestershire Worked All Brittain Group c/o D. Brooks 28 Avon Vale Road, Loughborough LE11 2AA
G4 LAD Stafford & Districts ARS c/o M. Howes Yarnbury Rufc, Brownberrie Lane, Leeds LS18 5HB
G4 LAE C. Wordley Whispering Winds, 7 Fulcher Avenue, Chelmsford CM2 6QN
G4 LAF R. Brodrick 16 Wallenge Drive, Paulton, Bristol BS39 7PX
G4 LAI C. Fone 12 Chiltern Rise, Ashby-de-la-Zouch LE65 1EU
G4 LAJ R. Hackett 4 Ryton Grove, Birmingham B34 7RS
G4 LAK R. Procter 83 Twickenham Road, Newton Abbot TQ12 4JG
G4 LAM R. Lamberton 28A Newtown Road, Raunds, Wellingborough NN9 6LX
G4 LAN P. Conway 14 Leahall Lane, Rugeley WS15 1JE
GM4 LAO A. Waddell 13 Auchenglen Road, Braidwood, Carluke ML8 5PH
G4 LAU Nunsfield House c/o C. Stevens 9 Newbury Avenue, Melton Mowbray LE13 0SR
G4 LAW F. Craven 2 Barn Owl Way, Stoke Gifford, Bristol BS34 8RZ
G4 LAY G. Dobbs Chaka, Grimsby Road, Market Rasen LN8 6DH
G4 LBC P. Rusling 20 Packman Lane, Kirk Ella, Hull HU10 7TL
GM4 LBE A. Tait 12 Greenwell, Gott, Shetland ZE2 9UL
G4 LBH R. Giles 33 Sowerby Avenue, Luton LU2 8AF
G4 LBJ L. Gurney Bluebridge Coach House, Colchester Road, Halstead CO9 1QG
G4 LBM RAF Waddington ARC c/o I. Jackson 5 Vivien Close, Chessington KT9 2DE
G4 LBQ J. Philipson Clifton Farm House, Pullover Road, King's Lynn PE34 3LS
G4 LBS King's Lynn ARC c/o K. Groom 2 Ruins Barn Road, Tunstall, Sittingbourne ME10 4HS
G4 LBT R. Harmer-Knight 3 Grendon Drive, Sutton Coldfield B73 6QA
G4 LBU E. Kersey 98 Campbell Road, Ipswich IP3 9RE
G4 LBY S. Wright 22 Crown St., Mansfield NG18 3JL
G4 LCB Dr M. Goldman 19 Myddelton Park, London N20 0HT
G4 LCE N. Watson 14 Mill Lane, Whittlesford, Cambridge CB22 4NE
GW4 LCF G. Williams 2 The Paddocks, Lodge Hill, Newport NP18 3BZ
G4 LCH M. Gregory Highleys Farm, 375 Tanworth Lane, Shirley, Solihull B90 4DX
G4 LCL E. Beardmore Kilaguni, The Avenue, Stoke-on-Trent ST9 9LW
G4 LCM P. Allsopp 32 Linden Close, Prestbury, Cheltenham GL52 3DU
G4 LCU M. Brownlow The Croft, 1 Byne Close, Pulborough RH20 4BS
GW4 LDA R. Lawrence Jabulani, Harbour Master'S Office, Penarth Portway, , Penarth CF64 1TQ
G4 LDB T. Kendall 86 Rockford Close, Redditch B98 7YL
G4 LDC A. Wallis 4 Trevose Close, Chandler'S Ford, Eastleigh SO53 3EB
G4 LDD P. Harling Pimlico House, Gisburn Road, Clitheroe BB7 4ES
G4 LDJ F. Gabell 25 Woodland Way, Crowborough TN6 3BQ
G4 LDL Bury St Edmunds ARS c/o A. Bettley 1 Dovetrees, Covingham, Swindon SN3 5AX
GI4 LDN S. Mcquaid Mullaghrodden, Dungannon, Co Tyrone BT70 3LU
GW4 LDP I. Dobby 43 Chestnut Avenue, West Cross, Swansea SA3 5NL
G4 LDR N. Underwood Blandings, Yarmley Lane, Winterslow, Salisbury SP5 1RB
G4 LDS C. Baker 14 Clarendon Road, Morecambe LA4 4HS
G4 LDT D. Holland 29 Lily Crescent, Sunderland SR6 7HN
G4 LDW M. Morris 11 Kingswood Avenue, Hampton TW12 3AU
GM4 LDX M. Mcforsyth Haltoun, Eddleston, Peebles EH45 8PW
G4 LED A. Wood 67 Bay View Road, Duporth, St. Austell PL26 6BN
G4 LEG P. Brent 14 Stagelands, Crawley RH11 7PE
G4 LEM Newton Le Willows ARC c/o E. Goodman 83 Avondale Road, Kettering NN16 8PL
G4 LEN A. Kendall 18 Chivenor Way Kingsway Quedgeley, Gloucester GL2 2BH
G4 LEP C. Jacobs 16 Woodyard Close, Mulbarton, Norwich NR14 8AS
GM4 LER T. Goodlad 72 North Lochside, Lerwick, Shetland ZE1 0PJ
G4 LES L. Macvean 27 Babs Field, Bentley, Farnham GU10 5LS
G4 LEV C. Veitch 108 Racecourse Road, Rd2, Otane 4277 New Zealand
G4 LEX G. Train 29 Waggoners Way, Morton, Bourne PE10 0XR
GM4 LFE R. Broom 1 Byron Court, Banff AB451FB
GW4 LFF Dr J. Devonshire 19 Voss Park Drive, Llantwit Major CF61 1YD
G4 LFG M. Davis 53 St. Georges Avenue, South Shields NE33 3EH
GM4 LFK L. Mclean Lower Hatton Cottage, Dunkeld PH8 0ET
GM4 LFL J. Rennie 1 The Banks, Brechin DD9 6JD
GW4 LFO Highfield ARC c/o D. Jenkins 44 Kensington Drive, Bridgend CF31 4QS
G4 LFQ J. Holloway Flat 1, 66 Unthank Road, Norwich NR2 2RN
G4 LFS R. Draycott 25 Flat Lane, Whiston, Rotherham S60 4EF
G4 LFT A. Busby 16A High Street, Sutton-On-Trent, Newark NG23 6QA
GW4 LFV B. Crow Lindisfarne, Pen Y Waun, Pentyrch, Cardiff CF15 9SJ
GW4 LFW T. Cross 50 Ty-Newydd, Whitchurch, Cardiff CF14 1NQ
G4 LGB B. Graham 19 Cannerby Lane, Sprowston, Norwich NR7 8NQ
G4 LGH K. Garside 191 Kenton Road, Newcastle upon Tyne NE3 4NR
GM4 LGM J. Mcgregor 26 Engels Street, Alexandria G83 0RZ
G4 LGO N. Thomas 24 Victoria Road, Teddington TW11 0BG
GI4 LGP S. Mccracken 29 Norwood Gardens, Belfast BT4 2DX
G4 LGU W. Hills Alperton, Rowhill Road, Dartford DA2 7QQ
G4 LGX J. Hall 30 Chatsworth Road, Harrogate HG1 5HS
G4 LGY P. Harber 28 Regent Road, Epping CM16 5DL
G4 LHA G. Reoch 350 Mavanelle Cove, Hempstead 77445 United States
G4 LHE J. Lee 41A Orchard Road, Seer Green, Beaconsfield HP9 2XH
G4 LHF S. Moffat 14 Churchill Rise, Burstwick, Hull HU12 9HP
G4 LHI P. Rosamond 13 Newnham Close Hartford, Huntingdon PE29 1RP
GM4 LHJ J. Campbell 23 Napier Avenue, Bathgate EH48 1DF
GW4 LHL S. Edwards 16 Maes Crugiau, Rhydyfelin, Aberystwyth SY23 4PP
G4 LHO P. Gillen 622 Glenwood Dr., Oxnard 93030 United States
G4 LHP G. Griffiths 21 Spring Lane, Olney MK46 5HT
GM4 LHQ W. Herron 21 Southfield Avenue, Paisley PA2 8BY
G4 LHR E. Williams 10 Eastbourne Close, Ingol, Preston PR2 3YR
G4 LHT J. Mcsorley 117 Park Avenue, Ruislip HA4 7UL
GM4 LHW S. Burnett 17 Crusader Drive, Roslin EH25 9NP
G4 LIA J. Gordon 36 Warbeck Close, Newcastle upon Tyne NE3 2FG
G4 LIC J. Graham 14 Ashbourne Grove, London W4 2JH
GI4 LIF R. Goligher Mountjoy East, County Tyrone BT79 7JJ
G4 LIG P. Hesketh 5 Beeches Close, Ixworth, Bury St. Edmunds IP31 2EW
G4 LIJ R. Nutt 4 Mercers Drive, Bradville, Milton Keynes MK13 7AY
G4 LIL C. Brown Sandysike Cottage, Sandysike, Carlisle CA6 5SS
G4 LIM G. Moody 37 Pine Street, Stockton-on-Tees TS20 2SP
G4 LIO J. Marshman 12 Neelands Grove, Cosham, Portsmouth PO6 4QL
G4 LIQ P. Williams 54 High St., Yelling, St. Neots PE19 6SD
G4 LIR P. Taylor 64 Walford Road, Rolleston-On-Dove, Burton-on-Trent DE13 9AR
GM4 LIS D. Wilkes 11 Trinity Crescent, Beith KA15 2HG
G4 LIX G. Greenwood 11 James Street, Holywell Green, Halifax HX4 9AS
G4 LIY C. Ware 4 Highfield Terrace, Lower Bentham, Lancaster LA2 7EP
G4 LJB J. Wild 6 Chestnut End, Headley, Bordon GU35 8NA
GU4 LJC Cmdr. B. Le Lievre Calabar Forest Road, Forest GY8 0AB Guernsey
G4 LJF Capt. I. Shepherd Hutts Farm, Blagrove Lane, Wokingham RG41 4AX
G4 LJK R. Mckee 5 Moorcroft, Ossett WF5 9JL
G4 LJN R. Bartlett 37 Church Road, Ferndown BH22 9ES
G4 LJR G. Garden 9 Gateway Avenue, Smithville L0R 2A0 Canada
GW4 LJS P. Harding Harbour Light, Five Roads, Llanelli SA15 5AQ
G4 LJT W. Hayward 19 Woodlands Coxheath, Maidstone ME17 4EE
G4 LJU C. Howell 43 Copsleigh Close, Salfords, Redhill RH1 5BJ
GW4 LJW J. Jenkins Pantycelyn, Llanwnnen, Lampeter SA48 7LW
G4 LJY J. Warren Clifden Farm, Quenchwell Carnon Downs, Truro TR3 6LN
G4 LKD J. Spurgeon Whitgift House, Whitgift, Goole DN14 8HL
GI4 LKG V. Tait 30 Corby Drive, Lisburn BT28 3HG

G4 LKM G. Clarke 33 Mulberry Avenue, Penwortham, Preston PR1 0LL
G4 LKP Dr K. Craven 8 Melander Close, York YO26 5RP
GW4 LKS W. Evans Dan Y Craig, Craig Road, Swansea SA7 9HS
G4 LKT P. Goodman 34 Fullers Road, South Woodford, London E18 2QA
G4 LKU D. Hill 8 Lingfield Walk, Corby NN18 9JS
G4 LKW P. Head 36A Ashacre Lane, Worthing BN13 2DH
G4 LKX D. Hepworth 2 Granby Crescent, Doncaster DN2 6AN
G4 LKZ C. Shuttleworth 17 Stirling Close, Clitheroe BB7 2QW
G4 LLG P. West 5 Stonehill Close, Appleton, Warrington WA4 5QD
G4 LLI G. Matthews 101 Trafalgar Road, Horsham RH12 2QL
G4 LLL Dr N. Rudgewick-Brown 10 Windsor Park, Dereham NR19 2SU
G4 LLM T. Barnes 20 Mayes Close, Warlingham CR6 9LB
G4 LLN R. Connolly Newfane, Temple Way, Slough SL2 3HE
G4 LLQ Northern Fells CG c/o A. Leeming 52 Kingfisher Drive, Pickering YO18 8TA.
G4 LLZ A. Barr 28 Roundway, Honley, Holmfirth HD9 6DD
G4 LMA J. Baylis 41 Ailesbury Way, Burbage, Marlborough SN8 3TD
G4 LMF R. Harrison 18 Gunners Lane, Studley B80 7LX
GM4 LMG D. Heasman 41 Honeyberry Drive, Rattray, Blairgowrie PH10 7RB
G4 LMK J. Morris 17 Overbrook Grange, Nuneaton CV11 6BQ
G4 LML W. Turner 11 Field View Close, Exhall, Coventry CV7 9BJ
G4 LMM P. Stears 127 Hughenden Avenue, High Wycombe HP13 5SS
G4 LMN D. Piper 45A Spielplatz, Lye Lane, Bricket Wood, St. Albans AL2 3TD
G4 LMR Inmarsat ARC c/o G. Sims 85 Surrey Street, Glossop SK13 7AJ
G4 LMV W. Loxley 92 Needlers End Lane, Balsall Common, Coventry CV7 7AB
G4 LMW R. Thomson Shire Jee Neevas, Cold Ash Hill, Cold Ash, Thatcham RG18 9PH
G4 LMX S. Crosson Smith The Old Pump House, Engine Road Ten Mile Bank, Downham Market PE38 0EN
G4 LMY J. Piggott 30 Farleigh Fields, Orton Wistow, Peterborough PE2 6YB
G4 LNC A. Friis 22 Garthwaite Crescent Shenley Brook End, Milton Keynes MK5 7AX
G4 LNE N. Howorth 42 Fairfield Avenue, Rossendale BB4 9TQ
G4 LNG F. Hollis 97 Manor Road, Chesterfield S40 1HZ
G4 LNM D. Brown 26 The Brucks, Wateringbury, Maidstone ME18 5PX
GM4 LNN C. Foden 4 Coastguard Houses, Cromwell Road, Kirkwall KW15 1LN
G4 LNQ K. Marshall 44 Rosemary Drive, Alvaston, Derby DE24 0TA
G4 LNR L. Miles 130 Well Lane, Willerby, Hull HU10 6HS
G4 LNT B. Thompson 113 Gordon Road, Stanford-le-Hope SS17 7QZ
G4 LNY A. Thurgood 19 Froment Way, Milton, Cambridge CB24 6DT
G4 LNZ G. Langford 15 Ambleside Drive, Hereford HR4 0LP
G4 LOB A. Major 33 Borough Road, Bridlington YO16 4HN
GW4 LOD D. Parrott 39 Groves Road, Newport NP20 3SP
G4 LOE G. Tuppeny 5 Ashlawn Crescent, Solihull B91 1PR
G4 LOF M. Adams 7 Finningley Drive, Allestree, Derby DE22 2XP
G4 LOG R. Farley Linden Lea, Close Hill, Redruth TR15 1EW
G4 LOH T. Fern South Boderwennack Farm, Trevenen Bal, Helston TR13 0PR
G4 LOI B. Howell 13 Westfield, Plympton, Plymouth PL7 2DY
G4 LOJ C. Black Charisma, Church Road, Norwich NR14 7PB
G4 LOM J. Boult Findon Lodge, Hartside, Durham DH1 5RJ
G4 LON J. Berg 25 Larch Close, Billinge, Wigan WN5 7PX
G4 LOO D. Ross 3 Little Lane, Clophill, Bedford MK45 4BG
G4 LOP C. Hannah 63 Chauntry Road, Alford LN13 9HJ
G4 LOR W. Mooney 21 Windsor Court, Poulton-le-Fylde FY6 7UX
G4 LOV J. Sutherland 31 Kensington Road, Sandiacre, Nottingham NG10 5PD
G4 LOX D. Morton 9 Metford Grove, Bristol BS6 7LG
G4 LOY B. Carr Spring House, Station Road, Grimsby DN36 5QS
G4 LPD R. Mills 3 Whitfield Close, Wilford, Nottingham NG11 7AU
G4 LPE I. Sill 36 Snowshill Drive, Cheswick Green, Solihull B90 4JT
G4 LPF S. Swain 9 Brickyard, Stanley Common, Ilkeston DE7 6FR
GM4 LPG W. Maslen Broomie Knowe, Skye Of Curr Road, Grantown-on-Spey PH26 3PA
GM4 LPJ Dr G. Kolbe Riccarton Farm, Newcastleton TD9 0SN
G4 LPL I. Davis 728 Ridge Rd #30, Lantana 33462 United States
G4 LPO Dr J. Hampson Ivy Lodge, Tor Side, Rossendale BB4 4AJ
G4 LPP Dr P. Holt Ellon House, Church Road, Sutton, Norwich NR12 9SG
GM4 LPT J. Hopkins 19 Cairnport Road, Stranraer DG9 8BQ
GW4 LPU J. Jones 9 Aelybryn, Ceinws, Machynlleth SY20 9EZ
G4 LPW C. Clarke 5 The Cottages, Low Road North Tuddenham, Dereham NR20 3DG
G4 LPY J. Carter 147 Maidenway Road, Paignton TQ3 2PT
G4 LPZ R. Doran 1 Maple Drive, Chellaston, Derby DE73 6RD
G4 LQD T. Alderman 8 Melrose Road, Weybridge KT13 8UP
G4 LQE N. Bishop 33 Pollards Green, Chelmsford CM2 6UH
G4 LQF N. Field 14 Regent Road, Harborne, Birmingham B17 9JU
G4 LQG C. Richardson 25 Hookstone Drive, Harrogate HG2 8PR
G4 LQH R. Sharpe Owl Cottage, Royal Oak Lane, Lincoln LN5 9DT
G4 LQI J. Cavanagh 133 Brox Road Ottershaw, Chertsey KT16 0LG
G4 LQL D. Lander 1 Colby Close, Forest Town, Mansfield NG19 0LS
G4 LQM T. Mccrimmon The Square, Newbiggin, Heads Nook, Brampton CA8 9DH
G4 LQP I. Plant 6 Randol Close, Mansfield NG18 5HY
GM4 LQR J. Reid 80 Bellside Road, Cleland, Motherwell ML1 5NU
G4 LQW North West ARC c/o D. Mcniel C/O G Mcniel, Stable Cottage, Alresford SO24 0HP
G4 LQX R. Coleman 35 Meadowside Road, Upminster RM14 3YT
G4 LRB K. Geen 34 Kensington Road, Ipswich IP1 4LD
G4 LRD D. Holt 241 New Hey Road, Oakes, Huddersfield HD3 4GH
G4 LRG J. West 9 Bainbridge Court St. Helen Auckland, Bishop Auckland DL14 9EJ
G4 LRH G. Obermaier 9 Milton Park Avenue, Southsea PO4 8JG
G4 LRL P. Wilkins 12 Chadcote Way Catshill, Bromsgrove B61 0JT
G4 LRN N. Barker 3 Silesbourne Close, Birmingham B36 9ST
G4 LRO R. Talbott 33 Highfield Street, Anstey, Leicester LE7 7DU
G4 LRP A. Boyd 5 Walmer Close, Southwater, Horsham RH13 9XY
G4 LRT S. Berry Hillview, Stanford Close, Cold Ashby, Northampton NN6 6EW
GM4 LRU T. Hood 29 Thomson Crescent Port Seton, Prestonpans EH32 0AN
G4 LRV N. Bundle Whiteway Cottage, Moorside, Sturminster Newton DT10 1HQ
G4 LSA J. Bell Byanna Cottage, Sturbridge, Stafford ST21 6LE
G4 LSE K. Darton 18 Highfield Avenue, Bishop's Stortford CM23 5LS
G4 LSG S. Smith 1 Parkland Crescent, Norwich NR6 7RQ
G4 LSK A. Sate 2 Lynns Hall Close, Great Waldingfield, Sudbury CO10 0FH
G4 LSL B. Lawrence 42 Cross St Crowle, Scunthorpe DN17 4LH
G4 LSQ P. Elmer 6 Elmers Lane, Kesgrave, Ipswich IP5 2GW
G4 LSU A. Burnett 72 Ightham Road, Erith DA8 1LU
G4 LSV C. Herrett 61 Mansfield Road, Alfreton DE55 7JN
G4 LSX G. Pearce 13 Walnut Close, Nailsea, Bristol BS48 4YH
G4 LTC J. Diment 16 Riverside Walk, Isleworth TW7 6HW
G4 LTH J. Allan 13 Vincent Close, Corringham, Stanford-le-Hope SS17 7QL
G4 LTI M. Coverdale 1A Halton Chase, Westhead, Ormskirk L40 6JR
G4 LTK P. Hinks 1 Richard Joy Close, Holbrooks, Coventry CV6 4EY
GM4 LTL N. Hyde 18 Mansefield, Methlick, Ellon AB41 7DF
G4 LTS B. Packington 83 Fitzroy Road, Whitstable CT5 2LE
G4 LTT A. Willetts 43 Galloway Avenue, Birmingham B34 6JL
G4 LTZ J. Peake 8 Surrey Drive, Congleton CW12 1NU
G4 LUA R. Gathergood 37 Hawkley Drive, Tadley RG26 3YH
G4 LUB D. Waldron 1 Galbraithe Close, Bilston WV14 8HX
GM4 LUD R. Bannerman 20 Post Box Road, Birkhill, Dundee DD2 5PX
G4 LUE E. Bailey 8 Hild Avenue Cudworth, Barnsley S72 8RN
G4 LUF R. Irish 15 Tenter Hill, Wooler NE71 6DB
G4 LUN A. Stickland 3 Kivernell Road, Milford On Sea, Lymington SO41 0PP
G4 LUO C. Morgans Merlewood, Maidstone Road, Sittingbourne ME9 7QA
G4 LUP D. Purslow Baylands, Pinkham, Cleobury Mortimer, Kidderminster DY14 8QE
G4 LUQ M. Tust 21 Laneside Close, Chapel En le Frith SK23 0TS
GM4 LUS S. Smith 80 Deanburn Park, Linlithgow EH49 6HA
G4 LUT W. Terry Morston, 121 Lodge Lane, Grays RM17 5SF
G4 LUW E. Johnson 29 Watering Lane, Collingtree, Northampton NN4 0NJ
GW4 LUX P. Biddle 14 Crossways Park, Howey, Llandrindod Wells LD1 5RD
G4 LUY D. Chubb 11 Pelham Close, Bembridge PO35 5TS
G4 LVA A. Lucas 4 Hewell Close, Kingswinford DY6 7RQ
G4 LVD B. Durrant 140 Fletcher Road, Ipswich IP30LA
G4 LVG D. Halls 7 Raeburn Road, Ipswich IP3 0EW
G4 LVI A. Entwistle 68 Sandy Lane, Stretford, Manchester M32 9BX
G4 LVK A. Kelly 40 Housman Park, Bromsgrove B60 1AZ
G4 LVO V. Stretch 5 Ledwych Road, Droitwich WR9 9LA
G4 LVR Dr B. Roe 7 Abbey Fields, Crewe CW2 8HJ
G4 LVV A. Hanson 1 Church Street, Kempsey, Worcester WR5 3JG
GM4 LVW Dr M. Bowman 5 Whinfield Gardens, Prestwick KA9 2PW
G4 LWB P. Smith 2A Kirby Lane, Kirby Lodge, Melton Mowbray LE13 0BY
G4 LWC L. Collins 44 Hollybush Lane, Penn, Wolverhampton WV4 4JJ
GW4 LWD W. Chandler 19 Cilhaul Terrace, Mountain Ash CF45 3ND
G4 LWF P. Green 1 Haddon Croft, Hayley Green, Halesowen B63 1JQ
G4 LWG I. Lambert 21 East View Terrace, Barnoldswick BB18 5NW
GW4 LWL K. Edwards 25 Gareth Close, Thornhill, Cardiff CF14 9AF
G4 LWN R. Nock 83 Coles Lane, West Bromwich B71 2QW
G4 LWQ G. Simpson Floral Cottage, 11 Summerwood Road, Street BA16 0RL
G4 LWU R. Moore 45 Lime Kiln Way, Salisbury SP2 8RN
G4 LWV E. Videan 40 Guessens Grove, Welwyn Garden City AL8 6RF
G4 LWY J. Bryce 6A Cawley Avenue, Culcheth, Warrington WA3 4DF
GW4 LWZ Tewkesbury Rynt c/o S. Trott 6 Mounton Drive, Chepstow NP16 5EH
G4 LXA D. Gibson 10 Church Close, Braybrooke, Market Harborough LE16 8LD
G4 LXC P. Johnson 148 Broadmead, Tunbridge Wells TN2 5NN
G4 LXD Dr F. De Bass Hawthorns, 10 Melville Road, Thetford IP24 1NG
G4 LXH D. Jones 34 Alpha Grove, Isle Of Dogs, London E14 8LH
G4 LXJ C. Phillips Bangla, Silchester Road, Tadley RG26 5EP
GM4 LXM G. Low 23 Bellfield Road, North Kessock, Inverness IV1 3XU
GW4 LXO Dr J. Eastment 211 Pantbach Road, Rhiwbina, Cardiff CF14 6AE
G4 LXR R. Hooper 88 Ninehams Road, Caterham CR3 5LJ
G4 LXU C. Lennox Lowfield House, Low Street, York YO61 4QA
G4 LXV A. Rose 40 Wilson Drive, Outwood, Wakefield WF1 3DN
G4 LXW A. Trousdale 65 Low Moor Side, New Farnley, Leeds LS12 5EA
G4 LXX I. Taylor Heatherlands, Felixstowe Road, Ipswich IP10 0DE
G4 LXY D. Millin Flat G12A, Elizabeth Court, Bournemouth BH1 3DX
G4 LYB C. KitchenerHamrest 5, Mill Road, , Cromer NR27 0BG
G4 LYC P. Collett 7 Saxon Rise, Earls Barton, Northampton NN6 0NY
G4 LYD D. Palmer 123 Bucklesham Road, Kirton, Ipswich IP10 0PF
G4 LYE N. Pilling 22 Templar Way, Selby YO8 9XH
G4 LYF Cross Border CG c/o Dr A. Webb 11 Crowland Road, Haverhill CB9 9LE
G4 LYG J. Lavis Brea Croft, St. Just TR19 7RN
G4 LYJ R. Carter 11 Ash Close, Shrewsbury SY2 6HU
G4 LYL H. Bonnor 1 Christchurch Road, Winchester SO23 9SR
G4 LYM Trowbridge And District ARC c/o G. Schiffeldrin 68 The Fairway, Alwoodley, Leeds LS17 7PD
G4 LYU B. Gauntlett 4 Sandbanks Gardens, Hailsham BN27 3TL
GM4 LYV W. Hattie 47 Border Way, Kirkintilloch, Glasgow G66 2BD
G4 LYX J. Wylie 15 Semley Road, Hassocks BN6 8PD
G4 LYY J. Schoolar 140 Slades Road, Golcar, Huddersfield HD7 4JR
G4 LZD G. Reading 73 Mayflower Close, Dartmouth TQ6 9JN
G4 LZE C. Lugard 5 Woodland Gardens, South Croydon CR2 8PH
G4 LZJ P. Garnett Drewen Garth, Church St, Hull HU11 4RN
G4 LZK R. Broughton 18 Elim Court Gardens, Crowborough TN6 1BS
GM4 LZO G. Mcdonald Ellrigg, Ballencrieff Toll, Bathgate EH48 4LD
GW4 LZP R. Smith 4 Glan Ysgethin, Talybont LL43 2BB
G4 LZQ G. Williams 2 Whitminster Lane, Frampton On Severn, Gloucester GL2 7HR
GI4 LZR W. Turner 31 Thiepval Avenue, Belfast BT6 9JF
GI4 LZS J. Smyth 12 Cleland Park Central, Bangor BT20 3EP
G4 LZT R. Brown 40 Pegholme Drive, Otley LS21 3NZ
G4 LZU E. Hayden Firbank, 1 Watery Lane, Taunton TA3 5BX
G4 LZV K. Brazington 38 Tamworth Road, Amington, Tamworth B77 3BT
G4 LZZ A. Siemieniago 3 Skye Close, Highworth, Swindon SN6 7HR
G4 MAB M. Barry 19 Trinity Road, Northampton NN3 3FA
GI4 MAC Dr M. Mckinney 117 Downpatrick Road, Crossgar, Downpatrick BT30 9EH
G4 MAG D. Lucas 23 Rectory Close, Wistaston, Crewe CW2 8HG
GM4 MAI Mid Warks Raynet Group c/o Dr A. Mcwilliam Lochaber, Braehead, Avoch IV9 8QL
GI4 MAJ W. Mcclintock 37 Belfast Road, Larne BT40 2PH
G4 MAK J. Gregg 11 Talbot Terrace , Rothwell, Leeds LS26 0DR
G4 MAR C. Rowe 29 Lucknow Road, Willenhall WV12 4QF
G4 MAS C. Day 59 Hoe Lane, Ware SG12 9LS
G4 MAU Dr D. Birchall 6 Hillmorton Road, Knowle, Solihull B93 9JL
GI4 MAZ P. Canny 7 Drumachose Park, Limavady BT49 0NY
G4 MB J. Bowes 20 Broomfield Road, Bexleyheath DA6 7PA
G4 MBA A. Cowsill 21 Manor Close, Bromham, Bedford MK43 8JA

G4 MBC Mid-Beds Contest Assoc c/o F. Handscombe Sandholm, Bridge End Road, Red Lodge, Bury St. Edmunds IP28 8LQ
G4 MBD I. Moth 145 Carisbrooke Road, Newport PO30 1DG
G4 MBE R. Scargill 17 Springfield Lane, Morley, Leeds LS27 9PL
GM4 MBG I. Simpson 1 Knockhall Road, Newburgh, Ellon AB41 6BJ
G4 MBH A. Embleton 4 Daventry Close, Mickleover, Derby DE3 0QT
G4 MBJ R. Hyett 18 Escley Drive, Hereford HR2 7LU
G4 MBK J. Broadbent Buttercross Cottage, Low Road, Gainsborough DN21 4ER
GW4 MBL S. Elmore Eirianfron, Llangoed, Beaumaris LL58 8PG
GI4 MBM S. Mcconnell 8 Carnesure Drive, Comber, Newtownards BT23 5LP
GI4 MBQ C. Black 5 Woodbrook Park, Warrenpoint, Newry BT34 3HL
G4 MBZ P. Taylor 32 Grasmere Road, Lightwater GU18 5TJ
G4 MCA J. Hanton 5 St. Davids Drive, Thorpe End, Norwich NR13 5HR
G4 MCE A. Audcent 9 Woodlands, Axbridge BS26 2AX
G4 MCF C. Begg 11 Lilac Road, Normanby, Middlesbrough TS6 0BS
GI4 MCH N. Crymble 60 Princes Drive, Newtownabbey BT37 0AZ
G4 MCM D. Hadaway 22 The Delph, Reading RG6 3AN
G4 MCQ S. Bailey 50 Quantock Close, Warmley, Bristol BS30 8UT
GD4 MCR D. Cannon 44 Derby Road, Peel, Isle of Man IM5 1HP
G4 MCU G. Stow 15 Hawthorne Gardens, Hockley SS5 4SW
GM4 MCV A. Patterson Wayside, 37 Abbotsford Road, Galashiels TD1 3HW
GI4 MCW G. Edgar 51 Kempe Stones Road, Newtownards BT23 4SQ
G4 MD South Bristol ARC c/o P. Howett 2, Parkfield Road, Stourbridge DY8 1HD
G4 MDB R. Tokley 9 Peel Road, Springfield, Chelmsford CM2 6AQ
G4 MDC J. Divall 2 Brockswood Lane, Welwyn Garden City AL8 7BG
GI4 MDD I. Gibson 4 Ilford Avenue, Belfast BT6 9SF
G4 MDE K. Hodkinson 13 Clovelly Road, Edenthorpe, Doncaster DN3 2PE
G4 MDF Chelsea Pensioners RC c/o B. Bristow Club Jays Lodge, Reading RG8 7QG
G4 MDH G. Feary 76 Parsons Way, Royal Wootton Bassett, Royal Wootton Bassett SN4 8DJ
G4 MDJ A. Smith 48 Milltown Way, Leek ST13 5SZ
G4 MDK G. Winfield 328 Stone Road, Stoke-on-Trent ST4 8NJ
G4 MDM P. Brassington 42 Dartmouth Avenue, Newcastle ST5 3NY
G4 MDN D. Fowler The Dees, Cross Lanes, Gerrards Cross SL9 0LR
GI4 MDO S. Hewitt 23 Drumard Road, Portadown, Craigavon BT62 4HP
G4 MDR Dengie Hundred ARS c/o A. Farmer 42 Sunridge Close, Newport Pagnell MK16 0LT
G4 MDT G. Fitton 29 Okus Grove, Upper Stratton, Swindon SN2 7QA
G4 MDU Dr J. Gudgeon Shillingsworth Cottage, Leckhampstead Road, Wicken MK19 6BY
GD4 MDY S. Keenan Fenella Villa, Peveril Road, Peel IM5 1PJ Isle of Man
G4 MDZ S. Cline 24 Petrel Way, Hawkinge, Folkestone CT18 7GZ
G4 MEA R. Hutchings 16 Le Marchant Road, Frimley, Camberley GU16 8RW
G4 MEB Leeds&Dist ARC c/o P. Green 1 Haddon Croft, Hayley Green, Halesowen B63 1JQ
G4 MEE D. Mobbs 39 Bramwell Road, Freckleton, Preston PR4 1SS
G4 MEF B. Rudkin 18 Beechfield Avenue, Birstall, Leicester LE4 4DA
G4 MEH M. Hughes 8 Elm Beds Road, Poynton, Stockport SK12 1TG
GW4 MEI I. Williams 27 Y Glyn, Caernarfon LL55 1HF
G4 MEK C. Chappell 6 Brayside Avenue, Cowcliffe, Huddersfield HD2 2PQ
G4 MEM M. David Ridgeway, Grainbeck Lane, Killinghall, Harrogate HG3 2AA
G4 MEO B. Elliott 13 Spring Grove, Sandy SG19 1EU
GI4 MEQ Rose & Crown RC c/o M. Kelly 6 Beechdene Gardens, Lisburn BT28 3JH
G4 MES J. Willis 1 Cedar Crescent, Royston SG8 5BP
G4 MET E. Robinson 60 Huntsmans Drive, Hereford HR4 0PN
G4 MEX M. Care The Stables, Hollerday Drive, Lynton EX35 6HQ
GM4 MFB C. Bowden West Reidford, Drumoak, Banchory AB31 5AU
G4 MFD W. Dean Bowerdene, Staplehay Trull, Taunton TA3 7HH
G4 MFE R. Kaiser Blackcoombe Farm, Henwood, Liskeard PL14 5BW
G4 MFI D. Roberts 88 Woodhouse Road, Urmston, Manchester M41 7WX
G4 MFK G. Dennick 29 Old Post Office Lane, Badsey, Evesham WR11 7XF
G4 MFN M. Jones 67 Dosthill Road, Two Gates, Tamworth B77 1JD
GM4 MFO M. Mackenzie Ash Lodge, 8 Brookend Brae, Helensburgh G84 0QZ
G4 MFP W. Woollen Greensward, Townsend, Didcot OX11 0DX
G4 MFQ R. Dunstan 100 Trevithick Road, St. Austell PL25 4RJ
G4 MFR C. Shanks 225 Freshfield Road, Brighton BN2 9YE
G4 MFS M. Smith 54B Kinson Grove, Bournemouth BH10 7JL
GI4 MFT Magherafelt ARS c/o H. Evans Oville House 404 Foreglen Road, Dungiven, Londonderry BT47 4PN
G4 MFV J. Marshall 278 Derby Road, Bramcote, Nottingham NG9 3JN
G4 MFW B. Fletcher 53 Onslow Gardens, London SW7 3QF
G4 MFX G. Davis Orchard House Peartree Avenue, Martham, Great Yarmouth NR29 4RJ
G4 MGB H. Mayor Lock House, Canal Bank, Preston PR4 6HD
G4 MGD T. Sear 14 Crowberry Drive, Scunthorpe DN16 3DE
G4 MGG Kingston Immortals CG c/o S. Esposito 21A Spencefield Lane, Leicester LE5 6PT
G4 MGH R. Hampson 30 Witts Lane, Purton, Swindon SN5 4EX
G4 MGI P. Kemmis Coton Clanford Farm Coton Clanford, Stafford ST18 9PE
G4 MGN M. Norman 7 Kingsway, Seaford BN25 2NE
G4 MGO J. Newman 20 Marshmoor Mobile Home Park, Wallow Lane, Ipswich IP7 7BZ
G4 MGP H. Boddy Greyholme, West Lane, Scarborough YO13 9AR
G4 MGQ R. Boddy Greyholme, West Lane, Scarborough YO13 9AR
G4 MGR Wirral & District ARS c/o D. Jones 39 Pensby Road Heswall, Wirral CH60 7RA
G4 MGV R. Pass 6 High Street, Hanslope, Milton Keynes MK19 7LQ
G4 MGW A. Lodge 92 Cheltenham Road, Gloucester GL2 0LX
G4 MGX J. Freeman 5A Beech Avenue, Briar Bank Park, Bedford MK45 3WE
G4 MGY J. Gibbs 27 East Hill Park, Knatts Valley, Sevenoaks TN15 6YF
G4 MHA P. Wright 2 Windsor Mews, Stanley DH9 8UH
G4 MHC Malvern Hills Radio Amateurs Club c/o D. Hobro 60 Linksview Crescent, Worcester WR5 1JJ
GI4 MHD G. Quaite 4 Drakes Bridge Road, Crossgar, Downpatrick BT30 9EW
G4 MHE R. Musto The Thatch, Mountain Castle, Dungarvan X35 A073 Ireland
G4 MHF J. Marshall Chaseborough House, Village Hall Lane, Wimborne BH21 6SG
G4 MHJ R. Hewitt 38 Eastry Road, Erith DA8 1NN
G4 MHK T. Fougere 48 Longland Road, Eastbourne BN20 8HY
G4 MHQ A. Bell 22 Ryde Place, Lee-on-the-Solent PO13 9AU
GW4 MHR 610 Sqn City of Chester Air Cadets Amateur Radio C c/o C. Norman 31 Cae Braenar, Holyhead LL65 2PN
G4 MHS N. Naish 85 Wear Bay Road, Folkestone CT19 6PR
G4 MHX B. Smith 6 Howbeck Crescent, Wybunbury, Nantwich CW5 7NX
G4 MIB D. Senior Court Barton, Bull Street, Creech St. Michael, Taunton TA3 5PW
G4 MID E. Pratt 65 Barton Road, Thurston, Bury St. Edmunds IP31 3PD
GM4 MIG I. Giffen 57 Glengarry Crescent, Falkirk FK1 5UE
G4 MIH D. Fenton Waverley, Warrington Road, Northwich CW8 2LW
GW4 MII P. Jenkins 2 Gwynfi Street, Treboeth, Swansea SA5 7DW
G4 MIJ R. Hunt 21 Springwell, Ingleton, Darlington DL2 3JJ
G4 MIK Exmouth ARC c/o Dr M. Bull Toad In The Hole, 25 Prospect Park, Tunbridge Wells TN4 0EQ
GM4 MIM Rvd. I. Morrison 53 Eastcroft Drive, Polmont, Falkirk FK2 0SU
G4 MIO P. Davies 9 Place Albert 1E, la Hulpe 1310 Belgium
GW4 MIP P. Phillips Woodreefe, Amroth, Narberth SA67 8NR
G4 MIS N. Allen 17 Winfield Road, Sedbergh LA10 5AZ
G4 MIT G. Hurst The Hollies, Derby DE6 6NB
G4 MIV K. Gibson 179 Ratcliffe Road, Sileby, Loughborough LE12 7PX
G4 MIX M. Howland 1 Swanton Farm Cottages, Lydden, Dover CT15 7JN
G4 MJA M. Swift 4 Embleton Drive, Chester le Street DH2 3JS
G4 MJC F. Jul-Christensen 66 Bushey Lodge Cottages, Firle, Lewes BN8 6LS
GI4 MJD M. Dunne 26 Duncreggan Road, Londonderry BT48 0AD
G4 MJF M. Hill 42 Oaklands Drive, Northampton NN3 3JL
G4 MJI T. Sturmey 35 Lane Court, Boscobel Crescent, Wolverhampton WV1 1QH
G4 MJT F. Harrison 98 The Stray, South Cave, Brough HU15 2AL
G4 MJU E. Smith 256 Stone Road, Stoke-on-Trent ST4 8NJ
G4 MJW S. Carey 27 Kilmar Way, St. Cleer, Liskeard PL14 5LU
G4 MJX G. Fisher 87 Ethersall Road, Nelson BB9 0RP
G4 MKD W. Kendal 9 The Glade, Furnace Green, Crawley RH10 6JS
G4 MKE A. Plaice 10 Stockhill Road, Chilcompton, Radstock BA3 4JL
G4 MKF Dr M. Franks 13 Fifth Road, Newbury RG14 6DN
G4 MKG P. Opie Timbers, Stockers Hill Road Rodmersham, Sittingbourne ME9 0PL
G4 MKI D. Bray 180 Greenhill Road, Herne Bay CT6 7RS
G4 MKP T. Burbidge 11 The Drift Little Gransden, Sandy SG19 3DX
G4 MKQ K. Barnes 2 Zeus Lane, Waterlooville PO7 8AG
G4 MKR R. Byford 46 Sutton Mill Road, Potton, Sandy SG19 2QG
G4 MKT B. Jackson Meadow Top Farm, Edgeside Lane, Rossendale BB4 9SD
GM4 MKU J. Flett 40 Commerce Street, Lossiemouth IV31 6QH
G4 MKW P. Bowden 12 Honeywood Road, Horsham RH13 6AE
G4 MKX C. Gericke Pear Tree House, Water Street, Bristol BS40 6AD
G4 MLB R. Padmore 3 Uldale Close, Nelson BB9 0ST
G4 MLG A. Denyer 94 Wood Lane, Chippenham SN15 3DZ
G4 MLI B. Mitchell 16 Perhaver Park, Gorran Haven, St. Austell PL26 6NZ
G4 MLO G. Rae 62 Brunel Drive, Upton, Northampton NN5 4AJ
G4 MLQ J. Lamont 9 Deepdale Croft, Barugh Green, Barnsley S75 1QG
G4 MLR J. Norris Freshfield House, Freshfield Lane, Haywards Heath RH17 7HE
G4 MLV L. Gaunt 31 Moat Hill, Birstall, Batley WF17 0DX
G4 MLW Dr I. Jones 114 Tennent Road, York YO24 3HG
G4 MLY I. Vincent 40 Treetops Close, London SE2 0DN
G4 MLZ J. Harrison 15 Wold Road, Pocklington, York YO42 2QG
GW4 MM Box 25 Contest Club c/o T. Kirby Parsonage House, St. Nicholas, Goodwick SA64 0LG
G4 MMA K. Barnard 89 Kings Road, Harrow HA2 9LD
G4 MMG A. Beecher 27 Normandale, Bexhill-on-Sea TN39 3LU
G4 MMH M. Evans Corners, Howbourne Lane, Uckfield TN22 4QB
G4 MMI R. Hodge 20 Linden Grove, Roydon, Diss IP22 4GJ
GI4 MMJ L. Kirk 26 Wallace Hill Road, Downpatrick BT30 9BU
G4 MMT A. Haley 9 Minster Avenue, Bude EX23 8RY
G4 MNA L. Meale 57 Chestnut Drive, Newton Abbot TQ12 4JZ
G4 MNE J. White 25 Fulwith Drive, Harrogate HG2 8HW
GI4 MNF N. Foote 4 Bushfield Road Moira, Craigavon BT67 0JB
G4 MNI W. Loucks 155 Brentwood Rd N, Toronto, Ontario M8X 2C8 Canada
GI4 MNN R. Barr 13 Fairhill Walk, Belfast BT15 4GR
G4 MNP M. Ward 14 Grayling Mead, Fishlake Meadows, Romsey SO51 7RU
G4 MNT R. Calkin 5 Bergen Court, Maldon CM9 6UH
G4 MNX C. Leece 101 Wellstead Way, Hedge End, Southampton SO30 2BH
G4 MOC P. Fawkes 118 Rhoon Road, Terrington St. Clement, King's Lynn PE34 4HZ
G4 MOE S. Noke 48 Hoadley Green, Salisbury SP1 3HS
GW4 MOG C. Tombs 14 Heol Merioneth, Boverton, Llantwit Major CF61 2GS
G4 MOH K. Moran 23 Dunlin Close Quedgeley, Gloucester GL2 4GS
G4 MOI D. Bone 69 Pick Hill, Waltham Abbey EN9 3LD
GW4 MOK V. Cashmore 31 Maes Y Dyffryn, Greenfield, Holywell CH8 7QR
G4 MOL K. Highley 3 West Hill Drive, Hythe, Southampton SO45 6DL
G4 MOP H. Williams 15 Hiawatha, Wellingborough NN8 3SH
G4 MOT K. Watson 12 Regency Park Grove, Pudsey LS28 8QD
G4 MOV E. Durey 71 Orchard Road, Maldon CM9 6EW
G4 MOY D. Bristow Flat 88, Millwood, Sycamore Avenue, Bingley BD16 1HW
GW4 MOZ J. Upstone 31 Broadway, Llanblethian, Cowbridge CF71 7EX
G4 MPA G. Squibb 36 Frognal Gardens, Teynham, Sittingbourne ME9 9HU
GM4 MPC D. Smith 13 Fernie Place, Dunfermline KY12 9BX
G4 MPG P. Grace 3 Warwick Grange, Solihull B91 1DD
G4 MPH N. Simmonds 77 Main Street, Long Whatton, Loughborough LE12 5DF
G4 MPI W. Sharples Foxgrove, Bonchurch Shute, Ventnor PO38 1NX
G4 MPJ M. Whitfield 26 Kingsclere Drive, Bishops Cleeve, Cheltenham GL52 8TG
G4 MPK S. Foster 25 Thorne Crescent, Bexhill-on-Sea TN39 5JH
G4 MPL T. Grimbleby 109 Downfield Avenue, Hull HU6 7XE
G4 MPO C. Duffy 87 Allenby Drive, Leeds LS11 5RX
G4 MPQ K. Clark 33 Landers Reach, Lytchett Matravers, Poole BH16 6NB
GM4 MPR D. Miller Old School, Ackergill, Wick KW1 4RG
G4 MPT D. Abbott 42 Rosebery Avenue, Blackpool FY4 1LB
G4 MPW M. Corbett Braemar, Heath Mill Lane, Guildford GU3 3PR
GM4 MPY A. Bell 48 Greenlaw Crescent, Paisley PA1 3RT
GI4 MQA M. Bradley 28 Church Road, Moneyrea, Newtownards BT23 6BB
G4 MQF A. Ramsey 51 Queens Road, Warmley, Bristol BS30 8EJ
G4 MQG C. Winters 45 Blackbush Spring, Harlow CM20 3DY
G4 MQK C. Cubitt The Flint House, Ostend Place, Flat 2, Walcott, Norwich NR12 0NJ
G4 MQL R. Cuddington The Barn, Vatch Lane, Stroud GL67LE
G4 MQP City of Belfast Radio Amateur Society c/o P. Crowe 22 Ringsbury Close, Purton, Swindon SN5 4DE

G4 MQQ S. Jones The Old Granary, Broomsmead, Lapford EX17 6NA
G4 MQR Dr G. Blower 30 The Glebe, Cumnor, Oxford OX2 9QA
G4 MQS P. Elliott 153 Glenhills Boulevard, Leicester LE2 8UH
G4 MQV J. Sanderson 5 Babbacombe Drive, Ferryhill DL17 8DA
G4 MQW R. Richardson Hazeldene, Sutton Road Fovant, Salisbury SP3 5LF
G4 MRB J. Feeley 177 Rock Street, Sheffield S3 9JF
G4 MRD Dr C. Scrase The Holt, Main Road Woolverstone, Ipswich IP9 1AR
G4 MRK J. Veitch 14 Dunmore Avenue, Sunderland SR6 8ET
G4 MRL N. Thomas 31 Gloucester Street, London SW1V 2DB
GI4 MRN J. Mccrea 14 Fairfield Park, Bangor BT20 4TX
G4 MRQ R. Marchington 78 Buxton Road, Dove Holes, Buxton SK17 8DW
G4 MRS Martlesham RS c/o A. Cook The Old Vicarage, High Road, Ipswich IP6 9LP
G4 MRU P. Sables 76 Sandyfields View, Carcroft, Doncaster DN6 8JQ
G4 MRW R. Westmeckett 3 Alton Grove, Portchester, Fareham PO16 9NJ
G4 MRX J. Cooch 1 Ash Grove, Wrea Green, Preston PR4 2NY
GI4 MRZ E. Smith 114 Bloomfield Road South, Bangor BT19 7HR
G4 MSA D. Price Peacehaven House, Chalbury , Wimborne. BH21 7EZ
G4 MSE J. Sivapragasam 1 Treve Avenue, Harrow HA1 4AL
GW4 MSI P. Needham Cil Y Sarn, Llanegryn, Tywyn LL36 9SB
G4 MSJ T. Moan 23 Laurel Grove, Sunderland SR2 9EE
G4 MSK W. Wilkinson 24 Greenway, Bromley BR2 8EY
GM4 MSL G. Wallace 21 Rosslyn Court, Rosslyn Avenue, Perth PH2 0GY
G4 MSN R. Slator 27 The Drive, Alwoodley, Leeds LS17 7QB
G4 MSP P. Shaw 1, Broadgate, Halifax HX4 9HZ
G4 MSQ Solway DX Group c/o F. Watson Syne Hurst Cottage, Kimbolton Road, Bedford MK44 2EW
G4 MSW L. Morgan 22 Stonelea Road, Hemel Hempstead HP3 9JY
G4 MSY R. Naylor 89 Pelham Avenue, Grimsby DN33 3NG
GW4 MTD H. Mcmurray 14 Hopkin St., Brynhyfryd, Swansea SA5 9HN
GW4 MTE R. Smith 6 Lavender Court, Brackla, Bridgend CF31 2ND
G4 MTF G. Williams 8 Blythe Close, Newport Pagnell MK16 9DN
G4 MTG J. Sillitoe 42 Marsham Road, Kings Heath, Birmingham B14 5HD
G4 MTH A. Smith 25 Lindsay Close, Stanwell, Staines TW19 7LF
GM4 MTI D. Spence Royal Fern, Dunollie Road, Oban PA34 5JQ
G4 MTP J. Theodorson 7 Kingfisher Court, Overstone Lakes, Ecton Lane, Northampton NN6 0BD
G4 MTR D. Coulter 9 Taylors Way, Whitehaven CA28 9PD
G4 MTW F. Cook 2 Burford Gardens, Sunderland SR3 1LX
GI4 MTZ G. Downs 19 Mullaghboy Road, Islandmagee, Larne BT40 3TT
G4 MUA A. Kingdon 4 Castlemead Walk, Northwich CW9 8GP
GI4 MUE J. Gwilt 207 Clandeboye Road, Bangor BT19 1AA
G4 MUI D. Brown 114 Telford Way, High Wycombe HP13 5TA
GW4 MUJ R. Parker Llanedw, Rhulen, Builth Wells LD2 3UU
G4 MUL D. Mayo 6 Leigh Avenue, Marple, Stockport SK6 6DF
GI4 MUN A. Lennon 5 The Drumlins, Ballynahinch BT24 8HW
G4 MUP G. Rouse 43 Oakwood Drive, Prenton CH43 7NX
G4 MUQ C. Ashlin 13 Brantfell Grove, Bolton BL2 5LY
G4 MUS D. Elwell 18 Padgetts Way, Hullbridge, Hockley SS56LR
G4 MUT Dr T. Hackwill 6 Ramsbury Drive Earley, Reading RG6 7RT
G4 MUU F. Westall 4 Francesca Lodge, Somerford Way, Christchurch BH23 3QN
G4 MUV S. Nicolle 17 Allensmore Close Matchborough, Redditch B98 0AS
G4 MUW G. Weaver 60 Crispin Road, Winchcombe, Cheltenham GL54 5JX
GM4 MUZ H. Angus South Grange Care Centre, Grange Road, Dundee DD5 4HT
GW4 MVA G. Burhouse 18 Leopard Moth Road, Sealand, Deeside CH5 2FX
G4 MVB A. Berrow 657 Old Lode Lane, Solihull B92 8NB
G4 MVD S. Cook 44 Ettrick Drive Sinfin, Derby DE24 3EA
G4 MVE D. Casey 18 Sandholme Drive, Ossett WF5 8QP
G4 MVP J. Paskins 190 Gore Road, New Milton BH25 5NQ
GI4 MVQ D. Mccluney 49 Upper Cairncastle Road, Larne BT40 2EG
G4 MVS G. Mellett Weston, Byways, Selsey, Chichester PO20 0HY
G4 MVX M. Gardiner 60 Rochford Way, Walton on the Naze CO14 8SR
GW4 MVY D. Davies 48 Bryn Eglur Road, Morriston, Swansea SA6 7PQ
G4 MVZ R. Pepper 4 Marine Avenue, Skegness PE25 3ER
GI4 MWA Dr F. Ruddell 16 Beechfield Manor, Aghalee, Craigavon BT67 0GB
G4 MWD I. Shaw 33 Park Farm Close, Horsham RH12 5EU
G4 MWF P. Wilkinson 14 Grasmere Close, Penistone, Sheffield S36 8HP
G4 MWG Dr I. Grant Middle Paradise, Garsdale, Sedbergh LA10 5PH
G4 MWH R. Blythe 4 Ashlea Close, Selby YO8 4NY
G4 MWJ R. Featherstone 3 Fairfield, Coningsby, Lincoln LN4 4SP
G4 MWL A. Keyworth 14 Robinson Road, Sheffield S2 5QW
G4 MWO P. Gaskell 131 Greenfield Road, Dentons Green, St. Helens WA10 6SH
G4 MWP T. Underhill 5 Lyndhurst Croft, Eastern Green, Coventry CV5 7QE
G4 MWQ C. Weir 1 Ashfield Place, Ilkley Road, Otley LS21 3PN
G4 MWR R. Sansoni 10 Charles Moor, Stockton Lane, York YO31 1BE
G4 MWS Macclesfield & District ARS c/o G. Acton 39 Craig Road, Macclesfield SK11 7YH
G4 MWW J. Mcleod 156 Brockhurst Road, Gosport PO12 3BB
G4 MWX L. Payne 40 Westmorland Drive, Costhorpe, Worksop S81 9JT
G4 MXE R. Swinnerton 8 Maple Close, Brereton Green, Sandbach CW11 1SQ
G4 MXF R. Wallace 161 Alma Avenue, Hornchurch RM12 6AT
G4 MXI K. Bruntlett Cronk-Ny-Mona, King Street, Louth LN11 0PN
G4 MXM W. Hawkridge 7 Langdale Gardens, Leeds LS6 3HB
G4 MXP D. Aunger 2 Leyland Lodge, Morval, Looe PL13 1PN
GI4 MXV M. Mckee 4 Loran Parade, Larne BT40 2DF
GI4 MXW D. Mckinney 13 Lynden Gate, Portadown, Craigavon BT63 5YH
G4 MXY B. Sowerby 3 Goughs Lane, Bracknell RG12 2JR
G4 MYB C. Barham 10 Little Brook Road, Sale M33 4WG
G4 MYD R. Clay 7 Dipper Close, Kilkhampton EX23 9RE
G4 MYE B. Chase 9 Claremont Drive, Taunton TA1 4JE
G4 MYN J. Thomas 42 Allington Drive, Billingham TS23 3UA
G4 MYQ G. Pettican 39 Kabin Road, Norwich NR5 0LW
G4 MYS A. Sillence 74 Atherley Road, Southampton SO15 5DS
GI4 MYT W. Stewart 11 Fairway Gardens, Castlereagh, Belfast BT5 7PS
G4 MYU A. Summers 6 Rothesay Road, Brierfield, Nelson BB9 5RS
G4 MYW B. Mallinson 7 Barnes Wallis Way, Churchdown, Gloucester GL3 2TR
G4 MYY E. BallAshleigh 111 Boslowick Road, Falmouth TR11 2ER
G4 MYZ R. Roscoe 4 Cobham Villas, Longden, Shrewsbury SY5 8EP
GW4 MZB R. Mills Heather Lea, Oaklands, Welshpool SY21 8HL
G4 MZC B. Horsman 53 Meadow Drive, Bembridge PO35 5XU
G4 MZF D. Earnshaw Po Box 1098, Sedona 86339 United States
G4 MZH K. Grimshire 8 St. Peters Close West Buckland, Barnstaple EX32 0TX
G4 MZI G. Dunn 20 The Grange, Wombourne, Wolverhampton WV5 9HX
G4 MZK P. Ansell 46 Rochford Way, Croydon CR0 3AD
G4 MZL E. Ailsby 15 Norman Close, Bridport DT6 4ET
G4 MZM P. Harper 9 The Orchard, Market Deeping, Peterborough PE6 8JS
G4 MZN M. Huntsman Pear Tree Cottage, Hildersham, Cambridge CB21 6BU
G4 MZQ R. Bickley 12 Cemetery Road, Market Drayton TF9 3BD
G4 MZU C. Harrison 6 Woodlands Close, Chandler'S Ford, Eastleigh SO53 5AT
G4 MZV R. Privett 2 Stevenson Court, Eaton Ford, St. Neots PE19 7LF
G4 MZY D. Plater 58 Mead Fields, Bridport DT6 5RF
G4 MZZ J. Powell 40 Kent Road, Formby, Liverpool L37 6BQ
G4 NAC D. Bosworth 13 Burns Road, Kettering NN16 9LA
GI4 NAE D. Jackson 1 Cloughey Road, Portaferry, Newtownards BT22 1ND
G4 NAJ N. Ashdown Cobwebs, Wilderness Lane, Uckfield TN22 4HT
G4 NAK R. Morey 1 Bradfield Cottages Queens Road, Freshwater PO40 9HB
G4 NAQ C. Maby 173 Clevedon Road Tickenham, Clevedon BS21 6RG
G4 NAV G. Quayle 10 Abbotsford Grove, Timperley, Altrincham WA14 5AZ
G4 NBC M. Hoare Chy Noweth, Seworgan, Falmouth TR11 5QN
G4 NBF A. Topsfield Wild Willow Cottage, Hancock Lane, Truro TR2 5DD
G4 NBG Dr C. Budd 12 Chedworth Close Claverton Down, Bath BA2 7AF
G4 NBH W. Cockshaw 14 Shropshire Road, Leicester LE2 8HW
G4 NBI L. Everton 18 Markham Road, Sutton Coldfield B73 6QR
GW4 NBM M. Jones Rhos Eithin, Brynsiencyn, Llanfairpwllgwyngyll LL61 6TZ
G4 NBN A. Bergman Flat 6, Grange Court, Grange Court Road, Bristol BS9 4DW
GI4 NBO G. Barr 51 Hillhead Road, Dundonald, Belfast BT16 1XD
G4 NBP D. Coggins 23 Fernleaze Coalpit Heath, Bristol BS36 2SB
G4 NBQ R. Hassell 2 Regnum Close, Eastbourne BN22 0XH
G4 NBS A. Collett 10 Quince Road, Hardwick, Cambridge CB23 7XJ
G4 NBW J. Alford 86 Grindleford Road, Great Barr, Birmingham B42 2SQ
GW4 NBY K. Barrett Gorbio House, 47 West Road, Bridgend CF31 4HD
G4 NCA P. Cook 38 Oak Road, Kettering NN15 7AP
G4 NCB K. Wooffindin Viewlands, Milford Road, Leeds LS25 6AF
G4 NCD K. Morey Iona, Colwell Road, Totland Bay PO39 0AH
G4 NCI R. Smith 25 Sisters Way, Birkenhead CH41 4FF
G4 NCJ J. Short Allybere, Marhamchurch, Bude EX23 0HY
G4 NCK P. Shapero 3 Princess Court, Leeds LS17 8BY
G4 NCP M. Carver 45 Harvester Way, Crowland, Peterborough PE6 0DG
G4 NCS C. Angove 9A Wanstead Road, Bromley BR1 3BL
G4 NCU M. Hewitt Hillcrest Bungalow, Middle Street, Crewkerne TA18 8LY
G4 NCV L. Hollingworth 55 Glenfield Avenue, Nuneaton CV10 0DZ
G4 NCY I. Woomans 223 Umberslade Road, Selly Oak, Birmingham B29 7SG
G4 NCZ J. Ramsay 79 Humphrey Lane, Urmston, Manchester M41 9PT
G4 NDC D. Mason 133 Bath Road, Atworth, Melksham SN12 8LA
G4 NDD J. Lloyd 72 Thornyville Villas, Plymouth PL9 7LD
G4 NDG N. Green 36 Chatsworth Road, Ellesmere Park M30 9DY
G4 NDL R. Davies 2 Torrhill Cottages Godwell Lane, Ivybridge PL21 0LT
G4 NDM R. Carter 49 Cambridge Road, West Bridgford, Nottingham. NG2 5NA
G4 NDP J. Burnett 42 Wentworth Drive, South Kirkby, Pontefract WF9 3RY
G4 NDR L. Lewis 653 Main Road, Dovercourt, Harwich CO12 4NF
G4 NDT Dr R. Bramley 8 Ivy Bank Park, Bath BA2 5NF
G4 NDU Prof. A. Bramley 8 Ivy Bank Park, Bath BA2 5NF
GM4 NDV W. Rattray 17 Brownside Road, Cambuslang, Glasgow G72 8NL
G4 NEA S. Rice 13 Wigram Way, Stevenage SG2 9TP
G4 NEE D. Foreman 1 Stour Valley Close, Upstreet, Canterbury CT3 4DB
G4 NEG W. Glew Carinya, Beltoft Belton DN9 1MB
G4 NEH J. Hankin Millfield Torpenhow, Wigton CA7 1JF
GW4 NEI K. Hodge 16A Mold Road, Mynydd Isa, Mold CH7 6TD
G4 NEJ K. Jackson 7533 Park Spring Circle, Orlando FL32815 United States
G4 NEL Kirkless Raynet c/o D. Bird 154 Cherrydown Avenue, Chingford, London E4 8DZ
G4 NEO C. Digby 7 Dagnall Road, Olney MK46 5BJ
G4 NEQ R. Welsh Holme View, Farleton, Lancaster LA2 9LF
G4 NER P. Lidbetter 1 Moor Lane, Westfield, Hastings TN35 4QU
G4 NEY J. Jarvis 116 Balland Field, Willingham, Cambridge CB24 5JU
G4 NFA F. Austin 45 Southdown Crescent, Cheadle Hulme, Cheadle SK8 6EQ
G4 NFE J. Edwards 5 Windmill Rise, York YO26 4TU
G4 NFF S. Frisby 19 Woodcock Close, Norwich NR3 3TB
GI4 NFH R. Jennings 117 Belsize Road, Lisburn BT27 4BS
GM4 NFI D. Leckie 6 Galloway Place, Fort William PH33 6UH
G4 NFL C. Peel The Ferns, Park Wood Drive, Newcastle ST5 5EU
G4 NFP M. Saunby Teachmore, Jacobstowe, Okehampton EX20 3AJ
G4 NFR R. Tyson 49 Strathaird Avenue Walney, Barrow-in-Furness LA14 3DE
G4 NFS N. Sharples 47 New Fosseway Road, Bristol BS14 9LW
G4 NFT G. Mcavoy Flat 5, Maple Lodge, Douglas Close, Poole BH16 5HE
G4 NFV J. Clark 12 Ogle Avenue, Morpeth NE61 2PN
GI4 NFW J. Hegarty 1 Cookstown Road, Moneymore, Magherafelt BT45 7QF
G4 NFY A. Clarke Ravenswood, Gull Road, Wisbech PE13 4ER
G4 NGB B. Gliddon Apartment 1, Idle Shores, Springfield Rd, Woolacombe EX34 7BX
G4 NGD M. Hadnum 79 Mowbray Road, Bedford MK42 9UX
G4 NGF M. Chapman Millway, Dunton Lane, Lutterworth LE17 5HX
GM4 NGJ C. Bridges Highfield, Ballinluig, Pitlochry PH9 0LG
G4 NGL J. Gass 2A, Orchard Way, Breachwood Green, Hitchin SG4 8NT
GI4 NGP D. Paul 4 Draperstown Road, Tobermore, Magherafelt BT45 5QG
G4 NGR C. Webber 107 Northfields Lane, Brixham TQ5 8RN
G4 NGS Dr G. Towler 77 Worrin Road, Shenfield, Brentwood CM15 8JL
G4 NGV A. Whittaker 31 Rydal Road, Haslingden, Rossendale BB4 4EE
G4 NHA P. Hastilow 18 Broadway Avenue, Croydon CR0 2LP
GW4 NHB P. Boyce 28 Alfreda Road, Cardiff CF14 2EH
G4 NHD M. Brightman Patriot'S Arms, 6 New Road, Swindon SN4 0LU
G4 NHE C. Evans 4 Fryers Copse, Wimborne BH21 2HR
G4 NHF A. Jones 51 Wiclif Way, Nuneaton CV10 8NH
GW4 NHH Dr M. Buck 23 Velindre Road, Cardiff CF14 2TE
GM4 NHI J. Cramond Robson'S Croft, Dunecht, Aberdeenshire AB32 7EQ
G4 NHL T. Dixon 30 Green Lane, Stamford PE9 1HF
G4 NHN C. Rowsell 31 Kingsfield Gardens, Bursledon, Southampton SO31 8AY
G4 NHO J. Pattemore 2 Edes Cottages Ottways Lane, Ashtead KT21 2PG

G4 NHP S. Porter 16 Alexander Close, Sidcup DA15 8QY
G4 NHQ A. Mcmackin 33 Poynder Place, Hilmarton, Calne SN11 8SQ
G4 NHR I. Turner 74 Diban Avenue, Elm Park Estate, Hornchurch RM12 4YF
G4 NHT Moorlands&Dis A c/o C. Beesley 15 Byron Close, Cheadle, Stoke-on-Trent ST10 1XB
G4 NHW D. Collins 22 Stalyhill Drive, Stalybridge SK15 2TR
GM4 NHX G. Brooks The Old Post Office, Scotscalder, Caithness KW12 6XJ
G4 NID G. Bromley 46 Independent Hill, Alfreton DE55 7DG
G4 NIF D. Lee 22 Woodland Rise, Parkend, Lydney GL15 4JX
G4 NIJ K. Sheldon Whitehaven, May Tree Road, Pershore WR10 2NY
G4 NIL R. Henshall St. Anns, Staplehay, Taunton TA3 7HB
G4 NIP D. Lewis 76 Reading Road, Finchampstead, Wokingham RG40 4RA
G4 NIU M. Sharples 9 Lower Lune Street, Fleetwood FY7 6DA
G4 NIV N. Rowcroft 110 Linceslade Grove, Loughton, Milton Keynes MK5 8BL
G4 NIX G. Cooke 7 Lime Grove, Royston SG8 7DJ
G4 NIY S. Cooke 5 Honey Way, Royston SG8 7ES
G4 NIZ R. King 20 Woodside East, Thurlby, Bourne PE10 0HT
G4 NJA R. Hewson 6 Talisman Drive, Bottesford, Scunthorpe DN16 3SW
G4 NJB K. Hepke 25 Victoria Avenue, Willerby, Hull HU10 6DD
G4 NJI A. Corker 59 Foljambe Road, Rotherham S65 2UA
G4 NJJ P. Cousins Roman Lodge Hall Road Clenchwarton, King's Lynn PE34 4DA
G4 NJK R. Elliott Stables End Cottage, Fatherford Farm, Okehampton EX20 1QQ
GW4 NJL D. Gerrard Salem Chapel, Llanfairtalhaiarn, Abergele LL22 8SS
G4 NJN A. Bowman 18 Essex Avenue, Isleworth TW7 6LF
GI4 NJQ P. Igo 84 Glebetown Drive, Downpatrick BT30 6PZ
G4 NJR M. Doe 2 Summerfields, Yarnfield, Stone ST15 0RH
G4 NJT R. Smith 2131 - 21St Ne, Salmon Arm, Bc V1E 2DN
G4 NJW G. Lowes 26 Huttoft Road, Sutton-On-Sea, Mablethorpe LN12 2QY
GI4 NKB F. Hunter Flat 9 50 Edenvale Crescent, Belfast BT4 2BH
G4 NKC M. Jones Racecourse Farm, Church Lane, Bridgnorth WV16 4NW
G4 NKI J. Shaw 1A Greenways, The Stookes, Chesterfield S40 3HF
GI4 NKK K. Planck 4 Westland Drive, Ballywalter, Newtownards BT22 2TH
G4 NKP D. Mellin 1 Whitfield Close, Warminster BA12 9HX
GW4 NKR G. Lewis The Old, Post Office Cottage, Brecon LD3 0UR
G4 NKU D. Dunn 37 Ridgemead, Calne SN11 9EW
G4 NKW G. Hilton 8 Sandwich Close, St. Ives PE27 3DQ
G4 NKX P. Digby 3 Copperview Mews, Cowplain, Waterlooville PO8 8BW
GI4 NKY W. Campbell 68 Richmond Court, Lisburn BT27 4QX
G4 NLA G. Goodrich 29 Cresswells, Corsham SN13 9NJ
G4 NLB B. Horne 77 Surrey Hills Residential Park, Boxhill Road, Tadworth KT20 7LZ
G4 NLC P. Hedison 85 Moorhouse Lane, Whiston, Rotherham S60 4NH
GW4 NLD P. Frost 1 Chester Street, Rhyl LL18 3ER
G4 NLG F. Askew 9 The Hall Spinney, Howden, Goole DN14 7FD
G4 NLH D. Haydon 50 Ward Close, Stratton, Bude EX23 9BB
G4 NLI R. Scott 10 Middle St., North Perrott, Crewkerne TA18 7SG
GM4 NLJ J. Martin Whitehill Foot Farm, Kelso TD5 8LB
G4 NLK R. Southall 7 The Willows, Brereton, Rugeley WS15 1EP
G4 NLL D. Bell 5 Byron Court, Dalton On Tees, Darlington DL2 2PX
G4 NLO K. Butcher 18 Windmill Street Whittlesey, Peterborough PE7 1HJ
GI4 NLQ P. Burns 41 Lambeg Road, Lambeg, Lisburn BT27 4QA
G4 NLU R. Williams 50 Hemerdon Heights, Plympton, Plymouth PL7 2EY
G4 NLW J. Wilding 8 Millbrook Way, Brierley Hill DY5 3YY
G4 NMC D. Willis 41 Chadbrook Crest, Richmond Hill Road, Birmingham B15 3RL
G4 NMD Rvd. G. Smith 6 Birtley Rise, Bramley, Guildford GU5 0HZ
G4 NMF W. Squire 7 Essex Crescent, Seaham SR7 8DZ
G4 NMK K. Petre 24 Harrogate Terrace, Murton, Seaham SR7 9PQ
G4 NMO J. Gallagher 1 Wythburn Mews, Langdale Road, Woodlesford, Leeds LS26 8FJ
G4 NMP B. Dudhill Westfield, Mablethorpe L N12
G4 NMR S. Harvey 2 Draycote Close, Worcester WR5 3SY
G4 NMS S. Burgess Borne House, Romsey Road, Stockbridge SO20 6PR
G4 NMT A. Godsiff 59 Vinson Close, Orpington BR6 0EQ
G4 NMU R. Jones Flat 4, Russell Court, Southport PR9 8NY
G4 NMV J. Baxter 16 Avon Close, Weston-Super-Mare BS23 4QS
G4 NMY P. Bennett 45 Ravenbank Road, Luton LU2 8EJ
G4 NNB G. Mackie 8 The Avenue, Biggleswade SG18 0PS
GM4 NNH J. Barber 156 Jamphlars Road, Cardenden, Lochgelly KY5 0ND
G4 NNI E. Bailey 213 Ashby Road, Hinckley LE10 1SJ
G4 NNJ A. Forster 31, Par Four Lane, Lydney GL15 5GB
GM4 NNK D. Harkness 22 Brockwood Crescent, Blackburn, Aberdeen AB21 0JZ
GW4 NNL M. Jones Jodanare, 72B Princes Drive, Colwyn Bay LL29 8PW
GI4 NNM J. Mcgillian 48 Millfield, Ballymena BT43 6PB
G4 NNN C. Winterflood 12 Bourne Road, Colchester CO2 7LQ
GW4 NNO T. Hadley Pastures Green Llandderfel Ll23 7Rf, Bala LL23 7RF
G4 NNP D. Andrews 10 Brantwood Drive, Paignton TQ4 5HZ
G4 NNS B. Coleman Woodlands, Redenham, Andover SP11 9AN
G4 NNX D. Ward 18 Henders, Stony Stratford, Milton Keynes MK11 1RB
G4 NNY D. Ransford 52 Loughborough Road, Bunny, Nottingham NG11 6QD
G4 NNZ G. Martorano 81 Sapcote Drive, Melton Mowbray LE13 1HG
G4 NOB R. Burbeck 20 St. Johns Road, Smalley, Ilkeston DE7 6EG
G4 NOC N. Black 7 Woodland Crescent, Bracknell RG42 2LH
G4 NOE M. Hollinghurst 30 Hall Road, Cheltenham GL53 0HE
G4 NOL B. Robinson 9 Henderson Road Norwich Nr4 7Jw, Norwich NR4 7JW
GW4 NOO M. Oliver 6 Flemish Close, St. Florence, Tenby SA70 8LT
G4 NOP R. Simpson 14 Cumberland Way, Dibden, Southampton SO45 5TW
G4 NOR C. Heaps 12 Oak Tree Close, Mansfield NG18 3EN
GW4 NOS R. Hopkins 8 Shady Road, Gelli, Pentre CF41 7UG
G4 NOT C. Sturgeon Windyridge, Linkside West, Hindhead GU26 6PA
G4 NOU R. Wigmore Little Landguard, Whitecross Lane, Shanklin PO37 7EJ
G4 NOW D. Shaw 14 Strongbow Road, Eltham, London SE9 1DT
G4 NOX K. Campbell 4 Orchard Close, Rowlands Gill NE39 1EQ
G4 NOY J. Mills 103 Irby Road, Wirral CH61 6UZ
G4 NPA A. Abbot 4 Nursery Drive, Birmingham B30 1DR
G4 NPB Dr S. Abbot 4 Nursery Drive, Birmingham B30 1DR
GW4 NPC R. Blayney 42 Ty Draw, Church Village, Pontypridd CF38 1UF
G4 NPD G. Chamberlain 13 Mayford Close, Beckenham BR3 4XS
G4 NPG P. Duffy 15 The Glade, Sheldon, Birmingham B26 3PW
G4 NPH J. Arnold 2 Duck Lane, Haddenham, Ely CB6 3UE
G4 NPN D. Westwood 1 Elmfield Road, Hartlebury, Kidderminster DY11 7LA
G4 NPS J. Woolliss Wharfdale, 245 Scartho Road, Grimsby DN33 2EA
G4 NPU P. Williams 122 Longhurst Lane, Mellor, Stockport SK6 5PG
G4 NPY C. Read 3 Wyrley Close, Lichfield WS14 9DA
G4 NQB S. Mould 53 Wolverley Avenue, Stourbridge DY8 3PJ
G4 NQC J. Neal 60 Balsdean Road, Brighton BN2 6PF
G4 NQI R. Atterbury Lomas De San Jose 20, la Vinuela 29712 Spain
G4 NQJ C. Brown Kingsdown Cottage, Fron, Montgomery SY15 6SB
G4 NQL M. Churms1 123 Urb Buenavista, Guadamar 3140 Spain
G4 NQQ N. Hemmings 3 Church Close, Shapwick, Bridgwater TA7 9LS
G4 NQS R. Young 79 Cradge Bank, Spalding PE11 3AF
G4 NQW P. Perrins 9 Merrick Close, Hayley Green, Halesowen B63 1JY
G4 NQZ R. Riley 103 St. Nicolas Park Drive, Nuneaton CV11 6DZ
G4 NRA J. Tisdale 2 Murley Grange, Bishopsteignton, Teignmouth TQ14 9TX
G4 NRC Raynet-UK c/o F. Mossop 4 Brookdale Way, Waverton, Chester CH3 7NT
G4 NRD A. Lindsay 21 Willow Road, Four Pools Industrial Estate, Evesham WR11 1YW
G4 NRE W. Ward 88 Central Road, Cromer NR279BW
G4 NRF H. Westwood 67 Bedford Close, Featherstone, Pontefract WF7 5LH
G4 NRG R. Greengrass Shanaclune, Dunhill X91 Y0H3 Ireland
G4 NRH D. Whitehead 50 Southey Lane, Kingskerswell, Newton Abbot TQ12 5JG
G4 NRI M. Dhami 118 Havelock Drive, Brampton, Ontario L6W 4E3 Canada
G4 NRK Newborough Radio Klub c/o P. Staton 52 School Road, Newborough, Peterborough PE6 7RG
G4 NRO J. Coupe 3 Longfield Avenue Coppull, Chorley PR7 4NT
G4 NRP J. Fish Kennels Cottage, Manor Road, Upper Bentley, Redditch B97 5TB
G4 NRQ Statford-Upon-Avon & District RS c/o J. Hamill 67 Windsor Avenue, Coleraine BT52 2DR Ireland
G4 NRR N. Astbury-Rollason Fern Lodge, The Parade, Newton Abbot TQ13 0JH
G4 NRS P. Goding 29 Caistor Drive, Hartlepool TS25 2QG
G4 NRT Cmdr. D. Bondy 328 Wilson Avenue, Rochester ME1 2ST
G4 NRV A. Diplock 24 Billings Hill Shaw, Hartley, Longfield DA3 8EU
G4 NRX S. Mantell 50 Coleshill St., Fazeley, Tamworth B78 3RA
G4 NRY I. Mantell 34 Piccadilly, Tamworth B78 2EP
G4 NRZ K. Moody 5 Moore Road, Mapperley, Nottingham NG3 6EF
G4 NS J. Hudson 22 Essex Gardens, Marsden, South Shields NE34 7JQ
G4 NSA D. Morgan 12 Rosalind Avenue, Bebington, Wirral CH63 5JR
G4 NSB Dr J. Winterburn 164 Malthouse Lane Earlswood, Solihull B94 5SD
G4 NSC W. Weatherspoon 12 Greenacres Close, Crawcrook, Ryton NE40 4TD
G4 NSD I. Mitchell Greenway Cottage, Greenway, Tatsfield, Westerham TN16 2BT
G4 NSE J. Rank 9 Fairfield Crescent, Scarborough YO12 6TL
G4 NSH S. Robinson Upper Hambleton Hill Farm, Wainstalls, Halifax HX2 7TX
G4 NSJ G. Heffer 35 Henty Road, Worthing BN14 7HE
GM4 NSL G. Greenlees 22 Hunters Grove, Hunters Quay, Dunoon PA23 8LQ
G4 NSM A. Bruce 5 Orchard Croft, Epworth DN91LL
G4 NSN M. Craft 8 Juniper Road, Farnborough GU14 9XU
G4 NSO S. Auckland 6 Lombard Crescent, Darfield, Barnsley S73 9PP
GI4 NSS L. Robinson 52 Ballymacormick Avenue, Bangor BT19 6AY
G4 NST S. Thorpe 11 Grove Road, Hethersett, Norwich NR9 3JP
G4 NSW F. Lacey 27 Howards Gardens, New Balderton, Newark NG24 3FJ
G4 NSZ M. Stanway 72 Sheldons Court, Winchcombe Street, Cheltenham GL52 2NR
G4 NTA P. Allan 2 Park View, Queensbury, Bradford BD13 1PL
G4 NTC D. Henderson 4 Vincent Street, Bolton BL1 4SA
G4 NTG J. Williamson 5 Rochester Close, Headless Cross, Redditch B97 5FP
G4 NTJ A. Rick 9 Sheldon Close, Loughborough LE11 5EZ
GM4 NTL J. Mcdermott Milking Green Gate, Eliock, Sanquhar DG4 6LD
GD4 NTR G. Kelly 5 Tynwald Close, Peel, Isle of Man IM5 1JJ
G4 NTV A. Wilkes 34 Tideswell Road, Great Barr, Birmingham B42 2DT
G4 NTW W. Douglas 2 Rockville, Sunderland SR6 9EL
GM4 NTX K. Elliott Northfield Cottage, Denny FK6 6RB
G4 NTY J. Higson 5 Primrose Avenue Worsley, Manchester M28 0TP
G4 NUA E. Williams 45 Bayford Place, Cambridge CB4 2UF
G4 NUB M. Tyler 27 Shakespeare Drive, Dinnington, Sheffield S25 2RP
G4 NUF Silverthorn RC c/o C. Muller 118 Park Lane, Northampton NN5 6PZ
G4 NUG R. Needs 13 Greenway, Great Horwood, Milton Keynes MK17 0QR
G4 NUJ K. Symonds 30 Fairlea Crescent Northam, Bideford EX39 1BD
G4 NUK J. Brown 33 Balmoral Drive, Leicester LE3 3AD
GM4 NUN G. Mackenzie 1 Walnut Grove, Blairgowrie PH10 6TH
G4 NUO A. Mackenzie Ivy House, 145 High Street, Marske By The Sea, Redcar TS11 6JX
G4 NUS A. Layland 16 Park Road, Quarry Bank, Brierley Hill DY5 2DA
GM4 NUU H. Park Carndearg, Upper Steelend, Dunfermline KY12 9LP
G4 NUV R. Richmond 7 Bishopdale Drive, Watnall, Nottingham NG16 1LE
G4 NUX C. Smith Wealdon Cottage, Dunsfold Road, Billingshurst RH14 0PJ
G4 NUY E. Wharton Vandling, Well Bank, Well, Bedale DL8 2QF
G4 NVA J. Dyke 2 Brooklands Drive, Goostrey, Crewe CW4 8JB
G4 NVH G. Boull 80 Ascot Road, Baswich, Stafford ST17 0AQ
GM4 NVI D. Chapman 9 Baillieswells Terrace, Bieldside, Aberdeen AB15 9AR
G4 NVL T. Copeman 1 Chestnut Avenue, Welney, Wisbech PE14 9RG
G4 NVM J. Duddridge 19 Ridgeway, Hurst Green, Etchingham TN19 7PJ
G4 NVN G. Harper Pathways, 41 Somerset Close, Congleton CW12 1SE
G4 NVP K. Kett 24 Deancourt Drive, New Duston, Northampton NN5 6PY
G4 NVQ S. Shirley 93 Alfred Road, Hastings TN35 5HZ
G4 NVT M. Musgrave 49 Vowler Road, Langdon Hills, Basildon SS16 6AQ
G4 NVV R. English 124 Hillside Road, Portishead, Bristol BS20 8LG
G4 NVY M. Juffs 32 Brooklands Park, Longlevens, Gloucester GL2 0DP
GM4 NWK T. Hill 3 Swift Crescent, Glasgow G13 4QN
G4 NWM M. Talbott 44 Tamworth Road, Amington, Tamworth B77 3BT
G4 NWN W. Talbott 44 Tamworth Road, Amington, Tamworth B77 3BT
G4 NWO K. Chaplin 1 Beechwood Crescent, Amington, Tamworth B77 3JH
G4 NWR North Wiltshire Raynet c/o A. Sharman 3 Deben Crescent Haydonwick, Swindon SN25 3QB
G4 NWS A. Wheeler 11 Barley Way, Rothley, Leicester LE7 7RL
G4 NWW J. Edgeley 12 The Glade, Horsham RH13 6DD
G4 NXA D. Kirk Apartment 28, Riverside, 8 Loxley Park, Sheffield S6 4TF
G4 NXB J. Evans 11 Columbia Close, Selston, Nottingham NG16 6GP
GW4 NXD R. Johns 12 Woodfield Road, New Inn, Pontypool NP4 0PT
G4 NXG A. Birch 6 Crescent Road, Wallasey CH44 0BQ
G4 NXI R. Light 72 Badger Rise, Portishead, Bristol BS20 8AX

GI4 NXJ M. Mcfall 4 Riverdale Close, Ballyclare BT39 9WE
G4 NXL M. Street 41 Shaw Lane, Holbrook, Belper DE56 0TG
G4 NXO Sheppey Western c/o A. Collett 10 Quince Road, Hardwick, Cambridge CB23 7XJ
G4 NXP D. Taylor The Acreage, View Farm, 11 Malvern Road, Worcester WR2 4SF
G4 NXR P. Rose Cuptree Cottage, 17 The Cross, Colchester CO7 9QQ
G4 NXS M. Davis 47 Meadow View, Holmewood, Chesterfield S42 5UL
GM4 NXT W. Davidson 7A South Street, Aberchirder, Huntly AB54 7XR
G4 NXV D. Gadsden 37 Cambridge Street, Wymington, Rushden NN10 9LG
G4 NXW K. Chesters 4 Kirton Crescent, Lytham St. Annes FY8 4BJ
G4 NXX D. Harbottle 33 Robin Crescent, Lyme Green, Macclesfield SK11 0LJ
G4 NYA R. Hyams 6 Gaudick Road, Eastbourne BN20 7QE
G4 NYB R. Knighton Merry Ways, The Green, Derby DE72 2BJ
G4 NYC G. Jannetta 14 Banks Lane, Heckington, Sleaford NG34 9QY
G4 NYD I. Watling 5 Claylands Court Bishops Waltham, Southampton SO32 1JS
G4 NYG D. Routledge 7 Littleton Croft, Solihull B91 3XR
G4 NYJ C. Webb 65 Littlebeck Drive, Darlington DL1 2TU
G4 NYK Prof. R. Williams 67A Sea Mills Lane, Bristol BS9 1DR
G4 NYL S. Brown 2 Windsor Close, Read, Burnley BB12 7QH
GU4 NYT N. Le Page Heathwick, Les Martins, Guernsey GY4 6QJ
G4 NYV M. Katzmann 6654 Barnaby Street Nw, Washington Dc 20015 United States
G4 NYW R. Schoales 4 Sandringham Place, Ravenfield, Rotherham S65 4NP
G4 NYY P. Ernster 36 Forest End, Fleet GU52 7XE
G4 NYZ J. Battle-Welch 25 Moorcroft Close, Callow Hill, Redditch B97 5WB
G4 NZB P. Neville 66 Oak Lodge Avenue, Chigwell IG7 5HZ
G4 NZC B. Manchett 12 Old Parsonage Court, Otterbourne, Winchester SO21 2EP
G4 NZE E. Third 1 Roman Road, Corby NN18 8FZ
G4 NZG S. Parsons Trehunsey Cottage Quethiock, Liskeard PL14 3SG
G4 NZK B. Laniosh 47 Barley Mow Lane, Catshill, Bromsgrove B61 0LU
G4 NZN K. Lowe Springwood, Priesthorpe Road, Pudsey LS28 5RE
G4 NZO G. Frederick 98 Coleridge Park Drive, Winnipeg MB R3K 0B5 Canada
G4 NZQ P. Brooks 7 Lindford Drive, Eaton, Norwich NR4 6LT
G4 NZU R. Wilson 9 Greythorn Drive, West Bridgford, Nottingham NG2 7GG
G4 NZX D. Cooper 6 Swinburne Close, Barnby Dun, Doncaster DN3 1BS
G4 NZY D. Coles 91 Grove Lane, Harborne, Birmingham B17 0QT
G4 NZZ B. Coulson 19 St. Lukes Close, Kettering NN15 5HD
G4 OAB M. Clutton 8 Ash Grove, Runcorn WA7 5LR
G4 OAE D. Crisp 20 Crawford Close, Earley, Reading RG6 7PE
G4 OAG A. Dymott St. Huberts, Ashey Road, Ryde PO33 4BB
G4 OAI H. Richter 84 Roehampton Drive, Wigston LE18 1HU
G4 OAK S. Richards Little Piece 1 Stocks Mead, Washington RH20 4AU
G4 OAN L. Wild The Drey, Penny Lane, Bedlington NE22 6HD
G4 OAR N. Mclaren Ingleside, Waterloo, Whixall, Whitchurch SY13 2PX
GM4 OAS G. Liddle Ashbeck, Morar, Mallaig PH40 4PD
G4 OAU G. Austin 38 Willow Crescent, Hatfield Peverel, Chelmsford CM3 2LJ
G4 OAV S. Ames 21 Common Lane, Harpenden AL5 5BT
G4 OAX W. Joiner 15 Laxton Grove, Great Holland, Frinton-on-Sea CO13 0SF
GW4 OBA R. Jarvis 3 Church View Llanwenog, Llanybydder SA40 9UU
G4 OBB Dr D. Kaylor 27 Fair Green, Glemsford, Sudbury CO10 7PH
G4 OBC M. Taylor 5 Hawford Avenue, Kidderminster DY10 3BH
GM4 OBD G. Sangster 36 St. Marys Drive, Ellon AB41 9LW
G4 OBE R. Snary 12 Borden Avenue, Enfield EN1 2BZ
G4 OBK P. Catterall 54 Westlands, Pickering YO18 7HJ
G4 OBN S. Harding 21 Abbey Road, Medstead, Alton GU34 5PB
G4 OBT N. Day Tanhouse Farm, Rusper Road, Dorking RH5 5BX
G4 OBV E. Day 42 Grosvenor Close, Bishop's Stortford CM23 4JP
G4 OBX R. Dobson 40 Dipton Gardens, Sunderland SR3 1AN
GM4 OCA P. Windsor Hillside Overbrae, Fisherie, Turriff AB53 5QP
G4 OCC J. Johnston 6482 Doctor Blair Crescent, North Gower, Ontario K0A 2T0 Canada
G4 OCF D. Clayton Flat 6 Clifton Court, Clevedon BS21 6QS
G4 OCH K. Dickens The Old Post Office, Cleobury Road, Ground Floor Flat, Bewdley DY12 2QG
G4 OCJ Dr E. Mullock 14 Cottage Walk Shawclough, Rochdale OL12 6DZ
GI4 OCK J. Mackay 12 Lynne Road, Bangor BT19 1NT
GI4 OCL R. Mccurry 82 Cumberland Road, Dundonald, Belfast BT16 2BB
G4 OCQ I. Blackman 69 Thorntons Close, Pelton, Chester le Street DH2 1QH
G4 OCR Dr M. Butler 41 Neale Road, Chorlton Cum Hardy, Manchester M21 9DP
G4 OCU D. Gipp 7 Edmunds Road, Cranwell Village, Sleaford NG34 8EL
G4 OCX L. Jewell Ashbrook Farm, Mill Hill, Bedford MK44 2HP
G4 OCZ C. Richardson 149 Old Fort Road, Shoreham-by-Sea BN43 5HL
G4 ODA B. Tatnall Poplar House, Delgate Bank, Weston Hills, Spalding PE12 6DH
G4 ODD M. Mathers Rose Cottage, Kirton Road Egmanton, Newark NG22 0HF
G4 ODE Dr N. Dovaston 53 Elmway, Chester le Street DH2 2LX
G4 ODF D. Faulkner Amber Croft, Dale Close, Mansfield NG20 9EB
G4 ODG V. Cawthron 8 Clay Hill Road, New Quarrington, Sleaford NG34 7TF
G4 ODI A. Dyer 34 Oakley Road, Chinnor OX39 4HB
G4 ODM C. Mott-Gotobed 14 Copse Road, New Milton BH25 6ES
GW4 ODN A. Whitticombe 160 Haven Drive, Hakin, Milford Haven SA73 3HN
G4 ODR Enfield CG c/o J. Young 30 Crofton Way, Enfield EN2 8HS
GI4 ODT W. Barker 96 Highlands Road, Limavady BT49 9LY
GM4 ODW D. Maclean Gramaiche Donavourd, Pitlochry PH16 5JS
GJ4 ODX S. Langlois L'Amarrage, La Route Orange, St. Brelade JE3 8GP Jersey
GD4 OEA C. Gerrard 6 Rheast Lane, Peel IM5 1BE Isle of Man
G4 OEB P. Grant 24 Dowlands Road, Bournemouth BH10 5LG
G4 OEC E. Mcpheat Dyche Old School House Holford, Bridgwater TA5 1SF
G4 OED G. Perry 12 Boydell Close, Shaw, Swindon SN5 5QT
G4 OEF D. Bayliss 38 Yarborough Crescent, Lincoln LN1 3LU
G4 OEH N. Rumble 24 Firle Road, North Lancing, Lancing BN15 0NZ
GW4 OEJ A. England 9 Priory Road, Milford Haven SA73 2DS
G4 OEK R. Jones 17B Plumpton Park Road, Doncaster DN4 6SQ
G4 OEM G. Hooker 42A Nether Hall Road, Doncaster DN1 2PZ
G4 OEP Dr A. Smith 15 Dyrham Close, Henleaze, Bristol BS9 4TF
G4 OEQ C. Thomas 69 Quakers Road, Downend, Bristol BS16 6JG
G4 OER D. Warner 28 Jameson Bridge Street, Market Rasen LN8 3EW
GW4 OES A. Pickard 89 Ael Y Bryn, Llanedeyrn, Cardiff CF3 7LL
G4 OEU Q. Campbell 8 Cookson Close, Corbridge NE45 5HB
G4 OEX G. Jones 3 Kings Mews, Bedford Street, Warrington WA4 6GY
G4 OEY Dr T. Sanderson Backershagenlaan 32, Wassenaar 2243 AD Netherlands
GM4 OEZ W. Taylor 2 Jubilee Terrace, Findochty, Buckie AB56 4QA
G4 OFA B. Millican 24 Wellington Terrace, Bramley, Leeds LS13 2LH
GM4 OFC J. Snelgrove Mill House, South Bridgend, Crieff PH7 4DH
GM4 OFI J. Robertson Springwell House, Auckengill, Wick KW1 4XP
G4 OFN P. Edmonds 170 Halton Road, Sutton Coldfield B73 6NZ
G4 OFO N. Baynes 3 Charles Babbage Close, Chessington KT9 2SA
G4 OFP J. Knowles 6 Seaway Gardens, Paignton TQ3 2PE
G4 OFU R. Burdess 33 The Green, Dartford DA2 6JS
G4 OGB L. Elliott Elm Lodge, 2 Hood Croft, Doncaster DN9 2FB
G4 OGG J. West 2 Sanders Close, Kempston, Bedford MK42 8RX
G4 OGL the Shortwave Shop c/o J. O'Brien 5 Highfields Mead, East Hanningfield, Chelmsford CM3 8XA
GM4 OGM S. Mather Pentland View, Limekiln Road, West Linton EH46 7BA
GW4 OGO S. Williams 31 Kensington Park, Magor, Caldicot NP26 3QG
GI4 OGQ W. Kernohan 3 Camphill Park, Ballymena BT42 2DQ
G4 OGW D. Thomas Handley Cross Cottage, Harewood End, Hereford HR2 8JT
G4 OGZ North Anglia Raynet c/o M. Walker 52 Derwent Road, Harpenden AL5 3NX
GW4 OH A. David 45 Amanwy, Llanelli SA14 9AH
G4 OHA L. Lux Hyde Brae, Hyde Hill Chalford, Stroud GL6 8NY
G4 OHB P. Taylor Flat 11, Romsley Hill Grange, Farley Lane, Romsley, Halesowen B62 0LN
G4 OHC R. Poore 8 Ainsdale Close, Worthing BN13 2QX
G4 OHF G. Bean 141 Narborough Road, Leicester LE3 0PB
GI4 OHH D. Cox 32 Kilmaconnell Road, Castleroe, Coleraine BT51 3QZ
G4 OHJ J. Porter 77 Westholme Road, Bidford-On-Avon, Warwickshire B50 4AN
G4 OHM Echelford DX Group c/o J. Parkin 18 Bradnock Close, Birmingham B13 0DL
G4 OHP L. Sharps 68 Vicarage Lane, Elworth, Sandbach CW11 3BU
G4 OHQ J. Reade 7 Wilmar Close, Hayes UB4 8ET
GM4 OHT T. Mitchell 12 Dalriada Place, Kilmichael Glassary, Lochgilphead PA31 8QA
G4 OHV C. Addison-Lees 18 Langley Avenue, Somercotes, Alfreton DE55 4LT
GI4 OHW N. Bell Rocklyn, 16 Dromore Road, Omagh BT78 1QZ
GM4 OHY R. Cameron 88 Little Vennel, Cromarty IV11 8XF
G4 OIA S. Hartgroves 54 Kensey Valley Meadow, Launceston PL15 9TJ
G4 OID J. Storry 99 Swineshead Road, Wyberton Fen, Boston PE21 7JG
G4 OIE R. Neale Field House, Recreation Road, Mansfield NG19 8TL
G4 OIG G. Peck 45 Bentley Close, Northampton NN3 5JS
G4 OII M. Morley Padagi, Town Road, Grimsby DN36 5JE
GM4 OIJ B. Robertson Frian, Westerton, Dalcross, Inverness IV2 7JL
G4 OIK J. Price 4 Housman Walk, Kidderminster DY10 3XL
G4 OIL J. Price 26 Hales Park, Bewdley DY12 2HT
G4 OIM P. Marchant 29 Hilldrop Road, Bromley BR1 4DB
G4 OIN A. Reeley Gibraltar House, 53 Pegasus Gardens, Gloucester GL2 4NP
G4 OIQ P. Storey 4 Sorrel Close, Wootton Bassett, Swindon SN4 7JG
G4 OIR N. Robinson 14 Glendale, Orton Wistow, Peterborough PE2 6YL
G4 OIS G. Reid 65 Rowelfield, Luton LU2 9HL
G4 OIV D. Mavin 20 Rowlington Terrace, Ashington NE63 0LZ
G4 OIW Dr C. Pine 2 Grange Drive, Stokesley, Middlesbrough TS9 5PQ
G4 OJD M. Guy 73 Penn Meadows, Brixham TQ5 9PF
G4 OJF G. Ball 12 Kelstern Close, Northwich CW9 5QR
G4 OJG J. Glass 70 Canterbury Road, Lydden, Dover CT15 7ES
G4 OJH A. Giles 11 Stanmore Close, Clacton-on-Sea CO16 7HQ
G4 OJI D. Schofield 18 Berrow Walk, Bristol BS3 5ES
G4 OJJ V. Holyoake 281 Causeway, Green Road, West Midlands B68 8LT
G4 OJK H. Baxendale 72B De Villiers Avenue, Liverpool L23 2XF
G4 OJL Mid Severn Vall c/o J. Bevan 5 Selsdon Close, Wythall, Birmingham B47 6HP
G4 OJN A. Semark 11 Fir Tree Close, Thorpe Willoughby, Selby YO8 9PF
G4 OJP R. Prosser 17 Lloyd Street, Hereford HR1 2HB
G4 OJQ A. Rowland Mole Cottage, Chapel Close, Morwenstow EX23 9JR
G4 OJR A. Stone 67 Bluebell Way, Preston PR5 6XQ
G4 OJS J. Rowlands 70 Braces Lane, Marlbrook, Bromsgrove B60 1DY
G4 OJU C. Lock 11 Stockton Close, Bristol BS14 0DS
G4 OJV A. Picton 5 Tuttles Lane East, Wymondham NR18 0EN
G4 OJW L. Szondy 6 Stanhope Gardens, London N6 5TS
G4 OJY A. Wright 2 Wards End Cottages, Tow Law, Bishop Auckland DL13 4JS
G4 OKA N. Crook 3 College Road, Reading RG6 1QE
G4 OKB B. Bloomfield 2 Walstead Manor Cottages, Scaynes Hill Road, Haywards Heath RH16 2QG
G4 OKC A. Gardner 19 Lower Rea Road, Brixham TQ5 9UD
G4 OKD Saltash Dist Ar c/o A. Forryan 21 Blakesley Road, The Meadows, Wigston LE18 3WD
G4 OKE M. Gould 10 Canterbury Close, Pelsall, Walsall WS3 4PB
GW4 OKF P. Granby 104 Priory Road, Milford Haven SA73 2ED
G4 OKH M. Fisher Witham Lodge, Fen Road, Newton-In-The-Isle, Wisbech PE13 5HT
G4 OKM M. Smith 7 Russley Green, Wokingham RG40 3HT
G4 OKS M. Porter Penlee, 11 Penwithick Road, Penwithick, St. Austell PL26 8UQ
GI4 OKU T. Patton 29 Greystone Park, Limavady BT49 0EQ
G4 OKW C. Trayner 2 Herisson Close, Pickering YO18 7HB
G4 OKY R. Wilkinson 10 Mildenhall Road, Loughborough LE11 4SN
G4 OKZ D. Wilson1 20 The Square, Worsthorne, Burnley BB10 3NG
G4 OLA R. Mccubbin 20 Wellesley Park, Wellington TA21 8PY
GM4 OLH I. Walsh 43 Sandyhill Road, Tayport DD6 9NX
G4 OLK A. Mackay 2 Highcliffe Grove, New Marske, Redcar TS11 8DU
G4 OLL D. Wilson 10 Winchester Drive, Stourbridge DY8 2LH
G4 OLO G. Spencer 5 Pitchcroft Lane, Church Aston, Newport TF10 9AQ
G4 OLP R. Parker 2 Laurel Road, Norwich NR7 9LL
G4 OLS J. Lloyd 16 Gilbanks Road, Stourbridge DY8 4RN
G4 OLU D. Steward 30 Riffhams Drive, Great Baddow, Chelmsford CM2 7DD
G4 OLY C. Morgan 316 Middle Road, Southampton SO19 8NT
G4 OLZ I. Sirley The Barn, Littleworth Lane, Horsham RH13 8JF
G4 OMD D. Wilson 1 Witchford Close, Lincoln LN6 0SS
G4 OMG C. Prescott 15 Sarabeth Drive, Tunley, Bath BA2 0EA
G4 OMI Dr A. Proudler 17 Sunnyside Road, Ketley Bank, Telford TF2 0DT
G4 OMJ G. Yarnall Blue Lias, Cropwell Road, Langar, Nottingham NG13 9HD
GI4 OMK P. Murphy 11 Danesfort Apartments, Belfast BT9 5QL
G4 OMN D. Thorndike 48 Cressingham Road, Reading RG2 7JR
G4 OMP M. Nyman 26 Silverstone Court, River Brook Drive, Birmingham B30 2SH
G4 OMS R. Reynolds 90 Manchester Road, Blackpool FY3 8DP

G4 OMT M. Taylor 320 Duffield Road, Derby DE22 1EQ
G4 OMV D. Marlow 53 The Lawns, Corby NN18 0TA
G4 OMZ L. Welding 67 Sunningdale Close, Burtonwood, Warrington WA5 4NS
G4 ONC E. Westcott 8 Portal Place, Ivybridge PL21 9BT
G4 ONF P. Sergent 6 Gurney Close, New Costessey, Norwich NR5 0HB
G4 ONG W. Lowe 34 Ridgeway, Lowton, Warrington WA3 2QL
G4 ONH C. Weller Mole End, Brent Hall Road, Braintree CM7 4JZ
GW4 ONI D. Mears 7 Tanydarren, Cilmaengwyn, Pontardawe, Swansea SA8 4QT
G4 ONJ A. Lightfoot 13 Midhurst Close, Ifield, Crawley RH11 0BS
G4 ONP Loughton & Epping Forest ARS c/o J. Mulye 83 Forest Drive East, London E11 1JX
G4 ONS M. Slade 5 Pedder Road, Clevedon BS21 5HB
G4 ONV G. Parker 49 Newlands, Dawlish EX7 0EA
G4 ONZ P. Tebbutt 115 Whitfield Mill, Meadow Road, Bradford BD10 0LP
G4 OO Spalding And District ARS c/o A. Hensman 24 Belchmire Lane, Gosberton, Spalding PE11 4HG
G4 OOB J. Westerman 7 Gascoigne Court, Barwick In Elmet, Leeds LS15 4NY
G4 OOC B. Simister 3 Beech Tree Road, Featherstone, Pontefract WF7 5EB
G4 OOE A. Langmead 10 The Copse, Scarborough YO12 5HG
G4 OOH S. Parker Flat 4, 72 Springfield Mount, Leeds LS18 5QE
G4 OOI C. Parker 18 Langdale Gardens, Leeds LS6 3HB
G4 OOJ C. Rose 3 Harley Drive, Leeds LS13 4QY
G4 OOK S. Stobbs 78 Hershall Drive, Town Farm, Middlesbrough TS3 8NX
G4 OOL Dr J. Yeandel Fairfield Farm, Penhallow, Truro TR4 9LT
G4 OOQ N. Ward 8 Meadowview Road, Kempston, Bedford MK42 7BE
GM4 OOU J. Forsyth1 10 Rowallan Crescent, Prestwick KA9 2HE
G4 OOX L. Harvey 27 Guernsey Drive, Birmingham B36 0PB
G4 OOY D. Bird 13 Kilvington Road, Arnold, Nottingham NG5 7HQ
G4 OPB N. Hydes 2 Stable Court, Martlesham Heath, Ipswich IP5 3UQ
G4 OPD A. Blissett 26 Cherry Orchard, Holt Heath, Worcester WR6 6ND
G4 OPE M. Hodges 40 Ennersdale Road, Coleshill, Birmingham B46 1EP
GI4 OPH T. Crawford 50 Thornleigh Gardens, Bangor BT20 4NP
G4 OPI A. Easom 1 Station Close, West Ayton, Scarborough YO13 9JQ
G4 OPK D. Carrett 80 Rotherfield Way, Emmer Green, Reading RG4 8PL
G4 OPL T. Bayliss 4 Sycamore Close, Polgooth, St. Austell PL26 7BW
G4 OPN W. Broxup 9 Kingsway, Hapton, Burnley BB11 5RB
G4 OPO C. Haddrell 9 Counterpool Road, Kingswood, Bristol BS15 8DQ
G4 OPQ Dr M. Holli 6 Morley Close Staple Hill, Bristol BS16 4QE
G4 OPR R. Hayward Sunnyfields, Lighthouse Road, Dover CT15 6EJ
G4 OPT A. Kemsley Newfield Lodge Rest Home, 93-99 St. Andrews Road South, Lytham St. Annes FY8 1PU
GM4 OPU J. Houston 26 Clerk Drive, Corpach, Fort William PH33 7LE
G4 OPV J. Jackson 15 Jackson Crescent, Stourport-on-Severn DY13 0EW
GW4 OPW B. Jones 12 Ashbourne Court, Aberdare CF44 8HA
G4 OPY S. Balmer 101 Marsh Lane, Shepley, Huddersfield HD8 8AP
G4 OQ G. Lidstone 76 Thames Drive, Leigh-on-Sea SS9 2XD
GW4 OQB A. Greatrex Clydfan, Dinas Cross, Newport SA42 0XS
G4 OQG Prof. M. Ayres 3 Wicks Drive, Chippenham SN15 3EL
G4 OQH Rvd. H. Callaghan 5 Manor Park View, Manor Park Road, Glossop SK13 7TL
G4 OQJ I. James 4 Lancaster Gardens, Earley, Reading RG6 7PA
G4 OQK R. Alderton 1 Comfrey Way, Thetford IP24 2UU
G4 OQL Dr C. Bowley Plum Tree House, Walk Close, Derby DE72 3PN
G4 OQN M. Barr 30 Hounslow Road, Twickenham TW2 7EX
G4 OQP J. Gerrity 14 Lostock Avenue, Hazel Grove, Stockport SK7 5JN
G4 OQR M. Huddart Buckminster Gliding Club Ltd Saltby Airfield Sproxton Road Skillington, Grantham NG33 5FE
G4 OQU J. Davenport 1 Lowfields, Staveley, Chesterfield S43 3QB
G4 OQV R. Beecham 7 Crummock Close, Coventry CV6 6GY
G4 OQX G. Cooper 61 Fallowfield Road, Hasbury, Halesowen B63 1BZ
GI4 OQY K. Chambers 44 Ballywillin Road, Portrush BT56 8JN
G4 OQZ B. Dawson Isca, 12 Lestock Way, Fleet GU51 3EB
G4 ORB G. Busby The Terrace, Terrace Road South, Binfield, , Bracknell RG42 4DS
G4 ORC Oldham Am RC c/o G. Oliver 158 High Barn St., Royton, Oldham OL2 6RW
G4 ORE A. Charles 14 Chorleywood Bottom, Chorleywood, Rickmansworth WD3 5JD
G4 ORI J. Hamill 67 Windsor Avenue, Coleraine BT52 2DR Ireland
G4 ORJ A. Jones Fairview, Frenchs Road, Wisbech PE14 7JF
G4 ORP M. Parsons 15 Sherbourne Road, Hangleton, Hove BN3 8BA
G4 ORQ A. Walker 4A Winston Drive, Eston, Middlesbrough TS6 9LY
G4 ORS W. Ragg 14 Mocatta Way, Burgess Hill RH15 8UR
G4 ORU G. Wadwell 7 Barkhart Drive, Wokingham RG40 1TW
G4 ORV D. Whatmough Flat 3, 170 Buxton Road, Stockport SK2 6HA
G4 ORW A. Atherley 2 Haydock Close, Dosthill, Tamworth B77 1QR
G4 ORX P. Baggett 33 Foxglove Way, Thatcham RG18 4DL
G4 ORY C. Bates Rivieri, Coventry Road, Kingsbury B78 2LH
G4 OSB T. Arris 7 Rowan Road, North Hykeham, Lincoln LN6 8LY
GI4 OSF J. Mcniece 14 Ballybracken Road, Doagh, Ballyclare BT39 0SE
GI4 OSG D. Robinson 17 Dalton Glen, Comber, Newtownards BT23 5RJ
G4 OSH A. Nevison 10 Birch Way, Tunbridge Wells TN2 3DA
G4 OSI D. Whitehouse 10 Felstead Street, Stoke-on-Trent ST2 7HJ
G4 OSJ P. Brewer 2 Mill Close, Wing, Oakham LE15 8RH
G4 OSK K. Hall 21 Eardulph Avenue, Chester le Street DH3 3PR
G4 OSO E. Binns Fieldview, 5 Moorside, Cleckheaton BD19 6JH
G4 OSP M. Binns Fieldview, 5 Moorside, Cleckheaton BD19 6JH
G4 OSR S. Roberts 1 Lakeside Crescent, Long Eaton, Nottingham NG10 3GH
GM4 OSS S. Campbell 14 Hillhouse Place, Stewarton, Kilmarnock KA3 3HT
G4 OST P. Cabban Ivydene, Upper Tockington Road, Tockington, Bristol BS32 4LQ
G4 OSU M. Dixey 50 Sandon Road, Ford Houses, Wolverhampton WV10 6EN
GM4 OSV C. Dunn 66 Glen Doll Road, Neilston, Glasgow G78 3QP
G4 OSX J. Griffiths 15 Victoria Road, Cambridge CB4 3BW
G4 OSY D. Gascoigne 2 Thorncliffes, Chapel Lane, Pontefract WF9 3NJ
G4 OTB Dr N. Hailes Yew Tree Cottage, The Hollies Common, Stafford ST20 0JD
G4 OTC South London Raynet c/o P. Gagen 16 Melbourne Road Sidemoor, Bromsgrove B61 8PE
G4 OTD C. Taylor 2 Kismet Avenue, Highbury 5089 Australia
G4 OTE M. Grayson One Elm, 58 Kaye Lane, Huddersfield HD5 8XU
GI4 OTG A. Mcneice 148 Doagh Road, Newtownabbey BT36 6BA
G4 OTI Dr P. Stockbridge 11 Fairways, Frodsham WA6 7RU
G4 OTJ J. Witchell Pennyquick Cottage, Broomhill Lane, Bristol BS39 5SA
G4 OTL M. Baker 49 Grove Lane, Gomersal Cleckheaton BD19 4JT
G4 OTS G. Eccleston 24 Orton Lane, Wombourne, Wolverhampton WV5 9AW
G4 OTU D. Fagan 3 Oxenham Green, Torquay TQ2 6DX
G4 OTV D. Green St. Annes Poundfield Road, Crowborough TN6 2BG
G4 OTX G. Hibberd 2 Carr Bank, Oakamoor, Stoke-on-Trent ST10 3EA
G4 OUB J. Whetstone 5 Zetland Road, Barnard Castle DL12 8LA
G4 OUG C. Beesley 15 Byron Close, Cheadle, Stoke-on-Trent ST10 1XB
G4 OUH J. Brockway 16 Dawlish Road, Dudley DY1 4LU
G4 OUI M. Brockway 16 Dawlish Road, Dudley DY1 4LU
G4 OUJ S. Carrigan 1 Milford Crescent, Littleborough OL15 9EF
G4 OUM T. Bolton 25 Woodfield Drive, Lichfield WS14 9HH
GI4 OUN D. Fulton 120 Dunnalong Road, Bready, Strabane BT82 0DP
GI4 OUP S. Henderson 47 Donaghedy Road, Bready, Strabane BT820DB
G4 OUS D. Bean 18 Witham Court, Higham, Barnsley S75 1PX
G4 OUT I. Cornes 17 Chilwell Avenue, Little Haywood, Stafford ST18 0QZ
GW4 OUU D. Grace The Laurels, Gwbert Road, Cardigan SA43 1AF
G4 OUZ L. Day 3 Harris Way, Lee Mill Bridge, Ivybridge PL21 9EU
G4 OVD G. Rugen 24 Highgate Road, Lydiate, Liverpool L31 0DA
GI4 OVE J. Mcelvanna 26 Lissummon Road, Newry BT35 6NA
G4 OVF P. Sampson 34 Solway Road, Moresby Parks, Whitehaven CA28 8XJ
G4 OVG J. Thompson 29 Arun, East Tilbury, Tilbury RM18 8SX
GW4 OVH H. Owen 5 Arllwyn Cefn Road, Bwlchgwyn, Wrexham LL11 5YF
G4 OVJ R. Read 29 Imber Road, Shaftesbury SP7 8RX
G4 OVL M. Allen 18 Philip Garth, Wakefield WF1 2LS
G4 OVM C. Barnes 13 Waterworks Road, Farlington, Portsmouth PO6 1NG
GI4 OVN S. Dawson 31 Rock Hill Warren Road, Donaghadee BT210FB
G4 OVO J. Featherstone Garden Close, Greenway Road, Torquay TQ1 4NJ
G4 OVR D. Fillingham 6 Kings Chase, Rothwell, Leeds LS26 0HL
G4 OVS F. Goddard 4 St. Peters Close, Barnburgh, Doncaster DN5 7EN
G4 OVT J. Grant The Bungalow, Drury Lane, Horsforth, Leeds LS18 4RL
G4 OVV N. Howard 9 Snowshill Drive Highfield, Wigan WN3 6AD
G4 OVW B. Jempson Flat 1, 3 Dacre, Scarborough YO11 2SP
G4 OVX M. Kennett Toms Cottage, Kendal Lane, York YO26 7QN
GI4 OWA G. Elliott 4 Fernbrae Gardens, Londonderry BT47 5XS
GI4 OWB J. Fallows 22 Meadow Way, Ballygowan, Newtownards BT23 5TQ
G4 OWH G. Gregor 41 Stonebridge Drive, Frome BA11 2TN
G4 OWK T. Pearsall 6 Vernon Close, Martley, Worcester WR6 6QX
G4 OWL C. Payne 7 Ellis Avenue, Onslow Village, Guildford GU2 7SR
G4 OWN A. Turner 1 Milton Road, Flitwick, Bedford MK45 1QA
GW4 OWQ D. Scott Isyfoel, Morfa Bychan, Porthmadog LL49 9YD
G4 OWS N. Cunliffe 44 Shore Road, Hesketh Bank, Preston PR4 6RB
G4 OWT S. Harwood 24 Firle Crescent, Lewes BN7 1QG
G4 OWY R. Howes 8 Kitchener Road, Weymouth DT4 0LN
G4 OXD T. Rose 41 Keats Way, Hitchin SG4 0DP
G4 OXG N. Wood Mentmore House, Woodside Hill Chalfont St. Peter, Gerrards Cross SL9 9TD
G4 OXK R. Waygood 2 Brookside Close, Bransgore, Christchurch BH23 8BT
GW4 OXL W. Smith 11 Connacht Way, Pembroke Dock SA72 6FB
GI4 OXO W. Fitzsimons 83 Boghill Road, Newtownabbey BT36 4QT
G4 OXR C. Mortimer Meadow Bank, Dowlish Wake, Ilminster TA19 0NZ
G4 OXU Dr D. Whan 1 Hillclose Avenue, Darlington DL3 8BH
GI4 OYG M. Black 38 Town Park, Carrickfergus BT38 8FG
G4 OYH A. Bowyer 37 St. Mark Drive, Colchester CO4 0LP
GI4 OYI D. Chambers 238 Donaghadee Road, Newtownards BT23 7QP
GI4 OYL J. Cuthbert 19 Antrim Road, Ballymena BT42 2BJ
GI4 OYM W. Elliott 23 Castle View Park, Portrush BT56 8AS
G4 OYN A. Fuller 17 Brington Drive, Barton Seagrave, Kettering NN15 6UW
G4 OYO A. Bramley 13 Moorland Avenue, Stapleford, Nottingham NG9 7FY
G4 OYP G. Savin 19 Hailey Avenue, Loughborough LE11 4QW
G4 OYR N. Lee Silverstone, Alverstone Road, Apse Heath, Sandown PO36 0LH
G4 OYT H. Moss 101 Barnford Crescent, Oldbury B68 8PR
G4 OYX D. Porter 8 Stanton Drive, Ludlow SY8 2PH
G4 OYZ M. Spencer 29 Kliffen Place, Halifax HX3 0AL
G4 OZC W. Smith 15 Henbury Drive, Woodley, Stockport SK6 1PY
G4 OZD R. Woolley 7 Geveze Way, Broughton Astley, Leicester LE9 6HJ
G4 OZG E. Haskett 23 Gloucester Road, King's Lynn PE30 4AB
GI4 OZI A. Kinghan 14 Sunningdale Park North, Belfast BT14 6RZ
GI4 OZJ G. Allen 6 Lougherne Road, Annahilt, Hillsborough BT26 6BX
G4 OZL P. Ingram Rosehill Cottage, Mount Carmel Road, Andover SP11 7ER
G4 OZM D. Bradberry 6 The Close, Easton On The Hill, Stamford PE9 3NA
G4 OZN P. Donaldson-Badger Heckdyke Cottage, Heckdyke, Doncaster DN10 4BE
G4 OZP S. Corrigan 11 Pear Tree Close, Bromham, Bedford MK43 8PR
G4 OZQ G. Fisher 9 Shrubbery Road, Drakes Broughton, Pershore WR10 2AX
GW4 OZU P. Hyams Tricklewood, Pembroke Dyfed SA71 5HY
G4 OZX C. Goble 12 Longfield Road, Emsworth PO10 7TR
G4 OZY A. Osborn 35 Griston Road, Watton, Thetford IP25 6DN
G4 PAA P. Booth 22 Charters Lane, Brandesburton, Driffield YO25 8QJ
GW4 PAF J. Thomas 2 Tudor Way, Llantwit Fardre, Pontypridd CF38 2NH
G4 PAH Dr M. Rollason Ash Ridge, Clint, Harrogate HG3 3DS
G4 PAI P. Wells 15 Apple Tree Grove, Ferndown BH22 9LA
G4 PAS P. Searles 63 Whalley Road, Ramsbottom, Bury BL0 0DP
G4 PAT J. Thirsk 47 Chestnut Avenue, Euxton, Chorley PR7 6BP
G4 PAV G. Brutnall 57 Wollaston Road, Irchester, Wellingborough NN29 7DA
G4 PBC J. Kilroy 119 Station Road Brimington, Chesterfield S43 1LJ
G4 PBD Capt. R. Hughes 8 Frinton Court, The Esplanade, Frinton-on-Sea CO13 9DW
G4 PBF P. Baker 23 Orde Close, Crawley RH10 3NG
G4 PBJ Axholme RC c/o B. Oakley 6 Windmill Way, Haxby, York YO32 3NL
G4 PBN J. Vivian 3 Station Road, Gunnislake PL18 9DX
G4 PBO D. Smith 21 Sydney Road, Benfleet SS7 5RD
G4 PBR D. Suttenwood White Lodge, Mersea Road, Colchester CO5 7LJ
GI4 PBS T. Wilson 39 Woburn Road, Millisle, Newtownards BT22 2HY
GI4 PBT T. Wilson Brambly Hedge, 39 Woburn Road, Newtownards BT22 2HY
G4 PBY B. Jones 13 Albert Street, Cheltenham GL50 4HS
G4 PBZ Weald ARS c/o T. Ashton 90 Secker Avenue, Warrington WA4 2RE
G4 PCB A. Cox 12 Merrymeet, Whitestone, Exeter EX4 2JP
G4 PCD M. Dally 11 Wrightson Terrace, Doncaster DN5 9ST
G4 PCE Rsars - Royal Signals ARS c/o R. Collins 389 Lode Lane, Solihull B92 8NN
G4 PCF P. Goodson 46 Southwold, Bracknell RG12 8XY
GW4 PCJ R. Belcher Parciau, Bronwydd Arms, Carmarthen SA33 6BN

G4 PCK B. James Rivendell, Kingsgate Close, Torquay TQ2 8QA
G4 PCL B. Walker 22 Peveril Road, Tibshelf, Alfreton DE55 5LQ
G4 PCN C. Lambert 23 Palmars Cross Hill, Rough Common, Canterbury CT2 9BL
GW4 PCO P. Mogford 27 Ynysymaerdy Road, Briton Ferry, Neath SA11 2TE
G4 PCP C. Shelton 18 Beaconsfield Drive, Coddington, Newark NG24 2RX
GI4 PCQ J. Quinn 86 Knocknacarry Road, Cushendun, Ballymena BT44 0NS
G4 PCR J. O'Hara 12 Ray Avenue, Nantwich CW5 6HJ
GM4 PCT A. Gordon 2 Duchray St., Riddrie, Glasgow G33 2DD
G4 PCW A. Tucker 3 Eston Close, Mabe Burnthouse, Penryn TR10 9JW
GW4 PCX R. Price 2 Grassholm Place, Broadway, Haverfordwest SA62 3HX
G4 PCZ D. St Quintin 16 Cromwell Road, Sprowston, Norwich NR7 8XH
G4 PDC P. Carter 16 Alexandra Park Paulton, Bristol BS39 7QS
G4 PDD F. Bibby 14 St. Clare Terrace, Chorley New Road, Bolton BL6 4AZ
G4 PDE R. Bradshaw 44 Hawthorn St., Derby DE24 8BD
G4 PDG B. Hillard Farmlea, Hele Lane, South Petherton TA13 5AP
G4 PDI B. Kenzie 9 Goodliffe Avenue, Balsham, Cambridge CB21 4AD
G4 PDK R. Davies 11 Tamar Green, Corby NN17 2LA
G4 PDQ J. Clayton 217 Prestbury Road, Cheltenham GL52 3ES
G4 PDR D. Hughes 19 Burnsall Close, Farnborough GU14 8NN
G4 PDU C. Carrington 3 Jeake Drive, Rye TN31 7FH
G4 PDY J. Brandhuber 3 Brigham Place, Felpham, Bognor Regis PO22 7NW
G4 PEA R. Flanders 51 Rookwood Court, Guildford GU2 4EL
G4 PED J. Hinde 12A Station Parade, Ockham Road South, Leatherhead KT24 6QN
G4 PEF Prof. W. Ingram 141 Churchill Road Willesden Green, London NW2 5EH
G4 PEK L. Dymond 13 Thornlea Avenue Fremington, Barnstaple EX31 3DA
G4 PEL W. Threapleton Cobbs Nook Farm, Newstead Lane, Stamford PE9 4JJ
G4 PEN R. Potts 180 Brook Hill, Thorpe Hesley, Rotherham S61 2PZ
G4 PEO J. Pitty 12 St. Leonards Road, Horsham RH13 6EJ
GI4 PES N. Robinson 3 Moorland Drive, Lisburn BT28 2XU
G4 PET J. Smith Pasturefields House, Pasturefields Lane, Stafford ST18 0RD
G4 PEU K. Smith1 Pasturefields House, Pasturefields Lane, Stafford ST18 0RD
G4 PEW R. Wood Abbots Croft, Abbey Road, St. Bees CA27 0EG
GW4 PEX W. Williams 168 Mumbles Road, West Cross, Swansea SA3 5AN
G4 PEY R. Wilmot 1 Retreat Cottages, Church Lane, Horsham RH12 3ND
G4 PFA P. Wheeler 21 Browns Road, Holmer Green, High Wycombe HP15 6SL
G4 PFE J. Laverick 5 York Crescent, Newton Hall Estate, Durham DH1 5PU
G4 PFF J. Potter 8 Mill Field Close, Burton Joyce, Nottingham NG14 5AA
G4 PFG M. Spooner 6 Cross Road, Starston, Harleston IP20 9NQ
G4 PFJ J. Backus 2 Southview Villas, Dunmow Road, Takeley, Bishop's Stortford CM22 6SW
G4 PFK G. Gifford 184 Chantrey Crescent Great Barr, Birmingham B43 7PG
GW4 PFL P. Freestone 8 Harcourt Road, Llandudno LL30 1TU
G4 PFO J. Gregory 22 Tower View Road, Great Wyrley, Walsall WS6 6HE
G4 PFQ Northwest Durham Raynet c/o T. Hanratty 12 Clarendon Street, Consett DH8 5LS
G4 PFR J. Harding 19 Carrington Crescent Wendover, Aylesbury HP22 6AW
G4 PFT J. Harris 45 Redehall Road, Smallfield, Horley RH6 9QA
G4 PFU D. Blunt 12 Mallard Place, East Grinstead RH19 4TF
G4 PFW H. Palmer 5 Hurst Close, Crawley RH11 8LQ
G4 PFX D. Palmer 14 Garibaldi Road, Redhill RH1 6PB
G4 PFY Dr J. O'Hagan 13 Chapel Road, Stanford In The Vale, Faringdon SN7 8LE
G4 PFZ J. Aspland 6 Trilithon Close, Hellesdon, Norwich NR6 5EP
G4 PGA B. Gage 2 Wellsbourne Road, Stone Cross, Pevensey BN24 5QX
G4 PGB P. Hayward 22 Falconers Park, Sawbridgeworth CM21 0AU
G4 PGD R. Hall 18 Park View, Truro TR1 2BW
G4 PGG P. Beesley 15 Byron Close, Cheadle, Stoke-on-Trent ST10 1XB
GI4 PGH J. Crawford 2 Holywood Road, Newtownards BT23 4TQ
G4 PGJ D. Ward 48 Moat Bank, Bretby, Burton-on-Trent DE15 0QJ
GM4 PGM P. Brash 4 Union Street, Lossiemouth IV31 6BA
GI4 PGN J. Bailie 4 Quarry Road, Greyabbey, Newtownards BT22 2QF
G4 PGO D. Fernant 2 Lonsboro Road, Wallasey CH44 9BR
G4 PGQ D. Harrison 517 Atherton Road, Hindley Green, Wigan WN2 4QF
GM4 PGV P. Lawless 37 Oaklands Avenue, Irvine KA12 0SE
G4 PGW N. Puttick 33 Alder Hill Drive, Totton, Southampton SO40 8JB
G4 PGX M. Williams 114 Ferry Street, Burton-on-Trent DE15 9EY
G4 PGY Dr R. White Beech Hill, Northampton NN7 4LL
GW4 PHB W. Vickers Creigiau, Penrhyndeudraeth LL48 6LS
G4 PHC G. Stearn 31, Regents Way, Minehead TA24 5HS
G4 PHK P. Colbeck 76 Church Road, Winterbourne Down, Bristol BS36 1BY
G4 PHL P. GreenDanewalk, North Road, Brotherton, Knottingley WF11 9ED
G4 PHP D. Foster 120 Green Lane, Cookridge, Leeds LS16 7HF
G4 PHR T. Clough 37 Park Avenue, Mirfield WF14 9PB
GW4 PHT D. Dalling 308 Townhill Road Mayhill, Swansea SA1 6PD
G4 PHV G. Bennison 35 Ermine Street, Thundridge, Ware SG12 0SY
G4 PHX T. Jackson 28 John Lee Road, Ledbury HR8 2FE
G4 PIA L. Roberts 18 Turret Grove, London SW4 0ET
GI4 PID B. Little 8 Ballynoe Road, Antrim BT41 2QT
G4 PIE D. Tyler 12 Bernards Way, Flackwell Heath, High Wycombe HP10 9EQ
G4 PIJ J. Goodman 4 Maloren Way, West Moors, Ferndown BH22 0BQ
G4 PIP C. Bottoms Treboro House, Ullenhall, West Midlands B95 5NN
G4 PIQ A. Cook The Old Vicarage, High Road, Ipswich IP6 9LP
G4 PIR J. Child 12 Beachill Road, Havercroft, Wakefield WF4 2EJ
G4 PJD H. Hoare Farvardale, The Street, Bishop's Stortford CM22 7LT
G4 PJE R. Kershaw 13 Silver Hill, Milnrow, Rochdale OL16 3UJ
G4 PJJ British Naturist ARS c/o Dr N. Garbutt Tudor Cottage, Main Road, Gloucester GL2 8JP
G4 PJK R. Mosedale 21 Druids Avenue, Aldridge, Walsall WS9 8LA
G4 PJL R. Bailey 5 Braemar Road, Doncaster DN2 5HN
G4 PJP M. Clay 24 Begonia Drive, Burbage, Hinckley LE10 2SW
GM4 PJR N. Yarrow 10 Coxburn Brae, Bridge Of Allan, Stirling FK9 4PS
G4 PJS P. Shields 81 Flaxton, Skelmersdale WN8 6PE
G4 PJT S. Schofield 18 Ascot Close, Mexborough S64 0JG
G4 PJZ J. Towle 46 Querneby Road, Nottingham NG3 5HY
G4 PKE R. Badham Caedman, Terrace Road North, Bracknell RG42 5JG
G4 PKF E. Wood 68 Baswich Crest, Stafford ST17 0HJ
GM4 PKJ D. Smith Haremuir Bungalow, Benholm, Montrose DD10 0HX
G4 PKK Dr S. Juden 17A Astonville St., Southfield, London SW18 5AN
G4 PKM J. Derrick 37 Admiralty Street, Keyham, Plymouth PL2 2BR
G4 PKO D. French 14 Linden Close, Prestbury, Cheltenham GL52 3DU
G4 PKP J. Jones Jason Photographic, New Moss Farm, Liverpool L37 0AH
G4 PKT Group Two c/o D. Lewin 14A Warwick New Road, Leamington Spa CV32 5JG
G4 PKV South Normanton, Alfreton & District ARC c/o D. Griffiths 61 The Drive, North Harrow, Harrow HA2 7EJ
G4 PKW C. Gerard 7 Parkwood Road Sidemoor, Bromsgrove B61 8UA
G4 PKX F. Gallimore 3 Wilson Crescent, Lostock Gralam, Northwich CW9 7QH
G4 PKZ R. Richardson Manor Farm, Manor Lane, Oakham LE15 7JL
G4 PLH R. Hughes 73 Upland Road, Sutton SM2 5JA
GM4 PLI J. Nellis 64 Kirkwood Avenue, Clydebank G81 2ST
G4 PLK S. Lewis 189 Ashburton Road, Hugglescote, Coalville LE67 2HE
G4 PLL I. Thomas 15 Wakefield Road, Fitzwilliam, Pontefract WF9 5AJ
G4 PLS A. Haigh White Horse Cottage, Maypole, Canterbury CT3 4LN
G4 PLU J. Bates 63 Sunny Blunts, Peterlee SR8 1LP
G4 PLV M. Seton 12 Chatsworth St., Roundthorn, Oldham OL4 5LF
G4 PLX A. Salata 64 Wildwood Road, London NW11 6UP
G4 PLY V. Morris 21 Cranhill Road, Street BA16 0BY
G4 PLZ P. Connors Manor Cottage, Mill Road, Banningham, Norwich NR11 7DT
G4 PMA D. Pearson 42 Church Street, Stapleford, Nottingham NG9 8DJ
G4 PMB F. Thompson 28 Lordsmead, Cranfield MK43 0HP
G4 PMG M. Green Huntley, Chesham Road, Wigginton, Tring HP23 6HH
GM4 PMH S. Dunn 4 Mid Street, Rosehearty, Fraserburgh AB43 7JS
G4 PMJ A. Santos 17 Elm Garth, Roos, Baytree, Hull HU12 0HH
GM4 PMK R. Blackwell Willowbank, Pennyghael, Isle of Mull PA70 6HB
G4 PMM R. Williams 2 Keepers Close, Bestwood Village, Nottingham NG6 8XE
GI4 PMP H. Smith 1A Taylor Park, Limavady BT49 0NT
G4 PMS P. Steele 107 Lower Shelton Road, Marston Moretaine, Bedford MK43 0LW
GM4 PMT A. Ross 16 Burnside Drive, Bridge Of Don, Aberdeen AB23 8PL
G4 PMV K. Grime 13 Runnymede Court, Jackson Street, Bolton BL3 5HX
G4 PMW R. Dunn 117 All Saints Way, West Bromwich B71 1RU
G4 PMY G. Bell Linden Lea, Crewe Road, Sandbach CW11 4RE
G4 PMZ P. Butcher 9 Little Platt, Guildford GU2 8JU
G4 PNB A. Bathurst 64 Oakfields, Guildford GU3 3AU
G4 PNC D. Hood 32 Bishops Wood, Nantwich CW5 7QD
G4 PND A. Daniel 10 Tamarisk Close, Hatch Warren, Basingstoke RG22 4UX
G4 PNH G. Aungiers 17 Broadwood Drive, Fulwood, Preston PR2 9SS
G4 PNI R. Bishop 29 Windsor Court, Poulton-le-Fylde FY6 7UX
G4 PNK T. Crosland Park Farm House, Park Road, Bedford MK43 7QF
G4 PNL A. Coe 112 Harborough Road, Desborough, Kettering NN14 2QY
GM4 PNM A. Wixon Riverview Cottage, Melrose TD6 9JB
G4 PNP B. Deak 57 Arundel Road, Peacehaven BN10 8RP
G4 PNQ R. Dhami 3327 Smoke Tree Road, Mississauga, Ontario L5N 7M5 Canada
G4 PNT A. Hellewell 41 Woodlea Grove, Armthorpe, Doncaster DN3 2HN
G4 PNX D. Painter 93 Oxclose Lane Arnold, Nottingham NG5 6FN
G4 POB T. Hutchings 9 Little Dell, Welwyn Garden City AL8 7HZ
GI4 POC R. Drain 5 Ravelstone Avenue, Bangor BT19 1EQ
G4 POD J. Gould 30 Weymouth Avenue, Middlesbrough TS8 9AB
G4 POF J. Hart 1 Meadow Court, Fordingbridge SP6 1LW
G4 POG W. Evans 3 Coastline Village, Ostend Road, Norwich NR12 0NE
G4 POI D. Lambert 8 Stretton Road, Barnsley S71 1XQ
G4 POL B. Robertson 12 Green Lane, Woodstock OX20 1JY
G4 POP T. Genes 28 Hillside Road, Burnham on Crouch CM0 8EY
G4 POR B. Banks 30 Hospital Road, Burntwood WS7 0ED
G4 POT D. Girling 20 Fore Street, Praze An Beeble, Camborne TR14 0JX
G4 POU P. Dyer 36 Margate Road, Ipswich IP3 9DE
G4 POW A. Owen 60 Brighton Avenue, Elson, Gosport PO12 4BX
G4 POY R. Kent Talstr 4, Eriskirch 88097 Germany
G4 PPB E. Marshall 75 Acacia Crescent, Wigan WN6 8NJ
G4 PPC B. Lowe 19 Wolverhampton Road, Bloxwich, Walsall WS3 2EZ
G4 PPE M. Bell 55 Park Road, Hampton Hill, Hampton TW12 1HX
G4 PPG J. O'Sullivan 40 Sheldon Avenue, Standish, Wigan WN6 0LW
G4 PPH A. Betts 11 Brisbane Close Mansfield Woodhouse, Mansfield NG19 8QZ
G4 PPJ S. Bone 6 Manor Road, Folksworth, Peterborough PE7 3SU
G4 PPK C. Everley 5 Firs Close, Hazlemere, Hazlemere HP15 7TF
G4 PPL P. Fisher 10 Magdalene Court, Seaham SR7 7DJ
G4 PPN D. Chapman 22 Horsley Drive, Kingston upon Thames KT2 5GG
G4 PPP R. Bailey 10 Epping Close, Walsall WS3 1TT
G4 PPR M. Spencer 67 Holmley Lane, Dronfield S18 2HQ
G4 PPS Dr D. Herbert 48 Furrow Grange, Middlesbrough TS5 8DP
GM4 PPT R. Hodge 34 Craig View Coylton, Ayr KA6 6LB
G4 PPU M. Roy 17 Elgar Avenue, Tolworth, Surbiton KT5 9JH
G4 PPW A. Keech 2 Mountfield Road, Irthlingborough, Wellingborough NN9 5SY
G4 PPZ J. Young Nonsuch, Oxbridge, Bridport DT6 3UB
G4 PQB D. Mathers Dovedale Lodge, Bourton On The Hill, Moreton-in-Marsh GL56 9TE
G4 PQI J. Raybould 2 Woodland Avenue, Brierley Hill DY5 1EQ
G4 PQM E. James 59 Queensway, Euxton, Chorley PR7 6PN
G4 PQP P. Malme Newhaven, Mill Lane East Runton, , Cromer, NR27 9PH
G4 PQS W. Tedbury Tyting House, Exeter Road, Honiton EX14 1AX
G4 PQU A. Harwood 108 Tudor Green, Jaywick, Clacton-on-Sea CO15 2PE
GI4 PQV T. Pollock 33 Seahill, Donaghadee BT21 0SH
G4 PQW M. Davis 478 Eastern Avenue, Gants Hill, Ilford IG2 6EQ
G4 PQX D. Taylor 28 Main Street, Broadmayne, Dorchester DT2 8EB
G4 PQY A. Williams 7 Bower Hall Drive, Steeple Bumpstead, Haverhill CB9 7ED
G4 PRB P. Ball 21 Doonamana Road, Dun Laoghaire A96 W6K3 Ireland
G4 PRD P. Dakin 12 Spinney Close, Kidderminster DY11 6DQ
G4 PRF S. Brown 27 The Court, Anderby Creek, Skegness PE24 5YQ
GI4 PRH D. Simpson 31 Beech Green, Doagh, Ballyclare BT39 0QB
G4 PRJ M. Worsfold 5 Turner Close, Langney, Eastbourne BN23 7PF
G4 PRL R. Hunt 13 Westlake Rise, Heybrook Bay, Plymouth PL9 0DS
GM4 PRO T. Oneil 187 Main St., Chapelhall, Airdrie ML6 8SF
GW4 PRP S. Lane 12 Carlos St., Port Talbot SA13 1YD
G4 PRQ I. Hooper Flat 5, Wenlock House, 41 Stanstead Road, London SE23 1HG
G4 PRS Poole RS c/o P. Ciotti 6 Bascott Road, Bournemouth BH11 8RH
G4 PRW P. Whitten 2 Eastmead, Woking GU21 3BP
G4 PSE M. Grime 10 East Park Avenue, Darwen BB3 2SQ
G4 PSG 1st Prestwood Scout Group c/o A. Willson Hilltop, Cryers Hill Road, Cryers Hill, High Wycombe HP15 6LJ

G4 PSI C. Franks 11 Orchard Close, Crook DL15 8QU
GM4 PSJ R. Stroud 24 Cullen St., Portsoy, Banff AB45 2PJ
GM4 PSL T. Grice 35 Approach Row, East Wemyss, Kirkcaldy KY1 4LB
G4 PSO A. Little 20 Vicarage Close, Shillington, Hitchin SG5 3LS
G4 PSP S. Gardner 191 Charlton Park Midsomer Norton, Radstock BA3 4BR
G4 PSR C. Sartorius 39 Althorne Gardens, London E18 2DA
G4 PSS S. Black 71 Bellerby Drive, Urpeth Grange, Ouston, Chester le Street DH2 1UF
G4 PST D. Turner Hurdletree Bank Farm, Hurdletree Bank, Spalding PE12 8QQ
G4 PSU A. Davidson 5 Hanover Parc, Indian Queens, St. Columb TR9 6ER
G4 PTE Dr K. Lown Maurice House, Callis Court Road, Broadstairs CT10 3AH
G4 PTF C. Keeping 12 St. Francis Avenue, Southampton SO18 5QJ
G4 PTK G. Mason 120 Scalford Road, Melton Mowbray LE13 1JZ
G4 PTM J. Fyrth 2 Merton Gardens, Farsley, Pudsey LS28 5DZ
GM4 PTQ Rvd. S. Bennie 13A Scotland Street, Stornoway HS1 2JN
G4 PTR G. Pinter 29 The Mill, The Boulevard, Horsham RH12 1GR
G4 PTU Blandford Garrison CW And DX Group c/o R. Carter 12 Glebe Close, Abbotsbury, Weymouth DT3 4LD
GD4 PTV B. Brough 4 The Bretney, Jurby IM7 3BL Isle of Man
G4 PTW A. Brunning 6 Newstead Road, Barnwood, Gloucester GL4 3TQ
G4 PTZ J. Topley 27 Inveraray Close, Sinfin, Derby DE24 3JA
G4 PUB Coalhouse Fort c/o S. Wensley 7 Bradshaw Close, Windsor SL4 5PS
GW4 PUC R. Rees 16 Brynheulog, Llanelli SA14 8AE
G4 PUD B. Langdon 80 Glen Rise, Birmingham B13 0EJ
G4 PUM R. Brookes Broadeaves, Bridgemere Lane, Hunsterson, , Nantwich CW5 7PN
G4 PUO W. Stumpf 18 Saxhorn Road, Lane End, High Wycombe HP14 3JN
G4 PUP B. Philipp 2 Red Lion Park, Denbigh Road, Battle TN33 9ET
G4 PUQ P. Mcewen Southerly, Church Road, Halesworth IP19 0EA
GM4 PUS Dr J. Murray Ose Farm House, Isle of Skye IV56 8FJ
GW4 PUX B. Gayther Coed Park, Penisarwaun, Caernarfon LL55 3PW
G4 PUZ N. Barrington 18 Brambleside, Thrapston, Kettering NN14 4PY
G4 PVC A. Smith 21, Darwin Crescent, , Morley 6062 Australia
G4 PVM P. Tittensor 47 St. Johns Road, Chelmsford CM2 0TY
G4 PVN A. Taylor 3 Mond Crescent, Billingham TS23 1DL
G4 PVP P. Painter 80 Willowsbrook Road, Hurst Green, Halesowen B62 9RF
GM4 PVQ D. Ross 11 Edinview Gardens, Stonehaven AB39 2EG
G4 PVS J. Maude Anthony Fold Farm, Bury Old Rd, Ramsbottom Lancs BL0 0RY
GW4 PVU A. Wilkinson 1 Langley Close, Penrhyn Bay, Llandudno LL30 3LN
G4 PVX A. Daws 9 Wellow Mead, Peasedown St. John, Bath BA2 8SA
G4 PVY R. Limb 3 Canford Heights, Western Road, Poole BH13 7BE
G4 PVZ G. Loach 39 Park Road West, Wolverhampton WV1 4PL
G4 PWA P. Dane Oakhill Lodge Hewelsfield, Lydney GL15 6UN
G4 PWB G. Smith 9A Lansdowne Drive, Rayleigh SS6 9AL
G4 PWD M. Mchale 41 Sheringham Drive, Etchinghill, Rugeley WS15 2YG
G4 PWE J. Veness 59 St. Helens Down, Hastings TN34 2BG
G4 PWF R. Harries The Mill, Mill Lane, Middle Rasen LN8 3LE
G4 PWG J. Hubbard 2 Carlton Road, Portchester, Fareham PO16 8JW
G4 PWI P. Heredge 118 Oxford Crescent, Didcot OX11 7AX
G4 PWM D. East 39 Chapel Lane Navenby, Lincoln LN5 0ER
G4 PWP D. Blackwell 58 Bleadon Hill, Weston Super Mare BS24 9JW
GM4 PWQ J. Foster 185 Sea Road, Methil, Leven KY8 2EQ
GM4 PWR A. Coutts Ballymeanoch Barn, Kildonan, Isle of Arran KA27 8SF
G4 PWS S. Keen 34 Unwin Road, Isleworth TW7 6HX
G4 PWV D. Howton 4 Stonepine Close, Wildwood, Stafford ST17 4QS
G4 PWY P. Hennessy 5 Smedley Court, Egginton, Derby DE65 6HD
GW4 PWZ W. Evans Windyridge Bungalow, Mount View, Merthyr Tydfil CF47 0UX
GM4 PXB M. Gale East Teuchan Cruden Bay, Peterhead AB42 0PP
G4 PXC A. Lord 16 Lark Valley Drive, Fornham St. Martin, Bury St. Edmunds IP28 6UG
G4 PXE A. Brend 42 West Garth Road, Exeter EX4 5AJ
G4 PXF M. Harries 63 Oakhill Road, Dronfield S18 2EL
GM4 PXG T. Worthington 5 Clairmont Place, Lerwick, Shetland ZE1 0BR
G4 PXH E. Southwell 60 Solent Breezes, Hook Lane, Southampton SO31 9HG
G4 PXJ J. Peet 66 Barry Road, Northampton NN1 5JS
GI4 PXM W. Nelson 111 Rutherglen Street, Belfast BT13 3LR
G4 PXN C. Sissons 9 Mount Pleasant, Goldenbank, Falmouth TR11 5BW
G4 PXR T. Geldart Langdale, Coast Road, Ulverston LA12 9QZ
G4 PXX P. Styles 23 Mereweather Avenue, Frankstow, Victoria 3199 Australia
G4 PXY G. Bloomfield 15 Beaulieu Drive, Pinner HA5 1NB
G4 PYA A. Ledger 32 St. Augustines Crescent, Whitstable CT5 2NW
G4 PYD C. Johnson 51 Newstead Avenue, Holton-Le-Clay, Grimsby DN36 5BQ
G4 PYG I. Bennett Collins Green, School Road, Colchester CO5 9TH
G4 PYH D. Jackson 9 Stour Close, Altrincham WA14 4UE
G4 PYI B. Burman 53 Field Avenue, Hatton, Derby DE65 5ER
GM4 PYJ J. Balfour 36 Causewayhead Road, Stirling FK9 5EU
G4 PYQ A. Hill 37 Rock St., Gee Cross, Hyde SK14 5JX
G4 PYS E. Devereux 15 Severn Close, Paulsgrove, Portsmouth PO6 4BB
G4 PYU S. Harding 6 Rooks Way, Tiverton EX16 6XJ
G4 PYV S. Parkin 4 Acacia Close, Worksop S80 3RD
G4 PYW M. Zubrzycki 4 Falklands Court, Easington, Hull HU12 0QE
G4 PZJ C. Christopher 15 Inman Road, Earlsfield, London SW18 3BB
G4 PZL Cheshunt & District A.R.S c/o R. Pain Hegerston, Long Road West, Colchester CO7 6ES
G4 PZN C. Poulson 12 Glebe Close, Appleby CA16 6RS
G4 PZU F. Day 27 Prince Charles Road, Lewes BN7 2HY
G4 PZV A. Terry 6 Seaton Close, Stubbington, Fareham PO14 2PX
G4 PZW R. Proctor 6 North Street, Burwell, Cambridge CB25 0BA
G4 PZX A. Tracey Well 'N' Garden, Abberton Road, Colchester CO5 7AS
G4 RAA M. Brown 12 Mead Way, Slough SL1 6HD
G4 RAB D. Ellis 1 Showering Close, Bristol BS14 8DY
G4 RAC J. Cooper 134 Jordan Avenue, Stretton, Burton-on-Trent DE13 0JD
G4 RAE R. Laney 7 Downfield Close, Alveston, Bristol BS35 3NJ
GD4 RAG J. Martin Tradewinds, Mount Gawne Road, Port St. Mary, Isle of Man IM9 5LX
GM4 RAH P. Robertson 32 Crosswood Crescent, Balerno EH14 7HS
GM4 RAI R. Shand 12 Bexley Terrace, Wick KW1 5HQ
G4 RAJ J. Shaw 31 Dartmouth Avenue, Almondbury, Huddersfield HD5 8UP
G4 RAK J. Hornby 21 West Wools, Portland DT5 2EA
G4 RAP V. Seaman 30 Dukes Orchard Nicholas Close, Writtle, Chelmsford CM1 3JZ
G4 RAR P. Clemens 18 Ladylea Road, Horsley, Derby DE21 5BN
G4 RAV P. Evans 2 Terence Airey Court, Harleston IP20 9JP
G4 RAY G. Pickering 1 Warrens Yard, Wells-next-the-Sea NR23 1PA
GM4 RAZ B. Smith 8 Mosside Drive, Portlethen, Aberdeen AB12 4NY
G4 RB Dr R. Baldwin 20 Lynn Road, Shouldham, King's Lynn PE33 0BT
G4 RBC C. Hawkridge11 2 Windward Close, Littlehampton BN17 6QX
G4 RBH T. Farmer 12 Rose Avenue, Mitcham CR4 3JS
G4 RBP R. Purdy 4 York Road, Brookenby, Market Rasen LN8 6EX
G4 RBQ D. Love Highridge, South Road, Haywards Heath RH17 7QS
G4 RBR C. Randall 38 Kilmorey Gardens, St. Margaret'S, Twickenham TW1 1PY
G4 RBU T. Crookes 167 Willow Drive, Handsworth Hill, Sheffield S9 4AU
G4 RBZ C. Dervin 24 Willow Park Way, Aston-On-Trent, Derby DE72 2DF
G4 RCB D. Thorp 24 Garnstone Drive, Weobley HR4 8TH
G4 RCC Caravan & Camping ARC c/o A. Wright 34 Webbs Way, Stoney Stanton, Leicester LE9 4BW
G4 RCD M. Capstick 186 Forest Lane, Harrogate HG2 7EE
G4 RCE Dr M. Capstick Gladbachstrasse 19, Zurich 8006 Switzerland
G4 RCF J. O'Dell 5 Further Ends Road, Freckleton, Preston PR4 1RL
G4 RCG J. Muzyka 2 Engine Fold Off Lindale Lane Kirkhamgate, Wakefield WF2 0PP
G4 RCH S. Thompson 2 Allenby Drive, Leeds LS11 5RP
G4 RCJ D. Underwood The Coach House, Todmorden OL14 8EP
GI4 RCK W. Mcmillen 26 Maymount St., Belfast BT6 8BH
GW4 RCM W. Williams 154 Llysfaen Road, Old Colwyn, Colwyn Bay LL29 9HP
GM4 RCN J. Young 13 Craig Crescent, Causewayhead, Stirling FK9 5LR
G4 RCP C. Marriott 19 Beechey Close, Denver, Downham Market PE38 0DH
G4 RCR P. Starley 40 Wordsworth Avenue, Warwick CV34 6JD
G4 RCY A. Beglin 3 The Mead, Shipham, Winscombe BS25 1TR
G4 RCZ I. Dempster 54 Fashoda Road, Selly Park, Birmingham B29 7QJ
G4 RDC J. Gumb 16 Raggleswood Close, Earley, Reading RG6 7LH
G4 RDG Dr G. Murray 176 Golfwood Drive, Hamilton L9C 7B8 Canada
G4 RDH M. Wirthner 51 College Road, Upper Beeding, Steyning BN44 3TB
GM4 RDI J. White 1 Banknowe Road, Tayport DD6 9LG
G4 RDL Leeds Raynet Group c/o G. Belt 3 Prospect Hill, Whitby YO21 1QE
G4 RDM P. Rouget 7 Palmer Close, Wellingborough NN8 5NX
G4 RDS B. Wood 100 Lower Road, Hullbridge, Hockley SS5 6DD
GW4 RDW L. Jones 6, Norton Terrace, Glyncorrwg, Port Talbot SA13 3AN
G4 RDY L. Harrison 48 Bleasdale Avenue, Thornton-Cleveleys FY5 3RQ
G4 REA A. Wilson 101 Watmore Lane, Winnersh, Wokingham RG41 5LG
G4 REC A. Marrows 7 Victoria Close, Yeadon, Leeds LS19 7AU
G4 REE P. Miller 2 The Pavilions End, Camberley GU15 2LD
GM4 REF W. Mclean 1 Richmond Court, Rutherglen, Glasgow G73 3BG
G4 REG A. Boocock 2 Vine Garth, Clifton, Brighouse HD6 4JZ
G4 REH R. England 5 Weir Road, Congresbury, Bristol BS49 5HL
GW4 REI D. May 19 Sycamore Street, Pembroke Dock SA72 6QN
G4 REK J. Tylee 40 Luna Road, Thornton Heath CR7 8NY
GM4 REN B. Strathdee 85 Weavers Knowe Crescent, Currie EH14 5PP
GW4 RER C. Carini 6 Whitewell Drive, Llantwit Major CF61 1TA
G4 REU J. Taylor 18 Fackley Way, Stanton Hill, Sutton-in-Ashfield NG17 3HT
GW4 REX P. Hassmann 11 Oak Close, Bulwark, Chepstow NP16 5RL
G4 RFA M. Tew 25 Broad Oak Lane, Penwortham, Preston PR1 0UX
G4 RFC S. Fletcher 90 Westcombe Park Road, Blackheath, London SE3 7QS
G4 RFF South Cheshire ARS c/o T. Mundiya Apt 5, 35 Mercer Street, New York 10013 United States
G4 RFI R. Linden 24 Hartland Drive, Edgware HA8 8RH
GD4 RFK M. Dodd Ellan Geay, Ballayockey Lane, , Ramsey IM7 3HP Isle of Man
G4 RFN A. Robey 54 Jarrett Avenue, Wainscott, Rochester ME2 4NL
G4 RFO B. Wood 11 Oakdale Avenue, Wibsey, Bradford BD6 1RP
G4 RFP A. Goodall 10 Beacon Close, Everton, Lymington SO41 0LQ
G4 RFR Flight Refueling ARS c/o R. Hawkins Forest Edge, Deer Park, Milton Abbas, Blandford Forum DT11 0AY
G4 RFU D. Abbott 21 Leckhampton Road, Cheltenham GL53 0AZ
G4 RFV B. Adams 14 Foxcroft Drive, Wimborne BH21 2JZ
G4 RGA J. Dunnett 43 Oakfield Park, Wellington TA21 8EX
G4 RGB M. Rogers 4 Hill Corner, Ledgemoor, Hereford HR4 8QG
G4 RGE N. Lovely Dolphin Cottage, Upper Green Road, Ryde PO33 1XE
G4 RGF P. Mccall 11 Elworthy Drive, Wellington TA21 9AT
G4 RGH D. Mclaughlin 1 Bosville Close, Ravenfield, Rotherham S65 4NF
GW4 RGI W. Baker 4 Connaught Place, Pembroke Dock SA72 6EZ
G4 RGM Greater Manchester Raynet c/o N. Czernuszka 12 Durham Drive, Ashton-under-Lyne OL6 8BP
G4 RGO J. Crocker 8 Oakwood Avenue, Havant PO9 3RA
G4 RGP E. Hall 93 Sthbourne Coast, Bournemouth BH6 4DX
GD4 RGR K. Grattan 41 Carrick Park, Sulby, Ramsey, Isle of Man IM7 2EY
GM4 RGS R. Smith 8 Mosside Drive, Portlethen, Aberdeen AB12 4NY
GM4 RGU D. Nicolson Silver Birches, Blebo Craigs, Cupar KY15 5UF
G4 RGY Burnham ARC c/o M. Lang 52 Gloucester Road, Burnham-on-Sea TA8 1JA
G4 RHB M. Bailey 10 Greenwood Avenue, Bolton Le Sands, Carnforth LA5 8AW
G4 RHC M. Kellett 1 Spa Cottages, Gilsland, Brampton CA8 7AL
G4 RHJ P. Vickers 2 Firbank Drive, Woking GU21 7QT
G4 RHK L. Woodcock 2 Poolhay Close Corse Lawn, Gloucester GL19 4NY
G4 RHL R. Langdon 15 St. Cuthberts Way, Sherburn Village, Durham DH6 1RH
G4 RHR K. Backhouse 113 Bucklesham Road, Kirton, Ipswich IP10 0PF
G4 RHX A. Moore 1 St. Andrews Road, New Marske, Redcar TS11 8AU
G4 RHY M. Pratt The Bays, Back Lane, Doncaster DN9 3AJ
G4 RHZ B. Coupe 9 School Lane, Auckley, Doncaster DN9 3JR
GW4 RIB D. Stoole Brookside Farm, Baltic Terrace, Cwmbran NP44 7AH
G4 RIE D. Littler 16 Lee Bank, Westhoughton, Bolton BL5 3HQ
G4 RIH N. Thorne 61 Horsham Avenue, London N12 9BG
G4 RIK R. Kirkwood 42 Porters Hill, Harpenden AL5 5HR
G4 RIM A. Day 3 Harris Way, Lee Mill Bridge, Ivybridge PL21 9EU
G4 RIO S. Williams 18 The Leas, Barkston, Grantham NG32 2PD
G4 RIP J. Creaseyy 8 Church Street Billingborough, Sleaford NG34 0QG
G4 RIQ Aylesbury Raynet Grp c/o L. Rushforth 90 Brearley Avenue, New Whittington, Chesterfield S43 2DZ
G4 RIS B. Didmom 45 Millstrood Road, Whitstable CT5 1QF
G4 RIU R. Jones 67 Plover Road, Larkfield, Aylesford ME20 6LA
GM4 RIV Wigtownshire ARC c/o A. Gaston 9 Lochans Mill Avenue, Lochans, Stranraer DG9 9BZ
G4 RJA I. Wilkinson 24 Isis Way, Hilton, Derby DE65 5LP
G4 RJD K. Ward 3 Levetts Hollow, Hednesford, Cannock WS12 2AW

GM4 RJF J. Weatherer 20 Gilloch Crescent, Dumfries DG1 4DW
G4 RJG I. Toon 18 Barrowfield Road, Stroud GL5 4DF
G4 RJM G. Heward 4 Hillside Drive, Little Haywood, Stafford ST18 0NN
G4 RJO B. Robertson 28 Heath Lane, Blackfordby, Swadlincote DE11 8AA
G4 RJQ C. Johnston 299 Constable Avenue, Clacton-on-Sea CO16 8YU
G4 RJS R. Shaw 1B Battle Close, Lindholme, Doncaster DN7 6DA
GM4 RJX J. Hatton 64 Abercromby Crescent, Helensburgh G84 9DN
G4 RJY C. Sidney 10 Colville Close, Bampton OX18 2NN
G4 RJZ A. Phillpott Southways, Stombers Lane, Folkestone CT18 7AP
G4 RKB J. Conlon 24 Goldcrest Close, Colchester CO4 3FN
GI4 RKC S. Jennings 34 Palmer Avenue, Lisburn BT28 3QB
G4 RKD T. Clarke 19 Ratby Lane, Markfield LE67 9RJ
G4 RKF B. Peart 7A Grundy Close, Abingdon OX14 3SD
G4 RKG J. Whiting 19 Watermore Close, Frampton Cotterell, Bristol BS36 2NQ
GM4 RKH T. Llewellyn The Shepherds Cottage, Buckies Farm, Thurso KW14 7XH
GW4 RKI K. Perryman 17 Fford Maes Gwilym Ffos Las Carway Kidwelly, Carmathenshire SA17 4AX
G4 RKK I. Welford Mistletoe House, Watton Road, Thetford IP24 1PB
G4 RKL W. Welford Bowling Green House, Griffin Lane, Attleborough NR17 2AD
GM4 RKM T. Cassidy 34 Torr-Na-Faire, Lochaline, Oban PA80 5XS
G4 RKO B. Cooper 20 The Paddock, Alconbury, Huntingdon PE28 4WS
G4 RKP R. Groom Tryst, Rackhams Corner, Lowestoft NR32 5LB
G4 RKR D. Geddes 9 Rosenella Close, Northampton NN4 8RX
G4 RKV L. Adams 50 Selsea Avenue, Herne Bay CT6 8SD
GW4 RKX G. Cook 22 Northlands Park, Bishopston, Swansea SA3 3JW
GW4 RKZ R. Cleverley 33 Tylchawen Crescent, Tonyrefail, Porth CF39 8AL
G4 RLA C. Butcher 7 Lascelles Hall Road, Kirkheaton, Huddersfield HD5 0AT
G4 RLC T. Isom 64 Cuffling Drive, Leicester LE3 6NF
G4 RLF M. Wright 24 Wessex Road, Wilton, Salisbury SP2 0LW
G4 RLL J. Woods 4 Wheatfield Drive, Burton Latimer, Kettering NN15 5YL
G4 RLM J. Hiscock 62 East Borough, Wimborne BH21 1PL
G4 RLN D. Rosevear 37 Sharaman Close, St. Austell PL25 3DH
GW4 RLO G. Seal Roda Villa, 14 Uchel Dre, Kerry, Newtown SY16 4PS
GW4 RLP T. Varney 29 Ffordd Eryri, Caernarfon LL55 2UR
G4 RLR J. Hogan 13 Strawberry Fields, Great Barford, Bedford MK44 3BQ
G4 RLU P. Quickfall Quickfall, 14 Lade Fort Crescent, Romney Marsh TN29 9YF
GM4 RLV W. Duguid Villach, 7 Hawthorn Place, Ballater AB35 5QH
G4 RLX H. Cave 3 Grace Meadow, Whitfield, Dover CT16 3HA
GI4 RMA L. Mccullough 9 Castleward Road Strangford, Downpatrick BT30 7LY
G4 RMC D. Marsden 67 Fourth Avenue, Watford WD25 9QH
G4 RMD J. Cobley 4 Briars Close, Hatfield AL10 8DQ
G4 RMG E. Guy 2 Broad Street, St. Columb Major TR9 6AS
G4 RMJ A. Cattani Rozel, Marsh Road, Spalding PE12 9PJ
GW4 RML G. Davies 101 Westlands, Port Talbot SA12 7DE
G4 RMN M. Hogan 16 Freshfield Close, West Earlham, Norwich NR5 8RA
G4 RMQ J. Dudley 2 Heathcote Grove, London E4 6RT
G4 RMT P. Johnson 4 High Beech, Lowestoft NR32 2RY
G4 RMV M. Buckle 3 Tilesford Park Tilesford, Pershore WR10 2LA
G4 RMX J. Phelps 33 Thirlmere Drive, North Anston, Sheffield S25 4JP
G4 RNA P. Dronfield 1 Kestrel Rise Swallownest, Sheffield S26 4SD
G4 RNC C. Blezard 26 Welford Avenue, Lowton, Warrington WA3 2RN
G4 RND C. Hawkins 57 Links Road, Knott End-On-Sea, Poulton-le-Fylde FY6 0DF
G4 RNF J. Handley Flat 31, Croft Manor Mason Close, Freckleton, Preston PR4 1RG
G4 RNK R. Dodson 22 Southgate Crescent Rodborough, Stroud GL5 3TS
GI4 RNP V. Mcfarland 16 Saint Annes Crescent, Carnmoney, Newtownabbey BT36 5JZ
G4 RNR J. Maunder Boat Inn Farm, Shipley Gate Eastwood, Nottingham NG16 3JE
G4 RNT D. Thorpe 10 Stoke Road, Taunton TA1 3EJ
G4 RNW M. Stewart 29 Elstree Road, Bushey Heath, Bushey WD23 4GH
G4 RNX A. Walker 5 Christchurch Road, Malvern WR14 3BH
G4 RNZ K. Page 51 Bournville Road, Weston-Super-Mare BS23 3RR
G4 ROA A. Chamberlain 16 Okehampton Road, Stivichall, Coventry CV3 5AU
G4 ROB R. Taylor 18 Spruce Avenue, Selston, Nottingham NG16 6DX
G4 ROC L. Odell 30 St. Hybalds Grove, Scawby, Brigg DN20 9DG
G4 ROH W. Smith 10 Playford Road, Rushmere St. Andrew, Ipswich IP4 5RH
G4 ROI S. Kiernan 60 Riverview Road, Epsom KT19 0LB
G4 ROJ R. Stafford 21 Kittiwake Drive, Kidderminster DY10 4RS
G4 ROK G. Sambrook 73 Hayes Drive, Barnton, Northwich CW8 4JX
G4 ROM M. Ellis Field Cottage, Hole Lane, Farnham GU10 5LP
G4 ROP C. White 18 Ashton Gardens, Old Tupton, Chesterfield S42 6JF
G4 ROR D. Harrison 55 Hudson Close, Worcester WR2 4DP
G4 ROU W. Maudsley 42 Crawford St., Clock Face, St. Helens WA9 4XH
GW4 ROV P. Weaver 24 Montclaire Avenue, Blackwood NP12 1EE
G4 ROX A. Capel 33 Romney Avenue, Bristol BS7 9ST
G4 RPA D. Court 4 Rucrofts Close, Bognor Regis PO21 3SL
G4 RPC R. Cassling 14 Canada Way, Lower Wick, Worcester WR2 4DJ
G4 RPD A. Else 77 Sherwood Street, Mansfield Woodhouse, Mansfield NG19 9NB
GM4 RPE J. Mccabe 109 Weirwood Avenue, Baillieston, Glasgow G69 6LQ
G4 RPI C. Parrish 16 Charter Close, Boston PE21 9PD
G4 RPJ I. Flaherty 10 Highfield Park, Heaton Mersey, Stockport SK4 3HD
G4 RPK J. Kaine 74 Camden Mews, London NW1 9BX
G4 RPL D. Ingham Auchengray, 51 Helena Street, Mexborough S64 9PF
GM4 RPO T. Gemmell 30 Goldie Crescent, Lochside, Dumfries DG2 0AJ
G4 RPP G. Kyte 25 Brasted Close, Bexleyheath DA6 8HU
G4 RPT M. Edis 28 High Street, Broughton, Kettering NN14 1NG
G4 RPV K. Baker 153 Long Nuke Road, Birmingham B31 1DX
G4 RQA T. Mills 14 Oxford Close, Padiham, Burnley BB12 7DB
G4 RQF R. Langer Elms Bungalow, Queens Road, Sheffield S20 1AW
G4 RQG S. Baggaley 35 Hayner Grove, Weston Coyney, Stoke-on-Trent ST3 6PQ
G4 RQI D. Warr 5 Monckton Drive, Castleford WF10 3HT
G4 RQJ R. Hannan 87 Plymouth Street, Walney, Barrow-in-Furness LA14 3AN
G4 RQK A. Johnson 14 Highfield, Duddington, Stamford PE9 3QD
G4 RQL M. Wilson The Old Chapel, Poulshot Road, Devizes SN10 1RW
G4 RQN N. Bartzeliotis 74B Ebrington Street, Plymouth PL4 9AQ
G4 RQO J. Pulford 68 York Avenue, Droitwich WR9 7DQ
G4 RQP G. Hallett 9 Dolcroft Road, Rookley, Ventnor PO38 3NT
GW4 RQQ T. Jones 19 Penlon, Menai Bridge LL59 5LR
GW4 RQS J. Leighton 12 Morley Avenue, Connah'S Quay, Deeside CH5 4RE
G4 RQU D. Young 9 Mercedes Avenue, Hunstanton PE36 5EJ
G4 RQW A. Mcewen Apartment 9, 10 Lismore Place, Carlisle CA1 1LX
G4 RRA P. Pasquet Honey Blossom Cottage Spreyton, Crediton EX17 5AL
G4 RRD I. Reynolds 4 Chappel Hill, Fakenham NR21 9HW
G4 RRH J. Green 83A High Street, Ramsey, Huntingdon PE26 1BZ
G4 RRJ C. Pemberton Flat 1, 14 Holmfield Road, Blackpool FY2 9TB
GW4 RRL Betws Y Coed RC c/o J. Williams Cartref, Capel Garmon, Llanrwst LL26 0RG
G4 RRM P. Walker 11 Flixton Drive, Crewe. CW2 8AP
G4 RRN C. Harrold Boundary Farm, Felbrigg Road, Norwich NR11 8PD
GM4 RRP R. Morris The Gables, Highfield, Muir of Ord IV6 7XN
G4 RRQ I. Wright 39 Queen Anne Gardens, Falmouth TR11 4SW
G4 RRR T. Bunce Pear Tree House, Greaves Lane, Malpas SY14 7AR
G4 RRU R. Crooks 6 Whylands Avenue, Worthing BN13 3HG
G4 RRX R. Saxton 7 Huxley Road, Old Lakenham, Norwich NR1 2JR
G4 RSC Reading School ARC c/o T. Walter 11 Silver Street Congresbury, Bristol BS49 5EY
G4 RSD D. Bones Flint Cottage, Ipswich Road, Woodbridge IP13 7PP
G4 RSF M. Booth 45 Park Avenue, Thackley, Bradford BD10 0RJ
GI4 RSI K. Allen 25 Knockgreenan Avenue, Omagh BT79 0EB
G4 RSL C. Bagley 47 Meadow View Road, Weymouth DT3 5PB
G4 RSN R. Burman Woodlands Vale, Calthorpe Road, Ryde PO33 1PR
G4 RSP D. Sandy The Chestnuts, Dumbs Lane, Norwich NR10 3BH
G4 RSS J. Upton 24 Heritage Drive, Clowne, Chesterfield S43 4ST
G4 RST D. Martin 111 Arkwright Road, Irchester, Wellingborough NN29 7EE
G4 RSU P. Winnett 148 Green Lanes, Epsom KT19 9UL
G4 RSW A. Bairstow 63 Barnes Road, Stafford ST17 9RL
G4 RSX M. Dean 117 Waltham Way, Chingford, London E4 8HD
G4 RTA K. Mellor 2 Clune St., Clowne, Chesterfield S43 4NJ
G4 RTC X. Iona 13 Vicars Close, Enfield EN1 3DW
G4 RTH R. Hamstead 1A The Close, North Walsham NR28 9HS
G4 RTI E. Handy 80 Watwood Road, Shirley, Solihull B90 2HY
G4 RTJ C. Howe 113 Fatfield Park, Washington NE38 8BP
GM4 RTN T. Morton 15 Craig Crescent, Causewayhead, Stirling FK9 5LR
G4 RTO G. Calkin 54 Pattermead Crescent, Ottawa, Ontario K1V 0G2 Canada
G4 RTP Dr A. Shattock The Stone House, Westport Road, Co. Galway 0 Ireland
G4 RTQ I. Whitehead 3 Botany Close, Thatcham RG19 4GJ
G4 RTS W. Bateson 10 Priestfield Avenue, Colne BB8 9QJ
G4 RTV C. Tucker 4 Kelsey Park Road, Beckenham BR3 6LJ
G4 RTW G. Rolf Flat 4, Hawksworth House, 73 St. Johns Road, Sandown PO36 8HE
G4 RTX G. Kingdon Flat 12, Courtfields, Lancing BN15 8PA
G4 RTY R. Hayward Alverstone, 28 Chatsworth Avenue, Shanklin PO37 7NZ
G4 RUA R. Medcalf 21 Greenbank, Falmouth, Cornwall TR11 2SW
G4 RUE I. Worsdale 10 Manton Road, Lincoln LN2 2JL
G4 RUI A. Keeble 9 Horsley Avenue, Shiremoor, Newcastle upon Tyne NE27 0UF
G4 RUJ P. Evans 706 St. Johns Road, Clacton-on-Sea CO16 8BN
G4 RUL A. Turner 42 Brassey Avenue, Hampden Park, Eastbourne BN22 9QG
G4 RUN M. Beesley 60 Ainsbury Road, Canley Gardens, Coventry CV5 6BB
GM4 RUP Rvd. J. Campbell 96 Boghead Road, Lenzie, Glasgow G66 4EN
G4 RUR M. Baker 2 St. Peters Yard, Wold Street, Norton, Malton YO17 9FH
G4 RUT H. Edwards 19 Cameron Road, Burpengary East, Queensland 4505 Australia
G4 RUW Lothians RS c/o R. Daniel 4 Gloucester Road, Newbury RG14 5JP
GW4 RUX A. Jones Forest Lodge, Glynhafod Street, Aberdare CF44 6LD
GW4 RVA T. Nicholas 15 Maes Llewelyn, Carmarthen SA31 1JJ
G4 RVE C. Andrews 29 Dell Drive, Angmering, Littlehampton BN16 4HE
GI4 RVF J. Burke 45 Shorelands, Greenisland, Carrickfergus BT38 8FB
G4 RVG I. Binding 40 Parklands, South Molton EX36 4EW
G4 RVH D. Hird 27 Red Beck Park, Cleator Moor CA25 5EU
G4 RVJ D. Jones 6 Priory Close, Pilton, Barnstaple EX31 1QX
G4 RVK D. Bentley 106 Pargeter Street, Walsall WS2 8RR
G4 RVL D. Gentle 1 Sunny Hill, Milford, Belper DE56 0QR
G4 RVO T. Collinson 8 Brownberrie Drive, Horsforth, Leeds LS18 5PP
G4 RVP S. O'Donnell 2 Derowen Drive, Hayle TR27 4JN
GD4 RVQ J. Wornham 64 Seafield Close, Onchan, Isle of Man IM3 3BU
G4 RVS A. Rodgers 278 Norton Lane, Norton, Sheffield S8 8HE
GI4 RVT R. Jenkins 11 Willowvale Crescent, Islandmagee, Larne BT40 3SQ
G4 RVU P. Wigley 7 Cavendish Close, Duffield, Belper DE56 4DF
G4 RVV M. Stoneham 139 Hever Avenue, West Kingsdown, Sevenoaks TN15 6DT
G4 RVY G. Richardson 69 O'Neill Drive, Peterlee SR8 5UD
G4 RWD K. Cheetham 71 Westmead Road, Barton Under Needwood, Burton-on-Trent DE13 8JR
GM4 RWE D. Brown Willow Crook Turin, Forfar DD8 2UZ
G4 RWF M. Piecha Oaklea, Gordons Close, Taunton TA1 3DA
G4 RWG R. Guest 67 Hanbury Road, Dorridge, Solihull B93 8DN
G4 RWH R. Williams 1 The Meadows, Newhall Green, Coventry CV7 8BF
G4 RWI N. Spear 33 Elmham Road, Beetley, Dereham NR20 4BW
G4 RWK W. Tolman Pulland Cottage, West Down, Ilfracombe EX34 8NH
G4 RWM P. Titherington 5 Hayland Green, Hailsham BN27 1SR
G4 RWN F. Rowan 1 Massey Walk, Wythenshawe, Manchester M22 5JY
G4 RWQ B. Wilkes 3 Alsop Crest, Acton Trussell, Stafford ST17 0SJ
GW4 RWR R. Thomas Ystrad Isa, Ystrad, Denbigh LL16 4RL
G4 RWS S. Valentine 65 Holland Street, Bolton BL1 8PA
G4 RWV P. Paling 15 Longfellow Road, Banbury OX16 9LB
G4 RWW P. Glaisher The Firs, 279 Addiscombe Road, Croydon CR0 7HY
G4 RXB K. Hawkings Church Barn, Dingle Lane, Sandbach CW11 1FW
GM4 RXD R. Gasken 3 Hameravirin, Glendale, Isle of Skye IV55 8WL
G4 RXF G. Bence 10 Valley Road, Mangotsfield, Bristol BS16 9HN
G4 RXG A. Mumford Upper Cross Farm, Thornton Lane, Sandwich CT13 0EU
G4 RXH H. Fowler 164 Rectory Road, Deal CT14 9NP
G4 RXK P. Mcmullan 6 Deepdale Road, Blackpool FY4 4UD
GI4 RXM T. Stitt 51 Lakeland Road, Hillsborough BT26 6PW
GW4 RXO P. Alexander 23 Maengwynne, Llanelli SA15 4NL
G4 RXQ R. Buck 2 Talbot Cottages, Birtley, Chester le Street DH3 1AR
G4 RXR R. Raine 47 Buckingham Road, Peterlee SR8 2DT
GI4 RXS R. Burnside 19 Hilton Park, Portglenone, Ballymena BT44 8HH
GM4 RXW N. Webster Meric, 7 Woodmuir Crescent, Newport-on-Tay DD6 8HL
GI4 RXX S. Tweedie 12 Glencraig Close, Newtownabbey BT36 5GZ
G4 RYB J. Baldwin 31 Beech Road, Branston, Lincoln LN4 1PG
G4 RYE D. Cocker 34 Beechfieldá, Leeds LS12 5QS

G4 RYH F. Appleby 10 Buckingham Orchard, Chudleigh Knighton, Newton Abbot TQ13 0EW
G4 RYI D. Ashcroft 9 Aldermere Crescent, Urmston, Manchester M41 8UE
GW4 RYJ A. Salisbury Heddwch, Ash Grove, Flint CH6 5RX
G4 RYK A. Richards Castell Forwyn, Abermule, Powys SY15 6JH
GI4 RYL M. Mccallan 16 Abbey Crescent, Newtownabbey BT37 9PD
G4 RYM I. Spalding 2 Briery Lands, Heath End, Stratford-upon-Avon CV37 0PP
G4 RYO P. Allan Ivy Cottage, Lee Mill Bridge, Ivybridge PL21 9EF
GI4 RYP J. Ferguson Drumbee-More, Armagh BT60 1HP
GW4 RYQ K. Edwards 10 Bala Drive, Rogerstone, Newport NP10 9HN
G4 RYS N. Black 42 Stonegate Way, Leeds LS17 6FD
G4 RYT J. Pickup 274 Mauldeth Road West, Chorlton Cum Hardy, Manchester M21 7TG
G4 RYV D. Rumbold 15 Lodge Grove, Yateley GU46 7AD
G4 RZC L. Ingerslev 20 Stoney Run Lane, Marion 2738 United States
G4 RZD L. Bradley 138 Templeton Road, Birmingham B44 9BY
GW4 RZE A. Taylor 5 Wyebank Rise, Tutshill, Chepstow NP16 7DS
G4 RZF G. May 5 The Burlongs, Glebe Road, Swindon SN4 7DR
G4 RZI M. Nagle The Woodlands, Pilton West, Barnstaple EX31 4JQ
G4 RZM B. Williams Warren Cottage, Polyphant, Launceston PL15 7PS
G4 RZN R. Leeds 1A Clare Road, Cromer NR27 0DD
G4 RZQ K. Russell Courtiles, Main Road, Ventnor PO38 3NH
G4 RZR R. Tooth 25 Northgate, Beccles NR34 9AS
GM4 RZW D. Taylor 42 Craiglockhart Road, Edinburgh EH14 1HG
G4 RZY L. Baker The Novers Park Community Centre, Rear Of 122-124, Bristol BS4 1RN
G4 RZZ I. Griffin 15 Hesselyn Drive, Rainham RM13 7EJ
G4 SAB C. Leat 8 White Point Court, Whitby YO21 3UR
G4 SAC S. Collings 4 Glamis Close, Waterlooville PO7 8JN
G4 SAJ C. Green 76 Dibleys, Blewbury, Didcot OX11 9PU
GI4 SAM S. Noble 19 New Line, Dundonald, Belfast BT16 1UU
G4 SAS R. Jones 42 Fastmoor Oval, Birmingham B33 0NR
G4 SAT Inmarsat ARC c/o D. John 27 Churchfields, Dartmouth TQ6 9HJ
G4 SAV F. Hepworth 5 Snydale Avenue, Normanton WF6 1SS
G4 SAW M. Arbon 106 The Tideway, Rochester ME1 2NN
GI4 SBA K. Branagh 17 Rathmoyle Park West, Carrickfergus BT38 7NG
G4 SBB Dr C. Fay Driftwood, Middle Road Sway, Lymington SO41 6BB
G4 SBC I. Bateman 29 Bradgate Croft Hasland, Chesterfield S41 0XZ
G4 SBD G. Bax 8 Hockeredge Gardens, Westgate-on-Sea CT8 8AN
G4 SBE K. Bowden 14 Pool Hey Lane, Scarisbrick, Southport PR8 5HS
G4 SBF P. Fry 54 Studley Avenue, Holbury, Southampton SO45 2PP
G4 SBG H. Crossland 107 Fairway, Normanton WF6 1SN
G4 SBM R. Harding 12 Keswick Avenue, Loughborough LE11 3RL
G4 SBN J. Ayers 3 Sovereign Way, Ryde PO33 3DL
GM4 SBP G. Allan 31 Jubilee Grove, Glenrothes KY6 1HW
G4 SBQ M. Rushton 14 Acorn Close, Leyland PR25 3AF
G4 SBS R. Phillips 4 Cumberland Drive, Fazeley, Tamworth B78 3YA
G4 SBU B. Gundry 37 Stoneham Park, Petersfield GU32 3BT
G4 SBW T. Carberry 10 Honeymeade Close, Stanton, Bury St. Edmunds IP31 2EF
G4 SCB M. Sargent 19 Pine Tree Close Cowes, Isle Of Wight Po31 8Dx, Cowes PO31 8DX
G4 SCE Dr P. Whitehead Carrick View, Bank End, Carlisle CA5 6QW
G4 SCG C. Curson 3 Cranmer Road, Edgware HA8 8UA
G4 SCJ D. Meakins 19 Booth Lane North, Northampton NN3 6JQ
GW4 SCK R. Hancock 1 Nevills Close, Gowerton, Swansea SA4 3BG
G4 SCL M. Starkey Cutlers Forth Farm, Radley Road, Newark NG22 8AP
G4 SCM J. Claxton Camino Del Perpen 25, Catral 3158 Spain
G4 SCO N. Drury 3 Northam Close, Marshside, Southport PR9 9GA
G4 SCS A. Fletcher Stonehouse Stores, West Rd, Ormesby NR29 3RJ
G4 SCV I. Gammon 1 Poppy Close, Willand, Cullompton EX15 2SX
G4 SCY P. Bradbury 52 Moss Park Avenue, Werrington, Stoke-on-Trent ST9 0EP
G4 SDI L. Footring 26 Ernest Road, Wivenhoe, Colchester CO7 9LG
G4 SDJ R. Freeman Flat 2, Russett Court, 15 Kirtleton Avenue, Weymouth DT4 7PS
G4 SDL B. Dorricott 6 Knowsley Avenue, Urmston, Manchester M41 7BT
GW4 SDO D. Phillips Trem Y Fammau, Tri Thy, Tir Y Fron Lane, Mold CH7 4TU
GW4 SDT S. Lansdown 11 Redbrook Road, Newport NP20 5AA
G4 SDU P. Smart 6 Nobold Close, Baschurch, Shrewsbury SY4 2EH
G4 SDX G. Townend 9 Warren Park Close, Hove Edge, , Brighouse HD6 2RU
G4 SDZ M. Gayler 39 Holmfield Avenue West Leicester Forest East, Leicester LE3 3FF
G4 SEA R. Seabridge 7 Heritage Avenue, Frankston South, Melbourne Australia
G4 SEB S. Swallow 16 Quarry Lane, Chesterfield S40 3AS
G4 SEF R. Jenkinson 4 Apple Croft Skidby, Cottingham HU16 5UG
G4 SEG A. Clayton 448 Gisburn Road, Blacko, Nelson BB9 6LZ
G4 SEJ B. Vane 3 Charnwood, Chestfield, Whitstable CT5 3QD
G4 SEK K. Aylwin 9 Hockeredge Gardens, Westgate-on-Sea CT8 8AN
G4 SEL G. Wilkes 49 Charlemont Road, Walsall WS5 3NQ
G4 SEN N. Whitham The Cottage, Castle Gate Nancledra, Penzance TR20 8BQ
G4 SEP C. Turner Saxavord, Humberston Road, Grimsby DN36 5NJ
G4 SEQ D. Vickers 48 Bromley Road, Hanging Heaton, Batley WF17 6EH
G4 SET C. Hall 8 Sharps Court, Exmouth EX8 1DT
G4 SEU J. Russell 9 Batten Close, Christchurch BH23 3BJ
G4 SEV N. Le Gresley 32 Churchill Road, Welton NN11 2JH
G4 SEW K. Weir Les Ginestes Appt 231, 28 Av Dr Gerhardt, Peymeinade 6530 France
G4 SEZ C. Greenland 21 Penleigh Close, Corsham SN13 9LE
GM4 SFA A. Keenan Darwin, Coalhall, Ayr KA6 6ND
G4 SFB D. Knowler Apartedo 1009, 8670 - 999, Aljezur Portugal
G4 SFD D. Birks 3 Wansdyke Drive, Calne SN11 0EW
GI4 SFE J. Mccullough 12 Bramble Grange, Newtownabbey BT37 0XH
G4 SFG P. O'Connor 36 Heron Road, Oldbury B68 8AQ
G4 SFH N. Richardson 22 Bramshott Drive, Hook RG27 9EY
G4 SFJ S. Stott 20 Lingfield Crescent, Wigan WN6 8QA
G4 SFN J. Tetlow 14 Fountains Crescent, Hebburn NE31 2HT
G4 SFP J. Nash 259 Weald Drive Furnace Green, Crawley RH10 6PN
G4 SFQ S. Fletcher The Bakery, Keswick Road, Norwich NR12 0HF
G4 SFS P. Grosjean Garden House, West Horrington, Wells BA5 3ED
GM4 SFT D. Mcalonan Glenonan House Cromlech St, Dunoon PA23 8PQ
GM4 SFW J. Stuart Tigh Na Coille, Bishop Kinkell, Dingwall IV7 8AW
G4 SFY R. Baker 69 Northfield Road Mundesley, Norwich NR11 8JN
GI4 SFZ C. Hought 37 Oldpark Avenue, Ballymena BT42 1AX
G4 SGA G. Barnes 3 Blandford Avenue, Castle Bromwich, Birmingham B36 9HX
G4 SGD S. Simpson 17 Astley Way Ashby De La Zouch, Leicester LE651LY
G4 SGE B. Hughes 86 Lewis Avenue, Wolverhampton WV1 2AR
G4 SGF Dr K. Ruiz Flat 12, The Woodlands, 39 Shore Lane, , Sheffield S10 3BU
G4 SGG D. Earp 88 Linton Rise, Nottingham NG3 7BY
G4 SGI S. Collings 46 St. Michaels Road, Cheltenham GL51 3RR
G4 SGJ J. Campbell 9 Blackdown Close, Dibden Purlieu, Southampton SO45 5QS
G4 SGN P. Playle 6 Walnut Tree Close, Cheshunt, Waltham Cross EN8 8NH
GW4 SGQ P. Hruza 18 Withy Avenue, Forden, Welshpool SY21 8NJ
GW4 SGR Bushvalley ARC c/o R. Magwood 13 Inverness Place, Cardiff CF24 4RU
G4 SGU G. Gilbertson 6 The Stray, South Cave, Brough HU15 2AL
G4 SGV K. Jones 228 Evesham Road, Headless Cross, Redditch B97 5EP
G4 SGW W. Prouse 1 Springfield Cottages, Bishops Tawton, Barnstaple EX32 0DF
G4 SGX I. Haywood 5 Pump Corner, Marsham, Norwich NR10 5PW
G4 SGY J. Winters 94 Wharncliffe Road, Loughborough LE11 1SN
G4 SHA M. Webb 9 Steele Close, Devizes SN10 3SL
G4 SHB P. Sheridan 17 Boakes Drive, Stonehouse GL10 3QW
G4 SHC R. Bentham 12 Tanners Way, Nantwich CW5 7FL
GW4 SHF S. Purser Penbrey, Llanfair Caereinion, Welshpool SY21 0DG
G4 SHH P. Brooking 49 Binstead Lodge Road, Ryde PO33 3TL
G4 SHJ N. Douglas 87 Hutton Avenue, Hartlepool TS26 9PR
G4 SHK G. Clifford Turnpike Road, Blunsdon, Swindon SN26 7EA
G4 SHM M. Baker 92 Moy Avenue, Eastbourne BN22 8UQ
G4 SHN J. Pitts 4 Tannery Close, Dagenham RM10 7EX
G4 SHO F. Dibden 127 Mayola Road, Clapton, London E5 0RG
G4 SHQ D. West 24 Oxfield Drive, Gorefield, Wisbech PE13 4LX
G4 SHY L. Afford 3 Kem Street, Nuneaton CV11 4LH
GM4 SID S. Will 53 Bishop Forbes Crescent, Blackburn, Aberdeen AB21 0TW
G4 SIE R. Mason 35 Princes Gardens, Blyth NE24 5HL
G4 SIF R. Rowsell 61 Barrack Road, Bexhill-on-Sea TN40 2AZ
GW4 SII P. Garston 85 Wood Lane, Hawarden, Deeside CH5 3JG
G4 SIJ B. Hammond 10 Grampian Close, Sleaford NG34 7WA
G4 SIL E. Tubman 54 Summerfield Avenue, Whitstable CT5 1NS
GI4 SIP J. Mckavanagh 28 Thompsons Grange, Carryduff, Belfast BT8 8TG
G4 SIR P. Rogers Pikk 36-46, Tallinn 10123 Estonia
G4 SIS R. Keefe 28 Burstead Drive, South Green, Billericay CM11 2QN
GI4 SIW Antrim & Dis Ar c/o Dr W. Hutchinson 40 Oldstone Hill, Muckamore, Antrim BT41 4SB
GI4 SIZ T. Thompson 135 Glenhead Road, Limavady BT49 9LR
GM4 SJB J. Bruce Kinnaird, Brora KW9 6NN
G4 SJD S. Davis 33 Pollard Close, Plymstock, Plymouth PL9 9RR
G4 SJG G. Upton 18 Cranthorne Drive, Nottingham NG3 7HD
G4 SJH B. Lewis 23 Lightwater Meadow, Lightwater GU18 5XH
G4 SJI A. Harris 10 Egroms Lane, Withernsea HU19 2LZ
G4 SJL S. Thomson 11 Beverley Road, London W4 2LL
G4 SJM J. Reeves 5 Arrows Crescent, Boroughbridge, York YO51 9LP
G4 SJN B. Hunt Tralee, Oakridge Lynch, Stroud GL6 7NY
GW4 SJO M. Edwards 1 Maes Yr Efail, Penparc, Cardigan SA43 1FB
G4 SJP S. Prior East Brantwood, Manor Road, Barnstaple EX32 0JN
GI4 SJQ G. Frazer 20 Old Rectory Park, Portadown, Craigavon BT62 3QH
G4 SJU Colburn & Richmondshire District ARS c/o S. Browne 38 Aldrin Road, Pennsylvania, Exeter EX4 5DN
G4 SJV D. Chapman 24 Broad Lane, Moulton, Spalding PE12 6PN
G4 SJW Bridges RC Hampshire c/o M. Troy 22 Jackie Wigg Gardens Totton, Totton, Southampton SO409LZ
GW4 SKA J. Barber 49 Blackmill Road, Bryncethin, Bridgend CF32 9YN
GM4 SKB M. Whyatt Backburn Cottage, Castleton Road, Auchterarder PH3 1JS
G4 SKM Maltby And District ARS c/o P. Archer 31 Stoney Bank Drive, Kiveton Park, Sheffield S26 6SJ
G4 SKO M. Brooke 6 Sun Street, Stanningley, Pudsey LS28 6DJ
GW4 SKP A. Clark 27 Heol Sant Bridget, St. Brides Major, Bridgend CF32 0SL
G4 SKU J. Blain 1 Handley Court, Bunyan Road, Sandy SG19 1BJ
G4 SLG K. Oliver 42 Minster Drive, Cherry Willingham, Lincoln LN3 4NA
GW4 SLI J. Plumley 34 Graigwen Crescent, Abertridwr, Caerphilly CF83 4BN
G4 SLL J. Buckland 245 Saunders Lane Mayford, Woking GU22 ONU
GI4 SLQ K. Boyd 29 Benburb Road, Moy, Dungannon BT71 7SQ
GM4 SLV J. Pumford-Green Greenmeadow, Clousta, Bixter, Shetland ZE2 9LX
G4 SLW D. Dudkowski 7 White Street, Brighton BN2 0JH
GM4 SLY J. Bell 13 Corrie Place, Troon KA10 6TZ
G4 SMA M. Goode Meadowgreen, Batch Valley, Church Stretton SY6 6JW
G4 SMB Capt. M. Briggs 70 Auchinleck Close Kellythorpe, Driffield YO25 9HE
G4 SMD D. Blackwell 2 Courtry Cottages, Bridgehampton, Yeovil BA22 8HF
G4 SME Norfolk Coast ARS c/o J. Rogers 186 Beavers Lane, Birleywood, Skelmersdale WN8 9BP
G4 SMK K. Wilson 15 Woodside Avenue, Cottingley, Bingley BD16 1RB
GW4 SML T. Morgan 37 Fan Heulog Talbot Green, Pontyclun CF72 8HQ
G4 SMM M. Manley Rolleston, Parkgate Road, Chester CH1 6JS
G4 SMQ D. Mcdonald 3 Cloverton Drive, Bridgwater TA6 4HQ
G4 SMT J. Short 42 Alt Road, Formby, Liverpool L37 6DF
G4 SMX J. Wilson 168 Elms Vale Road, Dover CT17 9PN
GI4 SNA D. Ross 127 Pond Park Road, Lisburn BT28 3RE
G4 SND M. Newey 148 High Street, Pensford, Bristol BS39 4BH
G4 SNI H. Farley 11 College Lane, Hatfield AL10 9PB
G4 SNJ C. Frenzel Butlers Hall, Butlers Hall Lane, Bishop's Stortford CM23 4BL
G4 SNL I. Dunworth 51 The Crest, Sawbridgeworth CM21 0ER
G4 SNN N. Booth-Isherwood 65 Burnaby St., Alvaston, Derby DE24 8RN
G4 SNO A. Duggins Dowles Bungalow, Dowles Road, Bewdley DY12 3AA
G4 SNQ T. Wadsworth 20 Rook Wood Way, Little Kingshill, Great Missenden HP16 0DF
G4 SNR E. Meekers 14 Croft Road, Christchurch BH23 3QQ
G4 SNU R. Hardie 12 Hopland Close, Longwell Green, Bristol BS30 9XB
G4 SNV G. Eastgate 103 Western Road, Leigh-on-Sea SS9 2PB
GM4 SNW M. Donnelly 23 St John Street, Creetown, Newton Stewart DG8 7JB
G4 SOA P. Instone 19 Dickenson Road, Swindon SN25 1WG
G4 SOB W. Hammond 27 Gainsborough Road, Colchester CO3 4QN
GW4 SOC V. Shaw 130 Aberthaw Road, Ringland, Newport NP19 9QS
G4 SOF J. Blight Lowbell, Handy Cross, Bideford EX39 3ET
G4 SOG M. Pridham 61 Bridge Street, Kington HR5 3DJ

G4 SOH M. Spence 5 St. Helens Avenue, Benson, Wallingford OX10 6RY
G4 SOI M. Wray The Old Croft, Top Street, Retford DN22 0LG
G4 SOK R. Hollow The Beeches, Grove Lane, Penzance TR20 9HN
G4 SOL I. Bale 2B Holes Lane, Knottingley WF11 8LH
G4 SOM J. Spiteri 27 Hillcote Close, Sheffield S10 3PT
G4 SOP K. Lillingstone 4 Old Bakery Court, Coltishall, Norwich NR12 7DQ
G4 SOQ B. Lyons 16 Ashdale Close, Sawtry, Huntingdon PE28 5SN
G4 SOR A. Collins 19 Cavendish Road, Skegness PE25 2QZ
G4 SOT D. Goodwin Route De Samatan, Frontignan Saves 31230 France
GI4 SOY Rvd. M. Dornan 8 The Hermitage, Dunmurry, Belfast BT61 8AH
G4 SOZ D. Herd Capricorn Cottage, Low Common, Norwich NR14 7BU
G4 SPA Antrim & District ARS c/o R. Marchington 78 Buxton Road, Dove Holes, Buxton SK17 8DW
G4 SPC B. Escreet 198 Front Street, Sowerby, Thirsk YO7 1JN
G4 SPD N. Jarvis 25 St. Augustines Close, Aldershot GU12 4SF
G4 SPE G. Callaghan 1 Wessex Close, Semington, Trowbridge BA14 6SA
GW4 SPL P. Ace 116 Gellionen Road, Clydach, Swansea SA6 5HF
G4 SPR F. Rattray 4 Winton Manor Court, Winton, Kirkby Stephen CA17 4HR
G4 SPS N. Beggs 11 Orion Close, Fareham PO14 2SQ
G4 SPT H. Golding Barany Uyca 2, Hodmezovasarhely 6800 Hungary
GI4 SPU N. Alcock 22 Chippendale Avenue, Bangor BT20 4PT
G4 SPV Hucknall Rolls Royce ARC c/o A. Adamson 520 York Road, Stevenage SG1 4EP
G4 SPW T. Devlin 12 Hawthorn Drive, , Yeadon Leeds LS19 7XB
G4 SPY A. Kay 36 Yorks Wood Drive, Birmingham B37 6DL
G4 SPZ P. Harris 22 Bramley Way, Bewdley DY12 2PU
G4 SQA D. Yeoman 1 Cartmell Court Bridge Street Deeping St James Pe6 8El, Peterborough PE6 8EL
G4 SQG R. Nicholson 7 Half Mile Gardens, Leeds LS13 1BL
G4 SQI G. Gulliford 18 Purslane, Abingdon OX14 3TR
G4 SQJ G. Turner 21 Barlow Road, Chichester PO19 3LD
G4 SQK G. Middleton 212 East Markham Avenue, Durham 27701 United States
GI4 SQL S. Adrain 10 Highgate Drive, Newtownabbey BT36 4WQ
GM4 SQM D. Anderson 34 Culzean Crescent, Kilmarnock KA3 7DT
GM4 SQO R. Riddiough 1 Cedar Road, Ayr KA7 3PE
G4 SQQ C. Hollister Rosemead, 326 Passage Road, Bristol BS10 7TE
G4 SQV J. Hart 88 Breckhill Road, Woodthorpe, Nottingham NG5 4GQ
G4 SRD R. Sealy 10 Mallard Close, Bowerhill, Melksham SN12 6TQ
GW4 SRE J. Reeves 5 Lon Y Bryn, Glynneath, Neath SA11 5BG
G4 SRF C. Radcliffe 85 Brian Avenue, Cleethorpes DN35 9DE
GW4 SRI A. Gray 24 Waunarlwydd Road, Cockett, Swansea SA2 0GB
GM4 SRL R. Cowan 85 Eastwoodmains Road, Clarkston, Glasgow G76 7HG
G4 SRP G. Brierley 6 Yeo Drive, Appledore, Bideford EX39 1RD
GI4 SRQ W. Mchugh 47 Main St., Hamiltonsbawn, Armagh BT60 1LP
GW4 SRR S. Richards 62 Glyn Bedw, Llanbradach, Caerphilly CF83 3PG
G4 SRV D. Darby 28 Coleshill Close, Hunt End, Redditch B97 5UN
G4 SRX G. Sampson 47 Netherthorpe Way, North Anston, Sheffield S25 4FL
GM4 SSA H. Hassel Sumra, Eshaness, Shetland ZE2 9RS
G4 SSC Dr A. Taylor 38 Summershades Lane, Grasscroft, Oldham OL4 4ED
G4 SSD South Devon RC c/o J. May 6 Hodson Close, Paignton TQ3 3NU
G4 SSE J. Sartin East Barn, Gubblecote, Tring HP23 4QG
GI4 SSF S. Craig 6 Kingswood Park, Belfast BT5 7EZ
G4 SSH R. Clayton 9 Green Island, Irton, Scarborough YO12 4RN
G4 SSJ R. Bolton 83 Sandicroft Close, Birchwood, Warrington WA3 7LY
G4 SSL W. Lancashire 111 Pentire Avenue, Newquay TR7 1PF
G4 SSP K. Waghorne 23 Bramley Hill, Mere, Warminster BA12 6JX
G4 SSV S. Smith 1 Buckfast Road, Lincoln LN1 3JS
G4 SSW J. Walker 15 Hillfield Road, Bilton, Rugby CV22 7EW
G4 SSY F. Druppel Suite 9559 - 488 Unit 9 Skyport Drive, West Drayton UB7 0LB
G4 SSZ D. Fox 49 Turnpike Hill, Hythe CT21 4SE
G4 STB P. Lock 82 Rownhams Road, Throop, Bournemouth BH8 0NL
G4 STD B. West 13 Tanglewood Close, Upper Shirley, Croydon CR0 5HX
G4 STE S. Farr 37A Bromsgrove Road, Studley B80 7PG
G4 STH G. Timbrell Crossing Cottage, Lamyatt, Shepton Mallet BA4 6NG
G4 STI M. Pinder 36 West Ridge, Allesley, Coventry CV5 9LN
G4 STK E. Brown 16 Springfield, Ovington, Prudhoe NE42 6EH
G4 STO P. Rose Pinchbeck Farmhouse, Mill Lane, Sturton By Stow, Lincoln LN1 2AS
G4 STP T. Mangles 46 Cedar Crescent, Willington, Crook DL15 0DA
G4 STW A. Buckley 5 Russell Walk, Messingham, Scunthorpe DN17 3TU
G4 STZ T. Bromsgrove 11 Moelwyn Drive, Ellesmere Port CH66 1TY
G4 SUA B. Common 59 Thornbera Gardens, Bishop's Stortford CM23 3NP
GW4 SUD K. Jones 111 Ewenny Road, Bridgend CF31 3LN
GW4 SUE M. Hill 13 Maesglas Grove, Newport NP20 3DJ
GM4 SUF P. Gane Ardmore Lodge, Station Road, Edderton, Tain IV19 1LA
G4 SUK M. Kent 304 Reculver Road, Herne Bay CT6 6SR
GW4 SUN B. Le Carpentier 43 Abernant Road, Aberdare CF44 0PY
G4 SUO P. Barwick Brook Cottage, 94 Ambleside Road, Lightwater GU18 5UJ
GM4 SUR R. Aitken 2 Eskdale Drive, Bonnyrigg EH19 2LD
G4 SUS S. Morgan Hollaway, Northbourne Road, Deal CT14 0LA
G4 SUU M. Nairn 13 Hanover Court, Lacey Street, Ipswich IP4 2PJ
G4 SUX R. Payne 13 Bowden Hill, Lacock, Chippenham SN15 2PW
G4 SVA J. Appleton 66 Bolton Road West, Ramsbottom, Bury BL0 9ND
G4 SVB A. Gatrell Sunnyside, Muddles Green, Lewes BN8 6HW
G4 SVC T. Axtell 146 Olivers Battery Road South, Winchester SO22 4LF
GD4 SVD A. Ames 20 Sunnybank Avenue, Onchan IM3 3BW Isle of Man
G4 SVE J. Hewett 84 Dunsgreen, Ponteland, Newcastle upon Tyne NE20 9EJ
G4 SVG S. Wensley 7 Bradshaw Close, Windsor SL4 5PS
G4 SVI C. Ames Heatherdene, 58 The Street, Norwich NR8 6AB
G4 SVL I. Hewitt 35 Birmingham Road, Alvechurch, Birmingham B48 7TB
GM4 SVM G. Hudson 17 Drylaw Crescent, Edinburgh EH4 2AU
G4 SVQ P. Haffenden 113 Pavilion Road, Worthing BN14 7 EG
G4 SVR W. Baddeley 12 Stockport Road, Altrincham WA15 8ET
G4 SVS D. Bush 8 Oldbury Chase, Bristol BS30 6DY
G4 SVV D. Lees 1, Davies Ave, Cheadle SK8 3PF
G4 SVY J. Perez 9 Rectory Road, Shanklin PO37 6NX
G4 SWA S. Authers 9 Conway Avenue, Birmingham B32 1DR
G4 SWH R. Jones Blakeney, Fore Street Weston, Hitchin SG4 7AS
G4 SWO S. Griffiths 25 Hanks Close, Malmesbury SN16 9UA
G4 SWQ R. Torence-Smith Birch Lodge, Flax Lane, Sudbury CO10 7RS
G4 SWR I. Hill 7 Cosford Close, Matchborough East, Redditch B98 0BH
G4 SWY D. Turner 15 Brooke Close, Bushey WD23 1FB
GW4 SXA P. Ap-Dafydd 37 Heol Nant Cwmdare, Aberdare CF44 8TE
G4 SXE B. Holden 76 The Lawns, Rolleston-On-Dove, Burton-on-Trent DE13 9DE
G4 SXG D. Fraser 63 Vicars Hall Gardens, Worsley, Manchester M28 1HW
G4 SXH L. Fletcher 27 Stanford Rise, Sway, Lymington SO41 6DW
GM4 SXJ R. Malcolm 43 Kinghorne St., Hospitalfield, Arbroath DD11 2LZ
G4 SXK A. Leighs 16 Spode Close, Cheadle, Stoke-on-Trent ST10 1DT
GU4 SXM P. Bannier 10 Le Bouet, Longstore, St. Peter Port GY1 2BA Guernsey
G4 SXQ D. Lempriere Harewarren Lodge, Salisbury SP2 0NF
G4 SXT D. Ayers 22 Holders Road, Amesbury, Salisbury SP4 7PP
GI4 SXV E. Barker 39 Birchwood, Omagh BT79 7RA
G4 SXX J. Key 29 Scots Court, Hook RG27 9QJ
G4 SXY G. Haines 25 Hook Hill, South Croydon CR2 0LB
G4 SXZ R. Hiles 19 Station Road, Kirton Lindsey, Gainsborough DN21 4BB
G4 SYA C. Allen 11 Chandos Court, Martlesham, Woodbridge IP12 4SU
G4 SYB P. Loveland 25 White Acres Road, Mytchett, Camberley GU16 6JJ
G4 SYC G. Lomas 2 Linney Road, Bramhall, Stockport SK7 3JW
G4 SYD H. Cook 24 Front St., Sherburn Hill, Durham DH6 1PA
G4 SYE M. Wilson The Old Post Office, Knowl Hill, Reading RG10 9YD
GM4 SYF N. Wallace 2 Mansefield, Leitholm, Coldstream TD12 4JQ
G4 SYG P. Tattersall Sun House, The Street, Nacton, Ipswich IP10 0EU
GW4 SYI L. Wilder 19 Cambrian Drive, Rhos On Sea, Colwyn Bay LL28 4SL
G4 SYL J. Frost 68 Wessex Road, Didcot OX11 8BP
GI4 SYM W. Donaldson 44 Drumman Hill, Armagh BT61 8RW
GW4 SYO H. Thomas 1 Cambrian Terrace, Llwynypia, Tonypandy CF40 2HN
GU4 SYQ L. Le Page Heathwick, Les Martins, Guernsey GY4 6QJ
G4 SYR S. Field 4 Lyndale, Kelvedon Hatch, Brentwood CM15 0BQ
G4 SYT D. Chambers 26 Drummond Gardens, Christ Church Mount, Epsom KT19 8RP
G4 SYV R. Ackroyd 14 Isis Avenue, Bicester OX26 2GS
G4 SYW A. Hargreaves 13 Linworth Road, Bishops Cleeve, Cheltenham GL52 8PA
G4 SZA I. Donaldson 25 Alwyn Road, Maidenhead SL6 5EG
G4 SZB E. Flannigan 14 Westbank Avenue, Blackpool FY4 5BT
GM4 SZG J. Freeland 48 Elgin Place, Shawhead, Coatbridge ML5 4JQ
G4 SZI A. Parry 18 Spinney Lane, Rabley Heath, Welwyn AL6 9TF
G4 SZO K. Painter 8 Fairview Road, Broadstone BH18 9AX
GI4 SZP N. Hughes 32 Kinedale Park, Ballynahinch BT24 8YS
GI4 SZQ K. Murphy 8 Rosscolban Meadows, Kesh, Enniskillen BT93 1UH
G4 SZS T. Harber 27 Yarlington Close, Norton Fitzwarren, Taunton TA2 6RR
GI4 SZU J. Mccurry 30 Carrowdoon Road, Dunloy, Ballymena BT44 9DL
GW4 SZV Humber Fortress DX ARC c/o G. Carruthers Henllys Farm, Cardigan SA43 2HR
GI4 SZW M. Keenan 30 Ballynabee Road Camlough, Newry BT35 7HD
G4 SZX M. Stockton 7 The Croft, Thorne, Doncaster DN8 5TL
GI4 SZY R. Wilson 57 Mill Green, Doagh, Ballyclare BT39 0PH
G4 TAD M. Wooltorton Willow End, 4 Hall Lane, Oulton, Lowestoft NR32 5DJ
G4 TAG T. Gammage 4 Grey Heights View, Chorley PR6 0TN
G4 TAH I. Conibear 5 Brunswick Street, Redfield, Bristol BS5 9QN
GI4 TAJ J. Bingham 35 Rathmena Drive, Ballyclare BT39 9HZ
G4 TAK J. Hancock 29 Convent Close, Aughton, Ormskirk L39 4XP
GM4 TAL A. Blyth 73 Glassel Park Road, Longniddry EH32 0TA
G4 TAM B. Chambers 93 Main Road, Hoo, Rochester ME3 9EU
G4 TAO S. Rogers Gaythorpe, Blacketts Wood Drive, Rickmansworth WD3 5QQ
GI4 TAP S. Mccabe 27 Baronscourt Road, Carryduff, Belfast BT8 8BQ
G4 TAT D. Ruth 1 Derwent Drive, Appley, Ryde PO33 1NT
GW4 TAU E. Davies Gwelydon, Lon Brynteg, Ynys Mon LL59 5UA
GI4 TAV M. Doherty 20 Drumcairn Close, Belfast BT8 8HQ
G4 TAW N. Perrott 56 Park Avenue, , Chatswood 2067 Australia
G4 TAZ F. Rewaj 5 Spring Gardens, Grizebeck, Kirkby-in-Furness LA17 7XJ
G4 TBF E. Popham 741 Landbase Australia, Locked Bag 25, Gosford 2250 Australia
G4 TBG D. Smith 34 Grays Lane Downley High Wycombe Hp13 5Tz, High Wycombe HP13 5TZ
G4 TBI P. Cornell 1 Orient Drive, Winchester SO22 6NZ
G4 TBJ R. Smith 184 Solihull Road, Shirley, Solihull B90 3LG
G4 TBK D. Nix 75 Mayfield Road, Chaddesden, Derby DE21 6FX
G4 TBM M. Tester 6 Harvard Close, Lewes BN7 2EJ
G4 TBN C. Anderson 10 School Lane, Everdon, Daventry NN11 3BW
G4 TBO C. Harris 33 Brent Street Brent Knoll, Highbridge TA9 4DT
G4 TBQ P. Harlow 36 Ravens Way, Burton-on-Trent DE14 2JS
G4 TCA R. Hawthorn Weirsmeet Bungalow, Mill Lane, Derby DE74 2EJ
G4 TCB P. Holland 9 Garmont Road, Leeds LS7 3LY
G4 TCC A. Hawkins 5 Hussey Road, Norton Canes WS11 9TP
G4 TCE P. Wade 356 Shirehall Road, Sheffield S5 0JP
G4 TCG R. Tams 4 Langdale Close, Fryston, Castleford WF10 2RB
G4 TCI M. Soars 8 Nomis Park Congresbury, Bristol BS49 5HB
G4 TCK N. Nicholls 15 Millennium Way, Westward Ho, Bideford EX39 1XN
G4 TCM W. Smith 65 Larchwood Road, Yew Tree Estate, Walsall WS5 4HE
G4 TCO D. Preece Tyning House, Westrip, Stroud GL6 6EY
G4 TCP J. Caddick 3 Church Walk, Avonwick, South Brent TQ10 9EJ
GI4 TCR Dr A. Jackson 9 Cove Crescent, Groomsport, Bangor BT19 6HW
GI4 TCS W. Jackson Shantara, 21 Carnreagh, Hillsborough BT26 6LJ
G4 TCT P. Johnson 5 Moorside Drive, Drighlington, Bradford BD11 1HE
G4 TDB D. Waterhouse 19 Finsbury Drive, Brierley Hill DY5 3NY
G4 TDC L. Wilson Eastwood, Common Road, Tadcaster LS24 9PQ
G4 TDF R. Copsey 7 Musson Close, Marston Green, Solihull B37 7HS
G4 TDG R. Dowson 14 The Warren, Tuffley, Gloucester GL4 0TT
G4 TDI D. Day 1 Kings Paddock, Ossett WF5 8EN
G4 TDO B. Fereday 16 Glentworth Gardens, Wolverhampton WV6 0SF
G4 TDP S. Bowden 62 Manor House Road, Wednesbury WS10 9PH
G4 TDQ L. Ashton Bacton Wood Mill, Spa Common, North Walsham NR28 9SH
G4 TDR C. Jay Hill House, Badgers Orchard, Pershore WR10 3HJ
G4 TDU B. Brandon 8 Moor Park Avenue, Castleton, Rochdale OL11 3JG
G4 TDV R. Stokes Sunnybank, The Arch, Exeter EX5 1LL
G4 TDW R. Maskew 23 Daventry Road, Rochdale OL11 2LN
G4 TDZ A. Jones Riverbank, Main Street, Newark NG22 0PP
G4 TEB P. Larbalestier 81 Gosfield Lake Park Church Road Gosfield, Halstead CO9 1UG

GI4 TED K. Doherty 77 Drumflugh Road, Benburb, Dungannon BT71 7QF
GM4 TEF A. Chalmers Mayfield, Malcolm Road, Peterculter AB14 0NX
G4 TEK J. Churchill Yew Tree Cottage, Grove Lane, Salisbury SP5 2NR
GD4 TEM J. Freestone C/O Kololi Holiday Services, Po Box 335, Douglas IM99 2QF Isle of Man
G4 TEN J. Burch 1 South Farm Close, Tarrant Hinton, Blandford Forum DT11 8JY
G4 TEP L. Kennedy 69 Drayton Road, Borehamwood WD6 2DA
GW4 TEQ J. Mattocks Brynheulog, Bronygarth Road, Oswestry SY10 7RQ
G4 TEU Avon Scouts ARC c/o J. Burr 4 The Fleet, Royston SG8 5BB
G4 TEW R. Cooper 4 Battle Close, Boroughbridge, York YO51 9GN
G4 TEZ J. Lever 30 Radcliffe Road, Winsford CW7 1RE
G4 TFB B. Gray 29 Verity Walk, Stourbridge DY8 4XS
G4 TFC D. Gray 29 Verity Walk, Stourbridge DY8 4XS
G4 TFD B. Pickard 3 Lodore Road, Bradford BD2 4HY
G4 TFF P. Ayre 1 Spring Gardens, Broadmayne, Dorchester DT2 8PP
G4 TFG B. Curtis 64 Market Avenue, Wickford SS12 0AB
G4 TFH P. Mach 17 Moorlands Avenue, Ossett WF5 9PR
G4 TFI W. White 9 Saffron Way, Tiptree, Colchester CO5 0AY
GM4 TFJ E. Wallace Lochiel Villa, Corpach, Fort William PH33 7LR
GW4 TFM Rvd. A. Davis 51 Gungrog Hill, Welshpool SY21 7UL
G4 TFO A. Coaton 138 Ratby Road, Groby, Leicester LE6 0BT
G4 TFP P. Herman Cranmer 38 Barbrook Lane, Tiptree, Colchester CO5 0EF
GW4 TFS A. Jones 6 Gower View, Llanelli SA15 3SN
G4 TFT C. Greenwood 3 Moorfield Drive, Oakworth, Keighley BD22 7EX
G4 TFU Dr A. Gerrard 2 Dudley Road, Timperley, Altrincham WA15 6UE
G4 TFV F. Kelly 2 Victoria Terrace, Kirkstall, Leeds LS5 3HX
G4 TFW H. Guy 24 The Mall, Binstead, Ryde PO33 3SF
GW4 TFX B. James 9 Brangwyn Close, Morriston, Swansea SA6 6AS
G4 TFZ D. Jarvis 21 Ashurst Place, Stannington, Sheffield S6 5LN
GW4 TGA S. Marvelley 36 Muirfield Drive, Mayals, Swansea SA3 5HS
G4 TGB D. Meadows 39 Sylvester Street, Mansfield NG18 5QS
G4 TGE J. Pullen 71 Barrow Road, Barton-upon-Humber DN18 6AE
G4 TGG G. Sifford 25 Kingsley Court, Fraddon, St. Columb TR9 6PD
G4 TGJ R. Tomlinson 25 Beverley Rise, Ilkley LS29 9DB
G4 TGK W. Wimble 87 Rolfe Lane, New Romney TN28 8JL
GW4 TGL W. Protheroe-Thomas Golwg Y Llan, Carmarthen SA32 8PR
G4 TGM A. Sherratt Anlyn, Norbury Drive, Brierley Hill DY5 3DP
G4 TGP R. Barling Maranello House, Pay Street, Folkestone CT18 7DZ
GI4 TGR T. Greer Ticino, 82 Purdysburn Hill, Belfast BT8 8JZ
G4 TGS D. Swarbrook 6 Westview Close, Leek ST13 8ES
GW4 TGT T. Threlfall 14 Clarence Street, Pembroke Dock SA72 6JP
G4 TGV I. Soaft 55 The Close, , Thurleigh MK44 2DT
G4 TGW D. Starmer 20 Garners Way, Harpole, Northampton NN7 4DN
G4 THA M. Crook 28 Porter Street, Preston PR1 6QN
G4 THC M. Arnison 57 Heywood Road, Cinderford GL14 2QU
G4 THF B. Smith 63 Hitchin Road, Stotfold, Hitchin SG5 4HT
G4 THI A. Robson 5 Wetton Lane, Tibshelf, Alfreton DE55 5NA
GW4 THK W. Moore 2 Heol Cae Glas, Sarn, Bridgend CF32 9UG
G4 THN M. Anthony1 Middlewood House, Blacksmiths Lane, Forward Green, Stowmarket IP14 5ET
G4 THP D. Last 16 South Street, Hockwold, Thetford IP26 4JG
G4 THU J. Read 35 Maytree Hill, Droitwich WR9 7QU
G4 THV R. Biddlecombe 24 West Avenue, Riverview Park, Althorne CM3 6DF
G4 THX G. Donoughue 1 Kings Mount, Leeds LS17 5NS
G4 THY K. Cornes 6 Haywood Heights, Little Haywood, Stafford ST18 0UR
G4 TIA S. Jarvis 1 Wakenslade Cottages, School House, Chard TA20 4PJ
G4 TIC B. Helman The Dingle, Redway, Minehead TA24 8QF
G4 TID D. Hall 6 St. Augustine Drive, Droitwich WR9 8QR
G4 TIF M. Jones 5 Congreve Close, Warwick CV34 5RQ
G4 TIG W. Pope 4 Salston Barton, Strawberry Lane, Salston, Ottery St. Mary EX11 1RG
G4 TIH C. Kay 26 Clare Close, Elstree, Borehamwood WD6 3NJ
G4 TIM G. Clementson 74 Pentley Park, Welwyn Garden City AL8 7SG
G4 TIQ D. Edwards Rosedene, Trewint Estate, Liskeard PL14 3RL
G4 TIV C. Roper 12 Canford Drive, Allerton, Bradford BD15 7AU
GW4 TIW D. Wood 159 Liswerry Road, Newport NP19 9QR
G4 TIX H. Woolrych 20 Meadow Drive, Devizes SN10 3BJ
GW4 TIZ P. Wyles The Lawns, Halkyn Road, Holywell CH8 7SJ
G4 TJA P. Prosser 47 Devereux Drive, Watford WD17 3DD
GM4 TJD M. Maclennan Tree Tops, Ruilick, Beauly IV4 7EY
G4 TJI R. Slim 58 Fairways Drive, Harrogate HG2 7ER
G4 TJK M. Porter 17 Lancaster Avenue, Guildford GU1 3JR
GM4 TJL J. Hebborn Elysian Fields, Spean Bridge PH34 4EX
G4 TJM E. Mordas 6 Walsingham Gate, High Wycombe HP11 1PA
GW4 TJN G. Smallwood Rushbrit, The Spinney Old Road, Bwlchgwyn, Wrexham LL11 5UF
GW4 TJQ J. Wallis 17 Pantbach Place, Whitchurch, Cardiff CF14 1UN
G4 TJS R. Dent 4 Woodgreen Close, Callow Hill, Redditch B975YR
G4 TJU F. Richards 9 Dales Grove, Worsley, Manchester M28 7JW
G4 TJY L. Barker 75 Hills Road, Saham Hills, Thetford IP25 7EW
G4 TKF P. Tuck 178 St. Ediths Marsh, Bromham, Chippenham SN15 2DJ
G4 TKO J. Sharman 102 Commercial Road, Skelmanthorpe, Huddersfield HD8 9DS
G4 TKP R. Peel 3 Martins Hill Lane, Burton, Christchurch BH23 7NJ
G4 TKS J. Clancy 22 Audrey Needham House, Victoria Grove, Newbury RG14 7RB
G4 TKW D. Hamilton 38 Gosport Road, Lee-on-the-Solent PO13 9EN
G4 TLE H. Kennard Chestnut Cottage, Main Street, Rye TN31 6UL
G4 TLG G. Holloway Rose Cottage, Deblins Green, Callow End, Worcester WR2 4UE
G4 TLL J. Jones Amusement Depot, Station Road, Cullompton EX15 1BQ
G4 TLM B. Jennings 33 Queen Margarets Drive, Brotherton, Knottingley WF11 9HR
G4 TLO P. Johnson Flat 5, Mulberry Lodge, 26 New Brighton Road, Emsworth PO10 7EW
G4 TLR B. Richards 37 Salisbury Grove, Sutton Coldfield B72 1YE
G4 TLS J. Norton 2 Hill Rise, Great Rollright, Chipping Norton OX7 5SW
G4 TLT A. Price 20 Feast Field, Wooliston NN29 7QG
G4 TLW H. Allen 425 Broadway, Chadderton, Oldham OL9 8AP
G4 TLY E. Holmes 36 Corn Gastons, Malmesbury SN16 0DR
G4 TMA P. Fielding 22 Meadow Crescent, Poulton-le-Fylde FY6 7QX
GI4 TMB M. Beggs 15 Marsham Court, Cotswold Drive, Bangor BT20 4RS
G4 TMC P. Barnett 8 Parsonage Road, Horsham RH12 4AR
G4 TMD L. Downes 357 Stone Road, Stafford ST16 1LD
G4 TMF P. Aisthorpe-Buckley 17 Pine View Road, Verwood BH31 6LQ
G4 TMG C. Sherwood 14 Amberley Road, Rustington, Littlehampton BN16 2EF
G4 TMI P. Johnson 3A Railway Street, Tow Law, Bishop Auckland DL13 4DU
G4 TML B. Parr 5 Ashes Lane, Almondbury, Huddersfield HD4 6TE
G4 TMQ J. Martin 22 Wansbeck Court, Front Street East, Bedlington NE22 5BU
G4 TMR S. Lacy 26 Sterndale Drive, Newcastle under Lyme ST5 4HS
G4 TMV E. Gale 4 Waingap Crescent, Whitworth, Rochdale OL12 8PX
G4 TMX W. Armstrong 121 Bede Street, Sunderland SR6 0NT
G4 TMY N. Hounslow 18 Crompton Place, Blackburn BB2 6LW
G4 TMZ D. Gillott 132 Racecommon Road, Barnsley S70 6JY
G4 TNA K. Pope 305 Hulton Lane, Bolton BL3 4LF
G4 TNB P. Dollery 22 Barley Mead, Danbury, Chelmsford CM3 4RP
G4 TND D. Jenkins The Hayloft, Week, Tavistock PL19 0NL
G4 TNE D. Horsman 33 Chanters Hill, Barnstaple EX32 8DN
GW4 TNF Windmill Amateur Radio DX Group c/o T. Jones 9 Hinsley Drive, Wrexham LL13 9QH
GM4 TNJ R. Milenkovic 10 Loganbarns Road, Dumfries DG1 4BU
GM4 TNP Northampton DX c/o J. Burke 25 Duncan Road, Auchmuty, Glenrothes KY7 4HS
G4 TNU A. Scott 79 Westwood Drive, Amersham HP6 6RR
G4 TNY D. Womack 58 Nelson Road North, Great Yarmouth NR30 2AT
GM4 TOE B. Horning Cemetery Lodge, Colleonard Road, Banff AB45 1DZ
G4 TOG B. Grainger 23 Heath Road, Hordle, Lymington SO41 0GG
G4 TOH R. Russell 23 Milfoil Avenue, Conniburrow, Milton Keynes MK14 7DY
G4 TOI P. Andrews 12 Cedarwood Grove, Sunderland SR2 9EJ
G4 TOM T. Turbert 200 Salisbury Terrace, York YO26 4XP
G4 TOO M. Final 3 Borda Close, Chelmsford CM1 4JY
GM4 TOQ A. Stewart Three Acres, Cochno Road, Clydebank G81 6PX
GI4 TOR A. Kincaid 63 Carolhill Park, Ballymena BT42 2DG
G4 TOT R. James Brantholme, Hasty Brow Road, Lancaster LA2 6AG
G4 TOX J. Glover Burrows Farm, Toot Hill Road, Ongar CM5 9QW
G4 TOY R. Handstock 38 Watson Close, Upavon, Upavon SN9 6AE
G4 TOZ S. Marchini Flat 9, Crofthill Court, Rochdale OL12 9UX
GW4 TPG M. Evans1 14 Heol Dewi, Hengoed CF82 7NP
G4 TPH T. Brockman 57 Ramsbury Drive, Hungerford RG17 0SG
GI4 TPI Dr G. Anderson 13 Ashley Park, Bangor BT20 5RQ
G4 TPJ R. Mepham 36 Bramble Close, Hildenborough, Tonbridge TN11 9HQ
G4 TPK P. Phillips 83 Arundel Road, Benfleet SS7 4EE
G4 TPM A. Malcher 30 Perham Road, London W14 9ST
G4 TPO S. Mcculloch 125 Comptons Lane, Horsham RH13 5NZ
GM4 TPQ W. Milligan 7 Girvan Road, , Turnberry KA26 9LP
GM4 TPR J. Mitchell 65 Robb Place, Castle Douglas DG7 1LW
G4 TPS P. Stinton 37 Harriet Close, Sutton Bridge, Spalding PE12 9QU
G4 TPV G. Jones 397 Fishponds Road, Eastville, Bristol BS5 6RJ
G4 TPW H. Igglesden Treeways, Littleworth Lane, Horsham RH13 8ER
GM4 TPX K. Gerard 9 Overdale Crescent, Prestwick KA9 2DB
GI4 TPY K. Boag 12 Plantation Road, Bangor BT19 6AF
G4 TQB Dr P. Grannell 1 Long Meadow, Newcastle-under-Lyme ST5 4HY
G4 TQC M. Anson 15 Clover Ridge, Cheslyn Hay, Walsall WS6 7DP
GW4 TQD J. Gulley Lawnfields, Brynhoffnant, Llandysul SA44 6EA
G4 TQL K. Prince Room 36, Millfield, Bury New Road, Heywood OL10 4RQ
G4 TQO P. Fowler 7 Ormonde Avenue, Orpington BR6 8JP
G4 TQR R. Wilkes 47 Richmond Drive, Glen Parva, Leicester LE2 9TJ
G4 TQS A. Wallis 27 Church Farm Road, Upchurch, Sittingbourne ME9 7AG
G4 TQT I. Waller 3 Reneville Close Moorgate, Rotherham S60 2AT
G4 TQY M. Addison 6 Hanley Orchard, Hanley Swan, Worcester WR8 0DS
G4 TQZ P. Foster 18 Stokesay Way, Sutton Hill, Telford TF7 4QE
G4 TRA S. Redway Hill House Grange Lane, Rodbourne SN160ES
G4 TRD R. Dafter 49 Balmoral Road, Salisbury SP1 3PZ
G4 TRE B. Boon Orchards, School Lane, Woodbridge IP13 0ES
G4 TRF I. Boon 327 Broomfield Road, Chelmsford CM1 4DU
G4 TRG I. Willmer 30 Portland Road, East Grinstead RH19 4EA
GM4 TRH A. Macdonald Tigh Na Mara, 6 Halistra, Hallin, Isle of Skye IV55 8GL
G4 TRI S. March 23 Pebworth Close, Church Hill North, Redditch B98 9JX
G4 TRM S. Burgess Muston Farm, Winterborne Muston, Blandford Forum DT11 9BU
G4 TRN J. Everingham 17 Collingwood Road, Redland, Bristol BS6 6PD
G4 TRP M. Fell 5 Sandown Close, Goring-By-Sea, Worthing BN12 4QA
G4 TRQ A. Westmorland Lilly Rose, Rosevear, Bugle, St. Austell PL26 8RJ
G4 TRR Gorleston ARS c/o P. Rose St. Margarets, Westrop, Swindon SN6 7HJ
GM4 TRS A. Pierce Mains Of Auchreddie, New Deer, Aberdeenshire AB53 6SL
G4 TRU A. Thompson 47 The Signals, Feniton, Honiton EX14 3UP
G4 TRV R. Pears 24 Westbourne Drive, St. Austell PL25 5EA
G4 TRW K. Prior 14 Bincombe Rise, Weymouth DT3 6AS
G4 TRY A. Moriarty 7 Meadow Drive, Bolton Le Sands, Carnforth LA5 8HA
GM4 TRZ T. Mcleod 1 Lochside Cottages, Otterston, Burntisland Fife KY3 0RZ
G4 TSA R. Turley Sunnybank, Matlock Road Kelstedge, Ashover, Chesterfield S45 0DX
G4 TSB Dr R. Cooper 8 Hollyfield Drive, Barnt Green, Birmingham B45 8HP
G4 TSD R. Edwards 8 Boney Hay Road, Burntwood WS7 9AB
G4 TSF P. Scholefield 10 Gainsborough Avenue, Leeds LS16 7PG
GW4 TSG J. Williams Cartref, Capel Garmon, Llanrwst LL26 0RG
GI4 TSK J. Skillen 3 Copeland Drive, Comber, Newtownards BT23 5JJ
G4 TSN J. Lee 46 Little Lane, Huthwaite, Sutton-in-Ashfield NG17 2RA
G4 TSQ M. Levett 5 Park Road, Yapton, Arundel BN18 0JE
G4 TST D. Richardson L'Ancresse, Uplands Road, Waterlooville PO7 6HE
G4 TSV J. Robinson 2 Bridge Mill Road, Nelson BB9 7BD
G4 TSW Tiverton South West ARC c/o T. Hugill West Whitnole House, Stoodleigh, Tiverton EX16 9QH
GW4 TTA Dragon ARC c/o J. Parry Ar-Allt, 19 Lon Hedydd, Llanfairpwllgwyngyll LL61 5JY
G4 TTB A. Gordon Flat 3, Woodlands, Acer Grove, Ipswich IP8 3RR
GM4 TTC P. Howes 43 Tanzieknowe Road, Cambuslang, Glasgow G72 8RD
GM4 TTD N. Loughrey 47 Obsdale Road, Alness IV17 0TU
G4 TTF Bishop Auckland Radio Amateurs Club c/o P. Haygarth 5 Forth Close, Peterlee SR8 1DG
G4 TTG H. Bryant 141 Shakespeare Road, Fleetwood FY7 7HH
G4 TTJ J. Lee 2 Rudgard Way, Liphook GU30 7GW

GI4 TTL M. Corcoran 20 Ringbuoy Cove, Cloughey, Newtownards BT22 1LL
G4 TTM J. Betts The Cottage, Meaford, Stone ST15 0PX
G4 TTN G. Redgewell 121 Gubbins Lane, Harold Wood, Romford RM3 0DL
G4 TTO D. Sandland 54 Bishopdale Drive, Rainhill, Prescot L35 4QH
G4 TTQ R. Philpot 38 Mountview Road, Clacton on Sea CO15 6LN
G4 TTS C. Harrison The Mobile Home, Langar Airfield, Nottingham NG13 9HY
G4 TTX R. Smith 405 Windmill Avenue, Kettering NN15 6PS
G4 TTY E. Macdonald 7 Alder Close, Crawley Down, Crawley RH10 4UL
G4 TTZ R. Margolis 12A Wyndham Close, Yateley GU46 7TT
G4 TUA T. Higgs 1 Merchants Row, Faraday Road, Kirkby Stephen CA17 4AU
GW4 TUD I. Williams 2 Church Street, Llanbadarn Fawr, Aberystwyth SY23 3QZ
G4 TUF A. Wilday 12 Duke Street, Bamber Bridge, Preston PR5 6FT
G4 TUH S. Elsdon Church View, Chapel Lane Sharnford, Hinckley LE10 3PE
GI4 TUJ W. Konos 27 Hillhead Road, Ballynahinch BT24 8LB
G4 TUK R. Scarfe Owlswood, Dereham Road, Shipdham, Thetford IP25 7NJ
G4 TUM J. Speakman 33 Leyburn Avenue, Bispham, Blackpool FY2 9AQ
G4 TUO E. Whitworth 129A Broomhill, Downham Market PE38 9QU
G4 TUP D. Norris 26 Freckleton Road, Southport PR9 9XE
GI4 TUV R. Bailie 26 Moatview Park, Dundonald, Belfast BT16 2BE
G4 TUX J. Baines 12B Tall Trees Park, Old Mill Lane, Mansfield NG19 0JP
G4 TVB C. Burnet 15 The Close, Aberdeenkirkham PR2 4UL
G4 TVD G. Hector 6 Benford Close, Bristol BS16 2UD
GW4 TVE S. Edwards 1 Maes Yr Efail, Penparc, Cardigan SA43 1FB
G4 TVJ A. Johns 5 Oakfields, Loddon, Norwich NR14 6UT
G4 TVN B. Yates 39 Moss Lane, Garstang, Preston PR3 1PD
GW4 TVQ R. Thomas 3 Tor Y Mynydd, Baglan, Port Talbot SA12 8LE
G4 TVT G. Spencer 322 Colchester Road, Ipswich IP4 4QN
GW4 TVU V. Sedgebeer 40 Pen Y Bryn, Croeserw Cymmer, Port Talbot SA13 3SD
G4 TVW R. Stone 51 Elaine Avenue, Rochester ME2 2YW
G4 TVX R. Lamb 27 The Ridgeway, Braintree CM7 1EB
G4 TWC D. Powell 8 Cranbrook Drive, Sittingbourne ME10 1RF
G4 TWG S. Greenwood Carter Place Farm, Hall Park, Rossendale BB4 5BQ
G4 TWH G. Wood-Hill 26 Bramerton Road, Hockley SS5 4PJ
G4 TWK H. Hart 20 Cowdray Drive, Goring-By-Sea, Worthing BN12 4LH
G4 TWL T. Lee 19A Imperial Avenue, Mayland, Chelmsford CM3 6AQ
G4 TWP L. Miles 21 Chaucer Walk, Eastbourne BN23 7QT
G4 TWS S. Holmes 7 Parkland Crescent, Old Catton, Norwich NR6 7RQ
G4 TWT H. Holmes 7 Parkland Crescent, Old Catton, Norwich NR6 7RQ
G4 TWW T. Bevan 30 Hawthorne Close, Stanton Hill NG17 3NQ
G4 TXA D. Mccartney 19 Friswell Road, Banbury OX16 9NW
G4 TXD M. Robbins 2 Tolview Terrace, Hayle TR27 4AG
G4 TXE A. Goode Tudor House, Chenhalls Road, Cornwall TR27 6HJ
G4 TXF C. White 7 Woodward Road, Pershore WR10 1LW
G4 TXG N. Hamilton North View, Chawston Lane, Bedford MK44 3BH
G4 TXK T. Stanley 35 Moorgate Road, Kippax, Leeds LS25 7ET
G4 TXL A. Stevenson Szabadsßg U. 32, Veresegyhßz 2112 Hungary
G4 TXM G. Porter 20 Fitzwilliam Drive, Barton Seagrave, Kettering NN15 6RG
GM4 TXN A. Newlands 21 Castle Crescent, Inverbervie, Montrose DD10 0SB
G4 TXO J. Middleton 8 Cullen Close, Newark NG24 1DF
G4 TXT D. Wales 105 Olney Road, Lavendon, Olney MK46 4ER
G4 TXV A. Turner 11 Holmcroft Road, Kidderminster DY10 3AQ
G4 TYA C. Carter 12 Grove Street, Leamington Spa CV32 5AJ
G4 TYD A. Kelly Brook House, Tremar Coombe, Liskeard PL14 5EN
GW4 TYH R. Roberts Bryn Gwyfan, Hiraddug Road, Rhyl LL18 6HS
G4 TYL M. Tyldesley 118 Rutland Street, Grimsby DN32 7NF
G4 TYN D. Bell 12 Parker Gardens, Stapleford, Nottingham NG9 8QG
G4 TYO G. Lilley 100 Trentham Drive, Nottingham NG8 3NE
G4 TYP K. Ward 9 Porlock Close, Long Eaton, Nottingham NG10 4NZ
G4 TYR C. Miles 23 Redacre Road, Sutton Coldfield B73 5EA
G4 TYT A. Hunt 141 Pickhurst Lane, Bromley BR2 7HU
G4 TYW R. Wilson 95 Longfield Road, Todmorden OL14 6ND
G4 TYY J. Worley 37 Fall Road, Heanor DE75 7PQ
G4 TZA C. Read 58 Somerset Road, Chiswick, London W4 5DN
G4 TZF C. Toby 32 Swallow Road, Langley Green, Crawley RH11 7RF
G4 TZG S. Mellor 52 Tetbury Drive, Bolton BL2 5NS
G4 TZK G. Prater 297 Highfield Road North, Chorley PR7 1PH
G4 TZL K. Rogers 7 Buckleigh Road, Wath-Upon-Dearne, Rotherham S63 7JB
G4 TZM I. Paterson 21 Beech Grove, , Little Oakley CO12 5NN
G4 TZO P. Pledger Mas Tracbuch, , Brunyola 17441 Spain
G4 TZQ D. Rouse 14 Kestrel Close, Downley, High Wycombe HP13 5JN
G4 TZR R. Stringfellow 18 Cline Court, Crownhill, Milton Keynes MK8 0DB
G4 TZV E. Cronin 10 Wellington Grove, Pudsey LS28 8DG
G4 TZX G. Everest 20 Seaway Road, St. Marys Bay, Romney Marsh TN29 0RU
G4 UAA J. Gaffney 77 South Street, Pennington, Lymington SO41 8DY
G4 UAF J. Higgins 124 Cromwell Road, South Kensington, London SW7 4ET
G4 UAI P. Cockman 29 Kensington Road, Southend-on-Sea SS1 2SX
GW4 UAJ E. Allwood 10 Fairfield Close, Aberdare CF44 0PF
G4 UAL J. Guffogg 8 Lincoln Road, Washingborough, Lincoln LN4 1EQ
G4 UAM A. Gould 3 Clarkson Road, Lingwood, Norwich NR13 4BA
G4 UAO J. Fisher 10 Belsay Gardens St. Gabriels, Sunderland SR4 7SZ
G4 UAQ I. Weston 53 Dickens Road, Maidstone ME14 2QR
G4 UAT A. Thomas 10 Brisco Avenue, Loughborough LE11 5HB
G4 UAU J. Parish 50 Far Hey Close, Radcliffe, Manchester M26 3GL
G4 UAV A. Waltham 100 Middleton Road, London E8 4LN
G4 UAY D. Grant 115 Clayton Road, Newcastle ST5 3EW
G4 UBB Heads of the Valleys ARC c/o J. Brown 21 Coulsdon Road, Sidmouth EX10 9JJ
G4 UBC K. Durrant 26 Dozule Close, Leonard Stanley, Stonehouse GL10 3NL
GM4 UBF A. Pontiero 1 Dalmeny Road, Hamilton ML3 6PP
G4 UBI A. Priddy 44 Frys Hill, Kingswood, Bristol BS15 4QJ
GM4 UBJ W. Tracey 65 Kirkland Street, Motherwell ML1 3JW
G4 UBK K. Martin 19 Rosevale Gardens, Luxulyan, Bodmin PL30 5EP
G4 UBM R. Bryant Plover, Hareby Road, Spilsby PE23 4JB
GW4 UBQ C. Powles 14 Willow Close, Four Crosses, Llanymynech SY22 6NF
G4 UBR P. Richardson 18 New Road, Heage, Belper DE56 2BA
G4 UBT K. Stone 63 Banks Road, Pound Hill, Crawley RH10 7BS
G4 UCC W. Trinder 354 Livesey Branch, Road, Blackburn BB2 4QJ
G4 UCE B. Davies 9 Paisley Avenue, Eastham, Wirral CH62 8DL
G4 UCF L. Bott 39 Auxerre Avenue, Redditch B98 7QW
G4 UCJ S. Gilbert 3 Shenley Road, Whaddon, Milton Keynes MK17 0LW
GW4 UCK G. Jones Frondeg, 18 Chapel Rd. Three Crosses, Swansea, Sa4 3Pu., Swansea SA4 3PU
G4 UCL A. Fallows 72 Soutergate, Ulverston LA12 7ES
G4 UCT A. Cooke 7 School Lane, Warmingham, Sandbach CW11 3QL
G4 UCU S. Hebel 18 Castle Road, Colne BB8 7AR
GW4 UCV I. Hynes Bryn, Llanddona, Beaumaris LL58 8UE
G4 UCX P. Johnson 38 Bristol Road, Ipswich IP4 4LP
G4 UCY C. Laird 31 Foxlease, Bedford MK41 8AP
G4 UCZ M. Kirk 2 Denton Gardens, East Cowes PO32 6EJ
G4 UDB C. Fay 36 Shooters Hill Close, Southampton SO19 1FW
G4 UDD S. Chapman 2 Birds Croft, Great Livermere, Bury St. Edmunds IP31 1JJ
GW4 UDE M. Ellis Seren Arian, Maesbury Hall Mill, Oswestry SY10 8BB
G4 UDF I. Fox 8 Priestfields, Leigh WN7 2RG
G4 UDG C. Fawkes 24 Moreton Close, Kidsgrove, Stoke-on-Trent ST7 4HP
G4 UDH P. Harley 6 Huntsbank Drive, Newcastle ST5 7TB
GI4 UDI J. Mccullagh 53 Fernagh Road, Omagh BT79 0PL
G4 UDK R. Wood 102 Ombersley Close, Redditch B98 7UT
G4 UDN C. Peake 279 Mansfield Road, Skegby, Sutton-in-Ashfield NG17 3AP
G4 UDU P. Godbold 13 Dawn Crescent, Upper Beeding, Steyning BN44 3WH
G4 UDW P. Hersey Cranbrook, Waghorns Lane, Uckfield TN22 4JA
G4 UDY B. Moorecroft 4 St. Davids Road, Locks Heath, Southampton SO31 6EP
G4 UDZ S. Tyler 28 Rushen Drive, Hertford Heath, Hertford SG13 7RB
G4 UEA P. Robinson 21 Waddow View, Waddington, Clitheroe BB7 3HJ
G4 UED G. Henstridge 21 John Gay Road, Amesbury, Salisbury SP4 7NN
G4 UEF K. Dalton 17 Shute Avenue Watchfield, Swindon SN6 8SX
GM4 UEH Rvd. A. Ford 14 Corsankell Wynd, Saltcoats KA21 6HY
GW4 UEJ M. Charman Noddfa, Dwrbach, Fishguard SA65 9RL
G4 UEL G. Hollebon Flat 24, Providence Place, Abbey Street, Farnham GU9 7RQ
G4 UEN K. Foskett 2 Ambleside Gardens, Southampton SO19 8EY
G4 UEO D. Stewart The Paddock, Allendale, Hexham NE47 9EL
GW4 UEP J. Morgan Arosfa, Upper Bridge Street, Newport SA42 0PL
G4 UET J. Rolfe 56 Elmhurst Road, Thatcham RG18 3DH
G4 UFC P. Fretwell 5 Main St, Brinsley NG16 5BG
GM4 UFD R. Gall 49A Ugie Street, Peterhead AB42 1NX
G4 UFG A. Johnson 27 Walden Avenue, Oldham OL4 2PW
G4 UFJ N. Taylor The Olde Barn, 369A Leymoor Road, Huddersfield HD7 4QQ
G4 UFK A. Watts 23 St. Marys Close, Torrington EX38 8AS
G4 UFL N. Wood 244 Leymoor Road, Golcar, Huddersfield HD7 4QP
GM4 UFP C. Ross The Old Cottages, Middlestead, Selkirk TD7 5EY
GW4 UFQ B. Jackson Bryn Tirion, Maes Y Waen, Bala LL23 7SF
G4 UFR E. Horsfield 13 St. Leonards Way, Ardsley, Barnsley S71 5BS
G4 UFS D. Pearson 48 Nuneham Grove, Westcroft, Milton Keynes MK4 4DH
G4 UFU B. Steen 30 Shady Grove, Alsager, Stoke on Trent ST7 2NH
G4 UFX D. Blackwell Rosegarth, 31 Main Street, Ilkeston DE7 6AU
G4 UFZ R. Greenwood 128 Towngate, Netherthong, Holmfirth HD9 3XZ
G4 UGB R. Bracegirdle 3 Westover Grove Warton, Carnforth LA5 9QR
G4 UGD I. Clover 80 Old Chester Road, Helsby, Helsby WA6 9PQ
GM4 UGF D. Duff Felcanty, Monikie, Broughty Ferry, Dundee DD5 3QN
GW4 UGI R. Crowley 15 Rudry Street, Penarth CF64 2TZ
G4 UGK C. Cattrall 57 Stonebridge, Orton Malborne, Peterborough PE2 5NT
G4 UGM D. Wade 28 Hazel Road, Altrincham WA14 1JL
GM4 UGN D. Duckworth 16 Kennedy Court, Caol, Fort William PH33 7PF
G4 UGO T. Farmer York House, Old Gloucester Road, Bristol BS35 3LQ
G4 UGQ Dr J. Davies 42 Boxworth Road, Elsworth, Cambridge CB23 4JQ
G4 UGR T. Burke 10 Goodwood Road, Lancaster LA1 4LZ
G4 UGT D. Hockin 18 Lower Down Road, Portishead, Bristol BS20 6PF
G4 UGU R. Hall 19 Buckingham Place, Downend, Bristol BS16 5TN
G4 UGV M. Hurrell 74 Southcote Road, Bournemouth BH1 3SS
G4 UGW P. Jones 46 Sergeants Lane, Whitefield, Manchester M45 7TS
GI4 UHA J. Maguire 4 Lawnakilla Park, Enniskillen BT74 7JN
GD4 UHB J. Parslow Traie Vane, Lhergy Dhoo, Peel IM5 2AE
G4 UHI D. Westby 55 Tarn Road, Thornton-Cleveleys FY5 5AY
G4 UHJ D. Lee 1 West View Cottage The Level, Pillowell, Lydney GL15 4QD
GW4 UHK J. Newell 8 Belgrave Road, Abergavenny NP7 7AL
G4 UHM S. Parsons 20 Fountayne Road, Hunmanby, Filey YO14 0LU
G4 UHQ B. Carr 23 Belford Drive, Bramley, Rotherham S66 3YW
G4 UHR D. Reekie 37 Harvey Way, Saffron Walden CB10 2AP
G4 UHS R. Rowlands 18 Green Crescent, Rowner, Gosport PO13 0DP
G4 UHT W. O'Reilly 12 Singledge Avenue, Whitfield, Dover CT16 3LQ
G4 UHU M. Rumens 18 De Legh Grove, West Allington, Bridport DT6 5QY
G4 UHZ D. Goulsbra Delfour, 3 Chapel Street, Market Rasen LN8 3AG
GW4 UIE S. Williams 17 Brettenham St., Llanelli SA15 3ED
G4 UIF A. Marston Stormsfield, Station Road, Limerick Ireland
G4 UIH S. Woolgar 8 Bowerland Avenue, Torquay TQ2 8QH
G4 UII D. Woolgar 8 Bowerland Avenue, Torquay TQ2 8QH
GW4 UIL R. Lewis Tal Engan, Tudweiliog, Pwllheli LL53 8ND
G4 UIO A. Cochrane 136 Osward, Courtwood Lane, Croydon CR0 9HE
G4 UIQ J. Greenhough 58 Gorsey Bank, Wirksworth, Matlock DE4 4AD
GW4 UIR J. Patterson Fairhaven, Tai Terfyn, Caerwys Road, Cwm Dyserth, Rhyl LL18 6HT
G4 UIT D. Geraghty Roselidden House Trevenen Bal, Helston TR13 0PT
G4 UIW P. Wood 61 Stoke Road, Bromsgrove B60 3EP
G4 UIY S. Hamilton 89 The Paddocks, Old Catton, Norwich NR6 7HE
G4 UJA J. Adshead 2 Gainsborough Avenue, Lostock Hall, Preston PR5 5JG
GW4 UJF M. Finnigan 3 Frances Avenue, Rhyl LL18 2LW
G4 UJI E. Cowperthwaite Woodlands, Garstang Road, Lancaster LA2 0EG
G4 UJJ D. Bodman 56A Martins Road, Keevil, Trowbridge BA14 6NA
G4 UJL B. Poole 1 Hungerford Piece Studley, Calne SN11 9LR
G4 UJO M. Greer The Pines, 5A Leek Road, Congleton CW12 3HS
G4 UJP B. Smith 17 Thornley Road, Wirral CH46 6HB
G4 UJS R. Harrison Green Lane House, Whixall, Whitchurch SY13 2PT
G4 UJV J. Fenn 40 Mildenhall Road, Fordham, Ely CB7 5NR
G4 UJW C. Elliott 52 Wellfield Road, Alrewas, Burton-on-Trent DE13 7EZ
G4 UKA C. Hawkridge 57 Wilkes Wood, Creswell, Stafford ST18 9QR
G4 UKD B. Gibson 161 Torbay Road, Harrow HA2 9QF

GM4 UKG M. Manekshaw 32 Inchcolm Drive, North Queensferry, Inverkeithing KY11 1LD
G4 UKO N. Hill 40A Hampden Road, Ashford TN23 6JL
G4 UKR S. Blayer 6 Lord Close, Poole BH17 8QW
GW4 UKU P. Jones 8 Tyn Y Pwll Estate, Llanbedrog, Pwllheli LL53 7PG
G4 UKV I. Leonard 11 St. Leonards Drive, Timperley, Altrincham WA15 7RS
G4 UKW K. Wevill 6 Henacre Wood Court, Queensbury, Bradford BD13 2LJ
G4 UKX R. Miller 31 Gladstone Road, Corton, Lowestoft NR32 5HJ
G4 UKZ R. Rounce Field House Farm, Blakeney Road, Fakenham NR21 0BU
G4 ULD R. Todd 616 Cowdin Drive, Glenwood Springs, Colorado 81601 United States
G4 ULG J. Rawlings Welwyn, Church Walk, Lydney GL15 4NY
G4 ULM J. Martin 25 Mcnish Court, Grenville Way, St. Neots PE19 8PE
G4 ULN I. Purdy 76 Lea Road, Dronfield S18 1SD
G4 ULP D. Pritchard 1 Sandstone Cottages, Walton-In-Gordano, Clevedon BS21 7AJ
G4 ULQ G. Judd 1 Mayfield Way, Ferndown BH22 9HP
G4 ULT L. Walker Oaklea, 13 Manor Road, Sandown PO36 9JA
G4 ULV D. Woodman 30 Sheridan Way, Longwell Green, Bristol BS30 9UE
G4 ULZ R. Ottway 9 Grove Road, Burgess Hill RH15 8LE
GM4 UMA S. Maclennan Tree Tops, Ruilick, Beauly IV4 7EY
G4 UMB P. Howard 63 West Bradford Road, Waddington, Clitheroe BB7 3JD
G4 UME H. Park 11A Morecambe Road, Morecambe LA3 3AA
G4 UMG M. Brassington 42 Dartmouth Avenue, Newcastle ST5 3NY
G4 UMJ R. Carslake 38 Loppets Road, Tilgate, Crawley RH10 5DW
G4 UMM A. Curran 9 Forton Road, Newport TF10 7JP
G4 UMP G. Daisley 10 Arundel Road, Benfleet SS7 4EF
G4 UMS M. Kinger 5 Fore Street, Gunnislake PL18 9BN
G4 UMT G. Elliott 10 Farningham Close, Spondon, Derby DE21 7DZ
G4 UMV P. Johnson 52 Evesham Road, Cookhill, Alcester B49 5LJ
G4 UMW R. Browning 28 Mowbray Close Bromham, Bedford MK43 8LF
G4 UMY M. Strong 92 Cobham Road, Halesowen B63 3JX
G4 UNB D. Williams 13 Hendon Road, Nelson BB9 9JL
G4 UNE S. Sharples 18 Singleton Crescent, Ferring, Worthing BN12 5DG
G4 UNF K. East 39 Chapel Lane, Navenby, Lincoln LN5 0ER
G4 UNH A. Pyne 414 Beacon Road, Bradford BD6 3DJ
G4 UNI T. Hepple 18 King Charles Walk, London SW19 6JA
G4 UNJ B. Walters 17 Oakway, Birkenshaw, Bradford BD11 2PG
G4 UNL R. Charlesworth Po Box 841, 9506 Koronadal City, South Cotobato 0 Philippines
G4 UNM R. Bushell 12 Sandham Close, Sandown PO36 9DS
G4 UNO J. Dobson 27 Darkfield Way, Woolavington, Bridgwater TA7 8JB
G4 UNS D. Brown 8 Gaynes Court, Upminster RM14 2JH
G4 UNW P. Everard The Bungalow, Toynton Fenside Road, Spilsby PE23 5DB
G4 UNX J. Fry 81 South Street, Lewes BN7 2BU
GM4 UOD L. Drake-Brockman 59 Sunnyside, Culloden Moor, Inverness IV2 5ES
G4 UOI R. Butterfield 33 Orchard Square, Wormley, Broxbourne EN10 6JA
G4 UON P. Prowse 9 Fairway, Carlyon Bay, St. Austell PL25 3QE
G4 UOO J. Bleaney 58 Jeans Way, Dunstable LU5 4PW
G4 UOR C. Bourke 36 The Drive, Fareham PO16 7NL
G4 UOS G. Newton 5 Southend Gardens, Highbridge TA9 3LD
G4 UOW J. Cosgrove 8 Wandsworth Road, Newcastle upon Tyne NE6 5AD
G4 UOZ E. Ball 50 Keldgate, Beverley HU17 8HY
G4 UPA J. Poxon 22 Sandhills Road, Bolsover, Chesterfield S44 6EY
GM4 UPB G. Read 58 Limepark Crescent, Kelty KY4 0FH
GI4 UPC W. Millar 121 Ballypollard Road, Magheramorne, Larne BT40 3JG
G4 UPD M. Parks 240 Stainbeck Road, Leeds LS7 2NN
GW4 UPG V. Jackson 24 Bishop Road, Ammanford SA18 3HA
G4 UPI J. Green St. Annes, Poundfield Road, Crowborough TN6 2BG
G4 UPK D. Thompson 112 Lexton Drive, Churchtown, Southport PR9 8QW
GM4 UPN P. Ingram Flat 3 5 Munro Street, Alexandria G83 0PU
G4 UPR J. Dickson 33 Ringwood Grove, Weston-Super-Mare BS23 2UA
G4 UPU R. Ainsworth 14 Edge Fold Crescent, Worsley, Manchester M28 7EX
GM4 UPX I. Wilson 18 High Street, Jedburgh TD8 6AG
G4 UPY A. Hodge 116A Broad Road, Eastbourne BN20 9RD
G4 UQA M. Goodman Randoms, Holt Road, Holt NR25 7UA
GM4 UQD A. Boyd 86 Ravenswood Rise, Livingston EH54 6PG
G4 UQE T. Fitzgerald 11 Hillcrest Road, Camberley GU15 1LF
G4 UQF M. Sole 17 Hyholmes, Bretton, Peterborough PE3 8LG
GM4 UQG R. Aitkenhead 11 Elm Court, Quarter, Hamilton ML3 7FB
G4 UQI A. Lether 16 The Dingle, Fulwood, Preston PR2 3EX
GM4 UQK J. Roberts 1 East Mains Lodge, Kirkinner, Newton Stewart DG8 9AQ
G4 UQM D. Grainger 25 Westwood Heath Road, Leek ST13 8LN
G4 UQN K. Stockley 19 The Lawns, Wisbech PE13 1SW
GD4 UQO P. Parker 46 Ballaquane Park, Peel IM5 1PX Isle of Man
G4 UQR J. Gibbs 3 Holts Green, Great Brickhill, Milton Keynes MK17 9AJ
G4 UQU D. Smith 2 Niton Road, Weddington, Nuneaton CV10 0BX
G4 UQW D. Beckett 433 New St., Biddulph Moor, Stoke on Trent ST8 7NG
G4 UQY M. Regan 36 Moor Park Gardens, Leigh-on-Sea SS9 4PY
G4 URA A. Haynes Thorrington Nurseries, Tenpenny Hill, Colchester CO7 8JB
GW4 URB R. Teesdale 22 Cwmgelli Drive, Treboeth, Swansea SA5 9BS
G4 URD R. Caira 12 West Hill Road, Herne Bay CT6 8HG
GW4 URG S. Richardson Awel Y Ddol, Glanrafon, Corwen LL21 0HA
G4 URM P. Butler Tanglewood, Elms Lane, Wolverhampton WV10 7JS
G4 URN M. Turvey 106 Foxwell St., Worcester, Worcester WR5 2ET
G4 URP R. Powell 57 Bartons Drive, Yateley GU46 6DW
G4 URS J. Osborne 64 Old Warren, Taverham, Norwich NR8 6GA
G4 URT P. Hutchison 1 Barkers Green, Wem., Shrewsbury SY4 5JN
G4 URV Dr W. Peel 34 Carlyn Avenue, Sale M33 2EA
G4 URW J. Allison 17 Gordon Terrace, Stakeford, Choppington NE62 5UE
G4 URX T. Robinson 26 Keeble Drive, Washingborough, Lincoln LN4 1DZ
G4 USC K. Appleton 5 Hart Hill Crescent, Full Sutton, York YO41 1LX
G4 USD Clydebank Cadet Centre c/o D. Brill 25 Boulevard Barbes, Paris 75018 France
G4 USK B. Finlay 4 Henden Mews, Maidenhead SL6 4GY
G4 USN M. Havard 61 Northwood End Road, Haynes, Bedford MK45 3QB
G4 USP S. Hall 22 Leam Road, Lighthorne Heath, Leamington Spa CV33 9TE
G4 USQ T. Hodgetts 14 St. Peters Road, Portishead, Bristol BS20 6QY
G4 UST A. Forbes Field Cottages, 44 Walkmills, Church Stretton SY6 6NJ
G4 USW W. Jenkins 5 Seatoller Place, Barrow-in-Furness LA14 4NH
G4 USX E. Pritchard 18 New Ridd Rise, Hyde SK14 5DD
G4 UTE M. Chaudhry 613 Service Road, G 10/4, Islamabad ZZ7 9PO Pakistan
G4 UTF A. Cockman 31 Kensington Road, Southend-on-Sea SS1 2SX
G4 UTG F. Collins 31 Mount Pleasant Road, Poole BH15 1TU
G4 UTJ J. Gorton 17 Oxford Road, Colchester CO3 3HW
G4 UTK D. James 3 El Mirador Del Embalse, Los Romanes, Malaga 29713 Spain
G4 UTM B. Dennis Thistledown Yallands Hill, Monkton Heathfield, Taunton TA2 8NA
G4 UTN G. Bromley 46 Independent Hill, Alfreton DE55 7DG
G4 UTQ M. Adamson 13 Towers Close, Bedlington NE22 5ER
G4 UTR M. Alder 342 Church St., Edmonton, London N9 9HP
GW4 UTS E. Bracey 3 Dyffryn Road, Waunlwyd, Ebbw Vale NP23 6UA
G4 UTV A. Cockerill 90 Stockton Road, Middlesbrough TS5 4AJ
G4 UUA M. Robinson 2 Bridge Mill Road, Nelson BB9 7BD
G4 UUB M. Lemin Mill House, Lingwood Road, Blofield, Norwich NR13 4AH
G4 UUE Dr L. Mcgrogan 239 Haslingden Road, Rossendale BB4 6RX
G4 UUF N. Kelly 3 The Terrace, Gawcott, Buckingham MK18 4HL
G4 UUG D. Payne 23 Laburnum Avenue, Newbold Verdon, Leicester LE9 9LQ
G4 UUH S. Rogers 31 Coleridge Road, Ottery St. Mary EX11 1TD
G4 UUI S. Rooker 67 Hawks Way, Ashford TN23 5UW
G4 UUJ E. Jeffery 11 Furze Hill Road, Shanklin PO37 7PA
G4 UUQ T. Tallis 16 Matlock Road, Wessington, Alfreton DE55 6DS
G4 UUT A. Turner 30 Wheatlands Road, Paignton TQ4 5HU
G4 UUU C. Clayton 17 Meadow Dene, East Ayton, Scarborough YO13 9EL
G4 UUW D. Williams 12 Springfield Road, Exmouth EX8 3JX
G4 UVA P. Money Meadow View, Podmore Lane, Dereham NR19 2NS
G4 UVB P. Gibson 9 Mallard Close Aughton, Ormskirk L39 5QJ
G4 UVD D. Asquith 516 Old Bedford Road, Luton LU2 7BY
G4 UVF J. Taylor 6 The Stray, South Cave, Brough HU15 2AL
G4 UVG D. Stewart 4 Towles Pastures, Castle Donington, Derby DE74 2RX
GW4 UVN J. Travers 40 Birchgrove Road, Birchgrove, Swansea SA7 9JR
G4 UVV D. Pike 22 Stable Court, Gatchell Oaks, Taunton TA3 7EG
G4 UVW E. Underhill 5 Lyndhurst Croft, Eastern Green, Coventry CV5 7QE
G4 UVX P. Lee 51 Ashford Road, Faversham ME13 8XN
G4 UVZ A. Whatmore Hollybank, Sellicks Green, Taunton TA3 7SD
G4 UWA M. Styne 11 Paisley Walk, Church Gresley, Swadlincote DE11 9FF
G4 UWF M. Kebbell 56 King Edward Avenue, Hastings TN34 2NQ
G4 UWG N. Dunn 4 Swyneghyll, Temple Sowerby, Penrith CA10 2AW
G4 UWM J. Mckenna 18 Frobisher Close, Goring-By-Sea, Worthing BN12 6EY
GM4 UWN R. Kane 39 Tollohill Drive, Aberdeen AB12 5DQ
G4 UWP L. Flynn 20 Heather Lea Place, Sheffield S17 3DN
GW4 UWR V. Thomas 6 Hillside Court, Pontnewydd, Cwmbran NP44 1LS
G4 UWS A. Walker 53 Parkstone Avenue, Parkstone, Poole BH14 9LW
G4 UWW A. Prior 17 Millfield, Castleton Way, Eye IP23 7DE
GM4 UWX J. Rennie 19 Harbour Place, Portknockie, Buckie AB56 4NR
G4 UXB R. Ball 144 Broad Lane, Hampton TW12 3BW
G4 UXC Newton Le Willows ARC c/o M. Butler Field Farm, Evesham WR11 7RP
G4 UXD D. Brandon 1 Woodlands Road, Saltney, Chester CH4 8LB
G4 UXG J. Dew 8 Silverbeck Way, Stanwell Moor, Staines-upon-Thames TW19 6BT
G4 UXH C. Wilkinson 14 Ryleyfield Road, Milnthorpe LA7 7PT
G4 UXJ T. Ager 5 Matthews Close, Bedhampton, Havant PO9 3NJ
G4 UXL K. Jones 10 Whetstone Hey, Great Sutton, Ellesmere Port CH66 3PH
G4 UXO N. Emson 9 Sands Close, Pattishall, Towcester NN12 8LU
G4 UXP M. Huxham 34 The Close, Brixham TQ5 8RF
G4 UXV C. Osborn 19 Maple Drive, Huntingdon PE29 7JE
GM4 UXX A. Hood 4 Murray Road, Law, Carluke ML8 5HR
G4 UXY C. Boulter 17 Forelands Way, Chesham HP5 1QP
GM4 UYE H. Martin 11 Ewing Court, Broomridge, Stirling FK7 0QP
G4 UYF L. Aldhous 5 Banks Lane, Heckington, Sleaford NG34 9QY
G4 UYJ B. Crow 690 Walmersley Road, Bury BL9 6RN
GM4 UYK J. Caddis 30 Newlands Drive, Kilmarnock KA3 2DW
G4 UYM Dr F. Roberts 5 Manor Farm Close, Broughton, Kettering NN14 1SL
GM4 UYP J. Smith 10 Witchknowe Avenue, Caprington, Kilmarnock KA1 4LQ
G4 UYR R. Noble Fallowfield, Chandler Road Stoke Holy Cross, Norwich NR14 8RG
GW4 UYT R. Jenkins 1 Lon Y Bryn, Glynneath, Neath SA11 5BG
GM4 UYZ R. Glasgow 7 Castle Terrace, Port Seton, Prestonpans, East Lothian EH32 0EE
GW4 UZC D. Ralph Tal-Y-Maes, Llanbedr, Powys NP8 1SY
G4 UZE C. Mason 145 Park Avenue, Ruislip HA4 7UN
G4 UZF B. Matthews 12 School Road, Thurston, Bury St. Edmunds IP31 3SP
G4 UZG G. Price 28 Leewood Close, Brampton Bierlow, Rotherham S63 6ET
G4 UZN A. Quest 86 Buckstone Avenue, Leeds LS17 5ET
G4 UZO K. Richards Trigg Court, Trewetha, Port Isaac PL29 3RU
GM4 UZR J. Low 4 Smith Avenue, Inverness IV3 5ES
GM4 VAC S. Murray Taigh Nam Moireach, Croy, Inverness IV2 5PN
G4 VAF N. Dorrington Im Buergel 4, Woertham 63939 Germany
GW4 VAG H. Green 2 Whitchurch Road, Bangor On Dee, Wrexham LL13 0AY
G4 VAH P. Hudson 15 Fellows Road, Farnborough GU14 6NU
G4 VAL V. Pellowe 191 Preston New Road, Blackpool FY3 9TN
G4 VAM P. Harrison 2 Allington Close, Bainton, Stamford PE9 3AG
G4 VAO M. Jordan Petalouda, Low Street, Ketteringham, Wymondham NR18 9RY
G4 VAP I. Kenyon 57 Gloucester Avenue, Lancaster LA1 4EF
G4 VAS E. Cooper 39 Violet Road, Southampton SO16 3GZ
G4 VAV A. Brooks 17 Grosvenor Avenue, Carshalton SM5 3EJ
G4 VAX S. Cope 24 Metcalf Road, Newthorpe, , Nottingham, NG16 3NL
GM4 VAY A. Newlands 7 Muir Close, Stewarton, Kilmarnock KA3 3HG
GD4 VBA Wigan & Dist Ar c/o R. Harrison 108 Ballacriy Park, Colby, Isle of Man IM9 4NB
G4 VBD S. Mcadam 92 Armstrong Close, Birchwood, Warrington WA3 6DJ
GM4 VBE R. Fairholm 28 Queensberry Avenue, Clarkston, Glasgow G76 7DU
G4 VBI R. Harte 32 Kingsgate Avenue, Kingsgate, Broadstairs CT10 3QP
G4 VBJ B. Kay 19 Langham Grove, Timperley, Altrincham WA15 6DY
G4 VBK R. Deeprose 70 Hollington Old Lane, St. Leonards on Sea TN38 9DP
GW4 VBM C. Leighton 12 Morley Avenue, Connah'S Quay, Deeside CH5 4RE
G4 VBO J. Mattock 1633 Dufferin Cres, Nanaimo V9S 5T4 Canada
G4 VBQ S. Largent 31 Penzance Road, Kesgrave, Ipswich IP5 1LU
G4 VBS P. Chapman 10 School Cottages, Hargrave, Bury St. Edmunds IP29 5HR
GW4 VBV R. Beckers 13 Taplow Terrace, Pentrechwyth, Swansea SA1 7AD
G4 VBX A. Currell 33 The Oval, Saham Toney, Thetford IP25 7HW
G4 VCA G. Mason Lilac Cottage, Tremar Coombe, Liskeard PL14 5EL
G4 VCB C. Melvin 5808 Sterling Trl, Mckinney 75071 United States
G4 VCE S. Sewell Medway, The Rosery, Norwich NR14 8AL

G4 VCJ R. Percival 6 Bulmer Place, Hartlepool TS24 9BQ
GW4 VCL R. Parry 2 Campanula Drive, Rogerstone, Newport NP10 9JG
G4 VCN A. Soars 118 Braddon Road, Loughborough LE11 5YZ
G4 VCO D. Seddon Zante, 31 Pembridge Road, Hemel Hempstead HP3 0QN
G4 VCP C. Smith 24 Watling Way, Whiston, Prescot L35 7NG
G4 VCQ R. Hogan 10 Lisle Place, Wotton-under-Edge GL12 7BJ
G4 VCX M. Beaumont 71 Lime Tree Avenue, Coventry CV4 9EZ
GI4 VCZ P. Donnelly 64 Aghnagar Road, Garvaghy, Dungannon BT70 2EL
G4 VDB D. Brocklehurst 73 Ridgeway, Clowne, Chesterfield S43 4BD
G4 VDF A. Palmer 1 Rosary Gardens, Yateley GU46 6JT
GM4 VDG J. Rankin 64 Forrest Walk, Uphall, Broxburn EH52 5PN
G4 VDH W. Peak 10 The Oval, Scarborough YO11 3AP
G4 VDJ B. Lee 61 Pendleway, Pendlebury, Manchester M27 8QS
GW4 VDP D. James Brig Y Gwynt, Penrhyn Geiriol, Trearddur Bay, Holyhead LL65 2YW
G4 VDX J. Menguy 6 Laurel Grove Lowton, Warrington WA3 2EE
GW4 VEB D. Lintern 108 Pontygwindy Road, Caerphilly CF83 3HF
G4 VEC M. Elliott 20 Haysel, Sittingbourne ME10 4QE
G4 VEH L. Skinner 15 Ridge Close, Portishead, Bristol BS20 8RQ
GW4 VEI C. Barrett 30 Brecon Road, Hirwaun, Aberdare CF44 9ND
G4 VEL J. Smith 43 Ash Close, Thetford IP24 3HQ
G4 VEO G. Barratt Charnwood, Great North Road, Retford DN22 8NL
GW4 VEQ T. Jones Penrhiw Bach, Bryngwran, Holyhead LL65 3RD
G4 VER Verulam ARC c/o P. King 96 Mancroft Road, Caddington, Luton LU1 4EN
G4 VET N. Greaves Forty Shilling Cottage, Oghill, Co. Kildare 0 Ireland
G4 VEW G. Davies 40 Derby Road, Talke, Stoke-on-Trent ST7 1SG
G4 VEY F. Havard Whitehalgh Farm, Whitehalgh Lane, Blackburn BB6 8ET
G4 VFC D. Monnery 8 Reeds Lane, Southwater, Horsham RH13 9DQ
GW4 VFE C. Davies 3 Bryn Onnen, Flint CH6 5QB
G4 VFG P. Lewis 18 Bittaford Wood Bittaford, Ivybridge PL21 0ET
G4 VFH G. Fuller 61 The Underwood, London SE9 3EP
G4 VFJ S. Shenton 36 Walleys Drive, Newcastle ST5 0NG
G4 VFK C. Archer 118 Cator Lane, Beeston, Nottingham NG9 4BB
G4 VFL Ipswich Rad Club c/o A. Holland 12 Riverside Drive, Egremont CA22 2EH
G4 VFR K. Hackwell 15 Standish Avenue, Billinge, Wigan WN5 7TF
G4 VFU C. White 4 Trenton Drive, Bradford BD8 7SZ
G4 VFX C. Perkins 32 Empshott Road, Southsea PO4 8AU
GW4 VGB R. Harper 114 Pantbach Road, Cardiff CF14 1UE
G4 VGL S. Luckhaus Wingertstrasse 5, Kleinwallstadt 63839 Germany
G4 VGM E. Grindel Bischofsheimer, Platz 24, Frankfurt D 60326 Germany
G4 VGN V. Havran Kurt-Schumacher, Ring 31, Dreieich D-63303 Germany
GM4 VGR J. Buchanan 114 Glasgow Road, Whins Of Milton, Stirling FK7 0LJ
GM4 VGU A. Lyttle 23 Heathfield Drive, Kirkmuirhill, Lanark ML11 9SR
G4 VGY D. Cline 68 Frenchgate, Richmond DL10 7AG
G4 VHB C. Averill-Elias 12 Bubwith Close, Chard TA20 2BL
G4 VHE R. Haase 674 Valley View Lane, Strafford 19087 United States
G4 VHG J. Fowler 6 Cridlake, Axminster EX13 5BS
G4 VHI M. Sawyers 20 Fairways, Ferndown BH22 8BA
G4 VHJ J. Taylor 29 Meadow Walk, Ewell, Epsom KT17 2EF
G4 VHK Prof. R. Leslie Tranquil, Rectory Lane Kingston, Cambridge CB23 2NL
G4 VHL T. Langford 11 The Grove Blackawton, Totnes TQ9 7BA
G4 VHM M. Hindley 12 Tremayne Avenue, Brough HU15 1BL
GI4 VHO D. Calderwood 43 Rathview Park Mullybritt, Lisbellaw, Enniskillen BT94 5EW
GW4 VHP S. Murdoch 55 Pendre Avenue, Prestatyn LL19 9SH
GW4 VHS E. Williams Newhaven, Kinmel Way, Abergele LL22 9NE
G4 VHV A. Sims 38 Giffard Drive, Welland, Malvern WR13 6SE
G4 VHW C. Thompson 2 Essen Lane, Kilsby, Rugby CV23 8XQ
G4 VHX J. Allen 18 Horsley Close, Chesterfield S40 4XD
GM4 VHZ N. Brown 7 Mid Road, Beith KA15 2AJ
G4 VIA J. Mcsherry 1 Station Houses, Corkickle, Whitehaven CA28 7XG
G4 VIF R. Watts 116 Hassall Road, Sandbach CW11 4HL
G4 VII J. Lawrence 25 Sylvia Crescent, Totton, Southampton SO40 3LP
GM4 VIK T. Irwin 6 Inverarnie Park, Inverarnie, Inverness IV2 6AX
G4 VIL J. Fielding 35 Amos Avenue, Litherland, Liverpool L21 7QH
G4 VIM B. Pulfrey 21 Emfield Road, Grimsby DN33 3BW
G4 VIO A. Greenbank 3 Cooperative Terrace, Stanley, Crook DL15 9SE
GI4 VIP the Sth Belfast c/o P. Murphy 11 Danesfort Apartments, Belfast BT9 5QL
G4 VIQ P. Brushwood 2 High Trees, Waterlooville PO7 7XP
GM4 VIS H. Cameron 14 Queen St., Castle Douglas DG7 1HX
G4 VIT M. Wood 42 Buckingham Drive, Willenhall WV12 5TD
G4 VIX the East Coast Vhf Group c/o D. Bartlett Apartamentos Las Adelfas 8A, C/ San Borondon 111, Playa Honda 35509 Spain
GI4 VIZ M. Jamieson 59 Curragh Road, Coleraine BT51 3RZ
G4 VJB V. Bloor 22 Regency Close, Talke Pits, Stoke-on-Trent ST7 1RH
GM4 VJH A. Douglas 12 Leaburn Drive, Hawick TD9 9NZ
G4 VJL J. Oldfield 4 The Elms, Taunton TA4 4AE
G4 VJN S. Matthews 30 Broadgate Lane, Deeping St. James, Peterborough PE6 8NW
G4 VJT K. Farmer 61 Queens, Beckenham, Kent BR3 4JJ
GI4 VJZ T. Wilson Wilden, 18 Ahoghill Road, Antrim BT41 3BJ
G4 VKC D. Lawrence 6 Hollycombe Close, Liphook GU30 7HR
G4 VKE R. Pearce 1A Green Lane, Dalton-in-Furness LA15 8LZ
GW4 VKG W. Weston 3 Factory Terrace, Aberkenfig, Bridgend CF32 9AF
GM4 VKI M. Kavanagh 4 Old Auchans View, Dundonald, Kilmarnock KA2 9EX
G4 VKJ T. Grant 81 Hillworth Road, Devizes SN10 5HD
G4 VKO J. Whittock 18 Westons Brake, Emersons Green, Bristol BS16 7BP
GI4 VKS A. Mccallion 3 Lisky Road, Strabane BT82 8NW
G4 VKV T. Linacre 69 Elizabeth Road, Fazakerley, Liverpool L10 4XL
G4 VKX I. Wade 59 St. Annes Road, Kettering NN15 5EQ
G4 VLA R. Trudgill The Retreat, Kiln Lane, Bedford MK45 4DA
G4 VLH B. Pash Dales Stores, Station Road, Cheltenham GL54 4HP
G4 VLI R. Allen 39 Deerpark, Co Meath Ireland
G4 VLK D. Haslehurst 23 Yew Tree Drive, Shirebrook, Mansfield NG20 8QH
G4 VLL C. Denham 3 Glenmore Close, Flackwell Heath, High Wycombe HP10 9DF
G4 VLN M. Evans 2A Moreton Road, Worcester Park KT4 8EZ
G4 VLP H. Knatchbull 19 Riverside Road, West Moors, Ferndown BH22 0LG
G4 VLS P. Turnham 71 Theobald Road, Norwich NR1 2NX
G4 VLT C. Tunna 52 Shaftoe Road, Springwell, Sunderland SR3 4EZ
GW4 VLU M. Hatwood Calgary, Denbigh Circle, Rhyl LL18 5HW
G4 VLV A. Flint 4 Churchill Way, Painswick, Stroud GL6 6RQ
G4 VLW R. Davey 35 The Pines, Faringdon SN7 8AT
GM4 VLX J. Brown 33 Gartmore Road, Paisley PA1 3NG
G4 VLZ M. Nettleship 141 Hollybank Drive, Sheffield S12 2BU
G4 VMA M. Anderson 1 Ridgeway Square, Knottingley WF11 0JY
G4 VMB D. Stoddart 16 Market Place, Long Buckby, Northampton NN6 7RR
G4 VMC P. Coates 20 The Flashes, Gnosall, Stafford ST20 0HL
G4 VMD C. Hackney Mar Azul 9, Apt.20 2 Fase, Alicante 03710 CALPE Spain
G4 VME R. Youell 1 Greenside Waterbeach, Cambridge CB25 9HW
G4 VMF S. Schofield 28 Rush Meadow Road, Cranbrook, Exeter EX5 7GB
G4 VMG J. Holmes 10 Chapel Road, Morley St. Botolph, Wymondham NR18 9TF
G4 VMI M. Pickworth 46 St. Andrews, Grantham NG31 9PE
G4 VMM S. Tidmarsh 4 The Grange, Earl Shilton, Leicester LE9 7GT
G4 VMO J. Harris 23 Brookvale Grove, Solihull B92 7JH
G4 VMQ A. Hyde 45 Camm Street, Sheffield S6 3TR
G4 VMR J. Watkins One Ash, Frogshall Lane, Haultwick, Ware SG11 1JH
GW4 VMT G. Williams 12 Heol Johnson, Talbot Green, Pontyclun CF72 8HR
G4 VMW J. Curtis 18 Carlidnack Close, Mawnan Smith, Falmouth TR11 5HF
G4 VMX A. Ritchie 24 Swift Close, Newport Pagnell MK16 8PP
G4 VMY A. Cooper 3 Marina Way, Ripon HG4 2LJ
G4 VMZ A. Jones 40 Alexandria Drive, Herne Bay CT6 8HX
G4 VNA R. Bell Long Meadows, Haddockstones, Harrogate HG3 3LA
G4 VNC M. Leak 32 Springdale Road, Market Weighton, York YO43 3JT
G4 VNE D. Hunt 233 Kingsley Road, Kingswinford DY6 9RP
G4 VNG R. Mccallum 9 Hardwick Close, Blackwell, Alfreton DE55 5LL
GW4 VNK L. Smith Blaenlluest, Cilcennin, Lampeter SA48 8RP
G4 VNM S. Frost 32 Hunters Lodge, Fareham PO15 5NE
G4 VNR R. Sharp 22 Sunderton Lane, Clanfield, Waterlooville PO8 0NU
GW4 VNS R. Sims 61 Constable Drive, Newport NP19 7QB
G4 VNX W. Wood 11 Walbert Avenue, Thurnscoe, Rotherham S63 0TN
G4 VOG A. Hepworth 9 Linden Grove, Kirkby-In-Ashfield, Nottingham NG17 8JJ
G4 VOJ A. Tennant 2 Chapel Hill Farm Cottage, Lower Lane, Preston PR3 3SL
G4 VOK P. Dresser 6 Acacia Avenue, Fencehouses, Houghton le Spring DH4 6JG
G4 VOU A. Pinkney 1 Hester Gardens. New Hartley. Whitley Bay., Whitley Bay. NE25 0SH
G4 VOV Skmrsdle & D Ar c/o D. Hallsworth Eastholme, Luddington Road, Scunthorpe DN17 4PP
G4 VOW B. Pluckrose 104 Edward Road, West Bridgford, Nottingham NG2 5GB
G4 VOY R. Powell Old School House, Broxwood, Leominster HR6 9JQ
G4 VOZ J. Jennings Mill Side, Mill Road, Lutterworth LE17 5DE
G4 VPA London Bbc Radio Group c/o J. Martindale The Old School House, Ipswich Way, Stowmarket IP14 6DJ
G4 VPC E. Ikin 30 Kelsborrow Way, Kelsall, Tarporley CW6 0NL
G4 VPD M. Pugh 44 Simms Lane, Hollywood & Wythall, Birmingham B47 5HY
G4 VPE Dr R. Derricott Birches, The Paddock, Stourbridge DY9 0RE
G4 VPF O. Davies 16 Central Way, Horninglow, Burton-on-Trent DE13 0UU
G4 VPI R. Riley 161 Botany Road, Kingsgate, Broadstairs CT10 3SD
G4 VPJ D. Bridgnell Penvale, 1 Tretherras Road, Newquay TR7 2RB
G4 VPL J. Villena Bota Santa Ana 74, Estartit, Gerona 17258 Spain
G4 VPM A. Stafford 233 Sparrow Branch Circle, Jacksonville 32259 United States
G4 VPS B. Lewis 6 Malt Dubs Close, Ingleton, Carnforth LA6 3DZ
G4 VPW P. Wilcock 12 Napier Road, Eccles, Manchester M30 8AG
GW4 VPX A. Jones Maes Y Llyn, Maesycrugiau, Pencader SA39 9DH
G4 VPZ A. Hill 36 Narrow Lane, Halesowen B62 9NQ
G4 VQE R. Spencer 6 Belland Drive, Charlton Kings, Cheltenham GL53 9HU
G4 VQF P. White 4 Barnett Lane, Wonersh, Guildford GU5 0SA
G4 VQH M. Clutton Cumberwell, Cumberland Lane, Whitchurch. SY13 2NJ
G4 VQI P. Hughes 92 Freshwater Drive, Paignton TQ4 7SD
G4 VQJ D. Northwood 5 Beech Grange, Landford, Salisbury SP5 2AL
GI4 VQK A. Ward 50 Derry Road, Strabane BT82 8LD
G4 VQL G. Dymond 28 The Green, Exmouth EX8 2QR
G4 VQP C. Smith Foxgloves, Northlew, Okehampton EX20 3PP
G4 VQR S. Reed 139 Potovens Lane, Outwood, Wakefield WF1 2LF
G4 VQS H. Jones 47 Penkett Road, Wallasey CH45 7QG
G4 VQT M. Pinnell 3 Inmans Lane, Petersfield GU32 2AN
GM4 VQY G. Leiper 76 Martin Drive, Stonehaven AB39 2LU
G4 VQZ J. Oakley 152 Little Breach, Chichester PO19 5UA
G4 VRB K. Raine 30 St. Andrews Gardens, Shepherdswell, Dover CT15 7LP
G4 VRC R. Doran 28 Buckingham Road, Petersfield GU32 3AZ
GM4 VRE P. Henderson 134 Gray Street, Aberdeen AB10 6JU
G4 VRG F. Margrave Templars, Peerley Road, East Wittering, Chichester PO20 8DW
G4 VRJ R. Clifft 11 Hambleton St., Wakefield WF1 3NW
G4 VRM A. Berry 148 Maple Way, Gillingham SP8 4RR
G4 VRN M. Blewett 32 Miltons Crescent, Godalming GU7 2NT
GW4 VRO P. Parsons 9 Military Road, Pennar, Pembroke Dock SA72 6SH
G4 VRP R. Porter 47 Milford Avenue, Wick, Bristol BS30 5PP
G4 VRS Aylesbury Vale RS c/o V. Gerhardi 24 Putnams Drive, Aston Clinton, Aylesbury HP22 5HH
G4 VRT B. Barker The Nook, Park Lane, Roughbirchworth, Oxspring, Sheffield S36 8WW
G4 VRU B. Stephenson 12 Claremont Terrace, York YO31 7EJ
G4 VRW K. Newbould 6 Pyenot Avenue, Cleckheaton BD19 5AY
G4 VRX Dr G. Brown 1 Dog Kennel Lane, Oldbury B68 9LU
G4 VSB A. Brown 6 The Firs, Rushbrooke Lane, Bury St. Edmunds IP33 2SY
G4 VSD P. Samuels 6 Miriam Close Caister-On-Sea, Great Yarmouth NR30 5PH
G4 VSI A. Stone 29 Nottingham Road, Belper DE56 1JG
G4 VSJ K. Drakeford Ballybrook, Eastfield Road, Firsby, , Spilsby PE23 5QZ
G4 VSK B. Skelton 8 Dunelm Drive, West Boldon, East Boldon NE36 0HJ
G4 VSL T. Watkins One Ash, Frogshall Lane, Haultwick, Ware SG11 1JH
G4 VSO R. Carter Wendover, Three Ahes HR2 8LU
G4 VSQ A. Bolton 17 Lomond Avenue, Caversham Reading RG4 6PL
G4 VSR S. Alston 21 Hilltop Road, Wingerworth, Chesterfield S42 6RX
G4 VSS M. Isherwood 32 Franklin Close Old Hall, Warrington WA5 8QL
G4 VSV G. Ingham Courthaven, South Duffield Road, Selby YO8 5HP
G4 VSW M. Taylor 2 Bickerton Drive, Hazel Grove, Stockport SK7 5QY
G4 VSX P. Reilly 40 Bollin Drive, Lymm WA13 9QA
G4 VSY K. Hutton 7 Roseveare Drive, Roseveare Park, Gothers, St. Austell PL26 8GY
G4 VTA J. Taylor 219 Mandarin Way, Cheltenham GL50 4SB

GM4 VTB M. Budas 20 Oak Avenue, Bearsden, Glasgow G61 3HD
G4 VTC A. Croydon Harvesters, Newdigate Road, Dorking RH5 4QB
G4 VTD I. Daniels 24 Ockley Lane, Keymer, Hassocks BN6 8BB
GW4 VTG E. Smith 21 St. Davids Road, Pembroke SA71 5JH
G4 VTM J. Hicks Cory House, Kilworth Road, Lutterworth LE17 6JW
G4 VTN J. Rodda 22 Balmoral Drive, Felling, Gateshead NE10 9TZ
G4 VTO P. Tanner Beechcroft, Station Hill, Newton Abbot TQ13 0EE
G4 VTQ D. Rainer Twin Oaks Knightstone Lane, Ottery St. Mary EX11 1PR
G4 VTU R. George 19 Apthorpe Street, Fulbourn, Cambridge CB21 5EY
G4 VUA A. Burton 26 Woffindin Close, Great Gonerby, Grantham NG31 8LP
G4 VUD J. Head 21 Reynell Avenue, Newton Abbot TQ12 4HE
G4 VUF C. Tugman 41 Chatsworth Road, Hunstanton PE36 5DJ
GM4 VUG C. Green 4 Gallowhill Gardens, Kinross KY13 8RT
GW4 VUH J. Washington Flat 9, Llys Canol, Holywell CH8 7XG
G4 VUI P. Sweeney 15 Alford Road, West Bridgford, Nottingham NG2 6GJ
G4 VUK L. Wolfson 7 Gilmore Drive, Prestwich, Manchester M25 1NB
G4 VUM D. Stocks 2 Newport Crescent, Mansfield NG19 6BY
G4 VUN P. Norris Thirn Farm Thirn, Ripon HG4 4AU
G4 VUP G. Newton 6 Yardley Way, Grimsby DN34 5UQ
G4 VUR I. Daniels 20A Stalham Road, Hoveton, Norwich NR12 8DG
G4 VUV S. Valori 7 Upton Close, Norwich NR4 7PD
G4 VUW R. Kemp 35 Rushett Drive, Dorking RH4 2NR
G4 VVD P. Taylor 8 High St., Clive, Shrewsbury SY4 3JL
G4 VVE E. Macmanus 41 Oldfield Crescent, Stainforth, Doncaster DN7 5PE
GW4 VVF N. Allen Lilac Cottage, Roddhurst, Presteigne LD8 2LH
G4 VVK B. Murray La Casa, 30 Middlegate Green, Rossendale BB4 8PY
G4 VVM A. Bennett 28 Kinglake Drive, Taunton TA1 3RR
G4 VVP B. Gillard Charmaine, Broadway, Chilcompton, Radstock BA3 4GT
G4 VVQ F. Shead 7 White Cottages, Fuller Street, Fairstead, Chelmsford CM3 2AY
G4 VVS J. Blanchard 41 Deane Drive, Galmington, Taunton TA1 5PQ
G4 VVT Lichfield Raynet Group c/o A. Moss 9 Summerfield Drive, Middleton, Manchester M24 2TQ
GM4 VVX C. O'Hennessy Savalbeg, Challenger Estate, Lairg IV27 4ED
GM4 VVY D. Davis 3 High Shore, Banff AB45 1DB
G4 VVZ C. Wilson 2 Bainton Close, Bradford-on-Avon BA15 1SE
G4 VWA S. Ward 88 Little Barn Lane, Mansfield NG18 3JJ
GI4 VWC G. Christie The Brambles, 9 Burnet Park, Newtownabbey BT37 0XY
G4 VWD A. Davies 85 Ridgewood Drive, St. Helens WA93XU
G4 VWE M. Courteney 36 Nursery Close, Hellesdon, Norwich NR6 5SJ
G4 VWF R. Hawkins 43 The Courtyard, Taylor Avenue, Northampton NN3 2DD
G4 VWG S. Vankassel 9 Tarragon Close, Swindon SN2 2SG
G4 VWI D. Hatton 9 Gregory Close, Thurmaston, Leicester LE4 8BP
G4 VWL J. Owen 7 Linear Park, Wirral CH46 6FL
GW4 VWO J. Bloodworth Sibrwd Yr Awel, Penrhyndeudraeth LL48 6AY
G4 VWS C. Davies Essex House, 42 Boxworth Road, Cambridge CB23 4JQ
G4 VWT B. Padgett 11 Crow Croft Road, Pilsley, Chesterfield S458HY
GM4 VWV R. Mcewan 12 Valleyfield Drive, Cumbernauld, Glasgow G68 9NW
G4 VWX A. Shone 7 Mayflower Close, Malvern WR14 2RH
GW4 VWY G. Whiteway 73 Victoria Avenue, Porthcawl CF363EY
GM4 VXA P. Williams 21 St. Clair Way, Ardrishaig, Lochgilphead PA30 8FB
G4 VXD R. King 1 Emmas Crescent, Stanstead Abbotts, Ware SG12 8AZ
GW4 VXE T. Kirby Parsonage House, St. Nicholas, Goodwick SA64 0LG
G4 VXG G. Buxton 12 Jute Road, York YO26 5EN
G4 VXH A. Rean 17 Mount Pleasant Road, Dawlish Warren, Dawlish EX7 0NA
GM4 VXM I. Munro1 12 Greenstone Place, Dundee DD2 4XB
G4 VXN C. Bennett 49 Keats Avenue, Redhill RH1 1AF
G4 VXP R. Want 19 Canterbury Road, Leyton, London E10 6EE
G4 VXU J. Haig 3 Hartland Court Gaping Lane, Hitchin SG5 2JH
G4 VXV R. Boulton 23 Stamford Bridge West, Stamford Bridge, York YO41 1AQ
G4 VXW R. Seddon 255 Westleigh Lane, Leigh WN7 5PN
G4 VXX S. Oakes 6 Wychwood Park, Weston, Crewe CW2 5GP
G4 VYA J. Jacobs 17 Cotswold Drive, Albrighton, Wolverhampton WV7 3DQ
G4 VYC V. Packman 241 Gurnard Pines Cockleton Lane, Cowes PO31 8RL
G4 VYE J. Harris 23 Balmoral Drive, Hednesford, Cannock WS12 4LT
G4 VYF K. Thomas 19 South End Hogsthorpe, Skegness PE24 5NE
G4 VYG B. Roberts 52 School Lane, Toft, Cambridge CB23 2RE
G4 VYH N. Baker 16 Boulderside Close, Norwich NR7 0JJ
G4 VYI M. Dalley 195 Marlcliffe Road, Sheffield S6 4AH
G4 VYJ J. O'Sullivan 30 Highbank Road, Kingsley, Frodsham WA6 8AE
G4 VYK R. Vaughan 73B Westward Drive, Pill, Bristol BS20 0JR
G4 VYL R. Reilly 4 Moreton Drive, Poulton-le-Fylde FY6 8ED
G4 VYN Dr J. Lawrence 1 Naples Close, Hopton, Great Yarmouth NR31 9SB
G4 VYP D. Rimmer 8A Mallee Avenue, Southport PR9 8NL
GM4 VYQ W. Harvey 32 Upper Glenfyne Park, Ardrishaig, Lochgilphead PA30 8HH
G4 VYR G. Mccartney 12 Timway Drive, West Derby, Liverpool L12 4YR
GM4 VYU O. Jackson Cossarshill Farm, Selkirk TD7 5JB
G4 VZB I. Ray Palmyra F2B-H8, , Vila Sol, 8125-307 QUARTEIRA Portugal
G4 VZC P. Stokes-Herbst 4 Thackeray Grove, Middlesbrough TS5 7QX
G4 VZH A. Hobkirk 216 Northwick Road, Worcester WR3 7EH
GM4 VZI W. Lawrie Sandana Old Woodhouselea, 2, Roslin EH25 9QJ
G4 VZK D. Perkins 10 The Foxes, Sutton Hill, Telford TF7 4NH
G4 VZL J. Caddick 5 Great Hay Drive Sutton Hill, Telford TF7 4DT
G4 VZR D. Cormack Lukes Orchard Far Green, Coaley GL11 5EL
G4 VZS L. Andrew 10 Front Street, Rookhope, Bishop Auckland DL13 2AY
G4 VZT P. Green 61 Gravel Hill, Wimborne BH21 3BJ
G4 VZW K. Rankine Alfredo L. Jones 34, Building Perez Roche 3Rd Floor, Apartment 306, Las Palmas 35008 Spain
GM4 VZY D. Deans 17 Montrose Way, Dunblane FK15 9JL
G4 WAB Worked All Britain Awards Group c/o K. Hale 58 St. Stephens Road, Saltash PL12 4BJ
G4 WAC Wythall RC c/o M. Pugh 44 Simms Lane, Hollywood & Wythall, Birmingham B47 5HY
G4 WAF A. Fewkes 21 Tong Road Bishops Wood, Stafford ST19 9AB
G4 WAG T. Morley 2 Hawthorn Place, Woodbridge IP12 4JZ
GI4 WAH J. Keenan 24 Leode Road, Hilltown, Newry BT34 5TJ
G4 WAK N. Rumbol 66 The Avenue, Hadleigh, Benfleet SS7 2DL
G4 WAL P. Walton 6 Gorse Grove, Longton, Preston PR4 5NP
G4 WAM M. Lockley 37 Farmside Lane, Biddulph Moor, Stoke-on-Trent ST8 7LY
G4 WAO J. Kimpton 721/30 Lantana Avenue, Narrabeen 2101 Australia
G4 WAP R. Southern 31 Burnsall Road, Brighouse HD6 3JS
G4 WAS K. Atack 29 High Hill, Essington, Wolverhampton WV11 2DW
G4 WAV A. Medland Trevilla, 93 Pengelly Road, Delabole PL33 9AT
G4 WAW South Bristol ARC c/o A. Jenner 24 The Willows, Nailsea, Bristol BS48 1JQ
G4 WAX J. Moon 25 Shotley Gardens, Gateshead NE9 5DP
G4 WAZ N. Mackinnon 49 Balmoral Way, Worle, Weston-Super-Mare BS22 9AL
G4 WBA B. Westbrook 14 Pickering Street, Maidstone ME15 9RS
G4 WBF P. Finney New Rivernook Farm, 10 Kinnersley Manor, Reigate RH2 8QJ
G4 WBG R. Dunn 13 Horton Gate, Giffard Park, Milton Keynes MK14 5JQ
G4 WBH P. Jackson 15 Bankside, Retford DN22 7UW
G4 WBI S. Haydon 58 Deanfield Road, Henley-on-Thames RG9 1UU
G4 WBO K. Johnson 23 Rotherhead Close, Horwich, Bolton BL6 5UG
G4 WBP D. Hatfield 29 Awbridge Road, Netherton, Dudley DY2 0HZ
GW4 WBT S. Clifton 15 Cae Clyd, Craig-Y-Don, Llandudno LL30 1BL
GM4 WBU W. Swinburne 29 Murray Place, Dollar FK14 7HP
G4 WBV A. Fry 128 Sylvan Way, Sea Mills, Bristol BS9 2LU
G4 WBW K. Odlum 17 Glebe Street, Talke, Stoke-on-Trent ST7 1NP
G4 WCD D. Longstaff 83 Spring Gardens, Anlaby Common, Hull HU4 7QG
GM4 WCE Dr P. Kirsop 2F1, 19 North West Circus Place, Edinburgh EH3 6SX
G4 WCH W. Houghton 28 Regent Street, Newton-le-Willows WA12 9LS
G4 WCK C. Baverstock 43 Tatnam Road, Poole BH15 2DW
G4 WCO D. Foy 37 Gorsey Croft, Eccleston Park, Prescot L34 2RS
G4 WCP S. Richardson 25 Kenmure Avenue, Patcham, Brighton BN1 8SH
G4 WCY H. Bottomley 8 Leyburn Place, Filey YO14 0DQ
G4 WDA J. Curtis 8 King Street, Wilton, Salisbury SP2 0AX
G4 WDC G. Cooke 106 Wirral Drive, Winstanley, Wigan WN3 6LD
G4 WDO A. Douglas 3 North Lodge Cottages, Ladykirk, Berwick-upon-Tweed TD15 1SU
G4 WDP D. Preston 77 Wensley Road, Woodthorpe, Nottingham NG5 4JX
G4 WDR West Devon Raynet c/o I. Harley 37 Crelake Close, Tavistock PL19 9AX
G4 WDS R. Silvera 10 White Hill, Kinver, Stourbridge DY7 6AD
G4 WDZ K. Bennett Lilac Cottage, St. Neots Road, Bedford MK44 2ER
G4 WEC G. Leesley Marsh View, Main Road, Maltby Le Marsh, Alford LN13 0JP
GM4 WED N. Tipping Millburn Cott, Weisdale ZE2 9LN
G4 WEE S. Leech 9 Parkside Drive, Old Catton, Norwich NR6 7DP
G4 WEH M. Pepper 56 Meadow Lane, Burgess Hill RH15 9JA
G4 WEL J. Bolton 110 Vale Road, Ash Vale, Aldershot GU12 5HS
G4 WEM A. Penney 110 Vale Road, Ash Vale, Aldershot GU12 5HS
G4 WEN K. Porter Thornfield, Mount Carmel Road, Andover SP11 7ES
G4 WEP W. Hewitt 101 Sunnyside Avenue, Ball Green, Stoke-on-Trent ST6 6DZ
G4 WET Tingley Moonbounce Society c/o M. Butler Field Farm, Evesham WR11 7RP
G4 WEV Dr A. Russell 6 Bartlemy Road, Newbury RG14 6JX
GM4 WEW C. Brown Glencraig, Ballantrae, Girvan KA26 0PA
G4 WEY B. Bush 45 Mimosa Avenue, Wimborne BH21 1TU
G4 WEZ K. Westley 29 The Limes, Sawston, Cambridge CB22 3DH
G4 WFC M. Morris Fieldhead Farm, Denholme, Bradford BD13 4LZ
GM4 WFE J. Longton 5 Allanshaw Grove, Hamilton ML3 8NZ
G4 WFF C. Mcguire11 40 Brook Gardens, Emsworth PO10 7LB
G4 WFK F. Seddon 20 Pinfold Lane, Bottesford, Nottingham NG13 0AR
G4 WFL P. Ford 24 Tonstall Road, Epsom KT19 9DP
GW4 WFM G. Moller 31 Wyngarth, Winch Wen, Swansea SA1 7EF
G4 WFR R. Cooper 53 Saturn Close, Southampton SO16 8BE
G4 WFT Amc RC c/o T. Kearsley 142 Avenue Road, Rushden NN10 0SW
GM4 WFV S. Duguid 22 Cairn Crescent, Ayr KA7 4PW
G4 WFW P. Edwards 1 Radley Avenue, Wickersley, Rotherham S66 2HZ
G4 WFZ P. Marsh Columbia, 28 Orcheston Road, Bournemouth BH8 8SR
G4 WGA R. Hall Hillside, Potten End Hill, Hemel Hempstead HP1 3BN
G4 WGB R. Vaughan 6 Dellside Grove, St. Helens WA9 5AR
GM4 WGC P. Naughton 16 Holton Crescent, Sauchie, Alloa FK10 3DZ
G4 WGD C. Pearse 77A Nutfield Road, Merstham, Redhill RH1 3ER
G4 WGE A. Cross 31 Mountcombe Close, Surbiton KT6 6LJ
G4 WGF G. Fairhurst 42 Chorley Road, Standish, Wigan WN1 2SS
G4 WGJ M. Collins 185 Church Road, Haydock, St. Helens WA11 0NB
G4 WGK G. Kemp 38 Merlin Way, Leckhampton, Cheltenham GL53 0LU
G4 WGN K. Wilson 102 Waddicar Lane, Melling, Liverpool L31 1DY
G4 WGR R. Gibson 52 Broomfields, Denton, Manchester M34 3TH
G4 WGT W. Taylor 27 Netherley Road Coppull, Chorley PR7 5EH
G4 WGU G. Tarry 8 Wareham Road, Blaby, Leicester LE8 4BE
G4 WGX North Bristol Amateurradio Club c/o B. Rivers Maybank, Athelney Bridge, Bridgwater TA7 0SB
G4 WGZ A. Brooker 18 Honeybourne Way, Petts Wood, Orpington BR5 1EZ
GM4 WHA G. Harper 15 Seaforth Park, Annan DG12 6HX
G4 WHF K. Wilson 111 Marple Road, Stockport SK2 5EP
G4 WHK R. Cann 39 Grafton Road, Harwich CO12 3BD
G4 WHL P. Callaghan 8 Abbey Road, Edwinstowe, Mansfield NG21 9LQ
G4 WHM P. Callaghan 15 John Barrow Close, Rainworth, Mansfield NG21 0GD
G4 WHT W. Tattersall 45 Russell Avenue, Alsager, Stoke-on-Trent ST7 2BN
G4 WHV M. Langdon 58 Upper Marsh Road, Warminster BA12 9PN
G4 WHY M. Foot Oakfield Farm, Horton Way, Verwood BH31 6JJ
G4 WHZ D. Cater 104 St. Johns Road, Clacton-on-Sea CO16 8DB
G4 WIA I. Whitmore Sunny Bank, Commercial Road, St. Keverne, Helston TR12 6LY
G4 WIG P. Lees 107 Balmoral Road, Wordsley, Stourbridge DY8 5JJ
G4 WIL J. Wilkinson 147 Alder Lane, Hindley Green, Wigan WN2 4ET
G4 WIM T. Forrester Dow Brook House, Brades Lane, Freckleton, Preston PR4 1HG
G4 WIN C. Mabbutt 17 Lakeside, Peterborough PE4 6QZ
G4 WIP A. Crickett 40 Ousden Close Cheshunt, Waltham Cross EN8 9RQ
G4 WIR I. Page 127 Whyke Lane, Chichester PO19 8AU
G4 WIS V. Gleek Fieldgate, The Warren, Radlett WD7 7DU
G4 WIY A. Clark Applecroft Care Home, Sanctuary Close, Chilton Way, Dover CT17 0ER
G4 WIZ D. Burleigh 12 Cadbury Road, Keynsham BS31 1JW
GM4 WJA J. Fraser Cherrybrae Croft, Aultmore, Keith AB55 6QU
G4 WJB R. Barratt 37 Cemetery Road, Whittlesey, Peterborough PE7 1RT
G4 WJE M. Fenelon 72 Fieldside, Epworth, Doncaster DN9 1DP
G4 WJG D. Nicolson 142 Shireburn Caravan Park, Edisford Road, Clitheroe BB7 3LB
G4 WJH P. Mathews 1 Erith Road, Belvedere DA17 6HB
G4 WJJ P. Short Flat 4, Orchard Court, Searle Street, Crediton EX17 2HA

G4 WJM W. Cooper 32 High St., Thurlby, Bourne PE10 0EE
G4 WJQ B. Blain 31 Crest Road, Bromley BR2 7JA
G4 WJR J. Singleton 11 Lound Place, Lound Street, Kendal LA9 7FE
G4 WJS W. Somerville Glendella, Wycombe Road Stokenchurch, High Wycombe HP14 3RP
G4 WJV J. Forrest 3 Martindale Park, Houghton le Spring DH5 8EX
G4 WJW T. Murphy 7 The Knapp, Templecombe BA8 0JP
G4 WJX M. Kessel 4 Harington Drive, Stoke-on-Trent ST3 5ST
G4 WJZ A. Kerr Braemoray, Dalditch Lane, Budleigh Salterton EX9 7AS
G4 WKB H. Poulton 1 Marnhull Close, Coventry CV2 2JS
G4 WKD K. Dunstan 41 Gravel Lane, Wilmslow SK9 6LS
G4 WKG G. Rowley Glassonby Lodge, Glassonby, Penrith CA10 1DT
GW4 WKQ L. Jones 8 Tyn Y Pwll Estate, Llanbedrog, Pwllheli LL53 7PG
G4 WKT N. Bleek 49 Lowhills Road, Peterlee SR8 2DJ
G4 WKW R. Benton 15 Polventon Parc, St. Keverne TR12 6PB
GW4 WKZ G. Fitch Tides Reach, 39 Hen Gei Llechi, Y Felinheli LL56 4PB
G4 WLA D. Dell Bushmead, 12 Penfield Gardens, Dawlish EX7 9NQ
G4 WLE N. Slater 17 Hall Park Drive, Lytham St. Annes FY8 4QR
G4 WLG K. Dunwell 8 Violet Grove, Thatcham RG18 4DQ
G4 WLI P. Nutt 1 Lime Gardens, Middleton, Manchester M24 4AE
G4 WLJ N. Bell 16 Amersham Close, Urmston, Manchester M41 7WH
G4 WLK M. Morgan 6 Blakeley Heath Drive, Wombourne, Wolverhampton WV5 0HW
G4 WLP S. Mccombe 14 Broadlands Avenue, Bournemouth BH6 4HQ
G4 WLS C. Smith 15 Bearsdown Close, Plymouth PL6 5TX
GW4 WLT J. Williams 7 Tynewydd, Nantybwch, Tredegar NP22 3SG
G4 WLV D. Gladwin Dorset House, St. Annes Road, Eastbourne BN21 2HR
G4 WMA Dr P. Haslam 55 Jesmond Park West, Newcastle upon Tyne NE7 7BX
G4 WMB W. Bell Rhydd Gardens, Worcester Road, Hanley Castle, Worcester WR8 0AB
GW4 WMD W. David Sirmione, Lawrenny Road, Kilgetty SA68 0SY
GI4 WME F. Hull 44 Killynether Walk, Belfast BT8 7DB
G4 WMF G. Blake Flat 5, 46 Marlborough Road, Ipswich IP4 5AX
GU4 WMG J. Gallienne Westward, Rue Des Marettes, St. Martin, Guernsey GY4 6JW
G4 WMH W. Hall 45 Dorchester Road, Solihull B91 1LN
G4 WMN J. Robb 3 Silver Dell, Watford WD24 5LT
G4 WMO P. Stainton Fairview, Mareham On The Hill, Horncastle LN9 6PQ
G4 WMP M. Bangle 21 Oakhill Road, Addlestone KT15 1DH
G4 WMQ A. Richardson 1 Silverton Terrace, Rothbury, Morpeth NE65 7QS
G4 WMV R. Bridge 93 Doxey Fields, Stafford ST16 1HH
G4 WMY G. Kay High Trees, Stockland Bristol, Bridgwater TA5 2PZ
G4 WMZ K. Law 50 Main Street, Little Downham CB6 2ST
G4 WNA H. Williams 37 Mickledales Drive, Marske-By-The-Sea, Redcar TS11 6DF
G4 WND R. Banks Crows Nest, Churchstoke, Montgomery SY15 6TP
G4 WNF F. Rhodes 248 Woolwich Road, London SE2 0DW
G4 WNG T. Furness 129 North Ridge, Bedlington NE22 6DF
GI4 WNH E. Loughran 6 Oaklea Road, Magherafelt BT45 6NH
G4 WNI J. Howarth 80 John F Kennedy Estate, Washington NE38 7AL
G4 WNP R. Tant 34 Manor Road, Wheathampstead, St. Albans AL4 8JD
GM4 WNQ P. Ramsey 1 Skye Place, Stevenston KA20 3DG
G4 WNU J. Smith George Bungalow, The Street, Axminster EX13 7RW
G4 WNV Dr S. Robinson 18 Headley Grove, Tadworth KT20 5JF
G4 WNW T. Almond Maranatha, Lumber Lane, Warrington WA5 4AX
G4 WNZ M. Watson 5A Clarendon Road, Shanklin PO37 7AG
G4 WOB J. Bazyk Heyford Cedars, Watling Street, Northampton NN7 4SB
G4 WOD J. Sheppard 37 Oakfield Road. Kingswood., Bristol BS15 8NT
G4 WOE D. Hudson 2 Muirfield Rise, St. Leonards-on-Sea TN38 0XL
G4 WOH P. Thwaytes 1 Sunningdale, Waltham, Grimsby DN37 0UA
G4 WOI R. Allen 115 Trerice Drive, Newquay TR7 2TE
G4 WOL R. Tenwolde 376 Buxton Road, Macclesfield SK11 7ES
G4 WOS D. Flello 1 St. Andrews Way, Tilmanstone, Deal CT14 0JH
GD4 WOW J. Jones Ballagarrow, Glen Auldyn, Ramsey IM7 2AF Isle of Man
GW4 WPA T. Leary 21 Gelli Glas Road, Morriston, Swansea SA6 7PS
G4 WPB P. Bruce Seascape, Surf Crescent, Sheerness ME12 4JU
G4 WPE M. Bland 18 Hill Street, Newhall, Swadlincote DE11 0JR
G4 WPG L. Hatton 51 Castner Avenue, Weston Point, Runcorn WA7 4EH
GW4 WPH S. Valentine Unit 21, Industrial Estate, Bala LL23 7NL
G4 WPI J. Fuller 22 Simmance Way, Amesbury, Salisbury SP4 7TD
G4 WPO D. Bevan 32 Thorley Park Road, Bishop's Stortford CM23 3NQ
G4 WPR D. Trotman 71 Bexley Street, Windsor SL4 5BX
G4 WPT D. Jackman Chapel Cottage, Naunton WR8 0PZ
G4 WQB K. Hamlyn 11 Bye Green, Weston Turville, Aylesbury HP22 5RU
GW4 WQC D. Williams 149 Rhiwr Ddar, Taffs Well, Cardiff CF4 7PD
G4 WQD J. Jocys 28 Vaudrey Drive, Timperley, Altrincham WA15 6HQ
GM4 WQH J. Naughton 126 Churchill Street, Alloa FK10 2JU
G4 WQL M. Bender Ivy Chimney Villa, Skinners Bottom, Redruth TR16 5DT
G4 WQO P. Truitt 2A Queens Gate Place, London SW7 5NS
G4 WQS N. Reading 30 Clifton Rise, Windsor SL4 5TD
G4 WQT R. Watson 158 Kingsfold Drive, Penwortham, Preston PR1 9EQ
G4 WQU P. Barrett 9 Mabena Close, St. Mabyn, Bodmin PL30 3BS
G4 WQZ J. Wiles 12A Ashling Gardens Denmead, Waterlooville PO7 6PR
G4 WRA Cambridge District ARC c/o S. Sands 33 High Street, Kinver, Stourbridge DY7 6HF
G4 WRB K. Beech 40 Star Street, Wolverhampton WV3 9BL
G4 WRC Carlisle & Dis. ARS c/o A. Wheeler 11 Barley Way, Rothley, Leicester LE7 7RL
G4 WRD N. Underwood 44 East View, Barnet EN5 5TN
GI4 WRJ R. Jennings 12 Garnerville Gardens, Belfast BT4 2PA
G4 WRK I. Edwards 9 Long Lane, Nr Wellington, Salop TF6 6HH
GU4 WRP D. Fletcher Celicia, 5 La Neuve Rue Estate, La Neuve Rue, , St. Peter Port GY1 1SF Guernsey
G4 WRQ D. Wring 8A Rectory Road, Easton-In-Gordano, Bristol BS20 0QB
GJ4 WRR F. Leighton 4 Victoria Village, Estate Trinity JE4 9VI Jersey
G4 WRX D. Cherrington 4 Bloomfield Close, Wombourne, Wolverhampton WV5 8HQ
G4 WSB A. Bowditch 28 Selby Crescent, Freshbrook, Swindon SN5 8PE
G4 WSE T. Saggerson 18 Ploughmans Way, Great Sutton, Ellesmere Port CH66 2YJ
G4 WSF R. Smith 37 Lyngford Road, Taunton TA2 7EF
G4 WSH F. Blaxland Seton Villa, Scorrier Road, Scorrier TR16 5AA
G4 WSI A. Brown 81 Ipswich Crescent, Great Barr, Birmingham B42 1LY
G4 WSL R. Cable 4A Ermine Close, St. Albans AL3 4JZ
G4 WSV I. Swancott 4 Ripple Close, Shrewsbury SY2 6LS
G4 WSW M. Heaver 24 Artillery Terrace, Guildford, Surrey GU1 4NL
G4 WTA M. Jones 57 Mountway Road, Bishops Hull, Taunton TA1 5DS
G4 WTD C. Flatman 36 Skoner Road, Bowthorpe Industrial Estate, Norwich NR5 9AX
G4 WTE M. Rye 33A Darnley Street, Gravesend DA11 0PH
GM4 WTK R. Fortune Stewarton Lodge, Eddleston, Peebles EH45 8PP
GU4 WTN A. Hamon 22 Mount Row, St. Peter Port GY1 1NT Guernsey
G4 WTQ N. Harvey 5 Harvey Gardens, Loughton IG10 2AD
G4 WTR A. Morris 18 Saxton Avenue, Doncaster DN4 7AX
GM4 WTS W. Stevenson 11 West Drive, Airdrie ML6 8BL
GI4 WTT T. Mcdonnell 52 Moira Road, Glenavy, Crumlin BT29 4JL
G4 WTU D. Pay Longmeadow House, Dunsford, Exeter EX6 7AD
G4 WTX V. Hansford Whitehouse Farm, Hardway, Bruton BA10 0RJ
G4 WTZ J. Hall 23 St. James Avenue, Congleton CW12 4DY
G4 WUA G. Brown 13 Francis Avenue, Moreton, Wirral CH46 6DH
G4 WUB D. Farr 10 Yeomanside Close, Whitchurch, Bristol BS14 0PZ
G4 WUG K. Medley 3 Beck Lane, Horsham St. Faith, Norwich NR10 3LD
G4 WUH I. Hopkins 1 Beauchamp Villas Kempley Green, Dymock, Newent GL18 2BW
G4 WUI J. Marr 11 Morley Crescent, Kelloe, Durham DH6 4NN
G4 WUJ N. Plant 73 Robert Burns Avenue, Cheltenham GL51 6NX
G4 WUK D. Dyer 64 Churchill Close, Sturminster Marshall, Wimborne BH21 4BH
G4 WUM F. Amos 53 Valley View, Jarrow NE32 5QT
G4 WUO P. Bullock 4 Yarmouth Road, Blofield, Norwich NR13 4JS
G4 WUQ P. Harding Flat D, 106 Bushey Hill Road, London SE5 8QQ
GM4 WUR C. Phillips Lonnie Sanday, Orkney KW17 2BA
G4 WUS W. Bingham 16 Carlin Park, Carlin How, Saltburn-by-the-Sea TS13 4DF
G4 WUU P. Williamson The Laurels, Norwich Road, Cawston NR10 4HA
G4 WUV C. Baker 48 Hazell Road, Farnham GU9 7BP
G4 WUW M. Baker 48 Hazell Road, Farnham GU9 7BP
G4 WUX Stafford Portable Operating Group c/o P. Bourne 6 Blythe Mount Park, Blythe Bridge, Stoke-on-Trent ST11 9PP
GW4 WVB J. Williams Arfryn, Windsor Road, Wrexham LL14 1ST
G4 WVC M. Jones 24 Whitford Road, Birkenhead CH42 7JA
G4 WVD M. Bundy 5 Dawe Crescent, Bodmin PL31 1PY
G4 WVF C. Farley 8 Church Road, Mellor, Stockport SK6 5PR
G4 WVH T. Hathaway 24 Oxford Meadow, Sible Hedingham, Halstead CO9 3QN
GW4 WVK D. Davies 20 Broadlands Way, Oswestry SY11 2YD
G4 WVM M. Keating 29 Pimm Road, Paignton TQ3 3XA
GI4 WVN Prof. H. Gilbody 5 The Plateau, Piney Hills, Belfast BT9 5QP
GW4 WVO Church Island ARG c/o P. Owen 13 Highland Close, Sarn, Bridgend CF32 9SB
G4 WVP F. Russell 56 Gatesgarth Road, Middleton, Manchester M24 4JJ
G4 WVQ T. Friesner 21 Knott Gardens Fishbourne, Chichester PO19 3FE
G4 WVR Sutton Coldfield RS c/o S. Day 14 The Crescent, Market Harborough LE16 7JJ
G4 WVT J. Stageman Sunray, Kennford, Exeter EX6 7XS
G4 WVW J. Lane 41 Ravenswood Crescent, Harrow HA2 9JL
G4 WVY J. Joynt Gurtymadden, , Loughrea COUNTY GALWAY Ireland
G4 WWB W. Bennett 61L Mansion Drive, Croxteth, Liverpool L11 9DP
GI4 WWF V. Fails 38 Fortsandel Avenue, Coleraine BT52 1TL
G4 WWG Dr A. Brown 44 Earlswood, Skelmersdale WN8 6AT
G4 WWH Dr P. Pavelin 7A Castletown, Portland DT5 1BD
G4 WWL I. Rowe 19 Poplar Avenue, Wetherby LS22 7RA
GW4 WWN M. Rowles 7 Gelli Deg, Bryncoch, Neath SA10 7PL
G4 WWP D. Barry Plough End, 8 Dell Lane, Bishop's Stortford CM22 7SJ
G4 WWR Three Counties ARC c/o D. Kamm Delabole Head, Week St. Mary, Holsworthy EX22 6UU
GM4 WWU R. Steel 4 Howford Road, Nairn IV12 5QP
G4 WWY P. Brown White Cottage, Woodside, Epping CM16 6LF
G4 WWZ A. Artt 5 Barnard Field, Amesbury, Salisbury SP4 7FF
G4 WXC S. Vaughan 8 Letcombe Avenue, Abingdon OX14 1EQ
G4 WXF R. Logan 52 Holbeach Drive Kingsway, Quedgeley, Gloucester GL2 2BF
G4 WXI F. Mckeown 1 Thirlmere Road, Preston PR1 5TR
G4 WXJ R. Harnett 41 Stepney Road, Scarborough YO12 5BT
G4 WXK G. Sears 36 Cedars Road, Exhall, Coventry CV7 9NJ
G4 WXO J. Pemberton Dunkirk Cottage, Dunkirk Lane, Chester CH1 6LU
GM4 WXQ W. Goudie 5 North Lochside, Lerwick, Shetland ZE1 0PA
G4 WXR B. Hayes 363 Watnall Road, Hucknall, Nottingham NG15 6EP
G4 WXT G. Shead 37 Shalford Road, Rayne, Braintree CM77 6BY
G4 WXX J. Charnock 20 Clifton Road, Ashton-In-Makerfield, Wigan WN4 0AZ
G4 WYC S. Powell 10 Foresters Square, Bracknell RG12 9ES
GI4 WYE P. Doran 143 Gransha Road, Bangor BT19 7RB
G4 WYF C. Ellison 31 Dudley Avenue, Blackpool FY2 0TU
G4 WYH A. Mcphail 300 Fletcher Road, Preston PR1 5HJ
G4 WYI A. Huff 4 Greding Walk, Brentwood Essex CM13 2UF
G4 WYL C. Levett 5 Park Road, Yapton, Arundel BN18 0JE
G4 WYN D. Harries 1 St. Michaels Close, Ashby-de-la-Zouch LE65 1ES
G4 WYO K. Brewer 14 Poplar Road, Kensworth, Dunstable LU6 3RS
GW4 WYX W. Thomas 24 Pontneathvaughan Road, Glynneath, Neath SA11 5NT
G4 WYZ M. Prescott 44 Glamis Drive, Chorley PR7 1LX
G4 WZA A. Nokes 24 Braces Lane, Marlbrook, Bromsgrove B60 1DY
G4 WZB H. Worley 22 Cross Road, Wellingborough NN8 4AT
GM4 WZD J. Nicholl Trees, 7 Holmisdale, Glendale, Isle of Skye IV55 8WS
GM4 WZG B. Mcintosh 14 River View, Dalgety Bay, Dunfermline KY11 9YE
G4 WZH A. Le Couteur Bisson 36 Gibson Way, Porthleven, Helston TR13 9AW
G4 WZJ M. Wiblin 60 Shepherds Lane, Bracknell RG42 2BT
G4 WZM S. Johnston Burn Moor End Farm, Wheathead Lane, Nelson BB9 6LD
GM4 WZP J. Gentles 8 Leadervale Terrace, Edinburgh EH16 6NX
G4 WZQ I. Smith 24 Sea View Road, Herne Bay CT6 6JA
GW4 WZS A. Glynn Cartref, Llanfachraeth, Holyhead LL65 4UY
G4 WZT N. Thomas Flat 18, Livability, Hereford HR4 9HP
G4 WZU L. Thompson 12 Long St., Great Gonerby, Grantham NG31 8LN
GM4 WZY B. Pennycook 3 Pirnie Mill, Forfar DD8 3ES
G4 WZZ B. Hunt 5 Osprey Close, Whitstable CT5 4DT
GI4 XAA C. Mccann Drumbally Hue House, Rock, Tyrone BT70 3JY
G4 XAB R. Hunt 3 Osprey Close, Whitstable CT5 4DT
G4 XAE A. Cole 13 Centre Close, Beccles NR34 9JJ
G4 XAG B. Mahany 3 Portland Road, Frome BA11 4JA
G4 XAH D. Mahany 3 Portland Road, Frome BA11 4JA
G4 XAL P. Lawrence 4 Monkshood Close, Wokingham RG40 5YE

G4 XAN C. Goddard 75 Downs Road, Slough SL3 7DA
G4 XAR M. Kearns 16 Fieldton Road, Liverpool L11 9AG
G4 XAT R. Evans 7 Westland Drive, Bromley BR2 7HE
GW4 XAU J. Rutkowski 7 Beach Road, Holyhead LL65 1ES
GM4 XAV J. Stevens 10 Tiel Path, Glenrothes KY7 5AX
GM4 XAW P. Nelson Croc Ard, Botany Street, Newton Stewart DG8 9JG
GW4 XAZ I. Mitchell 18-19 Hendre-Wen Road, Blaencwm, Treorchy CF42 5DR
G4 XBC A. Turner 19 Trelawney Road, St. Austell PL25 4JA
G4 XBD G. Nash 36 Lynton Avenue, Arlesey SG15 6TS
G4 XBF M. Ray Willow Mead House, Willow Mead, Godalming GU8 5NR
G4 XBG K. Murphy 34 Hawkenbury Way, Lewes BN7 1LT
G4 XBI D. Parslow 1 Willington Close, Harlescott, Shrewsbury SY1 3RH
G4 XBJ P. Kemp 9 Moorfield Way, Wilberfoss, York YO41 5PL
G4 XBS C. Smith 1 Langley Court, St. Ives PE27 5WX
G4 XBU J. Atkinson 8 Woodcock Road, Flamborough, Bridlington YO15 1LJ
G4 XBW R. Head The White House, School Lane, St. Austell PL25 3TJ
G4 XBX A. Wallman 8 Oakbank Avenue, Manchester M9 4EX
G4 XBZ Dr A. Roberts Apartment 17, Fleur De Lis Duttons Road, Romsey SO51 8LH
G4 XCB K. Rook 232 Wick Road, Brislington, Bristol BS4 4HN
G4 XCE A. Tamplin Browtop, Old Lane, Crowborough TN6 2AD
G4 XCK S. Boden 14 Potters Way, Ilkeston DE7 5EX
G4 XCM J. Tavener The Cube, North Drive, Wirral L60 0BD
G4 XCQ L. Prescott 58 Blenheim Road, Ashton-In-Makerfield, Wigan WN4 9JN
G4 XCR G. Wood The Old Corner Smithey, New Road, Dereham NR20 5TA
G4 XCV R. Barnett 7 Chapel Terrace, Pendeen, Penzance TR19 7SY
G4 XCX C. Clarke 33 James Road, Kidderminster DY10 2TP
G4 XCY F. Mills 14 Seagram Close, Aintree, Liverpool L9 0NA
G4 XDB A. Parry 189 Kimbolton Road, Bedford MK41 8DR
G4 XDC M. Taylor 26 Summerfield Place, Wenlock Road, Shrewsbury SY2 6JX
G4 XDE S. Crosskey 25 Meadow Gardens, Baddesley Ensor, Atherstone CV9 2DA
G4 XDG D. Humphreys 19 Meadow Close, Northwich CW8 2LZ
G4 XDH D. Humphreys 15 Burton Road, Warrington WA2 9AJ
G4 XDJ York ARS c/o B. Fields 64 Collins Street, Waikouaiti 9510 New Zealand
G4 XDK N. Heasman 5 Hackneys Corner Great Blakenham, Ipswich IP6 0JQ
G4 XDL M. Norman 52 Turkdean Road, Cheltenham GL51 6AL
G4 XDM Dr S. Comis 178 Lordswood Road, Birmingham B17 8QH
G4 XDP P. Knowles 18 Brookside, Pill, Bristol BS20 0JX
GW4 XDR A. Walker Stanley Cottage, Station Road, Wrexham LL13 0LJ
G4 XDT L. Booth 40 St. Georges Road, New Mills, High Peak SK22 4JT
G4 XDU D. Chislett Hilltops, 2A St. Marks Road, Maidenhead SL6 6DA
G4 XDV R. Hall 9 Stone Court, South Hiendley, Barnsley S72 9DL
G4 XDW A. Chidwick 4 Burgess Close, Whitfield, Dover CT16 3NP
G4 XDX S. Garbett 2 Redruth Court, Launceston Road, Wigston LE18 2FU
GU4 XEA P. Carre La Petite Miellette, La Miellette Lane, Vale GY3 5EN Guernsey
G4 XED S. Cowdell 6 Pearl St., Bedminster, Bristol BS3 3EA
G4 XEE D. Bate 15 Martins Drive, Ferndown BH22 9SG
GW4 XEF B. Passmore 16 Epworth Road, Rhyl LL18 2NU
G4 XEI D. Mcloughlin 23 St. Marys Court, Clayton Le Moors, Clayton le Moors, Accrington BB55LA
G4 XEJ A. Allen 86 Grayswood Park Road, Quinton, Birmingham B32 1HE
G4 XEL S. Evans 72 Sandown Road, Toton, Nottingham NG9 6JW
G4 XEO D. Holmes 32 Eastcheap, Rayleigh SS6 9JZ
GW4 XES D. Johns 16 Maes Yr Haf, Llansamlet, Swansea SA7 9ST
G4 XET J. Rawlinson Hollydene, Newbiggin, Penrith CA10 1TA
G4 XEW I. Rosenberg 11 Parkside Drive, Edgware HA8 8JU
G4 XEX P. Rivers 34 Coales Gardens, Market Harborough LE167NY
G4 XEZ A. Smith 3 The Fold, Penn, Wolverhampton WV4 5QY
G4 XFA R. Ball 4 The Brow Hesketh Bank, Preston PR4 6SJ
G4 XFC J. Fulton 8 Park View Legsby, Market Rasen LN8 3QP
GI4 XFE A. Calvin 20 Orangefield Crescent, Armagh BT60 1DS
G4 XFF J. Holdsworth 37 Harewood Crescent, Old Tupton, Chesterfield S42 6HS
G4 XFG N. Goodman 3 Fury Avenue, Grimoldby, Louth LN11 8UN
G4 XFM D. Steer 24 Manor Drive, Ivybridge PL21 9BD
GI4 XFN G. Smith 4 Conway Court, Belfast BT13 2DR
GI4 XFS C. Gilbody 5 The Plateau, Piney Hills, Belfast BT9 5QP
G4 XFT J. Tranter 275 Bosty Lane, Aldridge, Walsall WS9 0QE
GM4 XFU W. Davidson 31 Glenmuir Crescent, Logan, Cumnock KA18 3EY
G4 XFV A. Mckechnie 2 Batts Pond Lane, Dropping Holms, Henfield BN5 9YU
GI4 XFX R. Reid 6 Sperrin Heights, Townhill Road, Ballymena BT44 8AD
GI4 XFY E. Townley 27 Windmill Road, Kilkeel, Newry BT34 4LP
G4 XFZ R. Griffin 53 St. Johns Avenue, Warley, Brentwood CM14 5DG
G4 XGD P. Harman 25 Pitts Road, Slough SL1 3XG
G4 XGI R. Hales 239 Charlton Road, Shepperton TW17 0SH
G4 XGN P. Riggott 1 Mill Lane Queensbury, Bradford BD13 1LP
GI4 XGO G. Armstrong 45 Rathmena Drive, Ballyclare BT39 9HZ
G4 XGP M. Kelly Birkby Lodge, Brickley Park Road, Kent BR1 2AT
GI4 XGQ T. Devine 141 Longland Road, Dunamanagh, Strabane BT82 0PP
G4 XGR S. Clark 1 Holcroft, Orton Malborne, Peterborough PE2 5SL
G4 XGT J. Dawson 21 Church Street, Needingworth, St. Ives PE27 4TB
GM4 XGY G. Smith 1/2 1 Seres Court, Clarkston, Glasgow G76 7PL
G4 XHC F. Jackson 34 High Street, Blyton, Gainsborough DN21 3JY
G4 XHE R. Cook 7 New Road, Worthing BN13 3JG
GM4 XHH R. Franklin Flat 2/3 Dovecot Court, Peebles EH45 8FG
G4 XHK L. Soutter 2 Hyde Barton, Churchill Way, Northam, Bideford EX39 1NX
GI4 XHO F. Orr 29A Mccraes Brae, Whitehead, Carrickfergus BT38 9NZ
G4 XHP D. Daniels 40 Rennie Street, Dean Bank, Ferryhill DL17 8NG
GM4 XHQ G. Mcinnes 14 East Croft, Ratho, Newbridge EH28 8PD
G4 XHT E. Wilkinson 30 Old Bridge End, Roach Street, Blackford Bridge, Bury BL9 9TA
GM4 XHV G. Horsburgh 3 Dumyat Road, Alva FK12 5NN
G4 XHX M. Powers 12 Roman Avenue North, Stamford Bridge, York YO41 1DP
G4 XHZ F. Jolley 30 Oban Drive, Shadsworth, Blackburn BB1 2HY
G4 XIE R. Shard 76 Clipsley Lane, Haydock, St. Helens WA11 0UB
G4 XIL B. Hurst 25 Hoadly Road, Cambridge CB3 0HX
G4 XIM R. Bradfield 118 East Road, Langford, Biggleswade SG18 9QP
G4 XIN A. Henstock 16 The Coppice, Enfield EN2 7BY
G4 XIP J. Baker 11 London Road, Old Basing, Basingstoke RG24 7JE
GI4 XIR W. Bird 198 Ashmount Gardens, Lisburn BT27 5DB
GU4 XIT R. Bird Redroof, La Mare De Carteret, Guernsey GY5 7XD
G4 XIU M. Kelleway Greenhills, Newport Road, Ventnor PO38 2QW
G4 XIW S. Mackenzie 6 Bridge Farm Close, Grove, Wantage OX12 7QF
G4 XIX M. Purnell The Olde Cottage, Lewdown, Okehampton EX20 4DQ
G4 XIZ R. Heath 9 Woodside Lane, Leek ST13 7AN
GI4 XJD J. Doherty 75 Drumflugh Road, Benburb, Dungannon BT71 7QF
G4 XJE D. Brawn 16 Mansel Close Cosgrove, Milton Keynes MK19 7JQ
GM4 XJF J. Park The Stables, Whiting Bay, Isle of Arran KA27 8QH
G4 XJG R. Hirst 47A Rowley Lane, Fenay Bridge, Huddersfield HD8 0JG
GW4 XJK R. King 30 Railway Terrace, Llanelli SA15 2RH
G4 XJL D. Rogers Gabwell House, Stokeinteignhead, Newton Abbot TQ12 4QS
G4 XJN J. Williamson 12 Honeysuckle Road Widmer End, High Wycombe HP15 6BW
G4 XJS J. Smith 84 Oakwood Drive, St. Albans AL4 0XA
GM4 XJY D. Mcminn Crestholme, East Bay, Mallaig PH41 4QF
G4 XKA P. Adams 19 Thistledown, Tilehurst, Reading RG31 5WE
G4 XKC A. Sieroslawski 8 Poot Hall, Dewhirst Road, Rochdale OL12 0AS
G4 XKD K. Dixon 23 Dorking Walk, Corby NN18 9JL
G4 XKF D. Browne 67 Benfield Way, Portslade, Brighton BN41 2DN
G4 XKH S. Foster 4 The Mews, Rear Of 5 Warren Road, Torquay TQ2 5TQ
GI4 XKI J. O'Neill 225 Dungannon Road, Killeshill, Dungannon BT70 1TH
G4 XKK K. Burston Can Singala, Hope Corner Lane, Taunton TA2 7PB
G4 XKL R. Whetton 117 Tutbury Road, Burton-on-Trent DE13 0NU
G4 XKM P. Andrews 88 Connegar Leys, Blisworth, Northampton NN7 3DF
GM4 XKP K. Macgillivray 87 Castle St., Forfar DD8 3AG
G4 XKR R. Coward 10 Market Street, Hambleton, Poulton-le-Fylde FY6 9AP
G4 XKV G. Pigott 67 Mayplace Road West, Bexleyheath DA7 4JL
G4 XKZ M. Collins 39 Denver Road, Dartford DA1 3JU
G4 XLA T. Carruthers Flat 43, House 119, St. Petersberg 193024 Russian Federation
GI4 XLB Sunspots Rac c/o G. Curry 87 Burren Road, Ballynahinch BT24 8LF
G4 XLC E. Metcalfe 18 Kirkstone Drive, Morecambe LA4 5XP
G4 XLG M. Rollings 39 Summerleys, Edlesborough, Dunstable LU6 2HR
G4 XLM J. Todd Dorothy House, 127 Dorothy Avenue North, Peacehaven BN10 8DS
GM4 XLN J. Durrand 9 Breadalbane Crescent, Wick KW1 5AS
G4 XLO K. Tatlow 24 Princes Road West, Torquay TQ1 1PB
GM4 XLU E. Wallace 10 Gean Court, Cumbernauld, Glasgow G67 3LU
G4 XLY E. Grint 15 Ivythorn Road, Street BA16 0TE
G4 XMA B. Easton 8 Church View Road, Camborne TR14 8RQ
GM4 XMD W. Mcdicken 4 Baillie Drive, Logan, Cumnock KA18 3HS
G4 XME L. Shone 3 Ascot Drive, Dudley DY1 2SN
G4 XMJ G. Wiggins Cherry Trees Thorney Road, Emsworth PO10 8BN
G4 XML P. Glydon 24 Imperial Road, Knowle, Bristol BS14 9ED
G4 XMO S. Ashfield 28 Long Grove, Baughurst, Tadley RG26 5NY
G4 XMP C. Balderston 22A Maplecresent, Basingstoke RG215SS
G4 XMQ T. Cooling 17 Hawthorn Avenue Cherry Willingham, Lincoln LN3 4JS
GW4 XMR M. Richardson 4 Waterloo Road, Ammanford SA18 3SF
G4 XMS C. Munton 86 Amsbury Road, Hunton, Maidstone ME15 0QH
GW4 XMU D. Jones 30 Sorrell Drive, Penpedairheol, Hengoed CF82 8LA
G4 XMX L. Johnson 6 Hurst Court, Bunbury, Tarporley CW6 9QX
G4 XMY J. Colson 718 East Buckingham Drive, Lecanto 34461 United States
G4 XMZ P. Toms Longfleet, Shipton Lane, Bridport DT6 4NQ
G4 XNA A. Willis 16 Redwood Close, Lymington SO41 9LT
GM4 XND W. Clark 173 Dunnikier Road, Kirkcaldy KY2 5AD
G4 XNE F. Handy 429 Penn Road, Penn, Wolverhampton WV4 5LN
G4 XNF J. Cameron 23 Farley Crescent, Oakworth, Keighley BD22 7SH
G4 XNK H. Johnson 7, Deer Close, Walsall WS3 3EA
G4 XNO M. Goodearl Glenhurst, Wood Lane, Dartmouth TQ6 0DP
G4 XNP D. Rayner 69 Saracen Road, Hellesdon, Norwich NR6 6PB
GM4 XNQ D. Muir 28 Grange Crescent, Edinburgh EH9 2EH
G4 XNR P. Morris The Overlands, Church Minshull, Nantwich CW5 6DX
G4 XNS N. Speak Le Bois Trainard, Lizant 86400 France
G4 XNV D. Owen 39 Smithford Walk Tarbock Green, Prescot L35 1SF
G4 XNW J. Simmonds 19 Red Admiral Apartments, Worcester Street, Stourbridge DY8 1AJ
GD4 XOD W. Jones Ballanard Road, Onchan, Douglas IM4 5EA
G4 XOE M. Swaby 16 Daimler Avenue, Herne Bay CT6 8AE
G4 XOG C. Wood 9 Tamar Close, Walsall WS8 7LH
G4 XOH D. Blackwell 10 High Oaks Gardens, Bournemouth BH11 9LJ
G4 XOJ N. Wade 6 Aisthorpe, Capel St. Mary, Ipswich IP9 2HT
G4 XOL M. Osborne 27 Silverdale Road, Newton-le-Willows WA12 0JT
G4 XOM R. Egan 56 Walker Avenue, Stourbridge DY9 9EL
G4 XOP T. Cooper 55 Meadway, St. Austell PL25 4HT
G4 XOW D. Lomas Galmpton, Cannon Lane, Maidenhead SL6 3NR
G4 XPD K. Nicholls 2 Sherrier Way, Lutterworth LE17 4NW
G4 XPI P. O'Dea 5 Matthews Court, Blackpool FY4 2BT
G4 XPJ A. Gridley 13 Brockwell, Oakley, Bedford MK43 7TD
G4 XPP J. Davies-Bolton 39 Newholme Estate, Station Town, Wingate TS28 5EJ
G4 XPT A. Fernandez 2 Silverston Ave, Bognor Regis PO21 2RB
G4 XPU M. Bennett 7 Woburn Avenue, Firwood Ind Est, Bolton BL2 3AY
G4 XPV P. Maisey 155 Parkfield Drive, Birmingham B36 9TY
G4 XPY D. Fuller 51 Evenlode, Banbury OX16 1PQ
G4 XQA K. James 6 Holly Grove, Paddington, Warrington WA1 3HB
G4 XQB T. Rowe 1 Mere Road, Marston, Northwich CW9 6DR
G4 XQD D. Cast 100 Priory Court, Priory Park, Ipswich IP10 0JX
G4 XQE A. Turner 4 Taylor Road, Ashtead KT21 2HY
G4 XQF N. Clacher 3 Annan Crescent, Marton, Blackpool FY4 4RQ
G4 XQG K. Icke 85 Evendene Road, Evesham WR11 2QA
GM4 XQJ Waterlooville ARC c/o B. Waddell 3A Polmont Road, Laurieston, Falkirk FK2 9QQ
G4 XQQ J. Freeman 5A Beech Avenue, Briar Bank Park, Bedford MK45 3WE
G4 XQV T. Sismey Southland House, 1B West End Lane, Leeds LS18 5JP
G4 XQX D. Oliver 6 Kensington Road, Gosport PO12 1QY
G4 XQY A. Clack 4 Chestnut Grove, Withernsea HU19 2PH
G4 XQZ J. Fisher 6 Castle Way, Havant PO9 2RZ
G4 XRA the 5 Mhz Pioneers Club c/o R. Avery 64 Burnmill Road, Market Harborough LE16 7JF
G4 XRB J. Gagg 20 Stanstead Avenue, Tollerton, Nottingham NG12 4EA
G4 XRD G. Pope 16 Catchpole Close, Corby NN18 8DE
G4 XRG E. Godlieb 4 Tytherington Park Road, Macclesfield SK10 2EL

G4	XRJ	J. Mills Aquila, 4 Westhill Road South, Winchester SO21 3HP
G4	XRK	J. Lord 16 Lark Valley Drive, Fornham St. Martin, Bury St. Edmunds IP28 6UG
G4	XRM	B. Foster 38 Oakfield Road, Street BA16 0RE
G4	XRO	S. Hill 32 Hunters Croft, Haxey, Doncaster DN9 2NX
GM4	XRP	J. Porter 1 Loney Crescent, Denny FK6 5EG
G4	XRR	M. Willgoss 28 Southdown Avenue, Weymouth DT3 6HS
GM4	XRT	M. Taylor The Old Croft, Forse, Lybster KW3 6BX
G4	XRV	R. Bullock Putnams, Hawridge, Chesham HP5 2UQ
GW4	XRW	R. Wood Bwlcyn, Eifl Road, Caernarfon LL54 5HG
G4	XRX	R. Headland 18 Blucher Street, Liverpool L22 8QB
GM4	XRY	A. Rimmer 16 Johnston Drive, Barassie, Troon KA10 6SD
G4	XSA	A. Boniface 33 Caraway Place, Wallington SM6 7AG
G4	XSB	W. Bridgen 11 Turnesc Grove, Thurnscoe, Rotherham S63 0TY
G4	XSC	G. Trim 731 Dorchester Road, Weymouth DT3 5LF
GI4	XSF	M. Stevenson 69 Portaferry Road, Cloughey, Newtownards BT22 1HP
G4	XSG	S. Stuart 102 Mitton Road, Whalley, Clitheroe BB7 9JN
G4	XSI	J. Saunders Top Hill Farm, Woodside Green, Maidstone ME17 2ET
G4	XSM	G. Davey 49 Maltward Avenue, Bury St. Edmunds IP33 3XQ
G4	XSR	Burton On Trent Wireless Club c/o M. Ohara 36 Balance Hill, Uttoxeter ST14 8BT
G4	XST	P. Cheeseman 10 Limden Close, Stonegate, Wadhurst TN5 7EG
GW4	XSX	M. Tovey Frondewi, Aberaeron SA46 0JS
G4	XTA	P. Godolphin Shepherds Cottage, Flakebridge, Appleby-in-Westmorland CA16 6JZ
GI4	XTC	W. Armstrong 8 Killowen Crescent, Lisburn BT28 3DS
G4	XTE	J. Johnson Winterwood, West Lodge Crescent, Huddersfield HD2 2EH
G4	XTF	N. Hancocks Wesley House Allensmore, Hereford HR2 9BE
G4	XTG	I. Brown 453 Blackburn Road, Edgworth Turton, Bolton BL7 0PW
G4	XTK	A. Kurnatowski 24 Eversleigh Rise, Darley Bridge, Matlock DE4 2JW
G4	XTO	W. Reade 106 Wellington Road, Bollington, Macclesfield SK10 5HT
G4	XTR	N. Hearn Horsebrook Farm, South Brent TQ10 9EU
G4	XTS	J. Strutt Woodland, Gardiners Lane North, Billericay CM11 2XE
GD4	XTT	W. Brown Cleckheaton, Ballaragh, Laxey IM4 7PW Isle of Man
G4	XTU	J. Jones 3 Blackstope Lane, Retford DN22 6NW
G4	XTW	A. Bowes 3 Cameron Mews, Mill St., Bury St. Edmunds IP28 7DP
G4	XTX	C. Cooper 31 Beacon Park Drive, Skegness PE25 1HE
G4	XTZ	A. Taylor 36 Bodmin Avenue, Slough SL2 1SL
G4	XUA	R. Walton 275 Ridgacre Road, Quinton, Birmingham B32 1EG
GW4	XUE	D. Thomas 5 Parcydelyn, Carmarthen SA31 1TS
GW4	XUG	G. Patterson 10 Clettwr Terrace Pontsian, Llandysul SA44 4TU
G4	XUI	J. Gordon 54 Guibal Road, London SE12 9LX
GM4	XUJ	K. Traill 57 Ashfield Drive, Dumfries DG2 9BP
GW4	XUM	M. Platt Cobweb Cottage Pen Y Cefn Road Caerwys, Mold CH7 5BH
G4	XUQ	S. Winters 16 Rushton Grove, Harlow CM17 9PR
GM4	XUS	G. Smith 80 Deanburn Park, Linlithgow EH49 6HA
G4	XUV	D. Bevan 46 Park Lane, Hartford, Northwich CW8 1PY
G4	XUW	D. Hudson 54 Montfitchet Walk, Stevenage SG2 7DT
G4	XUZ	R. Chandler 29 Ravensbourne Avenue, Shoreham-by-Sea BN43 6AA
G4	XVE	J. Francis Pintail Cottage, St. Helena Westleton, Saxmundham IP17 3ED
G4	XVF	A. Henk 3 Well Road, Tweedmouth, Berwick-upon-Tweed TD15 2BB
G4	XVH	D. Van Haaren 7 Middle Boy, Abridge, Romford RM4 1DT
G4	XVI	J. Ames 58 The Street, Ringland, Norwich NR8 6AB
G4	XVM	M. Brett 25A First Avenue, Galley Hill, Waltham Abbey EN9 2AL
G4	XVO	Dr S. Bate 4 Jacksons Meadow, Bidford-On-Avon, Alcester B50 4HQ
G4	XVP	P. Hart 4 Kings Ride, Penn, High Wycombe HP10 8BL
G4	XVS	K. Hughes Annies Cottage, Gravel Walk, Malpas SY14 8JQ
G4	XVV	E. Davies 11 Herons Close, Fareham PO14 2HA
G4	XVW	I. Dobson Pine View, Forest Dale Road, Marlborough SN8 2AS
G4	XVY	D. Bastin 94 Clyfton Close, Broxbourne EN10 6NY
G4	XWA	M. Cohen 7 Northdale Park, Swanland, North Ferriby HU14 3RH
GW4	XWC	W. Crooks 52 St. Catherines Road, Baglan, Port Talbot SA12 8AS
G4	XWD	J. Cookson Arcadia. 11 St Winefrides Road, Littlehampton BN17 5HA
G4	XWE	L. Perrett 1 Churchill Close, Wells BA5 3HY
GD4	XWF	J. Harrison 33 Tynwald Close, St. Johns, Douglas IM4 3LZ Isle of Man
GM4	XWL	S. Gaw 10 Scotstoun Park, South Queensferry EH30 9PQ
G4	XWM	F. Walton Tollgate, 36 Cranfield Road, Wavendon, Milton Keynes MK17 8AW
GW4	XWN	H. Walker 46 Golden Grove, Rhyl LL18 2RS
G4	XWP	C. Boyce 41 Furlong Close, Buckfast, Buckfastleigh TQ11 0ER
G4	XWQ	D. Cottle The Brambles, Landkey Road, Barnstaple EX32 9BW
G4	XWR	P. Grainger 26 Beattie St., South Shields NE34 0NJ
GM4	XWS	D. Munro Eriskay, 4 Boswell Crescent, Inverness IV2 3ET
G4	XWT	F. Donachie 57 Avon Road West, Christchurch BH23 2DF
G4	XWW	G. Winyard 76 West Elloe Avenue, Spalding PE11 2BJ
G4	XWZ	D. Lerner 6 Willow Road, Kings Stanley, Stonehouse GL10 3HS
G4	XXA	F. Mills 66 Beeches Road, Charlton Kings, Cheltenham GL53 8NQ
G4	XXB	E. Mills 66 Beeches Road, Charlton Kings, Cheltenham GL53 8NQ
G4	XXD	S. White 15 Spurway Road, Canal Hill, Tiverton EX16 4ER
GW4	XXF	B. Morris 62 Gerllan, Tywyn LL36 9DE
G4	XXG	Stockton & District ARG c/o S. Bourne 1 Humewood Grove, Stockton-on-Tees TS20 1JU
G4	XXH	R. Miles Lone Oak, Clappers Farm Road, Silchester, Reading RG7 2LH
G4	XXI	G. Lee 5 Morton, Tadworth KT20 5UA
GW4	XXJ	N. Jones Hillesley, Montpellier Park, Llandrindod Wells LD1 5LW
G4	XXK	D. Hart 52 Scalwell Lane, Seaton EX12 2DJ
G4	XXM	D. Frederick 16 Phoenix Drive, Eastbourne BN23 5PG
GM4	XXO	I. Carbry 24 Craigenhill Road, Kilncadzow, Carluke ML8 4QT
GW4	XXP	J. Bancroft 101 Meliden Road, Prestatyn LL19 8LU
G4	XXS	G. Cooper 44 Nursery Close, Hucknall, Nottingham NG15 6DQ
G4	XXT	J. Cassidy 137 Heath Park Road, Gidea Park, Romford RM2 5XJ
G4	XXW	J. Groeger Waldweg 11, Schneverdingen D-29640 Germany
G4	XXZ	D. Palfreman 43 Southfield Close, Scraptoft, Leicester LE7 9UR
G4	XYB	M. Kingdon Watersmeet Cottage, Brewers Lane, Calne SN11 8EZ
G4	XYC	A. Reynolds 90 Windfield, Leatherhead KT22 8UJ
G4	XYD	R. Young 5 Edge Hill, Chellaston, Derby DE73 6RP
G4	XYG	S. Randall 129 Ryeland Way, Augusta Park, Andover SP11 6RH
G4	XYH	W. Stock The Cottage, Hollow Road, Winscombe BS25 1TG
GW4	XYI	P. Coombs 28 Cae Braenar, Holyhead LL65 2PN
G4	XYK	P. Mitchell 19 Ashbourne Avenue, Whetstone, London N20 0AL
G4	XYM	J. Hoskins 37 Green Close, Didcot OX11 8TE
G4	XYN	R. Savin 7 Bannard Road, Maidenhead SL6 4NG
G4	XYP	R. Jobes 11 Eglinton Street North, Sunderland SR5 1DY
G4	XYR	W. Clarkson 5A Sutton Avenue, Bradford BD2 1JP
G4	XYS	J. Mundy 19 Brickfield Grove, Halifax HX2 9AZ
G4	XYU	P. Davis 5 Well Close, New Milton BH25 6TA
G4	XYW	A. Pevy 2 Oaktree Way, Sandhurst GU47 8QS
G4	XYY	J. England 2 Clifford Road Bramham, Wetherby LS23 6RN
G4	XZA	E. Wardle 57 Brook View Drive, Keyworth, Nottingham NG12 5RA
G4	XZC	P. Gass 7 Chipperfield Close, Upminster RM14 3EA
G4	XZF	B. Irwin Rockside, Frog Lane, Braunton EX33 1BB
G4	XZG	P. Lawton 5 Belvedere Gardens, Leeds LS17 8BS
G4	XZI	G. Hall 22 Templenewsam View, Leeds LS15 0LW
GW4	XZJ	H. Jones Hafan, 7 Tan Y Bryn Street, Tywyn LL36 9UY
G4	XZK	S. Bond 12 Richmond Road, Farsley, Pudsey LS28 5DY
G4	XZM	K. Pickles 79 Mill Lane, Hanging Heaton, Batley WF17 6DZ
GM4	XZN	J. Macdonald 15 Muir Wood Drive, Currie EH14 5EZ
GW4	XZP	E. Wood Bwlcyn, Eifl Road, Caernarfon LL54 5HG
G4	XZS	R. Weston 2 Gill Park, Efford, Plymouth PL3 6LX
G4	YAB	J. Livesley 79 Mellor Road, New Mills, High Peak SK22 4DP
GM4	YAC	N. Howarth Kilbeg House Teangue, Isle of Skye IV44 8RQ
G4	YAF	A. Trudgen 14 Park An Pyth, Pendeen, Penzance TR19 7ET
G4	YAH	G. Warnes 90 High Haden Road, Cradley Heath B64 7PN
G4	YAJ	S. Woodhead 804 Huddersfield Road, Dewsbury WF13 3LZ
G4	YAK	C. Dobinson 37 Ladram Road, Thorpe Bay, Southend-on-Sea SS1 3PX
G4	YAL	S. Tuffin 21 Garraways, Royal Wootton Bassett, Swindon SN4 8NQ
G4	YAM	C. Atkin 23 Brewster Avenue, Immingham DN40 1DW
G4	YAN	R. Page 26 Colne Road, High Wycombe HP13 7XN
G4	YAP	G. South 9 Common Lane Royston, Barnsley S71 4JA
G4	YAQ	B. Setter Briarwood Alexandra Road, Crediton EX17 2DH
G4	YAS	E. Lucas-Davis 27 Cadbury Road, Sunbury-on-Thames TW16 7NA
GM4	YAT	T. Turner 5 Braemore Place, Fort William PH33 6HX
GM4	YAU	W. Scott Garden House, Fetternear, Inverurie AB51 5LY
G4	YAV	R. Dixon 91 Bondicar Terrace, Blyth NE24 2JR
G4	YAX	D. Diss 130 Beridge Road, Halstead CO9 1JU
G4	YAZ	H. Sheer Sea Echo, 53 Leonard Road, New Romney TN28 8RX
G4	YBA	G. Collis 13 Westbrook Close, Horsforth, Leeds LS18 5RQ
G4	YBD	G. Reed 6 Tentergate Close, Knaresborough HG5 9BJ
G4	YBG	A. White Northdale, Goughs Lane, Bracknell RG12 2RA
G4	YBH	B. Hawkins Andorra, Haw Lane, High Wycombe HP14 4JG
G4	YBI	P. Aust 28 The Green, Wennington, Rainham RM13 9DX
G4	YBJ	J. Davies 8 Randle Meadow Court, Great Sutton, Ellesmere Port CH66 2BL
GJ4	YBM	A. Alexandre Merryvale Cottage, La Vallee De St. Pierre, St. Lawrence JE3 1EZ Jersey
G4	YBN	I. Ansell 9 Sewell Harris Close, Harlow CM20 3HB
G4	YBP	P. Darcy Cherry Blossom Cottage, Hunts Lane, Netherseal DE12 8BJ
G4	YBS	Morecambe Bay ARS c/o D. Jackson 6 Bolton Lane, Carnforth LA5 8BL
G4	YBT	E. Tracey 100 Booth Close, Kingswinford DY6 8SP
GU4	YBW	P. Wadley Gironde, Lorier Lane, Vale, Guernsey GY3 5JG
G4	YBX	E. Scleparis 8 Devonshire Park, Reading RG2 7DX
G4	YCD	M. Lowe Crossley Farm, Bristol BS17 1RH
G4	YCE	L. Ball 16 Kelston View, Whiteway, Bath BA2 1NW
G4	YCG	C. Beeston 14 Valley Close, Waterlooville PO7 5DX
GW4	YCJ	Morecambe Bay ARS c/o A. Clift-Jones Coed Tew Mill, Nant Glas, Llandrindod Wells LD1 6PD
GW4	YCO	D. Gill 19 Rowling St., Williamstown, Tonypandy CF40 1QY
G4	YCP	G. Newman 2 Grange Road, Eldwick, Bingley BD16 3DH
G4	YCS	I. Carby Sunny Bank, North End Road Yapton, Ford, Arundel BN18 0DH
G4	YCV	C. Vickery 7 Higher Redgate, Tiverton EX16 6RJ
G4	YCW	C. Croxford Bodley Cottage, Parracombe, Barnstaple EX31 4PR
GI4	YCZ	J. Rainey 40 Cranny Lane, Portadown, Craigavon BT63 5SW
G4	YDB	F. Heald Brightling House, Alexandra Road, Heathfield TN21 8ED
GM4	YDC	S. Hunt 5 Highland Road, Crieff PH7 4LE
G4	YDD	W. Davidson 171 Ramsey Road, St. Ives PE27 3TZ
G4	YDE	E. Metcalf Beech Lee, Vicarage Lane, Alresford SO24 0DU
G4	YDH	G. Innes Stonehaven, Holwell Road, Hitchin SG5 3SL
G4	YDI	R. Benbow 54 Park Lea, Bradley Grange, Huddersfield HD2 1QH
G4	YDL	P. Martin 3 Switchback Road North, Maidenhead SL6 7UF
G4	YDM	J. Allsopp 30 Manor Park, Concord, Washington NE37 2BT
G4	YDO	Crossways CG c/o P. Boaler 21 Harts Close, Birmingham B17 9LE
GI4	YDP	G. Moore 12 Irish Green Street, Limavady BT49 9AD
G4	YDQ	D. Hannant 36 Coslany St., Norwich NR3 3DT
G4	YDR	D. Reed 14 Glenholt Road, Plymouth PL6 7JA
G4	YDT	J. Craddock 1 Brookes Road, Broseley TF12 5SB
G4	YDW	P. Grant 3 Craggwood Close, Horsforth, Leeds LS18 4RL
GW4	YDX	K. Gill 16 Hafodarthen Road, Llanhilleth, Abertillery NP13 2RY
G4	YDZ	M. Massen The Old School, Horning Rd, Norwich NR12 8JH
G4	YEB	D. Whitton 61 Greenacre Park, Gilberdyke, Brough HU15 2TY
G4	YEE	P. Hall Barnlea, Knapp Lane, Romsey SO51 9BT
G4	YEF	B. Renner 95 Reids Piece, Purton, Swindon SN5 4BA
G4	YEG	R. Paganuzzi 3 St. Johns Close, Hook RG27 9HW
G4	YEI	S. Masterman 5 Leggs Lane, Heyshott, Midhurst GU29 0DJ
G4	YEJ	A. Ayton 3 Links Avenue, Norwich NR6 5PE
G4	YEK	S. Clack 23 Cameron Grove, York YO23 1LE
G4	YEO	L. Gillain 1 Willow Avenue, Denham, Uxbridge UB9 4AG
GM4	YEQ	Gala & Dist ARS c/o J. Campbell 50 Glebe Place, Galashiels TD1 3JW
G4	YER	D. Davies 248 West Street, Hoyland Nr, Barnsley S749EE
G4	YES	J. Thompson 3 Newport Mount, Headingley, Leeds LS6 3DB
G4	YEX	J. Kennedy 60 Burnway, Albany, Washington NE37 1QQ
G4	YFC	T. Hill 11 Paget Cottages, Munden Road, Ware SG12 0NL
G4	YFF	E. Reynolds 4 Underwood Close, Stafford ST16 1TB
G4	YFI	R. Larter 12 Ashby Road, Hinckley LE10 1SL
G4	YFJ	N. Von Fircks 4 Park St., Salisbury SP1 3AU
G4	YFK	M. Aitchison 21 St. Pauls Road West, Dorking RH4 2HT
G4	YFO	G. Wardy 7 Yew Tree Road, Wistaston, Crewe CW2 8BN
G4	YFS	Leicestershire Worked All Brittain Group c/o S. Van Praag 12 Derby St., Darlington DL3 0NW
G4	YFT	B. Page 7 Muirfield Crescent Tividale, Oldbury B69 1PW

G4 YFU M. Parker 85 Elston Road, Aldershot GU12 4HZ
G4 YFV N. Banham Lodge Bungalow, Norwich Road, Diss IP21 4EE
G4 YFX P. Johnson 119 Riverstone Way, Northampton NN4 9QW
G4 YFZ B. Jones Jetza, Rose Avenue, Burton-on-Trent DE13 0DQ
G4 YGA S. Howarth 16 Bailey Close Old Catton, Norwich NR6 7FE
G4 YGB A. Graph 39 Poets Corner, Margate CT9 1TR
G4 YGD G. Aldred 212 Reepham Road, Norwich NR6 5SW
G4 YGE C. Oakley 9 Fitzroy Road, Landport, Lewes BN7 2UB
G4 YGH D. Hart 30 Dartford Avenue, Edmonton, London N9 8HD
G4 YGJ L. Rozentals 67 Linden Close, Eastbourne BN22 0TT
G4 YGL G. Reece 34 Priestley Gardens, Romford RM6 4SL
G4 YGM P. Reynolds 321 Sopwith Crescent Merley, Wimborne BH211XQ
GM4 YGN F. Albers 71 Old Evanton Road, Dingwall IV15 9RB
G4 YGP J. Vinson 11 Ripon Way, Carlton Miniott, Thirsk YO7 4LR
G4 YGQ M. Tate Stone Cottage, The Street, Norwich NR13 3PL
GM4 YGS G. Wells Squaredoch, Deskford, Buckie AB56 5YD
G4 YGT W. Waldron 16 Barke St., Highley, Bridgnorth WV16 6LQ
G4 YGU J. Hetherington 1 Downs Wood, Vigo, Gravesend DA13 0SQ
G4 YGV R. Barnes 7 Maycroft Close, Ipswich IP1 6RG
G4 YGW Washington ARC c/o F. Waters 96 Stockley Road, Barmston Village, Washington NE38 8DR
G4 YGY R. Fawke 36 Lavender Meadows, Corelli Close, Stratford-upon-Avon CV37 9FZ
G4 YGZ K. Ghillyer 54 Longmeadow Road, Saltash PL12 6DR
G4 YHG J. Hubner 30 Orchard Rise, Olveston, Bristol BS35 4DZ
G4 YHK H. Jackman 8 The Buchan, Camberley GU15 3XB
G4 YHN A. Gehammar The Lodge, 119 Ashdon Road, Saffron Walden CB10 2AJ
G4 YHP C. Jobling Joycliff, 20A Poplar Road, Grimsby DN41 7RD
GM4 YHS S. Grant 16 Netherton Place, Westmuir, Kirriemuir DD8 5LD
G4 YIA B. Robinson 196 Bristol Avenue, Farington, Leyland PR25 4QZ
G4 YIC A. Pitt The Long Barn, Millmans Farm, Southend, , Wotton under Edge GL12 7PD
GW4 YID M. James 14 Carmel Road, Winch Wen, Swansea SA1 7JY
G4 YIE C. Kelley Dunkeld, Bridge Street, Southam CV47 2XY
G4 YIF S. Lumbard 26 Waverleigh Road, Cranleigh GU6 8BZ
G4 YIG J. Cluley 24 Avon Green, Wyre Piddle, Pershore WR10 2JE
G4 YIH W. Maycey 21 Brook Drive, Wickford SS12 9EQ
G4 YIM J. Cameron 20 Fellmead, East Peckham, Tonbridge TN12 5EQ
G4 YIP V. Ulfik 31 Tarragon Way, East Hunsbury, Northampton NN4 0SF
G4 YIS J. O'Farrell Shannon, Gunville Road, Salisbury SP5 1PP
G4 YIT I. Toon 56 Cockbank, Turves, Peterborough PE7 2HN
G4 YIV B. Whyle 4 Britannia Gardens, Rowley Regis B65 8DT
G4 YIZ A. Mansfield 90 Vicarage Road, Mickleover, Derby DE3 0EE
G4 YJA B. Lambert Firbank, East Street, Rusper, Horsham RH12 4RE
G4 YJB R. Briggs 32 Waterside, Evesham WR11 1BU
G4 YJC G. Thornton 115 High Street, Studley B80 7HN
G4 YJD J. Donin 2 Crablands, Selsey, Chichester PO20 9AX
G4 YJH J. Hockey 11 Amulet Way, Shepton Mallet BA4 4TL
GW4 YJI T. Tilley 47 Stratton Way, Neath Abbey, Neath SA10 7BU
G4 YJK P. Duke 4 Doggets Lane, Fulbourn, Cambridge CB21 5BT
G4 YJM M. Leonard 7 Moorside Parade, Drighlington, Bradford BD11 1HR
G4 YJN M. Griggs Tudor Rose Cottage, Malting Green Layer-De-La-Haye, Colchester CO2 0JE
G4 YJP S. Stanton 10 Jenner Crescent, Northampton NN2 8NB
G4 YJQ D. Cutts 36 Lodge Road, Little Oakley, Harwich CO12 5EE
G4 YJS B. Parsons 65 Foster Street, Widnes WA8 6ET
G4 YJT C. Roberts 59A Wharncliffe Road, Loughborough LE11 1SL
G4 YJU C. Mount 6 Almond Close, Countesthorpe, Leicester LE8 5TG
G4 YJW G. Grundy 47 Northiam Road, Eastbourne BN20 8LP
G4 YJX J. Boyes 1 Fuller Close, Shepton Mallet BA4 5PX
G4 YJY W. Marsden 33 Kilton Crescent, Worksop S81 0AX
G4 YK B. Morrissey 50 Fingringhoe Road, Langenhoe, Colchester CO5 7LB
G4 YKB H. Ramsden 23 Nandywell, Little Lever, Bolton BL3 1JU
G4 YKE K. Howell 25 Shelleycotes Road, Brixworth, Northampton NN6 9NE
G4 YKG J. Pether Stead Farm, Quarry Land Lane, Nr Axbridge BS26 2QW
G4 YKH C. Littler 11 Richards Road, Stoke D'Abernon, Cobham KT11 2SX
G4 YKK A. Babbage 247-248 Molesey Avenue, West Molesey KT8 2ET
GW4 YKM K. Martich 25 Pentwyn Isaf, Energlyn, Caerphilly CF83 2NR
G4 YKQ P. Welford 11 Ridgeside, Bledlow Ridge, High Wycombe HP14 4JN
G4 YKR K. Ramsdale 770 Warrington Road, Risley, Warrington WA3 6AQ
G4 YKV K. Wood 257 Church Road, Haydock, St. Helens WA11 0LY
GW4 YKW A. Hopkins 30 Wavell Drive, Newport NP20 6QN
G4 YKX M. Blakeley Stowe House, Preston Gubbals Road, Shropshire SY4 3LY
G4 YKZ R. Harvey Richlyn House Cedar Road, Norwich NR9 3JY
G4 YLG E. Jacquemai 26 The Crescent, Brighton BN2 4TD
G4 YLI R. Barnes The Cottage, Hutton Row, Skelton, Penrith CA11 9TR
G4 YLK D. Adams 16 Centurion Close, Birchwood, Warrington WA3 6NE
GM4 YLN C. Grierson 6 Baberton Mains Court, Edinburgh EH14 3ER.
G4 YLO H. Timbrell Crossing Cottage, Lamyatt, Shepton Mallet BA4 6NG
GJ4 YLP C. Landor L'Auge, La Rue Du Puchot, Jersey JE3 6AS
G4 YLQ R. May Flat 4, Hyde Bank Court, Hyde Bank Road, New Mills SK22 4NE
G4 YLT J. Smith 43 Pagitt Street, Chatham ME4 6RE
G4 YLW P. Yaxley 10 Leybourne Close, Brighton BN2 4LU
G4 YMB M. Brass 11 Lealholm Way, Guisborough TS14 8LN
GM4 YMC J. Mccallum 86A Meadowfoot Road, Gladstone House, West Kilbride KA23 9BY
GM4 YMD W. Macdiarmid 167 Glasgow Road, Whins Of Milton, Stirling FK7 0LH
G4 YME M. Elsey Trekeek Farm, Camelford PL32 9UB
G4 YMF J. Cracklow 4 The Lawns, Sidcup DA14 4ET
G4 YMG P. Chorley 6 Conference Close, Warminster BA12 8TF
G4 YMH M. Harney 14 Druce Way, Thatcham RG19 3PF
GM4 YMI A. Dodds 9 Wellheads, Dunfermline KY11 3JG
G4 YMJ W. Fitzgerald 4 Sungold Place, Carterton OX18 1DN
GW4 YML E. Jones 15 St. Joseph Place, Llantarnam, Cwmbran NP44 3HH
GM4 YMM C. Dons 37 Ashley Drive, Edinburgh EH11 1RP
G4 YMQ A. Kimm 9 Tennis St., Burnley BB10 3AG
G4 YMT M. Taylor 64 Elmdene Road, Kenilworth CV8 2BX
G4 YMY R. Nicol 32 Mayfair Drive, Newbury RG14 6EE
G4 YMZ J. May Glanrhyd Uchaf, Stags Head, Llangeitho, Tregaron SY25 6QU
G4 YNG M. Garlick Church View, School Lane, Islip, Kettering NN14 3LQ
G4 YNH S. Payas 36 Tintern Close, Popley, Basingstoke RG24 9HE
G4 YNI H. Beckman 16 Wilton Road, Crumpsall, Manchester M8 4WQ
G4 YNK N. Fletcher 11 Parkgate Drive, Bolton BL1 8SD
G4 YNL the A Team CG c/o R. Banks Crows Nest, Churchstoke, Montgomery SY15 6TP
G4 YNM B. Spencer 33 New King Street, Bath BA1 2BL
G4 YNO P. Barnes 69 Southborne, Overcliff Drive, Bournemouth BH6 3NN
GW4 YNP D. Collins 34 Tone Road, Bettws, Newport NP20 7AW
G4 YNS S. Shakeshaft Belvedere Cottage, Wrexham Road, Pulford, Chester CH4 9DG
GW4 YNT M. Yallop 25, Hawthorn Avenue, Baglan, Neath, Port Talbot SA12 8PH
G4 YNU J. Scriven 1 Holgate Road, Pontefract WF8 4ND
G4 YNV H. Snaden 92 Avon Way, Portishead, Bristol BS20 6LU
G4 YNX B. Bassford 12 Little Brum, Grendon, Atherstone CV9 2ET
G4 YOA F. Harvey 137 Epping New Road, Buckhurst Hill IG9 5TZ
G4 YOC D. Gully 9 Shellards Road Longwell Green, Bristol BS30 9DU
G4 YOF G. Coomber 6 Birch Way, Birch, Colchester CO2 0NQ
G4 YOR R. Greenwood 26 Littlefield Walk, Bradford BD6 1UU
G4 YOS D. Corallini 8 Britannia, Puckeridge, Ware SG11 1TG
G4 YOT L. Zalicks 3 Retford Path, Harold Hill, Romford RM3 9NL
G4 YOV J. Metcalfe 3 Castle Close, Stockton-on-Tees TS19 0SL
GU4 YOX R. Beebe San Grato, Les Houguette, Castel, Guernsey GY5 7DZ Guernsey
G4 YOZ B. Starkey 52 Bermuda Road, Nuneaton CV10 7HP
G4 YPA J. Aisher 44 Cranleigh Road, Portchester, Fareham PO16 9DN
G4 YPC P. Croucher 66 Loop Road, Kingfield, Woking GU22 9BQ
G4 YPD F. Thorne 42 Roseworth Avenue, Liverpool L9 8HF
G4 YPE N. Hanney 62 Avonfield Avenue, Bradford-on-Avon BA15 1JF
G4 YPF W. Taylor 3 Westcroft, Leominster HR6 8HE
G4 YPG R. Mason Flat 4, Lysander House, Washington Road, Pulborough RH20 4RF
G4 YPH D. Rothwell 37 Eamont Avenue, Crossens, Southport PR9 9YX
G4 YPI A. Maires 26 Dunmow Road, Thelwall, Warrington WA4 2HQ
G4 YPK P. Knowles 6 Dorchester Close, Basingstoke RG23 8EX
GM4 YPL Dr R. Thompson Lochview West, 2 St. Ninians Avenue, Linlithgow EH49 7BP
G4 YPQ K. Simpson 5 Plover Fields, Madeley, Crewe CW3 9EG
GI4 YPR W. Swail 30 The Gables, Ballyphilip Road, Newtownards BT22 1RB
G4 YPS A. Bradley 22 Alexandra Crescent, Wigan WN5 9JP
G4 YPV D. Ramsden 76 Brigg Lane, Camblesforth, Selby YO8 8HD
G4 YQA M. Lawson 2 Low Lane, Embsay, Skipton BD23 6SD
G4 YQC P. Whiting 77 Melford Way, Felixstowe IP11 2UH
G4 YQD T. Mayfield 14 Wheatley Grange, Coleshill, Birmingham B46 3LZ
G4 YQG B. Hodgetts 3 Garsdale Road, Weston-Super-Mare BS22 8PT
G4 YQH J. Frampton 161 Longmead Avenue, Bristol BS7 8QG
G4 YQJ F. Collie 58 Waarem Avenue, Canvey Island SS8 9DZ
G4 YQK K. Taylor 22 Anderson, Dunholme, Lincoln LN2 3SR
G4 YQL R. Silvey 9 Kempe Road, Finchingfield, Braintree CM7 4LE
G4 YQP M. Simmens 1 Meaver Cottages, Meaver Road, Mullion, Helston TR12 7DN
G4 YQQ A. Booth 656 Southmead Road, Filton, Bristol BS34 7RD
G4 YQS T. White Rosewall Bungalow, Towednack Road, St. Ives TR26 3AL
G4 YQW K. Lawton 52 Gamble Lane, Leeds LS12 5LP
G4 YRA J. Hills 27 Wellington Road, Denton, Newhaven BN9 0RD
G4 YRC York RC c/o A. Palfrey 5 Ings View, York YO30 5XE
GM4 YRE E. Marcus Woodmyre House, Edzell, Brechin DD9 7UX
G4 YRF K. Amos 1 Byron Close, Upper Caldecote, Biggleswade SG18 9DF
G4 YRM R. Maynard Clarnard, 7 Phillipps Avenue, Exmouth EX8 3HY
GM4 YRO W. Patterson 11 Almond Place, Comrie, Crieff PH6 2BB
GI4 YRP T. Hutchinson 47 Ballylough Road, Donaghcloney, Craigavon BT66 7PQ
G4 YRT G. Cogger 40 The Crescent, Southwick, Brighton BN42 4LA
G4 YRV M. White 34 Pain'S Way, Amesbury, Salisbury SP4 7RG
G4 YRX J. Hilliard 44 Lerwick Way, Corby NN17 2DZ
G4 YRY M. Holloway 70 Baring Road, Southbourne, Bournemouth BH6 4DT
G4 YRZ R. Denton 48 Shireoaks Common, Shireoaks, Worksop S81 8PE
G4 YSB J. Leary 18A Chestnut Avenue, Andover SP10 2HE
G4 YSE Haven Valley Contest Club c/o G. Ring 31 Studland Park, Westbury BA13 3HQ
G4 YSF J. Stimpson 12 Fairhaven Court, Pittville Circus Road Roa, Cheltenham GL52 2QR
G4 YSG A. Cooper 85 Mansfield Road, Aston, Sheffield S26 2BR
G4 YSH C. Bowers Cornbury, Seymour Plain, Marlow SL7 3BZ
G4 YSJ P. Rowland 17 Hemel Hempstead Road, Redbourn, St. Albans AL3 7NL
GM4 YSN I. Brown Redland House, Westruther, Gordon TD3 6NF
G4 YSO M. Nixon 32 Gilbert Sutcliffe Court, Cleethorpes DN35 0SF
G4 YSP K. Metcalf 34 Framland Drive, Melton Mowbray LE13 1HY
G4 YSQ T. Rogers Lodge 12 Benson Waterfront, Benson, Wallingford, Oxford OX10 6SJ
G4 YSS J. Earnshaw Dunelm, Ayton Road, Irton, Scarborough YO12 4RQ
G4 YSZ R. Painton 17 Brookside, Pill, Bristol BS20 0JX
G4 YTB D. Slade B2 2A Brisas Del Mar, Formentera Del Segura, Formentera Del Segura 3179 Spain
G4 YTC M. Tyson 1 Warwick Road, Bude EX23 8EU
GM4 YTD T. Booth 41 Lochryan Street, Stranraer DG9 7HP
G4 YTF P. Godber 3 Chalvington Close, Evington, Leicester LE5 6XT
G4 YTG A. Gilbey 83 Chignal Road, Chelmsford CM1 2JA
G4 YTH T. Handford 20 Minehead Road, Knowle, Bristol BS4 1BN
G4 YTI N. Terry 2 Crosley House, Crosley Wood Road, Bingley BD16 4QD
G4 YTJ J. Pagett 26 Rednal Hill Lane, Rednal, Birmingham B45 9LR
G4 YTK S. Hopley 35 Norton Grange, Norton Canes, Cannock WS11 9QZ
G4 YTL Dr D. Hilton-Jones Home Farm, Lillingstone Lovell, Buckingham MK18 5BJ
G4 YTM I. Pettinger 266 West Street, Hoyland, Barnsley S74 9EQ
G4 YTN L. Thomas 31 Claude Avenue, Oldfield Park, Bath BA2 1AE
G4 YTO M. Yeomans 6 Badsey Close, Northfield, Birmingham B31 2EJ
G4 YTQ Colechester CG c/o D. Richardson Holmlea, Town Street, Immingham DN40 3DA
G4 YTR A. Hough 27 Doncaster Road, Askern, Doncaster DN6 0AL
G4 YTT M. Curran 40 Barnpark Road, Teignmouth TQ14 8PN
G4 YTU K. Maskell 2 Birkhall Close, Chatham ME5 7QD
G4 YTV R. Guttridge Ivy House, Rise Road, Skirlaugh Hull HU11 5BH
G4 YTY A. Dodd 109 Perinville Road, Babbacombe, Torquay TQ1 3PD
G4 YUA M. Rowland 27 Wilmot Close, Witney OX28 5NL

G4 YUF C. Sharon 7 Waverley Gardens, Barkingside, Ilford IG6 1PJ
G4 YUG C. Rogers 221 Dales Road, Ipswich IP1 4JY
G4 YUI R. Smith 27 Laburnum Road, Bournville, Birmingham B30 2BA
G4 YUK A. Ince 3. Craycroft Road. Westwoodside., Doncaster. DN9 2DG
G4 YUL R. Hudson 5 Common Lane, Hemingford Abbots, Huntingdon PE28 9AN
G4 YUN M. Fox 10 Alderhay Lane, Rookery, Stoke-on-Trent ST7 4RQ
G4 YUO E. Hodges 2 Joeys Field, Bishops Nympton, South Molton EX36 4PX
G4 YUR J. Saunders 40 Walkley Street, Sheffield S6 3RG
G4 YUV H. Boehner Not, Applicable, Resides OUTSIDE OF THE UK IN Germany
G4 YUZ I. Parker 26 Lowther Gardens, Whitehaven CA28 9LE
G4 YVA J. Guest 57 Park Road, Quarry Bank, Brierley Hill DY5 2HT
G4 YVB J. Finch Hillside House, Chapel Street, Camelford PL32 9UP
G4 YVD P. Challinor Las Ciguenas, 33 Hill Road, Telford TF2 8NA
G4 YVE C. Herwig 29 New Road, Cupernham, Romsey SO51 7LL
G4 YVF F. James 70 Broadway West, Walsall WS1 4DZ
G4 YVI Dr P. Rimmer 1 Pear Tree Close, Weaverham, Northwich CW8 3HD
G4 YVJ L. Beal 17 Park Street, Cleethorpes DN35 7NG
G4 YVK J. Felgate 31 Melbourne Road, Ipswich IP4 5PP
G4 YVM D. Perry 11 St. Lawrence Close, Stratford Sub Castle, Salisbury SP1 3LW
GW4 YVN G. Bertos Farallon, Beach Road, Pembroke Dock SA72 6TP
G4 YVO G. Howarth 79 Eden Avenue, Edenfield, Bury BL0 0LD
G4 YVQ R. Hargreaves Lawnswood, Lee Road, Blackpool FY4 4QS
G4 YVV P. Leetham 26 Petersham Drive, Alvaston, Derby DE24 0JU
G4 YVW P. Speed 52 Hunter Avenue Shenfield, Brentwood CM15 8PF
GW4 YVX J. Rafferty Pen Y Bont, Llanfachraeth, Holyhead LL65 4UY
G4 YVY T. Williams 31 Manor Road, Holbury, Southampton SO45 2NQ
G4 YVZ I. Cooling 1 Grosvenor Gardens, Southampton SO17 1RS
G4 YWA P. Cartwright Danae, Glen Road, Kingsdown, Deal CT14 8DD
G4 YWD W. Davies Davies Electrical Services (N W) Ltd, 104 Bromborough Village Road, Bromborough, Wirral CH62 7EX
G4 YWG D. Fowler 22 Larchwood Crescent, Leyland PR25 1RJ
GM4 YWI T. Ross 36 Appin Drive, Prestonpans EH32 9FB
G4 YWJ P. Ganley Holywell Farm House, Beltoft, Doncaster DN9 1NB
GW4 YWM W. Matthews Cornerways, William Street, Swansea SA9 1AT
G4 YWN G. Morris Rivendell, The Street, Braintree CM7 5HN
GM4 YWS G. Mckay Reay House, St. Vigeans, Arbroath DD11 4RA
GI4 YWT J. Crichton 10 Bann Drive, Londonderry BT47 2HW
GM4 YWU J. Bledowski 24 Arbirlot, Arbroath DD11 2NX
GM4 YWV W. Watson 21 Cameron Way, Bridge Of Don, Aberdeen AB23 8QD
G4 YWX A. Bell 43 Bigsby Road, Retford DN22 6SF
G4 YWZ C. Winning 14 Fairmead Way, Totton, Southampton SO40 7JH
G4 YXB A. Utley 384 Skipton Road, Utley, Keighley BD20 6HP
GM4 YXI Prof. K. Kerr East Loanhead, Auchnagatt, Ellon AB41 8YH
G4 YXJ T. Trethewey 8 Sunningdale Road, Saltash PL12 4BN
G4 YXR P. Waygood 89 George St., Wellington TA21 8HZ
G4 YXS F. Lake 77A Wood Lane, Chapmanslade, Westbury BA13 4AT
G4 YXU S. Rawcliffe 4 Rue Des Contamines, Gex 1170 France
G4 YXX N. Varnes Kelneath, West Hill, Wincanton BA9 9BZ
G4 YYC D. Craig 48 Fairholme, Bedford MK41 9DD
G4 YYD A. Birtwistle 6 Solness Street, Bury BL9 6PP
G4 YYE K. Smith 27 Fairleas, Branston, Lincoln LN4 1NW
G4 YYF S. Collings 44 Church Road Wootton Bridge, Ryde PO33 4PY
G4 YYG G. Plant 74 Elmwood Drive, Blythe Bridge, Stoke-on-Trent ST11 9NX
G4 YYH R. Blemings 1 Trethern Close, Troon, Camborne TR14 9ER
G4 YYI S. Tear 18 The Chase, Sinfin, Derby DE24 9PD
G4 YYL T. Anderton 12 Oaklands Close, Halvergate, Norwich NR13 3PP
G4 YYM M. Remnant Redwood Court, Tolcarne Road, Camborne TR14 9AA
G4 YYO P. Sutcliffe Rosemead, Cheadle Road, Stoke-on-Trent ST10 4BH
G4 YYP P. Plant 74 Elmwood Drive, Blythe Bridge, Stoke-on-Trent ST11 9NX
G4 YYR S. Gibbs 14 Castle Mead, Kings Stanley, Stonehouse GL10 3LD
G4 YYS S. Jones 2 Rutland Way, Southampton SO18 5PG
G4 YZA Salford University ARS c/o D. Brown 104 Kineton Green Road, Solihull SOLIHULL
G4 YZC E. Smith Willow Cottage, Mill Road, Louth LN11 9TF
G4 YZD F. Benstead 19 Davis Court, Eastland Road, Bristol BS35 1DP
G4 YZF B. Alston-Pottinger 16 Vincent Close, Great Yarmouth NR31 0HR
G4 YZH B. Calvert-Toulmin Brandesby House, 31 West End, Scunthorpe DN15 9NR
G4 YZK R. Marsh Hazatree, 71 Station Road, Worcester WR3 7UP
G4 YZL G. Woollams Pebbles, Pebsham Lane, Bexhill-on-Sea TN40 2NT
G4 YZM S. Green 4 Countess Drive, Walsall WS4 1HT
G4 YZN K. Chapman 10 Beck Lane, Collingham, Wetherby LS22 5BW
G4 YZP J. Mcmahon 15 Chatteris Park, Runcorn WA7 1XE
G4 YZR M. Baker 62 Court Farm Road, Whitchurch, Bristol BS14 0EG
GM4 YZT J. Simpson Flat 1/B, Isla Court, Perth PH2 7HJ
GD4 ZAB L. Taylor Burnt Mill House, Mount William, Summer Hill, Douglas, Isle of Man IM2 4PE
G4 ZAC M. Lewis Oak Fruit Farm, Devils Highway, Reading RG7 1XS
GW4 ZAG G. Woodworth 136 Wepre Park, Connah'S Quay, Deeside CH5 4HW
GI4 ZAH F. Anderson Flat 6, 25 Main Street, Coleraine BT51 4RA
G4 ZAI Milton Keynes ARS c/o A. Stevenson 11 Alexandra Road, Malvern WR14 1HA
G4 ZAL Central Lancs A R C c/o N. Head 11 Crowden Crescent, Tiverton EX16 4ET
G4 ZAM R. Lowe The Chimes, 4 Broadway, Swindon SN25 3BT
G4 ZAO D. Holmes 17 Green Lane, Scarborough YO12 6HL
G4 ZAP ARC of Nottm c/o C. Wilson 2 Bainton Close, Bradford-on-Avon BA15 1SE
GW4 ZAR D. Flanagan 13 Bryn Awelon, Flint CH6 5QA
G4 ZAS J. Searle 21 Chetwynd Drive, Southampton SO16 3HY
GW4 ZAW J. Aspinall 66 Lake Road East, Cardiff CF23 5NN
G4 ZAX S. Jones 71 Milford Road, Pennington, Lymington SO41 8DN
G4 ZAY J. Tench 20 Waterfield Meadows, North Walsham NR28 9LD
G4 ZBC L. Brazier 65 Cadman Crescent, Fallings Park, Wolverhampton WV10 0SH
G4 ZBE T. Anderson 38 Redwood Drive, Chase Terrace, Burntwood WS7 2AS
G4 ZBG T. Green Two 1 Conduit Road, Stamford PE9 1QQ
G4 ZBH A. Holder 47 Church Road, Gurnard, Cowes PO31 8JP
G4 ZBK J. Olsen Klintevej 218, Hjertebjerg, Stege DK 4780 Denmark
G4 ZBL S. Bateman 29 Nags Head Hill, St. George, Bristol BS5 8LN
GW4 ZBN L. Connery 37 Thomas St., Abertridwr, Caerphilly CF8 4AU
G4 ZBO R. Parker 34 Sandgate, Kendal LA9 6HT
G4 ZBQ J. Thomas 113 Southwood Drive, Coombe Dingle, Bristol BS9 2QR
G4 ZBS A. Appleyard 5 Rowan Close, Puriton, Bridgwater TA7 8AL
GW4 ZBU J. Johns 16 Maes Yr Haf, Llansamlet, Swansea SA7 9ST
G4 ZBW J. Sceal South Low, Lyth, Kendal LA8 8DJ
G4 ZBZ A. Cooper 26 Burlington Road, Skegness PE25 2EW
G4 ZCA R. Mccormick 22 Eric Road, Wallasey CH44 5RQ
G4 ZCG A. Ashworth 210 Liverpool Road, Hutton, Preston PR4 5HB
GI4 ZCH C. Harkin 39 Phillip Street, Derry City BT48 7PN
G4 ZCJ H. Cresswell 34 Kingsgate Avenue, Birstall, Leicester LE4 3HB
GW4 ZCL P. Jones 14 Fonmon Road Rhoose, Barry CF62 3DZ
GW4 ZCM K. Frowd 9 Heol Bryn Fab, Nelson CF46 6JF
G4 ZCN B. Grylls 22 Aldeburgh Close, Hartlepool TS25 2RG
G4 ZCP B. Roberts 70 Coombe Park Road, Coventry CV3 2PE
G4 ZCR C. Glenn 61 Ansell Road, Erdington, Birmingham B24 8LX
G4 ZCS C. Saunders Garlands, Malthouse Lane, Burgess Hill RH15 9XA
G4 ZCT C. Thomas 3 Poldice Terrace Poldice, St. Day, Redruth TR16 5QA
GM4 ZCV E. Prietzel 4 Cherry Lane, Cupar KY15 5DA
G4 ZCW D. Reed 4 Allwood Drive, Carlton, Nottingham NG4 3EH
GW4 ZCY F. Barwell Galahad, Penisarwaun, Caernarfon LL55 3BN
G4 ZDD B. Brookfield 17 St. Stephens Drive, Aston, Sheffield S26 2EP
G4 ZDE R. Boss 11 Penkridge Road, Church Gresley, Swadlincote DE11 9FH
G4 ZDF T. Langham 1 Chatsworth Avenue, Radcliffe-On-Trent, Nottingham NG12 1DG
G4 ZDG R. Mather 27 Bridgeacre Gardens, Coventry CV3 2NQ
G4 ZDH D. Hepworth Gander Green, Ings Lane Lastingham, York YO62 6TD
G4 ZDN G. Titterington 5 Stratford Crescent, Retford DN22 7NX
G4 ZDP S. Williams 18 Croft Road, Newbury RG14 7AL
G4 ZDQ A. Siddons 18 Earlswood Road, Evington, Leicester LE5 6JB
G4 ZDR A. Perrett 99 Welsford Avenue, Wells BA5 2HZ
GM4 ZDT D. Brodie Newton Of Muiresk, Turriff AB53 8AE
GW4 ZDU S. Kirkwood Glanrafon, Llanddeusant, Holyhead LL65 4AH
G4 ZDX A. Staniforth 2 Park View, Mapperley, Nottingham NG3 5FD
G4 ZDY R. Haining 2 Keswick Close, Kirby Cross, Frinton-on-Sea CO13 0TG
GW4 ZEA E. Hawkins 12 Marine Drive, Ogmore-By-Sea, Bridgend CF32 0PJ
G4 ZEB A. Richardson 117 Polgrean Place, St. Blazey, Par PL24 2LH
G4 ZEG E. Cross 15 Carisbrooke Crescent, Barrow-in-Furness LA13 0HU
G4 ZEJ R. Coombes Aaru - 15 Semington Strand, Swindon SN1 7DJ
G4 ZEL D. Rampton Chalemar, Eddeys Lane, Bordon GU35 8HU
G4 ZEN G. Gardner 10 Chestnut Close, Ventnor PO38 1DQ
G4 ZES R. Mills 5 Summerlands Road, Marshalswick, St. Albans AL4 9XB
G4 ZEU W. Dix 2 Churchdown Close, Boldon Colliery NE35 9HA
G4 ZEW D. Adams 4 St. Georges Close, Brampton, Huntingdon PE28 4US
GM4 ZEX G. Duncan Cromletvilla, South Road, Oldmeldrum, Inverurie AB51 0AB
G4 ZEY E. Gough 41 Matlock Green, Matlock DE4 3BT
G4 ZEZ J. Curwen 1 Oak Drive, Halton, Lancaster LA2 6QJ
G4 ZFC R. Marks 14 Carnation Road, Rochester ME2 2YE
G4 ZFD D. Roper Lunesdale, Halifax Road, Nelson BB9 0EG
G4 ZFE R. Everitt 6 Ormathwaites Corner, Warfield, Bracknell RG42 3XX
G4 ZFJ C. Roberts 122 Lower Road, Hullbridge, Hockley SS5 6BH
G4 ZFP P. Lewis 12 St. James Park, Tunbridge Wells TN1 2LH
G4 ZFQ A. Reeves 41 Nodes Road, Cowes PO31 8AD
G4 ZFR South Normanton And District ARC c/o P. Whiting 77 Melford Way, Felixstowe IP11 2UH
GM4 ZFS S. Graham 8 Kirkton Crescent, Dundee DD3 0BN
G4 ZFT M. Nurse 67 Grasleigh Way, Allerton, Bradford BD15 9BD
G4 ZFV D. Green 56 Southfields Road, Littlehampton BN17 6PA
G4 ZFX J. Blades 3 Briery Croft, Stainburn, Workington CA14 1XJ
G4 ZFY J. Bowes 8 Coxford Drove, Southampton SO16 5FD
G4 ZGE T. Stocks 1 Church Street Messingham, Scunthorpe DN17 3SB
G4 ZGG B. Storey 8-9 Chadley Lane, Godmanchester, Huntingdon PE29 2AL
G4 ZGM C. Macdonald 3 Shaftesbury Avenue, Doncaster DN2 6DT
G4 ZGP G. Pritchard 26 Anglesey Drive, Poynton, Stockport SK12 1BU
G4 ZGQ D. Richardson 55 Barton Road, Central Treviscoe, St. Austell PL26 7PT
G4 ZGZ S. Harris Pentillie Cottage Quethiock, Liskeard PL14 3SQ
G4 ZHA D. Raine 160 Aragon Road, Morden SM4 4QN
G4 ZHD P. Crofts 8 Sandown Avenue, Mickleover, Derby DE3 0QQ
G4 ZHE D. Cannon 69 Hayfield Road, Oxford OX2 6TX
G4 ZHG J. Nevin 26 Beech Avenue, Newark NG24 4DY
G4 ZHI J. Howell-Pryce Bwlch Teulu, Tynygraig, Ystrad Meurig SY25 6AJ
G4 ZHK D. Lennard 24 Southdown Road, Shoreham-by-Sea BN43 5AN
GM4 ZHL G. Graham 18 Hamarsgarth, Mossbank, Shetland ZE2 9TH
G4 ZHN D. Young The Gables, Aerodrome Road, Canterbury CT4 5EX
G4 ZHT Dr C. Strevens 15 Mere House, Frodsham WA6 0FN
G4 ZHX J. Burton 23 Dorchester Close, Dartford DA1 1ND
G4 ZHZ A. Nash 10 Broome Close, Yateley GU46 7SY
G4 ZIB A. Roberts 25 Sebright Road, Wolverley, Kidderminster DY11 5TZ
G4 ZID L. Chapman 6 Barholm Avenue, Lutton, Spalding PE12 9HS
G4 ZIF M. Taylor Holly House, Faussett Hill, Canterbury CT4 7AH
G4 ZIH R. West 51 Glen Avenue, Herne Bay CT6 6HU
G4 ZII A. Taylor 50 Long Hill Rise, Hucknall, Nottingham NG15 6GN
GM4 ZIL A. Brown Skellies Knowes East, Leswalt, Stranraer DG9 0RY
G4 ZIS R. Beech 131 Bounces Road, Lower Edmonton, London N9 8LJ
GM4 ZIT J. Brown 51 Braeside Park, Balloch, Inverness IV2 7HN
G4 ZIU M. O'Connell 5 Beckwith Close, Harrogate HG2 0BJ
G4 ZIW M. Hutchings 31 Newtown Road, Little Irchester, Wellingborough NN8 2DX
G4 ZIY M. Haddon 1 Victoria Place, Weston, Portland DT5 2AA
G4 ZIZ R. Barrett Langton House Gullet Lane, Kirby Muxloe LE9 2BL
G4 ZJC P. Berry 3 Village Farm Road, Preston, Hull HU12 8QH
G4 ZJD Dr L. Taylor 14 Spring Grove, Chiswick, London W4 3NH
G4 ZJE K. Faichney 57 Moorside Road, Brookhouse, Lancaster LA2 9PJ
G4 ZJH I. Tickle 7 Ashfords Close, Saxmundham IP17 1WB
GM4 ZJI C. Claydon 33 Craigievar Drive, Glenrothes KY7 4PH
G4 ZJK R. White 29 Princes Road, Clacton-on-Sea CO15 5LA
G4 ZJL D. Wood 29 Oakville Road, , Higher Heysham, Near Morecambe LA3 2TB
G4 ZJO H. Docherty 40 Norwood Drive, Morecambe LA4 6LU
G4 ZJP M. Ward Laurels, Eastergate Lane, Chichester PO20 3SJ
G4 ZJR E. Knibb The Cottage, Cold Newton Road, Leicester LE7 9DA
G4 ZKA J. Watson 158 Kingsfold Drive, Penwortham, Preston PR1 9EQ
G4 ZKD J. Moule Silver Dale, Callow Hill Rock, Kidderminster DY14 9DB
G4 ZKE M. Chapman 18 The Winter Knoll, Littlehampton BN17 6ND

G4 ZKG J. Corfield 5 Beasley Close, Great Sutton, Ellesmere Port CH66 2SX
G4 ZKH M. Curtis 11 Pentreath Terrace, Lanner, Redruth TR16 6HP
G4 ZKI M. Day 76 Freeman Road, Didcot OX11 7DB
G4 ZKJ W. Applebee 9 The Glade, Bucks Horn Oak, Farnham GU10 4LU
G4 ZKM W. Ingram 39 Ainsdale Drive, Peterborough PE4 6RL
G4 ZKN P. Robinson 1 Stennack, Troon, Camborne TR14 9JT
G4 ZKQ J. Garner Craythorne, Amberstone, Hailsham BN27 1PJ
G4 ZKR J. Olbrien 14 Ryecroft Close, Middlewich CW10 0PJ
G4 ZKS A. Howland Hollydene, Station Road, Colchester CO7 8LJ
G4 ZKT A. Hale 27 Danesbury Meadows, New Milton BH25 5GX
G4 ZKW T. Davies 7 Medway, Sturton By Stow, Lincoln LN1 2DY
GI4 ZLD G. Breslin 85 Whitehouse Park, Londonderry BT48 0QA
G4 ZLF L. Forde 3 Heather Way, Rosudgeon, Penzance TR20 9PT
G4 ZLI T. Schofield 25 Kingsfield, Ringwood BH24 1PH
G4 ZLJ P. Aspinall 20 Carr Lane, New Hall Hey, Rossendale BB4 6BE
G4 ZLK S. Pike 32 Rosewood Road, Dudley DY1 4DZ
G4 ZLN B. Phillips 2 Oriole Grove, Kidderminster DY10 4HG
G4 ZLP N. Crook 10 Shuttle Close, Rossington, Doncaster DN11 0FR
G4 ZLT R. Winkup 92 Barnes Crescent, Bournemouth BH10 5AW
G4 ZLU T. Clark Thaw House, Brunswick Street, Nelson BB9 0HZ
G4 ZLX A. Whillock 74 Chettell Way, Blandford St. Mary, Blandford Forum DT11 9PH
G4 ZMA J. Smith 7A The Green, East Leake, Loughborough LE12 6LD
G4 ZMB D. Sharples 11 Lina St., Accrington BB5 1SL
G4 ZMH G. Robinson 8 Fenlands Crescent, Lowestoft NR33 9AW
GM4 ZMK R. Coyle 216 Faifley Road, Clydebank G81 5EG
G4 ZML D. Coupe 14 Maltby Road, Thornton, Middlesbrough TS8 9BU
G4 ZMM J. Roberts 1 Grange Court, Hixon ST18 0GQ
G4 ZMN P. Shepherd 315 Daws Heath Road, Benfleet SS7 2TY
G4 ZMP D. Butler 42 Coombe Farm Avenue, Fareham PO16 0TR
G4 ZMR M. Reynolds Flat 2, Abbeyfield Court, 63 Abbey Foregate, Shrewsbury SY2 6BG
G4 ZMU Dr A. Vernon ? 29 Pinewood Avenue New Haw, Addlestone KT15 3AA
G4 ZMY A. Wakely 177 St.Hermans Estate, Hayling Island PO11 9NE
GM4 ZNC W. Findlay 46 Rowallan Drive, Kilmarnock KA3 1TU
G4 ZNI A. Wragg 11A Fall Road, Heanor DE75 7PQ
GM4 ZNS J. Callaghan 31 Hillview Road, Darvel KA17 0DQ
GM4 ZNX D. Stockton 13 Dunvegan Court, Crossford, Dunfermline KY12 8YL
G4 ZNY K. Brown 28 Blenheim Avenue, Stony Stratford, Milton Keynes MK11 1EX
G4 ZNZ D. Neilson 11 Craigs Way, Thirsk YO7 1UD
GM4 ZOA S. Mcgregor 35 Pentland Gardens, Edinburgh EH10 6NN
G4 ZOB P. Harris 47 North Park Grove, Roundhay, Leeds LS8 1EW
G4 ZOC J. Lawton Grenehurst, Pinewood Road, High Wycombe HP12 4DD
G4 ZOD J. Greenberg 12 Broadhurst Avenue, Edgware HA8 8TR
G4 ZOF A. Hughes Kemble Motors, Unit 9, Coundon BISHOP AUCKLAND
G4 ZOG D. Andrews 3 St. Davids Road, Thornbury, Bristol BS35 2JE
G4 ZOH C. Hetherington 10 Westway, Cowes PO31 8QP
G4 ZOI D. Hunsdale 1 Lesh Lane, Barrow-in-Furness LA13 9EA
G4 ZOQ J. Dennis 44 The Drive, Uckfield TN22 1BZ
G4 ZOR E. Wand 13 Weetmans Drive, Colchester CO4 9EA
GI4 ZOS W. Boyd 51 South Sperrin, Knock, Belfast BT5 7HW
G4 ZOT P. Varey 6 Ludlow Avenue Garforth, Leeds LS25 2LY
G4 ZOU D. Nuthall 29 Bloxham Crescent, Hampton TW12 2QG
G4 ZOX C. Moore Spion Cop, Blacksmiths Lane, Lincoln LN5 9SW
G4 ZOY D. Elliott 6 Linden Close, Stakeford, Choppington NE62 5LD
G4 ZPA B. Watling 18 Deverell Place, Waterlooville PO7 5ED
G4 ZPB R. Alexander 1 Locarno Road, Swanage BH19 1HY
G4 ZPC P. Collier 7 Cavendish Close, Bicton Heath, Shrewsbury SY3 5PG
G4 ZPH F. Machniak 18 Wyatt Road, Kempston, Bedford MK42 7EN
G4 ZPI D. Madden 17 Canberra Gardens, Birmingham B34 7LP
G4 ZPJ C. Marks 85 Madrona, Tamworth B77 4EJ
GW4 ZPL C. Barwell Galaghad, Penisarwaun, Caernarfon LL55 3BN
GW4 ZPM D. Thompson 1 West Kinmel Street, Rhyl LL18 1DA
G4 ZPN M. Brown 47 Threlfall Road, Blackpool FY1 6NW
G4 ZPO G. Belt 45.Prospect Road, Dorchester DT1 2PF
G4 ZPP B. Hope 19 Seaview Court, Hillfield Road, Chichester PO20 0JS
G4 ZPQ S. Drury 24 Mollison Road, Hull HU4 7HB
G4 ZPR R. Wilson 1 Larkfield Avenue, Harrow HA3 8NQ
G4 ZPW D. Mckie 16 Guys Close, Addison Square, Ringwood BH24 1PQ
G4 ZPZ I. Macpherson 18 Mountbatten Avenue, Dukinfield SK16 5BU
G4 ZQB G. Tapp 70 Charlesford Avenue, Kingswood, Maidstone ME17 3PH
G4 ZQC R. Alderson Old School House, Tattersett, King's Lynn PE31 8RS
G4 ZQF R. Worth 2 Orchard Drive, Otterton, Budleigh Salterton EX9 7JL
GM4 ZQH J. Howell1 26 Bonaly Crescent, Colinton, Edinburgh EH130EW
G4 ZQJ A. Mayes 103 Lionel Road, Canvey Island SS8 9DJ
G4 ZQL N. Higgins 7 Staveley Close, Middleton, Manchester M24 4RU
G4 ZQM J. Neary 29 Willow Avenue, Torquay TQ2 8DH
G4 ZQS L. Brown 17 Chaucer Walk, Langney, Eastbourne BN23 7QT
G4 ZQT J. Wright 85 Kingfisher Drive, Beacon Park Home Village, Skegness PE25 1TQ
GW4 ZQV I. Bradford The Meadows, Penyrheol, Pontypool NP4 5XS
GW4 ZQY M. Couch 37 Heol Rhosyn, Morriston, Swansea SA6 6ER
G4 ZRA G. Moffatt 30 Rose Walk, St. Albans AL4 9AF
G4 ZRB Dr W. Gerrard 9 St. Marys Gardens, Bagshot GU19 5JX
G4 ZRC M. Cole 25 Holly Gardens, West Drayton UB7 9PE
G4 ZRD B. Rosewarn 16 Stoke Park Close Bishops Cleeve, Cheltenham GL52 8UL
G4 ZRF T. Emery 23 Richmondfield Way, Barwick In Elmet, Leeds LS15 4HJ
GM4 ZRH A. Hutton 4 Linn Road, Stanley, Perth PH1 4QS
G4 ZRM D. Line 28 Wykeham Road, Higham Ferrers, Rushden NN10 8HU
GM4 ZRR I. Watt 21 Clerwood Way, Edinburgh EH12 8QA
G4 ZRT M. Johnson 5 Donigers Dell, Swanmore, Southampton SO32 2TL
G4 ZRV T. Russell 56 Curzon Avenue, Cleethorpes DN35 9HF
GW4 ZRW T. George 80 Yew Street, Troedyrhiw, Merthyr Tydfil CF48 4EE
GM4 ZRX J. Lindsay 6 Netherhouse Avenue, Lenzie, Glasgow G66 5NG
G4 ZRY A. Clemons 2 Cherry Tree Road, Rainham, Gillingham ME8 8JU
G4 ZRZ G. Baker 16 Tower Close, Emmer Green, Reading RG4 8UU
G4 ZSA A. Smith 16 Burley Close, South Milford, Leeds LS25 5BT
G4 ZSC A. Kent 9 Tolmers Gardens, Cuffley, Potters Bar EN6 4JE
G4 ZSD W. Guy 102 Bonington Road, Mansfield NG19 6QQ
G4 ZSG J. Savegar 39 Little Lane, Roundfield, Reading RG7 6RA
G4 ZSO N. Dakin 3 Tavistock Road, West Bridgford, Nottingham NG2 6FH
G4 ZSP G. Marriott 6 The Pastures, Barrow Upon Soar, Loughborough LE12 8LA
G4 ZSR D. Woollams Pebbles, Pebsham Lane, Bexhill-on-Sea TN40 2NT
G4 ZSS S. Simpson 8 Hallfield Avenue, Micklefield, Leeds LS25 4AU
G4 ZST D. Nuttall 92 Long Road, Lowestoft NR33 9DH
G4 ZSV J. Broughton 55 Webbs Close, Wolvercote, Oxford OX2 8PX
G4 ZSW P. Withall 19 Highfield Drive, Ewell, Epsom KT19 0AU
G4 ZSX P. Johnson 23 Ubbeston Way, Lowestoft NR33 7HG
G4 ZSY C. Ingram Woodwind, Kingstone, Hereford HR2 9HD
G4 ZSZ B. Watts 74 Westfield Road, Caversham, Reading RG4 8HJ
G4 ZTA J. Reed Easton Villa, Grangemoor Road, Morpeth NE61 5PU
G4 ZTC T. Cleghorn 12 Rennington Close, Stobhill Gate, Morpeth NE61 2TQ
G4 ZTD K. Wright 63 Dudley Avenue, Leicester LE5 2EF
G4 ZTF J. Scott Kemsley Street Cottage, Kemsley Street, Gillingham ME7 3LS
GW4 ZTG K. Ford Tan Y Bryn, Llanbedr LL45 2ND
G4 ZTM N. Rohsler 107 Quinton Lane, Quinton, Birmingham B32 2TT
GM4 ZTO J. Mcilwraith 54 Foreland, Ballantrae, Girvan KA26 0NQ
G4 ZTQ S. Chappell 67 Swanfield Drive, Chichester PO19 6GL
G4 ZTR J. Lemay Carlton House, White Hart Lane, Colchester CO6 3DB
G4 ZTS C. Thorne High Trees, Bradford On Tone, Taunton TA4 1EX
GI4 ZTU H. Morgan 42 Ardmore Road, Holywood BT18 0PJ
G4 ZTW C. Gallagher 9 St. Marys Road, New Romney TN28 8JB
G4 ZTX C. Hall 2 Brambling Lane, Wath-Upon-Dearne, Rotherham S63 7GT
G4 ZTY D. Dalton 22 Fernleigh Avenue, Mapperley, Nottingham NG3 6FL
G4 ZTZ K. Taylor 107 Trelowarren Street, Camborne TR14 8AW
GW4 ZUA W. Webb 62 Llewelyn Street Trecynon, Aberdare CF44 8LA
G4 ZUC G. Allin 1 Brookhill Court, Sutton-in-Ashfield NG17 1EP
G4 ZUD B. Perry Preswylfa, Carno, Caersws SY17 5JP
G4 ZUE R. Hopkins 259 Croft Road, Nuneaton CV10 7EE
G4 ZUH J. Rowles The Haven, 5 Honey Lane, Chatteris PE16 6LG
G4 ZUI P. Bevington 40 Carnarthen Street, Camborne TR14 8UP
GW4 ZUJ B. Willis 5 Park Street, Penrhiwceiber, Mountain Ash CF45 3YW
GM4 ZUK A. Duncan Barrhill House, Peterculter AB14 0LN
G4 ZUL S. Cocks 1 Church Ponds, Castle Hedingham, Halstead CO9 3BZ
G4 ZUN C. Gee 100 Plantation Hill, Worksop S81 0QN
GW4 ZUS G. Smith 8 Springfield Terrace, Llanhilleth, Abertillery NP13 2RQ
GD4 ZUU P. Chambers 15, Ramsey IM7 1HE Isle of Man
GW4 ZUW A. Hockley 44 Brookfields, Crickhowell NP8 1DJ
G4 ZUX N. Sparks 36 Tormynton Road, Worle, Weston-Super-Mare BS22 9HT
G4 ZVA Hinckley District Scouts c/o T. Webster 42 The Meadow, Mount Pleasant Residential Park, Crewe CW4 8JU
G4 ZVB G. Mantovani 74 Barnsley Road, South Kirkby, Pontefract WF9 3QE
G4 ZVD J. Birse 3 Main Road, Rathmell, Settle BD24 0LH
GM4 ZVF M. Sheriff Schoolhouse Dunmore, Kilberry Road, Tarbert PA29 6XY
G4 ZVK J. Taylor 123 Lancaster Road, Hindley, Wigan WN2 4JA
GW4 ZVL Riviera ARC c/o M. Higgins 8 Clos Rheidol, Caldicot NP26 4JD
G4 ZVN P. Baxter 20 Thorpe Street, Thorpe Hesley, Rotherham S61 2RP
GW4 ZVO Fenland Portable Group c/o B. Froley 20 Sandymeers, Porthcawl CF36 5LP
G4 ZVP B. Rhodes 13 Amanda Road, Harworth, Doncaster DN11 8HP
GW4 ZVQ S. Evans 6 Eastfield Way, Caerleon, Newport NP18 3EU
G4 ZVS C. Ford 19 Listowel Road, Kings Heath, Birmingham B14 6HH
G4 ZVU T. Chadwick 102 Feltham Road, Ashford TW15 1DP
GW4 ZVV B. Raby 4 Tyfica Road, Pontypridd CF37 2DA
G4 ZVX M. Russell 67 Dugard Road, Cleethorpes DN35 7SD
G4 ZVZ S. Josko 69 Newborough Road, Shirley, Solihull B90 2HB
G4 ZWA G. Johnson The Cottage, Mareham On The Hill, Horncastle LN9 6PQ
G4 ZWB R. Ayers 5 Cornflower Close, Leiston IP16 4UQ
G4 ZWD C. Spicer 27 Carden Crescent, Brighton BN1 8TQ
G4 ZWI F. Cooper 29 Mayfair Avenue, Mansfield NG18 4EQ
GM4 ZWJ S. Macfarlane 6 Edward Drive, Helensburgh G84 9QP
G4 ZWM H. Hoy The Meadows Cottage, Stow Heath Road, North Walsham NR28 0LR
GW4 ZWN T. Anziani 42 Tyn Rhos Estate, Penysarn LL69 9BZ
G4 ZWO A. Green Bryn Y Coed, Llanfair Road, Abergele LL22 8DH France
G4 ZWQ P. Smith 16 Church St., Owston Ferry, Doncaster DN9 1RG
G4 ZWR D. Edwards 2 Mason Close, Headless Cross, Redditch B97 5DF
G4 ZWX Dr P. Harrison 17 Kings Road, Ascot SL5 9AD
G4 ZWY S. Icke 11 Church Lane, Bromyard HR7 4DZ
G4 ZXA R. Smith 1 Hall Lane, Wolvey, Hinckley LE10 3LF
G4 ZXB A. Tomlins 44 Newlands, Balcombe, Haywards Heath RH17 6JA
G4 ZXF K. Kimber Greenacres, Garrigill, Alston CA9 3DY
GW4 ZXG L. Thomas Roughton, Corntown Road, Bridgend CF35 5BH
G4 ZXI N. Parnell 4 Forge Lane, Headcorn, Ashford TN27 9QQ
GM4 ZXJ J. Burns 7 Johns Road, Eyemouth TD14 5DX
G4 ZXN M. Ward 25 Margeson Close, Coventry CV2 5NU
G4 ZXP V. Legge 26 Goldcroft Avenue, Weymouth DT4 0ET
G4 ZXQ A. Mainwaring The Old Smoke House, Lodge Farm Barns, Hereford HR4 8NN
G4 ZXS E. Loach 99 Gorse Lane, Clacton-on-Sea CO15 4RJ
G4 ZXT M. Twigg 30 Valley Drive, Yarm TS15 9JQ
G4 ZXV W. Bailey 225 Holburne Road, London SE3 8HF
G4 ZXZ W. Johnson 12A Kings Road, Spalding PE11 1QB
G4 ZYH M. Timms 5 Lytchett Way, Nythe, Swindon SN3 3PJ
G4 ZYL J. Anderson 44 Overhill Road, Burntwood WS7 4SU
GW4 ZYM W. Williams Plum Tree Farm, Commonwood Road, Wrexham LL13 9TA
G4 ZYN M. Sherlock Flat 6, 34 Duke Street, Southport PR8 1JA
G4 ZYO B. Lawrance 14 Warren Close, Porthleven, Helston TR13 9BL
G4 ZYR H. Webber 6 Barn Ground, Highnam, Gloucester GL2 8LJ
GW4 ZYV J. Raymond 23 Castle Pill Crescent, Steynton, Milford Haven SA73 1HD
G4 ZYY G. Fildes 62 Higher Days Road, Swanage BH19 2LB
G4 ZYZ T. Seymour 21 Chainhouse Road, Needham Market, Ipswich IP6 8ER
G4 ZZD A. Hellier 64 Penlee Park, Torpoint PL11 2PZ
GM4 ZZH H. Firth Edan, Berstane Road, Kirkwall KW15 1NA
G4 ZZK B. Tutt 15 Alexandria Drive, Herne Bay CT6 8HX
G4 ZZL R. Lyons 15 Winston Avenue, Tiptree, Colchester CO5 0JU
GD4 ZZN A. Rickward 14 Ballakneale Avenue, Port Erin IM9 6ND
G4 ZZP K. Lock 5 Copthorne Crest, Shrewsbury SY3 8RU
G4 ZZS D. Bamber 3 Abbotts View, Sompting, Lancing BN15 0NG
G4 ZZV M. Singh-Gill 30 King Edwards Gardens, London W3 9RQ
GM4 ZZW R. Watts 1C Kirklands, 100 Greenock Road, Largs KA30 8PG

G4 ZZY T. Watts Carne Grey Cottage, Trethurgy, St. Austell PL26 8YE
G4 ZZZ T. Smith Lower Carniggey Farm, Greenbottom, Truro TR4 8QL

G5

G5 ADY A. Jones 67 Mandara Grove, Abbeydale., Gloucester GL4 5XT
G5 AIR R. Harwood 26 Barry Avenue, Bicester OX26 2DY
GI5 ALP Benbradagh Us Navcommsta c/o Dr E. Harper 404 Foreglen Road Dungiven, Londonderry BT47 4PN
GM5 ALX A. Johnstone 10 Earlspark Avenue, Bieldside, Aberdeen AB15 9BU
GW5 AMS Amplitude Modulation ARS c/o S. Taylor 43 Toronnen, Bangor LL57 4TG
G5 AOZ Burnham Beeches c/o U. Grunewald Nuptown Orchard, Nuptown, Warfield, Bracknell RG42 6HU
G5 ART A. Tait The Reddings, Dirty Lane, Beausale, Warwick CV35 7AQ
G5 AT RSGB Transatlantic Test Century Club No2 c/o N. Totterdell Moscar Cross House, Hollow Meadows, Sheffield S6 6GL
G5 ATT A. Horne 38 Bradley Road, Waltham, Grimsby DN37 0UZ
G5 AU A. Albinson Po Box 156, York WA 6302 Australia
GM5 AUG M. Topple Emmaus, 30 Robshill Court, , Newton Mearns G77 6UG
G5 BBC London Bbc Radio Group c/o S. Richards 6 Heathfield, Royston SG8 5BW
G5 BBL J. Verduyn 14 Ragleth Grove, Trowbridge BA14 7LE
G5 BCO P. Gautier-Lynham 95 Oxford Road, Marlow SL7 2PL
GM5 BDW R. Watson 37A Muirfield Crescent, Dundee DD3 8PY
GM5 BDX B. Corkindale 8 Rockland Park, Largs KA30 8HB
G5 BH M. Coleman Flat A, 53 De Parys Avenue, Bedford MK40 2TR
G5 BK Cheltenham A.R.A c/o A. Woolford 39 Apple Orchard, Prestbury, Cheltenham GL52 3EH
G5 BW W. Waugh 67 Cragside, Whitley Bay NE26 3EF
G5 CDC J. Kornreich 35 Charlotte Drive, Spring Valley United States
GI5 CEO G. Hill 10 Carntall Rise, Mossley, Newtownabbey BT36 5UF
GM5 CGA L. Landers 33 Newmanswalls Avenue, Montrose DD10 9DD
G5 CH Chester And District RS c/o P. Hughes 27 Hemsworth Avenue Little Sutton, Ellesmere Port CH66 4SG
G5 CTX C. Thorne 67 Devon Road, Cadishead, Manchester M44 5HB
GM5 CX R. Ferguson 19 Leighton Avenue, Dunblane FK15 0EB
GM5 DAV D. Strachan 30 Belhaven Park, Muirhead G69 9FB
G5 DJW A. Osmond 10 Deerhurst Park, Forest Row RH18 5GD
GM5 DNA Dr S. Henderson 10 Tor Gorm Road, North Kessock, Inverness IV1 3JJ
GW5 DOD T. Hoedjes C/O Borras Park 4 Townsend Avenue, Wrexham LL12 7UB
GW5 DRE D. Eade 16 Glynllifon Street, Blaenau Ffestiniog LL41 3AF
GM5 DVS Dv Scotland c/o I. Suart 37 Meldrum Mains, Glenmavis, Airdrie ML6 0QQ
G5 EDQ S. Hahn 26 Watling St., Gillingham ME7 2YH
G5 EDW E. Wright 2 Wulfrath Way, Ware SG12 0DN
G5 EFV M. Lauterborn 16 Bolden Street, London SE8 4JF
G5 FAS F. Szymanski 15 Greengate, Scarborough YO12 5NA
G5 FRG Farringdon Radio Group c/o D. Talaber 54 Southfield Park, North Harrow, Harrow HA2 6HE
G5 FZ Lincoln Short Wave Club c/o P. Rose Pinchbeck Farmhouse, Mill Lane, Sturton By Stow, Lincoln LN1 2AS
GM5 GAZ G. Lennox 49 Sorbie Road, Ardrossan KA22 8AP
G5 GIH G. Horobin 21 Welwyn Avenue, Allestree, Derby DE22 2JR
G5 GRG Genesis Radio Group c/o A. Royds 3A Fairfield Avenue, Rossendale BB4 9TG
G5 GWH G. Hunt 62 Pilkington Avenue, Sutton Coldfield B72 1LG
G5 GX J. Smith 4 Townend Villas, Humbleton, Hull HU11 4NR
G5 HAM Ham Radio Network c/o A. Whybrow 64 Church Road Sevington, Ashford TN24 0LF
G5 HBM H. Meyer 10 Cairnside, Ilfracombe EX34 8EW
G5 HI R. Birch 15 Chester Street, Cirencester GL7 1HF
G5 HOW Dr G. Howling Sysonby Knoll Hotel, Asfordby Road, Melton Mowbray LE13 0HP
G5 HY D. Wilkins 45 Seaward Avenue, Barton On Sea, New Milton BH25 7HN
G5 JBQ E. Lorenzoni 63 River Mill One, Station Road, London SE13 5FL
G5 JCB Last Post Museum c/o K. Hudson 20 Claude Street, Crumpsall, Manchester M8 5AW
GM5 JDG M. Evans 27 Greenacres Crescent, Peterhead AB42 3QH
GI5 JEB M. Houston 45 Castletown Court, Strabane BT82 9FZ
G5 JHE L. Batten Y Felin Barn, Llawr-Y-Glyn, Caersws SY17 5RH
G5 JIM J. Lang 7 Marion Grove, Liverpool L18 7HY
G5 JJ Taunton & District ARC c/o A. Walrond Leigh Hill Cottage, Lowton, Taunton TA3 7SU
G5 JON J. Housley Lesede Cottage, The Town, Carsington, Matlock DE4 4PX
G5 JPR J. Prince Wellfield Avenue, Luton LU33AT
G5 KC C. Quarton Flaxton Gatehouse, Flaxton, York YO60 7QT
GM5 KCC Straight Key Century Club c/o M. Topple Emmaus, 30 Robshill Court, , Newton Mearns G77 6UG
G5 KCP K. Pugh 4 Salt Boxes, Pinvin, Pinvin WR102LB
G5 KN Kettering & District ARS c/o K. Doswell 15A Queen Street, Desborough, Kettering NN14 2RE
G5 KUE I. Vernon 7 Seaton Road Wick, Littlehampton BN17 7LG
G5 KW the UK Six Metre Group c/o D. Toombs 1 Chalgrove, Welwyn Garden City AL7 2QJ
G5 LK Reigate Amateur Transmitting Society c/o P. Tribe Island View, 108 Portsmouth Road, Lee-on-the-Solent PO13 9AF
G5 LP L. Parker 128 Northampton Road, Wellingborough NN8 3PJ
G5 LSI Prof. L. Soares Indrusiak 6 Trafalgar House, Piccadilly, York YO1 9QP
GM5 LUK J. Macassey G/R 5 Dudhope Street Dundee, Dundee DD1 1JZ
G5 LUX J. Ratty 12 Challoners Close, East Molesey KT8 0DW
GM5 LWD G. Ledgerwood 28 Cannerton Park, Milton Of Campsie, Glasgow G66 8HR
GM5 MAJ J. Devlin 200 Second Avenue, Clydebank, Glasgow G81 3LE
G5 MAN R. Groves 34 Clover Way, Paddock Wood TN12 6BQ
G5 MHZ the 5 Mhz Pioneers Club c/o P. Gaskell 131 Greenfield Road, Dentons Green, St. Helens WA10 6SH
G5 MS Wab Awards Group c/o K. Hudson 20 Claude Street, Crumpsall, Manchester M8 5AW
G5 MUN H. Van Driel 20 Links Avenue Little Sutton, Ellesmere Port CH66 1QT
G5 MW Medway Amateur Receiving & Transmitting Society c/o J. Burton 22 Pear Tree Lane, Hempstead, Gillingham ME7 3PT
G5 MY H. Mee 268 Victoria Rd East, Leicester LE5 0LF
G5 NB N. Brown 3 Mulberry Tree Close, Filby, Great Yarmouth NR29 3HD
GW5 NF R. Ward Lower Ton-Y-Felin Farm, Croespenmaen, Crumlin, Newport NP11 3BE
GW5 NIC J. Nicholas Reservoir House, St. Lythan'S, Wenvoe CF5 6BQ
G5 NKW Key + Wire ARG c/o K. Walker 20 Thornhill Close, Bramcote, Nottingham NG9 3FS
G5 NLD D. Van Dijk 76 High Street, Tetsworth, Thame OX9 7AE
G5 OD Wey Valley ARG c/o A. Vine Hilden, Woodland Avenue, Cranleigh GU6 7HZ
G5 OLD T. Morgan 26 Farlers End, Nailsea BS48 4PG
G5 OW W. Wigg 7 Brendon Way, Long Eaton, Nottingham NG10 4JS
G5 OYY N. Weiner 3 Bluebell Court Lower Mardyke Avenue, Rainham RM13 8GF
G5 PAT P. Smyth 61 Bridge Street, Kington HR5 3DJ
GM5 PEB P. Bingham 129 Livingstone Terrace, Irvine KA12 9ER
G5 PI Caterham Radio Group c/o J. Wilson 20B High Green, Great Shelford, Cambridge CB22 5EG
G5 PJK P. Kiernan 4 Bradley Street, Southport PR9 9HW
G5 PMJ P. Stoneham The Croft, Blind Lane, Billericay CM12 9SN
G5 QK Pentland Firth Radio Hams c/o A. Radley 16 Kingsley Lane, Thundersley, Benfleet SS7 3TU
GM5 RAB R. Drummond 11 Firwood Drive, Bo'ness EH51 0NX
G5 RAH R. Hardy 4 Tideswell Close, Staveley, Chesterfield S43 3TE
G5 RC Skyline c/o A. Crespo 266 Trinity Road, London SW18 3RQ
GM5 RDX E. Smith 15 School Wynd, Quarrier's Village PA11 3NL
G5 RET G. Menendez 3 Maybrick Road, Bath BA2 3PT
G5 REV Rvd. B. Topham 2 Highgrove Gardens, Stamford PE9 2GR
G5 RFL Radio Fraternity Lodge (8040) c/o A. Boyd 5 Walmer Close, Southwater, Horsham RH13 9XY
G5 RGS G. Jordan 15 Ley Hill Road, Sutton Coldfield B75 6TF
G5 RJG R. Griffiths 31 Balcarres Road, Ashton-On-Ribble, Preston PR2 2BT
G5 RJH R. Harlow 28 Dovecliff Crescent, Stretton, Burton-on-Trent DE13 0JH
GM5 RP Vowhars c/o Dr I. White 2 Appleby Cottages, Whithorn, Newton Stewart DG8 8DQ
G5 RR Hucknall Rolls Royce A.R.C c/o S. Sorockyj 8 Bowden Avenue, Bestwood Village, Nottingham NG6 8XN
G5 RS Guildford CG c/o P. Croucher 66 Loop Road, Kingfield, Woking GU22 9BQ
G5 RSW R. Williams 44 King Street, Brierley Hill DY5 2DH
G5 RV Mid Sussex ARS c/o J. Gibson Four Oaks, Tylers Green, Haywards Heath RH17 5DZ
GI5 SBZ Prof. K. Zepf 33 Rannoch Road, Holywood BT18 0NB
G5 SHO S. Hodgkiss Estren, Station Road, Wendling, Dereham NR19 2NE
G5 SIX North London Six Metre Group c/o K. Rochester 22 Langford Road, Cockfosters, Barnet EN4 9DS
G5 SRC Swinton (Amateur) RC c/o T. Ward 173-175 Station Road, Pendlebury, Swinton, Manchester M27 6BU
G5 STU S. Green Meadow View, Ewleaze Farm, Tincleton, Dorchester DT2 8QR
G5 TAM the University of the Third Age c/o M. Meadows 8 Beeches Park Minchinhampton., Stroud GL6 9BA
G5 TCP G. Kerr 6 Cressages Close, Felsted, Dunmow CM6 3NW
GM5 TDX E. Blakeway 25 Allanton Grove, Wishaw ML2 7LL
G5 TEO T. Orzechowski 25 Lutterworth Road, Northampton NN1 5JR
GM5 TIM S. Doonan West Clanfin Farm Waterside, Kilmarnock KA3 6JQ
GI5 TKA Radio Security Service Memorial ARS c/o W. Bradley 14 Ardmore Grange, Ballygowan, Newtownards BT23 5TZ
G5 TO Sheffield & District Wireless Society c/o P. Day 38 Broomhill Road, Old Whittington, Chesterfield S41 9DA
GM5 TOA A. Carmichael 18 Waterside Gardens Carmunnock, Clarkston, Glasgow G76 9AL
G5 TRS T. Scothern 170 Wales Road, Kiveton Park, Sheffield S26 5RE
G5 TV B. Watson 173 Churchfield Lane, Darton, Barnsley S75 5EA
G5 UI R. Perkis 8 Mill Road, Cottingham, Market Harborough LE16 8XP
G5 UOS University of Suffolk RS c/o Dr K. Hotchkiss Hamilton House, Chapel Road, Mendlesham, Stowmarket IP14 5SQ
GM5 VG Obo Windy Yett CG c/o W. Miller 15 Glenalla Crescent, Ayr KA7 4DA
G5 VH P. Chapman 112 Sharpland, Leicester LE2 8UP
G5 VNX A. Wright 32 Temple Grove, Leeds LS15 0HT
G5 VO N. Clarke Brimham Lodge Farm, Brimham Rocks Road Burnt Yates, Harrogate HG3 3HE
GW5 VOG S. Cawsey 8 Hickman Road, Penarth CF64 2AJ
G5 VRA K. Cope 9 Amber Heights, Ripley DE5 3SP
G5 VZ C. Pearson 4 Brentwood Close, Thorpe Audlin, Pontefract WF8 3ES
G5 WQ I. Williams Alma Cottage , Old Vicarage Lane , South Marston, Swindon SN3 4SN
G5 WS RSGB Transatlantic Test Centenary Club c/o N. Totterdell Moscar Cross House, Hollow Meadows, Sheffield S6 6GL
G5 XDX C. Macleod 21 Halsetown, St. Ives TR26 3LY
G5 XGX P. Hughes 17 Worcester Park, Bath BA1 6QU
G5 XWH J. Horrocks 8 Moor View Close, Menston LS29 6RT
G5 YAX D. Wilkins Malt Hill Cottage, Malt Hill, Warfield, Bracknell RG42 6JG
G5 YC Icars c/o Dr S. Bunting 17 Sunnydene Avenue, Highams Park, London E4 9RE
G5 YL K. Haywood 126 Derby Street, Sheffield S2 3NF
G5 YSS D. Turner 21 Arnot Way, Higher Bebington, Wirral CH63 8LP
G5 YZI B. Gray 180 Buckingham Road, Hampton TW12 3JX
G5 ZG Bishops Stortford Ar Society c/o A. Judge 44 Thorley Lane, Bishop's Stortford CM23 4AD

G6

G6 AAB T. Sloane 42 Ashbury Drive, Blackwater, Camberley GU17 9HH
G6 AAC P. Mcgoldrick 23 Coleman Drive, Kemsley, Sittingbourne ME10 2EA
GW6 AAG F. Steadman 10 Oaktree Avenue, Sketty, Swansea SA2 8LL
GM6 AAJ G. Scattergood 14 Market Street, Forfar DD8 3EY
G6 AAK J. Smith 16 Cross Keys, Ossett WF5 9SJ
G6 AAR D. Bolingford Cobb Gate, School Lane, Pulborough RH20 4LL
G6 AAZ K. Woodward 19 Hazel Grove, Winchester SO22 4PQ
G6 ABA P. Dobson 16 Glenair Avenue, Parkstone, Poole BH14 8AD
G6 ABG D. Coldbeck 101 Westlands Road, Hull HU5 5NX
G6 ABJ M. Claydon 4 Sandringham Gardens, London N12 0NX
G6 ABM A. Chick The Rowans, Bourne View, Allington, Salisbury SP4 0AA
G6 ABO R. Campbell 207 Seabank Road, Wallasey CH45 1HD
G6 ABP C. Cave 31 Mill Road, Rearsby, Leicester LE7 4YN
G6 ACJ D. Frampton 28 Horsham Road, Owlsmoor, Sandhurst GU47 0YY

G6 AD A. Fairclough Forders House, Forders Lane, Marston Jabbett, Bedworth CV12 9SG
G6 ADD T. Hallam 98 Keppel Road, Sheffield S5 0TY
G6 ADG M. Kennedy 96 Kingsway, Boston PE21 0AU
G6 ADO S. Nicholas Greenbank, Chester High Road, Neston CH64 7TR
G6 AEB S. Neil 55 Colne Road, Brightlingsea, Colchester CO7 0DU
G6 AEC D. Nicholls 22 Yeo Way, Clevedon BS21 7UP
G6 AEK D. Molyneux 8 Ullswater Close, Hambleton, Poulton-le-Fylde FY6 9EE
GM6 AES M. Clark 12 Achaphubil, Fort William PH33 7AL
G6 AFA P. Paskin 36 Lewarne Road, Newquay TR7 3JT
GD6 AFB N. Bazley 77 Royal Park, Ramsey IM8 3UH Isle of Man
G6 AFE R. Plested 33 Hartbury Close, Cheltenham GL51 0NZ
G6 AFG A. Afford 2 Holly Court, Sandiway, Northwich CW8 2PP
G6 AFK J. Adams 6 Austen Road, Guildford GU1 3NP
G6 AFL P. Blay Treetops, Mount Pleasant, Crewkerne TA18 7AH
GW6 AFQ R. Bambrey Glangwili, Felinfach, Lampeter SA48 7PG
G6 AFS D. Bell 7 Chichester Drive, Cotgrave, Nottingham NG12 3JJ
G6 AFT A. Carr 4 Tansor Close, Corby NN17 2QP
G6 AFX A. Crickett 40 Ousden Close Cheshunt, Waltham Cross EN8 9RQ
G6 AGA G. Clark 2 Whitton Manor Road, Isleworth TW7 7NL
G6 AGN D. Darby 12 Laburnum Close, Clacton-on-Sea CO15 2DD
G6 AGO B. Bean 19 Coleshill Road, Sutton Coldfield B75 7AA
G6 AGP A. Patterson 10 Pear Tree Close, Wirral CH60 1YD
G6 AGR M. Taylor 8 Clifford Close, Long Eaton, Nottingham NG10 3BT
GW6 AGS R. Thomas 4 Duffryn Avenue, Cardiff CF23 6LF
G6 AGY A. Smith 103 Station Road, Seaham SR7 0BD
G6 AGZ G. Smith 103 Station Road, Seaham SR7 0BD
G6 AHC T. Snook 116 Rosemary Road, Poole BH12 3HE
G6 AHD M. Sumner Jaggen, Maldon Road, Chelmsford CM3 6LF
G6 AHE P. Young 8 The Slype, Wheathampstead, St. Albans AL4 8RY
G6 AHF C. Waterworth 16 Fountains Walk, Lowton, Warrington WA3 1EU
G6 AHH C. Walden The Briers, Scures Hill, Nately Scures, Hook RG27 9JS
G6 AHK C. Wallwork 40-44 Henwood Green Road, Pembury, Tunbridge Wells TN2 4LF
G6 AHN S. Reynolds 12 Lowlands Crescent, Great Kingshill, High Wycombe HP15 6EG
G6 AHO A. Oakes 12 Bridge Mill Court, Chorley PR6 9DU
G6 AHR R. Redpath Hillview, Tudballs, Exford, Minehead TA24 7PT
G6 AHV J. Spriggs Kreuzstr. 18, Tuerkenfeld D-82299 Germany
G6 AHX S. Evans 18 Hillview Lane, Twyning, Tewkesbury GL20 6JW
G6 AIB G. Farline Willow Cottage, Manor View Road, Scarborough YO11 3PB
G6 AIC R. Field Flat 27, Captain Webbs 161-165 Folkestone Road, Dover CT17 9SZ
G6 AIG H. Gibson 10 Trafalgar Street, Cambridge CB4 1ET
G6 AII R. George 27 Nocton Park Road, Nocton, Lincoln LN4 2BE
G6 AIK J. Gill Millside, Mill Road, Steyning BN44 3LN
G6 AIO P. Hillier 20 Firtree Road, Norwich NR7 9LG
G6 AIQ M. Homer 29 Holmefield Avenue, Fareham PO14 1EF
G6 AIU L. Harland 16 Burford Close, Dagenham RM8 3ST
G6 AIZ M. Holmes 15 Anderton Way, Garstang, Preston PR3 1RF
G6 AJ Barnsley & District ARC c/o D. Gillott 132 Racecommon Road, Barnsley S70 6JY
GM6 AJA M. Hunt Gaoith The Saorsa, 30 Kanachrine Place, Ullapool IV26 2TX
G6 AJC I. Hodgkins 2 Seagrave Road, Coventry CV1 2AA
G6 AJG T. Jenkins 134 Frankland Road, Croxley Green, Rickmansworth WD3 3AU
GW6 AJK P. Jones Ty?N Y Coed, Penley LL13 0LN
G6 AJS A. Sharp 17 Beechwood Avenue, Flanshaw, Wakefield WF2 9JZ
G6 AJT B. Kenneally 5 Havengore, Pitsea, Basildon SS13 1JU
G6 AJV G. Larcombe 55 Fairways International Caravan & Camping Park, Bath Road Bawdrip, Bridgwater TA7 8PP
G6 AJW D. Lucas 42 Falcon Way, Ashford TN23 5UR
G6 AJX S. Lampard 111 Whitworth Way, Wilstead, Bedford MK45 3EF
G6 AK J. Brister 49 Tiverton Road, Loughborough LE11 2RU
G6 AKG R. Ayley 1 Ballam Close, Upton, Poole BH16 5QT
G6 AKK P. Archer 26 Freshfield Drive, Macclesfield SK10 2TU
G6 AKN M. Bentley 9 Tinkers Castle Road, Seisdon, Wolverhampton WV5 7HF
GW6 AKS F. Barwell Galahad, Penisarwaun, Caernarfon LL55 3BN
G6 AKX T. Blackburn 42 Thames Drive, Biddulph, Stoke-on-Trent ST8 7HL
G6 ALB A. Burge 2 Lower End, Swaffham Prior, Cambridge CB25 0HT
G6 ALG N. Cutmore 3 Linden Close, Tadworth KT20 5UT
G6 ALJ T. Collins 11 Sutton Road, Maidstone ME15 9AE
G6 ALN G. Colclough 20 Pembroke Drive, Whitby, Ellesmere Port CH65 6TD
G6 ALR R. Delamare 28 Blandford Road, Plymouth PL3 5DU
G6 ALU Wessex CG ARS c/o S. Drury 25 Crosslands, Stantonbury, Milton Keynes MK14 6AY
G6 ALW B. Darby 96 Bassnage Road, Halesowen B63 4HG
G6 ALZ A. Davis 2 Wolverhampton Road, Essington, Wolverhampton WV11 2DB
G6 AMF B. Elliott 41 Henwick Lane, Thatcham RG18 3BN
GW6 AMK W. Needham Cnwc Y Rhedyn, Aberporth, Cardigan SA43 2DA
G6 AML J. Newcombe Elmcroft, Upper Basildon RG8 8LS
G6 AMV L. Mcconnell 12 Marlborough Drive, Weston-Super-Mare BS22 6DQ
G6 AMW C. Williams 133 Devon Drive, Chandler'S Ford, Eastleigh SO53 3GJ
G6 AMX P. Helm 74 Neston Road Walshaw, Bury BL8 3DB
G6 ANA P. Miller Flat 907, De Montfort House, Leicester LE1 5XR
GI6 ANC A. Murphy 53 Whitehouse Park, Newtownabbey BT37 9SH
G6 ANI J. Baverstock Meadow View, Newbridge, Cadnam, Southampton SO40 2NW
G6 ANJ C. Perrott 15 Chestnut Drive, Claverham, Bristol BS49 4LN
G6 ANO L. Goodwin Gallifrey, The Village, Chelmsford CM3 1AS
G6 ANR S. Garfirth 19 Ingleside Drive, Stevenage SG1 4RN
G6 ANV B. Gulliford 12 Hawthorn Road, Eynsham, Witney OX29 4NT
GM6 ANZ J. Howat 61 Carolside Avenue, Clarkston G76 7AD
G6 AOB A. O'Brien 25 Sands Road, Paignton TQ4 6EG
G6 AOF J. Henshaw 5 Charlton Beeches Charlton Marshall, Blandford Forum DT11 9NP
G6 AOH R. Hoblin 4 Portiswood Close, Pamber Heath, Tadley RG26 3UQ
GM6 AOJ W. Hay 11 Lovat Road, Glenrothes KY7 4RU
GM6 AOR G. Robertson 32 The Square, Ellon AB41 9JB
G6 AOS S. Pilbeam 74 Southbank Avenue, Blackpool FY4 5BX
G6 AOV C. White 20A The Beacon, Ilminster TA19 9AH
G6 APB G. Taylor 1 Haigh Street, Greetland, Halifax HX4 8JF
G6 APD L. Sawford 16 Queens Close, Lee-on-the-Solent PO13 9NA
G6 APE A. Schofield 16 Ken Jones Close, Lightmoor, Telford TF7 5QT
G6 APH C. Shiradski 69 Masefield Avenue, Borehamwood WD6 2HG
G6 APJ Rvd. G. Smith 6 Birtley Rise, Bramley, Guildford GU5 0HZ
GW6 APK G. Sinclair 4 Nant Y Mynydd, Seven Sisters, Neath SA10 9BU
G6 APQ F. Hill 12 Woodbine Walk, Chelmsley Wood, Birmingham B37 6SB
G6 APW T. Harvey 2 Trefor Jones Court, Brookfield Avenue, Dover CT16 2QP
G6 APX S. Handley 8 Nabb Close, St. Georges, Telford TF2 9PT
GM6 AQB A. Riddell 16 Lewis Drive, Old Kilpatrick, Glasgow G60 5LE
G6 AQI T. Smith 1 St. Jude Gardens, Colchester CO4 0QJ
GM6 AQL A. Ryan 6 Cumloden Court, Newton Stewart DG8 6AB
GM6 AQR Dr H. Wynne 103 New City Road, Glasgow G4 9JX
G6 AQW N. Wiltshire 66 Neville Road, Shirley, Solihull B90 2QW
G6 ARC Burnham Beeches RC c/o C. Keens Toad Hall, 69 Lillywhite Crescent, Andover SP10 5NA
GD6 ARJ C. Jennings 13 Barrule Drive, Ballasalla, Ballasalla IM9 2HA Isle of Man
G6 ARM N. Kett 10 Carrel Road, Great Yarmouth NR31 7RF
G6 ARO I. Kendall 65 Olive Grove, Swindon SN25 3DB
G6 ARR Cray Valley RS c/o S. Kimber 3 Gloucester Way, Glossop SK13 8RZ
G6 ART S. Langton Corner Cottage, Harlow Road Sheering, Bishop's Stortford CM22 7NB
G6 ASA Prof. N. Lipman Meadowcroft, Cotswold Road, Oxford OX2 9JG
G6 ASH N. Ash 16 St. Marys Road, Sawston, Cambridge CB22 3SP
G6 ASJ M. Bradley Flat 20, Crown Court, Crown Street, Portsmouth PO1 1QN
G6 ASK J. Matthews Moor View, Oldways End, East Anstey, Tiverton EX16 9JQ
GI6 ATD South Dorset Rs c/o G. Rodgers 23 Rathmore Park, Bangor BT19 1DQ
G6 ATK K. Austin 13 North End Grove, Portsmouth PO2 8NF
G6 ATS D. Bowen Flat 9, Moorfields Court, Silver Street, Bristol BS48 2AG
GW6 ATT M. Bryan 10 Woodlands Road, Barry CF63 4EF
G6 ATW R. Czajkowski 37 The Great Court, Royal Naval Hospital, Great Yarmouth NR30 3JU
GI6 ATZ G. Curry 87 Burren Road, Ballynahinch BT24 8LF
G6 AUC Prof. H. Whitfield Apartment 88, Rishworth Palace, Rishworth Mill Lane, Rishworth, Sowerby Bridge HX6 4RZ
G6 AUD S. Challis 38 Blacksmiths Lane, Wickham Bishops CM8 3NR
G6 AUE G. Cosham 85 Capsey Road, Ifield, Crawley RH11 0UF
GI6 AUI D. Doherty 42 Silverbrook Park, Newbuildings, Londonderry BT47 2RD
G6 AUO M. Graffham 106 Barford Road, Edgbaston, Birmingham B16 0EF
G6 AUP B. Goodyear 13 Moorland Avenue, Barnsley S70 6PQ
G6 AUR B. Golding 67 Milford Avenue, Wick, Bristol BS30 5PP
G6 AUW R. Howes 8 Kitchener Road, Weymouth DT4 0LN
G6 AUY D. Hawley The Old Dairy, Edgefield Hall Barns, Edgefield NR24 2RD
G6 AVI North Yorkshire ARG c/o R. Tucker Foxhall Cottage, Dukes Lane, Attleborough NR17 1BL
G6 AVK C. Thomson 160 Down Hall Road, Rayleigh SS6 9PD
G6 AVL H. Thompson 6 Alexandra Chase, Casa Rosa, Cramlington NE23 6AA
G6 AVN D. Shaw 33 The Fairway, Halifax HX2 9PZ
G6 AVP A. Rowe 2 Broom Mead, Bexleyheath DA6 7NY
G6 AVS J. Russell11 13 Stonebridge Lea, Orton Malborne, Peterborough PE2 5LY
G6 AVT G. Stanhope 39 Denham Close, Stubbington, Fareham PO14 2BQ
G6 AVY Tiverton South West ARC c/o D. Lane 230 Raeburn Avenue, Eastham, Wirral CH62 8BB
G6 AWF D. Miller 33 Springfield Park, Twyford, Reading RG10 9JG
G6 AWM C. Montgomery 70 Campbell Road, Twickenham TW2 5BY
G6 AWO R. Mansel Ashcroft House, Ashfield Road, Bury St. Edmunds IP30 9HJ
G6 AWP A. Mchardy The Haven, Hull Road, Hull HU12 0TE
G6 AWY N. Armstrong 27 High Street, Billinghay LN4 4AU
G6 AWZ P. Ashdown 1 Wheelers Patch, Emersons Green, Bristol BS16 7JL
G6 AXC R. Beaumont The New Hall, Fletchergate, Hull HU12 8ET
G6 AXE G. Broad 14 Albion Road, Westcliff-on-Sea SS0 7DR
G6 AXH P. Brothers 101 Bridgewater Drive, Northampton NN3 3AF
G6 AXK P. Butler 25 Orrishmere Road, Cheadle Hulme, Cheadle SK8 5HP
G6 AXV R. Baldock 12 Chippendale Close, Baughurst, Tadley RG26 5HF
G6 AXY P. Coombes Harleyford, Lower Wokingham Road, Crowthorne RG45 6BT
GM6 AXZ K. Cocks 60A Palmerston Place, Edinburgh EH12 5AY
G6 AY G. Kellaway 55 Ladbrooke Drive, Potters Bar EN6 1QW
G6 AYD D. Chorley Sunnylands, Sandpitts Hill, Langport TA10 0NG
G6 AYE S. Cotterill Arcadia, Leicester Lane, Leicester LE9 9JJ
G6 AYH K. Cooke 28 Curland Place, Longton, Stoke-on-Trent ST3 5JL
G6 AYI B. Collett 264A Lichfield Road, Sutton Coldfield B74 2UH
GW6 AYR R. Shearing Woodstock Fairview Old Winchfawr Road Clwydyfagwr, Merthyr Tydfil CF48 1HW
G6 AYS T. Ramsden 1A Fox Grove, Walton-on-Thames KT12 2AT
G6 AYU P. Rice 4 Council St., Walton, Peterborough PE4 6AQ
G6 AYX D. Robinson 3 King Edward Avenue, Wickham Market, Woodbridge IP13 0SL
G6 AYY T. Rumbold 23 Montague Road, Saltford, Bristol BS31 3LA
G6 AZE A. Roberts 9 Littlemoor Lane, Newton, Alfreton DE55 5TY
GW6 AZG S. Pearless Glanrafon, Dolgellau LL40 2AH
G6 AZL P. Tarmey 48 Merlin Crescent, Burton-on-Trent DE14 3JF
G6 AZP D. Glover 16 Cardigan Grove, Trentham, Stoke-on-Trent ST4 8XY
G6 AZR A. Granshaw 38 Tudor Gardens, Stony Stratford, Milton Keynes MK11 1HX
GW6 AZX R. Hughes 4 Brittania Terrace, Porthmadog LL49 9NB
G6 BAD C. Dovener The Old Barn, 39 Sun Street, Haworth, Keighley BD22 8BY
GW6 BAH G. Davis 2 New House, Ponthir Road, Gwent NP6 1PE
G6 BAL D. De La Haye 14 Palace Meadow, Chudleigh, Newton Abbot TQ13 0PJ
G6 BAM J. Draper 42 Pitt Street, Broadwaters, Kidderminster DY10 2UN
GM6 BAO A. Devine 12 Auchengate, Barassie, Troon KA10 6UG
G6 BAT D. Falstein 3 Gracefields, 121 The Avenue, Fareham PO14 3AA
G6 BAY C. Howes 1 Wharrage Road, Alcester B49 6QY
G6 BBD R. Hancock 16 Buttermere, Wellingborough NN8 3ZA
G6 BBG A. Harland 23 Shelley Drive, Stratford Sub Castle, Salisbury SP1 3JZ
G6 BBH N. Burton 32/11-19 Stirling Road, Claremont 6010 Australia
G6 BBI P. Ward 63 Salcombe Drive, Glenfield, Leicester LE3 8AG
G6 BBK S. Nelson 10 Wragg Drive, Newmarket CB8 7SD
G6 BBN J. Temple-Heald Shires, 28 West End, Cambridge CB22 4LX
G6 BBR M. Thomas 17 Rectory Park Avenue, Sutton Coldfield B75 7BL
G6 BBW J. Witts 35 Warton Road, Basingstoke RG21 5HL
G6 BCG R. Whitehouse 5 Parkland Drive, Darlington DL3 9DT

G6 BCL N. Miller Bc House, East Hanningfield Road, Chelmsford CM3 8EW
G6 BCM S. Ward 33 All Saints Way, Aston, Sheffield S26 2FJ
G6 BD M. Farmer 16 Beckside, Nettleham, Lincoln LN2 2PH
G6 BDA D. Harvey 42 Norbury Drive, Lower Heath, Congleton CW12 1NB
G6 BDH J. Kennard 52 Lavender Lane, Stourbridge DY8 3EF
GI6 BDI A. King 43 Orby Gardens, Belfast BT5 5HS
GW6 BDM C. Parker Dolifor Llanwrthwl, Llandrindod Wells LD1 6NU
GI6 BDN R. Larke 11 Ballymaconnell Road South, Bangor BT19 6DG
G6 BDW A. Sibley 25 Vesta Avenue, St. Albans AL1 2PG
G6 BDY R. Southern 208 Puxton Drive, Kidderminster DY11 5HJ
G6 BEB J. Lines Karen House, 11 Hill St., Brierley Hill DY5 2AY
G6 BEH K. Penaluna 5 Holkham Close, Rushmere St. Andrew, Ipswich IP4 5DW
G6 BEL S. Fairweather 65 Ambleside Avenue, Hornchurch RM12 5EU
G6 BEN UK Amateur Radio Discord c/o A. Burke 24 Wentworth Close, Farnham GU9 9HJ
G6 BER S. Boote The Shippen, Downgate, Callington PL17 8JX
GM6 BEY M. Craig 7 Hallyards Cottages, Kirkliston EH29 9DZ
G6 BFM A. Green 117 Acanthus Road, Liverpool L13 3DY
G6 BFP L. Humphrey Four Gables, Gilletts Lane, High Wycombe, Bucks HP12 4BB
G6 BGA K. Turvey St. Vincents Cottage, St. Vincents Lane, West Malling ME19 5BW
G6 BGH I. Macdiarmid 73 Stadium Avenue, Blackpool FY4 3QA
GM6 BGL K. Maclean 177 Lamond Drive, St. Andrews KY16 8JP
GM6 BGQ D. Small 50 Toll Court, Lundin Links, Leven KY8 6HH
G6 BGY J. Meek Flat 26, Wickham Court, Clevedon BS21 7TN
G6 BHA R. Smart 67 Corkland Road, Chorlton Cum Hardy, Manchester M21 8XT
G6 BHB J. Seager 9 Lodge Close, Brighstone, Newport PO30 4BX
G6 BHE V. Rogers 66 East Beach Park, 66 East Beach Park, Shoeburyness SS3 9SG
G6 BHH D. Palmer Firdene, Abbey Road, Alton GU34 5PB
G6 BHI A. Palmer Firdene, Abbey Road, Alton GU34 5PB
GW6 BHQ K. Williams 19 Narberth Crescent, Llanyravon, Cwmbran NP44 8RJ
GM6 BHR R. Warbrick 8 Bathurst Drive, Alloway, Ayr KA7 4QN
G6 BHS J. Watson 58 St. Georges Drive, Cheltenham GL51 8NX
G6 BHX C. Walker 19 Springfield Grove, Corby NN17 1EN
G6 BHY R. Vicarage 10 Fleming Avenue, Sidford, Sidmouth EX10 9NY
G6 BIA R. Thompson 39 Grotto Road, South Shields NE34 7AQ
GM6 BIG D. Anderson 20 Greenrig Road, Hawksland, Lesmahagow, Lanark ML11 9QA
G6 BIM J. Bowers 6 Fairview Park, Hetton-Le-Hole, Houghton le Spring DH5 0SE
G6 BIT Dr D. Crossley 25 Newhaven Close, Bury BL8 1XX
G6 BIU D. Carter 23 First Street, Low Moor, Bradford BD12 0JQ
G6 BIX E. Donbavand 6 Springmeadow, Charlesworth, Glossop SK13 5HP
G6 BJB Dr A. Forsyth 14 Highgrove Road, Lancaster LA1 5FS
G6 BJG I. Hancock 64 Swanswell Road, Solihull B92 7EY
G6 BJJ I. Harley 37 Crelake Close, Tavistock PL19 9AX
G6 BJL R. Harding 12 Aller Vale Close, Exeter EX2 5NH
G6 BJO P. Mctaggart 33 Manor Farm Close, Barton-Le-Clay, Bedford MK45 4TB
G6 BJR K. Hulbert 15 St. Germans Road, Forest Hill, London SE23 1RH
G6 BJY D. Vivash 16 Whitchurch Close, Maidenhead SL6 7TZ
GW6 BK Blackwood CG c/o R. Jones 2 Pen-Y-Cwarel Road, Wyllie, Blackwood NP12 2HP
G6 BKD J. Scotney 30 Trinity Road, Rothwell, Kettering NN14 6HY
G6 BKL P. Metcalfe 65 Saville Road, Whiston, Rotherham S60 4DZ
G6 BKY N. Arkwright Bay Hill, Woodhouse Lane, Heversham, Milnthorpe LA7 7EW
G6 BLA S. Woodford The Lord Nelson, 1 Hale Road, Thetford IP25 7RA
G6 BLC B. Conway 29 Mandeville Road, Southgate, London N14 7NJ
G6 BLK A. Johnson Edelweiss, Boxley Road, Chatham ME5 9JG
G6 BLU B. Nicholls 29 Wittmead Road, Mytchett, Camberley GU16 6ER
G6 BME D. Gibb 46 School Road, Charing, Ashford TN27 0JN
G6 BMG Ham Radio Builders Club c/o J. Hind 80 Forge Fields, Sandbach CW11 3RD
GM6 BML A. Ramsay 15 Dunalistair Gardens, Broughty Ferry, Dundee DD5 2RJ
GW6 BMP A. Roberts 16A High Street, Llangefni LL77 7NA
G6 BMY R. Satterthwaite 47 Aberford Road, Baguley, Manchester M23 1JY
G6 BMZ M. Williams 3 Teesdale Road, Nottingham NG5 1DA
GI6 BNI D. Mawhinney 14 Cayman Avenue, Bangor BT19 6XG
G6 BNO D. Dallaway 17 Bantams Close, Birmingham B33 0YL
GM6 BNS S. Lewis Eyin Helga, Evie, Orkney KW17 2PJ
G6 BNT M. Dronfield 1 Kestrel Rise Swallownest, Sheffield S26 4SD
G6 BNU Dr P. Entwistle Waverley Cottage, Sherfield Road, Bramley, Tadley RG26 5AG
G6 BNW Dr J. Garcia-Rodriguez St. Albans, Mill Lane, Dover CT15 4HR
G6 BOF G. Hollidge 10 Newfoundland Close, Worth Matravers, Swanage BH19 3LX
G6 BOK P. King 10 Heath Hey, Woolton, Liverpool L25 4TJ
G6 BOP A. Reid 19 Springfield Road, Wincanton BA9 9BL
G6 BOQ E. Parker Jasmine Cottage, Apperley, Gloucester GL19 4DE
G6 BOX S. Wilson 21 Plumian Way, Balsham CB21 4EG
G6 BPH F. Bennewitz 1 Millfield Avenue, Saxilby, Lincoln LN1 2QN
G6 BPK Dr S. Cook 50 Bath Road, Swindon SN1 4AY
G6 BPN R. Edmondson 91 Lewin Road, London SW16 6JX
G6 BPY W. Roe 39 Marlborough Road, Southwold IP18 6LR
G6 BQC M. Stuart 207 Saunders Lane, Mayford, Woking GU22 0NT
G6 BQE P. Tilley 22 Meadowsweet, Waterlooville PO7 8RS
G6 BQM P. Bentley Sandy Ridge, Church Street, Stoke-on-Trent ST7 4RS
G6 BQQ M. Barnes Drovers, Crampshaw Lane, Ashtead KT21 2UF
G6 BRA Bracknell ARC c/o I. Pawson 3 Orion, Roman Hill, Bracknell RG12 7YX
GW6 BRC Barry ARS c/o S. Trahearn 148 Gladstone Road, Barry CF62 8ND
G6 BRD W. Hammond 245 Broadoak Road, Ashton-under-Lyne OL6 8RP
G6 BRP P. Walter 2 Hallams Lane, Beeston, Nottingham NG9 5FH
GM6 BRU J. Steele 54 Myrtle Crescent, Bilston, Roslin EH25 9SB
G6 BRV R. Shelford 1 All Lands Cottages, Highcross, Rotherfield, Crowborough TN6 3QA
G6 BRW S. Sumner 27 High Street, Flore, Northampton NN7 4LL
G6 BRY C. Thomas 52 Derwent Road, Burton-on-Trent DE15 9FR
G6 BSP S. Guest Flat 2 7 Corscombe Close, Weymouth Dorset UK DT4 0UE
G6 BSS G. Higgs 68 Otterfield Road, Yiewsley, West Drayton UB7 8PF
G6 BTB C. Pringle 38 Priory Road, Littlemore, Oxford OX4 4NE
G6 BTC A. Layton 17 Maplehurst, Leatherhead KT22 9NB
G6 BTM N. Rice 31 Bold Street, Heysham, Morecambe LA3 1TS
G6 BTP E. Beswarick 2 Hurst, Beaminster DT8 3ES
G6 BTR M. Challis 18 Castlefield Close, Eastleaze, Swindon SN5 7EG
G6 BTX K. Holmes 313 Havering Road, Romford RM1 4BZ
G6 BUH J. Walsh 13 Byam Street, London SW6 2RB
G6 BUP Aldridge And Barr Beacon ARC c/o C. Chan 11 The Paddocks, Welwyn Garden City AL7 2BW
G6 BUT Harlow & District ARS c/o M. Simkins 37 St. Andrews Meadow, Harlow CM18 6BL
G6 BUU R. Costello 6 Qua Fen Common, Soham, Ely CB7 5DH
G6 BUV A. Cutts Highthorns Cottage, North Frodingham, Driffield YO25 8LS
GW6 BUW I. Davies Garthewyn, Caernarfon LL55 2RL
G6 BUY R. Gingell 23 Woodfarm Road, Malvern Wells, Malvern WR14 4PL
G6 BV South Birmingham RS c/o J. Storey 34 Austin Rise, Longbridge, Birmingham B31 4QN
G6 BVF Prof. N. Linge 21 Pennant Drive, Prestwich, Manchester M25 3BT
G6 BVO S. Green 3 Mulberry Close, March PE15 9FH
GI6 BVQ T. Finlay 4 Station Road, Eglinton, Londonderry BT47 3PR
G6 BVR R. Gammage 12 The Butts, Warwick CV34 4SS
GW6 BVS J. Hayman 22 Princess St., Abertillery NP3 1AR
G6 BWA C. Clarke 11 Eastmoor Villas, Epworth Road, Doncaster DN9 2LH
G6 BWE K. Edwards 289 Monks Walk, Buntingford SG9 9DZ
G6 BWJ J. Richards 44 Swain Street, Watchet TA23 0AG
G6 BWK T. Wallis 17 Alderbank, Wardle, Rochdale OL12 9NH
G6 BWM R. Smith 49 Aubourn Avenue, Lincoln LN2 2JW
G6 BWN J. Stewart 101 West Way, Lancing BN15 8LZ
G6 BWO J. Taberner 20 Stevenson Drive, Wirral CH63 9AH
G6 BWP D. Weaver 8 Strathmore Close, Worthing BN13 1PQ
G6 BWT A. Bajjon 35Ablackford Rd, , Shirely B90 4BU
G6 BXO C. Blackwell 20 Southworth Avenue, Blackpool FY4 3LH
G6 BXR R. Calvert 3A Panxworth Road, South Walsham, Norwich NR13 6DY
G6 BXS D. Ellison Riverside, Old Mill Drive, Colne BB8 0TX
G6 BXT M. Fry 61 Swift Road, Abbeydale, Gloucester GL4 4XH
GW6 BXU E. Hatherall 101 Park Crescent, Abergavenny NP7 5TL
G6 BXV D. Willis Rivendell, Shirnall Hill, Alton GU34 3EJ
G6 BYF C. Gomez The Gazebo, Military Road, Rye TN31 7NY
G6 BYK J. Parkes 65 Ferrier Road, Stevenage SG2 0NZ
G6 BYL D. Lycett 1 Saredon Close, Pelsall, Walsall WS3 4DH
G6 BYR T. Frangopulo Flat 11, Jack Edwards Court 5 Lapwing Lane, Manchester M20 2NT
G6 BZE M. James 9 Wyke Mark, Winchester SO22 5DJ
G6 BZG L. Green 37 Park Road, Northville, Bristol BS7 0RH
G6 BZL M. Adams The Vicarage, Intake Lane, Ormskirk L39 0HW
G6 BZQ G. Doubleday 1 St. Johns Avenue, Chelmsford CM2 0UA
G6 BZW E. Butt 97 Hawthorn Crescent, Yatton, Bristol BS49 4RG
G6 CAC J. Hallett 16 Streche Road, Swanage BH19 1NF
GI6 CAG W. Millar 9 Lynnehurst Drive, Comber, Newtownards BT23 5LN
G6 CAR Icom (UK) Ar c/o A. Baldwin Rathlin, Dromnea, Kilcrohane, Bantry P75 Y300 Ireland
G6 CAU M. Samuel 71 Brighton Road, South Croydon CR2 6EE
G6 CBB D. Beddow 34 Loweswater Road, Stourport-on-Severn DY13 8LP
G6 CBL D. Leslie 8 The Avenue, Swarland, Morpeth NE65 9JL
G6 CBP A. Pidgeon 106 Winchester Avenue, St Johns, Worcester WR2 4JQ
G6 CBY M. Jeeves 52 Castlefields, Istead Rise, Gravesend DA13 9EJ
G6 CCB A. Stonehouse 105 Humberston Avenue, Humberston, Grimsby DN36 4ST
G6 CCN L. Armstrong Deuteros House 1 Bank Top, Earsdon NE25 9JS
G6 CCQ R. Powell Manuela 25 Jack Haye Lane Light Oaks, Stoke-on-Trent ST2 7NG
G6 CDT G. Henshaw 18 Queens Avenue, Ilkeston DE7 4DL
G6 CDU G. Keeble 4 Bardfield Way, Frinton-on-Sea CO13 0AN
G6 CDV A. Morling 33 Russell Court, Chesham HP5 3JH
G6 CDW N. Miller 3 Upwood Gorse, Tupwood Lane, Caterham CR3 6DQ
G6 CEM E. Weir 10 St. Georges Crescent, Whitley Bay NE25 8BJ
G6 CEP A. Kneebone 34 Henver Road, Newquay TR7 3BN
G6 CEZ R. Brand 17 Park Road, Fordingbridge SP6 1EQ
G6 CFA J. Carrick Smith 15 The Vale, Oakley, Basingstoke RG23 7LB
G6 CFC G. Purchon 33 Lancaster Avenue, Hitchin SG5 1PA
G6 CFU N. Shaw Crane Hill, Oxenton, Cheltenham GL52 9SE
G6 CGC R. Sheppard 51 Marks Road, Wokingham RG41 1NR
G6 CGI M. Rowat 154 Hollingwood Lane, Bradford BD7 4DB
G6 CGO E. Parr 18 Arundel Close, Macclesfield SK10 2NS
G6 CGQ R. Hatch 4 Springfield Crescent, Parkstone, Poole BH14 0LL
G6 CGY Lids CW & Data Club c/o R. Percival 6 Bulmer Place, Hartlepool TS24 9BQ
G6 CHA E. Povey Hillcroft, Schoolfields, Henley-on-Thames RG9 4DH
G6 CHC V. Appleton 15 Pinewood Crescent, Ramsbottom, Bury BL0 9XE
G6 CHD P. Bridle Flat 25, Riverview Gardens, 289 Old Chester Road, Birkenhead CH42 3XQ
G6 CHI A. Bowley Plum Tree House, Walk Close, Derby DE72 3PN
G6 CHJ M. Carter 5 Orchard Brook, Long Melford, Sudbury CO10 9LF
G6 CHT M. Hall 31 Meendhurst Road, Cinderford GL14 2EF
G6 CHX P. Holland High Lea Cottage, Witchampton Lane, Wimborne BH21 5AF
G6 CIA T. Kenyon 31 Marble Hill Gardens, Twickenham TW1 3AU
G6 CID J. Andrews Top Flat, 211 Brighton Road, South Croydon CR2 6EJ
G6 CIE R. Townsend 3 Cranfield View, Darwen BB3 2HP
G6 CIF D. Taylor 8 Russell Drive, Wollaton, Nottingham NG8 2BH
G6 CII K. Sutton 9 Babbacombe Drive, Ferryhill DL17 8DA
G6 CIO J. Robinson 31 Church Road, Banks, Southport PR9 8ET
G6 CIP P. Ralston Laund House, 9 College Avenue, Liverpool L37 3JL
G6 CIT Grantham ARC c/o R. Young 143 Rodmell Avenue, Saltdean, Brighton BN2 8PH
G6 CJB P. White 8 Kingswood Court, Maidenhead SL6 1DD
GW6 CJJ Dr J. Alexander 2 Meadow Park, Burton, Milford Haven SA73 1NZ
G6 CJR S. Barber Homedale, St. Monicas Road, Tadworth KT20 6ET
G6 CJT B. Bradshaw 19 Wren Garth, Beeford, Driffield YO25 8FQ
G6 CKD L. Newbury 30 Sharon Close, Felmingham, North Walsham NR28 0LJ
G6 CKE C. Evans 21 Snowdrop Close, Crawley RH11 9EG
G6 CKH J. Muir 150 Thorntree Road, Thornaby, Stockton-on-Tees TS17 8LX
G6 CKJ D. Morris 255 Lichfield Road, Wolverhampton WV113EW
G6 CKK R. Martin 1 Rosemount Court, Rochester ME2 3NF
G6 CKL I. Martin 24 Heddington Close, Trowbridge BA14 0LH
G6 CKM D. Langdon 17 Forest Grove, Eccleston Park, Prescot L34 2RY
GM6 CKN R. Morrison 38 Burnfoot Road, Hawick TD9 8EN
G6 CKW R. Beattie 11 Pine Grove, Bricket Wood, St. Albans AL2 3ST

G6 CKY M. Gray 20 Ravenstone Street, London SW12 9SS
G6 CKZ P. Berwick 4 Brewer Road, Crawley RH10 6BP
G6 CLA G. Blacksell 152 Hawthorn Avenue, Colchester CO4 3YA
G6 CLC N. Carter 58 Meadow Road, Barlestone, Nuneaton CV13 0HQ
G6 CLD G. Coker 46 Clarendon Road, Ipplepen, , Newton Abbot TQ12 5QS
G6 CLK P. Carter 19 Felix Road, Walton-on-Thames KT12 2LB
G6 CLP J. Miller 7 Malvern Crescent, Ashby-de-la-Zouch LE65 2JZ
G6 CLU D. Lawes 8 High Beech Chalet Park, Battle Road, St. Leonards-on-Sea TN37 7BS
G6 CLW B. Lloyd 243 Stand Lane, Radcliffe, Manchester M26 1JA
G6 CLX D. Lloyd 31 Lever Park Avenue, Horwich, Bolton BL6 7LF
GI6 CMA R. Dawson 31 Clonmore Manor, Lisburn BT27 4EW
G6 CMB I. Dalton 10 St. Vincents Villas, Temple Hill, Dartford DA1 5HT
G6 CMD C. Driver 23 Mercers Row, St. Albans AL1 2QS
G6 CMF A. Daborn 49 Crescent Road, Locks Heath, Southampton SO31 6PE
G6 CML J. Sykes 1 Mulberry Lodge, Cudlow Gardens, Rustington BN16 2RN
G6 CMN A. Shaw 14 Delph Crescent, Clayton, Bradford BD14 6RY
GM6 CMQ D. Robson 35 Lady Nairne Road, Dunfermline KY12 9YD
G6 CMS M. Robertson 13 Orchard Cottages, Main Road, Boreham, Chelmsford CM3 3AD
G6 CMV D. Palmer Spidrift, Landsdown Road, Malvern WR14 1HX
G6 CMX J. Pell 33 Low Street, Winterton, Scunthorpe DN15 9RT
G6 CND J. Oliver 3 Savile Walk, Brierley, Barnsley S72 9HJ
G6 CNF J. Payne 71 Waarden Road, Canvey Island SS8 9AB
G6 CNK R. Freshwater 82 Sandford Road, Chelmsford CM2 6DH
G6 CNL P. Farnell 40 Thorney Lane, Luddendenfoot, Halifax HX2 6UX
G6 CNQ T. Genes 28 Hillside Road, Burnham on Crouch CM0 8EY
GW6 CNS Southdown ARS c/o J. Graham 23 Somerset Road, Barry CF62 8BL
G6 CNW J. Gibson Penrose Cottage, Carne, St. Austell PL26 8DB
G6 CNX J. Goodwin 10 Abingdon View, Worksop S81 7RT
G6 COB J. Hodkinson 3 Cypress Close, Market Drayton TF9 3HJ
G6 COE C. Hill Manor Farm Cottage, Portington, Goole DN14 7LZ
G6 COG D. Holdsworth Middle Pasture, Heath Lane, Halifax HX3 0AG
G6 COL Lincoln Shortwave c/o P. Rose Pinchbeck Farmhouse, Mill Lane, Sturton By Stow, Lincoln LN1 2AS
G6 COZ R. Turner 73 Digby Court, Nottingham NG7 1RG
G6 CP D. Cutter David Cutter Engineering, 34 Greengate Lane, Knaresborough HG5 9EL
G6 CPE K. Stanley 35 St. Blaize Road, Romsey SO51 7JY
G6 CPF J. Stephenson 16 Greenways, Driffield YO25 5HX
G6 CPO N. Wysocki 6 Rose Dene, Stourport-on-Severn DY13 8SU
G6 CPS A. Yates 12 Graham Drive, Middleton, King's Lynn PE32 1RL
G6 CPX M. Waples 24 Constable Drive, Wellingborough NN84UX
G6 CPY E. Whitham 72 Bole Hill, Treeton, Rotherham S60 5RE
G6 CQB M. Wilson 23 Claydown Way, Slip End, Luton LU1 4DU
G6 CQC A. Varty Wisteria, Hillcrest, Burnhope, Durham DH7 0BQ
G6 CQH J. Abbishaw Hastings House Farm, Littletown, Durham DH6 1QB
G6 CQR Coventry Raynet Group c/o C. Bailey 32 Ryland Road, Moulton, Northampton NN3 7RE
G6 CRC Cheshunt & District A.R.S c/o R. Gray 51 Wyatt Close, Ickleford, Hitchin SG53XY
G6 CRF T. Bailey 65 Edge Lane, Chorlton Cum Hardy, Manchester M21 9JU
G6 CRG B. Bowes 1 Rockall Close, Southampton SO16 8EH
G6 CRR R. Solomons 32 Church Road, Pembury, Tunbridge Wells TN2 4BT
GM6 CRX F. Mcleod-Stangroom 6 Leonach, Strathlachlan, Cairndow PA27 8DB
G6 CSC Fylde Coast Ray c/o W. Skidmore 29 The Meadows, Grisedale Road, Bakewell DE45 1TP
G6 CSK A. Beal 115 Maldon Road, Witham CM8 1HR
G6 CSL C. Redding 20 Bromley Street, Workington CA14 2TP
G6 CSN G. Chadwick 25 Passmonds Crescent, Rochdale OL11 5AW
G6 CSR H. Calloway 6 Franchise Gardens, Wednesbury WS10 9RQ
G6 CTA J. Davidson 12 Hanbury Close, Dronfield S18 1RF
G6 CTC Bridges RC Hampshire c/o J. Witt 67 Dillotford Avenue, Coventry CV3 5DS
G6 CTE L. Duffill 14 Leonard Street, Hull HU3 1SA
G6 CTH E. Dunne 16 Ulleswater Close, Little Lever, Bolton BL3 1UD
G6 CTP H. Wakefield 32 Mandene Gardens, Great Gransden, Sandy SG19 3AP
G6 CTV E. Eggs 62 Laurel Manor, 18 Devonshire Road, Sutton SM2 5EJ
G6 CTX M. Brown 31 Kingscote Road, Cowplain, Waterlooville PO8 8QD
G6 CTY C. Edwards 54 Thoroughgood Road, Clacton-on-Sea CO15 6DP
G6 CUA H. Erridge 15 Maurice Road, Southsea PO4 8HH
G6 CUE J. Frampton 54 Hudson Road, Bexleyheath DA7 4PG
G6 CUK A. Fisher 1 Elm Walk, Catterick, Richmond DL10 7PB
G6 CUQ N. Wedgbury 12 The Ridgeway, Astwood Bank, Redditch B96 6LT
GW6 CUR S. Williams 371 Coed-Y-Gores Llanedeyrn, Cardiff CF23 9NR
G6 CUT J. Whitehurst Serendipity, 97 Noke Common, Newport PO30 5TY
G6 CUV K. Wyeth 3 West Palace Gardens, Weybridge KT13 8PU
G6 CUY J. Wildsmith Lingmoor, 7 Lambert Road, Uttoxeter ST14 7QG
G6 CVB J. Taylor 12 Fairview Drive, Westcliff-on-Sea SS0 0NY
G6 CVD C. Thornley Sylvastone House, Herne Street, Herne CT6 7HG
G6 CVE R. Tanfield 8 Rede Close, Bedford MK41 7UH
G6 CVP D. Wilkins The Workshop, Rear Of 59 Jasmine Grove, London SE20 8JY
G6 CVR J. Geary 2 Eastgate Lane, Terrington St. Clement, King's Lynn PE34 4NU
G6 CVV M. Gumbrell 2A The Avenue, Carlby, Stamford PE9 4NA
G6 CVW W. Griffiths 6 Stanway Close, Middleton, Manchester M24 1HE
GW6 CVX R. Griffith The Hollies, Morton, Oswestry SY10 8AJ
G6 CVY H. Gibbons 27 Bentham Road Chesterfield, Chesterfield S40 4EZ
G6 CWF C. Hazell 18 Cleeve Hill, Downend, Bristol BS16 6HN
G6 CWH S. Harwood 24 Firle Crescent, Lewes BN7 1QG
G6 CWP D. Hartley 4 Park Gate, Euston Road Fakenham Magna, Thetford IP24 2QS
G6 CWW V. Holbrook 2 The Crescent, Derby DE21 6QB
GW6 CWZ D. Mccallum Glan Alaw Llanddeusant, Holyhead LL65 4AG
GI6 CXD R. Mcwhirter 200 Townhill Road, Portglenone, Ballymena BT44 8AR
G6 CXI A. Long 35 Heath Court, Grampian Way, Derby DE24 9NG
G6 CXN K. Lankshear 28 Monmouth Place, Newcastle-under-Lyme ST5 3DF
G6 CXO D. Lloyd 16 Kingsley Road, Brighton BN1 5NH
G6 CXV K. Phillips Stockings Barn, Whitbourne, Worcester WR6 5SR
G6 CXY R. Revan 50 Woodland Rise, Welwyn Garden City AL8 7LF
G6 CYA R. King 55 Coppins Road, Clacton-on-Sea CO15 3HS
G6 CYE A. Read 36 West St., Tollesbury, Maldon CM9 8RJ
G6 CYF D. Richards 433-435 Cronton Road, Widnes WA8 5QG
G6 CYH I. Roberts 32 Priory Drive, Plymouth PL7 1PU
G6 CYO I. Jarvis The Garden House, Walkley Wood, Stroud GL6 0RT
G6 CYT R. Kempton 14 Bloxam Gardens, Rugby CV22 7AP
G6 CYU M. Kendrick 157 Pinar De Gariata, La Nucia, Alicante 3530 Spain
G6 CYV P. Kirkham 9 Bluebell Close, Biddulph, Stoke-on-Trent ST8 6TJ
G6 CZB R. Poffley 3 Bowerhill Road, Salisbury SP1 3DN
G6 CZD M. Swanwick 45, Coach Way, Willington, Derby DE65 6ES
GW6 CZE C. Peacock 8 Heol Ewenny, Pencoed, Bridgend CF35 5QA
GM6 CZM I. Mcaulay 9 Randolph Cliff, Edinburgh EH3 7TZ
G6 CZO D. Mcghie 54 School Road, Newborough, Peterborough PE6 7RG
G6 CZX W. Aitchison 18 Kerensa Green, Falmouth TR11 2HE
G6 CZZ J. Abram 3 Frenchies View Denmead, Waterlooville PO7 6SH
G6 DAC M. Bounds 5 Ingleby Close, Heacham, King's Lynn PE31 7SA
G6 DAD D. Blagburn 10 Tottington Avenue, Springhead, Oldham OL4 4RY
G6 DAH D. Budd 81 Bohemia Chase, Leigh-on-Sea SS9 4PW
G6 DAI N. Brickwood 4 Vale Cottages, Shillingstone, Blandford Forum DT11 0SS
G6 DAN B. Daniels 113 Orchard Way, Wymondham NR18 0NZ
G6 DAO G. Bradbury 3 Westfield Bank, Barlborough, Chesterfield S43 4EG
G6 DAP J. Balmford Upper Brook Farm House, The Avenue, Aylesbury HP18 9LD
G6 DAQ A. Boonham 1 Oakleigh Drive, Sedgley, Dudley DY3 3LH
G6 DAU P. Bidwell 156 Elstree Park, Barnet Lane, Borehamwood WD6 2RP
G6 DAY M. Pemberton 37 Bardsley Close, Croydon CR0 5PS
G6 DBC A. Norfolk 18 Middle Lane, Amcotts, Scunthorpe DN17 4AT
GM6 DBJ J. Fairhurst 1. Ackmore Court, Kyleakin, Isle of Skye IV418PT
GW6 DBP J. Firmstone 1 Holly Grange, Rhoswiel, Weston Rhyn, Oswestry SY10 7TU
G6 DBQ D. Fryer Norwood, 105 Chester Road, Stockport SK7 6HG
G6 DBU R. Gambles 5 College Way, Horspath, Oxford OX33 1SQ
G6 DBX A. Grover 44 Stirling Court Road, Burgess Hill RH15 0PT
G6 DBY P. Gould Derna, Surrey Lane, Colchester CO5 0QT
G6 DBZ S. Griffin 50 Cherrybrook Drive, Broseley TF12 5SH
G6 DCH J. Molyneux 18 Bay Close, Horley RH6 8LF
G6 DCS G. Norris 1 Pear Tree Avenue, Newhall, Swadlincote DE11 0LZ
G6 DCT D. Littlewood 50 Industry Road, Sheffield S9 5FQ
GI6 DCX D. Lyons 64 Drumgavlin Road, Ballynahinch BT24 8QY
G6 DDA A. Moss 20 Black-A-Tree Court, Black-A-Tree Road, Nuneaton CV10 8BD
G6 DDC J. Leatherbarrow 10 Henley Drive, Southport PR9 7JU
GW6 DDF J. Morris 45 Ffordd Pentre Mynach, Barmouth LL42 1EN
G6 DDJ S. Pillinger Calle Sileno 10, Mailbox (Buzon) 470, Fortuna 30620 Spain
G6 DDN I. Oates 33 Evelyn Terrace, Mountain, Queensbury, Bradford BD13 1LF
G6 DDO R. Owen 36 Foley Road, Stourbridge DY9 0RT
G6 DDP R. Oakden 38 Brookfield Avenue, Hucknall, Nottingham NG15 6FF
G6 DDR L. Horne 8 Kingsway Avenue, Broughton, Preston PR3 5JN
G6 DDX M. Wild 46 All Saints Drive, North Wootton, King's Lynn PE30 3RY
G6 DEG T. Hampson 6 Rushmere Drive, Bury BL8 1DW
GW6 DEP M. Harris 11 Lower Rawlinson Terrace, Tredegar NP2 4JD
G6 DER K. Hewitt 6 Church Grove, Monk Bretton, Barnsley S71 2EY
G6 DET M. Heighton 3 Warner Road, Codsall, Wolverhampton WV8 1SA
G6 DEV D. Harris 15 Millwood Road, Orpington BR5 3LG
GI6 DEY F. Hunter Flat 9 50 Edenvale Crescent, Belfast BT4 2BH
G6 DFA C. Willies 17 Campion Way, Sheringham NR26 8UN
G6 DFB C. Smith 9 Barratts Close, Bewdley DY12 2ED
G6 DFC P. Johnson 3 Lance Drive, Chase Terrace, Burntwood WS7 1FA
G6 DFH J. Roberts 155 Langley Hall Road, Olton, , Solihull B92 7HB
G6 DFM J. Phelps Windy Dene, Green Lane, Chessington KT9 2DT
G6 DFR T. Parfitt 4 Back Street, Lakenheath, Brandon IP27 9HF
G6 DFV A. Parker 13 Hartley Street, Colne BB8 9DF
GW6 DFX D. James 5 Lon Y Parc, Cardiff CF14 6DF
G6 DFY Dr G. Joly 116 Hind Grove, London E14 6HP
G6 DFZ M. Jones 28 Winston Avenue, Colchester CO3 4NQ
G6 DGK G. Keegan 12 Allington Road, Newick, Lewes BN8 4NA
G6 DGQ J. Baines 2 Moor Close, Radcliffe, Manchester M26 4QF
G6 DGR N. Bean 19 Coleshill Road, Sutton Coldfield B75 7AA
GW6 DGU R. Britton Llwynon, 95 North Road, Cardigan SA43 1LT
G6 DGV C. Brock 37 Ashington Drive, Bury BL8 2TS
G6 DGW E. BallAshleigh 111 Boslowick Road, Falmouth TR11 2ER
G6 DGX J. Raby Cedar House, Coppenhall, Stafford ST18 9DA
G6 DHD A. Rollason Flat 4, 14 Woodland Terrace, Greenbank Road, Plymouth PL4 8NL
G6 DHI D. Kennedy 1 Lynton Road, Hindley, Wigan WN2 4EH
G6 DHT P. Chace Flat 14, Tudor Rose Court, South Parade, Southsea PO4 0DE
G6 DHU M. Chace 26 Stillwater Drive, Unit 9, Westbrook 4092 United States
G6 DHW I. Clayton 15 Ashbourne Drive, Desborough, Kettering NN14 2XG
G6 DIC P. Dickinson 7 Church Croft, Coton In The Elms, Swadlincote DE128HG
G6 DID J. Davis 38 Dover Close, Southwater, Horsham RH13 9XX
G6 DIE G. Drohan 23 Lindholme Drive, Rossington, Doncaster DN11 0UP
G6 DIF Rvd. V. Van Den Bergh St. Francis C Of E Church, Masefield Drive, Tamworth B79 8JB
G6 DIM T. Eves Banks Farm, Manor Road, Romford RM4 1NH
G6 DIO R. Everson Eversons Farm Bardfield Road Shalford, Braintree CM7 5HU
G6 DIQ J. Wilkins 14 Prospect Road, Shanklin PO37 6AE
G6 DIR M. Wray 18 Cleveland Street, Loftus TS13 4JB
G6 DIZ D. Feeley 177 Rock Street, Sheffield S3 9JF
G6 DJH D. Harvey 23 Sprules Road, Brockley, London SE4 2NL
G6 DJQ G. Tomlinson 10 Ashbourne Road, Underwood, Nottingham NG16 5EH
G6 DJS D. Sojkowski 7 Spenlow Drive, Chelmsford CM1 4UQ
G6 DJX M. Turner 24A Cedar Road, Balby, Doncaster DN4 9DT
G6 DJY W. Telford The Walnuts, Main Road, Boston PE20 2LQ
G6 DKE E. Reynolds 11 New St., Sudbury CO10 1JB
G6 DKF L. Marsh 18 Northgate, Hornsea HU18 1ES
G6 DKI R. Tew 11 Huson Road, Warfield, Bracknell RG42 2QX
G6 DKK S. Simes 53 Waterford Lane, Cherry Willingham, Lincoln LN3 4AN
G6 DKM L. Sandford 150 Tipton Road, Woodsetton, Dudley DY3 1AL
G6 DKS R. Saverton Flat 8 2 Christ Church Road, Surbiton KT5 8JJ
G6 DLJ P. Bridges Nutwood, Coldridge, , Crediton EX17 6AY
G6 DLM Q. Borthwick 106 Westpole Avenue, Cockfosters, Barnet EN4 0BB
G6 DLT J. Bartlett 5 Park View Gypsyville, Hull HU4 6NG
G6 DLZ P. Bosanquet-Bryant Flat 6, 54 West Avenue, Clacton-on-Sea CO15 1HA

G6 DMC D. Crabtree 145 Mendip Road, Yatton, Bristol BS49 4ER
G6 DMF D. Wilkins 124 Fullers Mead, Harlow CM17 9AU
G6 DMG S. Wellon 71 Toftdale Green, Lyppard Bourne, Worcester WR4 0PE
G6 DMM K. Webster 27 Glendale Close, Horsham RH12 4GR
G6 DMQ P. Singleton 37 Victoria Road, Harborne, Birmingham B17 0AQ
G6 DNA T. Cattermole 24 Cromwell Road, Colchester CO2 7EN
G6 DNH M. Carvell 12 Liskeard Drive, Allestree, Derby DE22 2GW
GI6 DNI D. Chapman 3 Brustin Lee, Ballygally, Larne BT40 2QA
G6 DNL K. Snellin 5 Kenley Close, Chislehurst BR7 6QT
G6 DNV G. Taylor 10 Scott Close, Hexham NE46 2QB
GW6 DOC R. Yarnold 47 Small Meadow Court, Caerphilly CF83 3RT
G6 DOD M. Wheeler 105 High Street, Wootton Bridge, Ryde PO33 4LU
G6 DOI C. Wigginton 4 Copes Haven, Shenley Brook End, Milton Keynes MK5 7HA
GW6 DOK C. Williams Caermai, Stad Pen Y Berth, Llanfairpwllgwyngyll LL61 5YT
G6 DON J. Walsh 7 Unicorn Place, Ball Green, Stoke-on-Trent ST6 6LX
G6 DOQ H. Davies 76 Brook Lane, Timperley, Altrincham WA15 6RS
G6 DOR D. Durrant 22 St. Martinsfield, Martinstown, Dorchester DT2 9JU
G6 DOW A. Deacon 1 Connaught Gardens, Crawley RH10 8NB
G6 DOZ L. Dell 205 Thelwall Lane, Warrington WA4 1NF
G6 DPE D. Evans 631 Chatsworth Road, Chesterfield S40 3NT
G6 DPH B. Flinn 65 Marina Avenue, Great Sankey, Warrington WA5 1JH
G6 DPL L. Green 76 Dibleys, Blewbury, Didcot OX11 9PU
G6 DPS M. Harrison 33 Campion Park, Up Hatherley, Cheltenham GL51 3WA
G6 DQA Radio Amateur Special Event Group c/o M. Morgan 125 Holymoor Road, Holymoorside, Chesterfield S42 7DR
GW6 DQB Dr J. Mitchell Y Graigwen, Cadnant Road, Menai Bridge LL59 5NG
GW6 DQH D. Moore 71 Woodlands Avenue, Talgarth, Brecon LD3 0AT
G6 DQK P. Mcbride 14 Hob Hill Meadows, Glossop SK13 8LW
G6 DQO I. Martin 6 Hollow Oak Lane, Cuddington, Northwich CW8 2XN
G6 DQT W. Lasbury Sonserra Flats, Flat 4 Fekruna St, St. Pauls Bay Malta
G6 DQU L. Goodison 16 Springfield, Sowerby Bridge HX6 1AD
G6 DQY J. Orrells Perry Willows, Yeaton, Shrewsbury SY4 2HY
G6 DQZ N. Perry 10 Carlyle Avenue, Kidderminster DY10 3QZ
G6 DRC D. Cooper 20 Simon De Montfort Drive, Evesham WR11 4NR
G6 DRG T. Place 73 Williams Street, Langold, Worksop S81 9NX
G6 DRH D. Hickton 27 Vanguard Road, Long Eaton, Nottingham NG10 1DX
GI6 DRK I. Humes 160 North Road, Belfast BT4 3DJ
G6 DRN P. Haylor 76 Beauchamp Road, Billesley Common, Birmingham B13 0NR
G6 DRP D. Hemmins 18 Burn Walk, Burnham, Slough SL1 7EW
G6 DSA R. Jeffery 7 Corfe Way, Winsford CW7 1LU
G6 DSD R. Jones 20 Bibsworth Avenue, Moseley, Birmingham B13 0BA
G6 DSG N. Austin 184 Tunstall Road, Knypersley, Stoke-on-Trent ST8 7AH
G6 DSP C. Addis 1 Newchurch Lane, Culcheth, Warrington WA3 5RW
G6 DTH A. Allnutt The Squirrels, Nutcombe Lane, Dorking RH4 3DZ
G6 DTN D. Crake Kentolop, Holyhead Road, Montford Bridge, Shrewsbury SY4 1EE
G6 DTT A. Campbell Eden Park, Den Cross, Edenbridge TN8 5PW
G6 DTW Dr P. Lee Links Corner Cottage Liks Road, Ashtead, Surrey, UK KT212EG
G6 DU G. Dunstan 29 Simon Street, Victoria Point Qld 4165 Australia
G6 DUC A. Rowlands Hill House, Bridge Road Leigh Woods, Bristol BS8 3PE
G6 DUH D. Crewe 71 Ladybalk Lane, Pontefract WF8 1LA
G6 DUI I. Castle 26 Lonsdale Drive, Sittingbourne ME10 1TS
G6 DUN R. Burrows 32 Frenchs Farm Road, Poole BH16 5RT
G6 DUT M. Bluck 26 Mayfield Avenue, Scarborough YO12 6DF
G6 DVE A. Redshaw 417 Marston Road, Marston, Oxford OX3 0JG
G6 DVO H. Warehand 79 Woodlands Road, Hertford SG13 7JF
G6 DVP R. Vickers 51 Charlecote Close, Redditch B98 0TQ
G6 DWM G. Sohal 15 Icknield Road, Luton LU3 2NY
G6 DWO R. Smart 52 Devonshire Avenue, Southsea PO4 9EF
G6 DWS N. Shearer 64 Balsall Heath Road, Edgbaston, Birmingham B5 7NE
GI6 DWZ T. Duffin 16 Park Lane, Newcastle BT33 0AR
G6 DXC C. Ellis 43 Epsom Walk, Hereford HR4 9NJ
G6 DXD Norfolk County Raynet c/o A. Edwards 35 Eldon Road, Cheltenham GL52 6TX
G6 DXP M. Gentry Maeldune, Orsett Road, Stanford-le-Hope SS17 8NS
G6 DXY T. Dix Willow Cottage 31 London Road, Woolmer Green SG3 6JE
G6 DYK S. Hicks 15 Chalice Close, Hampton Gardens, Peterborough PE7 8RL
G6 DYM G. Hudgell 18 Fellowes Lane, Colney Heath, St. Albans AL4 0QA
G6 DYR D. Bettie 54 Grendon Road, Polesworth, Tamworth B78 1NU
G6 DYU L. Horn 9 Musson Close, Irthlingborough, Wellingborough NN9 5XW
GW6 DYW C. Hughes Cym Lane, Rogerstone, Newport NP10 9EN
G6 DZI C. Kuss 20 Windermere Road, Haydock, St. Helens WA11 0ES
G6 DZJ S. Kitchener 101 Highfield Road., Tring HP23 4DS
G6 DZT D. Anstock 12 Raymoor Avenue, St. Marys Bay, Romney Marsh TN29 0RD
G6 DZX J. Beardmore 6 Essex Close, Congleton CW12 1SH
G6 EAH R. Carrington 45 Crompton Road, Pleasley, Mansfield NG19 7RG
G6 EAM J. Calder Wyrley Lodge, Hill Farm, Northwood Lane, Bewdley DY12 1AT
G6 EAR P. Dowler 21A Wash Lane, Clacton-on-Sea CO15 1UW
G6 EAX S. Hufschmied 99 Leverstock Green Road, Hemel Hempstead HP3 8PR
G6 EAY M. Hughes 15 Feckenham Road Headless Cross Redditch B97 5As, Redditch B97 5AS
G6 EAZ R. Hildebrand Meadow View, Cunningham Place, Bakewell DE45 1DD
G6 EBL M. Brundle 36 Campion Street, Derby DE22 3EF
G6 EBO B. Beckers 6 Patmore Way, Collier Row, Romford RM5 2HF
GI6 EBX S. Bird 70 Greencastle Road, Kilkeel, Newry BT34 4JJ
G6 ECG G. Bradley Greentree Cottage, Town End, Broadclyst, Exeter EX5 3HW
G6 ECN B. Clay 3 Sandy Close, Bollington, Macclesfield SK10 5DT
G6 ECS P. Buckingham Thrimley House, Thrimley Lane, Bishop's Stortford CM23 1HX
G6 ECT Trinity School RC c/o R. Close 208 Northampton Road, Wellingborough NN8 3PW
G6 EDC F. Davis 28 Western Drive, Claybrooke Parva, Lutterworth LE17 5AG
G6 EDD S. Donald 5 Windsor Road, Royston SG8 9JF
G6 EDF D. Evans 107 Bradbury Road, Solihull B92 8AL
G6 EDM D. Evans Caithness, Greenlands Road, Sevenoaks TN15 6PG
G6 EDR R. Fletcher 31 Snowdrop Close, Broadfield, Crawley RH11 9EG
G6 EDT M. Fletcher 9 The Causeway Carlton, Bedford MK43 7LT
G6 EDU M. Firth Kasamily, 73 Lions Lane, Ashley Heath, Ringwood BH24 2HH
G6 EEB W. Moodie 141 Wood Lane, Handsworth, Birmingham B20 2AQ
G6 EED N. Mockridge 2 Palm View, Waterloo Cross Caravan Park, Uffculme, Cullompton EX15 3ES
G6 EEE A. Mead 17 Beadle Way, Great Leighs, Chelmsford CM3 1RT
G6 EEF D. Malekout 59 Glebelands Avenue, Ilford IG2 7DL
GI6 EEH S. Mccullagh 18 Village Walk, Portadown, Craigavon BT63 5TL
G6 EER G. Middleton 37 Hamdon Close, Stoke-Sub-Hamdon TA14 6QN
G6 EES P. Morris 8 Millfield, Lambourn, Hungerford RG17 8YQ
G6 EET D. Monk 311 Birmingham Road, Lickey End, Bromsgrove B61 0ER
G6 EEU M. Meredith 55 New Barn Lane, Cheltenham GL52 3LB
GU6 EFB K. Le Boutillier Tiverton, Bailiffs Cross Road, St. Andrew GY6 8RT Guernsey
G6 EFE S. Weiss 7 Tennyson Avenue, Grays RM17 5RG
GJ6 EFW E. Walscharts Le Creux Country Park, La Route Orange, St. Brelade, Jersey JE3 8GQ
GI6 EGE R. Hadden 28 Belfast Road, Comber, Newtownards BT23 5EW
GI6 EGJ J. Potts 217 Donaghanie Road, Beragh, Omagh BT79 0RZ
G6 EGO D. Pink 31 The Fairway, Daventry NN11 4NW
G6 EGU B. Nixon 87 Field Avenue, Canterbury CT1 1TS
G6 EGY I. Niven Keepers Cottage, Sulby, Northampton NN6 6EZ
G6 EHE W. Ward 88 Central Road, Cromer NR279BW
G6 EHG R. Quiney 59 Malham Road, Stourport-on-Severn DY13 8NT
G6 EHJ J. Parker 16 Southland Road, Leicester LE2 3RJ
G6 EHL D. Partington 6 Celandine Avenue, Cowplain, Waterlooville PO8 9BE
G6 EIH R. Mccracken 16 Station Road, Rolleston-On-Dove, Burton-on-Trent DE13 9AA
G6 EIO A. Mitchell 85 Farriers Green, Monkton Heathfield, Taunton TA2 8PP
GI6 EIR D. Mullan 5 Mountfield Drive, Coleraine BT52 1TW
G6 EIU M. Parkins 6 Quantock Avenue Caversham, Reading RG4 6PY
G6 EIZ J. Austin 17 New Road, Ascot SL5 8QB
G6 EJD D. Bird 59 Speedwell Close, Melksham SN12 7TE
G6 EJF R. Amos 89 Stanstrete Field, Great Notley, Braintree CM77 7JW
G6 EJH J. Bradley 66 Belmont Road, Parkstone, Poole BH14 0DB
G6 EJI P. Barrett 4 Hazel Road, Middleton, Manchester M24 2WB
G6 EJM T. Burrows 11 Louis Close, Old Catton, Norwich NR6 7BG
G6 EJT J. Bibby 19 Richmond Crescent, Mossley, Ashton-under-Lyne OL5 9LQ
G6 EJU C. Biddles 129 Hallam Crescent East, Leicester LE3 1FG
GI6 EJW W. Mccormick 46 Gortlane Drive, Greenisland, Carrickfergus BT38 8SY
G6 EKM R. Perks 1 Tothill Court Shaldon, Teignmouth TQ14 0EJ
G6 EKS A. Stelfox 6 Surrey Street, Glossop SK13 7AH
G6 EKT Hornsea ARC c/o R. Guttridge Ivy House, Rise Road, Hull HU11 5BH
G6 ELG M. Wright 6 Tregalister Gardens, St. Germans, Saltash PL12 5NQ
G6 EMB G. Collins 33 West Hay Grove, Kemble, Cirencester GL7 6BE
G6 EML C. Exelby The Old Farm House Lower Denford, Hungerford RG17 0UN
G6 ENA M. Pyrah 53 St. Georges Road, Ramsgate CT11 7EF
G6 ENN D. Gordon 38 Deer Park Road, Langtoft, Peterborough PE6 9RB
G6 ENO B. Garrett 226 Rydal Drive, Bexleyheath DA7 5DG
GJ6 ENP J. Gready Avon Cottage, La Rue D'Elysee, St. Peter JE3 7DT Jersey
G6 ENQ G. Greenwood 27 Delph Mount, Great Harwood, Blackburn BB6 7QF
G6 ENR S. Grant The Vicarage, Downe Street, Driffield YO25 6DX
G6 ENS S. Gordon 20 Hawkins Crescent, Shoreham-by-Sea BN43 6TP
G6 ENT D. Gordon 20 Swinburne Court, 143 Brighton Road, Lancing BN15 8HX
G6 ENU I. Gordon 9 Park Road, Camberley GU15 2SP
G6 ENY N. Graham Millers Croft, Queens Road, Freshwater PO40 9ES
G6 ENZ G. Holmes 10 Birch Road, Stamford PE9 2FB
G6 EOK R. Lewis 12 Station Road, Wimborne BH21 1RG
G6 EON P. Martin 40 Carnarthen Street, Camborne TR14 8UP
G6 EOO R. Machin 236 Tamworth Road, Kettlebrook, Tamworth B77 1BY
G6 EOR W. Power 31 Darbys Hill Road, Tividale, Oldbury B69 1SE
G6 EPD J. Howe Flat 1, The Old Post Office, Market Thoroughfare, Bury St. Edmunds IP33 1DR
G6 EPL R. Jonas 49 Clarendon Road, Aylesham, Canterbury CT3 3AQ
G6 EPN P. Knight Hawkwind, Elcot Lane, Marlborough SN8 2AZ
G6 EPQ C. Kingston 3 Hill Street, Cheslyn Hay, Walsall WS6 7HR
GM6 EPU B. Sherman 240 Annan Road, Dumfries DG1 3HE
G6 EPX P. Shuttleworth 12 Oak Avenue, Penwortham, Preston PR1 0XQ
G6 EQB J. Singleton 48 Pennine Way, Ashby-de-la-Zouch LE65 1EW
G6 EQF R. Skinner 23 Woodstock Road, Worcester WR2 5ND
G6 EQI R. Smith 3 Mendip Edge, Weston-Super-Mare BS24 9JF
G6 EQL S. Thomas 64 Victoria Road, Aigburth, Liverpool L17 0DP
G6 EQP T. Thompson 7 West Bank, Dorking RH4 3BZ
G6 EQS J. Hastings 2 Coltsfoot Road, Rushden NN10 0GE
G6 EQT J. Aston 3 Valley Road, Darley, Harrogate HG3 2QE
G6 EQZ R. Bracken 72 Brampton Way, Portishead, Bristol BS20 6YT
G6 ERI R. Couch 54 Hill Park Road, Gosport PO12 3EB
G6 ERJ A. Croucher 73 Loxley Close, Church Hill North, Redditch B98 9JH
G6 ERK A. Cunliffe 28 Rosebank Close, Ainsworth, Bolton BL2 5QU
G6 ESJ P. Wookey 16 Danvers Way, Westbury BA13 3UE
G6 ESK D. Whittle 3 Mackley Way, Harbury, Leamington Spa CV33 9NP
G6 ESM D. Tankaria 23 Oakwood Avenue, Southall UB1 3QD
G6 ESQ P. Baker 12 College Close, Coltishall, Norwich NR12 7DT
G6 ETC J. Brown 44 Perowne Way, Sandown PO36 9BX
G6 ETL P. Cooke 55 Priory Road, Portbury, Bristol BS20 7TQ
G6 ETP J. Cookson Barker Fold Farm, Tockholes Road, Tockholes, Darwen BB3 0LU
GI6 ETQ A. Campbell 16 Parkwood, Lisburn BT27 4EF
G6 ETX M. Carter 22 John Morgan Close, Hook RG27 9RP
G6 ETZ C. Chalmers Flat 30, Luna Apartments, 272 Field End Road, Ruislip HA4 9DL
GM6 EUC D. Cruickshank 61 Woodside Road, Banchory AB31 4EN
G6 EUF P. Raynor 29 Kilvin Drive, Beverley HU17 9PG
G6 EUG C. Slater 70 Windsor Avenue, Ashton-On-Ribble, Preston PR2 1JD
G6 EUI C. Shaw 19 Church Road, Teversham, Cambridge CB1 9AZ
G6 EUO J. Slater 47 Broom Road, Lakenheath, Brandon IP27 9EZ
GW6 EUR P. Williams Llwyn, Manafon, Welshpool SY21 8BJ
GW6 EUS R. Williams Llwyn, Manafon, Welshpool SY21 8BJ
GW6 EUT A. Williams Brynfield, Kingswood Forden, Welshpool Powys SY21 8TS
G6 EUU A. Wilson Flat 5, Shelley House, London E2 0HE
G6 EUW A. Sheridan 6 Mill Road, Burnham-on-Crouch CM0 8PZ
G6 EUY W. Shadwell 2 Poppy Close, Yaxley, Peterborough PE7 3FA
G6 EVC C. Sleight Orchard House, School Hill, Napton, Southam CV47 8NN
G6 EVW S. Wem 10 Astley Crescent, Halesowen B62 9SX
G6 EVX A. Wood 54 Wilton Park Road, Shanklin PO37 7BU
G6 EVY H. Woolrych 20 Meadow Drive, Devizes SN10 3BJ
G6 EWH E. Turton 27 Langdale Avenue, Hesketh Bank, Preston PR4 6TD

G6 EWK D. Mason 15 Windmill Gardens, Prenton CH43 7YQ
GI6 EWO B. Davis 49 The Roddens, Larne BT40 1QL
GW6 EWQ C. Dormer 39 Eastmoor Road, Newport NP19 4NX
GW6 EWX N. Evans Abbey Dingle Care Home Abbey Road, Llangollen Denbighshire LL20 8DD
G6 EXC P. Gibson 9 Mallard Close Aughton, Ormskirk L39 5QJ
G6 EXE M. Graham 11 Robert Moffat, High Legh, Knutsford WA16 6PS
G6 EXG M. Gee 100 Plantation Hill, Worksop S81 0QN
G6 EXN E. Hall 9 Valance Avenue, Chingford, London E4 6DR
G6 EXU A. Jobber Church Hill, Kings North, Ashford TN23 3EG
G6 EXX B. Kent 4 Bedmond Road, Pimlico, Hemel Hempstead HP3 8SH
G6 EXZ A. Kent 166 Louth Road, Scartho, Grimsby DN33 2LG
G6 EYA P. Kershaw 8 Stoppers Hill, Brinkworth, Chippenham SN15 5AW
G6 EYI C. Moore Glen View, Fosseway, Radstock BA3 4BB
G6 EYJ D. Morton 27 Beechfield Way, Hazlemere, High Wycombe HP15 7TP
G6 EYS A. Morne 16 Warmden Avenue, Baxenden, Accrington BB5 2PR
G6 EZG I. Prince 21 Aberfield Drive, Crigglestone, Wakefield WF4 3PT
G6 EZH S. Pepper 149 The Hill, Glapwell, Chesterfield S44 5LU
G6 EZI J. Oldroyd 357 Commercial St #704, Boston 2109 United States
G6 EZM D. Winters 13A St. Catherines Road, Bournemouth BH6 4AE
G6 EZR F. Thompson 28 Lordsmead, Cranfield MK43 0HP
G6 EZY D. Powell 30 Fernley Road, Southport PR8 5AU
G6 FAF C. Narroway 26 Fern Way, Watford WD25 0HG
G6 FAH K. Lawrence 54 Sheldrake Road, Christchurch BH23 4BP
G6 FAL R. Stoneman 9 Winchester Road, Northampton NN4 8AZ
G6 FAX 5th Bham Rnt Gr c/o P. Brooks Flat 4, 8 Glencathara Road, Bognor Regis PO21 2SF
G6 FBA J. Butters 21 Erleigh Road, Reading RG1 5LR
G6 FBB R. Chidgey 14 Drury Road, Colchester CO2 7UX
G6 FBH G. Davis Westbury House, 3 Windermere, Tamworth B77 5TD
G6 FBJ J. Endicott 1 Elm Tree Park, Yealmpton, Plymouth PL8 2ED
GW6 FBV T. Howell 19 Uwchgwendraeth, Drefach, Llanelli SA14 7AR
G6 FCI C. Mcmahon 6 Layton Road, Blackpool FY3 8HS
G6 FCJ P. Magnus-Watson 95 Sutton Lane, Slough SL3 8AU
G6 FCL J. Mahoney Winton Dene, The Street, Sudbury CO10 8JP
G6 FCS D. Lane 10 Whylands Close, Worthing BN13 3HB
G6 FDD R. Pinchin 10 Epping Drive, Melton Mowbray LE13 1UH
G6 FDG I. Rivers 35 Cloverville Approach, Bradford BD6 1ET
G6 FDI B. Raymer 19 Caithness Drive, Crosby, Liverpool L23 0RG
G6 FDK S. Maskrey The Hayloft, Stamford Lane, Chester CH3 7QD
G6 FDO I. Moody 54 Lansdowne Road, Studley B80 7RD
G6 FDP S. Litobarski 7 Exeter Road, Southsea PO4 9PZ
GM6 FDQ G. Allan Corse Farm, Kininmonth, Peterhead AB42 4JU
G6 FDU R. Butterworth 49 Swandene, Pagham, Bognor Regis PO21 4UR
G6 FDX C. Bicknell Flat 25, Madderfields Court, London N11 2JL
GW6 FED D. Corsi 4 Horsley Drive, Wrexham LL12 8BE
G6 FEI D. Harris 53 Welwyn Drive, Salford M6 7PQ
G6 FEJ R. Hawkes 1 The Fairway, Wellingborough NN9 5YS
G6 FEM A. Harris April Cottage, Sheepcote Green, Clavering, Saffron Walden CB11 4SJ
GI6 FEN P. Irwin 6 Cairnburn Avenue, Belfast BT4 2HT
G6 FEQ P. Jolly 22 Wellhouse Road, Barnoldswick BB18 6DD
GW6 FES S. Jones 12 Meadow Croft, Cross Lanes, Wrexham LL13 0UJ
G6 FEX J. Sandford 23 South Lawn, Locking, Weston-Super-Mare BS24 8AD
G6 FFB D. Meaker 181 Dovecote, Yate, Bristol BS37 4PF
G6 FFH T. Sallis 54 West Way, Hove BN3 8LQ
G6 FFL T. Short Freemans Farm, Itchington, Alveston, Bristol BS35 3TL
G6 FFQ F. Bilton 50 Coldwell Road, Crossgates, Leeds LS15 7HA
G6 FFR B. Berry 7 Barlow Close, Telford TF3 2NQ
G6 FFU P. Coogan 39 Sycamore Crescent, Macclesfield SK11 8LW
G6 FGA R. Holyhead 42 Dockham Road, Cinderford GL14 2BH
G6 FGC C. Hawkins 80 Duston Wildes, Northampton NN5 6NR
G6 FGJ C. Tandy 7 The Swallows, Patrons Way West, Uxbridge UB9 5PB
G6 FGL T. Toulson 30 Old Park Avenue, Sheffield S8 7DR
G6 FGV J. Webber 5 Leda Mews, Achilles Close, Hemel Hempstead HP2 5WR
G6 FGW R. Weekes 84 Vera Road, Yardley, Birmingham B26 1TT
G6 FGY E. Westbrook 66 Nelson Close, Croydon CR0 3SW
G6 FHB M. Williams 20 Wignall Road, Stoke-on-Trent ST6 5LE
GI6 FHD A. Mcpartland 4 Clanbrassil Gardens, Portadown, Craigavon BT63 5YD
G6 FHK C. Leonard 138 Sundridge Drive, Chatham ME5 8JD
G6 FHM D. Sunderland 1 Allfield Cottages, Condover, Condover, Shrewsbury SY5 7AP
G6 FHR R. Plant 32 Buckland Road, Pen Mill Trading Estate, Yeovil BA21 5HA
G6 FIB T. Wicks 123 The Crescent, Andover SP10 3BN
GM6 FIK D. Stevenson Flat C, 2 Melbourne Court Braidpark Drive, Giffnock, Glasgow G46 6LA
G6 FIL D. Smith 323 Colchester Road, Ipswich IP4 4SF
G6 FIN A. Stevens 16 Tremlett Grove, Ipplepen, Newton Abbot TQ12 5BZ
G6 FIO J. Slater 154 Ralph Road, Shirley, Solihull B90 3JZ
G6 FIT A. Lewis 81 Ashton Avenue, Rainhill, Prescot L35 0QR
G6 FJA F. Aunger 2, Lowick Woodthorpe, York YO24 2RF
G6 FJE L. Plewa 174 Dorset Avenue, Chelmsford CM2 8YY
G6 FJG N. Pinkney 4 St. Hughs Road, Buckden, St. Neots PE19 5UB
G6 FJI M. Richards 1 Ashenden Close, Abingdon OX14 1QE
G6 FJL P. Standen 17 Canberra Road, Worthing BN13 3HH
G6 FJO S. Turner 14 The Poplars, Launton, Bicester OX26 5DW
G6 FJP North Cheshire RC c/o R. Wild 15 Cartridge Street, Heywood OL10 3AF
G6 FKB E. Taylor 19 Chester Road, Saltney Ferry, Chester CH4 0AQ
G6 FKE K. Redmond 8 George Street, Morecambe LA4 5SU
G6 FKL L. Taylor 127 Dundee Close, Fearnhead, Warrington WA2 0UJ
G6 FKN Dr M. Lee 55 Wodeland Avenue, Guildford GU2 4LA
GW6 FKP S. Moore 25 Overdale Avenue, Mynydd Isa, Mold CH7 6US
G6 FKR D. Roberts 10 Woodville Terrace, Darwen BB3 2JH
G6 FKS S. Robinson 4 Grayling Close, Cambridge CB4 1NP
G6 FKY T. Norris The Old Post Office, Arundel Road, Arundel BN18 0SD
G6 FLE N. Scott 33 Leaze Close, Thornbury, Thornbury BS35 2FH
G6 FLH J. Smith 25 Seafield Close, Seaford BN25 3JR
G6 FLK J. Walton 168 Park Road, Stanley DH9 7AJ
GM6 FLL A. Simpson-Fraser 430 Millcroft Road, Cumbernauld, Glasgow G67 2QW
G6 FLQ G. Smith 2 Wakeling Close, Southwell NG25 0JF
G6 FLR J. Smith 33 Hop Pole Green, Leigh Sinton, Malvern WR13 5DP
GW6 FLU C. Mock Homelea, Royal Oak Hill, Newport NP18 1JF
G6 FLW C. Thompson 27 Queensland Drive, Colchester CO2 8UD
G6 FLY H. Lee 26 Ratcliffe Avenue, Branston, Burton-on-Trent DE14 3DA
G6 FMF D. Timson 40 Rockwood Road, Calverley, Pudsey LS28 5AA
G6 FMN P. Rogers 12 St. Peters Rise, Headley Park, Bristol BS13 7LY
G6 FMS M. Peers Viewpoint, Wicker Lane Guilden Sutton, Chester CH3 7EL
G6 FMU I. Muir 7 Bovarde Avenue Kings Hill, West Malling ME19 4BS
GW6 FNB D. Morris 11 Ffordd-Y-Mynach, Pyle, Bridgend CF33 6HT
G6 FNJ R. Oglesby 1 High Street, Littleton Panell, Devizes SN10 4EL
G6 FNQ A. Smith 16 Hazel Way, Barwell, Leicester LE9 8GP
G6 FNY T. Strand 15 Sherwood Avenue, Melksham SN12 7HJ
G6 FOF C. Morris Flat Iv, Brummel Court, Worcester Road, Droitwich WR9 0DF
G6 FOI Dr A. Regan 153 Acre Lane, Cheadle Hulme, Cheadle SK8 7PB
GI6 FOR N. Lane 117A Hillhall Road, Lisburn BT27 5BT
GM6 FOT T. Armour 69 Hillend Road, Clarkston, Glasgow G76 7XT
G6 FOV D. Aldridge 17 Priory Close, Tavistock PL19 9DJ
G6 FOW P. Atkinson 30 Spital Terrace, Gainsborough DN21 2HQ
G6 FOX York ARC c/o T. Ray 1 Providence Lane, Leamore, Walsall WS3 2AQ
G6 FPC J. Body Flat 5 1 New Marchants Passage, Bath BA1 1AR
G6 FPF Dr G. Barnes Rockleigh, 17 Savile Park, Halifax HX1 3EA
G6 FPH M. Cole 45 Gainsborough Road, Tilgate, Crawley RH10 5LD
G6 FPK N. Cooper 53 Stanway Road, Benhall, Cheltenham GL51 6BU
G6 FPN Y. Dunn 117 All Saints Way, West Bromwich B71 1RU
G6 FPO G. Faulkner Flat 2, 1265 Melton Road, Syston, Leicester LE7 2EN
G6 FPP A. Ford 8 Merganser Drive, Bicester OX26 6UQ
G6 FPX S. Hill 7 Meadowcroft Court, Runcorn WA7 2NS
G6 FQL R. Heath Flat 2, Portland View, 62 High Street, Great Yarmouth NR31 6RQ
G6 FQP B. Kneebone 1 Chapel Terrace, Carnkie, Helston TR13 0DT
GI6 FQT D. Mcconville 28 Derrycor Lane, Derryadd, Craigavon BT66 6QW
G6 FRB J. Pearce 25 Boughton Street, St. Johns, Worcester WR2 4HE
G6 FRS Farnborough & District c/o M. Hearsey Halycon, Lawday Link, Farnham GU9 0BS
G6 FS 5th Reigate Scout Group c/o M. Danfer The Nook, Mill Common, Blaxhall, Woodbridge IP12 2ED
GM6 FSG W. Chamberlain Cwmmelyn, Kings Road, Whithorn, Newton Stewart DG8 8PP
G6 FSK D. Fisher 6 Small Holdings Road Clenchwarton, King's Lynn PE34 4DY
G6 FSP Ionspheric P Gr c/o D. Helliwell 1 Beechfield Avenue, Barton, Torquay TQ2 8HU
G6 FSU M. Apperly Chaundlers, Church Lane, Kingston, Cambridge CB23 2NG
G6 FTA M. Everall 17 Golden Park Avenue, Torquay TQ2 8LR
GM6 FTE P. Bondar 4 The Steadings Swinside Townhead Farm, Jedburgh TD8 6ND
G6 FTH D. Clark 43 Glenfield Crescent, Chesterfield S41 8SF
G6 FTJ P. Carter 145 Wakefield Road, Dewsbury WF12 8AJ
G6 FTL P. Dixon 37 Carlton Close, Parkgate, Neston CH64 6RB
GI6 FTM J. Dynes 30 Breagh Road, Portadown, Craigavon BT63 5LT
G6 FTR T. Fabbri 53 Langdale Road, Cheltenham GL51 3LX
G6 FTY R. Miles 60 Aylesham Way, Yateley GU46 6NT
G6 FUT I. Donn Charsleys, Weedon Hill Hyde Heath, Amersham HP6 5RN
GW6 FUY R. Fishwick The Elms, New Road, Brynteg, Wrexham LL11 6PD
G6 FVB R. Baker 23 Disraeli Road, London W5 5HS
G6 FVD J. Henville 5 Station Road, Hemyock, Cullompton EX15 3SE
G6 FVF P. Fenn 38 Harwood Close, Welwyn Garden City AL8 7SN
G6 FVJ A. James 82 Sandringham Drive, Spondon, Derby DE21 7QA
G6 FVL R. Young 134 Harport Road, Redditch B98 7PD
G6 FVM A. Williamson 31 Poulton Road, Southport PR9 7BE
G6 FVZ R. Munns 2 The Tyleshades, Romsey SO51 5RJ
G6 FWK D. Booth 54 Shaw Drive, Knutsford WA16 8JR
G6 FWO P. Dover 92 The Roundway, Claygate, Esher KT10 0DW
G6 FWU T. Pillar 14 Thoresby Mews, Bridlington YO16 7GZ
G6 FXE C. Walton 49 Blandford Drive, Walsgrave, Coventry CV2 2JD
G6 FXR D. Ainslie Brackendene, 17 Sandhurst Road, Crowthorne RG45 7HR
GI6 FXY E. Connolly 21 Clanrye Avenue, Newry BT35 6EH
GM6 FXZ A. Carnall 3 Main St., Glenluce, Newton Stewart DG8 0PN
G6 FYA C. Collins 29, Seaford BN25 1SP
G6 FYC J. Cowee 26 Arundel Road, Heatherside, Camberley GU15 1DL
G6 FYD D. Dale 81 Burton Manor Road, Stafford ST17 9PR
G6 FYE C. Das Neves Pedro The Leas, Bearwood, Leominster HR6 9EE
G6 FYL N. Harris 104 Blandford Drive, Walsgrave, Coventry CV2 2NE
G6 FYR A. Johnson 14 Norman Road, Newhaven BN9 9LJ
G6 FYT G. Whiting Glyn South View, Liskeard PL14 3EX
G6 FYU Sudbury And District Radio Amateurs c/o J. Walker 33 Erica Way, Horsham RH12 5XL
G6 FYX N. Holden 1 Brooklands Way, Lincoln LN6 0RH
G6 FZC D. Hall 1 Westfall, Wearhead, Bishop Auckland DL13 1JD
GI6 FZI G. Mcbriar 15 Ambleside Drive, Bangor BT20 4QB
G6 FZV W. Day 4 Queenswood Drive, Worcester WR5 3SZ
G6 FZW A. Eaves 3 Station Cottages, Station Road, Leighton Buzzard LU7 0SQ
G6 GA R. Rushton 53 Crossfield Avenue, Blythe Bridge, Stoke-on-Trent ST11 9PL
G6 GAB W. Honey 20 Pennor Drive, St. Austell PL25 4UW
G6 GAC G. Jenkins 75 Rectory Road, Coltishall, Norwich NR12 7HW
G6 GAF A. Franklin C/O 4 Princes Close, Seaford BN25 2EW
GI6 GAG N. Orr 405 Enniskeen, Drumgor, Craigavon BT65 4AB
GM6 GAH C. Mcdowell 3 Buchanan Street, Largs KA30 8PP
G6 GAK M. Tyrrell 189 Runcorn Road, Barnton, Northwich CW8 4HR
GI6 GAQ H. Warke 5 Meadow View, Ballymoney BT53 7AH
G6 GAW D. Peters 46 Sheridan Road, Worthing BN14 8ET
GI6 GBK J. Anderson 1 Claragh Hill Drive, Kilrea, Coleraine BT51 5YR
G6 GBL A. Abbott 164 Bath Road, Reading RG30 2HA
G6 GBT I. Cole 5 St. Milborough Close, Ludlow SY8 1XS
G6 GBU A. Dixon 33 Colgrove Road, Loughborough LE11 3NL
G6 GBY P. Pearce Criggion Mw Radio Station, Back Lane, Criggion, Shrewsbury SY5 9BE
G6 GCI C. Burnett 36 Mill Lane, Romsey SO51 8EQ
G6 GCJ J. Burnett 44 Bourne Vale, Hungerford RG17 0LL
GW6 GCK J. Cook St. Davids, Chepstow Road Langstone, Newport NP18 2JR
G6 GCM T. Collings 55 Flora Thompson Drive, Newport Pagnell MK16 8SR
G6 GCO D. Davies 79A Spenser Road, Bedford MK40 2BE

G6 GCW L. Wiltshire 4 Nether Close, Eastwood, Nottingham NG16 3DL
G6 GCY J. Robinson 84 Hereford Way, Middleton, Manchester M24 2NN
G6 GDH A. Froggatt 37 Charnock Drive, Sheffield S12 3HD
G6 GDI V. Gerhardi 24 Putnams Drive, Aston Clinton, Aylesbury HP22 5HH
G6 GDR C. Price-Gore 22 Oakham Close, Desborough, Kettering NN14 2FH
G6 GEK A. Elliott Knowle House, Hooke Road, Leatherhead KT24 5DY
G6 GEL K. Inman 15 Waterbridge Court, Appleton, Warrington WA4 3BJ
G6 GEN R. Ainsworth 18 Washington Drive, Slough SL1 5RE
G6 GEP A. Holdup Tunnel Farm, Tunnel Rd, Imbil (Po 155) 4570 Australia
G6 GES R. Haywood 16 The St., Kingston, Canterbury CT4 6JB
G6 GEV D. Ashton 12 Little Lees, Charlbury, Chipping Norton OX7 3HB
G6 GEX C. Farley 1 Wesley Cottages, Mutley, Plymouth PL3 4RB
G6 GFA P. Arscott 122 Woodlands Road, Ashurst, Southampton SO40 7AL
G6 GFC J. Burrows 4 Cavendish Crescent, Alsager, Stoke-on-Trent ST7 2EF
G6 GFG P. Cook 109 Crosthwaite Avenue, Wirral CH62 9DF
G6 GFJ H. Goozee 45 Brighton Road, Purley CR8 2LR
GM6 GFL Dr D. Begg 12 Broomhill Road, Penicuik EH26 9EE
G6 GFO N. Atrill 22 Lester Close, Hr Compton, Plymouth PL3 6PX
GM6 GFQ C. Barnard 122 Union Grove, Aberdeen AB10 6SB
G6 GFR T. Crook 21 Cleveland Close, Maidenhead SL6 1XE
G6 GGN M. Hoskin 7 Worrall Mews, Clifton, Bristol BS8 2HF
G6 GGT M. Huntley 81-82 The Avenue, Sunderland SR2 7EZ
G6 GGV B. Hollngworth 62 Illingworth Avenue, Halifax HX2 9JD
G6 GGW N. Gautrey 11 Bracken Close, Crawley RH10 8JR
G6 GGY L. Fitzwater The Old Cottage, Babylon Lane, Tadworth KT20 6XE
G6 GGZ M. Ferne 24 Essex Gardens, Leigh-on-Sea SS9 4HG
G6 GHE A. Rawdon 44 Southgate, Hornsea HU18 1AL
G6 GHU R. Wood 12 Roundhead Drive, Thame OX9 3DG
GI6 GIE J. Pinkerton 40 Seacon Park, Seacon, Ballymoney BT53 6QB
G6 GIF M. Oram 25 Jerome Close, Marlow SL7 1TX
G6 GIH P. Sewell 6 Hawthorne Avenue, Bedford MK40 4HJ
G6 GIU A. Stephens 4 Falcon House, Gurnell Grove, London W13 0AE
G6 GIV D. Newington 6 Walkdale Brow, Glossop SK13 6PX
G6 GJD C. Harper Flat 2, Dove Tree Court, Blackpool FY4 4NA
G6 GJN T. Biggs 3 Pentathlon Way, Cheltenham GL50 4SE
G6 GJV W. Willis 24 Old Hall Lane, Walton on the Naze CO14 8LE
GM6 GJW J. Leith Appiehouse, Stenness, Stromness KW16 3LB
G6 GJY S. Smith 36 Greenfields, Earith, Huntingdon PE28 3QH
G6 GKG W. Hodson 27 Belvedere Grove, London SW19 7RQ
G6 GKK A. Barton Orchard Bungalow, Westfield Road, Retford DN22 7BT
G6 GKL M. Borrow 189 Crofton Road, Orpington BR6 8JB
GW6 GKP J. Coyne 44 Brompton Avenue, Rhos On Sea, Colwyn Bay LL28 4TF
G6 GKT I. Houldridge 57 Heads Lane, Hessle HU13 0JH
G6 GLB G. Walker 141 Deerleap, Bretton, Peterborough PE3 9YD
G6 GLH Obo Manfield ARS c/o P. Burfield 33 St. Ediths Road, Kemsing, Sevenoaks TN15 6PT
G6 GLO Gloucestershire County Raynet c/o A. Webb 47 Granville Street, Gloucester GL1 5HL
G6 GLR Gt Lumley ARS c/o B. Corker 46 Danelaw, Great Lumley, Chester le Street DH3 4LU
G6 GLT Dr R. Bennett 11 Powys Close, Haslingden, Rossendale BB4 6TH
G6 GLW J. Fisher 4 Chancery Close, Lincoln LN6 8SD
G6 GLZ D. Clews 4 Chancellors Close, Coventry CV4 7ED
GW6 GMF M. Inness 6 Denning Road, Wrexham LL12 7UG
G6 GMH Lundy DX Group c/o G. Russell 57 Oak Tree Drive, Cutnall Green, Droitwich WR9 0QY
G6 GMR Greater Manchester Raynet c/o N. Czernuszka 12 Durham Drive, Ashton-under-Lyne OL6 8BP
G6 GMU J. Oldfield 30 Village Farm Caravan Site, Bilton Lane, Harrogate HG1 4DL
G6 GMW Thornton Cleveleys ARS (Tcars) c/o J. Rodway 9 York Avenue, Thornton-Cleveleys FY5 2UG
GM6 GMZ N. Saunders 6 Haughs Of Clinterty, Kinellar, Aberdeen AB21 0TZ
GI6 GNA H. Wright 2 Duncans Road, Lisburn BT28 3LP
G6 GNC J. Thornber 7 Buckland Close, Peterborough PE3 9UH
G6 GND R. Lambert 10 Ambleside, Rugby CV21 1JB
G6 GNE J. Sugden 3 Castle Keep, Hibaldstow, Brigg DN20 9JG
G6 GNO Civil Service ARS c/o B. Cooper 8 Stanley Road, Doncaster DN5 8RR
G6 GOG A. Kerr 10 Hillcrest Road, Crosby, Liverpool L23 9XS
G6 GOS M. Jones 58 Newton Road, Lewes BN7 2SH
G6 GOV B. Wood 8 Chichester Drive, Chelmsford CM1 7RY
G6 GOW R. Wheeler 2 Heather Close, Brereton, Rugeley WS15 1BB
G6 GOX L. Timbrell 3 Rushden Road, Wymington, Rushden NN10 9LN
G6 GPF J. Woodhouse 102 Newland Road, Worthing BN11 1LB
GM6 GPH J. Robertson 2 Cults Bungalow Cults, Cupar KY15 7TF
G6 GPR Dr D. Jefferies 96 Broad Street, Wood Street Village, Guildford GU3 3BE
G6 GPV K. Patching 1 Peak House 84 Trinity Street, Fareham PO16 7SJ
G6 GQF Dr G. Martin 9 Clarkes Avenue, Kenilworth CV8 1HX
G6 GQG I. Moston 19 Wegnalls Way, Leominster HR6 8TQ
G6 GQI R. Swann 3 Elizabeth Avenue, Newmarket CB8 0DJ
G6 GQJ J. Davies 1 Woodland Road, Halesowen B62 8JS
GI6 GRV J. Barnett 2 Donegall Park, Whitehead, Carrickfergus BT38 9ND
G6 GS Guildford & D Rd c/o A. Pevy 2 Oaktree Way, Sandhurst GU47 8QS
G6 GSF K. Edwards Whitehaven, High Street, Uckfield TN22 4JU
G6 GSG Hastings College RC c/o R. Grimley 11 Sewell Wontner Close, Kesgrave, Ipswich IP5 2GB
G6 GSI D. Millington 9 Roxburgh Croft, Leamington Spa CV32 7HT
GW6 GSR Holsworthy ARC c/o P. Williams 5 Whitewell Drive, Llantwit Major CF61 1TA
G6 GSV J. Williams 54 Fleminghouse Lane, Huddersfield HD5 8QG
G6 GTB J. Tracey 100 Booth Close, Kingswinford DY6 8SP
G6 GTC P. Willson 37 The Grove, Sidcup DA14 5NG
G6 GTH S. Leonard 231 Hale Road, Hale, Altrincham WA15 8DN
G6 GTJ L. Baldwin The Fields, Pinewood Road, Market Drayton TF9 4QE
GW6 GTS P. Dudman Chapel House, Pen Y Bryn, Wrexham LL14 1UA
G6 GTZ P. Wilson 162 Bowerdean Road, High Wycombe HP13 6XW
G6 GUC D. Ellis Field End, Northwood Green, Westbury-on-Severn GL14 1NB
G6 GUD M. Everley 5 Firs Close Hazlemere, High Wycombe HP15 7TF
G6 GUH P. Bonds The Gables, Crosslane Head, Bridgnorth WV16 4SJ
G6 GUT B. Turley 12 Legh Drive, Woodley, Stockport SK6 1PT
G6 GVF K. Waters 25 Edwin Road, Twickenham TW2 6SP
G6 GVH J. Marks 124 Stowey Road, Yatton, Bristol BS49 4EB
G6 GVI R. Wilkinson 84 Park Road, Bolton BL1 4RQ
G6 GVL M. Longley 78 Priory Road, Eastbourne BN23 7BE
G6 GVO M. Pearce 64 Goongarrie Drive, Wa 6169 Australia
G6 GVR G. Whittle 5 Chantry Close, Westhoughton, Bolton BL5 2LY
G6 GVS R. Wood 6 Timberlaine Road, Pevensey Bay, Pevensey BN24 6DE
G6 GVU S. Wood 30 Ramsay Way, Eastbourne BN23 6AL
G6 GVZ E. Rigby 12 Sorrel Avenue, Tean, Stoke-on-Trent ST10 4LY
GW6 GW Blackwood & District Amateur Radio Soc. c/o D. Lewis 23 Gelligroes Road, Pontllanfraith, Blackwood NP12 2JU
G6 GWE M. Ranger 13 Springfield Close, Crowborough TN6 2BN
G6 GWP J. Briggs Wood Lea, Bawtry Road, Doncaster DN10 5BS
G6 GWU P. Hopkinson 59 Mulberry Drive, Upton-Upon-Severn, Worcester WR8 0ET
G6 GWX C. Hore 45 Medrose Street, Delabole PL33 9BN
G6 GWY K. Dodd 1 Nansen St., Bulwell, Nottingham NG6 9JE
G6 GXE L. Jordan 20 Coniston Road, Folkestone CT19 5JF
G6 GXG D. Ridden 6 Maple Drive, Witham CM8 2LH
G6 GXK D. Wrigley 45 Norford Way, Rochdale OL11 5QS
G6 GXO C. Parks 29 Heighams, Harlow CM19 5NU
G6 GXS D. Mclean Quartier Les Tourres, Pourcieux 83470 France
G6 GXY M. White 8 Browning Avenue, Droylsden, Manchester M43 6QG
G6 GXZ B. Vaslet Heatherlea, Adbury Holt, Newbury RG20 9BN
G6 GYC D. Oultram 61 Bolton Road, Westhoughton, Bolton BL5 3DN
G6 GYF M. Marshman 12 Neelands Grove, Cosham, Portsmouth PO6 4QL
G6 GYG D. Langridge 4 The Puddledocks Puddledock Lane, Sutton Poyntz, Weymouth DT3 6LZ
G6 GYM D. Popely 24 Lawson Avenue, Stanground, Peterborough PE2 8PL
G6 GYN P. Price 67 Bennetts Road, Keresley End, Coventry CV7 8HY
G6 GYV Redditch Amateur Radio & CG c/o B. Thompson 17 Avenue Road, Askern, Doncaster DN6 0AR
G6 GYX J. Walton Copper Beech Lutton Gowts, Spalding PE12 9LQ
G6 GZC A. Jeffries 29 Amberleigh Close Appleton Thorn, Warrington WA4 4TD
G6 GZZ A. Hammond 23 St. Andrews Road, Fremington, Fremington EX313BS
G6 HAA A. Johns Glan Y Nant, Murcot, Broadway WR12 7HS
G6 HAT P. Calpin 36 Chatsworth Grove, Harrogate HG1 2AS
G6 HBF A. Bischtschuk 30 Livingstone Road, Wirral CH46 2QR
G6 HBJ T. Charman 1 Bowler Lea, Downley, High Wycombe HP13 5UD
G6 HBQ A. Ford 1 Hem Heath Cottage, Longton Road, Stoke on Trent ST4 8HP
G6 HBZ S. Jenkinson Field End, Castleton, Hope Valley S33 8WB
GD6 HCB A. Kennaugh 36 Seafield Close, Onchan, Douglas IM3 3BU Isle of Man
G6 HCF L. Carter Hattersbrick Farm, Lancaster Road, Preston PR3 6BN
G6 HCH C. Back The Gift Shop, 3 Albion Villas, Main Road, Wareham BH20 5RQ
G6 HCQ S. Crawford 71 Harewood Road, Bedford MK42 9TH
G6 HCT the Home Counties Atv Group c/o T. Grady 63 Bridport Close, Lower Earley, Reading RG6 3DG
G6 HCW D. Fieldsend 3 Rosehall Close, Redditch B98 7YD
G6 HDD P. Ingham 1411 Helderberg Avenue, Rotterdam 12306 United States
G6 HDF S. Kelly 4 Franklyn Close, Wolverhampton WV6 7SB
G6 HEB P. Ballance 6 Coronation Terrace, Knaresborough HG5 8JN
G6 HEF D. Bailey Ardmore, Chapel Lane, Bootle, Millom LA19 5UE
G6 HEJ G. Stewart 3 Harvest Crescent, Carterton OX18 1FF
G6 HFB A. Wedgwood Ardmore, Chapel Lane, Bootle, Millom LA19 5UE
G6 HFF G. Bates The Anvil, 4 Eastgate, North Newbald YO43 4SD
GM6 HFH I. Baker 31 Strathaven Road, Stonehouse, Larkhall ML9 3EN
G6 HFK L. Dutton 5 Beaver Close, Stoke-on-Trent ST4 6PR
G6 HFS B. Shaw 43 Egremont Road, Hardwick, Cambridge CB23 7XR
G6 HFW South Birmingham RS c/o J. Graham 142 Shakerley Lane, Atherton, Manchester M46 9TZ
G6 HFZ S. Homer 31 Shaftmoor Lane, Acocks Green, Birmingham B27 7RU
G6 HGE D. Heale 3 Evans Wharf, Hemel Hempstead HP3 9WU
G6 HGG R. Ireson 6 Walker Square, Wellingborough NN8 5PQ
G6 HGI P. Johnston 566 Woodchurch Road, Prenton CH43 0TT
G6 HGK E. Kesterton 24 Alexandra Road, Illogan, Redruth TR16 4DY
G6 HGM R. Buckle Bissom House, Parrotts Lane, Tring HP23 6NE
G6 HGR K. Potts 31 Sparnon Close, Redruth TR15 2RJ
GM6 HGW C. Topping 26 Crathes Close, Glenrothes KY7 4SS
G6 HGX B. Waterloo 55 Solent Road, Hill Head, Fareham PO14 3LB
G6 HH Hastings Electronics & RC c/o J. Tucker 17 Rye Industrial Park Harbour Road, Rye TN31 7TE
G6 HHE J. Avern 8 Napier Crescent, Fareham PO15 5BL
G6 HHH G. Dowse 60 Lower Mortimer Road, Southampton SO19 2HF
G6 HHK D. Birkbeck Low Farm Brotton, Saltburn-by-the-Sea TS12 2QX
G6 HIA A. Cook Woodlands House, Hempstead Road, Hemel Hempstead HP3 0DS
G6 HIB Hallam DX Group c/o C. Craven 24 Links Drive, Bexhill-on-Sea TN40 1TE
G6 HIE B. Edwards 28 Poppy Drive, Horam, Heathfield TN21 9BL
G6 HIG G. Edmonds Wellwood End, Waterworks Lane, Dover CT15 5JW
G6 HIO D. Ollerton 91 Church Road, Bickerstaffe, Ormskirk L39 0EB
G6 HIQ C. Lavis 88 Boundary Way, Glastonbury BA6 9PH
G6 HIU N. Lasher 1 Eton Heights 145 Whitehall Road, Woodford Green IG8 0FB
G6 HIV J. Martin 1 Marsh Street, Strood, Rochester ME2 4BB
G6 HIX J. O'Hagan Brubell, 13 Chapel Road, Faringdon SN7 8LE
G6 HJU J. Binns 2 Gawsworth Close, Poynton, Stockport SK12 1XB
G6 HJV J. Evill 54 Copsey Grove, Farlington, Portsmouth PO6 1NB
G6 HKA J. Moss 801 Manchester Road, Linthwaite HD7 5NF
GI6 HKE W. Leitch 7 Burghley Mews, Belfast BT5 7GX
G6 HKF R. Mew Tehig, 16 La Mustais, Sion Les Mines 44590 France
G6 HKH P. Randell 38 Hanover Drive, Brackley NN13 6JS
G6 HKL D. Martin 9 Twinberrow Lane, Woodmancote, Dursley GL11 4AP
G6 HKN W. Mccue 2 Downham Avenue, Culcheth, Warrington WA3 5RU
G6 HKP D. Merrington 31 North Road, Wellington, Telford TF1 3ED
G6 HKS R. Mason 11 Dyers Mews, Neath Hill, Milton Keynes MK14 6ER
G6 HKY W. Metcalfe 81 Westminster Drive, Bromborough, Wirral CH62 6AN
G6 HKZ R. Moses 80 Edgeworth, Yate, Bristol BS37 8YW
G6 HL Club Hut c/o W. Ward 88 Central Road, Cromer NR279BW
G6 HLL B. Allman 38 Whinchat Drive, Birchwood, Warrington WA3 6PB
G6 HLR G. Marshall 118 Heather Road, Small Heath, Birmingham B10 9TB
GM6 HLT J. Melville 6 Dixon Avenue Kirn, Dunoon PA23 8NA

G6 HLU T. Miller 23 Manchester Road, Altrincham WA14 4RQ
G6 HMA P. Matthews 47 Slyne Road, Morecambe LA4 6PD
G6 HMF R. Venison Brooklands, Sharnbrook Road, Souldrop, Bedford MK44 1EX
G6 HMG R. Trowsdale 422 Leatherhead Road, Chessington KT9 2NN
GW6 HMJ A. Upcott 67 Hunters Ridge, Brackla, Bridgend CF31 2LJ
G6 HMN R. Sunter 15, Wellhead, Winewall, Winewall Colne BB8 8BW
G6 HMS E. Veall 24 Meadow Drive, Tickhill, Doncaster DN11 9ET
G6 HMV R. Tilley 41 Rookery Road, Knowle, Bristol BS4 2DX
G6 HMX D. Tucker 2 Chardonnay Crescent, Thornton-Cleveleys FY5 3UH
G6 HNE E. Shirt 213 Carlton Road, Barnsley S71 2BL
G6 HNI D. Baker 16 Warners Bridge Chase, Rochford SS4 1JE
G6 HNN M. Bugg 39 Glencoe Road, Ipswich IP4 3PP
G6 HNP P. Beever 33 Masterton Road, Stamford PE9 1SN
G6 HNQ K. Blackburn 57 Hope Street, Leigh WN7 1NB
G6 HNR J. Ball 94 Marshall Lake Road, Shirley, Solihull B90 4PN
G6 HNS R. Ball 139 Bedford Road, Sutton Coldfield B75 6DB
G6 HOB D. Brebner 2 Oldborough Drive, Loxley, Warwick CV35 9HQ
G6 HOC A. Bird 95 Hundred Acre Road, Streetly, Sutton Coldfield B74 2BS
G6 HOR D. Padfield 9 Dunster Close, Minehead TA24 6BY
G6 HOS C. Playford 6 Nutberry Close, Teynham, Sittingbourne ME9 9SP
G6 HPE P. Simms 42 Bridgeside Close, Walsall WS87BN
G6 HPK D. Scott 8 Lynton Road, Chesham HP5 2BU
G6 HPL R. Seddon 255 Westleigh Lane, Leigh WN7 5PN
G6 HPR A. Swinburne 69 Northfield Avenue, Wigston LE18 1FX
G6 HPT D. Sumner 34 Japonica Close, Bicester OX26 3YB
G6 HQ P. Bushell The Fairway, Well Lane, Neston CH64 4AN
G6 HQX A. Cook 90 Ramsbury Walk, Trowbridge BA14 0UX
G6 HRA J. Chesterman 69 Heath Lane, Bladon, Woodstock OX20 1RZ
G6 HRM J. Wildman 14 Southorpe Close, Bridlington YO16 7QL
GW6 HRU R. Walker 54 Woolpitch Wood, Chepstow NP16 6DW
G6 HRX T. Whitehead 14 Somerset Road, Willenhall WV13 2RY
G6 HSC V. Williamson 28 Mill Park Drive, Braintree CM7 1XF
G6 HSD R. Willmott 85 Malthouse Lane, Earlswood, Solihull B94 5RZ
G6 HSG A. Walsh 14 The Rydings, Langho, Blackburn BB6 8BQ
G6 HSI S. Wallace 26 Parsons Drive, Glen Parva, Leicester LE2 9NS
G6 HSR J. Hague 33 East Rise, Royal Sutton Coldfield B75 7TH
G6 HSS P. Hardiman 12 Brempsons, Basildon SS14 2AZ
G6 HSW L. Hagger 48 Little Meadow Bar Hill, Cambridge CB23 8TD
G6 HTA P. Hartas 6 Newton St., Whitby YO21 1QX
G6 HTB E. Brodie 116 Pagham Road, Pagham, Bognor Regis PO21 4NN
G6 HTH C. Hall Oakenhill, North Pole Road, Maidstone ME16 9HH
G6 HTS R. Hooper 11 Joy Wood, Boughton Monchelsea, Maidstone ME17 4JY
G6 HTT G. Reece 9 Lambert Close, Framlingham, Woodbridge IP13 9TE
G6 HTY A. Rollitt St. Peters, 29 High Street, Lincoln LN5 0EE
G6 HTZ A. Rogers 11 Avebury Close, Curzon Park, Calne SN11 0EP
GW6 HUD R. Rees 5 Rhydyffynnon, Pontyates, Llanelli SA15 5UG
G6 HUI B. Tanner 2 Doveys Cottages Days Lane, Kington Langley SN15 5NT
G6 HUN A. Thompson Hillier Garden Centre, Priors Court Road, Thatcham RG18 9TG
G6 HUO J. Thompson Belmoor Lodge, Pilton Lane, Exeter EX1 3RA
G6 HUP M. Thompson 2 Cotman Road, Lincoln LN6 7PA
GW6 HUR N. Thursfield Highmoor, Llanymynech SY22 6HB
GW6 HVA M. Vernon 33 Ffordd Morfa, Llandudno LL30 1ES
G6 HVD N. Dunford 8 Fair Mead, Mountsorrel, Loughborough LE12 7BN
G6 HVE N. Martin 12 Coppice Side, Hull HU4 6XJ
G6 HVJ J. Durrant 35 Britford Avenue, Wigston LE18 2RF
G6 HVQ L. Dodson 24 Ashcombe Terrace, Tadworth KT20 5EW
G6 HVX D. Gladwish 36 All Saints Street, Hastings TN34 3BJ
GM6 HVY R. Goodwins 3 Westmost Close, Edinburgh EH6 4TE
G6 HWA F. Glover 11 Esk Valley, Grosmont, Whitby YO22 5BG
G6 HWI B. Wilson 102 Woodlands Road, Woodlands, Doncaster DN6 7JZ
G6 HWR M. Fern 8 Hackney Road Hackney, Matlock DE4 2PW
G6 HWT C. Freeman Mill House Great Bricett, Ipswich IP7 7DE
G6 HXB M. Aston Flat 51 Acton House 253 Horn Lane, Acton W3 9EJ
G6 HXI T. Ardern 7 North Drive, Harwell, Didcot OX11 0PE
G6 HXL D. Latham 89 Kestrel Park, Skelmersdale WN8 6TA
G6 HXU Dr E. Loader 13 Vale Road, Hartford, Northwich CW8 1PL
G6 HXW L. Leighton 177 Terringes Avenue, Worthing BN13 1JS
G6 HXZ P. Lovett Betula, High Halden, Ashford TN26 3LY
G6 HYD R. Ison 48 Staples Hill Partridge Green, Horsham RH13 8LF
G6 HYF C. Ironmonger 77 Boston Road, Spilsby PE23 5HH
G6 HYI P. Ingle 8 Slayleigh Delph, Sheffield S10 3RZ
G6 HYP N. Jones 7 Church Terrace, Church Road, Norwich NR16 2NA
G6 HZG K. Purser 6 Parkway, Ryde PO33 3UX
G6 HZH S. Prosser 53 Broadlands Rise, Lichfield WS14 9SF
GW6 HZJ D. Pemberton Yr Efail, Capel Coch, Llangefni LL77 7UR
G6 HZK G. Partridge 53 Acres Road, Brierley Hill DY5 2XY
G6 HZX R. Purdy 49 Mansfield Road, Eastwood, Nottingham NG16 3DY
G6 IAN I. Brooks 10 Windermere Close, Dunstable LU6 3DD
G6 IAO J. Sothcott Flat 4, 19 Auckland Road East, Southsea PO5 2HA
G6 IAT T. Bruce 17 Blaydon Road, Luton LU2 0RP
G6 IBD D. Bowles 23 Broughton Way, Rickmansworth WD3 8GW
GI6 IBL M. Barr 4 Sandelwood Avenue, Coleraine BT52 1JW
G6 IBN M. Bodill 24 Dalbeattie Close, Arnold, Nottingham NG5 8QX
G6 IBO S. Blay 68 Springvale, Gayton, King's Lynn PE32 1QZ
G6 IBP G. Burnett 314 Highcliffe, Spittal, Berwick-upon-Tweed TD15 2JN
G6 IBU M. Squance Flat 19, Brock House 2 Batter Street, Plymouth PL4 0EF
G6 IBW R. Savigar 95 Hillier Road, Devizes, Wiltshire SN10 2FB
G6 ICC I. Campbell 273 Crystal Palace Road, London SE22 9JH
G6 ICH R. Brothwood Amberley, Coombe Cross, Bovey Tracey, Newton Abbot TQ13 9EP
GD6 ICR M. Webb Coastguard House, 1 Mount Morrison, Peel IM5 1PN Isle of Man
G6 ICV I. Whittaker 20 Manor Drive, Leicester LE4 1BL
G6 ICZ Milton Keynes Amateur, RS c/o R. Waller 48 Lambfield Way, Ingleby Barwick, Stockton-on-Tees TS17 5BG
GM6 IDF Dr H. Stinton Lower Inchlumpie, Strathrusdale, Alness IV17 0YQ
G6 IDG C. Stringer Meadowbank, Station Road, Devon EX10 0ER
G6 IDL M. Waud 7 Chalkpit Lane, Candlesby, Spilsby PE23 5SE
G6 IDO A. Davies 8 Alexandra Road, Wednesbury WS10 9LH
G6 IDU I. Rose 144 Overton Road, Benfleet SS7 4DT
G6 IDW T. Roberts 9 Dixons Road, Market Deeping, Peterborough PE6 8AG
G6 IEE M. Elsley 25 Elmsdale Road, Wootton, Bedford MK43 9JW
G6 IEI P. Williams Peanjays, 4 Cutbush Close, Reading RG6 4XA
G6 IEQ P. Hawkridge 211 Goring Way, Worthing BN12 5BU
GI6 IES C. Hagan 15D Ballygalget Road, Portaferry, Newtownards BT22 1NE
G6 IET S. Hargreaves Dakyn Cottage Kirby Hill North Yorkshire, Richmond DL117JH
G6 IFE P. Holland 3 Manor Villas, Chilton Road, Chearsley, Aylesbury. HP18 0DN
G6 IFH J. Rimington 8 Harvesters, Tolleshunt D'Arcy, Maldon CM9 8UF
G6 IFN L. Rouse 69 Shackerdale Road, Wigston LE18 1BR
G6 IFQ S. Howcroft Warwick Cottage, 5 Ecclesgate Road, Blackpool FY4 5DW
G6 IFR G. Horwood 25 Briar Road, Shepperton TW17 0JB
G6 IFS N. Hollinshead 35 Parkside Drive, May Bank, Newcastle-under-Lyme ST5 0NL
G6 IFV J. Hunt 77 Scott Street, Burnley BB12 6NJ
G6 IGK L. Glasscock 37 Huntingfield Road, Bury St. Edmunds IP33 2JA
G6 IGO D. Gospel Sunnydene Farm, The Common, Fakenham NR21 9JB
G6 IGU A. Greenleaf The Lindens, Frating Road, Ardleigh, Colchester CO7 7SY
G6 IGV D. Gregson 8 Lennox Gate, Blackpool FY4 3JQ
G6 IGW N. Gutten 8 Chalfield Close, Crewe CW2 6TJ
GW6 IGY J. Mead 1 Tudor Court, Fagl Lane, Wrexham LL12 9PJ
G6 IHB F. Norton 62 Moorlands Drive, Shirley, Solihull B90 3RE
G6 IHD S. Maxwell 64 St. Georges Park, Wallasey CH45 9LW
G6 IHG H. Mitchell 17 Burners Close, Burgess Hill RH15 0QA
GI6 IHM R. Mcdowell 78 Brunswick Road, Bangor BT20 3DN
G6 IHO V. Marks 176 Middlemarch Road, Coventry CV6 3GL
G6 IHU J. Measom 41 Glebe Road, Thringstone, Coalville LE67 8NU
G6 IHW A. Mackinlay 26 Anderson Road, Erdington, Birmingham B23 6NN
G6 IIA A. Stansfield 22 Low Stobhill, Morpeth NE61 2SG
G6 IIF R. Sharpe 14 Dansie Court, Compton Road, Colchester CO4 0EA
G6 IIK D. Gill 79 Heather Walk, Bolton-Upon-Dearne, Rotherham S63 8BZ
G6 IIM P. Jones 2 Farmers Heath Great Sutton, Ellesmere Port CH66 2GX
G6 IIN P. Curtigan 5 Gayton Avenue, Wallasey CH45 9LJ
G6 IIP J. Clarke 19 Kensington Road Gaywood, King's Lynn PE30 4AT
G6 IIU R. Cooper 69 Vicarage Lane, Elworth, Sandbach CW11 3BU
G6 IIZ Dr J. Clark Brooklyn Cottage, Milton Combe, Yelverton PL20 6HP
G6 IJE P. Chapman 29 Ashwell Grove, Rotherham S65 1NF
G6 IJK A. Clayphon Downshire Lodge, Park Lane, Finchampstead, Wokingham RG40 4PT
G6 IJQ W. Cartwright 3 Masefield Rise, Halesowen B62 8SH
G6 IJW G. Stoelwinder Hampt Cottage, Middle Hampt Luckett, Callington PL17 8NR
G6 IKC S. Saunders 16 Hill Close, Pennsylvania, Exeter EX4 6HG
G6 IKE S. Lynch 21 Mill Lane, Bolton Le Sands, Carnforth LA5 8HR
G6 IKH A. Simpson 7 Hartington Street, Newcastle ST5 8DR
G6 IKM D. Tarbuck 23 Kingsway, Newton-le-Willows WA12 8LZ
GM6 IKN K. Towns Pluscarden Abbey, Pluscarden, Elgin IV30 8UA
G6 IKO M. Smith Mbe 34 Diamond Road, Watford WD24 5EW
G6 IKU J. Stockton 183 Eskdale, Tanhouse, Skelmersdale WN8 6ED
G6 ILC G. Sword Garden Cottage, Kiln Road, Salisbury SP5 2HT
G6 ILD P. Southgate Flat 1, The Old Yard, Mill Road, Holsworthy EX22 7RT
G6 ILH A. Davies 27 Whitecroft Lane, Mellor, Blackburn BB2 7HA
G6 ILN J. Dodge 5 Moat Way, Queenborough ME11 5BU
G6 ILT E. Elliston 117 Willbye Avenue, Diss IP22 4NP
G6 ILU E. Edmonds 1 Chalkhole Cottages, Flete Road, Margate CT9 4LL
G6 ILX B. Edward 27 Barford Close, Ainsdale, Southport PR8 2RS
GW6 ILY W. Evans Treetops, Whitchurch Road, Wrexham LL13 0BL
G6 ILZ I. Fullerton 54 Brashland Drive, Northampton NN4 0SS
G6 IMH R. Firth Kasamily, 73 Lions Lane, Ashley Heath, Ringwood BH24 2HH
G6 IMJ K. Walker 37 Willingdon Road, Liverpool L16 3NE
G6 IML I. Walsh 7 Winchester Avenue, Ashton-under-Lyne OL6 8BU
G6 IMN K. Wetherell 12 Parc Stephney, Budock Water, Falmouth TR11 5EJ
G6 IMQ J. Wild 20 Sandy Lane, Cholsey, Wallingford OX10 9PY
GW6 IMS T. Vernalls 5 Min Y Traeth, Minffordd, Penrhyndeudraeth LL48 6EG
G6 IMW T. Rogers Lodge 12 Benson Waterfront, Benson, Wallingford, Oxford OX10 6SJ
G6 INA P. Reidy Famagusta Avenue 45, House 2, Sotira 5390 Cyprus
GW6 INF J. Markham 4 Ty Arfon, Tywyn LL36 0TA
G6 ING S. Meigh Flat 94, Reginald Mitchell Court, Stubbs Lane, , Stoke on Trent ST1 3SP
G6 INI C. Mahony 20 Kenchester, Bancroft, Milton Keynes MK13 0QP
G6 INK E. Mcglen M.B.E. 22 Stratford Avenue, City of Sunderland SR2 8RX
G6 INL M. Nagle 47 Hawthorn Road, Minehead TA24 8EP
G6 INM K. O'Reilly 1 Parkfield, Crewe CW1 4TT
G6 INO R. Ottolini 154 Barwick Road, Leeds LS15 8SW
G6 INU D. Port 8 Betterton Drive, Sidcup DA14 4PS
G6 INV D. Pratley 2 Haseldine Meadows, Hatfield AL10 8HE
G6 INW M. Purnell 14 Cranleigh Close, Bournemouth BH6 5LD
G6 INX G. Pryke 38 Colne Drive, Walton-on-Thames KT12 3SQ
GW6 IOA C. Crow Lindisfarne, Pen Y Waun, Cardiff CF15 9SJ
G6 IOB P. Coghlan 96 Cambridge Road, Ely CB7 4HU
G6 IOE G. Crawford 4 Beverley Gardens, Gedling, Nottingham NG4 3LF
G6 IOM M. Cunningham 16 Cherry Waye, Eythorne, Dover CT15 4BT
G6 ION R. Civil 7 Sunnybanks, Hatt, Saltash PL12 6SA
G6 IOT P. Langley 40 Kingshayes Road, Aldridge, Walsall WS9 8RU
GI6 IOU M. Pollock 33 Seahill, Donaghadee BT21 0SH
G6 IOV P. Phelps 57 Southbrook Road, Havant PO9 1RL
G6 IOW D. Peachey 4 Windermere Drive, Great Notley, Braintree CM77 7UA
G6 IOX A. Pearce 49 Bishopswood Road, Tadley RG26 4HF
G6 IPB B. Doyle Ashwell Croft, Brunthwaite Lane, Keighley BD20 0ND
G6 IPC I. Downes 21 Caldbeck Court, Beeston, Nottingham NG9 5NH
G6 IPH R. Distin The Martletts, Broad Oak, Rye TN31 6DN
G6 IPN M. Davies 54 Helmside Road, Oxenholme, Kendal LA9 7HA
G6 IPQ R. Deakin 55 Pendeen Crescent, Plymouth PL6 6RE
GW6 IPR P. Drew 6 Clos Cae'R Wern, Caerphilly CF83 1SQ
G6 IPT S. Edwards 129 Rushden Road, Wymington, Rushden NN10 9LF
G6 IPW S. Featherstone 36 Denton Avenue, Grantham NG31 7JL
G6 IQC S. Leak 97 Lees Road, Ashton-under-Lyne OL6 8BQ
G6 IQF R. Harris Greve De Lec, 14A All Saints Lane, Clevedon BS21 6AY
GM6 IQH R. Wickenden 2 Buail-Bhan, Ballinluig, Pitlochry PH9 0NH

UK Callsigns

G6 IQI P. White 16 Charnwood Close, Chandler'S Ford, Eastleigh SO53 5QP
G6 IQM M. Wooding 5 Ware Orchard, Barby, Rugby CV23 8UF
G6 IQP D. Marriott 80 Andrew Avenue, Ilkeston DE7 5DW
G6 IQY J. Price 67 Broadway, Oldbury B68 9DP
G6 IRE R. Aynge 9 Sedgebrook Road, Blackheath, London SE3 8LR
G6 IRF S. Atwell 10 Belding Avenue, Manchester M40 3SE
G6 IRG M. Andrews 1 Garrick Close, Dudley DY1 3DF
G6 IRJ G. Andronov 90 Overbury Close, Northfield, Birmingham B31 2HD
GI6 IRL J. Agnew 23 Berwick Heights Moira, Craigavon BT67 0SZ
G6 IRP C. Gardner 7 Lesley Close, Bexley DA5 1LX
G6 IRU B. Green 23 Freemantle Avenue, Blackpool FY4 1SX
G6 IRW K. Holmes Gable Cottage, Low Hill, Helsby Warrington WA6 0NW
G6 IRX C. Holt 1 Vale View, Common Road, Wincanton BA9 9RB
G6 IRY R. Hobbs 120 Misbourne Road, Hillingdon, Uxbridge UB10 0HP
G6 IRZ W. Hughes 7 Richards Court, Bednall, Stafford ST17 0SP
G6 ISA J. Hutchins 18 Derby Road, Sale M33 5PR
G6 ISB the Guernsey Ar c/o A. Hunt 10 Sturton Street, Forest Fields, Nottingham NG7 6HU
G6 ISG P. Hancock 2 Gulistan Road, Leamington Spa CV32 5LU
G6 ISM J. Hancock 7 Hollies Close, Houghton-On-The-Hill, Leicester LE7 9GW
G6 ISY L. Hill 21 Liddiards Way, Purbrook, Waterlooville PO7 5QW
GW6 ITB J. Imperato 118 Heol Uchaf, Rhiwbina, Cardiff CF14 6SS
GW6 ITJ M. Jones 4 Plastirion Avenue, Prestatyn LL19 9DU
G6 ITM D. Jupp 26 Audley Avenue, Gillingham ME7 3AY
G6 ITU M. Bunn 45 Red Cat Lane, Burscough, Ormskirk L40 0RA
G6 ITV L. Parker 15 Savile Place, Mirfield WF14 0AJ
G6 ITW M. Blundell 68 Alton Road, Leicester LE2 8QA
G6 IUD Air Cadet ARS c/o R. Bracey 50 Harrow Way, Watford WD19 5ET
G6 IUF R. Brittain Sarenchel, St Cross, Harleston Norfolk IP20 0NY
GW6 IUK R. Bastable Gwynfryn, Ffordd Caergybi, Llanfairpwllgwyngyll LL61 5SZ
G6 IUQ M. Gaylard 66 Runnymede Road, Yeovil BA21 5SU
G6 IUS B. Gilbert 22 Oaklands Way, Hildenborough, Tonbridge TN11 9DA
G6 IVB A. Gornall 28 Woodward Close, Winnersh, Wokingham RG41 5NW
G6 IVC M. Griffiths 25 Lethbridge Road, Southport PR8 6JA
G6 IVD P. Guy 11 Ludlow Crescent, Redcar TS10 2LQ
GI6 IVJ R. Brown 157 Newtownards Road, Bangor BT20 4HS
G6 IVP J. Burton 22 Pear Tree Lane, Hempstead, Gillingham ME7 3PT
G6 IVR Itchen Valley ARC c/o K. Hastie 3 The Woodlands, Kings Worthy, Winchester SO23 7QQ
G6 IVW R. Balderson 15 Woodrush Way, Moulton, Northampton NN3 7HU
GW6 IVY C. Somerville 1 Glyn Isaf, Llandudno Junction LL31 9HF
GW6 IWC M. Saunders Rose Cottage, Pleasant Lane, Wrexham LL11 5DH
G6 IWD L. Sherratt Anlyn, Norbury Drive, Brierley Hill DY5 3DP
G6 IWK R. Skingley Kynance, Church Cove The Lizard, Helston TR12 7PQ
G6 IWT M. Rea Osmary, Station Road, Bishop's Stortford CM22 6LG
G6 IWU J. Rodwell 20 Nelson Street, King's Lynn PE30 5DY
G6 IWZ D. Jefferys 22 Cleveland Gardens, Cricklewood, London NW2 1DY
G6 IXE M. James 58 Spitfire Way, Hamble, Southampton SO31 4RT
G6 IXH D. Hodges 5 Greenlands, Leighton Buzzard LU7 3UJ
G6 IXM G. Hares 30 Copped Hall Drive, Camberley GU15 1NP
G6 IXN G. Hewitt 66 Portland Drive, Forsbrook, Stoke-on-Trent ST11 9AU
G6 IXP L. Holden 16 George Street, Clayton Le Moors, Accrington BB5 5QJ
GW6 IYA H. Woodnutt Latitude, Gannock Park West, Conwy LL31 9HQ
G6 IYD D. Noakes 117 Kingsmead Park, Allhallows, Rochester ME3 9TA
GM6 IYJ M. Plested Cregneash, Platcock Wynd, Fortrose IV10 8SQ
G6 IYM N. Perkins Heathcote Kents Road, Torquay TQ1 2NL
GW6 IYP R. Parry Glengarriff, Rhyl Road, Rhyl LL18 2TP
G6 IYS I. Porter 25 Wolds Retreat, Brigg Road, Fonaby, , Market Rasen LN7 6RU
G6 IYY R. Adamek The Old Engine House, 11 Top Road, Belaugh, Norwich NR12 8XB
G6 IZK A. Collier 2 The Hollies, Trerise Road, Camborne TR14 7HB
GM6 IZU K. Frame 102 High St., Galashiels TD1 1SQ
GW6 IZZ C. Evans 1 Ashdale Lane, Llangwm, Haverfordwest SA62 4NU
G6 JAC P. Dalley 32 Albert Road, Erdington, Birmingham B23 7LT
G6 JAF I. Evans 19 Grange Road, Stone ST15 8PR
G6 JAK S. Deacon 3 Blenheim Grove, Offord D'Arcy, St. Neots PE19 5RD
G6 JAL A. Dixon 23 Appleby Drive, Barrowford, Nelson BB9 6EX
G6 JAM Reading And District ARC c/o M. Dainty 5 Woodman Close, Wednesbury WS10 9UA
G6 JAP G. Denison Greengage Cottage 30 Long Street, Thirsk YO7 1AP
G6 JAR M. Drake 7 Orient Road, Paignton TQ3 2PB
G6 JAS A. Balding 8 Winston Way, Farcet, Peterborough PE7 3BU
G6 JAY R. Luckett 20 Leicester Villas, Hove BN3 5SQ
GM6 JBF D. Mardlin 35 Uist Road, Aberdeen AB16 6FN
G6 JBL G. Moore 2 Meadow Lea, Worksop S80 3QJ
GW6 JBN Nats & Caa RS c/o R. Thomas Post Office, Llanbedr LL45 2HH
G6 JBQ G. Taylor 54 Bowershott, Letchworth Garden City SG6 2EU
G6 JBY J. Bibby 24 Assarts Lane, Malvern WR14 4JR
G6 JCI W. Henson 1 Bonser Close Carlton, Nottingham NG4 1DP
GW6 JCK A. Harvey Cornerways, Mount Pleasant, Holyhead LL65 1SN
G6 JCM J. Hatfield Tenter Close, Husthwaite, York YO61 4PF
G6 JCV A. Haslehurst Westlands, Stinting Lane, Mansfield NG20 8EQ
G6 JCX F. Hewitt 12 Woodside, North Walsham NR28 9XA
G6 JCY B. Hedge Birchwood Lodge, Barnards Road, Norwich NR28 9RG
G6 JDC T. Kemp 85 Rosehill Road, Rawmarsh, Rotherham S62 7BX
GW6 JDF G. Walker Lluesty, Bryn Hyfryd Road, Tywyn LL36 9HG
G6 JDH G. Webster 153 Frogmore Lane, Waterlooville PO8 9RD
G6 JDO N. Wright 99 School Road, Saxon Street, Newmarket CB8 9RY
G6 JDP J. Mott-Gotobed 14 Copse Road, New Milton BH25 6ES
G6 JDW T. Scarfe 26 Norman Keep, Warfield, Bracknell RG42 7UY
G6 JEB J. Bailey 22 Hainfield Drive, Solihull B91 2PL
G6 JEF S. Wardley 5 Swindon Street, Bridlington YO16 4JD
G6 JEM P. Stoneman 111 Fletemoor Road, Plymouth PL5 1UL
GM6 JEP F. Cassidy Langside, Mill Farm Road, Strichen AB436RX
G6 JEU P. Chrysostomou 45 Leyborne Avenue, Ealing, London W13 9RA
G6 JEY Lord M. Cooper Flat 3, 32 Lansdowne Road, Worthing BN11 5HB
G6 JFE S. Fantom Cedar Ridge, The Dimple Fritchley, Belper DE56 2HP
G6 JFL D. Barsby 42 New Terrace, Pleasley, Mansfield NG19 7PY
G6 JFN P. Brazier Stud House, Mentmore, Leighton Buzzard LU7 0QE
GM6 JFP D. Brown 15 Eliots Park, Peebles EH45 8HB
G6 JFU A. Mayman Lingmell, Cedar Grove, Aldbrough HU11 4QH
GW6 JFV T. Morris 29 Heol Croes Faen, Nottage, Porthcawl CF36 3SW
GI6 JGB M. Mcninch 5 Bangor Road, Groomsport, Groomsport BT19 6JF
G6 JGF A. Morgan 8 Shaftesbury Road, Watford WD17 2RQ
GM6 JGH Dr J. Massheder 2B Pentland Park, Loanhead EH20 9PA
G6 JGP A. Lawrence Columbine Cottage, Ford, Salisbury SP4 6DJ
G6 JGR J. Richardson 30 Shaftesbury Avenue, Chandler'S Ford, Eastleigh SO53 3BS
G6 JGT A. Davis 7 Kennedy Crescent, Gosport PO12 2NL
G6 JHG J. Grieve 65 Royal Lane, Uxbridge UB8 3QU
GM6 JHH W. Gunn Tullochard, Scouriemore, Lairg IV27 4TG
GD6 JHP J. Pauley 32 King Edward Bay Apartments, Sea Cliff Road, Onchan IM3 2JF Isle of Man
G6 JHS A. Winrow 14 Green Lane, Bayston Hill, Shrewsbury SY3 0NS
G6 JHT M. Stannard 16 Rose Court Primrose Road, Dover CT17 0FP
GM6 JIC L. Paget 40 Davaar Drive, Kilmarnock KA3 2JG
G6 JIF J. Purdy 99 Ashford Road, Hastings TN34 2HY
GM6 JIL R. Mcmillan 17B Kingston Road, Neilston, Glasgow G78 3JA
G6 JIM J. King 4 Glenhurst Avenue, Ruislip HA4 7LZ
G6 JIR N. Brown 11 Tudor Green, Jaywick, Clacton-on-Sea CO15 2PA
G6 JJA N. Billingham 127 Atlantic Way, Westward Ho, Bideford EX39 1JG
G6 JJB B. Banks 16 Park Road, Burntwood WS7 0EE
G6 JJF C. Byrne 104 Ripon Hall Avenue, Ramsbottom, Bury BL0 9RE
G6 JJG K. Breakwell 91 Lynton Avenue, Claregate, Wolverhampton WV6 9NQ
G6 JJI A. Bromfield 11 Blackthorn Croft, Clayton-Le-Woods, Chorley PR6 7TZ
G6 JJK J. Bourne 91 Burwell Road, Exning, Newmarket CB8 7DU
GM6 JJN R. Berry Sylvan House, Glenmoriston, Inverness IV63 7YJ
G6 JJP J. Pinson 10 Kenelm Close, Clifton-On-Teme, Worcester WR6 6EB
GI6 JJR N. Loughrey 12 Billys Road, Newry BT34 2NA
G6 JJT E. Ferguson Willowmead, Church End, Ravensden, Bedford MK44 2RP
GW6 JJX R. Price Arfryn, Trecastle, Brecon LD3 8UP
G6 JKF M. Lowe 7 St. Michaels Close, Stafford ST17 0JA
G6 JKK G. Orchard 189 Sopwith Crescent, Wimborne BH21 1SR
GM6 JKU G. Henderson 34 Soutar Crescent, Perth PH1 1QB
G6 JKV R. Henneman Lydbury, Wistanswick, Market Drayton TF9 2BB
G6 JLI P. Dixey Rose Cottage, 197 Raikes Lane, Batley WF17 9QF
G6 JLL S. Douglas 1030 Shields Road, Walkerville, Newcastle upon Tyne NE6 4SR
G6 JLU A. Millar 8 Eisenhower Road, Shefford SG17 5UP
G6 JMB J. Mountain Thurlow House, Aldenham Avenue, Radlett WD7 8HJ
GW6 JMC D. Miller Maes Hyfryd, Llanfynydd, Wrexham LL11 5HH
GI6 JMD J. Moller 9 River Hill Lane, Newtownards BT23 7GQ
G6 JMG P. Parton 12 Duchess Drive, Bridgnorth WV16 4JD
G6 JMJ K. Renton 87 Shirley Gardens, Barking IG11 9XB
G6 JMO J. Page 9 Ascot Close, Elstree, Borehamwood WD6 3JH
G6 JMX B. Wendon 89 Palewell Park, London SW14 8JJ
GM6 JNJ D. Anderson 34 Culzean Crescent, Kilmarnock KA3 7DT
GM6 JNQ I. Cox 8 Traill St., Castletown, Thurso KW14 8UG
G6 JNS P. Crosland Sprackets Orchard, Curry Rivel, Langport TA10 0PP
G6 JNV M. Carter 17 Mcwilliam Road, Woodingdean, Brighton BN2 6BE
G6 JNW S. Carter 84 Barnett Road, Brighton BN1 7GH
G6 JNY I. Eliade 32 Meadow Sweet Road, Stratford-upon-Avon CV37 0TH
G6 JNZ W. Caine 53 Cromford Road, Crich, Matlock DE4 5DJ
GM6 JOA A. White Brodiescroft, Banff AB45 3BR
GM6 JOD T. Lawless 2 Lawers Place, Bourtreehill North, Irvine KA11 1LR
G6 JOL R. Young 53-55 Kirby Road, Leicester LE3 6BD
GI6 JOP A. Wallace 61 Locksley Park, Belfast BT10 0AS
G6 JOR D. Webb 1 Corelli Road, Basingstoke RG22 4NB
GM6 JOS K. Arrowsmith 2, Dornoch Road, Ness Castle, Inverness IV2 6EQ
GW6 JPC C. Jenkins 10 Marsh Court, Abergavenny NP7 5HQ
GU6 JPE A. Jephcott 12 Clos Du Beauvoir, Rue Cohu, Castel X X Guernsey
G6 JPG J. Gilliver 5 Yew Tree Park Homes Charing, Ashford TN27 0DD
G6 JPM S. Green 7 Brook View, Totnes TQ9 5FH
GI6 JPO H. Graham 104 Tattygare Road, Lisbellaw, Enniskillen BT94 5FB
G6 JPQ J. Gould 108 Newton Road, Burton-on-Trent DE15 0TT
G6 JPR Dr M. Glover 3 Sammons Way, Coventry CV4 9TD
G6 JPS J. Skertchly 132 Derby Road, Spondon, Derby DE21 7LX
G6 JPT D. Gleave 1 Fearnley Way, Newton-le-Willows WA12 8SQ
G6 JQD R. Skinner 23 Hardy Road, Greatstone, New Romney TN28 8SF
G6 JQE G. Tannahill 15 Bowfell Avenue, Newcastle upon Tyne NE5 3XB
GU6 JQF M. Trenchard Mont Gibel, 3 Clifton Stairs, St. Peter Port GY1 2PL Guernsey
G6 JQH J. Williams 18 The Leas, Barkston, Grantham NG32 2PD
G6 JRE S. Stanton 6 Trevor Road Beeston, Nottingham NG9 1GR
G6 JRI I. Wright1 3 Sykes Court, Wheldrake YO19 6GE
G6 JRL M. Bernard 1 Foxglove Way, Hambleton, Selby YO8 9UB
G6 JRM H. Bottomley Nerefield, Aylesbury Road, Aylesbury HP18 0BL
G6 JRS A. Cuthbertson 72 Bulford Road, Durrington, Salisbury SP4 8DJ
GM6 JRX D. Fraser Kylerhea, Harbour Road, Castletown, Thurso KW14 8TG
GI6 JRY Hastings & Rother Raynet Grp c/o R. Getty 6 Rocheville, Cookstown BT80 8QE
G6 JRZ S. Gunn 55 Station Road, West Byfleet KT14 6DT
G6 JSF M. Hayward 1 Station Road, Grateley, Andover SP11 8LG
G6 JSI A. Haswell 66 White Hart Lane, Fareham PO16 9BQ
GW6 JSJ D. James 6 Chave Terrace, Maesycwmmer, Hengoed CF82 7RZ
G6 JSN A. Sym 1 Beech Close, Spetisbury, Blandford Forum DT11 9HG
G6 JSR A. Mason 5 Birch Road, Kippax, Leeds LS25 7DY
G6 JTC M. Whiteley 9 Fernie Close, Barton Seagrave, Kettering NN15 6RE
G6 JTD D. Walker 100 Clifton Road, Kingston upon Thames KT2 6PN
G6 JTI H. Martin 80 Topcliffe Road, Sowerby, Thirsk YO7 1RT
G6 JTK R. Nokes 99 Harmers Hay Road, Hailsham BN27 1TW
G6 JTO the Gower/Gwyr Contesting Club c/o J. Parfrey 97 Gordon Road, Camberley GU15 2JQ
G6 JTT J. Trett 1 Moorland Way, Bridgwater TA6 4JL
G6 JTV R. Allen 65 Atherstone Road, Measham, Swadlincote DE12 7EG
G6 JTW H. Marshall Mardachrob, 4 Howell Road, Sleaford NG34 9RX
GW6 JTX E. Bielawski Rowanlea, Quarry Brow, Wrexham LL12 8SJ
GM6 JUA D. Brown 10 Culmore Place, Falkirk FK1 2RP
G6 JUE N. Cockayne 46 Canterbury Way, Stevenage SG1 4DQ

G6 JUI K. Dare One Bee, 1 Gloucester Road, Reading RG30 2TH
G6 JUP J. Sutton 252 Rawling Road, Gateshead NE8 4UH
G6 JUQ G. Williams 21 Arden Close, Southport PR8 2RR
G6 JUT J. Whiting Old Hall, Sykes Lane, Pickworth, Sleaford NG34 0TZ
G6 JVA J. Greevy 11A Norman Road, Walsall WS5 3QJ
GW6 JVB R. Griffiths 26 Brynglas, Gilwern, Abergavenny NP7 0BP
G6 JVI C. Hedges 398 Chemin De Lancement, Thezac 47370 France
G6 JVK M. Jeffery 14 Rosemary Avenue, Earley, Reading RG6 5YQ
G6 JVO M. Kidd 99 Ferry Road West, Scunthorpe DN15 8UG
G6 JVT C. Santer 51 Limbrick Lane, Goring-By-Sea, Worthing BN12 6AB
G6 JVX H. Schofield 15 Deerfield Road, March PE15 9AH
G6 JWD J. Davies Welfare, High Street, Borth SY24 5JD
GM6 JWF A. Paul 20 Upper Bridge Street, Alexandria G83 0LL
GM6 JWH D. Taylor 3 Abbotsgrange Road, Grangemouth FK3 9JD
GW6 JWL H. Roberts Pen Yr Erw, Graigfechan, Ruthin LL15 2EY
G6 JWM D. Le Grove Apartment 3, Beechwood, Ilkley LS29 8AH
G6 JWO A. Legg 2 Alkington Farm Lane Cottage, Heathfield, Berkeley GL13 9PL
G6 JXA K. Brown 165 Canterbury Road, Morden SM4 6QG
G6 JXC P. Chadbund 20 Northlands Road, Adstock, Buckingham MK18 2JH
GI6 JXG W. Collins 33 New Row, Kilrea, Coleraine BT51 5TA
G6 JXS W. Hughes 27 Winchester Close, Ashington NE63 9QJ
G6 JYB M. Niman 55 Harrow Way, Great Baddow, Chelmsford CM2 7AU
G6 JYN D. Watkinson 52 Tweed Close Haydon Wick, Swindon SN25 1PX
G6 JYO C. Allen F B S 20 Hollywood Lane, Hollywood, Birmingham B47 5PX
G6 JYR D. Benton 231 Prestwood Road, Wolverhampton WV11 1RF
G6 JYX R. Drew Derwent House, Landing Lane, Selby YO8 6RA
G6 JZE P. Graham 14 Carlaw Road, Birkenhead CH42 8QA
G6 JZN A. Ogden 6 Crossway, Thornton Cleveleys FY5 1LA
G6 JZV K. Lummis The Bothy, Upper Town Wetherden, Wetherden Upper Town, Stowmarket IP14 3NF
G6 JZW C. Muller 5 Ash Close, Flitwick, Bedford MK45 1JY
G6 KAE J. Bailey 54 Dimsdale Road, Northfield, Birmingham B31 5RD
G6 KAI M. Brighton 11 West Close, Norwich NR5 0NH
GM6 KAM A. Drummond Flat 4F, Crossfolds Crescent, Peterhead AB42 1RD
GW6 KAV Dr H. Hughes Hendre Bach, Cerrigydrudion, Corwen LL21 9TB
G6 KAW G. Instone 19 Dickenson Road, Swindon SN25 1WG
GM6 KAY C. Bates 10 Swanston Avenue, Edinburgh EH10 7BU
G6 KBC C. Philpot 17 Jervis Court, Ilkeston DE7 8PX
GW6 KBD D. Potts 11 Walmer Road, Newport NP19 8NU
G6 KBN R. Morby The Rectory, High Street, Edgmond, Newport TF10 8JR
G6 KBQ T. Williams 2 Hazelwood, Greasby, Wirral CH49 2RQ
G6 KBS J. Musgrave 57 Chiltern Road, Baldock SG7 6LT
GI6 KBX Rvd. J. Turner 45 Gloonan Hill, Ahoghill, Ballymena BT42 1PU
G6 KBZ N. Wilkins Norjen, Thorpe Market Road, Norfolk NR11 8NG
G6 KCG P. Sharpe 46 Beaumont Road, New Costessey, Norwich NR5 0HG
G6 KCJ D. Wynters 11 Heritage Lane, Ascott-Under-Wychwood, Chipping Norton OX7 6AD
G6 KCV P. Willmott C/O Oil Management Serv. Ltd., P.O. Box Hm 1751, Hamilton Hm Gx X X Bermuda
GM6 KDB K. Lee West Skares, Glens Of Foudland, Huntly AB54 6AT
GM6 KDD D. Scobbie 17 Roselea Drive, Brightons, Falkirk FK2 0TJ
GI6 KDN K. Mcinnes 39 St. Johns Place, Belfast BT7 3HA
G6 KDU Fists/International Morse Pre 'Soc' c/o M. Thorley 5 Burland Road, Newcastle ST5 7ST
G6 KDW G. Morton 42 Elwyn Road, March PE15 9DA
G6 KDY A. Perkins 3 Greenway Close, Radcliffe-On-Trent, Nottingham NG12 2BU
G6 KEH J. Golightly 1 Pannier Mews, Castle Street, Torrington EX38 8EE
G6 KEN K. Dasilva-Hill 5 Station Road, Charing, Ashford TN27 0JA
GM6 KEV D. Smith Mandala, Belhaven Road, Dunbar EH42 1NW
G6 KEX S. Smyth 18 Weston Road, Birmingham B19 1EH
G6 KEZ P. Pattison 18 Broadgate Lane, Deeping St. James, Peterborough PE6 8NW
G6 KFD P. Stockwell 62 Golden Cross Road, Ashingdon, Rochford SS4 3DQ
GM6 KFO G. Gordon 31 Stoneyhill Avenue, Musselburgh EH21 6SB
G6 KFR D. Jones 2 The Orchard Mill Lane, Kings Sutton, Banbury OX17 3RG
G6 KGA L. Coleman Lilac Cottage, Coley Lane, Stafford ST18 0XB
G6 KGK G. Gudgin 7 Merchant Place, Middleton, Milton Keynes MK10 9JL
G6 KGL G. Gudgin 890 W Iowa Ave, Sunnyvale 94086 United States
GW6 KGR M. Buck Upper Glaisfer, Llangynidr, Crickhowell NP8 1LN
G6 KGU Dr D. Craig Flat 5, The Chapel, The Plains, Totnes TQ9 5DW
G6 KHA T. Hyde 14 Wyley Road, Coventry CV6 1NW
G6 KHD K. Bierton 44 Stalmine Hall Park, Hall Gate Lane, Poulton-le-Fylde FY6 0LD
G6 KHG R. Champion 25 Congreve Road, Worthing BN14 8EL
G6 KHM L. Edwards 71 Gleneagles Road, Yardley, Birmingham B26 2HT
G6 KHN S. Harvey 53 Winleigh Road, Handsworth Wood, Birmingham B20 2HN
G6 KHW I. Bultitude 48 Forty Acres Road, Devizes SN10 3DG
G6 KIA C. Duckles 8 Railway Cottages, Skillings Lane, Brough HU15 1EN
G6 KIB P. Duesbury The Bungalow, Robins Lane, Cambridge CB23 8HH
G6 KIE D. Banks 145 Compton Crescent, Chessington KT9 2HG
G6 KIH P. Ball The Old Maggot Farm Stocking Lane, Knottingley WF11 8TH
G6 KIV S. Blythe 17 Ashlea Road, Wirral CH61 5UG
GW6 KIW M. Dennis 16 Gwel Y Llan, Llandegfan LL59 5YH
G6 KIZ M. Griffiths 70 Towcester Road, Far Cotton, Northampton NN4 8LQ
G6 KJA P. Hoyle Sunset Cottage, Water End Lane, Ayot St. Peter, Welwyn AL6 9BB
GI6 KJC Dr W. Abram 1 The Briggs, Groomsport, Bangor BT19 6HY
G6 KJE A. Dolby 27 Tucker Road , Ottershaw, Chertsey KT16 0HD
G6 KJH P. Horobin 12 Laurel Road, Blaby, Leicester LE8 4DL
G6 KJK J. Chappell 15 Edmund Avenue, Stafford ST17 9FT
G6 KJM J. Mirams 29 Martello Court, Jevington Gardens, Eastbourne BN21 4HR
G6 KJT S. Brabbins 8 Park Drive, Bingley BD16 3DF
G6 KJY L. Cartwright 18 High Causeway, Much Wenlock TF13 6BZ
G6 KKA J. Edmondson 6 Park Lea, Bradley, Huddersfield HD2 1QH
G6 KKN P. Clowes 14 Derek Drive, Sneyd Green, Stoke-on-Trent ST1 6BY
G6 KKW R. Rogers 31 Westgate Bay Avenue, Westgate on Sea CT88AH
GW6 KLC A. Morris Bodvel Hall, Llannor, Pwllheli LL53 6DW
G6 KLF A. Lythaby 25 Greenhill Road, Otford, Sevenoaks TN14 5RR
G6 KLH R. Taylor 57 Walnut Tree Road, Shepperton TW17 0RP
GW6 KLQ J. Laing Penyboncyn, Pen-Y-Garnedd, Oswestry SY10 0AN
G6 KMG I. Turnbull 45 Elton Road, Darlington DL3 8HU
GM6 KMK S. Windsor Hillside Overbrae, Fisherie, Turriff AB53 5QP
G6 KMQ C. Meadows 47 Widney Lane, Solihull B91 3LL
G6 KNE J. Wright Rhumbles, Station Approach, Dorking RH5 5HT
G6 KNK J. Solomon 11 Angle Close, Hillingdon, Uxbridge UB10 0BS
G6 KNM R. Suttenwood 51 High St., Rowhedge, Colchester CO5 7ET
G6 KNU Medway Raynet c/o H. Man 115 Northdown Park Road, Margate CT9 3PX
G6 KOB J. Sobanski 10 Robert Avenue, Barnsley S71 5RB
G6 KOE A. Reilly 14 Carleton Gardens, Carleton, Poulton-le-Fylde FY6 7PB
GM6 KON T. Wilkins Aelart, Lyth, Wick KW1 4UD
GM6 KOR K. Osborne 42 India St., Edinburgh EH3 6HB
G6 KPA G. Osborne 19 Orchard Close, Biggleswade SG18 0NE
G6 KPD J. Perrett 12 Horne Close, Stratton-On-The-Fosse, Radstock BA3 4SS
G6 KPJ P. Vaughan The Views, Bedford Road West, Northampton NN7 1HB
GM6 KPL A. Wilson 193 Irvine Road, Kilmarnock KA1 2LA
G6 KPT N. Woodley 20 St. Edwards Road, Cheddleton, Leek ST13 7JP
G6 KPW S. Taylor 17 Crays Hill, Leabrooks, Alfreton DE55 1LN
G6 KPX A. Thorne 31 Oak Farm Close, Sutton Coldfield B76 1PJ
G6 KQ K. Spicer Grove Cottage, Dallinghoo, Woodbridge IP13 0LR
G6 KQD G. Morris 20 Victoria Way, Stafford ST17 0NU
G6 KQJ H. Moon 14 Elmwood Road, Eaglescliffe, Stockton-on-Tees TS16 0AQ
G6 KQN G. Robertson 24 Begonia Avenue, Farnworth, Bolton BL4 0DS
G6 KQS J. Newton Shestnadeseta Street 6, Mindya 5044 Bulgaria
G6 KQZ B. Wiseman 307 Kempshott Lane, Basingstoke RG22 5LY
G6 KRG J. Freeman 18A Five Bells, Watchet TA23 0HZ
G6 KRJ J. Jones 52 Woodleigh Drive, Sutton-On-Hull, Hull HU7 4YZ
GW6 KRK E. Karklins Lonlas House, Lonlas, Neath SA10 6SD
G6 KRS N. Ashall 21 Buxton Lane, Droylsden, Manchester M43 6HL
G6 KRY C. Pieters 32 Olde Farm Drive, Blackwater, Camberley GU17 0DU
G6 KSK A. Hodgson 33 Higham Road, Wainscott, Rochester ME3 8BE
G6 KSO P. Lash 7 Park Road, Stockport SK4 4PY
G6 KSR F. Patman Northcote, 31 Church Road, Lincoln LN6 5UW
G6 KSV A. Sayers 9 Willow Way, Bottisham, Cambridge CB25 9BS
G6 KTB W. Curtis Innsbruck, Trevingey Crescent, Redruth TR15 3DF
G6 KTC Royal Naval ARS c/o W. Bramwell 15 Chadswell Heights, Lichfield WS13 6BH
G6 KTE D. Brunt 31 The Green, Kingsley, Stoke-on-Trent ST10 2AG
G6 KTG D. Clements 3 Tilefields, Hollingbourne, Maidstone ME17 1TZ
G6 KTK C. Handley Torcroft, 40 New Village Road, Little Weighton HU20 3XH
G6 KTN M. Lamerick 12 Woodhouse Close, Birchwood, Warrington WA3 6QP
G6 KTO J. Martyn-Clark 127 Blackpool Road North, Lytham St. Annes FY8 3DB
GM6 KTP K. Morrison 8 St. Helena Crescent, Hardgate, Clydebank G81 5PD
G6 KTR G. Birch 18 Hill Farm Way Boxted, Colchester CO4 5RD
G6 KTX A. King 2 Longstaff Gardens, Fareham PO16 7RR
G6 KUI P. Walker1 23 Denstone Drive, Alvaston, Derby DE24 0HZ
G6 KUJ F. Moulding 28 Woodbine Road, Bolton BL3 3JH
G6 KVA R. Dresser 6 Acacia Avenue, Fencehouses, Houghton le Spring DH4 6JG
G6 KVE C. Payne 4 George Street, Helpringham, Sleaford NG34 0RS
G6 KVG S. Reid 223 The Greenway, Epsom KT18 7JE
G6 KVI B. Gosling 15 Cherry Chase, Tiptree, Colchester CO5 0AE
G6 KVR P. Roberts 30 Baldwins Lane, Hall Green, Birmingham B28 0QX
GI6 KVS H. Porter 30 Twinburn Road, Newtownabbey BT37 0EL
G6 KVY S. Trotter 61 Trinity Road, Billericay CM11 2RY
G6 KWA D. King 20 Trinity Close, Haslingfield, Cambridge CB23 1LS
G6 KWH J. Dixon 16 Forest Lane, Martlesham Heath, Ipswich IP5 3ST
GW6 KWM P. Holt 5 Tudor Road, Treboeth, Swansea SA5 9HF
GM6 KWU R. Adamson 29 South Hermitage Street, Newcastleton TD9 0QE
G6 KWY D. Tipping 30 Bickerton Point, South Woodham Ferrers, Chelmsford CM3 5YG
G6 KWZ J. Manning 280 Ledbury Road, Hereford HR1 1QL
G6 KXB R. Linzey 29 Arkle Court, Alnwick NE66 1BS
G6 KXD M. Parker Hazel House, Talkin, Brampton CA8 1LE
G6 KXJ K. Turner 16 Orford Street, Liverpool L15 8HX
G6 KXN R. Perry 6 Morgan Close, Arley, Coventry CV7 8PR
GM6 KXP D. Flanagan Ryan Mar, Stair Drive, Stranraer DG9 8EY
G6 KXW A. Blair 35 South Court Avenue, Dorchester DT1 2BY
G6 KYE M. Davis 86 Upper Shaftesbury Avenue, Southampton SO17 3RT
G6 KZI R. Gregory 75 Station Road South, Belton, Great Yarmouth NR31 9LZ
G6 LAE J. Clifton 6 Chester Close, Newbury RG14 7RR
G6 LAO S. Parkinson Meadow Barn, Watery Lane, Astbury, Congleton CW12 4RR
G6 LAU D. Tanswell Highstead Farmhouse, Bradford, Holsworthy EX22 7AA
G6 LAW C. Rudge 1 Mill Lane, Alfington, Ottery St. Mary EX11 1PF
G6 LBE J. Massey 10 Rapley Avenue, Storrington, Pulborough RH20 4QL
G6 LBG N. Orgill 32 Upland Avenue, Chesham HP5 2EB
G6 LBJ P. Shadbolt 39 Ringstead Crescent, Weymouth DT3 6PT
G6 LBO K. Batty 19 Breckland Close, Stalybridge SK15 2QQ
G6 LBQ A. Hunter 22 Lynthorpe Avenue Cadishead, Manchester M44 5JQ
G6 LBR A. Ledger 9 Fox Wood, Westlea, Swindon SN5 7AW
G6 LCL T. Mallett 11 Caragh Road, Chester le Street DH2 3EA
G6 LCP P. Muzyka 2 Engine Fold Off Lindale Lane Kirkhamgate, Wakefield WF2 0PP
G6 LCS J. Mcneill 2 Greenwood Close, Weaverham, Northwich CW8 3RH
G6 LCU J. Retter 12 Palmerston Road, Grays RM20 4YR
G6 LCX M. Pomfret 17 Lovers Lane, Atherton, Manchester M46 0PG
G6 LD Denby Dale ARS c/o J. Stocks 96, North Street, Lockwood, , Huddersfield HD1 3SL
G6 LDA J. Round 53 Furlong Lane, Halesowen B63 2TB
GM6 LDG P. Clements Dhualton Cottage, Kirtomy, Thurso KW14 7TB
G6 LDJ R. Wilkinson 2 Conway Avenue, Billingham TS23 2HX
G6 LDM D. Shippen 11A Pear Tree Drive, Wincham, Northwich CW9 6EZ
G6 LDO C. Seeney 91 Dovehouse Close, Eynsham, Witney OX29 4EW
G6 LDP D. Scott 7 Greenfield Mount, Wrenthorpe, Wakefield WF2 0TJ
G6 LDW J. Tottle 327A Edificio, Calle Miguel Machado, Son Caliv, Calvia 07181, Spain
G6 LDY J. Seddon 11 Hilda St., Leigh WN7 5DG
G6 LEB T. Leader-Chew 10 Hawmead, Crawley Down, Crawley RH10 4XY
G6 LEI S. Meadwell 25 Redland Road, Oakham LE15 6PH
G6 LEK R. Mason 23 Fulmodeston Road, Stibbard, Fakenham NR21 0LT
G6 LEU D. Last Hillview, New Road, Bridport DT6 4NY
G6 LEY D. Miller 44 Long Lane, Ickenham, Uxbridge UB10 8TA
GM6 LEZ J. Mcdermott 12A Margaret Street, Greenock PA16 8AS
G6 LFA A. Machin 10 Buttlehide, Maple Cross, Rickmansworth WD3 9TZ
G6 LFC J. Mchale Glen Elg, The Green, Skipton BD23 4LB

G6 LFD J. Corderoy 1 Alandale Drive, Pinner HA5 3UP
G6 LFG J. Bradbury 281 Peter Street, Macclesfield SK11 8EX
G6 LFJ J. Aslan 16 Guildford Street, Brighton BN1 3LS
G6 LFQ R. Cross 84A Cranborne Avenue, Surbiton KT6 7JT
G6 LFR L. Carr 29 Hill Drive, Whaley Bridge, High Peak SK23 7BH
G6 LFT Dr G. Cooke 594 Preston Road, Clayton-Le-Woods, Chorley PR6 7EB
G6 LFW J. Ford 24 Tonstall Road, Epsom KT19 9DP
G6 LGM I. Rogers Tremayne Cottage, Calloose Lane, Hayle TR27 5ET
G6 LGR A. Picot 14 Ringshall Road, St. Pauls Cray, Orpington BR5 2LZ
G6 LHA F. Priestnall 56 Badger Gate, Threshfield, Skipton BD23 5EN
GW6 LHF D. Owen Grasmere, Pontycleifion, Cardigan SA43 1DW
G6 LHG B. O'Shea 37 Gardeners Road, Halstead CO9 2TA
G6 LHQ R. Harber 6 Westgate, Fareham PO14 2NY
G6 LI Lincolnshire Poachers CG c/o D. Johnson Deans Cottage, Dowsby Fen, , Fen Road, Dowsby, Bourne PE10 0TU
G6 LIB J. Baker 5 Larkspur Close, Bishop's Stortford CM23 4LL
G6 LIJ D. Chilton 6 Deneside Close, Yarm TS15 9NT
G6 LIK L. Clark 56 Rembrandt Avenue, South Shields NE34 8RU
GM6 LIN Essex Ham c/o J. Quinn 8 Cluny Drive, Newton Mearns, Glasgow G77 6YG
G6 LJC G. Weston Meadowside, Wardle Lane Light Oaks, Stoke-on-Trent ST2 7LP
GM6 LJE R. Waitt Orchard Cottage, Claygate, Canonbie DG14 0RZ
G6 LJF J. White 8 Well Side, Marks Tey, Colchester CO6 1XG
G6 LJH M. Wilson Hillside, Chapel Lane, Kettering NN14 4EA
G6 LJR Anglo-European School RC c/o D. Twyman 77 Essex Road, Maldon CM9 6JH
G6 LJU Nunsfield House ARG c/o J. Whitehouse The Paddock, Westmancote, Tewkesbury GL20 7EP
G6 LJX East Yorkshire Contest c/o S. Williams 187 London Road, Northwich CW9 8AR
G6 LKB D. Warburton 36 Bigland Drive, Ulverston LA12 9PD
G6 LKG R. Milne 9 Brunstath Close, Wirral CH60 1UH
G6 LKH C. Dunlop 32 Court Way, Twickenham TW2 7SN
G6 LKJ De Montfort University ARS c/o J. Depledge 37 Higher Bents Lane, Bredbury, Stockport SK6 1EE
G6 LKV G. Ashbee 6 The Green, Wimbledon, London SW19 5AZ
G6 LKW T. Ashbee Plough Heights, Main Road, Itchen Abbas, Winchester SO21 1BQ
G6 LKZ D. Bentley 302 Bordesley Green East Stechford, Birmingham B33 8ST
G6 LLD G. Bell 4 Dallymore Drive, Bowburn, Durham DH6 5ES
G6 LLF P. Bennett 1 The Briars, Newcastle ST5 9PU
G6 LLG M. Broadway 69 The Brambles, Crowthorne RG45 6EF
G6 LLL D. Burrows 32 Whitfield Cross, Glossop SK13 8NW
G6 LLP R. Farey 38 Trent Close, Yeovil BA21 5XQ
G6 LLU D. Setterfield 10 Birch Walk, Bride Street, Todmorden OL14 5ET
G6 LMB P. Steadman 41 The Linkway, Brighton BN1 7EJ
G6 LMC P. Webb 63 Trinity Road, Halstead CO9 1ED
GW6 LMI J. Evans 91 Queens Avenue, Flint CH6 5JP
G6 LMJ Dr G. Eardley 45 Little Moss Lane, Scholar Green, Stoke-on-Trent ST7 3BL
G6 LMR K. Fisher 26 Manila Street, Sunderland SR2 8RS
G6 LNA P. Cooper 2 Meadowsweet Way, Wimblebury, Cannock WS12 2GS
G6 LNF N. Clare 4 Arlington, Weymouth DT4 9SG
G6 LNL I. Dobson 73 Stanley Street, Seaham SR7 0AU
G6 LNS J. Duxbury Woodlands, Wallace Lane, Preston PR3 0BB
G6 LNU J. Durban 62 Westfield Way, Charlton, Wantage OX12 7EP
G6 LNV J. Cunliffe 142 Hall Road, Hull HU6 8SB
G6 LOC T. Stirrup 23 Round Wood, Penwortham, Preston PR1 0BN
G6 LOJ N. Pettit 10 Broom Road, Lakenheath, Brandon IP27 9ES
G6 LOR E. Wood 57D Halesowen St., Rowley Regis B65 0HF
G6 LPB R. Steele 27 Beasley Grove, Birmingham B43 7HG
G6 LPC A. Samways 61 Cooper Road, Rye TN31 7BG
G6 LPD A. Tucker 63 Oakes Road, Bury St. Edmunds IP32 6PU
G6 LPG Clyde Coast Contest Club c/o S. Taylor 76 Queensdown Gardens, Bristol BS4 3JF
G6 LPS T. Biddle 55 Barley Mow Lane, Bromsgrove B61 0LU
G6 LPT R. Bruckner Flat 2 Byron Court Fairfax Road, London NW6 4HB
G6 LPV D. Blackmore 2 Witten Gardens Northam, Bideford EX39 3RE
G6 LPX P. Brown 5 Fairview Close, Amington, Tamworth B77 3LA
G6 LQE D. Byrom 206 Didsbury Road, Stockport SK4 2AA
G6 LQG D. Bate 15 Martins Drive, Ferndown BH22 9SG
G6 LQI N. Bird 2 Manor Valley, Weston-Super-Mare BS23 2SY
G6 LQM G. Barker 99 Sheffield Road, Wymondham NR18 0HS
G6 LQP D. Brown Hillingswood, Acton, Newcastle ST5 4FD
G6 LQR B. Walker 5-6 Cochrane Terrace, Willington, Crook DL15 0HN
G6 LRT C. Johnson 52 Evesham Road, Cookhill, Alcester B49 5LJ
G6 LRU R. Jones 53 Wavertree Road, Blacon, Chester CH1 5AF
G6 LRY C. Kelland 11 The Meads, West Hanney, Wantage OX12 0LJ
G6 LSB N. Key Three Corner Cross, Rosecare, St. Gennys, Bude EX23 0BE
G6 LSC M. Kitchener 5 Whinbush Grove, Hitchin SG5 1PT
G6 LSD P. Kerry 35 Victoria Drive, Blackwell, Alfreton DE55 5JL
GW6 LSL S. Wood 2 Radyr Road, Llandaff North, Cardiff CF14 2FU
G6 LSO C. Wolf 35A Moorhouse Road, Carlisle CA2 7LU
G6 LST D. Rhodes 1 Tanpit Cottages, Winstanley, Wigan WN3 6JY
G6 LSW A. Stevenson 37 Hillside Road East, Bungay NR35 1JU
G6 LTB P. Townrow 64 Millham Road, Bishops Cleeve, Cheltenham GL52 8BG
G6 LTD P. Sutton-Atkins 57 Oakleigh Residential Park, Clacton Road, Weeley, , Colchester CO16 9DH
G6 LTK K. Wilson 44 Campbell Road, Caterham CR3 5JN
G6 LTN A. Wanford 4 Willows Close, Tydd St. Mary, Wisbech PE13 5QR
G6 LTR J. Warner 32 Rolleston Road, Wigston LE18 2EP
G6 LTT R. White 105 Engel Park, London NW7 2HN
G6 LUD A. Ryan 153 Sutton Road, Maidstone ME15 9AB
G6 LUE T. Yates 5 Manor Garth, Kellington, Goole DN14 0NW
G6 LUF A. Yates 59 Worden Lane, Leyland PR25 3BD
G6 LUJ R. Perry Thornaby, Queen Street, Colyton EX24 6JU
G6 LUK J. Russell 9 Batten Close, Christchurch BH23 3BJ
G6 LUM J. Papworth 339 Gayfield Avenue, Brierley Hill DY5 3JE
GW6 LUT A. Mills 1 Derwen Fawr, Llandybie, Ammanford SA18 2UY
G6 LUU A. Marshall 8 Barn Owl Close, Northampton NN40RQ
G6 LUY N. Mattey 27 Middleton Road, Daventry NN11 8BH
G6 LVB H. Long 40 Ashburn Place, London SW7 4JR
G6 LVC J. Shergold 35 Orchard Grove, New Milton BH25 6NZ
G6 LVG S. Normandale 5 The Beacon, Ilminster TA19 9AH
G6 LVI P. Hickey 36 Station Road, Alderholt, Fordingbridge SP6 3RB
G6 LVJ R. Hickey Mallorin, Blackfield Road, Southampton SO45 1EG
G6 LVM C. Holderness 7 Oakfield Avenue, Clayton Le Moors, Accrington BB5 5XG
G6 LVN R. Hope 35 Pinewood Gardens, North Cove, Beccles NR34 7PQ
G6 LVS D. Mallalieu Howard Flat 17, The London Well Street, Ryde PO33 2SS
G6 LVT C. Harvey 619 West Street, Crewe CW2 8SH
G6 LWA C. Hall 147 Gordon Avenue, Camberley GU15 2NR
G6 LWC R. Hardman 4 Alverstone Road, Wallasey CH44 9AA
G6 LWD P. Hurley 18 Pear Tree Lane, Wolverhampton WV11 1BD
G6 LWK M. Horsfield 13 St. Leonards Way, Ardsley, Barnsley S71 5BS
G6 LWT J. Mills Smiths Hill Petrockstow, Okehampton EX20 3EZ
G6 LWZ L. Miles 1 Wyndham Wood Close, Fradley, Lichfield WS13 8UZ
G6 LXE T. Doyle Leal House, Sparrow Pit, Buxton SK17 8ET
G6 LXF P. Duley 4 Brean Road, Stafford ST17 0PA
G6 LXL J. Ellis Goosters Green, Hope Bagot, Ludlow SY8 3AE
G6 LXP D. English 14 Elm Close, Ryde PO33 1ED
G6 LXU S. Westall 4 South View Great Harwood, Blackburn BB6 7NL
G6 LXV D. Woods 110 Sandy Lane, Warrington WA2 9JA
G6 LXW J. Weigh 167 Farm View Road, Rotherham S61 2BL
G6 LYA Dr P. Whysall 1 Greenlees Close, Fareham PO17 5GS
G6 LYD C. Weaver Linton Lodge, Pluckley Road, Ashford TN27 0AQ
G6 LYE G. Whiles 7 Thorndale Street, Hellifield, Skipton BD23 4JE
GM6 LYJ J. Young 1 Stevenson Place, Annan DG12 6BU
G6 LYM D. Miller 9 High Mead, Hockley SS5 4QG
G6 LZB P. Adams 4 Cherry Coft, Croxley Green, Rickmansworth WD3 3AL
G6 LZM G. Beddington Konrei, Tower Hill, Norwich NR8 5AX
G6 LZX B. Broad 1 Sussex Close, Laindon, Basildon SS15 6PR
G6 LZZ J. Bolton Huds House, Cowgill, Sedbergh LA10 5TQ
G6 MAA G. Bishop Oyston Lodge, Lynstone Road, Bude EX23 8LR
GW6 MAB J. Barwick 13 Greenfields Avenue, Bridgend CF31 4SR
G6 MAC B. Mcdonnell 68 Chaigley Road, Longridge, Preston PR3 3TQ
G6 MAJ A. Mulvaney 38 Ramwells Brow, Bromley Cross, Bolton BL7 9LL
G6 MAM K. Wright 63 Dudley Avenue, Leicester LE5 2EF
G6 MAR G. Wratten 6 Buckle Drive, Seaford BN25 2QN
G6 MAT A. Valentine 21 Naseby Road, Congleton CW12 4QX
G6 MAW R. Underwood 35 Greenfields, Langley Mill, Nottingham NG16 4GJ
G6 MAY D. Pool 6 Rivett Close, Clothall Common, Baldock SG7 6TW
G6 MBF D. Surgey 4 Down Lane Bathampton, Bath BA2 6UE
G6 MBH W. Stiling 11 Carrol Grove, Cheltenham GL51 0PP
G6 MBI D. Stainton Tilton House, 39 Redland Grove, Nottingham NG4 3ET
G6 MBL M. Snow 32 Orchard Avenue, Worthing BN14 7PY
G6 MBR Rnars London (Hms Belfast) Group c/o I. Mciver 3 Asgard Drive, Bedford MK41 0UP
G6 MBV C. Sutcliffe 1 St. James Street, Waterfoot, Rossendale BB4 7HN
G6 MC the University of the Third Age c/o S. Clarke Brimham Lodge Fm, Harrogate HG3 3HE
G6 MCB M. Baldry 10 Kingfisher Court, Lowestoft NR33 8PJ
G6 MCC L. Crompton 6 Moss Avenue, Ashton-On-Ribble, Preston PR21SH
G6 MCE P. Garde 21 Leicester Avenue, Timperley, Altrincham WA15 6HR
G6 MCG C. Garnham 1 Ennerdale Close, Felixstowe IP11 9SS
G6 MCQ J. Grane 15 Pinelands Way, Osbaldwick, York YO10 3QJ
GM6 MCV J. Mcvicar 2 Lilliardsedge Par, Mr Ancrum, Roxburghshire TD8 6TZ
G6 MCX P. Garland 6 Barn Piece, Chandler'S Ford, Eastleigh SO53 4HP
G6 MCY M. Goddard 65 Langley Hall Road, Solihull B92 7HE
GM6 MD Clyde Coast Contest Club c/o A. Dunn 50 Pemberton Valley, Ayr KA7 4UB
G6 MDC M. Green 9 Greencroft Avenue, Northowram, Halifax HX3 7EP
G6 MDG J. Tyson 1102 Rochdale Road, Blackley, Manchester M9 7EQ
G6 MDM W. Smyth 4 Dereham Road, Pudding Norton, Fakenham NR21 7NA
G6 MDN G. Tillett 43 Chippenham Road, Harold Hill, Romford RM3 8HJ
G6 MDR I. Stanley 6 Kennedy Avenue, Long Eaton, Nottingham NG10 3GF
G6 MDS A. Scott 3 Majestic Road, Hatch Warren, Basingstoke RG22 4XD
G6 MDX D. Davies 9 Lawnsfield Walk, Stafford ST16 1TS
G6 MED P. Cartmell 16 Churchfield Drive, York YO322FL
G6 MEH J. Turner 155 Bure Lane, Christchurch BH23 4HB
G6 MEI C. Thacker 1 Pine Grove, Chorley PR6 7BW
G6 MER M. Roberts 32 Orion, Bracknell RG12 7YX
G6 MEW A. Hunt 7 Wood Lodge, Calmore, Southampton SO40 2UP
G6 MFB R. Hodges 21A Preston Lane, Lyneham, Chippenham SN15 4AR
G6 MFU N. Cowley 126 Racecourse Road, Swinton, Mexborough S64 8DS
G6 MGA S. Cooper 27 Huntsmans Gate, Bretton, Peterborough PE3 9AU
G6 MGH M. Cooper 3 Marina Way, Ripon HG4 2LJ
G6 MGN D. Richardson 14 Wingfield Avenue, Lakenheath, Brandon IP27 9HS
G6 MGQ A. Reddish Wheelwrights House, Luckeys Corner, Hitcham, Ipswich IP7 7LR
G6 MGZ J. Middleton 187 Balcombe Road, Horley RH6 9EA
GM6 MHC Mid-Beds Contest Assoc c/o R. Mcnaught 5 Bonnymuir Crescent, Bonnybridge FK4 1GD
G6 MHF A. Marshall Thistledome, First Avenue, Watford WD25 9PS
G6 MHO I. Pomfret 20 Sandown Road, Bury BL9 8HN
G6 MHR J. Castelow 7 Langford Close, Burley In Wharfedale, Ilkley LS29 7NP
GW6 MHV B. Cooke 51 Celyn Avenue, Cardiff CF23 6EJ
G6 MHY Grimsby & Cleethorpes District Sas Radio Scouting Team c/o A. Carlile Top Flat 13 B Mill Road, Cleethorpes DN35 8HZ
G6 MIC M. Clayden 121 North Lane, East Preston, Littlehampton BN16 1HB
G6 MID P. Croft Exchange Buildings, Exchange Street, Normanton WF6 2AA
GW6 MIH M. Cleverley 33 Tylchawen Crescent, Tonyrefail, Porth CF39 8AL
G6 MIS S. Ransom 1 Bilberry Road, Clifton, Shefford SG17 5HB
G6 MIU W. Livesey 20 West Way, Little Hulton, Manchester M38 9GL
G6 MJA M. Addison Berrymead, Oxford Street, Great Missenden HP16 9JH
G6 MJB D. Lloyd Rangelands, Old Guildford Road, Camberley GU16 6PH
G6 MJM S. Parker 19 Sundour Crescent, Wolverhampton WV11 1AP
G6 MJT G. Painting 35 Selsdon Road, New Haw, Addlestone KT15 3HP
G6 MJW G. Davis 30 Bonny Wood Road, Hassocks BN6 8HR
GM6 MJY C. Donald 126 Newburgh Circle, Bridge Of Don, Aberdeen AB22 8XB
G6 MKD M. Douglass 20 Cadshaw Close Birchwood, Warrington WA3 7LR
G6 MKJ N. Ellis 140 Wollaston Road, Irchester, Wellingborough NN29 7DH
G6 MKL R. Ellis 4A Elmdale Road, Earl Shilton, Leicester LE9 7HQ
G6 MKO S. Everett 11 Chepstow Road, Felixstowe IP11 9BU

G6 MKQ G. Evans 31 Queen Elizabeth Crescent, Accrington BB5 2AS
GW6 MKR C. Foster Pentwyn House, Delfryn, Aberdare CF44 0TU
G6 MKZ P. Fisher 77 Swallow Street, Iver SL0 0ET
G6 MLH G. Marshall Fern House, Church Road, Newark NG23 7ED
GW6 MLI Dr D. Morgan Northwood Hotel, 47 Rhos Road, Colwyn Bay LL28 4RS
G6 MLJ T. Maker 25 Walthams Place, Pitsea, Basildon SS13 3PR
GW6 MLL B. Murphy 22 Deepglade Close, St. Thomas, Swansea SA1 8EJ
G6 MLS T. Abson 177 Meadowhall Road, Kimberworth, Rotherham S61 2JW
G6 MLV K. Barker 8 Shelley Gardens, Wembley HA0 3QG
G6 MMA Sunderland Raynet Group c/o M. Barlow 56 Pasturegreen Way, Irlam, Manchester M44 6TE
G6 MMB Bunkers On the Air c/o J. Bulbrook 33 Stonecross Way, March PE15 9DH
G6 MMD S. Burrows 2 Luscombe Farm Cottages, Heath End, Stratford-upon-Avon CV37 0PP
G6 MMG D. Brown 28 Bishop Drive, Whiston, Prescot L35 3JL
G6 MMJ P. Bromley 3 Georgia Avenue, Broadwater, Worthing BN14 8AZ
G6 MML V. Bates The Anvil, 4 Eastgate, North Newbald YO43 4SD
GW6 MMM J. Bowen 18 Admirals Walk, Sketty, Swansea SA2 8LQ
G6 MMR A. Salter 143 Eastwood Road North, Leigh-on-Sea SS9 4NB
G6 MMS J. Young 45 Eaves Lane, Chadderton, Oldham OL9 8RG
G6 MMT J. Ward 64 Gladstone Road, Ipswich IP3 8AT
G6 MNB M. Bulmer Highfield, 7 Fountain Avenue, Altrincham WA15 8LY
GW6 MNC W. Turner 37 Dan-Y-Bryn Avenue, Radyr, Cardiff CF15 8DD
G6 MND L. Davison 58 Priestley Court, South Shields NE34 9NQ
G6 MNI R. Andrews 10 Summerfield Close, Mevagissey, St. Austell PL26 6TZ
G6 MNJ P. Andrews 10 Summerfield Close, Mevagissey, St. Austell PL26 6TZ
G6 MNL A. Butler 45 Roewood Close, Holbury, Southampton SO45 2JT
G6 MOD P. Boden 54 Avill, Hockley, Tamworth B77 5QF
G6 MOI A. Bates 44 Chalfont Drive, Sileby, Loughborough LE12 7RQ
G6 MOT S. Kilmister 9 Wheal Jane Meadows, Threemilestone, Truro TR3 6EN
G6 MOZ P. Sealey 45 Haydon Way, Coughton, Alcester B49 5HY
G6 MPE J. Simmons 282 Bishopton Road West, Stockton-on-Tees TS19 7LY
G6 MPJ W. Smith Boleyn Service Station, 77 River Road, Barking IG11 0DS
G6 MPK T. Smith 87 Swanland Road, Hessle HU13 0NS
G6 MPN A. Shalders 29 Princess Drive, Sandbach CW11 1BS
G6 MPT P. Pritchard 5 Charlemont Road, Stone Cross, West Bromwich B71 3HX
GW6 MPX D. Prince 40 Ffordd Gryffydd, Llay, Wrexham LL12 0RT
G6 MQD P. Mullins 2A Knights Close, Great Brickhill, Milton Keynes MK17 9AW
G6 MQG C. Northrop Mayfield, 47B Hardhorn Road, Poulton-le-Fylde FY6 7SR
G6 MQH M. Pickard 9 Beatrice Way, Trowbridge BA14 7TX
G6 MQI G. Pointon 448 Stockport Road, Thelwall, Warrington WA4 2TR
G6 MQJ P. Racher 2 Heron Way, Horsham RH13 6DG
G6 MQK I. Stinton11 57 Wildfields, King's Lynn PE34 4DE
G6 MQN R. Wyatt 1 Ivy Villa, Kings Hill, Haverhill CB9 7NA
G6 MQP N. Roberts 81 Broad Lane, Coventry CV5 7AH
G6 MQU B. Plumtree Sunnyside, Station Road, Skegness PE24 5ES
G6 MQY P. Wilson Laurel Cottage, 43 Newnham Road, Ryde PO33 3TE
G6 MQZ T. Wilson Orchard House, Whitmoor Lane, Guildford GU4 7QB
G6 MRN N. Parr 24 Park Avenue, Awsworth, Nottingham NG16 2RA
GW6 MRO J. Reddaway Voltaire House, Ffordd Uchaf, Gwynfryn, Wrexham LL11 5UN
G6 MRP K. Playford 1 Cherwell Close, Abingdon OX14 3TD
G6 MRW P. Grant 117 Hazel Avenue, Farnborough GU14 0DW
G6 MRY C. Guy 78 Park Road, Bolton BL1 4RQ
G6 MSC T. Glover 70 Sandown Road, Toton, Nottingham NG9 6JW
G6 MSY Merseyside ARS c/o A. Birch 6 Crescent Road, Wallasey CH44 0BQ
G6 MTB D. Hesketh Flat 3, Redwood Court Plantation Terrace, Dawlish EX7 9FD
G6 MTE S. Heath 12 The Medway, Daventry NN11 4QU
G6 MTF T. Horn 9 Gipton Wood Avenue, Leeds LS8 2TA
G6 MTG D. Knight 119 Bracebridge Street, Nuneaton CV11 5PD
GI6 MTL M. Mccutcheon 10 Chestnut Brae, Gilford, Craigavon BT63 6FA
G6 MTV D. Lovell 5 Were Close, Warminster BA12 8TB
G6 MTY M. Matthews 30 Broadgate Lane, Deeping St. James, Peterborough PE6 8NW
G6 MUJ C. James 5 Arbroath Road, Luton LU3 3LA
GW6 MUP W. Jones Pen-Y-Berth, Pen-Y-Garth, Gwynedd LL55 1EY
G6 MUQ T. Jones 4 Anne Close, Christchurch BH23 2NW
G6 MUW B. Kent 6 Church Walk Mancetter, Atherstone CV91NX
G6 MUX A. Kelly 9 Cotswold Close, Dibden Purlieu, White SO455QW
GM6 MUZ Dr C. Duncan 12 Juniper Park Road, Juniper Green EH14 5DX
G6 MVF C. Stokes 7 St. Nicholas Close, Arnold, Nottingham NG5 6GU
G6 MVN R. Suckling 21 Warren Court, Meadowside, Dartford DA1 2RZ
G6 MVQ J. Simmonds Overbeck South, Stokesley Road, Guisborough TS14 8DL
G6 MVR D. Scott 20 Belmont View, Harwood, Bolton BL2 3QN
G6 MVS M. Sandler 21 Kenilworth Gardens, Hornchurch RM12 4SE
G6 MVW E. Sayer 27 Glenmere Park Avenue, Benfleet SS7 1SS
G6 MWB T. Gordon 101 Dorset Road, Bexhill-on-Sea TN40 2HU
G6 MWD Dr C. Goodhand 22 Somin Court, Doncaster DN4 8TN
G6 MWL D. Henderson 7 Glenhaven Avenue, Borehamwood WD6 1AY
G6 MWM B. Hall 32 Danby Road, Newton, , Hyde SK14 4DL
G6 MWQ D. Horne Flat 31, Homefleet House, Wellington Crescent, Ramsgate CT11 8JY
G6 MWS A. Hueck 9 Corden Avenue, Mickleover, Derby DE3 9AQ
G6 MXE R. Peeling 13 Greenview Crescent, Hildenborough, Tonbridge TN11 9DR
G6 MXL C. Redwood 53 Woodpecker Drive, Poole BH17 7SB
G6 MXV A. Poupard Woodlea, Crouch House Road, Edenbridge TN8 5EN
G6 MYH C. Mallory 11 Baymead Meadow, North Petherton, North Petherton TA6 6QW
G6 MYL C. Jones 63 Hockenhull Avenue, Tarvin, Chester CH3 8LR
G6 MYO B. Johnson 2 Plumtree Cottages, Hill Street, Swadlincote DE12 7PW
G6 MYT C. King 3 Huntingdon Gardens, Christchurch BH23 2TW
GW6 MYY D. Davies Penrallt, Abercaseg Road, Gerlan, Bangor LL57 3SP
G6 MYZ S. Doorey 11 Langley Gardens, Petts Wood, Orpington BR5 1AB
G6 MZF L. Cromar 17 Ipley Way, Hythe, Southampton SO45 3LG
G6 MZN M. Esser 10 Van Diemans Road, Wombourne, Wolverhampton WV5 0BQ
G6 MZT B. Fereday 16 Glentworth Gardens, Wolverhampton WV6 0SF
G6 MZV R. Titmuss 70 Mallards Rise, Church Langley, Harlow CM17 9PL
G6 MZW D. Speak 42 Penn Lea Road, Twerton, Bath BA1 3RB
G6 NAD M. Mcdermott 91 Hargwyne Street, London SW9 9RH
G6 NAG D. Lang 8 Church Hill, Cheddington, Leighton Buzzard LU7 0SY
G6 NAH P. Proudlove 14 Heath Avenue, Rode Heath, Stoke-on-Trent ST7 3RY
G6 NAJ Rvd. T. Leyland 4 Ellsmore Meadow, Lichfield WS13 6NJ
G6 NAL R. Pain 200 Butchers Lane Mereworth, Maidstone ME18 5QF
G6 NAP E. Lester 178 Newtown Road, Malvern WR14 1PJ
GI6 NAQ S. Mccullagh 7 Clanbrassil Gardens, Portadown, Craigavon BT63 5YD
G6 NAV R. Martin 110 Binscombe, Godalming GU7 3QJ
G6 NAX S. Moring 1 Burrows Cottages, Toot Hill Road, Ongar CM5 9QN
G6 NBF J. Baron 9 Milton Avenue, Doncaster DN5 8ER
G6 NBI J. Brett 127 Cranbrook Drive, Maidenhead SL6 6RY
G6 NBL R. Barnett 5 Overbrook, Evesham WR11 1DE
G6 NBM G. Bryant 37 Broad Leas Court, Broad Leas, St. Ives PE27 5XG
G6 NBP P. Blease 15 Shadybrook Lane, Weaverham, Northwich CW8 3PN
G6 NCE M. Craig 194 Elm Grove, Brighton BN2 3DA
G6 NCL R. Pickstone 33 Shore Mount, Littleborough OL15 8EN
GU6 NCZ P. Wild Honfleur, La Rue Du Le Hurel, Vale GY3 5AF Guernsey
G6 NDA C. Venn Stantor, High Road, Templecombe BA8 0DN
G6 NDH P. Walker 37 Cromwell Road, Grimsby DN31 2DN
G6 NDJ A. Wilson 23 Claydown Way, Slip End, Luton LU1 4DU
GI6 NDM G. O`Boyle 27A Drapersfield Road, Cookstown BT808RS
G6 NDS Northampton Sct c/o I. Rivett 30 Millside Close, Kingsthorpe, Northampton NN2 7TR
G6 NEA J. Dean 15 Park Close, Sonning Common, Reading RG4 9RY
G6 NEK P. Diss 130 Beridge Road, Halstead CO9 1JU
G6 NEZ R. Emerson 4 Freeford Gardens, Lichfield WS14 9RJ
G6 NFB T. Wright 182 Lansdowne Road, Oxton, Prenton CH43 7SQ
G6 NFC A. Young 1 Rapley Green, Bracknell RG12 7PS
G6 NFE R. White 10 Melbourne Rise Bicton Heath, Shrewsbury SY3 5DA
G6 NFJ J. Eden 23 Elm Green Close, Worcester WR5 3HD
GI6 NFK S. Ferguson 20 Old Road, Loughgall, Armagh BT61 8JD
G6 NFR R. Foden 1A Garden Cottages, Eaton Road, Liverpool L12 3HQ
G6 NGA J. Lamble 17 Willowvale, Lowestoft NR324UB
G6 NGF M. Tatlow 20 Windmill Way, Tysoe, Warwick CV35 0SB
G6 NGM S. Cross 7 April Place, Buckhurst Road, Bexhill-on-Sea TN40 1UE
G6 NGN D. Simpson The Hawthorns, Slacken Lane, Stoke-on-Trent ST7 1NQ
G6 NGR P. Thornton 3 Copy Cottages, Cliviger, Burnley BB10 4SZ
G6 NGV W. Taylor 29 Holmdale Road, Syston, Leicester LE7 2JN
G6 NHA M. Malyon 16 Tintern Road, Gosport PO123QN
GW6 NHB G. Mahoney 684 Beechley Drive, Pentrebane, Cardiff CF5 3SS
G6 NHG S. Marshall 25 Carlcroft, Wilnecote, Tamworth B77 4DL
G6 NHK Craighalbert RC c/o N. Martin Stonea House, Middle Road, March PE15 0AJ
GW6 NHL A. Mccallum Glan Alaw, Llanddeusant, Holyhead LL65 4AG
G6 NHO R. Smith 20 Dryden Way, Higham Ferrers, Rushden NN10 8DH
G6 NHU K. Maton 41 Bemerton Gardens, Kirby Cross CO13 0LQ
G6 NHV D. Meakins 19 Booth Lane North, Northampton NN3 6JQ
G6 NHW P. Minchin 122 Mildenhall Road, Great Barr, Birmingham B42 2PQ
G6 NHY K. Marriott 1 Holbeck Road, Hucknall, Nottingham NG15 7SR
GM6 NIA D. Mccall 11 Craiglockhart Dell Road, Edinburgh EH14 1JW
GM6 NIC J. Mcaulay 9 Randolph Cliff, Edinburgh EH3 7TZ
G6 NID J. Matthew 24 Southgate, Rochdale OL128UQ
G6 NIO M. Smith 39 Seliot Close, Poole BH15 2HQ
G6 NIW F. Shaw 43 Egremont Road, Hardwick, Cambridge CB23 7XR
G6 NIX E. Samuels 19 Glynleigh Drive, Polegate BN26 6LU
G6 NIZ A. Scott The Conifers, Back Lane, York YO30 2DF
G6 NJE J. Smith 414 Sparrowhawk Drive, Willow Grove Park, Poulton-le-Fylde FY6 0RS
G6 NJJ M. Swift Spa Cottage, Spa Lane, Ormskirk L40 6JQ
GM6 NJL M. Spittle Stablecleugh, Ewes, Langholm DG13 0HJ
G6 NJO P. Askham 1 Park House Cottage, Carr Lane, Thirsk YO7 3PF
G6 NJR P. Nikolic 29 Tern View, Market Drayton TF91DU
G6 NJT F. Neill 7 Bellevue Terrace, Southampton SO14 0LB
G6 NKI K. Brown 73 Church Avenue, Preston PR1 4UD
G6 NKL D. Baldock Deeside, Platts Lane, Bucknall, Woodhall Spa LN10 5DY
G6 NKS G. Pearn 178 Lloyds Avenue Kessingland, Lowestoft NR33 7TU
G6 NLC C. Rabe 79 Rectory Avenue, Corfe Mullen, Wimborne BH21 3EZ
G6 NLD A. Reed 6 Brancaster Close, Nottingham NG6 8SL
G6 NLE H. Roberts 3 Short Avenue, Allestree, Derby DE22 2EH
G6 NLG M. Ritchie Bruern Abbey, Bruern, Chipping Norton OX7 6QA
G6 NLN J. Burdass Cedarwood, High Road, Brightwell-Cum-Sotwell, Wallingford OX10 0PT
GW6 NLO T. Bott 13 Cheriton Road, Pennar, , Pembroke Dock SA72 6RN
GW6 NLP M. Bryant The Nook, Llanarmon Road, Wrexham LL11 5YP
G6 NLQ T. Fradley 40 Higher Green, Poulton-le-Fylde FY6 7BL
G6 NLS D. Budd Valhalla, 81 Bohemia Chase, Leigh-on-Sea SS9 4PW
G6 NLU S. Buxton 111 Digby Avenue, Nottingham NG3 6DT
G6 NLX G. Richardson 21 Broadlands Road, Hickling NR12 0YG
G6 NLZ M. Reynolds 24 Mill Road, Lydd, Romney Marsh TN29 9EJ
G6 NMA P. Ayers-Hunt 15 Kelvin Road, Leamington Spa CV32 7TF
G6 NME I. Griffiths 34 Dobbins Oak Road, Pedmore, Stourbridge DY9 9HX
G6 NMK M. Grimes 73 Ryston Road, Denver, Downham Market PE38 0DP
G6 NMQ G. Goodyer Flat, 54 Wyndham Road, Petworth GU28 0EQ
G6 NMU J. Greenhough 58 Gorsey Bank, Wirksworth, Matlock DE4 4AD
G6 NNA Dr A. Liggins 97 Vessel Crescent, Scarborough, Ontario M1C 5K5 Canada
GW6 NNB D. Owen Pen Y Bont Maerdy, Corwen LL21 0PE
G6 NNK P. Frampton 118 Ramnoth Road, Wisbech PE13 2JD
G6 NNO J. Evans 74 Trejon Road, Cradley Heath B64 7HJ
GI6 NNP V. Hagan 7 Emania Terrace, Armagh BT60 4AS
G6 NNS J. Hunt Honeytiles Culford, Bury St. Edmunds IP28 6DT
G6 NNU J. Archer 117 Heath Way, Erith DA8 3LZ
G6 NOI J. Whelan 8 Welland Road, Wirral CH63 2JU
G6 NOL J. Weir 17 Pasteur Drive, Apley, Telford TF1 6PQ
GM6 NOO C. Wood 5 Damhead Steading, Kinloss, Morayshire IV36 3UA
G6 NOW R. Beck 26 Cheshire Court, Ravenall Close, Birmingham B34 6PZ
G6 NPC R. Carlson 45 Firs Road, Milnthorpe LA7 7QF
G6 NPE A. Coates 2 Kipling Avenue, Burntwood WS7 2HS
G6 NPJ J. Copeland Little Cophall, Dowlands Lane, Crawley RH10 3HX
G6 NPW S. Caine 19 Turner Drive, Tingley, Wakefield WF3 1UD
G6 NPZ P. Chard 29 Nettle Gap Close, Wootton, Northampton NN4 6AH
G6 NQB S. Clements 39 Redland Close, Marlbrook, Bromsgrove B60 1DZ
G6 NQL J. Wilkinson The Old Joinery, Garsdale, Sedbergh LA10 5PJ
G6 NQM K. Whitchurch 65 Honey Hill Road, Kingswood, Bristol BS15 4HN

G6 NQQ A. Wilson 17 Rook Way, Horsham RH12 5FR
GW6 NQU J. Hoy 39 Blackbird Road, Caldicot NP26 5RE
G6 NQY R. Heath 222 Congleton Road, Talke, Stoke-on-Trent ST7 1LW
G6 NRH D. Hallifax 22 Wendover Way, Welling DA16 2BN
G6 NRK A. Hunt 39 Circular Road West, Liverpool L11 1AY
G6 NRL C. Hargreaves Viridis, Retford Road, Retford DN22 0BY
G6 NRM S. Hanscombe 24 St. Marks Drive, Wellington, Telford TF1 3GA
GW6 NSG J. Jones 26 Spring Road, Wrexham LL11 2LU
GW6 NSK A. Jones 4 Park View, Llanddew, Brecon LD3 9RL
G6 NSQ P. James 9 Smallholding, Tutbury Road, Burton-on-Trent DE13 0AL
G6 NSU P. Lewis 18 Bittaford Wood Bittaford, Ivybridge PL21 0ET
G6 NSZ D. Lawton 48 Woodlands Road, Woodlands, Doncaster DN6 7JZ
G6 NTE J. Lyons 8 Anstie Close, Devizes SN10 2EN
G6 NTM E. Murphy 25 Warrington Road, Ashton-In-Makerfield, Wigan WN4 9PJ
GI6 NTP D. Mcalpine 35 Carnamena Avenue, Belfast BT6 9PJ
G6 NTQ R. Morgan 1 Hillmeads Drive, Dudley DY2 7TS
G6 NTW K. Gosbee 64 Connaught Gardens, Palmers Green, London N13 5BS
G6 NTY B. Griffiths 18 Julius Drive, Coleshill, Birmingham B46 1HL
G6 NUI D. Chamberlain 44 Parsonage Chase, Minster On Sea, Sheerness ME12 3JX
GM6 NUL R. Crawford Glengarry, East Terrace, Kingussie PH21 1JS
G6 NUQ S. Coward 100 Lytham Road, Freckleton, Preston PR4 1XB
G6 NUS A. Croft 15 St. Marys Road, Bozeat, Wellingborough NN29 7JU
G6 NUX S. Clack Flat 9, The Grange, Emsworth PO10 7QP
G6 NUZ A. Charlton 26 Saundergate Lane, Wyberton, Boston PE21 7BZ
G6 NVC K. Castley Zone 2, (Opposite High School), Cagayan de Oro City 9000 Philippines
G6 NVD C. Webb 2 Rykhill, Chadwell St. Mary, Grays RM16 4RR
G6 NVF D. Mcglasson 19 Kennedy Street, Ulverston LA12 9EA
G6 NVH M. Morris 19 Gowy Close, Alsager, Stoke-on-Trent ST7 2HX
G6 NVI S. Mindel Longwood House, Arkley Lane, Barnet EN5 3JR
GW6 NVJ C. Marlow 27 Sandy Way, Connah'S Quay, Deeside CH5 4SH
G6 NVS P. Harrison 41 Chestnut Close, Handsacre, Rugeley WS15 4TH
G6 NVU Sheffield ARC c/o M. Haynes 10 Cypress Grove, Denton, Manchester M34 6EA
G6 NVW P. Higgins Jp 49 Milton Road, Hoyland, Barnsley S74 9AX
G6 NVY A. Hilbourne 24 Tamarisk Avenue, Reading RG2 8JB
G6 NWC P. Holley Wotton Farm, Buckfastleigh TQ11 0HB
G6 NWK M. Parker 31 Sandholme Drive, Burley In Wharfedale, Ilkley LS29 7RG
G6 NWN I. Poyser 24 Overstone Close, Sutton-in-Ashfield NG17 4NL
G6 NWS J. Hildreth 69 Mason Street, Sutton-in-Ashfield NG17 4HQ
G6 NWT W. Taylor 33 Lancaster Avenue, Dawley, Telford TF4 2HS
GM6 NX Stirling And District ARS c/o H. Martin 11 Ewing Court, Broomridge, Stirling FK7 0QP
GW6 NXH Stisted CG c/o W. Rees 1 St. Marys Close, Briton Ferry, Neath SA11 2JU
GW6 NXL T. Rees Ty Goleu, Llwyngwril LL37 2UZ
G6 NXM R. Rixon 11 The Ridings, Waltham Chase, Southampton SO32 2TR
G6 NXP S. Rafferty 57 Essex Crescent, Billingham TS23 4AW
G6 NXV M. Shannon 129 Hampton Lane, Blackfield, Southampton SO45 1WF
G6 NXW K. Sykes 68 Newtown Avenue, Cudworth, Barnsley S72 8DY
G6 NYF J. Aylward 7 Manygates Lane, Wakefield WF1 5NT
G6 NYG R. Adams 6 Worcester Road Wychbold, Droitwich WR9 7PE
G6 NYH G. Austin 21 St. Georges Place, Northampton NN2 6EP
G6 NYL P. Baylis 118 Eastgate, Deeping St. James, Peterborough PE6 8RD
GW6 NYR A. Davis 9 Taliesin Street, Llandudno LL30 2YE
GM6 NYT J. Danton 12 Laburnum Road, Methil, Leven KY8 2HA
G6 NZA M. Davenport 8 Cedar Avenue, Chesterfield S40 4ES
G6 NZG S. Edson Inglewood, Camelot Gardens, Mablethorpe LN12 2HP
G6 NZH J. Edwards 69 Eastford Road, Warrington WA4 6EY
G6 NZL P. Fletcher 43 Merlin Way, Woodville, Swadlincote DE11 7QU
G6 NZN G. Fowler 10 Ullswater Road, Wimborne BH21 1QT
G6 NZO M. Finney 49 Ashcroft Drive, Old Whittington, Chesterfield S41 9PA
G6 NZW D. Smith 90 Endhill Road, Kingstanding, Birmingham B44 9RP
G6 NZY B. Sparke 1 Norwich Street, Mundesley, Norwich NR11 8DN
G6 OAI S. Baverstock 43 Tatnam Road, Longfleet, Poole BH15 2DW
G6 OAN C. Bryan 113 Hoe View Road, Cropwell Bishop, Nottingham NG12 3DJ
G6 OAS A. Inglis 59 Chapel Street, Forsbrook, Stoke-on-Trent ST11 9DA
G6 OAU M. Jones 17 Puddingmoor, Beccles NR34 9PL
G6 OAV C. Jones 1 Stonehill Close, Leigh-on-Sea SS9 4AZ
GW6 OAW T. Jones Marvor, Madyn Road, Amlwch LL68 9DL
G6 OBA M. Kaznowski 85 St. Albans Road, Kingston upon Thames KT2 5HH
G6 OBB P. Kerr Burrow Farm, Burrowbridge, Bridgwater TA7 0RH
G6 OBD T. Keeling 1 New Lane, Brown Edge, Stoke-on-Trent ST6 8TQ
G6 OBE S. Kimblin 4 Horsham Close, Westhoughton, Bolton BL5 2GR
G6 OBG J. Kay 12 Williams Avenue, Newton-le-Willows WA12 0NN
G6 OBJ G. Webster Flat 3 Charlotte Broadwood, Vicarage Lane, Dorking RH5 5LL
G6 OBO L. Weiss 7 Tennyson Avenue, Grays RM17 5RG
G6 OBT H. Wilson 11 Palmerston Close, Haslington, Crewe CW1 5QE
G6 OBU S. Wright 21 Poplars Close, Watford WD25 7EW
G6 OCB D. Byers 11 Heath Road, Ashton-In-Makerfield, Wigan WN4 9DY
GI6 OCC K. Brennan 1 Ballyscullion Lane, Bellaghy, Magherafelt BT45 8NQ
G6 OCF R. Wallis 187 Langtons Meadow, Farnham Common, Slough SL2 3NT
G6 OCM E. Bunyan 1 Talman Close, Ifield, Crawley RH11 0RB
G6 OCO A. Bruce 48 Durham Road, Gillingham ME8 0JN
G6 ODA A. Bardy 67 Chase Side, Enfield EN2 6NQ
G6 ODF D. Adamson 22 Longacres, Cannock WS12 1LD
G6 ODT K. Lamford 27 School Lane, Irchester NN29 7AZ
G6 ODU R. Leong 55 Liverpool Road, Aughton, Ormskirk L39 5AP
G6 ODW L. Liffchak 6 Ashmore Grove, Welling DA16 2RU
G6 OEJ A. Barnard 36 St. Pauls Road, Walton Highway, Wisbech PE14 7DN
G6 OEM J. Bolland 18 Ward Avenue, Formby, Liverpool L37 2JD
G6 OER J. Topping 3 Dean Road, Handforth, Wilmslow SK9 3AF
G6 OES M. Smith Almond House, Waste Lane, Balsall Common, Coventry CV7 7GG
G6 OET M. Telford 11 Twyford Close, Swinton, Mexborough S64 8UH
G6 OEW C. Thorn 20 Kiln Road, Shaw, Newbury RG14 2HA
GM6 OFB J. Mcardle 40 Rodney Drive, Girvan KA26 9DZ
G6 OFD L. Morrison 29 Mead Hatchgate, Hook RG27 9PU
G6 OFM J. Cordial Flat 30, Homeview House, Seldown Road, Poole BH15 1TT
GM6 OFO M. Clark 38 Dunsinane Drive, Perth PH1 2DU
G6 OFV S. Crossland 16 Holland Road, High Green, Sheffield S35 4HF
G6 OFZ J. Roberts 8 Woodgate Close, Market Harborough LE16 8EX
GW6 OGD S. Dawber 7 Heol Y Pentir, Rhoose CF62 3LQ
GM6 OGN A. Sives 4 Fir Grove, Craigshill, Livingston EH54 5JP
G6 OGT A. Scholes 45 Howden Road, Blackley, Manchester M9 0RQ
G6 OGZ J. Mitchell 17 Spring Close, Lutterworth LE17 4DD
GM6 OHF D. Macliver 20 Lancaster Avenue, Beith KA15 1AR
G6 OHK P. Butler 219 Ridge Avenue, Burnley BB10 3JF
G6 OHM A. Dunham 28 Kingfisher Close, Chatteris PE16 6TP
G6 OHQ G. Finch 77 Furnivall Crescent, Lichfield WS13 6DB
G6 OHR R. Edwards 11 Litlington Court, Surrey Road, Seaford BN25 2NZ
G6 OIA P. Rattenbury 2 Main Road Upper Heyford, Northampton NN7 3LZ
G6 OIB J. Riley 11 Sutton Road, Mepal, Ely CB6 2AQ
G6 OIF A. Postans 62 Elm Grove, Bromsgrove B61 0DX
G6 OIH R. Phillips 1 Forge Close, Ashendon, Aylesbury HP18 0HJ
G6 OIN A. Talbot 159 Edgeside Lane, Rossendale BB4 9TR
GW6 OIO M. Thomas 13 Chestnut Grove, The Bryn, Blackwood NP12 2PU
G6 OIX J. Roberts 9 Tower Close, North Weald, Epping CM16 6HA
G6 OIY J. Roberts 6 Weavers Close, Braintree CM7 2WB
GI6 OJC O. Okane 39 Harberton Park, Ballymena BT43 6NF
G6 OJH S. Page 4 Trenley Close, Holbury, Southampton SO45 2HN
G6 OJK A. Mayall 8 Millennium Way, Bideford EX39 1XN
G6 OJN K. Michael Flat 21, Granville House, Victoria Parade, Ramsgate CT11 8DF
G6 OJV J. Greenley 22 Langley Drive, Norton, Malton YO17 9AR
G6 OJX E. Grayson Manor Gate, 2 Polsue Way, Truro TR2 4BE
G6 OJZ P. Anstock 12 Raymoor Avenue, St. Marys Bay, Romney Marsh TN29 0RD
G6 OKA C. Glover 16 Woodfield Road, Radlett WD7 8JD
G6 OKB R. Gilham Wren Cottage, Wayborough Hill, Ramsgate CT12 4HR
G6 OKC M. Gerrard 29 Forest Drive, Broughton, Chester CH4 0QT
G6 OKH P. Vanner 23 Logwell Court, Northampton NN3 9DJ
G6 OLJ Dr D. Hill 33 Cleveland Close, Thornbury, Bristol BS35 2YD
GM6 OLM R. Hendry 43 Barone Road, Rothesay, Isle of Bute PA20 0DY
G6 OLU P. Hobson 220 Station Road, Burton Latimer, Kettering NN15 5NT
G6 OLV A. White 20 Wyles St., Gillingham ME7 1ND
G6 OLY A. Williams 1 Leslie Drive, Leigh-on-Sea SS9 5NW
G6 OMH B. Staddon 311 Cheney Manor Road, Swindon SN2 2PE
G6 OMN G. Rogers 7 Flordon, Birch Green, Skelmersdale WN8 6PA
GW6 OMV P. Rea 159 Mill View Estate, Maesteg CF34 0DP
G6 ONE D. Williams 16 Church St., Owston Ferry, Doncaster DN9 1RG
G6 ONI B. Steponitis Flat 7, 6 Second Avenue, Hove BN3 2LH
G6 ONV M. Johnson 16 Gardner Close, Raunds, Wellingborough NN9 6HN
G6 ONW S. Jackson 256 Perry Road, Sherwood Rise, Nottingham NG5 1GP
GW6 ONZ P. Kinsey Glyn Elwy Allt Goch, St. Asaph LL17 0BP
GM6 OOA D. King Marionville, Donibristle, Cowdenbeath KY4 8EU
G6 OOH J. Stone 13 Winchester Close, Newport PO30 1DR
G6 OOK M. Stewart Fieldhead, New Barns Road, Arnside, Carnforth LA5 0BH
G6 OOP D. Ashton 25 Langdale Road Woodley, Stockport SK6 1BH
G6 OOT C. Atkins 11 Brambledown, West Mersea, Colchester CO5 8RY
G6 OPD L. Middleton 24 Townshend Road, Worle, Weston-Super-Mare BS22 7FW
G6 OPK M. Lee 23 Camford Close, Beggarwood, Basingstoke RG22 4UJ
G6 OPV E. Starkey 71 Elwick Drive, Liverpool L11 4UW
G6 OPY Coulsdon Amateur Transmitting Society c/o R. Van Cleak 19 Hanbury Road, Stoke Heath, Bromsgrove B60 4LS
G6 OQJ W. Castle 2 Wellington Close, Mundesley, Norwich NR11 8JF
GI6 OQL J. Craig 8 Muckamore View, Muckamore, Antrim BT41 2EU
GM6 OQN R. Campbell 32 Harvie Avenue Newton Mearns, Glasgow G77 6LQ
G6 OQO J. Douthwaite 38 Burnside Road, Newcastle upon Tyne NE3 2DU
G6 OQV B. Everitt The Hermitage, The Rookery, Galley Common, Nuneaton CV10 9PB
GW6 ORE R. Trangmar Ffynnon Bach Isaf, Tregarth, Bangor LL57 4PA
G6 ORH D. Wright 23 Oakenhall Avenue, Hucknall, Nottingham NG15 7TF
G6 ORJ A. Weller 104 Medina Avenue, Newport PO30 1HG
G6 ORL D. Woodhouse 5 Swallow Wood, Fareham PO168UF
G6 ORM S. Whiley 14 Clift Avenue, Chippenham SN15 1DA
G6 ORO G. Walsh 36 Westminster Street, Newtown, Wigan WN5 9BH
G6 ORS A. Bennett 39 West View, Parbold, Wigan WN8 7NT
G6 ORT L. Bailey 27 Birch Road, Congleton CW12 4NN
G6 OSH D. Ridley 37 Harewood Close, Whickham, Newcastle upon Tyne NE16 5SZ
G6 OSJ J. Roberts 1 Ollerdale Close, Allerton, Bradford BD15 9BT
G6 OSK E. Robinson 8 Barrier Mews, Stainforth, Doncaster DN7 5PT
G6 OSO D. Parr Lordings, Station Road, Pulborough RH20 1AH
G6 OSR P. Price 20 Froglands Way, Cheddar BS27 3NY
G6 OSV I. Woodward 20 Boyle Avenue, Warrington WA2 0EZ
GM6 OSZ B. Williams 29 St. Ternans Road, Newtonhill, Stonehaven AB39 3PF
GW6 OTD Hull And District ARS c/o P. Sizer Gambos End, Reynoldston, Swansea SA3 1BR
G6 OTE D. Shaw 19 Upper Moors Great Waltham Chelmsford., Chelmsford CM3 1RB
G6 OTL G. Blades 11 Willard Grove, Stanhope, Bishop Auckland DL13 2XY
G6 OTP M. Rainbow 38 Moselle Drive, Churchdown, Gloucester GL3 2RY
G6 OTQ T. Roddy 26 Chapeltown Road, Radcliffe, Manchester M26 1YF
G6 OTS W. Peck 5 Stirling Crescent, Horsforth, Leeds LS18 5SJ
G6 OTV A. Ricalton 33 Tintagel Close, Cramlington NE23 1NZ
G6 OTW A. Ricalton 84 Wansdyke, Morpeth NE61 3RA
G6 OTZ A. Shaw 4 Jones Lane, Burntwood WS7 9DS
G6 OUA Radio Society of Great Britain c/o R. Saunders 8 Norfolk Road, Luton LU2 0RE
G6 OUI M. Burnell 49 Ashfield Road, Carterton OX18 3QZ
G6 OUJ B. Bozman 33 Maple Road, Loughborough LE11 2JL
GM6 OUL N. Bowry 18 Mortonhall, Park Gardens, Edinburgh EH17 8SR
G6 OUM B. Breaden 6 Breydon Road, Sprowston, Norwich NR7 8EE
G6 OUO P. Burgess 232 Hightown Road, Luton LU2 0DN
G6 OUT D. Andrew 14 Westfield Grove, Morecambe LA4 4LQ
G6 OUX S. Smith 71 Rockford Close, Redditch B98 7SZ
G6 OVA N. Styne 2 Greenway, Burton-on-Trent DE15 0AR
G6 OVC B. Thurlow 1 Sheffield Way, Earls Barton, Northampton NN6 0PF
GW6 OVD M. Clee 29 Heol Uchaf Nelson, Treharris CF46 6NT
G6 OVL J. Hopkins 15 Wallace Drive, Wickford SS12 9NA
G6 OVX R. Hadfield 2 Bridge Street, Shaw, Oldham OL2 8BG
G6 OWB E. Hill 213A Leicester Road, Markfield LE67 9RF
G6 OWI P. Haworth 19 Arnside Crescent, Blackburn BB2 5DU

G6 OWS K. Jones 1 Chadlow Road, Liverpool L32 7QR
G6 OWT G. Kelly Brook House, Liskeard PL14 5EY
GD6 OXG J. Williams Brookfield, Douglas Road, Ballabeg, Isle of Man IM9 4EF
G6 OXI C. Webb 50 Ridgeway, Eynesbury, St. Neots PE192QY
G6 OXJ D. Webb 10 Nuns Meadow, Gosfield, Halstead CO9 1UB
GM6 OXL I. Wilkins 133 Gavin Street, Motherwell ML1 2RL
G6 OXN I. Walker 66 Wood Street, Kettering NN16 9SB
G6 OXQ A. Day 7 Seagers, Great Totham, Maldon CM9 8PB
G6 OXY E. Chinn 10 Ironstone Lane, Northampton NN4 8TR
G6 OXZ M. Charlton 20 Bailey Crescent, South Elmsall, Pontefract WF9 2TL
G6 OYF M. Matthews 22 Elm Drive Bradley Stafford St18 9Ds, Stafford UK ST18 9DS
G6 OYU R. Spinner 9 Lindholme Road, Lincoln LN6 3RQ
G6 OYV T. Silvers 15 Stanford Way, Walton, Chesterfield S42 7NH
G6 OZH D. Martin 2 Farm View Road, Kirkby-In-Ashfield, Nottingham NG17 7HF
G6 OZQ J. Woulfe Cherry Orchard Cottage, Chorley, Bridgnorth WV16 6PP
G6 OZT P. White 3 South View, Whitwell, Worksop S80 4NP
G6 OZU A. Wilson 67 Sandpits, Leominster HR6 8HT
G6 OZZ G. Llewellyn 3 Ilex Close, Pamber Heath, Tadley RG26 3DW
G6 PAA J. Brownsett 10 Great Aldens, Bedford MK41 8JS
G6 PAE R. Hillum 48 Lydiard Way, Trowbridge BA14 0UJ
G6 PAJ R. Green 1 Knightsbridge Road, Messingham, Scunthorpe DN17 3RA
G6 PAO M. Hale 32 Oakfield Grove, Biddulph, Stoke-on-Trent ST8 6UH
G6 PAP S. Hale 19 Nailers Drive, Burntwood WS7 0ES
G6 PAR P. Rhodes 1 Killinghall Avenue, Bradford BD2 4SA
GI6 PAZ W. Mcconnell 17 Beech Green, Doagh, Ballyclare BT39 0QB
G6 PBG N. Munnery 3 Monnington Lane, Poundbury, Dorchester DT1 3RJ
G6 PBI K. Partington 38 Queensgate Drive, Royton, Oldham OL2 5SD
G6 PBO J. Tobin 5 Ashley Close, Ringwood BH24 1QX
G6 PBW J. Wainwright 8 Common Lane, Cutthorpe, Chesterfield S42 7AN
G6 PBZ P. Wright 75 Preston Road, Abingdon OX14 5NG
G6 PCC R. Slade 2 South Lodge Drive, Fornham St. Genevieve, Bury St. Edmunds IP28 6TQ
G6 PCE R. Stamford 30 Craft Way, Steeple Morden, Royston SG8 0PF
G6 PCP J. Brown 1 Whitehouse Cottages, Woodham Walter, Maldon CM9 6LR
GM6 PCW P. Boyd 144 Brown Street, Paisley PA1 2JE
G6 PCX J. Beresford 1 Russell Place, Maltby, Rotherham S66 7HB
G6 PDA A. Beacher 9 Gleneagles Court, Normanton WF6 1WW
G6 PDE C. Irish 128 Rushmere Road, Ipswich IP4 4JX
G6 PDJ S. Jubb 4 Manor End, Worsbrough, Barnsley S70 5JB
G6 PDM S. Procter 1B York Villas, York Street, Colne BB8 0ND
GW6 PDR S. Riggs 3 Lawrence Terrace, Llanelli SA15 1SW
G6 PEG C. Price 42 Kipling Road, Kettering NN16 9JZ
G6 PEH A. Rands 20 Riby Road, Keelby, Grimsby DN41 8ER
G6 PEP Dr J. Morris 22 St. Amand Drive, Abingdon OX14 5RQ
G6 PFC P. Cooper The Lodge, Fairholme Road, Newton St. Faith, Norwich NR10 3LL
G6 PFF A. Willis Kilncroft, Broadlayings, Newbury RG20 9TS
GM6 PFJ G. Gott 21 Hamilton Avenue, Dumfries DG2 7LW
GW6 PFK L. Griffiths 36 Lonydd Glas, Llanharan, Pontyclun CF72 9FZ
G6 PFN A. Hewitt 29 Brabazon Road, Oadby, Leicester LE2 5HF
G6 PFP S. Hill 10 Honeycrft Drive, St. Albans AL4 0GE
G6 PFX I. Harris 1 Greenways, Mill Lane, Credenhill, Hereford HR4 7EH
G6 PFZ A. Holroyd 59 Southern Parade, Preston PR1 4NJ
G6 PGG Dr D. Jones Bramble Cottage, Newtown, Nantwich CW5 8BG
G6 PGJ J. Kyle 1A Lynmouth Gardens, Greenford UB6 7HR
G6 PGM D. Kaye The Pantiles, Bildeston Road, Stowmarket IP14 2JT
G6 PGN C. King 18812 Thornwood Circle, Huntington Beach, California United States
G6 PGO B. Key 65 Ravenhurst Road, Birmingham B17 9TB
G6 PGP S. Kinton 7 Ferndale Drive, Ratby, Leicester LE6 0LH
G6 PGQ M. Karaszy-Kulin Crossways Corner, Pulborough RH20 2QY
G6 PGT J. Chapman 43 Balas Drive, Sittingbourne ME10 5AS
G6 PGV N. Phillips 5A Nutter Road, Thornton-Cleveleys FY5 1BG
G6 PHC P. Dewick Corner House, High Street, Waddingham DN21 4SW
G6 PHF M. Dent 23 Spruce Avenue, Lancaster LA1 5LB
G6 PHH P. Dickens 2 Millfield Avenue, Marsh Gibbon, Bicester OX27 0HP
G6 PHM P. Durbin 2 Keswick Gardens, Pill, Bristol BS20 0DR
G6 PHT S. Fitzhugh 25 Bridge Meadow, Denton, Northampton NN7 1DA
G6 PHU Wirral & District ARC Wadarc c/o D. Ford 6 Bluebell Grove Needham Market, Ipswich IP6 8JH
G6 PHX C. Johnson 37 Oakfield Drive, Mirfield WF14 8PX
G6 PHZ P. Maddox 7 Keats Road, Flitwick, Bedford MK45 1QD
G6 PIB M. Livingston 22 Oak Avenue, Elloughton, Brough HU15 1LA
G6 PII D. Simpson 20 Belvoir Place, Balderton, Newark NG24 3HH
G6 PIM P. Lawford 44 Clarendon Road, Broadstone BH18 9HY
G6 PJC P. Brown1 13 Hillside Close, Biddulph Moor, Stoke-on-Trent ST8 7PF
G6 PJD M. Belshaw Tara Cottage, 11 Hectors Way, Blandford Forum DT11 9QP
G6 PJE D. Bull 2 School Road, St. Johns Fen End, Wisbech PE14 8JR
G6 PJL J. Blacker 74 Benomley Crescent, Almondbury, Huddersfield HD5 8LU
G6 PJP L. Bealing 18 Avon Road, Oakley, Basingstoke RG23 7DJ
G6 PKG R. Davis 72 Windyridge, Bisley, Stroud GL6 7DA
G6 PKM Dr J. Allen 27 Grafton Road, Whitley Bay NE26 2NR
GM6 PKP J. Allardyce 17 Hallglen Terrace, Glen Village, Falkirk FK1 2AP
G6 PKS B. Bean 46 Grand Drive, Herne Bay CT6 8JS
G6 PKV G. Branagan 434 Manchester Road West, Little Hulton, Manchester M38 9XU
G6 PKX P. Bishop 38 Parkside Gardens East Barnet, Barnet EN4 8JS
G6 PKY R. Bush 3 Charnwood Avenue Keyworth, Nottingham NG12 5JX
G6 PLF J. Smoker 9 Anson Way, Bicester OX26 4UH
GM6 PLG P. Sloan 24 Hythe View, Lossiemouth IV31 6TP
G6 PLL C. Leach 109 Congreve Road, Worthing BN14 8EN
GI6 PLO I. Bell 3 Stratford Drive, Bangor BT19 6ZW
G6 PLR L. Chandless 16 Crest Gardens, Ruislip HA4 9HD
G6 PLT E. Cheetham 172A Hesketh Lane, Tarleton, Preston PR4 6UD
G6 PLU A. Chenery 43 Wessex Estate, Ringwood BH24 1XD
GW6 PMC R. Evans 16 Monmouth Grove, Prestatyn LL19 8TS
G6 PMD C. Eagling 96 Regent Road Brightlingsea, Colchester CO7 0NZ
G6 PMF H. Langsley 39 Lavender Road, Basingstoke RG22 5NN
G6 PMJ S. Murphy 1 Orchard Cottage, Golden Valley, Newent GL18 1HN
G6 PMO I. Parker 27 St. Audries Road, Worcester WR5 2AL
G6 PMR P. Shaw 52 Belvedere Parade, Bramley, Rotherham S66 3WA
G6 PMW G. Goodier 3 The Paddock, Beckingham, Lincoln LN5 0FD
G6 PNB North Bristol ARC c/o R. Elford Prospects, Tormarton Road, Badminton GL9 1HP
G6 PNG P. Hill 28 Somerton Grove, Thatcham RG19 3XE
GM6 PNJ S. Hammond Graywalls, Denny FK6 5JF
G6 PNO P. Hill 33 The Pastures, South Beach, Blyth NE24 3HA
G6 POC R. Kinrade 23 Crofthill Road, Slough SL2 1HG
G6 POE J. Knott 3 Lords Wood, Welwyn Garden City AL72HF
G6 POG R. Williams 29 Bridle Close, Banbury OX16 9SZ
G6 POI J. Wright Chez Mon, Burton Road, Carnforth LA6 1QN
G6 POJ I. Worthy 7 The Paddocks, Pilsley, Chesterfield S45 8ET
GW6 POO R. Smallwood 12 Oak Close, Connah'S Quay, Deeside CH5 4GG
G6 POV M. Walker 232 Bideford Green, Leighton Buzzard LU7 2TS
G6 POW D. Pow 16 Ancaster Close, Trowbridge BA14 9DA
G6 POZ R. Farrall 7 The Meadows, South Cave, Brough HU15 2HR
G6 PPA T. Farmer 35 Ascot Drive, Dudley DY1 2SN
G6 PPD A. Morgan High Bow Cottage, Bow, Carlisle CA5 6EN
G6 PPU H. Chappell Oanley, East Lyng, Taunton TA3 5AU
G6 PPV P. Caswell 94 Dewsbury Road, Luton LU3 2HJ
G6 PPY S. Carter 3 Mary Street, Burnley BB10 4AJ
G6 PQI J. Finch The Croft, Dalby Road, Melton Mowbray LE14 3EX
G6 PQP P. Brooks Cherating, Hanning Road Horton, Ilminster TA19 9QH
G6 PRA J. Whittaker 6 Bradley Gardens, Burnley BB12 6JT
G6 PRE E. Snell 156 Brookdale Avenue South, Greasby, Wirral CH49 1SS
G6 PRL D. Brown 63A Great Northern Street, Huntingdon PE29 7HJ
G6 PRP W. Barker 297 Williamthorpe Road North Wingfield, Chesterfield S42 5NT
G6 PSA N. Turnham 153 Canterbury Road, Urmston, Manchester M41 0PY
G6 PSC M. Horn 3 Church Cottages, Pound Lane, Beccles NR34 0EX
G6 PSO I. Russell 24 Standard Avenue, Tile Hill, Coventry CV4 9BW
G6 PSQ R. Langton Dawn Cliffe, Goodwin Road, Dover CT15 6ED
G6 PSZ T. Shackleton 27 Court Crescent, Kingswinford DY6 9RJ
G6 PTF J. Wilson Belle Vue House, Common Side, Workington CA14 4PU
G6 PTT P. Hedley 7 Midhill Close, Langley Park, Durham DH7 9YA
GM6 PTX G. Gane 28 Queens Croft, Kelso TD5 7NN
G6 PUE M. Mackmin 15-3 Koumasion, Peyia 8560 Cyprus
G6 PUO A. Quantrill 105 High Street, Sawston, Cambridge CB22 3HJ
G6 PUR J. Howells 66 Rochester Avenue, Burntwood WS7 2DL
G6 PUV W. Holding 20 Lingfield Crescent, Wigan WN6 8QA
GM6 PVA M. Green Fulwood House, Coronation Road, Bellshill ML4 2RT
G6 PVC P. Coates Jacaranda, Cotswold Close, Staines TW18 2DD
GW6 PVK G. Jones 12 The Nurseries. Cymau, Nr Wrexham. LL11 5LE
G6 PVT L. Adamson 6 Castle View Estate, Derrington, Stafford ST18 9NF
G6 PVV C. Burt 1 Chapter Court, Vicarage Road, Egham TW20 8NL
G6 PVW G. Bayliffe 179 Breedon Street, Long Eaton, Nottingham NG10 4EW
G6 PVY G. Bennett 6 Jubilee Road, Holt NR25 6HH
G6 PWF S. Choules 43 Ashbrook Road, Old Windsor, Windsor SL4 2LT
G6 PWJ J. Chiddick Unioninkatu 45A10, Helsinki 170 Finland
G6 PWL R. Cloutman 35 Camlet Way, St. Albans AL3 4TL
G6 PWQ D. Dick 140 Chatham Street, Stockport SK3 9JU
G6 PWS R. Fuller 18 St. Leonards Crescent, Sandridge, St. Albans AL4 9EH
G6 PXJ A. Harrison Nirvana Cottage, 42 Bell Lane, Goole DN14 8RP
G6 PXN R. Bee 80 Hospital Road, Burntwood WS7 0EQ
G6 PXQ R. Boyce 3 Castleton Cottages, Westhide, Hereford HR1 3RF
GW6 PXW E. Davies 62 Heol Y Coedcae, Cwmllynfell, Swansea SA9 2FY
G6 PXX L. Dempsey 24 James Street, Great Harwood, Blackburn BB6 7JE
GM6 PYD A. Dunnett 11 Silverknowes View, Edinburgh EH4 5PY
G6 PYE Cambridge Reapeater Group c/o P. Nice 31 Elizabeth Drive, Chapel St. Leonards, Skegness PE24 5RS
G6 PYF D. Hills 9 Brook Gardens, Devizes SN10 2FX
G6 PYI M. Jones 15 Rowan Rise, Kingswinford DY6 8EE
G6 PYL P. Hatter 14 Morland Avenue, Bromborough, Wirral CH62 6BE
G6 PYM A. Hedges 25 The Lanes, Cheltenham GL53 0PU
GI6 PYP A. Gault 134 Leighan Road, Randalshough, Monea, Enniskillen BT93 7DN
G6 PYR H. Adams Hill Sixty, Happisburgh, Norwich NR12 0RB
G6 PZ P. Beecham The Haybarn, Church Street, , Bridgwater TA7 9AT
G6 PZE D. Jefferson 48 Neston Road, Walshaw, Bury BL8 3DB
G6 PZF P. James 18 Brackens Drive, Warley, Brentwood CM14 5UE
G6 PZH B. Hickman 7 Nina Close, Stourport-on-Severn DY13 9RZ
G6 PZN M. Mcloughlin 13 Old Manor Gardens, Wymondham, Melton Mowbray LE14 2AN
G6 PZS D. Carr 5 Church Meadow, Hyde SK14 4RT
GM6 PZY Dr Z. Yang 48 Foxglove Road, Newton Mearns, Glasgow G77 6FP
G6 QA L. Jopson 68 Greenmount Park Kearsley, Bolton BL4 8NT
G6 QM Southgate ARC c/o D. Berry 4 Holly Hill, Winchmore Hill, London N21 1NP
G6 QN T. Blakeman 31 Walsingham Gardens, Epsom KT19 0LS
G6 RAF Royal Air Force ARS c/o M. Garrett 489 Dorchester Road, Weymouth DT3 5BP
G6 RAH R. Hammond 126 Otley Drive, Ilford IG2 6QY
GM6 RAK D. Brown 14 Newton Crescent, Carnoustie DD7 6HW
GW6 RAO D. Griffiths 35 Greystones Crescent, Mardy, Abergavenny NP7 6JY
G6 RAQ S. Hayter Rookery Farm, Mill Road, Battisford, Stowmarket IP14 2LT
GW6 RAV W. Keeley 93 Park Crescent, Abergavenny NP7 5TL
G6 RAZ J. Paton 23 Courville Close, Alveston, Bristol BS35 3RR
GI6 RBD K. Brady 26 Kilbroney Road, Rostrevor, Newry BT34 3BJ
G6 RBM Luton Rep Grp c/o R. Jeffery 15 Greenway, Hulland Ward, Ashbourne DE6 3FE
G6 RBO W. Bennett 44 Wood Lane, Streetly, Sutton Coldfield B74 3LR
G6 RBP R. Pearsey 21 Ashwood Drive, Newbury RG14 2PN
G6 RBR M. Allen 1 Allens Yard, Chatteris PE16 6QE
GW6 RBZ R. Coombes 10 Goodrich Court, Llanyrafon, , Cwmbran, NP44 8RY
G6 RC Crawley ARC c/o R. Hadfield 45 Erica Way, Copthorne, Crawley RH10 3XG
G6 RCD P. Clark 166 Attenborough Lane, Attenborough, Nottingham NG9 6AB
GW6 RCK H. Fray 17 Homelands Road, Cardiff CF14 1UH
G6 RCT T. Stellar 27 Blackmore Chase, Wincanton BA9 9SB
GW6 RCX Red Brick Stables ARC c/o D. Smithies 26 Coed Mor, Penyffordd, Holywell CH8 9HY
G6 RCY D. Reed 11 Grenville Close, Corby NN17 2RP
G6 RDD I. Senter 33 King Coel Road, Colchester CO3 9AQ
G6 RDO A. Shaw 1 Chapel Lane, Clifford, Wetherby LS23 6HU

GW6 RDV B. Clarke Flat 3, Carling Court, Haig Place, Cardiff CF5 4PH
G6 REA J. Gilpin River Bank House 28A Ivel Road, Sandy SG19 1AX
G6 REC M. Hart 7 Ullswater Avenue, South Wootton, King's Lynn PE30 3NJ
GW6 REF D. Jones 16 New Road, Llandovery SA20 0ED
G6 REG A. Joyce Ashdene, Highworth Road, Swindon SN3 4SE
G6 REH J. Staplehurst 12 Trotter Way, Epsom KT19 7EW
G6 REM I. Atkinson 537753, 7 Headquarter Squadron, Bfpo 36 AA1 1AA
GW6 REQ W. Vize Cefn Rhos, Bethel, Caernarfon LL55 1YB
G6 REV J. Yam 4537 Mossburg Court, Marietta 30066 United States
G6 REW G. Seymour-Smith Glencarne, Bridgerule, Holsworthy EX22 7ED
G6 REY R. Mcminn 1C Bickley Avenue, Sutton Coldfield B74 4DY
G6 RFH D. Ruck 50 Kiel Walk, Corby NN18 9DE
G6 RFJ L. Waite The Towers, Castle St., Nottingham NG2 4AE
G6 RFL R. Rothery 12 Reevy Crescent, Bradford BD6 2BT
G6 RFM A. Warren 20 Wolverhampton Road, Stafford ST17 4BP
G6 RFR A. Brammer 5 The Green, Reepham, Lincoln LN3 4DH
G6 RFS A. Brown 22 Mount Wise Crescent, Plymouth PL1 4GQ
G6 RFU P. Csapo 87 Latchmere Road, Kingston upon Thames KT2 5TU
G6 RGA J. Ewen 26 Court Road, Eastbourne BN22 9EZ
GM6 RGD T. Murray 2 The Glebe, Edzell, Brechin DD9 7SZ
G6 RGI A. Shead 95 Sea Front, Hayling Island PO11 0AW
G6 RGN W. Stockley 4 Tothby Close, Alford LN13 0BG
GW6 RGT M. Morgan 17 Dunstable Road, Newport NP19 9NE
GM6 RGY W. Hardie 96 Carmuirs Avenue, Camelon, Falkirk FK1 4PB
G6 RHA S. Howes 46 High Street, Upwood, Huntingdon PE26 2QE
G6 RHB R. Howlett 4 Station Road, Willoughby, Alford LN13 9NG
G6 RHJ M. Swain 17 Sponnes Road, Towcester NN12 6ED
G6 RHK C. Spencer 21 Playford Road, Ipswich IP4 5QZ
G6 RHL P. West 6 Iveldale Drive, Shefford SG17 5AD
G6 RHN Dr D. Morris Rowley Farm, Rowley Lane, Borehamwood WD6 5PE
G6 RHT D. Owens 17 Priory Place, Greenham, Thatcham RG19 8XT
G6 RHV I. Smith 126A High Street, Teddington TW11 8JB
G6 RIC M. Ellis 28 High Meadows, Romiley, Stockport SK6 4PT
G6 RIG N. Golding Coppice View, 16 Littlewood Road, Walsall WS6 7EU
G6 RII M. Dodson Tree Tops, Badgeworth Lane, Cheltenham GL51 4UW
G6 RIJ R. Fletcher 33 Littlewood Lane, Cheslyn Hay, Walsall WS6 7EJ
G6 RIM C. Berry 258 Lowerhouse Lane, Burnley BB12 6NG
G6 RIQ G. Dunn 11 Ellesmere Rise, Grimsby DN34 5PE
G6 RIY A. Wilkinson 34 Coppice Lane, Hellifield, Skipton BD23 4JW
G6 RIZ B. Jones 96 Somerset Road, Farnborough GU14 6DS
G6 RJH J. Proffitt 38 Hockley Road, Poynton, Stockport SK12 1RW
G6 RJU M. Scaife 4 Arden Close, Wallsend NE28 9YB
G6 RJW R. Woodgate Roger Woodgate, C/O Ratana Paa, P.D.C., Turakina 4548 New Zealand
G6 RKF R. Taylor 20 Scraley Road, Heybridge, Maldon CM9 4BL
G6 RKG J. Walters The Gables, Lavenham Road, Sudbury CO10 0RN
G6 RKJ R. Butland 4 Park Close, Sonning Common, Reading RG4 9RY
G6 RKQ D. Hudson 19 Worcester Close, Lichfield WS13 7SP
G6 RKS R. Domville 12 Craig Terrace, Peterlee SR8 3AJ
G6 RLG J. Kerr 3 Lime Kiln Way, Salisbury SP2 8RN
G6 RLM R. Maddison Tom Butt, Hope Street, Elstead, Godalming GU8 6DE
G6 RMA D. Glover 14 Fitzgerald Avenue, Herne Bay CT6 8LN
G6 RMJ W. Clements 3 May Street, Durham DH1 4EN
GI6 RMO D. Johnston Olanda, Lisreagh, Co Fermanagh BT94 5BX
G6 RMV R. Sharp The Old School, Bridge Street, Bridport DT6 5LS
GW6 RNA T. Lovell 4 Maes Rathbone, Waen, St. Asaph LL17 0AD
G6 RNF R. Smith 12 East View, West Bridgford, Nottingham NG2 7QN
G6 RNR S. Allison Rudd Hall Cottage, East Appleton, Richmond DL10 7QD
G6 RNT G. Kingdon Wymering, Copley Drive, Barnstaple EX31 2BH
GW6 RNV C. Brewster 35 Ffordd Las, Sychdyn, Mold CH7 6DU
G6 RNZ D. Brook 50 Ashley Down Road, Bristol BS7 9JW
GI6 ROI J. Polson 6 Castlemara Drive, Carrickfergus BT38 7RB
G6 ROS J. Warwick 12 Oak St., Sutton in Ashfield NG17 3FF
G6 RPD R. Montford 394 Selbourne Road, Luton LU4 8NU
G6 RPH G. Tibbert 5 Upper Havelock Street, Wellingborough NN8 4PN
G6 RPK N. Waterton 1270 Killaby Drive, Mississauga, Ontario L5V 1B1 Canada
G6 RPW S. Andrews 1 Swynford Close, , Kempsford GL7 4HN
G6 RQA D. Nicholls 15 Poplar Avenue, Heacham, King's Lynn PE31 7EB
G6 RQJ R. Sherlock 34 St. Cecilias Road, Belle Vue, Doncaster DN4 5EG
GM6 RQU B. Hamilton 51 Grange Road, Grange, Edinburgh EH9 1UF
GM6 RQW T. Christie 4 Glebe Park, Bressay, Shetland ZE2 9ER
G6 RQZ B. Cripps 3 Sabre Court, Aldershot GU11 1YY
G6 RRJ R. Urwin 21 Winchester Avenue, Leicester LE3 1AX
G6 RRS D. Rotgans 18 Minter Avenue, Densole, Folkestone CT18 7DS
G6 RRV R. Weston The Old Dairy, Slate Cross, Bridgwater TA7 8QR
G6 RRY M. Dunbar 42 Wickham Way, Shepton Mallet BA4 5YG
G6 RSI L. Hart 25 Murcroft Road, Stourbridge DY9 9HT
G6 RSU R. Anstee 12 Ashmore Avenue, Stockport SK3 0QY
G6 RTD G. Small 6 Mary St., Longridge, Preston PR3 3WN
G6 RTE Capt. J. Menhinick Coburg, Barton Estate, East Cowes PO32 6NT
G6 RTG J. Sutton 29 Victory Avenue, Darlaston, Wednesbury WS10 7RR
G6 RTM R. Ashberry 30 Factory Lane Roydon, Diss IP22 4EG
GM6 RTN K. Bone School House Makerstoun, Kelso TD5 7PB
G6 RTY D. Johnson Penvern, Nacton, Ipswich IP10 0EW
GW6 RUE J. Newey Springwood Cottage, Tyntaldwyn Road Troedyrhiw, Merthyr Tydfil CF48 4NG
G6 RUM I. Nice 6 Malden Road, Sidmouth EX10 9LS
GW6 RUO J. Griffiths 4 Lon Elan, Meliden, Prestatyn LL19 8LP
G6 RUP J. Horner 43 Birch Close, Patchway, Bristol BS34 5SA
G6 RUU S. Burns 289 Wallasey Village, Wallasey CH45 3HA
G6 RUY J. Heaney 15 Perth Street, Nelson BB9 8EE
G6 RVH R. Jamieson 3 Waterpark Road, Prenton Park, Birkenhead CH42 9NZ
G6 RVP C. Pung 73 John Mace Road, Colchester CO2 8WW
G6 RVS R. Sohst 2 Shaftesbury Drive, Maidstone ME16 0JS
G6 RVZ D. Carruthers 168A Wanstead Park Avenue, London E12 5EF
GU6 RWD S. Hancock L'Hirondel, Hubits De Bas, St. Martin, Guernsey GY4 6NB
GW6 RWJ D. Silcox Troedyrhiw, Penparc, Cardigan SA43 2AE
G6 RXD M. Yirrell 66 Park Lane, Sandbach CW11 1EP
G6 RXF G. Priestley 7 Affleck Avenue, Radcliffe, Manchester M26 1HN
G6 RXK A. Orchard Kilimani, Cuilfail, Lewes BN7 2BE
G6 RXP S. Dwyer 10 Swan Street, Darwen BB3 2LW
GM6 RXQ A. Gordon 2, Merse Avenue, Kirkcudbright DG6 4RN
G6 RXV K. Keen 26 Brogden Close, Botley, Oxford OX2 9DS
G6 RXY D. Ferguson Aneataprint Four Ltd 3-5 Lord Street, Watford WD17 2LN
G6 RYM H. Wagg 43 Highfield Road, Birkenhead CH42 2BU
G6 RYW V. Smith 9 Pinewood Drive, Mansfield NG18 4PG
G6 RZR N. Stevens 151 Ferme Park Road, London N8 9BP
G6 RZS R. Wood 115 Anchorway Road, Green Lane, Coventry CV3 6JH
G6 RZY N. Harper 15 Epsom Close, Dosthill, Tamworth B77 1QT
GM6 SAA Dr P. Record 27 Westfield Road, Cupar KY15 5AR
G6 SAQ G. Hines 11 Montagu Gardens, Wallington SM6 8EP
GW6 SBD G. Davies 2 Ffordd Aled, Wrexham LL12 7PP
G6 SBG A. Lubrani Ranscombe Manor, Sherford, Kingsbridge TQ7 2DP
G6 SBI D. Smith 8 Corunna Drive, Horsham RH13 5HG
G6 SBN M. Searl Sinodun Road, Didcot OX118HW
GI6 SBW A. Alcock 22 Chippendale Avenue, Bangor BT20 4PT
G6 SCB A. Bean 66 Wickham Lane, London SE2 0XN
G6 SCG M. Lockwood 33 Elmtree Road, Calverton, Nottingham NG14 6QA
G6 SCM J. Webb Oakdene, 22 Meeting House Lane, Coventry CV7 7FX
G6 SDC B. Simmons Wootton Leas, 35 Benenden Green, Alresford SO24 9PE
G6 SDE A. Curley 21 Trinity Rise, Penton Mewsey, Andover SP11 0RE
G6 SDI P. Hall 28 Grangeway, Rushden NN10 9EZ
GM6 SDV J. More 51 Hilton Drive, Aberdeen AB24 4NJ
G6 SDW M. Rowan 1 Yanleigh Close, Bristol BS13 8AQ
G6 SDY R. Beecroft 28 Hall Garth Lane, West Ayton, Scarborough YO13 9JA
G6 SEE Z. Feast 2 Dyrham Close, Burnham-on-Sea TA8 2TT
G6 SEF J. Feast 2 Dyrham Close, Burnham-on-Sea TA8 2TT
G6 SEK P. Ovey 35 Lower Fairfield, St. Germans, Saltash PL12 5NH
GM6 SEL M. Munro 5 Corran Cismaol Horve, Isle of Barra HS9 5ZE
GM6 SEV I. Carr 36A Broomieknowe, Lasswade EH18 1LN
G6 SFC T. Foulds Deer Park, Detling Avenue, Broadstairs CT10 1SR
G6 SFE S. Al-Kattan 8 Little John Drive, Rainworth, Mansfield NG21 0JJ
G6 SFF C. Hewes 1 Broad Valley Drive Bestwood Village, Nottingham NG6 8XA
G6 SFH P. Barber 17 Wheelwright Avenue, Leeds LS12 4UW
GI6 SFO G. Miskimmin 332 Rathfriland Road, Dromara, Dromore BT25 2HN
G6 SFR Flt Refuelng ARS c/o A. Baker Highleaze, Deans Drove, Poole BH16 6EQ
G6 SFW P. Dodd 9 Rudge Croft, Kitts Green, Birmingham B33 9NZ
G6 SFY G. Barker 18 Penryn Close, Nuneaton CV11 6FF
G6 SGA S. Thornber 18 Lichfield Road Talke, Stoke on Trent ST7 1SQ
G6 SGD D. Carding Mill House, Walcot, Telford TF6 5ER
G6 SGE P. Bayliss 36 Slingates Road, Stratford-upon-Avon CV37 6ST
G6 SGM C. Macey 29 Burleigh Road, Sutton SM3 9NE
G6 SGU A. Aristotelous 8 Sudbury Drive, Huthwaite, Sutton-in-Ashfield NG17 2SB
G6 SGV M. Durkin Selsdon House, 23 Jameson Road, Bexhill-on-Sea TN40 1EG
G6 SGW J. Miller 6 Saunders Mews, Southsea PO4 9XZ
G6 SGY B. Sales 2 Highview, Hurley, Atherstone CV9 2RP
G6 SGZ J. Smith 1 Markby Close, Moorside, Sunderland SR3 2RG
G6 SHD G. Mcbrien 26 Lumb Carr Avenue, Ramsbottom, Bury BL0 9QG
G6 SHF M. Trolan Hilbre, East Taphouse, Liskeard PL14 4NJ
G6 SHQ M. Brown 6 Snell Hatch, West Green, Crawley RH11 7JB
G6 SHS K. Eldridge 44 Merley Gardens, Merley, Wimborne BH21 1TB
G6 SHZ S. Congrave 28 London Road, St. Ippolyts, Hitchin SG4 7NG
G6 SIG G. Timbrell Crossing Cottage, Lamyatt, Shepton Mallet BA4 6NG
G6 SIM J. Simarpi 6 Berryman Court Lethbridge Road, Wells BA5 2FF
G6 SIQ W. Whitcombe 11 The Elms, Deerton Street, Teynham, Sittingbourne ME9 9LH
GW6 SIX P. Macmillen Spinney Cottage, Sychnant Pass Road, Conwy LL32 8NS
G6 SJA W. Barnes 17 Greenhill Road, Long Buckby, Northampton NN6 7PU
G6 SJD A. Evans Spring Bank, Coventry Road Kingsbury, Tamworth B78 2LW
G6 SJG T. Hurton 4 Athlone Close, Enham Alamein, Andover SP11 6JY
G6 SJV P. Cliffe 63 Mill Lane, Upton, Chester CH2 1BS
G6 SKF L. Hopson 39A Fenside Road, Boston PE21 8HY
G6 SKK T. Parkin 8 Horsley Crescent, Holbrook, Belper DE56 0UB
G6 SKM W. Taylor 5 Gadbury Avenue, Atherton, Manchester M46 0LQ
G6 SKP A. Whitgreave 2 Oaklea Avenue, Hoole, Chester CH2 3RE
G6 SKR G. Walker 81 Normanshire Drive, Chingford, London E4 9HE
G6 SKS C. Learoyd Leofric House, 31 Leofric Avenue, Bourne PE10 9QT
G6 SKT W. Learoyd Leofric House, 31 Leofric Avenue, Bourne PE10 9QT
G6 SL C. Pettitt 12 Hennals Avenue, Redditch B97 5RX
G6 SLG A. Berry Emberley Leys, Ratley, Banbury OX15 6DS
G6 SLN C. Gleave 10 Henley Road, Neston CH64 0SG
GW6 SLO P. Charnley 9 Bryn Crescent, Rhuddlan, Rhyl LL18 5RF
G6 SLY D. Lewis 30 Printers Park, Hollingworth, Hyde SK14 8QH
G6 SLZ J. Mackenzie 11 Upper Heyshott, Petersfield GU31 4QA
G6 SMI Zycomm Elect Lt c/o G. Langstaff Flat 22 Benwell Close, Benwell Grange, Newcastle-upon-Tyne NE15 6RZ
G6 SMJ M. Pitts 30 Sandhurst Avenue, Surbiton KT5 9BS
G6 SMN N. Maslin 99 The Haywards, Thatcham RG18 4LY
GM6 SMW G. Harper 15 Seaforth Park, Annan DG12 6HX
G6 SNA D. Heather 65 Fairview Road, Headley Down, Bordon GU35 8HQ
G6 SND W. Jarvis Owlpen, 20 Park Road, Wiltshire SN10 4ED
G6 SNI G. Platt 15 Mount Close, Nantwich CW5 6JJ
G6 SNN D. Ramsey 11 Pendle Close, Basildon SS14 3NA
GJ6 SNQ A. Leighton 2 Le Petit Menage, Fountain Lane, St. Saviour JE2 7RL Jersey
G6 SNV R. Smith 3 Payne Road, Wootton, Bedford MK43 9JL
G6 SNZ I. Stones 1 Valdene Close, Farnworth, Bolton BL4 9NE
G6 SOA R. Wilday 4 Kenelm Road, Clifton-On-Teme, Worcester WR6 6DW
G6 SOO P. Williams 2 St. Christophers Drive, Romiley, Stockport SK6 3BE
G6 SOX J. Bradley 68 Rosedale Avenue, Alvaston, Derby DE24 0FJ
G6 SOY P. Berry Bundys Cottage, Colwood Lane, Haywards Heath RH17 5QQ
G6 SOZ A. Byrne Holly Cottage, Deacons Lane, Thatcham RG18 9RJ
G6 SPA P. Brown 7 Stoneberry Road, Bristol BS14 0JF
G6 SPB D. Corder 140 Edward Road, Somerford, Christchurch BH23 3EW
G6 SPG P. Cesnavicius 52 Boundary Road, Irlam, Manchester M44 6HD
G6 SPH J. Crookbain 11 Champlain Avenue, Canvey Island SS8 9QL
G6 SPI S. Carwood 34 Flemming Avenue, Ruislip HA4 9LF

G6 SPN R. Barton 82 Buckingham Road, South Woodford, London E18 2NJ
G6 SPQ W. Holmes 37 Barmpton Lane, Darlington DL1 3HH
G6 SQL A. Lowthian 38 Arthur Street, Ryde PO33 3BU
G6 SQS F. Sivyer 22 Boxley Road, Walderslade, Chatham ME5 9LF
G6 SQT C. Wall 151 Bisley Road, Stroud GL5 1HS
G6 SRE B. Stone Reindene, Faversham Road, Ashford TN25 4PQ
G6 SRJ A. Waring 2 Wroxton Close, Thornton-Cleveleys FY5 3EY
G6 SRS Stourbridge & District ARS c/o D. Scott Hyde Bungalow, The Hyde, Stourbridge DY7 6LS
G6 SRT D. Armstrong 103 Victoria Road, Oxford OX2 7QG
G6 SRU C. Alford 43 Cleeve, Glascote, Tamworth B77 2QD
G6 SRV R. Andrews 11 Holly Grove, Verwood BH31 6XA
G6 SRX Furness ARS c/o I. Aram 4 Severn Road, Chilton, Didcot OX11 0PW
G6 SRY S. Aram The Barn, Charleston Place, Eastbury, Hungerford RG17 7JN
G6 SSN K. Burton 2 Council House, Stainfield Road, Kirkby Underwood, Bourne PE10 0SG
G6 SSQ D. Bolton Huds House, Cowgill, Sedbergh LA10 5TQ
G6 STB G. Munn 1 Sovereign Close, Hastings TN34 2UB
G6 STD D. Macey Affric House, New Street, Banbury OX15 0SR
G6 STE B. Stevens 16 Fowey Close, Wellingborough NN8 5WW
G6 STF G. Smith Sunny Patch, Western Backway, Kingsbridge TQ7 1QB
G6 STI H. Staddon 45 Saxony Parade, Hayes UB3 2TQ
G6 STJ S. Smith 9 Beadon Drive, Salcombe TQ8 8NU
GW6 STK R. Sweet 9 Seafield Road, Colwyn Bay LL29 7HB
GW6 STS G. Jones The Bungalow, Castle Street, Pontypool NP4 9QL
G6 SUC J. Hudson 6 Brooks Close, Ringwood BH24 1NE
G6 SUK B. Trevor 39 Clayton Crescent, Brentford TW8 9PT
G6 SUR P. Thornsby 25 Kipling Way, Stowmarket IP14 1TS
G6 SUV J. Barlow 3 Shaw Brook Close, Rishton, Blackburn BB1 4ES
G6 SVH K. Henderson 42 Chartwell Avenue, Wingerworth, Chesterfield S42 6SP
G6 SVJ S. Harvey 148 Smithfield Road, Uttoxeter ST14 7LB
G6 SVL S. Harris 34 Butterfly Crescent, Nash Mills Wharf, Hemel Hempstead HP3 9GS
G6 SVV R. Gray Willett, 28 Hoe Lane, Romford RM4 1AX
G6 SW Cannock Chase ARS c/o M. Starkey 23 Arthur Street, Cannock WS11 5HD
G6 SWD A. Gibbings 3 Bonville Crescent, Tiverton EX16 4BN
G6 SWJ J. Askey The Maltings, Brewery Yard, Kettering NN14 3BT
G6 SWO M. Fincher Brickyard Farm, Lincoln Road, Horncastle LN9 5NW
G6 SWT S. French 47 Horn Lane, Woodford Green IG8 9AA
G6 SWW S. Ellin 7 Crawshaw Avenue, Beauchief, Sheffield S8 7DZ
G6 SWZ M. Davy1 22 Scott Gardens, Lincoln LN2 4LX
G6 SXB L. Dunham 5 King Street, Wimblington, March PE15 0QF
G6 SXC J. Mallichan 17 Napier Road, Gillingham ME7 4HB
G6 SXD E. Drinkwater 57 Ludlow Road, Bridgnorth WV16 5AH
G6 SXN G. Dixon 4 Yarborough Road, Keelby, Grimsby DN41 8HG
G6 SYA M. Mills 6 Bower Road, Hextable, Swanley BR8 7SE
G6 SYB J. Malcom 62 Linden Avenue, Ruislip HA4 8UA
G6 SYI P. Somerfield 27 Ormerod Street, Worsthorne, Burnley BB10 3NU
G6 SYW B. Bauly Poplar Farm Mendlesham, Stowmarket IP14 5SN
G6 SYX G. Brookes 47 Lucas Avenue, York YO30 6HL
G6 SZB J. Barton 76 Elvaston Road, North Wingfield, Chesterfield S42 5HH
GM6 SZJ M. Burke 3 Tarvit Avenue, Cupar KY15 5BW
G6 SZS R. Crook 26 Chapel Street Rishton, Blackburn BB14NP
G6 TAF P. Penny 79 Grove Avenue, New Costessey, Norwich NR5 0JA
G6 TAH D. Palmer 17 Atyeo Close, Burnham-on-Sea TA8 2EJ
G6 TAI J. Peel 9 Hillspring Road, Springhead, Oldham OL44SJ
G6 TAK P. Reay 26 Clifton Court, Workington CA14 3HR
G6 TAN M. Shread 21 The Strand, Mablethorpe LN12 1BQ
G6 TAP D. Squire Green Valley, Raleigh Road, Barnstaple EX31 4HY
G6 TAS R. Wroe 11 Malvern Close, Banbury OX16 9EL
GI6 TBC V. Loughran 10 Oakwood, Armagh BT60 1QR
GM6 TBE P. Lowrie 11 Berrymoss Court, Kelso TD5 7NP
G6 TBJ J. Larssen 228A Barnsole Road, Gillingham ME7 4JB
G6 TBV K. Cheers 112 Rickerscote Road, Stafford ST17 4HB
G6 TCV J. Halliday1 Calle Barrio 18 Ribera Baja, Alcala la Real 23691 Spain
G6 TDG K. Hodges 18 Leycester Close, Birmingham B31 4SS
G6 TDJ G. Ward 83 Buxton Road, Congleton CW12 2DX
G6 TDW J. Mead 12 Coltsfoot Close, Ixworth, Bury St. Edmunds IP31 2NJ
G6 TDX D. Yarrow 193 Ladygate Lane, Ruislip HA4 7RD
G6 TEB A. Varga 2 Yew Tree Lane, Malvern WR14 4LJ
G6 TEL S. Mayer 453 Wimborne Road, Poole BH15 3EE
GW6 TEO G. Smith 11 Sandy Leys, Castlemartin, Pembroke SA71 5HJ
G6 TEQ I. Stuckey Rancliffe, Higher Downs Road, Torquay TQ1 3LD
G6 TER R. Monksummers 29 Cloverfields, Peacemarsh, Gillingham SP8 4UP
G6 TET B. Smith 8 Devon Street, Leigh WN7 2NG
G6 TEX T. Speak 40 Orchard Close, Bolsover, Chesterfield S44 6DY
G6 TFE D. Standen Sunnybank, 54 Park Way, Hastings TN34 2PJ
GI6 TFF R. Symington 8 Thorndene Park, Carrickfergus BT38 9EA
G6 TFJ R. Smith 44 Yvonne Robertson House, Hastings Road, Bexhill-on-Sea TN40 2HQ
G6 TFL C. Sugars 6 Church Street, Denby Village DE5 8PQ
G6 TFP C. Mann Woodford, Listowel Co. Kerry 0 Ireland
G6 TFV D. Owen 18 Prescott Avenue, Atherton, Manchester M46 9LN
G6 TGB A. Pennells 56 Wilverley Place, Blackfield, Southampton SO45 1XW
G6 TGE G. Holman 5 Ingleton Road, Newsome, Huddersfield HD4 6QX
G6 TGJ J. Hirons Furlong House, Racecourse Lane Bicton Heath, Shrewsbury SY3 5BJ
G6 TGM W. Howard 2 Heather Drive, Rise Park, Romford RM1 4SP
G6 TGQ S. Houghton 259B St. Faiths Road, Norwich NR6 7BB
GW6 TGR G. Jones 3 Tan Y Buarth Estate, Bethel, Caernarfon LL55 1UP
G6 TGW R. Jones 49 Sycamore Drive, Huntingdon PE29 7JA
G6 THC M. Johnson Happy Valley, Highfield Road, Westerham TN16 3UX
G6 THM C. Smith 8 Terry Close, Stoke-on-Trent ST3 6NS
G6 THP L. Curwen 12 Garden Close, St. Ives PE27 3XZ
GM6 TIB I. Campbell 35 Thornwood Avenue, Lenzie, Glasgow G66 4EL
G6 TID M. Coleman 1 Burdon Drive, Bartestree, Hereford HR1 4DL
G6 TIQ A. Price Flat 2, South Elms, 69 Silverdale Road, Eastbourne BN20 7EU
G6 TIU M. Rainer 101 Gwydir Street, Cambridge CB1 2LG
G6 TIW D. Reeve 12 Lambourne Road, Birstall, Leicester LE4 4FU
G6 TJC S. Deville 39 Acre Close, Maltby, Rotherham S66 8BL
GM6 TJD J. Doull 52 Howburn Road, Thurso KW14 7ND
G6 TJE R. Dawson 10 St. Julien Close, New Duston, Northampton NN5 6QX
G6 TJJ R. Downham 11 Churchills Rise, High Street, Cullompton EX15 3AU
G6 TJK A. Dowell 54 Station St., Castle Gresley, Burton on Trent DE14 1BS
G6 TJY I. Randle 12 Cuckoo Avenue Hanwell, London W7 1BT
G6 TJZ P. Rendell 6 The Park, Bradley Stoke, Bristol BS32 0AP
G6 TKB K. Slaughter 652 Newchurch Road, Newchurch, Rossendale BB4 9HG
GU6 TKE C. Wild Honfleur, La Rue Du Le Hurel, Vale GY3 5AF Guernsey
G6 TKH J. Torring Ivy Cottage, Royal Oak, Filey YO14 9QE
G6 TKR E. Tratt 162 Stoddens Road, Burnham-on-Sea TA8 2EL
G6 TKV T. Tyrer 85 Swann Lane, Cheadle Hulme, Cheadle SK8 7HU
G6 TKW P. Tomlinson 158 Seamore Avenue, Benfleet SS7 4LA
G6 TKY Reading & District ARC c/o E. Caligari 209 Ormskirk Road, Upholland, Skelmersdale WN8 0AA
G6 TLA P. Curran-Bilbie 198 Birchwood Lane, Somercotes, Alfreton DE55 4NF
G6 TLB P. Curran 422 Carlton Road, Worksop S81 7QW
G6 TLN D. Allen 35 Fortescue Chase, Thorpe Bay, Southend-on-Sea SS1 3SS
G6 TLP D. Attree 36 Furze Road, Norwich NR7 0AS
G6 TLX C. Bull 35 Manor Road, Wokingham RG41 4AR
GM6 TMH D. Bell 11 Shebster Court, Thurso KW14 7ES
G6 TMN H. Bryan 2 Ashbrook Close, Hesketh Bank, Preston PR4 6LY
G6 TMQ L. Saagi 17 Broughton Close, Anstey, Leicester LE7 7EU
G6 TNA C. Walton 6 Gorse Grove, Longton, Preston PR4 5NP
G6 TNE G. Skulski 30 Eastfield Road, Laindon, Basildon SS15 4JE
G6 TNI B. Telford 18 Kirkstall Close, South Anston, Sheffield S25 5BA
G6 TNJ D. Thomas 97 The Gardens, Doddinghurst, Brentwood CM15 0LX
G6 TNK J. Turnbull 34 Bridge Avenue, Hanwell, London W7 3DJ
G6 TNP R. Brooks 8 Chesle Way, Portishead, Bristol BS20 8JB
G6 TNQ N. Bosanquet-Bryant 2 Dupont Close, Clacton-on-Sea CO16 8YD
G6 TNR D. Blackman 115 Ringwood, Bracknell RG12 8XU
G6 TNW I. Webb Cornerways Orchard Road, Eaton Ford, St. Neots PE19 7AN
G6 TOI A. Edgcombe 2 Providence Place, Fore Street, Loddiswell, Kingsbridge TQ7 4QP
G6 TOT A. White Flat A, 106 Palmerston Road, Chatham ME4 5SJ
GW6 TOX B. Taylor Swn-Y-Don, Beaumaris LL58 8RW
G6 TOY D. Williams Hollybank, Royston Road, Taunton TA3 7RE
GJ6 TPD E. Langlois L'Amarrage, La Route Orange, St. Brelade JE3 8GP Jersey
G6 TPE Gravesend And Bean DX Group c/o M. Long 28 Wentworth Drive, Wirral CH63 0JA
G6 TPG J. Littler 39 Wigan Road, Golborne, Warrington WA3 3TZ
G6 TPI C. Bryan 3 Hales Place, Longton, Stoke-on-Trent ST3 4NF
G6 TPO D. Bathe Moel Tryfan 37 Mallows Green, Harlow CM19 5SA
G6 TQC W. George 28 Melbourn Close, Duffield, Belper DE56 4FX
G6 TQF P. Game 15 Nightingale Close, Gosport PO12 3EU
GW6 TQH Winteringham Wireless Society c/o G. Giudice 31 Woodfield Cross, Tredegar NP22 4JG
G6 TQL T. Law 41 Lime Trees, Tonbridge TN12 0SS
G6 TQZ R. Andrews 10 Fourth Avenue, Havant PO9 2QX
G6 TRA D. Andrew The Willows, Nordelph, Downham Market PE38 0BY
G6 TRG Todmorden Ryt G c/o P. Rigg 1 Stones Hey Gate, Widdop Road, Hebden Bridge HX7 7HD
G6 TRM M. Bryant 104 Manor Road, Dover CT17 9JZ
G6 TRN D. Best 64 New Hey Road, Cheadle SK8 2AQ
G6 TRO B. Wilcox 1 Parklands, Stanwick, Wellingborough NN9 6QX
G6 TRQ J. Wright 9 Willow Close, Broadmeadows, Alfreton DE55 3AP
G6 TRW A. Toas 116 Rownhams Road, North Baddesley, Southampton SO52 9EU
G6 TRX J. Sugrue 124 Hall Lane, Upminster RM14 1AL
G6 TRY G. Smillie 5 Fleckers Drive, Up Hatherley, Cheltenham GL51 3BB
G6 TSC V. Simmons 88 Wellcome Avenue, Dartford DA1 5JW
G6 TSE A. Sullivan 20 Crockerne Drive, Pill, Bristol BS20 0LF
G6 TSF P. Shayler 38 Maryside, Slough SL3 7ET
G6 TSJ P. Hannam 7 Bodenham Close, Buckingham MK18 7HR
G6 TSL R. Hill Marclecote, Ledbury Road, Ross-on-Wye HR9 7BE
G6 TSM W. Hirst 8 Moss Road, Alderley Edge SK9 7HZ
G6 TSP Aberdeenshire CG c/o K. Hendry 23 Briscoe Way, Lakenheath, Brandon IP27 9SA
G6 TSX C. Heater 92 Salisbury Road, Farnborough GU14 7AE
G6 TSZ D. Hall 282 Dereham Road, Norwich NR2 3TL
GW6 TTA M. Harris 7 Washington Street Landore, Swansea SA1 2QE
G6 TTD R. Hewitt 11614 Waesche Drive, Mitchellville 20271 United States
G6 TTX W. Kenyon 13 Baskerfield Grove, Woughton On The Green, Milton Keynes MK6 3ES
GW6 TUD M. Prosser 18 Thornhill Way, Rogerstone, Newport NP10 9FT
GM6 TUE P. Mclaren Dalriada, Fogwatt, Moray IV30 8SY
G6 TUG I. Metcalfe 12 Clarence Way, Horley RH6 9GT
G6 TUS R. Page 53 The Brambles, Bar Hill, Cambridge CB23 8SZ
G6 TVA D. Biram 124 Keresforth Hill Road, Red Gables, Barnsley S70 6RG
G6 TVB R. Steele 98 Obelisk Rise, Boughton Green, Northampton NN2 8QU
G6 TVC D. Spooner Thorny How, Canon Pyon, Hereford HR4 8NT
GW6 TVD E. Sims Hafan, Engedi, Holyhead LL65 3RR
G6 TVE J. Tyreman 12 Richmond Close, Rochdale OL16 4RJ
G6 TVI G. Sculthorpe 50 Station Road, Dersingham, King's Lynn PE31 6PR
G6 TVJ I. Bennett 47 Bakers Ground, Stoke Gifford, Bristol BS34 8GD
G6 TVK A. Baker 10 Kingscroft Court, Northampton NN3 9BH
G6 TVP S. Burke 17 The Crescent, Wragby, Market Rasen LN8 5RF
GM6 TVR J. Black Solway View, Carlisle Road, Annan DG12 6QX
G6 TVX M. Collins Coburg Cottage Mount Road, East Cowes PO32 6NT
G6 TW South Cheshire ARS c/o P. Walker 11 Flixton Drive, Crewe. CW2 8AP
G6 TWA A. Woollard 30 John Grinter Way, Wellington TA21 9AR
G6 TWB South Cheshire ARS c/o B. Rigby 76 Woodland Road, Rode Heath, Stoke-on-Trent ST7 3TL
G6 TWD W. West Lectric, Alexandra Road, Crediton EX17 2DH
GD6 TWF C. Wood Deep Water, Glen Rushen Road, Peel IM5 3BA
G6 TWR K. Simmonds 1 Fallow Close, South Molton EX36 3FL
G6 TWX A. Tatterton 28 Kinloch Drive, Bolton BL1 4LZ
G6 TXB B. Thompson 12 Albion Road, Chatham ME5 8SR
G6 TXH R. Webb 90 Queens Road, Tunbridge Wells TN4 9JU

G6 TXP A. Cronk 65 Russell Court, Chatham ME4 5LE
G6 TXQ V. Covell-London 15 St. Nicholas Way, Potter Heigham, Great Yarmouth NR29 5LG
G6 TXV S. Calver 144A Smugglers Club Ground, Bridgemarsh Lane, Chelmsford CM3 6DQ
G6 TXY B. Coulstock 32 Climping Park, Bognor Road, Littlehampton BN17 5DW
G6 TYB J. Cooke 106 Wirral Drive, Winstanley, Wigan WN3 6LD
G6 TYE A. Decamps 11 Love Lane, Spalding DA5 1RJ
G6 TYF R. Duley Denova, Hornsby Lane, Grays RM16 3AU
GM6 TYL G. Davidson Flat 8, 8 Tait Wynd, Edinburgh EH15 2RJ
GW6 TYO B. Young 3 Bryn Road Pontlliw, Swansea SA4 9ED
G6 TYT S. White 8 Bilford Avenue, Worcester WR3 8PJ
GM6 TYX Dr C. Macleod Morven, Marybank, Isle of Lewis HS2 0DD
G6 TZE J. Richards 17 Chaffers Mead, Ashtead KT21 1NA
G6 TZO N. Roberts 11 Mallard Road, Barrow Upon Soar, Loughborough LE12 8BF
G6 TZT T. Rogers 36 Goodacre Road, Ullesthorpe, Lutterworth LE17 5DL
G6 UAJ P. Longstaff 27 Feather Wood, Westlea, Swindon SN5 7AG
G6 UAN M. Moseley Flat 40, Oceana Boulevard, Briton Street, Southampton SO14 3HU
G6 UAP E. Magnuszewski 49 Elvaston Road, Nottingham NG8 1JU
GW6 UAS J. Mcmurray 30 St. Martins Crescent, Llanishen, Cardiff CF14 5QA
G6 UAW M. Egerton 3 Boundary Close, Salisbury SP2 9FZ
G6 UBC J. Evans The Haven, Jubilee Gardens, Corfe Castle, Wareham BH20 5EN
G6 UBH M. Faithfull 99 Bramble Road, Hatfield AL10 9SB
G6 UBK W. Finkle 17 Newlands Ave, Sunderland SR3 1XW
G6 UCI G. Miller 11 Friars Avenue, Great Sankey, Warrington WA5 2AR
GM6 UCN W. Mckenzie 14 Bridgend, Dunblane FK15 9ES
G6 UCO G. Bee 80 Hospital Road, Burntwood WS7 0EQ
G6 UCQ M. Burgess 20 Norfolk Road, Luton LU2 0RE
G6 UCT P. Bowron 52 Eastcotes, Tile Hill, Coventry CV4 9AU
G6 UCW A. Brookes 8 Cedar Close, Ruskington, Sleaford NG34 9FH
GM6 UCX M. Bellerby Hamnavoe, Eabost West, Struan, Isle of Skye IV56 8FL
G6 UCY K. Porter 60 Spitfire Road, Wallington SM6 9GL
G6 UDA R. Pyrah Whispering Waves, The Shore, Poulton-le-Fylde FY6 9EA
G6 UDB F. Scott 15 Sunningdale Close, Kirkby-In-Ashfield, Nottingham NG17 8NW
G6 UDF P. Phelps 14 The Warren, Hazlemere, High Wycombe HP15 7ED
G6 UDG A. Price Brook House, Drury Lane, Shrewsbury SY4 1DT
G6 UDI S. Phillips 79 Selwyn St., Stoke, Stoke on Trent ST4 1ED
G6 UDX B. Oldford 16 Ludlow Drive, Stirchley, Telford TF3 1EG
G6 UED T. Lloyd 18 Coleville Road, Minworth, Sutton Coldfield B76 1XR
G6 UEG I. Norman Bridge Farm Snaffers Lane, Whaplode, Spalding PE12 6RX
G6 UEH E. Naylor 18 Mackenzie Crescent, Cheadle, Stoke-on-Trent ST10 1LU
G6 UEI P. Norman 20 Meadow Close, Budleigh Salterton EX9 6JN
G6 UEQ R. Rix Patterdale, Roe Downs Road, Alton GU34 5LG
G6 UER V. Rice 24 Harewood Close, Tuffley, Gloucester GL4 0SR
G6 UEV RSGB Contest Club c/o M. Piper 26 Hare Law Gardens, Stanley DH9 8DG
GW6 UFH S. Fry 10 Heaseland Place, Killay, Swansea SA2 7EQ
G6 UFI Rvd. P. Fanning 1 Naburn Grove, Moreton, Wirral CH46 0SN
G6 UFL S. Duckett 35 Fowlmere Road, Foxton, Cambridge CB22 6RT
GI6 UFO J. Fitzgerald 15 Bunnahesco Road, Bunnahesco, Enniskillen BT94 5HJ
GI6 UFU J. Campbell 22 Sheridan Drive, Helens Bay, Bangor BT19 1LB
G6 UFV C. Spencer 18 Coatsby Road Kimberley, Nottingham NG16 2TH
G6 UFZ K. Chamba 63 Patricia Avenue, Wolverhampton WV4 5AQ
G6 UGA M. Pinkney 169 Sandringham Road, Perry Barr, Birmingham B42 1PZ
GW6 UGC G. Phillips 83 Heol Y Llwynau, Trebanos, Swansea SA8 4DB
G6 UGE C. Power 296 Alderley, Digmoor, Skelmersdale WN8 9NB
G6 UGG A. Panton 35 Long Water Drive, Gosport PO12 2UP
G6 UGS M. Allison 6 Eden Road, Beverley HU17 7HD
G6 UGT T. Aherne 21 Burbage Place, Alvaston, Derby DE24 8NP
G6 UGW M. Bell 61 Oldbury Orchard, Churchdown, Gloucester GL3 2PU
G6 UGZ A. Scott 23 Wingfield Road, Great Barr, Birmingham B42 2QB
GM6 UHC A. Stewart Skerry Alvha, Torphins, Banchory AB31 4NB
G6 UHD B. Scott Linda Cottage, St. Giles-On-The-Heath, Launceston PL15 9RT
GM6 UHE A. Wilson Lochend, Ayrshire KA15 2LN
G6 UHL B. Ritchie 65 Ransome Avenue, Worcester WR5 3AL
G6 UHS A. Coe 22 St. Annes Way, Spalding PE11 3PN
G6 UIF I. Clark 41 Brook Close, Jarvis Brook, Crowborough TN6 2ET
G6 UIM S. Daniels 46 Freshwater Drive, Paignton TQ4 7SD
G6 UIP G. Dodd 6 Highfield Avenue Kirkby-In-Ashfield, Nottingham NG17 8GF
G6 UIT W. Dillon 49 Goring Way, Greenford UB6 9NN
G6 UJC C. Dukes 32 Greenacres Woolton Hill, Nr. Newbury RG20 9TA
GM6 UJG V. Simpson 43 Fortingall Place, Perth PH1 2NF
G6 UJI B. Staton 99 Linden Avenue, Prestbury, Cheltenham GL52 3DT
G6 UJJ N. Stoker 6 Beech Grove, Gateshead NE9 7RE
G6 UJR M. Severs 125 Hawthorne Way, Shelley, Huddersfield HD8 8QF
G6 UKC M. Brown 33 Stonegate, Cowbit, Spalding PE12 6AH
G6 UKM S. Brown 22 Asquith Close, Biddulph, Stoke-on-Trent ST8 7LN
G6 UKN W. Bailey 35 Elton Lane, Winterley, Sandbach CW11 4TN
GW6 UKO C. Barwell Galaghad, Penisarwaun, Caernarfon LL55 3BN
G6 UKQ J. Riley 56 Church St., Bignall End, Stoke on Trent ST7 8PE
G6 ULD R. Humphrys 10 St. Andrews Road, Bexhill-on-Sea TN40 2BQ
G6 ULJ R. Green Branford House, Valley Road, Tasburgh, Norwich NR15 1NG
G6 ULS P. Kent-Woolsey 32 Yaxham Road, Dereham NR19 1AJ
G6 UMH E. Harding 49 Compass Close, Murdishaw, Runcorn WA7 6DL
G6 UML T. Reader 76 West View Road, Dartford DA1 1TR
G6 UMN L. Gibson 27 Farm Street, Barrow-in-Furness LA14 2RX
G6 UMS C. Hadler Dowles Brook Lodge, Rock Cross Rock, Kidderminster DY14 9SF
G6 UMT A. Handcocks Woodpeckers, Chapel Lane, Southampton SO45 1YX
GW6 UMU A. Haigh Nant Fawr, Corwen LL21 9AA
G6 UMX J. Hibbert 125 Chase Hill Road, Arlesey SG15 6UF
G6 UNA W. Stoneman 5 Creaton Road, Hollowell, Northampton NN6 8RP
GM6 UNQ E. Leask 2/7 Barnton Avenue West, Edinburgh EH4 6EB
G6 UNR R. Rogers 97 Sutherland Avenue, Biggin Hill, Westerham TN16 3HH
G6 UNU A. Lunn 45 St. Anthonys Avenue, Eastbourne BN23 6LN
G6 UOH I. Thacker 3 Webster Way, Gonerby Hill Foot, Grantham NG31 8GH
G6 UOO D. Wilde 3 Canal Cottages, Buxworth, High Peak SK23 7NF
G6 UOX M. Walker 94 Lambert Road, Uttoxeter ST14 7QY
G6 UPA D. Wiseman 22 Queens Crescent, Clapham, Bedford MK41 6DA
G6 UPH M. Hackney Mar Azul 9, Apt.20 2 Fase, Alicante 03710 CALPE Spain
G6 UPI B. Hurrell 33 Meadow Way, Hellesdon, Norwich NR6 5NN
G6 UPL T. Hayhurst The Paddock, Crooklands, Milnthorpe LA7 7NL
G6 UPM A. Hayhurst11 Bank Field, Orton Road, Tebay, Penrith CA10 3TL
G6 UPQ D. Holloway 48 Wenrisc Drive, Minster Lovell, Witney OX29 0RQ
G6 UPR B. Hingston Hazelwood Farm Marldon, Paignton TQ3 1SQ
G6 UQ Stockport Rs c/o B. Naylor 47 Chester Road Poynton, Stockport SK12 1HA
G6 UQA S. Duckles 8 Railway Cottages, Skillings Lane, Brough HU15 1EN
G6 UQI E. Payne 4 Richmond Crescent, Barons Cross, Leominster HR6 8RX
G6 UQO D. Oliver 37 Milford Avenue, Elsecar, Barnsley S74 8DT
G6 UQZ A. Parkhurst 14 Church Street, Clare, Sudbury CO10 8PD
G6 URD J. Ratcliffe 63 Dickens Lane, Poynton, Stockport SK12 1NN
G6 URF A. Hartley 18 Smithy Close, Cronton, Widnes WA8 5BT
G6 URK J. Jennings 354 Williamthorpe Road, North Wingfield, Chesterfield S42 5NS
G6 URM Emf Hams c/o B. Johnson 6 Winston Avenue, Plymouth PL4 6AZ
GM6 URP R. Gray-Jones Flat C 7 Nelson Street, Aberdeen AB24 5EP
G6 URR I. Kirk 12 Edinbane Close, Rise Park, Nottingham NG5 5DU
G6 URT C. Kapoutsis 7A East Lane, Morton, Bourne PE10 0NW
G6 URU J. Keats 50 Ringway Road, Park Street, St. Albans AL2 2RD
G6 USA P. Love 2 Meadway, Dover CT17 0PS
G6 USD M. Matthews 213 Hucclecote Road, Gloucester GL3 3TZ
G6 USG G. Comer 27 Peckforton View, Kidsgrove, Stoke-on-Trent ST7 4TA
G6 USL B. Cowell 46 Gattison Lane, New Rossington, Doncaster DN11 0NQ
G6 USO P. Chamings 52 Crown Street, Redbourn, St. Albans AL3 7PF
G6 USR M. Davis Sunny Bank, Headcorn Road, Maidstone ME17 2AN
G6 UST P. Drury 5 Bede Place, Peterborough PE1 4EE
G6 USU T. Derbyshire 32 Hardie Avenue, Wirral CH46 6BJ
G6 USX W. Dennison 41 Tarbert Walk, Stepney, London E1 0EE
G6 USZ D. Deverell 23 Frankmarsh Park, Barnstaple EX32 7HN
G6 UTB H. Campbell 175 Thrupp Lane, Brimscombe, Stroud GL5 2RG
GW6 UTF Cambridge Consultants ARC c/o D. Foster 11 Dingle Road, Leeswood, Mold CH7 4SN
G6 UTK G. Fisher 16 Somerset Lane, Lansdown, Bath BA1 5SW
G6 UTL S. Foulser 32 Langhorn Road, Southampton SO16 3TN
G6 UTO N. Simpkin 2 Redland Close Beeston, Nottingham NG9 5LA
G6 UTT P. Sheppard Round Corners, 7 First Avenue, Bognor Regis PO22 6ED
GI6 UUC J. Thompson 21 Watch Hill Road, Ballyclare BT39 9QW
G6 UUQ L. Sheward 7 Harlington Avenue, Grove, Wantage OX12 7NQ
G6 UUR S. Whitehead 94 Cranmore Boulevard Shirley, Solihull B90 4RU
GI6 UUT Dr W. Page 4 Glebe Manor, Hillsborough BT26 6NS
G6 UVB C. Wilson Rustleigh, 7 Stagbury Close, Coulsdon CR5 3PH
G6 UVL G. Hall 23 Monarch Close, Chatham ME5 7PD
G6 UVN M. Henman 4 Lyne Walk, Hackleton, Northampton NN7 2BW
G6 UVO C. Heritage 29 Hill Head, Glastonbury BA6 8AW
G6 UVS P. Hannington 21 Little Gate, Westhoughton, Bolton BL5 2SD
G6 UVU J. Handy 77 Abbeyfield Road, Wolverhampton WV10 8TH
G6 UW Cambridge Uws c/o Prof. J. Keeler 67 Perne Avenue, Cambridge CB1 3RY
GM6 UWF J. Allan 87 Needless Road, Perth PH2 0LD
G6 UWI N. Bradshaw 26 Suffolk Gardens, South Shields NE34 7JF
G6 UWK J. Barden 2 Pondhall Cotts. Bradfield Rd, Manningtree CO112SP
G6 UWO D. Bullock 1 Selby Close, Beeston, Nottingham NG9 6HS
G6 UWS M. Byles 108 Kingsway, Wellingborough NN8 2EN
GW6 UWW M. Williams-Davies Plas Penrhos, Llwyngwril LL37 2QB
G6 UWX I. Whiting 15 Highfield Court, Grace Way, Stevenage SG1 5EH
G6 UWY D. Williams 16 Blaydon Walk, Wellingborough NN8 5YU
G6 UX I. Truslove 36 Leicester Road, Hinckley LE10 1LS
G6 UXE R. Wheeler 36 Kimbolton Crescent, Stevenage SG2 8RJ
G6 UXF K. Young 8 Magnolia Close, Worcester WR5 3SJ
G6 UXG A. Webb 35 Hill House Drive, Minster, Ramsgate CT12 4BE
G6 UXK D. Wookey 3 Westland Close, Boscombe Down, Amesbury, Salisbury SP4 7QS
G6 UXM S. Vinnicombe 8A Cross Road, Cholsey, Wallingford OX10 9PE
G6 UXU C. Stanley 494 Blackburn Road, Darwen BB3 0AJ
G6 UXW P. Mundy 25 Lonsdale Avenue, Cosham, Portsmouth PO6 2PU
G6 UXX P. Leese 4 Harefield, Harlow CM20 3EF
G6 UXY A. Lightly 8 Smithville Close, St. Briavels, Lydney GL15 6TN
G6 UYJ A. Page 35 Acorn Close, Christchurch 8023 New Zealand
G6 UYK P. Russell 1 Larch Grove, Kendal LA9 6AU
G6 UYM D. Richards 25-27 Burnivale, Malmesbury SN16 0BL
G6 UYN A. Rumney Church House Farm Cottage, Cheltenham GL51 0TW
G6 UZA A. Kotowicz 47 Portree Drive, Rise Park, Nottingham NG5 5DT
G6 UZG P. Ashby 26 Van Diemens Lane, Bath BA1 5TW
G6 UZJ J. Austen 13 Coverdale, Whitwick, Coalville LE67 5BP
G6 UZL P. Bunn Yew Trees Main Road, Little Haywood, Stafford ST18 0TS
G6 UZM S. Byford 21 Clarke Drive, Shaw, Swindon SN5 5SH
G6 UZO M. Brunsdon 7 Oldberg Gardens, Brighton Hill, Basingstoke RG22 4NP
G6 UZR A. Brown Badgers Way, Holton, Wincanton BA9 8AL
G6 UZY M. Owen 49 Southdale Drive Carlton, Nottingham NG4 1DA
G6 VAA G. Perks 55 Andrew Road, Tipton DY4 0AJ
G6 VAD P. Purdy 4 Hethersett Road, East Carleton, Norwich NR14 8HX
G6 VAE T. White 117A Western Road, Southall UB2 5HN
G6 VAL A. Oughton 16 Thompson Way, Hertford SG13 8FX
G6 VAR Obo March & Dist ARS, British Legion Club, Rookswood Road c/o J. Smith 6 Hollams Road, Tewkesbury GL20 5DG
G6 VAW C. Soars 118 Braddon Road, Loughborough LE11 5YZ
G6 VAX R. Saunders 93 Oaks Avenue, Worcester Park KT4 8XG
G6 VAZ D. Thomas 25 Lime Close, Mildenhall, Bury St. Edmunds IP28 7PR
G6 VBA D. Townend 25 De Trafford St., Huddersfield HD4 5DR
G6 VBD J. Savage 2 Alvecote Cottages, Alvecote Lane, Tamworth B79 0DJ
G6 VBE R. Ransom 1 Bilberry Road, Clifton, Shefford SG17 5HB
G6 VBJ P. Tasker Oaktree Cottage, Bunkers Hill, Ridley, Sevenoaks TN15 7EY
G6 VBK D. Hatton 24 Langdale Road, Leyland PR25 3AR
GW6 VBN M. Hunt 23 Swansea Road, Pontardawe, Swansea SA8 4AL
G6 VBQ A. Haddock 1 Heron Way, St. Ives PE27 6SS
GW6 VBR L. Cowley 3 Pleasant Villas, Pontarddulais, Swansea SA4 8QF
G6 VCF P. Brackstone 3 Wentworth Close, Beverley HU17 8XB
GI6 VCG J. Brownlees 8 Cairnbeg Park, Larne BT40 1UB
GI6 VCL K. Cunningham 4 Garvaghy Road, Portglenone, Ballymena BT44 8EF

G6 VCR H. Eden 142 Ringway, Thornton-Cleveleys FY5 2NW
G6 VDA A. Sutton 3 Cornflower Close, Willand, Cullompton EX15 2TT
G6 VDD P. Waddington 20 Littlington Court Surrey Road, Seaford BN25 2NZ
G6 VDK P. Lutas 616 Queens Drive, Swindon SN3 1AZ
G6 VDW R. Olliver 39 Nutshalling Avenue, Rownhams, Southampton SO16 8AY
G6 VDX I. Ogilvie 8 Devonshire Road, Prenton CH43 4UL
G6 VDY H. Jeffery-Wright 55 Burland Avenue, Wolverhampton WV6 9JJ
GW6 VED R. Straughan 1 Crossroads, Gilwern, Abergavenny NP7 0DX
G6 VEG T. Gray 8 Holystone Grange, Holystone, Newcastle upon Tyne NE27 0UX
GW6 VEI D. Pierce Coed Duon, Tremeirchion, St. Asaph LL17 0UH
G6 VEJ F. Stone 51 The Glen, Yate, Bristol BS37 5PJ
GW6 VEN A. Rose 4 Llys Clwyd, Kinmel Bay, Rhyl LL18 5EW
GW6 VET J. Goodson 22 Pant Gwyn, Bridgend CF31 5BA
G6 VEY I. Haver 4 Campion Way, Bourne PE10 0QE
G6 VEZ G. Helm 31 Faringdon Avenue, South Shore, Helmsman Electronics Ltd, Blackpool FY4 3QQ
G6 VF S. Illman 66 Frieth Road, Marlow SL7 2QU
G6 VFA M. Hine Tall Trees, Lime Lane, Derby DE21 4RF
G6 VFB W. Hogan 279 Halliwell Road, Bolton BL1 3PE
G6 VFC D. Hooton 80 Portland Road, Rushden NN10 0DJ
G6 VFF S. Jackson 47 Gurnard Pines, Cockleton Lane, Cowes PO31 8RF
GW6 VFH R. Jenkins 29 Pemberton St., Llanelli SA15 2RB
G6 VFI A. Jones The Studio, Fullers Vale Headley Down, Bordon GU35 8NR
G6 VFO J. Stokes 109 Hollyhedge Road, West Bromwich B71 3BT
G6 VGA C. Mccall Flat 6, Kent House, Park Cottages, Hawkhurst, Cranbrook TN18 4JH
G6 VGC Int. Morse Preservation Society c/o R. Woolley 82 Pennycroft Road, Uttoxeter ST14 7ET
G6 VGG Bromsgrove + District ARC c/o A. Kelly 40 Housman Park, Bromsgrove B60 1AZ
G6 VGH I. Allen 14 Bettridge Place, Wellesbourne, Warwick CV35 9LY
G6 VGO M. Barrett 1 Walterstead Cottage, Ladykirk, Berwick upon Tweed TD15 1XW
G6 VGS M. Bradbury 55 Crowthorp Road, Northampton NN3 5EY
G6 VGT D. Bowlas 38 Senneleys Park Road, Northfield, Birmingham B31 1AL
G6 VGV I. Craig 1 Whitton Drive, Chester CH2 1HF
G6 VGZ D. Cheriton 5 Cornwall Close, Warwick CV34 5HX
GM6 VHA M. Deverill Flannan House, Aird Uig Timsgarry, Isle of Lewis HS2 9JA
G6 VHE Dr M. Entwistle 34 Webbs Court, Lyneham, Chippenham SN15 4TR
G6 VHG A. Foster 35 Gloucester Place, Peterlee SR8 2HB
GW6 VIC V. Jones Gwel Y Mor, Porth Y Felin Road, Holyhead LL65 1BG
G6 VIF B. Morris 21 Loxley Gardens, Southdown, Bath BA2 1HS
G6 VIK I. King 11 Cockhall Close, Litlington, Royston SG8 0RB
G6 VIN J. Walker 44 Albany Road, Kilnhurst, Mexborough S64 5UG
G6 VIQ M. Watson Salt Pie Farm, Birdsedge, Huddersfield HD8 8XP
GM6 VIU A. Wilson 1 High Street, Dysart KY12TS
G6 VIY A. Wood 12 Bishops Meadow, Sutton Coldfield B75 5PQ
G6 VJA I. Taylor 97 George St., Cleethorpes DN35 8PL
G6 VJC J. Taylor 17 Aintree Way Milking Bank, Dudley DY1 2SL
G6 VJK A. Maslin 2 Clarks Cottages, White Horse Road, Colchester CO7 6TX
G6 VJM D. Lynch 30 Whitecroft View, Baxenden, Accrington BB5 2QP
G6 VJP D. Pilkington 45 High Meadows, Midsomer Norton, Radstock BA3 2RZ
G6 VJR D. Reading 23 Elwy Circle, Ash Green, Coventry CV7 9AU
G6 VKA C. Thompson Fourwinds, Walton Hill, Gloucester GL19 4BT
GW6 VKI C. Richardson Moorcroft, Kinnerley, Oswestry SY10 8DW
G6 VKL D. Mayers 15 Oakfield Road, Poynton, Stockport SK12 1AR
G6 VKP J. Littlewood 5 Laburnum Grove, Harrogate HG1 4EH
G6 VKS Dr I. Morgan Leigh House, 64 Widney Road, Solihull B93 9AW
G6 VKX M. Webber 23 Ramsey Close, Horley RH6 8RE
GW6 VKY A. White 86 Derlwyn, Dunvant, Swansea SA2 7QE
G6 VLC S. Paxton 11 Synderford Close, Didcot OX11 7UT
G6 VLL J. White 16 Lyon Walk, Newton Aycliffe DL5 5LZ
G6 VLT J. Higgins 190 Little Glen Road, Glen Parva, Leicester LE2 9TT
G6 VLV D. Colman 22 Peerley Close, East Wittering, Chichester PO20 8PB
GI6 VLY Dr J. Earle 25 Carnesure Park, Comber, Newtownards BT23 5LT
G6 VMB C. Gibson 103 Lydalls Road, Didcot OX11 7DT
G6 VMF R. Hope 30 Greendale Gardens, Hetton-Le-Hole, Houghton le Spring DH5 0EF
G6 VMI D. Milne 22 Eastnor Road, Reigate RH2 8NE
G6 VMR M. Adams 122 Green Lane Castle Bromwich, Birmingham B36 0BX
G6 VMV S. Brocklehurst Bank View, Reades Lane, Congleton CW12 3LL
G6 VNC R. Davies 27 Smiths Way, Water Orton, Birmingham B46 1TW
G6 VNI G. Duggan 28 Higher Rads End, Eversholt, Milton Keynes MK17 9ED
G6 VNO N. Hanson 100 Bassett Green Road, Southampton SO16 3EF
G6 VNW B. Major 3 Tithebarn Grove Wavertree, Liverpool L15 6TG
G6 VOE D. Simpkins 34 Rose Avenue, Weldon, Corby NN17 3HB
G6 VOV R. Leavold 8 Wilkinson Way, North Walsham NR28 9BB
G6 VPH R. Gorton 3 Pickford Avenue, Little Lever, Bolton BL3 1PN
G6 VPJ G. Hall 54 Townfields, Sandbach CW11 4PQ
G6 VPL J. Hopkinson 4 Marwood Croft, Streetly, Sutton Coldfield B74 3JU
G6 VPN B. Jameson 42 Eastgate, Fleet, Spalding PE12 8NA
G6 VPU S. Mulligan 406 St. Helens Road, Leigh WN7 3PQ
G6 VPV J. Wake 15 Deepdale Way, Darlington DL1 2TA
G6 VPW R. Stoate 19 Jean Road, Brislington Bs4 4Jt, Bristol BS4 4JT
G6 VQC A. Read Huenibachstrasse 75, Huenibach CH-3626 Switzerland
G6 VQN A. Morris 67 Broad Oak Way, Cheltenham GL51 3LL
G6 VQV S. Shenfield 3 Blackberry Grove, Bradwell-On-Sea, Southminster CM0 7QE
G6 VQW R. Seaton Wisteria Cottage, Welsh Road, Leamington Spa CV33 9AQ
GM6 VRC J. Brown Inchbeag Cottage, Inchcoonans, Perth PH2 7RB
G6 VRF B. Crowther 10 Askrigg Close, Marton Moss, Blackpool FY4 5RE
G6 VRI G. Eccleshare 22 Barley Close, Herne Bay CT6 7XG
GW6 VRN M. Jones 66 Brondeg, Heolgerrig, Merthyr Tydfil CF48 1TP
GM6 VRU G. Giles Monachan, Culrain IV24 3DW
G6 VSE A. Bansal Fernley, 2 Seaview Cotts, Chideock DT6 6JE
G6 VSM J. Morris 10 Batt Hall Kitchen Hill, Bulmer, Sudbury CO10 7EZ
G6 VSQ F. Whitehurst Roselands, Clarke Lane, Macclesfield SK10 5AH
G6 VSY C. Sheppard 42 Freeman Road, Didcot OX11 7DD
G6 VTA M. Fisher 46 Hedgerow Walk, Andover SP11 6FD
G6 VTE S. Chambers 1 Tatling Grove, Walnut Tree, Milton Keynes MK7 7EG
G6 VTH R. Carney 29 Hayton Close, Sunderland SR5 2BU
G6 VTN P. Green 79 The Spinney, Bar Hill, Cambridge CB23 8SU
G6 VTX M. Brindley 53B Seabridge Road, Newcastle ST5 2HU
GW6 VTZ D. Campbell 61 Maes Y Crofft, Morganstown, Melbourne CF15 8FE
G6 VUE S. Butler 231 Newman Road, Wincobank, Sheffield S9 1LU
G6 VUF F. Caulfield Kerney 47 Freemans Close, Stoke Poges, Stoke Poges SL2 4ER
G6 VUG R. Collins 37 Warwick Road, Twickenham TW2 6SW
G6 VUJ J. Davis 69 Bryanston Road, Solihull B91 1BS
G6 VUN M. Hodson 17 Marshfield Close, Redditch B98 8RW
G6 VUX S. Challoner Grosvenor Farm, Holme Street, Chester CH3 8EQ
G6 VVE Nunsfield House - ARG c/o S. Banks 29 Froxmere Close, Crowle, Worcester WR7 4AP
GM6 VVG G. Caldwell 10 Craigmath, Dalbeattie DG5 4EB
G6 VVL K. Hotchen 6 Nourse Close, Leckhampton, Cheltenham GL53 0NQ
G6 VVS R. Jackson 46 Ashford Road, Maidstone ME14 5BH
G6 VVU Dr S. Adkins 8 Kensey Valley Meadow, Launceston PL15 9NB
G6 VVZ T. Butler 103 Spring Gardens, Anlaby Common, Hull HU4 7QH
G6 VWF J. Holbrook 1 Segrave Grove, Hull HU5 5DJ
G6 VWI C. Kowcun 27 Mill Crescent, Kingsbury, Tamworth B78 2LX
GI6 VWS J. Quigg 9 Springhill Terrace, Limavady BT49 9BS
G6 VWV S. Cresswell 7 Japonica Drive, Nottingham NG6 8PU
G6 VXC R. Callaghan 5A Chapel Garth, West Ayton, Scarborough YO13 9HH
G6 VXL T. Buck 178 Rover Drive, Castle Bromwich, Birmingham B36 9LL
G6 VXN A. Hart 10 Walsgrave Close, Solihull B92 9PQ
G6 VXR H. Metcalf Beech Lee, Vicarage Lane, Alresford SO24 0DU
G6 VXZ A. Sorab Woodgaston Cottage, Woodgaston Lane, Hayling Island PO11 0RL
G6 VYK E. Williams 2 Ennerdale Court, Bridlington YO16 6HL
GM6 VYY A. Mcminn Siarardh, Mallaig PH41 4QY
GM6 VYZ W. Mcminn Glengyle, East Bay, Mallaig PH41 4QF
GW6 VZB M. Lennox 17 Coed Y Fron, Holywell CH8 7UJ
G6 VZF A. Dawes 1A Lower Olland Street, Bungay NR35 1BY
G6 VZG M. Frosdick 48 Woodfield, Briston, Melton Constable NR24 2JY
G6 VZM J. Johnson 62 Julien Road, Ealing, London W5 4XA
G6 VZS D. Goodall 94 Camp Mount, Pontefract WF8 4BX
G6 VZU C. Hunt 23 Beccles Road, Gorleston, Great Yarmouth NR31 0PW
G6 VZZ J. Hackett 18 Brow Edge, Rossendale BB4 7TT
GW6 WAG D. Jones Bradford House, The Square, Corwen LL21 0DL
G6 WAM P. Woodyard Sunny Nook Chapel Lane Scaleby Hill, Carlisle CA6 4LY
G6 WAN R. Rayner 12 Weedon Way, King's Lynn PE30 4YY
G6 WAO N. Austin 8 Chandler'S Court, Norwich NR4 6EY
G6 WAR Mid Warwickshire ARS c/o Q. Wright 9 Browning Avenue, Warwick CV34 6JQ
G6 WAS D. Carpenter 2 Milton Road, Little Irchester, Wellingborough NN8 2DY
G6 WAU Dr S. Connor 1 Tallis Walk, Grange Park, Swindon SN5 6BQ
G6 WAY J. Randall 3 Steins Lane, Humberstone, Leicester LE5 1ED
GM6 WAZ A. Ronnie 7 Beechwood Avenue, Stranraer DG9 0AU
G6 WBG P. Smith Juniper Cottage, Palestine SP11 7ER
G6 WBS S. Siggins 5 Arrow Lane Halton, Lancaster LA26QW
G6 WBT I. Thorp Pinelodge, Carleton Green, Pontefract WF8 3NJ
G6 WBX Strood Kent CG c/o P. Yorke 27 Luard Court, Havant PO9 2TN
G6 WCI M. Richards 72 Carlton Avenue, Westcliff-on-Sea SS0 0QL
G6 WCW M. Smith 234 Big Meadow Road, Wirral CH49 9AW
G6 WCX D. Mardle 22 Wayfield Link, Avery Hill, London SE9 2LP
G6 WDC L. Baldwin 26A Cheney Hill, Heacham, King's Lynn PE31 7BS
G6 WDH A. Hutchison 28 Allee De Quiberon, Colomiers 31770 France
GJ6 WDK M. Monteil Kalimera, Six Rues Villas 1 La Rue, St. Lawrence, Jersey JE3 1GL
G6 WDM J. Million 11 Derwent Mews, Blackhill, Consett DH8 8TU
G6 WDR A. Tett 19 Park Road, Shoreham-by-Sea BN43 6PF
G6 WDS Dr F. Deravi 22 Clifton Gardens, Canterbury CT2 8DR
G6 WEH R. Burrows 6 Frensham Drive, Hitchin SG4 0QP
GM6 WEI G. Cockcroft The Old Schoolhouse, Scatwell, Strathconon, Muir of Ord IV6 7QG
G6 WEL J. Reynolds 4 Rosewood Drive, Winsford CW7 2UW
GW6 WEU K. Turner 115 Newton Road, Newton, Swansea SA3 4SW
G6 WEW J. Fitzsimons 63 School Lane, Chapel House, Skelmersdale WN8 8EN
G6 WFF G. Solkow 12A Manor Court, Penkhull, Staffs ST4 5DW
G6 WFM K. Farr 3 Sheppard Drive, Chelmsford CM2 6QE
GM6 WFP S. Pollok School House Watten, Wick KW1 5YJ
G6 WFS G. Quantrill 47 Lambeth Road, Leigh-on-Sea SS9 5XR
GW6 WFW A. Humphreys 45 Cwm Place, Llandudno LL30 1LP
GI6 WFX R. Johnston 51 Kennedy Drive, Lisburn BT27 4JA
G6 WGA A. Swift 56 Birch Hall Avenue, Darwen BB3 0JB
G6 WGE N. Riding 15 Church Lane, Dewsbury Moor, Dewsbury WF13 4EN
G6 WGM M. Reilly Flat 5, 57 Cheriton Road, Folkestone CT20 1DF
G6 WGY R. Clague 11 Trebor Avenue, Bryntirion Park, Flintshire CH66DP
G6 WGZ Boothferry ARS c/o D. Collier 133 Woodstock Road, Moston, Manchester M40 0DG
G6 WHH S. Martin 19 Old Manor Rd, Rustington BN163QU
G6 WHS N. Read 296 Westdale Lane, Mapperley, Nottingham NG3 6EU
G6 WHT K. Willard 5 Waltham Way, Frinton-on-Sea CO13 9JE
G6 WHY K. Daniels Greenacre, 71 Little Yeldham Road, Halstead CO9 4LN
GI6 WHZ R. Freeburn 6 Killycurragh Road, Cookstown BT80 9LB
G6 WIG G. Crowton 64 Atlantic Road, Birmingham B44 8LQ
G6 WIL G. Wilden 39 The Laurels, Morris Avenue, Jaywick CO15 2JN
G6 WIO B. Mchugh 63 Three Butt Lane, Liverpool L12 7HE
G6 WIT G. Anderton 12 Oaklands Close, Halvergate, Norwich NR13 3PP
G6 WIW U. Harding 11 Hare Crescent, Watford WD25 7EE
G6 WJD J. Dobson 13 Elgin Close, Bedlington NE22 5HJ
G6 WJJ A. Kendal 3 Benbeck Grove, Tipton DY4 8AJ
G6 WJW H. Hutton Cassiobury, The Street, Diss IP22 2PS
G6 WJX E. Jackson Melford, 36 Ickleton Road, Cambridge CB22 4RT
G6 WKI R. Lewis 42 Launceston Close, Romford RM3 8HQ
G6 WKN C. Reed 14 Fletcher Drive, Wickford SS12 9FA
G6 WKO J. Richards 8 Westminster Crescent, Burn Bridge, Harrogate HG3 1LY
G6 WKQ P. Rowe 131 Cambridge Road, Great Shelford, Cambridge CB22 5JJ
GW6 WKU W. Walker 18 Parc Sychnant, Conwy LL32 8SB
G6 WKZ M. Jaques 3 The Rowans, Baldock SG7 6HJ
G6 WLE R. Bailey The Malt House, Great Shefford, Hungerford RG17 7ED
GM6 WLJ D. Milne 30 Bruceland Road, Elgin IV30 1SF
G6 WLM S. Simmonds 3 Robert Cramb Avenue, Tile Hill, Coventry CV4 9LA

G6 WLP G. Smith High Croft, 91 Rannerdale Drive, Whitehaven CA28 6JZ
G6 WLQ M. Smith 10 Riffams Court, Riffams Drive, Basildon SS13 1BQ
G6 WLX A. Davey Highdale, 82 Silver Street, Nailsea BS48 2DS
GM6 WMA D. Elam Achnacree, 38 Hunter Avenue, Loanhead EH20 9SN
G6 WME B. Gray 21 Litester Close, North Walsham NR28 9JA
G6 WMG D. Hastings Westering, Norwich NR13 6RQ
G6 WML J. Barrasford 34 Barnard Avenue, Ludworth, Durham DH6 1LS
G6 WMR Leicester RS c/o J. Barnett 11 Ridge Street, Stourbridge DY8 4QF
G6 WMT B. Roper 3 Whites Close, St. Agnes TR5 0TU
G6 WMU D. Pearce 247 Wigston Lane, Aylestone, Leicester LE2 8DJ
GJ6 WMZ M. L'Amy Tamarind, Le Mont De St. Anastase, St. Peter JE3 7ES Jersey
G6 WNB G. Bennett 6 Danescroft, Bridlington YO16 7PZ
G6 WNG J. Haines The Westlands, Wilcott, Shrewsbury SY4 1BJ
G6 WNN M. Farrimond Dragonfly Barn, Hall Farm, Stokesby, Great Yarmouth NR29 3EP
GM6 WNX D. Mitchell 65 Robb Place, Castle Douglas DG7 1LW
GW6 WOB H. Stevens Parc Y Dilfa, Talley, Llandeilo SA19 7YT
GM6 WOF A. Firth Edan, Berstane Road, Kirkwall KW15 1NA
G6 WOI G. Flint 782 College Road, Birmingham B44 0AL
G6 WOT D. Fishlock 93 Shackstead Lane, Godalming GU7 1RL
G6 WPD V. Demicoli 37 Elm Rd, Birmingham B30 2AX
G6 WPE S. Mason 46 Frankton Close, Redditch B98 0HJ
G6 WPJ M. Phillips Woodside, Bures CO8 5BN
G6 WPK J. Puttock 8 Millfield, St. Margarets-At-Cliffe, Dover CT15 6JL
G6 WPL S. Lawson 33 Country Meadows, Market Drayton TF9 3LP
G6 WPO A. Brislin Greengage, Plough Road, Droitwich WR9 7NL
G6 WPR D. Fleetwood The Watch House, Cadgwith, Ruan Minor, Helston TR12 7JX
G6 WQH J. Wilson 127 James Reckitt Avenue, Hull HU8 7TJ
GW6 WQJ A. Tidswell 9 Dewi Avenue, Holywell CH8 7UG
G6 WQN W. Convery 20 Grove Road, Hethersett, Norwich NR9 3JP
G6 WRC Warrington ARC c/o J. Lang 7 Marion Grove, Liverpool L18 7HY
GM6 WRY G. Smith 41 Glebe Place, Galashiels TD1 3JW
G6 WSF M. Strickland 25 Coniston Drive, Aylesham, Canterbury CT3 3HZ
G6 WSN D. Westgate 72 Bosworth Street, Leicester LE3 5RA
G6 WSX W. Carter 49 The Oval, Holmfirth HD9 3ET
G6 WSZ J. O'Hara 4 Lower Mill Close, Goldthorpe, Rotherham S63 9BY
G6 WTD R. Kenward The Bungalow, 20 Church Road, Coventry CV8 3ET
GW6 WTK B. Wiegold 8 Nant Ddu, Caerphilly CF83 3BU
G6 WTM M. Higlett 3 Clover Way, Killinghall, Harrogate HG3 2WE
GM6 WTT D. Anderson 11 Longside Road, Mintlaw, Peterhead AB42 5EJ
G6 WUD R. Green 10 Torwood Court, Cramlington NE23 2BZ
G6 WUR T. Price 54 Medeway, Lake, Sandown PO36 9HQ
GW6 WVD J. Williams 48 Belvedere Drive, Plas Coch, Wrexham LL11 2BG
G6 WVL J. Parr 114 Ashton Road, Golborne, Warrington WA3 3UX
G6 WVM D. Harrison 22 Oswin Grove, Coventry CV2 5GJ
G6 WVO Dr C. Hunt Greenfield House, Heapham, Gainsborough DN21 5PT
G6 WVR S. Worner 1 Tynedale, Hull HU7 6EL
G6 WVS P. Child 36 Crosslands, Caddington, Luton LU1 4ER
G6 WVV A. Ferencz 3 Batley Avenue, Hawthorndene 5051 Australia
G6 WWA T. Banham 28 Norwood Avenue, High Lane, Stockport SK6 8BJ
G6 WWR Three Counties ARC c/o D. Kamm Delabole Head, Week St. Mary, Holsworthy EX22 6UU
G6 WWS B. Smith 17 Thornley Road, Wirral CH46 6HB
G6 WWV P. Mann 9 Holcombe Road, Blackpool FY2 0SR
G6 WWY G. Miller Silvermine, Cooks Lane, Axminster EX13 5SQ
G6 WXI G. Ball Ciss Green Farm, Watery Lane, Congleton CW12 4RS
G6 WXJ L. Bagnall 15 Ypres Road, Allestree, Derby DE22 2NA
G6 WXK I. Buckie 156 Greenfield Crescent, Horndean, Waterlooville PO8 9EW
G6 WXN C. Bennett 12 Sherwood Road, Winnersh, Wokingham RG41 5NJ
G6 WXS R. Archer 40 Caroline Street Preston, Preston PR1 5UY
G6 WXZ A. Collier 2 Viceroy Court Gordon Road, Horndon-On-The-Hill, Stanford-le-Hope SS17 8NL
G6 WYD S. Chambers 52 Chapel Lane, Spondon, Derby DE21 7JW
G6 WYE A. Clack 4 Chestnut Grove, Withernsea HU19 2PH
G6 WYF R. Cook Arnel Ltd, Arnel House, 1 Peerglow Centre, Ware SG12 9QL
G6 WYH N. Daniels 2 Homelye Lane, Dunmow CM6 3AW
G6 WYL J. Williamson 5 Frensham Close, Stanway, Colchester CO3 0HP
G6 WYQ S. Quade 60 Carlton Mews, Birmingham B36 0AD
G6 WYS W. Patching 7 Bursledon Road, Hedge End, Southampton SO30 0BP
G6 WZA D. Wickens Auchensail, 3 Bews Lane, Chard TA20 1JU
G6 WZC D. Slatter 5 Opendale Road, Burnham, , Slough SL1 7LY
G6 WZD C. Sillence 104 Coleford Bridge Road, Mytchett, Camberley GU16 6DT
G6 WZE P. Robinson 108 Station Road, Mickleover, Derby DE3 9FP
G6 WZL B. Walker 81 Stacey Avenue, Wolverton, Milton Keynes MK12 5DN
G6 WZM H. Collinson 28 Tadcaster Avenue, Leicester LE2 9GA
G6 WZN M. Hodges 2 Coral Avenue, Westward Ho, Bideford EX39 1UW
G6 WZP D. Rogers No. 10 The Saltings, Seaton EX12 2XW
G6 WZY A. Gerrard 51 Sheringham Drive, Crewe CW1 3XJ
G6 WZZ B. Gibson 55 Ledward Street, Winsford CW7 3EN
G6 XAG A. Higgs Neatsfold, Hilton, Blandford Forum DT11 0DQ
G6 XAK C. Harding 15 The Stampers, Tovil, Maidstone ME15 6FF
G6 XAN S. Harding 29 Wey Barton, Byfleet, West Byfleet KT14 7EF
G6 XAR L. Hall 170 Macers Lane, Wormley, Broxbourne EN10 6EE
G6 XAT D. Armstrong 69 Station Crescent, Rayleigh SS6 8AR
G6 XAV G. Lawrence 5 Longwood View, Furnace Green, Crawley RH10 6PB
G6 XAW D. Lawrence 2205 Southwest 44Th Terrace Cape Coral, Florida, 33904 33904
G6 XBD Buxton ARS c/o S. Meakin 25 Derby Road, London E18 2PZ
G6 XBG J. Lines 6 Hawthorn Road, Denmead, Waterlooville PO7 6LJ
G6 XBS J. Newman 21 Stains Close, Cheshunt, Waltham Cross EN8 9JJ
GW6 XBV K. Simpson 6 New Market Street, Usk NP15 1AT
G6 XCC J. Sayer 19 Arras Boulevard, Hampton Magna, Warwick CV35 8TY
G6 XCD E. Ashworth 232 Clifton Road, Darlington DL1 5EA
G6 XCK S. Bishop 1 Walsh Close, Hitchin SG5 2HP
G6 XCO R. Piper 3 The Haven, Langley Park, Durham DH7 9UW
G6 XCU R. Willis 10 Nayling Road, Braintree CM7 2RZ
G6 XD J. Taylor 14 Woodway Close, Teignmouth TQ14 8QG
G6 XDB J. Woodnutt 17 Hill Farm Road, Chalfont St. Peter, Gerrards Cross SL9 0DD
G6 XDI C. Packman 4 Angel Lane, Hayes UB3 2QX
G6 XDK V. Oag Parkside, Stratton Park, Biggleswade SG18 8QS
G6 XDY K. Gibson-Ford 123 Hawthorn Crescent, Cosham, Portsmouth PO6 2TJ
G6 XDZ N. Glover 21A Jason Close, Bridlington YO16 6JA
G6 XEB D. Green 47 Siston Common, Bristol BS15 4PA
G6 XEF P. Hammond 31 Honey Way, Royston SG8 7ES
G6 XEL D. Hawkins Travellers Lodge, Bere Road, Wareham BH20 7PA
G6 XEN R. Hill 114 Moorside Crescent, Sinfin, Derby DE24 9PT
G6 XEX A. Croft Exchange Buildings, Exchange Street, Normanton WF6 2AA
G6 XFB B. Roe 11 Abbotts Way, Louth LN11 8BS
G6 XFR F. Fielder 103 Acworth Court, Acworth Crescent, Luton LU4 9JE
G6 XFU A. Edge 1 Newquay Drive, Macclesfield SK10 3NQ
GW6 XGA Barry ARS c/o D. Collins 12 Penybedd, Pembrey, Burry Port SA16 0HJ
G6 XGF D. Cadman 32 Breedon Hill Road, Derby DE23 6TG
G6 XGJ J. Davis 47 Meadow View, Holmewood, Chesterfield S42 5UL
G6 XGK M. Drinkall 11 Rossefield Gardens, Bramley, Leeds LS13 3RQ
G6 XGT M. Thornsby 2 Shelley Way, Bacton, Stowmarket IP14 4TP
G6 XGV M. Valenti 545 Gander Green Lane, North Cheam, Sutton SM3 9RF
G6 XHF S. Richards 58 Holm Lane, Oxton, Prenton CH43 2HS
GD6 XHG E. Rixon 65 Friary Park Road Ballabeg, Castletown, Isle of Man IM9 4EP
G6 XHI K. Ridgwell 23 Peter Bruff Avenue, Clacton-on-Sea CO16 8UA
G6 XHJ P. Raxworthy 32 St. Marys Avenue, Alverstoke, Gosport PO12 2HX
G6 XHK I. Roper 109 Birstall Pk Ct., Birstall WF17 9DL
G6 XID S. Mann 2 Fowey Close, Nailsea, Bristol BS48 2UR
G6 XIF C. Milton 31 Morley Road, Tiptree, Colchester CO5 0AA
G6 XII J. Miller The Oast House, Houghton Green Lane, Rye TN31 7PJ
G6 XIR M. Bennett Ravenswood, The Shires, Southampton SO3 4BA
G6 XIW D. Bye 2 Valentine Mansions, London N21 1BA
G6 XJB D. Wratten 42 North Road, Petersfield GU32 2AX
G6 XJC L. Whitehead Flat 2, Masons Court, Clacton-on-Sea CO15 3SE
G6 XJD D. Whysall Christ Church Vicarage, 587 Nuthall Road, Nottingham NG8 6AD
G6 XJE Rvd. J. Whysall Christ Church Vicarage, 587 Nuthall Road, Nottingham NG8 6AD
G6 XJF A. Webb 255 Bambury Street, Stoke-on-Trent ST3 5QY
G6 XJI D. Wiblin 98 Pemberton Road, Slough SL2 2JY
G6 XJJ S. Mckay 11 Brough Meadows, Catterick, Richmond DL10 7LQ
G6 XJN G. Valenti 31 Stratton Court, Bognor Regis PO22 8DP
G6 XJT D. Ramsden 76 Brigg Lane, Camblesforth, Selby YO8 8HD
G6 XJZ D. Rowe 5 Kelburn Close, Chandler'S Ford, Chandlers's Ford SO53 2PU
G6 XKE H. Papworth 339 Gayfield Avenue, Brierley Hill DY5 3JE
G6 XKF A. Parfitt 242 Hook Road, Chessington KT9 1PL
G6 XKJ I. Pinkard 10 Westminster Green, Handbridge, Chester CH4 7LE
G6 XKK Borden Gram Sch c/o H. Parrott 3 Fox Gardens, Lymm WA13 9EY
G6 XKO R. Mclellan 74 Mount Ambrose, Redruth TR15 1QR
G6 XKV D. Bodenham High Dale, Besbury, Minchinhampton, Stroud GL6 9EP
G6 XKX R. Newell 57 Evendene Road, Evesham WR11 2QA
G6 XKY G. Ogden 10 Hartington Drive, Standish, Wigan WN6 0UA
G6 XLB R. Morris 10 Danetre Drive, Daventry NN11 4GY
G6 XLC J. Mills 6 Borrowdale Road, Halfway, Sheffield S20 4HL
G6 XLG P. Pulley 7 St. Peters Close, Pirton, Worcester WR8 9EH
G6 XLR S. Nightingale Cefn Ydfa, Bartwood Lane, Ross-on-Wye HR9 5TA
G6 XMA S. Butler 45 Roewood Close, Holbury, Southampton SO45 2JT
G6 XMB T. Betts 3 Burns Avenue, Mansfield Woodhouse, Mansfield NG19 9JR
G6 XML W. Barnes 49 Sunningdale Road, Haydon Wick, Swindon SN25 3AZ
G6 XMM T. Bugg Gravel Hill, Nayland, Colchester CO6 4BJ
G6 XMT M. Samson 115A Far Gosford Street, Coventry CV1 5EA
G6 XMU G. Smith 71 Mount Pleasant Road, Wisbech PE13 3NQ
G6 XN Wey Valley ARG c/o A. Vine Hilden, Woodland Avenue, Cranleigh GU6 7HZ
G6 XND P. Smith 6 Nuthatch, Longfield DA3 7NS
G6 XNI A. Taylor 20 Mythop Road, Marton, Blackpool FY4 4UZ
G6 XNJ J. Taylor 21 Crestfield Crescent, Elland HX5 0LS
G6 XNK J. Theedom 5 Rodbridge Drive, Southend-on-Sea SS1 3DF
G6 XNN E. Townsend 10 Little Oak Avenue, Kirkby-In-Ashfield, Nottingham NG17 9BG
G6 XNP A. Trett 236 Avondale, Ash Vale, Aldershot GU12 5NQ
G6 XNQ R. Taylor 53 Hutton Park, Hutton Moor Lane, Weston-Super-Mare BS24 8RZ
G6 XNU V. Williams 24 Sunny Bank Avenue, Blackpool FY2 9EQ
G6 XOD E. Whitby 37 Regeneration Way, Beeston, Nottingham NG9 1NJ
G6 XOE F. Whitby 37 Regeneration Way, Beeston, Nottingham NG9 1NJ
G6 XOG C. Wells Troutbeck, Arthington Lane, Pool In Wharfedale, Otley LS21 1JZ
G6 XOR D. Winfield 1 Underhill Close, Derby DE23 1RH
G6 XOU H. Yeldham 19 Wade Reach, Walton on the Naze CO14 8RG
G6 XOX A. Patrick 22 Falcon Way, Dinnington, Sheffield S25 2NY
G6 XPB R. Partner 22 Moordale Avenue, Priestwood, Bracknell RG42 1RT
G6 XPF Three Counties ARC c/o G. Love 8 Scotts Way, Tunbridge Wells TN2 5RG
G6 XPY R. Chappell 17 Redcar Avenue, Hereford HR4 9TJ
G6 XPZ A. Carter 28A Smithwell Lane, Heptonstall, Hebden Bridge HX7 7NX
G6 XQB R. Carter 56 Main Road, Naphill, High Wycombe HP14 4QB
G6 XQO P. Gait 6 Martindale Road, Churchdown, Gloucester GL3 2DW
G6 XQP G. Garner Tredore, Haugh Road Banham, Norwich NR16 2DE
G6 XQR F. Gizzi 19 Kings Field, Bursledon, Southampton SO31 8EN
G6 XQT N. Godwin 9 Broadway, Barnsley S70 6QQ
G6 XQY J. Griffin 6 Heathfield, Royston SG8 5BW
GW6 XRE B. Helsdon Bryn Hedd, Cyffylliog LL15 2DW
G6 XRF J. Hicks Flat 213, Enterprise House, 112 Kings Head Hill, London E4 7ND
G6 XRH J. Hoare 8 Sunnyheath, Havant PO9 3BW
G6 XRI S. Hobbs 19 Ashfield Road, Kenilworth CV8 2BE
G6 XRK M. Huggins Black Firs, Pinewood Road, Iver SL0 0NJ
G6 XRS Leicester RS c/o P. Taylor 104 Winstanley Drive, Leicester LE3 1PA
G6 XRY G. Kobiela 61 Earith Road Willingham, Cambridge CB24 5LS
G6 XSB M. Dower 19 Fullwell Court, Fullwell Avenue, Ilford IG5 0RZ
G6 XSC I. Denison 5 Hazelwood Close, Cheltenham GL51 5RX
G6 XSK E. Firth 2 Gladstone Close, Littlemoor, Weymouth DT3 6RH
G6 XSL C. Franklin Troy Cottage, Hyde Heath, Amersham HP6 5RW
G6 XSS B. Gell 27 Park Road, Barnstone, Nottingham NG13 9JF
G6 XSY J. Goodey 62 Rose Hill, Binfield, Bracknell RG42 5LG
G6 XSZ D. Graham 127 Shephall View, Stevenage SG1 1RP
G6 XTC A. Tripp 3 Ash Close, Oathills, Malpas SY14 8JB
G6 XTD R. Hallsworth 27 Westfield Avenue, Heanor DE75 7BN
G6 XTG B. Haynes 6 Epping Walk, Furnace Green, Crawley RH10 6LX
G6 XTJ R. Harris 88 Earles Meadow, Horsham RH12 4HR

G6 XTK D. Harris Claws Cottage, Crablands, Chichester PO20 9AY
G6 XTT R. Holgate 5 Exley Gardens, Halifax HX3 9EE
G6 XTZ Skyline c/o T. Jarvis 1 Whitehall Avenue, Mirfield WF14 0AQ
G6 XUD S. Justin Garth, Park View Road, Pinner HA5 3YF
G6 XUV D. Lee 188 Manstone Ave, Sidmouth EX10 9TJ
G6 XUX R. Mettam 12 School Lane, Marsh Lane, Sheffield S21 5RS
G6 XVH G. Allen 68 Hawthorn Avenue, Armthorpe, Doncaster DN3 2ET
G6 XVQ P. Braybrooke 6 Tubbenden Lane, Orpington BR6 9PN
G6 XVY T. Barker 2 The Beeches Chapel Lane, Overton, Morecambe LA3 3HU
G6 XVZ A. Barker 33 Willoughby Avenue, Kenilworth CV8 1DG
GM6 XW A. Winton 2 Castlehill Cottages, Brisbane Glen Road, Largs KA30 8SN
G6 XWD C. Breckons Low Wood Farm, Lamonby, Penrith CA11 9SS
G6 XWM R. King 52 Ford Road, Tiverton EX16 4BE
G6 XWY D. Clarke 10 Dorchester End, Colchester CO2 8AR
G6 XX RSGB Contest Club c/o N. Totterdell Moscar Cross House, Hollow Meadows, Sheffield S6 6GL
G6 XXB D. Cook Stepping Stones, 31 Vicarage Hill, Paignton TQ3 1NH
G6 XXE S. Crowther 17 Carr Gate Crescent, Carr Gate, Wakefield WF2 0QR
G6 XXJ D. Clubley 37 Appleton Road, Beeston, Nottingham NG9 1NE
G6 XXL G. Carter Rivendell, North Reston, Louth LN11 8JD
G6 XXN A. Clarke 138 High Street, Barwell, Leicester LE9 8DR
GW6 XXY K. Dobson 152 Foryd Road, Kinmel Bay, Rhyl LL18 5LS
G6 XYD K. Elsworth 88 Mungo Park Way, Orpington BR5 4EQ
G6 XYF R. Ediss 5 Stirling Crescent, Totton, Southampton SO40 3BN
G6 XYL J. Luxton 2 Trinity Court, Westward Ho, Bideford EX39 1LT
G6 XYO J. Fazey 90 Beecher Road, Halesowen B63 2DW
G6 XYR J. Scothern 24 Cavendish Crescent, Kirkby-In-Ashfield, Nottingham NG17 9BN
G6 XYS A. Searle 22 Crowther Close, Southampton SO19 1BX
G6 XYU J. Stanton Waters & Stanton Plc, 22 Main Road, Hockley SS5 4QS
G6 XYV E. Strode 26 Churchill Close, Congleton CW12 4QU
G6 XYX B. Slater 47 Broom Road, Lakenheath, Brandon IP27 9EZ
G6 XZA M. Scott 28 Penwarden Way, Bosham, Chichester PO18 8LF
G6 XZC C. Shaw 17 South Street, Pilsley, Chesterfield S45 8BQ
G6 XZM C. Smith 104 Warren Road, Banstead SM7 1LB
G6 XZP R. Sammons 42 Woodcote Avenue, Wallington SM6 0QY
G6 XZS J. Thorn 20 Kiln Road, Shaw, Newbury RG14 2HA
G6 YAH C. Wheeler 11 Brooklands Way, Redhill RH1 2BN
G6 YAI I. Wilson 2 Kingswood Close, Owlthorpe, Sheffield S20 6SD
G6 YAK P. Willetts 49 Summervale Road, Hagley, Stourbridge DY9 0LX
G6 YAQ F. Barker 13 Ashbourne Road, Eccles, Manchester M30 0HW
G6 YAR R. Porteus 22 North View, Meadowfield, Durham DH7 8SQ
G6 YAS M. Brashill 42 Bannister Street, Withernsea HU19 2DT
G6 YB City Bristol Gr c/o D. Bailey 70A Park Road, Staple Hill, Bristol BS16 5LG
G6 YBC D. Anderson 142 Tyldesley Road, Atherton, Manchester M46 9AB
G6 YBH A. White 85 Goddard Way, Saffron Walden CB10 2EB
G6 YBV S. Hunt 33 Rutland Street, Ashton-under-Lyne OL6 6TX
G6 YCA J. Cooper 17C Suttons Lane, Deeping Gate, Peterborough PE6 9AA
G6 YCE A. Brooke 14 Counting House Road, Disley, Stockport SK12 2DB
G6 YCF M. Bartlett 206 Victoria Road, Romford RM1 2NP
G6 YCG A. Bennett 10 Burleaze, Chippenham SN15 2AY
G6 YCI M. Buck 178 Rover Drive, Castle Bromwich, Birmingham B36 9LL
G6 YCL M. Banner 7 Lowdham Road, Gedling, Nottingham NG4 4JP
G6 YCM R. Brookes 52 Larch Grove, Kendal LA9 6AU
G6 YCN R. Brassington Above Park Farm, Leek Road, Stoke on Trent ST10 2PT
G6 YCO J. Baddeley 52 Stephens Way Bignall End, Stoke-on-Trent ST7 8PL
GW6 YCT M. Le Ves Conte 74 Glan Road, Aberdare CF44 8BW
G6 YCW B. Lancaster 1 Belgrave Close, Dodleston, Chester CH4 9NU
G6 YCZ J. Massey 10 Rapley Avenue, Pulborough RH20 4QL
G6 YDN J. Mountain 15 Eldon Close, Chapel En le Frith SK23 0PX
G6 YDO F. Mirams 5 Shaftesbury Avenue, Cheadle Hulme, Cheadle SK8 7DB
G6 YDP A. Gallagher 1A Wynsome Street, Southwick, Trowbridge BA14 9RB
GW6 YDT E. Gittins 40 Melyd Avenue, Prestatyn LL19 8RN
G6 YEA N. Guy 43 Hereford Road, Bolton BL1 4NJ
G6 YEK D. Heard 103 Moorland Road, Weston-Super-Mare BS23 4HU
G6 YEY R. Hope 26 Chaucer Avenue, Andover SP10 3DS
G6 YFF G. Hunter 57 The Cedars, Hailsham BN27 1TU
G6 YFG S. Lles 3 Petersway Gardens, , Bristol BS5 8TA
G6 YFH R. Ingle 48 Barlborough Road Clowne, Chesterfield S43 4RF
G6 YFL H. Jones 15 Bonchurch Walk, Manchester M18 8BP
G6 YFY I. Pitfield 27 Winchester Crescent, Fulwood, Sheffield S10 4ED
G6 YFZ D. Paul Enfield, Gunton Road, Wymondham NR18 0QP
G6 YGB R. Preston 188 Dumers Lane, Radcliffe, Manchester M26 2GF
G6 YGC A. Pilkington 1 Woodshaw Grove Worsley, Manchester M28 7XX
G6 YGH A. Richardson 9 Webbers Way, Puriton, Bridgwater TA7 8AS
GW6 YGI B. Rogers Fronucha, Rhewl, Oswestry SY10 7AS
G6 YGJ R. Robinson 2 Badminton Close, Sewerby, , Bridlington, YO16 6GD
G6 YGP D. Lee 4 Blythe Cottages, Blythe Lane, Ormskirk L40 5UA
G6 YGV M. Lane Harewood Villa, Harewood Place, Halifax HX2 7PN
GM6 YGW B. Finch Anchor Cottage, Lybster KW3 6AS
G6 YHE G. Mander 70 Copthall Way, New Haw, Addlestone KT15 3TU
G6 YHF R. Marchant 12 Poplar Close, Huntingdon PE29 7BP
G6 YHK T. Miller 6 Captains Walk, Falmouth TR11 4HR
G6 YHL C. Miller 5 Lodge Lane, Bewsey, Warrington WA5 0AG
G6 YHP C. Molyneux 23 Kemp Close, Chatham ME5 9SP
G6 YHW G. Murly 1128 Route Du Trieux, Vigne Redonde 24360 France
G6 YIE S. Forbes 8 Nutmeg Close, Earley, Reading RG6 5GX
G6 YII K. Everington 1 Norfolk Road, Wigston LE18 4WH
G6 YIJ J. Elford 7 Cunliffe Road, Stoneleigh, Epsom KT19 0RJ
G6 YIK N. Drury 444 Upper Shoreham Road, Shoreham-by-Sea BN43 5NE
G6 YIO T. Chapman 17 Trevor Road, Swinton, Manchester M27 0YH
G6 YIP A. Cohen 9 Terrace Rd, 9 Terrace Rd, Plymouth Meeting 19462 United States
G6 YIQ J. Dixon 8 East View, St. Ippolyts, Hitchin SG4 7PD
G6 YIS R. Chell 3 Elderberry Close, Stourport-on-Severn DY13 8TF
G6 YIU P. Dawson Ivy Dene, Middle Lane, Wolverhampton WV8 2BE
G6 YIW W. Gilroy Little Harewood Farm, Clamgoose Lane, Kingsley, Stoke-on-Trent ST10 2EG
G6 YJD J. Govier 111 Pearson Crescent, Wombwell, Barnsley S73 8SF
G6 YJH A. Haills The Cherries, Main Road, Chelmsford CM3 1NR
G6 YJJ P. Hambly 22C Windsor Road, London W5 5PD
G6 YJO D. Arscott 20 Orchid Vale, Kingsteignton, Newton Abbot TQ12 3YS
G6 YJR J. Angel 33 Grovewood Close, Chorleywood, Rickmansworth WD3 5PX
G6 YLA J. Howard 11 Lightwood, Crown Wood, Bracknell RG12 0TR
G6 YLB G. Howse 1 Sutherland Close, Woodloes Park, Warwick CV34 5UJ
G6 YLD G. Hope 17 Church Road, Sutton At Hone, Dartford DA4 9EX
G6 YLN M. Hobbs 22 Swan Place, Reading RG1 6QD
G6 YLO P. Hizzey Borde Neuve, Maurens 31540 France
G6 YLQ D. Harrop C/Mariano Aguilo 2A, Edificio Formentera 1, Mallorca 7181 Spain
G6 YLR K. Harris 20 Rose Walk, Wicken Green Village, Fakenham NR21 7QE
G6 YLV J. Cromack 45 Chelsea Road, Aylesbury HP19 7BG
G6 YLW T. Cannon 36 St. Margarets Drive, Wigmore, Gillingham ME8 0NR
G6 YLX A. Crabtree 15 Richmond Gardens, Redhill, Nottingham NG5 8JS
G6 YLZ P. Cornes 46 Newland Avenue, Stafford ST16 1NL
GI6 YM the City of Belfast Ymca RC c/o W. Mcaleer 90 Gortin Park, Belfast BT5 7EQ
G6 YMA N. Clark 2 Barleycroft, Stevenage SG2 9NP
G6 YMD M. Cooke 22 Durham Close, Grantham NG31 8RL
G6 YMH C. Hughes 85 Benson Gardens, Wortley, Leeds LS12 4LA
G6 YMI A. Harris 10 Egroms Lane, Withernsea HU19 2LZ
GW6 YML D. James 68 Orchard Park, St. Mellons, Cardiff CF3 0AQ
GW6 YMS P. Humphreys Tyn Llan, Bodffordd, Llangefni LL77 7DZ
G6 YMU D. Hutchings 2 Burghley Avenue, Bishop's Stortford CM23 4PD
G6 YMY P. Jacques Caprius, The Parks, Evesham WR11 8JP
G6 YNL R. Perry Straight Mile Cottage, Gloucester Road, Rudgeway, Bristol BS35 3SB
G6 YNT S. Pentecost 3 Delamare Road Cheshunt, Waltham Cross EN8 9AP
G6 YNV S. Raddy 32 Berry Park, Saltash PL12 6EN
G6 YNW M. Reeves 17 Newark Avenue, Putnoe, Bedford MK41 8NX
G6 YOG M. Rutt 15 Salmons Road, Chessington KT9 2JE
G6 YOP P. Harding 54 Manor Road, Stretford, Manchester M32 9JB
G6 YOR J. Gillott 132 Racecommon Road, Barnsley S70 6JY
G6 YOZ J. Addison 20 Wychwood Rise, Great Missenden HP16 0HB
GW6 YPA M. Attfield 16 Rhodfar Eos, Cwmrhydyceirw, Swansea SA6 6TF
G6 YPD S. Aldridge Flat 7, Rosie Court, Newnham Street, Ely CB7 4PQ
G6 YPF J. Armstrong 14 Rickwood Park, Horsham Road, Dorking RH5 4PP
G6 YPJ J. Brown 30 The Avenue Brookville, Thetford IP26 4RF
G6 YPK A. Bradbury 20 Adrden Close, Warwick CV34 5SN
G6 YPM J. Willats 17 Purcell Road, Crawley RH11 8XJ
G6 YPY S. Davis 30 Bonny Wood Road, Hassocks BN6 8HR
GM6 YQA C. Davies 2 Sweyn Road, Thurso KW14 7NW
G6 YQI E. Fletcher-Cowen 18 Buckingham Avenue Horwich, Horwich, Bolton BL6 6NR
G6 YQJ D. Fisher 86 Parsons Lane, Littleport, Ely CB6 1JS
G6 YQN N. Fox 32 Westmorland Avenue, Kidsgrove, Stoke-on-Trent ST7 1AT
G6 YQT S. Forbes 11 Henfield View, Warborough, Wallingford OX10 7DB
G6 YQU R. Fuller The New House, Main Street, Lutterworth LE17 6NT
G6 YQW J. Taylor 7 Caddick Road, Birmingham B42 2RL
G6 YRB J. Stewart 107 Turnberry, Skelmersdale WN8 8EG
G6 YRC A. Smith 4 Wesley Grove, Burnley BB12 0JJ
GM6 YRH A. Smith Robsland, Strathaven Road, Lanark ML11 0HY
G6 YRI S. Sizmur 38 Longbourne Way, Chertsey KT16 9ED
G6 YRJ T. Simmons 3 West Hill Place, Brighton BN1 3RU
GM6 YRN A. Stewart 6 Lawers Place, Aberfeldy PH15 2BE
G6 YRV D. Bedford 28 Durfold Drive, Reigate RH2 0QA
G6 YRY R. Bearchell 81 Leaves Green Road, Keston BR2 6DG
G6 YSB J. Bates 16 Harewood Avenue, Great Barr, Birmingham B43 6QE
G6 YSJ J. Bicknell 52 Fieldcourt Gardens, Quedgeley, Gloucester GL2 4UD
G6 YSL S. Watts 15 Churchill Way, Northam, Bideford EX39 1DF
G6 YSN K. Ward 8 Hinckley Road, St. Helens WA11 9HU
G6 YSO P. Wayer 4 Chatburn Avenue, Waterlooville PO8 8UB
G6 YSQ S. Tricker 1 Drewitt Court, 75 Godstow Road, Oxford OX2 8PE
G6 YSZ P. Tonge 17 Thomas Street, Glossop SK13 8QN
G6 YTB R. Watts 41 Watford Road, Crick, Northampton NN6 7TT
G6 YTO Radio Society of Great Britain c/o R. Cassidy 9 Langham Way, Ely CB6 1DZ
G6 YTR R. Broughton Brookside, Blagdon Terrace, Newcastle upon Tyne NE13 6EY
G6 YTV A. Black Redholme, The Street, Thetford IP25 6NL
G6 YTW A. Bennett 29 Kennington Road, Kennington, Oxford OX1 5NZ
G6 YTX R. Burnett 46 Dorset Waye, Heston, Hounslow TW5 0ND
G6 YTY A. Bournes 115 Abbotts Ann Down, Andover SP11 7BX
GW6 YUC E. Brooksbank 22 King St., Carmarthen SA31 1BS
G6 YUX B. Clough Ashby Powerboat School, 31 Countess Road, Salisbury SP4 7AS
G6 YUY F. Crockford 41 Coram Green, Hutton, Brentwood CM13 1LW
G6 YVD G. Wood Tethers End, Angarrack Lane, Hayle TR27 5JF
G6 YVJ C. Ward 416A Portsmouth Road, Southampton SO19 9AT
G6 YVS J. Wilson 36 North Warren Road, Gainsborough DN21 2TU
G6 YWL A. Griffiths 45 Clarence Road, Bilston WV14 6NZ
G6 YWN P. Groom 2 Alms Road, Doveridge, Ashbourne DE6 5JZ
G6 YWU D. Harding 20 D'Arcy Road, Tiptree, Colchester CO5 0RP
G6 YWV M. Harrison Barn Cottage, Parwich, Ashbourne DE6 1QB
G6 YWZ E. Heath-Coleman 1 Longmead Oakford, Tiverton EX16 9DW
G6 YXB D. Hewson Woodwells, 52 Elmham Road, Beetley, Dereham NR20 4BW
G6 YXO S. Fisher 37 Elmlands Grove, York YO31 1ED
G6 YXT B. Evans C/O Angel 18, Xerta Tarragona 43592 Spain
G6 YXV K. Faulkner 5 Tregarrick, West Looe, Looe PL13 2SD
G6 YXW P. Foulkes 23 Callowbrook Lane, Rubery, Birmingham B45 9HW
G6 YXX N. Frederick 72 Cheltenham Street, Barrow-in-Furness LA14 5HW
G6 YXY C. Edwards Seymore, Greenhill Park Road, Evesham WR11 4NL
G6 YYN K. Mccann Treverven, Back Lane, Hemingbrough, Selby YO8 6QP
G6 YYU A. Mutimer 52 Sycamore Avenue, Wymondham NR18 0HX
G6 YZB L. Nunn 103 Bladindon Drive, Bexley DA5 3BT
G6 YZF S. Alston-Pottinger 86 Main Street, Walton, Street BA16 9QN
G6 YZR M. Smith 5 Derwent Close, North Anston, Sheffield S25 4GD
G6 YZU L. Nixon 87 Field Avenue, Canterbury CT1 1TS
G6 ZAA J. Wellard 19 South Motto, Kingsnorth, Ashford TN23 3NJ
G6 ZAC A. Wilson 21 Lakes Close, Chilworth, Guildford GU4 8LL
G6 ZAF D. Walker 27 Daltons Close, Langley Mill, Nottingham NG16 4GP
GM6 ZAK A. Sutton 22 St. Michaels Drive, Cupar KY15 5BS

G6 ZAL S. Ward 125 Heys Lane, Blackburn BB2 4NG
G6 ZAM P. Waldron 12 Lady Drive, Pengegon, Camborne TR14 7UF
G6 ZAX R. Hollick 7 Grenfell Road, Bournemouth BH9 2UD
G6 ZAY R. Hope 129 Lunedale Road, Dartford DA2 6JX
G6 ZBL K. Jones 18 Goldhurst Drive, Tean, Stoke-on-Trent ST10 4LS
G6 ZBO M. Julians 29 Trentdale Road, Carlton, Nottingham NG4 1BU
G6 ZBT D. Green 6 Garth Villas, Rimswell, Withernsea HU19 2DB
G6 ZBV A. Higham 12 Lakenheath Drive, Sharples, Bolton BL1 7RJ
GW6 ZCR J. Phillips 39 Bryn Glas, Rhosllanerchrugog, Wrexham LL14 2EA
GM6 ZCX M. Rochester Eadar Da' Sloc, Achmelvich, Lairg IV27 4JB
GM6 ZCY M. Rochester Eadar Da' Sloc, Achmelvich, Lairg IV27 4JB
G6 ZDB G. Reddington 2 South St., Newton, Alfreton DE55 5TT
G6 ZDE D. Ellingworth 3 Leighton Park West, Westbury BA13 3RW
GW6 ZDH R. Roberts All-Y-Coed, Sychnant Pass Road, Conwy LL32 8EU
G6 ZDP K. Baum 25 Lakers Meadow, Billinghurst RH14 9NP
G6 ZDS R. Baldwin 24 Keys Court, Banbury OX16 2AZ
G6 ZDV A. Beales Broomhill Bungalow, Mappleton Road, Hull HU11 4UW
G6 ZDY R. Bethell 5 Dorset Way, Maidstone ME15 7EL
G6 ZEM Flaxton Moor CG c/o J. Hollerbach 119 Mead End, Biggleswade SG18 8JU
G6 ZEN J. Homan 55 Ark Royal, Bilton, Hull HU11 4BN
G6 ZEQ R. Hubert 62 Bluebell Woods, Shalloak Road, Broad Oak, Canterbury CT2 0QB
G6 ZET D. Jackson 6 Bolton Lane, Carnforth LA5 8BL
G6 ZEW M. Jennings 6 Broomroyd Worsbrough, Barnsley S70 5DU
G6 ZEY Powys ARC c/o K. Johnson 66 Godwin Way, Cambridge CB1 8QR
G6 ZEZ C. Jones 709 Bath Road, Taplow, Maidenhead SL6 0PB
G6 ZFA J. Justice 6 Stanley Terrace, Devizes SN10 5AJ
G6 ZFG P. Wood 31 Larches Lane, Tettenhall, Wolverhampton WV3 9PX
GM6 ZFI D. Smith 39 High Croft, Kelso TD5 7NB
G6 ZFK E. Toohey No6 Block E, Peabody Avenue, London SW1V 4AS
G6 ZFO D. Tate 73 Sparth Avenue, Clayton Le Moors, Accrington BB5 5QH
G6 ZFT H. Smithey 9 Moverley Flatts, Pontefract WF8 2BX
G6 ZFU C. Stephen 12 Beaufort Close, Leegomery, Telford TF1 6XU
G6 ZFV T. Wynne-Jones 62 Moor Lane, Rickmansworth WD3 1LQ
G6 ZFX P. Senior 13 St. Michaels Avenue, Swinton, Mexborough S64 8NX
G6 ZFZ M. Turner 1 Bakers Gardens Codsall, Wolverhampton WV8 1HA
G6 ZG Bristol Raynet c/o B. Alston-Pottinger 16 Vincent Close, Great Yarmouth NR31 0HR
G6 ZGA M. Smith 6 Norton Crescent, Towcester NN12 6DN
G6 ZGB C. Salmon 20 Lime Close, Sandbach CW11 1BZ
G6 ZGC M. Tann 11 St. Margarets Grove, Redcar TS10 2HW
G6 ZGF D. Simpson 18 Croft House View, Morley, Leeds LS27 8NS
G6 ZGH R. York 8 The Rookery Brogden Street, Ulverston LA12 0DB
G6 ZGI C. Butler 8 Douglas Walk, Chelmsford CM2 9XQ
G6 ZGK G. Weston 2 Gill Park, Efford, Plymouth PL3 6LX
G6 ZGU J. Brown 10 Cherry Trail, Coldwater, Ontario L0K 1E0 Canada
GW6 ZGY R. Bennett 88 Coychurch Road, Pencoed, Bridgend CF35 5NA
G6 ZHB J. Booth 9 New Street, Abingdon OX14 3PE
G6 ZHF S. Bailey Silverthorne House, North Piddle, Worcester WR7 4PR
G6 ZHJ G. Butler 23 Roman Meadow, Downton, Salisbury SP5 3LB
G6 ZHL M. Leack 68 Dale Street, Lancaster LA1 3AW
GW6 ZHM R. Lannon 16 Heol Mabon, Rhiwbina, Cardiff CF14 6RL
G6 ZHO G. Lattin 5 Seymour Road, Broadfield, Crawley RH11 9ES
G6 ZHS K. Lupton Oak Tree Cottage, Post Office Lane, Frodsham WA6 8JJ
G6 ZHU P. Lightfoot 7 Fearns Avenue, Newcastle ST5 8ND
GW6 ZHY A. Mayers 2 Wyndham Gardens, Wrexham LL13 9LY
G6 ZIC J. Mccomb Bridge End Cottage, Bridge End, Hexham NE48 2RY
G6 ZIO N. Dessau 20 Coventry Circle, Mahopac 10541 United States
GI6 ZIR A. Duffy 81A Arney Road, Bellanaleck, Enniskillen BT92 2DL
G6 ZIY A. Fairhurst 16 Waverley Road, Hindley, Wigan WN2 3BN
G6 ZJD E. Fensome 77 Church Green Road, Bletchley, Milton Keynes MK3 6BY
G6 ZJI A. Washby 57 Cromwell Road, Hedon, Hull HU12 8GF
G6 ZJK F. Webster 1 Fir Tree Cottages, Lower Ansford, Castle Cary BA7 7JY
G6 ZJM D. Leese 22 Elm Road, Abram, Wigan WN2 5XG
G6 ZJN R. Williams 220 Euston Grove, Morecambe LA4 5LJ
G6 ZJS L. Wright 17 Drayton St., Alumwell Estate, Walsall WS2 9QB
G6 ZJV A. Winterbottom 38 Heaton Avenue Earlsheaton, Dewsbury WF12 8AQ
G6 ZKC D. Usher 26 Meneth, Gweek, Helston TR12 6UW
G6 ZKM J. Cornell 10 Craneswater Park, Southsea PO4 0NT
G6 ZKS M. Staniland 2 Epsom Road, Cantley, Doncaster DN4 6HX
G6 ZKU B. Sawyers 36 Frome Road, Bath BA2 2QB
G6 ZKX the University of Greenwich c/o R. Smith Smith Farms, Herne Lane, Dereham NR19 1QE
G6 ZKY E. Stebbings 1 Coupland Road, Wootton, Abingdon OX13 6DU
G6 ZKZ V. Smith 40 Princess Gardens, Blackburn BB2 5EJ
G6 ZLD P. Bent 7 Bandon Rise, Wallington SM6 8PT
G6 ZLJ M. Adams 62 Woodlands Road, Holmcroft, Stafford ST16 1QP
G6 ZLS G. Ashbee 34 Manorgate Road, Kingston upon Thames KT2 7AL
GM6 ZLY D. Brasenell 18 Whitelaw Avenue, Castle Douglas DG7 1GB
G6 ZMD S. Roberts 36 Hill Crescent, Dudleston Heath, Ellesmere SY12 9NA
G6 ZME Telford And District ARS c/o J. Humphreys 15 Colemere Drive, Wellington, Telford TF1 3HH
G6 ZMG G. Mills 57 Holborough Road, Snodland ME6 5PA
GW6 ZMN W. Mcdowall 36 Adenfield Way, Rhoose, Barry CF62 3EA
G6 ZMO J. Mooney 2 Madford Lane, Launceston PL15 9EB
G6 ZMU G. Randall 15 Bincombe Rise, Weymouth DT3 6AS
G6 ZMX A. O'Shaughnessy Southby, Buckland, Faringdon SN7 8QR
G6 ZNJ A. Reeve 188 Dorset Avenue, Great Baddow, Chelmsford CM2 8YY
G6 ZNT I. Cross 5 Upper Crescent, Minster Lovell, Witney OX29 0RT
G6 ZNW S. Cascino 3 Connaught Road, Folkestone CT20 1DA
G6 ZOB A. Crowther 16 Linden Avenue, Tuxford, Newark NG22 0JR
G6 ZOE C. English 124 Hillside Road, Portishead, Bristol BS20 8LG
G6 ZOI A. Bond 24 Mill Lane, Camblesforth, Selby YO8 8HW
G6 ZOJ A. Buchan 5 Copythorne Close, Brixham TQ5 8QG
G6 ZOL P. Lancaster 134 Wigan Road, Euxton, Chorley PR7 6JW
G6 ZOT J. Leary 24 Howard Drive, Old Whittington, Chesterfield S41 9JU
G6 ZPL P. Manning 21 Whitethorn Way, Oxford OX4 6ER
G6 ZPO K. Miller 47 Hermitage Green Hermitage, Thatcham RG18 9SL
G6 ZPR A. Morris 32 New Road, Wonersh, Guildford GU5 0SE
G6 ZPV N. Mansfield 2 Little Halt Portishead, Bristol BS20 8JQ
G6 ZQA G. Nolan 94 St. Andrews Road, Burgess Hill RH15 0PH
G6 ZQJ A. Doughty 42 Thornton Road, Ilford IG1 2ER
G6 ZQS M. Charlton 104 Foundry Street, Horncastle LN9 6AF
G6 ZQT S. Clapson 123 Norwich Road, Ipswich IP1 2PR
G6 ZQU D. Crook Sherington Nurseries, Bedford Road, Sherington, Newport Pagnell MK16 9NQ
G6 ZRL K. Sellens 7 The Knapps, Semington, Trowbridge BA14 6JG
G6 ZRO B. Stoner Montrose, Wesley Road, Robin Hoods Bay, , Whitby YO22 4RW
G6 ZRS B. Starr 121 Pretoria Road, Patchway, Bristol BS34 5PY
G6 ZRV P. Stainton 9 Park Lane Reepham, Norwich NR10 4JZ
G6 ZSF Dr D. Neely 3 Sidestrand Road, Newbury RG14 6HP
G6 ZSG L. Onions 8 Prince Charles Close, Rubery, Birmingham B45 0NB
G6 ZSH P. Owen 288 Chervil Rise, Wolverhampton WV10 0HR
G6 ZSQ J. Pepper Flat 55, Seaward Court, West Street, Bognor Regis PO21 1XJ
G6 ZSU P. Payton 11 Hexham Way, Dudley DY1 2UN
G6 ZTD B. Robinson 23 Croft Drive, Millhouse Green, Sheffield S36 9NE
G6 ZTF A. Bowler 12 Wrenbury Drive, Coventry CV6 6JZ
G6 ZTH M. Richardson Better View, Back Lane, Sutton-in-Ashfield NG17 2LL
G6 ZTL B. Rogers 24 Marmion Road, Coningsby, Lincoln LN4 4RG
G6 ZTP G. Down 8A Abbeville Close, Exeter EX2 4SJ
G6 ZTR S. Davies 111 Southend, Garsington, Oxford OX44 9DL
G6 ZTT O.B.O. Mid-Cheshire CG c/o M. Baguley 2 Kensington Way, Northwich CW9 8GG
G6 ZTZ S. French 22 Amity Street, Reading RG13LP
G6 ZUE W. Edwards 31 Cumberland Avenue, Benfleet SS7 5NU
G6 ZUO D. Gibson 14 Lowfield Road, Dewsbury Moor, Dewsbury WF13 3SR
GW6 ZUS J. Gray 36 Heol Pentre Felen, Llangyfelach, Swansea SA6 6BY
G6 ZUV J. Griffin 35 Cottage Street, Kingswinford DY6 7QE
G6 ZUZ J. Hampshire 14 Fellows Road, Cowes PO31 7JN
G6 ZVB K. Harris123 14 Dunstall Close, St. Marys Bay, Romney Marsh TN29 0QX
G6 ZVD M. Hicken 8, Calle Francisco Villaespesa, Oria 4810 Spain
G6 ZVL M. Hoskins 22 Rosedale Gardens, Thatcham RG19 3LE
G6 ZVO K. Howarth 79 Eden Avenue, Edenfield, Bury BL0 0LD
G6 ZVR D. Inskip 40 Burway Meadow Alrewas, Burton-on-Trent DE13 7EB
G6 ZVU S. Hughes 50 Albany Road, Dalton, Huddersfield HD5 9UW
G6 ZVV N. Hull 60 Portreath Place, Chelmsford CM1 4DN
G6 ZWC C. Brown 16 Old Croft Close, Good Easter, Chelmsford CM1 4SJ
G6 ZWL C. Wright 19 Redwood Glen, Chapeltown, Sheffield S35 1EA
G6 ZWM R. Wade 104 Brookehowse Road, London SE6 3TW
G6 ZXO J. Crowe 15 Lambert Road, Kendray, Barnsley S70 3AA
GW6 ZYI B. Jones 10 Hughes Street, Penygraig, Nr Tonypandy CF40 1LX
G6 ZYL P. Jones 274 Cannock Road, Westcroft, Wolverhampton WV10 8QG
G6 ZYM K. Keeble Hall Cottage, Hardwick Road, Harleston IP20 9PU
G6 ZYX G. Spruce 158 Wolverhampton Street, Wednesbury WS10 8UB
G6 ZYZ P. Skerritt Room 1. 157 Hagley Road, Edgbaston, Birmingham B16 8UQ
G6 ZZ RSGB Transatlantic Test Centenary Club c/o N. Totterdell Moscar Cross House, Hollow Meadows, Sheffield S6 6GL
G6 ZZE P. Read 11 Fairview Avenue Whetstone, Leicester LE8 6JQ
GW6 ZZF P. Thomas 42 Wyndham Road, Abergavenny NP7 6AF
G6 ZZR D. Wiltshire 19 Heron Way, Basingstoke RG22 5QF
G6 ZZS D. Watts 176 Blatchcombe Road, Paignton TQ3 2JP

G7

G7 AAI Dr A. Hickey 11 Barker Road, Wirral CH61 3XH
GM7 AAJ P. Mcmanus 59 Mauchline Road, Hurlford, Kilmarnock KA1 5AB
G7 AAR P. Comben Undine Cottage, Sedrup, Hartwell, Aylesbury HP17 8QN
G7 AAS D. Hawkins 8 Braybrook Street, East Acton, London W12 0AP
GW7 AAU H. Studdart 33 Linden Avenue, Connah'S Quay, Deeside CH5 4SN
GW7 AAV S. Studdart 33 Linden Avenue, Connah'S Quay, Deeside CH5 4SN
G7 AAY Scottish Isles DX Group c/o K. Richardson 514 Obelisk Rise, Northampton NN2 8SX
G7 ABE C. Duberley 2 The Grove, Greenford UB6 9BY
G7 ABF K. Austin 6 Boothey Close, Biggleswade SG18 0DG
G7 ABO Dr D. Bescoby 40 High Street, Wickham Market, Woodbridge IP13 0QS
G7 ABQ D. Ferns 18 Sandelswood End, Beaconsfield HP9 2AE
G7 ABR A. Clark Sheldon House, Sheldon, Bakewell DE45 1QS
G7 ABT D. Hepworth 1 Greengate Crescent Epworth, Doncaster DN9 1HA
G7 ABZ M. Bromage 14 Rhuddlan Way, Kidderminster DY10 1YH
G7 ACA B. Pearce 39 Fairholme Park, Ollerton, Newark NG22 9AS
G7 ACD R. Cariss Eaton Manor , Eaton-Under-Heywood, Church Stretton SY6 7DH
G7 ACG J. Baker 19 Green Lane, Rugeley WS15 2AR
G7 ACJ G. Mantle 6 North Green, Wolverhampton WV4 4RQ
G7 ACK G. Bromfield 63 Herondale Road, Mossley Hill, Liverpool L18 1JZ
G7 ACM S. Pinkney 39 Butterley Drive, Loughborough LE11 4PX
G7 ACN C. Hayes 9 Grenville Way, Thetford IP24 2JH
G7 ACO R. Horton 1 Stonehill Rise, Doncaster DN5 9HD
G7 ACR P. Blakemore 50 Longley Farm View, Sheffield S5 7JX
G7 ADF I. Bradbury 11 St. Stephens Avenue, Wigan WN1 3UQ
G7 ADH G. Williams 18 Luther Road, Bournemouth BH9 1LH
G7 ADP C. Baker 17 Coronation Road, Illogan, Redruth TR16 4SG
G7 ADS A. Cresswell 31 New Street, Doddington, March PE15 0SP
GM7 ADU Dr L. Morrison 22 Lodge Park, Kilmacolm PA13 4PY
G7 ADW G. Laycock 18 Montague Crescent, Garforth, Leeds LS25 2EP
GM7 ADY Dr M. Morrison 22 Lodge Park, Kilmacolm PA13 4PY
G7 AEA Gloucestershire County Raynet c/o R. Large 5 Jasmine Close, Abbeydale, Gloucester GL4 5FJ
G7 AEC Cheltenham Rynt c/o P. Kent 82 Despenser Road, Tewkesbury GL20 5TW
G7 AEE Tewkesbury Rynt c/o C. Davis 38 Courtney Close, Tewkesbury GL20 5FB
G7 AEF Frst of Dean Ry c/o G. Harden 13 Greenfield Road, Coleford GL16 8BY
G7 AEG Gloucestershire County c/o A. Ayres Bryn Hyffryd, Phocle Green, Ross-on-Wye HR9 7TW
G7 AEH Cotswold Raynet c/o G. Hayter 19 Austin Road, Cirencester GL7 1BT
G7 AEQ R. Murphy 17 Valley View Road, Paulton, Bristol BS39 7QB
G7 AES P. Crook 40 St. Aubins Avenue Brislington, Bristol BS4 4NX
G7 AEY D. Martin 12 The Willows, Kemsley, Sittingbourne ME102TE

GW7 AFC M. Grant 11 Golwg Yr Eglwys, Pontarddulais, Swansea SA4 8EE
GM7 AFE A. Erwood Lunna House, Lunna, Vidlin ZE2 9QF
G7 AFL A. Fountaine 19 Metcalfe Grove, Blakelands, Milton Keynes MK14 5JY
G7 AFO K. Hollingsworth 34 Marconi Drive, Yaxley, Peterborough PE7 3ZR
G7 AFQ K. Marlow Computer Science, Po Box 363, Edgbaston Birmingha B15 2TT
G7 AFS M. Sheldon 5 Runnymede Mews, Faversham ME13 8RU
G7 AFT K. Brazier 6 Sadlers Lane, Dibden Purlieu, Southampton SO45 4LZ
G7 AFV M. Weston 10 Flete Avenue, Newton Abbot TQ12 4EH
G7 AFW the Bedford And District ARC c/o M. Towers 44 Ravenscroft Drive, Chaddesden, Derby DE21 6NX
G7 AFZ D. Payea 10 Royal Drive, Seaford BN25 2XW
G7 AGA K. Askew 3A Craven Drive, Broadheath, Altrincham WA14 5JF
G7 AGB P. Rothwell 20 Henbury Close, Corfe Mullen, Wimborne BH21 3TF
G7 AGC M. Collis 2 Westwood Avenue, Urmston, Manchester M41 9NG
GW7 AGG Gravesend ARS c/o R. Ricketts 2 Brynystwyth, Penparcau, Aberystwyth SY23 1SS
G7 AGI D. De Silva 22 Bishop Road, Bristol BS7 8LT
G7 AGO J. Lee 188 Manstone Avenue, Sidmouth EX10 9TJ
G7 AGR Aylesbury Ryn Group c/o R. Clark 9 Conigre, Chinnor OX39 4JY
GM7 AHA V. Turnbull 18 Easterfield Court, Livingston Village, Livingston EH54 7BZ
G7 AHB T. Green 12 Springfields, Ambrosden, Bicester OX25 2AH
G7 AHO P. Russell 27 Main Street, Haconby, Bourne PE10 0UR
G7 AHP S. Crask 14 Southfield Road, Paignton TQ3 2SW
GW7 AHR M. Brett Bryn Bela, Llanfair Road, Abergele LL22 8PD
G7 AIB N. Denton 41 Monks Dale, Yeovil BA21 3JB
G7 AIC V. Newman 35 Netherton Road, Yeovil BA21 5NY
G7 AIF D. Grevatt 17 Foxdale Drive, Angmering, Littlehampton BN16 4HF
G7 AIH R. Whitenstall 4 Monksmead, Borehamwood WD6 2LQ
GW7 AIY V. Lamb 19 Pemba Drive, Buckley CH7 2HQ
G7 AIZ P. Pugh 9 Red Kite Close, Gateford, Worksop S81 8WA
G7 AJE F. Salt 6 Bodycoats Road, Chandler'S Ford, Eastleigh SO53 2GX
G7 AJG T. Ellis 29 St. Annes Road, Clacton-on-Sea CO15 3NF
G7 AJJ A. Hammond 5 Durness Close, Kettering NN15 5BN
G7 AJK A. Knowler 385 Capstone Road, Gillingham ME7 3JE
G7 AJN H. Lister 68 Spring Avenue, Gildersome, Leeds LS27 7BT
G7 AJP B. Staniforth 1 Pylon Cottages, Donington-On-Bain, Louth LN11 9RQ
G7 AJR S. Godfrey 7 Laburnum Close, North Baddesley, Southampton SO52 9JT
G7 AJS T. Green 34 Thorn Close, Kettering NN16 9BU
G7 AJT M. Carter 17 Ash Crescent, Higham, Rochester ME3 7BA
G7 AJX R. King 31 Lambert Road, Sprowston, Norwich NR7 8AA
G7 AKI P. Shambrook 7 The Close, Cheltenham GL53 0PQ
G7 AKJ W. Wrench 2 Maunders Place, Otterton, Budleigh Salterton EX9 7JE
G7 AKM D. Pearson 8 Walnut Way, Swanley Kent BR8 7TW
G7 AKP Y. Branch 38 Kynaston Road, Didcot OX11 8HD
G7 AKV E. Grantham 18 Fen End Lane, Spalding PE12 6AD
G7 ALC L. Civita 53 Rockhurst Drive, Eastbourne BN20 8XD
GI7 ALH J. Bailie 42D John Street Lane, Newtownards BT23 4LY
GM7 ALI A. Crighton Tighnacreag, Pacemuir Road, Kilmacolm PA13 4JJ
G7 ALR P. Goodman 85 Rantree Fold, Basildon SS16 5TW
G7 AMD G. Blakemore 6 Pine Tree Close Hednesford, Cannock WS12 4JT
G7 AMQ J. Morstatt 32 Elwy Circle, Ash Green, Coventry CV7 9AU
GW7 AMS A. Steel 74 Caradoc Road, Prestatyn LL19 7PF
G7 AMW P. Jackson 4 Abbotsbury, Orton Malborne, Peterborough PE2 5PS
G7 ANA R. Jackson 5 Home Farm Court, Hooton Pagnell, Doncaster DN5 7BL
G7 ANB D. Ball 16 Kelston View, Whiteway, Bath BA2 1NW
GM7 ANE W. Jamieson 20 Cairn Road, Cumnock KA18 1HN
G7 ANG A. Santagata 25 Swyncombe Avenue, London W5 4DR
G7 ANH F. Pattinson 4 Carlisle Road, Brampton CA8 1SR
G7 ANK C. Bryan Beau Rivage, Seaholme Road, Mablethorpe LN12 2DF
G7 ANO T. Hyde 10 Castleton Avenue, Riddings, Alfreton DE55 4AG
G7 ANQ J. Hedges 31 Meadow Road, Hartshill, Nuneaton CV10 0NL
G7 ANV S. O'Malley 140 Allerburn Lea, Alnwick NE66 2QP
G7 ANY A. King 31 Springhill, Pennycross, Plymouth PL2 3QZ
G7 AOA P. Gash 2 Betjeman Walk, Yateley GU46 6YP
GW7 AOE G. Williams 10 Strawberry Place Morriston, Morriston SA6 7AG
GJ7 AOG C. Eve Flat 17, Le Petit Hurel, Queen'S Road, St. Helier, Jersey JE2 3SY
G7 AOK R. Swynford-Lain 19 Kenton Road, Earley, Reading RG6 7LQ
GM7 AOM J. Curr 56 Drygate Street, Larkhall ML9 2DA
G7 AOQ J. Johnson 32 Bradlea Rise, Rotherham S625QJ
G7 AOU S. Wilkins 15 Roundway, Egham TW20 8BS
GU7 APA P. Ash Trigale House, Alderney, Guernsey GY9 3TZ
G7 APD Rugby Am Tra So c/o S. Tompsett 9 Ashlawn Road, Rugby CV22 5ET
G7 API S. Garlick 37 Edith Road, Kettering NN16 0QB
G7 APL S. Bonham 4 St. Martins Avenue, Studley B80 7JJ
GW7 APM C. Mockford Denamby Cottage, King Street, Leeswood, Coed-Llai CH7 4SB
G7 APO C. Monckton 219 Chaldon Road, Chaldon, Caterham CR3 5XN
GW7 APP I. Capon Pentre Garreg Bach, Marianglas LL73 8PP
G7 APQ A. Jones 6 Heatherbreea Gardens, Rushden NN10 6EH
G7 APU L. Young 7 Tudor Rose, 28 Northgate, Hunstanton PE36 6AP
G7 AQA M. Hawkshaw 14 Croft Drive, Menston, Ilkley LS29 6LX
G7 AQD W. Williams 2 Lightfoot Lane Fulwood, Preston PR2 3LP
G7 AQF A. Gregory 13 Combe Avenue, Portishead, Bristol BS20 6JR
G7 AQK N. Mcgrath 48 Willersley Avenue, Kent BR6 9RS
G7 AQL B. Burbage 9 Westerdale Drive, Frimley, Camberley GU16 9RB
G7 AQN I. Cooper Ceylon, 70 Carshalton Park Road, Carshalton SM5 3SW
GI7 AQO R. Todd 14 Glencroft Road, Newtownabbey BT36 5GD
G7 AQV A. Russell 73 Seymour Road, Newton Abbot TQ12 2PX
G7 ARF S. Wright 63 Cambridge Road, St. Albans AL1 5LF
G7 ARJ J. Baber 130 Lumsden Road, Southsea PO4 9LR
G7 ARK A. Wright 3 Wyborn Close, Hayling Island PO11 9HY
G7 ARP R. Orchard 31 Chiswick House, Bell Barn Road, Birmingham B15 2AA
GD7 ARS W. Wrigley 20 Fairy Hill Close, Ballafesson, Port Erin IM9 6TJ Isle of Man
G7 ART M. Blake 20 Triumphal Crescent, Plymouth PL7 4RW
G7 ASF Coventry ARS c/o J. Beech 124 Belgrave Road, Coventry CV2 5BH
GW7 ASL P. Jones 27 Hawthorn Road East, Llandaff North, Cardiff CF14 2LR
G7 ASY P. Matkin 2 Raven Close, Huntington, Cannock WS12 4TQ
G7 ASZ Dr N. Blair 19 Church Street, Bourn, Cambridge CB23 2SJ
G7 ATJ R. Williams 10 Barton Close, Whippingham, East Cowes PO32 6LS
G7 ATW G. Johnson Wood Nook, Blennerhasset, Wigton CA7 3RJ
G7 AUE B. Oubridge 54 Cantle Avenue, Downs Barn, Milton Keynes MK14 7QS
G7 AUF K. Gebhardt 15 Jubilee Road, Corfe Mullen, Wimborne BH21 3NH
G7 AUP A. White Tioram, Garthends Lane, Selby YO8 6QW
GW7 AUQ N. Smith 7 Lili Mai, Barry CF63 1DW
G7 AUR S. Davis 7 Kennedy Crescent, Gosport PO12 2NL
G7 AUU R. Selwood 33 Chandlers, Sherborne DT9 3RT
GM7 AUW J. Milne 24 Lorne Street, Edinburgh EH6 8QP
GM7 AUX E. Ramsay Tighnduin, 2 Queen Street, Dundee DD5 4HG
GI7 AUY I. Potts 46 Richmond Park, Omagh BT79 7SJ
GW7 AVB D. Daymond 15 Constance Street, Newport NP19 7DB
G7 AVF P. Honeybone 30 The Hordens, Barns Green, Horsham RH13 0PJ
G7 AVU R. Fisk 25 Cromwell Street, Gainsborough DN21 1DH
G7 AVZ L. Collings 37 Armstrong Road, Mansfield NG19 6HZ
G7 AWG V. Watson 3 Anderton Rise, Millbrook, Torpoint PL10 1DA
GM7 AWK D. Easton 86 Dryburn Road, Kelloholm, Sanquhar DG4 6SN
G7 AWP T. Noyes 59 Abbots Leys Road, Winchcombe, Cheltenham GL54 5QX
G7 AWS D. Henderson 19 Stuart Road, York YO24 3AX
G7 AWW M. Gynane 164 Stockbridge Lane, Huyton, Liverpool L36 8EH
G7 AXL A. Marchington 30 Warwick Avenue, Golcar, Huddersfield HD7 4BX
G7 AXM J. Smith 7 Ainsworth Court, Cameron Close, Freshwater PO40 9JH
G7 AXN R. Moxham 8 Dunroyal Close, Helperby, York YO61 2NH
G7 AXW Dr A. Parkes 19 Malt Mill Lane, Halesowen B62 8JA
G7 AYA G. Jessup 25 Harrier Green, Holbury, Southampton SO45 2EY
G7 AYB R. Upton 35 Weston Street, Swadlincote DE11 9AT
G7 AYE R. Phipps 39 Perrinsfield, Lechlade GL7 3SD
G7 AYI L. Faragher 4 Kirloe Avenue, Leicester Forest East, Leicester LE3 3LA
G7 AYL P. Thompson Flat 11, Old Gaol, 16 Grove Street, Bath BA2 6PJ
G7 AYO L. Hutt Fenwick Crossing House, Fenwick Lane, Doncaster DN6 0EZ
G7 AYP A. Gregory 9 Fordbridge Road, Ashford TW15 2TD
G7 AYQ A. Tregay 53 Haverscroft Close, Taverham, Norwich NR8 6LT
G7 AYS D. Webster11 5 Eastfield Road, Princes Risborough HP27 0JA
GM7 AYW J. Hunter 1 Mitchell Drive, Rutherglen, Glasgow G73 3QP
G7 AZA J. Cash 89 Peacocks, Harlow CM19 5NZ
G7 AZC A. Rogers Yoke Farm, Upper Hill, Leominster HR6 0JZ
G7 AZH V. Barber 8 Hollyberry Close, Redditch B98 0QT
G7 AZJ B. Sayers 13 Mulberry Close, Cambridge CB4 2AS
G7 AZT V. Mills 18 Muir Road, Maidstone ME15 6PX
G7 AZV R. Quick Halfway House, Upton Scudamore, Warminster BA12 0AE
G7 AZW M. Ney 4 Rathen Road, Withington, Manchester M20 4GH
G7 BAB W. Foden 209 Lord Lane, Failsworth, Manchester M35 0PX
G7 BAC M. Gohl 4 Yewtree Drive, Hull HU5 5YH
G7 BAE T. Searle Haven Orchard, Exwick Lane, Exeter EX4 2AP
GM7 BAS W. Hunter 2 Wallace Cottages, Southend, Campbeltown PA28 6RX
G7 BAV A. Roberts 4 Rocky Park Road, Plymouth PL9 7DQ
G7 BBC London Bbc Radio Group c/o J. Lee 44 Howard Road, Nuneaton CV10 7ES
G7 BBD C. Hartley 102A Bedford Road, Cranfield, Bedford MK43 0HA
G7 BBJ B. Jenkinson 14 Sandhill Way, Harrogate HG1 4JN
G7 BBN E. Crookall 17 Dundee St., Moorlands, Lancaster LA1 3DS
G7 BBT R. Pitman 21 Old Church Road, St. Leonards on Sea TN38 9HB
G7 BBU Dr C. Sharp Dept. Of Astronomy, University Of Arizona, Tuscon 85721 United States
GW7 BBY M. Jones Awelfa, Llangeler, Llandysul SA44 5EP
GM7 BCC R. Sutherland Tigh - Na - Coille, Mill Road, Nairn IV12 5EW
G7 BCI V. White 6 Laburnum Close, South Anston, Sheffield S25 5GL
G7 BCK N. Gillies 5 Pickmere Terrace, Dukinfield SK16 4JJ
G7 BCO A. Adey 8 Spinners Court, Telford TF5 0PG
GM7 BDD Northamptonshire Grammar School ARC c/o A. Mankin 10 Kerloch Crescent, Banchory AB31 5ZF
G7 BDK P. Blackett 32 Woodstock Road, Carshalton SM5 3DZ
G7 BDR A. Davis 5 Ludlow Close, Loughborough LE11 3TB
G7 BDS J. Mills 42 Maple Grove, Welwyn Garden City AL7 1NL
G7 BEJ W. Young 24 Peebles Road, Newark NG24 4RW
G7 BEP M. Van Der Steeg Horsebrook Farm, South Brent TQ10 9EU
GI7 BET R. Griffin 19 Jubilee Park, Cookstown BT80 8LJ
G7 BFE C. Broom 74A Newton Road, Torquay TQ2 7BN
G7 BFH N. Lambert 23 Knightlands Road, Irthlingborough, Wellingborough NN9 5SU
G7 BGM D. Allison 57 Algarth Road, Pocklington, York YO42 2HJ
G7 BGO P. Toll 83 Pepper St., Lymm WA13 0JT
G7 BGT R. Pykett The Lilacs, Station Road, Thurgarton, Nottingham NG14 7HD
G7 BGV C. Barker 5 Crossgates Wadworth, Doncaster DN11 9TE
G7 BGY M. Bellaby 21 Sprydon Walk, Nottingham NG11 9ET
G7 BGZ M. Hickman 84 Woodward Lakes & Lodges, Holme Wood Lane, Armthorpe, Doncaster DN3 3EH
G7 BHE P. Gledhill 36 Tylers Ride, South Woodham Ferrers, Chelmsford CM3 5ZT
G7 BHG R. Gill 24 Larkfield Crescent, Rawdon, Leeds LS19 6EH
G7 BHR D. Coombes 2 Ormesby Drive, Potters Bar EN6 3DZ
G7 BHU T. Mayfield 184 Wharf Road, Pinxton, Nottingham NG16 6LQ
G7 BHW J. Wilson 61 New Lane, Hilcote, Alfreton DE55 5HT
G7 BHY P. Bailey 21 Westhall Road, Mickleover, Derby DE3 0PA
G7 BIK D. Clark 33 Landers Reach, Lytchett Matravers, Poole BH16 6NB
GW7 BIL W. Knox 9 Harrow Close, Caerleon, Newport NP18 3EF
G7 BIM S. Beazley 10 Barnecut Close, St. Cleer, Liskeard PL14 5RU
G7 BIP G. Roffey 31 Saxville Road, Orpington BR5 3AN
G7 BIQ K. Lloyd 9 Hornbeam Walk, Witham CM8 2SZ
G7 BIV Kilmarnock And Loudoun ARC c/o R. Hudson 12 Magnus Drive, Colchester CO4 9WQ
G7 BIX T. Houlihane Blackgate House, Scotland Road, Dry Drayton, Cambridge CB23 8BX
G7 BIY B. Ansell 26 Stubby Lane, Wolverhampton WV11 3NL
G7 BJB C. Foster 10 Handel Street, Derby DE24 8AZ
G7 BJC M. Wiggins 158 Prince Charles, Avenue, , Derby DE3 4LQ
G7 BJD R. Ward 12 Meadow Lea Knighton Fields, Worksop S80 3QJ
G7 BJE Rvd. I. Godlington 46 Bren Way Hilton, Derby DE65 5HP
G7 BJG I. Harvey 367 Stone Road, Stafford ST16 1LD
G7 BJN J. Barlow 38 St. Pauls Road, Newcastle ST5 2PQ
G7 BJR G. Mitchell 8 Addison Road, Mexborough S64 0DJ
G7 BKJ C. Watney 23 The Wad, West Wittering, Chichester PO20 8AH

G7 BKL B. Moulton 70 St. Georges Avenue, Westhoughton, Bolton BL5 2EU
G7 BKN R. Hatcher 61 Holland Road, Oxted RH8 9AU
G7 BLD C. Holden 26 Valebridge Drive, Burgess Hill RH15 0RW
G7 BLJ R. Maytum 62 Coronation Close, Great Wakering, Southend-on-Sea SS3 0JG
G7 BLK T. Dicks 4 Nicholas Drive, Reydon, Southwold IP18 6RE
G7 BLL B. Weaver 13 Atlay Street, Hereford HR4 9PF
G7 BLT W. Manthorp 49 Cassell Road, Bristol BS16 5DE
G7 BLX M. Rowe 31 Thornhill Avenue, Thornhill, Southampton SO19 6PS
G7 BMC P. Hollis 2 Falcon Drive, Whittington, Lichfield WS14 9PF
G7 BME D. Watts 9 Filwood Drive, Kingswood, Bristol BS15 4HT
G7 BMM S. Hallam 125 Charnwood Road, Shepshed, Loughborough LE12 9NL
G7 BMP N. Gough Oak Den, Park View, Swynnerton, Stone ST15 0QG
G7 BMT S. Hodges 15 Middlewich Street, Crewe CW1 4BS
G7 BMY J. Moore Fairview, 28 Bulmer Lane, Great Yarmouth NR29 4AF
G7 BNB S. Reynolds Dowgill Head House, North Stainmore CA17 4EX
GW7 BNC J. Beadle The Coach House, Cwmdauddwr, Rhayader LD6 5HA
G7 BND B. Welthy 8 Du Cane Place, Witham CM8 2UQ
GM7 BNF M. Harrington Mount Pleasant House, North Road, Wick KW1 4DN
G7 BNI N. Pope 21 Moultrie Road, Rugby CV21 3BD
G7 BNK D. Wood 16 Church Road, Pelsall, Walsall WS3 4QN
G7 BNL A. Creek Westmoor House, Wisbech Road, Ely CB6 1RQ
G7 BNM A. Ison 32 Station Road Lode, Cambridge CB25 9HB
G7 BNN K. Sheppard Woodlands Bungalow, Gunby Road, Skegness PE24 5HT
G7 BNO C. Sheppard The Old Mineral Water Works, Pinfold Lane, , Irby in the Marsh PE245DH
G7 BNS T. Healey 5 St. Johns Crescent, Huddersfield HD1 5DY
G7 BNW G. Ingmire 93 Havelock Road, Luton LU2 7PP
G7 BNZ H. Williams 20 Barham Mews, Teston ME185BL
G7 BOH P. Craig 3 Rothsbury Drive, Eastleigh SO53 4QQ
GM7 BOW R. King 118 Boswell Road, Inverness IV2 3EW
GW7 BOY B. Hodgkinson 16 Swain Avenue, Buckley CH7 3BR
GM7 BOZ A. Bowie 374A High Street, Leslie, Glenrothes KY6 3AX
G7 BPF D. Rose 8 Ambrose Avenue, Hatfield, Doncaster DN7 6QQ
G7 BPG D. Dixon 5 Denbigh Close, Newcastle under Lyme ST5 3DL
G7 BPI C. Thompson 13 Wentworth Avenue, Luton LU4 9EN
G7 BPM N. Hemingway 24 Ealees Road, Littleborough OL15 0HQ
G7 BPN K. Pentecost 46 Austen Way, Crook DL15 9UT
G7 BPQ M. Meadows 4 The Grove, Wharncliffe Side, Sheffield S35 0EA
G7 BPR S. Winlove-Smith 7 Maughan Street, Shildon DL4 1AP
G7 BPX Oville Amateur Radio Portable Club c/o A. Gifford 56 Seymour Road, Gloucester GL1 5QD
G7 BPZ J. May-Golding 9 St. Barts Road, Sandwich CT13 0BG
G7 BQA R. Golding 9 St. Barts Road, Sandwich CT13 0BG
G7 BQM Dr E. Turk Sunny Meadow, Three Bridges Road, Long Buckby Wharf, Long Buckby, Northampton NN6 7PP
G7 BQO J. Hirons Furlong House, Racecourse Lane, Bicton Heath, Shrewsbury SY3 5BJ
G7 BQS M. Rodgers 14 North Street, Rawmarsh, Rotherham S62 5NH
G7 BQT Wirral ARS c/o B. Miller 22B Avondale Road, Fleet GU51 3BS
G7 BQU G. Daynes 25 Redwood Close, Keighley BD21 4YG
G7 BQY A. Brighton 22 Langport Drive Vicars Cross, Chester CH3 5LY
G7 BRA Dr J. Riley 132 Barrs Road, , Cradley Heath B64 7EZ
G7 BRB J. Marsh 39 Palace Gate, Odiham, Hook RG29 1JZ
G7 BRF D. Oliver 10. Stepping Stones, Bidford on Avon B50 4PH
G7 BRJ J. Mills 70 Crescent Road, Rochdale OL11 3LG
GM7 BRL Oldham Am RC c/o D. O'Donnell 188 Warriston Street, Glasgow G33 2LD
G7 BRM Hereford ARS c/o G. Jeffery 48 Minnis Lane, River, Dover CT17 0PR
G7 BRP J. Biggs 5 Churchgate St., Soham, Ely CB7 5DS
G7 BRS J. Rushton 391 Rossendale Road, Burnley BB11 5HP
G7 BRU J. Rhodes Little Meadow, Roke, Wareham BH20 7JF
G7 BRX R. Bell 10 Old Mill Way, Weston-Super-Mare BS24 7AS
G7 BRZ C. Gaunt 39 Sonja Crest, Immingham DN40 2EQ
GW7 BSC R. Snelling 91 Oakfield Road, Newport NP20 4LP
G7 BSF A. Lewis 20 Annes Walk, Caterham CR3 5EL
G7 BSG Bemerton Sc Grp c/o A. Carter 28 Springfield Road, Wellington TA21 8LG
G7 BSK J. Ingram 201 Ocean Drive, #06-11, the Azure Singapore 98584 Singapore
G7 BSL P. Bedford 48 Trentham Drive, Bridlington YO16 6EZ
G7 BSO A. Noble 42 Upper Street, Salisbury SP2 8LY
G7 BSP S. Farmer Horton Brook Cottage, Horton, Wem, Shrewsbury SY4 5NB
GW7 BTC Heckington And District Radio Group c/o R. Magwood 13 Inverness Place, Cardiff CF24 4RU
G7 BTI Madley ARG c/o N. Prosser 35 Holmfirth Close, Belmont, Hereford HR2 7UG
G7 BTP P. Jensen 50 Steynburg Street, Hull HU9 2PF
G7 BUF P. Parkin 2 The Knoll, Dronfield S18 2EH
G7 BUK D. Brinnen 134 Victoria Road, Mablethorpe LN12 2AJ
G7 BUL E. Williamson-Brown Meadowside, Green Lane, Stowmarket IP14 5DS
G7 BUN H. Ellis Home Cottage, Drury Square, King's Lynn PE32 2NA
G7 BUR G. Robinson 228 Bradford Road, Riddlesden, Keighley BD20 5JT
G7 BUS G. Payne 155 Camping Hill, Stiffkey, Wells-next-the-Sea NR23 1QL
G7 BVH M. Gurr Elan, Sandown Road, Sandwich CT13 9NY
G7 BVL K. Lambert 1 Langton Road, Chichester PO19 3LY
G7 BVS N. Hill123 16 Bittern Avenue, Abbeydale, Gloucester GL4 4WA
G7 BVZ B. Allen 84 Holland Road, Little Clacton, Clacton-on-Sea CO16 9RS
G7 BWE E. Gibbons 19 Queens Park Road, Caterham CR3 5RB
G7 BWF A. Gray 147 Kirby Road, Walton on the Naze CO14 8RL
G7 BWI J. White 11 Rowdown, Upper Lambourn, Hungerford RG17 8RF
G7 BWO D. Morgan 26 Lyndhurst Road, Exmouth EX8 3DT
G7 BWV R. Fosbraey 122 East St., Sittingbourne ME10 4RX
G7 BWW M. Widdows 27 Market Close, Barnham, Bognor Regis PO22 0LH
G7 BXA P. Austin 24 Fairfield Terrace, Bramley, Leeds LS13 3DH
G7 BXG F. Clarke 37 Cambridge Road , Rainworth, Mansfield NG21 0AX
G7 BXJ P. Turner 260 New Lane, Huntington, York YO32 9LY
G7 BXL M. Bickerton 54 Swinnel Brook Park, Grane Road, Rossendale BB4 4FN
G7 BXS R. Wake 55 Bearsdown Road, Eggbuckland, Plymouth PL6 5TR
G7 BXT R. Cross 9 Chesham House Leyburn Crescent, Romford RM3 8RU
G7 BXU S. Welton 18 Coningham Road, Reading RG2 8QP
GM7 BYB A. Mcintyre 18 Seal Craig Gardens, Altens, Aberdeen AB12 3SH
G7 BYE J. Hersom 10 Young St., Gilesgate, Durham DH1 2JU
G7 BYG A. Found 18 Mead Fields, Bridport DT6 5RF
G7 BYI M. Baldry 10 Kingfisher Court, Lowestoft NR33 8PJ
G7 BYK R. Barry 14 Home Mead Creswicke Road, Bristol BS4 1UQ
G7 BYN D. Bendrey 73 Kestrel Close, Chipping Sodbury, Bristol BS37 6XB
G7 BYS J. Pollard 25 Heath Avenue, Ramsbottom, Bury BL0 9UN
G7 BYU W. Mcguffie 1 Norbury Drive, Marple, Stockport SK6 6LL
G7 BYV T. Connolly 7 Springfield Crescent, Sherborne DT9 6DN
G7 BYW C. Stone 60 Staddon Park Road, Plymouth PL9 9HJ
G7 BZC R. Pickett 1 The Cottage, Hospital Road, Wingland, Spalding PE12 9YR
G7 BZD P. Yates Kingsomborne The Broadway, Totland Bay PO39 0BL
G7 BZE W. Gillott 14 Oakham Place, Barnsley S75 2ND
G7 BZM R. Brown Apartment 313 2 Austin Way, Birmingham B31 3GG
G7 BZQ G. Reynolds 187 Steelhouse Lane, Wolverhampton WV2 2AU
GW7 BZR J. Shurmer 126 Gaerwen Uchaf Estate, Gaerwen LL60 6JW
G7 BZU A. Mccoll 105 North East Road, Southampton SO19 8AF
GW7 BZY P. Mcfarland 20 Caer Sarn, Caernarfon LL54 7TW
G7 CAA N. Royle 62 Cynthia Close, Poole BH12 3JW
G7 CAF S. Bond 38 Hampsfell Drive, Morecambe LA4 4TU
G7 CAG A. Malpass 48 Geoffrey Barbour Road, Abingdon OX14 2ES
GW7 CAH M. Astley 30 Lon Ceirios, Newtown SY16 1PR
G7 CAS W. Welburn 31 West Bank, Scarborough YO12 4DX
G7 CAT G. Cutter 48 Hyndley Road, Bolsover, Chesterfield S44 6RX
G7 CBI M. Higgins 2 Walden Road, Keynsham, Bristol BS31 1QW
GW7 CBU J. Griffiths 143 Brynglas, Hollybush, Cwmbran NP44 7LL
G7 CBW S. Duffield 21 Calley Close, Tipton DY4 8XY
G7 CBY L. Cotton 6 Blacksmith Row, Lytham St. Annes FY8 4UE
G7 CBZ S. Cotton 6 Blacksmith Row, Lytham St. Annes FY8 4UE
G7 CCL S. Cullingworth 19 Springbank Garforth, Leeds LS25 1DD
GW7 CCR Flintshire Raynet c/o M. Ellett 14 Canon Drive, Bagillt CH6 6LS
G7 CCS G. Oliver 17 Jack Stephens Estate, Penzance TR18 2QE
G7 CCV P. Upton 73 Allington Close, Taunton TA1 2NA
G7 CDI P. Matthews 128 Thealby Gardens, Doncaster DN4 7EG
G7 CDO A. Corps 6A Salisbury Road, Leigh-on-Sea SS9 2JX
GW7 CEA D. Smith 62 Cheshire View. Brymbo., Wrexham LL11 5AW
G7 CEB J. Redpath 41 Sandford Green, Banbury OX16 0SB
G7 CEC J. Evans 28, Kenneth Gamble Court, West Avenue, Wigston LE18 2FP
G7 CED K. Barnett 126 Oldham St., Latchford, Warrington WA4 1EX
G7 CEN T. Martin 10 Hardy Close, Galley Common, Nuneaton CV10 9SG
GW7 CEQ W. Jones 26 Cwm Silyn, The Park, Caernarfon LL55 2AG
G7 CER C. Riley-Moxon 51 Nuttall Lane, Ramsbottom, Bury BL0 9JX
G7 CEW D. Everard 6 Leith Hill Green, St. Pauls Cray, Orpington BR5 2SB
G7 CEY P. Copeland 7 Stuart Avenue, Draycott, Stoke-on-Trent ST11 9AA
G7 CFC A. Chamberlain 11 Woden Crescent, Wolverhampton WV11 1PR
G7 CFS D. Halliday 33 Brentnall Close, Great Sankey, Warrington WA5 1XN
G7 CFT G. Taylor 48 Westwood Heath Road, Leek ST13 8LL
G7 CFW K. Morrison 7 Turnstone End, Yateley GU46 6PE
G7 CFX A. Newell 4 Dexter Square, Cricketers Way., Andover SP10 5DB
G7 CGC P. Oliver 32 Pearmain Way, Stanway, Colchester CO3 0NP
G7 CGI C. Honey 1 Brimley Court, Lower Brimley Road, Teignmouth TQ14 8LW
G7 CGN S. Hodkinson 17 Thorn Well, Westhoughton, Bolton BL5 2PJ
G7 CGT A. Day 8 The Garth, Ash, Aldershot GU12 6QN
G7 CHB Lord R. Montague 71 Middlethorpe Road, Cleethorpes DN35 9PP
G7 CHC J. Anderson111 Tre Noce Cottage, 75 St. Michaels Road, Paignton TQ4 5NA
G7 CHG R. Gould 16 Oakleaf Rise, Far Forest, Kidderminster DY14 9AE
G7 CHW Mid Severn Valley Raynet c/o A. Saunders 4, Severn Way, Bewdley DY12 2JQ
G7 CIA Dr J. Adams 14 Kedleston Close, Northampton NN4 0WF
G7 CIH C. Hayes 2 Castleford House, Castle Road, Okehampton EX20 1HZ
G7 CIK M. Kielthy 35 Alexandra Road, Sheringham NR26 8HU
G7 CIQ A. Wiseman 61 Hilton Avenue, Horwich, Bolton BL6 5RH
G7 CIT T. Mcguigan 18 Whalton Close, Gateshead NE10 8SW
G7 CIU P. Burbury 43 Locomotive Street, Darlington DL1 2QF
G7 CIV B. Perrin 8 Station Road, Gretton, Corby NN17 3BU
G7 CJC J. Hughes Hollies Bungalow, Valeswood, Shrewsbury SY4 2LH
G7 CJD C. Dyer 6 Witcombe, Yate, Bristol BS37 8SA
G7 CJG G. Stringer 13 Garfield St, Kettering NN15 7HX
G7 CJO J. Graves 7 Matthews Chase, Binfield, Bracknell RG42 4UR
G7 CJS D. Evans 16 Cruden Road, Gravesend DA12 4HD
G7 CJW J. Wardle 9 Leefield Road, Chapel-En-Le-Frith, High Peak SK23 0LF
G7 CKE D. Scriven 59 Breck Lane Dinnington, Sheffield S25 2LJ
G7 CKG R. Varley 23 Manor Court, Bingley BD16 1QD
G7 CKL B. Taylor 32 Marples Avenue Mansfield Woodhouse, Mansfield NG19 9HA
G7 CKP J. Surman 122 Burwell Meadow, Witney OX28 5JQ
G7 CKQ Taunton & Somerset Raynet Group c/o K. Hartley 39 Raleigh Road, Sunderland SR5 5RD
G7 CKS D. Davies 2 London Road, Battle TN33 0EU
G7 CLG P. Ord 52 Hillside Road, Stockton on Tees TS20 1JQ
G7 CLH D. Smith 7 Dunlin Close, Norton, Stockton-on-Tees TS20 1SJ
GM7 CLM J. Harrington Mount Pleasant House, North Road, Wick KW1 4DN
G7 CLO C. Gaukroger Meadow View, Lansallos, Looe PL13 2PU
G7 CLX K. Marsden 3 Lane Head, Heptonstall, Hebden Bridge HX7 7PB
G7 CLY J. Hill 55 The Oval, Welton, Brough HU15 1DA
G7 CMB D. Kennett 6 The Holdings, Hatfield AL95HQ
GI7 CMC N. Moore 94 Orby Drive, Belfast BT5 6AG
GU7 CMH G. Simon 3 Mahaut Villas, Collings Road, St. Peter Port GY1 1FP Guernsey
G7 CMI Prof. B. Birch 4 Kynnesworth Gardens, Higham Ferrers, Rushden NN10 8NH
GW7 CMM G. Jones 13 Palace Close, Flint CH6 5YE
G7 CMN Cheshire County Raynet c/o B. Williams 3 Welton Close, Wilmslow SK9 6HD
G7 CMP B. Fielding The Copse Charmouth Road, Axminster EX13 5SZ
G7 CNC D. Gray Flat 57, Wesley Court, 1 Millbay Road, Plymouth PL1 3LB
G7 CND J. Bell 2 Rake Lane, Milford, Godalming GU8 5AB
GU7 CNI M. Elliston La Guillard Lane, St. Andrew GY6 8YJ Guernsey
G7 CNM D. Clark 33 Antrim Road, Lincoln LN5 8TF
G7 CNP H. Weatherhead 39 Meadow Park, Dawlish EX7 9BU
GM7 CNW G. Dryburgh 86 Normand Road, Dysart, Kirkcaldy KY1 2XP
G7 CNX P. Hammond 23 Peppers Close, Brandon IP270PU
G7 CNZ K. Ford 8 Blakedon Road, Wednesbury WS10 7HY
G7 COA R. Johnson 7 West Parade, Warminster BA12 8LY
GW7 COB S. Coburn 54 Queensway, Hope, Wrexham LL12 9PE

G7 COC N. Whelan 54 Boroughbridge Road, Northallerton DL7 8BN
G7 COD A. Kitchen 4 Dairy Cottages, Bank Newton, Skipton BD23 3NT
G7 COG Coleraine & Dis c/o J. Lister 6 Fordlands Crescent, Fulford, York YO19 4QQ
G7 COP P. Payton 3 Astor Close, Winnersh, Wokingham RG41 5JZ
G7 COQ K. Raxworthy 9 Harrow Drive, Edmonton, London N9 9EQ
G7 COY C. Keevil 82 Chatham Grove, Chatham ME4 6LY
GM7 CPJ G. Currie The Old Post Office, Cuminestown AB53 5TQ
GM7 CPL C. Scott 115 Tarvit Terrace, Springfield, Cupar KY15 5SE
G7 CPN S. Burgess 59 Back Lane, Congleton CW12 4PY
G7 CPQ C. Ambrose 47 Whitton Close, Swavesey, Cambridge CB24 4RT
GM7 CPR J. Wright 9 Meadowhead Road, Plains, Airdrie ML6 7JF
GM7 CPY S. Leggat Ailach, St. Aethans Road, Elgin IV30 2YR
G7 CQA R. Ginn 91 High Street, Shoeburyness, Southend-on-Sea SS3 9AR
GW7 CQB D. Locock Bank House, Selattyn, Oswestry SY10 7DX
G7 CQG Madley ARG c/o R. Bradshaw 11 Glebe Avenue, Orton Waterville, Peterborough PE2 5EN
G7 CQH M. Smith 14B Witham Bank West, Boston PE21 8PU
G7 CQK Capt. I. Phillips Goldsworthy Farm, Stony Lane, Gunnislake PL18 9BL
GU7 CQN J. Gardner The Ferns, Rue De La Girouette, St. Saviour, Guernsey GY7 9NN
GM7 CQQ A. Donaldson 36 Rothes Park, Leslie, Glenrothes KY6 3LH
G7 CQW A. Riley 4 Birtle Drive Astley, Tyldesley, Manchester M29 7RE
G7 CQX D. Read 31 Grace Gardens, Bishop's Stortford CM23 3EU
G7 CQZ E. Courtnell 38 Woodchurch Lane, Birkenhead CH42 9PH
G7 CRA I. Brelsford 78 Borough Road, Redcar TS10 2EQ
G7 CRK S. Halbertsma 65 Gareth Grove, Bromley BR1 5EG
G7 CRM D. Stump 9 Shipton Grove, Swindon SN3 1BZ
G7 CRN R. Phin 35 Parkland Close, Newquay TR7 3EB
G7 CRQ A. Haywood Flat 21, Atholl House, 178 Woodcote Road, Wallington SM60PB
G7 CRR J. Wheeler 8 Slimbridge Close, Worcester WR5 3SH
G7 CRS Sos Radio Group c/o M. Hall 35 Bunns Lane, Dudley DY2 7RA
G7 CRU P. Wallis 6 Lancelott Court, Pershore WR10 1RE
G7 CRV B. Burnside The Vicarage, Baxtons Road, Helmsley, York YO62 5HT
G7 CRY J. Bain 7 Wrights Lane, Sutton Bridge, Spalding PE12 9RH
G7 CSD N. Whittle 31 St. Georges Avenue, Southall UB1 1PZ
G7 CSE R. Rowthorn 25 Branscombe Drive, Wootton Bassett, Swindon SN4 8HS
G7 CSF J. Plowright 11 Coddington Street, Newport United States
G7 CSI M. Morris 20 Bracken Way, Chobham, Woking GU24 8PR
G7 CSJ K. Chapman 19 St. Johns Rise, Woking GU21 7PN
GW7 CSK D. Winter 25 Pembroke St., Thomastown, Porth CF39 8DU
G7 CSL R. Dent 9 Dovey Close, St. Ives, Huntingdon PE27 6HW
G7 CSM R. Knight 32 Linnet Close Abbeydale, Gloucester GL4 4UA
G7 CSS M. Budd 103 Old Charlton Road, Shepperton TW17 8BT
G7 CST A. Clarke 86 Roebuck Road, Walsall WS3 1AL
G7 CSV Spen Valley ARS c/o T. Clough 37 Park Avenue, Mirfield WF14 9PB
G7 CSX A. Keen 20 Horam Park Close, Horam, Heathfield TN21 0HW
G7 CTE N. Farmer 4 Frank Hughes Avenue, Sandbach CW11 3TA
G7 CTG E. Ives 15 Northlands, Adwick-Le-Street, Doncaster DN6 7AX
G7 CTN M. Puddephatt Church Barn, Kirklington Road Eakring, Newark NG22 0DA
GM7 CTQ J. Dovaston 7 Mulloch View Dalry, Castle Douglas DG7 3UJ
G7 CTT S. Lock 17 Elgar Crescent, Droitwich WR9 7SP
GM7 CTV A. Smith 13 Park Terrace, Markinch, Glenrothes KY7 6BN
GI7 CTW E. Regan 4 Lecumpher Road, Desertmartin, Magherafelt BT45 5LY
G7 CUA R. Cookson Briarswood, Snow Hill Lane, Preston PR3 1BA
G7 CUB D. Price Summer Fields, Mayfield Road, Oxford OX2 7EN
G7 CUD J. Richardson 68 Place Farm Way, Monks Risborough, Princes Risborough HP27 9JY
G7 CUF G. Adrian 140 Shawfield Road Ash, Aldershot GU12 6SG
G7 CUL R. Bourn 7 Clitheroes Lane, Freckleton, Preston PR4 1SD
G7 CUO Devon Data Group c/o J. Folland 41 Rydal Avenue, Billingham TS23 1HX
G7 CUP P. Ingle 8 Burnstone Gardens, Moulton, Spalding PE12 6PS
G7 CUU D. Stainforth -Small 10 Balland Park, Ashburton, Newton Abbot TQ13 7BS
G7 CUW B. Minish Raheens, Castlebar Ireland
G7 CUY Dr K. Martin 8 Taylors Close, Meppershall, Shefford SG17 5NH
G7 CVA E. Curnow 9 Moreton Bay, Bilton, Hull HU11 4ER
G7 CVC M. Porter 19 Thistle Downs, Northway, Tewkesbury GL20 8RE
G7 CVF S. Evans 34 Kent Road, Southport PR8 4BJ
G7 CVM D. Illman 27 Blackborough Road, Reigate RH2 7BS
G7 CVY K. Helgesen 13 Mara Court, White Road, Chatham ME4 5TW
G7 CVZ A. Bevins 12 Wheatstone Road, Formby, Liverpool L37 6BF
G7 CWE D. Ishmael 38 Greenford Close, Orrell, Wigan WN5 8RH
G7 CWI I. Green 25 Riley Avenue, Lytham St. Annes FY8 1HZ
G7 CWM J. Denton 48 Seas End Road, Surfleet, Spalding PE11 4DQ
G7 CWN F. Merchant 3 Main Road, Shortwood, Mangotsfield BS16 9NH
G7 CWO D. Ward 107 Oundle Road, Birmingham B44 8ER
G7 CWT Prof. M. Joy Cheddon Corner, Cheddon Fitzpaine, Taunton TA2 8LB
G7 CWX C. Tarling 36 Lancaster Road, North Weald CM16 6JA
G7 CXB T. Mcinnes 7 Hilary Drive, Merry Hill, Wolverhampton WV3 7NJ
G7 CXO V. Cassar 51 Aylesford Avenue, Beckenham BR3 3SB
G7 CXT S. Haywood 7 Stamford Gardens, Dagenham RM9 4ET
G7 CXU S. Power 8 Green Lane, Chislehurst BR7 6AG
G7 CYD A. Jenkins 15 Tilstone Avenue, Eton Wick, Windsor SL4 6NF
G7 CYF C. Wardill 85 Station Road, Chellaston, Derby DE73 5SU
G7 CYN C. Casper 22 Eshton Road, Gargrave, Fell View, Skipton BD23 3SE
G7 CYQ E. Hornby 14 Essex Road, Stevenage SG1 3EZ
GW7 CYT D. Phillips 15 Herbert Street, Treorchy CF42 6AW
GM7 CZC R. Johnson 3 Hamilton Gardens, Edinburgh EH15 1NH
G7 CZF J. Jenkins 42 Butlers Court Road, Beaconsfield HP9 1SG
G7 CZL G. Miles 7 Dobbin Close Rawtenstall, Rossendale BB4 7TH
GM7 CZU J. Mclaughlan 2 Donaldson Drive, Irvine KA12 0QG
G7 DAB Holy Island ARC c/o I. Weeks 19 St. Michaels Road, Tunbridge Wells TN4 9JG
G7 DAH C. Moore 168 Church End Lane, Runwell, Wickford SS11 7DN
GU7 DAI J. Smith Ravello, 35 Le Villocq Estate, Le Villocq, Castel, Guernsey GY5 7SQ Guernsey
GM7 DAJ R. Hepburn 44 Macindoe Crescent, Kirkcaldy KY1 2JG
G7 DAL J. Ratigan 81 Cunningham Drive, Unsworth, Bury BL9 8PD
GM7 DAP A. Lord 5 Windsor Terrace, Brechin DD9 6SD
G7 DAR R. Charlton 13 Hollywood Avenue, Walkerville, Newcastle upon Tyne NE6 4TN
G7 DAZ J. Watson Flat 1, 53 Castle Street, Bolton BL2 1AD
G7 DBN E. Turner Rectory Cottage, Little Marsh, Marsh Gibbon, Bicester OX27 0AP
G7 DBO I. Leaver 2 Marnhull Close, Coventry CV2 2JS
G7 DBT R. Claridge 3 Wentworth Avenue, Leagrave, Luton LU4 9EN
G7 DBV D. Ritson C/O 12 Tudor Grange, Easington Village, Peterlee SR8 3DF
GI7 DBZ W. Hollinger 51 Collin Road, Ballyclare BT39 9JS
G7 DCF A. Mclennan 2 Phillip St, Beachmere 4510 Australia
G7 DCJ S. Warner 96 Walter Nash Road East, Kidderminster DY11 7BY
G7 DCM L. Bant 58 Severn Way, Cressage, Shrewsbury SY5 6DS
G7 DCT A. Horsfall 2 Temple Walk, Halton, Leeds LS15 7SQ
G7 DDF P. Johnson 6 Rugby Road, Lilbourne CV23 0SP
G7 DDN C. Rolinson 534 Haslucks Green Road, Shirley, Solihull B90 1DS
G7 DDQ W. Roberts 12 Camberley Drive, Penn, Wolverhampton WV4 5RP
G7 DDR R. Murray 8 Church Lane, Kirk Langley, Ashbourne DE6 4NG
G7 DDV C. James Apartment 78, Seckle House Steepleton Court, Cirencester Road, Tetbury GL8 8FQ
G7 DEC A. Grundy 647 Preston Old Road, Feniscowles, Blackburn BB2 5ER
G7 DEE K. Denniss 4 Aysgarth Road, Sheffield S6 1HU
G7 DEF J. Kingsley 3 The Orchard, Swarland, Morpeth NE65 9NB
G7 DEG R. Williams 6 Ralfland View, Shap, Penrith CA10 3PF
G7 DEH H. Carpenter 44 Bowbridge Road, Newark NG24 4BZ
G7 DEI S. Courtney-Crowe 28 Brymore Close, Prestbury, Cheltenham GL52 3DY
G7 DEU G. Renton 58 St. Christophers Road, Humberston, Grimsby DN36 4EA
G7 DEY P. Knowles 35 Raby Park Road, Neston. CH64 9SW
G7 DFC J. Worsnop 1217 Thornton Road, Thornton, Bradford BD13 3BE
G7 DFP J. Fitzpatrick 22 Ferry Road Surlingham, Norwich NR14 7AR
G7 DFV G. Jelley 28 Blanches Road, Partridge Green, Horsham RH13 8HZ
G7 DFW A. Crisp 17 Gaitskell House, Howard Drive, Borehamwood WD6 2PB
G7 DFX G. Allan Kent House, 106 Kent Road, Sheffield S8 9RL
G7 DGC M. Lewis 21 Woodlands Road, Ashton-under-Lyne OL6 9DU
G7 DGD B. Walton 28 Durham Terrace, Durham DH1 5EH
G7 DGE A. Biggin 14 Coultas Avenue, Deepcar, Sheffield S36 2PT
G7 DGF D. Coupe 22 West Street, South Normanton, Alfreton DE55 2AJ
G7 DGP B. Barrass 7 The Crescent, Easton On The Hill, Stamford PE9 3LZ
GM7 DHA K. Pugh 1 Barrington Gardens, Beith KA15 2BA
G7 DHD D. Dyson 5 Warwick Street, Church, Accrington BB5 4AL
GW7 DHG A. Gardner 28 Usk Court, Thornhill, Cwmbran NP44 5UN
G7 DHJ M. Harding 79 Beaumont Walk, Leicester LE4 0PP
G7 DHQ G. Davies 17 Remington Road, Walsall WS2 7EJ
G7 DHW D. Neale Greyhills Farm, Diptford, Totnes TQ9 7NQ
G7 DIB A. Finon Radford House, Hall Lane, Wymondham NR18 9TB
G7 DIE S. Salmon 35 Westgate Road, Lytham St. Annes FY8 2SG
G7 DIG R. Dee 10 Sanderson Street, Coxhoe, Durham DH6 4DG
GW7 DIL P. Davis The Willows, Park View, Cwmbran NP44 1RB
G7 DIO C. Harper 132 Park Avenue East, Dallas, Ga 30157 United States
G7 DIR A. Brinton 136 Efford Road, Plymouth PL3 6NQ
G7 DIS C. Baker Moffat House, Church Road, Broughton Moor, Maryport CA15 7SS
GI7 DIT D. Roberts 6 Plantation Road, Bangor BT19 6AF
G7 DIW A. Saul 18 Elm Bank Close, Cubbington, Leamington Spa CV32 6LR
G7 DIZ M. Beatrup Ivella, Recreation Street, Dudley DY2 9EU
G7 DJA M. Rayner 38A Fen Road Milton, Cambridge CB24 6AD
GW7 DJL T. Davis 127 Lon Glanyrafon, Newtown SY16 1QT
G7 DJN A. Coates 19 Bretton Avenue, Bolsover, Chesterfield S44 6XN
G7 DJT D. Tinley 2 Rosemount Close, Loose, Maidstone ME15 0AJ
G7 DJY N. Monkman 59 Fairway Approach, Normanton WF6 2LX
G7 DKB D. Simons 65 Dolphin Court Road, Paignton TQ3 1AB
G7 DKY G. Rowntree 20 Lancaster Lane, Leyland PR25 5SN
G7 DKZ J. Stearn Half Acre, Hatch Green, Taunton TA3 6TN
GW7 DLD R. Hilton 9 Waterloo Fields, Kingswood, Welshpool SY21 8LF
G7 DLE J. Hodges 70 Chestnut Drive, Brixham TQ5 0DD
GM7 DLY J. Whitcomb 1/1 30 Highburgh Road, Glasgow G12 9DZ
G7 DME R. Gornall 58 Homelawn House Brookfield Road, Bexhill on Sea TN401PN
G7 DMG W. Hetherington The Laurels 8 Kings Lane, Yelvertoft NN6 6LX
G7 DMH S. Hetherington 8 Kings Lane Yelvertoft, Northampton NN6 6LX
G7 DMK P. Drage Cranford House, 167 Rockingham Road, Kettering NN16 9JA
GM7 DMN Y. Benting Suthainn, Askernish, Isle of South Uist HS8 5SY
G7 DMP J. Barnes 26 Fairthorn Road, Sheffield S5 6LX
G7 DMQ S. Rafferty 22 Hengist Close, Horsham RH12 1SB
G7 DMS M. Horsfall 8 Greenbrook Road, Burnley BB12 6NZ
G7 DMX N. Taylor 5 Miranda Road, Preston, Paignton TQ3 1LE
G7 DMZ B. Knight Conchardrin, Tibberton, Gloucester GL2 8EB
G7 DNF T. Haye Woodside, Sutton Wood Lane, Alresford SO24 9SG
G7 DNG I. Casey 38 Wordsworth Road, Salisbury SP1 3BH
GJ7 DNI S. Mcadams Gemaur, St. Clements Road, St. Helier JE2 4PX Jersey
GJ7 DNJ I. Meade Etape De Base, La Rue Des Platons, Trinity, Jersey JE3 5AA
G7 DNM E. Ashworth 10 Wisteria Drive, Lower Darwen, Darwen BB3 0QY
G7 DNP A. Patel 11716 Vista Meadow Ln, Frisco 75035 United States
G7 DNQ S. Howard 5 Grummock Avenue, Ramsgate CT11 0RR
G7 DNR K. Crookes 64 Heron Drive, Audenshaw, Manchester M34 5QX
G7 DNT K. Handscombe 6 Abbeyfield House Market Cross, Malmesbury SN16 9AS
G7 DNV L. Carter 3 Cleviscroft, Stevenage SG1 1UJ
G7 DNX M. Morris 4 Meadow Brook Road, Northfield, Birmingham B31 1NE
G7 DOA D. Morris Flat 3, 748 Melton Road, Leicester LE4 8BD
G7 DOD S. Mann 55 Taunton Street, Wavertree, Liverpool L15 4ND
G7 DOF K. Mount 4 Hermitage Road, Abingdon OX14 5RN
G7 DOL Royal Naval ARS c/o J. Kirk 111 Stockbridge Road, Chichester PO19 8QR
G7 DOR Dorking And District RS c/o C. Berry 60 Copthorne Road, Leatherhead KT22 7EE
G7 DOS Prudential ARS c/o B. Smith 7 School Walk, Chase Terrace, Burntwood WS7 1NQ
G7 DOW G. Smith 59 Radipole Lane, Weymouth DT4 9RR
G7 DOY J. Baddeley 22 Scott Road, Denton, Manchester M34 6FT
G7 DPE J. Glass 8 Hazlemere View, Hazlemere, High Wycombe HP157BY
G7 DPF G. Brightman 5 Meadow Rise, Lacey Green, Princes Risborough HP27 0QY
GD7 DPG J. Wrigley 20 Fairy Hill Close, Ballafesson, Port Erin IM9 6TJ Isle of Man
GM7 DPI P. Blacklaw Flat 15, Servite House 21A High Street, Monifieth, Dundee DD5 4AA
G7 DPR F. Overbury 47 The Maltings, Dunmow CM6 1BY
G7 DPV A. Marks 57A Bulcote Drive Burton Joyce, Nottingham NG14 5AZ

G7 DPW A. Butterworth 3 Fir Tree Avenue, Worsley, Manchester M28 1LP
G7 DPZ E. Curd 11 Ashkirk Close, Waldridge, Chester le Street DH2 3HY
G7 DQA J. Hallin 12 Church Park Road, Plymouth PL6 7SA
G7 DQE M. Meerman University Of Surre, Dept Elec. Eng., Surrey GU2 5XH
G7 DQL P. Perkins 29 Parkhill, Middleton, King's Lynn PE32 1RJ
G7 DQQ Herts County C/ c/o M. Lampett 130 Clopton Road, Birmingham B33 0RL
G7 DQZ D. Keen 5 London Road, Uckfield TN22 1HU
G7 DRD T. Platt 30 Great Ellshams, Banstead SM7 2BA
G7 DRG R. Moseley 307 Archer Road, Stevenage SG1 5HF
G7 DRO W. Webber Springfield Lodge, Broadway, Winscombe BS25 1UE
G7 DRR D. Bellinger Holly Cottage, Deacons Lane, Thatcham RG18 9RJ
G7 DRT M. Dickinson 1 Tregaron Avenue, Cosham, Portsmouth PO6 2JU
G7 DRU A. Tink 13 The Wicketts, Filton, Bristol BS7 0SR
G7 DRW V. Wynn 12 Holly Road, Orpington BR6 6BE
G7 DRX J. Stelmasiak Golden Rocks, Coed Lane, Montgomery SY15 6AB
GM7 DRY S. Graham 15 Stone Crescent, Mayfield, Dalkeith EH22 5DT
G7 DSA S. Jeffery 7 Corfe Way, Winsford CW7 1LU
GU7 DSB P. Blampied 9 Rue Des Grons Estate, St. Martin GY4 6JT Guernsey
G7 DSO P. Yarnold 162 Harwill Crescent, Aspley, Nottingham NG8 5LF
G7 DSQ R. Roberts 16 Poplar Drive, Pucklechurch, Bristol BS16 9QF
G7 DST Dr D. Thomson 96 Rydal Avenue, Loughborough LE11 3RX
G7 DSU C. Tong 24 James Road Cuxton, Rochester ME2 1DJ
G7 DSV D. Crinson 16 Lomax Street, Geeat Harwood BB67DJ
GJ7 DTA A. Lange Les Bois, La Rue De La Pointe, St. Peter JE3 7AQ Jersey
GW7 DTB R. Dixon 19 The Burrows, Porthcawl CF36 5AJ
GM7 DTC J. Arthur 15 St. Andrews Place, Beith KA15 1JE
G7 DTG S. Le Poer Trench Brown Holy Mill, Longville TF13 6ED
G7 DTK T. Timms 16 Claverdon Road, Coventry CV5 7HP
G7 DTR M. Penny 79 Grove Avenue, New Costessey, Norwich NR5 0JA
G7 DTS A. Lowe 33 Dandies Chase, Eastwood, Leigh-on-Sea SS9 5RF
G7 DTT A. Reeves 13 Maple Way, Leavenheath, Colchester CO6 4PQ
G7 DTV H. Partridge 11 Elm Close, Stourbridge DY8 3JH
G7 DUB P. Spencer 11 Two Trees Estate, Wadebridge PL27 7PG
G7 DUC Basildon Dist Rd c/o B. Tonkin 9 Penhallick Road, Carn Brea, Redruth TR15 3YJ
G7 DUE W. Davis 6 Bushy Mead, Waterlooville PO7 5DY
GW7 DUI C. Teague 27 Pond Mawr, Maesteg CF34 0NG
G7 DUK M. Clarke 12 Moat Court, Shaw Close, Chertsey KT16 0PH
GJ7 DUX S. Raynes Mon Plaisir, La Rue De Samares, St. Clement JE2 6LZ Jersey
G7 DUY R. Jones 14 Lunsford Road, Liverpool L14 0NU
GD7 DUZ S. Kelly 16 Close Rushen, Castletown IM9 1NL
GW7 DVJ D. Maxted 33 Bryn Dryslwyn, Bridgend CF31 5BT
G7 DVO T. Spearing 139 Holt Road, Hellesdon, Norwich NR6 6UA
GI7 DWF J. Murphy 19 Kilburn Park, Armagh BT61 9HA
G7 DWH E. Shaw 5 Charlock Grove, Cannock WS11 7FR
G7 DWI A. Davies 23 Holly Place, Eastbourne BN22 0UT
G7 DWM C. Hadjigeorgiou 26 Priory Gardens, Hampton TW12 2PZ
G7 DWN A. Keen 29 Churchman Close Melton, Woodbridge IP12 1RN
G7 DWO S. Prisk 86 Wycliffe Grove, Werrington, Peterborough PE4 5DF
G7 DWQ P. Donovan 35 Ash Drive, Wardley, , Swinton M27 9QP
G7 DWU M. Stapleton 15 Haviland Way, Cambridge CB4 2RA
G7 DWV Dr K. Webster 9 King John Avenue, Fareham PO16 9AP
G7 DWY I. Barraclough Maru, 25 Blaithroyd Lane, Halifax HX3 9PS
G7 DXB L. Watson 9 Croft Close Cumwhinton, Carlisle CA4 8FG
G7 DXC J. Maxwell Tysties, Tile Barn, Newbury RG20 9UY
GM7 DXE A. Todd Waterfurrows, Breakachy, Beauly IV4 7AE
G7 DXH R. Ife 43 Bracknell Crescent, Nottingham NG8 5EU
G7 DXN P. Chapman 1 Bader Close, Watton, Thetford IP25 6FF
G7 DXQ K. Salt 4 Burghwood Road, Ormesby, Great Yarmouth NR29 3LT
GM7 DXT J. Macleod 59 Fife Street, Keith AB55 5EG
G7 DXV P. Shepherd 25 Tomkins Close, Stanford-le-Hope SS17 8QU
G7 DXX J. Walker 121 Park Drive, Upminster RM14 3AU
G7 DYB A. Rudling 1 St. Anthonys Close, Ottery St. Mary EX11 1EN
G7 DYD M. Green 59 Brand End Road Butterwick, Boston PE22 0JD
G7 DZD Dr K. Quinlan Polidoris Cottage, Polidoris Lane, High Wycombe HP15 6XD
GI7 DZE S. Thompson 19 Windsor Heights, Larne BT40 1UL
GM7 DZK J. Malone 8 St. Margarets Crescent, Polmont, Falkirk FK2 0UP
G7 DZR G. Shand 5 Bromyard Drive, Chellaston, Derby DE73 6PF
G7 DZY B. Daniel Tamar Bay Rd, Freshwater Bay, Isle of Wight PO40 9QS
G7 EAA D. Mills 3 Brookside Glasbury, Hereford HR3 5NF
G7 EAH Dr P. Stewart 34 North Park Road, Bramhall, Stockport SK7 3JS
G7 EAQ D. Potten 151 Sherborne Road, Yeovil BA21 4HF
G7 EAR Echelford ARS c/o P. Gray 2 Bryan Close, Sunbury Oppn Thames TW16 7UA
G7 EAT Dr J. Hatfield 22 Blackhorse Crescent, Amersham HP6 6HP
G7 EBF P. Langfield 88 Counthill Road, Oldham OL4 2PE
G7 EBI M. Evans 87 Tintagel Close, Andover SP10 4DB
G7 EBL G. Orlebar 21 Field Lane, Willersey, Broadway WR12 7QB
GI7 EBM A. Stewart Flat17 Forest Glen Mark Street, Glenarm BT44 0AN
G7 EBR M'hd & E.Berk Rd c/o R. Mclachlan Heathersett, Lightlands Lane Cookham, Cookham SL6 9DH
G7 ECA C. Price 18 Armley Park Road, Leeds LS12 2PG
G7 ECE Dr S. Holmes 31 Brightside Avenue, Staines TW18 1NE
G7 ECG K. Thompson 32 The Crescent Pattishall, Towcester NN12 8NA
G7 ECQ N. Murray East End House, Oak Lane, Sheerness ME12 3QR
G7 ECU Eastbourne & Wealden Raynet c/o J. Eade High Shaw Sandrock Hill, Sedlescombe TN33 0QR
G7 EDA R. Woodcock 5 Walton Road, Sidcup DA14 4LJ
G7 EDF D. Hall 29 Airedale Avenue, Tickhill, Doncaster DN11 9UH
G7 EDK I. Barkley 39 Fulbeck Avenue, Wigan WN3 5QN
G7 EDZ M. Hedges 17 Fairview Road, Dudley DY1 2RT
G7 EED N. Winter 66 Priory Wharf, Birkenhead, Wirral CH41 5LD
G7 EEE A. Mothew 7 Ashfields, Loughton IG10 1SB
G7 EEG T. Cole Walden, The Common, Lydney GL15 6NT
G7 EEJ D. Swift 63 Guiness Trust Buildings, Fulham Palace Road, London W6 8BD
G7 EEN S. Paget 2 Willow Cottages, Huxley Lane, Chester CH3 9BE
G7 EFA C. Horsfield Flat 2, Rosemary Court 53 Chantrey Road, Sheffield S8 8QU
G7 EFG M. Marston Apt 12B La Palmeras De Benavista, Calle Virgo, Malaga 29680 Spain
G7 EFL A. Crooks 32 March Road, Wimblington, March PE15 0RN
G7 EFV W. Waterton 23 Mill Drive, Leven, Beverley HU17 5NR
GM7 EGQ I. Harrop South Millburn Cottage, Benslie, Kilwinning KA13 7QY
G7 EGU Kettering & District A.R.S c/o P. Adams 12 The Birches, Benfleet SS7 4NT
G7 EGX M. Miller 12 Leighfields Avenue, Leigh-on-Sea SS9 5NN
GW7 EHD D. Cotton 135 Main Road, Bryncoch, Neath SA10 7TW
GM7 EHN J. Baird 26 Bearside Road, Stirling FK7 9BY
G7 EHR R. Watson 2 Stanley Cottages, Fair View Lane, Colyford, Colyton EX24 6QZ
G7 EHS Amplitude Modulation ARS c/o I. Tooley L'Eree, Burnham Road, Chelmsford CM3 6DP
G7 EHU B. Walley 52 Main St., Rosliston, Swadlincote DE12 8JW
G7 EHY D. Robinson 130 Magnolia Drive, Colchester CO4 3LX
G7 EIA S. Ralph 70 Mickleburgh Hill, Herne Bay CT6 6DX
G7 EIE T. Jacobs 43 Winfields, Pitsea, Basildon SS13 1HA
G7 EIK G. Edlin 2 Ashby Road, Donisthorpe, Swadlincote DE12 7QG
G7 EIS E. Shaddick 6 Haylands, Portland DT5 2JZ
G7 EJH J. Tyerman 7 Veronica Close, Branston, Lincoln LN4 1PU
G7 EJK J. Turner Shuckstone Lodge, Shuckstone Lane, Tansley, Matlock DE4 5GT
G7 EJO I. Oxley 29 The Gables, Newhall, Swadlincote DE11 0TG
G7 EKC A. Taylor 106 Raeburn Avenue, Surbiton KT5 9EA
G7 EKG P. Ainscow 10 Rectory Road, Felling, Gateshead NE10 9DH
G7 EKH S. Rishton 14A The Green, Settle BD24 9HL
G7 EKJ Kent Active Radio Amateurs c/o S. Cox 137 Perry Walk, Blackrock Road, Birmingham B23 7XL
G7 EKL N. Denker Nafferton, Killerton Road, Bude EX23 8EN
G7 EKM K. Mcgeough 57 Stonehouse Park, Thursby, Carlisle CA5 6NS
G7 EKT G. Aucott 21 Riverway, Wednesbury WS10 0DN
G7 EKW Coventry Tech ARC c/o S. Foxall 25 Western Road, Sutton Coldfield B73 5SP
G7 ELA M. Dunn 45 Chaddock Lane, Worsley, Manchester M28 1DE
G7 ELC N. Fountain The Venture, Green Lane, Upton, Huntingdon PE28 5YE
G7 ELE L. Dean Apperley Bridge Marina, Waterfront Mews, Bradford BD10 0UR
G7 ELG A. Scarisbrick 106 Edward Street, Grantham NG31 6JG
G7 ELH K. Graham 44A Adelaide Drive, Colchester CO2 8UB
G7 ELS A. Bartram 47 Temple Gate Crescent, Leeds LS15 0EZ
G7 ELV P. Lockwood 61 Beverley Road, Whitley Bay NE25 8JQ
G7 ELX J. Northfield 48 Gleanings Drive, Halifax HX2 0PA
G7 ELZ T. Leahy Flat 15 Old Brewery House 294 London Road, Wallington SM6 7DD
G7 EME Worcs Mnbounce c/o D. Palmer Spidrift, Landsdown Road, Malvern WR14 1HX
G7 EMH A. Adem 38 Cantley Gardens, Ilford IG2 6QA
GW7 EMO D. Brough 19 Cameron Street, Cardiff CF24 2NW
GW7 EMV Downland Radio Group c/o M. O'Reilly 40 St. Anthony Road, Heath, Cardiff CF14 4DJ
G7 EMZ I. Mellor 124 Ryknield Road, Kilburn, Belper DE56 0PF
G7 ENA D. Neal 33 Swallow Drive, Louth LN11 0DN
G7 ENC D. Flatters 7 Cornwall Crescent, Diggle, Oldham OL3 5PW
GM7 ENM S. Yates 67 Closeburn, Thornhill DG3 5HR
G7 ENQ W. Donald 15 Kingsland Parade, Portobello, Dublin Ireland
G7 ENR C. Davies 138 Cannons Gate, Clevedon BS21 5HN
G7 ENS N. Swift 19 Carlton Road, Caversham, Reading RG4 7NT
G7 ENT S. Alexander 18 Southbourne Grove, Hockley SS5 5EE
G7 EOA G. Harrold Birches, 2 Mill Hill Lane, Wymondham NR18 9DD
G7 EOC C. Hopkins 16 Maypole Road, Gravesend DA12 2LP
G7 EOD A. Capone North Lodge Farm Cottage Fixby, Huddersfield HD2 2EW
G7 EOE E. David 105 Kingsdown Park, Whitstable CT5 2DH
G7 EOG C. Flynn 2 Trafalgar Avenue, Grimsby DN34 5RE
G7 EOH G. Newnham 22 Warren Place Calmore, Southampton SO40 2SD
G7 EOK Bracknell ARC c/o P. Wilson 18 Riversdale Road, Halton, Runcorn WA7 2AP
G7 EPE D. Chamberlain 33 Drake Close, New Milton BH25 5JG
G7 EPF S. Arnold 8 Manor Lane, Sunbury-on-Thames TW16 5ED
G7 EPL L. Wong 320 Wilbraham Road, Chorlton Cum Hardy, Manchester M21 0UX
G7 EPM Lerwick RC c/o J. Enever 21 Waldegrave Way, Lawford, Manningtree CO11 2DT
G7 EPN J. Webb 36 Westfield Drive, Knutsford WA16 0BN
G7 EPR R. Jeeves 172 Aldwick Road, Bognor Regis PO21 2YQ
G7 EPX B. Coffin 3 Berkeley Close, Melksham SN12 6AZ
G7 EPY W. Wright 53 Forshaw Avenue, Grange Park Estate, Blackpool FY3 7PW
G7 EQG J. Sturman 41 Jay Close, Haverhill CB9 0JR
G7 EQK A. Grime 7 Egremont Road Milnrow, Rochdale Ol16 4Ep, Rochdale OL16 4EP
G7 EQO M. Ryan 13 Normandy Road, Heavitree, Exeter EX1 2SR
G7 EQR D. Gammans 35 Chute Avenue, High Salvington, Worthing BN13 3DS
G7 EQX P. Wickers 4 Little Foxburrows, Colchester CO2 7UG
G7 ERC Dorset & Wilts Wing ARC c/o G. Hodge 8 Stainsby Street, St. Leonards-on-Sea TN37 6LA
GW7 ERI A. Brown 3 Wyebank View, Tutshill, Chepstow NP16 7DR
G7 ERS J. Hauton 15 Bourne Close, Lincoln LN6 7DR
G7 ESE C. Tully 19 Glyn Place East Melbury, Shaftesbury SP7 0DP
GW7 ESF T. Ford 14 Hillsnook Road, Ely, Cardiff CF5 5DD
G7 ESI S. Gregory 73 Princess Way, The Walshes, Stourport-on-Severn DY13 0EL
G7 ESL M. Kemble 41 Princes Way Bletchley, Milton Keynes MK2 2FB
GM7 ESM J. Grundey 6 Ternemny Villas, Knock, Huntly AB54 7LR
G7 ESO K. Reynolds 20 Wenwood Gardens, Plymouth PL6 8TD
GD7 ESU C. Ellis Ballahams, 3 Glen Road, Laxey IM4 7AP Isle of Man
G7 ESX S. Bowman 39 Pearson Street, Spennymoor DL166HP
G7 ESY I. Bowman 22 Bryan Street, Spennymoor DL16 6DW
G7 ESZ D. Foster 5 Newman Road, Plymouth PL5 2DX
G7 ETC S. Abel 121 Angela Road, Horsford, Norwich NR10 3HF
G7 ETK S. Yates 14 Wright Street, Horwich, Bolton BL6 7HZ
G7 ETM A. Sherratt 22 Lane Green Avenue, Codsall, Wolverhampton WV8 2JT
G7 ETS S. Hambleton 6 Wylde Green Road, Sutton Coldfield B72 1HB
G7 EUB R. Tabor 9 Liskeard Road, Saltash PL12 4HE
G7 EUF G. Rhodes 54 Chell Green Avenue, Stoke-on-Trent ST6 7JY
G7 EUJ D. Waterworth 74 Oakdale Close, Halifax HX3 5RP
G7 EUT D. Richards Orchard Cottage, Ashbourne Road, Ashbourne DE6 4NJ
G7 EVC P. Stone 2 The Russets, Lees Close, Ashford TN25 6RW
G7 EVF G. Keene 28 Little Hoddington, Upton Grey, Basingstoke RG25 2RN
GW7 EVG G. Nicholas 37 Lon Y Berllan, Abergele LL22 7JF
G7 EVI J. Galbraith 13 Simeons Walk, Quarry Bank, Brierley Hill DY5 2EL

G7 EVK A. Wade 3 Ashendene Grove, Stoke-on-Trent ST4 8NW
G7 EVP M. Pritchard 155 Elliott Road, March PE15 8HF
G7 EVQ M. Jordan 139 Camping Hill, Stiffkey, Wells-next-the-Sea NR23 1QL
G7 EVR S. Sprint 14 Monterey Drive, Allerton, Bradford BD15 9LP
G7 EVT C. Mills 16 Broom Close, Wath-Upon-Dearne, Rotherham S63 7JU
G7 EVY G. Lawton 125 Glovers Way Burscough, Ormskirk L40 5AA
G7 EWA A. Hampton 5 Willow Crescent, Worthing BN13 2SU
GW7 EWD D. Siviter Cilgeraint Farm, St. Anns Bethesda, Bangor LL57 4AX
G7 EWH D. Hughes Balshult, Eriksmala 361 94 Sweden
G7 EWL S. Hogarth 34 High Street, Irthlingborough, Wellingborough NN9 5TN
G7 EWS P. Breck Little Orchard, Chambers Green Road, Pluckley, Ashford TN27 0RJ
G7 EWV P. Ridge 8 Hazel Coppice, Hook RG27 9RH
G7 EWX N. Price 245 Anchor Road, Longton, Stoke-on-Trent ST3 5DX
G7 EWY P. Kember 2 Sandhills Crescent, Wool, Wareham BH20 6HB
G7 EXD R. Fletcher 160 Barnsley Road, Denby Dale, Huddersfield HD8 8QW
GW7 EXH B. Mee Anncott, Hylas Lane, Rhuddlan, Rhyl LL18 5AG
G7 EXO K. Brown 15 Gloucester Road, Aldershot GU11 3SL
GW7 EXQ R. Morris 40 Maes Pedr, Carmarthen SA31 3BR
G7 EXT H. Jarvis Dovecote Farm, Patmans Lane, Boston PE22 8QJ
G7 EXX E. Gwilliam 15 Sheppard Way, Minchinhampton, Stroud GL6 9BZ
G7 EXZ S. Hobbs 15 The Valley, Salisbury SP2 9EJ
G7 EYA R. Hatcher 53 Lancaster Drive, Broadstone BH18 9EH
G7 EYE S. Finnegan 25 Westcliff Gardens, Margate CT9 5DT
G7 EYL M. Dixon 19 Stanford Way, Broadbridge Heath, Horsham RH12 3LH
G7 EYM M. Parkyn Brookfield, Clee St. Margaret, Craven Arms SY7 9DX
GW7 EYP M. Jenkins 23 Guenever Close, Thornhill, Cardiff CF14 9AH
G7 EYR P. Wiles 16 Churchill Road, Broadheath, Altrincham WA14 5LT
G7 EYS C. Chance 19 White Beam Rise, Clanfield, Waterlooville PO8 0LQ
G7 EYV A. Little 444 Dunsbury Way, Leigh Park, Havant PO9 5BJ
G7 EZE A. Byrne 23 The Deansway, Kidderminster DY10 2RH
G7 EZH P. Higginson 93 Oakfield Road, Wollescote, Stourbridge DY9 9DE
G7 FAD V. Ritson 24 Chapel Road, Pawlett, Bridgwater TA6 4SH
G7 FAG P. Lunn 1 Burman Road, Wath-Upon-Dearne, Rotherham S63 7NE
G7 FAQ E. Cloude 2 Bramshot Cottages, Cove Road, Fleet GU51 2RT
G7 FAR RAF Waddington ARC c/o M. Farmer 16 Beckside, Nettleham, Lincoln LN2 2PH
GM7 FAS A. Warnock 21B Melbost Point, Isle of Lewis HS2 0BG
G7 FAZ A. Mould 8, Foxwood, Brierley Hill DY5 2PH
G7 FBE E. Dare 17 Montgomery Drive, Spencers Wood, Reading RG7 1BQ
G7 FBT E. Woodhouse 7 Cow Heys, Dalton, Huddersfield HD5 9RG
GW7 FBV S. Hathaway 9 Mirehouse Place, Angle, Pembroke SA71 5BD
G7 FBY R. Furniss 7 Elizabeth Road, Sutton Coldfield B73 5AR
G7 FCC D. Bent 21 Loughborough Avenue, Nottingham NG2 4LN
G7 FCJ P. Honeywell 7 Hopton Gardens Hopton, Great Yarmouth NR31 9DF
G7 FCL D. Hames 10 Downs Close, East Studdal, Dover CT15 5BY
GI7 FCM S. Fleming 15 Castle Green, Ballynure, Ballyclare BT39 9GN
G7 FCO G. Howat 44 Castle Street Thornbury, Bristol BS35 1HB
GI7 FCP J. Mccormick 14 Ballyoran Park, Portadown, Craigavon BT62 1JN
G7 FCR Fylde Coast Ray c/o R. Knighton 262 Victoria Road West, Thornton-Cleveleys FY5 3QB
G7 FCU F. Smith 3 Downside Close, Findon Valley, Worthing BN14 0EZ
GI7 FCW P. Quinn 53 Dernanaught Road, Dungannon BT70 3BU
G7 FDD W. Cooper 33 Elm Drive Cherry Burton, Beverley HU17 7RJ
G7 FDS T. Whitehead 23 Rossendale Road, Lytham St. Annes FY8 3HY
G7 FDW C. Milburn 9 Woodhall Avenue, Bradford BD3 7BY
G7 FEA M. Codling 18 Ash Grove Pinehurst, Swindon SN2 1RX
G7 FED E. Wells 1A Brocklewood Avenue, Poulton-le-Fylde FY6 8BZ
G7 FEE F. Harvey 39 Simonside Terrace, Heaton, Newcastle upon Tyne NE6 5JY
G7 FEF J. Pipkin 46 Charles Avenue, Albrighton, Wolverhampton WV7 3LF
G7 FEG I. Kilkenny 23 Hazelhurst Road, Stalybridge SK15 1HD
G7 FEL J. Slater 313 Southend Road, Stanford-le-Hope SS17 8HL
G7 FEP Dover ARC c/o D. Birt 3 South Brent Close, Brent Knoll, Highbridge TA9 4BS
G7 FEQ B. Shipton 4 School Close, Kilcott Road, Wotton-under-Edge GL12 7RH
G7 FFB D. Egleton 16A Frescade Crescent, Basingstoke RG21 3NF
G7 FFC V. Prall 20 Marlowe Close, Basingstoke RG24 9DD
G7 FFI K. Harding 21 Doulton Way, Ashingdon, Rochford SS4 3BX
G7 FFK A. James 66 Rydal Crescent, Worsley, Manchester M28 7JD
G7 FFM P. Malpass 30 Countisbury Road, Norton, Stockton-on-Tees TS20 1PZ
G7 FFR J. Rutherford 270 Milburn Road, Ashington NE63 0PL
G7 FFS A. Pike 63 Mill Lane, Bentley Heath, , Solihull B93 8NN
G7 FFV I. Miller 5 Avenue Terrace, Sunderland SR2 7HB
G7 FFW S. Lonsdale 16 Hinkler St., Cleethorpes DN35 8PR
G7 FFZ J. Humphries 23 Sycamore Drive, Lutterworth LE17 4TR
G7 FGA D. Moreland 179 Carr Lane, York YO26 5HQ
G7 FGD J. Brown St. Winnolls House, St. Winnolls, Torpoint PL11 3DX
GM7 FGH J. Bartolo 84 Calderbraes Avenue, Uddingston, Glasgow G71 6ED
GI7 FGQ P. Faulkner 40 Glenariff Drive, Comber, Newtownards BT23 5HA
G7 FGR G. Cluley 1 Shepherds Lane, Greetham, Oakham LE15 7NX
GJ7 FGS J. Bette- Bennett 2 Aspley Villas, Bagatelle Road, Jersey JE2 7TA
G7 FGZ D. Mitchell 55 Halewick Lane, Sompting, Lancing BN15 0ND
G7 FHA C. Brailsford 65 Cherry Orchard, Codford, Warminster BA12 0PW
G7 FHU I. Davies 12 Kaye Avenue, Culcheth, Warrington WA3 5SA
G7 FHV T. Beeching 11 Kents Road, Haywards Heath RH16 4HL
GI7 FHZ E. Mccrystal 33 Richmond Park, Omagh BT79 7SJ
G7 FIA P. Page 42 Alamein Gardens, Stone, Dartford DA2 6BN
GM7 FIE I. Stevenson 68 Silverknowes Eastway, Edinburgh EH4 5NE
G7 FIJ B. Parsons 20 High Park Road, Halesowen B63 2JA
G7 FIK Braintree & District ARS c/o S. Regan 92-94 Lytham Road, Blackpool FY1 6DZ
GM7 FIS J. Russell 15 Glen View, Cumbernauld, Glasgow G67 2DA
G7 FJC P. Fellingham Flat 6, Duncan House, Collingwood Close, Peacehaven BN10 8BA
G7 FJU C. Ovenden 2 Firemans Cottage, Fortis Green, London N10 3PB
GI7 FJY N. Gamble 21 Anderson Crescent, Waterside, Londonderry BT47 2BY
G7 FJZ P. Selley 2 Coronation St., Barnstaple EX32 7AY
G7 FKF R. Phillips 24 Harris Lane, Wistow, Huntingdon PE28 2QG
G7 FKJ C. Holloway 23 Ryecroft Road, Stretford, Manchester M32 9BS
G7 FKP D. Henderson 2 Beverley Court, Beverley Road, York YO43 3NB
G7 FKS H. Ellis 10 Gardens Quay, Pitwines Close, Poole BH15 1XL
G7 FKX C. Wood1 2 Plain Cottages, Plain Road, Marden, Tonbridge TN12 9LS
G7 FKZ R. Bowden 35 Glebelands, Biddenden, Ashford TN27 8EA
GM7 FLG D. Pegg 11 Glenward Avenue, Lennoxtown, Glasgow G66 7EP
G7 FLI J. Moyse 2 Kestle Drive, Truro TR1 3PT
G7 FLS K. Lawson 60 Minster Avenue, Beverley HU17 0ND
G7 FLX B. Timms 74 Park Gwyn, St. Stephen, St. Austell PL26 7PN
GM7 FLZ E. Chesters Blackhill, Blackhill Road, Kirkwall KW15 1FP
G7 FMB R. Burns 43 Gibson St., Bickershaw, Wigan WN2 5TF
G7 FMI Sheppey Western c/o M. Kensall 40 Eskdale Avenue, Ramsgate CT11 0PB
G7 FMJ G. Mitchell 7 Buxton Close, Whetstone, Leicester LE8 6NT
G7 FML P. Clarke 93 Commercial Road, Spalding PE11 2YU
G7 FMQ J. Sutton 10 Cathcart Road, Stourbridge DY8 3UZ
G7 FMU D. Appleby Unterrohr 194, , Rohr Bei Hartberg AT-8284 Austria
G7 FMV D. Sweet 50 Mereside, Soham, Ely CB7 5XE
G7 FMW P. Olson 23 Dennett Close, Liverpool L31 5PD
G7 FND E. Millership 16 Bramble Way, Wirral CH46 7UP
G7 FNM J. Walmsley Bank Field, 5 Dimples Lane, Preston PR3 1RD
G7 FNN A. Morley 87 Epsom Drive, Ipswich IP1 6SS
GI7 FNP H. Massey 156 Killaughey Road, Donaghadee BT21 0BQ
GW7 FNQ D. Jones Noddfa, High Street, Bodorgan LL62 5AS
G7 FNT S. Taylor 39 Hookhills Road, Paignton TQ4 7LR
G7 FNU R. Beaumont 49 Vincent Close, Broadstairs CT10 2ND
GI7 FOD P. Mccollam 32 Robinson Way, Bangor BT19 6NR
G7 FOT W. Sanger Tregonning Lea, Laddenvean, Helston TR12 6QD
G7 FOX Melton Mowbray ARS c/o G. Mason 120 Scalford Road, Melton Mowbray LE13 1JZ
G7 FPJ J. Marks Flat 112 Weavers Quay, 51 Old Mill Street, Manchester M4 6GB
G7 FPM K. Pontin Flat 41 Park Farm The Street, Moredon Swindon SN25 3ES
GM7 FPN A. Mcpherson 3 Fulmar Road, Elgin IV30 4HL
G7 FPR G. Flanagan Flat 9, West Cliff Court, 25 Portarlington Road, Bournemouth BH4 8BX
G7 FPS S. Harrison 3 Smallbridge Close, Worsley, Manchester M28 7XS
G7 FPU C. Wissun 111 Berkeley Vale Park, Berkeley GL13 9TQ
G7 FPW P. Howard 3 Hollies Close, Shepton Mallet BA4 5LG
GM7 FPX A. Stewart 8 Crinan Place, Ardrossan KA22 7PT
G7 FPZ D. Foster 29 Harrowfield Road, Stechford, Birmingham B33 9BU
G7 FQE J. Whiffen 5 Sharpthorpe Close, Lower Earley, , Reading RG6 4DB
GW7 FQP S. Earle Clydfan Nanternis, New Quay SA45 9RP
G7 FQY G. Spark Park Lodge, Cheltenham Drive, Sale M33 2DQ
GM7 FQZ J. Brooks The Steading, West Mosstown, Lonmay, Fraserburgh AB43 8RU
GM7 FRC Fife Raynet Group c/o J. Burke 25 Duncan Road, Auchmuty, Glenrothes KY7 4HS
G7 FRH A. Russ 21 Francis Road, St. Pauls Cray, Orpington BR5 3LY
G7 FRR M. Samson 277 Upton Lane, London E7 9PR
G7 FRW K. Mccaffery 34 Ringwood Road, Luton LU2 7BG
G7 FSA R. Colclough 8 Parker Jervis Road, Stoke-on-Trent ST3 5RP
G7 FSC K. Davis 26 Mendip Drive, Nuneaton CV10 8PT
G7 FSD A. Holles 20 Sapcote Road, Burbage, Hinckley LE10 2AU
G7 FSH F. Pearson 15 York Road, Driffield YO25 5AT
G7 FSJ P. O'Connor 149 Roxeth Green Avenue, Harrow HA2 0QJ
G7 FSR A. Wyard 85 Swaledale, Bracknell RG12 7ET
G7 FTA M. Collins 15 Trevethan Rise, Falmouth TR11 2DX
G7 FTD K. Taber 110 Uplands, Peterborough PE4 5AF
GM7 FTK M. Long 52 Stirling Road, Milnathort, Kinross KY13 9XG
G7 FTM S. Clayton 22 Orchard Avenue, North Anston, Sheffield S25 4BW
G7 FTS O. Whiteside 9 Beech Grove House, Beech Grove, Harrogate HG2 0ES
G7 FUM N. Seath 6 Harvester Way, Sibsey, Boston PE22 0YD
G7 FUQ A. Vincent 12 Spelman Road, Norwich NR2 3NJ
G7 FUV I. Marsh 56B Oliver Crescent, Farningham, Dartford DA4 0BE
G7 FUW J. Birch 15 Adstone Grove, Birmingham B31 4AU
G7 FVH R. Barrick Orchard Bungalow, Pasture Lane, Middlesbrough TS6 8EH
G7 FVR C. Heywood 29 Smallwood Mews, Wirral CH60 6TE
GM7 FWA A. Pratt 129 Brodie Court, Glenrothes KY7 4UE
G7 FWD B. Nicholls 6 Jubilee Road, Devizes SN10 3FE
G7 FWE Strabane ARS c/o A. Smith 12 Northgate, Beccles NR34 9AS
G7 FXO P. Werba 47 Ulwell Road, Swanage BH19 1LG
G7 FXW P. Whitworth 67 Staddiscombe Road, Staddiscombe, Plymouth PL9 9LU
GW7 FXX B. Harries 12 Panteg, Llanelli SA15 3TF
G7 FXY P. Hallett 30 Summerdown Walk, Trowbridge BA14 0LJ
G7 FXZ G. Hodgetts 2 Friars Gorse, Stourton, Stourbridge DY7 6SP
GM7 FYB D. Wemyss 24 Brucklay Court, Peterhead AB42 2UF
GW7 FYG C. Wright 12 Bryn Teg, Arddleen, Llanymynech SY22 6PZ
G7 FZB Dr G. Ridgeway 4 Russell Avenue, Alsager ST7 2BL
G7 FZJ M. Whatley Woodside West, Wood Lane, Halifax HX3 8HB
G7 FZN P. Du Plessis 42 La Providence, Rochester ME1 1NB
GW7 FZW M. Heel 27 Englefield Drive Oakenholt, Flint CH6 5SB
G7 GAB R. Hagues 40 Barton Road, Rugby CV22 7PT
GM7 GAE I. Mackenzie 52/5 Craighall Road, Edinburgh EH6 4RU
G7 GAG J. O'Neill 24 Lily Lane, Bamfurlong, Wigan WN2 5JN
GW7 GAH R. Dore Maespoeth, Corris, Machynlleth SY20 9RD
G7 GAK D. Garbutt 8 Yorkshire Road, Partington, Manchester M31 4GW
G7 GAP J. Cartwright 109 Kneller Road, Twickenham TW2 7DT
G7 GAZ B. Kerrison 45 Bramley Crescent, Bearsted, Maidstone ME15 8JZ
GM7 GBD G. Macgregor 6 Kincaidfield Milton Of Campsie, Glasgow G66 8ER
G7 GBE Midland ARS c/o S. Burgoine 47 Squirrel Close, Hounslow TW4 7NU
G7 GBJ J. Kaczmarek 2 Westgate Terrace, London SW10 9BJ
G7 GBN P. Baird 168 Plumberow Avenue, Hockley SS5 5AT
G7 GBZ P. Leach 21 Abbess Close, Chelmsford CM1 2SE
G7 GCD S. Lee 51 Ireland Crescent, Red Deer T4R 3K8 Canada
G7 GCF P. Kell Flat25, Church Court, Church Rd., Haywards Heath RH16 3UE
G7 GCI M. Collett 54 Hollywood Way, Woodford Green IG8 9LQ
G7 GCU M. Edge 2 Yew Tree Place, Walsall WS3 3DG
G7 GCW Dr P. Andrew 3 Grayway Close, Highfields Caldecote, Cambridge CB23 7UZ
G7 GDA T. Wootton 1 Lingfield Drive, Walsall WS6 6LS
G7 GDC A. Gosden 10 Radcliffe Way, Northolt UB5 6HP
GM7 GDE A. Hood 26 Annan Avenue, East Kilbride, Glasgow G75 8XT
G7 GDV A. Porteous 73 Dowgate Close, Tonbridge TN9 2EJ
G7 GEA J. Broadfoot 65A Swan Meadow, Pewsey SN9 5HP

G7 GEE J. Gee 51 Hattons Lane, Childwall, Liverpool L16 7QR
G7 GEF G. Duthie 15 Wagtail Close, Twyford, Reading RG10 9ED
G7 GEI Keighley ARS c/o M. Arliss 22 Rowena Avenue, Edenthorpe, Doncaster DN3 2JF
G7 GEL J. Mansell 8 Himley Gardens, The Straits, Dudley DY3 3AS
G7 GEP C. Danks Neuadd Las, Llanddewi Brefi, Tregaron SY25 6NY
G7 GES B. Norcott 5 The Shrubbery, Upminster RM14 3AH
G7 GEU V. Bruntnell 4 Cypress Avenue, Dudley DY3 2JF
G7 GEX N. Potter 4 Eastleigh Drive, Mickleover, Derby DE3 9HZ
G7 GFC D. Mullock 18 Tewkesbury Close, Upton, Chester CH2 1NF
G7 GFH W. Baker 41 Kenwood Park Road, Sheffield S7 1NE
G7 GFK K. Percival 1608 Scant Row, Chorley Old Road, Bolton BL6 6PZ
G7 GFM J. Hunt 43 Felton Close, Redditch B98 0AG
G7 GFP I. Bishop 115 Burman Road Shirley, Solihull B90 2BQ
G7 GFQ M. Charlwood 60 Alfred Road, Feltham TW13 5DJ
G7 GFR B. Clifford 8 Caldbeck Place, North Anston, Sheffield S25 4JY
G7 GFX P. Everard 56 Hawkins Crescent, Shoreham-by-Sea BN43 6TP
G7 GGA N. Dawson 4 Bathurst Close, Staplehurst, Tonbridge TN12 0NA
G7 GGF C. Martin 073299, Bfpo 5442 BF19DB
G7 GGG G. Richardson 11 Queensway, Forest Town, Mansfield NG19 0BX
G7 GGH G. Hurrell 13 Hinton Road, Newport PO30 5QZ
G7 GGJ A. Edwards 45 Chilton Grove, Yeovil BA21 4AW
G7 GGM D. Thomalla 14 Walkers Lane, Penketh, Warrington WA5 2PA
G7 GGN J. Williams 41 Cote Green Lane, Marple Bridge, Stockport SK6 5EB
G7 GGT S. Mullins 549 Bromford Lane, Washwood Heath, Birmingham B8 2EA
GI7 GHC T. Lyons 3 Clanbrassil Gardens, Portadown, Craigavon BT63 5YD
GW7 GHE D. Pearson Warren Cottage Pontfadog, Llangollen LL20 7AT
G7 GHH I. Wraith 7 Bowman Close, Sheffield S12 3LR
G7 GHP G. Fellows 34 The Ridings, Bexhill-on-Sea TN39 5HU
G7 GHT M. Alexander 51 Park Lane, Bootle L20 6DJ
GM7 GIF K. Juner 56 Queens Gardens, East Calder, Livingston EH53 0EG
G7 GIG R. Vincent 14 Trevenson Street, Camborne TR14 8JB
G7 GIJ J. Barnett 20 Springford Gardens, Southampton SO16 5SW
G7 GIK D. Keene Firemark Cottage West Street Odiham, Hook RG29 1NT
GM7 GIO W. Mackinnon 31 Kirk Bauk, Symington, Biggar ML12 6LB
GM7 GIS M. Glendinning 148 Gala Park, Galashiels TD1 1HD
G7 GJA P. Cockayne 7A Wrekin Drive, Bradmore, Wolverhampton WV3 7HZ
G7 GJI D. Sager 29 Station Road, Mickleover, Derby DE3 9GH
GM7 GJM C. Unsworth 3 Thorfinn Place St. Margarets Hope, St. Margaret's Hope KW17 2TR
G7 GJN A. Khachaturian 377 Watford Road, St. Albans AL2 3DD
G7 GJO P. Morris 117 Lonsdale Avenue, Doncaster DN2 6HF
G7 GJS N. Cheesewright 5 Duberly Close, Perry, Huntingdon PE28 0BP
G7 GJT W. Everton Fencott, Fen Road, Lincoln LN4 1AE
G7 GJU G. Darby 5 Lumsden Terrace, Catchgate, Stanley DH9 8EQ
G7 GJV A. Gordon 1 Surrey Street, Hetton-Le-Hole, Houghton le Spring DH5 9LX
GI7 GJX N. Simmons 9 Rhanbuoy Park, Carrickfergus BT38 8BS
G7 GJY J. Chapman 77A Carnforth Gardens, Elm Park, Hornchurch RM12 5DR
G7 GJZ C. Brown 73 Ringstone, West Huntspill, Highbridge TA9 3RF
G7 GKC I. Boyd 43 Lower Landedmans, Westhoughton., Bolton BL5 2QL
G7 GKD L. Tryhorn 46 Mill Green Road, Amesbury, Salisbury SP4 7RE
GW7 GKN S. Gordon 6 Oakridge Acres, Tenby SA70 8DB
G7 GKQ L. Measures 163 Huddersfield Road, Meltham, Holmfirth HD9 4AJ
GM7 GKT R. Smith 27 Elm Lane, Foresters Lodge, Glenrothes KY7 5TD
GW7 GKX J. Rough 10 Beaconsfield Road, Shotton, Deeside CH5 1EZ
G7 GLA J. Mitchinson 93 Hinckley Road Leicester Forest East, Leicester LE3 3GN
GM7 GLJ A. Potter 42 Pender Gardens, Rumford, Falkirk FK2 0BJ
G7 GLL L. Carlile 26 The Bungalows, Stonebroom, Alfreton DE55 6LH
G7 GLP J. Walton 14 Chapel Street, Stanhope, Bishop Auckland DL13 2NB
G7 GLQ D. Cottrell 2 Foss Court, Summerhill Road, Bristol BS5 8HF
G7 GLR Wisbech Ar & Electronics Club c/o I. Jackson 5 Vivien Close, Chessington KT9 2DE
G7 GLS J. Pinna 31 Bowness Road, Little Lever, Bolton BL3 1UB
G7 GLW R. Cains 58 Sunnydale Road, Lee, London SE12 8JN
G7 GLZ R. Hourston 12 The Warren, Chesham HP5 2RY
G7 GMB J. Craig 1 Eldon Road, Eastbourne BN21 1UD
G7 GMD M. Ollerton 1 Hammy Way, Shoreham-by-Sea BN43 6GH
G7 GMQ D. Smith1 65 St. Anthonys Road, Kettering NN15 5JB
G7 GMR B. Golland 15 Turpin Close, Gainsborough DN21 1PA
G7 GMU G. Lamb Parisfield, Headcorn Road, Tonbridge TN12 0BT
G7 GMZ W. Newton 7 Moss Close, Bridgwater TA6 4NA
G7 GNA L. Smith 13 Eagle Avenue, Waterlooville PO8 9UB
GM7 GNO Huddersfield Technical College ARS c/o N. Goodall 26 Greenbank Loan, Edinburgh EH10 5SJ
G7 GNS J. Olive 2 Wyke Cottage Wotton Road, Rangeworthy, Bristol BS37 7NA
G7 GNU P. Brayshaw 38 Chilfrome Close, Canford Heath, Poole BH17 9WE
G7 GOA S. Constable 18 Salvington Gardens, Worthing BN13 2BH
GM7 GOE M. Doig 18 Gotterstone Drive, Broughty Ferry, Dundee DD5 1QW
G7 GOK N. Breckell Barn Hill Lodge, Barn Hill Road, Broadwell, , Coleford GL16 7BL
G7 GOV M. Hill 31 Brocklesby Avenue, Immingham DN40 2AS
GM7 GPG A. Jakowuik 167 Magdala Terrace, Galashiels TD1 2HZ
G7 GPI A. Baily 13 Longleigh Lane, Bexleyheath DA7 5SL
G7 GPJ R. Banks Highview, New Road, Sturminster Newton DT10 2HF
G7 GPL N. Giles 6 Bridgewater Mews, London Road, Warrington WA4 6LF
G7 GPU A. Sharman 9 Silver Close, Minety, Malmesbury SN16 9QT
G7 GQA A. Doswell 14 Carisbrooke Drive, Charlton Kings, Cheltenham GL52 6YA
G7 GQB M. Woodhouse 18 Soame Close, Aylsham, Norwich NR11 6JF
G7 GQC G. Beckingham Furcot'S Lair, 20 Baptist Close, Abbeymead, , Gloucester GL4 5GD
G7 GQD D. Pearce 2 Mell Avenue, Hoyland, Barnsley S74 9HF
G7 GQH R. Hannemann 112 Northern Road, Aylesbury HP19 9QY
G7 GQL J. Sutton 15 Lowther Street, Penrith CA11 7UW
G7 GQM E. Sutton 15 Lowther Street, Penrith CA11 7UW
G7 GQO R. Harman The Briars, Brambleberry Lane, Skegness PE24 5DQ
G7 GQW D. Williams 28 Mill Lane, Great Sutton, Ellesmere Port CH66 3PF
G7 GQX D. Howard 6 Draycote Close, Solihull B92 9PT
G7 GRC Grantham ARC c/o K. Burton 2 Council House, Stainfield Road, Kirkby Underwood, Bourne PE10 0SG
GM7 GRH N. Hardie 38 Sentry Knowe, Selkirk TD7 4BG
G7 GRM J. Booth 2 Fairfax Mews, London E161TY
G7 GRO D. Stimpson 19 Moss Bank, Winsford CW7 2ED
G7 GRR S. Everett 4 Ilkley Place, Newcastle ST5 6QP
G7 GRU M. Lucas 22 Ferny Brow Road, Wirral CH49 8EE
GI7 GRY S. Gordon 138 Mullalelish Road, Richhill, Armagh BT61 9LT
GI7 GSB A. Wiese 105 Milltown Avenue, Lisburn BT28 3TR
G7 GSC N. Godden 23 Rapsons Road, Willingdon, Eastbourne BN20 9RJ
G7 GSD K. Osborn 2A Sullington Gardens, Worthing BN14 0HR
G7 GSF S. Blandford Flat 18, Avro House, 5 Boulevard Drive, London NW9 5HF
G7 GSM D. Hebron 12 Scholars Gate, Guisborough TS14 8LT
G7 GSR J. Shrubsall 54 Park Avenue, Sittingbourne ME10 1QY
G7 GSX C. Penfold 149 Shuttlewood Road Bolsover, Chesterfield S44 6NX
G7 GTG A. Hyndman Norman House, Railway Terrace, Herts WD4 8JE
G7 GTH A. Marriott Norman House, Railway Terrace, Herts WD4 8JE
GM7 GTS C. Richman 18 Nigel Rise, Livingston EH54 6LT
G7 GTU G. Sharples 24 Kelboro Avenue, Audenshaw, Manchester M34 5UH
GM7 GTX M. Kaye 146 Newlands Road, Grangemouth FK3 8NZ
G7 GUA A. Dennis Chy An Ros, Riverside, Hayle TR27 5JD
G7 GUB A. Alderton 7 Bigland Drive, Ulverston LA12 9NU
G7 GUG G. Wales 1 Ryton Fold, North Anston S25 4AG
GM7 GUL C. Jordan 3 Birch Avenue, Rosemount, Blairgowrie PH10 6XE
G7 GUO S. Falconer 6 Ogilvie Road, High Wycombe HP12 3DS
GI7 GUT D. Watt 51 Rashee Road, Ballyclare BT39 9HT
GM7 GVD D. Innes 6 Mamore Terrace, Inverness IV3 8PF
GI7 GVI T. Henderson 7 Legaloy Road, Ballyclare BT39 9PS
G7 GVJ S. Fletcher Fernleigh, Ash Lane, Gloucester GL2 9PS
G7 GVP C. Price 16 Woodlands Drive, Warton, Preston PR4 1UQ
G7 GWA A. Jakins 29 Burchnall Close, Deeping St. James, Peterborough PE68QJ
GW7 GWO S. Evans 1 Brynheulwen Blaenannerch, Cardigan SA43 2AH
G7 GWT G. Taylor 33 Heol Aberwennol, Borth SY24 5NP
GM7 GWW S. Gardiner Kyendigaet, Whiteness, Shetland ZE2 9GJ
G7 GXE P. Kitson 15 Louvain Road, Derby DE23 6DA
GM7 GXI G. Cowan 15 Waterhaughs Grove, Glasgow G33 1RS
G7 GXR B. Clewes 19 Church Mews, Denton, Manchester M34 3GL
GI7 GXZ S. Dornan 3 Hampton Lane, Bangor BT19 7GB
G7 GYN C. Barlow 2/5 Hospital Steps, Gibraltar GX11 1AA Gibraltar
G7 GYR K. Wade Eccleston Hall, Lydiate Lane, Chorley PR7 6LY
G7 GZB C. Davies 84 Hob Hey Lane, Culcheth, Warrington WA3 4NW
G7 GZC D. Coles 20 James Darby House 11 Mereway Road, Twickenham TW2 6SA
G7 GZJ K. Oliver 155 Old Road, East Cowes PO32 6AX
G7 GZK I. Croft 34 Laburnum Drive, Armthorpe, Doncaster DN3 3HE
G7 GZU S. Selwyn 65 Porterhouse Road, Ripley DE5 3FL
G7 GZV H. Houldershaw The First Bungalow, Fen Road, Boston PE22 8EX
G7 GZZ E. Gaffney 1 White Hart Lane, Wistaston Green, Crewe CW2 8EX
GW7 HAE C. Davies Afallon, 3 Penygraig, Aberystwyth SY23 2JA
G7 HAF W. Hunton 60A Bondgate, Helmsley, York YO62 5EZ
G7 HAH Finningley ARS c/o M. Hotchin 122 Buckingham Avenue, Scunthorpe DN15 8NS
G7 HAR B. Ferris 5 Guildway, Todwick, Sheffield S26 1JN
G7 HAS A. Newton Rockburn, Victoria Road, Malvern WR14 2TE
G7 HBN P. Osborne 11 Galston Road, Luton LU3 3JZ
G7 HBO R. Cornell 161 Tuckers Road, Loughborough LE11 2PH
G7 HBU T. Hickling 6 Harrold Road, Bozeat, Wellingborough NN29 7LP
G7 HBV C. Heard 42 Hallowell Down South Woodham Ferrers, Chelmsford CM3 5FS
G7 HCB B. Atterbury 7 Ross Court, Stevenage SG2 0HD
G7 HCC D. Jones 120 Heathfield Road, Keston BR2 6BF
G7 HCJ A. Parr 52B Trent Boulevard, West Bridgford, Nottingham NG2 5BD
G7 HCL P. Good 80 Meredith Road, Stevenage SG1 5QS
G7 HCN A. Jones 179 Blandford Road, Efford, Plymouth PL3 6JZ
G7 HCO N. Lambert Bradfields Farm, Burntmills Road, Wickford SS12 9JX
G7 HCQ D. Browne 293 St. Albans Road, Hemel Hempstead HP2 4RP
G7 HCR G. Richardson The Homestead Washway Road, Holbeach, Spalding PE12 7PP
G7 HCT K. Moore 8 Lilac Close, Toftwood, Dereham NR19 1JY
GW7 HDC A. Rowe 5 Church Street, Knighton LD7 1AG
G7 HDR D. Horder 77 Grove Avenue, Harpenden AL5 1EZ
GW7 HDS S. Barker 11 Prosser Street, Treharris CF46 5LN
G7 HDU J. Tombs Mariedown, Bustards Lane, Wisbech PE14 7PQ
G7 HDW J. Bigger 128 Trueway Drive South, Shepshed, Loughborough LE12 9DY
G7 HEJ G. Atkinson 23 Fielding Road, Blackpool FY1 2QL
G7 HEK A. Owen 57 Melrose Avenue, Vicars Cross, Chester CH3 5JB
G7 HEN M. Priestley 29 Birchlands Avenue, Wilsden, Bradford BD15 0HB
G7 HEP A. Ellis Eikly Tregada, Launceston PL15 9NA
G7 HEY L. Morrell-Cross Delta Lodge, 14 Rushton Crescent, Bournemouth BH3 7AF
G7 HEZ J. French 25 Twickenham Court, Stourbridge DY8 4QG
G7 HFE S. Hitches 7 Church Close, Chedgrave, Norwich NR14 6NH
G7 HFL C. Elphick 2 Vine Way, Brentwood CM14 4UU
G7 HFP C. Castle 2 Wellington Close, Mundesley, Norwich NR11 8JF
G7 HFS I. Harling 71 De Cham Road, St. Leonards on Sea TN37 6HF
G7 HFW A. Wood Flat 413, Manchester M1 2FA
GW7 HFZ A. Strachan 16 Clos Y Wiwer, Llantwit Major CF61 2SG
G7 HGB J. Dunn 10 Endsleigh Close, Upton, Chester CH2 1LX
G7 HGD P. Allott 1 Abbey Court, Abbey Road, Knaresborough HG5 8HX
G7 HGF I. Simpson Honeysuckle Cottage, 39 Chewton Street, Nottingham NG16 3GY
G7 HGI R. Roberts 13 Tudor Way, Wickford SS12 0HS
G7 HGQ D. Horwood 12 Curtis Close, Mill End, Rickmansworth WD3 8QA
G7 HGT P. Stimpson 93 Chaucer Road, Farnborough GU14 8SR
GW7 HGU M. Howard 64 Lawrenny St., Neyland, Milford Haven SA73 1TB
GM7 HHB J. Brown 133 Meadowbank Road, Kirknewton EH27 8BH
G7 HHI S. Curry Barnabus Communications, Barnabus Cottage, Egley Road, Mayford, Mayford, Woking GU220NQ
G7 HHK R. Johnson Honeysuckle Cottage, Front Street, Northallerton DL6 2AA
G7 HHM L. Dring 22 Castle Street, Eastwood, Nottingham NG16 3GW
G7 HHN K. Glover 7 Mill Lane, Cressing, Braintree CM77 8HN
G7 HHQ R. Saunders The Grange, High Road, Wisbech PE13 4RG
G7 HHT M. Gotts 23 Beechcroft Avenue, Croxley Green, Rickmansworth WD3 3EG
G7 HHU A. Edwards 3 Simonside Close, Morpeth NE61 2XY

G7 HHW G. Phillips 14 Orchard Close, Plymouth PL7 2GT
G7 HHZ J. Whelan 1 Chevin Road, Milford, Belper DE56 0QH
G7 HIC K. Bow 16 Brook Road, Ivybridge PL21 0AX
G7 HID M. Burgess 63 Chalvey Park, Slough SL1 2HX
G7 HIH R. Pedro 65 Glebe Crescent, Harrow HA3 9LB
G7 HII D. Lloyd No.5 The Close, Burton Gardens, Hereford HR4 8RQ
G7 HIJ J. Gunia 21 Campbell Avenue, Leek ST13 5RR
G7 HIK J. Doherty 101 Padacre Road, Torquay TQ2 8QQ
G7 HIN P. Riddell 4 Pear Tree Road, Addlestone KT15 1SR
G7 HIO W. Austin 53 Giantswood Lane, Congleton CW12 2HQ
G7 HIQ J. Hickey 53 Norwood Avenue, Hasland, Chesterfield S41 0NN
GM7 HIR A. Pert 56 Lochiel Drive, Milton Of Campsie, Glasgow G66 8ET
G7 HIT P. Chambers 7 Redland Close, Beeston, Nottingham NG9 5LA
G7 HIU R. Hurst 33 Northern Road, Aylesbury HP19 9QT
G7 HIX R. Gray 12 St. Francis Close, Deal CT14 9LS
G7 HIY T. Jefford 7 Bellevue Street, Folkestone CT20 1HY
G7 HJD G. Holland 11 Swanton Drive, Dereham NR20 4DW
G7 HJG R. Blewitt Fazeley Mill Marina, Coleshill Road, Fazeley, Tamworth B78 3SE
G7 HJJ H. Holman 62 The Ridge, Kennington, Ashford TN24 9EU
G7 HJK R. Kearnes 25 Epsom Close, Clacton-on-Sea CO16 8FE
GW7 HJN S. Tweed Y Tardis 257 Penybanc Road, Ammanford SA18 3QW
G7 HJQ M. Erber 75 St. Andrews Road North, Lytham St. Annes FY8 2JF
G7 HJR T. Rudderham 24 Casswell House, Grimsby DN32 7SB
G7 HJT T. Reynard 12 Acorn Close, Selsey, Chichester PO20 9HL
G7 HKN P. Walsh 2 Elm Road, Winwick, Warrington WA2 9TW
G7 HKQ I. Tideswell 2 Pangbourne Avenue, Urmston, Manchester M41 0GF
G7 HKT C. Fowle 70 The Parade, Greatstone, New Romney TN28 8RE
G7 HKU J. Turner 7 Highfield Crescent, Baildon, Shipley BD17 5NR
G7 HKZ T. Allen 15 Manning Road, Cotford St. Luke, Taunton TA4 1NY
G7 HLB A. Coombs Featherlands, Holsworthy Beacon, Holsworthy EX22 7NH
G7 HLG B. Morrell-Tourle 77 Mallard Road, Bournemouth BH8 9PJ
G7 HLP K. Baldock 66 Port Road, Northampton NN5 6NL
G7 HLU V. Meads May Tree Barn, Main Street Upper Benefield, Peterborough PE8 5AN
G7 HLV J. Jordan 18 Sunningdale Crescent, Cullingworth, Bradford BD13 5BA
G7 HLW G. Burn 4 Goston Gardens, Thornton Heath CR7 7NQ
GW7 HLZ R. Davies 8 Deri Road, Abergavenny NP75SY
G7 HMA I. Smith 4 Stour Road, Grays RM16 4BS
G7 HMB G. Bull 48 Spragg House Lane, Stoke-on-Trent ST6 8DX
G7 HMF Technical Experimenters Group c/o E. Last 134 New Queens Road, Sudbury CO10 1PJ
G7 HMI R. Shelford 3 Browning Chase, Littleport, Ely CB6 1FH
G7 HMK A. Baldwin B M Box 6902, London WC1N 3XX
G7 HMN C. Boutell 6 Willow Way, Harwich CO12 4HR
G7 HMQ B. Boult 20 Perry Road, Long Ashton, Bristol BS41 9FE
G7 HMS Rnars London (Hms Belfast) Group c/o C. Read 58 Somerset Road, Chiswick, London W4 5DN
G7 HMU J. Stratton Hillview, Churchfoot Lane, Hazelbury Bryan, Sturminster Newton DT10 2DS
G7 HMV M. Wood 26 Parkfield Crescent Kimpton, Hitchin SG4 8EQ
G7 HMW W. Knight 30 Stretford Road, Urmston, Manchester M41 9JZ
G7 HMZ A. Murfin 31 Kings Road, St. Neots PE19 1LD
G7 HNF J. Baldwin 71 Norfolk Road, Littlehampton BN17 5HE
G7 HNG A. Lord 59 Monks Orchard Road, Beckenham BR3 3BJ
G7 HNL M. Carter Meadowview Bransford Road Rushwick, Worcester WR2 5SJ
G7 HNM G. Greatrix West Cottage, Main Road, Boston PE20 3PZ
G7 HNR A. Ball 39 Deepdale Avenue, Birmingham B26 3EL
G7 HOA Widnes & Runcorn ARC c/o D. Wilson 12 New Street, Elworth, Sandbach, CW11 3JF
GW7 HOC D. Warburton 71 Richards Terrace, Cardiff CF24 1RW
G7 HOE P. Goode 23 Byworth Road, Farnham GU9 7BT
G7 HOK P. Kellingley 290 Calmore Road, Calmore, Southampton SO40 2RF
G7 HOL D. Martin Alken, The Covert, Orpington BR6 0BT
GW7 HOM V. Cole 77 Parc Castell Y Mynach, Creigiau, Cardiff CF15 9NZ
G7 HON S. Martin Broad Oak House, Pheasant Lane, Maidstone ME15 9QR
G7 HOT J. Scott 16 Hawton Road, Newark NG24 4QB
G7 HPI C. Vance 64 Caulfield Road, Swindon SN2 8BT
G7 HQC I. Sorrell 67 Northfield Drive, Pontefract WF8 2DJ
G7 HQF P. Smith 17 Beverley Avenue, Canvey Island SS8 0DN
G7 HQJ A. Baker 34 Clare Street, Stoke-on-Trent ST4 6ED
GW7 HQL J. Caswell 31 Pontalun Close, Barry CF63 1QJ
G7 HQP J. Woods 1 Dean Road, Cosham, Portsmouth PO6 3DG
GM7 HQW B. Currie Fawn House, Abriachan, Inverness IV3 8LB
G7 HQY K. Walton Springfield, Green Lane, Uckfield TN22 5LA
G7 HRB P. Briggs Mistleberry Cottage, Newtown, Sixpenny Handley, , Salisbury SP5 5PF
G7 HRF S. Crutchley 8 Cloverland Drive, Hemsby, Great Yarmouth NR29 4JY
G7 HRH R. Conway 9 Whitworth Lane, Loughton, Milton Keynes MK5 8EB
G7 HRJ T. Jackson 18 Shepherds Close East Runton, Cromer NR27 9PQ
G7 HRL T. Turner 21 Spurgate, Hutton, Brentwood CM13 2LA
G7 HRM S. Baker 35 Cosmo Street, Westerly 2891 United States
G7 HRP I. Booth 16 Sandstone Drive, Leeds LS12 5SU
G7 HRR Hucknall Rolls Royce ARC c/o S. Sorockyj 8 Bowden Avenue, Bestwood Village, Nottingham NG6 8XN
G7 HRZ S. Haynes 10 Cypress Grove, Denton, Manchester M34 6EA
G7 HSA A. Cramp 7 St Margarets Road, Ludlow SY8 1XN
G7 HSB A. Green Moss View, Southport Road, Ormskirk L39 7JU
G7 HSL T. Reddish 72 Edgmond Close, Redditch B98 0JQ
G7 HSN J. Calder Grassington, Station Road, Bedale DL8 1SX
G7 HSO D. Hardinges 4 The Close, Eastcote, Pinner HA5 1PH
G7 HSS J. East 102 Westfield Lane, Wyke, Bradford BD12 9LS
GW7 HSW I. Edgington 108 Ger Y Llan, Penrhyncoch, Aberystwyth SY23 3TR
G7 HSY B. Stanton 111 Beaconside, South Shields NE34 7PT
GD7 HTG S. Hill 54 Wybourn Drive Onchan, Isle of Man IM3 4AT
G7 HTI Hilderstone RS c/o I. Lowe 54 College Road, Margate CT9 2SW
G7 HTN P. Seitz 6 Meadow Rise Iwade, Sittingbourne ME9 8SB
GW7 HTU P. Jenkins 28 King Edward Road, Brynmawr, Ebbw Vale NP23 4SD
GJ7 HTV A. Mourant Little Mead, Claremont Road, St. Saviour JE2 7RT Jersey
G7 HUC M. Fiorentini Crompton, Draffin Lane, Camber, Rye TN31 7RA
G7 HUG N. Markley 79 Peake Close, Peterborough PE2 9JE
G7 HUJ S. Telford 44 Northcote Crescent, Leeds LS11 6NN
G7 HUK P. Hart 104 St. Austell Drive, Wilford, Nottingham NG11 7BQ
G7 HUO Scarborough ARS c/o C. Terry Sandfields, Long Lane, Newbury RG14 2TH
G7 HUP M. Terry Sandfields, Long Lane, Newbury RG14 2TH
GW7 HVA N. Callan 24A Ynysmeurig Road, Abercynon, Mountain Ash CF45 4SY
GI7 HVC T. Kennedy 1 Inverleith Drive, Belfast BT4 1RJ
G7 HVF S. Garlick 20 Hill St, Upper Gornal DY3 2DF
G7 HVL C. Spires 31 East Street, Warminster BA12 9BY
G7 HVO R. Gerrard 12 Goldrill Gardens, Bolton BL2 5NL
G7 HVU S. Lamb 18 The Green, North Burlingham, Norwich NR13 4SZ
G7 HWM A. Brookes 8 Peppersgate, Lower Beeding, Horsham RH13 6ND
G7 HXF Cornwall Raynet Group c/o J. Newton 18 Siddeley Close, Broughton, Chester CH4 0SG
G7 HXI I. Duffin Mirabella, Bush Drive, Bush Estate, Norwich NR12 0SF
G7 HYG R. Barber 180 Beechfield, Hoddesdon EN11 9QN
G7 HYM N. Singer 11 Langley Road, Beckenham BR3 4AE
G7 HYS D. Germaney 22 Westbrook Road, Weston-Super-Mare BS22 8JX
G7 HYZ M. Thompson 23 Hare Park Lane, Crofton, Wakefield WF4 1HS
G7 HZQ S. Breen 20 Goodwood Close, Clophill, Bedford MK45 4FE
G7 HZS G. West West Cottage, Eaudyke Road, Boston PE22 8RU
G7 HZU A. Bateman 4 Fair Meadows, High Street, Rugeley WS15 3LD
G7 HZV R. Baal 10 Rosemary St, Exeter EX4 1QX
G7 HZZ A. Clayton 6 Albert Road, Bunny, Nottingham NG11 6QE
G7 IAE R. Lindley 23 Quadrant Close, Murdishaw, Runcorn WA7 6DW
GW7 IAK C. Hughes 16 Morgraig Avenue, Newport NP10 8UP
G7 IAM M. Chrzanowski 53 Lamb Street, Kidsgrove, Stoke-on-Trent ST7 4AL
G7 IAS R. Bell 3 Haywards Heath Road, Balcombe, Haywards Heath RH17 6NG
GW7 IAT Lt. Cmdr. M. Howells 34 Cobden Street, Cross Keys, Newport NP11 7PF
G7 IAU C. Penney 9 Elm Lane, Minster On Sea, Sheerness ME12 3SQ
G7 IAW C. Walsh 4 Musbury Crescent, Rossendale BB4 6AY
G7 IBD S. Beaumont 6 Coral Way, Aughton, Sheffield S26 3RE
G7 IBF R. Waller 6 Pitchcombe, Yate, Bristol BS37 4JX
G7 IBH K. Ashton 13 Laceys Avenue, Leverton, Boston PE22 0BG
G7 IBL M. Whatley 2 Thompsons Hill, Sherston, Malmesbury SN16 0PZ
GM7 IBM M. Robertson Woodside House, Feabuie, Inverness IV2 5EQ
G7 IBN F. Goodes 17 Ashmead Close, Lords Wood, Chatham ME5 8NY
GW7 IBT A. Earp 42 Tudor Gardens, Neath SA10 7RX
G7 IBU D. Nicholls 19 Kimmeridge, Crown Wood, Bracknell RG12 0UD
G7 IBX V. Finlayson 92 Herlington, Orton Malborne, Peterborough PE2 5PR
G7 ICD J. Mcdowall 19 Plaistow Court, Hallwood Park, Runcorn WA7 2GR
G7 ICE R. Catlow 137 Haven Lane, Oldham OL4 2QQ
G7 IDE I. Evans 18 Plemstall Way, Mickle Trafford, Chester CH2 4QJ
G7 IDH M. Newton 5 Granville Avenue, Newcastle ST5 1JH
G7 IEB R. Emberton 10 Lodway Close, Pill, Bristol BS20 0DE
G7 IED R. Martin 45 Quail Holme Road, Knott End-On-Sea, Poulton-le-Fylde FY6 0BT
G7 IEF D. Roadnight 14 Newquay Crescent, Harrow HA2 9LJ
GD7 IEH M. Blackburn 63 Westbourne Drive, Douglas IM1 4BB Isle of Man
G7 IEO A. Cook 26 Worcester Road, Stourport-on-Severn DY13 9PB
G7 IER A. Bent Three Gables, Craggs Hill, Carnforth LA6 1DJ
G7 IET Dr A. Dunlop High View, Milton Avenue, Sevenoaks TN14 7AU
GM7 IEU A. Steele 20 Stewart Way, Alford AB33 8UB
G7 IEY D. Chenery 25 Aldreth Road, Haddenham, Ely CB6 3PW
GI7 IEZ T. Mc Geown 1 Drumcairn Road, Armagh BT61 7SA
G7 IFB S. Thompson 39 Greenbank Road, Ambleside LA22 9BD
G7 IFI B. Jones 25 Milton Drive, Wistaston Green, Crewe CW2 8BT
G7 IFJ R. Page 68 The Ridgeway, St. Albans AL4 9PS
G7 IFL P. King 1 Rue Du Canelots, Saint Frajou 31230 France
G7 IFM J. Hewitt 16 Sandsdale Avenue, Fulwood, Preston PR2 9AZ
G7 IFO N. Rigby 2 Mill Lane, Sutton Manor, St. Helens WA9 4HW
G7 IFR T. Chibnell - Smith Nursery Cottage, Whitney-On-Wye, Hereford HR3 6HT
G7 IFU M. Mccartney 1 Tollemache Close, Manston, Ramsgate CT12 5LX
GI7 IFW S. Boskett 314 Shore Crescent, Belfast BT15 4JU
GM7 IFX J. Barnett 72 Cameron Toll Gardens, Edinburgh EH16 4TG
G7 IGF I. Fields Boarzell Cottage, London Road, Etchingham TN19 7QY
G7 IGR C. Crowhurst 143 Drayton High Road, Drayton, Norwich NR8 6BD
G7 IGU L. Evans 58 Westminster Drive, Bromborough, Wirral CH62 6AW
G7 IHD J. Bolsover 18 Millers Ford Low Bentham, Lancaster LA2 7BF
G7 IHE H. Robinson 16 Coniston Avenue, Ashton-In-Makerfield, Wigan WN4 8AY
GM7 IHH A. Tolson 4 Albert Place, Langholm DG13 0AT
GM7 IHJ M. Alexander 38 The Wynd, Dalgety Bay, Dunfermline KY11 9SJ
G7 IHN S. Harvey Gabled Cottage, Shipton Oliffe, Cheltenham GL54 4HZ
G7 IHP K. Weston 2 Beech Grove, Somerton TA11 6LG
GM7 IHR R. Brodie Midgeloch Cottage, Arbuthnott, Scotland AB3 1NX
G7 IHV G. Havell Flat 13, Waldron House, London SW2 1PA
G7 IHX J. Allan 60 Godfrey Road, Halifax HX3 0SU
GM7 IHZ Dr G. Hayes Flat 6, 87 London Road, Edinburgh EH7 5TT
G7 IIB P. Shields 3 Hawthorn Road, Tavistock PL19 9DL
G7 IIC S. Oliphant Homeside, Compton, Paignton TQ3 1TD
G7 IID E. Caunt 5 Littledale, Pickering YO18 8PS
G7 IIF A. Barrett 17 Wimborne Avenue, Wirral CH61 7UL
G7 IIH J. Bond 1 Old House Courtyard Southover High Street, Lewes BN7 1HT
G7 IIN M. Hewitt 2 Hill View, Worstead, North Walsham NR28 9SD
G7 IIO B. Bellamy 71 High Road, Benfleet SS7 5LH
G7 IIS C. Beatrup Bon Air, 34 Springfield Drive, Halesowen B62 8EU
G7 IIZ G. Dooley 93 Springfields, Walsall WS4 1JX
G7 IJC D. Wells 21 Kings Road, Barnetby DN38 6HF
G7 IJI D. Gibbs Flat 1 74 Priory Road Hall Green, Birmingham B28 0TE
G7 IJL D. Mcclew 135 Hermitage Street Rishton, Blackburn BB1 4ND
G7 IJW Dr B. Rushton Cherrydene, New Road, Windermere LA23 2LA
G7 IJY B. Evans 51 Katrina Grove, Featherstone, Pontefract WF7 5LW
GM7 IKB G. Riddell Lawhead Croft, Tarbrax, West Calder EH55 8LW
G7 IKG A. Thynne 1 Earlston Way, Birmingham B43 5JR
G7 IKM W. Willan 31 St. Oswalds Lane, Bootle L30 5QD
G7 IKS A. Raistrick 10 Orchard Way, Chinnor OX39 4UD
G7 ILA T. Brennan 9 Mill Lane, Felixstowe IP11 7RL
G7 ILD P. Brown 4 Solent Drive, Barton On Sea, New Milton BH25 7AW
G7 ILG I. Glossop 1 Harborough Hill Cottages, Birmingham Road, Kidderminster DY10 3LH

G7 ILI A. Page 29 Lambourne Close, Fareham PO14 1SL
G7 ILJ Vintage Radio Group c/o B. Thornton 21 Valley Road, Banbury OX16 9BQ
G7 ILL P. Atherton Findern Lane, Willington, Derby DE65 6DW
G7 ILP K. Naylor 3 Windrush Close, Bicester OX26 2AR
G7 ILS I. Warrilow 84 Marple Road, Stockport SK2 5RN
G7 ILX R. Voges 43 Eastgate, Fulwood, Preston PR2 3HS
G7 ILY D. Barber 2 St. Jamess Mews, Church, Accrington BB5 4JR
G7 IMB S. Jeffcoate 25A Northampton Road, Lavendon, Olney MK46 4EY
G7 IMD A. Spittlehouse 7 Fernbank, Battle Green, Doncaster DN9 1LJ
G7 IMH M. Fortescue 98 Campbell Road, Oxford OX4 3NU
G7 IMO L. Handley 11 Brook Close, Blythe Bridge, Stoke-on-Trent ST11 9PX
G7 IMQ P. Bannister 222 Haslucks Green Road, Shirley, Solihull B90 2LN
G7 IMR M. Taft 44 Langcomb Road, Shirley, Solihull B90 2PR
G7 IMT D. Gerard 15 Nyetimber Lane, Bognor Regis PO21 3HQ
GI7 IMU A. Reid 18 Orby Grove, Belfast BT5 6AL
G7 IMV R. King Old Orchard, South Milton, Kingsbridge TQ7 3JZ
G7 IMY S. Kemp1 16 Douglas Road, Aylesbury HP20 1HW
G7 IMZ G. Smith 19 Parker Road, Humberston, Grimsby DN36 4TT
G7 INC G. Bacon 36 Warnadene Road, Sutton-in-Ashfield NG17 5BD
G7 ING M. Darbyshire 50 Gaythorne Avenue, Preston PR1 5TA
GI7 INR A. Greer 86 Ballygowan Road, Banbridge BT32 3QX
G7 INY J. Coady Sunset, Station Road Wisbech St. Mary, Wisbech PE13 4RT
G7 IOB C. Knowlson 28 Hill Drive, Handforth, Wilmslow SK9 3AR
G7 IOC R. Mitchell 2 Corbar Road, Stockport SK2 6EP
G7 IOF K. Roebuck 20 Ryecroft Close, Outwood, Wakefield WF1 2LW
G7 IOI N. Telford 18 Kirkstall Close, South Anston, Sheffield S25 5BA
G7 ION M. Tennant 64 Aldenham Road, Kemplah Park, Guisborough TS14 8LD
G7 IOO P. Horton 408 Woodcrest Way, Forney 75126 United States
G7 IPA M. Clements 23 Pudding Lane, Gadebridge, Hemel Hempstead HP1 3JU
G7 IPH P. Baker 6 Firework Close, Kingswood, Bristol BS15 4LT
G7 IPI P. Crane 16 Lansbury Drive, Cannock WS11 4BH
GI7 IPO H. Stokes 32 Islay Street, Antrim BT41 2TS
G7 IPR N. Daniels 136 Womersley Road, Knottingley WF11 0DQ
GW7 IPS S. Hamlyn 6, New Road, Newcastle Emlyn SA38 9BA
G7 IPX C. Bowden 36 Aspin Drive, Knaresborough HG5 8HQ
G7 IQD R. Cook 17 Beech Gardens, , Ludlow SY8 1UT
G7 IQM P. Jaggs 218 New Road, London E4 9SJ
G7 IQO D. Flatters 87 Albert Promenade, Loughborough LE11 1RD
G7 IQZ R. Norman 87 Edenfield Gardens, Worcester Park KT4 7DX
GW7 IRD T. Jones 3 Woodlands Close, St. Arvans, Chepstow NP16 6EF
G7 IRF A. Saunders 24 Hallam Close Midsomer Norton, Radstock BA3 2FG
G7 IRG G. Wisbey 4 Avenue Road, Streatham, London SW16 4HL
G7 IRH W. Moth 145 Carisbrooke Road, Newport PO30 1DG
GM7 IRI K. Baxter Flat 2, 29B Corbiehill Road, Edinburgh EH4 5BQ
GI7 IRJ P. Mcateer 36 Ballyquillan Road, Aldergrove, Crumlin BT29 4RH
G7 IRK D. Deacon 14 Dukes Road, Braintree CM7 5UE
G7 IRL W. Thomas 6 Chestnut Rd Astwood Bank, Redditch B96 6AF
G7 IRN M. Scott 3 Summerhill, Ticehurst, Wadhurst TN5 7JA
G7 IRP D. Williams 66 Gover Road, Hanham, Bristol BS15 3JZ
G7 IRU C. Hosegood 4 The Orchard, Sixpenny Handley, Salisbury SP5 5QL
G7 ISD Dr C. Rizzo Downside Downs Road Funtington, Chichester PO18 9LS
G7 ISE G. Walters 12 Portstone Close, Northampton NN5 6QP
G7 ISR G. Lines 11A Gloucester Road North, Bristol BS7 0SG
GI7 ISX S. Butler 25 Chippendale Avenue, Bangor BT20 4PX
G7 ITB D. Davies Greyroofs Albert Place, Washington NE38 7BW
G7 ITM G. Clarkson 40 Wharf Road, Ash Vale, Aldershot GU12 5AY
G7 ITO M. De Banks 56 Blackwater Drive, Aylesbury HP21 9RX
G7 ITS M. Fasham 29 Granville Avenue, Ramsgate CT12 6DX
G7 ITT S. Import The Old Rectory, Dufton, Appleby-in-Westmorland CA16 6DA
G7 ITU S. Marlow 14 Lightgate Road, South Petherton TA13 5AJ
G7 ITW D. Fennelly 23 Trent View Gardens, Radcliffe-On-Trent, Nottingham NG12 1AY
G7 ITX T. Emblem-English 4 Mark Avenue, London E4 7NR
G7 ITZ B. Calvert Wall To Wall Communications, Unilink House, 21 Lewis Road, Sutton SM1 4BR
G7 IUB L. Porter 324-326 Lillie Road, London SW6 7PP
G7 IUE G. Clem 25 Alexander Close, Waterlooville PO7 5TB
GM7 IUF W. Howie 24 Newfield Drive, Dundonald, Kilmarnock KA2 9EW
G7 IUI Prof. W. Fagan 32 Eastfield Road Thurmaston, Thurmaston, Leicester LE48FP
G7 IVF C. Flux 35 Oaklyn Gardens, Shanklin PO37 7DF
G7 IVG K. Graham 10 Summerfields, Dalston, Carlisle CA5 7NW
G7 IVN P. Jagdev 10 St. Johns Road, Southall UB2 5AN
G7 IVU R. Walker 16 Norman Drive, Stilton, Peterborough PE7 3RS
GI7 IVX R. Connolly 21 Eleastan Park, Kilkeel, Newry BT34 4DA
G7 IWA A. Maunder 2 Downhouse Road, Waterlooville PO8 0TX
G7 IWEá T. Haggie 6 Rose Villas Middleburg Street, Hull HU9 2QR
G7 IWK G. Blackburn 10 Lodge Close, Redhill, Nottingham NG5 8NZ
G7 IWM S. Gusterson 19 Blackberry Close, Higham Ferrers, Rushden NN10 8FJ
G7 IWU H. Judge 8 Fontenoy Road, Balham, London SW12 9LU
G7 IWV A. Godley 177 Cheriton Road, Folkestone CT19 5HG
G7 IWW R. Gibbs 32 Beswick Avenue, Ensbury Park, Bournemouth BH10 4EY
G7 IWZ R. Murray 92 North Lane, East Preston, Littlehampton BN16 1HE
G7 IXC R. Barkley 39 Fulbeck Avenue, Wigan WN3 5QN
G7 IXG D. Doermann 19 Jackman Close, Fradley, Lichfield WS13 8PW
G7 IXH P. Lawton 207 Eachelhurst Road, Sutton Coldfield B76 1EA
G7 IXK G. Smith 12 Middle Greeve, Wootton, , Northampton NN4 6BB
G7 IXP P. Hammersley 30 Bonner Grove, Aldridge, Walsall WS9 0DU
G7 IYA B. Whittock 12 Hillside Crescent, Midsomer Norton, Radstock BA3 2NB
G7 IYG N. Hobbs 24 Falkland Road, Southport PR8 6LG
G7 IYH Dr L. Hobbs 24 Falkland Road, Southport PR8 6LG
G7 IYK D. Trewren 184 High Street Oldland Common, Bristol BS30 9QQ
G7 IYM T. Mann 34 Crows Grove, Bradley Stoke BS32 0DA
G7 IYN A. Attack 8 Lewis Road, Hornchurch RM11 2AJ
G7 IYQ K. Fulcher Derventio, High Street, Gainsborough DN21 5LY
G7 IYX R. Dodds 33 Westgate, Warley, Oldbury B69 1BA
G7 IYY P. Hollyoake 79 Borough Cres, Oldbury B69 1AJ
GW7 IZA G. Griffiths Newcastle Court, Evancoyd, Presteigne LD8 2PA
G7 IZC R. Phillips 25 Carlile Hill, Hemlington, Middlesbrough TS8 9SL
G7 IZE K. Palmer 6 Parklands Close, Arnold, Nottingham NG5 9QU
G7 IZM F. Lucas Ivella, Recreation Street, Dudley DY2 9EU
G7 IZN M. Dunn 39 Gainsbrook Crescent, Norton Canes, Cannock WS11 9TN
G7 IZU A. Smith 7 Hartley Avenue, Plymouth PL3 5HW
G7 IZV C. Funnell 61 Blackwatch Road, Coventry CV6 3GS
G7 IZW F. Chilton 127 Nicholls Field, Harlow CM18 6EB
G7 JAE C. Pritchard 11 Willow Green, Needingworth, St. Ives PE27 4SW
G7 JAF A. Lambert 50 Clarendon Road, Sheffield S10 3TR
GI7 JAM Dr K. Gibson 4 Ilford Avenue, Belfast BT6 9SF
G7 JAN J. Martyn Aspiration, Queens Road, Crowborough TN6 1QQ
G7 JAO C. King 39 West Street, Huntingdon PE29 1WT
G7 JAQ R. Adam 8 Lexington Court, Purley CR8 1JA
G7 JAS D. Harris 68 Tomlinson Avenue, Luton LU4 0QW
G7 JAX D. Hall The Crow'S Nest 9A Cheveley Road, Newmarket CB8 8AD
G7 JBD C. Storrie 3 Stocken Hall Mews, Stretton, Oakham LE15 7RL
G7 JBW P. Hoath 1 Red Lodge Drive, Bilton, Rugby CV22 7TT
G7 JBZ R. Cone 6 Renault Drive, Bracebridge Heath, Lincoln LN4 2QG
G7 JCD M. Jones 4 Bell Street, Tipton DY4 8HZ
G7 JCF S. Beamish The Old Vicarage, Vicarage Road, Woodbridge IP13 8DT
G7 JCQ G. Blunt 36 Whitefield Way, Liverpool L6 2NB
G7 JCX J. Price 37 The Court, Anderby Creek, Skegness PE24 5YQ
G7 JDA A. Roberts 5 Colmar Close, Daventry NN11 9BT
G7 JDB J. Blackburn 2 Heath Drive, Sutton SM2 5RP
G7 JDF J. Hope 48 Holbeck, Bracknell RG12 8XE
G7 JDH A. Nevill 46 Skipwth Gardens , New Rossington, Doncaster DN11 0TU
G7 JDI D. Carslake 21 Kestrel Drive, Bingham, Nottingham NG13 8QD
G7 JDK R. Rothwell 11 St Marks Road, Stourbridge DY9 7DT
G7 JDN M. Collins 8 Newfield Road, Marlow SL7 1JW
G7 JDR C. Bennett The Old Cottage, Waterside Road, Southminster CM0 7QT
GM7 JDS B. Reid 10 Badenoch Road Kirkintilloch, Glasgow G66 3NX
GW7 JDX Dr M. Ghassempoory 102 Colchester Avenue, Penylan, Cardiff CF23 9AZ
GI7 JEB M. Gibson 1 Downshire Park, Bangor BT20 3TP
GM7 JED I. Macdonald 3 Anderson Road, Stornoway HS1 2PG
G7 JEJ T. Hyder 83 Beam Hill Road, Burton on Trent DE13 0AD
GI7 JEM D. Branagh 146 Craigs Road, Carrickfergus BT38 9XA
G7 JFI T. Steeper 16 High Street, Eagle, Lincoln LN6 9DH
G7 JFM S. Smith 26 Broadsands Avenue, Paignton TQ4 6JN
GM7 JFN K. Maclean 10B Knockaird, Port Of Ness, Isle of Lewis HS2 0XF
G7 JFU R. Evison Sandpiper Fibbards Road, Brockenhurst SO42 7RD
G7 JGB A. Kemp 7 Hartley Way, Bishopdown, Salisbury SP1 3WS
G7 JGE C. Hobson 28 Withering Road, Swindon SN1 4GU
GW7 JGF T. Froggatt Gwyndy, Moelfre, Anglesey LL72 8LN
GM7 JGH A. Bruce 20 Weir Crescent, Milton, Wick KW1 5SS
G7 JGQ A. Greenland 19 The Ridgeway, Potton, Sandy SG19 2PS
GM7 JGR Dr J. Howie 29, Coates Gardens, Edinburgh EH12 5LG
G7 JGS G. Swindells 52 Western Avenue, Blacon, Chester CH1 5PP
GI7 JGT M. Mc Namee 22 St. Patricks Park, Rosslea, Enniskillen BT92 7QY
G7 JGW W. Holroyd 8 Carr Dene Court, Preston Street, Preston PR4 2XA
G7 JGY P. Smith 174 Willerby Road, Hull HU5 5JW
G7 JGZ R. Brooks 8 Chichester Place, Tiverton EX16 4BW
GW7 JHC T. Christie 7 Hayes View, Oswestry SY11 1TP
G7 JHE G. Beckett 34 Bradwall Road, Sandbach CW11 1GF
GW7 JHK P. Brettle 27 Neath Road, Resolven, Neath SA11 4AA
G7 JHM J. Mccollin 17 Lamsey Road, Hemel Hempstead HP3 9HB
G7 JHU S. Birchall 83 Wilton Avenue, Chapel St. Leonards, Skegness PE24 5YN
G7 JHV D. Gervais Seven Gables Lodge, Buckingham Road, Buckingham MK18 3NA
G7 JHW R. Johnson 30 Thorpe Downs Road, Church Gresley, Swadlincote DE11 9FB
G7 JHX Dr J. Williams 40 Tythe Barn Lane, Shirley, Solihull B90 1RW
G7 JHZ D. Randles 20 Felix Road, London W13 0NT
G7 JIB L. Evans Polvellan, School Hill, St. Austell PL26 6TG
G7 JIF S. Ruffell 2 Beulah Cottage, Church Street, West Stour SP8 5RL
G7 JIM W. Barton 4 Hawthorn Flats, Hawthorn Road, Dorchester DT1 2PE
G7 JIN C. Willis 9 Avington Close, Sedgley, Dudley DY3 3LN
G7 JJC P. Gerrard 6 Ellabank Road, Heanor DE75 7HF
G7 JJG K. Watts 68 Kentwood Hill, Tilehurst, Reading RG31 6DE
G7 JJJ C. Marshall Gladstan House, 70 Chester Road, Runcorn WA7 3DY
G7 JJP L. Towler 8 Stowehill Road, Peterborough PE4 7PY
G7 JJW S. Coffin 5 Colt Close, Streetly, Sutton Coldfield B74 2EA
G7 JJX R. Wallace 31 Salts Road, West Walton, Wisbech PE14 7EJ
G7 JKD M. Coward 7, Brackenrigg, Armathwaite, Carlisle CA4 9PX
G7 JKH C. Hyde 42 Fern Road, Whitby, Ellesmere Port CH65 6PB
GW7 JKK J. Mossman 13 Tynrhos Estate, Caergeiliog, Holyhead LL65 3HS
GI7 JKM S. Glendinning 2 Scotts Road, Moneymore, Magherafelt BT45 7TW
G7 JKW S. Avery Wilding Farm Cottage Cinder Hill, Lewes BN8 4HP
G7 JKY Dr S. Smith 73, Station Street, Rippingale, Bourne PE10 0SX
G7 JLC A. Edwards 34 Albion Road, Malvern Link, Malvern WR14 1PU
GI7 JLD J. Hunter 29 Mullaghacall Road, Portstewart BT55 7EG
G7 JLF R. Pike 57 Bishopstone Road, Stone, Aylesbury HP17 8QR
GW7 JLG A. Williams 2 Nant Y Berllan, Llanfairfechan LL33 0SN
G7 JLK R. Elliott 39 Amanda Way, Pensilva, Liskeard PL14 5RA
G7 JLO N. Townend 124 Rylands Road, Southend-on-Sea SS2 4LJ
G7 JLS D. Bryant Knowle Barns, Knowle, Broadhempston, Totnes TQ9 6DA
G7 JLT K. Bryant 18 Loundyes Close, Thatcham RG18 3EB
G7 JMB J. Baker Green Lane Farmhouse, Rugeley WS15 2AR
G7 JME P. Good 11 Moorland Road, Didsbury, Manchester M20 6BB
G7 JMQ M. Tidmarsh 16 Castleton Road, Mitcham CR4 1NY
G7 JMU D. Butterworth 27 Royds Avenue, Linthwaite, Huddersfield HD7 5QU
G7 JMW A. Weaver 116 Maldon Road, Tiptree, Colchester CO5 0BN
G7 JMZ J. Bache 62 Whittingham Road, Halesowen B63 3TP
G7 JNM A. White 6 Greenbank, Hadfield, Glossop SK13 1PD
G7 JNS S. Mclennan 179 King John Avenue, Bear Wood, Bournemouth BH11 9SJ
G7 JOA Rsc of Cheshire c/o C. Rickerby Brownsea, 113 Cliftonville Road, Woolston, Warrington WA1 4BJ
G7 JOW J. Ashbee 49 Sandwich Road, Whitfield, Dover CT16 3LT
G7 JPN M. Bateman 22 Bowling Green Lane, Albrighton, Wolverhampton WV7 3HL
G7 JQF W. Booth 8 Park Crescent, Bacup OL13 9RL
GD7 JQI A. Kissack 30 High View Road, Douglas IM2 5BH Isle of Man
G7 JQT E. Barry 8 Astley Crescent, Scotter, Gainsborough DN21 3SL

G7 JQW H. Derrick 28 Great Parks, Holt, Trowbridge BA14 6QP
G7 JQZ D. Beadle 4 Harlaxton Drive, Lincoln LN6 3NR
G7 JRC D. Smith 26 Mill Fields Todwick, Sheffield S26 1JS
G7 JRD T. Alwyn-Clark 1 Blackfriars Road, Lincoln LN2 4WS
GI7 JRG A. Mcnerlin 27 Roeview Park, Limavady BT49 9BQ
G7 JRJ C. Wainwright 31 Queens Road, Leytonstone, London E11 1BA
G7 JRK P. Dixon 7 Pincey Mead, Basildon SS13 3EW
G7 JRM C. Hinton 65 South Street, Tarring, Worthing BN14 7NE
G7 JRP T. Pratley 28 Charles Avenue, Watton, Thetford IP25 6BZ
GW7 JRT J. Tonge Bracken Brae, Gwalchmai, Holyhead LL65 4SL
G7 JRU A. Martin 36 Saxon Road, Lowestoft NR33 7BT
G7 JSC R. Brotherton 167 Pershore Road Hampton, Evesham WR11 2NB
G7 JSE S. Almond 2 King Street, Swinton, Mexborough S64 8ND
G7 JSG R. Maynard 8 Badgers Walk Pool Lane, Clows Top, Kidderminster DY14 9NT
GW7 JSH J. Field Dan-Y-Coed, North Beach Road, Aberystwyth SY23 3DT
G7 JSQ P. Domachowski 39 Wycliffe Road West, Coventry CV2 3DX
G7 JSS C. Watson 26 Jupiter Gate, Stevenage SG2 7ST
G7 JST Causeway RC c/o J. Wheatley 8 Winchester Close, Feniton, Honiton EX14 3EX
G7 JSV W. Mcareavey 3 Hall Farm Cottage, East Heckington, Boston PE20 3QG
G7 JSW R. Steward 2 Glenister House, 238 Avondale Drive, Hayes UB3 3PP
G7 JTB R. Pluck The Garden House, St. Leonards Avenue, Blandford Forum DT11 7PA
G7 JTD D. Lockett 10 Cornwall Drive, Bayston Hill, Shrewsbury SY3 0ER
G7 JTF A. Harvey Rose House, Rose Grove, Doncaster DN3 3AJ
G7 JTH J. Carter 30 Swift Way, Sandal, Wakefield WF2 6SR
G7 JTI G. Cuskin 57 Aln Street, Hebburn NE31 1XT
G7 JTK S. Bell 6 Broom Wood Court, Prudhoe NE42 6RB
G7 JTR D. Lock Pelican House, Chilton Candover, Alresford SO24 9TX
G7 JTV J. Caswell 3 Birch Road, Finchampstead, Wokingham RG40 3LB
G7 JTZ R. Smith 17 Julian Road, Spixworth, Norwich NR10 3QA
GW7 JUB T. Jones 37 Bro'R Dderwen, Clynderwen SA66 7NR
G7 JUC K. Marsh 21 Edward Road, Eynesbury, St. Neots PE19 2QF
GI7 JUH T. Cox 13 Shrewsbury Gardens, Belfast BT9 6PJ
G7 JUJ P. Moss 23 Lees Row, Padfield, Glossop SK13 1EN
G7 JUL A. Whitcher 12 Battersby Street, Bury BL9 7SG
G7 JUN M. Steadman 26 Walkers Green, Marden, Hereford HR1 3DU
G7 JUP J. Beckingham 20 Baptist Close, Abbeymead, , Gloucester GL4 5GD
G7 JUR P. Lock 1 Carters Walk, Farnham GU9 9AY
GW7 JUV C. Broadbent 12 Aelybryn Ceinws, Machynlleth SY20 9EZ
GM7 JUX W. Dyer 24 Southfield Road, Cumbernauld, Glasgow G68 9DZ
G7 JUZ R. Shams-Nia 1090 Eastern Avenue, Ilford IG2 7SF
G7 JVB P. Wade 41 Prospect Avenue, Stanford-le-Hope SS17 0NH
G7 JVC M. Hewitt 1 Harpswell Hill Park, Hemswell, Gainsborough DN21 5UT
G7 JVE N. Cook 35 Glanville Road, Hadleigh, Ipswich IP7 5SQ
G7 JVF S. Mobley 2 Lingham Close, Solihull B92 9NW
G7 JVG A. White 19 Haswell Close, Wardley, Gateshead NE10 8UE
G7 JVJ E. Peacock Octon Lodge, Langtoft, Driffield YO25 3BJ
G7 JVK R. Hardie 12 Hopland Close, Longwell Green, Bristol BS30 9XB
G7 JVN D. Greywolf 3 Denham Close, St. Leonards-on-Sea TN38 9RS
G7 JVO K. Saxby 184 Brodrick Road, Eastbourne BN22 9RH
G7 JVQ F. Sparks 36 High View Road, Guildford GU2 7RT
G7 JWD T. Place 34 Holcroft, Orton Malborne, Peterborough PE2 5SL
G7 JWE A. Liddell 4 Russet Court, Kingswood, Wotton-under-Edge GL12 8SG
G7 JWH A. Butler 22 Willow Park, Minsterley, Shrewsbury SY5 0EH
G7 JWI G. Harrison 58 Hollywall Lane, Stoke-on-Trent ST6 5PP
G7 JWJ E. Hickman Eriska, 33 Romany Way, Stourbridge DY8 3JR
G7 JWL E. Oakes 30 Linden Avenue, Stourport-on-Severn DY13 0EQ
G7 JWO R. Allcock 44 Newmount Road, Stoke-on-Trent ST4 3HQ
G7 JWQ B. Priestley Priorswood Cottage, Tyndale Road, Gloucester GL2 7DJ
G7 JWV R. Ebbetts Markway House, Blackbush Road, Lymington SO41 0PB
G7 JWW S. Charters Beechgrove, Haselor Lane Hinton-On-The-Green, Evesham WR11 2QZ
G7 JWX B. Maley 10 Wolsey Place, 49-51 London Road, Hailsham BN27 3FU
G7 JWY C. Ainley 23 Forresters Close, Norton, Doncaster DN6 9HX
G7 JXB K. Cox 16 Henty Close, Walberton, Arundel BN18 0PW
G7 JXD J. Pritchard 22 Osborne Way, Haslingden, Rossendale BB4 4DZ
G7 JXF M. Forknell 24 Sherbourne Avenue, Nuneaton CV10 9JH
G7 JXJ C. Smith 30 Rookery Close, St. Ives PE27 5FX
G7 JXL D. Kerridge 7 Haslers Place, Haslers Lane, Dunmow CM6 1AJ
G7 JXR G. Wiseman 7 Barton Road, Woodbridge IP12 1JQ
G7 JXT I. Ballantyne 2 Dunvegan Close Manea, March PE15 0LU
G7 JXU M. Barker 103 Friarswood Road, Newcastle ST5 2EF
G7 JXX I. Thaiss 4A Union Street, Market Rasen LN8 3AA
G7 JXY Windy Hill CG c/o I. Guffick 13 Alderwood Close, Hartlepool TS27 3QR
G7 JYG H. Odd Verona, Harrow Road, Sevenoaks TN14 7JU
GW7 JYJ T. Gittoes Oak Farm, Builth Wells LD2 3EN
GI7 JYK P. Lowrie 13 Carwood Park, Newtownabbey BT36 5JU
G7 JYL J. Sage 8 Foxwood Road, Bean, Dartford DA2 8BH
G7 JYQ T. Dabbs 4 Caverleigh, Cadogan Road, Surbiton KT6 4DH
GM7 JYW P. Lawrence Gateside Smithy, Munlochy IV8 8PA
G7 JYY M. Penn 5 Angus Close, Kenilworth CV8 2XH
G7 JYZ S. Turley 22 Powlers Close, Stourbridge DY9 9HH
G7 JZC A. Upchurch 68 Lindleys Lane, Kirkby-In-Ashfield, Nottingham NG17 8AD
G7 JZI W. Hilton 8 Ashfield Avenue, Hindley Green, Wigan WN2 4RG
G7 JZJ M. Doyle 133A Pope Lane, Penwortham, Preston PR1 9DD
G7 JZK W. Hancox Flat 34, Millbank Court, Barlows Lane, Liverpool L9 9HQ
G7 JZM G. Priestley 24 Saxton Avenue, Bradford BD6 3SW
G7 JZS M. Budd 37 Cheyne Walk, Hornsea HU18 1BX
G7 JZY K. Long Manor Farm, 27 Church Street, Hull HU11 4RN
G7 KAK I. Clewley 31 Kenilworth Road, Basingstoke RG23 8JF
GD7 KAM A. Swearman 56 Garth Avenue, Surby, Isle of Man IM9 6QU
G7 KAO D. Clarke 2 Wilmot Road, Dartford DA1 3BA
G7 KAV N. Stemp 3 Loxwood, East Preston, Littlehampton BN16 1DT
G7 KBD A. Carlton 32 Culver Road, Bradford-on-Avon BA15 1HZ
G7 KBE B. Mcintyre West Abbey Nursing Home, Stourton Way, Yeovil BA21 3UA
G7 KBH K. Wainwright 25 Tithebarn Road, Rugeley WS15 2QW
GW7 KBI G. Dreiling Picton Farm, Holywell CH8 9JQ
GM7 KBK E. Pratt 46 Sheddocksley Drive, Aberdeen AB16 6NX
G7 KBR P. Phillips 10 Byron Grove, East Grinstead RH19 1SG
G7 KBZ S. Hutchinson 32 Uppleby, Easingwold, York YO61 3BB
G7 KCC J. Durdin 16 Barnwood Close, Kingswood, Bristol BS15 4JA
G7 KCE J. Hannaford 22 Barn Park, Stoke Gabriel, Totnes TQ9 6SR
G7 KCK P. Langley 321 Maidstone Road, Rochester ME1 3EF
G7 KCN B. Elcoate 9 Parsonage Lane, Laindon Laindon, Basildon SS15 5YN
G7 KDG P. Edmondson 20 Mill Road, Impington, Cambridge CB24 9PE
G7 KDH D. Edmondson 4 Elm View, Steeton, Keighley BD20 6SZ
GW7 KDI P. Stevenson Nant Fach Cerrigydrudion, Corwen LL21 0SB
G7 KDJ A. Chadwick 2 Auden Place, Longton, Stoke-on-Trent ST3 1SJ
G7 KDM C. Campbell 21 Sellwood Drive, Carterton OX18 3AZ
G7 KDN A. Thomas 49 Tristan Close, Calshot, Southampton SO45 1BN
G7 KDQ K. Roan 133 Woodhouse Lane, Beighton, Sheffield S20 1AD
G7 KDR B. Hopkins 14 Falkenham Rise, Basildon SS14 2JQ
GW7 KDU M. Lewis 111 Willowbrook Gardens, St. Mellons, Cardiff CF3 0BY
G7 KDX R. Bell 5 Byron Avenue, Blyth NE24 5RN
G7 KEA R. Chapman Flat 3, Goda Court, Littlehampton BN17 6AS
GI7 KEC J. Stafford 31 Shimna Close, Belfast BT6 0DZ
G7 KEE B. Daw 19 Rowan Close, Yarnfield, Stone ST15 0EP
G7 KEI B. Edgley 2 Queens Close, Hyde SK14 5RE
G7 KEK R. Horsfall 7 Lytham Close, Doncaster DN4 6UT
G7 KEP A. Reeve 97 Mendip Vale, Coleford, Radstock BA3 5PP
G7 KFM I. Hasman Fleetway, The Spinney, Newark NG24 2NT
G7 KFN C. Hasman Fleetway, The Spinney, Newark NG24 2NT
G7 KFP G. Spicer 19 Byfield Way, Bury St. Edmunds IP33 2SN
G7 KFQ N. Camp 1 Higher Tresillian Cottages, Tresillian., Newquay TR8 4PL
GM7 KFS A. Wood Seaward, Toward, Dunoon PA23 7UA
G7 KFZ R. May 153 Station Road, Winsford CW7 3DE
GW7 KGD H. Wrighton 43 Bryn Celyn, Colwyn Bay LL29 6DH
G7 KGH M. Forder 157 Kennington Road, Kennington, Oxford OX1 5PE
G7 KGI E. Gould 53 Green Road, Kidlington OX5 2EU
G7 KGP J. Chisholm 162 Ardington Road, Northampton NN1 5LT
G7 KGR C. Saunders 148 Downs Barn Boulevard, Downs Barn, Milton Keynes MK14 7RR
G7 KGV I. Lewis Whitehill Lodge, Hextalls Lane, Redhill RH1 4QT
GM7 KHA S. Grant 2 Clayton Avenue, Irvine KA12 0TR
G7 KHE M. Knowlson 23 Hawthorne Avenue, Shipley BD18 2JB
G7 KHF S. Bates Chez Nous, Seaton Lane, Seaton, Seaham SR7 0LS
G7 KHL S. Smith 287 Campkin Road, Cambridge CB4 2LD
GI7 KHR W. Smyth 35 Davarr Avenue, Dundonald, Belfast BT16 2NT
G7 KHT A. Haw 16 Sunnybank Crescent, Yeadon, Leeds LS19 7TE
G7 KHV R. Irvine 1 Nutana Avenue, Hornsea HU18 1JU
G7 KHW D. Nock 112 Helmsley Close, Bewsey, Warrington WA5 0GB
G7 KHZ R. Hobbs 3 Duncombe Close, Bridgwater TA6 4UT
G7 KID Dr C. Baily 25 Rocks Park Road, Uckfield TN22 2AT
G7 KIE N. Kirkman 4 Woodhall Crescent, Saxilby, Lincoln LN1 2HZ
G7 KIF C. Davis 91 Station Road, Barton Under Needwood, Burton-on-Trent DE13 8DS
G7 KII M. Chilcott 16 Mount Gould Avenue St. Judes, Plymouth PL4 9EZ
G7 KIL C. Hunt 39 Withdean Crescent, Brighton BN1 6WG
G7 KIN B. Kinsella 8 Sherwood Park Road, Sutton SM1 2SQ
GW7 KIO G. Hawthorn-Slater Ty Croes, Garndolbenmaen LL51 9UJ
G7 KIQ P. Hyde 10 Highfield Crescent, Taunton TA1 5JH
GW7 KIS B. Latta 17 Park Lane, Holywell CH8 7UR
G7 KIT D. Hogg 26 Grenville Drive, Church Crookham, Fleet GU51 5NR
GW7 KIV R. Gadney 6 Dan Yr Eppynt Tirabad, Llangammarch Wells LD4 4DR
G7 KIW R. Henery 117 Marlborough Road, Swindon SN3 1NJ
GM7 KIY J. Webster 31 Harperland Drive, Kilmarnock KA1 1UH
G7 KJA R. Early 11 Wenlock Drive, Newport TF10 7HH
G7 KJD J. Smallwood 6 Thatchers Croft, Copmanthorpe, York YO23 3YD
G7 KJE A. Wilkes 51 Shrewsbury Drive, Newcastle ST5 7RQ
G7 KJI D. Lean 6 Orchard Close Edmondsham, Wimborne BH21 5RQ
GW7 KJO M. Wray Dinas Bran, Ceidio, Pwllheli LL53 8UG
G7 KJP R. Mcmahon 8 Meadow Close, Holburn Estate, Ryton NE40 3RU
G7 KJR C. Baxter 6 Merrington Close, Kirk Merrington, Spennymoor DL16 7HU
G7 KJT S. Mills 49 Temple Gate Crescent, Leeds LS15 0EZ
G7 KJV M. Litchman 26 Oak Tree Close, Loughton IG10 2RE
G7 KJW P. Haylock 25 Whitehouse Road, Sawtry, Huntingdon PE28 5UA
G7 KJX R. Tebbutt 37 Christchurch Drive, Daventry NN11 4RX
G7 KKW J. Marsden 11 Firethorn Drive, Hyde SK14 3SN
G7 KLJ S. Kerr 2 Shaw Cross, Kennington, Ashford TN24 9JY
G7 KLN J. Abbey 4 Northway, Curzon Park, Chester CH4 8BB
G7 KLP C. Hartigan Doonagore, Doolin Ireland
G7 KLR L. Pooley Grayson Green Farmhouse, Midtown, High Harrington, Workington CA14 5RE
G7 KLS A. Macaulay 14 Shipcote Lane, Gateshead NE8 4JA
G7 KLT T. Hassall 5 Ashworth Street, Bacup OL13 9LS
G7 KLV G. Lovegrove 64 Vicarage Lane, Great Baddow, Chelmsford CM2 8HY
G7 KLZ J. Fowler Quinnhaven, Banton Shard, Bridport DT6 3EB
G7 KMA S. Balkham 49 St. Georges Road, Hastings TN34 3NH
GI7 KMC J. Magee 2 Gilbourne Court, Belfast BT5 7JB
GW7 KMD N. Hilton 9 Waterloo Fields, Kingswood, Welshpool SY21 8LF
G7 KME D. Silverton 49 Brighton Road Holland-On-Sea, Clacton-on-Sea CO15 5SR
G7 KMF R. Lythall 6 Belmont Crescent, Little Houghton, Barnsley S72 0HT
G7 KMH S. Smith 12 Holgate Close, Malton YO17 7YP
G7 KMK N. Deacon 159 Waterworks Road, Coalville LE67 4HZ
G7 KMM S. Linksted Chestnut Lodge, Apleyhead Wood, Babworth DN22 8HQ
G7 KMO P. Butler 15 Roxby Close, Bessacarr, Doncaster DN4 7JH
G7 KMP J. Davies 14 Cullen View, Probus, Truro TR2 4NY
G7 KMT P. Jones 24 Valley Lane, Lichfield WS13 6SU
G7 KMW A. Brown 4 Kimberley Close, Redditch B98 8RL
G7 KNA ltv West RC c/o A. Jenner 24 The Willows, Nailsea, Bristol BS48 1JQ
G7 KNK H. Arrowsmith 15 Hermitage Close, Frimley, Camberley GU16 8LP
G7 KNM M. Giles 9 Bower Green, Lords Wood, Chatham ME5 8TN
GW7 KNN B. Jones Rivendell, Heol Llewelyn, Coedpoeth, Wrexham LL11 3PB
G7 KNQ C. Martin 27 Sheepfold Lane, Ruddington NG11 6NS
G7 KNR P. Grimshaw 12 Field Maple Drive, Ribbleton, Preston PR2 6EU

G7 KNS G. Bubb Clearways Hadlow Stair, Tonbridge TN10 4HD
G7 KNU P. Davis 29 Wiltshire Drive, Trowbridge BA14 0RX
G7 KNW D. Hatfield Ballyheane, Castlebar F23KT92 Ireland
G7 KOF L. Barr 7 Southwold Gardens, New Silksworth, Sunderland SR3 1LG
G7 KOI G. Russ 12 Marconi Road, Chelmsford CM1 1QB
G7 KON W. Humphreys 43 Arundel Street, Bolton BL1 6RR
G7 KOS S. Mccormick 22 Eric Road, Wallasey CH44 5RQ
GM7 KPE J. Reid 10 Fernhill Gardens, Windygates, Leven KY8 5DZ
G7 KPH M. Wood 2 Ridings Lane, New Mill Road, Huddersfield HD7 2SQ
G7 KPM J. Haywood 5 Canada Lane Caistor, Caistor LN7 6RN
G7 KQL D. Walker 34 Kingsford Street, Salford M5 5HX
GW7 KQN C. Parsons Little Foxes, Craig Penllyn CF71 7LE
G7 KQT S. Schrier 163 West Lane, Hayling Island PO11 0JW
G7 KRB S. Wells 55 Staverton Road, Daventry NN11 4EY
G7 KRC Cumbria Raynet Group c/o K. Conlon 4 Hill Crest Drive, Slack Head, Milnthorpe LA7 7BB
G7 KRE T. Benjamin 24 Moat Farm Drive Hillmorton, Rugby CV21 4HG
G7 KRG Keighley Ray Gr c/o T. Binns Cross Farm, 1. Cross Lane Oxenhope, Keighley BD22 9LE
G7 KRH T. Hurley 18 Manewas Way, Newquay TR7 3AJ
G7 KRI J. Tilley 40 Marlborough Road, Stretford, Manchester M32 0AN
G7 KRM B. Walker 3 Moorlands Drive, Mayfield, Ashbourne DE6 2LP
G7 KRO D. Willis 5 St. Andrews Place, Brightlingsea, Colchester CO7 0RH
GM7 KRQ H. Gordon The Cedars, Methlick, Ellon AB41 7DU
G7 KRS Kettering & District A.R.S c/o C. Woodward 32 Bryant Road, Kettering NN15 6JG
G7 KRT H. Leong 38 Woodland Road, Sawston, Cambridge CB22 3DU
GW7 KRY A. Ryall 1 Vine Tree, Rumble Street, Usk NP15 1QG
G7 KRZ S. Pountain 21 Hayfield Road, Chapel-En-Le-Frith, High Peak SK23 0JF
GM7 KSA R. Vennard 4 Braehead, Girdle Toll, Irvine KA11 1BD
G7 KSE A. Hill 53 Fairladies, St. Bees CA27 0AR
G7 KSH C. Coleman 16 Greyhound Road, Glemsford, Sudbury CO10 7SJ
G7 KSP G. Hampson 11 Gladstone Grove, Stockport SK4 4BX
G7 KSQ S. Little 25 Thrift Wood, Bicknacre, Chelmsford CM3 4HT
G7 KSS M. Watts 70 Kentwood Hill Tilehurst, Reading RG31 6DE
G7 KSV D. Pickering 15 Primrose Close, Purley On Thames, Reading RG8 8DG
G7 KTH G. Pargeter 2 Mayfair Drive Kingsmead, Northwich CW9 8GF
G7 KTP T. Daniels Three Yew Trees, Newton St. Margarets, Hereford HR2 0QG
G7 KTQ J. Klunder 58 Windsor Drive, Brinscall, Chorley PR6 8PX
G7 KTR A. Slinn Santon, Pound Lane, Sevenoaks TN14 7NA
GI7 KTU P. Donnelly 18 Marcella Park, Newtownards BT23 4SF
GM7 KTY P. May 6 Hillpark Way, Edinburgh EH4 7BJ
G7 KUB R. Warrell Rose Cottage, Brookbottom, High Peak SK22 3AY
G7 KUG Dr D. Rutherford 9 College Drive, Ruislip HA4 8SD
G7 KUM A. Yorke 45 Ling Road, Chesterfield S40 3HT
GM7 KUN C. Schofield Airidh Ghrianach Knock, Carloway, Isle of Lewis HS2 9AU
G7 KUR P. Rennison 30 Millfield Road, Chorley PR7 1RE
G7 KUU K. Bates Newhaven Cottage, Star Green, Stroud GL6 6AD
GM7 KVB A. Whyte 3 Glenfield Road, Cowdenbeath KY4 9EP
GI7 KVR P. Mcdonald 13 Heathfield, Culmore, Londonderry BT48 8JD
G7 KVT B. Moorey 132 Queensway, Hereford HR1 1HQ
GM7 KVU G. Kilgour 2/1 6 Thornwood Place, Glasgow G11 7PP
G7 KVZ J. Ashmore 46 Mease Close, Measham, Swadlincote DE12 7NA
G7 KWA J. Billam 46 Rugby Road, Rainworth, Mansfield NG21 0AU
G7 KWD Norwich North Scouts Fellowship ARC c/o M. Savin Flat 3, 30 Thurso Close, Reading RG30 4YJ
G7 KWF A. Richards 18 Orchard Way, Lower Kingswood, Tadworth KT20 7AD
G7 KWM R. Saunders 3 Curtismill Close, Orpington BR5 2JX
G7 KWN A. Dance 8 Eversley Road, Arborfield Cross, Reading RG2 9PU
G7 KWO J. Lewis 6 Abbots Way, Beckenham BR3 3RL
G7 KWP G. Lewis 7 Hollam Drive, Dulverton TA22 9EL
G7 KWQ T. Holliday 131 Skinburness Road, Silloth, Wigton CA7 4QH
G7 KWS S. Riches 5 Norfolk St., Forest Gate, London E7 0HN
G7 KWT J. Ruddock 13A Murray Road, Northwood HA6 2YP
G7 KXN M. Bonser 24 Meend Garden Terrace, Cinderford GL14 2EB
G7 KXS P. Adam 50 Lower Edge Road, Rastrick, Brighouse HD6 3LD
G7 KXT G. Belt 3 Prospect Hill, Whitby YO21 1QE
G7 KXV I. Eastham 51 Chapman Road, Fulwood, Preston PR2 8NY
G7 KXZ C. Holdford 23 Willow Close, Newbury RG14 7FX
G7 KYD S. Walker-Kier 45 Anstey Road, Peckham, London SE15 4JX
G7 KYG J. Hope 29 Horner Road, Taunton TA2 8DZ
G7 KYH J. Mann Hyatts Mead, East End, Banbury OX15 5LH
G7 KYI Rvd. M. Wilcockson Queen'S House 16 High Street, Linton, Cambridge CB21 4HS
G7 KYJ A. Clark 10 Garfield Close, Lincoln LN1 3QP
G7 KYL M. Lack 39 Riverview, Church Laneham, Retford DN22 0FL
G7 KYW C. Mellings 4 Kiln Lane, Horley RH6 8JG
G7 KYX G. Stones Ropercroft, Chapel Road, Boston PE22 9PW
G7 KZG C. Cain Rydal House, Audley Road, Newport TF10 7DT
G7 KZJ M. Woodland 8 Berkeley Crescent, Stourport-on-Severn DY13 0HJ
GM7 KZL J. Mawson 5 Forth View, Kirknewton EH27 8AN
G7 KZV C. Dodson 64 Stoneleigh Road, Solihull B91 1DQ
G7 KZY L. Stirrup 16 Berwyn Grove, St. Helens WA9 2AR
GM7 LAC P. Green Clochcan School Cottage, Auchnagatt, Ellon AB41 8UJ
G7 LAF M. Kidman 465 Grove Green Road, London E11 4AA
G7 LAK C. Wilkinson 9 Cheddar Close, Rainworth, Mansfield NG21 0HX
G7 LAL I. Mazura 45 Bolingbroke Road, Scunthorpe DN17 2NQ
G7 LAN D. Halsey 67 Watling St., Rochester ME2 3JH
GD7 LAV A. Gawne Keristal House, Marine Drive, Douglas IM4 1BJ Isle of Man
G7 LAW J. Danner 16 Batemans Acre South, Coventry CV6 1BE
G7 LAX W. Keeys 9 Broomfield Avenue, Rayleigh SS6 9EJ
G7 LBD L. Lewis 29 Sefton Avenue, Hove Edge, Brighouse HD6 2NA
G7 LBH Dr A. Champion 5 Airedale Cliff, Leeds LS13 1EA
GW7 LBI M. Durdin 52 Norton Road, Penygroes, Llanelli SA14 7RS
G7 LBL Dr A. Batey 9 Rampton Drift, Northstowe, Cambridge CB24 3EH
G7 LBM T. Howard 21 Church Lane, Thornhill, Dewsbury WF12 0JZ
G7 LBO D. Wilson 109 Nightingale Drive, Taverham, Norwich NR8 6TR
G7 LBP M. Akiki 103 Main Street, Tupper Lake 12986 United States
G7 LCD A. Sermons 18 Crispin Way, Uxbridge UB8 3WS
G7 LCK J. Berry Roseneath, Walcote Road, Lutterworth LE17 6EQ
GI7 LCQ C. Serplus 14 Claggan Park, Aghadowey, Colraine BT51 4BD
G7 LCS A. Daniels Wiscombe, Cleveland Road, Worcester Park KT4 7JQ
G7 LCV M. Sims 4 Arran Close, Stapleford, Nottingham NG9 8LT
G7 LCW C. Simpson 124 Tattershall Road, Boston PE21 9LR
G7 LDD R. Newton 114 Kingston Road, Taunton TA2 7SP
GW7 LDP P. Martin 19 Clos Bevan, Gowerton, Swansea SA4 3GY
G7 LDR E. Woolfenden 20 Belvedere Avenue Atherton, Manchester M46 9LQ
GM7 LDU W. Adie 16 Gordon Crescent, Methlick, Ellon AB41 7DH
G7 LEB Southgate ARC c/o F. Stevens 4 Pennine Road, Bedford MK41 9AS
G7 LED D. Miles 2 Barrington Road, Solihull B92 8DP
G7 LEL D. Hawkins 93 Buxton Drive, Bexhill-on-Sea TN39 4AS
G7 LEN Hillbillies CG c/o A. Clark Emergency Planning Department, Fire Brigade Headquarters, Lincoln LN5 8EL
G7 LET Barry ARS c/o I. Maughan 95 York Road, Swindon SN1 2JR
G7 LEX S. Wilkes The Coach House, Astley Abbotts, Bridgnorth WV16 4SP
G7 LEY International Police Association c/o G. Lloyd 9 Hornbeam Walk, Witham CM8 2SZ
G7 LFC D. Hughes 86 Colinmander Gardens, Ormskirk L39 4TF
G7 LFL S. Botterill 7 Plumtree Road, Cotgrave, Nottingham NG12 3HT
G7 LFM A. Cocker123 30 Shaw Road, Rochdale OL16 4SH
G7 LFQ J. White 56A Clarendon Street, Herne Bay CT6 8LZ
GM7 LFT A. Monk 36 North Road, Saline, Dunfermline KY12 9UQ
GM7 LFX A. Wilson 8 Grahamsdyke Road, Bo'ness EH51 9EG
G7 LFZ W. Mumford Agden Green Farm, The Green, Great Staughton, St. Neots PE19 5DQ
G7 LGI G. Bryce 135 Fairbridge Road, Upper Holloway, London N19 3HF
G7 LGS N. Green 2 Whittaker Mews, High Street, Rocester, Uttoxeter ST14 5JU
G7 LGV M. Smith 3 Tithe Barn, Merton, Bicester OX25 2NF
G7 LGY H. Abbott 1 St. Lawrence Close, Heanor DE75 7AN
G7 LHK T. Reynolds 11 Duncalfe Drive, Sutton Coldfield B75 5EX
G7 LHS S. Gray 26 Hatfield Gardens, Appleton, Warrington WA4 5QJ
G7 LHT F. Wilson 3A Vernon Road, Kirkby-In-Ashfield, Nottingham NG17 8EJ
G7 LHV G. Beaumont 16 Chelburn View, Littleborough OL15 9QQ
G7 LID G. Eddies 68 Little Harlescott Lane, Shrewsbury SY1 3PZ
G7 LIE B. Lovatt 12 Nelson Street, Leek ST13 6BB
G7 LIH S. Warren 41 Barton Road, Rugby CV22 7PT
G7 LII J. Eccles 30 The Stour, Daventry NN11 4PR
G7 LIK C. Tunbridge 12 Burnham Road, Latchingdon, Chelmsford CM3 6EU
G7 LIT G. Blaxall 27 St. Davids Road, Hextable, Swanley BR8 7RJ
G7 LIW D. Kent 10 Goldgarth, Grimsby DN32 8QS
G7 LJA P. Gibson 62 Glen Park Pensilva, Liskeard PL14 5PW
G7 LJB C. Mott-Gotobed 5 Cotswold Close, Basingstoke RG22 5BA
GM7 LJE J. Freer 30 Kilmarnock Drive, Cruden Bay, Peterhead AB42 0NG
GJ7 LJJ N. Utting Oberon, Bagatelle Road JE2 7TX Jersey
G7 LJL S. Murton 10 James Allchin Gardens, Kennington TN24 9SD
G7 LJQ G. Roser 26 Willow Road, Larkfield, Aylesford ME20 6QZ
G7 LKC J. Radtke 22 Spinney Drive, Banbury OX16 9TA
G7 LKI D. Connor Green Pastures, Stratton Road, Holcombe, Radstock BA3 5ED
G7 LKL S. Titterington 33 Victoria Road, Urmston, Manchester M41 5BZ
G7 LKR A. Maciver 55 Nordale Park, Rochdale OL12 7RT
G7 LKV R. Spray 132 Mansfield St., Sherwood, Nottingham NG5 4BD
G7 LKY D. Parkinson 36 Henley Road, Ipswich IP1 3SA
G7 LKZ C. Bowden 20 Parc Peneglos, Mylor Bridge, Falmouth TR11 5SL
G7 LLD M. Bewley 75 Sugden Road, Worthing BN11 2JG
G7 LLN W. Rodgers 28 Church Meadows, Calow, Chesterfield S44 5BP
G7 LLY J. Wharton 74 Brompton Park, Brompton On Swale, Richmond DL10 7JP
G7 LMH J. Davey 10 Kingsley Close, Harrogate HG1 4RA
G7 LMI J. Hollowood 10 Rossendale Close, Shaw, Oldham OL2 8JJ
G7 LMR K. Lavin 35 Manor Bend, Galmpton, Brixham TQ5 0PB
G7 LMT D. Pantrey 10 Columbine Close, East Malling, West Malling ME19 6ES
G7 LNB A. West 142 The Street, Kingston, Canterbury CT4 6JQ
G7 LND R. Williams 45 Station Road, Westbury BA13 3JW
G7 LNG J. Tucker 2 Ivydene Road, Ivybridge PL21 9BH
G7 LNI S. Czarnota 11 Spring Park, Chapel Road, Ipswich IP6 9NX
G7 LNJ R. Woolridge 8 Alastair Drive, Yeovil BA21 3BT
G7 LNK P. Knox 24 Bannister Drive, Banbury OX16 1GQ
G7 LNM D. Gilham 53 The Close, Bradwell, Great Yarmouth NR31 8DR
GM7 LNO G. Cash 3 Hallydown Crescent, Eyemouth TD14 5TB
G7 LNP A. Jones 1 Abbey Way, Rushden NN10 9HF
G7 LNT P. Cundall 40 Union Court, Otley LS21 3NW
G7 LNU N. Sparrow 46 Thomas Bell Road, Earls Colne, Colchester CO6 2PF
G7 LNV N. Turland 2 Ludlow Close, Beeston, Nottingham NG9 3BY
G7 LNY H. Mascall 37 Carnival Close, Ilminster TA19 9DG
G7 LOA L. Fisher 195 Malvern Road, Billingham TS23 2PJ
G7 LOE J. Bhogal 36 Titford Road, Warley, Oldbury B69 4QA
G7 LOG T. Smallwood 51 Barlow Road, Barlow, Blaydon-on-Tyne NE21 6JU
GM7 LOK D. Barr 17 Ballantrae, East Kilbride, Glasgow G74 4TZ
G7 LOV E. Farrar 23 Grovehill Road, Filey YO14 9NL
G7 LOW M. Bosberry 31 St Lukes Road, Gosport PO12 3JN
G7 LOY A. Powell 76 Glendale Avenue, Washington NE37 2JS
G7 LOZ K. Blackham 86 Heather Road, Small Heath, Birmingham B10 9TA
G7 LPB A. Sellick 15 Thorpe Street, Raunds, Wellingborough NN9 6LS
G7 LPD P. Wilkinson 43 Polperro Drive, Freckleton, Preston PR4 1YD
G7 LPF C. Hewitt 16 Sandsdale Avenue, Fulwood, Preston PR2 9AZ
G7 LPG P. Garcia 52 Pilot Road, Hastings TN34 2AN
G7 LPK R. Hilliard 8 Cromwell Crescent, Sleaford NG34 7HW
GW7 LPM L. La Traille 33 Festival Crescent, New Inn, Pontypool NP4 0NB
G7 LPN T. Snape 4 Back Street, Abbotsbury, Weymouth DT3 4JP
G7 LPO A. Perry 63A Brookland Road, Huish Episcopi, Langport TA10 9TH
G7 LPP F. Rice 42 Donegal Road, Knowle, Bristol BS4 1PL
G7 LPT A. Page The Farmhouse, Budges Shop, Trerulefoot, Saltash PL12 5DA
G7 LPV G. Soden 21 Bracknell Drive, Alvaston, Derby DE24 0BP
G7 LPW K. Sharples 11 West Drove North, Walpole St. Peter, Wisbech PE14 7HU
G7 LPZ D. Williams 7 Hampton Drive, Great Sankey, Warrington WA5 1JF
G7 LQD M. Baguley 2 Kensington Way, Northwich CW9 8GG
G7 LQK R. Dunn 12 Roseberry St., Beamish, Stanley DH9 0QR

G7 LQN C. King 33 Alexandra Road, Swallownest, Sheffield S26 4TA
G7 LQO Dr L. Brown 19 Stephen Drive, Sheffield S10 5NX
G7 LRB P. Stevens 62 Lansdowne Road, Bayston Hill, Shrewsbury SY3 0JG
G7 LSB L. Brown 4 Loraine Gardens, Ashtead KT21 1PD
G7 LSD P. Wainwright 3 Ashridge Close, Nuneaton CV11 4XG
G7 LSF J. Blain 91 Deanfield Road, Henley-on-Thames RG9 1UU
G7 LSG Dr P. Campbell 13 Springfield Close, Marden, Hereford HR1 3EH
GM7 LSI J. Stuart 3 Pringle Road, Elgin IV30 4HN
G7 LSP P. Harness 16 Norfolk Street, Boston PE21 6PW
G7 LSZ Dr M. Foreman Vallgatan 10, Vara 534 31 Sweden
G7 LTG P. Savage 60 Colonial Road, Bordesley Green, Birmingham B9 5NG
G7 LTO M. Milns 3 Merlin Court, Batley WF17 0RG
G7 LTP P. Sawyer 96 Violet Lane, Croydon CR0 4HG
G7 LTR D. Ingham 19 Recreation Avenue, Ashton-In-Makerfield, Wigan WN4 8SU
G7 LTT M. Phillips 2 Hemwood Road, Windsor SL4 4YU
G7 LTU G. Smith 36 Sandalwood Road, Loughborough LE11 3PS
G7 LTW T. Metcalfe 39 Chobham Road, Frimley, Camberley GU16 8PS
GM7 LTX A. Warner 41 Gaynor Avenue, Loanhead EH20 9LU
G7 LUB J. Broome Henbant Fach, Penuwch, Tregaron SY25 6QZ
G7 LUF G. Whitehouse 27 Kings End Road, Powick, Worcester WR2 4RB
G7 LUK P. Preston 45 Saxons Heath, Long Wittenham, Abingdon OX14 4PU
G7 LUL F. Russell 61A Fleet Street, Plymouth PL2 2BU
GM7 LUN J. Keddie Garrion, Bowland Road, Clovenfords, Galashiels TD1 3ND
G7 LUO N. Head 12 Heston Walk, Redhill RH1 5JB
G7 LUR J. Nolan 33 Cambridge Road, Langford, Biggleswade SG18 9PS
G7 LVA S. Sorrell 19 College Road, Hockwold, Thetford IP26 4LD
G7 LVE D. Wright Flat 1 The Annexe, Uxbridge UB9 5HJ
G7 LVG J. Ashton-Jones Kiddley Kopse, Mordiford, Hereford HR1 4LR
G7 LVM J. Maule 12 Edith Cavell Way, Steeple Bumpstead, Haverhill CB9 7EE
G7 LVN M. Odam 10 The Orchards, Meare, Glastonbury BA6 9PU
G7 LVS M. Unsworth 41 Aylesbury Crescent, Hindley Green, Wigan WN2 4TY
GM7 LWA S. Leith 3 County Houses, Roseisle, Elgin IV30 5YE
G7 LWF J. Totten 28 Newman Road, Devizes SN10 5LE
G7 LWH E. Dalley 5 Anstey Mill Close, Alton GU34 2QT
G7 LWT D. Storer 1616 Scant Row, Chorley Old Road, Horwich, Bolton BL6 6PZ
G7 LWU S. Porter 1 Belt Drove, Elm, Wisbech PE14 0BA
G7 LWY D. Northeast 11 Repton Road, Earley, Reading RG6 7LJ
G7 LXA C. Staff Uphill Road South, Weston Super Mare BS23 4TU
G7 LXB W. Roberts 36 Wray Court, Emerson Valley, Milton Keynes MK4 2GF
G7 LXH D. Hayzen 79 Swinburne Avenue, Hitchin SG5 2QZ
GW7 LXI J. Baines Pentre Clawdd Cottage, Gobowen, Oswestry SY10 7AE
G7 LXV N. Hobbs 224 Belchers Lane, Bordesley Green, Birmingham B9 5RY
G7 LXY J. Hopkins 7 Montgomery Close, Coventry CV3 4FS
G7 LYB R. Brown 61 Paddockhurst Road, Gossops Green, Crawley RH11 8EU
G7 LYH J. Briggs 16 Belmont Place, Colchester CO1 2HU
G7 LYL C. Nixon 52 Gloucester Drive, Basingstoke RG22 4PH
G7 LYN S. Laugher Jasmine Cottage, Healey, Ripon HG4 4LH
G7 LYS C. Ameigh 45 Manley Road, , Ilkley LS29 8QP
G7 LZB A. Howat 6 Richmond Road, London N2 8JT
G7 LZM L. Mountain 45 Westway Gardens, Redhill RH1 2JB
G7 LZY W. Eatwell 45 Admirals Walk, Minster On Sea, Sheerness ME12 3BB
G7 MAB M. Dodson 64 Stoneleigh Road, Solihull B91 1DQ
GM7 MAG P. Budgen 12 Boggs Holdings, Pencaitland, Tranent EH34 5BB
G7 MAR J. Rivers Wind In The Willows 1 Hazelwood Close, Ryde PO33 2UP
G7 MAT K. Hinton 10 Hillview Road, Basingstoke RG22 6BQ
G7 MAV A. Goodall 21 Sladburys Lane, Clacton-on-Sea CO15 6NX
GM7 MBB L. Millar 34 Brora Drive, Renfrew PA4 0XA
G7 MBH M. Davis 3 Thornley Close Ushaw Moor, Durham DH7 7NN
GI7 MBP W. Kane 21 Mount Coole Gardens, Belfast BT14 8JY
G7 MBU R. Marks 23 Parfitt Way, Dover CT16 2QW
G7 MBY D. Richards 6 Kingley Close, Wickford SS12 0EN
G7 MCE D. Wilkinson 56 Cobden Street, Dalton-in-Furness LA15 8SE
G7 MCK K. Singleton Spring Cottage, Barcombe Lane, Paignton TQ3 2QS
G7 MCS B. Mcshea 5 Frensham Avenue, Fleet GU51 3EL
G7 MCT C. Taylor 36 Harewood Road, Shaw, Oldham OL2 8EA
G7 MDI G. Hawkes 17 Beacon Hill, Burnham Market, King's Lynn PE31 8ET
GI7 MDJ S. Clarke 86 Roddens Crescent, Castlereagh, Belfast BT5 7JP
GI7 MDK D. Robinson 4 Ballylesson Road, Magheramorne, Larne BT40 3HL
G7 MDM S. Wilkins 5, Blackthorn Close, Gainsborough DN21 1WB
GI7 MDP S. Mcilvenna 10 Sycamore Court, Drumaness, Ballynahinch BT24 8QZ
G7 MDT D. Limb 34 Elmwood Avenue, Boston PE21 7RU
G7 MDV C. Prowse 125, Hill Road, Portchester, Fareham PO16 8JY
G7 MEA R. Thomas 25 Heath Hill, Heathfield, Newton Abbot TQ12 6SP
G7 MEE A. Wood 14 Anatase Close, Sittingbourne ME10 5AN
G7 MEG D. Cash 3 Marsh Lane, Wolverhampton WV10 6RU
G7 MER C. Hurst 28 Hengistbury Road, Barton-on-Sea BH25 7LU
G7 MES M. Stevens Autumn Cottage, Silver Street, Horncastle LN9 5NH
G7 MEU D. Hughes 25 Highfield Road, Carnforth LA5 9BE
G7 MEX Mexborough & District ARS c/o J. Saiger 10 Markham Avenue, Armthorpe, Doncaster. DN3 2AZ
G7 MEZ J. Arter 18 Essex Road, Westgate-on-Sea CT8 8AP
G7 MFA A. Sejwacz 20 Wellington Gardens, Newton-le-Willows WA12 9LT
G7 MFE S. Morris The Grange, Downash Farm, Rosemary Lane, Wadhurst TN5 7PS
G7 MFH B. Fifield Clyro, Lower Coombses, Chard TA20 2SX
G7 MFN I. Douglas 13 Castlereagh Street, New Silksworth, Sunderland SR3 1HJ
G7 MFO R. Parkes 7 Main Street, Preston, Hull HU12 8UB
G7 MFP A. Dresser 7 Torcross Grove, Calcot, Reading RG31 7AT
G7 MFR C. Jenkins-Powell 43 Cambridge Road, Lee-on-the-Solent PO13 9DH
G7 MFW D. Burdett 17 Brambledown, Chatham ME5 0DY
G7 MFX P. March 39 Rochford Garden Way, Rochford SS4 1QH
G7 MFY A. Wakeling The Willows, Litcham Road, King's Lynn PE32 2LJ
G7 MFZ M. Sherratt 21 Tweedale Close, Mursley, Milton Keynes MK17 0SB
G7 MGA R. Thorley 9 Birchendale Close, Tean, Stoke-on-Trent ST10 4LT
G7 MGC T. Gerrard 41 Auberson Road, Bolton BL3 3AU
G7 MGG R. Shirley 1 St. Richards Court, Bellingham Crescent, Hove BN3 7FW
G7 MGM D. Barnes 36 Westbrook Crescent, Cockfosters, Barnet EN4 9AS
G7 MGQ M. Ball 11 Plantation Road, Thorne, Doncaster DN8 5EA
G7 MGT P. Cox 17 Hyde Lane, Upper Beeding, Steyning BN44 3WJ
G7 MGV C. Chadburn 31 Darwin Close, Top Valley, Nottingham NG5 9LN
GW7 MGW E. Palmer 10 Maes Gwyn, Llanfair Caereinion, Welshpool SY21 0BD
G7 MGX P. Asbury 67 Orchard Way, Measham, Swadlincote DE12 7JZ
G7 MGY S. Welger 55 Burford Avenue, Swindon SN3 1BX
G7 MGZ S. Plant 99 Pegwell Road, Ramsgate CT11 0ND
GW7 MHB M. Burt 44 Overton Close, Buckley CH7 2AX
G7 MHD A. Thorp 34 Third Avenue, Hightown, Liversedge WF15 8JU
GW7 MHF R. Johnston Gledrid Cottage, Oaklands Road, Wrexham LL14 5DW
G7 MHL J. Britton Salters Rest, Salters Mill Northwood, Shrewsbury SY4 5NW
G7 MHO S. Fell 14 Rectory Avenue, Corfe Mullen, Wimborne BH21 3EZ
G7 MHQ G. Taylor 21 New Road, Kirkheaton, Huddersfield HD5 0JB
G7 MHV S. Stillwell 130 London Road, Chatteris PE16 6SF
G7 MID A. Haydon 9 Ash Close, Newport PO30 5UR
G7 MIE S. Hudson 20 Churchill Road, Gravesend DA11 7AQ
G7 MIF B. Dickenson 22 Ford Close, Herne Bay CT6 8AN
G7 MII D. Burgin 7 Bramble Close, Halliford, Shepperton TW17 8RR
G7 MIM T. Wheeler 60 Bredhurst Road, Gillingham ME8 0PE
G7 MIN A. Jones 17 Maybush Drive, Chidham, Chichester PO18 8SR
G7 MIP Sir H. Kneale 57 Danforth Close, Framlingham, Woodbridge IP13 9HP
G7 MIS A. Trott 8A Wyatt Road, Kempston, Bedford MK42 7EH
G7 MIT T. Good 11 Moorland Road, Didsbury, Manchester M20 6BB
G7 MIZ J. Locker Delamere, 8 Concordia Avenue, Wirral CH49 6JD
G7 MJD Eccles And District ARS c/o A. Bruring 5 Church Lane, Hartford, Huntingdon PE29 1XP
G7 MJI P. Sayers 23 Roseveare Road, Eastbourne BN22 8RS
G7 MJJ R. Delves 66 Palmeira Road, Bexleyheath DA7 4UX
G7 MJP C. Edwards 16 Martin Street, Normanton WF6 1DA
G7 MJS Dr G. Davies 78 Chatsworth Road, Southport PR8 2QF
G7 MJV A. Watts 32 Hedley Davis Court, Cherry Orchard Lane, Salisbury SP2 7UE
G7 MJX D. Hanson 64 Laxfield Way, Lowestoft NR33 7HH
G7 MKB J. Humphries 25 Wrekenton Row, Wrekenton, Gateshead NE9 7JD
G7 MKF B. Stracey 31 Westfield Road, Margate CT9 5PA
G7 MKG P. Bradbury 40 Titty Ho, Raunds, Wellingborough NN9 6DF
G7 MKJ N. Austin 30 Cardinal Avenue, Borehamwood WD6 1EP
G7 MKP A. Brooks 7 Lindford Drive, Norwich NR4 6LT
G7 MKQ A. Airey 2 Rossmere, Greenways Estate, Spennymoor DL16 6TZ
G7 MKV P. Nicholls 53 Fastolff Avenue Gorleston, Great Yarmouth NR31 7ND
G7 MLC G. Bunn 1 Twelve Acre Road, Norwich NR2 3PZ
G7 MLJ P. Skinner 84 Beresford Avenue, Tolworth, Surbiton KT5 9LW
G7 MLK D. Rose 87 Second Avenue, Sudbury CO10 1QX
G7 MLL M. Birtles 65 Hemsworth Road, Sheffield S8 8LJ
GW7 MLN J. Corcoran 23 Tan Yr Allt, Abercrave, Swansea SA9 1XF
G7 MLO J. Large 5 Raynsford Rise, Stanningfield Road, Great Whelnetham, Bury St. Edmunds IP30 0TS
G7 MLT A. Armstrong-Bednall 63 Wellington Street, Heanor DE75 7FW
G7 MLU S. Murray 26 Alfreda Avenue, Hullbridge, Hockley SS5 6LT
G7 MLW G. Kilbey 38 Midland Road, Stonehouse GL10 2DH
G7 MLX G. Crisp Hoppers Farm, Great Kingshill, High Wycombe HP15 6EY
G7 MMC S. Reed 32 Plantation Road, Amersham HP6 6HL
G7 MME E. Hughes-Lai 18 Ramillies Avenue, Plymouth PL5 2NU
GW7 MMG P. Pike 19 Hillrise Park, Clydach, Swansea SA6 5DX
GW7 MMH E. Cooke 32 Chapel Road, Three Crosses, Swansea SA4 3PU
GM7 MMI J. Wilson 32 Silverburn Road, Bridge Of Don, Aberdeen AB22 8RW
G7 MMJ S. Pratt 57 Regency Court, Bradford BD8 9EX
G7 MMK D. Baggaley 6 Bylands Place, Newcastle ST5 3PQ
G7 MMV K. Foster 5 Newman Road, Plymouth PL5 2DX
G7 MMW P. Francis 14 Fulmar Place, Meir Park , Stoke on Trent ST3 7QF
G7 MND W. South Dufonis, Dorchester Road, Wareham BH20 6EQ
G7 MNE B. Altman 5 Ridgemount Gardens, Enfield EN2 8QL
G7 MNG M. Whale 499 Maidstone Road, Wigmore, Gillingham ME8 0JX
G7 MNK J. Lambe 4 St. Georges Road, Enfield EN1 4TX
G7 MNL South Devon Raynet Group c/o C. Coker 46 Clarendon Road, Ipplepen, , Newton Abbot TQ12 5QS
G7 MNO R. Nightingale 58 Nutfield Grove, Filton, Bristol BS34 7LJ
G7 MNP G. Turner 23 Withycombe Road, Penketh, Warrington WA5 2QL
G7 MNQ P. Bland 19 Sookholme Drive, Warsop, Mansfield NG20 0DN
G7 MNS A. Cartwright 118 High Road West, Felixstowe IP11 9AL
G7 MNT B. Woods 64 Yarningale Road, Coventry CV3 3EQ
G7 MNZ C. Gaskin 2 The Briars, West Kingsdown, Sevenoaks TN15 6EZ
G7 MOB 59 Degrees North ARG c/o P. Thain 26 Haston Lee Avenue, Blackburn BB1 9QT
G7 MOD D. Rust 26 Mill Road, Wiggenhall St. Germans, King's Lynn PE34 3HL
G7 MOH E. Middleton Fairwinds, Southella Road, Yelverton PL20 6AT
G7 MOK N. Rieger-Ridd 3 Rockland Close, Swaffham PE37 7SP
G7 MOW K. Starnes 17 Boughey Place, Lewes BN7 2EN
G7 MOX W. Jones 62 Mallings Drive, Bearsted, Maidstone ME14 4HG
G7 MOY B. Jenkins 27 Glendale, South Woodham Ferrers, Chelmsford CM3 5TS
G7 MPF R. Ransome High Winds, High Town Green, Bury St. Edmunds IP30 0SZ
G7 MPH R. Cole 18 Borrowdale Close, Benfleet SS7 3HE
G7 MPJ J. Tweedy 59 St. Aloysius View, Hebburn NE31 1RH
G7 MPV J. Woods Rozel, Bigbury Road, Canterbury CT4 7ND
G7 MPZ C. Atkins 278 Walderslade Road, Chatham ME5 9AA
G7 MQC C. Thomas 1 George Gent Close, Steeple Bumpstead, Haverhill CB9 7EW
GW7 MQE D. Smith 11 Cymau Lane Caergwrle, Wrexham LL12 9DH
G7 MQF A. Kirkham 49 Macclesfield Road, Leek ST13 8LD
G7 MQP S. Richardson 73 Primrose Copse, Horsham RH12 5PZ
G7 MQQ H. Griffiths 11 Gensing Road, St. Leonards-on-Sea TN38 0ER
G7 MQU D. Sandever 57 Hayes Lane, Wimborne BH21 2JB
G7 MQW R. Carroll 71 Pelham St., Manton, Worksop S80 2TT
G7 MRF M. Farmer 3 Brackenberry, Cross Heath, Newcastle ST5 9PS
G7 MRH E. Cole 11 Ainsworth House, Wellington Road, Brighton BN2 3BG
G7 MRJ E. Stead Cranford, Church Road, Walpole St. Peter, Wisbech PE14 7NS
G7 MRL N. Williams 1 Dorset Close, Whitehaven CA28 8JP
G7 MRO B. Bowker 205 Smallshaw Lane, Ashton-under-Lyne OL6 8RJ
G7 MRZ R. Thompson 4 Hill Top Road, Birdwell, Barnsley S70 5QZ
G7 MSC B. Knight 3 Burgess Cottages, Mongeham Road, Deal CT14 8JW
G7 MSF K. Sanderson 45 Bygrove, New Addington, Croydon CR0 9DG
G7 MSG M. Olivant 2 Vicarage Gardens, Flamstead AL3 8EF

G7 MSH H. Samwells 2 Dudley Walk, Macclesfield SK11 8SD
G7 MSK T. Mann 21 Glastonbury Court, Yeovil BA21 3TW
G7 MSN D. Stead 15 Reeves Close, Porthleven, Helston TR13 9PB
G7 MSQ G. Shelley 41 Thornley Road, Stoke-on-Trent ST6 7AL
G7 MSS D. Forward 4B Cowper Road, Deal CT14 9TW
G7 MST T. Bennett Rose Cottage, High Street, Rotherham S62 6LN
G7 MTA C. Parr 13 Peartree Avenue, Southampton SO19 7JN
G7 MTE M. Coote 22 Tennyson Close, Boston PE21 8DL
G7 MTF T. Foley Flat 38, Windmill Court, Uxbridge Road, Swindon SN5 8RT
G7 MTG A. Blakeston 8 Victor Street, Cutsyke, Castleford WF105HB
G7 MTI F. Paley 68 Dennil Road, Leeds LS15 8SD
G7 MTJ C. Chase Asholt, Ermine Street, Scunthorpe DN15 0AD
G7 MTQ S. Saunders 3 The Terrace, High Street, Cavendish, Sudbury CO10 8AS
G7 MTV M. Bourne 100 Dimsdale View West, Newcastle ST5 8EL
G7 MTW R. Powell 4 Diana Close, Spencers Wood, Reading RG7 1HP
G7 MUB R. Harcourt 7 Lightfoot Close, Newark NG24 2HT
G7 MUD Christchurch ARS c/o D. Layne 5 Howe Close, Christchurch BH23 3JA
G7 MUE S. Roper 1 Holywell Road, Kilnhurst, Mexborough S64 5UQ
GM7 MUN J. Smith 14 John Collins Crescent, Galashiels TD1 2FA
G7 MUT T. Cannon 5 Barn Close, Upton, Poole BH16 5RX
G7 MUY A. Sadler 19 Lyndhurst Park Home Estate Sea Lane, Ingoldmells PE25 1PD
G7 MVE M. Cotton 2 Redhill View, Castleford WF10 4QL
GW7 MVG H. Millington Arran Clayton Road, Mold CH7 1SU
G7 MVU N. Brown 6 Hundon Place, Haverhill CB9 0AP
G7 MVX G. Trudgill 61 Lansdowne Road, Coxhoe, Durham DH6 4DN
G7 MVY R. Stockley 4 Tothby Close, Alford LN13 0BG
GI7 MWA Marches ARS c/o S. Stewart 3 Killyfaddy Road, Magherafelt BT45 6EX
G7 MWB W. Bone 217 Bensham Road, Gateshead NE8 1US
G7 MWC R. Moss 6 Adelaide Gardens, Stonehouse GL10 2PZ
G7 MWH P. Cross Churchill House, Churchill Road, Louth LN11 7QW
G7 MWI L. Hansen 19 Market St., Appledore, Bideford EX39 1PW
G7 MWJ A. Holloway 31 Gays Road, Hanham, Bristol BS15 3JR
G7 MWK J. Arkle 16 Sea View, Ashington NE63 0XH
GM7 MWL H. Murray 23 Denmore Gardens, Bridge Of Don, Aberdeen AB22 8LJ
G7 MWM A. Scott 22 Planters Grove, Lowestoft NR33 9QL
G7 MWS P. Parrish 5 Kestrel Lane, Cheadle, Stoke-on-Trent ST10 1RU
G7 MWU G. Haswell 16 Hither Green, Jarrow NE32 4LP
G7 MWW J. Smith 19 The Crescent, Mitcheldean GL17 0SB
GM7 MWX R. Raynor La Pergola, Kilmuir, Inverness IV1 1XG
G7 MXL T. Grange 7 The Cherries, Canvey Island SS8 0BB
G7 MXM C. Turner 55 Fordfield Road, Ford Estate, Sunderland SR4 6XG
G7 MXN L. Orchard 678 Devonshire Road, Blackpool FY2 0AW
G7 MXQ B. Gilbraith 19 Bullcote Green, Royton, Oldham OL2 6NJ
G7 MXS A. Harrison Midhope Lodge Midhopestones, Sheffield S36 4GW
G7 MXT D. Harris 12 Turner Avenue, Billingshurst RH14 9PU
GU7 MXZ B. Heath 21 Clos De Bas, Green Lanes, St. Peter Port, Guernsey GY1 1TS
GW7 MYD P. Williams 5 Bright St., Cross Keys, Newport NP1 7PB
GM7 MYF C. Dennett Lyn-Ard, Smollett Street, Alexandria G83 0DW
G7 MYI A. Stride 3 Barnfield Cottages, Edmondsham, Wimborne BH21 5RD
G7 MYJ R. Ball 5 Miller Fold Avenue, Accrington BB5 0NT
G7 MYM D. Roberts Chatterbox, 3A, Station Road, Pershore WR10 1NQ
G7 MYN C. George 5 Turing Gardens, Shefford SG17 5ZS
G7 MYO C. Mciver 2 Abbey Meadows, Chertsey KT16 8RA
G7 MYT P. Hilton 40 Megstone Avenue, Whitelea Chase, Cramlington NE23 6TU
G7 MYY L. Fuller 78C Seal Road, Sevenoaks TN14 5AT
G7 MZA R. Loukes Fagus, The Street, Staple, Canterbury CT3 1LL
G7 MZE T. Ingle 68 Wooldale Drive, Filey YO14 9ER
G7 MZJ I. Mitchell 87 Bluebell Avenue, Penistone, Sheffield S36 6AF
G7 MZK D. Mitchell 28 Southgate, Penistone, Sheffield S36 6EA
G7 MZL N. Baker 43 Little Park, Wadhurst TN5 6DL
G7 MZS L. Terry 8 Carters Close, Slyfield, Guildford GU1 1FR
G7 MZW A. Calvert 122 Grampian Way, Thorne, Doncaster DN8 5YW
G7 MZX Yorkshire DX Club c/o R. Barron 15 Fernhill Close, Poole BH17 8SQ
G7 MZY I. Sharp 6 Ullswater Drive, Bath BA1 6NP
GM7 MZZ G. Whiting 21 Leckethill Court, Cumbernauld, Glasgow G68 9EG
GM7 NAA I. Skeoch 1 Castleton Crescent, Grangemouth FK3 0BH
G7 NAE A. Westwood 3 Walgrave Close, Belper DE56 1UF
G7 NAI J. Stock 31 Grange Road, Wickham Bishops, Witham CM8 3LT
G7 NAL H. Wood 31 Goring Avenue, Gorton, Manchester M18 8WW
G7 NAO I. Langmuir 2 Nelson Road, Newport PO30 1QT
G7 NAP D. Gee 28 Rein Road, Morley, Leeds LS27 0JA
G7 NBE M. Goodwin 23 Saxon Way, Ashby-de-la-Zouch LE65 2JR
G7 NBF A. Sadler 23 Wolsey Road, Moor Park, Northwood HA6 2HN
G7 NBG L. Mcguire 200 Wellingborough Road, Rushden NN10 9SX
G7 NBI P. Webb 42 Holland Road, Ampthill, Bedford MK45 2RS
G7 NBJ D. Corfield 177 Hurst Rise, Matlock DE4 3EU
G7 NBL C. King Broadlea, Honey Hill, Fen Drayton, Cambridge CB24 4SF
G7 NBQ S. Ambrose 3 Three Mile Pond, Sawbridgeworth CM21 9ED
G7 NBR W. Hayward 15 Whitehouse Road, South Woodham Ferrers, Chelmsford CM3 5PF
G7 NBU K. Bassett Manor Farm, Marsh Green, Exeter EX5 2EX
G7 NBV F. Young 6 Birchvale Court, Desborough, Kettering NN14 2UY
G7 NBZ P. Millerchip 6 Washbrook View, Ottery St. Mary EX11 1EP
G7 NCD T. Willis 15 Cedar Court, Congleton CW12 3JP
G7 NCE K. Derbidge 1 Batch View, Grange Avenue, Street BA16 9PE
G7 NCG E. Turner 16 The Rowans, Doddington, March PE15 0SE
G7 NCP J. Merrington Cartref, Ball Lane, Frodsham WA6 8HP
G7 NCV K. Hobbs 61 Fairway, Waltham DN37 0NB
G7 NCW G. Hinton 25 Linden Close, Prestbury, Cheltenham GL52 3DX
GU7 NCZ Dr N. Turner Camellia Lodge, L'Aumone, Castel GY5 7RT Guernsey
G7 NDB K. Marshall Doveysmead, Chapel Street, Basingstoke RG25 2BZ
G7 NDC P. Hirst 47A Rowley Lane, Fenay Bridge, Huddersfield HD8 0JG
G7 NDI D. Bunney 3 Hilmanton, Lower Earley, Reading RG6 4HN
G7 NDN S. Ward Russet House, Beech Road, Haslemere GU27 2BX
G7 NDO H. Blackburn 4 Hawkridge, Furzton, Milton Keynes MK4 1BQ
G7 NDQ G. Bettyes 44 Springfield Road, Oundle, Peterborough PE8 4LT
G7 NDS M. Fry 14 The Lawns Collingham, Newark NG23 7NT
G7 NDT R. Walker 46 Lodge Road, Little Houghton, Northampton NN7 1AE
GI7 NEB J. Conlon 30 Drumglass Way, Dungannon BT71 4AG
G7 NEC C. Watson Westwood, Laneside, Queensbury BD13 1NE
G7 NED G. Gini Tenuta Buzzoletto Nuovo 6, Garbagna Novarese 28070 Italy
G7 NEE E. Maloney 56 Westonfields Drive, Longton, Stoke-on-Trent ST3 5JA
G7 NEG R. Smith 32 Water Lane, Wootton, Northampton NN4 6HE
G7 NEH G. Pemberton 2 Hockenhull Avenue, Tarvin, Chester CH3 8LP
G7 NEM J. Barber 2 Gresley Way, March PE15 8QA
G7 NER T. Stokes 33 Talbot Drive, Euxton, Chorley PR7 6PD
GI7 NET K. Nolan 34 Lisgoole Park, Drumgallan, Enniskillen BT74 5ND
GI7 NFB M. Robinson 92A Dromore Road, Hillsborough BT26 6HU
GM7 NFF P. Salmon Mill House, Monreith, Newton Stewart DG8 9LJ
G7 NFG M. Holdsworth 9 Beaumont Close, Bowburn, Durham DH6 5QA
G7 NFK J. Watmough 2 Burbage Heights, Buxton SK17 6YU
GW7 NFM E. Jones Maes Y Coed, Vownog Road, Mold CH7 6ED
G7 NFN R. Bailey 29 Priory Close, Bath BA2 5AL
G7 NFO M. Hall 30 Kingsley Avenue, Rugby CV21 4JY
G7 NFR J. Thomas 2 Alexandra Road, Uxbridge UB8 2PQ
GW7 NFT J. Parry Charlbury, Usk Road, Newport NP18 1LP
GW7 NFY M. Mee Anncott, Hylas Lane, Rhyl LL18 5AG
G7 NGB J. Alger Church Hill Cottage Church Hill, Caterham CR36SA
G7 NGF D. Hatcher 8 Churchfield, Monks Eleigh, Ipswich IP7 7JH
G7 NGI J. Price 32 Wiltshire Drive, Trowbridge BA14 0RE
G7 NGN R. Williams 73 Quedgeley Park, Greenhill Drive, Gloucester GL2 5NZ
G7 NGQ A. Bauer 5 Horse Fayre Fields, Spalding PE11 3FA
GW7 NGU J. Vaughan Montrose, 5 Trewarren Drive, Haverfordwest SA62 3TR
G7 NGX A. Mckenna 12 Sunnyside Road, Beeston, Nottingham NG9 4FH
G7 NHB R. Griffiths 4 Wolrige Way, Plympton, Plymouth PL7 2RU
G7 NHC D. Scholes 71 Pelham St, Ashton under Lyne OL70DU
G7 NHD T. Carroll 32 Marfords Avenue, Wirral CH63 0JW
G7 NHE J. Turnbull 32 Haydon, Washington NE38 8PF
G7 NHF K. Williams 1 St. Ives Way, Halewood, Liverpool L26 7YW
G7 NHL Appledore And District ARC c/o K. Mitchell 12 Lon Lafant, Llandudno Junction SK17 9PL
G7 NHQ K. Cotterill 1 Molineux Avenue, Broadgreen, Liverpool L14 3LT
G7 NHR P. Dunlop 4 Birket Avenue Moreton, Wirral CH46 1QZ
GM7 NHS Dr R. Johnson 3 Hopetoun Green, Bucksburn, Aberdeen AB21 9QX
GM7 NHU G. Devereux 44 Greenhead Road, Dumbarton G82 2PN
G7 NHV K. Johnson 43 Glencoe Road, Great Sutton, Ellesmere Port CH66 4NA
G7 NHW K. Mckane 60 Hazelwood Road, Callington PL17 7EU
GU7 NHX A. Dorrian Le Petit Jardin, Clos Des Emrais, Castel GY5 7YB Guernsey
G7 NHY V. Brooker Flat 6, 46 Foxglove Way, Wallington SM6 7JU
G7 NHZ A. Greatbatch 46 Java Crescent, Trentham, Stoke-on-Trent ST4 8RT
G7 NIA H. Seldon 22 Downside Avenue, Plymouth PL6 5SD
G7 NIB I. Clark 1 Hayward Parade, Oakengates, Telford TF2 6EZ
G7 NID T. Thorpe 7A Rosalind Close, Colchester CO4 3JH
G7 NIH A. Davies 1 Fire Station Yard, Rochdale OL11 1DT
G7 NII Prof. R. Kalawsky 23 Brook Lane, Loughborough LE11 3RA
G7 NIL J. Chin 198 Bermondsey Wall East, London SE16 4TT
G7 NIN D. Townsend 40 Popes Lane, Sturry, Canterbury CT2 0JZ
G7 NIR L. Jones 53 Ennisdale Drive, Wirral CH48 9UF
G7 NIU S. Tanner 31 Four Acres, Portland DT5 2JG
GW7 NIW G. Durno Lothlorien, Upper Denbigh Road, St. Asaph LL17 0BH
G7 NIX C. Shurety O Fran Villa, Camp Road, Norwich NR8 6LD
G7 NIZ A. Bowers Flat 2, Jevington House, Upperton Road, Eastbourne BN21 1LW
G7 NJB G. Harris 58 The Leas, Minster On Sea, Sheerness ME12 2NL
G7 NJD J. Stewart 22 Garden Road, Kendal LA9 7ED
G7 NJE C. White 13 Peel St., Heywood OL10 4QD
G7 NJG D. Godwin 2 Barncroft Drive, Hempstead, Gillingham ME7 3TJ
GW7 NJM P. Martin 2.Gwarllyn.Tudweiliog.Pwllheli, Pwllheli.Gwynedd LL538NG
G7 NJP M. Neal 3 Nursery Way, Grimston, King's Lynn PE32 1DQ
GW7 NJQ S. Richardson Holmleigh, Broughton, Cowbridge CF71 7QR
GW7 NJT J. Jones 8 Manor Court, Ewenny, Bridgend CF35 5RH
G7 NJW G. Bullen1 Flat 3, Hill Hook House, Clarence Road, Sutton Coldfield B74 4DX
G7 NJX S. Mullen 18 Helens Road, Sandford BS25 5PD
G7 NJZ J. O'Rourke 39 Rutherglen Road, Corby NN17 1ER
G7 NKH D. Smith 7 Salisbury Road, Carshalton SM5 3HA
G7 NKI A. Davin 7 Paynes Park, Hitchin SG5 1EH
G7 NKJ T. Westbrook 5 Newlands, Northallerton DL6 1SJ
GM7 NKN M. Dean 4 Beatrice Drive, Holytown, Motherwell ML1 4UT
G7 NKS Dr J. Cowburn 26 Birch Close, Broom, Biggleswade SG18 9NR
G7 NKU C. Prout 1 Westbrook Lustrells Vale, Saltdean, Brighton BN2 8EZ
G7 NKV A. Taylor Moonlight Cottage, 4 Alderley Road, Macclesfield SK11 9AP
G7 NKZ K. Bradley 161 Mortimer Road, Southampton SO19 2HJ
G7 NLA C. Milburn Greenleas, Furlongs Lane, Horncastle LN9 6LD
G7 NLF G. Bandara 26 Undine Street, London SW17 8PR
G7 NLJ R. Watts 33 Rockside View, Matlock DE4 3GP
G7 NLP J. Greenacre 30 Ramsey Grove, Bury BL8 2RE
G7 NLR Crawley ARC c/o D. Andrew 14 Westfield Grove, Morecambe LA4 4LQ
G7 NLY J. James 14 Fairview Drive, Bayston Hill, Shrewsbury SY3 0LE
G7 NLZ R. Bennett 16 Graham Close, Portslade, Mile Oak, Brighton BN41 2YE
G7 NMB International Shortwave League c/o K. Bennett 38 Northumberland St., Workington CA14 3EY
G7 NME B. Caldicott 1 Naish Road, Burnham-on-Sea TA8 2LE
G7 NMI K. Osborne 42 Barbrook Lane, Tiptree, Colchester CO5 0EF
GI7 NMK L. Breadon 32 Ashley Crescent, Millisle, Newtownards BT22 2BG
G7 NMT M. Beach 11 Lawday Link, Farnham GU9 0BS
GW7 NNA A. Watkin Ba Hnd Aarburg, Windsor Close, Oswestry SY11 2UA
GW7 NNM J. Day Tynwtra, Bwlch-Y-Ffridd, Newtown SY16 3HX
G7 NNR B. Hughes Dorfstrasse 71, Waldfeucht 52525 Germany
GM7 NNS A. Strachan Mormond View, New Leeds, Peterhead AB42 4HX
G7 NNU J. Wood 19 Arbour Crescent, Macclesfield SK10 2JB
G7 NNZ D. Dukeson Wadsley Nook, Far Lane, Sheffield S6 4FD
G7 NOI G. Evans 20 Bleasdale Court, Longridge, Preston PR3 3TX
G7 NOQ J. Howarth 61 Poplar Drive, Lamaleach Park, Lamaleach Drive, Freckleton, , Preston PR4 1EG
G7 NOR M. Bowers Uprising, Shottendane Road, Margate CT9 4NE
GI7 NOW R. Mcmaster 40 Woodlands, Ballycarry, Carrickfergus BT38 9JD

G7 NPL C. Daniel 24 Canterbury Road, Dewsbury WF12 7LA
GM7 NPR G. White 2 Keill Cottage, Isle of Gigha PA41 7AD
G7 NPT Farnborough CG c/o J. Boyd 26 Pear Tree Place, Warrington WA4 1AX
G7 NQJ B. Silcocks 2 Derham Road, Bristol BS13 7SA
GM7 NQP G. Kinnell 61 Hallforest Avenue, Kintore, Inverurie AB51 0TF
G7 NQU J. Varnham 1 Burgin Road, Anstey, Leicester LE7 7FA
G7 NQW A. Symon Flat3, 66A Clyde Road, Croydon CR06SW
G7 NQX J. Bailes 48 Harlech Close, Eston, Middlesbrough TS6 9SZ
G7 NQZ D. Harrison 18 The Crescent, Eaglescliffe, Stockton-on-Tees TS16 0JB
G7 NRB L. Kirk 10 Dyers Mews, Neath Hill, Milton Keynes MK14 6ER
G7 NRG P. Atkinson 8 Thanet Terrace, Appleby-in-Westmorland CA16 6TU
G7 NRO C. Flanagan 2 Wynyerd House, Durham Road, Wolviston, Billingham TS22 5LP
G7 NRP R. Crossley 52 Richard Road, Darton, Barnsley S75 5NP
G7 NRR P. Morris Antler Cottage, High Street, Scaldwell, Northampton NN6 9JS
G7 NRS A. Saunders Christ Church Vicarage Schofield Street, Leigh WN7 4HT
G7 NRU L. Carley The Limes, Grantham Road, Hough On The Hill, Grantham NG32 2BQ
G7 NRV R. Wheeldon 10 Mill View Court, School Lane, St. Neots PE19 8GJ
G7 NSJ S. Shufflebotham Upper House, Woodbank Abdon, Craven Arms SY7 9HX
G7 NSK P. Blunden 20 Fiskerton Road, Reepham, Lincoln LN3 4EB
G7 NSN J. Vinters 106 Halifax Road, Ripponden, Sowerby Bridge HX6 4AG
GW7 NTA T. Blunsdon 3 Railway Terrace, Aberbeeg, Abertillery NP13 2AD
G7 NTG J. Smith 54 Greenfield Avenue, Kettering NN15 7LL
G7 NTI A. Wood 23 Cross Ryecroft Street, Ossett WF5 9EW
G7 NTO D. Brown 5 Ash Close, Watlington OX49 5LW
GW7 NTP P. Banks Ysgol Emrys Ap Iwan, Rhuddlan Road, Abergele LL22 7HE
G7 NTQ C. Thomson 17 Tilley Crescent, Ryton NE40 4FE
G7 NUC D. Poulet 3 Barton Cottages, Newton St. Cyres, Exeter EX5 5DA
G7 NUE A. Isted 22 Tavy Road, Worthing BN13 3PG
G7 NUG T. Brown 125 Godinton Road, Ashford TN23 1LN
GW7 NUL J. Page 39 Rest Bay Close, Porthcawl CF36 3UN
G7 NUM M. Exton Thorn Cottage, 42 High Street, Bourne PE10 0SR
G7 NUN S. Latham 4 Shaston Road, Stourpaine, Blandford Forum DT11 8TA
GM7 NUQ C. Mair 23 Strathburn Gardens, Inverurie AB51 4RY
G7 NUT I. Nutley Czechers, Potten End Hill Water End, Hemel Hempstead HP1 3BN
GW7 NUU R. Clegg The Cottage At Fron Isaf, Pentrecelyn LL15 2HR
G7 NVB K. Fowler Flat 10, Westwood House Edinburgh Road, Norwich NR2 3RL
GM7 NVG C. Park Flat 4, Corrow, Cairndow PA24 8AD
G7 NVI P. Lancaster 2 North Farm Road, Lancing BN15 9BS
GW7 NVM R. Skelton 7 Whitethorn Place, Sketty, Swansea SA2 8HR
G7 NVS D. Poulton 93 Pretoria Road, Ibstock LE67 6LP
G7 NVZ G. Moore 1 Sibland Way, Thornbury, Bristol BS35 2EJ
G7 NWR North Wiltshire Raynet Group c/o A. Sharman 3 Deben Crescent Haydonwick, Swindon SN25 3QB
G7 NXV D. Ross 37 Cartmell Drive, Leeds LS15 0NQ
GM7 NYB N. Macfarlane 3 Kilmore Terrace, Devaig Isle of Mull PA75 6GN
G7 NYD B. Collinge 4 Ash Grove, Preesall, Poulton-le-Fylde FY6 0EW
G7 NYF P. Ridley 11 Thorney Close, Fareham PO14 3AF
GW7 NYP Gloucester Repeater Group c/o Prof. N. Negus Llain, Llanycefn, Clynderwen SA66 7XT
GM7 NZI R. Simpson 2/1 53 Jedworth Avenue, Glasgow G15 7QE
G7 NZM G. Clifton 21 Park Road, Featherstone, Wolverhampton WV10 7HS
G7 NZO S. Bate 5 Turnpike Way, Ashbourne DE6 1UD
G7 NZR A. Haslam The Goldings, Hayton, Brampton CA8 9JA
G7 NZU L. Elliott 62 Holmsley Lane Woodlesford, Leeds LS26 8RY
G7 NZV R. Easting 3 Ellistons Yard, Ballingdon Street, Sudbury CO10 2BU
G7 NZY C. Wells 6 Craister Court, Cambridge CB4 2SH
G7 NZZ D. Poole 239 Forest Road, Fishponds, Bristol BS16 3QY
G7 OAA J. Davies 78 Chatsworth Road, Southport PR8 2QF
GM7 OAF E. Capstick 24 Dalmore Crescent, Helensburgh G84 8JP
G7 OAH K. Adams Queena, Bicton, Liskeard PL14 5RF
G7 OAI J. Cannell 53 Thimble Close, Rochdale OL12 9QP
G7 OAJ S. Walker 3 Sefton Villas Spook Hill, North Holmwood, Dorking RH5 4JW
G7 OAL P. Knapper 17 Rothwell Street, Failsworth, Manchester M35 0FX
G7 OAS A. White Dovedale, Main Street, Wormington WR12 7NL
G7 OAV Dr A. Holohan 8 School House Terrace, Kirk Deighton, Wetherby LS22 4EH
G7 OAX K. Morrison 25 Holywell Close, Poole BH17 9BG
G7 OBC R. Hudman 27 Egerton Road, Streetly, Sutton Coldfield B74 3PQ
G7 OBD M. Peach 48 Melrose Avenue Portslade, Brighton BN41 2LS
G7 OBF J. Aston 9 Beaufort Close, Reigate RH2 9DG
GM7 OBM D. Macpherson 138 Broomhill Crescent, Alexandria G83 9QL
G7 OBP G. Turner 11 Royds Crescent, Rhodesia, Worksop S80 3HF
G7 OBR M. Biddles 338 High Road, Whaplode, Spalding PE12 6TG
G7 OBS M. Simkins 37 St. Andrews Meadow, Harlow CM18 6BL
G7 OBX N. Finbow 17 Bartholomew Close Bardney, Lincoln LN35XT
G7 OCB S. Fox 249 Warminster Road, Sheffield S8 8PR
G7 OCC A. Graham 19 Talbot Road, Rushden NN10 9NS
G7 OCH J. Doy 11A Shrubland Avenue, Ipswich IP1 5EA
G7 OCK B. Phillipson 27 Victoria Avenue., Crook DL15 9DB
G7 OCQ W. Horwood 2A Bellrope Lane, Roydon, Diss IP22 5RG
GM7 OCU G. Rule 105/19 Causewayside, Edinburgh EH9 1QG
G7 OCX J. Turton 60 Shafton Lane, Leeds LS11 9RE
G7 OCY D. Norton 52 Letchworth Road, Leicester LE3 6FG
G7 ODB R. Evans 18 Lilac Close, Keyworth, Nottingham NG12 5DN
G7 ODG R. Beadle 8 Erica Gardens, Croydon CR0 8LG
G7 ODM G. Wane 1A Rickyard Close, Polesworth, Tamworth B78 1DE
G7 ODN L. Nicoletti 6 Laverock Close, Kimberley, Nottingham NG16 2QX
GW7 ODP M. Jones Flat 2 Block 15 Heol Eifion Gorseinon, Swansea SA4 4PH
G7 ODR D. Cartwright 6 Peveril Road, Castleton, Hope Valley S33 8UA
G7 ODT C. Wright Top Farm Bungalow, Ermine Street, Huntingdon PE28 4EW
G7 ODV G. Bagley Woodcroft, Gatton Bottom, Redhill RH1 3BH
G7 ODZ B. Fowler 4 Langley Street, Derby DE22 3GL
G7 OEA P. Foulkes 60 Hornby Boulevard, Litherland, Liverpool L21 8HG
G7 OED R. Stanley 58 Wells Gardens, Basildon SS14 3QS
G7 OES W. Robinson 5 North View, Newfield, Chester le Street DH2 2SD
G7 OET L. Hitchen 40 Methuen Avenue, Fulwood, Preston PR2 9QX
G7 OEW S. Yohn Little Cottage, Hale, Milnthorpe LA7 7BL
G7 OEY G. Hodges 12 Linwal Avenue, Houghton-On-The-Hill, Leicester LE7 9HD
G7 OFI Capt. P. Smith 11A Springwell Close, Maltby, Rotherham S66 7HG
G7 OFM R. Squires 2 Fishergreen, Ripon HG4 1NW
G7 OFU N. Patterson 63 Squires Wood, Fulwood, Preston PR2 9QA
G7 OFV T. Wordsworth 61 Crane Road, Kimberworth, Rotherham S61 3HN
G7 OGL D. Parker 9 Warwick Gardens, Thrapston, Kettering NN14 4XB
G7 OGN D. Arnold 18 Pheasant Way, Spring Park, Northampton NN2 8BJ
G7 OGO K. Mann 89 Wootton Village, Boars Hill, Oxford OX1 5HW
G7 OGR D. Arthurs 32 Lowfield Avenue, Rotherham S61 4PD
GM7 OGS D. Rushmer 1A Low St., New Pitsligo, Fraserburgh AB43 6NQ
G7 OGT T. Mcdonald 3 Widden Close, Sway, Lymington SO41 6AX
G7 OHD P. Martindale 4 The Crayke, Bridlington YO16 6YP
G7 OHM R. Jarvis 5 Caldecote Avenue, Cockermouth CA13 9EQ
G7 OHO J. Hislop 10 Park Wood Close, Broadstairs CT10 2XN
G7 OHW L. Blanchard 1 Dibden Lane, Alderton, Tewkesbury GL20 8NT
G7 OIA M. Larcombe 52 Orchard Road, Burgess Hill RH15 9PL
G7 OIB S. Johnson 85 Bradley Road, Trowbridge BA14 0QS
G7 OIE S. Viney 5 Hawthorne Grove, Dudley DY3 2QQ
GW7 OIK D. Todd Tyrcae, Gwernogle, Carmarthen SA32 7SA
GM7 OIN J. Cowan 1 Treebank Crescent, Ayr KA7 3NF
G7 OIR Sutherland And District ARC c/o A. Grundy 21 Ribston Close, Shenley, Radlett WD7 9JW
G7 OIT I. Guest 3 Great Cliff, Dawlish EX7 9EX
G7 OJA P. Mann 11 New Mills Road, Hayfield SK22 2JG
GM7 OJJ J. Alexander Newton Of Kinmundy Cottage, Kinmundy, Peterhead AB42 5AY
G7 OJO R. Brown 34 Fallowfield Road, Solihull B92 9HH
GW7 OJT Dr E. Wolfenden 18 Edison Crescent, Clydach, Swansea SA6 5JF
G7 OJU F. Dixon 9 Lincoln Road, Fenton, Lincoln LN1 2EP
G7 OJX K. Trigg 41 Veasey Road, Hartford, Huntingdon PE29 1TA
G7 OJY V. Holyoake 14 Maudlin Court, De Cham Road, St. Leonards-on-Sea TN37 6JY
G7 OJZ J. Neale 20 Oakfield Road, Wollescote, Stourbridge DY9 9DL
G7 OKF M. Hawkins 294 Norton Lane, Earlswood, Solihull B94 5LP
G7 OKI W. Cornish 21 Centaur Street, Portsmouth PO2 7HB
G7 OKO F. Webb 50 Hassam Avenue, Newcastle ST5 9ET
G7 OKR J. Campbell Room 2, 94 Liscard Road, Wallasey CH44 8AB
G7 OKT S. Smith 17 Thackers Way, Deeping St. James, Deeping St. James , Nr Peterborough PE68HP
G7 OKV K. Porter 47 Pick Hill, Waltham Abbey EN9 3LD
GM7 OKX G. Chesworth Auchinway, Skares, Cumnock KA18 2RE
G7 OKY D. Schofield 26 The Chase, Coulsdon CR5 2EG
G7 OLC P. Broadhead 45 Priory Close, Dudley DY1 3ED
G7 OLF W. Levick 50 Wintern Court, Lea Road, Gainsborough DN21 1NA
G7 OLG M. Hodge 271 Marsh Lane, Bootle L20 5BG
G7 OLH A. Barnett 20 Mortlake Drive, Mitcham CR4 3RQ
G7 OLT I. Edwards 31 The Twistle, Byfield, Daventry NN11 6UR
G7 OLU W. Wood The Alley Off Of Gajdoru St, Xaghra, Gozo XRA 104 Malta
G7 OLW P. Newton 22 Barrow Rise, Weymouth DT4 9HJ
G7 OLX J. Chambers 46 Violet Road, West Bridgford, Nottingham NG2 5HA
G7 OMA M. Heales 11 Cardinals Walk, Hampton TW12 2TR
G7 OMF K. Gill 358 Moor End Road, Halifax HX2 0RH
G7 OMI J. Patel 1 The Glade, Furnace Green, Crawley RH10 6JS
G7 OMM R. Stroud 22 Marvell Close, Crawley RH10 3AL
G7 OMN J. Eyes 31 Langdale Road Wistaston, Crewe CW2 8RS
G7 OMQ J. Gillman 4 Yeosfield, Riseley, Reading RG7 1SG
GM7 OMU S. Macmillan 74 Canberra Avenue, Clydebank G81 4LN
GI7 OMY D. O'Buitigh 11 Rossnareen Avenue, Belfast BT11 8LP
G7 ONB C. Robinson Jordan, The Green, Stowmarket IP14 3AB
G7 ONE R. Jacobs Rose Mount, Grove Road, Ventnor PO38 1TH
G7 ONF M. Whitley Apple Tree Cottage, Ratten Row, Langtoft, Driffield YO25 3TJ
G7 ONI J. Churchill 30 Brigade Place, Caterham CR3 5ZU
GM7 ONJ A. Martin The Cairn, Duntrune, by Dundee DD4 0PP
G7 ONL R. Ramsey 17 Derby Road, Guisborough TS14 7DP
G7 ONR P. Green 23 Singleton Court, Patrington, Hull HU12 0SF
G7 ONV D. Das 4 Farcliff, Sprotbrough, Doncaster DN5 7RE
G7 OOB K. Barnes The Old School House, 32 Church Street, Peterborough PE6 8DA
G7 OOE J. Bone Flat 11, Homebreeze House, Beach Street, Morecambe LA4 6BT
G7 OOF E. Cottle 3 Mainstone, Romsey SO51 8HG
G7 OOH K. Mcallister Willow Croft, 109A Kaye Lane, Huddersfield HD5 8XT
G7 OOI R. Phillipson 22 Bagmere Close, Brereton, Sandbach CW11 1SG
GI7 OOM D. Magowan 35 Princeton Avenue, Lurgan, Craigavon BT66 8LW
G7 OOO Invicta CG c/o R. Clayton 9 Green Island, Irton, Scarborough YO12 4RN
G7 OOP A. Constantine1 Fairways, Birchington Close, Bexhill-on-Sea TN39 3TF
G7 OOS J. Smalley 88 Gledhow Wood Road, Leeds LS8 4DH
G7 OOT S. Godrich 11 The Ringway, Queniborough, Leicester LE7 3DN
G7 OOU Dr D. Witts Heelands, Heelands, Milton Keynes MK13 7PZ
G7 OOV R. Nunn 49 Lulworth Drive, Roborough, Plymouth PL6 7DT
G7 OPB A. Adams 44 Berkeley Vale Park, Berkeley GL13 9TG
G7 OPD P. Hundy 101 Goodway Road, Great Barr, Birmingham B44 8RS
G7 OPG R. Palmer 3 Oldfield Close, Maidstone ME15 8DY
G7 OPI D. Pink 87 Lillybrook Estate, Lyneham, Chippenham SN15 4AS
G7 OPJ J. Buttery 38 Wigmore Gardens, Worle, Weston-Super-Mare BS22 9AQ
GM7 OPN W. Cairns 74 Jean Armour Drive, Mauchline KA5 6DT
G7 OPS University of Kent c/o P. Whiting 77 Melford Way, Felixstowe IP11 2UH
G7 OPY L. Kelly 8 Solent Hill, Freshwater PO40 9TG
G7 OQB E. Dockray 2 The Gardens, Farsley, Pudsey LS28 5HW
GM7 OQE G. Murray 10 Mcgregor Court, Crossgates, , Cowdenbeath KY4 8ER
G7 OQG S. Williams 7 Wilton Crescent, Macclesfield SK11 8TH
G7 OQL C. Broad St. Eval Kart Circuit, St. Eval, Wadebridge PL27 7UN
G7 OQO A. Thomas 5 Elm Way, Bistall, Batley WF17 0EQ
G7 OQQ S. Ayers 20 Wytham View, Eynsham, Witney OX29 4LU
G7 OQT K. Peacock 9, Lake View Pool View Caravan Park, Buildwas, Telford TF8 7BS
G7 OQU Dr M. Jones 30 Redruth Street, Manchester M14 7PX
GW7 ORB D. Cullen 16 Chapel Street, Upper Brynamman, Ammanford SA18 1AD
G7 ORE Essex Raynet c/o N. Smith Clare Cottage, White Ash Green, Halstead CO9 1PD
G7 ORG R. Gunner White House, The Whiteway, Cirencester GL7 7BA
GM7 ORJ A. Ross 16 Croft Road, Kiltarlity, Beauly IV4 7HZ

G7 ORK D. Love Woodland View, Lower Street, Shepton Mallet BA4 6BB
G7 ORN S. Tideswell 35 Didcot Drive Marchington, Uttoxeter ST14 8LT
G7 ORS J. Lorenzen 40 Boundary Road, Ramsgate CT11 7NW
G7 ORT D. Buckley 22B Anerley Grove, Kingstanding, Birmingham B44 9QH
G7 ORV S. Edmonds 170 Halton Road, Sutton Coldfield B73 6NZ
G7 ORW R. Garnett-Frizelle 17 Bridport Avenue, New Moston, Manchester M40 3WP
GM7 ORX J. Lee 2/5 Heriot Bridge, Edinburgh EH1 2HR
G7 OSB C. Baker 22 Court Park, Thurlestone, Kingsbridge TQ7 3LX
G7 OSJ A. Ramm 17 Sharrington Road, Bale, Fakenham NR21 0QX
G7 OSK D. Ramm 24 Rowan Way, Holt NR25 6TZ
G7 OSO A. Kinnersley Weathertop, Barthomley Road, Stoke-on-Trent ST7 8HU
GM7 OSQ D. Clark Benmhor, Baluachrach, Tarbert PA29 6TF
G7 OST S. Timms 7 Portway Drive, High Wycombe HP12 4AU
G7 OTE S. Barlow 16 Arundel Avenue, Urmston, Manchester M41 6NQ
GW7 OTQ A. Dibbins 2 Edwards Close, Briggs Lane, Oswestry SY10 8PS
GM7 OTT I. Alexander Newton Of Kinmundy Cottage, Kinmundy, Peterhead AB42 5AY
G7 OUG K. Gatfield 76 Barnwood Avenue, Gloucester GL4 3AJ
G7 OUT M. Reynolds 15 Foxfield Drive, Stanford-le-Hope SS17 8HH
G7 OUZ B. Donkin 13 Saddlebow Road, King's Lynn PE30 5BQ
G7 OVB G. Hutton 17 Fonteyn Place, Stanley DH9 6XE
G7 OVE D. Brown 9 Lancaster Way, East Winch, King's Lynn PE32 1NY
G7 OVK C. Carson 47 Stratford Close, Cramlington NE23 8HW
G7 OVM N. Ward 79 Ulwell Road, Swanage BH19 1QU
G7 OVS J. Dilks Handley Farm Bungalow, Brant, Beckingham LN5 0RN
G7 OWB K. Moorcroft 58 Oakley Road, Dovercourt, Harwich CO12 4QU
G7 OWP J. Oliphant 16 Sylvias Close, Amble, Morpeth NE65 0GB
G7 OWQ B. Clifton 3 Kirton Road, Cosham, Portsmouth PO6 2ES
GM7 OWU B. Brander 3 Spartleton Place, Dundee DD4 0UJ
G7 OWV M. Baines 21 Acre Moss Lane, Kendal LA9 5QE
G7 OWX D. Allen 130 Seamer Road, Scarborough YO12 4EY
G7 OWZ D. Payne 147 Upper Marehay Road, Marehay, Ripley DE5 8JG
G7 OXA D. Giles 73 Barsby Drive, Loughborough LE11 5UJ
G7 OXB S. Richardson 6 Dane Ghyll, Barrow-in-Furness LA14 4PZ
G7 OXH R. Wilkins 85 St. Richards Road, Otley LS21 2AL
G7 OXK J. Leach 2 Andover Close, Feltham TW14 9XG
G7 OXN B. Burdis Toledillo 19, Malaga 29570 Spain
G7 OXP R. Singleton 91 Robins Lane, St. Helens WA9 3NF
G7 OXV R. Blott Chateau Perigord Ii Bloc E Apt 5, 6 Lacets Saint Leon, Monaco MC98000 Monaco
G7 OXY J. Mathew 139 Dowthorpe Hill, Earls Barton, Northampton NN6 0PX
G7 OYD P. Thompson Stanville Cowick Road Snaith, Goole DN14 9JG
G7 OYF B. Gilbert 3 Williams Way, West Row, Bury St. Edmunds IP28 8QB
G7 OYP S. Hollis 89 Longfield Lane, Cheshunt, Waltham Cross EN7 6AN
GU7 OYU A. Stoaling Carando, La Petite Mare De Lis Clos, La Rocquette, Castel GY5 7BN Guernsey
G7 OYX B. Dorey 8 Richmond Road, Swanage BH19 2PZ
G7 OZA G. Johnston 94 Abercorn Crescent, Harrow HA2 0PU
G7 OZE South Birmingham RS c/o K. Baldry 160 Rover Drive, Castle Bromwich, Birmingham B36 9LL
G7 OZH D. Albury 40 Mulberry Gardens, Fordingbridge SP6 1BP
G7 OZI A. Morley 6 Millway, Chudleigh, Newton Abbot TQ13 0JN
G7 OZJ S. Morley 7 Bank Avenue, Mitcham CR4 3DW
GW7 OZP J. Barrett Tree Tops, Comins Coch, Aberystwyth SY23 3BL
G7 OZQ T. Pluck 29 Templegate View, Leeds LS150HQ
G7 OZU Wenvoe ARC c/o M. Knight 30 Mountbatten Drive, Biggleswade SG18 0JJ
G7 PAC C. Shadlock Fransham, 15 Queen Street, Spooner Row, Spooner Row NR18 9JU
G7 PAE C. Bailes 2 Church Road, Catworth PE28 0PA
G7 PAF R. Scaife 50 Springbank Road, Gildersome, Leeds LS27 7DJ
G7 PAG P. Gould 10 Heron Park, Lychpit, Basingstoke RG24 8UJ
G7 PAK A. Smith 153 Seymour Way, Sunbury-on-Thames TW16 7NL
G7 PAN Wisbech Amateur Radio & Electronics Club c/o I. Leather 44 Newlands Road, Intake, Sheffield S12 2FZ
G7 PAY C. Wilson 107 Hamilton Avenue, Uttoxeter ST14 7FE
GM7 PBB J. Gray 5 Noth Dell, Isle of Lewis HS20SW
G7 PBC P. Sherburn 70 Briarwood Road, Stoneleigh Park, Epsom KT17 2NG
G7 PBH H. Parrish 5 Kestrel Lane, Cheadle, Stoke-on-Trent ST10 1RU
G7 PBK J. Oliver 27 Rosamund Avenue, Pickering YO18 7HF
G7 PBO M. Sewell 4 Cherfield, Minehead TA24 5TD
GW7 PBP V. Roberts 44 Mount Crescent, Morriston, Swansea SA6 6AP
GI7 PBQ R. Young 8 Glenside Avenue, Drumbo, Lisburn BT27 5LQ
G7 PBT R. Spirrell 32 Churchfield Drive, Castle Cary BA7 7LA
G7 PBV J. Mason 56 Skegby Road, Sutton-in-Ashfield NG17 4EZ
G7 PCE K. Toop 10 Hunt Road, Blandford Forum DT11 7LZ
G7 PCF J. Banfield Highbury, 81 Clophill Road, Bedford MK45 2AD
G7 PCG M. Bartlett 6 St Vincent Chase, Braintree CM7 9UJ
G7 PCT P. Treadwell 22 Meynell Close, Melton Mowbray LE13 0RA
G7 PCV A. Newman 115 Wolverhampton Road, Cannock WS11 1AR
G7 PCW C. Naylor 25 Tavistock Way, Wakefield WF2 7QS
GW7 PCX G. Bellis 70 Osborne St., Rhos, Wrexham LL14 2HT
G7 PDH J. Swallow 26 Balmoral Road, Abbots Langley WD5 0ST
G7 PDO M. Galea 17 Waterloo Road, Horsham St. Faith, Norwich NR10 3HS
G7 PDR J. Martin 45 Quail Holme Road, Knott End-On-Sea, Poulton-le-Fylde FY6 0BT
G7 PDU S. Willis 180 Thisselt Road, Canvey Island SS8 9BL
G7 PEB J. Edwards1 37 The Orchard, Swanley BR8 7UR
G7 PEC Essex Raynet c/o G. Tiller 15 Woodlands Gardens, Romsey SO51 7TE
G7 PEE T. Griffiths 19 James Copse Road, Waterlooville PO8 9RG
G7 PEH P. Gater 166 Rolls Ave, Crewe CW1 3QD
G7 PEN E. Penn 53 Manfield Avenue, Walsgrave, Coventry CV2 2QF
GW7 PEO P. Bennett 14 Harlech Crescent, Prestatyn LL19 8DG
G7 PER D. Boughton 59 Redland Drive, Kirk Ella, Hull HU10 7UX
G7 PEU R. Chapman 12 Lynton Road, Chesham HP5 2BU
GW7 PEX K. Barker 12 Swn Yr Afon Kenfig Hill, Bridgend CF33 6AJ
G7 PFD A. Thompson Forge House, Newark, Newark NG22 0PN
G7 PFG J. Smith 6 Aspen Close Ss89Jj, Canvey Island Essex SS89JJ
G7 PFI M. Green Runnymede, Aston Common, Aston, Sheffield S26 2AD
GW7 PFK E. Gillet 4 Camrose Court, Caldy Close, Barry CF62 9DR
G7 PFL M. Dormer 5 Kipling Walk, Basingstoke RG22 6BN
GM7 PFQ D. Chadwick 19 Lochslin Place, Balintore, Tain IV20 1UP
G7 PFT J. Richardson Unit F, Tollgate Business Centre, Stafford ST16 3HS
G7 PFY J. Ellis 11 Moorland Crescent Guiseley, Leeds LS20 9EF
G7 PGH M. Card 11 Manifold Road, Eastbourne BN22 8EH
G7 PGY G. Sleeman 21 Millbank, Kintbury, Hungerford RG17 9UW
G7 PHB S. Beesley 15 Byron Close, Cheadle, Stoke-on-Trent ST10 1XB
G7 PHC M. Porter 16 The Oval, Scarborough YO11 3AP
G7 PHD I. Connor 34 Mace Road Mildenhall, Bury St. Edmunds IP28 7FP
G7 PHE A. Lassman 69 St. Ladoc Road, Keynsham, Bristol BS31 2EQ
G7 PHF Dr G. Lloyd 256 Penns Lane, Sutton Coldfield B76 1LQ
G7 PHG D. Harbron 48 Sheridan Road, South Shields NE34 9JJ
G7 PHI T. Larsen 47 Lizard Lane Whitburn Village, Sunderland SR6 7AL
G7 PHK J. Wilkes 229 Merland Rise, Tadworth KT20 5JQ
G7 PHL R. Marshall 3 Lawrence Crescent, Sutton-in-Ashfield NG17 4HX
G7 PHR W. Stewart 1 Laing Close, Bardney, Lincoln LN3 5XS
G7 PHT K. Dennis 4 Ash Grove, Sheringham NR26 8PT
G7 PHW S. Dobson 166 Lynfield Drive, Bradford BD9 6EZ
G7 PHY D. Dobson 166 Lynfield Drive, Bradford BD9 6EZ
GW7 PIB J. Challenger 33 Blossom Close, Langstone, Newport NP18 2LT
G7 PIG C. Wood 2 Longfield Avenue, Heald Green, Cheadle SK8 3NH
G7 PIJ M. Beeson 1 Tamar Grove, Cheadle, Stoke-on-Trent ST10 1QQ
G7 PIK B. Bashford 51 Broadwater Road, Worthing BN14 8AH
GW7 PIN W. Griffiths 147 High Street, Tonyrefail, Porth CF39 8PL
G7 PIP R. Oswald 17 Dunclutha Road, Hastings TN34 2JA
G7 PIR J. Briggs 8 Premier Court, 100 Monyhull Hall Road, Birmingham B30 3QJ
G7 PIX D. Atkins 99 Mariners Way, Maldon CM9 6YX
GI7 PIZ S. Hewitt 11 Erindee Close, Donaghadee BT21 0NS
G7 PJD R. Howes 58 Mayfield Avenue, Orpington BR6 0AQ
GI7 PJF Callington ARS c/o R. Stewart 1 Portmore Hall, Ballydonaghy Road, Crumlin BT29 4WT
G7 PJG D. Gohill Flat 71, Hatton Place, Luton LU2 0FD
GI7 PJU C. Robinson 19D Divis Tower, Belfast BT12 4QB
G7 PKD R. Brown 1 Octavian Close, Hatch Warren, Basingstoke RG22 4TY
G7 PKG B. Jenkins 3 Wisteria Avenue, Branston, Lincoln LN4 1QB
G7 PKH P. Precious 99 Sherwood Avenue, St. Albans AL4 9PW
G7 PKJ D. Alway 79 Landseer Avenue, Bristol BS7 9YW
G7 PKK A. Sharp 170 Kingshill Road, Swindon SN1 4LL
G7 PKP J. Constance 1 Grayling Close, Grimsby DN37 9HA
G7 PKQ P. Troll 18 Bowness Road, Millom LA18 4LS
GM7 PKT R. Morrison Corran Gardens, Corran Gardens, Onich, Fort William PH33 6SJ
G7 PKY E. Plant 22 Bournville Road, London SE6 4RN
G7 PLE J. Goodliffe 25 Stansgate Avenue, Cambridge CB2 0QZ
G7 PLP I. Brown 9 Larford Walk, Stourport-on-Severn DY13 0HE
G7 PLS N. Partridge 13 Alderney Way, Immingham DN40 1RB
G7 PLV A. Miller 10 Limerick Close, Ipswich IP1 5LR
GW7 PMA J. Beach 11 St Annes, Western Lane, Swansea SA3 4EW
G7 PMB R. Cooke 2 Harvey Court, Warrington WA2 9SD
G7 PMF V. Lennox 64 Oak Avenue, Blidworth, Mansfield NG21 0TL
G7 PMG S. Rose 84 Woodhill Park, , Pembury TN2 4NP
G7 PMI D. Williams 6 Raven Crescent, Billericay CM12 0JF
G7 PMK W. Harrison 67 Connaught Gardens, Shoeburyness, Southend-on-Sea SS3 9LR
G7 PMQ P. Slight 4 Field Close Welton, Lincoln LN2 3TT
G7 PMU S. Lawrence 85 West Avenue, Clacton-on-Sea CO15 1HB
G7 PMV G. Gimber 10 Harrowdene Gardens, Teddington TW11 0DH
G7 PMW G. Wall Westerland, Sandhill Road, Buckingham MK18 2LZ
G7 PMX M. Firth 5 Courtenays, Seacroft, Leeds LS14 6JZ
G7 PMY B. Whittington Flat 5, Anjou Court, 8 Hereward Road, Eastbourne BN23 6TQ
G7 PNE D. Head 76 Dryden Crescent, Stevenage SG2 0JH
G7 PNF M. Cleverley 43 Friesian Gardens, Newcastle ST5 6BB
G7 PNG R. Owen 53 Huntingdon Drive, Castle Donington, Derby DE74 2SR
G7 PNM P. Smith 41A Thornhill Road Middlestown, Wakefield WF4 4RU
G7 PNP P. Collins Summerley, Hollow Road, Widdington, Saffron Walden CB11 3SL
GM7 PNX N. Armstrong 4 Arboretum Road, Edinburgh EH3 5PD
G7 POA Dr F. Greaves Ratooragh, Schull WEST CORK Ireland
G7 POC N. Phillips 9 Symonds Close, Chandler'S Ford, Eastleigh SO53 3TP
G7 POG M. Mountford 1 Bowater House, Moor St., West Bromwich B70 7AZ
G7 POI L. Selway Rua Do Rochio No 7, Poco Redondo, Tomar Portugal
G7 POQ R. Erdinc 811 York Road, Leeds LS14 6AA
G7 POS D. Ford 44 Cardinal Square, Beeston, Leeds LS11 8HR
G7 POT S. Eastwood Cross Trods Spellowgate, Driffield YO25 5UP
G7 POV A. Lickley 18 Byron Street, Macclesfield SK11 7PL
G7 POW J. Sanderson 40 Sheldon Close, Bransholme, Hull HU7 4RU
G7 PPC B. Lowe 19 Wolverhampton Road, Bloxwich, Walsall WS3 2EZ
G7 PPL A. Pardivalla 2 Mcdowell Way, Narborough, Leicester LE19 2RA
GM7 PPN H. Waugh 93 Denholm Road, Musselburgh EH21 6TU
G7 PPS H. Lodge 69 Helena Road, Rayleigh SS6 8LQ
G7 PQB K. Hunter 30 Loxley Road, Lowestoft NR33 9PG
G7 PQD B. Harrison 145 St. Leonard St., Hendon, Sunderland SR2 8QB
G7 PQL M. Robinson 1 Selby Close, Baxenden, Accrington BB5 2TQ
G7 PQM R. Buchan 16 Lomond Drive, Kettering NN15 5DE
G7 PQP D. Little 20 Vicarage Close, Shillington, Hitchin SG5 3LS
GW7 PQS M. Waite 4 Clos Bryngwyn, Garden Village, Swansea SA4 4BJ
G7 PQW N. Wills 18 Hollis Way, Southwick, Trowbridge BA14 9PH
G7 PQX P. Smith 4 Stone Lodge Lane, Ipswich IP2 9PA
G7 PRB W. Ross 8 Mayall Court, Waddington, Lincoln LN5 9PY
G7 PRC S. Newstead Rectory Cottage, Church Road, Norwich NR12 8YL
G7 PRD R. Ware 23 Harmsworth Drive, Stockport SK4 4RP
G7 PRI P. Newton 5 Sandford Road, Winscombe BS25 1HD
GW7 PRK R. Zeal 5 Llanthewy Road, Newport NP20 4JR
G7 PRO P. Julian 45 Rectory Avenue, Corfe Mullen, Wimborne BH21 3EZ
GW7 PRW M. Price Graig Hill Chapel, Skenfrith, Abergavenny NP7 8UF
G7 PRZ B. Reed Blackthorn, Wanborough Lane, Cranleigh GU6 7DS
G7 PSC P. Rivers 39 Ashton Road, Birmingham B25 8NZ
G7 PSF J. Block 40 Cromwell Road, Cambridge CB1 3EF
GM7 PSH A. Stevens 69 Polwarth Terrace, Prestonpans EH32 9PX
G7 PSK N. Kingsley-Lewis Forge Cottage, Rudham Road Helhoughton, Fakenham NR21 7BY

G7 PSL M. Lamb 52 Crookham Grove, Morpeth NE61 2XF
G7 PSS Wincanton Ladies CG c/o P. Allnutt 37 Moss Mead, Chippenham SN14 0TN
G7 PST T. Ellis 19 Cavendish Avenue, Colchester CO2 8BP
G7 PSU A. Drummond 10-12 Tottington Road, Turton, Bolton BL7 0HS
G7 PSV M. Bennett 83 Middlethorpe Road, Cleethorpes DN35 9PP
G7 PSW D. Wieloch 43 Northampton Grove, Langdon Hills, Basildon SS16 6ED
G7 PSZ A. Laurence Brookvale, Nooklands, Preston PR2 8XN
G7 PTA M. Masterman 7 Pond Bank, Blisworth, Northampton NN7 3EL
G7 PTB R. Player 49 St. Johns Road, Tilney St. Lawrence, King's Lynn PE34 4QJ
G7 PTC M. Watson 12 Milk Thistle Close, Stainton, Middlesbrough TS8 9FQ
G7 PTD J. Midwood 8 Thompson Avenue, Holt NR25 6EN
G7 PTH R. Graham 8 Pecche Place, Chineham, Basingstoke RG24 8AA
G7 PTM J. Taylor 8 Worsley Avenue, Blackpool FY4 2DH
G7 PTT A. Myers 24 Milburn Street, Crook DL15 9DY
G7 PTV M. Howse 28 Courtiers Drive, Bishops Cleeve, Cheltenham GL52 8NU
G7 PTX P. Castle 26 Chestnut Walk, Pulborough RH20 1AW
G7 PTZ R. Mold 134 Kipling Avenue, Brighton BN2 6UE
G7 PUA J. Campbell 48 Renforth Street, Gateshead NE11 9BE
GI7 PUG G. Clegg 45 Strandburn Drive, Belfast BT4 1NA
G7 PUK D. Glass 2 Thorne Square, Sunderland SR3 4PA
G7 PUL R. Hoggard 1 Whiphill Close, Bessacarr, Doncaster DN4 6DX
G7 PUN D. Russon 123 Queens Drive, Newton-le-Willows WA12 0LN
G7 PUP A. Hurd 15 Ashgate Court Mews, Fairfield Road, Chesterfield S40 4TU
G7 PUW S. Frizzell 58 Crisp Road, Lewes BN7 2TX
G7 PUZ L. Martyn 1 Canewdon Hall Close, Canewdon, Rochford SS4 3PY
G7 PVE G. Eddy 102 Springfield Close, Andover SP10 2QT
G7 PVF M. Folland 14 High St., Shoreham, Sevenoaks TN14 7TD
G7 PVG B. Fox 10 Materman Road, Stockwood, Bristol BS14 8SS
GU7 PVI M. Major East Liberty, Gibauderie, St. Peter Port GY1 1XJ Guernsey
G7 PVL C. Watts 8 Clark Close, Wraxall, Bristol BS48 1JL
G7 PVQ S. Denton 9 High Meadow, Tollerton, Nottingham NG12 4DZ
G7 PVU G. Rouse 18 Westfield Avenue, Woking GU22 9PH
G7 PVY D. Driver 53 Brabant Way, Westbury BA13 3UW
G7 PVZ W. Sefton 1 Old Station Close, Stockbridge SO20 6DG
G7 PWA N. Padley 12A Wey Close, Ash, Aldershot GU12 6LY
G7 PWI C. Thornton 1 Elizabeth Drive, Tring HP23 5HL
G7 PWJ J. Churchill 68 Anthony Road, London SE25 5HB
G7 PWK M. Bibb 9 Nelson Court, Old Nelson Street, Lowestoft NR32 1EH
G7 PWL M. Turnbull 11 Waverley Avenue, Whitley Bay NE25 8AU
GI7 PWQ G. Mccormick 24 Warren Park Drive, Lisburn BT28 1HF
G7 PWS C. Collins 32 St. Martins Road, New Romney TN28 8JY
G7 PWU H. Tomlinson 42 Gawsworth Avenue, Crewe CW2 8PB
G7 PWV P. Woolhouse 21 Coombe Wood Hill, Purley CR8 1JQ
GM7 PXJ L. Michie 5 Torridon Place, Rosyth, Dunfermline KY11 2EZ
GM7 PXL D. Warner Torran, Letterfinlay, Spean Bridge PH34 4DZ
G7 PXM J. Cordell 45 Queens Gardens, Dartford DA26HZ
G7 PXR P. Gilbert 4 Ruby Street, Bristol BS3 3DY
G7 PXS G. Mape 34 Amberwood Drive, Manchester M23 9NZ
G7 PXX E. Spires 8 Regent Terrace Barrow Road, New Holland, Barrow-upon-Humber DN19 7QB
G7 PYB J. Bailey 52 Longmeadow, Frimley, Camberley GU16 8RR
G7 PYN A. Wentworth 5 York Avenue, Prestwich, Manchester M25 0FZ
G7 PYQ N. Gell 1 Lawton Road, Rushden NN10 0DX
G7 PYR V. Tuff 8 Millcroft Court, Blyth NE24 3JG
G7 PYT G. Coleman 23 Graham Avenue, Patcham, Brighton BN1 8HA
G7 PYV A. Turner 20 Kipling Gardens, Upper Stratton, Swindon SN2 7LJ
G7 PYW K. Houghton 42 Pear Tree Avenue, Coppull, Chorley PR7 4NL
G7 PZB R. Dewsbery 8 Westfield Close, Market Harborough LE16 9DX
G7 PZE F. Eastham 4 Dunkirk Avenue, Fulwood, Preston PR2 3RY
G7 PZF W. Bailey 15 Norfolk Road, Congleton CW12 1NY
GM7 PZH M. Drennan 6 Hillpark Way, Edinburgh EH4 7BJ
G7 PZL A. Morton 54 Rose Farm Approach, Normanton WF6 2RZ
G7 PZM S. Carter 6 Bramalea Close, London N6 4QD
G7 PZQ P. Breese 6 Wheatley Close, Birmingham B75 5EJ
G7 PZT J. Keen 30 Fielding Crescent, Blackburn BB2 4TD
G7 PZU A. Haworth 8 Boulsworth Crescent, Nelson BB9 8DF
G7 RAB D. Evans 31 Kinsbourne Way, Thornhill, Southampton SO19 6HB
G7 RAE J. Kirkwood 6 Trinity Court, Rothwell, Kettering NN14 6YQ
G7 RAF Royal Air Force ARS (Rafars) c/o K. Sellens 7 The Knapps, Semington, Trowbridge BA14 6JG
G7 RAG Keighley Raynet Group c/o J. Dennis The Old Chapel House, Alford LN13 9PH
GI7 RAH T. Tweedie 17 Schomberg Park, Belfast BT4 2HH
G7 RAI M. Moorhouse 11 Hazel Grove, Huddersfield HD2 2JP
G7 RAJ D. Eggett Copper Beeches, 68 Forest Lane, Kirklevington, Yarm TS15 9ND
GM7 RAK J. Boyd 102 Provost Milne Grove, South Queensferry EH30 9PL
G7 RAL Loughborough & Distrct ARC c/o I. Hewitt 26 Outwoods Drive, Loughborough LE11 3LT
GI7 RAM J. Christie 3 Victoria Drive, Sydenham, Belfast BT4 1QT
G7 RAT Reigate Amateur Transmitting Society c/o P. Tribe Island View, 108 Portsmouth Road, Lee-on-the-Solent PO13 9AF
G7 RAU D. Edwards Blue Stones, 9-10 Mile End. Lizard, Helston TR12 7AS
G7 RAY B. Kagelmacher Rauhe Höge 5, Wismar 23966 Germany
G7 RAZ M. Wager 115 Queensway, Taunton TA1 4NL
G7 RBA M. Sims 23 Winding Way, Alwoodley, Leeds LS17 7RB
G7 RBB A. Perkins 9 St. Martins Close, Canterbury CT1 1QG
G7 RBC Stourbridge & District ARS c/o I. Rodgers 89 Braemar Road, Worcester Park KT4 8SN
G7 RBL C. Johnson 23 St Georges Avenue, Stoke-on-Trent ST6 7JR
G7 RBQ Bolton School ARC c/o P. Dodman 15 Goscote Close, Redditch B97 6UF
G7 RBR J. Pavia 67 Browns Rock Road, Burnt Hill , Rd1, Oxford 7495 New Zealand
G7 RBS A. Sercombe 28 Strumpshaw Road, Brundall, Norwich NR13 5PA
GM7 RBY M. Flett Braeside, Clachan, Tarbert PA29 6XL
G7 RBZ J. Hudson Atholgarth House, St. Johns Lane, Bewdley DY12 2QZ
G7 RCC R. Wendes 24 Whitehead Crescent Wootton Bridge, Ryde PO33 4JF
GI7 RCH G. Walker 16 Stormount Crescent, Belfast BT5 4NT
G7 RCK S. Motala 28 Fishwick View, Preston PR1 4YB
G7 RCL R. Abbott 2 Leybourne Drive, Springfield, Chelmsford CM1 6TX
G7 RCP D. Baines 157 Hall Green Road, West Bromwich B71 2DY
G7 RCS J. Harris 172 Shenstone Avenue, Stourbridge DY8 3DZ
G7 RCU A. Walker 37 East Road, Brinsford, Wolverhampton WV10 7NP
G7 RCW J. Matthews 126 Ingrave Road, Brentwood CM13 2AG
G7 RDA P. Brownsett 10 Great Aldens, Bedford MK41 8JS
GM7 RDH R. Spence Leyan Harray, Orkney KW17 2LQ
G7 RDJ R. Middleton 32 West Busk Lane, Otley LS21 3LW
G7 RDP B. Gawthorpe 19 Tower Hill, Clitheroe BB7 1PD
G7 RDQ T. Rochford 1C Oak Villa, Moors Avenue, Hartlebury, Kidderminster DY11 7YL
G7 RDT Dorset Raynet c/o S. Hawkins Forest Edge, Deer Park, Milton Abbas, Blandford Forum DT11 0AY
GM7 RDY J. Mowat Nether Bigging, Shapinsay, Orkney KW17 2EB
G7 REC D. Allison 52 Boyn Valley Road, Maidenhead SL6 4ED
GM7 REF Lv21 Lightship Museum c/o M. Harrington Mount Pleasant House, North Road, Wick KW1 4DN
GM7 REG J. Robertson 13 Swanston View, Edinburgh EH10 7DG
G7 REH P. Evans 45 Chiltern Drive, Charvil, Reading RG10 9QF
G7 REJ S. Hutchinson 42 Greenham Mill, Mill Lane, Newbury RG14 5QW
G7 REV T. Jones 7 Sedum Close, Huntington, Chester CH3 6BL
GM7 REY J. Macdonald 27 Melantee, Fort William PH33 6PY
GW7 RFA A. Lord The Mount, Trefecca, Brecon LD3 0PW
G7 RFC Coventry ARS c/o G. Farrell 95 Washington Road, Maldon CM9 6JF
G7 RFD P. Johnson Sixpenny Cottage, Farthings Fold, Bourne PE10 0RN
G7 RFE R. Johnson 8 Merlin Close, Bourne PE10 0BZ
G7 RFH J. Fearns 23 Homestead St., Stoke on Trent ST2 0RQ
G7 RFM G. Hunt 7 Kevington Drive, St. Pauls Cray, Orpington BR5 2NT
G7 RFO R. Thomson 123 Oak Avenue, Todmorden OL14 5PE
GW7 RFP P. Barry 44 Heol Onen, Brynmawr, Ebbw Vale NP23 4TS
G7 RFS K. Abeynayake 25 Anderson Avenue, Earley, Reading RG6 1HD
G7 RFT K. Whittle 26 Beachs Drive, Chelmsford CM1 2NJ
G7 RFX R. Oxlade 3 Thyme Court, Northampton NN3 8HY
G7 RFY G. Annett 8 Wheatsheaf Way, Linton, Cambridge CB21 4XB
G7 RFZ Dr J. Bilmen 145 The Maples, Harlow CM19 4RD
G7 RGA P. Cattanach Grosse Neugasse 5/4, Wien 1040 Austria
G7 RGG S. Emmett 14 Ernle Road, Calne SN11 9BT
G7 RGI H. Jones Four Acres, Bungalow Farm Winwick Gated Road, West Haddon NN6 7BH
G7 RGJ P. Musselwhite 80 Craven Road, Orpington BR6 7RT
G7 RGO E. Allan 282 Bilton Road, Rugby CV22 7EG
G7 RGR R. Jones 21, Northumberland Avenue, , Bishop Auckland DL14 6LW
G7 RGU R. Oxlade 3 Thyme Court, Lumbertubs, Northampton NN3 8HY
G7 RGV H. Jump 4 Bankwood, Shevington, Wigan WN6 8EY
G7 RHD M. Clarke 6 Oldbrook Fold, Timperley, Altrincham WA15 7PA
G7 RHE S. Payne 19 Weavers Lane, Sevenoaks TN14 5BT
G7 RHF A. Richards 3 Marsh Gate, Clee St. Margaret, Craven Arms SY7 9DU
G7 RHI A. Mclocklin 43 Forbes Avenue, Potters Bar EN6 5NB
G7 RHM K. Pang 30 Barnwood Avenue, Gloucester GL4 3AH
G7 RHT P. Bennett 47 Bakers Ground, Stoke Gifford, Bristol BS34 8GD
G7 RIA M. Cook 19 Reigate Avenue, Clacton-on-Sea CO16 8FB
GW7 RIB P. Nicholls 11 Ifor Hael Road, Rogerstone, Newport NP10 9FB
G7 RIE J. Glenn School House, 70 Norwich Road, Norwich NR12 7EG
G7 RIJ E. Devine 23 Radley Avenue, Wickersley, Rotherham S66 2HZ
G7 RIO W. Care 29 Wheal Gorland Road, St. Day, Redruth TR16 5LT
G7 RIS I. Higton 25 Shady Grove Hilton, Derby DE65 5FX
G7 RIU D. Kirk 19 The Meads, Hildersley, Ross-on-Wye HR9 7NF
GM7 RJG A. Forbes 28 Innes Street, Inverness IV1 1NS
G7 RJO C. Compton 55 Lulot Gardens, London N19 5TR
G7 RJR D. Crossley 33 Town Lane , Castle Acre, Kings Lynn PE322AU
G7 RJW D. Lamden 6 Ashbourne Way, Thatcham RG19 3SH
GW7 RKC A. Hall 29 Ely Street, Tonypandy CF40 1BY
G7 RKE J. Bottomley Grove House, 2 Woodlane, Falmouth TR11 4RG
G7 RKJ C. Hindmarsh 5 Jackman Drive, Horsforth, Leeds LS18 4HS
GW7 RKQ S. Rudge 1 Marl Mews, Marl View Terrace, Conwy LL31 9BJ
G7 RKT P. Jones 14 Westerleigh Road, Clevedon BS21 7US
G7 RKU P. Dickinson Haven, The Row, Bury St. Edmunds IP29 4DL
G7 RKV D. Wilson 32 Laurel Bank, The Highlands, Whitehaven CA28 6SW
G7 RKW J. Hart 16 Leabrook Close, Bury St. Edmunds IP32 7JH
G7 RKX C. Hill-Smith Top Flat, The Warehouse, West St., Newton Abbot TQ13 7DU
G7 RLK S. Drury 5 Hawthorn Close, Healing, Grimsby DN41 7SR
G7 RLO L. Van Beers Cob Cottage, Tram Inn, Hereford HR2 9AN
G7 RLQ T. Winton 84 Pembroke Road, Clifton, Bristol BS8 3EG
GW7 RLS Belfast Royal Academy Amateur c/o J. Gray City And County Of Swansea, Emergency Planning Unit, Swansea SA1 3SN
G7 RLV C. Pitchford 84 New Road, Rubery, Birmingham B45 9HY
G7 RLX G. Coleman 120 Kidderminster Road South, Hagley, Stourbridge DY9 0JH
GW7 RLZ G. Roberts 4 Fawnog Wen, Penrhyndeudraeth LL48 6PS
G7 RMD D. Devlin 5 Kelsall Avenue, Sutton Manor, St. Helens WA9 4DQ
G7 RME M. Buckley 8 Highthorne Street, Armley, Leeds LS12 3LB
GM7 RMF E. Walker 38 Greenbank Gardens, Edinburgh EH10 5SN
G7 RMG G. Chapman 15 Greenhays Rise, Wimborne BH21 1HZ
G7 RMJ M. Amies Home Farm, Hulme Walfield, Congleton CW12 2JJ
G7 RMQ R. Scarce 7 Mussidan Place, Theatre Street, Woodbridge IP12 4NN
G7 RMW RAF Waddington ARC c/o R. Medcalf 19 All Saints Road, Warwick CV34 5NL
G7 RMX N. Taylor West Mede, Exeter Road, Honiton EX14 1AX
G7 RMZ East Chehire Raynet Group c/o B. Williams 3 Welton Close, Wilmslow SK9 6HD
G7 RNA Obo North Anglia Raynet c/o K. Kent 5 Jubilee Road, Heacham, King's Lynn PE31 7AR
G7 RNB S. Bieber Tonkins Quay, Mixtow, Fowey PL23 1NB
GW7 RNC T. Heywood-Bell 4 Aberthaw Close, Newport NP19 9QA
G7 RNF T. Roberts 6 Petworth Road, Southport PR8 2QL
GM7 RNJ M. Dennis 47 Viewfield Road, Aberdeen AB15 7XP
G7 RNN North Norfolk Raynet c/o A. Farrow 18 The Green Trimingham, Norwich NR11 8ED
G7 RNQ R. Young 12 Elmwood Close, Stokesley, Middlesbrough TS9 5HX
G7 RNX A. Linney 5 Elliscales Avenue, Dalton-in-Furness LA15 8BW
GI7 ROB R. Degossely 82 Knightsbridge, Lisburn BT28 3DG
G7 ROC J. Armstrong 15B Lamberton, Berwick-upon-Tweed TD15 1XB

G7 ROI J. Naylor 46 Loxley Drive, Mansfield NG18 4FB
G7 ROM A. Boardman 147 Musgrave Road, Bolton BL1 4HW
G7 ROP R. Sykes 46 Crescent Road, Netherton, Dudley DY2 0NW
G7 ROY R. Clayton 9 Green Island, Irton, Scarborough YO12 4RN
G7 RPJ J. Barnard 39 Ecclestone Close, Bradwell, Great Yarmouth NR31 8RG
G7 RPK L. Goffin The Hollies, Belaugh Green Lane, Norwich NR12 7AJ
G7 RPP I. Gurney 28 Barrington Drive, Basingstoke RG24 9RS
GM7 RPT D. Hutchison 55 Springfield Road, Tarbolton, Mauchline KA5 5QU
G7 RPW S. Pike 1 Barley Garth, Burton Pidsea, Hull HU12 9AF
G7 RQD M. Folkes 3 Colindale Road, Ferring, Worthing BN12 5JF
GW7 RQI D. Pearson 142 Heol Bryngwili, Cross Hands, Llanelli SA14 6LY
G7 RQK S. Skidmore Kildemarken 223, Ringsted 4100 Denmark
G7 RQO B. Wylie 54 Cromwell Street, Lincoln LN2 5LP
GW7 RQV N. Jenkins 41 Park Street, Bridgend CF31 4AX
G7 RRD G. Smith 12 Oakwood Glade, Holbeach, Spalding PE12 7JS
G7 RRJ A. Mcconnachie 16 Poplar Avenue, Wyre Piddle, Pershore WR10 2RJ
GW7 RRM Dr S. Whitehouse 6 Dol Y Dderwen, Ammanford SA18 2GA
G7 RRO G. Gardner 47 Old Road, Stanningley, Pudsey LS28 6BG
GW7 RRS A. Davis 5 Jubilee Road, Bridgend CF31 3BA
G7 RRY M. Saltmer 12 Beechings Mews, Whitby YO21 3DW
G7 RSA C. Hawkes 103 Station Road, Roydon, King's Lynn PE32 1AW
GW7 RSE L. Clarke 83 Lancaster Street, Blaina, Abertillery NP13 3EQ
G7 RSK A. Scott 62 Berry Meade, Ashtead KT21 1SG
G7 RSM Dr R. Bloor 7 Highfield Court, Clayton Road, Newcastle ST5 3LT
G7 RTA S. Harding 39 Clayton Road, Lidget Green, Bradford BD7 2LX
GI7 RTB P. Mccrory 24 Drumcoo Green, Dungannon BT71 4AJ
G7 RTC G. Darby 69 Churchill Road, Earls Barton, Northampton NN6 0PQ
G7 RTI the Ham Fellowship c/o K. Werner 85 Brecon Way, Downley, High Wycombe HP13 5NW
G7 RTJ D. Bransby 7 West Cliff Avenue, Whitby YO21 3JB
G7 RTL Blaenau Gwent Radio Group c/o M. Pell 7 Churchfleet Lane, Gosberton, Spalding PE11 4NE
G7 RTN J. Burrows 37 Braydeston Crescent Brundall, Norwich NR135LD
G7 RTO B. Theaker 25 Pinewood Drive, Plymouth PL6 7SP
G7 RTQ M. Cowley 72 Warley Road, Warley, Oldbury B68 9TB
G7 RTR D. Freeman 59 Grange Road, Somersham, Huntingdon PE28 3JT
G7 RTX K. Brookes Kohima, Spout Lane, Stoke-on-Trent ST2 7LR
G7 RUC D. Millen 75 Mill Hill Lane Winshill, Burton upon Trent DE15 0BA
G7 RUH R. Peggram Starcroft Janes Close Blackfield, Southampton SO45 1WJ
G7 RUJ D. Brain 3 Mill Lane, Skipsea, Driffield YO25 8SP
G7 RUN M. Graves 20 Stace Way, Worth, Crawley RH10 7YW
G7 RUP J. Shamash 45 Elder Street, Lincoln LN5 8QX
G7 RUQ L. Murphy Flat 3, Evelyn Court, 187 South Coast Road, Peacehaven BN10 8NS
G7 RUR S. Todorovic 36.Trentley Road, Stoke on Trent ST4 8PJ
G7 RUS R. Parkin 25 Kent House Lane, Beckenham BR3 1LE
G7 RUX J. Gardner 24 Challin Street, London SE20 8LW
G7 RUY D. Ager 11 Tilbury Close, St. Pauls Cray, Orpington BR5 2JR
G7 RVC P. Sutherland 9 Lely Close, Bedford MK41 7LS
G7 RVG P. Doble High Street, Stoke-Sub-Hamdon TA146PT
G7 RVI T. Hankins Cawdor House, Cawdor, Ross-on-Wye HR9 7DN
GD7 RVP S. Rand 3 Yn Aittin Vooar, Bretney Road, , Jurby IM7 3EU Isle of Man
GM7 RVR N. Moir 34 Souter Drive, Inverness IV2 4XJ
G7 RVT T. Smith 9 Crofters Way, Westlands, Droitwich WR9 9HU
G7 RVW R. Crofts Little Isle, Woodgate Green, Tenbury Wells WR15 8LX
G7 RVY H. Branch 326 Springfield Road, Chelmsford CM2 6BA
G7 RWC C. Halbert 3 Third Row, Ellington, Morpeth NE61 5HF
G7 RWF J. Buck 14 Crosstree Walk The Willows, Colchester CO2 8QF
G7 RWN D. Taylor 48 Southcroft Road, Gosport PO12 3LD
GJ7 RWT A. Cutland Little Gables, La Route Orange, St. Brelade JE3 8GQ Jersey
G7 RWW J. Kirkham 100 Prince Charles Avenue, Derby DE22 4FL
G7 RWY B. Sankey 121 Green Lane, Coventry CV3 6EB
G7 RXB N. Larson 90 Lingfield Ash, Coulby Newham, Middlesbrough TS8 0SU
G7 RXE V. Donald 20 Parkhill Road, Barnby Dun, Doncaster DN3 1DP
G7 RXI V. Ball 30 Park Drive, Worlingham, Beccles NR34 7DJ
G7 RXJ J. Dyson 21 Highmoor, Morpeth NE61 2AS
G7 RXK R. Thompson 7 Rufford Close, Sutton-in-Ashfield NG17 4BX
GM7 RXL D. Winton 273 Hilton Drive, Aberdeen AB24 4NT
G7 RXO N. Larsen 6 Shrewsbury Close, Barwell, Leicester LE9 8JX
G7 RXW M. Lockitt 19 Roundway Down, Perton, Wolverhampton WV6 7SX
G7 RXX S. Cooper 30 Pinta Drive, Stourport-on-Severn DY13 9RY
G7 RXZ D. Cooper 1A Kent Street, Dudley DY3 1UU
G7 RYA D. Tomlin 154 Court Lane, Erdington, Birmingham B23 5RG
GM7 RYK G. Pollard 127 Braeside Park, Mid Calder, Livingston EH53 0TE
G7 RYL D. Sheridan 78 Oaklands Park, Buckfastleigh TQ11 0BP
G7 RYM R. Pugh 41 East Beach Park, Shoeburyness, Southend-on-Sea SS3 9SG
G7 RYN D. Proctor 11 Bedford Rise, Winsford CW7 1NE
G7 RYO K. Turner 34 Amherst Road, Kenilworth CV8 1AH
GM7 RYT D. Weller 66 Dolphin Road, Currie EH14 5SA
G7 RYW Ashford ARC c/o F. Trainer 23 Woodend Avenue, Hunts Cross, Liverpool L25 0NY
GM7 RZE Dr D. Bremner 'Braedine', Johnshill, , Lochwinnoch PA12 4EL
GW7 RZN E. Taylor 8 First Avenue, Prestatyn LL19 7LP
G7 RZQ N. Waterman 1 Wood Lane Close, Sonning Common, Reading RG4 9SP
G7 RZW A. Davies 16 Sutton Road, Bolton BL3 4QR
G7 SAC Sutton & Cheam RS c/o J. Puttock Sutton & Cheam Rs, 53 Alexandra Avenue, Sutton SM1 2PA
G7 SAI Scarborough Seg c/o E. Birt 10 Wilden Lane, Stourport on Severn DY13 9LR
GM7 SAK A. Jardine 17 Louisa Drive, Girvan KA26 9AH
GW7 SAQ R. Turner 60 Clos-Y-Deri, Porthcawl CF36 3PR
G7 SAT D. Marritt 96 Kynaston Road, Orpington BR5 4JZ
G7 SAX R. Newman 31 Oval Gardens, Alverstoke, Gosport PO12 2RA
GI7 SBF Dr J. Henderson 76 Tirmacspird Road, Crimlin, Ederney, Enniskillen BT93 0FB
GW7 SBJ E. Birtwistle 29 Church View, Pentre, Deeside CH5 2DP
G7 SBK D. Hunt 298 Cavendish Road, Carlton, Nottingham NG4 3QH
G7 SBN M. Royal 3 Bethany Place, St Just, Penzance TR19 7HB
GW7 SBO R. Thomas 25 Lon Lwyd Isaf, Pentraeth LL75 8LN
G7 SBP N. Hancocks 9A St. Philip Street, Penzance TR18 2DN
G7 SBZ M. Newton 24 Chestnut Avenue, York YO31 1BR
G7 SCE P. Farman 298 Laburnum Grove, Portsmouth PO2 0EX
GM7 SCJ G. Deas 81 Speirs Road, Bearsden, Glasgow G61 2LT
G7 SCL J. Robinson 4 Gardner Close, Loughborough LE11 5YB
G7 SCN P. Brotherton 73 Thorneywood Rise, Nottingham NG3 2PE
G7 SCO D. Brooke 34 Park Road, Burwell, Cambridge CB25 0ES
G7 SCP D. Wain 51 Foxstone Way, Eckington, Sheffield S21 4JX
G7 SCR Cambridge Reapeater Group c/o R. Keen 13 Mill View Close, Woodbridge IP12 4HR
G7 SCT G. Rutherford 24 Chestnut Avenue, Hedon, Hull HU12 8NH
G7 SCU D. Ibrahim 14 Dunvegan Road, London SE9 1SA
G7 SCV J. Straughan 16 Garner Close, Chapel Park, Newcastle upon Tyne NE5 1SQ
G7 SCX P. O'Rourke 186 Cottingham Road, Corby NN17 1SY
G7 SCZ D. Kiteley 13 Chiltern Close, Astley Cross, Stourport-on-Severn DY13 0NU
G7 SDC D. Coe 105 Raynham Road, Bury St. Edmunds IP32 6ED
G7 SDD M. Smith 38 Vestry Road, Street BA16 0HX
GW7 SDE I. Jones 14 Clare Court, Loughor, Swansea SA4 6UH
G7 SDG R. Martin 8 Short Lane, Bricket Wood, St. Albans AL2 3SE
G7 SDM G. Davies 11 Ninfield Close, Carlton Colville, Lowestoft NR33 8SD
GM7 SDP D. Ryan 53 Unity Terrace, Perth PH1 2BG
G7 SDQ M. Smith The Old Commercial Inn, Clanage Street, Bishopsteignton, Teignmouth TQ14 9QS
G7 SEG A. Harrison 44 Rosslyn Road, Whitwick, Coalville LE67 5PT
G7 SEJ M. Baskeyfield 69 Mill Lane, Attleborough NR17 2NW
G7 SEK R. Newham 26A Kilwardby Street, Ashby-de-la-Zouch LE65 2FQ
G7 SEO R. Plant 22 The Woodlands, Wokingham RG41 4UY
G7 SER Sutton Coldfield & Dist Raynet c/o J. Trickey 59 Shelley Drive, Sutton Coldfield B74 4YD
G7 SEU Dr E. Kershaw 83 Foxhunter Drive, Oadby, Leicester LE2 5FH
G7 SEY P. Simpson The Conifers, Woodhouse Lane, Telford TF4 3BJ
G7 SFA M. Stevens Flat 7, 1A Woodstock Road, Croydon CR0 1JS
G7 SFD M. King 4 Keith Avenue, Ramsgate CT12 6JQ
GM7 SFE R. Lawrie 84 Redlawood Road, Cambuslang, Glasgow G72 7TP
G7 SFF D. Hartshorn 21 Hucklow Avenue, Chesterfield S40 2LT
G7 SFI S. Merrifield 2 Larkspur Glade, Telford TF3 2AQ
G7 SFJ S. Pratt 15 Springwell Close, Cowling, Keighley BD22 0AP
G7 SFL M. James 8 Swan Quay Bath Lane, Fareham PO16 0DX
G7 SFM R. Wiltshire 8 Hilltop Lane, Heswall, Wirral CH60 2TT
G7 SFS L. Banner 7 Lowdham Road, Gedling, Nottingham NG4 4JP
G7 SFY B. Purkiss 99 Westland Road, Yeovil BA20 2AZ
G7 SGH B. Williams Hillside, Wigmore Lane, Eythorne, Dover CT15 4AW
G7 SGK R. Ward 16 Southgate Crossgates, Scarborough YO12 4NB
G7 SGM R. Gifford 100 Gadebridge Road, Hemel Hempstead HP1 3EW
G7 SGO R. Percival 145 Queen Street, Whitehaven CA28 7AW
G7 SGR South Gloucestershire Raynet c/o T. Humphreys 93 Cornwall Crescent Yate, Bristol BS37 7RU
G7 SHI C. Conce 35 Mortimer Drive, Sandbach CW11 4HS
G7 SJD S. Fitzpatrick 21 Corn Close, South Normanton, Alfreton DE55 2JD
G7 SJK T. Masson Apple Tree Cottage, Neath Gardens, Reading RG3 4UL
G7 SJS P. Roberts 5 Snelston Crescent, Littleover, Derby DE23 6BL
G7 SJX B. Shields 20 Gresley Court, Grantham NG31 7RH
G7 SKA P. Burnett 4 Lavendon Court, Barton Seagrave, Kettering NN15 6QH
GM7 SKB Dr D. Fortune 26 Newton Grove, Newton Mearns, Glasgow G77 5QJ
GW7 SKC Radio Group O c/o J. Gray City And County Of Swansea, Emergency Planning Unit, Swansea SA1 3SN
G7 SKH G. Murray Brookside, Thirlby, Thirsk YO7 2DJ
G7 SKL J. Koops 64 Winchester Avenue, Nuneaton CV10 0DW
G7 SKR D. Tarbatt 9 Dashwood Close, Warrington WA4 3JA
G7 SKV D. Graham 11 Hibernia St., Deane, Bolton BL3 5PQ
G7 SKW B. Mcinnes Thistledome, Tumby Road, Coningsby, Lincoln LN4 4RQ
G7 SKX A. Wilkinson 21 Solbys Road, Basingstoke RG21 7TG
G7 SLB R. Machin 11 Victoria Street, Irthlingborough, Wellingborough NN9 5TR
G7 SLJ D. Lloyd-Jones 2 Leyside, Rayne, Braintree CM77 6DE
G7 SLL G. Peach 120 Craven Road, Newbury RG14 5NR
GI7 SLN G. Mcafee 12 Skerryview, Craigahullier, Portrush BT56 8NJ
G7 SLP P. Hardcastle Rowantree, Woodburn Drive, Leyburn DL8 5HU
GJ7 SLU C. Whittaker Coeur Joyeux, La Rue Des Sapins, St. Peter JE3 7AD Jersey
G7 SLV R. Walker 210 London Road, Worcester WR5 2JT
G7 SLY P. Taylor 46 Ralph Road, Staveley, Chesterfield S43 3PY
G7 SLZ T. Gill 2 Church View Clatworthy, Taunton TA4 2EQ
G7 SMC G. Jameson 17 Lansbury Avenue, Mastin Moor, Chesterfield S43 3AG
G7 SME P. Helliwell 1 Beechfield Avenue, Barton, Torquay TQ2 8HU
G7 SMH G. Newton 8 Lynch Mead, Winscombe BS25 1AT
G7 SMN A. Holden 1 Rose Cottage, Little Bramford Lane, Ipswich IP1 2PH
G7 SMQ B. Cottee 41 Colesbourne Road, Clifton, Nottingham NG11 8JG
GW7 SMV L. Ashford 13 Cefn Court, Rogerstone, Newport NP10 9AH
G7 SMW W. Broyd 40 Hall Farm Road, Melton IP121PJ
G7 SMZ R. Walker 24 Colin St., Alfreton DE55 7HT
G7 SNB O. Newland 22A Cromwell Road, Basingstoke RG21 5NR
G7 SNC I. Palmer 182 Salhouse Road, Norwich NR7 9AD
G7 SNJ R. Chaytor 19 Granville Avenue, Hartlepool TS26 8ND
G7 SNP K. Jordan 7 Park Avenue, Bedlington NE22 7EH
G7 SNQ S. Taylforth 1 Clough Terrace, Barnoldswick BB18 5PD
G7 SNR Mid Norton Raynet Group c/o S. Brodie Waterloo Cottage, Tanners Green, Norwich NR9 4QS
G7 SNT B. Jordan 40 High Street, Coltishall, Norwich NR12 7HD
G7 SNW J. Ward 40 Lancaster Drive, Long Sutton, Spalding PE12 9BD
G7 SNX M. Pearce 42 Pine Close, Rudloe, Corsham SN13 0LB
GI7 SOB K. Elgin 50 Ballinteer Road, Macosquin, Coleraine BT51 4LZ
G7 SOE Rvd. M. Howard East Dean House, East End Langtoft, Peterborough PE6 9LP
G7 SOH C. Brown 9 Marjorie Street, Rhodesia, Worksop S80 3HR
G7 SOV C. Howarth 5 West Mount, Orrell, Wigan WN5 8LX
G7 SOZ S. Jude 9 Winchfield, Great Gransden, Sandy SG19 3AN
GM7 SPA J. Brown 11 Oak Gardens, Oak Drive, Lenzie, Glasgow G66 4BF
GM7 SPB M. Garrington South Orrock Bungalow, Balmedie, Aberdeen AB23 8XY
G7 SPE R. Keep 14 Foster Road, Kempston, Bedford MK42 8BU
G7 SPL D. Pomfret 52 Warwick Close, Bury BL8 1RT
G7 SPM C. Jones Nb Guanche Bradford On Avon Marina, Widbrook Bradford-on-Avon BA15 1UD

G7 SPN S. Townsley 222 Prince Consort Road, Gateshead NE8 4DX
G7 SPP H. Conrad 22 Low Stobhill, Morpeth NE61 2SG
G7 SPZ R. Brown 19 Comberton Road, Toft, Cambridge CB23 2RY
G7 SQH G. Chew 45 Brackley, Weybridge KT13 0BL
G7 SQM N. Crawford 20 Fearnley Crescent, Kempston, Bedford MK42 8NL
G7 SQW A. Woods 10 Radcliffe Road, Drayton, Norwich NR8 6XZ
G7 SQY D. Colton 9 Thornemead, Peterborough PE4 7ZD
G7 SRA Sudbury And District Radio Amateurs c/o A. Harman 107 Kempson Drive, Great Cornard, Sudbury CO10 0YF
G7 SRB D. Shorten 32 Stoneleigh Drive, Carterton OX18 1ED
G7 SRC Worthing And District Amateur Club c/o N. Hull C/O 95 Washington Road, Maldon CM9 6JF
G7 SRH M. Harper 31 Lorland Road, Cheadle Heath, Stockport SK3 0JJ
G7 SRI M. Lowe 2 White Post Bungalows, North Leverton, Retford DN22 0AS
GM7 SRJ S. Jones Smiddy Cottage, Auchencrow, Eyemouth TD14 5LS
G7 SRK R. Carder 45 Chalklands, Linton, Cambridge CB21 4JQ
G7 SRL A. Gallichan 4 Wigston Road, Rugby CV21 4LT
G7 SRV P. Everett 26 Tennyson Close, Horsham RH12 5PN
G7 SSA M. Addicott Orchardleigh, The Street, Radstock BA3 4HG
G7 SSB D. Jones 429 Redmires Road, Sheffield S10 4LF
G7 SSG J. Smye 24 Eastfield Road, Wincanton BA9 9LT
G7 SSJ D. Sutton 32 Queensway, Euxton, Chorley PR7 6PW
G7 SSK R. Walton Easingmoor House, Thorncliffe Road, Leek ST13 7LW
GW7 SSN N. Cole 40 Primrose Court, Ty Canol, Cwmbran NP44 6JJ
GW7 SSQ P. Cole 18 Juniper Crescent Henllys, Cwmbran NP44 6EH
G7 SSU J. Stewart 43A Independent Place, London E8 2HE
G7 SSW J. Haywood 7 Anna Walk, Stoke-on-Trent ST6 3BX
G7 STC K. Gater 110 Byrds Lane, Uttoxeter ST14 7NB
G7 STD L. Goodridge 110 Quarrendon Road, Amersham HP7 9EP
G7 STF D. Taylor 147A Callington Road, Saltash PL12 6JA
G7 STG B. Spavins 8 Berkeley Avenue, Briar Bank Park, Bedford MK45 3WH
GM7 STI Blacksheep Contest + DX Group c/o I. Pearce 1 Mount Farm Cottage, Cupar KY15 4NA
G7 STL M. Anderson 94 Tolworth Road, Surbiton KT6 7SZ
G7 STM M. Wyatt 8 St Mary'S Drive, Sutterton PE20 2LU
G7 STQ M. Oura The Quoins, Gloucester Road, Bath BA1 8AD
G7 SUA D. Wiseman 12 Hamilton Way, Acomb, York YO24 4LE
G7 SUM G. Fewings 22 Watcombe Road, West Southbourne, Bournemouth BH6 3LU
G7 SUQ A. Jobson 7 Dunlin Close, Norton, Stockton-on-Tees TS20 1SJ
G7 SUS R. Biss 1 Fairey Crescent, Gillingham SP8 4PE
G7 SUT C. James (James), Lower Kenneggy Farm, Lower Kenneggy, Rosudgeon, Penzance TR20 9AR
G7 SUU R. Wolk Calle Zaragoza 48, Castalla 3420 Spain
G7 SUV J. Patterson 28 Woodridge Avenue, Thornton-Cleveleys FY5 1PR
G7 SVE A. Jackson 14 West Field Gardens, Sandy SG19 1HF
G7 SVF K. Ingram 15 Kent Avenue, East Cowes PO326QN
G7 SVI C. Lambert-Hutchinson 63 Chalbury Close, Canford Heath, Poole BH17 8BP
G7 SVM D. Bradley 22 Grosvenor Road, Ettingshall Park, Wolverhampton WV4 6QY
G7 SVQ R. Holmes 18 Dresden Close, Mickleover, Derby DE3 0RD
G7 SVT D. Bultitude 48 Forty Acres Road, Devizes SN10 3DG
G7 SVU N. Hinchliffe 19 Grange Road, Blidworth, Mansfield NG21 0RN
G7 SWB Dr R. Bambrey 76 Boundary Way, Glastonbury BA6 9PH
G7 SWE F. Rowbotham 56 Farnborough Road, Clifton, Nottingham NG11 8GF
G7 SWH A. Howell 35 Melton Road, Wakefield WF2 7PR
G7 SWQ Cardiff And District ARC c/o I. Wild 153 Alexandra Road, Sheffield S2 3EH
G7 SWR M. Prentice 26 Meir View, Stoke-on-Trent ST3 6AH
G7 SWS A. Bird 18 Welbeck Drive, Spalding PE11 1PD
G7 SWV Dr C. Smith 11 Woods Close, Haskayne, Ormskirk L39 7JL
G7 SWW R. Jones Flat 32 Tottenhoe Court Colville Road Cherry Hinton, Cambridge CB5 8SE
GM7 SWX D. Curran 104 Mcpherson Crescent, Chapelhall, Airdrie ML6 8XL
G7 SWZ J. Halliday 14 Heath Gardens, Halifax HX3 0BD
G7 SXB D. Phillips 10 Broadstone Hall Road South, Stockport SK5 7DQ
G7 SXG D. Dean 17 Drayton Close, Runcorn WA7 4TW
GM7 SXI A. Williams Gardeners Cottage, Ardrossan KA22 8PH
G7 SXJ J. Farrow 74 The Droveway, St. Margarets Bay, Nr. Dover CT15 6DD
GW7 SXN D. Davies 35 Ty Llwyd Parc Estate, Quakers Yard, Treharris CF46 5LA
GW7 SXU I. Harries Gwastad, Maenygroes, New Quay SA45 9RJ
G7 SYC W. Jarvill 66 Gloucester Road, Newbury RG14 5JN
G7 SYD S. Applegate 180 Logan Street, Nottingham NG6 9FU
G7 SYE T. Laskey 72 Windermere Avenue, Ramsgate CT11 0PL
G7 SYI R. Hutchinson 10 Clifton Avenue, Eaglescliffe, Stockton-on-Tees TS16 9BA
G7 SYJ M. Hogg 55 Ardenfield Drive Wythenshawe, Manchester M22 5DJ
G7 SYL G. Thaxter Leymoon, Gayton Road East Winch, King's Lynn PE32 1NW
G7 SYQ A. Orchiston 16 Windsor Close, Collingham, Newark NG23 7PR
G7 SYS R. Baxter 107 Kendale Road, Bridgwater TA6 3QE
G7 SYT C. Denman 12 Woodland Close, Northampton NN5 6NH
G7 SYU Dr D. Bowers 88 Stamford Avenue, Springfield, Milton Keynes MK6 3LQ
G7 SYY S. Howarth 14 Eaves Lane, Chorley PR6 0PY
GM7 SZA S. Mussell Dunelm, Thornhill Road, Cuminestown, Turriff AB53 5WH
G7 SZB N. Kendal-Ward 29 Denmark Street, Gateshead NE8 1NQ
G7 SZF N. Hartley 66 Broad Lane, Norris Green, Liverpool L11 1AN
G7 SZG K. Gardner 27 Lindon Drive Alvaston, Derby DE24 0LP
G7 SZO R. Collinson 56 Orchard Valley, Hythe CT21 4EA
GI7 SZV A. Maclaine 172 Moylagh Road, Seskanore, Omagh BT78 2PN
G7 SZW D. Green 43 James Street, Selsey, Chichester PO20 0JG
G7 SZZ R. Roberts Connemara, High Hesket, Carlisle CA4 0JF
G7 TAE S. Wersby Oak Barn, 6 Timothys Field Abbotts Ann, Andover SP11 7AT
G7 TAF M. Hawes 78 Martyns Way, Bexhill-on-Sea TN40 2SH
G7 TAT J. Moye 33 Prince Charles Road, Colchester CO2 8NS
G7 TAV S. Houghton 28 Heron Way, Mayland, Chelmsford CM3 6TP
G7 TAX F. Roullier 19 Terling Road, Dagenham RM8 1DS
G7 TBC P. Stockdale-Woodhead 77 Fort Hill Road, Sheffield S9 1BA
G7 TBF N. Smith 47 Kiveton Lane, Todwick, Sheffield S26 1HJ
G7 TBJ J. Kewn 31 Trescoe Road, Long Rock, Penzance TR20 8JY
G7 TBO L. Shergold 8 The Moors, Lydiard Millicent, Swindon SN5 3LE
G7 TBP R. Gill 17 Old Hall Close, Calverton, Nottingham NG14 6PU
G7 TBU S. Fitzjohn 10 Samsons Close, Brightlingsea, Colchester CO7 0RP
G7 TBW T. Polain 22 Hilltop Avenue, Hullbridge, Hockley SS5 6BN
G7 TBX A. Siddle 5 Neneside, Benwick, March PE15 0YF
G7 TCB P. Hubberstey 10 Dove Avenue, Penwortham, Preston PR1 9RP
G7 TCD G. Ward 162 Greenbank Road, Darlington DL3 6ES
G7 TCH Burton On Trent Wireless Club c/o D. Grandfield Hastings College, Arts & Technology, St. Leonards on Sea TN38 0HX
G7 TCQ S. Preston 18 Station Road, Great Wyrley, Walsall WS6 6LQ
G7 TCW C. Haslewood 66 Hunter Road, Cannock WS11 0AF
GI7 TDA J. Mckeever 19 Corrycroar Road, Pomeroy, Dungannon BT70 3DY
G7 TDN A. Baines 60 Norton Drive, Halifax HX2 7RB
GW7 TDQ D. Banister 41 Tynycoed Road, Great Orme, Llandudno LL30 2QA
G7 TDR R. Smith 47 Kiveton Lane, Todwick, Sheffield S26 1HJ
G7 TEA A. Goddard 50 Ardmore Walk, Manchester M22 5QG
GI7 TEB J. Mathers 14 Castlewood Avenue, Coleraine BT52 1JR
G7 TEG G. Fletcher 171 Obelisk Rise, Northampton NN2 8TX
GW7 TEO P. Taylor 8 First Avenue, Prestatyn LL19 7LP
G7 TEP K. Blain 27 Prospect Road, Ash Vale, Aldershot GU12 5ED
G7 TET I. Mowbray 23 Rhodes Avenue, Bishop's Stortford CM23 3JN
G7 TEZ G. Masters 85 Petersham Road, Creekmoor, Poole BH17 7DW
G7 TFA B. Wrampling 18D May Avenue, Canvey Island SS8 7EE
G7 TFG H. Orchel Gildertofts, Ingleby Greenhow, Middlesborough TS9 6JF
GI7 TFK S. Mccormick 74 Belsize Road, Lisburn BT27 4BH
G7 TFL S. Dodds 4 Claremont Road, Wisbech PE13 2JR
GM7 TFN C. Paton 4 Abbeyhill, Dhailling Road, Dunoon PA23 8FG
G7 TFU B. George 43 Claverton Road West, Saltford, Bristol BS31 3DU
G7 TFX J. Patterson 11 Elmway, Chester le Street DH2 2LD
G7 TFY J. Pollard 1 Sawston Close, Ipswich IP2 9DQ
G7 TFZ D. Thomas 27 Kingsbury Road, Coventry CV6 1PW
GW7 TGB C. Bristow 18 Clarendon Close, Chepstow NP16 5TL
G7 TGF A. Gibson 100 Top Row, Darton, Barnsley S75 5JQ
G7 TGG C. Preston 34 Forrester Street Precinct, Walsall WS2 8RE
GI7 TGJ G. Heaney 38 Derryvore Lane, Portadown, Craigavon BT63 5RS
G7 TGK C. Coombe 123, Farleigh Road, Pershore WR10 1JY
G7 TGN P. Dawson 1 Eastfield Road, Bridlington YO16 7DZ
G7 THF M. Thompson 4 Saxony Way, Donington, Spalding PE11 4YA
GI7 THH T. White Shallamar, 3A Park Road, Strabane BT82 8EL
G7 THI F. Gillespie Low Fold, Hoff, Appleby-in-Westmorland CA16 6TA
G7 THJ B. Mills 37 Ashley Road, Hildenborough, Tonbridge TN11 9ED
G7 THK K. Grover 6 Wren Court, Battle TN33 0DU
GI7 THY R. Larimer 131 Carnalea Road, Seskanore, Omagh BT78 2PP
G7 THZ J. Reid 12 Marlay Grove, Crownhill, Milton Keynes MK8 0AT
G7 TIB D. Cross 91 Ilges Lane, Cholsey, Wallingford OX10 9PA
G7 TIE I. Chamberlain 14 High House Avenue, Wymondham NR18 0HY
G7 TIK C. Mcqueen 20 Willow Lane, Stanion NN14 1DT
G7 TIM G. Jones 42 Everard Road, Southport PR8 6NA
G7 TIN R. Martin 2 East View, North Walsham Road, Trunch NR28 0PJ
G7 TIR D. Thomas 5 Minster Drive, Urmston, Manchester M41 5HA
G7 TIV J. Askew 22 Cowslip Grove, Calne SN11 9QQ
GW7 TIX D. Price Sabrina, Pool Road, Newtown SY16 1DW
G7 TIY D. Miller 92 Eton Drive, Wirral CH63 1JS
G7 TJD M. Crosfill Polmennor Farmhouse, Heamoor, Penzance TR20 8UL
GW7 TJM M. Roberts 2 Donnen Street, Port Talbot SA13 1NE
G7 TJQ C. Shaw 30 Southern Way, Stoke-on-Trent ST6 1PX
G7 TJV C. Ho Po Box 900, Fanling Post Office, Hong Kong 0 Hong Kong
G7 TJZ I. Smith 39 Hollingsworth Road, Lowestoft NR32 4AU
G7 TKB F. Coles Le Bouillo, Estampes 32170 France
G7 TKG B. Mersi 4 Westdown Road, Bournemouth BH11 9EQ
G7 TKI R. Pettett 2 Windmill Close, Great Dunmow, Dunmow CM6 3AX
G7 TKM M. Hewitt 17 Farquhar Road, Maltby, Rotherham S66 7PD
G7 TKO M. Smith 11 Martigny Road, Melksham SN12 7PG
G7 TKP M. Hewitt 3 Orchard Rise, Bourne Lane, Reading RG7 5NS
G7 TKT Dr P. Ashford 3 Valley Road, Cheadle SK8 1HY
G7 TKW M. Peppiatt 31E Llverton St, Kentish Town, London NW5 2PE
G7 TLC D. Benton Hawthorn Cottage, Penrose, Wadebridge PL27 7TB
G7 TLD M. Clare 43 Birchfield Close Blackbird Leys Cowley, Oxford OX4 6DL
G7 TLK K. Hemsil Silkcot, Burngullow Lane High Street, St. Austell PL26 7TQ
G7 TLL H. Hodson 1 Chevin Avenue, Borrowash, Derby DE72 3HR
G7 TLR K. Marshall 28 Deerness Grove, Esh Winning, Durham DH7 9LY
G7 TMC M. Conlon 3 Selside, Brownsover, Rugby CV21 1PG
G7 TMF T. Foster 98 Station Road, Carlton, Nottingham NG4 3DA
G7 TMH A. Hunt 63A Toms Lane, Kings Langley WD4 8NJ
G7 TMM A. Kirkham Flat 6, The Laurels, 14 Marlborough Road, Buxton SK17 6RD
G7 TMO P. Foster 218 Stoops Lane, Bessacarr, Doncaster DN4 7JQ
GI7 TMQ J. Bell 72 Coleraine Road, Portrush BT56 8HN
G7 TMR R. Nelson 15 Poplars Close, Burgess Hill RH15 9SZ
G7 TMU V. Swanwick 43 Hormare Crescent, Storrington, Pulborough RH20 4QX
G7 TNO D. Lunn 23 Moynton Close, Crossways, Dorchester DT2 8TX
G7 TNQ the A Team CG c/o M. Mrzyglod 8 Beech Road, Shillingford Hill, Wallingford OX10 8LU
GW7 TNS M. Davies 33 Hazel Mead, Brynmenyn, Bridgend CF32 9AQ
G7 TNT B. Scarsbrook Salix, 96 Moss Lane, Alderley Edge SK9 7HW
G7 TNU P. Sparke 18 Gordon Road, Haywards Heath RH16 1EJ
G7 TNZ M. Wells 37 Water Meadows, Worksop S80 3DF
G7 TOA S. Haigh 2 Locker Avenue, Warrington WA2 9PS
G7 TOB R. Wardell 1 Enfield Close, Norden, Rochdale OL11 5RT
G7 TOF I. Pardington 36 Rivermeads Avenue, Twickenham TW2 5JJ
G7 TOI P. Goodayle 2 Downs Road, Seaford BN25 4QL
G7 TOO P. Crabtree 106 Sagecroft Road, Thatcham RG18 3BF
G7 TOU Mid-Thames Radio Direction-Finding Club c/o M. Mussard 35 Oakfield Gardens, Beckenham BR3 3AY
G7 TOY A. Peet 95 Recreation St., Mansfield NG18 2HP
G7 TOZ J. Whytock 48 Lythe Fell Avenue, Halton, Lancaster LA2 6NL
G7 TPB J. Kilminster 499 Hagley Road West, Quinton, Birmingham B32 2AA
G7 TPD T. Morton 28 Turnfields, Ickford, Aylesbury HP18 9HP
G7 TPG B. Barber 114 Scrogg Road, Newcastle upon Tyne NE6 4HA
G7 TPH R. Hand 70 Flansham Lane, Bognor Regis PO22 6AH
GI7 TPO G. Hodgkinson 675 Crumlin Road, Belfast BT14 7GD
G7 TPS B. Seed 10 South Place, Calne SN11 0JA

G7 TPW A. Grigor 48 Valebridge Drive, Burgess Hill RH15 0RW
G7 TQA D. Legge 28 Dresser Road, Prestwood, Great Missenden HP16 0NA
G7 TQC C. Banister York Avenue, East Cowes PO32 6JT
G7 TQE T. Brown 138 Holmesdale Road, South Norwood, London SE25 6HY
G7 TQT R. Denton 37 Tenby Road, Cheadle Heath, Stockport SK3 0UN
G7 TQU J. Orritt Pound Farm, Church Lane, Rudford, Gloucester GL2 8DT
GU7 TQX E. Grisley Les Clercs, Contree Des Clercs, St. Pierre Du Bois GY7 9DA Guernsey
G7 TRB P. Stevenson 50 Field Lane, Beeston, Nottingham NG9 5FJ
G7 TRG K. Liddle 36 Vicarage Lane, Grasby, Barnetby DN38 6AU
G7 TRL D. Perry-Wright 1 Marlfield Close, Preston PR2 7AL
G7 TRM K. White 20 Agnes Close, Bude EX23 8SB
G7 TRV D. Glover 59 Shelley Close, Abingdon OX14 1PP
G7 TSB E. Jones 26 Wood End, Bluntisham, Huntingdon PE28 3LE
G7 TSO K. Jones 87 Meade Street, Glen Innes 2370 Australia
G7 TSP C. Leman 92 Queens Crescent, Eastbourne BN23 6JP
G7 TSQ J. Stafford 89 Mossley Road, Ashton-under-Lyne OL6 9RH
G7 TTH G. Quint 4 Gibson Grove, Malvern WR14 1NX
GI7 TTO D. Dunlop 63 Cloyfin Road, Coleraine BT52 2NY
G7 TTP R. Hughes 5 Cornwallis Close, Bromham, Bedford MK43 8LG
GM7 TTU R. Emmott 5 Thorter Loan, Dundee DD1 3AW
GW7 TTX M. Tahla Penrhiw, Ffestiniog, Blaenau Ffestiniog LL41 4PN
G7 TTY A. Hubbard 40 Field Drive, Shirebrook NG20 0BP
GM7 TUD J. Pedley 4 Tinwald View Back Road, Locharbriggs, Dumfries DG1 1RT
G7 TUG N. Mitchell 49 Kersey Road, Felixstowe IP11 2UL
G7 TUH P. Ferguson 152 Chestnut Drive, Sale M33 4HR
G7 TUK J. Steel 10 Green Courts, Winterton-On-Sea, Great Yarmouth NR29 4AQ
G7 TUM J. Moore Waveney, Abbotts Way, Bush Estate, Norwich NR12 0TA
G7 TUP R. Irwin 7 Hameau Des Peupliers, Rue Du Vert Pre, Lys Lez Lannoy 59390 France
GW7 TUQ B. Forhead Autumn Villa, Chapel Street Newbridge, Wrexham LL14 3JH
G7 TUS R. Munden 2 Hain Villa, Forest Road, Ruardean GL17 9XR
G7 TUV J. Hewitt 6 Crawley Walk, Warley, Cradley Heath B64 5EX
G7 TVL E. Roberts 800 Walsall Road, Great Barr, Birmingham B42 1EU
G7 TVQ J. Gilbert Mills Caravan, Garland Cross Kings Nympton, Umberleigh EX37 9TT
G7 TVT L. Whiteside 8 The Orchards, Eaton Bray, Dunstable LU6 2DD
GI7 TVV A. Mccready 25 Glendun Park, Bangor BT20 4UX
G7 TWA D. Bullard 20 Chaney Road, Wivenhoe, Colchester CO7 9QZ
G7 TWC M. Ruttenberg 90 Heath View, London N2 0QB
G7 TWJ P. Edwards Cleveland, Blackberry Road, Lingfield RH7 6NQ
GM7 TWM I. Hipkin 1 Maclennan Place, Dufftown, Keith AB55 4EF
G7 TWU F. Clarkson 313 Normanby Road, Middlesbrough TS6 0BQ
G7 TWW C. Papaioannou 2 Temple Lane, Temple, Marlow SL7 1SA
G7 TXF A. Scott 60 Lowndes Park, Driffield YO25 5BG
G7 TXR S. Loyd Maple House, Pangbourne Road, Reading RG8 8LN
G7 TXV C. Wood 20 Tedworth Close, Guisborough TS14 7PR
G7 TXW M. Oliver 14 Harwood Road, Gosport PO13 0TT
G7 TXX D. Williams 57 Hillside Avenue, Kidsgrove, Stoke-on-Trent ST7 4LW
G7 TYB J. Hawley 89 Mansfield Avenue, Denton, Manchester M34 3NS
G7 TYH S. Furminger 9 Amberley Gardens, Wokingham RG41 1LN
G7 TYJ J. Pennington 6 Penny Hapenny Crt, Atherstone CV9 2AA
G7 TYO S. Birtwhistle 16 Woodley Street, Bury BL9 9HZ
G7 TYP B. Cook 40 Preston Avenue, Alfreton DE55 7JY
G7 TYR Sutton & Cheam RS c/o D. Collins 71 Trench Road, Tonbridge TN10 3HG
G7 TYT M. Claxton 9 Thompson Avenue, Beverley HU17 0BG
G7 TZB D. Vincent 6 Nathan Gardens, Poole BH15 4JZ
G7 TZD P. Jones 361 Wellingborough Road, Rushden NN10 6BA
GW7 TZG P. Kelly 33 Yeo St., Resolven, Neath SA11 4HS
GW7 TZI M. Tonkin 185 Pentregethin Road, Cwmbwrla, Swansea SA5 8AU
G7 TZN S. Buckingham 8 Tedder Avenue, Buxton SK17 9JU
G7 TZO C. Turner 308 North Road, Yate, Bristol BS37 7LL
G7 TZQ R. Darby 25 Bramley Road, Marsh Lane, Sheffield S21 5RD
G7 TZU T. Stalker 172 Kirkby Road, Barwell, Leicester LE9 8FS
G7 TZV G. Broughton 111 Broadway, Manchester M40 3NL
G7 TZW R. Wheatley 288 Bennett Street, Long Eaton, Nottingham NG10 4JA
G7 TZX D. Johnson 12 Heron Close, Broughton, Chester CH4 0RL
G7 TZZ J. Eyre 41 Wood Street, Kettering NN14 1BG
GM7 UAC E. Edwards 12 Highfield Place, Girdle Toll, Irvine KA11 1BW
G7 UAI UK DX c/o P. Blizzard 12 Hilton Road, Malvern WR14 3NP
G7 UAK S. Hunter 30 Adelaide Street, Barrow-in-Furness LA14 5TX
G7 UAL D. Witherall 221 Poynters Road, Dunstable LU5 4SH
G7 UAT J. Johnson 10 Croft Avenue, Newcastle ST5 8EY
G7 UAV I. Morris 60 Moorland Avenue, Lincoln LN6 7RD
G7 UAY D. Pickering 28 Keepers Wood Way, Chorley PR7 2FU
G7 UBB E. Knight 170 Roderick Avenue North, Peacehaven BN10 8AW
G7 UBD T. Thomas 166 Bluebell Road, Southampton SO16 3LP
G7 UBK G. Reddecliffe 5 Stanley Close, Dymchurch, Romney Marsh TN29 0TY
G7 UBO J. Pearson 22 Ashburnham Close, Norton, Doncaster DN6 9HJ
G7 UBP C. Howard 144 Fairfield Road, Heysham, Morecambe LA3 1LR
G7 UBQ T. Bray The Dell, Burnlee Road, Holmfirth HD9 2LF
G7 UBX P. Pleydell 6 The Croft, Meriden, Coventry CV7 7NQ
GI7 UBY C. Lunnon 3 Parkfield, Crumlin BT29 4SG
G7 UCB P. Hudson 47 Hall Farm Road, Duffield, Belper DE56 4FJ
G7 UCG J. Woodward 108 Tamworth Road, Sutton Coldfield B75 6DH
G7 UCL S. Dixon 5 Swanmore Road, Havant PO9 4LG
G7 UCN A. Allport 55 Byrds Lane, Uttoxeter ST14 7NF
G7 UCO D. Reed1 8 Wolverstone Drive Hollingdean, Brighton BN17FB
G7 UCP D. Hornby 7 Shawfield Grove, Rochdale OL12 7SU
G7 UCR K. Yeo 48 Great Goodwin Drive, Guildford GU1 2TY
GI7 UCS M. Grainger 1 Knocknamoe Bungalows, Omagh BT79 7LA
G7 UCT B. Lord 13 Park Ave, Norden, Timperley WA14 5AQ
G7 UCZ D. Evans 3 Dalkeith Close, Bransholme, Hull HU7 5AS
G7 UDE D. Clark 32 Laburnum Grove, Burstead Close, Brighton BN1 7HX
G7 UDJ C. Edwards 6 Blacksmiths Close, Nether Broughton, Melton Mowbray LE14 3EW
G7 UDM D. Bonfield 49 Linden Grove, Chandler'S Ford, Eastleigh SO53 1LE
G7 UDU J. Selwyn 5 Main Road, Billockby, Great Yarmouth NR29 3BG
G7 UDX C. Harris 8 Trelawney Rise, Callington PL17 7PT
G7 UEC D. Denyer 85 Highlands Road, Horsham RH13 5ND
G7 UEI D. Longhurst Burston, Wood Road, Hindhead GU26 6PZ
G7 UEJ S. Kitchen 344 Windward Way, Castle Bromwich, Birmingham B36 0UH
G7 UEK A. Jones 60 Heywood Drive, Starcross, Exeter EX6 8SD
G7 UEL R. Dean Sandshadow, Stow Road, King's Lynn PE34 3PF
G7 UET A. Levy 29 Ferndale Avenue, Reading RG30 3NQ
G7 UEV C. Fox 3 Manor Drive, Wragby, Market Rasen LN8 5SL
G7 UEX P. Cardwell 2 Hayfield Place, Sheffield S12 4XH
G7 UFF L. Brackstone 276 Ladyshot, Harlow CM20 3EY
G7 UFI B. Courtenay 251 Smeeth Road Marshland St. James, Wisbech PE14 8ES
GM7 UFN T. Graham 265 Gilmartin Road, Linwood, Paisley PA3 3SU
G7 UFT R. Elliott 16 Prince Philip Road, Colchester CO2 8PA
G7 UFV D. Riseborough 2 The Barn, Grigsons Wood, Norwich NR16 2LW
G7 UFW N. Brickwood 4 Patterson Close, Northampton NN3 3PE
G7 UGA M. Turner 14 The Rookery, Barrow Upon Soar, Loughborough LE12 8JZ
G7 UGB M. Howard Netherwood, Shortthorn Road, Stratton Strawless, Norwich NR10 5NU
G7 UGC A. Fellows 343 Wake Green Road, Moseley, Birmingham B13 0BH
G7 UGR D. Barnett 81 Bankside West Lynn, King's Lynn PE34 3JH
G7 UGW J. Smith 32 Aberdeen Street, Hull HU9 3JU
G7 UGY N. Thornley 7 Hisehope Close, Startforth, Barnard Castle DL12 9BZ
G7 UHE G. Tiller 15 Woodlands Gardens, Romsey SO51 7TE
G7 UHG D. Tropman 91 Reindeer Road, Fazeley, Tamworth B78 3SW
G7 UHL S. Yuill 20 Buttercup Court Deeping St. James, Peterborough PE6 8TF
G7 UHS C. Jewell 43 Rannoch Road, Bristol BS7 0SA
G7 UHT C. Griffiths 33 Westwood Road, Newport PO30 1TD
G7 UHW M. Mcdermott 4 Tolcairn Court, 28 Lessness Park, Belvedere DA17 5BT
G7 UHX B. Anderson 22 The Drive, Clacton-on-Sea CO15 4NN
G7 UHY R. Blewitt 62 Vicarage Road West, Dudley DY1 4NP
G7 UID S. Clarke 75 Beaumont Street, Netherton, Huddersfield HD4 7HE
G7 UII C. Savage 24 Park Hill, Awsworth, Nottingham NG16 2RD
G7 UIO N. Johnson 64 Rotten Row, Pinchbeck, Spalding PE11 3RH
GI7 UIP K. O'Reilly 52 Ballydoolagh Road, Enniskillen BT744JZ
GJ7 UIT C. Totty Flat 4, Beech Court, Woodlands Apartments, La Rue Des Cotils, Grouville, Jersey JE3 9AY
G7 UIU S. Palmer 54 Hawthorn Road, Exeter EX2 6EA
GW7 UIZ J. Hughes Maes Y Ffynnon, 7 Meadow Gardens, Llandudno LL30 1UW
G7 UJC G. Taylor 34 Hockley Road, Poynton, Stockport SK12 1RW
GM7 UJJ J. Scott 1 Carrick Knowe Drive, Edinburgh EH12 7EB
GM7 UJO S. Maxwell 24 Castle Drive, Airth, Falkirk FK2 8GD
G7 UJS C. Sheffield 1 Benson Close Perton, Wolverhampton WV6 7LU
G7 UJT S. Dransfield Gardener Ground House, West End, Goole DN14 8RW
G7 UJY M. Poole 184 Woodgates Lane, Swanland, North Ferriby HU14 3PR
G7 UKA T. Collier 23 The Riggs, Brandon, Durham DH7 8PQ
G7 UKF M. Ellis 64 Coppice Drive, Dordon, Tamworth B78 1QZ
G7 UKK A. Firth 59 Station Road, Shepley, Huddersfield HD8 8DS
G7 UKN K. Riley 27 Limewood Close, Blythe Bridge, Stoke-on-Trent ST11 9NZ
G7 UKR M. Blackburn 36 Mardale Grove, Barrow-in-Furness LA13 9QG
G7 ULC C. Probert 25 Elizabethan Way, Rugeley WS15 2EE
GI7 ULG S. Murdoch 78 Landgarve Manor, Crumlin BT29 4SF
G7 ULJ P. White 1 Hazel Hill Place, Nottingham NG5 5FA
G7 ULL P. Craig 6 Marsham Close, Chislehurst BR7 6JD
G7 ULN J. Grundy 47 Northiam Road, Eastbourne BN20 8LP
G7 ULS K. Hurst 94 East Park, Harlow CM17 0SB
G7 ULW S. Widdowson 45 Limes Avenue, Staincross, Barnsley S75 6JP
G7 UMA N. Tindall 87 The Grove, Marton-In-Cleveland, Middlesbrough TS7 8AN
G7 UMF D. Griffiths Home Farm House Cottage, Leebotwood, Church Stretton SY6 6LX
GW7 UMS K. Keepin 45 Heol Y Groes, Cwmbran NP44 7LT
GW7 UMW A. Banner 14 Carlson Drive, Wrexham LL11 2YF
G7 UMY D. Rockliffe 3 Hewell Lane, Barnt Green, Birmingham B45 8NZ
G7 UNB A. Bevington 54 Pheasant Road, Smethwick B67 5PD
G7 UNU N. Davies 16 St. Leonards Close, Scole, Diss IP21 4DW
GW7 UNV E. Jones Crungoed Farm Llanbister Road, Llandrindod Wells LD1 5UR
G7 UNW N. Othen 234A Regents Park Road, London N3 3HP
G7 UNZ W. Scott Rose Brae, Lazonby, Penrith CA10 1AJ
G7 UOD H. Golding 238 Greenkeepers Road, Great Denham, Bedford MK40 4GW
GW7 UOH S. Lupton Egryn, Ffordd Dewi Sant Nefyn, Pwllheli LL53 6EA
G7 UOL R. Bennion 3 Dorrington Close, Ruskington, Sleaford NG34 9EQ
G7 UOQ N. Birt 60 Church Road Woodley, Reading RG5 4QB
G7 UOS B. Yates 2 Larchwood, Countesthorpe, Leicester LE8 5RH
G7 UOU A. Colville Bachefield House, Kimbolton HR6 0EP
GM7 UPD C. Edwards The Bennachie Craft Centre Chapel Of Garioch, Inverurie AB51 5HE
G7 UPL S. Northeast 143 Henderson Road, Southsea PO4 9JE
G7 UPN C. Jackson 2 Northway, Guildford GU2 9SB
G7 UPP R. Hall Greenviews, Lower Kingsbury, Sherborne DT9 5ED
GI7 UPQ M. Cunningham 4 Garvaghy Road, Portglenone, Ballymena BT44 8EF
GI7 UPU F. Gillespie 33 Clonliffe Park, Londonderry BT48 8NT
G7 UPZ I. Sansom 26 Finedon Road, Wellingborough NN8 4EB
G7 UQA D. Haigh 29 Victoria Grove, Wakefield WF2 8UP
GW7 UQJ M. Mee Cerrig Gwynion, Penisarwaun, Caernarfon LL55 3PW
GM7 UQM M. Horne 10 Blair Place, Kirkcaldy KY2 5SQ
G7 UQQ T. Wakeling Coreopsis, Church Road, Peldon, Colchester CO5 7PT
G7 UQS S. Hill 24 Mason Road Worcestershire, Redditch B97 5DA
G7 UQV Friskney + East Lincolnshire Communications Club c/o M. Willoughby 30 Kipling Road, Ipswich IP1 6EW
GI7 UQW B. Neill 81 Orangefield Road, Belfast BT5 6DD
GI7 URC A. Brown 3 Clara Road, Belfast BT5 6FN
G7 URG C. Gaunt 15 Bar House Lane Utley, Keighley BD20 6HA
G7 URJ J. O'Brien 45 Rossall Promenade, Thornton-Cleveleys FY5 1LP
G7 URL D. Foster Pentlow, Crowle Bank Road, Scunthorpe DN17 3HZ
G7 URM T. Heartfield 69 Great Thrift, Petts Wood, Orpington BR5 1NF
G7 URP D. Palmer Edison House, Bow Street, Great Ellingham, Attleborough NR17 1JB
G7 URR S. Easter Flat 11, Saxon Court, Hitchin SG4 9TB
G7 URS R. Bird 9 Orchard Lane, Wembdon, Bridgwater TA6 7QY
G7 URT C. Langham 9 Laurence Close, Shurdington, Cheltenham GL51 4SZ

G7 URW N. Tucker 15 Mount Pleasant Road, Dawlish Warren, , Dawlish, EX7 0NA
GI7 USA A. Niblock Apartment 1, 6 Glenburn Court, Glynn, Larne BT40 3FF
G7 USB J. Ainsworth 42 Buttfield Road, Hessle HU13 0AS
GM7 USC G. Mckelvie 37 Carskeoch Drive, Patna KA67LR
G7 USG J. Sutherland 4 Cherbury Close, Bracknell RG12 9HT
G7 USI R. Hayselden 400 Heath End Road, Nuneaton CV10 7HG
G7 USJ T. Cogan 11 Highgrove Walk, Weston-Super-Mare BS24 7EF
G7 USM K. Reavill 11 Clarence Road, Beeston, Nottingham NG9 5HY
G7 USO N. Brodt-Savage 5 Granville Road, Westfield, Woking GU22 9ND
G7 USQ B. Siddall 6 Delside Avenue, Manchester M40 9LF
G7 USV Dr D. Atkins 14 Ryde Place, Lee-on-the-Solent PO13 9AU
G7 USX M. Woollard Barnside, Colchester Road, Colchester CO7 7EG
G7 UTB H. Scott-Telford 9 Squires Close, Rochester ME2 2TZ
G7 UTC M. Bean Ashmore, Belle Vue Road, Sudbury CO10 2PP
GM7 UTD D. Forrest 15 Invergarry Avenue, Thornliebank, Glasgow G46 8UR
G7 UTE B. Spencer 80 Horncastle Road, Boston PE21 9HY
G7 UTG J. Dodds 84 Borrowdale Avenue, Walkerdene, Newcastle upon Tyne NE6 4HL
G7 UTH R. Banks 50 Vale Road, Portslade, Brighton BN41 1GG
G7 UTI G. Ducros 21 Wardlow Gardens, Plymouth PL6 5PU
G7 UTR G. Kelsall 3 Raven Street, Bingley BD16 4LB
G7 UTS M. James 7 Greenfield Park, Portishead, Bristol BS20 6RG
G7 UTY C. Lewis 3 Jacobs Close, Stantonbury, Milton Keynes MK14 6EJ
G7 UUA P. Matthews 6 West Road, Halstead CO9 1EH
G7 UUB F. Gibbs 62 Wenvoe Avenue, Bexleyheath DA7 5BT
G7 UUC M. West 69 Frampton Crescent, Bristol BS16 4JD
G7 UUD K. Matthews 6 West Road, Halstead CO9 1EH
G7 UUG N. Griffiths 125 Coleridge Way, Crewe CW1 5LF
GW7 UUH A. Hughes 30 Liddell Drive, Llandudno LL30 1UH
G7 UUK M. Hooper 12 Meare, Dunster Crescent, Weston-Super-Mare BS24 9DY
G7 UUL A. Riggs Lower House, Stockland Bristol, Bridgwater TA5 2PY
G7 UUN R. Wood 7 Lilac Grove, Luston, Leominster HR6 0EF
G7 UUO J. Harbidge Low Balk Farm, Finkle Street, Bishop Burton, Beverley HU17 8QP
G7 UUP J. Chapman 8 Oakfield Court, Stanley Common, Ilkeston DE7 6XB
G7 UUQ R. Moorhouse 66 Wicor Mill Lane, Fareham PO16 9EG
G7 UUT A. Wilson 36 Davey Crescent, Great Shelford, , Cambridge CB225JF
G7 UUW S. Hearn 28 Neithrop Avenue, Banbury OX16 2NF
G7 UVB D. Anstie 20 Keyes Road, Norwich NR1 2JX
G7 UVF G. Cheetham 35 South Park Grove, New Malden KT3 5BZ
G7 UVL D. Croot 58 Dixie Street, Jacksdale, Nottingham NG16 5JZ
G7 UVN I. Cross 25 Yatesbury Avenue, Blakelaw, Newcastle upon Tyne NE5 3SZ
GW7 UVO R. Moss 10 Llys Eleanor Shotton Lane, Shotton, Deeside CH5 1EH
G7 UVP Luton Vhf Group c/o J. Swanwick Ramblers, Clarks Farm Road, Chelmsford CM3 4PH
GM7 UVS J. Graham 265 Gilmartin Road, Linwood, Paisley PA3 3SU
G7 UVV I. Perry Meadow Cottage, Mill Lane, Halstead CO9 2NW
G7 UVW Dr D. Mills 11 Northfield Road, Dagenham RM9 5XH
G7 UVY C. Carr 10 Bonds Road, Hemblington, Norwich NR13 4QF
G7 UWB B. Wright 2 Butterfly Gardens, Rushmere St. Andrew, Ipswich IP4 5TF
G7 UWC C. Wright 32 The Pastures, Rushmere St. Andrew, Ipswich IP4 5UQ
G7 UWE P. Smith 12A Sandicroft Place, Preesall, Poulton-le-Fylde FY6 0PB
G7 UWG R. Hancox 13 Regnum Close, Eastbourne BN22 0XH
G7 UWI M. Jones 41 Milton Brow, Weston-Super-Mare BS22 8DD
GM7 UWL D. Cottage Greystones, Wick KW1 4XP
G7 UWO G. Holland 15 Rollis Park Road, Oreston, Plymouth PL9 7LU
G7 UWP P. Groves Flat 3, County Chambers Station Road, Gloucester GL1 1DH
G7 UWS D. Brunt 91 Shaftesbury Avenue, Feltham TW14 9LW
G7 UWV I. Brown 35 Lees Terrace, Bilston WV14 8EL
G7 UWW C. Harding 1 Saddleton Grove, Saddleton Road, Whitstable CT5 4LY
G7 UWZ D. Pooley 25 Wharncliffe Road, Highcliffe, Christchurch BH23 5DB
G7 UXD R. Wade The Limes, Hunston, Bury St. Edmunds IP31 3EL
G7 UXH E. Gaunt 71 Hebron Road, Stokesley TS9 5DF
G7 UXK S. Hedges 25 Rudland Close, Thatcham RG19 3XW
G7 UXO P. Stokes Pl17 8Fd, Callington PL17 8FD
G7 UXQ J. Manwaring 17 Pleasant View, Burnhope, Durham DH7 0BA
G7 UXR M. Taylor 27 Lincoln Road, Newark NG24 2BU
G7 UXU H. Andrews 24 Belvoir Road, Widnes WA8 6HR
GW7 UXY R. Williams 16 Tir Dafydd, Pontyates, Llanelli SA15 5TP
G7 UYB C. Major 17 Jubilee Cottages, Station Road, Bedford MK43 0PN
G7 UYI B. Williamson 12 Middleton Close, Southampton SO18 2FP
G7 UYJ J. Jardine 41 Charles Drive, Anstey, Leicester LE7 7BH
G7 UYT J. O'Toole 4 Lindisfarne Road, Dagenham RM8 2RA
G7 UYW J. Kitchener 101 Highfield Road, Tring HP23 4DS
G7 UZA J. Rodinson 21 Graylands Road, Liverpool L4 9UG
G7 UZG A. Mcwilliam 43 Hylder Close, Swindon SN2 2SL
G7 UZI P. Pullen 12 Kimpton Road, Sutton SM3 9QJ
G7 UZN D. Dawson 12 Thurlow Terrace, Kentish Town, London NW5 4JB
G7 UZO D. Lock 1 Heaton Avenue, Huddersfield HD5 0LJ
G7 UZS N. Thompson 30 Dene View, Ashington NE63 8JF
G7 UZY A. Musther 2 Fakenham Close, Lower Earley, Reading RG6 4AB
G7 VAB M. Richards Sunnymead, Ardley End, Bishop's Stortford CM22 7AJ
G7 VAD M. Beeney Oakville Farm, Lewes Road, Uckfield TN22 5JH
G7 VAE J. Beeney 37 Coppice Avenue, Eastbourne BN20 9PP
G7 VAG G. Podmore 9 Pendlebury Street, Warrington WA4 1TU
G7 VAS M. Kay 12 The Crescent, Ashton-On-Ribble, Preston PR2 1JP
G7 VAY R. Dingle 87 Eighth Avenue, Bridlington YO15 2NA
G7 VBD M. Ennis 1 Nairn Road, Cramlington NE23 1RQ
GW7 VBE D. Harris 29 Queen Street, Blaengarw, Bridgend CF32 8AH
G7 VBF J. Barwell Flat 9, Corinth House 33 Barley Lane, Ilford IG3 8XE
G7 VBJ D. Wager 162 Harvest Fields Way, Sutton Coldfield B75 5TJ
G7 VBL J. Munday 20 Highcroft, Wood Road, Hindhead GU26 6PW
G7 VBN B. Richards 52 Ripon Way Carlton Miniott, Thirsk YO7 4LR
G7 VBU D. Firth 5, Birchfield Grove, Skelmanthorpe, Huddersfield HD8 9BS
GW7 VBY Dr D. Morrison-Smith 1 Neptune House, Upper Corris, Machynlleth SY20 9BQ
G7 VBZ P. Bunce 66 Berry Park, Saltash PL12 6EN
G7 VCB L. Tooze Flat 1, 91 Harbour Road, Seaton EX12 2NJ
G7 VCE M. Flack 31 Harebell Close, Cambridge CB1 9YL
G7 VCF J. O'Donnell 3 Linden Avenue, Altrincham WA15 8HA
G7 VCG I. Firby 19 St. Georges Drive, Manchester M40 5HL
G7 VCJ C. Hansford 14 Parsonage Crescent, Castle Cary BA7 7LT
G7 VCK M. Robertson 67 Oatland Gardens, Leeds LS7 1SL
G7 VCN R. Bawley 52 Pitville Avenue, Liverpool L18 7JG
G7 VCP P. Stubbs 2 Cynthia Road, Runcorn WA7 4TX
GI7 VCR S. Robertson 32 Castle Meadows, Carrowdore, Newtownards BT22 2TZ
GM7 VCV A. Brown 96 Barony Terrace, Kilbirnie KA25 6DB
G7 VCY D. Seymour 24 Farley Dell, Coleford, Radstock BA3 5PJ
G7 VCZ Sqdn. Ldr. P. Weaver 42 Erithway Road, Coventry CV3 6JT
G7 VDA I. Singer 197 Rosalind Street, Ashington NE63 9BB
G7 VDD P. Mcgowan Rua Oliva De La Frontera, 29, 1Esq, , Caldas Da Rainha 2500-886 Portugal
G7 VDH J. Brook 2 New Laithe Bank, Holmfirth HD9 1HL
G7 VDI D. Norris Flat 11, 10 Cromartie Road, London N19 3SJ
G7 VDJ Dr S. Henry Hertford College, Catte Street, Oxford OX1 3BW
G7 VDK G. Taylor 4 Brown Crescent, Eighton Banks, Gateshead NE9 7EX
GM7 VDL W. Steele 35 Devlin Court, Whins Of Milton, Stirling FK7 0NP
G7 VDN H. May 18 Pennant Hills, Bedhampton, Havant PO9 3JZ
G7 VDQ D. Butterworth 6 Fir Grove, Weaverham, Northwich CW8 3JD
G7 VDS M. Hudson 5 Berkeley Court High Street, Cheltenham GL526DA
G7 VDT G. Baines 20 Whitehall Rise, Wakefield WF1 2AL
G7 VDU A. Marston 111 Averil Road, Leicester LE5 2DE
G7 VDV N. Keech 14 Simpson Court, Ashington NE63 9SD
G7 VDX S. Taverner 8 The Rye Lea, Droitwich WR9 8SS
G7 VEB D. Wilson 210 Stanks Lane South, Swarcliffe, Leeds LS14 5PD
GM7 VEC S. Fearn 2 Carabhat, Carinish HS6 5HR
G7 VEE A. Saunders 25 Southern Drive, South Woodham Ferrers, Chelmsford CM3 5NY
G7 VEF R. Parkin 17 Roberts Road, Watford WD18 0AY
GW7 VEH J. Humphrey Bryn Ebbw, Beaufort Hill, Ebbw Vale NP23 5QR
G7 VEI A. Stripp 87 Elthorne Park Road, London W7 2JH
G7 VEL C. Lee 59 Quarry Gardens, Ludlow SY8 1RE
G7 VEX N. Hindle 19 Barkway Road, Royston SG8 9EA
G7 VEY J. Martin 3 The Rise, Calne SN11 0LQ
G7 VFA J. Juggins 5 Charter Close, Helston TR13 8SR
G7 VFC Paisley ARC c/o O. Dewberry The Stables, Barrack Street, Manningtree CO11 2RB
G7 VFE R. Wills 14 Penwood Heights, Penwood, Highclere, Newbury RG20 9EY
GW7 VFJ S. Magee Lle Da, Cefn Bychan Road, Mold CH7 5EL
G7 VFL K. Sherman 12 Portland Drive, Stourbridge DY9 0SD
G7 VFQ A. Latham 22 Moorland Road, Leek ST13 5BW
GM7 VFR J. Smith 28 Tollerton Drive, Irvine KA12 0QE
G7 VFU A. Mullord 296 City Way, Rochester ME1 2BL
G7 VFV G. Somers 16 Button Drive, Newquay TR7 3FB
G7 VFX R. Watts-Read 43 Whyteleafe Hill, Whyteleafe CR3 0AJ
G7 VFY S. Walters 16 North Lodge, 46 Somerset Road, New Barnet, Barnet EN5 1RJ
G7 VGA S. Bonney 22 Gordon Drive, Abingdon OX14 3SW
GW7 VGB D. Beynon 129 Eureka Place, Ebbw Vale NP23 6LN
G7 VGC B. Goody Flat 31, Homeweave House, Robinsbridge Road, Colchester CO6 1UL
G7 VGE R. Teague 18 Aspen Close, Great Blakenham, Ipswich IP6 0HQ
G7 VGH M. Smith 46 Bentham Way, Ely CB6 1BS
G7 VGJ A. Cole 58 Stradbroke Drive, Chigwell IG7 5QZ
G7 VGK W. Parrett 5 Coniston Close, Walton, Liverpool L9 0NG
G7 VGL D. Pemberton 12 Victor Road, Thatcham RG19 4LX
G7 VGM S. Taylor 22 Raby Square, Hartlepool TS24 8HH
G7 VGN A. Chamberlain 7 Mccalmont Way, Newmarket CB8 8HU
G7 VGO Key Performance Indicators c/o A. Hurst 12 Spilsby Close, Hartlepool TS25 2RD
GI7 VGR C. Dorrian 47 Albany Drive, Carrickfergus BT38 8BF
G7 VGT P. Schranz 42 South Townside Road, North Frodingham, Driffield YO25 8LE
G7 VGX H. Dolman 28 The Downs, Middleton, Manchester M24 1TJ
G7 VGY the Royal Naval (Ulster) ARC c/o D. Childs 7 Grange Road, East Cowes PO32 6EA
G7 VHC C. Spires 5 Springhead, Sutton Veny, Warminster BA12 7AG
GW7 VHD A. Jones 57 Dinerth Road, Rhos On Sea, Colwyn Bay LL28 4YG
G7 VHF East Anglian Six Meter Group c/o A. Garry-Durrant Casa Santosa Llano Del Espino, Albox 4800 Spain
G7 VHG S. Bryan 18 Whalley Crescent, Wroughton, Swindon SN4 9EP
G7 VHJ P. Gow 11 Rodley Square, Lydney GL15 5AZ
G7 VHN Mid-Beds Raynet Group c/o J. Hart 35 Aintree Close, Uxbridge UB8 3HS
G7 VHO Wordsley ARC c/o B. Hart 35 Aintree Close, Uxbridge UB8 3HS
GM7 VHQ I. Helie 22 Mcclue Road, Renfrew PA4 9BL
G7 VHS J. Mclaughlin 16 East Street, Batley WF17 5QY
G7 VHU D. Beastall 11 Hopwood Bank, Horsforth, Leeds LS18 5AW
G7 VHX J. Golding 65 Longworth Avenue, Tilehurst, Reading RG31 5JU
G7 VHZ E. Gilowski 126 Owlsmoor Road, Owlsmoor, Sandhurst GU47 0ST
G7 VIB D. Airs Cornerways Cottage, Poffley End, Witney OX29 9UW
G7 VIE J. Cooper 9 Highfield Crescent, Halesowen B632BD
G7 VIG A. Smith 19 Gibsons Gardens, North Somercotes, Louth LN11 7QH
G7 VIH P. Wilson 117 Naseby Road, Kettering NN16 0LL
G7 VIK N. Higgins 6 Larksfield Avenue, Bournemouth BH9 3LP
G7 VIL G. Mason 6 Willowtree Avenue, Gilesgate Moor, Durham DH1 1EB
G7 VIP F. Marston 1 Weaver Road, Leicester LE5 2RL
G7 VIR A. James 19 Coach Lane, Redruth TR15 2TP
G7 VIV A. Nussey 9 Brent Street, Brent Knoll, Highbridge TA9 4DU
GI7 VIW Lincolnshire Poachers CG c/o A. Harvey 5 Kilmaine Road, Bangor BT19 6DT
G7 VIX I. Rogers 8 Long Copse Lane, Emsworth PO10 7UL
G7 VIY A. Harper 3 Eskdale Crescent, Blackburn BB2 5DT
G7 VJA K. Sharman 1 The Greenwoods, Hartland, Bideford EX39 6JA
G7 VJD D. Skidmore Weavers, Kingsdale Road, Berkhamsted HP4 3BS
G7 VJE C. Rohrer Alpenrose, , Bishop's Stortford CM22 7TP
G7 VJG P. Veitch 58 Beaulieu Close, Toothill, Swindon SN5 8AQ
G7 VJH T. Scanlon 11 Caterhouse Road Framwellgate Moor, Durham DH1 5HP
G7 VJI G. Dawes 19 Benbridge Avenue, Bournemouth BH11 9HN
G7 VJJ T. Wood 39 Baker Road, Bournemouth BH11 9JD
GW7 VJK N. Cole Tycoch, Llandovery SA20 0UP
G7 VJM C. Margetts 16 Lahn Drive, Droitwich WR9 8TQ

G7 VJQ C. Radford 12 Homewood Drive, Kirkby-In-Ashfield, , Nottingham NG17 8QB
G7 VJT S. Tibbetts 113 Highfield Crescent, Halesowen B63 2AY
G7 VJU R. Jones 20 Carnoustie Close, Southport PR8 2FB
G7 VJY T. Hill 15 Catkin Walk, Rugeley WS15 2NS
G7 VKA K. Wandless 1 Lee Moor Cottages, Rennington, Alnwick NE66 3RL
G7 VKB A. Cunnington 131 Colson Road, Loughton IG10 3QY
G7 VKG M. Gibson 6 Harrison Road, Mansfield NG18 5RG
G7 VKJ C. Buckley 14 Sunny Drive, Prestwich, Manchester M25 3JJ
GM7 VKN R. Beharie Isengard, Norseman Village, Firth KW17 2NY
G7 VKY B. Shrimpling 45 Fairmont Road, Grimsby DN32 8DZ
G7 VLA M. Sandham 7 Mill Close, Caverswall, Stoke-on-Trent ST11 9HA
G7 VLB S. Kirkbright 48 Plant Crescent, Stafford ST17 4EH
GM7 VLC A. Haines 164 North High St., Musselburgh EH21 6AR
G7 VLD K. Howard 43 Hazeldell, Watton At Stone, Hertford SG14 3SN
G7 VLF N. Faiz 48 Cox House, Field Road, London W6 8HN
G7 VLH C. Hill 14 Blenheim Close, Chandler'S Ford, Eastleigh SO53 4LD
G7 VLJ E. Baker 29 Ashcroft Road, Ipswich IP1 6AB
G7 VLL J. Woodhouse 5 Dolphin Villas, Hazlerigg, Newcastle upon Tyne NE13 7NG
G7 VLR Vintage & Military ARS c/o A. Holmes 5 Launde Park, Market Harborough LE16 8BH
GM7 VLZ A. Pearce 105 Gyle Park Gardens, Edinburgh EH12 8NQ
G7 VME P. Schofield 22 Atherton Court, Meadow Lane, Windsor SL4 6BN
G7 VML L. Alden 20 Kings Walk, Shoreham-by-Sea BN43 5LG
G7 VMO S. Fawcett 45 Forresters Close, Norton, Doncaster DN6 9HX
G7 VMQ T. Jones Ockton House, 24 Station Road, Okehampton EX20 1EA
GW7 VMT E. Wetherall 38 Argyle Street, Pembroke Dock SA72 6HL
G7 VNC C. Cave Little Meadow, Brewham Road, Bruton BA10 0JD
G7 VND Dr G. Morris 17 Bradshaw Road, Inkersall, Chesterfield S43 3HJ
G7 VNG A. Elmes Pookeezows, 10 Farnham Avenue, Hassocks BN6 8NS
G7 VNJ M. Swain Cottage Farm, Langley Marsh, Wiveliscombe, Taunton TA4 2UL
G7 VNK R. Cronshaw Flat 2, 28 Adelaide Terrace, Blackburn BB2 6ET
G7 VNL C. Lambert 43 Church Road, Guildford GU1 4NQ
G7 VNM A. Melham 4 Constantine Road, North Bitchburn, Crook DL15 8AG
G7 VNN C. Backhouse Elm House, The Green Saxlingham Nethergate, Norwich NR15 1TH
G7 VNO C. Brown 6 Ellesmere Avenue, Derby DE24 8WD
G7 VNP K. Knights 13 Millfield Place, Wilburton,.Ely CB6 3SA
G7 VNQ K. Staddon 1 Aller Grove, Whimple, Exeter EX5 2TJ
G7 VOA D. Hughes 60 Martingale Place, Downs Barn, Milton Keynes MK14 7QN
G7 VOH S. Holland 49 Oxland Road, Illogan, Redruth TR16 4SH
G7 VOI T. Nicholas Talmont, Chester Road, Tarporley CW6 0SD
G7 VOK A. Bracey 42 Lampton Grove, Bristol BS13 0QA
G7 VOM J. Snelgrove 22 Plains Avenue, Maidstone ME15 7AU
G7 VON A. Snelgrove 22 Plains Avenue, Maidstone ME15 7AU
GW7 VOO Welland Valley ARS c/o P. Hockey 98 Meadow Rise, Brynna, Pontyclun CF72 9TF
G7 VOQ S. Casey 5 Willow Road, Leyland PR26 8NP
G7 VOT A. Moseley 15 Hawthorn Avenue, Billingham TS23 1EE
G7 VOX M. Ingram Foxhill, Lower Daggons, Fordingbridge SP6 3EE
G7 VPA N. Muncey 2 Ladysmith Avenue, Whittlesey, Peterborough PE7 1XX
G7 VPD R. Friend High Hedges, Church Road, Norwich NR12 8YL
G7 VPN A. Berry 13 Collimer Close, Chelmondiston, Ipswich IP9 1HX
G7 VPQ J. Bishop 27 Southway, Blacon, Chester CH1 5NW
G7 VPS D. Barwood 6 Hemington Close, King's Lynn PE30 3YB
GM7 VPT D. Leask Avonmuir, The Loan, Muiravonside, Eh496Lw, Linlithgow EH49 6LW
G7 VPU P. Ansell White Hatch, Uvedale Road, Oxted RH8 0EW
GW7 VQA N. Parker 47 Rosehill Road, Rhyl LL18 4TN
GM7 VQB T. Roy 1 Rose Terrace, Leven KY8 4DF
G7 VQC D. Driver 27 Cricketers Way, Chatteris PE16 6UR
G7 VQE D. Frost 24 Woodland Close, Northampton NN5 6NH
G7 VQI G. Burch 6 The Barracks, Parkend, Lydney GL15 4HR
G7 VQJ W. Willmott Walton House Sandwich Road, Eastry CT13 0DP
G7 VQL M. Endean 17 Dryden Place, Tilbury RM18 8HQ
G7 VQM C. Davis 8 Mulberry Grove, Wallasey CH44 6PZ
G7 VQO W. Good 53 Harrow Lane, St. Leonards-on-Sea TN37 7JY
G7 VQR P. Mawdsley 7 Aldebert Terrace, London SW8 1BH
G7 VQW P. Jessup 2 Tile Lodge Cottages, Hoath Road, Canterbury CT3 4JN
G7 VQX J. Hunter 7 Berry Hill, Nunney, Frome BA11 4NR
G7 VRJ M. Holland 31 The Avenue, Andover SP10 3EP
G7 VRK S. Balding 13 Church Close, Colby Road Banningham, Norwich NR11 7DY
G7 VRX R. Croft Wallbury Lodge, Dell Lane, Bishop's Stortford CM22 7SQ
G7 VRY P. Bambridge 8 Temple Lane, Tonwell, Ware SG12 0HP
GM7 VSB J. O'Neill Douglas Cottage, Kames, Tighnabruaich PA21 2BH
G7 VSE D. Briggs 15 Orkney Close, Manchester M23 2AT
GW7 VSF W. Thomas 2 Ffordd Trecastell, Llanharry, Pontyclun CF72 9ND
G7 VSJ D. Jones 6 Eastville, Bath BA1 6QN
G7 VSL R. Taylor 11 Ranscombe Close, Brixham TQ5 9UR
G7 VSM J. Skinner 85 Main St., Barton Under Needwood, Burton on Trent DE13 8AB
G7 VSN L. Franklin 4 Rossington Close, Metheringham, Lincoln LN4 3DS
GW7 VSO D. Lewis 45 Llewellyn St., Pontygwaith, Ferndale CF43 3LF
G7 VSP S. Wooster 44 King Johns Road, North Warnborough, Hook RG29 1EJ
GW7 VST G. Davies 41 Woodlands Road, Barry CF63 4EF
G7 VSW S. Weston 11 Friars Road, Abbey Hulton, Stoke-on-Trent ST2 8DQ
G7 VTC L. Dodd Mulberry Lodge, Hallaze Road, Hallaze, , St. Austell PL26 8YW
G7 VTE G. Forster 33 Deer Valley Road, Holsworthy EX22 6DA
G7 VTH D. Reacher 33 Cator Crescent, New Addington, Croydon CR0 0BL
G7 VTJ T. Scott 50 Davison Avenue, Whitley Bay NE26 1SH
G7 VTL C. Davis 38 Courtney Close, Tewkesbury GL20 5FB
G7 VTN P. Mccaulay 33 Millmoor Way, North Hykeham, Lincoln LN6 9PJ
G7 VTQ Prof. D. Parsons 1 Kent Drive, Congleton CW12 1SD
G7 VTR C. Taylor 1105 Evesham Road, Astwood Bank, Redditch B96 6EB
G7 VTS P. Green 81 Victoria Road, Farnborough GU14 7PP
G7 VTT J. King Portland Manor Care Home, Thornhill Road, Newcastle upon Tyne NE20 9PZ
G7 VTW D. Jacques 4 Cedar Close, Thorpe Willoughby, Selby YO8 9QL
G7 VUH C. Squire 19 Southfield Road, Burley In Wharfedale, Ilkley LS29 7PA
G7 VUL J. Cook 32 Ash Bank Road, Stoke-on-Trent ST2 9DR
G7 VUM D. Riches 118 Drayton Road, Norwich NR3 2DL
G7 VUP J. Milner Sweetcroft Brentor, Tavistock PL19 0NJ
G7 VUU K. Coe 5 George St., Enderby, Leicester LE19 4NQ
G7 VVF D. Rossiter 37 Meadway, Enfield EN3 6NT
G7 VVK P. Bradley 22 Cavalier Close, Romford RM6 5EJ
G7 VVL N. Quest 21 Neave Crescent, Romford RM3 8HN
G7 VVO I. Anderson 18 St. Anthonys Drive, Wick, Bristol BS30 5PW
G7 VVX A. Renton 18 Stoneworks Garth, Crosby Ravensworth, Penrith CA10 3JE
G7 VWA D. Lever 35 Carteret Road, Luton LU2 9JZ
G7 VWG R. Evans 113 Highbridge Road, Burnham-on-Sea TA8 1LW
G7 VWM J. Hazell 7 Higher Road, Woolavington, Bridgwater TA7 8EA
G7 VWN J. Blackwell 7 Church Road, Darley Dale, Matlock DE4 2GG
G7 VWO D. Lisle Kent Ii, Broadmoor Hospital, Crowthorne RG45 7EG
G7 VWW J. Brook 45 Colonial Court, Senoia 30276 United States
GI7 VXC A. Crozier 38 Hawthorn Hill, Dromara, Dromore BT25 2HY
G7 VXK B. Amare Po Box 30464, Addis Ababa Ethiopia
G7 VXQ C. Elcombe 10 Northport Drive, Wareham BH20 4DR
GM7 VXR P. Crankshaw 3 North Neuk, Troon KA10 6TT
G7 VXS G. Burchell 23B Luff Meadow, Stowmarket Road, Ipswich IP6 8DP
G7 VYB R. Patel 30 Buckingham Drive, Luton LU2 9RA
G7 VYF K. Tadesse Po Box 60229, Addis Ababa Ethiopia
G7 VYI R. Smith 82 Long Row, Shrewsbury SY1 4DD
G7 VYN M. Jarman 143 Rotherham Road, Barnsley S71 2LL
G7 VYQ I. Holman 7 The Silent Woman Park, Tavistock PL19 9LQ
GM7 VYR I. Findlay 2 Bothwell Road, Uddingston, Glasgow G71 7ET
G7 VYT J. Graver 15 Cartwright Road, Charlton, Banbury OX17 3DG
G7 VYW W. Moreton 17 Hadley Road, Bilston WV14 6RX
G7 VYY S. Middleton 22 Hall Villa Lane, Toll Bar, Doncaster DN5 0LH
G7 VYZ M. Lancastle 31 Ridgeside, Kirk Merrington, Spennymoor DL16 7HF
G7 VZD J. Payne 15 Belmont Road, Tiverton EX166AR
G7 VZI H. Charles 6 Bridewell Street, Wymondham NR18 0AR
G7 VZL D. Forster 15 Bracondale, Norwich NR1 2AL
G7 VZM M. Blacklock 39 Birtwistle Avenue, Colne BB8 9RS
G7 VZQ D. Polley 33 Wye Close, Crawley RH11 9QZ
G7 VZR C. Gain 14 Battens Avenue, Overton, Basingstoke RG25 3NL
G7 VZS A. Walsh 21 Rydal Avenue, Darwen BB3 2SA
GM7 VZV D. Henry 106 Whinhill Gate, Aberdeen AB11 7WF
G7 VZY Dr M. Page-Jones 2 Chestnut Close, Romsey SO51 5SP
G7 WAA E. Donaghy Mendips, 36 Attwood Road, Salisbury SP1 3PR
G7 WAB Worked All Britain Awards Group c/o K. Hale 58 St. Stephens Road, Saltash PL12 4BJ
G7 WAC Hull And District ARS c/o L. Volante Richmond House, Icknield Street, Birmingham B38 0EP
G7 WAE T. Ward 173-175 Station Road, Pendlebury, Swinton, Manchester M27 6BU
G7 WAF D. Keeble 71 St. Lawrence Avenue, Bolsover, Chesterfield S44 6HS
G7 WAQ A. Moss 21 Shrubbery Lane, Weymouth DT4 9LY
G7 WAS S. Staines 6 The Quantocks, Flitwick, Bedford MK45 1TQ
G7 WAW D. Thompson 12 Dam Head Road, Barnoldswick BB18 5NH
G7 WAY S. Foster 137 Cheltenham Road, Longlevens, Gloucester GL2 0JH
G7 WBA R. Grandshaw Treehaven, South Lane, Salisbury SP5 2BZ
G7 WBE D. Welch 38 Little Sammons Chilthorne Domer, Yeovil BA22 8RB
G7 WBH D. Small 17 Claygate Road, Wimblebury, Cannock WS12 2RN
G7 WBJ W. Naylor 5 Burman Close, Shirley, Solihull B90 2DR
G7 WBL J. Wheeler 92 Holford Road, Bridgwater TA6 7NZ
G7 WBM Dr P. Longhurst Burston, Wood Road, Hindhead GU26 6PZ
G7 WBO M. Stenning 56 Hampshire Court, Upper St. James'S Street, Brighton BN2 1JZ
GM7 WBP F. Kelly 50 Farm Road, Blantyre, Glasgow G72 9DT
G7 WBR Yeovil ARC c/o E. Davis 33 Truggers, Handcross, Haywards Heath RH17 6DQ
G7 WBU A. Hopley 124 Lonnen Road, Wimborne BH21 7AZ
G7 WBW A. Millward 26 Osprey Close, Scotton, Catterick Garrison DL9 3RA
G7 WBY B. Mulder 8 Chapel Close, Little Gaddesden, Berkhamsted HP4 1QG
G7 WBZ M. Malone 11 Pine Close, Rishton, Blackburn BB1 4JX
G7 WCB A. Bennett 12 Barns Close, Walsall WS9 9BD
G7 WCF C. Rose The Barn, Lower Killigorrick, St Keyne, Liskeard PL14 4QP
G7 WCG D. Seabrook 44 Village Centre, Richmond Letcombe Centre, Letcombe Regis OX12 9RG
G7 WCN K. Packer 47 Sheppard Road, Basingstoke RG21 3JH
G7 WCP C. Sharpe 30 Mardale Way, Loughborough LE11 3SS
GW7 WCR J. Pitkin 29 Dolwerdd Estate, Pen Y Parc, Cardigan SA43 1RF
GI7 WCS J. Stitt 199 Gobbins Road, Islandmagee, Larne BT40 3TX
G7 WDC M. Kiteley 13 Chiltern Close, Astley Cross, Stourport-on-Severn DY13 0NU
G7 WDD B. Bird 4 Berkeley Crescent, Frimley, Camberley GU16 8YN
G7 WDG P. Wyatt 6 Bridge Road, Coalville LE67 3PW
G7 WDM N. Feetham 154 Magdalen Lane, Hedon, Hull HU12 8LB
G7 WDN A. Goodridge 9 Radcliffe Way, Oundle, Peterborough PE8 4QE
G7 WDO C. Barker 15 Epping Green, Hemel Hempstead HP2 7JP
G7 WDS A. James 36 Lemon Hill, Mylor Bridge, Falmouth TR11 5NA
G7 WEB D. Raxter 2 Lower Croft, Cropthorne, Pershore WR10 3NA
GM7 WED R. Feilen 131 Croftend Avenue, Glasgow G44 5PF
G7 WEK C. Taylforth 1 Clough Terrace, Barnoldswick BB18 5PD
G7 WEM T. Hewitt 6 Mayfield, Catforth Road, Preston PR4 0HH
G7 WEN J. Marron 30 York Road, Nunthorpe, Middlesbrough TS7 0EZ
G7 WEP M. Williams 59 Thistledene, Thames Ditton KT7 0YH
G7 WER P. Fisher 21 Charlotte Close, Mount Hawke, Truro TR4 8TS
G7 WEW A. Cossey 17 Hazel Close, Norwich NR8 6YE
G7 WFC R. Smith 35 Montsale, Pitsea, Basildon SS13 1JL
G7 WFD M. Dockerty 16 Valley Way, Stalybridge SK15 2QZ
G7 WFH T. Ford 3 Greenmore Road, Bristol BS4 2LA
G7 WFK G. Tew 5 Hill Top Avenue, Tamworth B79 8QB
G7 WFQ J. Stevens Springfield Cottage, 57 Brindley Street, Stourport-on-Severn DY13 8JG
GM7 WFT G. Edwards 19 Howe Park, Edinburgh EH10 7HF
G7 WFV J. Kendall 3 High Street, Cottenham, Cambridge CB24 8SA
G7 WFZ R. Stone 222 Dedworth Road, Windsor SL4 4JP
G7 WGA J. Potter 198 Battle Road, St. Leonards-on-Sea TN37 7AL
G7 WGD D. Price 199 Central Drive, Bilston WV14 8JE
G7 WGE T. Forster 20 Bryant Avenue, Slough SL2 1LG
G7 WGI J. Gordon 19 Heywood Gardens, Havant PO9 4HR

G7 WGL K. Firth Chimneys, 30 Kingscroft, King's Lynn PE31 6QN
G7 WGN J. Hutton 42 Priory Road, Cottingham HU16 4SA
G7 WGO G. Bradshaw 3 Falmouth Avenue, Haslingden, Rossendale BB4 6QN
G7 WGP A. Brook 163 Station Road, Mickleover, Derby DE3 9FL
G7 WGX D. Sayles 82 Molineaux Road, Shiregreen, Sheffield S5 0JY
G7 WGY D. Wilson 75 Gainsborough Road, Scotter, Gainsborough DN21 3RU
G7 WGZ E. Morley 91 Allerton Road, Stoke-on-Trent ST4 8PQ
G7 WHA O. Morley 91 Allerton Road, Stoke-on-Trent ST4 8PQ
G7 WHI D. Page 55 Hinckley Road, Stoney Stanton, Leicester LE9 4LL
G7 WHM A. Howgate 7 Caledonian Way, Belton, Great Yarmouth NR31 9PQ
G7 WHP W. Jones 7 Hampstead Gardens, Hockley SS5 5HN
GM7 WHQ S. Gray-Thompson Vagastie, Lairg IV27 4AD
G7 WHU M. Nock Mesquida, Lorraine Road, Newhaven BN9 9QB
G7 WHX J. Bodle 48 Bolsover Road, Hove BN3 5HP
G7 WHZ T. Crane 15 Belchamps Way, Hawkwell, Hockley SS5 4NT
G7 WIC G. Probyn 24 Woollaton Close, Grange Park, Swindon SN5 6BB
G7 WID G. White 21 Tollfield Road, Boston PE21 9PN
G7 WIG R. Bilsland 56 Cowleigh Bank, Malvern WR14 1PH
G7 WIQ S. Pack 245A Beacon Road, Loughborough LE11 2QZ
G7 WIY M. Downing 12 Martindale Road, Woking GU21 3PJ
G7 WJC B. Webster 50 Blackburn Road, Rishton, Blackburn BB1 4BH
G7 WJE H. Coots 40 Essex Close, Romford RM7 8BD
G7 WJJ R. Morton 29 Lanmoor Estate, Lanner, Redruth TR16 6HN
G7 WJK J. Stephens 19 Aspen Fold, Oswaldtwistle, Accrington BB5 4PH
GM7 WJP A. Anderson 232 Annan Road, Dumfries DG1 3HE
G7 WJV R. Stroud 55 Haymeads Lane, Bishop's Stortford CM23 5JJ
G7 WJW G. Cripps 52 Cleveland, Tunbridge Wells TN2 3NQ
G7 WJZ P. Clarke 21 Long Furlong Road, Sunningwell, Abingdon OX13 6BL
G7 WKC T. Hasted Springfield House, Birds End, Bury St. Edmunds IP29 5HE
G7 WKG A. Roche Flat 22, Trident Court, Birmingham B20 2NX
G7 WKH Calderdale Raynet ARG c/o P. Clark 21 Sandfield Road, Arnold, Nottingham NG5 6QA
G7 WKP A. Andrew Thrift, Madles Lane, Ingatestone CM4 9QA
G7 WKV B. Jewell 49 Spinney Road, Burton Latimer, Kettering NN15 5ND
G7 WKW M. Davis Cherry Pie Bay Road, Freshwater PO40 9QS
GI7 WLA D. Calvin 65 Tannaghmore Road, Markethill, Armagh BT60 1TW
G7 WLC D. Evans 1 Brigadier House Captain Gardens, Colchester CO2 7LD
G7 WLL I. Irlam 31 Wyatt Road, Dartford DA1 4SN
G7 WLM J. Tamlyn Hedge Rise, Sidmouth Road, Exeter EX2 5QJ
GM7 WLO J. Burt Olivet, Lanton Road, Jedburgh TD8 6SD
G7 WLV G. Southall 6 Dudley Wood Avenue, Dudley DY2 0DG
G7 WLY R. Bagwell 20 School Approach, South Shields NE34 6DP
G7 WRS T. Misselbrook City Fields, Wakefield WF3 4GD
G7 WSH R. Munt Box 166, Se-177 23 Jarfalla SWEDEN Sweden
G7 WWW B. Cole 6 Parkstone Parade, Hastings TN34 2PS
G7 XPC P. Chorley Boone Hill House, Mount Boone Hill, Dartmouth TQ6 9NZ
G7 YAB A. Brewerton The Vicarage, Highthorn Road, Kilnhurst, Mexborough S64 5TX
G7 ZMS M. Larcombe 65 Western Road, Burgess Hill RH15 8QW
G7 ZRT R. Thayne 213 Carlton Road, , Boston PE21 8NG
G7 ZZY P. Pile Apartment 836, Lagos 8600 Portugal

G8

G8 AA M. Keilty 25 Lathom Avenue Wallasey, Wirral CH44 5UH
G8 AAC J. Billingham 14 St. Matthews Court, Sutherland Road, Brighton BN2 2EX
G8 AAD B. Blight 43 North Street, Oxon OX9 3BJ
G8 AAE D. Phillips 2 Walkers Close, Chelmsford CM1 6UW
GW8 AAF F. Blake 3 Morfa Gaseg, Llanfrothen, Penrhyndeudraeth LL48 6BH
G8 AAI M. Bues 7A Alice Parkins Close, Hadleigh, Ipswich IP7 6FE
G8 AAR F. May Quatre Vents, Church Road, Sudbury CO10 0QP
G8 AAT R. Pye 7 Meadow View, Potterspury, Towcester NN12 7PH
G8 AAU N. Stanners 22 Brands Hill Avenue, High Wycombe HP13 5QA
G8 ABB G. Rogers 10 The Laurels, Bletchley, Milton Keynes MK1 1BL
G8 ABX G. Catling 3 The Tene, Baldock SG7 6DG
G8 ACL H. Cosford 3 Applewood, Park Gate, Southampton SO31 7HQ
G8 ACQ R. Whattam The Aviary No1, Arkwright Rd, Beds MK44 1SE
G8 ADA J. Robinson 7 Rhyl St., Liverpool L8 6QL
G8 ADC J. Haile 145 Dunstable Road, Caddington, Luton LU1 4AN
G8 ADD B. Carter 51 Smirrells Road, Birmingham B28 0LA
GM8 ADK M. Ritchie 11 Cromwell Road, Aberdeen AB15 4UH
G8 ADQ J. Taylor 21 Launcestone Close, Earley, Reading RG6 5RY
G8 ADX E. Lawley 3 Barnicott Close, Newton Ferrers, Plymouth PL8 1BP
G8 ADY P. Harrison 2 The Barns, Bridge End, Carlton, Bedfordshire MK43 7LP
G8 ADZ N. Shepherd 7 High St., Kelvedon, Colchester CO5 9AG
G8 AEN P. Helm 74 Neston Road Walshaw, Bury BL8 3DB
G8 AER J. Tanner Merlins Mill, Toadsmoor Road, Stroud GL5 2UG
G8 AEU J. Nightingale 6 Aubrey Close, Chelmsford CM1 4EJ
G8 AFA C. Atkins 2 Eastlands, Yetminster, Sherborne DT9 6NQ
G8 AFI P. Funnell 25 Broadyates Road, Yardley, Birmingham B25 8JF
G8 AFN P. Cleall 139 Preston Grove, Yeovil BA20 2DB
G8 AFQ T. Hall 24 Church Street, Kelvedon, Colchester CO5 9AH
GI8 AFS M. Granville 33 Dunfield Terrace, Londonderry BT47 2ES
G8 AFU P. Gilby 191 Send Road, Send, Woking GU23 7ET
G8 AGJ J. Evans 1 Grosvenor Close, Hatch Warren, Basingstoke RG22 4RQ
GM8 AGM M. Collar Shoemakers Croft, Hatton, Peterhead AB42 0TB
G8 AGN Dr B. Chambers 5 The Ridge, Sheffield S10 4LL
G8 AGQ A. Strong 3 Ellorslie Drive, Stocksbridge, Sheffield S36 2BB
G8 AGR Dr S. Craddock 38 Briardene, Lanchester, Durham DH7 0QD
GW8 AHB P. Swinbank 13 Mundy Place, Cardiff CF24 4BZ
G8 AHE L. Arnold 402 Bournville Gardens, 49 Bristol Road South, Birmingham B31 2FT
G8 AHK University of Surrey Ears c/o N. Jones 61 Sutherland Close, Whitehill, Bordon GU35 9RE
G8 AHN J. Barnes 2 Mappins Road, Catcliffe, Rotherham S60 5TH
G8 AHR P. Rushworth 2 Aberdeen Close, Coventry CV5 7NE
G8 AIE P. Willcocks 27 Manor Road, Barnet EN5 2LE
G8 AIM F. Tarver 14 Southview Road, Leamington Spa CV32 7JD
G8 AIP M. Osment Flat 2, Weavers Court, Shoreham-by-Sea BN43 5ES
GI8 AIR W. Parkes 15 Bushfoot Park, Portballintrae, Bushmills BT57 8YX
GW8 AJA D. Hardy 7 Coed Y Go Cottages, Coed Y Go, Oswestry SY10 9AU
G8 AJM C. Payne 14 Watts Lane, Louth LN11 9DG
G8 AJP J. Eade High Shaw Sandrock Hill, Sedlescombe TN33 0QR
G8 AJZ R. Boardall 9 Oxford Street, Bury BL9 7EL
G8 AKA T. Wiltshire Bramblings, Pelican Road, Tadley RG26 3EL
G8 AKC C. Bell Croftner, Mary Tavy, Tavistock PL19 9QD
G8 AKE J. Warrington 26 Lynton Road, Melton Mowbray LE13 0NN
G8 AKF J. Ballantyne Brookeside, Ashwellthorpe Road, Wreningham, Norwich NR16 1AW
G8 AKL G. Ashcroft Wood End House, Wood End , Bluntisham, Huntingdon PE28 3LE
G8 AKM G. Roper 19 Normay Rise, Newbury RG14 6RY
G8 AKP P. Mcquade The Old Swan, Holt Road Sharrington, Melton Constable NR24 2PH
G8 AKQ S. Birkill 38538 Sky Pilot Drive, Squamish V8B 0T6 Canada
G8 AKU B. Willson Hilltop, Cryers Hill Road, Cryers Hill, High Wycombe HP15 6LJ
G8 AKX M. Perry 216 Marlpool Lane, Kidderminster DY11 5DL
G8 ALD M. Lunt 18 Longhurst Road, Hindley Green, Wigan WN2 4PL
G8 ALE M. Brereton Gleaston Water Mill, Gleaston, Ulverston LA12 0QH
G8 ALQ A. Whitlock 23 Daly Way, Aylesbury HP20 1JW
G8 ALR J. Cull 2 Drybrook Cottages, Amesbury Road, Cholderton, Salisbury SP4 0ER
G8 ALS M. Stevenson 15 Wall Hill Road, Allesley, Coventry CV5 9EN
G8 AMC Amc RC c/o D. Millward 77A Meadowcroft, St. Albans AL1 1UG
G8 AMD H. Bate 88 Darnick Road, Sutton Coldfield B73 6PG
G8 AMG M. Foster 9 Norman Way, Irchester, Wellingborough NN29 7AT
G8 AMJ D. Woolley Tweddell'S Garth, West End, Leyburn DL8 3HN
G8 AMK L. Parry 13 Cannon Hill, Bracknell RG12 7QA
G8 AMU C. Saveker 23 Southlands Avenue, Horley RH6 8BS
G8 ANN G. Townsend 61 Richmond Park Road, London SW14 8JU
G8 ANO D. Lawton Grenehurst, Pinewood Road, High Wycombe HP12 4DD
G8 ANT S. Holland 14 The Vineries, Eastbourne BN23 7TP
GD8 ANU C. Howard 5 Ballure Grove, Ramsey IM8 1NF Isle of Man
GM8 AOB J. Briscoe 2 Peebles Place, Fort William PH33 6UG
G8 AOE B. Duffell 7 Potto Close, Yarm TS15 9RZ
G8 AOG M. Browne 143 Thatch Leach Lane, Whitefield, , Manchester M45 6EP
G8 AOI T. Knight 3 Eaton Close, Rainworth, Mansfield NG21 0AR
G8 AOJ G. Smith Forest View Cottage, Gorsty Knoll, Coleford GL16 7LR
G8 AOK A. Porch 17 Purcell Close, Brighton Hill, Basingstoke RG22 4EL
G8 AOO B. Hills 3 Frithmead Close, Basingstoke RG21 3JW
G8 AOZ P. Hughes 247 High Greave, Sheffield S5 9GS
G8 APB C. Plummer Barley House Farm, Newtown, Stoke-on-Trent ST8 7SW
G8 APF D. Chaplin 4 Blenheim Close, Loughborough LE11 4SA
G8 APL G. Parsons 21 Wild Ridings, Fareham PO14 3BS
G8 APM G. White 1 Drakes Close, Hythe, Southampton SO45 5BP
G8 APW D. Taylor 87 Grasmere Road, Chester le Street DH2 3EU
G8 APY J. Bond Folly House, The Reddings, Cheltenham GL51 6RL
G8 APZ S. Lucas 84 Woodman Road, Warley, Brentwood CM14 5AZ
G8 AQA P. Nickalls Holy Mill, Longville, Much Wenlock TF13 6ED
G8 AQB Tamworth & Lichfield Raynet c/o M. Ballance 24 Western Road, Wolverton, Milton Keynes MK12 5BE
G8 AQH R. Hine 147/149 Bolton Hall Road Bolton Woods, Bradford BD2 1BQ
G8 AQN A. Hibberd 20 Barby Lane, Rugby CV22 5QJ
G8 AQO A. Copperwaite 71 Gladbeck Way, Enfield EN2 7EL
G8 AQP S. Warner 14 Andrews Way, Aylesbury HP19 8WA
G8 ARA B. King 15 Newstead Road, West Southbourne, Bournemouth BH6 3HJ
GW8 ARC Dr A. Craggs 15 Pen-Y-Groes Avenue, Cardiff CF14 4SP
G8 ARF L. Thompson 44 Tillmouth Avenue, Holywell, Whitley Bay NE25 0NP
G8 ARH N. Blackmore 35 Weyhill Gardens, Weyhill, Andover SP11 0QT
G8 ARM B. Pickrell Perrans, Ludgvan, Penzance TR20 8AJ
GW8 ARR P. Edwards Trevland, Felindre, Knighton LD7 1YL
GW8 ASA G. Wyatt 3 Creidiol Road, Mayhill, Swansea SA1 6TZ
G8 ASC P. Richards 134 Downhills Park Road, Tottenham, London N17 6BP
GW8 ASD A. Pugh Willcroft, Mold Road, Gwersyllt LL11 4AF
G8 ASG M. Farrell Hobberley House, Hobberley Lane, Leeds LS17 8LX
G8 ASJ G. Swan Morogar, Post Office Lane Kempsey, Worcester WR5 3NX
G8 ASP I. Gurton 28 Bloomfield Road, Harpenden AL5 4DB
G8 ASV D. Skinner Latch Cottage, Nursery Lane, Blackboys, Uckfield TN22 4EU
G8 ASW R. Warrender 102 Turnberry Road, Great Barr, Birmingham B42 2HT
G8 ASX A. Hoggan 25 Clingan Road, Bournemouth BH6 5PY
GM8 AT W. Beattie Alastrean House, Tarland, Aboyne AB34 4TA
G8 ATC Dr R. Gayton 20 Barton Close, Exton, Exeter EX3 0PE
G8 ATD A. Barter 503 Northdown Road, Margate CT9 3HD
G8 ATE R. Turlington 2 Laithwaite Close, Leicester LE4 1BX
G8 ATG M. Williamson 120 Warbreck Hill Road, Blackpool FY2 0TR
G8 ATK M. Hearsey Halycon, Lawday Link, Farnham GU9 0BS
G8 ATL M. Lankester 154 Gorse Lane, Clacton-on-Sea CO15 4RJ
G8 ATP K. Mintern 71 Crafts End, Chilton, Didcot OX11 0SB
G8 AUL P. Buck 41 Marion St., Brighouse HD6 2BJ
G8 AUN R. Chiddick 87 Aylsham Road, Norwich NR3 2HW
G8 AUU C. Partridge 6 Blagdon Walk, Teddington TW11 9LN
G8 AVB C. Dickson 5 Arrow View, Ledbury HR8 2FR
G8 AVC R. Evans Mansfield, 1 Horsehead Lane, Chesterfield S44 6HU
G8 AVK R. Kimberley 8 Nutwell Road, Weston-Super-Mare BS22 6EN
GM8 AVM I. Macdonald Benvoir, Lightlands Avenue, Wigtown, Newton Stewart DG8 9EE
G8 AVO J. Wainwright 33 Station View, Nantwich CW5 7BJ
G8 AVQ J. Florentin 17 Campden Hill Gardens, London W8 7AX
G8 AVZ M. Keeping 8 Calderdale Close, Southgate West, Crawley RH11 8SQ
G8 AWB R. Lawrence 16 Westover Road, Callington PL17 7HD
G8 AWE M. Wellspring 21 Rue De La Gendarmerie, Aigre 16140 France
G8 AWI C. Smith 129 Earls Road, Nuneaton CV11 5HP
GW8 AWM F. Evans Ty Cryr, Chepstow Road, Usk NP15 1HN
G8 AWN B. Procter 28 Holme Grove, Burley In Wharfedale, Ilkley LS29 7QB
G8 AWY J. Ward 44 Rugby Road, Barby CV238UB
G8 AXN C. Amery 9 View Close, Biggin Hill, Westerham TN16 3XE
G8 AXO A. Nunn 9 Elmhurst Court, Hamblin Road, Woodbridge IP12 1HB
G8 AXR R. Moore 22 Cardan Drive, Ilkley LS29 8PH
G8 AXV K. Shail Veeda Glenta, Blackmore Park Road, Malvern WR13 6NN
G8 AYC N. Walker 36 Meyrick Drive, Wash Common, Newbury RG14 6SX

G8 AYJ J. Hanson 22 Church Way, Falmouth TR11 4SG
G8 AYM N. Pritchard 108 Kynaston Avenue, Aylesbury HP21 9DS
G8 AYV J. Lewis Newnham House, Shurton, Bridgwater TA5 1QG
G8 AZB S. Smith 11 Grayshott Laurels, Lindford, Bordon GU35 0QB
G8 AZM D. Johnson 195 Staplers Road, Newport PO30 2DP
G8 AZN R. Barnes 18 Battle Road, Tewkesbury GL20 5TZ
G8 AZR J. Dimmock 93 Barton Road, Harlington, Dunstable LU5 6LG
G8 AZT J. Jones 9 Queens Walk, Thornbury, Bristol BS35 1SR
G8 AZZ G. Craddock Westlands, Bowden Hill, Yealmpton, Plymouth PL8 2JX
G8 BAD 1466 Holmfirth c/o D. Donati Rust Oaks, Emms Lane, Brooks Green, Horsham RH13 0TR
G8 BAG T. Rowley 7 Hall Farm Close, Castle Donington, Derby DE74 2NG
G8 BAJ P. Southby 51 Teddington Park, Teddington TW11 8DE
G8 BAK P. Knight 4 Dimmock Road, Wootton, Bedford MK43 9DW
G8 BAL C. Robinson 17 Fairview Close Hythe, Southampton SO45 5EX
G8 BAQ B. Kneller Mystic Flight, Brackenhill Road, Eastlound, Doncaster DN9 2LR
G8 BAS D. Gardiner 31 Alexander Drive, Cirencester GL7 1UG
G8 BAZ P. Talbot 19 Bladen Valley, Briantspuddle, Dorchester DT2 7HP
G8 BBK R. Nelson 10 Wragg Drive, Newmarket CB8 7SD
G8 BBV J. Goulty 1 Larksway, Felixstowe IP11 2PN
G8 BBZ P. Barker 3 Hudson Fold, Heptonstall, Hebden Bridge HX7 7PH
G8 BCA R. Chambers 11 Thetford Road, Mildenhall, Bury St. Edmunds IP28 7HX
G8 BCF G. Podmore Crownfield, Kings Lane, Faringdon SN7 7SS
G8 BCG P. Taylor Merrivale, Portuan Road, Looe PL13 2DW
G8 BCI E. Rowlands Wychanger Cottage, Luccombe, Minehead TA24 8TA
G8 BCJ A. Unsworth Meadow View, Clockhouse Lane, North Stifford Rm165Ur., Grays RM16 5UR
GW8 BCL H. Bottomley Llwyn Y Berllan, Battle, Nr Brecon LD3 9RN
G8 BCO C. Boys 34 Firacre Road, Ash Vale, Aldershot GU125JT
G8 BDF J. Hanney 16 Parsonage Barn Lane, Ringwood BH24 1PX
G8 BDM J. Adams 1 Powell Close, Creech St. Michael, Taunton TA3 5TE
G8 BDQ G. Hedley 260E 100 Sts, Raymond, Alberta TOK 250 Canada
GM8 BDX A. Scott 20 Treaty Park, Birgham, Coldstream TD12 4NG
G8 BDZ K. Cowdell 6 Pearl Street, Bristol BS3 3EA
G8 BEH D. Hill Care Uk, Prince George House, 102 Mansbrook Boulevard, Ipswich IP3 9GY
G8 BEK C. Dunn 75 Waddington Avenue, Burnley BB10 4LA
G8 BEQ K. Greenough 2 Bexley Close, Glossop SK13 7BG
G8 BFA S. Davis 21 Cordville Close, Chaddesden, Derby DE21 6WX
G8 BFC P. Johnson 15 Elvaston Lane, Alvaston, Derby DE24 0PX
G8 BFH J. Marriott 104 Whinbush Road, Hitchin SG5 1PN
G8 BFK S. Ballard 26 Crafts End, Chilton, Didcot OX11 0SA
G8 BFL B. Jayne 38 Townfields, Lichfield WS13 8AA
G8 BFM A. Whittaker 6 Kingsbridge Way, Bramcote, Nottingham NG9 3LW
GW8 BFO R. Hayter Glanyrafon, Talywern, Machynlleth SY20 8NY
G8 BFV D. Edwards 34 Campkin Road, Wells BA52DG
G8 BGI B. Hepburn 52 Hibiscus Grove, Bordon GU35 0XA
G8 BGL R. Gilliatt 21 Main St., Thorpe On The Hill, Lincoln LN6 9BG
G8 BGM M. Lee 32 Fernham Road, Faringdon SN7 7LB
G8 BGT A. Dermont 7 Pool Close, Little Comberton, Pershore WR10 3EL
G8 BGV P. Selwood 43 Keene Way, Galleywood, Chelmsford CM2 8NT
G8 BHC J. Richmond-Hardy 45 Burnt House Lane, Kirton, Ipswich IP10 0PZ
G8 BHE N. Gutteridge 68 Max Road, Quinton, Birmingham B32 1LB
G8 BHK J. Vickers 242B High Road, Trimley St. Martin, Felixstowe IP11 0RG
GM8 BHR G. Pearson 2 Hamilton Terrace, Edinburgh EH15 1NB
G8 BHX M. Berry111 27 Greenway Road, Heald Green, Cheadle SK8 3NR
G8 BHY A. Heath 7 Coral Close, Coventry CV5 7AD
G8 BIG M. Stebbings 15 St. Helena Way, Horsford, Norwich NR10 3EA
G8 BIH J. Akam 10 Apple Tree Road, Alderholt, Fordingbridge SP6 3EW
G8 BII B. Hunt 53 The Sands, Milton-Under-Wychwood, Chipping Norton OX7 6ER
G8 BIR R. Harris 35 Freemantle Road, Eastville, Bristol BS5 6SY
G8 BIS P. Lyon Frogs Hall, Cannon Street, New Romney TN28 8BJ
G8 BIW R. Booth 16 Darwynn Avenue, Swinton, Mexborough S64 8DU
G8 BIX A. Parcell Birdies Barn, Minions, Liskeard PL14 5LE
G8 BJA D. Couchy 8 Chapel St., Wincham, Northwich CW9 6DA
G8 BJB G. King 62 Heathfield Road, Sholing, Southampton SO19 1DP
GM8 BJF Dr B. Flynn 15 Riselaw Crescent, Edinburgh EH10 6HN
GM8 BJJ A. Morton 4 Mountstuart St., Millport KA28 0DP
G8 BJO J. Barfoot 21 Richard Crampton Road, Beccles NR34 9HN
G8 BJQ L. Case 58 Brookdale, Widnes WA8 4TB
G8 BKD P. Scotney 30 Trinity Road, Rothwell NN14 6HY
G8 BKE C. Towns 21 Seafield Close, Barton On Sea, New Milton BH25 7HR
G8 BKG D. Wright 61 Potton Road, St. Neots PE19 2NN
G8 BKH G. Shepherd 64 Dawley Road, Arleston, Telford TF1 2JF
G8 BKL E. Danks 18 Lichfield Street, Stourport-on-Severn DY13 9EU
G8 BKQ C. Clark 21A Headland Park Road, Paignton TQ3 2EN
G8 BLB P. Blakeney 45 Hampden Avenue, Chesham HP5 2HL
G8 BLD J. Draper 11 Parva Close, Little Barningham, Norwich NR11 7NJ
G8 BLK Sheffield Hf DX Group c/o M. Keightley 20 Longrood Road, Rugby CV22 7RG
G8 BLP C. Bond 5 Rushley Close, Sheffield S17 3EG
G8 BME F. Burrow 51 Stanhope Avenue, Morecambe LA3 3AJ
G8 BMH J. Parry 29 Heath Road Upton, Chester CH2 1HT
G8 BMI G. Theasby 115 Bevercotes Road, Sheffield S5 6HB
G8 BMP M. Taylor 96 Woodhouses Road, Burntwood WS7 9EJ
G8 BMQ B. Cedar 29 Velsheda Court, Hythe Marina Village, Southampton SO45 6DW
G8 BMZ P. Cowling 94 Welholme Road, Grimsby DN32 0NG
G8 BNB R. Gibbs 15 Gosford Hill Court, Bicester Road, Kidlington OX5 2XP
GI8 BNC J. Mccann 61 Glengawna Road, Glengawna, Omagh BT79 7WJ
G8 BNE R. Kendall Random Stones, Arkendale Road, Knaresborough HG5 0QA
G8 BNG A. Green 37 Bramcote Lane, Nottingham NG8 2NA
GM8 BNH I. Gall Cluaran, Bridge Of Don, Aberdeen AB23 8BD
G8 BNK P. Banbury 16 Gloucester Road, Whitstable CT5 2DS
G8 BNR R. Wells 279 Hatfield Road, St. Albans AL4 0DH
G8 BOB A. Robinson 29 Thomas Manning Road, Diss IP22 4HL
G8 BOI M. Simpson 9 Brock House, 2 Batter Street, Plymouth PL4 0EF
G8 BOJ K. Agombar 54 Julien Road, London W5 4XA
G8 BOP M. Palmer 109 Longfellow Road, Dudley DY3 3EF
G8 BOQ K. Phillips 1140 Riverberry Drive, Reno Nv 89509 United States
G8 BOS B. Saunders 88 Bramwoods Road, Chelmsford CM2 7LT
G8 BPH J. Rome 1 Bridge Cottages, Downhall Road, Hatfield Heath CM22 7AS
G8 BPN G. Wilkerson Hill House, Newton, Leominster HR6 0PF
G8 BPQ J. Wiseman 147 Hilton Road, Nottingham NG3 6AR
G8 BPS C. Booth 11 High St., Haxey, Doncaster DN9 2HX
G8 BPU H. Skelhorn 9 Moss Lane, Bollington, Macclesfield SK10 5HJ
G8 BPW A. Stoker 35A Church End Lane, Runwell, Wickford SS11 7JE
G8 BPY P. Hollis 5 Salisbury Road, New Malden KT3 3HZ
G8 BQF A. Dixon 2 Yorkdale Drive, Hambleton, Selby YO8 9YB
G8 BQH M. Marsden Hunters Moon, Buckingham Road Hardwick, Aylesbury HP22 4EF
GW8 BQK G. Oatway 21 Victoria Park, Colwyn Bay LL29 7AX
G8 BQT I. Hudson Flat 32, Three Crowns House, King's Lynn PE30 5DT
G8 BQZ P. Plunkett 30 Broadlands Avenue, Shepperton TW17 9DQ
G8 BRD Dr C. Dawson 33 Rough Common Road, Rough Common, Canterbury CT2 9DL
G8 BRG P. Mitchell 3 Goodwin Court, Farnsfield, Newark NG22 8LU
G8 BRL Shropshire Raynet c/o B. Ward 10 Upper Moorfield Road, Woodbridge IP12 4JW
G8 BRU G. Gallamore 30 Orchard Avenue, Partington, Manchester M31 4DL
G8 BSD J. Ceresole 7 Stokes Bay Home Park, Stokes Bay Road, Gosport PO12 2QU
G8 BSP A. Wicks 1 Castle Hill Close, Shaftesbury SP7 8LQ
GM8 BSQ A. Shepherd 2 Westwood Place, Skene, Westhill AB32 6WS
GM8 BSU A. Weller 18 Froghall Road, Aberdeen AB24 3JL
G8 BTC B. Fenwick 16 Pine Walk, Uckfield TN22 1TU
G8 BTD P. Sladen 2 Burlea Close, Crewe CW2 8SZ
G8 BTL H. Futcher Sarum, 12 Thursby Road, Woking GU21 3NZ
G8 BTU J. Dowson The Granary, St. Peters Road, Arnesby, Leicester LE8 5WJ
G8 BTV P. Marlow 1 Vineries Close, Leckhampton, Cheltenham GL53 0NU
GW8 BTX T. Storeton-West Tan Y Banc, Blaenpennal, Aberystwyth SY23 4TT
G8 BTY M. Dennis Thistledown, Yallands Hill, Taunton TA2 8NA
G8 BUB B. Goodall 10 Westoby Close, Shepshed, Loughborough LE12 9SS
GD8 BUE I. Rae 65 Lezayre Park, Ramsey IM8 2PT
G8 BUF Prudential ARS c/o M. Higgins 59 Clinton Crescent, Ilford IG6 3AH
G8 BUI Dr C. Nowikow 10 Windmill Road, Whitstable CT5 4NL
G8 BUV C. Chapman 6 Pickhurst Green, Hayes, Bromley BR2 7QT
G8 BUX Buxton Radio Amateurs c/o D. Carson 21 Harris Road, Harpur Hill, Buxton SK17 9JS
G8 BUZ J. Paine 1 Elm Close, London SW20 9HX
G8 BVB P. Power 8 The Fairway, Camberley GU15 1EF
G8 BVF J. Wearing 122 Dixon Drive, Chelford, Macclesfield SK11 9BX
G8 BVL M. Porter Birklands, 16 The Oval, Scarborough YO11 3AP
G8 BVQ R. Straker 26 Constance Crescent, Hayes, Bromley BR2 7QJ
G8 BVR G. Oddy 2 Manor Farm, Chard TA20 2EB
G8 BVU P. Reilly 19 Maunders Court, Liverpool L23 9YU
G8 BVY G. Spinks 40 Ferndale Avenue Walthamstow, London E17 9EH
G8 BWA M. Pollard 3 Highfield Road, Chertsey KT16 8BU
G8 BWH R. Robinson 1 John Dixon Lane, Darlington DL1 1HG
G8 BWK J. Harper 2 Wolves Mere Woolmer Green, Knebworth SG3 6JW
G8 BWP C. Jones 2 Windmill Crescent, Wolverhampton WV3 8HY
GW8 BWX A. Hancock 38 High Street, Pontycymer, Bridgend CF32 8HY
G8 BXA A. Nicol 18 Lower End, Swaffham Prior, Cambridge CB25 0HT
G8 BXC R. Clark 41 Avenue Road, Bexleyheath DA7 4
G8 BXD Dr R. Edgecombe 48 Birchwood Road, Woolaston, Lydney GL15 6PE
G8 BXH J. Pryke 52 Oaklands Avenue, Watford WD19 4LW
G8 BXJ A. Pullen 22700 Gault Street, West Hills, Ca 91307-2306 United States
G8 BXO J. Stacey 3 West Park, South Molton EX36 4HJ
G8 BXQ T. Hordley 9 Newtown, Charlton Marshall, Blandford Forum DT11 9NN
G8 BYB A. Hebden 1 Ringwood Avenue, Newbold, Chesterfield S41 8RA
G8 BYC C. Keen Brighton Road, Radio Relay, Lewes BN7 3JL
G8 BYI R. Burrows 76 Southfield, Southwick, Trowbridge BA14 9PW
G8 BZJ A. Matheson 1 St. Edmunds Close, Bromeswell, Woodbridge IP12 2PL
G8 BZL G. Lindsay 4 Downs View, Hove BN3 8EN
GW8 BZN D. Goadby Ty Mawr, Bryncroes, Pwllheli LL53 8EH
GM8 BZP D. Joiner 8 Damask Crescent, Newmachar, Aberdeen AB21 0NG
G8 BZR P. Clark 6 Littlefield Lane, Sixpenny Handley SP5 5NP
G8 BZT D. Allen 156 Middlecotes, Tile Hill, Coventry CV4 9AZ
G8 CA Derby And District ARS c/o P. Cross Balls Farm Cottage, Musbury Road, Axminster EX13 8TT
G8 CAA C. Broomfield 8 Woodview Crescent, Hildenborough, Tonbridge TN11 9HD
G8 CAB J. Sawford 68 Harlyn Drive, Pinner HA5 2DA
G8 CAF R. Price Flat 3, Dippons House Dippons Drive, Wolverhampton WV6 8HJ
G8 CAH A. Parsons 153 Denman Drive, Ashford TW15 2AP
GW8 CAK P. Kenyon The Elvins, Norton, Presteigne LD8 2EP
G8 CAM I. Foster 22 Margetts Place, Lower Upnor, Rochester ME2 4XF
G8 CAU J. Borradaile 25 Inglewood Crescent, Carlisle CA2 6JJ
G8 CAV C. Isenman Bracklinn House, Broadlands Road, Brockenhurst SO42 7PB
G8 CBA G. Tipler Scotts House, Chorley, Bridgnorth WV16 6PR
G8 CBE K. Quarman 127 Highfield Lane, Hemel Hempstead HP2 5JG
G8 CBO K. Smith 6 Hermitage Close, North Mundham, Chichester PO20 1JZ
G8 CBU R. Aldous 23 Aldhous Close, Luton LU3 2LZ
G8 CCD J. Hodge 71 Rawcliffe Road Walton, Liverpool L9 1AN
G8 CCF S. Hall Knackershole Barn, Dulverton TA22 9RU
G8 CCJ D. Petri 42 Lucas Road, Snodland ME6 5PY
G8 CCL J. White 18 Sawyers Road, Tolleshunt Major, Maldon CM9 8NE
G8 CCN R. Read 76 School Road, Downham, Billericay CM11 1QN
G8 CCO J. Hess 3 Havana Court, Eastbourne BN23 5UH
G8 CCV M. O'Donnell 40 Mercers Drive Bradville, Milton Keynes MK13 7AY
G8 CDA M. Richards Copperknobs, High Street, Stockbridge SO20 6HE
G8 CDB P. Strudwick 20 New Road, Broomfield CM1 7AN
G8 CDC Capt. P. Jones March House, Burnthurst Lane, Princethorpe, Rugby CV23 9QA
G8 CDD R. Leman Crundalls Farmhouse Gedges Hill Matfield, Tonbridge TN12 7EA
G8 CDG N. Broadbent 2 Market Hill, Clare, Nr Sudbury CO10 8NN
G8 CDV T. Jeacock 9 Parkwood Rise, Barnby Dun, Doncaster DN3 1LY
GM8 CEA R. Spencer Pitagown House, Cluny, Newtonmore PH20 1BS
G8 CEE Yorkshire Dales CG c/o C. Carr 6 Jervaulx Road, Morton On Swale, Northallerton DL7 9RA

G8 CEP D. Clough 165 Pilgrims Way, Andover SP10 5HT
G8 CET W. Marsden 163 Buxton Old Road, Disley, Stockport SK12 2AY
G8 CEX B. Turner 50 Bosworth Road, Leigh-on-Sea SS9 5AB
GJ8 CEY A. Hearne Hearnes Hastle, 2 Teighmore Park, La Chevre Rue, Grouville, Jersey JE3 9EF
G8 CEZ R. Fuller 35 Chichester Walk Merley, Wimborne BH21 1SL
G8 CFD R. Rimmer 6 The Dene, Blackburn BB2 7QS
G8 CGM P. Raybould 115 Curlew Crescent, Bedford MK41 7HY
G8 CGW J. Elliott 92 Hinckley Road, Barwell, Leicester LE9 8DN
G8 CHA N. Blackburn 158 Dyas Road, Great Barr, Birmingham B44 8SW
G8 CHC B. King 32 Mayfield, Buckden, St. Neots PE19 5SZ
G8 CHI A. Tidder 3 Fernway Close, Wimborne BH21 2ST
G8 CHK R. King 28 Jenkinson Road, Towcester NN12 6AW
G8 CHN G. Barber 666 Bradford Road, Birkenshaw, Bradford BD11 2EE
G8 CHO S. Humm 235 Felmongers, Harlow CM20 3DP
G8 CHY K. Twort 39 Mile End Lane, Stockport SK2 6BN
GM8 CIF D. Macdonald 22 Drummie Road, Devonside, Tillicoultry FK13 6HT
G8 CIG P. Tester Gable Crest, Longburton, Sherborne DT9 5PD
G8 CIJ F. Fyfe 28 Whitton Close, Greatworth, Banbury OX17 2EH
G8 CIT W. Mckillop 2 Moores Green, Wokingham RG40 1QG
G8 CIX M. Maynard 41 Liverpool Avenue, The Pyramid, Southport PR8 3NP
G8 CJA Dr M. Dowson The Granary, St. Peters Road, Leicester LE8 5WJ
G8 CJD C. Hutton 25 Fiddlers Lane, East Bergholt, Colchester CO7 6SJ
GM8 CJG R. Kirsch Milntack House Laurieston, Castle Douglas DG72PW
G8 CJH D. Fletcher 17 Durley Chine Road South, Bournemouth BH2 5JT
G8 CJL A. Dorling 4 The Pastures, Rushmere St. Andrew, Ipswich IP4 5UQ
G8 CJM A. Croft 15 Blenheim Avenue, Chatham ME4 6UU
G8 CJQ R. Barnes 3 Ivy Cottages, Church Lane, Knutsford WA16 7RD
G8 CJT C. Coles 15 Somerdale Avenue, Bath BA2 2PG
GM8 CJW J. West Of Stow Stow Mill, Stow, Galashiels TD1 2RB
G8 CKB P. Ebsworth Olamyra 20, Forland, Steinsland 5379 Norway
GW8 CKJ A. Williams 54 St. Augustine Road Griffithstown, Griffithstown Pontypool NP4 5EZ
G8 CKK Amberley Radio Group c/o A. Zerafa 2 Furnwood, St. George, Bristol BS5 8ST
G8 CKN R. Powers The Dell, Hussell Lane, Alton GU34 5PF
G8 CKS J. Sargent The Coach House, Speltham Hill, Waterlooville PO7 4RU
G8 CKV S. Dale 30 Almond Road, Peterborough PE1 4LT
G8 CLI D. Hall 32 Fernwood, Marple Bridge, Stockport SK6 5BE
G8 CLJ I. Richmond 58 Cauldron Barn Road, Swanage BH19 1QF
G8 CLK K. Woollven 7 Heatherstone Avenue, Dibden Purlieu, Southampton SO45 4LR
G8 CLW J. Griffin 185 Eastcote Avenue, West Molesey KT8 2EX
G8 CLY J. Lythgoe 18 Ranleigh Walk, Harpenden AL5 1SR
G8 CLZ Qrz ARG of Sussex c/o J. Eade High Shaw Sandrock Hill, Sedlescombe TN33 0QR
G8 CMD A. Ashford 56 Guarlford Road, Malvern WR14 3QP
G8 CMG R. Williams 18 Woodford Crescent, Plymouth PL7 4QY
G8 CMK W. Blankley 16 Charles Road, St. Leonards-on-Sea TN38 0QA
G8 CMO R. Grounds 101 Honeysuckle Way, Witham CM8 2XQ
G8 CMP C. Heymans 10 Rushmore Drive, Widnes WA8 9QB
G8 CMU M. Adcock Rudhall Farm, Phocle Green, Ross-on-Wye HR9 7TL
GW8 CNF S. Biddiscombe 20 Arlington Close, Malpas, Newport NP20 6QF
GW8 CNS W. Mathias Grenan Bungalow, Highland Avenue, Bridgend CF32 9YH
G8 CON J. Beith 18 Avenue Road, New Milton BH25 5JP
G8 COR G. Peters 156 Preston Road, Whittle-Le-Woods, Chorley PR6 7HE
G8 CPA J. Vizor 31 Somerset Road, Swindon SN2 1NE
G8 CPF M. Edwards 14 Cheyney Walk, Westbury BA13 3UH
G8 CPJ I. Lever 23 Anton Road, Andover SP10 2EN
G8 CPK D. Hibbin 95A Thorpe Acre Road, Loughborough LE11 4LF
G8 CPM C. Mortlock 27 Baldwin Road, Greatstone, New Romney TN28 8SY
G8 CPN J. Hawkins Westhay Farm, Higher Clovelly, Bideford EX39 5SH
G8 CPQ V. Humphrey 5 Wistow Road, Luton LU3 2UR
G8 CQG P. Cornell 1 Orient Drive, Winchester SO22 6NZ
G8 CQH Dr P. Best 21 Greening Drive, Edgbaston, Birmingham B15 2XA
G8 CQQ A. Paterson 36 Bracadale Road, Nottingham NG5 5EE
G8 CQV W. Hunter 2 Green Acre, Goosnargh, Preston PR3 2BQ
G8 CQX J. Hawes 193 Leckhampton Road, Cheltenham GL530AD
G8 CQZ C. Powlesland The Ferns, Broad Street, Gloucester GL19 3BN
G8 CRB S. Blunt 53 Butt Lane Milton, Cambridge CB24 6DG
G8 CRC C. Callegari 16 Rustington Court, St. Johns Road, Eastbourne BN20 7HS
GW8 CRH I. Troughton Rhiwbina, Pentre Lane, Cwmbran NP44 3AP
G8 CRM P. Watson Tall Oak, 6 New Road, Bury St. Edmunds IP29 5QL
G8 CRV J. Christian 5 Towers Way Corfe Mullen, Wimborne BH21 3UA
G8 CRX S. Winford Mayflower, South Hanningfield Road, Chelmsford CM3 8HJ
G8 CRZ P. Hunt 17 Selfridge Avenue, Southbourne, Bournemouth BH6 4NB
G8 CSA Silverthorn RC c/o L. Butterfields 22 Horsley Road, London E4 7HX
GM8 CSE H. Hogarth 32 Broomhall Park, Edinburgh EH12 7PU
G8 CSK S. Browning 12 Sunderland Close, Woodley, Reading RG5 4XR
G8 CSQ P. Benson Ashbank Bungalow, Bentham, Lancaster LA2 7HX
G8 CSR J. Credland Lieu-Dit Cornier, Prayssas 47360 France
G8 CTB K. Chambers 24 Primrose Close, Flitwick, Bedford MK45 1PJ
G8 CTD A. Tait Birch Glen, 71 Twemlows Avenue, Whitchurch SY13 2HD
G8 CTJ M. Maxey 28 Herald Way, Burbage, Hinckley LE10 2NX
G8 CTR Dr D. Upton Polwin, Budock Water, Falmouth TR11 5DT
G8 CTX C. Havercroft 28 Anglers Way, Cambridge CB4 1TZ
G8 CUA R. Boittier 5 The Crescent, Harlow CM17 0HN
G8 CUB R. Ray Little Mallards, Mallard Way, Brentwood CM13 2NF
G8 CUG P. Cockram 14 Langshott Close, Woodham, Addlestone KT15 3SE
G8 CUL M. Stevens 67 New Road, East Hagbourne, Didcot OX11 9JX
G8 CUN G. Rawlings 109 The Upway, Basildon SS14 2JD
G8 CUW S. Thackery 19 Pyne Point, Clevedon BS21 7RL
G8 CUX D. Stanton 106 Scrapsgate Road, Minster On Sea, Sheerness ME12 2DJ
G8 CVF P. Dobson 3 Wallingford Road, Wirral CH49 6PW
GM8 CVN J. Struthers 79 Woodfield Park, Colinton, Edinburgh EH13 0RA
G8 CVP R. Perry 49 Harwich Road, Little Clacton, Clacton-on-Sea CO16 9NE
G8 CVQ A. Parr 8 Kingston Avenue, North Cheam, Sutton SM3 9TZ
G8 CVS J. Jenkinson 26 Blenheim Drive, Oxford OX2 8DG
G8 CW Essex CW Contest Club c/o P. Tittensor 47 St. Johns Road, Chelmsford CM2 0TY
G8 CWE T. Cook 141 Station Road, Watlington, King's Lynn PE33 0JG
G8 CWJ J. Abbott 20 Highbury Avenue, Salisbury SP2 7EX
G8 CWQ G. Horsfall Lancaster New Road, Garstang, Preston PR3 1AD
G8 CXA D. Froggatt 2 Cobden Avenue, Mexborough S64 0AD
G8 CXF J. Lucas 48 Sycamore Drive, Ash Vale, Aldershot GU12 5PR
G8 CXI D. Phillips 13 Bowford Avenue, Bexleyheath DA7 4ST
G8 CXK G. Peck 45 Bentley Close, Northampton NN3 5JS
G8 CXT D. Coxhill 82 Williams Close, Hanslope, Milton Keynes MK19 7BT
G8 CXV R. Brown 19C Arlington Drive, Mapperley Park, Nottingham NG3 5EN
G8 CXW P. Appleby 23 Oban Drive, Ashton-In-Makerfield, Wigan WN4 0SJ
G8 CXZ M. Mills 145 Park St., Haydock, St. Helens WA11 0BL
G8 CYA N. Parker 10 Lockhart Close, Kenilworth CV8 1RB
G8 CYE S. Cook 24 Beaufort Court, Beaufort Road, Richmond TW10 7YG
G8 CYF M. Bucknall Driftaway, White Oak Green, Hailey, Witney OX29 9XP
G8 CYG W. Steer Downside, Membury, Axminster EX13 7AF
G8 CYK W. Poel Hockham Hill, Spring Elms Lane, Little Baddow, Chelmsford CM3 4SD
G8 CYL P. Smith Andelain, Drift Road Whitehill, Bordon GU35 9DZ
G8 CYT F. White 12 Burcombe Road, Bournemouth BH10 5JT
G8 CYU P. York-Jones 18 Solway Road, Cheltenham GL51 0LZ
G8 CYW S. Wisher 17 Kenmore Crescent Greenside, Ryton NE40 4QY
G8 CYX D. Storey 43 Harwood Close, Welwyn Garden City AL8 7ST
G8 CZE F. Beesley 9 Northway, Droylsden, Manchester M43 6EF
G8 CZG Colchester CG c/o Lord D. Bell 25A Mill Lane, , Great Harwood, Blackburn BB6 7UQ
G8 CZI D. Paterson 3 Shawcroft Close, Shaw, Oldham OL2 7DA
G8 CZJ J. Meredith 25 Frankel Avenue, Redhouse, Swindon SN25 2NJ
G8 CZM K. Jones 3 Webb Avenue, Perton, Wolverhampton WV6 7YH
G8 CZP V. Maund 24 Elliott Crescent, Bedford MK41 0HL
G8 CZQ I. Bayliss West Common Lodge, West Common Close, Gerrards Cross L9 7QR
GM8 CZU I. Davidson 43 Whitehall Avenue, Cardenden, Lochgelly KY5 0PH
G8 CZW A. Buxton 31 Sandringham Road, Macclesfield SK10 1QB
G8 DAI A. Justin Garth, Park View Road, Pinner HA5 3YF
G8 DAM D. Goodway 35 South Avenue, Buxton SK17 6NQ
G8 DAY G. Teague Birkenweg 6, Eiselfing Bachmehring 835459 Germany
G8 DBD R. Taylor 54 Portsmouth Road, Lee-on-the-Solent PO13 9AG
G8 DBH C. Wallwork Honeywicke Cottage, Honeywick Lane, Dunstable LU6 2BJ
G8 DBK P. Barker 24 Main Street, South Croxton, Leicester LE7 3RJ
G8 DBO K. Smith Wilson Hall Farm, Slade Lane, Wilson, Derby DE73 8AG
G8 DBP J. Mills 93 Gays Road, Hanham, Bristol BS15 3JX
G8 DBU N. Greensted High View Oust Care Home, Poulton Lane, Canterbury CT3 2NH
G8 DCD J. Durrant 27 Trafford Road, Willerby, Hull HU10 6AJ
G8 DCJ P. Mcquail 3 Post Office Lane, Draycott, Moreton-in-Marsh GL56 9JZ
G8 DCX R. Sangster 10 Addison Road, Banbury OX16 9DH
G8 DD South Notts ARC c/o D. Hill 86 The Downs, Nottingham NG11 7EB
G8 DDC Dunstable Dwn Rd c/o C. Asquith 36 Sunningdale, Luton LU2 7TE
G8 DDH M. Lelliott Well Lane Corner, Lower Froyle, Alton GU34 4LJ
G8 DDN P. Bennett Whitelands, Common Mead Lane, Gillingham SP8 4RB
G8 DDR S. Collinge 11 Glendale Grove, Spital CH63 9FP
G8 DDY P. Thompson 1A Downside Avenue, Niton, Ventnor PO38 2DE
G8 DEC A. Malcolm 68 Old Birmingham Road, Lickey End, Bromsgrove B60 1DG
G8 DEJ T. Ray 1 Providence Lane, Leamore, Walsall WS3 2AQ
G8 DEL D. Coppen 100 Atbara Road, Teddington TW11 9PD
G8 DEM B. Willetts 11 Albert Road, Warley, Oldbury B68 0NA
G8 DER R. Richardson Hazeldene, Sutton Road Fovant, Salisbury SP3 5LF
G8 DET J. Bowen 6 Bishops Court Gardens, Chelmsford CM2 6AZ
G8 DEX J. Hosking 21 Yeo Valley Way, Wraxall, Bristol BS48 1PS
G8 DEY D. Parr 58 Ritson St., Toxteth, Liverpool L8 0UF
GM8 DFC Dr R. Cliff 32 Lochardil Road, Inverness IV2 4LD
G8 DFI B. Oliver 6 Catherton Road, Cleobury Mortimer, Kidderminster DY14 8EB
GM8 DFX Rvd. J. Lincoln 59 Obsdale Park, Alness IV17 0TR
GI8 DGB B. Moore 34A Feumore Road, Ballinderry Upper, Lisburn BT28 2LH
G8 DGC S. Hall 3 Sleepers Delle Gardens, Winchester SO22 4NU
G8 DGH E. Townsend The Manor House, Leicester LE8 0AP
G8 DGR R. Smallwood The Island, Hyde End Lane, Brimpton RG74TH
G8 DGW M. Wickham 8 Verlands Close Niton, Ventnor PO38 2BG
G8 DHA D. Bishop Oyston Lodge, Lynstone Road, Bude EX23 8LR
G8 DHE G. Mather 72 Cranleigh Road, Worthing BN14 7QW
G8 DHF S. Matthews 213 Hucclecote Road, Gloucester GL3 3TZ
G8 DHI G. Roberts 56 Horse Shoes Lane, Birmingham B26 3HY
G8 DHJ C. Pickering 28 George V Avenue, Margate CT9 5QA
G8 DHQ D. Digby 73 Bedford Street, Crewe CW2 6JB
GW8 DHT J. Clifford Dippers Barn Coppice Lane Pool Quay, Welshpool SY21 9JY
G8 DHU M. Baxter 11B The Leys, Roade, Northampton NN7 2NR
G8 DHV N. Eaton 3 Thirslet Drive, Heybridge, Maldon CM9 4YN
GI8 DHW J. Hendron 9 Drumahiskey Road, Bendooragh, Ballymoney BT53 7QL
G8 DIQ T. Hall 7 Sweetlake Cottage, Nobold, Shrewsbury SY5 8NH
G8 DIR K. Walker 12 Willow Park, Minsterley, Shrewsbury SY5 0EH
G8 DIU B. Cannon 52 Goodhew Close, Yapton, Arundel BN18 0JA
G8 DIY P. Geeson 35 Brewhouse Lane, Soham, Ely CB7 5JD
G8 DJF A. Dickson 7 Sandford Gardens, High Wycombe HP11 1QT
G8 DJL J. Renaut 33 Ellesfield Drive, West Parley, Ferndown BH22 8QN
G8 DJO M. Adcock 37 Ashpole Road, Braintree CM75LW
G8 DJT G. Platts 1 Blacksmiths Court, Kingham, Chipping Norton OX7 6GE
G8 DJU J. Frisby 66 Clear Crescent, Melbourn, Royston SG8 6JD
G8 DJW G. Membury 21 Webbers Piece, Maiden Newton, Dorchester DT2 0AQ
GM8 DKB E. Taynton 42 Craigmount Park, Edinburgh EH12 8EE
G8 DKD C. Weale Boston Farm, Boston Lane, Hinton-On-The-Green, Evesham WR11 2RD
GM8 DKG Dr C. Pegrum 4 Northampton Drive, Glasgow G12 0LE
G8 DKI D. Lucas The Old Barn, The Street, Malmesbury SN16 9DL
G8 DKK B. Harber 45 Brandles Road, Letchworth Garden City SG6 2JA
G8 DKV M. Coldicott The Old Cottage, Church Lane, Morley, Ilkeston DE7 6DE
G8 DKW M. Solomons 389 B Alexandra Avenue, Harrow HA2 9EF
G8 DLH A. Hall 19 Crewkerne Road, Chard TA20 1EZ
G8 DLL M. Monro 6 Yew Tree Road, Hayling Island PO11 0QE
G8 DLP R. Baker Royal Oak House, Crich, Derbyshire DE4 5BH
G8 DLX M. Crampton 55 Gilbert Ave, Bilton, Rugby CV22 7BZ

G8 DLZ P. Lea 7 Cressex Road, High Wycombe HP12 4PG
G8 DML J. Hughes 12 Plough Garth, Kellington, Goole DN14 0PD
G8 DMN S. Rundle 4 Bridge Close, Evercreech, Shepton Mallet BA4 6LZ
G8 DMT M. Caley 40 Spenser Way, Jaywick, Clacton-on-Sea CO15 2QT
G8 DMU A. Frazer 11A, Leadhall Way Keld House, Harrogate HG29PG
G8 DNH J. Webber 21 Highfield Court, Wigton CA7 9DR
G8 DNL K. Smith 19 Westfield Avenue, South Croydon CR2 9JY
G8 DNP P. Donoghue Hillcrest, The Green, Harlow CM17 0QR
GW8 DOA G. Pollard 3 Carey Walk, Neath SA10 7DD
G8 DOB I. Stuart 87 Redgrove Park Hatherley Lane, Cheltenham GL51 6QZ
G8 DOF P. White 6 Curzon Court, Curzon Street, Chester CH4 8PA
G8 DOH Dr A. Seeds 114 Beaufort Street, London SW3 6BU
GM8 DOR A. Barrett Mains Of Glasclune Farm, , Blairgowrie PH10 6SF
G8 DOW B. Lee 19 Lizard Head, Littlehampton BN17 6RY
G8 DOY R. Elliott Flat 7, Queen Mother Court, 151 Sellywood Road, Birmingham B30 1TH
G8 DPE V. Brooks 19 Malham Avenue, Wigan WN3 5PR
G8 DPH T. Booth 155 Oxford Road, Windsor SL4 5DX
G8 DPQ D. Hendon 2 Ellis Avenue, Onslow Village, Guildford GU2 7SR
GM8 DPV J. Hunting 77 Califer Road, Forres IV36 1JB
G8 DPW D. Holden 63 High St., Queenborough ME11 5AG
G8 DQD T. Taylor 15 Kennard Road, Bristol BS15 8AA
G8 DQE R. Lees 6 Library Road, Ferndown BH22 9JP
G8 DQF L. Johnston 9 Tunbridge Close, Burwell, Cambridge CB25 0EL
G8 DQK A. Symonds 19 Danby Terrace, Exmouth EX8 1QS
G8 DQN N. Hunter 33 Chapel Court, Billericay CM12 9LX
G8 DQP J. Peden 51A Bewdley Road, Kidderminster DY11 6RL
GM8 DRA R. Macleod 9 Croftcroighn Gate, Glasgow G33 5JJ
G8 DRB K. Slee 4 Dibbinview Grove, Wirral CH63 9FW
G8 DRE D. Atkinson Colne House, Robinson Road, Brightlingsea CO7 0ST
G8 DRK R. Vince 5 Bay Tree Road, Bath BA1 6NA
G8 DRQ Dr R. Cochrane 134 Moor Lane South, Ravenfield, Rotherham S65 4QR
G8 DSG W. Jones Elm Hurst, Station Road, Shrewsbury SY4 2BB
G8 DSM J. Witherspoon 109 Bromsgrove Road, Redditch B97 4RL
GW8 DSO C. Warwick 33 Ceri Road, Townhill, Swansea SA1 6LS
G8 DST G. Smith 23 Whaggs Lane, Whickham, Newcastle upon Tyne NE16 4PF
G8 DSU R. Gill 61 Cross Deep Gardens, Twickenham TW1 4QZ
G8 DTA A. Parsons 20 Paddocks Lane, Prestbury, Cheltenham GL50 4NX
G8 DTE M. Pusey 6 Blagdon Close, Martinstown, Dorchester DT2 9JT
G8 DTF R. Price 29 Birchfield Drive Worsley, Manchester M28 1ND
G8 DTM F. Partington 21 East Road, Wymeswold, Loughborough LE12 6ST
G8 DTQ B. Petifer 14 Wood Lane, Caterham CR3 5RT
G8 DTS B. Norcliffe 2 Alexander Drive, Heswall, Wirral CH61 6XT
G8 DTT W. Moore 26 Richard Moon St., Crewe CW1 3AX
G8 DTX I. Sanderson 15 Gorse Road, Huddersfield HD3 4BN
G8 DUF R. Bird 129 Park Road, Formby, Liverpool L37 6AD
G8 DUI D. Cox 52 Avill Crescent, Taunton TA1 2PL
G8 DUO I. Casewell 7 Pine Drive, Finchampstead, Wokingham RG40 3LD
GW8 DUP R. Harris 64 Frederick Place, Llansamlet, Swansea SA7 9SX
G8 DUT H. Orgel 1 Taunton Grove, Whitefield, Manchester M45 6TJ
G8 DUV Dr C. Zammit 9 Sandbanks Drive, Hatch Warren, Basingstoke RG22 4UL
G8 DUW I. Redfern 8 Lilac Grove, Stourport-on-Severn DY13 8SR
GW8 DUY C. Davies 14 Twynpandy, Pontrhydyfen, Port Talbot SA12 9TW
G8 DVB J. Sandon 461 Archer Road, Pin Green, Stevenage SG1 5QP
G8 DVF T. Jones 5 Blue Hatch, Frodsham WA6 7QJ
G8 DVJ G. Wilks 8 Chestnut Grove, East Barnet, Barnet EN4 8PU
G8 DVK Letterston ARC c/o D. Aram 4 Severn Road, Chilton, Didcot OX11 0PW
G8 DVN D. Smith 2 Burnor Pool, Calverton, Nottingham NG14 6FL
G8 DVS A. Sterry 9 Finch Avenue, Wakefield WF2 6SE
G8 DVU R. West 55 Burney Bit, Pamber Heath, Tadley RG26 3TL
G8 DVW R. Leadbeater The Birches, Torpenhow, Wigton CA7 1JF
G8 DWF N. Earl 162 Winchmore Hill Road, London N21 1QP
G8 DWP P. Lee 223 Chelmsford Road, Shenfield, Brentwood CM15 8SA
G8 DWW C. Garcia 8 Lyme Road, Bath BA1 3LN
G8 DWX G. Haslip 1 Sea Cottages, 28 Steyne Road, Seaford BN25 1QF
GW8 DX J. White Keepers Lodge Pumpsaint, Llanwrda SA19 8DX
G8 DXF C. Tarran Woodlands, School Road, Romsey SO51 6AR
G8 DXI Huntingdonshire ARS c/o W. O'Connor 3 Sterndale Close, Desborough, Kettering NN14 2XL
G8 DXM C. Taylor 45 Greenfield St., Shrewsbury SY1 2PY
G8 DXO R. Humble 3 Plover Close, Milborne Court, Sherborne DT9 5DD
G8 DXP A. Cheasley 25 Normanhurst Road, Walton-on-Thames KT12 3EQ
G8 DXU B. Pollard-Wilkins Seacall Limited, 16 Seabeach Lane, Eastbourne BN22 7JG
G8 DXV H. King 11 Priory Mead, Doddinghurst, Brentwood CM15 0NB
G8 DXZ J. Sandys 28 The Maultway, Camberley GU15 1PS
G8 DYA C. West 14 Ashleigh Gardens, Wymondham NR18 0EX
G8 DYG M. Marshallsay 2 Prospect Cottages, Lime Street, Gloucester GL19 4NX
G8 DYI K. Holdway 18 Pennymore Close, Stoke-on-Trent ST4 8YQ
GM8 DYT J. Hotchin 2 Moorfield Place, Gatehead, Kilmarnock KA2 0AX
G8 DZC P. Martin 58 Hearn Road, Woodley, Reading RG5 3QG
G8 DZH J. Ray 7 Barn Mead, Theydon Bois, Epping CM16 7ET
G8 DZJ G. Booth 68 Tarragon Drive, Meir Heath, Stoke-on-Trent ST3 7YE
G8 DZN B. Bird 7 Old Kingsdown Close, Broadstairs CT10 2HG
G8 DZW R. Brookes 29 Ripley Road, Liversedge WF156QE
G8 EAD M. Hutchings 109 Longlands Way, Heatherside, Camberley GU15 1RU
G8 EAH I. Carress 1 Riplingham Road, Skidby, Cottingham HU16 5TR
G8 EAJ Prof. P. Cannon Field Cottage, Mathon Road, Malvern WR13 6ER
G8 EAM John Newton Memorial RC c/o R. Newton 8 Old Farm Road, Minehead TA24 8AS
G8 EAN J. Cunningham 62 Kings Hill, Beech, Alton GU34 4AN
G8 EAX S. Herod 8 Deben Way, Felixstowe IP11 2NS
G8 EBD G. Welch 18 Alderdale, Wolverhampton WV3 9JF
G8 EBM S. Haseldine Newton House, Bretby Lane, Newton Solney, Burton-on-Trent DE15 0RY
G8 EBQ R. Martin 10 Westways, Stoneleigh, Epsom KT19 0PQ
G8 EBT R. Lees Hurlands, Hurlands Lane, Godalming GU8 4NT
G8 EBX P. Starling 14 Merton Place, Littlebury, Saffron Walden CB11 4TH
G8 ECG K. Montgomery The Old Village Post Office, High St., Oxford OX44 9HP
G8 ECI D. Brown 8 Waddingham Place, New Waltham DN36 4QY
G8 ECR P. Jago 39 Royal Avenue, Flat 2, London SW3 4QE
G8 ECZ Wyre ARG c/o P. Barker 14 Elsworth Green, Newcastle upon Tyne NE5 3YB
G8 EDN T. Gallagher 35 Wilhelmina Avenue, Coulsdon CR5 1NL
G8 EDQ C. Soundy 16 Crane Cottages, West Cranmore, Shepton Mallet BA4 4QN
G8 EDS W. Hind 3 Birds Hill, Letchworth Garden City SG6 1PH
G8 EDX C. Vitiello 1 North Street, Rothersthorpe, Northampton NN7 3JB
G8 EEA D. Hill 872 Oldham Road, Rochdale OL11 2BN
G8 EEK B. Bruce Three Ways, Wisbech Road, Wisbech PE14 9RF
G8 EEM C. Gill 77 Main Road Hambleton, Selby YO8 9HW
G8 EEY A. Mobbs 149 The Paddocks, Old Catton, Norwich NR6 7HR
G8 EFG M. Vaughan 69 Seamore Avenue, Benfleet SS7 4EZ
G8 EFK E. Carter 44 Plattes Close, Shaw, Swindon SN5 5SA
G8 EFU C. Bloxidge 33 Rosemary Hill Road, Sutton Coldfield B74 4HL
G8 EGE J. Denton 32 Highfields Mead East Hanningfield, Chelmsford CM3 8XA
G8 EGG D. Hemingway Conygore Farm, Howell Hill, Yeovil BA22 7QZ
G8 EGL C. Burton 13 Newells Terrace, Misterton, Doncaster DN10 4DP
G8 EGM M. Booth 16 Falcon Drive, Birdwell, Barnsley S70 5SN
G8 EGU M. Smith 35 Queen Street, Balderton, Newark NG24 3NS
G8 EHD P. Brenton 40 Furneaux Road Milehouse, Plymouth PL2 3ET
G8 EHE K. Emerson Stone Gables, Upper Minety, Malmesbury SN16 9PR
G8 EHF J. Healen 12 Primrose Lane, Standish, Wigan WN6 0NR
G8 EHM Sir E. Vavasour 15 Mill Lane, Earl Shilton, Leicester LE9 7AW
GW8 EHO N. Holms 22 Heol Isaf Radyr, Cardiff CF15 8AL
GW8 EHQ J. Brown 106 Marlborough Road, Penylan, Cardiff CF23 5BY
G8 EHS Dr A. Fletcher 40, Dereham Avenue, Ipswich IP3 0QB
G8 EHX M. Melbourne 42 Pasture Road, Stapleford, Nottingham NG9 8GL
G8 EIE R. Forster 7 Western Way, Alverstoke, Gosport PO12 2NE
G8 EII M. Smith 3 Arnwood Drive, Bransgore, Christchurch BH23 8FH
G8 EIN N. Shepherd 166 Chaldon Way, Coulsdon CR5 1DF
G8 EJC R. Drew 9 Sona Merg Close Heamoor, Penzance TR18 3QL
G8 EJQ P. Vaughan 15 Humber Gardens, Wellingborough NN8 5WE
G8 EKD M. Nilson 9 Middlemead, Folkestone CT19 5UB
GM8 EKF F. Benson 53 Warriston Drive, Edinburgh EH3 5NA
G8 EKG G. Newstead 97 Hawthorn Crescent, Burton-on-Trent DE15 9QN
G8 EKN M. Biltcliffe 4 Fleming Close, Bicester OX26 2YA
G8 EKW G. Thornton 4 Fir Tree Close, Exmouth EX8 4EU
G8 EKZ A. Jones 97A Bakers Ground, Stoke Gifford, Bristol BS34 8GD
G8 ELG E. Joyce 34 Milton Avenue, Eaton Ford, St. Neots PE19 7LE
G8 ELH D. Fisher 17 Thrushel Close, Swindon SN25 3PP
G8 ELP A. Stockley Blacksole House, The Boulevard, Herne Bay CT6 6GZ
G8 ELW R. Straker 15 Rue Robert Garnier, le Mans 72000 France
G8 EMA D. Pedley Penn Gere, Hindon Road, Dinton, Salisbury SP3 5HW
G8 EMB W. Tickell 26 Shear Brow, Blackburn BB1 7EX
G8 EMH D. Roebuck 7 Elm Tree Close, North Anston, Sheffield S25 4FG
G8 EMU J. Wheeler 9 Elmer Close, Malmesbury SN16 9UE
G8 EMX G. Hankins 92 Sunningdale Road, Birmingham B11 3QJ
G8 EMY Mike-Whiskey DX Group c/o K. Britain Blenheim Cottage, Falkenham, Ipswich IP10 0QU
G8 ENA E. Fellows 343 Wake Green Road, Birmingham B13 0BH
G8 ENB R. Whitby 138 Browns Lane, Stanton-On-The-Wolds, Nottingham NG12 5BN
G8 END I. Bodie Thye Linney, Higher Poldown Farm, Truro TR13 0FB
G8 ENM A. Hall Rose Cottage, Burstall, Ipswich IP8 3DX
G8 ENS J. Morris 6 Barrowby Gate, Grantham NG31 7LT
G8 ENW P. Baker Top Of The Hill, Post Office Lane, Cheltenham GL52 3PS
G8 ENY R. Hersey 7 Tower Close, Brandon IP27 0LJ
G8 EOH G. Simpkins 12 Eastwood End, Wimblington PE15 0QJ
G8 EOJ E. March 23 Pebworth Close, Redditch B98 9JX
G8 EOM D. Garrard 48 Shorefields, Benfleet SS7 5BQ
G8 EOV B. Cross 3 The Meads, Haslemere GU27 1LA
G8 EOZ K. Waight 13 Kilda Road, Highworth, Swindon SN6 7HS
G8 EPC M. Dyke Cortijo Las Marrojas, Buzon 48 Palancar, 1820 Granada Spain
G8 EPH C. Kilvington 53 Hall St.Skegby, Sutton-in-Ashfield NG17 3EJ
G8 EPK D. Skye 16 Lulworth Avenue, Poole BH15 4DQ
G8 EPQ R. Prew 16 Stokenchurch Place, Bradwell Common, Milton Keynes MK13 8AT
G8 EPS G. Phelan 113 Albert Road, Epsom KT17 4EN
G8 EPZ C. Ward 4 The Hawthorns, Charvil, Reading RG10 9TS
G8 EQC D. Cliffe Common Farm, Riley Hill, Lichfield WS13 8JE
G8 EQD D. Wright 22 West Hill, Rotherham S61 2HB
GW8 EQI J. Fellows 8 The Links Gwernaffield, Mold CH7 5DZ
G8 EQO B. Tyler 842 Handsworth Road, North Vancouver V7R ZA2 Canada
G8 EQY F. Butler 511 Fulbridge Road, Peterborough PE4 6SB
G8 EQZ C. Reynolds 49 Westborough Way, Anlaby Common, Hull HU4 7SW
GW8 ERA M. Voss 9 Chapel Close, Garndiffaith, Pontypool NP4 7QS
G8 ERN R. Walker 12 Foldyard Close, Sutton Coldfield B76 1QZ
G8 ERV K. Blackman 7 Deanery Close, Ripley DE5 3TR
G8 ESK B. Kermode 7 Midgeham Grove, Harden, Bingley BD16 1DA
G8 ESL W. Miller 5 Givendale Close, Bridlington YO16 6GQ
G8 ESW W. Brade 51 Coventry Gardens, Herne Bay CT6 6SB
G8 ETD T. Rumble 5 Mulberry Way, Spalding PE11 2QJ
G8 ETI N. Foggin 12 Linnetsdene, Covingham, Swindon SN3 5AG
GM8 ETJ K. Mccartney Greystones, Eskdaill Street, Langholm DG13 0BG
G8 ETN S. Last 72 Humber Road, Chelmsford CM1 7PG
G8 ETP M. Furnival The Ballroom, Stokeley Manor, Stokenham, Kingsbridge TQ7 2SE
G8 ETR R. Cooke 4 New Forest Close, Far Forest, Kidderminster DY14 9TJ
G8 ETS D. Swale 369 Scalby Road, Scarborough YO12 6TG
G8 ETU A. Metcalf 10 Manor Bend, Galmpton, Brixham TQ5 0PB
G8 ETV P. Richardson 3 Butlers Close, Amersham HP6 5PY
G8 EUE M. Gasper The Barn, Back Road, Halesworth UK IP19 9DZ
G8 EUF C. Hall The Orchard, Arkholme, Carnforth LA6 1AX
GM8 EUG N. Robertson 10 Warrenpark Road, Largs KA30 8EF
GD8 EUH D. Pickard Mont Y Mer. St. Georges Crescent., Port Erin IM9 6HR
G8 EUV C. Fenton-Coopland 14 Chevril Court, Wickersley, Rotherham S66 2BN
G8 EVD T. Cartwright 132 Mere Road, Wigston LE18 3RL
G8 EVI Dr A. Clark 3 North Street, Owston Ferry, Doncaster DN9 1RT
G8 EVR K. Taylor 39 Bowerfield Crescent, Hazel Grove, Stockport SK7 6JB
G8 EVY Cambridge & District ARC c/o B. Davies 16 Pearmains Close, Orwell, Royston SG8 5QY

G8 EWC A. Rouse Clinton, Church Road, Colchester CO7 8HS
G8 EWD M. Smith 47 Salisbury Road, Market Drayton TF9 1AR
G8 EWF B. Gilbert 1 Wilmington Drive, Sutton-On-Sea, Mablethorpe LN12 2JU
G8 EWL C. Burgess Jalna, 12 Foley Close, Ashford TN24 0XA
G8 EWN D. Edmonds Great House Cottage, Great House Lane, Ripponden, , Sowerby Bridge HX6 4LQ
G8 EWT G. Diacon 45 Woodpecker Way, , Witney OX28 6NN
G8 EXF R. Slatter Ashwell House, Stratford Road, Oversley Green, Alcester B49 6PG
GD8 EXI Dr S. Baker Ballanarran House, Surby Road, Ballafesson, Port Erin IM9 6TE Isle of Man
G8 EXJ B. Jones 38 Wyresdale Road, Lancaster LA1 3DU
G8 EXK M. Hatch 6 Portland Street, Blyth NE24 1NP
G8 EXN C. Briggs 22 Woodlesford Crescent, Halifax HX2 0RB
G8 EXQ T. Connell 28 Tasman Close Corringham, Stanford-le-Hope SS17 7LD
G8 EXS P. Atherton 10 Cheriton Drive, Ravenshead, Nottingham NG15 9DG
GM8 EXU J. Steven Andor, Skitten, Wick KW1 4RX
G8 EXZ S. Warren 269 Upper Weston Lane, Southampton SO19 9HY
G8 EYA W. Rimmer 79 Brookhurst Avenue, Wirral CH63 0LA
G8 EYM N. Kearey 73 Wellesley Drive, Crowthorne RG45 6AL
G8 EYP Dr A. Faulkner 79A West Drive, Highfields Caldecote, Cambridge CB23 7RY
G8 EYQ J. Clee 34 Knebworth Road, Bexhill-on-Sea TN39 4JJ
G8 EYY M. Hancock 12 Mellor Road, Hillmorton, Rugby CV21 4BP
G8 EZB M. Whitlock 85 Antrobus Road, Sutton Coldfield B73 5EL
G8 EZD A. Gifford Broncroft, Rock Green Bank, Ludlow SY8 2DT
G8 EZE Mid Lanark ARS c/o P. Swallow 1 Auden Crescent, Ledbury HR8 2UU
G8 EZG A. Pybus Elm Bank Care Home 81, Northampton Road, Kettering. Northant?S, Kettering NN16 7JZ
G8 EZL T. Lambert 40 Deepdale Road, North Shields NE30 3AN
G8 EZR K. James 67 Drakes Way, Portishead, Bristol BS20 6LD
G8 EZT R. Elgy 130 Stebbing House, Queensdale Crescent, London W11 4TG
G8 EZU K. Darbyshire 24 Neston Road, Walshaw, Bury BL8 3DB
G8 EZV G. White 94 Wingate Road, Luton LU4 8PY
G8 EZZ R. Chambers 15 Barnfield Close, Braunton EX33 2HL
G8 FAB Southampton ARC c/o M. Troy 22 Jackie Wigg Gardens Totton, Totton, Southampton SO409LZ
G8 FAD W. Chown 7840 Sw 136Th Avenue, Beaverton 97008 United States
G8 FAK S. Sherratt 21 Tweedale Close, Mursley, Milton Keynes MK17 0SB
G8 FAR R. Elms Fernside, Great Burches Road, Benfleet SS7 3NA
G8 FAS S. Hotham 54 Devon Drive, Westbury BA133XQ
G8 FAT B. Haines 20 Westfield Gardens, Harrow HA3 9EJ
G8 FAX E. Bye 117 Bull Lane, Rayleigh SS6 8LZ
G8 FBA M. Hearne Thicket Cottage, Crawley Down Road Felbridge, East Grinstead RH19 2PS
G8 FBF D. Fellows 10 Benning Way, Wokingham RG40 1XX
G8 FBK L. West-Knights Kc 5 Bede House, Manor Fields, , London SW15 3LT
G8 FBM M. Bates 11 The Rise Partridge Green, Horsham RH13 8JB
G8 FBQ B. Corker 46 Danelaw, Great Lumley, Chester le Street DH3 4LU
G8 FBW A. Williams 16 Hillside Road, Penn, High Wycombe HP10 8JJ
G8 FC Royal Air Force ARS c/o R. Hyde 25 The Pastures, Cottesmore, Oakham LE15 7DZ
G8 FCO G. Onions 3 Tower Rise, Tividale, Oldbury B69 1NP
G8 FCQ M. Lister 246 Wigston Lane, Aylestone, Leicester LE2 8DH
G8 FCT R. Chadwick Ithaca, Heck Lane, Goole DN14 0RD
G8 FDE B. Mcmanus 6 Rowley Road, St. Neots PE19 1UF
G8 FDF J. Bastable 94 Baymead Lane, North Petherton, Bridgwater TA6 6RN
GW8 FDI G. Felton 10 Penbodeistedd, Llanfechell, Gwynedd LL68 0RE
G8 FDR M. Bingham 18 Ladywell Gate, Welton, Brough HU15 1NL
G8 FDZ D. Targett 10 Thames Mews, Poole BH15 1JY
G8 FEJ M. Woudstra Flat 1, 2 Upper Park Road, St. Leonards-on-Sea TN37 6SJ
G8 FEK E. Gawthorpe 35 Highfield Way, North Ferriby HU14 3BG
G8 FET J. Guppy 16 Barnfield Close, Hastings TN34 1TS
G8 FEZ F. Stuart 70 Peartree Road, Herne Bay CT6 7EQ
G8 FFA E. Davis 24 Redcar Avenue, Hereford HR4 9TJ
G8 FFC C. Mcmanus 6 Rowley Road, St. Neots PE19 1UF
G8 FFF C. Player 28 Darwin Walk Withersfield, Haverhill CB9 7ST
GM8 FFH Halkyn Radio Group c/o Dr D. Brown 14 Barloan Place, Dumbarton G82 3QW
GM8 FFK G. George 13 Balmoral Terrace, Elgin IV30 4JH
G8 FFM B. Jackson 23 Rylands Heath, Luton LU2 8TZ
G8 FFU C. Burrows 6 Brook Way, Lower Somersham, Ipswich IP8 4PE
GW8 FFW P. Rycroft Glyn, Llanerfyl, Welshpool SY21 0JD
GM8 FFX G. Knight 6 Findon Road, Findon, Aberdeen AB12 3RN
G8 FFZ P. Ewington 26 Dickens Road, Rugby CV22 5RW
G8 FGB S. Whitehead 74 Manchester Road, Haslingden, Rossendale BB4 5TE
G8 FGN H. Elstob 25 Lindfield Avenue, Seaford BN25 4DU
G8 FGQ H. Brittan Meadowhurst Cottage, Woodcock Heath, Uttoxeter ST14 8QS
G8 FGR K. Balch 10 Heming Place, Stoke-on-Trent ST2 9DF
G8 FGY P. Griffiths 5 Chestnut Crescent, Carlton Colville, Lowestoft NR33 8BQ
G8 FGZ C. Boon Corner Cottage, Brackenhill, Nottingham NG14 7EF
G8 FHC M. Passam Birchenbower, Birchendale, Stoke-on-Trent ST10 4HL
G8 FHI M. Clarke 2 The Grove Penton Grafton, Andover SP11 0RS
GM8 FHK J. Gallacher 23 East Avenue, Carluke ML8 5TS
G8 FHL Dr D. Green 83 Wigan Lower Road, Standish Lower Ground, Wigan WN6 8LJ
G8 FIE Dr N. Mcfetridge 16 Blagrove Lane, Wokingham RG41 4BA
G8 FIF Dr D. Howlett 11 Barfleur Rise, Lyme Regis DT7 3QY
G8 FIG C. Cole 157 Cherry Tree Road, Beaconsfield HP9 1BD
G8 FJA P. Webster 3 Eden Avenue, Morecambe LA4 6QL
G8 FJG R. Shoulder 264 Wennington Road, Rainham RM13 9UU
G8 FJR D. Jowett 59 Old Road, Thornton, Bradford BD13 3DQ
G8 FKF C. Sargeant Northview, 20 South Marsh Road, Grimsby DN41 8AN
G8 FKH Marconi Radio Group c/o D. Balharrie 27 Norfolk Road, Uxbridge UB8 1BL
GM8 FKL G. Twibell Mambeg, Dervaig, Tobermory, Isle of Mull PA75 6QN
G8 FLL D. Roseaman 101 Westbrook, Bromham, Chippenham SN15 2EE
G8 FLS I. Maciver 160 Marsden Road, Burnley BB10 2QP
G8 FLV A. Nicholson 29 Quaker Lane, Northallerton DL6 1EE
G8 FMA E. Sillars 34 Sandown Road, Stevenage SG1 5SF
G8 FMC D. Keston 8 Copse Gate Winslow, Buckingham MK18 3HX
G8 FMD C. Wells 5 Hepplewhite Close, Baughurst, Tadley RG26 5HD
G8 FME A. Hilton 28 Eastern Esplanade, Broadstairs CT10 1DR
G8 FMI F. Steed 19 Chancery Lane, Debenham, Stowmarket IP14 6RN
GM8 FMR D. Taylor 14 Fenton Street, Alloa FK10 2DT
G8 FMT P. March Devonholme, Bedford Road, Hitchin SG5 3RX
G8 FMW R. Whitehouse 92 Willenhall Road, Bilston WV14 6NP
G8 FMX D. Beard 9 Bowgate, Gosberton, Spalding, PE11 4ND.
G8 FMZ P. Mcnamara Sunnybank Cottage, Lower Swell, Cheltenham GL54 1LG
G8 FNG P. Robinson 52 Lea Court, New Road, Crewe CW3 9DN
G8 FNH M. Nash 12 Ruston Park, Rustington, Rustington, Littlehampton BN16 2AB
GW8 FNO R. Gregory 5 Bryn Castell, Radyr, Cardiff CF15 8RA
G8 FNR D. Stone 165 Wellington Hill West, Bristol BS9 4QW
GW8 FOL G. Spencer Tyn Cae, Llanfwrog, Anglesey LL65 4YL
G8 FOT B. Butterworth 21 Higher Drive, Purley CR8 2HQ
GW8 FOY L. Oakes Flat 2 Brython, 54-56 Lloyd St., Llandudno LL30 2YP
G8 FPA D. Hoult 1 East Mill Gate, Cherry Willingham LN3 4BZ
G8 FPG S. Banner Oedhofstrasse 18, Amstetten 3300 Austria
G8 FPU R. Hutton 5 Tollemache Road, Prenton CH43 8SU
G8 FPW F. Brown The Bungalow, Oxcroft Bank, Shepeau Stow, Spalding PE12 0TY
G8 FQN R. Schneider 15 Hope Lane, Upper Hale, Farnham GU9 0HY
G8 FQS Dr P. Simpson 17 Reynard Close, Horsham RH12 4GX
G8 FQZ C. Stocker 8 Brook Drive, Astley, Manchester M29 7HS
G8 FRH P. Lyall 10 Winston Court Old Road, Frinton on Sea CO13 9DG
G8 FRI J. Lucas 42 Westerleigh Road, Bath BA2 5JE
G8 FRJ P. Hayes 28 Rochester Road, Barnsley S71 2NJ
G8 FRS K. Gurr 119 Vaisey Road, Stratton, Cirencester GL7 2JW
G8 FRY N. Friday 140 Chapel Point Village, , Skegness PE245UZ
G8 FSJ R. Page 39 Carlton Street, Kettering NN16 8EB
G8 FSL A. Benham 141 The Close, Salisbury SP1 2EY
GW8 FSN B. Steadman 8 Machno Place, Denbigh LL16 3YA
GU8 FSU V. Rees Le Chene Lodge, Le Chene Hill, Forest GY8 0AJ Guernsey
G8 FSV A. Mason Ulitsa Svoboda 20, Banevo 8125 Bulgaria
G8 FTE R. Cowley 20 Mill Road, Willingham, Cambridge CB24 5UU
G8 FTP P. Jarrett 15 Groveside, East Rudham, King's Lynn PE31 8RL
G8 FTW R. Goodchild 48 Coral Drive, Ipswich IP1 5HS
G8 FTX D. Gotch The Bungalow, West Lane, Shipley BD17 5DW
G8 FTY C. Gray 135 Long Drive, Ruislip HA4 0HL
G8 FUB L. Jones 52 New Lane, Aughton, Ormskirk L39 4UD
G8 FUH S. Melling 15 Woodbridge Hill Gardens, Guildford GU2 8AR
G8 FUI W. Raybould 33 Roberts Green Road, Dudley DY3 2BB
G8 FUL J. Masterton 15 Maylins Drive, Sawbridgeworth CM21 9HG
G8 FUO R. Britton 12 Bulkeley Avenue, Windsor SL4 3LP
G8 FVC Farnborough CG c/o D. Mclay 6 Burton Road, Castle Gresley, Swadlincote DE11 9HD
G8 FVE K. Lake 79 Sherrards Way, Barnet EN5 2BP
GW8 FVI C. Reeves 37 Arnold Gardens, Kinmel Bay, Rhyl LL18 5NH
G8 FVJ D. Still 133A Feltham Road, Ashford TW15 1AB
G8 FVK Prof. P. Marks Ovinswell House, Low Street, Lastingham, York YO62 6TJ
G8 FVM P. Peake 34 Blackhalve Lane, Wolverhampton WV11 1BH
GM8 FVN G. Adams Heath Court, Morven Way, Ballater AB35 5SF
G8 FVT D. Bainton 86 Holywell Avenue, Whitley Bay NE26 3AD
G8 FWA J. Errington The Woodlands, Station Road, Leicester LE8 9FP
G8 FWC D. Sharp 2 Millbank Street, Goole DN14 5XF
G8 FWD T. Mckee 19 Wall Lane Terrace, Cheddleton, Leek ST13 7ED
G8 FWE J. Maidment Down Along, Cherry Lane, Barrow-upon-Humber DN19 7AX
G8 FWH J. Hill 21 Somersby Road, Mapperley, Nottingham NG3 5QB
G8 FWK J. Cranfield 65 Broome Manor Lane, Swindon SN3 1NB
G8 FWY P. Russell 57 Norburn Park, Witton Gilbert, Durham DH7 6SG
G8 FXA G. Griffiths 51 Bempton Road, Liverpool L17 5DB
G8 FXC M. Bradford 3 Veysey Close, Hemel Hempstead HP1 1XQ
G8 FXG N. Lay Orchard Cottage, Parkham, Bideford EX39 5PL
G8 FXL A. Patterson 139 Lowther Road, Bournemouth BH8 8NP
G8 FXM D. Toombs 1 Chalgrove, Welwyn Garden City AL7 2QJ
G8 FXN R. Thackeray 9 Jubilee Road, Formby L37 3HN
G8 FXU D. Percival Trebakken, 11 Lamborne Close, Sandhurst GU47 8JL
G8 FXV Dr M. White 2 Mill Close, Denmead, Waterlooville PO7 6PE
G8 FXX R. Limb Charnwood House, Station Road, Henley-on-Thames RG9 3JS
G8 FYK K. Payne Flat 4, Monton Bridge Court, Eccles M30 8UW
G8 FYX N. Fensch Glen Cottage, Bowl Road, Ashford TN27 0HB
G8 FZI M. Logsdon Pilgrims Cottage, Langford Budville, Wellington TA21 0RH
G8 FZT T. Unsworth Heathview, 15 Fenton Road, Huntingdon PE28 2SD
G8 FZV D. Ryan Turners Oak, Barrs Lane, Woking GU21 2JN
G8 FZW J. Brown 16 Greenwood Close, Moulton, Northampton NN3 7RD
G8 GAR H. Taylor 21 Windermere Road, Coulsdon CR5 2JF
G8 GAT Dr M. Smith 241 Sandbanks Road, Poole BH14 8EY
GM8 GAX P. Howson 1 Howetown Fishcross, Alloa FK10 3AW
G8 GBE P. Richardson 50 Amberley Road, Gosport PO12 4EW
G8 GBM R. Head 29 Kingslea Road, Solihull B91 1TQ
G8 GBP C. Fawdon 21 Bevan Close, Southampton SO19 9PE
G8 GBU D. Barker 311 Uttoxeter Road, Mickleover, Derby DE3 9AH
G8 GBY Hull And District ARS c/o B. Atkinson 165 Alliance Avenue, Hull HU3 6QY
G8 GCC Cannock Chase ARS c/o B. Gallear 5 Oak Avenue, Cannock WS12 4QA
G8 GCK G. Croome 10 Axford Close Gedling, Nottingham NG4 4BB
G8 GCM City & County of Swansea c/o J. Price 6 Kernick Road, Penryn TR10 8NX
G8 GCO N. Wall 9 North Close, Ipswich IP4 2TL
G8 GCS C. Coker 46 Clarendon Road, Ipplepen, , Newton Abbot TQ12 5QS
G8 GDC R. Laver 40 Middleton Close, Tysoe, Warwick CV35 0SS
G8 GDH D. Brown 56 Paddock Road, Staincross, Barnsley S75 6LE
GM8 GDN M. Brunton 2 Easter Place, Portlethen, Aberdeen AB12 4XL
G8 GDZ R. Thompson 23 Fox Hill, Selly Oak, Birmingham B29 4AG
G8 GEA K. Warriner Windover, 16 The Ridgeway, Eastbourne BN20 0EU
G8 GEB S. Rowsby 10 Echells Close, Bromsgrove B61 7EB
G8 GEE R. Sherwood 19 Norton Drive, Warwick CV34 5FE
G8 GET J. Shepherd 6 The Jordans, Coventry CV5 9JT
G8 GEV Dr J. Moore 17 Kings Grove, Barton, Cambridge CB23 7AZ
G8 GEZ L. Wooller 4 Old Court Close, Brighton BN1 8HF
G8 GFA R. Marshall The Village School, Upleatham, Redcar TS11 8AG
G8 GFB C. Jones 1 Primrose Hill Road Euxton, Chorley PR7 6BA
G8 GFF N. Sanderson 54 Kelvedon Close, Chelmsford CM1 4DG
G8 GFQ E. Hill 22 Botham Grove, Stoke-on-Trent ST6 5NX

G8 GFS M. Winiberg Summerhill, Smallhythe Road, Tenterden TN30 7NB
G8 GFW J. Douglas 1030 Shields Road, Newcastle upon Tyne NE6 4SR
G8 GFY D. King 108 Huddersfield Road, Meltham, Holmfirth HD9 4AG
G8 GFZ T. Cockram The Bungalow, Dyke Hill, Chard TA20 2PY
G8 GGI R. Geddes 107 Dukes Avenue, New Malden KT3 4HR
G8 GGM P. Burfoot 18 Ember Road, Langley, Slough SL3 8ED
G8 GGO P. Carson 16 Gaynes Park Road, Upminster RM14 2HJ
G8 GGR C. Coleman 19 Megdale, Matlock DE4 3JW
G8 GGS M. Clarke Ruskin, Ashurst Drive, Tadworth KT20 7LS
GW8 GGW N. Dudman Chapel House, Pen Y Bryn, Wrexham LL14 1UA
G8 GHH C. Gibbs 32 Gresham Avenue, Margate CT9 5EH
G8 GHK W. White 60 Parklands, Rochford SS4 1SH
G8 GHL S. Garland 53 The Crescent, Horsham RH12 1NA
G8 GHO J. Wood 17 Yew Tree Park Road Cheadle Hulme, Cheadle SK8 7EP
G8 GHP R. Whyte 64 Ash Tree Drive, West Kingsdown, Sevenoaks TN15 6LF
GM8 GHQ P. Laverock 3/6 Hopetoun Crescent, Edinburgh EH7 4AY
G8 GHR B. Farey 5 Ivel View, Sandy SG19 1AU
GM8 GHV W. Sherriffs Hillcrest, Disblair, Aberdeen AB21 0RJ
G8 GIF K. Turner 20 Rawdon Way, Faringdon SN7 7YT
G8 GIG A. Patterson Rose Cottage, Ingatestone Road, Ingatestone CM4 0RS
G8 GIH K. Foster 52 Bottesford Avenue, Scunthorpe DN16 3EN
G8 GIK J. Hart 28A Dunton Road Stewkley, Leighton Buzzard LU7 0HZ
G8 GIL M. Dimmock 44 Diddington Close, Bletchley, Milton Keynes MK2 3EB
GM8 GIQ C. Wearing 16 Campbell Drive, Troon KA10 6XE
G8 GIU Rugby Amateur Transmitting Society c/o J. Harman 13 Linthorpe Court, South Shields NE34 9BU
G8 GIZ Riverway ARS c/o D. Ollerhead 15 Kingsley Road, Chester CH3 5RR
G8 GJA P. Reeves 77 Cale Way, Wincanton BA9 9BS
G8 GJC J. Tillin 76 Holbrook Road, Belper DE56 1PB
G8 GJG N. Giltrow 7 The Square, Milton-Under-Wychwood, Chipping Norton OX7 6JN
GM8 GJI Suffolk Coastal Raynet c/o Dr E. Smith The Steading, Craigmyle, Banchory AB31 4LS
G8 GJM R. Harwood 9 Cornwall Close, Woosehill, Wokingham RG41 3AG
G8 GJO A. Heasman 170 Plum Lane, London SE18 3HF
G8 GJU M. Bernard 33 Station Road, Over, Cambridge CB24 5NJ
G8 GJV T. England 30 Sparrow Way, Burgess Hill RH15 9UL
G8 GJW C. Drouet Ash Trees, Hallfield, Dalston, Carlisle CA5 7QH
G8 GKC C. Ridley 39 Lancelot Road, Welling DA16 2HX
G8 GKH R. Hadley 36 Folly Lane, Cheltenham GL50 4BY
G8 GKL C. Rauch 40 Russett Close, King's Lynn PE30 3HB
G8 GKR M. Fellows 343 Wake Green Road, Birmingham B13 0BH
G8 GKX D. Nicholson Avenida EspaÒ□A, Edificio Sorrento, Malaga 29793 Spain
G8 GLB P. Brown 4 King Edgar Close, Ely CB6 1DP
G8 GLC I. Cooper 77A Benhill Wood Road, Sutton SM1 3SL
G8 GLD M. Bounford 62 High Street, Wood Lane, Stoke-on-Trent ST7 8PB
G8 GLI J. Husk Brandhu, Common Moor, Liskeard PL14 6EP
G8 GLL J. Woolford Daedalian, Trewoon Road Mullion, Helston TR12 7DS
G8 GLP B. Barnett 21 Primrose Walk, Maldon CM9 5JJ
G8 GLS G. Wimlett Yew Tree Lodge White Horse Lane Barton, Preston PR3 5AH
G8 GLV A. Brown 23 Vincents Way, Naphill, High Wycombe HP14 4RA
G8 GLY A. Higgins 86B Cranleigh Road, Bournemouth BH6 5JL
G8 GLZ A. Findlay 7 Market Square, , Winslow, MK18 3AB
G8 GMA D. Elliott 56 Lincoln Avenue, Willenhall WV13 1JQ
G8 GMB S. Bradshaw 82 Arden Way, Market Harborough LE16 7DD
G8 GML P. Melbourne 2 Jubilee Cottages, Jubilee Lane, Colchester CO7 7RY
G8 GMU B. Leathley-Andrew 4 Robinson Road, Bedworth CV12 0EL
G8 GNI Dr A. Thomas The Stone Barn, 1 Home Farm Close, Chesterton OX26 1TZ
G8 GNO S. Ferdenzi 4 Ashworth Road, Rossendale BB4 9JE
G8 GNX J. Bartholomew 33 Manor Way, Woodmansterne, Banstead SM7 3PN
G8 GNZ Newborough Radio Klub c/o G. Blake 22 Cannon Leys, Galleywood, Chelmsford CM2 8PD
GW8 GOC M. Black Mediascene Ltd, Unit A-D, Bowen Industrial Estate, Bargoed CF81 9AB
G8 GOM A. Ireson 32 The Avenue, Wellingborough NN8 4ET
G8 GON A. Jefford 37 Marions Way, Exmouth EX8 4LF
GW8 GOO P. Nelson 15 Hill Street Gerlan, Bethesda, Bangor LL57 3TD
G8 GOR A. Pearce 153 Henver Road, Newquay TR7 3EJ
G8 GOS K. Roche 96 Porter Road, Basingstoke RG22 4JR
G8 GOT D. Parkin 252 Standbridge Lane, Crigglestone, Wakefield WF4 3JA
G8 GPF D. Clark 6 Bradley Park Road, Torquay TQ1 4RD
G8 GPO Ofrac Baldock c/o D. Thorpe 70 Willow Way, , Ampthill MK45 2SP
GW8 GQE J. Moore Oak House, Falcondale Drive, Lampeter SA48 7SB
G8 GQF I. Mcenteggart 46 Bissley Drive, Maidenhead SL6 3UZ
G8 GQG J. Crow 58 Cooden Drive, Bexhill-on-Sea TN39 3AX
G8 GQJ R. Clark 9 Conigre, Chinnor OX39 4JY
G8 GQS B. Summers 9 Prior Croft Close, Camberley GU15 1DE
G8 GRB R. Day 20 Linacre Road, Torquay TQ2 8LF
G8 GRC J. Drakeley Rowan Cottage, Four Crosses Lane, Cannock WS11 1RU
G8 GRD L. Hetherington 5 Withey Close West, Bristol BS9 3SX
GD8 GRE C. Wilkinson The Grange 14 Montreux Court, Douglas IM2 6AF Isle of Man
G8 GRL K. Edwards 11 Foxes Road, Ashen, Sudbury CO10 8JS
G8 GRO R. Nicholls 10 Polmeere Road, Penzance TR18 3PD
G8 GRP E. Poole Ramillies Hall School, Ramillies Avenue, Cheadle SK8 7AJ
G8 GRQ A. Plail 46 Hayling Rise, Worthing BN13 3AG
G8 GRS Malvern Hills Radio Amateurs Club c/o R. Woodward 158 Highridge Road, Bishopsworth, Bristol BS13 8HU
G8 GRT R. Oakley 17 Windmill Close, Ellington, Huntingdon PE28 0AJ
G8 GSL I. Liston-Brown 20 Chatterton Avenue, Lichfield WS13 8EF
G8 GSU R. Wade The Hermitage 15 Heath Hill Road North, Crowthorne RG45 7PD
G8 GSY G. Young 6 The Maypole Thaxted, Dunmow CM6 2QZ
GW8 GT F. Clare Glen View, Newport Road, Caldicot NP26 3BZ
G8 GTD S. Porter 20 Newbridge Road, Ambergate, Belper DE56 2GR
G8 GTI K. Barnes 75 Southmeade, Liverpool L31 8EG
G8 GTR Dr D. Murray 27 Station Avenue, Walton-on-Thames KT12 1NF
G8 GTU P. Stephens 3 Inett Way, Droitwich WR9 0DN
G8 GTV B. Raby 10 Bulverton Park, Sidmouth EX10 9EW
G8 GTZ N. Matthews 12 Petrel Croft, Basingstoke RG22 5JY
G8 GUA R. Wood 339 Horse Road, Hilperton Marsh, Trowbridge BA14 7PE
G8 GUH G. Ohara 107 Castlesteads Drive, Carlisle CA2 7XD
GW8 GUJ J. Stubbs 14 The Glen, Langstone, Newport NP18 2NR
G8 GUN H. Parker 7 The Hollies, Clee Hill, Ludlow SY8 3NZ
G8 GUS M. Board 48 Skipper Way, Lee-on-the-Solent PO13 9EY
GM8 GUX J. Thomson 2 Wilton Hill, Hawick TD9 8BA
GW8 GVI M. Ritchie Bryngolman Farm, Llangolman, Clynderwen SA66 7QL
G8 GVL K. Woods 7 Ives Close, West Bridgford, Nottingham NG2 7LU
G8 GVN E. Shield 14 Wellwood Street, Amble NE65 0EL
G8 GVO C. Byard 32 Castlefields, Leominster HR6 8BJ
G8 GVV P. Richmond 57 The Fairway, Daventry NN11 4NW
G8 GVW P. Shillito Little Orchard, Thorney Road Kingsbury Episcopi, Martock TA12 6BG
G8 GVZ L. Sullivan 89 Richmond Crescent, Mossley, Ashton-under-Lyne OL5 9LQ
G8 GWB N. Awcock 16A Ongar Road, Writtle, Chelmsford CM1 3NU
G8 GWJ J. Vincent 12 Spelman Road, Norwich NR2 3NJ
G8 GWM N. Hay 20 The Ridgeway, Fetcham, Leatherhead KT22 9AZ
G8 GWP G. Atkinson Parkland Stables, Landmere Lane, Nottingham NG11 6ND
G8 GWR Dr S. Linney 164 Kineton Green Road, Solihull B92 7ES
G8 GWX R. Howells 52 Upton Park, Upton, Chester CH2 1DG
G8 GXF J. Ashmore 3 The Cedars, Stockwell Road, Wolverhampton WV6 9AZ
G8 GXN K. Raynor 17 Kirkstone Walk, Nuneaton CV11 6EZ
G8 GXO P. Rogers 26 Hall Lane, Sutton, Macclesfield SK11 0EP
G8 GXS Rvd. J. Hadjioannou The Vicarage, Wakefield Road, Pontefract WF9 5BX
G8 GYB V. Vesma Durvale, 5 Jonas Drive, Wadhurst TN5 6RJ
G8 GYI G. Stanley 133 Park Lane, Kidderminster DY11 6TE
G8 GYK G. Carter 19 Wych Elms, Park Street, St. Albans AL2 2AR
G8 GYL I. Bishop 5 Trent Close, Tolpuddle, Dorchester DT2 7HA
G8 GYM R. Claridge 124 Pemdevon Road, Croydon CR0 3QP
G8 GYP V. Holmes 104 York Avenue, Hayes UB3 2TP
G8 GYS P. Wright 1 Lambourne Way, Thruxton, Andover SP11 8NE
G8 GYV B. Simms 17 Peregrine Close, Weston-Super-Mare BS22 8UY
G8 GYX T. Ellinor 53 Hillside, Banstead SM7 1HG
G8 GYY C. Gregory 5 Fox Close, Wigginton, Tring HP23 6ED
G8 GZC G. Tew 73 King Cerdic Close, Chard TA20 2JB
GI8 GZM J. Mawhinney 12 Shane Park, Lurgan, Craigavon BT66 7HD
G8 GZN A. Hill 1 Greenways, Highcliffe, Christchurch BH23 5BA
G8 GZV R. Duke 5 Pembroke Close, Billericay CM12 0PF
G8 GZW Rvd. A. Davis 8 Roberts Road, Greatstone, New Romney TN28 8RL
G8 GZX J. Eadie 5 Silver Street Cublington, Leighton Buzzard LU7 0LJ
GW8 HAG T. Regan 2 Park View Gardens, Bassaleg, Newport NP10 8JZ
G8 HAM S. Collins 2 St Teresa Drive, Chippenham SN152BD
G8 HAU R. Lambarth 38 Kirkley Park Road, Lowestoft NR33 0LG
G8 HBQ P. Davies 24 Upland Grove, Leeds LS8 2SX
GM8 HBY C. Ross 16 Glebe Crescent, Airdrie ML6 7DH
G8 HBZ S. Stephenson 6 Livingstone Close Rothwell, Kettering NN14 6HT
G8 HCC N. Hunt F88 Trinidad House 3 Rossmore Road, Poole BH12 3ND
G8 HCK A. Rutter The Uplands, Castle Howard Road, Malton YO17 6NJ
G8 HCL V. Menday Huf House, Horseshoe Ridge, Weybridge KT13 0NR
G8 HCS H. Stratton 26 Marjorie Road, Chaddesden, Derby DE21 4HQ
G8 HCW C. Morgan 24 High Mead Royal Wootton Bassett, Sn4 8Lw, Swindon SN4 8LW
G8 HCZ Dr P. Iredale Mayfield, Woodlands Road, Raydon, Ipswich IP7 5LJ
GW8 HDH J. Dowdall 56 Goetre Bellaf Road, Dunvant, Swansea SA2 7RP
G8 HDJ P. Muxlow 17 Station Road, Grasby, Barnetby DN38 6AP
G8 HDL M. Connell 38 White Close, High Wycombe HP13 5NG
G8 HDM I. Arnold 44 Elwick Avenue, Acklam, Middlesbrough TS5 8NT
G8 HDP R. Jenkins 10 Ulstan Close, Woldingham, Caterham CR3 7EH
GW8 HEB T. Brady 8 Cefn Hawys, Red Bank, Welshpool SY21 7RH
G8 HER A. Lambert 2 Huxley Close, Locks Heath, Southampton SO31 6RR
G8 HEU P. Whitehead 7 Vulcan Road, Freckleton, Preston PR4 1JN
GW8 HF D. Phillips 34 Graig Terrace, Graig, Pontypridd CF37 1NH
G8 HFL L. Caine 25 Smallbrook Road, Broadway WR12 7EP
G8 HFW T. Hall 23 Burcott Gardens, Addlestone KT15 2DE
G8 HGG P. Abernethy Enfield House, Halford, Shipston-on-Stour CV36 5DA
G8 HGI M. Warriner Rapps Lodge, Rapps, Ilminster TA19 9LG
G8 HGL D. Lambert 4 Tamworth Road, Bedford MK41 8QY
G8 HGM K. Ellis 11 Ringwood Close, Eastbourne BN22 8UH
G8 HGN R. Harrison 59 Grange Road, Billericay CM11 2RQ
GM8 HHC P. Dick Napier House, 8 Colinton Road, Edinburgh EH10 5DS
G8 HHO M. Strange 60A Manor Road, Dersingham, King's Lynn PE31 6LH
G8 HHR J. Bardell 239 Meadow Road, Droitwich WR9 9BZ
G8 HI K. Burnitt 15 St. Bedes, East Boldon NE36 0LE
G8 HIG L. Cole 151 Carshalton Park Road, Carshalton SM5 3SF
G8 HIO T. Ellis Hollybush House, Hawley Green, Blackwater, Camberley GU17 9BP
G8 HIQ S. Whitehouse 2 Lindholme Drive, Rossington, Doncaster DN11 0UR
GW8 HJC C. Harper Deunant, Capel Curig, Betws-Y-Coed LL24 0DS
G8 HJD C. Tubis Rockleaze Mews, Rockleaze Avenue, Bristol BS9 1NG
G8 HJF C. Williams Kingsclere House, Fox'S Lane, Newbury RG20 5SL
G8 HJG C. Williams Kingsclere House, Fox'S Lane, Newbury RG20 5SL
G8 HJH M. Norton 179B Kimbolton Road, Bedford MK41 8DR
G8 HJK P. Hunt 40 Leighton Road, Toddington, Dunstable LU5 6AL
G8 HKK M. North 194 North Road, Combe Down, Bath BA25DN
G8 HKN Vision Youth Centre (Bovington) c/o R. Meakins 335 Court Road, Orpington BR6 9BZ
G8 HKP E. Jakins 2 South View Place, Midsomer Norton, Bath BA3 2AX
G8 HKS Guildford CG c/o M. Booth 30 Manor Green, Harwell, Didcot OX11 0DQ
G8 HLE R. Marshall 54 Tudor Avenue, Maidstone ME14 5HJ
G8 HLH Wellesley House RC c/o R. Wheeler 14 Robins Lane, St. Helens WA9 3NF
G8 HLJ E. Edwards Flat 27, Barncroft, Wirral CH61 6YH
G8 HLM S. Catlin 3 Manor Lane, Langham, Oakham LE15 7JL
G8 HLQ E. Birch 17 Canalside Cottages Chester Road, Preston Brook, Runcorn WA7 3AQ
G8 HMA R. Smith 5 Newton Close, Loughborough LE11 5UU
G8 HMG P. Walker 12 Brownlow Road, Redhill RH1 6AW
G8 HMJ M. Kellett Wistow Gate, Glen Road, Leicester LE8 9FH
G8 HMV J. Nicholas 4 Lion Lane, Clee Hill, Ludlow SY8 3NJ
G8 HMZ P. Cheseldine 6 Lissett Close, Lincoln LN6 0SY
G8 HNA S. Clark 1 Roman Road, Broadstone BH18 9DF

G8 HNM R. Parker 1 Whitmore Orchard, Whitmore Lane, Taunton TA2 6SR
G8 HNS R. Stanleigh 15 Lansdown Terrace Golden Hill, Bristol BS6 7YW
G8 HNT T. Thompson 25 Meadow Avenue, Codnor, Ripley DE5 9QN
G8 HOI R. Warner Barley Hill Farm, Combe St. Nicholas, Chard TA20 3HJ
G8 HOR E. York Combe Brune, Prayssac 46220 France
GW8 HOS V. Mander Meadowlands, Severn Lane, Welshpool SY21 7BB
G8 HOU H. Cox 21 North Avenue, Hayes UB3 2JE
G8 HPF Dr I. Mclenaghan 82 Cheam Road, Epsom KT17 1QP
G8 HPJ P. Beaumont 1 Byron Road, Mexborough S64 0DG
GW8 HPL W. Taylor Bywell, Chester Road, Rossett, Wrexham LL12 0HN
G8 HPN G. Staniewicz Flat 1, Jubilee Farm, Gillingham SP8 5SJ
G8 HPS A. Hancock 9 Elmside, Willand, Cullompton EX15 2RN
G8 HPV D. Green 67 Coombe Park Road, Binley, Coventry CV3 2NW
G8 HPW Newry High School RC c/o M. Hanaghan 9 Goole Road, Grindon Broadway, Sunderland SR4 8HT
G8 HPY A. Mander 18 Bridge Avenue, Otley LS21 2AA
GW8 HQM Raynet-UK c/o S. Bastow Bryn Goleu, Rhosgadfan, Caernarfon LL54 7LB
G8 HQO N. Johnson 56 Clarkson Avenue, Wisbech PE13 2EG
G8 HQW P. Kirby 2 Kneeton Park, Middleton Tyas, Richmond DL10 6SB
G8 HRA C. Ryalls 15 Belmont Way, South Elmsall, Pontefract WF9 2BT
G8 HRC Havering & District ARC c/o D. Nuttall 92 Long Road, Lowestoft NR33 9DH
G8 HRF K. Dodman 10 Newark Road, Lowestoft NR33 0LY
G8 HRW S. Watkin 9 Longden Close, Haynes, Bedford MK45 3PJ
G8 HSI J. Carey 7 Church Road, Walton on the Naze CO14 8DF
G8 HSR E. Warren 37 Kingston Drive, Mangotsfield, Bristol BS16 9BQ
G8 HSS M. Saxon 4 The Coppice, Impington, Cambridge CB24 9PP
G8 HST M. Sanders 19 Brunswick Gardens, Hainault, Ilford IG6 2QU
G8 HSV S. Otter Redlands Back Lane Bilsby, Alford LN13 9PT
GM8 HSY Dr H. Reekie 5 Golf Course Road, Bonnyrigg EH19 2EU
G8 HTA K. Parker 20 River Avenue, Hoddesdon EN11 0JS
G8 HTB A. Barker Bank Royd Barn, Bank Royd Lane, Halifax HX4 0EW
G8 HTF D. Fletcher 40 Bentham Drive, Liverpool L16 5EU
G8 HTM J. Taylor Perry House, 188 Walstead Road, Walsall WS5 4DN
G8 HTN A. Kettley 106 Denton Road, Audenshaw, Manchester M34 5BD
G8 HTO A. Farrell 206 London Road, Northampton NN4 8AU
G8 HTW P. Allen 46 Slade Avenue, Burntwood WS7 2EL
G8 HTZ S. Druitt 25 Holcroft, Orton Malborne, Peterborough PE2 5SL
GI8 HUD T. Huddleston 29 North Parade, Belfast BT7 2GF
G8 HUF S. Carpenter Fernlea, Fernhill Lane, Camberley GU17 9HA
G8 HUG I. Coulson 56 Potterdale Drive Little Weighton, Cottingham HU20 3UX
G8 HUH T. Rabbitts Laurel Cottage, Wick Lane, Highbridge TA9 4BU
G8 HUO D. Sharpe 37 Oulton Avenue, Bramley, Rotherham S66 2SS
G8 HUR N. Mills 3 Whitfield Close, Wilford, Nottingham NG11 7AU
GW8 HUS A. Mead 12 Wyelands View, Mathern, Chepstow NP16 6HN
G8 HUT N. Onions Windy Ridge, Dunmow Road, Dunmow CM6 3PJ
G8 HUV M. Rowlands 3 Littledown View, Great Durnford, Salisbury SP4 6AU
G8 HUY J. Hill 24 Hunters Hill Close, Guisborough TS14 7FH
G8 HVF C. Billson Knotts End, Bateman Road, Loughborough LE12 6NN
G8 HVT M. Evans 25 Walnut Close, Nailsea, Bristol BS48 4YH
G8 HVV C. Goadby Heligan, 12 School Road, Newmarket CB8 9RX
G8 HVX A. Staniforth 25 Brown Ct, East Brunswick 8816 United States
G8 HVZ R. Anderson 23 Heritage Court, Magdalene Street, Glastonbury BA6 9ER
G8 HWI J. Simons 7A Walton Way, Stone ST15 0JF
G8 HWJ E. Smith Brickyard House, Wainfleet Road, Irby-In-The-Marsh, Skegness PE24 5AT
GW8 HWL P. Jenkins 20 Dimbath Avenue, Blackmill, Bridgend CF35 6ED
GW8 HWS J. Mills 13 Egerton Street, Cardiff CF5 1RF
G8 HXD M. Ledger 58 Mount Pleasant Close, Lightwater GU18 5TR
G8 HXE K. Haywood 6 Lydney Road, Urmston, Manchester M41 8RN
G8 HXR M. Brooke 70 Wootton Avenue, Peterborough PE2 9EG
G8 HXW A. Sargent 22 Duckmill Crescent, Duckmill Lane, Bedford MK42 0AF
GW8 HYI E. Whitfield Erw Mor, Nebo, Amlwch LL68 9NE
G8 HYK J. Brockwell 10 Tregony Rise, Lichfield WS14 9SN
G8 HYL H. Tuff 4 Battery Terrace, Mevagissey, St. Austell PL26 6QS
G8 HYM S. Bradley 247 Filey Road, Scarborough YO11 3AE
G8 HYP M. Peers 45 Carlton Crescent, East Leake, Loughborough LE12 6JF
GW8 HYT P. Madden Ty Gwyn, Llandovery SA20 0NT
G8 HZI Wythall RC c/o M. Holdsworth 2 Newman Drive, Branston, Burton-on-Trent DE14 3DZ
G8 HZJ R. Ingamells Moor View, Small Banks, Moorside Ilkley LS29 0QQ
GW8 HZL D. Wildman 7 Alder Way, West Cross, , Swansea SA3 5PD
G8 HZN R. Orchard Tredinneck Moor, Newmill, Penzance TR20 8XT
G8 HZQ P. Healy 93 Tile Kiln Lane, Leverstock Green, Hemel Hempstead HP3 8NW
G8 HZS T. Storey 50 Longfield Road, Darlington DL3 0HX
GW8 IAD S. Bowen 6 Pen-Y-Waun Close St Dials, Cwmbran NP44 4JZ
G8 IAJ J. Richardson 43 Front St., Leadgate, Consett DH8 7SB
G8 IAK R. Thomas 88 Parkway, London SW20 9HG
GW8 IAM Dalridia ARC c/o S. Lloyd 4 Cwmdu Court, Cwmdu, Crickhowell NP8 1RU
G8 IAN Gwent Raynet Group c/o M. Lees 175 Overdale Road, Romiley, Stockport SK6 3EN
G8 IAR P. Smith 12 Old Library Mews, Norwich NR1 1ET
G8 IBC D. Herke 24 The Lawns, Farnborough GU14 0RF
G8 IBE R. Bailey 6 Kestrel Close, Horsham RH12 5WD
G8 IBK M. Murray Heads Nook Hall, Heads Nook, Brampton CA8 9AA
G8 IBL H. Hallybone Birch Bassett, 52 Busbridge Lane, Godalming GU7 1QQ
G8 IBO T. Gill 21 Winn Road, London SE12 9EX
G8 IBP R. May 10 Lime Close, Wokingham RG41 4AW
G8 IBR N. Davies 1 Helens Close, The Street, Redgrave, Diss IP22 1RW
G8 IC M. Dawson 60 Ashenhurst Road, Todmorden OL14 8DS
GM8 ICC A. Campbell 2 Cairndhu Cottage, Cairnbaan, Lochgilphead PA31 8SQ
GW8 ICT C. Hopley Clayton Cottage, Alltami Road, Mold CH7 6RW
G8 IDE J. Pimlott 40 Queens Road, Higher St. Budeaux, Plymouth PL5 2NW
G8 IDJ I. Judd 33 Coles Mede, Otterbourne, Winchester SO21 2EG
G8 IDK R. Voisey 2 Chester Place, Malvern WR14 1RQ
G8 IDL D. Smith The Old Forge, High Street, Brinkley, Newmarket CB8 0SE
G8 IEA S. Parham 132 Wrotham Road, Gravesend DA11 7LB
G8 IEI J. Mckillop 2 Moores Green, Wokingham RG40 1QG
G8 IEL R. Tust 28 Osprey Close, Beechwood, Runcorn WA7 3JH
GM8 IEM M. Hall 199 Clashmore, Lochinver, Lairg IV27 4JQ
G8 IER P. Nice 31 Elizabeth Drive, Chapel St. Leonards, Skegness PE24 5RS
G8 IEV B. Guy Hawthorn Folly, Cul De Sac, Boston PE22 8EY
G8 IEW C. Davies Applecroft, St. Johns Road, Gloucester GL2 7DF
G8 IEZ C. Moss 11 Sheepfold Crescent Barrow, Clitheroe BB7 9XR
G8 IFF N. Gunn 1865 El Camino Drive, Xenia 45385 United States
G8 IFH K. Thomas 1 Byways, Yateley GU46 6NE
G8 IFN N. Hinderwell 1 Bower Grove, West Mersea CO5 8GJ
G8 IFT I. Gordon 40 Grange Crescent, Rubery, Birmingham B45 9XB
G8 IHA J. Gregory 2 Abbey Dale Close, Kilburn, Belper DE56 0PY
G8 IHC S. Styler 85 Fairoaks Drive, Great Wyrley, Walsall WS6 6HA
G8 IHF D. Cochrane 18 Russell Avenue, Dunchurch, Rugby CV22 6PX
G8 IHT S. Chambers 7 Mowbray Road, Northallerton DL6 1QT
GM8 IID N. Paterson 4 Cambridge Road, Renfrew PA4 0SL
G8 IIG G. Punter 18 Lodge Road, Sharnbrook, Bedford MK44 1JP
GM8 IIH W. Jarvie Berryhill Farm, Tak-Ma-Doon Road, Kilsyth, Glasgow G65 0RY
G8 III L. Roberts 32 Orion, Bracknell RG12 7YX
G8 IIK D. Hooker 2 Fernlea Court, Lydd Road, Camber, Rye TN31 7RS
G8 IIS B. Heaney 19 Ormonde Drive, Liverpool L31 7AN
G8 IIZ W. Rush 17 Hagden Lane, Watford WD18 0HQ
G8 IJC Dr C. Phillipson 24 Wyatt Close, Martin, Lincoln LN4 3RN
G8 IJE B. Laxton 1 Stoney Lane, Walsall WS3 3RF
G8 IJG D. Adams 77 Chestnut Crescent, Shinfield, Reading RG2 9HA
G8 IJI K. Williamson 4 Lynwood Drive, Wakefield WF2 7EF
G8 IJM H. Wallington 5 Glebe Road, Royal Wootton Bassett, Swindon SN4 7DU
G8 IJS R. Sayer Vignouse, Paimpont 35380 France
GW8 IJT M. Cawood 51 Mayflower Drive, Marford, Wrexham LL12 8LD
G8 IK V. Morse 42 Kingscote Road, Dorridge, Solihull B93 8RA
G8 IKA D. Poll 66 Southlands Avenue, Orpington BR6 9NF
G8 IKG K. Raper 26 Lancaster Way, Scalby, Scarborough YO13 0QH
GW8 IKH R. Rolley Glas Cwm, Dyffryn Crawnon, Crickhowell NP8 1NU
G8 IKK J. Channon 49 Fallow Road, Helston TR13 8WH
G8 IKS D. Warwick Orchard Cottage, Colber Lane, Bishop Thornton, Harrogate HG3 3JR
G8 IKW P. Nutt 1 Lime Gardens, Middleton, Manchester M24 4AE
G8 ILB N. Allinson 42 Rook Lane, Stockton-on-Tees TS20 1SB
G8 ILD R. Barrow 50 Redhill Drive, Bredbury, Stockport SK6 2HQ
G8 ILG J. Law 29 Brackenwood, Orton Wistow., Peterborough PE2 6YP
G8 ILJ S. Nutt 23A Hesketh Drive, Southport PR9 7JX
G8 ILN M. Grindrod 20 Castle Mead, Kings Stanley, Stonehouse GL10 3LD
G8 ILP T. Voller 179 High Street, Harriseahead, Stoke-on-Trent ST7 4JU
G8 ILU J. Parker 15Burton Road, Heckington NG34 9QR
G8 ILW D. Couse 6 Reading Drive, Sale M33 5DL
G8 ILZ I. Walker 113 Whitlock Drive, Wimbledon, London SW19 6SH
G8 IMB M. Stubbs Crofters, Harry Stoke Road, Bristol BS34 8QH
G8 IMH M. Fereday 35 Manor House Park, Codsall, Wolverhampton WV8 1ES
G8 IMI C. KitchenerHamrest 5, Mill Road, , Cromer NR27 0BG
G8 IMJ R. Head 21 Church Street, Fleetwood FY7 6JR
G8 IMM R. Keeley 6 Standings Rise, Whitehaven CA28 6SX
G8 IMS M. Stroud 39 Brocks Drive, Guildford GU3 3NE
G8 IMZ A. Palfrey 5 Ings View, York YO30 5XE
G8 INA South Birmingham RS c/o D. Harris 102 Greatmeadow, Northampton NN3 8DF
G8 INC K. Davenport 10 Woodend Lane, Hyde SK14 1DT
G8 INL B. Miller 1 The Meadows, Monk Fryston, Leeds LS25 5PJ
G8 INO A. Brown 25 Birch Lane, Haxby, York YO32 3RP
G8 INS P. Williams 2 Rosamund Road, Crawley RH10 6QF
G8 INZ T. Prentice 36 Ives Close, Yateley GU46 7RD
GW8 IOA P. Crockford 4B Esperanto Way, Newport NP19 0RD
G8 IOJ D. Martin 54 The Crossway, Portchester, Fareham PO16 8PB
G8 IOK J. Noden 1 Ashley Court, Providence Hill, Southampton SO31 8AT
GM8 IOL R. Thomson Middlerig Farm, Bathgate EH482HH
G8 ION Dr J. Hollis 19 Burlington Grove, Sheffield S17 3PH
G8 IOS K. Evans 12 Moxhull Drive, Sutton Coldfield B76 1LZ
G8 IOW P. Wright 70 Hardy Barn, Shipley, Heanor DE75 7LY
G8 IPA A. Powell 8 Penzoy Avenue, Bridgwater TA6 5BT
G8 IPF H. Billingham Tanglewood, Brookside Orchard, Pulborough RH20 3BD
G8 IPG A. Shaw 92 Freemantle Road, Romsey SO510AX
G8 IPK C. Knight The Lodge, 16A Cromptons Lane, Liverpool L18 3EX
G8 IPN C. Foote 3 Mere Road, Weybridge KT13 9NU
G8 IPQ A. Badcock 7 Heathfield Road, Chandler'S Ford, Eastleigh SO53 5RP
G8 IPT P. Hughes 27 Hemsworth Avenue Little Sutton, Ellesmere Port CH66 4SG
G8 IPY East Anglian CG c/o B. Hewitt 177 Avery Hill Road, London SE9 2EX
GW8 IQC 49f Son Air Cadet RC c/o M. White 5 Marlowe Close, Rogerstone, Newport NP100BT
G8 IQF C. Newell 16A Pembroke Road, Framlingham, Woodbridge IP13 9HA
G8 IQK D. Bryan 4 Kingfisher Close, Barrow Upon Soar, Loughborough LE12 8AX
G8 IQT T. Spicer 3 Parkers Fields, Quorn, Loughborough LE12 8EJ
G8 IQX M. Dixon 57 Northease Drive, Hove BN3 8PP
G8 IRC D. De Fraine Block 8 Lot 3, Rosalina Village 1, Upper Libby Road, Davao City 8023 Philippines
G8 IRL Dr K. Brown 56 Haydock Close, Alton GU34 2TL
G8 IRM M. Emery 45 Old Pasture Road, Frimley, Camberley GU16 8RT
G8 IRN A. Telford 9 Fellside, Tower Wood, Windermere LA23 3PW
G8 IRS J. Wiles 12A Ashling Gardens Denmead, Waterlooville PO7 6PR
G8 ISE G. Sharp 46 Coronation St., Monk Bretton, Barnsley S71 2ES
G8 ISI F. Breame 68 Church Road, Bramshott, Liphook GU30 7SH
G8 ISJ Riviera ARC c/o J. Witt 67 Dillotford Avenue, Coventry CV3 5DS
G8 ISM P. Goldsmith 5 Old School Court, Rock Hill Road, Ashford TN27 9DW
G8 ITB R. Perzyna 29 Lakeside Drive, Bromley BR2 8QQ
GI8 ITD T. Davidson 26 Lower Parklands, Dungannon BT71 7JN
GU8 ITE D. Eaton Glenfield, Le Foulon, St. Andrew GY6 8UF Guernsey
G8 ITG P. Levitt 21 Station Road, Firsby, Spilsby PE23 5PX
GW8 ITI J. Evans Rosegarth, Woodbine Road, Blackwood NP12 1QH
G8 ITJ M. Admans 8 Webb Street, Nuneaton CV10 8JQ
G8 ITU P. Wragg 7 St. Johns Mount, Thirsk Road, Easingwold, York YO61 3HG
G8 ITX O. Williams Meadow View, Irthington, Carlisle CA6 4NN
G8 IUB Birmingham ARS c/o D. Cottam 14 Barnard Close, Rednal, Birmingham B45 9SZ

G8 IUC R. Glover 8 Woodberry Way, Chingford, London E4 7DX
G8 IUD B. Sermons 17 Well Side, Marks Tey, Colchester CO6 1XG
G8 IUG P. Tewkesbury 267 York Road, Stevenage SG1 4HD
GW8 IUM M. Richardson Sunrise, Trefonen SY10 9DQ
G8 IUN S. Tolputt Walnut Lodge, Annings Lane, Bridport DT6 4QN
G8 IUP I. Walukiewicz Louise Cottage, Branksome Avenue, Stockbridge SO20 6AH
G8 IUQ M. Wareing 20 Middlesex Avenue, Burnley BB12 6AA
G8 IVB P. Samson 49 Crest View Drive, Petts Wood, Orpington BR5 1BZ
G8 IVO R. Hartland Three Gables, Crozens Lane, Hereford HR1 1XY
GM8 IVR P. Ager 50 Pitbauchlie Bank, Dunfermline KY11 8DP
G8 IWB A. Parker 33 Colerne Drive, Hucclecote, Gloucester GL3 3SX
G8 IWE R. Thomas 6 Copeland Drive, Poole BH14 8NW
G8 IWF T. Bierney 5318 N 106 Avenue, Glendale 85307 United States
G8 IWI J. Pearce 34 Fleetwood Avenue, Westcliff-on-Sea SS0 9RA
G8 IWJ G. Strange 12 Bronington Avenue, Bromborough, Wirral CH62 6DT
GM8 IWL Dr C. Sutherland 15 Stanley Drive, Brookfield, Johnstone PA5 8UF
G8 IWO N. Jones 14 Salcombe Grove, Swindon SN3 1ER
G8 IWQ A. Jacques 17 Pyrethrum Way, Willingham, Cambridge CB24 5UX
GM8 IWR T. Campbell 5 Coastguard Station, Heugh Road, Portpatrick, Portpatrick DG9 8TF
G8 IWT R. Shears 15 Hale Pit Road, Great Bookham, Leatherhead KT23 4BS
G8 IWX B. Homer 116 Shorncliffe Road, Folkestone CT20 2PQ
G8 IXC L. Prior 64 Montfort Road, Walderslade, Chatham ME5 9HA
G8 IXK B. Owen 21 Marlborough Road, Luton LU3 1EF
G8 IXL P. Baker Doules Mead, Heath Lane, Farnham GU10 5PA
G8 IXM R. Pickett 126-127 Cuckoofield Lane, Mulbarton, Norwich NR14 8BA
G8 IXN K. Watkins 23 Mount Ambrose, Redruth TR15 1NX
G8 IXP R. Lister 8 Carlton Avenue, Wilmslow SK9 4EP
G8 IXX J. Brister 49 Tiverton Road, Loughborough LE11 2RU
GM8 IXZ A. Legood 25 Frankfield Place, Dalgety Bay, Dunfermline KY11 9LR
G8 IYD D. Hancock 4 Elmside, Willand, Cullompton EX15 2RN
G8 IYE P. Shore 1 Whatsill, Hopton Wafers, Kidderminster DY14 0QB
G8 IYH A. Bevington Malthouse, Hoggs Lane, Purton, Swindon SN5 4HQ
G8 IYJ C. Buckland 7 The Maltings, Royal Wootton Bassett, Swindon SN4 7EZ
G8 IYK R. Sayers 3 Riversdale Cottages, The Staithe, Stalham, Norwich NR12 9BY
G8 IYN C. Marsh 6 De Burgh Hill, Dover CT17 0BS
G8 IYS J. Simkins 18 Riding Hill, South Croydon CR2 9LN
G8 IYZ A. Barker 8 Manor Avenue, Attenborough, Nottingham NG9 6BP
GI8 IZB Beaneaters DX Group c/o J. Crawford-Baker 131 Gobbins Road, Islandmagee, Larne BT40 3TX
G8 IZR P. Higginson 18 Park Meadow, Westhoughton, Bolton BL5 3UZ
G8 IZW P. Cain 22 Ditton Green, Luton LU2 8RU
G8 IZY S. Eldridge 6 Cobbles Crescent, Crawley RH10 8HA
G8 JAB A. Berriman Meadowside, Little-In-Sight, St. Ives TR26 1AX
G8 JAC A. Jackson 59 Leas Road, Warlingham CR6 9LP
G8 JAD J. Townsend 56 Seymour Road, Northfleet, Gravesend DA11 7BN
G8 JAG D. Williams Flat 7, Yewbarrow Lodge, Main Street, Grange-over-Sands LA11 6EB
G8 JAI A. Livesley Gates Garth, Barbon, Carnforth LA6 2LJ
G8 JAN P. Biggadike 49 Willow Road, Downham Market PE38 9PG
G8 JAQ J. Walker 2 Morris Drive, Stafford ST16 3YE
G8 JAW B. Heed 3 Woodcote Green Downley, High Wycombe HP13 5UN
G8 JAY A. Jay Jasper, The Reddings, Cheltenham GL51 6RT
G8 JBC C. Jervis 8 Portobello Close, Willenhall WV13 3QA
G8 JBD P. Godfrey 3 Lowry Way, Lowestoft NR32 4LW
GM8 JBJ J. Berry Willowburn, Kirkton, Hawick TD9 8QJ
G8 JBM Dr S. Wood The Mews, 18A Grange Road, Shanklin PO37 6NN
G8 JBP G. Head 34 Balds Lane, Stourbridge DY9 8SG
G8 JBQ R. Hughes Court Church View, South Perrott, Beaminster DT8 3HU
G8 JBT D. Bellingham 22 Princes Drive, Codsall, Wolverhampton WV8 2DJ
G8 JBV D. Dawe 7 Princes Road, Romford RM1 2SR
G8 JCB P. Pullinger 1 Sycamore Cottages, Upper Wield, Alresford SO24 9RP
G8 JCC Chesterfield & District ARS c/o C. Purchase 35 Pasture Way, Bridport DT6 4DW
G8 JCD M. Northey Achill Mist House, Kilmeaney, Listowel V31 VX95 Ireland
GM8 JCF P. Carnegie 29 Castle Terrace, Cullen, Buckie AB56 4SD
G8 JCL J. Essex 40 Lincoln Walk, Heywood OL10 3JB
G8 JCN W. Allen 143 Cherry Crescent, Rawtenstall, Rossendale BB4 6DS
G8 JCS A. Bunting Manor House, Market Place, Binbrook LN8 6DE
G8 JCV P. Hewitt 28 Amersham Avenue, Langdon Hills, Basildon SS16 6SJ
GW8 JDB V. Grayson Willow Lodge, Croeslan, Llandysul SA44 4SJ
G8 JDC Aberdeen ARS c/o T. Robinson 17 Balliol Road, Brackley NN13 6LY
G8 JDD R. Kelsall The Cottage, Denford Road, Stoke-on-Trent ST9 9QG
G8 JDN B. Deefholts Yew Tree House, 28 St Marys Road, Meare, Glastonbury BA6 9SP
G8 JDQ K. Few 35 Whitton Close, Swavesey, Cambridge CB24 4RT
GW8 JEI N. Cross Glan Alaw, Llanddeusant, Holyhead LL65 4AG
G8 JEM E. Cheer 15 Stibbs Way, Bransgore, Christchurch BH23 8HG
G8 JET D. Higginson 43 North Street West Butterwick, Scunthorpe DN17 3JR
G8 JFC F. Wilmott 2 Manor Close, Misson, Doncaster DN10 6HE
G8 JFL D. Crough 32 Roundaway Road, Ilford IG5 0NP
G8 JFT N. Hewitt 36 Princes Terrace, Kemp Town, Brighton BN2 5JS
G8 JFX T. Simmons11 Cedar House 5, Blueberry Close, Maidwell NN6 9XL
GM8 JGB W. Fleming 65 Dundonald Park, Cardenden, Lochgelly KY5 0DG
G8 JGF P. Walters 3 Inkerman Street, Selston, Nottingham NG16 6BQ
G8 JGL N. Owen 59 Fernwood Drive, Leek ST13 8JA
G8 JGM J. Martin 19C Willow Tree Road, Altrincham WA14 2EQ
G8 JGQ M. Thomas Pantglas, Llanddewi Brefi, Tregaron SY25 6PE
G8 JGU R. Hallam 37 Dingle Avenue Appley Bridge, Wigan WN6-9LF
G8 JHA T. White 24 Chapel Street, Tingley, Wakefield WF3 1RE
G8 JHC I. Whitworth 104 The Dormers, Highworth, Swindon SN6 7PD
G8 JHE M. Brogan 31 Rempstone Road, East Leake, Loughborough LE12 6PW
G8 JHG Windscale ARS c/o Dr J. Collins Hill Crest, The Hill, Millom LA18 5HB
G8 JHH M. Baugh 71 Hatch Lane, Old Basing, Basingstoke RG24 7EF
G8 JHL J. Lovell 2 Moran Close, Wilmslow SK9 3UF
G8 JHM I. Carney 39 Blenheim Crescent, Luton LU3 1HB
G8 JHO P. Evans 5 Hunters Close, Bilston WV14 7BN
G8 JIE C. Riding 14 The Coppice, Clayton Le Moors, Accrington BB5 5RU
G8 JIP G. Miller 39 Scrivens Mead, Thatcham RG19 4FQ
G8 JIS T. Macey Whitegates, Histons Hill, Wolverhampton WV8 2HA
G8 JIT J. Mckinnon 142 Hughes Street, Bolton BL1 3EZ
G8 JIU P. Dunham 19 The Lunds, Kirk Ella, Hull HU10 7JJ
G8 JJF North Kent RS c/o J. Puddifoot Rua De Mira Tamega, 6080, Marco de Canaveses 4635-563 Portugal
G8 JJK T. Barrett Flat 38, Queens Court, Cheltenham GL50 2LU
GM8 JJN J. Pryde 7 The Engine Green, Fishcross, Alloa FK10 3JN
GW8 JJP P. Tabberer 8 Wynnstay Road, Old Colwyn, Colwyn Bay LL29 9DS
G8 JJR K. Mcmahon 27 Marlborough Avenue, Doncaster DN5 8EH
GW8 JJZ R. Merrick-Jenkins 17 Marlais Park, Carmel SA14 7UF
G8 JKB C. Hemmings 11 Brookside, Desborough, Kettering NN14 2UD
GW8 JKC J. Kendall 26 Bryn Seiri Road, Conwy LL32 8NR
G8 JKD C. Littman 70 Orbel Street, London SW11 3NY
G8 JKV D. Leary Blackers Hill Farm, Lowndes Drove, Needingworth, St. Ives PE27 4NE
G8 JLA K. Turner 13 Stanhope Street, Saltburn-by-the-Sea TS12 1AL
G8 JLB B. Silver 280 Britten Road, Brighton Hill, Basingstoke RG22 4HR
G8 JLD J. Garters Sun Patch, Garfield Road, Hailsham BN27 2BT
G8 JLM P. Higham 56 Coopers Avenue, Heybridge, Maldon CM9 4YX
G8 JLY L. Leach 2 Nightingale Place, Droitwich WR9 7HG
G8 JMB J. Button 16 Meadow Rise, Broadstone BH18 9ED
G8 JMG J. Gartland 175 Talbot Street, Whitwick, Coalville LE67 5AY
G8 JMK D. Butler 144 Longridge Way, Weston-Super-Mare BS24 7HS
G8 JMO B. Justin 1704 Cottontown, Forest Virginia 24551 United States
G8 JMP D. Beech 8 Copthorne Drive, Lightwater GU18 5TE
G8 JMS S. Miller 44 Greenway Road, Galmpton, Brixham TQ5 0LZ
G8 JMU J. Potter 15 Alterton Close, Goldsworth Park, Woking GU21 3DD
G8 JMY D. Hugman 6 Barley Close, Henley-in-Arden B95 5HU
G8 JNI D. Bookham 1 Monks Rise, Fleet GU51 4HB
G8 JNO S. Munday 25 Southend Road, Weston-Super-Mare BS23 4JY
G8 JNR R. Hedderley 17 Linford Close, Handsacre, Rugeley WS15 4EF
G8 JNZ K. Crowder 15 Fleetwood Close, Minster On Sea, Sheerness ME12 3LN
GI8 JOA D. Thompson 16 Lynden Gate Park, Portadown, Craigavon BT63 5YJ
G8 JOC Dr E. Powell 49 Normanby Road, Worsley, Manchester M28 7TS
G8 JOX J. Dobson Home Farm, Mansmore Lane, Charlton On Otmoor, Kidlington OX5 2US
GW8 JOY T. Bowen 7 Bedford Close, Greenmeadow, Cwmbran NP44 5HN
G8 JPA J. Hunt Woodstone Farm, High Common Road, Diss IP22 2HS
GI8 JPF T. Phillips 52 Belfast Road, Bangor BT20 3PU
G8 JPJ D. Jones 17 Stanmore Close, Clacton-on-Sea CO16 7HQ
G8 JPU D. Potts 25 Southlands Road, Congleton CW12 3JY
G8 JPV M. Parnell 101 Ridgeway, Wellingborough NN8 4RZ
G8 JPW J. Abbott 11 Red House Road, Bodicote, Banbury OX15 4BB
G8 JQG J. Hough 77 Pennine Court, Macclesfield SK10 2RN
G8 JQH P. Wright 33B Slack Lane, Crofton, Wakefield WF4 1HX
G8 JQS G. Greensmith Japonica, Hawthorne Avenue, Biggin Hill, Westerham TN16 3SG
G8 JQV D. Marchant 11 Derehams Lane, Loudwater, High Wycombe HP10 9RH
G8 JQW R. Thomas 8 Sherwood Court, Sherwood Street, Bolsover, Chesterfield S44 6GF
GI8 JRE J. Donnelly 9 Lomond Heights, Cookstown BT80 8XW
G8 JRF M. Willson 19 The Willows, Highworth, Swindon SN6 7PG
G8 JRI P. Aldridge 2 Beldams Close, Thorpe-Le-Soken, Clacton-on-Sea CO16 0NT
G8 JRN R. Stockdale 53 Brightwalton, Newbury RG20 7BT
G8 JRW M. Austin The House, Four Seasons Village, Winkleigh EX19 8DP
G8 JRZ A. Mills 42 Mona Avenue, Chadderton, Oldham OL9 0EJ
G8 JSC K. Austin 139 Sewall Highway, Coventry CV2 3NG
G8 JSE F. Cowlin 9 Zealand Close, Hinckley LE10 1TJ
G8 JSF R. Williams 35 Broadhurst Grove, Lychpit, Basingstoke RG24 8SB
G8 JSL P. Smith 13 Manor Garth, Pakenham, Bury St. Edmunds IP31 2LB
G8 JSM C. Wood 57 Holly Crescent, Rainford, St. Helens WA11 8ER
G8 JSN P. Bailey 50 Amis Avenue, New Haw, Addlestone KT15 3ET
G8 JSR V. Hinksman 1 Shaw Lane, East Woodburn, Hexham NE48 2SL
G8 JSS D. Mobbs 59 Victoria Road, Bude EX23 8RH
G8 JTD Otley ARS c/o J. Castelow 7 Langford Close, Burley In Wharfedale, Ilkley LS29 7NP
G8 JTG E. Spanton 14 Days Lane, Sidcup DA15 8JN
G8 JTL M. Davies 25 Walker Avenue, Quarry Bank, Brierley Hill DY5 2LY
G8 JUC J. Wheatley 44 Kingswood Close, Boldon Colliery NE35 9LG
G8 JUG N. Spenceley 18 Rectory Road, Broadmayne, Dorchester DT2 8EG
G8 JUK B. Storeton-West Nazdar, Camps Heath Oulton, Lowestoft NR32 5DW
G8 JUS T. Gale 58 Westwood Road, Newbury RG14 7TL
G8 JUT S. Linton 41 Long Close., Bristol BS162UF
G8 JUV S. York 10 Beechwood Avenue, Wallasey CH45 8NX
GM8 JUY R. Mcmillan 12 Parkthorn View, Dundonald, Kilmarnock KA2 9EZ
G8 JVE M. Rowe 97 Old Worthing Road, East Preston, Littlehampton BN16 1DU
G8 JVI A. Hicks 10 Evans Close, Eynsham, Witney OX29 4QY
G8 JVM R. Bown Park View, Chapel Street, Telford TF4 3DD
G8 JVS Rvd. M. Fairey 10 Mallard Close, York YO10 3BS
G8 JVU A. Johnson Clematis Cottage, Wheatlow Brooks, Stafford ST18 0EW
G8 JVV Foyle And District ARC c/o J. Burchell Gooseleys Farm, Harrow Hill, Halstead CO9 4LX
G8 JVW P. Boswell 10 The Grange, Wombourne, Wolverhampton WV5 9HX
GM8 JVZ Dr M. Nimmo The Court, 6 Farington Street, Dundee DD2 1PJ
G8 JWC D. Luscombe 31 Tewkesbury Drive, Prestwich, Manchester M25 0HR
G8 JWD I. Rees Knowle Cottage, Whittonditch Road, Marlborough SN8 2PX
G8 JWE J. Hickman 41 Field Road, Ramsey, Huntingdon PE26 1JP
G8 JWK R. Staveley 52 New Road, Wootton Bassett, Swindon SN4 7DG
GW8 JWL G. Smith 13 Lapwing Close, Penarth CF64 5GA
GW8 JWP J. Griffiths Llygad Yr Haul, Ferwig, Cardigan SA43 1PX
GM8 JWQ K. Faloon Moss-Side Croft, 6 Rothiemay, Huntly AB54 5NY
G8 JWT R. Trett Low Barn, Norwich Road, Woodton, Bungay NR35 2LP
G8 JWX J. Wright 7 Basket Gardens, London SE9 6QP
G8 JXG J. Dean 6 Greenleas Pembury, Pembury, Tunbridge Wells TN2 4NS
G8 JXK S. Blew 24 Batts Park, Taunton TA1 4RE
G8 JXP D. Mccabe 78 Oakleigh Road, Stratford-upon-Avon CV37 0DN
G8 JXS M. Stephenson 6 Cedar Road, Tewkesbury GL20 8PX

G8 JXU C. West-Bulford 25 Sunnyside Close, Heacham, King's Lynn PE31 7DX
G8 JXV T. Trew Stockers Lodge, Bere Farm Lane, Fareham PO17 6JJ
G8 JYN Basingstoke ARC c/o P. Cresswell 108 Hawthorn Way, Basingstoke RG23 8NH
G8 JYS M. Fletcher 20 Leahurst Close, Norton, Malton YO17 9DF
G8 JYX P. Johnson 42 College Gardens, London E4 7LG
G8 JZI P. Smith 4 Fellstone Vale, Withnell, Chorley PR6 8UE
G8 JZO J. Gibbs 6 Southampton Close, Blackwater, Camberley GU17 0HB
G8 JZT S. Osborn 67 Chessington Avenue, Bexleyheath DA7 5NP
G8 JZX C. Stephenson Armanby, Main Street, Selby YO8 8QT
G8 JZZ R. Taylor Higher Priestacott, Belstone, Okehampton EX20 1QX
G8 KAE R. Bushell 102 Winchester Gardens, Northfield, Birmingham B31 2QB
G8 KAM J. Hurnandies 70 Orchard Rise West, Sidcup DA15 8SZ
G8 KAP D. Patrick Quarryside, Stockdalewath, Dalston, Carlisle CA5 7DP
G8 KAS B. Buschl 27 The Drive, Court Farm Road, Newhaven BN9 9DJ
G8 KB P. Johnson 55 Rodney Hill, Loxley, Sheffield S6 6SG
G8 KBB D. Roberts 32 Woodbridge Close, Appleton, Warrington WA4 5RD
G8 KBG A. Price 5 Stoneybrook Leys, Wombourne, Wolverhampton WV5 8JE
G8 KBH D. Ward 3 Sherbourne Close, Poulton-le-Fylde FY6 7UB
GW8 KBK A. Short 70 Tremains Court, Brackla, Bridgend CF31 2SS
G8 KBM I. Whiting 11 Woodsend Close, Burton Joyce, Nottingham NG14 5DY
G8 KCB J. Nally 313 Wyndhurst Road, Stechford, Birmingham B33 9DL
GW8 KCH K. Houston 6 Ashgrove, Llanellen, Abergavenny NP7 9HP
GW8 KCY M. Bover Glynfach Bungalow, Pontyates, Llanelli SA15 5TF
G8 KDD D. Coton 17 Flambards Close, Meldreth, Royston SG8 6JX
G8 KDF Dr M. Sach Old School, Cambridge Road, St. Neots PE19 6ST
G8 KDM Lifeboat ARS c/o A. Smith 2A Chesterfield Road, Barlborough, Chesterfield S43 4TR
G8 KDO Dr P. Topham 5 Kings Road, Cambridge CB3 9DY
G8 KDU R. Eager 45 Fleetwood Avenue, Herne Bay CT6 8QW
G8 KEA M. Sutton 178 Cole Lane, Borrowash, Derby DE72 3GN
G8 KED C. Mullineaux 27 Ashfield Avenue, Lancaster LA1 5EB
G8 KEJ M. Johnson 23 The Crest, Surbiton KT5 8JZ
G8 KEK P. Wilson 5 Mons Close, Harpenden AL5 1TD
G8 KEO Chesterfield & District Scouts A R C c/o P. Dickinson Halshanger Farm, Ashburton, Newton Abbot TQ13 7HY
GI8 KEP K. Bones 54 Derryvolgie Park, Lisburn BT27 4DA
GW8 KEV K. Shafto 3 Harding Close, Boverton, Llantwit Major CF61 1GX
G8 KFD R. Gwynn 227 Sketchley Road, Burbage, Hinckley LE10 2DY
G8 KFF R. Parker 17 Valley Road, Streetly, Sutton Coldfield B74 2JE
GI8 KFG P. Douglas 21 Hillhead Road, Ballycarry, Carrickfergus BT38 9HE
G8 KFJ D. Greig 23 Parsons Walk, Walberton, Arundel BN18 0PA
G8 KFK P. Loten 15 Hornsea Burton Road, Hornsea HU18 1TP
G8 KFN R. Heron 46 Bradvue Crescent, Bradville, Milton Keynes MK13 7AJ
G8 KFS B. Russell 56 Kingsmead Avenue, Surbiton KT6 7PP
G8 KGE S. Bailey 50 Quantock Close, Warmley, Bristol BS30 8UT
G8 KGG Dunstable Dwn Rd c/o A. Ward 49 Spielplatz, Lye Lane, St. Albans AL2 3TD
G8 KGH M. Cooper 35 Cote Lea Park, Bristol BS9 4AH
G8 KGR R. Tidswell Helloplane, Clubhurn Lane, Spalding PE11 4BQ
G8 KGS C. Suslowicz 2366 Coventry Road Sheldon, Birmingham B26 3LS
G8 KGV P. Jessop 84 Common Road Kensworth, Dunstable LU6 3RG
G8 KHF J. Dove 33 The Haystack, Daventry NN11 0NZ
G8 KHH C. Young 26 Horsham Avenue, Peacehaven BN10 8HX
G8 KHI R. Partridge 5 Beck Close, Mundesley NR11 8QL
G8 KHU Stirling And District ARS c/o D. Fielding 216 Andover Road, Newbury RG14 6PY
G8 KHV Brannock High Radio Group c/o R. Evans 6 Park End, Lichfield WS14 9US
G8 KIG P. Winwood 2 The Warren, Abingdon OX14 3XB
G8 KIH J. Sargent 9 Lee Woottens Lane, Basildon SS16 5HD
G8 KIJ C. Followell 9 Mayflower Gardens, Nailsea, Bristol BS48 1QW
G8 KIK D. Bland 17 Knowles Close, Kirklevington, Yarm TS15 9NL
GM8 KIQ J. Harper 11 Cathburn Holding, Cathburn Road, Wishaw ML2 9QL
G8 KIW Kings College ARS c/o H. Muller 118 Park Lane, Northampton NN5 6PZ
G8 KIZ S. Morris 23 Ellesmere Way, Carlisle CA2 6LZ
G8 KJI RAF Digby ARC c/o J. Richardson 4 Torrington Lane, East Barkwith, Market Rasen LN8 5RY
G8 KJJ L. Haywood 9 Canberra Crescent, West Bridgford, Nottingham NG2 7FL
GW8 KJK G. Park 34 Delafield Road, Abergavenny NP7 7AW
GM8 KJO D. Moodie 1 Lageonan Road, Grandtully, Aberfeldy PH15 2QY
G8 KJP P. King 25 Lamellyn Drive, Truro TR1 3JR
G8 KJT R. Burgess 3 Deeside Avenue, Chichester PO19 3QF
G8 KKA B. Stevens 2 Hawthorn Crescent, Shepton Mallet BA4 5XR
G8 KKG M. Elliott 9 Moyleen Rise, Marlow SL7 2DP
G8 KKH C. Hills 8 Blackdale, Cheshunt, Waltham Cross EN7 6DF
G8 KKN D. Mcfarlane 10 Green Lane, Vicars Cross, Chester CH3 5LA
G8 KKU J. Walker 21 Garden Hedge, Leighton Buzzard LU7 1DJ
G8 KLA C. Bonner 6 Camley Park Drive, Maidenhead SL6 6QF
G8 KLC P. Webber 37 Rasen Road, Market Rasen LN8 3XL
G8 KLT P. Valteris 25 Copthill Way, Houghton le Spring DH4 4FB
G8 KMK Kirkless Raynet c/o G. Edinburgh 77 Westerley Lane, Shelley, Huddersfield HD8 8HP
G8 KMM J. Bryant 12 Dale Tree Road, Barrow, Bury St. Edmunds IP29 5AD
G8 KMP M. Pollock 25 Meadow Lane, Burgess Hill RH15 9HZ
G8 KMR M. Davis 8 Mead Close, Leckhampton, Cheltenham GL53 7DX
G8 KNC S. Wheatley The Gables, Green Road, Wivelsfield Green, Haywards Heath RH17 7QA
G8 KNF D. Hawkins 109 Elphinstone Road Walthamstow, London E17 5EY
GW8 KNJ J. Blinco 2 Agnes Hunt Drive, Park Hall, Oswestry SY11 4FE
G8 KNN Capel Battery Radio Group c/o J. Bigwood 133 Gilbert Road, Cambridge CB4 3PA
G8 KNS M. Jelfs Adams Acre, Chapel Lane, Wimborne BH21 3SL
G8 KNU R. Jacobs 1 Coverdale, Northampton NN2 8UU
G8 KOC R. Backham 15 Rushmead Close, South Wootton, King's Lynn PE30 3LY
G8 KOD R. Adams Swinneys, Station Road, Carterton OX18 3PR
GM8 KOF D. Mcnaughton 6 Wilderhaugh Court, Galashiels TD1 1QL
G8 KOL D. Slocombe 7 Talbot Avenue, Herne Bay CT6 8AD
G8 KOM D. Hanson 42 Choseley Road, Knowl Hill, Reading RG10 9YT
G8 KOQ N. Morris 88 Tynesbank, Worsley, Manchester M28 0SL
G8 KOS S. Head 3 Ripon Gardens, Waterlooville PO7 8ND
G8 KOZ A. Clarke 42 Chamberlain Way, Biddulph, Stoke-on-Trent ST8 7BB
G8 KPD B. Fothergill 53 Meadow Court Ponteland, Newcastle upon Tyne NE20 9RA
G8 KPE E. Howard 15 Amherst Road, Bexhill on Sea TN40 1QH
G8 KPG G. Wright 58 Lifton Croft, Kingswinford DY6 8RZ
GM8 KPH M. Hobson 17 Well Brae, Pitlochry PH16 5HH
G8 KPL S. Williams Flat 7, Yewbarrow Lodge, Main Street, Grange-over-Sands LA11 6EB
G8 KPV G. Hickman Pine Tree Cottage, Calverton Road, Blidworth, Mansfield NG21 0NW
G8 KPY D. Pratt 77 Hayfield Road, St. Mary Cray, Orpington BR5 2DL
G8 KQA R. Laslett Dinnages, Street End Lane, Heathfield TN21 8SA
G8 KQB the Raynet Association c/o S. Prior East Brantwood, Manor Road, Barnstaple EX32 0JN
G8 KQV Dr S. Evans 4 Holcot Lane, Anchorage Park, Portsmouth PO3 5TR
G8 KQZ G. Dawkins 8 Chancery Lane, Eye, Peterborough PE6 7YF
G8 KRB K. Barnes Fairseat Close, Totnes TQ9 5AN
G8 KRG Prof. C. Harrison 11 Ringley Park, Whitefield, Manchester M45 7NT
G8 KRV J. Cottier 83 Elizabeth Drive, Tamworth B79 8DE
G8 KSA W. Hall 67 Selwyn Drive, Stockton-on-Tees TS19 8XF
G8 KSC D. Goodwin 41 Newpool Road, Knypersley, Stoke-on-Trent ST8 6NT
G8 KSD A. Hewett 1 Mountside, Westfield Lane, Folkestone CT18 8BY
GW8 KSE W. Salisbury 28 Dyke Street, Brymbo, Wrexham LL11 5AH
GW8 KSF A. Salisbury 28 Dyke St., Brymbo, Wrexham LL11 5AH
G8 KSH A. Wilkins 2 Beechfield Crescent, Banbury OX16 9AR
GM8 KSJ D. Cowie 8 Centre Street, Kelty KY4 0EQ
GW8 KSL R. Cleaver 61 Llewellyn Park Drive Morriston, Swansea SA6 8PF
G8 KSM R. Beament Midlands Farm, Horndon, Mary Tavy, Tavistock PL19 9NQ
G8 KST T. Mayer 61 Rawley Crescent, New Duston, Northampton NN5 6PU
G8 KSW J. Wood 38 Beech Lane, West Hallam, Ilkeston DE7 6GU
G8 KSX A. Thompson Carloway, Turner Lane, Addingham LS29 0LE
G8 KSZ I. Newbold 40 Heath Close, Stonnall, Staffordshire WS9 9HU
G8 KTA P. Thomas 76 Church Road, Braunston, Daventry NN11 7HQ
G8 KTC M. Rhys 2 Sun Lane, Teignmouth TQ14 8EF
G8 KTG D. Smith 76 Reigate Road, Brighton BN1 5AG
G8 KTX M. Butler 7 Bassett Road, Coventry CV6 1LF
G8 KUA C. Bridgland 10 Eastlands Grove, Stafford ST17 9BE
G8 KUV A. Simonds 3 Links Close, Seaford BN25 4NU
G8 KUZ J. Wiggins 35 Downing Avenue, Newcastle ST5 0LB
G8 KVN A. Nelson 29 Coxford Road, Southampton SO16 5FG
G8 KVO C. Miller Broomwood, South Park, Sevenoaks TN13 1EL
G8 KVU Ham Radio Builders Club c/o Rvd. G. Smith 6 Birtley Rise, Bramley, Guildford GU5 0HZ
G8 KW R. Shears 15 Hale Pit Road, Great Bookham, Leatherhead KT23 4BS
G8 KWD G. Bettley 1 Dovetrees, Covingham, Swindon SN3 5AX
G8 KWH N. Liddle 67 Algarth Road, Pocklington, York YO42 2HJ
G8 KWJ D. Barnwell Bernagh, Duncombe Street, Kingsbridge TQ7 1LR
G8 KWN R. Bryant 81 Dukes Drive, Halesworth IP19 8TJ
G8 KWP A. Darragh The Gables, Belle Vue Lane, Chester CH3 7EJ
G8 KWV J. Bailey 20 Smitham Downs Road, Purley CR8 4NB
GM8 KXF G. Robb 3 Doonholm Park, Ayr KA6 6BH
G8 KXO B. Gamble 79 Humphries House, Lindon Drive, Walsall WS8 6DL
GW8 KXW J. Watts 6 Castle View, Haverfordwest SA61 2JA
GI8 KYI T. Carlisle 46 Middle Road, Carrickfergus BT38 9DN
G8 KYK C. Keens 3 Kirk Gardens, Totton, Southampton SO40 9UZ
G8 KYP J. Buckley 11 Salisbury Grove Giffard Park, Milton Keynes MK14 5QA
GW8 KZA J. Wells 30 St. Andrews Road, Barry CF62 8BR
G8 KZG P. Delaney 6 East View Close, Wargrave, Reading RG10 8BJ
G8 KZJ E. Lockyear 140 Andover Road, Orpington BR6 8BL
G8 KZN W. Clinton 5 Moorland Crescent, Castleside, Consett DH8 9RF
G8 KZO R. Edgeley 6 Hearne Gardens, Shirrell Heath, Southampton SO32 2NR
G8 KZY C. Denison 40 Leysholme Drive, Leeds LS12 4HQ
G8 LAB R. Harste 2 Park Drive, Ingatestone CM4 9DT
G8 LAM R. Lambley 31 Ridgeway Road, Redhill RH1 6PQ
G8 LAU D. Peck 3 Dearnford Avenue, Wirral CH62 6DX
G8 LAY E. Hibbett Trumps Lodge, Broad Street, Ottery St. Mary EX11 1BY
GM8 LBC C. Dalziel 2 Alder Avenue, Hamilton ML3 7LL
G8 LBG J. Cook Highlands, Littledown, Shaftesbury SP7 9HD
G8 LBR M. Rogers 30 Neasden Avenue, Clacton-on-Sea CO16 7HG
G8 LBS C. Ranson 281 Hawthorn Drive, Ipswich IP2 0QG
G8 LBT M. Rigby 16 Juniper Way, Penrith CA11 8UF
G8 LCA J. Scott 123 Cotswold Way, Tilehurst, Reading RG31 6SR
G8 LCE M. Perrett 21 Tredova Crescent, Falmouth TR11 4EQ
G8 LCI A. Goode 42 Fourth Street, Black Rock 3193 Australia
GI8 LCJ D. Craig 40 Chilton Road, Carrickfergus BT38 7JT
G8 LCK L. Reynolds 1845 Van Buren Road, Caswell 4750 United States
G8 LCL S. Tames 21 Lind Close, Earley, Reading RG6 5QX
G8 LCM K. Day Powys Lodge, 6 Court Road, Worcester WR8 9LP
G8 LCP N. Jamieson 1 Langdale Place, Newton Aycliffe DL5 7DX
G8 LCS J. Monte 11 Woodfield Avenue, Hyde SK14 5BB
G8 LCU A. Frankling 5 Chadborn Avenue Gotham, Nottingham NG11 0HT
G8 LCZ J. Sellick 24 Windsor Road, Lytham St. Annes FY8 1ET
G8 LDB K. Oldham 165 Mountsorrel Lane, Rothley, Leicester LE7 7PU
G8 LDC Dr J. Salthouse 10 Ramillies Avenue, Cheadle Hulme, Cheadle SK8 7AL
G8 LDJ C. Douglas 22 Connaught Road, Sittingbourne ME10 1EH
G8 LDU Dr G. Noble 9 Sunrise Way Kings Hill, West Malling ME19 4DL
G8 LDV B. Harrad 32 Woodfield Avenue, Northfleet, Gravesend DA11 7QG
G8 LDW P. Harness 7 Castlegate, Gipsey Bridge, Boston PE22 7BS
G8 LDY R. Tompkins 16 Garden Close, Watford WD17 3DP
GM8 LEA N. Adam Bridaig Villa, Gladstone Avenue, Dingwall IV15 9PG
G8 LEB R. Hill Rose Lodge, 35 Colne Fields, Huntingdon PE28 3DL
G8 LED Northampton RC c/o J. Cockrill 28 Northampton Road, Harpole, Northampton NN7 4DD
G8 LEG K. Hardy 2 Forest Hill, Maidstone ME15 6UU
G8 LEM R. Griffith 9 Devonshire Road, West Kirby, Wirral CH48 7HR
G8 LES M. Sanders 39 Telegraph Lane, Four Marks, Alton GU34 5AX
G8 LF E. Byrne 40 Wentworth Avenue, Ascot SL5 8HQ
GM8 LFB J. Rabbitts 38 Murchison Street, Wick KW1 5HW
GM8 LFI Prof. M. Cartmell 33 Orrok Park, Edinburgh EH16 5UW

GI8 LFY A. Penn 9 Milltown Road, Donaghcloney, Craigavon BT66 7NE
G8 LGA R. Ward 1 Horton, Downswood, Maidstone ME15 8TN
G8 LGE P. Devine 3 The Hawthorns, Outwood, Wakefield WF1 3TL
G8 LGM R. Field 20 Hill Road, Watlington OX49 5AD
G8 LGP K. Harris 20 Westminster Close, Devizes SN10 1BF
G8 LGS P. Chitty 109 Bannings Vale, Saltdean, Brighton BN2 8DH
G8 LGT D. Blakemore 20 Derwent Road, Coventry CV6 2HB
G8 LGU R. Milliken 15 Lee Grove, Chigwell IG7 6AD
G8 LGW R. Liddiard 26 Dowgate Close, Tonbridge TN9 2EL
G8 LGY R. Tyson 18 Blackthorn Close, Gedling, Nottingham NG4 4AU
G8 LHD D. Allen 21 Goldings Close, Haverhill CB9 0EQ
G8 LHF P. Earl Holly Cottage, Popes Lane, Chappel, Colchester CO6 2DZ
G8 LHI M. Levy 22 Boyne Road, Lewisham, London SE13 5AW
G8 LHP Dr A. Milne 49 Cleevemount Road, Cheltenham GL52 3HF
G8 LHQ M. Tuffrey 50 Lynette Avenue, London SW4 9HD
G8 LHS D. Waters 84 Littlehaven Lane, Horsham RH12 4JB
G8 LHT I. Harwood 38 Spring Crescent, Sprotbrough, Doncaster DN5 7QF
G8 LHZ P. Avon 81 Parsonage Barn Lane, Ringwood BH24 1PU
G8 LID N. Dowler 1 Cottage Walk, Clacton-on-Sea CO16 8DG
G8 LIE N. Borrell Chapel Cottage Cotherstone, Barnard Castle DL12 9PQ
G8 LIH G. Storey 27 Dyche Road, Sheffield S8 8DQ
G8 LII J. Lee 225 Avenue Road, Rushden NN10 0SN
G8 LIK S. Hurst Fareview, Woodhead Road, Holmfirth HD9 2PX
G8 LIP B. Greenbeck 10 Campbell Avenue, Bottesford, Scunthorpe DN16 3SA
G8 LIU N. Clyne 78 Halford Road, Ickenham, Uxbridge UB10 8QA
G8 LIX R. Keates 35 Walsh Grove, Birmingham B23 5XE
GW8 LJJ E. Edwards 11 Old Village Road, Barry CF62 6RA
G8 LJO W. Ricketts 22 Westfield Close Durrington, Salisbury SP4 8BY
G8 LJQ C. Asquith 142B Newbegin, Hornsea HU18 1PB
G8 LJU J. Spicer 6 Avenue Road, Worcester WR2 4ES
G8 LJY A. Griffiths 17 Ferenberge Close, Farmborough, Bath BA2 0DH
G8 LKA S. Whitehead 4 Colleton Crescent, Exeter EX2 4DG
G8 LKB Woofferton Transmitting Station c/o I. Rabson 50 Burwell Meadow, Witney OX28 5JQ
G8 LKK R. Horsford 2 Old Mill, Mill Lane, Chard TA20 2ND
GM8 LKL A. Hogg 43 Muir Wood Road, Currie EH14 5JN
G8 LKP J. Duchscherer 36 Hamdon Close, Stoke-Sub-Hamdon TA14 6QN
G8 LKQ D. Falkner 45 Westwood Carleton, Skipton BD23 3DW
G8 LKS D. Burton 48 West Beeches Road, Crowborough TN6 2AG
G8 LKW H. Colville Flat 33, Hamilton Court 165 Northfield Road, Birmingham B30 1DU
GW8 LKX Obo Armagh And Dungannon DARC c/o M. Corrigan 3 Heathway, Heath, Cardiff CF14 4JQ
G8 LLD P. Pritchard 15 Hilldene Close Flitwick, Bedford MK45 1AQ
G8 LLJ M. Tutt 9 Russell Drive, Dunbridge, Romsey SO51 0RA
G8 LLS S. Perkins 17 Lime Tree Avenue, Malvern WR14 4XE
G8 LLV G. Christie 9 / 49 Democrat Drive, The Basin, Victoria 3154, Australia, Melbourne 3154 Australia
G8 LLZ S. Hickling Shamba Slade Lane Galmpton, Brixham TQ50PE
G8 LM J. Jennings Mill Side, Mill Road, Lutterworth LE17 5DE
G8 LMC R. Lovell 16 North View, Staple Hill, Bristol BS16 5RU
G8 LMF P. Rigby 92 Albany Road, Ansdell, Lytham St. Annes FY8 4AR
G8 LMI D. Morgan 23 Banstead Road, Caterham CR3 5QH
G8 LMW C. Smith 73 Desford Road, Newbold Verdon, Leicester LE9 9LG
G8 LMY D. Sweetland 15 Wasdale Close, Owlsmoor, Sandhurst GU47 0YQ
G8 LNC D. Golding 27 Wesermarsch Road, Cowplain, Waterlooville PO8 8JJ
G8 LNG D. Severn 20 Somerton Avenue, Wilford, Nottingham NG11 7FD
GM8 LNH R. Pascal 19 Clach Na Strom, Whiteness, Shetland ZE2 9LG
G8 LNQ C. Tindill The Old School, Bellerby, Leyburn DL8 5QN
G8 LNU L. Tucker 10 The Meadow, Waterlooville PO7 6YJ
G8 LOF S. Champion 4 Oldcastle Croft, Tattenhoe, Milton Keynes MK4 3EN
G8 LOJ S. Dorrington-Ward Higher Dairy, Stoke Abbott, Beaminster DT8 3JT
GM8 LON B. Bruce 10 John Huband Drive, , Birkhill DD2 5RY
G8 LOP P. Coomber 10 Streeton Way, Earls Barton, Northampton NN6 0HX
G8 LOU Dr P. Mattos Olive House, Rock Road Rock, Wadebridge PL27 6NW
G8 LOZ J. Ramsay Strathmore, 5 Parkhurst Road, Guildford GU2 8AP
G8 LPA N. Hilbery 16 Albert Road, Ashford TW15 2LU
G8 LPC R. Cawley 59 The Horseshoe, Hemel Hempstead HP3 8QS
G8 LPI R. Bray 2 Hill Park, Walsall Wood, Walsall WS9 9RD
G8 LPN K. Edwards 22 Claverton Estate, Stoulton, Worcester WR7 4RH
G8 LPX C. Morgan 43 Ferndown Road, Manchester M23 9AW
G8 LQB W. Morrison 14 Browns Grove, Kesgrave, Ipswich IP5 2GP
G8 LQF J. Pettifor 12 Windmill Road, Atherstone CV9 1HP
GW8 LQH M. White Penrhos, Llanddewi, Llandrindod Wells LD1 6SL
GM8 LQL W. Cowell High Clachaig, Kilmory, Isle of Arran KA27 8PG
G8 LQM P. Green Nut House, 2 Warren Barns, Warren Lane, Bedford MK45 4AS
G8 LQN G. Bryce 6A Kingfisher Drive, Whitby YO22 4DY
G8 LQO C. Mckenzie Apartment 4 8 Maryport Drive, Timperley, Altrincham WA15 7NS
G8 LQP R. Lines11 5 Dowling Drive, Pershore WR10 3EF
G8 LQZ R. Banfield 2 Laleham Close, Eastbourne, East Sussex BN21 2LQ
G8 LRD P. Hutchings 59 Braemor Road, Calne SN11 9DU
GW8 LRO A. Williams 1 Glyncoch Terrace, Pontypridd CF37 3BW
G8 LRS D. Massey 28 Rufus Close, Rownhams, Southampton SO16 8LR
G8 LSA H. Potter Burwood House, Salisbury Road, Woking GU22 7UR
G8 LSC P. Wheeler 3 Oatfield Road, Orpington BR6 0ER
G8 LSD A. Wyatt 75 Millbrook Road, Crowborough TN6 2SB
G8 LSH D. Oakley 136 Chanctonbury Way, London N12 7AD
G8 LSI R. Dungan 2 Lamorna Close, Orpington BR6 0TD
G8 LSS A. Tompson 15 Plumbley Meadows, Winterborne Kingston, Blandford Forum DT11 9BY
GI8 LTB R. Mcwilliams 4 Wheatfield Drive, Coleraine BT51 3RD
G8 LTC R. Hore 19 Warrenside Close, Ramsgreave, Blackburn BB1 9PE
G8 LTD S. Vaslet 4 Coniston Crescent, Redmarshall, Stockton-on-Tees TS21 1HT
G8 LTN A. Brown Casita, The Ridge, Cold Ash, Thatcham RG18 9HT
GW8 LTV G. Snellgrove 142 Arail St., Six Bells, Abertillery NP13 2NQ
G8 LTY A. Harman 107 Kempson Drive, Great Cornard, Sudbury CO10 0YF
G8 LUL R. Myers 33 Withenfield Road, Manchester M23 9BT
G8 LUP A. Semark 11 Fir Tree Close, Thorpe Willoughby, Selby YO8 9PF
GI8 LUR A. Hewitt 18 Knockview Avenue, Newtownabbey BT36 6TZ
G8 LUV G. Fairbrass 230 Kirkby Road, Barwell, Leicester LE9 8FS
G8 LUZ A. Szczepanek Harling House, Calcraft Road, Corfe Castle, Wareham BH20 5EL
G8 LVC P. Johnson 54 Beechwood Close, Chandler'S Ford, Eastleigh SO53 5PB
G8 LVF A. Sierota 20 Marder Road, London W13 9EN
G8 LVL D. Holmes 30 Roydale Close, Loughborough LE11 5UW
G8 LVM A. Holmes 5 Launde Park, Market Harborough LE16 8BH
G8 LVQ White Rose ARS c/o E. Hannaby 170A Weston Drive Otley, Leeds LS21 2DT
G8 LVU E. Brown 37 Lynton Way Sawston, Cambridge CB22 3EA
G8 LVW C. Snell 138 Main Road, Great Leighs, Chelmsford CM3 1NP
G8 LWA D. Tyler Wayside View, Orsett Road, Stanford-le-Hope SS17 8PN
G8 LWC Dr J. Stuart 13 Gloucester Close, Petersfield GU32 3AX
G8 LWO F. Merritt 17 Blakes Way Eaton Socon, St. Neots PE19 8PU
G8 LWQ S. Wood Lucerne, Berrycroft, Soham, Ely CB7 5BL
G8 LWS Edinburgh Hacklab c/o G. Rowlands C/O Gareth Rowlands, Engineering Pigeon Holes, Acton W3 0RP
G8 LXS G. Pascoe Newhaye, Broadhempston, Totnes TQ9 6DB
G8 LXY S. Clarke 128 Putteridge Road, Luton LU2 8HQ
G8 LYB S. Tompsett 9 Ashlawn Road, Rugby CV22 5ET
G8 LYG W. Leach 15 Beech Lea, Blunsdon, Swindon SN26 7DE
GM8 LYO P. Mahood 4 Irvine Court, Glasgow G40 3LE
GM8 LYQ I. Lindsay 10/4 Mertoun Place, Edinburgh EH11 1JZ
G8 LYV K. Kearns 79 Church Road, Hatfield Peverel, Chelmsford CM3 2LB
G8 LYW B. Theedom 5 Rodbridge Drive, Southend-on-Sea SS1 3DF
G8 LZG G. Allen 21 Dale Road, Welton, Brough HU15 1PE
G8 LZK M. Ball 46A Daniels Crescent, Long Sutton, Spalding PE12 9DS
G8 LZO J. Hibbert 80 High Street, Newchapel, Stoke-on-Trent ST7 4PT
G8 LZS P. Martin 35 Martineau Lane, Hurst, Reading RG10 0SF
G8 LZV A. Fulcher Keepers, The Street, Ashford TN27 0QF
GW8 LZY S. Brown Maes Yr Haidd, 8 Glanceulan, Aberystwyth SY23 3HF
G8 MAA G. Chaplin 8 Manor House Drive, Northwood HA6 2UJ
G8 MAD P. Tostevin 70 Robson Road, Goring-By-Sea, Worthing BN12 4EN
G8 MAF T. Beckham 2 Sandbanks Place, Ersham Road, Hailsham BN27 3LJ
G8 MAR M. Sibley 10 Ainley Close, Huddersfield HD3 3RJ
G8 MAV P. Lewis Westbank, 46 Weyside Road, Guildford GU1 1HX
G8 MAW K. Maw 12 Strickland Way, Wimborne BH21 2GF
G8 MAY A. Lake 9 Grafton Close, King's Lynn PE30 3EZ
G8 MBE S. Fouracres Grove House, Castle Street, Bampton, Tiverton EX16 9NS
G8 MBJ J. Parsons 34 Mill Hill, Brancaster, King's Lynn PE31 8AQ
G8 MBK P. Bland 17 Knowles Close, Kirklevington, Yarm TS15 9NL
G8 MBM C. Proctor 15 Chiltern Street, Aylesbury HP21 8BN
G8 MBQ R. Jones 46 Wilmington Close, Woodley, Reading RG5 4LR
G8 MBS R. Vitiello 6 Meeting Oak Lane, Winslow MK18 3JU
G8 MBU R. Williams 8 Gurnard Heights, Gurnard PO31 8EF
G8 MBV I. Wood Tessian Lodge, Oak Hill, Swingfield, Dover CT15 7HF
G8 MCA G. Bryan 34 Shelbury Close, Sidcup DA14 4BE
G8 MCC C. Divall 12 Latchmount Gardens, Axminster EX13 5JT
G8 MCJ B. Pritchard 14 Rugby Way, Croxley Green, Rickmansworth WD3 3PH
G8 MCR V. Eagles 3 Church Road, Buckhurst Hill IG9 5RU
G8 MCT C. Bate 28 Argyle Avenue, Tamworth B77 3PH
G8 MCW P. Elkins 615 Blandford Road, Upton, Poole BH16 5ED
G8 MCY M. Dannatt 46 Laburnham Road, Biggleswade SG18 0NX
G8 MDG D. Shaw 35 Tinshill Lane, Leeds LS16 6BU
G8 MEA C. Wilson 2 Blackthorn Garth, Beverley HU17 8FZ
G8 MEC D. Uttley 1 Edgeside, Great Harwood, Blackburn BB6 7JS
G8 MED P. Shirtliff 2 Birch Avenue, Newton, Preston PR4 3TX
G8 MEE K. Patman 5 Lime Grove, Holbeach, Spalding PE12 7NG
G8 MEH L. Steele Caprice, Woodville Road, Bude EX23 9JA
G8 MEI R. Whitby 24 Macaulay Avenue, Great Shelford, Cambridge CB22 5AE
G8 MEM A. Lillywhite 1 Roblin Close, Aylesbury HP21 9DT
GW8 MER M. Busson 14 Squires Gate, Rogerstone, Newport NP10 0BP
G8 MEX I. Glenn 257 Wimpole Road, Barton, Cambridge CB23 7AE
G8 MFF Pembs RS c/o R. Hedley 20 Spencer Drive, Tiverton EX16 4PY
G8 MFH R. Lake 2A Hyde Lane, Bovingdon, Hemel Hempstead HP3 0EG
G8 MFI S. Mcguigan 16 Queen Street, Middleton Cheney, Banbury OX17 2NP
G8 MFM R. Wood 36 New England Road, Haywards Heath RH16 3JS
G8 MFO T. Sorensen 22 The Cottrells, Angmering, Littlehampton BN16 4AF
GW8 MFQ A. John 79 Harding Close, Boverton, Llantwit Major CF61 1GX
G8 MFR R. Irwin Copperfield, 97 Offerton Lane, Stockport SK2 5BS
G8 MFU D. Parry 19 Norton Lane, Great Wyrley, Walsall WS6 6PE
G8 MFV R. Hickmott 8 Dyke Road, Folkestone CT19 6BS
GM8 MFZ Dr N. Kennedy Deveron, North Deeside Road, Pitfodels, Aberdeen AB15 9PL
G8 MGD D. Marshall 7 Aesops Orchard, Woodmancote, Cheltenham GL52 9TZ
G8 MGE G. Young 30 Degenhardt Streett, South Australia 5545 Australia
GW8 MGF J. Tait 2 Bron-Y-Coed, Coed-Y-Glyn, Wrexham LL13 7QJ
G8 MGG W. Whiteside 5 Church Close Levens, Kendal LA8 8QE
G8 MGK J. Dosher 40 Bromfield Road, Redditch B97 4PN
G8 MGO J. Marshall 34 Derwent Drive, Swindon SN2 7NJ
G8 MGP A. Hill 5 Lilac Walk, Kempston, Bedford MK42 7PE
G8 MGQ D. Garwood 13 Market Street, Bradford-on-Avon BA15 1LL
G8 MGZ P. Haynes 2 The Chase, Furnace Green, Crawley RH10 6HW
G8 MHA L. Humphrey 1 Falkenham Road, Kirton IP100NP
G8 MHD C. Cooper 16 Paulton Drive, Bishopston, Bristol BS7 8JJ
G8 MHE G. Cross 117 Broadway, Eccleston, St. Helens WA10 5PB
G8 MHI K. Russell 12 Evans Close, Greenhithe DA9 9PG
G8 MHL D. Shaw 51 Emmanuel Road, Southport PR9 9RP
G8 MHN S. Scrase 5 Clinton Road, Retired KT22 8NU
G8 MHO A. Fraser 184 Old Road, Harlow CM17 0HQ
G8 MHT G. Dallaway Flat 2, School Court, Meyer Street, Stockport SK3 8JE
GM8 MHU I. Fraser 12 Auchlea Place, Aberdeen AB16 6PD
G8 MIA Hastings Electronics & RC c/o A. Malbon The Lodge, Blithbury Road, Rugeley WS15 3HJ
G8 MIC M. Williams Flat 2, High Point, Highgate N6 4BA
G8 MIE K. Croucher 140 Dane Road, Coventry CV2 4JW
G8 MIF F. Golding 16 Lessness Park, Belvedere DA17 5BG
G8 MIH R. Green 33 Bulkington Avenue, Worthing BN14 7HH
G8 MII T. Ashton 30 Highfields Road, Chasetown, Burntwood WS7 4QU
G8 MIN R. Welsh 14 Drayton Close, High Halstow, Rochester ME3 8DW

G8 MIT C. Wyatt 273 Nuthurst Road, Birmingham B31 4TQ
GI8 MIV G. Hutchinson 40 Oldstone Hill, Muckamore, Antrim BT41 4SB
G8 MIW J. West 21 Gardenia Crescent, Mapperley, Nottingham NG3 6JA
G8 MJE R. Farman 19 Mayfield Park, Cheltenham Road, Bagendon, Cirencester GL7 7BH
G8 MJF K. Bottomley Whispering Winds, 15 Marvell Rise, Harrogate HG1 3LT
G8 MJH P. Harrison 154 Cherrydown Avenue, Chingford, London E4 8DZ
GM8 MJV T. Melvin Blue House, Remote, Pathhead EH37 5UP
G8 MJX D. Coomber 1 Brympton Road, Coventry CV3 1GW
G8 MKC Eden Valley RS c/o D. White 1 Whaddon Road, Shenley Brook End, Milton Keynes MK5 7AF
G8 MKE C. Rose 45 Clent Road, Warley, Oldbury B68 9ES
G8 MKG G. Barraclough 1 Meadowcroft Road, Middlesbrough TS6 0JD
G8 MKN I. Wager 106 Turner Road, Colchester CO4 5JT
G8 MKO R. Pocock 3 Brewery Cottages, Netherley Road, Prescot L35 1QG
G8 MKQ A. Bullock 35 Parkstone Avenue, Thornton-Cleveleys FY5 5AE
G8 MKS P. Moore 3A High Street, Mow Cop, Stoke on Trent ST7 3ND
G8 MKW J. Green Huntley, Chesham Road, Tring HP23 6HH
G8 MKX J. Donnithorne 6 Bulbourne Court, Tring HP23 4TP
G8 MLA P. Richardson 11 Overstone Road Coldham, Wisbech PE14 0ND
G8 MLB N. Bourner 11 Richborough Road, Sandwich CT13 9JE
G8 MLD M. Warren 17 Bolehill Park, Hove Edge, Brighouse HD6 2RS
G8 MLK J. Owen The Old Coach House, Callow Hill, Virginia Water GU25 4LD
G8 MLW D. Carr 39 Fallowfield Road, Walsall WS5 3DH
GM8 MMA W. Williamson Leeskol, Camb, Shetland ZE2 9DA
G8 MMF P. Dorrington 57 Ferring Lane, Ferring, Worthing BN12 6QS
G8 MMG D. Bentley 55 Saddlers Road, Quedgeley, Gloucester GL2 4SY
G8 MMM G. Nicholas Greenbank, Chester High Road, Neston CH64 7TR
G8 MMN Dr M. Holmes 8 High Street, Norley, Frodsham WA6 8JS
G8 MMP M. Swain 38 Longdale Lane, Ravenshead, Nottingham NG15 9AD
G8 MMV N. Goodwin 17 The Acres, Lower Pilsley, Chesterfield S45 8DT
GM8 MMW W. Dick 58 Kirkland Road, Glengarnock, Beith KA14 3AJ
G8 MNC M. Bilkey Pebble Flek, The Green, St. Austell PL25 5TA
GM8 MNG C. Raine Broomhill Edgehead, Pathhead EH37 5RS
G8 MNL P. Carruthers 16 Wivenhoe Close, Rainham, Gillingham ME8 7QB
GM8 MNM R. Hood Milton Of Auchindoir House, Rhynie, Huntly AB54 4JB
G8 MNO W. Stewart 9 Ashley Road, Marnhull, Sturminster Newton DT10 1LQ
GM8 MNR D. Jenkins 16 Bentinck Street, Galston KA4 8HT
G8 MNY J. Stockley 27 Campden Road, South Croydon CR2 7ER
G8 MOF F. Bellamy 3 Manor Road, Crowle, Scunthorpe DN17 4ET
G8 MOG D. Dale Blackwood Hall, Felton, Morpeth NE65 9QW
GM8 MOI C. Stirling 20 Craigford Drive, Bannockburn, Stirling FK7 8NQ
G8 MOL P. Marshall 134 Gladbeck Way, Enfield EN2 7EN
G8 MOS A. Reale 20 Wickham Close, Alton GU34 1RR
GI8 MOV F. Warwick 34 Shaws Wood Cullybackey, Ballymena BT42 1SB
GW8 MOZ G. Elliott 9 Hove Avenue, St. Julians, Newport NP19 7QP
G8 MPG G. Rigby 1 Route Danton, Petit Caudos, , Mios 33380 France
G8 MPM W. Brock 15 Picketleaze, Chippenham SN14 0DN
G8 MQF M. Cooper Woodstock, Snow Hill, Crawley RH10 3EG
G8 MQK J. Lindley 17 Leyfield Bank, Holmfirth HD9 1XU
G8 MQT T. Smith 416 Charminster Road, Bournemouth BH8 9SG
G8 MQX R. Eccles 6 Queens Drive, Barnsley S75 2QJ
G8 MQY B. Densham 47 High Street, Paulerspury, Towcester NN12 7NA
G8 MRI R. Davey 23 Campbell Close, Hunstanton PE36 5PJ
G8 MRN M. Watch 9 High Drive, Gosport PO13 0QS
GM8 MST Dr G. Kelly 36 Craigleith Drive, Edinburgh EH4 3JU
G8 MSY J. Wilkinson 11 Wigmore Road, Tadley RG26 4HH
G8 MTB M. Greenfield 8 The Spinney, Clayton, Newcastle ST5 4DA
G8 MTI M. Dibsdall 28 Court Farm Avenue, Epsom KT19 0HF
G8 MTV J. Wood Coach House, Croft On Tees, Darlington DL2 2SL
G8 MTW D. Sutherland 44 Old Barrack Road, Woodbridge IP12 4ET
G8 MUF J. Ames 16 Vere Gardens, Henley Road, Ipswich IP1 4NZ
G8 MUV B. Clarke Flat 26, Compton Grange, Whitehall Road, Cradley Heath B64 5BG
G8 MUX J. Mottram Church View, New Road, High Peak SK23 7NH
G8 MVC R. Westlake Flat 9, Grosvenor Court, 135-139 The Grove, London W5 3SL
G8 MVD K. Wilks 72 Grasmere Road, Bradford BD2 4HX
G8 MVH J. Armstrong 30A Abbey Fields, Faversham ME13 8JA
G8 MVJ C. Chambers Hollybank, Back Street, Driffield YO25 3TD
G8 MVS N. Fuller 48 White Street, Easterton, Devizes SN10 4PA
G8 MVY E. Phillips 2 Primrose Cottage, The Street, Reading RG7 1QY
G8 MWA Medway Amateur Receiving & Transmitting Society c/o J. Burton 22 Pear Tree Lane, Hempstead, Gillingham ME7 3PT
G8 MWD D. Lewing 94 Carville Crescent, , Brentford TW8 9RD
G8 MWE K. Knight 54 Vicarage Lane, Water Orton, Birmingham B46 1RU
G8 MWN K. Harris Bella Vista, Station Road, Yelverton PL20 7JS
G8 MWU P. Stafford 5 Westmead Drive, Newbury RG14 7DJ
G8 MWW W. Westlake West Park, Clawton, Holsworthy EX22 6QN
G8 MWX A. Priestley 55 Derwent Avenue, Garforth, Leeds LS25 1HN
G8 MXD G. York 13 Cherwell Close, Thornbury, Bristol BS35 2DN
G8 MXQ A. Taylor 311 Smeeth Road, Marshland St. James, Wisbech PE14 8ES
G8 MXR W. Pitt 1 Windy Ridge, James Street, Stourbridge DY7 6ED
G8 MXT L. Mansfield 25 Carlton Road, Derby DE23 6HB
G8 MXV K. Ayriss 6 Langstons, Trimley St. Mary, Felixstowe IP11 0XL
G8 MXW C. Down 100 Lynwood Drive, Merley, Wimborne BH21 1UQ
G8 MYF M. Johnson 42 Marlborough Road, Ryde PO33 1AB
G8 MYG C. Hunt Rowan Bank, 2 Cranston Rise, Bexhill-on-Sea TN39 3NJ
G8 MYJ C. Drewe 37 Baker Street, Chelmsford CM2 0SA
G8 MYK A. Rowley Holly Cottage, 368 Highters Heath Lane, Birmingham B14 4TE
GM8 MYO C. Tyler 26 Sinclair Way, Knightsridge, Livingston EH54 8HW
G8 MYV D. Webster 35 Raymond Road, Maidenhead SL6 6DF
G8 MZA D. Garrett Brookside Farm, Tonge, Derby DE73 8BD
G8 MZD P. Diggins 8 Gloucester Gardens, Bagshot GU19 5NU
G8 MZQ Dr W. Katz The Beacon, Goathland, Whitby YO22 5AN
GW8 MZR R. Harris 15 Quarry Rise, Undy, Caldicot NP26 3JU
G8 MZW S. Adams 20 Queens Drive, Skegness PE25 1RE
G8 MZY Prof. D. Cushman 50 St. Peters St., Syston, Leicester LE7 1HJ
G8 MZZ P. Boam 36 Copeland Drive, Stone ST15 8YP
GW8 NAC K. Davies 45 Castle View, Simpson Cross, Haverfordwest SA62 6EN
G8 NAG M. Smith 18 Manor Lane, Verwood BH31 6HX
G8 NAI J. Lazzari 3 Terson Way, Weston Coyney., Stoke-on-Trent ST3 5RQ
GM8 NAL P. Corbishley Tweedbank House, Cardrona, Peebles EH45 9HX
G8 NAM P. Buttress 18 Taffrail Gardens, South Woodham Ferrers, Chelmsford CM3 5WH
G8 NAP P. Beacon 67 St. Helena Road, Polesworth, Tamworth B78 1NJ
G8 NAU C. Searle 55 Chenies Close, Tunbridge Wells TN2 5LN
G8 NAV N. Vernon 11 Alexandra Road, Birchington CT7 0DX
GW8 NBF C. Morgan 84 Treowen Road Newbridge, Newport NP11 3DP
G8 NBI A. Buxton 95 Cantelupe Road, Ilkeston DE7 5HT
G8 NBO L. Phillips 14 Heal Park Crescent, Fremington, Barnstaple EX31 3AP
GM8 NBV C. Davies 35 Laverock Avenue, Hamilton ML3 7DD
G8 NCK N. Brown 9 Redhill Close, Tamworth B79 8EJ
GW8 NCN Edinburgh And District ARC c/o D. Sanford 35 Summerfield Avenue, Cardiff CF14 3QA
G8 NCS M. Green 21 Hill View Rise, Northwich CW8 4XA
G8 NCU J. Watkins Llwynteg, Glanwern, Borth SY24 5LT
G8 NDB Dr G. Jarrett 1 Church Street, Twycross, Atherstone CV9 3PJ
G8 NDE J. Turner 9 Clifton Avenue Culcheth, Warrington WA3 4PD
G8 NDF D. Simpson 10 Buckingham Way, Byram, Knottingley WF11 9NN
G8 NDK K. Lindley 25 Lindsey Court, Epworth, Doncaster DN9 1SD
G8 NDN C. Keens Toad Hall, 69 Lillywhite Crescent, Andover SP10 5NA
G8 NDP R. Callan The Tower, Seaside Road, Easington, Hull HU12 0TY
G8 NDR N. Burridge 8 Cedar Close, Ware SG12 9PG
G8 NDT N. Thompson 125 Summer Road, Erdington, , Birmingham. B23 6DX
G8 NDV P. Fay 42 Roberts Road, Salisbury SP2 9BY
G8 NED Wisbech Amateur Radio & Electronics Club c/o A. Bridgeland 17 Oldfield Lane, Wisbech PE132RJ
G8 NEF Dr R. Peel 76 Cypress Grove, Ash Vale, Aldershot GU12 5QW
G8 NEI K. Marsh 1 Parr Close, Exeter EX1 2BG
G8 NEL S. Nightingale 75 Gordon Road, Herne Bay CT6 5QX
G8 NEO D. Edwards 3 Murton Close, Burwell, Cambridge CB25 0DT
GM8 NET A. Fraser 50 Tannin Crescent, East Kilbride G75 9FS
G8 NEY D. Millard Weavern House, Hartham Lane, Biddestone, Chippenham SN14 7EA
G8 NFD K. Gardiner 8 Foxlands Drive, Sutton Coldfield B72 1YZ
GM8 NFG J. Aitken Hamabo, 10 Lynnpark, Kirkwall KW15 1SL
G8 NFM F. Turner 46 Main Street, Kings Newton, Derby DE73 8BX
G8 NFP A. Crockett 57 Upland Road, Sutton SM2 5HW
GM8 NFT N. Leitch 4 Dullatur Road, Dullatur, Glasgow G68 0AF
G8 NFZ S. Sims 35 Garfield Road, Hailsham BN27 2BU
G8 NGE K. Ebborn 18 St. Marys Park, Ottery St. Mary EX11 1JA
G8 NGF D. Stone Aston Hill Cottage, 5 Aston Hill, Westbury, SY5 9JS
G8 NGJ P. Richardson Domaine De Calcat, 47150 La Sauvetat Sur Lede, Lot Et Garonne, France, la Sauvetat Sur Lede 47150 France
G8 NGM N. King 42 Constance Close, Witham CM8 1XY
G8 NGR M. Roberts 62 Oakwood Road, Bricket Wood, St. Albans AL2 3QA
G8 NGZ South Gloucestershire Raynet c/o V. Edwards 33 Eyrescroft, Bretton, Peterborough PE3 8ES
G8 NHD P. Mart 6 Rose Creek Gardens, Great Sankey, Warrington WA5 3TT
G8 NHG R. Wilkins New Copse House, Fishbourne Lane, Ryde PO33 4EZ
G8 NHO J. Austin 5 Mercia Road, Baldock SG7 6RZ
G8 NIE D. Sharpe 5 Drydales, Kirk Ella, Hull HU10 7JU
G8 NIL D. Bales 30 Railway Road, Wisbech PE13 2QA
GU8 NIS Guernsey ARS c/o D. Eaton Glenfield, Le Foulon, St. Andrew GY6 8UF Guernsey
G8 NIU R. Whiting Heather Bank Glebelands, Minehead TA24 8DH
G8 NJA Gloucestershire County Raynet c/o D. Webber 43 Lime Tree Walk, Milber, Newton Abbot TQ12 4LF
G8 NJI P. Woodhead 24 North Bar Without, Beverley HU17 7AB
G8 NKJ L. Reid 26 Mansion Avenue, Whitefield, Manchester M45 7SS
G8 NKM A. O'Donovan 2 Mackenzie Road, Beckenham BR3 4RU
G8 NKN S. Gorwits 29 Howitt Drive, Bradville, Milton Keynes MK13 7DY
GI8 NLF I. Munro 54 Recreation Road, Larne BT40 1EN
G8 NLK M. Bennett 68 Meadow Hill Road, Birmingham B38 8DA
G8 NLS S. O'Brien Flat 2, 99A Howard Street, North Shields NE30 1NA
G8 NMH B. O'Regan 10 School Hill, Little Sandhurst, Sandhurst GU47 8LD
G8 NMK C. Eccles 18 Infinity View, Stockton-on-Tees TS18 2FN
G8 NMM C. Reid 138/17, Bang Saray, Pattaya 20230 Thailand
G8 NMO D. Pechey Jays Lodge, Crays Pond, Reading RG8 7QG
G8 NMT J. Hicks 7 North Croft, High Wycombe HP10 0BP
G8 NNA B. Crellin 60 College Fields, Woodhead Drive, Cambridge CB4 1YZ
GW8 NNF R. Galpin 23 Heol Y Delyn, Lisvane, Cardiff CF4 5SR
G8 NNL J. Hurley 64 Carleton Road, Chorley PR6 8UB
G8 NNP B. Gower 132 Goldsworthy Way, Slough SL1 6AY
G8 NNS G. Stamp 41 Willoughby Road, Wallasey CH44 3DZ
G8 NNU T. Rowe 68 Cobourg Road, Montpelier, Bristol BS6 5HX
G8 NNX M. Cohen 41 South Station Road, Liverpool L25 3QE
G8 NNZ N. Carrington 22 Parrs Wood Avenue, Manchester M20 5ND
G8 NOB N. Bean 33 Badger Close, Guildford GU2 9PJ
G8 NOD M. Stamford 1 Canon Drive, Norton Canon, Hereford HR4 7BJ
G8 NOF R. Holt Tile House, Vicarage Hill, Solihull B94 5EB
G8 NOP P. Price Calwich View, Dove Street, Ellastone, Ashbourne DE6 2GY
G8 NOS A. Swallow 67A Strines Road, Marple, Stockport SK6 7DT
GW8 NP Highfields ARC c/o S. Williams 371 Coed-Y-Gores Llanedeyrn, Cardiff CF23 9NR
G8 NPD J. Hodnett 126 Northwood Lane, Newcastle ST5 4BN
G8 NPH A. Arnold 2 Duck Lane, Haddenham, Ely CB6 3UE
G8 NPP A. Brown Dunlop Hiflex Powerbend, Pennywell Industrial Estate, Sunderland SR4 9EN
G8 NPR J. Blackshaw 23 Cherry Orchard, Oakington, Cambridge CB24 3AY
G8 NPT A. Work Faleriolaan 253, Hillegom 2182 TM Netherlands
G8 NPZ P. Whiteman 22 Hartsbourne Road, Earley, Reading RG6 5PY
G8 NQC P. Manser 61 Galsworthy Drive, Caversham, Reading RG4 6QB
G8 NQI J. Gartside 12 Starfield Avenue, Hollingworth Lake, Littleborough OL15 0NG
G8 NQK G. English 25 Powell Gardens, Newhaven BN9 0PS
G8 NQN M. Bancroft Westfield, Towngate Southowram, Halifax HX3 9QZ
G8 NQO A. Whyatt 11 The Perrings, Nailsea, Bristol BS48 4YD

G8 NQY W. Lea 20 Gloucester Road, Walsall WS5 3PN
G8 NRC S. Deighton 132 Horsehead Lane Bolsover, Chesterfield S44 6XH
G8 NRF G. Wood 3 Cleveleys Road, Great Sankey, Warrington WA5 2SR
G8 NRP Exeter ARS c/o M. Andrew 80 Hamble Drive, Abingdon OX14 3TE
G8 NRR R. Bambrook 26 Croft Road, Thame OX9 3JF
G8 NRU D. Carr 78 Kingsleigh Road, Heaton Mersey, Stockport SK4 3PG
G8 NSD F. Taylor 96 Elvaston Road, North Wingfield, Chesterfield S42 5HH
G8 NSE F. Wood 96 Manchester Road, Astley, Manchester M29 7EJ
G8 NSK J. Barnes 23 Spenser Road, King's Lynn PE30 3DP
G8 NSO S. Fleetham Hill House Leckhampton Hill, Cheltenham GL53 9QG
G8 NST J. Leek 30 Casuarina Road, Bucklands Beach, Auckland 1706 New Zealand
G8 NSV R. Mcnair 43 Hastings Gardens, New Hartley, Whitley Bay NE25 0RN
G8 NSX R. Miller 89 Moorside, Spennymoor DL16 7DZ
G8 NSZ M. Stanway 72 Sheldons Court, Winchcombe Street, Cheltenham GL52 2NR
G8 NTD K. Johnson 24 Capers Close, Enderby, Leicester LE19 4QD
G8 NTG W. Howell 6 Unity Avenue, Sneyd Green, Stoke-on-Trent ST1 6DE
G8 NTH A. Hewat 41 Summersbury Drive, , Shalford GU4 8JG
G8 NTJ K. Hand 75 Hill Street, Hednesford, Cannock WS12 2DW
G8 NTQ M. Roper 6 Ilmington Close, Hatton Park, Warwick CV35 7TL
G8 NTR Dr J. Williams 133A Wiltshire Lane, Pinner HA5 2NB
G8 NTS J. Smart Greystone, High Street, Swindon SN26 7AR
G8 NTY C. Mallows 9 Chestnut Drive, Shenstone, Lichfield WS14 0JH
G8 NTZ D. Kowalczyk 5 Priestthorpe Lane, Bingley BD16 4ED
G8 NVB N. Brown 9A Decoy Drive, Eastbourne BN22 0AB
G8 NVC B. Ellis 7 Highmoor Close, Corfe Mullen, Wimborne BH21 3PU
GM8 NVE D. Watters 28 Bruce Road, Crossgates, Cowdenbeath KY4 8AZ
GM8 NVG Wallen Antennae RC c/o A. Wilson Lochend, Beith KA15 2LN
G8 NVH S. Reynolds 242 Butchers Lane, Mereworth, Maidstone ME18 5QH
G8 NVI A. Stevens 67 New Road, East Hagbourne, Didcot OX11 9JX
G8 NVS S. Hindle Innisfree, Barton Terrace, Dawlish EX7 9QH
G8 NVT R. Hatfield 1 Slade Close, Ottery St. Mary EX11 1SY
G8 NVX Cumbria Emergency Planning Unit c/o M. Moss 24 Magna Lane, Dalton, Rotherham S65 4HH
G8 NVZ Dr G. Evans 4 Holcot Lane, Anchorage Park, Portsmouth PO3 5TR
G8 NWC G. Boor 27 Welbeck Drive., Spalding PE11 1PD
G8 NWI J. Vine 117 Betterton Road, Rainham RM13 8ND
G8 NWL J. Mason 46 Bradford Street, Chelmsford CM2 0FJ
G8 NWM V. Maxfield 50 Hanthorpe Road, Morton, Bourne PE10 0NT
G8 NWU M. Wright 69 Wroxham Drive, Nottingham NG8 2QR
G8 NWZ M. Percy 73 Ridgeway, Wellingborough NN8 4RY
G8 NXA T. Ehlen 58B Warriner Gardens, London SW11 4DU
G8 NXB N. Borrett 44 Oakhill Road, Ashtead KT21 2JG
G8 NXD Horsham ARC c/o M. Waterfall 12A Boskenna Road, Four Lanes, Redruth TR16 6LS
G8 NXE S. Eyles 2 Salisbury Close, Lichfield WS13 7SN
G8 NXG R. Burn 3 Barnsite Gardens, Rustington, Littlehampton BN16 3QG
G8 NXJ I. Livesey 26 Hilltop Road, Twyford, Reading RG10 9BN
GW8 NXK G. Garner 4 Laurel Drive, Woodland Park, Waunarlwydd, Swansea SA5 4QH
G8 NXQ North Pennines Radio Group c/o W. Povey 31 Baddlesmere Road, Whitstable CT5 2LB
G8 NXS D. Stevenson 86 Kingston Road, Luton LU2 7SA
G8 NXY W. Godwin Heathwood House, Burton Road, Duddon Heath, Duddon, Tarporley CW6 0GJ
G8 NYB D. Reed 59 Cowley Avenue, Chertsey KT16 9JJ
G8 NYC J. Primmer 45 Roman Way, Felixstowe IP11 9NP
G8 NYD M. Perry 23 Victors Crescent Hutton, Brentwood CM13 2HZ
G8 NYH R. Adams 2 Longwill Avenue, Melton Mowbray LE13 1UR
G8 NYJ I. Gibbs 3 Badger Drive, Lightwater GU18 5TS
G8 NYK M. Nicholson Flat 14, Whyke Court, Chichester PO19 8TP
G8 NYM M. Lister 97 Hightown Road, Liversedge WF15 8DG
G8 NYR B. Rabey 36 Park Way, St. Austell PL25 4HR
GM8 NYV S. Richardson Rowan Bank, Melvich, Thurso KW14 7YJ
G8 NYZ N. Cooper 4 Mossfield Crescent, Kidsgrove, Stoke-on-Trent ST7 4YA
G8 NZA P. Baker 12 Hockeredge Gardens, Westgate-on-Sea CT8 8AN
G8 NZB B. Durrant 16 Merrymeet, Whitestone, Exeter EX4 2JP
G8 NZC K. Edmunds 44 Antonia Circuit, Hallett Cove SA 5158 Australia
G8 NZD C. Atkinson 8 Southwood Road, Dunstable LU5 4EA
G8 NZK N. O'Hagan 5 Bankside, Finchampstead, Wokingham RG40 3QB
GM8 NZL E. Hogg 43 Muir Wood Road, Currie EH14 5JN
GW8 NZN D. Roberts 12 Erw'R Llan, Nannerch, Mold CH7 5RF
G8 NZO J. Crozier 43 Shepherds Way, Birmingham B23 5XR
G8 NZR K. Pullan 18 Heathfield, Mirfield WF14 9BJ
G8 OAD G. Baxter 4 Deeping Road, Baston, Peterborough PE6 9NP
GM8 OAH W. Easton 21 Cameron Avenue, Bishopton PA7 5ES
G8 OBB T. Hooker Inglewood, Woodside Road, Luton LU1 4DJ
GM8 OBO R. Thomson 43 Ashgrove Street, Ayr KA73BG
G8 OBP D. Payne 11 Welbeck Close, Blaby, Leicester LE8 4HF
G8 OBR J. Readle 17 Rossendale View, Todmorden OL14 6HN
G8 OBT T. Graham 22 Locker Park, Wirral CH49 2RZ
G8 OCA J. Astle River View, Brough, Kirkby Stephen CA17 4BZ
G8 OCE J. Hardy 42 Fir Tree Drive, Wales, Sheffield S26 5LZ
G8 OCF R. Harris 25 Pear Tree Drive, Chard TA20 2FP
G8 OCM E. Dubbins 2 Elizabeth Avenue, Rose Green, Bognor Regis PO21 3EL
G8 OCO M. Hughes 49 Reedings Road, Barrowby, Grantham NG321AU
GI8 OCR J. Mcilveen 31 Edenaveys Crescent, Armagh BT60 1NT
G8 OCS D. Simpson 6 St. Martins Close, Stratford-upon-Avon CV37 9QW
G8 OCT S. Terry 207 Lakeside Drive, Walhalla Sc 29691 United States
G8 OCV C. Smart Old Queens Head, Ipswich Road, Diss IP21 4XP
G8 ODK R. Varley 41 Lang Lane, West Kirby, Wirral CH48 5HQ
GM8 OEG Chichester ARC c/o A. Swiffin Glebe House, Kellas, Dundee DD5 3PD
G8 OEJ E. Bray Rothesay, 6 Empshott Road, Southsea PO4 8AU
G8 OEK P. Brown Estate Yard House, Beverley HU17 7PN
G8 OEO J. Thompson 4 The Grove, Ponteland, Newcastle upon Tyne NE20 9HQ
G8 OEU T. Hipwood 3 Camview, Paulton, Bristol BS39 7XA
G8 OFA Dr M. Cranage Corris House, West Gomeldon, Salisbury SP4 6LS
G8 OFI G. Radivan 15 Agecroft Road West, Prestwich, Manchester M25 9RE
G8 OFN R. Pashley 24 Essendine Crescent, Sheffield S8 8PB
G8 OFO R. Short Langtree House, Castle Hill, Fordingbridge SP6 2AX
GM8 OFQ G. Dobson Ocean View, Stromness KW16 3PQ
G8 OFR R. Coole Courtyard Cottage, Horse Fair Lane, Swindon SN6 6BN
G8 OFX A. Nelson 37 Brook Way, Romsey SO51 7JZ
G8 OFZ I. Mcgowan Meld House, Hawthorn Road, Shrewsbury SY3 7NB
G8 OGP S. Martin Aldon, The Hayes, Cheddar BS27 3HS
G8 OGR J. Holton 24 Great Austins, Farnham GU9 8JQ
G8 OHC G. Scholes 14 Braemar Road, Bulwell, Nottingham NG6 9HN
G8 OHG J. Myall 52 Princethorpe Way, Binley, Coventry CV3 2HF
G8 OHH J. Morgan 41 Lingen Avenue, Hereford HR1 1BY
G8 OHM East Chehire Raynet Group c/o N. Gutteridge 68 Max Road, Quinton, Birmingham B32 1LB
G8 OHP C. Cainsford-Betty 19 Reads Street, Stretham, Ely CB6 3JT
G8 OHS M. Emery 25 Bradgate Drive, Sutton Coldfield B74 4XG
GW8 OIJ A. Stark 5 Ladyhill Road, Newport NP19 9RY
G8 OIU Dr J. Cave 4 Grove Road, Newbury RG14 1UH
G8 OIY Dr S. Robertson 249 Ware Road, Hertford SG13 7EJ
G8 OJK V. Willett 20 The Green, Sharlston Common, Wakefield WF4 1EF
G8 OJR W. Pearce 160 Philip Lane, Tottenham, London N15 4JN
G8 OKB R. Mccann Goss House, Clark Street, Stourbridge DY8 3UF
G8 OKD M. Bailey 28 St. Pauls Hill Road, Hyde SK14 2SW
G8 OKE R. Brown 8 Grassmere Way, Waterlooville PO7 8QD
G8 OKI L. Mather 8 Carnoustie Avenue, Chesterfield S40 3NN
G8 OKN M. Gallagher 50 Warwick Road, Southam CV47 0HW
GW8 OKR B. Kirkpatrick 88 Gaer Park Drive, Newport NP20 3NR
G8 OKS B. Dawson 9 Redmayne Close, Billingham TS23 3HG
G8 OKZ D. Shillington 6 Moss Close, Willaston, Neston CH64 2XQ
G8 OLA R. Elwood 28 Manor Lane, Selsey, Chichester PO20 0NX
GI8 OLH T. Lavery 21 Mussenden Grange, Articlave, Coleraine BT51 4US
G8 OLK P. Smith 21A Meadow Way, Bracknell RG42 1UE
G8 OLL R. Porter 9 School Avenue Brownhills, Walsall WS8 6AG
G8 OLP M. Matthews 19 Perrylands, Charlwood, Horley RH6 0BL
G8 OLY D. Curwell 9 St. Georges Road, Aldershot GU12 4LD
G8 OMB D. Parker 146 Merlin Avenue, Nuneaton CV10 9QJ
G8 OMC D. Smith 71 Ashbourne Avenue, Aspull, Wigan WN2 1HW
G8 OMK C. Robins 49 Brockenhurst Gardens, London NW7 2JY
G8 OMQ D. Bliss 7 Silver Lane, West Wickham BR4 0SG
G8 OMW F. Rowan 91 St. Nicholas Road, Littlemore, Oxford OX4 4PW
G8 ONH J. Sager Well Cottage, Fenn Lane, Woodbridge IP12 4NZ
GW8 ONP Dr J. Eastwood 30 Gerddi Rheidol Trefechan, Aberystwyth SY23 1DB
G8 ONR M. Loader 20 Edgcumbe Drive, Tavistock PL19 0ET
G8 ONS K. Creighton 10 Oram Close, Allery Banks, Morpeth NE61 1XF
G8 ONY B. Goodhew 101 Brier Road, Sittingbourne ME10 1YL
G8 OO A. Holdsworth Millfield, The Green, Dereham NR20 5LL
G8 OOC J. Morecroft 217A Longhurst Lane, Mellor, Stockport SK6 5PN
G8 OOF Dr G. Ellison 36 Park Hill, Clapham, London SW4 9PB
G8 OOQ Dr M. Barton 23 Caledonia Place, Bristol BS8 4DL
G8 OOS M. Reeson 18 Hazelnut Way, Louth LN11 7BZ
G8 OPA P. Barry 32 Rutland Avenue, Sidcup DA15 9DZ
G8 OPC D. Crawley 9 Gwynns Walk, Hertford SG13 8AD
G8 OPE M. De Rouffignac 2 Westgate, Old Malton, Malton YO17 7HE
G8 OPI J. Spooner 67 Covert Mead, Handcross, Haywards Heath RH17 6DL
G8 OPO G. Bartels 37 Faircross Avenue, Romford RM5 3SX
G8 OPP J. Birkett 13 The Strait, Lincoln LN2 1JD
G8 OPX T. Willford 15 Foxglove Close, Broughton Astley, Leicester LE9 6YU
G8 OPY G. Winston 8 Linnet Close, Shoeburyness, Southend-on-Sea SS3 9YE
G8 OQC J. Kent 106 Victoria Road, Barnet EN4 9PA
G8 OQG P. Jobbins 35 Keys Avenue Horfield, Bristol BS7 0HQ
G8 OQK A. Taylor Flo & Elsies Cottage, 77 Bedford Road, Marston Moretaine, Bedford MK43 0LA
G8 OQP T. Spacagna 3A Station Road, Romsey SO51 8DP
G8 OQR M. Crossman 127 The Grove, Southend-on-Sea SS2 4DA
G8 OQT J. Lambert 125 Tudor Way, Mill End, Rickmansworth WD3 8HT
G8 ORM M. Baguley 42 Kendall Avenue, Shipley BD18 4DY
G8 ORO D. Coulter 9 Taylors Way, Whitehaven CA28 9PD
G8 ORR C. Brown 179 Bournville Lane, Birmingham B30 1LY
G8 ORX K. Rashleigh 43 Oxshott Way, Cobham KT11 2RU
G8 OSG N. Mcalpine 15 Sparrows Herne, Basildon SS16 5JH
G8 OSH N. Hubbard 31 Bridlington Crescent, Monkston, Milton Keynes MK10 9HG
G8 OSJ D. Halliwell 9 Berkeley Avenue, Alsager, Stoke-on-Trent ST7 2BW
G8 OSX K. Dawson 3 The Green, Bonehill, Tamworth B78 3HW
G8 OSZ S. Ashley 12 Dene Close, Wellingborough NN8 5QP
G8 OTA H. O'Tani 23 Bennett Street, Bath BA1 2QL
G8 OTC C. Anderson 42 Elizabethan Way, Rugeley WS15 2EE
G8 OTD S. Ballard 11 Laburnum Gardens Quedgeley, Gloucester GL2 4WF
G8 OTG R. Cannon 111 Brangbourne Road, Bromley BR1 4LP
G8 OTH C. Churchill 87 Bradley Crescent, Shirehampton, Bristol BS11 9SR
GM8 OTI Dr J. Cooke 6/2 Greenbank Terrace, Edinburgh EH10 6ER
G8 OTZ D. Logan 33 Foxwood Drive, Kirkham, Preston PR4 2DS
G8 OUG L. Gray 20 Turner Way, Clevedon BS21 7YN
G8 OUH I. Harfield White Gates, Crofton Avenue, Lee on the Solent PO13 9NJ
G8 OUI D. Baines 1 Carole Close, Sutton Leach, St. Helens WA9 4PW
GW8 OUM D. Briggs 43 Monmouth Walk, Markham, Blackwood NP12 0QR
G8 OUS S. Greendale 15 Bosworth Road, Cambridge CB1 8RG
G8 OUY D. Smith 41 Mitcham Road, Camberley GU15 4AR
G8 OVO N. Lihou 47 Lichfield Drive, Brixham TQ5 8DG
G8 OVZ S. Gosby 20 Woodland Mount, Hertford SG13 7JD
G8 OWA Dr R. Lewin 24 Gilliver Close, Burton-on-Trent DE14 2FL
G8 OWG Dr K. Rowe 43 Knighton Close, Duston, Northampton NN5 6NE
G8 OWS J. Greenall 6 Neasham Drive, Darlington DL1 4LG
G8 OWV Prof. J. Everard 4 Lloyd Close Heslington, York YO10 5EU
G8 OWZ O. Cockram 446 Holdenhurst Road, Bournemouth BH8 9AE
G8 OXD P. Brown 41 School St., Castleford WF10 2SB
G8 OXE M. Brooks 3 Wood Side, Wood Street, March PE15 0SB
G8 OXG N. Powell 42 Sheraton Drive, Kidderminster DY10 3QR
G8 OXI E. Mannix La Vieille Scierie, Les Allues 73550 France
G8 OXS A. Lambert Electronic Media Services Ltd, Building 51, Whitehill & Bordon Enterprise Park, Budds Lane, Bordon GU35 0FJ
G8 OXU C. Madge 89 Heron Gardens, Rayleigh SS6 9TU

G8 OXW P. Bettinson 22 Parkway Gardens Chandlers Ford, Eastleigh SO532EN
G8 OXX P. Bailey 236 Sandy Lane, Droylsden, Manchester M43 7JX
G8 OYA S. Mckinty Sands House 5 Nightingale Gardens, Moreton-in-Marsh GL56 0FX
G8 OYB K. Armstrong 2 Blackthorn Grove, Shawbirch, Telford TF5 0LL
G8 OYF J. Popplewell 6 Roseleigh Avenue, Manchester M19 2NP
G8 OYL W. Shave 26 Hessle Avenue, Boston PE21 8DA
G8 OYM P. Taylor 15 De Montfort Road, Lewes BN7 1SP
G8 OYQ M. Everitt 48 Rant Meadow, Hemel Hempstead HP3 8EQ
GW8 OYT B. Jones 6 Rhodfa Maes Hir, Rhyl LL18 4JF
G8 OYX C. Blanchard 19 Cuckfield Avenue, Ipswich IP3 8RZ
G8 OYY J. Bishop No1 Billhurst Cottage, Plaistow Street, Lingfield RH7 6EY
G8 OZD A. Batty 23 Sandyshot Walk, Peel Hall, Wythenshawe, Manchester M22 5AQ
G8 OZH J. Burrell 6 Blenheim Croft, Brackley NN13 7ET
G8 OZP R. Platts 43 Iron Walls Lane, Tutbury, Burton-on-Trent DE13 9NH
G8 OZQ S. Pallett 6 Lancaster Close, Coalville LE67 4TG
G8 OZT N. Morley Mazongill, Orton, Penrith CA10 3RZ
G8 OZY P. Harrison 91 Obelisk Rise, Northampton NN2 8QU
G8 PAB M. Humphries 20 Taunton Street, Swindon SN1 5EE
G8 PAD J. Field Flat 24, Dibden Court, Honeywall, Stoke-on-Trent ST4 1PB
G8 PAE H. Jackson 22 Harkness Drive, Waterlooville PO7 8SH
G8 PAG M. Rose 20 Broad Piece, Soham, Ely CB7 5EL
GM8 PAH D. Schofield 166 Harbour Place, Dalgety Bay, Dunfermline KY11 9AA
G8 PAI D. Rout Two Akers, Wrabness Road, Harwich CO12 5NE
G8 PAK C. Hollier The Sheiling, Ryall Road, Ryall, Upton-Upon-Severn, Worcester WR8 0RH
G8 PAL P. Hankinson 37 Victoria Avenue, Whitefield, Manchester M45 6DP
G8 PAN S. Day 14 The Crescent, Market Harborough LE16 7JJ
G8 PAT P. Mcguinness 9 Farmdale Road, Carshalton Beeches SM5 3NG
G8 PBH Dr A. Kent 46 Russley Road, Bramcote, Nottingham NG9 3JE
G8 PBI I. Murphy The Nurseries, Carnon Crease, Truro TR3 6LJ
G8 PBM A. Royston 4 Cedars Walk, Dunstable Street, Ampthill, Bedford MK45 2JY
G8 PBY G. Coles Dairy Farmhouse, West Winterslow, Salisbury SP5 1RE
GJ8 PCY P. Falle 2 Greystones, Gorey Village Main Road, Grouville JE3 9EP Jersey
G8 PDE W. Burin 35 Beaconsfield Road Low Fell, Gateshead NE9 5EU
GI8 PDK Dr D. Courtney 79 Fort Road, Belfast BT8 8LX
G8 PDM C. Commander 8 Cannon Place, Hampstead, London NW3 1EJ
G8 PDP R. Hinchliffe 34 Oaklea, Ash Vale, Aldershot GU12 5HP
G8 PDY S. Procter 8 Pond End Road, Sonning Common, Reading RG4 9SA
G8 PEA K. Wibberley 5A Marston Road, Croft, Leicester LE9 3GX
GM8 PEB Dr P. Fineron The Courtyard, Crauchie, East Linton EH40 3EB
G8 PEN C. Vernon 48 Long Beach, Hemsby, Great Yarmouth NR29 4JD
G8 PEW A. Pewsey Pineholm, High Close, Bovey Tracey, Newton Abbot TQ13 9EX
G8 PFL J. Turner 32 Petunia Crescent, Chelmsford CM1 6YP
G8 PFR M. Gibson Eccleswall Farm Bromsash, Ross on Wye HR9 7PW
GW8 PFT P. Hinson 7 Awel Tywi, Llangunnor, Carmarthen SA31 2NL
G8 PFZ R. Harrison Badgers Oak, Redbrook Street, Ashford TN26 3QU
G8 PGA March Army Cadet Force c/o R. Martinus 62 High Oaks Close, Locks Heath, Southampton SO31 6SX
G8 PGE Silcoates School Ar & Elec.Club c/o D. Sinclair 12A Sunnydown Road, Winchester SO22 4LD
G8 PGF A. Price 11 Gatcombe Gardens, Titchfield, Fareham PO14 3DR
G8 PGH K. James 44 The Oakfield, Littledean Hill Road, Cinderford GL14 2DE
G8 PGI P. Lord Beechwood House, Main Street, Lutterworth LE17 5QA
GI8 PGJ D. Campbell 18 The Counties, Mark Street, Portrush BT56 8QA
G8 PGO D. Carter 49 Hinckley Road, , Sapcote LE9 4LG
G8 PHB P. Mckenzie 21 Arnside Walk, Chapel House, Newcastle upon Tyne NE5 1BT
G8 PHG Dr B. Cook Hinsley Mill House, Hinsley Mill Lane, Market Drayton TF9 1HP
G8 PHJ M. Palmer 38 Windermere Close, Dartford DA1 2TX
G8 PHM M. Kent Meadow Bank, Rye Lane, Sevenoaks TN14 5JF
G8 PHQ Aries ARC c/o C. Challender 9 Blick Close, West Winch, King's Lynn PE33 0UA
G8 PHS E. Campbell 2 Russell Avenue, March PE15 8EL
G8 PHV R. Wetton 8 St. Moritz Close, Northwick, Worcester WR3 7ND
G8 PIC C. Pomphrett 47 North Leas Avenue, Scarborough YO12 6LJ
G8 PIN West Glamorgan Raynet c/o R. Bannister 14 Amery Close, Worcester WR5 2HL
G8 PIO O. Futter 25 Amhurst Gardens Belton, Great Yarmouth NR31 9PH
G8 PIP P. Elwell 4 Richmond Grove, Wollaston, Stourbridge DY8 4SF
G8 PIR Lowestoft District and Pye ARC c/o R. Stone 63 Sands Lane, Lowestoft NR32 3ER
GM8 PIV E. Souter 3/2 10 James Gray Street, Glasgow G41 3BS
G8 PIY D. Clifton 10 Scotney Road, Basingstoke RG21 5SR
G8 PJD P. Deffee 18 Poplar Road, Kensworth, Dunstable LU6 3RS
GW8 PJE P. Evans 10 Cae Perllan Brackla, Bridgend CF31 2HL
G8 PJF E. Summers 161 North Cray Road, Sidcup DA14 5LT
G8 PJQ C. Cole 70 Throgmorton Road, Yateley GU46 6FA
G8 PK B. Wilson 58 High Street, Saxilby, Lincoln LN1 2HA
GW8 PKB L. Rudge 8 Penrallt Estate, Llanystumdwy, Criccieth LL52 0SR
G8 PKG I. Bosworth 7 Sandbourne Court, 54-56 West Overcliff Drive, Bournemouth BH4 8AB
G8 PKJ G. Rowland 18 Heights Way, Leeds LS12 3SN
G8 PKM C. Mitchell 25 Belper Road, Ashbourne DE6 1BB
GW8 PKV M. James 28 Bloomfield Gardens, Narberth SA67 7EZ
G8 PL A. Garry-Durrant Casa Santosa Llano Del Espino, Albox 4800 Spain
G8 PLI N. Vranic 30 Mitchell Street, Sheffield S3 7NL
G8 PLJ J. Bailey 8 Hild Avenue Cudworth, Barnsley S72 8RN
G8 PLO R. Clubley Church Hill House, High Street, Braintree CM7 4BY
GI8 PMA L. Pennell 21 Dundrum Road, Clough, Downpatrick BT30 8SH
G8 PMJ D. Hughes 18 Bailey Close, Pewsey SN9 5HU
G8 PMR L. Morris 17 Kestrel Close, Hornchurch RM12 5LS
GW8 PNE P. Griffiths 13 Wesley Court Warren Street, Tenby SA70 7JT
G8 PNM N. Cocking 15 Spring View Road, Sheffield S10 1LS
G8 PNN Hippings Methodist Primary School ARC c/o G. Emmerson 72 The Gables, Widdrington, Morpeth NE61 5RB
G8 POE J. Phillips 235 Barn Mead, Harlow CM18 6ST
G8 POG P. Wood 23 Shipley Avenue, Newcastle upon Tyne NE4 9QY
G8 POI D. Evans 14 Lower Wharf, Wallingford OX10 9AA
G8 POK G. West 6 Willerton Close, Chidswell, Dewsbury WF12 7SQ
G8 POL M. Williams 22 Charlecote Drive, Nottingham NG8 2SB
G8 POO S. Robinson 23 Jameson Drive, Corbridge NE45 5EX
G8 POP R. Mundy 12 Cantors Way, Minety, Malmesbury SN16 9QZ
G8 POQ B. Salter 19 Garnier Park, Wickham, Fareham PO17 5LD
G8 POS A. Axon 7 Tudor Grove, Groby, Leicester LE6 0YL
G8 PPA S. Ornstein Sandy Lodge, Trusthorpe Road, Sutton-On-Sea, Mablethorpe LN12 2LN
G8 PPD A. Saunders Suffolk House, Main Road, Ipswich IP9 1DX
G8 PPN A. Osmond The Moorings, 10 North Road, Shanklin PO37 6DB
G8 PPQ G. Boakes 16 Terminus Drive, Herne Bay CT6 6PP
GD8 PPU G. Brookes 44 Magherchirrym, Port Erin, Isle of Man IM9 6DB
G8 PQA A. Gapper 9 St. Annes Drive Wick, Bristol BS30 5PN
G8 PQB G. Grantham 18 Fen End Lane, Spalding PE12 6AD
G8 PQH F. Rowsell 4 Coxswain Way, Selsey, Chichester PO20 0UA
G8 PQJ J. Robinson Rose Cottage, Wavering Lane West, Gillingham SP8 4NR
G8 PQN M. Henshaw 23A Bedford Road, Northill, Biggleswade SG18 9AH
G8 PQZ G. Collier 6 Copse Close, Tilehurst, Reading RG31 6RH
G8 PRC 59 Degrees North ARG c/o P. Connor 20 Longfield, Lutton, Ivybridge PL21 9SN
G8 PRH A. Hartley 16 Old Thorne Road, Hatfield, Doncaster DN7 6ER
G8 PRJ S. Sanders 19 Brunswick Gardens, Hainault, Ilford IG6 2QU
G8 PRK R. Holmwood 3 Stanstead Road, Caterham CR3 6AD
G8 PRN R. Morley 21 Meadow View Skelmanthorpe, Huddersfield HD8 9ET
G8 PRP B. Youster 24 Sunningdale Road, Weston-Super-Mare BS22 6XP
G8 PSC J. Benoy 46 Bickham Road, Plymouth PL5 1SB
G8 PSF A. Ball 20 Inverness Avenue, Enfield EN1 3NT
GW8 PSJ L. Finch Hafan Deg, Waungilwen, Felindre, Llandysul SA44 5YG
G8 PSO R. Gould 20 Southwood Drive, Coombe Dingle, Bristol BS9 2QU
G8 PSS J. Meldrum 18 Oatlands Way, Pity Me, Durham DH1 5GL
GM8 PSV B. Thomson 51 Main Street, Newmill, Keith AB55 6UR
G8 PSZ C. Wood 5 Farm Close, Market Drayton TF9 3UH
G8 PTF R. Duddin 16 Gateley Road, Warley, Oldbury B68 0NU
G8 PTH A. Emmerson 71 Falcutt Way, Northampton NN2 8PH
G8 PTL M. Fleming 16 Church Street, Chasetown, Burntwood WS7 3QL
G8 PTN D. Stoney 7 Sandwell Close, Long Eaton, Nottingham NG10 3RG
GW8 PTS W. Leddington 4 Cherry Walk, Monmouth NP25 5DE
G8 PTW C. Wallace Windy Ridge, Langley Priory, Derby DE74 2QQ
G8 PTY P. Thornton-Evison Greyfriars, Townsend, Wantage OX12 0AT
G8 PUB S. Lucas 84 Woodman Road, Warley, Brentwood CM14 5AZ
G8 PUE J. Taylor The Jays, 5 Watling Close, Bourne PE10 9XL
G8 PUH B. Merrell 16 Box Close, Broadfield, Crawley RH11 9QT
G8 PUK S. Mann 1 Blackthorn Avenue Bramley, Rotherham S66 2LU
G8 PUN J. Keleher 9 Broadwell Drive, Leigh WN7 3NE
G8 PUR T. Rose 41 Keats Way, Hitchin SG4 0DP
G8 PUT C. Townsend 2 Netherfield Drive, Netherthong, Holmfirth HD9 3ES
G8 PUY N. Dowsett 4 Pankhurst Avenue, London E16 1UT
G8 PVG D. Hobbs 46 Gloucester Road, Bridgwater TA6 6DZ
G8 PVK R. Still The Manor House, 5 Beechwood Avenue, Bournemouth BH5 1LY
GJ8 PVL P. Bertram Roz-Den, La Rue De La Guilleaumerie, St. Saviour JE2 7HQ Jersey
G8 PVR J. Riggs 21 Janeva Court, Liskeard Road, Saltash PL12 4FD
G8 PVV M. Walton 12 Davison Street, Newcastle upon Tyne NE15 8NB
G8 PWE I. Ashford 28 Gorsey Lane, Great Wyrley, Walsall WS6 6JA
G8 PWK M. Forsey 84 Garner Road, Walthamstow, London E17 4HH
G8 PWO J. Thwaites 15 Spring Head Road, Kemsing, Sevenoaks TN15 6QL
G8 PWQ N. Wilson 114 Northgate Way, Terrington St. Clement, King's Lynn PE34 4LH
G8 PWT L. Cook 16 Florence Road, Maidstone ME16 8EN
G8 PWU J. Crossland 1 Carter Lane, Flamborough, Bridlington YO15 1LW
G8 PX Oxford And District ARS c/o G. Diacon 45 Woodpecker Way, , Witney OX28 6NN
G8 PXI M. Parker 65 Shoreham Drive, Penketh, Warrington WA5 2HY
G8 PXM K. Townsend 11 Caws Avenue, Seaview PO34 5JS
G8 PXO G. Murray Old Court, Westleigh, Tiverton EX16 7HT
G8 PXU B. Gascoigne 108 Blandford Avenue, Castle Bromwich, Birmingham B36 9JD
G8 PXW A. Stevens 23 Millers Green Drive, Kingswinford DY6 0DA
G8 PYD G. Farrell 95 Washington Road, Maldon CM9 6JF
G8 PYE P. Barrett 30 Rosslyn Park Road, Plymouth PL3 4LN
G8 PYU I. Walton 153 Vicarage Crescent, Redditch B97 4RP
G8 PZD T. Wills 66 Kipling Road, St. Marks, Cheltenham GL51 7DQ
G8 PZF B. Simpson 14 Priestthorpe Lane, Bingley BD16 4EE
G8 PZI W. Nolan South Lawn, 77 Reigate Road, Reigate RH2 0RE
GW8 PZS J. Coady 21 Garth Wen, Llanfaes, Beaumaris LL58 8PT
G8 PZX F. Gunn 8 College Gardens, Hornsea HU18 1EF
G8 QM V. Flowers Eothen Homes Ltd, 45 Elmfield Road, Newcastle upon Tyne NE3 4BB
G8 QZ D. Sager 29 Station Road, Mickleover, Derby DE3 9GH
G8 RAC J. Maines Brick House Farm, Marden, Hereford HR1 3ET
G8 RAF Royal Air Force ARS c/o S. Mckinnon 145 Enville Road, Kinver, Stourbridge DY7 6BN
G8 RAJ R. Shore 42 King George Avenue, Bournemouth BH9 1TX
G8 RAN K. Reeman 4 Alfreda Avenue, Hullbridge, Hockley SS5 6LT
G8 RAO A. Yates 87 Princess Road, Warley, Oldbury B68 9PW
GW8 RAS R. Neville 11 Heol Urban, Llandaff, Cardiff CF5 2QP
G8 RAU R. Lewis 66A Derek Gdns, Southend on Sea SS26QY
G8 RAV K. Jackson 45 The Crescent Bilsthorpe, Newark NG22 8QX
G8 RAX C. Hill 183 Manchester Road, Swinton, Manchester M27 4FA
G8 RBI C. Allen 8 Shoulbard, Fleckney, Leicester LE8 8TX
G8 RBK I. Brown 56 Church Lane, Darley Abbey, Derby DE22 1EY
GM8 RBR W. Egerton Croft House, Upper Breakish, Isle of Skye IV42 8PY
G8 RBS P. Bickersteth Tregarth, Fernsplatt, Truro TR4 8RJ
G8 RBU T. Dewey 93 Calverton Road, Arnold, Nottingham NG5 8FQ
G8 RBV D. Deighton 3 Bluebell Way, Huncoat, Accrington BB5 6TD
G8 RBW C. Ellison 29 Ashton Road, Clay Cross, Chesterfield S45 9FA
G8 RBX L. Fitzpatrick-Browne 24 Beechmount Avenue, Hanwell, London W7 3AG
G8 RBY P. Hodson 43 Thorpe Road, Melton Mowbray LE13 1SE
G8 RCE K. Shergold 47 Moorcroft Gardens, Redditch B97 5WG
G8 RCK R. Woollard 68 Trunk Furlong, Aspley Guise, Milton Keynes MK17 8HX
G8 RCL G. Whiston 86 Elmsett Close, Great Sankey, Warrington WA5 3RX
G8 RCO D. Russell 53 The Campions, Borehamwood WD6 5QE

G8 RCZ G. Fermor 26 Byron Road, Exeter EX2 5QN
G8 RDA K. Forster 10 Springfield Oval, Witney OX28 6EG
G8 RDB R. George Juniper Cottage, Hillesden, Buckingham MK18 4BX
G8 RDG R. Maltby Meadow Croft, Bishop Lane, Henfield BN5 9DG
GW8 RDI R. Colclough Login Fach, Waunarlwydd, Swansea SA5 4NJ
G8 RDJ J. Davies 11 Bromley Road, Macclesfield SK10 3LN
G8 RDK L. Mayhew 47 Beeches Avenue, Worthing BN14 9JE
G8 RDN T. Sale 20 Redwood Drive, Chase Terrace, Burntwood WS7 2AS
G8 RDP J. Webb 6 Chatsworth Avenue, Fleetwood FY7 8EG
G8 RDQ P. Williams 30 Duchess Drive, Bridgnorth WV16 4JD
G8 RDT K. Williams Jasmine Cottage, Evesham Road, Dodwell, Stratford-upon-Avon CV37 9ST
G8 REF P. Ellis 15 Alexander Close, Bognor Regis PO21 4PS
GM8 REG R. Bell Fairview, Main Street, Huntly AB54 7SY
G8 REO R. Mitchell 4 Friendly Fold Road, Halifax HX3 5QF
G8 REQ F. Robinson 13 Dorset Drive, Wirral CH61 8SX
G8 RER J. Fothergill 53 Meadow Court, Ponteland, Newcastle upon Tyne NE20 9RA
G8 RES M. Howard Lodge Cottage, Stody Estate, Stody, Melton Constable NR24 2EW
GW8 REV C. Marsh 50 Timothy Rees Close, Cardiff CF5 2AU
G8 REX G. Stephens 71 Quicks Road, London SW19 1EX
G8 RF F. Raby 20 Lime Tree Road, Codsall, Wolverhampton WV8 1NT
G8 RFC R. Cassell 1 St. Saviour Close, Colchester CO4 0PW
GW8 RFD P. Short 6 Broadmead Pontllanfraith, Blackwood NP12 2NL
G8 RFE M. Wallace 26 Parsons Drive, Glen Parva, Leicester LE2 9NS
G8 RFF K. Richardson 12 Badger Hill, Brighouse HD6 3QX
G8 RFL D. Robinson 25 Angelica, Amington, Tamworth B77 3JZ
G8 RFP D. Clarke 57 Stubley Drive, Dronfield S18 8QY
G8 RFV R. Bulmer 4 Valerian Drive, Stafford ST16 1FJ
G8 RFW G. Sharpe Holborn House 28 Sheffield Road, Dronfield S18 2GG
G8 RFY C. Garner Flat 10, Chapel Court, Leicester Road, Narborough, Leicester LE19 2FX
G8 RFZ D. Cooke 27 Goosehill Court, Balby, Doncaster DN4 8SX
G8 RGN J. Harling 6 Fontwell Road, Little Lever, Bolton BL3 1TE
GM8 RGO M. Robson Whistlefield Cottage, Loch Eck, Dunoon PA23 8SG
G8 RGU M. Burt Hartcliff Farm, Okeford Fitzpaine, Blandford Forum DT11 0EF
G8 RHC J. Cranage Corris House, West Gomeldon, Salisbury SP4 6LS
G8 RHM K. Hoggett 14 Wyld Court, Allesley, Coventry CV5 9LQ
G8 RHN M. Kirkham 60 Westminster Drive, Grimsby DN34 4TY
GW8 RHP Rvd. J. Williams Monte Vista, Llandyrnog, Denbigh LL16 4HH
G8 RHQ S. Ormondroyd 15 Meadowlands, Blundeston, Lowestoft NR32 5AS
G8 RHU E. Carvill 61 Midhurst Drive, Ferring, Worthing BN12 5BQ
G8 RHZ A. Robertson 22 Court Way, Twickenham TW2 7SN
G8 RIB P. Fallon 17 Blundell Road, Widnes WA8 8SS
G8 RIC K. Murphy 79 Torkington Road, Hazel Grove, Stockport SK7 6NR
G8 RIK R. Milner Lyndene, Holyhead Road, Shrewsbury SY4 1EE
G8 RIM J. Boardman 81 Shipston Road, Stratford-upon-Avon CV37 7LW
G8 RIP M. Walmsley 27 Russell Avenue, Preston PR1 5TP
G8 RIR S. Newbury 19 Kings Meadow Drive, Winkleigh EX19 8HD
G8 RIS R. Merriman 8 Abbots Meadow, Chittlehampton, Umberleigh EX37 9QE
G8 RIW B. Harvey 56 Oakwood Drive, Grimsby DN37 9RN
G8 RJB R. Bridgwater 31 Pembroke Avenue, Worthing BN11 5QS
G8 RJF K. Freer 54A High Lane East, West Hallam, Ilkeston DE7 6HW
G8 RJM S. Reap The Staddles, Romsey Road, Stockbridge SO20 8DB
G8 RJQ M. Corke 6 Dhow Street, Sun Valley, Cape Town 7975 South Africa
G8 RJZ M. Wills 9 Allerdale Close, Thirsk YO7 1FW
G8 RKG D. Peck Flat 1, Shrewsbury Court, 21-23 Manor Road, Worthing BN11 3RU
G8 RKH L. Hunt 15 Oxford Street, Cowes PO31 8PT
G8 RKO J. Butler 36 Park Road, Bracknell RG12 2LU
G8 RKX A. Titley 6 Spring View, Luddendenfoot, Halifax HX2 6EX
G8 RLD R. Dowdell 2 Coles Courtyard, 5 Black Horse Way, Horsham RH12 1NU
GI8 RLE J. Ashe 49 Deans Walk, Richhill, Armagh BT61 9LD
G8 RLF R. Dickerson 7 Sixpenny Close, Titchfield Common, Fareham PO14 4SY
GI8 RLG H. Emerson Little Castle Dillon 55 Portadown Road, Armagh BT61 9HJ
GW8 RLI V. Banfield New Haven, Main Road, Pontypridd CF38 1RY
G8 RLN J. Barnett 11 Ridge Street, Stourbridge DY8 4QF
G8 RLW G. Woodward 6 Lang Road, Huntington, York YO32 9SD
G8 RMI S. Blake 10 Dimore Close, Hardwicke, Gloucester GL2 4QQ
G8 RML M. Juby Silver Birch, High Road, Diss IP22 5RU
G8 RMP J. Bond 19 Compton Avenue, Mannamead, Plymouth PL3 5DA
GM8 RMR E. Scott 81 Rosehaugh Road, Inverness IV3 8SR
GI8 RNG W. Smyth 11 Alexander Park, Armagh BT61 7JB
G8 RNM I. Lucking 32 Nolton Place, Edgware HA8 6DL
G8 RNT P. Walkling Flat 36, Highlands House, Wharncliffe Road, Southampton SO19 7GG
G8 RNU I. Strange Holly Lane, Tansley, Matlock DE4 5FF
G8 RNV N. Jefferies 8 Cambridge Green, Fareham PO14 4QX
G8 ROC Radio Operators Cornwall c/o C. Macleod 21 Halsetown, St. Ives TR26 3LY
G8 ROF C. Taylor 26 Bernard Road, Brighton BN2 3EQ
G8 ROG Dr A. Johnston 12 Whitby Court Caversham, Reading RG4 6SF
G8 RON R. Eyes 6 Bakers Lane, Southport PR9 9RN
G8 ROS R. Platt 40 Solent Drive, Darcy Lever, Bolton BL3 1RN
G8 ROU D. Hardy 394 Slade Road, Birmingham B23 7LG
G8 RPA K. Mendum 30 Hollow Lane, Hitchin SG4 9SD
G8 RPD J. Fennell Broad Park Cottage, Stanbury Copse, Ilfracombe EX34 8DW
GM8 RPE J. Robinson 18 Craigshannoch Road, Wormit, Newport-on-Tay DD6 8ND
G8 RPI G. Atkinson 25 Appletrees, Bar Hill, Cambridge CB23 8SJ
GI8 RPP M. Elder 44 Learmount Road, Claudy, Londonderry BT47 4AQ
GI8 RPT N. Copeland 34 Glenkyle Park, Newtownabbey BT36 6SP
G8 RQF J. Duffy 5 Birch Court, Prudhoe NE42 6PZ
G8 RQH Bournemouth RS c/o I. Sherer 1A Appleyard Drive, Barton-upon-Humber DN18 5TD
GI8 RQI Coleg Menai RC c/o D. Allen 40 Bramblewood Drive, Banbridge BT32 4RA
G8 RQN P. Needham 2 Woodridge Close, Bracknell RG12 9QX
GM8 RQW J. Kennedy 6 Ryedale Terrace, Dumfries DG2 7DL
G8 RRC P. Sharpe 27 Record Road, Emsworth PO10 7NS
G8 RRN M. Jones 26 Cliff Road, Felixstowe IP11 9PJ
GJ8 RRP J. Parry 2 Thornley, Bagatelle Road, Jersey JE2 7TZ
G8 RRR H. Potter 134 Ifield Road, Crawley RH11 7BW
G8 RRS M. Ellison 22 Cotebrook Drive, Upton, Chester CH2 1RD
G8 RSA Dr S. Hasko 105 High Street Brampton, Huntingdon PE28 4TQ
GM8 RSC J. Chinnock 3B Dundee Street, Letham, Forfar DD8 2PQ
G8 RSE J. Murphy 15 Loders Close, Poole BH17 9BF
G8 RSI G. Whitney 4 Sefton Crescent, Sale M33 7EN
G8 RSK P. Tyrell 14 Park Farm Road, Horsham RH12 5EW
G8 RSQ L. Williamson 17 Ray Bond Way, Aylsham, Norwich NR11 6UT
G8 RSV S. Staniforth 4 Moses View, Shireoaks, Worksop S81 8NH
G8 RSX T. Beck 10 Rookery Close, Hatfield Peverel, Chelmsford CM3 2DF
G8 RTB North Hertfordshire Raynet Assoc. c/o R. Breeze 119 Sundorne Road, Shrewsbury SY1 4RP
GM8 RTI Dr J. Grieve 1 Orchard Way, Inchture, Perth PH14 9QB
G8 RTK L. Man 4 Back Lane, Yeadon, Leeds LS19 7SQ
G8 RTN G. Smith 1 Abbey Road Goldington, Bedford MK41 9LG
G8 RUX R. Gough 3 Meadowlands, Havant PO9 2RP
G8 RVB A. Augustus 35 Cedar Grove, Wolverhampton WV3 7EB
G8 RVO Dr A. Parsons 8 Queen Annes Gardens, Ealing, London W5 5QD
GJ8 RVT Jersey ARS c/o M. Turner 4 Le Clos Sara, St. Lawrence JE3 1GT Jersey
G8 RVY P. Lee 8 Sandringham Gardens, Ellesmere Port CH65 9EY
G8 RVZ I. Martin Maycroft, Hatchett Lane, Edingale B79 9JG
G8 RW P. Standley 9 Capelands, New Ash Green, Longfield DA3 8LG
G8 RWG N. Montanana 91 Coulsdon Road, Coulsdon CR5 2LD
G8 RWH I. Jackson 5 Vivien Close, Chessington KT9 2DE
G8 RWJ R. Adams Ground Floor Flat, 91 Mount Pleasant Road, Hastings TN34 3SL
G8 RWM F. Box 11 Cook Avenue, Newport PO30 2LL
G8 RWN Radio-Tele Lincolnshire Group c/o M. Mckenzie Flat 40, Broomfield House, Huddersfield HD3 4RS
G8 RWU Glasgow University Wireless Society c/o A. Capron 28 Windmill Road, Hemel Hempstead HP2 4BN
G8 RWZ K. Hodson 117 High Lane, Brown Edge, Stoke-on-Trent ST6 8RT
G8 RXY G. Alcock 61 Henshall Hall Drive, Congleton CW12 3TY
G8 RXZ P. Allgood 11 Dover Drive, Leegomery, Telford TF1 6TD
G8 RYE D. Cope 29 Desford Road, Newbold Verdon, Leicester LE9 9LG
G8 RYJ P. Tegg Glendale, 36 Wrecclesham Hill, Farnham GU10 4JW
G8 RYK R. Taylor 18 Spruce Avenue, Selston, Nottingham NG16 6DX
G8 RYL I. Smith 12 Windmill Lane, Fulbourn, Cambridge CB21 5DT
G8 RYO P. Sargent Hill View, North Road, Tetford, Horncastle LN9 6QH
G8 RYX S. Jones The Old Granary, Broomsmead, Lapford EX17 6NA
GM8 RYZ I. Mclaren 2 Cleeve Drive, Perth PH1 1HH
G8 RZ P. Webster 15 Napier Street, Workington CA14 2PT
G8 RZL T. Claydon 20 Ivy Lane, Royston SG8 9DQ
G8 RZN D. Dunn Valarms, The Pigeons, Wisbech PE13 4JU
G8 RZS E. Humpston 2 The Glebe, Hildersley, Ross-on-Wye HR9 5BL
G8 RZZ K. Colman 10 South Rise, North Walsham NR28 0EE
G8 SAA I. Adamson 5 Rose Meadows Somersham, Huntingdon PE28 3YU
G8 SAL Saltash Dist Ar c/o K. Hale 58 St. Stephens Road, Saltash PL12 4BJ
G8 SAN Dr R. Charlton 31 Meriden Road, Hampton-In-Arden, Solihull B92 0BS
GM8 SAP D. Cooper 4 Ruskie Avenue, Callander FK17 8LA
G8 SAR M. Elliott 54 Bankhouse Road, Trentham, Stoke-on-Trent ST4 8EL
GM8 SAU B. Titmarsh Caberfeidh, Clachan Na Luib HS6 5HD
G8 SAX P. Wilkinson 60 Whalley Drive, Aughton, Ormskirk L39 6RF
G8 SBH H. Cromack 6 West Park, South Molton EX36 4HJ
G8 SBJ T. Blankley 16 Charles Road, St. Leonards-on-Sea TN38 0QA
GW8 SBK L. Cleak Danetre, Newport Road, Cwmbran NP44 3AE
GW8 SBN J. Kemp Poldhu 259 Delffordd, Rhos, Swansea SA8 3EP
GW8 SBO P. Sibert Chibembe Lodge Lamborough Lane, Clarbeston Road SA63 4XD
G8 SBQ D. Wooller 95 Havant Road Drayton, Portsmouth PO6 2JE
G8 SBS J. Westlake 41D Shirley Road, Southampton SO15 3EW
G8 SCG J. Downes 6 Lagonda, Glascote, Tamworth B77 2RY
G8 SCI M. Bynorth 374 Bloxwich Road, Walsall WS2 7BG
G8 SCO Street Club Operators c/o D. Barber 3 Vestry Road, Street BA16 0HY
G8 SCT B. Cater 50 Woodside Darras Hall, Ponteland NE20 9JB
G8 SCY C. Rosewall 12 Treloggan Lane, Newquay TR7 2JN
G8 SDE R. Pitts 84 Prospect Avenue, Pye Nest, Halifax HX2 7HP
G8 SDN E. Mciver 31 Hartshill, Bedford MK41 9AL
G8 SDS M'hd & E.Berk Rd c/o W. Barton 4 Hawthorn Flats, Hawthorn Road, Dorchester DT1 2PE
G8 SDU R. Clayton 1 Raymond Road, Hellesdon, Norwich NR6 6PL
G8 SDX S. Dale 25 Copeland Avenue, Tittensor, Stoke-on-Trent ST12 9JA
G8 SEA Capt. J. Pearce Clematis Cottage, 52 Cheselbourne, Dorchester DT2 7NP
G8 SED P. Starling 5 Ash Close, Bacton, Stowmarket IP14 4NR
G8 SEE R. Stone 10 Rosemullion Gardens, Tolvaddon, Camborne TR14 0EY
G8 SEI J. Snell 18 Green Lane, Grendon, Atherstone CV9 2PL
G8 SEK C. Watts 5 Pennycress Way, Bridgwater TA5 2FQ
G8 SEQ J. Beech 124 Belgrave Road, Coventry CV2 5BH
G8 SEV P. Matthews 10 Norway Close, Corby NN18 9EG
G8 SEW N. Humphreys 13 Gordon Drive, East Boldon NE36 0TD
G8 SEY A. Graver 8 Avenue Road, Bishop's Stortford CM23 5NU
G8 SFA S. Milsom 30 Beechwood Drive, Prudhoe NE42 5PN
G8 SFD C. Williams 14 Milton Place, Bideford EX39 3BN
G8 SFF M. Watson 7 Grange Lane, Willingham By Stow, Gainsborough DN21 5LB
G8 SFI S. Firth 8 Lyndale Avenue, Osbaldwick, York YO10 3QB
G8 SFM K. Saunders 4 Rue Des Camelias, Villalier 11600 France
G8 SFO T. Robson The Old Post Office, Winchester Road, Burghclere, Newbury RG20 9EQ
G8 SFQ T. Mcnamara 12 Scarsdale Road, Great Barr, Birmingham B42 2JW
GW8 SFT D. Mansell 57 Ger Y Llan Penrhyncoch, Aberystwyth SY23 3HQ
G8 SGB P. Houseago 11 Arnstones Close, Colchester CO4 3AS
G8 SGF P. Gilliland11 34 Cavan Drive, St. Albans AL3 6HP
G8 SGH P. Marshall 123 Rochford Garden Way, Rochford SS4 1QJ
G8 SGI S. Pascoe 34 Ravens View, Witham St Hughs, Lincoln LN6 9JE
G8 SGM A. Boyce 22 Peggotty Close, Chelmsford CM1 4XU
G8 SGP G. Wheeler 14 Sparkmill Terrace, Beverley HU17 0PA
G8 SGV M. Williams The Hideaway, 39 Terrys Avenue, Victoria 3160 Australia
G8 SGX S. Saltmer 12 Beechings Mews, Whitby YO21 3DW
G8 SH J. Storey 34 Austin Rise, Longbridge, Birmingham B31 4QN
G8 SHC P. Hammond 35 West Green, Barrington, Cambridge CB22 7RZ
G8 SHE R. Shears 20 Regency Gardens, Grantham NG31 9JW

G8 SHF C. Scrase 2 Park View, Seaton Junction, Axminster EX13 7PS
G8 SHR P. Goodfellow 10 St. Agnes Walk, Knowle, Bristol BS4 2DL
GW8 SIE R. Stark Roneragh, Llanrhaeadr, Denbigh LL16 4NN
G8 SIG A. Jeffery 14 Holly Mount Shavington, Crewe CW2 5AZ
G8 SIK A. Sturt 6 Kenley Road, Kingston upon Thames KT1 3RW
G8 SIM J. Green 7 Russell Road, Runcorn WA7 4BG
G8 SIN D. Dobson 93 Old Road, Headington, Oxford OX3 8SX
GW8 SIT M. Shewring 2 Glan Hafan, Trefechan, Aberystwyth SY23 1AT
G8 SIU D. Stillwell 2B Lesley Owen Way, Shrewsbury SY1 4RB
G8 SJA P. Farrar 17 Clough Lane, Halifax HX2 8SG
G8 SJK D. Monk Manor Farmhouse, Willen Road, Milton Keynes Village, Milton Keynes MK10 9AF
G8 SJO S. Ootam 9 Harewood Road, Isleworth TW7 5HB
GI8 SJS R. Hoey 28 Hanwood Heights, Dundonald, Belfast BT16 1XU
G8 SKA R. Holden 5 Lawrence Grove, Kidderminster DY11 7DR
G8 SKG L. Challis 30 London Road, Kirton, Boston PE20 1JA
GI8 SKN D. Reid 2 New Line, Carrickfergus BT38 9DL
GI8 SKR G. Bannister 65 Osborne Drive, Belfast BT9 6LJ
G8 SLB P. Lockwood 36 Davington Road, Dagenham RM8 2LR
G8 SLE G. Leake Flat 3, Edward May Court, Bournemouth BH11 8AW
G8 SLP J. Barry 18 Hough Green, Chester CH4 8JG
G8 SLU M. Hack Anmee The Ride, Ifold, Billinghurst RH14 0TF
G8 SMA C. Ward 25 Blewbury Drive Tilehurst, Reading RG31 5HJ
G8 SMH K. Hempsall 69 Wantage Road, Didcot OX11 0AE
G8 SMZ C. Shaw 1 Guilford Cottages, East Langdon, Dover CT15 5JD
GM8 SNB G. Allan 13 Mitchell Drive, Rutherglen, Glasgow G73 3QP
G8 SND W. Hoskins 89 Boston Road, Lytham St. Annes FY8 3PS
GM8 SNE P. Walter 4 Rose Gardens Cairneyhill, Dunfermline KY12 8QS
G8 SNF I. Hewitt 26 Outwoods Drive, Loughborough LE11 3LT
G8 SNJ P. Townsend 7E Back Lane, Holmfirth HD9 1HQ
G8 SNQ R. Knock 18 The Hawthorns, Eccleston, Chorley PR7 5QW
G8 SNR D. Mangnall 23 Blundell Road, Lytham St. Annes FY8 3AG
G8 SNV Dr M. Dey Windy Lodge, 18 Ripley Road, Hampton TW12 2JH
G8 SOI D. Carter 35 Upland Road, West Mersea, Colchester CO5 8DR
GM8 SOK J. Sturrock 4 Ann Street, Edinburgh EH4 1PJ
G8 SOU R. Topping 47 Celtic Road, Deal. CT14 9EF
G8 SPC R. White Flat 1 Oaklands, Elton Road, Clevedon BS21 7QZ
G8 SPD M. Beevers Pool House Cottage, Astley, Stourport-on-Severn DY13 0RH
G8 SPE R. Armstrong Flat 8, University Court, Buckingham Close., London W5 1TZ
G8 SPG G. Wood 40 Larch Grove, Peterborough PE1 4JY
G8 SPM Dr P. Armitage 5 Park Court, Heath Road, Brixham TQ5 9AX
G8 SPP C. Parkinson 77 Lime Grove, Doddinghurst, Brentwood CM15 0QX
G8 SPU R. Doughty 47 Red Lion Close, Tividale, Oldbury B69 1TP
G8 SPW A. Rouse 49 Ghyllside Avenue, Hastings TN34 2QB
G8 SQA T. Povey 24 Townley Way, Earls Barton, Northampton NN6 0HR
G8 SQH D. Hutchinson Ryton Villa, Horsecroft Lane, Dymock GL18 2EJ
G8 SQK L. Mcmahon 25 Belmont Avenue, Warrington WA4 1LY
G8 SQP R. Marquiss 66 Oakwood Rise, Tunbridge Wells TN2 3HF
G8 SQY S. Cade 2 Grosvenor Crescent, Louth LN11 0BD
G8 SQZ S. Westlake 11 Mount Road, Evesham WR11 6BE
G8 SRC Gloucester Repeater Group c/o D. Forrest 166 Meadowcroft, Swindon SN2 7LE
G8 SRN E. Day 10 Carlrayne Lane, Menston, Ilkley LS29 6HH
G8 SRS Stockport Rs c/o B. Naylor 47 Chester Road Poynton, Stockport SK12 1HA
G8 SRV Clacton RC c/o A. Ashe 34 College Avenue, Maidenhead SL6 6AX
G8 SRZ B. Atkinson 3 Sandy Close, Whitwell, Worksop S80 4PY
G8 SSE K. Lawrence 7 Canada Way, Pak House, Worcester WR2 4DJ
G8 SSL A. Marwood 65 Castleton Avenue, Arnold, Nottingham NG5 6NH
G8 SSP C. Horswell Hungereckstrasse 60/3, Vienna A-1230 Austria
G8 SSS Exmoor RC c/o J. Stacey 3 West Park, South Molton EX36 4HJ
G8 SSX D. Baker 99 Repton Road, Wigston LE18 1GD
G8 SSY E. Davies 5 Cheapside, Horsell, Woking GU21 4JG
G8 STD Burnham ARC c/o E. John Obo St. Dunstans Ars, 52 Broadway Avenue, Wallasey CH45 6TD
G8 STF T. Woods 12 Primrose Court, Egerton Street, Wallasey CH45 2PE
G8 STI J. Maiden 42 Timberdine Avenue, Worcester WR5 2BD
G8 STJ D. Carter 4 Sandale Close, Gamston, Nottingham NG2 6QG
G8 STM A. Bilton 34 Wimberley Way, South Witham, Grantham NG33 5PU
G8 STR B. Beestin 45 Pinehill Road, Crowthorne RG45 7JP
G8 STW J. Ferguson Meridian, The Street, Hepworth, Diss IP22 2PX
G8 STY J. Holmes 45 College Avenue, Gillingham ME7 5HY
G8 SUG G. Peterson 15 Hindhead Green, Watford WD19 6TR
G8 SUJ W. Shambrook 49 Beaufort Road, Church Crookham, Fleet GU526AY
G8 SUM K. Smith 11 Church Street, Earl Shilton, Leicester LE9 7DA
G8 SUN S. Williams 11 Cotman Drive, Hinckley LE10 0GB
G8 SUQ J. Corbidge 11 Berkeley Close, Folkestone CT19 5NA
G8 SUV B. Pont 56 Ravenhill Road, Bristol BS3 5BT
G8 SUW N. Pont Maisemoor, 17 Vicarage Lane, Bridgwater TA7 9LR
G8 SVD M. Bridle 28 Blakes Farm Road, Southwater, Horsham RH13 9GJ
G8 SVR J. Allart 54 Urban Gardens, Concord, Washington NE37 3DE
G8 SVT T. Ellis 58 Greenland Drive, Sheffield S9 5GJ
G8 SWC H. Moyle 9 Park Approach, Welling DA16 2AW
G8 SWK D. Taylor 19 Armley Grange Oval, Leeds LS12 3QJ
G8 SWL E. Theodorson 7 Kingfisher Court, Overstone Lakes, Ecton Lane, Northampton NN6 0BD
G8 SWO C. Watts 42 Truscott Avenue, Bournemouth BH9 1DB
G8 SWW P. Copeman 1 Chestnut Avenue, Welney, Wisbech PE14 9RG
G8 SXA J. Davies Ballards Piece, Forest Hill, Marlborough SN8 3HN
G8 SXB D. Mullenger 6 Churchfields, Kingsley, Bordon GU35 9PJ
G8 SXD B. Davies Ballards Piece, Forest Hill, Marlborough SN8 3HN
GW8 SXI Rvd. G. Howells 53 Abbey Road, Rhos On Sea, Colwyn Bay LL28 4NR
G8 SXJ F. Hutchings 21 School Lane, St. Ives, Ringwood BH24 2PF
GM8 SXQ A. Leigh Duaig, Lochavich, Taynuilt PA35 1HJ
G8 SXU J. Simmons 167 Bourne Vale, Hayes, Bromley BR2 7LX
G8 SYA K. Parker 3 Cross Roads, East Stour, Gillingham SP8 5LW
G8 SYD J. Thomson 11 Uranus Road, Hemel Hempstead HP2 5QF
G8 SYE N. Trotman1 38 Oldbury Road, Nuneaton CV10 0TD
G8 SYM D. Whittle 16 Garner Drive, Astley, Manchester M29 7RT
G8 SYV J. Morgan Linden Lea, Fakes Road, Great Yarmouth NR29 4JL
GW8 SZC P. Henry 1 Afan Valley Road, Neath SA11 3SS
G8 SZG C. Just 2 The Old Rectory, The High Road Felmersham, Bedford MK43 7HN
GW8 SZL D. Phillips 9 Baldwin Street, Newport NP20 2LT
G8 SZP D. Price 196 Monument Court, Stevenage SG1 3BT
G8 SZR M. Matthews 11 Church End, Ashdon, Saffron Walden CB10 2HG
GM8 SZS B. Mccaffrey 5 The Old Orchard, Limekilns, Dunfermline KY11 3HS
G8 SZZ S. Jackson 18 Blakesware Gardens, Edmonton, London N9 9HU
G8 TAE G. Wooltorton 155 El Alamein Way, Bradwell, Great Yarmouth NR31 8SX
G8 TAQ A. Dyce 26 Forest Road, Winford, Sandown PO36 0JY
G8 TAU A. Fisher 2 Hillside Mansions, Barnet Hill, Barnet EN5 5RH
GI8 TAX R. Mcloughlin 27 The Manor, Portadown, Craigavon BT62 3QU
G8 TAY R. Manning North Waterhouse Farm, Runnon Moor Lane, Hatherleigh, Okehampton EX20 3PL
G8 TBB J. O'Meara 117 Little Sutton Lane, Sutton Coldfield B75 6SN
G8 TBF R. Jenkins 11 Westfield Drive, Worksop S81 0JS
GW8 TBG M. Terry 265 Delffordd, Rhos, Swansea SA8 3EP
G8 TBL N. Mosedale Flat 113, Altitude Apartments 9 Altyre Road, Croydon CR0 5BP
G8 TBU N. Doe Longclose, Langtree, Torrington EX38 8NR
G8 TBV A. Kyle 6 Mill Hill Drive, Halesworth IP19 8DB
G8 TBW R. Crathorne 340 Farnborough Road, Castle Vale, Birmingham B35 7PD
G8 TBX Wales Digital Radio Group c/o S. Pybus 26 White Laithe Court, Leeds LS14 2EQ
GW8 TBY D. Padley 31 South Drive, Rhyl LL18 4SU
GM8 TCG J. Blackie Drumcharry, Montrose Road, Auchterarder PH3 1BZ
GM8 TCH M. Bell 1 Dow Brae, Town Yetholm, Kelso TD5 8SA
G8 TCP C. Dawe 21 Portmellon Park, Mevagissey, St. Austell PL26 6XD
G8 TCQ A. Dening 42 Grove Avenue, Yeovil BA20 2BD
G8 TDP D. Cooke 19 St. Aldwyn Road, Seaham SR7 0AN
G8 TEB D. Clarke Bugbrooke Marina, The Wharf, Bugbrooke, Northampton NN7 3QB
G8 TEC G. Cook Flat 3, Southdown Court, Southdown Road, Winchester SO21 2BX
G8 TED Wimbledon & District ARS c/o T. Coote 25 Somersby Green, Boston PE21 9PH
G8 TEF A. Crute Shangri La, Winsor Estate, Looe PL13 2JY
G8 TEK K. Worley 33 Lynbrook Close, Netherton, Dudley DY2 9HE
G8 TEL P. Hynes 3 Holt Park Gardens, Leeds LS16 7RB
G8 TEO B. Jay 13 Oakhurst Road, West Moors, Ferndown BH22 0DW
G8 TEQ Dr D. Linsdall 2B Linkswood Road, Burnham, Slough SL1 8AT
G8 TFB S. Haywood 12 Elm Terrace, Tividale, Oldbury B69 1UD
G8 TFR S. Cottis 61 Oaken Grove, Maidenhead SL6 6HN
G8 TFU P. Simpson 1 Emmit Field Close, Chesterfield S40 2UH
G8 TFW L. Stamp 41 Willoughby Road, Wallasey CH44 3DZ
G8 TFY N. Richards 38 Parsons Road, Irchester, Wellingborough NN29 7EA
G8 TGB M. Verrall 1 Speedwell Avenue, Weedswood, Chatham ME5 0SB
G8 TGD D. Troop 10 Mellowdew Road, Coventry CV2 5GL
G8 TGH B. Wilmott 27 Apple Grove, Bognor Regis PO21 4NB
GW8 TGS Dr W. Williams 17 Llys Yr Onnen, Coity, Bridgend CF35 6FA
G8 THE Dr R. Hill 12 Winchelsea Lane, Hastings TN35 4LG
G8 THH D. Baker 5 Larkspur Close, Bishop's Stortford CM23 4LL
GW8 THL Dr F. Morgan Glantowy Lodge, Capel Dewi Road, Carmarthen SA32 8AA
GW8 THM M. Griffin 3 Pritchard Close, Danescourt, Cardiff CF5 2QS
G8 THR P. Crossley Firpark Farm, Fir Park, Market Rasen LN8 3YL
G8 THZ A. Tipper 24 Waverley Road Hoylake, Wirral CH47 3DD
G8 TIA D. Trickett 25 Spring Street, Halesowen B63 2SY
G8 TIO P. Dear Oakdale, Gadbridge Lane Ewhurst, Cranleigh GU6 7RW
G8 TIU A. Brierley Thorndale House, Brockhurst Farm, Watersfield, Pulborough RH20 1NX
GW8 TIX G. May 19 Mclaren Cottages, Abertysswg, Rhymney, Tredegar NP22 5BH
G8 TJG F. Starkey 13 Thorncliffe Drive, Darwen BB3 3QA
G8 TJI M. Oldfield Willows, Stablebridge Road, Buckinghamshire HP22 5ND
G8 TJR C. Colebrook 14 Yiewsley Drive, Darlington, Dl3 9Xs, Darlington DL3 9XS
G8 TKD D. Hensby 28 Moorland Crescent, Whitworth, Rochdale OL12 8SU
G8 TKQ J. Ackerley 24 Macaulay Road, Lutterworth LE17 4XB
G8 TKY T. Bootyman 14 Vale View, Ackworth, Pontefract WF7 7HQ
G8 TLC S. Parker 22 Lincoln Drive, Syston, Leicester LE7 2JW
G8 TLH R. Rogers 14 Coningsby Drive, Franche, Kidderminster DY11 5LU
G8 TLL L. Stewart The Spinney, Holmes Lane, Scunthorpe DN15 9QY
G8 TLP R. Pearce Doral, The Grove, Sevenoaks TN15 6JJ
G8 TLT J. Beveridge 18 Sir Christopher Court, Hythe SO45 6JR
G8 TMA H. Colborn Orchard Cottage, Blakeney Hill, Blakeney GL15 4BS
G8 TMD T. Clint 11 Home Lea, Rothwell, Leeds LS26 0PP
GI8 TME J. Campbell 115 Dromore Road, Ballynahinch BT24 8HU
G8 TMH H. Caseley 7 Lacy Way, Leominster HR6 9AY
G8 TMJ P. Faulkner 8 Parkfield Road, Cheadle Hulme, Cheadle SK8 6EX
G8 TML J. Foster 14 Braemar Grove, Heywood OL10 3RR
G8 TMM E. Gilbert 34 School Lane, Harpole, Northampton NN7 4DR
G8 TMQ P. Stevens 17 Weaver Close, Brierley Hill DY5 4QN
G8 TMR P. Taylor 22 Windermere Drive, Rainford, St. Helens WA11 7LD
G8 TMV C. Tuckley 98 Woodland Road, Cambridge CB22 3DU
G8 TNA S. Thompson Hollies, Chapel Hill, Sticker, St. Austell PL26 7HG
G8 TNB P. Thompson Lyndhurst Cottage, Main Street, Newark NG23 6ST
G8 TND C. Schiffman 14 Caspian Way, Purfleet RM19 1LE
G8 TNE D. Pickford 80 Hollowood Avenue, Littleover, Derby DE23 6JD
G8 TNH P. Jeffries 22 Ingrams Way, Hailsham BN27 3NP
G8 TNK North Kent RS c/o S. Osborn 67 Chessington Avenue, Bexleyheath DA7 5NP
G8 TNS S. Ward 2 Nursery Road, Rugeley WS15 1EZ
G8 TNU A. Lambert 69 Anvil Crescent, Broadstone BH18 9DZ
G8 TOI R. Hempstead 21 Lymington Avenue, Clacton-on-Sea CO15 4PJ
G8 TOP N. Huggins 4 Kelmscott Close, Goldings, Northampton NN3 8XN
G8 TOQ J. Jackson 1 Dolly Garth, Arkengarthdale Road, Richmond DL11 6QX
G8 TOT D. Lodge 29 Lea Lane, Netherton, Huddersfield HD4 7DP
GW8 TOX K. Taylor Swn Y Don, Llanfaes, Beaumaris LL58 8RG
G8 TPC B. Taylor 161 Sidegate Lane, Ipswich IP4 4JN
G8 TPF N. Sanvoisin 11 Ter Rue Lecoq, St. Nom la Breteche 78860 France
G8 TPM N. Wellsbury 15 Woodlands Road, Cookley Village, Kidderminster DY10 3TL
G8 TPP M. Strudwick 65 Neave Crescent, Harold Hill, Romford RM3 8HN
G8 TQH A. Mcmullin Fair View, Rickham, East Portlemouth, Near Salcombe TQ8 8PJ
G8 TQI P. Herod 4 St. James Road, Little Paxton, St. Neots PE19 6QW

G8 TQJ R. Markfort 105 Woodlands Way, Southwater, Horsham RH13 9TF
G8 TQK A. Mayhew 51 Upland Road, Sutton SM2 5HW
G8 TQO G. Harrison 1 Wellis Gardens, St. Leonards-on-Sea TN38 0UX
G8 TQP R. Healey 14 Jardine Drive Bishops Cleeve, Cheltenham GL52 8XQ
G8 TQV R. Tuckett 89 Hillbrook Road, Tooting, London SW17 8SF
G8 TQZ B. Woods 84 Beauly Way, Rise Park, Romford RM1 4XR
G8 TRF Invicta CG c/o I. Hope 5 The Crescent, Northfleet, Gravesend DA11 7EB
G8 TRG Museum of Communication ARS c/o R. Green 2 Ragley Walk, Rowley Regis B65 9NT
GW8 TRO K. Prosser 12 Sycamore Court, Woodfieldside, Blackwood NP12 0DA
G8 TRQ M. Walker 20 Littlewood Lane, Cheslyn Hay, Walsall WS6 7EJ
G8 TRR W. Pickard 19 Canham Close, Kimpton, Hitchin SG4 8SD
G8 TRS Tamworth ARS c/o T. Robertson 10 Athelstan Way, Tamworth B79 8LB
G8 TRU S. Lynch 4 Tanglewood, Welwyn AL6 0RU
G8 TRY Sight Matters ARC c/o G. Scott 19 Penkett Road, Wallasey CH45 7QF
G8 TSC J. Collins 12 Malham Road, Thatcham RG19 3XB
G8 TSG D. Johansen 45 Marfords Avenue, Bromborough, Wirral CH63 0JJ
GI8 TSI I. Raine 48 Gardners Road, Lisburn BT27 5PD
G8 TSV G. Rowlands 39 Nelthorpe Street, Lincoln LN5 7SJ
G8 TSZ A. Twyford 70 Ellesfield Drive, West Parley, Ferndown BH22 8QW
G8 TTD Dr P. Palin Flat 2, Springfield Heversham, Heversham, Milnthorpe LA7 7EJ
G8 TTE Dr E. Thomas Fairfield, St. Marys Road, Oakham LE15 8SU
G8 TTI D. Kearns 14 Draycot Cerne, Chippenham SN15 5LD
G8 TTJ J. Holland 2 Hadfield Close, Staunton, Gloucester GL19 3QY
G8 TTP R. Nicholls 57 Mandalay Court, London Road, Brighton BN1 8QW
G8 TTU C. Smithson 4 Calder Avenue, Littleborough OL15 9JE
G8 TTX K. Sugg 28 Well Close, Winscombe BS25 1HQ
G8 TUH D. George East Winch Road, Blackborough End, Kings Lynn PE32 1SF
G8 TUN C. Denton 34 Brook Lane, Ormskirk L39 4RE
G8 TUU R. Boyce 9 Kestrel Close, Bexhill-on-Sea TN40 1UG
G8 TVC T. Webb 25 Wheatfield Drive, Ramsey, Huntingdon PE26 1SH
G8 TVM K. Biggs 33 Blanford Gardens, West Bridgford, Nottingham NG2 7UQ
G8 TVU A. Cole 14 Ellesmere Grove, Stainforth, Doncaster DN7 5BS
GM8 TVV N. Coote 130 Castle Gardens, Paisley PA2 9RD
G8 TVW D. Young 58 Furzefield Road, Welwyn Garden City AL7 3RJ
GW8 TVX R. Hope 75 Priors Way, Dunvant, Swansea SA2 7UH
G8 TVZ R. Midgeley 2 Digswell Park Cottages, Digswell Park Road, Welwyn Garden City AL8 7NN
G8 TWA G. Jasper Clann Farm, Clann Lane, Bodmin PL30 5HD
GI8 TWB B. Mitchell 7 Crea Road, Randalstown, Antrim BT41 3DX
G8 TWN G. Darlington The Raven, Northfield Lane, over Stratton, South Petherton TA13 5LJ
G8 TWR J. Evans 49 Inverness Avenue, Enfield. EN1 3NU
G8 TWS M. Corbett 32 Bibury Road, Cheltenham GL51 6BA
G8 TWT B. Fisher 24 Vessey Road, Worksop S81 7PG
G8 TWZ I. Goodman 271 Alcester Road, Hollywood, Birmingham B47 5HJ
G8 TXA R. Heeley 4 Cherry Tree Lane, Halesowen B63 1DU
GM8 TXC J. Hedley 3/2 48 White Street, Glasgow G11 5EA
G8 TXJ D. Shaw Cotehouse, Bleatarn, Appleby-in-Westmorland CA16 6PX
G8 TXK P. Shulver Skanes House Redmoor, Bodmin PL30 5AT
G8 TXL J. Sillitoe 42 Marsham Road, Kings Heath, Birmingham B14 5HD
G8 TXT B. Linkins Curlew Cottage, Higher Wringworthy Farm, Looe PL13 1PR
G8 TXW G. Sutcliffe 41 Rose Avenue, Irlam, Manchester M44 6AQ
G8 TXX J. Taylor The Jays, 5 Watling Close, Bourne PE10 9XL
G8 TYB R. Uccellini 49 Slaithwaite Road, Meltham, Holmfirth HD9 5PG
G8 TYD D. Prouse Flat 7, Waterfront House, 211 Lower Bristol Road, Bath BA2 3DQ
G8 TYF H. Sasse Flat 8 The Beeches, 43 Queens Road, Leicester LE2 1WQ
G8 TYH R. Marsh 58 Statham Avenue, Lymm WA13 9NL
G8 TYX J. Hodgson 21 Portcullis Drive, Wallingford OX10 9LY
G8 TYY Dr T. Hopkins 5 Rochester Close, Bacup OL13 8RN
G8 TZE D. Pritt 387 London Road, Clanfield, Waterlooville PO8 0PJ
G8 TZJ A. Sellers 2 Dunkenshaw Crescent, Lancaster LA1 4LQ
G8 TZN R. North 5 George Road, Guildford GU1 4NP
G8 TZU R. Collis 17 Belvedere Close, Guildford GU2 9NP
G8 TZW G. Dunn 29 Sundridge Road, Kingstanding, Birmingham B44 9NY
G8 UAD R. White 72 Green Lane, Bournemouth BH10 5LF
G8 UAE D. Scott Hyde Bungalow, The Hyde, Stourbridge DY7 6LS
G8 UAF Trowbridge And District ARC c/o G. Scargill 98 Southleigh Road, Leeds LS11 5SG
G8 UAI D. Lockwood 61 Green Lane, Tickton, Beverley HU17 9RH
GW8 UAM Burton & District RS c/o L. Wright 19 Lon Y Fran, Caerphilly CF83 2RX
GW8 UAP T. Williamson Fforch Farm House, Cemetery Road, Treorchy CF42 6TF
G8 UAY J. Earley 36 Camelot, Cornish Avenue, Weltevredenpark 46 South Africa
G8 UBD A. Baker 19 Rockington Way, Crowborough TN6 2NJ
G8 UBF G. Beadle 8 Woodland View, Southwell NG25 0AG
G8 UBJ R. Lester 71 Ronelean Road, Surbiton KT6 7LL
G8 UBN G. Hodgson East Cottage, Chineham Lane, Basingstoke RG24 9LR
G8 UBP A. Jacketts Flat 7, Jenneth Court, 44 Mauldeth Road, Stockport SK4 3NB
G8 UBU R. Jarvis 39 Moy Road, Colchester CO2 8NZ
G8 UBX R. Manning 2 Reydon Close, Haverhill CB9 7WG
G8 UCC I. Bradley 8 Hunt Avenue, Heanor DE75 7QB
G8 UCK T. Colligan 53 Datchworth Turn, Hemel Hempstead HP2 4PB
G8 UCL N. Cook 61 Redland Road, Malvern WR14 1LY
G8 UCN A. Crookes 23 Helliwell Lane, Deepcar, Sheffield S36 2NH
G8 UCP M. Culling 101 Orchard Drive, Park Street, St. Albans AL2 2QL
G8 UCR D. Davis 2 Bowen Road, Rotherham S65 1LH
GI8 UCS A. Edwards 6 Carwood Park, Newtownabbey BT36 5JU
G8 UCY D. Walker 43 Wimborne Road, Corfe Mullen, Wimborne BH21 3DS
G8 UCZ C. Wren 24 Willow Way, Martham, Great Yarmouth NR29 4SH
G8 UDA B. Watson 3 Anderton Rise, Millbrook, Torpoint PL10 1DA
G8 UDD S. James 42 Wilshire Avenue Springfield, Chelmsford CM2 6QW
G8 UDG D. Roberts 25 Metcalfe Road, Cambridge CB4 2DB
G8 UDI P. Newman 4 Old Barn Court, Ludford, Market Rasen LN8 6AZ
GW8 UDJ Selex Galileo Sports & Leisure Club c/o M. Loach Rhiwgwraidd, Llanilar SY23 4SQ
G8 UDS J. Farrant Partida Barrancs 16, Orba, Alicante 3790 Spain
G8 UDV A. Frost 10 Ramsden Square, Cambridge CB4 2BJ
G8 UDZ A. Gilbertson The Corn Store, Manor Farm, Old Alresford, Alresford SO24 9DH
G8 UED W. Harris 1 Marsh Cottages, Middle Street, Yeovil BA22 7AP
G8 UEE S. Melvin 2 Belsay Gardens, Newcastle upon Tyne NE3 2AU
G8 UEF P. Mobberley The Willows, Hobro, Kidderminster DY11 5ST
G8 UEI A. Howells 27 Swallow Lane, Aylesbury HP19 7HW
GW8 UEK N. Hosker Laburnum Cottage, Church Lane, Old Aston Hill, Ewloe CH5 3BF
G8 UEU J. Thornburn 19 Pingate Lane South, Cheadle Hulme, Cheadle SK8 7NP
G8 UEY J. Rice 2 Medalls Path, Stevenage SG2 9DX
G8 UEZ C. Southall 40 Tathall End, Hanslope, Milton Keynes MK19 7NF
G8 UFO E. Charlton 2 Bullock Road, Washingley, Peterborough PE7 3SH
G8 UFX I. Fowler 1 Mayfields, Shefford SG17 5AU
G8 UGK P. Warburton 4384 Henneberyy Road, Manilus 13104 United States
GM8 UGO W. Kay 22 Linton Terrace, Perth PH1 1LE
G8 UGS P. Marks 106 Darlton Drive, Arnold, Nottingham NG5 7LW
G8 UHJ P. Couch 6 Quantock Gardens, Ramsgate CT12 6SW
G8 UHK A. Rolls 9 Mareschal Road, Guildford GU2 4JF
G8 UHL J. Rangeley Tanglewood Buckden, Skipton BD23 5JA
G8 UHM W. Rosser Fanshawgate House, Fanshaw Gate Lane, Holmesfield, Dronfield S18 7WA
G8 UHO D. Reay 78 Wyresdale Road, Lancaster LA1 3DY
G8 UHT P. Shaw Poole Bank Cottage, Poole, Nantwich CW5 6AL
G8 UHW C. Mobbs 5 Garth Avenue, Leeds LS17 5BH
G8 UID R. Nock 83 Coles Lane, West Bromwich B71 2QW
G8 UIG M. Chedzoy Helena, Picts Hill, Langport TA10 9EZ
G8 UIL J. Gale Barn End, Highampton, Beaworthy EX21 5LT
G8 UIO D. Salter 9 Old Milverton Road, Leamington Spa CV32 6BA
GI8 UIU Dr P. Moore 13 Ballygallum Road, Downpatrick BT30 7DA
G8 UIV Dr P. Morton-Thurtle 23 Wife Of Bath Hill, Canterbury CT2 8PQ
G8 UIW S. Threlfall-Rogers 43 Nanpantan Road, Loughborough LE11 3ST
G8 UJF D. Headland Hazelwood, Haywards Lane, Cheltenham GL52 6RF
G8 UJO B. Leveton Orman House, 17A Grove Avenue, Norwich NR5 0JD
G8 UJS A. Mcdermott-Roe 100 Dorset Ave, West Parley, Ferndown, Ferndown BH22 8DZ
G8 UJV M. Johnson 9 South Road, Brampton, Huntingdon PE28 4PX
G8 UKH L. Luck 1313 Rodney Lane, Winchester K0C 2K0 Canada
G8 UKI Echelford ARS c/o T. Gawn 15 Barradon Close, Torquay TQ2 8QE
G8 UKO P. Coupe 10A West End Avenue, Brundall, Norwich NR13 5RF
G8 UKV M. Vincent 9 Sleapford, Long Lane, Telford TF6 6HQ
G8 UKY R. Mills 3 Hallam Moor, Liden, Swindon SN3 6LS
GW8 UKZ Dr A. Walker Glandenys, Cross Inn, Llanon SY23 5NA
G8 ULH J. Wilson 2 Reston Court, Cleethorpes DN35 0JQ
G8 ULJ G. Sowter 21 Seawell Road, Bude EX23 8PD
G8 ULL M. Williams 9 Fir Tree Drive, West Winch, King's Lynn PE33 0PR
G8 ULM D. Petty 93 High Street, Hinxton, Saffron Walden CB10 1QY
G8 ULQ C. Copsey Flat 17, Cherrywood Court, Moordown Avenue, Solihull B92 8QS
G8 UMA J. Spencer 78 Copse Avenue, Farnham GU9 9EA
G8 UMB R. Kennedy Little Thatch, Canterbury Road, Dover CT15 7HJ
G8 UML M. Spencer 79 Salisbury Close, Alton GU34 2TP
GM8 UMN J. Norrie 13 Pentland Crescent, Dundee DD2 2BU
G8 UMO M. Walker Winterfell, Fen Road, Boston PE22 8HA
GI8 UMV P. Melly 1 Bessfield Park, Carrickfergus BT38 7BY
G8 UMY P. Mahoney 2 Elm Lodge, Elm Avenue, Ruislip HA4 8PH
G8 UND J. Warburton 164 High Street, Lewes BN7 1XU
G8 UNO R. Clarke 58 Orpin Road, Merstham, Redhill RH1 3EY
G8 UNP R. Burningham 27, Sunningdale Gardens North Bersted, Bognor Regis PO22 9LE
G8 UOJ A. Abrahams 69 Culverhouse Road, Luton LU3 1PY
G8 UOL B. Cooper 2 Vernon Way, Plot 58 - Poltair Vale, Penryn TR10 8SJ
G8 UOZ M. Freestone 12 St. Martins Approach, Ruislip HA4 7QD
GM8 UPC A. Wells 64 Denwell Road, Insch AB52 6LH
G8 UPD R. Payne 1 Mill Hill, Horning, Norwich NR12 8LQ
G8 UPF K. Hutchinson 8 Innage Crescent, Bridgnorth WV16 4HU
GM8 UPI D. Mcalpin Birchwood, Belzies, Lockerbie DG11 1SA
GW8 UPJ Rvd. P. Nunn 1 Talwrn Court, Coedpoeth, Wrexham LL11 3NN
G8 UPK D. Akester 19 Bracken Park, Bingley BD16 3LG
G8 UPO D. Steele 43 Lancastria Mews, Boyndon Road, Maidenhead SL6 4SA
G8 UPX M. Westwood 59 Woodland Drive, St. Albans AL4 0EN
GW8 UQC Dr D. Bolton Mill Farm, Manorowen, Fishguard SA65 9PT
G8 UQR T. Riley 3 Hoefield Crescent, Nottingham NG6 8AY
G8 UQV F. Hall 478 Darwen Road, Bromley Cross, Bolton BL7 9DX
GM8 UQX S. Roberts Blaruaine, Dunoon PA23 8TH
G8 UQY G. Waywell 14 Causeway, Great Harwood, Blackburn BB6 7HU
G8 URB F. Mcgilp High Beech Ashley, Tiverton EX16 5PA
G8 URG P. Ridgeon 1 Fiennes Road Herstmonceux, Hailsham BN27 4LN
G8 URI G. Cross 11 Highfield App/Ch, Billericay, Essex CM11 2PD
G8 URZ N. Bird 15 Bramley Close, Powick, Worcester WR2 4SR
G8 USA M. Dawes 20 Alden Street, Danvers 1923 United States
G8 UST M. Freeman Sunnyside, Long Lane, Mansfield NG20 8AZ
G8 UTH G. Sunderland 28 Tillotson Avenue, Sowerby Bridge HX6 1BX
GW8 UTK B. Davies Rhosyr, Llanfair Pg LL61 5JB
G8 UTQ W. Vander Byl 45 Scotby Road, Scotby, Carlisle CA4 8BD
G8 UTW H. Mirams 58 Ing Head Terrace, Shelf, Halifax HX3 7LB
G8 UTY J. Wardle Spring Cottage, Chapel Road, Hayle TR27 6BA
G8 UUC G. York 10 Beechwood Avenue Wallasey Village, Wallasey CH45 8NX
GI8 UUN Betws Y Coed RC c/o G. Curtis-Smith 19A Glenavy Road, Lisburn BT28 3UT
G8 UUO D. Drizen 3 Crittall Close, Silver End, Witham CM8 3SY
G8 UUR P. Dicken 99 Stanton Road, Burton-on-Trent DE15 9SE
G8 UUS P. Owen 2 Plantation Road, Wollaton, Nottingham NG8 2ER
G8 UUV M. Copsey 1 Cooper Terrace, Dereham Road, Dereham NR19 2BJ
GM8 UUW C. Fyfe 20 Larch Grove, Galashiels TD1 2LB
G8 UVF T. Cassell 3 Rose Hill, Waterlooville PO8 9QU
G8 UVG C. Parker-Larkin 40 Head Street Goldhanger, Maldon CM9 8AZ
G8 UVN B. Rigby 76 Woodland Road, Rode Heath, Stoke-on-Trent ST7 3TL
G8 UVU M. Soble Whitethorn Farm, Carey, Hereford HR2 6NG
G8 UVY Southgate ARC c/o J. Haste 11 Corporation Road, Chelmsford CM1 2AR
G8 UVZ B. Hart 63 Newcastle Road, Congleton CW12 4HL
G8 UWD R. Hornby 63 Shearwater Drive, Bicester OX26 6YR
G8 UWE M. Jefford 37 Marions Way, Exmouth EX8 4LF
G8 UWG A. Kay 19 Chesnut Grove, Higher Tranmere, Birkenhead CH42 0LB

G8 UWI A. Downing 21 Firfield Road, Thundersley, Benfleet SS7 3UU
G8 UWL D. Cooper 18 Stockfield Road, Stoke on Trent ST37AP
G8 UWM M. Crossley 3 Derby Street, Stockport SK3 9HF
GW8 UWP N. Elgood Rhyd Y Paith Rhydyfelin, Aberystwyth SY23 4PU
G8 UXB B. Payne 45 Kellaway Avenue, Westbury Park, Bristol BS6 7XS
G8 UXD A. Gill 4 Cornubia Close, Hayle TR27 4RL
GW8 UXL P. Nicholson 85 Llay New Road, Llay, Wrexham LL12 0PS
G8 UXW J. Benton 2 Appleton Road, Fareham PO15 5QH
G8 UXX K. Brazington 38 Tamworth Road, Amington, Tamworth B77 3BT
G8 UYB K. Reece Winsford Grange Nursing Home, Station Road By Pass, Winsford CW7 3NG
G8 UYF R. Ritchie Flat 2, Sundon Park Parade, Luton LU3 3BH
G8 UYK F. Rowntree 5 Cornel House, Osborne Road, Windsor SL4 3SQ
G8 UYL R. Rumbelow The Spinney, The Chase, Leatherhead KT22 0HR
G8 UYM J. Sanderson 5 Babbacombe Drive Rudd Hill Estate, Ferryhill DL17 8DA
G8 UYR B. Smith 3 Harwin Close, Wolverhampton WV6 9LF
G8 UYW T. Carrig 12 Longmoor Drive, Liphook GU30 7XA
G8 UYY M. Daish 27 Westbourne Road, Portsmouth PO2 7LB
G8 UZM Dr R. Jefferson 23 Valerian Avenue, Heddon-On-The-Wall, Newcastle upon Tyne NE15 0EA
G8 UZQ S. Haywood 19 Crich Way, Newhall, Swadlincote DE11 0UU
G8 UZW J. Durrant 114 Rosebank Avenue, Hornchurch RM12 5QS
G8 UZY D. Fisher 8 Beech Road, Stibb Cross, Torrington EX38 8HZ
G8 VAD J. Goodings 133 Lache Lane, Chester CH4 7LU
G8 VAE A. Griffiths Tranby Croft, Matt Pits Lane, Wainfleet, Skegness PE24 4LY
G8 VAF K. Chittenden Ponders, Hall Road, Colchester CO6 3DX
GM8 VAM G. Brazier 117A East Clyde Street, Helensburgh G84 7PL
G8 VAN M. Goodwin 15 Meadow Close, Repton, Derby DE65 6GT
G8 VAR B. Harper 51 Cross Lane, Scarborough YO12 6DQ
G8 VAT G. Denton 90 Willow Lane, Knottingley WF11 8AJ
G8 VBA R. Webb 78 Station Road, Rolleston-On-Dove, Burton-on-Trent DE13 9AB
G8 VBC R. Timms 20 Driftside, Blackfordby, Swadlincote DE11 8BD
G8 VBE C. Thomas Apartment 231, Bournville Gardens Village, Birmingham B31 2FS
G8 VBI D. Sproson 1 The Old Orchard, Whitehall, South Petherton TA13 5AQ
G8 VBK R. Penver 56 Cottesmore Avenue, Ilford IG5 0TG
G8 VBW M. Corbett 86 Jordan Avenue, Stretton, Burton-on-Trent DE13 0JD
GM8 VBX D. Coulthart 23 Larchfield Road, Dumfries DG1 4HU
GW8 VCA D. Dyer Ty Newydd, 24D Fforest Hill, Neath SA10 8HD
G8 VCH J. Grevatt 17 Foxdale Drive, Angmering, Littlehampton BN16 4HF
G8 VCI G. Gwynne 5 Stanstead Avenue, Nottingham NG5 5BL
G8 VCJ C. Gunn 21 Epworth Close, Truro TR1 1UP
G8 VCL R. German 29 Glenthorne Gardens, Sutton SM3 9NL
G8 VCN M. Newport 53 Southfiels Drive, Yeovil BA213FJ
G8 VCO B. Nicholls 35 Lynn Road, Downham Market PE38 9NJ
G8 VCQ W. Norman 27 Newport Road, Barnstaple EX32 9BG
G8 VCU S. Morgan 5 Parklands, Ufford, Woodbridge IP13 6ES
G8 VDJ R. Pauley Lamplight, Casterton Lane, Tinwell, Stamford PE9 3UQ
G8 VDP K. Roberts 35A Rockley Avenue, Birdwell, Barnsley S70 5QY
G8 VDQ C. Parnell 213A Northfield Avenue, London W13 9QU
GW8 VEE G. Blore Ty Newydd, Cymau, Wrexham LL11 5EU
G8 VEM A. Challis 12 Moorland Crescent, Boultham Moor, Lincoln LN6 7NL
G8 VEN H. Chapman 24 Croft Road, Cosby, Leicester LE9 1SE
G8 VEQ A. Stone 47 Oakford Villas, North Molton, South Molton EX36 3HJ
G8 VER Verulam ARC c/o P. King 96 Mancroft Road, Caddington, Luton LU1 4EN
G8 VEZ T. Wagg 15 Barncroft Way, Havant PO9 3AA
GW8 VFF A. Wilkins 11 Redhouse Road, Ely, Cardiff CF5 4FG
G8 VFI D. Franklin 49 Hope Road, Benfleet SS7 5JQ
G8 VFM J. Callaghan 271 Belvoir Road, Coalville LE67 3PL
G8 VFP D. Evans 2 Cottage Lane, Marlbrook, Bromsgrove B60 1DW
GW8 VFQ R. Elliott 19 Pencoed, Dunvant, Swansea SA2 7PQ
G8 VG A. Windle Flat 11, Parham House 15 King George'S Drive, Liphook GU30 7GB
GW8 VGB R. Morgan 4 Underhill Lane, Horton, Swansea SA3 1LB
G8 VGI A. Lilly 47 Horton Street, Frome BA11 3DP
G8 VGQ P. Andrews 1 Waite Meads Close, Purton, Swindon SN5 4ET
G8 VGY E. Davis 850 Lantana Rd, Lantana 33462 USA
G8 VHB M. Fitzgibbons 8 Lundhill Close, Wombwell, Barnsley S73 0RW
G8 VHF Reading & District ARC c/o J. Goodier 20 Poleacre Lane, Woodley, Stockport SK6 1PG
G8 VHG I. Gower 10 Homethorpe, Hull HU6 9EU
G8 VHI Clyde Valley DX Group c/o R. Woolley 103 Mancetter Road, Nuneaton CV10 0HP
G8 VHK M. Stanford Fircroft, Mutton Hall Lane, Heathfield TN21 8NR
G8 VHL S. Price 35 Western Road, Goole DN14 6QW
G8 VHO S. Pratt 4 Bussex Square, Westonzoyland, Bridgwater TA7 0HD
G8 VHX G. Shering 66 Oliver Street, Ampthill, Bedford MK45 2QL
G8 VIB J. Sim 22 Dene View, Ashington NE63 8JT
G8 VIC M. Rump 24 Stoneleigh Avenue, Brighton BN1 8NP
G8 VIV K. Bilke 8 Ibworth Lane, Fleet GU51 1AU
G8 VJG K. Halls Little Rema, Cray Road, Crockenhill, Swanley BR8 8LP
G8 VJO C. Green 12 Spenser Grove, Great Harwood, Blackburn BB6 7JU
G8 VJP L. French 14 Manor Farm Court, Thrybergh, Rotherham S65 4NZ
G8 VJR D. Fowles 52 Lucy Lane South, Stanway, Colchester CO3 0HY
G8 VJU K. Earl 210 Churchill Avenue, Chatham ME5 0JS
G8 VJW G. Davey 147 Deeds Grove, High Wycombe HP12 3PA
G8 VJY A. Crafer 155 Upham Road, Swindon SN3 1DR
GI8 VKA R. Coulter11 34 Toberdowney Valley, Ballynure, Ballyclare BT39 9TS
GM8 VKN M. Tarr 1 Methven Drive, Dunfermline KY12 0AH
G8 VKO G. Tandy 24 Coast Road, Hopton Great Yarmouth NR31 9BT
G8 VKQ C. Sparke 47 Jobes, Balcombe, Haywards Heath RH17 6AF
G8 VKR A. Williams 10 Catalina Close, Woodley, Reading RG5 4UG
GW8 VKS Dr J. Williams Glenview, Penrhos, Usk NP15 2LF
GM8 VL Gmdx Group c/o R. Ferguson 19 Leighton Avenue, Dunblane FK15 0EB
G8 VLL A. Kett 476 Earlham _Road, Norwich NR4 7HP
G8 VLP M. Clark 6 Shalcross Drive, Cheshunt, Waltham Cross EN8 8UX
G8 VLR R. Law 19 Central Drive, Bramhall, Stockport SK73JU
G8 VLS D. Leeder 2 Robin Court, East Rainton, Houghton le Spring DH5 9RW
G8 VLY R. Macbeth 9 Woodside, Stroud GL5 1PL
G8 VMF M. Clayton 54 Banks Road Golcar, Huddersfield HD7 4RE
G8 VML L. Gibson 57A Heritage Park, Hatch Warren, Basingstoke RG22 4XT
G8 VMP K. Webster 5 Ridgmont Rd, St. Albans AL1 3AG
G8 VMQ Dr A. Parker 78 Whitbarrow Road, Lymm WA13 9BA
G8 VMY D. Whitfield Framingham, Manor Road, Hayling Island PO11 0QR
G8 VMZ M. Walpole 9 The Paddocks, Brandon IP27 0DX
G8 VNF B. Benson 31 Helston Close Brookvale, Runcorn WA7 6AA
G8 VNL J. Darlington 111 Maas Road, Northfield, Birmingham B31 2PP
G8 VNN A. Dowsett 70 Warren Drive, Broughton, Chester CH4 0PT
G8 VNO G. Edmonds 29A Manor Park, Woolsery, Bideford EX39 5RH
G8 VNP R. Elden 124 Larchcroft Road, Ipswich IP1 6PQ
G8 VNX J. Dixon 19 Pheasant Drive, Wincham, Northwich CW9 6PX
G8 VOB K. Fisher 50 Queen Street, Henley-on-Thames RG9 1AP
G8 VOH Dr P. Renshaw Hayes Pond Cottage, Hayes Plat, Rye TN31 6HQ
G8 VOI R. Reeves 4 Elmwood Avenue, Waterlooville PO7 7LG
G8 VOJ J. Read 99 Blackheath Road, Lowestoft NR33 7JF
G8 VOQ R. Rogalewski 47 Partridge Crescent, Dewsbury WF12 0HT
G8 VOY Hilderstone Radio And Electronic Society c/o J. Mcparlane 47 Aragon Road, Kingston upon Thames KT2 5QB
G8 VPD I. Morley 2 Livingstone Walk, Park Wood, Maidstone ME15 9JB
G8 VPE J. Noy 14 Poplar Drive, Filby, Great Yarmouth NR29 3HU
G8 VPG S. O'Sullivan 15 Witney Close, Saltford, Bristol BS31 3DX
G8 VPH B. Aveling 6 Brambling, Wilnecote, Tamworth B77 5PQ
G8 VPO J. Chalmers 10 Hornbeam Close, Wokingham RG41 4UR
GW8 VPP S. Croft 43 York Road, Colwyn Bay LL29 7EY
G8 VPR Surrey Radio Contact Club c/o B. Gallear 5 Oak Avenue, Cannock WS12 4QA
G8 VPX A. Davies 14 Primrose Grove, Keighley BD21 4NP
G8 VPY R. Davies 1 Seaham Close, Norton, Stockton-on-Tees TS20 1RT
G8 VQA S. Foulser 9 Oak Coppice Close, Eastleigh SO50 8PH
G8 VQE B. Haylett 160 Hookfield, Harlow CM18 6QN
G8 VQH M. Russ 71 Farriers Close, Martlesham Heath, Ipswich IP5 3SN
G8 VQJ M. Otterson 161 Tollgate Lane, Bury St. Edmunds IP32 6DF
G8 VQK M. Nightingale 31 Cradge Bank, Spalding PE11 3AB
G8 VQN R. Martin Suite 248, 548-550 Elder House Elder Gate, Milton Keynes MK9 1LR
G8 VQQ J. Locke 2 Norton Close, Daventry NN11 4GW
G8 VQX A. Hyde 5-7 St. Nicholas Drive, Caister-On-Sea, Great Yarmouth NR30 5QT
G8 VR K. Rochester 22 Langford Road, Cockfosters, Barnet EN4 9DS
G8 VRN D. Sutton 14 Brocklesby Close, Gainsborough DN21 1TT
GW8 VRS Dr D. Fone 29 South Rise, Cardiff CF14 0RF
G8 VRV G. Dyer Chez Nous, 11 Fore Street, St. Austell PL26 7NN
G8 VRW M. Davis 4 Joel Close Earley, Reading RG6 5SN
G8 VSF J. Williams Staithe Marsh House, The Staithe, Norwich NR12 9DA
G8 VSH P. Taylor 62 Westfield Avenue, Ashchurch, Tewkesbury GL20 8QP
G8 VSI M. Sutcliffe 3 Sunningdale Way, Neston CH64 0UY
G8 VSM A. Swenson Bankholme, Ashleigh Road, Arnside, Carnforth LA5 0HE
G8 VSN G. Tennant 85 Coronation Drive, South Normanton, Alfreton DE55 2HS
G8 VSO C. Tompkinson 24 Rodney Close, Exmouth EX8 2RP
G8 VSR J. Rowley 10 Friars Close, Cheadle, Stoke-on-Trent ST10 1AT
G8 VSV D. Petty 7 Luscombe Close, Ipplepen, Newton Abbot TQ12 5QJ
G8 VSX R. Hammond 52A Mill Road, Cleethorpes DN35 8JA
GI8 VTK A. Boston 14 Galloway Point, Donaghadee BT21 0ES
G8 VTV A. Hart 78 Shepherds Way, Rickmansworth WD3 7NR
G8 VTX I. King Mill Ghyll, Low Lane, Kendal LA8 8AT
G8 VTY N. Knight 36A Abbey Street, Rugby CV21 3LH
G8 VTZ M. Keay 492 Falmer Road, Brighton BN2 6LH
G8 VU D. Blair 121 Longstomps Avenue, Chelmsford CM2 9BZ
GW8 VUG I. Wilkinson 6 Cwm Teg, Old Colwyn, Colwyn Bay LL29 8ZA
G8 VUK A. Palmer 24 Vellum Drive, Carshalton SM5 2TL
G8 VUM P. Reade 30 Hayleigh House, Silcox Road, Bristol BS13 0JG
G8 VUN R. Roberts 5480 Laburnum Ave., British Columbia V8A 4MB Canada
G8 VUO D. Roberts 32 Central Road, Gloucester GL1 5BY
G8 VUS A. Branton 20 Sling Lane, Malvern WR14 2TU
G8 VUU R. Clifford 1 Darell Croft, Sutton Coldfield B76 1HU
GW8 VUV A. Gravell 49 Rehoboth Road, Five Roads, Llanelli SA15 5DJ
G8 VVB C. Heath 3 Trebellan Drive, Hemel Hempstead HP2 5EL
G8 VVC C. Haver 31 Edenham Road, Hanthorpe, Bourne PE10 0RB
G8 VVG S. Hackett 11 Fairfield Avenue, Upminster RM14 3AZ
G8 VVM D. Merrick 259 Wigmore Road, Gillingham ME8 0LZ
G8 VVP P. North 84A Park Road, Great Sankey, Warrington WA5 3ET
G8 VVR C. Key 23 Oxford Road, Kesgrave, Ipswich IP5 1EL
G8 VVU M. Strange 15 Waterside, Willesborough, , Ashford TN24 0AX
GW8 VVX R. Williams The Basement, 9 Charlton St., Llandudno LL30 2AA
G8 VVY R. Shelley 69 Avenue De Gien, Malmesbury SN16 9GX
G8 VVZ A. Stephens 12 Sutherland Walk, Aylesbury HP21 7NS
G8 VWH S. Hall 31 Somerton Gardens, Earley, Reading RG6 5XG
G8 VWJ M. Hoare 45 Tilehurst Road, Reading RG1 7TT
G8 VWU J. Marriott 9 Albany Walk, Peterborough PE2 9JN
G8 VWV A. Ball White Cottage, Barracks Lane, Reading RG7 1BB
G8 VXB D. Young 66 Porchester Road, Kingston upon Thames KT1 3PS
G8 VXR C. Hunt 41 Maylands Way, Harold Wood, Romford RM3 0BQ
G8 VXU D. Llewelyn 56 Marlpit Lane, Seaton EX12 2HN
G8 VXY P. Nicol 38 Mitten Avenue, Rubery, Birmingham B45 0JB
G8 VYK West Lincs Rynt c/o M. Purser 17 Firecrest Road, Chelmsford CM2 9SN
G8 VYO A. Swain 6 Abbots Grove, Belper DE56 1BX
G8 VYP C. Syms 24 Warmdene Road, Patcham, Brighton BN1 8NL
G8 VYQ G. Todd 100 Avebury Drive, Washington NE38 7DB
G8 VYT D. Tombs 24 Ferndale Road, Northville, Bristol BS7 0RP
GM8 VYZ A. Raine 10 Castle View, Airth, Falkirk FK2 8GE
G8 VZB A. Poole 6 Rutland Avenue, Willsbridge, Bristol BS30 6EZ
G8 VZD C. Ramsey 10 Lindley Road, London E10 6QT
G8 VZI M. Warren 39 St. Marks Road, Weston-Super-Mare BS22 7PF
G8 VZJ M. Webb 24 College Avenue, Grays RM17 5UW
G8 VZR N. Giddings Colliford Lake Park St. Neot, Liskeard PL14 6PZ
G8 VZS I. Chapman 188 Goodhart Way, West Wickham BR4 0HA
G8 VZT D. Hall 4 Steventon Road, Wellington, Telford TF1 2AS
G8 VZY B. Levie 51 Budges Road, Wokingham RG40 1PL
G8 VZZ P. Marks Flat 3, 47 The Thoroughfare, Woodbridge IP12 1AH
G8 WAE P. Vella Chez Tillet, Rougnac 16320 France
G8 WAJ N. Treanor 23 Norton Avenue, Penketh, Warrington WA5 2RB

G8 WAL Norfolk County Raynet Group c/o P. Taylor 14 Cedern Avenue, Elborough, Weston-Super-Mare BS24 8PA
G8 WAM G. Weeks Forge Cottage, The Bury, Hook RG29 1ND
G8 WAP R. Warren 22 Tyndale, North Wootton, King's Lynn PE30 3XD
G8 WAV C. Jacobs 133 Fordham Road, Isleham, Ely CB7 5QX
G8 WAW I. Howard 85 Mollington Avenue, Liverpool L11 3BQ
G8 WBG S. Netherton 33 Bethel Road, St. Austell PL25 3HB
G8 WBK A. Maufe 28 Dale View, Ilkley LS29 9BP
G8 WBL T. Mole 20 Horns Park, Bishopsteignton, Teignmouth TQ14 9RP
G8 WBN D. Neale 24 Addison Road, Reading RG1 8EN
G8 WBO S. Holley 6 Middle Ground, Fovant SP35LP
G8 WBP P. Humphreys 910 High Lane, Stoke-on-Trent ST6 6HE
G8 WBT A. Farnborough 9 Mitchelmore Road, Yeovil BA21 4BA
G8 WBU A. Greenall Flat 11, Sutherland House Royal Herbert Pavilions, Gilbert Close, London SE18 4PS
G8 WBY J. Osborne Little Martins Langham, Colchester CO4 5PY
GI8 WBZ A. Smith 12 Sandringham Heights, Carrickfergus BT38 9EG
GW8 WCA K. Winter Derwen, Hillside, Monmouth NP25 4LY
G8 WCD A. Smith Lyndhurst House, 19 Fleet End Close, Havant PO9 5ED
G8 WCH R. Shepherd 299 West Wycombe Road, High Wycombe HP12 4AA
G8 WCQ V. Mcclure 43 Roman Way, Seaton EX12 2NT
G8 WCT A. Grindrod 54 Priestley Drive, Pudsey LS28 9NQ
G8 WCX A. Essex 32 Crossfield Drive, Skellow, Doncaster DN6 8RJ
G8 WDC Otley ARS c/o G. Scott 19 Penkett Road, Wallasey CH45 7QF
G8 WDX R. Lamkin 3 Homestead Close, Upton, Aylesbury HP17 8XQ
G8 WEM M. O'Neill Coolrake, Moone, Co Kildare 0 Ireland
GW8 WEY T. Jones 80 Taff Embankment, Cardiff CF11 7BG
GI8 WFA W. Harvey 25 Shanes Hill Road, Kilwaughter, Larne BT40 2PA
G8 WFP C. Kershaw 50 Wellgarth, Halifax HX1 2BJ
GW8 WFS J. Lawson-Reay The Nook, Conway Road, Llandudno LL30 1PY
G8 WGD P. Randall-Cook 3 Wellmeadow, Staunton, Coleford GL16 8PQ
G8 WGE I. Robinson 26 Wick Road, Teddington TW11 9DW
G8 WGN J. Marks Stam 69, Huizen 1275 CG Netherlands
G8 WGP Gilwell Park Scout RC c/o S. Barber Homedale, St. Monicas Road, Tadworth KT20 6ET
G8 WGQ D. Onione 19 Chapman Close, Kempston, Bedford MK42 8RU
GM8 WGU A. Irving 23 Woodlea Park, Sauchie, Alloa FK10 3BG
G8 WHB H. Couchman Pond Cottage, Woodside Green, Maidstone ME17 2EU
G8 WHD P. Whittington 7 Bowden Rise, Seaford BN25 2HZ
GI8 WHP S. Craig 8 Andrew Avenue, Larne BT40 1EB
G8 WHR S. Wood 90 Plymyard Avenue, Bromborough, Wirral CH62 6BR
G8 WIJ B. Magrath 14 Strines Road, Marple SK67BT
G8 WIM Backpackers Radio Activity Group c/o G. Cripps 115 Bushey Road, Raynes Park, London SW20 0JN
G8 WIR J. Vousden 44 Castle Road, Tankerton, Whitstable CT5 2DY
GI8 WIU S. Douthart 75 Market St., Ballycastle BT54 6DS
G8 WIW T. Elsey Flat 4, 32 Pembridge Square, London W2 4DT
G8 WJB J. Geer 31 The Beeches, Salisbury SP1 2JH
GM8 WJK J. Nicolson Clickhimin, Serrigar, Orkney KW17 2RL
GI8 WJN A. Humphreys 20 Ballyreagh Road, Tempo., Enniskillen BT94 3EH
G8 WJY M. Garton 13 Damaskfield, Worcester WR4 0HY
G8 WKA R. Reich Cob Barn, Northlew, Okehampton EX20 3NR
G8 WKE J. Bloxham 15 Windmill Road, Breachwood Green, Hitchin SG4 8PG
G8 WKH I. Jones 2 Castle Keep Mews, Newcastle ST5 2SD
G8 WKK M. Daniels 6 Middlemead Stratton-On-The-Fosse, Radstock BA3 4QH
G8 WKL Downside Scl Ar c/o M. Daniels 6 Middlemead Stratton-On-The-Fosse, Radstock BA3 4QH
G8 WKX R. Denton 18 Sealand Court, Esplanade, Rochester ME1 1QH
G8 WKZ K. Spragg 88 Low Lane, Middlesbrough TS5 8EB
G8 WLB S. Austen Shiralee, The Plain Road, Ashford TN25 6RA
G8 WLD W. Parrott 11 St. Georges, Chester CH1 3HG
G8 WLL S. Lown 50 Fall Birch Road, Lostock, Bolton BL6 4LG
G8 WLV R. Barber 10 St. Leonards Close, Upper Minety, Malmesbury SN16 9QB
G8 WLY J. List 41 Westbury Crescent, Dover CT17 9QQ
G8 WMC E. Holman Weavers Cottage, The Shoe, Chippenham SN14 8SA
G8 WMF A. Dawe Highcroft, Upper House Lane, Guildford GU5 0SX
G8 WMG J. Bassnett 105 Edgemoor Drive, Crosby, Liverpool L23 9UF
G8 WMK G. Bessant 4 Sleigh Road, Sturry, Canterbury CT2 0HR
G8 WMW A. Rowell 25 Headcorn Gardens, Cliftonville, Margate CT9 3ES
GW8 WNB K. Phillips Lluest Y Coed, 39 Llwyn Ynn, Talybont LL43 2AG
GW8 WNK J. Davies Fronallt Llanbedrog, Pwllheli LL53 7PB
G8 WNQ R. Harrison Margaty, Pencoys, Redruth TR16 6LR
G8 WOX A. Hartland 16 Hillgrove Crescent, Kidderminster DY10 3AP
G8 WOZ D. Hopkins 14 Abraham Drive, Silver End, Witham CM8 3SP
G8 WPA Dr B. Lloyd 238 Brecknock Road, London N19 5BQ
G8 WPF A. Middleton 13 Ragleth Road, Church Stretton SY6 7BN
G8 WPU I. Rivett 30 Millside Close, Kingsthorpe, Northampton NN2 7TR
G8 WPV A. Reason 71 Cavendish Road, Hazel Grove, Stockport SK7 6HU
G8 WPX G. Ratcliffe 68 Priory Close, Tavistock PL19 9DG
G8 WQ Weymouth And District Short Wave Club c/o G. Watts 3 Maple Grove Knightsdale Road, Weymouth DT4 0FE
G8 WQC J. Maxworthy 3 Hoylake Close, Slough SL1 5UR
G8 WQE A. Vaughan 12 Kingsley Road, Frodsham WA6 6SG
G8 WQT T. Rickard 137 Hugin Avenue, Broadstairs CT10 3HN
G8 WQW J. Shergold 35 Orchard Grove, New Milton BH25 6NZ
G8 WQZ D. Mead 9 Abraham Drive, Silver End, Witham CM8 3SP
G8 WRB Epping Forest Raynet Group c/o Dr D. Kirkby Stokes Hall Lodge, Burnham Road, Chelmsford CM3 6DT
GW8 WRC T. Harston Ogilvie House, St. Ishmaels, Haverfordwest SA62 3TD
G8 WRG Martlesham RS c/o D. Salter 9 Old Milverton Road, Leamington Spa CV32 6BA
G8 WRI W. Lawrence 15 Rissington Road, Tuffley, Gloucester GL4 0HP
G8 WRL R. Williamson 35 Villiers Avenue, Twickenham TW2 6BL
G8 WRV R. Bygrave 35 East St., St. Neots, Huntingdon PE19 1JU
G8 WRY G. Brock 54 Lord Haddon Road, Ilkeston DE7 8AW
G8 WSB M. Beardsley 121 Wood Road, Lower Gornal, Dudley DY3 2LR
G8 WSC R. Burg 20 Rowan Way, Witham CM8 2LJ
G8 WSF F. Price 26 Teviot Gardens, Pensnett, Brierley Hill DY5 4QL
G8 WSH M. French 32 St. Michaels Road, Long Stratton, Norwich NR15 2PH
G8 WSM Weston Super Mare RS c/o D. Dyer 26 Locking Road, Weston-Super-Mare BS23 3DF
G8 WSP P. Arup Alma House, Broadway Road, Windlesham GU20 6BU
G8 WSQ A. Beeston 8 Meadow Close, Repton, Derby DE65 6GT
G8 WSR Wirral Schools RC c/o S. Wood 90 Plymyard Avenue, Bromborough, Wirral CH62 6BR
G8 WSS M. Blair 12 Medoc Close, Pitsea, Basildon SS13 1NR
G8 WSU J. Hoggarth Cotherstone, Rockingham Paddocks, Kettering NN16 9JR
G8 WSV M. Cartwright 9 Montgomery Close, Kettering NN15 5BY
G8 WSW R. Carter 46 Arterial Road, Leigh-on-Sea SS9 4DA
G8 WSX DV Scotland c/o G. Goodyer Flat, 54 Wyndham Road, Petworth GU28 0EQ
G8 WSY P. Bloor 216 Waterloo St., Burton on Trent DE14 2NB
G8 WSZ J. Foster 1 Thorn Court, Four Marks, Alton GU34 5BY
G8 WTB D. Crowe 27 Hartfield Court, Collett Road, Ware SG12 7LT
G8 WTM R. Britt 2 Lindisfarne Court, Maldon CM9 6UQ
G8 WTN J. Capon 24 Furness Close, Chadwell St. Mary, Grays RM16 4JB
G8 WTZ D. Holland 29 Lily Crescent, Sunderland SR6 7HN
G8 WUF D. Legg 2 Birkbeck Road, Wimbledon, London SW19 8NZ
G8 WUG R. Spence 8 Stoneleigh, Sawbridgeworth CM21 0BT
GW8 WUM H. Matthews 24 Clos Y Berllan, Rhuddlan, Rhyl LL18 2UL
G8 WUO K. Baker 57 Hedingham Road, Hornchurch RM11 3QH
G8 WUR S. Browning 360 Aureole Walk, Newmarket CB8 7AZ
G8 WUS P. Besley Higher Minzies Down Farm, Bolventor, Launceston PL15 7TT
G8 WUU J. Cooper 156 Church Road, Benfleet SS7 4EN
G8 WVB S. Ayer 335 Ings Road, Kingston Upon Hull, Hull HU7 4UY
G8 WVH J. Bull 12 Eastfield Crescent, Laughton, Sheffield S25 1YT
G8 WVO P. Dawson 35 Crofton Road, Ipswich IP4 4QP
G8 WVU D. Carr 59 Belvoir Road, Bristol BS6 5DQ
G8 WVZ C. Edwards 9 Bradworth Close, Osgodby, Scarborough YO11 3PZ
G8 WWC G. Ludlow 48 Clifford Avenue, Walton Cardiff, Tewkesbury GL20 7RW
G8 WWD G. Hunter 151 Norwich Drive, Wirral CH49 4GD
G8 WWF P. O'Ryan 12 Minton Close, Congleton CW12 3TD
G8 WWI P. Leverington 28 Burymead, Stevenage SG1 4AY
G8 WWJ J. Kirton 13 Saltersford Road, Grantham NG31 7HH
G8 WWM A. Morgan 316 Middle Road, Southampton SO19 8NT
G8 WWO J. Jackson 12 Lower Laith Avenue, Todmorden OL14 5RU
G8 WWW M. Harrington 17 Church Road, Penponds, Camborne TR14 0QE
GM8 WWY W. Kemp 35 Quarry Drive, Kirkintilloch, Glasgow G66 3RY
GW8 WXP R. Hadland 41 Colby Road, Burry Port SA16 0RH
G8 WXV A. Faulkner 8 Wayside Trull, Taunton TA3 7HS
G8 WYI P. Herring 52 Mellowship Road, Eastern Green, Coventry CV5 7BY
G8 WYR Verulam ARC c/o M. Howes Yarnbury Rufc, Brownberrie Lane, Leeds LS18 5HB
GW8 WYW C. Burn Ynysfallen, Church Street, Wrexham LL14 2RL
G8 WZJ A. Collier 44 Cockington Close, Leigham, Plymouth PL6 8RQ
G8 WZK D. Collins 71 Trench Road, Tonbridge TN10 3HG
G8 WZO P. Evans 63 Broadfield Road, London SE6 1NQ
GW8 WZR D. Gale 50 Pickle Line Road, Newport NP19 4DL
G8 WZW K. Aspden Langriggs, Goose House Lane, Darwen BB3 0EH
G8 XAA Bristol Raynet c/o A. Williams 38 Seneca Street, Bristol BS5 8DX
G8 XAJ T. Sherman 4 St Margarets Close, Seasalter CT5 4ST
G8 XAK C. Porteous 1 Earlsfield, Stonards Brow, Shamley Green, Guildford GU5 0UY
G8 XAO G. Woodman 8 Westfield, Loughton IG10 4EB
GW8 XAS G. Evans Wynona, Esplanade, Penmaenmawr LL346LY
G8 XAX K. Tully 225 Main Road, Harwich CO12 3PL
G8 XBY P. Allwood 4 Wightmans Orchard, Piddletrenthide, Dorchester DT2 7QQ
G8 XCE D. Baker 48 Elmwood Street, Burnley BB11 4BP
G8 XCJ I. Coton 77 Lockesfield Place, London E14 3AJ
G8 XCL I. Davis 28 Sycamore Close, Lydd, Romney Marsh TN29 9LE
G8 XCW R. Thomson Shire Jee Neevas, Cold Ash Hill, Cold Ash, Thatcham RG18 9PH
G8 XCY A. Worsfold 5 Turner Close, Langney, Eastbourne BN23 7PF
G8 XDD D. Lucas 43 Larcombe Road, Petersfield GU32 3LS
G8 XDL R. Medcalf 19 All Saints Road, Warwick CV34 5NL
G8 XDM P. Mutter 129 Demesne Road, Wallington SM6 8EW
G8 XDR C. Johnstone 24 Elibank Road, Eltham, London SE9 1QH
G8 XDV S. Huyton 33 Hide Gardens Rustington, Littlehampton BN16 3NP
G8 XDY K. Lloyd 18 Heather Court, Broughton, Chester CH3 5SN
G8 XEF M. Mciver 31 Hartshill, Bedford MK41 9AL
G8 XEI V. Nolan 127 Martins Lane, Blakehall, Skelmersdale WN8 9BQ
G8 XEN H. Hughes 101 Mousehold Avenue, Norwich NR3 4RX
G8 XER J. Smith 32 Station Crescent, Lidlington, Bedford MK43 0SD
G8 XET C. Street Russets, Isle Brewers, Taunton TA3 6QN
G8 XEU R. Stephens 21 St. James Avenue, Lancing BN15 0NN
G8 XEZ M. Ward 11 Rogate Gardens, Portchester, Fareham PO16 8DS
G8 XFK R. Young 34 Wharfedale Drive, Bridlington YO16 6FB
G8 XFY I. Downie 17 Clyfton Crescent, Immingham DN40 2AZ
G8 XGB K. Dickson 29 Sunnyfields Drive, Minster On Sea, Sheerness ME12 3DH
G8 XGG S. Gwilliam 40 Falcon Close, Droitwich WR9 7HF
G8 XGK P. Manford Smithy Hay, Hay Lane, Rugeley WS15 4QG
G8 XGO P. Mckellow 155 Pittmans Field, Harlow CM20 3LE
G8 XGS J. Hindmarsh Roseworth Cottage, Roseworth Cottage West, Hexham Road, Newcastle upon Tyne NE15 9EB
G8 XGT M. Saul 23 Rockingham Road, Bury St. Edmunds IP33 2SA
G8 XGV J. Schofield Wildwinds, St. Johns Road, Wroxall PO38 3EH
G8 XGW N. Shearing 51 Mill Lane, Huthwaite, Sutton-in-Ashfield NG17 2SJ
G8 XHD P. Riebold 7 Clitsome View, Roadwater, Watchet TA23 0RH
G8 XHK K. Prior 9 Tangmere Road, Crawley RH11 0JJ
G8 XHN R. Harman Dirleton Cottage, Church Hill, Godshill, Godshill Ventnor PO38 3HY
G8 XHU G. Arrowsmith 2 Orchard Drive, Bishops Hull, Taunton TA1 5ES
G8 XIM I. Churchill 12 Wyedale Avenue, Coombe Dingle, Bristol BS9 2QQ
G8 XIN M. Chapman 4 Amberley Court, Sidcup DA14 6JT
G8 XIR K. Church 11 Cambria Crescent, Gravesend DA12 4NJ
G8 XIY A. Tee 136 Burstellars, St. Ives PE27 3TJ
G8 XIZ H. Tillotson 30 St. Laurence Road, Northfield, Birmingham B31 2AX
G8 XJB B. Simmons 88 Wellcome Avenue, Dartford DA1 5JW
GW8 XJC R. Smith 6 Lavender Court, Brackla, Bridgend CF31 2ND
G8 XJE J. Williams 32 Fair St., Broadstairs CT10 2JL

G8 XJJ G. Hayes 58 Church Road, Woodley, Reading RG5 4QB
G8 XJL M. Halford 20 Fulwood Drive, Longeaton, Nottingham NG10 3RF
G8 XJN W. Hefferman 74 Balmoral Drive, Borehamwood WD6 2RB
G8 XJO S. Hedicker 1 Hares Close Cottages, Selborne Road, Liss GU33 6HG
G8 XKD Aldridge & Barr Beacon ARC c/o R. Dance 402 Wimborne Road East, Ferndown BH22 9NB
G8 XKH W. Flood 3 March Meadow, Wavendon Gate, Milton Keynes MK7 7TB
G8 XKI G. Fowler 26 Laburnum Drive, Armthorpe, Doncaster DN3 3HE
G8 XKT D. Last 77 Brunswick Road, Ipswich IP4 4BS
GM8 XKW J. Ness Fenway, Dalbeattie Road, Dumfries DG2 8LN
G8 XLA L. Mayes Stone House, Goathland, Whitby YO22 5AN
G8 XLB J. Martin Thatched Cottage, Thaxted Road, Saffron Walden CB11 3BJ
G8 XLE W. Metcalf 30 Rosemary Road, Waterbeach, Cambridge CB25 9NB
G8 XLG C. Proctor 69 Goodrington Road, Paignton TQ4 7HZ
G8 XLH A. Ralph 15 Portchester Close, Stanground, Peterborough PE2 8UP
G8 XLI J. Rigby 93 Birch Grove, Ashton-In-Makerfield, Wigan WN4 0QX
GW8 XLL R. Stubbs 35 Laburnum Drive, Rhyl LL18 4JH
G8 XLZ K. Riley 122 Dryden Road, Gateshead NE9 5TX
G8 XMH D. Higgins 80 Hill Morton Road, Sutton Coldfield B74 4SG
G8 XML J. Hopper 21 Knowles Avenue, Crowthorne RG45 6DU
G8 XMO H. Houghton 21 John Gwynn House, Newport St., Worcester WR1 3NY
G8 XMS L. Sellar Kiawah Ringfield Drive Fownhope Hereford, Fownhope, Hereford HR1 4PR
G8 XMU E. Jones 3 Byland Close, Boston Spa, Wetherby LS23 6PU
GW8 XMW D. Jones 7 Llys Y Godian, Trimsaran, Kidwelly SA17 4BQ
G8 XMZ R. Linton 11 Keats Lane Wincham, Northwich CW9 6PP
G8 XNA J. Lane 12 Penarwyn Woods, St. Blazey Gate, Par PL24 2DG
G8 XNB R. Lelliott Smugglers Cottage, Oreham Common, Henfield BN5 9SB
G8 XNC R. Lacey 12 Melville Avenue, Frimley, Camberley GU16 8NA
G8 XND D. Lucas 6 Holborns Site, Main Road, Spalding PE12 9PF
G8 XNL J. Rigby 43A Corser Street, Stourbridge DY8 2DE
G8 XNN H. Vadgama 20 Hollies Walk, Wootton, Bedford MK43 9LB
G8 XNO P. Lambert 92 Winterslow Drive, Leigh Park, Havant PO9 5DZ
G8 XOB P. Ashcroft Fendley Corner, Common Lane, Harpenden AL5 5DW
G8 XOC D. Bird 119 Brandon Road, Watton, Thetford IP25 6LL
G8 XOE B. Baker Linden Lea, Fivehead, Taunton TA3 6PU
G8 XOM P. Cook Orchard Cottage New Road, Elmswell IP30 9BS
G8 XOR D. Sparrow 23 Tranmere Grove, Ipswich IP1 6DU
G8 XOU I. Spinks 30 Lime Tree Walk, Watton, Thetford IP25 6EU
G8 XOV L. Sedgwick 28 Fairhaven Road, Redhill RH1 2LA
G8 XOX R. Sneath 16 Wavish Park, Torpoint PL11 2HJ
G8 XPB K. Chadwick 5 Mason Close, Great Sutton, Ellesmere Port CH66 2GU
G8 XPD M. Dawkins11 3 Prestwick Terrace, Whitminster, Gloucester GL2 7PA
G8 XPQ B. Whitehead 17A Home Close, Histon, Cambridge CB24 9JL
G8 XPZ S. Lovell 98B Baker Road, Newthorpe, Nottingham NG16 2DP
G8 XQA P. Lineham 10 Streetsbrook Road, Shirley, Solihull B90 3PL
G8 XQB M. Lees 127 Mayfield Gardens, London W7 3RA
G8 XQD T. Miller 35 Caudle Avenue, Lakenheath, Brandon IP27 9AU
G8 XQH E. Massey 21 Arlington Drive, Macclesfield SK11 8QL
G8 XQL J. Alcock Shirley Cottage, Welland Road, Worcester WR8 0SJ
G8 XQN A. Cleave 37 Alledge Drive, Woodford, Kettering NN14 4JQ
G8 XQS M. Chapple 10 Alderley Heights, Lancaster LA1 2HR
G8 XQT C. Dodds Cornhill, Magpie Close, Flackwell Heath, High Wycombe HP10 9DZ
G8 XQZ Dr G. Farmer 39 Plough Rise, Upminster RM14 1XR
G8 XRG R. Margetts Mowbray, Arbor Road, Leicester LE9 3GE
G8 XRL R. Mills 131 High Road East, Felixstowe IP11 9PS
G8 XRP R. Pryor 27 Hollickwood Avenue, London N12 0LS
GW8 XRR J. Nicholson 85 Llay New Road, Llay, , Wrexham, LL12 0PS
G8 XRS G. Nuttall 120 Cleevelands Avenue, Cheltenham GL50 4PX
G8 XRW D. Owen 18 Bushey Close, Capel St. Mary, Ipswich IP9 2HW
G8 XRX A. O'Kavanagh 24 Greentrees Crescent, Sompting BN159SY
G8 XSA W. Ash 53 Waxland Road, Halesowen B63 3DN
GI8 XSB F. Aughey 239 Bridge Street, Portadown, Craigavon BT63 5AR
G8 XSD J. Atkinson 8 Grove End, Luton LU1 5PF
G8 XSF M. Ainley 152 Bourne View Road, Huddersfield HD4 7JS
G8 XST W. Butchers 12 Church Road, St. Marychurch, Torquay TQ1 4QY
G8 XSU M. Bond 58 St. Pauls Street, Clitheroe BB7 2LS
GI8 XSY K. Steenson 108 Morgans Hill Road, Cookstown BT80 8BW
G8 XTD R. Cavendish 66 Coachmans Drive, Liverpool L12 0HX
G8 XTE P. Connor 20 Longfield, Lutton, Ivybridge PL21 9SN
G8 XTJ J. Fitzgerald 21 Honor Road, Prestwood, Great Missenden HP16 0NJ
G8 XTO R. Evans 53 Dolphin Court Road, Paignton TQ3 1AG
G8 XTR P. Emmans 16 Foresters Close, Rags Lane, Waltham Cross EN7 6TF
G8 XTU M. Fowler 28 St. Hildas Road, Doncaster DN4 5EE
G8 XTW P. Seaford 14 Nevis Close, Leighton Buzzard LU7 2XD
G8 XUB N. Reddish 15 Drakes Close, Redditch B97 5NG
G8 XUE L. Radcliffe 25 Oakleigh Drive, Codsall, Wolverhampton WV8 1JP
G8 XUH J. Pearson 14 Gorse Close, Brampton Bierlow, Rotherham S63 6HW
GM8 XUK D. King 59 South Knowe, Crossgates, Cowdenbeath KY4 8AW
G8 XUL D. James 19 Estuary Drive, Felixstowe IP11 9TL
GW8 XUM P. Jeavons Manora, Penisarwaun, Caernarfon LL55 3PW
G8 XUN M. Hickman 24 Calverley Road, Kings Norton, Birmingham B38 8PW
G8 XUU E. White 97 Fillongley Road, Meriden, Coventry CV7 7LW
G8 XUW D. Shields 54 Wildmoor Lane, Catshill, Bromsgrove B61 0PA
G8 XVJ E. Gedvilas 23 Pennington Drive Newton Le Willows, Newton-le-Willows WA12 8BA
G8 XVO C. Hetherington 23 Falkland Court, Braintree CM7 9LL
G8 XWH C. Langham 27 Fyfield Avenue, Swindon SN2 5ED
G8 XWR M. Izzard 17 Greenfields Avenue, Alton GU34 2ED
G8 XXA J. Harrison 10 Gaia Lane, Lichfield WS13 7LW
G8 XXC P. Prince 21 Ash Close, Appley Bridge, Wigan WN6 9HU
G8 XXG S. Richardson 52 Nailsea Park, Nailsea, Bristol BS48 1BB
G8 XXI J. Akines 105 Sutcliffe Avenue, Grimsby DN33 1EZ
G8 XXJ J. Allchin 40 Vale Road, Seaford BN25 3EZ
G8 XXM C. Beecher 6 Brices Meadow Shenley Brook End, Milton Keynes MK5 7HB
G8 XXU M. Caulton 115 Delves Green Road, Walsall WS5 4NH
G8 XXV G. Clarke 28 Little Potters, Bushey WD23 4QT
G8 XXZ P. Grace 6 Davis Grove, Yardley, Birmingham B25 8LQ
G8 XYA N. Southorn 20 Bratton Avenue, Devizes SN10 5BA
G8 XYJ M. Porter 20, Southfield Road, Much Wenlock TF13 6AX
G8 XYQ D. Stanford Laurel House, Top Road, Woodbridge IP13 6JF
G8 XYR R. Tiller Wayside, Ockley Lane, Hassocks BN6 8NU
G8 XYS R. Travett 39 Amwell Road, Cambridge CB4 2UH
G8 XZB J. Payne 25 Ringwood Road, Bath BA2 3JL
G8 XZC A. Pinder 2 Eleanor Road, Woodlands, Harrogate HG2 7AJ
G8 XZQ M. Fowler 1 Mayfields, Shefford SG17 5AU
G8 XZX J. Tyler 30 Northons Lane, , Holbeach PE12 7PZ
G8 XZZ J. Tabberer 22 Ogden View Close, Halifax HX2 9LY
G8 YAE C. Wenn 11 Bysouth Close, Ilford IG5 0XN
GM8 YAQ R. Wroblewski 1 Normandy Place, Rosyth, Dunfermline KY11 2HJ
G8 YAS A. Miller 113 West Front Road, Bognor Regis PO21 4TB
G8 YAT I. Naylor 8 Churchill Close, Uttoxeter ST14 8BB
G8 YAU R. Newton Cascades, Top Road, Brigg DN20 0NN
G8 YAZ G. Oates 21 Churchill Mansions, Cooper Street, Runcorn WA7 1DH
G8 YBC S. Ford 14 Merryfield, Fareham PO14 4SF
G8 YBH A. Bristow 2 Nursery Cottages, Staplehurst Road, Tonbridge TN12 9BS
G8 YBO R. Colebrook 21 Hillclose Avenue, Darlington DL3 8BH
G8 YBR I. Davidson 1 Mooracre Lane, Bolsover, Chesterfield S44 6ER
G8 YBT N. Dilley 26 Linhey Close, Kingsbridge TQ7 1LL
GI8 YBU M. Dunne 26 Duncreggan Road, Londonderry BT48 0AD
G8 YBZ M. Hampson 7 Merryfield Close, Bransgore, Christchurch BH23 8BS
G8 YCI A. Lewis 8 Arundel Road, Hartford, Huntingdon PE29 1YW
G8 YCJ G. Vickery 56 Hilden Park Road Hildenborough, Tonbridge TN11 9BL
G8 YCK K. Tomlinson 27 Brackens Lane, Alvaston, Derby DE24 0AQ
G8 YCL S. Turner-Smith 26 Ash Church Road Ash, Aldershot GU12 6LX
G8 YCP J. Sergeant 5 Jedburgh Close, North Shields NE29 9NU
G8 YCQ N. Storey 15 Tower Avenue, Upton, Pontefract WF9 1ED
G8 YDB R. Merry Havenwood, Oak Farm Lane, Sevenoaks TN15 7JU
G8 YDC J. Jebb 30 Runnymede, Nunthorpe, Middlesbrough TS7 0QL
G8 YDE S. Inns 11 Hodds Wood Road, Chesham HP5 1SQ
G8 YDJ M. Alexander 101 Richmond Street, Stoke-on-Trent ST4 7DZ
GW8 YDR D. Catleugh 48 Tyn Y Celyn, Glan Conwy, Colwyn Bay LL28 5NN
GM8 YEC P. Eunson Sandwick Cottage, Bridge End, Burra Isle ZE2 9LD
G8 YEF A. Eaton 16 Wood Road, Godalming GU7 3NN
G8 YEJ J. Glover1 5 Meadows Rise, Wymondham LE14 2AP
G8 YEN M. Stevens 29 Luscombe Close, Ipplepen, Newton Abbot TQ12 5QJ
G8 YEO Yeovil ARC c/o R. Spirrell 32 Churchfield Drive, Castle Cary BA7 7LA
G8 YEP B. Meyer 6 Barrington Road, Sutton SM3 9PP
G8 YEQ N. Littleboy 22 Sylvaner Court, Vyne Road, Basingstoke RG21 5NZ
G8 YFA A. Regnart 3 Preston Avenue, North Shields NE30 2BW
G8 YFH D. Oliver 30 Lipscombe Rise, Alton GU34 2HP
G8 YFK J. Mason 80 Swallow Drive, Milford On Sea, Lymington SO41 0XG
G8 YFP J. Wells 24 Tomlinson Way Ruskington, Sleaford NG34 9TW
GI8 YGG P. Foley 5 Woodland Drive, Cookstown BT80 8PL
GM8 YGI P. Sime 29 Huntingtower Road, Baillieston, Glasgow G69 7BH
G8 YGK W. Standing 72 Ivydore Avenue, Durrington, Worthing BN13 3JD
G8 YGM D. Southward 3A Carnoustie Close, West Derby, Liverpool L12 9NE
G8 YGO G. Tarr 40 The Garth, Coniston LA21 8EQ
G8 YGT B. Senior 1 Bedale Close, Coalville LE67 3BE
G8 YHF S. Kenyon 8 Dunedin Gardens, Ferndown BH22 9EQ
G8 YHO R. Hodge 4 Hidcote Road, Kenilworth CV8 2PP
G8 YIG C. Fawcett 11 Hurst Close, Glossop SK13 8UF
GM8 YIK A. Robson 69 Redburn Road, Prestonpans EH32 9NA
G8 YIN S. Wood 246 Rush Green Road, Romford RM7 0LA
GI8 YJD R. Perver 6 Gransha Road, Bangor BT20 4TG
GI8 YJF D. Roxburgh 5 Forestbrook Park, Rostrevor, Newry BT34 3DX
G8 YJL C. Pogmore Sunnybanks, West End, Barlborough, Chesterfield S43 4HE
GW8 YJN A. Price 45 Baring Gould Way, Haverfordwest SA61 2SB
G8 YJQ P. Holt Flat 13, Norbiton Hall, Kingston upon Thames KT2 6RA
G8 YJS G. Hammond 21 Cawston Road, Reepham, Norwich NR10 4LU
G8 YJT C. Jarvis 516 Kingsbury Road, Erdington, Birmingham B24 9NF
GI8 YJV P. Lloyd 18 Demesne Road, Holywood BT18 9NB
G8 YJZ Rvd. P. Rayson 2 Exmouth Gardens Horton Heath, Eastleigh SO50 7LL
G8 YKE C. Andrew 17 St. James Close, Kettering NN15 5HB
G8 YKG M. Armour 22 Langcliffe Close, Culcheth, Warrington WA3 4LR
G8 YKM A. Browne 140 Tongham Road, Aldershot GU12 4AT
G8 YKO S. Bardsley 73 Highlands, Royton, Oldham OL2 5HL
G8 YKS D. Barton Brook House, Hemington Road, Polebrook, Peterborough PE8 5LS
GM8 YKT E. Brumby 141 Morriston Road, Elgin IV30 4NB
G8 YKV A. Cragg 28 Damian Way, Hassocks BN6 8BJ
G8 YKY D. Canham 82 Rugby Road, Binley Woods, Coventry CV3 2AX
G8 YLA R. Cato Orrell House, Winterpit Lane, Horsham RH13 6LZ
G8 YLC R. Cura 105A Bexley Road, Erith DA8 3SN
GW8 YLK B. Evans Mynyddmelin, Pontfaen, Fishguard SA65 9SL
G8 YLM M. Farnworth 16 Lees Court, Ribble Avenue, Darwen BB3 0HW
G8 YLR Meopham Parish RC c/o R. Foss 4 Sandy Close, Wimborne BH21 2NG
G8 YLS D. Fox 4 Lacey Grove, Annesley, Nottingham NG15 0EG
G8 YMD J. Carins 26 Roman Way, St. Margarets-At-Cliffe, Dover CT15 6AH
G8 YMM P. Stevenson 6 Dighton Gate, Stoke Gifford, Bristol BS34 8XA
G8 YMN M. Shorter 10 Lodgefield Road, Chestfield, Whitstable CT5 3RF
G8 YMR A. Snow 28 The Maltings Station Street, Tewkesbury GL20 5NN
G8 YMS P. Swarbrook 14 The Willows, Leek ST13 8XF
G8 YMT Llanelli ARC c/o D. Smith 7 Peterdale Road, Brimington, Chesterfield S43 1JA
G8 YMU L. Shaw 108 Brookvale Road, Solihull B92 7JA
G8 YMW A. Sneath 21 Garrick Close, Lincoln LN5 8TG
G8 YMZ J. Trent The Hollies Bourne Road, West Bergholt, Colchester CO6 3EP
G8 YNB A. Taylor 4 Tarrant Drive, Harpenden AL5 1RP
G8 YNC P. Tuck 30 Brownlow Road, New Southgate, London N11 2DE
G8 YNE S. Horner 15 Newhouse Road, Huddersfield HD2 1ED
G8 YNF G. Holman 62 The Ridge, Kennington, Ashford TN24 9EU
G8 YNG A. Hall 33 Deanwood Road, Dover CT17 0NT
G8 YNH M. Hall 20 Cubitt House, Black Bull Road, Folkestone CT19 5SH
G8 YNK M. Higton 12 Chestnut Avenue, Mickleover, Derby DE3 9FT
G8 YOC M. Witchard 110 Bradley Road, Huddersfield HD2 1QY
G8 YOE C. Victory Pennros, Treworgans, Cuberts TR8 5HH
G8 YOG J. Woodard 213 Leicester Road, Ibstock LE67 6HP

G8 YOK J. Ward 3 Sherbourne Close, Poulton-le-Fylde FY6 7UB
G8 YOX A. Munday 77 Postland Road, Crowland, Peterborough PE6 0JB
G8 YOY M. Maxwell 962 Bury Road, Bolton BL2 6NX
G8 YPH T. Mcknight 31 Cavendish Road, Eccles, Manchester M30 9EE
G8 YPK V. Maddex 140A Kents Hill Road, Benfleet SS7 5PH
G8 YPL P. Martin 3 Grange Avenue, Southport PR9 9AH
G8 YPN P. Lutman 47 Conan Drive, Richmond DL10 4PQ
G8 YPQ M. Waring Woodside Cottage, Mansfield Road, Ollerton, Newark NG22 9DX
GW8 YPR R. Williams 54 Woodlands Avenue, Talgarth, Brecon LD3 0AT
G8 YPV G. Williams 54 Greenacre, Wembdon, Bridgwater TA6 7RF
G8 YPY D. Wilson 120 Poulton Road, Fleetwood FY7 7AR
G8 YQA D. Arnold 10 Shaw Place, Leek ST13 6ES
G8 YQC M. Beetlestone 19 Tenbury Road, Birmingham B14 6AD
G8 YQH V. Carter 69 Angela Crescent, Horsford, Norwich NR10 3HE
G8 YQN P. Gebbie 1 Burniston Road, Scarborough YO12 6PG
G8 YQO D. Henderson Reverie Pennys Lane, Margaretting, Essex CM4 0HA
G8 YQU G. Lenihan 28 Paddock Crescent, Sheffield S2 2AR
GM8 YRE J. Firth 6 Upper Burnside Drive, Thurso KW14 7XB
G8 YRF R. Foxley 20B Alder Copse, Horsham RH12 1LD
G8 YRG G. Williams 2 Windsor Court Watton, Thetford IP25 6XB
G8 YRL B. Trim Endon Cottage, 63B Rose Street, Wokingham RG40 1XS
GM8 YRT W. Stewart 20 Corrie Place, Scone, Perth PH2 6QE
G8 YRW R. Williams 29 Woodfield Road, Bude EX23 8JB
GM8 YRX E. Saxon 73 Upper Burnside Drive, Thurso KW14 7XB
G8 YSA Polish ARC c/o P. Powers 5 Bracken Close, Hugglescote, Coalville LE67 2GP
G8 YSH L. Jannetta 1 Lake Road, Hadston, Morpeth NE65 9TF
G8 YSJ W. Bannerman 3 The Cornfield, Langham, Buttercups, Holt NR25 7DQ
G8 YTF G. Mcgowan 281 Ashgate Road, Chesterfield S40 4DB
GI8 YTH S. Moore 7 Cyprus Avenue, Belfast BT5 5NT
GW8 YTO A. Ham 46 Celtic Way, Rhoose, Barry CF62 3FT
G8 YTP S. Holgate 91 Valley Road, Stockport SK4 2DB
G8 YTR S. Higgs 5 Lawnswood Close, Cowplain, Waterlooville PO8 8RU
G8 YTU F. Adams 27 Challenger Close, Malvern WR14 2NN
G8 YTX K. Bagshaw 36 St. Peters Road, Buxton SK17 7DX
G8 YTZ J. Cockett 2a Priory Avenue, Orpington BR5 1JF
G8 YUC M. Forrester 7 Cottage Gardens, Bamber Bridge, Preston PR5 6AG
GM8 YUI G. Mcclintock 13 St. Andrews Drive, Gourock PA19 1HY
GW8 YUJ J. Milburn Orme View, Anglesey LL73 8PE
G8 YUK A. White 10 Stott Drive, Urmston, Manchester M41 6WA
GM8 YUM G. Walker 24 George Street, Cellardyke, Anstruther KY10 3AU
G8 YUO M. Taylor 4 Yew Tree Court, Botley Road, Swanwick, Southampton SO31 1EA
G8 YUP West Wales Radio Group c/o B. Stevens 77 Dean Lane, Hazel Grove, Stockport SK7 6EJ
G8 YUR M. Robelou 12 Cooks Drove, Earith, Huntingdon PE28 3QG
G8 YVC M. Smith 31 Burringham Road, Scunthorpe DN17 2BD
G8 YVM N. Matthes 24 Albany Road, Fleet GU51 3LY
G8 YVP M. Nicholson 33 Painshawfield Road, Stocksfield NE43 7PX
G8 YVQ Dr C. Harper Chusan, Farley Court, Church Road, Reading RG7 1TT
G8 YVS R. Hillan 128A Bridge Street, Deeping St. James, Peterborough PE6 8EH
G8 YVW C. Stacey 157 Ormond Road, Sheffield S8 8FT
GI8 YWE M. Anderson 17 Leydene Court, Lisburn BT28 3LL
G8 YWJ A. Frost 76 Tregrea Estate, Beacon, Camborne TR14 7SU
G8 YWK W. Gleave 6 Sidlaw Avenue, Chester le Street DH2 3DD
G8 YWL G. Pitt 17 Penfound Gardens, Bude EX23 8FF
G8 YWQ D. Petch 12 Avon Way, Portishead, Bristol BS20 6JQ
G8 YXI D. Shemeld 13 Arran Road, Sheffield S10 1WQ
G8 YXJ R. Skells 31 Perry Road, Leverington, Wisbech PE13 5AE
G8 YXQ D. Chatterton 27 Victoria Road, Folkestone CT19 5AT
G8 YXR E. Ferris Karravas, Osborne Road, Deal CT14 8BT
G8 YXZ R. Dominy 8 Meadow Road, Claygate, Esher KT10 0RZ
G8 YYA H. Duesbury 4 Harbour View Close, Poole BH14 0PF
G8 YYC G. Miller 93 Shepherds Grove Park Stanton, Bury St. Edmunds IP31 2BN
GW8 YYF K. Jones 3 Penffordd, Pentyrch, Cardiff CF15 9TJ
G8 YYL Lady G. Johnson Kilmurry House, Kilmurry Estate, Fermoy BLANK Ireland
GI8 YYM I. Ferris 48 Abbey Gardens, Belfast BT5 7HL
G8 YYW M. Freeman 2 Poolthorne Farm Cottage, Cadney, Brigg DN20 9HU
G8 YYX A. Layton 7 Higher Saxifield, Harle Syke, Burnley BB10 2HB
G8 YZA K. Sawday 15 Moorland View, Buckfastleigh TQ11 0AF
G8 YZC R. Smith 86 Manor Road, Borrowash, Derby DE72 3LN
G8 YZF M. Bishop 6 Tiverton Close, Kingswinford DY6 8PD
G8 YZL P. Thackeray 19 Moneyfly Road, Verwood BH31 6BL
G8 YZT P. Wing 14 Huntingdon Road, Kempston MK42 7EX
G8 YZY D. Spencer 28 Watery Lane, Minehead TA24 5NZ
G8 ZAD R. Mantle 37 Willis Road, Stockport SK3 8HQ
G8 ZAJ C. French 26 Wood Street, Ash Vale, Aldershot GU12 5JG
GM8 ZAK H. Gemmell 53 Southesk Avenue, Bishopbriggs, Glasgow G64 3AD
G8 ZAT J. Haslip 18 Downsview Drive, Wivelsfield Green, Haywards Heath RH17 7RW
G8 ZAU D. Hoodless 21 Meadow Close, Eastwood, Nottingham NG16 3DQ
G8 ZAX R. Rees 69 Pewley Way, Guildford GU1 3PZ
G8 ZBC C. Lucas 8 Hawker Close, Broughton, Chester CH4 0SQ
G8 ZBJ W. Sheldon 15 Hawthorn Place, Walsall WS2 0HZ
G8 ZBN T. Nye 18 Kingsway, Chandler'S Ford, Eastleigh SO53 2FE
G8 ZCJ J. Skidmore 55 Elmsleigh Road, Heald Green, Cheadle SK8 3UD
G8 ZCK C. Wilson 19 Chace Avenue, Potters Bar EN6 5LX
GM8 ZCS A. Westerman 5 Eldon Gardens, Bishopbriggs G64 2EU
GI8 ZDB R. Logue 46 Brunswick Road, Londonderry BT47 5SZ
G8 ZDS P. Hocking 10 South Terrace, Camborne TR14 8ST
G8 ZDT P. Langford 1 Kingaby Gardens, Rainham RM13 7PH
G8 ZEE A. Hudson 1 Laburnum Court, Cheltenham GL51 0XE
GW8 ZEI E. Whitham 44 Tyddyn Isaf, Menai Bridge LL59 5DA
G8 ZEK P. Jacobi Highbury, Furzehill, Wimborne BH21 4HD
GM8 ZEQ M. Smith Haremuir Bungalow, Benholm, Montrose DD10 0HX
G8 ZES P. Street 50 Dickson House, Ridgway Road, Stoke on Trent ST1 3BA
G8 ZEV C. Hartt 9 Laura Grove, Paignton TQ3 2LR
G8 ZEW A. Joy 15 Wymersley Close, Great Houghton, Northampton NN4 7PT
G8 ZEX S. Lauritson 42 Woodstock Road, Kingswood, Bristol BS15 9UE
G8 ZFD C. Askin 54 York Road, Hull HU6 9RA
G8 ZFI P. Bryant 21 Devonshire Close, Stevenage SG2 8RY
G8 ZFL A. Butcher 4 Maple Close, Bristol BS30 9PX
G8 ZFQ M. Kanelis 57 Ringwood Avenue, Redhill RH1 2DY
G8 ZFS P. Wiley 36 Hungate Lane, Hunmanby, Filey YO14 0NN
G8 ZFT R. Thompson 329 Prestbury Road, Prestbury, Cheltenham GL52 3DF
G8 ZFU G. Taylor 8 Ullathorne Road, Streatham, London SW16 1SN
GM8 ZFW J. Morris 1 Wealthyton Cottages Keig, Alford AB33 8BH
G8 ZFX P. Blake 35 Kings Court 71-76 Wright Street, Hull HU2 8JR
GI8 ZFZ D. Alexander 33 Greenan Road, Newry BT34 2PJ
GM8 ZGC C. Dowers 31 Jenny Gray Place, Lochgelly, Fife, Lochgelly KY5 9BF
G8 ZGF R. Mackrell 17 Townfield Avenue, Worsthorne, Burnley BB10 3JG
G8 ZGK A. Mockford 58 Wendover Heights, Wendover HP22 6PH
G8 ZGM S. Berks 14 Austen Way, Hastings TN35 4JH
G8 ZGQ A. Longuet 10 Severnmead, Grovehill, Hemel Hempstead HP2 6DX
G8 ZGS J. Holden 128 Greenways, Norwich NR4 6HA
G8 ZGY R. Bareham 49 Wharf Road, Crowle, Scunthorpe DN17 4HU
G8 ZHA R. Morrall 32 Broadstone Avenue, Walsall WS3 1EW
G8 ZHN P. Gibbons 13 Canon Park, Berkeley GL13 9DF
G8 ZHR N. Lawes 87 Glebelands, Crayford, Dartford DA1 4RY
G8 ZHS P. Lester 1C Eastwood Road, London E18 1BN
GI8 ZHW J. Mcdonnell 6 Sandhurst Park, Bangor BT20 5NU
G8 ZIA A. Bowman Evergreen, Durham Road, Stockton-on-Tees TS21 3LT
G8 ZIC C. Harrison 2 Bridgemere Close, Radcliffe, Manchester M26 4FS
G8 ZIH J. Eady Pytchley Lodge, Pytchley, Kettering NN14 1EE
G8 ZIK E. Serwa 102 Cornwall Road, Wolverhampton WV6 8UZ
GW8 ZIL I. Bell 102 Ewenny Road, Bridgend CF31 3LN
G8 ZIP K. Lake 22 Chapmans Close Stirchley, Telford TF3 1ED
G8 ZIW G. Ludar-Smith 2 Springmead, Queenborough Lane, Braintree CM77 7PX
G8 ZIY P. Eyre 27 Holborn View, Codnor, Ripley DE5 9RB
G8 ZJH B. Mccourt 3 Littlemore Close, Upton, Wirral CH49 4GS
G8 ZJK R. Cole Flat 6, Barton Court, 19 Southwood Road, Hayling Island PO11 9PS
G8 ZJO S. Tomschey 36, Lord Lytton Ave, Coventry CV2 5JW
G8 ZK West Wight RS c/o C. Archer 118 Cator Lane, Beeston, Nottingham NG9 4BB
GM8 ZKF D. Robson 6 Ladywood Estate, Milngavie, Glasgow G62 8BE
GM8 ZKN I. Diment 22 Academy Place, Bathgate EH48 1AS
GM8 ZKU S. Hawley Hill Of Ardiffery, Hatton, Peterhead AB42 0TB
G8 ZLF T. Gilleard 3 Paul Crescent, Humberston, Grimsby DN36 4DF
G8 ZLL I. Thomas 30 Alcot Close, Crowthorne RG45 7NE
G8 ZLN P. Thompson 81 Ashmead Road, Banbury OX16 1AA
GW8 ZLT M. Chambers 3 Manod Road, Blaenau Ffestiniog LL41 4DD
G8 ZLU M. Wright 17 Colwyn Crescent, Stockport SK5 7LL
G8 ZMC A. Mccalden 127 Kings Road, Godalming GU7 3EU
G8 ZME M. O'Toole Daffodil Cottage, Dunsmore, Aylesbury HP22 6QH
GM8 ZMF M. Osborn 3 Lovers Lane, South Queensferry EH30 9UP
G8 ZMG S. Watson 61 Glenview Road, Shipley BD18 4AR
G8 ZMH K. Robinson 33 Cranford Road, Northampton NN2 7QU
G8 ZML B. Ewart 36 Sycamore Rise, Holmfirth HD9 7TJ
G8 ZMM R. Bunney 35 Grayling Mead, Romsey SO51 7RU
G8 ZMQ P. Burnley 45 Ashwell Road, Heaton, Bradford BD9 4AX
G8 ZNB A. Harris 55 Frenchgate, Richmond DL10 7AE
G8 ZNK G. Barnes 18, Wellesley Avenue, Goring-By-Sea, Worthing, West Sussex BN12 4PN
GW8 ZOE S. Trott 6 Mounton Drive, Chepstow NP16 5EH
G8 ZOJ G. Barrett The Old Chapel, 5 Tappers Lane, Bridgwater TA6 6SJ
G8 ZOO J. Molinghen 16 Dumpers Lane, Chew Magna, Bristol BS40 8SS
G8 ZOV Dr R. Nicholson 24 Barnmead, Haywards Heath RH16 1UZ
GM8 ZOW P. Oram 24 John Smith Place, Kelty KY4 0NL
G8 ZOY G. Page 1A Montagu Gardens, Wallington SM6 8EP
G8 ZPD P. Davies 46 Spring Street, Colley Gate, Halesowen B63 2SZ
G8 ZPE P. Cooper The Bungalow, Clopton, Kettering NN14 3DZ
G8 ZPH D. Bucknell 46 Heath Row, Bishop's Stortford CM23 5DE
G8 ZPO R. Blackwell 46 Wyatts Drive, Thorpe Bay, Southend-on-Sea SS1 3DG
G8 ZPW A. Martin 23 Portfield Road, Christchurch BH23 2AF
G8 ZQA P. Stonebridge 207 Henley Road, Ipswich IP1 6RL
G8 ZQB J. Smith 7 Mill Hill Close, Whetstone, Leicester LE8 6NF
G8 ZQG S. Wood 8A Glendale Avenue, Glenfield, Leicester LE3 8GF
G8 ZQJ D. Young 9 Larchfield House, Highbury Estate, London N5 2DE
G8 ZQM K. Pascoe 21 Cotswold Avenue, Sticker, St. Austell PL26 7ER
GM8 ZQY S. Frey 2 Balgeddie Gardens, Glenrothes KY6 3QR
G8 ZRD I. Gilzean 35 Pieces Terrace, Waterbeach, Cambridge CB25 9NE
G8 ZRE D. Hewitt 31 Broadmead, Vicars Cross, Chester CH3 5PT
G8 ZRM R. Myers 9 Romney Road, Rottingdean, Brighton BN2 7GG
G8 ZRN G. John 29 Park Road, Northville, Bristol BS7 0RH
G8 ZRQ R. Knight 2 Stag Place, Pocklington, York YO42 2FN
G8 ZRU D. Moger 47 Powys Grove, Banbury OX16 0UG
G8 ZRV G. Sargant 9 Orchard Way, Reigate RH2 8DS
G8 ZSD I. Worthington 7 Bowness Close, Gamston, Nottingham NG2 6PE
G8 ZSK A. Allcock 30 Clyde Grove, Crewe CW2 8NA
G8 ZSM L. Barlow 4 Bucknell Place, Thornton-Cleveleys FY5 3HZ
G8 ZSP A. Blanchard 41 Deane Drive, Galmington, Taunton TA1 5PQ
G8 ZSZ I. Dickinson 16 Heathfield Grove, Beeston, Nottingham NG9 5EB
G8 ZTB S. Fenn 21 Waarem Avenue, Canvey Island SS8 9DS
G8 ZTD J. Francis 9 Holland Close, Bognor Regis PO21 5TW
G8 ZTF J. Hargraves 321 Northway, Maghull, Liverpool L31 0BW
G8 ZTG J. Harman 20 Sunview Avenue, Peacehaven BN10 8PJ
G8 ZTM N. Ledeux 14 Jubilee Close, Cam, Dursley GL11 5JQ
G8 ZTN F. Lock Monks Rest, The Street, Charmouth, Bridport DT6 6PE
G8 ZTR J. Macdonald 74 Bradford Road, Boston PE21 8BJ
GM8 ZTV F. Millar 13 Edzell Park, Kirkcaldy KY2 6YB
G8 ZUF K. Rogers 36 Goodacre Road, Ullesthorpe, Lutterworth LE17 5DL
G8 ZUL R. Yates 16 Arnold Grove, Shirley, Solihull B90 3JR
G8 ZUU M. Smith 2 Newbury Close, Mapperley, Nottingham NG3 5QW
G8 ZUZ D. Unwin 1 Bentinck Close, Nuncargate, Nottingham NG17 9ET
G8 ZVI L. Hart 28A Dunton Road, Stewkley, Leighton Buzzard LU7 0HZ
G8 ZVK B. Ackroyd 91 Bulford, Wellington TA21 8DH
G8 ZVM M. Atkinson Menamber Farm, Trenear, Helston TR13 0HE
G8 ZVS R. Bird 80 Clearmount Road, Weymouth DT4 9LE
G8 ZVX A. Breeds 26 Heighton Road, Newhaven BN9 0JU
G8 ZVZ I. Collins Knapp Cottage, Pixley, Ledbury HR8 2QB

G8 ZWA P. Collins 40 Shacklegate Lane, Teddington TW11 8SH
G8 ZWC L. Curtis 34 Gaisford Road, Worthing BN14 7HW
G8 ZWF R. Cowling 20 Claremont Hill, Shrewsbury SY1 1RD
G8 ZWN M. Davies Sunningdale, Sulhamstead Hill, Reading RG7 4DE
G8 ZWU K. Graham 670 Stafford Road, Ford Houses, Wolverhampton WV10 6NW
G8 ZXI R. Nixon 18 Normill Terrace, Aylesbury Road, Aston Clinton, Aylesbury HP22 5AG
G8 ZXL Lord R. Pretty 52 Queens Road, Hersham, Walton-on-Thames KT12 5LW
GM8 ZXQ J. Mcdermott Milking Green Gate, Eliock, Sanquhar DG4 6LD
G8 ZXT J. Marshall 58 Sandbed Court, Leeds LS15 8JJ
G8 ZXU P. Mcguinness 83 Beaconsfield, Telford TF3 1NH
G8 ZXY W. Mason 365 Heath Road South, Birmingham B31 2BJ
G8 ZXZ D. Holmes 17 Spring Hall Close, Shelf, Halifax HX3 7NE
G8 ZYC Zycomm Elect Lt c/o M. Sneap Ivy Farm Bungalow, Farm Close, Pentrich, Ripley DE5 3RR
G8 ZYH E. Hitch 35 Hawthorndene Road, Hayes, Bromley BR2 7DY
G8 ZYI N. Hitch 1B Greenlands, Platt, Sevenoaks TN15 8LL
G8 ZYM I. Hammond 1 Old Rectory Close, Barham, Ipswich IP6 0PY
G8 ZYR P. Hodgkinson 25 Polisken Way, St. Erme, Truro TR4 9RB
G8 ZYT S. Higlett 28 Oak Crescent, Potton, Sandy SG19 2PY
G8 ZZB D. Kellet April Cottage, 10 Yorkdale Drive, Selby YO8 9YB
G8 ZZF A. Leatherbarrow 33 Chester Avenue, Sale M33 4NS
G8 ZZG T. Lock 40 Chertsey Road, Ashford Common, Ashford TW15 1SQ
G8 ZZK D. Lee 14 Woodview Close, West Kingsdown, Sevenoaks TN15 6HP
G8 ZZL P. Lake 166 Burrs Road, Clacton on Sea CO15 4LH
G8 ZZR P. Vince 19, Links Road, Ashtead KT21 2HB
G8 ZZS D. Vaughan Orchard Farm House, Framsden, Stowmarket IP14 6HD
G8 ZZT J. Tonks Flat, 3 Greystone Passage, Dudley DY1 1SL
G8 ZZV A. Tye 3 Parkwood Court, Forest Park, Nottingham NG6 9FB
G8 ZZW I. Shepherd 12 Grains Road, Delph, Oldham OL3 5DS
G8 ZZY A. Smart 101 Bardon Road, Coalville LE67 4BF

M0

M0 AAA Reading And District ARC c/o V. Robinson 4 Hilltop Road, Caversham, Reading RG4 7HR
M0 AAC P. Bergin 15 Monks Way, Harmondsworth, West Drayton UB7 0LE
M0 AAD M. Stockton 37 Ney Street, Ashton-under-Lyne OL7 9NL
M0 AAF D. Hodgson 1B Court Farm Avenue, Epsom KT19 0HD
M0 AAG A. Salt 1 Chantry Close, Harrow HA3 9QZ
M0 AAK M. Pearson 56 Parkwood Green, Parkwood, Gillingham ME8 9PP
M0 AAM R. Armstrong 71 Bradshaw View, Queensbury, 71 Bradshaw View, Queensbury, Bradford, BD13 2FF
M0 AAN W. Glover 21 West End Way, Lancing BN15 8RL
M0 AAP I. Parker 23 Southdown Road, Benham Hill, Thatcham RG19 3BF
M0 AAR J. Kemp 394 Great Thornton Street, Hull HU3 2LT
M0 AAS J. Whittaker 10 Pownall Court, Wilmslow SK9 5QE
M0 AAV S. Bates 6 Foxdell, Northwood HA6 2BU
MI0 AAW S. Blakley 123 Mount Merrion Avenue, Belfast BT6 0FN
MI0 AAZ J. Anderson 1 Claragh Hill Drive, Kilrea, Coleraine BT51 5YR
M0 ABA T. Hackett 16 Eagle Way, Shoeburyness SS3 9RJ
MM0 ABB Nuneaton & District ARC c/o C. Kane 46 Hillmoss, Kilmaurs, Kilmarnock KA3 2RS
M0 ABC D. Cracknell 120 Woodhill, London SE18 5JL
MI0 ABD J. Mccarrison 11 Boretree Island Park, Newtownards BT23 7BW
M0 ABF K. Molyneux 220 Woodlands Holiday Homes Pk, Dowles Road, Bewdley DY12 3AE
M0 ABG A. Powell Crosstrees, Main Road, Theberton, Leiston IP16 4RX
M0 ABH R. Tyler 343 Broadwater Crescent, Stevenage SG2 8EZ
M0 ABI M. Lennon 12 Byron Road, Barton On Sea, New Milton BH25 7NX
MM0 ABJ C. Ewart 13 Princes St., Innerleithen EH44 6JT
M0 ABK M. Gray 19 Marsh View Newton, Preston PR4 3SX
MI0 ABN N. Crawford 10 White Mountain Road, Lisburn BT28 3QY
M0 ABO J. Valle Espin 203 Broadway, Horsforth, Leeds LS18 4HL
M0 ABP J. Barker Karma, 6 Acredykes, Bridlington YO15 1LY
M0 ABQ Sir W. Couse 68/29 Moo 3 Rattanapron Village, Tambon Khungkong, Chiang Mai 50230 Thailand
M0 ABT S. Little 46 Marine Drive, Seaford BN25 2RU
M0 ABU K. Simkin The Flat, Cinque Ports, 49 High Street, Seaford BN25 1PP
MW0 ABV P. Plummer Hill Road, Neath Abbey SA108ND
M0 ABY A. Soane 24 Nurseries Road, Wheathampstead, St. Albans AL4 8TP
M0 ABZ A. Allbright Greenacre, Carne Road, Newlyn, , Penzance TR18 5QA
M0 ACA E. Morley 91 Allerton Road, Stoke-on-Trent ST4 8PQ
M0 ACB E. Mcdonald 32 Butterwick Road, Messingham, Scunthorpe DN17 3PB
M0 ACC A. Dixon 17 Coppice Court, Weymouth DT3 5SA
M0 ACI Finningley ARS c/o S. Stacey Trehill, Trekenner, Launceston PL15 9PH
M0 ACK M. Jackson 121 Kiln Lane, Eccleston, St. Helens WA10 4RH
M0 ACM D. Forrest 166 Meadowcroft, Swindon SN2 7LE
MM0 ACN J. Green Tigh Callum, Culkein Drumbeg, Lairg IV27 4NL
MM0 ACR L. Skinner Manse Hall, Drumoak, Banchory AB31 5HA
MM0 ACT R. Skinner Manse Hall, Drumoak, Banchory AB31 5HA
M0 ACU M. Eddyvean 41 Liddell Road, Cowley, Oxford OX4 3QU
M0 ACV T. Bevan 6 Buttermere Grove West Auckland, Bishop Auckland DL14 9LG
M0 ACW Over the Hill DX Group c/o R. Williams Dyffryn Coed, Union Road, Coleford GL16 7QB
M0 ADA C. Rule Marconi House, Meaver Road, Helston TR12 7AH
M0 ADB N. Pringle 21 Petersmiths Drive, New Ollerton, Newark NG22 9RZ
M0 ADD A. Saville 1 Bellingham Close, Shaw, Oldham OL2 7UU
M0 ADG D. Morris 86 Richardson St., Carlisle CA2 6AG
M0 ADH North Warks Raynet Group c/o M. Mettam 190 Scotter Road, Scunthorpe DN15 7EQ
M0 ADL A. Jermaks 52 Laburnum Crescent, Allestree, Derby DE22 2GR
M0 ADO D. Kennard 16 Cantilupe Crescent, Aston, Sheffield S26 2AT
M0 ADR G. Galbraith 24 Airedale, Hadrian Lodge West, Wallsend NE28 8TL
M0 ADW R. Latham 47 Oldfield Park, Westbury BA13 3LQ
M0 ADY A. Grundy 21 Ribston Close, Shenley, Radlett WD7 9JW
M0 ADZ A. Gall 26 West Acre Drive, Norwich NR6 7HX
M0 AEC Dr S. Roper 15 St. Gerards Road, Solihull B91 1TZ
M0 AEE M. Clarke 19 Hodroyd Cottages, Brierley, Barnsley S72 9JA
M0 AEJ V. Trend 64 Shutlock Lane Moseley, Birmingham B13 8NZ
M0 AEK J. Sloan Flat 39, Colonel Stevens Court, 10A Granville Road, Eastbourne BN20 7HD
MW0 AEL S. Townsend 42 Burns Crescent, Bridgend CF31 4PY
M0 AEN M. Austen 11 Corn Avill Close, Abingdon OX14 2ND
M0 AEP G. Dawes 11 Ferriby Road, Barton-upon-Humber DN18 5LE
M0 AEQ M. Bardell 47 Calverleigh Crescent, Furzton, Milton Keynes MK4 1HY
M0 AEU F. Heritage 50 Laurel Close, North Warnborough, Hook RG29 1BH
MW0 AEV E. Jones 18 Madryn Terrace, Llanbedrog, Pwllheli LL53 7PF
MI0 AEX J. Smith 54A Blackstaff Road, Kircubbin, Newtownards BT22 1AF
M0 AEZ M. Herpe 25 Gordon Street Sutton In Craven, Keighley BD20 7EU
M0 AFC T. Boon 27 Meadowside Avenue, Clayton Le Moors, Accrington BB5 5XF
MW0 AFD S. Edwards 59 St. Andrews Road, Colwyn Bay LL29 6DL
M0 AFF F. Hallsworth Flat 2 Derwent Court Salt Ayre Lane, Lancaster LA1 5JP
M0 AFJ T. Hague 2 Winns Row, Godolphin Road, Helston TR13 8QH
M0 AFQ B. Eagleton 12 Park Court, Hadley, Telford TF1 6AD
M0 AFR P. Walker 1 Vicarage Lane, Fordington, Dorchester DT1 1LH
M0 AFS P. Whiteley 53 Sharp Lane, Almondbury, Huddersfield HD4 6SS
MI0 AFT J. Stewart 23 Swifts Quay, Carrickfergus BT38 8BQ
M0 AFV R. Rippin Lyndhurst, Myatts Field Harvington, Evesham WR11 8NG
M0 AFW C. Parkinson 4 Campion Drive, Killamarsh, Sheffield S21 1TG
M0 AFX D. Waters Station House, Station Road, Manningtree CO11 2LH
M0 AFY R. Ford 70 Jubilee Road, Darnall, Sheffield S9 5EH
M0 AFZ P. Nairne 137 Barden Road, Tonbridge TN9 1UX
M0 AGA K. Gunstone 67 Woodside, Sutton in Ashfield NG17 3EB
MW0 AGE J. Chinnock 22 Mill Road, Pyle, Bridgend CF33 6AP
M0 AGJ A. Bowker 120 Broomhouse Lane, Doncaster DN4 9DB
M0 AGP M. Weber Tall Chimneys, Malacca Farm, West Clandon, West Clandon GU4 7UG
M0 AGR M. Bray 2 Camborne Drive, Fixby, Huddersfield HD2 2NF
M0 AGS E. Smeaton 27 Sandringham Avenue, Burton-on-Trent DE15 9BJ
M0 AGT R. Markham 8 Railway Cuttings, Ilminster TA19 9FG
M0 AGU J. Shorthouse 84 Mount Pleasant, Ackworth, Pontefract WF7 7HU
MM0 AGV T. Mcguigan 42 Fidra Avenue, Burntisland KY3 0AZ
M0 AGW W. Mason 104 Chester Road, Poynton, Stockport SK12 1HG
M0 AGY M. Griffin 15 Victoria Road, St. Austell PL25 4QF
MM0 AHC M. Collins Redwoods, Barcaldine, Oban PA37 1SG
M0 AHD M. Carter 47 Grantham Green, Middlesbrough TS4 3QS
M0 AHF G. May 14 Tennyson Avenue, Dukinfield SK16 5DP
MI0 AHH C. Doris 9 Gortalowry Park, Cookstown BT80 8JH
MI0 AHI J. Doris 92 Coolnafranky Park, Cookstown BT80 8PW
M0 AHJ C. John 5 Highfield Gardens, Aldershot GU11 3DB
M0 AHS M. Nicholas 4 Chesterfield Mews, Chesterfield Road, Ashford TW15 3PF
M0 AHT W. Burt 5 Aged Miners Homes, Springwell Terrace, Hetton-Le-Hole, Houghton le Spring DH5 0BA
M0 AHV H. Banks 104 Viking Road, Bridlington YO16 6TB
M0 AHY G. Parsons Gull Cottage, Briar Close, Hastings TN35 4DP
M0 AHZ R. Brown 17 Ridgeway, North Seaton, Ashington NE63 9TJ
M0 AIB S. Budd 19 Queen Street, Worthing BN14 7BL
M0 AIC S. Deary 43 Old Road, Tintwistle, Glossop SK13 1LH
M0 AID K. Marsh Highgrove, Creech Heathfield, Taunton TA3 5EW
MW0 AIE R. Duncombe 1 Pennar Court, Pembroke Dock SA72 6NW
MI0 AIH D. Martin 34 Lower Kildress Road, Cookstown BT80 9RN
M0 AIJ C. Blake 30 Pine Tree Walk, Poole BH17 7EH
MM0 AIK Scottish DX Contest Club c/o B. Devlin 112 Benview, Bannockburn, Stirling FK7 0HJ
M0 AIQ B. Fryett 9 Trethiggey Crescent, Quintrell Downs, Newquay TR8 4LF
M0 AIS A. Benns 7 Brooklands Road, Burnley BB11 3PR
M0 AIT R. Holt 41 Garden Avenue, Ilkeston DE7 4DF
M0 AIY R. Carter 16 Holts Lane, Clayton, Bradford BD14 6BL
MW0 AIZ R. Ramm Mor Welir, Sarnau, Llandysul SA44 6QY
M0 AJB the North West 320 DX Club c/o A. Birch 6 Crescent Road, Wallasey CH44 0BQ
M0 AJC M. Mcinally Flat 7, 32-33 Edgar Road, Margate CT9 2EJ
M0 AJD M. Saxton Heathercrest, 7 Boston Road, , Horncastle LN9 6EY
MW0 AJH J. Donnell 42 Wentworth Crescent, Mayals, Swansea SA3 5HT
M0 AJI S. Nursey 9 Tydd Low Road, Long Sutton, Spalding PE12 9AR
M0 AJJ P. Olson 23 Dennett Close, Liverpool L31 5PD
MM0 AJQ J. Stone 1 Seafield Crescent, Bilston, Roslin EH25 9TD
M0 AJT C. Towle 116 Stainton Drive, Grimsby DN33 1JB
M0 AJX G. Jones 7 Hardwick View, Skegby, Sutton-in-Ashfield NG17 3BW
M0 AKD Dr G. Dublon 25 Carr Lane, Sandal, Wakefield WF2 6HJ
M0 AKE R. Johnson 24 Balmoral Avenue, Stanford-le-Hope SS17 7BD
M0 AKF M. Temblett 42 Westward Road, Bristol BS13 8DB
M0 AKI E. Woollen 6 Back Lane, Kington Magna, Gillingham SP8 5EL
M0 AKJ R. Hunt 8 Spicer Close, Cullompton EX15 1QD
M0 AKK S. Elden 124 Larchcroft Road, Ipswich IP1 6PQ
MM0 AKM J. Hood 88/2 Craighouse Gardens, Edinburgh EH10 5LW
M0 AKQ R. Gawan 39 The Filberts, Fulwood, Preston PR2 3YS
M0 AKR K. Daniels 122 Furzehatt Road, Plymstock, Plymouth PL9 9JT
M0 AKS R. Lusty 483 Bacup Road, Rossendale BB4 7JA
MI0 AKU Foyle And District ARC c/o T. Campbell 27 Silverbrook Park, Newbuildings, Londonderry BT47 2RD
MM0 AKX Oxford And District ARS c/o J. Ramsay 150 City Road, Dundee DD2 2PW
M0 AKY T. Money 119 Twyford Way, Canford Heath, Poole BH17 8SR
M0 AKZ R. Taylor 46 Crescent Road, Netherton, Dudley DY2 0NW
M0 ALB N. Hixson Flat 35, Milward Court, Reading RG2 7BG
M0 ALD J. Britten 10 Broadgate Avenue, Horsforth, Leeds LS18 5DT
M0 ALE P. Johnson 91 Highlands Road, Andover SP10 2PZ
M0 ALF R. Faithfull 5 Hadleigh Road, Portsmouth PO6 3RD
MW0 ALG D. Burge Ucheldir, Maenygroes, New Quay SA45 9TH
M0 ALH S. Case 5 Haldon Grove, Birmingham B31 4LN
M0 ALK R. Cook 3 Mill Close, Hartford, Huntingdon PE29 1YL
MM0 ALM D. Wood West Raedykes, Rickarton, Stonehaven AB39 3SY
MW0 ALN A. Lane 14 Hertford Place, Newport NP19 7SN
M0 ALO D. Hooper 21 High Street, Great Linford, Milton Keynes MK14 5AX
M0 ALQ G. Denby 14 Talman Grove, Stanmore HA7 4UQ

M0 ALR T. Knight 117 Ennerdale Road, Cleator Moor CA25 5LR
MI0 ALS E. Stanford 33 Glenview Gardens, Belfast BT5 7LY
M0 ALT I. Halliwell 61 Cliffe Road, Shepley, Huddersfield HD8 8AG
M0 ALX M. Feasey 4 Abbeydale, Carlton Colville, Lowestoft NR33 8WJ
MM0 ALY A. Brown 4 Averon Park, Blackburn, Aberdeen AB21 0LH
M0 ALZ T. Thompson 23 Oaklands, Paulton, Bristol BS39 7RP
M0 AMB B. Metcalfe 5 Oakdale Avenue, Bradford BD6 1RP
M0 AME D. Draper 36 Highfield Gardens, Combe Martin, Ilfracombe EX34 0HQ
M0 AMF R. Jefferies 38 Towbury Close, Redditch B98 7YZ
MW0 AMI R. Hall 33 Heol Y Garreg Las, Llandeilo SA19 6EB
MW0 AMJ L. Carter 13 Maes Dolau, Idole, Carmarthen SA32 8DQ
M0 AMM G. Smith East Lodge, Woodlands Drive, Bradford BD10 0NX
MW0 AMN G. Thomas Stonehall Mill Farm, Wolfscastle, Haverfordwest SA62 5NT
M0 AMP A. Davies 27 Foxley Grove, Bicton Heath, Shrewsbury SY3 5DF
MW0 AMQ G. Thomas Stonehall Mill Farm, Wolfscastle, Haverfordwest SA62 5NT
M0 AMS M. Burke 8 Childwall Gardens, Ellesmere Port CH66 1RL
MM0 AMV R. Moodie 18 Bennecourt Drive, Coldstream TD12 4BY
MM0 AMW D. Gillies 10 Killeonan, Campbeltown PA28 6PL
M0 AMX J. Howell Orchard House, Blennerhasset, Wigton CA7 3QX
M0 AMZ J. Williams 61 Longfield Road, South Woodham Ferrers, Chelmsford CM3 5JJ
M0 ANC Straight Key Century Club c/o R. Jones 31 Main Street, Awsworth, Nottingham NG16 2RH
M0 ANH J. Waller 56 Daventry Road, Dunchurch, Rugby CV22 6NS
M0 ANK S. Cotterill 320 Hamstead Road, Great Barr, Birmingham B43 5EH
M0 ANN G. Wardale 25 The Crescent, Huyton, Liverpool L36 6ER
M0 ANO R. Spencer 19 Trafalgar Road, Cirencester GL7 2EJ
M0 ANP Medway Amateur Receiving & Transmitting Society c/o N. Crooks 3 Grove Court, Settle BD24 9QR
M0 ANQ J. Chadwick 6 Harper Fold Road, Radcliffe M26 3RU
M0 ANS A. Rawlings 57 High Street, Nash, Milton Keynes MK17 0EP
M0 ANU G. Coolledge 49A Enfield Avenue, New Waltham, Grimsby DN36 4RB
MW0 ANV R. Davies 4 Maes Derlwyn, Llanberis, Caernarfon LL55 4TW
MW0 ANX J. Jensen Pistyll Canol Farm, Llandeilo Road, Ammanford SA18 2LQ
M0 AOA D. Young 47 Horseshoe Crescent Pocklington, York YO42 1UN
M0 AOB J. Allen 149 Penistone Road, Waterloo, Huddersfield HD5 8RP
M0 AOD D. Kay-Newman Kay-Spray, Pottery Road, Ilminster TA19 9QN
MM0 AOF D. Henry 106 Whinhill Gate, Aberdeen AB11 7WF
M0 AOG G. Dyson 32 Farleigh Fields, Orton Wistow, Peterborough PE2 6YB
M0 AOH J. Barber 7 Thomas Street, Carlisle CA2 5DZ
M0 AOI J. Russell 46 Eastleigh Drive, Tingley, Wakefield WF3 1PF
M0 AOJ A. Elliott 26 Watery Lane, Minehead TA24 5NZ
M0 AOK S. Millar 4 Broomfield, Benfleet SS7 2ST
MM0 AOL R. Bloomfield 35 Shaw Street, Dunfermline KY11 4AX
M0 AOM M. Goodrich Urb Les Basetes B3, Adsubia, Alicante 3786 Spain
MM0 AOQ C. Greig 5 Mitchell Place, Stuartfield, Peterhead AB42 5WE
M0 AOT M. Stanley 16 Fenton, Keswick CA12 4AZ
MM0 AOY D. Stephen 16 The Square, Portlethen, Aberdeen AB12 4QA
M0 AOZ M. Boothman Ballysax, Curragh R56 PX47 Ireland
M0 APC M. Brown 6 Rose Court, Garforth, Leeds LS25 1NS
M0 APD J. Udall 4 Church Lane Chilcote, Swadlincote DE12 8DL
MM0 APF Inverclyde CG c/o J. Fisher High Birches, Culbokie, Dingwall IV7 8JS
M0 APH A. Gilbert 79A Station Road, Brimington, Chesterfield S43 1LJ
M0 APK D. Allen 162 Wood Lane, Newhall, Swadlincote DE11 0LY
M0 APL B. Tucker 2 Hundall Court, Grasscroft Close, Chesterfield S40 4HN
M0 APN A. Nelson 29 Coxford Road, Southampton SO16 5FG
M0 APY R. Arey 4 Iveson Lawn, Leeds LS16 6NA
M0 APZ F. Piper 6 Russell Street, Little Hulton, Manchester M38 0LW
M0 AQA G. Shaw 6 Bromstone Road, Broadstairs CT10 2HA
M0 AQE E. Entwistle 43 Brock Road, Chorley PR6 0DB
M0 AQF T. Davies 20 The Coppice, Impington, Cambridge CB24 9PP
M0 AQH E. Blackburn 2 Stockwell Drive, Knaresborough HG5 0LW
MJ0 AQJ N. Jones 1 Cornucopia Court, Le Mont Pinel, St. Helier JE2 4RS Jersey
M0 AQK K. Hesketh 13 Elm Road, St. Helens WA10 3NE
M0 AQO G. Willson 40 Grace Gardens, Bishop's Stortford CM23 3EX
M0 AQP A. Bellamy 8 Dorothy Road, Kettering NN16 0PH
M0 AQQ K. Evans 5 Garswood Avenue, Rainford, St. Helens WA11 8JW
MW0 AQT E. Lucocq 96 Carisbrooke Way, Cardiff CF23 9HX
M0 AQW City Bristol Gr c/o M. Storkey 9 Waterman Court, Acomb, York YO24 3FB
MI0 AQX J. May 8 Oak Vale Avenue, Newry BT34 2BQ
MW0 AQZ L. Griffiths Tros Y Garreg Plas Road, Holyhead LL65 2LU
M0 ARA J. Layton 6 Granby Road, Cheadle Hulme, Cheadle SK8 6LS
M0 ARC East Yorkshire Contest c/o V. Lindgren 143 Hull Road, Anlaby., Hull HU106ST
MW0 ARD A. Davies 19 Maes-Y-Dderwen, Dinas Cross, Newport SA42 0XF
M0 ARH N. Ravilious 17 Halls Green, Weston, Hitchin SG4 7DR
M0 ARK B. Shepherd 17 Huntock Place, Brighouse HD6 2NW
MW0 ARL H. Davies Garth Wen, 1, Bryn Siriol, Coedpoeth, Wrecsam LL11 3PZ
M0 ARO J. Parsons 36 Gainsborough, Milborne Port, Sherborne DT9 5BD
M0 ARQ J. Churchill 59 Highfield, Letchworth Garden City SG6 3PY
MW0 ARV M. Thomas 12 School Road, Rhosllanerchrugog, Wrexham LL14 1BB
M0 ARX M. Richardson 39 Wilson Avenue, Deal CT14 9NL
M0 ARY M. O'Rourke Brookside Farm, Walpole, Halesworth IP19 9BH
M0 ARZ S. Hurst 25 Florence Road Abington, Northampton NN1 4NA
MM0 ASB R. Barbour 40 Mannerston Holdings, Linlithgow EH49 7ND
M0 ASC A. Clayton 7 Salisbury Avenue, Broadstairs CT10 2DT
M0 ASD A. Gallichan 4 Wigston Road, Rugby CV21 4LT
M0 ASG C. Nehmzow 26 Woodlands, Colchester CO4 3JA
M0 ASI N. Johns 85 South Hill, Hooe, Plymouth PL9 9PT
M0 ASJ S. Griggs 16 Sharleston Drive Stainforth, Doncaster DN7 5PU
MW0 ASL J. Phillips 57 Ffordd Llanerch, Penycae, Wrexham LL14 2ND
M0 ASN J. Marron 190 Cotswold Crescent, Billingham TS23 2QH
M0 ASR D. Campanario 3 Foxearth, Leek Road, Werrington, Stoke on Trent ST9 0DG
M0 AST M. Kulac 59 Gore Road Raynes Park, London SW20 8JN
M0 ASU K. Hamer Flat 7, Red Court, 66 Upper Park Road, Salford M7 4JA
MI0 ASV G. Best 1 Bensons Road, Lisburn BT28 3QX
M0 ASY S. Werner 4225 Place Sainte-Helene, Laval H7W 1P3 Canada
M0 ATA A. Rundle 9 Windsor Terrace, East Herrington, Sunderland SR3 3SF
M0 ATB R. Hutton 57 Sandy Lane, Upton, Poole BH16 5EJ
M0 ATC Dorset & Wilts Wing ARC c/o R. Ley 23 Heronbridge Close, Westlea, Swindon SN5 7DR
M0 ATD A. Holzapfel Flat 1-4, 4 Tanner Street, London SE1 3LD
MW0 ATG H. Thomas 15 Coronation Terrace, Pontypridd CF37 4DP
MW0 ATI G. Roberts Glanafon, Drefach, Llanybydder SA40 9YB
MW0 ATK S. Brewer 16 Oxwich Close, Cefn Hengoed, Hengoed CF82 7JB
M0 ATL P. Nash 110 Cranborne Road, Potters Bar EN6 3AJ
M0 ATQ J. Torry 41 Nevill Road, Rottingdean, Brighton BN2 7HH
MW0 ATR D. Williams 17 Brynawelon, Llanelli SA14 8PU
M0 ATS Monitoring Monthly c/o J. Ammundsen 62 Linden Avenue, Broadstairs CT10 1HR
MW0 ATT E. Cooke Anelog, Rhewl Fawr Road, Holywell CH8 9HJ
M0 ATV A. Reilly 19 The Ridgway, Romiley, Stockport SK6 3EE
M0 ATX E. Williams Dyffryn Coed, Union Road, Coleford GL16 7QB
M0 ATY C. Kirkland 9 Holland Way, Newport Pagnell MK16 0LL
M0 ATZ C. Hardy 56 Vyner Road, Wallasey CH45 6TF
M0 AUA D. Hazel 7 Arley Close, Upton, Chester CH2 1NW
M0 AUF N. Handforth 26 Appleby Crescent, Mobberley WA16 7GB
M0 AUG G. Ashton 95 Willow Drive, Lamaleach Park, Lamaleach Drive, Freckleton, Preston PR4 1DF
M0 AUK J. Sporton 199 Glaisdale Drive West, Nottingham NG8 4GY
M0 AUR A. Taylor 18 Chestnut Road, Glemsford, Sudbury CO10 7PS
M0 AUS A. Dowie 12 Malvern Drive, Gonerby Hill Foot, Grantham NG31 8GA
M0 AUT D. Randles 12 Wain Court. Rakeway, Saughall. Chester CH16BF
M0 AUW R. Hull 1 Northfield Cottage, Withington Road, Cheltenham GL54 4LL
M0 AUY L. Jeffries 37 Woodfield Road, Bournemouth BH11 9EU
M0 AVA D. Salsbury 1 Somerset Avenue, Tyldesley, Manchester M29 8LQ
M0 AVF A. De Araujo 13 Fifth Avenue Shaws Trailer Park, Knaresborough Road, Harrogate HG2 7NJ
M0 AVH D. Eaton-Watts 129 Blake Road, West Bridgford, Nottingham NG2 5LA
MI0 AVI Newry High School RC c/o G. Millar 1 Mullybrannon Road, Dungannon BT71 7ER
M0 AVK D. Swift 8 Grove Lane, Buxton SK17 9HG
M0 AVL D. Meakin 83 Kingsley Street Pleck Walsall, Walsall WS2 9QZ
M0 AVN A. Oatey Robin Hill, Blackpost Lane, Totnes TQ9 5RF
M0 AVP A. Baughan Camino Tigalate No. 31-33, Villa De Mazo, St. Cruz de Tenerife 38730 Spain
M0 AVQ J. Worthington 23 Sefton Avenue, Congleton CW12 3DB
M0 AVS V. Saundercock 14 Rashleigh Avenue, Plymouth PL7 4DA
M0 AVU M. Scott 36 Glebe Crescent, Newcastle upon Tyne NE12 7JR
M0 AVW C. Spence 32 Woodford Walk, Thornaby, Stockton-on-Tees TS17 0LT
M0 AVY A. Johnson 131 Rylands Road, Kennington, Ashford TN24 9LU
M0 AVZ D. Clutterbuck 2 Spring Valley Drive, Leeds LS13 4RN
M0 AWB A. Boom Oakthorpe House, 8A Peterborough Road, Peterborough PE6 0BA
M0 AWD M. Mansfield Piso 4 (Izq), Avda Jaime I - 14, Altea (Alicante) 3590 Spain
M0 AWE A. Ellis 50 Taylors Crescent, Cranleigh GU6 7EN
M0 AWH P. Bush 144 Stoke Lane, Westbury-On-Trym, Bristol BS9 3RN
M0 AWI J. Ross 13 Wensleydale Crescent, Oakridge Park, Milton Keynes MK14 6GX
MM0 AWJ K. Gray 18 Greenmantle Place, Glenrothes KY6 3QQ
MI0 AWL A. Smith 12 Sandringham Heights, Carrickfergus BT38 9EG
M0 AWN C. Gladman 24 Priory Road, Chessington KT9 1EF
MW0 AWO S. Jones 6 Heol Will Hopkin, Llangynwyd, Maesteg CF34 9ST
M0 AWP P. Oliver 21 Charlotte Close, Mount Hawke, Truro TR4 8TS
MM0 AWU G. Moffat 16/1 Laichpark Loan, Edinburgh EH14 1UH
M0 AWX G. Schoof 6 Canal Row, Haigh, Wigan WN2 1NA
M0 AWY D. Ashdown Cartwheels, 4 Honeysuckle Close, Hailsham BN27 3TP
MW0 AXA W. Townsend 133 Hazeldene Avenue, Brackla, Bridgend CF31 2JR
M0 AXC D. Russell 8 Norburton, Burton Bradstock, Bridport DT6 4QL
M0 AXE D. Marshall 34 Brentwood Road, Sheffield S11 9BU
M0 AXG K. Wheeler 26 Melverton Avenue, Wolverhampton WV10 9HN
M0 AXJ A. Clay 22 Park Street, Wallasey CH44 1AT
M0 AXL J. Cook 36 Kotuku Street, Coffs Harbour NSW 2450 Australia
M0 AXN J. Davis 60 West Bar Street, Banbury OX16 9RZ
M0 AXO G. Morris 7 Rowley View, Bilston WV14 8DE
MM0 AXR T. Rees 23 Doune Road, Dunblane FK15 9AT
M0 AXV M. Amplett 44A Darby Road, Coalbrookdale, Telford TF8 7EW
M0 AXW J. Davies 243A Bradford Road, Winsley, Bradford-on-Avon BA15 2HL
M0 AXX E. Moody The Apiaries, Rufford Lane, Newark NG22 9DG
M0 AXZ P. Morgan 20 Bishops Way, Buckden, St. Neots PE19 5TZ
M0 AYA S. Sellman Inselhof, Banbury Road Gaydon, Warwick CV35 0HH
M0 AYB T. Davies 95 High Brigham, Cockermouth CA13 0TJ
M0 AYC J. Soakell 162 Manor Road, New Milton BH25 5ED
MM0 AYE J. Welsh 7 South Cathkin Cottage, Rutherglen, Glasgow G73 5RG
M0 AYF D. Kostryca 12 High Street, Upton, Gainsborough DN21 5NL
M0 AYG M. Wood Weavers, Kingsdale Road, Berkhamsted HP4 3BS
M0 AYI G. Waring 7 Tynedale Terrace, Stanley DH9 7TZ
M0 AYO H. Parker 21 Mayfield Street, Hull HU3 1NS
M0 AYS C. Pocock 4 Broadfields, Harpenden AL5 2HJ
M0 AYU I. Gibson 7 Peverells Wood Close, Chandler'S Ford, Eastleigh SO53 2FY
M0 AYV A. Eyles 25 Chatham Road, Winchester SO22 4EE
M0 AYX A. King 28 King Street, Cirencester GL7 1JT
M0 AYY R. Corfield 35 Taplings Road, Winchester SO22 6HE
M0 AZB R. Goddard 6 Upper Ley Dell, Chapeltown, Sheffield S35 1AL
M0 AZC P. Niel 19 Fountains Close, Whitby YO21 1JS
M0 AZE M. Surplice 8 Mainwaring Drive, Sutton Coldfield B75 5QA
M0 AZG J. Kisiel Wayside Cottage, South Stoke Road, Reading RG8 0PL
M0 AZJ G. Gould 32 Archer Road, Kenilworth CV8 1DJ
M0 AZK D. Hill 35 Bridle Lane, Sutton Coldfield B74 3QE
MW0 AZN R. Cullis 20 Larch Close, New Inn, Pontypool NP4 0RT
M0 AZP P. Lemasonry 7 Eastwood Road, Sittingbourne ME10 2LZ
M0 AZR S. Gale 39 Thorley Park Road, Bishop's Stortford CM23 3NG
M0 AZS R. Buckle 25 Portsmouth Close, Rochester ME2 2QY
M0 AZT M. Thomas 36 Seaview Avenue, Peacehaven BN10 8SA
M0 AZV N. Devine 46 Tytton Lane West, Wyberton, Boston PE21 7HL
M0 AZW K. Coe 5 George St., Enderby, Leicester LE19 4NQ
M0 AZY F. Willis 99 Kenilworth Court, Coventry CV3 6JB
M0 AZZ A. Bond 6 Meadway, Knebworth SG3 6DN
MW0 BAA Blacksheep Contest + DX Group c/o S. Purser Penbrey, Llanfair Caereinion, Welshpool SY21 0DG
MM0 BAC C. Mackay 27 Barleyknowe Terrace, Gorebridge EH23 4EQ

M0 BAD T. Brusel De La Torre Flat 5, Douglas Hall, 3 Victoria Street, Preston PR1 7QR
M0 BAE A. Radford 25 Priory Avenue, Kirkby-In-Ashfield, Nottingham NG17 9BU
MM0 BAG G. Craig 9 Green Drive, Inverness IV2 4EX
M0 BAH A. Tyler Bracken 15 Chanctonbury Close, Washington RH20 4AR
M0 BAI P. Byrne 9 Glenluce Road, Liverpool L19 9BX
M0 BAJ D. Nelson 46 The Cunnery, Kirk Langley, Ashbourne DE6 4LP
M0 BAK K. Williams Cranesbie, 6 Dore Road, Sheffield S17 3NB
M0 BAL F. Johnson 7 Pharos Court, Pharos St., Fleetwood FY7 6BG
M0 BAM A. Tobin 17 Brockhampton, Cheltenham GL54 5XH
M0 BAO A. Edwards 75 Combe Park, Yeovil BA21 3BE
M0 BAP W. Stewart Hillfield Bungalow, Grunt Lane, Stroud GL6 6PH
M0 BAR B. Bartley 5 Cookes Wood Broompark, Durham DH7 7RL
MI0 BAT S. Gilmore 8 Fortfield, Dromore BT25 1DD
M0 BAU G. Hoyle 39 Randle Meadow, Great Sutton, Ellesmere Port CH66 2BG
M0 BAV L. Evans 16 Kynaston Drive, Wem, Wem SY4 5DE
M0 BAW D. Rose 31 Mount Crescent, Warley, Brentwood CM14 5DB
M0 BAY G. Bilson Fieldgate, 55 Littlemoor Lane, Alfreton DE55 5TY
M0 BAZ J. Waterfield 287 Turves Green, Birmingham B31 4BS
M0 BB Burnham Beeches RC c/o U. Grunewald Nuptown Orchard, Nuptown, Warfield, Bracknell RG42 6HU
M0 BBE J. Hayward 76 Lincoln Road, Skegness PE25 2EE
MI0 BBF D. Doherty 175 Bridge Road, Glarryford, Ballymena BT44 9QA
M0 BBH M. Redman 19 Richmond Road, Rugby CV21 3AB
M0 BBK J. Meakin White House Farm, Osmotherley, Northallerton DL6 3QA
MW0 BBL A. Hadden 164 Derwen Fawr Road, Sketty, Swansea SA2 8DP
MW0 BBM B. Meredith 27 Hyde Place, Llanhilleth, Abertillery NP13 2RT
M0 BBO S. Woodford 31 Seaborough View, Crewkerne TA18 8JB
M0 BBQ K. Taylor 241 Daventry Road, Cheylesmore, Coventry CV3 5HH
M0 BBT T. Pirrie Walnut Thatch, Tysoe Road, Warwick CV35 0UE
MW0 BBU S. Lloyd 41 Coombs Drive, Milford Haven SA73 2NU
M0 BBV R. Lyford 4 Wrentham Estate, Old Tiverton Road, Exeter EX4 6ND
M0 BBW S. Gleadall 59 Old Chapel Road, Warley, Smethwick B67 6HU
M0 BCC R. Clapp 11 Kensington Gardens, Ilkeston DE7 5NZ
M0 BCE Dr W. Johnstone 67 Station Lane, Birkenshaw, Bradford BD11 2JE
M0 BCF R. Cranwell 7 Central Drive, Elston, Newark NG23 5NT
M0 BCG I. Williams Alma Cottage , Old Vicarage Lane , South Marston, Swindon SN3 4SN
M0 BCH C. Chadburn 31 Darwin Close, Top Valley, Nottingham NG5 9LN
M0 BCI N. Armstrong 112 Chandos Street, Netherfield, Nottingham NG4 2LW
M0 BCJ G. Lewis 42 Ladywood Road, Ilkeston DE7 4NE
M0 BCK K. Bell 71 Wheatfield Road, Stanway, Colchester CO3 0YA
M0 BCL P. Williams 37 Winyards View, Crewkerne TA18 8JA
M0 BCN D. James 54 Woolacombe Lodge Road, Birmingham B29 6PX
MM0 BCR L. Haynes 29 Invercauld Road, Aberdeen AB16 5RP
M0 BCT M. Danfer The Nook, Mill Common, Blaxhall, Woodbridge IP12 2ED
M0 BCV S. Graham 4 Oakland Avenue, Ellenborough, Maryport CA15 7BU
M0 BCW P. Mason 15 Granton Avenue, Clifton, Nottingham NG11 9AL
M0 BCZ R. Burton 1 Avenue Court, Mount Avenue, London W5 1PY
MM0 BDA Dr R. August Smiddyhill House, Stracathro, Brechin DD9 7QE
M0 BDB R. Taylor 86-88 Hillside Crescent, Leigh-on-Sea SS9 1HQ
M0 BDD D. Webster 210 Walesby Lane, New Ollerton, Newark NG22 9UU
M0 BDE B. Thorburn 3 Victoria Road, Bexhill-on-Sea TN39 3PD
M0 BDF J. Reid Rosebury, Soldridge Road, Alton GU34 5JF
M0 BDH P. Fisher 21 Charlotte Close, Mount Hawke, Truro TR4 8TS
MM0 BDJ R. Hawkins 7 Ola Drive, Scrabster, Thurso KW14 7UE
M0 BDL D. Ferris 167 Lonsdale Avenue, Doncaster DN2 6HF
M0 BDQ K. Kisselev 37 Stanley Av., Barking, Essex IG11 0LD
M0 BDS Dr G. Butler 11 Keppel Drive, Bridlington YO16 6ZD
M0 BDU D. Goodwin 15 Tennyson Road, Bentley, Doncaster DN5 0EG
M0 BDW P. Hayes 1 Stile Plantation, Royston SG8 9HP
MI0 BDX A. Patterson 33 Marlborough Park, Carryduff, Belfast BT8 8NL
MI0 BDZ M. Chancellor 55 Brae Hill Park, Belfast BT14 8FP
MW0 BEA C. Lewis 60 Caeau Gleision, Rhiwlas, Bangor LL57 4UA
M0 BEC R. Millerchip 16 Kennedy Crescent, Gosport PO12 2NN
MM0 BED J. Macdonald 22 New Parliament Place, Campbeltown PA28 6GY
M0 BEE Waterside ARS c/o N. Williams 1 Dorset Close, Whitehaven CA28 8JP
M0 BEH P. Mutter 129 Demesne Road, Wallington SM6 8EW
M0 BEJ G. Moody 25 Norbiton Common Road, Kingston upon Thames KT1 3QB
M0 BEK C. Dunn 75 Waddington Avenue, Burnley BB10 4LA
MW0 BEL A. Owen Rafael Fawr House, The Fraich, Fishguard SA65 9QJ
M0 BEM M. Taperell 16 Parkhall Croft, Birmingham B34 7BU
M0 BEO K. Anderson 53 Priory Grove, Hull HU4 6LU
M0 BEQ H. Walsh 38 Potter Hill, Greasbrough, Rotherham S61 4PA
MW0 BER D. Jones Hafan Mynydd Bodafon, Llanerchymedd LL71 8BG
MI0 BES J. May 8 Oak Vale Avenue, Newry BT34 2BQ
MW0 BET V. Hughes Manley, 1 Garden Drive, Llandudno LL30 3LL
M0 BEV R. Knapp 85 Eastern Avenue, Liskeard PL14 3TD
M0 BEX J. Hrycan 40 Marina Drive, Marple, Stockport SK6 6JL
M0 BFA D. Wilson 30 Little Avenue, Swindon SN2 1NL
M0 BFB K. Francks 63 Parc Godrevy, Newquay TR7 1TY
M0 BFM S. Jones 69 Colville Street, Liverpool L15 4JX
M0 BFT S. Smith 225 South Drove, Lutton Marsh, Spalding PE12 9NT
M0 BFV L. Papazoglou 37A Eaton Road, West Kirby, Wirral CH48 3HE
M0 BGE T. Parker 24 Burrows Close, Lawford, Manningtree CO11 2HE
MM0 BGO R. Herd 4 Smithy Lane, Balmullo, St. Andrews KY16 0FG
M0 BGR H. Howard 10 Lawnside, London SE3 9HL
M0 BGS G. Steedman 5 Allerton Grange Gardens, Leeds LS17 6LL
M0 BGT G. Dyson-Bawley Grahil, 33 Ridgetor Road, Liverpool L25 6DG
M0 BGU J. Moran 8 Doffcocker Lane, Bolton. BL1 5RG
MM0 BGW A. Munro 2 Woodlands View, Inshes Wood, Inverness IV2 5AQ
M0 BHA C. Baker 1 Astley Green Darleyhall, Luton LU2 8TS
M0 BHE M. Sadler Hill View, Horton, Ilminster TA19 9QU
M0 BHH D. Newing 13 Maxwell Road, Broadstone BH18 9JG
M0 BHJ P. Worlledge 181 Roselands Drive, Paignton TQ4 7RN
M0 BHK G. Robertson 22 Carlton Villas, Hatt, Saltash PL12 6PS
M0 BHM the Sth Belfast c/o A. Whitehouse 19 Cleeve Road, Marlcliff, Alcester B50 4NX
M0 BHN P. Jarvis 26 Nally Drive, Woodcross, Bilston WV14 9UT
M0 BHO S. Coe 113 Highfield Road, Yeovil BA21 4RJ
M0 BHP G. King 12 Bracken Close, Blackburn BB2 5AH
M0 BHQ F. Lugg 4 Newbury Close, Walsall WS6 6DF
M0 BHR G. King 8 Oak Lane, Burghill, Hereford HR4 7QP
M0 BHV J. Cocks 12 Birch Pond Road, Plymouth PL9 7PG
M0 BHW G. Jeckells Hogals End, Mill St., Thetford IP25 7QN
MI0 BHX T. Costford 19 Cornacully Road, Meenarainy, Belcoo, Enniskillen BT93 5BR
M0 BIC J. Brocklebank Springfield House, Sixhills Lane, Market Rasen LN8 6AN
M0 BIH R. Deakin 40 Brussels Road, Stockport SK3 9QG
M0 BII S. Keightley 11 Sandringham Avenue, Wisbech PE13 3ED
M0 BIJ C. Harden 19 Nutshalling Avenue, Rownhams, Southampton SO16 8AY
M0 BIT P. Smith Karinya, Rectory Road Haddiscoe, Norwich NR14 6PG
MM0 BIX E. Cameron 5 King Street Ferryden, Montrose DD10 9RR
M0 BIZ M. Thomas 39 Treworder Road, Truro TR1 2JZ
M0 BJD B. Duffy 27 Kinloch Close Halewood, Liverpool L26 9XZ
M0 BJE A. Cockram 70 Arlington Drive, Marston, Oxford OX3 0SJ
M0 BJJ S. Miyake Hakata Radio, P.O.Box 232, Hakata-North 812-8799 Japan
M0 BJK G. Scothorn School House, Kirk Balk, Barnsley S74 9HU
M0 BJL S. Jarvis Kellow, Old Lyndhurst Road, Southampton SO40 2NL
MD0 BJM M. Rodgers 1 Kings Court, Ramsey, Isle of Man IM8 1LJ
M0 BJN F. Humphris 169 Bloxham Road, Banbury OX16 9JU
M0 BJO D. Greenway 37 Primrose Hill Park Homes, Primrose Hill, Somerton TA11 7AP
M0 BJP R. Pearce Kolner, 86 Thrupp Lane, Stroud GL5 2DG
M0 BJR M. Brown 3 The Vines, Kelsale, Saxmundham IP17 2PU
M0 BJS P. Hall 25 Ham Green, Pill, Bristol BS20 0EY
M0 BJT K. Davison 2 Sitwell Close, Spondon, Derby DE21 7GT
MJ0 BJU A. Mourant Little Mead, Claremont Road, St. Saviour JE2 7RT Jersey
M0 BJX R. Glover 89 Cambridge Road, Linthorpe, Middlesbrough TS55LD
M0 BKA J. Slough 2554 Hamilton Rd., Lebanon Oh 45036 United States
M0 BKD P. Mccormack 3 Greenway Close, Torquay TQ2 8EF
M0 BKF C. Brodrick 6 Carlton St., Hartlepool TS26 9ES
M0 BKG G. Rundle 15 Sandown Road, Paignton TQ4 7RL
M0 BKJ P. Mulliner 3 Watts Close, Cogenhoe NN7 1PD
M0 BKK Dr J. Rowe The Old Rectory, Wickenby, Lincoln LN3 5AB
M0 BKL S. Passmore 35 David Road, Paignton TQ3 2QF
M0 BKN S. Sherwin 1 Nursery Close, Wroughton, Swindon SN4 9DR
M0 BKS K. Sim 3 Thorngate Close Penwortham, Preston PR1 0XN
M0 BKV D. Kamm Delabole Head, Week St. Mary, Holsworthy EX22 6UU
M0 BLD S. Tsuzuki Flat 4, Lancing House, Watford WD24 4RL
M0 BLF D. Smith 67 Lambs Lane, Cottenham, Cambridge CB24 8TB
M0 BLH S. Laddiman 27 Morgans Way Hevingham, Norwich NR10 5PD
M0 BLI J. Crangle 43 Scarfell Close, Peterlee SR8 5PF
M0 BLN K. Hopps 100 Etherley Lane, Bishop Auckland DL14 6TU
M0 BLO R. Jackson 4 Hornbrook Gardens, Plymouth PL6 6LS
M0 BLR A. Cresswell 31 New Street, Doddington, March PE15 0SP
M0 BLS G. Hickford Sanclare, 94 Manor Road, Fleetwood FY7 7HY
M0 BLT M. Waldron 32 Windmill Street, Upper Gornal, Dudley DY3 2DQ
MW0 BLU W. Jepson 3 Marchog, Holyhead LL65 2HD
M0 BLV G. Peacock 7 Pensclose, Witney OX28 2EG
MM0 BLW A. Grant 4 Bellswood Crescent, Banchory AB31 5TE
M0 BLY S. Young 126 Stevens Road, Dagenham RM8 2QL
MM0 BLZ A. Blackburn 6 Dunadd View, Kilmichael Glassary, Lochgilphead PA31 8QA
MM0 BMA W. Irwin 15 Inchcolm Place, East Kilbride, Glasgow G74 1DR
M0 BMB B. Bentham 89 Westborough Way, Hull HU4 7SW
M0 BMD J. Green 788 The Ridge, St. Leonards-on-Sea TN37 7PS
MI0 BME P. Maile 3 Cairnmore Avenue, Lisburn BT28 2DW
M0 BMF D. Anger 17 Dell Road, Andover SP10 3JT
MM0 BMG N. Stewart 160 Carrick Knowe Drive, Edinburgh EH12 7EW
M0 BMJ I. Singer 197 Rosalind Street, Ashington NE63 9BB
MI0 BML P. Doris 92 Coolnafranky Park, Cookstown BT80 8PW
MI0 BMM B. Moore 8 Orken Lane, Aghalee, Craigavon BT67 0ED
M0 BMN P. Webb 41 Lancaster Gardens, Wolverhampton WV4 4DN
M0 BMR P. Pirrazzo 30 Coronation Road, Middlewich CW10 0DL
M0 BMT D. Bunting 6 Mill Gardens, Worksop S80 3QG
M0 BMU J. Moritz Carillon, 6 Bell Lane, Hatfield AL9 7AY
M0 BMW K. Wrack 18 Carrs Road, Cheadle SK8 2EE
M0 BMX M. Fitchett Barkenroy, Ludgvan, Penzance TR20 8AJ
M0 BMY L. Bilson Fieldgate, 55 Littlemoor Lane, Alfreton DE55 5TY
M0 BMZ A. Martin 11 The Mount, Worcester Park KT4 8UD
MW0 BNB T. Rogers The Willows, 48 Hillock Lane, Wrexham LL12 8YL
M0 BNC N. Clark 28 Thickthorn Close, Kenilworth CV8 2AF
M0 BND A. Sherman 45 Norfolk Road, Weymouth DT4 0PW
M0 BNF I. Copping 54 Hartley Road, Kirkby-In-Ashfield, Nottingham NG17 8DP
M0 BNO G. Ryder 11 Claremont Gardens, Farsley, Pudsey LS28 5BF
M0 BNP J. Taylor 3 Inhams Close, Murrow, Wisbech PE13 4HS
M0 BNR N. Rodley 5 Ward Avenue, Bilton, Hull HU11 4EE
M0 BNS British Naturist ARS c/o C. Beesley-Reynolds Kaos Roams, Palmerston Close, Kibworth Beauchamp LE8 0JJ
M0 BNZ D. Brooks The Elms, Trewoon Road, Helston TR12 7DS
M0 BOA Bunkers On the Air c/o W. Tinnion 3 Brayton Road, Aspatria, Wigton CA7 3DJ
M0 BOB R. Adlington 7 Kynance Close, Romford RM3 7LB
M0 BOH S. Saiger 10 Markham Avenue, Armthorpe, Doncaster DN3 2AZ
M0 BOI B. Johnson 10 Saffron Road, Tickhill, Doncaster DN11 9PW
MI0 BOK P. Connolly 94 North Parade, Belfast BT7 2GJ
M0 BOL R. Rose-Round 2 Lee Road, Blackpool FY4 4QS
M0 BOM R. Wilson 22 Leadhills Way, Bransholme, Hull HU7 4ZA
M0 BOQ R. Slater 79A Ainsworth Road, Radcliffe, Manchester M26 4FA
MI0 BOU J. Orr 17 Argyll View, Larne BT40 2JR
M0 BOX S. Wilson 21 Plumian Way, Balsham CB21 4EG
M0 BOY L. Sanduly 1 Archford Croft, Emerson Valley, Milton Keynes MK4 2EZ
MI0 BPB A. Mulholland 83 Tullyrain Road, Donaghcloney, Craigavon BT66 7PP
M0 BPC K. Hunt 4 Oak Avenue, Willington, Crook DL15 0BJ
MM0 BPF R. Armstrong 98 Burnbank Road, Ayr KA7 3QJ
M0 BPM D. Barclay 13 Avondale Terrace, Chester le Street DH3 3ED
M0 BPN N. Taplin 149 Frindsbury Road, Strood, Rochester ME2 4JD
MM0 BPP K. Lindsay 34 Park Crescent, Newtown St. Boswells, Melrose TD6 0QS
M0 BPQ Dr S. Bunting 17 Sunnydene Avenue, Highams Park, London E4 9RE
M0 BPS D. Lawrence 42 Upper Packington Road, Ashby-de-la-Zouch LE65 1UL
M0 BPT R. Walker Fists The International Morse Preservation Society, Po Box 6743, Tipton DY4 4AU

M0 BPU T. Lyne The Ark, 10 Blackfields Avenue, Bexhill-on-Sea TN39 4JL
MM0 BPV A. Finlayson 10B Flesherin, Isle of Lewis HS2 0HE
M0 BPW A. Shelswell Waterway, Dock Lane, Melton, Woodbridge IP12 1PE
MM0 BPX Dr K. Scott Kirklands, Craigend Road, Galashiels TD1 2RJ
M0 BPY P. Henson 35 Westbrook Drive, Rainworth, Mansfield NG21 0FB
M0 BQB T. Lee 91 Old Vicarage Park, Narborough, King's Lynn PE32 1TG
M0 BQC D. Bradley 1 Traddles Court, Chelmsford CM1 4XZ
M0 BQD E. Lee 16 Phoenix Chase, North Shields NE29 8SS
M0 BQE C. Margetts 16 Lahn Drive, Droitwich WR9 8TQ
M0 BQF M. Preece 51 Drancy Avenue, Willenhall WV12 5RD
M0 BQH K. Goodacre 22 New Ferry Road, Wirral CH62 1BJ
MM0 BQI J. Martin 3 Lismore Avenue, Edinburgh EH8 7DW
MM0 BQJ T. Cassidy 44 Wellpark Road, Saltcoats KA21 5LJ
MM0 BQL A. Cromack 10 Manse Road, Ardersier, Inverness IV2 7SR
M0 BQO U. Rose 45 Ringstead Crescent, Weymouth DT3 6PT
M0 BQT S. Smith 55 Market Street, Ilkeston DE7 5RB
M0 BQZ J. Romanis 23 Old Farm Lane, Stubbington, Fareham PO14 2BZ
M0 BRA Reigate Amateur Transmitting Society c/o G. Leonard 65 Qualitas, Bracknell RG12 7QG
M0 BRE P. Mann 38 Green Lane, Halesowen B629LP
MM0 BRG R. Harman 2 Dornoch Court, Kilwinning KA13 6QN
M0 BRH T. Linham 44 Vestry Road, Street BA16 0HX
M0 BRI P. Walker 20 Arbury Banks, Chipping Warden, Banbury OX17 1LU
M0 BRL T. Wells Flat Above Londis, 150 High Street, Boston Spa LS23 6BW
M0 BRM E. Turner 16 The Rowans, Doddington, March PE15 0SE
MW0 BRO M. Lewis 8 Rose Close, Pembroke SA71 4TR
M0 BRP M. Wastie 5 Pensclose, Witney OX28 2EG
M0 BRU W. Griffin 48 Wardle Way, Kidderminster DY11 5UJ
M0 BSB W. Carr 37 Keats Road, Greenmount, Bury BL8 4EP
M0 BSC P. Bonsey Wood View, Lower Kelly, Calstock PL18 9RY
M0 BSD B. Daly 10 Parfitts Close, Farnham GU9 7DH
M0 BSF N. Rigazzi-Tarling 3 The Planes, Bridge Road, Chertsey KT16 8LE
M0 BSH W. Kerslake 42 Silverdale Road, Newcastle ST5 2TB
M0 BSI L. Preston Hillside, Trewennack, Helston TR13 0PQ
M0 BSJ P. Poore 42 Kibblewhite Crescent, Twyford, Reading RG10 9AX
M0 BSL M. Pell 7 Churchfleet Lane, Gosberton, Spalding PE11 4NE
MM0 BSM S. Mcquillian 68 Strathmore Drive, Cornton, Stirling FK9 5BE
M0 BSP E. Doyle 33 Bodenham Road, Birmingham B31 5DP
M0 BSQ J. Doyle Orchard Corner, Piddington Road, Ludgershall, Aylesbury HP189PJ
MI0 BSU D. Stanley 59 Gransha Road, Kircubbin, Newtownards BT22 1AJ
M0 BSV A. Ilett 34 Westbrook Park Road, Woodston, Peterborough PE2 9JG
MM0 BSX G. Scattergood 14 Market Street, Forfar DD8 3EY
M0 BSZ A. Wanford 4 Willows Close, Tydd St. Mary, Wisbech PE13 5QR
MM0 BTD J. Ganson 1 Beechwood Terrace West, St. Fort, Newport-on-Tay DD6 8JH
M0 BTG G. Barrett 114 William Street, Long Eaton, Nottingham NG10 4GD
MW0 BTI E. Thomas Craig-Y-Don, Conwy LL27 0JJ
M0 BTJ J. Ingram 58 Belben Road, Poole BH12 4PJ
MI0 BTK A. Cobb 62 Katesbridge Road, Dromara, Dromore BT25 2PN
M0 BTL A. Withers 5 Tintern Road, Skelton-In-Cleveland, Saltburn-by-the-Sea TS12 2YN
MI0 BTM G. Duffy Carrageen Cottage, Enniskillen BT74 6ET
M0 BTN J. Escreet Colfield, Carlton Lane, Hull HU11 4RA
M0 BTO S. Collins 69 Walkers Heath Road, Kings Norton, Birmingham B38 0AL
M0 BTP E. Edmunds 5 Nelsons Quay, St. Helens, Ryde PO33 1TA
M0 BTR J. Springett 31 Mountbatten Court, Andover Road, Winchester SO22 6BA
MW0 BTU G. Rowlands Dalar Wen, Rhosmeirch, Llangefni LL77 7SJ
M0 BTX R. Elliott 8 Bridge Place, Amersham HP6 6JF
M0 BTZ R. Harrison 55 Stapleford Close, Romsey SO51 7HU
M0 BUA R. Miller 1 Mews Cottages, Penview Crescent, Helston TR13 8RX
M0 BUC P. Buck 6 Bedford Road 2 Bracebridge House, Sutton Coldfield B756AA
M0 BUE J. Page Cherry View, 28 Ellerslie Lane, Bexhill-on-Sea TN39 4LJ
M0 BUF E. Smith Lamorna, Broadmead Road, Woking GU23 7AD
M0 BUG N. Jamieson 1 Langdale Place, Newton Aycliffe DL5 7DX
MM0 BUH Capt. N. Smith Nether Dallachy Farmhouse, Boyndie, Banff AB45 2JT
M0 BUI R. Styles 4 Cameron Close, Bude EX23 8SP
M0 BUR N. Armer 17 Keswick Road, Lancaster LA1 3HJ
M0 BUT T. Gale 33 Watson Close Upavon, Pewsey SN9 6AF
M0 BUV A. Zerafa 2 Furnwood, St. George, Bristol BS5 8ST
M0 BUY J. Swann 1 Sunnindale Drive, Tollerton, Nottingham NG12 4ES
M0 BVD C. Turner North Holme, Drain Lane, York YO43 4DQ
M0 BVF G. Kapranos 89 Spohr Terrace, South Shields NE33 3LQ
MI0 BVG T. Wedlock 13 Drumawhey Road, Newtownards BT23 8RS
M0 BVI Dr P. Rogers Flat 4 Holmdale, 2 Osborne Road, Poole BH14 8SD
M0 BVM D. Hancock 4 Elmside, Willand, Cullompton EX15 2RN
M0 BVN A. Tavender 38 Hesley Grove, Chapeltown, Sheffield S35 1TX
M0 BVO J. Cull 2 Drybrook Cottages, Amesbury Road, Cholderton, Salisbury SP4 0ER
M0 BVQ G. Stokes 9 The Haven, Harwich CO12 4LA
M0 BVT J. Colles The Orangery, Ufford Place Ufford, Woodbridge IP13 6DP
M0 BVU S. Noble 30 Flude Road, Coventry CV7 9AQ
M0 BVV F. Hibberd 58 Stoke Green, Coventry CV3 1AN
M0 BVW B. Cox 13 The Graylands, Coventry CV3 6EW
M0 BVX D. Poulton 14 George Street, Gun Hill, Coventry CV7 8HL
M0 BVY N. Marshall Key Cottage, Haynall, Little Hereford, Nr. Ludlow SY8 4AY
M0 BVZ P. Poulton 14 George Street, Gun Hill, Coventry CV7 8HL
M0 BWB RSARS - Royal Signals ARS c/o J. Ridout 136 Church Hill Road, Cheam, Sutton SM3 8NA
M0 BWC J. Barton 27 Francis Way, Salisbury SP2 8EF
M0 BWF D. Surman 27 Stanley Road, Hinckley LE10 0HP
M0 BWH S. Jones 34 Bury Green, Little Downham, Ely CB6 2UH
M0 BWI W. Wilson 8 Nora Street, South Shields NE34 0RA
M0 BWL S. Vinnicombe 8A Cross Road, Cholsey, Wallingford OX10 9PE
MW0 BWM P. Lane 51 Maesgwyn, Cwmdare, Aberdare CF44 8TH
M0 BWN H. Seldon 22 Downside Avenue, Plymouth PL6 5SD
M0 BWO B. Watts 24 Carisbrooke Way, Redcar TS10 2LJ
M0 BWP R. Bebbington 64 Stafford Road, Toll Bar, St. Helens WA10 3JH
M0 BWQ H. Robinson 5 Coppice Close, Haxby, York YO32 3RR
M0 BWS D. Pacheco Iii 35 Baker Close, Caversfield, Bicester OX27 8FQ
M0 BWU A. Hall 16 Brushfield Avenue, Sileby, Loughborough LE12 7NX
M0 BWW K. Allen 55 Brandish Crescent, Clifton, Nottingham NG11 9JZ
M0 BWY D. Hill 86 The Downs, Nottingham NG11 7EB
M0 BWZ A. Foad 42 Catkin Way, New Balderton, Newark NG24 3DT
M0 BXA L. Jensen 17 Middleton Close, Tysoe, Warwick CV35 0SS
M0 BXB E. Fuller 36 North Road, Hull HU4 6LJ
M0 BXC S. Coulston 15 West View, Clitheroe BB7 1DG
M0 BXD C. Robinson Flat 1, St. Michaels Court, Worcester WR2 5QR
M0 BXF P. Pickup 13 Siddows Avenue, Clitheroe BB7 2NX
M0 BXG J. King 155 Barcombe Avenue, London SW2 3BQ
M0 BXM D. Grimshaw Apartment 52 Albion Mill, Blackburn BB2 4LX
M0 BXP P. Bramberg St. Margarets School House, Margery Lane, Durham DH1 4QJ
M0 BXU P. White 61 North St., Pewsey SN9 5ES
MM0 BYB A. Lee Burrowgate, , Stronsay KW17 2AN
M0 BYI D. Moorey 35 Broadway Manor, The Broadway, Hull HU9 3PN
M0 BYJ R. Stoddart 4 Belmont Road, Rednal, Birmingham B45 9LW
M0 BYK B. Kurtcebe Apartment 22, Rose Court, Baltic Avenue, Brentford TW8 0FU
M0 BYL British Young Ladies Club c/o J. Jones 21 The Maltings, Warminster BA12 8JR
M0 BYM J. Robson Flat 6, 57 Lewisham Park, London SE13 6QP
MI0 BYR D. Christie 5 Moneydig Park Garvagh, Coleraine BT51 5JP
MW0 BYS W. Reed 2 St. Marys Park, Jordanston, Milford Haven SA73 1HR
MW0 BYT G. Roscoe 45 Bro Infryn, Glasinfryn, Bangor LL57 4UR
M0 BYU C. Dodshon 62 Moor Road, Melsonby, Richmond DL10 5PE
M0 BYV J. Goldsbrough 63 Aske Road, Redcar TS10 2BP
M0 BYY J. Ellner 21 Cranmer Road, Hampton Hill, Hampton TW12 1DW
M0 BYZ K. Crossman 24 Coxs Drive, Baltonsborough, Glastonbury BA6 8RG
M0 BZA R. Stoddart 163 Flatts Lane, Middlesbrough TS6 0PP
M0 BZB S. Thirlaway 10A Sea View, Blackhall Colliery, Hartlepool TS27 4AX
M0 BZC A. Phillips 39 Stonechat Road, Billericay CM11 2NZ
M0 BZE Dr B. Allen 1 St. Marys Close, Mursley, Milton Keynes MK17 0HP
M0 BZH M. Wilkinson 124 Doncaster Road Darfield, Barnsley S73 9JA
M0 BZI F. Western 12 St. Oswalds Crescent, Brereton, Sandbach CW11 1RW
M0 BZK D. Mapeley 6 Green Lane, Wolverton, Milton Keynes MK12 5HB
M0 BZN L. Evans 184 West St., Dunstable LU6 1NX
M0 BZO G. Major 17 Jubilee Cottages, Station Road, Bedford MK43 0PN
M0 BZQ J. Doxey 34 Lime Tree Crescent, New Rossington, Doncaster DN11 0BT
M0 BZR M. Taylor 38 Manor Road, Slyne, Lancaster LA2 6LB
M0 BZS R. Cornthwaite 18 Slaidburn Drive, Accrington BB5 0JJ
M0 BZU N. Burkill 4 Giles Street, Cleethorpes DN35 8EA
M0 BZV P. Wade 11 Hillside Crescent, Puriton, Bridgwater TA7 8AP
M0 BZX S. Turner 75 Keir Hardie Avenue, Stanley DH9 6JU
M0 BZY M. Haywood 4 Wentworth Gate, Birmingham B17 9EB
M0 BZZ R. Bunker 12 Bedford Avenue, Birkenhead CH42 4QX
MW0 CAB Dr I. Jones 4 Crowhill, Haverfordwest SA61 2HL
MI0 CAC H. Mcgoldrick 2 Carsdale, Mullanahoe Road, Dungannon BT71 5GA
M0 CAD B. Lovatt 12 Nelson Street, Leek ST13 6BB
MM0 CAE J. Gauson 112A High Street, New Pitsligo, Aberdeenshire AB43 6NN
M0 CAG I. Solly 49 Dorothy Drive, Ramsgate CT12 6TL
M0 CAJ D. Tysoe 21 Burnt Close, Luton LU3 3SU
M0 CAM Granta CG c/o M. Marsden Mill Cottage, Shrowle, East Harptree, Bristol BS40 6BJ
M0 CAN J. Wallis 8 Kennedy House Hainworth Lane, Keighley BD21 5BD
M0 CAR S. Robertson 5 Sear Hills Close, Balsall Common, Coventry CV7 7QL
M0 CAS M. Cassidy 6 Mosley Street, Blackburn BB2 3ST
M0 CAT J. Isaksson 4 Horse Chestnut Gardens, Chellaston, Derby DE73 6SB
M0 CAV J. Pearson 82 Devoke Avenue, Worsley, Manchester M28 7EN
M0 CAX D. Deakin Restholme Cottage, Mosham Road, Doncaster DN9 3BA
M0 CAZ A. Dahalay The Manor, 80 Beach Road, Weston Super Mare BS22 9UU
M0 CBA G. Gunn The Old Rectory, Coreley, Ludlow SY8 3AW
M0 CBD W. Eldridge Minafon, Llangeitho, Tregaron SY25 6TT
M0 CBF N. Lock 57 Western Way, Basingstoke RG22 6DF
M0 CBG A. Godney 38 Botley Drive, Havant PO9 4QY
MM0 CBI J. Isaacs 5 Main Road, Glencraig KY5 8AL
M0 CBK H. Purves 2 Bourtree Close, Wallsend NE28 9AA
MM0 CBL G. Black 9 Mcculloch Road, Girvan KA26 0EF
M0 CBM R. Newman 11 Pine View Close, Woodfalls, Salisbury SP5 2LR
M0 CBN R. Liversidge 17 Millbank Close, High Green, Sheffield S35 4NS
M0 CBP B. Muizelaar Birtley Cb Services, Phoenix Communications, 33 Penshaw View, Portobello Road, Birtley DH3 2JL
M0 CBT L. Betts 33 Four Wells Drive, Sheffield S12 4JB
MI0 CBX R. Graham 21 Meadowvale Crescent, Bangor BT19 1HQ
M0 CCA S. Ramsden 24 Western College Road, Plymouth PL4 7AG
MM0 CCC J. Maclean 15 Muirpark Terrace, Tranent EH33 2AS
M0 CCD J. Newsome 241 Skellow Road, Skellow, Doncaster DN6 8JL
M0 CCF Firepower Museum c/o M. Buckley Springfield, 12 Ranmore Avenue, Croydon CR0 5QA
M0 CCG N. Griffiths 85 Foljambe Road, Chesterfield S40 1NJ
MW0 CCK K. Dancer 63 Romilly Crescent, Cardiff CF11 9NQ
MW0 CCL K. Dancer 118 Fairwater Grove West, Cardiff CF5 2JR
MW0 CCN J. Jones 4 Westfield Road, Rhyl LL18 4PN
M0 CCQ P. Burgess 27 Watergate Street, Ellesmere SY12 0EX
MW0 CCS A. Patterson 3 Trem Y Foryd, Kinmel Bay, Rhyl LL18 5JE
M0 CCU J. Buckley 14 Eastfield, Foxholes, Driffield YO25 3QW
M0 CCV L. Parsons Gull Cottage, Briar Close, Hastings TN35 4DP
M0 CCW M. Thornton Bramble Lodge, Youngers Lane, Skegness PE24 5JQ
M0 CCZ A. Cubitt 132 Cauldwell Hall Road, Ipswich IP4 5BP
M0 CDB T. Faris 7 Mount Close, Fetcham, Leatherhead KT22 9EF
M0 CDF M. Brooks 9 The Green, Blaby, Leicester LE8 4FQ
MW0 CDG C. Gozzard Craig Dulas, Rhydyfoel Road, Llanddulas, Abergele LL22 8EG
M0 CDJ S. Spevack Pips Hill, Garden Close, Leatherhead KT22 8LR
MM0 CDK D. Dickson 115 Hatton Gardens, Glasgow G52 3PU
M0 CDL J. Griffin 35 Cottage Street, Kingswinford DY6 7QE
M0 CDN N. Bullough 29 Redfern Road, Stone ST15 0LF
MW0 CDO P. Tomlinson 17 Heol Yr Orsedd, Port Talbot SA13 2HL
M0 CDQ J. Grant 15A Elmdon Road, Marston Green, Birmingham B37 7BU
M0 CDS S. Sanders 52 Hazelwood Road, Callington PL17 7EU
M0 CDU E. Linley 40 Belvoir Road, Cleethorpes DN35 0SE
MM0 CDW A. Bryce 23 Primrose Avenue, Inverkip, Greenock PA16 0DS
M0 CDY J. Elsworth 15 Elm Avenue, Christchurch BH23 2HJ

M0 CDZ K. Mountford 22 Hollington Drive, Oxford, Stoke-on-Trent ST6 6TZ
M0 CEB M. Bridge 100 Pennine Road, Bacup OL13 9PH
M0 CEC T. Ditchfield 13 Alexandra Road, Waterloo, Liverpool L22 1RJ
M0 CEG A. Shipp 12 Rainbow Court Paston Ridings, Peterborough PE4 7UP
M0 CEO B. Lewin 68 Brackley Square, Woodford Green IG8 7LS
M0 CEQ N. Taylor 18 Chestnut Road, Glemsford, Sudbury CO10 7PS
M0 CER T. Oka 65 Curzon Street, London W1J 8PE
M0 CES D. Hadden 52 Brant Road, Lincoln LN5 8SH
M0 CEW J. Lehane 174 Stamfordham Drive, Liverpool L19 6PZ
M0 CEX P. Newell Thwaites Bank, Spring Avenue, Keighley BD21 4TD
MM0 CEZ P. Moran 3 Dunottar Avenue, Coatbridge ML5 4LL
M0 CFB A. Bennett 11A Ratcliffe Road Haydon Bridge, Hexham NE47 6ER
M0 CFD N. Hould 53 Laurel Avenue, Forest Town, Mansfield NG19 0DW
MM0 CFE S. Campbell 9 Arbuthnott Place, Stonehaven AB39 2JA
M0 CFF Dr R. Fukuda 2-6-17-218 Mita, Meguro-Ku, Tokyo 153-0062 Japan
M0 CFH A. Hewett 1 Mountside, Westfield Lane, Folkestone CT18 8BY
MW0 CFQ P. Brennan 1 Gerddi Mair, St. Clears, Carmarthen SA33 4ET
M0 CFR R. Gould 22 Vereker Drive, Sunbury-on-Thames TW16 6HF
M0 CFT K. Pye 5 Teme Avenue, Wellington, Telford TF1 3HU
M0 CFZ B. Farrington Flat 5 4 Southfield Rise, Paignton TQ3 2NE
M0 CGA A. Desoer 53 Highfield Road South, Chorley PR7 1RH
M0 CGB A. Wiseman 28 Inchfield, Worsthorne, Burnley BB10 3PS
M0 CGE R. Corneloues 21 Brook Street, Manningtree CO11 1DL
M0 CGF I. Schofield 100 Purley Downs Road, Sanderstead CR2 0RB
M0 CGO R. Pratt 11 Park Road, Ryde PO33 2BG
MW0 CGP T. Peters 37 Lon Coed Bran, Cockett, Swansea SA2 0YD
M0 CGR J. Clarey 2 Sebastopol Cottages, Redmere, Ely CB7 4SS
M0 CGS S. Hancock Monrad, Back Street, Gainsborough DN21 3DL
M0 CGT G. Thompson 160 Rempstone Road, Wimborne BH21 1SX
MI0 CGV Benbradagh US Navcommsta c/o T. Keery 51 School Road, Ballyroney, Banbridge BT32 5JF
M0 CGW C. Selwyn-Smith Miranda, East Molesey KT8 9AN
MM0 CGZ C. Cregan 4 Fowlers Court, Prestonpans EH32 9AT
M0 CHD J. Neale 20 Oakfield Road, Wollescote, Stourbridge DY9 9DL
M0 CHE D. Letton 21 Westfield, Bradninch, Exeter EX5 4QU
MW0 CHI M. Broxton 4 Owen St, Orange Gardens, Pembroke SA71 4EP
M0 CHJ S. Birbeck 3 Accrington Road Hapton, Burnley BB11 5QL
M0 CHL D. Fitzpatrick 21 Stambridge Road, Clacton-on-Sea CO15 3JR
MU0 CHN J. Gardner The Ferns, Rue De La Girouette, St. Saviour, Guernsey GY7 9NN
M0 CHO G. Burton 41 Etchingham Road, Langney, Eastbourne BN23 7DS
M0 CHP S. Nicholls 50 Gorleston Road, Lowestoft NR32 3AQ
M0 CHR D. Whitelock-Wainwright 21 Whelan Gardens, St. Helens WA9 5TD
M0 CHS M. Stanley 9 Clare Way, Clacton on Sea CO168BX
M0 CHU T. Hodby 1 Hawksworth Close, Rotherham S65 3JX
MM0 CHV W. Adamson 47 Clarinda Gardens, Dalkeith EH22 2LW
MM0 CHX G. Mckenna 19 Gordon Terrace, Ayr KA8 0EF
MW0 CIA N. Lemon 75 Tai Llwyd Road, Neath SA10 7DY
MI0 CIB P. Bell 1Knockbracken Drive, Coleraine BT521WN
M0 CIC R. Partington 47 Sandmoor Road, New Marske, Redcar TS11 8DJ
M0 CIE M. Coles 1 Church Street, Taunton TA1 3JE
M0 CIF D. Rosewarn 16 Charles Crescent, Taunton TA1 2XN
MW0 CIH J. Jones 14 Heol Tywysog, Pentre Halkyn, Holywell CH8 8HA
MM0 CIK D. Cox Seye Hethen, Sanquhar DG4 6JZ
M0 CIO R. Wyatt 8 Millbrook Road, Bushey WD23 2BU
M0 CIP L. Thompson 9 Elmwood Drive, Ponteland, Newcastle upon Tyne NE20 9QQ
M0 CIR C. Ryalls 15 Belmont Way, South Elmsall, Pontefract WF9 2BT
MW0 CIS A. Bray Coedcelyn, Talley, Llandeilo SA19 7YR
MW0 CIT V. Bray Coedcelyn, Talley, Llandeilo SA19 7YR
M0 CIW H. Delafield 205 South Avenue, Abingdon OX14 1QU
MW0 CJB D. Newton-Goverd 2 Blaen Y Morfa, Morfa, Llanelli SA15 2BG
M0 CJC G. Fowle 12 Lytham Road, Broadstone BH18 8JS
M0 CJD C. Rhodes Wayside, Hardstoft, Pilsley, Chesterfield S45 8AH
M0 CJE N. Murphy 9 Eliot Gardens, Newquay TR7 2QE
MM0 CJF J. Smith 41 Dickie Drive, Peterhead AB42 1HB
M0 CJG D. Jesinger C/O B Jesinger, 29 Breakspeare Close, Watford WD24 6DA
MM0 CJH N. Fowler 59 Milnefield Avenue, Elgin IV30 6EJ
M0 CJI R. Turner 53 Queens Drive, Sandbach CW11 1BN
M0 CJJ R. Brown 1 Octavian Close, Hatch Warren, Basingstoke RG22 4TY
M0 CJK S. Coulthard 90 Rochester Crescent, Crewe CW1 5YQ
M0 CJM N. Toombes 46 The Vale, Oakley, Basingstoke RG23 7LD
M0 CJN J. Gilmore 30 Fairhaven Road, Caversfield, Bicester OX27 8TU
M0 CJO A. Kay Pear Tree Cottage, Hale House Lane, Farnham GU10 2JG
M0 CJR N. Porter 27 Severn Road, Aveley, South Ockendon RM15 4NR
M0 CJS M. Elliott 42 Ryelands Crescent, Stoke Golding CV13 6EP
MM0 CJT A. Mctaggart 14/1 Lady Nairne Place, Edinburgh EH8 7LZ
M0 CJY I. James 56A Ridgeway, East Herringthorpe, Rotherham S65 3NN
M0 CJZ G. Dobson 35 Parkhill Road, Doncaster DN3 1DP
M0 CKA P. Webb 42 Holland Road, Ampthill, Bedford MK45 2RS
M0 CKB K. Prior The Annex, 26 St David?S Road, Norwich NR93DH
M0 CKC P. Wakelam 6 Frankland Road, Durham DH1 5HZ
M0 CKE J. Balls 70 Risegate Road, Gosberton, Spalding PE11 4EY
MM0 CKF J. Lewis 9 Cessnock Road, Troon KA10 6NJ
M0 CKG D. Kilburn Shepherds Cottage, Burradon, Morpeth NE65 7HF
M0 CKI S. Edwards 5 Chalk Lane, Sutton Bridge, Spalding PE12 9YF
MM0 CKK A. Mcluckie 25 Churchill Avenue, Kilwinning KA13 7JN
M0 CKL G. Sell 135 Northfields, Norwich NR4 7ET
M0 CKM K. More 18 Douglas Close, Ford, Arundel BN18 0TG
M0 CKN W. Dodge 78 Littleworth Road, Dover 03820-4331 United States
M0 CKO S. Westall 4 South View Great Harwood, Blackburn BB6 7NL
M0 CKP D. Warner 6 Desborough Road, Rushton, Kettering NN14 1RG
M0 CKS R. Cockings 4 Freeman Gardens, High Green, Sheffield S35 4NT
MW0 CKT C. Campbell-Moore Maes Celyn Henllan Street, Denbigh LL16 3PE
M0 CKU W. Meisenbach 5 Byland Court, Whitby YO21 1JJ
M0 CKV R. Wiltshire 8 Hilltop Lane, Heswall, Wirral CH60 2TT
M0 CKX F. Graseley 11 Wilberforce Road, South Anston, Sheffield S25 5EG
MW0 CLB M. Youlden 33 Treseifion, Porthdafarch Road, Holyhead LL65 2NN
M0 CLD I. Arnold 4 Larkspur Close, Tanfield Lea, Stanley DH9 9UH
M0 CLE A. Pascoe 53 Priory, Bovey Tracey, Newton Abbot TQ13 9HP
M0 CLG G. Gundry 181 Stoneleigh Avenue, Worcester Park KT4 8YA
M0 CLI E. Donaghy Mendips, 36 Attwood Road, Salisbury SP1 3PR
M0 CLJ J. Townsend 28 West Street, Darfield, Barnsley S73 9NF
M0 CLK P. Fitzpatrick 16 Lichens Crescent, Oldham OL8 2NS
M0 CLL J. Nixon 25 Grove Court, Alsager, Stoke-on-Trent ST7 2DS
M0 CLM M. Bromley Chartley, Norton Lea, Warwick CV35 8JX
M0 CLN P. Tose Flat 4, 33 Rohilla Close, Whitby YO22 4BU
M0 CLO K. Farthing 86 Coldnailhurst Avenue, Braintree CM7 5PY
MI0 CLP M. Hunter 11A Maydown Road, Armagh BT61 8BU
M0 CLR W. Huddleston 17 Moorside Road, Brookhouse, Lancaster LA2 9PJ
M0 CLW S. Pearson 8 The Pastures, Dunstable LU6 2HL
M0 CMC M. Veary 4 Millennium Way, Stone ST15 8ZQ
M0 CME D. Viney 5 Waters Edge, Bognor Regis PO21 4AW
M0 CMF T. Beaumont 61 Mitcham Road, Camberley GU154AR
M0 CMH M. Hemmings 117 Kingston Hill Avenue, Romford RM6 5QP
M0 CMK L. Taylor 34 Bradegate Drive, , Peterborough PE1 4SP
M0 CMN B. Pearce 12 Bean Avenue, Bracebridge, Worksop S80 2EW
MM0 CMO E. Skea Craigard, Craigton, Inverness IV1 3YG
M0 CMP J. Miller 23 Wetherby Gardens, Farnborough GU14 6BW
M0 CMQ E. Brendish Flat 1, 19 Blake Hall Road, London E11 2QQ
M0 CMS J. Lewis 3 Jacobs Close, Stantonbury, Milton Keynes MK14 6EJ
M0 CMT J. Donald 67 Cradley, Widnes WA8 7PL
M0 CMW J. Hudson 3 Vogan Avenue, Crosby, Liverpool L23 0SG
M0 CMZ I. Okanoue 142-1 Yoshida, Okoh-Cho, Kochi 783-0045 Japan
MW0 CNA M. Evans 322 Heol Gwyrosydd, Penlan, Swansea SA5 7BR
MW0 CNB B. Polgreen 344 Heol Gwyrosydd, Penlan, Swansea SA5 7BP
MW0 CNC T. Symons Brynchwyth Farm, Fairyland Road, Neath SA11 3QE
MW0 CND M. Davies 10 Torrington Road, Gendros, Swansea SA5 8DU
M0 CNE P. Black 43 Malvern Avenue, Rugby CV22 5JN
MM0 CNF J. Clark 16 Bonnyton Avenue, Drongan, Ayr KA6 7DG
M0 CNG D. Pearce 2 Mell Avenue, Hoyland, Barnsley S74 9HF
M0 CNH B. Cockerill 4 Foxglove Close, Rugby CV23 0TS
MI0 CNI W. Hamilton-Sturdy 243 The Woods, Larne BT40 1BD
M0 CNK J. Serrano 42 Fleetwood Walk, Murdishaw, Runcorn WA7 6DZ
M0 CNL P. Glover 14 Holbrook Close, Clacton-on-Sea CO16 8TH
M0 CNM A. Williams 18 Sturdee Close, Thetford IP24 2LF
M0 CNN S. Mcnally 14 Cornelius Drive, Pensby, Wirral CH61 9PR
M0 CNP D. Edwards 3 Murton Close, Burwell, Cambridge CB25 0DT
M0 CNS Essex DX Group c/o T. Hackett 16 Eagle Way, Shoeburyness SS3 9RJ
M0 CNU T. Norman 113 Dracaena Avenue, Falmouth TR11 2ER
MM0 CNV W. Aitken 63 Newlands Road, Grangemouth FK3 8NT
M0 CNW H. Martin 23 St. Marks Road, Gorefield, Wisbech PE13 4QQ
M0 CNX P. Frampton 118 Ramnoth Road, Wisbech PE13 2JD
M0 CNY S. Hayes 36 Mayfield Road, Chaddesden, Derby DE21 6FW
M0 CNZ P. Enrico Trengrouse House, Polhorman Lane, Helston TR12 7JD
M0 COA Hertfordshire Peak Assault (Scouts) c/o C. Egginton Flat 117, Seaward Tower, Gosport PO12 1HH
MW0 COB R. Cobb 107 High Street, Neyland, Milford Haven SA73 1TR
M0 COC R. Dean 10 Livingstone Road, Ellesmere Port CH65 2BE
MW0 COD C. Queeley 63 Tonna Road, Caerau, Maesteg CF34 0RU
MW0 COE R. Thomas 11 Heol-Y-Parc, North Cornelly, Bridgend CF33 4LT
MW0 COF L. Thomas 11 Heol-Y-Parc, North Cornelly, Bridgend CF33 4LT
M0 COI N. Burgess 12 Glenfield Road, Grimsby DN37 9EE
M0 COJ D. Clenshaw 4 Spring Meadow, Glemsford, Sudbury CO10 7PN
M0 COM D. Fryer 16 Elston Place, Aldershot GU12 4HY
M0 CON J. Bell 5 Burntwood Close, London SW18 3JU
M0 COO M. Goulbourne 9 Hawksworth Close, Liverpool L37 7EX
M0 COP P. Wesley Stalden, Ludlow Road, Church Stretton SY6 6RB
M0 COQ Dr C. Cane 69 Stoughton Drive North, Leicester LE5 5UD
M0 COT J. Pears 8 The Hawthorns, Ellesmere SY12 9ER
M0 COV V. Fairhurst 44 Harold Road, Coventry CV2 5LG
MW0 COZ J. Gearey 46 Priory St., Carmarthen SA31 1NN
M0 CPB D. Tucker 137 Seaford Road, London W13 9HS
M0 CPC C. Cook Hilbry Cottage, Douglas Road, Crowborough TN6 3QT
MW0 CPD P. Jennings 5 Pantydwr, Nantybwch, Tredegar NP22 3RZ
M0 CPE H. Broyles 16 Clifton Road, Shefford SG17 5AE
M0 CPF K. Payne 20 Laburnum Road, Exeter EX2 6EG
M0 CPG Capel Battery Preservation Group c/o J. Button 4 Capstone Ridge, Hempstead, Gillingham ME7 3AQ
M0 CPK A. Childs 5 Barnes Wallis Drive, Leegomery, Telford TF1 6XT
M0 CPL W. Harrison 2 Mount House Road, Formby, Liverpool L37 3LB
MW0 CPN M. Watkins Highwinds, Bryn Pydew, Llandudno Junction LL31 9QF
MM0 CPS Cockenzie & Port Seton ARC c/o R. Glasgow 7 Castle Terrace, Port Seton, Prestonpans, East Lothian EH32 0EE
M0 CPT P. Smith 3 Glenfield Square, Farnworth, Bolton BL4 7TG
M0 CPU N. Bartlett 61 Uplands Road, West Moors BH220BU
M0 CPW P. Thompson 73 Erlstoke Close, Plymouth PL6 5QN
MM0 CPZ R. Roberts 16 Swanston Avenue, Edinburgh EH10 7BX
M0 CQF C. Warhurst 26 Cherry Tree Close, Ilkeston DE7 4HQ
M0 CQH I. Hirst 30 Lincoln Road, Skellingthorpe, Lincoln LN6 5UU
M0 CQK J. Palmer 1 Titley Cottages, West Monkton, Taunton TA2 8NN
M0 CQL A. Young 33 Springbank Close, Barnsley S71 3HZ
M0 CQO P. Hunter 31 Itchen Grove Perton, Wolverhampton WV6 7QY
M0 CQQ D. Newman 89 Sea Place, Goring-By-Sea, Worthing BN12 4BH
MW0 CQR D. Conde 13 Broxton Road, Wrexham LL13 9BA
MM0 CQT S. Forsyth West Park, Innes Road, Fochabers IV32 7NL
M0 CQV J. Hubbard 99 Tuckers Road, Loughborough LE11 2PH
M0 CQW J. Drummond 60 Park Lane, Exeter EX4 9HP
MW0 CRA M. Clarke Freshwinds Halkyn, Holywell CH88ES
M0 CRD P. Mayne 17 School Street, Cottingley, Bingley BD16 1QB
M0 CRG Calderdale Raynet ARG c/o M. Cox Haugh Shaw Hall Haugh Shaw Road, Halifax HX1 3LE
M0 CRH A. Roberts 23 Church St., Great Harwood, Blackburn BB6 7NF
MW0 CRI D. Marston Rosario, High Street, Cardigan SA43 3EF
M0 CRJ R. Jones 8 Downing Avenue, Newcastle ST5 0JY
M0 CRM N. Williams 1 Dorset Close, Whitehaven CA28 8JP
M0 CRN UK DX c/o S. Corbett 80 Helmsdale Lane, Great Sankey, Warrington WA5 1SY
M0 CRO G. Mansell 87 Bifield Road, Stockwood, Bristol BS14 8TT

M0 CRP A. Harradine 4 Hesketh Drive, Lostock Gralam, Northwich CW9 7QJ
MI0 CRQ K. Mcauley Layde View, 19 Rathlin Avenue, Ballycastle BT54 6DQ
MI0 CRR P. Quinn 11 Blackpark Road, Ballyvoy, Ballycastle BT54 6QZ
M0 CRT N. Hall 38 Cottongrass Gardens, Dinnington, Sheffield S25 2DF
M0 CRU J. Bradley 47 Brunswick Park Road, Wednesbury WS10 9HH
M0 CRW J. Roebuck 2 Royston Close, Walton, Chesterfield S42 7NE
M0 CRZ S. Crabtree 107 Rochdale Road, Shaw, Oldham OL2 7JT
M0 CSA C. Shepherdson 149 Scarborough Road, Norton, Malton YO17 8AD
M0 CSB M. Tait 2 Wilsford Close, Walsall WS4 1QP
MW0 CSC C. Osborn Lowfield, Station Road, Usk NP15 2EP
M0 CSD T. Quinn 11 Meadowfield Stokesley, Middlesbrough TS9 5EL
M0 CSE J. Burdett 27 Hayfield Road, Woolston, Warrington WA1 4PE
M0 CSF A. Prandoczky 18 Kestrel Court Birtley, Chester le Street DH3 2PT
M0 CSG City And Sticks ARG c/o A. Hosking 30 Edrick Road, Edgware HA8 9JD
M0 CSO C. Mckenzie 15 Belmont Drive, Saltney Ferry, Chester CH4 0AL
M0 CSP E. Watson 49 Beechfield Road, Bolton BL1 6HZ
M0 CSQ R. Bates 51 Boyton Road, Ipswich IP3 9PD
M0 CSR J. Gardiner 18 Granville Terrace, Guiseley, Leeds LS20 9DY
M0 CST A. Holbrook 24 Victoria Street, Sheerness ME121YA
M0 CSU M. Deacon 26 Brecon Chase, Sheerness ME12 2HX
M0 CSV R. Tavener 3 Mill Close Roxwell Chelmsford Essex Cm1 4Pg 3 Mill Close, Roxwell, Chelmsford CM1 4PG
M0 CSZ S. Clarey 36 Birchin Bank, Elsecar, Barnsley S74 8DP
M0 CTC P. Ridgeon 1 Fiennes Road Herstmonceux, Hailsham BN27 4LN
M0 CTF D. Kaye 20 Tryfan Close, Redbridge, Ilford IG4 5JX
M0 CTI B. Johnson 268 Badsley Moor Lane, Rotherham S65 2QP
M0 CTJ G. Tepper 5 Herbert Street, Mexborough S64 0JZ
M0 CTK M. Stocking 125 Bassnage Road, Halesowen B63 4HD
M0 CTL S. Ball 64 Stallington Road, Blythe Bridge, Stoke-on-Trent ST11 9PD
M0 CTM R. Pearson 21 West Parade, Spalding PE11 1HD
M0 CTN D. Haughton 31 Holmes Carr Road, New Rossington, Doncaster DN11 0QF
M0 CTP G. Hyde 4 Albright Close, Pocklington, York, York YO42 2PE
M0 CTQ C. Greaves 2 Attlee Avenue, New Rossington, Doncaster DN11 0QX
M0 CTR A. Smith 25 Hill Corner Road, Chippenham SN15 1DW
MM0 CTT D. Stewart 45 Kilwinning Road, Irvine KA12 8RZ
MM0 CTU C. Stewart 45 Kilwinning Road, Irvine KA12 8RZ
MW0 CTX B. Jones 87 Heol Llanelli, Pontyates, Llanelli SA15 5UB
MW0 CUA D. Herbert 50 Clos Cilsaig, Dafen, Llanelli SA14 8QU
M0 CUD S. Marsh 6 Mayland Drive, Streetly, Sutton Coldfield B74 2DG
M0 CUF S. Hall 28 Pugneys Road, Wakefield WF2 7JT
MM0 CUG G. Grant 11 Auchriny Circle, Bucksburn, Aberdeen AB21 9JJ
M0 CUH D. Coisson 21 Medalls Path, Stevenage SG2 9DX
M0 CUI K. Lucas 7 Wales Road, Kiveton Park, Sheffield S26 6RA
M0 CUK M. Slade 7 Glebe Field, Chaddleworth, Newbury RG20 7EZ
M0 CUL M. Stevens 67 New Road, East Hagbourne, Didcot OX11 9JX
MI0 CUN P. Alexander 59A Lismurn Park, Ahoghill, Ballymena BT42 1JW
M0 CUP M. Phillips Roseneath, 4C Valley Road, Kenley CR8 5DG
M0 CUS G. Mack 1085 Evesham Road, Astwood Bank, Redditch B96 6EB
M0 CUT S. Cruise 18 Morris Court Close, Bapchild, Sittingbourne ME9 9PL
M0 CUU C. Wardell 8 Lower End, Bricklehampton, Pershore WR10 3HL
M0 CUY S. Odell 41 Pevensey Park Road, Westham, Pevensey BN24 5HW
M0 CVA V. Dewey 100D Bromley High Street, London E3 3EG
M0 CVB J. Sherbourne 4 Chelston Terrace, Chelston, Wellington TA21 9HT
M0 CVC C. Blackburne 1 Watson Road, Leeds LS14 6AE
M0 CVG B. Watmough 28 Aspin Oval, Knaresborough HG5 8EL
M0 CVH I. Cowie 25 Penrice Close, Colchester CO4 3XN
M0 CVJ C. Dale The Common, Alsager, Stoke on Trent ST7 2TQ
M0 CVK R. Henshall 19 Townson Road, Ashmore Park, Wolverhampton WV11 2PP
M0 CVN A. Fulcher Keepers, The Street, Ashford TN27 0QF
M0 CVO N. Booth 68 Vernon Road Kirkby-In-Ashfield, Nottingham NG17 8ED
M0 CVP B. Sutherland 9 Park Drive South Hoole, Chester CH2 3JT
M0 CVR P. Gurney Perhams Green, Plymtree, Cullompton EX15 2LW
M0 CVS F. Sadler 12 Yokecliffe Drive, Wirksworth, Matlock DE4 4EX
MW0 CVT R. Evans The Brae, Coed-Cae-Ddu Road, Pontllanfraith Blackwood NP12 2DA
M0 CVU A. Forrest 2 Otterburn Grove, Blyth NE24 4QP
MW0 CVW P. Wright 145 Park Avenue, Bryn-Y-Baal, Mold CH7 6TR
M0 CVZ D. Haynes 113 The Glade, Croydon CR0 7QP
MM0 CWB J. Benson 15 Hawkhill Place, Stevenston KA20 4HN
MW0 CWF S. Beer 4 Churchfields, Barry CF63 1FP
MM0 CWI C. Mcgowan 82 Normand Road, Dysart, Kirkcaldy KY1 2XP
MM0 CWJ J. Cameron 407 Smerclate, Isle of South Uist HS8 5TU
M0 CWN J. Thorne Willow Lodge, The Street, Bury St. Edmunds IP29 5AP
MW0 CWS B. Chapman Gwyddfan, Mountain Road, Kidwelly SA17 4EY
M0 CWT M. Kozakowski 18 Town Barn Road, Crawley RH11 7EB
M0 CWX Prof. A. Morris Shenlea, Leasowes Lane, Halesowen B62 8QE
M0 CWY G. Andrews 18 Vyne Road, Sherborne St. John, Basingstoke RG24 9HX
M0 CWZ K. Summers 30 David Street, Kirkby-In-Ashfield, Nottingham NG17 7JW
MM0 CXA A. Burns 49 Stoneycroft Lane, Arbroath DD11 1PX
MM0 CXB K. Gillen Flat 1/R, 12 Fergus Drive, Glasgow G20 6AG
MI0 CXE G. Rea 50 Culrevog Road, Dungannon BT71 7PY
MW0 CXH P. Evans 35 Trinity Road, Llanelli SA15 2AB
M0 CXL D. Findlay 78 South View Road, Bradford BD4 6PJ
M0 CXO A. Preece 1 Springfield Close, Thirsk YO7 1FH
M0 CXQ C. Zdziech 4 Pine Tree Mews, Barlby YO8 5GZ
MW0 CXW I. Gray 4 Llundain Fach, Felinfoel, Llanelli SA15 4PF
M0 CXY E. Threadingham 2 Cullum Close, Chichester PO19 6GG
MM0 CXZ V. Madden 38 Hunters Avenue, Dumbarton G82 2RZ
M0 CYB Radio Gaga CG c/o T. Cannon 35 Loddon Bridge Road, Woodley, Reading RG5 4AP
M0 CYD S. Bourne 1 Humewood Grove, Stockton-on-Tees TS20 1JU
M0 CYE J. Taylor 46 Lever House Lane, Leyland PR25 4XL
M0 CYG P. Davies Devildchies, West Bay Road, Bridport DT6 4EH
M0 CYI T. Witherspoon Po Box 2, Swannanoa 28778 United States
M0 CYJ Dr D. Nicole 53 Cobbett Road, Southampton SO18 1HJ
M0 CYM Southampton ARC c/o E. Tewsley 20 Crookhorn Lane, Waterlooville PO7 5QF
M0 CYR Dr P. Palin Flat 2, Springfield Heversham, Heversham, Milnthorpe LA7 7EJ
M0 CYT Prof. S. Warrillow Po Box 452, Canterbury, Victoria 3126 Australia
M0 CYU D. Holdroyd 2 Vicarage Lane, Naburn, York YO19 4RS
M0 CYX G. Yoxall 74 Summerson Lodge, 94 Alverstone Road, Southsea PO4 8GS
M0 CZA Cheshire County Raynet c/o A. Moss Lime Kiln Basin, Whitebridge Estate, Stone ST15 8LQ
M0 CZB J. Webber 5 Leda Mews, Achilles Close, Hemel Hempstead HP2 5WR
M0 CZC M. Wade 42 New Road, Burnham-on-Crouch CM0 8EH
M0 CZE M. Garton 13 Damaskfield, Worcester WR4 0HY
MI0 CZF W. Belshaw 9 Ashmount Gardens, Lisburn BT27 5BZ
MM0 CZH D. Mitchell 3 Lade Crescent, Bucksburn, Aberdeen AB21 9HJ
MM0 CZK C. Rogers 18 Primrose Lane Rosyth, Dunfermline KY11 2SL
MM0 CZM A. Stewart G/1 290 Dumbarton Road, Old Kilpatrick, Glasgow G60 5LJ
M0 CZN W. Martin 17 Dickens Road, Maidstone ME14 2QW
M0 CZO J. Dixon 83 Portmeads Rise, Chester le Street DH3 2NW
M0 CZP T. Ostley 30 Ashley Way Brighstone, Newport PO30 4HH
M0 CZR B. Richtering 20 Shaftesbury Avenue, Hornsea HU18 1LX
M0 CZT D. Ward 42 Fern Grove, Cherry Willingham, Lincoln LN3 4BG
M0 CZU D. Gregson 15 Alice Street, Oswaldtwistle BB5 3BL
M0 CZX F. Boele 301 Cell Barnes Lane, St. Albans AL1 5QB
MM0 DAA S. Mackie Kierycraigs Lodge, Blairadam, Kelty KY4 0JF
M0 DAB D. Bowles 23 Broughton Way, Rickmansworth WD3 8GW
M0 DAC D. Tanner 55 Arundel Drive, Bramcote, Nottingham NG9 3FN
M0 DAD D. Roper 84 Tynedale Drive, Cowpen, Blyth NE24 4DS
M0 DAE M. Haladij 50 Liberty Drive, Duston, Northampton NN5 6TU
M0 DAG D. Godden 94 The Common, South Normanton, Alfreton DE55 2EP
M0 DAH S. Brown 4 Foundry Lane, Manchester M4 5LB
M0 DAL A. Pounder 6 Barbondale Grove, Knaresborough HG5 0DX
M0 DAN D. Black 8 Cornwood Close, Finchley, London N2 0HP
MW0 DAR P. Drew 14 Longfield Court, Hirwaun, Aberdare CF44 9NG
M0 DAS A. Cockburn Grangewood, Blenheim Road, Littlestone, New Romney TN28 8RD
MM0 DAT D. Thain 20 Spey Street, Fochabers IV32 7EH
M0 DAW D. Wilcox Medlar Cottage, Faringdon Road, Swindon SN6 8AJ
MW0 DAX C. Evans Y Rhosfa, High Street, Llanfyllin SY22 5AF
M0 DAZ D. Drake 38 Frinton Road Broxtowe, Nottingham NG8 6GQ
M0 DBA Dr M. Brown 2 Beacon Road, Marazion TR17 0HF
M0 DBB J. Hayhurst 18 Ducket Close, Richmond DL10 5QD
MM0 DBC D. Brown 10 Culmore Place, Falkirk FK1 2RP
M0 DBD L. Dean 24 Spennithorne Avenue, Leeds LS16 6JA
MM0 DBF W. Callanan 3 Eden Place, Aberdeen AB25 2YF
M0 DBG G. Hodges 12 Linwal Avenue, Houghton-On-The-Hill, Leicester LE7 9HD
M0 DBH S. Palik 10 Clare Road, Northborough, Peterborough PE6 9DN
M0 DBI Dr A. Wylie 15 Worcester Gardens, Greenford UB6 0BH
M0 DBJ O. Peters 3 Churchill Close, Sutton, Ely CB6 2QF
MI0 DBK A. Quinn 11 Derrynaught Road, Collone, Armagh BT60 1LZ
M0 DBM C. Bloxidge 33 Rosemary Hill Road, Sutton Coldfield B74 4HL
MM0 DBR A. Butler 68 Laws Road, Aberdeen AB12 5LJ
M0 DBT C. Gregson 11 Coupe Green, Hoghton, Preston PR5 0JR
M0 DBX R. Batten 118 Marryat Road, New Milton BH25 5JF
M0 DBY A. Burton Normanby View, Otby Lane, Market Rasen LN8 3UT
M0 DCB D. Brown 7 Limber Hill, Cheltenham GL50 4RJ
MM0 DCC R. Rutherford 3 Stevenson Street, Oban PA34 5NA
M0 DCD A. Ripley 3 Linden Way Wetherby, Leeds LS22 7QU
M0 DCG Dr T. Spence 38 Burtonwood Road, Great Sankey, Warrington WA5 3AJ
MW0 DCM D. Maydew 40 Penrhys Road, Tylorstown, Tylorstown CF43 3BD
M0 DCO B. Tutty 14 Nursery Walk, Canterbury CT2 7TF
M0 DCP Y. Ochiai 1-8-7 Tamagawa Den-En-Chofu, Setagaya-Ku, Tokyo 1580085 Japan
MW0 DCQ W. Ashton 44 Bryn Awel, Bettws, Bridgend CF32 8SA
M0 DCS D. Taylor Garth Farm, Hull Road, Selby YO8 6NH
MW0 DCT P. Grace 14 Langcliff, Swansea SA3 4JF
M0 DCU North of Scotland CG c/o S. Quantrill 7 Clare Road, Kessingland, Lowestoft NR33 7PS
M0 DCV P. Howell 18 High Street Foxton, Cambridge CB22 6SP
M0 DCW D. Eggleton 79 Hazel Close, Twickenham TW2 7NP
M0 DCY G. Tarr 40 The Garth, Coniston LA21 8EQ
M0 DCZ J. Farrer 3 Pitt Garth, Haggs Lane, Grange-over-Sands LA11 6PH
M0 DDA B. Waterloo 55 Solent Road, Hill Head, Fareham PO14 3LB
M0 DDB C. Parr 13 Peartree Avenue, Southampton SO19 7JN
M0 DDC A. Copperwaite 71 Gladbeck Way, Enfield EN2 7EL
MW0 DDE A. Bullock 3 Cae Celyn, Berriew, Welshpool SY21 8BT
M0 DDI P. Cassidy 34 Belgrave Avenue, Penwortham, Preston PR1 0BH
M0 DDK R. Milne 8 Connaught Walk, Rayleigh SS6 8UY
M0 DDT Dr C. Potter 12 Beech Road, Headington, Oxford OX3 7RR
M0 DDU L. Shepherd 334 Copnor Road, Portsmouth PO3 5EL
M0 DDV M. Watts 51 Chester Road, Sidcup DA15 8RX
MI0 DDW S. Cooper 31 Kilbroney Valley, Rostrevor, Newry BT34 3SR
M0 DDY J. Todd 108 Clee Road, Grimsby DN32 8NX
M0 DEA N. Jacovides 11 Stravonos Street, Nicosia 2335 Cyprus
M0 DEB Mango c/o E. Lugmayer 17 Borough End, Beccles NR34 9YW
MM0 DEC I. Street 5 Calder Road, Bellsquarry, Livingston EH54 9AA
M0 DEF G. Thacker 4 Duffy Place, Rugby CV21 4EF
M0 DEI Hull & District ARS c/o R. Liversidge 17 Millbank Close, High Green, Sheffield S35 4NS
M0 DEJ A. Neumann 5 Denfield Avenue, Halifax HX3 5NL
M0 DEK D. Dyde 10 Essex Close, Worcester WR2 5RW
M0 DEL A. Lindley 2 Mickle Hill Farm Cottage, Mickle Hill Road, Blackhall Colliery, Hartlepool TS27 4DF
M0 DEN D. Miller Brookhill Gardens Pinxton, Nottingham NG16 6JX
M0 DEO C. Wheldon 39 Felton Avenue, South Shields NE34 6RY
M0 DEP R. Speed 48 Stony Lane, Burton, Christchurch BH23 7LE
M0 DEQ M. Clarke 21 Sycamore Road, Greenstead Estate, Colchester CO4 3NF
M0 DER R. Hannigan 4 Westlands Avenue, Tetney, Grimsby DN36 5LP
M0 DES R. Beck 26 Corcoran Street, Duncraig, Perth 6023 Australia
M0 DEV M. Twells Camels, Annscroft, Shrewsbury SY5 8AN
MW0 DEW D. Walters 17 Heol Islwyn, Llanrhystud SY23 5BW
M0 DEX J. Constable 438 Old Road, Clacton-on-Sea CO15 3SB
M0 DEY H. Watt 40 Long Wood Road, Bristol BS16 1FD
M0 DFA D. Crake Kentolop, Holyhead Road, Montford Bridge, Shrewsbury SY4 1EE
M0 DFD S. Sparkes Flat 1, 14 Hall Road, Wilmslow SK9 5BN
M0 DFF M. Bywater 16 Grove Road, Cromer NR27 0BY

M0 DFH C. Miles 26 Meadowside, Grindleton, Grindleton, Clitheroe, Lancs BB7 4RR
M0 DFL C. Houghton 20 St. Peters Way, Thurston, Bury St. Edmunds IP31 3RZ
MW0 DFN D. Thomas 48 Gilbert Road, Llanelli SA15 3RA
MI0 DFO R. Kennedy 83 Craigstown Road, Randalstown, Antrim BT41 2PN
M0 DFQ P. Dickman 1 Old Hall Close, Henley, Ipswich IP6 0RJ
M0 DFW D. Wright 61 Potton Road, St. Neots PE19 2NN
M0 DFX D. Riley 9 Century Avenue, Mansfield NG18 5EE
M0 DFY M. Watts 14 Beverley Close, Lowestoft NR33 8QQ
M0 DGA A. Bevins 12 Wheatstone Road, Formby, Liverpool L37 6BF
M0 DGB D. Balharrie 27 Norfolk Road, Uxbridge UB8 1BL
MM0 DGI S. Spence Halley, Deerness, Orkney KW17 2QL
M0 DGJ G. Hannam 26 Cornflower Way, Melksham SN12 7SW
M0 DGK M. Luby 19 Robin Lane Bentham, Lancaster LA2 7AB
M0 DGQ B. Zarucki 26 Heathfield Road, Kings Heath, Birmingham B14 7DB
MM0 DGR Hubnet Amateur Radio & Digital Online Network c/o J. Phunkner 7 Plenshin Court, Glasgow G53 6QW
M0 DGT P. Kelly 4 The Crescent, Guildford GU2 8AL
M0 DGU C. Thompson 80 Aston Road, Willerby, Hull HU10 6SG
MI0 DGX N. Mccully 12 Cargygray Road, Hillsborough BT26 6BL
M0 DHE J. Coxon Links View Farm, Fairy Lane, Sale M33 2JT
MW0 DHF P. King 11 Lord Street, Penarth CF64 1DD
M0 DHI B. Phillips 79 Leen Valley Drive, Shirebrook, Mansfield NG20 8BJ
M0 DHM E. Dean 26 Silverdale, Maidstone ME16 9JG
M0 DHN G. Kirkpatrick 23 Hornhatch, Chilworth, Guildford GU4 8AY
M0 DHO D. Honey Bluebell Cottage, Crondall Road, Fleet GU51 5SU
M0 DHP R. Benitez 4 Raphael Drive, Thames Ditton KT7 0BL
MM0 DHQ A. Clark 20 Church Street, Kilwinning KA13 6BE
M0 DHU M. Tachibana 81334076949, Tokyo 1070062 Japan
M0 DHX H. Wood 402 Chemin De Peyrebelle, Valbonne 6560 France
MM0 DHY A. Hart Tigh Na Coille, Daviot, Inverness IV2 5EP
M0 DID R. Macgregor 125 Spring Lane, Birmingham B24 9BY
M0 DIG J. O'Mahoney 36 Scotton Gardens, Catterick Garrison DL9 4HX
M0 DIJ J. Allen 2 Chichester Walk, Chichester Road, Ramsgate CT12 6NX
M0 DIL S. Lowe 31 Court Farm Road, Bristol BS14 0EH
M0 DIN Dr A. Bottrill 4 St Luke'S Mews Gilesgate, Durham DH11JA
M0 DIQ R. Mullen 18 Chandlers Ridge, Nunthorpe, Middlesbrough TS7 0JL
MM0 DIS I. Elder 24 Birniehill Avenue, Bathgate EH48 2RR
M0 DIT J. Tildesley 16 Calver Grove, Keighley BD21 2RX
M0 DIW D. Walls 15 Brant Road, Lincoln LN5 8RL
M0 DJA A. Garner The Chestnuts, Surfleet, Spalding PE11 4BA
M0 DJB A. White 368 Hall Lane, Whitwick, Coalville LE67 5PF
M0 DJD D. Gould 11 Secret Garden, Hunslet, Leeds LS9 8FB
M0 DJF D. Fradley 33 Churchill Way Riseing Brook Stafford, Stafford ST17 9NZ
M0 DJH D. Hale 51 Lynhurst Avenue, Sticklepath, Barnstaple EX31 2HY
M0 DJI J. Ford 1 Cherry Road, Enfield EN3 5SE
M0 DJQ B. Cushing 51 Anderida Road, Eastbourne BN22 0PZ
M0 DJT D. Turner 47 Cliveden Road, London SW19 3RD
M0 DJW Dr D. Westland 8 Faris Barn Drive, Woodham, Addlestone KT15 3DZ
M0 DKD Dr D. Bains 2 Arundel Road, Brighton BN2 5TD
M0 DKJ I. Rose 26 Beckford Road, Cowes PO31 7SG
M0 DKL E. Whittle 15 Weavers Court, Scorton, Preston PR3 1NQ
M0 DKN D. Turney 2 Beult Meadow, Cage Lane, Ashford TN27 8PZ
M0 DKP J. Jones 3 St. Judes Walk, Cheltenham GL53 7RU
M0 DKR D. Austin 16 Southfold Place, Lytham St. Annes FY8 4PZ
M0 DKS D. Kees 23 Walsingham Way, Billericay CM12 0YE
M0 DKT D. Knott 80 Melville Court, Chatham ME4 4XJ
M0 DKU P. Giles Fenimora Cottage, Paines Hill, Bicester OX25 4SQ
M0 DKV D. Bowers Oak House, Church Stile Lane, Exeter EX5 1HP
M0 DKX J. Horsfield 91 Harlington Road, Mexborough S64 0DT
M0 DLB S. Orange 20 Borrowdale Avenue, Fleetwood FY7 7LF
M0 DLC P. Bartlett 43 Chamberlain Way, Pinner HA5 2AU
M0 DLE R. Tingay 1 Ullswater Road, Sompting, Lancing BN15 9UF
M0 DLG J. Conway 27 Victoria Road, Gorleston, Great Yarmouth NR31 6EF
MM0 DLH A. Dunsmore 21 East Croft, Ratho, Newbridge EH28 8PD
M0 DLI Wellesley House RC c/o J. Hislop 10 Park Wood Close, Broadstairs CT10 2XN
M0 DLL D. Gray 68 Sixth Cross Road, Twickenham TW2 5PD
M0 DLM D. Mann Hunters Lodge, Grange Road, Bedford MK44 3NT
M0 DLP D. Birch 18 Ronhill Lane, Cleobury Mortimer, Kidderminster DY14 8AU
M0 DLR K. Jones 21 Wharton Crescent, Beeston, Nottingham NG9 1RJ
M0 DLS M. Bell 56 Boyd Road, Wallsend NE28 7SQ
M0 DLX D. Cleal 9 Ladymere Place, Ockford Road, Godalming GU7 1AH
M0 DLY C. Kerrison 18 Parks Road, Dunscroft, Doncaster DN7 4AH
M0 DLZ P. Parry 31 Swinburne Way, Daybrook, Nottingham NG5 6BX
M0 DMA D. Marshall 75 North Hill Road, Sheffield S5 8DT
M0 DMB C. Simpson 6 Dalton Close, Driffield YO25 6YE
M0 DMD M. Coe 2 Burnslack Road, Ribbleton, Preston PR2 6EX
M0 DME A. Pell 5 Gallery Close, Northampton NN3 5NT
M0 DMF C. Price 27 Naunton Way, Cheltenham GL53 7BQ
M0 DMI J. Enderby 42 Claremont Avenue, Chorley PR7 2HL
M0 DMJ C. Peters 9 Evelyn Close, Twickenham TW2 7BL
MM0 DMK D. Mckenzie 17 Alexander Drive, Livingston EH54 6DB
M0 DMR D. Rolf 40 Gunton Drive, Lowestoft NR32 4QB
M0 DMS D. Schofield 9 Manor View, Shafton, Barnsley S72 8NQ
MI0 DMT J. Hyndman 4 Larchfield Gardens Kilrea, Coleraine BT51 5SB
MM0 DMU J. Stewart 76 Caroline Terrace, Edinburgh EH12 8QU
M0 DMX B. Donnachie 3 Downs View Close, Scaynes Hill, Haywards Heath RH17 7EQ
M0 DMY C. Hicken 76 Peaksfield Avenue, Grimsby DN32 9QG
M0 DMZ M. Hughes 9 North Street, Owston Ferry DN9 1RT
MI0 DNB E. Hyndman 4 Larchfield Gardens Kilrea, Coleraine BT51 5SB
M0 DND D. Brown 97 Hewett Road North End, Portsmouth PO2 0QS
MW0 DNF C. Owen 58 Bedwellty Road, Cefn Fforest, Blackwood NP12 3HB
MM0 DNH Dr B. Shippey Binn Ericht, 438 Perth Road, Dundee DD2 1JT
M0 DNJ D. Cook Flat 6, 7 Heath Court, Felixstowe IP11 0YQ
MW0 DNK R. Law 12 Gwel Y Llan Llandegfan, Menai Bridge LL59 5YH
MI0 DNM W. Wilson 68 Ballygowan Park, Banbridge BT32 3AW
M0 DNN A. Wharton 184 Surbiton Road, Stockton-on-Tees TS19 7SH
M0 DNO C. Anderson Flora, 61 Merrilees Crescent, Clacton-on-Sea CO15 5XY
M0 DNR R. Dickson 88 Station Road Drayton, Portsmouth PO6 1PL
M0 DNU A. Bateman 3 Church Croft, Bramshall, Uttoxeter ST14 5DE
M0 DNV A. Truman 92 Spring Meadow, Sutton Hill, Telford TF7 4AQ
M0 DNW S. Dyson 18 Coniston Road, Chorley PR7 2JA
MM0 DNX D. Barrett 88 Camp Rd Baillieston, Glasgow G69 6QP
M0 DNY P. Crump 41 Vernon Way, Guildford GU2 8DE
M0 DNZ D. Pauley Charente, Westerfield Road, Ipswich IP6 9AJ
M0 DOA D. Brace 10 Sawles Road, St. Austell PL25 4UD
M0 DOB S. Dobbs Firbeck House Farm Cottage, Steetley, Worksop S80 3EB
M0 DOC J. Jones 16 Laurel Avenue, Darwen BB3 3AG
M0 DOD T. Bodily Flat 5, Denbeigh House, Rushden NN10 0AT
M0 DOH R. Golsby 19 Glascote Close, Shirley, Solihull B902TA
M0 DOK A. Abrams The Cottage, Victoria Road, Rushden NN10 0AS
M0 DOM S. Cassidy 71 Kensington Avenue, Penwortham, Preston PR1 0EE
M0 DON Scottish DX Contest Club c/o J. Theodorson 7 Kingfisher Court, Overstone Lakes, Ecton Lane, Northampton NN6 0BD
M0 DOP A. Hoskins 38 Tasmania Close, Basingstoke RG24 9PQ
MW0 DOR E. Roberts 6 Trem Y Moelwyn, Tanygrisiau, Blaenau Ffestiniog LL41 3SS
M0 DOS A. Yearp 29 Humber Road, Ferndown BH22 8XN
MM0 DOT D. Macnaughton 24 Kepplehills Drive, Bucksburn, Aberdeen AB21 9PQ
M0 DOW A. Holden 21 East View Meadowfield, Durham DH7 8RY
M0 DOY R. Kendrick Forest View, 126 Ameysford Road, Ferndown BH22 9QE
MW0 DOZ Heads of the Valleys ARC c/o A. Sneddon 3 Marigold Close, Gurnos, Merthyr Tydfil CF47 9DA
M0 DPF S. Thompson Flat 3, 225 Westmacott Drive, Feltham TW14 9XB
MD0 DPG P. Taylor 14 Royal Park, Ramsey IM8 3UF Isle of Man
M0 DPH P. Tullock 17 Owthorne Walk, Bridlington YO16 7GB
M0 DPJ B. Crawshaw 112 Waller Road, Sheffield S6 5DQ
M0 DPK P. Tucker 2 Kemps Field, Cranbrook, Exeter EX5 7AZ
M0 DPQ R. Meadley 2 Lower Hollacombe Cottages, Torquay Road, Paignton TQ3 2DP
M0 DPS C. Seager 77 Stonewood, Bean, Dartford DA2 8BZ
M0 DPV A. Clark 195 Ivyhouse Road, Dagenham RM9 5RS
M0 DPW D. Wakefield Rosebank, 109 Spring Road, Stoke-on-Trent ST3 7JA
M0 DPY K. Wilks 22 Uppercroft, Haxby, York YO32 3GD
M0 DQB J. Brown Ladythorn, Cleeve Hill, Cheltenham GL52 3QB
M0 DQD P. Mcgonigal 4 West View, Evenwood, Bishop Auckland DL14 9QH
M0 DQH G. Waugh 5008 Spartanburg Cove, Austin, Texas 78730 United States
M0 DQK H. Beckett 5 Chatsworth, Benfleet SS7 3BB
M0 DQL D. Porteus 2 Piers Close, Malvern WR14 3JH
M0 DQN R. Richards Broad Oak Bascombe Road, Churston, Brixham TQ5 0JZ
M0 DQO C. Bloy 6 Cornfield, Fareham PO16 8UE
MM0 DQP J. Mckay 28 Lumsden Crescent, St. Andrews KY16 9NQ
M0 DQS P. Mccleaft Gas House Farm, Shavington Park, Market Drayton TF9 3SY
MM0 DRA A. Cattanach 8 Auchencairn Place, Monifieth, Monifieth, Dundee DD5 4TS
M0 DRB F. Dawson Silverthorne, Lower Sea Lane, Bridport DT6 6LR
M0 DRD Dr D. Homer 118 Brampton Way, Portishead BS20 6YT
M0 DRE J. Barnett 5 Manknell Road, Chesterfield S41 8LZ
M0 DRF D. Falkner 45 Westwood Carleton, Skipton BD23 3DW
M0 DRG D. Green 39 Heritage Park, Hatch Warren, Basingstoke RG22 4XT
M0 DRI A. Lyons 20A Russell Street Devonport, Auckland 624 New Zealand
M0 DRK D. Carman 30 Parsonage Close, Burwell, Cambridge CB25 0ER
M0 DRL D. Lane The Coltmoor, Peterchurch, Hereford HR2 0SW
M0 DRM Rose & Crown RC c/o J. Mahoney 27 Linby Drive, Bircotes, Harworth, Doncaster DN11 8PF
M0 DRN A. Small 28 Capel Street Capel-Le-Ferne, Folkestone CT18 7LZ
M0 DRO Capt. R. Bell Sandgate House, Gough Road, Sandgate, Folkestone CT20 3BE
M0 DRQ M. Munn 54 Longbeech Park, Canterbury Road, Ashford TN27 0HA
M0 DRS Dr S. Sampathkumar 52 Crowstone Road, Westcliff-on-Sea SS0 8BD
MM0 DRT D. Taylor Hillcrest, Woodside Place, Banchory AB31 5XW
MW0 DRU D. Underwood 891 Heol Y Ffynon, Penrhys, Ferndale CF43 3RN
M0 DSC D. Westwood 60 Selwyn Drive, Stockton-on-Tees TS19 8XF
M0 DSF Dr D. Fenna 84 High Park Road, Ryde PO33 1BX
M0 DSI D. Shaw 35 Tinshill Lane, Leeds LS16 6BU
M0 DSL P. Zatylny 100 Larkspur Way, West Ewell, Epsom KT19 9LU
MM0 DSM E. Mcneill 3 Sunnybrae Terrace, Maddiston, Falkirk FK2 0LP
M0 DSN D. Tomlinson 3 Holgate Road, Nottingham NG2 2EB
M0 DSO R. De Ieso 57 Kingfisher Road, Thrapston NN14 4GN
MM0 DSP C. Addison 8 Winding Brae, Longside, Peterhead AB42 4XQ
M0 DSR N. Passam 177 Uttoxeter Road, Blythe Bridge, Stoke-on-Trent ST11 9HQ
M0 DSS D. Smith Church House Farm, Church Terrace, Alnwick NE66 2YD
M0 DSW J. Wright 31 Cherry Orchard, Marlborough SN8 4AS
M0 DSX D. Mir 2 Chadhurst Cottages, Coldharbour Lane, Dorking RH4 3JH
M0 DSY D. Bailey 28C Cliff Road Dovercourt, Harwich CO12 3PP
MW0 DSZ D. Church Elm Cottage, Pant, Oswestry SY10 9RB
M0 DTA R. Cheverall 1 Clarkwood Cottages, Twitty Fee, Chelmsford CM3 4PG
M0 DTB T. Belton Flat 6, 28 First Avenue, Hove BN3 2FF
MW0 DTD R. Edwards 59 Allt-Yr-Yn Close, Newport NP20 5EE
MW0 DTH R. Howes 8 Birbeck Road, Caldicot NP26 4DX
M0 DTJ Dr J. Holloway 76 Sir Thomas Whites Road, Coventry CV5 8DR
M0 DTK D. Cox 45 Victoria Road, Walderslade, Chatham ME5 9HB
MM0 DTL Aberdeenshire CG c/o I. Ross Idlewildeá, Inverurie AB51 0XA
M0 DTR D. Anderson 76 South Road, Northfield, Birmingham B31 2QY
M0 DTS R. Swinbank Oxhill Farm, Hilton, Yarm TS15 9LB
MM0 DTW R. Clark 52 Loons Road, Dundee DD3 6AQ
M0 DUB D. Wood 3 Ripley Close, Wakefield WF3 2FG
M0 DUP S. Cox 63 Netherfield Avenue, Eastbourne BN23 7BT
M0 DUQ A. Bridgeland 17 Oldfield Lane, Wisbech PE132RJ
MM0 DUR M. Mckay Tryggo, Sarclet, Wick KW1 5TU
M0 DUT J. Barley 2 Little Hobbyvines, Duckend, Stebbing CM6 3BP
M0 DUU D. Allen 20 Evenley Road, Northampton NN2 8JR
M0 DUV D. Tootill Apartment 18, Ley Gardens, Lawley Close, Church Stretton SY6 6GA
M0 DUY D. Bates Bold Gate Lodge, Praze, Camborne TR14 0NQ
MM0 DVB A. Paton 17 Union Terrace, Keith AB55 5EQ
M0 DVD P. Barnes 20 Benbow Drive, South Woodham Ferrers, Chelmsford CM3 5FP
M0 DVF I. Dummer 29 Chisholm Close, Southampton SO16 8GU
M0 DVG W. Bosworth 87 Tregorrick Road, Exhall, Coventry CV7 9FH
MI0 DVH M. Mclaughlin 22 Duncreggan Road, Londonderry BT48 0AD
M0 DVK S. Christodoulou 27 Manor Place, Cambridge CB1 1LE
MW0 DVM D. Morris 14 Church Terrace, Porth CF39 0ET

M0 DVQ A. Yates 12 Rowsley Avenue, Derby DE23 6JY
M0 DVR P. King 130 Langer Lane, Chesterfield S40 2JJ
M0 DVT J. Adlington 23 Newstead Road, Abbey Hulton, Stoke-on-Trent ST2 8HU
M0 DVU T. Wilf Flat 10, Church View Church Walk, Bourne PE10 9UQ
M0 DVW G. Mallinson 6 Deerplay Court, Bacup OL13 8GE
MM0 DVZ J. Craig Borrowhill Farmhouse, Strichen, Fraserburgh AB43 6TJ
M0 DWB M. Tidman 3 Stannet Way, Wallington SM6 8BE
M0 DWC D. Wraight 2 Silver Mead, Congresbury, Bristol BS49 5EX
MI0 DWD D. Hamilton 7 Bolea Park, Limavady BT49 0SH
MI0 DWE D. Eames 108 Rabbitburrow Road, Farnamullan, Lisbellaw, Enniskillen BT94 5FL
MM0 DWF Dr L. Boehme 26 Sandylands Road, Cupar KY15 5JS
M0 DWG Dr R. Gooch 14 Cotterill Road, Surbiton KT6 7UN
M0 DWK Dr C. Jewell 4 Springfield, Bentham, Lancaster LA2 7BA
M0 DWM D. Cartlidge 2 Walton Road, Walsall WS9 8HN
M0 DWP D. Peters 5 Riverside, Buntingford SG9 9HJ
M0 DWS D. Summerwill 52 Lanmoor Estate, Lanner, Redruth TR16 6HN
M0 DWT D. Todd 20 Wasdale Close, Plymouth PL6 8TL
MW0 DWU A. Fleming 30 Corbett Close, Tywyn LL36 0BL
M0 DWW M. Giffin 15 Chalk Lane, Sutton Bridge PE12 9YF
M0 DWX D. Mcintosh 14 Anne Crescent, Barnstaple EX31 3AF
M0 DWZ R. Webster 74 Bescar Brow Lane, Scarisbrick, Ormskirk L40 9QG
MW0 DX Burnham Beeches RC c/o T. Clapp Crunns Farm. Coxhill, Narberth SA67 8EH
M0 DXA T. Vasdaris 86 Hall Cross Road, Huddersfield HD5 8LD
MM0 DXC C. Stevenson 8 Carlaverock Grove, Tranent EH33 2EB
MM0 DXD J. Wilson 20 Ballumbie Gardens, Dundee DD4 0NR
M0 DXF P. Binswanger 115 Hawthorn Bank, Spalding PE11 1JQ
MM0 DXH J. Hume 51 Vogrie Road, Gorebridge EH23 4HL
M0 DXJ C. Humphris Glebe House, School Lane, Spilsby PE23 4AU
M0 DXM G. Pesch Nikolaus-Jansen-St. 10, Simmerath D52152 Germany
M0 DXN T. Green 9 Craiglands Park, Ilkley LS29 8SX
M0 DXP D. Poole 17 London St., Chertsey KT16 8AP
M0 DXQ P. Dougherty 79 Beverly Road, West Caldwell, New Jersey 07006-6532 United States
M0 DXR M. Haynes Yew Tree Cottage, Brick End CM6 2BL
M0 DXS D. Sheppard 75 St. Nicholas Road, Littlestone, New Romney TN28 8QA
M0 DXT W. Tinnion 3 Brayton Road, Aspatria, Wigton CA7 3DJ
M0 DXV M. Whitehead 29 Coulsons Road, Bristol BS14 0NN
MW0 DXX S. Pettipher Min Yr Afon, Llandysul SA44 5AT
M0 DXZ J. Arnell Jersey Farm, Little London, Bishop's Stortford CM23 1BD
M0 DYA O. Haselden 15 Broadmeadow Close, Totton, Southampton SO40 8WB
M0 DYB Dr S. Kemp 5 Oakdale Avenue, Frodsham WA6 6PY
M0 DYG P. Swan 44 Gillingham Road, Gillingham ME7 4RR
M0 DYH P. Lockley 2 Valley View, Bowling Green, Constantine TR11 5AP
M0 DYO M. Blair 20 Hazel Drive, Burn Bridge, Harrogate HG3 1NY
M0 DYQ P. Davis 9 Chartwell Road, Kirkby-In-Ashfield, Nottingham NG17 7HB
M0 DYR S. Durham 24 Morgan Close, Yaxley, Peterborough PE7 3GE
MW0 DYS A. Davies 17 Queens Road, Merthyr Tydfil CF47 0NB
M0 DYU Rvd. W. Walker 9 Malthouse Lane, Dorchester-On-Thames, Wallingford OX10 7LF
M0 DYV R. Hitchins 1 The Villas, Romansleigh, South Molton EX36 4JW
M0 DYW D. Donnelly 11 Anton Street, London E8 2AD
MM0 DYX D. Francis 2 Morlich Crescent, Dalgety Bay, Dunfermline KY11 9UW
MW0 DYZ R. Jeffery 66 Redhouse Road, Cardiff CF5 4FH
M0 DZA M. Brinnen 82 Victoria Road, Mablethorpe LN12 2AJ
M0 DZB K. Johnson 16 Lamberts Close, Weasenham, King's Lynn PE32 2TE
M0 DZC R. Barton 57 Croxteth Drive, Rainford, St. Helens WA11 8LA
M0 DZD M. Smith 16 Thornton Drive, Brierley Hill DY5 2BS
M0 DZG B. Banks 1 Worthington Road, Dunstable LU6 1PN
M0 DZH P. Holloway 43 Little Sammons, Chilthorne Domer BA228RB
M0 DZL M. Fitzpatrick 3 Orchard Close, Yealmpton, Plymouth PL8 2JQ
M0 DZM J. Enright Flat 11/A, Cold Springs Farm, Manchester Road, Buxton SK17 6ST
M0 DZO A. Hambidge 25 Lock Drive, Stechford, Birmingham B33 8AB
M0 DZT R. Clark 10 Clarendon Road, Bournemouth BH4 8AL
M0 DZV J. Maynard 5 Farm Close, Crowthorne RG45 6SE
M0 DZW I. Massey 129 Church Street, Milnthorpe LA7 7DZ
M0 DZX A. Nolan 41 Taylor Street, Rochdale OL12 0HX
M0 EAB S. Pochojka 38 West Cotton Close, Northampton NN4 8BY
M0 EAD J. Hart 1 Cuckoo Nest, Harden, Bingley BD16 1BD
M0 EAE K. Kotarba 14 North Park, Bristol BS15 1UW
M0 EAF R. Astbury 14 Thornberry Drive Dy12Pl, Dudley DY1 2PL
MM0 EAI D. Jamieson Drumrae, Barbour Road, Helensburgh G84 0JN
M0 EAK F. Cappleman 8 The Woodlands, Lilleshall, Newport TF10 9EN
M0 EAL G. Gauld 251 Fleetwood Road South, Thornton-Cleveleys FY5 5EA
M0 EAM G. Bourne 72 Cornish Way, Royton, Oldham OL2 6JY
MW0 EAN R. Westcott 8 Pen Y Bigyn, Llanelli SA15 1PB
M0 EAO C. Mcdonnell Sunholme, Witham Bank West, Boston PE21 8PU
M0 EAQ Dr T. Gale 38 Lantree Crescent, Trumpington, Cambridge CB2 9NJ
MM0 EAR Mid Severn Valley Raynet c/o K. Carroll 32E Meadowburn Place, Campbeltown PA28 6ST
M0 EAS P. Humphrey 19 Donnay Close, Gerrards Cross SL9 7PZ
MW0 EAT A. Phillips 85 Gorseinon Road, Penllergaer, Swansea SA4 9AB
M0 EAU D. Harrison 21 Springfield Estate, Scopwick, Lincoln LN4 3NP
MM0 EAX D. Thomson Boat House, Finstown, Isle of Orkney KW17 2EH
M0 EAY C. Knight Shamrock, Stow Lane, Wisbech PE13 2JU
M0 EAZ R. Bennett Hawthorn Cottage, Mill Lane, Norwich NR12 8HP
M0 EBD R. Chew 1 Exeter Close, Aintree, Liverpool L10 8LU
M0 EBG P. Good 80 Meredith Road, Stevenage SG1 5QS
M0 EBI Prof. N. Billingham 19 Tumulus Road, Saltdean, Brighton BN2 8FR
M0 EBJ S. Rope 12 Edrich Close, Norwich NR134JD
M0 EBN S. Bywater Birch Wood, Norwich Road, Cromer NR27 0HG
M0 EBO I. Bryant 17 Kent Road, Southampton SO17 2LJ
M0 EBP G. Joyce 8 Christ Church Street, Preston PR1 8PJ
M0 EBQ A. Medhurst 44 Battle Road, Hailsham BN27 1DS
M0 EBR M. Shuttleworth 19 Edgerton Drive, Tadcaster LS24 9QW
MW0 EBT J. Larden Pavilla, 2 Craignant, Nantmel, Llandrindod Wells LD1 6EW
M0 EBU K. Naylor 35 Whiston Road, Northampton NN2 7RR
M0 EBV H. Hepworth The Leas, Main Street Bishop Wilton, York YO42 1RX
M0 EBX M. Tween 79 West Close, Fernhurst, Haslemere GU27 3JS
M0 ECC E. Cassidy 3 The Elms, Great Chesterford, Saffron Walden CB10 1QD
MW0 ECF D. Edwards 25 Bryn Coed, Gwersyllt, Wrexham LL11 4UE
M0 ECK Hms Cavaller RC c/o B. Lucas 8 Gilbert Close, Hempstead, Gillingham ME7 3QQ
M0 ECL E. Lerpiniere The Windmill, Millwrights, Colchester CO5 0LQ
M0 ECM C. Martin 14 Freeston Terrace, St. Georges, Telford TF2 9HD
M0 ECP K. Fujita Flat A, 33/F, Block 5, 10 Sheung Ning Road, Tseung Kwan O, Kowloon, Hong Kong, Kowloon 0 Hong Kong
M0 ECQ I. Nutt Green Acres, Chapple Road, Newton Abbot TQ13 9JY
M0 ECR East Cheshire Radio Group c/o G. Hannan 20 Arlington Drive, Stockport SK2 7EB
M0 ECT M. Mrozinski 6 Grange Road, Northampton NN3 2AZ
MM0 ECV J. Sleet Keppel Gate, Forglen, Turriff AB53 4LX
M0 ECW the City of Belfast Ymca RC c/o T. Watts 26 Woodger Close, Guildford GU4 7XR
M0 ECX Dr N. Rothwell Hughes Cefn Glas, Clyro, Hereford HR3 5JT
MW0 ECY H. Fingerhut-Holland Osnok, 1 South Cliff Street, Tenby SA70 7EB
M0 ECZ E. Williams 22 Sherwood Avenue, Melksham SN12 7HL
M0 EDA S. Hemmings Leylands, Leigh Road, Frome BA11 3LR
M0 EDE A. Holmes 10 Doe Park, York YO30 4UQ
MI0 EDF P. Hawthorne 77 Pollock Drive, Lurgan, Craigavon BT66 8JP
MU0 EDN B. Gray 21 Auderville, Alderney GY9 3XE Guernsey
M0 EDO S. Williams 6 Grasmere Road, Dewsbury WF12 7PU
M0 EDP J. Barr 41 Salisbury Road, London E12 6AA
MW0 EDQ N. Heyne New House, Hope, Welshpool SY21 8JD
M0 EDR S. Pritchard 18 Cumberland Avenue, Basingstoke RG22 4BG
MW0 EDS G. Edwards Douglas Arms Hotel, Ogwen Terrace, High Street, Bethesda, Bangor LL57 3AY
M0 EDU M. Wade Watts Palace Cottage, Chitcombe Road, Rye TN31 6EX
MW0 EDX A. Koval 287 Heol Y Coleg, Newtown SY16 1RA
M0 EEB M. Brady 3 Bransdale, Worksop S81 0XY
M0 EEG Dr C. Pomfrett 17 Manifold Close, Sandbach CW11 1XP
M0 EEH P. Halpin 50 Celtic Road, Deal CT14 9EF
M0 EEK D. Edgar 31 Albany Villas, Hove BN3 2RT
M0 EEL S. Connelly 79 Pettycot Crescent, Gosport PO13 0SJ
M0 EEP J. Nethercott 30 Goldcrest Road, Chipping Sodbury, Bristol BS37 6XF
MM0 EFI F. Wenseth 2 Sunnybank Cottage, Logie Coldstone, Aboyne AB34 5PQ
MM0 EFJ Cray Valley RS c/o M. Donnachie Roselea, Cluny Road, Dingwall IV15 9NJ
MI0 EFM E. Mulligan 27 Hillside Park, Belfast BT9 5EL
MU0 EFR D. Robert Nos Treis 7 Liberation Drive, Route Des Clos Landais, St. Saviour, Guernsey GY7 9PH
M0 EGA R. Price 2 Wordsworth Avenue, Easington Lane, Houghton le Spring DH5 0NR
M0 EGC East of Greenwich RAC c/o D. Green 6 Garth Villas, Rimswell, Withernsea HU19 2DB
M0 EGG E. Gurler Flat 302 Trafford House Cherrydown East, Basildon SS16 5FW
M0 EGL C. Dunne 48 Drury Road, Harrow HA1 4BW
M0 EGN C. Lewis 3 Sovereign Way, Calcot, Reading RG31 4US
M0 EGV I. Robinson 5 The Meadows, Bempton, Bridlington YO15 1LU
M0 EHA G. Beam 35A Moor Lane, York YO24 2QX
M0 EHF Essex Raynet c/o G. Tiller 15 Woodlands Gardens, Romsey SO51 7TE
M0 EHL M. Longbottom 32 Ann'S Hill Road, Gosport PO12 3JY
M0 EHS P. Shaw 16 Sutherland Road, Cradley Heath B64 6EA
M0 EIJ P. Micuda 9 Manor Court, Manorgate Road, Kingston upon Thames KT2 7AN
M0 EIR S. Mcmickan Glenevry, Hillquarter, Coosan, Athlone N37R8Y2 Ireland
M0 EIW J. Walsh 35 Carisbrooke Road, Bushbury, Wolverhampton WV10 8AB
M0 EJB D. Banks 9 Woodbank, Egremont CA22 2RL
M0 EJF C. Briggs 21 Peak View Road, Chesterfield S40 4NW
M0 EJG J. Freeman High Meadow, Martens Lane, Colchester CO6 5AG
M0 EJL P. Kendall 3 Hurstwood Close, Lincoln LN2 4TX
M0 EJW Dr M. Bishop 22 Herrington Road, Dorchester DT1 2BS
M0 EKB G. Patrick Athena, 121 Ringmer Road, Worthing BN13 1DX
M0 ELA A. Mckenzie Little Wishmore, Whitbourne, Worcester WR6 5SR
M0 ELC L. Chesters 5 Lingley Fields, Frizington CA26 3RU
MM0 ELF W. Mccue 188 Redburn, Alexandria G83 9BU
M0 ELO C. Bootz 10 Davenport Avenue, Nantwich CW5 5QJ
MM0 ELP C. Maxwell 29 Ambleside Rise, Hamilton ML3 7HJ
M0 ELS J. Randall 32 Holm Oak Close, Canterbury CT1 3JL
MM0 EMC E. Mcpherson 12 Lambourn, Wolfhill, Perth PH2 6TQ
M0 EMD E. Deeley 26 Eversley Crescent, Isleworth TW7 4LS
M0 EME P. Tomlinson 217 Old Hall Road, Tapton, Chesterfield S40 1HQ
M0 EMM D. Martin 14 Freeston Terrace, St. Georges, Telford TF2 9HD
M0 EMR C. Wright 8 Kendal Road, Sheffield S6 4QG
M0 EMW E. Wheeler 3 Praze Road, Porthleven, Helston TR13 9LR
MI0 ENR R. Mcfadden 36 Trinity Drive, Ballymoney BT53 6EQ
M0 EOT B. Podmore 78 Ridge Road, Stoke-on-Trent ST6 5LP
M0 EOU J. Hinds 2C Telegraph Street, Stafford ST17 4AT
MM0 EPC Eaaro ARC c/o J. Phunkner 7 Plenshin Court, Glasgow G53 6QW
M0 EPR E. Rippon 319 Beechdale Road, Nottingham NG8 3FF
M0 EPX L. Stone 32 Watermeadow Lane, Storrington, Pulborough RH20 3GU
MM0 EQE A. Thompson 22 Lochend Road, Carnoustie DD7 7QF
MW0 EQL J. Sneddon 3 Marigold Close, Gurnos, Merthyr Tydfil CF47 9DA
M0 EQM R. Agacy 23 Highgate Lane, Bolton-Upon-Dearne, Rotherham S63 8HR
M0 EQY M. Campbell 377 Bushbury Lane, Wolverhampton WV10 8JZ
M0 ERG Eagle Radio Group c/o T. Stow 38 The Strand, Mablethorpe LN12 1BQ
M0 ERJ E. Jones 37 Sluice Road, Denver, Downham Market PE38 0DY
MM0 ERK B. Murray Sherwood Cottage, Farnell, Brechin DD9 6UH
M0 ERN E. Coleby 13 Farm Close, Sunniside, Newcastle upon Tyne NE16 5PP
M0 ERS J. Hauton 15 Bourne Close, Lincoln LN6 7DR
M0 ERY M. Samborskyy St Johns College, St Johns Street, Cambridge CB2 1TP
M0 ESB Wessex Aw Club c/o D. Baldwin 356 Uppingham Road, Leicester LE5 2BF
M0 ESP Ilera c/o L. Marobin Flat 60, Tudor Court, King Henrys Walk, London N1 4NU
M0 ESR Cheltenham Hackspace ARS c/o P. Phillips 2 Millstream Close, Goostrey, Crewe CW4 8JG
M0 ESU M. Bown 47 Ullswater Crescent, Weymouth DT3 5HF
M0 ESW E. Swanepoel Bridge Cottage, Tinhay, Lifton PL16 0AH
M0 ESZ M. Owens 66 Woodlands Road, Bishop Auckland DL14 7LZ
M0 ETA G. Andrews 158 Latchmere Road, Kingston upon Thames KT2 5TU

M0 ETE R. Home Beech Cottage, The Green, Huntingdon PE28 9NA
M0 ETP J. Bolton 2 Patterdale Street, Hetton-Le-Hole, Houghton le Spring DH5 0BH
M0 ETQ D. Bolton 2 Patterdale Street, Hetton-Le-Hole, Houghton le Spring DH5 0BH
M0 ETS C. Lyon 3 Doodstone Avenue, Lostock Hall, Preston PR5 5TY
M0 ETY S. Hindle 35 Heyhead Street, Brierfield, Nelson BB9 5BN
M0 EUI G. Plant 11 Shinwell Grove, Stoke-on-Trent ST3 7UG
M0 EUK G. Stoker 32 Willow Way, Ponteland, Newcastle upon Tyne NE20 9RF
MW0 EUS A. Jones Pant Y Beddau, Bethania, Llanon SY235NN
M0 EUY A. Swan 47 Warren Close, Whitehill, Bordon GU35 9EX
M0 EVE P. Beier 20 Markham Avenue, Armthorpe, Doncaster DN3 2AZ
M0 EVI A. Butler 88 Manor Road, New Milton BH25 5EJ
M0 EVK R. Kidd 4 Oakfields Close, Norwich NR4 6XH
M0 EVT E. Haralampiev 13 South Hill Grove, Harrow HA1 3PR
M0 EWG B. Read 111 Fitzpain Road, West Parley, Ferndown BH22 8SF
M0 EWW R. Moreton 25 Holyoake Place, Rugeley WS15 2NP
M0 EXM B. Wheeler 2 Rose Street, Houghton le Spring DH4 5BB
MW0 EYE Dr C. Varghese 15 Crestacre Close Newton, Swansea SA3 4UR
M0 EYT P. Marsh 10 Pardys Hill, Corfe Mullen, Wimborne BH21 3HW
M0 EZO R. Griffin 15 Donside Close, Boldon Colliery NE35 9BS
M0 EZP D. Brewerton 10 Porter Avenue, York YO19 4AG
M0 FAK R. Chick 15 Bonfire Close, Chard TA20 2EG
MU0 FAL C. Fallaize Lorbert, Pleinheaume Road, Vale GY6 8NR Guernsey
M0 FAR A. Rawson 64 Dukes Mead, Fleet GU51 4HE
M0 FAT A. Moffatt 10 Fenn Road, Barnsley S75 3DE
M0 FAZ D. Fower 31 Hillswood Avenue, Leek ST13 8EQ
M0 FBB S. Smale 230 Wareham Road, Corfe Mullen, Wimborne BH21 3LW
M0 FBM A. Lock 2 Knutscroft Lane, Thurloxton, Taunton TA28RL
M0 FCA W. Mannerfelt 16 Suffolk Road, London SW13 9NB
M0 FCB F. Brunt 74 Bardley Crescent, Tarbock Green, Prescot L35 1RJ
M0 FCD M. Christieson September Cottage, Rushlake Green, Heathfield TN21 9PP
M0 FCI D. Houghton 106 Lawn Avenue Woodlands, Doncaster DN6 7TT
MM0 FCM C. Sheridan 8 South Park Grove, Biggar ML12 6GJ
M0 FCP F. Parsons 17 Hannah More Close, Wrington, Bristol BS40 5QG
M0 FCR Swindon ARC c/o A. Crespo 266 Trinity Road, London SW18 3RQ
M0 FCT P. Ryder 6 Lodge Drive, Moulton, Northwich CW9 8RQ
M0 FCW M. Ballard 41 Middlefield Avenue, Halesowen B62 9QJ
M0 FCY D. Thornton 63 Houghtonside, Houghton le Spring DH4 4BW
MW0 FDG J. Myszka 65 Heol-Y-Frenhines, Bridgend CF31 4RN
M0 FDX G. Marsden 71 Sedgley Avenue, Rochdale OL16 4TY
M0 FEU M. Pullan Hauptstrasse 13/2, St. Radegund Bei Graz 8061 Austria
M0 FEY E. Fey 56 Bathurst Close, Burnham on Sea TA8 2SZ
MM0 FFC I. Douglas1 15 Henderson Crescent, Broxburn EH52 6HA
M0 FFX T. Wootten Trinity Hall, Cambridge CB2 1TJ
M0 FGA D. Mccarty Po Box :4910, the Woodlands Tx 77387 United States
M0 FGB F. Buck 89 Marsh St., Barrow in Furness LA14 2AD
M0 FGC T. Seed 1A, Purok , Brngy Arenas. Arayat, Pampanga NONE Philippines
M0 FIL P. Graham 29 Lancaster Street, Colne BB8 9AZ
M0 FIS P. Fisher 12 St. Anns Avenue, Grimsby DN34 4PW
MD0 FIX N. Wallace 85 Erin Vale, Lezayre Park, Ramsey IM8 3PU
M0 FJM J. Hudson 62 Old Road, Churwell, Morley, Leeds LS27 7RT
M0 FJS F. Stevenson Andalsnes 33 Highfield Close Amersham Hp6 6Hg, Bucks HP6 6HG
M0 FLC I. Pollard Ilderton Glebe Cottage, Ilderton, Alnwick NE66 4YD
M0 FLF C. Wilson 12 Desmond Avenue, Hornsea HU18 1AF
M0 FMT P. March Devonholme, Bedford Road, Hitchin SG5 3RX
M0 FMY J. Mallichan 17 Napier Road, Gillingham ME7 4HB
M0 FOG N. Brereton 10 Coverdale Close, Stoke-on-Trent ST3 7RZ
M0 FOR G. Mcinnes 32 Bis Rue D'Ezy, 32 Bis Rue D'Ezy, Ivry la Bataille 27540 France
M0 FOX P. Leicester 30 Knighton Street, North Wingfield, Chesterfield S42 5JA
M0 FPA R. Etchells 6 Woodbank Court, Canterbury Road, Manchester M41 7DY
M0 FQZ D. Garner 1 Rowland Avenue, Field Street, Hull HU9 1HR
M0 FRA T. Fray 20 St. Catherines Road, Blackwell, Bromsgrove B60 1BN
M0 FRC Southdown ARS c/o R. Topliss 12 Dorothy Avenue, Skegness PE25 2BP
M0 FRD M. Taylor 10 Piers Road, Glenfield, Leicester LE3 8BN
M0 FRG A. Howard 4 Woodgarth Avenue, Manchester M40 1QE
M0 FRH I. Fraser Flat 2, 77 Bayford Road, Littlehampton BN17 5HN
M0 FRS C. Boys 34 Firacre Road, Ash Vale, Aldershot GU125JT
MW0 FRY R. Fry Old Police Station, Parkmill, Swansea SA3 2EQ
M0 FSH N. Harris 23 Winchester Avenue, Chatham ME5 9AR
M0 FSK F. Kennedy 3 Brookfield View, Bolton Le Sands, Carnforth LA5 8DJ
M0 FSN P. Wells 7 Kings Meadow, Overton, Basingstoke RG25 3HP
M0 FTL R. Metcalfe 33 Midland Terrace, Hellifield, Skipton BD23 4HJ
M0 FTR Box 25 Contest Club c/o J. Adlington 23 Newstead Road, Abbey Hulton, Stoke-on-Trent ST2 8HU
MW0 FUN B. Fisher 93 Heol Llanelli, Pontyates, Llanelli SA15 5UH
M0 FVD M. Kittika 51 Overlea Drive, Burnage, Manchester M19 1QY
M0 FVV A. Hanner 8 Countryside Farm Park, Church Lane, Steyning BN44 3HF
MM0 FWG R. Gaisford 9 Rattray St., Boness EH51 9PE
M0 FWM F. Mifflin Windsor House, Harras Road, Whitehaven CA28 6SG
M0 FWO J. Thomson 51 Birch Avenue, Cuerden Residential Park, Leyland PR25 5PD
M0 FXB A. Macrides 16 Newtons Road, Kewstoke, Weston-Super-Mare BS22 9LG
M0 FXX R. Limb Charnwood House, Station Road, Henley-on-Thames RG9 3JS
M0 FYA A. Young 39 Thornton Drive, Hoghton, Preston PR5 0LX
M0 FZR R. Wickenden Selwood Cottage, Moor Lane, Wincanton BA9 9EJ
M0 FZU C. Walcott 28 Balfour Road, London W13 9TN
MM0 FZV G. Bourhill 30C Salters Road, Wallyford, Musselburgh EH21 8AA
M0 FZW N. Crampton 7 Barneveld Avenue, Canvey Island SS8 8NZ
M0 FZX S. Norman 27 Ashburton Road, Ickburgh, Thetford IP26 5JA
M0 GAC G. A'Court Three Oaks, Greenhill Lane, Winscombe BS25 5PE
M0 GAE G. Errington 22 Willoughby Drive, Whitley Bay NE26 3DY
M0 GAG R. Burrell 38 Standhill Crescent, Barnsley S71 1SU
M0 GAH A. Cunningham 23 Heathgate Close, Birstall, Leicester LE4 3GW
MM0 GAI T. Spencer 3 Tramore Crescent, Prestwick KA9 1LT
MM0 GAL J. Vilar Ameijeiras 76 Willowbank Road, Aberdeen AB11 6XL
M0 GAN P. Street 50 Dickson House, Ridgway Road, Stoke on Trent ST1 3BA
M0 GAQ K. Ingram 15 Kent Avenue, East Cowes PO326QN
M0 GAV A. Burton 51, Wilcox Road Foxhill, Sheffield S6 1BQ
M0 GAX C. Ponder 12 Wood Street, Doddington, March PE15 0SA
M0 GBA G. Allison 16 Copse Road, Plymouth PL7 1PZ
M0 GBB Winteringham Wireless Society c/o D. Ogg 36 Cliff Road, Winteringham, Scunthorpe DN15 9NQ
M0 GBC W. Jefferies 26 Norcutt Road, Twickenham TW2 6SR
M0 GBF B. Findler 1 Gordon Avenue, Stoke-on-Trent ST6 2LY
M0 GBH C. Johnston-Stuart 35 Robbins Close, Bradley Stoke, Bristol BS32 8AS
M0 GBK S. Nash 6 Berry Road, Meltham, Holmfirth HD9 5PL
M0 GBO J. Zemlicka 37 Ascot Gardens, Southall UB1 2SA
MI0 GBU Causeway RC c/o N. Bolt 32 Bush Gardens, Bushmills BT57 8AE
MW0 GBW B. Collins 58 Brockhill Way, Penarth CF64 5QD
M0 GBZ E. Mcpherson 138 Shephall View, Stevenage SG1 1RR
M0 GCA T. Sheridan 4 Stane Close, Bishop's Stortford CM23 2HU
M0 GCB T. Kim 9 Temeraire Heights, Folkestone CT20 3TL
M0 GCC G. Cook 61 Mortomley Lane, High Green, Sheffield S35 3HS
MM0 GCF J. Brown 78 Egilsay St., Glasgow G22 7RG
M0 GCH G. Holmes 11 Claudeen Close, Southampton SO18 2HQ
M0 GCI M. Rowley Flat 23, Minstrel Court, 170 High Street, Harrow HA3 7AX
M0 GCN D. Bourne 6 The Maples, Abbeymead, Gloucester GL4 5WQ
M0 GCQ G. Soteriou Whiteleys Cottages, 32 Mornington Avenue, London W14 8UW
M0 GCR G. Rumsey 13 Greenhills Road, Northampton NN2 8EL
MW0 GCS L. Powell 2 Gelliderw Pontardawe, Swansea SA8 4NB
MW0 GCT S. Tweddle 3 Bron Ffinan, Pentraeth LL75 8UT
M0 GCU Eshaness RC c/o J. Jordan The Cottage, Papcastle, Cockermouth CA13 0LA
MI0 GCV Mid-Thames RDF c/o T. Conlon 7 Waringfield Gardens, Moira, Craigavon BT67 0FQ
M0 GCX Scarborough Seg c/o V. Narinian 15 Headley Gardens, Great Shelford, Cambridge CB22 5JZ
MM0 GCY C. Bewley Fodderlee Dell, Hawick TD9 8JE
M0 GDC R. Bakken Flat 6, Marlow House, Abbey Street, London SE1 3DW
MM0 GDG Northampton Sct c/o Dr A. Curlis 94 Kirkhill Road, Aberdeen AB11 8FX
M0 GDH G. Hogg 7 Elbra Farm Close, Ellenborough, Maryport CA15 7RG
MM0 GDI Gloucestershire County Raynet c/o B. Massie Flat 4/R, 74 Commercial Street, Dundee DD1 2AP
M0 GDJ London Raynet G c/o B. Holt 36 Tenter Hill Lane, Sheepridge, Huddersfield HD2 1EJ
MM0 GDL D. Lindsay 114 Strathblane Road, Milngavie, Glasgow G62 8HD
MW0 GDM Horsham Scout ARG c/o P. De Mengel Fern Cottage, 1 The Gail, Haverfordwest SA62 4HJ
M0 GDP R. Parkinson 18 Lime Tree Gardens Lowdham, Nottingham NG14 7DJ
M0 GDS G. De Sousa 29 Slippers Hill, Hemel Hempstead HP2 5XT
M0 GDT D. Holland 13 Linley Drive, Boston PE21 7EJ
M0 GDU R. North 24 Gadesden Road, Epsom KT19 9LB
M0 GDV D. Harbron 6 West View, Penshaw, Houghton le Spring DH4 7HP
M0 GDX D. Hayes 43 Linden Avenue, Sheffield S8 0GA
M0 GEB G. Beesley Stone Barn, Black Dog, Crediton EX17 4QX
M0 GEC G. Clennell 69 Seventh Row, Ashington NE63 8HX
M0 GED F. Holt 8 Pleasant View, Coppull, Chorley PR7 4PH
M0 GEF G. Freeman 11 Westward Road, Malvern WR14 1JX
MW0 GEI S. Walmsley 29 Shelley Court Machen, Caerphilly CF83 8TT
M0 GEK L. Gange 15 Ham Close, Worthing BN11 2QE
M0 GEL S. Attwood 60 Underwood Avenue, Ash, Aldershot GU12 6PL
M0 GEN G. Barusevicus 6 Middlebrook Crescent, Bradford BD8 0EN
MM0 GEO G. Muir 51 Lindsay Way, Livingston EH54 8LQ
M0 GEP A. Holdup Tunnel Farm, Tunnel Rd, Imbil (Po 155) 4570 Australia
M0 GEU A. Nicholson 14 Rossinyol, Los Arcos, Alicante 3530 Spain
M0 GEX C. Farley 1 Wesley Cottages, Mutley, Plymouth PL3 4RB
M0 GEY Vintage Operating Group c/o M. Spinks 26 Church Hill, Royston, Barnsley S71 4NH
MM0 GFA Stevenage & District A.R.S c/o P. Mcbride 1 Hillside, Croy, Glasgow G65 9HJ
M0 GFD J. Smith 3 Glamis Road, London E1W 3EE
MI0 GFE Antrim & District ARS c/o R. Robinson 31 Brantwood Gardens, Antrim BT41 1HP
M0 GFF B. Courtney 44 Uxbridge Road, Rickmansworth WD3 7AR
M0 GFJ D. Russell Halfway House, Holbrook Road, Ipswich IP9 1BP
M0 GFK A. Ochot 99 Canterbury Avenue, Slough SL2 1DY
M0 GFM G. Dawson Bramwell, Winchester Road, Southampton SO32 2LG
M0 GFN J. Bell 50 Colchester Terrace, Sunderland SR4 7RY
M0 GFO R. Mcdermott 2 Monument Close, Wellington TA21 9AL
MM0 GFP A. Stewart 21 Mansfield Avenue, Newtongrange, Dalkeith EH22 4SJ
MM0 GFR G. Forster 4 Kirk Brae, Morvern, Oban PA80 5XW
M0 GFX P. Hull 1 Sawpits Close, Stogumber, Taunton TA4 3TX
MI0 GGB S. Quigg 100 Whispering Pines, Limavady BT49 0UF
MM0 GGD G. Duncan 5 Jarvis Place, Carnoustie DD7 7BR
MM0 GGG D. Banks 60 Leander Crescent, Bellshill ML4 1JB
M0 GGH J. Gubric 279 Nottingham Road, Eastwood, Nottingham NG16 2AP
M0 GGK D. Lawson 30 Meadowcroft, St. Helens WA9 3XQ
M0 GGL C. Lester 21 Barwell Way, Witham CM8 2TY
M0 GGM G. Markey Trebrown Farm, Horningtops, Liskeard PL14 3PU
M0 GGO C. Loughran 8 Douglas Road, Dover CT17 0BD
M0 GGP Angel of the North ARC c/o S. Townsley 222 Prince Consort Road, Gateshead NE8 4DX
M0 GGQ A. Shaw 2 Dowber Court, Thirsk YO7 1SP
M0 GGT B. Ashton 31 Home Close Renhold, Bedford MK41 0LB
M0 GGU G. Medlicott Lower Medlicott Farm, Wentnor, Bishops Castle SY9 5EL
M0 GGW G. Milsom 31 Chichester Close, Bowerdean Road, High Wycombe HP13 6AU
M0 GGX J. Patient 4 Bucklebury Heath, South Woodham Ferrers, Chelmsford CM3 5ZU
M0 GGZ A. Chaplin 33 The Crofts, Little Wakering, Southend-on-Sea SS3 0JS
M0 GHA M. Dudley 8 Woolpack Meadows, North Somercotes, Louth LN11 7QG
M0 GHC M. Wojcik 43 Connaught Road, London W13 0TF
M0 GHE L. Nordgren 41 Forest Road, London E7 0DN
MI0 GHI A. Murphy 13 Torrens Park, Lislagan Upper, Ballymoney BT53 7DE
M0 GHK T. Lee 54 Shielfield Terrace, Tweedmouth, Berwick-upon-Tweed TD15 2EE
MM0 GHM G. Cochrane 33 Portland Road, Galston KA4 8EA
MM0 GHN N. Inglis Orchard House 35 Portland Park, Hamilton ML3 7JY
M0 GHO G. Hopkins 27 The Templars, Worthing BN14 9JT
M0 GHR I. Millar 3A South Street, Wiveliscombe, Wiveliscombe TA4 2LZ
MM0 GHT K. Brown 21 Strain Crescent, Airdrie ML6 9ND

M0 GHV S. Young 27A Norton Road, London E10 7LQ
M0 GHX C. Painter 45 Meadow Lane, Beeston, Nottingham NG9 5AE
M0 GHY P. Hollas 46 Askham Fields Lane, Askham Bryan, York YO23 3PS
M0 GHZ D. Millard Weavern House, Hartham Lane, Biddestone, Chippenham SN14 7EA
M0 GIA S. Amesbury 92 Range Court, Macclesfield SK10 2RR
M0 GIB D. Gibbons 4 Ivychurch Mews, Runcorn WA7 5AR
M0 GID G. Dunne Swan Hotel, Market Place, Sturminster Newton DT10 1AR
M0 GIE P. Ellis 40 Grasmere Road Royton, Oldham OL2 6SR
M0 GIF R. Manser Flat 38, St. Johns Court, Portsmouth PO2 8NA
M0 GIG D. Wharlley 15 Crampton Court, Grosvenor Road, Broadstairs CT10 2XU
MI0 GIJ J. Thompson 119 Rathkyle, Antrim BT41 1LN
M0 GIL G. Wildman 55 Hill Street, Bradley, Wolverhampton WV14 8SB
M0 GIM Dr G. Mitchener Cabins, Wenham Road, Ipswich IP8 3EY
MW0 GIN S. Peel 28 Dan Yr Allt, Llanelli SA14 8AT
M0 GIP W. James 3 Midfield Close, Gillow Heath, Stoke-on-Trent ST8 6RD
M0 GIQ G. Griffiths 16 Back Lane, Winteringham, Scunthorpe DN15 9NW
M0 GIU P. Tier 16A Burcombe Road, Bournemouth BH10 5JT
M0 GIW D. Ryan Goosehill House, West Street, Thorne, Doncaster DN8 5QU
M0 GIY Dr P. Swansbury 119A Trelowarren Street, Camborne TR14 8AW
M0 GIZ C. Melia 45 Sheriff Street, Hartlepool TS26 8EZ
M0 GJA K. Nyquist 25 Marsh View, Newton, Preston PR4 3SX
MM0 GJC G. Costa 54 High Street, Dollar FK14 7BA
M0 GJD J. Farrant Orchard Cottage, Claycastle, Crewkerne TA18 7PB
M0 GJH A. Vine Hilden, Woodland Avenue, Cranleigh GU6 7HZ
MM0 GJJ G. Johnson Speur Mor, Gifford Road, Longformacus, Duns TD11 3NZ
M0 GJK G. Knight 20 Crossway, Welwyn Garden City AL8 7EE
M0 GJL R. Brodie 8 West View Terrace, Main Road, Salcombe TQ8 8AB
MI0 GJN M. Edwards 15 Highgrove Road, Carrickfergus BT38 9AG
M0 GJS G. Boyd 26 Bluebell Drive, Littlehampton BN17 6UL
M0 GJU J. Freeman 38 City Road, Cambridge CB1 1DP
M0 GJV E. Jones 43 Wesley Road, Wimborne BH21 2QB
M0 GJX A. Jordan 21 Madison Avenue, Exeter EX1 3AH
M0 GKA S. Len 13 Griffiths Gardens, Caversfield, Bicester OX27 8FL
MM0 GKB K. Mackintosh Chez Nous, Scatwell, Strathconon, Muir of Ord IV6 7QG
MW0 GKD A. Mockford 35 Lingfield Gardens, Whitland SA34 0BJ
M0 GKG R. Ley 23 Heronbridge Close, Westlea, Swindon SN5 7DR
M0 GKJ R. Frencham 1 Eggerslack Cottages, Windermere Road, Grange-over-Sands LA11 6EX
M0 GKK Birmingham ARS c/o D. Fishlock 93 Shackstead Lane, Godalming GU7 1RL
MI0 GKL Bushvalley ARC c/o J. Traynor 8 Roeville Terrace, Limavady BT49 0BH
MM0 GKN J. Hogg 31 Woodlea Court, Crosshouse, Kilmarnock KA2 0ES
M0 GKO G. Thorpe 81 Knoll Drive, Coventry CV3 5PJ
M0 GKP Dr P. Haydn Smith 3 King Henrys Road, Lewes BN7 1BT
M0 GKR A. Steel 78 Water Meadows, Worksop S80 3DB
MM0 GKT Amsat-UK c/o Dr D. Bushby Coach House, Dalginross, Crieff PH6 2HB
MM0 GKU T. Mccall 119 Claremont, Alloa FK10 2ER
MW0 GKV M. Williams 30 Elm Drive, Risca, Newport NP11 6HJ
MW0 GKW M. Sullivan 14 Clerks Court, Mill Lane, Welshpool SY21 7JA
M0 GLE G. Tomkins 12 Jarrah Court, Bathurst 2795 Australia
M0 GLF Dr J. Stanford 1941 Ute Creek Drive, Longmont, Co 80504 United States
MI0 GLG T. Currie 26 High Street, Portaferry, Newtownards BT22 1QT
M0 GLI M. Shasby 19 Crawshaw Grange, Crawshawbooth, Rossendale BB4 8LY
MD0 GLK Sth Manchestr Rd c/o A. Dorman 1 Sprucewood Rise, Foxdale, Douglas IM4 3JP Isle of Man
M0 GLL C. Smith 2 Blankney Close, Fareham PO14 3RX
M0 GLP Thames ARG c/o G. Parker 420 Meadow Lane, Nottingham NG2 3GD
M0 GLQ C. Senior 13 Oak Crescent, Woolaston, Lydney GL15 6PF
MW0 GLS S. Day 2 Cae Job, Piercefield Lane, Aberystwyth SY23 1RJ
M0 GLT M. Rosenbrand 12 Greville Road, Cambridge CB1 3QL
M0 GLU A. Vincz 9 St. Brelades Road, Crawley RH11 9RQ
M0 GLV M. Jusko 13 Ellerby Grove, Hull HU9 3PR
MM0 GLX B. Burt 182 Old Inverkip Road, Greenock PA16 9JG
M0 GMA G. May 95 Moorfield Avenue, Denton, Manchester M34 7TX
M0 GMC G. Collier 6 Copse Close, Tilehurst, Reading RG31 6RH
M0 GMD D. Gray 68 Endeavour Way, Hythe Marina Village, Southampton SO45 6LA
M0 GME G. Ellis 46 The Uplands, Scarborough YO12 5HX
M0 GMG R. Bell 92 Dean Drive, Wilmslow SK9 2EY
MW0 GMH O. Williams 39 Camden Road, Maes-Y-Coed, Brecon LD3 7RT
M0 GMI C. Woodbridge 53 Baffins Road, Portsmouth PO3 6BE
M0 GMK C. Dawson 9 Mulberry Close, Poringland, Norwich NR14 7WF
M0 GMN W. Owen 8 Sandhurst Avenue, Lytham St. Annes FY8 2DA
M0 GMO P. Cheshire 29 Madison Avenue, Exeter EX1 3AH
M0 GMQ P. Hall 13 Sheard Avenue, Ashton-under-Lyne OL6 8DS
M0 GMS Rvd. S. Smith 5 Melhuish Close, Witheridge, Tiverton EX16 8AZ
M0 GMT D. Clapp 150 Brougham Road, Worthing BN11 2PH
M0 GMU P. Sweatman 62 Barncroft Way, Waterlooville PO93AQ
M0 GMW C. Watts 10 Kemble Gardens, Bristol BS11 9RY
MW0 GMZ H. Hughes 21 Maes Geraint, Pentraeth LL75 8UR
M0 GNA A. Shaw 21 Laburnum Road, Prenton CH43 5RP
M0 GNB M. Dreszer 47 Lydgate Court, Nuneaton CV11 5RR
M0 GNC A. Ellison 24 The Grove, Brentwood CM14 5NS
M0 GND Eureka RC c/o R. Johnson 30 Thorpe Downs Road, Church Gresley, Swadlincote DE11 9FB
MW0 GNF C. Hughes 26 Tan Y Bryn, Valley, Holyhead LL65 3ES
M0 GNG S. Gee 20 Nesham Place, Houghton le Spring DH5 8AG
MM0 GNH K. Foreman 16 Beveridge Place, Kinross KY13 8QY
M0 GNI G. Inman 11 Sylvan Way, Gillingham SP8 4EQ
M0 GNJ D. Coventry 1 Seacrest Avenue, Fleetwood FY7 6FG
M0 GNK R. Jennings 8769 Greengrass Way, Parker 80134 United States
M0 GNL Dr C. Price Byways, Taylors Lane, Chichester PO18 8QQ
M0 GNM Dr H. Donnelly 6 Famet Walk, Purley CR8 2DY
M0 GNO E. Whiten 17 Scott Close, Ashby-de-la-Zouch LE65 1HT
M0 GNP D. Salter 142 Brays Road, Birmingham B26 2PP
MM0 GNS C. Stewart 9 Rousay Wynd, Kilmarnock KA3 2GP
M0 GNU D. White 3 West Street, South Normanton, Alfreton DE55 2AJ
MM0 GNX A. Messner 6 Elistoun Drive, Tillicoultry FK13 6NT
M0 GNY M. Zlobinski 16 Birkdale Avenue Atherton, Manchester M46 9PY
M0 GOA J. Goacher 41 Clay Hill, Two Mile Ash, Milton Keynes MK8 8AY
M0 GOB B. Holland 11 Silverlands Park, Buxton SK17 6QX
M0 GOC T. Ward 1 Darrismere Villas, Edinburgh Street, Hull HU3 5AS
MM0 GOF J. Mcculloch 2 Riverbank Wynd, Gatehouse Of Fleet, Castle Douglas DG7 2EA
MM0 GOG North Riding Rafars c/o D. Baillie 126 Main St., Fauldhouse, Bathgate EH47 9BW
M0 GOH P. Preston 49 Cowpasture Lane, Sutton-in-Ashfield NG17 5AR
M0 GOI K. Hornby 39 Parkland View, Barnsley S71 5LG
M0 GOK D. Richards 73 Greenfields Avenue, Alton GU34 2EW
M0 GOL T. Goldsmith 37 Cowdray Road, Sunderland SR5 3PG
MW0 GOM J. Roissetter 2 The Willows Usk Road, Caerleon, Newport NP18 1JB
MM0 GON G. Craig 1 Butt Avenue, Helensburgh G84 9DA
M0 GOO J. Brook The Clock Tower, Rectory Lane, Chichester PO20 9DT
M0 GOP G. Oliver 17 Jack Stephens Estate, Penzance TR18 2QE
M0 GOQ J. Barbieri 20 Gilbard Court, Chineham, Basingstoke RG24 8RG
M0 GOT S. Martin 3 Houndsmill, Horsington, Templecombe BA8 0ED
MW0 GOV C. Davis 132 Steynton Road, Steynton, Milford Haven SA73 1AN
M0 GOW C. Gowing 5 Curson Road, Tasburgh, Norwich NR15 1NH
MW0 GOX J. Davies Penralt, Abercaseg Road, Bangor LL57 3SP
M0 GOY R. Klima 87 Campion Avenue, Hull HU4 7AR
MI0 GOZ V. Maksimavicius 14 Leckagh Walk, Magherafelt BT45 6JU
MI0 GPB G. Bunting 8 Moor Park Avenue, Belfast BT10 0QE
M0 GPC S. Withnall 2 Lansdown Close, Cheltenham GL51 6QP
M0 GPD J. Fuller Bramble Cottage, Leggatt Hill, Lodsworth, Petworth GU28 9DP
M0 GPE T. Loker 24 St. Albans Hill, Hemel Hempstead HP3 9NG
M0 GPG G. Dyson 111 Chester Road, Ellesmere Port CH65 6SB
M0 GPH T. Hall 18 Common Lane New Haw 18 Comman Lane New Haw, Addlestone KT153LH
M0 GPJ D. Waite 1 Naseby Court, Bradville, Milton Keynes MK13 7EP
M0 GPK Dr W. Gasser 76 Empress Road, Derby DE23 6TE
MM0 GPL C. Jones Croy Lodge, Shandon, Helensburgh G84 8NN
M0 GPO G. Otter 3 Glen Park Avenue, Glenfield, Leicester LE3 8GH
MW0 GPP B. Doyle 3 Bryn Road, Flint CH6 5HU
M0 GPQ A. Wieckowski 40 Askham Court, Askham Road, London W12 0NX
M0 GPU A. Norrie 45 Eastern Way, Ponteland, Newcastle upon Tyne NE20 9RD
M0 GPV W. Tommasini 49 Taverner Close, Poole BH15 1UP
M0 GPW P. Andrew 13 Luke Road, Aldershot GU11 3BW
M0 GPX B. Wagiel 116 Hurworth Avenue, Slough SL3 7FQ
M0 GPY Prof. H. Schmidt 88 Candlemas Lane, Beaconsfield HP9 1AE
MM0 GPZ G. Paterson 33 Parkneuk Road, Blantyre, Glasgow G72 0TR
M0 GQB M. Cox Haugh Shaw Hall Haugh Shaw Road, Halifax HX1 3LE
M0 GQD J. Parrett 6 Shelley Road, East Grinstead RH19 1TA
M0 GQE G. Moss 10 Thistlegreen Road, Dudley DY2 9JT
MM0 GQF Z. Biorka 150 St. Michaels Road, Newtonhill, Stonehaven AB39 3XW
MI0 GQG B. Crozier 33 Cullentragh Road, Poyntzpass, Newry BT35 6SD
MI0 GQI M. Crozier 33 Cullentragh Road, Poyntzpass, Newry BT35 6SD
M0 GQJ D. Downer 19 Watergate Road, Newport PO30 1XN
M0 GQM T. Kidwell 2 Batts Farmyard, Wilton, Marlborough SN8 3SS
M0 GQP B. Sims 4 New Cottages, Cranwich Road, Thetford IP26 5EQ
M0 GQR A. Matheson 21 Warren Hill Road, Woodbridge IP12 4DU
M0 GQS D. Roguszczak 4 Home Farm Close, Reading RG2 7TD
M0 GQT S. Hubin 10 Coneygeare Court, Eynesbury, St. Neots PE19 2UL
M0 GQU M. Burzynski 2 Somerset Avenue, Luton LU2 0PJ
M0 GQV P. Langabeer 1 Newfield Crescent, Middlesbrough TS5 8RE
M0 GQW E. Tart Sunnybank Farm, Wattlesborough Heath, Shrewsbury SY5 9EG
M0 GRA G. Hickford 56 Alexander Close, Abingdon OX14 1XB
M0 GRB N. Harris 45 Sleigh Road, Sturry, Canterbury CT2 0HT
M0 GRE W. Greenall 356 Warrington Road, Abram, Wigan WN2 5XA
M0 GRF D. Blyth 45 Clarence Road, Bilston WV14 6NZ
MI0 GRG M. Mcgrory 10 Bramley Court, Red Lion Road, Kilmore, Armagh BT61 8ND
M0 GRH G. Hart 55 Runswick Drive, Nottingham NG8 1JE
M0 GRI R. Ingham 84 Marian Court, Gateshead NE8 2JB
MW0 GRJ G. Jones 12 Field Close, Flint CH6 5RQ
MI0 GRN A. Cartin 64 Ashgrove Park, Magherafelt BT45 6DN
M0 GRP G. Priestley 53 Millfield Gardens, Crowland, Peterborough PE6 0HA
M0 GRR S. Turner 12 Park Street, Morecambe LA4 6BN
M0 GRT N. Rotari 81 School Road, Dagenham RM10 9QD
M0 GRU A. Webb 17 Dickins Way, Horsham RH13 6BQ
M0 GRW C. Gibbs 112 Barkham Ride, Finchampstead, Wokingham RG40 4EN
M0 GRX Kings Lynn ARC c/o E. Roberts 117 Walstead Road, Walsall WS5 4LU
M0 GRY G. Collis 16 Hill Grove, Barrow Hill, Chesterfield S43 2NW
MW0 GRZ G. Woloszun 44 Cowbridge Road, Bridgend CF31 3DA
M0 GSC Rvd. M. Bracci 12 Bowling Green Close, Bognor Regis PO21 4HB
M0 GSI C. Nelmes 119 Exeter Road, Dawlish EX7 0AN
M0 GSK M. Silver 52 Park Crescent, Elstree, Borehamwood WD6 3PU
M0 GSL G. Walker 6 Tenbury Drive, Shrewsbury SY2 5YB
M0 GSN P. Newton 61 Ashbourne Crescent, Taunton TA1 2RA
M0 GSO R. Harris 1 Hollinhey Close, Bootle L30 7RN
M0 GSP S. Palmer 58 Highlands Way, Whiteparish, Salisbury SP5 2SZ
MM0 GSQ A. Young 4/4 Prestonfield Terrace, Edinburgh EH165EE
MW0 GSR S. Poyser Glandwr, Snowdon Street, Porthmadog LL49 9DF
MM0 GSS G. Smith 40 Pirleyhill Drive, Shieldhill, Falkirk FK1 2EA
MM0 GSW I. Wishart 7 Cairngorm Crescent, Kirkcaldy KY2 5RF
M0 GSX P. Stocking 6 Royal Oak Road, Rowley Regis B65 8NX
M0 GSZ G. Starling 4 Three Corner Drive, Norwich NR6 7HA
M0 GTE P. Allen 6 Hobson Close, Wigston LE18 2JH
M0 GTH R. Killen 3 Great Charles Close, St. Stephen, St. Austell PL26 7PW
MI0 GTI South Dorset Rs c/o A. Jamison 11 Richmond Gardens, Newtownabbey BT36 5LA
M0 GTJ R. Henderson 14 Oxford Avenue, St. Albans AL1 5NS
M0 GTL K. Cossey 34 Pinewood Road, Hordle, Lymington SO41 0GP
MI0 GTM J. Sills 145 Ballycolman Estate, Strabane BT82 9AJ
MW0 GTN Prof. D. Embrey 21 Rockfield Glade, Parc Seymour, Penhow, Caldicot NP26 3JF
M0 GTO J. Bateman 26 Thackeray Road, East Ham, London E6 3BW
M0 GTP G. Perry 173 Litchfield Road, Wednesfield WV11 3HX
M0 GTQ N. Bennett 16 Dickens Road, Worksop S81 0DP

M0 GTR P. Henderson 24 Farrow Road Whaplode Drove, Spalding PE12 0TS
M0 GTT R. Wilcox 107 Somerset Avenue Yate Bristol Bs377Sj, Bristol BS377SJ
MM0 GTU A. Cumming 15 Cockburn Crescent Whitecross, Linlithgow EH49 6JT
M0 GTV D. Suplatowicz 93 Huntingtower Road, Grantham NG31 7AZ
MM0 GTX B. Thomson 2 Auchencorvie Cottages, Campbeltown PA28 6PH
MW0 GTY G. Jones 182 Pontardulais Road, Tycroes, Ammanford SA18 3RD
M0 GTZ Derwent Valley Radio Group c/o K. Sanderson 39 Kirkland Street Pocklington, York YO42 2BX
M0 GUC M. Elkington 32 The Knoll, Kingswinford DY68JT
M0 GUD G. Gash 61 Beaconsfield Road, Rotherham S60 3HB
MM0 GUE J. Mcmorland 382 Maryhill Road, Glasgow G20 7YQ
M0 GUF G. Jones 57 Oxford Road, Banbury OX16 9AJ
M0 GUG E. Govan 9 Willowbank, Sandwich CT13 9QA
M0 GUH E. Cobb 8 Manor Avenue, Poole BH12 4LD
M0 GUJ D. Tarrant 17 Orchard Close, Corfe Mullen, Wimborne BH21 3TW
M0 GUL C. Repton Hp6 6Qu, Amersham HP6 6QU
M0 GUM D. Adshead 16 Moat Way, Swavesey, Cambridge CB24 4TR
M0 GUO P. Fry 4 Stretham Road, Wicken, Ely CB7 5XH
M0 GUR J. Harris 2 Elm Cottages, Boreham Lane, Wartling, Hailsham BN27 1RS
M0 GUU G. Moore 40 Main Street, South Rauceby, Sleaford NG34 8QG
MM0 GUW M. Mccabe 15 Laggan Road, Glasgow G43 2SY
MM0 GUX M. Potts 14 Constarry Road, Croy, Kilsyth, Glasgow G65 9HF
M0 GVB W. Brennan Flat 34 Compton Court Shopfiel Close, Rustington BN163JQ
MI0 GVC Castlerock ARS c/o J. Mcpeake 47 Dunclug Gardens, Ballymena BT43 6NN
M0 GVE M. Kentell 24 Kendal Court, Congleton CW12 4JN
M0 GVI D. Capon 144 Stow Road, Magdalen, King's Lynn PE34 3BD
M0 GVK D. Lyon-Mckeil 1372 Turnstone Way, Sunnyvale, Ca 94087-3736 United States
M0 GVL P. Smith 24 Bede Crescent, Benington, Boston PE22 0DZ
M0 GVN P. Keane Pembroke Lodge, Byes Lane, Reading RG7 2QB
M0 GVP Lv21 Lightship Museum c/o C. Turner 84 Gravel Hill Way Dovercourt, Harwich CO12 4XN
M0 GVQ A. Sibley 27 Sherwood Road, Tetbury GL8 8BU
M0 GVT C. Lee Grove House, Harrowbarrow, Callington PL17 8JN
M0 GVW M. Wibberley 5 The Row, Broadwell, Rugby CV23 8HF
M0 GVX A. Farrar 8 Wensley Street, Thurnscoe, Rotherham S63 0PX
M0 GVY M. Hall 29 The Spinney, Wokingham RG40 4UN
M0 GVZ C. Turton 32 Northfield Crescent, Driffield YO25 5ES
M0 GWA G. Rodmell 2 Meadow Way, Walkington, Beverley HU17 8SD
M0 GWB M. Baker 39 Exham Close, Warwick CV34 5UL
M0 GWC G. Chaloner 9 Fairthorne Rise, Old Basing, Basingstoke RG24 7EH
M0 GWD M. Ponsford 83 Grant Road, Farlington, Portsmouth PO6 1DU
M0 GWE K. Graffham 15 Hayes Road, Clacton-on-Sea CO15 1TX
M0 GWF A. Randles 62 Brookside Avenue, Poynton, Stockport SK12 1PW
M0 GWH D. Iveson 11 Newport Road, North Cave, Brough HU15 2NU
M0 GWK N. Green 11 Wythburn Way, Rugby CV21 1PZ
MW0 GWL J. Gwilliam 39 Wyndham Street, Glynfach, Porth CF39 9HT
M0 GWM R. Poole 57 Loxley Avenue, Shirley, Solihull B90 2QF
MM0 GWO H. Storie 33 Harbour Street, Plockton IV52 8TN
M0 GWQ J. Baker Birdsong, Princes Close Redlynch, Salisbury SP5 2HQ
M0 GWR J. Akinin 70 Valley Road, West Bridgford, Nottingham NG2 6HQ
MW0 GWT G. Thomas 20 Ael Y Bryn, Caerau, Maesteg CF34 0YG
M0 GWU W. Probst Flat 19, Lindsay Park, 16 Lindsay Road, Poole BH13 6AU
MW0 GWV A. Warner 35 Lon Y Berllan, Abergele LL22 7JF
MW0 GWY Dr I. Williams 19 Stryd Y Brython, Ruthin LL15 1JA
M0 GXB G. Bichard 9 Kelburne Close, Winnersh, Wokingham RG41 5JG
MW0 GXC Thanet Radio And Electronics Club c/o G. Zaza 42 Borras Road, Wrexham LL12 7EP
MW0 GXE T. Banks 18 Leicester Road, Newport NP19 7ER
M0 GXH J. Hayward 91 Hoyland Road, Hoyland S74 0AP
M0 GXK J. Rodriguez Cemillan Flat 51, Waxham, London NW3 2JJ
M0 GXM Leeds & Dis ARS c/o Dr M. Roe 68 Argyle Street, Cambridge CB1 3LR
M0 GXN S. Woodmore 66 Imperial Way, Chislehurst BR7 6JR
M0 GXO I. Sheppard 13 Meynell Close, Chesterfield S40 3BL
MM0 GXQ Club of Friendship c/o G. Milne 139 Rannoch Drive, Cumbernauld, Glasgow G67 4ES
MM0 GXU G. Sutherland 7 Abbotsgrange Road, Grangemouth FK3 9JD
M0 GXV C. Berry 60 Copthorne Road, Leatherhead KT22 7EE
M0 GXW R. Lee 39 Pickering Close Cramlington, Whitley Bay NE23 6QB
MM0 GXY St Dunstans ARS c/o R. Mannifield 2 Plewlands Avenue, Edinburgh EH10 5JY
M0 GYA Yate CG c/o R. Moody 372 Walsall Road, Perry Barr, Birmingham B42 2LX
M0 GYB M. Peterson 29 Warwick Close, Saxilby LN1 2FT
M0 GYC D. Fletcher 10 Russell Place, Plymouth PL4 6NJ
MM0 GYD A. Young 21 Corrour Road, Glasgow G43 2DY
MW0 GYF 610 Sqn City of Chester ATC ARC c/o M. Buxton 25 Pen Y Bryn, Sychdyn, Mold CH7 6EE
MM0 GYG A. Fletcher 164 Mayfield Road, Edinburgh EH9 3AR
M0 GYH M. Pearce 38 Salisbury Road, Beaconsfield Upper, Victoria 3808 Australia
M0 GYI Colchester Radio Amateurs Club c/o D. Leigh 39 Hill Chase, Chatham ME5 9HE
M0 GYK A. Roberts 14 Cowper Close, Newport Pagnell MK16 8PG
M0 GYL Dr M. Redman 4 Deauville Avenue, Cowes PO31 7GA
M0 GYM P. Mcewen 7 Springfield, Longhoughton, Alnwick NE66 3NT
M0 GYN Angel of the North ARC c/o K. Hulme Sutherland Road, Longsdon, Stoke-on-Trent ST9 9QD
M0 GYO D. Parker 53 Brisbane Way, Cannock WS12 2GR
M0 GYP Ariel Rad Group c/o S. Gillard 1 Chevening Close, Stoke Gifford, Bristol BS34 8NJ
M0 GYS Flintshire Raynet c/o D. Garner Flat 4, Joseph Nye Court, Portsmouth PO1 3RD
M0 GYU L. Karchev 6 Croombs Road, London E16 3RY
MW0 GYV P. Oseland 6 Oaklands Close, Bridgend CF31 4SJ
MM0 GYX I. Watson 10 Christie Place, Elgin IV30 4HX
MW0 GYY G. Lewis Bryn Cottage, Clydach, Abergavenny NP7 0LL
MM0 GZA S. Hargreaves 4 Oxenfoord Avenue, Pathhead EH37 5QD
M0 GZB A. Armitage 6 Rosebery Avenue Hythe, Southampton SO45 3HJ
M0 GZC the Moors CG c/o I. Coulson 2 Marl Hurst, Edenbridge TN8 6LN
M0 GZD Sherwood ARC c/o E. Rippon 319 Beechdale Road, Nottingham NG8 3FF
M0 GZE P. Slup 1 The Meadow, Copthorne, Crawley RH10 3RG
M0 GZF Bolton Wireless Club c/o M. Lamport Bartwood, Dancing Green, Ross-on-Wye HR9 5TE
M0 GZH Kings Lynn ARC c/o N. Smith 2 Norton Villas, Vicarage Road, Maidstone ME18 6DX
M0 GZI Alderley Explorer Scout ARU c/o M. Sharp Pentober, Firmingers Road, Orpington BR6 7QG
M0 GZK Bar-Packers CG c/o C. Yung Flat 8, Bridge Court, London E10 7JS
MW0 GZL A. Burleton 11 Orchard Close, Caldecott NP26 4BH
M0 GZM J. Oldman High Waters, Bentfield Green, Stansted CM24 8HX
M0 GZW GMDX Group c/o T. Lorn 152 Brougham Court, Peterlee SR8 1PZ
MM0 GZZ D. Taylor 1 Mayfield Farm Cottages, Reston, Eyemouth TD14 5LG
MW0 HAB M. Mainwaring 36 Oak Street, Gilfach Goch, Porth CF39 8UG
MW0 HAC C. Thomas 2 Ffordd Donaldson, Copper Quarter, Swansea SA1 7FJ
M0 HAD H. Tomlinson 5 Lynam Way, Madeley, Crewe CW3 9HX
M0 HAF R. Hambly 144 Station Road, Irchester, Wellingborough NN29 7EW
M0 HAG G. Hill-Adams 6 Broadleaze Way, Winscombe BS25 1JX
M0 HAH M. Summers 21 Quantock Avenue, Caversham, Reading RG4 6PY
M0 HAL P. Musselwhite 80 Craven Road, Orpington BR6 7RT
M0 HAM C. Bays 116 Rochester Road, Durham DH1 5PN
M0 HAN P. Maennel 4 Central Buildngs, Market Place, York YO61 3AB
M0 HAO M. Martin De La Fuente 35 Silchester Road, Reading RG30 3EJ
MW0 HAP A. Phillips 3 Pen Y Llys, Rhyl LL18 4EH
MM0 HAR H. Stuart 31 Robertson Road Lhanbryde, Elgin IV30 8PE
MW0 HAT R. Hatfield 35 Victoria Road, Penarth CF64 3HY
M0 HAU Cannock Chase ARS c/o J. Goodale 82 Farnborough Road, Farnborough GU14 6TH
M0 HAW Harpenden And Wheathampstead District Scout Group c/o M. Wood 26 Parkfield Crescent Kimpton, Hitchin SG4 8EQ
M0 HAZ A. Freeman 34 Marmion Road, Coningsby, Lincoln LN4 4RG
M0 HBC B. Broad 22 Minchin Acres Hedge End, Southampton SO30 2BJ
M0 HBE Grantham ARC c/o M. Robins 17 Old Turnpike, Fareham PO16 7HB
M0 HBH E. Mathieson 30 Lynfield Road, Frome BA11 4JB
M0 HBJ S. Blaikie 22 Juno Close, Goring-By-Sea, Worthing BN12 4UB
M0 HBL Prof. P. Richmond 7 Softley Drive, Norwich NR4 7SE
M0 HBM Dr B. Denyer-Green Dunsley South, Park Road, Forest Row RH18 5BX
M0 HBN Strathclyde Regional Raynet Groups c/o J. Bell 255 Willington Street, Maidstone ME15 8EP
M0 HBO K. Such 38 Hornby Grove, Hull HU9 4PG
M0 HBT Poole Radio Scouts (Prs) c/o D. Nelson 110 Chandag Road, Keynsham, Bristol BS31 1QF
M0 HBU I. Duffie Trebeighan Farm, Saltash PL12 5AE
M0 HBV D. Ingrey 1 Ponders Road, Fordham, Colchester CO6 3LX
M0 HBX Dr J. Pelham 5 The Crescent, Shortstown MK42 0UJ
M0 HBY Obo Windy Yett CG c/o P. Watkins 135 Lodge Road, Writtle, Chelmsford CM1 3JB
MW0 HCA F. Price 2 Bryniau Duon Estate, Llandegfan, Menai Bridge LL59 5PP
MW0 HCC C. Dumitrescu Aelfryn, Pen Y Cefn Road, Caerwys CH75BE
M0 HCE M. De Jong Grachtstraat 64, Oirsbeek 6938 HP Netherlands
M0 HCI G. Burton 26 Church View, Egremont CA22 2DT
M0 HCM L. Michalowski 136 St. Bernards Road, Newcastle ST5 6HL
M0 HCN Guisborough & District ARC c/o D. Mills 261 West Wycombe Road, High Wycombe HP12 3AS
MM0 HCO S. Mckenzie 0/2 69 Glenkirk Drive, Glasgow G15 6AU
M0 HCP Exmoor RC c/o J. Hunt 14 Nevill Close, Hanslope, Milton Keynes MK19 7NY
M0 HCR North Anglia Raynet c/o K. Kent 5 Jubilee Road, Heacham, King's Lynn PE31 7AR
M0 HCT M. Fitzjohn 96 Nightingale Gardens, Nailsea, Bristol BS48 2BN
M0 HCV Woodpecker CG c/o C. Wallace 16 Morley Square, Bristol BS7 9DW
MW0 HCW J. Morgan 3 Maes Yr Hebog, Penrhyn Bay, Llandudno LL30 3EY
M0 HCY Blackwater Radio CG c/o A. Copperwaite 71 Gladbeck Way, Enfield EN2 7EL
M0 HCZ C. Lycett 2 Royce Avenue, Hucknall, Hucknall, Nottingham NG15 6FU
MM0 HDA R. Fairfull Blackford Farm Gartocharn, Alexandria G83 8SD
M0 HDC Farmors School RC c/o S. Lucas 31 Lilian Close, Norwich NR6 6RZ
M0 HDE A. Morris 71 Lurdin Lane, Standish, Wigan WN6 0AQ
M0 HDG Hallam DX Group c/o N. Totterdell Moscar Cross House, Hollow Meadows, Sheffield S6 6GL
M0 HDJ D. Hall 8 Colston Close, Bristol BS16 4PQ
M0 HDK E. Erbes 488 Birkfield Drive, Ipswich IP2 9JE
M0 HDN B. Richter C/O Dr Steffen Grant, Wolfson College, Oxford OX2 6UD
M0 HDP P. Bolton 2 Alexander Court Chute Lane, Gorran Haven, St. Austell PL26 6NU
M0 HDQ G. Van Breemen 58 Horseshoe Lane, Bromley Cross, Bolton BL7 9RR
M0 HDR Royal Air Force ARS c/o R. Scholey Barleycroft, Lower Road, Ipswich IP6 9AR
M0 HDS Hinckley District Scouts c/o M. Smith 14B Witham Bank West, Boston PE21 8PU
M0 HDT Ramsbury Amateur Radio DX Group c/o D. Simpson 50 Castle Hill, Berkhamsted HP4 1HF
M0 HDU J. Legrain 17 Route De La Cote, St. Laurent Sur Gorre 87310 France
M0 HDV Glenrothes & District ARC c/o D. Cowling 11 Shakespeare Avenue, Scunthorpe DN17 1SA
MM0 HDW Welland Valley ARS c/o J. Duncan 36 Bank Row, Wick KW1 5EY
M0 HEJ White Rose ARS c/o G. Hatt 4H Colman House, Earlham Road, Norwich NR4 7TJ
M0 HEM J. O'Toole 4 Lindisfarne Road, Dagenham RM8 2RA
M0 HEN T. Kindts Northwood, Marhamchurch, Bude EX23 0HH
M0 HEP G. Zorzi 3B Ambleside Avenue, Telscombe Cliffs, Peacehaven BN10 7LS
M0 HET Gloucestershire County c/o S. Cordner 29 Buxton Road, Aylsham, Norwich NR11 6JD
M0 HEW T. Johnson 15 Tennyson Road, Creswell, Worksop S80 4DW
M0 HEX J. Ash 47 Stein Road, Emsworth PO10 8LB
M0 HEY M. Hickford 56 Alexander Close, Abingdon OX14 1XB
M0 HFA Vowhars c/o A. Birkett 67A Branston Road, Burton-on-Trent DE14 3BY
M0 HFB P. Szewczyk 514 Whaddon Way, Bletchley, Milton Keynes MK3 7LD
M0 HFC Humber Fortress DX ARC c/o J. Cunliffe 142 Hall Road, Hull HU6 8SB
M0 HFE Barnsley And District ARC c/o J. Sobanski 10 Robert Avenue, Barnsley S71 5RB
M0 HFF E. Bray 28, Henshall Avenue, Latchford, , Warrington WA4 1PY
M0 HFH J. Rowden 7 Regents Close, Thornbury, Bristol BS35 1HX
M0 HFI Clan Maclean ARS c/o J. Mclean 24 Durham Drive, Oswaldtwistle BB5 3AT

M0 HFO M. Jessop Department Of Electronic Engineering, Claverton Down BA2 7AY
M0 HFQ 1st Ringmer Scout Group c/o C. Lai Storeys Way, Cambridge CB3 0DG
MM0 HFU Rhondda ARS c/o E. Horn 3 Mckay Place, Newton Mearns, Glasgow G77 6UZ
M0 HFW De Havilland Heritage Radio Group c/o J. Newton 6413 Hillegass Avenue, Oakland 94618 United States
M0 HFX A. Walker 17 Carr House Road, Halifax HX3 7QY
M0 HFY B. Eames 22 Ashgrove Close, Hardwicke, Gloucester GL2 4RT
M0 HFZ B. Cox 7 Wolsey Avenue, London E6 6HG
M0 HGA D. Corless 3 Barn Close, Clifford Chambers CV37 8HJ
M0 HGD D. Molloy 187 Babylon Lane, Heath Charnock, Chorley PR6 9ET
M0 HGG Dr C. Regan 1 Fairways, Birkenhead CH42 8JZ
MW0 HGK W. Tse Marino Room, Fulton Houose, Swansea SA2 8PP
MW0 HGM A. Pritchard 12 Llys Le Breos, Mayals, Swansea SA3 5DL
MM0 HGN D. Higgins 1 Meggatland Farm Cottage, Inchture PH14 9QL
M0 HGO S. Westwood 118 Abbey Lane, Leigh WN7 5NU
M0 HGS M. Shepherd North Waver Cottage, Bells Road Belchamp Walter, Sudbury CO10 7AR
M0 HGV R. Dodds West Villa, The Green, Wallsend NE28 7PG
M0 HGY J. Read 49 Bransdale Way, Macclesfield SK11 8QT
M0 HHA M. Meehan 14 Grosvenor Road, Walton, Liverpool L4 5RB
M0 HHB G. Willard 4 Varrier Jones Place, Papworth Everard, Cambridge CB23 3XP
M0 HHC K. Jackson 4 Milfoil Close, Marton-In-Cleveland, Middlesbrough TS7 8SE
M0 HHD P. Rogers 16 Begonia Close, Basingstoke RG22 5RA
M0 HHE Dr G. Panico 7 Hollybush Lane, Orpington BR6 7QN
M0 HHF C. Greenwood 1 Bentinck Close, Boughton, Newark NG22 9HP
M0 HHG G. Aldridge Greenridge, Fore Street Bishopsteignton, Teignmouth TQ14 9QR
MD0 HHH Pontefract And District ARS c/o H. Dorman 1 Sprucewood Rise, Foxdale, Douglas IM4 3JP Isle of Man
M0 HHI J. Hughes Milestone House, Easole Street, Nonington Dover CT15 4HE
M0 HHM W. Roberts 218 Kristiansand Way, Letchworth Garden City SG6 1TU
M0 HHP M. Kasprzyk 80 Ederline Avenue, London SW16 4SA
M0 HHR M. Lee 34 Astley Road, Liverpool L36 8DA
MI0 HHU R. Benko 23 Six Mile Water Mill Drive, Antrim BT41 4FG
MI0 HHV B. Craney 8A Drumhoy Drive, Carrickfergus BT38 8NN
M0 HHX A. Currie 31 Launceston Road, Bodmin SO50 6AY
M0 HIB Cobham Marlow ARG c/o M. Passler The Chelton Centre Fourth Avenue, Marlow SL7 1TF
M0 HIC W Scotland ARS c/o M. De Silva 31 Rosemary Avenue, Hounslow TW4 7JQ
MW0 HID S. Leo 37 The Coldra, Newport NP18 2LS
M0 HIG B. Hultquist 37 New Road, Tiptree, Colchester CO5 0HN
M0 HIH Birmingham ARS c/o K. Manos 102 Goodwood Avenue, Sale M33 4QL
M0 HIL D. Hill 11 Paddock Lane, Metheringham, Lincoln LN4 3YG
M0 HIM P. Newsome 47 Bramhall Drive, Washington NE38 9DE
M0 HIN Hinckley Sea Cadets c/o V. Hopkins 6 Daimler Road, Coventry CV6 3GD
M0 HIO D. Wood 10 Tadgedale Avenue, Market Drayton TF94DD
M0 HIP Hippings Methodist Primary School Amateur Radio Cl c/o J. Mclean 24 Durham Drive, Oswaldtwistle BB5 3AT
M0 HIQ D. Cotton 1 Fieldfare Close, Penwortham, Preston PR1 9NG
M0 HIW P. Jones Tallonhouse, Mill Lane, Pulham St. Mary, Diss IP21 4QY
M0 HIX A. Holmes 49 Elm Grove South, Barnham, Bognor Regis PO22 0EJ
M0 HIY A. Thomas 1 Millers Close, Ruardean Hill, Drybrook GL17 9AU
M0 HIZ W. Easdown 11 Mulcaster Avenue, Kidlington OX5 2HG
M0 HJB M. Stillman 58 Highfield Road, Bognor Regis PO22 8PH
MM0 HJC Clydebank Cadet Centre c/o K. Brown 10 Richmond Street, Clydebank G81 1RF
M0 HJD D. Harbron 6 West View, Penshaw, Houghton le Spring DH4 7HP
M0 HJE P. Frost 11 Church Road, Swainsthorpe, Norwich NR14 8PH
M0 HJF H. Felstead Rosmede, Windmill Drive, Rustington, Littlehampton BN16 3HW
MW0 HJG N. Williams 27 Meadway Rogiet, Caldicot NP26 3SA
M0 HJI R. Havart 2 Holly Farm Road, Reedham, Norwich NR13 3TH
M0 HJJ A. Wierdis 39 Milton Road, London SW19 8SF
M0 HJL R. Taylor 27 The Holt, Hailsham BN27 3ND
M0 HJN W. Tomczyk 3L, 76 Berry Street, New York 11249 United States
M0 HJO J. Brooks Treven House, Treven, Tintagel PL34 0DT
M0 HJQ P. Garrett 21 Wychbury Road, Wolverhampton WV3 8DN
M0 HJR D. Vale 21 Chelston Road, Ruislip HA4 9SA
M0 HJW I. Ftaiha 8 Parkside, London NW7 2LH
M0 HJY R. Green 44 Aldwyn Place, Larchwood Drive, Egham TW20 0RZ
MW0 HKA M. Day 11 Troedrhiw-Trwyn, Pontypridd CF37 2SE
M0 HKB K. Brunning 45 Dover Road, Ipswich IP3 8JQ
M0 HKC K. Cullum 7 Gate Farm Road, Shotley Gate, Shotley, Ipswich IP9 1QH
M0 HKE A. Mullin 111 Arps Road, Codsall, Wolverhampton WV8 1SG
M0 HKG M. Clarke 40 Fingringhoe Road, Langenhoe CO5 5AD
M0 HKH Dr A. Fronters Flat 54, Central Quay North, Bristol BS1 4AU
M0 HKI L. Todman 17 Hall Road, St. Dennis, St. Austell PL26 8BE
M0 HKK Dr A. Doe 26 Beachfield Road, Bembridge PO35 5TN
M0 HKL G. Alberti Josef-Retzer-Strasse 48, M³nchen 81241 Germany
M0 HKM R. Richardson 11 Packman Green, Countesthorpe, Leicester LE8 5WS
M0 HKP Dr D. Potts 90 Albemarle Road, Willesborough, Ashford TN24 0HN
M0 HKS M. Booth 30 Manor Green, Harwell, Didcot OX11 0DQ
M0 HKT O.B.O. Tayside Raynet c/o A. Farrar 8 Wensley Street, Thurnscoe, Rotherham S63 0PX
MM0 HKU E. Duncan 3 George Street, Banff AB45 1HS
M0 HKV P. Bull 87 Braemor Road, Calne SN11 9DU
M0 HKW A. Brand 6 Walnut Close, Milton, Cambridge CB24 6ET
M0 HLB D. Slater 13 Longford Close, Rainham, Gillingham ME8 8EW
M0 HLC C. Taylor 1 Jasmine Gardens, Warrington WA5 1GU
M0 HLD D. Hicks 36 Middlesex Road, Maidstone ME15 7PL
M0 HLI D. Hyde 136 Station Road, Woodmancote, Cheltenham GL52 9HN
M0 HLM K. Schmidt Church, Corner, Mareham-le-Fen PE22 7RA
MM0 HLN H. Mason 20 David'S Crescent, Kilwinning KA13 6JJ
M0 HLP G. Bunting 31 Hardwick Avenue, Allestree, Derby DE22 2LN
MM0 HLQ G. Gilmour 3 Campsie Drive Milngavie, Glasgow G62 8HX
M0 HLR A. Brown 3 Alston Road, New Hartley, Whitley Bay NE25 0ST
M0 HLS C. Murphy 17 Shepherd Street, Littleover, Derby DE23 6GA
MM0 HLU I. Konstas 25/4 Milton Street, Edinburgh EH8 8HA
M0 HLV C. Hicks 11 Patch Street, Bath BA25BN
MW0 HLW W. Maxwell 25 Laurel Drive, Buckley CH7 2QP
M0 HLX D. Bailey 2B Queens Road, Enfield EN1 1NE
M0 HLZ M. Meachen 20 Wilkinson Road, Rackheath, Norwich NR13 6SG
M0 HMB the North West 320 DX Club c/o R. Stratford 32A Priory Avenue, High Wycombe HP13 6SW
M0 HME R. Campbell 17 Elgar Road, Southampton SO19 0JG
M0 HMF M. Smith Ii The White House, Old Avenue, West Byfleet KT14 6AE
M0 HMI G. Tamir Service Now, Strata, 1 Bridge Street, Staines-upon-Thames TW18 4TW
M0 HMJ D. Sadauskas 18 Flat, Newport House Newport Street, Tiverton EX16 6FJ
M0 HMO Dr H. Nickalls Holy Mill, Longville, Much Wenlock TF13 6ED
M0 HMR C. Harmer Spring Corner, Rockness Hill, Nailsworth, Stroud GL6 0PJ
M0 HMS E. Purvis 36 Birchington Avenue, Middlesbrough TS6 7EZ
M0 HMU Fleetwood Radio Enthusiasts Group c/o J. Earnshaw 128 Shakespeare Road, Fleetwood FY7 7HJ
MW0 HMV C. Josey 726 Llangyfelach Road, Treboeth, Swansea, SA5 9EL
M0 HMX R. Sykes 11 Lodwells Orchard, North Curry, Taunton TA3 6DX
MI0 HMY A. Hamill 15 Maythorn Avenue, Coleraine BT52 2EU
M0 HMZ P. Iljin 15 The Green, Newton Burgoland LE67 2SS
M0 HNA Southern Microwave Group c/o D. Austen Tudorlands, Silchester Road, Bramley, Tadley RG26 5DG
M0 HNC A. Ribeiro 38A Galpins Road, Thornton Heath CR7 6EB
M0 HND Spixworth Scout Radio Group c/o B. Smith 73 Devon Street, Hull HU4 6PL
M0 HNE R. Ashley 15 Wimbourne Drive, Gillingham ME8 9EN
M0 HNF P. Dickson 49 Signal Road, Grantham NG31 9BL
M0 HNG A. Douglas Gobbins Cottage, Sandy Lane, Lathom, Ormskirk L40 5TU
M0 HNH A. Reason 1 Iles Cottages, St. Marys, Stroud GL6 8NX
M0 HNI R. Weaver 15 Sharps Field, Headcorn, Ashford TN27 9UF
M0 HNJ P. Evans 67 Grenville Street, Stokport SK3 9ER
M0 HNK Dr R. Tofts Elmcroft, Redhill Road, Ross-on-Wye HR9 5AU
M0 HNL Dr R. Campbell 2 Hesketh Bank, York YO10 5HH
MW0 HNM L. Coleman Felin Newydd, Ciliau Aeron, Lampeter SA48 7PX
M0 HNN T. Walsh 6 Brass Thill, Durham DH1 4DS
M0 HNO H. Nishio 28 New Lane, Havant PO9 2NQ
MI0 HNQ Hilltop ARC Co.Down c/o A. Mcgarvey 66A Scaddy Road, Downpatrick BT30 9BS
M0 HNT A. Angus 51 Osprey Drive, Blyth NE24 3QS
M0 HNX R. Raeburn 145 Paddock Road, Basingstoke RG22 6QQ
M0 HOB G. Brotherhood 17 Baldwin Close, Forest Town, Mansfield NG19 0LR
M0 HOF B. Ajeti 88 Bushfield Crescent, Edgware HA8 8XJ
M0 HOI Dr S. Musgrave Orchard Cottage, Stalmine, Poulton le Fylde FY6 0LZ
M0 HOJ F. Costa 7 Aylesborough Close, Cambridge CB4 2HH
M0 HOK L. Carberry 23 Greens Beck Road, Stockton-on-Tees TS18 5AR
MM0 HOL C. King 19 Gleneagles Way, Deans, Livingston EH54 8EW
M0 HOM M. Hotchin 122 Buckingham Avenue, Scunthorpe DN15 8NS
M0 HOO A. Hodgeon 30 Rock Bank, Buxton SK17 9JF
M0 HOP Sir A. Hopson 1 Hall Lane, Leicester LE2 8SF
M0 HOQ J. Johnson 49 Beach Priory Gardens, Southport PR8 2SA
M0 HOT H. Rose 10 St. Vincents Close, Girton, Cambridge CB3 0PE
M0 HOU H. Atifeh 57 Lincoln Drive, Rugby CV23 1BS
M0 HOV A. Hristov Flat A, 71 Beckenham Lane, Bromley BR2 0DN
M0 HOY S. Curtis 354 St. Helens Road, Leigh WN7 3PQ
MI0 HOZ M. Na BpYob 94 Curlyhill Road, Strabane BT82 8LS
M0 HPB D. Bisbey 17 Benson Close, Lichfield WS13 6DA
MI0 HPE P. Dorris 29 Eia Street, Belfast BT14 6BT
M0 HPF G. Cheeran 37 Farnol Road, Dartford DA1 5NG
M0 HPG Cumbria Raynet Group c/o P. Woodburn 21 The Row, Silverdale, Carnforth LA5 0UG
MW0 HPH Ystrad Mynach College c/o P. Jones 23 Pinecroft Avenue, Aberdare CF44 0HY
M0 HPJ J. Whiteside The Old Antique Shop, Bank Street Pulham Market, Diss IP21 4TG
M0 HPL R. Masshedar 6 Hutton Avenue, Hartlepool TS26 9PN
MM0 HPP G. Fleming Tarbat View, Achavandra Muir, Dornoch IV25 3JB
M0 HPR H. Richardson 7 Regent Road, Leyland PR25 2LJ
M0 HPS Dr H. Powell 6 Sowbury Park, Chieveley, Newbury RG20 8TZ
M0 HPT John Newton Memorial RC c/o D. Prior 10 Birley Close, Appley Bridge, Wigan WN6 9JL
M0 HPU D. Rudling Rose Cottage, Ludwells Lane, Southampton SO32 2NP
M0 HPV D. Green 67 Coombe Park Road, Binley, Coventry CV3 2NW
M0 HPW M. Phillips 59 Bradeley Road Haslington Crewe, Crewe CW1 5PX
MI0 HPX A. Sweeney 117 Lisnablagh Road, Coleraine BT52 2HD
M0 HPZ L. Edmonds 3 Waterlow Road, London N19 5NJ
M0 HQA P. Massolt 26 Redgate Heights, Hunstanton PE36 5EA
M0 HQB K. Kulpinski 85 Severn Drive, Taunton TA1 2PW
M0 HQC M. Copse 3 The Limes, Market Overton, Oakham LE15 7PX
MM0 HQD D. Searle 48 George Court, Hamilton ML3 9HG
M0 HQG Sutton Coldfield & Dist Raynet c/o J. Mason 56 Skegby Road, Sutton-in-Ashfield NG17 4EZ
MM0 HQI L. Richings 2 St. Margarets Place, Edinburgh EH9 1AY
M0 HQJ H. Quigg 60 Oak Road, Ripon HG4 2NB
M0 HQL S. Mitchell 42 Fairfield Avenue, Felixstowe IP11 9JJ
M0 HQM R. Givens 13 Wakehurst Drive, Crawley RH10 6DL
M0 HQO P. Freeman 57 Ruffa Lane, Pickering YO18 7HN
M0 HQP N. Marley Penstemons, Chapel Lane Pen Selwood, Wincanton BA9 8LY
M0 HQQ P. Taylor 54 Church Road, Stanley, Liverpool L13 2BA
M0 HQR P. Naik 82 Misbourne Road, Uxbridge UB10 0HW
M0 HQU C. Lombao 82 Cirrus Drive, Shinfield, Reading RG2 9FL
M0 HQZ J. Widdowson 26 Woodville Gardens West, Boston PE21 8BW
M0 HRA H. Alderson 66 Houghtonside Estate, Houghton le Spring DH4 4BW
M0 HRC W. Nicholas 16 Withymoor Road, Netherton, Dudley DY2 9LA
MW0 HRD C. Hughes 88 Derlwyn Street, Phillipstown, New Tredegar NP24 6BA
MI0 HRG Hill Top Radio Group c/o B. Vaughan 32 Claremore Road, Castlederg BT81 7RF
M0 HRH C. Morley 191 Purbrook Way, Havant PO9 3RS
MM0 HRI I. Candy 6 Provost Milne Gardens, Arbroath DD11 5FG
MM0 HRL I. Gourlay 76 Largo Road, St. Andrews KY16 8NJ
M0 HRM C. Greenwood 21 Valley Drive, Thornhill Dewsbury WF120HE
MI0 HRO Team Thunderbox c/o C. Stockdale 3 Hightown Drive, Newtownabbey BT36 7TG

M0 HRP R. Huelin 15 Hill Chase, Walderslade, Chatham ME5 9HE
M0 HRT R. Bryan 1 White Cottage, Old Warwick Road, Lapworth, Solihull B94 6LN
MI0 HRV T. Darrah 42 Pinewood Avenue, Carrickfergus BT38 8EW
M0 HRW A. Parker 9 Milecastle Court West Denton, Newcastle-upon-Tyne NE5 2PA
M0 HRY S. Wheeler 98 Charterhouse Road, Orpington BR6 9EW
M0 HRZ D. Irving 8 Durlston Road, Swanage BH19 2DL
MM0 HSB W. Forrester 149 Whyterose Terrace, Methil, Leven KY8 3AR
M0 HSC Northeast ARS c/o G. Cockburn 20 Hexham Avenue, Hebburn NE31 2HN
M0 HSG P. Scrimshaw 126 Nelson Road, Leighton Buzzard LU7 3EG
M0 HSH Dr J. Brooks 44 Rowan Drive, Seaton EX12 2UH
MW0 HSI Clwud Portable Operating Group c/o M. Allington Merton Place Nursing Home, 8 Pwllycrochan Avenue, Colwyn Bay LL29 7BU
M0 HSJ Cockenzie & Port Seton ARC c/o H. Jones 116 Dark Lane, Bedworth CV12 0JH
M0 HSQ I. Tsimperidis Flat 410, Birch Court, Howlands, Welwyn Garden City AL7 4LR
MM0 HSR Brannock High Radio Group c/o P. Bainbridge 49 Hare Moss View, Bathgate EH47 0DN
M0 HSS A. Perrow 16 Bannister Walk, Cowling, Keighley BD22 0NU
M0 HSU S. Challis 73 Rivenhall Way, Hoo, Rochester ME3 9GF
MM0 HSV K. Baird 24 Main Street, Sorn KA5 6HU
M0 HSW H. Scott Whittle 7 Skyline House, Dickens Yard, Longfield Avenue, London W5 2BJ
M0 HSX M. Josi 10 Robert Close, Billericay CM12 9DS
M0 HSZ J. Merritt 41 Great Grove, Bushey WD23 3BQ
M0 HTA I. Cooke 11 Farriers Gate, Chatteris PE16 6AY
M0 HTB H. Banasiak 16 St Christophers Close, Bath BA2 6RG
M0 HTE J. Taylor 90 Village Road, Gosport PO12 2LG
M0 HTF C. Rose 132 Golf Green Road, Jaywick, Clacton-on-Sea CO15 2RW
MW0 HTG G. Edwards 17 Glan Y Mor Road, Penrhyn Bay, Llandudno LL30 3NL
M0 HTI S. Storey 10 Amble Way, Trimdon Station TS29 6DZ
M0 HTJ Hamtests.Co.UK c/o P. Gibson 60 Raglan Road, Bromley BR2 9NW
M0 HTK H. Kassier 26 Higher Port View, Saltash PL12 4BX
MM0 HTL D. Hegarty 13 Gaitschaw Lane, Selkirk TD7 4HS
MW0 HTO Summerfields ARC c/o D. Bowen 25 Maendu Terrace, Brecon LD3 9HH
M0 HTQ K. Numata 28 Guildhouse Street, London SW1V 1JJ
M0 HTR Ashton In Makerfield ARC c/o P. Williams 35 Cansfield Grove, Ashton In Makerfield, Wigan WN4 9SE
M0 HTS C. Mellor 104 Rocky Lane, Eccles, Manchester M30 9LY
M0 HTU J. Stokoe 77 High Street Market Deeping, Peterborough PE6 8ED
M0 HTV M. Gregson 10 Eden Avenue, Consett DH8 6EZ
M0 HTW H. Chan Flat 20A, Pine Manor, 61 Waterloo Road, Mongkok 852 Hong Kong
M0 HTX D. Anderson 53 Collywell Bay Road Seaton Sluice, Whitley Bay NE264RG
M0 HTY M. Tointon 13 Ridgeway, Broadstone BH18 8DY
M0 HUA A. Brown 63 Pound Green Lane, Shipdham, Thetford IP25 7LH
MM0 HUF S. Harvey 63 Darley Road, Cumbernauld, Glasgow G68 0JR
M0 HUG S. Eyre St. Michael Mead, The Common Barton Turf, Norwich NR12 8BA
M0 HUH P. Tan Cambridge, CB3 0BN
M0 HUI I. Magness Orchard House Ravenswood Drive, Camberley GU15 2BU
M0 HUL A. Molloy Rue Apesenia 12, Urrugne 64122 France
M0 HUN J. Hunt Flat 1, Strand House, 16 Wells View Drive, Bromley BR2 9UL
MM0 HUQ M. Flaws West Voe, Sumburgh, Shetland ZE3 9JN
M0 HUS H. Steers 39 Upwood Road, London SE12 8AE
MW0 HUU M. Pope 4 Croft Villas, Narberth SA67 7DY
M0 HUV E. Valdez 65 Broken Cross, Charminster, Dorchester DT2 9QB
M0 HUW G. Gentile Via B, Vecchia, Preganziol 31022 Italy
MW0 HUY K. Saltmarsh 15 Colbourne Road, Beddau, Pontypridd CF38 2LN
M0 HUZ C. Warwick 104 Church Road, Formby, Liverpool L37 3NH
MM0 HVA J. Hawkins 113 Meadowpark Avenue, Bathgate EH48 2ST
MW0 HVB H. Bancroft Stop And Call, Goodwick SA64 0EX
M0 HVC R. Barnard 3 Heaths Close, Enfield EN1 3UP
M0 HVD D. King 25 Church Road, Worthing BN13 1ET
M0 HVE L. Sargent 18 Lyndhurst, Maghull L316DY
MM0 HVF Z. Bak 62/6 North Gyle Loan, Edinburgh EH12 8LD
M0 HVI M. Kurczab 159 Huddersfield Road, Halifax HX3 0AH
M0 HVK D. Ackrill 62 Lapsley Drive, Banbury OX16 1EW
MW0 HVL E. England 2 Luton Street, Blaenllechau, Ferndale CF43 4PB
M0 HVM J. Vollbrecht Reinsdorf, Steingasse 3, Nebra (Unstrut) 6642 Germany
M0 HVN D. Connolly 2 Layton Close, Birchwood, Warrington WA3 6PT
M0 HVO I. Bailey 8 Willow Drive, Ringwood BH24 3BE
M0 HVP Hucknall Rolls Royce A.R.C c/o N. Tindall Royds Mount, Linthwaite, Huddersfield HD7 5QX
M0 HVQ D. Holland 22 Morris Croft, Cottingham HU16 5GU
M0 HVR J. Brawn 5 Downs Cote View, Westbury On Trym, Bristol BS9 3TU
M0 HVS N. Bethell 4 Magazine Road, Wirral CH62 3LH
MM0 HVU D. Smith 25 High Academy Street, Armadale, Bathgate EH48 3HG
M0 HVV M. Hickman 13 Millfields Avenue, Rugby CV21 4HJ
MM0 HVW D. Plummer 39 St. Nicholas Drive, Banchory AB31 5YG
M0 HVX C. Pattison-Hart 133 Park Road, Bingley BD16 4EJ
M0 HWC Hadley Wood CG c/o M. Ruttenberg 90 Heath View, London N2 0QB
M0 HWD D. Levy Flat 36, Claydon House, London NW4 1LS
MI0 HWG P. Moore 32 Kinnegar Rocks, Donaghadee BT21 0EZ
M0 HWH K. Quigley 12 Silver Lane, Billingshurst RH14 9RJ
M0 HWI A. Trzepietowski 8 Parsons Nook, Coventry CV2 4QY
M0 HWJ Dr A. Boldireff Strzeminski 47 Waters Edge, Canterbury CT1 1WX
M0 HWL R. Riches Flat 21, Hawthornden, 84 Bradford Road, Otley LS21 3LE
M0 HWM S. Baddeley 50 Western Esplanade, Herne Bay CT6 8JA
M0 HWN E. Wilcockson Flat 17, Dale Court, Seymour Road, Slough SL1 2NU
M0 HWO G. Phillips 73 Gotham Road, Wirral CH63 9NG
M0 HWP J. Tarrant 70 Sunnymead, Midsomer Norton, Radstock BA3 2SD
M0 HWQ P. Browne 151 North Road, St. Andrews, Bristol BS6 5AH
M0 HWS G. Hewis 10 Albert Road, New Malden KT3 6BS
M0 HWT G. Mutch 94 Abbotswood Road, Brockworth, Gloucester GL3 4PF
MW0 HWU Raynet Pembrokeshire c/o I. Baker 24 Donovan Reed Gardens, Pembroke Dock SA72 6EW
M0 HWV P. Heiney 12 Church Lane, Walberswick IP186UZ
M0 HWW B. Vemic 7A Selby Road, London SE20 8SF
M0 HWZ D. Mccann 41 Weighton Road, Harrow HA3 6HY
M0 HXA A. Crosland 18 Duncan Crescent, Bovington, Wareham BH20 6NN
MI0 HXB T. Browne 7 Hawthorn Park Greysteel, Londonderry BT47 3YE
MW0 HXC F. Kroon Flat 10, Court Rise, Hoggan Park, Brecon LD3 9SZ
M0 HXE R. Hill 108 Hitchin Close, Romford RM3 7EQ
M0 HXF Dr R. Thorpe 6 Millthorpe, Sleaford NG34 0LD
M0 HXH J. Matthewson 20 Yew Tree Road, New Ollerton, Newark NG22 9UL
M0 HXI C. Baker 29 Green Lane, Bristol BS11 9JD
M0 HXK H. Carrythers 31 Baden Street, Hartlepool TS26 9BJ
M0 HXM D. Estevez Oceano Atlantico, 38, Tres Cantos 28760 Spain
M0 HXN J. Orme 42 Dovecote, Newport Pagnell MK16 8BB
M0 HXO J. Neal 48 Mansfield Road, South Normanton, Alfreton DE55 2ER
M0 HXS E. Haener 110 Great Stone Road, Manchester M16 0HD
M0 HXV A. Yeomans 17 Hollyfield Close, Tring HP23 5PL
MW0 HXX D. Machon 22 Albert Street, Caerau, Maesteg CF34 0UF
M0 HXZ M. Mallette Not, Applicable, Resides OUTSIDE OF THE UK IN Germany
MW0 HYA J. Starbuck 28 Plas Panteidal, Aberdyfi LL35 0RF
M0 HYC T. Rutt Granthorpe, Hull Road, Hull HU11 5RN
M0 HYD F. Hyde 10 Devonshire Drive, Barnsley S75 1EE
M0 HYE T. Byers 1 Hazelwood Avenue, Sunderland SR5 5AH
M0 HYG H. Hope 51 Margravine Gardens, London W6 8RN
M0 HYH C. Glass The Old Homestead, Havikil Lane, Knaresborough HG5 9HN
M0 HYI A. Brocking 9 Glenhaven Avenue, Urmston, Manchester M41 5BN
M0 HYJ A. Rand 17 Fairways Drive, Harrogate HG2 7ES
MW0 HYK K. Dutfield-Cooke Tan Yr Efail, Segurinside, Llandudno Junction LL31 9QE
M0 HYL A. Robnett 38B Woodmere Avenue, Watford WD24 7LN
MM0 HYM W. Jackson 3 Annick Road, Dreghorn, Irvine KA11 4EY
M0 HYN B. Davies 12 Scalebor Gardens, Burley In Wharfedale, Ilkley LS29 7BX
MW0 HYP D. Thomas 67 Crynallt Road, Neath SA11 3RN
MI0 HYQ A. Zakrzewski 11 Millbrook Gardens, Kilrea, Coleraine BT51 5RZ
M0 HYX South And West Yorkshire Wing ATC c/o P. Marchant 16 Melrose Drive, Peterborough PE2 9DN
M0 HZ Horizontal Net c/o A. Gravell 21 Wickridge Close, Stroud GL5 1ST
M0 HZA G. Charlesworth 6 Eastfield Close, Sutterton, Boston PE20 2JF
M0 HZB Z. Yao Room 4C, Unit 12-2, Zhonghaibanshanxigu Garden, No.15 Zhongqing Rd., Yantian District, Shenzhen 518000 China
M0 HZC V. Perovic Staudenb³hlstrasse 126, Zurich 8052 Switzerland
MI0 HZD K. Mikicki 17 Glenhoy Drive, Belfast BT5 5LB
M0 HZE P. Preston 12 Backney View, Greytree, Ross-on-Wye HR9 7JP
M0 HZF G. Craioveanu Flat 8, Apex House, Burch Road, Northfleet, Gravesend DA11 9FF
MM0 HZI T. Johnston The Old Schoolhouse, Luggate Burn, Haddington EH41 4QA
MM0 HZJ S. Thomas Macsherry, Glenachulish, Ballachulish PH49 4JZ
M0 HZK D. Pearson 37 Elmridge, Leigh WN7 1HN
MM0 HZL H. Mckay 3 Flemington Gardens, Whitburn, Bathgate EH47 0NS
M0 HZM D. Greenland 1 Hilltop, Tuesley Lane, Godalming GU7 1SB
MM0 HZO N. Clark 27 Deansloch Crescent, Aberdeen AB16 5UY
M0 HZP D. Morrow 57 Bispham Road, Poulton-le-Fylde FY6 7PE
M0 HZR N. Barker 17 Pippin Walk, Hardwick, Cambridge CB23 7QD
M0 HZT J. Jones 21 The Maltings, Warminster BA12 8JR
M0 HZU D. Eate 69 Dunyeats Road, Broadstone BH18 8AE
M0 HZV M. Mirchev 82 Collingwood Road, Uxbridge UB8 3EL
M0 HZW W. Dawkins 2 Nativity Close, Sittingbourne ME10 1ET
M0 HZX M. Stephens 52 Waterslea Drive, Bolton BL1 5FJ
M0 HZY J. Strandberg Apartment 201, Satin House, 15 Piazza Walk, London E1 8PW
M0 IAA I. Astley 6 Shay Court, Crofton, , Wakefield WF4 1SL
M0 IAD I. Macdonald Broomhill Mill Lane, Worthing BN13 3DH
M0 IAE Medway Raynet c/o M. Adcock 37 Ashpole Road, Braintree CM75LW
M0 IAF I. Fletcher 19 Church Street, St. Day, Redruth TR16 5JY
M0 IAG S. Donath 12A Comerford Road, London SE4 2AX
M0 IAH I. Pryke 9 Charles Avenue, Grundisburgh, Woodbridge IP13 6TH
M0 IAJ I. Jones 21 Kennet Green, Worcester WR5 1JQ
M0 IAK I. Karbhari Flat B, 226 Westbourne Park Road, London W11 1EP
MM0 IAL I. Lindsay 265 Stirling Street, Denny FK6 6QJ
M0 IAM C. Collins 31 Warren Road, Godalming GU7 3SH
M0 IAS G. Reywer 1 Tiverton Close, Houghton le Spring DH4 4XR
M0 IAT I. Chick 7 Furzegood Marldon, Paignton TQ3 1PH
M0 IAX M. Bumstead Middle Leys, Leys Farm, Withypool, Minehead TA24 7RU
M0 IAZ R. Dykes Gorse Cliff, West Hill, Heybrook Bay, Plymouth PL9 0BB
M0 IBD W. Thiele 50B, The Highway, London E1W 2BG
MM0 IBE P. Woods 92 Preston Crescent, Prestonpans EH32 9RD
MW0 IBH K. Davies 16 Lon Ffawydd, Abergele LL22 7DU
MW0 IBI Taff Vale ARC c/o A. Burns 34 Lakeside Gardens, Merthyr Tydfil CF48 1EN
M0 IBK H. Schr÷der Hamptstrabe 8, Dieblich 56332 Germany
MM0 IBL C. Niven The Coachmans Cottage, Balmullo Farm, Balmullo, St. Andrews KY16 0AQ
M0 IBN W. Parish 44 Fitzroy Drive, Lee-on-the-Solent PO13 8LZ
MM0 IBO J. Moreno 1-19 Albion Street, Glasgow G1 1LH
M0 IBQ M. Savage 3 Marlborough Close, Cheltenham GL53 7RY
M0 IBR B. Clayton 26 Wood Walk, Mexborough S64 9SG
MW0 IBT D. Jones 44 Stryd Y Wennol, Ruthin LL15 1QN
M0 IBW Dr S. Harrison 8 St. Michaels Close, Buckland Dinham, Frome BA11 2QD
M0 IBX A. Beacham 25 Baldock Road, Canterbury CT1 1XH
M0 IBY Utc Sheffield ARC c/o M. Rigby 75 Manchester Road, Deepcar, Sheffield 36 2QX
MW0 IBZ I. Baker 24 Donovan Reed Gardens, Pembroke Dock SA72 6EW
M0 ICA P. Rushby 16 Foxhill Lane, Selby YO8 9AR
M0 ICB I. Buchner 33 Blakewell Gardens Tweedmouth, Berwick-upon-Tweed TD15 2HJ
MW0 ICE C. Evans 25 Beech Drive, Hengoed CF82 7JP
M0 ICG G. Tagg Tinkers Cottage, Nevendon Road, Wickford SS12 0QB
M0 ICI J. Le Roux 20 Varsity Drive, Twickenham TW1 1AG
M0 ICJ Warks Raynet Gr c/o L. Zywicki 18 Springbank, Brigg DN20 8PW
M0 ICK M. Heywood 16 Edinburgh Drive, Hindley Green, Wigan WN2 4HL
M0 ICL J. Salek 19 Eskmont Ridge, London SE19 3PZ
MW0 ICO P. Jones 76 Pengwern, Llangollen LL20 8AS
M0 ICP I. Pass 69 Cotswold Road, Bath BA2 2DL
MD0 ICS C. Schofield Rockside, Dreemskerry Road, Maughold IM7 1BL Isle of Man
M0 ICT M. Gascoyne 31 Dale View, Hemsworth, Pontefract WF9 4TA
M0 ICZ D. Jones Drove Farm, Sheepdrove, Lambourn, Hungerford RG17 7UN

M0 IDC J. Clark 27 The Gabriels, Newbury RG14 6PZ
M0 IDE C. Laycock 35A High Street, Henlow SG16 6AA
M0 IDG I. Garrard 33 Uplands Road, Hockley SS5 4DL
M0 IDI P. Askew 6 Claremont Avenue, Newcastle upon Tyne NE15 7LB
M0 IDJ S. Lo 12 Kingston Wharf, Kingston Street, Hull HU1 2ES
M0 IDK I. King 7, Greenacres Avenue, Blythe Bridge, Stoke on Trent ST11 9HU
M0 IDL G. Stockley Flat 1, The Pentagon, 94 Stanley Green Road, Poole BH15 3AG
M0 IDM A. Ferriroli 142 Hillbury Road, Warlingham CR6 9TD
M0 IDR I. Reeve 36 Stone Pippin Orchard, Badsey, Evesham WR11 7AA
MW0 IDT I. Booth 18 Clos Y Wiwer, Pentre Cwrt, Llantwit Major CF61 2SG
M0 IDW D. Byrne Suite 214, 1 Hanley Street, Nottingham NG1 5BL
M0 IDY S. Gorski 2 Samphire Close, Didcot OX11 6HP
M0 IEA C. Bowler 42A Honor Road Prestwood, Great Missenden HP16 0NL
M0 IEB Dr C. Bridges 23 Bramley Vale, Cranleigh GU6 7FY
M0 IED A. Wedge 30 Primrose Way Locks Heath, Southampton SO31 6WX
MW0 IEH D. Lockyer 19B Drury Lane, Buckley CH7 3DU
MM0 IEJ J. Macdonald 24 St. Pauls Drive, Armadale, Bathgate EH48 2LT
M0 IEK P. Phipps Meakers Cottage, Long Load, Langport TA10 9JX
MM0 IEL I. Lee Roehill, Crossroads, Keith AB55 6LQ
M0 IEM M. Feast 30 Peveril Drive, Riddings, Alfreton DE55 4AP
M0 IEO M. Sanderson 2 East Crescent, Canvey Island SS8 9HL
M0 IEP V. Williams 11 Priory Green Highworth, Swindon SN6 7NU
M0 IEQ M. Priest 35 Albert Road, Chaddesden, Derby DE21 6SJ
M0 IER A. Coote 148 Clarendon Street, Dover CT17 9RB
M0 IES M. Reaney Odessa Marine, Little London, Newport PO30 5BS
M0 IET C. Blount 55 Silverthorne Drive, Caversham, Reading RG4 7NR
MJ0 IEW Dr C. Poole Kayalami, La Ruelle Du Clos Du Parcq, St Brelade, Jersey J E3 8AQ
M0 IEY C. Ryder Sunnymead, Well Head Road Newchurch-In-Pendle, Burnley BB12 9LW
M0 IEZ N. Cohen 8 Henry Gepp Close, Adderbury, Banbury OX17 3FE
M0 IFB D. Endean 11 Forrester Drive, Brackley NN13 6NE
M0 IFC A. Ward 70 Southworth Avenue, Warrington WA5 0DU
MI0 IFG Blacks Hillbillies ARC c/o P. Hosey 13 Glenelly Gardens, Omagh BT79 7XG
M0 IFH I. Harley 1 Portland Crescent, Meden Vale, Mansfield NG20 9PJ
M0 IFP M. Byzdra 3 Brownlow Street, Whitchurch SY13 1QW
M0 IFT R. Hodson 99 Alcester Road, Hollywood, Birmingham B47 5NR
M0 IGB I. Bennett 44 Haig Avenue, Whitley Bay NE25 8JG
M0 IGF M. Lane Greenacres, Bickington Road, Barnstaple EX31 2JG
M0 IGG S. Wright 23 Kitchener Street, Walney, Barrow-in-Furness LA14 3QW
M0 IGJ Tintagel And District RAC c/o J. Brooks Treven House, Treven, Tintagel PL34 0DT
MI0 IGL D. Neill 8 Castle Meadows Carrowdore, Newtownards BT22 2TZ
M0 IGM R. Powley 8 Treadgold Avenue Great Gonerby, Grantham NG31 8PD
MM0 IGO W. Mcbain 56 Scotstoun Park, South Queensferry EH30 9PQ
M0 IGP E. Coles 41 Venn Court Brixton, Plymouth PL82AX
M0 IGU D. Harrop 2 St. Marys Hill, Chester CH1 2DW
M0 IGW A. Griffiths 67 Griffiths Rd, Upwey 3158 Australia
M0 IGX A. Griffiths 67 Griffiths Rd, Upwey 3158 Australia
M0 IGY J. Hardman 45 Doncaster Avenue, Manchester M20 1DH
M0 IHB R. Monaghan Reservoir Cottage, Tavistock Road Roborough, Plymouth PL6 7BD
MM0 IHE I. Hepworth Bronte Cottage, Inverugie, Peterhead AB42 3DN
M0 IHJ M. Rose 149 Claremont Road, Blackpool FY1 2QJ
M0 IHM I. Millman 3 Oyster Mews, 1-3 Forest Road, Poole BH13 6EN
M0 IHN K. Missenden 47 Roseacre Drive, Elswick, Preston PR4 3UQ
M0 IHR E. Squires Forest Care Village, 10-20 Cardinal Avenue, Borehamwood WD6 1EP
M0 IHT H. Lennertz 18 Church Terrace, Exeter EX2 5DU
M0 IHU P. Farley 1 Holders Road Amesbury, Salisbury SP4 7PW
MW0 IHW G. Reason 454 Cowbridge Road West, Cardiff CF5 5BZ
M0 IHX M. Allinson 60 King Street Aspatria, Wigton CA7 3AH
M0 IHY B. Page 12 Hitchens Close, Hemel Hempstead HP1 2PP
M0 IIA A. Bullard 15 Rowan Drive, Lutterworth LE17 4SP
M0 IIE J. Bennet 53 Haven Road, Barton-upon-Humber DN18 5BS
MU0 IIF S. Bougourd La Petite Folie, Folie Lane, Guernsey GY3 5SE
MI0 IIG I. Gibb 1 Shankill Road, Garvary, Enniskillen BT94 3DB
M0 IIM Horizontal Net c/o C. Gordon 90 Sunholme Drive, Wallsend NE28 9YW
M0 IIQ A. Birkett 21 Cedar Drive Wyke, Bradford BD12 9HL
M0 IIU J. Fogg 17 Lyppiatt Road, Bristol BS5 9HW
M0 IIZ D. Saunders 6 Cherry Tree Drive Sedgefield, Sedgefield TS21 3DN
M0 IJA C. Heal 2 Thorn Gardens, Ramsgate CT11 7AS
M0 IJC P. Drobny Flat 5, Busby Court Ceylon Place, Eastbourne BN21 3JG
MW0 IJE Dr J. Woore 10 Haverfordwest Road, Letterston, Haverfordwest SA62 5UA
M0 IJG I. Mcmullan Glenayr, Glen Cuileann, Kilquade, Greystones, A63 XH90 Ireland
M0 IJI M. Harper 15 Furrows End, Drayton, Abingdon OX14 4GN
M0 IJJ M. Cox 17A Church End, Weston Colville, Cambridge CB21 5PE
M0 IJL T. Dix Ebb House Downhouse Lane Higher Eype, Bridport DT6 6AH
M0 IJO Dr E. Melikyan 39 St. Evox Close Rownhams, Southampton SO16 8FS
M0 IJP A. Jamieson 48 Shackleton Road, Crawley RH10 5BX
M0 IJR P. Jones 446 Winchester Road, Southampton SO16 7DG
M0 IJT J. Rushby 2 Arthurs Bridge Farm Cottages, Evercreech, Shepton Mallet BA4 6NE
M0 IJU A. Wakefield Kuleana, 4 Barnrigg Barbon, Carnforth LA6 2LJ
M0 IJY K. Scotney 5 Candler Drive, Stone ST15 0WA
M0 IKA G. Mills Bowness, Moortown Lane Brighstone, Newport PO30 4AN
M0 IKB A. Young 15 Shelton Avenue, East Ayton, Scarborough YO13 9HB
M0 IKC G. Bryant 11 Cadwallon Road, London SE9 3PX
M0 IKD M. Draper 160 Chanctonbury Road, Burgess Hill RH15 9HA
M0 IKE D. Bilson 31 Middleton Drive, Inkersall, Chesterfield S43 3HS
MM0 IKG J. Dick 85 Mellerstain Road, Kirkcaldy KY2 6UD
MW0 IKH C. Eyre Croes Yr Onen, Chapel Street Newbridge, Wrexham LL14 3JH
M0 IKI J. Swann 5 Lanark Close, Hazel Grove, Stockport SK7 4RU
M0 IKM T. Palmer 70 Channel View Road, Eastbourne BN227LL
M0 IKN Dr F. Kuttikkate 34 Shetland Crescent, Rochford SS4 3FJ
M0 IKS J. Homersham 35 Meeting Street, Ramsgate CT11 9RT
M0 IKT D. Capstick 3 Andrew Close Dibden Purlieu, Southampton SO45 4LS
M0 IKV P. Day 1 Pine Close, Lutterworth LE17 4UT
M0 IKW G. Cannon 30 Main Street, Flixton, , Scarborough YO11 3UB
M0 ILA D. Mielczarek 7 Bryant Close, Camberley GU16 8AD
MM0 ILC S. Jordaan 16 Hillside Terrace, Selkirk TD7 4LT
M0 ILF B. Devon 8 Bruckner Place, Claremont Meadows 2747 Australia
MW0 ILG J. Lewis 26 Penlan Crescent, Swansea SA2 0RL
MI0 ILH A. Storey 31 Willowbrook Park, Bangor BT19 7GY
M0 ILI B. Neville 14 Alabury House Birch Close, Huntington, York YO31 9PP
MI0 ILJ D. Fisher 11 Summer Street, Belfast BT14 6ES
M0 ILK D. Critoph 26 Pevensey Close, Aylesbury HP21 9UB
M0 ILM M. Miller Barn Cottage, Wingfield Hall, Manor Road, Alfreton DE55 7NH
M0 ILN D. Bee 25 Blatcher Close, Minster On Sea, Sheerness ME12 3PG
M0 ILO M. Noblet 1 Lingdale Road, Wirral CH48 5DG
M0 ILT P. Evans 12 Cottage Corner, Ilton, Ilminster TA19 9ER
M0 ILU J. Taylor 41 Waters View, Yarwell Mill, Yarwell, Peterborough PE8 6EU
M0 ILX D. Vaughan 1 Boulnois Avenue, Poole BH14 9NX
M0 ILY A. Trueman 10 Mountbatten Ave, Dukinfield SK165BU
M0 ILZ P. Moore Barrio Los Ventorros 16, Comares 29195 Spain
M0 IME M. Scobie Suncourt, Meadfoot Sea Road, Torquay TQ1 2LQ
M0 IMF G. Brown 29 Park Close, Little Eaton, Derby DE21 5DY
M0 IML Glanford Electronics c/o B. Vile 24 Hudson Close, Dover CT16 2SG
M0 IMM S. Imms 26 Greenwood Avenue, Rowley Regis B65 9NJ
M0 IMP I. Pollard 24 Terminus Road, Littlehampton BN17 5BX
M0 IMQ A. Tomaszewski Tw34Ad, London TW34AD
M0 IMS M. Sims 5 Sandy Leaze, Bradford-on-Avon BA15 1LX
M0 IMT I. Turner 1 Elmwood Rise, Sedgley DY3 3QJ
MM0 IMU K. Swierczynski 103 Hendry Road, Kirkcaldy KY2 5DB
M0 IMV Kimbolton School ARC c/o D. Brattle 34 York Street, Bedford MK40 3RJ
M0 IMW I. Walker 24 Hawthorn Road, Norwich NR5 0LP
M0 IMY O. Fienko 42 Fenside Road, Boston PE21 8JH
M0 IMZ E. Preda 34A Church Street, Willingham, Willingham CB24 5HT
M0 INB I. Barraclough Maru, 25 Blaithroyd Lane, Halifax HX3 9PS
MW0 INC I. Curnock 62 Heol Y Banc, Bancffosfelen, Llanelli SA15 5DL
M0 IND P. Ind 30 Thompson Road, Stroud GL5 1SY
MM0 INE P. Thompson Windward Honeyfield Road, Jedburgh TD8 6JW
M0 INF Dr A. Fugard Flat 136, Willowbrook House, Coster Avenue, London N4 2ZT
MM0 INH W. Goodfellow 1 Yester Place., Haddington EH41 3BE
M0 INI M. Smith Church Farm, Mucklestone, Market Drayton TF9 4DN
MI0 INL O. Hart 18 Sandringham Court Portadown, Craigavon BT63 5BF
M0 INM L. Dettman 55 Canada Drive, Cottingham, Hull HU16 5EH
M0 INP I. Popgueorguiev 239 Westborough Road, Westcliff-on-Sea SS0 9PR
MM0 INS C. Ralph 37 Seaview Terrace, Edinburgh EH15 2HE
MM0 INT Outer Hebrides Iota Group c/o C. Mcgowan 21 Franchi Drive, Stenhousemuir, Larbert FK5 4DX
M0 INU Chelsea Pensioners RC c/o R. Petrie Royal Hospital Chelsea, Royal Hospital Road, London SW3 4SR
MW0 INW J. Evans 9 Cynfi Terrace Deiniolen, Caernarfon LL55 3LG
M0 INX S. Griffiths 14 Vicarage Lane, Oxford OX1 4RQ
M0 INY I. Davis Top Pub Brown Edge, Hill Top, Stoke-on-Trent ST6 8TX
M0 IOA the Isle of Avalon ARC c/o C. Heritage 29 Hill Head, Glastonbury BA6 8AW
MM0 IOB A. Macleod 2 Eoligarry, Isle of Barra HS9 5YD
M0 IOC I. O'Connor 7 Grove Court, Shotton Colliery, Durham DH6 2QD
M0 IOE K. Malinowski 46, Bennett Court, 2 Pitcher Lane, Ashford TW15 2BN
M0 IOI Dr S. Leask 1 Collington Street, Beeston, Nottingham NG9 1FJ
M0 IOK D. Proctor 4 The Green, Sproatley, Hull HU11 4XF
MM0 IOL A. Morrison 6A Upper Barvas, Isle of Lewis HS2 0QX
MD0 IOM M. Perry 48 Anagh Coar Road, Douglas, Isle of Man IM2 2AR
M0 IOO M. Kowalczyk 10 Hillberry Close, Narborough LE193EW
M0 IOS B. Stone 8 Gissons Lane, Kennford, , Exeter, EX6 7UB
M0 IOT C. Norris 115 Sutton Road, Walpole Cross Keys PE344HE
MI0 IOU T. Herbison 22 Dernaveagh Road, Ballymena BT43 6SX
M0 IOV Dr K. Singh Flat 401 Wolsey House, Princess Street, Ipswich IP1 1RS
M0 IOW B. Cant 15 Mountbatten Drive, , Newport, Iow. PO30 5SG
M0 IOZ Dr K. Singh Flat 401 Wolsey House, Princess Street, Ipswich IP1 1RS
M0 IPB S. Buckley 1 Columbia Street Fairview, Cheltenham GL52 2JR
MM0 IPD L. Flis 30 Bruce Avenue, Inverness IV3 5HE
M0 IPE R. Fulcher 1 Edwards Close Hutton, Brentwood CM13 1BU
M0 IPH A. Grzegorek Basement Flat 1, 17 Fitzroy Street, Bristol BS4 3BY
M0 IPI R. Hall 1 The Willows Brent Knoll, Highbridge TA9 4EJ
M0 IPJ B. Mcnamara Ham Hatches, 1 Recreation Road, Amesbury, Salisbury SP4 7BB
MW0 IPL Men's Shed RC c/o C. Beech 9 Wesley Court, Pembroke Dock SA72 6NE
M0 IPO B. Davies 16 Pearmains Close, Orwell, Royston SG8 5QY
M0 IPQ R. Zielinski 33 Ellis Road, Cambridge CB2 9BG
M0 IPR I. Ridings 25 Mond Road, Irlam, Manchester M44 6QA
M0 IPS A. Hollings 39 Rendham Road, Saxmundham IP17 1EA
M0 IPV D. Steel 18 Sandal Hall Mews, Wakefield WF2 6ED
M0 IPZ A. Griffin 4 Bramley Close Kingswood, Wotton-under-Edge GL12 8SF
M0 IQC R. Hall 47 Woodlands Drive Skelmanthorpe, Huddersfield HD8 9DB
M0 IQF M. Blazejewski 73 Byron Road, Luton LU4 0HX
M0 IQG G. Robinson 91 Tilstock Crescent, Shrewsbury SY2 6HH
M0 IQH D. Pemberton 142 Norfolk Road, Huntingdon PE29 1RH
MM0 IQK I. Andrews 29 The Riggs Auchtermuchty, Cupar KY14 7DX
M0 IQL H. Taylor 25 Northolme Avenue, Nottingham NG6 9AP
M0 IQM A. Bailey 52 Berkeley Road, Shirley, Solihull B90 2HT
MM0 IQN D. Petrovic 206 Mayfield Drive Armadale, Bathgate EH48 2JL
M0 IQQ T. Bell 28 Inwood Drive, Coleford GL168EZ
M0 IQR C. Wrobel 33 Harsnett Road, Colchester CO1 2HS
M0 IQX A. Emmerson 8 Weston Close, Heath Hayes, Cannock WS11 7YX
M0 IQY C. Cromie 140 Whalley Road Wilpshire, Blackburn BB1 9LJ
MW0 IQZ G. Cattle 12 Claerwen, Gelligaer, Hengoed CF82 8EW
M0 IRB S. Gandy 54 Saxty Way, Sowerby, Thirsk YO7 1SB
MM0 IRC C. Fraser Rockside, Locheport, Isle of North Uist HS6 5EU
M0 IRD Genesis Radio Group c/o I. Day 137 Tuffley Lane, Tuffley, Gloucester GL4 0NZ
M0 IRF Radio Group O c/o S. Constable 29 Sheppeys, Haywards Heath RH16 4NP
M0 IRG W. Linton 29 Lancaster Park, Richmond TW10 6AB
M0 IRH P. Geary 51 Heathermount Drive, Crowthorne RG456HJ
M0 IRI R. Trelease 23 Torridon Close, Woking GU21 3DB
MM0 IRJ I. Johnstone 14 Carledubs Crescent, Uphall, Broxburn EH52 6TH

M0 IRK P. Holmes 1 Leonards Place, Bingley. BD16 1AD
M0 IRL G. Maddden 192 Ardilaun, Portmarnock D13 FA03 Ireland
M0 IRO J. Ramirez Gonzalez 107 Maidstone Road, London N11 2JS
M0 IRP I. Pipe 9 Sherlock Hoy Close, Broseley, Telford TF125JB
M0 IRS D. Spinks 15 Brunlees Drive, Telford TF3 2NH
M0 IRT I. Thomas 47 Salisbury Avenue, Coventry CV3 5DA
M0 IRU A. Whybrow 64 Church Road Sevington, Ashford TN24 0LF
M0 IRV L. Jones 116 Captain Fold Road Little Hulton, Manchester M38 9UB
MI0 IRX A. Mcguigan 34 Ardymagh Road, Ballyclare BT39 9TJ
MI0 IRZ D. Gregg 9 Willowfield, Tandragee, Craigavon BT62 2EJ
MW0 ISC S. Charles Tower Side, Pantymwyn, Mold CH7 5HY
M0 ISE W. Su No.27, Ln. 800, Yingde St. Qianzhen Dist., , Kaohsiung 80651 Taiwan, Province of China
MW0 ISF C. Astbury Old Police House, Llanegryn, Tywyn LL36 9SL
M0 ISG F. Rundle 3 Winchester Court, Jarrow NE32 4TN
M0 ISI A. Cucchiara 172 Gonville Crescent, Stevenage SG2 9LZ
M0 ISJ Windmill Amateur Radio DX Group c/o J. Connor 28 Church Street, Hungerford RG17 0JE
M0 ISK G. Howard 8 Paddock Road, Woodford, Kettering NN14 4FL
M0 ISL A. Paulick Wormbacher Weg 27, Berlin 12207 Germany
M0 ISN Funny Contest G c/o E. Entwistle 43 Brock Road, Chorley PR6 0DB
M0 ISQ R. Lilley 3 Coultshead Avenue, Billinge, Wigan WN5 7HS
M0 IST N. Lasseter 7 Huntington Mews, York YO31 8JB
M0 ISU H. Baker 300 Lowerhouse Lane, Burnley BB12 6LZ
M0 ISW I. Singlehurst-Ward 39 Nadder Close, Tisbury, Salisbury SP3 6JL
MI0 ISY C. Maguire 1 Churchill Street Antrim Road, Belfast BT15 2BP
M0 ISZ B. Hardy 10 Spring Farm Road, Burton-on-Trent DE15 9BN
M0 ITA R. Ritossa 182, Rue Chateau Des Rentiers, Paris 75013 France
M0 ITB M. Brown 99 Apprentice Drive, Colchester CO45SE
M0 ITC R. Fearn 1 Broadheath Close, Droitwich WR9 7SW
M0 ITG N. Crudgington Appledore Blackness Lane, Keston BR2 6HL
M0 ITI M. Bruce 28 Pheasants Way, Rickmansworth WD3 7ES
MI0 ITS P. Ford 25 Carnhill, Londonderry BT48 8BA
M0 ITT J. Griffiths 83 Golborne Road, Ashton-In-Makerfield, Wigan WN4 8XA
M0 ITX N. White 99 The Common, Mellis, Eye IP23 8EF
M0 ITY J. Culak 2 Linkside Road, Bishop's Stortford CM23 5LP
M0 ITZ S. Proverbs 14 Spring Lane, Shepshed, Loughborough LE12 9JE
M0 IUA Dr C. Oxley Appletree House Halam Road, Southwell NG25 0AH
M0 IUB J. Fisher 4 Orchard Close, East Leake, Loughborough LE12 6PL
M0 IUC M. Curtis 9 Oaklands Crescent, Holt NR25 6UD
M0 IUE D. Cassidy 30 Oakland Road, Botley, Southampton SO32 2SX
M0 IUG S. Campbell 50 Northwood Road, Whitstable CT5 2ES
M0 IUI J. Pinto 7 Wright'S Way, Colchester CO6 4NS
M0 IUK D. Grayson 79 Errington Avenue, Sheffield S2 2EA
M0 IUM G. Clark 65 Chyvelah Vale, Gloweth, Truro TR1 3YJ
MW0 IUN I. Jones 21 Albert Street, Maesteg CF34 0UF
M0 IUQ G. Hinson The Leys, Brierley Hill DY5 3UJ
M0 IUR E. Vrentzos 12 Floyer Close, Queens Road, Richmond TW10 6HS
M0 IUT R. Williams 10 Bramley Close, Twickenham TW2 7EU
M0 IUU I. Stephens 14 Vardon Close Kingston Hill, Stafford ST16 3YW
M0 IUV Nel Rynt L/B Wt c/o S. Helm 10 St. Annes Avenue, Middlewich CW10 0AE
M0 IUZ D. Waterhouse 6 Sands Lane, South Ferriby, Barton-upon-Humber DN18 6JS
M0 IVC I. Marques 7A Kings Park Road, Southampton SO15 2AS
M0 IVE I. Valkov Flat 10, Warlingham House, London SE16 3DQ
M0 IVJ J. Raybould 33 Lincoln Road, Dorrington, Lincoln LN4 3PT
M0 IVK A. Parkin 22 Glorney Mead, Badshot Lea, Farnham GU9 9NL
M0 IVL P. O'Shea 37 Barclay Court, Ilkeston DE7 9HJ
M0 IVM A. Nel 12 Wain Court, Rakeway, Saughall, Chester CH1 6BF
M0 IVN P. Mocker Flat 24, The Gardens, Clapton Common, London E5 9AZ
M0 IVO W. Gissing 2 Yeo Moor, Clevedon BS21 6UQ
M0 IVS M. Mann Woodlands, Coston, Norwich NR9 4DT
M0 IVU M. Holland 32 Saltersgate Drive, Birstall, Leicester LE4 3FF
M0 IVW D. Barrett 183 Wilson Avenue, Brighton BN2 5PD
M0 IVZ C. Cumming 2 Alexander Close, New Milton BH25 5NS
M0 IWA D. Brooks 61 Carisbrooke High St., Newport PO30 1NR
M0 IWB I. Bunting 14 Mill Pightle, Aylsham, Norwich NR11 6LX
M0 IWG M. Tyrrell 47 Woodley Hill, Chesham HP5 1SL
M0 IWI D. Gilbertson 6 Lewens Lane, Wimborne BH21 1LE
M0 IWJ J. Siviter 96 Byne Road, London SE26 5JD
M0 IWN P. Taylor 138 Paulhan Street, Bolton BL33DT
M0 IWO S. Noller 3 Thor Road, Norwich NR7 0JS
M0 IWQ T. Warner 6 Cotswold Court, Skelmersdale Road, Clacton on Sea CO15 6EN
MM0 IWS J. Greene Ingleneuk, Beltie Road, Torphins AB31 4JU
M0 IWT A. Gale 22 Graham Street, Swindon SN1 2EY
MW0 IWU Eryri DX c/o J. Pritchard 1 Tan Y Coed, Maesgeirchen, , Bangor LL57 1LU
M0 IWV R. Molyneux 23 Birks Drive, Bury BL8 1JA
M0 IWW R. Molyneux 23 Birks Drive, Bury BL8 1JA
MW0 IWX Blaenau Gwent Radio Group c/o A. Lewis 33 Heol Helig, Brynmawr, Ebbw Vale NP23 4TY
M0 IWZ A. Hanna 35 Orchard Drive, Mayland, Chelmsford CM3 6EP
M0 IXC Dr A. Mazur 1 Oak Trees New Road, Stoborough BH20 5BB
M0 IXF A. Smythson 64 Blinco Rd, Rushden NN10 0EA
M0 IXG I. Omenaca-Gavin 2 Mayflower Court Highbridge Wharf, Reading RG1 3AJ
MI0 IXH M. Mcquillan 21 Glen River Park Glenavy, Crumlin BT29 4FX
MM0 IXJ K. Brown 10 Richmond Street, Clydebank G81 1RF
M0 IXL M. Norfield 122 Huntingdon Road, Upwood, Ramsey, Huntingdon PE26 2QQ
M0 IXM R. Gaskell Cottonwood California, Baldock SG7 6NU
M0 IXP D. Hall 31 Edinburgh Drive, Didcot OX11 7HS
M0 IXQ O. Reyes Salazar P.O Box 117, Heidelberg 3084 Australia
MM0 IXT S. Kelso 98 Highmains Avenue, Dumbarton G82 2QB
M0 IXU A. Foote Flat One, Kimber'S Close Kennet Road, Newbury RG14 5JF
MW0 IXV W. Groves 3 Tetbury Close, Newport NP20 5HX
M0 IXW A. Royds 3A Fairfield Avenue, Rossendale BB4 9TG
M0 IXX S. Colquhoun-Lynn 76 Chestnut Drive, Sale M33 4HL
M0 IXY A. Atkinson 21 Dennington Crescent, Basildon SS14 2FF
M0 IXZ P. Rollason 68 Heathfield Lane West, Wednesbury WS10 8QP
M0 IYC Basingstoke Makerspace c/o K. Roche 96 Porter Road, Basingstoke RG22 4JR
MM0 IYD G. Mann Halcyon, Strachan, Banchory AB31 6NL
M0 IYE A. Gibbons 305 Monks Road, Lincoln LN2 5LB
M0 IYF M. Christensen 40 Aireview Terrace, Broughton Road, Skipton BD23 1RX
M0 IYG D. Stevenson 37 Larkhill, Skelmersdale WN8 6TE
M0 IYH T. Machii 3-6-17-311 Nakacho, Musashino City, Tokyo 180-0006 Japan
M0 IYJ P. Reeves 11 Quines Hill Roadá Forest Towná, Mansfield Notts NG190NW
M0 IYK C. Walson 30 West Crescent, Duckmanton, Chesterfield S44 5HE
M0 IYM B. Clark 8 Langdale Close, Farnborough GU14 0LQ
M0 IYN G. Moss 125 Lavender Avenue, Mitcham CR4 3RS
M0 IYP G. Wood 28 Marford Crescent, Sale M33 4DH.
M0 IYQ D. Buchan 40 Bradstone Road, Winterbourne, Bristol BS36 1HQ
MM0 IYT M. Mercanti 45 Brueacre Drive, Wemyss Bay PA18 6HA
M0 IYV D. Tennant 9 Beaumaris Road, Hindley Green, Wigan WN2 4NB
M0 IYZ Dr P. Duvoisin 6A Newland Street, Witham CM8 2AQ
M0 IZA A. Nikitits 270A The Ridgeway, St. Albans AL4 9XQ
M0 IZF A. Robeson 19 Cameron Drive Woodlands, Ivybridge PL21 9TS
M0 IZG A. Evans 8 Plowden Close, Bolton BL3 3NU
M0 IZM P. Stuart 5 Welbeck Gardens, Woodthorpe, Nottingham NG5 4NX
MI0 IZN Orchard County DX Club c/o E. Simpson 10 Woodview Park, Tandragee, Craigavon BT62 2DD
MI0 IZP J. Henshaw 80 Ballystrudder Road, Islandmagee, Larne BT40 3SJ
M0 IZR D. Butler Westbrook, Lower Farm Road, Ringshall, Stowmarket IP14 2JE
M0 IZS D. Sexton 16 Rufus Isaacs Road, Caversham, Reading RG4 6DD
M0 IZV P. Gardner 7 Park Avenue, Shipley BD18 3LW
M0 IZW P. Raftery Ballybride, , Roscommon, F42TK49 Ireland
MW0 IZX K. Earnshaw 5 Castle Mews George Street, Pontypool NP4 6BU
M0 JAD P. Holland 30 Knighton Park Road, London SE26 5RJ
M0 JAE J. Allen 20 Spa Hill, Kirton Lindsey, Gainsborough DN21 4BA
M0 JAF J. King 22 Latchmere Gardens, Leeds LS16 5DN
M0 JAG A. Pegg 18 Blythe Way, Shanklin PO37 7NJ
M0 JAI P. Gaur 34 Queensberry Avenue, Copford, Colchester CO6 1YN
M0 JAJ J. Stedman 60 Sandown Road, Ipswich IP1 6RE
M0 JAK J. Swain 84 Sunnymead Drive, Waterlooville PO7 6BX
M0 JAM J. Mortimer 4 Nethercliffe Crescent, Guiseley, Leeds LS20 9HN
MW0 JAN J. Day 20 St. Johns Drive Pencoed, Bridgend CF35 5NF
M0 JAO J. Sansom 4 Vicarage Road, Eastbourne BN20 8AU
M0 JAP Dr D. James Bramble Cottage, Tray Lane, Atherington, Umberleigh EX37 9HY
M0 JAQ J. Malia 55 West Farm Avenue Longbenton, Newcastle upon Tyne NE12 8LS
MI0 JAR J. Rice 42 The Crescent, Ballymoney BT53 6ES
MI0 JAT J. Mcgoldrick 23 Lettercarn Road, Clare, Castlederg BT81 7QY
M0 JAV Dr J. Rogers Impala 81 Hady Hill, Chesterfield S41 0EE
MW0 JAW J. Stevens 8 St. Josephs Court, Llanelli SA15 1NR
M0 JAX J. Edwards 45 Bramshaw Gardens, Bournemouth BH8 0BT
MI0 JAY W. Graham 19 Margaret Square, Ballymoney BT53 6BZ
M0 JAZ J. Sadler 10 Spindle Warren, Havant PO9 2PU
M0 JBA J. Baines 7 Willerby Low Road, Cottingham HU16 5JD
M0 JBC J. Crank 38 Harley Avenue, Harwood, Bolton BL2 4NU
M0 JBD J. Day 124 Radstock Road, Southampton SO19 2HU
M0 JBF J. Cobb 32 Dellmont Road, Houghton Regis, Dunstable LU5 5HU
MW0 JBH A. Jozwik 160 Broad Mead Park, Newport NP19 4PF
M0 JBI J. Booth-Isherwood 59 Brackens Lane, Alvaston, Derby DE24 0AQ
MI0 JBK J. Mackenzie 30 Dalriada Gardens, Ballycastle BT54 6DZ
MM0 JBS J. Summers 1 Main Road, Fairlie, Largs KA29 0DP
MI0 JBT J. Traynor 8 Roeville Terrace, Limavady BT49 0BH
M0 JBV M. Harvey May Tree Cottage, Kelvedon Road, Tiptree, Colchester CO5 0LJ
M0 JBW J. Mclaughlin 34 Cambridge Road, Birstall, Batley WF17 9JF
M0 JBY E. Riddle 37B Stubbs Lane, Braintree CM7 3NR
M0 JBZ J. Chalmers 19 Brettenham Crescent, Ipswich IP4 2UB
M0 JCC I. Jefferson 19 Orchard Way, Flitwick, Bedford MK45 1LF
M0 JCD J. Dalgliesh 61 Clonners Field, Stapeley, Nantwich CW5 7GU
M0 JCE J. Crewe 22 Myrtle Tree Crescent Sandbay, Weston-Super-Mare BS22 9UL
M0 JCG C. Grant 27 Bulrush Close, Chatham ME5 9BN
M0 JCH J. Paul East Park, Church Road, Cowes PO31 8HA
M0 JCK E. Beechill Belleroyd Farm Blackshaw Head, Hebden Bridge HX7 7JP
M0 JCL J. Plant 67 Kenley Road, London SW19 3JJ
M0 JCM J. Murray 2 The Cuttings Hampstead Norreys, Thatcham RG18 0RR
M0 JCP J. Haynes 16 Mountsfield, Frome BA11 5AR
M0 JCQ J. Stevens 51 Cheddington Road, Pitstone, Leighton Buzzard LU7 9AQ
M0 JCR J. Reynolds 15 Chestnut Mead, Oxford Road, Redhill RH1 1DR
M0 JCS J. Stevenson 18 Drakehouse Laneá, Sheffield S20 1FW
M0 JCT J. Townsend 47 Main St., Wolston, Coventry CV8 3HH
M0 JCV GNRS c/o R. Moseley 20 Bleakley Avenue, Notton, Wakefield WF4 2NT
M0 JCY R. Manciu 3 Delta Court, Standard Road, Hounslow TW4 7AN
M0 JCZ M. Ciechan 109 Rose Vale, Liverpool L5 3PD
M0 JDA J. Dale Corydon, Church Street, Sevenoaks TN14 7SW
M0 JDB J. Pollard 72 Windy Arbour, Kenilworth CV8 2BB
M0 JDD J. Dedier 10 Lowry Close, Haverhill CB9 7GH
M0 JDE D. Foxall 1 Doe Hey Grove, Farnworth, Bolton BL4 7HS
M0 JDF Ncc Group RS c/o P. Hodges 191 Broadstone Road, Stockport SK4 5HP
M0 JDL J. Dowdeswell 18 Lechlade Gardens, Fareham PO15 6HF
M0 JDP J. Page 5 Riddimore Avenue, Hereford HR2 7LJ
M0 JDS J. Sweatman 14 Clover Court, Jasmine Grove, Waterlooville PO7 8BP
M0 JDT J. Tyrrell 28 Park Avenue, Princes Avenue, Hull HU5 3ER
M0 JDU J. Dowson 4 Thorntree Close, Goole DN14 6HJ
MW0 JDW J. Williams 1 Nant Terrace, Pentraeth LL75 8YE
M0 JDY Z. Zhao Christs College, Cambridge CB2 3BU
M0 JDZ R. Griffin 15 Donside Close, Boldon Colliery NE35 9BS
M0 JEA M. Augustus 3 Heathend Cottages Heathend, Wotton-under-Edge GL12 8AS
M0 JEB J. Braithwaite 129 Cherry Tree Gardens, Blackpool FY4 4PY
M0 JEC N. Bland 63 Swindon Road, Wroughton, Swindon SN4 9AG
M0 JEG B. Mcdonald-Watson 39 Heald Way, Nantwich CW5 6SQ
M0 JEH D. Mackie 41 Fairfield Avenue, Sandbach CW11 4BP
M0 JEI O. Phillips 54 Marshall Road, Cambridge CB1 7TY
M0 JEJ S. Hunt 7 Cokefield Avenue, Nuthall, Nottingham NG16 1AU
M0 JEK A. Skarzynski 1 River View Moorings, Bridge Road, Stoke Ferry, King's Lynn PE33 9TS
M0 JEL J. Loughrey 54 Marshall Drive, Bramcote, Nottingham NG9 3LD
M0 JEM J. Mayo 134 Bromsgrove Road, Redditch B97 4SP

M0 JEO R. Bishop 12A Goseley Avenue, Hartshorne, Swadlincote DE11 7EZ
M0 JEP J. Price Oxted Place East, Broadham Green Road, Oxted RH8 9PF
M0 JEQ D. Tynan 2 Horsham Grove Whelley, Wigan/Greater Manchester WN2 1AP
MJ0 JER R. Taylor 21 Samares Avenue La Grande Route De St. Clement, St. Clement JE2 6NY Jersey
M0 JES J. Saunders 105 Wembdon Hill, Wembdon, Bridgwater TA6 7QB
MM0 JET Antrim & Dis AR c/o B. Burt 182 Old Inverkip Road, Greenock PA16 9JG
M0 JEU R. Tadhunter 44 Queensway, Melksham SN12 7LD
M0 JEZ J. Powell 46 Woodmancote, Yate, Bristol BS37 4LL
M0 JFB J. Button 1 Amber Close, Rainworth, Mansfield NG21 0FU
M0 JFD J. Dixon 23 Dee Way, Winsford CW7 3JB
M0 JFE J. Earnshaw 128 Shakespeare Road, Fleetwood FY7 7HJ
M0 JFH D. Flood 506 Preston Old Road, Blackburn BB2 5LY
M0 JFI S. Kennett 8 Red Lion Lane, Chobham, Woking GU24 8RH
M0 JFJ Y. Alghurair 12 Second Avenue, Bath BA2 3NN
M0 JFK P. Waine 14 Grace Street, Sutton, St. Helens WA93NG
M0 JFM J. Marsh 14 Eyam Road, Hazel Grove, Stockport SK7 6HP
M0 JFO J. Owen 8 Highridge Crescent, Bristol BS13 8HN
M0 JFR L. Vencel 12 Alston Road, Ipswich IP3 8EU
M0 JFW J. Wheeler 428 Bromsgrove Road, Hunnington, Halesowen B62 0JL
MW0 JFX K. Boxley 9 Llys-Y-Pentre, Afonwen CH7 5UY
M0 JGB J. Greenway-Brown 207 Lowe Avenue, Wednesbury WS10 8NS
M0 JGC Royal Signals Swindon c/o Dr R. Ashman 44 Conan Doyle Walk, Swindon SN3 6JB
MW0 JGE N. Lewis 1 Clyne Drive, Blackpill, Swansea SA3 5BU
M0 JGF P. King 14 Glenhurst Drive, Whickham, Newcastle upon Tyne NE16 5SJ
M0 JGH J. Hunt 15 Greenway, London SW20 9BQ
M0 JGK A. Day 15 Hillview Crescent East Preston, Worthing BN16 1RD
M0 JGM J. Marlett 6 Delamere Avenue Sutton Manor, St. Helens WA9 4AP
MM0 JGP J. Pirie Millhouse, Watermill, Fraserburgh AB43 7ED
M0 JGR J. Glover 12 Willow Street, London E4 7EG
M0 JGS J. Seaton 88 Stanley Road, Cambridge CB5 8LB
M0 JGU P. Shaw 32 Hardwick Road East, Worksop S80 2NT
M0 JHB Eureka RC c/o J. Butcher 3 Basket Gardens, London SE9 6QP
MW0 JHC J. Clarke 29 Sisial Y Mor, Rhosneigr LL64 5XB
M0 JHD J. Hardy Lambda House, Seanor Lane, Chesterfield S45 8DH
M0 JHF J. Foster Westfield House, 23 High Street, Cumnor OX2 9PE
M0 JHG Leeds Radio Amateurs c/o J. Ginever 66 London Road, Maidstone ME16 8QU
MM0 JHL J. Hutchinson Hawthorn Cottage, Muirhall, West Calder EH55 8NL
M0 JHM J. Rymer 23 Chetwode Road, Tadworth KT20 5PS
M0 JHN M. Rees Hillcrest, Cumeragh Lane, Whittingham, Preston PR3 2AL
M0 JHO B. Hodgson 28 Grove Drive, Woodhall Spa LN10 6RT
M0 JHP H. Alava Moreira 185 Riverdale Road, Erith DA8 1PZ
M0 JHV G. Dulcu 49 Pearson Road, Crawley RH10 7AJ
M0 JHW J. Wheeldon 11 Stathern Walk, Grantham NG31 7XG
M0 JIB Hilltop ARC Co.Down c/o J. Bickers 3 The Old Brickyard, West Haddon, Northampton NN6 7GP
M0 JIC M. Matusiewicz 6 Westminster Crescent, Doncaster DN2 6JQ
M0 JID T. Gore 22 Stoppard Road, Burnham-on-Sea TA8 1QB
M0 JIE K. Reynolds 20 Wentwood Gardens, Plymouth PL6 8TD
M0 JIJ J. Matthews 2 Farm Close, Bungay NR35 1JG
M0 JIL G. Heyes 5 Ashgarth Way, Harrogate HG2 9LD
M0 JIM J. Heinen The Coach House, St. Mary Bourne, Andover SP11 6EW
M0 JIO A. Edwards 23 Brittany Avenue, Ashby-de-la-Zouch LE65 2QY
M0 JIP J. Pownall 75 Park Barn Drive, Guildford GU2 8ER
M0 JIQ D. Samarin Long Gables, Gorse Avenue, Kingston Gorse, East Preston BN16 1SQ
MJ0 JIS Jersey Scouts ARC c/o C. Totty Flat 4, Beech Court, Woodlands Apartments, La Rue Des Cotils, Grouville, Jersey JE3 9AY
M0 JIU J. Uren 4 Killivose Road, Camborne TR14 7RN
M0 JIX Dr C. Fletcher 7 Highfield Crescent, Baildon, Shipley BD17 5NR
M0 JJA R. James 51 The Lampreys, Gloucester GL4 6QU
M0 JJB J. Barton 93 Cardigan Road, Bridlington YO15 3JU
M0 JJC A. Coghlan 1 Bell View Cross Houses, Shrewsbury SY5 6JR
M0 JJD J. Dignan Plum Tree Cottage, Station Road, Marsh Gibbon, Bicester OX27 0HN
M0 JJE M. Chisholm 19 Tudor Close, Newtoft, Market Rasen LN8 3NQ
M0 JJH G. Cavie Dawn, Maypole Road, Colchester CO5 0EN
M0 JJI A. Shakesby 14 Dawnay Road, Bilton, Hull HU11 4HB
M0 JJK J. King 18 Ross Road, Wallington SM6 8QB
M0 JJM D. Oliver 20 Five Oaks Close, Malvern WR14 2SW
M0 JJN J. Nicholls 6 Marasca End, Holt Drive, Colchester CO2 0DL
M0 JJP J. Walker Flat A 14 Elswick Road, London SE13 7SR
MM0 JJQ M. Butterworth 97A Dunearn Drive, Kirkcaldy KY2 6AL
M0 JJR J. Reilly 22 Charlecote Gardens, Sydenham, Leamington Spa CV31 1GE
MM0 JJU M. Love 17 Lindsay Road, East Kilbride, Glasgow G74 4HZ
MM0 JJV J. Vennard 4 Braehead, Girdle Toll, Irvine KA11 1BD
M0 JJX D. Jarman 62 Combe Drive, Dunstable LU6 2AE
MM0 JJZ K. VielhaberNether Littlefold, Crieff PH7 3NY
M0 JKB S. Lucas 31 Lilian Close, Norwich NR6 6RZ
M0 JKE J. Howarth 19 Farnham Croft, Leeds LS14 2HR
M0 JKF J. Ferrol 29 Westlands, Haltwhistle NE49 9BS
M0 JKG J. Gaskin Badgers Barn, Canterbury Road, Folkestone CT18 8DF
MM0 JKH D. Smith 9 Ceum A Bhealaich, Stornoway HS1 2UB
M0 JKI A. Munns Peverels, Stoke Farm Drive, Battisford, Battisford Tye, Stowmarket IP14 2NA
M0 JKM J. Lugsden 21 Overhill Way, Beckenham BR3 6SN
M0 JKN J. Wilson Flat 5, Blake House, London SE1 7DX
M0 JKP M. Howden 11 Marsh Lane Gardens, Kellington, Goole DN14 0PG
M0 JKQ C. Poulson 9 Scattergate Green, Appleby-in-Westmorland CA16 6SP
M0 JKS Dr D. Pegler September Cottage, East Bank, Winster DE42DT
MW0 JKU C. Taylor 23 Heol Derw, Brynmawr, Ebbw Vale NP23 4TT
M0 JKV A. Allan The Oxford Health Co Ltd, Unit 4, Longlands Road, Bicester OX26 5AH
M0 JKW J. Whitehead 20 Seamill Park Crescent, Worthing BN11 2PN
M0 JKX Invicta CG c/o I. Hope 5 The Crescent, Northfleet, Gravesend DA11 7EB
MM0 JKY A. Chambers 35 Echline Grove, South Queensferry EH30 9RU
MI0 JLC S. Mulligan 70 Lisnoe Walk, Lisburn BT28 1QD
M0 JLE J. Cafe Flat 6, Smiths Court, 73 East Borough, Wimborne BH21 1PJ
M0 JLI I. Bertea 102 Clarence Way, Buckles Lane, London RM15 6QE
MW0 JLN N. Howells1 21 Coed Bach, Pencoed, Bridgend CF35 6TF
M0 JLP J. Powell 23 Park Road, Norton, Malton YO17 9DZ
M0 JLR J. Redhead 28 Sandfields, Frodsham WA6 6PT
M0 JLS A. Stevens 16 Hunters Chase, March PE15 9EL
M0 JLT L. Taylor 33 Priestley Avenue, Darton, Barnsley S75 5LG
M0 JLW J. Welford 26 Templewood, Welwyn Garden City AL8 7HX
M0 JLY A. Page 207 Brooklyn Road, Cheltenham GL51 8DZ
MM0 JMB J. Brown 11 Fairway Avenue, Elgin IV30 6XF
M0 JMC J. Mccutcheon 15 Maytrees, Hitchin SG4 9LT
M0 JME J. Blundell 21 Walmsley Street, Fleetwood FY7 6LJ
M0 JMF J. Foster 41 Coinagehall Street, Helston TR13 8ER
MM0 JMI Prof. J. Davies 5 County Council Houses, Kingston, North Berwick EH39 5JE
M0 JMJ J. Jukes 22 Hazelmere Road, Creswell, Worksop S80 4HS
MM0 JMK J. Mckechnie 8 Waulker Avenue, Stirling FK8 1SA
MI0 JML J. Mccaw 62 High Street, Ballymena BT43 6DT
M0 JMN R. Andrews Owls Rest, Park Lane, Worcester WR2 6PQ
M0 JMP J. Poole 18 Grosvenor Avenue, Kidderminster DY10 1SS
M0 JMQ M. Ireland 45 Pheasant Rise, Bar Hill, Cambridge CB23 8SA
M0 JMS J. Stimpson 149 Hungate Street, Aylsham, Norwich NR11 6JZ
M0 JMT J. Tanner 4 Pool Meadow Close, Birmingham B13 9YP
M0 JMU Dr J. Hunt 10 Couzens Close, Chippenham SN15 1US
M0 JMV D. Vanstone 1 Deben Drive, Sudbury CO10 2QH
M0 JMY J. Mcmullan 17 Banbury Close, Accrington BB5 4BZ
M0 JNF B. Coley 17 Livingstone Road Handsworth, Birmingham B20 3LS
M0 JNJ J. Butcher 14 Park Road, Lowestoft NR32 1SW
MM0 JNL G. Crawford La Quinta Blackhill, Kirriemuir DD8 5NX
M0 JNP J. Perry 7 Lancaster House Belle Vue Road, Paignton TQ4 6HD
M0 JNQ S. Rhenius Baythorne Cottage Baythorne End, Halstead CO9 4AB
M0 JNS J. Shatford 31 Pinner Park Avenue, Harrow HA2 6LG
M0 JNU J. Nuttall 73 Severn Drive, Walton-Le-Dale, Preston PR5 4TD
M0 JNV S. Pomeroy 55 Woodmancote Vale, Woodmancote, Cheltenham GL52 9RJ
M0 JNX J. Hall 1 Nash Close, Earley, Reading RG6 5SL
MM0 JNY Dr E. Newman Yewbank, Bamff Road, Alyth, Blairgowrie PH11 8DR
M0 JOB J. O'Brien 76 Berkeleys Mead, Bradley Stoke, Bristol BS32 8AU
M0 JOC J. O'Connell 23 Halstead Road, Gosfield CO9 1PG
M0 JOD J. Preece 51 Drancy Avenue, Willenhall WV12 5RD
M0 JOH J. Godfrey 17 Lichfield Road, Sneinton, Nottingham NG2 4GF
M0 JOI P. Edwards 4 Stodart Road, London SE20 8ET
M0 JOJ J. Green 26 Foxhill Road Burton Joyce, Nottingham NG14 5DB
MM0 JOK J. Burgoyne 5 Shankston Crescent, Cumnock KA18 1HA
M0 JOL M. O'Leary 4 Park Farm Close, Martinstown, Dorchester DT2 9TW
MM0 JOM J. Mann 10 Brimmond Walk, Westhill AB32 6XH
M0 JOO A. Williams 12 St. Wilfrids Crescent, Brayton, Selby YO8 9EU
MW0 JOP R. Bowen 16 Lon Hywel, Whitland SA34 0BE
M0 JOR J. Orr 13 Haldane Close, Brierley, Barnsley S72 9LL
M0 JOW D. Hamilton 16 Conifer Crest, Newbury RG14 6RT
MM0 JOX P. O'Hara 13/3 135 Kirkton Avenue, Glasgow G13 3EP
M0 JOY J. Bilson 31 Middleton Drive, Inkersall, Chesterfield S43 3HS
M0 JPA J. Wake 60 Cloverville Approach, Bradford BD6 1ET
M0 JPB J. Bull 91 Lime Road, Wednesbury WS10 9NF
MI0 JPC J. Cosgrove 91 Church Street, Newtownards BT23 4AN
MI0 JPD J. Doyle 25 Parkmore Road, Magherafelt BT45 6PF
M0 JPE J. Poole 61 Lower Vickers Street, Miles Platting, Manchester M40 7LX
M0 JPG J. Gleeson 124 Rushes Mead, Harlow CM18 6QE
M0 JPK G. Anestopoulos 60 Jaguar Lane, Bracknell RG12 9PE
MI0 JPL J. Jones 107 Belfast Road, Whitehead, Carrickfergus BT38 9SU
M0 JPM J. Meijer Birchwood East End, Gooderstone, King's Lynn PE33 9DB
M0 JPN T. Nakagawa 7 Milton Street, Barrowford, Nelson BB9 6HE
MI0 JPO J. Alexander 24 Alexandra Avenue, Ballymoney BT53 6EX
M0 JPP S. Clark 20 Herbert Street, Loughborough LE11 1NX
M0 JPR C. Kwok 27 Blackberry Lane, Stockport SK5 8JZ
M0 JPS J. Styles 42 Brook Street, Woodbridge IP12 1BE
M0 JPT Hadrabs Cont Gr c/o J. Tan 22, Jalan Geikie, , Miri, 98000 Malaysia
M0 JPW J. Woolvin 62 Whitewood Park, Liverpool L9 7LG
M0 JPX J. Parkes 24 Kenilworth Road, Lichfield WS14 9DP
M0 JQB J. Bovey 17 North Bank Road, Bingley BD16 1UH
M0 JQD D. Start The Rise, Valley Lane Swaby, Alford LN13 0BH
M0 JQJ N. Brierley Whitewalls Farm, Warleggan, Mount, Bodmin PL30 4HF
M0 JQK T. Firth 126 Tombridge Crescent, Kinsley, Pontefract WF9 5HE
M0 JQL M. Dhur 3 Sharpes Corner, Lakenheath, Brandon IP27 9LA
M0 JQO Daventry Amateur Radio Repeater Group c/o A. Baker 10 Kingscroft Court, Northampton NN3 9BH
M0 JQP P. Kilby 1 Home Farm Barns, South End, Milton Bryan, Milton Keynes MK17 9HS
MI0 JQS Tri County ARC c/o V. Mcfarland 16 Saint Annes Crescent, Carnmoney, Newtownabbey BT36 5JZ
M0 JQV G. Huff 135 Westminster Road Toothill, Swindon SN5 8JE
M0 JQW Mf Propagation Research Group c/o J. Wang Flat 31, 74 Arlington Avenue, London N1 7AY
MI0 JQY Rvd. M. Donald Station Road, Garvagh BT51 5LA
M0 JRA A. Jessop 4 Katherine Street, Thurcroft, Rotherham S66 9LG
M0 JRE J. Eaton1 38 Litchford Road, New Milton BH25 5BQ
MM0 JRF Altrincham Grammar School for Boys c/o J. Fyfe 53A Ware Road, Glasgow G34 9AR
M0 JRI J. Dyson 77 Grantham Road, Southport PR8 4LT
M0 JRJ J. Jenkins 31 Pendrell Street, London SE18 2PH
M0 JRL J. Lynn 72 Badger Close, Guildford GU2 9WA
M0 JRN D. Higton 24 Holly Avenue, Bradwell NR31 8NL
M0 JRO J. Roberts 'Meadowside', Keswick Lane, Bardsey, Leeds LS17 9AD
M0 JRP R. Pearson 14 Highfields, Lakenheath, Brandon IP27 9DZ
MM0 JRR J. Rayne 8 Bankton Grove, Livingston EH54 9DW
M0 JRU C. Denton 100 Lincoln Road Deeping Gate, Peterborough PE6 9BA
M0 JRW J. Wilson 1 Locarno Avenue, Runwell, Wickford SS11 7HX
MW0 JRX O. Bross 8 Queens Drive, Buckley CH7 2LJ
M0 JRZ J. Robb 37 Wroxham Road, Woodley, Reading RG5 3AX
M0 JSA A. Jones 5 Meadowlands, Kirton, Ipswich IP10 0PP

M0 JSD J. Delaney 33 Deepdale Close, Ibstock LE67 6LW
M0 JSE M. Egan 32 Hawksworth Avenue Guiseley, Leeds LS20 8EJ
MM0 JSG J. Galloway 144 Strathkinnes Road, Kirkcaldy KY2 5PZ
M0 JSH J. Yan St Edmund'S College, Mount Pleasant, Cambridgeshire CB3 0BN
MI0 JSJ J. Smith 54A Blackstaff Road, Kircubbin, Newtownards BT22 1AF
M0 JSK G. Booton 69 The Street, Deal CT14 0AJ
M0 JSL J. Swales 90 Earlswood Road, Dorridge, Solihull B93 8RN
M0 JSN Severn Valley Rd c/o J. Sowman 53 Newton Wood Road, Ashtead KT21 1NN
M0 JSP J. Swiffen Dragonelle, Marina Drive, Shireoaks S81 8NQ
M0 JSR J. Street 22 Roman Acre, Wick, Littlehampton BN17 7HN
MM0 JSU J. Smit Annabells Cottage, Deirdre, Connel, Oban PA37 1PL
M0 JSW J. Woodland 14 Kelham Green, Nottingham NG3 2LP
M0 JSX J. Sawyer 27 Croft Road, Wallingford OX10 0HN
M0 JSY B. Josyfon 25 Norrice Lea, London N2 0RD
M0 JSZ F. Jackson 5 Chalmers Avenue, Haversham, Milton Keynes MK19 7AG
M0 JTB J. Brown 55 Barrington Road, Rubery, Birmingham B45 9EU
M0 JTD D. Baker 65 Madison Street, Tunstall ST6 5HS
MI0 JTE J. Elliott 183 Kilraughts Road, Ballymoney BT53 8NL
MW0 JTG B. Williams Bryn Celyn, Hendre Road, Conwy LL32 8RJ
M0 JTH J. Thomas 77 Hawthorn Avenue, Lowestoft NR33 9BB
M0 JTJ J. Talbot-Jones 21 Downsview Drive, Wivelsfield Green, Haywards Heath RH17 7RN
M0 JTL I. Bardell 17 Stanton Avenue, Bradville, Milton Keynes MK13 7AR
M0 JTM G. Ridley 12 Garforth Avenue, Steeton, Keighley BD20 6SP
M0 JTN M. Chivers 4 Hunters Lodge, Fareham PO15 5NF
M0 JTQ J. Cook 28 Wenny Estate, Chatteris PE16 6UX
MM0 JTX Dr J. Wills 2 Old Dalmore Gardens Auchendinny, Penicuik EH26 0RR
M0 JUA J. Thirlwell 6A Brook Street, Warminster BA12 8DN
MM0 JUC R. Cook Flat 6, Howards Court, Caledonian Road, Perth PH1 5NJ
M0 JUI P. Stavropoulos 60 Trent Valley Road, Stoke-on-Trent ST4 5JA
M0 JUJ M. Smith 24 Fifth Avenue, Portsmouth PO6 3PE
M0 JUK J. Allison 72 Chantry Croft, Kinsley, Pontefract WF9 5JL
MM0 JUL R. Mitchell Broadhills, Isle of Coll PA78 6TB
M0 JUN J. Unwin 39 Whinfell Drive, Normanby, Middlesbrough TS6 0BG
M0 JUP Rvd. S. Perry 8 Collingwood Avenue, March PE15 9EF
M0 JUQ Capel Battery Radio Group c/o I. Hope 5 The Crescent, Northfleet, Gravesend DA11 7EB
M0 JUT D. Heron 36A Forest Road, Hartwell, Northampton NN7 2HE
M0 JUU R. Bazant 1807 128Th St, College Point 11356 United States
M0 JUV D. Howard 31 White Mullein Drive, Redlodge IP288XP
M0 JUX A. Seedhouse 29 Great Northern Road, Dunstable LU5 4BN
M0 JVC K. Francis 203 Colchester Road Lawford, Manningtree CO11 2BU
M0 JVE K. Hedges 28 Hill Park, Congresbury, Bristol BS49 5BT
M0 JVG J. Geisau Huelchratherstr 37, Koeln D-50670 Germany
M0 JVL R. Scott 34 Moorfield Road, Birmingham B34 6QY
M0 JVM D. Turton 8 Lightwoods Road, Warley, Smethwick B67 5AY
M0 JVR D. Galloway Riverview Barn, Ferry Lane, Twyford, Barrow-On-Trent, Derby DE73 7AA
M0 JVT J. Turner White Haven 7 Lawrence Hill Gardens, Dartford DA1 3AP
M0 JVU J. Connolly 2 Waring Avenue, St. Helens WA9 2QG
M0 JVV J. Waddy 70 Linden Avenue, Prestbury, Cheltenham GL52 3DS
M0 JVW J. Wild1 11 Whitefield Avenue, Newton-le-Willows WA12 8BY
M0 JVX D. Hudson 30 Sarmatian Fold, Ribchester, Preston PR3 3YG
M0 JVY P. Smythe 15 Falcon Drive, Trowbridge BA14 7GE
M0 JVZ L. Bailey Flat 14 Walter Hull Court King Edward Road Loughborough, Loughborough LE11 1SU
M0 JWA Folkestone & District ARS c/o J. Arrow 29 Billington Gardens, Hedge End, Southampton SO30 2AX
M0 JWC H. Griffiths The Old Church Hall, Christmas Common, Watlington OX49 5HL
M0 JWD J. Wright 32 Carlton Road, Nottingham NG10 3LF
M0 JWE J. Williamson 64 Sandy Lane, Irlam, Manchester M44 6WJ
MM0 JWH J. Hosea 61 John Street, Helensburgh G84 9JZ
M0 JWJ J. Jordan 31 Rotherham Road, Dinnington, Sheffield S25 3RG
M0 JWL M. Lee Up To Date House, Shore Road, Boston PE22 0NA
M0 JWM J. Mccown 2105 Viking Drive, Colorado Springs 80910 United States
M0 JWN Royal Engineers Swindon c/o Dr R. Ashman 44 Conan Doyle Walk, Swindon SN3 6JB
MW0 JWP J. Pritchard 1 Tan Y Coed, Maesgeirchen, , Bangor LL57 1LU
M0 JWQ M. Krol 74 Boakes Drive, Stonehouse GL10 3QW
M0 JWR J. Richardson 6 Clarence Road Scorton, Richmond DL10 6EE
M0 JWT J. Wieczorek 1 Wylie Gardens, Basingstoke RG24 9TU
M0 JWU R. Jones St. Fillans, The Warren, East Horsley, Leatherhead KT24 5RH
M0 JWV P. Preston 4 Barn Close, Torriano Avenue, London NW5 2SY
MW0 JWW J. Williams 12 Pantycelyn, Fishguard SA65 9EH
M0 JWX P. Preston 4 Barn Close, Torriano Avenue, London NW5 2SY
MI0 JWY Newry And District ARC c/o M. Keenan 30 Ballynabee Road Camlough, Newry BT35 7HD
M0 JWZ J. Blakemore 16 Brentwood Avenue, Newbiggin-by-the-Sea NE64 6JH
M0 JXD J. Drinkell 11 Valley Walk, Kettering NN16 0LY
M0 JXE D. Ennion 347 Parkgate Road, Chester CH1 4BE
M0 JXF B. Power 20 Marlpool Place, Kidderminster DY11 5BB
M0 JXG J. Guess Flat 257, Helen Gladstone House Nelson Square, London SE1 0QB
MM0 JXI J. Innes 33 Monktonhall Place, Musselburgh EH21 6RR
M0 JXM D. Easterling 14 Brunswick Close, Biggleswade SG18 0DA
M0 JXR B. Glen 105 Heron Gardens, Portishead, Bristol BS20 7BN
M0 JXS C. Ember 35 Mattock Lane, Ealing, London W5 5BH
M0 JXY Red Brick Stables ARC c/o J. Hall 4 Dorking Crescent, Clacton-on-Sea CO16 8FQ
MW0 JYC M. Coleman Felin Newydd Ciliau Aeron Lampeter Sa48 7Px, Lampeter SA48 7PX
M0 JYF D. Gaskell 26 Stoneyland Drive, New Mills, High Peak SK22 3DL
M0 JYG E. Mccormick 526 Watling Street Road, Ribbleton, Preston PR2 6TU
M0 JYJ J. Lundrigan 9 Richards Close, March PE15 8UH
M0 JYM J. Sangster 139 Higher Road, Liverpool L26 1UN
M0 JYN G. Kermeen 31 Coppice Close, Sedgley, Dudley DY3 3NP
M0 JYP K. Sale 19 Mayfield Grove, Stockport SK5 7JB
M0 JYQ B. Sadler 56 Hayden Lane, Hucknall, Nottingham NG15 8BS
MM0 JYR J. Rae 4 Hillside Crescent, Langholm DG13 0EE
M0 JYS B. Sadler 56 Hayden Lane, Hucknall, Nottingham NG15 8BS
MM0 JYV S. Spoor The Old School, Main Street, Whitsome, Duns TD11 3NB
MW0 JYW G. Uvner The Stratford House B&B 8 Craig Y Don Parade, Llandudno LL30 1BG
M0 JYZ N. Honey 29 Mill House Lane, Winterton, Scunthorpe DN15 9QP
MM0 JZB J. Brown 60 Laburnum Lea, Hamilton ML3 7LZ
MW0 JZE A. David 45 Amanwy, Llanelli SA14 9AH
M0 JZF I. Johnston 4 Fieldway Crescent, Cowes PO31 8AJ
M0 JZG A. Loukes 14 Batchwood View, St. Albans AL3 5TD
MM0 JZI A. Mclean 10 Seaforth Gardens, Annan, Dumfries And Galloway DG12 6UH
M0 JZK J. Howlett 29 Little London, Heytesbury, Warminster BA12 0ES
MW0 JZM S. Kedward 9 Lawrence Avenue, Aberdare CF44 9EW
M0 JZN J. Nichols 4 Elmlee Close, Chislehurst BR7 5DU
M0 JZO B. Rusu 47 Elmgrove Crescent, Harrow HA1 2QT
M0 JZS K. Mueller 33 Rylands Road, Southend-on-Sea SS2 4LW
M0 JZT M. Hunt 189 Gibbins Road, Birmingham B29 6NH
M0 JZU K. Mueller 33 Rylands Road, Southend-on-Sea SS2 4LW
M0 JZV Z. Nikolic 6 Warbank Lane, Kingston upon Thames KT2 7ES
MM0 JZW J. Burton 6 Kilfinnan Lodges, Spean Bridge PH34 4EB
MI0 JZZ C. Mclelland 14 Clifton Park, Coleraine BT52 2HW
M0 KAB K. Bull 12 Hinksford Mobile Home Park, Kingswinford DY6 0BG
M0 KAC Spalding And District ARS c/o R. Walker 35 Romany Close, Letchworth Garden City SG6 4LA
M0 KAD K. Allen 48 Beaumont Rise, Worksop S80 1YG
M0 KAE K. Ely 6 Baxter Square, Town End Farm Estate, Sunderland SR5 4ND
MI0 KAG E. Rantin 8A Buchanans Road, Newry BT35 6NS
M0 KAI K. Boothman 5 Millwood Road, Balby, Doncaster DN4 9DA
M0 KAJ D. Ramsell 36 West Street, Burton-on-Trent DE15 0BW
M0 KAK D. Knight Flat 9, Loddon House, London Road, Ruscombe, Reading RG10 9BW
MI0 KAM K. Mullan 19 Parklea, Portstewart BT55 7HA
M0 KAN Pontefract & District ARS c/o K. Nicholson 11 Lancaster Way, Skellingthorpe, Lincoln LN6 5UF
M0 KAO K. Jeffery 9 Gordon Road, Tunbridge Wells TN4 9BL
M0 KAP P. Smith 3 Watts Road, Colchester CO2 9DZ
M0 KAR K. Lott 6 Centurion Close, Sandhurst GU470HH
M0 KAU K. Sharpe 18 Dudhill Road, Rowley Regis B65 8HT
M0 KAW K. Wells 2 Holmefield, Farndon, Newark NG24 3TZ
MW0 KAX M. Reynolds 2 Scales Row, Aberdare CF44 0PW
M0 KBA A. Bhakoo 4 Bryden Cottages, High Street, Uxbridge UB8 2NY
M0 KBB Norfolk County Raynet Group c/o K. Bright 20 Radley Road, Bristol BS16 3TL
M0 KBC K. Aird Kemberwood, Canterbury Road, Folkestone CT18 7EQ
M0 KBD P. Smiths 12 Lambton Road, Stockton-on-Tees TS19 0ER
M0 KBG V. Hayes 68 Billingsley Road, Birmingham B26 2EA
M0 KBH N. Kimber 127 Beaconsfield, Withernsea HU19 2EW
M0 KBJ S. Harvey 110 Vicarage Road, Wednesfield, Wolverhampton WV11 1SF
M0 KBK Prof. R. Houlston 51 Adelaide Road, Surbiton KT6 4SR
MI0 KBL P. Brennan 23 Ardchrois, Donaghmore, Dungannon BT70 3LB
MW0 KBN Holy Island ARC c/o C. Thorley Helston, The Mountain, Holyhead LL65 1YR
M0 KBP B. Al-Rawi Flat 17, Harrow Lodge, London NW8 8HR
M0 KBR W. Townsend 8 Beech Avenue Kirkham, Preston PR4 2UE
MM0 KBT D. Bisset Jordieland Cottage, Kirkcudbright DG6 4XT
M0 KBU C. Norris Heather View, Forest Front, Hythe, Southampton SO45 3RJ
M0 KBW M. Cerveny Dvorska 1871/90, Blansko 67801 Czech Republic
M0 KBX K. Hewson 48 Ruskin Road, Belvedere DA17 5BB
M0 KBY M. Criscione 46 South Molton Street, London W1K5RX
M0 KBZ K. Tietz 112 Over Lane, Belper DE56 0HN
M0 KCA J. Cater 5 Shady Grove, Hilton, Derby DE65 5FX
M0 KCC K. Cartwright 53 Sedgley Road, Dudley DY1 4NE
MM0 KCD K. Davies 1 Myreton Way, Falkirk FK1 5NZ
M0 KCE T. Peterson 59 Fleet Street, Holbeach, Slane Lodge, Spalding PE12 7AU
M0 KCF G. Mccaffery 7 Cliffe Court, Sunderland SR6 9NT
M0 KCI Ashtead RC c/o Dr P. Lee Links Corner Cottage Liks Road, Ashtead, Surrey, UK KT212EG
M0 KCJ C. Kennedy 81 Corporation Road, Audenshaw, Manchester M34 5LZ
M0 KCO K. Cornmell 19 Forest Road, Chandler'S Ford, Eastleigh SO53 1NA
M0 KCP R. Hutton 22A Victoria Road, Maldon CM9 5HF
MM0 KCS M. Brunsdon 25 Buckstone Lea, Edinburgh EH10 6XE
M0 KCV K. Conlon 136 Chart Downs, Dorking RH5 4DG
M0 KCW C. Wade 31 Melton Green, Wath-Upon-Dearne, Rotherham S63 6AA
M0 KCY J. Crabtree 5408 Oaklawn Avenue, Edina 55424 United States
M0 KCZ R. Robinson 22 Riddings Court, Timperley, Altrincham WA15 6BG
M0 KDA Wings Museum c/o S. Bolton 201 Lime Tree Avenue, Crewe CW1 4HZ
M0 KDE A. Lee 14 Bernice Avenue, Chadderton, Oldham OL9 8QJ
M0 KDH D. Hughes 75 Suncote Avenue, Dunstable LU6 1BN
MW0 KDL K. Lewis 21 Wheatley Place, Merthyr Tydfil CF47 0TA
M0 KDM Swinton (Amateur) RC c/o P. Sephton 11 Moss Avenue, Leigh WN7 2HH
M0 KDO J. Saxon 134 Sherwood Drive, Wigan WN5 9RS
M0 KDR K. Roberts 1 Tregerles Barns, Newquay TR8 4PW
M0 KDT J. Hobbs 2 Eccles Road, Wittering, Peterborough PE8 6AU
M0 KDU Cunninghame & District Amt RC c/o M. Shingler 4 Church Lane, Checkley, Stoke-on-Trent ST10 4NJ
M0 KDV D. Craven 69 Markham Avenue, Rawdon, Leeds LS19 6NE
M0 KDX P. Martin 26 Chart Lane, Reigate RH2 7DY
MM0 KDY D. Latto 8 Aspen Avenue, Glenrothes KY7 5TA
M0 KEB K. Legg Bennetts, High Street, Thorpe-Le-Soken, Clacton-on-Sea CO16 0EG
M0 KED A. Keddie 6 Vulcan Crescent, North Hykeham, , Lincoln, LN6 9SB
M0 KEE J. Charlton Hillside House, Ham Lane, Bristol BS41 8JA
M0 KEF P. Munson 8 Longley Lane, Spondon, Derby DE21 7AT
M0 KEG I. Bain 45 Larpool Crescent, Whitby YO22 4JD
M0 KEJ K. Borszlak 37 Bank Lane, Little Hulton, Manchester M38 9UH
M0 KEL K. Gale 4 Field Court, Sea Road, East Preston BN16 1JS
M0 KEO Prof. I. Neal 8 Rushey Gill, Brandon, Durham DH7 8BL
M0 KEP T. Keep Coombe Cottage, Coombe Lane, Cradley, Malvern WR13 5JF
MW0 KEQ K. Mogford 49 Cefn Road, Rogerstone, Newport NP10 9AQ
MM0 KES E. Wigram Inglenook, North Road, Lerwick, Shetland ZE1 0PR
M0 KEU T. Barris Vogelgartenstr. 41/1, Eislingen/Fils 73054 Germany
MW0 KEV K. Dawson 57 Fair View, Blackwood NP12 3NR

M0 KEW D. Askew 19 Oliver Road Staplehurst, Tonbridge TN12 0TE
M0 KEY B. Wagstaff 4 Fleets Road, Sturton By Stow, Lincoln LN1 2BU
M0 KFA A. Ferenc 2A Rosedene Avenue, London SW16 2LT
M0 KFB K. Bell 11 Mill Lane, Hogsthorpe, Skegness PE24 5NF
M0 KFF M. Harrington Tanglewood House, Station Road, , Tilbrook PE28 0JY
M0 KFH I. Hemmens The Laurels, Barnstaple Road, South Molton EX36 3RD
M0 KFI S. Gilbert 34 Ramsey Road, St. Ives PE27 5RD
M0 KFJ J. Moyler 38 Golden Hill, Whitstable CT5 3AR
M0 KFK D. Westwood 28 Weybridge Mead, Yateley GU46 7UY
MW0 KFL Dr E. Flikkema 7 St. James Mews, Great Darkgate Street, Aberystwyth SY23 1DW
MM0 KFP Dr M. Sutcliffe 11 Low Borland Way, Eaglesham, Glasgow G76 0BP
MM0 KFR K. Fisher 9 Haymarket Crescent, Livingston EH54 8AP
M0 KFU D. Sweeney 3 Robin Hood Close, Woking GU21 8SS
M0 KFV K. O'Connell 63 Hazelton Road, Colchester CO4 3DS
M0 KFW K. Whittaker 32 Ashleigh Mount Road, Exeter EX4 1SW
M0 KFY J. Gilvarry 16 Fox Walk, Sheffield S6 3QZ
M0 KGA S. Pollak 22 Gloucester Road, Avonmouth, Bristol BS11 9AD
M0 KGE Dr D. Rodriguez 23 Thorpe Way, Cambridge CB5 8UJ
MW0 KGG D. Owen Tanrallt, Blaenpennal, Aberystwyth SY23 4TP
M0 KGI S. Byatt High Trees, High Street, Pavenham, Bedford MK43 7NJ
MM0 KGK D. Munro 47A Garrabost, Isle of Lewis HS2 0PW
M0 KGM K. Graham 22 Repton Avenue, Oldham OL8 4JB
MW0 KGP Dr P. Kelly Arosfa, Westminster Road, Moss, Wrexham LL11 6DN
MM0 KGS G. Sinclair 33 Keptie Road, Arbroath DD11 3EF
MW0 KGU P. Bennett 2 Tan Yr Wylfa, Abergele LL22 7DX
M0 KGV C. Moulding 28 Queens Avenue, Highworth, Swindon SN6 7BA
M0 KGW Dr B. Minnis 16 Dene Tye, Crawley RH10 7TS
MW0 KGY K. Yearsley Garth Lea, Lon St. Ffraid, Holyhead LL65 2YH
M0 KHA C. Wale 6 West Howe Close, Bournemouth BH11 8AE
M0 KHB K. Bridge 22 Oaksfield, Darwen BB3 0AU
M0 KHD S. Gaspar 54 Herberts Park Road, Wednesbury WS10 8QH
MM0 KHG Craighalbert RC c/o J. Brown 78 Egilsay St., Glasgow G22 7RG
M0 KHK C. Cheung 23 Mosley Road, Timperley, Altrincham WA15 7TF
M0 KHM K. Maddy 56 Coachwell Close, Telford TF3 2JB
M0 KHN J. Riches 4 Kings Road, Chalfont St. Giles HP8 4HU
M0 KHO D. Pluright 21 Buscot Drive, Abingdon OX14 2BJ
M0 KHQ Swinton RC c/o T. Ward 173-175 Station Road, Pendlebury, Swinton, Manchester M27 6BU
M0 KHS K. Shuttleworth 27 Queens Drive, Fulwood, Preston PR2 9YJ
M0 KHW Montrose Amateur Radio Station c/o K. Wright 12 Bushmead Road, Luton LU2 7EU
M0 KHX M. Cooper 49 Bolton Lane, Ipswich IP4 2BX
M0 KHZ K. Wheatley 1 Braithwaite Court, Egremont CA22 2DN
M0 KIB C. Mckinney Swallow Cottage, Chickney Road, Henham, Bishop's Stortford CM22 6BG
M0 KIC N. Jaggs 15 St. Anthonys Road, Kettering NN15 5HT
M0 KID N. Fairbairn 15 Hewitt Road, Dover CT16 1TH
M0 KIF M. Chalkley 36 Cowper Road, Bournemouth BH9 2UJ
M0 KIG K. Gamble 67 Queen Street, Burntwood WS7 4QQ
MW0 KIJ N. Sugg 12 Caerleon Grove Castle Park Cf48 1Jh, Merthyr Tydfil CF48 1JH
M0 KIL K. Lockstone Oceana, The Parade Pevensey Bay, Pevensey BN24 6LX
M0 KIN A. Clarke 57 Welland Avenue, Grimsby DN34 5JP
M0 KIO Kingston Immortals CG c/o N. Newby 167 Watersplash Road, Shepperton TW17 0EN
M0 KIQ S. Dowling 34 Kit Hill Avenue Walderslade, Chatham ME5 9EX
M0 KIR M. Kirkman 8 Ashington Drive, Arnold, Nottingham NG5 8GH
M0 KIW A. Thorn 12 Hardy Drive Hardingstone, Northampton NN4 6UX
M0 KIZ D. Allen 19 Brooklands Close, Uttoxeter ST148UH
M0 KJC K. Cole 4 Marsham Road Hazel Grove, Stockport SK7 5JB
M0 KJE V. Loshakov 109 Booth Road, Colindale, London NW9 5JU
MM0 KJG K. Glacken 14 Hailes Avenue, Edinburgh EH13 0NA
M0 KJH C. Field 8 Twelve Acres, Braintree CM7 3RN
MM0 KJJ T. Goodenough 83 Craufurdland Road, Kilmarnock KA3 2HU
M0 KJK K. Kolesnik 15 Steer Road, Swanage BH19 2RU
MM0 KJM K. Martin 4 Hunter Crescent, Troon KA10 7AH
MW0 KJN K. Nedin 28 Llys-Y-Coed Birchgrove, Swansea SA7 9PR
M0 KJO N. Cranston 42 Curling Vale, Guildford GU27PH
M0 KJT the UK Six Metre Group c/o K. Todman 12 Winscombe, Bracknell RG12 8UD
M0 KJU S. Rigato Flat 16, Saffron Court, 2A Maryland Park, London E15 1HU
M0 KJZ S. Bowen Flat 6, Wyndley House, 9 Welshmans Hill, Sutton Coldfield B73 6RY
M0 KKA C. Wheatley-Hince Jasmine Cottage, Main Street, Chaddleworth, Newbury RG20 7EH
M0 KKB S. King 23 Queens Avenue, Canterbury CT2 8BA
MM0 KKC D. Yeaman 7 Brimmond Crescent, Westhill AB32 6RD
MI0 KKD K. Dorman 25 Blackthorn Road, Newtownabbey BT37 0GH
M0 KKE M. Bender Ivy Chimney Villa, Skinners Bottom, Redruth TR16 5DT
MM0 KKF W. Mackenzie 7 Urquhart Grove, Elgin IV30 8TB
MD0 KKG Sight Matters ARC c/o D. Wilson 23 Snugborough Avenue, Union Mills, Braddan, Isle of Man IM4 4LT
M0 KKH N. Haigh 10 Moor Park Gardens, Dewsbury WF12 7AS
MM0 KKL J. Mclaren 33 Foulford Street, Cowdenbeath KY4 9NB
M0 KKM K. Minett Rosedene, Honey Hill, Wimbotsham, King's Lynn PE34 3QD
M0 KKN PI RC c/o M. Gosztyla 50 Lingmell Avenue, St. Helens WA11 7AZ
M0 KKO T. Stack 31A Chester Road South, Kidderminster DY10 1XJ
M0 KKZ S. Stothard 84 Guntons Close, Soham CB7 5DN
M0 KLA S. Collins 3 Sedgemoor Road, Camps Bay South Africa
M0 KLB E. Gomez Lozano 29 Wykeham Crescent, Oxford OX4 3SD
M0 KLG R. Searby 5 Bosvean Gardens, Truro TR1 3NQ
M0 KLH K. Hawes 837 Garratt Lane, London SW17 0PG
M0 KLJ J. Athersmith 20 Fell Close, Ulverston LA12 0AE
M0 KLK W. Foster 55 Drake Avenue Minster On Sea, Sheerness ME12 3SA
M0 KLL K. Lloyd 1 Fordenbridge Square, Sunderland SR4 0BA
M0 KLM B. Whiteley 2A Beechfield Close, Thorpe Willoughby, Selby YO8 9QJ
M0 KLN K. Nevins 4 Ubbanford, Norham, Berwick-upon-Tweed TD15 2LA
MM0 KLR Kilmarnock And Loudoun ARC c/o A. Clark 20 Church Street, Kilwinning KA13 6BE
M0 KLT G. Clark 28 Manor Road, Woolton, Liverpool L25 8QG
MW0 KLW A. Jones 2 Erw Terrace, Bethel, Caernarfon. LL55 1YT
M0 KLX R. Harper 19 Tennyson Avenue, King's Lynn PE30 2QG
M0 KMB A. Bailey 58 Billy Buns Lane, Wombourne, Wolverhampton WV5 9BP
M0 KMI K. Mills 6 West Coombe, Bristol BS9 2BA
MI0 KMJ A. Mcguinness 42 Downshire Road, Carrickfergus BT38 7LD
MM0 KMK C. Murphy 3 Bolestyle Crescent, , Kirkmichael - Maybole KA19 7PW
M0 KML K. Leesmith 31 Dogger Lane, Wells-next-the-Sea NR23 1BE
M0 KMN C. Falas Christs College, Cambridge CB2 3BU
MM0 KMQ I. Morrison Coach House, Olivers Brae, Stornoway HS12SX
M0 KMR Medway Raynet c/o R. Sohst 2 Shaftesbury Drive, Maidstone ME16 0JS
MW0 KMS K. Smith 62 Waterloo Road, Talywain, Pontypool NP4 7HJ
M0 KMT M. Sadler 12 John Corbett Drive, Amblecote, Stourbridge DY8 4BW
M0 KMU K. Sanderson 29 Ingleton Road, Stockton-on-Tees TS19 8EE
M0 KMW M. Ballard 41 Middlefield Ave, Halesowen B62 9QJ
M0 KMX K. Murphy 120 Elmway, Chester le Street DH2 2LQ
M0 KNB M. Ronan 49 Dorset Street, Nottingham NG8 1PU
MM0 KNE Ripon & District ARS c/o T. Kane 40B Brisbane Street, Greenock PA16 8NP
M0 KNF C. Bolton 1A Salterns Terrace, Bideford EX39 4AG
M0 KNH K. Holman 39 Trellech Court, Yeovil BA21 3TE
M0 KNI St Peter-In-Thanet Junior School ARC c/o J. Hislop 10 Park Wood Close, Broadstairs CT10 2XN
M0 KNK K. Kariraman 17 Kipling Close, Galley Common, Nuneaton CV10 9SJ
M0 KNL T. Wong 24 Chiswick Staithe, Hartington Road, , London W4 3TP
M0 KNM M. Hemmings 62 Spencer Way, Stevenage SG2 8GD
MM0 KNN C. Kennedy 16 New Garrabost, Isle of Lewis HS2 0PL
M0 KNP C. Burls 86 All Saints Road, Kings Heath, Birmingham B14 7LN
M0 KNQ P. Scarratt 4 Sandy Lonning, Maryport CA157LW
MM0 KNS R. Barlow Iphs, Westray KW17 2DW
M0 KNV K. Taylor 44 Main Street, Willoughby, Rugby CV23 8BH
M0 KNW M. Burge 24 Oakdale Road, Sheffield S254EY
M0 KNX M. Mattiello Via Luigi Gaudio, 21, Faedis, Udine 33040 Italy
M0 KOB A. Prestwich Highfield, Exminster, Exeter EX6 8AT
M0 KOE A. Sampson1 15 Laburnum Avenue, Crook DL15 9LH
M0 KOG C. Haw 72 Mayflower Road, Boston PE21 0EZ
M0 KOH K. Hough 15 Moorside Road Endmoor, Kendal LA8 0EN
M0 KOI P. Burrows 37 Dorrington Close, Murdishaw, Runcorn WA7 6JR
M0 KOM L. Wilson 6 Marrick Road, Middlesbrough TS3 7RX
M0 KON Taunton & Somerset Raynet Group c/o D. Smith 47 Laburnum St., Taunton TA1 1LB
M0 KOO K. Milone The Granary, Bilton Grange, Tockwith Lane, Bilton-In-Ainsty, York YO26 7NZ
M0 KOQ J. Visona The Old Wilds, Eastcote Road, Gayton NN7 3HQ
M0 KOR D. Korzeniewicz 6 Barley Down Drive, Winchester SO22 4LS
M0 KOT N. Baulf 1 Lower Chart Cottages, Brasted Chart, Westerham TN16 1LS
M0 KOV F. Limbert Knowlecroft Little Ribston, Wetherby LS22 4ET
M0 KOW Taunton & Somerset Raynet Group c/o D. Smith 47 Laburnum St., Taunton TA1 1LB
M0 KOY Taunton & Somerset Raynet Group c/o D. Smith 47 Laburnum St., Taunton TA1 1LB
MM0 KOZ R. Thomson 30 Sealstrand, Dalgety Bay, Dunfermline KY11 9NG
MI0 KPA S. Frazer 2 Cavanballaghy Road, Killylea, Armagh BT60 4NZ
M0 KPB K. Blanshard 30 Torquay Crescent, Symonds Green, Stevenage SG1 2RS
M0 KPC P. Gagliardi 7 Saxon Way, Jarrow NE32 3QA
M0 KPD R. Simpson 48 Weatherhill Road, Horley RH6 9LY
M0 KPE F. Keane 14 Pollard Court, Beverley Road, Hull HU5 2TP
M0 KPF I. Shelton 6 Kingfisher Way, Sutton-in-Ashfield NG17 4PR
M0 KPI R. Mackay 7 Darwin Close, Lee-on-the-Solent PO13 8LS
M0 KPJ M. Barrett 57 Marlborough Avenue, Hornsea HU18 1UA
M0 KPK D. Smith 333A, Forton Road, Gosport PO12 3HF
M0 KPO S. Warren 7 Crich Way, Newhall, Swadlincote DE11 0UU
M0 KPP P. Baker 69 Cavendish Road, Walsall WS27HH
M0 KPQ Taunton & Somerset Raynet Group c/o D. Smith 47 Laburnum St., Taunton TA1 1LB
M0 KPT K. Towler 3 Elm Grove, Barnham, Bognor Regis PO22 0HF
M0 KPU C. Weight 166 Prince Henry Road, Charlton SE7 8PJ
MM0 KPV GM Contest Club c/o J. Phunkner 7 Plenshin Court, Glasgow G53 6QW
M0 KPW C. Leviston 1 Great Carrs Close, Askam-in-Furness LA16 7FL
M0 KPX M. Rollings 107 Upper Sherwood Road, Seaford BN25 3EA
M0 KPY D. Lee 31 Lower Adeyfield Road, Hemel Hempstead HP2 5DG
MM0 KQC UK Travellers Club c/o J. Phunkner 7 Plenshin Court, Glasgow G53 6QW
MM0 KQG C. Burkinshaw 12 Haas Grove, Lockerbie DG11 2QP
M0 KQI University of Exeter Amateur Radio c/o C. Leaver 20 Sylvania Drive, Exeter EX4 5DT
M0 KQJ F. Bila-Nicola 48 Shepherds Pool, Evesham WR11 4JG
M0 KQK J. Green Belvoir Vale Cottage, Barrowby Stenwith, Grantham NG32 2HE
M0 KQN J. Murphy 20 Shakespeare Gardens, London N2 9LJ
M0 KQR P. Ionescu 1 Hare Close, Buckingham MK18 7EN
MI0 KQU S. Homer 10 Bann Drive, Londonderry BT47 2HW
M0 KQV I. Briley 12 Stirling Avenue, Waterlooville PO7 7NH
M0 KQW D. Wilding 404 Howlands, Welwyn Garden City AL7 4HB
M0 KRA S. King Box 14, West Union South Carolina 296960014 United States
MM0 KRC K. Carroll 32E Meadowburn Place, Campbeltown PA28 6ST
M0 KRD D. Niggemann 35 Holm Court, Twycross Road, Godalming GU7 2QT
M0 KRE K. Knight 11 Sweetbriar Lane Holcombe, Dawlish EX7 0JZ
M0 KRH C. Pierce 42 Staplehurst Road, Reigate RH2 7PY
M0 KRK Dr D. De-Cogan 52 Gurney Road, New Costessey, Norwich NR5 0HL
M0 KRL K. Fell 5 Henry Road, Wath-Upon-Dearne, Rotherham S63 7NF
M0 KRM Medway Raynet c/o R. Sohst 2 Shaftesbury Drive, Maidstone ME16 0JS
M0 KRO Kent Active Radio Amateurs c/o K. Richardson 35 Vidgeon Avenue, Hoo, Rochester ME3 9DE
M0 KRP R. Bullen 2 Redlands Cottages, East Coker, Yeovil BA22 9HF
M0 KRR A. Kerr 23 Manor Park, Duloe, Liskeard PL14 4PT
MW0 KRS C. Young 34 Penlan Crescent, Uplands, Swansea SA2 0RL
M0 KRU P. Crewe 31 Seas End Road Moulton Seas End, Spalding PE12 6LD
M0 KRW I. Bricknell 171 Springthorpe Road, Birmingham B24 0SN
M0 KRX K. Rosema Apartment 801, 25 Goswell Road, London EC1M 7AJ
MM0 KRZ Sutton & Cheam c/o K. Suchomski Flat 4, 6 Nigg Kirk Road, Aberdeen AB12 3DF

M0 KSA R. Love 48 Langland Drive, Dudley DY3 3TH
M0 KSB Kingsmead School c/o J. Matthews Moor View, Oldways End, East Anstey, Tiverton EX16 9JQ
M0 KSE S. Shields 28 Bank Field Westhoughton, Bolton BL5 2QG
M0 KSG M. Wood 26 Parkfield Crescent Kimpton, Hitchin SG4 8EQ
M0 KSL K. Lack 21 Lexington Close, Borehamwood WD6 1XA
M0 KSO the Kings School RS c/o D. Lee 188 Manstone Ave, Sidmouth EX10 9TJ
M0 KSR K. Mcinnes 5A Northbrook Road, London N22 8YQ
MM0 KSS K. Scott 70 Raeden Crescent, Aberdeen AB15 5WJ
MW0 KST S. Davies 5 Maldwyn Street, Cardiff CF11 9JR
M0 KSX F. Stepien 11A Wirral Gardens, Wirral CH63 3BD
MM0 KTE K. Taylor 77 Queen Margaret Fauld, Dunfermline KY12 0RL
MM0 KTL K. Lane 23 Mayfield Avenue, Tillicoultry FK13 6HB
M0 KTN A. Snell-Pym 50 Newton Avenue, Gloucester GL4 4NU
M0 KTR A. Rowbottom 8 Hedge Drive, Colchester CO2 9DT
M0 KTT C. Jacobs Flat 33, The Lodge, Lavender Road, Waterlooville PO7 8BX
M0 KTV Warrington ARC c/o B. Chauhan 45 Burnham Drive, Whetstone, Leicester LE8 6HY
M0 KTY G. Furlonger 7 St. Nicholas Close, Sturry, Canterbury CT2 0NT
M0 KUH Hull & District ARS c/o R. Pike 44 Falkirk Close Bransholme, Hull HU7 5BX
MI0 KUJ T. Calka 71 Willowfield Street, Belfast BT6 9AW
M0 KUK J. Kowalski 280 Old Farm Avenue, Sidcup DA15 8AR
M0 KUL K. Richards 207 Beaulieu Gardens, Blackwater, Camberley GU17 0LG
M0 KUP A. Anderson 89A Malmesbury Park Road, Bournemouth BH8 8PS
M0 KUR S. Campion Flat 6, Carousel Steps, 10 Hawtree Close, Southend-on-Sea SS1 2TZ
M0 KUV I. Barnes 12 Sunderton Lane, Waterlooville PO8 0NU
M0 KUY G. Michilin 46 Via N.Sauro, Preganziol 31022 Italy
M0 KVA A. Dokic 28 Tudor Gardens, Shoeburyness, Southend-on-Sea SS3 9JG
M0 KVB A. Taylor 60 Wood Ride, Petts Wood, Orpington BR5 1PY
M0 KVF K. Emery-Ford 48 Welham Grove, Retford DN22 6TS
M0 KVK K. Sim 49 St. Julians Wells, Kirk Ella, Hull HU10 7AF
M0 KVM K. Mills 6A Schofield Street, Mexborough S64 9NH
M0 KVN K. Finn 132 Lansdowne Grove, Wigston LE18 4LY
M0 KVR A. Burfield 4 Eastern Crescent, Chelmsford CM1 4JQ
M0 KWA Headcorn Aerodrome c/o P. Blunt 17 Offens Drive, Staplehurst, Tonbridge TN12 0LR
M0 KWB K. Bradd Flat 2, Montrose Court, London NW9 5BS
M0 KWK D. Hall 1 Pendreth Place, Cleethorpes DN35 7UR
M0 KWM Yorkshire Net UK c/o D. Wright 203 Winn Street, Lincoln LN2 5EY
M0 KWP J. Simon 14 Le Peyrefus, Daignac 33420 France
M0 KWR K. Royce 11 Church Lane, Stibbington, Peterborough PE8 6LP
M0 KWS S. Kendrick 29 Waterside Silsden, Keighley BD20 0LQ
M0 KWV M. Evans 16 Colville Grove, Sale M33 4FW
M0 KWW K. Willson Ludpit Cottage, Ludpit Lane, Etchingham TN19 7DB
M0 KWY D. Wells 27 Victoria Avenue, Camberley GU15 3HT
M0 KXD D. Rivron Spring Cottage, Mill Lane, Bellerby, Leyburn DL8 5QN
M0 KXK R. Britt Thoroughfare House, South Burlingham Road, Norwich NR13 4FA
M0 KXL K. Laing 16 Cherrywood Drive Gonerby Hill Footá, Grantham NG31 8QL
MW0 KXN K. Nicholls 35 Partridge Road, Cardiff CF24 3QW
M0 KXQ P. Hardacre 13 St. Johns Street, Bridlington YO16 7NL
MI0 KYD G. Starling 63 Clowney Street, Belfast BT12 7LZ
M0 KYI K. Armstrong 29 Thorntree Avenue, Crofton, Wakefield WF4 1NU
M0 KYL A. Kyle Greengates, Ainsworth Street, Ulverston LA12 7EU
M0 KYR K. Orfanidis Flat 36, Cumberland Court, London W1H 7DP
M0 KYX I. Morgan 30 Farm Road, Hutton, Weston-Super-Mare BS24 9RH
MM0 KZA A. Marques Gomes 34 Campbell Close, Hamilton ML3 6BF
M0 KZB E. Arkinstall 79 Sundorne Road, Shrewsbury SY1 4RU
M0 KZC M. Clack 42A Provost Street, Fordingbridge SP6 1AY
M0 KZH D. Taylor 49 Boggart Hill Gardens, Seacroft, Leeds LS14 1LJ
MM0 KZJ A. Thomson 5 Gib Grove, Dunfermline KY11 8DH
M0 KZM M. Osborne 9 Sunningdale Court, Jupps Lane, Goring-By-Sea, Worthing BN12 4TU
M0 KZP N. Simmonds 3 Noneley Hall Barns, Noneley, Wem, Shrewsbury SY4 5SL
M0 KZT L. Kinzett 52 Thistle Drive, Peterborough PE2 8HX
M0 LAA S. Jefferson 145 Duke St Fenton, Stoke-on-Trent ST4 3NR
M0 LAB M. Labourn 6 Healey Drive, Ossett WF5 8NA
M0 LAE Rvd. L. Clark 226 Philip Lane Tottenham, London N15 4HH
M0 LAF N. Smith 40 Fairdale Drive, Newthorpe, Nottingham NG16 2FG
M0 LAG D. Whelan 431 Leeds Road, Huddersfield HD2 1XT
M0 LAH S. Chng Cambridge, CB3 9BB
M0 LAI P. Tolcher 15 Langstone Close, Torquay TQ1 3TX
M0 LAL C. Mole 6 Clements Road, Chorleywood, Rickmansworth WD3 5JT
MW0 LAO A. Powell 31 Highmead, Pontllanfraith, Blackwood NP12 2PF
M0 LAR G. Jackson 14 Southdown Crescent, Cheadle Hulme, Cheadle SK8 6EQ
M0 LAS L. Milford 82B Oakley Lane, Oakley, Basingstoke RG23 7JX
M0 LAT A. Laity 9 Haverhill Road, Stapleford, Cambridge CB22 5BX
M0 LAW M. Martin 2821 Bissonnet St, Houston 77005-4014 United States
M0 LAY A. Brooks 19 Hawthorn Close, Knodishall, Saxmundham IP17 1XW
M0 LAZ A. Burton 12 Hanover Court Salisbury Road, Bath BA1 6QX
MI0 LBA Church Island ARG c/o T. Flanagan 18 Hunters Park, Bellaghy, Magherafelt BT45 8JE
MW0 LBB B. Bush 2 Penrhiw Cottages, Brynithel, Abertillery NP13 2AU
M0 LBD D. Lagan 24 Milvain Close, Gateshead NE8 3RS
MM0 LBF R. Bertram 46 Main Street Main Street, Pathhead EH37 5QB
M0 LBJ B. Jenson 10 Tintern Close, Paulsgrove, Portsmouth PO6 4LS
M0 LBK L. Karthauser 17 Manor Close Abbotts Ann, Andover SP11 7BJ
M0 LBL Marine Radio Museum Society c/o W. Cross 31 Joshua Close, Liverpool L5 0TD
M0 LBM P. Matthews 15 Tennyson Way, Melton Mowbray LE13 1LJ
MW0 LBR Dr F. Labrosse 72 Ger Y Llan Penrhyncoch, Aberystwyth SY23 3HQ
MI0 LBS L. O'Sullivan 24 Swifts Quay, Carrickfergus BT38 8BQ
MM0 LBX J. Cattigan Lunan Home Farm Cottage, Lunan Bay, Arbroath DD11 5ST
M0 LBY L. Lay 17 Herbert Road, Hornchurch RM11 3LD
M0 LCA E. Taylor 32 Knoll Drive, Warwick CV34 5YQ
M0 LCC Lymington Community Association RC c/o K. Cromar 74 Shakespeare Drive, Totton, Southampton SO40 3NS
MW0 LCH L. Holder 133 Maple Drive, Brackla, Bridgend CF31 2PR
MW0 LCK M. Heathcote Ingledown, Trelogan, Holywell CH8 9BZ
M0 LCM L. Micallef 1 Brockholme Mews, Great Cambourne, Cambourne CB23 6GU
M0 LCR V. Lynch 16 Okehampton Crescent, Sale M33 5HR
M0 LCS A. Selwood 3 Warborough Cottages Warborough Road, Letcombe Regis, Wantage OX12 9LE
M0 LCW Croham Callers c/o A. Hill 53 Fairladies, St. Bees CA27 0AR
M0 LCX Dr L. Clift 150 Greenstead Road, Colchester CO1 2SN
M0 LCY M. Taylor 18 Mallow Walk, Westgate, Morecambe LA3 3QA
M0 LDC L. Spriggs 19 Mackenzie Square, Shephall, Stevenage SG2 9TT
MW0 LDD M. Ladd 50 Brynmelyn Avenue, Llanelli SA15 3RT
M0 LDE G. Matts 34 Barry Road, Leicester LE5 1FA
M0 LDG West Lincs Ry G c/o J. Edmunds Caroline Cottage, New Passage Road, Bristol BS35 4LZ
M0 LDH L. Hawkins 121 Selmeston Road, Eastbourne BN21 2TL
MW0 LDJ L. Jessup 23 Penallt Estate Llanelly Hill, Abergavenny NP7 0RA
M0 LDK S. Alexander 13 Padgate, Thorpe End, Norwich NR13 5DG
M0 LDP D. Parker 51 Parsonage Road, Henfield BN5 9HZ
M0 LDQ D. Arnold The Chase, Rectory Road, Penzance TR19 6BB
M0 LDR L. Reynolds 12 Providence Crescent, Boundary Way, Hull HU4 6EF
M0 LDV D. Curtis 7 Neale Close, Aylsham, Norwich NR11 6DJ
M0 LDX S. Cape 92 Davison Avenue, Whitley Bay NE26 3SY
M0 LDY J. Davies 5 Beauchamp Road, Kenilworth CV8 1GH
M0 LDZ L. Cook 43 Midge Hall Drive, Rochdale OL11 4AX
MW0 LEA P. Price Stable Cottage , Pendre, Cardigan SA43 1JU
M0 LEB Prof. A. Sadka 8 Wilson Close, Upper Heyford, Bicester OX25 5BE
M0 LED L. Dixon 23 Gipsy Lane, Buckfastleigh TQ11 0DL
M0 LEE L. Jones 439 Tangmere Drive, Birmingham B35 6PZ
MW0 LEF M. Verardi 21 Snowdon Street, Y Felinheli LL56 4HQ
M0 LEH M. Reynolds 24 Burton Close, Corringham, Stanford-le-Hope SS17 7SB
M0 LEK Leek And District ARC c/o S. Jefferson 145 Duke St Fenton, Stoke-on-Trent ST4 3NR
MJ0 LEL L. Langlois Brookfield, La Rue D'Empierre, Trinity JE3 5QF Jersey
MM0 LEN L. Cochrane 2 Muir Terrace, Paisley PA3 4LT
M0 LEO L. Boberschmidt 3928 Denfeld Court, Maryland United States
M0 LEP Sir R. Hewett 17 Westfield, Dursley GL11 4EP
MM0 LER M. Dickeson 44 Mossmill Park, Mosstodloch, Fochabers IV32 7JY
M0 LET R. Collis 20 Little Meadows, Haxby, York YO32 3YY
M0 LEV M. Leveridge 17 Gladstone Court, Dewsbury WF13 4DQ
MW0 LEW L. Thomas 2 Goytre Crescent, Goytre, Port Talbot SA13 2YD
M0 LEX R. Styles Padcroft, Weir OL13 8QL
M0 LEY C. Kirby 8 Reynolds Way, Swindon SN25 4GF
M0 LEZ L. Robinson 15 Seldown Lane, Poole BH15 1UA
M0 LFC Friskney + East Lincolnshire Communications Club c/o B. Derbin-Sykes 1 Lentons Lane, Friskney, Boston PE22 8RR
M0 LFS Dr L. Spacek 33 Wellesley Road, Colchester CO3 3HE
M0 LGA O. Parker 37 Springbank Crescent, Gildersome, Leeds LS27 7DN
M0 LGB G. Benson 2 Guisborough Road, Nunthorpe, Middlesbrough TS7 0LB
M0 LGC Essex DX Group c/o M. Russell 107 Cambridge Road, Hitchin SG4 0JH
MI0 LGD Lord G. Drummond 41 Calhame Road, Ballyclare BT39 9NA
MW0 LGE R. Samphire Magor Court, Newport Road, Magor, Caldicot NP26 3BZ
M0 LGL L. Layland 3 Thirlmere Road, Golborne, Warrington WA3 3HH
M0 LGN B. Fitzgerald-O'Connor 14 Goldwing Close Custom House, London E16 3EQ
M0 LGP L. Porter 134 Grimshaw Lane, Middleton, Manchester M24 2AH
MM0 LGR D. Boden 42, Kirkwynd, Maybole KA197AE
MM0 LGS Maxpak c/o J. Gorczynski 54 Lindsay Gardens, Bathgate EH48 1DU
MM0 LGT M. Mckay 75 Cardross Road, Dumbarton G82 4JL
M0 LGW W. Grocott 12 Marigold Road, Stratford-upon-Avon CV37 7DW
M0 LHA M. Burton 139 Avenue Road, Erith DA8 3BA
M0 LHB L. Bramley 140 Nevill Road, Hove BN3 7QB
M0 LHC A. Walrond Leigh Hill Cottage, Lowton, Taunton TA3 7SU
M0 LHF Fenland Portable Group c/o C. Day 14 Windsor Drive, Ramsey Forty Foot, Ramsey, Huntingdon PE26 2XX
M0 LHK L. King The Old School, Coombe Cross Bovey Tracey, Newton Abbot TQ13 9EP
M0 LHR P. Lonsdale 77 Burtons Road, Hampton Hill, Hampton TW12 1DE
M0 LHS R. Silcox 103 Oakdale Road, Downend, Bristol BS16 6EG
M0 LIE E. St Quinton Mill Cottage, The Thorofare, Woodbridge IP13 8BB
M0 LIF Sos Radio Group c/o D. Hughes 86 Colinmander Gardens, Ormskirk L39 4TF
M0 LIJ D. Smout Sunrays, Warbage Lane, Bromsgrove B61 9BH
M0 LIO Northwest ARC c/o A. Hull 1 Occupation Lane, New Bolingbroke, Boston PE22 7LW
M0 LIS E. Buckland 11 Veronica Close, Basingstoke RG22 5NW
MM0 LIV Livingstone District ARS c/o N. Morris 23 Sedgebank, Sedgebank, Livingston EH54 6HE
MM0 LJA Dr L. Auchterlonie 2 Lawhead Road East, St. Andrews KY16 9ND
M0 LJC Aeroventure ARS c/o L. Carpenter Shardlow Marina, London Road, Derby DE72 2GL
M0 LJD L. Goldsmith Hunters Cottage, 61 Fengate Drove, Weeting, Brandon IP27 0PW
M0 LJH L. Hopgood 62 Briarwood Drive, Blackpool FY2 0EB
M0 LJK L. Marriott 94 Lyndhurst Road, Worthing BN11 2DW
M0 LJL L. Lewis 28 Brow Hey, Bamber Bridge, Preston PR5 8DS
M0 LJT A. Tranter 122 Summerhill Road, Bristol BS5 8JU
M0 LKD L. Kahlbau 7 Hamilton Court, De La Warr Road, Lymington SO41 0PR
M0 LKE L. Kett 52 Northgate, Hornsea HU18 1EU
M0 LKJ K. Mccarthy 34 Shawley Way, Epsom KT18 5PB
M0 LKL L. Brigham 42 Cayley Close, Clifton, York YO30 5PT
M0 LKN C. Emery 50 Hobkirk Drive, Sinfin, Derby DE24 3DT
M0 LKR R. King 12 South Road, Marden TN12 9EN
M0 LKS R. Boruch 4 Hafton Road, Salford M7 3TF
M0 LKT L. Taylor Marralomeda 7 Gatekeeper Close, Great Park, Newcastle upon Tyne NE13 9EH
MW0 LKX W. Dabrowski 2 Underhill Crescent, Knighton LD7 1DG
M0 LKY A. Lee 27 Victoria Avenue, Camberley GU15 3HT
MI0 LLG S. Horner 10 Meadow Court, Bushmills BT57 8SD
MW0 LLK C. Tanner Pen Y Gogarth Llaneilian, Amlwch LL68 9NH
MI0 LLM R. Hetherington 112 Screeby Road, Fivemiletown BT75 0LG
MW0 LLO C. Hill 9 Oliver Road, Newport NP19 0HU

M0 LLS I. Powell 9 Cardinal Crescent, Bromsgrove B61 7PR
M0 LLW A. Bostock 26 Ingham Road, Bawtry, Doncaster DN10 6NW
M0 LLX L. Andrews 72 Grange Road, Alresford SO24 9HF
MW0 LLY R. Martin Llan Owen, Rhulen, Builth Wells LD2 3UY
M0 LMB B. Savage Rufford, Barnes Lane, Milford On Sea, Lymington SO41 0RR
MM0 LMC L. Chan 5 Lansdowne Drive, Cumbernauld, Glasgow G68 0JB
M0 LMG P. Jones Sussex Cottage, The Limes, Felbridge, East Grinstead RH19 2QY
M0 LMH L. Hudson 68 Eleanor Road, Harrogate HG2 7AJ
M0 LMI L. Mikolka Flat, 1 Scotney Court, Romney Marsh TN29 9JP
M0 LMN D. Robinson Height End Farm, Kirk Hill Road, Haslingden, Rossendale BB4 8TZ
M0 LMO M. Moody 16 Tyersal Court, Bradford BD4 8EN
M0 LMQ R. Lewis 20 Hillary Road, Rugby CV22 6EU
M0 LMR D. Stanley 58 Wells Gardens, Basildon SS14 3QS
M0 LMS D. Partridge 44 Trumpet Terrace, Cleator CA23 3DY
MW0 LMW D. Jenkins 36 Brynawel Road, Gorseinon SA4 4UX
M0 LNE D. Bruce Nunfield House, Bull Lane, Sittingbourne ME9 7SL
M0 LNK Herts Repeater Group (Linking Call) c/o M. Wood 26 Parkfield Crescent Kimpton, Hitchin SG4 8EQ
MI0 LNL M. Robinson 84 Windsor Crescent, Cookstown BT80 8EZ
M0 LNO C. Sjostedt 14 New Road, Marlow SL7 3NG
M0 LNY M. Mccoy The Oaks, Liverton, Newton Abbot TQ12 6EZ
M0 LOB M. Garry 34 Conway Road, Paignton TQ4 5LH
M0 LOC D. Lock 20 Jasmine Close Trimley St. Martin, Felixstowe IP11 0UY
M0 LOF D. Loftus 5 Knowle Mount, Burley, Leeds LS4 2PP
MM0 LOG W. Stuart Meikleton, Aith, Bixter, Shetland Isles ZE2 9NE
M0 LOH P. Rasmussen 7 Portal Drive North, Upper Heyford, Bicester OX25 5TH
M0 LOL R. Marshall 29 Yew Tree Road, Witley, Godalming GU8 5RQ
M0 LOU D. Cave 22 Longsight Road, Mapplewell, Barnsley S75 6HB
M0 LOW D. Barnes 11 Yewside, Gosport PO13 0ZD
M0 LOX S. Cilliers 61 Heathwood Gardens, London SE7 8ET
MM0 LOZ D. Leech The Croft House, 9, Ruilick, Beauly IV4 7AB
M0 LPA N. Hilbery 16 Albert Road, Ashford TW15 2LU
M0 LPB L. Brown 28 Farley Way, Stockport SK5 6JD
M0 LPF B. Hall 30 Worcester Road, Dudley Westmidlands DY29LN
MW0 LPG L. Parsons Ty Crwn, Rhosgadfan, Caernarfon LL54 7HU
M0 LPK M. Napieralski Flat 2, Reeves House, Crawley RH10 7SW
M0 LPL G. Roberts 25 Chalfont Way, Liverpool L28 3QB
M0 LPM A. Monaghan Briar Patch 117 Elm High Rd, Wisbech PE14 0DN
MI0 LPO J. Mcerlean 24 Mccorley Road, Toomebridge, Antrim BT41 3NH
M0 LPR Dulverton Junior School c/o J. Matthews Moor View, Oldways End, East Anstey, Tiverton EX16 9JQ
M0 LPT P. Perreas 108 Lakonikis St., Kalamata 24133 Greece
M0 LPW L. Walker 7 Stroudley Close, Ashford TN240TY
M0 LPZ G. Harmath 101 Whitton Avenue East, Greenford UB6 0QE
MM0 LQF J. Bennet 24/11 Greenpark, Edinburgh EH177TA
M0 LQR Wearside Electronics And ARS c/o T. Longmore 3 Dairy Farm Cottages, Northlands Road, Gainsborough DN21 5DN
M0 LQW D. Michalczyk 155 Ewart Road, Nottingham NG7 6HG
M0 LQY I. Gilmore 19 Green Sward Lane, Redditch B98 0EN
M0 LRA Leeds Radio Amateurs c/o S. Priestley 49 Victoria Crescent, Leeds LS28 7SS
M0 LRB R. Baechle Dr.-Schuhwerk-Strasse 32A, St. Blasien 79837 Germany
MI0 LRC S. Davison 60 Cornation Place, Craigavon BT66 7AN
M0 LRD R. Taylor 89 St. Johns Road, Pelsall, Walsall WS3 4EZ
M0 LRG Essex Packet Gr c/o F. Barkhouse 312 Humberstone Lane, Leicester LE4 9JP
M0 LRH Dr R. Lecybyl 49 Thorndon Close, Orpington BR5 2SH
MD0 LRK C. Larkham Monte Rosa , 7 Ballaughton Close, , Douglas IM2 1JE
MW0 LRO M. Street The Lamb Inn, Hermon, Glogue SA36 0DS
M0 LRR R. Sansom 72 Wannock Lane, Eastbourne BN20 9SQ
M0 LRS L. Smith 20 St. Loyes St., Bedford MK40 1ZL
M0 LRY L. Bown 4 Oak Leys, Brewood, Stafford ST19 9EH
M0 LRZ S. Carr 24 Park Road, Blyth NE24 3DH
M0 LSA L. Mossop 4 Brookdale Way, Waverton, Chester CH3 7NT
M0 LSG G. Holland 6 Moorfield Road, Widnes WA8 3JE
M0 LSH M. Humphreys 12 Pixiefields, Cradley, Malvern WR13 5ND
M0 LSI N. Highfield 298 Mersea Road, Colchester CO2 8QY
M0 LSL A. Wood 4 St. Andrews, The Common, Cranleigh GU6 8NX
MM0 LSM A. Halcrow Da Cro, Branchiclate, Burra Isle ZE2 9LA
M0 LSN Rvd. D. Harding 9, Gilbert Street, Blenheim 7201 New Zealand
M0 LSO G. Coltman The Oaks Rushden, Buntingford SG9 0SN
M0 LSS L. Storry The Chimes, Madeira Drive, Bude EX23 0AJ
M0 LSV R. Derham Netherwood, Copse Lane, Long Sutton, Hook RG29 1SX
M0 LSW South Lancashire ARC c/o D. Horner 21 Ainsworth Road, Little Lever, Bolton BL3 1RG
M0 LSX A. Dale 37 Bussey Road, Norwich NR6 6JF
M0 LSY Y. Li Song 8 Birkin Court, Welwyn Garden City AL7 3FA
M0 LSZ S. Deakin 20 Riccat Lane, Stevenage SG1 3XY
M0 LTA A. Tokely 17 Sycamore Avenue, Horsham RH12 4TP
M0 LTD C. Brink 138 Brookside, Burbage, Hinckley LE10 2TN
M0 LTE T. Fanning 26 Mandeville Close, Tilehurst, Reading RG30 4JT
M0 LTG M. Champion 155 Walton Road, Walton on the Naze CO14 8NF
M0 LTK J. Forsyth 2 Littleworth, Oxford OX33 1TR
M0 LTL K. Leung Apartment 915, Metropolitan House, 1 Hagley Road, Birmingham B16 8HU
M0 LTN A. Rademaker 26 Elm Park Close, Houghton Regis LU5 5PN
M0 LTP Z. Lukasz 28 Heather Lane, West Drayton UB7 8AW
M0 LTS L. Stant Ears, University Of Surrey Student'S Union, Guildford GU2 7XH
M0 LTT M. Lovatt 3 Withington Close, Atherton, Manchester M46 0EZ
M0 LTW L. Farrant 21 Lime Tree Walk, Newton Abbot TQ12 4LF
MI0 LTX I. Kondrashenkov 44 Forge Manor, Magheralin, Craigavon BT67 0XP
M0 LUC A. Gee 14 Pipistrelle Drive, Onehouse IP14 1GS
M0 LUD G. Stanley 11 Marlborough Street, Ossett WF5 8JW
M0 LUF A. Clack 3 Darwin Close, Swindon SN3 3NF
M0 LUH N. Porter 114 Kingston Avenue, Worcester WR3 8PP
MW0 LUK C. Moreton 20 Millbrook Court, Little Mill, Pontypool NP4 0HT
MM0 LUP A. Young Broomloan, Staffin, Portree IV51 9JX
M0 LUS Royal Air Force ARS c/o C. Campbell 5 Ryebank, Holmfirth HD9 1EU
M0 LUT A. Lutley Springfield, Rookery Hill, Ashtead KT21 1HY
M0 LUY L. Isaac Sneath 21 Garrick Close, Lincoln LN5 8TG
M0 LVL S. Kiel 32 Weavers Avenue, Frizington CA26 3AT
M0 LVR Dr O. De Peyer Flat 5, Molasses House, Clove Hitch Quay, London SW11 3TN
M0 LVW L. Van Wezel 1 Waveney Road, Felixstowe IP11 2NT
M0 LWA A. Wale 17 Roundhouse Close, Welford, Northampton NN6 6NN
M0 LWC Long Wave Club c/o Dr R. Ashman 44 Conan Doyle Walk, Swindon SN3 6JB
M0 LWM B. Blackstone 2 Sneating Hall Cottages, Sneating Hall Lane, Frinton-on-Sea CO13 0EW
MM0 LWS M. Strachan 62 Charleston Drive, Dundee DD2 2EZ
M0 LWT C. Jenkins 25 Longmeadow Grove West Heath, Birmingham B31 4SU
M0 LWZ Royal Signals Museum c/o G. Budden 7 Ashburton Gardens, Bournemouth BH10 4HP
M0 LXA C. Staff Uphill Road South, Weston Super Mare BS23 4TU
M0 LXD D. Sawyer 46 Brendon Gardens, Fair Oak, Eastleigh SO50 7GG
M0 LXG C. Whatmough 11 Blackchapel Drive, Rochdale OL16 4QU
M0 LXI E. Pestano 38 Third Avenue, Bexhill-on-Sea TN40 2PA
M0 LXS C. Claxton Forest Edge, Deer Park, Milton Abbas, Blandford Forum DT11 0AY
M0 LXV D. Edwards 212, Eastern Avenue North Kingsthorpe, Northampton NN2 7AT
M0 LXY R. Thompson Croft Michael Farm, Croft Mitchell, Troon, Camborne TR14 9JJ
M0 LYB P. Lyba 6 Ambassadors Way, North Shields NE29 8ST
M0 LYD R. Beck Moorings, Pleasance Road Central, Romney Marsh TN29 9NP
M0 LYF K. Fletcher Beverley Hotel, 55 Old Brumby Street, Scunthorpe DN16 2AJ
M0 LYI M. Lynn 12 Mowbray Close, Crook DL15 9GH
M0 LYN G. Molyneaux 3 Wilson Close, Thelwall, Warrington WA4 2ET
M0 LYQ E. Gudziunas 66 Packwood Close, Bentley Heath, Solihull B93 8AW
M0 LYR Lowestoft And Gt Yarmouth Repeater Group c/o J. Crawford 4 Church Road, Woodton NR35 2NB
M0 LZH L. Herman 28 Millwood Court, Bury BL9 9SD
M0 LZI I. Duxbury 334 Linnet Drive, Chelmsford CM2 8AL
M0 LZM P. Radford 43 Bells Lane, Nottingham NG8 6EX
M0 LZN J. Marshall 6 Foster Walk, Sherburn in Elmet LS25 6EU
M0 LZQ D. Bate 14 Bromley Drive, Leigh WN7 5NA
M0 LZS J. Kelly Whiteley Bank Lodge, Canteen Road, Whiteley Bank, Ventnor PO38 3AF
M0 LZU K. Kostov 47 Wapshott Road, Staines-upon-Thames TW18 3HB
M0 LZX J. Mutter 27 Snowdonia Way, Huntingdon PE29 6XP
MW0 LZZ C. Stubbs 50 Laburnum Close, Rogerstone, Newport NP10 9JQ
M0 MAC J. Mcgowan 72B Adelphi Crescent, Hornchurch RM12 4JZ
M0 MAF M. Milne Flambards, Manor Road, Dunmow CM6 2JR
M0 MAG M. Tinnion 3 Hillhead Road, Newcastle upon Tyne NE5 5AP
MW0 MAH M. Arnett 5 Ffordd Cerrig Mawr, Caergeiliog, Holyhead LL65 3LU
M0 MAI M. Mahoney Hurdle Cottage, Mannington, Wimborne BH21 7JZ
M0 MAJ M. Jones 20 Chelsea Drive, Sutton Coldfield B74 4UG
M0 MAL M. Elliott 4 Maple Close, Keelby, Grimsby DN41 8EL
MD0 MAN R. Cunningham 3 Kellets Cottage, Lhergy Cripperty, Union Mills IM4 4NF Isle of Man
M0 MAO M. Boland 24 Hallam Close, Moulton, Northampton NN3 7LB
MI0 MAP J. Phillips 23 Edenderry Gardens, Banbridge BT32 3BQ
M0 MAQ E. Macgurk 10 Elmore Road, Lee on Solent PO139DU
M0 MAR A. El Khalidi 66 Trevor Crescent, Ruislip HA4 6ND
M0 MAT M. Jeffery 25A Stockwell Drive, Mangotsfield, Bristol BS16 9DW
MW0 MAU M. Uphill 1 Brynview Avenue, Ystrad Mynach, Hengoed CF82 7DB
M0 MAW N. Cheesewright 5 Duberly Close, Perry, Huntingdon PE28 0BP
M0 MAX M. Skinley 69 George Street, Wellington TA21 8HZ
MW0 MAY M. Stokes 23 Goetre Fawr Road Killay, Swansea SA2 7QS
M0 MAZ M. Stevenson 127 Walton Road, Chesterfield S40 3BX
M0 MBA Dr Z. Derzsi 217 Bensham Road, Bensham, Gateshead NE8 1US
M0 MBB M. Bowell 28 Jubilee Close, Byfield, Daventry NN11 6UZ
MM0 MBC J. Curtis 11 Haston Crescent, Perth PH2 7XD
M0 MBD D. De La Haye 4 Nicola Mews, Ilford IG6 2QE
M0 MBE S. Holt 14 Fir Street, Cadishead, Manchester M44 5AU
M0 MBG M. Cooper 9 Conway Close, Crewe CW1 3XN
M0 MBH Dr M. Holbrook 140 High Street, Steyning BN44 3LH
M0 MBI A. Fox 53 Shakespeare Way, Taverham, Norwich NR8 6SL
M0 MBM Devon County Council Emergency Plng c/o M. Bridgehouse 43 Age Croft, Oldham OL8 2HG
M0 MBO M. Bay 10 Mitchinson Street Steeple Claydon, Buckingham MK18 2GS
M0 MBR M. Mutkin 13 The Grove, Radlett WD7 7NF
M0 MBS M. Hyman 6 Belvedere Court, St. Anns Road, Manchester M25 9LB
M0 MBT M. Mckenna 54 Whickham Road, Hebburn NE31 1QU
M0 MBV M. Hennessey 3 Northgate Cottage, Falmer Road, Rottingdean, Brighton BN2 7DT
M0 MBW F. Smith 32 Amesbury Drive, London E4 7PZ
M0 MBZ M. Bray 13 Rosebay Close, Hartlepool TS26 0ZL
M0 MCA A. Howden 7 West Vale, Filey YO14 9AY
MI0 MCB J. Mcbride 21 Mosside Gardens, Mosside, Ballymoney BT53 8QQ
MI0 MCC C. Mcclelland 2 Stuart Park, Ballymoney BT53 7BE
M0 MCE D. Mcewan 29 St. Andrews Avenue, Weymouth DT3 5JS
M0 MCG the Moors CG c/o R. Edgar 45 Exeter Road, Dawlish EX7 0AB
M0 MCH M. Chapman 3 Whitton Close, Doncaster DN4 7RB
M0 MCI P. Illidge 55 East Park Road, Spofforth, Harrogate HG3 1BH
M0 MCL K. Winton 130 George V Avenue, Worthing BN11 5RX
M0 MCO M. Tinsell-Stanton 38 Comberton Road, Kidderminster DY10 3DT
M0 MCP R. Van-Der-Wijst 6 Willow Street, Romford RM7 7LJ
M0 MCT M. Bradbury 104 Lilly Hall Road Maltby, Rotherham S66 8AT
M0 MCV R. Treacher 93 Elibank Road, London SE9 1QJ
M0 MCW K. Phillips 31B Waterloo Close, Blackburn BB2 4RQ
M0 MCY M. Rolph 17 Moorlands Park, Sinope LE67 3BD
M0 MDC M. Clements 21 Mallard Place, Twickenham TW1 4SW
M0 MDE D. Edmondson 21 Hawthorne Close, Heathfield TN21 8HP
M0 MDG Middlesex DX Group c/o S. Smith 12 Stoneleigh Avenue, Hordle SO41 0GS
MM0 MDH M. Herridge The Hollies, Petticoat Lane, Orkney KW17 2RP
MW0 MDJ M. Johns 151 Somerset Street, Abertillery NP13 1DR
M0 MDO Obo Wigan Deanery High School c/o D. Mcauslan Casa Arco Iris, Via Variante Nascente, 8005-491 SANTA BARBARA DE NEX Portugal
M0 MDP P. Murphy 41 Tower Street, Sunderland SR2 8NF
MW0 MDT M. Griffiths Mandalay, Bromfield Street, Wrexham LL14 1NF
MW0 MDV D. Morgan 28 Harbour Village, Goodwick SA64 0DY

M0 MDZ M. Day 14 Windsor Drive, Ramsey Forty Foot, Ramsey, Huntingdon PE26 2XX
M0 MEA M. Attlesey 1 The Landway, Borough Green, Sevenoaks TN15 8RG
M0 MED D. Creighton 8 Stockton Road West, Hawthorn, Seaham SR7 8RS
M0 MEF M. Jasiorkowski 275 Hartland Ave, Sheffield S20 2PZ
M0 MEG A. Stephenson 1 Northrop Close, Sunnybrow, Crook DL15 0NS
M0 MEH M. Horton 12 Kelburn Close, Chandler'S Ford, Eastleigh SO53 2PU
M0 MEI Shefford & District ARS c/o M. Eilec 5 Balata Way, Basingstoke RG24 9YP
M0 MEL M. Kirk 128 Perry Hill Road, Oldbury B68 0BJ
M0 MEN M. Norris 35 Sudbrooke Road, London SW12 8TQ
M0 MEO Mexborough & District ARS c/o S. Saiger 10 Markham Avenue, Armthorpe, Doncaster DN3 2AZ
MI0 MEV D. Lynas 7 Regency Avenue, Dollingstown, Craigavon BT66 7TY
M0 MEW T. Garcia-Quismondo 11 Half Moon Lane, Worthing BN13 2EN
MW0 MEX G. Johnson 47 Heol Fawr, Penyrheol, Caerphilly CF83 2JU
M0 MEY T. Hart 17A Meyrick Park Crescent, Bournemouth BH3 7AG
M0 MFA F. Alfrey 16 Walls Road, Bembridge PO35 5RA
MW0 MFB Dr M. Brown Bryn Siriol Mount Road, St. Asaph LL170DB
M0 MFC M. Chester 84 Edinburgh Drive, Spalding PE11 2RT
M0 MFH M. Huson 21 Fairfield Avenue, Stoke-on-Trent ST3 4NU
MI0 MFI E. Taylor 17 Rutherglen Street, Belfast BT13 3LR
M0 MFL K. Bantock 22 Deepdale Drive, Consett DH8 7EH
M0 MFP C. Reed 2 Drapers Lane, Hedon, Hull HU12 8BG
M0 MFS A. Brook Inkerman House 113 Clovelly Road, Bideford EX39 3BY
M0 MFT Lincoln Short Wave Club c/o M. Tweedie 14 St. Cuthbert Drive Romanby, Northallerton DL7 8JF
M0 MFX Dr M. Fox 24 The Avenue, Sandy SG19 1ER
M0 MGA M. Smyth 111 Forest Road, Whitehill, Bordon GU35 9BA
MM0 MGB A. Britton 15 Glenbrook, Balerno EH14 7JE
M0 MGF J. Gower 10 Dann Court, Hedon HU12 8GT
M0 MGI M. Isbell 20 Woodland Crescent, Wolverhampton WV3 8AS
MI0 MGJ M. James 6 Portaferry Road, Newtownards BT23 8NN
M0 MGK G. Marley 144 Moreland Road, South Shields NE34 8NJ
MM0 MGM M. Macfarlane 9 Dreghorn Park, Colinton, Edinburgh EH13 9PH
M0 MGP G. Champion 34 Greenfields, Edenside, Kirby Cross., Frinton-on-Sea CO13 0SW
M0 MGS M. Smith 313 Stourbridge Road, Dudley DY1 2EF
M0 MGW S. Dean 154 Broad Lane, Walsall WS3 2TQ
M0 MHG G. Hamilton 5 Felixstowe Road, Sunderland SR4 0BF
M0 MHJ I. Jabegu 51 South Crescent, Blandford Camp, Blandford Forum DT11 8AJ
M0 MHO B. Silveira 6 Amesbury, Waltham Abbey EN9 3LQ
MM0 MHP H. Percival 3/2 21 Prince Albert Road, Glasgow G12 9JU
M0 MHQ Braunstone Troop Military Radio Group c/o E. Scott 3 School Drive, Coalville LE67 4AN
M0 MHT M. Thompson 4 Jubilee Court Ravenscroft, Holmes Chapel, Crewe CW4 7HA
M0 MHU M. Hubbard 14 Parkfield Crescent, Kimpton, Hitchin SG4 8EQ
M0 MHW G. Hardman 12 Fernleigh Chorley New Road, Horwich, Bolton BL6 6HD
M0 MHY J. Mahoney 27 Linby Drive, Bircotes, Harworth, Doncaster DN11 8PF
MM0 MHZ Backpackers RAG c/o P. Thompson 31 St. Marys Drive, Perth PH2 7BY
M0 MIB Strathclyde R G c/o V. Ball 24 Carr Lane, Warsop, Mansfield NG20 0BN
M0 MID P. Staite Chestnut Farm, Eastville, Boston PE22 8LX
MW0 MIE M. Ireland Pen Y Gadlas, Ffordd Bryniau, Prestatyn LL19 8RD
M0 MIG M. Navarro 22 Lark Crescencent Harford, Huntingdon PE29 1YN
MM0 MIJ J. Smith 15E Afton Road, Cumbernauld, Glasgow G67 2DW
M0 MIM M. Pearce 1 Briars Wood, Horley RH6 9UE
M0 MIQ M. Iqbal 22 Rupert Avenue, High Wycombe HP12 3NG
M0 MIR M. Lesniowski 3 Woodland Avenue, Worksop S80 2RB
M0 MIT B. Mitchell 34 St. Marys Avenue, Gosport PO12 2HX
M0 MIZ S. Stebbings 4 Coltsfoot Lane, Bull'S Green, Knebworth SG3 6SB
MW0 MJB M. Lee 4 Hill Street, Haverfordwest SA61 1QF
M0 MJD M. Davis The Innings, Cricketts Lane, Chippenham SN15 3EG
M0 MJF M. Firth 209 High Street, Wickham Market, Woodbridge IP13 0RQ
M0 MJG M. Garrett 489 Dorchester Road, Weymouth DT3 5BP
M0 MJH M. Hickford 3 Ashen Road, Clare, Sudbury CO10 8LQ
M0 MJK M. Keyte 3 Lower High St., Mow Cop, Stoke on Trent ST7 3PB
M0 MJS M. Sykes Hope Cottage, 8 Brookside Road, Wimborne BH21 2BL
M0 MJT N. Tyerman 44 Hawkstone Close, Guisborough TS14 7PE
M0 MJW M. Whitfield 10 Bede Haven Close, Bude EX23 8QF
M0 MJX M. Johnson 54 Birchwood Drive, Ulverston LA12 9PN
MM0 MJY M. Yarrow Lomond Villa, Downies Village, Aberdeen AB12 4QX
M0 MKE King Edward VII School c/o P. Treadwell 22 Meynell Close, Melton Mowbray LE13 0RA
MW0 MKG M. Gray 15 The Circle, Two Locks NP44 7JP
M0 MKH M. Hadfield 22 Mansfield Road, Clowne, Chesterfield S43 4DH
M0 MKO M. O'Connor 28 Cardigan Road, Southport PR8 4SF
M0 MKR Milton Keynes Raynet c/o J. Breen 68 Honeysuckle Way, Bedford MK41 0TF
M0 MKV A. Crawford 4 Trimpley Drive, Kidderminster DY11 5LB
M0 MKY J. Wilkinson 8 Hunters Point, Chinnor OX39 4TG
MM0 MLD W. Lawson 60 Inglis Avenue, Port Seton EH32 0AQ
M0 MLE J. Statham Oakwoods, School Lane Upper Basildon, Reading RG8 8LT
M0 MLG M. Goff 27 Harley Road, Oxford OX2 0HS
M0 MLH F. Peters 56 Moresdale Lane, Leeds LS14 6SY
M0 MLJ R. Tattersall 70 Selwyn Street, Hillstown, Bolsover, Chesterfield S44 6LR
M0 MLK M. Kipling 12 Jolly Brows, Harwood, Bolton BL2 4LZ
M0 MLM M. Millen Flat 9, Sussex Court, Park Road, Bognor Regis PO21 2PY
MM0 MLO M. Olesen 19 Holmston Crescent, Ayr KA7 3JJ
M0 MLS A. Stevenson 252 Perry Wood Road, Great Barr, Birmingham B42 2BH
M0 MLT M. Titcombe 1A Langdale Avenue, Harpenden AL5 5QU
M0 MLV T. Jones Flat 92, Berkeley Court, Marylebone Road, , London NW1 5ND
M0 MLW M. Wren Church Farm, Pointon Fen, Sleaford NG34 0LF
M0 MLY J. Malley 18 Park View, Seaton Delaval, Whitley Bay NE25 0AL
M0 MLZ M. Mills 17 Hornby Street, Plymouth PL2 1JD
M0 MMC K. Sansom 23 Victoria Crescent, Poole BH12 2JQ
M0 MMD M. Moore 52 Limefield Street, Accrington BB5 2AF
MM0 MMG M. Gourlay 14 Holmes Holdings, Broxburn EH52 5NS
M0 MMJ M. Majhail 3 Poynders Hill, Hemel Hempstead HP2 4PQ
MM0 MMN I. Campbell Fossoway Lodge, Kinross KY13 0PD
M0 MMO J. Summerhill 43 Rangers Walk, Bristol BS15 3PW
M0 MMP Y. Weng 20 Kendal Grove, Leeds LS3 1NS
M0 MMR S. Quinn 7 Poppleton Court, Tingley, Wakefield WF3 1UY
M0 MMS Dr M. Mattingley-Scott Bergheimer Strasse 28, Heidelberg 69115 Germany
M0 MMT M. Johnson 3 Bersted Mews Bersted Street, Bognor Regis PO22 9RR
M0 MMU A. Moss Winstons, Mayfield Lane Durgates, Wadhurst TN5 6DG
M0 MMX D. Watt 4 Spring Gardens Terrace, Padiham, Burnley BB12 8JB
M0 MNB J. Davidson 78 Old Heath, Shrewsbury SY1 4SE
M0 MNG E. Spicer 3 Golden Avenue Close, East Preston, Littlehampton BN16 1QS
M0 MNO D. Edge 122 Aldbanks, Dunstable LU6 1AJ
MM0 MNS M. Stuart 3 Arran View, Largs KA30 9ER
M0 MNU P. Richardson 14 Portland Street, Worksop S80 1RZ
M0 MNV P. Bienko 94 Foster Street, Lincoln LN5 7QF
MW0 MNX Artie Moore ARS c/o K. Dawson 57 Fair View, Blackwood NP12 3NR
M0 MNZ L. Bennett 19 Campion Crescent, Cranbrook TN17 3QJ
MM0 MOB M. Overthrow 63 Primrose Avenue, Larkhall ML9 1JX
MM0 MOC Museum of Communication ARS c/o K. Horne 10 Blair Place, Kirkcaldy KY2 5SQ
MI0 MOD T. Thompson 8 Knockburn Avenue, Lisburn BT28 2QF
M0 MOI S. Pettitt 11 Derling Drive, Raunds, Wellingborough NN9 6LF
MM0 MOK L. Matchett 9 Colcoon Park, Gorebridge EH23 4RS
M0 MOL G. Mollard 1 Barnard Street, Barrow-in-Furness LA13 9TD
MW0 MON D. Williams 10 Bronllys, Gaerwen LL60 6JN
M0 MOR A. Turner 29 Welling Road, Orsett, Grays RM16 3DW
M0 MOS M. Beckett 59 Broadacre, Caton, Lancaster LA2 9NH
MM0 MOT A. Smith 56 Ayr Road Douglas, Lanark ML11 0QA
M0 MOW J. Fautley 71 Pullman Lane, Godalming GU7 1YB
MM0 MPA D. Panton 64 Dochart Crescent, Polmont, Falkirk FK2 0RE
M0 MPB M. Budd 28 Ladymeadow Court Middleton, Milton Keynes MK10 9HZ
MW0 MPD D. Jenkins 44 Kensington Drive, Bridgend CF31 4QS
M0 MPF M. Finn Atzenbach 38, Idar-Oberstein 54473 Germany
M0 MPI M. Ibbett 13 Lely Close, Bedford MK41 7LS
M0 MPM M. Meerman 24 Horseshoe Crescent, Burghfield Common, Reading RG7 3XW
M0 MPQ J. Dales 6 Woodfield Drive, Sawtry, Huntingdon PE28 5TZ
M0 MPS B. Hopkins 28 Dean Lodge Grange Road, Southbourne, Bournemouth BH6 3ND
M0 MPT M. Travis Cherrydayle, 6 Lingwood Close, Southampton SO16 7GJ
MM0 MPW F. Dove 5 Cairnorchies, Mintlaw, Peterhead AB42 4LH
M0 MPY S. Gray 2 Gloucester Road, Pilgrims Hatch, Brentwood CM15 9ND
M0 MRB Dr M. Brickley The Shearings, Milborne Port Road, Charlton Horethorne, Sherborne DT9 4NH
M0 MRC C. Robinson 9 Chatsworth Avenue, Culcheth, Warrington WA3 4LD
MI0 MRG Marconi Radio Group c/o P. Quinn 11 Blackpark Road, Ballyvoy, Ballycastle BT54 6QZ
M0 MRH A. Hawksworth 87 Bradeley Road, Haslington, Crewe CW1 5PX
M0 MRI A. Titmus The Old Police House, Arundel Road, Fontwell, Arundel BN18 0SX
M0 MRJ M. Jebbett 16 Hastings Meadow Close, Kirby Muxloe, Leicester LE9 2DR
M0 MRK M. Newton Hall Farm Bungalow, Holbeck, Worksop S80 3NF
M0 MRL D. Weight 15 Bourn Road, Caxton, Cambridge CB23 3PP
MM0 MRM A. Moe 43 Huntlyburn Terrace, Melrose TD6 9BH
MM0 MRO C. Munro 11 Craigleith Hill Green, Edinburgh EH4 2ND
M0 MRP M. Phillips 71 Stour View Gardens, Corfe Mullen, Wimborne BH21 3TL
M0 MRQ Dr P. Lawrence 20 Pond Lane, Drayton, Norwich NR8 6PP
MW0 MRS Marches ARS c/o M. Bobby Hafan, Church Street, Penycae LL14 2RL
M0 MRT K. Turner 3 Park Close, Pinxton, Nottingham NG16 6QQ
MI0 MRV M. Kashkoush 41 Dunamallaght Road, Ballycastle BT54 6PF
M0 MRW M. Williams 124 Ringwood Road, Christchurch BH23 5RF
M0 MRX M. Roper 26 Malpas Close, Bransholme, Hull HU7 4HH
M0 MRY J. Mullins 61 St. Johns Road, Slough SL2 5EZ
M0 MSA Mid Somerset ARC c/o C. Lavis 88 Boundary Way, Glastonbury BA6 9PH
MI0 MSB B. Campbell 3, B Crewe Road, Ballinderry Upper, Lisburn BT28 2PL
M0 MSE M. Edmonds 60 Shenstone Road, Maypole, Birmingham B14 4TJ
M0 MSF T. Reed Seafield, Charing Hill, Charing, Ashford TN27 0NG
M0 MSG M. Gibbons 117 Ettingshall Road, Bilston WV14 9XF
MM0 MSH B. Mccosh 5 Muckhart Road, Dollar FK14 7AE
MI0 MSM D. Dellett 5 Larchfield Gardens, Kilrea, Coleraine BT51 5SB
M0 MSN M. Boisriveau-Mitchell 16 Pochard Close, Quedgeley, Gloucester GL2 4LL
MI0 MSO Data - DXers c/o P. Hosey 13 Glenelly Gardens, Omagh BT79 7XG
M0 MSS M. Simpson 7 New Hall Farm, Cowling, Keighley BD22 0JQ
M0 MSV S. Morozov Haroldene, Towpath, Shepperton TW17 9LL
M0 MSX M. Smith 6 Neeps Terrace, Middle Drove, Wisbech PE14 8JT
M0 MSY Merseyside ARS c/o J. Woolvin 62 Whitewood Park, Liverpool L9 7LG
M0 MSZ M. Strange 101 Southbroom Road, Devizes SN10 1LY
M0 MTA M. Atfield 42 Pauls Croft Cricklade, Swindon SN6 6AJ
M0 MTC Banbury Amateur Radio c/o G. Brown 13 Francis Avenue, Moreton, Wirral CH46 6DH
M0 MTD T. Davidson 2 Ridgeway, Rotherham S65 3PQ
M0 MTF M. Falkowski 23 Casson Street, Crewe CW1 3EG
M0 MTG G. Clarke 2 Gort Lomie, Clonlara V94 H96H Ireland
M0 MTI M. Juvonen 23 Harwood Road East Hagbourne, Didcot OX11 9LX
M0 MTJ M. Smith 6 Peverill Road, Perton WV6 7PH
M0 MTN C. Martin 14 Campbell Road, Eastleigh SO50 5AD
MM0 MTO R. Foulds 83 Croftfoot Road, Glasgow G44 5JU
M0 MTQ D. Hartley 102 Moss Lane, Sale M33 5BE
MW0 MTR M. Roynon 16 Greenwood Avenue, Pontnewydd, Cwmbran NP44 5JE
M0 MTS C. Small Riddings Barn, Hope Bagot, Ludlow SY8 3AE
M0 MTW J. Bailey 22 Wilford Drive, Ely CB6 1TL
M0 MTX A. Price 23 Greenway, Wingerworth, Chesterfield S42 6NP
M0 MUC Ystrad Mynach College c/o M. Wolfson 4 Crabmill Lane, Easingwold, York YO61 3DE
M0 MUI M. Tsun Dyson'S Farm, Long Row, Tibenham NR16 1PD
MM0 MUL A. Jackson Union Farm, Craigrothie, Cupar KY15 5PJ
MW0 MUM A. Sneddon 3 Marigold Close, Gurnos, Merthyr Tydfil CF47 9DA
MM0 MUN Newton-Le-Willows Raynet c/o E. Munro 55 Abergeldie Road, Aberdeen AB10 6ED
MM0 MUR G. Murray The Barn House, Springfield Farm, Carluke ML8 4QZ
M0 MUU S. Bunting Highfield, Three Ashes HR2 8LU
M0 MUZ M. Hickman 40 Tredington Grove, Caldecotte, Milton Keynes MK7 8LR
M0 MVB S. Norman 38 The Croft, Christchurch, Wisbech PE14 9PU

M0 MVE M. Lambev 2 Keats Way, Hitchin SG4 0DR
M0 MVL M. Lloyd 17 Williams Mead, Bartestree, Hereford HR1 4BT
MW0 MVM A. Holt Tyshoni, New Street, Llandrindod Wells LD1 6BU
M0 MVO R. Zakrzewski 31 Kingston Crescent, Chelmsford CM2 6DN
M0 MVS Maritime Volunteer Service RC c/o L. Miller 28 Arthur Road Cliftonville, Margate CT9 2EN
MM0 MVX D. Wilson Rivendell Lodge, Glenkindie, Alford AB33 8RN
MW0 MWA A. Nisbet 8 Pen Dinas, Tonypandy CF40 1JD
M0 MWB S. Brown 18 Goring Ave. Gorton., Manchester M18 8WW
MW0 MWJ J. Jillings Cane Garden, Dolau, Llandrindod Wells LD1 5TE
MW0 MWL D. Mead 35 Holly Street, Rhydyfelin, Pontypridd CF37 5DA
M0 MWN M. Singer 1 Bentley Road, Slough SL1 5BB
M0 MWR M. Redstall 56 Westmorland Road, Felixstowe IP11 9TJ
M0 MWS M. Smith Ashcroft, Black Horse Lane, Winterbourne Earls, Salisbury SP4 6HW
M0 MWT R. Wilkes 157 Saltwells Road, Dudley DY2 0BN
MM0 MWW Orkney ARC c/o E. Holt Ashwell, St. Ola, Kirkwall KW15 1SX
MW0 MWX Mike-Whiskey DX Group c/o C. Morris Hideaway Bettws Cedewain, Newtown, Powys SY16 3DS
M0 MXC M. Craven 78 Connaught Road, Brookwood, Woking GU24 0HF
MU0 MXF O. Borisov The Palms La Couture St Peter Port, Guernsey GY1 2DZ
M0 MXN R. Lovell Formby, Formby, Livepool L37 4BP
M0 MXX M. Day 33 Ryndle Walk, Scarborough YO12 6JT
M0 MYA D. Passey 5 The Croftings, Felton Close, Ludlow SY8 1DS
M0 MYB H. Ibbitson Tor View, Whitstone, Holsworthy EX22 6TB
M0 MYC R. Browne 2 Martham Close, London SE28 8NF
M0 MYE D. Myers 39 Rowan Avenue, Shildon DL4 2AS
M0 MYG M. Young 72 Goddard Way, Saffron Walden CB10 2EB
M0 MYJ A. Frost 23 Judeland, Chorley PR7 1XJ
M0 MYK M. Knowles 86 West Shore Road, Walney, Barrow-in-Furness LA14 3UD
MM0 MYL C. Williamson 31 Medrox Gardens, Cumbernauld, Glasgow G67 4AJ
M0 MYM D. Crane 3 Middlemead Close West Hanningfield, Chelmsford CM2 8UR
M0 MYN C. George 5 Turing Gardens, Shefford SG17 5ZS
M0 MZC M. Carpenter The Retreat, High Lane Manaccan, Helston TR12 6HT
M0 MZD B. Greenwood 13 Mayflower Street, Blackburn BB2 2RX
M0 MZE D. Pilkington 197 Saltings Road, Snodland ME6 5HP
M0 MZN P. Hilton Shankly Cottage 161 Highgate, Jennings Yard, Kendal LA9 4EN
M0 MZX A. Watts 12 Duchy Close, Dorchester DT1 2EL
M0 NAA Denby Dale ARS c/o G. Porter Higher Bramble, Trusham, Newton Abbot TQ13 0NW
MW0 NAB N. Hockenhull 6 Bryn Gannock, Deganwy, Conwy LL31 9UG
M0 NAE Peterborough And District ARC c/o B. Smith Maple Lodge, Burtoft Lane South, Boston PE20 2PF
M0 NAG N. Parry 125 Lawsons Road, Thornton-Cleveleys FY5 4PL
M0 NAI R. Neufeld 19 Douai Grove, Hampton TW12 2SR
MI0 NAJ P. Thompson 57 Ballyduff Road, Newtownabbey BT36 6PA
M0 NAK C. Marshall 51 Hedgerow Close, Redditch B98 7QF
M0 NAL Dr P. Shaw 25 Headcorn Road, Platts Heath, Maidstone ME17 2NH
M0 NAM N. Matthes 24 Albany Road, Fleet GU51 3LY
M0 NAO N. Tateishi 4-24-14-402 Taihei, Sumida-Ku, Tokyo 1300012 Japan
M0 NAP A. Newell 17 Southlands Grove, Thornton, Bradford BD13 3BG
M0 NAQ R. Smith 15 Hollybush Road, North Walsham NR28 9XT
M0 NAR Dorset Raynet c/o R. Seddon 255 Westleigh Lane, Leigh WN7 5PN
M0 NAS N. Smith Clare Cottage, White Ash Green, Halstead CO9 1PD
M0 NAU N. Coventry 1 Seacrest Avenue, Fleetwood FY7 6FG
M0 NAW N. Carey 28 Tremayne Road, St. Austell PL25 4NE
M0 NAX G. Mosner Zum Roehrbrunnen 16, Dreieich 63303 Germany
M0 NAY C. Pegrum 3 Bretland Road, Tunbridge Wells TN4 8PS
M0 NAZ A. Davies 4 Capella Path, Hailsham BN27 2JY
M0 NBA B. Chalmers 19 Brettenham Crescent, Ipswich IP4 2UB
M0 NBC North Bristol ARC c/o P. Stevenson 6 Dighton Gate, Stoke Gifford, Bristol BS34 8XA
M0 NBD Norfolk County Raynet Group c/o A. Mobbs 149 The Paddocks, Old Catton, Norwich NR6 7HR
M0 NBJ N. Jones 26B Wellington Road, Wallasey CH45 2NG
M0 NBK G. Hicks 8 Mill Lane, Stockton-on-Tees TS20 1LG
M0 NBL F. Noble 1045, 45Th Street Apartment A, California 94608 United States
M0 NCA Norfolk Coast ARS c/o R. Leeds 1A Clare Road, Cromer NR27 0DD
M0 NCC Northampton DX c/o G. Bansil 15 Abington Close, Crewe CW13TL
M0 NCE N. Irvine 100 Cavendish Road, Sunbury-on-Thames TW16 7PL
M0 NCG M. Dumpleton 24 Barley Close, Newton St. Faith, Norwich NR10 3GY
M0 NCI Lifeboat ARS c/o D. Hughes 86 Colinmander Gardens, Ormskirk L39 4TF
M0 NCJ C. Nicholson 97 Station Road, Burgess Hill RH15 9ED
M0 NCK N. Jewitt 10 Gorse Lane, Oadby, Leicester LE2 4RQ
M0 NCL M. Seabrook 1 The Covers, Morpeth NE61 2RU
M0 NCN Rvd. M. Gillingham 14 Nethergreen Gardens, Killamarsh S21 1FX
M0 NCR Norfolk County Raynet Group c/o A. Mobbs 149 The Paddocks, Old Catton, Norwich NR6 7HR
M0 NCV G. Killpack 20 Fisher Close, Banbury OX16 3ZW
M0 NCZ N. Czernuszka 12 Durham Drive, Ashton-under-Lyne OL6 8BP
M0 NDA Nuneaton & District ARC c/o D. Parker 146 Merlin Avenue, Nuneaton CV10 9QJ
M0 NDC N. Caulfield Smallbrook Road, Whitchurch SY13 1BS
MW0 NDE N. Evans 16 Oakwood Road, Y Rhyl LL18 4BH
M0 NDF D. Finlay 23 Glen Way, Oadby, Leicester LE2 5YF
M0 NDJ D. Noe 21 Gale Crescent, Banstead SM7 2HZ
MI0 NDK N. Jameson 15A Ednagee Road, Castlederg BT81 7QF
M0 NDL S. Emary Mallards, Fishers Lane, Mark, Highbridge TA9 4LZ
MM0 NDM N. Maclucas Lochnell Lodge, Benderloch, Oban PA37 1QS
M0 NDO A. Nall 22 Park Grove, Swillington, Leeds LS26 8UN
M0 NDP N. Plunkett 11 Stoneleigh Gardens, Grappenhall, Warrington WA4 3LE
M0 NDT D. Farrant 13 Bramham Down, Guisborough TS14 7BY
M0 NDU J. Knight 30 Ash Meadow, Lea, Preston PR2 1RX
MM0 NDX C. Mcgowan 21 Franchi Drive, Stenhousemuir, Larbert FK5 4DX
M0 NDY R. Potter 8 Hansard Way, Kirton, Boston PE20 1QN
M0 NDZ T. Daskalov 40 Elm Road, Norton Canes, Cannock WS11 9QN
M0 NEC N. Eccles 55 New Street, Lymington SO41 9BP
M0 NEG K. Metcalfe 33 Corsican Drive, Hednesford, Hednesford WS12 4SS
M0 NEH Dr N. Hoare 5 Kelsey Head, Port Solent, Portsmouth PO6 4TA
M0 NEM P. Sanders 6 Primrose Hill, Warwick CV34 5HW
MM0 NEO N. Thomson Four Winds, Holland Bush Hightae, Lockerbie DG11 1JL
M0 NER A. Fraser 18 Donside Close, Boldon Colliery NE35 9BS
M0 NEU D. Newgas 4207 3Rd Ave Nw, Seattle 98107 United States
M0 NEV L. Creek 382 Ripon Road, Stevenage SG1 4NQ
MM0 NEW B. Newcombe 9 Calder Road, Bellsquarry, Livingston EH54 9AA
M0 NEX L. Jepson 143 Walnut Avenue Weaverham, Northwich CW8 3DX
M0 NFB Derwent Valley Radio Group c/o N. Bisiker 31 Lansdowne Avenue, Waterlooville PO7 5BL
M0 NFD GB7ADX CSG/Cymru CG c/o C. Davies 28 Neville Road, Darlington DL3 8HY
M0 NFI N. Mooney 60 Rhyddings Street, Oswaldtwistle., Accrington BB5 3EY
M0 NFN P. Dernikos P.O Box 599, Ashburton 3147 Australia
M0 NFR 59 Squadron Air c/o R. Ferguson 31 Barton Court Road, New Milton BH25 6NW
M0 NFY N. Young 139 Northumberland Street, Norwich NR2 4EH
M0 NFZ M. Levett 16 Firs Avenue, Waterlooville PO8 8RS
M0 NGB N. O'Brien-Bird 18 Milsted Close, Sunderland SR3 2RF
M0 NGC C. Austin 12 Laburnum Grove, Newbury RG14 1LF
M0 NGH C. Ng 45 Langdown Road, Southampton SO45 6EX
M0 NGI P. Strachan-Buckley 9 Short Street, Aldershot GU11 1HA
M0 NGK K. Charazinski 114 Marmet Avenue, Letchworth Garden City SG6 4QF
M0 NGL N. Nash Roann, Bedmond Road, Pimlico, Hemel Hempstead HP3 8SH
M0 NGN N. Green 44 Rushyford Drive, Chilton, Ferryhill DL17 0EQ
M0 NGS Brede Steam ARS c/o R. Tickle 5 Bramley Court, Harrold, Bedford MK43 7BG
M0 NGY J. Fautley 71 Pullman Lane, Godalming GU7 1YB
M0 NGZ S. Lawrance 69 Athelstan Gardens, Wickford SS11 7EF
M0 NHK Newbury And District Hackspace c/o N. Bland 63 Swindon Road, Wroughton, Swindon SN4 9AG
MM0 NHM N. Morris 23 Sedgebank, Sedgebank, Livingston EH54 6HE
M0 NHY M. Pittas 89 Seddon Road, Morden SM4 6ED
M0 NIC N. Bellamy 23 Poskett Way, Charfield, Wotton under Edge GL12 8FF
MI0 NID N.I DXer's Group c/o S. Barnes 191 Marlacoo Road, Portadown, Craigavon BT62 3TD
M0 NIE B. Niewiadomski 41 The Crescent, Keresley End, Coventry CV7 8LB
M0 NIF G. Calder 41 Wood End Way, Chandler'S Ford, Eastleigh SO53 4LN
M0 NIG N. Howe 45 Kettering Road, Islip, Kettering NN14 3JT
M0 NIL Dr R. Blackwell Vikings Hall, Baylham, Ipswich IP6 8JS
M0 NIW N. Wiseman 23 Kingsway, Langley Park, Durham DH7 9TB
M0 NJD J. Davies 35 Semley Road, Hassocks BN6 8PD
M0 NJE N. Eustice 22 Lower Wear Road, Exeter EX2 7BQ
M0 NJI N. Isherwood 41 Livingstone Road, Blackburn BB2 6NE
M0 NJJ N. Pipkin 46 Charles Avenue, Albrighton, Wolverhampton WV7 3LF
M0 NJP N. Pettefar 44 Duck Lane, Laverstock, Salisbury SP1 1PU
MM0 NJS N. Sheridan Cemetery Lodge, Lochmaben, Lockerbie DG11 1RL
M0 NJW N. Wears 25 Topcliffe Mews, Morley, Morley LS27 8UL
M0 NJX Dr M. Nassau 4A London Road, Liphook GU30 7AN
M0 NKE N. Yorke 30 Bramdene Avenue, Nuneaton CV10 0DH
M0 NKR A. Goldsmith 61 Fengate Drove, Weeting, Brandon IP27 0PW
M0 NKS B. Maggs 44 Coldharbour Road, Hungerford RG17 0AZ
MD0 NKX D. Wilson 23 Snugborough Avenue, Union Mills, Braddan, Isle of Man IM4 4LT
M0 NKY J. Atkinson 17 Agricola Gardens Hadrian Park, Wallsend NE28 9RX
M0 NLI Sos Radio Group c/o D. Hughes 86 Colinmander Gardens, Ormskirk L39 4TF
M0 NLP C. Bowman 26 Albany Hill, Tunbridge Wells TN2 3RX
M0 NLR Dr A. Clark 3 North Street, Owston Ferry, Doncaster DN9 1RT
M0 NLT P. Austin Bush Farmhouse Clee St. Margaret, Craven Arms SY7 9DT
M0 NLW East Anglian Six Meter Group c/o P. Williams 2 Sycamore Avenue, Newton-le-Willows WA12 8LT
MI0 NLY S. Carlin 9 Mullandra Park, Kilcoo, Newry BT34 5LS
M0 NMA S. Marsh 31A Broad Street, Stamford PE9 1PJ
M0 NMC N. Mcintyre 27 Chapel Close, St Ann'S Chapel, Gunnislake PL18 9JB
M0 NMD N. Davison 1 Retford Close Breadsall Estate, Derby DE21 4DX
M0 NMH N. Hilton 20 Darbyshire Close, Deeping St. James, Peterborough PE6 8SF
M0 NMI D. Blake Pound Farm, Swan Lane, Leigh, Swindon SN6 6RD
M0 NMM I. Ezhov 133A Birdwood Road, Cambridge CB1 3TB
M0 NMO RAF Coningsby ARC c/o K. Nice 19 Southill Road, Poole BH12 3AW
MM0 NNA Dr V. Vyshemirsky 2103 Great Western Road, Glasgow G13 2XX
M0 NNB A. Hopper 7 Holmesdale Villas, Swallow Lane, Dorking RH5 4EY
M0 NNE D. Hanwell 28 Chipperfield Road, Norwich NR7 9RR
M0 NNH G. Bansil 15 Abington Close, Crewe CW13TL
M0 NNL Dr N. Lutte Oak House, Brandy Hole Lane, Chichester PO19 5RX
MW0 NNX R. Cotterell 49 Graham Court, Caerphilly CF83 1RF
M0 NOA A. De Broise 113 Conisbrough Grove, Leeds LS25 2QB
M0 NOC P. Bolton 1 Acorn Rise, Hollesley, Woodbridge IP12 3JT
M0 NOE M. Hodgson 10A Myrtle Grove, Enfield EN2 0DZ
M0 NOI Key + Wire ARG c/o C. Birkin 16 Marystow Close, Allesley, Coventry CV5 9EA
M0 NOK B. Handley 68 Northfield Avenue, Rothwell, Leeds LS260SW
M0 NOL M. Hodgkinson 116 Bidwell Hill, Houghton Regis, Dunstable LU5 5EP
M0 NOM M. Wickens Haven Lea, Queens Drive, Windermere LA23 2EL
MI0 NOR N. Mckee 54 Castlemore Park, Belfast BT6 9RP
M0 NOS A. Hayward 54 Eastern Avenue, Mitcheldean GL17 0DF
M0 NOV E. Lane 50 Oakhurst Close, Belper DE56 2TR
M0 NOW N. Walker 79 Arklow Drive, Hale Village, Liverpool L24 5RR
M0 NOY D. Noyek 34 The Spinney, Sidcup DA14 5NF
M0 NOZ J. Norrington 32 Fulfen Way, Saffron Walden CB11 4DW
M0 NPA N. Aleksander 3 Elm Walk, London NW3 7UP
M0 NPB N. Prater 100 Pitfold Road, London SE12 9HY
MW0 NPC Letterston ARC c/o H. Bancroft Stop And Call, Goodwick SA64 0EX
M0 NPD N. Du Pre Honeysuckle, Donald Way, Winchelsea Beach, Winchelsea TN36 4HF
M0 NPG North Pennines Radio Group c/o B. Lenton 32 Forstersteads, Allendale NE47 9AS
M0 NPH N. Holden Plum Cottage, Avon Dassett, Southam CV47 2AP
M0 NPL N. Livingstone 5 Drayton Court, High Street, Polesworth, Tamworth B78 1EX
M0 NPQ Northampton RC c/o N. Ubonis 8 Burleigh Close, Great Yarmouth NR30 2RU
MI0 NPR N. Prentice 26 Claranagh Road, Claranagh, Enniskillen BT94 3FJ
M0 NQB N. Berrie 12 Packenham Road, Basingstoke RG21 8XT
M0 NQU E. Wagner 3 Sarre Road, London NW2 3SN
MM0 NQY P. Davis 4 Daisy Park, Baltasound, Unst, Shetland ZE2 9EA

M0 NRC Newton Le Willows ARC c/o K. Horsfield 59 Queens Drive, Newton-le-Willows WA12 0LY
M0 NRG N. Grigsby 67 Abshot Road, Titchfield Common, Fareham PO14 4NB
M0 NRH N. Hickson 27 Cressing Road, Witham CM8 2NP
M0 NRJ N. Johnson Belair, Western Road, Crediton EX17 3NB
M0 NRK E. Musselle 2 Rectory Crescent, Middle Barton, Chipping Norton OX7 7BP
M0 NRP A. Andrew 80 Hamble Drive, Abingdon OX14 3TE
M0 NRS N. Stoker 11 Hewley Crescent, Throckley, Newcastle upon Tyne NE15 9AT
M0 NRW M. Reeve 9 Kingfisher Walk, Loddon, Norwich NR14 6FB
M0 NRY L. Brackstone 276 Ladyshot, Harlow CM20 3EY
MW0 NSC Neath & District Sea Cadet Unit c/o J. Mason 2 Golwg-Y-Bryn, Off Woodland Road, Skewen, Neath SA10 6SP
M0 NSI B. Taylor 15 Gledhall Street, Stalybridge SK15 1LE
M0 NSK N. Skermer 17 Grange Farm Close, Sutton in Ashfield NG17 1NJ
M0 NSP G. Jurgaitis 70B Ingleby Road, Ilford IG1 4RY
M0 NSR Norfolk Scout Radio c/o C. Rolph 21 South End Hogsthorpe, Chapel St. Leonard PE245NE
MD0 NSS N. Smith 4 Cooil Farrane, Douglas IM2 1NX
M0 NTA T. Brookes 94 Newton Road, Lowton, Warrington WA3 1DG
M0 NTC G. Bull 9 Kilburn Place, Dudley DY2 8HP
M0 NTG Simpson ARS c/o J. Spurgeon Whitgift House, Whitgift, Goole DN14 8HL
M0 NTH D. Higgs 4 Rowsley Road, Stretford, Manchester M32 9QA
M0 NTI T. Ward 173-175 Station Road, Pendlebury, Swinton, Manchester M27 6BU
M0 NTK J. Carrington 15 Astley Court, Newcastle upon Tyne NE12 6YR
M0 NTN N. Norris 6 Tell Grove, London SE22 8RH
M0 NTT J. Naylor 15 Cawder Road, Skipton BD23 2QE
M0 NTY C. Shane 21 Avon Walk, Leighton Buzzard LU7 3DE
M0 NTZ M. Montgomery 4 Merryhill Country Park, Telegraph Hill, Norwich NR9 5AT
M0 NUC Brede Steam ARS c/o D. Adkin 31 Fieldway, Broad Oak, Rye TN31 6DL
M0 NUD N. Cull 8 Eaton Road, Norwich NR4 6PY
M0 NUG N. Lewis 81 Long Lane, Upton, Chester CH2 1JG
MM0 NUO Dr C. Brown 4 Damselfly View, Edinburgh EH17 8XH
M0 NUT D. Mainwaring 1 Buckingham Close, Didcot OX11 8TX
M0 NUX J. Horn 8 Princess Close, Watton IP25 6XA
M0 NUZ A. Charlton 26 Saundergate Lane, Wyberton, Boston PE21 7BZ
M0 NVJ C. Drury 129 Greenhill Road, Mossley Hill, Liverpool L18 7HQ
M0 NVK E. Ridoutt The Bungalow, Main Road, Porchfield, Newport PO30 4LP
M0 NVQ R. Lynch 2 Launceston Close, Oldham OL8 2XE
M0 NVT N. Tennant 22 The Lizard, Wymondham NR18 9BH
MW0 NVY Mid Sussex ARS c/o W. Oliver Pwllmeyric, Chepstow NP16 6LE
MI0 NWA J. Baker 324 Clonmeen, Drumgor, Craigavon BT65 4AT
M0 NWC North West ARC c/o J. Mcinnes-Boylan 10 Faith Street, Leigh WN7 4TS
M0 NWE A. Wood 85 Love Lane, Rayleigh SS6 7DX
MI0 NWG North West ARC c/o D. Keys 71 Madison Avenue, Eglinton, Londonderry BT47 3PW
M0 NWI P. Martin 5 Shropshire Drive, Wilpshire, Blackburn. BB1 9NF
MW0 NWJ Dr N. Jones 54 Glanrhyd, Coed Eva, Cwmbran NP44 6TY
MW0 NWM S. Taylor 43 Toronnen, Bangor LL57 4TG
MI0 NWO E.C A.R.C. c/o D. Adams 65 Rose Park, Limavady BT49 0BF
M0 NWT J. Turner 2 South Drive, Padiham, Burnley BB12 8SH
M0 NWW N. Warner 12 Bay Road, Harwich CO12 3JZ
M0 NWY S. Newhouse 28 Hillmorton Lane, Lilbourne, Rugby CV23 0SS
M0 NXA D. Pounder 15 Eldon Grove, Hartlepool TS26 9LY
M0 NXB N. Beck 2 Killerton View, Silverton, Exeter EX5 4JZ
M0 NXD N. Downes 17 Knightswood Close, Rosliston, Swadlincote DE12 8JJ
M0 NXF W. Rowland 5 Gwinear Downs, Leedstown, Hayle TR27 6DJ
M0 NXP M. Whitaker 5 Horns Drove, Rownhams, Southampton SO16 8AH
MI0 NYC J. Murdie 9 Henderson Park, Bangor BT19 1NS
M0 NYG N. Cox 182 North Tenth Street, Milton Keynes MK9 3AY
M0 NYM Guisborough & District ARC c/o R. Dutton 3 Kilkenny Road, Guisborough TS14 7LE
M0 NYP S. Vzor 40 Henlow Road, Birmingham B14 5DS
M0 NYX Dr J. Hyde The Grove, 7 Mill Lane, Kidderminster DY10 3ND
M0 NYY N. Woodruffe 139 North Home Road, Cirencester GL7 1DY
M0 NZA V. Vesma Durvale, 5 Jonas Drive, Wadhurst TN5 6RJ
M0 NZL D. Foster 42 Cranborne Avenue, Eastbourne BN20 7TT
M0 NZR D. Bond 4 Alfred Road, Haydock, St. Helens WA11 0QD
M0 OAA Ashtead RC c/o J. Chatterton 6 Bayliss Road, Wargrave, Reading RG10 8DR
M0 OAB B. Hodson 176 Carters Mead, Harlow CM17 9EU
M0 OAC Harwell ARS c/o D. Ion 78 Blackmore Street, Derby DE23 8AX
M0 OAD S. Davis 74 The Driveway, Canvey Island SS8 0AD
M0 OAE Q. Wright 9 Browning Avenue, Warwick CV34 6JQ
M0 OAJ A. Johnson 19 Meadow Vale, Bristol BS5 7RG
M0 OAL A. Weller 169 Ashgate Road, Chesterfield S40 4AN
M0 OAT G. Walker 2 Cliffe Bank Cottages, Piercebridge DL2 3SX
M0 OAU L. Boylan 30 Pembrey Way, Liverpool L25 9SN
MI0 OBC D. Best 13 Cranley Green, Bangor BT19 7FE
M0 OBD M. Hopkins 27 Girtford Crescent, Sandy SG19 1HR
MI0 OBE J. Watt 23 Riverview Park, Ballymoney BT53 7QS
M0 OBK D. Cull 6 Compass Way, Bromsgrove B60 3GP
M0 OBL M. Orbell 21 Reedings Road, Barrowby, Grantham NG32 1AU
M0 OBM D. Peacock 41 Oxford Meadow, Sible Hedingham, Halstead CO9 3QW
MI0 OBR A. Savage 469 Old Belfast Road, Bangor BT19 1RQ
MM0 OBT R. Hutcheon 12 Denbecan, Alloa FK10 1QZ
M0 OBU J. Haddleton 3, Sawley Avenue, Whitefield M45 8PP
M0 OBW D. Wilson 12 New Street, Elworth, Sandbach, CW11 3JF
M0 OBY D. Clavey 32 Apollo Close, Dunstable LU5 4AQ
M0 OBZ J. Mcinnes-Boylan 10 Faith Street, Leigh WN7 4TS
M0 OCC Oxo Contest Club c/o C. Wilmott 60 Church Hill, Royston, Barnsley S71 4NG
M0 OCE P. Clarke 14 Wellfield Road, Culcheth, Warrington WA3 4JP
M0 OCJ C. Overson Studio Flat, 6 Grenville Street, Bideford EX39 2EA
M0 OCK A. Mock 60 Effra Road Wimbledon, London SW19 8PP
M0 OCL L. Hendry 109 Grove Avenue, New Costessey, Norwich NR5 0HZ
M0 OCM I. Johnson 9 Brook Road, Pontesbury, Shrewsbury SY5 0QZ
M0 OCP D. Drynski Flat 22, Nicholas Court, Corney Reach Way, Chiswick, London W4 2TS
M0 OCV N. Powis 24 Rosemullion Close, Exhall, Coventry CV7 9NQ
M0 ODB G. Rodriguez 32 Mount Pleasant, Prestwich, Manchester M25 2SD
M0 ODD I. Rotheram 60 Whitewood Park, Liverpool L9 7LG
M0 ODE S. Hawkins Forest Edge, Deer Park, Milton Abbas, Blandford Forum DT11 0AY
MM0 ODI R. Kelly 11 Kelvin Drive, Chryston, Glasgow G69 0LZ
MM0 ODL F. Gordon Crofts Of Torrancroy, Strathdon AB36 8UJ
M0 ODM Preston ARS c/o D. Merridale The Granary, Falledge Lane, Upper Denby HD8 8YH
MW0 ODQ Ci Bach DX Group c/o J. Williams Cartref, Capel Garmon, Llanrwst LL26 0RG
M0 ODS D. Eccles 123 New Inn Lane, Stoke-on-Trent ST4 8HA
MM0 ODX R. Johnstone 15 Barassiebank Lane, Troon KA10 6SH
M0 ODZ G. Fenton 40 High Street, Easington Lane, Houghton le Spring DH5 0JN
M0 OEB C. Brennan 19 The Furrow, Littleport, Ely CB6 1GL
M0 OED G. Marfell Springfields Bungalow, Drybrrok GL17 9BW
M0 OEG S. Finch 25 Bluebell Avenue, Wigan WN6 8NS
M0 OEK C. Taylor 51 Barnes Close, Sturminster Newton DT10 1BN
M0 OER Dr M. Cianni 121 Springfield Park Avenue, Chelmsford CM2 6EW
M0 OES I. Jones 33 Cobham Avenue, Liverpool L9 3BP
M0 OET B. Mead 8 Wordsworth Road, Kettering NN16 9LB
M0 OFE H. Bond Flat 4, Athrington Court, First Avenue, Felpham, Bognor Regis PO22 7LB
M0 OFF R. Withers 50 Coneygear Road, Hartford, Huntingdon PE29 1QL
M0 OFL D. Gunn 40 The Pastures, Oadby, Leicester LE2 4QD
M0 OFM P. Joyce 2 Harold Road, Cuxton, Rochester ME2 1EE
M0 OGD G. Davies 6 Bayleys Close, Empingham, Oakham LE15 8PJ
M0 OGI M. Shopland 128 Whitewood Park, Liverpool L9 7LG
M0 OGS M. Jordan 10 Wilmot Green, Great Warley, Brentwood CM13 3DD
M0 OGX K. Fujita 3-21 Denenchofu-Honcho, Ota-Ku, Tokyo 1450072 Japan
M0 OGZ G. Singleton 2 Rome Avenue, Burnley BB115LQ
M0 OHA Ormiston Horizon Academy RC c/o L. Preece 29 Elliott Street, Newcastle under Lyme ST5 1JL
M0 OHI G. Chaffey 63 Underwood Road, Eastleigh SO50 6FX
M0 OIA P. Rutkowski 17 Beaminster Gardens, Ilford IG6 2BN
M0 OIC B. Downes 6 Greenland Crescent, Beeston, Nottingham NG9 5LB
MI0 OIM M. Edwards 58 Rosemount Park, Jordanstown BT37 0NL
M0 OIO R. Hubbard Southbroom School House, Estcourt Street, Devizes SN10 1LW
M0 OJC C. Curry 41 Bargate, Richmond DL10 4QY
M0 OJG J. Canning The Mount, Birmingham Road, Alcester B49 5EG
MM0 OJJ J. Jarvie Berryhill Farm, Tak-Ma-Doon Road, Kilsyth, Glasgow G65 0RY
M0 OJO N. Hudson Woodpecker Cottage Red Lane, Aldermaston, Reading RG7 4PA
M0 OJS O. Spurway 6 Alder Grove, Crewkerne TA18 7DJ
M0 OJX T. Yamamoto 1141-5, Kozukue, Kouhoku, Kanagawa 222-0036 Japan
M0 OKB B. Bailur 27 Russell Road, Felixstowe IP11 2BG
M0 OKD D. Shelsher Gables, Colchester Road, Ardleigh, Colchester CO7 7PQ
MM0 OKG Dr J. Bowes 1 Greendyke Cottage, Falkirk FK2 8PP
M0 OKK G. Cooper Holmfield, Chelmorton, Buxton SK17 9SG
M0 OKL A. Gates 46 Gloucester Place, Littlehampton BN17 7AL
M0 OKQ M. Broum 137 Culvers Avenue, Carshalton SM5 2BA
M0 OKS S. Mookerjee 6 Dipper Drive, Altrincham WA14 5YF
M0 OKT Flight Refuelling ARS c/o C. Law 23 Yeldersley Close, Chesterfield S40 4LG
MM0 OKY J. Williamson Clunie Cottage, Tullibardine Road, Auchterarder PH3 1LX
M0 OLD S. Old Firtrees, Main Street, Scarborough YO11 3UD
MW0 OLE O. Thomas Garth Celyn, St. Davids Road, Aberystwyth SY23 1EU
M0 OLG M. Hirst 69 Potton Road, St. Neots PE19 2NN
M0 OLM G. Dewey West Riding Five Ash Down, Uckfield TN22 3AP
M0 OLS A. Lee 1 Lower Stoke, Limpley Stoke, Bath BA2 7FU
M0 OLW Rvd. N. Cooper Frindsbury Vicarage, 4 Parsonage Lane, Rochester ME2 4UR
M0 OMA Dr M. Hocking 52 Ashbourne Drive, Newcastle ST5 6RL
MW0 OMB R. Rimmer Dwyfor, Heol Las, Llantrisant, Pontyclun CF72 8EG
M0 OMC Holsworthy ARC c/o D. Roomes View Field, Milton Damerel, Holsworthy EX22 7NY
M0 OMD D. Haigh 80 Saddlers Road, Quedgeley, Gloucester GL2 4SY
MM0 OMG G. Robinson Asgard, 12 Upper Waston Road, Burray KW172TT
M0 OMI J. Jones 1 Knebworth Road, Bexhill-on-Sea TN39 4JH
MW0 OMK C. Morris Hideaway Bettws Cedewain, Newtown, Powys SY16 3DS
MM0 OMS M. Scullion 24 Langmuir Road, Kirkintilloch, Glasgow G66 2QE
M0 OMT S. Thompson 30 Southport Parade, Hebburn NE31 2AQ
M0 OMV J. Barton 37 Lytton Road, Sheffield S5 8AX
MW0 OMZ N. Shepherd Prospect, Newchapel, Llanidloes SY18 6JY
M0 ONB Eaaro ARC c/o J. Williams 66 Oakfield Avenue, Hitchin SG4 9JD
M0 OND J. Isherwood 11 Manor Crescent, Chesterfield S40 1HU
M0 ONH A. Brown 51 Towncroft, Chelmsford CM1 4JX
M0 ONI G. Swift 43 Storth Lane, Kiveton Park, Sheffield S26 5QS
M0 ONL the Online RC c/o A. Amos Willow Tree House, Deers Green, Clavering, Saffron Walden CB11 4PX
M0 ONO R. Vermeulen 7 Butterfly Crescent, Evesham WR11 1BP
M0 ONQ A. Richardson The Chalet, Lincoln Road, Lincoln LN4 2EX
M0 ONS C. Stone 26 Chesham Road North, Weston-Super-Mare BS22 8AD
MM0 ONX R. Walker 10 Westpark Gate, Saline, Dunfermline KY12 9US
M0 ONY P. Bown 19 Victory Villas, Hatherop Road, Fairford GL7 4JU
M0 ONZ A. Cross 12 Appleby Drive, Langdon Hills, Basildon SS16 6NU
M0 OOD Capt. J. Newman Reeds, The Street, Cranbrook TN17 4DB
MM0 OOF A. Bennett 8 Buckstone Loan East, Edinburgh EH10 6XD
MD0 OOH A. Elliott Round Table House, Ronague, Castletown, Isle of Man IM9 4HJ
MM0 OOI H. Chen Iq Fountainbridge Room 118E 114 Dundee Street, Edinburgh EH11 1AD
M0 OOO A. White 1 The Red House, Old Gallamore Lane, Market Rasen LN8 3US
M0 OOR D. Thomas 8 Cedar Avenue, Weston-Super-Mare BS22 8HL
M0 OOS A. Norton 9 The Common, West Tytherley, Salisbury SP5 1NS
M0 OOT Obo Jubilee Sailing Trust (ARS) c/o J. Logan Whetstead, Grange Road, Gillingham ME7 2UN
M0 OOW G. Cameron 33 Railway Terrace, York YO24 4BN
M0 OOZ Sir G. Cowne 8 Arthur Road, New Malden KT36LX
MI0 OPC Oville Amateur Radio Portable Club c/o H. Evans Oville House 404 Foreglen Road, Dungiven, Londonderry BT47 4PN
M0 OPG O. Griffiths 272 Worcester Road, Malvern WR14 1BD
M0 OPK P. Kirby 11 Bembridge Court, Crowthorne RG45 6BN
M0 OPL P. Lounton 107 Browning Hill Coxhoe, Durham DH6 4SA
MI0 OPM D. Kirkwood 1 Rural Cottages, Front Road, Lisburn BT27 5LF
MW0 OPS H. Willott 14 Warwick Close, Chepstow NP16 5BU

MW0 OPY D. Pingel 7 Duffryn Close Bassaleg, Newport NP10 8PD
M0 OQO G. Brindle 25 Chedworth Drive, Witney OX28 5FS
MM0 OQR P. Taylor 2 Laurel Grove, Aberdeen AB22 8YJ
M0 ORC J. King Plum Tree Cottage, Royston Place, Barton On Sea, New Milton BH25 7AJ
M0 ORE G. Moore 2 Spinacre, Barton On Sea, New Milton BH25 7DF
M0 ORI D. Dart 164A Salterton Road, Exmouth EX8 2PA
MM0 ORK 59 Degrees North ARG c/o E. Holt Ashwell, St. Ola, Kirkwall KW15 1SX
M0 ORM Quantum Amateur Radio & Technology Society c/o D. Hughes 86 Colinmander Gardens, Ormskirk L39 4TF
M0 ORN S. Pafrey 75 New Queen St., Bristol BS15 1DE
M0 ORR C. Hale 24 Wolverhampton Road, Kidderminster DY10 2UT
M0 ORS D. Holman 38 Polyear Close, Polgooth, St. Austell PL26 7BH
M0 ORZ Dr H. Umemoto Hashiba 2-16-15-308, Taito-Ku 111-0023 Japan
M0 OSB G. Webster 15 Bridge Road, Chichester PO19 7NW
M0 OSE R. Allan 44 Elderdene, Chinnor OX39 4EJ
M0 OSH W. Rogalski 22 Sadler Road, Walsall WS8 6BG
M0 OSI S. Haigh 17 Glebe Street, Swadlincote DE11 9BW
M0 OSL S. Latimer 40 Petersham Road, Long Eaton, Nottingham NG10 4DD
M0 OSM C. Penfold 14 Romney Road, Tetbury GL8 8JU
M0 OSO C. Dinu Apartment 8, 16 Abbey Road, London NW8 9GS
M0 OSX A. Logan 23 Cherry Tree Rise, Walkern, Stevenage SG2 7JL
M0 OSY S. Houssart Flat 3, Virginia Court, London SE16 6PU
M0 OTA G. Hutchinson 128 Crescent Drive North, Woodingdean, Brighton BN2 6SF
MI0 OTC B. Emerson 67 Castlemore Avenue, Belfast BT6 9RH
M0 OTE D. Barlow 21 Yellow Brook Close, Aspull, Wigan WN2 1ZH
MW0 OTG D. James Coombe House, Coombe, Presteigne LD8 2HL
MW0 OTH A. Hodgson Browns Holiday Park Towyn Road, Conwy LL22 9HD
M0 OTJ Dr M. Ibison 40 Regent Drive, Fulwood, Preston PR2 3JB
M0 OTL G. Humphrey 324 Snarlton Lane, Melksham SN12 7QW
M0 OTO M. Pesendorfer 13 Blake Road, London N11 2AD
M0 OTS Pontefract + District ARS c/o R. Bateman 81 Stanton St., Derby DE23 6NF
M0 OTT Dr C. Darby Brookfield, Forest Green, Dorking RH5 5SG
MW0 OUC J. Bidwell 26 Lone Road, Clydach, Swansea SA6 5HR
M0 OUS J. Holley Lookers, Blenheim Road Littlestone, New Romney TN28 8PR
MI0 OUT P. Gibson 118 Coleraine Road, Portstewart BT55 7HS
MM0 OUU C. Mcconnochie 72 Duddingston Avenue, Kilwinning KA13 6RS
M0 OVB Bolton Wireless Club c/o R. Gowers 43 Tungstone Way, Market Harborough LE16 9GA
MM0 OVD D. Adamson 5 Central Quadrant, Ardrossan KA22 7DY
M0 OVI O. Popa 5 Lanark Close, Horsham RH13 5RY
MM0 OVK M. Mcallister 36 Girvan Crescent, Newmilns KA16 9HZ
MM0 OVV A. Bernard 200 Carden Avenue, Cardenden, Lochgelly KY5 0EN
M0 OVW G. Sandell 1 St. Margaret Road, Ludlow SY8 1XN
M0 OWG D. Burgfeld 20 Wilson Row, Crowthorne RG45 6WE
MM0 OWL C. Barclay 57 Belhaven Park, Muirhead, Glasgow G69 9FA
MW0 OWM Capt. R. Cooke 6 The Forestry, Trecastle, Brecon LD3 8YA
M0 OWO A. Roworth 17 Davidson Avenue, Congleton CW12 2EQ
M0 OWS P. Stocks 12 Bredbury Drive Farnworth, Bolton BL4 7QD
M0 OXD C. Romocea 21 Hurst Lane, Cumnor, Oxford OX2 9PR
M0 OXO C. Wilmott 60 Church Hill, Royston, Barnsley S71 4NG
M0 OXR S. Wyatt 55 Ridgefield Road, Oxford OX4 3BX
M0 OXW D. Ellis 4 Vane Terrace, Darlington DL3 7AT
MM0 OXX A. Berry 8 Hill Street Striling Fk7 0Dh, Striling FK7 0DH
M0 OXZ P. Rodley 27 Tollgate Close, Northampton NN2 6RP
M0 OYH C. Staples 32 Browns Lane, Netherton, Bootle L30 5RW
M0 OYQ S. Long 14 Petley Close, Flitwick, Bedford MK45 1XP
M0 OYR S. Roberts 77 Lambwath Road, Hull HU8 0HB
M0 OYZ G. Brierley 35 Ochrewell Avenue, Deighton, Huddersfield HD2 1LL
MW0 OZI C. Osborne Thornleigh, Tremont Road, Llandrindod Wells LD1 5BH
M0 OZJ B. Aicheler Rose Cottage, Chilsworthy, Holsworthy EX22 7BQ
M0 OZO I. Taylor-Hayward Roslyn House, 57 Sheriff Highway, Hedon, Hull HU12 8HA
MM0 OZY D. Leiper 29 Thornton Avenue, Bonnybridge FK4 1AR
M0 PAA Dr P. Thompson 9 Thames Mews, Poole BH15 1JY
M0 PAC P. De Camps 22 Osier Road, Spalding PE11 1UU
M0 PAF G. Batty 85 Cobcar Lane, Elsecar, Barnsley S74 8BW
M0 PAG T. Pagden 199 Woad Farm Road, Boston PE21 0EN
M0 PAI A. Dodd 14 Davies Street, Macclesfield SK10 1GE
M0 PAJ P. Alley 21 Orchard Court, Thorney PE6 0QW
M0 PAL B. Beck 21 Winston Grove, Retford DN22 6SQ
M0 PAM A. Martins 32 Godwin Road, Canterbury CT1 3UF
M0 PAO P. Mcfadden Maple Cottage, Leighton Buzzard LU7 9DZ
M0 PAQ Worksop Amateur RS c/o P. Meerman 24 Horseshoe Crescent, Burghfield Common, Reading RG7 3XW
M0 PAR A. Holland 18 Mason Close, Malvern WR14 2NF
M0 PAV S. Richards 18 Lowfields Staxton, Scarborough YO12 4SR
M0 PAW K. Pawley 24 Chatsworth Road, Torquay TQ1 3BL
M0 PAX Mexborough & District ARS c/o G. Levine 65 Clitheroe Road, Romford RM5 2SL
M0 PAY J. Houghton 7 West View, Skeffling, Hull HU12 0US
MM0 PAZ S. Mckinnon 8 Rowanlea Avenue, Paisley PA2 0RP
M0 PBB P. Burt 56 Winslade Road, Sidmouth EX10 9EX
M0 PBD C. Smith 21 Earl Spencer Court, Peterborough PE2 9PQ
M0 PBF P. Bone 11 Fox Hill Drive, Stalybridge SK15 2RP
M0 PBN P. Biggin Galadean, Farriers Way, Newport PO30 3JP
M0 PBO M. Waistell 23 Halton Court, Sheffield S12 4ND
M0 PBR P. Rawlinson 15 Elmbourne Drive, Belvedere DA17 6JE
M0 PBT Essex CW Contest Club c/o P. Burgess 61 Grosvenor Avenue, Torquay TQ2 7JX
M0 PBX P. Batson 71 North Parade, Falmouth TR11 2TE
M0 PBZ P. Bond 16 Little Avenue, Swindon SN2 1NL
M0 PCA P. Asbury 67 Orchard Way, Measham, Swadlincote DE12 7JZ
M0 PCB I. Kelly 261 Bodiam Avenue, Tuffley, Gloucester GL4 0XW
M0 PCE P. Crema Not, Applicable, Resides OUTSIDE THE UK IN Italy
M0 PCG P. Gay Ryecroft, Finningham Road, Rickinghall, Diss IP22 1LT
M0 PCH C. Morgan 28 Tewther Road, Bristol BS13 0NL
MI0 PCJ P. Hume 2 Seabourne Parade, Belfast BT15 3NP
M0 PCK P. Clay Sylvesterweg 35, Viktring 9073 Austria
M0 PCN Pelican Radio Group c/o D. Perry 11 St. Lawrence Close, Stratford Sub Castle, Salisbury SP1 3LW
M0 PCO P. James 44 Narbonne Avenue Ellesmere Park Eccles, Manchester M30 9DL
M0 PCR P. Rudd 27 Loxwood Avenue, Worthing BN14 7QY
M0 PCS P. Sefton 27 Donovan Avenue, London N10 2JU
MW0 PCT Bournemouth RS c/o S. Gau Disgwylfa, The Downs, Cardiff CF5 6SB
MI0 PCW Dr R. Bishop 2 Alexander Park, Carrickfergus BT38 7LL
M0 PCX P. Chronopoulos Flat 11, Eton Hall, London NW3 2DW
M0 PCZ P. Colyer 23 Florida Road, Torquay TQ1 1JY
M0 PDA P. Stallibrass 12 Sheerwater Close, Bury St. Edmunds IP32 7HR
M0 PDB B. Beed 72 Looseleigh Lane, Plymouth PL6 5HH
M0 PDC P. Collins 59 Portman Road, Scunthorpe DN15 8PE
MM0 PDD S. Burnside Woodend Farm, Buchlyvie, Stirling FK8 3PD
M0 PDE Dr C. Clark 2 Orchard Close, Elmstead, Colchester CO7 7AS
M0 PDF Dr A. Clark 2 Orchard Close, Elmstead Market, Colchester CO7 7AS
M0 PDG J. Nicholls 2 Karen Rise, Arnold, Nottingham NG5 8GE
M0 PDH P. Hardwick 2 Cliffe Cottages, Sandy Lane, Rake GU33 7JE
M0 PDL S. Symonds 301 North Fairlee Farm, Fairlee Road, Newport PO30 2JU
M0 PDN N. Dinsdale 3 Pond Cottages, Faulkland, Radstock BA3 5XB
M0 PDP S. Martin 77 Chatford Drive, Shrewsbury SY3 9PH
M0 PDQ C. Almey 152 Queensgate, Bridlington YO16 6RW
MW0 PDR P. Randall 24 Ffordd-Y-Goedwig, Pyle, Bridgend CF33 6HY
M0 PDS P. Schoenmaker 24 Greenheys Drive, London E18 2HB
M0 PDU L. Fuller Rosemar Lodge Westford, Wellington TA21 0DX
MW0 PDV P. Devlin Brynteg, Fron Bache, Llangollen LL20 7BP
M0 PDW P. Whiteley Grantham House, Grantham Road, Halifax HX3 6PL
M0 PDX N. Pagdin 74 Thelwall New Road, Thelwall, Warrington WA4 2HY
M0 PDY P. Dyer 19 Church Road, Evesham WR11 2NE
M0 PDZ P. Harper 36 Barrow Close, Marlborough SN8 2BD
M0 PEA G. Pearson 41 Myrica Grove, Hoole, Chester CH2 3EW
M0 PEB P. Burke 38 Bosworth Square, Rochdale OL11 3QG
M0 PEG P. Grainger 36 Orchard Road, Wigton CA7 9JL
MW0 PEH B. Sellers 86 St. John Street, Ogmore Vale CF32 7BB
M0 PEI I. Daraban 10 Holly Blue Close, Little Paxton, Saint Neots PE19 6TD
M0 PEM C. Rawlin 5 Japonica Hill, Immingham DN40 1LT
M0 PEQ S. Grammenos Flat 2B, Chelverton Road, Putney, London SW15 1RH
M0 PER A. Perkins 3 Intake Close Willaston, Neston CH64 2XG
M0 PES R. Blacker 1 Ashwindham Court, Woking GU21 8AW
M0 PET P. Gough 7 The Crossings, Cannock WS110EZ
M0 PEW P. Woolley 84 Bowthorpe Road, Norwich NR2 3TP
M0 PEX D. Munn 36 Moor Lea, Braunton EX33 2PE
M0 PFC Coalvile RC c/o D. Poulton 93 Pretoria Road, Ibstock LE67 6LP
M0 PFF G. Peters Curlew Court, Guys Head Road, Sutton Bridge, Spalding PE12 9QQ
MM0 PFH Pentland Firth Radio Hams c/o D. Morrison 4 West Murkle, Murkle, Thurso KW14 8YT
M0 PFJ Splitters c/o K. Baker 64 Pendle Drive, Basildon SS14 3LZ
MM0 PFN P. Darmady Breckster Upper Camster, Lybster KW3 6BD
M0 PFO P. Noble 14 Park Street, Swallownest, Sheffield S26 4UP
M0 PFT A. Perfect 3 Chelmarsh Close, Chellaston, Derby DE73 6PB
M0 PFW P. Borer 88 Beechings Way, Gillingham ME8 6LX
M0 PGC P. Corley 90 Hill Road, Benfleet SS7 1AL
M0 PGD P. Dann 14 Sedgebourne Way, Northfield, Birmingham B31 5HQ
M0 PGH UK Travellers Club c/o G. Hart 11 Sadlers Ride, West Molesey KT8 1SU
M0 PGI G. Hartless 32 Long Acre, Mablethorpe LN12 1JF
M0 PGL A. Md Ali 67 Thorley Lane Timperley, Altrincham WA15 7BA
M0 PGM P. Meadows 6A College Road, Maidenhead SL6 6BE
M0 PGN P. Niewiadomski Flat1 79A Dartmouth Road, London SE23 3HT
M0 PGO S. Mansfield The Old Piggery, Ham Lane, Compton Dundon, Somerton TA11 6PQ
M0 PGS P. Smith The Dingle, 27 Habberley Road, Bewdley DY12 1JH
M0 PGW P. Whiffing 38 Green Close, Stannington, Morpeth NE61 6PE
M0 PGX P. Graham 28 Newburgh House, Highworth, Swindon SN6 7DW
M0 PHB P. Bartlett 58 Ashdown Road, Chandler'S Ford, Eastleigh SO53 5QJ
MM0 PHD P. Dutton 20/9 Craighall Crescent, Edinburgh EH6 4RZ.
M0 PHL P. Stephens 100A Foxglove Road, Eastbourne BN23 8BX
M0 PHM P. Matthews Sky Reach, Appletree Farm, St.Austell PL268RT
M0 PHN Dr P. Mason 29 Grantham Road, Bristol BS15 1JR
M0 PHO P. Honey 3 Peterswood, Harlow CM18 7RJ
M0 PHP C. Rodway 26 Redesdale Avenue, Newcastle upon Tyne NE3 3PP
M0 PHV Glenrothes & District ARC c/o P. Velzeboer The Highnings, Coneygree Fold, Chipping Campden GL55 6JL
M0 PHX Phoenix Radio Group c/o A. Clayton 6 Albert Road, Bunny, Nottingham NG11 6QE
M0 PIA C. Kolderman 6 Flanders Close, Kemsley, Sittingbourne ME10 2PX
M0 PIB P. Badley 37 Martins Lane, Dorchester-On-Thames, Wallingford OX10 7JE
MW0 PIC R. Miles 63 Phillip Street, Mountain Ash CF45 4BG
M0 PIE R. Cockroft 8 Lumb Lane, Huddersfield HD4 6SZ
M0 PIK B. Pike 19 Cardigan Gardens, Reading RG1 5QP
M0 PIP C. Sidey Inish Mor 19 Greenwood Court, Bideford EX39 3SF
M0 PIR J. Clark 4 Exeter Street, North Tawton EX20 2HB
M0 PIT P. Hayes 4 London Road, Roade, Northampton NN7 2NL
M0 PIX R. Bibby 40 Morval Crescent, Runcorn WA7 2QS
M0 PIZ D. Street 18 Meadow Gardens, Bath BA1 3RY
M0 PJA P. Archer 31 Stoney Bank Drive, Kiveton Park, Sheffield S26 6SJ
M0 PJC P. Crabtree 106 Sagecroft Road, Thatcham RG18 3BF
M0 PJD P. Davies 53 Lammas Road, Cheddington, Leighton Buzzard LU7 0RY
M0 PJF P. Franklin 1 Aberdeen Court, Newcastle upon Tyne NE3 2XU
M0 PJG P. Galer 62 Court Mount, Canterbury Road, Birchington CT7 0BT
MW0 PJJ P. Jones 23 Pinecroft Avenue, Aberdare CF44 0HY
M0 PJK P. Knappett Hope Cottage, The Green, Clacton-on-Sea CO16 0BU
MI0 PJL P. Letters 24 Old Grange Avenue, Carrickfergus BT38 7UE
M0 PJM P. Mcmillan 82 Front Street, Tudhoe Colliery, Spennymoor DL16 6TJ
M0 PJP P. Pearson 22 Norris Street, Darwen BB3 3DR
MM0 PJQ P. Quinn 24 Highfield Avenue, Paisley PA2 8LG
MW0 PJR P. Rees 8 Pencae Terrace, Llanelli SA15 1NZ
MI0 PJS P. Smiley 100 Lislaban Road, Cloughmills, Ballymena BT44 9HZ
M0 PJT P. Tomlinson 11 Haynes Close, Clifton, Nottingham NG11 8JN
M0 PJX D. Dickson 6 St. Johns View, Old Hutton, Kendal LA8 0NG

M0 PJY P. Yarwood 3 Old Blundells Court, Station Road, Tiverton EX16 4LF
M0 PKE P. Walsh 181 Hermes Close, Hull HU9 4DR
M0 PKH P. Halloway 82 Northwall Road, Deal CT14 6PP
M0 PKL the North Norfolk Arg c/o C. Wilson-Shah 42 Glenthorne Road, London N11 3HJ
M0 PKV P. Slade End Cottage, Monyash Road, Bakewell DE45 1FG
M0 PKW P. Watson 10 Whitelands Crescent, Baildon, Shipley BD17 6NN
M0 PLB P. Bromley Broadwood Treovis Upton Cross, Liskeard PL14 5BQ
MI0 PLC R. Thomson 1 Litchfield Park, Coleraine BT51 3TN
M0 PLE C. Pryke 50 Raglan Gardens, Watford WD19 4LL
M0 PLG P. Gyngell 54 Association Walk, Rochester ME1 2XD
M0 PLH P. Hamnett 13 Breakwater Court West, Berry Head Road, Brixham TQ5 9AG
M0 PLN J. Kendrick 29 Waterside, Silsden, Keighley BD20 0LQ
M0 PLO P. Lesiecki 38 Pendock Road, Bristol BS16 2PN
M0 PLP P. Hallson 4 Cranbrook Drive, Esher KT10 8DL
M0 PLR D. White 10 Meaux Road, Wawne, Hull HU7 5XD
M0 PLS J. Walczak 18 Heathfield, Chippenham SN15 1BQ
M0 PLT Coalvile RC c/o G. Myers 6 Ullswater Close, Biggleswade SG18 8LX
M0 PLV P. Le Vallois 14 London Row, Arlesey SG15 6RX
M0 PLW P. Latham 135 Ashgate Road, Chesterfield S40 4AN
M0 PLX J. Telecki 61 Wistaria Road, Wisbech PE13 3RH
M0 PLY R. Austen Holly Tree Cottage, Station Road, Immingham DN40 3AX
MJ0 PMA Dulverton Junior School c/o P. Ahier 1 Mahara Mews, Upper Kings Cliff, La Pouquelaye, St. Helier, Jersey JE2 3GP Jersey
M0 PMC P. Curnow 13A Warden Close, Maidstone ME16 0JL
M0 PMH P. Holmquest 6 Rhyme Hall Mews, Fawley, Southampton SO45 1FX
M0 PMI P. Mccormick Fieldview, Crown East Lane, Lower Broadheath, Worcester WR2 6RH
M0 PMJ P. Mullen 14 Anderson Road, Hemswell Cliff, Gainsborough DN21 5XP
M0 PMM P. Levetsky 9 Moat Walk, Pound Hill, Crawley RH10 7ED
MD0 PMN P. Best 5 The Willows, Ballasalla IM9 2EW Isle of Man
M0 PMO C. Breen 64 Linkstor Road Woolton, Liverpool L25 6DH
M0 PMR S. Macauley 1 Moricambe Crescent, Anthorn, Wigton CA7 5AS
M0 PMV P. Mansfield 27 Popplechurch Drive, Swindon SN3 5DE
MM0 PMW M. Mclauchlan 8 Craigie St., Ballingry, Lochgelly KY5 8NS
MI0 PMX G. Todd 27 Ardreagh Road, Aghadowey, Coleraine BT51 4DN
M0 PMY P. May 33 Minsterley Avenue, Shepperton TW17 8QS
M0 PNA P. Fulbrook 167 Droitwich Road, Fernhill Heath WR3 7TZ
M0 PNB P. Bozikis 336 Higham Hill Road, London E17 5RG
MW0 PNC M. Bloore Halfway House, Hyfrydle Road, Talysarn, Caernarfon LL54 6HG
M0 PNN P. Bowen 12 Powell Place, Newport TF10 7BS
M0 PNZ R. Maddock 48 Collygree Parc, Goldsithney, Penzance TR20 9LY
M0 POA A. Polesel 33 The Maltings, Leighton Buzzard LU7 4BS
MW0 POB J. Lewis 2 Tymaen Crescent, Cwmavon, Port Talbot SA12 9EA
MM0 POD A. Conlon Kilrae, Barrpath, Glasgow G65 0EX
M0 POE A. Chlebikova St. Catharine'S College, Cambridge CB2 1RL
M0 POG Kettering & District ARS c/o P. Wilkes 8 Cloverdale, Stafford ST17 4QJ
M0 POI Grtr Lndn Rayne c/o M. Boland 24 Hallam Close, Moulton, Northampton NN3 7LB
M0 POQ R. Finch 19B Kiln Road, Newbury RG14 2LS
M0 POS G. Postlethwaite 20 Birkett Drive, Ulverston LA12 9LS
M0 PPC Central Radio Amateur Circle c/o M. Hallard 77 Banklands Road, Dudley DY2 8BT
M0 PPF S. Harding Acres Hill, Jacobs Lane, Alhampton BA4 6PZ
M0 PPG G. Beacher 22 Trowbridge Gardens, Luton LU2 7JY
MW0 PPM F. Miers Pentre Isaf, Bryneglwys, Corwen LL21 9NA
MW0 PPO V. Frostick Clawdd Llwyd, Ceunant, Caernarfon LL55 4RR
M0 PPP G. Batty 85 Cobcar Lane, Elsecar, Barnsley S74 8BW
M0 PPR P. Rozenek 10 South Road, Portsmouth PO1 5QT
M0 PPS I. Underwood 12 Forge Lane, Gillingham ME7 1UG
MI0 PPW J. Macfarlane 5 Sunbeam Terrace, Lisnaskea, Lisnaskea BT92 0LL
M0 PPX A. James 36 Westcote Close, Solihull B92 8PL
M0 PPY J. Martin 20 Hall Green Road, West Bromwich B71 3LA
M0 PPZ P. Zanek Mulberry Hill, Violet Lane, Tadley RG26 5JX
M0 PQA P. Casado Arias 24 Oldridge Road, London SW12 8PJ
M0 PQI T. Pearsall 16 Langdale Road, Leyland PR25 3AR
MM0 PQN Tall Trees CG c/o R. Tripney 7 Sunnyside St., Camelon, Falkirk FK1 4BJ
MI0 PQR Greenisland Electronics ARS c/o B. Mckeen 11 Fairymount Terrace Taylors Avenue, Carrickfergus BT38 7HN
M0 PRA D. Runyard Tilecroft, Shortheath Crest, Farnham GU9 8SA
MM0 PRB P. Bacon 12 The Greens, Maddiston, Falkirk FK2 0FN
MW0 PRC P. Randall 24 Ffordd-Y-Goedwig, Pyle, Bridgend CF33 6HY
M0 PRD P. Denham Arlyn House, 10 Prince Alfred Avenue, Skegness PE25 2UH
M0 PRF the Online RC c/o J. Petch-Harrison 13 Church Lane, Shepton Mallet BA4 5LE
MW0 PRI D. Price 2 Heol Tyn-Y-Fron, Penparcau, Aberystwyth SY23 3RP
MI0 PRM E. Simpson 10 Woodview Park, Tandragee, Craigavon BT62 2DD
M0 PRN P. Norman Mellon, Hincaster, Milnthorpe LA7 7ND
MW0 PRO J. White Keepers Lodge Pumpsaint, Llanwrda SA19 8DX
MW0 PRP P. Pugh 27 Bank Street, Tonypandy CF40 1PJ
M0 PRT B. Dey 15 Bradenham Road, Grange Park, Swindon SN5 6EB
M0 PRV D. Parvin 11 Stanhope Way, Sevenoaks TN13 2DZ
M0 PRW P. West Stour House, Church Street, West Stour, Gillingham SP8 5RL
MM0 PSA P. Smith 13 Newmills Grove, Balerno EH14 5SY
M0 PSB J. Bell 8 Firsleigh Park, Roche, St. Austell PL26 8JN
M0 PSC P. Croxford 1 Meteor Close, Bicester OX26 4YA
M0 PSD P. Davies 2 Lynfords Drive, Runwell, Wickford SS11 7PP
M0 PSE P. Glandfield Flat 5, 7 Cargate Avenue, Aldershot GU11 3EP
MW0 PSG 1st Pencoed Scout Group c/o L. Ward 11 Verlands Way, Pencoed, Bridgend CF35 6TY
M0 PSI Dr A. Al-Azzawi 33 Clare Mead, Rowledge, Farnham GU10 4BJ
M0 PSK Dr C. Gibson 1 Ryelands Orchard, Leominster HR6 8QQ
MM0 PSM S. Milne 5 Moriston Court, Grangemouth FK3 0JJ
M0 PSR East Cheshire Radio Group c/o R. Tickle 5 Bramley Court, Harrold, Bedford MK43 7BG
M0 PSS P. Shaw 10 Godbold Road, London E15 3AL
M0 PST P. Stevens 19 Elmfield Place, Newton Aycliffe DL5 7BD
M0 PSW J. Godfrey 4 Cherry Close, Houghton Conquest, Bedford MK45 3LQ
M0 PSY D. Shuttleworth 27 Union St., Egerton, Bolton BL7 9SP
M0 PSZ S. Macdonald Woodside Cottage, Horton Way, Verwood BH31 6JJ
M0 PTA LSWC Portable Group c/o S. Mcbain 13A St. Lukes Close, Cherry Willingham, Lincoln LN3 4LY
M0 PTB P. Singleton 9 Sherbourne Road, Middleton, Manchester M24 6FF
MM0 PTE P. Bainbridge 49 Hare Moss View, Bathgate EH47 0DN
M0 PTG P. Threakall 83 Gregory Avenue, Birmingham B29 5DG
M0 PTO M. Reeve 9 Kingfisher Walk, Loddon, Norwich NR14 6FB
M0 PTR P. Clifford 90 Sherwood Avenue, Poole BH14 8DL
M0 PTS P. Boultwood 32 Makepiece Road, Bracknell RG42 2HJ
M0 PTT R. Winthrop 98 Creighton Avenue, Carlisle CA2 7DW
MM0 PTX A. Morgan Broomrig House, Harviestoun Road, Dollar FK14 7PT
MW0 PTY P. Matthews The Chateau, Wynnstay Hall Estate, Ruabon, Wrexham LL14 6LA
M0 PTZ Prof. P. Curtis Cotswold, Salisbury Road, Abbotts Ann, Andover SP11 7NX
M0 PUC Dr J. Woods Haycocks Farm, Haycocks Lane, Colchester CO5 8SS
M0 PUD A. Eyre St. Michael Mead, The Common, Norwich NR12 8BA
M0 PUH Dr M. Foster 58 New Terrace, Staverton, Trowbridge BA14 6NY
M0 PUK P. Wilson 90 Kenwood Crescent, Ingleby Barwick, Ingelby Barwick TS17 5BS
M0 PUR R. Janezko 10A Lens Road, Allestree, Derby DE22 2NB
M0 PUS B. Beard 8 Monks Close, Newcastle-under-Lyme ST5 3QU
M0 PUT K. Puttock 12 Beechfields, School Lane, Petworth GU28 9DH
M0 PVI P. Handley 97 Applegarth Avenue, Guildford GU2 8LX
M0 PVN P. Nicholls 23 Bishops Gate, Birmingham B31 4AJ
M0 PVO K. Chippindall-Higgin 173 St. Augustine Road, Southsea PO4 9AB
M0 PVP Baron B. Mendham 252 Gregson Lane, Hoghton, Preston PR5 0LA
M0 PVS University College London ARS c/o V. Sterea 67 Lamberhead Road, Wigan WN5 9TU
M0 PVU R. Coomer 30 Torquay Road, Kingskerswell, Newton Abbot TQ12 5EZ
MW0 PVW P. Witts 23 Aelfryn, Llanharry, Pontyclun CF729LQ
M0 PWB P. Booth 12 Heathgate, Wickham Bishops, Witham CM8 3NZ
M0 PWC P. Clark 4 Chestnut Crescent, Blythe Bridge, Stoke-on-Trent ST11 9NH
M0 PWD P. Woodburn 21 The Row, Silverdale, Carnforth LA5 0UG
M0 PWF W. Feather Long House Farm, Ellers Road, Keighley BD20 7BH
MD0 PWI C. Ingles 1 Hillberry View, Onchan IM3 3GB Isle of Man
M0 PWL M. Mynn The Town House, Parsons Field, St. Mary'S, Hugh Town TR21 0JJ
MM0 PWM P. Mackie 8 Letham Avenue, Pumpherston, Livingston EH53 0NG
M0 PWS P. Snelson 103 Queens Road, Vicars Cross, Chester CH3 5HF
M0 PWT Dr D. Garnett Hill View, Snailbeach, Shrewsbury SY5 0NS
M0 PWV P. Wells 6 Westmead Road, Chichester PO19 3JD
MW0 PWY L. Jones Ty'R Ysgol, Holland Street, Ebbw Vale NP23 6HT
M0 PXD P. Donaghy 67 Brockenhurst Way, Bicknacre, Chelmsford CM3 4XN
M0 PXI L. Evans 1 Leave Acre, New Buildings, Sandford, Crediton EX17 4PL
M0 PXL J. Dyson 5 Welton Park, Daventry NN11 2JW
M0 PXM Yorkshire Resistors c/o P. Matthew 24 Jubilee Close, Pamber Heath, Tadley RG26 3HP
M0 PXO P. Offord 1 Adare Close, Dunmow CM6 2GR
M0 PXP J. Maudsley Knight Stainforth Hall, Little Stainforth, Settle BD24 0DP
M0 PXR D. Eggett 5 Winifred Road, Poole BH15 3PU
M0 PXS P. Simpson 7 Hawthorn Close, Wootton DN39 6RB
M0 PXY A. Schofield 4 St. Guthlacs Close, Crowland, Crowland PE6 0ES
M0 PXZ South Anglia Raynet c/o Dr C. Fox 45 Park Road, Wivenhoe, Colchester CO7 9LS
M0 PYA W. Allen 109 Barston Road, Oldbury B68 0PU
M0 PYB P. Bunting 29 Marion Avenue, Alverthorpe, Wakefield WF2 0BJ
M0 PYC D. Polley 6 Coneygear Road, Hartford, Huntingdon PE29 1QL
M0 PYE N. Atkins The Old Rectory, Church Lane, Chippenham SN14 6DE
M0 PYG G. Fielding Chapel Court, Chapel Lane, Cradley, Malvern WR13 5HX
MI0 PYN S. Pynappels 38 Dora Avenue, Newry BT34 1JW
MM0 PYR Elderslie ARS c/o P. Temple 23 Ramsay Place, Johnstone PA5 0EX
MM0 PYS Swinton RC c/o K. Gillen Flat 1/R, 12 Fergus Drive, Glasgow G20 6AG
M0 PYT P. Kimberlee 24 Jacey Road, Shirley, Solihull B90 3LJ
M0 PZC Dr C. Hulbert Nutmeg Cottage, 7 St. Leonard'S Street, Stamford PE9 2HU
M0 PZD A. Jakstas 36 Boxridge Avenue, Purley CR8 3AQ
M0 PZR P. Hanman 7 Tremenheere Road, Penzance TR18 2AH
M0 RAB Dr C. Finnegan 249 Winchester Road, Basingstoke RG22 6EP
M0 RAC R. Cochrane 7 Lawn Terrace, Blackheath, London SE3 9LJ
M0 RAD Avon Valley Ara c/o P. Badham 201 York Avenue, East Cowes PO32 6BH
MM0 RAG J. Hutson 13 Greenan Road, Ayr KA7 4ET
MM0 RAI T. Vanderydt 33 Portland Road, Galston KA4 8EA
MM0 RAM D. Stevenson 51 Shannon Drive, Falkirk FK1 5HU
M0 RAN M. Moran 27 Burnet Close, Padgate, Warrington WA2 0UH
M0 RAP R. Peech 17 Chestnut Avenue, Crossgates, Leeds LS15 8ED
M0 RAR N. Booth Greenfield, Westmancote, Tewkesbury GL20 7EP
M0 RAT N. Mcwilliam 13 Rawlins Street, Liverpool L7 0JE
M0 RAU R. Johnson 15 Magazine Close, Wisbech PE13 1LH
M0 RAW R. Wyeth 112 Main Road, Crockenhill, Swanley BR8 8JL
M0 RAX M. Pike 5 Rowan Drive, Heybridge, Maldon CM9 4BW
M0 RAZ Dr R. Bowman 48 Eliot Drive St. Germans, Saltash PL12 5NL
MW0 RBA R. Arnould 13 Laurel Place, Sketty, Swansea SA2 8JL
M0 RBB R. Brier 48 Burton Rise, Kirkby-In-Ashfield, Nottingham NG17 9BR
M0 RBC R. Colman 197 Coppins Road, Clacton-on-Sea CO15 3LA
M0 RBD M. Czerski 3 Ladymead Close, Whaddon, Milton Keynes MK17 0LL
M0 RBE R. Smith 4 London Road, Lindal La12 0LI, Ulverston LA12 0LL
M0 RBF R. Ferguson 31 Barton Court Road, New Milton BH25 6NW
M0 RBG R. Blandford 60, Benomley Road, Almondbury, Huddersfield HD5 8LS
M0 RBH R. Hookham 20 White Lodge Park Portishead, Bristol BS20 7HH
M0 RBI C. Bennett The Old Cottage, Waterside Road, Southminster CM0 7QT
M0 RBJ N. Roberts 40 Armour Road, Tilehurst, Reading RG31 6HN
M0 RBK Believe In Better Cg c/o Dr R. Bleaney 40 Broadstone Road, Harpenden AL5 1RF
MW0 RBL R. Lovesey 33 Ty Isaf Park Avenue, Risca, Newport NP11 6NB
M0 RBM R. Medland 5 Bay Tree Cottages, Hospital Road, Bude EX23 9BP
MM0 RBN R. Corkey 11 Golf View Cardenden, Lochgelly KY5 0NW
M0 RBQ R. Simpson 22 Kenworthy Road, Stocksbridge, Sheffield S36 1BZ
MM0 RBR R. Hutton 2 Watson Place, Dunfermline KY12 0DR
M0 RBT Dr R. Hunt 7 Knotley Hall Cottages, Chiddingstone Causeway, Tonbridge TN11 8JH
M0 RBU D. Perrin 54 High Street, West Wratting, Cambridge CB21 5LU
MW0 RBV R. Briant Talarvor, Llanon SY23 5HG

M0 RBX R. Buckland 34 Beechwood Drive, Meopham, Gravesend DA13 0TX
M0 RBY R. Hall Concorde Cottage, Ellingstring, Ripon HG4 4PW
M0 RCC R. Chadwick 4 Gleneagles Drive, Haydock, St. Helens WA11 0YS
M0 RCD E. Capstick 130 Ship Lane, Farnborough GU14 8BJ
M0 RCE G. Morse 34 Headford Road, Bristol BS4 1QE
MI0 RCF Lough Erne ARC c/o H. Graham 104 Tattygare Road, Lisbellaw, Enniskillen BT94 5FB
MW0 RCH Taunton & Somerset Raynet Group c/o S. Smith 102 Gresford Road, Llay, Wrexham LL12 0NW
M0 RCI R. Chappell 11 Highfields Road, Darton, Barnsley S75 5ER
M0 RCK R. Wells 27 Victoria Avenue, Camberley GU15 3HT
M0 RCL Dr A. Jackson 40 Richardby Crescent, Durham DH1 3TY
M0 RCM R. Moxham 8 Dunroyal Close, Helperby, York YO61 2NH
M0 RCN R. Hart 4 Glade Mews, Guildford GU1 2FB
M0 RCP Dr R. Peterson 9 Moseley Wood View, Leeds LS16 7ES
M0 RCR R. Room 197 Newbridge Road, Bath BA1 3HH
M0 RCT R. Tomkinson 24 Beech Drive, Wistaston Green, Crewe CW2 8RE
M0 RCV South East Hampshire Raynet c/o P. Raxworthy 32 St. Marys Avenue, Alverstoke, Gosport PO12 2HX
M0 RCX R. Rawson 30 Harty Road, Haydock, St. Helens WA11 0YY
M0 RCY D. Osborne 12 Sandringham Close, Brackley NN13 6JQ
MW0 RCZ R. Shipman 1 Lledfair Place, Heol Pentrerhedyn, Machynlleth SY20 8DL
M0 RDA A. Rivers 34 Brookfield, Mawdesley, Ormskirk L40 2QJ
M0 RDB Capt. R. De Savigny-Bower 55 Fenwick Close, Woking GU21 3BZ
M0 RDC R. Cooke 1 Fiona Walk, Fazakerley, Liverpool L10 4YW
MM0 RDD R. Duncan 12 Douglas Loan, Kirkwall KW15 1FU
M0 RDI A. Riddick 30 Britannia Road, Banbury OX16 5DW
M0 RDK Phase Array DX Group c/o S. Gadsby 30 Woodside Close, Knaphill, Woking GU21 2DD
M0 RDP R. Parker 53 Tunstall Road, Canterbury CT2 7BX
M0 RDR R. Rawlinson 17 Walmer Place, Winsford CW7 1HA
M0 RDS R. Staniland The Cottage, Martin Moor, Lincoln LN4 3BQ
MM0 RDT R. Tripney 7 Sunnyside St., Camelon, Falkirk FK1 4BJ
M0 RDV R. Vincent 141 Timberleys, Littlehampton BN17 6QD
M0 RDX J. Scott 443 Ford Green Road, Stoke-on-Trent ST6 8LX
M0 RDY R. Hawkins 3 Fairways Drive, Harrogate HG2 7ES
M0 RDZ R. Dell 18 Greenacres, Fulwood, Preston PR2 7DA
M0 REA A. Duffield 4 Crabmill Lane, Easingwold, York YO61 3DE
M0 REB D. Hook 28 Rifford Road, Exeter EX2 5JT
M0 REC R. Clare 26 Hall Park, Swanland, North Ferriby HU14 3NL
M0 RED G. Sharp 18 Whitbred Road, Salisbury SP2 9PE
M0 REG Worthing Radio Events Group c/o M. Folkes 3 Colindale Road, Ferring, Worthing BN12 5JF
MW0 REH R. Harlow Swyn-Y-Mor, Penrallt Road, Holyhead LL65 2UG
MI0 REI R. Bestek 111 Cloughwater Road, Ballymena BT43 6SZ
M0 REJ R. Jones 39 Dalton Lane, Barrow-in-Furness LA14 4LE
M0 REM M. King 6 Aylesbury Close, Hockley Heath, Solihull B94 6PA
M0 REP Dr A. Wallman Oakwood, 30 Elmsway, Bramhall, Stockport SK7 2AE
M0 REQ E. Stammers 40 Tillingbourne Road, Shalford, Guildford GU4 8EY
M0 REV Rvd. J. Drake 37 Weston Lane, Southampton SO19 9GN
M0 REX Midland CG c/o R. Duffy 45 Chatham Road, Winchester SO22 4EE
M0 REZ Dragon ARC c/o S. Brown 4 Dorado Gardens, Orpington BR6 7TD
MM0 RFA R. Aird 26 Underbank, Largs KA30 8SS
M0 RFK D. Gardner 122 All Saints Avenue, Maidenhead SL6 6LT
M0 RFM R. Mannock Craybourne, Higher Brill, Falmouth TR11 5QG
M0 RFU J. Byrne 316 Turncroft Lane, Stockport SK1 4BP
M0 RFW Dr R. White 2 Uplands Cottages, Rattle Road, Pevensey BN24 5DT
M0 RFY D. Murray 9 Woodend, Sutton SM1 3LW
M0 RGB G. Baker Ling Cottage, Crag Foot, Carnforth LA5 9SA
M0 RGC G. Henshall 43 Cumberworth Road, Skelmanthorpe, Huddersfield HD8 9AB
M0 RGD R. Dale 17 Spencer Gardens, Brackley NN13 6AQ
M0 RGE Lowestoft District And Pye ARC c/o R. Edwards 46 Lavers Oak, Martock TA12 6HG
M0 RGF A. Mobbs 149 The Paddocks, Old Catton, Norwich NR6 7HR
M0 RGH R. Smith Five Elms, Lullington Road Edingale, Tamworth B79 9JA
M0 RGI I. Reichenfeld 7 Hazelbank Close, Liphook GU30 7BY
M0 RGL Gloucestershire County Raynet c/o M. White 7 Overthwart Crescent, Worcester WR4 0JW
M0 RGN P. Williams 35 Cansfield Grove, Ashton In Makerfield, Wigan WN4 9SE
M0 RGO D. Robertson 53 Moor Lane, Weston-Super-Mare BS22 6RA
M0 RGP Redditch Amateur Radio & CG c/o R. Jones 6 Wychwood Drive Hunt End, Redditch B97 5NW
MJ0 RGR R. Bisson A122, Le Capelain House, Castle Quay, La Rue De L'Etau, St. Helier JE2 3EA Jersey
M0 RGS the Kings School RS c/o D. Saul 78 Ingleton Drive, Lancaster LA1 4QZ
M0 RGV G. Green 90 Princes Way, Fleetwood FY78DX
MI0 RGX R. Gilmore 11 Abbots Gardens, Newtownabbey, Belfast BT379QZ
M0 RGY A. Mobbs 149 The Paddocks, Old Catton, Norwich NR6 7HR
M0 RHA P. Van Staveren 14 Fortune Green Road Flat 3, London NW6 1UE
M0 RHB R. Burton 23 Freston, Paston, Peterborough PE4 7EN
M0 RHC Dr R. Hopper 44 Ditton Walk, Cambridge CB5 8QE
MW0 RHD R. Hughes-Burton 6 Troed Y Garn Llangybi, Pwllheli LL53 6DQ
M0 RHE R. Head 24 Beaufort Road, Church Crookham, Fleet GU52 6AZ
M0 RHG D. Bines 39 School Close, Bretton, Peterborough PE3 9FS
M0 RHI R. Hunt 4 Rue Josy Printz, Hesperange L-5841 Luxembourg
MM0 RHL Edinburgh Hacklab c/o T. Hawes 83 Dinmont Drive, Edinburgh EH16 5RY
M0 RHO R. Morgan 14 Ash Road, Ashurst, Southampton SO40 7AT
M0 RHQ J. Middleton 8 Cullen Close, Newark NG24 1DF
M0 RHR A. Thomas 1 Millers Close, Ruardean Hill, Drybrook GL17 9AU
M0 RHS R. Hawkins Forest Edge, Deer Park, Milton Abbas, Blandford Forum DT11 0AY
MW0 RHT R. Titcombe 82 Liverpool Road, Buckley CH7 3NB
MM0 RHU R. Humphrey 11 Pearce Grove, Edinburgh EH12 8SP
M0 RHW W. Westlake 2 Chegwin Court, Newquay TR7 2DE
M0 RHX R. Haynes 28 Ridgeway View, Montgomery SY15 6BF
M0 RIA T. Lea 1 Roseland Close, Keyworth, Nottingham NG12 5LQ
MI0 RIB W. Nicholl 58 Dunnalong Road, Bready, Strabane BT82 0DW
M0 RIC D. Moore 379 Main Road, Harwich CO12 4DW
M0 RIF A. Barrett 4 Wood Cottages, Cummings Cross, Liverton, Newton Abbot TQ12 6HJ
M0 RIG A. Rigg 12 Bydales Drive, Marske-By-The-Sea, Redcar TS11 7HJ
M0 RIK F. Woodhams Greenways, Mill Lane, Fareham PO15 5DU
M0 RIS M. Baines 21 Acre Moss Lane, Kendal LA9 5QE
M0 RIU D. Simmons 8 Lower Grange, Huddersfield HD2 1RU
M0 RIV Riviera ARC c/o S. Crask 14 Southfield Road, Paignton TQ3 2SW
M0 RJB R. Blaney 16 Pages Close, Wymondham NR18 0TU
M0 RJE R. Edwards 23 Queens Walk, Ruislip HA4 0LX
M0 RJH Dr J. Reynolds 38 Spring Lane, Hockley Heath, Solihull B94 6QY
MM0 RJJ R. James The Garret, Alyth, Blairgowrie PH11 8HQ
M0 RJK R. Kavanagh Chatslea, 57 Mill Lane, Littlehampton BN16 3JP
M0 RJM R. Millington Quaintways, The Avenue, Tarporley CW6 0BA
MI0 RJN R. Neill 15 Hawthorn Place, Coleraine BT52 2ES
M0 RJO R. Odell 41 Pevensey Park Road, Westham, Pevensey BN24 5HW
MM0 RJR South Tyneside ARS c/o R. Renshaw Smithy House, Scotscalder, Halkirk KW12 6XJ
M0 RJS R. Somerville Roberts Bank House, 7 Mill Lane, Stoke-on-Trent ST7 3LD
M0 RJT R. Gilbert 61 Coltstead, New Ash Green, Longfield DA3 8LN
MI0 RJW R. Wylie 9A Kinnegar Drive, Holywood BT18 9JQ
M0 RJX R. Harrison 18-20 Hall Lane, Kirkburton, Huddersfield HD8 0QW
MI0 RJY J. Young 7 Dunmore Close, Cookstown BT80 8AS
M0 RJZ N. Noda 14 Widmer Court, Vicarage Farm Road, Hounslow TW3 4NL
M0 RKA M. Hawker Popes Corner Marina, Holt Fen, Little Thetford, Ely CB6 3HR
MW0 RKB M. Brady Ty Mawr Uchaf, Dulas LL70 9DQ
MW0 RKD R. Hark 5 Victoria Park, Bagillt CH6 6JS
M0 RKE D. Roake 9 Falcondale Walk, Westbury On Trym, Bristol BS93JG
M0 RKF C. Whitelaw 18 Marine Drive, Bishopstone, Seaford BN25 2RT
M0 RKH R. Hayes 59 Elm Road, Folksworth, Peterborough PE7 3SX
MD0 RKI R. Kijak 13 Falcon Cliff Court, Douglas, Isle of Man IM2 4AH
M0 RKK R. Pike 44 Falkirk Close Bransholme, Hull HU7 5BX
M0 RKM R. Kershaw 40 The Butts, Frome BA11 4AA
MM0 RKN C. Welsh 28 Peacock Wynd, Motherwell ML1 4ZL
M0 RKR G. Hirst 94 Upper Brighton Road Sompting, Lancing BN15 0LB
MM0 RKT R. Towers 34 South Park Road, Hamilton ML3 6PN
M0 RKW R. Watson 5 Angrove Gardens, Sunderland SR4 7TB
M0 RKX M. Hemming 11 Blackberry Way, Evesham WR11 2AH
M0 RKY R. Brown 24 Malthouse Court, Wellington, Telford TF1 1QJ
M0 RLC Dr R. Coley 165 Westerfield Road, Ipswich IP4 3AB
MW0 RLD B. Shelley Sunray, Pendine, Carmarthen SA33 4PD
M0 RLF R. Beardmore 28 Broadway, Ilkeston DE7 8TD
M0 RLI D. Thomas 130 Norwich Avenue, Southend-on-Sea SS2 4DH
MW0 RLJ R. Johns 1 Llanferran St. Nicholas, Goodwick SA64 0LL
M0 RLM R. Lima Matos 37 Palace Road, London N8 8QL
MM0 RLN D. Nixon 17 Semple Place, Linwood, Paisley PA3 3RT
M0 RLP R. Le Piez 279 Oakley Road, Southampton SO16 4NR
MD0 RLS R. Smith 3 Rheast Barrule, Castletown IM9 1HW
M0 RLV R. Oliver 3 Histon Road Cottenham, Cambridge CB24 8UF
M0 RLW R. Williams 16 Irving Road, Norwich NR4 6RA
MM0 RMD R. Nelson 8 South Street Cambus Fk102Pa, Stirling FK102PA
M0 RMF G. Coupe 112 Greenwood Drive, Kirkby-In-Ashfield, Nottingham NG17 8GH
M0 RMG G. Chapman 15 Greenhays Rise, Wimborne BH21 1HZ
M0 RMH R. Halliwell99 38 Larch Grove, Kendal LA9 6AU
M0 RMI R. Miller 23 Clarendon Road, Sevenoaks TN13 1EU
M0 RMJ R. Jeffs 45 Forest Road, Bingham, Nottingham NG13 8RL
MI0 RMK A. Mackenzie 30 Dalriada Gardens, Ballycastle BT54 6DZ
M0 RML Barnsley & District ARC c/o P. Greenhalgh 13 Primrose Avenue, Urmston M41 0TY
M0 RMM R. May 8 Boscaswell Estate, Pendeen, Penzance TR19 7EU
M0 RMN M. Watmough 6 Blair Park, Knaresborough HG5 0TH
M0 RMO C. Larner 98 Allandale, Hemel Hempstead HP2 5AT
M0 RMP R. Perkin 26 Hall Avenue, Leek ST13 6BU
M0 RMT J. Gorlinski Flat 2, 5 Motcombe Lane, Eastbourne BN21 1PS
M0 RMU A. Murayama 248-4 Nakano-Cho, Ise-City 516-0034 Japan
M0 RMW R. Williams 11 Solent Way, Selsey PO20 0JR
M0 RMY T. Rowlands 7 Northfield Crescent, Beeston, Nottingham NG9 5GR
M0 RMZ R. Mansfield 8 Haysoms Drive, Greenham, Thatcham RG19 8EY
M0 RNC M. Brigham 21 Overdale Close, York YO24 2RT
M0 RND A. Greig 17 Begbroke Lane, Kidlington OX5 1RN
M0 RNH I. Justice 4 Saxon Street, Droylsden, Manchester M43 7FR
M0 RNI D. Harris 2 Horsecroft, , Ewyas Harold HR20EQ
M0 RNP P. Rainer 6 Highland Close, Folkestone CT20 3SA
M0 RNR B. Pickup 3 Mews Court, Houghton le Spring DH5 8GB
M0 RNU M. Nutt 110 Birkinstyle Lane, Shirland, Alfreton DE55 6BT
M0 RNW R. Wellsted 127 Goldthorn Hill, Wolverhampton WV2 4PS
M0 RNX J. Faulks 11 Fishguard Spur, Slough SL1 1TS
M0 RNZ N. Cooper Romer Cottage, Long Reach, Ockham, Woking GU23 6PF
M0 ROC M. Russell 107 Cambridge Road, Hitchin SG4 0JH
M0 ROJ R. Reeves Goldford House, Goldford Lane, Malpas SY14 8LL
M0 ROL A. De Mora 36 West Park, Minehead TA24 8AN
M0 ROM R. Pasika 192 Longfield Lane, Cheshunt, Waltham Cross EN7 6AQ
M0 RON A. Eustace 3 Linworth Road, Bishops Cleeve, Cheltenham GL52 8PF
M0 ROO R. Smith South Cottage, Radley Green Road, Chelmsford CM1 4NW
MM0 ROR I. Learmonth 14 Deansloch Terrace, Aberdeen AB16 5SN
MM0 ROV M. Gerrard 10 Whinhill Gardens, Aberdeen AB11 7WD
M0 ROW Chepstow + District ARS c/o P. Ormsby Wood Cottage, Little Heck, Goole DN14 0BU
M0 ROX Dr R. Pediani Old School House Great Coxwell, Faringdon SN7 7NB
M0 ROY R. Henson 2 Byron Street, Shirebrook, Mansfield NG20 8PJ
MW0 RPB R. Bennett 28 Neyland Path Fairwater, Cwmbran NP44 4PX
M0 RPD I. Handley Rosedale, Chapman Street, Market Rasen LN8 3DS
MW0 RPE Newbury And District Hackspace c/o R. Evans Brackenwood, 30 Northop Country Park, Mold CH7 6WD
M0 RPF R. Fullagar 6 Locke Way, Stafford ST16 3RE
M0 RPI D. Akerman The Brick Barn Old Gore, Ross-on-Wye HR9 7QW
M0 RPJ Second Class Operators Club (UK) c/o Dr R. Jepsen 20 The Mount, Aspley Guise, Milton Keynes MK17 8EA
M0 RPK R. King 188 Providence Road, Sheffield S6 5BE
M0 RPO R. Powell 26 Fenwick Avenue, South Shields NE34 9AJ
M0 RPR M. Roper 13 St. Cuthbert Street, Worksop S80 2HN

MI0 RPT R. Tomalin 22 Drumfad Road, Millisle, Newtownards BT22 2JQ
M0 RPZ R. Whiteway 53 Tavistock Road, Weston-Super-Mare BS22 6NX
M0 RQD S. Cownley 5 Pumphouse Lane, East Cowes PO32 6FJ
M0 RQK B. Dare 1 St. Johns Villas, Sivell Place, Exeter EX2 5ES
M0 RQN J. Foster 23 High Street, Cumnor, Oxford OX2 9PE
M0 RQQ D. Cockram 86 Tamworth Road, Two Gates, Tamworth B77 1EG
M0 RQX L. Coneley 4 Primrose Close, Gosport PO13 0WP
M0 RRC Bromsgrove And District ARC c/o S. Williams 32 Waterdell Lane, St. Ippolyts, Hitchin SG4 7QZ
MI0 RRE R. Rantin 8A Buchanans Road, Newry BT35 6NS
M0 RRF I. Sharpe 4 Low Dowfold, Crook DL15 9AE
M0 RRG Richmond Raynet Group c/o J. Kirby 2 Kneeton Park, Middleton Tyas, Richmond DL10 6SB
M0 RRL K. Woodhams 83 Langdale Place, Newton Aycliffe DL5 7DY
MM0 RRM R. Murray The Barn House, Springfield Farm, Carluke ML8 4QZ
M0 RRN D. Barker 12 The Weavers, Denstone, Uttoxeter ST14 5DP
M0 RRR R. Reeves 15 Higher Albert Street, Chesterfield S41 7QE
M0 RRX R. Ridge Roskellan House, Maenlay, Helston TR12 7QR
MW0 RRY Guildford & D Rd c/o G. Sherry 22 York Street, Oswestry SY11 1LX
M0 RSA D. Silburn 34 Northfields, Strensall, York YO32 5XW
M0 RSC Chesterfield & District Scouts ARC c/o K. Greatorex 54 Lilac Grove, Glapwell, Chesterfield S44 5NG
M0 RSD K. Winwood 146 Chapel Street, Pensnett, Brierley Hill DY5 4EQ
M0 RSE Radio Society of Great Britain c/o S. Thomas 2 Myrtle Cottages, Sandy Lane, Saxmundham IP17 1HR
M0 RSF C. Darlow 418 Broad Lane, Bramley, Bramley, Leeds LS13 3DF
M0 RSG E. Flint The Bell House, Kingston Deverill, Warminster BA12 7HE
M0 RSH R. Hansford 17 Dolver Close, Corby NN18 8NB
MM0 RSI R. Inglis 13 Princes Street, California, Falkirk FK1 2BX
M0 RSJ R. Dunstan 53 Church View Road, Camborne TR14 8RQ
MI0 RSN R. Robinson 30 Trasnagh Drive, Newtownards BT23 4PD
MI0 RSO S. Mcclean 28A Ashfield Court, Donaghadee BT20 0BF
M0 RSP Dr R. Paden 21 The Rookery, Balsham, Cambridge CB21 4EU
MW0 RSR A. Noakes Westra Holt, Westra, Dinas Powys CF64 4HA
M0 RST C. Taylor 48 Northdown Park Road, Cliftonville, Margate CT9 3PT
M0 RSU R. Suchocki 8 High Street, Bluntisham, Huntingdon PE28 3LD
MW0 RSV Jubilee Sailing Trust(ARS) c/o G. Jones 31 Liverpool Road, Buckley CH7 3LH
M0 RSW R. Weatherup Sidney Sussex College, Cambridge CB2 3HU
M0 RSX R. Stone 63 Sands Lane, Lowestoft NR32 3ER
M0 RSY A. Davies Penthouse Caravan, Shutt Green Lane, Stafford ST19 9LX
MM0 RTD D. Robertson 17 Keswick Drive, Hamilton ML3 7HN
M0 RTE A. Green 40 Claines Road, Northfield, Birmingham B31 2EE
M0 RTH R. Duffield 4 Crabmill Lane, Easingwold, York YO61 3DE
M0 RTK Dr R. Tempo 35 Warminster Road, Bath BA2 6XG
M0 RTL A. Garthwaite 278 Carlton Road, Barnsley S71 2BA
M0 RTM C. Bell 4 Main Street, Newbold, Rugby CV21 1HW
M0 RTP R. Paster 8 Rachaels Lake View, Warfield, Bracknell RG42 3XU
M0 RTQ K. Jones 4 Smithwick Place, Tregenver Road, Falmouth TR11 2QE
MM0 RTT R. Turpie 11 Ashkirk Place, Dundee DD4 0TN
M0 RTV R. Coombs 55 Highfield Road, Hemsworth, Pontefract WF9 4EA
M0 RTW C. Colless 128 Ditton Lane, Fen Ditton, Cambridge CB5 8SS
MI0 RTY M. Strawbridge 9 Wheatfield Crescent, Coleraine BT51 3RA
M0 RUB D. Golding Windrush Cottage 84-85 Bradenstoke, Chippenham SN15 4EL
MI0 RUC N. Bolt 32 Bush Gardens, Bushmills BT57 8AE
M0 RUG J. Fry Deal Cottage, Ipswich Road, Long Stratton, Norwich NR15 2TF
MW0 RUH D. Thomas 23 Merthyr Dyfan Road, Barry CF62 9TG
M0 RUK C. Lote 8 Warren Place, Walsall WS8 6BY
M0 RUM M. Symmonds 24 Woodville Grove, Stockport SK5 7HU
MI0 RUR D. Boyd 11 Abbey Gardens, Belfast BT5 7HL
M0 RUT J. Wright 4 Sweeters Field Road, Alfold GU6 8UD
M0 RUX J. Dunn 3 Hobbs Way, Rustington, Littlehampton BN16 2QU
M0 RUZ R. Brierley 39 Hatfield Road, Alvaston, Derby DE24 0BU
M0 RVB Dr J. Harmer 1 Wynford Rise, Leeds LS16 6HX
MW0 RVC R. Gripp 23 Edmond Locard Court, Chepstow NP16 6FA
M0 RVD R. Dewes 31 Woodlea Avenue, Lutterworth LE17 4TU
MI0 RVH T. Nelson 25 Monaghan Road Annashanco Rosslea, Belfast BT927PT
M0 RVJ Rvd. J. Goodman 34 Hillside Close, Chalfont St. Peter, Gerrards Cross SL9 0HN
M0 RVK V. Rotaru 15 Leicester Drive, Glossop SK13 8SH
M0 RVN R. Vern 104 High Street, Netheravon, Salisbury SP4 9PJ
M0 RWA R. Anderson 56A Cheriton Avenue, Adwick-Le-Street, Doncaster DN6 7BT
M0 RWB R. Broadbridge 102 Bailey Crescent, Poole BH15 3HB
M0 RWD D. Eastwood 13 Riverwood Drive, Halifax HX3 0TH
M0 RWG R. Grout 5 Branton Close, Great Ouseburn, York YO26 9SF
M0 RWH R. Hornby 61 Fulwood Heights, Fulwood, Preston PR2 9AW
MM0 RWJ R. Welsh 28 Peacock Wynd, Motherwell ML1 4ZL
M0 RWK West Kent Raynet c/o D. Parvin 11 Stanhope Way, Sevenoaks TN13 2DZ
M0 RWL R. Lane 9 Hartoft Road, Hull HU5 4JZ
M0 RWM R. Mayfield 75 Cartwright Street, Loughborough LE11 1JW
M0 RWN R. Nock 83 Coles Lane, West Bromwich B71 2QW
M0 RWR Riverway ARS c/o E. Reynolds 4 Underwood Close, Stafford ST16 1TB
M0 RWS R. Stokes 20 Elizabeth Road, Waterlooville PO7 7LY
M0 RWT P. Field 63 Hartford Road, Davenham, Northwich CW9 8JE
M0 RWV S. Forshaw 32 Fox Way, Eastfield YO11 3PH
M0 RWW R. Wells 57 The Ridings, Paddock Wood, Tonbridge TN12 6YA
M0 RXA P. Costall 3 Gaynsford Place, Little Canfield, Dunmow CM6 1WB
M0 RXB R. Badami Flat F 373 Camden Road, London N7 0SH
MW0 RXD R. Dutton Burn Naze, Old Mill Road, Penmaenmawr LL34 6TE
M0 RXG M. Robson 270 Calder Road, Lincoln LN5 9TL
M0 RXM A. Matynka 2 Chatham Road, Sunderland SR5 3QA
M0 RXN R. Abeykoon 5 Challney Gardens, Luton LU4 8QQ
M0 RXS J. Brewer 1 Bentley Road, Forncett St. Peter, Norwich NR16 1LH
M0 RXV R. Mansell 1412 Warwick Road, Knowle, Solihull B93 9LG
M0 RXX S. Southern 37 Conway Road, Calcot, Reading RG31 4XP
M0 RXZ B. Lupton 124 Wolsey Crescent, New Addington, Croydon CR0 0PF
M0 RYA J. Kay Uplands Farm, Dallington, Heathfield TN21 9NG
M0 RYB Dr P. Lock The Firs, The Butts, Norwich NR16 2EQ
MM0 RYE R. Rothon 112 Ravenswood Rise, Livingston EH54 6PG
M0 RYG Ramsbury Amateur Radio DX Group c/o J. Connor 28 Church Street, Hungerford RG17 0JE
M0 RYK M. Granatt 16 Culverden Avenue, Royal Tunbridge Wells TN4 9RF
MI0 RYM R. Murphy 40 Stoneypath, Londonderry BT47 2AF
M0 RYO R. Makioka 77-38 Kawai-Honcho, Asahi-Ku, Yokohama-City 2410803 Japan
MM0 RYR R. Clow 25 Scott Street, Newcastleton TD9 0QQ
M0 RYS R. Sayre 8 Lorne Road, Richmond TW10 6DS
MW0 RZC G. Bodley 34 Claremont Road Newbridge, Newport NP11 5DL
M0 RZD R. Luscombe 7 Orchard Close, Tideford, Saltash PL12 5HR
M0 RZE I. Laidler 5 South Street, West Rainton, Houghton le Spring DH4 6PA
M0 RZF R. Brown 16 Meadow Edge, Waterlooville PO7 5AZ
M0 RZO A. Brzosko 32 Lower Park Street, Cambridge CB5 8AR
MW0 RZS M. Beasley Ffynnon Wen, Bontnewydd, Aberystwyth SY23 4JJ
M0 RZX Manchester & District ARS c/o B. Forrest 32 Idonia Road, Wolverhampton WV6 7NQ
M0 RZY E. Moore 33 Avon Drive, Congleton CW12 3RQ
M0 SAA B. Matthews 30 Oaklands Drive, Brandon IP27 0NR
M0 SAB A. Brackstone 3 Petunia Close, Basingstoke RG22 5NX
M0 SAC M. Couchman 140 The Tideway, Rochester ME1 3QE
M0 SAD D. Platt 50 Poplars Road, Stalybridge SK15 3EN
MM0 SAH S. Henderson 13 Dunnottar Place, Kirkcaldy KY2 5YX
MI0 SAI S. Barnes 191 Marlacoo Road, Portadown, Craigavon BT62 3TD
MM0 SAJ S. Smith 10 Munro Street, Stenhousemuir, Larbert FK5 4QF
MM0 SAK A. Jardine 17 Louisa Drive, Girvan KA26 9AH
M0 SAL D. Salter 142 Brays Road, Birmingham B26 2PP
MI0 SAM S. Christie 17 Kilburn, Belfast BT12 6JS
M0 SAO A. Dossa 24 Warwick Drive, Cheshunt, Waltham Cross EN8 0BW
MI0 SAP S. Murray 117 Knockview Drive, Tandragee, Craigavon BT62 2BL
M0 SAQ D. Astley 34 Church Terrace, Glossop SK13 7RL
M0 SAR S. Roy 28 Kingston Rise, New Haw, Addlestone KT15 3EY
M0 SAV A. Smithies 35 Dialstone Lane, Stockport SK2 6AA
MM0 SAX G. Sproul 25 Mulben Place, Glasgow G53 7UP
M0 SAY D. Sayles 82 Molineaux Road, Shiregreen, Sheffield S5 0JY
M0 SAZ M. Parker Ridgeways, Mill Common, Westhall, Halesworth IP19 8RQ
M0 SBA S. Walker 33 Parkside Somercotes, Alfreton DE55 4LA
M0 SBB A. Southwell 56 Lambrook Road, Taunton TA1 2AF
M0 SBC K. Smith 7 Rosebery Avenue, Morecambe LA4 5RU
M0 SBD M. Denut 17 Quillet Road, Newlyn, Penzance TR18 5QR
M0 SBF S. Larkins 4 Water Lane, Greenham, Thatcham RG19 8SS
M0 SBH Dr S. Nandalan 22 The Common, Parbold, Wigan WN8 7DA
MW0 SBJ D. John 29 Eleanor Street, Tonypandy CF40 1DW
M0 SBK S. Johnson 2 North Square, Edlington, Doncaster DN12 1ED
M0 SBL P. Trembath 48 Treveneth Crescent, Newlyn, Penzance TR18 5NG
MM0 SBO S. Boyd 1 St. Marks Lane, Edinburgh EH15 2PX
M0 SBR H. Hamilton Flat B, 9 Cambridge Drive, London SE12 8AG
M0 SBT S. Burtsal 69A Pewley Way, Guildford GU1 3PZ
M0 SBY S. Bassett 3 Lower Merryfield, Anchor Road, Radstock BA3 5PG
M0 SBZ D. Smith 105 Princes Street, Dunstable LU6 3AS
M0 SCA S. Light 16 Cabot Road, Yeovil BA21 5FQ
M0 SCB T. Bacon Norreum, Church Road, Reading RG7 1TJ
M0 SCE D. Hagan 8 Charles Close, Westcliff-on-Sea SS0 0EU
M0 SCG Sands Amateur Radio Communications Group c/o B. Watson 7 Branksome Drive, Morecambe LA4 5UJ
M0 SCN S. Nuttall 17 Redgate, Northwich CW8 4TQ
M0 SCO S. Court 16 Worcester Road, Woodthorpe, Nottingham NG5 4HY
M0 SCP D. Purbrick 94A Polperro Way, Hucknall, Nottingham NG15 6JW
M0 SCR Cornwall Raynet Group c/o K. Harris 8 Trelawney Rise, Callington PL17 7PT
M0 SCS West Kent Raynet c/o S. Smith 113 Deaconsfield Road, Hemel Hempstead HP3 9JA
M0 SCT G. Rutherford 24 Chestnut Avenue, Hedon, Hull HU12 8NH
M0 SCU S. Culshaw 37 Netherby Road, Wigan WN6 7PU
M0 SCW S. Warren 7 Crich Way, Swadlincote DE11 0UU
M0 SCX Dr S. Wing 107 Highlands Boulevard, Leigh-on-Sea SS9 3TH
M0 SCY Sandringham School ARC c/o A. Gray 5 Meadow Close, Marshalswick, St. Albans AL4 9TG
M0 SDA E. Gedvilas 23 Pennington Drive Newton Le Willows, Newton-le-Willows WA12 8BA
M0 SDB D. Bower 4 Winsford Road, Sheffield S6 1HT
M0 SDC 7m CG c/o C. Wilson 57 James Andrew Crescent, Sheffield S8 7RJ
MW0 SDD Willow CG c/o J. Bidwell 26 Lone Road, Clydach, Swansea SA6 5HR
M0 SDE Farringdon Radio Group c/o S. Preston The Chapel, Robson St., Shildon DL4 1EB
M0 SDF P. Ashton 32 Sycamore Road, New Ollerton Nr Newark, Notts NG22 9PS
M0 SDG M. Torrington 4 Aylesby Gardens, Grimsby DN33 1SB
MM0 SDK M. Bartlett 93 Lumsden Crescent, Almondbank PH1 3UA
M0 SDM S. Mason 3 Grange Cottages, Low Road, Barrowby, Grantham NG32 1DL
M0 SDP S. Plows York House, York Close, Measham, Swadlincote DE12 7JH
MI0 SDR D. Reid 13 Gelvin Grange, Londonderry BT47 2LD
M0 SDS S. Stocker 2 Peveril Avenue, Borrowash, Derby DE72 3JJ
M0 SDT S. Theaker 10 Grange Fields Mount, Leeds LS10 4QN
M0 SDU L. Soldan 35 Lingfield Gate, Leeds LS17 6DB
M0 SDW Dr S. Willoughby 33 Viscount Close, Diss IP22 4GL
M0 SDY P. Cattermole Blaxhall Hall Crossing, Little Glemham, Woodbridge IP13 0BP
M0 SEA A. Newns 3 Fox'S Yard, Harbour Village, Penryn TR10 8GF
M0 SEB S. Banach 13 Hilton Crescent Worsley, Manchester M28 1FY
MW0 SEC L. Hayward Cefn Gribyn Carmel, Llannerch-Y-Medd LL71 7BU
M0 SED D. Cockburn 4 Tranmere Avenue, Heysham, Morecambe LA3 2BB
M0 SEJ S. Adams 1 Byford Way, Winslow, Buckingham MK18 3RJ
MM0 SEK J. Mcphillips 86 Glenburn Avenue, Motherwell ML1 5EF
M0 SEL S. Elliott 50 West End Road, Mortimer Common, Reading RG7 3TH
M0 SEM M. Skinner 5 Sycamore Avenue, Upminster RM14 2HR
M0 SEO Dr R. Seo Flat 130, Oslo Court, London NW8 7EP
M0 SER C. Lewis 9 Chatsworth Gardens, Sydenham, Leamington Spa CV31 1WA
M0 SET P. Harvey 35 Isaac Street, Liverpool L8 4TH
M0 SEV P. Holmes 53 Bishops Hull Road, Bishops Hull, Taunton TA1 5EP
M0 SEW J. Sewell 8 Anglesmede Crescent, Pinner HA5 5SP
MM0 SEY N. Rogers 108 Beechwood Road, Cumbernauld, Glasgow G67 2NP
MW0 SEZ PI RC c/o S. Ezard 59 Station Farm, Croesyceiliog, Cwmbran NP44 2JW

M0 SFA S. Astbury 131 Denton Avenue, Grantham NG31 7JG
M0 SFD Dr F. Derry 13 Fraucup Close, Ford, Aylesbury HP17 8XU
M0 SFI F. Sidzhimov 51 Barnwood Road, Guildford GU2 8JD
MM0 SFM S. Forrest-Mcneill 27 King Edwards Way, Kirkliston EH29 9DS
M0 SFR Dr A. Shafarenko Church House, Kimbolton Road, Bolnhurst, Bedford MK44 2ES
M0 SFT D. Swift 31 Meadow Lane, Westbury BA13 3AE
M0 SGA A. Suttle 61 Albert Street, Shildon DL4 2DN
MW0 SGD S. Doherty 104 Cromwell Road, Milford Haven SA73 2EN
M0 SGF S. Francis 17 Garden Close, Rough Common, Canterbury CT2 9BP
M0 SGG A. Rowan 14 Craven Lea, Liverpool L12 0NF
M0 SGH S. Hall Orchard End, Asenby, Thirsk YO7 3QR
M0 SGJ S. James 94 North Road, Hull HU19 2AY
M0 SGK S. Knott 24 John Street, Leek ST13 8BL
MM0 SGQ S. Gill 5 Ramornie Place, Kingskettle, Cupar KY15 7PT
M0 SGS S. Priestley 49 Victoria Crescent, Leeds LS28 7SS
MW0 SGU C. Cowling 14 Shelley Drive, Bridgend CF31 4QA
M0 SGV S. Vanstone 2 Walker Crescent, Weymouth DT4 9AU
M0 SGW S. Whalley 1 Cambridge Road, Gatley, Cheadle, Stockport SK8 4AE
MW0 SGX J. Bidwell 26 Lone Road, Clydach, Swansea SA6 5HR
M0 SGZ J. Alincastre 14919 Hope Hills Lane, Cypress, Texas 77433 United States
M0 SHA Surbiton Heritage Amateur Radio c/o T. Fell 24 Ardmay Gardens, Surbiton KT6 4SW
MM0 SHB S. Boyd 6 Schaw Road, Prestonpans EH32 9HA
M0 SHD S. Hyde-Dryden 90 Broadoaks Grange, Carlisle CA1 2TA
M0 SHF N. Newman 1 Hadham Park Cottages, Cradle End, Ware SG11 2EH
M0 SHH P. Val 72 Devonport Road, London W12 8NU
M0 SHI M. Joshi 14 Doyle Close, Erith DA8 3QT
M0 SHK K. Holloway 6 Britons Lane Close, Beeston Regis, Sheringham NR26 8SH
M0 SHM S. Marriott 4 Stone Cross Gardens, Catterall, Preston PR3 1YQ
M0 SHN A. Al-Shakarchi 17 Fairfax Place, London NW6 4EJ
M0 SHO S. Howarth 19 Farnham Croft, Leeds LS14 2HR
M0 SHP S. Shepherd 24 Brayton Road, Whitehaven CA28 6EF
M0 SHQ S. Hedgecock 37 Tennyson Road, Maldon CM9 6BE
M0 SHR Hornsea ARC c/o P. Gaskell 131 Greenfield Road, Dentons Green, St. Helens WA10 6SH
M0 SHV Thales ARC c/o A. Sidhu-Brar White Gates, Main Road, Northampton NN7 3NA
M0 SHY S. Wildman 55 Hill Street, Bradley, Bilston WV14 8SB
M0 SHZ P. Shires 30 Philip Garth, Wakefield WF1 2LS
MM0 SIA Scottish Isles DX Group c/o B. Titmarsh Caberfeidh, Clachan Na Luib HS6 5HD
M0 SIH S. Hammond Ellsworth, Thrigby Road, Filby, Great Yarmouth NR29 3HJ
M0 SII Gilwell Park Scout RC c/o V. Lynch 16 Okehampton Crescent, Sale M33 5HR
MM0 SIL J. Connelly 60 Frankfield Street, Glasgow G33 1BU
M0 SIN T. Brundrett 45 Talbot Crescent, Whitchurch SY13 1PH
MW0 SIP A. Ferguson The Mount Stables, Salem, Llandeilo SA19 7HD
MJ0 SIT S. Whitfield Ceylon Cottage, Journeaux Street, St. Helier JE2 3XQ Jersey
M0 SIY S. Shaul 31 Chatterton Avenue Ermine West, Lincoln LN1 3SZ
M0 SJD S. Davies 2 Greenways, Hyde Lea, Stafford ST18 9BD
M0 SJG S. Goodwin 9 Downsview, Warminster BA12 9DU
M0 SJJ S. Jones 39 Dalton Lane, Barrow-in-Furness LA14 4LE
M0 SJK S. Kearley 36 Priory Road, Wirral CH48 7EU
M0 SJL S. Low 11 Bitterley Close, Ludlow SY8 1XP
M0 SJR S. Roberts 7 Alberta Grove, Prescot L34 1PX
M0 SJS S. Stanhope 61 Heathfield St., Manchester M40 1LF
M0 SJV S. Viney 5 Hawthorne Grove, Dudley DY3 2QQ
M0 SJW S. Whitehead 55 Crombie Road, Sidcup DA15 8AT
M0 SKA D. Bennett 2 Broadway, Blackburn BB1 8QZ
M0 SKC S. Clay Akers Lodge, 6 Penn Way, Rickmansworth WD3 5HQ
MW0 SKD E. Edwards 6 Kerslake Terrace, Tonypandy CF40 1EQ
M0 SKF S. Keating-Fry 18 Hewitt Avenue, London N22 6QD
M0 SKG Strood Kent CG c/o B. Howard 8 St. Margarets Close, Seasalter, Whitstable CT5 4ST
M0 SKI Dr J. Skittrall 14 Tamarin Gardens, Cambridge CB1 9GH
M0 SKM S. Marshall 96 Bidwell Hill, Houghton Regis, Dunstable LU5 5EP
M0 SKN Dr M. Romenskyy 11A Kensington Park Road, London W11 3BY
M0 SKO A. Skolik 13 Locks Meadow, Owls, Dormansland RH7 6AW
M0 SKP R. Ibbotson 33 St. Peters Avenue, Caversham, Reading RG4 7DH
M0 SKR South Kesteven ARS c/o S. Mason 3 Grange Cottages, Low Road, Barrowby, Grantham NG32 1DL
M0 SKT D. Webb 52 Simpkin Close, Eaton Socon, St. Neots PE19 8PD
M0 SKV M. Sherrey 25 Aspen House, Stratton-On-The-Fosse, Radstock BA3 4RW
MM0 SKX S. Kirkbride 18 North Roundall, Limekilns, Dunfermline KY11 3JY
M0 SKY L. Sparks 9 Hawk Place, Moresby Parks, Whitehaven CA28 8YG
MM0 SLB Stromness Academy ARC c/o D. Smith 23 Torness, Kirkwall KW15 1UU
M0 SLC K. Molnar 201 Fold Croft, Harlow CM20 1SW
M0 SLD S. Lovell 20 Courtenay Walk, Weston-Super-Mare BS22 7TQ
MI0 SLE D. Tarnowski 11 Millbrook Gardens, Kilrea, Coleraine BT51 5RZ
M0 SLF S. Farnell 16 Lily Way, Lowestoft NR33 8NN
M0 SLG D. Logan Cedar House, Reading Road North, Fleet GU51 4AQ
MW0 SLH Plymouth Radio Community c/o J. Hewitt 1 Highfield, Gloucester Road, Chepstow NP16 7DF
M0 SLL D. Firth 5 Mowhay Gardens, Hatherleigh EX20 3FE
M0 SLO Dr A. Saje 72 Bedworth Road, Bulkington, Bedworth CV12 9LL
M0 SLP S. Richardson 89 Mead End, Biggleswade SG18 8JR
M0 SLY S. Lyon 10 Sycamore Close, Preston, Hull HU12 8TZ
M0 SMA B. Beckett 38A Whinney Banks Road, Middlesbrough TS5 4HG
MM0 SMB B. Mcsherry 3 Taylor Road, Whitburn, Bathgate EH47 0NL
M0 SMC S. Mcgregor 2 Castleheath Close, Wirral CH46 3SL
MM0 SMD J. Nicol 18 Tininver Street, Dufftown, Keith AB55 4AZ
M0 SME 5xx Group Daventry c/o G. Bystryakov 20 Elmhurst Gardens, Leeds LS17 8BG
M0 SMF C. Smith 21 Mill House Drive, Cheltenham GL50 4RG
M0 SMG A. Booth 16 Coronation Street, Wessington, Alfreton DE55 6DX
M0 SMH S. Hassan 69 Waltham Close, West Bridgford, Nottingham NG2 6LD
M0 SMJ M. Seaward 7 St. Olafs Road, Stratton, Bude EX23 9AF
MW0 SML Sparks At the Shed c/o A. Adams 1 Nant Y Ffynnon, Letterston, Haverfordwest SA62 5SX
M0 SMN Dr D. Richardson 89A Bean Oak Road, Wokingham RG40 1RJ
M0 SMP S. Peel 21 Fairfield Avenue, Ormesby, Middlesbrough TS7 9BB
M0 SMQ S. Mansfield 545 Portswood Road, Southampton SO17 3SA
M0 SMS W. Pinkhardt 43 Cambrian Way, Calcot, Reading RG31 7DD
M0 SMT S. Tasker 94 Davenport Drive, Cleethorpes DN35 9JR
MI0 SMV S. Mcveigh 28 Waringfield Avenue, Moira, Craigavon BT67 0FA
MI0 SMY S. Dallas 101 Coagh Road Stewartstown, Dungannon BT71 5JL
M0 SMZ O. Carp Rothera Research Station, Stanley FIQQ 1ZZ Antarctica
M0 SNA P. Gacek 66C Church Street, Berwick upon Tweed TD15 1DU
M0 SNB Secret Nuclear Bunker CG c/o Dr G. Smart 30 Cornmills Road, Soham CB7 5AT
M0 SND J. Popple 11, Chapel Close, Waterbeach, Cambridge CB25 9JW
MI0 SNG S. Gilmour 14G Malcolm Road, Lurgan, Craigavon BT66 8DF
M0 SNJ N. Jones 61 Sutherland Close, Whitehill, Bordon GU35 9RE
MM0 SNK J. Dow 58 Beatty Crescent, Kirkcaldy KY1 2HS
MI0 SNY G. English 15 Murrays Hollows, Ballyroney, Banbridge BT32 5ES
M0 SNZ S. Dodd Bradden Lane, Gaddesden Row, Hemel Hempstead HP2 6JB
M0 SOA G. Mccourty The Orchard, Eaton, Tarporley CW6 9AJ
M0 SOC Second Class Operators Club (UK) c/o R. Pike 57 Bishopstone Road, Stone, Aylesbury HP17 8QR
M0 SOE B. Macmillan 27 Fiveways Rise, Deal CT14 9QN
M0 SOL Solway DX Group c/o C. Wolf 35A Moorhouse Road, Carlisle CA2 7LU
M0 SOO J. Barker Pearl Bungalow, Killerby Cliff, Cayton Bay, Scarborough YO11 3NR
M0 SOR S. Orchard 30 Wilkes Court, Ipswich IP5 2EQ
M0 SOT A. Cowan 217 South Park Road, Wimbledon, London SW19 8RY
M0 SOU J. Lovelock Sea Spray, Lighthouse Road The Lizard, Helston TR12 7NU
M0 SOV S. Riesenberg 30 Aberdare Gardens, London NW6 3QA
M0 SOX G. Galliver 29 Archery Fields, Odiham, Hook RG29 1AE
M0 SPA Staffordshire ARC c/o N. Briggs 20 Broad Lane, Pelsall, Walsall WS4 1AP
M0 SPB R. Evans 21 Quilter Close, Bilston WV14 9AX
M0 SPC C. Smith 59 Moneyfly Road, Verwood BH31 6BL
M0 SPD S. Davies 10 Knutsford Green, Wirral CH46 8TT
M0 SPH S. Hodkinson 17 Thorn Well, Westhoughton, Bolton BL5 2PJ
M0 SPJ P. Snook 7 Sandhurst Avenue, Kwazulu Natal 3610 South Africa
M0 SPK P. Susa 3 Ainsdale Drive, Whitworth, Rochdale OL12 8QB
MM0 SPL S. Ling Leadburnlea Leadburn, West Linton EH46 7BE
M0 SPN S. Netting 39 Poulton Street, Swindon SN2 1BH
M0 SPS A. Hutley 90 Main Road, Crick, Northampton NN6 7TX
M0 SPV A. Vitiello 8 Pegasus Road, Leighton Buzzard LU7 3NJ
M0 SPX Spixworth Scout Radio Group c/o P. Burgess 26 William Peck Road, Spixworth, Norwich NR10 3QB
M0 SPZ Dr S. Pearce 15 Hillfield Court Road, Gloucester GL1 3QS
M0 SQC Polish ARC c/o B. Jakubowski 120 Chandos Street, Coventry CV2 4HT
M0 SRA Simpson ARS c/o I. Ridings 25 Mond Road, Irlam, Manchester M44 6QA
M0 SRB S. Britten 10 Second Avenue, Wolverhampton WV10 9PP
M0 SRJ R. Shenton 14 , Leighton Close, Uttoxeter ST148SJ
MI0 SRM S. Mccormick 8 New Close, Portavogie, Newtownards BT22 1DZ
M0 SRN P. Holland 2 Blythorpe, Hull HU6 9HG
M0 SRO D. Coupe 6 Berry Avenue, Kirkby-In-Ashfield, Nottingham NG17 8GE
M0 SRP P. Skidmore 36 Princes Drive, Harrow HA1 1XH
MI0 SRR D. Poots 18 Upper Quilly Road, Dromore BT25 1NP
MM0 SRX Yeovil ARC c/o T. Kane 40B Brisbane Street, Greenock PA16 8NP
M0 SRZ S. Robottom-Scott 73 St. Bernards Road, Solihull B92 7DF
MW0 SSB I. Rowlands 22 Maes William Williams Vc, Amlwch LL68 9DS
M0 SSD G. Birkby 8 Kestrel Drive, Dalton-in-Furness LA15 8QA
M0 SSE J. Mossman 12 Cheviot Crescent, Hadston, Morpeth NE65 9SP
M0 SSF Canon P. Midwood 4 Larch Crescent, Holt NR25 6TU
MM0 SSG C. Haldane 6A Earls Gate, Bothwell, Glasgow G71 8BP
M0 SSH Dr S. Herman Barbary House, California Lane, Bushey WD23 1EX
M0 SSJ P. Dekkers 21 Nodens Way, Lydney GL15 5NP
M0 SSK K. Baker 64 Pendle Drive, Basildon SS14 3LZ
M0 SSM South Notts ARC c/o S. Mcmurtrie 5 Hill Road, Carshalton SM5 3RA
M0 SSN B. Woods 28 Delph Drive, Burscough, Ormskirk L40 5BE
M0 SSO M. Slater 8 Flaxman Rise, Oldham OL1 4QB
M0 SSP 759 Sqn(Beccles)ATC c/o R. Shillabeer 29 Newlease Road, Waterlooville PO7 7BX
M0 SSR S. Stewart 2 Broadlands, Clinton Way, Fairlight TN35 4DL
M0 SST Thames ARG c/o R. Finch 12 Simcox Street, Hednesford, Cannock WS12 1BG
M0 SSV S. Vickers 22 Thistle Green, Birmingham B38 9TT
M0 SSW Silcoates School Ar & Elec.Club c/o N. Wears 25 Topcliffe Mews, Morley, Morley LS27 8UL
M0 SSX Sussex 4x4 Response c/o A. Brierley Thorndale House, Brockhurst Farm, Watersfield, Pulborough RH20 1NX
M0 SSY R. Moss 27 Ravens Close, Bignall End, Stoke-on-Trent ST7 8QE
M0 STA R. Stafford 133 Essex Road, Stamford PE9 1LA
M0 STC S. Cheal 12 Heather Lane, Worthing BN13 3BU
M0 STF S. Binns 174 Enfield Chase, Guisborough TS14 7LQ
M0 STI D. Smith Heath Farm, Heath Road, Woolpit, Bury St. Edmunds IP30 9RL
M0 STJ C. Cherry 12 Scarisbrick New Road, Southport PR8 6PY
M0 STK P. Roberts 121 Hartshill Road, Stoke-on-Trent ST4 7LU
M0 STL A. Palmer 18 Windsor Avenue, Great Yarmouth NR30 4EA
M0 STN S. Neale 48 Five Acres Fold, Northampton NN4 8TQ
M0 STS G. Sowden The Grange Lodge, Rodley Lane, Calverley, Pudsey LS28 5QH
M0 STT S. Gordon 6 Aspinall Grove, Hailsham BN27 3GP
MM0 STU S. Macpherson 41 Mull Avenue, Port Glasgow PA14 6DP
M0 STV S. Ridgeon 7 Southlands, Haxby, York YO32 2PB
M0 SUD S. Griffiths 37 Stourton Close, Knowle, Solihull B93 9NP
M0 SUF S. Batley 2 Boulge Road, Hasketon, Woodbridge IP13 6LA
M0 SUG D. Eastlake 148 Pursey Drive, Bradley Stoke, Bristol BS32 8DP
M0 SUI R. Isnenghi Flat 24, Weller Court, Melvill Road, Falmouth TR11 4ES
M0 SUN C. Tate 11A Nether Lea, Cranage, Crewe CW4 8HX
M0 SUR Flt Refuelng ARS c/o P. Dickson 49 Signal Road, Grantham NG31 9BL
MM0 SUS I. Macdonald The Cottage, High Craigton, Glasgow G62 7HA
M0 SUU Spen Valley ARS c/o W. Malcolm-Brown Flat 11, Chiltern Court, Harpenden AL5 5LY
M0 SUY D. Valaris 18 Charnwood Gardens, Gateshead NE9 5SB
M0 SUZ Vulture Squadron CG c/o S. Coombes 33 Clarence Park Road, Bournemouth BH7 6LF

M0 SVA T. Papadopoulos 77 Cottrell Road, Bristol BS5 6TN
M0 SVB S. Bell 1 Cherwell Road, Aylesbury HP21 8TW
MM0 SVE S. Shaw 2 Highfield, Dalry KA24 4HP
M0 SVH Dr S. Hill 36 The Woodlands, Market Harborough LE16 7BW
M0 SVN S. Farnsworth 26 Burrows Close, Headington, Oxford OX3 8AN
M0 SVP G. Mylonas 68 Lancaster Gate, Cambourne CB236AT
M0 SVR S. Ring 35 Sturmer Close Yate, Bristol BS37 5UR
M0 SVT Dr A. Dickson 1 Roebuck Drive Baldwins Gate, Newcastle-under-Lyme ST5 5FE
M0 SVV S. Wade 42 Beauclerk Green, Winchfield, Hook RG27 8BF
M0 SVX A. Thorpe 8 Syke Avenue, Tingley, Wakefield WF3 1LU
MW0 SWB A. Jones 69 Hendre Gwilym, Tonypandy CF40 1HF
M0 SWC D. Brough 38 Tynedale Avenue, Crewe CW2 7NY
M0 SWE M. Sweeney 3 Orchard Cottages, Asenby, Thirsk YO7 3QW
M0 SWF S. Fry Mount Pleasant Farm, Thorpe Fendykes, Skegness PE24 4QR
M0 SWH S. Heard 29 Grange Farm Road, Yatton, Bristol BS49 4RB
M0 SWI R. Dixon Coach House, Chapel Lane, Ellel, Lancaster LA2 0PN
M0 SWL B. Bosson 80 White Horse Road, Marlborough SN8 2FE
MW0 SWN P. Barnes 55 Erw Werdd, Birchgrove, Swansea SA7 0HF
M0 SWO A. Sword Plevna 100 Eaton Road, Norwich NR4 6PS
MW0 SWR G. Waters 47 Close Yr Eryr, Bridgend CF35 6HE
M0 SWT M. Colpman 20 Rochford Road, Basingstoke RG21 7TQ
M0 SWV B. Elms-Lester Ferndale House Kerry'S Gate, Hereford HR2 0AH
M0 SWZ I. Swindells 69 Danby Close, Newton Moor, Hyde SK14 4AF
M0 SXA Essex Ham c/o P. Sipple 52 Fillebrook Avenue, Leigh-on-Sea SS9 3NT
M0 SXE Downland Radio Group c/o G. Keegan 12 Allington Road, Newick, Lewes BN8 4NA
M0 SXH S. Hunter 9 Gelt Burn, Didcot OX11 7TZ
M0 SXM S. Morris 23 De Courtenai Close, Bournemouth BH11 9PG
M0 SXN S. Neale 43 Crompton Road Pleasley, Mansfield NG19 7RG
M0 SYG A. Sygerycz 75 O'Brien Road, Cheltenham GL51 0UP
M0 SYJ P. Krzeminski Flat 46, Polden House, Bristol BS3 4LG
M0 SYL S. Krolak 179 Beatrice Street, Swindon SN2 1BD
M0 SYM S. Ludlam Lower Wolves, Bosham Hoe, Bosham, Chichester PO18 8EU
MI0 SYN J. Bradshaw 51 Albany Drive, Carrickfergus BT38 8BF
M0 SYS S. Strange 94 Digby Avenue, Nottingham NG3 6DY
M0 SYW S. Ward Wellbeck, Wheel Road, Alpington, Norwich NR14 7NH
M0 SYY C. Gibson 3 Conway Drive, Billinge, Wigan WN5 7LH
MM0 SZC S. Gohl 40 Tantallon Gardens, Bellsquarry, Livingston EH54 9AT
M0 SZD S. Denman 12 Dyke Vale Road, Sheffield S12 4ER
M0 SZL Z. Szot 45 Ealing Park Gardens, London W5 4EX
M0 SZQ S. Jones Marvin House, Ryhill Pits Lane, Cold Hiendley, Wakefield WF4 2DU
MW0 SZR A. Lamb 30 Dale Road, Queensferry, Deeside CH5 1XE
M0 TAA E. Slevin Woodcock Hall, Cobbs Brow Lane, Newburgh WN8 7NB
M0 TAB A. Brotherhood 5 Longcliffe Road Shepshed, Loughborough LE12 9LW
M0 TAD B. Catchpoole 8 Buckland Avenue, Basingstoke RG22 6JL
MW0 TAF E. Brookes 40 Llancayo Street, Bargod, Bargoed CF81 8TG
M0 TAJ T. Kemp 30 Tawny Sedge, King's Lynn PE30 3PW
M0 TAK T. Cooper Flat 6, Smiths Court, 73 East Borough, Wimborne BH21 1PJ
M0 TAL C. Travis 4 Kingsdale, Worksop S81 0XJ
M0 TAM M. Cox Havenview Stables, Main Road, Havenstreet, Ryde PO33 4DR
M0 TAN T. Nichols 12 Ivy Grove, Shipley BD18 4JZ
M0 TAO Dr O. Bock Okenstr. 34, Jena D-07745 Germany
M0 TAP W. Cooper 20 Staple Close, Waterlooville PO7 6AH
M0 TAQ F. Clements 40 Ellison Fold Terrace, Darwen BB3 3EB
M0 TAR T. Balls Rostan, 99 Front Road, Murrow, Wisbech PE13 4JQ
M0 TAT F. Felix Flat 41, King Edward Court, Wembley HA9 7DQ
M0 TAV V. Hopkins 6 Daimler Road, Coventry CV6 3GD
M0 TAW T. Woodhouse The Old Granary, 12 Limekiln Lane, Newport TF10 9EZ
M0 TAX E. Underhill 61 Goldthorne Avenue, Sheldon, Birmingham B26 3LA
M0 TAY A. Ayres Bryn Hyffryd, Phocle Green, Ross-on-Wye HR9 7TW
M0 TAZ D. Cutts 38 Berkeley Drive, Hornchurch RM11 3PY
M0 TBA A. Baker 97 Pond Lane, Wolverhampton WV2 1HG
MW0 TBB C. Morris 17 Percy Road, Wrexham LL13 7EA
MI0 TBD A. Kelly 19 Union Street Mews, Coleraine BT52 1EN
MI0 TBE E. Hill 24 Whitehouse Park, Newtownabbey BT37 9SQ
M0 TBG Team Thunderbox c/o C. Moulding 28 Queens Avenue, Highworth, Swindon SN6 7BA
MM0 TBH J. Kelly 41 Glenshee Street, Glasgow G31 4RT
M0 TBI S. Smith 12 Stoneleigh Avenue, Hordle SO41 0GS
M0 TBJ T. Buck 6 Lynn Road, Terrington St. Clement, King's Lynn PE34 4JX
M0 TBK E. Cree 24 Old Lincoln Road, Caythorpe, Grantham NG32 3EJ
MI0 TBN S. Donnelly 14 Derryloste Road, Derrytrasna, Craigavon BT66 6PS
M0 TBQ D. Nicholls 62 Queen Elizabeth Way, Telford TF3 2JW
M0 TBR R. Thorogood 4 Deerhurst Close, Calcot, Reading RG31 7RX
M0 TBS T. Tiesdell-Smith 4 Godwin Close, Epsom KT19 9LD
MI0 TBV T. Mckee 4 Earlford Heights, Newtownabbey BT36 5WZ
M0 TBW R. East 6 Ashley Road, Worcester WR5 3AY
MM0 TBY S. Turnbull 15 Woodruff Gait, Dunfermline KY12 0NL
M0 TBZ Dr C. Cowen Rosita, White Street Green, Sudbury CO10 5JN
M0 TC T. Palmer Edison House, Bow Street, Great Ellingham NR17 1JB
M0 TCA T. Codner-Armstrong 22 Thoresby Road, Rainworth, Mansfield NG21 0DS
M0 TCB D. Howarth 6 Lyndhurst Close, Norton, Doncaster DN6 9PY
M0 TCC Dr T. Choy 39 Netherton Road, Manchester M14 7FN
M0 TCD A. Allen Milverton, Mill Road West Chiltington, Pulborough RH20 2PZ
M0 TCE Dr C. Eaglen 46 Sark Close, Hounslow TW5 0PZ
M0 TCF L. Allen 481 Topsham Road, Exeter EX2 7AQ
MW0 TCJ T. Jones 4 Llys Y Nant Glais, Swansea SA7 9JB
M0 TCL D. Mort 7 Sheldon Avenue, Congleton CW12 3LD
M0 TCM the Home Counties Atv Group c/o A. Nightingale 42 Spilsby Road, Horncastle LN9 6AW
M0 TCN C. Lyne 4 Bridge Close, Catterick Garrison DL9 4PG
M0 TCO T. Corcoran 191 Queensway, Rochdale OL11 2NA
MM0 TCP K. Brown 33 Windsor Gardens, Largs KA30 9DN
MM0 TCQ T. Campbell 10 Barra Gardens, Old Kilpatrick, Glasgow G60 5HR
M0 TCR T. Rozier 124 Deansfield Road, Wolverhampton WV1 2LD
M0 TCT T. Collins 11 Joseph Gardens, Silver End, Witham CM8 3SN
MM0 TDB D. Gartshore 85 Springhill Street, Douglas, Lanark ML11 0NZ
M0 TDC R. Stevenson 97 Queen Street, Crewe CW1 4AL
M0 TDD M. Xian 39 Belson Road, London SE18 5PU
MW0 TDF W. Welch Kenilworth, School Lane, Gobowen, Oswestry SY11 3LD
M0 TDG T. Grant 51A Hursley Road, Eastleigh SO53 2FS
M0 TDK A. Tyrwhitt-Drake Holly Cottage, Church Lane, Beccles NR34 0AU
M0 TDM R. Hydes 60 Handsworth Grange Road, Sheffield S13 9HH
M0 TDP A. Pickett 4 Trembel Road, Mullion, Helston TR12 7DY
MW0 TDQ J. Grzywaczewski 16 Dyffryn Road, Port Talbot SA132UG
MM0 TDS D. Scott Farewell, Arnage, Auchnagatt, Ellon AB41 8UW
MW0 TDV R. Smith Hendafarn, Sarnau, Llanymynech SY22 6QJ
M0 TDW B. Woodroffe 2 Little Mead, Shalbourne, Marlborough SN8 3QB
M0 TDY A. Askam 8 The Pastures, Weston-On-Trent, Derby DE72 2DQ
MW0 TDZ T. Clapp Crunns Farm. Coxhill, Narberth SA67 8EH
M0 TEA A. Goddard 50 Ardmore Walk, Manchester M22 5QG
M0 TEB M. Bell 36 Schneider Road, Barrow-in-Furness LA14 5DW
M0 TEF A. Smith 101 Chaucer Drive, Lincoln LN2 4LT
M0 TEG D. Horner 21 Ainsworth Road, Little Lever, Bolton BL3 1RG
MM0 TEI Worcester Radio Amateurs Association c/o A. Wright 17A James Court, 493 Lawnmarket, Edinburgh EH1 2PB
M0 TEK E. Moore 44 Bridge Street, Oxford OX2 0BB
M0 TEN E. Williams 15 Tenth St., Peterlee SR8 4NE
M0 TER B. Ashcroft 16 Edge Lane, Thornton, Crosby., Liverpool L23 9XE
M0 TES C. Brown Town End House, Ulverston Road, Gleaston, Ulverston LA12 0PZ
M0 TET A. Ford 5 Prenede, Roches 23270 France
M0 TEX R. Rushlow 94 Tennyson Street, Guiseley, Leeds LS20 9LW
M0 TEZ T. Mullaney 8 Westerham Close, Macclesfield SK10 3BG
M0 TFB B. Titmus 68 Hobart Road, Cambridge CB1 3PT
M0 TFC Edgware & District RS c/o P. Kirkden 22 Leas Green, Broadstairs CT10 2PL
M0 TFF R. Cichocki 12 Crossland Crescent, Wolverhampton WV6 9JY
M0 TFH T. Hull 12 Durley Road, , Gosport PO12 4RT
MI0 TFK R. Vage 80 Chinauley Park, Banbridge BT32 4JL
M0 TFN T. Nolan 2 Shore Road, Cowes PO31 8LB
M0 TFO R. Styles 52 Vernham Grove, Bath BA2 2TB
M0 TFS T. Smith 43 Hall Street, St. Helens WA94XN
MM0 TFU I. Macalister 33 King Street Crosshill, Maybole KA19 7RE
M0 TFX T. Fisk 2 Hall Farm Cottage, Caston Road, Caston, Attleborough NR17 1BW
M0 TFY D. Butler Church Cottage, Church Road, Badminton GL9 1HT
MM0 TGB T. Brown 11 Approach Row, East Wemyss, Kirkcaldy KY1 4LB
M0 TGC J. Hay Hudson House, Fremington Road, Seaton EX12 2HX
M0 TGE T. Gentry 8 Bowldown Cottages, Bowldown Road, Westonbirt GL8 8UD
M0 TGF M. Leach 64 Grove Street, Wantage OX12 7BG
MM0 TGG G. Jamieson 6 Maryville Park, Aberdeen AB15 6DU
MI0 TGL R. Greer 11 , Mullaghcarton Road, Lisburn BT28 2TE
M0 TGM D. Pask Apartment 403, 1314 Tower Road, Halifax NS B3H 4S7 Canada
M0 TGN D. Trudgian 18 Hart Close, Wootton Bassett, Swindon SN4 7FN
M0 TGS Altrincham Grammar School for Boys c/o G. Binns 21, Rydal Close, Winsford CW7 2SE
M0 TGT S. Faulkner Mount Pleasant, Elkstones, Buxton SK17 0LU
M0 TGV G. Cooke 12 Marcus Road, Dartford DA1 3JX
M0 TGW M. Rigby 75 Manchester Road, Deepcar, Sheffield S36 2QX
M0 TGX T. Green 35 Park Road, Allington, Grantham NG32 2EB
M0 TGY T. Guy 16 Cogdeane Road, Poole BH17 9AS
M0 THA T. Hurren 4 Grove Bungalows, Upper Street, Horning, Norwich NR12 8NF
M0 THB Dr T. Barratt Flat 41, The Panoramic, 30 Park Row, Bristol BS1 5LS
MM0 THE A. Lang 202 Devonside Road, Carmichael, Biggar ML12 6PQ
M0 THF Tawney Hall Farm ARC c/o R. Clark 41 Avenue Road, Bexleyheath DA7 4
M0 THJ A. Howell-Jones 11 Staffick Close, Kenton, , Exeter EX6 8NS
M0 THM T. Mcconnell 51 Langney Road, Eastbourne BN21 3QD
M0 THN Castle House School ARC c/o R. Blane Redfield, Buckingham Road, Buckingham MK18 3LZ
M0 THO A. Boato Via A Diaz 20, Marcon, Venezia 30020 Italy
MJ0 THP M. Thorpe Dolphin Cottage, Union Road, Grouville, Jersey JE3 9ER
M0 THT Laser ATC Rac (South) c/o T. Toon 9 Boundstone Lane, Sompting, Lancing BN15 9QL
M0 THY H. Tang 14 Bridle Close, Borehamwood WD6 5QA
M0 THZ G. Mattocks 70 Doncaster Road, Hatfield, Doncaster DN7 6AT
MM0 TIE R. Adamson 6 Camdean Crescent, Rosyth, Dunfermline KY11 2TJ
M0 TIF J. Housego 16 Ligo Avenue, Stoke Mandeville HP225TX
M0 TIK C. Pascoe Treleven, Primrose Hill, Goldsithney, Penzance TR20 9JR
M0 TIL Coalhouse Fort c/o J. Parker 76 Elm Road, Grays RM17 6LD
M0 TIN D. Le Grove Apartment 3, Beechwood, Ilkley LS29 8AH
MI0 TIP W. Thompson 25 Darby Road, Carrickfergus BT38 7XU
MM0 TIR M. Apaza Machaca 12 Donald Street, Dunfermline KY12 0BY
M0 TIU A. Beaumont 112 Whittington Road, Hutton, Brentwood CM13 1JZ
M0 TIW A. Thornton 78 Wellington Road, Ryde PO33 3QJ
M0 TIX R. Cowles Bonnie Rock, 76 Fordham Road, Ely CB7 5AL
M0 TIZ A. Tyrrell Flat 1, 61 Vicarage Road, Eastbourne BN20 8AH
M0 TJB T. Barnes Flat 38, Mill Court, Edinburgh Gate, Harlow CM20 2JG
MW0 TJD T. Davies 58A Ynyswen Road, Treorchy CF42 6ED
M0 TJJ T. Jinkerson 104 Foxcote, Finchampstead, Wokingham RG40 3PE
M0 TJL T. Leavold 129 Aylsham Road, Norwich NR3 2AD
MI0 TJM Elderslie ARS c/o T. Mulholland 215 Finaghy Road North, Belfast BT11 9ED
M0 TJN T. Newton 4 Manor Close, Bradford Abbas, Sherborne DT9 6RN
MM0 TJR T. Thorne Top Flat, 26 Mary Elmslie Court, Aberdeen AB24 5BE
MW0 TJS T. Scott 44 Croft Road Broad Haven, Haverfordwest SA62 3HY
MM0 TJT the Jaggy Thistles c/o W. Findlay 46 Rowallan Drive, Kilmarnock KA3 1TU
M0 TJU E. Duffield 92 Crosby Street, Stockport SK2 6SP
M0 TJV C. Vernon 29 Alice St., Deane, Bolton BL3 5PJ
M0 TJW T. Beardwood Flat 9, Alma House, Ripon HG4 1NG
M0 TJX T. Humphreys 93 Cornwall Crescent Yate, Bristol BS37 7RU
M0 TKA T. Kay 64 Cowcliffe Hill Road, Huddersfield HD2 2PE
M0 TKC T. Canning Whitegates, Mayfield Avenue New Haw, Addlestone KT15 3AG
M0 TKD K. Raistrick 2 Greenacres Grove, Shelf, Halifax HX3 7RN
MM0 TKE Prof. T. Kerby 1 St. Mark'S Lane, Edinburgh EH15 2PX
M0 TKF R. Nicholson 27 Bishopton Way, Hexham NE46 2LR
M0 TKG Dr T. Bishop 2 Jennings Street, Stockport SK3 9BX
M0 TKM M. Gosi 49 Elms Drive, Marston, Oxford OX3 0NW
M0 TKR C. Barker 11 Long Meadows, Chorley PR7 2YA
M0 TKS Dr T. Kyriacou 54 Sutton Avenue, Silverdale, Newcastle-under-Lyme ST5 6TB

M0 TKT R. Bradshaw 272 Councillor Lane, Cheadle Hulme, Cheadle SK8 5PN
M0 TKU D. Carter 3 Queens Road, Brentwood CM14 4HE
M0 TKW T. Kelly 32 The Foxgloves, Hedge End SO30 0UG
M0 TKX Qrz ARG of Sussex c/o A. Sood Parima, Sewardstone Road, London E4 7RA
M0 TLC R. Ainsworth 181 Carlton Road, Boston PE21 8NG
MI0 TLF T. Flanagan 18 Hunters Park, Bellaghy, Magherafelt BT45 8JE
MI0 TLG R. Greer 11 , Mullaghcarton Road, Lisburn BT28 2TE
M0 TLJ R. Mcwilliam 3 Fountains Close, Riccall, York YO19 6QN
M0 TLL B. Smith 24 Oakwood Park, Leeds LS8 2PJ
M0 TLM British Young Ladies Club c/o I. Williams 36 Telford Road, Tamworth B79 8EY
M0 TLN S. Moissejev 41 Queens Road, Caversham, Reading RG4 8DN
M0 TLO R. Hunter 3 Sandy Way, Croyde, Braunton EX33 1PP
M0 TLR G. Taylor Thorndale, Oakfield Drive, Boxted, Colchester CO4 5RX
M0 TLX D. Burdsall 37 Fulmar Walk, Whitburn, Sunderland SR6 7BW
M0 TLY M. Casey 7 Cobham Avenue, Manchester M40 5QW
M0 TMA T. Moncaster 24 Minster Yard, Lincoln LN2 1PY
M0 TMB R. Cook 6 Aster Road, Ipswich IP2 0NQ
M0 TMC E. Newby 22 Acton Road, Liverpool L32 0TT
M0 TMD H. Melhuish 22 Mayflower Close, Glossop SK13 8UD
M0 TMF A. Fullwood 16 Hollands Place, Walsall WS3 3AU
MM0 TMG K. Cussick 15A Finlow Terrace, Dundee DD4 9ND
MW0 TMH T. Mitchell 9 Rhiw Grange, Colwyn Bay LL29 7TT
MW0 TMI Kingsmead School c/o D. Willis 51 Fforchaman Road, Cwmaman, Aberdare CF44 6NG
MW0 TMJ T. James Penrallt, Mountain, Holyhead LL65 1YR
M0 TMM M. Elliott 60A Forest Street, Shepshed, Loughborough LE12 9DA
M0 TMN T. Nguyen 9 Green Street, Cambridge CB2 3JU
M0 TMO K. Chadwick 17 Nettlebed Nursery, New Road, Shaftesbury SP7 8QS
MI0 TMP T. Mcelwee Orchard Bank, 1A Dunover Road, Ballywalter, Newtownards BT22 2LE
M0 TMS T. Schwabe 112 Clarkson Court, Hatfield AL10 9GW
M0 TMT R. Aldridge 37 Vincent Road, Luton LU4 9AN
M0 TMX D. Mcglone 32 Shipley Mill Close, Kingsnorth, Ashford TN23 3NR
MM0 TMZ A. Miles 9 Buchanan Drive, Lenzie, Kirkintilloch, Glasgow G66 5HS
M0 TNB M. Jakubowski 75 Ashcombe Road, London SW19 8JP
M0 TNC A. Burton 12 Munden Grove, Watford WD24 7EE
M0 TNE T. Newman 10 Dereham Road, Garvestone, Norwich NR9 4AD
M0 TNF G. Barnard Red Ridges Ghyll Road, Crowborough TN6 1SU
M0 TNG Havering & District ARC c/o S. Adaway 20 Foundry Street, Barnsley S70 1PL
M0 TNL V. Behal 21 Bromley Road, Walthamstow, London E17 4PR
M0 TNT A. Roberts Chy Kerenza, Parc Morrep, Penzance TR20 9TE
M0 TNV M. Clough 8 Skeldyke Road, Kirton, Boston PE20 1LR
M0 TNX K. Haworth 11 Petersfield Close, Bootle L30 1SG
MM0 TOB T. Burnett 8 Kyles View, Colintraive PA22 3AS
M0 TOF J. Morgan Glas Y Dorlan, Pontrhydfendigaid, Ystrad Meurig SY25 6EJ
M0 TOG D. White Woodpeckers, Top Green, Romsey SO51 0JP
M0 TOL Northampton RC c/o A. Rixon 17 Brimmers Way, Aylesbury HP19 7HR
M0 TOM T. Wilcox 7 Jacob Court, Billinge, Wigan WN5 7GE
M0 TOP A. Topsfield Wild Willow Cottage, Hancock Lane, Truro TR2 5DD
M0 TOR J. Dearden 7 Wadworth Street, Denaby Main, Doncaster DN12 4EN
M0 TOZ G. Agostinelli 16 George Street, Shefford SG17 5BS
M0 TPA A. Patrick 2 Beacon Grange, Malvern WR14 3EU
M0 TPE T. Pearce 101 Primrose Hill, Widmer End, , High Wycombe HP156NT
M0 TPG A. Gravell 21 Wickridge Close, Stroud GL5 1ST
M0 TPH G. Emsden Flat 47, Cedar Court, London N10 1EG
M0 TPJ T. Mallaband 29 Ferndale Road, Burgess Hill RH15 0HB
M0 TPL A. Capon 1 Windermere Way, North Common BS305XN
M0 TPT D. Millward 77A Meadowcroft, St. Albans AL1 1UG
M0 TPW T. Winyard 48 Windsor Drive, Yate, Bristol BS37 5DY
M0 TPY T. Pollett 10 Bridport Road, Poole BH12 4BS
MM0 TQH R. Hay Roddach Cottage East, Cummingston, Burghead, Elgin IV30 5XY
M0 TQV R. Tuckett 89 Hillbrook Road, Tooting, London SW17 8SF
M0 TRB T. Bowden Carmel, Swallowcliffe, Salisbury SP3 5PW
MI0 TRC P. Clarke 26 Derryhale Lane, Portadown, Craigavon BT62 4HL
MW0 TRE Dr A. Kicman Jasmay, Grimpo, West Felton, Oswestry SY11 4HG
M0 TRF T. Knox 21 St. Annes Avenue, Bournemouth BH6 3JR
M0 TRK R. Tarling Moonrakers, Ashley, Box, Corsham SN13 8AN
M0 TRN T. Horsten Kastelsvej 4, 2.Tv, Copenhagen E 2100 Denmark
MW0 TRO A. Roberts 14 Beech Road, Monmouth NP25 5DA
M0 TRP A. Pursglove 78 Alfreton Road, Westhouses, Alfreton DE55 5AJ
MM0 TRS S. Troscheit 20 James Street, St. Andrews KY16 8YA
M0 TRU A. Hartwell 41 Orchard Grove, Brixham TQ5 9RH
M0 TRV T. Hammett 9 Coral Close, Aughton, Sheffield S26 3RB
M0 TRW T. Wormald 12 Church Lane, Owermoigne, Dorchester DT2 8HS
M0 TRX Dr M. Cooper Apartment 33, St. Peters Court, 2 St. Peters Street, Worcester WR1 2PJ
M0 TRY R. Barnes 1 Moira Close, Chaddesden, Derby DE21 4RL
M0 TRZ S. Jordan 8 Averham Close, Swadlincote DE11 9SG
M0 TSA I. Macfarlane 70 Ashby Drive, Rushden NN10 9HH
MM0 TSB J. Morris 42B Church Street, Borve, Isle of Lewis HS2 0RT
M0 TSD S. Smith 103 Comberford Road, Tamworth B79 8PE
M0 TSM M. Bull Sunrise, Ram Lane, Norwich NR15 2DG
M0 TSN M. Lee 46 Little Lane, Huthwaite, Sutton-in-Ashfield NG17 2RA
M0 TSW T. Walker 11 Banburies Close, Bletchley, Milton Keynes MK3 6JP
M0 TSZ T. Skorzewski 40 Hedges Way, Luton LU4 9FD
M0 TTB A. Bright 86 Fourth Avenue, Watford WD25 9QQ
M0 TTE S. Fairbourn 17 Perry'S Lane, Wroughton, Swindon SN4 9AX
MW0 TTF D. D'Mellow 2 Twynpandy Pontrhydyfen, Port Talbot SA12 9TW
M0 TTG Greenisland Electronics ARS c/o B. Gale Tall Trees Farm, Noah'S Ark Lane, Great Warford WA16 7AX
M0 TTH T. Haley 3 Akeman Rise, Ramsden, Chipping Norton OX7 3BJ
M0 TTI Dr S. White Upton Farm, Upper Strode, Bristol BS40 8BG
MW0 TTK M. Buxton 25 Pen Y Bryn, Sychdyn, Mold CH7 6EE
M0 TTL A. Dickinson 18 Cullingworth Avenue, Hull HU6 7DD
M0 TTN R. Colman-Whaley 37 Suters Drive, Taverham, Norwich NR8 6UU
M0 TTO G. Grant 15 Watson Close, Rugeley WS15 2PE
MW0 TTU M. Evans The Brae, Coed-Cae-Ddu Road, Blackwood NP12 2DA
M0 TTX G. Watkins 21 Comberton Av, Kidderminster DY10 3EG
M0 TTY D. Spence 30 Chestnut Drive, Shirebrook, Mansfield NG20 8NH
MI0 TUB D. Given 15 Middle Road, Lisburn BT27 6UU
M0 TUI H. Drysdale Waunycaerau Ffynonn Gynnydd, Hereford HR3 5ND
M0 TUK P. Ray 136 Haselbury Road, London N18 1QD
M0 TUM R. Robinson 4 Limetree Court, Taverham, Norwich NR8 6QY
M0 TUN G. Bergeret 20 Rue Labrouste, Paris 75015 France
M0 TUR D. Biyikli Basement, 300 Portobello Road, London W10 5TA
M0 TUT S. Prescott 210 Inver Road, Blackpool FY2 0LW
M0 TUV A. Dingwall 48 Village Farm Caravan Site, Bilton Lane, Harrogate HG1 4DL
M0 TUW R. Harris 7 Fosse Lane, Shepton Mallet BA4 4PS
M0 TUX B. Sutton 4 Church Cottages, Carriers Road, Cranbrook TN17 3JR
M0 TVA C. Beresford 13 Chaseside Avenue, Twyford, Reading RG10 9BT
M0 TVG M. Shurley 43 Charles Close, Wroxham, Norwich NR12 8TU
M0 TVL C. Mackay 665A Edenfield Road, Rochdale OL11 5XE
M0 TVR T. Parker 100 Horsebridge Hill, Newport PO30 5TL
M0 TVS A. Travis 2 Home Farm Cottage, Ossington, Newark NG23 6LH
M0 TVT K. Wilson 26 Mill Field, Sutton, Ely CB6 2QB
M0 TVU GNRS c/o P. Swingewood 9 Goodall Grove, Great Barr, Birmingham B43 7PQ
M0 TVV M. Hillman Flat 5, 32 South Terrace, Littlehampton BN17 5NU
M0 TVX R. Taylor 6 Appleton Drive, Belper DE56 1FQ
M0 TWC Travelling Wave CG c/o K. Haywood 6 Lydney Road, Urmston, Manchester M41 8RN
M0 TWG P. Hallewell 32 Shaldon Grove, Aston, Sheffield S26 2DH
M0 TWJ W. Twemlow Flat 6, 27 Marmion Road, Liverpool L17 8TT
MM0 TWK C. Hall 3 Academy Street, Tain IV19 1ED
M0 TWM J. Nethercott 15 French Gardens, Cobham KT11 2AJ
M0 TWO P. Dunn 13 Stanton Avenue, Newsham Farm Estate, Blyth NE24 4PL
M0 TWS T. Wood 44 Wincobank Lane, Sheffield S4 8AA
M0 TWW T. Larman 861B London Road, Westcliff on Sea SS0 9SZ
MM0 TWX Dr P. Calvi-Parisetti 1 Aytoun Road, Glasgow G41 5RL
M0 TXA D. Harvey 5 Tithe Barn Drive Bray, Maidenhead SL6 2DF
M0 TXB C. Howell 47 Birch Park Coalway, Coleford GL16 7RU
M0 TXD G. Rowberry 32 Tiree Avenue, Worcester WR5 3UA
M0 TXH H. Gruen Allfrey House, Herstmonceux, Hailsham BN27 4RS
M0 TXK M. Fletcher Wray 70, Crook O'Lune Caravan Park, Caton Road, Lancaster LA29HP
M0 TXL P. Dunnicliffe 19 Woodland Road, Chelmsford CM1 2AT
MI0 TXM A. Mcgarvey 66A Scaddy Road, Downpatrick BT30 9BS
M0 TXN S. Moore 5795 Jefferson Commons Way, Apt. 203, Kalamazoo, Michigan 49009 United States
MM0 TXO Radio Amateur Invalid And Blind Club c/o A. Reid Johnston Farm, Leslie, Insch AB52 6PD
M0 TXP P. Cassells 49 Dodds Lane, Maghull, Liverpool L31 0BD
M0 TXS M. Townsend 25 Barton Road, Bedford MK42 0NA
M0 TXX G. Acton 39 Craig Road, Macclesfield SK11 7YH
M0 TYG D. Moore Camara, 379 Main Road, Harwich CO12 4DW
M0 TYH T. Hill 14 Hunters Mead, Motcombe, Shaftesbury SP7 9QG
M0 TYK J. Shephard 19 Duffryn, Telford TF3 2BU
M0 TYN G. Cockburn 20 Hexham Avenue, Hebburn NE31 2HN
MM0 TYR C. Taylor 47 Seaview, Knock, Isle of Lewis HS2 0PD
M0 TYW W. Hibberd 169 Highbury Grove, Cosham, Portsmouth PO6 2RL
M0 TZD A. Maiden 79 Green End Road, Manchester M19 1LE
M0 TZM T. Menzies 4 Meadow Road, Muxton, Telford TF2 8JH
M0 TZO P. Gibson 60 Raglan Road, Bromley BR2 9NW
M0 TZR P. Haygarth 5 Forth Close, Peterlee SR8 1DG
M0 TZT S. Emmett Middle Farm, East Side, North Littleton, Evesham WR11 8QW
M0 TZY S. Crabb 1 Council Houses, Hall Lane, Crostwick, Norwich NR12 7BB
M0 TZZ P. Moore 24 Plough Road, Dormansland, Lingfield RH7 6PS
MW0 UAA D. Bowen 25 Maendu Terrace, Brecon LD3 9HH
M0 UAC D. Carter Bungalow 7, Higher Ingsdon Quarry, Liverton, Newton Abbot TQ12 6JA
M0 UAR UK Amateur Radio Discord c/o D. Keene Firemark Cottage West Street Odiham, Hook RG29 1NT
M0 UAS D. Williams 2 Tyning Road, Peasedown St. John, Bath BA2 8HT
M0 UAT I. Marsh 56B Oliver Crescent, Farningham, Dartford DA4 0BE
M0 UAV T. Ward Flat 26, Bassett Court, Bassett Avenue, Southampton SO16 7DR
MI0 UBE L. Calderwood 43 Rathview Park, Mullybritt, Enniskillen BT94 5EW
M0 UCD J. Turner 17 Beechwood Road, Dronfield S18 1PW
M0 UCH C. Howard 1 Beale Road, Cheltenham GL51 0JN
M0 UCK A. Manning 10 The Quarry, Kidderminster DY10 2QD
MW0 UCL K. Ucele 54 Sebastopol Street, St. Thomas, Swansea SA1 8BL
M0 UDA A. Cattell 2 St. James Close, Ruscombe, Reading RG10 9LJ
M0 UDB D. Beard 1 Bond Close, Leonard Stanley, Stonehouse GL10 3GQ
MM0 UDI R. Duncan South Backieley, Turriff AB53 4GS
M0 UDK D. Cater 6 Mere Oak Road Perton, Wolverhampton WV6 7NB
M0 UDL D. Woodhams 83 Langdale Place, Newton Aycliffe DL5 7DY
M0 UEE R. Goldsack 5 Parc Dellen, Croft Farm Park, Luxulyan, Bodmin PL30 5EW
M0 UEH Dr S. Smith 557, Riverside Island Marina, Isleham, , Ely, CB7 5SL.
M0 UEI I. Dukes 36 Red Barn Road Brightlingsea, Colchester CO7 0SJ
M0 UET R. Taylor 68 Charter Road, Chippenham SN15 2RA
M0 UEZ R. Scholefield 4 Minnie Street, Haworth, Keighley BD22 8PR
M0 UFA M. Atherton 4 Bakers Park Saltney, Chester CH4 8FB
M0 UFC M. Bryant 284 Brantingham Road Chorlton Cum Hardy, Manchester M21 0QU
MI0 UFL A. Davidson 8 Rashee Court, Ballyclare BT39 9SE
MI0 UFT J. Martin 23 Winters Gardens, Omagh BT79 0DZ
M0 UGD D. Underwood 24 Wheatcroft Road, Rawmarsh, Rotherham S62 5ED
M0 UGE R. Evans 11 Swane Road Stockwood, Bristol BS14 8NQ
M0 UGG Dr R. Weller 67 Uplands Way, Diss IP22 4DF
M0 UGH A. Stevenson Antonienstrasse 40, Berlin 13403 Germany
M0 UGM G. Mountain 34 Albert Road, Warlingham CR6 9EP
MI0 UGP Rushyhill RS c/o T. Thompson 8 Knockburn Avenue, Lisburn BT28 2QF
M0 UGR C. Luckett 257 Folkestone Road, Dover CT17 9LL
M0 UGX M. Street 7 Salisbury Street, Sowerby Bridge HX6 1EE
M0 UHH E. Hanna 35 Orchard Drive Mayland, Chelmsford CM3 6EP
MM0 UIG M. Mackinnon 17 Valtos, Miavaig, Isle of Lewis HS2 9HR
M0 UJD K. Colman 10 South Rise, North Walsham NR28 0EE
M0 UKA Wellington School RS c/o R. Wood 7 Wishart Green, Old Farm Park, Milton Keynes MK7 8QB

M0 UKF F. Hennigan 18 Friary Gardens, Newport Pagnell MK16 0ZJZ
M0 UKI UK Islands Group c/o C. Wilmott 60 Church Hill, Royston, Barnsley S71 4NG
M0 UKM M. Busch Dammstrabe 4, Neuwied 56564 Germany
M0 UKN A. Mcdonald 11 Micklegate Murdishaw, Runcorn WA76HT
M0 UKO A. Shaw 9 Stamford Lane, Warmington, Peterborough PE8 6TW
M0 UKS J. Banham Timandra, Mill Road, Hardwick, Norwich NR15 2ST
MM0 UKW C. Houston 8 Gray Crescent, Irvine KA12 8HS
MW0 UKX G. Ralls 31 Greenfield Terrace, Penydarren, Merthyr Tydfil CF47 9HN
M0 ULB T. Tate 339 West Dyke Road, Redcar TS10 4PS
M0 ULC W. Biernacki 2A Tewkesbury Terrace, London N11 2LT
M0 ULD M. Elford 10 Meadowlands, Lymington SO41 9LB
M0 ULE M. Da Silva Casquilho Suite 9559-488 Unit 9, Skyport Drive, West Drayton UB7 0LB
MI0 ULK S. Morrow 769 Farranseer Park, Macosquin, Coleraine BT51 4NB
MM0 ULL K. Mackenzie Alderwood, Braes, Ullapool IV26 2TB
M0 ULR J. Gromadzki 13 Merrill Heights, Maidenhall Approach, Ipswich IP2 8GA
M0 UMG M. Ribbands Dyson'S Farm, Long Row, Tibenham NR16 1PD
MM0 UMH L. Mitchell Hynd Smithy House, Bruichladdich, Isle of Islay PA49 7UN
M0 UMM A. Insarov Broadway Chambers, 20 Hammersmith Broadway, London W6 7AF
M0 UMS M. Salt 1 Chantry Close, Harrow HA3 9QZ
MI0 UNA U. Murray 80 Canterbury Park, Londonderry BT47 6DU
M0 UNI G. Rigby Gas House Farm, Shavington Park, Market Drayton TF9 3SY
M0 UNJ A. Perek 28 Sefton Avenue, Plymouth PL4 7HB
M0 UNM Blackpool Amateur Radio Network c/o L. Hopgood 62 Briarwood Drive, Blackpool FY2 0EB
M0 UNN S. Jukna 85 St. Davids Crescent, Aspull, Wigan WN2 1SZ
M0 UNU H. Ilie Sos. Iancului Nr. 33 Bloc 105A Scara B Apt 63, Bucharest 21717 Romania
M0 UOE Harpenden And Wheathampstead District Scout Group c/o U. Nehmzow University Of Essex, Department Of Biological Sciences, Colchester CO4 3SQ
M0 UOG the University of Greenwich c/o P. Smith University Of Greenwich, Park Row, Greenwich SE10 9LS
M0 UOK B. Eddy 345 Moo 2, Phon Thong., Chaiyaphum 36000 Thailand
M0 UOO R. Bone 6 Danehurst Place, Locks Heath, Southampton SO31 6PP
M0 UOY University of York ARC c/o M. Walker 19 Highbury Place, Headingley, Leeds LS6 4HD
M0 UPA J. Van Der Elsen 6 Kent Close, Churchdown, Gloucester GL3 2HQ
MW0 UPH A. Williams 8 Old Tanymanod Terrace, Blaenau Ffestiniog LL41 4BU
M0 UPL C. Panaitescu 131 Stafford Road, Croydon CR0 4NN
M0 UPS V. Vaznais 35 Lynwood Drive, London KT4 7AA
M0 UPU A. Stirk 59 West Avenue, Lightcliffe, Halifax HX3 8TJ
M0 UQB T. Chan 29 Marthall Drive, Sale M33 2XP
M0 URB G. Urban 33 High Meadow, Hathern, Loughborough LE12 5HW
M0 URF A. Vincent Post Office House, Station Road, Andoversford, Cheltenham GL54 4HP
M0 URJ S. Spencer 18 Goseley Avenue, Hartshorne, Swadlincote DE11 7EZ
M0 URL P. Gavin 11 Campbell Close, Yateley GU46 6GZ
MM0 URN I. Quinnell Acarsaid, Kinlochbervie, Lairg IV27 4RP
M0 URX T. Beaumont 83 Limbrick Avenue, Tile Hill, Coventry CV4 9EX
M0 USB R. Davies 24 Evesham Avenue, Whitley Bay NE26 1QR
M0 USC S. Chaney 54 Clementine Avenue, Seaford BN25 2XG
MW0 USK C. Burke 84 Elgam Avenue, Blaenavon, Pontypool NP4 9QU
M0 USM 5th Reigate Scout Group c/o G. Mountain 34 Albert Road, Warlingham CR6 9EP
MI0 UST S. Dockery 70 Main Street, Greyabbey BT22 2NG
M0 USV D. Soames 40 Woodland Drive, North Anston, Sheffield S25 4EP
M0 USY P. Shields 34 Dryden Close, Grantham NG31 9QS
M0 UTA A. Emmerson 31 Culver Road, Stockport SK3 8PG
M0 UTD M. Jones 110 Becconsall Drive Leighton, Crewe CW1 4RP
M0 UTG J. Dodds 84 Borrowdale Avenue, Walkerdene, Newcastle upon Tyne NE6 4HL
M0 UTH G. Guinan 5A Temple Lane, Silver End, Witham CM8 3QY
M0 UTK S. Kembrey 101 Yew Tree Drive, Bristol BS15 4UF
MW0 UTT B. Bull Swan Cottage, Swan Road, Welshpool SY21 0RH
M0 UTX J. Swift Raf Holmpton Rysome Lane, Holmpton HU19 2RG
MI0 UTY D. Cartin 6 Grange Avenue, Magherafelt BT45 5RP
M0 UUU A. Sheard 15 Bent Lanes, Urmston, Manchester M41 8PB
M0 UVZ Glanford Electronics c/o G. Cowling Laissez Faire, Reedness, Goole DN14 8ET
M0 UWD G. Deacon 32 Gloucester Road, Exwick, Exeter EX4 2EF
M0 UWS I. Lindsay 17 Middleforth Green, Penwortham, Preston PR1 9TB
M0 UXB D. Coomber 14 Francis Green Lane, Penkridge, Stafford ST19 5HF
M0 UXO M. Gritton 53 Brinkburn Grove, Banbury OX16 3WX
M0 UXS C. Dutton Twillingate Farm, Tiptoe, Lymington SO41 6EJ
MI0 UYD P. Page 259 Bridge Street, Portadown, Craigavon BT63 5AR
M0 UYR R. Brooker 18 Honeybourne Way, Petts Wood, Orpington BR5 1EZ
MW0 UZO D. White 222 St. Fagans Road, Cardiff CF5 3EW
M0 VAA G. Mcgowan 281 Ashgate Road, Chesterfield S40 4DB
MI0 VAC V. Crothers 5 Thornleigh Park, Ballymoney BT53 7BX
M0 VAD Mid Warwickshire ARS c/o D. Cook 44 Statfold Lane, Fradley, Lichfield WS13 8NY
M0 VAG A. Grant 26 Fountains Avenue, Boston Spa, Wetherby LS23 6PX
M0 VAH E. Whitehouse 16 Rue Gaston De Caillavet, Paris 75015 France
M0 VAI Crystal Palace Radio & Electronics Club c/o D. Vainas 51 Magister Road, Bowerhill, Melksham SN12 6FD
M0 VAM M. Medcalf 47 Paddock Drive, Chelmsford CM1 6UX
M0 VAP M. Alcaino Pizani Flat 45, Brian Redhead Court, 123 Jackson Crescent, Manchester M15 5RR
M0 VAR Belvoir Vale AR c/o B. Hiley 9 Pinfold Lane, Harby, Melton Mowbray LE14 4BU
M0 VAS V. Papanikolaou 104 West Drive Gardens, Soham, Ely CB7 5EX
M0 VAT A. Rodgers 123 Mill Lane, Northfield, Birmingham B31 2RP
M0 VAU Dr M. Vaughan Hillside Tregorrick, St. Austell PL26 7AG
M0 VAW V. Werrett 3 Hardingham Drive, Sheringham NR26 8YE
M0 VAY R. Bewick 357 Franklands Village, Haywards Heath RH16 3RP
M0 VBE B. Barwise 10 Morland Road, London E17 7JB
M0 VBR J. Baughan Chestnut Farm, Eastville, Boston PE22 8LX
MW0 VBT M. Kveksas 4 Llys Coed Derw, Llantwit Fardre, Pontypridd CF38 2JB
M0 VBV V. Be`Dard 53 Cottingley Crescent, Leeds LS11 0HZ
M0 VBW B. Whall 3 Farrow Close, Great Moulton, Norwich NR15 2HR
M0 VBY D. Potter 30 Mersham Gardens, Goring-By-Sea, Worthing BN12 4TQ
M0 VCA J. Davis 29 Willow Tree Rise, Bournemouth BH11 8EE
M0 VCB C. Fox Millstone Cottage, Prior Wath Road, Scarborough YO13 0AZ
MW0 VCC H. Morris The Heights Hotel, 74 High Street, Llanberis, Caernarfon LL55 4HB
M0 VCE N. Baker 56 Chalklands, Bourne End SL8 5TJ
M0 VCP S. Pryke 12 Seaward Avenue, Leiston IP16 4BB
M0 VCR P. Woodhouse 8 Greenhill Road, Halesowen B62 8EZ
M0 VCS V. Stocker 25 Davies Drive, Uttoxeter ST14 7EQ
MW0 VCV P. Lang 10 Coed Ty Maen, Bridgend CF31 4TG
M0 VCX J. Woods Invicta Cottage, Carbrooke Road, Thetford IP25 6SD
M0 VDO A. Collier 65 Sandpits, Leominster HR6 8HT
M0 VDQ A. Bolster 45 Headlands Drive, Hessle HU13 0JP
M0 VDU R. Tree 9 Heather Close, Worcester WR5 3LX
MW0 VDX Group Two c/o J. White Keepers Lodge Pumpsaint, Llanwrda SA19 8DX
M0 VEC R. Trevan 35 Oaktree Drive, Hook RG27 9RA
M0 VEL Dr S. Favell Lodge Farm, Moor Lane, Reepham, Lincoln LN3 4EE
M0 VES M. Thompson 10 Whitsun Grove, Cottingham HU16 4BX
MW0 VET M. Williams 10 Clettwr Terrace Pontsian, Llandysul SA44 4TU
M0 VEX T. Cope 47 Kings Avenue, Atherstone CV9 1JY
M0 VEY P. Sidwell 7 Spring Field Close Sigglesthorne, Hull HU11 5QP
M0 VEZ S. Debenham 14 Aldergrove Close, Lincoln LN6 0SL
M0 VFC R. Chipperfield 13 Harlestones Road, Cottenham, Cambridge CB24 8TR
M0 VFG P. Hawkins 12 Bittaford Wood, Bittaford, Ivybridge PL21 0ET
M0 VFR Hill Top Radio Group c/o S. Tomlinson 7 Springwell Close, Crewe CW2 6TX
M0 VGA D. Silkstone 169 Otley Road, Harrogate HG2 0DA
M0 VGC R. West 557 East Bank Road, Sheffield S2 2AG
M0 VGG J. Tricklebank 1 Hewell Road, Barnt Green, Birmingham B45 8NG
M0 VGH R. Ashworth 10 Mulberry Close, Wigan WN5 9QL
M0 VGK A. Green 28 Queensway, Old Dalby, Melton Mowbray LE14 3QH
M0 VGL A. Lunn 57 Greets Green Road, West Bromwich B70 9ES
M0 VGT R. Newell 44 New Street, Chagford, Newton Abbot TQ13 8BB
M0 VGV G. Venugopalan 3 Southwater Close, London E14 7TE
M0 VHC T. Oliver 17 East Lea, Newbiggin-by-the-Sea NE64 6BQ
M0 VHG V. Greatwood 11 The Green, Long Preston, Long Preston, Skipton BD23 4PQ
M0 VIG A. Smith 19 Gibsons Gardens, North Somercotes, Louth LN11 7QH
M0 VII A. Price Barn Owl Roost, Astwith, Chesterfield S45 8AN
MM0 VIK E. Crawford Keldi, Houss, Bridge End, Shetland ZE2 9LE
M0 VIN C. Vincent 64 Park End Road, Romford RM1 4AU
M0 VIR D. Smith 48 Shirley Gardens, Tunbridge Wells TN4 8TH
M0 VIS R. Morris 45 St. Kildas Road, Bath BA2 3QL
M0 VIT J. Franks 14 The Hamlet, Slades Hill, Templecombe BA8 0HJ
M0 VJR Cambridge District ARC c/o S. Rhenius Baythorne Cottage Baythorne End, Halstead CO9 4AB
M0 VJX B. Walker 255 Packington Avenue, Birmingham B34 7RU
M0 VKB R. Bearcroft 45 Broad Marston Road, Pebworth, Stratford-upon-Avon CV37 8XT
M0 VKC N. Williams 17 Sunnyside, Malpas SY14 7AA
M0 VKG A. Smith 305, 1414 5Th St Sw, Calgary T2R 0Y8 Canada
M0 VKJ A. Yiangou 153 Hoppers Road, London N21 3LP
M0 VKK R. Cresswell Meadow View, Hulver Road, Beccles NR34 7UW
M0 VKP C. Schroth Flat 3, 15B Cavendish Road, Bournemouth BH1 1QX
M0 VKR L. Bullen11 5 West View, Long Sutton TA10 9LT
M0 VKS D. Vickers 178 Bakewell Road, Matlock DE4 3BA
M0 VKU D. Chadwick 19 Regent Crescent, Failsworth, Manchester M35 0LR
M0 VKX R. Routledge 42 Eighth Row, Ashington NE63 8JX
M0 VKY S. Billingham Kewell House, Wombourne Road, Swindon, Dudley DY3 4NF
M0 VLA A. Howsen Oakland Villa, Seaton Road, Maryport CA15 8ST
M0 VLC F. Harwood 1, South Highall Cottage, Lincolnshire LN10 6UR
M0 VLI South East Hampshire Raynet c/o W. Toher The Chapel, Station Road, Darlington DL2 1JG
M0 VLK V. Lucock 34 Wentworth Drive, Ipswich IP8 3RX
M0 VLL V. Leppard 39 Queensland Drive, Colchester CO2 8UD
MW0 VLO B. Henley Rhewin Glas, Llandysul SA44 6PR
M0 VLP Radio Millenium Lodge c/o P. Barville Felucca, Pinesfield Lane, West Malling ME19 5EN
M0 VLT A. Macdonald Woodside Cottage, Horton Way, Verwood BH31 6JJ
M0 VLX D. Cassidy 172 Lyde Road, Yeovil BA21 5PN
M0 VMC D. Burkin 26 Rampton Road, Cottenham, Cambridge CB24 8UL
MD0 VMD Jersey Scouts ARC c/o P. Birchall 7 Richmond Close, Douglas IM2 6HR Isle of Man
M0 VMH V. Hocking 80 Barton Tors, Bideford EX39 4HA
M0 VMV R. Vale 611 College Road, Birmingham B44 0AY
M0 VMW Vintage & Military ARS c/o S. Mckinnon 145 Enville Road, Kinver, Stourbridge DY7 6BN
M0 VMX M. Seaney 56 Winnham Drive, Fareham PO16 8QG
M0 VNG M. White 7 Overthwart Crescent, Worcester WR4 0JW
M0 VNO D. Harwood 36 Seaview Drive, Great Wakering, Southend-on-Sea SS3 0BE
M0 VNR N. Ramsey Dalestones, Lansdown Road, Bath BA1 5TB
M0 VNY V. Brindle 185 Brunshaw Road, Burnley BB10 4DL
MU0 VOE Dr H. Voehrs 50 High Street, Alderney GY9 3 TG Guernsey
M0 VOG Vintage Operating Group c/o M. Buckley Springfield, 12 Ranmore Avenue, Croydon CR0 5QA
MM0 VOK R. Remnant Hillock O Leys, New Pitsligo, Fraserburgh AB43 6QN
M0 VOL C. Brayshaw Wayside Cottage, Main Street, Harpham, Driffield YO25 4QY
M0 VOM N. Curran 8 Daneswood Close, Whitworth, Rochdale OL12 8UX
M0 VON G. Smith 92 Brighton Road, Banstead SM7 1BU
M0 VOS S. Devos Applecross Cottage, Main Road, Newark NG23 7HR
M0 VOZ M. Crockford Centre Cottage Kelk, Kelk YO258HL
M0 VPC J. Elstone 54 Oakfield, Woking GU21 3QS
M0 VPE I. Stirzaker Avenida De Huelva 7, Conjunto Latinos 44, Rojales 3170 Spain
MM0 VPF H. Phillips Maplebank, Leithen Road, Innerleithen EH446NJ
M0 VPG R. Killington 5 Ladymead Close, Maidenbower, Crawley RH10 7JH
M0 VPI V. Prystaj 69 Kingsdale Crescent, Bradford BD2 4DP
M0 VPK M. Smith 18 Hawthorn Road. Old Leake, Boston PE22 9NY
M0 VPL M. Wells 23 Eastmead, Bognor Regis PO21 4QT
MM0 VPM A. Cowan 32 Esk Valley Terrace, Dalkeith EH22 3FT
MM0 VPR P. Rice 255 Eskhill, Penicuik EH26 8DF

M0 VQJ RAF Holmpton Ara c/o J. Swift Raf Holmpton Rysome Lane, Holmpton HU19 2RG
M0 VQP A. Majoch 66 Boughton Green Road, Northampton NN2 7SP
M0 VRD R. Drage 4 Bruce'S Close Conington, Peterborough PE73QW
M0 VRG Vintage Radio Group c/o A. Clayton 6 Albert Road, Bunny, Nottingham NG11 6QE
M0 VRI J. Groves 350 Middle Deal Road, Deal CT14 9SN
M0 VRP I. Bell 164 The Broadway, Herne Bay CT6 8HY
MW0 VRQ S. Trahearn 148 Gladstone Road, Barry CF62 8ND
M0 VRS J. Strange Culloden, Ulting Road Hatfield Peverel, Chelmsford CM3 2LU
M0 VRT L. Hummerstone 70 Salisbury Road, Plymouth PL4 8TA
M0 VRW P. Wilson1 45 Newquay Close, Hartlepool TS26 0XG
M0 VSD Dr L. Kirkcaldy 19 Itchen Avenue, Bishopstoke, Eastleigh SO50 8JW
M0 VSE P. Taylor 104 Winstanley Drive, Leicester LE3 1PA
M0 VSP N. Briggs 20 Broad Lane, Pelsall, Walsall WS4 1AP
M0 VSQ Vulture Squadron CG c/o I. Kelly 261 Bodiam Avenue, Tuffley, Gloucester GL4 0XW
M0 VSR T. Wai Ming 313 Devizes Road, Salisbury SP2 9LU
MM0 VSU L. Bradley Amon Sul, Kiltarlity, Beauly IV4 7HT
M0 VSW S. Whall 17 Vicarage Road, Deopham, Deopham NR18 9DR
MM0 VTA Dr D. Mcnicholl 42 Dean Street, Edinburgh EH4 1LW
M0 VTC P. Robins 20 Saffron Close, Chineham, Basingstoke RG24 8XQ
M0 VTD S. Iles Bigbury Bay Holiday Park, Challaborough, Kingsbridge TQ7 4HS
M0 VTE A. Gallop 75 Shearmans, Fullers Slade, Milton Keynes MK11 2BQ
M0 VTG D. Howlett 21 Chandlers, Orton Brimbles, Peterborough PE2 5YW
M0 VTJ T. Scott 50 Davison Avenue, Whitley Bay NE26 1SH
MW0 VTK J. Martin 78 Llwyn Ynn, Talybont LL43 2AG
M0 VTR M. Newell 55 Station Road Brimington, Chesterfield S43 1JU
M0 VTS P. Wilkes 8 Cloverdale, Stafford ST17 4QJ
MM0 VTV R. Farrer 23 Upper Craigour, Edinburgh EH17 7SE
M0 VUB S. Daley 1 North Green, Calverton, Nottingham NG14 6NT
M0 VUE C. Suddell Lynhurst, Littleworth Lane Partridge Green, Horsham RH13 8JX
MM0 VUV R. Fraser 72 Ferguson Drive, Denny FK6 5AG
M0 VUW C. Smith 2 Burley Gardens, Street BA16 0SN
M0 VVA A. Amos Willow Tree House, Deers Green, Clavering, Saffron Walden CB11 4PX
M0 VVC M. Walker 6 Broadoak, Tadley RG26 3UZ
M0 VVG Elkstones ARS c/o R. King 8 Rydal Court, Congleton CW12 4JL
M0 VVM T. Aldred 31 Cock Road, Bristol BS15 9SH
MW0 VVO S. Barry 1 Pearson Cottages, St. Brides, Haverfordwest SA62 3BN
M0 VVQ N. Ham 4 Heighes Drive, Alton GU34 2FJ
M0 VVR C. Chak P.O. Box 691 Tuen Mun Central Post Office, Hong Kong 999077 Hong Kong
M0 VVT M. Mcgregor 141 Herne Road, Ramsey St. Marys, , Huntingdon PE26 2SY
M0 VVV J. Worthington The Old Hundred, Farm Lane, Farnham GU10 5QE
M0 VVX J. Platt 12 Tawny Grove Four Marks, Alton GU34 5DU
MW0 VWC W. Wiggans Bronysgawen, Llanboidy, Whitland SA34 0EX
M0 VWD V. Downes 55 Ashfield Road Bromborough, Wirral CH62 7EE
M0 VWK M. Poole 15 Roberts Place, Dorchester DT1 2JJ
M0 VWP R. Russell 7 Holt Place, Coach House Mews, Ferndown BH22 9UX
MM0 VWR D. Green 35 Douglas Avenue, Brightons, Falkirk FK2 0HB
M0 VWS M. Smith The Lawns Tylers Green, High Wycombe HP108BH
M0 VWT C. Poole 15 Devon Close, Macclesfield SK10 3HB
M0 VWW J. Bielen 5 Long Hale, Pitstone LU7 9GF
M0 VXC V. Cepraga 9 Kynon Close, Gosport PO12 4LW
M0 VXD S. Finlayson 41 Low Catton Road, Stamford Bridge, York YO41 1DZ
M0 VXX T. Quiney 20 Britannia Gardens, Stourport-on-Severn DY13 9NZ
M0 VXY K. Poulton 21 East View, London E4 9JA
M0 VYB Vision Youth Centre (Bovington) c/o S. Hawkins Forest Edge, Deer Park, Milton Abbas, Blandford Forum DT11 0AY
M0 VYC A. Handley 4 Southwood Drive, Thorne, Doncaster DN8 5QS
M0 VYW A. Willsher 1 Tolputt Court Gladstone Road, Folkestone CT19 5NE
M0 VZA Stisted CG c/o G. Nurse Orchard Bungalow, Priors Green, Stisted, Braintree CM77 8BP
M0 VZR C. Gain 14 Battens Avenue, Overton, Basingstoke RG25 3NL
M0 VZS D. Clewer 45 Ashfield Road, Andover SP10 3PE
M0 VZT R. Clay 75 Trinity View, Ketley Bank, Telford TF2 0DY
M0 VZV M. O'Donovan 10 Rockfield Close, Teignmouth TQ14 8TS
M0 WAB W. Baxter 19 Westbury Road, Nottingham NG5 1EP
M0 WAC M. Friesch Stag Hill Court 39-E, University Campus, Guildford GU2 7JG
M0 WAD A. Waddington 8 Redbrook Close, Bromborough, Wirral CH62 6EA
M0 WAE L. Severe 19324 Paddock View Drive, California 33647 United States
M0 WAF P. Marchant 16 Melrose Drive, Peterborough PE2 9DN
M0 WAG O. Prin 19 The Colliers, Heybridge Basin, Maldon CM9 4SE
M0 WAH W. Horsewell 15 Highcroft Lane, Waterlooville PO8 9NX
M0 WAI C. Lam 58 Sparrow Hill, Loughborough LE11 1BU
M0 WAJ A. Hagland 11 Coppice View, Heathfield TN21 8YS
MM0 WAK W. Laurie 306 Lanark Road West, Currie EH14 5RR
M0 WAM D. Beet 1 Shottesford Avenue, Blandford Forum DT11 7XU
M0 WAO B. Walstra 138 Tanhouse Farm Road, Solihull B92 9EY
MM0 WAP F. Pudsey 21/2 Bathfield, Edinburgh EH6 4DU
M0 WAQ M. Welland 76 Lovel Road, Chalfont St. Peter, Gerrards Cross SL9 9NX
M0 WAR D. Warwick 3 The Pines, Stoke Ferry, King's Lynn PE33 9XW
M0 WAS O. Staines 52 James Place, Flitwick MK45 1GW
M0 WAU Derbys Worked All Britain ARC c/o J. Lynch Beechway, Raddel Lane, Warrington WA4 4EE
M0 WAV A. Snelson 6 Rayleigh Close, Braintree CM7 9TX
MM0 WAX Fareham And District ARC c/o B. Hendry 9 Glen Aray View, Inveraray PA32 8TW
M0 WAY W. Thomas 5 Thornley Road, Wolverhampton WV11 2HR
M0 WAZ W. Payne 1 Niton Road, Rookley PO38 3NP
M0 WBB W. Brown 126 Alexandra Road, Ashington NE63 9LU
M0 WBC J. Phillips 56 Rosemary Avenue, Hounslow TW4 7JG
M0 WBD D. Blake The Bramleys, Gaysfield Road, Fishtoft, Boston PE21 0SF
M0 WBF W. Millington 93 Feiashill Road, Trysull, Wolverhampton WV5 7HT
M0 WBG N. Challis 48 Brunsfield Close, Wirral CH46 6HE
M0 WBJ B. Webb 88 Stanley Road, Cambridge CB5 8LB
M0 WBK W. Knapp 32 Turner Close, Shoeburyness, Southend-on-Sea SS3 9TL
M0 WBM A. Birch 22 Ullswater Road, Burnley BB10 4HX
M0 WBR R. Walker 1A Winifred Way, Caister-On-Sea, Great Yarmouth NR30 5AB
M0 WBS W. Bennison 21 Ashdene Close, Chadderton, Oldham OL1 2QG
M0 WBY Tower Radio Group c/o J. Willby 10 Sunbury Road, Birmingham B31 4LJ
M0 WBZ K. Hunt 11 De Marnham Close, West Bromwich B70 6RJ
M0 WCA M. Bostock 86 Beauvale Drive, Ilkest on DE7 8SJ
M0 WCB Northumbria ARC c/o D. Trudgian 18 Hart Close, Wootton Bassett, Swindon SN4 7FN
MM0 WCD C. Docherty 23 The Maltings, Haddington EH41 4EF
M0 WCE C. Conghos 11A Rowans Way, Leavenheath, Colchester CO6 4UU
MM0 WCG Woodpecker CG c/o R. Fraser Hopefield Cottage, Gladsmuir, Tranent EH33 2AL
M0 WCH C. Haynes 4 Thorn Close, Rugby CV21 1JN
M0 WCK C. Kakoutas Trinity College, Trinity Street, Cambridge CB2 1TQ
M0 WCL C. Lonie Jr 41 De La Hay Avenue, Plymouth PL3 4HS
M0 WCM W. Maddox 28A Redcar Avenue, Ingol, Preston PR2 3YY
MW0 WCP Dr E. Harries Ty Traeth, Caerwedros, Llandysul SA44 6BS
M0 WCR M. Mcsherry 5 Briery Croft, Stainburn, Workington CA14 1XJ
M0 WCS D. Sewell 19 St. Leonards Way, Ashley Heath, Ringwood BH24 2HS
MM0 WCT T. Woods Marsden, Lochard Road, Aberfoyle, Stirling FK8 3SZ
M0 WCW C. Wise 32 Commercial Street, Willington DL15 0AD
M0 WCZ D. Wells 96 Tennyson Avenue, Rugby CV22 6JF
M0 WDD D. Mcarthur 7 Gore Avenue, Salford M5 5LF
M0 WDE F. Windridge 84 Queens Road, Skegness PE25 2JE
M0 WDG D. Wressell 6 St. Josephs Close, Bishopdown, Salisbury SP1 3FX
M0 WDJ D. Watson 56 Lambton Avenue, Delves Lane Industrial Estate, Consett DH8 7JE
M0 WDK P. Barrows 5A Magdalen Road, Willoughby, Rugby CV23 8BJ
M0 WDL D. Lee The Old Barn, Hatch Beauchamp, Taunton TA3 6AE
M0 WDO T. Smith Chy Crowshensy, Clifton Road, Redruth TR15 3UD
M0 WDP W. Phillips 55 Kilton Crescent, Worksop S81 0AX
M0 WDU D. Walsh 8 Prestwold Way, Aylesbury HP19 8GZ
M0 WDZ S. Horne 29 Shaftesbury Street, Fordingbridge, Hampshire SP6 1JF
M0 WEB B. Munro-Smith 8 Billings Way, Cheltenham GL50 2RD
M0 WEC P. Wagstaff 49 The Paddock, Earlsheaton, Dewsbury WF12 8BY
MW0 WEE A. Brown Oakridge, 6 Bro Hafan, Llandysul SA44 6NQ
MM0 WEI Stockport RS c/o E. Ireland The Steading, Blairmains, Shotts ML7 5TJ
M0 WEL D. Wells 40 Barnham Broom Road, Wymondham NR18 0DF
M0 WEN C. Owen Garden Cottage, Holbeck Woodhouse, Worksop S80 3NQ
M0 WEO D. Hart 7 Penrose Road, Ferndown BH22 9JF
M0 WES H. Partridge 19 Dickens Drive, Melton Mowbray LE13 1HZ
M0 WET T. Clarke 80 Bendall Road, Birmingham B44 0SN
M0 WEV J. Wedge 9 Claremont Mews, Wolverhampton WV3 0EB
M0 WEW A. Mcewen 4 The Pantyles, Nightingale Lane, Sevenoaks TN14 6BX
M0 WFA A. Walker 14 Maritime Avenue, Hartlepool TS24 0XF
MW0 WFB L. Bowman Chanrick, Penderyn Road, Aberdare CF44 9RU
M0 WFC C. Cooper 25 Waterside Close, Loughborough LE11 1LP
M0 WFI C. Tam Flat 33, Signature Studio Living, 1 Humber Avenue, Coventry CV3 1BB
M0 WFK P. Ashton 14 Poppy Close, Boston PE21 7TJ
M0 WFM M. Deeley Unit 4 Beechwood Business Park, Cannock WS11 7GB
M0 WFN W. Newton 7 Moss Close, Bridgwater TA6 4NA
M0 WFO S. Harris 1 Eastbank Drive, Worcester WR3 7BH
M0 WFR F. Waller 19 Wortley Avenue S738Sb, Wombwell S738SB
M0 WFX C. Bolton 201 Lime Tree Avenue, Crewe CW1 4HZ
M0 WGA R. Mahorney Walnut Cottage, Church Lane, Wallingford OX10 0SD
M0 WGB G. Beale 34 Teville Road, Worthing BN11 1UG
M0 WGC C. Watkins 25 Citadilla Close, Gatherley Road, Richmond DL10 7JE
M0 WGF E. Lewis 105 Wards Hill Road, Minster On Sea, Sheerness ME12 2LH
M0 WGI S. Sugihara Southfield, Park Lane, Wokingham RG40 4PY
MI0 WGL W. Leonard 57 Old Coach Road Mullanavehy, Enniskillen BT92 2EW
MI0 WGM G. Mccusker 10 Birchdale, Lurgan, Craigavon BT66 7TR
M0 WGO I. Paterson 11 Ocho Rios Mews, Eastbourne BN23 5UB
MW0 WGR West Glamorgan Raynet c/o W. Britten-Jones 101 Mill View Estate, Maesteg CF34 0DE
M0 WGS Wings Museum c/o B. Bloomfield 2 Walstead Manor Cottages, Scaynes Hill Road, Haywards Heath RH16 2QG
MI0 WGW E. Kyle 2 Wattstown Crescent, Coleraine BT52 1SP
M0 WGY R. Williams 4 Bluebell Close, Barlborough, Barlborough S43 4WT
MM0 WHA W. Anderton 15 Queens Crescent, Lockerbie DG11 2BA
M0 WHB W. Bray 46 Alexandra Road, Lostock BL6 4BB
M0 WHC W. Clayton 403 Queens Drive Walton, Liverpool L4 8TY
MI0 WHG the Isle of Avalon ARC c/o S. Frazer 2 Cavanballaghy Road, Killylea, Armagh BT60 4NZ
M0 WHJ W. Hoar 46 Pendean Avenue, Liskeard PL14 6DA
M0 WHK K. White 4 Top Birches, St. Neots PE19 6BD
M0 WHL A. Lancefield 19 Tawny Sedge, King's Lynn PE30 3PW
MM0 WHM J. Scott Mid Henshilwood Farm, Braehead Forth, Lanark ML11 8HB
M0 WHO M. Sims 133 Canterbury Road, Hawkinge, Folkestone CT18 7BS
M0 WHP R. Hoppe 1 Grove Road Houghton Regis, Dunstable LU5 5PD
M0 WHQ Norfolk County Raynet c/o A. Mobbs 149 The Paddocks, Old Catton, Norwich NR6 7HR
M0 WHR D. Williams Flat 3 The Old Council House, Market Street, Atherstone CV9 1ET
M0 WHT K. Snipe 5 Draycott Road, Chiseldon, Swindon SN4 0LT
MI0 WHX S. Gibson 22 Station Road, Bangor BT19 1HD
M0 WIA W. Armes 11 Rutland Road, Broadheath, Altrincham WA14 4HW
MM0 WIC C. Aitken Windybraes, Upper Gills, Canisbay, Caithness, Canisbay KW1 4YB
M0 WIG S. Biggs 81 St Abbs Drive, Bradford BD6 1EJ
M0 WIK Dr K. Morris 44 Leamington Road, Weymouth DT4 0EZ
MW0 WIL W. Howe 78 Coychurch Road, Pencoed, Bridgend CF35 5NA
M0 WIN O. Prosser 2 Caroline Close, Ventonleague, Hayle TR27 4EX
M0 WIO D. Owen 8 Crag Bank Crescent, Carnforth LA5 9EQ
M0 WIT D. Whitley 10 Kenmore Drive, Cleckheaton BD19 3EJ
M0 WIV R. Laidler Swallow Cottage, Wiverton, Plympton, Plymouth PL7 5AA
M0 WIZ I. Moore Sun House, 33 Church Lane, Trowbridge BA14 0TE
M0 WJA D. Redmayne 10 The Square, Kington HR5 3BA
MI0 WJC W. Campbell 9 Rochester Court, Coleraine BT52 2JL
M0 WJG W. Garvey 254 Bury Road, Tottington, Bury BL8 3DT
M0 WJH W. Haddock 5 Bradley Close, Middlewich CW10 0PF

M0 WJL G. Hayers 87 Bradleigh Avenue, Grays RM17 5RH
MI0 WJM W. Murray 80 Canterbury Park, Londonderry BT47 6DU
MM0 WJP J. Pike Mosside Farmhouse, Culsalmond, Insch AB52 6TU
M0 WJT W. Taylor 99 St. Marys Close, Littlehampton BN17 5QQ
M0 WJW Tamworth ARS c/o W. Wellington 57 Hillcrest, Whitley Bay NE25 9AF
M0 WKG K. Latham 45 Hollybank Close, Northwich CW8 4GS
MM0 WKJ W. Jenkins 3A Manse Grove, Stoneyburn, Bathgate EH47 8EW
M0 WKL W. Lam 29 Marthall Drive, Sale M33 2XP
M0 WKO P. Holton 66 Mill Road, Gillingham ME7 1JB
M0 WKR N. Clarke Brimham Lodge Farm, Brimham Rocks Road Burnt Yates, Harrogate HG3 3HE
M0 WKT G. Williams 18 Elmsleigh Road, Farnborough GU14 0ET
MM0 WKY R. Mackay 12 Robertson Square, Wick KW1 5NF
M0 WLA Royal Air Force ARS c/o R. Wynne South Graceholme, High Lorton, Cockermouth CA13 9UQ
M0 WLB L. Barlow 51 Hare Hill Road, Littleborough OL15 9HE
M0 WLD B. Wild 1 Sunnymount, Midsomer Norton, Radstock BA3 2AS
M0 WLF I. Prater 470 Bishport Avenue, Bristol BS13 0HS
M0 WLH Prof. W. Lionheart Marsham, Start Lane, Whaley Bridge, High Peak SK23 7BP
M0 WLK R. Readman 1 Millside Close, Kilham, Driffield YO25 4SF
MM0 WLL W. Fleming 65 Dundonald Park, Cardenden, Lochgelly KY5 0DG
MW0 WLP P. Williams 127 Penchwintan Road, Bangor LL57 2YG
M0 WLS W. Le Serve 120 Cheam Road, Cheam, Sutton SM1 2EB
MU0 WLV A. Prosser Woodlands, La Vassalerie, , St. Andrew GY6 8XL Guernsey
M0 WLY A. Omar 84 Beaumont Hill, Darlington DL1 3ND
M0 WMB M. Chanter 7 Woodford Crescent, Plymouth PL7 4QY
M0 WMG R. Shepherd 19 Elford Avenue, Newcastle upon Tyne NE13 9AP
MW0 WML R. Davison 2 Marlow Terrace, Mold CH7 1HH
M0 WMO D. Tordoff 49 Dale Edge, Eastfield, Scarborough YO11 3EP
M0 WMR W. Ross 62 Derwent Drive, Tewkesbury GL20 8BB
M0 WMS M. Smith 78 Shiregate, Metheringham, Lincoln LN4 3DR
M0 WMT M. Lawrence 83 Gresham Drive, Northampton NN4 9SB
MW0 WMW M. Suddaby Bryn Eiddion, Rhydymain, Dolgellau LL40 2AS
M0 WMX D. Colver 85 Whitemoor Lane, Belper DE56 0HD
M0 WNA A. Toal 29 Highland Drive, Oakley, Basingstoke RG23 7LF
M0 WNF N. Fellingham 23 Brooklands, Colchester CO1 2WA
M0 WNI R. Karpinski 55 Cambridge Avenue, New Malden KT3 4LD
MW0 WNL R. Bartrum 1A Bakers Way, Bryncethin, Bridgend CF32 9RJ
M0 WNT A. Taylor 90 Coppice Avenue, Eastbourne BN20 9QJ
M0 WNV T. Higginson 109 London Road, Biggleswade SG18 8EE
MM0 WNW N. White 2 Appleby Cottages, Whithorn, Newton Stewart DG8 8DQ
MM0 WOA Prof. G. Woan 6 Sandpiper Road, Lochwinnoch PA12 4NB
M0 WOB D. Bowden 58 Southville, Yeovil BA21 4JF
M0 WOD G. Norgrove The Rockings Alcester Road Burcot, Bromsgrove B60 1PJ
M0 WOF Woofferton Transmitting Station c/o M. Porter 20, Southfield Road, Much Wenlock TF13 6AX
M0 WOJ A. Landless 2 Aspen Way, Banstead SM7 1LE
M0 WOM D. Shingleton 6 Newsham Walk, Manchester M12 5QB
M0 WON H. Moore 52 Limefield Street, Accrington BB52AF
M0 WOS W. Barnes Cushendall, Lyngate Road, North Walsham NR28 0DH
M0 WOT D. Watts 3 Witney Road, Crawley RH10 6GJ
M0 WOW D. Dunne 1 Burton Gardens, Brierfield, Nelson BB9 5DR
M0 WPA S. Robbins Sugar Mouse Luxury Confectionary Ltd, 2 Central Buildings, Market Place, Easingwold, York YO61 3AB
M0 WPI T. Hobson-Smith 15 Henconner Lane, Chapel Allerton, Leeds LS7 3NX
M0 WPJ P. Joyner 3 Barton Road, Canterbury CT1 1YG
M0 WPL P. Loda St. Albans Court, Sandwich Road Nonington, Dover CT15 4HH
M0 WPN W. Nichols Newcourt Farmhouse, Silverton, Exeter EX5 4HT
M0 WPR W. Rees 67 Chine Walk, West Parley, Ferndown BH22 8PS
M0 WPS W. Phillips 36 Beeches Road, Great Barr, Birmingham B42 2HF
M0 WPT J. Rosa Flat 7, Wick Hall, Abingdon OX14 3NF
M0 WPX Data - DXers c/o K. Holloway 6 Britons Lane Close, Beeston Regis, Sheringham NR26 8SH
M0 WPY R. Hampson 10 Lindisfarne Close, Sandy SG19 1TT
M0 WQK D. Blackie 8 Kingswood Road, Manchester M14 6SB
M0 WQR V. Keeley Hawthorns, Cowbit Drove, West Pinchbeck, Spalding PE11 3TG
M0 WRA B. Wray 10 Winstanley Road, Sale M33 2AR
M0 WRD D. Whitehouse 6 Larch Close, Heathfield TN21 8YW
M0 WRE G. Norris 12 Westfield Placescholes, Bradford BD19 6DU
M0 WRI P. Hodge 141 Linden Place, Newton Aycliffe DL5 7BQ
M0 WRJ Dr Y. Suzuki Eng. 5-203, Shizuoka University, 3-5-1 Johoku, Hamamatsu-Naka 432-8561 Japan
MD0 WRK R. Kissack 6 Falcon Cliff Court, Douglas, Isle of Man IM2 4AQ
MM0 WRO M. Lorenowicz 10F Forrester Park Grove, Edinburgh EH12 9AJ
MW0 WRP W. Powell 40 Heol Ty Newydd, Cilgerran, Cardigan SA43 2RT
MW0 WRQ Tonypandy Scout Group c/o B. Jones 10 Hughes Street, Penygraig, Nr Tonypandy CF40 1LX
M0 WRR Waveney Wireless c/o L. Cropley San Ferryann, Brundish Road, Wilby, Eye IP21 5LS
M0 WRS W. Smith Flat 48, Winehala Court, 50A Sandbeds Road, Willenhall WV12 4GA
M0 WRU W. Rudge 33 Wyrley Rd, Wolverhampton WV11 3NY
MM0 WRX K. Mccormick 51 Millford Drive, Paisley PA3 3EJ
MW0 WRY J. Richardson 15 Calland Street, Plasmarl, Swansea SA6 8LE
M0 WSA Reading School ARC c/o S. Williams Flat 35, Winterton House, London E1 2QR
M0 WSB A. Craddock 58 Vicarage Road, Mickleover, Derby DE3 0ED
M0 WSC T. Kwok 313 Devizes Road, Salisbury SP2 9LU
MW0 WSD N. Orchard 152 Garden Suburbs, Trimsaran SA174AF
M0 WSE A. Champion 4 Oldcastle Croft, Tattenhoe, Milton Keynes MK4 3EN
MM0 WSG Glasgow University Wireless Society c/o T. Storkey 1 Hartington Gardens, Edinburgh EH10 4LD
MW0 WSH Dr J. Blaxland 7 Maes Y Sarn, Pentyrch, Cardiff CF15 9QQ
MM0 WSK J. Muchowski 71 The Braes, Tullibody, Alloa FK10 2TT
M0 WSN R. Swinburne 32 Hollywell Road, Birmingham B26 3BX
M0 WSP S. Peare 15 Clydesdale Gardens, Bognor Regis PO22 9BE
M0 WSR B. Harrison 43A Rumbridge Street, Totton, Southampton SO40 9DR
MI0 WST D. Caiden 14 Church Street, Greyabbey, Newtownards BT22 2NQ
M0 WSW Reading Scouts Radio c/o S. Whittaker 25 Cleveleys Road, Blackburn BB2 3JS
M0 WSX Wessex Ham c/o R. Thomas 28 Clarks Meadow, Shepton Mallet BA4 4FD
M0 WSZ J. Hocking 26 Musket Road, Heathfield, Newton Abbot TQ12 6SB
M0 WTC J. Paradas 1A Brocks Ghyll, Eastbourne BN20 9RQ
M0 WTG D. Cooper Little Heath, Bradfield Common, North Walsham NR28 0QR
M0 WTJ T. Weston 4 The Pightle, Peasemore, Newbury RG20 7JS
MW0 WTK P. Iles 150 Pen-Y-Bryn, Caerphilly CF83 2LA
M0 WTL O. Fallon 26 Central Avenue, Corfe Mullen, Wimborne BH21 3JD
M0 WTN J. Withington 20 Bond Way, Hednesford, Cannock WS12 4SN
M0 WTO M. Roberts Flat 6, 463 Brighton Road, Lancing BN15 8LF
M0 WTW W. Walker 247 Forest Road, Tunbridge Wells TN2 5HT
M0 WTX S. Jackson 64 Main Road, Moulton, Northwich CW9 8PB
M0 WTY R. Clare Kimberley, Boston Road, Bicker, Boston PE20 3AP
MM0 WTZ K. Waitz C/O Waitz-Rainey, 16 Inverkeithing Road, Aberdour, Burntisland KY3 0RS
M0 WUL W. Stewart 5 St. Catherines Close, Uttoxeter ST14 8EF
M0 WUS S. Burns 22 Pendle Close, Peterlee SR8 2JS
MW0 WVA Wales Digital Radio Group c/o D. D'Mellow 2 Twynpandy Pontrhydyfen, Port Talbot SA12 9TW
M0 WVE M. Huggett 12 West View Cottages Lewes Road, Lindfield, Haywards Heath RH16 2LJ
M0 WVL C. Lacey 2 Purbeck Cottages Acton, Langton Matravers, Swanage BH19 3LU
M0 WVQ J. Rogers 288 Wareham Street, Middleborough 23462906 United States
MI0 WWB W. Bradley 14 Ardmore Grange, Ballygowan, Newtownards BT23 5TZ
M0 WWD J. Godfrey 24 Walton Way, Barnstaple EX32 8AE
M0 WWE D. Higginbottom 137 Ridgehill Avenue, Sheffield S12 2GN
MI0 WWF S. Nash 5 Drumard Cottages , Dans Road, Ballymena BT42 2PX
MM0 WWH N. Robertson Ladyburn, Port William, Newton Stewart DG8 9QN
MM0 WWM R. Jowett Fearnoch Ardentallen, Oban PA34 4SF
MW0 WWR West Wales Radio Group c/o I. Gray 4 Llundain Fach, Felinfoel, Llanelli SA15 4PF
M0 WWV A. Norden 10 School Lane, Watton At Stone, Hertford SG14 3SF
M0 WWX Warwickshire Wireless Society c/o M. Whitehall 114 Nuneaton Road, Bedworth CV12 8AR
MM0 WXD Dr D. Fisher 1 Inverleith Row, Edinburgh EH3 5LP
MM0 WXE A. Barclay 21 Netherlea, Scone, Perth PH2 6QA
M0 WXF A. England 12 Bronte Court Swinburne Road, Wellingborough NN8 3BF
M0 WXO M. Shibata 18-41 Moegino Aobaku, Yokohama 156-0045 Japan
M0 WXP S. Martin 6 Cherry Tree Drive, Sedgefield, Stockton-on-Tees TS21 3DN
MM0 WXS A. Mccall Shielhill, Greengairs, Airdrie ML6 7TJ
M0 WXU P. Elsey 62B Coleraine Road, London SE3 7PE
M0 WXX D. Sharpen 52 Woodsend Road, Urmston, Manchester M41 8QT
M0 WXY S. Baldwin 143 Oxford Road, Swindon SN3 4JA
M0 WYB J. Scully 10 Eckweek Road, Peasedown St. John, Bath BA2 8EQ
M0 WYC the RC c/o D. Jones Marvin House, Ryhill Pits Lane Cold Hiendley, Wakefield WF4 2DU
M0 WYE H. Burnham 13 The Close, Wye, Ashford TN25 5BD
M0 WYH D. Lester 171 Glenavon Road, Birmingham B14 5BT
M0 WYM C. Ivermee 58 Home Meadow, Totnes TQ9 5XY
MW0 WYN D. Davies 2 Hendre Ddu, Manod, Blaenau Ffestiniog LL41 4BH
M0 WYP J. Marvel Dean House Farm, Nordan, Leominster HR6 0AW
M0 WYR Wyre ARG c/o K. Haworth 11 Petersfield Close, Bootle L30 1SG
M0 WYT T. Webster 1 Fen Close, Newton, Alfreton DE55 5TD
M0 WYW Y. Wong 15B Block 27, City One Shatin N.T. . Hong Kong
M0 WYZ K. Winwood 146 Chapel Street, Pensnett, Brierley Hill DY5 4EQ
M0 WZM M. Kitt 18 Brickmakers Road, Colden Common, Winchester SO21 1TT
M0 WZT M. Fulbrook 2 Cob Place, Westbury BA13 3GS
MW0 WZX C. Davies 23 Nottingham Street, Cardiff CF5 1JP
MM0 WZZ W. Ramsay 1 Northburn Road, Eyemouth TD14 5AU
MM0 XAB A. Lark 20 Lawfield, Coldingham, Eyemouth TD14 5PB
M0 XAC G. Dean 62 Baptist Close, Abbeymead, Gloucester GL4 5GD
MW0 XAD S. Gordon 8 Maesteg, Cymau, Wrexham LL11 5EP
M0 XAE B. Smith 8 Mill Field Close, South Kilworth, Lutterworth LE17 6FE
M0 XAJ A. Calvert 5 Pond Cottages Butts Pond, Sturminster Newton DT10 1BE
M0 XAK Monmouth School Amateur School Society c/o A. Kent 4 Sellerdale Drive, Wyke, Bradford BD12 9DA
M0 XAM A. Morgan 18 Keysworth Drive, Wareham BH20 7BD
MM0 XAO Dr A. Onken 19/7 Damside, Edinburgh EH4 3BB
M0 XAR S. Halliday 8 Newby Farm Road, Scarborough YO12 6UN
M0 XAS A. Shepherd 33 Meadow Way, Turton, Bolton BL7 0DE
M0 XAT M. Harwood 36 Coronation Avenue, Seaton, Workington CA14 1DW
MM0 XAU Obo Strathmore ARC c/o H. Stoeteknuel C/O Marriott Parkview, Dunrossness, Shetland ZE2 9JG
M0 XAV J. Hazeltine 21 Hassock Way, Wimblington, March PE15 0PJ
M0 XAW R. Watson 8 Bourne Close, Warminster BA12 9PT
M0 XBA G. Harvey 55 Trelawney Road Hainault, Ilford IG6 2NJ
M0 XBB R. Boland 10 Kenilworth Drive, Kidderminster DY10 1YD
MM0 XBD Surrey Space Centre c/o B. Donnelly 19 Douglas Drive, Dunfermline KY12 9YG
M0 XBI A. Romanov 10 Gloucester Walk, Westbury BA13 3XG
M0 XBM C. Atkinson 7 Hamilton Road, Grantham NG31 9QG
M0 XBN B. Johnson 6 Trevor Road, Swinton, Manchester M27 0YH
M0 XBQ Dr F. Sedgemore 12 The Brambles, Royston SG8 9NQ
M0 XBR A. Brade áSand Gap, Bursea Lane, York YO43 4DF
M0 XBW B. Woollett 23 Kinglake House, Southall UB24FZ
M0 XBY Bromley & District ARS c/o R. Perzyna 29 Lakeside Drive, Bromley BR2 8QQ
M0 XCA C. Ashley 22 Pasture Close, Lower Earley, Reading RG6 4UY
M0 XCC Cross Counties ARC c/o S. Forshaw 32 Fox Way, Eastfield YO11 3PH
MD0 XCE A. Elliott Ballacannell, Earstane, Colby, Isle of Man IM9 4HN
M0 XCF I. Titchener 18 King Edgar Close, Ely CB6 1DP
M0 XCH C. Harding 27 Eston Avenue, Malvern WR14 2SR
M0 XCJ C. Jackson 84 Ogley Road, Walsall WS8 6BB
M0 XCL C. Loud 24 Harrington Avenue, Lowestoft NR32 4JU
M0 XCO M. Macdonell 54 Cinque Foil, Peacehaven BN10 8DZ
MM0 XCP J. Reid 69 Limekiln Wynd, Mossblown KA6 5BE
M0 XCR C. Raphael 86 Main Street, South Rauceby, Sleaford NG34 8QQ
MM0 XCS C. Sharp 12 Manse Place, Inverkeithing KY111AZ
M0 XCT Hilderstone RS c/o D. Aldred 14 The Meadows, Radcliffe, Manchester M26 4NS
M0 XCX P. Houghton 6 Olivers Court, Calne SN11 0FL

M0 XCZ M. Hartley 24 Burnham Avenue, Bognor Regis PO21 2JU
M0 XDA D. Sullivan 16 Wadley Close Tiptree, Colchester CO5 0SL
M0 XDC D. Copsey Fairview, Mill Lane, Hook End, Brentwood CM15 0PP
M0 XDF D. Ferrington 20 Innings Lane, Warfield, Bracknell RG42 3TR
M0 XDJ K. Gribben 33 Strawberry Close, Birchwood, Warrington WA3 7NT
M0 XDK C. Griffiths 12 Bank View, Northampton NN4 0RS
M0 XDL Cambridge & District ARC c/o R. Coles 10 Littlemoor Road, Weymouth DT3 6AA
MW0 XDN Dr A. Newsome Dros-Dro, Station Road, Letterston SA62 5RY
M0 XDS D. Sharp 8 Beechfield, Hoddesdon EN11 9QH
MW0 XDT R. Snape Bodlondeb, North Road, Whitland SA34 0AX
M0 XDV A. Dalla-Volta 88 Claygate Lane, Esher KT10 0BJ
M0 XDX P. Dumpleton 20 Cambridge Road North, Mablethorpe LN12 1QR
MM0 XEA J. Dunlop 5 Loudon Road, Glasgow G33 6NJ
M0 XEE M. Toher The Chapel, Station Road, Darlington DL2 1JG
M0 XEF C. Tacon 2 Pavillion Court, 74-76 Northlands Road, Southampton SO15 2NN
M0 XEK E. Kemp 4 Foundry Flats, Foundry Square, Hayle TR27 4AE
M0 XEM S. Christie 124 Bickershaw Lane, Abram, Wigan WN2 5PP
M0 XER L. Bodnar 47 Alchester Court, Towcester NN12 6RL
M0 XEY C. Blain 8 Fern Way, Weaverham, Northwich CW8 3EZ
M0 XFG F. Grande Restrup Engvej 26, Aalborg 9000 Denmark
M0 XFL C. Coles 2 Fern Square, Chickerell, Weymouth DT3 4NZ
MW0 XFU C. Jenkins Flat 5, The Lawns, Usk NP15 1BA
M0 XFX J. Hawkes 53 Mill Hill, Derby DE24 5AF
M0 XG Braintree & District ARS c/o G. Nurse Orchard Bungalow, Priors Green, Stisted, Braintree CM77 8BP
M0 XGB K. Dickson 29 Sunnyfields Drive, Minster On Sea, Sheerness ME12 3DH
M0 XGG M. Bayliff 80 The Meadows, Leominster HR6 8RE
M0 XGK J. Haig 1 Vallibus Close, Lowestoft NR32 3DS
M0 XGL G. Lund 1 Thrush Close, Gloucester GL4 4WZ
M0 XGN G. Norbury 3 Sherard Croft, Birmingham B36 0LS
M0 XGR W. Smith 81 Hazelgrove Residential Park Milton Street, Saltburn-by-the-Sea TS12 1FE
M0 XGS G. Stanley 95 Old Vicarage, Westhoughton, Bolton BL5 2EG
M0 XGT Dr T. Thomas 55 Bath Street, Southampton SO14 6GR
M0 XGW G. Whall 10 Hillcrest Court, Ipswich Road, Diss IP21 4YJ
M0 XGX G. Round 128 Leicester Road, Shepshed, Loughborough LE12 9DH
M0 XHD I. Hammond 88 Great Innings North, Watton At Stone SG14 3TD
M0 XHN Hubnet Amateur Radio & Digital Online Network c/o S. Curtis 354 St. Helens Road, Leigh WN7 3PQ
M0 XIA I. Alderman 107 Manton Drive, Luton LU2 7DL
M0 XIC M. Savage 5 Mere Hall Barns, Mere Lane, Enville, Stourbridge DY7 5JL
M0 XID G. Hurst 173 Halling Hill, Harlow CM20 3JP
M0 XIG J. Wakefield Oakhurst, Lower Common Road, Romsey SO51 6BT
M0 XIK J. Lambert Hurley 64 Henry Road, West Bridgford, Nottingham NG2 7ND
M0 XJG J. Goodman 70 Bradford Road Eccles, Manchester M30 9FT
M0 XJL W. Pickles 31 Longfield Road, South Woodham Ferrers CM3 5JL
M0 XJM J. Meek Rose Tree Cottage, Charlbury OX7 3RX
M0 XJN J. Neal 5 Shelley Close, Huntingdon PE29 1NF
M0 XJT P. Hodson 21 Green Hill, London Road, Worcester WR5 2AA
MM0 XKA C. Robertson 5 Broomlands Place, Irvine KA12 0DU
M0 XKD L. Booth 8 Rowthorne Close, Northampton NN5 4WB
MW0 XKL R. Jones Flat 2, Tan Y Geraint, 33 Princess Street, Llangollen LL20 8RD
M0 XKM M. Koster Stalworthy Manor Farm, Suton Lane, Suton, Wymondham NR18 9JG
M0 XKO P. Goodridge 22 Horefield, Porton, Salisbury SP4 0LE
M0 XKW K. Williams 35 Lord Street, Coventry CV5 8DA
M0 XKX Kent County Raynet c/o M. Granatt 16 Culverden Avenue, Royal Tunbridge Wells TN4 9RF
M0 XLA J. Tompkins 3 Hartwell Road, Portsmouth PO3 5TN
M0 XLB S. Borrell Rose Cottage, Colchester Main Road, Colchester CO7 8DD
M0 XLH L. Hollingworth 43 Wingfield Road, Hull HU9 4PR
MI0 XLK Ayrshire Raynet Group c/o S. Baird 11 Laral Park, Newtownabbey BT37 0LH
M0 XLR D. Roberts 19 East Avenue, Warrington WA2 8AD
M0 XLT K. Jackson 7 River Place, Gargrave, Skipton BD23 3RY
M0 XLX H. Knight 10 Welford Road, Barton, Alcester B50 4NP
M0 XLY M. Oxley 49 Dalton Crescent, Shildon DL4 2LE
M0 XMB P. Foster 100 Howe Road, Norton, Malton YO17 9BL
M0 XMC M. Coad 227D Woodham Lane, New Haw, Addlestone KT15 3NR
M0 XMD M. Davidson 19 Mason Street, Workington CA14 3EH
MW0 XMG P. Provis Dingle Gardens, Croesbychan, Aberdare CF44 0EJ
M0 XMH M. Hopewell 4 Cotes Crescent Bicton Heath, Shrewsbury SY3 5AS
MW0 XMI R. Lacey 7 Oak Tree Drive, Cefn Hengoed, Hengoed CF82 8FN
M0 XMK M. Rose 19 Hawthorn Street, Peterlee SR8 3LY
M0 XML Ex-Military Land Rover Assoc. c/o J. Butcher Mount Pleasant, Trampers Lane, North Boarhunt, Fareham PO17 6DG
M0 XMP M. Pope Drove Lodge 39 The Drove Barroway Drove, Downham Market PE38 0AJ
MM0 XMQ L. Anderson 6 St. Keiran Crescent, Stonehaven AB39 2GQ
M0 XMS M. Smith Higher Alterhay, Combe St. Nicholas, Chard TA20 3LT
M0 XMT P. Setter 199 Southbourne Grove, Westcliff-on-Sea SS0 0AN
M0 XMX M. Lewis 1 Kingsmead Stretton, Burton on Trent DE13 0FQ
M0 XMZ Dr D. Boocock 47 Long Close, Anstey, Leicester LE7 7QG
MM0 XNC R. Biggart Lodgebush Cottage, Craigie, Kilmarnock KA1 5NA
M0 XNG J. Blackwell 164 York Road Haxby, York YO32 3EL
M0 XNR N. Reeve 16 Hospital Drove, Little Sutton, Long Sutton, Spalding PE12 9EL
M0 XNW North West Radio Group c/o D. Roberts 19 East Avenue, Warrington WA2 8AD
M0 XOB J. Dominy 19 Church Hill Avenue, Warton, Carnforth LA5 9NU
M0 XOC M. Cox 55 Malvern Crescent, Ince, Wigan WN3 4QA
M0 XOD G. Smith 47 Percy Road, Carlisle CA2 6ER
MI0 XOJ G. Travers 88 Palmerston Road, Belfast BT4 1QD
M0 XOK D. Mitchell 3 Ivy Cottage, Main Road, Theberton, Leiston IP16 4RX
M0 XOL T. Brownen 43 Great Rea Road., Brixham. TQ5 9SW
M0 XOM R. Smith 21 Canal Road Crossflatts, Bingley BD16 2SR
M0 XON K. Handscombe 6 Abbeyfield House Market Cross, Malmesbury SN16 9AS
M0 XOR M. Hauser 23 Prince William Way Sawston, Cambridge CB22 3SZ
M0 XOS O. Snowdon Churchill College, Cambridge CB3 0DS
MW0 XOT J. Messenger 34 Goylands Close, Llandrindod Wells LD15RB
M0 XOU I. Spinks 30 Lime Tree Walk, Watton, Thetford IP25 6EU
M0 XOX J. Hill 45 Venus Street, Congresbury, Bristol BS49 5HA
M0 XPA P. Hekman 28 Beechcroft Avenue, Crewe CW2 6SQ
M0 XPB North West ARC c/o P. Bannon 73 London Road, Worcester WR5 2DU
M0 XPD Dr P. Darlington 8 Uplands Road, Urmston, Manchester M41 6PU
M0 XPH P. Hennessey 11 Monmouth Drive, Eaglescliffe, Stockton-on-Tees TS16 9HU
M0 XPJ J. Parfitt 12 Jodrell Place, Selsey, Chichester PO20 0FQ
M0 XPK P. Davies 2 Lynfords Drive, Runwell, Wickford SS11 7PP
M0 XPL C. Lawrence Croft House, Station Road, Lancaster LA2 8ER
M0 XPM P. Mullen 12 Poplar Grove, Conisbrough, Doncaster DN12 2JG
MM0 XPP L. Johnstone 67 Belfast Quay, Irvine KA12 8PR
M0 XPS P. Standley 9 Capelands, New Ash Green, Longfield DA3 8LG
MM0 XPT C. Watkinson Hare House, Perth Road, Birnam, Dunkeld PH8 0AA
MM0 XPZ S. Groves 1/1 99 Belville Street, Greenock PA15 4SX
M0 XQB C. Louie 29 Marthall Drive, Sale M33 2XP
M0 XQS P. Jarvis 50 Northcott, Bracknell RG12 7WR
M0 XQX T. Tzvetkov 52 Clearwell Gardens, Cheltenham GL52 5GH
M0 XRA J. Batsman 141 Bury Street, Ruislip HA4 7TQ
M0 XRC Banbury Ray Grp c/o J. Fennell Bajamar House, Belton Road, Doncaster DN9 1JL
M0 XRF J. Margarson 26 David Street, Grimsby DN32 9NL
M0 XRH Eccentricity Radio Hams Club c/o J. Rolfe 56 Elmhurst Road, Thatcham RG18 3DH
MM0 XRI R. Irvine 9 Pearce Grove, Edinburgh EH12 8SP
M0 XRL R. Langford 103 Huddersfield Road, Elland HX5 0EE
M0 XRM D. Bingham 33 Sheffield Road, Creswell, Worksop S80 4HN
MW0 XRT D. Burt 2 Cae Masarn, Pentre Halkyn, Holywell CH8 8JY
MW0 XRU T. Nutbeem 6 Morris Rise, Blaenavon, Pontypool NP4 9PA
M0 XRV J. Coates 2 Holstein Drive, Scunthorpe DN16 3TT
M0 XRW A. Diaz 29, Parkside Gardens, Widdrington, Morpeth NE61 5RP
MM0 XRZ M. Nicholls Grahams Onsett Farm, Newcastleton TD9 0TT
M0 XSD Southampton ARC c/o C. Catlin 27 Main Street, Frizington CA26 3SA
M0 XSG C. Braisby 4 Langmans Way, Woking GU21 3QY
M0 XSH D. Shaw 81 Chesterfield Road Tibshelf, Alfreton DE55 5NJ
M0 XSJ Northwich Repeater Group c/o S. Jackson 64 Main Road, Moulton, Northwich CW9 8PB
M0 XTD C. Morgan 5 Montgomery Avenue Hampton-On-The-Hill, Warwick CV35 8QP
M0 XTG C. Day 14 Windsor Drive, Ramsey Forty Foot, Ramsey, Huntingdon PE26 2XX
MW0 XTK P. Hoath 8 Liverpool Terrace, Llithfaen, Pwllheli LL53 6NN
MM0 XTW A. Wallis Pine Cottage, South Smallburn, Peterhead AB42 5BL
M0 XTX A. Cristofoletti 39 Jessop Street, Codnor Ripley DE5 9RN
MW0 XTZ M. Digby 40 Waterloo Road, Ammanford SA18 3SF
M0 XUA Dr M. Sinclair 40 Grotto Road, South Shields NE34 7AH
M0 XUB Hamtests.Co.UK c/o B. Xu 138 King'S College, Cambridge CB2 1ST
M0 XUH Dr G. Thomas 3 The Croft, Wilton, Egremont CA22 2PW
M0 XUI J. Ribbands Dyson'S Farm, Long Row, Tibenham NR16 1PD
M0 XUM W. Webb 84 Bruce Street, Swindon SN2 2EN
M0 XUU R. Gopan 84 Hilmanton, Lower Earley, Reading RG6 4HN
M0 XVF J. Smith 8 Mayfields, Spennymoor DL16 6RN
M0 XVI M. Collins Chilterns, Little Frieth, Henley-on-Thames RG9 6NR
M0 XVK R. Redmond 28 Common Lane, Polesworth, Tamworth B78 1LS
M0 XVL G. Elsigan Traunuferstr 143 A, Haid A-4053 Austria
M0 XVN M. Tarrant Holly Cottage Nanstallon, Bodmin PL30 5JZ
M0 XVX Dr A. Smith Borrowdale House, 10 Borrowdale Road, , Malvern WR14 2DS
M0 XWD J. Brown 8 Chatsworth Street, Sutton-in-Ashfield NG17 4GG
M0 XWP M. Meyer 22 Orchard Grove, Newton Abbot TQ12 1FZ
M0 XWS M. Swann 56 Mansel Drive Old Catton, Norwich NR6 7NB
M0 XXC Thames ARG c/o A. Atkinson 21 Dennington Crescent, Basildon SS14 2FF
M0 XXD J. Cadman 1 Kittiwake Road, Chorley PR6 9BA
M0 XXI Abergavenny RS c/o S. Lindberg 6 Meadow View, Belper DE56 1UT
M0 XXJ J. Creaser 1 Trotsford Meadow, Blackwater GU17 0PD
M0 XXK K. Mitchell 1 Denstroude Cottages, Denstroude Lane, Canterbury CT2 9JX
M0 XXL D. Tinn 28 South Road, Kirkby Stephen CA174SN
M0 XXM M. Jennings Springfield Farm, The Causeway, Stow Bridge, King's Lynn PE34 3PP
MM0 XXP A. Pitkethley 99 Margaretvale Drive, Larkhall ML9 1EH
M0 XXT Lapworth RC c/o C. Mccormick 65 Glendon Way, Dorridge, Solihull B93 8SY
M0 XXV Yorkshire DX Club c/o C. Smith 199A Richardshaw Lane, Stanningley, Pudsey LS28 6AA
MM0 XXW M. Whyte 147/2 Lower Granton Road, Edinburgh EH5 1EX
M0 XXX H. Taylor Sunnyside Well, Chaingate Lane, Bristol BS37 9XN
M0 XYA P. Hodges 191 Broadstone Road, Stockport SK4 5HP
M0 XYL M. Turner 2 Oakleigh Road, Droitwich WR9 0RP
M0 XYM K. Polston 10 Marsh Farm Road, South Woodham Ferrers, Chelmsford CM3 5WP
M0 XYN Yorkshire Net UK c/o L. Riley 51 Brier Lane, Havercroft, Wakefield WF4 2AT
M0 XYT West Wight RS c/o C. Oliver 3 Windsor Drive, Freshwater PO40 9GB
M0 XYX A. Loyd Maple House, Pangbourne Road, Reading RG8 8LN
M0 XYZ A. Clark 10 Garfield Close, Lincoln LN1 3QP
M0 XZG Dr G. Welch Amazonas, Sandy Lane, Hightown, Liverpool L38 3RP
M0 XZS A. Coetzee 6 Covent Gardens, Colwall, Malvern WR13 6FA
M0 XZW C. Set Hughes Hall, Wollaston Road, Cambridge CB1 2EW
M0 XZX S. Lowe 14 Windmill Rise, York YO26 4TX
MM0 YAB C. Phillips 8 The Square, Newtongrange, Dalkeith EH22 4QD
MW0 YAC K. Smith 3 Pendoylan Walk, Cwmbran NP44 7JX
MW0 YAD Cwmbran ARS c/o K. Smith 3 Pendoylan Walk, Cwmbran NP44 7JX
MW0 YAE G. Thatcher-Sharp 20 Dilys Street, Blaencwm, Treorchy CF42 5DT
MW0 YAG A. Graham 2 Heol Undeb, Beddau, Pontypridd CF38 2LB
M0 YAH Laser ATC Rac (South) c/o W. Coburn 42 Hinton Wood Avenue, Christchurch BH23 5AH
M0 YAL W. Dowkes Woodlea, Gillamoor Road, York YO62 6EL
MI0 YAM D. Foley 14 Chestnut Hall Court, Moira, Maghaberry BT67 0GJ
MD0 YAU D. Smith Alwyn, Four Roads Port St. Mary, Isle of Man IM9 5LH
M0 YAV W. Jones 8 Oakbrook Close, Ewyas Harold, Hereford HR2 0NX
M0 YAW D. Cooper Seinna Bryne Lane, Padbury MK18 2AL
M0 YAY D. Young 35 Limber Hill, Cheltenham GL50 4RJ
M0 YBC D. Croft 33 Roughaw Road, Skipton BD23 2PY
M0 YBD P. Trew 5 Gifford Close, Fareham PO15 6PJ
MI0 YBH J. Mccourt 70A Sessiagh Scott Road Rock, Dungannon BT70 3JU

M0 YBT M. Harrison 2 Rosemount Court, Holly Bank Road, York YO24 4EG
MW0 YBZ P. Smith 29 Heol Cwarrel Clark, Caerphilly CF83 2NE
M0 YCB C. Button 30 Southfall Close, Ranskill, Retford DN22 8NE
M0 YCG Yorkshire Dales CG c/o D. Falkner 45 Westwood Carleton, Skipton BD23 3DW
M0 YCH Dr C. Haws 5 Mallow Close, Locks Heath, Southampton SO31 6XF
MM0 YCJ Prof. C. Jones 11B Ettrick Road, Edinburgh EH10 5BJ
MI0 YCK D. Mayock 27 Woodvale, Bessbrook BT35 7FD
M0 YCQ S. Burgess 10B Scotland Street, Ellesmere SY12 0EG
M0 YCS C. Steuwe Moor Barn Farm, Madingley Road, Cambridge CB23 7PG
M0 YCX G. Bevan 2 Geddes Street, Blackburn BB2 5LQ
M0 YCY A. Capitan 41 Cunningham Drive, Runcorn WA7 4DL
M0 YDB D. Breed 8 Tudor Street, New Rossington, Doncaster DN11 0JG
M0 YDF D. Fowler 5 Highfield Cottages, Everingham, York YO42 4JG
M0 YDH D. Holman 20 Green Drive, Wolverhampton WV10 6DW
M0 YDJ D. Jackson 3 Laburnum Road, Cadishead, Manchester M44 5AS
M0 YDK A. Morrell 19 Nairn Road, Stamford PE9 2YR
M0 YDM Dr D. Maldoom Hyde Manor, The Street, Kingston, Lewes BN7 3PB
M0 YDP S. Allington 57 Lightfoot Drive, Carlisle CA1 3BP
M0 YDW D. White 9 Wyatts Lane, Tavistock PL19 0EU
MW0 YDX B. Dallimore 4 Llys Dyffryn, St. Asaph LL17 0SX
M0 YEE A. Chapman 24 Oakford Park, Halnaker, Chichester PO18 0BF
MM0 YEK East Kilbride ARS c/o A. Hood 26 Annan Avenue, East Kilbride, Glasgow G75 8XT
M0 YEP D. Stinson 1 The Croft, Earls Colne, Colchester CO6 2NH
MM0 YEQ G. Pearce 2 Kirkriggs, Forfar DD8 2AT
M0 YES P. Shaw 32 Hardwick Road East, Worksop S80 2NT
MM0 YET G. Burnett 1B Craig Road, Troon KA10 6DA
M0 YFT A. Murdoch 128 Whalley Road, Langho, Blackburn BB6 8DD
M0 YGB A. Birch 3 Partridge Way, High Wycombe HP13 5JX
M0 YGG A. Mansfield 20 High Street, Broughton, Kettering NN14 1NG
MW0 YGJ Greater Manchester Raynet c/o G. Owen 14 Bideford Road, Newport NP20 3BJ
M0 YGT T. Kadleck 22 Oaklands Road, Liskeard PL14 3TX
M0 YGW J. Newton 6413 Hillegass Avenue, Oakland 94618 United States
M0 YHA Aries ARC c/o A. Clayton 6 Albert Road, Bunny, Nottingham NG11 6QE
MM0 YIA A. Maclennan 5 Baluachrach, Culbokie, Dingwall IV7 8FP
M0 YIG G. Coleman 15 Redwood Drive, Ormskirk L39 3NS
M0 YIJ G. Chesters11 50 Primrose Chase, Goostrey, Crewe CW4 8LJ
M0 YIM M. Ellwood 27 Bath Meadow, Halesowen B63 2XH
M0 YIT D. Gresty 81 Old Hall Road, Sale M33 2HU
M0 YJD J. Davidson 4 Willows Avenue, Alfreton DE55 7ER
M0 YJL C. Liu 7 Nocks Avenue, Birmingham B24 9NB
M0 YJO Newquay & District ARS c/o J. Hatton 49 Buxton Street, Morecambe LA4 5SR
M0 YJT C. Jarvis 516 Kingsbury Road, Erdington, Birmingham B24 9NF
M0 YJW J. Williams 66 Oakfield Avenue, Hitchin SG4 9JD
M0 YKB D. Yakub 42 Swift Close, Blackburn BB1 6LF
M0 YKC D. Forshaw 14 Hope Carr Road, Leigh WN7 3ET
M0 YKR Yorkshire Resistors c/o G. Bystryakov 20 Elmhurst Gardens, Leeds LS17 8BG
M0 YKS S. Davison 5 Denby Drive, Baildon, Shipley BD17 7PQ
M0 YLA R. Cato Orrell House, Winterpit Lane, Horsham RH13 6LZ
M0 YLG G. Roddis 61 Everton Road, Potton, Sandy SG19 2PB
MW0 YLS S. Smith 102 Gresford Road, Llay, Wrexham LL12 0NW
MI0 YLT S. Mccormick 46 Lany Road, Moira, Craigavon BT67 0NZ
M0 YLY S. Myland 2 Willhays Close, Kingsteignton, Newton Abbot TQ12 3YT
M0 YMA A. Banks 2 Holt Close, Farnborough GU14 8DG
MW0 YMB Dr W. Dickson The Rowans, Pwllmeyric, Chepstow NP16 6LA
MI0 YMF M. Foley 44 Gallows Street, Dromore BT25 1BD
MM0 YMG M. Gibson 18 Pentland View, Edinburgh EH10 6PS
M0 YMJ P. Coppin 3 Firtree Close, Rough Common, Canterbury CT2 9DB
M0 YMM N. Trangmar 8 Maxstoke Close, Meriden, Coventry CV7 7NB
M0 YNK Y. Watkins 1 St. Saviour Close, Colchester CO4 0PW
M0 YNY N. Oldrid 4 Bar Lane Mapplewell, Barnsley S75 6DQ
M0 YOB D. Ferguson 39 Fouracres, Maghull, Liverpool L317BP
M0 YOJ J. Boone Amberley, Pinewood Road, High Wycombe HP12 4DA
M0 YOL D. Parker 16 Aldborough Road, Dagenham RM10 8AS
M0 YOM J. Thresher Quarry Grange, Nuneaton Road Over Whitacre, Coleshill, Birmingham B46 2NH
MM0 YOS O. Sturm Lochinvar House, Dalry, Castle Douglas DG7 3XJ
M0 YOT J. Partington 56 Rutherford Drive, Bolton BL5 1DL
M0 YPJ P. Kirby 30 New Street, Eccleston, Chorley PR7 5TW
M0 YPS Hunmanby Primary ARC c/o C. Fox Millstone Cottage, Prior Wath Road, Scarborough YO13 0AZ
M0 YPW Dr P. Woodfin 14 Broomcroft Road, Bognor Regis PO22 7NJ
M0 YQB R. Suda 3-11-5-206, Hiyoshi-Honcho, Kohoku-Ku, Yokohama-City, Kanagawa-Pref 223-0062 Japan
M0 YQC D. Berry 27 Harcourt Terrace, Headington, Oxford OX3 7QF
M0 YQQ S. Evans 30 Dalglish Creasent Radbrook, Shrewsbury SY3 9FW
M0 YRC Mango c/o M. Hodgkinson 34 Pennine Way, Brierfield, Nelson BB9 5DT
M0 YRF Yorkshire Radio Friends c/o R. Potter 30 High Street, Doncaster DN7 6RY
M0 YRG A. Tring 12 Ainsdale Close, Orpington BR6 8DJ
M0 YRM M. Radulov 14 Grove Road, Chatham ME4 5HS
M0 YSE K. Moysey 109 Langbrook Cottages, Langbrook, Ivybridge PL21 9JX
M0 YSG St George's School c/o M. Hubbard 14 Parkfield Crescent, Kimpton, Hitchin SG4 8EQ
MM0 YSK S. Ram 28 Craigievar Gardens, Kirkcaldy KY2 5SD
M0 YSR R. Moys 12A Palmerston Avenue, Fareham PO16 7DP
MM0 YTA J. Anderson The Grange, Leslie Road, Scotlandwell, Kinross, KY13 9JE
M0 YUG Sherwood ARC c/o G. Bates 230 Brook Street, Erith DA8 1DZ
M0 YUX S. Fabris Via Marchesan 43, Treviso 31100 Italy
M0 YVG D. Cordes 22 Holywell Avenue, Newcastle upon Tyne NE6 3RY
MW0 YVK E. Howells 72 Holywell Crescent, Abergavenny NP7 5LG
MW0 YVT C. Williams 1 South View, Freeholdland Road, Pontypool NP4 8LL
M0 YVX B. Tomlinson 7 Springwell Close, Crewe CW2 6TX
M0 YWA M. Pack 59 West End Falls, Nafferton YO25 4QA
M0 YWO M. Bruyneel 14 Riversmead, St. Neots PE19 1HA
M0 YXR C. Richardson Heathercroft, Kirkby Mills, Kirkbymoorside, York YO62 6NN
M0 YYA G. Bridge 16 Victoria Street, Ramsbottom, Bury BL0 9ED
M0 YYT S. Bennett 7 Holme Park, High Bentham LA2 7ND
M0 YYV M. Lomax 7 Planetree Road, West Derby, Liverpool L12 6RE
M0 YYY Mid Somerset ARC c/o H. Taylor Sunnyside Well, Chaingate Lane, Bristol BS37 9XN
MM0 YZE West Scotland Air Cadet RC c/o K. Brown 10 Richmond Street, Clydebank G81 1RF
M0 YZF C. Preece 14 Dock Street, Widnes WA8 0QX
M0 YZG G. Castledine 11 Cam Road, Cheltenham GL52 5QS
M0 YZT Dr H. Orridge Stonecross Cottage, Wadworth Hall Lane, Wadworth, Doncaster DN11 9BH
M0 YZV L. Sadler 2 Birkdale Avenue, Dinnington, Sheffield S25 2SX
M0 YZY S. Collins 5 Fernleigh Gardens, Stafford ST16 1HA
M0 ZAA J. Wellard 19 South Motto, Kingsnorth, Ashford TN23 3NJ
MW0 ZAB A. Bubb Ground Floor Flat 5 Temple Street, Newport NP202GJ
M0 ZAE H. Ehm 17 Stuart Road, Kempston, Bedford MK42 8HS
M0 ZAF R. Barter 8 Orchard Close, Newton Abbot TQ12 3DF
M0 ZAI Border ARS c/o M. Grice 48 St. Ives Road, Coventry CV2 5FZ
M0 ZAK J. Steel 6 Central Avenue, Shepshed, Loughborough LE12 9HP
MM0 ZAL B. Keiller Da Cro, Branchiclate, Burra Isle ZE2 9LA
M0 ZAM G. Campbell 10 Welbeck Road, Rochdale OL16 4XP
M0 ZAN 1st Prestwood Scout Group c/o Dr E. Nieuwoudt Not, Applicable, Outside of UK South Africa
MW0 ZAP J. Davies Rose Villa, Creigiau, Cardiff CF15 9NN
MW0 ZAQ C. Rayment Brambles, Alltami Road, Mold CH7 6RT
M0 ZAR Dr S. Smith Po Box 446, Clanwilliam, Clanwilliam 8135 South Africa
M0 ZAV R. Amos 6 Eccles Road, Wittering, Peterborough PE8 6AU
MM0 ZAW A. Woodford Nordkette, Evnabrek, Levenwick, Shetland ZE2 9GY
M0 ZAY H. Jones 17 Doves Yard, London N1 0HQ
MM0 ZBD D. Brown 181/1 (Gf) Gorgie Road, Edinburgh EH11 1TT
MM0 ZBH Worked All Britain Awards Group c/o P. Mclaren 1 Morayvale, Aberdour, Burntisland KY3 0XE
M0 ZBT S. Green 6 Garth Villas, Rimswell, Withernsea HU19 2DB
M0 ZBZ M. Carvell 10 Burns Close, Stevenage SG2 0JN
MW0 ZCE M. Douglas 486 Malpas Road, Newport NP20 6NB
MM0 ZCG Shetland CG c/o H. Stoeteknuel C/O Marriott Parkview, Dunrossness, Shetland ZE2 9JG
M0 ZCJ C. Jonas 198 High Street, Chesterton, Cambridge CB4 1NX
M0 ZCM A. Adams 45 Four Oaks Road Tedburn St. Mary, Exeter EX6 6AP
M0 ZCO C. O Broin Dewhurst, Orchard End, Weybridge KT13 9LS
M0 ZCP C. Parker The Grange, Watercombe Cornwood, Nr. Ivybridge PL21 9RB
MM0 ZCT Dr C. Thompson 32 Damselfly Road, Edinburgh EH17 8XG
M0 ZCW P. Smith 101 Brunel Avenue, Newthorpe, Nottingham NG16 3RE
M0 ZDB D. Brownsea 47 Southill Road, Bournemouth BH9 1SH
M0 ZDC Dr D. Cooke Apartment 9, 27 Sheldon Square, London W2 6DW
M0 ZDD A. Henderson Clocktower Lodge, Cragside, Rothbury, Morpeth NE65 7PU
M0 ZDE D. Kirkden 57 Crow Hill Road, Margate CT9 5PF
M0 ZDG D. Griffin 101 Kingsway, Duxford CB22 4QN
M0 ZDH D. Hardwick 30 Halfcot Avenue, Stourbridge DY9 0YB
M0 ZDJ S. Scott 13 Silver Close, Harrow HA36JT
M0 ZDM D. Mason 94 Guessburn, Stocksfield NE43 7QR
M0 ZDO A. Hipkiss 24 Newington Road, Birmingham B37 7RW
M0 ZDU A. Hawkes Coachmans Emsworth Road, Lymington SO41 9BL
MW0 ZDX G. Western 5 Meredith Close, Acrefair LL14 3GB
M0 ZEB D. Featherby 14 Station Road, Sutton, Ely CB6 2RL
M0 ZED P. Phillips 2 Millstream Close, Goostrey, Crewe CW4 8JG
M0 ZEE J. Rufes Flat 1 & 3-8 12 Smyrna Road, London NW6 4LY
M0 ZEH S. Hendy Flat 2, 33 Kingston Road, , Leatherhead KT22 7SL
M0 ZEL S. James 35 Prospect Road, Dronfield S18 2EA
M0 ZEM R. Donaldson 10 Berry Avenue, Trimdon Grange, Trimdon Station TS29 6EE
M0 ZEN A. Bolton 2 Temperance Cottages, Lower Cross, Clearwell, Coleford GL16 8LD
MM0 ZEQ F. Trigg Levante, Bankend, Dumfries DG1 4RP
MM0 ZET Blacks Hillbillies ARC c/o H. Hassel Sumra, Eshaness, Shetland ZE2 9RS
M0 ZEY N. Horton 51 Walsingham Gardens, Epsom KT19 0LS
M0 ZFF D. Almond 55 Forde Park, Yeovil BA21 3QP
MM0 ZFG S. Street Tangaroa, Fairfield Gardens, Kilcreggan, Kilcreggan G84 0HS
M0 ZGB J. Hobbs 82 Perry'S Lane, Wroughton, Swindon SN4 9AP
M0 ZGT E. Ayre 1 Spring Gardens, Broadmayne, Dorchester DT2 8PP
M0 ZID S. Frampton 20 Winslow Close, Boldon Colliery NE35 9LR
MM0 ZIF M. Hazel-Mcgown 9 Barra Wynd, Irvine KA11 1DB
M0 ZIG J. Stoppard 15 South Lodge Court, Old Road, Chesterfield S40 3QG
M0 ZIM M. Raynor 68 Cambridge Street, South Elmsall, Pontefract WF9 2AR
M0 ZIP Cross Border CG c/o K. Pritchard 9 Golf Close, Pyrford., Woking GU22 8PE
M0 ZJB M. Collier 32 London Road, Dereham NR19 1AW
M0 ZJO Obo North Anglia Raynet c/o J. Rawlinson Westfield Farm, Risden Lane, Cranbrook TN18 5DU
M0 ZJQ A. Rawlinson Westfield Farm, Risden Lane, Hawkhurst, Sandhurst, Cranbrook TN18 5DU
M0 ZJV S. De Vries 46 Chaulden House Gardens, Hemel Hempstead HP1 2BP
M0 ZKA F. Hruszka 20 Winchester Avenue, Leicester LE3 1AU
M0 ZKK M. Bayman Marsworth, Tring HP23 4LX
MI0 ZKX S. Mcnulty 13 Syenite Place, Rostrevor, Newry BT34 3EP
M0 ZLE UK Islands Group c/o M. Holmes 6 Wells Court, Saxilby, Lincoln LN1 2GY
M0 ZLF T. Lovell 7 Victoria Wharf Victoria Street North, Grimsby DN31 1PQ
M0 ZLH A. Stabler 11 Lincolns Avenue, Gedney Hill, Spalding PE12 0PQ
M0 ZLI D. Challis 3 West Leaze Place, Bradley Stoke, Bristol BS32 8AF
M0 ZLK C. Forber 32 Larch Avenue, Newton-le-Willows WA12 8JF
M0 ZLP R. Edmondson Kamway, Stanhoe Road, King's Lynn PE31 8NJ
M0 ZMB P. Smart 142 Finch Road, Chipping Sodbury, Bristol BS37 6JB
M0 ZMM Dr R. Hodgkinson 39 Oxford Road, Carlton-In-Lindrick, Worksop S81 9BD
M0 ZMO L. Whitfield 49 Northview Road, Dunstable LU55HB
M0 ZMS M. Strickland Ancoats, Piercy End, York YO62 6DQ
M0 ZMT Herts Repeater Group (Linking Call) c/o M. Thompson 133 Redford Avenue, Horsham RH12 2HH
M0 ZMX M. Hardingham Prospect House, High Street Isle Of Grain, Rochester ME3 0BS
M0 ZNP C. Gray 2 Gloucester Road, Pilgrims Hatch, Brentwood CM15 9ND
M0 ZNZ Pelican Radio Group c/o Dr G. Richardson Berwick Cottage, Bailes Lane, Guildford GU3 2AX

M0 ZOE Z. Dunne 1 Burton Gardens, Brierfield, Nelson BB9 5DR
MM0 ZOG P. Riddle Carngeal, Pitlochry PH16 5JL
M0 ZOM C. Langdon 6 Glebe Close, Stockton, Southam CV47 8LG
M0 ZOO Worcester Radio Amateurs Association c/o P. Badham 201 York Avenue, East Cowes PO32 6BH
M0 ZOR B. Trayhurn 15 Wight Drive, Caister-On-Sea, Great Yarmouth NR30 5UN
M0 ZOV J. Renmans 17 Cartmel Crescent, Chadderton, Oldham OL9 8DA
M0 ZPA P. Davies 68 Sidmouth Avenue, Stafford ST17 0HF
M0 ZPC E. Krebser P.O.Box 621 Port Edward, Kwazulu, Natal, South Africa, 4295, Port Edward 4295 South Africa
M0 ZPG P. Morris 10 Haslam Avenue, Sutton SM3 9ND
M0 ZPL D. Janowicz 20 Salisbury Road, St. Leonards-on-Sea TN37 6RX
M0 ZPM P. Mcgillewie 64 Caradoc View, Hanwood, Shrewsbury SY5 8ND
M0 ZPU R. Compton 18 Drove Road, Gamlingay SG193NY
M0 ZPZ C. Parry 27 Tynedale Close, Stockport SK5 7NA
MM0 ZRC R. Chroston North Stonganess, Cullivoe, Shetland ZE2 9DD
M0 ZRD D. Hind 19 Ellington Road, Arnold, Nottingham NG5 8SJ
M0 ZRF R. Fidler 44 Windermere Avenue, Ramsgate CT11 0PF
M0 ZRG G. Reynolds 43 Orchard Drive, Watford WD17 3DX
M0 ZRR T. Cooper 9 Websters Close, Shepshed, Loughborough LE12 9AT
M0 ZRS Artie Moore ARS c/o R. Styles 4 Coningsby Close, Gainsborough DN21 1SS
M0 ZRX K. Bianchini 10 St. Leonards Road, Headington, , Oxford OX3 8AA
MI0 ZSC J. Sinclair 29 Tern Crescent, Carrickfergus BT38 7RU
M0 ZSJ J. Gibson 22 Woodburn Drive Chapeltown, Sheffield S351YS
M0 ZSM S. Sissens 20 Fallow Drive, Eaton Socon, St. Neots PE19 8QL
M0 ZSS M. Chamberlain 30 Roxton Rd, Great Barford MK44 3 LR
M0 ZTD Dr T. Digman 74 Baddlesmere Road, Whitstable CT5 2LA
M0 ZTE S. Broom 128 Springhill Road, Wolverhampton WV11 3AQ
M0 ZTG A. Hill 5 Park Road, Thurnscoe, Rotherham S63 0TG
M0 ZUB D. Zubrzycki 4 Falklands Court, Easington, Hull HU12 0QE
M0 ZUI D. Bill 5 Kennington Road, Wolverhampton WV10 9RJ
MW0 ZUS T. Lewis 34 Erw Goch, Ruthin LL15 1RR
M0 ZVB P. Carpenter 11 Lakeside, Beckenham BR3 6LX
M0 ZVF R. Baxter 30 Croft Gate, Bolton BL2 3JJ
M0 ZVR B. Bateman 27 Imperial Avenue, Kidderminster DY10 2RA
MU0 ZVV J. Bligh The Bounty, Salines Lane, St. Sampson GY2 4FL Guernsey
MW0 ZWR G. Spicer 6 Cromwell Road, Neath SA10 8DR
M0 ZWT I. Lonsdale 23 Hunts Field Clayton-Le-Woods, Chorley PR6 7TT
M0 ZWW Dr W. Warwicker 28 Porters Wood, Petteridge Lane, Matfield, Tonbridge TN12 7LR
M0 ZXG G. Carless Silver Cottage, Silver Street, South Petherton TA13 5BY
MM0 ZXI J. Stewart 6 Sutherland Way, East Kilbride, Glasgow G74 3DL
M0 ZXJ J. Wildsmith 15 Harebell Close, Hartley Wintney RG27 8TW
M0 ZXQ I. Talbot 41 Elmwood Close, Cannock WS11 6LX
M0 ZXW D. Bambrough 7 Barnwell View, Herrington Burn, Houghton le Spring DH4 7FB
MW0 ZXY A. Pugh 27 Cae'R Pandy, Penrhyn-Coch, Aberystwyth SY23 3FT
M0 ZYD D. Moger 23 Elmsleigh Road, Paignton TQ4 5AX
M0 ZYF Taunton & District ARC c/o Dr T. Langdon 58 Upper Marsh Road, Warminster BA12 9PN
MM0 ZYT J. Mccoll 25 Nelson Ave, Coatbridge, Ml5 5Lr, Coatbridge ML5 5LR
M0 ZYZ F. Wagstaff 19 Grange Park Avenue, Bedlington NE22 7EF
M0 ZZA A. De Maillet Brock Cottage, The Park, Lower Brailes, Banbury OX15 5JB
M0 ZZE S. Sims 2 Kintbury, Duxford, Cambridge CB22 4RR
M0 ZZI A. Downing Kyriacou Tziapra 25, Alexandria Building 9, Bungalow Number 6, , Larnaca 7520 Cyprus
MM0 ZZO J. Mcginty 9 Powgree Crescent, Beith KA15 1ES
M0 ZZT S. Webber 59 Mincinglake Road, Stoke Hill, Exeter EX4 7DY

M1

M1 AAC L. Healy 31 Roach Road, Sheffield S11 8UA
MW1 AAH D. Creber 8 George Manning Way, Gowerton, Swansea SA4 3HB
M1 AAS A. Jackson 29 Bramble Avenue, Birkenhead CH41 0AX
M1 AAY Dr N. Thomson 11 School Close Stamford Bridge, York YO41 1PT
MM1 ABA Whitehaven ARC (T.S.Bee) c/o I. Hopley 53 Redmoss Road, Aberdeen AB12 3JJ
M1 ABC B. Johnson 20 Valleyside, Hemel Hempstead HP1 2LN
M1 ABF K. Perkin 25 Rownall View, Leek ST13 8JN
M1 ABG H. Mallin Riverside House, Rope Walk, Southampton SO31 4HD
M1 ABM M. Chapman Woodcroft, Windmill Green, Pevensey BN24 5DY
MW1 ABT R. Macleod 24 Heol Powis, Gungrog Hill, Welshpool SY21 7TP
M1 ABU S. Norman 53 Saddlers Park, Eynsford, Dartford DA4 0HA
M1 ABV B. Breet 23 Mitchell Street, Eccles, Manchester M30 8AJ
M1 ABX S. Painting Claytons, Inkpen, Newbury RG17 9QE
M1 ABY A. Highams Rua Das Americas, 17, Fim Da...Rua America Do Norte, Portal Das Americas, , Morretes 83350-000 Brazil
MW1 ABZ I. Jones 11 Glamorgan Street Canton, Cardiff CF5 1QS
M1 ACA G. Barnett 63 Sandcroft, Sutton Hill, Telford TF7 4AB
M1 ACB Radio Security Service Memorial ARS c/o S. Thomas 2 Myrtle Cottages, Sandy Lane, Saxmundham IP17 1HR
M1 ACC J. Chambers 9 Farnborough Road, Swindon SN3 2DR
M1 ACE P. Musk 38 Glenwood, Welwyn Garden City AL7 2JS
M1 ACF Rad Soc Harrow c/o M. Buckley Springfield, 12 Ranmore Avenue, Croydon CR0 5QA
M1 ACJ S. Shearing 42 Meadow Park, Wesham, Preston PR4 3DN
M1 ACK R. Mackay Conifers, The Street, Guildford GU4 7TJ
M1 ACL Bradley Wood Scout Radio Group c/o K. Green 42 Dartmouth St., Stoke on Trent ST6 1HB
M1 ACN M. Goom 47 Sandringham Court, Slough SL1 6JU
M1 ACO R. Hankin Solway View, Whitrigg, Torpenhow, Wigton CA7 1JG
M1 ACQ S. Thomas 111 Jersey Avenue, Bristol BS4 4QX
M1 ACT C. Leeds 12 Northfields, Norwich NR4 7EU
M1 ADK G. Pratt 2 Houghton Road, Newbottle, Houghton le Spring DH4 4EF
M1 ADN T. Whiting 28 Legarde Avenue, Hull HU4 6AP
M1 ADP K. Eastwood 35 Finisterre Rise, Great Yarmouth NR30 5TT
M1 ADT R. Vickerstaff 16 Sewell Wontner Close, Kesgrave, Ipswich IP5 2GB
M1 ADV I. Owen 104 Hall Farm Road, Melton, Woodbridge, IP12 1RW
M1 ADX J. Towns 72 Longfields Road, Norwich NR7 0NA
M1 ADZ N. Davis 71 Brettenham Road, London E17 5AZ
M1 AEA Spalding And District ARS c/o M. Waldron 32 Windmill Street, Upper Gornal, Dudley DY3 2DQ
M1 AEB G. Haughie 100 Henton Road, Edwinstowe, Mansfield NG21 9LE
M1 AED M. Cattell 12 Fairway Road, Warley, Oldbury B68 8BE
M1 AEG A. Green Michigan, North Road, Whitemoor, Nanpean, St. Austell PL26 7XN
M1 AEH D. Cossey 11 Halden Avenue, Norwich NR6 6UX
M1 AEI D. Bowyer Little Meadow, Station Road, Bow, , Crediton, EX17 6HY
M1 AEJ A. Benjamin Fieldview, Field Common Lane, Walton-on-Thames KT12 3QH
M1 AEK D. Mulliner 23 Nostell Way, Bridlington YO16 6FY
MM1 AEL C. Haswell 6 Lochlann Road, Culloden, Inverness IV2 7HB
M1 AEO R. East 27 Caddywell Meadow, Torrington EX38 7NZ
M1 AEP M. Griffin 76 Waylands, Swanley BR8 8TN
M1 AEQ F. Allenby 9 Church View, Holme-On-Spalding-Moor, York YO43 4BG
M1 AEV A. Aiello Llamedos, Walnut Road, Wisbech PE14 7NP
M1 AEX R. Pyman 16 Bramshott Close, Allington ME16 0RX
MM1 AEZ R. Barrett 59 Dunrobin Road, Kirkcaldy KY2 5YT
M1 AFF R. Dyer 4 Downleaze, Durrington, Salisbury SP4 8AB
M1 AFP P. Jefford 61 Willow Way, Flitwick, Bedford MK45 1LN
M1 AFQ A. Brooks 86 Violet Road, Norwich NR3 4TS
M1 AFU S. Derwin 5 Hawthorne Grove, Yarm TS15 9EZ
M1 AFV S. Wells 1 Neath Gardens, Leeds LS9 6RG
MW1 AFW C. Jones Arosfa House, 7 Wood Street, Bargoed CF81 8NW
M1 AFX D. Hill 3 Morcar Road, Stamford Bridge, York YO41 1PR
M1 AFY S. Jones Dairy Cottage, Menith Wood, Worcester WR6 6UB
M1 AFZ C. Grizzell 17 Barley Close Newton St Faith, Norwich NR10 3GY
M1 AGA R. Taylor 5 Thirlmere Drive, Bury BL9 9QE
M1 AGE D. Thorley 53.The Spires, Moreton on Lugg HR4 8FJ
M1 AGH D. Hirst 2 The Birches, Marlborough Road, Swindon SN3 1PT
M1 AGK Sth Tottenhm Ar c/o R. Large 5 Jasmine Close, Abbeydale, Gloucester GL4 5FJ
M1 AGP J. Davies 10 Gorselands, Hollesley, Woodbridge IP12 3QL
M1 AGR M. Skinner 44 Westminster Crescent, Sheffield S10 4EX
M1 AGW S. Whitehouse 47 Mulberry Road, Bloxwich, Walsall WS3 2NG
M1 AGY J. Cuddy 4 Thames Gardens, Plymouth PL3 6HD
M1 AHA S. Lefevre 9 Old Barn Crescent, Hambledon, Waterlooville PO7 4SW
M1 AHF M. Byatt 15 Lower Farm Road, Plympton, Plymouth PL7 1JJ
M1 AHJ P. Clarke 36 Eldred Drive, Orpington BR5 4PF
MM1 AHL D. Mcarthur 12 Laburnum Grove, Lenzie, Kirkintilloch, Glasgow G66 4DF
M1 AHN G. Boyce 10 Quarry Close, Ross-on-Wye HR9 7DR
M1 AHR K. Peters 82 Blackmoor Road, Moortown, Leeds LS17 5JP
M1 AHT L. Russell 106 Stambridge Road, Rochford SS4 1DP
MW1 AHU Dr G. Armstrong 61 Victoria Drive, Llandudno Junction LL31 9PF
M1 AHY M. Rhodes 1 Chetwode, Overthorpe, Banbury OX17 2AB
M1 AIB P. Lewis 58 Ocean Close, Fareham PO15 6QP
M1 AIK T. Bardgett 49 St. James St., South Petherton TA13 5BN
M1 AIM A. Moore Silver Trees, Woodlands Lane, Pulborough RH20 3HG
M1 AIN D. White 19A Gravenhurst Road, Campton, Shefford SG17 5NY
M1 AIX W. Greenall 356 Warrington Road, Abram, Wigan WN2 5XA
M1 AIY M. Bastin 14 Golvers Hill Road, Kingsteignton, Newton Abbot TQ12 3BP
M1 AJA Dr L. Mason 50 Portholme Drive, Selby YO8 4QF
M1 AJG M. Ramskill 7 Hobart Road, Dewsbury WF12 7LS
M1 AJM S. Hoskins 21 Wicken House, London Road, Maidstone ME16 8QP
M1 AJQ W. Clarke 10 Athol Close, Sinfin, Derby DE24 9LZ
M1 AJT P. Amos 16 Eastry Road, Erith DA8 1NN
M1 AJU S. Bradford 28 Downs Road, , Deal CT14 7SY
M1 AKF P. Wilson 9 The Brooklands, Wrea Green, Preston PR4 2NQ
M1 AKH D. Mulvana 17 Wildlake Orton Malborne, Peterborough PE2 5PG
M1 AKL V. Parsons 20 Old Top Road, Hastings TN35 5DJ
M1 AKN R. Day 20 Gardiner Close, Bury St. Edmunds IP33 2UB
M1 AKQ S. Hutton Astrid, Main Street, Reedness DN14 8ER
M1 AKT D. Thomas 10 The Willows Oxspring, Sheffield S36 8ZZ
M1 AKV R. Kowalski 2 Newcross Park, Kingsteignton, Newton Abbot TQ12 3TJ
M1 AKZ M. Dean 164 Ambleside Road, Lightwater GU18 5UW
M1 ALA D. Mawson 84 Walnut Avenue, Weaverham, Northwich CW8 3DX
M1 ALE A. Ratcliff 71 Spring Gardens, Leek ST13 8DD
M1 ALF R. Coston 272 Warley Road, Blackpool FY2 0UG
M1 ALG J. Sindall 9 James Street, Rotherham S60 1JU
M1 ALH K. Whinney 20 Reedcutters Avenue, Brundall, Norwich NR13 5RZ
M1 ALM R. Hodgkins 38 Byron Road, Gillingham ME7 5QH
M1 ALO P. Rogers Flat 11, Green Court, Lewes BN7 1HY
M1 ALR C. Moore Colchester House, Farrington Road, Bristol BS39 7LW
M1 ALT M. Oram 43 Peverell Avenue West, Poundbury, Dorchester DT1 3SU
M1 ALU T. Summers Caravan 26, Waitegates Caravan Site, Doncaster DN7 4EJ
M1 ALX D. Philip 83 Helman Tor View, Bodmin PL31 1RF
M1 AMA A. Day 9 Arundel Road, Tewkesbury GL20 8AS
M1 AMB D. Snow 7 Aynsley Close, Cheadle, Stoke-on-Trent ST10 1DP
M1 AMI D. Chambers 94 Hawthorn Avenue, Colchester CO4 3JR
M1 AMJ D. Bonnett 254 Norwich Road, Wisbech PE13 3UT
M1 AMP H. Withers 23 Fernie Road, Guisborough TS14 7LZ
M1 AMW C. Whitehead 18 Victoria Quay, Ashton-On-Ribble, Preston PR2 2YW
M1 AMZ K. Birch 16 Brentwood Avenue, Thornton-Cleveleys FY5 3QR
M1 ANC C. Mclean 18 Chatfield Road, Gosport PO13 0TN
MW1 AND M. Griffiths 32 Hill Street, Aberdare CF44 6YG
M1 ANK R. Taylor 86-88 Hillside Crescent, Leigh-on-Sea SS9 1HQ
M1 ANL D. Clapp Wenlock Edge, Park Hill, Shepton Mallet BA4 4AZ
M1 ANN A. Webb 20 The Fleet, Stoney Stanton LE9 4DY
M1 ANO C. Worlledge 181 Roselands Drive, Paignton TQ4 7RN
MM1 ANP Dr J. Tobias Gowanpark House, Gowanpark, Cumnock KA18 2NZ
M1 ANQ D. Hirst 66 Turncroft Lane, Stockport SK1 4AB
M1 ANR R. Hall 12 West Lane, Edwinstowe, Mansfield NG21 9QT
M1 ANT D. Simcock 51 Broadway, Stockport SK2 5SF
M1 AOB R. Pentney 11 Beech Park Holsworthy Beacon, Holsworthy EX22 7NB
M1 AOD V. Bolger Little Annaside, Bootle, Millom LA19 5XL
M1 AOF A. Wainwright111 8 Mount Drive, Purdis Farm, Ipswich IP3 8UU
M1 AOG S. Moriarty 31 Guernsey Way, Banbury OX16 1UE
M1 AOL J. Pepper 16 Chartwood, Loggerheads, Market Drayton TF9 4RJ

M1 AOR S. Darrigan 70 Somerset West Kirby, Wirral CH486EJ
M1 AOU R. Pinchen 9 Orwell Close, Swindon SN25 3LZ
M1 AOX J. O'Neill 4 Heathlands Road, Little Sutton, Ellesmere Port CH66 5PB
M1 APB S. Sanders 52 Hazelwood Road, Callington PL17 7EU
M1 APC J. Hulme 2 East Road Menheniot, Liskeard PL14 3RR
M1 APF J. Thompson 18 Winder Way, Micklefield, Leeds LS25 4FX
M1 APH Wigtownshire ARC c/o P. Hildebrand 82 Reed Drive, Redhill RH1 6TB
M1 APL A. Speakman 28 Eden Gardens, Leeds LS4 2TQ
M1 APQ M. Hodson 93 Deer Park Road, Fazeley, Tamworth B78 3SZ
MM1 APS C. Stuart 5 Deloraine Court, Hawick TD9 7QE
M1 APT I. Waterhouse Little Bracken, 9 Willingdon Drove, Eastbourne BN23 8AL
M1 APV G. Andrews 18 Elms Close, Whitefield, Manchester M45 8XR
M1 AQC J. Jedrzejewski 7 Shedfield Way, Northampton NN40SD
M1 AQI M. Sanderson 14 Hazelwood Avenue, York YO10 3PD
M1 AQJ P. Jackson 55 Queens Road, Hazel Grove, Stockport SK7 4HZ
M1 AQN P. Wilson 14 Jerome Way, Shipton-On-Cherwell, Kidlington OX5 1JT
M1 AQP C. Chapman 9 Edinburgh Avenue, Sawston, Cambridge CB22 3DW
M1 AQR A. Cooper 61 Greenhaze Lane, Great Cambourne, Cambridge CB23 5EF
M1 AQX A. Pell 4 Mill Close, Braunston, Daventry NN11 7HY
M1 AQY G. Cottam 26 Ayton Court, Bedlington NE22 6NS
M1 ARF D. Riley 9 Century Avenue, Mansfield NG18 5EE
M1 ARH W. Mountford 3 Spurstow Close, Prenton CH43 2NQ
M1 ARL D. Edwards 28 Solingen Estate, Blyth NE24 3EP
MW1 ARM D. Rees Y Coed, Tan Lan Hill, Holywell CH8 9JB
M1 ARS H. Cawley 11 Cleveland Way, Winsford CW7 1QL
M1 ART F. Melhuish Alverdean, Mile End Road, Coleford GL16 7QD
M1 ARU R. Stanley 113 Upper Brents, Faversham ME13 7DL
M1 ARX S. Russell 13 Burdett Road, Crowborough TN6 2EN
MI1 ASN P. Mcmahon 26 Ballycraigy Road, Newtownabbey BT36 5ST
M1 ASR G. Jefferies Fenland Lodge, Landing Lane, Broomfleet, Brough HU15 1RE
M1 ASV A. Evans 14A New Road, Tiptree, Colchester CO5 0HJ
M1 ATA A. Betts 31 Porters Wood, Petteridge Lane, Matfield, Tonbridge TN12 7LR
M1 ATB G. Gale 2 Manston Crescent, Crossgates, Leeds LS15 8QZ
M1 ATC Reading And West Berkshire Raynet c/o R. Courtney 5 Bute Close, Highworth, Swindon SN6 7HN
M1 ATI R. Proctor 11 Bedford Rise, Winsford CW7 1NE
M1 ATJ P. Whitby 90 Manor Road, Martlesham Heath, Ipswich IP5 3SY
M1 ATP A. Plitsch 64 Oxford Road, Lowestoft NR32 1TP
MM1 ATR L. Robinson 17 Burn Brae Avenue, Westhill, Inverness IV2 5RG
M1 ATU M. Bloss 20 Barry Walk, Brighton BN2 0HP
MM1 ATY D. Shirley 17 Carlaverock Terrace, Tranent EH33 2PL
MM1 AUF C. Mcclintock 13 St. Andrews Drive, Gourock PA19 1HY
MM1 AUG M. Mcclintock 30 Findhorn Road, Inverkip, Greenock PA16 0HX
M1 AUH A. Easton 1 Wood Street, Warrington WA1 3AY
MI1 AUI V. Hughes 7 Craiglands Manor, Newtownabbey BT36 5FG
M1 AUK B. Pittaway 66 Montrose Avenue, Leamington Spa CV32 7DY
M1 AUN J. Yarnall 85 Wombourne Park, Wombourne, Wolverhampton WV5 0LX
M1 AUO Rvd. D. Eady The Rectory, Rectory Lane, Cheltenham GL51 9RD
M1 AUP A. Bailey 47 Whiteridge Road, Kidsgrove, Stoke-on-Trent ST7 4TH
M1 AUR M. Garlick 59 Foundry Avenue, Leeds LS9 6BY
MW1 AUV P. Davies 4 Caradog Place, Townhill, Swansea SA1 6NH
M1 AUW T. Roberts 109 Gordon Avenue, Norwich NR7 0DS
M1 AUZ G. Wright School House Farm, The Gravel, Mere Brow, Preston PR4 6JX
M1 AVB J. Pidgeon 6 Marlborough Road, Musbury, Axminster EX13 8AH
M1 AVM S. Cooper 93 Langton Road, Norton, Malton YO17 9AE
MM1 AVR S. Mciver 9 Balvicar Road, Oban PA34 4RP
M1 AVU G. Purrier Archways, Forge Hill, Lydbrook GL17 9QS
M1 AVV S. Linney 3 Severn Road, Walney, Barrow-in-Furness LA14 3TS
M1 AVW R. Hedges 10 The Green, Goldenbank, , Falmouth TR11 5PR
M1 AWC M. Worrall 15 Whitegate Drive, Bolton BL1 8SF
MI1 AWM D. Reid 179 Melmount Road, Sion Mills, Strabane BT82 9LA
M1 AWN J. Sanders 135 Windmill Avenue, Kettering NN15 7DZ
M1 AWS A. Jones 35 St. Marys Close, Aspull, Wigan WN2 1RL
MW1 AWT P. Mccarthy 43 Bodnant Road, Llandudno LL30 1LT
MM1 AWV R. Lynch 21 Carnoustie Avenue, Gourock PA19 1HF
M1 AWX S. Yendell 35 Chester Road, Newquay TR7 2RH
M1 AXD P. Gartell 1 Springfields, Richards Castle, Ludlow SY8 4EP
M1 AXE G. Ecclestone 11 Longrood Road, Rugby CV22 7RG
M1 AXG D. Tucker 2 New Cottages, Bill Hill, Wokingham RG40 5QU
M1 AXM D. Russell 39 Mowries Court, Somerton TA11 6NF
M1 AXP B. Cornall Fernholm, Taylors Lane, Preston PR3 6AB
M1 AYA P. Booth 61 Coalpit Lane, Rugeley WS15 1EW
M1 AYC A. Booth 35 Gillamore Drive, Whitwick, Coalville LE67 5PA
M1 AYG R. Sanders 17 Shelley Lane, Kirkburton, Huddersfield HD8 0SW
MI1 AYL B. Byrne 6 Holymount Road, Gilford, Craigavon BT63 6AT
M1 AYN J. Hewlett 28 Coombs Road, Coleford GL16 8AY
M1 AYR K. Miller 15A Holly Close, Cherry Willingham, Lincoln LN3 4BH
M1 AYU R. Freeman 6 Sutton Road, Leverington, Wisbech PE13 5DW
M1 AZA M. Gould 36 Wistaria Road, Wisbech PE13 3RH
M1 AZB J. Titterton Longfield Farm, Ifield Road, Horley RH6 0DR
M1 AZF S. Arbuckle 39 South Road Pe24 5Tl, Chapel Saint Leonards PE24 5TL
M1 AZG A. Ruston 42 The Straits, Dudley DY3 3BH
MW1 AZI S. Dunlop 34 Dan-Y-Coed Clydach, Abergavenny NP7 0LS
M1 AZJ A. Bottrell 36 Tremodrett Road, Roche, St. Austell PL26 8JA
M1 AZM P. Jefferson 20 Buckstone Grove, Leeds LS17 5HW
M1 AZO K. Austen 12B Downs Road, Folkestone CT19 5PW
M1 AZQ A. Beale Flat 2, Brabstone House, Medway Drive, Greenford UB6 8LN
MW1 AZR Oxo Contest Club c/o R. Snelling 91 Oakfield Road, Newport NP20 4LP
M1 AZV R. Aston 2 Brockwood Crescent, Keyworth, Nottingham NG12 5HQ
M1 AZY J. Drummond 60 Park Lane, Exeter EX4 9HP
M1 BAA M. Crow 180 Bois Moor Road, Chesham HP5 1SS
M1 BAC J. Booth 35 Gillamore Drive, Whitwick, Coalville LE67 5PA
M1 BAD D. Redding 11 Camley Gardens, Maidenhead SL6 5JW
M1 BAI A. Saunders 128 Foxcroft Drive, Wimborne BH21 2LA
MW1 BAJ Jersey ARS c/o J. Alexander 19 Rhes Brickyard Row, Llanelli SA152DZ
M1 BAN Dr T. Baldwin 17 Warn Crescent, Oakham LE15 6LZ
M1 BAR Bar-Packers CG c/o N. Roscoe 35 Kenilworth Road, Cheadle Heath, Stockport SK3 0QL
M1 BAS C. Bastin 42 Peterborough Road, Exwick, Exeter EX4 2EG
M1 BAV G. Noble 96 Foxroyd Lane Estate, Dewsbury WF12 0BD
M1 BBB P. Marshall 1 Prospect Cottages, St. Anns Chapel, Gunnislake PL18 9HH
M1 BBH C. Tan 62A Jalan Sepah Puteri 5/6, Kota Damansara, Selangor Malaysia
M1 BBR M. Deglos 405 Armidale Place, Bristol BS6 5BQ
M1 BBS Dr D. Hawkes The Bank East Northdown Close, Margate CT9 3YA
M1 BBU G. Price 118 Broadstone Road, Heaton Chapel, Stockport SK4 5HS
M1 BCB D. Ball 14 Ayot Path, Borehamwood WD6 5BJ
M1 BCM New Forest ARS c/o J. Worthing 27 Mayfield Close, Shrewsbury SY1 4BF
M1 BCR A. Richards 3 Beeston Close, Watford WD19 6LF
M1 BCU A. Howe 6 Crabapple Close, Wymondham NR18 0XT
M1 BCY T. Keeler 72 Grafton Road, Selsey, Chichester PO20 0JB
M1 BCZ R. Craggs 1 Hylton Court, Bowmonts Road, Tadley RG26 3SH
M1 BDD E. Reynolds Cloonagh, Ballinagore Ireland
M1 BDH T. Woods Lower Courtyard Rooms, Swan House Tarrington, Hereford HR1 4EU
M1 BDJ G. Hamlin Down Farm Bungalow, Stockbridge SO20 8EA
M1 BDL A. Statham 24 Fulton Close, High Wycombe HP13 5SP
M1 BDO W. Lodeweegs 68 Totterdown Lane, Weston-Super-Mare BS24 9NJ
M1 BDR Lough Erne ARC c/o G. Farrell 95 Washington Road, Maldon CM9 6JF
M1 BDS P. Colwell 56 Hamelin Road, Gillingham ME7 3EX
MW1 BDV W. Davis Cartref, Blaenannerch, Cardigan SA43 1SN
M1 BDY V. Beard 13 Mayesford Road, Romford RM6 4NU
M1 BED Bedford And District ARC c/o H. Ehm 17 Stuart Road, Kempston, Bedford MK42 8HS
M1 BEO D. Tatlow Mulberry House, Bettys Grave, Cirencester GL7 5ST
M1 BEP A. Amos 118 Mount Hill Road, Bristol BS15 8QR
MW1 BEQ A. Strange 8 Tregarn Close, Langstone, Newport NP18 2JL
MW1 BEW C. Sampson Stable Cottage, Lower House Farm, Welshpool SY21 8LA
M1 BEX North Cheshire RC c/o G. Olsen 36 Bluebell Way, South Shields NE34 0BZ
MM1 BFE J. Murray 27 Wellpark Road, Banknock, Bonnybridge FK4 1TP
M1 BFF A. Buckley 191 Broadway, Stoke-on-Trent ST3 5PW
M1 BFG H. Tribe 20 Penrith Road, Bournemouth BH5 1LT
M1 BFI Z. Billington 27 Saxon Gardens, Caister-On-Sea, Great Yarmouth NR30 5AH
M1 BFO P. Aplin 30 Cheviot Drive, Charvil, Reading RG10 9QD
M1 BFR A. Day 26 Windyridge Gardens, Cheltenham GL50 4SX
M1 BFV F. Richardson 3 Carl Moult House, Regent St., Swadlincote DE11 9PH
M1 BFX J. Belcher 101 Colne Drive, Romford RM3 9LA
M1 BFY A. Davey 34 Monkswood, Littleport, Ely CB6 1JD
M1 BGF M. Shearman 4 Mcdonough Close, Fitton Hill, Oldham OL8 2PD
M1 BGK J. Lewis 12 Eastleigh Road, Staple Hill, Bristol BS16 4SQ
M1 BGL G. Langford 22 Kensington Way, Oakengates, Telford TF2 6NA
M1 BGS M. Elvers 3 St. Michaels Road, Maidstone ME16 8BS
M1 BGT R. Williams 19 Venice Close, Chellaston, Derby DE73 5BX
M1 BGY I. Townson 53 Brompton Road, Bradford BD4 7JD
M1 BHC M. Lee 11 Sturrocks, Vange, Basildon SS16 4PQ
M1 BHE B. Vickers 17 Linden Close, Dewsbury WF12 8PL
M1 BHN J. Chambers 16 Wood Way, Huntington, York YO32 9QG
MM1 BHO R. Hopkins 15 Station Drive, Dalbeattie DG5 4FA
M1 BHP D. Hartley 10 Tamworth Grove, Clifton, Nottingham NG11 8JA
M1 BHZ T. Shepherd 40 Pheasant Way, Cirencester GL7 1BL
M1 BIB P. Brooke 34 Park Road, Burwell, Cambridge CB25 0ES
M1 BIG T. Read 57 Ollard Avenue, Wisbech PE13 3HF
M1 BIK C. Blackmur 9 Cameron Close, Chatham ME5 0DD
M1 BIL C. Pugh 16 Park Gate, Somerhill Road, Hove BN3 1RL
M1 BIX G. Perrins 1 Cornhill Gardens, Leek ST13 5PZ
M1 BIY J. Laker 7 Station Court, West Lane, Hayling Island PO11 0FP
MW1 BJB S. Mandal 1 High Ridge Drive, Bersham Road, Wrexham LL14 4JD
M1 BJC P. Marshall 75 Drewstead Road, London SW16 1AA
M1 BJE S. Robinson 140 The Street, Kirtling, Newmarket CB8 9PD
MM1 BJG O. Ferula 4 Castledyke Road, Carstairs, Lanark ML11 8SU
MM1 BJO R. Lowe 9 South Quarry Gardens, Gorebridge EH23 4GX
M1 BJS G. Ausher 94 New Road, Ditton, Aylesford ME20 6AE
MM1 BJT D. Smith 1076 Aikenhead Road, Glasgow G44 4TJ
MM1 BJZ P. Fraser 8 Devon Walk, Cumbernauld, Cumbernauld G68 9NT
M1 BKE P. Hewlett 28 Coombs Road, Coleford GL16 8AY
M1 BKF W. Hill 492 Earlham Road, Norwich NR4 7HP
M1 BKI J. Carins 26 Roman Way, St. Margarets-At-Cliffe, Dover CT15 6AH
M1 BKL P. Coddington 2 Canal View Chemistry, Whitchurch SY13 1BZ
M1 BKQ T. Beadman 6 Gaiafields Road, Lichfield WS13 7LT
M1 BKS M. Kevern Wheal Bal, Trewellard, Penzance TR19 7SP
M1 BKW S. Plant 17 New Road, Driffield YO25 5DJ
MW1 BLE C. Beech 9 Wesley Court, Pembroke Dock SA72 6NE
M1 BLJ S. Brion 165 Kings Head Hill, London E4 7JG
M1 BLO P. Hoggard 41 Malpas Close, Bransholme, Hull HU7 4HH
M1 BLW E. Banks 165 Burstall Hill, Bridlington YO16 7NH
M1 BLX P. Buxton Bower House, Thornham Road, Eye IP23 8HP
MI1 BLZ D. Kyle Sea Breezes, Rathlin Island, Ballycastle BT54 6RT
M1 BMC K. Termie 14 Hollins Lane, Marple Bridge, Stockport SK6 5BB
M1 BMD W. Curtis 5 Cambridge Road, Kesgrave, Ipswich IP5 1EN
MM1 BMK Norfolk County Raynet c/o E. Mitchell Carradale Yondertonhill, Hatton, Peterhead AB42 0RE
M1 BML M. Carroll 712 Hoover Street, Nelson VIL 4X4 Canada
M1 BMQ J. Waters 71 Sixth Avenue, Blyth NE24 2SU
M1 BMR C. Robinson 76 Ingrams Way, Hailsham BN27 3NX
M1 BMU E. Woodward 6 Lang Road, Huntington, York YO32 9SD
M1 BMV J. Fox Stonecroft, Horley, Banbury OX15 6BJ
M1 BMW J. Burrill 3 Town Farm Close, Pinchbeck, Spalding PE11 3SG
M1 BNG R. Smith 45 The Avenue, Mortimer Common, Reading RG7 3QU
M1 BNH P. Walton 135 Starling Road, Bury, Greater Manchester BL8 2HF
M1 BNI M. Clarke 1 Burbury Close, Bedworth CV12 8DU
M1 BNK A. Wood 1 Minch Road, Hartlepool TS25 3QY
MI1 BNO B. Bonnar 49 Cullycapple Road, Aghadowey, Coleraine BT51 4AR
M1 BNR Holsworthy Community College c/o G. Forster 33 Deer Valley Road, Holsworthy EX22 6DA
MW1 BNY B. Fitzpatrick 1 Pistyll Newydd Mynyddygarreg, Kidwelly SA17 4NW
M1 BOA G. Heard The Saddlers, The Street, Thurgarton, Norwich NR11 7PD
M1 BOB R. Allen 43 Vowell Close, Bristol BS13 9HS

M1 BOD P. Hanfrey 49 Allotment Road, Niton, Ventnor PO38 2DZ
MI1 BOE A. Prenter 5 Knockview Gardens, Newtownabbey BT36 6UA
M1 BOL D. Harding Po Box 11755 Apo, Grand Cayman Cayman Islands
M1 BOP M. Riley 3 Foxley Close, Ipswich IP3 8BW
M1 BOZ A. Thompson 26 Balmoral Avenue, Clitheroe BB7 2QH
M1 BPA R. Hill Margarey, Exeter Road, Newton Abbot TQ12 2SF
M1 BPD A. Collins Old Orchards, Romsey Road, Stockbridge SO20 6PR
M1 BPK J. Bloor Hillview House, Whitegates, Bromyard HR7 4ES
M1 BPN A. Burchell 25 Cherbury Close, London SE28 8PG
M1 BPS Chesham & District ARS c/o A. Wellman Lyndale, Northleach, Cheltenham GL54 3JJ
M1 BPU B. Lake 1 Eunice Grove, Chesham HP5 1RL
M1 BPW P. Williams 26 Downs View Road, Westbury BA13 3AQ
M1 BPY D. Dixey 102 Blackberry Road, Stanway, Colchester CO3 0RZ
M1 BQC the Polish Scout ARC - London c/o S. Sparks 36 Tormynton Road, Worle, Weston-Super-Mare BS22 9HT
M1 BQD K. Sparks 29 Pennycress, Weston-Super-Mare BS22 8QH
M1 BQE C. Sparks 36 Tormynton Road, Worle, Weston-Super-Mare BS22 9HT
M1 BQF J. Stewart 45 The Cross, Wivenhoe, Colchester CO7 9QH
M1 BQM G. Preedy 12A Grange Court, Prescot Road, Stourbridge DY9 7LA
MW1 BQO E. Buckley 3 Cae Eithin, Minffordd, Penrhyndeudraeth LL48 6EF
M1 BQS F. Gibson 125 Chelveston Drive, Corby NN17 2QJ
M1 BQT F. Lloyd 8 Balfour Cottages Burcot, Abingdon OX14 3DR
M1 BQU S. Daniels 9 Beechwoods, Burgess Hill RH15 0DE
M1 BQW Dr V. Edwards Elder Cottage, Skeyton Common, Skeyton NR10 5AU
M1 BQY Trowbridge And District ARC c/o D. Birch 32 Union Street, Trowbridge BA14 8RY
M1 BQZ W. Rowley 6 Sea King Crescent, Colchester CO4 9RJ
MI1 BRA Belfast Royal Academy Amateur c/o N. Moore 94 Orby Drive, Belfast BT5 6AG
MI1 BRS R. Dickey 8 Coachmans Way, Hillsborough BT26 6HQ
M1 BRU D. Pope 32 Barn Crescent, Newbury RG14 6HD
M1 BRX D. Seddon 22 Newton Road, Lowton, Warrington WA3 1EB
M1 BRY J. Fisher 18 The Smooting, Tealby, Market Rasen LN8 3XZ
M1 BRZ D. Lee 53 Portmellon Park, Mevagissey, St. Austell PL26 6XD
M1 BSB W. Charman Chaplains Lodge, St. Marys Church Close, Southampton SO18 2ST
M1 BSE J. Wharton Flat 12, Brent Court, Leicester LE3 2XQ
M1 BSF S. Ogden 39 Levens Drive, Heysham LA3 1JN
M1 BSI C. Bowen 45 Morris Road, Nottingham NG8 6NE
M1 BSM G. Meyer 447-449 Manchester Road, Stockport SK4 5DJ
M1 BSN J. Riley Peacehaven, Great Street, Stoke-Sub-Hamdon TA14 6SH
M1 BSO B. Ford 7 Courtwick Road, Wick, Littlehampton BN17 7NE
M1 BSP M. Ford 27 The Street, Rustington, Littlehampton BN16 3PA
M1 BSU T. Osborne 134 Merridale Road, Wolverhampton WV3 9RJ
M1 BSV M. Black 37 Castle Street, Lancashire BB9 0TW
M1 BSX Dr A. May 22 Hillside, Abbotts Ann, Andover SP11 7DF
M1 BSY W. Ginger Wenick House, 152 Hawks Road, Hailsham BN27 1NA
MW1 BTA G. Haines 12 Tyn Rhos Estate, Gaerwen LL60 6HL
M1 BTD D. Wilkinson 89 The Northern Road, Liverpool L23 2RD
MW1 BTM C. Jones 17 Grove House Court, Pontygwaith, Ferndale CF43 3LJ
M1 BTO N. Martin 28 Churchmead Close, Lavant, Chichester PO18 0AY
M1 BTR J. Charles Ash Tree, Priory Lane Grimoldby, Louth LN11 8SP
M1 BTU P. Mason 23 Glade Close, Little Billing, Northampton NN3 9SN
M1 BUC A. Benson 12 Longfellow Road, Caister-On-Sea, Great Yarmouth NR30 5RH
M1 BUG M. Dugdale 57 Macauley Avenue, Blackpool FY4 4YF
M1 BUJ I. Lewis 15 Margaret Close, Thurmaston, Leicester LE4 8GL
MW1 BUN D. Luke 56 Maerdy Park, Pencoed, Bridgend CF35 5HX
M1 BUP N. Foyen Sherborne Valley Kennels, Sherborne DT9 4SZ
M1 BUQ I. Houghton 39 Fiskerton Way, Oakwood, Derby DE21 2HY
M1 BUR B. Murray 7 Herdwick Place, Middlewich CW10 9QY
M1 BUU C. Evans 10 Holme Park Bentham, Lancaster LA2 7ND
M1 BUX S. Leaker 166 Beckett Road, Doncaster DN2 4BB
M1 BVI K. Worrall 66 Elm St., Hollingwood, Chesterfield S43 2LH
M1 BVP M. Taylor 56 Newgate Lane, Mansfield NG18 2LQ
M1 BVT M. Beardsley 121 Wood Road, Lower Gornal, Dudley DY3 2LR
MM1 BVW A. Gordon The Paddock, Greenhead Farm, West Saltoun EH34 5EH
M1 BVX S. Simpson 20 Staveley Grove, Keighley BD22 7DH
M1 BWH F. Laycock 8 Melling Way, Liverpool L32 1TP
M1 BWJ J. Bottle 15B Elizabeth House, Alexandra Street, Maidstone ME14 2BX
M1 BWN S. Jarrett 17 Wolmers Hey, Great Waltham, Chelmsford CM3 1DA
M1 BWR Dr E. Oakley Brooklands Lodge, Park View Close, Ventnor PO38 3EQ
M1 BWS Dr A. Kerr 14 Glamorgan Close, St. Helens WA10 3XT
M1 BWZ P. Quick 70 Trent Avenue, Maghull, Liverpool L31 9DE
M1 BXC A. Blakeney 7 Gayton Road, Eastcote, Towcester NN12 8NG
M1 BXD M. Cross 51 Edgehill Drive Lang Farm, Daventry NN11 0GR
MM1 BXF G. Nesbitt 12 Wester Boghead Crosshill Road, Lenzie G66 4SR
M1 BXJ M. Ellis 62 Peterborough Road, Crowland, Peterborough PE6 0BA
M1 BXM M. Forster 10 Weaver Valley Road, Winsford CW7 3JU
M1 BXO D. Ellis 67 Ambersham Crescent, East Preston, Littlehampton BN16 1AJ
M1 BXQ J. Squire 57 The Avenue, Chinnor OX39 4PE
M1 BXU D. Napper 47 Mallard Walk, Sidcup DA14 6SG
MW1 BXX M. Broxton 4 Owen St, Orange Gardens, Pembroke SA71 4EP
M1 BYG A. Chatel 17 Star Holme Court, Star Street, Ware SG12 7EA
M1 BYH A. Moss Festina Lente Macclesfield Canal Centre Brook Street, Macclesfield SK11 7AW
M1 BYI P. Stockley 41 Fairway Court, Cleethorpes DN35 0NN
M1 BYQ R. Josephs 113 Patrick Street, Grimsby DN32 9PQ
M1 BYT H. Bloomfield 49 Oak Crescent, Garforth, Leeds LS25 1PW
M1 BZD D. Atkins 63 Elmwood Road, Keighley BD22 7DW
M1 BZF S. Gore 10 Cambridge Street, Guiseley, Leeds LS20 9AU
M1 BZG L. Raybould 7 Tenbury House, Highfield Lane, Halesowen B63 4RN
M1 BZH J. Rose 200 Burntwood Road Norton Canes, Cannock WS11 9UR
M1 BZI F. Lee 26 Lache Hall Crescent, Chester CH4 7NF
M1 BZJ P. Buer 71 Belvedere Road, Ashton-In-Makerfield, Wigan WN4 8RX
M1 BZK D. Riches 15 Ashton Way, Saltash PL12 6JE
M1 BZR D. Wright 18 Allensway, Stanford-le-Hope SS17 7HE
M1 BZZ P. Lonsdale 14 Donne Close, Wirral CH63 9YJ
MM1 CAC G. Mathers 46 Castle Street, Fraserburgh AB43 9DH
M1 CAE R. Naylor 93 Woodland Road Halton, Leeds LS15 7DN
M1 CAH C. Bennet 32 Angelica Avenue, Stotfold, Hitchin SG5 4HH
M1 CAO R. Cameron 23 Ravenscroft, Hook RG27 9NP
M1 CAR R. Griffiths 22 Quarry Road, Hereford HR1 1SS
M1 CAV J. Morrison West View, Sunk Island Road, Ottringham, Hull HU12 0DX
M1 CAX K. Biggs 22 Wallingford Close, Bracknell RG12 9JE
M1 CAY M. Harris Weathercock Cottage, East Mersea Road, Colchester CO5 8SL
M1 CBC P. Wainwright 5 Pulcroft Road, Hessle HU13 0ND
M1 CBG A. Brown 12 Masefield Drive, Rushden NN10 6BH
M1 CBH J. Tomlins 3 Turnstone Crescent, Mansfield NG18 3SP
M1 CBK R. Scott 198 Slade Green Road, Erith DA8 2JG
M1 CBO R. Appleby 3 St. Judes Way, Burton-on-Trent DE13 0LR
M1 CBT C. Taylor 48 Northdown Park Road, Cliftonville, Margate CT9 3PT
M1 CBU R. Stansfield Sundene, 157 Hollin Lane, Wakefield WF4 3EG
M1 CBV G. Leeder 89 Chesterton Avenue, Harpenden AL5 5ST
M1 CBY D. Howse 24 Sandown Road, Bishops Cleeve, Cheltenham GL52 8BY
M1 CBZ A. Howse 24 Sandown Road, Bishops Cleeve, Cheltenham GL52 8BY
M1 CCA B. Thomas Hazel Mount, Lockhams Road, Southampton SO32 2BD
M1 CCF Blackwood CG c/o M. Buckley Springfield, 12 Ranmore Avenue, Croydon CR0 5QA
M1 CCG G. Smales 6 Chestercourt Cottages, Camblesforth, Selby YO8 8HZ
M1 CCL R. Chantler 68 Chilton Lane, Ramsgate CT11 0LQ
M1 CCN P. Terry 13 Lamsey Lane, Heacham, King's Lynn PE31 7LA
M1 CCQ A. Bantoft 110 St. Peters Road, Wiggenhall St. Peter, King's Lynn PE34 3HF
MM1 CCR A. Annan Easter Cottage Blairlogie, Stirling FK9 5PX
MI1 CCT B. Vaughan 32 Claremore Road, Castlederg BT81 7RF
MI1 CCU I. Morrow 90 Bracky Road Sixmilecross, Omagh. BT79 9PH
M1 CCX Lord P. Tully 3 Doe Crag Houses, Otterburn Camp, Otterburn, Newcastle upon Tyne NE19 1NX
M1 CCY R. Coxon 7 Elworthy Road, Longhoughton, Alnwick NE66 3LS
M1 CDJ F. Waite 91 Priors Hill, Wroughton, Swindon SN4 0RL
M1 CDL D. Hall 4 Burns Close, Peterborough PE1 3JJ
M1 CDP A. Mcdade 20 Westbury Walk, Corby NN18 0AE
M1 CDQ R. Damm 18 Mayfair Crescent, Waltham, Grimsby DN37 0EE
M1 CDT C. Behan St Chads Close, Hornjnglow, Burton upon Trent DE13 0ND
M1 CDV A. Mcewen 23 Cobholm Road, Great Yarmouth NR31 0BU
M1 CDX K. Leach 6 Tewkesbury Avenue, Blackpool FY4 2NF
M1 CEA T. Gladman 14 Chalford Close, West Molesey KT8 2QL
M1 CEM B. Harper 36 Percy Street, Oswaldtwistle, Accrington BB5 4LY
M1 CEW M. Trueblood 44 Wallgate Road, Liverpool L25 1PR
M1 CEY J. Page 5 Grassington Place, Thatcham RG19 3XD
MW1 CFA K. Thorley Helston, Mountain, Holyhead LL65 1YR
MM1 CFC D. Goodfellow 4 West Grange Street, Monifieth, Dundee DD5 4LD
MW1 CFE Dundee ARC c/o A. Evans Flat 5, Nanthir Lodge, Nanthir Road, Bridgend CF32 8BL
M1 CFG S. Appleby Westholme, Asterby Lane, Asterby, Louth LN11 9UE
M1 CFS J. Mac Lyndale, Torbryan, Newton Abbot TQ12 5UR
M1 CFW R. Powell 151 Bury Hill Close, Anna Valley, Andover SP11 7LL
M1 CFZ A. Moore 23 Ashgrove Court, Oakwood, Derby DE21 2LH
M1 CGF L. Griffiths 91 Worrall Road Wadsley, Sheffield S6 4BA
M1 CGI A. Fishwick Causeway House Farm, Coppice Lane, Heapey, Chorley PR6 9DA
M1 CGJ S. Fishwick Causeway House Farm, Coppice Lane, Chorley PR6 9DA
M1 CGM K. Foster 48 The Street, Newbourne IP12 4NY
M1 CGO N. Hewgill 40 Lime Tree Place, Stowmarket IP14 1BT
M1 CGP D. Doyle 100 New Road, South Darenth, Dartford DA4 9AR
M1 CGQ N. Mulryan Flat 31, Chatsworth Lodge, Buxton SK17 6XX
M1 CGR P. Bowles 25 North Down, Staplehurst, Tonbridge TN12 0PG
M1 CHF B. Crossley 19 Westwick Close, Walsall WS9 9EA
M1 CHM M. Bailey 10 Argyll Avenue, Doncaster DN2 6LG
MM1 CHQ D. Wildridge 1 Glamis Gardens, Dalgety Bay, Dunfermline KY11 9TD
M1 CHS J. Arundale 29 Deepdale Avenue, Scarborough YO11 2UQ
M1 CHU W. Llewellyn 105 Sandford Avenue, Church Stretton SY6 7AB
M1 CIE J. Morgan 15 Town Head, Dearham, Maryport CA15 7JW
M1 CIG S. Spurr 12 Rushmoor Close, Rickmansworth WD3 1NA
M1 CIJ J. Turk 25 Berkeley Road, Newbury RG14 5JE
M1 CIM B. Kidane Po Box 10130, Addis Ababa Ethiopia
MM1 CIR P. Merckel 1 Mortimer Court, Dalgety Bay, Dunfermline KY11 9UQ
M1 CIS P. Jameson 1 White Acres Road, Mytchett, Camberley GU16 6EY
M1 CIZ S. Jones 34 Marlborough Road, Rugby CV22 6DD
M1 CJB L. Holyer 45 Crabble Hill, Dover CT17 0RX
M1 CJE A. Eastland 4 Bergamot Close, Manton, Marlborough SN8 4HT
M1 CJF R. Barrett Upland, Tidings Hill, Halstead CO9 1BJ
M1 CJM R. Wallis 26 Heather Bank, Osbaldwick, York YO10 3QH
M1 CJN S. Ogiela 107 Osbaldwick Lane, York YO10 3AY
M1 CJT A. Brotherhood 5 Longcliffe Road Shepshed, Loughborough LE12 9LW
M1 CJX T. Cotterell 52 The Crofts, Hatch Warren, Basingstoke RG22 4RF
M1 CJZ A. Roberts 3 Jaynes Close, Banbury OX16 9ES
M1 CKJ G. Reeds 26 Holme Leaze, Steeple Ashton, Trowbridge BA14 6EH
MW1 CKK S. Lowe 2 Bryn Eglwys, Llanfachreth, Dolgellau LL40 2EF
M1 CKO S. Chapman 9 Edinburgh Avenue, Sawston, Cambridge CB22 3DW
M1 CKQ B. Roth 46 Newport Road, Saffron Walden CB11 4BS
M1 CKU M. Brooks 13 Weatherside, Blaydon-on-Tyne NE21 5QL
MM1 CKW A. Johnson 20 Falkirk Road Glen Village, Falkirk FK1 2AG
M1 CLI M. Poulter 26 West Crescent, Duckmanton, Chesterfield S44 5HE
M1 CLO G. Capon 24 Beech Drive, Brackley NN13 6JH
MM1 CLR R. Vause 100 Carmuirs Avenue, Camelon, Falkirk FK1 4PB
M1 CLW E. Brown 76 West Park Drive, Swallownest, Sheffield S26 4UY
M1 CLX A. Robinson 11 The Crescent, Whalley, Clitheroe BB7 9JW
M1 CLZ J. Taylor 307 Birmingham Road, Lickey End, Bromsgrove B61 0ER
M1 CML K. Grout 36 Churchill Road, Exmouth EX8 4DN
M1 CMM J. Timmis Upper House, Abdon, Craven Arms SY7 9HX
M1 CMN M. Curtis 20 Alder Road, Folkstone CT19 5BZ
M1 CMR B. Jarvis 26 Longhouse Road, Halifax HX2 8RE
MM1 CMU J. Freeland 12 Mccathie Drive, Newtongrange, Dalkeith EH22 4BW
M1 CMW R. Cottington 3 Dickens Drive, East Malling, West Malling ME19 6SJ
M1 CMX M. Hurley 5 Borough Crescent, Stourbridge DY8 3UT
MJ1 CNB N. Fryer 25 Walter Benest Court, La Route Des Quennavais, St. Brelade JE3 8NS Jersey

M1 CND T. Wooldridge 12 Redwood Avenue, Leyland PR25 1RN
M1 CNE M. Howard Paddock House, School Lane, Aby LN13 0DL
M1 CNH J. Gibb 25 Ferndown Gardens, Cobham KT11 2BH
M1 CNI G. Lambley Jasmic, Main Road, Spilsby PE23 4BE
M1 CNJ F. Manley 37 Goodrington Close, Banbury OX16 0DB
M1 CNK P. Wilton 217 Chamberlayne Road, Eastleigh SO50 5HZ
M1 CNL R. Tew 66 St. Nicholas Estate, Baddesley Ensor, Atherstone CV9 2EZ
MW1 CNN D. Hayward 1 Elidyr Road, Newbridge, Newport NP11 3EE
M1 CNP C. Robinson 9 Chatsworth Avenue, Culcheth, Warrington WA3 4LD
M1 CNS E. Mathias 17 St. Johns Terrace, Lewes BN7 2DL
M1 CNX S. Russell 11 Lowgate Avenue, Bicker, Boston PE20 3DF
M1 CNY K. Wilson1 12 New Street, Elworth, Sandbach CW11 3JF
MW1 COB K. Sands 5 Ynysgau Street, Ystrad, Pentre CF41 7UE
M1 COE R. Alford 225 N Main Box 174, Gas 66742 United States
MW1 COJ A. Jones 22 Wendover Avenue, Towyn, Abergele LL22 9LP
M1 COL Colchester Radio Amateurs Club c/o H. Yeldham 19 Wade Reach, Walton on the Naze CO14 8RG
MJ1 COO D. Gallichan 20 Cranham Court La Rue Des Chenes, St. Helier, Jersey J E2 4RY
M1 COQ M. Panton 116 Kings Road Glemsford, Sudbury CO10 7QZ
MM1 COS S. Laepong 29D Hill St., Montrose DD10 8AZ
M1 COV S. Rollinson College Farm Cottage, Humber Lane, Hull HU12 0UX
MW1 COY S. Beer 4 Churchfields, Barry CF63 1FP
M1 CPB K. Ellison 33 Priory Grove, Sunderland SR4 7SU
M1 CPC F. Moy 86 Manningford Road, Birmingham B14 5LX
M1 CPD F. Marrai 19 Hind Close, Chigwell IG7 4EA
M1 CPL R. Ramsay Fairview, Briar Close, Hastings TN35 4DP
M1 CPP I. Mcleary 11 Malcolm Court, Whitley Bay NE25 8NN
M1 CQC S. Eglinton 2 Victoria Road, Saltash PL12 4DL
M1 CQF M. Barnes 58 Prince St., Dalton in Furness LA15 8EU
M1 CQI A. Oxlade-Gotobed 22 St. Peters Road, Basingstoke RG22 6TD
M1 CQK N. Ore Willowdene, Rode Lane, Norwich NR16 1NW
M1 CQL A. Ore Willowdene, Rode Lane, Norwich NR16 1NW
M1 CQM B. Suyat 24 Lynmouth Road, London E17 8AF
M1 CQN B. Adkins 4 Orion Close, Ward End, Birmingham B8 2AU
MW1 CQP K. Blackwell-Chambers 69 Burford Gardens, Cardiff CF11 0AP
M1 CQQ M. Sheanon 3 Bridge Road, Little Sutton, Long Sutton, Spalding PE12 9EG
M1 CQR R. Field 3 Waveney Drive Belton, Great Yarmouth NR31 9JU
M1 CQS G. Browne 30 Dereham Road Easton, Norwich NR9 5EJ
M1 CQT A. Wheeler 60 Bredhurst Road, Gillingham ME8 0PE
M1 CQU N. Anderson 19 Berrylands, Liss GU33 7DB
M1 CQX M. Jones 44 Purley Road, Sunderland SR3 1QS
M1 CRA Wacral (World Association of Christian Radio Amate c/o P. Jackson 8 Buttree Court, South Kirkby, Pontefract WF9 3NB
MW1 CRE S. Woolley 139 Prince Of Wales Avenue, Flint CH6 5JU
M1 CRF N. Ovenden 12 Wooburn House, Boundary Road, Loudwater, High Wycombe HP10 9EG
M1 CRL J. Edwards 44 Hunter Road, Norwich NR3 3PY
M1 CRO Rustyradios A.R.C.G. c/o J. Lemay Carlton House, White Hart Lane, Colchester CO6 3DB
M1 CRP P. Booth 39 New Close, Eyam, Hope Valley S32 5QX
M1 CRQ J. Nuttall 114 Plumstead Road, Norwich NR1 4JX
M1 CRZ A. Tudge 7 Moreton Avenue, Whitefield, Manchester M45 8GG
MI1 CSA J. Higgins 43 Temple Road, Garvagh, Coleraine BT51 5BJ
M1 CSC S. Swancutt 5 Olde Hall Lane, Great Wyrley, Walsall WS6 6LL
M1 CSE C. Childs 43 Eastdale Road, Burgess Hill RH15 0NJ
M1 CSG G. Millsott 11 Kingsthorn Road, Poundbury, Dorchester DT1 3RR
M1 CSI D. Avery 38 Junction Road, Burgess Hill RH15 0JN
M1 CSL D. Cooke 125 Glenhills Boulevard, Leicester LE2 8UH
M1 CSU I. Bliss 3 Ford Road, Ashford TW15 2RF
M1 CSZ S. Eggleton 5 Ladywood Grange, Lady Margaret Road, Ascot SL5 9QH
M1 CTB C. Dale 3 Ivatt Close, Bawtry, Doncaster DN10 6QF
M1 CTG M. Hopkins The Black Swan, Burn Bridge Road, Harrogate HG3 1PB
MW1 CTJ T. Joyes 4 The Smithy, Devauden, Chepstow NP16 6QA
M1 CTK Cobham Marlow ARG c/o D. Hunt 4 Warmdene Road, Brighton BN1 8NL
M1 CTM A. Mcmullen 70 Sylvan Avenue, Timperley, Altrincham WA15 6AB
M1 CTO L. Chung 104 Penland Road, Haywards Heath RH16 1PH
MI1 CTQ J. Murphy 19 Fernagh Road, Omagh BT79 0HX
M1 CTY M. Gray 8 Middleton Terrace Cross Street, Cowes PO31 7TE
M1 CUB K. Hope 37 Hollyhurst Road, Darlington DL3 6HT
M1 CUC J. Mcgowan 72B Adelphi Crescent, Hornchurch RM12 4JZ
MI1 CUS J. Woods 18 Mullaghdrin Road, Dromara, Dromore BT25 2AF
M1 CUX R. Matthews 18 Hawkins Close, Daventry NN11 4JQ
M1 CUY J. Wood 3 Harold Collins Place, Colchester CO1 2GQ
M1 CVB P. Bull 8 Mayfield Lane, Martlesham Heath, Ipswich IP5 3TZ
M1 CVF C. Block 13 Beatrice Road, Capel-Le-Ferne, Folkestone CT18 7LH
M1 CVG D. Wood 17 St. Peters Close, Henley, Ipswich IP6 0RH
M1 CVH S. Blewitt 9 Durlston Close, Amington, Tamworth B77 3QG
M1 CVK K. Bennett 34 Shrubbery Close, Barnstaple EX32 9DG
M1 CVL M. Crossley 196 Middleton Road Hopwood, Heywood OL10 2LH
MW1 CVM D. Giles 9 Ty Newydd Court, Pontnewydd, Cwmbran NP44 1LJ
M1 CVT A. Mallett 8 Shaws Close, Prestwood, Great Missenden HP16 0SL
M1 CVU K. Smolkovic 26 Keeling Way, Attleborough NR17 1YF
M1 CVX S. Taylor 17 York Close, Clayton Le Moors, Accrington BB5 5RB
M1 CWA Prof. J. Gough 2 Brunswick Cottages, Cambridge CB58DL
M1 CWB I. Bryant 17 Kent Road, Southampton SO17 2LJ
M1 CWD D. Taberer 4 Hillfields Road, Brierley Hill DY5 2NG
M1 CWG T. Foreman 39 Railway Street, North Fleet DA11 9DU
M1 CWN B. Bailey 8 Alderton Road, Grittleton, Chippenham SN14 6AN
M1 CWR North Wiltshire Raynet c/o A. Sharman 3 Deben Crescent Haydonwick, Swindon SN25 3QB
M1 CWV A. Dykes 149 Mayfield Road, Chaddesden, Derby DE21 6FZ
M1 CWY O. White 35 Drage Street, Derby DE1 3RW
M1 CXA J. Marcus 115 Kimberley Road, Solihull B92 8QA
M1 CXK A. Cordier 49 Laburnum Avenue, Dartford DA1 2QN
M1 CXN H. Neal 141 Manor Road, Erith DA8 2AQ
MM1 CXO J. Donnelly 21 Mcdonald Drive, Irvine KA12 0QS
M1 CXP R. Gill 45 Biggin Lane, Ramsey, Huntingdon PE26 1NB
M1 CXV G. Allison 16 Copse Road, Plymouth PL7 1PZ
M1 CXW D. Hines 31 Clegge Street, Warrington WA2 7AT
MJ1 CYD C. Paland 19 Maison St. Louis, St. Saviour JE2 7LX Jersey
M1 CYJ G. Jenkinson 4 Brundish House, Braithwell Road, Rotherham S66 8JT
M1 CYK E. Doran 24 Maple Leaf Close, Ingol, Preston PR2 7DZ
M1 CYL K. Langhamer 26 Maple Drive, Burgess Hill RH15 8AW
M1 CYM M. Husband 31 Crescent Road, Colwall, Malvern WR13 6QW
M1 CYN W. Hamilton 22 Rayford Close, Dartford DA1 3AJ
M1 CYP B. Millard 11A Fourways, Tetney, Grimsby DN36 5NF
M1 CYR K. Scott 362 Cannock Road, Heath Hayes, Cannock WS12 3HA
M1 CYT S. Davey 2 Staveley Road, Dunstable LU6 3QQ
M1 CYX D. Bryan 14 Fairfield Way, Totland Bay PO39 0EF
M1 CZA K. Churchill 76 Preston Drive, Bexleyheath DA7 4UE
M1 CZF M. Lewis The Manor House, The Green, Banbury OX17 1BU
M1 CZI I. Johnson 2 Latton Close, Didcot OX11 0SU
M1 CZL S. Bond Powell Cottage, 1 Powell Close, Leamington Spa CV33 9PX
M1 CZM J. Stocks 10 Hollycroft Road, Emneth, Wisbech PE14 8AY
M1 CZO D. Charles 29 Acacia Gardens, Upminster RM14 1HT
M1 CZX T. Ruane Ventnor, High Lane, Haslemere GU27 1AZ
M1 CZY D. Mageehan 37 Gosbecks Road, Colchester CO2 9JR
M1 CZZ Coventry ARS c/o D. Gallier 86 Pine Tree Road, Oldham OL8 3LQ
M1 DAB M. Mcphail 126 Welbeck St., Creswell, Worksop S80 4AN
M1 DAH J. Saiger 10 Markham Avenue, Armthorpe, Doncaster. DN3 2AZ
MM1 DAK I. Mcdonald 5 Well Street, Rosehearty, Fraserburgh AB43 7NW
MW1 DAM A. Cartwright 7 Pen Parc, Malltraeth, Bodorgan LL62 5BG
M1 DAN D. Black 8 Cornwood Close, Finchley, London N2 0HP
M1 DAP M. Purcell 14 Adelaide Road, Blacon, Chester CH1 5SY
M1 DAS D. Nicolson Woodbridge House, Wembworthy, Chulmleigh EX18 7SN
MW1 DAU C. Cater 40 Frances Avenue, Wrexham LL12 8BN
MI1 DAW R. Bamber 15 Ladybrook Parade, Belfast BT119ER
MW1 DBA P. Kinley 18 Larchwood Road, Wrexham LL12 7SG
M1 DBC R. Carroll 27 Sheraton Grange, Stourbridge DY8 2BE
M1 DBF G. Jones 12 Birks Holt Drive, Maltby, Rotherham S66 7JZ
M1 DBK M. Lawrance 18 The Green Road, Sawston, Cambridge CB22 3LP
M1 DBM B. Barrett Kite Hill Camping Park, Firestone Copse Road, Wootton Bridge PO33 4LQ
M1 DBW P. Roberts 7 Boscombe Road., Swindon SN25 3EZ
M1 DCE P. Rollinson 4 Turmarr Villas, Easington, Hull HU12 0TJ
MW1 DCF G. Hopkins 132 Laurel Road, Bassaleg, Newport NP10 8PT
M1 DCH Burnham Beches RC c/o P. Wakefield Flat 3, 21 Priests Road, Swanage BH19 2RG
MW1 DCI P. Bevan 61 Dinas St., Plasmarl, Swansea SA6 8LQ
M1 DCK W. Curry-Peace 14 Springfield Road, Stoke-on-Trent ST4 6RU
M1 DCV M. Mcbride 127 Leicester Causeway, Coventry CV1 4HL
M1 DCX T. Dore 53 Queens Avenue, Shipston on Stour CV36 4DJ
M1 DCY A. Dore 7 Cavendish Drive, Kidderminster DY10 2SX
M1 DDB T. Smith 69 Sunningdale, Grantham NG31 9PF
M1 DDF O. Spevack 16 Ranmore Road, Dorking RH4 1HD
M1 DDI K. Skidmore 239 Alfreton Road Blackwell, Alfreton DE55 5JN
M1 DDR D. Carter 30 Swift Way, Sandal, Wakefield WF2 6SR
M1 DDW J. Reed 9 Mercer Drive, Harrietsham, Maidstone ME17 1AY
M1 DDY T. Reed Seafield, Charing Hill, Charing, Ashford TN27 0NG
MM1 DEA G. Leadbetter 8 Tomtain Brae, Cumbernauld, Glasgow G68 9ER
MM1 DEE G. Hall 0/2 1 Thistle St Kirkintilloch, Glasgow G661NU
M1 DEG R. Smith 41 Middle Deal Road, Deal CT14 9RG
M1 DEJ M. Hibbert 5 Cliff View Road, Cliffsend, Ramsgate CT12 5ED
M1 DER A. Cooper 11 Pendlebury Grove, Hoyland, Barnsley S74 0LL
M1 DEY K. Armstrong 8 Caxton Garth, Threshfield, Skipton BD23 5EZ
MI1 DEZ T. Reid 15 Gillistown Road, Ahoghill, Ballymena BT42 2RJ
M1 DFB A. Dunster 113 Canterbury Road, Folkestone CT19 5NR
M1 DFC S. Bell 14 Charlotte Avenue, Wickford SS12 0DX
M1 DFK C. Ansell 51 East Road, Brinsford, Wolverhampton WV10 7NP
M1 DFM K. Davies 58 Popes Lane Sturry, Canterbury CT2 0LA
M1 DFO A. Bruce 4 Drayton Manor 507 Parrswood Road, Manchester M20 5GJ
MW1 DFQ B. Howard 64 Lawrenny St., Neyland, Milford Haven SA73 1TB
M1 DFW Dr K. Prakash 14 Masham Road, Harrogate HG2 8QF
M1 DGE D. Cockayne 32 Shaw Close, Garforth, Leeds LS25 2HA
M1 DGK J. Kirkham Flat 31, 123 St. Anns Road, London W11 4BT
M1 DGL R. Walsh 10 Standen Road Bungalows, Clitheroe BB7 1LA
M1 DGP C. Anderson 15 John Gunn Close, Chard TA20 1DG
M1 DGQ P. Talbot 5 Stones Walk, Burghfield Common, Reading RG7 3JA
MW1 DGR C. Smith 22 Berth Y Glyd Road, Old Colwyn, Colwyn Bay LL29 9HT
M1 DGS D. Snell 154 Oaks Cross, Stevenage SG2 8NA
M1 DGW M. Wharton 9 Orchard View, Linton Colliery, Morpeth NE61 5SP
M1 DGX J. Griffiths 10 Cote Road, Telford TF5 0NQ
M1 DGY H. Shemming 6 Smiths Place, Kesgrave, Ipswich IP5 2YR
M1 DHA A. Davis 19 Grange Street, Barnoldswick BB18 5LB
M1 DHC A. Killing 102 Coquet Grove, Newcastle upon Tyne NE15 9LH
M1 DHG M. Hilton 40 Megstone Avenue, Whitelea Chase, Cramlington NE23 6TU
M1 DHI J. Edwards Willows, Sunray Avenue, Whitstable CT5 4EQ
M1 DHJ I. Maltas 20 Suddaby Close, Hull HU9 3RG
M1 DHM R. Fraser 12 Birchen Road, Halewood, Liverpool L26 9TL
M1 DHO R. Bloxam 39 Claremont Drive, Ravenstone, Coalville LE67 2ND
M1 DHT D. Shackleton 29 Windmill Green, Ditchingham, Bungay NR35 2QP
MM1 DHU P. Mcbride 1 Hillside, Croy, Glasgow G65 9HJ
M1 DHV G. Turner 35 Horncastle Road, Wragby, Market Rasen LN8 5RB
M1 DHW J. Simlat 7 Coventry Close, Wroughton, Swindon SN4 9BB
M1 DHY D. Sandell 29 Manor Road, Herne Bay CT6 6RF
M1 DIB D. Beck 25C Lickless Gardens, Horsforth, Leeds LS18 5QU
M1 DIE J. Halsall 83 Poole Road, Leeds LS15 7HD
M1 DIL B. Jones 39 Rosewood Avenue, Burnham-on-Sea TA8 1HE
M1 DIM K. Ingram 1 Hazel Close, Brackley NN13 6PE
M1 DIN D. Blythe 4 Laburnum Terrace, Hexham NE47 7EB
M1 DIR R. Duncan Lake View, 2 Upper New Road, Cheddar BS27 3DH
M1 DJA A. Garner The Chestnuts, Surfleet, Spalding PE11 4BA
M1 DJB C. Meakin 56 Coronation Walk, Gedling, Nottingham NG4 4AQ
M1 DJC G. Griffiths 8 Grays Lane, Paulerspury, Towcester NN12 7NW
M1 DJG K. Miller 52 Stanway Road, Shirley, Solihull B90 3JE
M1 DJI A. Waddington 5 Glenview Avenue, Bradford BD9 5PA

MM1 DJJ G. Waddington Wester Lathallan, Leven KY8 5QP
M1 DJN R. Francis 50 Parsonage Chase, Minster On Sea, Sheerness ME12 3JX
M1 DJO B. Kynaston 76 Thorncliffe Avenue, Dukinfield SK16 4UD
M1 DJP A. Saltmarsh 55 Wentworth Grove, Winsford CW7 2LJ
M1 DJS I. Turk 11 Medway Crescent, North Hykeham, Lincoln LN6 8UB
MI1 DJW J. Campbell 2 Lakeview, Crumlin BT29 4YA
M1 DJX A. Baker 27 Liney Road, Westonzoyland TA70EU
M1 DKA A. Lewis 111A Cheltenham Road, Longlevens, Gloucester GL2 0JG
M1 DKF R. Saward Vigeland, 61A Old Main Road, Boston PE20 2BU
M1 DKK N. Hall 378/113 B410 Diamond Suites Condo, Moo 10, Soi 15, Thappraya Road, Nongprue 20150 Thailand
M1 DKL Strathclyde Emergency Planning Unit c/o D. Liddard 81 Tattersall Gardens, Leigh-on-Sea Essex SS9 2QS
MW1 DKM A. Williams 25 Tre Rhosyr, Newborough, Llanfairpwllgwyngyll LL61 6TG
M1 DKP A. Maylin 221 Branksome Avenue, Stanford-le-Hope SS17 8DD
M1 DKW R. Illman 35 Courtenay Park, South Brent TQ10 9BT
M1 DKY H. Sanders Copper Coins, Deans Drove, Poole BH16 6EQ
M1 DKZ D. Forster 3 West View, Middleton, Ludlow SY8 3ED
M1 DLE M. Gifford 22 St. Agnes Way, Kesgrave, Ipswich IP5 1JZ
M1 DLG D. Smith 19 Victoria Grove, Newbury RG14 7RA
M1 DLM D. Russell 6 Deansgate Lane North, Liverpool L37 7ER
M1 DLR R. Wood Bank View, Bilham Road, Huddersfield HD8 9PA
M1 DLS J. Wharton 18 Queen Street, Newbiggin-by-the-Sea NE64 6DE
M1 DLX Blackwood & District Amateur Radio Soc. c/o J. Griffiths 10 Cote Road, Telford TF5 0NQ
M1 DMB S. Atton A1 Manor Park, Happisburgh, Norfolk NR12 0PW
MM1 DME R. Fuggle 27 Ewart Crescent, Hamilton ML3 8LX
M1 DMH J. Hardman 2 Well Orchard, Bamber Bridge, Preston PR5 8HJ
M1 DMN R. Gilbert 8 Church Road, West Kingsdown, Sevenoaks TN15 6LL
M1 DMR B. Pilcher 283 London Road, Portsmouth PO2 9HE
M1 DMT J. Hull 68 Meadow Avenue, West Bromwich B71 3EE
MM1 DMU D. Mccann 69 Davies Drive, Lomond Industrial Estate, Alexandria G83 0UF
M1 DNA D. Bruce 15 St. Richards Road, Deal CT14 9JR
M1 DNC L. Hall 15 Fullwood Avenue, Newhaven BN9 9SP
M1 DNE S. Moppett 59 Piccadilly, Tamworth B78 2ER
M1 DNG R. Steward Long Meadow, Seven Acres Lane, Southwold IP18 6UL
M1 DNJ D. Houbart 10 Lancelot Close, Rochester ME2 2YT
M1 DNQ J. Golding Flat 6, Daver Court, London SW3 3TS
MD1 DNT D. Hughes 13 Julian Road, Douglas IM2 6HW Isle of Man
MW1 DNY D. Reid 11 Caer Delyn, Bodffordd, Llangefni LL77 7EJ
M1 DNZ D. Herridge 93 Freshbrook Road, Lancing BN15 8DE
M1 DOA S. Jefferson 145 Duke St Fenton, Stoke-on-Trent ST4 3NR
MI1 DOG S. Mcauley Layde View, 19 Rathlin Avenue, Ballycastle BT54 6DQ
MW1 DOO V. Lee 40 Willins Coed Eva, Cwmbran NP44 4TJ
M1 DOR D. Stuart 58 Woodplace Lane, Coulsdon CR5 1NF
M1 DOS C. Smith 13 West Winds Road, Winterton, Scunthorpe DN15 9RU
M1 DOT S. Myall 71 Barnes Avenue, Fearnhead, Warrington WA2 0BL
MW1 DOU M. Jones 3 St. Catherines Close, Llanfaes, Beaumaris LL58 8LH
M1 DOX N. Price 162 Stamshaw Road, Portsmouth PO2 8LX
M1 DOZ D. Sampson Spirits Hall Cottage, Mountains Road, Great Totham CM9 8BY
MM1 DPC M. Mclauchlan 8 Craigie St., Ballingry, Lochgelly KY5 8NS
M1 DPE L. Stockwell 167 Hathaway Road, Grays RM17 5LW
MM1 DPH J. Crichton 51 Obree Avenue, Prestwick KA9 2NN
M1 DPI E. Thompson 9 Elmwood Drive, Ponteland, Newcastle upon Tyne NE20 9QQ
M1 DPJ A. Bonner Flat 15, Enderleigh House, Havant PO9 1LQ
MI1 DPL J. Stewart 45 Mull Road, Antrim BT41 2TR
M1 DPO J. Gould 14 Homestead Road, Orpington BR6 6HW
M1 DPQ Dr P. Orr 74 Amalfi Tower, Lakeside Village, Sunderland SR3 3AL
M1 DPU P. Cain 108 Spencer Road, Norwich NR6 6DG
M1 DPW P. Whiffing 38 Green Close, Stannington, Morpeth NE61 6PE
M1 DPX D. Collins 1 New Street Close, Stradbroke, Eye IP21 5JH
M1 DPY J. Bowes 40 Nursery Road, Angmering, Littlehampton BN16 4FH
MI1 DQB C. Gardner 50 Kirkliston Park, Belfast BT5 6ED
M1 DQE I. Fletcher Priory House, 56 Fairfield Road, Saxmundham IP17 1BA
M1 DQG R. Kennedy-Bright Sandiacre, Orchard Lane, Hanwood Sy58Le, Shewsbury SY58LE
M1 DQH A. Duffield-Dyche 1A The Hawthorns, Brockton, Shrewsbury SY5 9JY
M1 DQI M. Jones 35 Pendle Way, Meole Brace, Shrewsbury SY3 9QS
M1 DQQ S. Stanley 35 Statham Close, Lymm WA13 9NN
M1 DQU A. Bedford 44 Kirtling Place, Haverhill CB9 0AU
MM1 DQW Scottish-Russian ARS c/o G. Steven 36 Springhill Terrace, Springside, Irvine KA11 3AL
M1 DQX P. Ormerod 14 Fanny Moor Crescent, Huddersfield HD4 6PL
M1 DRB K. Glover 14 Crawley Crescent, Eastbourne BN22 9RN
M1 DRK A. Thomson 33 Finchley Close, Hull HU8 0AN
M1 DRL D. Luff 12 Swan Lane, Sellindge, Ashford TN25 6EP
M1 DRM M. Bird 33 Corbyns Hall Road, Brierley Hill DY5 4QY
MI1 DRP P. Mcdaid 66 Laurel Drive, Strabane BT82 9PN
M1 DRZ J. Stevens Springfield Cottages, 57 Brindley Street, Stourport-on-Severn DY13 8JG
MM1 DSD G. Duncan 5 Jarvis Place, Carnoustie DD7 7BR
M1 DSE P. Gibson 77 Sunlea Avenue, North Shields NE30 3DT
M1 DSI D. Palmer 2 Merrill Road, Thurnscoe, Rotherham S63 0NN
M1 DSQ N. Taylor 36 Bodmin Avenue, Slough SL2 1SL
M1 DSU J. Henderson 12 Chathill Terrace, Newcastle upon Tyne NE6 3BB
M1 DSV A. Newell Thwaites Bank, Spring Avenue, Keighley BD21 4TD
MM1 DSX J. Spiers 29B Carlyle Gardens, Haddington EH41 3LS
M1 DSZ S. Turner 27 Huntspill Road, Highbridge TA9 3DQ
M1 DTG M. Whitchurch 94 Hundred Acres Lane, Amersham HP7 9BN
MM1 DTN W. Gray 1 Regent Court, Regent Street, Keith AB55 5ED
M1 DTO T. Jones 40 Chester Road South, Kidderminster DY10 1XJ
M1 DTS E. Cawte Woodpeckers, Rectory Gardens, Church Stretton SY6 6DP
MW1 DTT S. Walters 31 John Street, Pentre CF41 7JT
MM1 DTU J. Weddell 10 High Street, Eyemouth TD14 5EU
M1 DUA D. Riley Flat 2, Crown Crest Court, Sevenoaks TN14 5AS
M1 DUB D. Stalley 42 Gadby Road, Sittingbourne ME10 1TJ
M1 DUC J. Parker 76 Elm Road, Grays RM17 6LD
M1 DUD R. Burrows-Ellis 49 Highland Drive, Worlingham, Beccles NR34 7AR
MW1 DUJ D. Jones 13 Pontardulais Road, Cross Hands, Llanelli SA14 6NT
M1 DUO R. Easthope 8 Gilbert House, 6 Mill Park, Cambridge CB1 2FJ
M1 DUV S. Crook 24 Gainsborough Drive, Northfleet, Gravesend DA11 8NH
MM1 DVC I. Hendry 47 Fraser Place, Keith AB55 5EB
M1 DVJ C. Wood 4 Waterdale Farm Park, Kingskerswell TQ12 5EX
M1 DVO R. Waters Romosco, Mill Lane, Bradfield, Manningtree CO11 2QP
M1 DVV P. Wise 13 Waltham Road, Newton Abbot TQ12 1LH
M1 DWQ I. Lowcock Sunflower Cottage Loddiswell, Kingsbridge TQ7 4QJ
M1 DWT D. Hamilton 120 Hall Road, Hull HU6 8SB
MM1 DWU G. Mcvittie 46 Mote Hill Road, Girvan KA26 0EB
M1 DWV Dr M. Stitson 91 St. Judes Road Englefield Green, Egham TW20 0DF
M1 DWW W. Bennett Ore House School Road Tunstall, Woodbridge IP12 2JF
M1 DXA P. Beech Grange Farm North Road, Atwick YO25 8DW
M1 DXB B. Smith 39B Palace Avenue, Paignton TQ3 3EQ
M1 DXG R. Williamson Beverley, Swineshead Road, Boston PE20 1SG
M1 DXL C. Churchward 36 Drake Road, Broadheath, Altrincham WA14 5LN
M1 DXO N. Onions 34 Redwing Court, Southsea PO4 8PB
M1 DXQ M. Rhead 11 Shelley Road, Stoke-on-Trent ST2 8JN
MM1 DXU M. Richards Rowan Lea, Weyland Gait, Kirkwall KW15 1QR
MD1 DXW W. Griffiths 7 Cooyrt Shellagh, Ballasalla, Isle of Man IM9 2EU
M1 DYC J. Guilford 2 Lacey Close, Ilkeston DE7 9LF
M1 DYD F. Frost Flat 34, Kingsley Court, 21 Pincott Road, Bexleyheath DA6 7LA
M1 DYE D. Ejugue Po Box 62449, Addis Ababa Ethiopia
M1 DYF A. Teffera Po Box 819, Addis Abba Ethiopia
M1 DYG N. Teklehaimanot Po Box 21866, Addis Abba Ethiopia
M1 DYH M. Belete Po Box 181922, Addis Ababa Ethiopia
M1 DYI E. Melaku 77 Chaucer Drive, Lincoln LN2 4LT
M1 DYJ A. Williams Alwent Farm, Staindrop, Darlington DL2 3NS
M1 DYK H. Davison 15 High St., Rippingale, Bourne PE10 0SR
M1 DYL D. Davison 15 High St., Rippingale, Bourne PE10 0SR
M1 DYO A. Cruise Badgers Holt, Park Grove, Chalfont St. Giles HP8 4BG
M1 DYS R. Broadbridge 102 Bailey Crescent, Poole BH15 3HB
M1 DYU G. Rogers 55 Upholland Road, Billinge, Wigan WN5 7JA
M1 DYW O. Barnes 14 Caroline Close, Wivenhoe, Colchester CO7 9SD
M1 DZM J. Nichols 51 Ashbourne Road, Barnsley S71 3QD
M1 DZR R. Daglish Beck Lea, Pasture Road, Frizington CA26 3XN
MM1 DZW R. Heath Roadside Cottage, , Glenkindie, Alford AB33 8SH
MW1 EAA G. Tucker 18 Plymouth Road, Penarth CF64 3DH
M1 EAB A. Thornton 11 St. Nicholas Close, Richmond DL10 7SP
M1 EAG K. Asher 48 Laburnum Crescent, Allestree, Derby DE22 2GQ
M1 EAI A. Beale 6 Meadow View, , Belper DE56 1UT
M1 EAJ Y. Bessell-Baldwin 10 Tudor Close, Barton-Le-Clay, Bedford MK45 4NE
M1 EAK C. Day 35 Rochford Road St. Osyth, Clacton-on-Sea CO16 8PH
M1 EAL A. Nottage 11 Chequers Drive, Horley RH6 8DR
M1 EAN B. Authers 91 Hay Green Lane, Bournville, Birmingham B30 1RF
M1 EAS J. Schofield 19 Shottery Walks, Bredbury, Stockport SK6 2HR
M1 EAW K. Gallacher 63 Holst Avenue, Basildon SS15 5RH
M1 EAZ J. Barker 53 Derby Street, Colne BB8 9AA
M1 EBC J. Best Longview, Central Road, Maryport CA15 7ER
M1 EBD D. Best Longview, Central Road, Maryport CA15 7ER
M1 EBH L. Emmerson Flat 18, 12 Merley Lane, Wimborne BH21 1RX
M1 EBI P. Bird 10A Shackleton Road, Bloxwich, Walsall WS3 3BZ
M1 EBK M. Rowley 20 Long Leasow, Selly Oak, Birmingham B29 4LT
M1 EBL C. Venables Deepdene, Rickford, Guildford GU3 3PQ
M1 EBS P. Beckwith 4 Hunters Yard, , Riseley MK44 1EN
M1 EBU W. Mitchell 8 Woodland Crescent, Burgess Hill RH15 0LJ
M1 EBW I. Merrill 26 Catkin Drive, Giltbrook, Nottingham NG16 2UB
M1 EBY P. Clark Flat Six Haynes House Booker Place, High Wycombe HP12 4QD
M1 ECB K. Cronin East Cottage Westviile Rd, Thornton le Fen LN4 4YJ
M1 ECC D. Wright 74 Witchards, Basildon SS16 5BN
M1 ECD North-Northants Raynet Group c/o M. Wright 27 Willow Road, Kettering NN15 7BA
M1 ECH S. Kirby 2 Kneeton Park, Middleton Tyas, Richmond DL10 6SB
M1 ECI A. Funnell 15 Hendham Road, London SW17 7DH
M1 ECM M. White 100 Burnham Road, Coventry CV3 4BQ
M1 ECQ C. Hamilton 101 Gipsy Lane, Swindon SN2 8DL
M1 ECT M. Procter 141 Ruddington Lane, Nottingham NG11 7BY
M1 ECV D. Bould 38 Curlew Grove, Bridlington YO15 3NX
M1 ECW N. Mcmahon 23 St. James Close, Bramley, Tadley RG26 5XH
M1 EDA S. Fabian 3 Manor Cottages Manor Orchard Horley, Banbury OX15 6DZ
M1 EDF G. Powell Sycamore Cottage, Church Lane, Tamworth B79 0LD
M1 EDL A. Wolverson 28 Thorness Close Alvaston, Derby DE24 0UY
M1 EDO J. Hurst 13 Peregrine Road, Hainault, Ilford IG6 3SR
M1 EDW P. Tinkler 27 Cavendish Drive, Carlton, Nottingham NG4 3DX
MM1 EDY J. Goldstraw 26 Craigmill Gardens, Carnoustie DD7 6HT
M1 EEN G. Butterfield Pasadena, St. Helens Road, Walcott, Norwich NR12 0LU
M1 EEP J. Nethercott 30 Goldcrest Road, Chipping Sodbury, Bristol BS37 6XF
M1 EEQ A. Waite 221 Hasler Road, Poole BH17 9AH
M1 EER G. Ball 11 Jersey Close, Congleton CW12 3TW
M1 EEW A. Beckwith 19 Westmorland Avenue, Dukinfield SK16 5JA
M1 EEY N. Beckley 76 Keir Hardie Way, Barking IG11 9NY
M1 EEZ A. Kypriadis 119 Whitfield Villas, South Shields NE33 5NH
M1 EFP J. Carter 5 Hastings Avenue, Seaford BN25 3LB
M1 EFT P. Swanton 54 South Avenue, Warrington WA2 8BQ
M1 EGC G. Hancock 24 Kings Avenue, Corsham SN13 0EG
M1 EGD S. Sykes 12 Banksville, Holmfirth HD9 1XP
M1 EGG P. Bird 37 Beachwood Avenue, Kingswinford DY6 0HL
M1 EGL P. Ritchley 34 Chesildene Avenue, Throop, Bournemouth BH8 0DS
M1 EGM B. Ritchley 25 Branwell Close, Christchurch BH23 2NP
M1 EGN J. Eyres 13 Newburn Crescent, Swindon SN1 5ES
M1 EGP R. Argent 48 Church Green, Staplehurst, Tonbridge TN12 0BE
MM1 EGS T. May 34 Dee Place, East Kilbride, Glasgow G75 8RZ
M1 EGV C. Mills 6 Levisham Gardens, Bewsey, Warrington WA5 0GD
M1 EGW D. Green 5 St. Benedicts Close, Cranwell Village, Sleaford NG34 8DB
M1 EGX M. Beddard 18 Hyacinth Way, Burbage LE10 2UH
M1 EGZ W. Gravestock 23A Murrell Road, Ash, Aldershot GU12 6ST
M1 EHB J. Nixon Willow Lodge, North Townside Road, North Frodingham, Driffield YO25 8LB

M1 EHD P. Leadill 7 Keldale, Haxby, York YO32 3GG
M1 EHF G. Gore-Thorne 19 Poplar Way, Ringwood BH24 1UY
M1 EHI I. Croasdale 16 Buttermere Avenue, Chorley PR7 2JG
M1 EHJ R. Roychoudhuri 62A Parkway, Eastbourne BN20 9DY
MM1 EHO S. Mcneil 35 Sutors Avenue, Nairn IV12 5AZ
M1 EHV C. Aram 1 Snuggs Lane, East Hanney, Wantage OX12 0HU
M1 EHZ J. Brown 3 Malton Close, Blyth NE24 5AS
M1 EIE S. Stephenson 31 Sherbrooke Avenue, Hull HU5 4AG
MI1 EIH G. Brennan 15 Kinnegar Rocks, Donaghadee BT21 0EZ
M1 EIJ S. Sweetlove 36 Park Avenue, Corsham SN13 0JT
M1 EIO A. Rixon 17 Brimmers Way, Aylesbury HP19 7HR
M1 EIR E. Fishbourne 8 Somers Walk, Tupsley, Hereford HR1 1QX
M1 EIU C. Martin Flat 3, Dodds House, Vicarage Lane, Tarporley CW6 9BP
M1 EIW G. Kinney 1 Eden Park, Brixham TQ5 9LS
M1 EIZ L. Rutherford 197 Rosalind Street, Ashington NE63 9BB
M1 EJD D. Pickering 28 Fosse Way Nailsea, Bristol BS48 2BG
M1 EJE D. Clarke 1 Combe Bank, Lindthorpe Way, Brixham TQ5 8PB
M1 EJG J. Clarke 21 Mill Lane, Blakedown, Kidderminster DY10 3ND
M1 EJI B. Hunt 2A Golf Road, Radcliffe-On-Trent, Nottingham NG12 2GA
M1 EJJ M. Laurie 2 The Steps, Phocle Green, Ross-on-Wye HR9 7TW
M1 EJL P. Langfield 21 Amy Johnson Court, Great Passage Street, Hull HU1 2AJ
M1 EJO P. Matthews The Stables, Alkham Road, Dover CT16 3EE
M1 EJQ M. Cross 7 Hallside Road, Enfield EN1 4AD
M1 EJS S. Birchall 11 Rosebery Road, Felixstowe IP11 7JR
M1 EJX M. Heley 22 St. Lawrence Road Dunscroft, Doncaster DN7 4AS
M1 EKA B. Parker 38 Cross St., Thurcroft, Rotherham S66 9NJ
M1 EKB K. Baldwin 5 Rosedale Close, Belmont, Hereford HR2 7ZD
M1 EKD S. Pounder 4 Otley Mount, East Morton, Keighley BD20 5TD
M1 EKH D. Bowker 54 Edward Street, Middleton, Manchester M24 6BN
M1 EKK K. George 20 Oakleigh Avenue, Clayton, Bradford BD14 6QE
M1 EKM G. Harden 13 Greenfield Road, Coleford GL16 8BY
M1 EKP D. Meeds 5 Lenton Way Frampton, Boston PE20 1AU
M1 EKU P. Lancaster 16 Wiltshire Close, Bury BL9 9EY
M1 ELB C. Mitchell A 3, Pakkalanrinne 14, Vantaa 1510 Finland
MM1 ELE D. Marshall 10 Spencer Crescent, Carnoustie DD7 6DQ
M1 ELI D. Welch 41 Mersey Way, Bletchley, Milton Keynes MK3 7PS
M1 ELM A. Dent 5 Loader Close, Bournemouth BH9 1LR
M1 ELN M. Pike 1 Sevelm, Up Hatherley, Cheltenham GL51 3RZ
M1 ELQ P. Houghton 19 Gilthwaites Lane, Denby Dale, Huddersfield HD8 8SG
M1 ELR C. Davies 94 Alnwick Drive, Moreton, Wirral CH46 6ET
M1 ELS J. Matias 15 Ennerdale Gardens, Wembley HA9 8QY
M1 ELW H. Watson 5 Arbroath, Ouston, Chester le Street DH2 1QY
M1 EMB Cornish RAC c/o S. Buckley 1 Chouler Gardens, Stevenage SG1 4TB
M1 EMC K. Heselton 3 Winterslow Road, Penhill, Swindon SN2 5JJ
M1 EMG D. Woodcroft 23 Wilkin Walk, Cottenham, Cambridge CB24 8TS
M1 EMO P. Warden 12A Landscape View, Saffron Walden CB11 4AU
M1 EMP N. Douglas 2 Huntingdon Close, Fareham PO14 4JP
M1 EMR J. Dixon 5 Laburnum Avenue, Moorends, Doncaster DN8 4SF
M1 EMU P. Gilmore 2 Ridgeway, Billericay CM12 9NT
M1 EMX H. Harvey 153 Stradbroke Grove, Clayhall, Ilford IG5 0DL
M1 ENA J. Long 1 Tangway, Chineham, Basingstoke RG24 8SU
M1 ENE S. Marcot 5 The Crescent, West Wickham BR4 0HB
M1 ENJ C. Berry 29 Marlborough Crescent, Long Hanborough, Witney OX29 8JP
M1 ENK J. Berry 29 Marlborough Crescent, Long Hanborough, Witney OX29 8JP
M1 ENQ R. Townsend 56 Seymour Road, Northfleet, Gravesend DA11 7BN
M1 ENX S. Caine Magnolia House, The Larches, East Grinstead RH19 3QL
M1 ENZ R. Rouse 15 Mulimbah Street, Eleebana 2282 Australia
MW1 EOK P. Davies 9 Cramer Court, Rhyl LL18 2BX
MW1 EOO H. Matthews 24 Clos Y Berllan, Rhuddlan, Rhyl LL18 2UL
M1 EOP M. Jurkiewicz 21 Porlock Avenue, Stafford ST170HS
MW1 EOR M. Edwards 2 The Twyn, Fleur De Lis, Blackwood NP12 3UL
M1 EOU T. Nadin 4 Firtree Rise, Chapeltown, Sheffield S35 1QG
M1 EOV P. Spurgeon 15 Ketts Close, Wymondham NR18 0NB
M1 EOZ C. Cartwright 8 Hudson Road, Blackpool FY1 6LY
MJ1 EPG B. Allchin Pont Marquet Cottage, La Rue Des Mans, St. Brelade JE3 8BL Jersey
MW1 EPI D. Williams Rhilin, Rhydwyn, Holyhead LL65 4EA
M1 EPK K. Hichisson 2 Tithe Farm Close, Langford, Biggleswade SG18 9NE
M1 EPN R. Shaddick 5 Shrewsbury Bow, Weston-Super-Mare BS24 7SB
M1 EPR A. Wilson 22 Ormesby Road, Raf Coltishall, Norwich NR10 5JY
M1 EPU South Devon Raynet c/o G. Coker 46 Clarendon Road, Ipplepen, , Newton Abbot TQ12 5QS
M1 EPX M. Clark 8 Willow Close, Clevedon BS21 6HR
M1 EQA N. Trewin 70 Trelowen Drive, Penryn, Cornwall TR10 9WS
M1 EQB G. Trouse 1 Amanda Close, Bexhill-on-Sea TN40 2TB
M1 EQD P. Burton 99 Western Avenue, Blacon, Chester CH1 5QX
MM1 EQE A. Thompson 22 Lochend Road, Carnoustie DD7 7QF
MI1 EQI C. Blake 9 Mullaghbrack Road, Hamiltonsbawn, Armagh BT60 1JU
M1 EQO C. Hayward 12 Trouvere Park, Hemel Hempstead HP1 3HY
M1 EQV M. Edwards 36 Mount Street, Coventry CV5 8DE
M1 EQW S. Harrison 20 Wisewood Avenue, Wisewood, Sheffield S6 4WG
M1 ERA Orkney Wireless Museum c/o S. Trimble Pentreath, Cury Cross Lanes, Helston TR12 7BJ
M1 ERD A. Trimble Pentreath, Cury Cross Lanes, Helston TR12 7BJ
M1 ERF F. Tatlow 45 Pasture Road, Stapleford, Nottingham NG9 8HR
M1 ERH S. Bird 12 Commercial Road, Shepton Mallet BA4 5DH
M1 ERJ P. Chandler 94 Shrubland St., Leamington Spa CV31 3BD
MI1 ERL P. Cranston 135 Saintfield Road, Lisburn BT27 6YW
M1 ERN J. Baugh 172 Pontefract Road, Cudworth, Barnsley S72 8BE
M1 ERO D. Eastope 9 St. Davids House, Willow Way, Redditch B97 6PG
M1 ERP S. Carruthers 13 Belah Road, Carlisle CA3 9RE
M1 ERS S. Webster 402 Windmill Lane, Sheffield S5 6FZ
M1 ERU S. Carrington 137 Richmond Park Road, Bournemouth BH8 8UA
M1 ERV S. Chambers 1 Northleaze, Corsham SN13 0QW
M1 ERY K. Sylvester 8 Beacon Park Close, Skegness PE25 1HQ
M1 ESH W. Inch 1 Blencowe Crescent Steeple Claydon, Buckingham MK18 2GY
M1 ESI D. Crane 132 Windermere Drive, Warndon, Worcester WR4 9JD
M1 ESM L. Wallace 20 Radworthy, Furzton, Milton Keynes MK4 1JH
M1 ESV R. Scotland 11 Edwards Court, Slough SL1 2HY
M1 ETC M. Ribton 80 Trafalgar St., Gillingham ME7 4RN
M1 ETM A. Hayward 67 Pinecroft, Carlisle CA3 0DB
M1 ETN D. Allen Ash Tree Farm, Small End, Friskney, Boston PE22 8PF
M1 ETS E. Coleby 13 Farm Close, Sunniside, Newcastle upon Tyne NE16 5PP
M1 ETT T. Hindson 73D Leigh Road, Wimborne BH21 2AA
M1 ETW A. Talabi 1 Crealock Grove, Woodford Green IG8 9QZ
M1 EUB A. Wearne 58 Mainstone Avenue, Princerock, Plymouth PL4 9NB
M1 EUE J. Underwood 27 Woodville Road, London E17 7ER
M1 EUF J. Cunningham 16 Welbeck Road, Doncaster DN4 5EY
M1 EUL B. Fielding 16 The Horseshoe, York YO24 1LX
M1 EUM P. Thorne 6 Wordsworth Terrace, Penrith CA11 7QT
M1 EUN J. Fletcher 66 Deightonby Street, Thurnscoe, Rotherham S63 0JA
M1 EUR P. Allen 32 Milward Road, Loscoe, Heanor DE75 7JX
M1 EUX C. Bartlett 25 Westfields, Buckingham MK18 1DZ
MI1 EVD T. Carlisle 12 Drumawhey Gardens, Bangor BT19 1SR
M1 EVF S. Larden1 Flat 42, Worcester House, Hill Street, Halesowen B63 4TJ
M1 EVH K. Wright 60 Ashley Road, Walsall WS3 2QF
MM1 EVJ J. Conway 26 Kerse Avenue, Dalry KA24 4DJ
M1 EVN J. Haughey 31 Woodfield Drive, West Mersea, Colchester CO5 8PX
M1 EVZ S. Punch 39 Wilson Road Stalham, Norfolk NR12 9FL
M1 EWB P. Dabell 21 Winchcombe Drive, Burton-on-Trent DE15 9EN
M1 EWD T. Beevers 4 Cloud Avenue, Stapleford, Nottingham NG9 8BN
M1 EWF R. Read 111 Fitzpain Road, West Parley, Ferndown BH22 8SF
MW1 EWJ E. Williams 82 Maes Llwyn, Amlwch LL68 9BG
M1 EWM A. O'Hea 12 De Vitre Place, Grove, Wantage OX12 0DA
M1 EWP B. Lester 37 Cormorant Drive, St. Austell PL25 3BB
M1 EWT C. Berry Berriscot, 7 Gloweth Villas, Truro TR1 3LU
M1 EWV K. Trench 10 Victoria Road, Morley, Leeds LS27 9DS
M1 EXJ M. Wohlgemuth 39 Great Mead, Denmead, Waterlooville PO7 6HH
M1 EXO S. Andrews 64 Bradgate Road, Markfield LE67 9SN
M1 EXQ M. Peters Yare House, Thuxton, Norwich NR9 4QJ
M1 EXS G. Burton 2 Derwent Street, Darwen BB3 1EF
M1 EXW M. Gardner 21 Tiptree, Castlehaven Road, London NW1 8TL
M1 EYA R. Neale-Gardner 72 Queensway Barwell, Leicester LE9 8AP
M1 EYG R. Barrow Fern House, Ripponden Old Lane, Sowerby Bridge HX6 4PA
MW1 EYH F. Bailey 28 Coopers Field, St. Martins, Oswestry SY11 3BU
MM1 EYI N. Kingon-Rouse 148 Oldwood Place, Livingston EH54 6UX
M1 EYL A. Banks 12 Taylor Court, Weston-Super-Mare BS22 7LU
M1 EYO A. Poxon 34 Conduit St., Tintwistle, Glossop SK13 1LR
M1 EYP Todmorden Ryt G c/o T. Read 31 Merebrook Road, Macclesfield SK11 8RH
M1 EYQ B. Davis 104 Lever House Lane, Leyland PR25 4XP
M1 EYS E. Shears 22 Richborough Drive, Charlton, Andover SP10 4EZ
M1 EYT A. Kok 14 Throop Road, Templecombe BA8 0HR
MW1 EYU D. Kok 14 Castle View, Fron Goch, Caernarvon LL55 4LE
MM1 EYZ D. Bruce Carnichal, Maud, Peterhead AB42 4QG
M1 EZB C. Coonick 36 Magdalen Way, Weston-Super-Mare BS22 7PG
M1 EZC G. Sawyer 101 Southern Drive, Loughton IG10 3BY
M1 EZD G. Wesson 12 Lea Close, Alcester B49 6AP
M1 EZE S. Davies 35 Queensland Crescent, Chelmsford CM1 2DZ
M1 EZG M. Lane 21 Winterbourne Road, Poole BH15 2ES
M1 EZH L. Copley Elmford, Mount Pleasant South, Whitby YO22 4RQ
M1 EZJ M. Skelton Annexe, 1 Muxton Lane, Muxton, Telford TF2 8PB
M1 EZK K. Sheehan 43 Central Avenue, Beverley HU17 8LL
M1 EZL J. Anderson 16 Hanham Road, Corfe Mullen, Wimborne BH21 3PZ
M1 EZP M. White 1 Nursery Close, Wroughton, Swindon SN4 9DR
M1 EZR D. Birkenshaw 29 St. Peters Drive, Thornton, Coalville LE67 1AX
M1 EZT M. Marston The Oaks, Nether Compton, Sherborne DT9 4PZ
M1 EZU D. Bedwell 12 Towcester Close, Chippenham SN14 0XX
M1 EZX D. Bridle 8 Cowleaze Road, Broadmayne, Dorchester DT2 8EW
MI1 EZZ R. Bradley 41 Graymount Road, Newtownabbey BT36 7DR
M1 FAA T. Atkinson 10 Invicta Close, Chislehurst BR7 6SJ
M1 FAF T. Jones 354 Bridgeman St., Bolton BL3 6SJ
M1 FAI S. Taylor Cherry Tree Cottage, Heslington, York YO10 5DX
M1 FAJ D. Green 7 Greenside Court, Mickleover, Derby DE3 0RG
MI1 FAR A. Bryce 123 Newtownards Road, Comber, Newtownards BT23 5LD
MM1 FAS R. Krawczyk 7 Anderson Crescent, Bishopmill, Elgin IV30 4HJ
MW1 FAT C. Jayne 65A Park Crescent, Abergavenny NP7 5TL
M1 FAX D. Jones 87 Forster St., Warrington WA2 7AX
M1 FAY K. Rowsell Elmtree Cottage, Chilworthy, Chard TA20 3BH
M1 FBF A. Wilson 2 Briar Close, Newhall, Swadlincote DE11 0RX
M1 FBI S. Plews 70 Baulkham Hills, Penshaw, Houghton le Spring DH4 7RZ
M1 FBL J. Rowe Hunters Brook, Fine Lane, Newport PO30 3JY
M1 FBN I. Dalton 10 Durkar Fields, Durkar, Wakefield WF4 3BY
M1 FBS D. Faul Ph 1113, 1200 The Esplanade North, Ontario L1V6V3 Canada
M1 FBW J. Machalski 28 Longdales Road, Lincoln LN2 2JU
MI1 FCB T. Kilgore. Mbe 8 Slieve Shannagh Park, Newcastle BT33 0HW
M1 FCC C. Chuter 35 Longford Road, Bognor Regis PO21 1AB
M1 FCE D. Sampson 32 The Gannets, Stubbington, Fareham PO14 3SY
M1 FCF D. Patrick 2 The Orchard, Old Totnes Road, Buckfastleigh TQ11 0FG
M1 FCG R. Boyns 65 Alma Road, Plymouth PL3 4HE
M1 FCH T. Johnson Orchard House, Tollerton, York YO61 1PS
MI1 FCQ P. Mccauley 5 Brookmount Rise, Omagh BT78 5AL
M1 FCV C. Roberts 57 Chandos Road, Lightpill, Stroud GL5 3QT
M1 FCW M. Yeomans11 7 Oak Avenue, Elloughton, Brough HU15 1LA
M1 FCX A. Rudnicki 1 Willoughby Court London Colney, St. Albans AL2 1HL
M1 FCZ E. Snowdon 22 Twizziegill View, Easington, Saltburn-by-the-Sea TS13 4NX
MM1 FDF S. Taylor Sunnybrae, Kinellar, Aberdeen AB21 0TY
M1 FDH F. Howker 65 Boxley Drive, West Bridgford, Nottingham NG2 7GN
M1 FDK G. Charnock Oldhouse, Newton St Margarets, Hereford HR2 0QR
MW1 FDN D. Morgan 6 Oxwich Grove, Newport NP10 8HR
M1 FDO C. Heading 27 Broadlands Avenue, Eastleigh SO50 4PP
M1 FDY L. Browning 5 Redstart Avenue, Kidderminster DY10 4JR
M1 FEK D. Stewart 1 Twelve Acres, Welwyn Garden City AL7 4TG
M1 FEM F. Priborsky Fohrenstrasse 49, Tuttlingen 78532 Germany
MM1 FEO P. Gaskin Tft Electronics, Unit 1, Skeld Industrial Estate, Skeld ZE2 9NL
M1 FEQ K. Watson Bolehall Manor Club, Amington Road, Tamworth B77 3LH
M1 FER G. Watson 85 Thomas Street, Tamworth B77 3PP

UK Callsigns

M1 FET E. Dodd 2 Chichester Crescent, Chadderton, Oldham OL9 0RW
MW1 FEU M. Williams 68 Hengoed Road, Penpedairheol, Hengoed CF82 8BR
M1 FEW A. Franklin 11 Harden Hills, Shaw, Oldham OL2 8NE
M1 FEX C. Wells 37 Water Meadows, Worksop S80 3DF
M1 FEY I. Trail 10 Hillary Drive, Crowthorne RG45 6QE
M1 FEZ S. West 142 Burrowmoor Road, Cambridgeshire PE15 9SS
M1 FFA Dr D. Byfield 23 New Cross, Longburton, Sherborne DT9 6EJ
M1 FFC P. Francis Unit 15 Wellington Road, Bridgwater TA6 5HA
MW1 FFE D. Evans 5 Brunel Road, Fairwater, Cwmbran NP44 4QT
M1 FFF D. Leech 9 Little Gransden Lane, Great Gransden, Sandy SG19 3BA
M1 FFG G. Siviter Flat 106, Lancaster House, Rowley Regis B65 0QE
M1 FFM M. Dawson 140A Healey Road, Scunthorpe DN16 1HT
M1 FFN J. Willingham 3 Cherry Road, Nailsea, Bristol BS48 2EE
M1 FFO E. Sanderson Flat, Old Post Office, High Street, Yeovil BA22 7NQ
M1 FFP B. White 7 Park Street, Castle Cary BA7 7EH
M1 FFR K. Singam 44 Forty Lane, Wembley HA9 9HA
M1 FFS J. Bates Flat 7, Colyer House, London SE2 0AJ
M1 FFV J. Cobb 17G Church Lane, Saxilby, Lincoln LN1 2PE
M1 FFX C. Burgess 13 Glyne Drive, Pebsham, Bexhill-on-Sea TN40 2PW
MW1 FFY R. Davies 100 Lewis Road, Neath SA11 1DQ
MW1 FGB R. Tremelling 45 Bryntawe Road, Ynystawe, Swansea SA6 5AD
M1 FGH Dr M. James Oak Tree House St. Matthews Terrace, Leyburn DL8 5EL
M1 FGM T. Beswick 37 Grovewood Road, Misterton, Doncaster DN10 4EF
M1 FGO Dr A. Kerr-Munslow 28 Swallow Court, St. Neots PE19 1NP
MW1 FGV J. Rowe Flat 27 Rhodfa Frank, Ammanford SA18 2QE
M1 FHA T. Earp 14 Drakes Avenue, Devizes SN10 5AZ
M1 FHB R. Earp 1 Yew Tree Cottages, Erlestoke, Devizes SN10 5UD
MI1 FHE K. Lunney 20 Inniskeen Close, Enniskillen BT74 6HD
M1 FHJ J. Bolton 11 Forest Drive, Lytham St. Annes FY8 4PF
MM1 FHL J. Crockett 70 Inchview Terrace, Edinburgh EH7 6TH
MM1 FHO L. Norman 16 Cotton Street, Balfron, Glasgow G63 0PF
M1 FHP E. Parr 36 Ridley Drive, Great Sankey, Warrington WA5 1HP
M1 FHQ I. Anderson 11 Rays Drive, Lancaster LA1 4NT
MM1 FHR G. Hunt 1 Love St., Kilwinning KA13 7LQ
MM1 FHS N. Sampson 47 Muirend Road, Perth PH1 1JD
M1 FHT A. Di Domenico 11 Bracken Drive, Freckleton, Preston PR4 1TH
M1 FHX D. Jordan 38 Weston Lane, Otley LS21 2DB
MM1 FHZ Dr D. Bickle Lon Mhor, 10 Grean, Isle of Barra HS9 5XU
M1 FIB P. Heathcote Flat 6, Balmoral House, 12 Balmoral Road, Westcliff-on-Sea SS0 7AZ
M1 FIE Dr A. Phelan Calle Misericordia 14, Jimena de la Frontera 11330 Spain
M1 FIG S. Hunt 1 Trefusis Cottages, Flushing, Falmouth TR11 5TE
M1 FII C. Parker 3 Lyon Close, Abingdon OX14 1PT
M1 FIL P. Smith 185 Bradwell Lane, Newcastle ST5 8JB
M1 FIP F. O'Sullivan 10 Hampton Close, London NW6 5LR
M1 FIR J. Coles 45 Common Lane, Titchfield, Fareham PO14 4BX
MI1 FIS D. Stewart 16 Weavers Lodge, Donaghcloney, Craigavon BT66 7LE
M1 FIV W. Wailes 8 Staindrop Terrace, Stanley DH9 8JW
M1 FJA R. Clifford-Smith 60 Petworth Gardens, Southampton SO16 8EF
M1 FJB S. Hunt Maxxwave House Hill Lane Business Park, Markfield LE67 9PY
M1 FJC J. Soltysik 24 Cottage Close, Hednesford, Cannock WS12 1BS
M1 FJD Newry And District ARC c/o J. Coleman 52 Brook Street, Colchester CO1 2UT
M1 FJF P. Griffiths Probation Hostel Haworth House, Blackburn BB2 2HL
M1 FJG A. Roberts 39 Colsterdale Carlton Colville, Lowestoft NR33 8TN
M1 FJH F. Bate 10Alder Close, Worcester WR3 8QH
M1 FJJ A. Parkinson 21 Lambton Street, Bolton BL3 3LG
MW1 FJK K. Hughes 33 Brynglas, Penygroes, Llanelli SA14 7PY
M1 FJL R. Collins Perch Hill Cottage, Perch Hill, Wells BA5 1JA
MM1 FJM R. Moore Ard Na Mara, Gulberwick, Shetland ZE2 9TX
M1 FJP P. Blackman 73 St. Marks Road, Chester CH4 8DE
M1 FJQ D. Tinker 116 Longley Avenue West, Sheffield S5 8WF
MW1 FLY R. Eyre 9 Poplar Close, Rogiet, Caldicot NP26 3TL
M1 FMC A. Bailey 2 Kingswood Road, Shrewsbury SY3 8UX
M1 FNE G. Scott 19 Witton Gardens, Jarrow NE32 5YJ
M1 FRB F. Barnes 4 Pound Close, Ducklington, Witney OX29 7TH
MI1 FRM Dr F. Montgomery 4 Thornbrook, Lisburn BT27 5LW
M1 FTA J. Fernandes 11 Ferndown Road, Watford WD19 6HU
M1 FUR Coulsdon Amateur Transmitting Society c/o A. Briers 33 Deans Walk, Coulsdon. CR5 1HR
M1 FWD S. Davies 17 Taw View Fremington, Barnstaple EX31 2NJ
MM1 FZR S. Gillies 49 Meadowside Road, Queenzieburn, Glasgow G65 9EJ
M1 GAP A. Prince 29 St. Stephens Road, West Bromwich B71 4LR
M1 GAR A. Gardner 39 Court Close, Twickenham TW2 5JH
MM1 GAS P. Cheeseman 12 Christie Grange, Bucksburn, Aberdeen AB21 9SE
MM1 GBS D. James 9 Dunbar Lane, Duffus, Elgin IV30 5QN
M1 GCS G. Steedman 61 Granville Street, Barnsley S75 2TQ
M1 GDB G. Bell 4 Fairley Way, Cheshunt, Waltham Cross EN7 6LG
M1 GDE G. Edgar 61 Winchester Avenue, Lancaster LA1 4HX
M1 GDH D. Hayward 16 Heathway, Dagenham RM10 9PP
M1 GEO Dr G. Smart 30 Cornmills Road, Soham CB7 5AT
M1 GFE F. Erridge 17 Head Street, Goldhanger, Maldon CM9 8AY
M1 GGG G. Ma 26 Church Lane, Chalgrove, Oxford OX44 7TA
M1 GHT D. Buckerfield 62 Springfield Crescent, Somercotes, Alfreton DE55 4LH
M1 GIZ S. Bridger 80 Springhill Crescent, Madeley, Telford TF7 4DP
MW1 GLD G. Davies 77 Tydraw St., Port Talbot SA13 1BR
M1 GMO M. Hastry 56 Kilsyth Close, Fearnhead, Warrington WA2 0SQ
M1 GOH R. Horry 1 Council House Nidds Lane, Kirton, Boston PE20 1LZ
M1 GPC G. Carpenter 67 Angela Crescent, Horsford, Norwich NR10 3HE
M1 GPE G. Emmerson 29 Dulsie Road, Talbot Woods, Bournemouth BH3 7DY
M1 GRA G. Stephens 25 Fore Street Langtree, Torrington EX38 8NG
M1 GRP G. Pound 19 Victory Green, Portsmouth PO2 8RH
M1 GSM S. Watson 6 Mount Pleasant, Stanley, Crook DL15 9SF
M1 GSX I. Greenfield 1 Dale Road, Shrewsbury SY2 5TE
M1 GTI D. Burgin 15 Birch Grove, Chippenham SN15 1DD
M1 GUR P. Gurney 12 Church Street, Wymeswold, Loughborough., Leicestershire LE12 6TX
M1 GUS J. Batchelor 5 Gladden Fields, South Woodham Ferrers, Chelmsford CM3 7AH
M1 GWA G. Warburton 50 Clarendon Road, Sheffield S10 3TR
M1 GWZ Dr P. Miller Tate 19 Esher Avenue, Walton-on-Thames KT12 2SZ
M1 GXL T. Higginson 109 London Road, Biggleswade SG18 8EE
M1 HEK Heckington And District Radio Group c/o G. Jannetta 14 Banks Lane, Heckington, Sleaford NG34 9QY
M1 HFM M. Poole 18 Lockway, Drayton, Abingdon OX14 4LG
M1 HFX R. Ayers 1 Handley Park, Sixpenny Handley SP5-5PL
M1 HGV M. Rule 23 Rue Du Puits Doux, Sacy, St. Christophe A Berry 2290 France
M1 HHL B. Miller Oakhurst, 1 Southern Oaks Barton On Sea, New Milton BH25 7JT
M1 HJE S. Elliott11 7 Manor Close Harston, Cambridge CB22 7QF
M1 HLG H. Glover 14 Crawley Crescent, Eastbourne BN22 9RN
M1 HLL S. Batchelor 2 Belmont Avenue, Atherton, Manchester M46 9RR
M1 HMP P. Grech 108 Hind Grove, London E14 6HU
MM1 HMV B. Shearer Latheron, 113 Auchamore Road, Dunoon PA23 7JJ
MM1 HMZ B. Allison 5 Mayfield Drive, Howwood, Johnstone PA9 1BJ
M1 HOP Sir A. Hopson 1 Hall Lane, Leicester LE2 8SF
M1 HPR I. Hooper 25 Honey Lane, Buntingford SG9 9BQ
M1 HQX Inverclyde CG c/o W. Hammond 28 Fengate Mobile Home Park, Peterborough PE1 5XD
M1 HVJ A. Jefferiss 27 Sherbourne Drive, Maidenhead SL6 3EP
MM1 HWB P. Oldham 2 Old Bar Road, Nairn IV12 5BX
M1 HZZ A. Barbour 36 Roseacre Drive, Elswick, Preston PR4 3UQ
M1 IAN I. Tennent Flat 2, 97 Rydens Road, Walton-on-Thames KT12 3AW
MM1 ICE A. Somerville 8 Craiglockhart Park, Edinburgh EH14 1HE
M1 ICL I. Leather White Cottage, Aston Lane, Aston, Runcorn WA7 3BU
M1 IFT Dr A. Bartle 10 Holme Dene, Haxey, Doncaster DN9 2JX
M1 IHM D. Jones 19 Queens Road, Donnington, Telford TF2 8DB
M1 IKE M. Collins 3 Beacon View, Grayrigg, Kendal LA8 9BT
M1 IMC I. Mcleod 12 Seathwaite Avenue , Marton, Blackpool FY4 4RL
M1 IOS J. Goody 9 Garrison Lane, St. Mary'S, Isles of Scilly TR21 0JD
M1 IOW P. Legg 20 Arthur Moody Drive, Newport PO30 5JR
M1 IRB I. Bush 17 Queens Place, Shoreham-by-Sea BN43 5AA
M1 IRM P. Rowley 6 Duesbury Green, Stoke-on-Trent ST3 2RZ
MM1 JAC J. Campbell 119 Campbell Avenue, Dumbarton G82 3PB
M1 JAK A. Hanson-Brown 35 York Avenue, Bedworth CV12 9EL
MW1 JAN J. Carfoot 24 Marble Church Grove, , Bodelwyddan LL18 5UP
MM1 JAS J. Shankland 2 Strathdoon Place, Ayr KA7 4PB
M1 JCB T. Wightman Laithbutts Farm, Cowan Bridge, Carnforth LA6 2JL
M1 JCL J. Plant 67 Kenley Road, London SW19 3JJ
M1 JCS C. Starr 64 Green Lane, Lambley, Nottingham NG4 4QE
M1 JDH J. Hammond 8 Rowntree Way, Saffron Walden CB11 4DG
M1 JDW J. Mitchell Sheet Hill Farmhouse, Winfield Lane, Sevenoaks TN15 0LZ
M1 JEC J. Cook 35 Holdbrook Way, Romford RM3 0JD
M1 JES J. Gilbert Thr Oaktree, Ellenbrook Lane, Hatfield AL10 9NT
M1 JHF J. Fielding 39 Wicklow Avenue, Melton Mowbray LE13 1DY
M1 JHG J. Green 33 Edenvale Crescent, Lancaster LA1 2NW
M1 JHL S. Jouhal 35 Cherrywood Gardens, Nottingham NG3 6LR
M1 JIM J. O'Hea 12 De Vitre Place, Grove, Wantage OX12 0DA
M1 JJN J. Nicholson 6 Mill Gardens, West End, Southampton SO18 3AG
M1 JJS P. Springate 10 Pipers Close, Burnham, Slough SL1 8AW
M1 JLM B. Murfitt 21 Priors Drive, Norwich NR6 7LJ
M1 JMB J. Bevan 1 Condor Close, Weston-Super-Mare BS22 8SE
M1 JMM Duke J. PÚrez SolÝs C/ San Cecilio 2 P03 K, Loja - Granada 18300 Spain
MI1 JOE W. Murray 80 Canterbury Park, Londonderry BT47 6DU
M1 JON J. Godding 58 Dukeswood Road, Longtown, Carlisle CA6 5UJ
M1 JPS J. Patterson 39 Coquet Drive, Ellington, Morpeth NE61 5LN
M1 JSS J. Smout 34 Kidderminster Road, Bromsgrove B61 7JT
M1 JTA J. Tyers Hillcrest, Papermill Lane, Evedon, Sleaford NG34 9PD
MM1 JWF J. Frame 24 Douglas Crescent, Erskine PA8 6BJ
M1 JWM J. Machin 42 Woodstock Road, Loxley, Sheffield S6 6TG
M1 JWR J. Rutherford Nook On Lyne, Longtown, Carlisle CA6 5TS
M1 JWS J. Smith 23 Horsham Lane, Upchurch, Sittingbourne ME9 7AN
M1 KAZ Dr A. Forrest 261 East End Road, London N2 8AY
M1 KCB K. Crank 319 Manchester Road, Clifton, Manchester M27 6PT
M1 KDH K. Harvey 29 The Hobbins, Bridgnorth WV15 5HH
M1 KDJ K. Jowett The Gables, Brinkworth Road, Royal Wootton Bassett, Swindon SN4 8DT
MW1 KDP M. Heron Heron House, Park Road, Gwynedd LL42 1PL
M1 KEJ R. Jeffs 45 Forest Road, Bingham, Nottingham NG13 8RL
M1 KES M. Oconnor 13 Ashburnham Road, Southend-on-Sea SS1 1QB
M1 KEV K. Mahoney 11 Leyland Walk, Bristol BS13 8PY
M1 KEY M. Hardy 4 Kirk Balk Hoyland, Barnsley S74 9HU
M1 KGL K. Large 6 Sylvden Drive, Wisbech PE13 3UD
M1 KIP L. Kipling 16 Northampton Close, Bracknell RG12 9EF
M1 KIQ D. Rosser P.O. Box 268, Enmore 2042 Australia
M1 KMC A. Coathup 54 Rydal Road, Kendal LA9 6LB
MM1 KML D. Line 55 Burn Brae Westhill, Inverness IV2 5RH
M1 KMT K. Tomlinson 5 Lynam Way, Madeley, Crewe CW3 9HX
M1 KOS Dr K. Tsioumparakis 10 Lavender Close, Leatherhead KT22 8LZ
M1 KPW K. Whitmarsh 7 Foxs Furlong, Chineham, Basingstoke RG24 8WN
M1 KPZ T. Carnegie 27 St. Andrews Road, Montpelier, Bristol BS6 5EG
M1 KSB K. Best 42 Falmer Avenue, Goring-By-Sea, Worthing BN12 4TD
M1 KTA R. Baines 34 Bury Road, Stapleford, Cambridge CB22 5BP
M1 KTY K. Mallows 57 Top Road, Kingsley, Frodsham WA6 8DA
M1 KVN K. Finn 132 Lansdowne Grove, Wigston LE18 4LY
M1 KWH K. Hargreaves Langton Lodge, Fordcombe Road, Tunbridge Wells TN3 0RB
M1 LAN A. Worsley 10 Millfield View, Worksop S80 3QB
M1 LAP L. Pollard 45 Nanny Marr Road, Darfield, Barnsley S73 9AB
MM1 LBA A. Bulloch 4 Cartleburn Gardens, Kilwinning KA13 7ND
M1 LCL A. Kvilums 78 Wagon Lane, Solihull B92 7PN
MM1 LDR A. Sloan 36 Paterson Avenue, Irvine KA12 9JJ
M1 LEO A. Bennett 16 Manor Avenue, Crewe CW2 8BD
M1 LES L. Rodger 19 South Walk, West Wickham BR4 9JA
M1 LIP A. Lippett 2 Ralph Court, Stafford ST17 9FR
MM1 LJB C. Newman Upper Flat, 3 Lindsay Gardens, Alexandria G83 0US
MW1 LLL M. Greatorex Cwm Pennant, Moel View Road, Prestatyn LL19 9SU
M1 LMJ L. Jones 47 Pine Crescent, Chandler'S Ford, Eastleigh SO53 1LN
M1 LMO N. Waring 3 Sampson St., Eastoft, Scunthorpe DN17 4PQ

M1 LOL A. King 8 Rydal Court, Congleton CW12 4JL
M1 LOU R. Cave 295 Sackup Lane, Staincross, Barnsley S75 5BG
M1 LRX D. Horwood 42 Southlands Drive, Timsbury, Bath BA2 0HB
M1 LSD L. Dawes 52 Ridley Road, Carlisle CA2 4LD
MW1 LSG A. Jenkins 25 Maes Hyfryd, Flint CH6 5LN
M1 LTS L. Stone 32 Watermeadow Lane, Storrington, Pulborough RH20 3GU
M1 LXM A. May 7 Stanton Close, Blandford Forum DT11 7RT
M1 LYE F. Lye 5 New Road, Hextable, Swanley BR8 7LS
M1 LYN L. Asbury 67 Orchard Way, Measham, Swadlincote DE12 7JZ
M1 MAB I. Patrick 17 Stamford Way, Fair Oak, Eastleigh SO50 7JJ
M1 MAD M. Cottrell 9 Woodland Terrace, Kingswood, Bristol BS15 9PU
M1 MAJ Dr M. Johnson 2A St. Margarets Road, Girton, Cambridge CB3 0LT
M1 MAL M. Cadman Flat 17, Harmon House, London SE8 3AS
M1 MCL Dr M. Lawson 9 Headingley Mews, Wakefield WF1 3AB
M1 MCW S. Lace 19 Methuen Street, Walney, Barrow-in-Furness LA14 3PS
M1 MDE D. Elwood High Farm Cottage 7, Newport Road, Market Drayton TF9 2TH
M1 MDP M. Palmer 57 Bemersley Road, Stoke-on-Trent ST6 8JF
MW1 MFY D. Lee 13 Yr Efail, Treoes, Bridgend CF35 5EG
MM1 MHD P. Overton Cluanie, Cairnballoch, Alford AB33 8HQ
M1 MIC M. Hodgson French Farm Bungalow, French Drove, Thorney, Peterborough PE6 0PQ
M1 MIJ W. Waddington 117 Dominion St., Walney, Barrow in Furness LA14 3BP
M1 MIT T. Haynes 30A London Road, Wymondham NR18 9JD
M1 MKL M. Livesey 33 Carrington Close, Birchwood, Warrington WA3 7QA
M1 MLM A. Lomas 32 Crestway Road, Baddeley Green, Stoke-on-Trent ST2 7LD
M1 MNR Mid Norton Raynet Group c/o L. Knighton 5 Quidenham Road, East Harling, Norwich NR16 2JD
M1 MOB M. Dimambro 26 Fetcham Court, Bank Top, Newcastle upon Tyne NE3 2UL
M1 MOD A. Straw Flat 17, Poseidon Court, Homer Drive, London E14 3UG
M1 MOG M. Taylor 136 Lenthall Avenue, Grays RM17 5AB
MM1 MOY J. Dye Allt Na Slanaichd, Moy, Inverness IV13 7YE
M1 MPA M. Allgar 13 Deacon Avenue, Kempston, Bedford MK42 7DU
M1 MPB M. Burfield 11 Myrtle Grove, Wallasey CH44 6QA
M1 MPK M. Kassai 6 Cranhill Close, Littleover, Derby DE23 3XU
M1 MRB M. Butler 210 Green Wrythe Lane, Carshalton SM5 2SP
M1 MRS R. Shepperley Flat F, London N3 1QL
M1 MSF M. Forster 4 West Street, Top Flat, Horncastle LN9 5JF
M1 MST C. Walker Fairfield House, Brimfield Cross, Ludlow SY8 4ND
M1 MTV A. Mason 76 Burlington Way Mickleover, Derby DE3 9BD
M1 MUM P. Worlledge 181 Roselands Drive, Paignton TQ4 7RN
M1 MUS M. Suleyman 26 Old Park Road South, Enfield EN2 7DB
M1 MVX Itchen Valley ARC c/o A. Webb 5 Highfield Avenue, Bristol BS15 3RA
M1 NAD D. Barratt Chapel Cottage, Briantspuddle, Dorchester DT2 7HX
M1 NAS N. Swann 11 Erdyngton Road, Leicester LE3 1JF
M1 NCC N. Cordell 392 Laceby Road, Grimsby DN34 5LX
M1 NER A. Taylor 16 Bellmans Road Whittlesey, Peterborough PE7 1TY
M1 NEW K. Mason 12 Vicarage Close, Weston-Super-Mare BS22 7PA
M1 NHR Radio Operators Cornwall c/o K. Edwards 289 Monks Walk, Buntingford SG9 9DZ
M1 NIS B. Ashcroft 16 Edge Lane, Thornton, Crosby., Liverpool L23 9XE
M1 NIZ M. Austin 11A Beverley Road, Ipswich IP4 4BU
M1 NMG N. Gunnell 8 Upper Dingle Madeley Madeley, Telford TF7 5RX
M1 NNN J. Earnshaw Dunelm, Ayton Road, Irton, Scarborough YO12 4RQ
M1 NPH N. Holland 40 Marlborough Road, Castle Bromwich, Birmingham B36 0EH
M1 NTV N. Tartt 47 Leatham Park Road, Featherstone, Pontefract WF7 5DP
M1 NTY D. Basson Penrhyn, Stonehouse Road, Sevenoaks TN14 7HW
M1 NUS R. Cartwright 14 Bromsgrove Avenue, Eccles, Manchester M30 8WB
M1 NXX J. Lynch 14 The Pastures, Cayton, Scarborough YO11 3UU
M1 OBR D. O'Brien 14 Lower Bettesworth Road, Ryde PO33 3EL
M1 OCN C. Wilson Houseboat Wolf Chandlers Quay, Maldon CM9 4LF
M1 OJS O. Smith 106 Middle Street, Blackhall Colliery, Hartlepool TS27 4EB
M1 ONE C. Chambers 75A Main St., Sedbergh LA10 5AB
MI1 OPM D. Kirkwood 1 Rural Cottages, Front Road, Lisburn BT27 5LF
M1 OXR A. Garthwaite 278 Carlton Road, Barnsley S71 2BA
M1 PAB P. Bush 1A Sherwood Close, Kennington, Ashford TN24 9PT
M1 PAC P. Cole 6 Back Street, Bramham LS23 6BR
M1 PAF P. Fletcher 7 Gatesyde Place, Eskdale, Holmrook CA19 1UD
M1 PAH D. Hardy 1 Upper Steeping, Desborough, Kettering NN14 2SQ
M1 PAM P. Rodger 19 South Walk, West Wickham BR4 9JA
M1 PAS W. James 3 Midfield Close, Gillow Heath, Stoke-on-Trent ST8 6RD
M1 PEL D. Pelling 8 Fowler Close Earley, Reading RG6 7SS
M1 PFS P. Symonds 2 Gorling Close, Ifield RH11 0TJ
M1 PGH P. Howell 16 Everard Road, Bedford MK41 9LD
M1 PGT P. Tomlin 10 The Martins, Thatcham RG19 4FD
M1 PJB P. Backx 43 Tindale Avenue, Cramlington NE23 2BP
M1 PJH P. Hill Flat 12, Danecourt 14 St. Peters Road, Poole BH14 0PA
M1 PKB P. Booth 19 Gunville Crescent, Bournemouth BH9 3PZ
M1 PKW P. Wilton Downsview Cottage, Wappingthorn Farm Lane, Steyning BN44 3AG
M1 PLC M. Woods 76 Kings Road, Evesham WR11 3BS
M1 PMR A. Marshall 53 Birch Close, Corfe Mullen, Wimborne BH21 3TB
M1 PRC Peterborough & District ARC c/o T. Ralph 15 Portchester Close, Stanground, Peterborough PE2 8UP
M1 PTE P. Merrick 12 Wilkinson Road, Wednesbury WS10 8SH
M1 PTR P. Ridley 6 Elm Close, Poynton, Stockport SK12 1QH
MM1 PTT C. Morrison 28 Solway Road Bishopbriggs, Glasgow G64 1QW
M1 PUW Dr P. Rogers 12 Cook'S Folly Road, Sneyd Park, Bristol BS9 1PL
M1 PVC P. Craven Hithe, Chuck Hatch Lane, Colemans Hatch, Hartfield TN7 4EN
M1 PVF P. Flavell 26 Hulles Way, North Baddesley, Southampton SO52 9NS
M1 PWT P. Telco 7 Brockswood Lane, Welwyn Garden City AL8 7BA
M1 PXB J. Hopkins Millinder House, Westerdale, Whitby YO21 2DE
M1 PYE J. Pye 19 Nellan Crescent, Stoke-on-Trent ST6 1PS
M1 RAD D. Clapp 150 Brougham Road, Worthing BN11 2PH
MM1 RAH R. Hemesley Overton Farm House, Kirknewton EH27 8DD
M1 RAL R. Leach 4 Honey Pot Drive, Baildon, Shipley BD17 5TJ
MI1 RAY R. Maguire 139 Carrowshee Park, Drumhaw, Enniskillen BT92 0FS
MI1 RDR R. Ross 13 Eureka Drive, Belfast BT12 5NR
M1 RDX A. Casson 5 Hazeltree Road, Ulverston LA12 9JP
M1 REC G. Doughty 1A The Crescent, Ketton, Stamford PE9 3SY
M1 REJ R. Jacklin 63 Ventnor Rise, Heathfield, Nottingham NG5 1NW
M1 REK R. King 8 Rydal Court, Congleton CW12 4JL
MW1 RES F. Garrett 11 Third Avenue, Flint CH6 5LT
MI1 RGL R. Leigh 42 Comber Road Killinchy, Newtownards BT23 6PB
M1 RGW R. Warren 23 Bramshaw Close, Winchester SO22 6LT
M1 RIC R. Booth 108 Newlands Gardens, Workington CA14 3PE
M1 RIG D. Gleadell 10 Swires Terrace, Halifax HX1 2EP
MM1 RIK R. Irvine 83 Glenacre Road, Cumbernauld, Glasgow G67 2NT
M1 RJG R. Gooderham 3 School Meadow, Barnby, Beccles NR34 7QL
M1 RJJ J. Fowler 48 Grangefields Road, Jacob'S Well, Guildford GU4 7NP
M1 RJL Rvd. R. Lapwood Rose Cottage, Challow Road, Wantage OX12 9DN
M1 RJS R. Speak Flat 1, Queens Court, East Road, Colchester CO5 8EB
M1 RKB R. Browning 575 Rayleigh Road, Leigh-on-Sea SS9 5HR
M1 RKY M. Harriott 20 Old Road, Tean, Stoke-on-Trent ST10 4EG
M1 RMO R. Oakley 41 Cyrano Way, Grimsby DN37 9SQ
MM1 RMS R. Scott Morven, Isle of Lewis HS2 0QX
M1 RMW R. Warry 51 South Street, Crewkerne TA18 8DB
M1 ROD R. Jones 7 Warner Avenue, North Cheam, Sutton SM3 9RH
M1 ROE A. Roebuck Holford Farm, Chester Road, Knutsford WA16 0TZ
MD1 RPC R. Woolley Moaralyn, King Williams Road, Castletown IM9 1BL Isle of Man
M1 RSB R. Bain Oak Tree Cottage, Long Barn Road, Sevenoaks TN14 6NH
M1 RST R. Dixon 20B Ratcliffe Road, Hedge End, Southampton SO30 4HA
M1 RTT R. Thong 10 Lowry Close, Haverhill CB9 7GH
M1 RWB R. Blears The Pump House, Warrington Road, Chester CH2 4DQ
M1 SAB S. Armatage Hayleazes, Lincoln Hill, Hexham NE46 4BE
M1 SAC S. Clark Flat 11, Saxon House Aylward Drive, Stevenage SG2 8UY
M1 SAM S. Evans 44 Edwards Drive, Plymouth PL7 2SU
M1 SAN R. Sanders Magnolia Cottage, Lanreath, Looe PL13 2NX
MW1 SAS A. Jones 4 Crowhill, Haverfordwest SA61 2HL
M1 SAZ S. Haynes 4 Monkend Terrace, Croft On Tees, Darlington DL2 2SQ
M1 SCS S. Smith 113 Deaconsfield Road, Hemel Hempstead HP3 9JA
M1 SCW S. Whiteman 109 Gordon Road, West Bridgford, Nottingham NG2 5LX
M1 SDE A. Raine Stable Cottage, St. Martins, Richmond DL10 4SJ
M1 SEM S. Rear 18 Brook Lane Cottages Sellindge, Ashford TN25 6HG
M1 SFS Summerfields ARC c/o Rvd. R. Lapwood Rose Cottage, Challow Road, Wantage OX12 9DN
M1 SGA S. Ash 1 Twyford Mill, Pig Lane, Bishops Stortford CM22 7PA
M1 SHA S. Creber 4 Joyce Close, Swindon SN25 4GX
M1 SHE J. Little 7 Deerfern Close, Great Linford, Milton Keynes MK14 5BZ
M1 SIM C. Simcock 51 Broadway, Stockport SK2 5SF
M1 SIN T. Brundrett 45 Talbot Crescent, Whitchurch SY13 1PH
M1 SJA A. Stockwell 28 Scholars Walk, Chalfont St. Peter, Gerrards Cross SL9 0EJ
M1 SJH S. Harrison 64 Douglas Road, Leigh WN7 5HG
M1 SKA D. Spencer 4 Douglas Road, Portsmouth PO3 6AU.
M1 SKI A. Grabianski 29 Lismore Road, Highworth, Swindon SN6 7HU
M1 SKY A. Johnston 3 Troutbeck Gardens, Barrow-in-Furness LA14 4LR
M1 SLH K. Taylor 23 The Chestnuts, Abingdon OX14 3YN
M1 SMF S. Flanagan 33 Ullswater Road, Chorley PR7 2JB
M1 SNM M. Wilson 3 Brookhurst Close, Chelmsford CM2 6DX
M1 SPW N. Walker 79 Arklow Drive, Hale Village, Liverpool L24 5RR
M1 SPY S. Pybus Grewgrass Farm, Grewgrass Lane, Redcar TS118EB
M1 SRC Surrey Raynet c/o T. Dabbs 4 Caverleigh, Cadogan Road, Surbiton KT6 4DH
M1 SRH S. Howlett 20 Long Perry, Capel St. Mary, Ipswich IP9 2XD
M1 SRP S. Pickin 138 Boreham Field, Warminster BA12 9EF
M1 SSB S. Bygrave 10 Spinney Close, North Cove, Beccles NR34 7PT
M1 STI B. Hall 5 Perche Court, Midhurst GU29 9TE
MM1 STK T. Storkey 1 Hartington Gardens, Edinburgh EH10 4LD
M1 SUE S. Gunn 5 School Villas, Broxted, Dunmow CM6 2BS
M1 SUM D. Sumner 30E Malvern Avenue, Ellesmere Port CH65 5AD
M1 SWB S. Bainbridge 6 Sandyville Grove, Liverpool L4 8UL
M1 SWL International Shortwave League c/o A. Kinson 6 Uplands Park, Broad Oak, Heathfield TN21 8SJ
M1 SWR S. Rigby 2 Tong Forge Lizard Lane Shifnal Shropshire Tf11 8Qd, Shifnal TF11 8QD
M1 SWS S. Southworth 157 Birkwood Avenue, Cudworth, Barnsley S72 8JB
MM1 SYD S. Mccance 34 Calside, Paisley PA2 6DB
M1 TAD T. Denby 22 Okehampton Crescent, Welling DA16 1DE
MW1 TAF M. Williams 1 Gomer Court, Abergele LL22 7UU
M1 TAP A. Prince 1 Woodside, Inglesbatch, Bath BA2 9DZ
M1 TAT J. Pymm Larkfield, Goxhill Road, Barrow-upon-Humber DN19 7EE
M1 TAZ C. Curtis 69 Lerwick Croft, Bicester OX26 4XX
M1 TCI T. Cheeseman 1 Queens Avenue, Birchington CT7 9QN
M1 TCP P. Braidwood 77 Pheasant Way, Cirencester GL7 1BJ
M1 TCQ J. Cheese 43 Moorside Road, Brookhouse, Lancaster LA2 9PJ
M1 TCR T. Rozier 124 Deansfield Road, Wolverhampton WV1 2LD
M1 TDD Lord T. Denham 20 Kirby Road, Basildon SS14 1RX
M1 TES J. Crawford 4 Church Road, Woodton NR35 2NB
M1 TET C. Causby 5 Blenheim Drive Higher Folds, Leigh WN7 2YR
M1 TMF A. Fullwood 16 Hollands Place, Walsall WS3 3AU
M1 TOD R. Todd 26 Dene Road, Guildford GU1 4DD
M1 TOM T. Boardman 9 Elm Grove, Farnworth, Bolton BL4 0AY
M1 TRC C. Marren 7 Mill Lane, East Ardsley, Wakefield WF32BL
M1 TSU I. James 21 Evelyn Way, Irchester, Wellingborough NN29 7AP
M1 TUG J. Wilson 36A Havelock Road, Maidenhead SL6 5BJ
M1 TVR M. Bryan 16 Walesmoor Avenue, Kiveton Park, Sheffield S26 5RG
M1 TXT P. Dimambro 26 Fetcham Court, Bank Top, Newcastle upon Tyne NE3 2UL
M1 UKC B. Welland 171 Hillcrest Road, Newhaven BN9 9EZ
M1 ULD G. Auld Victoria Villa, Sheepwash, Choppington NE62 5NG
MI1 UNA U. Murray 80 Canterbury Park, Londonderry BT47 6DU
MW1 USK Usk Side RC c/o D. Collins 34 Tone Road, Bettws, Newport NP20 7AW
MW1 VCD J. Roberts 13 Maes-Y-Coed, Gwersyllt, Wrexham LL11 4PF
M1 VGH A. Hutchinson 112 York Road, Haxby, York YO32 3EG
M1 VHF B. Burden 18 Challenor Close, Finchampstead, Wokingham RG40 4UJ
M1 VHT K. Morrison 31 Simonside Crescent, Hadston, Morpeth NE65 9YA
M1 VIP A. Evans Apartment 28, Stocks Court, 2 Harriet Street, Manchester M28 3JW
M1 VLS B. Wilson 131 Denmark Road, Beccles NR34 9DW
M1 VPL A. Westland 8 Faris Barn Drive, Woodham, Addlestone KT15 3DZ
M1 VPN D. Coombes The Old School, Combe Raleigh, Honiton EX14 4UL

M1 VRC R. Broadley The Smithy, Elstronwick, Hull HU12 9BP
M1 VSR C. Worthington 11 Manor Hill Road, Marple, Stockport SK6 6LP
MM1 VTB C. Budas 20 Oak Avenue, Bearsden, Glasgow G61 3HD
M1 WAW W. Waller 4 New Road, Sheerness ME12 1BW
M1 WAZ S. Eggleton 12 Cedar Grove, Trowbridge BA14 0HS
M1 WDK S. Papworth1 Spring Cottage, Gringley Road, Walkeringham, Doncaster DN10 4HT
M1 WDX F. Buck 89 Marsh St., Barrow in Furness LA14 2AD
M1 WEH M. Harrold 29 Barnford Crescent, Warley, Oldbury B68 8PP
MW1 WEJ W. Jones 53 Bro Enddwyn, Dyffryn Ardudwy LL44 2BG
M1 WHO S. Calver 4 George Place, St. Neots PE19 2QG
M1 WIN C. Higgins 52 Pittsfield, Cricklade, Swindon SN6 6AW
MM1 WKD A. Cartledge Ard Shonas, Rogart IV28 3XE
M1 WRX J. Mallows 57 Top Road, Kingsley, Frodsham WA6 8DA
M1 WVS E. Brown Rose Cottage, Grindlow, Buxton SK17 8RJ
MI1 WWG R. Mccann 96 Carncullagh Road, Stranocum, Ballymoney BT53 8PS
M1 WWW A. Varley Bank Farm, Matlock Road, Spitewinter, Chesterfield S45 0LL
MW1 WYN W. Britten-Jones 101 Mill View Estate, Maesteg CF34 0DE
M1 XCG C. Gaskell Folly Cottage, Watermillock, Penrith CA11 0LS
MM1 XJS K. Brown 21 Strain Crescent, Airdrie ML6 9ND
M1 XPS W. Chu 3 Windflower Close, Langley, Maidstone ME17 3YP
M1 XRC T. Crellin 2 Senlac Green, Uckfield TN22 1NN
M1 XXT C. Willetts The Retreat, Wood Lane, Colchester CO3 9TR
M1 XZG R. Mckenzie 48 Fuller Close, Swindon SN2 7TN
MM1 YAM C. Allanson Five Acres, Lairg IV27 4DG
M1 YOW D. Bland 16 Tennyson Avenue, Grays RM17 5RG
M1 ZAR J. Wilkinson 26 Hazelwood Grove, Sanderstead, South Croydon CR2 9DU
M1 ZEM J. Ogden 14 Bishops Close, Little Downham, Ely CB6 2TQ
M1 ZXG N. Moffat 22 Churchill Way, Acklington, Morpeth NE65 9DB
M1 ZXZ P. Clarke 31 Northern Rise, Great Sutton, Ellesmere Port CH66 4QY
M1 ZZA C. Thomas 55 High Street, Aylburton, Lydney GL15 6BZ
M1 ZZY T. Quinn 43 Stirtingale Road, Bath BA2 2NG

M3

MD3 AAI M. Rish 3 Lime St., Port St. Mary IM9 5ED
M3 AAQ L. Handley 11 Brook Close, Blythe Bridge, Stoke-on-Trent ST11 9PX
M3 AAS P. Rixon 52 New Road, Hatfield Peverel, Chelmsford CM3 2JA
M3 AAY N. Sanderson 54 Kelvedon Close, Chelmsford CM1 4DG
M3 ABQ J. Sharp Hunrosa, Crowlas, Penzance TR20 8DS
M3 ABY J. Turnbull 32 Haydon, Washington NE38 8PF
M3 ACA G. Reeds 26 Holme Leaze, Steeple Ashton, Trowbridge BA14 6EH
M3 ACF M. Buckley Springfield, 12 Ranmore Avenue, Croydon CR0 5QA
M3 ACY J. Bailey 13 Newark Road, Mexborough S64 9EZ
M3 ADB Dr A. Bottrill 4 St Luke'S Mews Gilesgate, Durham DH11JA
M3 ADJ B. Waterloo 55 Solent Road, Hill Head, Fareham PO14 3LB
M3 ADL P. Jefford 61 Willow Way, Flitwick, Bedford MK45 1LN
MM3 ADM A. Mair 8 Cockburn Crescent Whitecross, Linlithgow EH49 6JT
M3 ADT R. Vickerstaff 16 Sewell Wontner Close, Kesgrave, Ipswich IP5 2GB
M3 AEA P. Ewington 26 Dickens Road, Rugby CV22 5RW
M3 AEE A. Buckman 116 Ashling Park Road, Waterlooville PO7 6EG
M3 AEJ A. Pearce 8 Carworgie Way, St. Columb Road, St. Columb TR9 6PT
M3 AEZ S. Hall 122 Norwich Road, New Costessey, Norwich NR5 0EH
M3 AFF A. Barnes 3 Sparks Villas, Black Torrington, Beaworthy EX21 5PX
MW3 AFR K. Barry Flat 1, 24 Vale Street, Denbigh LL16 3BE
M3 AFS I. Dwyer 39 Berry Road, Newquay TR7 1AS
MW3 AFX L. Cook 8 Swn Yr Afon Flats, Cefn Coed, Merthyr Tydfil CF48 2SA
M3 AGA P. Bore 29 Edgerton Road, Lowestoft NR33 9BG
M3 AGB A. Bennett 16 Manor Avenue, Crewe CW2 8BD
M3 AGE D. Doroba Flat 3, 305A London Road South, Lowestoft NR33 0DX
M3 AGH M. Arnold 27 Poplar Avenue, Bentley, Walsall WS2 0NT
M3 AGI M. Taylor 29 Ferndale Avenue, Reading RG30 3NQ
MI3 AGR P. Mcdaid 66 Laurel Drive, Strabane BT82 9PN
M3 AHJ R. Selwood 33 Chandlers, Sherborne DT9 3RT
M3 AHL D. Furness 9 Ouzel Drive, Bradford BD6 3YN
M3 AHO M. Coulson 64 Craddock Street, Spennymoor DL16 7TA
M3 AHQ J. Maloney 196 Finchale Road, Hebburn NE31 2BW
M3 AHR T. Hamilton 16 Weardale Street, Spennymoor DL16 6ER
M3 AHS A. Stevenson 97 Queen Street, Crewe CW1 4AL
M3 AHU G. Taylor 63 Millbrook Towers, Stockport SK1 3NL
M3 AHZ A. Hardman 47 Oatlands Road, Manchester M22 1AH
M3 AIE C. Gibson 11 Parkside Avenue, Queensbury, Bradford BD13 2HQ
M3 AIG K. Rendell 20 Milford Park, Yeovil BA21 4QD
M3 AIL G. Langdon 43 Daniel St., Ryde PO33 2BH
MI3 AIN R. Martin 23 Scaddy Road, Downpatrick BT30 9BW
M3 AIR P. Robinson 16 Bartlett Close, Liverpool L31 8BZ
M3 AIS S. Martin 25 Wellswood Road, Ellesmere Port CH66 1JX
M3 AIZ J. Smith 44 Knapp Way, Malvern WR14 1SG
M3 AJA J. Crowhurst 5 Hampshire Road, Canterbury CT1 1SJ
M3 AJC A. Charbit 65 Bourne Street, London SW1W 8JW
M3 AJD I. Humberstone 20 Kingswood Road, Colchester CO4 5JX
MI3 AJK D. Poots 18 Upper Quilly Road, Dromore BT25 1NP
M3 AJN S. Panczel 11 Chauncy Road, Manchester M40 3GG
M3 AJU R. Smith 41 Middle Deal Road, Deal CT14 9RG
M3 AJV A. Brown 28 Parkland Drive Wingerworth, Chesterfield S42 6UU
M3 AJW A. Watmough 37 Heath Park Road, Buxton SK17 6NY
M3 AKE R. Hoey 225 King Avenue, Bootle L20 0BY
M3 AKH A. Hanson Pilgrim Cottage, South Road, Truro TR3 7AD
M3 AKQ R. Vickerstaff 16 Sewell Wontner Close, Kesgrave, Ipswich IP5 2GB
M3 AKR A. Beers The Beeches/New Bridge, Wisbech PE14 9DW
M3 AKW K. Pentney 17 Hilly Park, Norton Fitzwarren, Taunton TA2 6RH
M3 ALB A. Borda 4 Bow Arrow Lane, Dartford DA1 1YY
MM3 ALG J. Freeland 12 Mccathie Drive, Newtongrange, Dalkeith EH22 4BW
M3 ALX B. Harrison 15 Helmington Terrace, Hunwick, Crook DL15 0LQ
M3 ALZ S. Matley 67 Alexandra Road, Chandlers Ford SO53 2BP
M3 AMA A. Ayling 58 Lower Derby Road, Portsmouth PO2 8EX
M3 AMB S. Hegarty 6 Wymbush Crescent, Bristol BS13 0BB
M3 ANE R. Odle 24 Longfellow Road, Gillingham ME7 5QG
M3 ANH G. Gore-Thorne 19 Poplar Way, Ringwood BH24 1UY
M3 ANW A. Dennis Menlo Park, Salisbury Road, Marlborough SN8 3RP
M3 AOC D. Fawcett 6 Wand Hill, Boosbeck, Saltburn-by-the-Sea TS12 3AW
M3 AOM R. Adkins 91 Fernbank Road, Birmingham B8 3LL
M3 AOP B. Smith 45 Branson Avenue, Stoke-on-Trent ST3 5LA
M3 AOQ I. Sephton 131 Smeaton Road, Upton, Pontefract WF9 1LG
M3 APA A. Sims 127 Cooks Spinney, Harlow CM20 3BW
M3 APE D. Baines 157 Hall Green Road, West Bromwich B71 2DY
M3 APO J. Watson 38 Ryelands Road, Lancaster LA1 2QW
M3 APQ D. Hawken The Crest, Cuttinglye Road, Crawley Down RH10 4LR
M3 AQF J. Paul 67 Fleet Avenue, Upminster RM14 1PZ
M3 AQG P. Harrison 55 Hudson Close, Worcester WR2 4DP
M3 AQJ L. Jesson 17 Omaha Drive, Hinckley LE10 0WU
M3 AQK P. O'Shea 37 Barclay Court, Ilkeston DE7 9HJ
MM3 AQM W. Fullerton 28 Kerr Avenue, Saltcoats KA21 5PS
M3 AQN G. Macauley 1 Moricambe Crescent, Anthorn, Wigton CA7 5AS
M3 AQP J. Burns 34 Fleswick Avenue, Whitehaven CA28 9PB
MM3 AQW C. Maclean 16 Glamis Avenue, Elderslie, Johnstone PA5 9NR
M3 ARB A. Bronze 215 Ivyhouse Road, Dagenham RM9 5RS
MW3 ARM D. Rees Y Coed, Tan Lan Hill, Holywell CH8 9JB
M3 ARS A. Simpson 36 Little Sammons, Chilthorne Domer, Yeovil BA22 8RB
M3 ARU A. Smith 130 Watkin Street, Warrington WA2 7DN
M3 ASC D. Goldsbrough 45 Tithe Barn Road, Stockton-on-Tees TS19 8SZ
MW3 ASG J. James 77 Cypress Crescent St. Mellons, Cardiff CF3 2WL
MI3 ASH A. Nicholl 58 Dunnalong Road, Bready, Strabane BT82 0DW
M3 ASN D. Hartless 2 Brendon, Wilnecote, Tamworth B77 4JW
M3 ASZ B. Gilligan 4 Orion Close, Ward End, Birmingham B8 2AU
M3 ATB T. Brierley 6 Bridle Avenue, Wallasey CH44 7BJ
M3 ATC I. Kilkenny 23 Hazelhurst Road, Stalybridge SK15 1HD
MM3 ATI B. Rodger 95A Main Street, Coaltown, Glenrothes KY7 6HX
MI3 ATT D. Cooke 7 Killyclooney Road, Dunamanagh, Strabane BT82 0LZ
M3 AUB D. Albison 6 Rossendale Way, Shaw, Oldham OL2 7TX
M3 AUC D. Edney 49 Burns Road, Loughborough LE11 4ND
M3 AUF R. Cassell 10 Palmer Close, Branston, Burton-on-Trent DE14 3DY
M3 AUK M. Blake 125 Ludlow Road, Portsmouth PO6 4AF
M3 AUL K. Rickard 7 Thorngate Close, Penwortham, Preston PR1 0XN
M3 AUN L. Rowley 10 Derby Place, Newcastle ST5 3DX
M3 AUP S. Burns 22 Pendle Close, Peterlee SR8 2JS
MM3 AUX J. Milne 24 Lorne Street, Edinburgh EH6 8QP
M3 AVF J. Hancox 3 Hillfoot Road, Liverpool L25 7UJ
MI3 AVJ D. Johnson 6 Sand Pits, Fenaghy Road, Ballymena BT42 1JL
MW3 AVU R. Taylor 48 Clark Avenue, Pontnewydd, Cwmbran NP44 1RZ
M3 AVZ R. Henshall 14 Greenway, Congleton CW12 4PS
MM3 AWC Dr J. Harrington 1 Kynoch Terrace, Keith AB55 5FX
M3 AWD S. Mcleman 55 Anxey Way, Haddenham, Aylesbury HP17 8DJ
MW3 AWI D. Davies Penrallt, Abercaseg Road, Gerlan, Bangor LL57 3SP
M3 AWN B. Procter 28 Holme Grove, Burley In Wharfedale, Ilkley LS29 7QB
M3 AWQ M. Stables 58 Cedar Street, Derby DE22 1GE
M3 AWS A. Smith 87 Willian Way, Letchworth Garden City SG6 2HY
M3 AXB S. Balkham 49 St. Georges Road, Hastings TN34 3NH
M3 AXT S. Plant Columbell, 43 Fairfield Road, Bromsgrove B61 9JW
MW3 AXW D. Evans 7 Bryn Piod, Llanfachreth, Dolgellau LL40 2EE
M3 AXX D. Monnington 77 Stanley Gardens, Paignton TQ3 3NX
M3 AXZ K. Wood Willow Brook, Stapley, Taunton TA3 7QB
M3 AYC T. Symonds 68 Manor Crescent, Pan, Newport PO30 2BH
M3 AYJ A. Parkman 27 Arctic Road, Cowes PO31 7PE
M3 AYP A. Allen 43 Marriott Road, Dudley DY2 0JY
M3 AYQ D. Payne 31 Cockering Road, Canterbury CT1 3UP
MM3 AYS G. Mcgann Strathview, 43 Main Street, Fintry, Glasgow G63 0XE
M3 AYT W. Wilson 35 Darbishire Road, Fleetwood FY7 6QA
M3 AZE P. Harris 18 The Broadway, Wombourne, Wolverhampton WV5 0HY
M3 AZF D. Hastings 43 Delmar Avenue, Leverstock Green, Hemel Hempstead HP2 4LZ
M3 AZH P. Beresford 58 Plumptre Way, Eastwood, Nottingham NG16 3LR
M3 AZK A. Collier 65 Sandpits, Leominster HR6 8HT
M3 AZP R. Butland 4 Park Close, Sonning Common, Reading RG4 9RY
M3 AZR D. Jones 31 Summerhill Drive, Liverpool L31 3DN
M3 BAA M. Eggleton 78 Toronto Avenue, Blackpool FY2 0PD
M3 BAH B. Holmes 11 Deerness Road, Bishop Auckland DL14 6UB
M3 BAL A. Mccarthy Lockinuar House, Treswell Road, Retford DN22 0HU
M3 BAN J. Sergeant 15 Tennyson Place, Walton-Le-Dale, Preston PR5 4TT
M3 BAO L. Walker Little Chapple, Skilgate, Taunton TA4 2DP
M3 BAS B. Squance 4 Glenholt Road, Plymouth PL6 7JA
M3 BAT P. Batty 134 Plymouth Road, Scunthorpe DN17 1TS
M3 BAW B. Welthy 8 Du Cane Place, Witham CM8 2UQ
M3 BBA R. Banfield 2 Laleham Close, Eastbourne, East Sussex BN21 2LQ
M3 BBB M. Goodwin 23 Saxon Way, Ashby-de-la-Zouch LE65 2JR
M3 BBC J. Holme 17 Oxlea Grove, Westhoughton, Bolton BL5 2AF
M3 BBF J. Mackett 49 Tennyson Road, Cowes PO31 7PY
M3 BBL B. Blackham 5 Reedham Drive, Bramley, Rotherham S66 2SW
MW3 BBQ G. Richards 3 Pen Y Mynydd, Bettws, Bridgend CF32 8SE
M3 BBS R. Pilgrim 36 Wessex Gardens, Twyford, Reading RG10 0AY
M3 BBY R. Taylor 93 Blue Dolphin Park, Reculver Lane, Herne Bay CT6 6SS
MM3 BCA A. Macinnes 377 South Boisdale, Isle of South Uist HS8 5TE
MM3 BCC R. Sutherland Tigh - Na - Coille, Mill Road, Nairn IV12 5EW
M3 BCH B. Heirene 9 Ryecroft Crescent, Barnet EN5 3BP
MI3 BCM A. Birkhead 21 Carson Villas, Upperlands Maghera BT46 5SH
M3 BCQ M. Shepherd 47 Ripley Grove, Barnsley S75 2RX
MI3 BCR T. Washbourne 15 Apsley Street, Belfast BT7 1BL
M3 BCS B. Smith 98 Orange Hill Road, Burnt Oak, Edgware HA8 0TW
M3 BCW S. Peake 3 Marigold Walk, Bermuda Park, Nuneaton CV10 7SW
M3 BDA A. Yates Kingsomborne, Broadway, Totland Bay PO39 0BL
M3 BDC C. Ciotti 6 Bascott Road, Bournemouth BH11 8RH
M3 BDH T. Woods Lower Courtyard Rooms, Swan House Tarrington, Hereford HR1 4EU
M3 BDQ J. Harvey Flat 15, Gloucester House, The Walk., Felixstowe IP11 9DE
M3 BEE P. Sykes 2 Thornton Villas, Barrow Road, Barrow-upon-Humber DN19 7QG
MI3 BEG A. Nicholl 58 Dunnalong Road, Bready, Strabane BT82 0DW

M3 BEK R. Cowlishaw 23 Aldrich Drive, Willen, Milton Keynes MK15 9HP
M3 BER S. Berry 4 Newlands Park Way, Newick, Lewes BN8 4PG
M3 BET A. Waters 12 Anvil Court, Whittonstall, Consett DH8 9JU
M3 BFB M. Bellamy 23 Hazelwood, Benfleet SS7 4NW
M3 BFG R. Hogwood 22 Queen Elizabeth Drive Easington Lane, Houghton le Spring DH5 0NW
M3 BFJ E. Harvey 62 Archibald Road, Romford RM3 0RH
M3 BFU L. Buttriss 5 Church Close, Upper Sheringham, Sheringham NR26 8UB
M3 BFX M. Jones 138 Brompton Farm Road, Rochester ME2 3RE
M3 BFY D. Bowler 43 Stirtingale Road, Bath BA2 2NG
M3 BGE R. Allen 77 Highfield Road, Stroud GL5 1ES
M3 BGT J. Ahmed 59 Ramsgate, Lofthouse, Wakefield WF3 3PX
MM3 BHD R. Freeland 12 Mccathie Drive, Newtongrange, Dalkeith EH22 4BW
MM3 BHG E. Hay 11 Lovat Road, Glenrothes KY7 4RU
M3 BHI D. Wilkinson 139 Church Road, Jackfield, Telford TF8 7ND
M3 BHK S. Prinnett 29 Ford Park Road, Plymouth PL4 6RD
M3 BHP D. Rugen 19 Jacksons Close, Haskayne, Ormskirk L39 7LD
M3 BIB A. Mcgoff 55 Knights End Road, March PE15 9QA
M3 BIC T. Humphries 10 Cropthorne Avenue, Leicester LE5 4QL
MI3 BIE D. Hamilton 7 Bolea Park, Limavady BT49 0SH
M3 BIK D. Newell 24 Melchester Grove, Stoke-on-Trent ST3 7FW
M3 BIO A. Ness 15 Scotforth Road, Lancaster LA1 4TS
M3 BIR D. Cupit Hazelnut Cottage, Weston Road, Newark NG22 0HB
M3 BIZ M. Turner 2 Higher Farm Cottages, Swallowcliffe, Salisbury SP3 5PE
M3 BJB A. Berry 4 Newlands Park Way, Newick, Lewes BN8 4PG
M3 BJE R. Thorpe 129 Fairham Road, Stretton, Burton-on-Trent DE13 0BT
M3 BJH B. Hill 26 Providence Close, Leamore, Walsall WS3 2AL
M3 BJJ B. Jenkinson 7 Chestnut Avenue, Thorngumbald, Hull HU12 9LD
M3 BJL W. Lewis 3 St. Martins Road, Folkestone CT20 3LA
M3 BJV B. Adkins 4 Orion Close, Ward End, Birmingham B8 2AU
M3 BJW B. Moseley 232 West Bromwich Road, Walsall WS1 3HL
M3 BJZ D. Koch 83 Springfield Park, Maidenhead SL6 2YU
MI3 BKA J. Calvert 29 Recreation Road, Larne BT40 1EW
M3 BKI M. Davenport 44 Vicarage Road, Hastings TN34 3LY
M3 BKJ T. Davenport 44 Vicarage Road, Hastings TN34 3LY
M3 BKU P. Bridger 2 Marline Avenue, St. Leonards-on-Sea TN38 9HP
M3 BKV K. Blanch 2C Brunswick Terrace, North St., Sandown PO36 8BG
M3 BLF D. Saunders 42 Peascroft, Long Crendon, Aylesbury HP18 9AU
M3 BLG L. Grainger 81 Windmill Rise, Tadcaster LS24 9HR
MI3 BLN W. Coates 3 Thornleigh Park, Bangor BT20 4NN
M3 BLO M. Herdman 23 Marshall Avenue, Huncoat, Accrington BB5 6NB
M3 BLR D. Griffiths 48 Conygar View, Dunster, Minehead TA24 6PW
M3 BLX M. Mina 15 Manor Drive, Manchester M21 7QG
M3 BMH F. Patrovits 30 St. Ronans Drive, Seaton Sluice, Whitley Bay NE26 4JQ
M3 BMI D. Richmond 105 Lord Street, Crewe CW2 7DP
M3 BMN M. Smith 8 Devon Street, Leigh WN7 2NG
M3 BMQ J. Nicholls 55 Moat Avenue, Coventry CV3 6BT
M3 BMU R. Scott 157 Fairview Road, Stevenage SG1 2NE
M3 BMV G. Brown 3 Willow Lane, Goostrey, Crewe CW4 8PP
M3 BMW B. Ashcroft 16 Edge Lane, Thornton, Crosby., Liverpool L23 9XE
M3 BNU M. Hall 20 Cubitt House, Black Bull Road, Folkestone CT19 5SH
M3 BOB F. Kennedy 19 High Street, East Hoathly, Lewes BN8 6DR
MW3 BOC F. Hart 60 Heol Bryncwils, Sarn, Bridgend CF32 9UE
MM3 BOJ D. Boden 42 Kirkwynd, Maybole KA19 7AE
M3 BON N. Peters Cemetery Lodge, 12 Mount Pleasant, Crewkerne TA18 7AH
MW3 BOO O. Williams 39 Camden Road, Maes-Y-Coed, Brecon LD3 7RT
MW3 BOP K. Stimson 10 Heol Twyn Du, Merthyr Tydfil CF48 1LU
M3 BOQ M. Pettifer 28 Hurst Close, Staplehurst, Tonbridge TN12 0BX
M3 BOR R. Parfitt 7 Water Lane Close, Barnstaple EX32 9JX
M3 BOU M. Mullis 18 Springfield Grove, Southam CV47 0ES
M3 BOV D. Waters 17 Lower Herne Road, Herne Bay CT6 7NA
M3 BPF L. Gaspar 18 North Hill Close, Burton Bradstock, Bridport DT6 4RY
M3 BPG P. Robertson 64 Castle St., Frome BA11 3DY
M3 BPL A. Browell 67 Stadium Avenue, Blackpool FY4 3QA
M3 BPN M. Newell 55 Station Road Brimington, Chesterfield S43 1JU
MM3 BPR S. Mclaughlin 21 Shirrel Road, Motherwell ML1 4RD
MI3 BPS G. Brennan 15 Kinnegar Rocks, Donaghadee BT21 0EZ
M3 BPY P. Hollis 5 Salisbury Road, New Malden KT3 3HZ
MW3 BPZ N. Heyne New House, Hope, Welshpool SY21 8JD
MM3 BQK M. Cleland 85 Carfin St., Motherwell ML1 4JL
M3 BQN R. Coleman 77 Millstrood Road, Whitstable CT5 1QF
M3 BQT J. Rabbitt 66 Parkfield Avenue, Delapre, Northampton NN4 8QB
MI3 BRJ S. Molloy 6 Glenloch Park, Coleraine BT52 1TY
M3 BRL P. Long 17 Wellesley Way, Newport PO30 2GA
M3 BRQ T. Leaver 132 Old London Road, Hastings TN35 5LZ
MM3 BRR R. Hall 13 Cleat, Castlebay, Isle of Barra HS9 5XX
M3 BRT B. Ingham 19 Recreation Avenue, Ashton-In-Makerfield, Wigan WN4 8SU
M3 BRU M. Brunsdon 7 Oldberg Gardens, Brighton Hill, Basingstoke RG22 4NP
M3 BRV D. Prout 2 Pine Crest Way, Bream, Lydney GL15 6HG
M3 BRW P. Buttery 103 Elsie Street, Goole DN14 6DY
MI3 BRX H. Mairs 9 Eureka Drive, Belfast BT12 5NR
MM3 BSC B. Clark 6 The Links, Cumbernauld, Glasgow G68 0EP
M3 BSF G. Mate 200 Chesterholm, Carlisle CA2 7XY
M3 BSH B. Hunt 15 High Street North, West Mersea, Colchester CO5 8JU
M3 BSI A. Storer 16 Eastfields, Braunston, Daventry NN11 7JN
MW3 BSJ B. Jones 4 Old Tanymanod, Blaenau Ffestiniog LL41 4BU
M3 BSM L. Cookman The Flat Above Cobblers Corner, Ewhurst Road, Cranleigh GU6 7AA
MI3 BSN J. Murphy 19 Kilburn Park, Armagh BT61 9HA
M3 BTG G. Simpson 17 Temple Crescent, Leeds LS11 8BG
M3 BTI J. Roddam Birches Farm, Button Street, Preston PR3 2LH
M3 BTJ J. Meredith 42 Hollins Crescent, Talke, Stoke-on-Trent ST7 1JY
MW3 BTN M. Luxton 3 The Paddocks, Newgate Street, Brecon LD3 8DJ
M3 BTZ P. Bull 8 Mayfield Lane, Martlesham Heath, Ipswich IP5 3TZ
M3 BUA J. Bull 8 Mayfield Lane, Martlesham Heath, Ipswich IP5 3TZ
M3 BUH A. Brownley 5 Skipton Road, Sheffield S4 7DD
MI3 BUT K. Martin 19 North St., Ballycastle BT54 6BW
M3 BUU J. Olive 2 Wyke Cottage Wotton Road, Rangeworthy, Bristol BS37 7NA
MM3 BUZ B. Cameron 6/4 Parkgrove Green, Edinburgh EH4 7RQ
M3 BVA P. Booth 19 Gunville Crescent, Bournemouth BH9 3PZ
M3 BVK E. Shirley Ham House, Ham Lane, Shepton Mallet BA4 5JW
M3 BVL G. Sowden The Grange Lodge, Rodley Lane, Calverley, Pudsey LS28 5QH
M3 BVM C. Timm 14 Little Copse Chase, Chineham, Basingstoke RG24 8GL
M3 BVP R. Hirst 105 Haregate Road, Leek ST13 6PX
M3 BVQ A. Waller 2 Barkis Mead, Owlsmoor, Sandhurst GU47 0GT
M3 BVX J. Pickard 39 Eafield Avenue, Milnrow, Rochdale OL16 3UN
M3 BVY A. Spinks 10 Foxley Close, Norwich NR58DQ
M3 BWF P. Cook 5 Home Farm, Highworth, Swindon SN6 7EG
M3 BWT B. Williams Dunsmoir Cottage, Chapel Hill Road Wreay, Carlisle CA4 0RP
MM3 BWV A. Smith 4 The Terrace, Lhanbryde, Elgin IV30 8NY
M3 BWZ K. Scully 2 St. Michaels & All Angels Church, Canada Road, Deal CT14 7BL
M3 BXC W. Bray 46 Alexandra Road, Lostock BL6 4BB
M3 BXE D. Morris 17 Mallory Way, Daventry NN11 0UN
M3 BXG H. Chick 15 Bonfire Close, Chard TA20 2EG
M3 BXH C. Wheeler 17 Orchard Way, Timberscombe, Minehead TA24 7UL
M3 BXN B. Holland 23 White Avenue, Langold, Worksop S81 9PT
MI3 BXQ B. Hannigan 4 Silverhill Road, Strabane BT82 0AE
M3 BXS R. Wake 55 Bearsdown Road, Eggbuckland, Plymouth PL6 5TR
M3 BXW B. Akiens 444 Groby Road, Leicester LE3 9QB
M3 BXX J. Smith 8 Albert Gardens, Halifax HX2 0HT
M3 BXY K. Walker 77 Blackwood Grove, Halifax HX1 4QG
M3 BYA D. Passey 5 The Croftings, Felton Close, Ludlow SY8 1DS
M3 BYF J. Milne 9 Roman Road, Colchester CO1 1UR
MI3 BYJ D. Bell 1 Knockbracken Drive, Coleraine BT52 1WN
M3 BYL B. Lewis 20 Annes Walk, Caterham CR3 5EL
MI3 BYQ J. Martin 19 North St., Ballycastle BT54 6BW
M3 BYS T. Scott 157 Fairview Road, Stevenage SG1 2NE
M3 BYX N. Clayton 280A Loughborough Road, Leicester LE4 5LH
MD3 BZA A. Radcliffe Cronk-Vue, Ballayockey, Andreas IM7 3HP
M3 BZC S. Galloway 1 Mount Pleasant, Leeds LS10 3TB
M3 BZQ R. Bostock 5 Hethersett Way, New Rossington, Doncaster DN11 0RZ
M3 CAA T. Groves 43 Grasmere Road Kennington, Ashford TN249BQ
MI3 CAB P. Gibson 118 Coleraine Road, Portstewart BT55 7HS
M3 CAD D. Coubrough 16 Celandine Close, Billericay CM12 0SU
M3 CAE J. Chantler 1 Glencoe Road, Margate CT9 2SL
M3 CAP P. Barusevicus 6 Middlebrook Crescent, Bradford BD8 0EN
M3 CAQ C. Hosegood 4 The Orchard, Sixpenny Handley, Salisbury SP5 5QL
M3 CAW S. Court 16 Worcester Road, Woodthorpe, Nottingham NG5 4HY
M3 CAZ C. Mitchell-Watson 144 Shakespeare Crescent, Dronfield S18 1ND
M3 CBC P. Eden 22 Greenside, Stoke Prior, Bromsgrove B60 4EB
M3 CBH B. Kenny 37 Coningswath Road, Carlton, Nottingham NG4 3SF
MI3 CBJ N. Mckittrick The Coach House, 74 Lyle Road, Bangor BT20 5LT
MI3 CBL S. Wylie 38 Elmfield Park, Donaghadee BT21 0AX
M3 CBN K. Hughes High Lane Cottage, Congleton Road, Macclesfield SK11 9RR
MM3 CBO J. Mccash 11 Tintagel Gardens, Chryston, Glasgow G69 0PH
MW3 CBS K. Hulme 13 Lime Street, Gorseinon, Swansea SA4 4AD
M3 CBV B. Robertshaw 14 Lawrence Avenue, Mansfield Woodhouse, Mansfield NG19 8DJ
M3 CBW A. Edwards 6 St. Pauls Road, Nuneaton CV10 8HL
MW3 CBX K. Roberts 8 Bronant, Lixwm, Holywell CH8 8NG
M3 CBY D. Smith 58 Marriott Road, Leicester LE2 6NT
MI3 CCA G. Wright 38 Fern Grove, Bangor BT19 1FG
M3 CCB A. Gunn 6 St. Pauls Road, Nuneaton CV10 8HL
MW3 CCE C. Evans 25 Beech Drive, Hengoed CF82 7JP
M3 CCF T. Toon 9 Boundstone Lane, Sompting, Lancing BN15 9QL
M3 CCJ S. Grainger 62 Meadowleaze, Longlevens, Gloucester GL2 0PS
MI3 CCN R. Reilly 220 Ardmore Road, Londonderry BT47 3TE
M3 CCP P. Smith 56 Darnhall Crescent, Nottingham NG8 4PZ
M3 CCS A. Sutton 11 Arnhill Road, Gretton, Corby NN17 3DN
MI3 CCT T. Vaughan 20 Killen Park, Killen, Castlederg BT81 7TJ
M3 CCY J. White 18 Sawyers Road, Tolleshunt Major, Maldon CM9 8NE
MI3 CDA C. Adjey 1 Foyle Park, Portstewart BT55 7DL
M3 CDE S. Martins 46 Ruskin Road, Mansfield NG19 7LX
M3 CDI C. Pearson Greystones Farm Cottage, Richmond DL11 7AJ
MW3 CDL J. Roberts 1 Seymour Drive, Rhuddlan, Rhyl LL18 5PP
MJ3 CDP C. Paland 19 Maison St. Louis, St. Saviour JE2 7LX Jersey
M3 CDV A. Willoughby 25 Maple Close, Louth LN11 0DW
M3 CDY J. Bennett 39 West View, Parbold, Wigan WN8 7NT
M3 CEB B. Cooke 2 Harvey Place, Andover SP10 2BU
MI3 CEM C. Murray 80 Canterbury Park, Londonderry BT47 6DU
M3 CEN C. Nowell 39 Brockfield Park Drive, York YO31 9EL
M3 CER M. Clews 16 Chestnut Street, Worcester WR1 1PA
M3 CET M. Tibbits 8 Holly Road, Northampton NN1 4QR
M3 CEZ B. Stratford 5 The Sycamores, Peacehaven BN10 8AB
M3 CFI C. Isherwood 32 Franklin Close, Old Hall, Warrington WA5 8QL
M3 CFJ S. Taylor 12 Minehead Avenue, Burnley BB10 2NP
M3 CFM K. Sawyers 27 Dukeswood Road, Longtown, Carlisle CA6 5UJ
M3 CFP C. Penfold 149 Shuttlewood Road Bolsover, Chesterfield S44 6NX
M3 CFU P. Flook 17 Valentine Close, Bristol BS14 9ND
MI3 CGA R. Doherty 48 Drumard Park, Londonderry BT48 0RL
M3 CGC J. Stanway Flat 17, Charlton Court London Road, Gloucester GL1 3QH
M3 CGH C. Horne 56 Pilkington Road Braunstone, Leicester LE3 1RA
M3 CGI D. Hunter 39 Nicholas Avenue, Rudheath, Northwich CW9 7LD
M3 CGJ J. Nenova 18 Longtown Road, Romford RM3 7QL
M3 CGM S. Allen 9 Tiled House Lane, Brierley Hill DY5 4LG
M3 CGO C. Oliver 25 Mary Peters Drive, Greenford UB6 0SS
M3 CGP B. Townsend 21 Royds Drive, New Mill, Holmfirth HD9 1LH
M3 CGS C. Shirley Ham Houes, Ham Lane, Shepton Mallet BA4 5JW
MI3 CGT C. Brown 33 Broadlands Drive, Carrickfergus BT38 7DJ
MI3 CGU A. Brown 33 Broadlands Drive, Carrickfergus BT38 7DJ
MI3 CGZ S. Buchanan 162 Victoria Road, Bready, Strabane BT82 0DZ
M3 CHE D. Letton 21 Westfield, Bradninch, Exeter EX5 4QU
M3 CHU Dr J. Waterhouse 31 Donnington Place, Wantage OX12 9YE
MW3 CHZ M. Bowen 7 Ael Y Bryn, Beddau, Pontypridd CF38 2AL
MI3 CID C. Morrow 46 Tullyard Way, Belfast BT6 9NU
M3 CIE C. Barker 14 Hall Road, Wilmslow SK9 5BN

M3 CIG J. Smith 9 Trafalgar Road, Newport PO30 1QD
M3 CIJ C. Martin 92 Thrupp Lane, Thrupp, Stroud GL5 2DG
M3 CIO T. Brookes 370 Broxtowe Lane, Nottingham NG8 5ND
M3 CIP C. Powis 28 Kington Gardens, Birmingham B37 5HS
M3 CIS C. Smith 6 Birtley Rise, Bramley, Guildford GU5 0HZ
MI3 CIV M. Semple 58 Green Drive, Larne BT40 2ER
MI3 CIW D. Ritchie 58 Green Drive, Larne BT40 2ER
MI3 CIZ C. Mccord 23 Blackthorn Green, Larne BT40 2JE
M3 CJA C. Alexander 25 Diamedes Avenue, Stanwell, Staines-upon-Thames TW19 7JE
M3 CJE R. Wilberforce 106 Marlborough Road, Slough SL3 7JY
M3 CJH C. Houghton 23 Carpenters Way, Badshot Lea, Farnham GU9 9FT
M3 CJI D. Sharp 11 Dovedale Gardens, Leeds LS15 8UP
MM3 CJP C. Paton 4 Abbeyhill, Dhailling Road, Dunoon PA23 8FG
M3 CJX C. Cross 131 Arnold Lane, Gedling, Nottingham NG4 4HF
M3 CKD P. Loomes 107 Main Street, Sedgeberrow, Evesham WR11 7UE
MI3 CKF D. Hamilton 1 Meadow Bank, Ballysally Road, Coleraine BT52 2QA
M3 CKH C. Hammett 63 Treffry Road, Truro TR1 1WL
M3 CKM S. Plumb 94 Wellington Street, New Whittington, Chesterfield S43 2BG
M3 CKO C. Plumb 94 Wellington Street, New Whittington, Chesterfield S43 2BG
MM3 CKP S. Aron 2/3, 14 Woodend Road, Glasgow G73 4DX
M3 CKU G. Moorhouse 29 Dee Road, Walsall WS3 1NW
MM3 CLA C. Henderson 22 Bowmont Place, East Kilbride, Glasgow G75 8YG
M3 CLO K. Williams 36 Castle St., Tiverton EX16 6RG
M3 CLP C. Palmer 21 Ibbett Close, Kempston, Bedford MK43 9BT
M3 CLT C. Turner 1B Amberbanks Grove, Blackpool FY1 6DW
MJ3 CMB C. Boudier Flat 141 La Tour Heron, Le Clos De Samares, St. Clement, Jersey JE2 6GG Jersey
MW3 CMG C. Griffiths 43 Cysgod Y Graig, Denbigh LL16 3TD
M3 CMI H. Dickinson 1 Larch Close, Kirkheaton, Huddersfield HD5 0NJ
M3 CMK A. Saville 4 Shannon Court, Downs Barn, Milton Keynes MK14 7PP
M3 CMM B. Perryman 55 Alderfield, Penwortham, Preston PR1 9HD
M3 CMP C. Pidd 36 Hunster Close, Doncaster DN4 6RE
M3 CMW C. Willimot 5 Green Lane, Upton, Huntingdon PE28 5YE
M3 CMX J. Roberts 52 School Lane, Toft, Cambridge CB23 2RE
M3 CNC A. Wood 29 Hornbeam Drive, Wingerworth, Chesterfield S42 6FY
M3 CND S. Bradley 6 Downing Street, South Normanton, Alfreton DE55 2HE
MW3 CNL D. Emanuel 27 Gwysfa Road, Ynystawe, Swansea SA6 5AE
MM3 CNS G. Kennedy 1 Martins Buildings, High Street, Perth PH2 7QP
M3 CNV C. Parry 43 Castle Mount, Dewsbury WF12 0DW
M3 COB J. Hodkinson 3 Cypress Close, Market Drayton TF9 3HJ
M3 COE R. Alford 225 N Main Box 174, Gas 66742 United States
MI3 CON A. Mclernon 520 Carneety Terrace, Castlerock, Coleraine BT51 4SZ
M3 CPH C. Harrap Waverley Court, 22 Forth Avenue, Portishead, Bristol BS20 7NY
MD3 CPK C. Kelly 16 Viking Road, Douglas IM2 6PB Isle of Man
M3 CPN D. Dunford 151 Doncaster Road, Rotherham S65 2BY
M3 CPX N. Thorne Barford Stream, Churt Road, Farnham GU10 2QU
M3 CPY L. Goodby 21 Kiniths Way, Hurst Green, Halesowen B62 9HJ
MI3 CQB T. Elliott 183 Kilraughts Road, Ballymoney BT53 8NL
MI3 CQO R. Brennan 15 Kinnegar Rocks, Donaghadee BT21 0EZ
M3 CQP J. Gilbert Sweden End, Ambleside LA22 9EX
MI3 CQR R. Hendy 10 Captains Road Forkhill, Newry BT35 9RR
M3 CQT A. Loasby 89 Jubilee Crescent, Wellingborough NN8 2PQ
M3 CQW D. Wooding 3 Chichele Court, North St., Rushden NN10 6BU
MI3 CQX M. Finnegan 18 Springfarm Heights, Newry BT35 8XA
M3 CRD C. Matthews 18 Tennyson Gardens, Darlington DL1 5BJ
M3 CRL C. Finnis 44 Disraeli Road, Christchurch BH23 3NB
M3 CRV D. Priestner 4 Oak Street, Northwich CW9 5LJ
M3 CSF C. Finnis 44 Disraeli Road, Christchurch BH23 3NB
M3 CSI C. Massimo 2 Beattie Close, Great Bookham, Leatherhead KT23 3JF
M3 CSK C. Newton 7 Moss Close, Bridgwater TA6 4NA
M3 CSM M. Lawson 233 Southwell Road West, Mansfield NG18 4HF
M3 CSN J. Prichard 1 Polton Dale, Swindon SN3 5BN
M3 CSR D. Hunter 9 Sleigh Road, Sturry, Canterbury CT2 0HR
MI3 CSS R. Ennis 91 Main Street Carrowdore, Newtownards BT22 2HW
MI3 CST D. Browne 26 Brooklands Gardens, Dundonald, Belfast BT16 2PQ
M3 CSV P. Southern 32 Mayville Avenue, Scarborough YO12 7NP
MM3 CSX A. Thomson 5 Gib Grove, Dunfermline KY11 8DH
M3 CSZ D. Croft 33 Roughaw Road, Skipton BD23 2PY
M3 CTB M. Burnett 16 Church Lane, Reepham, Lincoln LN3 4DQ
M3 CTN M. Mcewen 31 Holmes Carr Road, New Rossington, Doncaster DN11 0QF
M3 CTO A. Kvilums 78 Wagon Lane, Solihull B92 7PN
M3 CTT M. Holroyd 9 Coniston Green, Aylesbury HP20 2AJ
M3 CUH S. Twynam 129 Poplar Drive, Herne Bay CT6 7QA
M3 CUU D. Taylor 10 Church Rd, , Northwich CW9 5NT
MM3 CVB C. Budas 20 Oak Avenue, Bearsden, Glasgow G61 3HD
M3 CVD C. Dewberry 2 Cleatham Villas, Cleatham, Gainsborough DN21 3HY
M3 CVH C. Holley 28 White Horse, Uffington, Faringdon SN7 7SE
M3 CVL C. Egan 20 Woodhatch Road, Brookvale, Runcorn WA7 6BJ
M3 CVM D. Healey 28 Witley Drive, Sale M33 5NQ
M3 CVO A. Beal 29 Bennetts Road North, Keresley End, Coventry CV7 8JX
M3 CVW R. Twose 10 Galingale Close, Bicester OX26 3FD
M3 CWA T. Watkins 6 Linnet Close, Waterlooville PO8 9UY
M3 CWC J. Harrison 12 Hillside Crescent, Skipton BD23 2LE
M3 CWH C. Harlow 12 Penhurst Court, Grove Road, Worthing BN14 9DG
MM3 CWO J. Mills 65 Strathkinnes Road, Kirkcaldy KY2 5PX
M3 CWV C. Harding 9 Westbourne Road, Middlesbrough TS5 5BN
M3 CWZ A. Robinson 30 Cope Street, Walsall WS3 2AT
MI3 CXD W. Duffy 4 Deramore Drive, Strathfoyle, Londonderry BT47 6XL
MI3 CXM P. Boyd 784 Shore Road, Newtownabbey BT36 7DG
M3 CXX A. Boyle 15 Slatter, Satchell Mead, London NW9 5UQ
M3 CYJ G. Jenkinson 4 Brundish House, Braithwell Road, Rotherham S66 8JT
MW3 CYM D. Taylor 7 Trellewelyn Close, Rhyl LL18 4NF
MW3 CYQ D. Hughes-Burton 6 Troed Y Garn, Llangybi, Pwllheli LL53 6DQ
M3 CYS Eagle Radio Group c/o K. Limbert 7 Acacia Avenue, Liverpool L36 5TL
MW3 CYU S. Hughes-Burton 6 Troed Y Garn, Llangybi, , Pwllheli LL53 6DQ
M3 CYW N. Duke 5 Hannay Close, Barrow-in-Furness LA14 1SZ
M3 CZB M. Hillary 145 Elizabeth Road, Waterlooville PO7 7NN
M3 CZE M. Gray 14 Florence Crescent, Gedling, Nottingham NG4 2QJ
M3 CZJ G. Whitear 19 Camelia Close, Littlehampton BN17 6UT
M3 CZL R. Morley 6 Ford Drive, Yarnfield, Stone ST15 0RP
M3 CZM J. Stocks 10 Hollycroft Road, Emneth, Wisbech PE14 8AY
M3 CZW C. Walton Old Drive, Sulby, Northampton NN6 6EZ
M3 CZX P. Godfrey Bramley End, Old Lane, Chester CH3 6QX
M3 CZY D. Mageehan 37 Gosbecks Road, Colchester CO2 9JR
M3 DAB D. Bambrook 18 Vervain Close, Bicester OX26 3SR
M3 DAE D. Edge 20 Parkers Court, Hallwood Park, Runcorn WA7 2FP
M3 DAF D. Fearn 15 Broome Acre, Broadmeadows, Alfreton DE55 3AW
M3 DAM A. Petts 19 Sandwell Avenue, Darlaston, Wednesbury WS10 7RH
M3 DAV R. Halsall 50 Leominster Drive, Manchester M22 5DH
M3 DBA K. Miller 8 Cooks Close, Ashburton, Newton Abbot TQ13 7AN
MI3 DBB D. Brown 17 Parkmore Drive, Strathfoyle, Londonderry BT47 6XA
MW3 DBF J. Mcconnell Ty Cerrig, Tremeirchion, St. Asaph LL17 0UP
M3 DBG B. Miller The Piglet, Saddleback Barn, Staverton TQ9 6AN
M3 DBS D. Baines 21 Vera Road, Norwich NR6 5HU
M3 DBU D. Burrows Pen Parc, 9 Clough Hall Road, Stoke-on-Trent ST7 1AR
M3 DBX D. Billinge 29 Stanley Avenue, Wallasey CH45 8JN
M3 DBY D. Pratt 4 Bussex Square, Westonzoyland, Bridgwater TA7 0HD
M3 DBZ S. Issatt 7 Birch Road, Doncaster DN4 6PD
M3 DCI T. Billington 31 June Avenue, Blackpool FY4 4LQ
M3 DCJ J. Whiffin 335 High Street, Eastleigh SO50 5NE
M3 DCL M. Phillips Orchards, Brains Green, Blakeney GL15 4AJ
MI3 DCM D. Christie 1 Marino Park, Ballymoney BT53 7BB
MM3 DCN A. Mclellan 57 Hunter Street, Kirn, Dunoon PA23 8JR
M3 DCP D. Palmer Edison House, Bow Street, Great Ellingham, Attleborough NR17 1JB
M3 DCS R. Cowles Bonnie Rock, 76 Fordham Road, Ely CB7 5AL
MI3 DDK E. Kashkoush 41 Dunamallaght Road, Ballycastle BT54 6PF
MM3 DDQ P. Lucas 69A Broomhill Crescent, Alexandria G83 9QT
MM3 DDS D. Scott Farewell, Arnage, Auchnagatt, Ellon AB41 8UW
M3 DDY M. Rote 71B Headland Crescent, Exeter EX1 3NP
M3 DDZ E. Perez-Mendez 56 Gaunt Close, Sheffield S14 1GD
M3 DEA D. Adams 9 Brancaster Avenue, Charlton, Andover SP10 4EN
M3 DEB D. Stanley Harby, Brightlingsea Road, Colchester CO7 8JH
MM3 DEC E. Mcneill 3 Sunnybrae Terrace, Maddiston, Falkirk FK2 0LP
M3 DEE D. Lewis 10 Addington Road, Bolton BL3 4QZ
M3 DEH M. Dollin 102 Rossall Road, Thornton-Cleveleys FY5 1HQ
MW3 DEI D. Jones Maes Y Llwyn, Tan Y Foel, Borth-Y-Gest, Porthmadog LL49 9UH
M3 DEJ D. Balls 7 Rowan Close, Holbeach, Spalding PE12 7BT
MW3 DEL A. Siddle-Ward 40 Wynn Avenue North, Old Colwyn, Colwyn Bay LL29 9RH
MW3 DEM M. Dancer 281 Fishguard Road, Llanishen, Cardiff CF14 5PW
M3 DFB A. Dunster 113 Canterbury Road, Folkestone CT19 5NR
M3 DFC D. Chambers 94 Hawthorn Avenue, Colchester CO4 3JR
MM3 DFG A. Graham 72 India St., Montrose DD10 8PW
M3 DFL D. Little 3 Swallow Dale, Thringstone, Coalville LE67 8LY
M3 DFM K. Davies 58 Popes Lane Sturry, Canterbury CT2 0LA
M3 DFP D. Wicks Friars Piece, Parham, Woodbridge IP13 9LY
MI3 DFR P. Clarke 26 Derryhale Lane, Portadown, Craigavon BT62 4HL
M3 DFS R. Mordaunt 330 Harborough Avenue, Sheffield S2 1UU
M3 DFU J. Mole 11 Branfill Road, Upminster RM14 2YX
M3 DFV G. Jelley 28 Blanches Road, Partridge Green, Horsham RH13 8HZ
M3 DFW D. Ferrow 1 Temple Avenue, Blyth NE24 5ET
MM3 DFZ A. Zimnowlocki 44 Dunbar Place, Kirkcaldy KY2 5SE
M3 DGD C. Densham 27 Lloyds Crescent, Exeter EX1 3JQ
M3 DGJ D. Gardiner 28 Winfield, Newent GL18 1QB
M3 DGN A. Lister 15 Elmwood Drive, Breadsall, Derby DE21 4GB
M3 DGR D. Riley 7 High St., Bolsover, Chesterfield S44 6HF
M3 DHA S. Webber 124A Exeter Road, Kingsteignton, Newton Abbot TQ12 3LY
MM3 DHE R. Murray 11 Castleview, Dundonald, Kilmarnock KA2 9HZ
MM3 DHG J. Blades 58 Hunter Road, Crosshouse, Kilmarnock KA2 0LD
MM3 DHN P. Traill 38 Burnside Road, Gorebridge EH23 4EU
MI3 DHR D. Richards 70 Cherryhill Avenue, Dundonald, Belfast BT16 1JD
M3 DHS D. Sherwin 5 North Road, Buxton SK17 7EA
M3 DHV S. Mccron 72 Yeoman Way, Trowbridge BA14 0QP
M3 DHW J. Mccron 72 Yeoman Way, Trowbridge BA14 0QP
M3 DIB K. Dibben 82 Lenthay Road, Sherborne DT9 6AF
M3 DIG K. Dignall 11 Mottershead Road, Widnes WA8 7LD
MW3 DIL D. Jenkins 10 Ty Fry Close, Brynmenyn, Bridgend CF32 8YB
M3 DIM H. Brace 56A Patching Hall Lane, Chelmsford CM1 4DA
M3 DIS R. Hensman 24 Belchmire Lane, Gosberton, Spalding PE11 4HG
M3 DIT C. Jacklin 69 Prince William Drive, Butterwick, Boston PE22 0JG
M3 DIU J. Neenan 11 Shaftesbury Square, West Bromwich B71 1DX
M3 DIW D. Brooks 61 Carisbrooke High St., Newport PO30 1NR
M3 DIY C. Ashworth 79 Stonehouse Road, Rugeley WS15 2LL
MM3 DIZ D. Mcarthur 12 Laburnum Grove, Lenzie, Kirkintilloch, Glasgow G66 4DF
M3 DJG K. Miller 52 Stanway Road, Shirley, Solihull B90 3JE
MI3 DJM J. Mcbride 22 Birchwood, Omagh BT79 7RA
MW3 DJV T. Lewis 22 Munro Place, Barry CF62 8BU
M3 DJW D. Wilson 158 Higher Lane, Rainford, St. Helens WA11 8BH
MW3 DJZ K. Lewis 22 Munro Place, Barry CF62 8BU
MM3 DKA J. Blair 47 Chapelhill Mount, Ardrossan KA22 7LU
M3 DKC L. Johnson 222 Norwich Road, Norwich NR5 0EZ
M3 DKG D. Morris 4 Carnarvon Road, Reading RG1 5SD
M3 DKK A. Milne 4 Bearsden Way, Broadbridge Heath RH12 3AQ
MI3 DKN H. Gillespie 30 Groarty Road, Rosemount, Londonderry BT48 0JX
M3 DKO L. Best 97 Pine Tree Avenue, Canterbury CT2 7TA
M3 DKT D. Knott 80 Melville Court, Chatham ME4 4XJ
M3 DKW C. Hemingway 78 Hildyard Street, Grimsby DN32 7NJ
M3 DKZ G. Tetley 21 Lowther Crescent, Leyland PR26 6QA
M3 DLB D. Beach 2 Millward Close, Telford TF2 8AR
M3 DLC D. Cole Amber Lights, Market Lane, Walpole St. Andrew, Wisbech PE14 7LT
M3 DLE D. Edwards 5 Chalk Lane, Sutton Bridge, Spalding PE12 9YF
M3 DLJ D. Mcdougall 15 Caldew Drive, Dalston, Carlisle CA5 7NS
MI3 DLO M. Harte 17 Main St., Carrickmore, Omagh BT79 9AY
M3 DLP C. Thulborn11 7 Damsire Close, Liverpool L9 9EJ
M3 DLT J. Hankin 271 Windrows, Skelmersdale WN8 8NP
M3 DLU M. Salt 84 Hayfield, Stevenage SG2 7JR

M3 DLZ E. Harvey 125 North Road, Clowne, Chesterfield S43 4PQ
M3 DMG D. Glazebrook 12 Kestrel Road, Haverhill CB9 0PH
M3 DMJ D. Jodrell 2 Charlesworth Street, Crewe CW1 4DE
MI3 DMM D. Mcauley Layde View, 19 Rathlin Avenue, Ballycastle BT54 6DQ
M3 DMW R. Weir 130 Alexander Square, Eastleigh SO50 4BX
M3 DMY A. Hoskins 38 Tasmania Close, Basingstoke RG24 9PQ
MM3 DMZ A. Perks The Lodge, Cemetery Drive, Dumbarton G82 5HD
M3 DNA D. Roberts 19 East Avenue, Warrington WA2 8AD
M3 DNB N. Brown 241 Bury Road, Tottington, Bury BL8 3DY
M3 DNC B. Brown 241 Bury Road, Tottington, Bury BL8 3DY
MI3 DNN W. Crozier 3 Carnhill Avenue, Newtownabbey BT36 6LE
M3 DNX J. Cox 5 Golden Avenue Close, East Preston, Littlehampton BN16 1QS
M3 DOA K. Hodder 12 Garden Crescent, Barnham, Bognor Regis PO22 0AR
MI3 DOD K. Campbell 11 Caldwell Drive, Portrush BT56 8ST
M3 DOM D. Pritchard 8 The Paddock, Dawlish EX7 0EJ
M3 DOO J. Dooley 29 The Drive, Alsagers Bank, Stoke-on-Trent ST7 8BB
MM3 DOP J. Shields 4 Rhindmuir Drive, Baillieston, Glasgow G69 6ND
M3 DOR G. Membury 21 Webbers Piece, Maiden Newton, Dorchester DT2 0AQ
M3 DOX A. Bajjon 35A Blackford Road, Shirley, Solihull B90 4BU
M3 DPB M. Oram Lancroft, West End Road, Boston PE21 7NQ
MW3 DPF D. Fitzgerald 60 The Links, Trevethin, Pontypool NP4 8DQ
M3 DPG D. Ganner 58 Kilngate, Lostock Hall, Preston PR5 5UW
M3 DPI J. Spencer 19 Meadowside Avenue, Bolton BL2 2SS
M3 DPP D. Pitchfork 57 Calvary Crescent, Bentilee, Stoke-on-Trent ST2 0AQ
M3 DPQ R. Palmer 34 Ditmas Avenue, Kempston, Bedford MK42 7DP
M3 DPS D. Sager 29 Station Road, Mickleover, Derby DE3 9GH
MW3 DPV D. Vaughan 40 Meadow Terrace, Phillipstown, New Tredegar NP24 6BW
M3 DPX A. Holden 21 East View Meadowfield, Durham DH7 8RY
M3 DPY W. Ellis 16 Furlong Drive, Tean, Stoke-on-Trent ST10 4LD
MW3 DQB R. Cotterell 49 Graham Court, Caerphilly CF83 1RF
M3 DQJ A. Barker 21 Raysmith Close, Southwell NG25 0BG
M3 DQQ D. Horsley1 1 Mead Close, Swanley BR8 8DQ
MM3 DQV B. Mcrae 29 Woodneuk Road, Gartcosh, Glasgow G69 8AG
MM3 DQX K. Mcrae 29 Woodneuk Road, Gartcosh, Glasgow G69 8AG
M3 DRA D. Abbott 29 Malvern Road, Peterborough PE4 7TT
M3 DRH D. Horner 38 Lyndhurst Drive, Bicknacre, Chelmsford CM3 4XL
M3 DRM D. Mostyn 3 Woodlands Avenue, Cheadle Hulme, Cheadle SK8 5DD
M3 DSA D. Bartlett 9 Macdonald Avenue, Hornchurch RM11 2NF
M3 DSE D. Essery Hilgay, First Avenue, Watford WD25 9PS
M3 DSI A. Clarke 3 Epps Road, Sittingbourne ME10 1JD
MI3 DSM D. Mccord 24 Craigstown Road, Moorfields, Ballymena BT42 3DF
M3 DSU S. Jenner 4 Christie Close, Chatham ME5 7NG
M3 DSW D. Wadey 15 Canberra Place, Tangmere, Chichester PO20 2WB
MW3 DTC R. Charge Llifon, Waunfawr, Caernarfon LL55 4YY
M3 DTD G. Asher Hillside, 23 Mill Road, Saxmundham IP17 1DP
M3 DTH R. Hart 11 Ivy Hall Road, Sheffield S5 0GX
MW3 DTO D. Owens 16 Blanche Street, Dowlais, Merthyr Tydfil CF48 3PE
M3 DUL R. Clark 10 Clarendon Road, Bournemouth BH4 8AL
M3 DUO D. Flaherty 24 Ansdell Drive, Eccleston, St. Helens WA10 5DW
M3 DUR M. Durrant 40 Wood Street, Mow Cop ST7 3PE
M3 DVA S. Maskrey The Hayloft, Stamford Lane, Chester CH3 7QD
M3 DVB J. Moreton 71 Bevendean Avenue, Saltdean, Brighton BN2 8PF
MM3 DVD P. Gazinski 14 Corstorphine Road, Edinburgh EH12 6HN
M3 DVG D. Green 89 Upper Ratton Drive, Eastbourne BN20 9DJ
M3 DVH D. Swaby 34 Lambourne Road, Barking IG11 9PS
M3 DVM J. Diston 15 Colletts Gardens, Broadway WR12 7AX
M3 DVN D. Noel 58 Easenhall Lane, Redditch B98 0BJ
M3 DVP A. Whyman 8 Staplers Close, Great Totham, Maldon CM9 8UN
M3 DVQ A. Cree 24 Old Lincoln Road, Caythorpe, Grantham NG32 3EJ
M3 DVU J. Binnell 146 Hales Crescent, Warley, Smethwick B67 6QX
M3 DWA B. Hollis 212 Rye Lane, Halifax HX2 0QP
M3 DWD D. Wraight 2 Silver Mead, Congresbury, Bristol BS49 5EX
MI3 DWQ D. Mcglone 10 O'Neill Terrace, Dromore, Omagh BT78 3AW
M3 DWV D. Simmons 39 Crosier Court, Upchurch, Sittingbourne ME9 7AS
MW3 DWZ B. Chapman Gwyddfan, Mountain Road, Kidwelly SA17 4EY
M3 DXD A. Clark Sheldon House, Sheldon, Bakewell DE45 1QS
M3 DXI T. Townson 4 Crawford Street, Bradford BD4 7JJ
M3 DXL L. Sucharyna Thomas Dame School House, 103 High Street, Milton Keynes MK11 1AT
M3 DXN B. Worthington 9 Greenway, Penwortham, Preston PR1 0TD
MD3 DXW W. Griffiths 7 Cooyrt Shellagh, Ballasalla, Isle of Man IM9 2EU
M3 DYF P. Beier 20 Markham Avenue, Armthorpe, Doncaster DN3 2AZ
M3 DYL W. Malin 729 Wellingborough Road, Northampton NN3 3JE
M3 DYR P. Burns 395 Hastilar Road South, Sheffield S13 8EH
MM3 DYT P. Mckenzie 76 East Bankton Place, Livingston EH54 9BZ
M3 DYU L. Horn 9 Musson Close, Irthlingborough, Wellingborough NN9 5XW
M3 DYY D. Ellison 48 Keenan Drive, Bootle L20 0AL
M3 DZC M. Blakely 2957 Wilderness Blvd, Florida 34219 United States
MI3 DZD W. Brown 16 Wallace Park, Rasharkin, Ballymena BT44 8QH
M3 DZK H. Cartwright 142 Ferness Road, Hinckley LE10 0SE
M3 DZN D. Newman 11 Pine View Close, Woodfalls, Salisbury SP5 2LR
M3 DZQ R. Sharpe 31 Carolyn House, Littlehampton Road, Worthing BN13 1RE
M3 DZT South East Hampshire Raynet c/o A. Dominy 58C Church St, Harwich CO12 3DS
MM3 DZW C. Graham 4 Dodridge Cottages, Pathhead EH37 5UJ
M3 EAE H. Carter 23 Brookdale Avenue, Marple, Stockport SK6 7HP
MW3 EAI A. Whitburn 14 Westgil Pen Ffordd, Blackwood NP12 3QS
MI3 EAQ S. Flanagan 18 Hunters Park, Bellaghy, Magherafelt BT45 8JE
M3 EAS A. Stacey 21 Edward Terrace, Stanley DH9 7JW
M3 EAT T. Bowskill 522 New St., Hilcote, Alfreton DE55 5HU
M3 EBA A. Barnett 53A Walkford Road, Walkford, Christchurch BH23 5QD
M3 EBF E. Fury 1 Wigley Drive, Wigley, Ludlow SY8 3DR
M3 EBG H. Crossley 196 Middleton Road, Heywood OL10 2LH
M3 EBK M. Hall 22 Leam Road, Lighthorne Heath, Leamington Spa CV33 9TE
M3 EBU E. Burrows 9 Clough Hall Road, Kidsgrove, Stoke-on-Trent ST7 1AR
M3 EBZ A. Cattermole Blaxhall Hall Crossing, Little Glemham, Woodbridge IP13 0BP
M3 ECD D. Snowdon Holly Cottage, Romsey Road, Romsey SO51 0HG
M3 ECF R. Maltby 15 Haglane Copse, Pennington, Lymington SO41 8DT
M3 ECJ P. Stringfellow 5 Cowslip Way, Romsey SO51 7RR
MM3 ECO W. Davidson 44 Abercromby Crescent, Helensburgh G84 9DX
M3 ECQ J. Ranger 9 Mitchells Close, Romsey SO51 8DY
M3 ECS R. Parkhouse 3 Thornfield Close, Seaton EX12 2SS
M3 ECU A. Caws 61 Kinver Close, Romsey SO51 7JW
M3 ECV D. Bould 38 Curlew Grove, Bridlington YO15 3NX
M3 ECW R. Parsons 145 Middlemarch Road, Coventry CV6 3GJ
MW3 ECZ M. Douglas 486 Malpas Road, Newport NP20 6NB
M3 EDC E. Cook 13 High Street, Cawston, Norwich NR10 4AE
M3 EDS S. Elliott 25 Staunton Heights, 27 Dunsbury Way, Havant PO9 5AR
M3 EDU M. Islam 158 Somerville Road, Chadwell Heath, Romford RM6 5AT
MM3 EDW C. Martin 82 Biggart Road, Prestwick KA9 2EQ
M3 EEJ J. Maguire 15 Brown Lane, Heald Green, Cheadle SK8 3RR
MD3 EEW E. Wood The Hawthorns, Droghadfayle Road, Port Erin, Isle of Man IM9 6EL
MU3 EFB K. Le Boutillier Tiverton, Bailiffs Cross Road, St. Andrew GY6 8RT Guernsey
M3 EFC P. France 238 Strand Road, Bootle L203HN
M3 EFL M. Arnold 334 Stourbridge Road, Halesowen B63 3QR
M3 EFQ D. Hillman 132 Vicarage Road, Oldbury B68 8HY
M3 EFV S. Leathes Harrogate Ladies' College, Clarence Drive, Harrogate HG1 2QG
M3 EFW C. Broadbent 9 Orchard Road, Bromley BR1 2PR
M3 EFX R. Lockyear 114 Wishaw Close, Redditch B98 7RF
MW3 EGB E. Bateman 32 Park Avenue, Bodelwyddan, Rhyl LL18 5TB
M3 EGC E. Colley 14 Hawthorne Close, Tyldesley, Manchester M29 8PH
M3 EGF R. Mahoney 1 Warner Avenue, Barnsley S75 2EQ
MI3 EGJ C. Hazlett 13 Faughanview Park, Claudy, Londonderry BT47 4HQ
M3 EGM T. Brain 47 Bankwood Crescent, New Rossington, Doncaster DN11 0PU
M3 EGU R. Knight 35 Bayswater Road, Headington, Oxford OX3 9PB
M3 EGV M. Johnson 5 Blackbird Close, Thurston, Bury St. Edmunds IP31 3PF
M3 EGY J. Johnson 5 Blackbird Close, Thurston, Bury St. Edmunds IP31 3PF
MW3 EGZ C. Cutcliffe 12 Heol Fargoed, Bargoed CF81 8PP
M3 EHF D. Austen Tudorlands, Silchester Road, Bramley, Tadley RG26 5DG
M3 EHH C. Mason Apartment 23, Burnside House, Carleton Road, Skipton BD23 2BE
M3 EHJ S. Harris 22 Westmoore Court, Westdale Lane West, Nottingham NG3 6EE
M3 EHK M. Ellaway 3 Lamb Close, Thatcham RG18 3UE
MM3 EHM R. Allan 30 Woodside Way, Glenrothes KY7 5DF
M3 EHP D. Moses 121 Badger Avenue, Crewe CW1 3JN
M3 EHY R. Cheesley 1 Lechlade Road, Inglesham, Swindon SN6 7RB
M3 EIA R. Southworth 37 Pound Close, Lyneham, Chippenham SN15 4PJ
M3 EIJ C. Bailey 13 Newark Road, Mexborough S64 9EZ
MM3 EJB J. Burgoyne 5 Shankston Crescent, Cumnock KA18 1HA
M3 EJH I. Haughton 17 Rivermead, Byfleet, West Byfleet KT14 7BZ
M3 EJL E. Lawrence 4 Malvern Road, Gillingham ME7 4BA
M3 EJM E. Mcgee 107 Fakes Road Hemsby, Great Yarmouth NR29 4JL
M3 EJR E. Trueman 2 Nursery Close, Saxilby, Lincoln LN1 2JD
MM3 EJV S. Waldron 11 Raithhill, Ayr KA7 4UF
M3 EJX J. Bennett 21 Scott Avenue, Sutton Manor, St. Helens WA9 4AN
M3 EKA E. Denman 12 Woodland Close, Northampton NN5 6NH
M3 EKC A. Taylor 106 Raeburn Avenue, Surbiton KT5 9EA
MM3 EKL R. Harrigan 7 Almond Crescent, Paisley PA2 0NG
M3 EKP J. Virdee 29 Larkfield Crescent, Houghton le Spring DH4 4PE
M3 EKR N. Harris 45 Sleigh Road, Sturry, Canterbury CT2 0HT
M3 EKU P. Lancaster 16 Wiltshire Close, Bury BL9 9EY
M3 EKY D. Osbourne 34 Lambourne Road, Barking IG11 9PS
M3 EKZ D. Smedley 27 Stirling Avenue, Loughborough LE11 4LJ
M3 ELD E. Dalton 120 Goodway Road, Birmingham B44 8RG
M3 ELN H. Van Schie 135 Mellish Court, Bletchley, Milton Keynes MK3 6PE
M3 ELP A. Ogburn 88 Castle Rise, Runcorn WA7 5XW
M3 ELS M. Skinner 5 Sycamore Avenue, Upminster RM14 2HR
MM3 ELT A. Kerr 14 Ellisland Drive, Dumfries DG2 9DZ
M3 ELV L. Tremble 7 Allerton Grove, Birkenhead CH42 5LR
M3 EMA E. Holmes 11 Deerness Road, Bishop Auckland DL14 6UB
M3 EMN K. Alexander 25 Diamedes Avenue, Stanwell, Staines TW19 7JE
M3 EMO E. O'Neal 22 Hill Lane, Birmingham B43 6NA
M3 EMS S. Calver 4 George Place, St. Neots PE19 2QG
M3 EMU R. Searle Hollies, 27 Cuckmere Rise, Heathfield TN21 8PG
M3 EMX P. Hardwick 2 Cliffe Cottages, Sandy Lane, Rake GU33 7JE
M3 ENE R. Evans 53 Faygate Road, Eastbourne BN22 9RR
M3 ENF T. Evans 53 Faygate Road, Eastbourne BN22 9RR
M3 ENJ C. Berry 29 Marlborough Crescent, Long Hanborough, Witney OX29 8JP
M3 ENO E. Cross 17 Nicholson Court, Tideswell, Buxton SK17 8PX
MM3 ENP W. Wilson Laurieston Farm, Hollybush, Ayr KA6 6HB
M3 ENS R. Nelson 10 Westmorland Avenue, Willington Quay, Wallsend NE28 6SN
M3 ENY N. Wootton 54 York Road, Harlescott, Shrewsbury SY1 3RA
MI3 EOD N. Crawford 10 White Mountain Road, Lisburn BT28 3QY
MI3 EOH B. Mccalmont 19 Drumsesk Place, Warrenpoint, Newry BT34 3NL
M3 EOL B. Jenson 10 Tintern Close, Paulsgrove, Portsmouth PO6 4LS
M3 EOQ D. Horton Glen View, New Road, Bude EX23 9LE
M3 EOT G. Gidman 8 Minerva Close, Knypersley, Stoke-on-Trent ST8 6SZ
M3 EOX J. Allen 2 Chichester Walk, Chichester Road, Ramsgate CT12 6NX
MW3 EOY S. Mclaughlin 7 Marine Terrace, Criccieth LL52 0EF
M3 EOZ A. Hammond 52 Esther Avenue, Wakefield WF2 8BX
M3 EPC N. Ball 12 Dixons Farm Mews, Clifton, Preston PR4 0PA
MW3 EPJ S. Mcdonald 56 Scotchwell View, Haverfordwest SA61 2RE
MW3 EPK G. Llewellyn Hazeldene, Abercrave, Swansea SA9 1SP
M3 EPQ D. Caldwell 44 Maxwell Road, Littlehampton BN17 7BW
M3 EPR A. Wilson 22 Ormesby Road, Raf Coltishall, Norwich NR10 5JY
MW3 EQE O. Richards 57 Maesgwyn, Aberdare CF44 8TL
M3 EQL S. Suresh 2 Amberley Walk, Kingsmead, Milton Keynes MK4 4AX
M3 EQP T. Thompson 7 West Bank, Dorking RH4 3BZ
M3 EQQ J. Laney 18 Dyrham Close, Thornbury, Bristol BS35 1SX
MI3 EQS T. Mcdonnell 52 Moira Road, Glenavy, Crumlin BT29 4JL
M3 EQW M. Savage 23 Queen Mary Road, Salisbury SP2 9LD
M3 EQY S. Heard 42 Hallowell Down South Woodham Ferrers, Chelmsford CM3 5FS
MM3 ERD E. Davidson 44 Abercromby Crescent, Helensburgh G84 9DX
MM3 ERP P. Smith 1 Hillside Cottages, Tillymorgan, Insch AB52 6UN
M3 ERR D. Barnes 11 Yewside, Gosport PO13 0ZD
MW3 ESE S. Reed 2 St. Marys Park, Jordanston, Milford Haven SA73 1HR
MW3 ESF S. Reed 2 St. Marys Park, Jordanston, Milford Haven SA73 1HR

M3 ESG A. Pickersgill 9 The Malthouse, Ashbury SN6 8NB
MW3 ESH E. Hoy 39 Blackbird Road, Caldicot NP26 5RE
M3 ESK K. Crane 15 Leighton Road, Ipswich IP3 0LJ
M3 ESN J. Carragher High Gorses, Henley Down, Battle TN33 9BP
M3 ESQ M. Peters 9 Evelyn Close, Twickenham TW2 7BL
M3 ESS M. Stirling 3 Rother Croft, New Tupton, Chesterfield S42 6BE
MW3 ETB F. Llewellyn 47 St. Teilos Road, Abergavenny NP7 6HB
M3 ETH J. Goodyear 30 Ashburton Road, Alresford SO24 9HH
M3 ETI D. Mcspadden 37 Halliday Crescent, Southsea PO4 9JU
M3 ETQ N. Greene 308 Cedar Road, Nuneaton CV10 9DY
M3 EUE J. Underwood 27 Woodville Road, London E17 7ER
M3 EUF W. Atherton 64 Dam Lane, Rixton, Warrington WA3 6LB
M3 EUM N. Miller 1 Alanbrooke Road, Colchester CO2 8EG
M3 EUP C. Cutler 18 Berkeley Road, Peterborough PE3 9PA
M3 EUR F. Watt 5 Brambling Road, Horsham RH13 6AX
M3 EUU D. Broad 34 Arderne Avenue, Crewe CW2 8NS
M3 EUW A. Lomas Scanderlands Farm, Gloves Lane, Alfreton DE55 5JJ
M3 EUY S. Swan 47 Warren Close, Whitehill, Bordon GU359EX
M3 EVB E. Munn 32A Brunswick Street, Wakefield WF1 4PW
M3 EVC M. Lorimer 1A Lingford Street, Hucknall, Nottingham NG15 7SJ
M3 EVE D. Watson 14 Gawber Road, Barnsley S75 2AF
M3 EVF D. Redmayne 10 The Square, Kington HR5 3BA
M3 EVI A. Bowron 11 Lealholme Grove, Fairfield, Stockton-on-Tees TS19 7AP
M3 EVJ B. Green 12 The Ridgeway, Coal Aston, Dronfield S18 3BY
M3 EVM S. Burnand 53 Sidley Road, Eastbourne BN22 7JL
MW3 EVN D. Evans 35 Caroline Road, Llandudno LL30 2TY
M3 EVP A. Munton 56 Jacklin Drive, Leicester LE4 7SU
M3 EVR P. Smith 77 Holymoor Road, Holymoorside, Chesterfield S42 7EA
M3 EVV M. Harrison 43 Erskine Road, South Shields NE33 2TH
MD3 EVY J. Phillips 1 Cronk Elfin, Ramsey IM8 2EX Isle of Man
MM3 EWI J. Woods 12 Westbank Terrace, Macmerry, Tranent EH33 1QE
M3 EWN E. Nevard Millinder House, Westerdale, Whitby YO21 2DE
M3 EWQ E. Quinn 20 Greenfield Road, Rotherham S65 3NX
MW3 EWR E. Roberts 10 Ael Y Bryn, Waunfawr, Caernarfon LL55 4AZ
M3 EWV J. Kooner 44 Headingley Road, Birmingham B21 9QD
M3 EWW S. Cooper 53 Queensway, Warton, Preston PR4 1XU
M3 EWY N. Kellow 17 Queensway, Warton, Preston PR4 1XT
M3 EWZ R. Dobson 1 Auster Crescent, Freckleton, Preston PR4 1JL
M3 EXJ L. Taylor 17 Lacy Street, Hemsworth, Pontefract WF9 4NW
M3 EXK J. Wilkes 47 Greenwood Park, Hednesford, Cannock WS12 4DQ
MM3 EXW G. Rotherham 43 Colliston Road, Dunfermline KY12 0XW
M3 EXY R. Shuttleworth 42 Buxton Lane, Marple, Stockport SK6 7QL
MI3 EYB P. Mckeown 7 Knockoneill Road, Maghera BT46 5NX
M3 EYH A. Carter 37 Seathorne, Withernsea HU19 2BB
M3 EYK J. Dodsworth 12 Fowlmere Road, Birmingham B42 2EA
MM3 EYM C. Somerville 39 Edgehead Village, Pathhead EH37 5RL
MM3 EYN R. Somerville 39 Edgehead Village, Pathhead EH37 5RL
M3 EYP J. Read 49 Bransdale Way, Macclesfield SK11 8QT
M3 EYR C. Greene 308 Cedar Road, Nuneaton CV10 9DY
M3 EYS C. Lewis 41 Hazelholt Drive, Havant PO9 3DL
M3 EYW S. Thornton 2 Sceptre Grove, New Rossington, Doncaster DN11 0RW
M3 EYX W. Thornton 2 Sceptre Grove, New Rossington, Doncaster DN11 0RW
M3 EYY E. Driver 99 Queens Road, North Weald, Epping CM16 6JQ
M3 EYZ C. Board Pinmoor, Moretonhampstead, Newton Abbot TQ13 8QA
M3 EZB G. Leake 154 Wareham Road, Lytchett Matravers, Poole BH16 6DT
MI3 EZF P. Rice 11 Kirkwood Park, Saintfield, Ballynahinch BT24 7DP
M3 EZH L. Copley Elmford, Mount Pleasant South, Whitby YO22 4RQ
M3 EZJ M. Pires 7 Felstead Close Earley, Reading RG6 5TP
MI3 EZK B. Flanagan 50 Towncastle Road, Strabane BT82 0AJ
M3 EZX W. Taylor 5 Council Bungalows, Churchtown, Belton, Doncaster DN9 1PD
M3 EZY M. James 82 Hill Crescent, Sutton-in-Ashfield NG17 4JA
M3 FAA D. Mcglone 32 Shipley Mill Close, Kingsnorth, Ashford TN23 3NR
M3 FAC C. Perkins Havasu, Treragin, Callington PL17 8BL
M3 FAE J. Modha 95 Stanway Road, Shirley, Solihull B90 3JF
M3 FAK M. O'Brien 4 Teal Close, Hawkinge, Folkestone CT18 7TG
M3 FAL F. Van Den Langenberg Flat 26, Yew Tree Court, Shifnal TF11 9BF
M3 FAY F. Eavis 61 Hitchmead Road, Biggleswade SG18 0NL
M3 FBG B. Jennings 6 The Bungalow, St. Johns Road, Wroxall, Ventnor PO38 3EL
M3 FBJ T. Mccann 21 Ladyseat, Longtown, Carlisle CA6 5XX
M3 FBP C. Leal 41 The Orchard, Croston, Leyland PR26 9HS
M3 FBR M. Morris 12 Redver Gardens, Newport PO30 5JJ
MI3 FBW P. Coulter 6 Skelton Close, Carrickfergus BT38 8GP
MI3 FBX C. Gardner 10 Abbington Manor, Bangor BT19 1ZQ
MI3 FCA S. Churchill 20 Killen Park, Killen, Castlederg BT81 7TJ
M3 FCB J. Mcdonald 43 Barley Close, St. Ives PE27 3AJ
MM3 FCG W. Mccue 188 Redburn, Alexandria G83 9BU
MI3 FCK J. Morgan 6 Gannet Way, Carrickfergus BT38 7RT
M3 FCN P. Norman 25 Hillswood Avenue, Leek ST13 8EQ
M3 FCO B. Dawson 29 Hillswood Avenue, Leek ST13 8EQ
M3 FCR L. Moreland 25 St. Georges Avenue, Bridlington YO15 2ED
M3 FCS R. Horne 1 Ireland Road, Ipswich IP3 0EJ
M3 FDB J. Johnson 5 Oakey Ley, Bradfield St. George, Bury St. Edmunds IP30 0AU
M3 FDM M. Goss 80 Merryhills Drive, Enfield EN2 7PD
MW3 FDO W. Corbett 27 Waunfawr Gardens. Crosskeys, Newport NP11 7AJ
M3 FDQ A. Proctor 3 The Courtyard, Tattingstone Park, Ipswich IP9 2NF
M3 FDV A. Goodwin 36 Cambridge Street, Bridlington YO16 4JZ
M3 FEA R. Morley 191 Purbrook Way, Havant PO9 3RS
M3 FEC F. Curry 22 Caernarvon Close, Towcester NN12 6UP
M3 FED F. Dunn 71 Redfield Road, Midsomer Norton BA3 2JH
M3 FEG J. Restall 1 Johndory, Dosthill, Tamworth B77 1NY
M3 FEL Dr E. Fellows 95 Arnold Road, Eastleigh SO50 5AS
MI3 FEO R. Robinson 92 Groomsport Road, Bangor BT20 5NT
M3 FES F. Shirley Ham House, Ham Lane, Shepton Mallet BA4 5JW
MM3 FET A. Galbraith 22 Jeffrey Street, Kilmarnock KA1 4EB
MI3 FEX Castel Contest Club c/o D. Rantin 8 Buchanans Road, Newry BT35 6NS
M3 FEY J. Brinnen 134 Victoria Road, Mablethorpe LN12 2AJ
M3 FFA T. Johnson 43 Cherry Orchard Avenue, Halesowen B63 3RZ
M3 FFE W. Johnson 43 Cherry Orchard Avenue, Halesowen B63 3RZ
M3 FFI D. Ross 27 The Meadows, Skegness PE25 2JA
M3 FFK D. Lythall 71 Bennett Street, Kimberworth, Rotherham S61 2JZ
MW3 FFL B. Kendrick 77 Heolddu Crescent, Bargoed CF81 8US
M3 FFO P. Hoe 12 Ashbridge Rise, Chandler'S Ford, Eastleigh SO53 1SA
M3 FFU D. Deakin 75 Dairyground Road, Bramhall, Stockport SK7 2QW
M3 FFV P. Lloyd 71 Grove Road, Stourbridge DY9 9AE
M3 FGG S. James 94 North Road, Hull HU19 2AY
MM3 FGH N. Macaulay 68 Lorn Road, Dunbeg, Oban PA37 1QQ
MM3 FGI C. Gillespie 18 Roslin Crescent, Rothesay, Isle of Bute PA20 9HT
MI3 FGK N. Craig 29 Oughtagh Road, Killaloo, Londonderry BT47 3TR
MM3 FGL A. Macdonald Manderley, Benvoullin Road, Oban PA34 5EF
M3 FGO J. Taylor 8 Orchard Grove, Dudley DY3 2UU
M3 FGQ I. Prior 81 Ladymeade, Ilminster TA19 0EA
M3 FGR D. Rootes 1 Shelfinch, Toothill, Swindon SN5 8AR
M3 FGU S. Fellows 24 Habberley Road, Rowley Regis B65 9QN
MW3 FGV J. Rowe Flat 27 Rhodfa Frank, Ammanford SA18 2QE
M3 FGX J. Wells 54 Queens Road, Everton, Liverpool L6 2NG
M3 FHK M. Tompkins 4 Prospect View, Rawtenstall, Rossendale BB4 8JG
MI3 FHM C. Patton 13 Oldpark Avenue, Ballymena BT42 1AX
M3 FHO G. Flack 20 The Pastures, Hardwick, Cambridge CB23 7XA
M3 FHP D. Haines 29 Parks Road, Mitcheldean GL17 0DQ
M3 FHQ C. Price 10 St. James Park, Lower Milkwall, Coleford GL16 7LG
M3 FHV B. Cahill 56 Dene Road, Headington, Oxford OX3 7EE
M3 FIA A. Smeed Flatt 8 9 Peir Terrace, Lowestoft NR330AB
M3 FIB S. Watling 1 Chediston Green, Chediston, Halesworth IP19 0BB
M3 FIH G. Street Flat 9, Weavers Cottages, Congleton CW12 1AG
M3 FIM K. Meredith 3 Abbots Road, Abbey Hulton, Stoke-on-Trent ST2 8DU
M3 FIP J. Shingler 19 Cherry Tree Avenue, Runcorn WA7 5JJ
M3 FIW J. Watson 32 Franklin Close, Old Hall, Warrington WA5 8QL
M3 FIX A. Thompson 51 Kempe Way, Weston-Super-Mare BS24 7DZ
M3 FIY E. Watson 32 Franklin Close, Old Hall, Warrington WA5 8QL
M3 FIZ A. Finn 105 Lynmouth Close, Biddulph, Stoke-on-Trent ST8 6LS
MM3 FJA G. Cull 2 Pitairlie Farm Cottages, Pitgaveny, Elgin IV30 5PQ
M3 FJB J. Bell 2 Rake Lane, Milford, Godalming GU8 5AB
M3 FJC B. Hinchliffe 272 South St., Rotherham S61 2NP
M3 FJD A. Lythall 6 Belmont Crescent, Little Houghton, Barnsley S72 0HT
M3 FJE M. Jones 6D Terrace Road, Walton-on-Thames KT12 2SU
M3 FJN A. Siebert Po Box 127, Nantwich CW5 8AQ
M3 FJP J. Park 18 Ladgate Grange, Middlesbrough TS3 7SL
M3 FJQ M. Rafique 21 Syddall Avenue, Heald Green, Cheadle SK8 3AA
M3 FJR L. Siebert Po Box 127, Nantwich CW5 8AQ
MW3 FJW F. Finch Porthgwyh, Lamb Road, Aberdare CF44 9JU
MM3 FJX J. O'Connor 23 Osborne Terrace, Cockenzie, Prestonpans EH32 0BY
M3 FKK J. Waddington 2 Heron Court, Daventry NN11 0XT
M3 FKL M. Rose 128 Boultham Park Road, Lincoln LN6 7TG
M3 FKM G. Rose 128 Boultham Park Road, Lincoln LN6 7TG
M3 FKN M. Mellish 302 Belvedere Road, Burton-on-Trent DE13 0RD
MM3 FKO C. Lorimer 70A Morningside Drive, Edinburgh EH10 5NU
M3 FKS B. Hoare 2 St. Peters Close, South Newington, Banbury OX15 4JL
M3 FKV R. Johnson 90 Regent Street Church Gresley, Swadlincote DE11 9PJ
M3 FKW S. Papworth Caravan 6, 4Hillside Minorca Lane Bugle, St. Austell PL26 8QN
MW3 FLA G. Backhouse De10 Isaf, Bryneglwys, Corwen LL21 9NP
M3 FLB A. Bean 25 Riverfield Grove, Bolehall, Tamworth B77 3NB
M3 FLC F. Hanmore 7 Tarbert Walk, London E1 0EE
M3 FLE D. Wallstone 128 Maltby Road, Mansfield NG18 3BL
MW3 FLI C. Bainbridge The Brindles, Primrose Hill, Connahs Quay CH5 4QA
M3 FLJ G. Jackson 5 Woodside Close, Siddington, Macclesfield SK11 9LQ
MW3 FLK G. Badham 103 Stanfield Street, Cwm, Ebbw Vale NP23 7TG
M3 FLL I. Hamilton 36 North Parade, Hoylake, Wirral CH47 3AJ
M3 FLP S. Dec 101 Cranford Road, Northampton NN2 7QY
MW3 FLU A. Yates 4 High St., Abergele LL22 7AR
M3 FLV J. Showell 14A Station Approach, Hayes, Bromley BR2 7EH
M3 FLZ M. Mccormick Sarnia, 73 Abelia, Tamworth B77 4EZ
MM3 FMB S. Markey 232 Main Street, Renton, Dumbarton G82 4QA
M3 FME D. Bennett 29 Margraten Avenue, Canvey Island SS8 7JD
M3 FMI N. Ashley 3 Nightingale Road, Trowbridge, Wiltshire BA14 9TP
M3 FMK A. Jones 36 Sutherland Drive, Newcastle ST5 3NZ
M3 FMP D. Gilhooly 50 Hillborough Crescent Houghton Regis, Dunstable LU5 5NX
M3 FMQ A. Ritchie 50 Hillborough Crescent Houghton Regis, Dunstable LU5 5NX
M3 FMV C. Hatter 14 Morland Avenue, Bromborough, Wirral CH62 6BE
MM3 FMY L. Dickenson 9 Naver Road, Thurso KW14 7QA
M3 FNA A. Brooks 93 Durham Road, Stockton-on-Tees TS19 0DE
M3 FNC F. Chance 128 Chapel St., Pensnett, Brierley Hill DY5 4EQ
M3 FNH W. Stopforth 52 Cypress Road, Southport PR8 6HF
M3 FNM P. Hewitt 166 Sheringham Avenue, London E12 5PQ
M3 FNO J. Scott-Brown 2 Haddon Close, Fareham PO14 1PH
M3 FNR S. Pitchford 419 Chell Heath Road, Stoke-on-Trent ST6 6PB
M3 FNT S. Williams 11 Hilda Street, Leigh WN7 5DG
M3 FNY J. Eagle 1B Kingsley Avenue, Daventry NN11 4AN
M3 FOD L. Newby 22 Acton Road, Liverpool L32 0TT
MM3 FOE S. Espie 70 Everard Rise, Livingston EH54 6JD
MI3 FOJ D. Kane 22 Rowan Road, Ballymoney BT53 7AQ
M3 FOK J. Old 33 Rookhill Road, Pontefract WF8 2BY
M3 FOQ J. Woods 3 Ingle Avenue Morley, Leeds LS27 9NP
M3 FOR M. Baldwin 52 Salisbury Road Me45Nn, Chatham ME4 5NN
M3 FOS I. Woods 3 Ingle Avenue, Morley, Leeds LS27 9NP
M3 FOV D. Read L'Eglise, Durley St., Southampton SO32 2AA
M3 FPA R. Etchells 6 Woodbank Court, Canterbury Road, Manchester M41 7DY
MI3 FPB I. Buchanan 162 Victoria Road, Bready, Strabane BT82 0DZ
MI3 FPE J. Rice 42 The Crescent, Ballymoney BT53 6ES
MW3 FPF J. Briers 117 Heath Mead, Cardiff CF14 3PL
M3 FPG A. Jennings 6 The Bungalow, St. Johns Road, Wroxall, Ventnor PO38 3EL
M3 FPH P. Harris Flat 33, Buckingham Court Shrubbs Drive, Bognor Regis PO22 7SE
M3 FPM L. Noel 58 Easenhall Lane, Redditch B98 0BJ
MI3 FPN C. Mcintyre 18 Glanroy Crescent, Newtownabbey BT37 9JZ
M3 FPS J. Reid 5 Hamlet Road, Fleetwood FY7 7HW
M3 FPT P. Turner 92 Lancashire Street, Leicester LE4 7AE
M3 FPU S. Cash 6 The Mariners, Valetta Way, Rochester ME1 1FB

M3 FPZ D. Turner 11 Weetwood Road, Congresbury, Bristol BS49 5BN
M3 FQA S. Wadsworth 47 Kilnhurst Road, Todmorden OL14 6AX
MI3 FQD J. Moore 7 Hillview Terrace, Castledawson BT45 8HJ
M3 FQG J. Heagren 14 Pepperbox Rise, Whaddon, Salisbury SP5 3BF
MM3 FQI G. Robinson 12 Hannahston Avenue, Drongan, Ayr KA6 7AU
M3 FQM J. Dunning 16 Shaggs Meadow, Lyndhurst SO43 7BN
M3 FQN S. Dunning 16 Shaggs Meadow, Lyndhurst SO43 7BN
M3 FQT C. Inwood 7 The Poplars, George Street, Mablethorpe LN12 2BP
M3 FQX J. Wadeson 75 Bedford Drive, Sutton Coldfield B75 6AX
M3 FRB D. Munday 29 Coombe Park, Wroxall, Ventnor PO38 3PH
M3 FRD F. Mcloughlin 128 Windrows, Church Farm, Skelmersdale WN8 8NW
M3 FRE J. French Eypes Mouth Country Hotel, Eypes, Bridport DT6 6AL
M3 FRJ R. Fearnley 31 Radburn Court, Dunstable LU6 1HW
M3 FRQ S. Smith 7 Rosebery Avenue, Morecambe LA4 5RU
M3 FRS B. Page 65 Meadow View, Charminster, Dorchester DT2 9RE
MM3 FRU G. Ison North Craighousesteads, Tundergarth, Lockerbie DG11 2QL
M3 FRX M. Breffit 10 Garrard Place, Ixworth, Bury St. Edmunds IP31 2EP
M3 FSB S. Babic 17 Ashwood Drive, Broadstone BH18 8LN
M3 FSC F. Creese 8A Brockman Road, Folkestone CT20 1DL
M3 FSD D. Babic 17 Ashwood Drive, Broadstone BH18 8LN
M3 FSE D. Edge Lymn Bank Cottage, Lymn Bank, Skegness PE24 4PJ
M3 FSQ D. Gornall 40 Welbrow Drive, Longridge, Preston PR3 3TB
MI3 FSR J. Brown 4 Stratford Gardens, Bangor BT19 6ZH
M3 FSS A. Goodchild Gravel Lane, Ringwood BH24 1LL
M3 FSU G. Lewis 57 Oakwood Road, Sutton Coldfield B73 5EH
M3 FSV P. Murray 2 Thurlow Gardens, Bishop Auckland DL14 7GH
MI3 FSW F. White 28 Lord Warden'S Parade, Bangor BT19 1YU
MI3 FSX T. Mulholland 215 Finaghy Road North, Belfast BT11 9ED
M3 FSY K. Doorbar 23 Oaktree Road, Rugeley WS15 1AD
M3 FTA M. Everall 17 Golden Park Avenue, Torquay TQ2 8LR
MW3 FTB S. Robson 16 Dunraven Road, Sketty, Swansea SA2 9LG
MW3 FTC D. Robson 16 Dunraven Road, Sketty, Swansea SA2 9LG
M3 FTE S. Manley Hill Cottage, Exminster Hill Exminster, Exeter EX6 8DW
M3 FTI F. Gibbs 62 Wenvoe Avenue, Bexleyheath DA7 5BT
M3 FTJ J. Lightly 8 Smithville Close, St. Briavels, Lydney GL15 6TN
M3 FTK C. Gale 51 Heron Way, Horsham RH13 6DW
MW3 FTP J. Mckenna 33 Low Islwyw, Prestatyn LL19 8HQ
M3 FTU P. Panayiotou 7 Aireville Rise, Bradford BD9 4ES
M3 FTV A. Dunham 28 Kingfisher Close, Chatteris PE16 6TP
M3 FTW C. Hughes 80 Ayrshire Close, Buckshaw Village, Chorley PR7 7DB
MW3 FTY R. Edwards Hillside, 3 Treforis, Ammanford SA18 2RA
M3 FTZ A. De Vries 21 Brean Down Road, Plymouth PL3 5PU
M3 FUB N. Phillips First Floor Flat, 116 Lodge Road, Croydon CR0 2PF
M3 FUD S. Merison Flat 2, 5 Church Lane, Banbury OX16 5LR
MM3 FUG W. Gillespie 33 Lochnell Road, Dunbeg, Oban PA37 1QJ
M3 FUH Dr C. Pomfrett 17 Manifold Close, Sandbach CW11 1XP
M3 FUQ S. Day The Lodge, Attleborough Fish Farm Norwich Road, Attleborough NR17 2LA
M3 FUR P. Henderson 214 Marsh Street, Barrow-in-Furness LA14 1BQ
M3 FUV P. Spowart Ruggs Hall, Clatterway Hill, Matlock DE4 2AH
M3 FVA S. Grimbleby 96 Waldeck Street, Reading RG1 2RE
MW3 FVC Rvd. J. Huntington 87 Dinerth Road, Rhos On Sea, Colwyn Bay LL28 4YH
M3 FVE D. Millard 114 Ainsdale Drive, Werrington, Peterborough PE4 6RP
MW3 FVH M. Hallett 24 Brynhyfrydst, Clydach Vale, Tonypandy CF40 2DZ
M3 FVJ C. Brown 44 Stanley Avenue, Inkersall, Chesterfield S43 3SY
MI3 FVW D. Shaw 9 The Ten Cottages, Newtownards Road, Donaghadee BT21 0PU
M3 FVX P. Sarratt Unit 20F, Brooke Business Park, Lowestoft NR33 9LZ
M3 FWA P. Matthew-Brown 57 The Limes Avenue, London N11 1RD
M3 FWJ C. Green 160 Ashbrook Road, London N19 3DJ
M3 FWO K. Beckett 95 Warrens Hall Road, Dudley DY2 8DH
M3 FWR D. Marsh 1 Caunts Crescent, Sutton-in-Ashfield NG17 2FH
M3 FWS J. Steele 70 The Crescent, Andover SP10 3BU
M3 FWT J. Cooper 1 Dearing Close, Lyndhurst SO43 7JP
M3 FWU R. Wilson 7 Cornwall Close, Kirton Lindsey, Gainsborough DN21 4DF
MI3 FXE J. Higginson 47 Ballycorr Road, Ballyclare BT39 9DD
M3 FXM D. Toombs 1 Chalgrove, Welwyn Garden City AL7 2QJ
M3 FXQ H. Gregory 178 Over Lane, Belper DE56 0HL
M3 FXU F. Siviter Flat 76, Lancaster House, Rowley Regis B65 0QE
M3 FXX V. Hocking 80 Barton Tors, Bideford EX39 4HA
MW3 FYA C. Alloway 9 Millands Park, Llanmaes, Llanwitmajor CF61 3XR
MM3 FYF J. Fyfe 5 Beaufort Avenue, Newlands, Glasgow G43 2YL
M3 FYM A. Reynolds 5 Broadlands , Broadmeadows , South Normanton, Alfreton DE55 3NW
MM3 FYN D. Innes 39 Mormond Place, Strichen, Fraserburgh AB43 6SY
MW3 FYR C. Johnson 50 Treberth Avenue, Newport NP19 9TA
M3 FYV B. Cairns 4 Spence Court, Great Ayton, Middlesbrough TS9 6DW
M3 FYZ A. Shellam 1 Trafalgar House, Nelson Drive, Cannock WS12 2GH
M3 FZB P. Paduch 291 Rochfords Gardens, Slough SL2 5XH
M3 FZC J. Charter 36 Northumberland Avenue, London E12 5HD
M3 FZE R. Humphreys 19 Monks Green, Fetcham, Leatherhead KT22 9TL
MM3 FZI R. Mcdonald 12 Queen Street, Tayport DD6 9NE
M3 FZJ L. Larkins 34 Guycroft, Otley LS21 3DS
M3 FZM C. Ramsdale 87 Mill Lane, Kirk Ella, Hull HU10 7JN
M3 FZO B. Shepherd 19 Washfield Lane, Treeton, Rotherham S60 5PU
M3 FZS A. Green 10 Howard Close, Teignmouth TQ14 9NW
M3 FZT R. Carpenter 218 Mansel Road West, Southampton SO16 9LR
M3 FZV M. Reed Channel Pool, Armathwaite, Carlisle CA4 9QY
MD3 GAB J. Espey 9A Hilltop View, Douglas IM2 2LA Isle of Man
M3 GAE J. Law 5 Sudbury Close, Chesterfield S40 4RS
M3 GAF G. Allen 39 Hallam Road, Newton Heath, Manchester M40 2SY
M3 GAG P. Gagliardi 7 Saxon Way, Jarrow NE32 3QA
MI3 GAM G. Moucka 5 Glebe Gardens, Moira, Craigavon BT67 0TU
M3 GAP G. Porter 65 Bartlett St., Wavertree, Liverpool L15 0HN
M3 GAV C. Tomlinson 9 Wells Close, Astley, Manchester M29 7WF
M3 GBA G. Barlow Ingleneuk, Hammersley Hayes Road, Stoke-on-Trent ST10 2DW
M3 GBB G. Bartley 19 South Avenue, , Shadforth, Durham DH61LB
M3 GBC M. Casey 7 Cobham Avenue, Manchester M40 5QW
MW3 GBD B. Dallimore 4 Llys Dyffryn, St. Asaph LL17 0SX
MJ3 GBJ S. Boudier 124 La Tour Dunlin, Le Clos De Samares, St. Clement JE2 6GE Jersey
M3 GCD E. Elsworth-Wilson 31 Douglas Avenue, Brixham TQ5 9EL
M3 GCH S. Shreeves 20 Selly Oak Road, Jordanthorpe, Sheffield S8 8DU
M3 GCJ G. Johnson 30 Trinidad Close, Basingstoke RG24 9PY
M3 GCM G. Masters 85 Petersham Road, Creekmoor, Poole BH17 7DW
M3 GCN G. Newton 18 Parks Road, Dunscroft, Doncaster DN7 4AH
M3 GCP G. Papworth 70 Edward Road, West Bridgford, Nottingham NG2 5GB
M3 GCR G. Watt 5 Brambling Road, Horsham RH13 6AX
M3 GCS G. Sadler 43 Laurel Grove, Stafford ST17 9EF
M3 GCT P. Wylie 40 Sheepwash Avenue, Choppington NE62 5NN
M3 GDE G. Edgar 61 Winchester Avenue, Lancaster LA1 4HX
M3 GDI T. Whittam 27 Dimples Lane, Garstang, Preston PR3 1RD
M3 GDK P. Weaver 1 Madeley Street, Newcastle ST5 6LS
MW3 GDL G. Jones 31 Parcy Mynach, Pontyberem, Llanelli SA15 5EN
M3 GDV D. Bearne 59 Foxhole Road, Foxhole Estate, Paignton TQ3 3TD
M3 GDX J. Ormerod 14 Fanny Moor Lane, Hall Bower, Huddersfield HD4 6PJ
M3 GDY V. Dealey 11 Ashcombe Close, Witney OX28 6NL
MI3 GEI G. Convery 12 Linen Grove, Belfast BT14 8PP
M3 GEK C. Poolman 5 Slessor Street, Waddington, Lincoln LN5 9NE
M3 GEN B. Jones 39 Dalton Lane, Barrow-in-Furness LA14 4LE
MI3 GER G. Reilly 220 Ardmore Road, Londonderry BT47 3TE
MM3 GEW R. Brooks 6 Amochrie Drive, Paisley PA2 0BE
M3 GEX E. Rogers Maes Gwersyll, Garthmyl, Montgomery SY15 6RS
M3 GEZ D. Fowles 2 The Red House, Gallamore Lane, Market Rasen LN8 3UB
MI3 GFA G. Freeman 817 Windyhall Park, Coleraine BT52 1TU
M3 GFE T. Dunn Rakers Rest, 31 Orleigh Avenue, Newton Abbot TQ12 2TP
MW3 GFG R. Lightly Carver 129 Carbonne Close, Monmouth NP25 5EH
M3 GFH M. Dunn 6 Hamilton Drive, Newton Abbot TQ12 2TL
M3 GFO N. Fox 15 Hawthorne Grove, Bentley, Doncaster DN5 0PQ
M3 GFW P. Seaman 18 Earlsford Road, Mellis, Eye IP23 8DY
M3 GFZ M. Hewson 27 Grange Crescent, Lincoln LN6 8BT
M3 GGA K. Mccarthy 260 Whalley Drive, Bletchley, Milton Keynes MK3 6PJ
M3 GGE G. Townsend 23 Lodgefield Park, Stafford ST17 0YE
M3 GGV 7m CG c/o T. White Watt House, Manor Road, Terrington St. Clement, King's Lynn PE34 4NF
M3 GHA G. Halls 16 Lovent Drive, Leighton Buzzard LU7 3LR
M3 GHD D. Bell 27 Kings Coombe Drive, Kingsteignton, Newton Abbot TQ12 3YU
M3 GHE M. Barnes 49 Harrowden Road, Bedford MK42 0RS
MW3 GHF G. Creed Great House Farm, Croesypant, Pontypool NP4 0JD
M3 GHG P. Walton 15 Arbour Close, Northwich CW9 7BF
M3 GHH G. Hazlewood 102 Throne Road, Rowley Regis B65 9JX
M3 GHI J. Haslam 25 Lulworth Road, Eccles, Manchester M30 8WP
M3 GHL G. Law 14 Sandpit Lane, Hilton, Bridgnorth WV15 5PH
M3 GHO G. Cliffe 5 Laurel Cottages, Ongar Hill Road, King's Lynn PE34 4JB
M3 GHR R. Gill 84 Leypark Road, Exeter EX1 3NT
M3 GHS S. Stanhope 61 Heathfield St., Manchester M40 1LF
MI3 GHW G. Mclernan Drumcor Hill, Enniskillen BT74 6BQ
MI3 GHY I. Gibb 1 Shankill Road, Garvary, Enniskillen BT94 3DB
M3 GIE R. Harper 19 Tennyson Avenue, King's Lynn PE30 2QG
M3 GIF E. Roberts 8 Skamacre Crescent, Lowestoft NR32 2QG
M3 GIH E. Peck 11 Blake Road, Stapleford, Nottingham NG9 7HN
M3 GIK M. Haughey 10 Sharp St., Hull HU5 2AB
MM3 GIR K. Gibson 136 Henrietta Street, Girvan KA26 0AE
M3 GIX A. Scrutton 35 Gainsborough Road, Warrington WA4 6DA
M3 GIY J. Eaton 10 Motcombe Farm Road, Heald Green, Cheadle SK8 3RW
MI3 GJG G. Mcgill 4 Grainan Park, Londonderry BT48 7UA
MI3 GJI P. Bingham 45 Gowanvale Drive, Banbridge BT323GD
M3 GJN P. Marriott 38 Westfields, Tilney St. Lawrence, King's Lynn PE34 4QS
M3 GJW G. Watson 2 Bow St., Mansfield Woodhouse, Mansfield NG19 9PJ
MW3 GKB D. Khan 20 Cae Penrallt, Trearddur Bay, Holyhead LL65 2WA
M3 GKE B. Luetchford 136 Summerwood Estate, Great Massingham, King's Lynn PE32 2HS
M3 GKG A. Boag 53 Castlewood Road, London N16 6DJ
M3 GKH G. Buxton 18 Savernake Close Rubery, Rednal, Birmingham B45 0DD
MW3 GKI N. Axon 30 Rating Row, Beaumaris LL58 8AF
M3 GKJ S. Willis 1 Whiffins Orchard, Coopersale Common, Epping CM16 7HT
M3 GKK M. Ahmed 75 Drove Road, Swindon SN1 3AE
M3 GKX J. Boot 110 Wallace Road, Bradley, Bilston WV14 8AU
M3 GKY A. Reed 32 Hollis Gardens, Cheltenham GL51 6JQ
M3 GLA G. Astbury 12 Southall Road, Ashmore Park, Wolverhampton WV11 2PZ
M3 GLC G. Colclough Little Hallands, Norton, Seaford BN25 2UN
MM3 GLH G. Bruce 60 Kingsmills, Elgin IV30 4BU
M3 GLM G. Parkins 73 Orwell View Road, Shotley, Ipswich IP9 1NW
M3 GLT G. Talbot 26 Chevalier Grove, Crownhill, Milton Keynes MK8 0EJ
M3 GMB S. Bradshaw 82 Arden Way, Market Harborough LE16 7DD
M3 GMG G. Mcgeough 57 Stonehouse Park, Thursby, Carlisle CA5 6NS
MI3 GMI M. Crozier 33 Cullentragh Road, Poyntzpass, Newry BT35 6SD
M3 GML G. Linfield 82 Claremont Road, Swanley BR8 7QT
MM3 GMP J. Williams Baptist Manse, Balemartine, Isle of Tiree PA77 6UA
M3 GMY A. Pickering 16 Chestnut Grove, Accrington BB5 0ND
MW3 GNB J. Carfoot 24 Marble Church Grove, , Bodelwyddan LL18 5UP
M3 GNM S. Mcloughlin 7 Wilmots Way, Pill, Bristol BS20 0JT
M3 GNN Addiscombe ARC c/o A. Glover 103A Latimer Street, Liverpool L5 2RF
M3 GNY A. Hunt Chesnut Cottage, Hine Town Lane, Shillingstone, Blandford Forum DT11 0SN
MM3 GOE T. Macdonald Main Road Farm, Balephuil, Isle of Tiree PA77 6UE
MM3 GOI S. Adam 231/1 Gogarloch Syke, Edinburgh EH12 9JF
MM3 GOT E. Griffiths Achnamara, Heanish, Isle of Tiree PA77 6UL
M3 GOV A. Ward 81 Northbrooks, Harlow CM19 4DB
MM3 GOX C. Williams Ormer, Kirkapol, Isle of Tiree PA77 6TW
MM3 GOY E. Williams Ormer Cottage, Kirkapol, Scarinish PA77 6TW
M3 GOZ N. Gostling 49 Roundhouse Road, Dudley DY3 2AX
MM3 GPB A. Williams Ormer Cottage, Kirkapol, Scarinish PA77 6TW
M3 GPC G. Cockram 28 Tabley Road, Bolton BL3 4BR
MW3 GPG S. Griffiths 8 Heol Cynwyd, Llangynwyd, Maesteg CF34 9TB
MW3 GPJ G. Jackson 40 Ellis Avenue, Old Colwyn, Colwyn Bay LL29 9LB
MM3 GPL G. Lawrie 4 Bank House Flats, Kingussie PH21 1AP

M3 GPM P. Mansfield 106 Field Lane, Burton-on-Trent DE13 0NN
M3 GPN R. Orton 38 Whitehill Avenue, Barnsley S70 6PP
M3 GPP A. Ault 89 Southbourne Coast Road, Bournemouth BH6 4DX
M3 GPR G. Richards 1 Maple Close, Seaton EX12 2TP
M3 GPX S. Russell 4 Swin Close, Swineshead, Boston PE20 3LD
M3 GQB J. Gyton 10 Longcroft, Southdown Road, Shoreham-by-Sea BN43 5AY
M3 GQD J. Reynolds 3 Ardleigh, Basildon SS16 5RA
MW3 GQE J. Doyle 18 The Paddocks, Tonna, Neath SA11 3FD
M3 GQI K. Cramp 20 Combeland Road, Minehead TA24 6BT
M3 GQL P. Ridgers 231 The Greenway, Epsom KT18 7JE
M3 GQM T. Sheppard 1 Waveney Walk, Crawley RH10 6RL
M3 GQP A. Hall 28 Tudor Road, Chester le Street DH3 3RY
MM3 GQR S. Mciver 9 Balvicar Road, Oban PA34 4RP
MW3 GQS C. Williams Pen Y Cae, Bodeiliog Road, Denbigh LL16 5PA
MM3 GQT G. Mc Gregor 34 Cairn Road, Cumnock KA18 1HN
M3 GQW E. Tart Sunnybank Farm, Wattlesborough Heath, Shrewsbury SY5 9EG
MM3 GQY H. Dineley Banks, Burray, Orkney KW17 2ST
MW3 GRC G. Coombes 25 Afan Valley Road, Cimla, Neath SA11 3SS
M3 GRF A. Grace 2 St. Peters Crescent, Bicester OX26 4XA
M3 GRI T. Griffiths 56 The Avenue, Totland Bay PO39 0DN
M3 GRY G. Crane 35 Betjeman Avenue, Wootton Bassett, Swindon SN4 8JY
M3 GSI C. Nelmes 119 Exeter Road, Dawlish EX7 0AN
MM3 GSL G. Shaw 1 Fir Park, Sorn, Mauchline KA5 6HY
M3 GSM P. Taylor 77 Ladstone Towers, Sowerby Bridge HX6 2QP
M3 GSQ W. Howarth 68 Valentine Road, Sheffield S5 0NZ
M3 GSR W. Chave 91 Newman Road, Exeter EX4 1PQ
MI3 GSW G. Heggan 19 Lough Road, Lisburn BT276TS
M3 GTA D. Langmead 38 Milton Grove, London N11 1AX
M3 GTB G. Bland 20 Brereton Close, Castlefields, Runcorn WA7 2LR
MM3 GTF F. Davidson 27 Gordon Way, Livingston EH54 8JG
M3 GTG R. Chisholm 162 Ardington Road, Northampton NN1 5LT
M3 GTH E. Jones 43 Wesley Road, Wimborne BH21 2QB
M3 GTK J. Walker 34 Vian Road, Waterlooville PO7 5TW
MW3 GTM G. Mainwaring 3 Elias St., Neath SA11 1PP
MI3 GTO G. Shaw 49 Cloughey Road, Portaferry, Newtownards BT22 1NQ
M3 GTQ G. Thompson 28 St. Georges Road, Atherstone CV9 3BP
M3 GTT G. Hines 126 Linacre Lane, Bootle L20 6ES
M3 GTV A. Burfield 4 Eastern Crescent, Chelmsford CM1 4JQ
M3 GUF R. Glossop 21 Elizabeth Avenue, Tattershall Bridge, Lincoln LN4 4JJ
M3 GUG K. Bell 12A Mill Lane, Carlton, Goole DN14 9NG
MW3 GUH S. Tolhurst Gwernrynydd Fach, Nantmel, Llandrindod Wells LD1 6EW
M3 GUJ A. Brimble-Brice 19 Rosebery Road, Exmout EX8 1SJ
M3 GUM K. Waterhouse 74 Clifford Road, West Bromwich B70 8JY
M3 GUO S. Shaw 4 Perry Hill, Chelmsford CM1 7RD
M3 GUQ M. Reynolds 15 Michelham Close, Eastbourne BN23 8JD
M3 GUU S. Whiting 56 Station Road, Branston, Lincoln LN4 1LH
M3 GVC P. Wayer 4 Chatburn Avenue, Waterlooville PO8 8UB
MM3 GVE C. Brown 9 Newton Crescent, Rosyth, Dunfermline KY11 2QW
MW3 GVF A. Roberts 9 Llys Hendre, Rhuddlan, Rhyl LL18 5YF
M3 GVJ L. Ballinger 9 Somerville Court, Cirencester GL7 1TG
M3 GVN J. Bacheta 32 Essex Road, London E12 6RE
M3 GVT G. Finney 78 Lockley St., Stoke on Trent ST1 6PQ
MW3 GVU J. Brennan 1 Gerddi Mair, St. Clears, Carmarthen SA33 4ET
M3 GWC S. Clarkson Carisbrooke, Poolhouse Road, Wolverhampton WV5 8AZ
MW3 GWH G. Haines 12 Tyn Rhos Estate, Gaerwen LL60 6HL
MI3 GWQ T. Cosgrove 301 Russell Court, Claremont Street, Belfast BT9 6JX
M3 GWW G. Wheelhouse 86 Severn Street, Hull HU8 8TQ
MW3 GWZ P. French 4 Acacia Avenue, Newport NP19 9AT
M3 GXG M. Sartorius Holmwood, Priory Road, Sunningdale, Ascot SL5 9RH
M3 GXI B. Saunders 4 Mendip Road, Worthing BN13 2LP
M3 GXX I. Bardell 17 Stanton Avenue, Bradville, Milton Keynes MK13 7AR
M3 GYA H. Denmead 47 Holland Road, Clevedon BS21 7YJ
M3 GYB M. Peterson 29 Warwick Close, Saxilby LN1 2FT
M3 GYH R. Blake Taita, Linnards Lane, Northwich CW9 6ED
M3 GYI L. Horton 36 Merevale Crescent, Morden SM4 6HL
MM3 GYU T. Bloomfield Midyard House, Carnwath, Lanark ML11 8LH
M3 GYY G. Clarke 233 Muirhead Avenue, Liverpool L13 0AY
M3 GZD D. Blackmore 197 Cotton Lane Wa75Jb, Runcorn WA7 5JB
M3 GZE S. Lee 154 Grangeway, Runcorn WA7 5JA
MM3 GZG K. Cunningham 11 Glendoune Street, Girvan KA26 0AA
M3 GZI A. Farmar Hawkes Place, Horslett Hill, Holsworthy EX22 6RS
M3 GZJ M. Strowger 88 Castle Rise, Runcorn WA7 5XW
M3 GZP I. Plain 18 Arundel Road, Bath BA1 6EF
M3 GZQ M. Parris 19B Milfoil Drive, Eastbourne BN23 8BR
M3 GZT A. Cain 55 Lytham Green, Muxton, Telford TF2 8SQ
M3 GZU R. Lang 89 Dodthorpe, Hull HU6 9HA
M3 GZW R. Shadbolt 58 Westfield Road, Manea, March PE15 0LN
M3 HAC J. Hilton 32 Dowry St., Fitton Hill, Oldham OL8 2LP
M3 HAD H. Rhymes 12 Reedling Drive, Southsea PO4 8UF
MW3 HAE C. Davies Afallon, 3 Penygraig, Aberystwyth SY23 2JA
MM3 HAF M. Hoskin 15 East Den Brae, Letham DD9 2PJ
M3 HAI G. Young 47 Birdhill Road, Woodhouse Eaves, Loughborough LE12 8RP
M3 HAJ I. Griffiths 147 Greenlawns, St. Marks Road, Tipton DY4 0SU
M3 HAK R. Siebert Po Box 127, Nantwich CW5 8AQ
M3 HAL S. Hall Orchard End, Asenby, Thirsk YO7 3QR
M3 HAM G. Roberts 25 Chalfont Way, Liverpool L28 3QB
MW3 HAQ C. Olding 10 Ty Nant, Caerphilly CF83 2RA
MW3 HAS H. Mustafa 17 Furness Close, Ely, Cardiff CF5 4PG
M3 HAT H. Kennedy 19 High Street, East Hoathly, Lewes BN8 6DR
M3 HAU W. Kent Long Spring Cottage Gracious Lane, Sevenoaks TN13 1TJ
M3 HAW S. Avery-Hawkins 41 Daniels Welch, Coffee Hall, Milton Keynes MK6 5DA
M3 HAZ H. Flower 17 Scott Grove, Morecambe LA4 4LN
MW3 HBC M. Lewis 96 Roundhouse Close, Nantyglo, Ebbw Vale NP23 4QY
MW3 HBF D. Williams 54 Howell Street, Pontypridd CF37 4NR
M3 HBM J. Baxter 10 Speedwell Close, Bedworth CV12 0NS
M3 HBP S. Bethell 35 Fulford Road, Bristol BS13 9RL
M3 HBS J. Bodie 69 Williton Crescent, Weston Super Mare BS234QZ
M3 HBT T. Richley 30 Chicheley Road, Harrow HA3 6QL
M3 HBX D. Glenn 84 Cambridge Street, Normanton WF6 1ER
M3 HCA E. Foster 12 Dunham Grove, Leigh WN7 3DS
M3 HCB H. Benton Emiviz, The Ridge, Salisbury SP5 2LQ
M3 HCE A. Macnauton 27A Lincoln Road, Poole BH12 2HT
M3 HCG C. Hawkins 118 Aldebury Road, Maidenhead SL6 7HE
M3 HCL C. Lott 55 Juniper Road, Farnborough GU14 9XU
M3 HCP D. Hounslow 3 Hengrave Green, Ivington, Leominster HR6 0JL
MW3 HCW P. Webb 7 Cherry Close, Prestatyn LL19 7DQ
M3 HDL R. Guess 69 Rowan Drive Kirkby-In-Ashfield, Nottingham NG17 8FP
M3 HDV B. Hampson 38 Parley Road, Bournemouth BH9 3BB
M3 HEE A. Totterdell Moscar Cross House, Hollow Meadows, Sheffield S6 6GL
M3 HEI S. Collett 81 Wycombe Road, Prestwood, Great Missenden HP16 0HW
M3 HEJ E. Haycock 55 Ashbourne Road, Rocester, Uttoxeter ST14 5LF
M3 HEO A. Fagan 77 Watling Street West, Towcester NN12 6AG
M3 HER F. Nation 1 Claydon Path Stoke Mandeville, Aylesbury HP21 9EF
M3 HET H. Thomas 15 Ashgrove Way, Bridgwater TA6 4UB
M3 HEV A. Sturgess Hawks Barn, Long Lane, Shaftesbury SP7 0BJ
M3 HFA E. Gainford 10 The Spinney, Ashford TN23 3LF
M3 HFH S. Overall Flat 74, Douglas Buildings Marshalsea Road, London SE1 1EL
MW3 HFO H. Foster 11 Rosedale Gardens, Rhyl LL18 4TY
M3 HFT E. Rogers Maes Gwersyll, Garthmyl, Montgomery SY15 6RS
MM3 HFU G. Johnson Speur Mor, Gifford Road, Longformacus, Duns TD11 3NZ
M3 HFX B. Mccann 21 Ladyseat, Longtown, Carlisle CA6 5XX
M3 HGA T. Winwood 2 The Warren, Abingdon OX14 3XB
M3 HGE A. Hitchens 16 Harrisons Place, Northwich CW8 1HX
M3 HGH K. Stewart 17 Delamere Street, Bury BL9 6NE
M3 HGL B. Peck 60 Richmond Road, Ipswich IP1 4DP
M3 HGM N. Peters 57 High Street, Collingtree, Northampton NN4 0NE
MW3 HGO R. Ward Beech Cottage, Saron Road, Goytre NP4 0BN
M3 HGP P. Bland 5 Pembroke Way, Winsford CW7 1QZ
MW3 HGR H. Roberts Hen Ddol, Northfield Road, Barmouth LL42 1PT
M3 HGT D. Leverton 21 Laburnum Grove, Killamarsh, Sheffield S21 1GR
M3 HGW M. Bancroft Oak Lodge 3 The Oaks Scothern, Lincoln LN2 2WB
M3 HGX D. Clark 12 Wilson Crescent, Lostock Gralam, Northwich CW9 7QH
M3 HGZ J. Sejwacz Flat 9, Mayrick Court, Newton le Willows WA12 9GB
M3 HHB M. Clarke 40 Fingringhoe Road, Langenhoe CO5 5AD
M3 HHC S. Crossley 29 Rycroft Avenue, Bingley BD16 1PU
M3 HHN A. Highfield 29 Blewitt Street, Brierley Hill DY5 4AW
M3 HHQ D. Sejwalz 4 Ash Avenue, Newton-le-Willows WA12 8HJ
M3 HHX M. Arnott 2 Hambleton Close, Elsecar, Barnsley S74 8DS
M3 HIE B. Edwards 28 Poppy Drive, Horam, Heathfield TN21 9BL
M3 HIG T. Higgins 15 Ellen Street, Warrington WA5 0LY
M3 HIM F. Doyle 1 Claydon Path Stoke Mandeville, Aylesbury HP21 9EF
M3 HIN A. Watkinson 34 Marble House, Felspar Close, London SE18 1LN
M3 HIO S. Jarvis 18 Savernake Close Rubery, Rednal, Birmingham B45 0DD
M3 HIP J. Dixon 6 Howland Close, Eastbourne BN23 5AJ
M3 HIT A. Bryan 16 Walesmoor Avenue, Kiveton Park, Sheffield S26 5RG
MW3 HIX P. Owen 24 Sirhowy Court, Green Meadow, Tredegar NP22 4PL
M3 HJB H. Beier 20 Markham Avenue, Armthorpe, Doncaster DN3 2AZ
MM3 HJC C. Hazle 6 Dalneigh Road, Inverness IV3 5AH
M3 HJE H. Evans Littlefield House, Bolney Road, Haywards Heath RH17 5AW
M3 HJF S. Gilchrist Kening, Ashleigh Crescent, Barnstaple EX32 8LA
M3 HJG A. Howe 18 Co-Operation Street, Crawshawbooth, Rossendale BB4 8AG
M3 HJJ S. Norris 15 East View, Choppington NE62 5UF
M3 HJU P. Webster Magoos Gweek, Helston TR12 6TX
M3 HJV C. Lishman 6 Clarence Road, Accrington BB5 0NA
M3 HJW R. Smith 21 Lambert Close, Waterlooville PO7 5XA
MM3 HKE L. Higgins 11 Strathyre Place, Broughty Ferry, Dundee DD5 3WN
MW3 HKH S. Crighton 12 Cwm Road, Waunlwyd, Ebbw Vale NP23 6TR
M3 HKM K. Monaghan The Bulstone Hotel, Branscombe, Seaton EX12 3BL
M3 HKV L. Flawn Autumns, Jacks Lane, Bishop's Stortford CM22 6NT
M3 HLA J. Meeks 64 Belford St., Burnley BB12 0DF
M3 HLD R. Metcalfe 33 Midland Terrace, Hellifield, Skipton BD23 4HJ
MM3 HLG S. Paul 10 Beechwood Gardens, Westhill AB32 6YE
M3 HLN H. Noel 58 Easenhall Lane, Redditch B98 0BJ
M3 HLP P. Hallas 37 Oakfield Road, Bromborough, Wirral CH62 7BA
M3 HLV J. Ferguson 41 Brunswick St., Burnley BB11 3NX
M3 HLX B. Buskin 6 Elgin Close, Bedlington NE22 5HJ
MJ3 HMA M. Haddon Balik Pulau, Bradford Ave, La Route Des Genets, St. Brelade JE3 8DP Jersey
M3 HME E. Cotton 98 Severn Street, Hull HU8 8TQ
M3 HMK H. Knighton 5 Quidenham Road, East Harling, Norwich NR16 2JD
M3 HML B. Just 2 The Old Rectory, The High Road, Bedford MK43 7HN
MM3 HMM D. Macmillan 2 Fladda Road, Oban PA34 4HZ
M3 HMT H. Tate 52 Marlborough Road, London N22 8NN
M3 HNE M. Ellis 58 Egghill Lane, Northfield, Birmingham B31 5NT
M3 HNK T. Lee 68 Wharton Drive, Springfield, Chelmsford CM1 6BF
M3 HNL J. Stone 27 The Heathlands, Warminster BA12 8BU
M3 HNM N. Evans 38 Cockster Road, Longton, Stoke-on-Trent ST3 2EG
MW3 HNP J. Nelson 31 Y Drim, Ponthenry, Llanelli SA15 5NY
M3 HNQ I. Rowland 45 Birks Road, Mansfield NG19 6JU
M3 HNU G. Sandell 20 Kirkby View, Sheffield S12 2NB
M3 HNV P. Breckell 45 Gordon Avenue, Mansfield NG18 3AZ
M3 HOD A. Hodson 22 Walmley Ash Road, Sutton Coldfield B76 1HY
M3 HOE A. Hoe 12 Ashbridge Rise, Chandler'S Ford, Eastleigh SO53 1SA
M3 HOM J. Homsey 105 Lynwood, Folkestone CT19 5DD
M3 HOQ N. Lambert 17 Starcross Road, Weston-Super-Mare BS22 6NY
M3 HOU E. Edwards 5 Brindley Road, Silsden, Keighley BD20 0LD
M3 HOV B. Brown 3 Swaledale, Worksop S810UY
MW3 HOY M. Hoy 39 Blackbird Road, Caldicot NP26 5RE
M3 HPM D. Woodward 140 Ewe Lamb Lane, Bramcote, Nottingham NG9 3JW
M3 HPN W. Hodgson 11 Tudor Court, Hitchin SG5 2BE
M3 HPO A. Riches 84 Elgar Drive, Shefford SG17 5RA
M3 HPT P. Taylor Fenway Farm, Ten Mile Bank, Downham Market PE38 0EU
M3 HPY K. Tokley 9 Peel Road, Springfield, Chelmsford CM2 6AQ
M3 HPZ A. Bidwell 134 Milton Road, Weston-Super-Mare BS23 2US
M3 HQB C. Smith 71 Connaught Road, Luton LU4 8ER
MM3 HQC J. Henry 7 Wenlock Road, Paisley PA2 6UJ

MW3 HQD R. Wilkes 9 Brynmawr Road, Ebbw Vale NP23 5FF
MM3 HQL G. Fuller 20 Drumellan Road, Ayr KA7 4XA
M3 HQN M. Haworth 26 Willowhey, Marshside, Southport PR9 9TW
M3 HQP J. Parrott 2 Boyd Close, Wirral CH46 1RX
M3 HQQ H. Dixon 45 Penkhull Terrace, Stoke-on-Trent ST4 5DH
M3 HQS M. Williamson 5 Fernbank Close, Crewe CW1 6ES
M3 HQU M. Copeman 1 Chestnut Avenue, Welney, Wisbech PE14 9RG
MW3 HQV E. Carter 34 Wrexham Road Brynteg, Wrexham LL11 6HR
M3 HQW A. Hillbeck 28 Darent Avenue, Walney, Barrow-in-Furness LA14 3NU
MW3 HRE S. Esp 34 Wrexham Road Brynteg, Wrexham LL11 6HR
M3 HRM D. Morgan 171 Town Road, London N9 0HJ
M3 HRN C. Fower 31 Hillswood Avenue, Leek ST13 8EQ
M3 HRT S. Brashill 42 Bannister Street, Withernsea HU19 2DT
M3 HRV D. Porter 39 Panama Road, Burton-on-Trent DE13 0SQ
M3 HRY P. Odle 24 Longfellow Road, Gillingham ME7 5QG
M3 HSC J. Taggart 250 Thomas Drive, Liverpool L14 3LF
M3 HSE C. Hoyle 43 Helme Drive, Kendal LA9 7JB
M3 HSH K. White 30 Nuneaton Road, Bedworth CV12 8AL
M3 HSI A. Mcgann 8 Hertford Close, Whitley Bay NE25 9XH
M3 HSJ D. Teasdale 43 Easington Road, Stockton-on-Tees TS19 8ES
M3 HSM C. Hayes 7 Harries Court, Waltham Abbey EN9 3NS
M3 HSR P. Hilton 14 Masefield Road, Thatcham RG18 3AF
M3 HSS J. Little 41 Sevenoaks Road, Portsmouth PO6 3JP
M3 HSV C. Srinivasan 2 Hall Drive, Burley In Wharfedale, Ilkley LS29 7LL
M3 HSW H. White 96 Clarendon Road, Luton LU2 7PJ
MW3 HSZ P. Bishop 76 Heol Homfray, Cardiff CF5 5SB
M3 HTA S. Fuller 29 Beckley Road, Wakefield WF2 9QB
M3 HTE E. Townley Beetham, Water Hill Lane, Halifax HX2 7SG
M3 HTF D. O'Flanagan 16 Corbett Road, London E11 2LD
M3 HTG A. Murphy 22 Shenley Fields Drive, Birmingham B31 1XH
M3 HTO P. Hardy 21 West Avenue, Boston Spa, Wetherby LS23 6EJ
M3 HTR S. Adlam 50 High St., Westtown, Dewsbury WF13 2QF
MM3 HTY D. Cunningham 105 Lower Bathville Armadale, Bathgate EH48 2JS
M3 HUB A. Hubbard 70 Little Morton Road North Wingfield, North Wingfield S42 5HN
M3 HUG R. Hughes 7 Willow Place, Darlington DL1 5LX
M3 HUN P. Hunter 7 Fairfield Avenue Hilcote, Alfreton DE55 5HL
M3 HUS N. Payne 19 Sid Park Road, Sidmouth EX10 9BW
M3 HUW H. Weatherhead 39 Meadow Park, Dawlish EX7 9BU
M3 HUX N. Gibson 19 Nene Side Close, Badby, Daventry NN11 3AD
M3 HUY D. Dolan 29 Byland Way, Monk Bretton, Barnsley S71 2JY
M3 HVA P. Robinson 8 Clewley Road, Branston, Burton-on-Trent DE14 3JE
M3 HVE R. Dolman 3 Cloonmore Avenue, Orpington BR6 9LE
M3 HVH M. Bell 25 Derwent Crescent, Barnsley S71 3QU
M3 HVL K. Phizacklea 23 High Duddon Close, Askam-in-Furness LA16 7EW
M3 HVN P. Harris 17 Seymour Avenue, Great Yarmouth NR30 4BB
M3 HVO G. Omar 140 Twickenham Road, Isleworth TW7 7DJ
M3 HVP T. Moss 3 Haddon Close, Macclesfield SK11 7YG
M3 HVS A. Hunt 14 Offranville Close, Leicester LE4 8NR
M3 HVU K. Jessop 61 Fountayne Street, Goole DN14 5HQ
M3 HVV K. Ozwell 109 Abbey Road, Grimsby DN32 0HN
M3 HVW J. Naylor 12 Princess Avenue, Wesham, Preston PR4 3BA
M3 HVX D. Proctor 58 Hornby Drive, Newton, Preston PR4 3SU
M3 HVY C. Hacker 49 Lamaleach Drive, Freckleton, Preston PR4 1AJ
MJ3 HWC H. Carrel 5 Belmont Road, St. Helier JE2 4SA Jersey
M3 HWN D. Wardman 45 The Grainger, North West Side, Gateshead NE8 2BG
M3 HWP D. Potts 19 Clay Street, Workington CA14 2XZ
M3 HWS J. Cleverley 4A Godfrey Street, Netherfield, Nottingham NG4 2JG
M3 HWV A. Hicks 30 Manna Drive, Elton, Chester CH2 4RP
M3 HWW H. Wilson 2 Railway Close, Burwell, Cambridge CB25 0DW
M3 HWX D. Beer 46 The Mailyns, Gillingham ME8 0DZ
M3 HWY G. Elsworthy 40 Moorfield Way, Wilberfoss, York YO41 5PL
M3 HXB B. Wheat 62 Havenwood Rise, Nottingham NG11 9HE
M3 HXF J. Merchant 186 Manor Hall Road, Southwick, Brighton BN42 4NH
M3 HXG D. Haestier 18 Midhurst Rise, Brighton BN1 8LP
M3 HXH J. Clarke 144 St. Johns Avenue, Kidderminster DY11 6AU
M3 HXM W. Cole The Spinney, Holmes Chapel Road, Congleton CW12 4SN
M3 HXO M. Shaw 10 Beechwood Avenue, Shevington, Wigan WN6 8EH
M3 HXQ T. Johnson 27 Fonthill Road, Bristol BS10 5SR
M3 HXS M. Campbell 71 Sages Lea Woodbury Salterton, Exeter EX5 1RA
M3 HXW K. Bouris 3 Suffolk Court, Vicarage Road, Maidenhead SL6 7DT
M3 HYD M. Douglas 13 Castlereagh Street, New Silksworth, Sunderland SR3 1HJ
M3 HYE D. Bruce 6 Princes Way, King's Lynn PE30 2QL
M3 HYF A. Sargent 15 Wilton Road, Balsall Common, Coventry CV7 7QW
MM3 HYG P. Mcarthur 22 Bridgeway Terrace, Kirkintilloch, Glasgow G66 3HJ
M3 HYI A. Reynolds Fairview, Coombe Way, Teignmouth TQ14 9QA
M3 HYO E. Hearne 6 Hillview Road, Basingstoke RG22 6BQ
M3 HYQ W. Oakley 1 Southern Avenue, Henlow SG16 6EY
M3 HYV K. Jones Court House Farm, Holmes Chapel Road, Crewe CW4 8AS
M3 HZA D. Pryor 10 Thornton Crescent, Church Langton, Market Harborough LE16 7TA
MW3 HZB K. Clark 56 Morris Avenue, Llanishen, Cardiff CF14 5JW
M3 HZC C. Chen Conville And Cains College, Trinity Street, Cambridge CB2 1TA
M3 HZD D. Evans 7 Bowerwood Road, Fordingbridge SP6 1BJ
M3 HZE D. Mannion 17 Balmoral Road, Haslingden, Rossendale BB4 4EA
M3 HZH D. Hubbard 99 Tuckers Road, Loughborough LE11 2PH
M3 HZK Dr R. Gooch 14 Cotterill Road, Surbiton KT6 7UN
M3 HZM M. Holden 26 Valebridge Drive, Burgess Hill RH15 0RW
M3 HZN R. Rippin 28 Ridgeway West, Market Harborough LE16 7LG
M3 HZO P. Sarll 81 Austendyke Road, Weston Hills, Spalding PE12 6BX
M3 HZP B. Holden 26 Valebridge Drive, Burgess Hill RH15 0RW
M3 HZW J. Trevarrow 17 Harland Road, Elloughton, Brough HU15 1JT
M3 IAA D. Rose Homeberry House, 13 Ashcroft Gardens, Cirencester GL7 1RU
M3 IAC I. Crabb 9A Lonsdale Road, Southend-on-Sea SS2 4LZ
M3 IAE S. Wilkinson 144 West End Road, Morecambe LA4 4EF
M3 IAF I. Firby 19 St. Georges Drive, Manchester M40 5HL
MM3 IAG I. Gerrard 10 Station Road, Ardersier, Inverness IV2 7ST
MI3 IAI T. Scott 25 Lisavon Drive, Belfast BT4 1LJ
M3 IAO A. Hirst 34 Woodhall Avenue, Bradford BD3 7BU
M3 IAP G. Evans 57 Lock Crescent, Kidlington OX5 1HF
M3 IAQ P. King Philinda, Carmen Street, Saffron Walden CB10 1NR
M3 IBE M. Hagan 186 Saltwell Road, Gateshead NE8 4XH
M3 IBJ D. Dickinson 8 East Grange Garth, Leeds LS10 3EJ
MM3 IBM C. Mckillop 7 Auchneagh Farm Lane, Greenock PA16 7BJ
M3 IBS F. Carter 26 Union Road, Shirley, Solihull B90 3DQ
M3 IBT J. Ferrol The Hayloft, Brunstock CA64QG
M3 IBY J. Leaman 40 Higher Budleigh Meadow, Newton Abbot TQ12 1UL
M3 IBZ D. Langridge 12 Battles Lane, Kesgrave, Ipswich IP5 2XF
M3 ICA A. Schmidt Church Corner, Fieldside, Boston PE22 7RA
MM3 ICD W. Ton 24 Craigmount Hill, Edinburgh EH4 8DL
M3 ICH M. Green 4 Tudor Court, Grimethorpe, Barnsley S72 7NA
M3 ICN A. Groat 23 Knightwake Road, New Mills, High Peak SK22 3DQ
M3 ICO L. Widdowson 11 Belmont Drive, Staveley, Chesterfield S43 3PQ
M3 IDA D. Cothey Summerley House, Skircoat Moor Road, Halifax HX3 0HA
M3 IDB N. Dallen 13, Epsom Court, 40 Upper High Street, Epsom KT17 4ER
M3 IDC F. Harris 51 Hillmans Road, Newton Abbot TQ12 1AA
M3 IDD D. Barker 21 Boundary Crescent Lower Gornal, Dudley DY3 2HJ
M3 IDF J. Kelly 1 Bramble Close, New Ollerton, Newark NG22 9TN
M3 IDH K. Reynolds 3 Lilac Close, Chelmsford CM2 9NY
M3 IDK D. Norman 22 Stirling St., Hull HU3 6SL
M3 IDO A. Thompson 3 Rufford Road, Long Eaton, Nottingham NG10 3FP
M3 IDQ A. Stevenson 2 Diddington Close, Bletchley, Milton Keynes MK2 3EB
MM3 IDR K. Tait 33 Bankton Avenue, Livingston EH54 9LD
M3 IDW J. Valle Espin 203 Broadway, Horsforth, Leeds LS18 4HL
M3 IDY J. Taylor Roseneath, 4C Valley Road, Kenley CR8 5DG
M3 IEA P. Jays 138 Lower Wear Road, Exeter EX2 7BD
MM3 IEC E. Cohen 234 Allison Street, Glasgow G42 8RT
M3 IEF S. Painting 15 Surrey Walk, Walsall WS9 8JG
M3 IEG M. Luxton 6 Tumbling Field Lane, Tiverton EX16 4LN
M3 IEL R. Tust 28 Osprey Close, Beechwood, Runcorn WA7 3JH
M3 IEM T. Rogers 45 Church Road, Westoning, Bedford MK45 5LP
M3 IEP N. Freeman 27 Montpelier Drive, Caversham, Reading RG4 6QA
M3 IEQ J. Wheeler 41 Winnards Park, Sarisbury Green, Southampton SO31 7BX
M3 IET C. Blount 55 Silverthorne Drive, Caversham, Reading RG4 7NR
M3 IEU T. Bougourd 1 Poplar Close, Newton Abbot TQ12 4PG
M3 IEV M. Blount 55 Silverthorne Drive, Caversham, Reading RG4 7NR
M3 IEW P. Mellish 302 Belvedere Road, Burton-on-Trent DE13 0RD
M3 IFA A. Harrison 16 Ingshead Avenue, Rawmarsh, Rotherham S62 5BH
M3 IFB D. Symonds 2 Montgomery Cottages, Stonham Road, Stowmarket IP14 5LS
M3 IFE M. Ferguson 80 Chester Road, Holmes Chapel, Crewe CW4 7DR
M3 IFF P. Webster 15 Napier Street, Workington CA14 2PT
M3 IFG F. Gatenby 6 Telford Close, Audenshaw, Manchester M34 5FB
MI3 IFI D. Sloan 15 Deramore Drive, Portadown, Craigavon BT62 3HH
M3 IFJ B. Royce 82 Ridge Lane, Watford WD17 4TA
M3 IFK K. Holloway 26 Aldgate Drive, Brierley Hill DY5 3NT
M3 IFM J. Lomako 23 Stretton Close Ackton, Pontefract WF7 6HT
MI3 IFO E. Mcclements 5 Eastbank, Strathfoyle, Londonderry BT47 6UW
MW3 IFZ N. Bruines 24 Trenel, Burry Port SA16 0UT
M3 IGA Dr L. Igali 22 Mile End Road, Norwich NR4 7QY
M3 IGN I. Nicholls 8 Northcroft Road, Corsham SN13 0LS
MI3 IGO A. Blythe 159 Victoria Road, Bready, Strabane BT82 0DZ
MW3 IGZ M. Matthias 18 Brynmally Park, Pentre Broughton, Wrexham LL11 6BP
M3 IHA S. Duffy 38 Well Lane, Newton, Chester CH2 2HL
MW3 IHB G. Jones Wern Lodge, Gobowen, Oswestry SY10 7JY
M3 IHC G. Ward 5 Allerton Road, Shrewsbury SY1 4QQ
MW3 IHD I. Davies 221 Trowbridge Green, Rumney, Cardiff CF3 1RE
M3 IHN R. Hargeaves 58 Horsewell Lane, Wigston LE18 2HQ
M3 IHO K. Brown 143 Princes Road, Ellesmere Port CH65 8EP
M3 IHQ T. Wall 50 Higham Gobion Road, Barton-Le-Clay, Bedford MK45 4LT
M3 IHR D. Tilley 15 Dowhills Park, Liverpool L23 8SS
M3 IHS C. Ivers 11 Twelve Acre Crescent, Farnborough GU14 9PW
M3 IHU J. Smith Wilson Hall Farm, Slade Lane, Derby DE73 1AG
M3 IHV Dr H. Donnelly 6 Famet Walk, Purley CR8 2DY
MW3 IHX D. Eacott-Palfrey 165 High Street, Blaina, Abertillery NP13 3AW
MI3 IHY S. Quigg 100 Whispering Pines, Limavady BT49 0UF
MI3 IHZ K. Mcdonald 37 Ardgarvan Cottages, Limavady BT49 0NF
M3 IIA R. Dale 17 Spencer Gardens, Brackley NN13 6AQ
MI3 IIH T. Mcconnell 41 Moyra Road, Doagh, Ballyclare BT39 0SQ
MW3 IIJ R. Parry 5 Accar Y Forwyn, Denbigh LL16 3PW
MI3 IIL C. Mcconnell 41 Moyra Road, Doagh, Ballyclare BT39 0SQ
M3 IIN B. Stokes 19 Hall Park, Barrow-On-Trent, Derby DE73 7HD
M3 IIP K. Hunt 13 Beaumaris Court, Spondon, Derby DE21 7RG
MM3 IIT G. Saunders Tower Guest House, 32 James Street, Stornoway HS1 2QN
M3 IIV J. Hurkett 9 Fair Field Park, Five Lanes, Launceston PL15 7RQ
M3 IJD J. Doyle 4 Columbia Road, Prescot L34 2SB
M3 IJE I. Ewen 26 Court Road, Eastbourne BN22 9EZ
M3 IJF E. Livesey Reevsmoor, Hollington, Ashbourne DE6 3AG
MM3 IJI M. Carmichael 39 Longsdale Crescent, Oban PA34 5JR
M3 IJO P. Chaney 246 Agar Road, Illogan Highway, Redruth TR15 3NJ
M3 IJS I. Sapstead 7 Shrubbery Grove, Royston SG8 9LJ
M3 IJT C. King 1 Victoria Court, Hadley, Telford TF1 5FL
M3 IJV J. Millichip 33 Lincoln Road, Stevenage SG1 4PJ
M3 IJZ C. Young 15 Shelton Avenue, East Ayton, Scarborough YO13 9HB
MW3 IKC J. Kenchington 36 Lando Road, Pembrey, Burry Port SA16 0UR
M3 IKD J. Hockedy 22 Victoria Road, Frome BA11 1RR
M3 IKE M. Murray Po Box 55, Calle San Jaime, Benijofar 3178 Spain
M3 IKI J. Batson 19 Seaview Road, Canvey Island SS8 7PB
M3 IKJ A. Knitter 15 Thompson Drive, Hatfield, Doncaster DN7 6JX
M3 IKM T. Palmer 70 Channel View Road, Eastbourne BN227LL
M3 IKN C. Staite 11 Lavender Lane, Great Denham MK40 4SB
M3 IKR D. Powis Fircroft, Pound Lane, Woodbridge IP13 0LN
MM3 IKS G. Frew 20 Achlonan, Taynuilt PA35 1JJ
M3 IKV E. Smith Grange Farm, Main Street, Newark NG23 5PX
M3 ILB N. Silveston 28 Quernstone Lane, Northampton NN48UN
M3 ILG B. Evans 12 The Mead Thaxted, Dunmow CM6 2PU
M3 ILJ E. Copper 238 Canterbury Road, Kennington, Ashford TN24 9QL
M3 ILM N. Hickson 27 Cressing Road, Witham CM8 2NP

M3 ILR I. Roper 1 Holywell Road, Kilnhurst, Mexborough S64 5UQ
M3 ILV P. Hinchliffe 21 Prospect Hill, Haslingden, Rossendale BB4 5EF
M3 ILY M. Andrews 27 Bramble Avenue, Norwich NR6 6LN
M3 ILZ B. Mcandrew 8 Springhill Walk, Morpeth NE61 2JT
M3 IMB I. Berry 4 Newlands Park Way, Newick, Lewes BN8 4PG
MM3 IMC I. Mccuaig 1 Ericht Bank Drive, Kirn, Dunoon PA23 8HB
M3 IMJ P. Coppin 3 Firtree Close, Rough Common, Canterbury CT2 9DB
MM3 IMK I. Mackinnon 6 Glencruitten Rise, Oban PA34 4RX
M3 IMM I. Margetts 48 Spetchley Road, Worcester WR5 2NL
MI3 IMO T. Mcnaughter 36 Elms Park, Coleraine BT52 2QE
M3 IMP I. Phillpott 14 Buttercup Close, Paddock Wood, Tonbridge TN12 6BG
M3 IMR A. Ashworth 22 Crow Lane, Ramsbottom, Bury BL0 9BR
M3 INC A. Nicholls 117 Hurligham Road, Kingstanding B440NG
M3 IND V. Lowe 35 Elm Place, Armthorpe, Doncaster DN3 2DE
M3 INH Dr E. Hayes 11 Ashleigh Wood, Monaleen Ireland
M3 INJ A. Hayes 11 Ashleigh Wood, Monaleen Ireland
M3 INL I. Lockyer 11 Lorina Road, Ramsgate CT12 6DD
M3 INO G. Patterson 28 Highcliffe, Spittal, Berwick-upon-Tweed TD15 2JH
M3 INQ Z. Ardern 6 Chaucer Avenue, Mablethorpe LN12 1DA
MI3 INS D. Quigg 9 Springhill Terrace, Limavady BT49 9BS
MM3 IOF M. Mitchell Easter Kilwhiss Farm, Ladybank, Cupar KY15 7UR
MI3 IOH W. Bradley 16 Mullaghanagh Road, Dungannon BT71 7AY
MJ3 IOJ N. Taylor 21 Samares Avenue, La Grande Route De St. Clement South, St. Clement JE2 6NY Jersey
M3 IOK R. Peel 34 Pagdin Drive, Styrrup, Doncaster DN11 8LU
MM3 IOM J. Grundey 6 Ternemny Villas, Knock, Huntly AB54 7LR
M3 IOQ B. Reynard 90 Barnsley Road, Darton, Barnsley S75 5NS
M3 IOT L. Bewley 21 Duloe Gardens, Pennycross, Plymouth PL2 3RS
M3 IOX S. Bridges 140 Highbridge Road, Burnham-on-Sea TA8 1LW
M3 IPD K. Barron 80 Primrose Crescent, Norwich NR7 0SF
MW3 IPK T. Vincent 88 Lake St., Ferndale CF43 4HE
M3 IPM S. Jackson 1 The Avenue, Burton-Upon-Stather, Scunthorpe DN15 9EX
M3 IPQ A. Badcock 7 Heathfield Road, Chandler'S Ford, Eastleigh SO53 5RP
M3 IPT W. Bull 117 Walton Road, Wednesbury WS10 0EU
M3 IPY L. Earnshaw 63 Manor Road, Fleetwood FY7 7LJ
M3 IPZ P. Wyles Casa De La Rosa, Torbay Road, Torquay TQ2 6RG
MM3 IQA G. Robbins 33 Moffat Court, Glenrothes KY6 1JR
MM3 IQD M. Gourlay 14 Holmes Holdings, Broxburn EH52 5NS
M3 IQF D. Green 12 Nostell Road, Ashton-In-Makerfield, Wigan WN4 9XD
M3 IQG S. Parris 3 Manor Close, Ringmer, Lewes BN8 5PA
M3 IQJ C. Evans Bridge Farm, Shrawley, Worcester WR6 6TQ
M3 IQM J. Fitzpatrick 21 Corn Close South Normanton, Alfreton DE55 2JD
M3 IQN M. Davis 63 Glebe Road, Barrington, Cambridge CB22 7RP
M3 IQP L. Bailey 18 Dudley Place, St. Helens WA9 1BL
M3 IQQ A. Harris 32 King Edward Road, Gillingham ME7 2RE
M3 IQS M. Cox 7 Wilson Close, Daventry NN11 9WH
MM3 IQU W. Curry 36 Banklands Newburgh, Cupar KY146DN
MW3 IQY S. Beer 49 Central Street, Pwllypant, Caerphilly CF83 2NJ
M3 IRF S. France 13 Cirencester Close, Little Hulton, Manchester M38 9HB
M3 IRH I. Harris 19 Holmrook Road, Carlisle CA2 7TB
M3 IRJ G. Rogers 25 Easton Road, Wirral CH62 1DR
M3 IRM J. Richardson 3 Aylesbury Avenue, Urmston, Manchester M41 0SB
M3 IRP P. Cannam 1 Field Close, Hinckley LE10 1TH
M3 IRQ S. Cannam 82 Barwell Lane, Hinckley LE10 1SS
M3 IRR A. Gregory 140 Alder Street, Newton-le-Willows WA12 8HP
M3 IRS D. Mountford 189 Lloyd St., Stockport SK4 1NH
M3 IRU P. Dawson 88 Urmson Road, Wallasey CH45 7LQ
MI3 IRV E. Mercer 8 Woodside Gardens, Portadown, Craigavon BT62 1EW
M3 IRX A. Ryan 60 Stanstead Road, Halstead CO9 1YB
MI3 IRY W. Cooney 30 Clanbrassil Park, Portadown, Craigavon BT63 5XT
MM3 ISA I. Mcdermid 52 Main Street, Pathhead EH37 5QB
MI3 ISC P. England 30 Kernan Grove, Portadown, Craigavon BT63 5RX
M3 ISG A. Evans 58 Lime Tree Avenue, Crewe CW1 4HL
M3 ISH S. Brooks 7 Mayfield Road, Northwich CW9 7AS
M3 ISI S. Russon 165 Billington Avenue, Newton-le-Willows WA12 0AU
M3 ISJ M. Turner 72 Neville Street, Newton-le-Willows WA12 9DB
M3 ISN A. Dickinson 77 Ullswater Avenue, Warrington WA2 0NQ
M3 ISO I. Butler 11 School Close, Basingstoke RG22 5FY
M3 ISQ R. Lilley 3 Coultshead Avenue, Billinge, Wigan WN5 7HS
MI3 ISX S. Butler 25 Chippendale Avenue, Bangor BT20 4PX
M3 ISY M. Edwards Rouse Farm, Normans Lane, Warrington WA4 4PY
MM3 ITA J. Veal 6 Morrison Avenue, Tranent EH33 2AR
M3 ITH P. Cowin 16 West Lane Shap, Penrith Cumbria CA10 3LT
M3 ITK D. Willson Flat 26, King Charles Place, Shoreham-by-Sea BN43 5JH
M3 ITL I. Buckton 67 Tennyson Avenue, Middlesbrough TS6 7ND
M3 ITM S. Garthwaite 278 Carlton Road, Barnsley S71 2BA
M3 ITT D. Jones 77 Brinkburn Grove, Banbury OX16 3WX
M3 ITU M. Stephenson 15 Springwood Road, Hoyland, Barnsley S74 0AZ
M3 ITZ S. Heywood 16 Edinburgh Drive, Hindley Green, Wigan WN2 4HL
M3 IUC M. Sturt1 2 Golden Villas Heathfield Road, Freshwater I.O.W PO40 9LQ
M3 IUH A. Morris 22 Dixon Avenue, Newton-le-Willows WA12 0NE
M3 IUK A. Mason 16 Newstead View, Fitzwilliam, Pontefract WF9 5DP
MW3 IUS I. Canterbury Brynllethryd Bungalow, Senghenydd, Caerphilly CF83 4HJ
M3 IUV P. Watson 32 Shrewsbury Way, Saltney, Chester CH4 8BY
M3 IUX R. Higham 17 Walkmill Gardens, Wellington, Seascale CA20 1EF
M3 IUZ B. Homer 7 King Street, Quarry Bank, Brierley Hill DY5 2DH
M3 IVA K. Yates 103 Raleigh Crescent, Stevenage SG2 0EB
M3 IVD A. Taplin 45 Sycamore Grove, Southend-on-Sea SS2 5HE
M3 IVI W. Ivison 11 Durham St., Fence Houses, Houghton le Spring DH4 6LA
M3 IVN G. Newman 5 Edward Crescent, Skegness PE25 3SA
M3 IVO W. Gissing 2 Yeo Moor, Clevedon BS21 6UQ
M3 IVV G. Merrington Cartref, Ball Lane, Frodsham WA6 8HP
M3 IVX R. Balm 250 Coppice Road, Arnold, Nottingham NG5 7HF
M3 IVY S. Dixon 5 Swanmore Road, Havant PO9 4LG
MW3 IWC M. Phillips 45 Lewis St., Aberbargoed, Bargoed CF81 9DZ
M3 IWG J. Pusey 29 Arthur Moody Drive, Carisbrooke, Newport PO30 5JR
M3 IWJ B. Larman Cornhill, Mount Bovers Lane, Hawkwell SS5 4JE
M3 IWK D. Judge 12 Heelas Road, Wokingham RG41 2TL
M3 IWN S. Reynolds 2 Lawson Court, Boldon Colliery NE35 9NH
M3 IWO W. Hall 16 Barrington Close, Chelmsford CM2 7AX
M3 IWR J. Chapman South View, Mill End Rushden, Buntingford SG9 0SU
M3 IWT M. Champness 10 Isaac Square, Great Baddow, Chelmsford CM2 7PP
M3 IWX S. Evans 7 Gloster Ropewalk, Dover CT17 9ES
M3 IWZ A. Hanna 35 Orchard Drive, Mayland, Chelmsford CM3 6EP
M3 IXC D. Watson 10 Gimson Close, Tuffley, Gloucester GL4 0YQ
M3 IXD D. Watson 45 Kennel Lane, Brockworth, Gloucester GL3 4NP
M3 IXE H. Hector 71 Edinburgh Drive, North Anston, Sheffield S25 4HB
M3 IXF M. Lucas 38 Hazel Avenue, Braunton EX33 2EZ
M3 IXH R. Maas 15 Pine Court, Attleborough NR17 2HU
MW3 IXJ A. Littleford 19 Llys Arthur, Towyn LL22 9PH
M3 IXK A. Pickett 4 Trembel Road, Mullion, Helston TR12 7DY
M3 IXM P. Bell 7 Greenwood Way, Norwich NR7 9HW
M3 IXO D. Lowe 5 Daisy St., Bury BL8 2QG
M3 IXT S. Widdowson 11 Belmont Drive, Staveley, Chesterfield S43 3PQ
M3 IXU D. Skinner 77 Rolleston Avenue, Petts Wood, Orpington BR5 1AL
M3 IXX D. Cattermole 39 Moor Lea, Braunton EX33 2PF
M3 IXY A. Guest 1 Green Meadows, Cannock WS12 3YA
MW3 IXZ H. Alban 15 Catherine Close, Abercanaid, Merthyr Tydfil CF48 1YY
M3 IYE P. Ellwood 13 Wensley Drive, Manchester M20 3DD
M3 IYG A. Dixon Flat 4, Farmer House, London SE16 4BY
MI3 IYH A. Wilson 108A Salia Avenue, Carrickfergus BT38 8NE
M3 IYO K. Peabody 23 Grange Mount, West Kirby, Wirral CH48 6ET
MI3 IYP S. Kelly 1 Ardnamoyle Park, Londonderry BT48 8HN
M3 IYX P. Chapman 4 The Street, Sutton, Pulborough RH20 1PS
M3 IYY J. Side Railway Crossing Cottage, Ash Road, Sandwich CT13 9JB
M3 IZB P. Lewis 37 Speedwell Close, Melksham SN12 7TE
M3 IZD G. Curtis 11 Bloomery Way, Maresfield, Uckfield TN22 2DP
M3 IZH J. Winson 41 Windsor Crescent, Ilkeston DE7 4HD
M3 IZI D. Groom 10 Sunnymead Road, Burntwood WS7 2LL
M3 IZJ J. Winson 12 Rydal Ave, Long Eaton NG10 4EB
M3 IZM K. Winson 41 Windsor Crescent, Ilkeston DE7 4HD
M3 IZN A. Lodge 45 Laneside Avenue, Sutton Coldfield B74 2BU
MM3 IZO E. Stuart 6D Dundee Street, Letham DD8 2PQ
M3 IZP B. Johnson 199 Lynwood, Folkestone CT19 5TA
M3 IZQ K. Coad 17 Dilly Lane, Barton On Sea, New Milton BH25 7DQ
MW3 IZV C. Sadler Brimaston Cottage, Hayscastle, Haverfordwest SA62 5PW
M3 IZW J. Reynolds Fairview, Coombe Way, Teignmouth TQ14 9QA
M3 JAB J. Butler 219 Ridge Avenue, Burnley BB10 3JF
M3 JAC J. Sefton 87 Lillibrooke Crescent, Maidenhead SL6 3XL
M3 JAL A. Lloyd 10 Makepeace Close, Vicars Cross, Chester CH3 5LU
MW3 JAP J. Phillips 39 Bryn Glas, Rhosllanerchrugog, Wrexham LL14 2EA
M3 JAZ J. Hudspeth 108 Fir Tree Lane, Burtonwood, Warrington WA5 4NE
M3 JBB J. Benson 51 Hollowood Avenue, Littleover, Derby DE23 6JD
M3 JBE J. Brocklesby 34 Sinnington End, Highwoods, Colchester CO4 9RE
M3 JBF N. Morphew 5 Canterbury Close, Canterbury Road, Folkestone CT19 5EL
M3 JBK S. Glass 16 Norman Way, Colchester CO3 4PS
M3 JBM M. Butchers 67 Keepers Coombe, Bracknell RG12 0TW
M3 JBQ J. Daniells Holly Villa, Foxholes, Wem, Shrewsbury SY4 5UJ
M3 JBW J. Wolohan 24 Granby Close, Corby NN18 0AG
M3 JBZ A. Bell 28 Haven Baulk Avenue, Littleover, Derby DE23 4BJ
M3 JCA C. Ashman 40 St. Matthews Road, Kettering NN15 5HE
MI3 JCB M. Crawford-Baker George'S Nest, 131 Gobbins Road, Larne BT40 3TX
M3 JCE T. Higgins 7 Nobles Close, Coates, Peterborough PE7 2BT
MW3 JCG J. Scott Winfield, Templeton, Narberth SA67 8SP
M3 JCH J. Hill Anglecroft, Somerford Booths, Congleton CW12 2JU
M3 JCQ J. Bond Oakley, 19 Poplar Road, Tenterden TN30 7NT
M3 JCS J. Sanderson 54 Kelvedon Close, Chelmsford CM1 4DG
M3 JCT L. Jenner 231 Greeswood Rord, Tunbridge Wells TN2 3HU
M3 JCU J. Turton 2 Elkstone Road, Chesterfield S40 4UT
M3 JCY J. Connolly 2 Waring Avenue, St. Helens WA9 2QG
MW3 JDA J. Ruck 286 Barry Road, Barry CF62 8HF
M3 JDF J. Feather 21 Cedar Avenue, Wickersley, Rotherham S66 2NT
M3 JDG J. Godding 58 Dukeswood Road, Longtown, Carlisle CA6 5UJ
M3 JDJ J. Handley 4 Manor Close, Draycott, Stoke-on-Trent ST11 9AZ
M3 JDN J. Dobson 17 Cadley Causeway, Fulwood, Preston PR2 3RU
MI3 JDQ J. Quigg 9 Springhill Terrace, Limavady BT49 9BS
M3 JDS J. Smith Clare Cottage, White Ash Green, Halstead CO9 1PD
M3 JDX J. Porteious 62B Church Close, Stilton, Peterborough PE7 3RG
M3 JEE J. Edmondson 11 Cedar Terrace, Fencehouses, Houghton le Spring DH4 5ND
M3 JEH J. Hutt 48 Hill Crest, Swillington, Leeds LS26 8DL
MW3 JEK C. Gilbert 11 Parc Y Deri, Neath SA10 6BQ
M3 JEM Isle of Man ARS c/o J. Carvill The Lodge, Oldbury Road, Worcester WR2 6AA
M3 JEP M. Clarke 138 Colne Road, Halstead CO9 2HJ
M3 JER J. Higgins 136 Rossmore Road, Poole BH12 2HL
M3 JFA P. Cowan 230 High St., Felixstowe IP11 9DS
M3 JFB J. Beezer 23 Milburn St., Sunderland SR4 6AU
M3 JFF J. Neighbour 112 Holyfields, West Allotment, Newcastle upon Tyne NE27 0EX
M3 JFP J. Payne Jomayne, Farm Lane, Evesham WR11 8TL
M3 JFS J. Skinner 12 Hanbury House, Cardy Close, Redditch B97 6LP
MM3 JFW A. Wilson 20 Ballumbie Gardens, Dundee DD4 0NR
M3 JGH J. Hewitt 166 Ormskirk Road, Rainford, St. Helens WA11 8SW
M3 JGI J. Ives 87 Sheepwalk, Paston, Peterborough PE4 7BJ
M3 JGJ A. Skinner Chevington, Carlton Road, Godstone RH9 8LD
M3 JGN R. Wild 90 Broadway East, Redcar TS10 5DP
M3 JGQ A. Greenland 19 The Ridgeway, Potton, Sandy SG19 2PS
MM3 JGR J. Gracie 34 Ayr Road, Dalmellington, Ayr KA6 7SJ
MD3 JGS Dr A. Foxon 39 Droghadfayle Road, Port Erin IM9 6EN Isle of Man
MM3 JGT J. Thomson 26 South Dean Road, Kilmarnock KA3 7RB
M3 JGU M. Connell 17 The Crescent, Stockport SK3 8SL
M3 JGW W. Holroyd 8 Carr Dene Court, Preston Street, Preston PR4 2XA
M3 JGX A. Hadfield 50 Eastbourne Road, Southport PR8 4DT
M3 JHC J. Clark 27 The Gabriels, Newbury RG14 6PZ
M3 JHJ J. Hickey 11 Greenfield Avenue, Hodthorpe, Worksop S80 4XT
M3 JHL J. Locke 2 Fairnley Road, Nottingham NG8 4AH
M3 JHR J. Richardson 44 Cross Tree Road, Wicken, Milton Keynes MK19 6BT
MM3 JHS J. Hume 51 Vogrie Road, Gorebridge EH23 4HL

M3 JHT J. Tarver 14 South View Road, Leamington Spa CV32 7JD
M3 JHV C. Smith 11 Chesterton Road, Thatcham RG18 3UH
M3 JHW J. Warren 1 Acre Close, Rochester ME1 2RE
M3 JIA J. Allanson Tresco, Hampton, Swindon SN6 7RL
M3 JIC P. Baker 25 Regency Court, Winsford CW7 1FE
M3 JID J. Douglas 26 Walker Drive, Bootle L20 6NN
M3 JIG L. Higgs 14 Lydgate Close Lawford, Manningtree CO11 2SU
M3 JIH D. Marsland 154 Moss Lane, Litherland, Liverpool L21 7NN
M3 JII A. Riley-Marsland 154 Moss Lane, Litherland, Liverpool L21 7NN
MW3 JIJ K. Powell 11 Terrig Street, Shotton, Deeside CH5 1XU
M3 JIK S. Sanders 3 Edmunds Square, Mickleover, Derby DE3 0DU
M3 JIL G. Hinsley 38 Swindon Lane, Prestbury, Cheltenham GL50 4NY
MM3 JIN J. Nicol 18 Tininver Street, Dufftown, Keith AB55 4AZ
M3 JIR S. Buckley 35 Marlborough Road, Irlam, Manchester M44 6HH
M3 JIT W. Carty 49 Princess Gardens, Blackburn BB25EJ
M3 JIU G. Rigby 106 Broadway Crescent, Binstead, Ryde PO33 3QS
M3 JIV R. French 19 Melstone Avenue, Stoke-on-Trent ST6 6EX
M3 JIW J. Biggin Galadean, Farriers Way, Newport PO30 3JP
M3 JJB J. Barton 93 Cardigan Road, Bridlington YO15 3JU
MM3 JJC Dr A. Curlis 94 Kirkhill Road, Aberdeen AB11 8FX
M3 JJH A. Williams 78 Hales Crescent, Smethwick B67 6QS
M3 JJM J. Martin 1 Collins Lane, West Harting, Petersfield GU31 5NZ
M3 JJN J. Nicholson 6 Mill Gardens, West End, Southampton SO18 3AG
M3 JJS J. Stanton Waters & Stanton Plc, 22 Main Road, Hockley SS5 4QS
M3 JJT D. Thompson 5 Vashon Drive, Droitwich WR97JP
M3 JJU M. Slee 88 West Avenue, Lightcliffe, Halifax HX3 8TJ
M3 JKA D. Moffatt 5 Florence Place, Decoy Road, Newton Abbot TQ12 1DX
MW3 JKB Dr J. Arkinstall Wrexham House, Domgay Road, Four Crosses, Llanymynech SY22 6SW
M3 JKG T. Shaughnessy 220 Ladybank Road, Mickleover, Derby DE3 0RS
M3 JKI D. Evans 69 Westbourne, Honeybourne, Evesham WR11 7PT
M3 JKJ S. Asling 18 Cecilia Grove, St. Peters, Broadstairs CT10 3DE
M3 JKM K. Mclaughlin 34 Cambridge Road, Birstall, Batley WF17 9JF
M3 JKP S. Asling 18 Cecilia Grove, St. Peters, Broadstairs CT10 3DE
M3 JKT J. Phillips The Manor, Blackwoods, York YO61 3ER
MM3 JKX J. Kirkpatrick Brims School House, Longhope, Stromness KW16 3NZ
M3 JKZ J. Hawkins 294 Norton Lane, Earlswood, Solihull B94 5LP
M3 JLA J. Astbury 12 Southall Road, Ashmore Park, Wolverhampton WV11 2PZ
M3 JLB J. Bevan 18 Martin Road, Diss IP22 4HR
M3 JLD J. Denny 9 Hawthorn Way, Macclesfield SK10 2DA
M3 JLE J. Isard-Brown 5 Grove Crescent, Croxley Green, Rickmansworth WD3 3JT
M3 JLF K. Bailey 55 Bridgend Park Brewery Road, Wooler NE716QG
M3 JLH J. Hood 17 Grays Close, Motcombe, Shaftesbury SP7 9QB
M3 JLI J. Breckon Greenbank, Chester High Road, Neston CH64 7TR
MW3 JLK J. Knowles 72 Uplands Avenue, Connah'S Quay, Deeside CH5 4LG
M3 JLR J. Ramsay Lane Cottage, Weymore Cottages, Bucknell SY7 0EP
MM3 JLS S. Bence 14 Stein Terrace, Ferniegair, Hamilton ML3 7FR
M3 JLV S. Matthews 60 Brook Street, Erith DA8 1JQ
M3 JLX S. Crouch 152 Thornhill Road, Brighouse HD6 3AH
MI3 JMC J. Mccaw 62 High Street, Ballymena BT43 6DT
M3 JMI J. Ioannou 22 Somerly Close, Binley, Coventry CV3 2LA
M3 JMJ J. Jenkinson 7 Chestnut Avenue, Thorngumbald, Hull HU12 9LD
M3 JMK J. Keegan The Cottage, 11 Condor Grove, Lytham St. Annes FY8 2HE
M3 JMQ D. Baines 3 Dunkirk Avenue, Houghton le Spring DH5 8HN
M3 JMU M. Crowley 133 Jessop Road, Stevenage SG1 5LH
M3 JMW J. Moxley-Wyles 7 Gidley Way, Horspath, Oxford OX33 1RQ
M3 JMX J. Moore 110 Pooles Lane, Willenhall WV12 5HW
M3 JMY J. Varty 4 St. Cuthberts Close, Burnfoot, Wigton CA7 9HQ
M3 JNB J. Kearney 12 Forshaw Lane, Burtonwood, Warrington WA5 4ES
M3 JND J. Dukes 79 Jubilee Avenue, Boston PE21 9LE
MM3 JNJ A. Campbell 3 North Shawbost, Isle of Lewis MS2 9BD
M3 JNQ J. Clark 1 Brooklime Road, Liverpool L11 2YH
M3 JNR K. Challoner 23 Chapel Lane, Queensbury, Bradford BD13 2QA
M3 JNT J. Foulds 7 Bridge Road, Little Sutton, Spalding PE12 9EG
M3 JNU S. Robinson 29 Grange Lane, Mountsorrel, Loughborough LE12 7HY
MW3 JNX M. January Blaenau Ucha Farm, Treudoyn, Flintshire CH7 4NS
M3 JNY J. Hutton Cassiobury, The Street, Diss IP22 2PS
M3 JOA N. Lambert 3 Nightingale Walk, Stockton-on-Tees TS20 1SZ
M3 JOF J. Miles 11 Enborne Gate, Newbury RG14 6AZ
M3 JOJ J. Smith 5 Manifold Gardens, Plymouth PL3 6HL
M3 JOM C. Loughran 8 Douglas Road, Dover CT17 0BD
M3 JOS D. Bown 34 Kings Gardens, Bedworth CV12 8JG
M3 JOW J. Fletcher 32 Chapel Lane, Barwick In Elmet, Leeds LS15 4EJ
MW3 JPF J. Freelove 12 Honeyborough Road, Neyland, Milford Haven SA73 1RE
M3 JPG Rvd. G. Bowen 93 Pelham Road, Bexleyheath DA7 4LY
M3 JPI J. Pickering Batemill, Batemill Lane, Macclesfield SK11 9BW
M3 JPM G. Meyer 37 Marina Village, Preston Brook, Runcorn WA7 3BH
M3 JPP H. Tonge 38 Colemeadow Road, Billesley Common, Birmingham B13 0JL
MM3 JPS J. Greig 100A Sheephousehill, Fauldhouse, Bathgate EH47 9EG
M3 JPU J. Patient 4 Bucklebury Heath, South Woodham Ferrers, Chelmsford CM3 5ZU
M3 JPW J. Wake 60 Cloverville Approach, Bradford BD6 1ET
MW3 JQC M. Breakwell 9 Llys Y Dderwen, New Quay SA45 9SY
MI3 JQD B. Young 63 Scarvagherin Road Spamount, Castlederg BT81 7NW
M3 JQF S. Walrond 17 Madam Lane, Weston-Super-Mare BS22 6PW
M3 JQG D. Johnson 11 Horseshoe Avenue, Dove Holes, Buxton SK17 8DP
M3 JQJ L. Berry 6 Warren Park Close, Brighouse HD6 2RU
MW3 JQK E. Williams Criafol, Upper Llandwrog, Caernarfon LL54 7PU
M3 JQM L. Ross 2 Bedford Street, Blackburn BB2 4EU
M3 JQN V. Greco 38 Horbling Lane, Stickney, Boston PE22 8DQ
M3 JQS J. Hellowell Upper Hole Head Farm, Ash Hall Lane, Soyland HX6 4NU
M3 JQT S. Watson 32 Bradley View, Holywell Green, Halifax HX4 9DN
M3 JQV P. Handy 30 Kingfisher Drive, Cheltenham GL51 0WN
M3 JQW D. Roscoe 28A Princess St., Chorley PR7 3AP
MW3 JQX T. Leaworthy 7 Maesderwen Rise, Stafford Road, Pontypool NP4 5SS
M3 JQY J. Johnson 30 Thorpe Downs Road, Church Gresley, Swadlincote DE11 9FB
M3 JRA A. Jessop 4 Katherine Street, Thurcroft, Rotherham S66 9LG
M3 JRF J. Francis 11 Middle Stream Close, Bridgwater TA6 6LF
M3 JRI J. Rowe 45 Durham Road, Wilpshire, Blackburn BB1 9NH
MM3 JRK G. Smith 40 Pirleyhill Drive, Shieldhill, Falkirk FK1 2EA
M3 JRM J. Marter 4 Meadow Way, Seaford BN25 4QT
M3 JRN J. Newman 25 Milebush Road, Southsea PO4 8NF
M3 JRQ L. Pearson 4 Brentwood Close, Thorpe Audlin, Pontefract WF8 3ES
M3 JRR J. Read 26 Chaucer Road, Walsall WS3 1DF
MM3 JSB J. Bence 5 Braeside Gardens, Hamilton ML3 7PN
M3 JSF J. Flores-Watson 10 Bramwell Gardens, Coventry CV6 6NB
MI3 JSH S. Hutchinson 21 Lord Warden'S Grange, Bangor BT19 1YN
M3 JSK J. Killian 7 Dankworth Road, Basingstoke RG22 4LJ
M3 JSM A. Mclaughlin 34 Cambridge Road, Birstall, Batley WF17 9JF
M3 JSO J. O'Shea 56 Crummock Gardens, London NW9 0DJ
M3 JSQ M. Woodruff 14 Primatt Crescent, Shenley Church End, Milton Keynes MK5 6AS
M3 JST J. Taylor 5 Pyman Close, Martham, Great Yarmouth NR29 4UR
MI3 JTB J. Black 17 Fairymount Terrace, Taylors Avenue, Carrickfergus BT38 7HN
MW3 JTJ J. Jones Bronydd, Blaenffos, Boncath SA37 0HZ
MI3 JTM J. Monteith 58 Bells Hill, Limavady BT49 0DQ
M3 JTO J. Bosworth 10 Aston Street, Leeds LS13 2BJ
MD3 JTT J. Talbot 43 Harcroft Meadow New Castletown Road, Douglas, Isle of Man IM2 1JT
M3 JTU A. Bailey 58 Billy Buns Lane, Wombourne, Wolverhampton WV5 9BP
M3 JTZ R. Smith 17 Julian Road, Spixworth, Norwich NR10 3QA
M3 JUC K. Marsh 21 Edward Road, Eynesbury, St. Neots PE19 2QF
M3 JUF Dr J. Schofield 6 Robin Royd Avenue, Mirfield WF14 0LF
M3 JUL J. Townsend 56 Seymour Road, Northfleet, Gravesend DA11 7BN
MW3 JUM R. Jones 10 Erw Wen Road, Colwyn Bay LL29 7SD
M3 JUO S. Jacklin 32 Edison Drive, Rugby CV21 1FB
M3 JUX C. Quilter 473 Sidcup Road, Mottingham, London SE9 4ET
M3 JUY R. Thomas 52 Victoria Road, Saltney, Chester CH4 8SS
M3 JUZ R. Shams-Nia 1090 Eastern Avenue, Ilford IG2 7SF
MW3 JVH E. Chell Mesen Fach, Llanybydder SA40 9TY
MI3 JVJ P. Hannigan 4 Silverhill Road, Strabane BT82 0AE
M3 JVK J. Kirkham 35 Central Avenue, Woodlands, Doncaster DN67NW
M3 JVP J. Papworth Flat 1, 70 Edward Road, Nottingham NG2 5GB
M3 JVR B. Gibbs Flat 6, Castleton Court, Southsea PO5 3AU
MI3 JVV K. Hannigan 4 Silverhill Road, Strabane BT82 0AE
M3 JVW J. Wheway 20 Radnor Street, Derby DE21 6DZ
M3 JVX E. Mcgowan Plovers, Moor Road, Walesby LN8 3UR
M3 JWJ D. Shields 42 Studland Park, Westbury BA13 3HL
M3 JWM J. Watts 10 Lacy Road, Ludlow SY8 2NS
M3 JWN J. Newell 7 Talbot, Tamworth B77 2RS
M3 JWQ W. Johnson 10 Archdale Road, Nottingham NG5 6EB
MW3 JWV L. Percival Blue Cedars, Gresford, Wrexham LL12 8RN
M3 JWW J. Wainwright 8 Common Lane, Cutthorpe, Chesterfield S42 7AN
M3 JWZ S. Sewell The Old Vicarage, Church Bank, Crewe CW4 8PG
MI3 JXA C. Matchett 28 Glendale Avenue East, Belfast BT8 6LF
M3 JXE D. Ennion 347 Parkgate Road, Chester CH1 4BE
MI3 JXG C. Birney 45 Sallys Wood, Irvinestown, Enniskillen BT94 1HQ
M3 JXI H. Southall 12 Prescot Close, Mickleover, Derby DE3 0TB
M3 JXN P. Jackson Langsmead Barn, Eastbourne Road, Blindley Heath, Lingfield RH7 6JX
MI3 JXO D. Burke 7 Edinburgh Villas, Omagh BT79 0DW
M3 JXV C. Trew Ringstone Lodge, 66 Oakwood Road, Horley RH6 7BX
M3 JXX J. Driver 99 Queens Road, North Weald, Epping CM16 6JQ
M3 JXY A. Gowans 38 Beech Way, Twickenham TW25JT
M3 JYA D. Kemp 7 Hillhurst Grove, Birmingham B36 9TS
M3 JYE B. Edwards 16 Whitland Close, Rednal, Birmingham B45 8SJ
M3 JYG D. Batty 168 Rotherhithe New Road, London SE16 2AP
M3 JYH G. Swain 3 Flaxfield Drive, Crewkerne TA18 8DF
M3 JYO D. Perks 6 Old School Gardens, Yatton Keynell, Chippenham SN14 7BB
M3 JYP C. Lester 21 Barwell Way, Witham CM8 2TY
M3 JYW X. Chen Harrogate Ladies' College, Clarence Drive, Harrogate HG1 2QG
M3 JYZ C. Williamson 53A High St., Whitwell, Hitchin SG4 8AJ
M3 JZA K. Armstrong 29 Thorntree Avenue, Crofton, Wakefield WF4 1NU
M3 JZD K. Hart 2 Springfield Cottages, Acton, Newcastle ST5 4EF
M3 JZE S. Peregrine 18 Gisborne Close, Mickleover, Derby DE3 9LU
M3 JZF D. Wall 96 Albert Street, Wigan WN5 9EF
M3 JZI M. Rolls 3 Gleneagles Crescent, New Holland, Barrow-upon-Humber DN19 7TL
M3 JZK M. Stinton 57 Wildfields Road, Clenchwarton, King's Lynn PE34 4DE
M3 JZL D. Atkins 20 Nappsbury Road, Luton LU4 9AL
M3 JZN M. Johnson 25 Rowan Drive, Kirkby-In-Ashfield, Nottingham NG17 8FU
M3 JZO S. Bacon 34 Fishers St., Kirkby In Ashfield, Nottingham NG17 9AH
M3 JZP A. Bacon 34 Fishers St., Kirkby In Ashfield, Nottingham NG17 9AH
M3 JZT W. Soffe 96 Urban Road, Doncaster DN4 0EP
MW3 JZV J. Jenkins 13 Birch Hill, Newport NP20 6JD
M3 JZX M. Cooper 69 Leicester Road, Kibworth Harcourt, Leicester LE8 0NP
M3 KAC D. Carr 19 Kingsmead Walk, Speedwell, Bristol BS5 7RL
M3 KAE B. Mcfarlane 11 Hill St., Barnsley S71 5AL
M3 KAK K. Harden 59 Violet Avenue, Edlington, Doncaster DN12 1NW
M3 KAL K. Lawton Meadowbank, Sutton St. Nicholas, Hereford HR1 3BJ
M3 KAN K. Hudson 3 Worcester Ave, Mansfield NG19 8QJ
MM3 KAQ P. Cotton 5 Carronhall Avenue, Carronshore, Falkirk FK2 8AN
M3 KAU H. Leaver 1 Litcham Close, Litcham, King's Lynn PE32 2QX
M3 KAX D. Larkin 19 Elizabeth Court, Hemsworth, Pontefract WF9 4TQ
M3 KAY K. Limbert 82 Kipling Avenue, Liverpool L36 0TZ
M3 KBB K. Barrow Fenway Farm, Ten Mile Bank, Downham Market PE38 0EU
M3 KBE J. Kelly 1 Bramble Close, New Ollerton, Newark NG22 9TN
M3 KBF C. Harley 1 Portland Crescent, Meden Vale, Mansfield NG20 9PJ
M3 KBG J. Crowther 16 Linden Avenue, Tuxford, Newark NG22 0JR
M3 KBL K. Lee 10 Queens Way, Pontefract WF8 2LX
MU3 KBP K. Pratt Avalon Le Clos Des Sablon, Sandy Lane, St. Sampson GY2 4RN Guernsey
M3 KBQ C. Daniells 7 Arlington Road, York YO30 5GF
M3 KBY M. Kirby 76 Burton Road, Overseal, Swadlincote DE12 6JJ
M3 KBZ R. Griffiths 7 Macnaghten Road, Tankersley, Barnsley S75 3DD
M3 KCA C. Jameson Flat 9, Britannia Court, Poole BH12 3HF

M3 KCC C. Evans 158 Delamore St., Liverpool L4 3SX
M3 KCG R. Jones 79 Turpins Rise, Stevenage SG2 8QZ
MW3 KCL M. Brennan 37 Marguerites Way, Cardiff CF5 4QW
M3 KCO K. Cornmell 19 Forest Road, Chandler'S Ford, Eastleigh SO53 1NA
M3 KCP K. Preen 12 Sandpit Lane, Hilton, Bridgnorth WV15 5PH
M3 KCQ T. Stansfield 40 Rushmore House, Rubery, Birmingham B45 9RU
MD3 KCT J. Kennaugh White Gables, 25 Kissack Road, Castletown IM9 1NW Isle of Man
M3 KCU M. Johnson 25 Rowan Drive, Kirkby-In-Ashfield, Nottingham NG17 8FU
M3 KDK S. Allington 137 Marshall Lane, Northwich CW8 1LA
M3 KDL C. Lote 8 Warren Place, Walsall WS8 6BY
M3 KDM D. Langley 2 Holly Bush Cottages Holmesdale Road, Sevenoaks TN13 3XN
MM3 KDN A. Traynor 2 Mains Of Carmyllie Farm Cottages Carmyllie, Arbroath DD11 2RJ
M3 KDO R. Smith 3 Vernon Road, Southport PR9 7EZ
MI3 KDR K. Dickson 66 Lisnabreeny Road, Belfast BT6 9SR
M3 KDV D. Craven 69 Markham Avenue, Rawdon, Leeds LS19 6NE
M3 KDY C. Lindsay 152 Dinmore Avenue, Blackpool FY3 7QS
M3 KEC S. Conlon 6 Cardigan St., Ashton On Ribble, Preston PR2 2AS
M3 KEF K. Forster Meadow View, Cracow Moss, Crewe CW3 9BS
M3 KEJ K. Jefferson 45 Rutherford Crescent, Leighton Buzzard LU7 3GE
M3 KEL K. Watwood 57 Cliveden Road, Stoke-on-Trent ST2 8LP
M3 KER M. Springett 31 Mountbatten Court, Andover Road, Winchester SO22 6BA
M3 KEV K. Graffham 15 Hayes Road, Clacton-on-Sea CO15 1TX
M3 KEW J. Taylor 90 Aldam Road, Doncaster DN4 9EL
M3 KEY K. Calderbank 6 Heathfield, Heath Charnock, Chorley PR6 9LA
M3 KEZ D. Brough 57 Francis Road, Ashford TN23 7UP
M3 KFE S. Evans 54 Stafford Crescent, Newcastle ST5 3EA
M3 KFH R. Drake 42 Sunningdale Close, Doncaster DN4 6UR
MI3 KFI T. Warmington 57 Carbet Road, Portadown, Craigavon BT63 5RJ
M3 KFL P. Hooper 4 Castlemead Close, Saltash PL12 4LF
M3 KFO W. Cromack 45 Southroyd Park, Pudsey LS28 8AX
M3 KFP J. Rouse Nettleden, Galane Close, Northampton NN4 9YR
M3 KFQ N. Camp 1 Higher Tresillian Cottages, Tresillian., Newquay TR8 4PL
M3 KFR F. Coles 8 Moore Close, Church Crookham, Fleet GU52 6JD
M3 KFT J. Ferrol 29 Westlands, Haltwhistle NE49 9BS
M3 KGE S. Angove 62 Trelissick Fields, Hayle TR27 6HZ
M3 KGG K. Gordon 308 Claremont Road, Swanley BR8 7QZ
M3 KGJ K. Weston 114 Morland Road, Ipswich IP3 0LZ
M3 KGO R. Dunn 15 Catkins Close, Catshill, Bromsgrove B61 0TT
MW3 KGP Dr P. Kelly Arosfa, Westminster Road, Moss, Wrexham LL11 6DN
MW3 KGQ J. Kelly Arosfa, Westminster Road, Wrexham LL11 6DN
M3 KGV C. Moulding 28 Queens Avenue, Highworth, Swindon SN6 7BA
M3 KHA A. Hughes 8 Bullens Green Lane Colney Heath, St. Albans AL4 0QS
MW3 KHC A. Vick Flat 8, Kingshill Court, Newport NP20 4DT
M3 KHE K. Bradley 9 Spruce Grove, Kirkby-In-Ashfield, Nottingham NG17 7QB
MW3 KHH A. Peake 24 Rhigos Gardens, Cardiff CF24 4LS
M3 KHI C. Webb 5 Pound Lane, Preston Bissett, Buckingham MK18 4LX
M3 KHJ C. Ashman 56 Farriers Close, Swindon SN1 2QT
MW3 KHK S. Rosser 25 Clos Tir Ypwll, Pantside, Newport NP11 5GE
M3 KHM L. Lebaldi 11 Artle Place, Lancaster LA1 2QP
M3 KHT H. Kwan Harrigate Ladies College, Clarence Drive, Harrogate HG1 2QG
M3 KHW K. Stocker 50 Mount Drive, Harrow HA2 7RP
MW3 KHY J. Robinson 41 Ashbrook, Brackla, Bridgend CF31 2AT
MI3 KIL P. Hill Flat 12 Kilcreggan Homes, Elizabeth Avenue, Carrickfergus BT38 7EP
M3 KIN D. Kinsey 161 Heath Road South, Weston, Runcorn WA7 4RP
M3 KIO E. Smith The Cabin, Nothe Parade, Weymouth DT4 8TX
M3 KIQ M. Lebaldi 11 Artle Place, Lancaster LA1 2QP
M3 KIR K. Wheeler 27 Elley Green, Neston, Corsham SN13 9TX
MW3 KIS C. Heaton Flat, St. Asaph Conservative Club, High Street, St. Asaph LL17 0RG
M3 KIT K. Cattermole Blaxhall Hall Crossing, Little Glemham, Woodbridge IP13 0BP
M3 KIU S. Moore 10 Strathcona Avenue, Hull HU5 4AD
M3 KIZ P. Lewis 16 Valley Road, St. Albans AL3 6LR
M3 KJB K. Brooks 24 Morris Drive, Weaverham, Northwich CW8 3LP
M3 KJC K. Cole 4 Marsham Road Hazel Grove, Stockport SK7 5JB
M3 KJD K. Davies 20A Hart Road, Wolverhampton WV11 3QJ
M3 KJE W. Taylor 99 St. Marys Close, Littlehampton BN17 5QQ
MM3 KJG Hms Cavaller RC c/o K. Glacken 14 Hailes Avenue, Edinburgh EH13 0NA
M3 KJK J. Mason 22 Eskdale Avenue, Halifax HX3 7NH
M3 KJM K. Marsh 11 Apollo Road, Stourbridge DY9 8YG
M3 KJS K. Sealey 8 Esplanade, Burnham-on-Sea TA8 1BE
M3 KJV D. Baker 65 Madison Street, Tunstall ST6 5HS
M3 KJY C. Atkins 87 Wentworth Road, Doncaster DN2 4DA
M3 KKA N. Holdridge 15 Ballam Avenue, Doncaster DN5 9DY
M3 KKB S. Silvers 39 Hickinwood Crescent, Clowne, Chesterfield S43 4AQ
M3 KKF R. Taylor 38 Edleston Road, Crewe CW2 7HD
M3 KKG K. Gledhill 19 Palmers Terrace, Treknow, Tintagel PL34 0EH
M3 KKI I. Todorovic 3 Braemar Close, Stoke-on-Trent ST2 8NL
M3 KKN G. Allen 60 Danefield Road, Cheshire CW95PX
M3 KKO M. O'Neill 15 School Road, Hockley Heath, Solihull B94 6QH
MI3 KKP A. Temple 27 Oakland'S Glenshane Rd Claudy Co Londonderry Bt47 4Ff, Londonderry BT47 4FF
M3 KKQ G. Thomas 15 Buckley Avenue, Byley, Middlewich CW10 9NW
M3 KKS B. Roaf 8 Weare Close, Portland DT5 1JP
M3 KKX A. Jones 27 Fishpond Lane, Holbeach, Spalding PE12 7DQ
M3 KKZ R. Kirby 44 Wilby Avenue, Little Lever, Bolton BL3 1QE
M3 KLB M. Crombie 12 Sir James Reckitt Haven, Hull HU8 8QR
MM3 KLO S. Dobie Flat 20, 3 Arneil Place, Edinburgh EH5 2GU
M3 KLS K. Symonds 68 Manor Crescent, Pan, Newport PO30 2BH
M3 KLT M. Rogers Jones Glanva, 20 Birchwood, Leyland PR26 7QJ
M3 KLU N. Andrews Temple View House, Shopland Road, Rochford SS4 1LH
M3 KLV P. Bailey 71 Norfolk Avenue, Leigh-on-Sea SS9 3HA
M3 KLY K. Lingham 102 Chancery Lane, St. Helens WA9 1SQ
MI3 KMB K. Barr 64 Owenreagh Drive, Strabane BT82 9DT
M3 KMH K. Haywood 6 Lydney Road, Urmston, Manchester M41 8RN
MW3 KML K. Iball Highcroft, 18 Tan Y Coed, Mold CH7 6TU
M3 KMN K. Macnauton 27A Lincoln Road, Poole BH12 2HT
M3 KMO K. Owen 10 Pitcher Lane, Leek ST13 5DB
M3 KMS K. Stanley 3 Hale Way, Colchester CO4 5BD
M3 KMT T. Ramsden 37 Hyde Abbey Road, Winchester SO23 7DA
MW3 KMU G. Cattle 12 Claerwen Gelligaer, Hengoed CF828EW
M3 KMW K. Ward 127 Lower Lime Road, Oldham OL8 3NP
MM3 KMX S. Mclachlan 531 Blair Avenue, Glenrothes KY7 4RF
MW3 KNE A. Mctaggart Brick Hall, Hundleton, Pembroke SA71 5QX
M3 KNF R. Curno 19 Beckwith Road, Yarm TS15 9TG
M3 KNK H. Arrowsmith 15 Hermitage Close, Frimley, Camberley GU16 8LP
MW3 KNR R. Seal 5 Millfield, Lisvane, Cardiff CF14 0RW
M3 KNT K. Royce 11 Church Lane, Stibbington, Peterborough PE8 6LP
M3 KNV A. Lebaldi 11 Artle Place, Lancaster LA1 2QP
MM3 KNY K. Brown 21 Strain Crescent, Airdrie ML6 9ND
M3 KOA J. Beecroft 48 Manor Road, Alton GU34 2PB
M3 KOF R. Johnson 30 Avenue Road, Coalville LE67 3PB
M3 KOJ K. Lowe 7 King Street, Creswell, Worksop S80 4ER
M3 KOR H. Ross 9 First Avenue, Edwinstowe, Mansfield NG21 9NZ
M3 KOU A. Rose 133 Petersmith Drive, New Ollerton, Newark NG22 9SG
M3 KPB K. Bromley 40 Winfrith Road, Fearnhead, Warrington WA2 0QE
M3 KPF J. Simmons 246 Ruskin Road, Crewe CW2 7JY
M3 KPG K. Stretton 6 Highfields, Hilltop Drive, Rye TN31 7HT
M3 KPL J. Kelly 1 Bramble Close, New Ollerton, Newark NG22 9TN
M3 KPO S. Warren 7 Crich Way, Newhall, Swadlincote DE11 0UU
M3 KPQ A. Jermyn 3 Tudor Walk, Carlton Colville, Lowestoft NR33 8NE
M3 KPU A. Parkes 59 Wellington Gardens, Battle TN33 0HD
M3 KPZ T. Wood 33 Somerdale Avenue, Bristol BS4 2XN
M3 KQB P. Broughton 23 Ivy Place, Tantobie, Stanley DH9 9PT
M3 KQC S. Chandler 4 Gladstone House, Horton Crescent, Epsom KT19 8BW
M3 KQD B. Minks 132 North Road, Clowne, Chesterfield S43 4PF
M3 KQF P. Woodcock 27, Knighton, Stafford ST20 0QH
MM3 KQI R. Houston 19 Deansloch Place, Aberdeen AB16 5SB
M3 KQP A. Eastwell 11 Middlebere Drive, Wareham BH20 4SD
M3 KQR B. Hall 126 Eton Road, Burton-on-Trent DE14 2SN
M3 KQS N. Ralph 24 Back Street, Laxton, Goole DN14 7TP
M3 KQT D. Berry Flat5, 106 Braybrooke Road, Hastings TN34 1TG
M3 KQW H. Malpas 148 Queen Street, Crewe CW1 4AU
M3 KQY R. Brisley 15 Elm Fields, Old Romney, Romney Marsh TN29 9SN
M3 KRB L. Kirby 76 Burton Road, Overseal, Swadlincote DE12 6JJ
M3 KRD K. Dukes 127 Carlton Road, Boston PE21 8LL
M3 KRE R. Jacobs 35 Edgar Road, Canterbury CT1 1NR
MI3 KRL K. Mccrystal 85 Tamlaght Road, Omagh BT78 5BB
M3 KRM S. Whitlock Railway Crossing Cottage, Ash Road, Sandwich CT13 9JB
MW3 KRN C. Richards 2 Castle Lodge Crescent, Caldicot NP26 4JL
M3 KRO G. Ticehurst 118 Old Roman Bank, Terrington St. Clement, King's Lynn PE34 4JP
M3 KRP K. Taylor 3 The Drive, Lichfield WS14 9QT
M3 KRQ N. Gadalla 26 South Parade, Boston PE21 7PN
MM3 KRR D. Corbett 1F1, 13 Learmonth Place, Edinburgh EH4 1AX
M3 KRS C. Cheverall 3 Egerton Road, Room 3, Egerton House, Bexhill-on-Sea TN39 3HH
M3 KRX D. Shires 1 West Close, High Coniscliffe, Darlington DL2 2LN
M3 KRY R. Dyson 4 Royston Lane, Royston, Barnsley S71 4NL
M3 KRZ B. Renowden Hilrowenick, Polwithen Drive, St. Ives TR26 2SP
MW3 KSE N. Lane 13 Traston Road, Newport NP19 4RQ
M3 KSG K. Gordon 308 Claremont Road, Swanley BR8 7QZ
M3 KSH A. Wilkins 2 Beechfield Crescent, Banbury OX16 9AR
MW3 KSI M. Giudice 31 Woodfield Cross, Tredegar NP22 4JG
M3 KSK M. Sheppard 107 Queen St., Swinton, Mexborough S64 8NF
MD3 KSN F. Kelly 5 Maynrys Castletown, Isle of Man IM9 1HP
M3 KSP S. Eldridge 20 Edendale Road, Melton Mowbray LE13 0EW
M3 KSS A. Eades Violet Bank, 18 Hillside Road, Leigh-on-Sea SS9 2DT
MM3 KSV P. Mccluskey 119 Tower Drive, Gourock PA19 1SG
M3 KTA R. Baines 34 Bury Road, Stapleford, Cambridge CB22 5BP
M3 KTD K. Davidson 5 Hanover Parc, Indian Queens, St. Columb TR9 6ER
M3 KTH K. Howard 5 St Nicholas Street, Dereham NR19 2BS
M3 KTT B. Gardner 38 Apley Rd, Stourbridge DY84PA
M3 KTV B. Bean 46 Grand Drive, Herne Bay CT6 8JS
M3 KUE B. Hall 65 Cavendish Road, Worksop S80 2ST
M3 KUG M. Amos 233B Abington Avenue, Northampton NN1 4PU
M3 KUH A. Grannon The Chestnuts, Church Lane, Hull HU11 4PR
M3 KUJ D. Russell 72 Langholm Drive, Cannock WS12 2EZ
M3 KUK J. Jones 15 Kinnaird Road, Sheffield S5 0NN
M3 KUM C. Sellors 73 Wilkinson Drive Kesgrave, Ipswich IP5 2DS
M3 KUN J. Wilson 20 Elgitha Drive, Thurcroft, Rotherham S66 9PD
M3 KUO D. Holden 2 Beacon Road, Bickershaw, Wigan WN2 4AF
M3 KUQ T. Bush 19 Spring Vale, Waterlooville PO8 9DA
M3 KUS P. Connelly 3 Finch Close, Weston-Super-Mare BS22 8XS
MM3 KUU G. White 119 Waggon Road, Brightons, Falkirk FK2 0EJ
M3 KUV P. Baines 4 Oxford Crescent, Hetton-Le-Hole, Houghton le Spring DH5 9LJ
M3 KUY S. Church The Willows, Warboys Road, Huntingdon PE28 3AH
M3 KUZ K. Mckeown 27 Lusty Glaze Road, Newquay TR7 3AE
M3 KVC A. Stacey 311 Hyde End Road, Spencers Wood, Reading RG7 1DD
M3 KVD D. Speed 137 Church Road North, Skegness PE25 2QQ
M3 KVG D. Hannon 20 High Street, Whittlebury, Towcester NN12 8XJ
M3 KVH K. Harrison 55 Hudson Close, Worcester WR2 4DP
M3 KVI D. Swann 37 Burgh Road, Skegness PE25 2RA
M3 KVJ K. King 16 Clare Way, Bexleyheath DA7 5JU
M3 KVK J. Mckeown 27 Lusty Glaze Road, Newquay TR7 3AE
M3 KVL K. Martin 19 Comrie Crescent, Burnley BB11 5HX
MM3 KVN K. Clark 38 Dunsinane Drive, Perth PH1 2DU
M3 KVR K. Robertson 189 Harrowby St., Farnworth, Bolton BL4 7DF
M3 KVU E. Smith 98 Chapel Fields Charterhouse Road, Godalming GU7 2AA
MM3 KVV W. Morrison 7 Knowehead Crescent, Kirriemuir DD8 5AB
M3 KVW M. Keelan 16 North Drive, Harwell, Didcot OX11 0PE
MM3 KVX G. Marshall West Inch Farm, Kinnordy, Kirriemuir DD8 5ET
MM3 KVY M. Mcconnell 6 Langlaw Road, Mayfield, Dalkeith EH22 5AX
M3 KWF S. Mainzer Lillypool House, Waldersea, Wisbech PE14 0NR
M3 KWR M. Sutty 14 Sedgwick Street, Cambridge CB1 3AJ
M3 KWS S. Dunn 64 Stucley Road, Bideford EX39 3EQ
M3 KWZ D. Leese 41 Woolston Avenue, Congleton CW12 3DZ

M3 KXB R. Walsh 8 Eyton Place, Dawley, Telford TF4 2DL
M3 KXD C. Collins 2 Kew Crescent, Sheffield S12 3LP
M3 KXE P. Beresford 23 High Lowe Avenue, Congleton CW12 2EP
M3 KXF J. Gregory 9 Longfields Crescent, Hoyland, Barnsley S74 9HZ
M3 KXG R. Robinson 18 O'Connell Road, Liverpool L3 6JF
M3 KXI C. Day 4 Marlborough Way, Market Harborough LE16 7LW
M3 KXS E. Booth 18 Maple Road, Kiveton Park, Sheffield S26 5PH
M3 KXV D. Hollinrake 4 Sandwood Avenue, Broughton, Chester CH4 0RJ
M3 KXY D. Preston Home View, Paradise Lane, Reading RG7 6NU
M3 KXZ P. Millis 26 Chalkland Rise, Brighton BN2 6RH
M3 KYG S. Humphreys Flat 1, Winslow Court 100 Fordwych Road, London NW2 3NN
M3 KYH J. List 41 Westbury Crescent, Dover CT17 9QQ
M3 KYK J. Hall 1 Nash Close, Earley, Reading RG6 5SL
MM3 KYO K. Rafferty 13 Robin Crescent, Buckhaven, Leven KY8 1EZ
M3 KYQ K. Rennison 49 Syston Avenue, St. Helens WA11 9JJ
M3 KYV S. Littlewood Townside Lodge, Townside, Immingham DN40 3PS
M3 KYZ D. Whitelock 22 Anne Crescent, Waterlooville PO7 7NA
M3 KZB C. Cascarino 87 Esther Grove, Wakefield WF2 8EX
M3 KZC M. Clack 42A Provost Street, Fordingbridge SP6 1AY
MM3 KZD S. Corstorphine 33 Springfield, West Barns, Dunbar EH42 1UF
M3 KZI A. Pateman 37 Hemans Road, Daventry NN11 9AL
M3 KZJ P. Lamb 13 Pool End, St. Helens WA9 3RE
M3 KZP K. Page 62 Farndon Avenue, Sutton Manor, St. Helens WA9 4DN
M3 KZR A. Bailey 9 Park View, Abram, Wigan WN2 5QR
M3 KZS P. Allen 4 The Links, Northam, Bideford EX39 1LS
M3 KZT D. Baugh 36 Well Hay Close, Plymouth PL9 8DT
M3 KZV P. Camplin 16 Green Street, Hoyland, Barnsley S74 9RF
M3 KZW S. Grainger 15 Carr House Lane, Wirral CH46 6EN
M3 LAG L. Gething 106 Westgate, Elland HX5 0BB
M3 LAJ L. Jarman 53 Enderby Crescent, Gainsborough DN21 1XQ
M3 LAP A. Clarke 57 Welland Avenue, Grimsby DN34 5JP
M3 LAQ T. Mynors 6 Walcott Avenue, Christchurch BH23 2NG
M3 LBC R. Kenton 65 Warren Drive, Broughton, Chester CH4 0PU
M3 LBD T. Donegan 19 Teesgate, Thornaby, Stockton-on-Tees TS17 9AN
M3 LBG D. Davis 36 Arbour Street, Southport PR8 6SQ
M3 LBJ I. Blundell 43 Ponsonby Place, London SW1P 4PS
M3 LBK L. Karthauser 17 Manor Close Abbotts Ann, Andover SP11 7BJ
M3 LBM N. Waller 16 Rother Croft, New Tupton, Chesterfield S42 6BE
M3 LBN G. Lowe 8 Markland Crescent, Clowne, Chesterfield S43 4NG
M3 LBP D. Clarke 30 Chelford Road, Bromley BR1 5QT
M3 LBQ M. Bradshaw 342 Manchester Road, Blackrod, Bolton BL6 5BG
M3 LBR S. Cross 31 Parkfields, Abram, Wigan WN2 5XR
M3 LBT R. Carter 43 Sheldon Avenue, Standish, Wigan WN6 0LW
MW3 LBX D. Roberts 118 Ffordd Ddyfrdwy, Mostyn, Holywell CH8 9PQ
M3 LBY P. Wallstone 3 Wilson Street, Mansfield NG19 7JW
M3 LBZ J. Fletcher Paradise Barn, Bounds Lane, Chard TA20 2TJ
MM3 LCC M. Pentland Castleview, Salterhill, Elgin IV30 5PT
M3 LCE R. Armstrong 26 Lancaster Road, Carnforth LA5 9LD
M3 LCF D. Lovell 109 Aylesbury Crescent, Plymouth PL5 4HX
M3 LCI M. Davidson 19 Mason Street, Workington CA14 3EH
M3 LCL C. Lewis 19 Elgar Close, Great Sutton, Ellesmere Port CH65 7AZ
M3 LCP L. Pentney 4 Caley Road, Tunbridge Wells TN2 3BL
M3 LCS M. Croxford Simmons 37 Queens Road, Askern, Doncaster DN6 0LU
M3 LCU P. Loose Tae Ping, Main Road, King's Lynn PE31 8BP
M3 LCZ R. Humpage 15 Middleton Road, Middleton Morecambe LA3 3JS
M3 LDC L. Cattermole Blaxhall Hall Crossing, Little Glemham, Woodbridge IP13 0BP
M3 LDE L. Davis Romano House, Gorefield Road, Wisbech PE13 5AS
M3 LDF S. Brown 21 Woborrow Road, Heysham, Morecambe LA3 2PW
M3 LDH L. Holmes 48 Woodpecker Close, Wirral CH49 4QP
M3 LDI J. Bircumshaw 39 Woodthorpe Lane, Sandal, Wakefield WF2 6JG
M3 LDJ J. Harvey 125 North Road, Clowne, Chesterfield S43 4PQ
MM3 LDK T. Mcfarlane 39 Beechwood Road, Tarbolton, Mauchline KA5 5RP
M3 LDL J. Parker 1 Schoose Caravan Park, Workington CA14 4JA
M3 LDM T. Craddock 12 Wold Road, Burton Latimer, Kettering NN15 5PN
MI3 LDO P. Logan 18 Castle Lane, Lisnaskea, Enniskillen BT92 0FW
M3 LDQ L. Paddon 21 Oak Park Drive, Havant PO9 2XE
MM3 LDR A. Sloan 36 Paterson Avenue, Irvine KA12 9JJ
M3 LDS R. Walker 17 Brookside Paulton, Bristol BS39 7NL
M3 LDT R. Taylor 7 Wall Street Blackpool, Blackpool FY1 2EG
MW3 LDY N. Cole Tycoch, Llandovery SA20 0UP
M3 LEB S. Bell 221 Horninglow Road, Sheffield S5 6SG
M3 LEF J. Peace 237A Mapperley Plains, Nottingham NG3 5RG
MD3 LEG H. Leslie 2 Close Y Lhergy, Union Mills, Isle of Man IM4 4LU
M3 LEK E. Little 7 Deerfern Close, Great Linford, Milton Keynes MK14 5BZ
M3 LEL L. Sargeant 99 Pot Kiln Road, Great Cornard, Sudbury CO10 0DX
M3 LEN L. Brackstone 276 Ladyshot, Harlow CM20 3EY
M3 LEO L. Flynn 2 Trafalgar Avenue, Grimsby DN34 5RE
MD3 LEP A. Le Prevost 58 Meadow Crescent, Douglas, Isle of Man IM2 1QX
MW3 LEW L. Jenkins 12 The Dell, Bryncethin, Bridgend CF32 9BJ
M3 LEX A. Green Croft House, Welbeck Road, Chesterfield S44 6DH
M3 LFC D. Hughes 86 Colinmander Gardens, Ormskirk L39 4TF
MI3 LFE A. Boylan 36 Callan Bridge Park, Armagh BT60 4BU
M3 LFG L. Gill 2 Loxton Court, Mickleover, Derby DE3 0PH
M3 LFH S. Gregory 1 St Martin Street, Atherton, Manchester M29 9DN
MM3 LFI D. Pomphrey Flat 2/9, 109 Bell Street, Glasgow G4 0TQ
MW3 LFL K. Morgan Gwel-Yr-Afon, Penrhyncoch Road, Aberystwyth SY23 3EA
M3 LFO S. Yeldham 19 Wade Reach, Walton on the Naze CO14 8RG
M3 LFP R. Brown 9 Bayleaf Crescent, Oakwood, Derby DE21 2UG
M3 LFQ W. Mcgill 49 Anthony Close, Colchester CO4 0LD
M3 LFU A. Heyes 528 Manchester Road, Paddington, Warrington WA1 3TZ
M3 LFV D. Mear 10 Peters Court, Hatton, Derby DE65 5JG
M3 LFX A. Lamb 9 Budworth Avenue Seaton Sluice, Whitley Bay NE26 4DB
M3 LFZ M. Haseldine 59 Brackley Road, Bedford MK42 9SH
M3 LGF D. Sporton 40 Eastwood Park Drive, Hasland, Chesterfield S41 0BD
M3 LGH S. Hallam 18A Market Street, Hoylake, Wirral CH47 2AE
M3 LGI I. Scott 13 Fairmount Road, Bexhill-on-Sea TN40 2HN
M3 LGJ P. Bishop 2 Hall Farm Bungalows, Illington, Thetford IP24 1RR
MI3 LGL L. Logue 21 Moyagh Road, Cullion, Londonderry BT47 2SL
M3 LGM M. Steeples 17 Windsor Avenue, Thurlstone, Sheffield S36 9RX
MW3 LGS S. Lewis 11 Treseder Way, Cardiff CF5 5NW
MM3 LGU R. Pennykid 50 Queen Street, Edinburgh EH2 3NS
M3 LGX G. Cash 28 Bramblewood Close, Prenton CH43 9YT
M3 LGY M. Cash 28 Bramblewood Close, Prenton CH43 9YT
MW3 LHA L. Hailstone 1 Hornbeam Close, Cimla, Neath SA11 3XA
M3 LHE J. Gammer 12 West Rise, Tonbridge TN9 2PG
M3 LHF D. Dunne 1 Burton Gardens, Brierfield, Nelson BB9 5DR
M3 LHG G. Gammer 12 West Rise, Tonbridge TN9 2PG
M3 LHI Z. Dunne 1 Burton Gardens, Brierfield, Nelson BB9 5DR
M3 LHM L. Marshall Thistledome, First Avenue, Watford WD25 9PS
M3 LHQ K. Brice 10A Nelson Drive, Exmouth EX8 2PU
M3 LHU C. Jenkins 5 Marlborough Close, Eastbourne BN23 8AN
M3 LHW A. England Beech House, Vicarage Gardens, Bradford BD11 2EF
M3 LHX A. Woodsford 5 Eliot Close, Wickford SS12 9ED
M3 LHZ S. Yates 14 Rushden Road, Sandon, Buntingford SG9 0QR
M3 LIB J. Bradburn 4 Lathkil Grove, Buxton SK17 7PH
M3 LIN L. Chesters 8 Conway Grove, Blacon, Chester CH1 5RU
M3 LIU I. Cartmell 10 Derwent Avenue, Burnley BB10 1HZ
M3 LIV Z. Bayliss 16 Oakmere Close, Sandbach CW11 1WN
M3 LIW J. Brown 8 Chatsworth Street, Sutton-in-Ashfield NG17 4GG
M3 LIX M. Knell 8 Wimborne Gardens, Kirby Cross, Frinton-on-Sea CO13 0TH
M3 LIY S. Mason 12 Blue Cedar Drive, Streetly, , Sutton Coldfield B74 2AE
M3 LJA L. Abel 121 Angela Road, Horsford, Norwich NR10 3HF
M3 LJB L. Bazley 18 Wellington Street , Radcliffe M26 2RB
M3 LJF Dr L. Gudgeon 5 Winkley Street, London E2 6PY
M3 LJI G. Billington 47 Smithy Leisure Park, Cabus Nook Lane, Preston PR3 1AA
M3 LJJ J. Lightfoot Apple Tree Cottage, Flowers Hill, Reading RG8 7BD
M3 LJK A. Smith 46 Mulberry Close, Goldthorpe, Rotherham S63 9LB
MI3 LJQ W. Phair 34 Sketrick Island Park, Newtownards BT23 7BN
MD3 LJS J. Keig 60 Garth Avenue, Surby, Port Erin IM9 6QZ Isle of Man
M3 LJX G. Jones 7 St. Ives Road, Weston-Super-Mare BS23 3XX
M3 LJZ L. Jennings 29 Mountbatten Drive, Newport PO30 5SJ
M3 LKD J. Dixon 23 Dee Way, Winsford CW7 3JB
M3 LKE E. Smith 30 Teignmouth Road, Torquay TQ1 4EA
M3 LKJ P. Manning 1 Waverley Gardens, Ash Vale, Aldershot GU12 5JP
M3 LKO K. Cope 64 Queen St., Pensnett, Brierley Hill DY5 4HA
M3 LKU A. Head 34 Balds Lane, Stourbridge DY9 8SG
MM3 LKV C. Duncan 131 Croftend Avenue, Glasgow G44 5PF
M3 LKY S. Smith 10 Parkway South, Doncaster DN2 4JS
M3 LLB L. D'Aubray-Butler Copse Edge 16 Francis Road, Frodsham WA6 7JR
M3 LLC L. Steele 14 Rowley View, West Bromwich B70 8QR
M3 LLK A. Soper 16 Queen Elizabeth Drive, Crediton EX17 2EJ
M3 LLM J. Proudman 61 Iffley Road, Oxford OX4 1EB
M3 LLN A. Sibley 27 Sherwood Road, Tetbury GL8 8BU
M3 LLQ J. Beards 175 Blackhalve Lane, Wolverhampton WV11 1AH
M3 LLT C. Moseley 15 Holden Crescent, Walsall WS3 1PY
MM3 LLU B. Gaudie Sunnyside, Harray, Orkney KW17 2JS
MW3 LLV H. Leonard 11 Newton Road Grangetown, Cardiff CF11 8AJ
M3 LLX J. Wright 2 Regent Road, Church, Accrington BB5 4AR
M3 LLZ P. Davies 53 Lammas Road, Cheddington, Leighton Buzzard LU7 0RY
M3 LMA L. Adkins 4 Orion Close, Ward End, Birmingham B8 2AU
M3 LMB P. Breslin 8A Mountfield, Tennyson Road, Yarmouth PO41 0PS
M3 LMC L. Mclaughlin 34 Cambridge Road, Birstall, Batley WF17 9JF
M3 LMD B. Dake 100 Lodge Road, West Bromwich B70 8PL
M3 LME G. Beardmore 9 Ashmore Drive, Gnosall, Stafford ST20 0RP
M3 LMH L. Holme 11 Oxlea Grove, Westhoughton, Bolton BL5 2AF
M3 LML Fyle Coast R.G. c/o M. Carney 2 Lilac Meadows, Lawley Village, Telford TF4 2NX
M3 LMQ R. Hines 7 Allendale Road, Caister On Sea, Great Yarmouth NR30 5ES
MI3 LMR R. Spence 299 Moyarget Road, Mosside, Ballymoney BT53 8DL
MW3 LMU E. Meek 7 High Tree Rise, Oakdale, Blackwood NP12 0DP
MW3 LMV A. Harris Flat 11, Tyn-Y-Coed, Cwmbran NP44 4PQ
M3 LMY F. Rowlands 45 Union Street, Market Rasen LN8 3AA
MW3 LMZ M. Williams 30 Elm Drive, Risca, Newport NP11 6HJ
MI3 LNC H. Davis 29 Fir Park, Broughshane, Ballymena BT42 4DH
M3 LNJ R. Woolridge 8 Alastair Drive, Yeovil BA21 3BT
M3 LNM S. Lawton 4 Astland Gardens, Tarleton, Preston PR4 6SX
M3 LNN J. Tomlinson 43 Haig Avenue, Southport PR8 6JY
M3 LNQ C. Woodruff 14 Primatt Crescent, Shenley Church End, Milton Keynes MK5 6AS
MM3 LNT M. Paterson 20 Loch Street, Rosehearty, Fraserburgh AB43 7JT
M3 LNU M. Durban 62 Westfield Way, Charlton, Wantage OX12 7EP
M3 LNY L. Goff 27 Harley Road, Oxford OX2 0HS
M3 LOA R. Loader Sunnyside, Main Street, Leyburn DL8 4LU
M3 LOE D. Gosling 19 Alcester Close, Plymouth PL2 1EA
MM3 LOF H. Paterson 20 Loch Street, Rosehearty, Fraserburgh AB43 7JT
MW3 LOI L. Pring 42 The Links, Trevethin, Pontypool NP4 8DQ
M3 LOT D. Filby 14 Jeffcut Road, Chelmsford CM2 6XN
M3 LOX D. Loxley 33 Longwood Road, Tingley, Wakefield WF3 1UG
M3 LPE J. Wright 26 Walmsley Close Church, Accrington BB5 4HL
MI3 LPF S. Hemmings Leylands, Leigh Road, Frome BA11 3LR
MI3 LPH L. Hesketh 61 Elmwood Cottages, Newtownabbey BT36 5WQ
M3 LPI R. Brierley 26 Jacobsen Avenue, Hyde SK14 4DW
M3 LPJ L. Renmans 70 Burman Road, Liverpool L19 6PW
M3 LPK G. Brierley 26 Jacobsen Avenue, Hyde SK14 4DW
M3 LPN B. Kersey 61 Crown Road, Portslade, Brighton BN41 1SJ
M3 LPR J. Main 15 Byron Road, Lydiate, Liverpool L31 0DB
M3 LPT L. Towler 8 Stowehill Road, Peterborough PE4 7PY
M3 LPU P. Higgins 9 Claremont Grove, Exmouth EX8 2JW
MD3 LPW L. Wernham Rogane Cottage, Church Lane, Santon IM4 1EZ Isle of Man
M3 LQA P. Ellis 40 Grasmere Road Royton, Oldham OL2 6SR
M3 LQB A. Foulds 4 Kropacz Court, South Street, Doncaster DN6 7JL
MW3 LQC J. Bhart Flat 1 90 Walter Road, Swansea SA1 4QF
M3 LQD S. Moakes 46 Parsonage St., Stockport SK4 1HZ
M3 LQE M. Clarke 48 Delves Wood Road, Huddersfield HD4 7AS
M3 LQI S. Briggs 56 Broadfields Road, Exeter EX2 5RG
M3 LQJ P. Langford 24 Asotte Way, Southam CV47 1GH

MM3 LQK W. Caithness 36 Wards Drive, Muir of Ord IV6 7PX
MI3 LQN H. Mcerlean 47 Barrack Road, Magherafelt BT45 6LY
M3 LQO R. Claydon 3 Birch Trees, Ambleside Road, Windermere LA23 1EU
M3 LQP D. Brown 21 Woborrow Road, Heysham, Morecambe LA3 2PW
M3 LQV P. Newton 61 Ashbourne Crescent, Taunton TA1 2RA
M3 LQW R. Edwards 46 Lavers Oak, Martock TA12 6HG
M3 LQX M. Hall 2 Hallcroft Road, Haxey, Doncaster DN9 2HP
M3 LQY R. Vigors 12 Sandfield Park, Lichfield Road, Brownhills WS8 6LN
M3 LRF P. Callaghan 41 Higher Ash Road, Talke, Stoke-on-Trent ST7 1JN
M3 LRJ A. Smith 101 Chaucer Drive, Lincoln LN2 4LT
M3 LRK M. Davey 67 Rotherham Baulk, Carlton-In-Lindrick, Worksop S81 9LE
M3 LRN A. Page 148 Waleton Acres, Carew Road, Wallington SM6 8PY
M3 LRP R. Plater Garsides, Keeling Street, Louth LN11 7QU
MI3 LRR W. Turtle 35 Buckna Road, Broughshane, Ballymena BT42 4NJ
M3 LRU J. Fitzpatrick 29 Delmar Road, Knutsford WA16 8BG
M3 LRW L. Wilson The Rectory, Church Close, Thetford IP25 7LX
M3 LRX D. Horwood 42 Southlands Drive, Timsbury, Bath BA2 0HB
M3 LRZ L. Reynolds 12 Providence Crescent, Boundary Way, Hull HU4 6EF
M3 LSE M. Newton 20 Orchard Way, Timberscombe TA24 7UL
M3 LSF L. Caslin 30D Holmewood, Holme, Peterborough PE7 3PG
MW3 LSG C. Goodridge 94 High Street, Nantyffyllon, Maesteg CF34 0BP
M3 LSK T. Newton 43 Hayfield Road, Minehead TA24 6AD
M3 LSL K. Summers 11 Jefferson Court, Franklin Street, Hull HU9 1JB
MM3 LSO S. Arnott 2 Nisbet Avenue, Eyemouth TD14 5BF
M3 LSS L. Stephenson 91 Hoyland Road, Hoyland Common, Barnsley S74 0AP
M3 LSU J. Evans 7 Westland Drive, Hayes, Bromley BR2 7HE
M3 LSX A. Dale 37 Bussey Road, Norwich NR6 6JF
M3 LTA L. Talbot 26 Chevalier Grove, Crownhill, Milton Keynes MK8 0EJ
M3 LTG L. Gear 26 Woodlands Way, Denaby Main, Doncaster DN12 4LR
M3 LTH M. Preston 53 Links Road, Birmingham B14 4TW
M3 LTP P. Whittall 165A High St., Brierley Hill DY5 3BU
M3 LTR D. Ingham 19 Recreation Avenue, Ashton-In-Makerfield, Wigan WN4 8SU
M3 LTT A. Old 51 Villiers Close, Plymouth PL9 7QP
MW3 LTU M. Brady Ty Mawr Uchaf, Dulas LL70 9DQ
M3 LTV A. Walker 76 Greenway, Birmingham B20 1EQ
M3 LTW J. Bennett 2 Victoria St., Pensnett, Brierley Hill DY5 4LB
M3 LUD P. Ludders 280 Hopewell Road, Bilton Grange, Hull HU9 4HH
M3 LUK L. Mandeville 6 Oxford Road, Benson, Wallingford OX10 6LX
M3 LUO D. Shoubridge 51 Pipers Field, Ridgewood TN225SD
M3 LUP S. Ingram Flat 1, Philip Howard Court, Glynne Street, Farnworth, Bolton BL4 7DQ
M3 LUU A. Gillett 441 Radipole Lane, Weymouth DT4 0QF
M3 LUW G. Stocks 62 Ridge Park Avenue, Plymouth PL4 6QA
M3 LUZ A. Bent 14 Pleasant Road, Eccles, Manchester M30 0FS
M3 LVA L. Adlington 21 Newstead Road, Stoke-on-Trent ST2 8HU
MW3 LVF R. Hawkins Nook Cottage, Common-Y-Coed, Caldicot NP26 3AX
M3 LVK C. Street Russetts, Isle Brewers, Taunton TA3 6QN
M3 LVM K. West 36 Watlington Road, Cowley, Oxford OX4 6SS
M3 LVP J. Gilbert 148 Purcell Road, Coventry CV6 7LB
M3 LVR E. Mills 76 Main St., Burton Joyce, Nottingham NG14 5EH
MM3 LVT J. Dock 75 Ferguslie Park Avenue, Paisley PA3 1BE
M3 LVX P. Hampton 4 Moorland View, Plymstock, Plymouth PL9 8NW
M3 LVY A. Kendrick 13 Queens Drive, Middlewich CW10 0DG
MI3 LVZ M. Mccloy 4 Audleys Park, Newtownards BT23 8UA
M3 LWG J. Arrowsmith 45 Hilderic Crescent, Dudley DY1 2EU
MM3 LWJ R. Bertram 46 Main Street Main Street, Pathhead EH37 5QB
MM3 LWO G. Mallolm 32 Holly Crescent, Dunfermline KY11 8BT
M3 LWP J. Clowes 52 Pennine Drive, St. Helens WA9 2BU
MD3 LWQ M. Corlett 18 Queens Drive, Peel IM5 1BQ Isle of Man
MM3 LWT S. Conway 26 Rennie St., Kilmarnock KA1 3AR
MI3 LWU A. Geary 114 Maddan Road, Armagh BT60 3LJ
M3 LWV M. Powell 37 Newnham Close, Mildenhall, Bury St. Edmunds IP28 7PD
M3 LWX W. Alder 21 Manor Gardens, London SW20 9AB
MM3 LWZ C. Coore 14 Craigs Drive, Edinburgh EH12 8UW
M3 LXA S. Jones 10 Litchborough Grove, Whiston, Prescot L35 7NE
M3 LXB J. Wynne 43 Lansdown Road, Broughton, Chester CH4 0NZ
MI3 LXE Dr I. Stevenson 55 Churchtown Road, Downpatrick BT30 7AZ
M3 LXF S. Mills 27 Boscow Crescent, St. Helens WA9 3SX
M3 LXH S. Wright Flat 2, 19 Cearns Road, Prenton CH43 2JL
MI3 LXJ B. Crozier 33 Cullentragh Road, Poyntzpass, Newry BT35 6SD
M3 LXK C. Wynne 43 Lansdown Road, Broughton, Chester CH4 0NZ
MI3 LXN D. Bryans 1 Meadowvale Avenue, Bangor BT19 1HG
M3 LXP P. Enfield 1 Horton Park, Blyth NE24 4JD
M3 LXR L. Dexter 27 Underwood Avenue, Worsbrough, Barnsley S70 4AU
M3 LXS M. Woolley 4 Robert Street, Warrington WA5 1TQ
M3 LXU T. Moscrop 64 Gresham Road, Norwich NR3 2NG
M3 LXV J. Thompson 51 Grafton Road, Oldbury B68 8BP
MI3 LXW M. Lewis 7 Liester Park, Ballyrobert, Ballyclare BT39 9RZ
MI3 LXZ Dr A. Mcdowell 10 Lord Wardens Vale, Bangor BT19 1GH
M3 LYA C. Cooper The Haven, Ipswich Road, Norwich NR15 2TA
M3 LYC L. Bentley 1 Cotswold Road, Lupset, Wakefield WF2 8EL
M3 LYG R. Dunkley 25 St. Andrews Crescent, Wellingborough NN8 2ES
MM3 LYH C. Rodger 23 Harrysmuir Road, Pumpherston, Livingston EH53 0NT
M3 LYP I. Hallatt 11 Cheshire St., Audlem, Crewe CW3 0AH
MW3 LYQ R. Lovesey 33 Ty Isaf Park Avenue, Risca, Newport NP11 6NB
M3 LYR T. Martin 46 Hayes Crescent, Frodsham WA6 7PG
MM3 LYS E. Smith 27 Elm Lane, Foresters Lodge, Glenrothes KY7 5TD
M3 LYU C. Arner 6 Welshampton Close, Great Sutton, Ellesmere Port CH66 2WL
M3 LYV C. Walsh 133 Belfield Road, Accrington BB5 2JD
M3 LYX D. Shaw 37 Smithwaite View, Normanton WF6 1AW
M3 LYZ F. Martin 1 Marsh Street, Strood, Rochester ME2 4BB
MI3 LZA T. Conway 203 Garrymore, Moyraverty, Craigavon BT65 5JF
MW3 LZC T. Rule 35 Pill St., Penarth CF64 2JS
MM3 LZD J. Dinning 1 South Brae Aiket Road, Dunlop, Kilmarnock KA3 4BP
MI3 LZF J. Steele 46 Circular Road, Newtownards BT23 4BN
M3 LZK J. Delves 14 Stanthorne Avenue, Crewe CW2 8NH
M3 LZL F. Shields 4 Occupation Lane, Earlsheaton, Dewsbury WF12 8PY
M3 LZR N. Finlay 50 Melchett Crescent, Rudheath, Northwich CW9 7EP
M3 LZT D. Young 20 Summerhouse, Tickenham, Clevedon BS21 6SN
MM3 LZU C. Tait 39 Baleshrae Crescent, Kilmarnock KA3 2GN
M3 MAA M. Ahmed 59 Ramsgate Lofthouse, Wakefield WF3 3PX
M3 MAC J. Mcintyre Hunters Moon, 34 Potter Street, Northwood HA6 1QE
MD3 MAN A. Espey 9A Hilltop View, Douglas, Isle of Man IM2 2LA
MM3 MAO M. Overthrow 63 Primrose Avenue, Larkhall ML9 1JX
M3 MAR M. Jeffery 14 Holly Mount Shavington, Crewe CW2 5AZ
M3 MBC M. Bridgeland 17 Oldfield Lane, Wisbech PE13 2RJ
M3 MBF A. Fryer 9 The Oval, Guildford GU2 7TS
M3 MBH D. Hemmings 1 Sunray Grove, Hucknall, Nottingham NG15 6RF
M3 MBI S. Stuart 46 Breach Road, Heanor. DE75 7NJ
MW3 MBL M. Watkins 1 Llys Melyn, Tregynon, Newtown SY16 3EE
MI3 MBM M. Buchanan 49 Glengiven Avenue, Limavady BT49 0RW
M3 MBQ M. Daniells 6 Carlton Road, Manchester M16 8BB
M3 MBR M. Roberts 13 St. Michaels Court, Stevenage SG1 5TB
M3 MBV H. Su 135 Devana Road, Leicester LE2 1PN
M3 MCA A. Matthews 70 Branksome Hall Drive, Darlington DL3 9SR
MD3 MCB B. Perrin 18 Bellevue Park, Peel IM5 1UF Isle of Man
M3 MCF M. Frame 23 Greenside Court, Sunderland SR3 4HS
MM3 MCG M. Gaston Ellena, Lochans Mill Avenue, Stranraer DG9 9BZ
M3 MCU J. Mccallum 48 Heber St., Bristol BS5 9JT
MM3 MDB M. Brunsdon 25 Buckstone Lea, Edinburgh EH10 6XE
M3 MDI J. Gleeson 19 Alston Avenue, Shaw, Oldham OL2 7SX
M3 MDK K. Sawyers 27 Dukeswood Road, Longtown, Carlisle CA6 5UJ
M3 MDN M. Dobson 17 Cadley Causeway, Fulwood, Preston PR2 3RU
M3 MDS A. Flemming 20 Chatham Hill, Chatham ME5 7AA
MI3 MDV A. Jess 25 Gransha Road, Dundonald, Belfast BT16 2HB
M3 MDW M. Webster 2 Brook Close, Nottingham NG6 8NL
M3 MDY M. De Young 97 Kingfisher Road, Larkfield, Aylesford ME20 6RE
M3 MEB M. Collins 73 Westholme Road, Bidford-On-Avon, Alcester B50 4AN
M3 MEE S. Parker 100 Horsebridge Hill, Newport PO30 5TL
M3 MEF M. Fry 46 Butt Parks, Crediton EX17 3HE
M3 MEG D. Cash 3 Marsh Lane, Wolverhampton WV10 6RU
MM3 MEH M. Mumford 13 Galloway Drive, Culloden, Inverness IV2 7ND
M3 MEI G. Childs 144 Sturdee Avenue, Gillingham ME7 2HL
M3 MEP M. Porteious 17 Church Walk, Yaxley, Peterborough PE7 3YD
M3 MER R. Fox 3 Cherry Blossom Close, Harlow CM17 0EX
M3 MES N. Messenger 7 Skinners Close, Swordy Park, Alnwick NE66 1EU
M3 MEU D. Cowman 23 Kirk Flatt, Great Urswick, Ulverston LA12 0TB
M3 MEW M. Wright 9 Dinmore Avenue, Blackpool FY3 7RR
MW3 MEY G. Alker Bryn Y Mor, Lon Ganol, Menai Bridge LL59 5YA
MI3 MFD F. Doherty 52 Madison Avenue Eglinton, Londonderry BT47 3PW
M3 MFE P. Penfold 2 The Leas, Essenden Road, St. Leonards-on-Sea TN38 0PU
M3 MFF M. Frohnsdorff 75 Alexander Drive, Faversham ME13 7TA
M3 MFG C. Goulty Seletar, 22 Western Avenue, Felixstowe IP11 9TS
MM3 MFN A. Harkess 7 Gardiner Road, Prestonpans EH32 9HF
MM3 MFR M. Mcminn 28 Dunlop Road, Dumfries DG2 9NN
M3 MFS S. Spencer 55 Witton Lane, West Bromwich B71 2AA
M3 MFT A. Hill 1 Rochester Close, Mountsorrel, Loughborough LE12 7UH
M3 MFU D. Mutlow Dunvegan, Wood Lane, Nuneaton CV13 0AU
M3 MFX C. Richardson 16 Church Road, Boreham, Chelmsford CM3 3EF
M3 MFZ J. Gosling 11 Pinfold Place, Harby, Melton Mowbray LE144BX
M3 MGD D. Riggs 37 Moot Gardens, Downton, Salisbury SP5 3LG
M3 MGI T. Rogers 18 Field Road, Bridlington YO16 4AU
M3 MGJ M. Minshull 12 Dunnett Close, Attleborough NR17 2NG
MM3 MGK S. Brown 21 Whiteford Avenue, Dumbarton G82 3JU
M3 MGL M. Talbot 26 Chevalier Grove, Crownhill, Milton Keynes MK8 0EJ
M3 MGN M. Naylor 96 New Meadows, Rawmarsh, Rotherham S62 7FE
M3 MGO M. Butler Wood Green, Astley, Stourport-on-Severn DY13 0RU
M3 MGP M. Champion 155 Walton Road, Walton on the Naze CO14 8NF
M3 MGQ J. Browne 29 Longbridge Close, Tring HP23 5HG
M3 MGU R. Gregory Town End, Kirkby Road, Askam-in-Furness LA16 7EY
M3 MGZ M. Curtis 10 Woodstock Gardens, Blackpool FY4 1JP
M3 MHD M. Downes 38 Queensway, Warton, Preston PR4 1XU
MW3 MHG G. Rees 47 Loftus Street, Cardiff CF5 1HL
M3 MHL A. Parrish 5 Kestrel Lane, Cheadle, Stoke-on-Trent ST10 1RU
M3 MHM M. Mountford 1 Bowater House, Moor St., West Bromwich B70 7AZ
M3 MHN M. Hurren 257 Norwich Road, Wroxham, Norwich NR12 8SL
M3 MHP E. Skinner 11 Finch Crescent, Leighton Buzzard LU7 2PE
MM3 MHQ A. Mccurdy 5 Kestrel Place, Greenock PA16 7BL
M3 MHR M. Reavell 85 Mccarthy Close, Birchwood, Warrington WA3 6RS
MM3 MHS M. Shearer 113 Auchamore Road, Dunoon PA23 7JJ
M3 MHV M. Vaughan 12 Kingsley Road, Frodsham WA6 6SG
M3 MHZ D. Lawrence 23 Parkmead Road, Wyke Regis, Weymouth DT4 9AL
MM3 MID K. Middleton 1 Campbell Court, Lochmaben, Lockerbie DG11 1NF
MI3 MIE Y. Wilson 59 Crew Road, Upperlands, Maghera BT46 5TU
M3 MIF J. Tranter 64 Geneva Drive, Newcastle ST5 2QH
M3 MII G. Elsworth 367 West Dyke Road, Redcar TS10 4PS
M3 MIJ J. Dean 2 Wellington Street, Chesterfield S43 2BJ
M3 MIN A. Jones 17 Maybush Drive, Chidham, Chichester PO18 8SR
M3 MIO A. Cox 11 Windmill Drive, Audlem, Crewe CW3 0BE
M3 MIP R. Parrish 5 Kestrel Lane, Cheadle, Stoke-on-Trent ST10 1RU
M3 MIQ M. Fradley 43 Grange Drive, Penketh, Warrington WA5 2JN
M3 MIR S. Carruthers 13 Belah Road, Carlisle CA3 9RE
M3 MIS J. Greatrix West Cottage, Main Road, Boston PE20 3PZ
M3 MIU F. Lie Harrogate Ladies' College, Clarence Drive, Harrogate HG1 2QG
M3 MIV T. Skinner Flat 4, 25-27 Bridge Street, Leighton Buzzard LU7 1AH
M3 MIX M. Cole 9 Troopers Drive, Romford RM3 9DE
M3 MJD M. Dennison 27 Chapel St., Cawston, Norwich NR10 4BG
M3 MJE M. Edwards 3 George Street, Bourne PE10 9HE
M3 MJH M. Hickford 3 Ashen Road, Clare, Sudbury CO10 8LQ
MI3 MJI R. Wylie 69 Rubane Road, Kircubbin, Newtownards BT22 1AU
M3 MJJ Cwmbran ARS c/o J. Miller Flat 1 Block 2, St. Phillips Place, Eastbourne BN22 8LW
M3 MJL M. Lee Up To Date House, Shore Road, Boston PE22 0NA
M3 MJM M. Marter 4 Meadow Way, Seaford BN25 4QT
M3 MJN M. Noon 97 Cherrycroft, Skelmersdale WN8 9EF
M3 MJV M. Verrechia 20 The Wyvern Grafham, Huntingdon PE28 0GG

M3 MJY M. Kirby 2 Morton Crescent, Bradwell, Great Yarmouth NR31 8NT
M3 MKB M. Baxter 5 Farnborough Street, Farnborough GU14 8AG
M3 MKD M. Davis 15 Farmcroft Road Mansfield Woodhouse, Mansfield NG19 8QU
M3 MKH M. Hall 10 Darwin Walk, Northampton NN5 6LR
M3 MKJ M. Allen 8 Green Close, South Wonston, Winchester SO21 3EE
M3 MKK M. Kilkenny 23 Hazelhurst Road, Stalybridge SK15 1HD
M3 MKM D. Kelly Clearance Crescent Whitley Bay, Newcastle NE262DZ
MW3 MKN M. Joseph 32 Charles Street, Trealaw, Tonypandy CF40 2UN
M3 MKO J. Duffield 4 Church Hill, Easingwold, York YO61 3JS
M3 MKV J. Beech 124 Belgrave Road, Coventry CV2 5BH
M3 MKZ J. Swift 52 North Street, Burwell, Cambridge CB25 0BB
M3 MLA P. Richardson 11 Overstone Road Coldham, Wisbech PE14 0ND
MD3 MLB M. Bazley 77 Royal Park, Ramsey IM8 3UH Isle of Man
M3 MLF M. Firth 126 Tombridge Crescent, Kinsley, Pontefract WF9 5HE
M3 MLG M. Goff 27 Harley Road, Oxford OX2 0HS
M3 MLI M. Litt-Wilson 14 Wastwater Rise, Seascale CA20 1LB
M3 MLK S. Ellison 23 Murphy Grove, St. Helens WA9 1QY
MM3 MMB M. Baird Creag Saval, Lairg IV27 4ED
MI3 MMC M. Mcclure 12 St. Patricks Park, Ballymoney BT53 6JG
M3 MMG M. Glover 96 Byron St., Macclesfield SK11 7QA
MM3 MMI F. Millar 13 Edzell Park, Kirkcaldy KY2 6YB
MW3 MMJ M. Jones 47 Maes Derw, Llandudno Junction LL31 9AN
M3 MML M. Lemer 1 Holmbush Court, Brent St., London NW4 2NS
M3 MMN G. Larrigan 9 Sandpiper Gardens, Chippenham SN14 6YH
MM3 MMO D. Frost Flat 6, 88 Albion Street, Glasgow G1 1NY
M3 MMP P. Evans 4 Havelock Court, Havelock Street, Aylesbury HP20 2NU
M3 MND A. Harrop 35 Langdale Crescent, Dalton-in-Furness LA15 8NR
MU3 MNG R. Bougourd 3 Bartholemew, Victoria Road, St. Peter Port GY1 1JB Guernsey
M3 MNQ B. Gould 26 Trembel Road, Helston TR12 7DY
M3 MNR J. Redrup 58 Shaftesbury Road, Bournemouth BH8 8ST
M3 MNT T. Williamson 286 Glynswood, Chard TA20 1BX
M3 MNU P. Richardson 14 Portland Street, Worksop S80 1RZ
MW3 MNV N. Hill 53 Broadmead, Pontllanfraith, Blackwood NP12 2NJ
M3 MNY K. King Chad Lane Farm, Chad Lane, St. Albans AL3 8HW
M3 MOB J. Howarth 5 Sydenham Building, Bath BA2 3BS
M3 MOF J. Jones The Studio, Ferney Hoolet, Hook RG27 8SW
M3 MOH G. Worrall 94 Scotia Road, Stoke-on-Trent ST6 4ET
MW3 MOJ K. Pitt 21 Maes-Yr-Onen, Nelson, Treharris CF46 6LF
M3 MOP A. Barden 38 Silver Close, Tonbridge TN9 2UY
M3 MOQ S. Hutchinson 28 Willow Road, New Balderton, Newark NG24 3DA
MI3 MOT A. Thompson 23 Causeway End Park, Lisburn BT28 2HX
M3 MOV M. Greenhow 39 Boston Avenue, Runcorn WA7 5XE
M3 MPC M. Coles 29 Sydney Road, Exeter EX2 9AH
MM3 MPK M. Kilday 3 Union Drive, Whitburn, Bathgate EH47 0AJ
MI3 MPL H. Currie 58 Duneden Park, Belfast BT14 7NF
M3 MPM M. Murrell 11 Ajax Close, Hull HU9 4BE
M3 MPT P. Denehy 17 Coverdale, Hull HU7 4AL
M3 MQA L. Woolley 4 Robert Street, Warrington WA5 1TQ
M3 MQB J. Hing 6 Peartree Walk, Billericay CM12 0PY
M3 MQC J. Taylor 6 Hawks Close, Walsall WS6 7LE
M3 MQH N. Hilton 20 Darbyshire Close, Deeping St. James, Peterborough PE6 8SF
M3 MQI A. Paxton Havencroft, Dalbury Lees, Ashbourne DE6 5BE
M3 MQJ G. Jackson Long Pools Farm, Marsh Lane, Market Drayton TF9 2TG
M3 MQM K. Rooney Red Cap Farm, Green Fairfield, Derbyshire SK17 7JF
M3 MQP M. Richards 57 Coronation Close, Broadstairs CT10 3DL
M3 MQR D. Rogers 9 Prospect Place, Stafford ST17 4HZ
M3 MQX R. Rowe 16 Orchard Road, Plymouth PL2 2QY
M3 MQY G. Robinson Crowstone Mews, Syke House Lane, Greetland HX4 8PA
M3 MRA N. Rogers 4 Lawson Court, Millfield Avenue, Market Harborough LE16 8XR
MW3 MRC M. Clayton 12 The Broadway, Abergele LL22 7DF
MI3 MRF T. Mccullough 23 Edenvale Park, Antrim BT41 1AY
MI3 MRG A. Colligan 8 Mourneview Crescent, Lisburn BT28 3HD
M3 MRJ J. Johnson 3 Rumbold Road, Hoddesdon EN11 0LP
MW3 MRK M. Knowles 72 Uplands Avenue, Connah'S Quay, Deeside CH5 4LG
MW3 MRL M. Williams 5A Derllwyn Close, Tondu, Bridgend CF32 9DH
M3 MRM J. Milner Stone Cottage, Wistanstow, Craven Arms SY7 8DG
M3 MRN J. Orange 20 Borrowdale Avenue, Fleetwood FY7 7LF
M3 MRO M. Ross 143 Rose Lane, Romford RM6 5NR
M3 MRQ N. Thain 24 Wilmington Road, Hastings TN34 2BT
M3 MRS L. Henley 5 Gosselin Street, Whitstable CT5 4LA
M3 MRU D. Edwards 10 Queens Avenue, Ilfracombe EX34 9LN
MM3 MRX G. Robertson 24 Tippet Knowes Court, Winchburgh, Broxburn EH52 6UW
M3 MRZ T. Hall 8 Vicarage Close, Billesdon, Leicester LE7 9AN
M3 MSB S. Bridge 59 Alder Hey Road, St. Helens WA10 4DN
M3 MSC L. Nicklin 59 Laurel Road, Armthorpe, Doncaster DN3 2ES
M3 MSH M. Hunt 57 Colsterdale, Worksop S81 0XH
M3 MSJ M. Stocker 1 Rectory Close, Carlton, Bedford MK43 7JT
M3 MSL M. Witter 62 Old Road, Churwell, Leeds LS27 7RT
M3 MSN M. Austin 38 Garden Road, Folkestone CT19 5RA
M3 MSQ E. Bartlett 86 Usk Road, Tilehurst, Reading RG30 4HU
M3 MST M. Wilson 234 Aylsham Drive Ickenham, Uxbridge UB10 8UF
M3 MSU L. Wardle 35 Woodland Close, Barnstaple EX32 0EG
M3 MSX M. Jenkins 1 Green End Road, Sawtry, Huntingdon PE28 5UX
M3 MSY M. Spicer 13 Strawberry Path, Oxford OX4 6RA
M3 MSZ M. Swift 8 Grove Lane, Buxton SK17 9HG
MW3 MTB M. Buxton 25 Pen Y Bryn, Sychdyn, Mold CH7 6EE
M3 MTC C. Mathewson 33 Thornton Road, Bootle L20 5AN
M3 MTL M. Price 9 Herbarth Close, Liverpool L9 1JZ
MM3 MTM M. Mcleary 146 Captains Road, Edinburgh EH17 8DX
M3 MTP M. Powell 2A Park Avenue, Uttoxeter ST14 7AX
M3 MTQ S. Saunders 3 The Terrace, High Street, Cavendish, Sudbury CO10 8AS
M3 MTR M. Bryan 16 Walesmoor Avenue, Kiveton Park, Sheffield S26 5RG
M3 MUA J. Anderson 121 Barton Road, Stretford, Manchester M32 9AF
M3 MUF N. Wilson 3 New Road Bovington, Wareham BH20 6JZ
M3 MUI M. Bromfield The Cottage, Huttoft LN13 9RF
M3 MUO P. Riddle 8 Lower Dingle, Oldham OL1 4PB
M3 MUP N. Speight Flat 14, Cranbrook, London NW1 0LJ
M3 MUQ W. Mitchell Flat 12, 1 Benwell Road, London N7 7AY
M3 MUX D. Baseden 27 Bayfield, Painters Forstal, Faversham ME13 0EF
M3 MVI M. Pick 290 Horsley Road, Washington NE38 8HS
M3 MVJ R. Jefferiss 21 East Street, Fritwell OX27 7PX
M3 MVK M. Egerton 43 New Street, Crewe CW15PN
M3 MVM J. Goulding 79 Dalston Drive, Manchester M20 5LQ
M3 MVO D. Campbell Beeches, Hammersley Lane, High Wycombe HP10 8HG
M3 MVR A. Hobbs 2 The Mead, Beaconsfield HP9 1AW
MW3 MVT R. Williams Royston, 18 Grove Street, Maesteg CF34 0HY
M3 MVV J. Mullarkey 41 Foyle Avenue, Chaddesden, Derby DE21 6TZ
MM3 MVY M. Gerrard 10 Whinhill Gardens, Aberdeen AB11 7WD
MI3 MWA B. Allen 48 Kevlin Gardens, Omagh BT78 1QS
M3 MWG M. Gosling 1 Zion Street, Plymouth PL1 2HX
M3 MWM M. Martin 80 Waveney Road, Hull HU8 9LY
MW3 MWO A. Rowlands 8 Cleveland Avenue, Tywyn LL36 9EG
M3 MWQ B. Hall 64 Synehurst Crescent, Badsey, Evesham WR11 7XX
M3 MWT R. Waters 38 Pasture Road, Kirkbymoorside, York YO62 6FH
M3 MWV M. Williams 6 Richmond Terrace, Barrow-in-Furness LA14 5LH
M3 MXA M. Anderson 5 Saffron Court, Wakefield WF2 0FQ
MW3 MXC J. Baker 43 Clyde St., Risca, Newport NP11 6BP
M3 MXF R. Mason 27 Meadway, Malvern WR14 1SB
M3 MXG S. Turner 28 Fox Lea, Kesgrave, Ipswich IP5 2YU
M3 MXH M. Price 43 Heckington Drive, Nottingham NG8 1LF
M3 MXJ P. Smith 1 Grappenhall Hall School House, Church Lane, Warrington WA4 3ES
M3 MXK M. Kendall 67 Slade Valley Road, Ilfracombe EX34 8LG
M3 MXM R. Hardy 35 Chilton Road, Ipswich IP3 8PD
MM3 MXN S. Reid 14 St. Marys, Monymusk, Inverurie AB51 7HH
M3 MXO R. Kemp 4 Scoones Close, Bapchild, Sittingbourne ME9 9SW
M3 MXP A. Macgregor 53 Napier Place, Orton Wistow, Peterborough PE2 6XN
MD3 MXU J. Forrester 107 Ballaquark, Douglas, Isle of Man IM2 2EN
M3 MXV K. Moulder 51A Aston Cantlow Road, Wilmcote, Stratford-upon-Avon CV37 9XN
M3 MXW D. Platt 50 Poplars Road, Stalybridge SK15 3EN
M3 MXX T. Chapman 16 Andover Place, Cannock WS11 6EH
M3 MXZ M. Rowe 15 Atlantic Close, Treknow, Tintagel PL34 0EL
M3 MYE D. Sykes 2 The Street, Claxton, Norwich NR14 7AS
M3 MYG K. Toner 17 Crooked End Place, Ruardean GL17 9YN
M3 MYI C. Toner 17 Crooked End Place, Ruardean GL17 9YN
M3 MYK M. Lees 28 Pleasant Avenue, Bolsover, Chesterfield S44 6LL
M3 MYM D. Cox 3 Besley Court, Lethbridge Road, Wells BA5 2FE
M3 MYQ R. Jenkins Glascoed, Garthmyl, Montgomery SY15 6RT
M3 MYT N. Groat 138 Freedom Road, Sheffield S6 2XE
M3 MYW M. Edmond 12 Yeoman Close, Worksop S80 2RR
M3 MYZ M. Jones 65 Montgomery Avenue, Bournemouth BH11 8BN
M3 MZA R. Nicholson 24 Barnmead, Haywards Heath RH16 1UZ
M3 MZC A. Nicholson 24 Barnmead, Haywards Heath RH16 1UZ
M3 MZG D. Butterfield 57 Holmes Road, Retford DN22 6QU
M3 MZN S. Hamilton 25 Keefe Close, Chatham ME5 9AG
MM3 MZO I. Graham 17 Royal Avenue, Stranraer DG9 8ET
M3 MZP T. Burcombe 49 Huntingdon Close, Mitcham CR4 1XJ
M3 MZR E. Moon Moon Marine, Rock Channel, Rye TN31 7HJ
M3 MZT L. Jones 13 Bracewell Close, Sutton, St. Helens WA9 3SH
M3 MZV A. Linton 10 Adelaide Square, Shoreham-by-Sea BN43 6LN
M3 MZW M. Weller 27 Rochester Avenue, Woodley, Reading RG5 4NA
MM3 MZX A. Bowers 4A Pine Street, Greenock PA15 4HW
MW3 NAE N. Edwards 17 Queensway, Garnlydan, Ebbw Vale NP23 5EE
M3 NAF N. Foster 18 Austen Ave, Sawley, Nottingham NG103GG
M3 NAH N. Higham-Hook 31 Ringwood, Bracknell RG12 8YG
M3 NAL C. Corbishley 15 High St., Hardingstone, Northampton NN4 7BT
M3 NAO D. Bradshaw 65 Lichfield Court, Sheen Road, Richmond TW9 1AX
MW3 NAQ S. Marles 4 Maes Y Llan, Conwy LL32 8NB
M3 NAR B. Johnson 15 Oak Avenue, Willington, Crook DL15 0BJ
M3 NAT N. Poate 15 Jackdaw Rise, Eastleigh SO50 9JT
M3 NAW N. White 10 Elm Crescent, Alderley Edge SK9 7PQ
M3 NBB N. Beith 18 Avenue Road, New Milton BH25 5JP
M3 NBD N. Draper 107 Arkwrights, Harlow CM20 3LY
M3 NBG L. Mcguire 200 Wellingborough Road, Rushden NN10 9SX
M3 NBH T. Dunne 23 Warstone Lane, Birmingham B18 6JQ
M3 NBI K. Fell 5 Henry Road, Wath-Upon-Dearne, Rotherham S63 7NF
M3 NBK A. Rooney 44 Heritage Drive, Gillingham ME7 3EH
M3 NBL N. Bland 63 Swindon Road, Wroughton, Swindon SN4 9AG
M3 NBQ S. Cross 138 Crow Lane West, Newton-le-Willows WA12 9YL
M3 NBU J. Shufflebotham 316 Stockport Road, Hyde SK14 5RU
M3 NBX V. Pomfrett 17 Manifold Close, Sandbach CW11 1XP
M3 NBZ N. Birnie 61 Pipers Croft, Dunstable LU6 3JZ
M3 NCB D. Lawson 30 Meadowcroft, St. Helens WA9 3XQ
MI3 NCC P. Haughey 10 Captains Road, Forkhill, Newry BT35 9RR
M3 NCD N. Welsh 18 Linkway, Runcorn WA7 5EJ
M3 NCE M. Palmer 22 Nightingale Drive, Poulton-le-Fylde FY6 7UQ
M3 NCG M. Dumpleton 24 Barley Close, Newton St. Faith, Norwich NR10 3GY
M3 NCH P. Blackie 30 Queens Avenue, Ilfracombe EX34 9LS
M3 NCL N. Lees 31 Cosford Drive, Dudley DY2 9JN
MM3 NCM N. Cunningham 11 Glendoune Street, Girvan KA26 0AA
M3 NCN Rvd. M. Gillingham 14 Nethergreen Gardens, Killamarsh S21 1FX
M3 NCO M. Collingswood School House, Norton Canes High School, Burntwood Road, Norton Canes, Cannock WS11 9SP
M3 NCP L. Steer 51 Kings Chase, East Molesey KT8 9DG
M3 NCQ M. Williams 136 Courtfield Road, Quedgeley, Gloucester GL2 4UF
MW3 NCS N. Sedgebeer 16 Metcafe Street Caerau, Maesteg CF340TB
M3 NCT N. Croft 22 King Edward Crescent, Leeds LS18 4BE
M3 NDC N. Crowley 133 Jessop Road, Stevenage SG1 5LH
M3 NDF D. Fard 187 Fleetwood Road South, Thornton-Cleveleys FY5 5NS
M3 NDJ C. Belham 1 Kenmare Bank, Northwich CW9 8BN
MW3 NDO J. Hoskins 18 Bryn Yr Onnen, Southsea, Wrexham LL11 6RG
M3 NDR N. Nash Roann, Bedmond Road, Pimlico, Hemel Hempstead HP3 8SH
MW3 NDU R. Williams Bryn Mawr, Gwalchmai, Holyhead LL65 4PY
M3 NDZ A. Humphriss 44 Bishops Close, Stratford-upon-Avon CV37 9ED
M3 NEA J. Rice 2 Medalls Path, Stevenage SG2 9DX

M3 NEC N. Chisholm 162 Ardington Road, Northampton NN1 5LT
M3 NEE M. Jay Web-Stile Farm, Coley Hill, Bristol BS39 5ED
M3 NEG P. Mcgarry 10 Douglas Avenue, Soothill, Batley WF17 6HG
MW3 NEI N. Mcloughlin 19 Byron Road, Newport NP20 3HJ
M3 NEL N. Snape 20 Marlow Court, Adlington, Chorley PR7 4LE
MI3 NEN N. Nicholl 34 Berryhill Road, Artigarvan, Strabane BT82 0HN
M3 NEP N. Payne 7 Lane House, Eastfield Close, Worcester WR3 7TT
M3 NFB S. Billingham Kewell House, Wombourne Road, Swindon, Dudley DY3 4NF
M3 NFE I. Jenkin 62 Burns Road, Wellingborough NN8 3RS
MW3 NFF B. Armstong 3 Walton Cres, Llandudno Junction LL31 9RR
M3 NFG N. Gonzalez 46 Whitton View, Rothbury, Morpeth NE65 7QN
M3 NFH E. Ashley The Chapel, Ashford, Ludlow SY8 4BX
M3 NFJ A. Williams 327 Locking Road, Weston-Super-Mare BS23 3LY
M3 NFK M. Watmough 39 Ripon Gardens, Buxton SK17 9PL
M3 NFQ S. Grainger 132 Honiton Way, Hartlepool TS25 2PY
M3 NFU A. Scarlett 87 Coronation Avenue, Shildon DL4 2AZ
M3 NFW M. Millward 48 Nightingale Close, Farnborough GU14 9QH
MM3 NFX I. Marshall 6 Broom Wynd, Shotts ML7 4HP
MW3 NFZ D. Matthews 42 College Road, Oswestry SY11 2SG
M3 NGC G. Champion 34 Greenfields, Edenside, Kirby Cross., Frinton-on-Sea CO13 0SW
M3 NGE N. Taggart 61 Well Lane, Curbridge, Witney OX29 7PB
M3 NGF R. Dickerson 68 Chestnut Avenue, Spixworth, Norwich NR10 3QQ
M3 NGG N. Clare 123 Cunningham Road, Tamerton Foliot, Plymouth PL5 4PU
MM3 NGJ A. Bernard 200 Carden Avenue, Cardenden, Lochgelly KY5 0EN
M3 NGK N. Kaye 33A Aggborough Crescent, Kidderminster DY10 1LQ
M3 NGM J. Gregory 17 Meadowgarth, Belford NE70 7PA
MW3 NGN B. Cook 218 Ffordd Pennant, Mostyn, Holywell CH8 9NZ
M3 NGO M. Hirst 34 Welldon Crescent, Harrow HA1 1QR
MW3 NGP A. Cook 218 Ffordd Pennant, Mostyn, Holywell CH8 9NZ
M3 NGU R. Bunting 9 Hammond Way, Attleborough NR17 2RQ
MM3 NGV M. Baird 28 Loch Road, Bridge of Weir PA11 3NB
M3 NGY A. Gromen-Hayes 95 Maypole Road, Ashurst Wood, East Grinstead RH19 3RB
M3 NGZ D. Boot 10 Madehurst Rise, Sheffield S2 3BJ
M3 NHA B. Ratcliff 27 Furlong Road, Manchester M22 1UD
MW3 NHC I. Meek 30 St. Peters Road, Penarth CF64 3PP
M3 NHD N. Harley 1 Portland Crescent, Meden Vale, Mansfield NG20 9PJ
M3 NHE J. Kelly 12 Park Road, Milford On Sea, Lymington SO41 0QU
M3 NHI S. Whitehead 55 Crombie Road, Sidcup DA15 8AT
MW3 NHN R. Williams 92 Bowleaze, Greenmeadow, Cwmbran NP44 4LF
M3 NHP N. Powell 4 Holme Court Avenue, Biggleswade SG18 8PF
M3 NHS N. Smith 248A South Street, Romford RM1 2AD
M3 NHU L. Gale 9 Ely Close, Worthing BN13 1BH
M3 NHV R. Gale 9 Ely Close, Worthing BN13 1BH
M3 NHW A. Hill 39 Lambs Row, Lychpit, Basingstoke RG24 8SL
M3 NHZ N. Hubbard 7 Creake Road, Syderstone, King's Lynn PE31 8SF
MW3 NIA H. Samuels 11 Bennions Road, Wrexham LL13 7AW
M3 NIC N. Rowland 11 Babylon Way, Eastbourne BN20 9DL
MI3 NIE C. Morton 29 Lackaboy View, Enniskillen BT74 4DY
M3 NIF F. Radford 3 Pierpoint Terrace, Brighton Road, Hassocks BN6 9TR
M3 NII C. Sims 7 Ainthorpe Lane, Ainthorpe, Whitby YO21 2JN
M3 NIT M. Paris 13 Butfield, Lavenham, Sudbury CO10 9SD
M3 NIZ G. Myers 25 Sherwood Road, North Bersted, Bognor Regis PO22 9DR
M3 NJA N. Atrill 22 Lester Close, Hr Compton, Plymouth PL3 6PX
M3 NJC N. Cook Idle Shores Springfield Road, Woolacombe EX34 7BX
M3 NJD N. Darby 60 Pine St., Grange Villa, Chester le Street DH2 3LX
M3 NJJ D. Wharlley 15 Crampton Court, Grosvenor Road, Broadstairs CT10 2XU
M3 NJK R. Mcallister 57 Wigan Road, Standish, Wigan WN6 0BE
M3 NJM N. Marsh 16 Daytona Quay, Eastbourne BN23 5BN
M3 NJO N. O'Hara 41 Exeter Street, Blackburn BB2 4AU
M3 NJQ C. Johnston 15 Queens Road, Haydock, St. Helens WA11 0RH
MI3 NJU P. Mcmahon 9 Sycamore Court, Drumaness, Ballynahinch BT24 8QZ
MM3 NJV P. Vernon 54 Brothock Way, Arbroath DD11 4BH
M3 NJY N. Young 5 Winslow Road, Boston PE21 0EJ
M3 NKB J. Banks 195 The Cornfields, Weston-Super-Mare BS22 9DZ
M3 NKC D. Ansell 30 Curzon Avenue, Horsham RH12 2LB
MW3 NKG R. Rooker 37 High Close, Nelson, Treharris CF46 6HJ
M3 NKL D. Toyne 19 Poachers Rest, Welton, Lincoln LN2 3TR
M3 NKN J. Hickman Ardoch, Harlestone Road, Northampton NN6 8AW
M3 NKO N. March 25 Emlyn Road, London W12 9TF
M3 NKP J. Keeble 2 Astley Cooper Place, Brooke, Norwich NR15 1JB
M3 NKU C. Prout 1 Westbrook Lustrells Vale, Saltdean, Brighton BN2 8EZ
M3 NKW J. Dale 37 Bussey Road, Norwich NR6 6JF
M3 NKX D. Doggett 82 Hurst Road, Kennington, Ashford TN24 9RS
M3 NKZ M. Ashton 31 Home Close Renhold, Bedford MK41 0LB
M3 NLA N. King 222 Prince Consort Road, Gateshead NE8 4DX
M3 NLF J. Grosvenor 10 Neves Close, Lingwood, Norwich NR13 4AW
MM3 NLH S. Anderson Newbigging Toll House, Drumsturdy Road, Dundee DD5 3RE
M3 NLI D. Bale 22 Highgrove Court, Rushden NN10 0DH
M3 NLJ N. Jeffery 7 Corfe Way, Winsford CW7 1LU
M3 NLM P. Maybin 16 Appleby Road Canning Town, London E16 1LQ
M3 NLN A. Paulizky 18 Cherry Tree Drive, London SW16 2PE
M3 NLP N. Pearson 34 Downside Road, Sutton SM2 5HP
M3 NLQ M. Thompson 4 Oat Hill Road, Towcester NN12 6EZ
M3 NLW L. Boull 80 Ascot Road, Baswich, Stafford ST17 0AQ
M3 NLX R. Mcdermott 2 Monument Close, Wellington TA21 9AL
M3 NMB N. Beech 94 Victoria Road, Runcorn WA7 5ST
MI3 NMG N. Mcgonigle 5 Woodend Road, Strabane BT82 8LF
M3 NMJ S. Martin 1 Buddleia Close, Weymouth DT3 6SG
M3 NMK A. Kemplay 8 Rue Du Lavoir, le Chillou 79600 France
M3 NMM J. Jones 40 Trembear Road, St. Austell PL25 5NY
M3 NMP M. Pedley 60 Ack Lane East, Bramhall, Stockport SK7 2BY
M3 NMR C. Purkiss Flat 6, 220 Greenheys Lane West, Manchester M15 5AF
M3 NMV L. Yates 108 Hawthorn Avenue, Lowestoft NR33 9BB
M3 NMX B. Scrivens 7 Normandy Way, Fordingbridge SP6 1NW
MW3 NNA A. Lydford 93 Hardwick Avenue, Chepstow NP16 5EB
M3 NNG R. Simms 3 The Byeway, London SW14 7NL
M3 NNH K. Bansil 65 Hervey Street, Northampton NN1 3QL
M3 NNI A. Mason 4 Quay Mill Walk, Great Yarmouth NR30 1JG
M3 NNJ J. Moore 2 Newsons Meadow, Lowestoft NR32 2NW
M3 NNM P. Morling 7 Hobill Close, Leicester Forest East, Leicester LE3 3PS
MM3 NNO C. Lewis 9 Cessnock Road, Troon KA10 6NJ
M3 NNQ J. Blamey 46 First Avenue, Canvey Island SS8 9LP
M3 NNV J. Paul East Park, Church Road, Cowes PO31 8HA
M3 NNY J. Brough 10 Linnet Close, Huntington, Cannock WS12 4TP
M3 NNZ B. James 19 Dukes Crescent, Sandbach CW11 1BL
M3 NOD N. Lightfoot 4 Prospect Close, Hatfield Peverel, Chelmsford CM3 2JE
M3 NOE D. Noe 21 Gale Crescent, Banstead SM7 2HZ
M3 NOF D. Donnelly 72 Bagots Oak, Stafford ST17 9SB
MI3 NOH N. O'Hagan 55 Meadowside, Antrim BT41 4HD
M3 NOJ J. Reynolds 4 Perriclose, Chelmsford CM1 6UJ
M3 NOM R. Smallman 128 Bevan Lee Road, Cannock WS11 4PT
M3 NON N. O'Sullivan 20 Pond Lane, Drayton, Norwich NR8 6PP
M3 NOR A. Norman 91 Church Road, Radley, Abingdon OX14 3QF
M3 NOS D. Leggett Fools Watering, London Road, Beccles NR34 8AQ
M3 NOW S. Wilson 55 Kent Road, Reading RG30 2EJ
M3 NOY S. James 124 Alcock Avenue, Mansfield NG18 2NF
M3 NPA A. Nicholson 7 Lingfoot Crescent, Sheffield S8 8DA
M3 NPC N. Collins Hill Farm, Broadheath, Tenbury Wells WR15 8QN
M3 NPE J. Nicholson 7 Lingfoot Crescent, Sheffield S8 8DA
MM3 NPG M. Maltman 30 Haldane Place, Dundee DD3 0JR
M3 NPH A. Arnold 2 Duck Lane, Haddenham, Ely CB6 3UE
M3 NPI M. Harrell 34 Nelson Drive, Cannock WS12 2GF
M3 NPK N. Kerner Headingley Cottage, Ryehurst Lane, Bracknell RG42 5QZ
M3 NPO D. Shuttleworth 27 Union St., Egerton, Bolton BL7 9SP
M3 NPS N. Sharpe 23 Cheney Road, Faversham ME13 8DG
M3 NPW P. Webster 39 Farndale Terrace, Leeds LS14 5BQ
M3 NPX S. Taylor-Mccormick North View Boarding Kennels, Skitham Lane, Preston PR3 6BD
M3 NPZ A. Paterson 35 Darlington Road, Richmond DL10 7BG
M3 NQA R. Clark 12 Ash Drive Haughton, Stafford ST18 9EU
MW3 NQE S. Walmsley 29 Shelley Court Machen, Caerphilly CF83 8TT
MW3 NQH S. Pope 11 Haman Place, Gelligaer, Hengoed CF82 8EG
M3 NQI P. Denham Arlyn House, 10 Prince Alfred Avenue, Skegness PE25 2UH
MW3 NQK J. Argent 7 Lloyds Hill, Buckley CH7 3ER
M3 NQL L. Kelsey 111-113 George Street, Mablethorpe LN12 2BS
M3 NQN K. Roberts 40 Portland Drive, Skegness PE25 1HF
M3 NQO A. Whyatt 11 The Perrings, Nailsea, Bristol BS48 4YD
M3 NQS N. Turner 18 The Green Road, Sawston, Cambridge CB22 3LP
MM3 NQT M. Simon 100 Findhorn Place, Edinburgh EH9 2NZ
M3 NQU Moorlands And District ARS c/o E. Wagner 3 Sarre Road, London NW2 3SN
M3 NQY G. Eklund 26-28 Zulu Road, Nottingham NG7 7DR
MI3 NRB N. Bolt 32 Bush Gardens, Bushmills BT57 8AE
M3 NRI A. Miles 5 Pershore Road, Basingstoke RG24 9BE
M3 NRJ N. Johnson 27 Redford Crescent, Bristol BS13 8SA
MW3 NRK N. Kind 32 Maesgarmon, Castle Caereinion, Welshpool SY21 9AN
M3 NRQ A. Nicholson 7 Lingfoot Crescent, Sheffield S8 8DA
M3 NRV D. Giering 39 High House Drive, Inkberrow, Worcester WR7 4EG
M3 NRW E. Cromwell 92 Hatch Road, Pilgrims Hatch, Brentwood CM15 9QA
M3 NRX J. Williams 39 St. Edmunds Close, Castleford WF10 3LL
M3 NSB A. Oatey Robin Hill, Blackpost Lane, Totnes TQ9 5RF
MI3 NSF R. Foley 6 Lislane Drive, Saintfield, Ballynahinch BT24 7HU
M3 NSG N. Garry 4 Fairstead, Skelmersdale WN8 6RD
M3 NSH S. Shelley 8 Harewood Close, Eastleigh SO50 4NZ
M3 NSJ K. Ingham 231 Regent Street, Nelson BB9 8SQ
M3 NSM G. Coyle 19 Mounsey Road, Bamber Bridge, Preston PR5 6LS
M3 NSO St Peter-In-Thanet Junior School ARC c/o N. Walch 52 Marsh House Road, Sheffield S11 9SP
M3 NSQ S. Beedham 27 Malpas Close Bransholme, Hull HU7 4HH
MI3 NSR G. Mccullough 32 Thistlemount Park, Lisburn BT28 2UN
M3 NSS M. Price 25 School Crescent, Lydney GL15 5TA
M3 NST N. Stirling 3 Rother Croft, New Tupton, Chesterfield S42 6BE
M3 NSX I. Thompson 33 Longsight Road, Mapplewell, Barnsley S75 6HD
MW3 NTE H. Fingerhut-Holland Osnok, 1 South Cliff Street, Tenby SA70 7EB
MU3 NTH N. Thomas 6 Tunstall Terrace, Gibauderie, St. Peter Port GY1 1XJ Guernsey
M3 NTI M. Gough 57 Ravenglass Road, Westlea, Swindon SN5 7BN
M3 NTJ N. Thompson 33 Longsight Road, Mapplewell, Barnsley S75 6HD
M3 NTQ M. Gough 58 Church Street, Brierley, Barnsley S72 9JG
M3 NTR M. Pratchett 119 Swindon Road, Wroughton, Swindon SN4 9AD
M3 NTW C. Northwood Apartment 50, 2 Munday Street, Manchester M4 7BB
MM3 NTX G. Askew 49 Kittlegairy Road, Peebles EH45 9LX
M3 NTZ S. Woodward 19 Beech Court, Spondon, Derby DE21 7TP
M3 NUB N. Vichitcheep1 6 Ormathwaites Corner, Warfield, Bracknell RG42 3XX
M3 NUE A. Lunn 57 Greets Green Road, West Bromwich B70 9ES
M3 NUH M. Gladders 2 Albion Mansions, Saltburn-by-the-Sea TS12 1JP
M3 NUI Rvd. S. Scotson 16 Merryfield Road, Dudley DY1 2PD
M3 NUL S. Heaton 13A Roche Avenue Bilton, Harrogate HG1 4ES
M3 NUM A. Ogle 22 Warwick Street, Daventry NN11 4AL
M3 NUO K. Holdt 18 Garrard Road, Banstead SM7 2ER
MW3 NUP M. Lewis 4 Coldwell Terrace, Pembroke SA71 4QL
M3 NUQ M. Dickenson 6 The Pavilions, Blandford Forum DT11 7GF
M3 NUU D. Copsey Fairview, Mill Lane, Hook End, Brentwood CM15 0PP
MI3 NUW A. Kincaid 428 Cushendall Road, Ballymena BT43 6QE
MW3 NUX D. James 10 Hafan Deg, Pencoed, Bridgend CF35 6YG
M3 NVA A. Caine 53 Cromford Road Crich, Matlock DE4 5DJ
M3 NVC P. Brown 13 Rydal Road, Weston-Super-Mare BS23 3RT
MM3 NVD D. Baillie 126 Main St., Fauldhouse, Bathgate EH47 9BW
M3 NVE K. Whiteley 14 Milton Street, Goole DN14 6EL
M3 NVF N. Fletcher 2 Handforth Road, Crewe CW2 8PL
M3 NVG J. Webster 72 Grosvenor Street, Derby DE248AT
M3 NVH J. Boull 80 Ascot Road Baswich, Stafford ST17 0AQ
M3 NVK L. Bolton 9 Nab Crescent, Meltham, Holmfirth HD9 5LT
M3 NVL J. Jones 5 Cranleigh Road, Liverpool L25 2RP
M3 NVO L. Clift 8 Kendal Road, Gloucester GL2 0NB
M3 NVP M. Weir 153 Tyndale Crescent, Birmingham B43 7HX

MW3 NVQ R. Wright1 12 Bryn Teg, Arddleen, Llanymynech SY22 6PZ
M3 NVR P. Gutteridge 75A Collingwood Drive, Birmingham B43 7JW
M3 NVS J. Caddick 135 Broadway, Dunscroft, Doncaster DN7 4HB
M3 NVV C. Elliot 106 Occupation Road, Corby NN17 1EG
MI3 NVX M. Mccay 2 Riverview, Spamount, Castlederg BT81 7NA
M3 NWD A. Broll1 17 Broadway, Farcet, Peterborough PE7 3AY
MM3 NWF K. Whyte 27 Queens Road, Inverbervie, Montrose DD10 0RY
M3 NWH F. Gear 251 Abington Avenue, Northampton NN3 2BU
M3 NWK S. Lofthouse 32 Westbrook Park Road, Peterborough PE2 9JG
MI3 NWO D. Adams 65 Rose Park, Limavady BT49 0BF
M3 NWQ F. Stone Rrt, Gosw, 2 Rivergate, Bristol BS1 6EH
MI3 NWU G. Mckeever 45 Blackthorn Court, Coleraine BT52 2EX
M3 NWY T. Bown 16 Sandringham Court, Queen Elizabeth Road, Nuneaton CV10 9AR
MM3 NWZ A. Smith 17 High Street, Stranraer DG9 7LL
M3 NXA A. Davenport 10 Woodend Lane, Hyde SK14 1DT
M3 NXC P. Waring 1 Fanshaw Road Eckington, Sheffield S21 4BW
MW3 NXD A. Downing 3A Pant Hirgoed, Pencoed, Bridgend CF35 6YD
M3 NXE S. Burrows 78A Coronation Road, Earl Shilton, Leicester LE9 7HJ
M3 NXF N. Forbes 55 The Henrys, Thatcham RG18 4LS
M3 NXH I. Rimell 51 Woodlands Avenue, Woodley, Reading RG5 3HF
M3 NXJ I. Livesey 26 Hilltop Road, Twyford, Reading RG10 9BN
M3 NXK S. Dudley 28 Walnut Lane, Wednesbury WS10 0BH
M3 NXO B. Bradford 3 Beverly Close, Thornton-Cleveleys FY5 5DR
M3 NXQ V. Stokes 52 Brantley Avenue, Wolverhampton WV3 9AR
MM3 NXY R. Hay 12 Mitchell Brae, Balmedie, Aberdeen AB23 8PW
M3 NXZ S. Robinson 1 Woodgate Road, Manchester M16 8LX
M3 NYA M. Hoggan 19 The Drive, Uckfield TN22 1BY
MI3 NYB N. Throne 12 Mason Road, Magheramason, Londonderry BT47 2RY
M3 NYF A. Probst 37 Devonshire Street, Skipton BD23 2ET
M3 NYG N. Carson Gov Office For The South West, 2 Rivergate, Bristol BS1 6EH
M3 NYI A. Parradine 76 Parsonage Road, Rainham RM13 9LF
M3 NYK K. Nye 32 Seafield Close, Chichester PO20 8DP
M3 NYM G. Whitehurst 28 Vicarage Fields, Warwick CV34 5NJ
M3 NYQ M. Hives 32 Partridge Way, Chadderton, Oldham OL9 0NS
M3 NYX J. Mock 29 Tavistock Place, Paignton TQ4 7NZ
MM3 NYY G. Cleary 2 Merlinford Avenue, Renfrew PA4 8XS
M3 NZG A. Wheeler 14 Sparkmill Terrace, Beverley HU17 0PA
M3 NZK J. Bayliss 39 Elms Avenue, Littleover, Derby DE23 6FB
M3 NZN C. Jones 20 Freeman Road, Wednesbury WS10 0HQ
M3 NZR D. Oddie 5 The Bridleway, Forest Town, Mansfield NG19 0QJ
M3 NZV L. Williams 2 The Tannery, Dol-Y-Bont, Borth SY24 5LX
M3 NZW A. Woods 42 St. Pauls Drive, Wellington, Telford TF1 3GD
MM3 NZX M. Mcdonald 17 Ramsay Mews, Strathaven ML10 6GN
MW3 NZZ I. Powe 7 Wellington Drive, Greenmeadow, Cwmbran NP44 5HH
M3 OAB A. Barker 43 Ploughmans Drive, Shepshed, Loughborough LE12 9SG
M3 OAC G. Dray 2 Mulberry Drive, Malvern WR14 4AT
M3 OAG W. Scott 8 Woodacre, Whalley Range, Manchester M16 8QQ
M3 OAJ S. Clay Akers Lodge, 6 Penn Way, Rickmansworth WD3 5HQ
M3 OAK D. Smith 98 Orange Hill Road, Burnt Oak, Edgware HA8 0TW
M3 OAL M. Edwards 30 Morrison Road, Tipton DY4 7PU
M3 OAM M. Parkin 51 Far Lane, Rotherham S65 2HQ
M3 OAQ S. Tingay 6 Butts Close, Farnley Tyas, Huddersfield HD4 6UR
M3 OAS I. Miller 64 Queens Road, Vicars Cross, Chester CH3 5HD
M3 OAT J. Judson 559 Colne Road, Burnley BB10 2LG
M3 OAX Hinckley Sea Cadets c/o F. Smith 32 Amesbury Drive, London E4 7PZ
M3 OAZ R. Mccarthy 11 Britain Street, Bury BL9 9PD
M3 OBB O. Boar 19 Blyford Road, Lowestoft NR32 4PZ
M3 OBD A. Dow 5 Verney Crescent, Liverpool L19 4UR
M3 OBM G. Scarr 35 Ilford Avenue, Cramlington NE23 3LE
M3 OBN L. Stott 70 Elizabeth St., Ashton under Lyne OL6 8SX
M3 OBO R. Goody 113 Kenneth Road, Basildon SS13 2BH
M3 OBS M. Simkins 37 St. Andrews Meadow, Harlow CM18 6BL
M3 OBU C. Camsey 1 Park Close, Milford On Sea, Lymington SO41 0QT
M3 OBX D. Sanderson 65 Holm Flatt Street, Parkgate, Rotherham S62 6HJ
M3 OBZ D. Thomas 51 Sandringham Avenue, Vicars Cross, Chester CH3 5JF
M3 OCA P. Bainbridge 6 Waterland Lane, St. Helens WA9 3AF
M3 OCC B. O'Connor 58 St Johns Way, Thetford IP24 3NP
M3 OCJ S. Rafter 30 Monmouth Grove, St. Helens WA9 1QB
M3 OCL H. Lister 68 Spring Avenue, Gildersome, Leeds LS27 7BT
M3 OCP M. Cave 295 Sackup Lane, Staincross, Barnsley S75 5BG
M3 OCQ R. Hall 5 Lea Hill Road, Birmingham B20 2AS
M3 OCR O. Crump 3 Vosper Road, Southampton SO19 9SS
M3 OCS A. Owen 5 Croft Close, Rowton, Chester CH3 7QQ
MM3 OCY D. Mcclelland 5 Cambusmoon Terrace, Gartocharn, Alexandria G83 8RU
M3 ODC P. Hatter 14 Morland Avenue, Bromborough, Wirral CH62 6BE
M3 ODH D. Hughes 75 Suncote Avenue, Dunstable LU6 1BN
M3 ODK L. Bedford 29 Kent Road, Brookenby, Market Rasen LN8 6EW
M3 ODL K. Bedford 29 Kent Road Brookenby, Binbrook, Market Rasen LN8 6EW
M3 ODO E. White 3 Davy Drive, Maltby, Rotherham S66 7EN
MM3 ODV A. Cairns 57 Miller St., Dumbarton G82 2JA
M3 OEB D. Norris 24 Northway, Fulwood, Preston PR2 9TP
MW3 OEC L. Williams1 18 Waun Wen Terrace, Nantymoel, Bridgend CF32 7NB
MD3 OED R. Britton 3 Meadowfield, Port Erin IM9 6PH Isle of Man
M3 OEE B. Hood 52 Kent St., Preston PR1 1RY
M3 OEF S. Tattum 20 Drew Close, Poole BH12 5ET
M3 OEG J. Cook 42 Pampas Close, Colchester CO4 9ST
M3 OEH A. Mclean 47 Tarn Drive, Bury BL9 9QB
MW3 OEJ M. Kay 3 Protheroe Avenue, Pen-Y-Fai, Bridgend CF31 4LU
M3 OEM B. Horton 44 Chamberlain St., St. Helens WA10 4NL
M3 OEN J. Edwards 36 Westerley Lane, Shelley, Huddersfield HD8 8HP
M3 OEO M. Hammersley 47 Shaftesbury Avenue, Timperley, Altrincham WA15 7NP
MI3 OEQ A. Jamison 11 Richmond Gardens, Newtownabbey BT36 5LA
M3 OER S. Warner 28 Jameson Bridge Street, Market Rasen LN8 3EW
M3 OEV M. Cuff 1 Ivy Close, St. Leonards, Ringwood BH24 2QZ
M3 OFA N. Shaw Greenacres Poultry, Three Lowes, Stoke on Trent ST10 3BW
M3 OFB J. Woodruff 10 Bailey Close, Blackburn BB2 4FT
M3 OFC S. Dunne 1 Burton Gardens, Brierfield, Nelson BB9 5DR
M3 OFD D. Dunne 1 Burton Gardens, Brierfield, Nelson BB9 5DR
MM3 OFE J. Jackson 25 Lomond Crescent, Alexandria G83 0RJ
M3 OFH J. Wright 277 Monkmoor Road, Shrewsbury SY2 5SS
M3 OFJ M. Jones 20 Chelsea Drive, Sutton Coldfield B74 4UG
M3 OFN C. Bamford 71 Trent Road, Shaw, Oldham OL2 7YQ
M3 OFS E. Marsh 16 Laurel Close, North Warnborough, Hook RG29 1BH
M3 OFU T. Smith 151 Halfords Lane, West Bromwich B71 4LQ
M3 OFV M. Kempson 67 Esther Avenue, Wakefield WF2 8BY
MI3 OFX T. Conlon 7 Waringfield Gardens, Moira, Craigavon BT67 0FQ
M3 OGC L. Li Harrogate Ladies' College, Clarence Drive, Harrogate HG1 2QG
M3 OGD R. Whiteside Hill Crest, Farley Hill, Farley, Matlock DE4 5LT
M3 OGL T. Woodhouse The Old Granary, 12 Limekiln Lane, Newport TF10 9EZ
M3 OGM B. Debenham 80 Stewart Road, Chelmsford CM2 9BD
MM3 OGS R. Keay 26 Cherrywood Drive, Beith KA15 2DZ
MM3 OGU T. Mussell Dunelm, Thornhill Road, Cuminestown, Turriff AB53 5WH
M3 OGV R. Bicknell-Thompson 4 Linden Court, Greenfrith Drive, Tonbridge TN10 3LW
M3 OHC M. Holgate 10 Brecon Crescent, Ashton-under-Lyne OL6 8UA
MM3 OHD D. Hague 13 North Dell, Ness, Isle of Lewis HS2 0SW
MI3 OHE E. Paterson 1 Sycamore Grove, Belfast BT4 2RB
MI3 OHF M. Graham 21 Meadowvale Crescent, Bangor BT19 1HQ
MI3 OHG N. Wylie 26 Lisnoe Walk, Lisburn BT28 1QD
M3 OHI G. Chaffey 63 Underwood Road, Eastleigh SO50 6FX
M3 OHJ B. Stanfield 24 Rowan Close, Kingsbury, Tamworth B78 2JR
M3 OHL B. Widdowson 28 Highfield Lane, Chesterfield S41 8AU
M3 OHN P. Martin Flat 9, Paddock Court, Graham Avenue, Portslade, Brighton BN41 2WU
M3 OHO M. Murray 2 Meadway, Penwortham, Preston PR1 0JL
MI3 OHP J. Leetch 30 Murob Park, Ballymena BT43 6JG
M3 OHQ W. Kong Clarence House, Harrogate Ladies College, North Yorkshire HG1 2QG
M3 OHR L. Gray 29 Longview Road, Liverpool L36 1TA
M3 OHX D. Taylor 58 Shenstone Road, Great Barr, Birmingham B43 5LN
M3 OHY H. Chan Harrogate Ladies' College, Clarence Drive, Harrogate HG1 2QG
M3 OHZ H. Fleming 29 Model Village, Creswell, Worksop S80 4BN
MI3 OIB D. Couser 5 Colinbrook Park, Dunmurry, Belfast BT17 0NZ
M3 OIC L. Crane 32 Clarke Avenue, Newark NG24 4NY
M3 OII T. Hoggan 9 Nursery Field, Buxted, Uckfield TN22 4NG
MW3 OIK W. Jaggard 2 Aled Drive, Rhos On Sea, Colwyn Bay LL28 4UU
M3 OIL O. Hutley 1 John Ray Street, Braintree CM7 9DZ
M3 OIN H. List 41 Westbury Crescent, Dover CT17 9QQ
MD3 OIS M. Wallace 61 Vernon Road, Ramsey IM8 2EG
MM3 OIX N. White 29 Forgie Crescent, Maddiston, Falkirk FK2 0LY
M3 OIY P. Yuen Harrogate Ladies' College, Clarence Drive, Harrogate HG1 2QG
MM3 OIZ J. Mcmonigle 19 Monach Gardens, Dreghorn, Irvine KA11 4EB
MM3 OJE T. Mcfarlane 15 West Edith Street, Darvel KA17 0EE
M3 OJJ A. Chance 24 Doddsfield Road, Slough SL2 2AD
M3 OJK N. Schall 39 Emery Avenue, Chorltonville, Manchester M21 7LE
M3 OJN P. Moule 30 Hillview Road, Chelmsford CM1 7RX
M3 OJP D. Moulding 5 Chalk Lane, Sutton Bridge, Spalding PE12 9YF
MM3 OJR J. Rae Bowhouse Farm Cottage, Auchtermuchty, Cupar KY14 7ES
M3 OJS P. Rowlands 314 Stourbridge Road Catshill, Bromsgrove B61 9LH
M3 OJU P. Watling 1 Chediston Green, Chediston, Halesworth IP19 0BB
MM3 OJV C. Mcgougan 6 Calder Place, Kilmarnock KA1 3QL
M3 OJW S. Earl 9A Florida Street, Daws Hill Lane, High Wycombe HP11 1QA
M3 OJX L. Parkman 9 Malim Way, Gonerby Hill Foot, Grantham NG31 8QF
M3 OJY B. Mason 4108 Hingston Avenue, Montreal H4A 2J7 Canada
M3 OJZ J. Neale 20 Oakfield Road, Wollescote, Stourbridge DY9 9DL
M3 OKC P. Barnesok 39 Prospect Avenue, Rochester ME2 3BZ
M3 OKE A. Ewence 9 Mount Pleasant, Bradford-on-Avon BA15 1SJ
MD3 OKG K. Glaister 42 Barrule Drive, Onchan, Douglas IM3 4NR Isle of Man
MD3 OKH B. Glaister 42 Barrule Drive, Onchan, Douglas IM3 4NR Isle of Man
M3 OKP R. Bancroft Ashrigg, Lazonby, Penrith CA10 1AT
M3 OKY R. Eglington 33 Bradley Lane, Bilston WV14 8EW
M3 OKZ A. Cammish 6 West Vale, Filey YO14 9AY
M3 OLD S. Old 49 Penrith Cresent, Castleford WF10 2RG
M3 OLE S. Jervis 45 Wyndham Road, Stoke-on-Trent ST3 3LX
M3 OLF I. Scott Croft Cottage, Cumwhinton, Carlisle CA4 8ER
M3 OLI O. Palmer 21 Ibbett Close, Kempston, Bedford MK43 9BT
MI3 OLM C. Doole 110 Moyagall Road, Knockloughrim, Magherafelt BT45 8PJ
M3 OLN E. Hall 5 The Paddocks, Thursby, Carlisle CA5 6PB
M3 OLQ P. Oliver 12 Walkmill Crescent, Carlisle CA1 2WF
MW3 OLT L. Thomas 15 Blaenwern, Newcastle Emlyn SA38 9BE
M3 OLU A. Shepherd 39 Minehead Road, Dudley DY1 2NZ
M3 OLW A. Mills 107 Blyth Ave, Southend-on-Sea SS3 9NJ
MW3 OLX W. Hance 27 Maesybont, Glanamman, Ammanford SA18 2AY
M3 OLZ L. Bullen11 5 West View, Long Sutton TA10 9LT
M3 OME Dr A. Sadanandam 23 Clare Avenue, Hoole, Chester CH2 3HT
M3 OMF O. Fury 1 Wigley Drive, Wigley, Ludlow SY8 3DR
MM3 OMI B. Bowman 15 Aboyne Road, Aberdeen AB10 7BS
M3 OMT T. Baggley 16 Seaton Road, Seaton, Workington CA14 1DT
M3 OMU J. Fletcher 217 Homefield Road, Sileby, Loughborough LE12 7TG
M3 OMX K. Boulton 17 Grange Avenue , Goodrington , Paignton.Tq4 7Jy, Paignton TQ4 7JY
M3 OMZ D. Jarvis 7 Leonard Road, Greatstone, New Romney TN28 8UJ
M3 ONB I. Woollen 33 The Oaks, Taunton TA1 2QX
M3 ONE C. Chambers 75A Main St., Sedbergh LA10 5AB
MW3 ONG J. Brydges 9 Twynygarreg, Treharris CF46 5RL
M3 ONH S. Fooks 52 Abbey Road, Darlington DL3 8ND
MM3 ONI O. Stein Library House, Stafford Street, Tain IV19 1AZ
M3 ONK R. Bray 49 Montacute Way, Wimborne BH21 1TZ
M3 ONM P. Connor 244 Gregory Avenue, Birmingham B29 5DR
MD3 ONP J. Taylor Ballafayle Cottage, Ballafayle, Ramsey IM7 1ED Isle of Man
M3 ONV J. Bonar Flat, 5A Friday Street, Minehead TA24 5UB
MM3 ONX L. Paget 40 Davaar Drive, Kilmarnock KA3 2JG
M3 OOA T. Durant 39 Snydale Road, Normanton WF6 1NY
M3 OOC B. Cooper 71 High Street, Birstall, Batley WF17 9RG
M3 OOE G. Kelly 141 Narbeth Drive, Aylesbury HP20 1PZ
M3 OOH S. Howroyd 7 Garendon Road, Loughborough LE11 4QB
M3 OOL T. Peterson 9 Moseley Wood View, Leeds LS16 7ES

M3 OOP J. Dietsch 21 Lake View Avenue, Chesterfield S40 3DR
M3 OOQ A. Shaw 2 Montrose Avenue, Montrose Street, Hull HU8 7RY
MM3 OOT J. Ferrans 77 Knockinlaw Road, Kilmarnock KA3 2AS
M3 OOU A. Hoskins 38 Tasmania Close, Basingstoke RG24 9PQ
M3 OOX F. Spencer 29 Kliffen Place, Halifax HX3 0AL
M3 OOY C. Spencer 29 Kliffen Place, Halifax HX3 0AL
M3 OPA N. Van-Den-Langenberg 2 Grove Crescent, Bridgnorth WV15 5BS
M3 OPB O. Blackburn 128 High St., Crigglestone, Wakefield WF4 3EF
M3 OPC J. Spencer 29 Kliffen Place, Halifax HX3 0AL
M3 OPD K. Spencer 29 Kliffen Place, Halifax HX3 0AL
M3 OPG P. Reed 32 London Road, Warmley, Bristol BS30 5JH
M3 OPM J. Newby 22 Acton Road, Liverpool L32 0TT
M3 OPN J. Shemwell 4 Darvel Close, Bolton BL2 6UD
M3 OPS A. Hindle 41 Seedfield, Staveley, Kendal LA8 9NJ
M3 OPT D. Wicks 148 Long Lane, Staines TW19 7AJ
M3 OPU K. Morris 80 Bridge Street, Chatteris PE16 6RN
M3 OPV K. Ashcroft 9 Aldermere Crescent, Urmston, Manchester M41 8UE
M3 OPW H. Foot 1 South View, Piddletrenthide, Dorchester DT2 7QS
M3 OPX A. Robinson 75 De La Pole Avenue, Hull HU3 6RD
M3 OQD H. Nehmzow 16 Goldington Avenueá, Bedford MK403BY
M3 OQG A. Chruscinski 39 Sherwood Rise, Mansfield Woodhouse, Mansfield NG19 7NP
M3 OQH D. Cooper 52 Meadow Lane, Birkenhead CH42 3YE
M3 OQI G. Woodward 5 Barnard Road, Chelmsford CM2 8RR
M3 OQJ B. Male 44 Lakefields, West Coker, Yeovil BA22 9BT
M3 OQK G. Williamson 26 Portland Mews, Bridlington YO16 4EH
M3 OQL S. Davin 40 Theynes Croft, Long Ashton, Bristol BS41 9NA
M3 OQQ T. Garvey 21 Oak Park Drive, Havant PO9 2XE
MM3 OQR P. Taylor 2 Laurel Grove, Aberdeen AB22 8YJ
M3 OQS B. Daley 129A Kingsway South, Warrington WA4 1RW
MM3 OQV J. Campbell 6 Dunard Court, Carluke ML8 5RX
M3 OQZ P. Hall 13 Sheard Avenue, Ashton-under-Lyne OL6 8DS
M3 ORB A. Goold 6 The Elms, Kempston, Bedford MK42 7JN
M3 ORE G. Beale 34 Teville Road, Worthing BN11 1UG
M3 ORL A. Price 23 Greenway, Wingerworth, Chesterfield S42 6NP
M3 ORN O. Newton 84 Ameysford Road, Ferndown BH22 9QB
MW3 ORP J. Marlow West Bulthy, Bulthy, Welshpool SY21 8ER
M3 ORQ M. Winch 2 Cranleigh Gardens, Cowes PO31 8AS
M3 ORT B. Williams 3 Welton Close, Wilmslow SK9 6HD
M3 ORU C. Baker 10 Kirton Close, Coventry CV6 2PG
MW3 ORY N. Lewis 1 Clyne Drive, Blackpill, Swansea SA3 5BU
M3 ORZ G. Marshall 12 Arthur Avenue, Caister-On-Sea, Great Yarmouth NR30 5PQ
M3 OSA J. Ross 142 Bridle Road, Croydon CR0 8HJ
M3 OSC E. Walker 2 The Green, Blencogo, Wigton CA7 0DF
M3 OSF J. Cobbold 2 The Green, Blencogo, Wigton CA7 0DF
MW3 OSI R. Johnson 25 Lon Tyrhaul Llansamlet, Swansea SA7 9SF
MM3 OSK A. Twort 17 Balallan, Isle of Lewis HS2 9PN
M3 OSP P. Moss 26 Woodlands Avenue, Farnham GU9 9EY
MW3 OSQ R. Phillips 188 Charston, Greenmeadow, Cwmbran NP44 4LD
M3 OSS C. Dell 18 Greenacres, Fulwood, Preston PR2 7DA
M3 OSU T. Pollard 35 Weatherall St. North, Salford M7 4TH
M3 OSW K. Sheldon 35 Weatherall St. North, Salford M7 4TH
M3 OSY W. Shiu Harrogate Ladies' College, Clarence Drive, Harrogate HG1 2QG
M3 OTG A. Hill 5 Park Road, Thurnscoe, Rotherham S63 0TG
M3 OTI D. Scotcher 17 St. Dominics Square, Luton LU4 0UN
M3 OTP R. Tailford 28 Paddock Wood, Prudhoe NE42 5BJ
M3 OTQ T. Chapman 17 Trevor Road, Swinton, Manchester M27 0YH
M3 OTR S. Taylor 49 Chestnut Avenue, West Drayton UB7 8BU
M3 OTS G. Rees 32 Glencroft Close, Burton-on-Trent DE14 3GJ
M3 OTU R. Woolgar 2 Solent Way, Milford On Sea, Lymington SO41 0TE
M3 OTZ M. Wright 8 St. Wilfrids Road, Oundle, Peterborough PE8 4NX
MW3 OUC J. Bidwell 26 Lone Road, Clydach, Swansea SA6 5HR
M3 OUF R. Crewe 12 Dimple Gardens, Ossett WF5 8LJ
M3 OUG C. Rickwood 7 Bromley Mount, Wakefield WF1 5LB
M3 OUH P. Loxton 32 Parkhill Crescent, Wakefield WF1 4EZ
M3 OUI P. Marsh 16 Laurel Close, North Warnborough, Hook RG29 1BH
M3 OUL G. Martin 12 Poolside, Phase 1, St. Joseph Trinidad and Tobago
MI3 OUN S. Fulton 120 Dunnalong Road, Bready, Strabane BT82 0DP
M3 OUQ W. Speak 20 Pear Tree Drive, Wincham, Northwich CW9 6EZ
M3 OUS E. Mcfalls 119 Park Avenue, Shelley, Huddersfield HD8 8JZ
M3 OUV S. Clarke 49 Torr View Avenue, Plymouth PL3 4QN
M3 OVA T. Durant 2 Tamar Way, North Hykeham, Lincoln LN6 8TZ
M3 OVC P. Goodburn 2 Sunray Cottages, Holt Street, Dover CT15 4HZ
M3 OVE M. Love 9 Firswood Drive, Swinton, Manchester M27 5QY
M3 OVF M. Clay-Burley 90 Huntington Terrace Road, Cannock WS11 5HB
M3 OVG T. Speak 20 Pear Tree Drive, Wincham, Northwich CW9 6EZ
M3 OVM K. Nelson 40 Staunton Road, Newark NG24 4EX
M3 OVO A. Rowe Southern Point, Grange View, Houghton le Spring DH4 4HU
MW3 OVT J. Jones 40 Ffordd Coed Marion, Caernarfon LL55 2EF
MM3 OVV S. Monaghan 13 Ballyhennan Crescent, Tarbet, Arrochar G83 7DB
M3 OVX J. Alexander 14 Barlow Road, Stretford, Manchester M32 0RG
M3 OWF A. Holter 42 Collingwood Close, Eastbourne BN23 6HW
M3 OWI A. Hodgeon 30 Rock Bank, Buxton SK17 9JF
M3 OWN O. Dixon 28 Manchester Road, Audenshaw, Manchester M34 5GB
M3 OWO M. Middleditch 56 Rowan Way, Yeovil BA20 2NR
M3 OWQ P. Mcspirit 26 Horridge Avenue, Newton-le-Willows WA12 0AS
M3 OWU D. Adlam 31 Coxons Close, Huntingdon PE29 1TS
M3 OWZ S. Waller 17 Vere Road, Peterborough PE1 3DZ
MM3 OXB B. Rodriguez Sprouston House, Newtown St. Boswells, Melrose TD6 0RY
M3 OXD C. Romocea 21 Hurst Lane, Cumnor, Oxford OX2 9PR
M3 OXN P. Jodrell 2 Greggs Avenue, Chapel-En-Le-Frith, High Peak SK23 9TU
MM3 OXQ S. Mckinnon 8 Rowanlea Avenue, Paisley PA2 0RP
MW3 OXV J. Evans 71 Bangor Road, Johnstown, Wrexham LL14 2SR
M3 OXY O. Bazar 1 Claremont Road, London NW2 1BP
MM3 OYB S. Morgan 23 Duncan Road, Glenrothes KY7 4HS
MW3 OYC S. Poyser Glandwr, Snowdon Street, Porthmadog LL49 9DF
M3 OYE Lady C. Windsor 44 Paragon Place, Norwich NR2 4BL
M3 OYJ C. Rose 32 Hobart Place, Thornton-Cleveleys FY5 3DQ
MM3 OYL A. Shearman 4 Millbrae Crescent, Clydebank G81 1EH
M3 OYN R. Watson 60 Beresford Avenue, Surbiton KT5 9LJ
MI3 OYP C. Brennan 1 Ballyscullion Lane, Bellaghy, Magherafelt BT45 8NQ
M3 OYQ N. Loughran 22 Edulf Road, Borehamwood WD6 5AD
M3 OYR A. Kirby 36 Baron Street, Darwen BB3 1NP
M3 OYS A. Parkes Oliver Court Bath Hill Terrace, Great Yarmouth NR30 2LF
M3 OYU N. Carr 6 Baldwin Avenue, Eastbourne BN21 1UJ
MM3 OYW J. Waugh 9 Kedar Bank, Mouswald, Dumfries DG1 4LU
M3 OYZ M. Ward 25 Highbury Crescent, Doncaster DN4 6AL
M3 OZB J. Robinson 34 High St., Dragonby, Scunthorpe DN15 0BE
M3 OZC H. Derbyshire 12 Trinity Homes, St. Clare Road, Deal CT14 7PX
M3 OZE J. Baldry 160 Rover Drive, Castle Bromwich, Birmingham B36 9LL
M3 OZI S. Bird 9 Almery Drive, Carlisle CA2 4EX
MI3 OZK J. Sills 145 Ballycolman Estate, Strabane BT82 9AJ
M3 OZN P. Davies 92 Thirlmere Road, Hinckley LE10 0PF
M3 OZP C. Rolfe 45 St Clements Court Wear Bay Crescent, Folkestone CT19 6BP
MI3 OZT R. Hepburn 34 Pinewood Crescent, Claudy, Londonderry BT47 4AD
MM3 OZU G. Craig 1 Butt Avenue, Helensburgh G84 9DA
MM3 OZW W. Pauley 43 Pringle Avenue, Tarves, Ellon AB41 7NZ
M3 OZY O. Morris 44 Leamington Road, Weymouth DT4 0EZ
MM3 PAE C. Hume Sundhopeburn, Yarrow, Selkirk TD7 5NF
M3 PAI M. Norton Springfield, Back Lane, Kingston, Sturminster Newton DT10 2DT
M3 PAP J. Parrott 2 Boyd Close, Wirral CH46 1RX
M3 PAU P. Laing 6 St. Vincent Street, Barrow-in-Furness LA14 2NR
M3 PAX N. Haigh 10 Moor Park Gardens, Dewsbury WF12 7AS
M3 PBA P. Alce 1/2 Arawa Street, Christchurch 8013 New Zealand
M3 PBB R. Bannon 18 Clavell Road, Liverpool L19 4TR
M3 PBE R. Rudd 11 Woodlands Way, Lepton, Huddersfield HD8 0JA
M3 PBK B. Kellner 95 Shakespeare Road Ipswich Suffolk, Ipswich IP16ET
M3 PBP D. Parsons Barn Owl Cottage, Stoke St. Mary, Taunton TA3 5BY
MM3 PBQ M. Hopkins 15 Station Drive, Dalbeattie DG5 4FA
M3 PBR T. Cumming 2 Ash Grove, Perth Street, Hull HU5 3PF
M3 PBU W. Wilkinson 35, Fitzgerald Court, Haughton Green M34 7LB
MW3 PBV S. Williams Flat 28, Llys Celyn Cedar Crescent, Tonteg, Pontypridd CF38 1LF
M3 PBW P. Roberts 7 Boscombe Road., Swindon SN25 3EZ
M3 PCC P. Crossley Firpark Farm, Fir Park, Market Rasen LN8 3YL
MI3 PCF P. Ford 25 Carnhill, Londonderry BT48 8BA
M3 PCP P. Papper 50 Lincoln Road, Stevenage SG1 4PL
M3 PCQ J. Campbell 22 Horsewhim Drive, Kelly Bray, Callington PL17 8GL
M3 PCW A. Maxwell Tysties, Tile Barn, Newbury RG20 9UY
MW3 PCX G. Bellis 70 Osborne St., Rhos, Wrexham LL14 2HT
M3 PDC P. Cooper 62 Fredericks Road, Beccles NR34 9UG
M3 PDD P. Bennett 94 Queensway, Taunton TA1 5QT
MW3 PDE P. Eckersley 14 Bro Nantfer, Gwaun Cae Gurwen, Ammanford SA18 1BR
M3 PDG E. Aitken 20 Plover Drive, Bury BL9 6JH
M3 PDK J. Fuller Rosemar Lodge, Westford, Wellington TA21 0DX
MI3 PDL Taunton & Somerset Raynet Group c/o P. Burns 25 Orchard Road, Strabane BT82 9QS
MM3 PDM P. Mckay 7 Buchanness Drive, Boddam, Peterhead AB42 3AT
MI3 PDN R. Neill 84 Carnreagh, Craigavon BT64 3AN
M3 PDP J. Clarkson 56 Edward Bailey Close, Binley, Coventry CV3 2LZ
M3 PDU L. Fuller Rosemar Lodge Westford, Wellington TA21 0DX
M3 PDY P. Dyer 19 Church Road, Evesham WR11 2NE
MW3 PEH B. Sellers 86 St. John Street, Ogmore Vale CF32 7BB
M3 PEQ M. Vincent 8 Waldemar Park, Norwich NR6 6TD
MD3 PER S. Perry 1 Cronk Grianagh Estate, Strang, Douglas IM4 4QP Isle of Man
MM3 PEV R. Stevenson 17 Springbank Gardens, Lawthorn, Irvine KA11 2BY
MM3 PEY D. Oates 14 Craighlaw Avenue, Eaglesham, Glasgow G76 0EU
MM3 PFA R. Maddock 24 Dalhousie Terrace, Montrose DD10 9BX
M3 PFE R. Laverick 55 Bondicar Terrace, Blyth NE24 2JW
M3 PFF A. Fisher 17 Spicers Way, Totton, Southampton SO40 9AX
M3 PFK J. Lennon 6 Piccadilly Square, Burnley BB11 4QG
M3 PFL K. Gallagher Flat 1, 59 Trinity Road, Bridlington YO15 2HF
M3 PFM P. Morris Canary Cottage, Eye Road, Eye IP23 7JX
M3 PFN J. Fisk The Cottage In The Croft, The Croft, Costessey, Norwich NR8 5DT
M3 PFU S. Silver 2 Brandon Close, Grange Park, Swindon SN5 6AA
M3 PFY P. Murthwaite 34 Cambridge St., Bridlington YO16 4JZ
M3 PGB D. Brierley 639 Borough Road, Birkenhead CH42 9QA
M3 PGD P. Danvers Heath Farm, Aylsham Road, North Walsham NR28 0JP
M3 PGH P. Howell 16 Everard Road, Bedford MK41 9LD
M3 PGI G. Pollard 24 Terminus Road, Littlehampton BN17 5BX
M3 PGK A. Farrar 8 Wensley Street, Thurnscoe, Rotherham S63 0PX
M3 PGL P. Lockwood 42 The Gables, Sedgefield TS21 3EU
MW3 PGN A. Gazi 51 Cyncoed Road, Cardiff CF23 5SB
M3 PGO P. Goodchild 577 Parrs Wood Road, East Didsbury, Manchester M20 5QS
M3 PGS P. Stevenson 6 Dighton Gate, Stoke Gifford, Bristol BS34 8XA
M3 PGU G. Farrar 174 Houghton Road, Thurnscoe, Rotherham S63 0SA
M3 PGY C. Graham 19 Pontop View, Rowlands Gill NE39 2JP
MM3 PHC T. Given 26 Campbell Court, Cumnock KA18 1NP
M3 PHF J. Austin 66 Homewood Avenue, Sittingbourne ME10 1XJ
M3 PHG P. Greenway 26 Coleridge Gardens, Burnham-on-Sea TA8 2QA
M3 PHO J. Wilson 448 Hythe Road, Willesborough, Ashford TN24 0JH
MM3 PHP P. Goodhall 12 Templand Road, Lhanbryde, Elgin IV30 8PP
M3 PHQ N. Benes 2 Leasowes Close, Watling Street North, Church Stretton SY6 7BB
M3 PHR P. Norman 22 Stirling St., Hull HU3 6SL
M3 PHS P. Saben Tredinneck Moor, Newmill, Penzance TR20 8XT
M3 PHX C. Duffill 181 Foden Road, Great Barr, Birmingham B42 2EH
M3 PHZ A. Billings 46 Thorley Drive, Cheadle, Stoke-on-Trent ST10 1SA
M3 PIA G. Mears 59 Hastoe Park, Aylesbury HP20 2AB
M3 PIH D. Huckle 1 Glebe Road, Biggleswade SG18 0PE
M3 PIK M. Gould 57 Fowler Close, Leicester LE4 0SF
M3 PIL R. Harvey 50 Warren Walk, Ferndown BH22 9LY
M3 PIO T. Nakagawa 7 Milton Street, Barrowford, Nelson BB9 6HE
M3 PIQ T. Bourne 100 Dimsdale View West, Newcastle ST5 8EL
M3 PIW J. Witchell 7 Watercombe Lane, Yeovil BA20 2ED
M3 PIY D. Dewsbury 62 Yew Tree Drive, Leicester LE3 6PL
M3 PJG P. Goodayle 2 Downs Road, Seaford BN25 4QL
M3 PJI P. Jones 86 Oaks Lane, Rotherham S61 3ND

M3 PJJ P. Johnson 7 Harrington Court, Hertford Heath, Hertford SG13 7QT
MI3 PJM P. Mccausland 31 Oakleigh Fold, North Street, Lurgan, Craigavon BT67 9BS
M3 PJN P. Northover 181 Mullway, Letchworth Garden City SG6 4BD
M3 PJS P. Seabrook 29 Gadby Road, Sittingbourne ME10 1TJ
MW3 PJU M. Sawford 62 Heol Briwnant, Cardiff CF14 6QH
MW3 PKC L. Hill 12 Heol Coedcae, Bargoed CF81 8QJ
M3 PKE R. North 11 Tintagel Close, Keynsham, Bristol BS31 2NL
M3 PKH C. Bell 12A Mill Lane, Carlton, Goole DN14 9NG
M3 PKL B. North 11 Tintagel Close, Keynsham, Bristol BS31 2NL
M3 PKM P. Bliss 6 Jubilee Gardens, Biggleswade SG18 0JW
M3 PKQ J. Blackburn 64 Marsh Lane, Birmingham B23 6PJ
M3 PKR Rsc of Cheshire c/o B. Parker 117 Corporation Road, Dudley DY2 7QT
M3 PKZ Beaneaters DX Group c/o M. Woolley 84 Bowthorpe Road, Norwich NR2 3TP
M3 PLB P. Beier 20 Markham Avenue, Armthorpe, Doncaster DN3 2AZ
M3 PLI P. Lister 73 Seabrook Court, Seabrook, Hythe CT21 5RY
M3 PLN P. Lewin The Hawthorns, Hawthorne Drive, Stafford ST19 9NQ
M3 PLP P. Price 4 Priory Avenue, North Ferriby HU14 3AE
M3 PLU T. Palmer Edison House, Bow Street, Great Ellingham NR17 1JB
M3 PLV C. Mccollum 25 Byron Road, Locking, Weston-Super-Mare BS24 8AG
M3 PMF R. Farman 19 Mayfield Park, Cheltenham Road, Bagendon, Cirencester GL7 7BH
M3 PMI J. Segrove 87 Henry Road, West Bridgford, Nottingham NG2 7ND
M3 PMK P. Kidd 78 Studfield Road, Sheffield S6 4SU
M3 PML P. Lines 56 Old Hall Close, Amblecote, Stourbridge DY8 4JQ
M3 PMN P. Nicholls 30 Nailbourne Court, Palm Tree Way, Lyminge CT18 8LX
M3 PMO P. Brindle 5 Showfield Close, Sherburn In Elmet, Leeds LS25 6LW
MI3 PMR R. Catney 32 Cairndore Avenue, Newtownards BT23 8RF
MW3 PMU D. Jones 34 Pen Y Bryn, Rassau, Ebbw Vale NP23 5AJ
MI3 PMW M. Pollock 5 St. Marys Terrace, Stream Street, Newry BT34 1HL
M3 PMX A. Marlow Yeomans Barn, Kingsbridge TQ7 3BH
M3 PMY T. Wright 11 Ash Close, Daventry NN11 0XH
M3 PNA N. Phillpott Escroft, Stombers Lane, Folkestone CT18 7AP
M3 PNB B. Crosswell 5 Harty Ferry View, Whitstable CT5 4TE
M3 PNF G. Taylor 31 Ashfurlong Crescent, Sutton Coldfield B75 6EN
M3 PNH N. Hoyle 34 The Drive, Halifax HX3 8NJ
M3 PNI E. Bishop 29 Windsor Court, Poulton-le-Fylde FY6 7UX
M3 PNO C. Turner 28 Fox Lea, Kesgrave, Ipswich IP5 2YU
MW3 PNR R. Williams Plaen Cottage, Bodfari, Denbigh LL16 4BS
M3 PNV J. Nelmes 118 Silcoates Lane, Wrenthorpe, Wakefield WF2 0PE
M3 PNY P. Robinson 19 St. Wilfrids Crescent, Brayton, Selby YO8 9EU
M3 PNZ R. Maddock 48 Collygree Parc, Goldsithney, Penzance TR20 9LY
MI3 POB P. O'Brien 71 Whitepark Road, Ballycastle BT54 6LP
M3 POH D. Dean 119 Queens Drive, Newton-le-Willows WA12 0LN
MM3 POI T. Penna North Windbreck Deerness, Orkney KW17 2QL
M3 POP J. Morris Lurdin Lodge, 71 Lurdin Lane, Wigan WN6 0AQ
M3 POQ R. Finch 19B Kiln Road, Newbury RG14 2LS
M3 POV N. Trangmar 8 Maxstoke Close, Meriden, Coventry CV7 7NB
M3 POW M. Davies 5 Twyford Avenue, Great Wakering, Southend-on-Sea SS3 0EZ
MM3 PPA S. Parsons Linksview, Barrock, Thurso KW14 8SY
MI3 PPD R. Reilly 220 Ardmore Road, Londonderry BT47 3TE
M3 PPG J. Godfrey 4 Cherry Close, Houghton Conquest, Bedford MK45 3LQ
MI3 PPI P. Pearson 1 Rock Cottages, Springwell Road, Bangor BT19 6LZ
M3 PPK N. Green 11 Wythburn Way, Rugby CV21 1PZ
M3 PPO A. Wade 40 Throxenby Lane, Scarborough YO12 5HW
MW3 PPQ P. Sharrock 23 Grosvenor Gardens, Wrexham LL11 1EF
M3 PPR P. Saving Room 1, Abbeyfield, The Glebe Field, Sevenoaks TN13 3DR
M3 PPU M. Whitten 75 Regent Street, Whitstable CT5 1JQ
M3 PPY J. Evans 21 Quilter Close, Bilston WV14 9AX
M3 PPZ R. Bullen 67 Abberley Road, Liverpool L25 9QY
M3 PQB D. Fagg 62 Hawkins Road, Folkestone CT19 4JA
MW3 PQE A. Paffey 1 St. Vincent Road, Newport NP19 0AN
MW3 PQF E. Paffey 1 St. Vincent Road, Newport NP19 0AN
M3 PQG J. Goldfinch 138 Palmerston Road, Chatham ME4 5SJ
M3 PQI M. Bhatia Swaynes House, Room 2 Flat 2, Wirenhoe Park CO4 3SQ
M3 PQJ J. Pratt 4 Bussex Square, Westonzoyland, Bridgwater TA7 0HD
M3 PQL A. Mellor 112 Allerton Road, Stoke-on-Trent ST4 8PL
MI3 PQM P. Millar 37 Thorncroft, Ahoghill, Ballymena BT42 1RX
M3 PQN A. Phillips 15 Hertford Close, Woolston, Warrington WA1 4EZ
M3 PQQ R. Fern 3 Park Road, Featherstone, Wolverhampton WV10 7HS
M3 PQT S. Broadbent 86 Inverness Road, Dukinfield SK16 5AB
M3 PQU A. Milton-Eldridge 2 Partridge Close, Didcot OX11 6AB
M3 PQV M. Nolan Bath Road Post Office, Post Restante. Bath Road, Devizes SN10 1QG
M3 PRN S. Martin 35 Hermitage Green, Hermitage, Thatcham RG18 9SL
M3 PRS G. Manchester 251 Osmaston Park Road, Allenton, Derby DE24 8DA
M3 PRU P. Broadbere Refail, Resthill Road, Wirral CH63 6HN
M3 PRY S. Mariott 4 Stone Cross Gardens, Catterall, Preston PR3 1YQ
M3 PRZ P. Radmall Appleford, Bowcombe Road, Kingsbridge TQ7 2DJ
M3 PSB S. Birch 6 Crescent Road, Wallasey CH44 0BQ
M3 PSC P. Cattel 21 School Hill, Chickerell, Weymouth DT3 4BA
M3 PSD P. Sheppard 107 Queen Street, Swinton, Mexborough S64 8NF
M3 PSE P. Elliott 11 Forgefields, Herne Bay CT6 7TB
M3 PSF R. Baroch 15 Salters Lane, Redditch B97 6JH
M3 PSI K. Morton 47 Trinity Court, Halstead CO9 1PY
MM3 PSL P. Leech The Croft House, 9 Ruilick, Beauly IV4 7AB
M3 PSM S. Maughan 17 Upper Dane, Desborough, , Kettering NN14 2LB
M3 PSO W. Weaver Challacombe House, Perrinpit Road, Bristol BS36 2AT
M3 PSR P. Roberts 43 Ashbourne Crescent, Sale M33 3LQ
M3 PSS P. Swanepoel Bridge Cottage, Tinhay, Lifton PL16 0AH
M3 PSU J. Davidson 5 Hanover Parc, Indian Queens, St. Columb TR9 6ER
M3 PSZ A. Laurence Brookvale, Nooklands, Preston PR2 8XN
M3 PTA P. Chambers 257 Kings Acre Road, Hereford HR4 0SR
M3 PTB T. Bishop 4 Walnut Grove, Worlington, Bury St. Edmunds IP28 8SF
M3 PTG T. Green Huntley, Chesham Road, Tring HP23 6HH
M3 PTI B. Parton 51 Marston Grove, Stoke-on-Trent ST1 6EF
M3 PTQ W. Hughes 6A Park Road, Melton Mowbray LE13 1TT
M3 PTR P. Dryden 27 Delaval Crescent, Blyth NE24 4AZ
M3 PTV B. Fitzakerley 38 Hazel Grove, Armthorpe, Doncaster DN3 3HG
M3 PTX C. Hunter 30 Glebelands, Pulborough RH20 2JJ
M3 PUB P. Swynford 6 The Rise, Cold Ash, Thatcham RG18 9PD
M3 PUE A. Hannon 8 Circular Road West, Liverpool L11 1AZ
MI3 PUH J. Dunlop 118 Ardenlee Avenue, Belfast BT6 0AD
M3 PUL P. Stead 36 Reeds Avenue East, Wirral CH46 1RQ
M3 PUN J. Rideout 35 Colmead Court, Northampton NN38QE
M3 PUQ J. Hunt 104 Hamilton Avenue, Cheam, Sutton SM3 9RL
M3 PUT C. Townsend 2 Netherfield Drive, Netherthong, Holmfirth HD9 3ES
MW3 PUU B. Hill 97 Maesglas Grove, Newport NP20 3DN
M3 PUZ S. Turford 1 Portland Crescent Bolsover, Chesterfield S446EG
M3 PVB T. Ireland 114 Alder Lane, Warrington WA2 8AW
MW3 PVC M. Cook 9 Drenewydd, Park Hall, Oswestry SY11 4AH
M3 PVI P. Handley 97 Applegarth Avenue, Guildford GU2 8LX
M3 PVP A. Botley Flat 1/B, 46 Trull Road, Taunton TA1 4QH
M3 PVQ E. Rhodes The Old Forge, Stoke Gabriel, Totnes TQ9 6RL
M3 PVU J. Watson 20 St. Marys Gardens, Hilperton Marsh, Trowbridge BA14 7PG
M3 PVV T. Crisp 6 Tumlins, All Cannings, Devizes SN10 3PQ
M3 PVX Dr D. Jones 22 Taylor Street, Hollingworth, Hyde SK14 8PA
M3 PWE P. Ward 69 Woodlands Avenue, Tadcaster LS24 9HP
M3 PWK S. Platts 59 Sea View Road, Drayton, Portsmouth PO6 1EW
M3 PWM P. Mitchell 13 Ashorne Close, Matchborough, Redditch B98 0EY
M3 PWO D. Robertson 53 Moor Lane, Weston-Super-Mare BS22 6RA
M3 PWS P. Sykes 2 Thornton Villas, Barrow Road, Barrow-upon-Humber DN19 7QG
M3 PWW P. Wright 16 Hainault Avenue, Giffard Park, Milton Keynes MK14 5PA
M3 PWZ B. Pearson 119 Tolkien Road, Eastourne BN23 7AQ
M3 PXE K. Peel 123 Cunningham Road, Tamerton Foliot, Plymouth PL5 4PU
M3 PXF T. Gabriel 57 West Down Road, Delabole PL33 9DT
MM3 PXG S. Simpson 11 Dighty Burn Court, Dundee DD4 8FE
M3 PXK R. Ellery 12 Sentry Close, St. Issey Wadebridge PL27 7QD
M3 PXL P. Houghton 37 Cedar Avenue, Cottingham HU16 4AL
MM3 PXO E. Mccook 6 Elms Place, Stevenston KA20 4EF
M3 PXP M. Williams 9 Clarence Place, Stonehouse, Plymouth PL1 3JN
M3 PXQ N. Kendall 19 Clowance Lane, Mount Wise, Plymouth PL1 4HU
M3 PXT P. Mutavdzic 1 Hawthorne Drive, Kingwood, Henley-on-Thames RG9 5WE
M3 PXU V. Parton 51 Marston Grove, Stoke-on-Trent ST1 6EF
M3 PXW B. Smith 19 Alexandra Square, Winsford CW7 2YR
M3 PXY C. Fox 45 Park Road, Wivenhoe, Colchester CO7 9LS
M3 PXZ Dr C. Fox 45 Park Road, Wivenhoe, Colchester CO7 9LS
M3 PYB D. Furlong 6A Glebe Avenue, Ruislip HA4 6QZ
M3 PYD M. Smith 6 Neeps Terrace, Middle Drove, Wisbech PE14 8JT
M3 PYG A. Webb 104 Birds Nest Avenue, Leicester LE3 9ND
M3 PYI W. Mcbain Willow Cottage, Gedney Broadgate, Spalding PE12 0DE
M3 PYJ S. Randall 23 Onslow Road, Plymouth PL2 3QG
M3 PYO D. Horner 21 Ainsworth Road, Little Lever, Bolton BL3 1RG
M3 PYR P. Rushby 16 Foxhill Lane, Selby YO8 9AR
M3 PYS E. Ransom 10 Gillercomb, Redcar TS10 4SG
M3 PYT C. Harris 16 Downfield Way, Plymouth PL7 2DU
M3 PYW B. Simmonds 55 Pepys Road, St. Neots PE19 2EN
MI3 PYX D. Caiden 14 Church Street, Greyabbey, Newtownards BT22 2NQ
MW3 PYY C. Thomas 22 Sea Road, Abergele LL22 7BU
MW3 PZC M. Marston The Retreat, The Catch, Holywell CH8 8DU
M3 PZF J. Bealey 17 Chelston Road, Newton Abbot TQ12 2NN
MM3 PZJ S. Ling Leadburnlea Leadburn, West Linton EH46 7BE
M3 PZL R. Mckenzie 4 Simpkin Street, Abram, Wigan WN2 5QD
M3 PZN L. Mckenzie 4 Simpkin Street, Abram, Wigan WN2 5QD
MW3 PZO S. Connor 8 Bro Arfon, Upper Llandwrog, Caernarfon LL54 7BH
MI3 PZV C. Stewart 18 Tullagh Dale, Ballymena BT42 2LQ
M3 PZX P. Seabrook 29 Gadby Road, Sittingbourne ME10 1TJ
M3 PZZ S. Cooke 1 White Meadow, Chilton Polden, Bridgwater TA7 9EZ
MW3 RAA A. Gordon 5 Parc Hendy, Mold CH7 1TH
M3 RAE J. Webb 6 Chatsworth Avenue, Fleetwood FY7 8EG
M3 RAK R. King 79 Holmside Avenue, Minster On Sea, Sheerness ME12 3EZ
MW3 RAU T. Rowlands 3, Pool, Llanfairfechan LL330TN
MM3 RBF D. Gemmell 36 Church St., Dumfries DG2 7AS
M3 RBI R. Gilbert 61 Coltstead, New Ash Green, Longfield DA3 8LN
MM3 RBJ B. Johnston 71 Upper Mastrick Way, Aberdeen AB16 5QG
MI3 RBM R. Abraham 9 Milfort Gardens, Waringstown, Craigavon BT66 7PD
M3 RBP R. Peacock 27 Greenside, Kendal LA9 5DU
M3 RBQ C. Boarer 37 The Martlets, Rustington, Littlehampton BN16 2UB
M3 RBT R. Kerr The Dower House, Church Square, Derby DE73 8JH
M3 RBU B. Upton 1 Sunningdale Close, Eastleigh SO50 8PU
M3 RBX R. Swynford 6 The Rise, Cold Ash, Thatcham RG18 9PD
M3 RCC N. Prescott 3 View Fields, Station Road, Doncaster DN9 3AE
M3 RCD R. Cooke 22 Shepperton Close, Great Billing, Northampton NN3 9NT
M3 RCE R. Edwards 15 Burghley Street, Bourne PE10 9NS
M3 RCG R. Gifford Mill House, Mill Road, Topcroft, Bungay NR35 2BW
M3 RCI K. Steele Flat 22, Bradgate Court, Staunton Avenue, Derby DE23 1PR
M3 RCQ M. Snowden Amber Lights, Market Lane, Walpole St. Andrew, Wisbech PE14 7LT
MM3 RCR J. Mcmartin 19 Bruce Street, Bannockburn, Stirling FK7 8UF
M3 RCS R. Swietlik 3 Tarvin Close, Sutton Manor, St. Helens WA9 4DL
M3 RCT M. Russell 107 Cambridge Road, Hitchin SG4 0JH
M3 RCV R. Treacher 93 Elibank Road, London SE9 1QJ
M3 RCW R. Wiggins 68 Beaconsfield Road, Burton-on-Trent DE13 0NT
MM3 RCX K. Carroll 32E Meadowburn Place, Campbeltown PA28 6ST
MM3 RCZ A. Conlon Kilrae, Barrpath, Glasgow G65 0EX
M3 RDA R. Astbury 12 Southall Road, Ashmore Park, Wolverhampton WV11 2PZ
M3 RDH R. Hastings 43 Delmar Avenue, Leverstock Green, Hemel Hempstead HP2 4LZ
MM3 RDP D. Moore 47 Lockhart Street, Germiston, Glasgow G21 2AP
MW3 RDS R. Smith Hendafarn, Sarnau, Llanymynech SY22 6QJ
M3 RDV R. Dewes 31 Woodlea Avenue, Lutterworth LE17 4TU
M3 RDW R. Wells 37 Water Meadows, Worksop S80 3DF
M3 RDY R. Young 4 Hammond Court, Mablethorpe LN12 2EL
M3 REL R. Lowis 53 Harewood Crescent, Louth LN11 0JD
M3 REM A. Kernick 40 Leyster Street, Morecambe LA4 5NF
M3 REP A. Wilkinson 6 Humbledon View, Sunderland SR2 7RX
M3 REQ A. Williamson 25 Manor Road, Rugby CV21 2SZ

M3 RET M. Parkes 12 Penderel Street, Walsall WS3 3DX
M3 REX S. Thompson Rutland, Quaker Lane, Wirral CH60 6RD
M3 REZ A. Adkins 91 Fernbank Road, Birmingham B8 3LL
M3 RFF P. Richardson 31 Castlefields Drive, Brighouse HD6 3XF
M3 RFG R. Gray Upper Bisterne Farmhouse, Bisterne, Ringwood BH24 3BP
M3 RFH R. Henderson 9 Green Mead, South Woodham Ferrers, Chelmsford CM3 5NL
M3 RFI K. Hilton 199A Sale Lane, Tyldesley, Manchester M29 8PG
M3 RFK Rvd. R. Eardley Bridge Cottage, Martin, Fordingbridge SP6 3LD
M3 RFO Dr J. Mccue 40 Bradbury Road, Stockton-on-Tees TS20 1LE
M3 RFR S. Fallows 23 Howard Street, Burnley BB11 4BJ
M3 RFW R. Ford 30 Cartmel Close, Worcester WR4 9NT
M3 RFX R. Ashworth 10 Mulberry Close, Wigan WN5 9QL
M3 RGC R. Cummings Juan Rodriguez El Cusques, No. 25 (Plot 20A), Alicante 0 Spain
MW3 RGD R. Hogben The Steppes, Presteigne Road, Knighton LD7 1HY
M3 RGE M. Tate 52 Marlborough Road, London N22 8NN
M3 RGG J. Brown 339 Manor Road, Brimington, Chesterfield S43 1NU
MM3 RGH R. Heath Roadside Cottage, , Glenkindie, Alford AB33 8SH
M3 RGJ R. Jamieson 3 Waterpark Road, Prenton Park, Birkenhead CH42 9NZ
M3 RGK K. Harley Care Of: Mr K Harley 9 Amyas Way, Northam, Bideford EX39 1UT
M3 RGN P. Frampton 118 Ramnoth Road, Wisbech PE13 2JD
M3 RGP R. Prangnell 124 St. Marys Road, Cowes PO31 7SR
M3 RGU A. Oxlade 27 Spenfield Court, Northampton NN3 8LZ
MM3 RGZ J. Cairney 5 James Street, Bannockburn, Stirling FK7 0NQ
MM3 RHA W. Hawthorn 8 Drummond Place, Stirling FK8 2JE
M3 RHB J. Palmer 2 Dagonet Road, Bromley BR1 5LR
M3 RHD R. Duffield 4 Crabmill Lane, Easingwold, York YO61 3DE
M3 RHG R. Greatrix 24 Berwick Drive, Cannock WS11 1NS
MW3 RHI R. Chalk 42 Erskine Road, Colwyn Bay LL29 8EU
M3 RHJ M. Dennis 10 Welland Court, Burton Latimer, Kettering NN15 5ST
M3 RHK S. Bharrich 8 Ferrers Ave, Tutbury DE13 9JR
M3 RHL R. Looker 165 Mollison Drive, Wallington SM6 9GX
M3 RHO R. Frylinck 46 Buckingham Road, Richmond TW10 7EQ
M3 RHP Lord R. Montague 71 Middlethorpe Road, Cleethorpes DN35 9PP
M3 RHR G. Kensett 12 Rustics Close, Calvert, Buckingham MK18 2FG
MM3 RHT G. Fyfe 7 Coralmount Gardens, Kirkintilloch, Glasgow G66 3JW
M3 RIE M. Hatton Elisha Cottage, St. Peters Walk, Hull HU7 5FB
MI3 RIF J. Smyth 37 Ardfreelin, Newry BT34 1JG
MI3 RIL L. Scott 98 Aghafad Road Dunamanagh, Strabane BT82 0QQ
M3 RIP M. Pearce 104 Sea Lane, Goring-By-Sea, Worthing BN12 4PU
M3 RIU M. Manser 17 Emperor Way, Kingsnorth, Ashford TN23 3QY
MI3 RIV S. Datchanamourty 77 Violet Hill Avenue, Newry BT35 6DS
MM3 RIX R. Guthrie 27 Meadowbank Road, Kirknewton EH27 8BH
M3 RJB R. Bird 78 Arden Road, Hockley, Tamworth B77 5JE
M3 RJF R. Fitzgerald 3 Sefton Lane Maghull, Liverpool L31 8AE
M3 RJH R. Hicks 31 Arundel Road, Great Yarmouth NR30 4LD
M3 RJI P. Cummings 24 Spindle Road, Malvern WR14 2WB
M3 RJK R. Kelso 55D Lewisham Hill, London SE13 7PL
M3 RKE K. Simmons 26 Red Hill Close, Studley B807BZ
MM3 RKF R. Mannifield 2 Plewlands Avenue, Edinburgh EH10 5JY
M3 RKJ J. Mccoll 6 Grenville Close, Bodmin PL31 2FB
M3 RKK E. Whiten 17 Scott Close, Ashby-de-la-Zouch LE65 1HT
M3 RKN R. Neville 4 Danson Gardens, Blackpool FY2 0XH
M3 RKR R. Rudd 43 Greenlands Road, East Cowes PO32 6HT
M3 RKV A. Gallop 75 Shearmans, Fullers Slade, Milton Keynes MK11 2BQ
M3 RKZ P. Lewin 42 Eastland Road, Yeovil BA21 4EX
MI3 RLA A. Holmes 5 Cambrai Cottages, Belfast BT13 3PS
M3 RLB A. Marlborough Maximillian Cottage, Manswood Common, Wimborne BH21 5BH
MM3 RLG M. Geldart 13B Greystone Place, Newtonhill, Stonehaven AB39 3UL
M3 RLH M. Thompson 9 Stratfield Place, New Milton BH25 5XE
M3 RLM M. Rose 71 Old Street, Ludlow SY8 1NS
M3 RLO C. Rodway 11 Cleveland Avenue, Bishop Auckland DL14 6AR
M3 RLT B. Tarpey 54 Lowforce, Wilnecote, Tamworth B77 4LU
M3 RLX G. Cox 6 Bullfinch Close, Poole BH17 7UP
M3 RMD M. Rout 2 Woods End Cottages, Kirby Bedon, Norwich NR14 7EB
M3 RMG G. Chapman 15 Greenhays Rise, Wimborne BH21 1HZ
M3 RMH R. Hunt 4 Rue Josy Printz, Hesperange L-5841 Luxembourg
M3 RMI J. Salmon 25 Helston Road, Chelmsford CM1 6JF
M3 RMQ J. Weston 25 Cambridge Road, Orrell, Wigan WN5 8PL
M3 RMS R. Stevenson 97 Queen Street, Crewe CW1 4AL
M3 RMU A. Teed 57 Lymington Road, Torquay TQ1 4BG
M3 RMV R. Ley 23 Heronbridge Close, Westlea, Swindon SN5 7DR
M3 RMX R. Moore 47 Darwin Road, Walsall WS2 7EN
M3 RMZ P. Randall 289 Wilson Avenue, Rochester ME1 2SS
M3 RNG P. Davies Bluebells, Station Road, Bere Alston, Yelverton PL20 7EP
M3 RNK V. Penprase 62 California Gardens, Plymouth PL3 6SZ
M3 RNM M. James 7 Pixey Place, Oxford OX2 8BB
MI3 RNN R. Nicholl 30 Foyle Crescent, Newbuildings, Londonderry BT47 2QR
M3 RNO World DX RC c/o R. Baker 12 Byland Road, Skelton-In-Cleveland, Saltburn-by-the-Sea TS12 2NJ
M3 RNS P. Rainey 27 School Road, Silver End, Witham CM8 3RZ
M3 RNU R. Hargate 79 Boundary Road, Beeston, Nottingham NG9 2QZ
M3 RNW R. Whitehead 1 Smithy Site, Farnborough, Wantage OX12 8NS
M3 RNY R. Kirk 5 Sidcup Court, Southgate Way, Chesterfield S43 2NR
M3 ROF D. Johnson 184 Howbeck Road, Arnold, Nottingham NG5 8QE
M3 ROI W. Chorlton 25 Ash Grove, Orrell, Wigan WN5 8NG
M3 ROQ R. Bird 12 Windsor Road, Loughborough LE11 4LL
M3 ROU P. Curnow 301 Hill Top Drive, Rochdale OL11 2AG
MM3 ROV D. Brown Courtyard Cottage, Letters Farm, Argyll PA27 8BX
M3 ROW F. Webley 2 Octavian Drive, Bancroft, Milton Keynes MK13 0PN
MW3 ROX I. Davies 2 Dinas Terrace, Aberystwyth SY23 1BT
M3 RPA O. Akanyeti Sq/H8/4/A University Quays, Lightship Way, Colchester CO2 8GY
M3 RPD I. Handley Rosedale, Chapman Street, Market Rasen LN8 3DS
M3 RPE R. Evenden 1 Castle Place, 2 Castle Street, Tonbridge TN9 1BN
M3 RPF R. Fullagar 6 Locke Way, Stafford ST16 3RE
M3 RPH R. Hales 8 Barton Close, Kingsbridge TQ7 1JU
M3 RPK H. Friberg 19 Holmcroft, Newbiggin-by-the-Sea NE64 6DQ
M3 RPQ J. Faulkner 3 Britannia Quay, 37 River Road, Littlehampton BN17 5DB
M3 RPR R. Roebuck 50 Henson Avenue, Blackpool FY4 3LY
M3 RPS P. Cooper 10 Headon Gardens, Exeter EX2 6LE
MW3 RPX A. Davies 24 Ash Lane, Mancot, Deeside CH5 2BR
MW3 RPZ M. Stokes 23 Goetre Fawr Road Killay, Swansea SA2 7QS
M3 RQB R. Wood 1 Kildare Garth, Kirkbymoorside, York YO62 6LN
MM3 RQC J. Livingstone 17 Livingstone Drive, Bo'ness EH51 0BQ
MM3 RQG H. Smith Ryefield, Windyknowe Road, Galashiels TD1 1RG
M3 RQJ M. Powell 2 Walton Avenue, Twyford, Banbury OX17 3LB
M3 RQO D. Rouse 46 Frensham Drive, Bradford BD7 4AS
MM3 RQP F. Pudsey 21/2 Bathfield, Edinburgh EH6 4DU
M3 RQQ Dr R. Baldwin 20 Lynn Road, Shouldham, King's Lynn PE33 0BT
M3 RQR L. Robinson 19 Adur Avenue, Shoreham-by-Sea BN43 5NN
MI3 RRE R. Rantin 8A Buchanans Road, Newry BT35 6NS
M3 RRJ J. Roughley 42 Thistledown Close, Wigan WN6 7PA
M3 RRN D. Dunstan 2 Trevarren Avenue, Four Lanes, Redruth TR16 6NH
MW3 RRU G. Clements 9 Esgair Y Gog, Bronllys, Brecon LD3 0HY
M3 RRV R. Taylor 2 Chadwick Road, Moorends, Doncaster DN8 4NG
MW3 RRW G. Tucker 18 Plymouth Road, Penarth CF64 3DH
M3 RRZ S. Garrett 44 Wardle Crescent, Leek ST13 5PW
M3 RSH Dr R. Hodgkinson 39 Oxford Road, Carlton-In-Lindrick, Worksop S81 9BD
M3 RSN C. Keeley 3A St. Marks Road, Huyton, Liverpool L36 0XA
MM3 RSR R. Rogerson 93 Auchencrieff Road, Locharbriggs, Dumfries DG1 1UZ
MI3 RST J. Donaldson 12 Drumcrow Road, Glenanne, Armagh BT60 2JQ
M3 RSX R. Shippey 43 Westbury Street, Bradford BD4 8PB
M3 RTE R. Turner 2 Gate House Cottages, Hunton Road, Tonbridge TN12 9SG
MM3 RTH W. Fitzsimons 34 Caledonian Road, Stevenston KA20 3LG
M3 RTI R. Brew 45 Stephenson Road, Braintree CM7 1DL
M3 RTP S. Tingay 6 Butts Close, Farnley Tyas, Huddersfield HD4 6UR
M3 RTR S. Davis 104 Cairo Avenue, Peacehaven BN10 7LA
MW3 RTU M. Kidner 4 Tonypistyll Road, Newbridge, Newport NP11 4HJ
M3 RUI R. Wang 86 Sunnyside Road, Beeston, Nottingham NG9 4FG
M3 RUK K. Moody 114 Acomb Road, York YO24 4EY
M3 RUL C. Rule 109 Carshalton Park Road, Carshalton SM5 3SJ
M3 RUO Hastings Electronics & RC c/o S. Mcguinness 64 Newshaw Lane, Hadfield, Glossop SK13 2AT
M3 RUR J. Stimpson 2 Church Avenue, Kings Sutton, Banbury OX17 3RJ
MI3 RUV D. Mccloskey 1 Dernaflaw Cottages Dernaflaw Road, Dungiven, Londonderry BT47 4PP
M3 RUW Eryri DX c/o D. Jaynes 81 Bude Crescent, Stevenage SG1 2QL
MM3 RUZ G. Ruzgar 22 Ochil Terrace, Dunfermline KY11 4BW
M3 RVE Cross Counties ARC c/o G. Thorpe 81 Knoll Drive, Coventry CV3 5PJ
M3 RVJ Chelmsford ARS c/o D. Purser 11 Barnards Close, Malvern WR14 3NJ
M3 RVK South Devon RC c/o G. Myall 418 Chester Road, Warrington WA4 6ES
M3 RVM South Devon Raynet c/o D. Morley 5 Pelham Close, Westham, Pevensey BN24 5NL
MW3 RVN S. Jones 15 Corn Hill, Porthmadog LL49 9AT
MW3 RVP R. Lasbury 57 Westbourne Road, Whitchurch, Cardiff CF14 2BR
M3 RVQ Taunton & Somerset Raynet Group c/o R. Lester 17 Clarence Road, Capel-Le-Ferne, Folkestone CT18 7LW
M3 RVS R. Sohst 2 Shaftesbury Drive, Maidstone ME16 0JS
M3 RVX M. Brandon 9 Holly Drive, Winsford CW7 1DZ
M3 RWC R. Cornwall 9 Bishop Close, Dunholme, Lincoln LN2 3US
M3 RWD R. Davidson 3 Eastridge Drive Bishopsworth, Bristol BS13 8HQ
M3 RWI Taunton & Somerset Raynet Group c/o J. Marshall 18 Dunnett Road, Folkestone CT19 4BX
M3 RWK Tony Pandy Scout Group c/o J. Lecaille Tezlan, Colton Road, Norwich NR9 5BB
M3 RWN R. Nock 83 Coles Lane, West Bromwich B71 2QW
M3 RWR D. Griffiths 1 Ballard Crescent, Dudley DY2 9EZ
M3 RWV R. Chown 7 Foden Walk, Wilmslow SK9 2HQ
M3 RXD Street Club Operators c/o R. Dryburgh 21 Glebe Close, Stow On The Wold, Cheltenham GL54 1DJ
MI3 RXF Worthing & District Scout ARC c/o I. Smyth 42 Mullintill Road, Claudy, Londonderry BT47 4JN
M3 RXG Bishop Auckland Radio Amateurs Club c/o R. Jolly 102 Swanstree Avenue, Sittingbourne ME10 4LF
MW3 RXH R. Rimmer Dwyfor, Heol Las, Llantrisant, Pontyclun CF72 8EG
MW3 RXK S. Merrifield 37 South View Drive Rumney, Cardiff CF3 3LX
MM3 RXM Chesham & District ARS c/o R. May 12 Clochbar Gardens, Milngavie, Glasgow G62 7JP
M3 RXO L. Milburn 87 Caburn Court, Crawley RH11 8SX
M3 RXP Dr R. Whittle 20 Marlbrook Lane, Marlbrook, Bromsgrove B60 1HN
M3 RXQ M. Milne Flambards, Manor Road, Dunmow CM6 2JR
M3 RXT J. Bridson 10 Clegg Street, Astley, Tyldesley, Manchester M29 7DB
MI3 RXU I. Ophert 5 Cloghboy Road, Bready, Strabane BT82 0DN
M3 RXW R. Webb 17 St. Marys Close, Chudleigh, Newton Abbot TQ13 0PL
M3 RYA R. Petts 19 Sandwell Avenue, Darlaston, Wednesbury WS10 7RH
MI3 RYD S. Davison 60 Cornation Place, Craigavon BT66 7AN
M3 RYG R. Hughes 117 Liverpool Road, Irlam, Manchester M44 6EH
M3 RYI S. Ashcroft 9 Aldermere Crescent, Urmston, Manchester M41 8UE
MI3 RYJ J. Hazlett 25 Gorteen Crescent, Limavady BT49 9EW
M3 RYN R. Fowler Ryland, Back Lane, Doncaster DN9 3AJ
M3 RYO M. Shasby 19 Crawshaw Grange, Crawshawbooth, Rossendale BB4 8LY
M3 RYR Scottish Tourist Board Radio c/o R. Farrar 41 Newtown Avenue, Cudworth, Barnsley S72 8DY
M3 RYT J. Abbott 22 Brent Close, Witham CM8 1TJ
M3 RYY R. Crowther 6 Kaliton, Church Street, Callington PL17 7GB
M3 RYZ A. Clunnie 19 Griffin Road, Warwick CV34 6QX
M3 RZB R. Brittain 159 Caledonia Road, Wolverhampton WV2 1JA
M3 RZE C. Russell 255 Leeds Road, Shipley BD18 1EH
M3 RZF S. Harris Cross House, Mill Lane, Preston PR3 2JX
M3 RZG J. Plant The Cottage, Back Springfield Road, Lytham St. Annes FY8 1TN
M3 RZI O. Rabbitt 20 Lysander Drive, Padgate, Warrington WA2 0GL
M3 RZJ B. Trayhurn 15 Wight Drive, Caister-On-Sea, Great Yarmouth NR30 5UN
M3 RZL A. Linden 12 Godstone House, Pardoner St., London SE1 4DT
M3 RZM G. Kingstone 17 Ullswater Drive, Leighton Buzzard LU7 2QR
M3 RZN I. Seaman 6 Aylsham Road, Buxton, Norwich NR10 5EX
M3 RZO N. Bardell 1 Walshs Manor, Stantonbury, Milton Keynes MK14 6BU

M3 RZP North Wiltshire Raynet c/o R. Powell 53 St. Marys Road, Adderbury, Banbury OX17 3HA
MI3 RZT J. Thompson 119 Rathkyle, Antrim BT41 1LN
M3 RZU U. Ekpe Cathedral Court, University Campus, Guildford GU2 7JH
M3 RZV R. Millington Quaintways, The Avenue, Tarporley CW6 0BA
MW3 RZW R. Lancaster 10 Railway Terrace, Tirphil, New Tredegar NP24 6EY
M3 RZX J. Shorey 47 Stanham Road, Dartford DA1 3AN
M3 RZY S. Trotter 62 Regent Street, Whitstable CT5 1JQ
M3 SAA S. Atkinson 10 Pond Lane, New Tupton, Chesterfield S42 6BG
M3 SAB S. Hughes 9 Melverton Avenue, Wolverhampton WV10 9HN
MW3 SAI R. Blore Ty Nwydd, Cymau, Wrexham LL11 5EU
MM3 SAK A. Mcneil 21 Dumbreck Terrace, Queenzieburn, Glasgow G65 9EA
M3 SAO A. Osmond 36 Knowles Road, Leicester LE3 6LU
M3 SAR S. Abraham 12 Graham Road, Halesowen B62 8LJ
M3 SAS G. Deakin 145 Duke Street, Stoke-on-Trent ST4 3NR
M3 SAY S. Yapp Dickers Farm, Beechy Road, Uckfield TN22 5JG
M3 SAZ S. Greatorex 54 Lilac Grove, Glapwell, Chesterfield S44 5NG
M3 SBA A. Savory 51 Catterick Way, Towcester NN12 6NX
M3 SBB B. Stoneley 44 Ilthorpe, Hull HU6 9ER
M3 SBE S. Edwards 5 Gorse Hill Road, Brickfields, Worcester WR4 9TU
M3 SBJ S. Inman 9 Colbert Avenue, Ilkley LS29 8LU
M3 SBP S. Palin Rose Tree Cottage, 17 Rowland Lane, Thornton-Cleveleys FY5 2QX
M3 SBQ K. Walsh Interval, Liverpool Marina, Liverpool L3 4BP
M3 SBS D. Green 144 Dilloways Lane, Willenhall WV13 3HJ
M3 SBT B. Lockley 35 High Street, Blackpool FY1 2BN
M3 SBY B. Young 25 Rombalds Drive, Skipton BD23 2SP
M3 SCA S. Ahmed 59 Ramsgate, Lofthouse, Wakefield WF3 3PX
M3 SCF H. Fish New House Peaton, Peaton, Craven Arms SY7 9DW
M3 SCH S. Holden 7 Macbeth Close, Colchester CO4 3SZ
M3 SCJ C. Short 35 Whitley Willows Lepton, Huddersfield HD8 0GD
MM3 SCO G. Macleod 12A Loyal Terrace, , Tongue IV27 4XQ
MM3 SCQ D. Macgregor The Tundra, Upper Lybster, Lybster KW3 6AT
M3 SCX S. Williamson 19 Alcester Close, Plymouth PL2 1EA
M3 SDB Braintree & District ARS c/o S. Bennett 17 Knox Close, Norwich NR1 4LN
M3 SDH S. Hodgson 31 Ullswater Avenue, Crewe CW2 8QQ
M3 SDJ S. Hackwood 8 Ronald Walk Dresden, Stoke-on-Trent ST34SN
M3 SDK J. Donald 11 Row Brow Park, Dearham, Maryport CA15 7JU
M3 SDN N. Hurst 74 Holden Road, Salterbeck, Workington CA14 5LZ
MM3 SDP I. Lipkowitz Chuccaby, Longhope, Stromness KW16 3PQ
M3 SDQ M. Behrooz-Kafshdooz 20 Byron Road, London W5 3LL
M3 SDV Dr J. Pelham 5 The Crescent, Shortstown MK42 0UJ
M3 SEE S. England 4 Ouse Close, Chandler'S Ford, Eastleigh SO53 4RW
M3 SEJ J. Shepherd 9 Wrea Head Close Scalby, Scarborough YO13 0RX
MI3 SEK R. Thomson 1 Litchfield Park, Coleraine BT51 3TN
MI3 SEO S. Murray 117 Knockview Drive, Tandragee, Craigavon BT62 2BL
MM3 SES S. Smart Cherrytrees, Top Street, Conon Bridge, Dingwall IV7 8BH
MW3 SET S. Taylor 43 Toronnen, Bangor LL57 4TG
MI3 SEV M. Severn 99 Crawfordsburn Road, Bangor BT19 1BJ
M3 SEY M. Howes 1 The Meadows, Herne Bay CT6 7XB
MW3 SEZ S. Ezard 59 Station Farm, Croesyceiliog, Cwmbran NP44 2JW
M3 SFC A. Woodward 19 Hazel Grove, Winchester SO22 4PQ
M3 SFJ F. Smith 9 Bramwell St., St. Helens WA9 2DP
M3 SFN A. Passey 3 The Yard, Bayton, Kidderminster DY14 9LH
MW3 SFP S. Parry Aukland Terrace, Crymych SA41 3QG
M3 SFZ North Bristol ARC c/o S. Free Mill Farm, Hargham Road, Attleborough NR17 1DT
M3 SGE C. Sargent Bradley, Holcombe Village, Dawlish EX7 0JT
M3 SGF S. Blount 55 Silverthorne Drive, Caversham, Reading RG4 7NR
M3 SGG N. Evans 70 Victoria Street, Chesterton, Newcastle ST5 7EW
M3 SGI S. Mallinson 11 Union Road, Liversedge WF15 7HW
M3 SGJ Rvd. J. Scott The Parsonage, 102A, Nutley Lane, Reigate RH2 9HA
MM3 SGQ S. Gill 5 Ramornie Place, Kingskettle, Cupar KY15 7PT
M3 SGS S. Salmon 35 Westgate Road, Lytham St. Annes FY8 2SG
M3 SGV R. Greaves 7 Eller Brook Close, Heath Charnock, Chorley PR6 9NQ
MW3 SGX J. Bidwell 26 Lone Road, Clydach, Swansea SA6 5HR
M3 SGZ J. Bentham 18 Cauldon Avenue, Swanage BH19 1PQ
M3 SHB S. Brown 6 Good Avenue, Trimdon Grange, Trimdon Station TS29 6EF
M3 SHI A. Shillabeer 29 Newlease Road, Waterlooville PO7 7BX
M3 SHJ S. Hughes 4 Cobden Court, Birkenhead CH42 3YH
M3 SHK Wales Digital Radio Group c/o R. Silversides 7 Earles Lane, Kelsall, Tarporley CW6 0QR
M3 SHN S. Neale 28 Needham Drive, Sutton St. James, Spalding PE12 0EG
M3 SHQ K. Browne 24 Oaktree Avenue, Cuerden Residential Park, Leyland PR25 5PJ
MM3 SHT Grajon Radio Group c/o D. Mcclure 10 Greystone Close, Strathaven ML10 6FW
M3 SHW K. Shaw 2 Montrose Avenue, Montrose Street, Hull HU8 7RY
M3 SHX A. Davies 13 The Close, Stalybridge SK15 1HU
M3 SHZ P. Bennett 1 Queens Road, Carterton OX18 3YB
M3 SII K. Borthwick 15 Thomas Close, Ixworth, Bury St. Edmunds IP31 2UQ
MI3 SIL S. Linton 68 Old Frosses Road, Cloughmills, Ballymena BT44 9NA
M3 SIM S. Lord 34 Alsop Street, Leek ST13 5NZ
M3 SIS L. Simmons 2 Blakemere Way, Sandbach CW11 1XU
M3 SIW J. Wellings 133 Griffins Brook Lane, Birmingham B30 1QN
M3 SIY S. Shaul 31 Chatterton Avenue Ermine West, Lincoln LN1 3SZ
M3 SIZ J. Easdown 38 North Street, Barming, Maidstone ME16 9HF
M3 SJD S. Darby 4 Whately Mews, Whately Road, Lymington SO41 0XS
M3 SJH S. Hewitt 4 Carrow Road, Dagenham RM9 4TJ
M3 SJK S. Kerrison 18 Parks Road, Dunscroft, Doncaster DN7 4AH
M3 SJL J. Lowe 46 Runshaw Avenue, Appley Bridge, Wigan WN6 9JN
M3 SJM S. Whitaker 34 Alder Grove, Poulton-le-Fylde FY6 8EH
M3 SJQ J. Cleaver 27 Lawton Crescent, Biddulph, Stoke-on-Trent ST8 6EH
M3 SJR S. Rigby 36 Richmond Road, Stoke-on-Trent ST4 8RH
M3 SJV P. Mccarthy 38 Lyndhurst Drive, Leyton, London E10 6JD
M3 SJW S. Wills 8 Frobisher Road, Yeovil BA21 5FP
M3 SJX B. Shields 20 Gresley Court, Grantham NG31 7RH
M3 SJY K. Young 14 Beechwood Avenue, Chatham ME5 7HH
M3 SKB S. Brown Rushbrook, Holly Grange Road, Lowestoft NR33 7RR
M3 SKC D. Bryan 3 George Street, Bourne PE10 9HE
M3 SKD S. Kidd 27 Hillswood Avenue, Leek ST13 8EQ
M3 SKN P. Probst 37 Devonshire Street, Skipton BD23 2ET
M3 SKQ P. Henderson 24 Farrow Road Whaplode Drove, Spalding PE12 0TS
M3 SKT S. Taylor-Toms 34 Larkspur Drive, Chandler'S Ford, Eastleigh SO53 4HU
M3 SKU V. Leddington 20 Bewell Head, Bromsgrove B61 8HY
M3 SKV J. Hawkes 53 Mill Hill, Derby DE24 5AF
MW3 SKW M. Barber 1 Gwernant, Cwmllynfell, Swansea SA9 2FT
M3 SKX B. Martin 111 The Avenue, Wallsend NE28 6SD
M3 SKY S. Keevil Gamekeepers Cottage, Snarehill, Thetford IP24 2QA
M3 SKZ J. Parfitt 5 Sheridan Road, Frimley, Camberley GU16 7DU
MM3 SLB S. Berry Willowburn, Kirkton, Hawick TD9 8QJ
MM3 SLD J. Bradley Tongue Of Bombie, Kirkudbright DG6 4QD
M3 SLF R. Cave 295 Sackup Lane, Staincross, Barnsley S75 5BG
MW3 SLI J. Backhouse De10 Isaf, Bryneglwys, Corwen LL21 9NP
MW3 SLL S. Lawton Bryngarw Lodge, Brynmenyn, Bridgend CF32 8UU
MW3 SLO D. Dash 36 Rockvilla Close, Varteg, Pontypool NP4 7QF
M3 SLQ N. Jones 10 Leamington Close, Cannock WS11 1PW
MI3 SLT S. Whitten 2 Springwell Manor, Castlederg BT81 7DR
M3 SLZ T. Gill 21 Trevor Smith Place, Taunton TA1 3RW
M3 SMD A. Davies 7 Windermere Grange, Edlington, Doncaster DN12 1NQ
MM3 SMI J. Smith 14 John Collins Crescent, Galashiels TD1 2FA
M3 SMK S. Mackimm 16 Stanneybrook Close, Rochdale OL16 2YH
M3 SML S. Lowe 59 Knight Avenue, Gillingham ME7 1UE
M3 SMM S. Mole 17A Marlborough, Seaham SR7 7SA
M3 SMN S. Kent 4 Arden Close, Chesterfield S40 4NE
M3 SMR R. Shepperley Flat F, London N3 1QL
M3 SMY S. Harkness 114 Morland Road, Ipswich IP3 0LZ
M3 SMZ S. Rdwards 59 Laburnum Road, Tipton DY4 9QS
MM3 SNB G. Mcgeouch 59 Torogay Street, Glasgow G22 7RA
M3 SNF I. Hewitt 26 Outwoods Drive, Loughborough LE11 3LT
MW3 SNH B. Jones Browerdd, Llangybi, Lampeter SA48 8NH
MW3 SNJ S. Jones 14 Lower Cross Road, Llanelli SA15 1NQ
M3 SNL R. James 50 Andrew Allan Road, Rockwell Green, Wellington TA21 9DY
M3 SNN N. Chapman 29 Johnson Road, Uppingham, Oakham LE15 9RY
M3 SNO R. Snowden 5 Eastfield Road, Wisbech PE13 3EJ
M3 SNQ L. Thornton 11 Polruan Road, Truro TR1 1QR
M3 SNR S. Rennalls 7 Hollybush Close, Sowerby Bridge HX6 1AH
MW3 SNW S. Williams Hillsboro Aberkenfig, Bridgend CF32 0EW
M3 SNX G. Rigden Corner House, Ashford Road, Kent TN27 0EE
M3 SNY R. Beardshall 41 Hill Crest, Hoyland, Barnsley S74 0BU
M3 SNZ S. Seath 9 Winchester Avenue, Morecambe LA4 6DX
MW3 SOC N. Davies 65 Walters Road, Neath SA11 2DW
M3 SOF S. Vaux 171 Foxon Lane, Caterham CR3 5SH
M3 SOG R. Stearn 18 Kings Avenue, Chippenham SN14 0UJ
M3 SOQ P. Swann 2 Little Walton, Eastry, Sandwich CT13 0DW
M3 SOT S. Gregory 11 Ribblesdale Avenue, Congleton CW12 2BS
M3 SOV P. Fernie 39 North Parade, Falmouth TR112TE
M3 SOY J. Rolph The Hollies, Back Lane, Norwich NR10 4HL
M3 SPA I. Beresford 16A Holbeck Hill, Scarborough YO11 2XD
M3 SPG S. Garthwaite 278 Carlton Road, Barnsley S71 2BA
M3 SPJ S. Hinds 69 Carshalton Grove, Wolverhampton WV2 2QZ
M3 SPL A. Ladell 25 Harwood Avenue, Thetford IP24 2LY
M3 SPP R. Penrose 41 Milton Road, Eastbourne BN21 1SH
M3 SPQ S. Schonborn 116 Hough Lane, Wombwell, Barnsley S73 0EF
M3 SPR S. Mclaughlin 34 Cambridge Road, Birstall, Batley WF17 9JF
M3 SPU P. Saunders 62 Parkfield Avenue, Eastbourne BN22 9SF
M3 SPY R. Gardner 6 Meade Road, Liverpool L13 9AA
M3 SQE D. Mcdonald 3 Lindley Street, Mansfield NG18 1QE
M3 SQG M. Breslin 15 Acorn Gardens, East Cowes PO32 6TD
M3 SQH A. Lawrence 11 Pembroke Court St. Johns Road, Chesterfield S41 8NX
M3 SQI N. Thake Easton House, Water Street, Berwick St. John, Shaftesbury SP7 0HS
MM3 SQJ J. Morris 10 Middlemas Road, Dunbar EH42 1GJ
MM3 SQM D. Mchardy 486 Kilmarnock Road, Glasgow G43 2BW
M3 SQO P. Burke 38 Bosworth Square, Rochdale OL11 3QG
M3 SQP S. Scotching 26 Newton Way, Leighton Buzzard LU7 4YU
M3 SQQ S. Oxenham 10 Arnside Close, Plymouth PL6 8UU
M3 SQS R. Garland 113 The Drive, Feltham TW14 0AH
M3 SQT C. Eyre 23 Nelson Street, Congleton CW12 4BS
M3 SQU M. Van Den Bergh The Parsonage, Masefield Drive, Tamworth B79 8JB
M3 SQV S. Sandford 11 Browning Close, Tamworth B79 8NB
M3 SQX I. Okorji 37 Lydwell Park Road, Torquay TQ1 3TQ
M3 SQZ N. Swift 59 Milton Avenue, Malton YO17 7LB
MM3 SRF R. Farrer 23 Upper Craigour, Edinburgh EH17 7SE
MI3 SRG E. Coates 148 Springwell Road, Groomsport, Bangor BT19 6LX
M3 SRH S. Hubball 24 Newstead Road, Stoke-on-Trent ST2 8HX
M3 SRI S. Issatt 69 St. Lawrence Avenue, Snaith, Goole DN14 9JH
MM3 SRK A. Ross 16 Croft Road, Kiltarlity, Beauly IV4 7HZ
MI3 SRL S. Rea 70 Raloo Road, Larne BT40 3DU
M3 SRQ J. Stoppard 15 South Lodge Court, Old Road, Chesterfield S40 3QG
M3 SRT S. Thompson 30 Southport Parade, Hebburn NE31 2AQ
M3 SRV J. Matthews 23 Elmhurst, Bridgnorth WV15 5DJ
M3 SRY P. Seymour 3 Larch Avenue Allington Gardens, Allington, Grantham NG32 2FG
M3 SSG A. Butler 12 South Bank Cottages South Stoke, Reading RG8 0HX
M3 SSI S. Bangalore 2 Amberley Walk, Kingsmead, Milton Keynes MK4 4AX
M3 SSL M. Belcher 52 Kynaston Road, Didcot OX11 8HD
M3 SSO B. Hawes 3 Orchard Close, Cassington, Witney OX29 4BU
M3 SSU E. Little 41 Sevenoaks Road, Portsmouth PO6 3JP
M3 STJ S. Jordan 10 Kirkby Avenue Garforth, Leeds LS252BN
M3 STQ S. Fox Hawthorns, School Lane, Martlesham, Woodbridge IP12 4RR
M3 STR N. Soltysik 24 Cottage Close, Hednesford, Cannock WS12 1BS
MI3 STW R. Bradley 41 Graymount Road, Newtownabbey BT36 7DR
MI3 STY S. Nicholl 89 Glenshane Road, Londonderry BT47 3SF
MW3 SUF N. Smith 7 Hawthorne Avenue, Connah'S Quay, Deeside CH5 4TF
M3 SUI S. Allen 33 Rookhill Road, Pontefract WF8 2BY
M3 SUJ J. Cook 86 Beechfield Road, Hemel Hempstead HP1 1PL
M3 SUK A. Holland 40 Sunnyside Road, Poole BH12 2LQ
MM3 SUS S. Holt Ashwell, Cannigall, Kirkwall KW15 1SX

MM3 SUV A. Mccaig 46 Patterson Drive, Law, Carluke ML8 5LT
M3 SUW P. Hopkins 40, Grange Close, Condover, Shrewsbury SY5 7AT
M3 SUY T. Van Den Bergh 19 Perrycrofts Crescent, Tamworth B79 8UA
M3 SVB S. Black 7 Harwood Close, Gosport PO13 0TY
M3 SVC S. Cox 19 Exbury Way, Andover SP10 3UH
M3 SVD M. Hewitt Redwood House, Adbury Holt, Newbury RG20 9BW
M3 SVF L. Leung Harrogate Ladies' College, Clarence Drive, Harrogate HG1 2QG
M3 SVH K. Henderson 42 Chartwell Avenue, Wingerworth, Chesterfield S42 6SP
M3 SVJ R. Gee Flat 1D, Quarmby Road, Huddersfield HD3 4HQ
M3 SVL S. Leech 9 Brocklehurst Mews, Macclesfield SK102GY
MI3 SVM S. Murray 80 Canterbury Park, Londonderry BT47 6DU
M3 SVN S. Adkins 4 Orion Close, Ward End, Birmingham B8 2AU
M3 SVO L. Birdsall 8 North Cote, Ossett WF5 9RE
M3 SVP A. Chapman 24 Eaton Grange Drive, Long Eaton, Nottingham NG10 3QE
M3 SVT S. Taylor 17 Mendip Drive, Bolton BL2 6LQ
M3 SVY L. Beardsley 1 Amber Villas, Sutton St. Nicholas, Hereford HR1 3DF
M3 SVZ G. Brierley 35 Ochrewell Avenue, Deighton, Huddersfield HD2 1LL
MM3 SWA S. Anderson 33 Dryden Avenue, Loanhead EH20 9JT
MI3 SWD S. Mcauley Layde View, 19 Rathlin Avenue, Ballycastle BT54 6DQ
M3 SWF R. Jenkinson Esperance, West End Road, Doncaster DN9 1LB
MM3 SWG S. Groves 1/1 99 Belville Street, Greenock PA15 4SX
M3 SWK S. Walker 64 Belmont Road, Rugby CV22 5NY
MW3 SWO S. Owen 19 Longtown Grove, Newport NP10 8HD
M3 SWQ A. Marks Grosvenor Hotel, 51 Grosvenor Road, Scarborough YO112LZ
M3 SWS S. Lowe 31 Court Farm Road, Bristol BS14 0EH
MM3 SWU S. Jenkins 66 Spruce Avenue, Johnstone PA5 9RG
M3 SWV J. Moppett 59 Piccadilly, Tamworth B78 2ER
MM3 SWW D. Elliot Thisleycrook, Torphins AB214NR
M3 SXF S. Forbes 55 The Henrys, Thatcham RG18 4LS
MI3 SXI I. Mckeown 19 Castlehill Comber, Newtownards BT23 5XA
MM3 SXJ P. Duckles 26 Meadowpark Road, Bathgate EH48 2SJ
M3 SXK C. Little 33 Wigmores, Telford TF7 5NB
MI3 SXM G. Hutton 13 Meadowbank, Sepatrick, Banbridge BT32 4PZ
M3 SXP S. Perring 16 Salford Road, Bolton BL5 1BL
MI3 SXQ C. Cunningham 1 Ballykeel Court, Ballymartin, Newry BT34 4XW
MI3 SXR A. Robb 10 Rosepark East, Belfast BT5 7RL
MM3 SXT S. Thorogood 38 Forres Drive, Glenrothes KY6 2JU
M3 SXU J. Jackson 90 Horne Street, Bury BL9 9HS
M3 SXV S. Simms Fairview, Naburn Lane, Deighton, York YO19 6HH
M3 SXZ D. George Flat 9, The Old Court House, Waterloo, Frome BA11 3FE
MM3 SYB D. Nicholson 4 Upper Barvas, Isle of Lewis HS2 0QX
M3 SYC A. Chaplin 33 The Crofts, Little Wakering, Southend-on-Sea SS3 0JS
MI3 SYF D. Bates 31 Drumard Park, Lisburn BT28 2HU
M3 SYH D. Horton 2 Brampton Way, Bulkington, Bedworth CV12 9PR
MI3 SYI B. Mcdonald 20 Aughan Park, Poyntzpass, Newry BT35 6TW
M3 SYL S. Stratford 23 The Fairway, Banbury OX16 0RR
M3 SYN S. Lanaway 1 Clovers Cottages, Faygate Lane, Faygate, Horsham RH12 4SH
MM3 SYO S. Angus 20 Norlands, Errol, Perth PH2 7QU
MM3 SYQ A. Clark 17 Glentilt Terrace, Perth PH2 0AE
MM3 SYU C. Higgins 65A Forthill Road, Broughty Ferry, Dundee DD5 3DQ
M3 SYV T. Symons Southgate, The Commons, Mullion TR12 7HZ
M3 SYW C. Voke 16 Exton Road, Chichester PO19 8BP
M3 SYY B. Sweeney 14 Eaves Lane, Chorley PR6 0PY
M3 SYZ S. Symonds 68 Manor Crescent, Pan, Newport PO30 2BH
M3 SZC S. Crabtree 107 Rochdale Road, Shaw, Oldham OL2 7JT
MW3 SZD M. Musgrave Hillside Cottage, Hiraddug Road, Rhyl LL18 6HS
MW3 SZF G. Williams Ty Newydd, Rhyd, Penrhyndeudraeth LL48 6ST
MJ3 SZI M. Brown 77 Andium Court, Langtry Gardens, St. Saviours Hill, St. Saviour, Jersey JE2 7AH
M3 SZK M. Gridley 41 Tower Way, Dunkeswell, Honiton EX14 4XH
MM3 SZM S. Macdonald 110 High Street Cuminestown, Turriff, AB53 5YH
M3 SZO R. Savery 75 Bramley Road, Tewkesbury GL20 8AQ
M3 SZQ S. Snelson 212 Dickson Road, Blackpool FY1 2JS
M3 SZS D. Forster 23 Field Street, Padiham, Burnley BB12 7AU
M3 SZT J. Smith 32 Youlgreave Drive, Sheffield S12 4SE
M3 SZY S. Holt 108 Blandford Avenue, Castle Bromwich, Birmingham B36 9JD
M3 TAE T. Eadon Chapel Cottage, Newcastle Road South, Sandbach CW11 1RS
MW3 TAF C. Williams 96 Bryn Road, Swansea SA2 0AT
M3 TAG A. Aldred 78 The Drive, Horley RH6 7NH
MM3 TAM T. Aitken 27 Beeches Avenue, Clydebank G81 6HX
M3 TAN S. Greenfield 4 Charlesworth Square, Gomersal, Cleckheaton BD19 4NX
MM3 TAV A. Mcconochie 15 Slains Crescent, Cruden Bay, Peterhead AB42 0PZ
M3 TAW T. Whittam 27 Dimples Lane, Garstang, Preston PR3 1RD
M3 TBF T. Ferguson 6 Binley Close, Birmingham B25 8NE
M3 TBG A. Archer 23 St. Ives Road, Somersham, Huntingdon PE28 3ER
M3 TBH T. Hobbs 2 The Lynch, West Stour, Gillingham SP8 5RN
M3 TBK E. Cree 24 Old Lincoln Road, Caythorpe, Grantham NG32 3EJ
MI3 TBL Dr T. Littler 15 Belmont Grove, Lisburn BT28 3YB
M3 TBP C. Parker Red Leas, 22 Bent Lane, Colne BB8 7AA
M3 TBQ I. Hill 74 Clarence Road, Torpoint PL11 2LT
M3 TBU T. Burnham Creedy Barn, Kennerleigh, Crediton EX17 4RU
M3 TBV D. Parker 16 Aldborough Road, Dagenham RM10 8AS
M3 TBW J. Humphrey Flat 1, Kingswood House, 10 Lewes Road, Eastbourne BN21 2BX
M3 TBZ D. Heathcote 154 High Street Harriseahead, Stoke-on-Trent ST7 4JX
M3 TCG T. Graham First Floor Flat, 43 Belgrave Crescent, Bath BA1 5JU
M3 TCR P. Ryall Windsor Lodge, Pantile Hill, Southminster CM0 7BA
M3 TCT S. Farrar 20 Cleveland Grove, Lupset, Wakefield WF2 8LD
M3 TCU G. Read Flat 3, Parkmead Court, Ryde PO33 2HD
M3 TCX T. Carroll 14 Glenpark Drive, Southport PR9 9FA
M3 TCY T. Earnshaw 63 Manor Road, Fleetwood FY7 7LJ
M3 TDB T. Berry Roseneath, Walcote Road, Lutterworth LE17 6EQ
M3 TDH T. Hewitt 6 Mayfield, Catforth Road, Preston PR4 0HH
M3 TDM T. Reddington 174 Home Farm Road, Wirral CH49 7LH
M3 TDP T. Packham Bradstow Lodge, 19 Crow Hill, Broadstairs CT10 1HN
M3 TDT A. Dockerill 8 Bennett Road, Swanton Morley, Dereham NR20 4LY
M3 TEE D. West 42 Scholars Green, Wigton CA7 9QW
M3 TEG T. Mason 31 Manor Park Road, Hailsham BN27 3AT
M3 TEI R. Mcknight Ardralla, Church Cross, Skibbereen P81 RK12 Ireland
M3 TEL T. Pink 11 Harmony Meadow, Roche, St. Austell PL26 8EJ
MI3 TEM S. Kirkwood 1 Rural Cottages, Front Road, Lisburn BT27 5LF
M3 TEN T. Newman Sometimes (The Workshop), South Pew, Dorchester DT2 9HZ
M3 TEP T. Payne 2 Greenleas, Waltham Abbey EN9 1SZ
MM3 TEQ A. Leiper 6 Inchyra Place, Grangemouth FK39EQ
M3 TEV O.B.O. University of Plymouth ARS c/o S. Ball 13 Yew Tree Road, Hayling Island PO11 0QE
M3 TEY Maltby And District ARS c/o J. Hartshorne 8 Ashbee Street, Bolton BL1 6NT
M3 TFA A. Tran Flat 4, Room 9, Rayleigh Tower, Colchester CO4 3SQ
M3 TFB South Lancashire ARC c/o D. Parkinson 4 Meadow View, Sherburn In Elmet, Leeds LS25 6BY
M3 TFE J. King Plum Tree Cottage, Royston Place, Barton On Sea, New Milton BH25 7AJ
MI3 TFF Gala & Dist ARS c/o F. Finlay 62 Slieveboy Road, Claudy, Claudy BT47 4AS
M3 TFG B. Jones 75 Mamble Road, Stourbridge DY8 3SY
M3 TFI Brighton & Dist c/o D. Woods 3 Brook Street, Port Sunlight, Wirral CH62 5DB
M3 TFK P. Hemsley 140 Greenhill Lane, Riddings, Alfreton DE55 4EX
M3 TFM S. Lytollis 91 Townfoot Park, Brampton CA8 1RZ
M3 TFO R. Styles 52 Vernham Grove, Bath BA2 2TB
M3 TFP T. Plummer 33 East St., Sudbury CO10 2TU
M3 TFS R. Shoubridge 2 Copestake Drive, Burgess Hill RH15 0LD
M3 TFW T. Harlow 10 Fraser Road, Poole BH12 5AY
M3 TFX T. Fisk 2 Hall Farm Cottage, Caston Road, Caston, Attleborough NR17 1BW
M3 TFZ E. Miller 8 Arthur Avenue, Caister-On-Sea, Great Yarmouth NR30 5PQ
M3 TGA G. Arnold 1 Louthe Way, Sawtry, Huntingdon PE28 5TR
M3 TGC G. Cooper 21 Thistle Bridge Road, Chivenor, Barnstaple EX31 4FL
M3 TGD J. Park 3 Flaxfield Drive, Crewkerne TA18 8DF
M3 TGE R. Lindon 134 Station Road, Sutton Coldfield B73 5LD
M3 TGJ J. Alexander 335 Canterbury Road, Birchington CT79TY
M3 TGK P. Wale Munstead Oaks, Hascombe Road, Godalming GU8 4AB
M3 TGL G. Thorne 72 Devonshire Road, London E16 3NJ
M3 TGO G. Hope 27 Clearmount Drive Charing, Ashford TN27 0LH
M3 TGP T. Porter 208 Clapgate Lane, Ipswich IP3 0RG
M3 TGT H. Doman 13 Cumbria Close, Maidenhead SL6 3DD
M3 TGW T. Willis 32 Sandover, Northampton NN4 0TS
M3 TGZ D. Lines 68 Rugby Place, Brighton BN2 5JA
M3 THE M. Peck 60 Riverside Drive, Tern Hill, Market Drayton TF9 3QH
MW3 THI M. Price 18 Rhiw Tremaen, Brackla, Bridgend CF31 2JA
M3 THJ B. Mills 37 Ashley Road, Hildenborough, Tonbridge TN11 9ED
M3 THN P. Cobley 58 John Street, Newhall, Swadlincote DE11 0SR
M3 THQ J. Peerless 503 Honeypot Lane, Stanmore HA7 1JH
M3 THY T. Lupton 81 Home Farm Lane, Bury St. Edmunds IP33 2QL
M3 TIC M. Tinsell-Stanton 38 Comberton Road, Kidderminster DY10 3DT
M3 TID B. Maddox 72 Church Road, Hartshill, Nuneaton CV10 0LY
M3 TIE L. Morrell-Cross Delta Lodge, 14 Rushton Crescent, Bournemouth BH3 7AF
M3 TIF C. Robertson 4 Pleasant Street, Walshaw BL8 3AU
M3 TII B. Aylward 53 Overdown Rise, Portslade, Brighton BN41 2YF
M3 TIJ J. Aylward 53 Overdown Rise, Portslade, Brighton BN41 2YF
M3 TIK M. Richardson 1 Cedar Drive, Lowestoft NR33 9HA
M3 TIL J. Tillson 23 The Fitches, Knodishall, Saxmundham IP17 1UX
M3 TIQ D. Perkins 56 Cliff Street, Rishton, Blackburn BB1 4EE
M3 TIY A. Green 65 Rosamond Road, Bedford MK40 3UG
M3 TIZ D. Marsh 16 Laurel Close, North Warnborough, Hook RG29 1BH
M3 TJD A. Di Domenico 11 Bracken Drive, Freckleton, Preston PR4 1TH
MW3 TJG T. Gwyther 15 Denbigh Court, Caerphilly CF83 2UN
M3 TJI T. Adams 11 St. Georges Crescent, Gravesend DA12 4AR
M3 TJJ J. Jones 19 Southbank Street, Leek ST13 5LS
MI3 TJK J. Mackenzie 30 Dalriada Gardens, Ballycastle BT54 6DZ
M3 TJL T. Lake 85 Clarkson Road, Norwich NR5 8ED
MI3 TJM T. Moore 43 Woodburn Park, Londonderry BT47 5PS
M3 TJO T. Jones 4 Anne Close, Christchurch BH23 2NW
M3 TJQ A. Newton 8 Trent Meadow, Taunton TA1 2NP
MI3 TJR T. Ruddell 30 Ballynacor Meadows, Portadown, Craigavon BT63 5UU
M3 TJT T. Ticehurst 10 Rushlake Road, Coldean, Brighton BN1 9AD
M3 TJU E. Duffield 92 Crosby Street, Stockport SK2 6SP
MM3 TKE R. Mckie 16 Silver Street, , Creetown, Newton Stewart DG8 7HU
MW3 TKI I. Hoyle-Jackson 21 Kimberley Close, Sketty, Swansea SA2 9DZ
M3 TKN M. Roche 1 Lancaster Close, London NW9 5RE
M3 TKO K. Smart 33 East Street, Littlehampton BN17 6AU
M3 TKQ J. Macey 62 Uttoxeter Road, Hill Ridware, Rugeley WS15 3QU
M3 TKT M. Turowski 57 Millwood Road, Orpington BR5 3LQ
M3 TKU P. Loveden 57 St. Marys Road, Rawmarsh, Rotherham S62 5BD
M3 TKV T. King 24 Royston Avenue, Basildon SS15 4EW
M3 TKW K. Rowley 10 Mount Close, Wombourne, Wolverhampton WV5 9ER
M3 TLB D. Weller 6 Aldervale, Fermor Road, Crowborough TN6 3BY
M3 TLD D. Tattersall 17 Badger Close, Durkar, Wakefield WF4 3QD
M3 TLG A. Dann 18 Salcombe Way, Ruislip HA4 6BA
MM3 TLH T. Holt Ashwell, Cannigall, Kirkwall KW15 1SX
M3 TLJ B. Hawkins 60 High Street Barwell, Leicester LE9 8DR
M3 TLL L. Nilon 5 Denby Drive Baildon, Shipley BD17 7PQ
M3 TLM T. Lockett 14 Tildsley Crescent, Weston, Runcorn WA7 4RN
M3 TLN C. Dean 119 Queens Drive, Newton-le-Willows WA12 0LN
M3 TLO D. Livings 30, Grenfell Avenue, Holland-On-Sea, Clacton-on-Sea CO15 5XH
MW3 TLP A. Buckley 53 Derlwyn St., Phillipstown, New Tredegar NP24 6AZ
MM3 TLQ D. Field 7 Admiralty Street, Portknockie, Buckie AB56 4NB
M3 TLT A. Mcgregor 41 Breedon Close, Corby NN18 9PG
M3 TLU M. Bailey 71 Somerfield Road, Walsall WS3 2EG
MI3 TLV M. Nicholl 34 Berryhill Road, Artigarvan, Strabane BT82 0HN
M3 TLW R. Crook 80 Kings Road, Biggin Hill, Westerham TN16 3XY
M3 TLX D. Burdsall 37 Fulmar Walk, Whitburn, Sunderland SR6 7BW
M3 TLY I. Bain 45 Larpool Crescent, Whitby YO22 4JD
M3 TLZ D. Pointon 1 Cross Cottages, Alsager Road, Audley, Stoke-on-Trent ST7 8JQ
M3 TME M. Trick 24 King St., Tiverton EX16 5JE
M3 TMG G. Thorpe Jasmine, Sutton Road, Mablethorpe LN12 2PT
M3 TMM T. Mckain 21 Oakhurst Grove, East Dulwich, London SE22 9AH
MI3 TMN A. Mcnulty 2 Devenish Crescent, Devenish, Enniskillen BT74 4RB
M3 TMQ Solihull ARS c/o M. Mutton 74 Alexandra Road, Sheerness ME12 2AT

MW3 TMR R. Thomas 35 Under Ffrydd Wood, Knighton LD7 1EF
M3 TMX J. Harrop 5 Edward Street, Newcastle ST5 0JE
M3 TMY T. Mulraney 3 Salvia Close Churchdown, Gloucester GL3 1LL
M3 TMZ A. Laister 23 Berry Street, Greenfield, Oldham OL3 7EF
M3 TNB M. Anthony Magpie Bungalow, Googumpas, St. Day, , Redruth TR16 5JL
M3 TND N. Tam Room 102, 30 Evelyn Gardens, London SW7 3BG
M3 TNE C. Pickford 80 Hollowood Avenue, Littleover, Derby DE23 6JD
MM3 TNF J. Odonnell 33 Broomward Drive, Johnstone PA5 8HR
MM3 TNG D. Stewart 45 Kilwinning Road, Irvine KA12 8RZ
M3 TNH K. Stockley 357 Clements Road, Ramsgate CT12 6UG
M3 TNJ J. Armstrong-Taylor Driftwood, Station Road, Yelverton PL20 7JS
M3 TNK N. Fong Harrogate Ladies' College, Clarence Drive, Harrogate HG1 2QG
M3 TNL P. Dawson 400 Ropery Road, Gainsborough DN21 2TH
M3 TNM T. Nguyen 9 Green Street, Cambridge CB2 3JU
M3 TNN T. Ellis 84 Revelstoke Road, London SW18 5PB
M3 TNO J. Dean 25 Chantry Avenue, Bexhill TN40 2EA
M3 TNV M. Clough 8 Skeldyke Road, Kirton, Boston PE20 1LR
M3 TNW P. Stanford 4 Barkway Road, Royston SG8 9EA
M3 TNY T. Limbert 7 Acacia Avenue, Liverpool L36 5TL
MW3 TOB H. Matthews 24 Clos Y Berllan, Rhuddlan, Rhyl LL18 2UL
M3 TOE A. Ward 39 Linley Close, Bridgwater TA6 4HL
M3 TOF D. Holyoake 281 Causeway, Green Road, West Midlands B68 8LT
MW3 TOI M. Francis 72 Bro Ednyfed, Llangefni LL77 7WD
M3 TOJ H. Clough 8 Skeldyke Road, Kirton, Boston PE20 1LR
M3 TOR C. Jones Po Box 293, Ford, Plymouth PL2 1WT
M3 TOT E. Brown Rose Cottage, Grindlow, Buxton SK17 8RJ
MM3 TOV E. Wilson The Old Schoolhouse, Fife KY15 4NB
M3 TOY A. Ashworth 79 Stonehouse Road, Rugeley WS15 2LL
M3 TPD T. Dooley 32, Coult Avenue, North Hykeham Coult Avenue, Lincoln LN6 9RG
MM3 TPF N. Mann Ramsburn Cottage, Knock, Huntly AB54 7LQ
M3 TPG T. Greenall Hall Lane Farm, Hall Lane, Warrington WA4 4AF
M3 TPH T. Hazel 84 Rodwell Avenue, Weymouth DT4 8SQ
M3 TPI A. Paxton 20F Green End, Granborough, Buckingham MK18 3NT
MW3 TPJ T. Price 5 Rhodfa'R Pant, Pant, Merthyr Tydfil CF48 2DG
M3 TPN T. Norrington 32 Fulfen Way, Saffron Walden CB11 4DW
MI3 TPR R. Thompson 14 Gelvin Grange, Londonderry BT47 2LD
M3 TPU R. Morgan'S 7 Tennyson Close, Braintree, Essex CM7 1AB
M3 TPW T. Wooldridge 12 Redwood Avenue, Leyland PR25 1RN
M3 TPY T. Purcell 18 Millberg Road, Seaford BN25 3ST
M3 TQA A. Madge Flat, 17 Newcomen Road, Dartmouth TQ6 9BN
M3 TQB D. Holmes 4 Council House, Nidds Lane, Boston PE20 1LZ
M3 TQD J. Lear 6 South View Green, Bentley, Ipswich IP9 2DR
M3 TQF C. Bailey 55 Bridgend Park Brewery Road, Wooler NE71 6QG
M3 TQG G. Joy Fair Oak, Higher Furzeham Road, Brixham TQ5 8QP
MM3 TQH R. Hay Roddach Cottage East, Cummingston, Burghead, Elgin IV30 5XY
MM3 TQI D. George D C George, 91 Regent Street, Keith AB55 5ED
M3 TQJ T. Kemp 30 Tawny Sedge, King's Lynn PE30 3PW
MM3 TQK J. Traill 31 Sherwood Place, Bonnyrigg EH19 3JY
M3 TQN N. Davies 156 Britannia Avenue, Dartmouth TQ6 9LQ
M3 TQP R. Messingham 2 The Lodge, Sotherington Lane, Blackmoor, Liss GU33 6DA
M3 TQU J. Bingham 31 Wyre Close, Paignton TQ4 7RU
M3 TQX A. Basterfield Red 3, Purn Holiday Park, Bridgwater Road, Weston-Super-Mare BS24 0AN
M3 TQY G. Rogers 26 Chaucer Close, Waterlooville PO7 6AQ
M3 TRC P. Lee 15 Talkin Drive, Middleton, Manchester M24 5LS
M3 TRH T. Hill 10 Parade View, Walsall WS8 7JA
M3 TRJ T. Jones 2 Holly Road, Penketh, Warrington WA5 2AG
M3 TRO S. Phillips 37 Wensley Road, Barnsley S71 1SB
M3 TRP S. Walker 26 The Warren, Hardingstone, Northampton NN4 6EW
MI3 TRR R. Elliott 183 Kilraughts Road, Ballymoney BT53 8NL
MM3 TRZ T. Reilly 21 North Street, Motherwell ML1 1LQ
M3 TSA T. Hickson 27 Cressing Road, Witham CM8 2NP
M3 TSE E. Hughes 74 Westmorland Road, Coventry CV2 5BT
M3 TSF P. Shayler 38 Maryside, Slough SL3 7ET
M3 TSG A. Ryder 4 Edgeway, Strelley, Nottingham NG8 6LY
M3 TSI P. Hewson 30 Princess Road, Kirton, Boston PE20 1JW
MW3 TSJ S. Trott 6 Mounton Drive, Chepstow NP16 5EH
M3 TSN M. Lee 46 Little Lane, Huthwaite, Sutton-in-Ashfield NG17 2RA
M3 TSO T. Owens 74 Tees Crescent, Stanley DH9 6JD
M3 TSV K. Reason 28 St. Marys Grove, Swindon SN2 1RQ
M3 TTA E. Davies 58 Popes Lane, Sturry, Canterbury CT2 0LA
M3 TTH T. Haley 3 Akeman Rise, Ramsden, Chipping Norton OX7 3BJ
M3 TTK S. Ridley 123 Lanercost Drive, Newcastle upon Tyne NE5 2DL
M3 TTS D. Brook 140 Dearne Hall Road, Barugh Green, Barnsley S75 1LX
MW3 TUB D. Butler 12, Bro Gwynfaen, Llandysul SA44 4ST
M3 TUC T. Gerrard 16 Haig Road, Carlisle CA1 3AS
M3 TUD C. Copeman 1 Chestnut Avenue, Welney, Wisbech PE14 9RG
M3 TUF J. Caswell 3 Pavillion Court, Roydon, Diss IP22 5SP
M3 TUH B. Tucker 12 Alpha Place, Appledore, Bideford EX39 1QY
M3 TUL A. Dodd 14 Davies Street, Macclesfield SK10 1GE
MW3 TUO A. Browning 11 Heather Close, Sirhowy, Gwent NP224PW
MM3 TUR P. Turner 99 Maitland Hog Lane, Edinburgh EH29 9DU
MI3 TUS W. Donnell 71 Niblock Oaks, Antrim BT41 2DP
M3 TUU G. Milsom 31 Chichester Close, Bowerdean Road, High Wycombe HP13 6AU
M3 TUW S. Tucker 11 Maple Drive, Killamarsh, Sheffield S21 1GA
MI3 TUZ A. Gault 7 Gardenmore Place, Larne BT40 1SE
M3 TVC R. Evans 7 Westland Drive, Hayes, Bromley BR2 7HE
M3 TVD D. Steele 22 Grindle Close, Thatcham RG18 3PD
M3 TVJ J. Evans 16 Longfield Place, Poulton-le-Fylde FY6 7DB
M3 TVK J. Hodgson 5 Clifton Place, Freckleton, Preston PR4 1RQ
M3 TVN T. Bunce 31 Kensington Avenue, Middlesbrough TS6 0QQ
M3 TVO N. Allen 52 Storth Lane, Kiveton Park, Sheffield S26 5QT
MM3 TVQ C. Smith 37 Glebe Road, Mosstodloch, Fochabers IV32 7JH
M3 TVV D. Render 4 Station Terrace, Allerton Bywater, Castleford WF10 2BS
M3 TVZ D. Darby 116 Middle Road, Southampton SO19 8FS
MM3 TWA I. Whiteford 54 Bilby Terrace, Irvine KA12 9DT
M3 TWB S. Bourdon 35 Main Road, Woolverstone, Ipswich IP9 1BA
MM3 TWG T. Galbraith 77 Netherwood Park, Deans, Livingston EH54 8RW
M3 TWK W. Tam Harrogate Ladies' College, Clarence Drive, Harrogate HG1 2QG
MI3 TWM C. Smallwoods 73 Knightsbridge, Londonderry BT47 6FE
M3 TWP D. Jenner 116 Trench Road, Tonbridge TN10 3HQ
MW3 TWQ P. Jones 36 Hopkin Street, Treherbert, Treorchy CF42 5HL
M3 TWS T. Stanford 27 Mill Gardens, Elmswell, Bury St. Edmunds IP30 9DQ
M3 TWV J. Benbow 20 Clifton Close, Thornton-Cleveleys FY5 4NG
MM3 TWW E. Wallace 57 Henderson Park, Windygates, Leven KY8 5DL
M3 TWY M. Sewell Flat 2-4, 6 Augusta Road, Ramsgate CT11 8JP
M3 TXA T. Ruddick Hazel Gill, Croglin, Carlisle CA4 9RR
M3 TXF H. Northcote 58 Warren Avenue, Wakefield WF2 7JN
M3 TXG D. Mardlin 13 Churchill Crescent Sonning Common, Reading RG4 9RU
M3 TXH A. Hensman 20 St. Marys Road, Braintree CM7 3JR
MI3 TXI R. Carlin 10 Top Of The Hill, Londonderry BT47 2HA
M3 TXJ R. Williams 198 Leverington Common, Leverington, Wisbech PE13 5BP
M3 TXL T. Graham Woodtown, Sampford Spiney, Yelverton PL20 6LJ
M3 TXM M. Bristow 9 Chadwick Drive, Harold Wood, Romford RM3 0ZA
M3 TXP A. Barker 37 Newbarns Road, Barrow-in-Furness LA13 9SF
M3 TXQ P. Holman 107 Eathorpe Close, Redditch B98 0HH
M3 TXR E. Smith 13 Eagle Avenue, Waterlooville PO8 9UB
M3 TXS H. Huckle 43 The Baulk, Biggleswade SG18 0PX
MI3 TXT M. Kashkoush 41 Dunamallaght Road, Ballycastle BT54 6PF
M3 TXU West Kent ARS c/o J. Fesel 3 Brook Street, Port Sunlight, Wirral CH62 5DB
M3 TXV R. Fuller 183 Nottingham Road, Alfreton DE55 7FL
MM3 TYA M. Anthoney 10 Cedar Road, Kilmarnock KA1 2HP
MW3 TYC G. Lewis Bryn Cottage, Clydach, Abergavenny NP7 0LL
M3 TYG Crawley ARC c/o M. Kyriacou 54 Sutton Avenue, Silverdale, Newcastle ST5 6TB
M3 TYI I. Morris 6 St. Nicholas Court, Gloucester GL1 2QZ
M3 TYL M. Tyler 40 Bullards Lane, Woodbridge IP12 4HE
M3 TYM T. Martin 14 Campbell Road, Eastleigh SO50 5AD
M3 TYO G. Aldridge Greenridge, Fore Street Bishopsteignton, Teignmouth TQ14 9QR
M3 TYQ D. Brown 9 Chisholm Close, Standish, Wigan WN6 0QP
M3 TYS K. Sanchez-Garci 74 Gorthorpe, Hull HU6 9EZ
M3 TYU M. Rawlings 3 Greenlands, Woolton Hill, Newbury RG20 9TB
M3 TYV D. Atkinson 133 Lingmoor Rise, Kendal LA9 7PL
M3 TYW D. Gray Flat 2 413 Weelsby Street, Grimsby DN328BJ
M3 TYX D. Mcgrath 48 Willersley Avenue, Orpington BR6 9RS
M3 TYZ D. Prince 29 St. Stephens Road, West Bromwich B71 4LR
M3 TZB D. Vincent 6 Nathan Gardens, Poole BH15 4JZ
M3 TZE R. Tse 1 Oaklands, , Gallows Lane, , Westham BN24 5AW
M3 TZF F. Wells 12 Portelet Place, Hedge End, Southampton SO30 0LZ
M3 TZI J. Cowell Mount Rivers, Bootle, Millom LA19 5XN
M3 TZN M. Robins 13 Sarum Way, Hungerford RG17 0LJ
MM3 TZP S. Wright 22 Beechwood, Linlithgow EH49 6SF
M3 TZQ G. Wright 2 Hillcrest Drive, Castleford WF103QN
M3 TZS Q. Mdlongwa 27 Faifax Avenue, Bierley, Bradford BD4 6JY
MW3 UAA H. Lee 3 Summerhill Park, Simpson Cross, Haverfordwest SA62 6EU
M3 UAE T. Baines 10 Croydon Avenue, Leigh WN7 1TP
M3 UAG D. Garland 8 Ladywell Gate, Welton, Brough HU15 1NL
M3 UAJ A. Neale 5 Millside, Wombourne, Wolverhampton WV5 8JJ
M3 UAK M. Brown 10 Terry Cooney Place, Newcastle upon Tyne NE5 2FA
M3 UAM C. Byrne 31 Graham Drive, Castleford WF10 3EY
M3 UAO M. Lewis 6 Remembrance Road, Newbury RG14 6BA
MW3 UAP G. Davies 45 Greensway, Abertysswg, Tredegar NP22 5AR
M3 UAQ A. Quinton 111 Spenser Road, Ipswich IP1 6HP
M3 UAR A. Riches 20 Western Gardens, Crowborough TN6 3EB
M3 UAV G. Preece 22 Crofters Green, Bradford BD10 8RZ
M3 UAW T. Arrow Crystalwood, Stonemans Hill, Newton Abbot TQ12 5PZ
M3 UAX D. Foyston 13 Hollinwood Road, Stoke on Trent ST71DQ
M3 UAY M. Scarr 15 Church Park, Overton, Morecambe LA3 3RA
MM3 UBB G. Goddard 17 Burnside, Flotta KW16 3NP
MM3 UBD D. Branson Derelochy, Kingsteps, Nairn IV12 5LF
M3 UBE A. Jones 26 Cloisters Avenue, Barrow-in-Furness LA13 0BA
M3 UBF J. Bonney 37 Watery Lane, Brackley NN13 7NJ
M3 UBG G. Chambers 26 Parkin Close, Cropwell Bishop, Nottingham NG12 3DG
M3 UBH C. Davis 1 Ashland Court, North Street, Crewkerne TA18 7AP
M3 UBK B. Oakley 8A Crooked Mile, Waltham Abbey EN9 1PS
M3 UBL D. Hutchinson 91, Pentland Avenue, Billingham TS23 2RF
MW3 UBQ J. Robinson 25 Stad Castellor, Cemaes Bay LL67 0NP
MM3 UBR J. Branson 129 Thornhill Road, Elgin IV30 6DX
M3 UBS D. Foord 28 Ferndale, Teversham, Cambridge CB1 9AL
M3 UBT C. Bowes 11 Burghwallis Lane, Sutton, Doncaster DN69JU
M3 UBU K. Kettle 19 St. Trinians Drive, Richmond DL10 7SS
MW3 UBY M. Davies 70 Heol Bryncwils, Sarn, Bridgend CF32 9UE
M3 UCA F. Bano 14 Norman Trollor Court, Cromer NR27 9RR
M3 UCC K. Wilton 71 Aston Clinton Road, Weston Tuville, Buckinghamshire HP22 5AB
M3 UCF T. Austin 51 Ashburnham Road, Ramsgate CT11 0BH
M3 UCH D. Francis 1905 London Road, Leigh-on-Sea SS9 2SY
MM3 UCI M. Mccallum 15 Quarry Road, Law, Carluke ML8 5HB
M3 UCJ C. Hewett 20 Cornwallis Avenue, Herne Bay CT6 6UQ
M3 UCL X. Liu Harrigate Ladies College, Clarence Drive, Harrogate HG1 2QG
M3 UCO D. Lawson 2 The Blossoms, Fulwood, Preston PR2 9RF
MI3 UCS M. Edwards 15 Highgrove Road, Carrickfergus BT38 9AG
M3 UCU S. Finch 25 Bluebell Avenue, Wigan WN68NS
M3 UCV T. Kilroy 55 Summerfield Crescent, Brimington, Chesterfield S43 1HB
M3 UCY L. Long 25 St. Matthias Road, Deepcar, Sheffield S36 2SG
M3 UCZ B. Smallbone 46 Lesters Road, Cookham, Maidenhead SL6 9LS
MW3 UDA G. Lloyd 2 Bryn Y Coed, Holywell CH8 7AU
MM3 UDB D. Brown 18 Louisa Drive, Girvan KA26 9AH
M3 UDC D. Moran 47 Radcliffe Park Road, Salford M6 7WP
M3 UDD M. Fisher 12 Abbey Mews, Pontefract WF8 1TD
M3 UDF D. Fagan 58 Main Street, Linton, Swadlincote DE12 6PZ
MM3 UDI P. Pirie Willowbank, Kirkton Of Tough, Alford AB33 8ER
M3 UDJ M. Pharoah 116 Chatsworth Street, Barrow-in-Furness LA14 5TP
M3 UDK J. Cunningham 56 Askam Avenue, Pontefract WF8 2PN
MM3 UDL B. Nielsen House Of Shannon Wester Templands, Fortrose IV10 8RA
M3 UDN Dr P. Thompson 9 Thames Mews, Poole BH15 1JY
MW3 UDO R. Oliver 6 Clevedon Avenue, Sully, Penarth CF64 5SX

MM3 UDQ J. Smith 10 High Street, Portknockie, Buckie AB56 4LD
M3 UDS W. Sun 10 Scholfield Way, , Eastbourne BN23 6HQ
M3 UDU S. Hadfield 56 Risborough Road, Bedford MK41 9QW
MM3 UDV D. Herd 4 West Fairbrae Drive, Edinburgh EH11 3SY
MW3 UDW M. Evans 9 St Teilos Close, Ebbw Vale, Blaenau Gwent NP23 6NE
M3 UDZ D. Mellor 28 Winster Road, Staveley, Chesterfield S43 3NJ
MM3 UEA E. Ewing Arisaig Priestland, Darvel KA17 0LP
M3 UED G. Henstridge 47 Churchfield Drive, Castle Cary BA7 7LB
M3 UEE J. Price 3 Perlethorpe Close, Gedling, Nottingham NG4 4GF
M3 UEF P. Marchant 16 Melrose Drive, Peterborough PE2 9DN
MW3 UEG A. Evans 9 St Teilos Close, Ebbw Vale, Blaenau Gwent NP23 6NE
M3 UEJ C. Liversidge Flat 3-8 Cromwell Terrace, Scarborough YO11 2DT
MW3 UEK N. Hosker Laburnum Cottage, Church Lane, Old Aston Hill, Ewloe CH5 3BF
M3 UEL S. Long 25 St. Matthias Road, Deepcar, Sheffield S36 2SG
M3 UEN A. Sammut The Quaker Cottage, Wainfleet Bank, Skegness PE24 4JP
M3 UEP A. Hudders 10 Waterton Grove, Wakefield WF28HR
M3 UEQ G. Lambert 17 Starcross Road, Weston-Super-Mare BS22 6NY
M3 UER P. Stone 2 Endeavour Close, Lower Stondon, Henlow SG166JR
MM3 UET M. Henderson 22 Bowmont Place, East Kilbride, Glasgow G75 8YG
MW3 UEU K. Rowney 14A Fagwr Road, Craig-Cefn-Parc, Swansea SA6 5TB
M3 UEW A. Allanson 5 Kingsley Avenue, Crofton, Wakefield WF4 1RN
M3 UEX D. Carr 36 Rookwood Mount, Leeds LS9 0LL
M3 UEY A. Holder 12A Evenlode Road, Gloucester GL4 0JT
M3 UFA S. Leggett 246 Delaware Road, Shoeburyness, Southend-on-Sea SS3 9NT
MI3 UFD R. Stirrup 211 Leckagh Drive, Magherafelt BT45 6ND
M3 UFE J. Chau Harrogate Ladies' College, Clarence Drive, Harrogate HG1 2QG
M3 UFF A. Mullen 81 Worcester St., Stourbridge DY8 1AX
M3 UFG D. Gunn 40 The Pastures, Oadby, Leicester LE2 4QD
MW3 UFH C. Livingstone-Lawn 72 Ty Mawr Avenue, Rumney, Cardiff CF3 3AG
M3 UFJ S. Jarvis 10 Wood Lane, Wolverhampton WV10 8HJ
M3 UFK N. Nash Flat 21, Farringdon House, Green Lane, Walsall WS2 8NP
M3 UFL W. Oakey11 2 Chemin Des Alleuds Moulin, Largeasse 79240 France
M3 UFT D. Redman 5 Dee Road, Lancaster LA1 2QX
M3 UFU K. Hillbeck 28 Darent Avenue, Walney, Barrow-in-Furness LA14 3NU
M3 UFW N. Mclean 21 Matlock Avenue, Wigston LE18 4NA
M3 UFX S. Blackwell 38 Peatburn Avenue, Heanor DE75 7RL
M3 UFY J. Poland 14 Malleson Road, Liverpool L139DF
M3 UFZ A. Saddington 7 Baker Court, Thrapston, Kettering NN14 4XA
M3 UGA T. Barnard 8 Argyle Road, Poulton-le-Fylde FY6 7EW
M3 UGB G. Boast 102 Abbeyfield Road, Birmingham B23 5LL
M3 UGD D. Underwood 33 Meadow Avenue, Rawmarsh, Rotherham S62 7EE
M3 UGF M. Sims 133 Canterbury Road, Hawkinge, Folkestone CT18 7BS
M3 UGH M. Cole 25 Freemans Road, Minster, Ramsgate CT12 4EL
MI3 UGI M. Downey 18 Castlevue Park, Moira, Craigavon BT67 0LN
M3 UGJ G. Douch 63 Greenaways Ebley, Stroud GL5 4UN
M3 UGK M. Rimmer 15 Brade Street, Southport PR9 8LS
MM3 UGL N. Rogers 108 Beechwood Road, Cumbernauld, Glasgow G67 2NP
M3 UGO D. Hill 19 Farren Road, Birmingham B31 5HH
M3 UGQ P. Woodyard 65 Raglan Street, Lowestoft NR32 2JS
M3 UGR C. Luckett 257 Folkestone Road, Dover CT17 9LL
M3 UGT G. Taylor Thorndale, Oakfield Drive, Boxted, Colchester CO4 5RX
M3 UGV S. Jawor 5 Cotteswold Rise, Stroud GL5 1HD
MW3 UGW G. Wilcock 42 Erskine Road, Colwyn Bay LL29 8EU
M3 UGX W. Owen 8 Sandhurst Avenue, Lytham St. Annes FY8 2DA
MD3 UGY M. Wernham Fair Isle, Lhoobs Road, Douglas IM4 3JB Isle of Man
M3 UGZ R. Dearden 218 South Street, Highfields, Doncaster DN6 7JQ
M3 UHB A. Hartley 47 Windways, Little Sutton, Ellesmere Port CH66 1JG
M3 UHC J. Bramley 17 Oakholme Rise, Worksop S81 7LJ
M3 UHG R. Smith 20A Waverley, Skelmersdale WN8 8BD
M3 UHH H. Hopkins 3 Colegrave Road, Bloxham, Banbury OX15 4NT
MI3 UHI E. Kyle 2 Wattstown Crescent, Coleraine BT52 1SP
M3 UHJ P. Hopkinson 28 Stockdove Way, Thornton-Cleveleys FY5 2AR
MM3 UHK W. Spiers 19/2 150 Charles Street, Glasgow G21 2QF
MI3 UHL W. Mark 82 Glenleslie Road, Clough, Ballymena BT44 9RH
M3 UHN P. Sherratt 39 Vimy Road, Leighton Buzzard LU7 1FQ
MW3 UHO B. Cayford 13 Harford Square, Newtown, Ebbw Vale NP23 5FA
M3 UHQ L. Richardson 127A Mount Gould Road, Plymouth PL4 7PY
M3 UHS I. Stephens 2 Boniface Walk, Burnham-on-Sea TA8 1RE
M3 UHV D. Lisi 60 Middlecroft Road, Staveley, Chesterfield S43 3XH
M3 UHW M. Troth 21 Willow Road, Bromsgrove B61 8PN
M3 UHX A. Anderson 89A Malmesbury Park Road, Bournemouth BH8 8PS
M3 UHY R. Bond 21 Coleridge Close, Bletchley, Milton Keynes MK3 5AF
MI3 UIA P. Dallas 12 Glendun Crescent, Coleraine BT52 1UJ
M3 UIB B. Wilde Seaways, Cliff Rise, Fowey PL23 1QQ
M3 UIC D. Simpson 140 Church Road, Redfield, Bristol BS5 9HN
M3 UIF M. Wheeler Homeleigh, Station Road, Rochester ME3 7RN
M3 UII S. Odonoghue 15 Chandlers Close, New Waltham, Grimsby DN36 4WH
M3 UIJ M. Rogers 41 Barton Hill Drive, Minster On Sea, Sheerness ME12 3NF
M3 UIK D. Mccrae 37 Burnside, Wigton CA7 9RE
M3 UIL T. Brice 10A Nelson Drive, Exmouth EX8 2PU
M3 UIM P. Mckeever 7 St Canices Park, Eglinton BT47 3AQ Ireland
MW3 UIN L. Collins 283 Graig Road, Godrergraig, Swansea SA9 2NZ
M3 UIP J. Dunkin 6 Kingsley Grove, Grimsby DN33 1NL
MW3 UIQ A. Henderson 45 Brynamman Road, Lower Brynamman, Ammanford SA18 1TR
M3 UIS A. Marshall 13 The Markhams, New Ollerton, Newark NG22 9QX
M3 UIU U. Ukommi 30 Oregano, Room 3, Hazel Farm, Surrey GU2 9TY
MI3 UIV V. Madden 2, Rathbeg, Limavady BT49 0AT
MI3 UIW J. Smyth 39 Whitehill Park, Limavady BT49 0QF
MM3 UIX C. Dyer 55 Duthie Road, Gourock PA19 1XS
MW3 UIY P. Henderson 45 Brynamman Road, Lower Brynamman, Ammanford SA18 1TR
M3 UJD K. Colman 10 South Rise, North Walsham NR28 0EE
M3 UJE M. Corbett 23 Heathfield Road, Fleetwood FY7 7LY
M3 UJF J. Finney 40 Tempest Avenue, Darfield, Barnsley S73 9BJ
M3 UJH J. Harbron 48 Sheridan Road, Biddick Hall, South Shields NE34 9JJ
M3 UJL J. Bailey 38 Barlow Drive, Sheffield S6 5HQ
M3 UJN C. Hall 10 First Street, Pont Bungalows, Consett DH8 6JG
M3 UJO J. Phillips Flat L 15, International House, Guildford Court, Guildford GU2 7JL
M3 UJP J. Fletcher 2 Sunflower Meadow, Irlam M44 6TD
MM3 UJQ A. Hackman The Herdsman Cottage, Brighouse Bay, Kirkcudbright DG6 4TT
M3 UJR J. Roberts 93 Earlshall Road, Eltham, London SE9 1PP
M3 UJS J. Suter 11 Summerdown Close, Durrington, Worthing BN13 3QG
M3 UJT I. Dransfield Gardener Ground House, West End, Goole DN14 8RW
M3 UJV G. Roth 121 St. Annes Road, Wolverhampton WV10 6SL
M3 UJX L. Parish 6 Courtwick Road, Wick, Littlehampton BN17 7NE
M3 UJY C. Meakin 102 Ryknield Road, Kilburn, Belper DE56 0PF
M3 UJZ J. Cooke 9 Nyetimber Crescent , Pagham, Bognor Regis PO21 3NN
M3 UKB C. Price 9 Arlington Avenue, Aston, Sheffield S26 2AA
M3 UKD Central Radio Amateur Circel c/o J. Parfrey 47 Ford Lane, Rainham RM13 7AS
M3 UKF P. Seaton Sunnymead, Newland, Barnstaple EX32 0ND
MM3 UKG D. Gilmour 35 Hailes Gardens, Edinburgh EH13 0JH
M3 UKH G. Nelson 812 Hessle Road, Hull HU4 6RD
M3 UKJ J. Boyd 14 James Street, Seaham SR7 7QW
MW3 UKK M. Davison Maes Y Neuadd Hotel, Talsarnau LL47 6YA
M3 UKN N. Hewitt Redwood House, Adbury Holt, Newbury RG20 9BW
M3 UKO N. Batchelor 15 Sufton Rise. Mordiford., Hereford HR1 4EN
M3 UKR A. Rogers 260 Griffiths Drive, Wolverhampton WV11 2JS
M3 UKU M. Qassim Winchester Road, Kings Somborne, Stockbridge SO20 6NY
M3 UKV T. Vincent 12 Lancaster Place, Kenilworth CV8 1GL
MI3 UKW M. Mcerlean 38A Culbane Road Portglenone, Portglenone BT448NZ
M3 UKX J. Cornish 2 Micklehill Drive, Shirley, Solihull B90 2PU
M3 UKY D. Gordon 32 Claremont Road, Stockport SK2 7AR
M3 ULB S. Owen 21 Market Place, Hingham, Norwich NR9 4AF
M3 ULE L. Lawrence Rookery Rise, 8 Woodside, Brede TN31 6DS
M3 ULI C. Dunn 13 Springfield Road, Leyland PR25 1AR
M3 ULL M. Bull 5 Roach Place, Rochdale OL16 2DD
M3 ULM I. Cottom 8 Bridgewater Rise, Brackley NN13 6DA
M3 ULN R. Fricker 2 Buttermere Drive, Allestree, Derby DE22 2SN
M3 ULO S. Clarke 2 Dawn Crescent, Upper Beeding, Steyning BN44 3WH
M3 ULQ J. Gromadzki 13 Merrill Heights, Maidenhall Approach, Ipswich IP2 8GA
M3 ULS J. Firth 36 Howley Grange Road, Halesowen B62 0HW
M3 ULT S. Dickinson 28 Ingleton Walk, Barnsley S70 2NE
M3 ULU A. Eizzard 41 Sycamore Drive, Waddington, Lincoln LN5 9DR
M3 ULW L. Weston 114 Morland Road, Ipswich IP3 0LZ
M3 ULX G. Quilter 83 Jameson Street, Wolverhampton WV6 0NT
M3 ULZ A. Nokes 3 Eastview, Ditcheat, Shepton Mallet BA4 6PN
M3 UMA M. Abel 121 Angela Road, Horsford, Norwich NR10 3HF
M3 UMB M. Bryans The Lodge, The Warren Croydon Road, Bromley BR2 7AL
MI3 UMC J. Mcauley 29 Ilse Court, Larne BT40 3NT
M3 UMD B. Walton 40 Princess Street, Mapplewell, Barnsley S75 6ET
M3 UMJ M. Hockin 7 Gourders Lane, Kingskerswell, Newton Abbot TQ12 5DZ
M3 UML M. Layton 16 Gwelmor, Camborne TR14 7BP
M3 UMM M. Case 6 Boldventure, Close, St. Austell PL25 3DY
MD3 UMN D. Kneale 4 Glashen Terrace, Ballasalla IM9 2ET Isle of Man
M3 UMR S. Rutt Granthorpe, Hull Road, Hull HU11 5RN
M3 UMV P. Shaves 33 Derwent Drive, Bletchley, Milton Keynes MK3 7BG
M3 UMW M. Wilkins 31 Stratton Audley Road, Fringford, Bicester OX27 8ED
M3 UMX R. Bates 61 Park View, Crowmarsh Gifford, Wallingford OX10 8BN
MM3 UMY C. Brogan 13 Mitchell Avenue, Cambuslang, Glasgow G72 7SQ
MM3 UMZ J. Bain 13 Mitchell Avenue, Cambuslang, Glasgow G72 7SQ
M3 UNB K. Puttock 12 Beechfields, School Lane, Petworth GU28 9DH
M3 UNF H. Coram 16B Park Road, Dawlish EX7 9LQ
M3 UNH A. Hitchcott 121 Oakhurst Road, Acocks Green, Birmingham B27 7PB
M3 UNI N. Bradshaw 291B Hull Road, Hull HU4 7RH
M3 UNK M. Wright 2 Regent Road, Church, Accrington BB5 4AR
M3 UNL P. Webb 104 Birds Nest Avenue, Leicester LE3 9ND
M3 UNP N. Patel 16 Camellia Court, 18 Copers Cope Road, Beckenham BR3 1NB
MM3 UNQ N. Stevenson 12 Glamis Place, Greenock PA16 7NB
M3 UNR A. Worlledge 181 Roselands Drive, Paignton TQ4 7RN
M3 UNT P. Morris 689 Tonge Moor Road, Bolton BL2 3BW
M3 UNY L. Bain 45 Larpool Crescent, Whitby YO22 4JD
MW3 UNZ S. Valentine 12 Cwrt Y Glyn, Carmel, Llanelli SA14 7SA
M3 UOB C. Bradley 6 Copeland Row, Evenwood, Bishop Auckland DL14 9PY
M3 UOC Dr J. Mohammed 2 Moffat Avenue, Ipswich IP4 3JH
M3 UOD J. Cooke 2 Church Street Close, Thurnscoe, Rotherham S63 0QT
MM3 UOE R. Wilson 12 Queen Road, Irvine KA12 0XA
M3 UOJ K. Barry 25 Delabole Road, Merstham, Redhill RH1 3PB
M3 UOK A. Abraham Flat 2, 41 Francis Road, Birmingham B33 8SL
M3 UOL L. Haworth 139 Manchester Road, Accrington BB5 2NY
M3 UOM M. Catterall 2 Cedar Road, Bishop Auckland DL14 6ET
M3 UON S. Kamal 184 Aycliffe Road, Borehamwood WD6 4EG
M3 UOO R. Bone 6 Danehurst Place, Locks Heath, Southampton SO31 6PP
MM3 UOR R. Paterson 1/R 162, Glasgow Street, Ardrossan KA228HA
MM3 UOS D. Young 81 Leaven Place, Irvine KA12 9PA
M3 UOX A. Tate 24 Brentingby Close, Melton Mowbray LE13 1ES
M3 UOY G. Fripp 35 Kiln Close, Bovey Tracey, Newton Abbot TQ13 9YL
M3 UOZ A. Knight 39 Thurnview Road, Evington, Leicester LE5 6HL
M3 UPB L. Banahan 18 Lynn Road, Ely CB6 1DA
M3 UPJ P. Harley 16 Clover Drive, Rushden NN10 0TZ
M3 UPK J. Addy 12 Wortley Avenue, Swinton, Mexborough S64 8PT
M3 UPL B. Banks 44 Manor Road, Swinton, Mexborough S64 8PY
M3 UPN P. Needham 9 Westwood, Broughton, Brigg DN20 0AU
M3 UPO D. Clarke 2 Dawn Crescent, Upper Beeding, Steyning BN44 3WH
M3 UPP L. Wilkes 24 Ilminster, Dunster Crescent, Weston-Super-Mare BS24 9EB
M3 UPQ M. Davis 3 Pollards Court, Rochford SS4 1GH
M3 UPT S. Thomas 103 Liverpool Road, Upton, Chester CH2 1BB
MW3 UPX M. Davies 11 High Street, Malltraeth, Bodorgan LL62 5AS
MM3 UPY S. Dunbar 48 Heatherpark, Seafield, Bathgate EH47 7BY
M3 UQA P. Allison Flat 11, Forest Court, 5-11 Salisbury Road, Fordingbridge SP6 1EG
M3 UQB C. Kerridge 2 Allerdale Close, Thirsk YO7 1FW
M3 UQD E. Sabbatella Riccardi 97 Harp Island Close, London NW10 0DQ
M3 UQE M. Burstow 40 Wyndham Road, Petworth GU28 0EQ
M3 UQF G. Hillbeck 28 Darent Avenue, Walney, Barrow-in-Furness LA14 3NU
M3 UQG H. Roxbrough 17 Stanwell Close, Sheffield S9 1PZ
M3 UQH J. Parrett 6 Shelley Road, East Grinstead RH19 1TA

M3 UQI G. Jones 57 Oxford Road, Banbury OX16 9AJ
M3 UQJ G. Dyson 111 Chester Road, Ellesmere Port CH65 6SB
M3 UQL D. Smith 131 Canons Walk, Thetford IP24 3PT
MM3 UQN D. Taylor 1 Mayfield Farm Cottages, Reston, Eyemouth TD14 5LG
M3 UQO G. Smith 92 Brighton Road, Banstead SM7 1BU
M3 UQP K. Eccleston 22 Lowes House, Rodney Drive, Woodley, Stockport SK6 1SL
MM3 UQT C. Page 15 Jackton View, East Kilbride, Glasgow G75 9NW
M3 UQY R. Cox 17 Hyde Lane, Upper Beeding, Steyning BN44 3WJ
M3 UQZ D. Hartley Flat 4, The Old Mill, Station Road, Ellesmere Port CH66 1NY
M3 URA M. Barber11 Homedale, St. Monicas Road, Kingswood, Tadworth KT20 6ET
M3 URD R. Dell 18 Greenacres, Fulwood, Preston PR2 7DA
M3 URE R. Eddy 15 Western Place, Penryn TR10 8HQ
M3 URF R. Freeman 9 Bramley Road, Wisbech PE13 3PA
MW3 URG R. Grayson Willow Lodge, Croeslan, Llandysul SA44 4SJ
M3 URH L. Harman 7 Loughborough Road, Walton On The Wolds, Loughborough LE12 8HT
MW3 URO D. Fullick 69 Shakespeare Road, St. Dials, Cwmbran NP44 4LW
MM3 URQ Shirehampton ARC c/o S. Urquhart 1 Hatchery Cottage, Station Road, Duns TD11 3HS
M3 URS R. Singh Stainburn House, Barrowby Lane, Harrogate HG3 1HY
M3 URT S. Mather 35 Neargates Charnock Richard, Chorley PR7 5EY
M3 URV R. Earnshaw 53 Blue Waters Drive, Paignton TQ4 6JF
M3 URW R. Woodford South Calvadnack, Carnmenellis, Redruth TR166PN
M3 URX Dr D. Craig Flat 5, The Chapel, The Plains, Totnes TQ9 5DW
M3 URZ C. Evans 1 Rialto Road, Mitcham CR4 2LT
M3 USB M. Perocevic 12 Ash Road, Crewe CW1 4DU
M3 USC S. Coleman 32 Southwell Road, Wisbech PE13 3LQ
M3 USF A. Willis 17 Ladypit Terrace, Whitehaven CA28 6AQ
M3 USH S. Hall 14 Nicholson Place, East Hanningfield, Chelmsford CM3 8UT
M3 USJ S. Hall 5 Ropery Lane, Barton-upon-Humber DN18 5TW
MW3 USK C. Burke 84 Elgam Avenue, Blaenavon, Pontypool NP4 9QU
M3 USN J. Challis 37 St. Ronans Drive, Seaton Sluice, Whitley Bay NE26 4HZ
MW3 USP S. Barwell 50 Mill Close, Caerphilly CF83 2LL
MW3 USS D. Provis Dingle Gardens, Croesbychan, Aberdare CF44 0EJ
M3 UST G. Evans 2 Tower Farm Cottages, Featherbed Lane, Hemel Hempstead HP3 0BT
MM3 USV A. Dunn 85 Neilston Road, Uplawmoor, Glasgow G78 4AG
M3 USW P. Jenkins 48 The Pantiles, Bexleyheath DA7 5HG
MW3 USX S. Tozer 110 Glanffornwg, Wildmill, Bridgend CF31 1RL
MM3 UTH P. Dower 1670 Maryhill Road, Glasgow G20 0HJ
M3 UTJ S. Crichton 27 Rosewood Ave, Stockport SK4 2DQ
M3 UTM A. Williams 1 Nimmings Close, Birmingham B31 4TA
M3 UTN K. Bailey The Firs, South Chard, Chard TA20 2RX
M3 UTP T. Pentz 30 Lindrick Way, Harrogate HG3 2SU
M3 UTQ D. Bell 18 Julius Hill, Warfield, Bracknell RG42 3UN
MM3 UTU M. Magee 30 Burnfield Drive, Mansewood, Glasgow G43 1BW
M3 UTV M. Compagno 18 Bromford Crescent, Birmingham B24 9RJ
MI3 UTY M. Regan 80 Killowen Drive, Magherafelt BT45 6DS
M3 UTZ S. Abdullah 24 The Grove, Walsall WS5 4BX
M3 UUE H. Roberts 71A Brook Street, Stourbridge DY8 3UX
M3 UUF R. Dicker 38 Inkerman Road, Southampton SO19 9DA
M3 UUG S. Lenton Badgers Rise, The Willows, Torquay TQ2 7TB
M3 UUL G. Rowe 10 Alexander Avenue, Selston, Nottingham NG16 6FW
M3 UUN T. Harding 7 Chiltern Avenue, Poulton-le-Fylde FY6 7DY
M3 UUO D. Lynch 12 Shipley Close, Blackpool FY3 7UJ
M3 UUS S. Mallinson 63 Celandine Avenue, Locks Heath, Southampton SO31 6WZ
MM3 UUT L. Thomson-Best 5 Gib Grove, Dunfermline KY11 8DH
M3 UUW J. Douch 63 Greenaways, Ebley, Stroud GL5 4UN
MW3 UUY D. Williams 10 Bronllys, Gaerwen LL60 6JN
M3 UUZ R. Chandler 26 Chalky Bank, Gravesend DA11 7NY
M3 UVA J. Grainger 32 Ellenfoot Drive, Maryport CA15 7DB
M3 UVC R. Milton 66 Hoo Marina Park, Vicarage Lane, Rochester ME3 9TG
M3 UVD D. Tregear 73 Alderley Lane, Leigh WN7 3DW
MM3 UVF W. Mcblain 50 Hamilton Crescent, Stevenston KA20 4JE
M3 UVJ J. Binfield 55 Gladstone Road, Broadstairs CT10 2HY
M3 UVK W. Penny 159 Coxford Road, Maybush, Southampton SO16 5JX
M3 UVM R. Murphy 23 Lowndes Close, Stockport SK2 6DW
M3 UVO A. Ruocco 16 Conyers Avenue, Grimsby DN33 2BY
M3 UVQ A. Barton 11 Grove Avenue, Beeston, Nottingham NG9 4ED
M3 UVS S. Thorne 72 Devonshire Road, London E16 3NJ
M3 UVT M. Garry 34 Conway Road, Paignton TQ4 5LH
M3 UVV A. Mackay 16 Vicarage Close, Chard TA20 2HH
M3 UVX D. Housden 12 Regent House, Cheltenham Gardens, Southampton SO30 2UD
M3 UWB M. Tromans 10 Crofters View, Little Wenlock, Telford TF6 5AU
M3 UWE D. Scott 7 Teal Close, Brookside, Telford TF3 1NY
M3 UWF S. Roberts Errys, Stunts Green, Herstmonceux, Hailsham BN27 4PP
M3 UWI K. Jones 41 Milton Brow, Weston-Super-Mare BS22 8DD
M3 UWJ R. Wilmot 41 Milton Brow, Weston-Super-Mare BS22 8DD
M3 UWM T. Kitto 4 Pennard Council Houses, St. Breock, Wadebridge PL27 7LL
M3 UWR S. Dawson 51 St. Edwards Road, Gosport PO12 1PW
M3 UWT A. Chapman 10 Derwent Road, Seaton Sluice, Whitley Bay NE26 4JH
M3 UWU M. Redfern 62 Meadow Road, Dudley DY1 3JU
M3 UWV C. King 21 Lowdham, Wilnecote, Tamworth B77 4LX
M3 UWW R. Massimino 115 Trelowarren Street, Camborne TR14 8AW
MU3 UWX M. Barker Rosee Terres, Les Effards Road, St. Sampson GY2 4YW Guernsey
M3 UWY T. Baddeley 44 Lowry Close, Willenhall WV13 3BD
MW3 UWZ J. Evans 2 Collfryn Cottages, Bethesda Bach, Caernarfon LL54 5SF
M3 UXC A. Spaxman 70 Park View, Shafton, Barnsley S72 8PY
M3 UXE E. Bruce 26 Queens Road, Wilbarston, Market Harborough LE16 8QJ
M3 UXF M. Bridgehouse 43 Age Croft, Oldham OL8 2HG
M3 UXG M. Leech 11 Westlake Close, Torpoint PL11 2BZ
M3 UXH K. Sowter 55 Ward Street, New Tupton, Chesterfield S42 6XR
M3 UXI L. Allanson 5 Kingsley Avenue, Crofton, Wakefield WF4 1RN
M3 UXK K. Jones 41 Tibbs Hill Road Abbott Langley, Watford WD5 0EE
M3 UXL A. Bond 21 Coleridge Close, Bletchley, Milton Keynes MK3 5AF
M3 UXM M. Bruce 26 Queens Road, Wilbarston, Market Harborough LE16 8QJ
M3 UXN P. Overton 39 Bridle Road, Madeley, Telford TF7 5HB
MM3 UXO L. Aitken 92B Belville Street, Greenock PA15 4TA
M3 UXR M. Griffiths 11 Frogwell Park, Chippenham SN14 0RB
M3 UXS C. Dutton Twillingate Farm, Tiptoe, Lymington SO41 6EJ
M3 UXU D. Almond 2 Farm Veiw, New Tupton, Derbyshire S426BD
M3 UXX M. Selvey 52 Coppice Close, Cheslyn Hay, Bella Casa, Walsall WS6 7EZ
M3 UYC D. Griffin 101 Kingsway, Duxford CB22 4QN
M3 UYF M. Heritage High View, Common Lane, Corley, Coventry CV7 8AQ
M3 UYG A. Goldsmith 61 Fengate Drove, Weeting, Brandon IP27 0PW
MW3 UYH G. Broadbent 7 James Close, Llanon SY23 5HP
MW3 UYJ J. Davies 19 Falcon Place, Blaenymaes, Swansea SA5 5NX
M3 UYK Y. Chiu Harrigate Ladies College, Clarence Drive, Harrogate HG1 2QG
M3 UYL G. Richards 1 Brisbane Road, Weymouth DT3 6RB
M3 UYO R. Dewis 6 St Nicolas Close, Pevensey BN245LB
M3 UYQ T. Hull 12 Durley Road, , Gosport PO12 4RT
M3 UYU B. Elderbrant 20 Loxley Road, Southport PR8 6NR
M3 UYV J. Rudd 5 St Andrews Close Blofield, Norwich NR13 4JX
M3 UYW P. Underhill 35 Windermere Road, Reading RG2 7HU
MW3 UYX A. Harvey 15 Pen Y Lan, Penclawdd, Swansea SA4 3LL
M3 UYY M. Dale 37 Bussey Road, Norwich NR6 6JF
M3 UZA P. Greenhalgh 33 Shepherds Lane, Chester CH2 2DH
M3 UZB S. Bailey 23 Maple Avenue, Tolladine, Worcester WR4 9RD
M3 UZE B. Monksummers 10 Breach Close, Bourton, Gillingham SP8 5BB
MW3 UZH J. Alfei 9 Brookside, Gowerton, Swansea SA4 3AY
M3 UZK A. Lancefield 19 Tawny Sedge, King's Lynn PE30 3PW
M3 UZL S. Clarke Brimham Lodge Fm, Harrogate HG3 3HE
M3 UZN R. Truelove 104 Malines Avenue, Peacehaven BN10 7RL
MW3 UZO D. White 222 St. Fagans Road, Cardiff CF5 3EW
MW3 UZP J. Young 10 Heol Fion, Gorseinon, Gorseinon SA4 4PN
M3 UZV J. Porter 18 Cornerway, Birmingham B38 9RH
M3 UZW J. Boag 60 Harebell, Amington, Tamworth B77 4NA
M3 UZZ G. Hamilton 11 The Spinney, West Lavington, Devizes SN10 4HP
M3 VAE C. Johns 6 Cranham Close Kingswood, Bristol BS15 4QB
M3 VAF K. Brown 41 Church Street, Swinton, Mexborough S64 8EF
M3 VAG B. Gittings 29 Highdown Way, Swindon SN25 4YD
MM3 VAH T. Hamilton 57/6 North Street, Bo'ness EH51 0AE
M3 VAM M. Medcalf 47 Paddock Drive, Chelmsford CM1 6UX
M3 VAQ J. Bowley 2 Cottage, Middle Battenhall Farm, Worcester WR5 2JL
M3 VAR A. Dokic 28 Tudor Gardens, Shoeburyness, Southend-on-Sea SS3 9JG
M3 VAS V. Papanikolaou 104 West Drive Gardens, Soham, Ely CB7 5EX
M3 VAT D. Fielding 2 Christchurch Road, Bradford-on-Avon BA15 1TB
MM3 VBF S. Campbell 78 Liddel Road, Cumbernauld, Glasgow G67 1JE
M3 VBG B. Garry 34 Conway Road, Paignton TQ4 5LH
M3 VBH A. Townsend 42 Grove Avenue, Yeovil BA20 2BD
M3 VBI J. Broadhurst Flat 2, Pennant Court, Rowley Regis B65 8DW
M3 VBL J. Williams 3 Corner Field, Kingsnorth, Ashford TN23 3NH
M3 VBM V. Maynard 34 Heath Farm Park, Barford St. Martin, Salisbury SP3 4BQ
M3 VBN K. Sloan Woodland Halt, Old Station Road, Winchester SO21 1BA
M3 VBP G. Bramham 1 Watson Avenue, Dewsbury WF12 8PZ
M3 VBT Ex-Military Land Rover Assoc. c/o V. Be-Dard 53 Cottingley Crescent, Leeds LS11 0HZ
M3 VBV G. Foster 18 Austen Avenue, Long Eaton, Nottingham NG10 3GG
M3 VBW B. Whall 3 Farrow Close, Great Moulton, Norwich NR15 2HR
M3 VBY D. Potter 30 Mersham Gardens, Goring-By-Sea, Worthing BN12 4TQ
M3 VBZ A. Bolton 26, St Margarets Avenue, Sutton SM3 9TT
M3 VCB C. Burbridge 9 Victoria Road, Stirchley, Birmingham B30 2LS
MI3 VCI Wacral (World Association of Christian Radio Amate c/o G. Lyttle 37 Cloyfin Park, Coleraine BT52 2BL
M3 VCK M. Brown 9 Warsop Road, Barnsley S71 3NR
M3 VCM C. Mallory 11 Baymead Meadow, North Petherton, North Petherton TA6 6QW
M3 VCO A. Knowles 260 Haunchwood Road, Nuneaton CV10 8DL
M3 VCP S. Pryke 12 Seaward Avenue, Leiston IP16 4BB
M3 VCQ S. Wilson 57 James Andrew Crescent , Greenhill, Sheffield S8 7RJ
M3 VCV L. Smith 11 Appleby Avenue, Timperley, Altrincham WA15 7HY
M3 VCW C. West 1 Willetts Mews, Hoddesdon EN11 9DX
M3 VCY G. Shakespeare 6 Waterworks Cottages, Clough Road, Hull HU6 7QB
M3 VCZ M. Rigby 75 Manchester Road, Deepcar, Sheffield S36 2QX
M3 VDA A. Crowther 11 Goodman Court, Central Drive, Chesterfield S44 5BA
M3 VDE R. Ferguson 31 Barton Court Road, New Milton BH25 6NW
M3 VDF G. Bertola 17 Caraway Drive Branston, Burton-on-Trent DE14 3FQ
M3 VDH D. Hind 116 Gowthorpe, Selby YO8 4HA
M3 VDL M. Gosling 12 The Grove, Studley B80 7QL
M3 VDN J. Owen 90 Granville Drive, Kingswinford DY6 8LW
M3 VDO D. Oliver 11 Crooked Creek Road, Rendlesham, Woodbridge IP12 2GL
M3 VDU B. Marston 111 Averil Road, Leicester LE5 2DE
M3 VDV G. Haggas 8 Retford Close, Stockton-on-Tees TS19 9EJ
M3 VDZ G. Yuill 14 Gardyn Croft, Taverham, Norwich NR8 6UZ
MW3 VEH D. Thomas 57 Brynhyfryd Street, Treorchy CF42 6DT
M3 VEJ R. Buckwell 75 Brookside Avenue, Polegate BN26 6DQ
MW3 VEL L. Wilson 48 The Woodlands, Brackla, Bridgend CF31 2JG
M3 VEM C. Vernon 80 Shirley Drive, Worthing BN14 9BB
MW3 VEN P. Hoy 39 Blackbird Road, Caldicot NP26 5RE
MI3 VEQ B. Mcconnell 14 Ballymacruise Park, Millisle, Newtownards BT22 2NW
M3 VEW A. Allanson 5 Kingsley Avenue, Crofton, Wakefield WF4 1RN
M3 VEX K. Fox 39 Felton Avenue, South Shields NE34 6RY
M3 VEY A. Selvey Bella Casa, 52 Coppice Close, Cheslyn Hay, Walsall WS6 7EZ
MW3 VFB B. Pugh Plas Newydd, 12 Fair Meadow Close, Milford Haven SA73 3TF
M3 VFC M. Derringer 19 Skylark Road, Trumpington, Cambridge CB2 9AQ
M3 VFE P. Goddard 62 Woodlands Drive, Thetford IP24 1JJ
MI3 VFF J. Dunlop 34 Keel Park, Moneyrea, Newtownards BT23 6DE
MI3 VFJ N. Orr 5 Manor Court, Donaghadee BT21 0NR
MM3 VFK J. Scally 15 Ardoch Court, Stevenston KA20 3PU
M3 VFL A. Senior 38 Haslemere, Way, Crewe CW1 4JZ
M3 VFM V. Millard 20 Droveway Gardens, St. Margarets Bay, Dover CT15 6BS
MW3 VFN E. Thomas 29 Maes Y Wern, Carway, Kidwelly SA17 4HF
M3 VFP N. Roberts 40 Armour Road, Tilehurst, Reading RG31 6HN
M3 VFS M. Toher The Chapel, Station Road, Darlington DL2 1JG
M3 VFU M. Whitehead 19 Wrose Brow Road, Shipley BD18 2NT
MI3 VFZ T. Currie 26 High Street, Portaferry, Newtownards BT22 1QT

M3 VGF A. Lewington 6 Brookhill Road, Darton, Barnsley S75 5EL
M3 VGH G. Hunter 5 Charlton Grove, Silsden, Keighley BD20 0QG
MW3 VGJ A. Lloyd 4 Gladstone Terrace, Miskin, Mountain Ash CF45 3BS
M3 VGK B. Lunn 204A Main Street, Horsley Woodhouse, Ilkeston DE7 6AX
M3 VGP P. Temple 136 Roborough Close, Bransholme, North Yorkshire HU7 4RP
M3 VGT Viscount A. Andover Bishoper Farmhouse, Brokenborough, Malmesbury SN16 9SR
M3 VGX M. Price 52 Newmarket Street, Norwich NR2 2DW
M3 VGZ D. Adshead 16 Moat Way, Swavesey, Cambridge CB24 4TR
M3 VHA R. Parker 29 Hill Lea Gardens, Cheddar BS27 3JH
M3 VHB S. Allen 27 Cottons Meadow, Kingstone, Hereford HR2 9EW
M3 VHC W. Ho Harrogate Ladies' College, Clarence Drive, Harrogate HG1 2QG
M3 VHE J. Heinonen Riittiontie 155, Vampula 32610 Finland
M3 VHH K. Foster 10 Bleaswood Road Oxenholme, Kendal LA9 7EY
M3 VHI R. Silcox 103 Oakdale Road, Downend, Bristol BS16 6EG
MM3 VHM V. Moran 31 Hermitage Crescent, Coatbridge ML5 4NE
M3 VHO F. Ho Clarence House, Harrogate Ladies College, North Yorkshire HG1 2QG
M3 VHQ J. Winson 35 Newington Avenue, Southend-on-Sea SS2 4RD
M3 VHU J. Redfearn 3 Taylor Hill Road, Huddersfield HD4 6HN
M3 VHV L. Kelly 9 Ham Lane, Farrington Gurney, Bristol BS39 6TW
MI3 VHW T. Boyd 40 Walnut Park, Larne BT40 2WF
M3 VHY A. Blake 53 Valley Road, Middlesbrough TS4 2RY
M3 VHZ S. Southern 37 Conway Road, Calcot, Reading RG31 4XP
M3 VIA N. Purkiss 357 Fair Oak Road, Eastleigh SO50 8AA
M3 VIB M. Michalak 100 Nursery Lane, Northampton NN2 7TJ
M3 VIG D. Potter 1 Wentworth Road, Rugby CV22 6BG
M3 VII A. Price Barn Owl Roost, Astwith, Chesterfield S45 8AN
M3 VIJ D. Williams 18 Lower Greave Road, Meltham, Holmfirth HD9 4DY
M3 VIO C. Skinner Beeston Marina Ltd, 1A The Quay, Beeston Marina, Riverside Road, Nottingham NG9 1NA
M3 VIR S. Kim 9 Magness Road, Deal CT14 9JF
MM3 VIS J. Dowson 19 Tweed Crescent, Wishaw ML2 8QR
M3 VIU J. Bettles 2 Ellfield Close, Bristol BS13 8EF
MM3 VIV V. Smith 10 Quilco, Dounby, Orkney KW17 2HW
M3 VIW C. Wall 26 Wallace Lane, Whelley, Wigan WN1 3XT
M3 VJC J. Cooper Gulean. Cross Common, the Lizard TR12 7PE
M3 VJE J. Edmunds 17 Stephens Road, Liskeard PL143SX
M3 VJI D. Mcevoy 33 Heathcote Drive, Hasland, Chesterfield S41 0BB
M3 VJJ V. Bowkett 9 Gwealmayowe Park, Helston TR13 0PE
MW3 VJL V. Lea Rose Cottage, Y Ffor, Pwllheli LL53 6UR
M3 VJM J. Ball 26 Verona Court, Yeo Vale Road, Barnstaple EX32 7EN
MW3 VJN T. Thomas Emlyn House Cawdor Terrace, Newcastle Emlyn SA38 9AS
M3 VJO J. Sawyer 27 Croft Road, Wallingford OX10 0HN
M3 VJS V. Sansom 70 Valley Road, West Bridgford, Nottingham NG2 6HQ
M3 VJW J. Whittington Meadowbank, Bridgerule, Holsworthy EX22 7EN
M3 VJX P. Buckley 2 Edward Lake Drive, Hoo, Rochester ME3 9UY
MW3 VKA A. Vincent 88 Lake Street, Ferndale CF43 4HE
M3 VKB V. Britton Badgers Hollow, Witt Road, Salisbury SP5 1PL
M3 VKF R. Mottershead 10 St. Mary Close, Blackpool FY3 7UB
M3 VKJ K. Jones Railway Crossing Cottage, Ash Road, Sandwich CT13 9JB
MW3 VKM R. Morgan 14 Woodland Road, Pontllanfraith, Blackwood NP12 2LS
M3 VKN I. Astley 6 Shay Court, Crofton, , Wakefield WF4 1SL
MM3 VKO S. Macdonald 366 Millcroft Road Cumbernauld, Glasgow G67 2QW
MM3 VKP C. Allan Ardroy, Kinclaven Road, Perth PH1 4EY
M3 VKS D. Vickers 178 Bakewell Road, Matlock DE4 3BA
M3 VKT E. Kottis Guildford Court Reception, University Campus, Guildford GU2 7JL
M3 VLG C. Periam 5 Elliott Walk, Preston PR1 7TP
M3 VLH P. Harlow 92 Eton Road, Burton-on-Trent DE14 2SW
M3 VLI W. Toher The Chapel, Station Road, Darlington DL2 1JG
MW3 VLJ L. Jones Ty'R Ysgol, Holland Street, Ebbw Vale NP23 6HT
M3 VLL L. Burbidge 33 Burcote Fields, Towcester NN12 6TH
M3 VLN V. Grimmer 48 Bingham Avenue, Sutton-in-Ashfield NG17 3AR
M3 VLO P. Todd 93 Derwent Drive, Tibshelf, Alfreton DE55 5LT
M3 VLT Oldham Am Rad C c/o P. Honey 3 Peterswood, Harlow CM18 7RJ
M3 VLX S. Mintram 25 Blunden Drive, Slough SL3 8WG
M3 VMA M. Collins 8 Pictor Grove, Buxton SK17 7TQ
MD3 VMD P. Birchall 7 Richmond Close, Douglas IM2 6HR Isle of Man
M3 VME C. Frost 6 Link Way, Arborfield Cross, Reading RG2 9PD
MD3 VMN V. Matthewman Monte Rosa, 7 Ballaughton Close, Isle of Man IM2 1JE
M3 VMQ J. Farrer 37 Priory Grove, Ditton, Aylesford ME20 6BB
M3 VMU J. Easterbrook 2 Warden Road, Eastchurch, Sheerness ME12 4EJ
M3 VMV R. Vale 611 College Road, Birmingham B44 0AY
MW3 VMY T. Peters 74 Maes Y Capel, Pembrey, Burry Port SA16 0EG
M3 VNG J. Hardy Lambda House, Seanor Lane, Chesterfield S45 8DH
M3 VNH N. Halford 20 Albany Avenue, Manchester M11 1HQ
M3 VNI J. Preece 25 Broadmead Catford, London SE6 3TG
M3 VNL C. Lockyear 26 Wentworth Gardens, Exeter EX4 1NH
M3 VNM A. Crawford 39 Fownhope Close, Redditch B98 0LA
M3 VNN V. Nikolaidis 35-46 Ernst Chain Road, Manor Park, Guildford GU2 7YW
M3 VNO D. Harwood 36 Seaview Drive, Great Wakering, Southend-on-Sea SS3 0BE
M3 VNP G. Simcock 11 Bannatyne Close, Manchester M40 3TD
M3 VNQ Raynet Pembrokeshire c/o T. Mcbride 53 Blackdown Grove, St. Helens WA9 2BD
MW3 VNR C. Mukans 2 Ffordd Cottages, Johnstown, Carmarthen SA33 5BL
M3 VNS Dr S. Sampathkumar 52 Crowstone Road, Westcliff-on-Sea SS0 8BD
MM3 VNT M. Robertson 1A Church Street, Lochgelly KY5 9JS
MM3 VNU R. Robb 8 Morven View, Tarland, Aboyne AB34 4UH
MW3 VNV M. Williams 25 Clos Sant Paul, Llanlli SA15 1HR
MM3 VNW A. Sim 44 Hillmoss, Kilmaurs, Kilmarnock KA3 2RS
M3 VNX L. Wolfe 90 Alderney Road, Erith DA8 2JD
MW3 VNZ R. Blackmore 96 Vachell Road, Cardiff CF5 4HJ
M3 VOA Dr A. Sartorius Holmwood, Priory Road, Sunningdale, Sunningdale, Ascot SL5 9RH
M3 VOB P. Sheargold 7 Mendip Close, Rough Hills, Wolverhampton WV2 2HF
M3 VOI R. Reeves 4 Elmwood Avenue, Waterlooville PO7 7LG
M3 VOJ A. Kreissl 382 Oldfield Road, Altrincham WA14 4QT
M3 VOL J. Garner 2 Coniston Grove, Haresfinch, St. Helens WA11 9NH
M3 VON Essex CW ARC c/o A. Ellerington 4 Wathcote Close, Richmond DL10 7DX
M3 VOR I. Mitchell 4 Walhouse Drive, Penkridge, Stafford ST19 5SP
M3 VOU M. Carr 25 Malvern Avenue, Fareham PO14 1QF
M3 VOW N. Pettefar 44 Duck Lane, Laverstock, Salisbury SP1 1PU
M3 VOY P. Williams 45 Blackgate Lane, Tarleton, Preston PR4 6US
M3 VOZ J. Morris 4 Pleasant Terrace, Lincoln LN5 8DA
M3 VPA A. Penfold 179 Byron Rd, Thornhill, Southampton SO19 6FB
M3 VPB T. Horsoo 5 Kelmarsh Court, Great Holm, Milton Keynes MK8 9EN
M3 VPD J. Bridges 2 Bridgeman Court, Sir John Moore Avenue, Hythe CT21 5EB
M3 VPH P. Hanson 34 20Th Avenue, Hull HU6 9JH
M3 VPJ P. Allen 21 Chase Vale, Burntwood WS7 3GD
MM3 VPK C. Hebenton 43 East Avenue, Uddingston, Glasgow G71 6LG
M3 VPN M. Thornton 72 Northfield Lane, Wickersley, Rotherham S66 2JA
MI3 VPO D. Smith 164 Ballygowan Road, Hillsborough BT26 6EG
M3 VPP L. Reeves 38 Eagle View, Aston, Sheffield S26 2GL
M3 VPQ S. Woods 30 Kenilworth Drive, Earby, Barnoldswick BB18 6NA
MW3 VPS M. Watts 17 The Stables, High Fawr Avenue, Oswestry SY11 1TG
MW3 VPT Apau CG c/o G. Taylorr 17, Llygad Yr Haul, Glynneath SA11 5RL
M3 VPX P. Dodds 22 Wheatear Lane, Ingleby Barwick, Stockton-on-Tees TS17 0TB
M3 VPY P. Mcdonough 91 Lever Street, Little Lever, Bolton BL3 1BA
M3 VQC A. Roberts 23 Seaton Avenue Houghton-Le-Spring, Sunderland DH5 8EQ
M3 VQD K. Morgan 43 Kenilworth Drive, Earby, Barnoldswick BB18 6NA
M3 VQF A. White 38 Hillcrest Road, Yeovil BA21 4RA
M3 VQG S. Walcot Chapel House, Hemington BA3 5XU
MI3 VQH J. Kane 5 Woodlawn Court, Carrickfergus BT38 8DP
M3 VQI J. Watts 6 Elm Court, Newhaven BN9 9NR
MW3 VQJ G. Ellis 3 Cae Bach, Talybont, Bangor LL57 3YJ
M3 VQL T. Morton 43 Hob Moor Drive, York YO24 4JU
M3 VQN B. Roberts 102 Brougham Road, Marsden, Huddersfield HD7 6BJ
M3 VQP A. Majoch 66 Boughton Green Road, Northampton NN2 7SP
M3 VQQ C. Unwin Mansells Farm, Mansells Lane, Hitchin SG4 8TJ
M3 VQS B. Jamieson 14 Ridgeway, Ashington NE63 9TJ
M3 VQV D. Page 12 Blacksmiths Meadow, Coads Green, Launceston PL15 7FF
M3 VQZ N. Long 25 Blendworth Lane, Southampton SO18 5GY
M3 VRA V. Tomlinson 180 Kendal Drive, Castleford WF10 3QZ
M3 VRD R. Elias Roganann, Dol Y Bont, Borth SY24 5LX
MM3 VRI M. Mccallum 15 Quarry Road, Law, Carluke ML8 5HB
M3 VRL Dr S. Sethuraman 9 Bramcote Close, Aylesbury HP20 1QE
M3 VRM M. Veal 2 Bernards Close, Christchurch BH23 2EH
M3 VRN R. Noake 44 Loxton Square, Bristol BS14 9SF
M3 VRP S. Broadhurst Flat 8 Pennant Court, Ross Heights, Rowley Regis B65 8DW
M3 VRU T. Baxendale 3 Greylands Close, Sale M33 6GS
M3 VRV M. Hemstock 6 Hucknall Crescent, Gedling, Nottingham NG4 4HZ
MM3 VRX N. Thomson Four Winds, Holland Bush Hightae, Lockerbie DG11 1JL
M3 VRY J. Docherty 208 Thornton Close, Newton Aycliffe DL5 7NP
M3 VSB K. Dunne 1 Burton Gardens, Brierfield, Nelson BB9 5DR
MM3 VSC S. Clark 75 Treeswoodhead Road, Kilmarnock KA1 4PB
M3 VSF B. Nicholls 5 Golden Miller Close, Newmarket CB8 7RT
MW3 VSG D. Riley-Kydd 35 Pensyflog, Porthmadog LL49 9LB
M3 VSH A. Freedman 276 Newchurch Road, Stacksteads, Bacup OL13 0TA
M3 VSL V. Cronin 4 Carnarvon Road, Reading RG1 5SD
M3 VSO R. Langmuir 24 Briar Road, Bexley DA5 2HN
M3 VSQ M. Wilde 18 Ledston Luck Cottages Kippax Leeds Ls257Bx, Leeds LS257BX
M3 VST F. Mcdermott 6 Bruce St., Swindon SN2 2EL
MM3 VSU S. Rodwell Bourtree, Kennethmont, Huntly AB54 4NN
M3 VSW S. Whall 17 Vicarage Road, Deopham, Deopham NR18 9DR
MM3 VSX L. Forbes Woodside, The Muirs, Huntly AB54 4GD
M3 VSZ L. Schofield 23 The Mount, Wrenthorpe, Wakefield WF2 0NZ
M3 VTA C. Duffield 32 Mount Close, Honiton EX14 1QZ
M3 VTH A. Ball 18 Seaview Crescent, Ostend Road, , Walcott NR120NL
MI3 VTJ T. Dorrian 29 Sperrin Road, Limavady BT49 0AS
MW3 VTK J. Martin 78 Llwyn Ynn, Talybont LL43 2AG
M3 VTL P. Wilson 2 Mary Rose Close, Cheslyn Hay, Walsall WS6 7BE
M3 VTN N. Mayall 10 South Rise, North Walsham NR28 0EE
M3 VTO S. Tanner 30 Christina Park, Totnes TQ9 5UR
M3 VTP M. Tuffs Izzyinn 87, Foxhills Road, Scunthorpe DN158LL
M3 VTQ E. Macgurk 10 Elmore Road, Lee on Solent PO139DU
M3 VTS P. Wilkes 8 Cloverdale, Stafford ST17 4QJ
M3 VTV A. Bedford 1 Carder Crescent, Bilston WV14 0JT
M3 VTX P. Hind 116 Gowthorpe, Selby YO8 4HA
M3 VUA J. Hunt 101 Kinoulton Court, Grantham NG31 7XR
M3 VUC N. Lay 32 School Road, Billericay, Essex CM129LH
M3 VUH S. Colman 22 Shearwater Way, Stowmarket IP14 5UG
M3 VUI S. Hefford 5 Fenners, Worle BS22 7DR
MW3 VUJ R. Williams 34 Maendu Terrace, Brecon LD3 9HH
M3 VUK B. Lewis 68 Irwin Avenue, Rednal, Birmingham B45 8QU
M3 VUN G. Diggins 7 Minterne Road, Bournemouth BH9 3EH
M3 VUO G. Twigg 4 Crossway, Widnes WA8 8SQ
M3 VUP A. Lowe 65 North Road, Clowne, Chesterfield S43 4PG
M3 VUQ G. Grimshaw 1 Hardy Close, Pinner HA5 1NL
M3 VUS V. Ban Christs College, Cambridge CB2 3BU
M3 VUV P. Larner Flat1A 53, West Street, Horncastle LN9 5JE
M3 VUX T. Loker 24 St. Albans Hill, Hemel Hempstead HP3 9NG
M3 VUY S. Heard The Annexe, 2 Church Walk, , Bideford EX39 2BP
M3 VUZ V. Ball 24 Carr Lane, Warsop, Mansfield NG20 0BN
M3 VVA K. Jones 4 Hawthorne Road, Castle Bromwich, Birmingham B36 0HH
M3 VVB L. Cunningham 96 Kingsleigh Drive, Castle Bromwich, Birmingham B36 9DY
M3 VVH G. Spencer 7 Squadron Close, Castle Vale, Birmingham B35 7PF
MM3 VVI J. Mason 27 Niddrie Marischal Gardens, Edinburgh EH16 4LX
MW3 VVJ W. Jones Bryn Golau, Mynytho, Pwllheli LL53 7RL
MW3 VVO S. Barry 1 Pearson Cottages, St. Brides, Haverfordwest SA62 3BN
M3 VVQ N. Mcdougall 15 Answell Avenue, Manchester M8 4GG
MM3 VVS I. Lindsay 265 Stirling Street, Denny FK6 6QJ
M3 VVW S. Smith 12 Stoneleigh Avenue, Hordle SO41 0GS
M3 VVY M. Soper 16 Queen Elizabeth Drive, Crediton EX17 2EJ
MM3 VVZ F. Coombes 44 Lochfield Road, Paisley PA2 7RL
M3 VWD G. Clarke 11 Blackfordby Lane, Moira, Swadlincote DE12 6EX
MW3 VWE E. Jones Afon Lodge Caravan Park, Parciau Bach, Carmarthen SA33 4LG

M3 VWF T. Fentiman 64 St. Nicholas Road, Faversham ME13 7PD
M3 VWG P. Brown 4 Heather'S Edge, Heather Lane, Hathersage, S32 1Dt, Hathersage S32 1DT
MI3 VWH R. Hunter 5 Castle Rise, Tandragee, Craigavon BT62 2NE
M3 VWJ P. Kavanagh 83 Imperial Avenue, Southampton SO15 8PT
M3 VWK M. Poole 15 Roberts Place, Dorchester DT1 2JJ
MW3 VWO D. Best 8 Carno Street, Rhymney, Tredegar NP22 5EA
M3 VWP C. Johnson 42 Reap Lane, Portland DT5 2JX
M3 VWR A. Bartlett 62 Kewstoke Road, Bath BA2 5PU
M3 VWW S. Hegarty 31 Beaconsfield Road, Deal CT14 7DA
M3 VWY N. Jewitt 10 Gorse Lane, Oadby, Leicester LE2 4RQ
M3 VXB B. Walker 18 Seals Green, Kings Norton, Birmingham B38 9UW
M3 VXC D. Stinson 1 The Croft, Earls Colne, Colchester CO6 2NH
M3 VXG G. Kavanagh 83 Imperial Avenue, Southampton SO15 8PT
M3 VXH S. Liles 62 Southwood Drive, Surbiton, Surbiton KT5 9PH
MI3 VXI S. Henry 105 Ramsey Park, Macosquin, Coleraine BT51 4NG
M3 VXK J. Diplock 8 Lodge Road, Messing, Colchester CO5 9TU
MM3 VXL Worcs Mnbounce c/o J. May 12 Clochbar Gardens, Milngavie, Glasgow G62 7JP
M3 VXM Bolton Raynet G c/o J. Stephenson 54 Elizabeth Road, Haydock, St. Helens WA11 0PP
M3 VXN M. Parkinson 4 Meadow View, Sherburn In Elmet, Leeds LS25 6BY
M3 VXO O. Carpenter-Beale 3 Linkden Cottages, Lomas Lane, Sandhurst TN18 5PU
MM3 VXP P. Mckay Bailur 1 Calligary, Ardvasar IV45 8RU
M3 VXX T. Quiney 20 Britannia Gardens, Stourport-on-Severn DY13 9NZ
M3 VXY S. Hill 1 Meadow Crescent, Wesham, Preston PR4 3BB
MM3 VYA A. Rodgers 13 Mill Street, Caldercruix, Airdrie ML6 7QB
M3 VYB Christleton High School A.R.C c/o P. Anstis 112 West Street, Hartland, Bideford EX39 6BQ
M3 VYD K. Lowcock 43 Larch Street, Nelson BB9 9RH
M3 VYE H. Ewer 89 Bridle Close, Enfield EN3 6EB
M3 VYF M. Ward 93 Sandsfield Lane, Gainsborough DN21 1BQ
M3 VYK L. Shallcross 6 Wimbrick Close, Wirral CH46 9RY
M3 VYM L. Clark 16 Kibblewhite Crescent, Twyford, Reading RG10 9AX
M3 VYN V. Roberts 17 Houldsworth Crescent, Coventry CV6 4HL
MM3 VYR I. Findlay 2 Bothwell Road, Uddingston, Glasgow G71 7ET
M3 VYS M. Clifford 32 Tiverton Way, Chessington KT92QS
M3 VYT S. Ferguson 12 Summerfields, Dalston, Carlisle CA5 7NW
MM3 VYU W. Young 26 Needle Green, Carluke ML8 4AF
M3 VYV J. Arblaster 22 Wood Lane, Carlton, Barnsley S71 3JJ
MM3 VYY T. Mason 11 St Serf Road, Glenrothes KY74EA
M3 VZC C. Mcnulty 91 Barn Hey Crescent, Meols CH47 9RW
M3 VZH S. Prichard 4 Morecambe Road, Scale Hall, Lancaster LA1 5JA
MM3 VZI D. O'Kane 0/1 51 Girvan Street, Glasgow G33 2DP
M3 VZL S. Haycock 51 South Crescent, Southend-on-Sea SS2 6TB
M3 VZP S. Arpino 24 Broad Lane, Rochdale OL16 4PG
M3 VZQ D. Riddick 36 Shadygrove Road, Carlisle CA2 7LD
M3 VZS D. Clewer 45 Ashfield Road, Andover SP10 3PE
M3 VZU A. Murphy 9 Harlton Close, Peterborough PE2 8LW
M3 VZV S. Thompson 64 Church Road, Fordham, Colchester CO6 3NJ
M3 VZZ C. Halls 2 Cock Fen Road, Lakesend, Wisbech PE14 9QE
M3 WAC W. Clarke 41 Upton Road, Atherton, Manchester M46 9RQ
M3 WAF W. Morgan Little Sandyhurst House, 186 Sandyhurst Lane, Ashford TN25 4NX
MW3 WAL S. Patchett Ty Ucha Farm, Nantyr, Llangollen LL20 7DD
M3 WAP A. Cosic 35 Betteridge Drive, Sutton Coldfield B76 1FN
M3 WAV A. Swain 1 George Street, Brimington, Chesterfield S43 1HG
M3 WAY P. Jones 22 Blair Road, Trowbridge BA14 9JZ
M3 WBA P. Allin 8 Kiln Close, Dove Holes SK178FQ
MD3 WBC J. Wernham Fair Isle, Lhoobs Road, Douglas IM4 3JB Isle of Man
M3 WBI W. Whitcher 17 Watermead, Stratton St. Margaret, Swindon SN3 4WE
M3 WBJ B. Williams 2 Pokas Cottages, Chelveston, Wellingborough NN9 6AL
M3 WBK W. Knapp 32 Turner Close, Shoeburyness, Southend-on-Sea SS3 9TL
MI3 WBL B. Lockhart 5 Lisnalee Park, Mountnorris, Armagh BT60 2UP
M3 WBM M. Leake 20 Witchampton Road, Broadstone BH18 8HZ
M3 WBN R. Cranston Hyrton House, Middle Street, Lincoln LN1 2RG
M3 WBQ W. Argyle 62 Yew Tree Drive, Leicester LE3 6PL
M3 WBR H. Lowthian West Brownrigg, Penrith CA11 9PF
M3 WBS S. Bridgeman 32 Brooklands, Brinkworth, Chippenham SN15 5BA
M3 WBT B. Weston 10 Clement Drive, Peterborough PE2 9RQ
MI3 WBU S. Rantin 8A Buchanans Road, Newry BT35 6NS
MM3 WBV S. Verth 23 The Quilts, Leith, Edinburgh EH65RY
M3 WCA W. Alexander 53 Woodlands Drive, Stanmore HA7 3PB
MW3 WCE J. Thomas-Jones 10 Greenwood Avenue, Gwersyllt, Wrexham LL11 4EB
M3 WCI A. Hand 150 Curtin Drive, Moxley, Wednesbury WS10 8RN
M3 WCM W. Cornish 21 Centaur Street, Portsmouth PO2 7HB
M3 WCO T. Purcell 28 Millberg Road, Seaford BN25 3ST
M3 WCQ J. Winter Flat 23, Knightlow Lodge Knightlow Avenue, Coventry CV3 3HH
M3 WCR C. Wilson 234 Aylsham Drive Ickenham, Uxbridge UB10 8UF
M3 WCX C. Whall 52 Spitfire Road, Upper Cambourne, Cambridge CB23 6FN
M3 WCZ M. Beckett 59 Broadacre, Caton, Lancaster LA2 9NH
M3 WDB C. Jones 34 Cadle Road, Wolverhampton WV10 9SJ
M3 WDC W. Carless 39 Harrison Road, Cannock WS11 0AQ
M3 WDH Dr W. Henderson 14 Highfield Road, Newcastle upon Tyne NE5 5HS
MI3 WDI D. Wiggins 12 Vauxhall Park, Belfast BT9 5GZ
M3 WDK P. Shaw 45 Wood End Road, Wolverhampton WV11 1NW
M3 WDN E. Temple 32 Lower Barresdale, Alnwick NE66 1DW
MI3 WDO D. Dallas 28 Kemp Park, Ballycastle BT54 6LE
MM3 WDT W. Thom 1 Bennan, Mossdale, Castle Douglas DG7 2NG
M3 WDU A. Stanmore 198 Heathfield Road, Southport PR8 3HE
M3 WDV D. Golding Windrush Cottage 84-85 Bradenstoke, Chippenham SN15 4EL
M3 WDY W. Duffield 26 Mount Close, Honiton EX14 1QZ
M3 WDZ W. Davies 17 Oakdale Avenue, Harrogate HG1 2JN
M3 WEA T. Bradley 1 Park Close, North Weald, Epping CM16 6BP
M3 WEF P. Brown 4 Parsonage Drive, St. Helens WA9 4ZW
MM3 WEI E. Ireland The Steading, Blairmains, Shotts ML7 5TJ
M3 WEJ J. Varley 54 Richmond Park Road, Kingston upon Thames KT2 6AH
M3 WEQ A. Schuler 6 Tatham Court, Taunton TA1 5QZ
MI3 WES W. Spence 8 Kilmahamogue Road, Moyarget, Ballycastle BT54 6JH
MM3 WEV G. Weir 95 White Street, Whitburn, Bathgate EH47 0BH
M3 WEZ L. Wayman Oak Tree Lodge, Redbridge Road, Dorchester DT2 8BG
M3 WFB D. Read 9 Meadow Road, Albrighton, Wolverhampton WV7 3DZ
M3 WFC J. Paradas 1A Brocks Ghyll, Eastbourne BN20 9RQ
M3 WFE A. Kitney 3 Wordsworth Close, Torquay TQ2 6EA
MW3 WFF R. Jones 5 Heol Llwyn Gollen, Merthyr Tydfil CF48 1LR
M3 WFG W. Griffiths 68 Altcar Lane, Formby, Liverpool L37 6AY
MW3 WFH W. Harries 18 Bro Teify, Alltyblacca, Llanybydder SA40 9SR
MD3 WFJ Maritime Volunteer Service RC c/o J. Hill 54 Wybourn Drive, Onchan IM3 4AT Isle of Man
M3 WFK P. Ashton 14 Poppy Close, Boston PE21 7TJ
M3 WFL M. Rimmer 7 Brookdale, Southport PR8 3UA
M3 WFO A. Jamieson-Colville 77 Salters Way, Dunstable LU6 1UG
MI3 WFT C. Mooney 12 Curragh Walk, Derry City BT48 8HX
MM3 WFU D. Green 1 The Square, Tomintoul, Ballindalloch AB37 9ET
M3 WFY C. Nelson 14 Windy Harbour Road, Southport PR8 3DU
M3 WGB K. Moore Flat 8, Lindis Court, Boston PE21 8SX
M3 WGI S. Sugihara Southfield, Park Lane, Wokingham RG40 4PY
M3 WGK L. Mobley 2 Boxhedge Road West, Banbury OX16 0BS
M3 WGM M. Brown 27 Greenfield Close, Dunstable LU6 1TS
M3 WGO I. Paterson 11 Ocho Rios Mews, Eastbourne BN23 5UB
M3 WGV V. Wright Beech Cottage, Baron Wood, Carlisle CA4 9TP
MM3 WGW G. Wallace 29 Dunlop Street, Stewarton, Kilmarnock KA3 5AT
M3 WGY D. Wilson 75 Gainsborough Road, Scotter, Gainsborough DN21 3RU
M3 WGZ G. Wells 22 Mill Road, Deal CT14 9AA
M3 WHA A. Ballinger 9 Somerville Court, Cirencester GL7 1TG
M3 WHB B. Balchin 301 New Hall Lane, Preston PR1 5XE
M3 WHG T. Tunstell 23 Swallow Crescent, Innsworth, Gloucester GL3 1BL
M3 WHH J. Cook 20 Huntingdon Close, Totton, Southampton SO40 3NX
M3 WHL D. Dawson 11 Aukland Grove, St. Helens WA9 5LR
M3 WHQ G. Woods 8 Wareham Road, Lytchett Matravers, Poole BH16 6DP
M3 WHR D. Williams Flat 3 The Old Council House, Market Street, Atherstone CV9 1ET
MM3 WHS D. Mclean 72 Bowfield Crescent, Glasgow G52 4HJ
M3 WHV S. Everson 41 Westminster Lane, Newport PO30 5ZF
M3 WHX H. Smith 6 Tynemouth Place, North Shields NE30 4BJ
M3 WHY G. Cahill 81 Albemarle Road, Willesborough, Ashford TN24 0HJ
M3 WIA W. Armes 11 Rutland Road, Broadheath, Altrincham WA14 4HW
M3 WIC R. Ashwick 98 Woodbury Avenue, East Grinstead RH19 3UX
M3 WID J. Free Flat 6, 60 Wyncroft Road, Widnes WA8 8QE
M3 WIJ H. Parkinson 61 Cinnamon Lane, Fearnhead, Warrington WA2 0AG
M3 WIT J. Withers 16 Tamworth Close, Etherley Dene, Bishop Auckland DL14 0RN
MW3 WIV L. Elston 11 Woodland Walk, Blaina, Abertillery NP13 3JS
M3 WIX M. Wilson 12 Gorsey Lane, Great Wyrley, Walsall WS6 6JA
M3 WJA A. Whitelam 107 Welholme Road, Grimsby DN32 0NQ
MM3 WJD D. Wishart Curcum, Swannay, Orkney KW17 2NS
M3 WJK J. Knowles 10 Grove Hill, Hessle HU13 0RT
M3 WJN J. Gorman 1 Patterdale Road, Ashton-In-Makerfield, Wigan WN4 0EF
MI3 WJO W. Jordan 24 Spelga Place, Newtownards BT23 4ND
MW3 WJP P. Watson Eirianfa Cwmduad, Carmarthen SA33 6XJ
M3 WJU J. Mowlam 46 Walpole Street, Weymouth DT4 7HQ
M3 WJV C. Cherry 12 Scarisbrick New Road, Southport PR8 6PY
M3 WJY A. Margaswamy 59 Grants Yard, Station Road, Burton-on-Trent DE14 1BW
MM3 WJZ I. Cogle 56 Dryburgh Avenue, Rutherglen, Glasgow G73 3EU
M3 WKC A. Campbell Gate House, The Bog, Shrewsbury SY5 0NG
MM3 WKF D. Goodwin Little Dens, Stuartfield, Peterhead AB42 5DG
M3 WKK R. Horgan 74 Inglewhite Road, Longridge, Preston PR3 2NA
M3 WKL C. Winch 2 Cranleigh Gardens, Cowes PO31 8AS
M3 WKM B. Howard 5 St Nicholas Street, Dereham NR19 2BS
M3 WKV K. Dale 26 Warwick Place, Langdon Hills, Basildon SS16 6DU
M3 WKZ B. Drury 6 Ellen Grove, Harrogate HG1 4RH
MM3 WLA M. Dickeson 44 Mossmill Park, Mosstodloch, Fochabers IV32 7JY
MW3 WLB L. Pearce 31 High Street, Abertridwr, Caerphilly CF83 4DD
M3 WLD W. Douglas 1 Sleetbeck Road, Roadhead, Carlisle CA6 6PA
M3 WLG D. Flynn Alberi, Manor Road, Chichester PO20 0SF
M3 WLL W. Garnett Starfish Cottage, Higher Clovelly, Bideford EX39 5ST
M3 WLO M. Anderson 38 Shellard Road, Filton, Bristol BS34 7LU
MW3 WLS D. Elias 31 Banc Y Gors, Upper Tumble, Llanelli SA14 6BR
M3 WLU J. Hepburn 32 Green Croft, Ashington NE63 8EF
M3 WLV D. Burden 16 Milnthorpe Lane, Wakefield WF2 7DE
MI3 WLW L. Elliott 19 Gosford Road, Collone, Armagh BT60 1LQ
M3 WLX M. Gooch Flat 16, Townfield Court, 32 Horsham Road, Dorking RH4 2JE
M3 WLY R. Readman 1 Millside Close, Kilham, Driffield YO25 4SF
M3 WMC W. Carr 18 Whiteway Close, Bristol BS5 7QZ
MW3 WMI N. Mitchell 37 Brookside, Glan Y Mor, Fairbourne LL38 2BX
MI3 WMK M. Mckeen 27 Old Grange, Carrickfergus BT38 7HQ
M3 WMO R. Laughlin 7 Catherine Hunt Way, Colchester CO2 9HN
MW3 WMP W. Phillips Annfield, Penrhyndeudraeth LL48 6LS
MM3 WMQ N. Davidson 25 Hopetoun Court, Bucksburn, Aberdeen AB21 9QS
M3 WMS W. Stone 79 Woodlands Road, Allestree, Derby DE22 2HH
M3 WMU S. Breese 11 Balham Grove, Birmingham B44 0NF
M3 WMV Tri County ARC c/o D. Messenger 18 Glebelands, Harlow CM20 2PA
M3 WNC S. Willmott Emborough Grove, Radstock BA34SF
MM3 WNH A. Maitland 6 Thorn Avenue, Coylton, Ayr KA6 6NL
M3 WNI R. Karpinski 55 Cambridge Avenue, New Malden KT3 4LD
M3 WNM N. Edwards 15 Penderry Rise, Catford, London SE6 1EZ
MM3 WNP S. Murray 25 Braefoot, Girdle Toll, Irvine KA11 1BY
M3 WNT T. Corker North Side, Wingerworth Hall Estate, Chesterfield S42 6PL
M3 WNV J. Giffard 5 Hazelwood Road, Oxted RH8 0JA
M3 WNZ J. Strawbridge 36 St. Dunstans Road, Salcombe TQ8 8AN
M3 WOC C. Bellis Cliffe Bungalow, Barnsley Road, Barnsley S72 9JX
M3 WOD D. Wood 27 St. Mildreds Avenue, Ramsgate CT11 0HT
M3 WOI R. Burlong 20 Jockey Mead, Horsham RH12 1LF
M3 WOK B. Burden 18 Challenor Close, Finchampstead, Wokingham RG40 4UJ
M3 WOL P. Bickley Smithy Cottage, Old Post Office Road, Bury St. Edmunds IP29 5RD
M3 WOQ K. Meyer 42 Sandcross Lane, Reigate RH2 8EL
M3 WOS W. Barnes Cushendall, Lyngate Road, North Walsham NR28 0DH
MW3 WOV H. Lyall 29 King Alfreds Road Sedbury, Chepstow NP16 7AQ
M3 WOW C. Constable 29 Sheppeys, Haywards Heath RH16 4NP

M3 WOX J. Ward 64 Laxey Road, Blackburn BB2 3LQ
M3 WOY C. Dunstan 67 Knights Way, Mount Ambrose, Redruth TR15 2BN
M3 WPC R. Brett 3 Rectory Close, Chingford, London E4 8BG
MW3 WPH S. Harrison 2 Hendre, Newtown, Ebbw Vale NP23 5FE
M3 WPI O. Prin 19 The Colliers, Heybridge Basin, Maldon CM9 4SE
M3 WPK H. Burch 46 School Lane, Horton Kirby, Dartford DA4 9DQ
M3 WPM M. Almeida 20 Gresley Court, Grantham NG31 7RH
MW3 WPN R. Blackett 10 Acton Gardens, Wrexham LL12 8DD
M3 WPO P. Woolley 84 Bowthorpe Road, Norwich NR2 3TP
M3 WPP D. Rayner 42 Chapelgate, Sutton St. James, Spalding PE12 0EE
M3 WPS W. Snowden 5 Eastfield Road, Wisbech PE13 3EJ
M3 WPU M. Mchugh 51 Rutland Street, Hyde SK14 4SY
M3 WPV M. Hardy 66 Exeter Road, Doncaster DN2 4LF
M3 WPW W. Whyatt 686 Whitchurch Lane, Whitchurch, Bristol BS14 0EJ
M3 WQA M. Herbert 31 Mayfield Avenue, New Haw, Addlestone KT15 3AQ
MI3 WQC S. Dillon 2 Otter Park, Strathfoyle, Londonderry BT47 6YU
MW3 WQE P. Pritchard 1A Pant Hirgoed, Pencoed, Bridgend CF35 6YD
M3 WQF D. Robinson 1 Common Piece, Swinefleet, Goole DN14 8DE
M3 WQG D. Pearson 49 Longford Road, Twickenham TW2 6EB
M3 WQI K. Squires 10 Markham Avenue, Armthorpe, Doncaster DN3 2AZ
M3 WQL H. Short 71 Lilac Crescent, Burnopfield, Newcastle upon Tyne NE16 6QF
M3 WQN J. Gordon 17 Cateran Way Collingwood Grange, Cramlington NE23 6EX
MM3 WQO E. Hughes 12 Cults Drive, Tomintoul, Ballindalloch AB37 9HW
MI3 WQT A. Mcbride 2 Glenbrook Cottage, Lugan, Craigavon BT66 8QT
MW3 WQV E. Morgan Holly Cottage, Old Racecourse, Oswestry SY10 7PQ
M3 WQX J. Neal 75 Park Lane, Castle Donington, Derby DE74 2JG
M3 WRA J. Turner 35 Horncastle Road, Wragby, Market Rasen LN8 5RB
M3 WRB R. Wheatley 46 Victory Road, Steeple Claydon, Buckingham MK18 2NY
MW3 WRH W. Hucker 14 Greenway Court, Barry CF63 2FE
M3 WRJ R. Waghorne 5 Freelands Drive, Church Crookham, Fleet GU52 0TE
M3 WRK S. Sarwar Redfern 57A, Uni Of Warwick, Coventry CV47AL
M3 WRL C. Lemin 44 Barton Road, Berrow, Burnham-on-Sea TA8 2LT
M3 WRM R. Dadge 14 North Roskear Village, Camborne TR14 0AS
M3 WRN J. Hills 67 Thornham Road, New Milton BH25 5AE
M3 WRO O. Woods 8 Fairway Close, Croydon CR0 7SH
M3 WRQ C. Smith 224 Hither Green Lane, London SE13 6RT
M3 WRS S. Webber 59 Mincinglake Road, Stoke Hill, Exeter EX4 7DY
M3 WRZ L. Coyne 75 Newtown Road, Worcester WR5 1HH
MW3 WSC G. Owen 8 Masshyfrmd, Garndolbenmarn, Gwymedd LL51 9SX
M3 WSE J. Cullen 22 Longlands Road, Beeston, Nottingham NG9 1LR
M3 WSH S. Holmes 11 Holford Rise, Bremilham Road, Malmesbury SN16 0EA
M3 WSI S. Warren 9 Warning Tongue Lane, Doncaster DN4 6TB
M3 WSJ J. Woodroof 37 Danefield Road, Northampton NN3 2LT
M3 WSO R. Pitman 10 Somerville Way, Bridgwater TA6 5SA
M3 WSQ L. Simpson 101 Toftwood Road, Sheffield S10 1SL
M3 WSR B. Harrison 43A Rumbridge Street, Totton, Southampton SO40 9DR
M3 WSS L. Shand 52 Ten Acre Way, Rainham, Gillingham ME8 8TL
M3 WSV Dr T. Kyriacou 54 Sutton Avenue, Silverdale, Newcastle-under-Lyme ST5 6TB
M3 WSW S. Ngai Harrogate Ladies' College, Clarence Drive, Harrogate HG1 2QG
M3 WTA D. Whitton Sea View, Baycliff, Ulverston LA12 9RL
M3 WTB B. Walden 59 Brook View Drive, Keyworth, Nottingham NG12 5RA
M3 WTC W. Caine 53 Cromford Road, Crich, Matlock DE4 5DJ
M3 WTD R. Burrow 1 Tinshill Crescent , Cookridge, Leeds LS16 7AS
M3 WTG D. Cooper Little Heath, Bradfield Common, North Walsham NR28 0QR
M3 WTN B. Watkin 48 Peel Park Crescent, Little Hulton, Manchester M38 0BU
M3 WTO S. Liu Harrogate Ladies' College, Clarence Drive, Harrogate HG1 2QG
M3 WTP E. Lowe 21 Sherwood Avenue, Creswell, Worksop S80 4DL
MW3 WTR W. Randall 3 Penygraig, Aberystwyth SY23 2JA
MI3 WTT A. Mcdonnell 52 Moira Road, Glenavy, Crumlin BT29 4JL
M3 WTU J. Townsend 124 Rough Common Road, Rough Common, Canterbury CT2 9BU
M3 WTY R. Clare Kimberley, Boston Road, Bicker, Boston PE20 3AP
M3 WUA A. Smith 20 South Terrace, Northampton NN1 5JY
M3 WUE D. Tidswell 1 Cherrytree Grove, Spalding PE11 2NA
M3 WUG J. Stainton 40-41 Dyke End Golcar, Huddersfield HD7 4LA
M3 WUH J. Middleton 16 Kyme Road, Boston PE21 8NQ
MM3 WUI P. Bingham 129 Livingstone Terrace, Irvine KA12 9ER
M3 WUJ J. Garner 30 Pendula Road, Wisbech PE13 3RR
M3 WUK W. Norwood Flat 5, 20 Upperton Gardens, Eastbourne BN21 2AH
M3 WUM R. Miles Haseley Lodge, Birmingham Road, Warwick CV35 7HF
M3 WUN I. Mcmahon 8 Thackeray Close, Liverpool L8 8NE
M3 WUO I. Taylor Flat 65, Kemsley, Lewisham Park, London SE13 6QW
MM3 WUP W. Steele 35 Devlin Court, Whins Of Milton, Stirling FK7 0NP
M3 WUQ I. Petropouleas 16 Amfissis Street, Holargos, Athens 155 62 Greece
M3 WUS M. Blenkinsop 23 Pilmoor Drive, Richmond DL10 5BJ
M3 WUV F. Harwood 1, South Highall Cottage, Lincolnshire LN10 6UR
M3 WUW M. Noakes 26 Box Lane, Pontefract WF8 2JW
M3 WUX K. Bailey 58 Billy Buns Lane, Wombourne, Wolverhampton WV5 9BP
M3 WVB P. George 4 Mandelbrote Drive, Littlemore, Oxford OX4 4XG
M3 WVC S. Howard 11 Jolly Gardeners Court, Norwich NR3 3HD
M3 WVD M. Bradshaw 118 Queens Road, Vicars Cross, Chester CH3 5HE
M3 WVF G. Bradshaw 118 Queens Road, Vicars Cross, Chester CH3 5HE
M3 WVG K. Pain 200 Manor Road, Mitcham CR4 1JF
M3 WVI J. Hanley 5 Timline Green, Bracknell RG12 2QP
M3 WVJ R. Brown 8 Eliot Walk, Kidderminster DY10 3XP
MI3 WVL J. Mooney 12 Curragh Walk, Londonderry BT48 8HX
MM3 WVN S. Clark 5B Ladykirk Road, Prestwick KA9 1JW
M3 WVO D. Willey 17 Bridge Place, Saxilby, Lincoln LN1 2QA
MM3 WVP R. Hutton 1 Grianairigh, Northton, Isle of Harris HS3 3JA
MM3 WVQ R. Robinson 12 Hannahston Avenue, Drongan, Ayr KA6 7AU
M3 WVT A. Bradshaw 118 Queens Road, Vicars Cross, Chester CH3 5HE
M3 WVX L. Gaynor 225 Watson Court, Stadium Way, Watford WD18 0FA
M3 WVY E. Pinvisase 225 Watson Court, Stadium Way, Watford WD18 0FA
M3 WWD L. Hornby 4 Shakespeare Road, Prestwich, Manchester M25 9GW
M3 WWH R. Fraser 7 Hawthorne Avenue, Fleetwood FY7 7PY
MI3 WWJ E. Simpson 10 Woodview Park, Tandragee, Craigavon BT62 2DD
MM3 WWM R. Jowett Fearnoch Ardentallen, Oban PA34 4SF
M3 WWN W. Northover 13 Dagenham Avenue, Dagenham RM9 6LD
MW3 WWO H. Golaszewski 16 Wingate Drive, Llanishen, Cardiff CF14 5LR
MM3 WWP C. Mccosh 5 Bridgend Gardens, Windygates, Leven KY8 5BP
MM3 WWQ A. Patrick 2/L 38Glasgow Street, Millport KA28 0DL
M3 WWR W. Witham 4 King George Road, Colchester CO2 7PE
M3 WWU G. Armitage Windmill Cottage, Greens Gardens, Nottingham NG2 4QD
MM3 WWV R. Kennedy 45 Rodney Road, Gourock PA19 1XG
M3 WWY G. Miller Silvermine, Cooks Lane, Axminster EX13 5SQ
M3 WWZ R. Alexander 118 Pepper Lane, Standish, Wigan WN6 0PW
M3 WXB B. Williamson 114 Radburn Road, New Rossington, Doncaster DN11 0SH
M3 WXD R. Fenn 78 Sapphire Road, Bishops Cleeve, Cheltenham GL52 7YU
M3 WXG G. Lewis 7 The Barns Church Aston, Newport TF10 9JJ
M3 WXH C. Renouf 27 Ashburton Road, Croydon CR0 6AP
M3 WXI C. Bossons 31 Hanbridge Avenue, Newcastle ST5 8HH
MW3 WXN B. Chandler Flat 11, Westwood Court, Stanwell Road, Penarth CF64 2EZ
M3 WXP D. Brookes 13 Princess Street, Woodlands, Doncaster DN6 7LX
MD3 WXS C. Ashworth Ravenscourt Lodge, Peel Road, Douglas IM1 5EQ Isle of Man
M3 WXU D. Grundy 25 Albert Street, Bignall End, Stoke-on-Trent ST7 8QB
M3 WXW C. Glitsun 152 St. Awdrys Road, Barking IG11 7QE
M3 WXX J. Child 12 Beachill Road, Havercroft, Wakefield WF4 2EJ
M3 WYA K. Willoughby 11 Hardistry Drive, Pontefract WF8 4BU
M3 WYG P. Engledow 62 Purland Road, Norwich NR7 9DZ
MM3 WYI Rugby Am Tra So c/o N. Stewart 220 Grieve Road, Greenock PA16 7AL
M3 WYJ B. Gale Barn End Burdon Lane, Highampton EX21 5LT
MM3 WYM M. Stewart 15 Fancy Farm Road, Greenock PA16 7LH
M3 WYQ P. Holton 66 Mill Road, Gillingham ME7 1JB
M3 WYR C. Smith 199A Richardshaw Lane, Stanningley, Pudsey LS28 6AA
M3 WYT R. White 2 Chambers Manor Cottages, Epping Upland, Epping CM16 6PJ
M3 WYV E. Scott 31 South Croft, Upper Denby, Huddersfield HD8 8UA
M3 WYZ C. Pearman Wicken Cottage Mill Hill, Edenbridge TN8 5DB
M3 WZF D. Walker 8 Wescoe Avenue, Great Houghton, Barnsley S72 0DW
M3 WZG W. Power 111 Woodlands Road, Ditton, Aylesford, Maidstone ME206EF
MM3 WZH S. Armstrong 85 Blantyre Court, Erskine PA8 6BP
M3 WZJ R. Davies 43 Woodfield Road, Holt NR25 6TX
MM3 WZL J. Scott 5 Barrwood Gate, Galston KA4 8NA
M3 WZN J. Mitchell 27 Tanager Close, Norwich NR3 3QD
M3 WZP M. Williams 59 Guilford Avenue, Dover CT16 3NG
M3 WZR P. Fry 76 Mount Pleasant Road, New Malden KT3 3LB
M3 WZS S. Thorne 2 Ellfield Close, Bristol BS13 8EF
M3 WZT M. Fulbrook 2 Cob Place, Westbury BA13 3GS
M3 WZV Fleetwood Radio Enthusiasts Group c/o S. Norman 44 Martival, Leicester LE5 0PH
M3 WZY J. Dunn 9 Wakefield Road, Stoke-on-Trent ST4 5PU
MW3 WZZ S. Bobby 56 Ffordd Offa, Rhosllanerchrugog, Wrexham LL14 2EY
M3 XAC A. Curry 30 Hillside Road, Norton, Stockton-on-Tees TS20 1JG
MM3 XAF A. Ferries Cairnbeathie, Lumphanan AB31 4QA
M3 XAG C. Tame 28 Tyrrells Way, Sutton Courtenay, Abingdon OX14 4DF
M3 XAH A. Holmes 614 City Road, Manor, Sheffield S2 1GH
M3 XAI H. Parfitt 5 Sheridan Road, Frimley, Camberley GU16 7DU
MW3 XAJ A. Morgan 46 Greensway, Abertysswg, Tredegar NP22 5AR
M3 XAK M. Gaunt 12 Glastonbury Abbey, Bedford MK41 0TX
M3 XAM A. Morgan 18 Keysworth Drive, Wareham BH20 7BD
M3 XAN A. Brooks 52 Houldsworth Drive, Chesterfield S41 0BS
M3 XAO A. Hughes 80C Royle Green Road, Manchester M22 4WB
M3 XAR M. Roberts 15 Pineside Avenue, Cannock Wood, Rugeley WS15 4RG
M3 XAS J. Saxon 134 Sherwood Drive, Wigan WN5 9RS
M3 XAU E. Landon 24 Larchwood Close, Sale M33 5RP
M3 XAV R. Hartle 7 Boggard Lane, Charlesworth, Glossop SK13 5HL
M3 XAY A. Yorkston 26 Hamilton Road, London NW10 1PA
M3 XBC B. Chamberlain 10 Scott Road, Bishop's Stortford CM23 3QH
MW3 XBE S. Best 38 Greensway Abertysswg, Rhymney, Tredegar NP22 5AR
M3 XBF M. Fisher 25 Tennyson Road, Diss IP22 4PY
M3 XBH B. Harrison 24 Alderton Road, Nottingham NG5 6DX
M3 XBL L. Rabone 6 Cranwell Grove, Kesgrave, Ipswich IP5 2YN
M3 XBN O. Popa 25 Wells Park Road, London SE26 6JQ
M3 XBO J. Kaby 3 Kexby Mill Close, North Hykeham, Lincoln LN6 9TB
M3 XBS O. Rogers 22 Robson Drive, Hoo, Rochester ME3 9EA
M3 XBT B. Totterdell Moscar Cross House, Hollow Meadows, , Sheffield S6 6GL
M3 XBZ M. Stead 38 Park Road, Bracknell RG12 2LU
M3 XCA A. Clarke 14 Tower Court, Haverhill CB9 9DD
M3 XCB J. Gardiner 146 Durley Drive, Prenton CH43 3BB
M3 XCF R. Cockayne 20 The Shrubbery, Rugeley WS15 1JJ
M3 XCH C. Howard 62 Bourne Valley Road, Poole BH12 1DU
M3 XCJ J. Fergusson 200 Tang Hall Lane, York YO10 3RA
M3 XCN C. Norton 34 The Grove, Little Aston, Sutton Coldfield B74 3UD
M3 XCP C. Parker The Grange, Watercombe Cornwood, Nr. Ivybridge PL21 9RB
M3 XCS B. Porter 124 Stella Nova, Washington Parade, Bootle L20 4TE
M3 XCU J. Callis 51 Pipistrelle Way, Oadby, Leicester LE2 4QA
M3 XCV M. Abberley 10 Cranesbill Close, Featherstone, Wolverhampton WV10 7TY
M3 XCX D. Michael 33 Garfield Close, Lincoln LN1 3QL
MW3 XDB D. Barnett 49 Parcyrhun, Ammanford SA18 3HD
M3 XDD D. Baseden Butt 29 Shearwater Way, Stowmarket IP14 5UG
M3 XDH D. Harris 23 Shearer Road, Portsmouth PO1 5LL
M3 XDI J. Peain 29 Wild Flower Way, Ditchingham, Bungay NR35 2SF
MM3 XDP D. Paterson 42 Third Avenue, Alexandria G83 9BJ
M3 XDQ A. Green 37 Fisher Close, Worsley Mesnes, Wigan WN3 5UT
MW3 XDT R. Snape Bodlondeb, North Road, Whitland SA34 0AX
M3 XDV J. Hall11 9 Stone Court, South Hiendley, Barnsley S72 9DL
MM3 XDW D. Woods 39 Northfield, Tranent EH33 1HU
M3 XDZ P. Neal 14 Hilltop Close, Desborough, Kettering NN14 2LQ
M3 XEA K. Bindley 56 Iona Close, Beaumont Leys, Leicester LE4 0QY
MI3 XEB R. Throne 12 Mason Road, Magheramason, Londonderry BT47 2RY
M3 XEF M. Goodwin Bramble Cottage, Well Hill Lane, Orpington BR6 7QJ
M3 XEG B. Pearce 4 Mary Chapman Close, Norwich NR7 0UD
M3 XEI W. Walther 57 A Lower Court Road, Epsom KT198SW
M3 XEJ E. Ellison 5 Darwin Terrace, Darwin Street, Shrewsbury SY3 8QQ
M3 XEL M. Morris 8 Millfield, Lambourn, Hungerford RG17 8YQ
M3 XEN J. Fautley 71 Pullman Lane, Godalming GU7 1YB
M3 XEO B. Hay Riverside Road, Great Yarmouth NR31 6PZ

M3 XEQ P. Whalan The Old Post Office, South Street, Faversham ME13 9NR
M3 XEX N. Farrow 30 Highdown, Southwick, Brighton BN42 4QS
MI3 XEY C. Cartin 64 Ashgrove Park, Magherafelt BT45 6DN
M3 XFA F. Alexis 44 Osborne Road, Enfield EN3 7RW
M3 XFB J. Emery 63 Warren Road, Orpington BR6 6JF
M3 XFC M. Calvert 8 Brixham Drive, Wigston LE18 1BH
M3 XFD N. Field 9 Shepherds Fold Drive, Winsford CW7 2UE
M3 XFH M. Ashton Lodge Farm Bungalow, Wattisham Road, Ipswich IP7 7LU
M3 XFI M. Radford 3 Cockshott Drive, Armley, Leeds LS12 2RL
MM3 XFM B. Burrows 27 Bughtknowes Drive, Bathgate EH48 4DP
M3 XFN K. Cross 31 Parkfields, Abram, Wigan WN2 5XR
MM3 XFP C. Lee-Marr 65 Barry Road, Carnoustie DD7 7QQ
M3 XFS J. Sewell 56 Victoria Court, Luddesdown Road, Swindon SN5 8HL
M3 XFT B. Dixon 21 Pankhurst Road, Hoo, Rochester ME3 9DF
MM3 XFX D. Sandilands Cuil Moss Cottage, Ardgour, Fort William PH33 7AB
M3 XFZ A. Coop 47 Amy Street, Rochdale OL12 7NJ
M3 XGA G. White 89 Kings Drive, Thingwall, Wirral CH61 9QA
M3 XGB I. Bennett 4 Frances Close, Wivenhoe, Colchester CO7 9RP
M3 XGC K. Emlay 195 Barrington St., Manchester M11 4FB
M3 XGD J. Challinor 69 Brimrod Lane, Rochdale OL11 4QF
M3 XGI E. Brook 30 Pitchstone Court, Farnley, Leeds LS12 5SZ
M3 XGL G. Clarke 28 Mayfield Way, Mendlesham, Stowmarket IP14 5SH
MM3 XGP G. Kelly 1 Bankfoot, Prestonpans EH32 9SG
MI3 XGR J. Doherty 62 Coolessan Walk, Limavady BT49 9EN
MM3 XGS G. Suttie 9B Pentland Crescent, Dundee DD2 2BU
M3 XGU C. Frizzell 85 Gibbon Road, Newhaven BN9 9ER
M3 XGV Rutherford Appleton Laboratory ARC c/o M. Wills 23 Moat Avenue, Coventry CV3 6BT
M3 XGW G. Whall 10 Hillcrest Court, Ipswich Road, Diss IP21 4YJ
M3 XGY S. Kay 476 North Drive, Thornton-Cleveleys FY5 2HX
M3 XGZ J. Murray 2 The Cuttings Hampstead Norreys, Thatcham RG18 0RR
M3 XHB J. Kelly 66 Denison Road, Feltham TW13 4QG
M3 XHC J. Akinin 70 Valley Road, West Bridgford, Nottingham NG2 6HQ
M3 XHG Royal Air-Force Air Cadets/Air Training Corps c/o M. Bolton 11 Silvia Way, Fleetwood FY7 7JF
M3 XHH S. Kiley 178 Kingfisher Drive, Woodley, Reading RG5 3LQ
M3 XHK P. Goodall 61 Turf Hill Road, Rochdale OL16 4XG
MW3 XHL J. Edwards 19 Bryntirion, Henllan, Denbigh LL16 5YL
M3 XHM P. Aitken 25 Clunbury Road, Northfield, Birmingham B31 3SY
M3 XHN S. Hutchinson 17 Monsom Lane, Repton, Derby DE65 6FX
M3 XHQ D. Jewitt 26 Sands Lane, Barmston, Driffield YO25 8PG
M3 XHT M. Shipham 1 The Farmhouse, Farmhouse Lane, Hemel Hempstead HP2 7AR
M3 XHU C. Hall 28 Tidebrook Place, Stoke-on-Trent ST6 6XF
M3 XHW H. Wright 168 Spinney Hill Road, Northampton NN3 6DN
M3 XHY B. Smith 7 Kestrel Avenue, Bransholme, Hull HU7 4ST
M3 XHZ M. Tickner 5 Tunnmeade, Ifield, Crawley RH11 0QR
MM3 XIA 93contest Group c/o I. Anderson Cantyhaugh, Ogscastle, Carnwath ML11 8NE
M3 XID J. Roberts 51 Bradfield Road, Broxtowe, Nottingham NG8 6GP
M3 XIE D. Robinson David Robinson 19 Meadow Lane Crossheath, Newcastle ST5 9AJ
M3 XIF J. Williams 41 Overton Lane, Hammerwich, Burntwood WS7 0LQ
M3 XIG S. Abberley 10 Cranesbill Close, Featherstone, Wolverhampton WV10 7TY
M3 XIH A. Williams 41 Overton Lane, Hammerwich, Burntwood WS7 0LQ
MW3 XIJ A. Pritchard 20 St. Malo Road, Cardiff CF14 4HN
M3 XIK J. Lambert Hurley 64 Henry Road, West Bridgford, Nottingham NG2 7ND
M3 XIL S. Hoy 114 Sheppey Beach Villas, Manor Way, Sheerness ME12 4QY
M3 XIM D. Hackling 20 Millers Lane, Norwich NR3 3LU
M3 XIO M. Stevenson 127 Walton Road, Chesterfield S40 3BX
M3 XIP A. Pomfrey-Jones 46 Hampton Road, Erdington, Birmingham B23 7JJ
MI3 XIU M. Sinton 65 Henderson Drive, Bangor BT19 1NP
M3 XIV R. Treherne 58 Cherry Orchard, Tewkesbury GL20 8PJ
MM3 XIW T. Hunter 1A Glen Avenue, Largs KA30 8RQ
M3 XIY B. Kerry 1, Churchway, Diss IP22 1RN
MM3 XIZ I. Hepworth Bronte Cottage, Inverugie, Peterhead AB42 3DN
MM3 XJA J. Arthur West Lodge, Murdoustoun, North & South Road, Motherwell ML1 5LB
M3 XJE D. Holdsworth 3 Briardale Road, Bradford BD9 6PU
M3 XJF J. Ferrington The Redwoods, 20 Innings Lane, Bracknell RG42 3TR
M3 XJG J. Gaskin Badgers Barn, Canterbury Road, Folkestone CT18 8DF
M3 XJH A. Hylton 3 Jubilee Cottages, Tring Road, Dunstable LU6 2JU
M3 XJK A. Weaver 77 East Acres, Widdrington, Morpeth NE61 5NT
M3 XJL J. Landless 2 Aspen Way, Banstead SM7 1LE
M3 XJM L. Faik Saarstr. 50, Dienheim 55276 Germany
M3 XJO J. Boids 2 Crown Street, Hoyland, Barnsley S74 9HS
M3 XJP J. Plows 187 Whitebeam Road, Birmingham B37 7PA
M3 XJQ R. Barter 8 Orchard Close, Newton Abbot TQ12 3DF
M3 XJW J. Wood Pear Tree Cottage, Agden, Whitchurch SY13 3UA
M3 XJX A. Reay 12 Victoria Avenue, South Hylton, Sunderland SR4 0QZ
M3 XJZ J. Gabriel 60 Goodwin Road, Ramsgate CT11 0JJ
M3 XKF N. Tideswell 19 Wish Court, Ingram Crescent West, Hove BN3 5NY
MM3 XKH K. Hail 70 Nobleston Estate, Alexandria G83 9DB
M3 XKI L. Taute 4 Mendelssohn Grove, Browns Wood, Milton Keynes MK7 8DH
M3 XKJ J. Lewis Millbrook, Church Street, Market Drayton TF9 2TF
MW3 XKL Hornsea ARC c/o R. Jones Flat 2, Tan Y Geraint, 33 Princess Street, Llangollen LL20 8RD
M3 XKM K. Mountford 7 Flaxman Close, Barlaston, Stoke-on-Trent ST12 9BD
M3 XKN L. Griffiths 90 Keats Road, Wolverhampton WV10 8NB
M3 XKO P. Goodridge 22 Horefield, Porton, Salisbury SP4 0LE
M3 XKP I. Pass 69 Cotswold Road, Bath BA2 2DL
M3 XKY G. Leverton 24 Saxton Avenue, Bradford BD6 3SW
M3 XLB P. Bailey Scald End Farm, Mill Road, Thurleigh, Bedford MK44 2DP
M3 XLC G. Auld 6 Pheabens Field, Bramley, Tadley RG26 5BX
M3 XLJ L. Justin Garth, Park View Road, Pinner HA5 3YF
M3 XLK R. Crerar 60 Gloucester Drive, London N4 2LN
M3 XLM L. Matthewman 2 St. Margaret Road, Ludlow SY8 1XN
MM3 XLO J. Nattress 44 Broadlands, Carnoustie DD7 6JY
MM3 XLQ S. Rennie 27 Whiting Road, Wemyss Bay PA18 6EB
MW3 XLR J. Percival Blue Cedars, Gresford, Wrexham LL12 8RN
M3 XLS L. Turner 16 Woodland Place, Scarborough YO12 6EP
M3 XLW K. Weston 44 Shelley Road, Wellingborough NN8 3DB
M3 XMA Dr M. Ali 4 The Crescent, Great Horkesley, Colchester CO6 4EH
M3 XMB M. Brittain 159 Caledonia Road, Wolverhampton WV2 1JA
MW3 XME M. Elmer 3 Maes Dolfor, Llanfairfechan LL33 0RP
MW3 XMG P. Provis Dingle Gardens, Croesbychan, Aberdare CF44 0EJ
M3 XMH M. Higham 6 Holbeck Avenue, Morecambe LA4 6NP
M3 XMJ University of Surrey Ears c/o S. Spicer 34 Hillcrest Avenue, Halesowen B63 2PR
M3 XMK M. Rose 19 Hawthorn Street, Peterlee SR8 3LY
M3 XMO A. Reed 94 Moor Lane, Loughborough LE11 1BA
M3 XMP G. Prasad 112 Eastlands Road, Rugby CV21 3RR
M3 XMQ D. White 10 Meaux Road, Wawne, Hull HU7 5XD
M3 XMS Dr F. Gavins 85 Dance Square, London EC1V 3AJ
MM3 XMT A. Stevenson Starwood Croft, Craigellachie, Aberlour AB38 9SQ
M3 XMU G. Hunter 58 Repps Road, Martham, Great Yarmouth NR29 4QT
MW3 XMY D. Worthington 22 Rhys Avenue, Kinmel Bay, Rhyl LL18 5NS
M3 XMZ Rushyhill RS c/o R. Honeybourne Flat 40, Napier Court West, Southend-on-Sea SS1 1NH
M3 XNA H. Ball Manor House, Tolgus Hill, Redruth TR15 1AX
M3 XNB N. Newby 22 Acton Road, Liverpool L32 0TT
M3 XNC G. Reywer 1 Tiverton Close, Houghton le Spring DH4 4XR
M3 XNE A. Warr 2 Fairfield Road, Bournheath, Bromsgrove B61 9JN
M3 XNK N. Price 22 Hanover Road, Warley, Rowley Regis B65 9DZ
M3 XNM I. Stevenson 79 Lunedale Road, Darlington DL3 9AT
M3 XNN D. Elliott 54 Grisedale Gardens, Gateshead NE9 6NP
M3 XNO J. Manning 9 Belmont Road, Taunton TA1 5NS
MM3 XNP N. Page 61 Calderglen Avenue, Blantyre, Glasgow G72 9UP
M3 XNR K. Batt 14 Milne Park West, New Addington, Croydon CR0 0DN
M3 XNT P. Johannessen 72 Duncombe Road South, Garston, Liverpool L19 1QJ
M3 XNU A. Hollis 9 St. Georges Road, Donnington, Telford TF2 7NP
M3 XNV D. Dunn 69 Broadwaters Drive, Kidderminster DY10 2RY
MW3 XNW N. Williams 1 Picton Terrace, Pontlottyn, Bargoed CF81 9PT
M3 XNX N. Stubbs 5 Newland Street, Wakefield WF1 5AH
M3 XOA W. Dunstan 57 Orchard Vale Flushing, Falmouth TR11 5TT
M3 XOD I. Donnelly 17 Jessop Close, Horncastle LN9 6RR
M3 XOE D. Coe 199 Newark Road, North Hykeham, Lincoln LN6 8QS
M3 XOH M. Finn Atzenbach 38, Idar-Oberstein 54473 Germany
MI3 XOI G. Gorman 11 Cootehall Road Crawfordsburn, Bangor BT19 1JA
M3 XOJ M. Benson 11 Hield Grove, Aston By Budworth, Northwich CW9 6LN
MM3 XOK B. Hughes 49 Marmion Drive, Kirkintilloch, Glasgow G66 2BH
M3 XOL P. Stone 60 Acorn Avenue, Braintree CM7 2LR
M3 XOQ D. Vale 221 Thurstone Road, Birmingham B31 4PA
M3 XOR M. Nelmes 119 Exeter Road, Dawlish EX7 0AN
M3 XOT E. Blake Ty Capel, Tynygraig, Ystrad Meurig SY25 6AE
M3 XOU 1127 (Kendal) Sqn Air Cadets c/o J. Howell 56 Prouds Lane, Bilston WV14 6PU
M3 XOV T. Harris 2 Elm Green, Dudley DY1 3RE
M3 XOW R. Woodworth 2 Harrington Court, Meltham, Holmfirth HD9 4ED
M3 XOY K. Chu Harrogate Ladies' College, Clarence Drive, Harrogate HG1 2QG
M3 XOZ W. Yung 17 York Road, Harrogate HG1 2QL
M3 XPF P. Collins 14 Roundel Way, Marden, Tonbridge TN12 9TW
M3 XPH P. Hennessey 11 Monmouth Drive, Eaglescliffe, Stockton-on-Tees TS16 9HU
M3 XPI A. Ackroyd Prospect House, Causeway, Weymouth DT4 9RX
M3 XPJ J. Parfitt 12 Jodrell Place, Selsey, Chichester PO20 0FQ
M3 XPK C. Park 197 Occupation Road, Albert Village, Swadlincote DE11 8HD
M3 XPL P. Rabone 6 Cranwell Grove, Kesgrave, Ipswich IP5 2YN
M3 XPN J. Shannon 16 Croft Drive, Tickhill, Doncaster DN11 9UL
M3 XPP P. Jarvis 24 St. Peters Gardens, Leeds LS13 3EH
M3 XPR P. Ryan 10 Inchwood, Birch Hill, Bracknell RG12 7ZX
M3 XPS P. Scarratt 339 Utting Avenue East, Norris Green, Liverpool L11 1DF
M3 XPU R. Wellburn 86 Granville Street, Grimsby DN32 9NU
M3 XPW J. Oglesby 22 Elm Drive, Finningley, Doncaster DN9 3EG
M3 XPY D. Chatzikos 53 Benbow Court, Shenley Church End, Milton Keynes MK5 6JE
M3 XQB M. Harbron 39 Raleigh Road, Sunderland SR5 5RD
MW3 XQE V. Wallace 10 Maes Llydan, Benllech, Tyn-Y-Gongl LL74 8RD
M3 XQG V. Littlewood 31 Herriot Drive, Chesterfield S40 2UR
M3 XQH J. Wilson 46 Redwood Drive, Maltby, Rotherham S66 8DL
MW3 XQJ P. Jones 76 Pengwern, Llangollen LL20 8AS
M3 XQK D. Goodfellow 60 Pickering Green, Gateshead NE9 7DX
M3 XQL M. Watson 5 Birchwood Avenue, Whickham, Newcastle upon Tyne NE16 5QS
M3 XQM A. Macrae Oak Lodge, Verwood Road, Three Legged Cross, Wimborne BH21 6RR
M3 XQO S. Lewis 40 Bridle Road, Burton Latimer, Kettering NN15 5QP
M3 XQQ D. Burman 6 Goodyers Avenue, Radlett WD7 8BA
M3 XQT J. Thompson 13 Wentworth Avenue, Luton LU4 9EN
M3 XQV R. Dutton 473 Manchester Road, Lostock Gralam, Northwich CW9 7QD
M3 XQW E. Slevin Woodcock Hall, Cobbs Brow Lane, Newburgh WN8 7NB
M3 XQX B. Cameron-Laker The School Room, Haughley Green, Stowmarket IP14 3RQ
M3 XQY J. Dutton 473 Manchester Road, Lostock Gralam, Northwich CW9 7QD
M3 XQZ J. Freeman 38 City Road, Cambridge CB1 1DP
M3 XRD D. Rogers 44 County Street, Oldham OL8 3RN
M3 XRG G. Duffy 34 Twentyfifth Avenue, Blyth NE24 2QW
M3 XRH S. Lawford 26 Venetian Crescent, Darfield, Barnsley S73 9PL
MW3 XRI J. Jones Greystones, Rhewl, Oswestry SY10 7AS
M3 XRO R. Doughty 1 Woodland Road, Wakefield WF2 9DR
M3 XRP R. Potter 1 Wentworth Road, Rugby CV22 6BG
M3 XRQ B. Benson Jnr 12 South Drive, Rudheath, Northwich CW9 7JQ
MI3 XRT L. Murray 80 Canterbury Park, Londonderry BT47 6DU
M3 XRV R. Peacock 21 Breamish Drive, Washington NE38 9HS
M3 XRW R. Wainwright 69 George A Green Road, Wakefield WF2 8HA
M3 XRY G. Jones 31 Cranage Close, Halton Lodge, Runcorn WA7 5YN
M3 XSA K. Sproates 33 Frome Road, Radstock BA3 3JZ
M3 XSD P. Davies 15 Kingsley Road, Chester CH3 5RR
MM3 XSF S. Turnbull 15 Woodruff Gait, Dunfermline KY12 0NL
M3 XSG D. Levey Heriots Wood, The Common, Stanmore HA7 3HT

UK Callsigns

M3 XSI S. I'Anson 5 Heather Gardens, Leeds LS14 3HU
M3 XSJ J. Wildsmith 15 Harebell Close, Hartley Wintney RG27 8TW
M3 XSK Dr J. Skittrall 14 Tamarin Gardens, Cambridge CB1 9GH
M3 XSN A. Thompson 7 Lammermoor Road, Liverpool L18 4QP
M3 XSP S. Purkiss 19 The Hurstings, Maidstone ME15 6YN
M3 XSR T. Came 15 Brookland Road, Langport TA10 9TA
M3 XST K. Wood 52 Trench Road, Tonbridge TN10 3HB
M3 XSU J. Martin 52 Maple Avenue, Farnborough GU14 9UR
M3 XSV M. Ridpath The Grange, Main Street, Hull HU12 0JF
M3 XSY C. Throup Willow House, O'Keys Lane, Worcester WR3 8RL
M3 XSZ Dr M. Bekara 9 Southwood Court Pine Grove, Weybridge KT13 9AT
M3 XTA D. Smith 28A Bagshot Green, Bagshot GU19 5JR
M3 XTE J. Sansom 4 Vicarage Road, Eastbourne BN20 8AU
MW3 XTF W. Welch Kenilworth, School Lane, Gobowen, Oswestry SY11 3LD
M3 XTG T. Greenaway 11 Gribben Close, Tregonissey, St. Austell PL254EA
M3 XTK P. Mortiboy 72 Uplands, Stevenage SG2 7DW
M3 XTL M. Porter 20, Southfield Road, Much Wenlock TF13 6AX
M3 XTM J. Maguire 14 Botha Road, St. Eval, Wadebridge PL27 7TS
M3 XTP K. Hemmings 11 Collenswood Road, Stevenage SG2 9ER
M3 XTR P. Britton 71 Upper Forster St., Walsall WS4 2AB
M3 XTT N. Pearson 116 The Stour, Daventry NN11 4PT
M3 XTV T. Benson 83 Glovers Road, Birmingham B10 0LE
MD3 XUA C. James 75 Silverburn Crescent, Ballasalla IM9 2DY Isle of Man
MI3 XUC J. Mccollum 26 Corkey Road, Loughgiel, Ballymena BT44 9JJ
M3 XUE S. Leadbetter 11 Cogos Park, Mylor Bridge, Falmouth TR11 5SF
M3 XUF F. Leung Harrogate Ladies' College, Clarence Drive, Harrogate HG1 2QG
M3 XUG R. Barnes 1 Moira Close, Chaddesden, Derby DE21 4RL
M3 XUH Dr G. Thomas 3 The Croft, Wilton, Egremont CA22 2PW
MM3 XUI G. Taylor 15 Ronaldsvoe, Kirkwall KW15 1XE
M3 XUJ P. Hodgkinson 122 Oulton Road, Stone ST15 8DY
M3 XUO J. Steven 1 Tree Terrace, Tree Road, Brampton CA8 1TY
M3 XUR P. Andrews 15 Park Lane, Bath BA1 2XH
MI3 XUS S. Barnes 191 Marlacoo Road, Portadown, Craigavon BT62 3TD
M3 XUT R. Udall 139 Leicester Road, Measham, Swadlincote DE12 7JG
M3 XUU R. Gopan 84 Hilmanton, Lower Earley, Reading RG6 4HN
M3 XUV E. Paddison 3 Westacre Gardens, Ormesby, Great Yarmouth NR29 3SP
M3 XUW G. Marsh 35 Wolverton Avenue, Bispham, Blackpool FY2 9NU
MM3 XUX J. Munro 5 Wallace Gait, Perth PH1 2NS
MM3 XUY G. Nicholson 4 John Street, Oban PA34 5NS
MW3 XVB D. Jones 26 Ffos Y Cerridden, Nelson, Treharris CF46 6HQ
M3 XVC Lord V. Couchman Flat 3, Arbroath Court, Sedlescombe Gardens, St. Leonards-on-Sea TN38 0TF
MM3 XVD I. Woods 12 Westbank Terrace, Macmerry, Tranent EH33 1QE
M3 XVF J. Smith 8 Mayfields, Spennymoor DL16 6RN
M3 XVJ M. Emmott 1 Swallow Close, Kendal LA9 7SN
M3 XVK L. Ward 20 North Street, Maryport CA15 6HR
M3 XVQ J. Mitchell 27 Watts Close, Southampton SO16 9WA
MW3 XVR A. Jones 11 Bigyn Road, Llanelli SA15 1NT
M3 XVU B. Lovius 10 Templemore Avenue, Liverpool L18 8AH
M3 XVW V. Walker Kirby Welch & Co, West View, Longlands Lane, Wetherby LS22 4BB
M3 XVZ P. Read 53 Hill Top Road, Oldbury B68 9DU
MM3 XWB J. Mccoll 25 Nelson Ave, Coatbridge, Ml5 5Lr, Coatbridge ML5 5LR
M3 XWC J. Conway 18 Headland Close, Welford On Avon, Stratford-upon-Avon CV37 8EU
MW3 XWE T. Banks 18 Leicester Road, Newport NP19 7ER
M3 XWF J. Coogan 1 Langsett Rise, Sheffield S6 2TY
M3 XWK L. Thompson 34 Broadway, Gateshead NE9 5PY
M3 XWM T. Sayers 12 Hutton Terrace, Willington, Crook DL15 0DS
M3 XWN C. Cartwright 8 Hawes Grove, Bradford BD5 9AN
M3 XWP A. Wright 149 Burton Road, Overseal, Swadlincote DE12 6JL
M3 XWR J. Halsall 8 Woodcock Street, Wakefield WF1 5LG
MM3 XWS W. Anderson 4 Brackendene, , Houston PA6 7DE
M3 XWV J. Barbieri 20 Gilbard Court, Chineham, Basingstoke RG24 8RG
MI3 XWW D. Bradley 33 Lilac Avenue, Limavady BT49 0HS
M3 XWX S. Kerslake 4 Guipavas Road, Callington PL17 7PL
M3 XWY E. Jennings 17 Manor Way, Worcester Park KT4 7PH
M3 XXA J. Smith 3 Glamis Road, London E1W 3EE
M3 XXB S. Bunce 15 Downs View Road, Bembridge PO35 5QS
M3 XXE N. Dwyer 82 Staunton Road, Kingston upon Thames KT2 5TL
M3 XXG G. Collins 110 Hawthorn Crescent Stapenhill Burton-On-Trent Staffs, Burton-on-Trent DE15 9QW
MM3 XXI J. Redmond 14 Bankfaulds Avenue, Kilbirnie KA25 6AB
M3 XXK K. Mitchell 1 Denstroude Cottages, Denstroude Lane, Canterbury CT2 9JX
M3 XXL S. Pearce 20 Barcote Walk, Plymouth PL6 5QE
M3 XXM M. Jennings Springfield Farm, The Causeway, Stow Bridge, King's Lynn PE34 3PP
MM3 XXO S. Rogan 21 Montrose Place, Selkirk TD7 5BH
MM3 XXP North Berwick DX Club c/o A. Pitkethley 99 Margaretvale Drive, Larkhall ML9 1EH
M3 XXS S. Mellor 11 Bolton Meadow, Leyland PR26 7AJ
M3 XXU A. Friswell 142 Aldermans Green Road, Coventry CV2 1PP
MW3 XXY K. Dobson 152 Foryd Road, Kinmel Bay, Rhyl LL18 5LS
M3 XYA R. Pasika 192 Longfield Lane, Cheshunt, Waltham Cross EN7 6AQ
MI3 XYB J. Throne 12 Mason Road, Magheramason, Londonderry BT47 2RY
M3 XYC M. Cowan Oak Haven, Smugglers Lane, Chichester PO18 8QW
M3 XYH C. Bell 60 East Vines, Sunderland SR1 2DP
M3 XYI W. Cooper 20 Staple Close, Waterlooville PO7 6AH
M3 XYJ S. Wright Emergency Planning Unit, Nycc County Hall, Northallerton DL7 8AD
M3 XYK A. Raby 209 Duke Of York Avenue, Wakefield WF2 7DH
M3 XYM M. Leonard 75 Skillings Lane, Brough HU15 1BA
M3 XYN L. Marsh 14 Herrick Road, Barnby Dun, Doncaster DN3 1AW
M3 XYO J. Douglas 14 Mountfields Walk, South Kirkby, Pontefract WF9 3SJ
M3 XYP G. Thompson 24 Fairmead Way, Sunderland SR4 0NA
M3 XYT M. Carter 17A Goodramgate, York YO1 7LW
M3 XYU C. Coverley 25 Fore Street, Milton Abbot, Tavistock PL19 0PA
MW3 XYW A. Rosser 25 Clos Tir-Y-Pwll Newbridge, Newport NP11 5GE
M3 XYX W. Slater 47 Broom Road, Lakenheath, Brandon IP27 9EZ
M3 XYZ Dr R. Hill 12 Winchelsea Lane, Hastings TN35 4LG
MW3 XZB S. Gardner 19 Crosscombe Terrace, Cwm, Ebbw Vale NP23 7SP
M3 XZD J. Kelly 2 Tamar Close, Higham, Barnsley S75 1PS
M3 XZE P. Webb 159 High Street, Arlesey SG15 6SZ
M3 XZF X. Fang Harrogate Ladies' College, Clarence Drive, Harrogate HG1 2QG
M3 XZG D. Mestel 1B Hayfield Road, Oxford OX2 6TX
M3 XZH P. Snook 7 Sandhurst Avenue, Kwazulu Natal 3610 South Africa
M3 XZJ L. Aldred 1 Eaton Grange Cottages, Eaton, Grantham NG32 1ET
M3 XZK A. Smith 20 Linden Road, Coxheath, Maidstone ME17 4QS
M3 XZN M. Robinson 10 Bramley Gardens, Poulton-le-Fylde FY6 7RD
MW3 XZP C. Maggs 15 Stuart Street, Treorchy CF42 6SN
M3 XZR D. Davis 9 Park Gate Mews, Upper Norwich Road, Bournemouth BH2 5RA
M3 XZS S. Greaves 9 Park Gate Mews, Upper Norwich Road, Bournemouth BH2 5RA
M3 XZT K. Lewinton-Smith 4 Old School Road, Barnstaple EX32 9DP
MW3 XZV W. Jones 2 Derwen Close, Connah'S Quay, Deeside CH5 4AU
M3 XZY M. Reilly Flat 59, The Keep, Stafford ST17 9TW
M3 YAD A. Rossant 4 Hands Lane, Abbotsbury, Weymouth DT3 4JW
MW3 YAE G. Thatcher-Sharp 20 Dilys Street, Blaencwm, Treorchy CF42 5DT
M3 YAI H. Litten 55 Downton View, Ludlow SY8 1JE
M3 YAJ A. Jay Jasper, The Reddings, Cheltenham GL51 6RT
M3 YAL B. Loughran 26 Squirrels Field, Mile End, Colchester CO4 5YA
M3 YAO E. Bicknell 12 Victory Road, Southampton SO15 8QZ
M3 YAS G. Cummings 18 Castleton Boulevard, Skegness PE25 2TX
M3 YAV P. Meredith Bottom Flat, 74 Earl Street, Grimsby DN31 2PW
M3 YAW S. Johnson 43 Terry Gardens Kesgrave, Ipswich IP5 2EP
M3 YAX L. Mason 9 Trenethick Avenue, Helston TR13 8LU
M3 YBB P. Ryalls 22 Carr Lane, Riddlesden, , Keighley BD20 5HN
MM3 YBD W. Doull 9 Mcdowall Avenue, Ardrossan KA22 7AJ
MW3 YBF P. Teszner 8 Tan Y Bryn, Lon Ty Llwyd, Llanfarian, Aberystwyth SY23 4UH
MM3 YBG C. Gerrard 10 Whinhill Gardens, Aberdeen AB11 7WD
MI3 YBI T. Quin 165 Marlacoo Road, Portadown, Craigavon BT62 3TD
M3 YBJ R. Marsh 56B Oliver Crescent, Farningham, Dartford DA4 0BE
MW3 YBK T. Jones 25 Pritchard Terrace, Phillipstown, New Tredegar NP24 6BS
M3 YBL B. Lace 19 Methuen Street, Walney, Barrow-in-Furness LA14 3PS
M3 YBN K. Stokes 10 Heal Park Crescent Fremington, Barnstaple EX313AP
MM3 YBQ K. Verrall 7 Roshven View, Arisaig PH39 4NX
M3 YBR A. Knight Flat 6, Brickworks, 9 Halstow Way, Ashford TN23 4EQ
M3 YBT M. Fearon 70 George Street, Heywood OL10 4PW
M3 YBU M. Shields 88 Thompson Avenue, Richmond TW9 4JN
M3 YBW B. Warner 15 Grosvenor Gardens, Shifnal TF11 8EB
MW3 YBX D. Smethurst 16 Falcon Road, Haverfordwest SA61 2UE
M3 YCB Kilmarnock And Loudoun ARC c/o C. Boston 53 Bullock Road, Terrington St. Clement, King's Lynn PE34 4PR
M3 YCD D. Clarke 57 Glanville Place, Kesgrave, Ipswich IP5 1NQ
MM3 YCG C. Graham 64 Forgewood Road, Motherwell ML1 3TH
MM3 YCI S. Burt 182 Old Inverkip Road, Greenock PA16 9JG
M3 YCJ V. Powell 35 Bramber Close, Banbury OX16 0XF
M3 YCK C. Fung Harrogate Ladies' College, Clarence Drive, Harrogate HG1 2QG
MW3 YCL C. Steer 1 Park Way, Park, Merthyr Tydfil CF47 8RH
M3 YCM M. Hemmings 62 Spencer Way, Stevenage SG2 8GD
M3 YCN W. Fowler 20 The Court, Anderby Creek, Skegness PE24 5YQ
M3 YCO D. Cope 138 Leycett Road, Scot Hay, Newcastle ST5 6AU
MW3 YCR C. Rayment Brambles, Alltami Road, Mold CH7 6RT
M3 YCS C. Sutton 56 Neatherd Road, Dereham NR20 4AY
M3 YCT P. Howells 20 Warwick Street, Stourport on Severn DY13 8JB
M3 YCU A. Wang Harrogate Ladies' College, Clarence Drive, Harrogate HG1 2QG
M3 YCV C. Watts 35 Coldharbour Lane, Salisbury SP2 7BY
M3 YCZ C. Landless 2 Aspen Way, Banstead SM7 1LE
M3 YDA A. Collins Robin Post House, Robin Post Lane, Hailsham BN27 3RA
M3 YDB D. Bush 19 Spring Vale, Waterlooville PO8 9DA
MI3 YDF D. Foley 14 Chestnut Hall Court, Moira, Maghaberry BT67 0GJ
MM3 YDH D. Hume Sundhopeburn, Yarrow, Selkirk TD7 5NF
M3 YDI M. Siddle 8 Coleridge Close, Oulton, Leeds LS26 8ET
M3 YDJ D. Wilkinson 6 Brambledown Road, South Croydon CR2 0BL
MM3 YDK D. Kilgour 56 Green Street, Rothes, Aberlour AB38 7BD
M3 YDL S. Thornton 29 Farrar Avenue, Mirfield WF14 9ED
M3 YDM T. Yardley 19 Elms Close, Shareshill, Wolverhampton WV10 7JT
MW3 YDS S. Morgan 38 Ffordd Cadfan, Bridgend CF31 2DP
MW3 YDT E. Gittins 40 Melyd Avenue, Prestatyn LL19 8RN
M3 YDV D. Brame 7 Roche Garden Countess Wear, Exeter EX26LS
M3 YDW D. White 9 Wyatts Lane, Tavistock PL19 0EU
M3 YDY D. Murfitt 10 Benefield Road, Moulton, Newmarket CB8 8SW
M3 YEA L. Pollard 11 Alfriston Road, Worthing BN14 7QU
MM3 YEC J. Dupont 11 Golf View Cardenden, Lochgelly KY5 0NW
M3 YEE H. Ngi Harrogate Ladies' College, Clarence Drive, Harrogate HG1 2QG
MW3 YEG J. Thorne 11 Dowland Road, Penarth CF64 3QX
M3 YEJ J. Marsh 31 Clay Street, Soham, Ely CB7 5HJ
M3 YEK A. Taylor 16 Bellmans Road Whittlesey, Peterborough PE7 1TY
M3 YEM Dr R. Browne 30 Cromwell Road, Southowram, Halifax HX3 9SE
MM3 YEQ G. Pearce 2 Kirkriggs, Forfar DD8 2AT
M3 YET A. Reilly-Cooper 40 Clough Lane, Northwich CW8 1JR
M3 YEU J. Mobbs 6 School View, Banbury OX16 4SD
M3 YEZ J. Dearden 218 South Street, Highfields, Doncaster DN6 7JQ
M3 YFG G. Walton 11 Redwood Close, Hoyland, Barnsley S74 0EJ
M3 YFH A. Sherman 31 Peartree Avenue, Kingsbury, Tamworth B78 2LG
M3 YFI T. Hall 18 Common Lane New Haw 18 Comman Lane New Haw, Addlestone KT153LH
M3 YFJ A. King 6 Dunsfold Close, Crawley RH11 8EY
M3 YFL J. Solomon 17 Chadwick Terrace, Macclesfield SK10 2DQ
M3 YFM A. Reader 7 Church Road, Swindon SN1 3HF
M3 YFN M. Green Fire Beacon Cottage, East Hill, Sidmouth EX10 0LR
M3 YFR D. Cockburn 30 Queensberry Road, Burnley BB11 4LH
MM3 YFT P. Finnie 20 St. Margarets Road, Ardrossan KA22 7ER
M3 YFW R. Stevens 53 Keeble Way, Braintree CM7 3JX
M3 YFX J. Loveridge 96 High Road, Islington, King's Lynn PE34 3BN
M3 YFY D. Rolfe 49 Hillrise Avenue, Sompting, Lancing BN15 0LU
M3 YFZ J. Blower 4 Lamorna Close, Luton LU3 2TH
M3 YGB A. Birch 3 Partridge Way, High Wycombe HP13 5JX
M3 YGC G. Cowley 2 Manor Close, Farcet, Peterborough PE7 3AA

M3 YGD G. Hyland-Davis 34 Melody Close, Warden, Sheerness ME12 4PU
M3 YGF G. Ferguson 31 Barton Court Road, New Milton BH25 6NW
MM3 YGI R. Stratton 18 Dunnock Park, Perth PH1 5FN
M3 YGK Y. Gopikrishna 29 Alandale Drive, Pinner HA5 3UP
M3 YGL D. Gribben 44 Fern Close, Birchwood, Warrington WA3 7NU
M3 YGO C. Chew 10 Bruce Drive, South Croydon CR2 8SL
M3 YGQ M. Smith 17 Dalton Close, Grantham NG31 8WS
M3 YGR J. Katz 8 Astor Drive, Birmingham B13 9QR
M3 YGS S. Dingmar 10 Kertland Street, Savile Town, Dewsbury WF12 9PU
M3 YGT G. Giannakopoulos 3 Vauxhall Quay, Plymouth PL4 0EZ
M3 YGU B. Chamberlain 2 Stocks Loke, Cawston, Norwich NR10 4BS
M3 YGV J. Harris 2 Elm Cottages, Boreham Lane, Wartling, Hailsham BN27 1RS
M3 YGY C. Lee-Koo Flat 5, 211 Sussex Gardens, London W2 2RJ
M3 YGZ I. Mccourt 181 Fircroft Road, Ipswich IP1 6PS
MM3 YHA D. Morrison 4 West Murkle, Murkle, Thurso KW14 8YT
MW3 YHC D. Jones Bryn Hyfryd, Pandy Tudur, Abergele LL22 8UL
M3 YHD K. Marshall 38 Staunton Road, Newark NG24 4EX
M3 YHF G. Hand Hollinhurst Farm, Park Lane, Stoke-on-Trent ST9 9JB
M3 YHG J. Stringer 31 Pipit Lane, Birchwood, Warrington WA3 6NY
M3 YHH B. Hand Hollinhurst Farm, Park Lane, Stoke-on-Trent ST9 9JB
M3 YHJ A. Hussain 789 Scarborough Street, Dewsbury WF12 9AY
M3 YHL H. Lee Harrogate Ladies' College, Clarence Drive, Harrogate HG1 2QG
M3 YHM W. Mok 17 York Road, Harrogate HG1 2QL
M3 YHN M. Roberts 7 Maxwell Place, Stoke-on-Trent ST4 6RE
M3 YHP H. Pentz 30 Lindrick Way, Harrogate HG3 2SU
M3 YHQ Dr R. Mason 9 Farmfields Rise, Woore, Crewe CW3 9SZ
M3 YHR A. Rolland Flat 29 Renfrew Court, Allfrey Road, Eastbourne BN227SZ
M3 YHT G. Winterbottom 35 Abingdon View, Worksop S81 7RT
M3 YHU C. Cleverley 4A Godfrey Street, Netherfield, Nottingham NG4 2JG
M3 YHV M. Chidgey 46 Station Road, Shirehampton, Bristol BS11 9TX
MW3 YHW J. Loughlin 453 Heol-Y-Waun, Penrhys, Ferndale CF43 3NW
M3 YHX L. Tedik 11 Bakers Lane, Shutlanger NN12 7RT
M3 YHY F. Noble 1045, 45Th Street Apartment A, California 94608 United States
M3 YHZ C. Amos 33 Douglas Road, Newcastle ST5 9BP
M3 YIC L. Crabtree 23 Ava Crescent, Richmond Hill, Ontario L4B 2X1 Canada
M3 YIE A. Fuller Flat 5 Maple House, 3 Fairfield Road, Havant PO9 1AG
M3 YIF N. Spooner 14 Glebe Road, Ongar CM5 9HW
MM3 YIG C. Doolan 56 Forfar Road, Greenock PA16 0YL
MM3 YIH C. Rodgers 3 Merrylee Avenue, Port Glasgow PA14 5UT
M3 YII A. Boag 60 Harebell, Amington, Tamworth B77 4NA
M3 YIL A. Bloor Cart Lodge, Cold Harbor Croft, Haywards Heath RH17 7RS
MM3 YIO B. Fullerton 55 Alexander Avenue, Stevenston KA20 4BG
MM3 YIQ R. Doolan 56 Forfar Road, Greenock PA16 0YL
M3 YIT T. Yip 6 Fulwith Grove, Harrogate HG2 8HN
M3 YIV A. Little Wisteria Cottage, 7 Shawfield Road, Havant PO9 2SY
M3 YIX M. Norris 35 Sudbrooke Road, London SW12 8TQ
M3 YIY R. Egan 26 Baldwin Road, Birmingham B30 3LG
M3 YJA J. Appleby 79 Glenwoods, Newport Pagnell MK16 0NG
M3 YJB J. Birch 6 Crescent Road, Wallasey CH44 0BQ
M3 YJD J. Dowdeswell 18 Lechlade Gardens, Fareham PO15 6HF
MI3 YJE J. Elliott 30 Moyle Road, Ballycastle BT54 6AN
M3 YJF D. Moran 33 Pilgrim Drive, Bere Alston, Yelverton PL20 7DB
M3 YJG R. Lawton 41 Almond Avenue, Armthorpe, Doncaster DN3 2HE
MM3 YJH J. Hume 2 Tinnis Farm Cottage, Yarrow, Selkirk TD7 5JZ
MW3 YJJ L. Gleed 9 Medlock Close, Bettws, Newport NP20 7EJ
M3 YJL J. Cairns1 17 Alfred Avenue, Worsley, Manchester M28 2TX
M3 YJM J. Patrick-Gleed 5 Glastonbury Drive, Lydney GL15 5TT
M3 YJN N. Allen 5 Limecroft View, Wingerworth, Chesterfield S42 6NR
M3 YJP G. Patrick-Gleed 6 Julius Way, Lydney GL15 5QS
M3 YJT H. Taylor 21 Charlecote Drive, Chandler'S Ford, Eastleigh SO53 1SF
M3 YJU P. Boreham 67 Brent Lane, Dartford DA1 1QT
M3 YJW J. Williams 66 Oakfield Avenue, Hitchin SG4 9JD
M3 YJY L. Bourne 9A Partridge Croft, Lichfield WS13 6SD
MW3 YKA A. Francis 18 Darent Close Bettes, Newport NP207SQ
M3 YKC K. Comben 9 West Lane, North Baddesley, Southampton SO52 9GB
M3 YKF K. Fung Harrogate Ladies' College, Clarence Drive, Harrogate HG1 2QG
M3 YKH A. Harding Sunnydene, Wellmead, Axminster EX13 7SQ
M3 YKI A. Sulieman 22 Warren Court, 80 Charlton Church Lane, London SE7 7AD
MW3 YKL C. Warburton 71 Richards Terrace, Cardiff CF24 1RW
M3 YKN D. Bright 103B Langer Road, Felixstowe IP11 2EA
M3 YKO T. Chan Harrigate Ladies College, Clarence Drive, Harrogate HG1 2QG
MM3 YKR R. Murray 18 Braids Road, Kirkcaldy KY2 6JE
M3 YKT L. Jones 16 Oxland Road, Illogan, Redruth TR16 4SH
M3 YKZ D. Warren 36 Milner Road, Heswall, Wirral CH60 5RZ
M3 YLB R. Beck 73 Crowborough Road, Southend-on-Sea SS2 6LW
M3 YLJ G. Whitehead 29 Coulsons Road, Bristol BS14 0NN
M3 YLK J. Swift 56 Leymoor Road, Huddersfield HD3 4SW
M3 YLL J. Swift 56 Leymoor Road, Huddersfield HD3 4SW
M3 YLM Y. Lam Harrogate Ladies' College, Clarence Drive, Harrogate HG1 2QG
M3 YLN C. Mills 118 Saxon Gardens, Shoeburyness, Southend-on-Sea SS3 9PX
MW3 YLO A. Powell 31 Highmead, Pontllanfraith, Blackwood NP12 2PF
MM3 YLP C. Edwards 18 Gelshfield, Halkirk KW12 6UZ
M3 YLQ J. Painter 58 Glenfield Road, Plymouth PL6 7LN
MW3 YLR D. Trevelyan 35 St. Kingsmark Avenue, Chepstow NP16 5LY
M3 YLT L. Timmins 83 Loxdale Sidings, Bilston WV14 0TN
MW3 YLV K. Bell 2 Hill Street, Risca, Newport NP11 6QH
MD3 YLX D. Cain 7 Cronk Y Berry Mews, Douglas IM2 6HQ Isle of Man
M3 YLZ Capt. M. Brewster Blackthorn Farm, Common Road, Dickleburgh IP21 4PH
M3 YMC M. Comben 9 West Lane, North Baddesley, Southampton SO52 9GB
M3 YMD M. Dudley 418 Sandon Road, Stoke-on-Trent ST3 7LH
MI3 YMF M. Foley 44 Gallows Street, Dromore BT25 1BD
M3 YMG M. Crawley 16 The Meadows, Herne Bay CT6 7XF
M3 YMH M. Hurst 20 Albany Avenue, Manchester M11 1HQ
MM3 YMM M. Holmes 1 Lauren Way, Paisley PA2 9JW
MM3 YMN Dr J. Henderson 7 Rowanhill Close, Port Seton, Prestonpans EH32 0SY
M3 YMS M. Statham 17 Nicholas Meadow, Higher Metherell, Callington PL17 8DE
MM3 YMU R. Morrison 4 West Murkle, Murkle, Thurso KW14 8YT
M3 YMX J. Lewis 4 Moor Park, Clevedon BS21 6EH
MW3 YMY M. Ireland Pen Y Gadlas, Ffordd Bryniau, Prestatyn LL19 8RD
MW3 YNA M. Seagrave Glanmarlaishome, Maes Piode, Llandybie, Ammanford SA18 3YS
M3 YNB N. Buttery 22 Mallard Road, Rowlands Castle PO9 6HN
M3 YNC C. Mackintosh 18 Park Avenue, Castleford WF10 4JT
M3 YND B. Higgins 41 Lower Meadow, Harlow CM18 7RE
M3 YNE J. Godfrey 36 Greenwich Road, Hailsham BN27 2PE
M3 YNH C. Arundel 54 Broadmead, Castleford WF10 4SE
M3 YNI R. Pike 66 Prowses Hemyock, Cullompton EX15 3QG
M3 YNJ A. Richards 5 Bloomfield Drive, Bracknell RG12 2JW
MW3 YNK G. Lane 32 Caellepa, Bangor LL57 1HF
M3 YNM R. Johnson 50 Barnaby Rudge, Chelmsford CM1 4YG
M3 YNN S. Linton 89 Ragpath Lane, Stockton-on-Tees TS19 9JS
M3 YNO A. Williamson 32 Beech Close Eastfield, Scarborough YO11 3QZ
M3 YNS M. Gibson 58 Byron Street, Barrow-in-Furness LA14 5RL
M3 YNX D. Brewer 15 Morella Road, London SW12 8UQ
M3 YNY N. Oldrid 4 Bar Lane Mapplewell, Barnsley S75 6DQ
M3 YOB P. Lake 19 Orchard Vale Midsomer Norton, Radstock BA3 2RA
MM3 YOC R. Munro 20 County Cottages, Piperhill, Nairn IV12 5SE
M3 YOE A. Yorke 33 Avon Crescent, Stratford-upon-Avon CV37 7EX
M3 YOG K. Cartledge Oysterber Farm, Burton Road, Lancaster LA2 7ET
M3 YOH L. Mason 432 Lichfield Road, Sutton Coldfield B74 4BL
MM3 YOL D. Davidson 44 Abercromby Crescent, Helensburgh G84 9DX
M3 YOM J. Thresher Quarry Grange, Nuneaton Road Over Whitacre, Coleshill, Birmingham B46 2NH
M3 YOO A. Williams 54 Longbridge, Willesborough, Ashford TN24 0TA
M3 YOP T. Court Eastgate Cottage, Perrys Lane, Norwich NR10 4HJ
M3 YOQ M. Harris Rodmarton House, Broad Town, Swindon SN4 7RG
M3 YOT J. O'Malley 140 Allerburn Lea, Alnwick NE66 2QP
M3 YOU M. Young 19 Tallents Close, Sutton At Hone, Dartford DA4 9HS
M3 YOW G. Eycott 1 Ham Road, Wanborough, Swindon SN4 0DF
M3 YOX W. Hopkins 30 Charles Darwin Road, Plymouth PL1 4GU
M3 YOZ Guernsey ARS c/o S. Hughes 117 Liverpool Road, Irlam, Manchester M44 6EH
M3 YPA S. Watts 29 Brook Drive, Corsham SN13 9AU
M3 YPB C. Bond Tryfan, Vicarage Lane, Neston CH64 5TJ
M3 YPD J. Scholz 1930 Grant Ave, Williamstown 8094 United States
M3 YPG W. Gratton Park House, Brimham Rocks Road, Harrogate HG3 3HE
MM3 YPH E. Bertram 20 Kyles View, Largs KA30 9ET
M3 YPI B. Deakin 8 Patey St., Manchester M12 5RP
M3 YPJ P. Kirby 30 New Street, Eccleston, Chorley PR7 5TW
MM3 YPN S. Hargreaves 4 Oxenfoord Avenue, Pathhead EH37 5QD
M3 YPP C. Snow 19 Salters Road, Haylands, Isle of Wight PO33 3HU
M3 YPR P. Ruocco 8 Chain Road, Manchester M9 6QX
M3 YPS L. Brady 9 Wordsworth Close, Wootton Bassett, Swindon SN4 8HJ
M3 YPU T. Galloway 1 Farley Close Shadoxhurst, Ashford TN26 1NB
M3 YPW R. Coleman 5 Meeting Lane, Burton Latimer, Kettering NN15 5LS
M3 YPX D. Parkhouse 5 Long Yard, Briston, Melton Constable NR24 2LB
M3 YQC P. Rogers 16 Begonia Close, Basingstoke RG22 5RA
M3 YQG A. Bandtock 28 Campion Road, Westoning, Bedford MK45 5LB
M3 YQH G. Smith 6 Grange Crescent, Childer Thornton, Ellesmere Port CH66 5NB
MW3 YQL J. Edwards 2 Maes Merddyn, Gaerwen LL60 6DG
M3 YQM D. Morbey 44 Browning Road, Plymouth PL2 3AP
MM3 YQP the Gliding Centre ARS c/o C. Pate 75 Castings Avenue, Falkirk FK2 7BJ
M3 YQQ Dr N. Lutte Oak House, Brandy Hole Lane, Chichester PO19 5RX
M3 YQR C. Hay Sea Cadets, Riverside Road, Great Yarmouth NR31 6PX
M3 YQT J. Best 24 Suggitts Lane, Cleethorpes DN35 7JJ
MW3 YQU J. Jones 55 London Road, Holyhead LL65 2NS
MM3 YQX K. Mcbride 1 Cowal Place, Gourock PA19 1EJ
M3 YRB D. Grantham 56 Knapp Avenue, Eastwood, Nottingham NG16 3JW
M3 YRC S. Evans 2 Firbeck Crescent, Langold, Worksop S81 9SB
M3 YRH R. Horner 21 Ainsworth Road Little Lever, Bolton BL3 1RG
M3 YRJ T. Lamont 30 Shackleton Close, Old Hall, Warrington WA5 9QE
M3 YRM R. Moles 14 Dorsett Road, Stourport-on-Severn DY13 8EL
M3 YRO P. Harvey 35 Isaac Street, Liverpool L8 4TH
M3 YRR C. Reid 128 Main Street, Hensingham, Whitehaven CA28 8PX
M3 YRS T. Godfrey 12 Beacon House, Chulsa Road, London SE26 6BP
M3 YRT R. Tasker 16 Lodge Hill Walk, Leeds LS12 5TP
M3 YRV S. Honywood 169 Primrose Lane, Croydon CR0 8YQ
M3 YRW J. Woods 21 Appleyard Crescent, Norwich NR3 2QN
M3 YRX M. Mckone 12 Hawkshead Road, Knott End-On-Sea, Poulton-le-Fylde FY6 0QE
M3 YRZ K. Lovell Wimbledon Court, 3, Miiddlesbrough TS5 5JP
M3 YSA J. Hallam 63B Poppleton Road, London E11 1LP
M3 YSC S. Croucher 17 Sundridge Road, Woking GU22 9AU
M3 YSD S. Dudley 365 Sandon Road, Stoke-on-Trent ST3 7LJ
M3 YSF S. Forrest 58 Victoria Road, Manchester M14 6BZ
MM3 YSJ J. Sinclair 21 Oxford Avenue, Gourock PA19 1XU
M3 YSL C. Brooks 61 Boxfield Green, Stevenage SG2 7DR
M3 YSM S. Marshall 43 Glenkerry House, 98 Burcham St., London E14 0SL
M3 YSN M. Beardsley 2 Wingrove Avenue, Newcastle upon Tyne NE4 9AL
M3 YSQ J. Powell 46 Woodmancote, Yate, Bristol BS37 4LL
M3 YSS S. Stewart 8 Craig Street, Peterborough PE1 2EJ
M3 YSU G. Cummings 10 Perth Close, Skegness PE25 2HY
M3 YSV D. Benwell 10 Harvester Way, Crowland, Peterborough PE6 0DA
M3 YSW G. Thrower 8 Upton Gardens, Worthing BN13 1DA
M3 YSY A. Zabalujevs 30 Miles Close, London SE28 0NJ
M3 YSZ C. Gao 17 York Road, Harrogate HG1 2QL
M3 YTA J. Marland 8 Dulverton Gardens, Edinburgh Road, Bolton BL3 1TR
MM3 YTB M. Martin Flat A, 11 Craigpark Street, Clydebank G81 5BS
M3 YTE A. Duffield 4 Crabmill Lane, Easingwold, York YO61 3DE
M3 YTF G. Garman 11 Rye Close, Norwich NR3 2LF
M3 YTG A. Clark 330 Stafford Road, Caterham CR3 6NJ
MI3 YTH A. Shilliday 26 Iskymeadow Road, Armagh BT60 3JS
MM3 YTI G. White 52 Union Road, Whitburn, Bathgate EH47 0AP
M3 YTL T. Lee Harrogate Ladies' College, Clarence Drive, Harrogate HG1 2QG
M3 YTQ J. Hughes 84 Rodwell Avenue, Weymouth DT4 8SQ
M3 YTV D. Telford 37 Swillington Lane, Swillington, Leeds LS26 8QF

M3 YTZ J. Chaplin 101 St. Cuthberts Drive, Gateshead NE10 9AB
M3 YUA M. Boyd 4 Crowton Cottages, Winsford Road, Cholmondeston, Winsford CW7 4DP
M3 YUB D. George 13 Cheltenham Way, Mablethorpe LN12 2AX
M3 YUC A. Waudby 7 Forest Grove, York YO31 1BL
M3 YUD S. Nutt 23A Hesketh Drive, Southport PR9 7JX
M3 YUH P. Lai Harrogate Ladies' College, Clarence Drive, Harrogate HG1 2QG
M3 YUK J. Burman 6 Goodyers Avenue, Radlett WD7 8BA
M3 YUN R. Spalding 7 Kingfisher Close, Scawby Brook, Brigg DN20 9FN
M3 YUP B. Stevens 77 Dean Lane, Hazel Grove, Stockport SK7 6EJ
MD3 YUQ D. Kelly 41 High Street, Port St. Mary IM95DN Isle of Man
M3 YUR C. Potter 4 Tomlinson Street, Stoke-on-Trent ST6 4NW
MM3 YUS P. Mccann 3 Exmouth Place, Gourock PA19 1JE
MI3 YUT D. Elliott 15 Derrychara Park, Enniskillen BT74 6JP
MM3 YUU M. Morrison 8 Garallan, Kilwinning KA13 6LU
M3 YUV L. Hudson 68 Eleanor Road, Harrogate HG2 7AJ
MM3 YUW D. Williamson 31 Medrox Gardens, Cumbernauld, Glasgow G67 4AJ
MM3 YUX C. Williamson 31 Medrox Gardens, Cumbernauld, Glasgow G67 4AJ
MM3 YUY D. Bendoris 35 St. Michael'S Wynd, Kilwinning KA13 6WH
MI3 YVB B. Craney 8A Drumhoy Drive, Carrickfergus BT38 8NN
M3 YVD T. Munro 71 Zig Zag Road, Liverpool L12 9EQ
M3 YVE Y. Neary 3 Wordsworth Close, Torquay TQ2 6EA
M3 YVF M. Lowe 22 Ryelands Close, Market Harborough LE16 7XE
M3 YVG M. Lowe 22 Ryelands Close, Market Harborough LE16 7XE
M3 YVJ I. Priest 11 Dunlin Close, Kingswinford DY6 8XP
MW3 YVK D. Bowen 25 Maendu Terrace, Brecon LD3 9HH
M3 YVN Dr R. Wall 30 Church Street, Skerries, Co. Dublin Ireland
M3 YVR L. Palir 116 Carville Crescent, Brentford TW8 9RD
M3 YVT J. Storey 3 Woodside Road, Poole BH14 9JH
MM3 YVU R. Corrieri 160 Telford Road, East Kilbride, Glasgow G75 0BX
M3 YVW L. Brisco 1 Bescot Way, Thornton-Cleveleys FY5 3QA
M3 YVY J. Hudd South Crofty Cottage North Pool Road, Redruth TR15 3JQ
M3 YVZ J. Smith 98 Dorset Road, Coventry CV1 4EB
MW3 YWC A. Johnston 44 Cradoc Road, Brecon LD3 9LH
M3 YWD W. Disney 98 Widney Lane, Solihull B91 3LL
M3 YWE J. Watkins 6 Glebelands, Biddenden, Ashford TN27 8EA
M3 YWF S. Griffiths 22 Manor Rise, Arleston, Telford TF1 2ND
M3 YWG W. Gradwell 9 Nottingham Drive, Bolton BL1 3RH
M3 YWH K. Skerry 18 Park Avenue, Cheadle, Stoke-on-Trent ST10 1LZ
M3 YWI F. Ingram 1 Hazel Close, Brackley NN13 6PE
M3 YWJ E. Mcmahon 120 New Ferry Road, Wirral CH62 1DY
M3 YWM P. Higgins 35 Whittington Street, , Plymouth PL3 4EG
M3 YWO K. Lovell 2 Beckingham Hall Cottages, Tolleshunt Major, Maldon CM9 8EH
MI3 YWT V. Crichton 10 Bann Drive, Londonderry BT47 2HW
M3 YWU J. Morris 96 Bradford Crescent, Durham DH1 1HW
M3 YXB R. Barrett 18 Bullstake Close, Oxford OX2 0HN
M3 YXC D. Jones 18 Maxwell Drive, Hazlemere, High Wycombe HP15 7BX
M3 YXD D. Wilson 24 Hallamshire Mews, Wakefield WF2 8YB
M3 YXE J. Peters 11 Clockhouse Lane, Ashford TW15 2EP
M3 YXF J. Barber 13 Dock Road, Sharpness, Berkeley GL13 9UA
M3 YXH G. Sweet 12 Old Harrow Road, St. Leonards-on-Sea TN37 7EG
M3 YXJ A. Mcgreish 36 Eastmoor Road, Oxborough, King's Lynn PE33 9PX
M3 YXK G. Mcgreish 36 Eastmoor Road, Oxborough, King's Lynn PE33 9PX
M3 YXL A. Needham 49 Macclesfield Road, Buxton SK17 9AG
M3 YXM J. Pick 178 Alcester Road South, Kings Heath, Birmingham B14 6DE
MM3 YXN D. Cowie 69 Broomfield Park, Portlethen, Aberdeen AB12 4XT
M3 YXP J. White 15 Norham Drive, Newcastle upon Tyne NE5 5PR
M3 YXQ S. Pye 23 Dene Way, Donnington, Newbury RG14 2JL
M3 YXR X. Ren Lincoln House, Clarence Drive, Harrogate HG1 2QD
M3 YXS L. Reddall 68 Broadhurst Green, Hednesford, Cannock WS12 4LF
M3 YXT N. Wall 6 Ashton Lane, Braithwell, Rotherham S66 7AJ
M3 YXU Felixstowe Dars c/o P. Dossett 92 Dale Valley Road, Southampton SO16 6QU
M3 YXV C. Reynolds 9 Skeyton Road, North Walsham NR28 0BS
M3 YXW D. Cox 9 Northbrook Copse, Bracknell RG12 0UA
MI3 YXX L. Bradley 4 Rathbeg Drive, Limavady BT49 0BB
M3 YYE A. Hinckley 114 Lawn Lane, Hemel Hempstead HP3 9HS
M3 YYG S. Senior 4 Flowers Meadow, Liverton, Newton Abbot TQ12 6UP
M3 YYJ J. Whitford-Robson 13 Perryman Close, Plymouth PL7 4BP
M3 YYK K. Yardley 4 Park Road, Hillton, Wolverhampton WV107HS
M3 YYM I. Mcpherson 86 Fletemoor Road, St. Budeaux, Plymouth PL5 1UH
M3 YYO M. Scott-Martin 14 Coles Gardens, Poole BH15 4DX
MW3 YYQ R. Smith 20 Fron Uchaf, Colwyn Bay LL29 6DS
M3 YYR N. Brown 85 Larkham Lane, Plympton, Plymouth PL7 4PL
M3 YYS S. Hurrell 4 Woodland Drive, Plympton, Plymouth PL7 1SN
MI3 YYT D. Bradley 33 Lilac Avenue, Limavady BT49 0HS
M3 YYU R. Woolley 84 Bowthorpe Road, Norwich NR2 3TP
M3 YYV S. Chiew 11 Svenskaby, Orton Wistow, Peterborough PE2 6YZ
M3 YYW J. Elsmore 142 St. Marks Road, Chester CH4 8DH
M3 YZA S. Russell 11 Morville Road, Dudley DY2 9HR
M3 YZC D. Matheson 21 Warren Hill Road, Woodbridge IP12 4DU
M3 YZE T. Knight 25 Allcot Road, Portsmouth PO3 5DE
M3 YZH J. Kelly Martins, Fairwarp, Uckfield TN22 3BE
M3 YZI B. Sims 4 New Cottages, Cranwich Road, Thetford IP26 5EQ
M3 YZJ J. King 2 Perys Court, Cracknore Hard Lane, Southampton SO40 4UT
M3 YZK D. Lisi 56 Gipsy Lane, Old Whittington, Chesterfield S41 9JB
M3 YZN J. Murray 6 Sheridan Court, St. Edmunds Road, Dartford DA1 5NF
M3 YZO M. Lewis Hillside, Caldy, Craven Arms SY7 8QR
M3 YZP M. Tydeman 6 Colin Close, Corfe Mullen, Wimborne BH21 3QG
M3 YZQ J. Thompson 32 Church Street, Warnham, Horsham RH12 3QR
M3 YZU I. Dennison 18 Tredington Grove, Caldecotte, Milton Keynes MK7 8LR
M3 YZV M. Hughes 19 Pendine Crescent, North Hykeham, Lincoln LN6 8UW
M3 ZAA M. Walker 4 Lemonroyd Marina, Fleet Lane, Leeds LS26 9AJ
M3 ZAE R. Giles 75 Bradstocks Way, Sutton Courtenay, Abingdon OX14 4DA
M3 ZAI J. Elderfield 2 Westwood Close, Amersham HP6 6RP
M3 ZAL J. Million 5 Passfield Square, Thornley, Durham DH6 3DB
M3 ZAM N. Tweed 42 Ophir Road, Worthing BN11 2SS
M3 ZAN S. Tokley 9 Peel Road, Springfield, Chelmsford CM2 6AQ
MW3 ZAQ Z. David 45 Amanwy, Llanelli SA14 9AH
M3 ZAR R. Whitehouse 29 Greswolde Road, Solihull B91 1DY
M3 ZAU A. Cunningham 96 Kingsleigh Drive, Birmingham B36 9DY
M3 ZAV A. Cooper 23 Ash Street, Manchester M9 5XY
M3 ZAW C. Beresford 13 Chaseside Avenue, Twyford, Reading RG10 9BT
M3 ZAY T. Hooper 71 Collins Parc, Stithians, Truro TR3 7RB
M3 ZAZ Z. Ripley 11 Gallows Hill Drive, Ripon HG4 1UP
M3 ZBF J. Evans 112 Seagrave Crescent, Sheffield S12 2JP
M3 ZBH B. Hudson 34 Eastwood Road, Bexhill-on-Sea TN39 3PS
M3 ZBI A. Yarnold 1 Ford Close, , Ivybridge PL21 9TQ
M3 ZBJ Z. Jennings 29 Mountbatten Drive, Newport PO30 5SJ
M3 ZBL R. Leese 22 Southlands Road, Congleton CW12 3JY
M3 ZBR R. Offord 16 Clive Avenue, Ipswich IP1 4LU
M3 ZBS A. Knight 37 Crispe House, 72 Dovehouse Mead, Barking IG11 7EB
M3 ZBV A. Higham 12 Lakenheath Drive, Sharples, Bolton BL1 7RJ
M3 ZBW R. Weaver 116 Carville Crescent, Brentford TW8 9RD
M3 ZBX J. Gleeson 124 Rushes Mead, Harlow CM18 6QE
M3 ZBZ M. Carvell 10 Burns Close, Stevenage SG2 0JN
M3 ZCA A. Bevan 15 Windsor Terrace, Haverhill CB9 9BE
M3 ZCB C. Blackmun 2A St. Margarets Road, Girton, Cambridge CB3 0LT
M3 ZCE R. Sutton 56 Neatherd Road, Dereham NR20 4AY
M3 ZCG C. Gregory 81 Fiskerton Way, Oakwood, Derby DE21 2HY
MW3 ZCI P. Probert 54 New Hall Road, Ruabon, Wrexham LL14 6AT
M3 ZCJ J. Ma Harrogate Ladies' College, Clarence Drive, Harrogate HG1 2QG
M3 ZCM D. Griffin 16A Kent Road, Fleet GU51 3AH
M3 ZCN M. Barnes 51 Lower Way, Great Brickhill, Milton Keynes MK17 9AG
MW3 ZCO C. Owen 97 Maesglas Grove, Newport NP20 3DN
M3 ZCR C. Zarucki 26 Heathfield Road, Kings Heath, Birmingham B14 7DB
MM3 ZCS S. Kirkbride 18 North Roundall, Limekilns, Dunfermline KY11 3JY
MW3 ZCU D. Cook 19 Almond Avenue Risca, Newport NP11 6PF
M3 ZCW C. Lewis 9 Chatsworth Gardens, Sydenham, Leamington Spa CV31 1WA
M3 ZCX D. Stringer 18 Townfield Close, Ravenglass CA18 1SL
MI3 ZCY C. Wang 32 Broadlands, Carrickfergus BT38 7BL
MM3 ZDG Z. Graham 15 Stone Crescent, Mayfield, Dalkeith EH22 5DT
MM3 ZDI R. Cook 5 Monkstadt, Linicro, Portree IV51 9YN
M3 ZDK D. Osmand Flat 5, Long Barn Rosevidney, Penzance TR20 9BX
MW3 ZDQ A. Davies 25 Llanfair Road, Tonypandy CF40 1TA
M3 ZDS D. Scott 33 Manor Crescent, Honiton EX14 2DF
M3 ZDT I. Arrow Crystalwood, Stonemans Hill, Newton Abbot TQ12 5PZ
M3 ZDV D. Vasey 22 Rickleton Village Centre, Washington NE38 9ET
M3 ZDW D. Whyatt 11 The Perrings, Nailsea, Bristol BS48 4YD
M3 ZEH S. Hendy Flat 2, 33 Kingston Road, , Leatherhead KT22 7SL
M3 ZEI D. Mctaggart 59 Gainsborough Road, Richmond TW9 2DZ
M3 ZEJ R. Fagg 62 Hawkins Road, Folkestone CT19 4JA
M3 ZER D. Nazer 20 College Road, Ringwood BH24 1NX
MM3 ZET H. Dally 3 Gremmasgaet, Lerwick, Shetland ZE1 0NE
M3 ZEV B. Dexter 237 Wordsworth Avenue, Sheffield S5 8NE
M3 ZEW E. Melman 177 Grantham Road, London E12 5NB
M3 ZEY M. Blagg 17 Flint Avenue, Forest Town, Mansfield NG19 0DS
M3 ZFB J. Creed 11 Athelstan Road, Faversham ME13 8QL
M3 ZFH S. Recht 1 Ireton Close, Chalgrove, Oxford OX44 7RZ
M3 ZFI C. Godfrey 97 Whalley Drive, Bletchley, Milton Keynes MK3 6HX
M3 ZFJ J. Tomlinson 14 Shorecliffe Rise, 27 Radcliffe New Road, Radcliffe, Manchester M26 1LE
MM3 ZFK W. Taylor Garth Wood, Fishers Brae, Eyemouth TD14 5NJ
M3 ZFL C. Rycott 7 Crescent Grove, London SW4 7AF
M3 ZFN D. Hoare 47 High Street, Chalgrove, Oxford OX44 7SJ
M3 ZFO D. Davies Marshlands, Low Road, Thurlton, Norwich NR14 6RL
M3 ZFS S. Fisher 4 Beaufont Gardens, Bawtry, Doncaster DN10 6RT
M3 ZFV P. Whiteley 1 Newton Close, Fareham PO14 3LF
M3 ZFW D. Tinn 28 South Road, Kirkby Stephen CA174SN
M3 ZFY R. Fye 201 North Wing The Residence, Kershaw Drive, Lancaster LA1 3SY
M3 ZGA G. Campbell 10 Welbeck Road, Rochdale OL16 4XP
M3 ZGC V. Jolliffe 54 Glendale Avenue, Wash Common, Newbury RG14 6RU
M3 ZGD G. Donaldson Moddershall House, Moddershall, Stone ST15 8TG
M3 ZGE C. Walton 6 Hilltop Road, Bearpark, Durham DH7 7DP
M3 ZGF P. Daniel 57 Jephson Road, Sutton-in-Ashfield NG17 5EH
M3 ZGG T. Turner 86 Bevan Close, Huntingdon PE29 1TJ
M3 ZGH G. Houlton 7 Bartletts Hillside Close, Chalfont St. Peter, Gerrards Cross SL9 0HH
M3 ZGI M. Jurczyszyn 115 The Twitchell, Sutton-in-Ashfield NG17 5AX
MM3 ZGK G. Munn 28 Broomfield Park, Portlethen, Aberdeen AB12 4XT
M3 ZGM G. Machin 7 Lansdown, Yate, Bristol BS37 4LS
M3 ZGO C. Ikeda-Chew 10 Bruce Drive, South Croydon CR2 8SL
MW3 ZGR G. Renshaw 46 Forge Close Caerleon, Newport NP18 3PW
M3 ZGS G. Somerville 22 Woolven Close, Burgess Hill RH15 9RR
M3 ZGT J. Throup Willow House, O'Keys Lane, Worcester WR3 8RL
M3 ZGU L. Retford 111 Lander Close, Old Hall, Warrington WA5 9PL
M3 ZGX P. Beltrami 15 Woodroffe Square, Calne SN11 8PW
M3 ZHC X. Beltrami 20 Brunel Way, Calne SN11 9FN
MD3 ZHD W. Callister 13 Fairfield Avenue, Onchan IM3 4BG Isle of Man
MU3 ZHF H. Fletcher Le Villocq House, Le Villocq, Castel GY5 7SA Guernsey
M3 ZHG C. Barnes 23 South Street, Crewe CW2 6HN
M3 ZHI J. Brookes 85 St. Johns Road, Rotherham S65 1LT
M3 ZHM L. Martin Little Acre, Swan Lane, Edenbridge TN8 6AJ
M3 ZHP D. Pastwik 451 Chorley Old Road, Bolton BL1 6AH
M3 ZHQ J. Martin Little Acre, Swan Lane, Edenbridge TN8 6AJ
MW3 ZHU M. Hilliar-Mills 1-3 Queens Road, Criccieth LL52 0EG
M3 ZHV T. Gorbutt 26 Whitethorn Avenue, Withernsea HU19 2LN
M3 ZHW T. Hulme 157 Birkinstyle Lane, Stonebroom, Alfreton DE55 6LD
M3 ZHX K. Baker 3 Manor Park, Duloe, Liskeard PL14 4PT
M3 ZHY J. Mason 9 Farmfields Rise, Woore, Crewe CW3 9SZ
M3 ZHZ L. Kay 20 Mytton Lane, Shawbury, Shrewsbury SY4 4JE
M3 ZIA Z. Ul Haq 7 Alden Walk, Stockport SK4 5NW
M3 ZID R. Farrington Sunny Brook, Broadway Rd, , Evesham WR11 7RN
M3 ZIE Wisbech Amateur Radio & Electronics Club c/o H. Ehm 17 Stuart Road, Kempston, Bedford MK42 8HS
M3 ZIF J. Whittick 91 Godfrey Way, Dunmow CM6 2SQ
M3 ZIH C. Warwick 104 Church Road, Formby, Liverpool L37 3NH

M3 ZII S. Savastano Pound Farm, Hawkchurch, Axminster EX13 5XN
M3 ZIL E. Greatorex 22 Marlborough Way, Uttoxeter ST14 7HL
M3 ZIM T. Fuller 49 Scotby Avenue, Chatham ME5 8ER
M3 ZIN P. Bruce Pilgrims, Broadmead, Lymington SO41 6DH
M3 ZIO M. Denon 3 Duke Street, Clowne, Chesterfield S43 4RZ
M3 ZIV I. Vickers 3 Nesbit Road, St. Marys Bay, Romney Marsh TN29 0SF
M3 ZIX S. Norris 95 Waterloo Road, Ashton-On-Ribble, Preston PR2 1BH
M3 ZIZ K. Druce Moonraker The Leas Kingsdown, Deal CT148ER
M3 ZJB M. Collier 32 London Road, Dereham NR19 1AW
M3 ZJD F. Stevens 191 Parkinson Drive, Chelmsford CM1 3GW
M3 ZJE M. Lake 17 Sunnyside View, Peasedown St. John, Bath BA2 8JN
M3 ZJF P. Hayward 14 Micklewright Avenue, Crewe CW1 4DF
M3 ZJG P. Gibbs 19 Lupin Close, Etherley Dene, Bishop Auckland DL14 0TP
M3 ZJH J. Herant 75 Victoria St., Chesterton, Newcastle ST5 7EP
M3 ZJJ G. Chiu Harrogate Ladies' College, Clarence Drive, Harrogate HG1 2QG
M3 ZJK S. Knight 18 Ludlow Close, Chippenham SN15 3UG
MM3 ZJL J. Logan 13 Hornel Road, Kirkcudbright DG6 4LH
M3 ZJM J. Leavesley 37 Western Road, Stourbridge DY8 3XU
MI3 ZJN G. Connolly 54 Granemore Park, Keady, Armagh BT60 2GP
M3 ZJO J. Rawlinson Westfield Farm, Risden Lane, Cranbrook TN18 5DU
M3 ZJQ A. Rawlinson Westfield Farm, Risden Lane, Hawkhurst, Sandhurst, Cranbrook TN18 5DU
M3 ZJS G. Stokes 24 Dayslondon Road, Waterlooville PO7 5NN
M3 ZJV G. Hewis 10 Albert Road, New Malden KT3 6BS
MM3 ZJY G. Boyter 36 Main Street, Springfield, Cupar KY15 5SQ
M3 ZKA K. Allgar 13 Deacon Avenue, Kempston, Bedford MK42 7DU
M3 ZKD K. Dungey 247 Park Road, Sittingbourne ME10 1ER
M3 ZKE K. Pemberton 38 Milford Drive, Bournemouth BH11 9HJ
M3 ZKF K. Franklin 1 Aberdeen Court, Newcastle upon Tyne NE3 2XU
M3 ZKI I. King 7 Shardlow Close, Haverhill CB9 7RF
M3 ZKL Dr I. Jutting 68 The Ridgeway, Tonbridge TN10 4NN
M3 ZKO L. Bodie Harrigate Ladies College, Clarence Drive, Harrogate HG1 2QG
M3 ZKT M. Sargeant 3 Old Springfields Padbury, Buckingham MK18 2AR
MW3 ZKW K. Watt 28 Llysgwyn, Morriston, Swansea SA6 6BJ
MW3 ZKX A. Griffiths 6 South View, Taffs Well, Cardiff CF15 7SE
MW3 ZKY J. Jones 3 Rhedyw Road, Llanllyfni, Caernarfon LL54 6SN
M3 ZLA G. Burrows Aubrietia, Malpas SY14 8AY
M3 ZLE L. Ellerington 120 Stagsden Road Bromham, Bedford MK43 8QJ
M3 ZLI P. Dee 3 Cherry Orchard, Upton-Upon-Severn, Worcester WR8 0LR
M3 ZLJ Axe Vale ARC c/o J. Priestman 198 Felmongers, Harlow CM20 3DW
M3 ZLL J. Giovinazzo 3 Eleanor Avenue, Epsom KT19 9HD
M3 ZLO M. Dring 1 Hannaford Close, St. Columb Road, St. Columb TR9 6FH
M3 ZLP Z. Gribben 44 Fern Close, Birchwood, Warrington WA3 7NU
M3 ZLR L. Rodrigues 61 Grayling Way, Boston PE21 8FS
M3 ZLS L. Sammut 16 Queen Marys Road, New Rossington, Doncaster DN11 0TS
M3 ZLV M. Ferenc 2A Rosedene Avenue, London SW16 2LT
M3 ZLW D. Wakefield 12 Goddard Court, Wokingham RG41 5HR
MW3 ZLX R. Miles 63 Phillip Street, Mountain Ash CF45 4BG
M3 ZLY T. Farr 125 Rochford Way, Walton on the Naze CO14 8SP
M3 ZLZ L. Wilmott 60 Church Hill, Royston, Barnsley S71 4NG
M3 ZMB M. Bateup 47 St Johns Ave, Burgess Hill RH15 8HJ
M3 ZME M. Davison 1 Chancel Way, Barnsley S71 2HS
MI3 ZMJ M. Johnston 52 Lansdowne Road, Newtownards BT23 4NT
M3 ZMM D. Mainwaring 1 Buckingham Close, Didcot OX11 8TX
M3 ZMN M. Nash 11 Frederick St., Warrington WA4 1HX
M3 ZMO A. Maguire 132 Wigan Road, Ormskirk L39 2BA
MI3 ZMP M. Paterson 121 Ballybunden Road Killinchy, Newtownards BT23 6RZ
M3 ZMQ M. Tew Willowell, Spring Valley Lane, Colchester CO7 7SD
M3 ZMR I. Nicholson 2 Broom Close, Leyland PR25 5RQ
M3 ZMS M. Strickland Ancoats, Piercy End, York YO62 6DQ
M3 ZMT M. Thompson 133 Redford Avenue, Horsham RH12 2HH
MW3 ZMU J. Taylor 60 Rockfield Way, Undy, Caldicot NP26 3FD
M3 ZMV P. Hayward 14 Micklewright Avenue, Crewe CW1 4DF
M3 ZMX I. Botham 12 Lairgill, Bentham, Lancaster LA2 7JZ
M3 ZMZ M. Bryant 26 Coronation Road, Melksham SN12 7PF
M3 ZNC D. Crosland Simmonds Green, Varley Road, Huddersfield HD7 5TY
M3 ZNF G. Panton Dale View, Thorpe Fendykes, Skegness PE24 4QN
M3 ZNG N. Galbraith 213 Queens Road, Portsmouth PO2 7LX
M3 ZNJ A. Harvey 28 Langdown Road, Hythe, Southampton SO45 6EW
M3 ZNL J. Ho 231 Rush Green Road, Romford RM7 0JP
M3 ZNM A. Sivyer 63 Sugden Road, Worthing BN11 2JG
MM3 ZNN Dr P. Holmes Maraval, Doune Road, Dunblane FK15 9AT
M3 ZNO T. Cao Harrogate Ladies' College, Clarence Drive, Harrogate HG1 2QG
M3 ZNP R. Young 26 Silent Woman Park Coldharbour, Wareham BH20 7PE
MM3 ZNQ W. Beaton 4 Moorfield Gardens, Springfield, Cupar KY15 5SH
M3 ZNR D. Belton 4 Sandown Road, Toton, Nottingham NG9 6GN
M3 ZNT I. Govan 9 Willowbank, Sandwich CT13 9QA
M3 ZNV N. Edwards 36 Joseph Luckman Road, Bedworth CV12 8BQ
M3 ZNX T. Hayward 11 Radnor Close, Bodmin PL31 2BZ
M3 ZNY P. Robinson Flat 2, 46 Cliff Road, Sheringham NR26 8BJ
M3 ZNZ L. Burke 24 Pinecliffe Avenue, Bournemouth BH6 3PZ
M3 ZOG J. Hyde The Grove, 7 Mill Lane, Kidderminster DY10 3ND
M3 ZOH N. Kerry 3 Edinburgh Cottages, West Newton, King's Lynn PE31 6AX
M3 ZOI J. Ogden 295 Church Road, St. Annes, Lytham St. Annes FY8 3NP
M3 ZON C. Holdt 18 Garrard Road, Banstead SM7 2ER
M3 ZOO M. Lucas 20 Collin Road, Kendal LA9 5HN
M3 ZOP C. Poulson 9 Scattergate Green, Appleby-in-Westmorland CA16 6SP
M3 ZOR Z. Rabbitt 66 Parkfield Avenue, Delapre, Northampton NN4 8QB
M3 ZOU E. Nicholson 24 Barnmead, Haywards Heath RH16 1UZ
M3 ZPB P. Burgess Tally Ho Cottage, High Street, Swindon SN4 0AE
M3 ZPE P. Evans 22 Northumberland Road, Wigston LE18 4WL
M3 ZPJ P. Smyth 91Beechcroft Ave, Darcy Lever, Bolton BL26HB
M3 ZPM P. May 95 Moorfield Avenue Denton, Manchester M34 7TX
M3 ZPO L. Gravel 16 North Street, Crowle, Scunthorpe DN17 4NB
M3 ZPR P. Rogers 11 Beech Crescent, Mexborough S64 9EH
M3 ZPT P. Tatham 54 High Street, Cleckheaton BD19 3PX
M3 ZPW South Glamorgan Raynet Group c/o B. Matthews 30 Oaklands Drive, Brandon IP27 0NR
M3 ZPY L. Dobson 8 Coronation Street, Darfield, Barnsley S73 9HA
M3 ZPZ P. Shurmer 1 The Glebe, East Harling, Norwich NR16 2SZ
M3 ZQA P. Stonebridge 207 Henley Road, Ipswich IP1 6RL
M3 ZQB M. Plummer 27 Bowness Court, Congleton CW12 4JR
M3 ZQC M. Hackett 26 Wantage Road, Didcot OX11 0BP
M3 ZQF M. Ashworth 123 Forest Road, Liss GU33 7BP
M3 ZQG D. Bates 80 St. Leonards View, Polesworth, Tamworth B78 1JY
M3 ZQJ B. Yarwood 12 Charminster Close, Waterlooville PO7 7RP
M3 ZQM K. Pascoe 21 Cotswold Avenue, Sticker, St. Austell PL26 7ER
M3 ZQN A. Wheeler 8 Elsworth Grove, Birmingham B25 8EJ
MM3 ZQP J. Docherty 23 Turret Drive, Polmont, Falkirk FK2 0QW
M3 ZQV L. Marshall 65 Bacons Lane, Chesterfield S40 2SX
MM3 ZQW C. Bryson 29 Roull Road, Edinburgh EH12 7JW
MM3 ZQX A. Falconer 61 Mountcastle Drive North, Edinburgh EH8 7SP
M3 ZRA R. Alford 1 School Lane, Winmarleigh, Preston PR3 0JY
M3 ZRB R. Brown 194 Wymersley Road, Hull HU5 5LN
MM3 ZRF R. Fletcher Balnacraig Cottage, Mid Balnacraig, Alness IV17 0XL
M3 ZRG R. Gladman 18 Willingdon, Ashford TN23 5YF
M3 ZRH R. Harrison 16 Curlew Rise, Morley, Leeds LS27 8US
MW3 ZRK D. Rowlands 1 Forge Lane, Bassaleg, Newport NP10 8NF
M3 ZRM R. Mcgregor 84 Churchill Way, Burton Latimer, Kettering NN15 5RS
M3 ZRO B. Stoner Montrose, Wesley Road, Robin Hoods Bay, , Whitby YO22 4RW
M3 ZRP E. Stooke 11 Westgate, Grantham NG31 6LT
M3 ZRQ G. Crane 33A Carlisle Gardens, Horncastle LN9 5LP
M3 ZRR C. Taylor 1 Jasmine Gardens, Warrington WA5 1GU
M3 ZRS R. Stokes 44 Broxhead Road, Havant PO9 5LA
M3 ZRV R. Allwood 1, 46, Guildford GU2 7JN
M3 ZRW D. Williams 15 Charwood Road, Wokingham RG40 1RY
M3 ZRX C. Waterworth 4 Mossdale Road, Ashton-In-Makerfield, Wigan WN4 0EQ
M3 ZRY L. Acton 39 Craig Road, Macclesfield SK11 7YH
MM3 ZRZ D. Hibberd 18 Whitestripes Path, Bridge Of Don, Aberdeen AB22 8WF
M3 ZSA S. Arthur 17 Bromeswell Road, Ipswich IP4 3AS
M3 ZSB S. Bannister 162 Dobcroft Road, Sheffield S11 9LH
M3 ZSC S. Colman 197 Coppins Road, Clacton-on-Sea CO15 3LA
M3 ZSD E. Worthington 7 Bowness Close, Gamston, Nottingham NG2 6PE
MM3 ZSF S. Faccenda 27 Dollar Avenue, Falkirk FK2 7LF
M3 ZSH S. Hampson 12 Flying Fields Drive, Macclesfield SK11 7GE
M3 ZSI S. Airs 6 The Willows, Culham, Abingdon OX14 4NN
M3 ZSJ J. Gibson 22 Woodburn Drive Chapeltown, Sheffield S351YS
MM3 ZSK C. Brash 4 Union Street, Lossiemouth IV31 6BA
M3 ZSL R. Hudson 35 Pear Tree Mead, Harlow CM18 7BY
MW3 ZSM P. Leyshon 34 South Street, Porth CF39 0EG
MW3 ZSO R. Madson 44 Manor Court, Church Village, Pontypridd CF38 1DW
M3 ZST C. Butters 4 The Dovecote, Pitsford, Northampton NN6 9SB
M3 ZSU Capt. P. Westwell Roden House, Dobsons Bridge, Whitchurch SY13 2QL
M3 ZSV T. Bendelow 3 St. Giles Close, Thirsk YO7 3BU
M3 ZSW S. White 1 The Red House, Old Gallamore Lane, Market Rasen LN8 3US
M3 ZSY D. Bendelow 17 Poppy Drive Sowerby, Thirsk YO7 3SJ
MW3 ZTB T. Beach 97 Van Road, Caerphilly CF83 1LA
M3 ZTD G. Berry 5 Oakholme Rise, Worksop S81 7LJ
MW3 ZTH J. Evans 8 Hengoed Crescent, Cefn Hengoed, Hengoed CF82 7HF
M3 ZTK T. King 215 Hartland Road, Reading RG2 8DN
M3 ZTL R. Hincks 79 Forest Street, Shepshed, Loughborough LE12 9BZ
M3 ZTN A. Romanov 132 Latchmere Drive, Leeds LS16 5DY
M3 ZTP J. Wilson 14 Elms Drive, Morecambe LA4 6DQ
M3 ZTR T. Reed Channel Pool, Armathwaite, Carlisle CA4 9QY
MM3 ZTS C. Owens 4 Union Road, Camelon, Falkirk FK1 4PG
M3 ZTT A. Jeffery 14 Holly Mount Shavington, Crewe CW2 5AZ
M3 ZTU D. Stewart 79 Eastfield Road, Driffield YO25 5EZ
M3 ZTW A. Mackenzie 91 Glenwood Drive, Romford RM2 5AR
M3 ZTX J. Thornhill 47 Hopton Lane, Mirfield WF14 8JP
M3 ZUA P. Coleman 18 Carr Road, Fleetwood FY7 6QJ
M3 ZUB D. Burrows 19 Fleming Avenue, Bottesford, Nottingham NG13 0ED
M3 ZUC A. O'Keeffe 1 Elmfield Road, Liverpool L9 3BL
M3 ZUD D. Marshall 117 Ogley Hay Road, Chase Terrace, Burntwood WS7 2HU
MW3 ZUF D. Harris 15 Mill Road, Pontllanfraith, Blackwood NP12 2GE
M3 ZUG D. Duncan 12 Carrington Avenue, Poppleton Road, York YO26 4SH
M3 ZUI C. Shaw 9 Pheasant Close, Mulbarton, Norwich NR14 8BL
M3 ZUJ A. Trudgett 103 Shandon Road, Worthing BN14 9EA
M3 ZUL D. Stinson 234 Pelsall Lane, Rushall, Walsall WS4 1NG
MM3 ZUP C. Webster 13 Harbour Road, Tayport DD6 9EX
M3 ZUS S. Skinner Chevington, Carlton Road, Godstone RH9 8LD
M3 ZUT P. Streatfield 66 Ockendon Road, London N1 3NW
M3 ZUU R. Duncan 12 Carrington Avenue, Poppleton Road, York YO26 4SH
M3 ZUW J. Caithness 6 Station Road, Catworth, Huntingdon PE28 0PE
M3 ZUY B. Pike 19 Cardigan Gardens, Reading RG1 5QP
M3 ZUZ A. Finn 29 Argyle Road, Weymouth DT4 7LX
M3 ZVA S. Toher The Chapel, Station Road, Darlington DL2 1JG
MW3 ZVB G. Spencer Erw Uchaf, Gorad, Valley, Holyhead LL65 3BT
MW3 ZVD D. Jones 7 Bryn Rhedyw, Llanllyfni, Caernarfon LL54 6SS
M3 ZVF B. Lane 26 Deben Valley Drive, Kesgrave, Ipswich IP5 2FB
MW3 ZVH C. Nicholls 26 Maes Geraint, Pentraeth LL75 8UR
M3 ZVI F. Lees 5 St. Winifred Road, Rainhill, Prescot L35 8PY
M3 ZVK V. King 215 Hartland Road, Reading RG2 8DN
M3 ZVS K. Brown 142 Moor Lane, Woodford, Stockport SK7 1PJ
M3 ZVT M. Joynson-Ellis 20 Morland Court, Skaters Way, Peterborough PE4 6GW
M3 ZVU M. Flanagan 804 New Hey Road, Huddersfield HD3 3YW
M3 ZVW J. Woods Invicta Cottage, Carbrooke Road, Thetford IP25 6SD
M3 ZVX J. Barker 26 Ardley Road, Fewcott, Bicester OX27 7PA
MM3 ZVY K. Bourhill Cherry Tree Cottage, 30C Salters Road, Wallyford EH21 8AA
M3 ZWB N. Watts 404 March Road, Turves, Peterborough PE7 2DW
M3 ZWD Z. Dixon 18 Norfolk Close, Plymouth PL3 6DB
M3 ZWF E. Watson 4 Moorland Avenue, Blackburn BB2 5EQ
M3 ZWG I. Garratt 2 Hayclose Crescent, Kendal LA9 7NT
M3 ZWH R. Simmons 10 Whitehall Road, Kingswinford DY6 9DY
M3 ZWI C. Riley 15 Keddington Crescent, Louth LN11 0AP
M3 ZWJ G. Hill-Adams 6 Broadleaze Way, Winscombe BS25 1JX
M3 ZWK C. Powell 37 Newnham Close, Mildenhall, Bury St. Edmunds IP28 7PD

M3 ZWL R. Amos 6 Eccles Road, Wittering, Peterborough PE8 6AU
M3 ZWM I. Coulson 2 Marl Hurst, Edenbridge TN8 6LN
M3 ZWN N. Woodstock 8 Fernheath Close, Bournemouth BH11 8SL
M3 ZWP K. Bainbridge 29 Bluebell Grove, Calne SN11 9QH
M3 ZWQ K. Toohey 197 Broad Oak Road, St. Helens WA9 2AQ
MW3 ZWR G. Spicer 6 Cromwell Road, Neath SA10 8DR
MW3 ZWS S. Todd 4 Uplands Road, Pontardawe, Swansea SA8 4AH
M3 ZWW Dr W. Warwicker 28 Porters Wood, Petteridge Lane, Matfield, Tonbridge TN12 7LR
M3 ZWX J. Turley 19 Ibstock Drive, Stourbridge DY8 1NW
M3 ZXA N. Gleaden 25 Ridgway Avenue, Darfield, Barnsley S73 9DU
M3 ZXC S. Makins 10 Lower Mill St., Ludlow SY8 1BH
M3 ZXE N. Egginton Emergency Planning Unit, Shropshire County Council, Shrewsbury SY2 6ND
M3 ZXG G. Carless Silver Cottage, Silver Street, South Petherton TA13 5BY
MM3 ZXL S. Fradley 30 Polmont Park, Polmont, Falkirk FK2 0XT
M3 ZXN A. Walton 65 Broadway East, Rotherham S65 2XA
M3 ZXQ I. Talbot 41 Elmwood Close, Cannock WS11 6LX
M3 ZXX K. Willets 32 Yates Way, Ketley Bank, Telford TF2 0AZ
MW3 ZYE P. Mason 20 Coronation Road, Six Bells, Abertillery NP13 2PJ
M3 ZYF H. Armstrong 83 Hillary Grove, Carlisle CA1 3JQ
M3 ZYG D. Zygadllo 56 Scarisbrick Crescent, Liverpool L11 7DW
M3 ZYI I. Ball 17 Homecroft Road, Goldthorpe, Rotherham S63 9DX
M3 ZYK M. Edge 19 Burton Av, Rushall, West Mids WS41NH
M3 ZYM D. Lavender 39 Albany Crescent, Bilston WV14 0HT
M3 ZYO B. Hall 6 Marshall Close, Parkgate, Rotherham S62 6DB
MW3 ZYQ M. Daymond 58 Holmesdale Street, Cardiff CF11 7HF
M3 ZYR A. Caulfield 1 Carleton Close, Amesbury, Salisbury SP4 7TU
MM3 ZYS S. Robertson 20 Knockard Place, Pitlochry PH16 5JF
M3 ZYT N. Bond 21 Coleridge Close, Bletchley, Milton Keynes MK3 5AF
MI3 ZYU A. Kelly 19 Union Street Mews, Coleraine BT52 1EN
M3 ZYV A. Brown 7 Whilton Crescent, West Hallam, Ilkeston DE7 6PE
M3 ZYW A. Pemberton 38 Pond Green Way, St. Helens WA9 3SD
M3 ZYX T. Parker 2 Kipling Close, Grantham NG31 9ND
M3 ZYY P. Day 46 Beatrice Avenue, Saltash PL12 4NG
M3 ZYZ C. Wilmott 60 Church Hill, Royston, Barnsley S71 4NG
M3 ZZA M. Daniels 10 Downland Avenue, Peacehaven BN10 8TH
M3 ZZD D. Smith 7 Kestrel Avenue Bransholme, Hull HU7 4ST
M3 ZZE E. Olver 41 Mount Tamar Close, Plymouth PL5 2AL
M3 ZZF D. Webb 18 Lavender Avenue, Minster On Sea, Sheerness ME12 3UA
M3 ZZH H. Metson Higher Churchtown Barn, North Hill, Launceston PL15 7PQ
M3 ZZI D. Brooker-Evans 379 Crownhill Road, Plymouth PL5 2LN
M3 ZZJ L. Haley 17 Oak Drive, Crownhill, Plymouth PL6 5TZ
M3 ZZL P. Stretton 38 Queens Way Melbourne, Derby DE73 8FG
M3 ZZN A. Hobson 8 Sycamore Road, Colchester CO4 3NF
M3 ZZQ J. Snape 2 Orchard Close, Fort Avenue, Preston PR3 3YS
M3 ZZS J. Skinner 36 Milton Road, Waterloo, Liverpool L22 4RF
MW3 ZZU E. Jones 39 Ger-Y-Llan, Velindre, Llandysul SA44 5YB
M3 ZZV G. Jones 39 Thurlow Way, Barrow-in-Furness LA14 5XP
M3 ZZW G. Smith 7 Kestrel Avenue, Bransholme, Hull HU7 4ST
M3 ZZX I. Curtress 7 Gwinnett Court, Shurdington, Cheltenham GL51 4GQ

M5

M5 AA Mf Propagation Research Group c/o R. Parsons Netherhall Barn, Hallmoor Road, Darley Dale, Matlock DE4 2HF
M5 AAT R. Nikolova 206 Eastcote Av West Molesey, Surrey KT82EX
M5 AAW C. Taylor 40 Chilton Drive, Watnall, Nottingham NG16 1HL
M5 ABC D. Last Hillview, New Road, Bridport DT6 4NY
M5 ABH D. Drew 14 Greensfields, Skegby, Nottingham NG17 3DN
M5 ABJ J. Bodle 48 Bolsover Road, Hove BN3 5HP
M5 ABN P. Herbert Flat 1, 7 Cockington Lane, Paignton TQ3 1EE
MW5 ABR P. Lovelock Garn Cottage, Aberbechan, Newtown SY16 3AY
M5 ABT R. Clark 36 Southfields, Stanley DH9 7PH
M5 ACD G. Steabler 1 Westhill Road, Grimsby DN34 4SG
M5 ACF Dr E. Mclusky 11 Ripon Road, Killinghall, Harrogate HG3 2DG
M5 ACJ R. Payne 58 Sheepcote Lane, Amington, Tamworth B77 3JW
M5 ACR A. Jakusz-Gostomski 15 Goodliffe Gardens, Tilehurst, Reading RG31 6FZ
M5 ACS M. Arnfield Cleabarrow, Plumley Moor Road, Knutsford WA16 0TU
M5 ACT E. Bluer 8 Cedar Close, Waterlooville PO7 7LN
M5 ACX M. Anderson 1 Thames Close, Ferndown BH22 8XA
M5 ADA Dr C. Goodhand 22 Somin Court, Doncaster DN4 8TN
MW5 ADD A. Dimmock Gwyndy, Llandegfan, Menai Bridge LL59 5PW
M5 ADE A. Deane Flat 1-6, 76 Church Street, Tewkesbury GL20 5RX
M5 ADF D. Hook 15 Wordsworth Avenue, Sutton-in-Ashfield NG17 2GG
M5 ADI D. Williams 4 Dunkirk Rise College Bank Way, Rochdale OL12 6UH
M5 ADL A. Lambert 69 Anvil Crescent, Broadstone BH18 9DZ
M5 ADM K. Marriott 99 Stapleford Lane, Toton, Nottingham NG9 6FZ
M5 ADQ B. Woolnough 99 Abbey Road, Leiston IP16 4TA
MW5 ADW P. Booth 68 Trem Eryri, Llanfairpwllgwyngyll LL61 5JF
M5 AEC M. Drinkwater Green Quarter, Ward Green Old Newton, Stowmarket IP14 4EZ
M5 AEE A. Steadman 4 Vineyard Way, Buckden, St. Neots PE19 5SR
M5 AEF R. Burrows-Ellis 49 Highland Drive, Worlingham, Beccles NR34 7AR
M5 AEH C. Shackleton 54A Blueleighs Park, Chalk Hill Lane, Great Blakenham, Ipswich IP6 0ND
M5 AEI E. Howe 22 Freston, Peterborough PE4 7EN
M5 AEO J. Kempster 4 Church Cottages, Great Gaddesden, , Hemel Hempstead HP1 3BU
MM5 AES J. Robertson 138 East Main St., Armadale, Bathgate EH48 2PB
M5 AFE W. Higgs 955 Oldham Road, Rochdale OL16 4SE
M5 AFG D. Hall 4 Steventon Road, Wellington, Telford TF1 2AS
M5 AFH J. Denmead 47 Holland Road, Clevedon BS21 7YJ
MI5 AFL I. Mccrum 12 Bishops Court Road, Kilclief, Downpatrick BT30 7NU
M5 AFV J. Jones 10 Huntington Close, Redditch B98 0NF
M5 AFX G. Rhodes 60A Somerset Road, Cinderford GL14 2HJ
M5 AFY S. Hall 58 Lower Meadow Court, Northampton NN3 8AX
M5 AGB A. Koeller 116 Parham Road, Gosport PO12 4UE
M5 AGG C. Ellis Broken Ridge, Fir Tree Close, Ringwood BH24 2QW
M5 AGI Rvd. J. Addison 20 St. Davids Drive, Callands, Warrington WA5 9SB
MM5 AGM C. Campbell 18 Parkview Avenue, Falkirk FK1 5JX
M5 AGR G. Macadam Acacia Lodge, 10 The Green, Herts SG8 7AD
M5 AGS Capt. J. Lightfoot Flat 18, The Cloister, Wokingham RG40 1AW
M5 AGV D. Adkins 196 High Road, North Weald, Epping CM16 6EF
M5 AGW W. Green 2 Irkdale Avenue, Enfield EN1 4BD
M5 AGY A. Dermont 7 Pool Close, Little Comberton, Pershore WR10 3EL
M5 AGZ M. Gill The Cottage, Barrowell Green, London N21 3AU
M5 AHF M. Latimer-Sufit Flat 615, Jacqueline House, 52 Fitzroy Road, London NW1 8UB
MI5 AHG J. Campbell 7A Desert Road, Mayobridge, Newry BT34 2JB
MM5 AHM G. Welch Beechlea, Yieldshields Road, Carluke ML8 4QY
MM5 AHO G. Crowley 3 Park View, Westfield, Bathgate EH48 3PP
M5 AIB H. Whiteoak 18, Gregory Springs Mount, Mirfield WF14 8LG
MM5 AII L. Mccay 4 South Mound, Houston, Johnstone PA6 7DX
M5 AIQ A. Bain 5 Norgrove Park, Gerrards Cross SL9 8QT
M5 AJB J. Button 1 Ross Cottages Southey Green, Sible Hedingham, Halstead CO9 3RN
MI5 AJH E. Holmes 7 Bamford Park, Dundrod, Crumlin BT29 4JW
M5 AJK T. Dawson 54 Graeme Road, Enfield EN1 3UT
MM5 AJN D. Gerrie 13 Martin Terrace Auchnagatt, Ellon AB41 8TF
M5 AJO Dr M. Woodhouse 5 Redhill Close, Bristol BS16 2AH
M5 AJP D. Hone 16 Newton Avenue, Aylesbury HP18 0BN
MM5 AJW D. Mckay Tryggo, Sarclet, Wick KW1 5TU
M5 AJZ Dr S. Coles 54 Brasslands Drive, Portslade, Brighton BN41 2PN
M5 AKT M. Ellis1 4 Magna Crescent, Great Hale, Sleaford NG34 9JX
M5 AKW L. Howell 18 High St., Foxton, Cambridge CB2 6SP
M5 AKY D. Vosper 5 Franklyn Terrace, Farrington Gurney, Bristol BS39 6UD
MM5 AKZ I. Mcclelland Parkburn, Dumfries DG1 1RB
M5 ALA A. Shone 50 Whitefield Avenue, Norden, Rochdale OL11 5YG
M5 ALC R. Hatcher 61 Holland Road, Oxted RH8 9AU
M5 ALG A. Levy 29 Ferndale Avenue, Reading RG30 3NQ
MI5 ALJ Falkirk & District ARS c/o C. Hannigan 4 Silverhill Road, Strabane BT82 0AE
MI5 ALO E. Clementson 84 Portaferry Road, Newtownards BT23 8SN
M5 ALS D. Munro 16 Gullimans Way, Leamington Spa CV31 1LA
M5 ALU A. Horsfield 45 Burchnall Close, Deeping St. James PE6 8QJ
MM5 ALX D. Ross 37 Shillinghill, Alness IV17 0SZ
MM5 AMM Dr P. Adams 32 The Woods, Milnathort, Kinross KY13 0RA
M5 AMN A. Waddington 41 Willow Way, Farnham GU9 0NU
MI5 AMO J. Houston 9 Lismore Drive, Dundonald, Belfast BT16 1SL
MM5 AMV H. Cardwell 26 South Road, Cupar KY15 5JF
M5 AOI N. Pollard 5 The Stackfield, Wirral CH48 9XS
MM5 AON R. Henry Woodyard House, Woodyard Road, Dumbarton G82 4BG
M5 ARC Wisbech Ar & Electronics Club c/o J. Balls 70 Risegate Road, Gosberton, Spalding PE11 4EY
M5 ASK W. Booker 3 Hollybank Avenue, Sheffield S12 2BL
M5 ASR J. Richardson 9 Hilton Avenue, Aylesbury HP20 2EX
M5 ATR T. Ralph 15 Portchester Close, Stanground, Peterborough PE2 8UP
M5 AXA I. Bassett 47 Queensdown Gardens, Brislington, Bristol BS4 3JD
M5 BAD Dr C. Leese 15 Ewden Way, Barnsley S75 2JW
M5 BAE B A E Systems Great Baddow ARC c/o P. Tittensor 47 St. Johns Road, Chelmsford CM2 0TY
M5 BAZ B. Carter 25 Chester Close, Chafford Hundred, Grays RM16 6ET
M5 BBB B. Bot 52 Kenilworth Avenue, Wimbledon SW19 7LW
M5 BFL S. Shenstone Primrose Cottage, Stubb Lane, St Michael South Elmham, Bungay NR35 1ND
M5 BGR D. Wrigley 32 Avon Road, Chadderton, Oldham OL9 0PH
M5 BIL W. Cooper 16 Beaumont Hill, Dunmow CM6 2AP
M5 BIR B. Jakubowski 120 Chandos Street, Coventry CV2 4HT
M5 BJC B. Crighton 2 The Lake House, Savage Cat Farm, Gillingham SP8 5QR
M5 BMW R. Read 111 Fitzpain Road, West Parley, Ferndown BH22 8SF
M5 BOP M. Riley 3 Foxley Close, Ipswich IP3 8BW
M5 BTB P. Brown 42 Foxglove Close, Burgess Hill RH15 8UY
M5 BUF R. Hanney 74 Avon Road, Bournemouth BH8 8SF
MI5 BUG G. Moucka 5 Glebe Gardens, Moira, Craigavon BT67 0TU
M5 BXB S. Burrows 33 Pettys Close, Cheshunt, Waltham Cross EN8 0EW
M5 CAB W. Daly 85 Lordens Road, Huyton, Liverpool L14 9PA
MW5 CAD G. Cadwaladr Madog Yacht Club, Pen Y Cei, Porthmadog LL49 9AT
M5 CAH C. Hunt 105 Cropston Road, Anstey, Leicester LE7 7BQ
M5 CBR A. Hutton 29 Manor Road, Ashford TW15 2SL
M5 CBS M. Beale 8 Blakeney Avenue, Swindon SN3 3NW
MM5 CFA C. Allan 35 Hamewith Court, Alford AB33 8QW
MI5 CFM L. Baine 50 Lawnbrook Drive, Newtownards BT23 8XD
M5 CHH C. Hollins 56 Lovell Road, Cambridge CB4 2QR
M5 CJH C. Hindle 10 Barrington Meadows, Bishop Auckland DL14 6NT
MW5 CKN S. Swinden 4 Uwch Y Maes, Dolgellau LL40 1GA
M5 CLO F. Jones 19 Moreton Street, Prees, Whitchurch SY13 2EG
M5 CMO B. Armstrong 35 Northfields Crescent, Settle BD24 9JP
M5 COL H. Craven 4 Amanda Drive, Louth LN11 0AZ
MW5 CYM St. Helens Raynet Group c/o I. Taylor 7 Trellewelyn Close, Rhyl LL18 4NF
MW5 DAD P. Stevenson Nant Fach Cerrigydrudion, Corwen LL21 0SB
M5 DAP D. Parker 12 Sedge Close, Ivybridge PL21 0WD
MI5 DAW R. Bamber 15 Ladybrook Parade, Belfast BT119ER
M5 DB Willow CG c/o A. Sheppard 81 Deer Way, Horsham RH12 1PX
M5 DHB D. Brattle 34 York Street, Bedford MK40 3RJ
M5 DIK R. King 10 Bucks Avenue, Watford WD19 4AS
M5 DJC M. Cressey 33 Parklands Drive, Harlaxton, Grantham NG32 1HX
MW5 DJO E. Owen Pant-Y-Fedwen, 39 Glanrafon Estate, Bontnewydd, Caernarfon LL55 2UW
M5 DLA R. Felds 93 Bancroft Lane, Mansfield NG18 5LL
M5 DND N. Read 3A Bull Street, Bishops Castle SY9 5DB
M5 DNK D. Kennedy Holmcroft, Lewis Road, Selsey PO20 0RG
MM5 DOG K. Macdonald 5 The Stances, Kilmichael Glassary, Lochgilphead PA31 8QA
MW5 DQQ M. Poreba 7 Waterloo Place, Brynmill, Swansea SA2 0DE
M5 DR D. Rugen 19 Jacksons Close, Haskayne, Ormskirk L39 7LD
M5 DRW J. Noel 58 Easenhall Lane, Redditch B98 0BJ
M5 DUO P. Richards 16 Fruiterers Arms Caravan Park, Uphampton Lane, Ombersley, Droitwich WR9 0JW

MM5 DWW D. Wishart Curcum, Swannay, Evie KW17 2NS
M5 DX Hallam DX Group c/o N. Totterdell Moscar Cross House, Hollow Meadows, Sheffield S6 6GL
M5 DXX R. Morton 16 Penshaw Avenue, Wigan WN3 5NG
M5 DZH J. Lynch 21 Worthington Avenue Hopwood, Heywood OL10 2LN
M5 EAY Christchurch ARS c/o G. Hobson 30 Leigh Road, Westbury BA13 3QL
M5 ECX T. Watts 6 Keynsham Walk, Swindon SN3 2AL
MI5 EEM W. Hesketh 49 Mount Michael Park, Belfast BT8 6JX
M5 EHG R. Bateman 81 Stanton St., Derby DE23 6NF
M5 EI Milton Keynes ARS c/o Dr P. Gould 152 High Street North, Stewkley, Leighton Buzzard LU7 0EP
M5 ENM R. Gurowich 1 St. Cuthberts Villas, Haybridge, Wells BA5 1AH
M5 ERN E. Coleby 13 Farm Close, Sunniside, Newcastle upon Tyne NE16 5PP
M5 ESA A. La Fauci 50 Gore Road, London SW20 8JL
M5 EXY M. Brashill 42 Bannister Street, Withernsea HU19 2DT
M5 FAB S. Sawyers Sleepy Meadow House, Cliburn, Penrith CA10 3AL
M5 FOX D. Ross 37 Cartmell Drive, Leeds LS15 0NQ
MM5 FWD S. Robertson Grove Cottage, 30 Commerce Street, Insch AB52 6HX
M5 GAC G. Pendrick 23 Hazel Drive, Spondon.De21 7Ds, Derby DE21 7DS
M5 GDK M. Gadeke 3 Ripon Close, Towcester NN12 6PL
M5 GHT G. Thompson 15 Caithness Road, Hylton Castle Estate, Sunderland SR5 3RE
M5 GJO G. Orlebar 21 Field Lane, Willersey, Broadway WR12 7QB
M5 GTA G. Tappenden 6 Longtail Rise, Herne Bay CT65PZ
M5 GUS R. Guscott 19 Springfield Way, Threemilestone, Truro TR3 6BJ
M5 GUY G. Austin 23 Ngunguru Heights Rise, Ngunguru, Whangarei 173 New Zealand
M5 GVY N. Drumm 8 Harpur Place, Thornhill, Egremont CA22 2SG
M5 GWH L. Preece 29 Elliott Street, Newcastle under Lyme ST5 1JL
M5 HDF Midland CG c/o M. Waldron 32 Windmill Street, Upper Gornal, Dudley DY3 2DQ
M5 HFJ J. Glover 22 Hampden Road, Birkenhead CH42 5LH
MI5 HIL Gt Lumley ARS c/o B. Hill 5 Heathers Close, Magheralin, Craigavon BT67 0RN
MI5 HNA C. Archibald 37 Jellicoe Drive, Belfast BT15 3LA
MW5 HOC D. Warburton 71 Richards Terrace, Cardiff CF24 1RW
M5 HOT M. Palmer 21 Ibbett Close, Kempston, Bedford MK43 9BT
M5 IC Invicta CG c/o I. Lowe 54 College Road, Margate CT9 2SW
M5 IGE D. Russell 90 Halleys Way, Houghton Regis, Dunstable LU5 5HZ
MI5 IMB D. Barnes 14 Old Station Road, Belleek BT93 3EZ
M5 IMI C. Wilson The Rectory, Church Close, Thetford IP25 7LX
MM5 ISS G. Milne 19 Fairview Crescent, Danestone, Aberdeen AB22 8ZB
M5 ITE P. Hayler 27 Birch Way, Heathfield TN21 8BB
M5 JAO J. Owen 10 Pitcher Lane, Leek ST13 5DB
M5 JB J. Booth 27 Moorlands Scholes, Holmfirth HD9 1SW
M5 JON J. Edmunds Caroline Cottage, New Passage Road, Bristol BS35 4LZ
M5 JPG J. Griffiths 9 Cauldale Close, Middleton, Manchester M24 5SU
M5 JWR J. Richardson 6 Clarence Road Scorton, Richmond DL10 6EE
M5 JWS J. Summers 3 Thatchers Close, Burgess Hill RH15 0QU
MI5 JYK P. Lowrie 13 Carwood Park, Newtownabbey BT36 5JU
M5 KEN J. Sharples 21 Alexandra Pavilions, Stanleyfield Close, Preston PR1 1QW
M5 KHH C. Young 26 Horsham Avenue, Peacehaven BN10 8HX
M5 KJM K. Murphy 79 Torkington Road, Hazel Grove, Stockport SK7 6NR
M5 KVK Blackpool Amateur Radio Network c/o G. Howell 19 Constable Avenue, Eaton Ford, St. Neots PE19 7RH
M5 KZI M. Hillary 45 Frances Road, Purbrook, Waterlooville PO7 5HH
M5 LAR S. Pounder 4 Otley Mount, East Morton, Keighley BD20 5TD
M5 LDF L. Farrington 130 Millway Road, Andover SP10 3AY
M5 LMG L. Griffiths 51 Linley Avenue, Pontesbury, Shrewsbury SY5 0TL
M5 LMY D. Sweetland 15 Wasdale Close, Owlsmoor, Sandhurst GU47 0YQ
M5 LRO C. Pickett 1 Woolwich Close, Chatham ME5 0HU
MI5 LYN E. Lynn 60 Lurgan Tarry, Lurgan, Craigavon BT67 9HN
M5 MC Fwe CG c/o A. Smith 12 Northgate, Beccles NR34 9AS
M5 MCH M. Heenan 15 Woodacre Green Bardsey, Leeds LS17 9AB
M5 MDH M. Hampton 126 The Crescent, Eastleigh SO50 9BH
M5 MDX Stockport RS c/o B. Naylor 47 Chester Road Poynton, Stockport SK12 1HA
M5 MK Milton Keynes Amateur, RS c/o T. Cowell 102 Stamford Avenue, Springfield, Milton Keynes MK6 3LQ
MI5 MTC M. Clarke 19 Ardlougher Road Irvinestown, Enniskillen BT94 1RN
M5 MUF M. Johnson 1 Ferndale Drive, Ratby, Leicester LE6 0LH
MW5 MWR M. Randall 15 Erw Wen, Pencoed, Bridgend CF35 6YF
M5 NEV N. Bridle 95 Kings Stone Avenue, Steyning BN44 3FJ
M5 NEX T. Jones 2 Holly Road, Penketh, Warrington WA5 2AG
M5 OMO M. Legg 3 The Avenue, Wincanton BA9 9HH
M5 OOO H. Clayton 3 Waterden Court, Queensdale Place, London W11 4SQ
M5 ORC Oldham ARC c/o G. Oliver 158 High Barn St., Royton, Oldham OL2 6RW
M5 OTA A. Hickson Randolph Cottage, France Lynch Chalford Hill, Stroud GL6 8LH
M5 PDL S. Symonds 32 Redhoave Road, Poole BH17 9DU
M5 PGC P. Challans Flat 4, Sandringham Court, 2 Chandos Square, Broadstairs CT10 1QN
M5 PIP Dr P. Nicolson 56 Serpentine Road, Harborne, Birmingham B17 9RE
M5 PM P. Molloy 4 Tilt Meadow, Cobham KT11 3AJ
M5 POO S. Robinson 23 Jameson Drive, Corbridge NE45 5EX
MM5 PSL P. Leybourne 13 Sanblister Place, Virkie, Shetland ZE3 9JX
M5 PSW P. Walking Flat 36, Highlands House, Wharncliffe Road, Southampton SO19 7GG
M5 PWR Buxton Radio Amateurs c/o P. Collins 50 Seacroft Esplanade, Skegness PE25 3BE
M5 PYE E. Boyd 7 Fritton Court, Haverhill CB9 8LX
M5 RAG R. Mullen 4 Bay View Grove, Barrow-in-Furness LA13 0EQ
M5 REG R. Barber 35 Lower Park Crescent, Bishop's Stortford CM23 3PU
M5 REM M. Remplakowski 11 De Montfort Road, Speen, Newbury RG14 1TA
M5 REV Rvd. R. Moll Penn Cottage, Green End, Buckingham MK18 3NT
M5 REW R. Walker 159 Cuckfield Road, Hurstpierpoint, Hassocks BN6 9RT
M5 RFD C. Wardale 18 Wolsey Way, Lincoln LN2 4QH
M5 RHG R. Gower 21 Saltings Crescent, West Mersea, Colchester CO5 8GG
M5 RHS M. Krupa 15 Pasture Avenue, Goole DN14 6LG
M5 RIC R. Brokenshaw Flat 11, Stonehill Court, 27 Westhill Road, Weymouth DT4 9NB
M5 RJC R. Crowther 14 Alice Close, Hereford HR1 1XQ
MI5 RJS R. Scott 98 Aghafad Road Dunamanagh, Strabane BT82 0QQ
M5 RMF R. Fisher 65 Kylemore Avenue, Mossley Hill, Liverpool L18 4PZ
M5 ROB R. Johnson 4 Ton Lane, Lowdham, Nottingham NG14 7AR
MI5 ROY R. Arthur 3 Stevenson Park Tullyally, Londonderry BT47 3QS
M5 RPT R. Tickle 5 Bramley Court, Harrold, Bedford MK43 7BG
M5 RST D. Warren 10 Meadow Bank, Meadow Lane, Alfreton DE55 2BR
M5 SB Stewart Bryant Radio Station c/o Dr S. Bryant 154 London Road North, Merstham, Redhill RH1 3AA
M5 SE R. Johnston 60 Green Lane, Seaforth, Liverpool L21 3UB
MM5 SHA Dr S. Stephenson 7 Charlotte Gardens, Aberdeen AB25 1LW
M5 SJM S. Mcbain 13A St. Lukes Close, Cherry Willingham, Lincoln LN3 4LY
M5 SJS J. Covel 111A Balmoral Drive, Southport PR9 8QH
M5 SKY Believe In Better Cg c/o A. Mace 14 High Street, Melbourn, Royston SG8 6EB
M5 SLC S. Collis 11 Luxton Way, Wiveliscombe, Taunton TA4 2BW
M5 SRE P. Scott 15 Victoria Drive, Blackwell, Alfreton DE55 5JL
M5 SSB D. Rogers 12 Archdale, High Wycombe HP11 2JP
M5 SUE S. Coombs 10 Horseshoe Walk, Widcombe, Bath BA2 6DE
M5 SW M. Williamson 5 John F Kennedy Walk, Tipton DY4 0SF
M5 TAM J. Bore 14 Westwood Avenue, Lowestoft NR33 9RH
MI5 TCC T. Campbell 27 Silverbrook Park, Newbuildings, Londonderry BT47 2RD
M5 TLA M. Robinson 19 St. Wilfrids Crescent, Brayton, Selby YO8 9EU
MW5 TLE T. Evans Gwauntrebeddau, Tregynon, Newtown SY16 3ER
M5 TMG T. Green 1 Forest Road, Blidworth, Mansfield NG21 0SJ
M5 TNT S. Purdy Wetheral House, Great Salkeld, Penrith CA11 9NA
M5 TRT Dr C. Hardmeier Robert Jacobsens Vej 80, 2.1., Copenhagen 2300 Denmark
M5 TT M. Knowles 11 Thorneycroft Avenue, Birkenhead CH41 8HJ
M5 TTT J. Chisholm 162 Ardington Road, Northampton NN1 5LT
M5 TUE N. Wadsworth Haygarth, Docker, Kendal LA8 0DF
MM5 TUW G. Collie Newton Cottage, Newton Avenue Elderslie, Elderslie, Johnstone PA5 9BE
M5 TWO C. Van Zuilen Stiermarkenweg 12, Alkmaar 1827 EK Netherlands
M5 TXJ D. Shaw Cotehouse, Bleatarn, Appleby-in-Westmorland CA16 6PX
M5 WAH W. Hetherington The Laurels 8 Kings Lane, Yelvertoft NN6 6LX
M5 WGD W. Dalzell 9 Pyms Lane, Crewe CW1 3PJ
MM5 WIG I. Macdonald Benvoir, Lightlands Avenue, Wigtown, Newton Stewart DG8 9EE
M5 WIZ R. Wiseman 12A Breadcroft Lane, Harpenden AL5 4TE
MI5 WJB W. Blanchflower 7 Casaeldona Park, Belfast BT6 9RB
M5 WJF W. Faulkner 49 Oakfield Road, Shrewsbury SY3 8AD
M5 WNS W. Sampson Rowena, Clifford Street, Chudleigh, Newton Abbot TQ13 0LH
MW5 WRG Wales Digital Radio Group c/o D. D'Mellow 2 Twynpandy Pontrhydyfen, Port Talbot SA12 9TW
M5 WSS Gloucester Amateur Radio & Electronics Society c/o P. Norman 3 The Gables, Waterloo Road, Wellington TA21 8JB
M5 XYZ C. Edgar 9 Winchester Avenue, Morecambe LA4 6DX
M5 YEX R. Hooperok 1 Yale Close, Washingborough, Lincoln LN4 1DX
MM5 YLO N. Marriott Parkview, Dunrossness, Shetland ZE2 9JG
M5 ZAP A. Morgan 153 Beanfield Avenue, Coventry CV3 6NY
M5 ZZZ S. Burke 17 The Crescent, Wragby, Market Rasen LN8 5RF

M6

MM6 AAA A. Martin Cathcart, Farr, Inverness IV2 6XJ
M6 AAC R. Johns 27 Stoneby Drive, Wallasey CH45 0LG
M6 AAD A. Mullinder 15 Withington Close, Oakengates, Telford TF2 6JR
M6 AAE A. Ellsom 21 High Street, Isle of Grain ME3 0BJ
M6 AAG A. Taylor 4 Oxford Street, Carnforth LA5 9LG
M6 AAJ J. O'Brien 16 Arnold Road, Darlington DL1 1JG
M6 AAK A. Comber 7 Quantock Close, Rushmere St. Andrew, Ipswich IP5 1AS
M6 AAL M. Castro 11 Monfa Avenue, Stockport SK2 7BH
MM6 AAM K. Mcclung 23 Maple Grove, Troon KA10 6QW
M6 AAO J. Holmes 15 Nash Close, Farnborough GU14 0HL
M6 AAP K. Metcalfe 33 Corsican Drive, Hednesford, Hednesford WS12 4SS
MM6 AAR R. Donaghy 66 Newton Road, Dundee DD3 0LT
M6 AAU C. Todd 35 Main Road, Uffington, Stamford PE9 4SN
MW6 AAW A. West Bryn Goleu, Gwalchmai, Holyhead LL65 4SW
M6 AAX A. Garman 50 Heys Road, Prestwich, Manchester M25 1JY
M6 ABB A. Bocutt 15 Peel Avenue, Frimley, Camberley GU16 8YT
M6 ABE A. Egan 4 Rutter Avenue, Warrington WA5 0HP
M6 ABG J. Larcombe 3 Archer Terrace, Plymouth PL1 5HD
M6 ABN P. Fulbrook 167 Droitwich Road, Fernhill Heath WR3 7TZ
MM6 ABO K. Balfour 55B Cockels Loan, Renfrew PA4 0NE
M6 ABQ R. Fiddy 19 Ingham Road Stalham, Norwich NR12 9DR
M6 ABR A. Cost 55 Whitcliffe Grange, Richmond DL10 4ET
M6 ABS A. Skarzynski 1 River View Moorings, Bridge Road, Stoke Ferry, King's Lynn PE33 9TS
M6 ABT E. Johnston 57 Bishop Ken Road, Harrow HA3 7HU
M6 ABV M. Magnall 18 Osprey Avenue, Westhoughton, Bolton BL5 2SL
M6 ABX S. Townsend 8 Heather Gardens, Belton, Great Yarmouth NR31 9PP
M6 ABZ A. Jones 41 Milton Brow, Weston-Super-Mare BS22 8DD
M6 ACA W. Shelley 91 Canterbury House, Stratfield Road, Borehamwood WD6 1NT
M6 ACB P. Flood 115 Court Farm Road, Newhaven BN9 9DY
M6 ACC A. Cheng Pastures Green, Firwood Road, Virginia Water GU25 4NG
M6 ACD T. Barker 93 Hughenden Road, St. Albans AL4 9QN
M6 ACE D. Murphy 23 Lowndes Close, Stockport SK2 6DW
M6 ACF A. Fulton 17 Rowan Garth, Skidby, Cottingham HU16 5TT
MM6 ACI D. Simpson Bridgefoot Of Ironside Cottage, New Deer, Turriff AB53 6UP
M6 ACJ A. Petrie 17 Brecon Close, Ashington NE63 0HT
MM6 ACM A. Crawford Hillview, Kintore, Inverurie AB51 0XX
M6 ACP G. Dale 26 Kensington Close, Houghton Regis, Dunstable LU5 5TJ
M6 ACQ M. Ward 22 Old Road, Leighton Buzzard LU7 2RE
MM6 ACV R. Moore 96 Queen Street, Castle Douglas DG7 1EG
MM6 ACW A. Woodford Nordkette, Evnabrek, Levenwick, Shetland ZE2 9GY
M6 ACX A. Clark 386 Wold Road, Hull HU5 5QG
MM6 ACY A. Connelly 40 Queen Street, Castle Douglas DG7 1HS
MW6 ACZ S. Holmes 26 Station Road, Llanrwst LL26 0EP
M6 ADB A. Banks 2 Holt Close, Farnborough GU14 8DG
M6 ADD A. Dade 15 Barton Road, Berrow, Burnham-on-Sea TA8 2LT
M6 ADE A. Smith 93 Sheriffs Highway, Gateshead NE9 6QN
MW6 ADF E. Pierce 26 Station Road, Aberconwy Aerials, Llanrwst LL26 0EP
M6 ADG A. Gladman 19 Colchester Road, Wymering, Portsmouth PO6 3RH

M6 ADI W. Knight 1 Caravan Site, Drakes Drive, St. Albans AL1 5AE
M6 ADJ H. Murray 39 Warneford Way, Leighton Buzzard LU7 4JG
M6 ADM A. Mccallum 15 Leabank, Newcastle upon Tyne NE15 7LN
MJ6 ADQ J. Crowder 90 Hue Court, Hue Street, Jersey JE2 3RX
MW6 ADS A. Scott 11 Clive Road, St. Athan, Barry CF62 4JD
M6 ADT A. Carman 26 Coronation Close, Happisburgh, Norwich NR12 0RL
MW6 ADU D. Jones 1 Brig Y Nant, Llangefni LL77 7QD
M6 ADX D. Richards Flat 40, Leander Court, Teignmouth TQ14 8AQ
M6 ADY A. Slim Troublesome Reach, Playford Road, Woodbridge IP13 6ND
MW6 ADZ A. Lewis 3 Aster View, Port Talbot SA12 7ED
MM6 AEB K. Hay 13 Braehead, Langholm DG13 0PS
M6 AEE S. Hedgecock 37 Tennyson Road, Maldon CM9 6BE
M6 AEG C. Holmes 8 Byron Way, Caister-On-Sea, Great Yarmouth NR30 5RW
M6 AEI S. Woodmore 66 Imperial Way, Chislehurst BR7 6JR
M6 AEJ T. Roberts 62 Ullswater Avenue, Jarrow NE32 4EY
M6 AEK J. Clark-Mcintyre 6 Belvedere Road, Blackburn BB1 9NS
M6 AEL J. Hofman Brookside, High Street, Stockbridge SO20 6EY
M6 AEM D. Barraclough Flat 18 89 Park Road, London SW19 2BD
MW6 AEN D. Field 5 Clos Crugiau, Rhydyfelin, Aberystwyth SY23 4RN
M6 AEQ B. Livesey 14 Sycamore Avenue, Tyldesley, Manchester M29 8WQ
M6 AER C. Fielden 44 Hylion Road, Leicester LE2 6JE
M6 AES A. Southwell 56 Lambrook Road, Taunton TA1 2AF
M6 AET G. Hunt 35 Outram Street, Sutton-in-Ashfield NG17 4BA
M6 AEU A. Seymour Home Farm, Lodge Lane, Northampton NN6 7PQ
M6 AEW A. Cockett 10 San Marcos Drive, Chafford Hundred, Grays RM16 6LT
MW6 AEX A. Bayliss 2 Hillside Terrace, Tonypandy CF40 2HJ
M6 AEY J. Smith 140 Lowmoor Road, Kirkby-In-Ashfield, Nottingham NG17 7JE
M6 AEZ R. Gore 4 Westaway Park, Yatton, Bristol BS49 4JU
M6 AFA A. Lambert 1 Glebelands, Lympstone, Exmouth EX8 5JD
M6 AFB A. Brown 45 Saffron Park, Kingsbridge TQ7 1RW
M6 AFC L. Hillier 34 Kittiwake Close, Herne Bay CT6 6JS
M6 AFF J. Hone 12 Marlborough Close, Exmouth EX8 4NA
M6 AFG C. Stacey 9 Nile Road, Southampton SO17 1PF
M6 AFH M. Hillam Common Farm, Swinefleet, Goole DN14 8DW
MI6 AFI G. Todd Dp1, Shanaghy, Enniskillen BT92 0EQ
MW6 AFK A. Studdart 24 Wepre Park, Connah'S Quay, Deeside CH5 4HN
M6 AFL A. Lawrence 39 Newbury Drive, Daventry NN11 0WQ
MI6 AFM J. Rice Shanaghy, Lisnaskea, Enniskillen BT92 0EQ
M6 AFN J. Hitchens 57 Batchelor Way, Downton, Salisbury SP5 3FN
MM6 AFQ J. Watt 10 Marshall Gardens, Luncarty, Perth PH1 3YX
M6 AFS P. Carter 18 Park Road, Allington, Grantham NG32 2EB
M6 AFU T. Turner 23 Pankhurst Drive, Bracknell RG12 9PS
M6 AFW S. Coombes Pantiles, South Crescent, Skegness PE24 5RQ
M6 AFX S. O'Donnell 26 Park Avenue, Skegness PE25 2TF
M6 AFY W. Haddock 5 Bradley Close, Middlewich CW10 0PF
M6 AGA G. Wright 21 Larkfield Road, Redditch B98 7PL
M6 AGD J. Salter 20 Burrow Road, Chigwell IG7 4HQ
MD6 AGF J. Kaighin 10 Kerrocruin, Kirk Michael, Isle of Man IM6 1AF
M6 AGG A. Greig 17 Begbroke Lane, Kidlington OX5 1RN
M6 AGH Dr A. Holt 36 The Maltings, Malmesbury SN16 0RN
M6 AGK R. Hambly 144 Station Road, Irchester, Wellingborough NN29 7EW
MM6 AGL R. Perkins 33 Station Road Newmachar, Aberdeen AB21 0NS
M6 AGO Dr B. Denyer-Green Dunsley South, Park Road, Forest Row RH18 5BX
M6 AGQ K. Jones 40 Sandrock Hill Road, Wrecclesham, Farnham GU10 4RJ
M6 AGR A. Rayner 12 Newhaven Drive, Lincoln LN5 9UF
MW6 AGS D. Evans 1 Heol Glyndwr, Fishguard SA65 9LN
M6 AGT R. Macdonald 31 Addison Drive, Stratford upon Avon CV377PL
MI6 AGV A. Hamilton 9 Slievenamaddy Avenue, Newcastle BT33 0DT
M6 AGY D. Pearson 73 Stackwood Avenue, Barrow-in-Furness LA13 9HJ
M6 AGZ J. Gore 88 Rowan Drive, Kirkby-In-Ashfield, Nottingham NG17 8FR
M6 AHA P. Whitmore 3 Crown Bank, Talke, Stoke-on-Trent ST7 1PT
MM6 AHB A. Hearty 42 Grange Avenue, Wishaw ML2 0AQ
M6 AHD A. Dossa 24 Warwick Drive, Cheshunt, Waltham Cross EN8 0BW
M6 AHE E. Yohn Ortner House Farm, Abbeystead, Lancaster LA2 9BD
M6 AHF W. Starkey Flat 13, Ramsden Court, Barrow-in-Furness LA14 2HH
M6 AHH A. Hall 14 Stanelow Crescent, Standon, Ware SG11 1QF
MM6 AHJ W. Noon 0/1 445 Royston Road, Glasgow G21 2DE
MM6 AHK R. Renshaw Smithy House, Scotscalder, Halkirk KW12 6XJ
M6 AHL R. Butterfield 35 Bede Crescent, Benington, Boston PE22 0DZ
MM6 AHN A. Macdougall 8 Mount Stuart Drive, Wemyss Bay PA18 6DX
MI6 AHO A. Ismay 21 Hillsborough Drive, Belfast BT6 9DS
M6 AHP A. Meekins 12 Myrtle Road, Kettering NN16 9TW
MW6 AHQ D. Williams 23 Parc Y Ffynnon, Ferryside SA17 5TQ
MI6 AHR T. Grzybek 75 Kilburn Street, Belfast BT12 6JT
M6 AHS A. Strong 55 Coley View, Halifax HX3 7EB
MW6 AHT O. Davis 18 Ty Gwyn Drive, Brackla, Bridgend CF31 2QF
M6 AHU K. Lambert 38 Whittleford Road, Nuneaton CV10 9HU
MW6 AHV D. Burton 11 Cwrt-Ucha Terrace, Port Talbot SA13 1LD
M6 AHW A. Whitehead 82 High Park Avenue, Stourbridge DY8 3NA
MM6 AHX M. Mason 101 Avalon Gardens, Linlithgow Bridge, Linlithgow EH49 7PL
MM6 AHY J. Liddell 49 Inchbrae Road, Glasgow G52 3HA
M6 AHZ P. Dawes 49 Altofts Lodge Drive, Altofts, Normanton WF6 2LB
M6 AIA A. Barton 51 Fieldhead Gardens, Dewsbury WF12 7SN
M6 AIB C. Durrant 12 Witney Close, Ipswich IP3 9QF
MW6 AIC Dr R. Morgan Spinning Wheel, Derwydd, Llandybie, Ammanford SA18 2LX
M6 AIE C. Ring 29 Shelley Close, Newport Pagnell MK16 8JB
M6 AIF K. Ball 29 Heather Grove, Wigan WN5 9PJ
M6 AIH A. Bailey 49 Grange Crescent, Gosport PO12 3DS
M6 AII B. Ansell 7 Bramley Green Road, Bramley, Tadley RG26 5UE
M6 AIJ R. Tarling Moonrakers, Ashley, Box, Corsham SN13 8AN
MM6 AIK D. Seivwright 35 Invercauld Gardens, Aberdeen AB16 5RR
M6 AIL M. Barba 5 Vetch Way, Andover, Hampshire SP11 6RR
M6 AIN C. Campbell 5 Ryebank, Holmfirth HD9 1EU
M6 AIO A. Hadfield 12 Manor Road, Caister-On-Sea, Great Yarmouth NR30 5HG
M6 AIP K. Mcfadden 3A Talbot Rd, Northampton NN1 4HZ
M6 AIQ L. Fletcher 110 West Street, North Creake, Fakenham NR21 9LH
M6 AIR M. Ward Flat 1, The Old Chapel, Chapel Street, Holsworthy EX22 6AY
M6 AIT J. Barker-Gunn 5 De Montfort Road, Lewes BN7 1SP
M6 AIU D. Crashley 14 Arran Close, Nuneaton CV10 7JX
M6 AIW A. Nutt 9 Hereford Road, Southport PR9 7DX
M6 AIZ M. Burbeck 5 Wouldham Terrace, Saxville Road, Orpington BR5 3AT
MI6 AJA J. Bingham 27 Carrickdale Gardens, Portadown, Craigavon BT62 3BN
M6 AJB A. Brown Ponsharden Cottage, High Offley Road, Stafford ST20 0LG
M6 AJC A. Cosham 1 The Orchard, Redisham, Beccles NR34 8PA
MW6 AJD R. Ware 147 Terrace Road, Swansea SA1 6HU
M6 AJF A. Fysh 74 Kingsway, King's Lynn PE30 2EL
M6 AJH A. Higham 30 Broome Road, Southport PR8 4EQ
MM6 AJI M. Mcgrorty 59 Craighall Street, Stirling FK8 1TA
M6 AJJ M. Baker 8 Higher Polsham Road, Paignton TQ3 2SY
M6 AJL A. Lashbrook 1 Fortescue Road, Exeter EX2 8LA
M6 AJM A. Mitchell 18 Holly Leys, Stevenage SG2 8JA
MI6 AJN A. Ruddell 16 Beechfield Manor, Aghalee, Craigavon BT67 0GB
MI6 AJO I. Hoey 58 Tullynamullan Road, Shankbridge, Ballymena BT42 2LR
M6 AJP A. Pilkington 26 Ryelands Close, Market Harborough LE16 7XE
M6 AKA G. Adgie Flat 1, 6 Grayfield Avenue, Birmingham B13 9AD
M6 AKE I. Brayley 9 Church View Close, Southampton SO19 8SJ
M6 AKF P. Edwards 8 Aughton Way, Broughton, Chester CH4 0QE
M6 AKG H. Turner 39 Court Crescent, Kingswinford DY6 9RJ
M6 AKH S. Ward 11 Blenheim Drive, Hawkinge, Folkestone CT18 7FA
M6 AKI B. Smith Maple Lodge, Burtoft Lane South, Boston PE20 2PF
M6 AKJ L. Lee 8 William Avenue, Margate CT9 3XT
M6 AKK A. Forsythe 24 The Welkin, Lindfield, Haywards Heath RH16 2PH
M6 AKO S. Clarke 27 Coronation Road, Callington PL17 7BX
MM6 AKQ S. Burns 3/R 170 Lochee Road, Dundee DD2 2NH
M6 AKT A. Millin 79 Court View, Stonehouse GL10 3PJ
M6 AKV A. Vivian 15 Kew Klavji, Rhind Street, Bodmin PL31 2FE
M6 AKW I. Duffie Trebeighan Farm, Saltash PL12 5AE
M6 AKY Dr R. Bleaney 40 Broadstone Road, Harpenden AL5 1RF
MM6 AKZ D. Henderson 10 Rye Crescent, Glasgow G21 3JS
M6 ALB A. Burns 76, 76, Morprth NE65 0TF
MW6 ALC A. Cotter 3 George Street, Aberdare CF44 6RY
M6 ALE D. Trollope 80 Azalea Drive, Trowbridge BA14 9GG
M6 ALG M. Yates 8 Johnsons Street, Ludham, Great Yarmouth NR29 5NZ
M6 ALH A. Higham 30 Broome Road, Southport PR8 4EQ
M6 ALJ A. Jones 23 Cranley Road, Hersham, Walton-on-Thames KT12 5BT
M6 ALK A. Kent 4 Sellerdale Drive, Wyke, Bradford BD12 9DA
MI6 ALL A. Armstrong 45 Rathmena Drive, Ballyclare BT39 9HZ
M6 ALM T. Hamilton Flat 14, Kingsmead Court, Dunstable LU6 1NQ
M6 ALO A. Nugorski 49 Buxton Street, Morecambe LA4 5SR
M6 ALP R. Mcleod 75 Davis Street, Stanley FIQQ 1ZZ Falkland Islands (Malvinas)
M6 ALQ S. Hassall 21 Bridgnorth Grove, Chesterton, , Newcastle under Lyme, ST5 7QP
M6 ALT D. Ingrey 1 Ponders Road, Fordham, Colchester CO6 3LX
M6 ALU J. Harrington 69 Lexden Road, Colchester CO3 3QE
M6 ALW J. Anderson 15 Dickson Road, London SE9 6RA
M6 ALX A. Jones 5 Meadowlands, Kirton, Ipswich IP10 0PP
M6 ALY A. Hall 254 Walton Road, Walton on the Naze CO14 8LT
MM6 AMA A. Campbell 1B Craig Road, Troon KA10 6DA
M6 AMC A. Cottrell 4 Oak Grove, Armthorpe, Doncaster DN3 2DJ
M6 AMD Dr A. Dingle 29 Castle View, Witton Le Wear, Bishop Auckland DL14 0DH
M6 AMG J. Smith 2 Sunfields Close, Polesworth, Tamworth B78 1LW
M6 AMI T. Gladman 39 Fairview Avenue, Rainham RM13 9RL
M6 AMJ B. Jones 9 St. James Close, Hanslope, Milton Keynes MK19 7LF
M6 AML A. Lennon 12 Lockgate East, Windmill Hill, Runcorn WA7 6LB
M6 AMN C. Johnson 58 Cheviot Road, London SE27 0LG
M6 AMO A. Fleming 39 Urswick Green, Barrow-in-Furness LA13 0BH
M6 AMR A. Riddick 30 Britannia Road, Banbury OX16 5DW
M6 AMT A. Taylor 11 Fillingfir Drive, Leeds LS16 5EG
M6 AMV G. Lyon 1 Eckersley Street, Wigan WN1 3PP
M6 AMW A. Webb 3 Blackmore Close, Thame OX9 3ZH
M6 AMX T. Brown 6 Wolstanholme Close, Congleton CW12 3RX
MM6 ANB P. Rice 255 Eskhill, Penicuik EH26 8DF
M6 AND A. Faulkner Northwood, Cranham, Gloucester GL4 8HB
M6 ANE D. Mansfield 1 Arun Court, Lower Street, Pulborough RH20 2DD
M6 ANI P. Hanman 7 Tremenheere Road, Penzance TR18 2AH
M6 ANM E. Seymour Home Farm, Lodge Lane, Northampton NN6 7PQ
M6 ANN E. Mann 6 London Road St. Georges, Telford TF2 9LQ
M6 ANO J. Seymour Home Farm, Lodge Lane, Northampton NN6 7PQ
M6 ANP A. Coulthard 47 Keston Crescent, Stockport SK5 8NQ
M6 ANV G. Parkinson 11 Curtis Road, Poole BH12 3AQ
M6 ANX P. Tillotson 9 Holker Street, Barrow-in-Furness LA14 5RQ
M6 ANY M. Tuffill 1 Madden Close, Nottingham NG5 5US
M6 AOB A. Back 34 Willow Road, Redhill RH1 6LW
MM6 AOD R. Gauld 20F Holland Street, Aberdeen AB25 3UL
MW6 AOE S. Parfitt 14 Heol Y Twyn, Rhymney, Tredegar NP22 5DW
M6 AOF S. Salisbury 6 Ryan Close, Leyland PR25 2XW
M6 AOG M. Whelan 134 Fields Farm Rd, Hyde.Cheshire. SK14 3QW
MM6 AOH S. Mckenzie 0/2 69 Glenkirk Drive, Glasgow G15 6AU
M6 AOJ J. Burnham 5 Shirley Court, Torwood Gardens Road, Torquay TQ1 1TZ
M6 AOK C. Macleod 21 Halsetown, St. Ives TR26 3LY
M6 AOL P. Butler 38 Oak Hill, Hollesley, Woodbridge IP12 3JY
MM6 AON A. Watson 10 Christie Place, Elgin IV30 4HX
M6 AOP O. Ernster 19 Rose Gardens, Farnborough GU14 0RW
M6 AOQ M. Keilty 17 Cliff Road, Wallasey CH44 3DJ
MI6 AOR A. O'Reilly 15 Killowen Point, Rostrevor, Newry BT34 3AN
M6 AOT Dr F. Fang 21 Loris Court, Cambridge CB1 9GF
M6 AOV R. Simpson 22 Kenworthy Road, Stocksbridge, Sheffield S36 1BZ
M6 AOW A. Edwards 2 Keystone Gardens, Steventon New Road, Ludlow SY8 1LE
MI6 AOX T. Darrah 42 Pinewood Avenue, Carrickfergus BT38 8EW
MW6 AOY J. Regan 30 Tynybedw Terrace, Treorchy CF42 6RL
MI6 AOZ D. Mudd 14 Bloomfield Road, Belfast BT5 5LT
M6 APA A. Armstrong 30 Tennyson Avenue, Hull HU5 3TW
M6 APB A. Bradshaw 130 Low Lane, Morecambe LA4 6PS
M6 APC A. Cook 12 National Drive, Manchester M5 3AZ
M6 APG N. Briggs 20 Broad Lane, Pelsall, Walsall WS4 1AP
M6 APM A. Munford 6 Column Mews, Alnwick NE66 1RZ

M6 APO A. Bais 9 Lemongrass Road, Bicester OX27 8BQ
M6 APR A. Ralph 3 The Leys, St. Albans AL4 9HD
M6 APS T. Bell 28 Inwood Drive, Coleford GL168EZ
M6 APU D. Bolton 88 Goldsborough, Wilnecote, Tamworth B77 4DF
M6 APW A. Woodhouse 4 Grafton Close, St. Albans AL4 0EX
M6 AQA A. Wright 69 Thomas Street, Tamworth B77 3PR
M6 AQD P. Hateley 44 Painters Croft, Coseley, Bilston WV14 8AP
M6 AQE M. Mohammed Shafi 575 Wood Lane, Dagenham, London RM8 1DR
M6 AQG P. Frost 86 Grantham Road, Sleaford NG34 7NW
M6 AQI C. Croasdale 76 Kingsway, Euxton, Chorley PR7 6PP
MW6 AQJ A. Hanley 11 Heol Pearetree, Rhoose, Barry CF62 3LB
M6 AQK T. Pearsall 16 Langdale Road, Leyland PR25 3AR
M6 AQL S. Melton 2A The Orchard, Bishopthorpe, York YO23 2RX
MM6 AQM C. Mcintyre 2/1, 6 Castle Street, Glasgow G81 4HH
M6 AQN E. Aksamit 14 Popplewell Gardens, Gateshead NE9 6TU
M6 AQO G. Foster 22 Bradley Cottages, Consett DH8 6JZ
M6 AQT L. Matthews 20 Harbridge Road, Broughton, Chester CH4 0FT
MW6 AQU R. Davison 2 Marlow Terrace, Mold CH7 1HH
M6 AQV V. Yotov Sanford House 81 Skipper Way, St. Neots PE19 6LT
M6 AQW T. Newman 39 Claremont Road, West Byfleet KT14 6DY
M6 AQY L. Gibbons 18 Langdale Road, Ribbleton, Preston PR2 6AN
M6 ARB A. Boucher 138 Ladysmith Road, Grimsby DN32 9SW
M6 ARC S. Froggatt 140 Greenlea Court, Huddersfield HD5 8QB
M6 ARD W. Peel 15 Brockhurst Close, Horsham RH12 1UY
M6 ARF M. Dailey 58 Waincliffe Mount, Leeds LS11 8AH
M6 ARH A. Hargreaves 27 Meadow Head Close, Blackburn BB2 4TY
M6 ARI B. Ansdell 8 Oakland Drive, Dudley DY3 2SH
M6 ARJ H. Clough 8 Skeldyke Road, Kirton, Boston PE20 1LR
M6 ARK M. Harris 29 Queen Street, Halesowen B63 3TZ
M6 ARL S. Skirving 1 Hallington Close, Bolton BL3 6YH
M6 ARM A. Martyn 54 North Side, Hepthorne Lane North Wingfield, Chesterfield S42 5HY
MM6 ARN A. O'Neill 39 Ardneil Court, Ardrossan KA22 7NQ
M6 ARP A. Holmes 49 Elm Grove South, Barnham, Bognor Regis PO22 0EJ
M6 ARQ J. Whitehead 20 Seamill Park Crescent, Worthing BN11 2PN
M6 ARR I. Clark 21 James Street, Epping CM16 6RR
M6 ARS B. Holland 11 Silverlands Park, Buxton SK17 6QX
MM6 ART A. Young 4/4 Prestonfield Terrace, Edinburgh EH165EE
M6 ARU J. Cherry 12 Scarisbrick New Road, Southport PR8 6PY
M6 ARV L. Williams Broomstreet Farm, Porlock, Minehead TA24 8JR
M6 ARW A. Winkley 77 Lechlade Road, Birmingham B43 5ND
MW6 ARX K. Bevan 76 Tennyson Close, Pontypridd CF37 5ER
M6 ARY H. Watts 696 Knowsley Lane, Knowsley, Prescot L34 9EH
M6 ASC A. Davies 68 Wood Street, Castleford WF10 1LN
M6 ASD A. Dhillon 13 Weston Road, Guildford GU2 8AU
M6 ASE A. Everett 16 Robyns Road, Beeston Regis, Sheringham NR26 8YJ
M6 ASF A. Foote Flat One, Kimber'S Close Kennet Road, Newbury RG14 5JF
M6 ASH D. Amos 86 Royal Military Avenue, Folkestone CT20 3EJ
M6 ASI T. Williams Broomstreet Farm, Porlock, Minehead TA24 8JR
M6 ASJ R. Bawden 43 Enys Road, Camborne TR14 8TW
M6 ASK A. Keatley 156 Earlswood Way, Colchester CO2 9NE
M6 ASM A. Mclachlan 4 Stratton Close, Bexleyheath DA7 4AJ
M6 ASO T. Roper 48 Rowthorne Lane, Glapwell, Chesterfield S44 5QD
M6 ASP T. Chapman 1 East Dean Road, Lockerley, Romsey SO51 0JL
M6 ASQ A. Hefford 9 Chamomile Gardens, Farnborough GU14 9XY
M6 AST B. Herrick Honing Road, Dilham NR28 9PL
M6 ASV C. Redmond 6 Apsley Road, Southsea PO4 8RH
M6 ASZ G. Wall Flat 2, Coniston House Holyoake Road, Worsley, Manchester M28 3DH
M6 ATA S. Waldock 102 Beaconsfield Road, London N15 4SQ
MW6 ATC A. Hare 243 Heritage Park, St. Mellons, Cardiff CF3 0DU
M6 ATD E. Gomez Lozano 29 Wykeham Crescent, Oxford OX4 3SD
M6 ATF N. Bisiker 31 Lansdowne Avenue, Waterlooville PO7 5BL
MW6 ATG A. Jones 3 Manor Way, Kinmel Bay, Rhyl LL18 5BP
M6 ATH A. Hicks The Granary, Vann Lake Road, Dorking RH5 5JB
M6 ATI A. Paricsi Vine House, Northwick Road Pilning, Bristol BS35 4HA
M6 ATJ M. Brasher 48 Eldertree Road Thorpe Hesley, Rotherham S61 2TQ
M6 ATK R. Atkins 2 Sandpiper Crescent, Malvern WR14 1UY
M6 ATL G. Hatt 4H Colman House, Earlham Road, Norwich NR4 7TJ
M6 ATM A. Malin 65A Coventry Road, Burbage, Hinckley LE10 2HL
M6 ATP D. Gibson 25 Middleham Close, Ouston, Chester le Street DH2 1TA
M6 ATQ I. Cooper 54 St Margarets Road, Lowestoft NR 324 HT
MM6 ATR A. Robertson 6C Fergusson Road, Cumbernauld, Glasgow G67 1LR
M6 ATS J. Watts 70 Castleway North, Leasowe, Wirral CH46 1RW
M6 ATU T. Mcbride 33 Somerville Square, Stafford ST17 9JU
M6 ATW A. Wilkinson 17 Stewkins, Audnam, Stourbridge DY8 4YW
M6 ATX G. Williams 19 Holden Walk, Wigan WN5 9JQ
M6 ATY F. Payne 45 Foxhill, Shaw, Oldham OL2 7NQ
M6 AUA P. Cooke 44 Brooklands Park, Craven Arms SY7 9RL
M6 AUB S. Drake-Brockman 13 St. Johns Place, Bury St. Edmunds IP33 1SW
M6 AUC S. Hall 3 Cedar Grove, Prestwich, Manchester M25 3DY
M6 AUE G. Carter 19 Brathay Crescent, Barrow-in-Furness LA14 2BG
M6 AUF E. Washbrook 47 Westgate, Leominster HR6 8SA
MM6 AUG B. Moerman 19/3 Pirniefield Bank, Edinburgh EH6 7QQ
M6 AUH H. Beaumont 1 Ashley Walk, Orleton, Ludlow SY8 4HD
MM6 AUI D. Cunningham 21 Constable Acre, Cupar KY15 4AE
MM6 AUJ F. Linn 14 Elphinstone Road, Tranent EH33 2HR
M6 AUL H. Hylton 214 School Road, Hall Green, Birmingham B28 8PF
M6 AUM Lord W. Saint Flat 74, Ferrier Point, London E16 1QW
M6 AUO J. Mcmahon 39 Hobmoor Croft, Birmingham B25 8TJ
M6 AUP T. Mundell 98 Westley Road, Bury St. Edmunds IP33 3SD
M6 AUQ P. Thearle 12 Grange Walk, Bury St. Edmunds IP33 2QB
M6 AUT M. Spurr 10 Ferncliffe Terrace, Leeds LS13 3PM
MI6 AUU M. Rushbrooke 22 Dublin Road, Omagh BT78 1ES
MW6 AUX D. Jones 17 Miners Row, Aberdare CF44 0TP
M6 AUZ S. Michaelis Fieldgate, Coltstaple Lane, Horsham RH13 9BB
M6 AVA C. Hughes 10 Langford Road, Stockport SK4 5BR
M6 AVD M. Darwen 31 Cleveland Road, Lytham FY8 5JH
MM6 AVE J. Rankin 17 Dippin Place, Saltcoats KA21 6AB
M6 AVF D. Bailey 92 Morrell Street, Maltby, Rotherham S66 7LH
M6 AVG L. Scott 28 Cavendish Place, New Silksworth, Sunderland SR3 1JW
MM6 AVH C. Lee Old House, 238 Garraheillie, Isle of South Uist HS8 5SX
M6 AVJ A. Johnson 3 Lindens Close, Thorney Toll, Wisbech PE13 4AR
MW6 AVK R. Gripp 23 Edmond Locard Court, Chepstow NP16 6FA
M6 AVL P. Smart 142 Finch Road, Chipping Sodbury, Bristol BS37 6JB
M6 AVM A. Hunter 9 Gelt Burn, Didcot OX11 7TZ
M6 AVO A. Craven 45 Benhams Drive, Horley RH6 8QT
M6 AVQ C. Brown 77 Grange Road, Ramsgate CT11 9LP
M6 AVR A. Mackay 11 Greatheed Road, Leamington Spa CV32 6ES
M6 AVS R. Paxman 11 Gibsons Gardens, North Somercotes, Louth LN11 7QH
M6 AVT R. Whalley 188 Astley Street, Astley, Tyldesley, Manchester M29 7AX
M6 AVU A. Pawlak 8 Healey Close, Crewe CW1 4RS
M6 AVV J. Noon 108 Cardinal Avenue, Morden SM4 4SX
M6 AVW A. Wilson 11 Headland Way, Alton, Stoke-on-Trent ST10 4AN
M6 AVY I. Clark 22 Rosemary Avenue, Grimsby DN34 4NJ
M6 AVZ K. Nicholson 11 Lancaster Way, Skellingthorpe, Lincoln LN6 5UF
M6 AWA A. Warnes 13 Warren Avenue, Fakenham NR21 8NP
M6 AWB A. Baker 2 Stileway, Meare, Glastonbury BA6 9SH
M6 AWC A. Coombes 78 Wilton Crescent, Southampton SO15 7QE
M6 AWE D. Barrett 208 Doncaster Road, Rotherham S65 2UE
M6 AWG M. Feltham Flat 5 Rebecca Court, 9, Beckenham BR3 1NN
M6 AWI B. Sturman 26 Howes Avenue, Thurston, Bury St. Edmunds IP31 3PY
M6 AWJ M. Harland Challenger Quay, Falmouth TR11 3YL
M6 AWL M. Shepherd North Waver Cottage, Bells Road Belchamp Walter, Sudbury CO10 7AR
M6 AWN C. Smith 44 Brooksfield, Bildeston, Ipswich IP7 7EJ
M6 AWO M. Nelson 30 Unicorn Place, Bury St. Edmunds IP33 1YP
M6 AWP L. Amunyela 1 Artillery Street, Colchester CO1 2JJ
M6 AWQ P. Vickers 37 Faraday Road, Ipswich IP4 1PU
M6 AWR A. Wright 52 Thorpe Way, Cambridge CB5 8UB
M6 AWS A. Stabler 11 Lincolns Avenue, Gedney Hill, Spalding PE12 0PQ
M6 AWU J. Harris 4 Burgh Old Road, Skegness PE25 2LN
MW6 AWV A. Williams 88 Heol Homfray, Cardiff CF5 5SB
M6 AWY J. Van-Boques-Tal 9 Stubbins Lane, Gazeley, Newmarket CB8 8RL
M6 AWZ A. Wilson 4 Oxford Street, Doe Lea, Chesterfield S44 5PH
M6 AXA M. Marsh 6 Carr Green, Lowton WA3 1EQ
M6 AXB A. Blamire 21 The Laurels, Banstead SM7 2HG
M6 AXC D. Howarth 6 Lyndhurst Close, Norton, Doncaster DN6 9PY
M6 AXD S. Denman 12 Dyke Vale Road, Sheffield S12 4ER
M6 AXF M. Atherton 53 Gillars Green Drive, Eccleston, St. Helens WA10 5AU
M6 AXG T. Winter 8 Thorpe Street, Hartlepool TS24 0DX
M6 AXH A. Hanson 14 Braithwaite Avenue, Keighley BD22 6EU
M6 AXI A. Chapman 24 Oakford Park, Halnaker, Chichester PO18 0BF
M6 AXJ S. Mccluskey 29 Hotspur Avenue, Bedlington NE22 5TD
M6 AXL A. Lowery 102 Brompton Park, Brompton On Swale, Richmond DL10 7JP
M6 AXM M. Rai Coldham Hall, Stanningfield, Bury St. Edmunds IP29 4SD
M6 AXP J. Christoforou 55 Wood Street, Taunton TA1 1UW
M6 AXR R. Williamson 2 Hobbs Road, Shepton Mallet BA4 4LS
M6 AXS N. Sutherland 34 Little Heath Road, Chobham, Woking GU24 8RL
MM6 AXT S. Glen 5/3 Renfrew Chambers, 136 Renfield Street, Glasgow G2 3AU
M6 AXU G. Briggs 54 Behind Berry, Somerton TA11 6JY
M6 AXW P. Hearnshaw Flat 10, 83 Swallows Meadow, Solihull B90 4PH
M6 AXX A. Marsden 38 Sandhill Road, Rawmarsh, Rotherham S62 5NT
MW6 AYA A. Young 14 Ramsons Way, Cardiff CF5 4QY
M6 AYC T. Jagger 50 North Street, Lower Hopton, Mirfield WF14 8PN
M6 AYE A. Yap Harrogate Ladies' College, Clarence Drive, Harrogate HG1 2QG
M6 AYG B. Leckey 76 Cardigan Lane, Leeds LS4 2LN
M6 AYH C. Wilson 87 Levensgarth Avenue, Fulwood, Preston PR2 9FP
M6 AYI R. Unsworth 22 Meadow House Park, Badcocks Lane, Tarporley CW6 9RT
M6 AYJ M. West Flat 1, 32 High Street, Dawlish EX7 9HP
M6 AYK D. Moger 23 Elmsleigh Road, Paignton TQ4 5AX
M6 AYL A. Sharam 30 Heywood Avenue, Maidenhead SL6 3JA
M6 AYM P. Armitage 250 Abbeydale Road South, Totley Rise, Sheffield S17 3LL
M6 AYN A. Taylor 14 The Lawns, Hinckley LE10 1DY
M6 AYP R. Styles 4 Coningsby Close, Gainsborough DN21 1SS
M6 AYQ J. Fudge 25 Virginia Orchard, Ruishton, Taunton TA3 5LP
MM6 AYR J. Mcmorland 382 Maryhill Road, Glasgow G20 7YQ
M6 AYS W. Abbott 12 Yew Tree Gardens, Birchington CT7 9AJ
M6 AYU C. Green 14 St. Andrews Road, Bletchley, Milton Keynes MK3 5DR
M6 AYV A. Lidster 23 William Street, Rotherham S60 2NG
M6 AYW K. Reeves 5 Westby Crescent, Whiston, Rotherham S60 4EA
M6 AYY C. Leviston 1 Great Carrs Close, Askam-in-Furness LA16 7FL
M6 AYZ M. Tarrant Wayside Cottage, Gabber Lane, Plymouth PL9 0AW
M6 AZA A. Adams 45 Four Oaks Road Tedburn St. Mary, Exeter EX6 6AP
M6 AZD A. Lines 26 Mcintyre Walk, Bury St. Edmunds IP32 6PF
M6 AZE A. Seedig 21 Ambleside Close, Mytchett GU16 6DG
M6 AZF S. Case 136 Old Basin, Bridgwater TA6 6LJ
M6 AZG P. Culverwell 29 Elland Road, Brierfield, Nelson BB9 5RX
M6 AZH E. Cross 12B Oakridge, Three Rivers Country Park, Clitheroe BB7 3JW
M6 AZK S. Tyler 11 Windmill Cottages, Dilmore Lane, Worcester WR3 7RX
M6 AZM P. Comley-Ross 48 High Street, Topsham, Exeter EX3 0DY
M6 AZN G. Villiers 88 Redwald Road, Rendlesham, Woodbridge IP12 2TE
MW6 AZP A. Dighton 84 Trefelin, Aberdare CF44 8LF
M6 AZQ M. Osband 22 Samian Crescent, Folkestone CT19 4JW
M6 AZR K. Braisher 7 Ormond Road, Thame OX9 3XN
M6 AZS P. Mcfadden Maple Cottage, Leighton Buzzard LU7 9DZ
M6 AZT S. Mason 3 Grange Cottages, Low Road, Barrowby, Grantham NG32 1DL
M6 AZV D. Simpson 50 Castle Hill, Berkhamsted HP4 1HF
M6 AZX N. Robinson Flat 20, Hanover Court, Old Vicarage Road, Dovercourt, Harwich CO12 4BU
M6 AZY W. Millington 93 Feiashill Road, Trysull, Wolverhampton WV5 7HT
M6 BAA B. Gutteridge 121 Station Road South, Walpole St. Andrew, Wisbech PE14 7LZ
M6 BAC B. Warner 41 Birling Road, Ashford TN24 8BD
M6 BAD B. Barnes-Martin 145 Farm Road, Barnsley S70 3DW
M6 BAG D. Roberts 27 Nairn Street, Jarrow NE32 4HX

M6 BAH S. Bassett 5 The Terrace, The Green, Stratford-upon-Avon CV37 0JD
MI6 BAI B. Baird 12 Manse Park, Newtownards BT23 4TN
MI6 BAJ Dr A. Bell 4 Mount Pleasant View, Newtownabbey BT37 0ZY
M6 BAK C. Baker 60 Belvedere Road, Danbury, Chelmsford CM3 4RB
M6 BAM B. Butler 42 Station Road, Stanbridge, Leighton Buzzard LU7 9JF
M6 BAN B. Banner 99 Swingate, Kimberley, Nottingham NG16 2PU
M6 BAQ B. Williamson 23 Tower Hamlets Street, Dover CT17 0DY
M6 BAR D. Barry Coasters, Station Road, Colchester CO7 8LH
MW6 BAS B. Sweet 14 Bryn Celyn, Colwyn Bay LL29 6DH
MW6 BAU K. Burgess 18 Fairmeadows, Maesteg CF34 9JL
M6 BAV D. Woodbine 29 Compass Tower, Munnings Road, Norwich NR7 9TW
M6 BAW J. Middleton 17 Woods Loke West, Lowestoft NR32 3DN
M6 BAX R. Baxter Rose Dene, Hornsby, Brampton CA8 9HF
M6 BAY K. Linklater 32 King Edward Road, Gillingham ME7 2RE
M6 BAZ B. Pike 36 Larkfield Avenue, Sittingbourne ME10 2DP
M6 BBA J. Alincastre 14919 Hope Hills Lane, Cypress, Texas 77433 United States
M6 BBC A. Bright 86 Fourth Avenue, Watford WD25 9QQ
M6 BBD P. Symonds 10 Cowper Court, Willunga 5172 Australia
M6 BBE E. Puttock 12 Beechfields, School Lane, Petworth GU28 9DH
M6 BBF G. Symonds 10 Cowper Ct, Willunga 5172 Australia
M6 BBG P. Holland 30 Knighton Park Road, London SE26 5RJ
M6 BBH M. Ramsey 21 Goldsmith Road, Eastleigh SO50 5EN
MM6 BBK P. Robertson 41 Hawthorn Street, Leven KY8 4QE
M6 BBL N. Stanley 253 Brownley Road Wythensawe, Manchester M22 9UX
M6 BBP A. Page 207 Brooklyn Road, Cheltenham GL51 8DZ
M6 BBQ A. Smith 311 Albion Street, Southwick, Brighton BN42 4AT
M6 BBR C. John 27 Berberis Walk, West Drayton UB7 7TZ
M6 BBS Dr P. Hyde Flat 4, 1 Longhorn Avenue, Gloucester GL1 2AR
M6 BBT M. Boswell 5 Woods Avenue, Marsden, Huddersfield HD7 6JX
M6 BBU R. Boardman 12 St. Margarets Road, Alderton, Tewkesbury GL20 8NN
MM6 BBX A. Wright 17A James Court, 493 Lawnmarket, Edinburgh EH1 2PB
M6 BCB R. Balmforth 33 Lees Hall Road, Dewsbury WF12 0RH
M6 BCC A. Boots 36A Church Street, Charlton Kings, Cheltenham GL53 8AR
MM6 BCF C. Flynn 6C White Street, Ayr KA8 9BW
M6 BCG A. Pickles 87A Laburnum Road, Waterlooville PO7 7EW
M6 BCH B. Chauhan 45 Burnham Drive, Whetstone, Leicester LE8 6HY
M6 BCJ J. Bullock 49 Gallimore Close, Stoke-on-Trent ST6 4DZ
M6 BCK T. Humphries 32 Bonds Meadow, Oulton Broad NR32 3QL
M6 BCN P. Kingston 2 Deepdale, Great Easton, Market Harborough LE16 8SS
M6 BCQ P. Chester 44 Kirtling Place, Haverhill CB9 0AU
M6 BCU B. Miles 35 Plantation Drive, Walkford, Christchurch BH23 5SG
M6 BCV M. Cooper 9 Conway Close, Crewe CW1 3XN
M6 BCW G. Fearnhead 27 Lukins Drive, Dunmow CM6 1XQ
M6 BCY J. Edwards 45 Bramshaw Gardens, Bournemouth BH8 0BT
M6 BCZ K. Percy 55 Buxton Avenue, Heanor DE75 7UN
MM6 BDA B. Adams 18 Bellfield Road, North Kessock, Inverness IV1 3XU
M6 BDD J. O'Brian 83 Bramdean Crescent, London SE12 0UJ
M6 BDG B. Greenberg 94 Rivermead Court, Ranelagh Gardens, London SW6 3SA
M6 BDI A. Bucknell 12 Cliveden Grove, Hereford HR4 0NE
MJ6 BDJ L. Langlois Brookfield, La Rue D'Empierre, Trinity JE3 5QF Jersey
M6 BDL B. Little 25 Thrift Wood, Bicknacre, Chelmsford CM3 4HT
M6 BDM D. Mead 32 Sherborne Road, Farnborough GU14 6JT
M6 BDQ D. Nelson 110 Chandag Road, Keynsham, Bristol BS31 1QF
M6 BDR B. Roberts 10 Morningside Way, Liverpool L11 1BD
MW6 BDS D. Bancroft Stop And Call, Goodwick SA64 0EX
M6 BDV B. Vile 24 Hudson Close, Dover CT16 2SG
M6 BDX T. Scott 39 Neil Avenue, Holt NR25 6TG
M6 BEB S. Smedley Spring Cottage, Frys Well, Radstock BA3 4HA
M6 BEC B. Beckett 21 Horseshoes Lane, Langley, Maidstone ME17 1SR
MW6 BED J. Bratchley 22 Zealand Park, Caergeiliog, Holyhead LL65 3PQ
M6 BEE B. Azzaro 5 Rye Hill Close, Bere Regis, Wareham BH20 7LU
MI6 BEF W. Tosh 38 Ballycastle Road, Coleraine BT52 2DY
MW6 BEG P. Sherwood High Croft Jeffreyston, Kilgetty SA68 0RG
M6 BEH B. Humphrey 45 Rose Avenue, Hazlemere, High Wycombe HP15 7PH
M6 BEI I. Whitlock 109 Sorrell Drive, Newport Pagnell MK16 8TZ
M6 BEJ J. Preston 25 Hamlet Road, Haverhill CB9 8EH
M6 BEK R. Clare Kimberley, Boston Road, Boston PE20 3AP
M6 BEL K. Whitelaw 18, Seaford BN252RT
M6 BEM M. Hooper 5 Pillowell Close, Cheltenham GL525GJ
M6 BEN B. Robinson 13 Dene View, Ellington, Morpeth NE61 5HQ
M6 BEQ C. Bradley 47 Long Meadow, Skipton BD23 1BP
M6 BEX R. Whitehead 29 Coulsons Road, Bristol BS14 0NN
M6 BEZ P. Bailey 81A Kings Parade Holland-On-Sea, Clacton-on-Sea CO15 5JF
M6 BFB V. Walsh 11 Coronation Drive Crosby, Liverpool L23 3BN
M6 BFC B. Spaxman 12 Stanhope Gardens, Barnsley S75 2QB
M6 BFF K. Rivett 89 Maidstone Road, Felixstowe IP11 9EE
M6 BFG C. Bowman 26 Albany Hill, Tunbridge Wells TN2 3RX
MM6 BFH B. Haynes 18 Drummond Street, Greenock PA16 9DN
M6 BFI A. Jobbling 3 Orchard Close, Cranfield, Bedford MK43 0HX
MI6 BFJ J. Crozier 9 Colinbrook Park, Dunmurry, Belfast BT17 0NZ
M6 BFK A. Eaton 40A Summer Street, Leighton Buzzard LU7 1HT
M6 BFL K. Goldsworthy Flat 2, Jordan House, Biggleswade SG18 8FS
M6 BFO C. Murphy 17 Shepherd Street, Littleover, Derby DE23 6GA
M6 BFP R. Roberts 25 Carlisle Avenue, Bootle L30 1PX
M6 BFQ T. Dodd 6 Victoria Terrace Prinsted Lane, Emsworth PO108HX
MM6 BFR P. Riddle Carngeal, Pitlochry PH16 5JL
M6 BFS N. Grudgings North Lodge, Templewood Lane, Slough SL2 3HW
M6 BFU A. Elwin 30 Kingsland Road, Aylesbury HP21 9SL
M6 BFV O. Gledhill 20 Curtis Way, Kesgrave, Kesgrave IP5 2FX
M6 BFX S. Cave Weymess Farm, Park Lane, Banbury OX17 2RX
M6 BFZ N. Edwards 11 Sandringham Road, Eccleston, Chorley PR7 5SN
M6 BGA K. Baker 27 St. Matthews Close, Cherry Willingham, Lincoln LN3 4LS
M6 BGB A. Ajmani Apartment 605, 7 Anchor Street, Ipswich IP3 0BW
M6 BGC J. Moody 16 Lingcrest, Gateshead NE9 6SN
MI6 BGD B. Gilliland 28 Baird Avenue, Donaghcloney, Craigavon BT66 7LP
M6 BGE S. Saunders 5 Park Court, Woking GU22 7NW
M6 BGF P. Martin 56 Devonshire Gardens, Bursledon, Southampton SO31 8HE
M6 BGG M. Chiesa 56 Waddon Road, Croydon CR0 4JD
M6 BGH B. Higgins 2 Bishops Yard, High Street, Huntingdon PE28 3JB
M6 BGJ A. Jones 22 Standish Road, Sheffield S5 8XU
M6 BGK A. Yates 1 Roberts Court, Whitwell, , Hitchin, SG4 8AF
M6 BGL K. Oliver 28 King Richards Hill, Earl Shilton, Leicester LE9 7EY
M6 BGM P. James 44 Narbonne Avenue Ellesmere Park Eccles, Manchester M30 9DL
MW6 BGO D. Lacaman 6 Falcon Road, Haverfordwest SA61 2UE
M6 BGP B. Peacock 17 Herril Ings, Tickhill, Doncaster DN11 9UE
M6 BGS A. Somerville 106 Bush Hill, Northampton NN3 2PG
M6 BGT J. Summerhill 43 Rangers Walk, Bristol BS15 3PW
MM6 BGV A. Timmins 16 Queens Crescent, Garelochhead, Helensburgh G84 0DW
M6 BGW M. Walters 110 Slade Road, Portishead, Bristol BS20 6BB
M6 BGZ T. Chapman 74 Kidderminster Road, Bewdley DY12 1BY
M6 BHA T. Harris 12 Maple Close, Stourport-on-Severn DY13 8TA
M6 BHB R. Smith Five Elms, Lullington Road Edingale, Tamworth B79 9JA
M6 BHC A. Tipler 27 Clumber Street, Hucknall, Nottingham NG15 7PJ
MW6 BHD D. Rees 49 Fair View, Hirwaun, Aberdare CF44 9SA
M6 BHE S. Verity 29 Patterdale Avenue, Fleetwood FY7 8NW
M6 BHF M. Atherton 4 Bakers Park Saltney, Chester CH4 8FB
M6 BHH G. Hartless 32 Long Acre, Mablethorpe LN12 1JF
M6 BHI Dr R. Smyth 46 Eagle Way, Abbeydale, Gloucester GL4 4WS
M6 BHJ B. Hall 1 Lytham Road, Leicester LE2 1YD
M6 BHK C. Lai Storeys Way, Cambridge CB3 0DG
M6 BHM M. Biadon 57 Fern Hill Road, Oxford OX4 2JW
M6 BHN J. Hunter 35 Inglefield, Hartlepool TS25 1RN
MW6 BHO A. Edwards Tir Brwyn, Rhydargaeau, Carmarthen SA33 6BL
M6 BHP R. Earland 11 Roseberry Avenue, Great Ayton, Middlesbrough TS9 6EN
M6 BHQ L. Copeland 78 Penderyn Crescent Ingleby Barwick, Stockton-on-Tees TS17 5HQ
MM6 BHS J. Read Knockenny Farm, Glamis, Forfar DD8 1UE
M6 BHT T. Scales 77 Upper Eastern Green Lane, Coventry CV5 7DA
M6 BHU P. Chamberlain Mulberry, Main Road, Edenbridge TN8 6HZ
M6 BHY M. Worley 12 Hall Farm Close, Melton, Woodbridge IP12 1RL
M6 BHZ P. Palmer 2 Dagonet Road, Bromley BR1 5LR
M6 BIA R. Kijewski 14 East Street, Heanor DE75 7NE
M6 BIB N. Foster Four Beeches, Gribthorpe, Goole DN14 7NT
M6 BID M. Knights 3 East View Cottages, Church Road, Woodbridge IP13 0AT
M6 BIE J. Lucas 64 Fitzroy Road, Whitstable CT5 2LE
M6 BIF D. Sobey Flat 2 73 Park Road, Blackpool FY1 4JQ
M6 BIG D. Robson 5 Rutherglen Drive, Hull HU9 3PF
MM6 BIH G. Fordyce 2 Church Street, East End, Earlston TD4 6HS
MI6 BII P. Floyd 25 Glenside, Omagh BT79 7GL
M6 BIJ I. Jarvis 6 Mullett Road, Wednesfield, Wolverhampton WV11 1DD
M6 BIM H. Salter 34 Hillside Villas, Millendreath Holiday Village, Millendreath, Millendreath Looe PL13 1PE
M6 BIN R. Granger 7 Webster Road, Winchester SO22 5NT
MM6 BIO N. Mackenzie 59 Plasterfield, Stornoway HS1 2UR
MM6 BIP A. Stanley 40 Coll, Isle of Lewis HS2 0LP
M6 BIR B. Jakubowski 120 Chandos Street, Coventry CV2 4HT
M6 BIU C. Andrews 34 Russell Street, Kettering NN16 0EL
M6 BIV O. Hall 2 Beverley Lodge, Paradise Road, Richmond TW9 1LL
M6 BIX J. Coward 15 Tyson Square, Ulverston LA12 0BG
MM6 BIY J. Corrigan 27 Stonecraig Road, Wishaw ML2 8BZ
M6 BJA B. Ardrey 13 Grebe Avenue, St. Helens WA10 3QL
M6 BJB K. Richards 5 Ramsay Close, Birchwood, Warrington WA3 6PS
M6 BJC M. Holding 8 Dunsop Close, Blackpool FY1 6NP
M6 BJE B. Emmerson 1 Tivydale Drive, Darton, Barnsley S75 5PG
M6 BJF B. Froggatt 11 Goldsmith Road, Walsall WS3 1DL
MI6 BJG R. Gilmore 11 Abbots Gardens, Newtownabbey, Belfast BT379QZ
M6 BJI I. Taylor 111 Kings Road, Lancing BN15 8EQ
M6 BJK K. Ashton 23 Wood Close, Wells BA5 2GA
M6 BJL M. Blackmore Timbers, Wolvershill Road, Banwell BS29 6DG
M6 BJM M. Bunting 31 Hardwick Avenue, Allestree, Derby DE22 2LN
M6 BJN B. Potter 55 Lindsworth Road, Kings Norton, Birmingham B30 3RP
M6 BJO K. Hawkins 2 Couford Grove, Huddersfield HD2 1TH
M6 BJP P. Blackmore Timbers, Wolvershill Road, Banwell BS29 6DG
M6 BJQ R. Stokes 3 Parham Walk, Grange Park, Swindon SN5 6EQ
M6 BJT B. Claydon 31 Riverside.Horley.Surrey, Horley RH67LN
M6 BJW B. White 53 Dacre Road, á, Brampton CA81BN
M6 BJX C. Thorpe-Morgan 31 Dingledery, Olney MK46 5ES
M6 BJY M. Walker 20 Fernhurst Road, Mirfield WF14 9LJ
MI6 BJZ P. Donnelly 43 Ashfield Gardens Fintona, Omagh BT78 2DD
M6 BKA M. Byrnes 61 Furze Close, Peatmoor, Swindon SN5 5DB
M6 BKC P. Glasper 2 Iris Close, Stockton TS18 1AX
MI6 BKD J. Tierney 48 Ashfield Gardens Fintona, Omagh BT78 2DD
M6 BKE C. Opie 354 Beaumont Road, Plymouth PL4 9EN
M6 BKF Dr C. Lane 47 Glenarm Road, London E5 0LY
M6 BKH D. Storer 13 The Square, Lower Burraton, Saltash PL12 4SH
M6 BKI S. Shaw 3 Wellington Terrace, Littleborough OL15 9DA
M6 BKJ P. Crow 31 Lakeside, Overstone Park, Northampton NN6 0QS
M6 BKK B. Tompkins 13 Magazine Close, Wisbech PE13 1LH
M6 BKL K. Jackson 4 Milfoil Close, Marton-In-Cleveland, Middlesbrough TS7 8SE
M6 BKM D. Turnbull 63 Brecklands, Mundford, Thetford IP26 5EG
M6 BKN S. Walker 73 Sunnybank Road, Halifax HX2 8RL
M6 BKO J. Legrain 17 Route De La Cote, St. Laurent Sur Gorre 87310 France
MM6 BKP R. Adamson 6 Camdean Crescent, Rosyth, Dunfermline KY11 2TJ
MW6 BKS N. Adam Tan Ffordd, Mynydd Llandygai, Bangor LL57 4LX
M6 BKT S. Keeble 38 Sandford Rise, Sandy SG19 1ED
M6 BKU B. Cox 7 Wolsey Avenue, London E6 6HG
M6 BKV M. Archer 4 The Bungalows, Mill Lane, Grays RM20 4YD
M6 BLA J. Forder 157 Kennington Road, Kennington, Oxford OX1 5PE
M6 BLB J. Sell 3 Powter Close, Elsenham CM226UT
M6 BLC B. Cripps 215 Bournemouth Road, Poole BH14 9HU
MM6 BLE S. Robertson 43 Kindar Drive, New Abbey, Dumfries DG2 8DA
M6 BLF M. Featherstone 62 Poles Hill, Chesham HP5 2QR
M6 BLG T. Clayton 4 Stable Mews, Lords Hill, Coleford GL16 8BJ
M6 BLH R. Pringle 14 Marjorie Street, Cramlington NE23 6XQ
M6 BLI S. Algar 17 Riseway Close, Norwich NR1 4NJ

M6 BLK P. Blake 70 Front Street, South Hetton, Durham DH6 2RG
M6 BLM K. Gallop Airedale House, Airedale Drive, Castleford WF10 2QA
M6 BLN M. Halliday 1 Everard Close, Bury St. Edmunds IP32 6RU
M6 BLO L. Monshall 7 Sweden Close, Harwich CO12 4JU
M6 BLP B. Pearn 230 Lloyds Avenue, Kessingland, Lowestoft NR33 7TU
M6 BLQ N. Fahey 5 Hillside, Felmingham, North Walsham NR28 0LE
M6 BLS T. Malik 83A Cleveland Road, Manchester M8 4GT
M6 BLT W. Thomson 72 Hurstwood Avenue, Bexleyheath DA7 6SG
M6 BLV J. Moore 53 The Boulevard, Great Sutton, Ellesmere Port CH65 7DX
MI6 BLW E. Macgra 197 Beraghvale, Skeoge Road, Londonderry BT48 8UJ
M6 BLX R. Bowles 7 Vineside, Gosport PO13 0ZU
MM6 BLY W. Robb 18 Buckie, Erskine PA8 6EE
M6 BLZ J. Blezard 10 North Row, Barrow-in-Furness LA13 0HE
M6 BMB Dr J. Bell 6 Highfields, Fetcham, Leatherhead KT22 9XA
MI6 BMC W. Mccormick 6 Church Street, Rosslea, Enniskillen BT92 7DD
MW6 BME P. Toal 17 Gronant Street, Rhyl LL18 1PN
M6 BMF A. Ledger 23 Wentworth Park, Freshbrook, Swindon SN5 8QX
M6 BMI P. Davies 2 Lynfords Drive, Runwell, Wickford SS11 7PP
M6 BMJ B. Adamson Flat 6 Eastgate Lodge 1 Eastgate Gardens, Guildford GU1 4AZ
M6 BMK P. Hardy 50 Harridge Road, Leigh-on-Sea SS9 4HA
MW6 BMM W. Murphy 148 Caergynydd Road, Waunarlwydd, Swansea SA5 4RE
M6 BMO B. Morris 131 Littlehampton Road, Worthing BN13 1QX
MM6 BMP A. Moerman 11 Cupar Road, Kettlebridge, Cupar KY15 7QD
MM6 BMQ J. Bannerman 77/2 Park Avenue, Edinburgh EH15 1JP
M6 BMT B. Thistlethwaite 29 Lamorna Drive, Callington PL17 7QH
M6 BMV R. Gocher 20 Mulberry Crescent, South Shields NE34 8DD
M6 BMW B. Martin 11 Alpha Street, Toll Bar, Doncaster DN50RA
M6 BMY S. Maqbool 69 Waltham Close, West Bridgford, Nottingham NG2 6LD
M6 BMZ J. Kay 36 Winnington Road. Marple, Stockport SK6 6PT
M6 BNA A. Piper 3 Bakers Court, Bakers Court Lane, Lynton EX35 6EW
M6 BNB J. Duffy 19 Lydford Gardens, Bolton BL2 6TU
M6 BNC B. Chambers 13 Cherry Tree Crescent, Walton, Wakefield WF2 6LQ
MW6 BNF S. Morgan Holly Cottage, Old Racecourse, Oswestry SY10 7PQ
M6 BNG R. Watson 8 Bourne Close, Warminster BA12 9PT
MW6 BNH G. Thomas 4 Oak Tree Close, Buckley CH7 3JU
M6 BNJ G. Hartley 1 Manor View, Shafton, Barnsley S72 8NQ
M6 BNK M. Wilsher 70 Norris Road, Blacon, Chester CH1 5DZ
M6 BNL D. Beck 94 Shaldon Crescent, Plymouth PL5 3RB
M6 BNM J. Sanders 76 Fullerton Road, Plymouth PL2 3AX
MM6 BNN O. Mckenzie 179 Gordons Mills Road, Aberdeen AB24 2XS
MM6 BNO K. Mckenzie 26 Gladstone Place, Woodside, Aberdeen AB24 2RP
M6 BNP D. Brookes 177 Charnwood Close, Rubery, Birmingham B45 0JY
M6 BNQ C. Marshall 51 Hedgerow Close, Redditch B98 7QF
MM6 BNS Prof. T. Donaldson Brawview Cottage, Maybole KA19 8EN
MM6 BNT L. Waller 1 Anne Arundel Court, Heathhall, Dumfries DG1 3SL
M6 BNU S. Ditchburn 18 St. Hilda Avenue, Waterlooville PO8 0JF
M6 BNV I. Balboa 7 Edenhall Close, Tilehurst, Reading RG31 6RR
MM6 BNX W. Evans Ninewar Farm, Duns TD11 3PP
M6 BNY J. Beeney 17 Norton Avenue, Herne Bay CT6 7TA
M6 BOA T. Cocks 9 Mountfield Way, Westgate-on-Sea CT8 8HR
M6 BOB R. Manning 15 Gurston Rise, Northampton NN3 5HY
MW6 BOC G. Williams 36 Park Street, Taibach, Port Talbot SA13 1TD
MM6 BOD D. Lightbody Glenorchy, Brownrigg Road, Falkirk FK1 3BA
M6 BOF B. Savage Rufford, Barnes Lane, Milford On Sea, Lymington SO41 0RR
M6 BOI K. Mecca 38 Abbots Road, Faversham ME13 8DE
M6 BOK V. Casambros 27 Oaks Road, Folkestone CT20 3JY
M6 BOR G. Wren Ashleigh, Yew Tree Hill, Holloway, , Matlock DE4 5AR
MM6 BOS A. Manson Clochcan Schoolhouse, Ellon AB41 8UJ
M6 BOT T. Gedvygas 9 The Mill, Kirton, Boston PE20 1LB
MW6 BOW T. Bowen 7A Heol Maes Y Cerrig, Loughor, Swansea SA4 6SW
M6 BOX C. Pegrum 3 Bretland Road, Tunbridge Wells TN4 8PS
M6 BOY C. Butler 210 Green Wrythe Lane, Carshalton SM5 2SP
MW6 BOZ E. Bosley Glandwr, Snowdon Street, Porthmadog LL49 9DF
M6 BPA A. Parnell-Brookes 14 Greens Close, Hullavington, Chippenham SN14 6EG
MW6 BPB K. Furlong 58A High Street Nantyffyllon, Maesteg CF34 0BT
M6 BPC G. Colson 3 Dartford Road, Dartford DA1 3EE
M6 BPD D. Taylor 143 Sandhurst Road, London SE6 1UR
M6 BPE P. Skinner 27 Westcots Drive, Winkleigh EX19 8JW
M6 BPG B. Patrick-Gleed 6 Julius Way, Lydney GL15 5QS
MW6 BPH B. Parsons 29 Hillside Crescent, Buckley CH7 2JS
M6 BPI W. Colquhoun 92 Byfield Road, Woodford Halse, Daventry NN11 3QS
M6 BPK J. Johnson 4 Wallace Close, Hullbridge, Hockley SS5 6NE
M6 BPL M. Lefton 35 Hawkstone Avenue, Whitefield, Manchester M45 7PR
MM6 BPM I. Wightman 13 Gillsland, Eyemouth TD14 5JF
M6 BPO R. Extrance 1 Morngate Caravan Park, Bridport Road, Dorchester DT2 9DS
M6 BPQ J. Chatterton 6 Bayliss Road, Wargrave, Reading RG10 8DR
M6 BPR K. Butt 15 Hamble Park, Fleet End Road, Southampton SO31 9JU
M6 BPS P. Dent 25 Clyde Avenue, Hebburn NE31 2JN
M6 BPU C. Shepherd 1 Holley Park, Okehampton EX20 1PL
M6 BPV J. Mcrobie 6 Southill Gardens, Bournemouth BH9 1SJ
M6 BPW B. Wang Harrogate Ladies' College, Clarence Drive, Harrogate HG1 2QG
M6 BPX J. Revell 37 Tennyson Street, Goole DN14 6EB
MM6 BPY L. Joseph 3/4 60 Wilson Street, Glasgow G1 1HD
MW6 BQA R. Staples 16 Wellington Street, Aberdare CF44 8EW
M6 BQB R. Babb 15 Sylvan Lane, Hamble, Southampton SO31 4QG
M6 BQC M. Powis 3 Greenacres, Ludlow SY8 1LU
M6 BQD H. Brewster 1 Pinewood Drive, Camblesforth, Selby YO8 8JU
M6 BQE M. Simpson 32 Underhill Lane, Wolverhampton WV10 8NS
M6 BQF G. Evans 13 Lydgate Road, Sale M33 3LW
MM6 BQG B. Wetton Tigh Air Achnoc, West Helmsdale, Helmsdale KW8 6HH
M6 BQJ R. Thompson 16 West Leys Court, Moulton, Northampton NN3 7UB
M6 BQK C. Holmes 10 Southampton Street, Farnborough GU14 6AX
MW6 BQL Dr J. Griffiths Llain Bach, Beach Road, Porthmadog LL49 9YA
MW6 BQM K. Williamson 1 Rhug Gardens, Corwen LL21 0EH
M6 BQN T. Gilmour 83 Billington Road, Leighton Buzzard LU7 4TG
M6 BQO W. Martin 54 Merritt Road, Greatstone, New Romney TN28 8SZ
M6 BQP J. Griffiths 7 Tynedale Road, Blackpool FY3 7UE
M6 BQQ S. Murray 79 Nightingale Road, Liverpool L12 0QN
M6 BQR T. Walton 72 Burleigh Road, Frimley, Camberley GU16 7EB
M6 BQS R. Brabazon 10 St. Dominic Close, St. Leonards-on-Sea TN38 0PH
M6 BQV A. Brackpool 79 St Leonards Road, Horsham RH13 6EH
M6 BQW K. Jeffery 9 Gordon Road, Tunbridge Wells TN4 9BL
M6 BQZ M. Iqbal 22 Rupert Avenue, High Wycombe HP12 3NG
M6 BRB L. Bethell 30 Finger Road, Dawley, Telford TF4 3LB
M6 BRH S. Plummer 2 Langdale Avenue, Outwood, Wakefield WF1 3TX
M6 BRJ R. Green 44 Aldwyn Place, Larchwood Drive, Egham TW20 0RZ
M6 BRK A. Freshwater 90A New Road, Minster On Sea, Sheerness ME12 3PT
M6 BRL M. Burrell 16 Atholl, Ouston, Chester le Street DH2 1RS
M6 BRN M. Brundrit 144 Reginald Road, Southsea PO4 9HP
M6 BRO J. Smith 38 The Vineries, Burgess Hill RH15 0NF
M6 BRP S. Fiske 12 Meliden Crescent, Bolton BL1 6AJ
M6 BRQ A. Clements 28 Durham Way Wyton, Huntingdon PE28 2EQ
M6 BRR S. Ray 28 Stenbury View, Wroxall, Ventnor PO38 3DB
M6 BRS D. Glover 11 Collbrook Avenue, Odsal, Bradford BD6 1HL
M6 BRT A. Williams 101 Horsebridge Hill, Newport PO30 5TL
M6 BRU A. Dean 14 Harvest Close, Worsbrough, Barnsley S70 5AY
MM6 BRV D. Drysder 37 Farburn Drive, Stonehaven AB39 2BZ
M6 BRY B. Allen 13 Woodgrove Road, Rotherham S65 3RW
M6 BRZ B. Ager 6 South Dibberford Farm, Beaminster DT8 3HD
M6 BSA C. Jacobs Flat 33, The Lodge, Lavender Road, Waterlooville PO7 8BX
M6 BSB S. Hamill 3 Pear Tree Croft, Brede, Rye TN31 6EJ
M6 BSC B. Sewell 12 Haylands Square, South Shields NE34 0JB
M6 BSF A. Rizzo 12 North Street, Wicken, Ely CB7 5XW
M6 BSG B. Garton 13 Damaskfield Road, Lyppard, Kettleby WR4 0HY
M6 BSI A. Butler 50 Scafell Way, West Bromwich B71 1DQ
MM6 BSK I. Bannerman 54 Broadleys Avenue, Bishopbriggs, Glasgow G64 3AQ
M6 BSL T. Bilsel 16 Hide Close, Sawston, Cambridge CB22 3UR
M6 BSO B. Scott 6 Congleton Edge Road, Congleton CW12 3JJ
M6 BSP B. Smith 8 Mill Field Close, South Kilworth, Lutterworth LE17 6FE
M6 BSQ P. Bull 87 Braemor Road, Calne SN11 9DU
MW6 BSR B. Sturgess 22 Heol Pant Y Deri, Cardiff CF5 5PL
M6 BSS J. Chambers 10 Derwent Street, Astley, Manchester M29 7AT
M6 BST T. Hughes 1 Sunnybank Road, Astley, Manchester M29 7BJ
MW6 BSU A. Buckley Cliff House, Sudbrook, Caldicot NP26 5TB
M6 BSV C. Hayes 7 Hadstock Close, Sandiacre, Nottingham NG10 5LQ
M6 BSW B. Stone 27 Mountbatten Close, Stretton, Burton-on-Trent DE13 0FD
M6 BSX J. Cummins 55 Rowley Street, Walsall WS1 2AZ
M6 BSY D. Carter 36 Sanderling Drive, Leigh WN7 1HU
M6 BSZ C. Batchelor 8 Howarde Court, Stevenage SG1 3DF
M6 BTA L. Brown 28 Farley Way, Stockport SK5 6JD
M6 BTB R. Dalley Birchley, Pine Avenue, Camberley GU15 2LY
MW6 BTC V. Kennedy Gerynant, Felin Ban Farm Estate, Cardigan SA43 1PG
M6 BTE B. Ellerton 4 Darfield Avenue, Owlthorpe, Sheffield S20 6SU
M6 BTF P. Likitplug Harrogate Ladies' College, Clarence Drive, Harrogate HG1 2QG
M6 BTG M. Mutkin 13 The Grove, Radlett WD7 7NF
M6 BTH D. D'Souza 3B Friend Street, London EC1V 7NS
M6 BTK J. Barton 8 Melton Lane, Sutton Bonington, Loughborough LE12 5RQ
M6 BTM B. Gillett 65 Kingsway, Wallasey CH45 4PN
M6 BTN R. Nicholson 57 Barnbridge, Tamworth B77 1DF
M6 BTP P. Wilmot 12 Brierholme Close, Hatfield, Doncaster DN7 6EH
M6 BTQ C. Manning 12 Whitehill Close, Camberley GU15 4JR
MM6 BTR J. Rayne 8 Bankton Grove, Livingston EH54 9DW
M6 BTS B. Seward 21 Chapel Close, Gunnislake PL18 9JB
M6 BTT P. Moye 13 Post Mill Gardens, Grundisburgh, Woodbridge IP13 6UP
M6 BTY P. Matthews The Old Chapel, High Street, Huntingdon PE28 0PF
M6 BUA H. Bennett 59 Scott Road, Bishop's Stortford CM23 3QN
M6 BUB R. Hughes 86 Colinmander Gardens, Ormskirk L39 4TF
M6 BUC J. Chambers 2 Farm Cottages, Magdalen Laver, Ongar CM5 0ES
M6 BUE J. Shepherd 169 Northbrooks, Harlow CM19 4DQ
MW6 BUF J. Hurst 1 Castle Cottage, Aberedw, Builth Wells LD2 3UL
MM6 BUG J. Millar The Kennels, Lanton, Jedburgh TD8 6SU
MW6 BUH I. Baylis 248 Trebanog Road, Porth CF39 9EL
M6 BUI D. Sewell 19 St. Leonards Way, Ashley Heath, Ringwood BH24 2HS
M6 BUJ G. Sheppard 32 Bramble Drive, Hailsham BN27 3EG
M6 BUL J. Ball Conifers, Main Road, Spilsby PE23 4BY
M6 BUR T. Skinner 13 Sawbrook, Fleckney, Leicester LE8 8TR
M6 BUU Dr D. Barrow 48 Main Street, Greysouthen, Cockermouth CA13 0UL
M6 BUW S. Bishton 18 Galloway Road, Poole BH15 4JX
M6 BUX P. Kerr 35 Coppice Gardens, Stone ST15 8BL
M6 BUY C. Marlow 59 Purford Green, Harlow CM18 6HN
M6 BVC C. Norwood 94 Waverley, Woodside, Telford TF7 5LU
M6 BVD R. Owen 4 Aldersleigh Drive, Stafford ST17 4RY
M6 BVE A. Hamilton 56 Wyvern, Telford TF7 5QH
MW6 BVG J. Campbell 1B Bush Road, Mountain Ash CF45 3BY
M6 BVK G. Kennedy-Brown 4 Seymour Gardens, Brockley, London SE4 2DN
M6 BVM S. Baynton 50 Briton Way, Wymondham NR18 0TT
MI6 BVN P. Mccullagh 6 Striff Lane, Omagh BT79 0WA
M6 BVP B. Hull 3 Cavalry Crescent, Eastbourne BN20 8NT
MW6 BVQ R. Unsworth 20 Arran Drive, Rhyl LL18 2NS
MW6 BVR J. Cowles 76 Pendine Close, Barry CF62 9DE
M6 BVS B. Watkins 41 Kingshill Avenue, St. Albans AL4 9QH
MW6 BVU M. Ryall 27 Clos Afon Twyi, Blackwood NP12 3FX
M6 BVV Dr J. Morgan Cedars, Springhill, Longworth, Abingdon OX13 5HL
M6 BVW D. Edwards 29 Larch Road, Maltby, Rotherham S66 8AZ
M6 BVX A. Booth 32 Acacia Avenue, Maltby, Rotherham S66 8DS
M6 BVZ R. Ede 14 Elm Close, Kidsgrove, Stoke-on-Trent ST7 4HR
M6 BWB W. Beston 79 Priestlands, Romsey SO51 8FJ
M6 BWC W. Tunstall 89 Lever Street, Little Lever, Bolton BL3 1BA
M6 BWE G. Brown 51 Arncliffe Drive, Knottingley WF11 8RH
M6 BWF R. Leverington 130 Osborn Road, Barton-Le-Clay, Bedford MK45 4NY
MW6 BWG A. Madden 51 Valley View, Cwmtillery, Abertillery NP13 1JE
M6 BWH W Cheshire Ray c/o Dr V. Venugopalan 6 Kennington Oval, Stoke-on-Trent ST4 8FX
M6 BWJ A. Wale 17 Roundhouse Close, Welford, Northampton NN6 6NN
M6 BWK S. Casey Flat 4, Belmont Court, Plymouth PL3 4DN
M6 BWL Z. Burningham 255 Welland Park Road, Market Harborough LE16 9DP

M6 BWN N. Bown 12 Sellwood Road, Abingdon OX14 1PE
M6 BWO M. Walters 65 Bannawell Street, Tavistock PL19 0DP
M6 BWP P. Walsh 181 Hermes Close, Hull HU9 4DR
M6 BWQ A. Carden Hazelgrove, South Allington, Kingsbridge TQ7 2NB
MW6 BWR P. Smith 19 Grandison Street, Neath SA11 2PG
M6 BWV A. Forber 32 Larch Avenue, Newton-le-Willows WA12 8JF
M6 BWW S. Laurence 23 West Street, Bridlington YO15 3DX
M6 BWZ S. Thirlwall 2 Crossfield Avenue Blythe Bridge, Stoke-on-Trent ST11 9PL
M6 BXA J. Lambert Birchwood Norwich Road, Cromer NR27 0HG
M6 BXB M. Stopper 29 Daisy Dale, Boston PE21 6DS
M6 BXD P. Hargreaves 9 Croston Road, Lostock Hall, Preston PR5 5LA
MM6 BXF D. Hook The Delvine, Amisfield, Dumfries DG1 3LH
MM6 BXH J. Curran 355D Charleston Drive, Dundee DD2 4HP
MW6 BXI R. Briant Talarvor, Llanon SY23 5HG
M6 BXJ M. Stillman 58 Highfield Road, Bognor Regis PO22 8PH
M6 BXK J. Street 22 Roman Acre, Wick, Littlehampton BN17 7HN
M6 BXL J. Lugsden 21 Overhill Way, Beckenham BR3 6SN
M6 BXM H. Phillips Flat 10, Fitch Court, 59-63 Effra Road, London SW2 1DD
M6 BXN C. Barthel 176 Lumb Lane, Droylsden, Manchester M43 7LJ
M6 BXO K. Cartwright 53 Sedgley Road, Dudley DY1 4NE
M6 BXP B. Pearson 9 Dunbar Close, Kidderminster DY10 3XS
MM6 BXQ A. Main 1 Waulkmill Loan, Currie EH14 5SS
MM6 BXR D. Reid 111 Oswald Road, Ayr KA8 8NX
M6 BXS A. Parker 9 Milecastle Court West Denton, Newcastle-upon-Tyne NE5 2PA
MM6 BXT B. Templeton 43 Elm Park, Ardrossan KA22 7BZ
M6 BXU B. Xu 138 King'S College, Cambridge CB2 1ST
MW6 BXV C. Astbury Old Police House, Llanegryn, Tywyn LL36 9SL
M6 BXW K. Morris 95 Murrayfield Drive, Wirral CH46 3RR
M6 BXX J. Baker 29 Ravensmoor Close, North Hykeham, Lincoln LN6 9AZ
M6 BXY P. Arnold 20 Upper Seagry, Chippenham SN15 5EX
M6 BXZ J. Walczak 18 Heathfield, Chippenham SN15 1BQ
M6 BYA K. Stowe 7 Glebe Way, Corsham SN13 9UL
M6 BYE T. Byers 1 Hazelwood Avenue, Sunderland SR5 5AH
M6 BYF D. Rogers 5 Semple Gardens, Chatham ME4 6QD
M6 BYH K. Broscomb Flat 34, White Willows, 70 Dyche Road, Sheffield S8 8DS
MM6 BYJ M. Stypka 28/9 Halmyre Street, Edinburgh EH6 8QD
M6 BYL W. Charlton 2 Mallory Street, Earl Shilton, Leicester LE9 7PH
M6 BYM T. Flynn 15 Dale Garth, Scarborough YO12 5NB
M6 BYN R. Jones 62 Nathaniel Road, Long Eaton, Long Eaton NG10 1GB
M6 BYO S. Downe 9 Danesway, Exeter EX4 9ES
M6 BYP P. Hoyle Flat 3, 84 Beverley Road, Hull HU3 1YD
M6 BYQ M. Abid 16 Cliff Gardens, Scunthorpe DN15 7PJ
M6 BYR J. Byrne 316 Turncroft Lane, Stockport SK1 4BP
M6 BYS A. Marriott 11 Downing Street, South Normanton DE55 2HE
MW6 BYT G. Pierce Ley Farm, Green Lane, Halton, Wrexham LL14 5BG
M6 BYU G. Webster 15 Bridge Road, Chichester PO19 7NW
MW6 BYV R. Potts 4 Maes Glan, Rhosllanerchrugog, Wrexham LL14 2DT
MW6 BYX N. Williams 60 Denbigh Close, Wrexham LL12 7TW
M6 BYY S. Bradley 9 Crofton Road Crofton Road, Southsea PO4 8NX
M6 BYZ G. Flinn 38 Fir Grove, Whitehill, Bordon GU35 9ED
M6 BZA D. Rudling Rose Cottage, Ludwells Lane, Southampton SO32 2NP
M6 BZB Y. Zhan 65A Mason Street, Edge Hill, Liverpool L7 3EN
MI6 BZC R. Benko 23 Six Mile Water Mill Drive, Antrim BT41 4FG
M6 BZD P. Elsey 62B Coleraine Road, London SE3 7PE
M6 BZE B. Evelyn 19 Laudsdale Road, Rotherham S65 3LG
M6 BZF K. Cunningham 18 Dovenby Fold, Ince, Wigan WN2 2PS
M6 BZG Thorn Emi ARC c/o R. Paddock 1 Harris Road, Bexleyheath DA7 4QD
M6 BZH R. Packman 62 Amsterdam Road, London E14 3JB
MI6 BZI C. Mccormick Flat 4, Legacorry House, Main Street, Armagh BT61 9RW
M6 BZK R. Boyes 63 Larch Road, New Ollerton, Newark NG22 9SX
M6 BZM K. Slater 63 29Th Avenue, Hull HU6 8DG
M6 BZN Capt. G. Slater 63 29Th Avenue, Hull HU6 8DG
MW6 BZP M. Clifford 27 Primrose Street, Tonypandy CF40 1BW
MM6 BZQ Rvd. A. Catterall Asta House, Scalloway, Shetland ZE1 0UQ
M6 BZT A. Smith 305, 1414 5Th St Sw, Calgary T2R 0Y8 Canada
M6 BZU E. St Quinton Mill Cottage, The Thorofare, Woodbridge IP13 8BB
M6 BZV S. Barker Strait Hey Farm, Stock Hey Lane, Todmorden OL14 6HB
M6 BZW F. Phillips 34 Park Drive, Maldon CM9 5JQ
MW6 BZX J. Palmer Clettwr Hall, Pontshaen, Llandysul SA44 4TU
M6 BZY I. Kazlauskaite 4 Spencer Way, Redhill RH1 5LY
M6 CAA C. Sparrey 166 Abberley Avenue, Stourport-on-Severn DY13 0LT
MI6 CAD A. Cahalan 131 The Meadows, Randalstown, Antrim BT41 2JD
MW6 CAE C. Catley 40 Rockvilla Close, Varteg, Pontypool NP4 7QF
M6 CAF C. Furlong 22 Swaisland Road, Dartford DA1 3DE
M6 CAG M. Sinclaire 21 Ford Court, Winsford CW7 1NJ
M6 CAH B. Heath 108 Cow Lane, Bramcote, Nottingham NG9 3BB
M6 CAI C. Ingamells 41 Princess Anne Road, Boston PE21 9AP
M6 CAJ P. Lewickyj 37 Maple Street, Lincoln LN5 8QS
MI6 CAK C. Doyle 13 Tobin Park, Cookstown BT80 0JL
MW6 CAN C. Davies 3 Penhydd Houses, Oakwood Avenue, Pontrhydyfen, Port Talbot SA12 9SE
M6 CAP C. Preece 14 Dock Street, Widnes WA8 0QX
M6 CAQ C. Allen 29 Cedar Road, Hythe, Southampton SO45 3PH
M6 CAR C. Taylor 212 Plantation Hill, Worksop S81 0HD
M6 CAS S. Hill 1 Beresford Road, Walsall WS3 1JX
M6 CAT D. Welford 24 Hawthorn Crescent Quarrington Hill, Durham DH6 4QW
MI6 CAV C. Mccammick 23 Atkinson Avenue, Portadown, Craigavon BT62 3HY
M6 CAW C. Waldron 2 The Bourne, Eastleach, Cirencester GL7 3NN
MI6 CAY M. Mullaney 21 Aghagay Meadows, Aghagay, Enniskillen BT92 8AE
M6 CBA T. Ingleby 27 Burley Wood Lane, Leeds LS4 2SU
M6 CBC C. Kent 19 Coppice Rise, Harrogate HG1 2DP
M6 CBD C. Arblaster 22 Wood Lane, Carlton, Barnsley S71 3JJ
MM6 CBE C. Ellison 1 Newton Road, St. Fergus, Peterhead AB42 3DD
MI6 CBG C. Gorman 5 Linden Gardens, Bangor BT19 6EB
MM6 CBI C. Addison 8D Thomson Street, Johnstone PA5 8RZ
M6 CBJ C. Jerome Kennel Cottage, Hobland Road, Great Yarmouth NR31 9AR
MW6 CBL S. Morgan 31 Church Street, Briton Ferry, Neath SA11 2JG
M6 CBM L. Clark 18 Cutters Close, Narborough, Leicester LE19 2FY
M6 CBN C. Nunnen 16 Meden Avenue, Warsop, Mansfield NG20 0PS
M6 CBO A. Carter 24 The Esplanade, Weymouth DT48DN
M6 CBP S. Bennett 154 Humberstone Road, Grimsby DN32 8HR
M6 CBQ A. Cranston 196 West End Costessey, Norwich NR8 5AW
M6 CBU W. Porter 92 Turner Road, Tonbridge TN10 4AJ
M6 CCA M. Landon 29 Portland Road, Hucknall, Nottingham NG15 7SL
M6 CCB G. Winchester 36 Crofters Green, Preston PR1 7UG
M6 CCE C. Etchells 7 Woodlands Drive, Sandford, Wareham BH20 7QA
M6 CCF C. Finbow Medlars Cottage, The Street, Woodbridge IP13 7JP
MW6 CCG S. Williams 1 Lawrence Terrace, Llanelli SA15 1SW
M6 CCH A. Fisher 63 Soloman Drive, Bideford EX39 5XY
M6 CCI F. Roberts 17 Ansisters Road, Ferring, Worthing BN12 5JG
M6 CCJ J. Davidge 45 Ferring Street, Ferring, Worthing BN12 5JW
M6 CCM D. James 123 Bruce Road, Woodley, Reading RG5 3DY
M6 CCN M. Tucker 182 Salisbury Road, Amesbury, Salisbury SP4 7HW
MM6 CCO W. Hannah 71 Herriot Avenue, Kilbirnie KA25 7JB
M6 CCR A. Bhakoo 4 Bryden Cottages, High Street, Uxbridge UB8 2NY
MM6 CCS C. Stewart 14 Stanley Road, Saltcoats KA21 5BB
MI6 CCU N. Morrow 90 Bracky Road, Sixmilecross, Omagh BT79 9PH
M6 CCV P. Marshall 21 Lymn Avenue, Gedling, Nottingham NG4 4EA
MM6 CCW J. Ransome Evanton, Novar, Dingwall IV16 9XH
M6 CCX M. Perkins 2 Buckingham Orchard, Chudleigh Knighton, Chudleigh, Newton Abbot TQ13 0EW
MM6 CCY A. Douglas 163 Glenlora Drive, Glasgow G53 6BL
M6 CDB A. Welch 18 Monk Close, Tipton DY4 7TP
M6 CDC C. Chalmers 2 Canterbury Road, Bracebridge Heath, Lincoln LN4 2TD
M6 CDD M. Wollaston 5 Clover Court, Shardlow, Derby DE72 2GE
M6 CDE G. Leedham Hawkes 103 Wood Lane, Hednesford, Cannock WS12 1BW
M6 CDG D. Chang Montefiore House, Wessex Lane, Southampton SO18 2NU
M6 CDH C. Hailstone 2 Thornfield Avenue, Thornton-Cleveleys FY5 5BH
M6 CDK J. Rowe 22 Treaty Road Glenfield, Leicester LE3 8LU
M6 CDL R. Winstanley 173 Wimborne Road, Poole BH15 2EF
M6 CDN D. Watkiss 54 Rolston Close, Plymouth PL6 6TN
M6 CDQ A. Camp 3 Acre Close, Witnesham, Ipswich IP6 9EU
M6 CDR M. Mason 7 Langland Close, Malvern WR14 2UY
M6 CDT J. Schleswick 9 Wick House Close Saltford, Bristol BS31 3BZ
M6 CDU E. Mortimer 6 Lanes End, Gastard, Corsham SN13 9QS
MW6 CDV K. House Liddington, Dehewydd Lane Llantwit Fardre, Pontypridd CF38 2EN
M6 CDW C. Wade 31 Melton Green, Wath-Upon-Dearne, Rotherham S63 6AA
MW6 CDZ G. Saltmarsh 15 Colbourne Road Beddau, Pontypridd CF38 2LN
M6 CEA M. Anderson 27 Laing Road, Colchester CO4 3UT
M6 CEB M. Bamber 10 Sedgeley Mews, Freckleton, Preston PR4 1PT
M6 CEE T. Massey 2 Cannon Heath Farm Cottages, Cannon Heath, Basingstoke RG25 3EJ
M6 CEF C. Davies 1 Meadow Close, Bagworth, Coalville LE67 1BR
MW6 CEG J. Martin 2 Pant Heulog, Dyffryn Ardudwy LL44 2BU
M6 CEH C. Halloway 82 Northwall Road, Deal CT14 6PP
M6 CEI K. Knapp 46 Robin Hood Road, St. Johns, Woking GU21 8SY
MI6 CEJ C. Jeffery 197 Finvola Park, Dungiven, Londonderry BT47 4ST
MM6 CEL R. Wood 42 Moorhouse Avenue, Paisley PA2 9NY
MM6 CEM C. Macdonald 2 Porterfield Road, Inverness IV2 3HW
M6 CEN R. Bird 431 Ombersley Road, Worcester WR3 7DQ
M6 CEO A. Humphries 3 Suffolk Drive, Worcester WR3 8QT
M6 CEP C. Pritchard 8 Hoon Avenue, Newcastle ST5 9NY
M6 CEQ D. Brown 2 Kibworth Grove, Stoke-on-Trent ST1 5QP
MM6 CES C. Canning 17 Blackadder Crescent Greenlaw, Duns TD10 6XN
M6 CET M. Knowles 86 West Shore Road, Walney, Barrow-in-Furness LA14 3UD
M6 CEU A. Burkitt 53 Westwood Drive, Bourne PE10 9PY
MM6 CEW C. Watson 20 Norlands, Errol, Perth PH2 7QU
MW6 CEX M. Brittle 11 Haig Place, Gendros, Swansea SA5 8BT
M6 CEY M. Green 26 Drake Crescent, Kidderminster DY11 6EE
M6 CEZ I. Hill 3 Beresford Rd, Walsall WS31JX
MM6 CFA S. Mackenzie 20 Meadowhouse Road, Edinburgh EH12 7HP
M6 CFD J. Seaton 3 Tunstall Street, Middlesbrough TS3 6PE
M6 CFE D. Cowdrey 264 Stamford Road, Brierley Hill DY5 2QF
MM6 CFH C. Fraser-Hopewell 2/1 70 Albert Road, Glasgow G42 8DW
M6 CFI K. Davis 26 Mendip Drive, Nuneaton CV10 8PT
M6 CFL M. Shortall 16 Darnaway Close, Birchwood, Warrington WA3 6TR
M6 CFN N. Layland 3 Thirlmere Road, Golborne, Warrington WA3 3HH
M6 CFO M. Summers 21 Quantock Avenue, Caversham, Reading RG4 6PY
M6 CFQ R. Smith 15 Hollybush Road, North Walsham NR28 9XT
M6 CFR J. Mather 33A Forest Road, Southport PR8 6JD
MM6 CFS F. Sturrock 16 Carlyle Crescent, Buckhaven, Leven KY8 1DW
M6 CFT C. Mole 6 Clements Road, Chorleywood, Rickmansworth WD3 5JT
M6 CFU S. Watson 8 Church Close, Overstrand, Cromer NR27 0NY
M6 CFV A. Stride 2 Bailey Close, Pewsey SN9 5HU
M6 CFW A. Comerford 21 New Cross Road, Headington, Oxford OX3 8LP
M6 CFX M. Lawrence 61 Jack Warren Green, Cambridge CB5 8US
M6 CFY A. Collison Bramble Lodge, Walkford Lane, New Milton BH25 5NL
M6 CFZ J. Gilpin 17 Roundstone Crescent, East Preston, Littlehampton BN16 1DG
M6 CGB C. Bradley 3 Trecarne Gardens, Delabole PL33 9DP
M6 CGC C. Cheadle 55 Ellison Street, Sheffield S3 7JH
M6 CGF S. Gateson Flat 6, Kenley House, Croydon CR0 6AQ
MW6 CGH M. Jones 102 Thomas Street, Tonypandy CF40 2AH
M6 CGJ R. Wheeler 12 Drake Avenue, Didcot OX11 0AD
M6 CGK Dr D. Richardson 89A Bean Oak Road, Wokingham RG40 1RJ
M6 CGL G. Livesey 48 Kingsway, Leyland PR25 1BL
M6 CGM C. Martin 63 Oversetts Road, Newhall, Swadlincote DE11 0SL
MI6 CGQ C. Glenn 40 Deanfield, Londonderry BT47 6HY
M6 CGS C. Smith 17 Sunningdale Avenue, Sale M33 2PJ
MW6 CGV C. Briant Talarvor, Llanon SY23 5HG
M6 CGW C. Watkins 25 Citadilla Close, Gatherley Road, Richmond DL10 7JE
M6 CGX R. Coy 45 Coton Grove, Shirley, Solihull B90 1BS
M6 CGY A. Lycett 2 Royce Avenue, Hucknall, Nottingham NG15 6FU
M6 CGZ A. Dalgliesh 62 Newbury Street, Wantage OX12 8DF
M6 CHC M. Corrigan 33 Westbourne Road, Knott End-On-Sea, Poulton-le-Fylde FY6 0BS

M6 CHD R. Kergozou Lilac Cottage, The Street, Cheddar BS27 3TH
M6 CHF Dr C. Ferguson Royd Moor, Royd Moor Lane, Badsworth, Pontefract WF9 1AZ
M6 CHG L. Hartley 206 Latimer Road, Eastbourne BN22 7JF
M6 CHH J. Wishart 19 Chepstow Close, Chippenham SN14 0XP
M6 CHI M. Boon 45 Carlton Road, Wickford SS11 7ND
M6 CHJ J. Chaplin-Madden 96 Salisbury Road, Great Yarmouth NR30 4LS
M6 CHL D. Cobbold 65 St. Olaves Road, Bury St. Edmunds IP32 6RR
MM6 CHM R. Russell 4 Valley Court, Patna, Ayr KA6 7LQ
MM6 CHN C. Denny Darnaway, Castleton Place, Braemar, Ballater AB35 5ZQ
M6 CHP M. Cooper 86 Grove Road, Tiptree, Colchester CO5 0JG
MW6 CHQ A. Bowlzer 3 Moreia Terrace, Harlech LL46 2YW
M6 CHS M. Drury 19 Cuffley Avenue, Watford WD25 9RB
M6 CHT A. Trowse 110 New Road, Hethersett, Norwich NR9 3HQ
M6 CHU R. Last 30 Abbot Road, Bury St. Edmunds IP33 3UB
MM6 CHV R. Gilchrist 5 Glenburn, Leven KY8 5BD
M6 CHW C. Wright 14 Orchard Close, Poughill, Bude EX23 9ES
M6 CHX M. Harrison 2A Arundel Road, Camberley GU15 1DL
MM6 CHY C. Hay 30 Kincardine Place, East Kilbride, Glasgow G74 3DN
M6 CHZ H. Eager 45 Fleetwood Avenue, Herne Bay CT6 8QW
MM6 CIA C. Anderson 11 Willowpark Place, Aberdeen AB16 6XY
M6 CIE W. Canavan 9 The Ridings, Deanshanger, Milton Keynes MK19 6JD
M6 CIG E. Black 34 White Bank Road, Oldham OL8 3JH
MM6 CIJ C. Thomson 54 Alderman Road, Glasgow G13 3YE
M6 CIK P. Turner 38 La Ferte Bernard Close, Louth LN11 0ZN
M6 CIP A. Rowson-Brown The Fox, Station Road, Baldock SG7 5RN
M6 CIQ M. Walls 38 Poplar Drive, Royston SG8 7ER
M6 CIS Dr B. Rush Springside, Uploders, Bridport DT6 4NU
M6 CIU G. Brotherhood 17 Baldwin Close, Forest Town, Mansfield NG19 0LR
M6 CIW C. Waite 69 Ellesmere Street Swinton, Manchester M27 0JT
M6 CIX A. Skinner 17 Hawkins Close, Harrow HA1 4DJ
MW6 CIY M. Bennett 24 Bryn Street, Merthyr Tydfil CF47 0TG
M6 CJA C. Apps 17 Kent Road, Formby L37 6BG
MM6 CJC J. Scanlan Aitnoch Farmhouse, Grantown-on-Spey PH26 3PX
M6 CJD C. Dawson 9 Mulberry Close, Poringland, Norwich NR14 7WF
M6 CJE D. Chidgey 42 Half Acre, Williton, Taunton TA4 4NZ
M6 CJF C. Foster 8 William Iliffe Street, Hinckley LE10 0LY
M6 CJG C. Groves 3 Hudson Davies Close, Pilley, Lymington SO41 5PA
MW6 CJH C. Hill 9 Oliver Road, Newport NP19 0HU
M6 CJI J. Power 43 Valley Road, Melton Mowbray LE13 0DU
M6 CJJ P. Spry 8 Maun Green, Newark NG24 2HA
M6 CJK R. Tuffin 133 Shirley Drive, Hove BN3 6UJ
M6 CJM C. Martin 2 Whitethorn Cottages, Dark Lane, Cheltenham GL51 9RW
M6 CJN C. Jenkins 5 Mayors Buildings, Bristol BS16 5AU
M6 CJP C. Petrie 14 Rotherfield Avenue, Eastbourne BN23 8JQ
M6 CJQ B. Hiley 9 Pinfold Lane, Harby, Melton Mowbray LE14 4BU
M6 CJR C. Rundle 1 Trezaise Close, Roche, St. Austell PL26 8HW
MW6 CJS C. Sweeney 67 Blandon Way, Cardiff CF14 1EH
M6 CJT C. Wright Home Farm, Bretford CV230LB
M6 CJU J. Brown 34 Salisbury Road, Worthing BN11 1RD
M6 CJV S. Clarke 20 Woodlands Way, Southwater, Horsham RH13 9HZ
M6 CJW C. Moss 19 Tozer Close, Wallisdown, Bournemouth BH11 8RB
M6 CJX I. Ftaiha 8 Parkside, London NW7 2LH
MM6 CJY C. Yohn 3F/C 567 George Street, Aberdeen AB25 3XX
MM6 CJZ J. Wright 43 Spey Court, Stirling FK7 7QZ
MM6 CKC M. Thomson 30 Pladda Road, Saltcoats KA21 6AQ
M6 CKD R. Basford 4 Renoir Close, St. Ives PE27 3HF
M6 CKE L. Hardy 221 Rookery Lane, Lincoln LN6 7PJ
M6 CKF C. Farmer 3 Laxton Way, Banbury OX17 1GJ
M6 CKI R. Baines 9 Deer Close, Walsall WS3 3EA
M6 CKJ E. Taylor Lanthorn Close, Broxbourne EN107NR
M6 CKL P. Ridley 218 Lichfield Road, Rushall, Walsall WS4 1SA
MW6 CKM J. Jones Isfryn Bungalow, Glan-Y-Nant, Llanidloes SY18 6PQ
M6 CKO E. Redmond 28 Common Lane, Polesworth, Tamworth B78 1LS
M6 CKP C. Prior 38 Windmill Road, Wombwell, Barnsley S73 8PP
M6 CKQ Dr A. Colman 5 Burn Heads Road, Hebburn NE31 2TB
M6 CKR M. Sadler 14 Woodlands Avenue, Water Orton, Birmingham B46 1SA
M6 CKT C. Jenkings 8 Tolworth Hall Road, Birmingham B24 9NE
M6 CKU P. Naik 82 Misbourne Road, Uxbridge UB10 0HW
M6 CKV P. Pain 12 Maple Drive, Bamber Bridge, Preston PR5 6RA
M6 CKY N. Gamulea Schwartz Flat 5, 127 Princes Avenue, Hull HU5 3HH
M6 CKZ D. Osborne 12 Sandringham Close, Brackley NN13 6JQ
M6 CLA S. Clark 43 Age Croft, Oldham OL8 2HG
M6 CLB E. Williams 35 Cansfield Grove, Ashton-In-Makerfield, Wigan WN4 9SE
MM6 CLC C. Collins Redwoods, Barcaldine, Oban PA37 1SG
M6 CLF C. Studdart 41 Queensgate Street, Hull HU3 2TT
M6 CLG J. Hurlbutt 55 Prospect Avenue, Seaton Delaval, Whitley Bay NE25 0EL
M6 CLI C. Herlingshaw 48 Keats Road, Normanby, Middlesbrough TS6 0RW
MW6 CLJ C. Jones 19 Crud Y Castell, Denbigh LL16 4PQ
M6 CLK M. Clark 34 Magdalene Road, Owlsmoor, Sandhurst GU47 0UT
M6 CLL A. Mccall 95 Newton Drive, Blackpool FY3 8LX
MM6 CLM C. Houston 115 Burnbank Road, Ayr KA7 3QH
M6 CLO P. Clough 101 Cathedral View, Houghton le Spring DH4 4HN
MW6 CLP R. Piper 16 Elm Rise, Bryncethin, Bridgend CF32 9SX
M6 CLQ M. Eccleston 15 Wood Road North, Manchester M16 9GQ
M6 CLV C. Vernau 62 Princethorpe Road, Ipswich IP3 8NX
M6 CLW A. Sockett 37 Windsor Road, Thorpe Hesley, Rotherham S61 2QS
M6 CLX P. Cooke 3 St. Stephens Court, Congleton CW12 1QW
M6 CLZ C. Sinclair 43 Newton Way, Leighton Buzzard LU7 4SU
M6 CMB L. Gorecki 6 Robinhood Lane, Winnersh, Wokingham RG41 5LX
M6 CMF B. Hill 41 Heath Road, Leighton Buzzard LU7 3AB
M6 CMG C. Goodhand 37 Westwick Gardens, Lincoln LN6 7RQ
M6 CMI D. Lyons 2 Goswick Farm Cottages, Berwick-upon-Tweed TD15 2RW
M6 CMK C. Kennedy 30 Tatton Close, Cheadle SK8 2LZ
MM6 CMM C. Mckinlay 19 Ash Grove, Blackburn, Bathgate EH47 7QJ
M6 CMO I. Steggles 4 Hawley Vale, Hawley Road, Dartford DA2 7RL
MI6 CMQ R. Donnan 71 Victoria Avenue, Newtownards BT23 7ED
M6 CMT C. Travis 4 Kingsdale, Worksop S81 0XJ
M6 CMW C. Walsh 8 Kemp Road, Leicester LE3 9PS
MM6 CMY C. Matheson 322 Millfield Hill, Erskine PA8 6JN
M6 CMZ M. Jones 89 Gloucester Road, Coleford GL16 8BN
M6 CNA M. Bradley 11 Pike Road, Coleford GL16 8DE
MM6 CNC C. Comrie 11 Glendoune Street, Girvan KA26 0AA
M6 CND D. Ward 65 Water Street, Accrington BB5 6QU
M6 CNG G. Southall 28 Manor Road, Woodford Halse, Daventry NN11 3QP
M6 CNI R. Johnson 24 Fairfields, Upper Denby, Huddersfield HD8 8UB
M6 CNK D. Stockton 78B Alderfield Penwortham, Preston PR1 9HA
M6 CNL H. Hope 51 Margravine Gardens, London W6 8RN
M6 CNN D. Bedford 163 Mortimer Road, South Shields NE34 0RR
MM6 CNO S. Leighton 4 Earn Court, Alloa FK10 1PT
M6 CNP M. Pratt 1 Ashvale Gardens, Romford RM5 3QA
M6 CNQ P. Phillips 49 Lower Northam Road, Hedge End, Southampton SO30 4HE
M6 CNR J. Taylor 90 Village Road, Gosport PO12 2LG
M6 CNS M. Phillips 66 Minden Way, Winchester SO22 4DU
MM6 CNV M. Jamieson 5 Straid Bheag, Barremman, Helensburgh G84 0QX
M6 CNX J. Mooney 107 Tedder Road, South Croydon CR2 8AR
M6 CNY R. East 6 Ashley Road, Worcester WR5 3AY
M6 CNZ J. Haddow 28 Church Marks Lane, East Hoathly, Lewes BN8 6EQ
M6 COB R. Williams 11 Trinity Buildings, Carlisle CA2 7DA
MW6 COD D. Codd Gwachal-Tagy Farm, Haverfordwest SA62 6HF
M6 COE C. Bassett 3 Downshire Terrace, Street Lane, Haywards Heath RH17 6UL
MM6 COF G. Steele 1/2 50 Motehill Road, Paisley PA3 4ST
M6 COH S. Parris-Hughes 23 Hobney Rise, Westham, Pevensey BN24 5NN
M6 COI M. Bernasinski 1 Elizabeth Place, Clyde Road, London N15 4LA
M6 COL C. Spicer 3 School View, Tunstall, Sittingbourne ME9 8DX
MI6 COM S. Carroll Drumguiff Lane, Enniskillen BT92 7HP
M6 CON C. Morgan 41 Torwood Mount, Old Torwood Road, Torquay TQ1 1PX
M6 COP L. Elphick 30 The Meadows, Heskin, Chorley PR7 5NR
M6 COU P. Coulsey 25 Weavers Close, Horsham St. Faith, Norwich NR10 3HY
M6 COV A. Hughes 86 Colinmander Gardens, Ormskirk L39 4TF
M6 COX A. Cox 20 Dunns Dale, Maltby, Rotherham S66 7NR
MW6 COY C. Burden 21 Salisbury Road, Barry CF62 6PB
M6 CPA C. Godding 11 Oakleaze Road, Thornbury, Bristol BS35 2LL
M6 CPC C. Cook 95 Station Road, Eccles, Manchester M30 0PZ
MM6 CPE C. English Easter Backlands, Roseisle, Elgin IV30 5YD
M6 CPH C. Holden 49 Whalley Road, Lancaster LA1 2HE
MW6 CPJ D. Martin Neuadd Hendidley, Beehive Lane, Newtown SY16 3NA
MM6 CPK E. Dungavel 9C Anderson Crescent, Ayr KA7 3RL
M6 CPN C. Norris 115 Sutton Road, Walpole Cross Keys PE344HE
M6 CPO S. Parker 10 Wheelwrights Close, Sixpenny Handley, Salisbury SP5 5SA
M6 CPP N. Barnard 10 Whites Lane Kessingland, Lowestoft NR33 7TF
M6 CPQ T. Walsh 6 Brass Thill, Durham DH1 4DS
M6 CPR G. Doxey 2 Nettlecroft, Barnsley S71 5SD
M6 CPS C. Stratford 15 Ferndale Road, Banbury OX16 0RZ
M6 CPT D. Baker 21 Dolman Road, Gosport PO12 1RB
MU6 CPV P. Martin 1 Hazeldene, Grandes Maisons Road, Guernsey GY2 4JS
M6 CPW P. Winkley 25 Grindley Avenue, Manchester M21 7NE
M6 CPX R. Gleave 52 Cranborne Avenue, Warrington WA4 6DE
M6 CPZ M. Spiers 99 Chapel Street, Tiverton EX16 6BU
M6 CQB I. Mckean 14 Maltings Close, Cranfield, Bedford MK43 0BY
M6 CQD A. Fitton 72 Newnham Court, Ipswich IP2 9UE
M6 CQE D. Bishop 62 Brindley Crescent, Hednesford, Cannock WS12 4DS
MJ6 CQF J. Crill Rocqueberg Farm, Rocqueberg Close, Jersey JE2 6LT
M6 CQH M. Fletcher Wray 70, Crook O'Lune Caravan Park, Caton Road, Lancaster LA29HP
M6 CQJ A. Tranter 122 Summerhill Road, Bristol BS5 8JU
MW6 CQL C. Summerfield 11 Woodland Park, Penderyn, Aberdare CF44 9TX
MW6 CQO C. Osborne Thornleigh, Tremont Road, Llandrindod Wells LD1 5BH
M6 CQR P. Dunn 9 Parkside Gardens, Widdrington, Morpeth NE61 5RP
MI6 CQS S. Allen 21 Derrymore Meadows, Bessbrook, Newry BT35 7GA
M6 CQV C. Norman 15 Maple Close, Sedbergh LA10 5JE
M6 CQX G. Davies 5 Rockall Drive, Hailsham BN27 3BG
M6 CRA C. Ancill 8 Ipley Way, Hythe, Southampton SO45 3LJ
M6 CRE C. Etches The Nook, Low Road Friskney, Boston PE22 8NQ
M6 CRF C. Faulkner Mount Pleasant, Elkstones, Buxton SK17 0LU
M6 CRG C. Gloess 5 Elmers Lane, Kesgrave, Ipswich IP5 2GW
M6 CRH C. Hogan 82 Abbott Road, Didcot OX11 8HY
M6 CRJ C. Jackson 30 Coronation Avenue, Mile Oak, Tamworth B78 3NW
M6 CRO C. Ordish 37 Mill Lane, Earl Shilton, Leicester LE9 7AY
M6 CRP C. Pulford 17 Canada Road, Cromer NR27 9AH
MM6 CRQ C. Welsh 28 Peacock Wynd, Motherwell ML1 4ZL
M6 CRR R. Flevill 62 Grosvenor Way, Horwich, Bolton BL6 6DJ
M6 CRT C. Thomas 17 The Green, Nether Heyford, Northampton NN7 3LE
MW6 CRU C. Uphill 167 Nantgarw Road, Caerphilly CF83 1AN
MM6 CRW C. Watkinson Hare House, Perth Road, Birnam, Dunkeld PH8 0AA
M6 CRX A. Haynes 16 Hills Crescent, Colchester CO3 4NU
M6 CRZ C. Roberts 25 Queens Lea, Willenhall WV12 4JA
M6 CSA S. Clark 1 Hanover Cottages, Lippen Lane, Southampton SO32 3LE
M6 CSC R. Metcalfe Little Shernden, Shernden Lane, Edenbridge TN8 5PS
M6 CSE C. Cartwright 8 Charles Close, Westcliff-on-Sea SS0 0EU
M6 CSF R. Powell 13E Rothsay Road, Bedford MK40 3PP
M6 CSG C. Greenwood 21 Valley Drive, Thornhill Dewsbury WF120HE
M6 CSI C. Morris 38 Beltony Drive, Crewe CW1 4TX
M6 CSJ B. Appleby 64 Lundy Close, Southend-on-Sea SS2 6HB
M6 CSK D. Polley 6 Coneygear Road, Hartford, Huntingdon PE29 1QL
M6 CSL C. Langmaid Flat 4, Woodlawn High Street, Partridge Green West Sussex RH13 8HR
M6 CSM C. Moppett 13 Piccadilly, Tamworth B78 2ER
M6 CSN C. Snelling-Nash 33 Church Lane, Kimpton, Hitchin SG4 8RR
M6 CSP C. Pritchard 9 Furnace Drive, Crawley RH10 6HZ
M6 CSR C. Robinson Watermill Farm Cottage, The Moor, Middleton, Saxmundham IP17 3LW
M6 CSS K. Latham 45 Hollybank Close, Northwich CW8 4GS
M6 CSU C. Sharif 38 King George Road, Ware SG12 7DT
M6 CSW J. Cheng Clarence Drive, Harrogate HG1 2QG
M6 CSX A. Askam 8 The Pastures, Weston-On-Trent, Derby DE72 2DQ
M6 CSZ A. Chaplin 10 St. Leonards Road, Malinslee, Telford TF4 2EB

M6 CTA P. Kerton 15 Barncroft Way, Havant PO9 3AA
MW6 CTD T. Cooper Dorset Cottage, Alltyblacca, Llanybydder SA40 9SU
MW6 CTE M. Sunderland 16 Heather Avenue, Cardiff CF5 5AH
MW6 CTG M. Adam Tan Ffordd, Mynydd Llandygai, Bangor LL57 4LX
MM6 CTH S. Fenton 19 Rowan Road, Girvan KA26 0BY
M6 CTI J. Roberts 4 Oaks Road, Staines-upon-Thames TW19 7LG
M6 CTJ C. Johnson 29 Linden Road Creswell, Worksop S80 4JT
M6 CTK C. Taylor 9 Pendeen Crescent, Threemilestone, Truro TR3 6SP
MM6 CTL C. Livingstone 391 Dyke Road, Glasgow G13 4QE
M6 CTN S. Knott 15 Meadowlands Drive, Haslemere GU27 2FD
M6 CTO D. Mcgleenan Greenfields, Ellonby, Penrith CA11 9SJ
MM6 CTQ A. Montgomery 13 Rathad A'Mhaoir, Plasterfield, Stornoway HS1 2UP
MW6 CTS D. Hooper Nant Y Dryslwyn Cottage, Ty Mawr, Llanybydder SA40 9RD
M6 CTT C. Rogers 65 Darwin Close, Taunton TA2 6TR
M6 CTV C. Leek 46 The Hollies, Holbeach, Spalding PE12 7JQ
M6 CTW C. Winnan 133 Deepcut Bridge Road, Deepcut, Camberley GU16 6SD
M6 CTX T. Rowlands 7 Northfield Crescent, Beeston, Nottingham NG9 5GR
M6 CTY P. Keane 18 Main Road, Cannington, Bridgwater TA5 2JN
M6 CTZ J. Richards 5 Dumfries Place, Weston-Super-Mare BS23 4LQ
MW6 CUA G. Young 104 Ystrad Road, Pentre CF41 7PW
M6 CUC R. Keast 7 The Finches, Newport PO30 5GU
M6 CUD S. Kneale 59 Mayflower Avenue, Saxmundham IP17 1BU
M6 CUE N. Connor 28 Church Street, Hungerford RG17 0JE
M6 CUJ J. Parsons Three Shire, Ross-on-Wye HR9 7PR
M6 CUL D. Richards 58 Holm Lane, Oxton, Prenton CH43 2HS
M6 CUN J. Colledge-Wiggins Raffles, Southcombe, Chipping Norton OX7 5QH
M6 CUP T. Cavanagh 186 Cole Valley Road, Birmingham B28 0DQ
M6 CUQ G. Fantom 6 Middle Street, Farcet, Peterborough PE7 3AX
MW6 CUR J. Troughton Rhiwbina, Pentre Lane, Cwmbran NP44 3AP
M6 CUS S. Williams 16 Pinewood Avenue, New Haw, Addlestone KT15 3AA
M6 CUU K. Boaler 12 Belmont, Slough SL2 1SU
M6 CUV A. Douglas Gobbins Cottage, Sandy Lane, Lathom, Ormskirk L40 5TU
M6 CUW A. Allan 13 Oakfield Road, East Cowes PO32 6DX
M6 CUX A. Stott 2 Douglas Drive, Maghull, Liverpool L31 9DG
MI6 CUZ G. Angus 59 Millisle Road, Donaghadee BT21 0HZ
M6 CVB J. Adler 1 Searles Meadow, Dry Drayton, Cambridge CB23 8BW
MW6 CVC C. Smith 29 Heol Cwarrel Clark, Caerphilly CF83 2NE
M6 CVE P. Herron 102 Garden City Villas, Ashington NE63 0EU
M6 CVF L. Bishop Franklin House, Canford School, Wimborne BH21 3AF
M6 CVJ J. Smith Flat 9 Mount Gardens, 19 Davenport Road, Coventry CV5 6QH
M6 CVK C. Saxton 23C Forest Way, Humberston, Grimsby DN36 4HQ
MW6 CVN M. Beecroft Hafod Y Wennol, Llanddoged, Llanrwst LL26 0TY
MI6 CVO R. Lambe 39 Tildarg Avenue, Belfast BT11 9LU
M6 CVP L. Stamper 22 Douglas Road, Workington CA14 2QY
M6 CVU Z. Szikszai 135 Devon Road, Newark NG24 4JL
M6 CVV J. Szikszai 135 Devon Road, Newark NG24 4JL
MI6 CVW A. Logan 15 Park Lane, Saintfield, Ballynahinch BT24 7PR
M6 CWA G. Moon 1 Brankenwall, Muncaster, Ravenglass CA18 1RG
MM6 CWB C. Beedie 21 Marywell Village, Arbroath DD11 5RH
MI6 CWC C. Campbell 21 Elms Park, Coleraine BT52 2QF
M6 CWD B. Ledson 16 Caton Close, Southport PR9 9XF
M6 CWF A. Mills 6 Kildare Drive, Peterborough PE3 9TS
M6 CWG D. Levy Flat 36, Claydon House, London NW4 1LS
M6 CWI S. Chapman 5 Bracton Drive, Nottingham NG3 2LN
M6 CWJ B. Harrison 104 Jeavons Lane, Great Cambourne, Cambridge CB23 5FN
M6 CWK C. Austin 1A Arundel Road, Peacehaven BN10 8TE
MW6 CWL K. Saltmarsh 15 Colbourne Road, Beddau, Pontypridd CF38 2LN
MM6 CWN C. Cowan 23 Park Road, Hamilton ML3 6PD
M6 CWO J. Cox 87 Richmond Road, Leighton Buzzard LU7 4RF
M6 CWP C. Hicks 6 Westlands Avenue, Weston-On-The-Green, Bicester OX25 3RD
M6 CWQ A. Toth 14 Runcorn Road, Sunderland SR5 5ET
M6 CWR C. Ralphson 20 Monsal Grove, Buxton SK17 7TF
M6 CWU J. Henningway 64 South Cliff, Bexhill-on-Sea TN39 3EE
M6 CWV N. Symes 23 Elm Road, Portslade, Brighton BN41 1SA
M6 CWW C. Warhurst 30 Nether Royd View, Silkstone Common, Barnsley S75 4QQ
M6 CWX Dr T. Digman 74 Baddlesmere Road, Whitstable CT5 2LA
M6 CWZ L. Starrett 50 Danes Road, Bicester OX26 2LP
M6 CXB C. Burnham 93 Colliers Break, Emersons Green, Bristol BS16 7EB
M6 CXC B. Burr Flat 87, Horatia House, Southsea PO5 4AL
MM6 CXD A. Donald 10 Fraser Road, Burghead, Elgin IV30 5YN
M6 CXH R. Pilkington 64 School Lane, Higher Bebington, Wirral CH63 2LW
M6 CXI G. Megson 147 Duke Of York Avenue, Wakefield WF2 7DA
MM6 CXJ R. Inglis The Shop, Roadside Skirza, Freswick, Wick KW1 4XX
M6 CXM J. Van De Vondel 50 Holmsley Lane, South Kirkby, Pontefract WF9 3JF
M6 CXN F. Strickland Hinds Cottage, Beverley Road, Driffield YO25 9PF
M6 CXO M. Hickey 100 Chester Road, Poynton, Stockport SK12 1HG
M6 CXP B. Fitzgerald-O'Connor 14 Goldwing Close Custom House, London E16 3EQ
M6 CXV A. Todd 16 Worcester Close, Bracebridge Heath, Lincoln LN4 2TY
M6 CXW A. Little 13 Belmont Grove, Burnley BB10 4NR
M6 CXZ M. Hack 28 Horsefield View, Melton Mowbray LE13 0TF
M6 CYA C. Rowe 1 Avondale Street, Stoke-on-Trent ST6 4NN
M6 CYB K. Williams 18 Bye Road, Lidlington, Bedford MK43 0RU
M6 CYC C. Carter 142 Hall Street, Briston, Melton Constable NR24 2LQ
M6 CYD A. Mcinnes 87 Sovereigns Quay, Bedford MK40 1TF
M6 CYE A. Forrester 16 East Street, Hebburn NE31 1HL
M6 CYF R. Silcock 18 Saxon Road, Southampton SO15 1JJ
M6 CYG C. Ashworth 40 Fairholme, Sedbergh LA10 5AY
M6 CYL Y. Chong Harrogate Ladies' College, Clarence Drive, Harrogate HG1 2QG
MW6 CYM G. Lewis Upper Cwm Farm, Llantilio Crossenny, Abergavenny NP7 8TG
M6 CYN G. Couzens Oak Tree Cottage Kite Hill Wootton Bridge, Ryde PO33 4LE
M6 CYO J. Mason 77 Albutts Road, British WS8 7ND
M6 CYP Dr S. Mouradian 1 Sandfield Road, St. Albans AL1 4JZ
MM6 CYQ G. Lewin Larch Cottage, Lein Road, Fochabers IV32 7NW
M6 CYS A. Owen 27 Granville Road, Chorley PR6 0HZ
M6 CYT D. Elias 12 Dagmar Terrace, London N1 2BN
MW6 CYU K. Evans Maesyronnen, Sarnau, Llanymynech SY22 6QL
M6 CYV K. Nicholson 68 Brunswick Street, Leigh WN7 2PL
M6 CYW J. Wood 17 Harrington Avenue, Lincoln LN6 7UP
MW6 CYX J. Shanahan 23 Pentre Gwyn, Trewern, Welshpool SY21 8DY
M6 CYY A. Brown 114 Tavistock Road, Birmingham B27 7LA
M6 CZA M. Singfield 8 Barnes Crescent, Sutton-in-Ashfield NG17 5BL
M6 CZB N. Shajan 19 Sturgess Avenue, London NW4 3TR
M6 CZC B. Crowhurst 33 Clarendon House, Clarendon Road, Hove BN3 3WW
M6 CZD T. Ayland 225 Stroud Road, Gloucester GL1 5JU
M6 CZF S. Amos 5 Curbar Close, Mansfield NG18 4XS
MM6 CZH S. Ram 28 Craigievar Gardens, Kirkcaldy KY2 5SD
M6 CZI J. Prodger Piper Cottage, Alde House Drive, Aldeburgh IP15 5EE
M6 CZJ J. Chalmers 19 Brettenham Crescent, Ipswich IP4 2UB
M6 CZL Dr T. Watson Montrose Farm, Bury Road, Diss IP22 2PY
M6 CZR R. Houghton 17 Coronet Drive, Ibstock LE67 6QG
MW6 CZS P. Hampton Caretakers Flat, T.A. Centre, Newport NP20 5XE
MW6 CZU L. Taylor 140 Cotswold Way, Risca, Newport NP11 6RG
MW6 CZV C. Parry 91 Grove Road, Risca, Newport NP11 6GL
M6 CZW T. Grayson 20 Harrison Street, Tow Law, Bishop Auckland, , Co. Durham DL13 4EE
M6 DAC D. Clough 44 Small Drove, Weston, Spalding PE12 6HS
M6 DAD J. Armstrong Sherrylea, Coventry Road, Coventry CV7 8BY
M6 DAG D. Walker Flat 5, Seward Court, 380-396 Lymington Road, Christchurch BH23 5HD
MW6 DAI D. Evans 29 Mount Pleasant, Bedlinog, Treharris CF46 6SD
M6 DAK C. Wood 18 Tufa Close, Chatham ME5 9LU
M6 DAL D. Cannon Hook 2 Sisters, Mautby Site Mautby, Great Yarmouth NR29 3JB
M6 DAM J. Malia 55 West Farm Avenue Longbenton, Newcastle upon Tyne NE12 8LS
M6 DAN D. Trudgian 18 Hart Close, Wootton Bassett, Swindon SN4 7FN
M6 DAP D. Pasika 192 Longfield Lane, Cheshunt, Waltham Cross EN7 6AQ
MM6 DAQ D. Rooney 19 Aurs Drive, Barrhead, Glasgow G78 2LR
MW6 DAR D. May 12 Marl Crescent, Llandudno Junction LL31 9HS
M6 DAS D. Starling Long Field Barn, Clarkes Lane, Beccles NR34 8HR
MM6 DAT I. Mclachlan Railway Cottage, Nether Falla, Peebles EH45 8QZ
M6 DAU D. Skates Flat 23, Orford Court Marsh Lane, Stanmore HA7 4TQ
M6 DAW D. Harris 2 Horsecroft, , Ewyas Harold HR20EQ
M6 DAX D. Blackhorse-Hull 1 Occupation Lane, New Bolingbroke, Boston PE22 7LW
MI6 DAY A. Galbraith 62 Millbrook Gardens, Castlederg BT81 7DF
M6 DAZ D. Coleman 9 Hogan Close, Newport PO30 5UF
M6 DBA D. Austrin 50 Lowestoft Road, Gorleston, Great Yarmouth NR31 6LZ
MW6 DBB S. Elias 20 Attlee Way, Cefn Golau, Tredegar NP22 3TA
MW6 DBD M. Williams 31 Waundeg, Nantybwch, Tredegar NP22 3SN
M6 DBE D. Bateson 19 Rothesay Road, Heysham, Morecambe LA3 2UR
M6 DBF P. Allen Flat 32, Whitworth Court, 9 Whitworth Road, Southampton SO18 1JR
M6 DBG D. Bentley 36 Byron Road, Mexborough S64 0DG
M6 DBH D. Coote 4 Hunters Oak Watton, Thetford IP25 6HL
M6 DBI R. Buckland 34 Beechwood Drive, Meopham, Gravesend DA13 0TX
MM6 DBJ A. Brown 17 Glamis Drive, Dundee DD2 1QN
M6 DBK D. Wells 27 Victoria Avenue, Camberley GU15 3HT
M6 DBL D. Smith 186 Weekes Drive, Slough SL1 2YR
MM6 DBN W. Brannan 16 Cairngorm Gardens Cumbernauld, Glasgow G68 9JD
M6 DBP D. Robinson 10 Pendragon Park, Glastonbury BA6 9PQ
M6 DBQ D. Baldwin 17 Teddy Gray Avenue Elworth, Sandbach CW11 3AR
M6 DBS S. Ward 22 St. Margarets Close, Horstead, Norwich NR12 7ER
MM6 DBT J. Ross 159 Grahams Road, Falkirk FK2 7BQ
M6 DBV A. Higgins 7 Waterloo Terrace, Bideford EX39 3DJ
M6 DBW D. Woodridge 8 Goold Close, Corston, Bath BA2 9AF
M6 DBX M. Hall 29 The Spinney, Wokingham RG40 4UN
M6 DCA D. Cotz 159 Ecclesfield Road, Sheffield S5 0DH
M6 DCB P. Jones Stonehead Farm Over Wyresdale, Lancaster LA2 9DL
M6 DCD G. Clarke 2 Beare Green Cottages, Horsham Road, Dorking RH5 4PE
MI6 DCH D. Hickey 83 Cairnmartin Road, Belfast BT13 3PQ
MW6 DCI D. Morgan Castle Cottage, Newbridge, Newport NP11 3NT
M6 DCJ C. Davies 83 Freeston Avenue, St. Georges, Telford TF2 9EN
M6 DCK D. Renshaw 25 Ashley Road, Worksop S81 7JS
M6 DCL D. Lane 46 Berkeley Vale Park, Berkeley GL13 9TQ
MM6 DCM D. Mifsud 25 Priory Road, Linlithgow EH49 6BP
MW6 DCN D. Davies 49 Heol Y Wal, Bradley, Wrexham LL11 4BY
M6 DCO D. Close 22 Station Road, Dodworth, Barnsley S75 3JE
M6 DCP A. Pearson 29 Broadoak Road, Langford, Bristol BS40 5HD
M6 DCQ D. Manning 153 Pavilion Road, Worthing BN14 7EG
M6 DCS D. Shephard 17 Grimsby Road, Laceby, Grimsby DN37 7DF
MM6 DCT D. Tourish 1 Langside Drive, Kilbarchan, Johnstone PA10 2EL
M6 DCU D. Matthews 81 Kipling Avenue, Goring-By-Sea, Worthing BN12 6LH
M6 DCW D. Wetton 2 Easton Grove, Hollywood, Birmingham B47 5LN
M6 DCZ D. Mooney 41 Queenhill Road, South Croydon CR2 8DW
M6 DDB D. Beavis 17 Kingsbere Crescent, Dorchester DT1 2DY
M6 DDC D. Chebsey 21 Shortlands Lane, Walsall WS3 4AG
M6 DDD V. Vesma Pound House, Market Square, Newent GL18 1PS
M6 DDF D. Poulton 115 Highview, Vigo, Gravesend DA13 0TQ
M6 DDH D. Muirhead 13 Berry Street, Skelmersdale WN8 8QZ
M6 DDI D. Barnett 49 Loundes Road, Unstone, Dronfield S18 4DE
M6 DDL P. Smith 156 Esther Grove, Wakefield WF2 8ET
M6 DDM D. Maycroft 100 Benwick Road, Doddington, March PE15 0UH
MM6 DDN D. Hornal 95 Traprain Crescent, Bathgate EH48 2BD
M6 DDO D. Daniels 128 Woodcock Road, Norwich NR3 3TD
M6 DDP D. Stallibrass 12 Sheerwater Close, Bury St. Edmunds IP32 7HR
M6 DDQ D. James 13 Lincoln Road, Fenton, Lincoln LN1 2EP
MW6 DDR D. Reeves 27 Beaufort Road, Pembroke SA71 4PX
M6 DDT D. Cross 4 Burns Avenue, Gloucester GL2 5BJ
M6 DDU D. Hind 17 Mill Way, Selby YO8 4FL
M6 DDV D. Borrett 84 Kingsway, Mapplewell, Barnsley S75 6EX
MM6 DDX M. Dougan 41D Balmerino Road, Dundee DD4 8RP
M6 DDY J. Betteridge 57 Wood Road, Chaddesden, Derby DE21 4LY
M6 DEC D. Hunter 9 Gelt Burn, Didcot OX11 7TZ
MI6 DED A. Mccann 6 Bowens Meadow, Lurgan, Craigavon BT66 7UT
M6 DEE D. Daniels 16 Bainbridge Road, Warsop, Mansfield NG20 0ND
M6 DEF A. Clarke 1 Riversdale Cottages, The Staithe, Stalham, Norwich NR12 9BY
M6 DEG D. Edge 18 Sandringham Avenue, Whitehaven CA28 6XL
M6 DEI D. Beresford 40 Fern Crescent, Congleton CW12 3HQ
M6 DEJ D. Thomas Stud Farm Bungalow, Stud Farm Drive, Tamworth B78 3HS

M6 DEK D. Knighton Holme Lea, Church Lane, Chesterfield S44 5AL
M6 DEL D. Millard 112 Avenue Road, Sandown PO36 8DZ
M6 DEM L. Demirkaya 11 Beech Park, Holsworthy Beacon, Holsworthy EX22 7NB
M6 DEO D. Sproston 22 Oakland Avenue, Haslington, Crewe CW1 5PB
M6 DER I. Dermondy 8 Glenwood Avenue, Baildon, Shipley BD17 5RL
M6 DES D. Kirkden 57 Crow Hill Road, Margate CT9 5PF
M6 DEV M. Raynor 68 Cambridge Street, South Elmsall, Pontefract WF9 2AR
MJ6 DEY J. Bryant 5 Louiseberg Court Queen'S Road, St. Helier JE2 3GQ Jersey
M6 DEZ J. Butters 7A Green Oak Avenue, Sheffield S17 4FT
MW6 DFC D. Gee 20 Davies Avenue, Brymbo, Wrexham LL11 5AS
M6 DFD D. Pennison 69 Caneland Court, Waltham Abbey EN9 3DS
M6 DFE J. Wood 3 Lion Lane, Haslemere GU27 1JF
M6 DFG D. Gardner 106 Willclare Road, Birmingham B26 2NY
M6 DFI E. Garry 34 Conway Road, Paignton TQ4 5LH
M6 DFJ T. Saunders 40 Southdown Avenue, Brixham TQ5 0AN
M6 DFK J. Bailey 22 Wilford Drive, Ely CB6 1TL
M6 DFL D. Loveys 3 Prestwood Court, Stafford ST17 4DY
M6 DFM D. Mcauslan Casa Arco Iris, Via Variante Nascente, 8005-491 SANTA BARBARA DE NEX Portugal
M6 DFP D. Parker 53 Brisbane Way, Cannock WS12 2GR
M6 DFQ A. Singh 19 Severn Crescent, Slough SL3 8UU
M6 DFS D. Slade 22 Oaklands Road, Mangotsfield, Bristol BS16 9EY
M6 DFT I. Lonsdale 23 Hunts Field Clayton-Le-Woods, Chorley PR6 7TT
M6 DFU K. Pownall 17 Horsebridge Road, Blackpool FY3 7BQ
M6 DFW D. Whitley 10 Kenmore Drive, Cleckheaton BD19 3EJ
MW6 DFX M. Lima Barbosa 65 Precelly Place, Milford Haven SA73 2BW
M6 DFY R. Medway 54 Peasland Road, Torquay TQ2 8PA
M6 DFZ D. Antcliffe 3 Wiltshire Mews, Cottam, Preston PR4 0NP
M6 DGA D. Ashford 56 Finch Close, Shepton Mallet BA4 5GL
MM6 DGC D. Baillie 77 Main Street, Fauldhouse, Bathgate EH47 9AZ
M6 DGD D. Bailey 2B Queens Road, Enfield EN1 1NE
M6 DGG M. Harrison 43 Second Avenue, Woodlands, Doncaster DN6 7QQ
M6 DGI P. Kirby 11 Bembridge Court, Crowthorne RG45 6BN
M6 DGJ D. Cole 4 Rosedale Road, Margate CT9 2TD
M6 DGM D. Mouland 2 Chafy Cottages, Holnest, Sherborne DT9 6HX
M6 DGN I. Coleman 24 Westwood Avenue, Plymouth PL6 7HS
M6 DGO A. Thacker 47 Hamilton Street, Walsall WS3 3EN
M6 DGP P. Dransfield Hill Top, Moss Croft Lane, Hatfield, Doncaster DN7 6BD
MW6 DGQ A. Evans 27 Oaklands Park Drive, Rhiwderin, Newport NP10 8RB
M6 DGR A. Deaves 46-48 Charlemont Drive, Manea, March PE15 0GA
M6 DGS D. Sole 10 Glyn Place, East Melbury, Shaftesbury SP7 0DP
M6 DGU Dr R. Jones Mole Corner, Red Shute Hill, Thatcham RG18 9QW
MW6 DGW D. Whitcombe 4 Fairview Road, Llangyfelach, Swansea SA5 7JJ
MM6 DGY D. Coyle 3 Claddyburn Terrace, Cairnryan, Stranraer DG9 8RD
MM6 DGZ D. Baillie 8 Cairnsmore Road, Palnure, Newton Stewart DG8 7AZ
M6 DHB A. Curd 44 Elbridge Avenue, Bognor Regis PO21 5AD
MM6 DHF D. Forsyth Solano, Stirling Road, Dumbarton G82 2PF
M6 DHG D. Lappage 37 Matlock Drive, Cannock WS11 6EN
MM6 DHI R. Buchan 5 Fairview Terrace, Danestone, Aberdeen AB22 8ZH
MW6 DHK D. Drake 56 Bryn Llidiard, Bridgend CF31 1QN
MM6 DHS K. Paterson 69 Middlepart Crescent, Saltcoats KA21 6LN
M6 DHT A. Herd 7 Water Lane, Greenham, Thatcham RG19 8SH
M6 DHU D. Maton 41 Bemerton Gardens, Kirby Cross, Frinton-on-Sea CO13 0LQ
M6 DHV D. Vincent 38 Methuen Street, Walney, Barrow-in-Furness LA14 3PR
M6 DHW D. Collins 50 Woodville, Barnstaple EX31 2HL
MM6 DHZ I. Smith 32 Kaimes Avenue, Kirknewton EH27 8AU
M6 DIB D. Bloxsome 74 Dunclair Park, Plymouth PL3 6DE
M6 DIF S. Goldsbrough 2 Stutte Close, Louth LN11 8YN
MW6 DIG D. Gozzard Craig Dulas, Rhydyfoel Road, Abergele LL22 8EG
M6 DIH A. Arnold 1 Fairford Gardens, Wordsley, Stourbridge DY8 5RF
M6 DII R. Wilson 16 Thorndon Hall, Thorndon Park, Ingrave, Brentwood CM13 3RJ
M6 DIJ D. Yates 16 Sunnyfield Road, Prestwich, Manchester M25 2RD
M6 DIL D. Yakub 42 Swift Close, Blackburn BB1 6LF
M6 DIM D. Vainas 51 Magister Road, Bowerhill, Melksham SN12 6FD
M6 DIO J. Cant 42 Copandale Road, Beverley HU17 7BW
M6 DIT N. Baulf 1 Lower Chart Cottages, Brasted Chart, Westerham TN16 1LS
M6 DIU P. Sewell 17 Chatham Close, Coventry CV3 1LY
M6 DIV J. Inwood Flat 13, Kelvestone House, 47 Park Road, Cannock WS11 1NZ
M6 DIZ D. Balsdon 25 Hanover Road, Plymouth PL3 6BY
M6 DJA D. Aspital 47 St. Augustines Park, Ramsgate CT11 0DF
MM6 DJC D. Caveney Westerhill, Clashnamuiach, Tain IV20 1XP
M6 DJG Z. Zmajkovic 67 Oakington Avenue, Little Chalfont, Amersham HP6 6SX
M6 DJI D. Oliver 20 Five Oaks Close, Malvern WR14 2SW
M6 DJJ B. Dodson 3 Bradley Road, Patchway, Bristol BS34 5LF
M6 DJK L. Oxberry 7 Kays Cottages Windy Nook, Gateshead NE10 9ST
M6 DJL D. Reader 190 Barnby Dun Road, Doncaster DN2 4RF
M6 DJN D. Neville 2 Oak View, Luddendenfoot, Halifax HX2 6HQ
M6 DJQ M. Hodges 4 Pond Fields Close, Little Hallingbury, Bishop's Stortford CM22 7FF
M6 DJS D. Smith 173 Leicester Road, Shepshed, Loughborough LE12 9DG
MW6 DJT D. Terrell 82 Baglan Street, Treherbert, Treorchy CF42 5AR
M6 DJW D. Wheeler 20 Hillwood Road Northfield, Birmingham B31 1DJ
M6 DJX D. Jones 102 Bryce Road, Brierley Hill DY5 4ND
M6 DKC D. Cowan 30 Crucian Way, Liverpool L12 0AW
M6 DKF B. Hodgson 28 Grove Drive, Woodhall Spa LN10 6RT
MM6 DKI S. Young 6 Ramsey Cottages, Bonnyrigg EH19 3JG
M6 DKK P. Weston 9 Clarendon Road, Smethwick B67 6DA
M6 DKL P. Shaw 21 Urmson Street, Oldham OL8 2AN
M6 DKM W. Molloy 32 Millers Barn Road, Jaywick, Clacton-on-Sea CO15 2QB
MM6 DKN N. Ford 11 Kenmore Way, Coatbridge ML5 4FN
M6 DKO J. Greenwood 18 Rookery Lane, Lincoln LN6 7PY
M6 DKR D. Reeves 37 The Cliff, Bryanston, Blandford Forum DT11 0PP
M6 DKS G. Hutton 8 Popples Drive, Halifax HX2 9SQ
M6 DKT D. Carter 18 Silent Woman Park, Coldharbour, Wareham BH20 7PE
M6 DKU B. Burgess 26 Bakers Way, Morton, Bourne PE10 0XW
M6 DKW C. Nicholl 36 Eylewood Road, London SE27 9NA
M6 DKY O. Silva Flat 71, Long Acre House, Pettacre Close, London SE28 0PB
M6 DLA D. Aldred 19 Birch Avenue, Bacton, Stowmarket IP14 4NT
M6 DLB P. Booth 7 Handley Crescent, East Rainton, Houghton le Spring DH5 9QX
M6 DLC D. Smith 106 Middle Street, Blackhall Colliery, Hartlepool TS27 4EB
M6 DLD H. Rasooli Nia 14 Larch Close, London N11 3NN
M6 DLH D. De La Haye 4 Nicola Mews, Ilford IG6 2QE
M6 DLI J. Walters 12 Leighton Road, Bilston WV14 8LN
M6 DLJ D. Caldicott 564 Fulbridge Road, Peterborough PE4 6SA
M6 DLL D. Hallsworth 29A Stephenson Court, Station Road, Stockport SK5 6LE
M6 DLM D. Le Mare The Sycamore, Church Bank, Barnard Castle DL12 0AH
M6 DLO R. Coles 10 Littlemoor Road, Weymouth DT3 6AA
M6 DLP M. Palmer New Haven, Stoneraise, Durdar, Carlisle CA5 7AX
M6 DLQ P. Leng The Barn, Gildersleets, Settle BD24 0AH
M6 DLR T. Knight 12 Vale Drive, Horsham RH12 2JX
M6 DLU S. Bland 61A Terminus Terrace, Southampton SO14 3FE
M6 DLV C. Tompkins 9 Billbrook Road, Hucclecote, Gloucester GL3 3QS
M6 DLW D. Endean 11 Forrester Drive, Brackley NN13 6NE
M6 DLX B. Tufnell 1 Moorlands Court Wath-Upon-Dearne, Rotherham S63 6DD
M6 DLY A. Omar 84 Beaumont Hill, Darlington DL1 3ND
M6 DMA D. Stockton 19 Chadwick Road, Middlewich CW10 0EA
M6 DMB D. Booth I3 Annandale Site Roman Bank, Skegness PE251HS
MM6 DMC D. Mccrae 4 Castramont Road, Gatehouse Of Fleet, Castle Douglas DG7 2JE
M6 DMD D. Mathieson-Dodd 1 Dag Lane, North Kilworth, Lutterworth LE17 6HD
M6 DME D. Elmy 2 Mill Road Drive, Purdis Farm, Ipswich IP3 8UT
M6 DMF J. Wilkins 124 Fullers Mead, Harlow CM17 9AU
M6 DMG D. Gillingham 5 Hillfield, St. Marks, Cheltenham GL51 7BQ
M6 DMN J. Robbins 5 South Close, Greatworth, Banbury OX17 2DZ
M6 DMO L. Wilburn 23 Sutton Road, Kirk Sandall, Doncaster DN3 1NY
M6 DMP D. Allen 9 The Crescent Cootham, Pulborough RH20 4JU
M6 DMQ D. Mcquirk 6 French Gardens Blackwater, Camberley GU17 9DP
M6 DMW D. Carey 78 Bentley Road, Bramley, Rotherham S66 1UH
M6 DMY A. Hoskins 38 Tasmania Close, Basingstoke RG24 9PQ
M6 DNB D. Eltham 47 Glenville Close, Royal Wootton Bassett, Swindon SN4 7EU
M6 DNC D. Swift 31 Meadow Lane, Westbury BA13 3AE
MI6 DND D. Fisher 11 Summer Street, Belfast BT14 6ES
M6 DNF D. Featherby 14 Station Road, Sutton, Ely CB6 2RL
M6 DNJ P. Hendry 19 Parsons Close, Portsmouth PO3 5LN
M6 DNL P. Swainson 11 Conway Drive, Banbury OX16 0QW
M6 DNM D. Mitchell 4 Millstream Close, Goostrey, Crewe CW4 8JG
M6 DNN D. Baker 117 Locks Road, Locks Heath, Southampton SO31 6LJ
M6 DNO C. Sheath 53 Manor Road, Trowbridge BA14 9HS
MW6 DNP Dr D. Morgan Ty Bettws, Kilgwrrwg, Chepstow NP16 6PN
M6 DNQ B. Southern 25 Chilgrove Avenue, Blackrod, Bolton BL6 5TR
M6 DNR L. Childs 354 Linnet Drive, Chelmsford CM2 8AL
M6 DNS M. Carter 61 Catherine Avenue, Swallownest, Sheffield S26 4RQ
M6 DNV G. Cater 7 Seymour Street, Chelmsford CM2 0RX
M6 DNX D. Garnham 16 Cherrytree Road, Great Cornard Sudbury CO10 0LJ
M6 DNZ D. Hart 163 Wakering Road, Shoeburyness, Southend-on-Sea SS3 9TN
M6 DOA D. Morris-Jones 121 Plas Dinas, Blacon, Chester CH1 5SW
M6 DOB A. Cowan 111 Oaks Drive, St. Leonards, Ringwood BH24 2QS
MM6 DOC Dr A. Davis 13 High Road, Auchtermuchty, Cupar KY14 7BE
MI6 DOD A. Henry 10 Drumbreda Crescent, Armagh BT61 7PE
M6 DOF A. Barker 49 Rockingham Avenue, York YO31 0TD
M6 DOG H. Godzisz 1 Jutland Place, Egham TW20 8ET
M6 DOJ A. Deery 25 Ribblesdale Place, Preston PR1 3NA
M6 DOM D. Chrumka 94 Clock House Road, Beckenham BR3 4JT
M6 DON D. Ashcroft Don Ashcroft, Iken House, 8Acer Rd, Rendlesham, Woodbridge IP12 2GA
M6 DOW J. Hogg 31 Roseberry Grove, York YO30 4SU
M6 DOX J. Armand 95 Derby Road, Golborne, Warrington WA3 3JJ
M6 DOZ D. Logan Cedar House, Reading Road North, Fleet GU51 4AQ
M6 DPA C. Cooke 42 Biddle Road, Leicester LE3 9HG
M6 DPF P. Fletcher 14 Snape Hill Crescent, Dronfield S18 2GQ
M6 DPG D. Glover 21 Monastery Road, Paignton TQ3 3BU
M6 DPH D. Hancock 2 Trevine Meadows, Indian Queens, St. Columb TR9 6NB
M6 DPI S. Hammond 89 Beeston Road, Sheringham NR26 8EJ
M6 DPJ P. Jones 50 Clay Lane, Doncaster DN2 4RJ
M6 DPL G. Payne Jay Close, Eastbourne BN23 7RW
M6 DPN R. Fripp 41 Sweyns Lease, East Boldre, Brockenhurst SO42 7WQ
MM6 DPO P. Beck Castle Clanyard Farm, Drummore, Stranraer DG9 9HF
M6 DPP D. Parkinson 24 New Road, Chatteris PE16 6BW
MM6 DPQ M. Mcmath 12 Townhead Crescent, Dalry, Castle Douglas DG7 3UR
M6 DPR D. Runyard Tilecroft, Shortheath Crest, Farnham GU9 8SA
M6 DPU C. Lombao 82 Cirrus Drive, Shinfield, Reading RG2 9FL
MM6 DPV D. Peacock Wyncum, Bridge Road, Castle Douglas DG7 1TN
M6 DPW D. Wilde 9 Redstone Park, Redhill RH1 4AS
M6 DPX A. Potts 28 Thistlebarrow Road, Salisbury SP1 3RT
M6 DPY D. Hansford 62 Bays Road, Pennington, Lymington SO41 8HN
M6 DPZ D. Stansfield 19 Cotman Fields, Norwich NR1 4EN
M6 DQB A. Deiana 41 Ashlar Drive, Eastfield, Scarborough YO11 3FP
M6 DQC Dr R. Blackwell Vikings Hall, Baylham, Ipswich IP6 8JS
M6 DQD J. Earye 28 Halls Drift, Kesgrave, Ipswich IP5 2DE
MM6 DQG A. Stewart 94 Dick Crescent, Burntisland KY3 0BT
M6 DQH B. Hartland 2 Brookland Close, Pevensey Bay, Pevensey BN24 6RT
M6 DQK A. Walker 4 Pretymen Crescent, New Waltham, Grimsby DN36 4NS
M6 DQN J. Davis 23 Blueberry Gardens, Andover SP10 3XD
M6 DQO D. Stocker 113 St. Marys Road, Bodmin PL31 1NH
M6 DQP Rvd. A. Lewis Four Winds Cottage, Main Street, Brough HU15 1RJ
MW6 DQQ Hunmanby Primary ARC c/o D. Jones 6 Frondeg, Tredegar NP22 3NT
M6 DQR G. Youll 4 Shaftsbury Court, Barnstaple Road, Scunthorpe DN17 1YB
M6 DQS Dr T. Meaker 1 Chemin Des Brugues, Roquevidal 81470 France
M6 DQT W. Martindale 57 Limefield Street, Accrington BB5 2AF
M6 DQV P. Patterson 3 Barnes Close, Southampton SO18 5FE
M6 DQW A. Hose 9 Cothey Way, Ryde PO33 1QY
MM6 DQY J. Dow 58 Beatty Crescent, Kirkcaldy KY1 2HS
M6 DQZ S. James 35 Prospect Road, Dronfield S18 2EA
M6 DRG G. Lythgoe 137 Dore Avenue, Fareham PO16 8DU
M6 DRH D. Hughes Warrenwood, Granville Rise, Totland Bay PO39 0DX
M6 DRI T. Mckinley 7 May Close, Godshill, Ventnor PO38 3HB
MM6 DRJ M. Cormack 2 Coghill Street, Wick KW1 4PN

UK Callsigns

M6 DRK D. Kurn 10 Dymoke Road, Mablethorpe LN12 2BF
MW6 DRM D. Machon 22 Albert Street, Caerau, Maesteg CF34 0UF
M6 DRO D. Drought 14 Mcminnis Avenue, St. Helens WA9 2PL
M6 DRP D. Rogers 24 Bramblewood Way, Halesworth IP19 8JT
M6 DRR D. Roderick 88 Broadway, Wakefield WF2 8LY
MW6 DRV D. Jones 59 Llewelyn Street, Aberdare CF44 8LA
M6 DRW D. Williams Lower House Farm Huntington, Kington HR53PU
M6 DRZ A. Potts 103 Etherstone Street, Leigh WN7 4HY
MM6 DSC K. Henderson 66 Rashierigg Place, Longridge, Bathgate EH47 8AT
MM6 DSD D. Mcmillan 45 Eton Avenue, Dunoon PA23 8DG
M6 DSE S. Clarke 23 Sinclair Court, Scarborough YO12 7SD
M6 DSH D. Sheehan 46 Darwin Close, London N11 1TA
M6 DSJ D. Johnson 5 Blackbird Close, Thurston, Bury St. Edmunds IP31 3PF
M6 DSO D. Osborne 1 Bramble Close, Wilnecote, Tamworth B77 5GG
MW6 DSP D. Pitman 5 Maesgwyn Street, Port Talbot SA12 6PF
M6 DSQ M. Bayman Marsworth, Tring HP23 4LX
M6 DSR D. Rhodes 66 Lindale Gardens, Blackpool FY4 3PQ
M6 DSS W. Stewart 43 Newlands Drive, Halesowen B62 9DX
M6 DST D. Stubbs 39 Torcross Way, Redcar TS10 2RU
MW6 DSU J. Wynne 8 Coed Artro, Llanbedr LL45 2LA
M6 DSV S. Townsend 13 Cornwall Drive, Bury BL9 9ET
M6 DSW D. Weight 15 Bourn Road, Caxton, Cambridge CB23 3PP
M6 DSY S. Waters 28C Cliff Road, Dovercourt, Harwich CO12 3PP
M6 DSZ D. Seedhouse 8 Levett Road, Tamworth B77 4AB
M6 DTA D. Easden 20 Brunel Way, Calne SN11 9FN
M6 DTC D. Cane 98 Lancaster Road, Northolt UB5 4TL
M6 DTD K. Worton 39 Staite Drive, Cookley, Kidderminster DY10 3UA
MI6 DTE D. Best 13 Cranley Green, Bangor BT19 7FE
M6 DTF M. Bell 36 Schneider Road, Barrow-in-Furness LA14 5DW
M6 DTH D. Hodgson Flat 21, John Norgate House, Two Rivers Way, Newbury RG14 5TF
M6 DTJ D. Brown 8 Tees Cresent, Spennymoor DL16 6QU
M6 DTL B. Jarvis 398 Aldermans Green Road, Coventry CV2 1NN
M6 DTM D. Mcgarrigle 40 The Glade, Waterlooville PO7 7PE
M6 DTN D. Thurman-Newell 95 Drake Avenue, Minster On Sea, Sheerness ME12 3RZ
M6 DTO G. Howe 3 Halton Close, Lincoln LN6 0YZ
M6 DTP M. Topham Scardale, Thorn Bank, Hebden Bridge HX7 5HS
M6 DTR S. Rhenius Baythorne Cottage, Baythorne End, Halstead CO9 4AB
M6 DTS D. Mason 15 Lake Street, Dudley DY3 2AU
M6 DTT D. Timbrell 15A Firgrove Crescent Yate, Bristol BS37 7AH
M6 DTU M. Woodruff 5 Regency Heights, Caversham, Reading RG4 7RH
MM6 DTV I. Campbell 1 Carbostbeg, Carbost, Isle of Skye IV47 8SH
M6 DTX P. Rose 10 Milton Street, Worthing BN11 3NE
M6 DTY C. Bell 28 New Forest Motel, 230 Hurn Road, Ringwood BH24 2BT
M6 DUB T. Price 40 East Street, Kidderminster DY10 1SE
M6 DUD M. Pugh 31 Cambridge Street, Reading RG1 7PA
M6 DUE D. Duell Manor House, 144 Stonhouse Street, London SW4 6BE
M6 DUF L. Duffy 46 Church Road, Worcester WR3 8NU
M6 DUH B. Brown 40 Stansfield Road, London SW9 9RZ
M6 DUI G. Walton 7 Burlington Street, Ulverston LA12 7JA
M6 DUJ A. Arnold 25 Church Lane, Saxilby, Lincoln LN1 2PE
M6 DUK S. House 12 Glebe Close, Abbotsbury, Weymouth DT3 4LD
MW6 DUL D. Davies 2 Hendre Ddu, Manod, Blaenau Ffestiniog LL41 4BH
M6 DUM J. Rowe 9 Corfield Close, Finchampstead, Wokingham RG40 4PA
M6 DUN A. Rigler 10 The Ball, Dunster, Minehead TA24 6SD
M6 DUO A. Richards 16 Fruiterers Arms Caravan Park, Uphampton Lane, Ombersley, Droitwich WR9 0JW
MI6 DUP R. Mcauley 37 Ladyhill Road, Antrim BT41 2RF
MI6 DUR A. Savage 469 Old Belfast Road, Bangor BT19 1RQ
M6 DUT T. Collins 56 Grasvenor Avenue, Barnet EN5 2DB
MM6 DUV D. Loughren 25 Cuiken Avenue, Penicuik EH26 0DR
M6 DUW T. Brooks 200 Kingsway, College Estate, Hereford HR1 1HE
M6 DUX S. Jones 3 Brockton, Lydbury North SY7 8BA
M6 DUZ B. Davies 12 Scalebor Gardens, Burley In Wharfedale, Ilkley LS29 7BX
M6 DVA J. Bligh-Wall 3 George Street, Elworth, Sandbach CW11 3BL
M6 DVE D. Thomson 20 Westlea, Clowne, Chesterfield S43 4QJ
M6 DVF K. George Wylye, Auberrow Wellington, Hereford HR4 8AN
M6 DVH D. Harrison 30 Horseshoe Drive, Cannock WS12 0FR
MM6 DVM D. Simpson 35 Westbourne Avenue Tillicoultry Clackmannanshire Fk136Pu, Tillicoultry FK136PU
MI6 DVN M. Devlin 17 Moninna Park, Newry BT35 8PP
MW6 DVP D. Price 2 Heol Tyn-Y-Fron, Penparcau, Aberystwyth SY23 3RP
MW6 DVQ A. Jones 31 Russell Terrace, Carmarthen SA31 1SZ
MM6 DVR R. Dover Brooklet, Dodside Road, Newton Mearns, Glasgow G77 6PZ
M6 DVS E. Johnson 3 Conifer Way, Dunmow CM6 1WU
M6 DVT D. Truscott 37 Langley, Chulmleigh EX18 7BQ
M6 DVU M. Hughes 183 Station Road, Hednesford, Cannock WS12 4DP
M6 DVV S. Storey 10 Amble Way, Trimdon Station TS29 6DZ
M6 DVW S. Challis 73 Rivenhall Way, Hoo, Rochester ME3 9GF
M6 DVY J. Sawyer 18 The Mead, Dunmow CM6 2PD
M6 DVZ J. Haywood 8 Cedar Close, Market Rasen LN8 3BE
M6 DWA Dr A. Dickson 1 Roebuck Drive Baldwins Gate, Newcastle-under-Lyme ST5 5FE
MM6 DWC K. Mclachlan 13 Dick Terrace, Penicuik EH26 8BW
M6 DWE D. Webb 52 Simpkin Close, Eaton Socon, St. Neots PE19 8PD
M6 DWF D. Waring 12 Mary Street, Farnhill, Keighley BD20 9AU
M6 DWG D. Gaskell 10 Freshford, St. Helens WA9 3WT
M6 DWI C. Hopper 31 Mary Road, Deal CT14 9HW
M6 DWJ D. Johnston 34 Coxley Crescent, Netherton, Wakefield WF4 4LR
M6 DWM D. Mitchard 49 Gladstone Road, Broadstairs CT10 2HY
M6 DWO A. Martin 97 The Maltings, Dunmow CM6 1BY
MM6 DWP D. Park 54 Coblecrook Gardens, Alva FK12 5BL
M6 DWS D. Saunders 17 Sandy Lane, Prestwich, Manchester M25 9RU
M6 DWT D. Potts 103 Etherstone Street, Leigh WN7 4HY
M6 DWV M. Tointon 13 Ridgeway, Broadstone BH18 8DY
M6 DWW D. Ward 11 Pearce Road, Sheffield S9 4JG
M6 DWY R. Friar 16 Upper Field Close, Redditch B98 9LE
M6 DWZ D. Whitaker 67 Oakington Avenue, Amersham HP6 6SX
M6 DX 350 DX c/o L. Marsland 154, Moss Lane, Litherland., Liverpool L21 7NN
M6 DXA A. Kendrick 29 Waterside, Silsden, Keighley BD20 0LQ
M6 DXC A. Crawford 18 Albatross Close, Standish, Wigan WN6 0WB
M6 DXG D. Gaffney 9 Tudor Road, Newton Abbot TQ12 1HT
MM6 DXH I. Craig 1/2 Drumlanrig Square, Hawick TD9 0AS
M6 DXI J. Mee 42 Potters Mead, Wick, Littlehampton BN17 7HY
MM6 DXJ S. Lewington 2 Coulliehare Cottages, Udny, Ellon AB41 7PH
M6 DXL M. Groves 15 Plains Lane, Littleport, Ely CB6 1RJ
M6 DXN D. Curtis 7 Neale Close, Aylsham, Norwich NR11 6DJ
M6 DXQ D. Baker 26 Lighthouse Close, Happisburgh, Norwich NR12 0QE
M6 DXS D. Martin 25 Broadmead Road, Blaby LE8 4AB
M6 DXV G. Frayne 98 De Lacy Court, New Ollerton, Newark NG22 9RW
M6 DXW D. Sawyer Realm House Heathfield Road, Bembridge PO35 5UQ
M6 DXZ P. Jones 18 Meadowbrook Road, Wirral CH46 0RS
M6 DYA D. Greaves 2 Willow Walk, Keynsham, Bristol BS31 2TR
M6 DYB D. Brough 38 Tynedale Avenue, Crewe CW2 7NY
M6 DYC P. Richards 1 Dorrs Drive, Watton, Thetford IP25 6HB
M6 DYD M. Jones 1 Elizabeth Court, Elizabeth Avenue, Norwich NR7 0GY
MW6 DYF J. Jones 40 Maes Mona, Amlwch LL68 9AT
M6 DYG S. Aspinall 3 Grasswood Road, Wirral CH49 7NT
MM6 DYH D. Houston The Knowe, Hardgate, Castle Douglas DG7 3LD
M6 DYI L. Edmonds 3 Waterlow Road, London N19 5NJ
MW6 DYL D. Witts 82 Park View, Llanharan, Pontyclun CF72 9SB
MD6 DYM J. Dunbar 3 Glenview Terrace, Port Erin, Isle of Man IM9 6HA
M6 DYN C. Hall 14 Marsh Road, Cowes PO31 8JQ
M6 DYO D. Osborne 3 Low Farm Road, Tunstall, Norwich NR13 3PU
M6 DYP R. Laver 20 Hall Street, Church Gresley, Swadlincote DE11 9QU
M6 DYR A. Hateley 26 Keelinge Street Tipton West Midlands Dy4 8Uq, Tipton DY4 8UQ
M6 DYU Dr H. Coghlan 1Bell View Cross Houses, Shrewsbury SY5 6JJ
M6 DYV R. Dunn 29 Hawe Lane, Sturry, Canterbury CT2 0LL
M6 DYW S. Jones 28A Park Road, Fowey PL23 1ED
M6 DYX M. Carter 15 Lashbrooks Road, Uckfield TN22 2AY
M6 DYY N. Christopher 161 Manor Road, Verwood BH31 6DX
M6 DZA D. Gilbert 34 Sullivan Way, Elstree, Borehamwood WD6 3DH
M6 DZB M. Bennetts 2 Chywoone Terrace, Newlyn, Penzance TR18 5NR
MM6 DZC T. Whyte Easter Unthank Farmhouse, Duffus IV30 5RN
M6 DZF A. Tiling 9 Coombe Street, Coventry CV3 1GG
M6 DZH H. Kassier 26 Higher Port View, Saltash PL12 4BX
M6 DZN A. Bertoneri Flat 1, 1 Albert Road, Nottingham NG9 2GU
M6 DZP M. Marsh 25 Southdown Road, Seaford BN25 4PD
M6 DZQ E. Zieba 12B Chingford Avenue, Chingford, London E4 6RP
M6 DZR D. Malins 36 Church Croft, Madley, Hereford HR2 9LT
M6 DZS Dr Z. Derzsi 217 Bensham Road, Bensham, Gateshead NE8 1US
M6 DZT J. Emery Mulberry Cottage, Quarry Lane, Combe St. Nicholas, Chard TA20 3PH
M6 DZV D. Abbott 3 Brewhouse Lane, Soham, Ely CB7 5JD
M6 DZX A. Norton 44 Jessamine Road, Southampton SO16 6AL
M6 DZY M. Parker 35 Prescelly Close, Nuneaton CV10 8QA
M6 DZZ D. Hunt 6 Lewisham Terrace, Newtown, Berkeley GL13 9NP
M6 EAA E. Johnston 67 Eversfield Road, Horsham RH13 5JS
MI6 EAC E. Luchi 68 Grovehill Gardens, Bangor BT20 4NS
MW6 EAD C. Jones 33 Graig Ebbw, Rassau, Ebbw Vale NP23 5SF
M6 EAE J. Collier 133 Woodstock Road, Moston, Manchester M40 0DG
MI6 EAF G. Monteith 66 Allen Park Dunamanagh, Strabane BT82 0PD
M6 EAH M. Bell 2 Hawthorne Road, Blyth NE24 3DT
MI6 EAI E. Taylor 17 Rutherglen Street, Belfast BT13 3LR
M6 EAJ J. Lavery Flat 2, Mentmore House, Mentmore Road, Ramsgate CT12 6RY
M6 EAL E. Westwood 17 Ennerdale Drive, Congleton CW12 4FR
M6 EAN I. Parsons 44 Hungerford Crescent, Bristol BS4 5HQ
MW6 EAO J. Johnson 1 Church View, Ruabon, Wrexham LL14 6TD
M6 EAP R. Taylor 18A Barnhall Road, Tolleshunt Knights, Maldon CM9 8HA
MI6 EAS Dr J. Henderson 76 Tirmacspird Road, Crimlin, Ederney, Enniskillen BT93 0FB
M6 EAT Forest of Dean ARG c/o R. Carr 38 Court Mead, Stone, Berkeley GL13 9LF
M6 EAW E. Whitehouse 16 Rue Gaston De Caillavet, Paris 75015 France
M6 EAX F. Davies 64 Main Road, Northwich CW9 8PB
M6 EBA A. Thompson 30 Birchwood Avenue, Lincoln LN6 0JB
M6 EBC M. Petchey 74 Avondale, Ellesmere Port CH65 6RW
MM6 EBD A. Connell 39 Glebe Crescent, Maybole KA19 7HZ
M6 EBE E. Fogarty Noke Farm, Hogscross Lane, Coulsdon CR5 3SJ
M6 EBG J. Thomas 128 Nuns Way, Cambridge CB4 2NS
M6 EBI C. Bakes Container City Building, 48 Trinity Buoy Wharf, London E14 0FN
MM6 EBJ K. Mccuish 6 Lighthouse Buildings, Birch Drive, Isle of Islay PA43 7HZ
M6 EBL R. Clarke 37B Northdown Park Road, Margate CT9 2NH
M6 EBO B. Clayton 26 Wood Walk, Mexborough S64 9SG
M6 EBP D. Morton 3 Pritchett Road, Birmingham B31 3NL
M6 EBQ D. Stanley 58 Wells Gardens, Basildon SS14 3QS
M6 EBR A. Buckland 21 Malton Close, Monkston, Milton Keynes MK10 9HR
MI6 EBS E. Mcknight 14 Marlacoo Beg Road, Portadown, Craigavon BT62 3TF
M6 EBU L. Wells 54 Giffords Cross Avenue, Corringham, Stanford-le-Hope SS17 7NH
M6 EBW C. Rowles 15 Rockhampton Walk, Colchester CO2 8UJ
M6 EBX S. Lucas Sunnyside, Church Road, Woodbridge IP13 7NU
M6 EBY C. Barrett 5 Oakapple Drive, Dereham NR19 2SR
MM6 EBZ E. Ball 32 Corton Lea, Ayr KA6 6GJ
MW6 ECB E. Brady 24 Gregory Avenue, Colwyn Bay LL29 7ND
MI6 ECC G. Haslem 124 Castle Rise, Tandragee, Craigavon BT62 2NF
M6 ECD C. Dennis 1 West Villa, Crathorne, Yarm TS15 0BA
M6 ECE G. Mutch 94 Abbotswood Road, Brockworth, Gloucester GL3 4PF
M6 ECH E. Hearn 41 Romney Avenue, Newcastle ST5 7JR
M6 ECJ A. Meszaros 9 Dunster Gardens, Cheltenham GL51 0QT
M6 ECM G. Stapylton 5 Coxcomb Walk, Crawley RH11 8BA
MM6 ECO V. Nelson 63 Melbourne Avenue, Clydebank G81 4QD
MW6 ECR K. Gulliford 5 The Square Abertridwr, , Caerphilly, Glamorgan CF83 4DH
M6 ECT E. Singleton 1 Ermin Park, Brockworth, Gloucester GL3 4BD
MI6 ECV C. Rafferty-Floyd 25 Glenside, Omagh BT79 7GL
M6 ECW C. Wilson 31 Violet Road, South Woodford, London E18 1DG
M6 ECX J. Steventon 11 Wyvern Way, Blandford Forum DT11 7XQ
M6 ECZ J. Hart Flat 4, 17 Trinity Gardens, Folkestone CT20 2RP
M6 EDA A. Bache 62 Whittingham Road, Halesowen B63 3TP

MM6 EDB J. Beck Brigadoon, Stair Street, Drummore, Stranraer DG9 9QE
M6 EDD E. Young 210 High Street, Gt Wakering SS3 0LS
M6 EDG N. Chamberlain 8 Southfields, Binbrook, Market Rasen LN8 6DX
M6 EDH E. Hitchins 50-52 Moorland Road, Weston-Super-Mare BS23 4HR
M6 EDI J. Cherry 12 Scarisbrick New Road, Southport PR8 6PY
M6 EDJ D. Jeffery 4 Sandhurst Drive, Beeston, Nottingham NG9 6NH
MI6 EDK K. Pearson 17, Knocknamoe Road, Omagh BT79 7LB
M6 EDL M. Arblaster 22 Wood Lane, Carlton, Barnsley S71 3JJ
M6 EDM P. Short 16 Limetree Avenue, , Glapwell, Chesterfield S44 5LE
M6 EDN A. Strange 38 Manor Road, Martlesham Heath, Ipswich IP5 3SY
M6 EDO D. Rasbarry 27 Royle Close, Romford RM2 5PS
MI6 EDP A. Connolly 68 Willowbank Gardens, Belfast BT15 5AJ
MW6 EDQ Dr J. Blaxland 7 Maes Y Sarn, Pentyrch, Cardiff CF15 9QQ
MW6 EDR E. Rees 23 Northlands Park, Bishopston, Swansea SA3 3JW
M6 EDS E. Kaye 119 St. Bernards Avenue, Louth LN11 8AS
M6 EDV P. Tranter Swn Y Gwynt, Trefeglwys, Caersws SY17 5PU
M6 EDW J. Tranter Swn Y Gwynt, Trefeglwys, Caersws SY17 5PU
M6 EDZ E. Twagirayezu 33 Morton Street, Stoke-on-Trent ST6 3PN
M6 EEA M. Rabl 14 Bramble Close, Durrington On Sea, , Worthing BN13 3HZ
M6 EEB J. Cameron 29 Webster Road, Stanford-le-Hope SS17 0BE
MI6 EEC S. Carlin 9 Mullandra Park, Kilcoo, Newry BT34 5LS
M6 EED C. Wright 24 Charlemont Crescent, West Bromwich B71 3DA
M6 EEF A. Al Busaidi 18 Sirdar Road, Southampton SO17 3SJ
M6 EEG A. Gow-Barber Plum Tree Villa, Butchers Lane, Ormskirk L39 6SY
M6 EEH C. Branch 62 Turnpike Road Connor Downs, Hayle TR27 5DT
MW6 EEJ J. Nicholas Reservoir House, St. Lythan'S, Wenvoe CF5 6BQ
M6 EEL L. Layland 3 Thirlmere Road, Golborne, Warrington WA3 3HH
M6 EEM D. Roberts 3 Heather Avenue, Melksham SN12 6FX
M6 EEN P. Farren 89 Fosterd Road, Newbold, Rugby CV21 1DE
M6 EEO S. Davey 38 Wordsworth Street, Barrow-in-Furness LA14 5SE
M6 EEP M. Bartlett 7 Thoresby, Tamworth B79 7SQ
MM6 EEQ G. Brown 37 Shandon Crescent, Bellshill ML4 1LE
MM6 EER E. Grant 6 Lindsay Place, Wick KW1 4PF
M6 EET P. Sladen 25 Linden Grove, Beeston, Nottingham NG9 2AD
M6 EEU A. Bullard 15 Rowan Drive, Lutterworth LE17 4SP
M6 EEW Prestatyn ARC c/o A. George 15 Ely Road, Croydon CR0 2LW
MM6 EEX L. Grant 6 Lindsay Place, Wick KW1 4PF
M6 EEZ P. Murphy 17 Birley Moor Avenue, Sheffield S12 3AQ
M6 EFC D. Shannon 68 Thornhill Road, Claydon, Ipswich IP6 0DZ
MI6 EFD E. Hudson 10 Kiloanin Crescent, Banbridge BT32 4NU
M6 EFE T. Barnden 27 Milton Road, Wokingham RG40 1DE
M6 EFF E. Phipps 12 Salisbury Close, Wokingham RG41 4AJ
M6 EFG J. Jones-Robinson 32 Verity Close, London W11 4HE
MM6 EFH P. Donachie 53 Seaforth Avenue, Wick KW1 5NE
M6 EFJ A. Davies 20 Hope Street, Halesowen B62 8LU
MW6 EFK A. Williams 27 Hick Street, Llanelli SA15 1AR
M6 EFL A. Miariti 19 Camelot Avenue, Nottingham NG5 1DW
M6 EFM E. Falomir Montanes 31 Albany Road, Birmingham B17 9JX
M6 EFP P. Williamson 22 Earls Road, Shavington, Crewe CW2 5EZ
M6 EFR D. Richman 42 Hathaway Road, Fleetwood FY7 7JH
M6 EFU S. Nyakabau 4 Vale View, Charvil, Reading RG10 9SJ
MW6 EFV J. Dyer Abergwawr House, Belmont Terrace, Aberdare CF44 6UW
M6 EFW C. Steele 40 Landor Road, Whitnash, Leamington Spa CV31 2JX
M6 EFY R. Bell 2 Hawthorne Road, Blyth NE24 3DT
M6 EGA V. Santhosh 119 Vaughan Road, Harrow HA1 4EF
M6 EGB I. Kendrick Whybank 10 Bankhouse Drive, Congleton CW12 2BH
MM6 EGC J. Flannigan 21 Kirkbean Avenue Rutherglen, Glasgow G73 4EA
MW6 EGD P. Jones 2 Fourth Avenue, Gwersyllt, Wrexham LL11 4EE
M6 EGE J. Powell 23 Park Road, Norton, Malton YO17 9DZ
M6 EGF H. Reeves 15 Mill Rise, Kidsgrove, Stoke-on-Trent ST7 4UR
M6 EGG M. Jackson 1 Kindred Barns, Ludlow SY8 4LF
MW6 EGH E. Barker 7 Upper High Street, Bedlinog CF46 6RY
M6 EGJ A. Chudasama 19 Walton Court, Bolton BL3 6QP
M6 EGK C. Suddell Lynhurst, Littleworth Lane Partridge Green, Horsham RH13 8JX
M6 EGL K. Hendricks Flat 2, The Halfway Cafe, Main Road, Darsham, Saxmundham IP17 3PL
M6 EGM L. Karaalp 45 Baird Grove, Kesgrave, Ipswich IP5 2DQ
M6 EGN A. Blackburn 81 Belvedere Road, Ipswich IP4 4AD
M6 EGO P. Bailey 103 Jarden, Letchworth Garden City SG6 2NZ
M6 EGP J. Churchill West Winds, Brandheath Lane, New End, Astwood Bank, Redditch B96 6NG
M6 EGS M. Sparrow 21 Langwell Crescent, Ashington NE63 8AB
M6 EGT J. Drake 38 Fawcett Road, Stevenage SG2 0EJ
M6 EGU A. Sharma 152 Ladybarn Lane, Manchester M14 6RW
M6 EGV C. Cousins 43 Avon Close, Little Dawley, Telford TF4 3HP
M6 EGW M. Delaney 24 Lonsdale Road, Manchester M19 3FL
M6 EGX M. Cross 11 Polyplatt Lane, Scampton, Lincoln LN1 2TL
M6 EGZ M. Rozier 5 Pond Piece, Brandeston, Woodbridge IP13 7AW
M6 EHA M. Philpott Garmisch, Hazel Road, Aldershot GU12 6HP
M6 EHB N. Livingstone 5 Drayton Court, High Street, Polesworth, Tamworth B78 1EX
M6 EHC E. Harris 147 Longdown Road, Congleton CW12 4QR
M6 EHD E. Delasalle 31 West Hill Road, Hoddesdon EN11 9DL
M6 EHE L. Hayward The Chimes, Farley Way, Hastings TN35 4AS
M6 EHH N. Hunt 21 Reams Close, Fishtoft, Boston PE21 0LL
M6 EHI D. Humm 15 Sherborne Road, Farnborough GU14 6JS
M6 EHK S. Buruma 40, 40, Groningen 9746PL Netherlands
MM6 EHL S. Nicol 9 Cowley Street, Methil, Leven KY8 3QG
M6 EHM J. Jackson 49 Leafield Rise, Two Mile Ash, Milton Keynes MK8 8BX
M6 EHO G. Gibbs 11 Fieldway Avenue, Leeds LS13 1ED
M6 EHP P. Garraway The Poplars, Crowell Road, Chinnor OX39 4HP
M6 EHQ P. Simmen 7 Thorpe Road, Thornton, Bradford BD133AT
M6 EHR P. Marshall 5 Thorntree Court Crofton, Wakefield WF4 1SP
M6 EHU D. Hodgson 11 Harmony Place, Mountain, Queensbury, Bradford BD13 1LD
M6 EHV W. Bradley 4 Forest View Avenue, London E10 6DX
M6 EHW P. Fellows Flat C1, 10 London Road, Boston PE21 8AG
MW6 EHX B. Bull Swan Cottage, Swan Road, Welshpool SY21 0RH
M6 EHY K. Brundle 17 The Paddocks, Hailsham BN27 3AQ
M6 EIA J. Rutter 19 Shaftesbury Avenue, Great Harwood, Blackburn BB6 7ST
MW6 EIB A. Timms Fistral, Red Bank, Welshpool SY21 7PN
M6 EIC S. Hammond 1 Elmer Close, Bognor Regis PO22 6JU
M6 EIE R. Rose 30 Brudenell Close, Cawston, Rugby CV22 7GN
M6 EIF K. Goss 57 Nursery Road, Leicester LE5 2HQ
M6 EIG R. Webb Norbury, Terrills Lane, Tenbury Wells WR15 8DD
M6 EIH K. Harris 78 Dovedale Road, Thurmaston, Leicester LE4 8NB
M6 EII J. Rodgers 5 Richil House 7 Ayston Road, Oakham LE15 9RL
M6 EIJ J. Iason 23 Baydon Grove, Calne SN11 9AT
MW6 EIK G. Edwards 17 Glan Y Mor Road, Penrhyn Bay, Llandudno LL30 3NL
M6 EIL S. Wilkes 24 Shelley Grove, Droylsden, Manchester M43 7YG
M6 EIM J. Darmont 114 Bowers Avenue, Norwich NR3 2PS
M6 EIR A. Mulcahy 85 Grifon Road, Chafford Hundred, Grays RM16 6NP
M6 EIS I. Mahoney 74 Radegund Road, Cambridge CB1 3RS
M6 EIU G. Farnbank 2 Market Hill, Foulsham, Dereham NR20 5RU
M6 EIW D. Blake Pound Farm, Swan Lane, Leigh, Swindon SN6 6RD
M6 EIY J. Lee 30 Wentworth, Yate, Bristol BS37 4DJ
M6 EIZ G. Denman 123 Links Avenue, Norwich NR6 5PQ
M6 EJA A. Ashton 46 Kingsland, Harlow CM18 6XL
M6 EJB P. State 53 Long Lane, Shirebrook, Mansfield NG20 8AZ
M6 EJC E. Crosby 16 Tweed Avenue, Ellington, Morpeth NE61 5ES
M6 EJF R. Glynn 106 Fairway Avenue, West Drayton UB7 7AP
MW6 EJG A. Zennadi 54 Llys Gwyrdd, Henllys, Cwmbran NP44 7LS
MW6 EJI R. Colson 1 Wildon Cottages, Twyn Allwys Road, Abergavenny NP7 9RS
M6 EJJ M. Corr Flat 7, Sovereign Court, 3 Winn Road, Southampton SO17 1EH
MI6 EJK J. Milligan 12 Rose Park, Limavady BT49 0BF
M6 EJL M. Leggett 10 Manor Road, Long Stratton, Norwich NR15 2XR
MM6 EJO B. Holderness 13 Glencairn Street, Camelon, Falkirk FK1 4LY
M6 EJP J. Siddle 7 Farebrother Street, Grimsby DN32 0NH
M6 EJR A. Graham 34 Balmoral Road, Stockport SK4 4EB
M6 EJS E. Scott 26 Fyfe Road, Corby NN172RD
MI6 EJT C. Rafferty 25 Glenside, Omagh BT79 7GL
M6 EJV A. Bowman The Glen, Vicarage Road, Bude EX23 8LN
M6 EJW E. Williamson 70 Douglas Drive, Stevenage SG1 5PH
M6 EJX R. Dickerson 61 Highfield Terrace, Queensbury, Bradford BD13 2BE
M6 EJY M. Boddy 4 Witton Close, Reedham, Norwich NR13 3HJ
M6 EKA E. Armstrong 30 Tennyson Avenue, Hull HU5 3TW
M6 EKB E. Beech 124 Evering Avenue, Poole BH12 4JH
M6 EKC S. Klymenko Flat 9, Dinerman Court, 38-42 Boundary Road, London NW8 0HQ
M6 EKD M. D'Arcy 1 Moorfield Avenue, Denton, Manchester M34 7TF
M6 EKE J. Welch 130 Sandy Lane, Upton, Poole BH16 5LY
M6 EKF G. Haime 5 Poets Way, Dorchester DT12FE
M6 EKI G. Lloyd 1 Holmside Terrace, Stanley DH9 6ET
M6 EKK A. Powell Crosstrees, Main Road, Theberton, Leiston IP16 4RX
M6 EKL S. Ross 3 Highlands, Lakenheath, Brandon IP27 9EU
M6 EKM M. Collyer 26 Beaumont Drive, Northampton NN3 8PS
M6 EKO S. Johnson Willow End Cottage, Willow Corner, Thetford IP25 6SS
M6 EKP S. Milner Pavilion House, School Lane, Ormskirk L40 3TG
M6 EKQ M. Isbell 20 Woodland Crescent, Wolverhampton WV3 8AS
M6 EKS J. Goodson 109 Millside, Stalham, Norwich NR12 9PB
MW6 EKT A. Price 141 Attlee Way, Cefn Golau, Tredegar NP22 3TE
M6 EKV A. Rimmer 9 Weymouth Avenue, St. Helens WA9 3QX
M6 EKW D. Rimmer 41 Ashburton Road, Wallasey CH44 5XB
M6 EKX C. Evenden 36 Castle View Road, Fareham PO16 9LA
M6 EKY A. Irwin 145 Coppermill Lane, London E17 7HD
M6 ELB W. Astill 14 Barlow Road, Exton, Oakham LE15 6BL
M6 ELC J. Parkes 24 Kenilworth Road, Lichfield WS14 9DP
M6 ELD A. Titmus The Old Police House, Arundel Road, Fontwell, Arundel BN18 0SX
M6 ELE E. Nichols 61B Norwich Common, Wymondham NR18 0SW
M6 ELF K. Windass 58 Nicholas Gardens, High Wycombe HP13 6JG
M6 ELH N. Henderson 14 Herbert Street, Carlisle CA1 2QE
M6 ELI R. Wells 27 Victoria Avenue, Camberley GU15 3HT
M6 ELJ S. Girdwood 8 Blaydon Walk, Wellingborough NN8 5YU
M6 ELL K. Ellison 69 Old Mill Road, Sunderland SR5 5TL
M6 ELN D. Shuttleworth 11 Bladen Close, Countesthorpe, Leicester LE8 5SB
M6 ELQ L. Cullip 11 Spiers Close, Tadley RG26 3SF
M6 ELR E. Reeve 12 Sime Street, Worksop S80 1TD
M6 ELS J. Fay 16 Foxhill, Whissendine, Oakham LE15 7HP
MM6 ELU G. Ross 5 Main Street, Alford AB33 8QA
MM6 ELW P. Thornley 13 Keose Lochs, Isle of Lewis HS2 9JT
MW6 ELX E. Davies 26 Heol Sant Gattwg, Llanspyddid, Brecon LD3 8PD
M6 ELY E. Cockett 2A Priory Avenue, Petts Wood, Orpington BR5 1JF
M6 ELZ E. Luckett 180 Clarendon Place, Dover CT17 9QF
M6 EMA E. Bansil 65 Hervey Street, Northampton NN1 3QL
M6 EMC J. Landless 2 Aspen Way, Banstead SM7 1LE
M6 EME E. Dunstan 57 Orchard Vale Flushing, Falmouth TR11 5TT
M6 EMF P. Cotton 29 Peake Avenue, Nuneaton CV11 6DW
M6 EMG D. Gray 68 Endeavour Way, Hythe Marina Village, Southampton SO45 6LA
M6 EMH N. Asling 9 First Avenue, Halifax HX3 0DL
MI6 EMI K. Mikicki 17 Glenhoy Drive, Belfast BT5 5LB
M6 EMJ E. Jubb 59 Buckingham Road Conisbrough, Doncaster DN12 3DG
MW6 EMK M. Klimaszewski 1 Woodwards Cottages, New Brighton, Wrexham LL11 3ED
M6 EML R. Copus 58 Colchester Rd, Holland on Sea CO15 5DG
M6 EMM E. Barton 86 Forge Lane, Kingswinford DY6 0LG
M6 EMP P. Holman 20 Green Drive, Wolverhampton WV10 6DW
M6 EMQ A. Johnson Avonworth, Grange Road, Bedford MK43 7HJ
M6 EMS C. Shaw 52 Margaret Avenue, Halesowen B63 4BX
M6 EMT K. Titmarsh 8 Bainbridge Close, North Walsham NR28 9UP
M6 EMV E. Varley Flat 8, Brookside Court, Liverpool L23 0TT
M6 EMW S. Hunter 9 Gelt Burn, Didcot OX11 7TZ
M6 EMX B. Di-Giulio 4 Highlands, Lakenheath, Brandon IP27 9EU
M6 EMY E. Porter 16 The Oval, Scarborough YO11 3AP
M6 EMZ E. Stubbs 39 Torcross Way, Redcar TS10 2RU
M6 ENB T. Gurney 24 Langley Way, Kettering NN15 6HL
M6 ENC F. Pauling Kingswood Farm House, Dalehouse Lane, Kenilworth CV8 2JZ
M6 END D. Kataria 87 Oakhill Road, Horsham RH13 5LH
M6 ENE V. Roets 7 Thorne Close, Harworth, Doncaster DN11 8SN
M6 ENF E. Briggs 20 Broad Lane, Pelsall, Walsall WS4 1AP

M6 ENH F. Baker 275 Bye Pass Road, Beeston, Nottingham NG9 5HS
M6 ENI C. Passey 1 Forest House, Baxworthy, Bideford EX39 5SF
M6 ENJ O. Karaalp 45 Baird Grove, Kesgrave, Ipswich IP5 2DQ
M6 ENK A. Seelig 9 Warren House Court 17 St Peters Avenue Caversham, Reading RG4 7RG
M6 ENM T. Smythe 19 Lime Close, Witham CM8 2PA
M6 ENN S. Burton 20 Flowerdown Avenue, Cranwell, Sleaford NG34 8HZ
M6 ENP K. Matthews St. Helens Cottage, Flimby, Maryport CA15 8RX
MW6 ENQ M. Mckenna 25 Heaton Place Norton Road, Rhos On Sea, Colwyn Bay LL284TL
MI6 ENR M. Carruthers 4 Kernan Avenue, Portadown, Craigavon BT63 5TB
M6 ENS D. Spencer 38 Town House Road, Nelson BB9 9LL
M6 ENU C. Gordon 9 Park Road, Camberley GU15 2SP
M6 ENV D. Lock Flat 33, Harrison Court, Harrison Close, Hitchin SG4 9SG
M6 ENW R. Ferreira 22 Vereker Road, London W14 9JS
M6 ENX M. Reid 17 Douglas Close, Hartford CW81SH
MW6 ENY K. Miles 25 Park Street, Penrhiwceiber, Mountain Ash CF45 3YW
M6 ENZ S. Margrave 30 Newtown Road, Bedworth CV12 8QU
M6 EOA C. Shaw 26 Grant Road, Spixworth, Norwich NR10 3NN
M6 EOB B. Bland 12 Park Lane, Pickmere, Knutsford WA16 0JX
M6 EOC P. Flanagan 71 Fellway, Pelton Fell, Chester le Street DH2 2BY
MM6 EOE A. Mather 38 Shandon Crescent, Balloch, Haldane G83 8EX
M6 EOK P. Wilson 45 Newquay Close, Hartlepool TS26 0XG
M6 EOM D. Kawayida Flat 7, Centre Point, London SE1 5NU
M6 EON I. Patrick Beech Cottage, Church Road, Friskney, Boston PE22 8RD
MW6 EOP L. Rowlands 14 Claerwen, Gelligaer, Hengoed CF82 8EW
M6 EOQ R. Edmonds 65 Green Lane, Redhill RH1 2DF
M6 EOR D. Kershaw 2 Croftlands, Neddy Hill, Carnforth LA6 1JE
M6 EOS A. Beresford 22 Tennyson Road, Rotherham S65 2LR
M6 EOT L. Seaton 4 Highfield Close, Empingham, Oakham LE15 8QB
MM6 EOU A. Gorman 36 Harestanes Road, Armadale, Bathgate EH48 3LA
M6 EOV L. Marsh 15 Buttercup Way, Southminster CM0 7RY
M6 EOW E. Taylor 11 Carlton Street, Featherstone, Pontefract WF7 6AA
M6 EOX I. Smith 4 Lammas Leas Road, Market Rasen LN8 3AP
M6 EOY I. Jones 90 Preston, Cirencester GL7 5PR
MM6 EOZ G. Headridge 79 Sheriffs Park, Linlithgow, West Lothian EH49 7SR
M6 EPA T. Bannister 6 Tanners Road, North Baddesley, Southampton SO52 9FD
M6 EPD D. O'Shea 37 Barclay Court, Ilkeston DE7 9HJ
MW6 EPE P. Plummer 26 Hill Road, Neath Abbey, Neath SA10 7NR
M6 EPF A. Cook 84 Clent View Road, Birmingham B32 4LW
M6 EPH D. Hollis 192 Rodbourne Road, Swindon SN2 2AF
M6 EPJ P. Jensen 36 Douglas Street, Derby DE23 8LH
M6 EPL N. Wachs 59 Broad Oak Way, Cheltenham GL51 3LL
M6 EPN D. Cocks 30 Great Arler Road, Leicester LE2 6FF
M6 EPO A. Smith 5 Boundary Close, , Holcombe BA3 5FP
M6 EPQ D. Tomlinson 74 Bradshaw Avenue, Riddings, Alfreton DE55 4AA
M6 EPS S. Cook 270 Heneage Road, Grimsby DN32 9NP
M6 EPT J. Rosa Flat 7, Wick Hall, Abingdon OX14 3NF
MM6 EPV A. Fraser Garden Cottage, Moy, Inverness IV13 7YQ
M6 EPW E. Waller 27 Main Street, Frizington CA26 3SA
MW6 EPX Dr D. Bentley 9 Pen-Y-Lan Place, Cardiff CF23 5HE
M6 EPZ N. Highfield 298 Mersea Road, Colchester CO2 8QY
MI6 EQA J. Mcgoldrick 45 Stewarts Road, Dromara, Dromore BT25 2AN
M6 EQB M. Prince Crantrock, Fourth Avenue, Greytree, Ross-on-Wye HR9 7HR
MI6 EQC E. Hamill 24 Beechvalley, Dungannon BT71 7BN
MI6 EQD A. Clearn 182A Clonmore Road, Dungannon BT71 6HX
M6 EQE K. Missenden 47 Roseacre Drive, Elswick, Preston PR4 3UQ
MW6 EQF P. Mcguinness The Coach House, Cwmdauddwr, Rhayader LD6 5HA
M6 EQG A. Nicholson 147 Watling Avenue, Seaham SR7 8JG
M6 EQJ A. Sherwin 10 Closes Side Lane, East Bridgford, Nottingham NG13 8NA
M6 EQK J. Tarrant 70 Sunnymead, Midsomer Norton, Radstock BA3 2SD
M6 EQL R. Paul Mayo, Raleigh Park, Barnstaple EX31 4JD
MM6 EQM D. Frost Lanfair, Kennethmont, Huntly AB54 4NN
M6 EQN D. Akerman The Brick Barn Old Gore, Ross-on-Wye HR9 7QW
M6 EQO A. Walton 11 Parkfield Road, Northwich CW9 7AR
M6 EQQ M. Stiles 13 Westover Close, Ivybridge PL21 9BA
M6 EQR D. Leonard 19 Nancy Street, Manchester M15 4FZ
M6 EQS J. Khan 123 Highfield Road, Hall Green, Birmingham B28 0HR
M6 EQV A. Gould 57 Northfield Road, Onehouse, Stowmarket IP14 3HE
MM6 EQW C. Hutton Langford, Broadfield, Symington, Biggar ML12 6JZ
MM6 EQY K. Archibald 31 Westwood Park, Deans, Livingston EH54 8QP
M6 EQZ S. Carlisle 103 Crossway Plympton, Plymouth PL7 4HZ
M6 ERA C. Jenkins 25 Longmeadow Grove West Heath, Birmingham B31 4SU
M6 ERC E. Thornes 11 Ringtale Place, Baldock SG7 6RX
M6 ERD R. Overy 62 Dykelands Road, Sunderland SR6 8ER
M6 ERF E. Felton 29 Pavitt Meadow, Galleywood, Chelmsford CM2 8RQ
M6 ERH E. Halford 69 Thirlmere Avenue, Astley, Manchester M29 7PZ
M6 ERI M. Konstantynowicz 12 Long Wall, Haddenham, Aylesbury HP17 8DL
M6 ERN R. Springall 18 Westbourne Park, Scarborough YO12 4AT
MM6 ERO Z. Bak 62/6 North Gyle Loan, Edinburgh EH12 8LD
M6 ERP S. Jones 36 Woodford Crescent, Burntwood WS7 9AE
M6 ERR N. Wood 10 Perriclose, Chelmsford CM1 6UJ
M6 ERS T. Glenday 52 Hollow Road, Bury St. Edmunds IP32 7AZ
M6 ERU M. Cook 194 Exeter Road, Kingsteignton, Newton Abbot TQ12 3NJ
M6 ERW I. Westby 5 Rosklyn Road, Chorley PR6 0NJ
MW6 ERZ E. Gibson 8 Llanthewy Close, Croesyceiliog, Cwmbran NP44 2PF
MW6 ESA S. Beavis 65 Old Road, Baglan, Port Talbot SA12 8TU
M6 ESH M. Young 4 Priestburn Close, Esh Winning, Durham DH7 9NF
MM6 ESL G. Robinson Asgard, 12 Upper Waston Road, Burray KW172TT
M6 ESO P. Melling 23 Appletree Close, Cottenham, Cambridge CB24 8UJ
M6 ESR A. Hunt 4 Weedon Road, Swindon SN3 4EE
M6 ESS M. Chamberlain 30 Roxton Rd, Great Barford MK44 3 LR
M6 EST R. Aldridge 37 Vincent Road, Luton LU4 9AN
M6 ESV P. Taylor 104 Winstanley Drive, Leicester LE3 1PA
MW6 ESW E. Sweeney 67 Blandon Way, Cardiff CF14 1EH
M6 ESY Dr S. Prior 17 Queens Walk, London NW9 8ES
M6 ESZ E. Palo 13 Welwyn Close, St. Helens WA9 5HL
M6 ETA D. Wall 227 Wayfield Road, Chatham ME5 0HJ
M6 ETC P. Hayes 4 London Road, Roade, Northampton NN7 2NL
M6 ETD R. Sindall 16 Chantrell Road, Wirral CH48 9XP
MI6 ETE W. Curry 7 Ballyversal Road, Coleraine BT52 2ND
M6 ETG P. Loades 126 Wealcroft, Gateshead NE10 8QS
M6 ETH E. Heath 63 Meadway, Dunstable LU6 3JT
M6 ETJ T. Widdowson 18 Newland Road, Banbury OX16 5HQ
M6 ETL D. Arthur 10 Lyndhurst Court, Lyndhurst Road, Hove BN3 6FZ
M6 ETN R. Bradshaw 272 Councillor Lane, Cheadle Hulme, Cheadle SK8 5PN
M6 ETP E. Price 47 Albany Road, Reading RG30 2UL
M6 ETR S. Guest 19 Ellesmere Avenue, Ashton-under-Lyne OL6 8UT
M6 ETS N. Luckett Flat 3, 25 Upton Park, Slough SL1 2DA
M6 ETU P. Gaur 34 Queensberry Avenue, Copford, Colchester CO6 1YN
MW6 ETW R. Atkins 88 Park Crescent, Abergavenny NP7 5TN
MW6 ETY D. Pesticcio Ty Ffynnon Farm, St. Mellons Road, Cardiff CF3 2TX
M6 ETZ M. Harvey-Ross Flat 4, Harley Court, Church Road, Southampton SO31 9GD
MM6 EUA C. Lockerbie 13 Granitehill Road, Aberdeen AB16 7AX
M6 EUB O. Wilson 99 Farnham Road, Durham DH1 5LN
M6 EUE N. Austerfield 22B Princes Avenue, Withernsea HU19 2JA
M6 EUH D. Levett 11 Love Lane, London SE25 4NG
M6 EUI D. Nolan Flat 7, Fonthill Court, Honor Oak Road, London SE23 3SJ
MW6 EUK S. Taylor 9 Crud Yr Awel, Prestatyn LL19 8YQ
M6 EUL N. Ramsey Dalestones, Lansdown Road, Bath BA1 5TB
MW6 EUO A. Jones 75 Hollybush Road, Cardiff CF23 6SZ
M6 EUP J. Wilson 125 Langroyd Road, Colne BB8 9ED
M6 EUU R. Scholey 2 Newfield Crescent, Wath-Upon-Dearne, Rotherham S63 6JN
M6 EUV A. Shitov Flat 4, 31 St. Leonards Road, Exeter EX2 4LR
M6 EUW M. Augustus 3 Heathend Cottages Heathend, Wotton-under-Edge GL12 8AS
M6 EUX R. Bradley 116B Old Hawne Lane, Halesowen B63 3ST
M6 EUZ R. Smith 2 Queen Street, Boston PE21 8XB
M6 EVA M. Knowles 12 Dalestorth Avenue, Mansfield NG19 6NT
M6 EVC J. Chapelle 7 Elizabeth Way, Stowmarket IP14 5AX
M6 EVD C. Harper 4 Bentley Avenue, Jaywick, Clacton-on-Sea CO15 2JW
M6 EVE B. Evered Ivy Cottage, Rookham Hill, Rookham, Wells BA5 3AL
M6 EVH H. Woodfin 8 Bank Hall Close, Bury. BL8 2UL
M6 EVI C. Charles 83 Dibdale Road, Dudley DY1 2RX
M6 EVK P. Hadley Narrow Boat Oregon The Moreings, Kinver DY7 6LG
M6 EVL A. Hoile 4 Lindale Mount, Wakefield WF2 0BH
M6 EVM Capt. J. Thompson Rose Cottage, Mickleton, Barnard Castle DL12 0JD
MM6 EVO J. Rickerby 12/10 Hermand Street, Edinburgh EH11 1LR
M6 EVR S. Rodgers 32 Merrimans Road Shirehampton Bristol, Bristol BS11 0AG
M6 EVU M. Marrs 43 Ely Close, Toothill, Swindon SN5 8DB
M6 EVW E. Woodward 309 Hartfields Manor, Hartfields, Hartlepool TS26 0NW
M6 EVX C. Haynes 25 Barnards Hill Lane, Seaton EX12 2EQ
M6 EVY R. Dalton 1 Ballard Estate, Four Lanes, Redruth TR16 6QL
M6 EVZ D. Lynch 7 Dollant Avenue, Canvey Island SS8 9EJ
M6 EWC P. Jenkins 137 Hawkhurst Road, Brighton BN1 9EB
M6 EWD G. Dobson 4 Durley Gardens, Orpington BR6 9LL
M6 EWH M. Lovering 16 Portland Avenue, Sittingbourne ME10 3QY
M6 EWI R. Farhall 2 Banks Cottages, Mountfield, Robertsbridge TN32 5JZ
M6 EWJ R. Scott 9 Burrows Close, Lawford, Manningtree CO11 2HE
M6 EWK S. Dempster 18 Salisbury Street, Swindon SN1 2AN
M6 EWL C. Donovan 130 Long Lynderswood, Basildon SS15 5BG
M6 EWN D. Carmichael 22 California Close Great Sankey, Warrington WA5 8WU
M6 EWO P. Chadwick 112 Sandy Lane, Warrington WA2 9JA
M6 EWQ A. Williams 5 Burleys Road, Crawley RH10 7DB
MM6 EWR A. Robson 100 Dawson Avenue, East Kilbride, Glasgow G75 8LH
M6 EWT Dr A. Grounds 49 Fletcher Way, Henfield BN5 9FR
M6 EWU L. Schofield 21 Wyche Close, Rudheath, Northwich CW9 7TY
M6 EWV S. Wagstaff 20 Hawfinch Road, Cheadle, Stoke-on-Trent ST10 1RX
M6 EWW E. Whitewood 2 Finn Farm Cottage, Finn Farm Road, Ashford TN23 3EX
MM6 EWX A. Talplacido 15 Park Avenue, Thurso KW14 8JP
MM6 EWY A. Kinnersley 5A Regent Terrace, Dunshalt, Cupar KY14 7HB
M6 EWZ E. Wali Zangana 78 Hivings Hill, Chesham HP5 2PG
M6 EXA S. Morris 5 Wensleydale Close, Stoke-on-Trent ST1 6XR
M6 EXB T. Thompson 14 Queen Street, Northwich CW9 5JL
M6 EXC A. Hawksworth 87 Bradeley Road, Haslington, Crewe CW1 5PX
M6 EXD M. Chapman 7 Dragons Lane, Shipley, Horsham RH13 8GD
M6 EXE C. Rose 55 Boddington Gardens, Biggleswade SG18 0PH
M6 EXF S. Brown 53 Prestwich Avenue, Worcester WR5 1QF
M6 EXG A. Holt 3 Cliff Mews, Hanging Heaton, Batley WF17 6FZ
M6 EXH D. Phelps 728 Sewall Highway, Coventry CV6 7JJ
M6 EXI J. Edgson 59 Gilmour Crescent, Worcester WR3 7PJ
M6 EXJ S. Baldwin 143 Oxford Road, Swindon SN3 4JA
MW6 EXL L. Mead 5 Hill Top, Ebbw Vale NP23 6PJ
M6 EXN A. Gaskin 44 Vale Road, Sutton SM1 1QH
M6 EXO T. Crocker 32 Godmanston Close, Poole BH17 8BU
MI6 EXP A. Boyd 27 The Meadows, Dungannon BT71 6PW
M6 EXQ D. Harrison 104 Gracemere Crescent, Birmingham B28 0TZ
M6 EXR M. Jackson 35 Marshall Road, Willenhall WV13 3PB
M6 EXS J. Harrison 69 The Cliff, Wallasey CH45 2NN
M6 EXT R. Murray 41 East Street, Rochdale OL16 2EG
MI6 EXU T. Reid Bamford 2 New Line, Carrickfergus BT38 9DL
MW6 EXV E. Mcmorrow 164 Maes Glas, Caerphilly CF83 1JW
M6 EXW T. Williams 5 Burleys Road, Crawley RH10 7DB
M6 EXX B. Jackson 2A Scrooby Street, Rotherham S61 4PL
M6 EXY S. Lake 85 Clarkson Road, Norwich NR5 8ED
M6 EXZ G. Cockerell 21 Coningsby, Bracknell RG12 7BE
M6 EYA M. Rasell 21 Avenue Sucy, Camberley GU15 3EB
M6 EYB S. Crouch 24 Church Road, Stanfree S44 6Aq, Chesterfield S44 6AQ
M6 EYD S. Houssart Flat 3, Virginia Court, London SE16 6PU
M6 EYF A. Screen Greenglade, Frith End, Bordon GU35 0RA
M6 EYG P. Cooper 3 Heath View Mill Lane, Grassmoor, Chesterfield S42 5AD
M6 EYH S. Brown 3 Grazebrook Croft, Birmingham B32 3NL
MW6 EYI B. Ravenhill-Lloyd 3 Bryn Awel, Conwy LL32 8WB
M6 EYJ R. Miller 23 Clarendon Road, Sevenoaks TN13 1EU
M6 EYK J. Peden 9 De Lacy Avenue Featherstone, Pontefract WF7 6AY
M6 EYL D. Humphrey 11 Colborne Close, Poole BH15 1UR
MW6 EYM D. Smith 3 Charles Close, Abergavenny NP7 6AP

M6 EYO R. Sansom 72 Wannock Lane, Eastbourne BN20 9SQ
M6 EYP S. Strange 94 Digby Avenue, Nottingham NG3 6DY
M6 EYS P. Smith 29 Ford Hayes Lane, Stoke-on-Trent ST2 0HB
M6 EYT Dr G. Thomas 6 Bankside, Headington, Oxford OX3 8LT
MW6 EYU E. Parry 22 Park Road, Tanyfron, Wrexham LL11 5SH
M6 EYV J. Allen 17 Inglemere Gardens Arnside, Carnforth LA5 0BX
M6 EYW G. Littlechild 85 Long Road, Canvey Island SS8 0JB
M6 EYY D. Wilkinson 20 Manchester Road, Barnoldswick BB18 5PR
M6 EYZ D. Mullard 46 Green Lane, Clanfield, Waterlooville PO8 0JX
M6 EZA G. Iredale Ship Cottage, Main Street, Maryport CA15 7DX
M6 EZC M. Bentley 5 Stokewell Road, Wath-Upon-Dearne, Rotherham S63 6EL
M6 EZE M. Wood 22 Prickett Road, Bridlington YO16 4AT
M6 EZF J. Holland 1 Ravenwood, Swadlincote DE11 9AQ
M6 EZG R. Zerafa 2 Furnwood, Bristol BS5 8ST
M6 EZM A. Metselaar 5A Canons Corner, Edgware HA8 8AE
M6 EZN A. Barrett-Sprot 1 Malting End, Wickhambrook, Newmarket CB8 8YH
MM6 EZO G. Irving 55 Gillbank Avenue, Carluke ML8 5UW
M6 EZP J. Crabb 23 Chaffinch Crescent, Billericay CM11 2YX
M6 EZR P. Harper 36 Barrow Close, Marlborough SN8 2BD
M6 EZS J. Marks Chantry End, Oak Hill, Epsom KT18 7BU
M6 EZT A. Wareham-Kirk Orchard Cottage, Church Street, Haverhill CB9 7SG
M6 EZV K. Gosney 76 Westgate Lane, Lofthouse, Wakefield WF3 3NS
M6 EZX D. Buchan 40 Bradstone Road, Winterbourne, Bristol BS36 1HQ
M6 EZY C. Cowley Rushbrook, Leafield Road, Chipping Norton OX7 6EA
M6 EZZ C. Garratt 19 Cherry Tree Place, St. Helens WA9 2AF
M6 FAB L. Rolinson 534 Haslucks Green Road, Shirley, Solihull B90 1DS
M6 FAC B. Whelan 147 Lawsons Road, Thornton-Cleveleys FY5 4PL
M6 FAE S. Moorcroft 44 Spencers Lane, Skelmersdale WN9 8JR
M6 FAF J. Taylor 19 Juniper Walk, Kempston, Bedford MK42 7SX
M6 FAH S. Wilson 63 Halford St, Tamworth B79 7RA
M6 FAJ T. Loveland 32 Whateley Lane, Whateley, Tamworth B78 2ET
M6 FAK A. Ladd 42 Plovers Way, Bury St. Edmunds IP33 2NJ
MW6 FAM A. Williams 211A New Road, Skewen, Neath SA10 6ET
MW6 FAN A. Burgess 18 Fairmeadows, Maesteg CF34 9JL
M6 FAS F. Southgate 52 Jeffrey Lane, Belton, Doncaster DN9 1LT
MI6 FAU W. Montgomery 56 Hazelbank Road, Drumahoe, Londonderry BT47 3NY
MW6 FAW S. Fawley 21 Broad View Pontnewydd, Cwmbran NP44 5JA
M6 FAX R. Stewart 20 Siddal Street, Halifax HX3 9BH
M6 FAY M. Furnivall 10 Wilwick Lane, Macclesfield SK11 8RS
MW6 FBA M. Young 104 Ystrad Road, Pentre CF41 7PW
M6 FBB M. Carr 51 Langton Road, Holton-Le-Clay, Grimsby DN36 5BH
M6 FBD D. Smith Heath Farm, Heath Road, Woolpit, Bury St. Edmunds IP30 9RL
M6 FBE M. Money 2 Lodge Farm Cottage, Black Horse Road, Norwich NR10 5DJ
M6 FBF A. Robson 6 Wood View, Maltby, Rotherham S66 7PA
M6 FBH F. Hatfull 16B Church Street, Easton On The Hill, Stamford PE9 3LL
MW6 FBJ B. Jordan 605 Monnow Way, Bettws, Newport NP20 7DJ
M6 FBM R. Proudman 7A Fithern Close, Dudley DY3 1YA
M6 FBO J. Borthwick 62A Rea Valley Drive, Birmingham B31 3XE
M6 FBP B. Preskey 14 Birchen Close, Chesterfield S40 4JT
M6 FBQ J. Hewitt 105 Princess Avenue, Pontefract WF9 2QX
M6 FBR C. Chew 9 Raven Park, Haslingden, Rossendale BB4 4HN
M6 FBT I. Long Flat 6, Kingfisher Court, Woking GU21 6DQ
MM6 FBU A. Miller 100 Kerrylamont Avenue, Glasgow G42 0DW
M6 FBV S. Cooney 56 Manor House Lane, Preston PR1 6HN
M6 FBW S. Richards 28 Lincoln Way, Thetford IP24 1DG
M6 FBY S. Chau Foley House, Heath Lane, Stourbridge DY8 1QX
M6 FBZ A. Raven 14 Paddock Close, Belton, Great Yarmouth NR31 9NT
MM6 FCA R. Mcdonald 135 Pappert, Alexandria G83 9LG
M6 FCC F. Cornes 18 Barke Street, Highley, Bridgnorth WV16 6LQ
M6 FCD N. Fisher 23 Matlock Road, Ferndown BH22 8QT
M6 FCE C. Smith 50 Oakwood Avenue, Wakefield WF2 9JS
M6 FCF B. Taylor 1 Whinchat Close, Stockport SK2 5UU
M6 FCG G. Miles 14 Woodlands Road, Great Shelford, Cambridge CB22 5LW
M6 FCI A. Renny 216 Malvern Road, Bournemouth BH9 3BX
M6 FCJ D. Vincelli 90 Broadbottom Road, Mottram SK14 6JA
M6 FCM C. Boyle 8 Westlees Close, North Holmwood, Dorking RH5 4TN
M6 FCN A. John Flat 10, Melbourne Court, 46 Seabourne Road, Bournemouth BH5 2HT
M6 FCQ D. Aissa 28 Sirdar Road, London N22 6RG
M6 FCR B. Scannell 60 Burnside Road, Dagenham RM8 1XD
M6 FCS F. Spickernell Stockstreet Farm, Mile Elm, Calne SN11 0NE
M6 FCT M. White 130 Main Street Walton, Street BA16 9QX
M6 FCV G. Allen 38 Edenside Kirby Cross, Frinton-on-Sea CO13 0TQ
M6 FCW L. Heaney 2 The Spinney, Eastleigh SO50 8PF
M6 FCX L. Palmer 39 Baird Grove, Kesgrave, Ipswich IP5 2DQ
M6 FCZ C. Cooper 25 Waterside Close, Loughborough LE11 1LP
M6 FDD B. Evans 2 Hastings Road, Eccles, Manchester M30 8JR
M6 FDF D. Sankey 30 Ballad Gardens, Manadon, Plymouth PL5 3FF
M6 FDG D. Chapman 27 Cuff Crescent, London SE9 5RF
M6 FDH K. Wells 45 Laburnum Avenue, Yaxley, Peterborough PE7 3YQ
M6 FDI W. Dawson 7 Field Close, Warboys, Huntingdon PE28 2UT
M6 FDK A. Patmore 5 Milton Close, Ramsey, Huntingdon PE26 1LU
M6 FDL P. Dixon 87, Ranger Drive, Wolverhampton WV10 6BZ
M6 FDN M. Martin 7 Carlton Drive, Priorslee, Telford TF2 9SH
M6 FDR A. Raine 91 Lulworth Avenue, Jarrow NE32 3SB
M6 FDS S. Jackson 5 Duchess Park Close, Shaw, Oldham OL2 7YN
M6 FDU K. Gilpin 181 Newport Road, Cowes PO317ER
M6 FDW Dr T. Krishnamurthy 64 Ingleby Road, Ilford IG1 4RY
M6 FDX D. Mcmorrin 24 Katherine Close, Addlestone KT151NX
M6 FDY C. Sparks 160 Wetmore Road, Burton-on-Trent DE14 1QS
M6 FEA Dr F. Agboma Flat 6, Mayfair Court, Edgware HA8 7UH
MI6 FEB F. Beattie 19 Derrin Road, Enniskillen BT74 6AZ
M6 FEC P. Mills 212 Carsic Road, Sutton-in-Ashfield NG17 2BS
M6 FED H. Clark 8 Snowberry Avenue, Belper DE56 1RE
M6 FEE F. Mather 98 Heathgate, Norwich NR3 1PN
MW6 FEF M. Lewis 6 Criccieth Close, Buckley CH7 3QF
MW6 FEH R. Harris 11 Maes Morgan, Llanrhaeadr Ym Mochnant, Oswestry SY10 0LH
M6 FEI W. Darvill 35 Allard Close, Northampton NN3 5LZ
M6 FEJ A. Croston Rosedene, Back Lane, North Molton, South Molton EX36 3HW
M6 FEK R. Fekete 22 St. Andrews Road, Ellesmere Port CH65 5DG
M6 FEL M. Bryant 19 Brooklands Road, Havant PO9 3NS
M6 FEM S. Simpson 48 Weatherhill Road Smallfield, Horley RH6 9LY
M6 FEN S. Morris 108 Templeside, Temple Ewell, Dover CT16 3BA
M6 FEO R. Hunter 3 Sandy Way, Croyde, Braunton EX33 1PP
M6 FEP R. Phillips 162 Glebelands, Pulborough RH20 2JL
M6 FEQ C. Ford 61 Gould Road, Barnstaple EX32 8ET
M6 FEU P. Gould 18 Tanfield Gardens, South Shields NE34 7DY
M6 FEZ T. Ryder 9A London Road, Slough SL3 7RL
M6 FFB M. Hall 67 Darlinghurst Grove, Leigh-on-Sea SS9 3LF
M6 FFD A. Ramsbottom 1 Barn Gill Close, Blackburn BB2 3HU
M6 FFE E. Durkin 30 Douglas Road West, Stafford ST16 3NX
M6 FFF West Midlands County Raynet c/o A. Beardsley 10 Moreton Close, Church Crookham, Fleet GU52 8NS
M6 FFI C. Lowe 45 Winster Road, Chaddesden, Derby DE21 4JY
M6 FFJ J. Preston 3 Essex Drive, Kidsgrove ST71HE
M6 FFK M. Ross 11 Queens Place, Otley LS21 3HY
M6 FFM S. Saville Little Cranebrook, St. Michaels Road, Verwood BH31 6JA
MW6 FFN G. Wilkins 7 Byron Road, Newport NP20 3HJ
M6 FFO Dr B. Issac 9B Poplar Grove, Stockport SK2 7JD
M6 FFP T. Skerton 60 Oakington Avenue, Harrow HA2 7JJ
MM6 FFQ D. Harvie 58 Esk Drive, Livingston EH54 5LE
M6 FFR R. Moys 12A Palmerston Avenue, Fareham PO16 7DP
M6 FFS R. Young 48 Sussex Street, Cleethorpes DN35 7NP
M6 FFT S. Merridale The Granary, Falledge Lane, Upper Denby HD8 8YH
M6 FFU L. Smith 7 Hereson Road, Ramsgate CT11 7DP
M6 FFV N. Calderley 48 Woodlands, Horbury, Wakefield WF4 5HH
M6 FFW Dr F. Harvey 14 Wilton Road, Hornsea HU18 1QU
M6 FFY M. Clarke 4 Mill Lane, Brant Broughton, Lincoln LN5 0RP
M6 FFZ R. Radley 20 Thorntondale Drive, Bridlington YO16 6GW
M6 FGA P. Meanwell 20 Crow Park Avenue, Sutton-On-Trent, Newark NG23 6QG
M6 FGC D. Newsome 1 Freemans Wharf, Plymouth PL1 3RN
M6 FGE M. Swan 35 Colston Close, Plymouth PL6 6AY
M6 FGG K. Legg Bennetts, High Street, Thorpe-Le-Soken, Clacton-on-Sea CO16 0EG
M6 FGH N. Speller 2 Hurst Rise Road, Oxford OX2 9HQ
M6 FGI K. Phillips 10 Canhaye Close, Plymouth PL7 1PG
MW6 FGJ M. Roberts 38 Gwilliam Court, Monkton, Pembroke SA71 4JL
M6 FGM B. Wilcock 32 Mallard Road Scotton, Catterick Garrison DL9 3NP
MW6 FGN N. Thomas 8 Western Terrace, Blaengwynfi, Port Talbot SA13 3YE
MW6 FGQ P. Passmore 127 High Street, Neyland, Milford Haven SA73 1TR
MM6 FGR A. Morrison 4 West Murkle, Murkle, Thurso KW14 8YT
M6 FGW D. Crouch 7 Tresco Road, Berkhamsted HP4 3JZ
M6 FGY B. Bestwick 185 Ashbourne Road, Turnditch, Belper DE56 2LH
M6 FGZ T. Crawley 41 Lynmouth Drive, Ruislip HA4 9BY
M6 FHB D. Bell 27 Parkfields Avenue, London NW9 7PG
M6 FHD D. Hartropp 185 Leighton Road, London NW5 2RD
M6 FHE Tolmers Scout Campsite c/o J. Hawbrook 7 Birkdale, Norwich NR4 6AF
M6 FHF A. Bailie 2 Loch Lane, Watton, Thetford IP25 6HE
MI6 FHG Dr P. Donaghy 41 Greenvale Manor Antrim, Antrim BT41 1SB
M6 FHH F. Stisted 19 Danehurst Street, London SW6 6SA
M6 FHI C. Colless 128 Ditton Lane, Fen Ditton, Cambridge CB5 8SS
MI6 FHJ N. Cully 42 Omerbane Road, Cloughmills, Ballymena BT44 9PE
MW6 FHK N. Williams 20 Railway Terrace, Pontyberem, Llanelli SA15 5HN
M6 FHM B. Hodgkinson 26 Daywell Rise, Rugeley WS15 2RE
M6 FHN I. Popovic 13 Walden Avenue, Arborfield, Reading RG2 9HR
M6 FHO K. Hunt Pound Farm, Gallows Hill, Diss IP22 1RZ
M6 FHQ J. Mount The Limes, The Green, Wickhambreaux, Canterbury CT3 1RQ
M6 FHR G. Daymond 30 Elizabeth Drive, Newcastle upon Tyne NE12 9QP
M6 FHS R. Goldup 57 Partridge Way, Old Sarum, Salisbury SP4 6PX
M6 FHV H. Peberdy 53 Far Lane, Normanton On Soar, Loughborough LE12 5HA
M6 FHX S. Powell 29 Coppice End Road, Derby DE22 2TA
M6 FHY D. Pearson 37 Elmridge, Leigh WN7 1HN
M6 FIA X. Zhang Harrogate Ladies' College, Clarence Drive, Harrogate HG1 2QG
M6 FIB M. Sutton 4 Centenary Close Broughton, Chester CH4 0FY
M6 FID P. Hack Catte Street, Oxford OX1 3BW
M6 FIG M. Fisher 96 Reepham Road, Norwich NR6 5PD
M6 FII J. Thurman Warbanks Farm, Cockfield, Bury St. Edmunds IP30 0JP
M6 FIJ Dr D. De-Cogan 52 Gurney Road, New Costessey, Norwich NR5 0HL
MI6 FIK D. Wilson 81 Parknasilla Way, Aghagallon, Craigavon BT67 0AU
M6 FIL P. Dunnicliffe 19 Woodland Road, Chelmsford CM1 2AT
MW6 FIN F. Toomey-Langford 8 Heol Dinas Isaf, Williamstown, Tonypandy CF40 1NG
M6 FIO D. Smith 62 Fulwoods Drive, Leadenhall, Milton Keynes MK6 5LB
M6 FIR M. Firth 209 High Street, Wickham Market, Woodbridge IP13 0RQ
M6 FIT M. Bryan 13 Elmwood Avenue, Sunderland SR5 5AW
MI6 FIU P. Reid 1 Nettlehill Mews, Lisburn BT28 3HN
M6 FIW R. Nicholson 1 Isis Close, Aylesbury HP21 9LY
MW6 FIX R. Hemming 21 New Road, Jersey Marine, Neath SA10 6JR
MW6 FIY S. Binnion 55 Dythel Park, Pen-Y-Mynydd, Llanelli SA15 4RR
M6 FIZ R. Faulkner 31 Faulkland View, Peasedown St. John, Bath BA2 8TG
M6 FJA A. Ferriroli 142 Hillbury Road, Warlingham CR6 9TD
M6 FJC C. Featherstone 3 Pogmoor Road, Barnsley S75 2EW
M6 FJD K. Dobson 1 Howarth Road, Ashton-On-Ribble, Preston PR2 2HH
M6 FJE C. Hinds 15 Logan Way, Hemyock, Cullompton EX15 3RD
M6 FJF J. Slater 1 Highfield Court, Swinton, Mexborough S64 8RF
M6 FJI J. Tonkyn 22 Glencarron Way, Southampton SO16 7EF
M6 FJJ F. Johnson 12 Hollins Road, Harrogate HG1 2JF
M6 FJL J. Robb 37 Wroxham Road, Woodley, Reading RG5 3AX
M6 FJM D. Marden 1 Stroma Gardens, Hailsham BN27 3AZ
M6 FJN M. Singer 1 Bentley Road, Slough SL1 5BB
M6 FJO I. Waddingham 102 Nethershire Lane, Sheffield S5 0QE
MW6 FJQ A. Lawson Hafryn, Dyserth Road, Rhyl LL18 5RB
MM6 FJS O. Sharp 18 Urquhart Court, Kirkcaldy KY2 5TX
MW6 FJT T. Cook 19 Cae Bach Aur Estate, Bodffordd, Llangefni LL77 7JS
M6 FJU P. Appleton The Old Rectory, Station Road, Grimsby DN36 5SQ
MM6 FJV W. Little Burnside, Main Street, Lochans, Stranraer DG9 9AW
M6 FJW S. Michael 191 Sutton House, Scunthorpe DN15 6SN
MW6 FJX C. Williams 18 Caer Delyn, Llannerch-Y-Medd LL71 8EJ

M6 FJZ S. Stewart-Sandeman 7 Bears Lane, Hingham, Norwich NR9 4LL
M6 FKA L. Hall 51 Bridgeacre Gardens, Coventry CV3 2NQ
MM6 FKB S. Broll 23 St. Medans, Monreith, Newton Stewart DG8 9LL
MW6 FKC G. Stevenson 3 Llwyn Rhosyn, Cardiff CF14 6NS
MM6 FKD E. Rattray Blashieburn Cottage, Blairingone, Dollar FK14 7NT
M6 FKE K. Collings 97 Jackmans Place, Letchworth Garden City SG6 1RF
M6 FKF A. Cucchiara 172 Gonville Crescent, Stevenage SG2 9LZ
M6 FKG J. Shears 161 Park Road, Keynsham, Bristol BS31 1AS
M6 FKH J. Fensom Foxdown House, Blandford Camp, Blandford Forum DT11 8BP
M6 FKI W. Tennison 85 Primrose Field, Harlow CM18 6QT
M6 FKJ F. Johnson 13 Honeymead Lane, Sturminster Newton DT10 1EW
M6 FKN A. Blackwell 27 Prince Charles Crescent, Farnborough GU14 8DJ
M6 FKO G. Carver 4 Andrews Road, Farnborough GU14 9RY
M6 FKP C. Robinson 3 Folly Hall Road, Bradford BD6 1UL
M6 FKQ A. Smith 116 Pilling Lane, Preesall, Poulton-le-Fylde FY6 0HG
M6 FKR G. Howard 8 Paddock Road, Woodford, Kettering NN14 4FL
M6 FKS T. Nash 32 Collington Rise, Bexhill-on-Sea TN39 3RS
M6 FKU L. Harris-Pugh 13 Myddelton Park, London N20 0HT
M6 FKW A. Metcalf 71 Harper Road, Coventry CV1 2AL
M6 FKZ C. Spencer 18 Coatsby Road Kimberley, Nottingham NG16 2TH
MW6 FLA Dr F. Labrosse 72 Ger Y Llan Penrhyncoch, Aberystwyth SY23 3HQ
M6 FLB F. Buss 44 Courtenay Road, Maidstone ME15 6UL
M6 FLC C. Horridge 6 Back Sreet, East Stockwith, Gainsborough DN21 3DL
M6 FLF C. Wilson 12 Desmond Avenue, Hornsea HU18 1AF
MM6 FLG M. Bradshaw 32 Greycraigs, Cairneyhill, Dunfermline KY12 8XL
M6 FLH D. Adkin 31 Fieldway, Broad Oak, Rye TN31 6DL
M6 FLJ S. Bumstead Middle Leys, Leys Farm, Withypool, Minehead TA24 7RU
M6 FLK J. Kelly 74 Hatfield Crescent, Bedford MK41 9RB
M6 FLL D. Dobson 1 Howarth Road, Ashton-On-Ribble, Preston PR2 2HH
M6 FLN J. Horn 8 Princess Close, Watton IP25 6XA
M6 FLQ K. Hobbs 75 Coronation Walk, Gedling, Nottingham NG4 4AS
M6 FLR J. Bicknell 10 Farnefold Road, Steyning BN44 3SN
M6 FLS D. Greenland 1 Hilltop, Tuesley Lane, Godalming GU7 1SB
M6 FLU J. Salter 103 Cootes Avenue, Horsham RH12 2AF
M6 FLW S. Miller 14 Queens Croft, Queens Street, Swinton, Mexborough S64 8NA
M6 FLX T. Clarke 20 Cliff Road, Felixstowe IP11 9PJ
M6 FLY J. Hall 9 Westland Drive, Fernwood, Newark NG24 5BJ
M6 FLZ R. Sutton 80 Fishbourne Lane, Ryde PO33 4EU
M6 FMC M. Straughan 21 Silcoates Avenue, Wrenthorpe, Wakefield WF2 0UP
M6 FME F. Moulsdale 63 Windermere Avenue, St. Helens WA11 7AG
M6 FMF C. Bassi Cherries, Hadham Road, Ware SG11 1LH
M6 FMG P. Booker 17 Colton Copse, Chandler'S Ford, Eastleigh SO53 4HQ
M6 FMI O. Earl 46 Rowan Way, Chelmsley Wood, Birmingham B37 7QT
M6 FMJ K. Browne 4 Darcey Drive, Brighton BN1 8LF
M6 FML F. Lowndes 46 Danes Road, Manchester M14 5JS
M6 FMN G. Morris 6 Eastcourt Road, Worthing BN14 7DB
M6 FMP F. Morson-Pate 3 Tudor Lawns, Carr Gate, Wakefield WF2 0UU
M6 FMS F. Mace 40 Quarry House Gardens, East Rainton, Houghton le Spring DH5 9RD
M6 FMT A. Webber 73 Greenwood Road, Yeovil BA21 3LF
M6 FMU J. D'Aubray-Butler Copse Edge 16 Francis Road, Frodsham WA6 7JR
MW6 FMV M. Plowman Ffynnonwen, Plwmp, Llandysul SA44 6EY
MW6 FMW C. Jones 32 Vale Street, Denbigh LL16 3BE
M6 FMY N. Mountford 78B Meeting House Lane, London SE15 2TX
M6 FMZ K. Bianchini 10 St. Leonards Road, Headington, , Oxford OX3 8AA
MW6 FNA Dr E. Flikkema 7 St. James Mews, Great Darkgate Street, Aberystwyth SY23 1DW
M6 FNB N. Berry The Mount, Deerfold, Bucknell SY7 0EF
M6 FNE R. Merrifield 6 Miersfield, High Wycombe HP11 1TX
M6 FNF M. Byrd 17A Castle Gates, Shrewsbury SY1 2AB
M6 FNG M. Grice 48 St. Ives Road, Coventry CV2 5FZ
MW6 FNK F. Williams 5 Tyn Rhos Estate, Gaerwen LL60 6HL
MM6 FNL G. Cunningham 17 Robb Place, Castle Douglas DG7 1LW
M6 FNM R. Bound 24 Buttington Road, Shrewsbury SY2 5TS
MI6 FNO C. Johnston 32 Maghereagh Road, Randalstown, Antrim BT41 4NS
MM6 FNQ K. Jones 6 Traill Street, Castletown, Thurso KW14 8UG
M6 FNR Braintree Raynet Group c/o F. Riches 4 Priory Close, Chelmsford CM1 2SY
MW6 FNT P. Taylor 11 Watford Close, Watford Park, Caerphilly CF83 1NQ
M6 FNU K. Cole 30 Wood Road, Rotherham S61 3RQ
MW6 FNV D. Robins 25 Clos Mancheldowne, Barry CF62 5AB
M6 FNX N. Farrington-Smith 3 Milton Road, Wokingham RG40 1DE
M6 FNY D. Chatterton 3 Hunt Close, South Wonston, Winchester SO21 3HY
MI6 FNZ R. Mckay 31 Squires Hill Crescent, Belfast BT14 8RE
M6 FOD C. Dempster 18 Concorde Way, Gloucester GL4 4PU
M6 FOE L. Downs 18 Burnopfield Gardens, Newcastle upon Tyne NE15 7DN
M6 FOJ D. Sutherland 22 Cherry Street, Wigston LE18 2BB
M6 FOK C. Fletcher 18 Kings Road, Barnetby DN38 6HJ
M6 FOL W. Bartle 5 Crosskeys Row. Chapel Milton, Chapel En le Frith SK23 0QQ
M6 FOR I. Forester 35 Thackeray Street, Sinfin, Derby DE24 9GY
M6 FOT S. Moore 170 Walsall Road, Aldridge WS90JT
M6 FOX S. Spence 13 First Street Crookhall, Consett DH8 7LX
M6 FOY F. Foy 4 The Square, East Rounton, Northallerton DL6 2LB
M6 FOZ A. Foster 47 Westmorland Drive, Desborough, Kettering NN14 2XB
M6 FPC J. Lord 5 Langworthy Avenue Little Hulton, Manchester M38 9GQ
M6 FPD L. Coveney 72 Bonney Road, Leicester LE3 9NH
M6 FPE R. Coveney 72 Bonney Road, Leicester LE3 9NH
M6 FPF F. Clifton Battery Road, Paull, Hull HU12 8FP
M6 FPG D. Worth 33 Lime Grove Close, Leicester LE4 0UG
MW6 FPH M. Clarke 48 Brynglas Road Llanbadarn Fawr, Aberystwyth SY23 3QR
MM6 FPI A. Mcclements 24 Glen Brae, Bridge of Weir PA11 3BH
M6 FPJ D. Lavell 51 Kingfisher Close, Newport PO30 5XS
MW6 FPL A. Muldoon 29 Eider Close, St. Mellons, Cardiff CF3 0DF
M6 FPM T. Smith* 1 Bents Close, Clapham, Bedford MK41 6DY
M6 FPN L. Varga 14. Tegla Str., Miskolc 3516 Hungary
M6 FPP J. Misztela 15 Severus House, 155 Varcoe Gardens, Hayes UB3 2FJ
M6 FPQ E. Turner 2 Shepherds Close, Fen Ditton, Cambridge CB5 8XJ
M6 FPU Rvd. E. Duell 1 Stumpacre, Bretton, Peterborough PE3 8HP
M6 FPV T. Van Den Bosch 8 Stopford Garth, Wakefield WF2 6RT
M6 FPW S. Bale 121 Washbrook Road, Rushden NN10 6UR
MM6 FPX T. Mcconnell 2 Stewartgill Place, Ashgill, Larkhall ML9 3BB
MM6 FPY I. Blackstock 149D Carlisle Road, Crawford, Biggar ML12 6TP
MW6 FQA E. Parry Blaen Gwenin, Llanrhystud SY23 5BZ
M6 FQC M. Roberts 41 Mcneill Avenue, Crewe CW1 3NW
MW6 FQD M. Roberts 20 Upper Robinson Street, Llanelli SA15 1TR
M6 FQE K. De La Hunty Bangers Whistle, Yarmouth Road, Newport PO30 4LZ
M6 FQF A. Goldstein 33 Hughenden Avenue, Harrow HA3 8HA
M6 FQH C. Goldsmith 15 Stephenson Road, Cowes PO31 7PP
M6 FQJ B. Boxall 10 Honor Avenue, Wolverhampton WV4 5HH
M6 FQK D. Briggs 5 Links Avenue, Cromer NR27 0EQ
M6 FQL G. Milner 17 Haragon Drive, Amesbury, Salisbury SP4 7FS
M6 FQM A. James 27A Lilliana Way, Bridgwater TA5 2GG
M6 FQN R. Vickers 3 Grenville Avenue, Goring-By-Sea, Worthing BN12 6JE
M6 FQO Dr A. Lipson Field House, The Haven, Cambridge CB21 5BG
M6 FQP A. Phillips 43 Jutland Road, Hartlepool TS25 1LP
M6 FQU R. Parker 53 Tunstall Road, Canterbury CT2 7BX
M6 FQW L. Jones 206 Lowe Avenue, Wednesbury WS10 8NS
M6 FQX M. Rowland 30 Tunnmeade, Harlow CM20 3HL
MI6 FQY B. Maguire 39 Carland Road, Dungannon BT71 4AA
M6 FQZ A. Williams Corner Cottage, South Brent TQ10 9JF
M6 FRC F. Clement 40 Ellison Fold Terrace, Darwen BB3 3EB
MM6 FRD F. Redden 31 Lawfield, Coldingham, Eyemouth TD14 5PB
M6 FRF R. Abel 43 Church Street, Didcot OX11 8DG
M6 FRG J. Marshall 1 Farriers Reach, Bishops Cleeve, Cheltenham GL52 7UZ
M6 FRI J. Hewart 14 Kestrel Close, Marple, Stockport SK6 7JS
M6 FRJ D. Janowicz 20 Salisbury Road, St. Leonards-on-Sea TN37 6RX
M6 FRK F. Waller 19 Wortley Avenue S738Sb, Wombwell S738SB
M6 FRO N. Froggatt 6 Beech Grove, , New Malden KT3 3HR
M6 FRP N. Williams The Bothy, Norton Bavant, Warminster BA12 7BB
MI6 FRQ J. Coleman 25 Kinnaird Street, Belfast BT146BE
M6 FRR J. Hingley-Hickson The White House, School Lane, Grantham NG32 2ES
M6 FRT J. Parsons 4 Saunders Close, Uckfield TN22 2BX
M6 FRV F. Raven-Vause 624 Hillbutts, Wimborne BH21 4DS
MW6 FRW A. Middleton 56 Caradoc Road, Prestatyn LL19 7PF
M6 FRY C. Fryer 14 Perks Road, Wolverhampton WV11 2ND
M6 FRZ J. Potter 21 Ulverston Crescent, Lytham St. Annes FY8 3RZ
M6 FSA S. Taylor 18 Cycle Street, York YO10 3LJ
MM6 FSB D. Kelly 21 Dhailling Road, Dunoon PA23 8EA
M6 FSC R. Cooper 86 Grove Road, Tiptree, Colchester CO5 0JG
M6 FSD M. Tayler-Grint 94 Dairymans Walk, Guildford GU4 7FF
M6 FSE L. Beaney Brookside, Walkers Lane, Shorwell, Newport PO30 3JZ
M6 FSG C. Moore 12 Melstock Road, Swindon SN25 1XF
M6 FSI D. Woolger 17 Wyvern Close, Tangmere, Chichester PO20 2GQ
M6 FSJ S. Burgess 101 Brendon Way, Nuneaton CV10 8NW
M6 FSM H. Morgan 18 Chrystel Close, Tipton St. John, Sidmouth EX10 0AY
M6 FSN A. Lillis 4 Sinclair Close, Gillingham ME8 9JQ
M6 FSO J. Forbes Weald Barkfold Farm, Plaistow, Billingshurst RH14 0PJ
MM6 FSP S. Paterson Springbank, 2 Mains Of Cuffurach, Clochan, Buckie AB56 5HP
MM6 FSQ C. Mackenzie 89C Needless Road, Perth PH2 0LD
MW6 FST W. Morris Fairview, Trefonen, Oswestry SY10 9DP
M6 FSU M. Cain 59 Bailey Road, Leigh-on-Sea SS9 3PJ
MM6 FSV P. Hadley 51 Victoria Road, Buckhaven, Leven KY8 1BG
M6 FSW P. Bowen Arbor Tree Bungalow, Mill Street, Craven Arms SY7 8EN
M6 FSZ D. Walker 22 Fir Terrace, Esh Winning, Durham DH7 9AG
MM6 FTA L. Pinkowski 69 Nelson Avenue, Livingston EH54 6BZ
MW6 FTC H. Lloyd Apartment 126, Woodlands Hayes Road, Sully, Penarth CF64 5QE
M6 FTF M. Goodman 80 Clay Street, Soham, Ely CB7 5HL
MM6 FTG F. Gorman 11 Maxwell Drive, Baillieston, Glasgow G69 6JB
M6 FTI T. Wetherill Upper House Cottage, Holme Marsh, Kington HR5 3JS
MW6 FTK M. Botham 17 Laburnum Drive, Oswestry SY11 2QW
M6 FTN D. Nathan 138 Anchor Lane, Hemel Hempstead HP1 1NS
M6 FTO G. Lawlor 14 Woodkirk Close, Seghill, Cramlington NE23 7TZ
M6 FTP R. Denham 179 Pandon Court, Shield Street, Newcastle upon Tyne NE2 1XY
M6 FTQ A. Street 110 Magdalen Street, Colchester CO1 2LF
M6 FTT M. Knowles 11 Thorneycroft Avenue, Birkenhead CH41 8HJ
M6 FTU R. Pownall 17 Horsebridge Road, Blackpool FY3 7BQ
M6 FTV S. Barber 31 Aysgarth Avenue, Crewe CW1 4QE
M6 FTW M. Bowman 26 Albany Hill, Tunbridge Wells TN2 3RX
M6 FTX V. Reeve 29 Benfield Way, Portslade, Brighton BN41 2DN
M6 FTY N. Baker 13 Elm Way, Melbourn, Royston SG8 6UH
M6 FUA C. Fear 13 Frome Road, Chipping Sodbury, Bristol BS37 6LD
M6 FUD A. Bentley 40 Wyndale Road, Leicester LE2 3WR
M6 FUE A. Davies 15 The Crescent, Woodside Park, Poulton-le-Fylde FY6 0QW
M6 FUF N. Irvine 100 Cavendish Road, Sunbury-on-Thames TW16 7PL
M6 FUH J. Reich 5 Wolsey Way, Cambridge CB1 3JQ
M6 FUJ H. Braithwaite 20 Richard Moon Street, Crewe CW1 3AX
M6 FUL M. Fuller Rose Cottage, Wickmoor, Stathe, Bridgwater TA7 0JR
M6 FUM Lord I. Kent 111 Sinclair Avenue, Banbury OX16 1BQ
MW6 FUN P. Hawkes Brodawel Cynghordy, Llandovery SA20 0LS
M6 FUQ A. Howard 24 Ladybower Lane, Poulton-le-Fylde FY6 7FY
M6 FUR D. Esdale The Bell Inn, Central Lydbrook, Lydbrook GL17 9SB
M6 FUT R. Harris 63 The Drive, High Barnet, Barnet EN5 4JG
M6 FUU S. Bloomfield 20 Farmers Way, Copmanthorpe, York YO23 3XU
M6 FUW C. Wilkinson Flat 5, Hometide House Beach Road, Lee-on-the-Solent PO13 9BP
M6 FUY A. Robnett 38B Woodmere Avenue, Watford WD24 7LN
MW6 FVA S. Matthews Aberbran Fawr, Aberbran, Brecon LD3 9NG
M6 FVC G. Gee 17 Portherras Villas, Pendeen, Penzance TR19 7TJ
MM6 FVD A. Mcdonald Woodbine Cottage, The Muirs, Rhynie, Huntly AB54 4GD
M6 FVE A. Day 47 Rembrandt Way, Bury St. Edmunds IP33 2LT
M6 FVH J. Mcmullen 281 The Broadway, Cullercoats, North Shields NE30 3LH
M6 FVJ P. Randerson 7 Roman Crescent, Swindon SN1 4HH
MM6 FVK R. Rothon 112 Ravenswood Rise, Livingston EH54 6PG
M6 FVL E. Smethurst 4 Lower New Row, Worsley, Manchester M28 1BE
M6 FVM J. Poriyath 48 Howard Close, Cambridge CB5 8QU
M6 FVN A. Storey 1 Mill Hill Road, Bingham, Nottingham NG13 8YR

M6 FVO M. Oates 63 Longfellow Drive, Rotherham S65 2LH
M6 FVP B. Straker 37 Beech Grove Avenue, Garforth, Leeds LS25 1EF
M6 FVR J. Rawley 33 Dorset Way, Billericay CM12 0UD
M6 FVS R. Sage 9 North Hall Farm Bungalows, Barley Road, Royston CB76AZ
MW6 FVT K. Young 29 High Street, Coedpoeth, Wrexham LL11 3RY
M6 FVV M. Roberts Flat 6, 463 Brighton Road, Lancing BN15 8LF
M6 FVW C. Fox Millstone Cottage, Prior Wath Road, Scarborough YO13 0AZ
M6 FVZ E. Preda 34A Church Street, Willingham, Willingham CB24 5HT
M6 FWB C. Andrews 10 Hartford Road Hartley Wintney, Hook RG27 8QW
M6 FWC H. Wilcox 3 Hutton Street Hutton Wandesley, York YO26 7ND
MI6 FWD J. Tipping 16 The Oaks, Portadown, Craigavon BT62 4HX
M6 FWE M. Scambell 8 South Bank Road, East Cowes PO32 6JE
M6 FWG S. Barclay 64 Deepdene Avenue, Dorking RH5 4AE
M6 FWL C. Pugsley 86 West Town Lane, Bristol BS4 5DZ
M6 FWM M. Tonkin 8 Artis Avenue, Wroughton, Swindon SN4 9BP
M6 FWO J. Capovila 30 Church Street, Barrow-in-Furness LA14 2JG
M6 FWP L. Churchill 1 Fowlswick Cottages, Allington, Chippenham SN14 6LU
M6 FWQ M. Duque 2 Spekehill, London SE9 3BN
M6 FWS M. Smith 151 Rowlands Road, Worthing BN11 3LE
M6 FWT M. Goslin 44 Teasdale Road, Walney, Barrow-in-Furness LA14 3SF
M6 FWV R. Sage 29 Rosewarne Park, Connor Downs, Hayle TR27 5LJ
M6 FWW D. Longson 1 Beckside, Plumpton, Penrith CA11 9PD
M6 FWX D. Jones Drove Farm, Sheepdrove, Lambourn, Hungerford RG17 7UN
M6 FWY D. Smith 7 Oakley Grove, Wolverhampton WV4 4LN
M6 FWZ Dr F. Da Dalt 13 Johnstone Street, Bath BA2 4DH
MW6 FXA S. Hodge 2 Parc Cemlyn, Prestatyn LL19 9NX
MU6 FXB R. Batiste Asile De Paix, Clos Des Sablons, Sandy Lane, Guernsey GY2 4RN
M6 FXE R. Hall 9 Deemers Stile, Telford TF2 9WR
M6 FXG T. Chapman 37 Pheasant Way, Cirencester GL7 1BJ
M6 FXI D. Thomas 51 Barrhill Avenue, Brighton BN1 8UE
M6 FXJ D. Parkes 4 Round Saw Croft, Rubery, Birmingham B45 9TT
M6 FXL T. Goodenough 12 Jerounds, Harlow CM19 4HE
MM6 FXM T. Johnston The Old Schoolhouse, Luggate Burn, Haddington EH41 4QA
M6 FXO K. Burton 16 Lupton Close, Glasshouses, Harrogate HG3 5QX
MM6 FXQ J. Anderson The Grange, Leslie Road, Scotlandwell, Kinross, KY13 9JE
M6 FXU R. Ball 7 Cliff Closes Road, Scunthorpe DN15 7HT
M6 FXX C. Freckelton 44 Hawton Road Newark, Newark NG24 4QB
M6 FXY L. Gregory 11 Roundhill Road, Castleford WF10 5AF
MM6 FXZ W. Goodfellow 1 Yester Place., Haddington EH41 3BE
M6 FYB P. Amond Treetops, Gedding, Bury St. Edmunds IP30 0QD
M6 FYD L. Clark 43 Grange View Harworth, Doncaster DN11 8QP
M6 FYE P. Escott 84 Salisbury Avenue, Bootle L30 1PZ
MM6 FYF T. Burnett 45 The Murrays Brae, Edinburgh EH17 8UF
M6 FYG T. Weston 4 The Pightle, Peasemore, Newbury RG20 7JS
M6 FYH C. Cosgrove 5 Kingswell Avenue, Wakefield WF1 3DY
M6 FYJ S. Lo 12 Kingston Wharf, Kingston Street, Hull HU1 2ES
M6 FYK Staffordshire ARC c/o E. Dutson 19 Maythorn Drive, Cheltenham GL51 0QH
M6 FYL M. Kipling 12 Jolly Brows, Harwood, Bolton BL2 4LZ
M6 FYM A. Armstrong View Firth, Brewery Brow, Parton, Whitehaven CA28 6PE
M6 FYN R. Gunson 3 Kenilworth Drive, Halifax HX3 8XP
MM6 FYR D. Fraser Applecross, Easterton Hatton, Peterhead AB42 0TQ
M6 FYT P. Jones 3 Smith Road, Walsall WS2 9DJ
M6 FYV V. Stanley 82 Sycamore Grove, Bracebridge Heath, Lincoln LN4 2RD
MM6 FYW A. Booth 29 Golf Terrace, Insch AB52 6JY
MW6 FYY C. Marchant 8 Morningside Walk, Barry CF62 9TE
M6 FZA R. Carr 14 Southwell Close, Kirkby-In-Ashfield, Nottingham NG17 8GP
M6 FZB P. Cox 25 Coronation Crescent, Margate CT9 5PN
M6 FZC I. Skeggs 24 Kendall Road, Beckenham BR3 4PZ
M6 FZE A. Korben 219 Leyland Road, Penwortham, Preston PR1 9SY
M6 FZF R. Waterson 43 Highland Road, Twerton, Bath BA2 1DY
M6 FZG J. Williamson 7 Dale Close, Wrecclesham, Farnham GU10 4PQ
M6 FZJ J. Foster Westfield House, 23 High Street, Cumnor OX2 9PE
M6 FZK C. Bowler 42A Honor Road Prestwood, Great Missenden HP16 0NL
M6 FZL D. Finlay 23 Glen Way, Oadby, Leicester LE2 5YF
M6 FZM I. Jones 8 Hengrove Avenue, Bristol BS14 9TB
M6 FZN C. Williams 31 Dent Close, South Ockendon RM15 5DS
M6 FZR C. Bowskill 5 Trent Walk, Daventry NN11 4QF
M6 FZS M. Holmes Lower Farm, Stony Moor, Newton-On-Rawcliffe, Pickering YO18 8QJ
MM6 FZT R. Clow 25 Scott Street, Newcastleton TD9 0QQ
M6 FZV N. Walker Wayfield Farm, Calstock Road, Gunnislake PL18 9BY
M6 FZW T. Crampton 7 Barneveld Avenue, Canvey Island SS8 8NZ
M6 FZX T. Cash 27 Robert Wynd, Bilston WV14 9SE
M6 FZY J. Langdon 6 Glebe Close, Stockton, Southam CV47 8LG
MW6 GAA G. Green 17 Danybanc, Llanelli SA15 4NS
M6 GAB G. Barnes 24 Gainsborough Court, Andover SP10 3SS
M6 GAD G. Briggs 5 Links Avenue, Cromer NR27 0EQ
M6 GAE Dr P. Shaw 25 Headcorn Road, Platts Heath, Maidstone ME17 2NH
M6 GAF G. Furlong 22 Swaisland Road, Dartford DA1 3DE
M6 GAH G. Hardill 107 Leicester Road Whitwick, Coalville LE67 5GN
M6 GAI G. Robinson 16 Stanley Road South, Rainham RM13 8AA
M6 GAK G. Cockburn 20 Hexham Avenue, Hebburn NE31 2HN
M6 GAL T. Kimberlee 24 Jacey Road, Shirley, Solihull B90 3LJ
M6 GAM A. Mawson 14 Pontop View, Delves Lane, Consett DH8 7JB
M6 GAN G. Rudley 37 Cherry Way, Shepperton TW17 8QQ
M6 GAO G. Webster 24 Martin Avenue, Barrow Upon Soar, Loughborough LE12 8LG
MI6 GAQ G. O'Reilly 20 Lower Clonard Street, Belfast BT12 4NH
M6 GAR G. Veale 8 Duchy Cottages, Stoke Climsland, Callington PL17 8PA
M6 GAS B. Hammond 35 Stratford Avenue, Newcastle ST5 0JS
MI6 GAU G. Ferguson 22 Cloneen Drive, Ballymoney BT53 6PT
MW6 GAV J. Ali 5 Darent Close, Bettws, Newport NP20 7SQ
M6 GAW G. Watson 70 Garden Hey Road, Moreton, Wirral CH46 5NE
M6 GAX G. Austin 4A Garlinge Road, Tunbridge Wells TN4 0NR
MW6 GBA A. Jones 2 Erw Terrace, Bethel, Caernarfon. LL55 1YT
M6 GBB D. Nock 33 Hale Road Bradenham, Thetford IP25 7RA
MI6 GBC G. Boyes 50 Halfpenny Gate Road, Moira, Craigavon BT67 0HW
M6 GBD G. Davage 17 Howlett Close, North Walsham NR280BF
M6 GBE H. Vecenans 155 Upper Dale Road, Derby DE23 8BP
M6 GBF G. Bouchier Flat 24, Hine House, St. Albans AL4 0EY
MU6 GBG G. Lanoe 1 Pinetrees Estate, Route De L'Islet, Guernsey GY2 4EX
M6 GBH I. Appleby 96 Cranbrook Road, Poole BH12 3BT
M6 GBI C. Gibson 17 Clyde Court, Grantham NG31 7RB
M6 GBJ G. Walker 25 Bear Close, Woodstock OX20 1JT
M6 GBL L. Ulvenmoe Flat 6, Turnstone Court, Greenfinch Way, Heysham, Morecambe LA3 2JF
M6 GBM N. Cunningham 115 Trafalgar Road, Washington NE37 3DJ
M6 GBN G. Barton 9 Tees Crescent, Stanley DH9 6HX
M6 GBO B. Goodman 6 Victoria Court, Penzance TR18 2EX
MM6 GBP G. Holford 5 Deerpark Cottages, Evanton, Dingwall IV16 9XH
M6 GBQ A. Self 10 Cecil Road, Hertford SG13 8HR
M6 GBR P. Chamberlain 22 Stanedge Grove, Wigan WN3 5PL
MM6 GBS G. Somerville 4 Kirkhill Way, Penicuik EH26 8HH
M6 GBU D. Carter Bungalow 7, Higher Ingsdon Quarry, Liverton, Newton Abbot TQ12 6JA
M6 GBV A. Finn 202 Northgate Road, Stockport SK3 9NJ
M6 GBW G. Wright Flat 27, Maudland House, Preston PR1 2YJ
MM6 GBX G. Mcfarlane 59 Millburn Road, Bathgate EH48 2AF
M6 GBY R. Hughes 96 Retallick Meadows, St. Austell PL25 3BZ
M6 GBZ D. Smith 48 Shirley Gardens, Tunbridge Wells TN4 8TH
M6 GCB G. Beynon-Fisher 21 Scopsley Green, Whitley, Dewsbury WF12 0NF
M6 GCC G. Clements 4 Hobart Square, Norwich NR1 3JB
M6 GCD G. Lees 16 Kingfisher Close, Congleton CW12 3FF
M6 GCE J. Berry 245A Eaves Lane, Chorley PR6 0AG
MU6 GCI J. Littlewood Wayland Les Martin, L'islet GY2 4XW
M6 GCJ G. Cooke 25 Avill, Hockley, Tamworth B77 5QE
M6 GCK G. Cornish 78 Kerry Avenue, Ipswich IP1 5LD
M6 GCM N. Baker 56 Chalklands, Bourne End SL8 5TJ
MW6 GCN G. Leedham 15 Vale Park, Rhyl LL18 2EN
M6 GCO S. Gilbert 34 Ramsey Road, St. Ives PE27 5RD
M6 GCP M. Poulter 32 Woburn Street, Hull HU3 5LW
MW6 GCQ G. Jones 6 Tre Ambrose, Holyhead LL65 1LR
M6 GCS G. Smith Glencairn, Gaston Lane, Malmesbury SN16 0LY
M6 GCT G. Taylor 24 Otter Close, Bletchley, Milton Keynes MK3 7QP
M6 GCV G. Adey 24 Burycroft, Welwyn Garden City AL8 7AW
M6 GCW G. Wager Dovecote, Turbary, Doncaster DN9 1DY
M6 GCX M. Parnham 9 The Close, Addington, West Malling ME19 5BL
M6 GCY B. Thomson 40 Northgate Road, Stockport SK3 0LQ
M6 GDA J. Kemp 9 Chequers Orchard, Stone Street, Canterbury CT4 5PN
M6 GDB G. Buckley 30 Saxon Road, Wheathampstead, St. Albans AL4 8NZ
M6 GDC D. Cowling 11 Shakespeare Avenue, Scunthorpe DN17 1SA
MI6 GDD Lord G. Drummond 41 Calhame Road, Ballyclare BT39 9NA
M6 GDI J. Williams 41 Overton Lane, Hammerwich, Burntwood WS7 0LQ
M6 GDJ G. Johnson 31 Hillcrest Avenue, Grays RM20 3DA
M6 GDL C. Foulkes 5 Kennedy Close, Chaddesden, Derby DE21 6LW
MI6 GDN G. Nelson 65 Dernawilt Road, Annagolgan, Enniskillen BT92 7FN
M6 GDO C. Stallard 8 Holmeside Avenue, Trimdon TS296HE
M6 GDP A. Parker 18 Lincoln Close, Keynsham, Bristol BS31 2LJ
M6 GDQ G. Whatmough Stud Bungalow, Wakefield Lodge Estate, Towcester NN12 7QX
M6 GDS G. Sole 16 Beech Crescent, Hythe, Southampton SO45 3QG
M6 GDT G. Thomas 14 Lower Meadow, Cheshunt, Waltham Cross EN8 0QU
M6 GDV G. Butterworth 11 St. Davids Road, Robin Hood, Wakefield WF3 3TG
MM6 GDY Lady D. Collins 6, Blackberry Farm, Throntonloch EH42 1QT
M6 GEA G. Abraham 18 Harpenden Close, Bedford MK41 9RG
M6 GEE B. Gee 23 Kenmore Close, Gateshead NE10 8WJ
M6 GEG A. Gleave 38 Barkus Way Stokenchurch, High Wycombe HP14 3RE
M6 GEJ G. Johnson 4 Wallace Close, Hullbridge, Hockley SS5 6NE
M6 GEK S. Sissens 20 Fallow Drive, Eaton Socon, St. Neots PE19 8QL
M6 GEP S. Ingledew 34 Sunningbrook Road, Tiverton EX16 6EB
M6 GEQ M. Emmett 8 Coates Drove, Isleham, Ely CB7 5SJ
M6 GEU G. Walker 43 Wordsworth Crescent, Kidderminster DY10 3EY
M6 GEV R. Heslop 7 Fieldfare Close, Clanfield, Waterlooville PO8 0NQ
M6 GEY A. Callaghan 46 Highfields, Great Yeldham, Halstead CO9 4QQ
M6 GFA C. Claypole 1 Richmond Crescent, Leominster HR6 8RX
M6 GFE G. Teale Flat 3, 89-91 Barrack Road, Christchurch BH23 2AJ
M6 GFF G. Gibbs 2 Salesfrith Cottages, Bicknacre Road, East Hanningfield, Chelmsford CM3 8AP
MM6 GFG D. Speirs 45 Elmbank Crescent, Arbroath DD11 4EZ
M6 GFH G. Molendijk 47 Lodge Road, Scunthorpe DN15 7EN
M6 GFM G. Cummins 10 Baugh Gardens Downend, Bristol BS16 6PN
MI6 GFO G. Craig 103 Moyle Parade, Larne BT40 1ET
MW6 GFP G. Stephens 3 Yr Hen Orsaf, Y Felinheli LL56 4AB
M6 GFQ J. Cooper 70 Rockhampton Close, Weymouth DT3 6NG
MU6 GFR D. Robert Nos Treis 7 Liberation Drive, Route Des Clos Landais, St. Saviour, Guernsey GY7 9PH
M6 GFS G. Williams 70 Southill Road, Poole BH12 3AS
M6 GFV P. Dorman 2 Moises Hall Road, Wombourne, Wolverhampton WV5 0LF
M6 GFW G. Watson The Dell, Nova Scotia Road, Great Yarmouth NR29 3QD
M6 GFY G. Waring 18 The Mead, Beaconsfield HP9 1AW
M6 GFZ G. Linnett 62 Melrose Avenue, Burtonwood, Warrington WA5 4NW
M6 GGD G. Durham 12 Beech Close, Sproatley, Hull HU11 4XB
MM6 GGE G. Semple 28 Newton Brae, Cambuslang, Glasgow G72 7UW
MI6 GGF A. Dowling 74 Ashmount Gardens, Lisburn BT27 5DA
M6 GGG D. Rosenschein 101 Christchurch Road, London SW14 7AT
M6 GGI G. Ridley 12 Garforth Avenue, Steeton, Keighley BD20 6SP
M6 GGJ G. Jacks 2 Corve View, Fishmore, Ludlow SY8 2QD
M6 GGL S. Cooke Flat 25, Attree Court, Brighton BN2 0FZ
M6 GGM A. Mansfield 20 High Street, Broughton, Kettering NN14 1NG
MI6 GGN G. Scullion 50 Tober Road, Pharis, Ballymoney BT53 8NY
M6 GGO L. Jones 44 Althorpe Drive, Loughborough LE11 4QU
M6 GGQ G. Wilson 28 Stanley Grove, Redcar TS10 3LN
M6 GGT D. Turford 51 Moorfield Avenue, Bolsover, Chesterfield S44 6EJ
M6 GGV G. Guest 90 Pearson Crescent Wombwell, Barnsley S73 8SG
M6 GGW G. Warnock 31 Greycote, Shortstown, Bedford MK42 0XD
M6 GGX C. Ogidih 89 West Road, Birmingham B43 5PG

M6 GGZ T. Holman 20 Green Drive, Wolverhampton WV10 6DW
MI6 GHA E. Forde 35 Torr Gardens, Larne BT40 2JH
M6 GHB G. Brooks 14 Chalton Crescent, Havant PO9 4PT
M6 GHC L. Arenas Martinez 22 Brickfield Road, Southampton SO17 3AE
MW6 GHD A. Burleton 11 Orchard Close, Caldecott NP26 4BH
MM6 GHF G. Farrer 23 Upper Craigour, Edinburgh EH17 7SE
MM6 GHH D. Livingstone The Bungalow, Stewarton, Campbeltown PA28 6PG
M6 GHJ J. O'Donnell 24 Foxhill, Watford WD24 6SY
M6 GHL G. Nicholls 26 The Headlands, Darlington DL3 8RP
M6 GHM G. Bell 34 Manor Road, Eastham, Wirral CH62 8BN
M6 GHN J. Oldham 5 Amersham Rise, Nottingham NG8 5QG
M6 GHP G. Pearson 41 Myrica Grove, Hoole, Chester CH2 3EW
M6 GHQ B. Gamble 196 Chase Road, Burntwood WS7 0DX
M6 GHR J. Greenfield 12 Firbeck Road, Nottingham NG8 2FB
M6 GHT G. Hart 11 Sadlers Ride, West Molesey KT8 1SU
M6 GHV J. Harris 108 Gresley Wood Road, Church Gresley, Swadlincote DE11 9QN
M6 GHW G. Harrison-Webb 2 Jubilee Close, Henlow SG16 6FD
MM6 GHX R. James The Garret, Alyth, Blairgowrie PH11 8HQ
MW6 GHY C. Beamish 15 Pen Yr Hwylfa, Harlech LL46 2UW
MW6 GIA D. Owen Tanrallt, Blaenpennal, Aberystwyth SY23 4TP
M6 GIB R. Gibbs 7 Thornhill, Eastfield, Scarborough YO11 3LY
M6 GIC A. Oliveira 13 A Lakefield Road, London N22 6RR
MM6 GID Leiston ARC c/o W. Elder 3 Mossgiel Place, Dundee DD4 8AP
MI6 GIF G. Clarke 12 Church Green, Dromore BT25 1LL
M6 GIH G. Hutchinson 128 Crescent Drive North, Woodingdean, Brighton BN2 6SF
M6 GII C. Baines 19 Kingfisher Road, Mountsorrel, Loughborough LE12 7FG
M6 GIL G. Coleman 5 Meeting Lane, Burton Latimer, Kettering NN15 5LS
MI6 GIN B. Emerson1 67 Castlemore Avenue, Belfast BT6 9RH
M6 GIQ D. Lynch M L & S Martin Lynch & Sons Ltd, Wessex House, Drake Avenue, Staines-upon-Thames TW18 2AP
MW6 GIU A. Sprott 61 Brynymor, Three Crosses, Swansea SA4 3PE
MW6 GIV A. Vowles 232 Pentregethin Road, Gendros, Swansea SA5 8AW
MM6 GIW C. Ferguson 18 The Braes, Lochgelly KY5 9QH
MW6 GIX A. Hodgson Browns Holiday Park Towyn Road, Conwy LL22 9HD
M6 GIY B. Smith 26 Epperstone Court Patrick Road, West Bridgford NG2 7QR
M6 GIZ M. Staley 164 Rotherham Road Maltby, Rotherham S66 8NA
M6 GJA A. Gallagher 174 Queensway, West Wickham BR4 9DZ
M6 GJB G. Baigent 15 Soyland Town Road, Sowerby Bridge HX6 4NB
M6 GJC G. Cregg 17 Rake Way, Aylesbury HP21 9AL
M6 GJD D. Toller Field Cottage, Ash Lane, Etwall, Derby DE65 6HT
M6 GJE G. Groves 5 Beech Road, Ashurst, Southampton SO40 7AY
M6 GJG S. Kelly 4 Berrells Court, Olney MK46 4AR
M6 GJH G. Henderson 4 Castell Road, Loughton IG10 2LT
M6 GJI T. Kelly 32 The Foxgloves, Hedge End SO30 0UG
M6 GJL J. Paine 28 Laurel Way, Bottesford, Nottingham NG13 0FP
M6 GJM G. Morris 7 Rowley View, Bilston WV14 8DE
M6 GJN G. Noble 6 Sturrocks, Basildon SS16 4PQ
M6 GJO G. Owens 6, Khyber Close, Torpoint PL11 2JB
M6 GJP G. Priest Lilac Cottage, School Lane, Tamworth B79 9JJ
M6 GJQ E. Wilkinson 40 Charmouth Road, St. Albans AL1 4SN
M6 GJS G. Shaw 6 Vickers Close, Woodley, Reading RG5 4PA
M6 GJT G. Tyler Crofton, Stoney Ley, Worcester WR6 5NG
M6 GJV A. Wesselby 16 Hudson Way, Grantham NG31 7BX
M6 GJX A. Hunt 76 Andrew Street, Bury BL9 7HB
M6 GJY T. Garrett 12 Poulders Gardens, Sandwich CT13 0BE
MI6 GKB Points of Historical Interest c/o G. Black 45 Meeting House Lane, Lisburn BT27 5BY
M6 GKC G. Kapfunde Gunnels Wood Road, Stevenage SG1 2AS
MM6 GKE L. Scott 5 Links Road, Saltcoats KA21 6BE
M6 GKF G. Morris-Roe 77 Bridlebank Way, Weymouth DT3 5RP
M6 GKG G. Gordon 40 Grange Crescent, Rubery, Birmingham B45 9XB
M6 GKH R. Curant 18 Ramley Road, Lymington SO41 8GQ
M6 GKI J. Mcgowan 2 Turnstone Drive, Liverpool L26 7WR
M6 GKK A. Maclean 10 Elizabeth Close, West Hallam, Ilkeston DE7 6LW
M6 GKM J. Harrison 20 Brafield Close, Belper DE56 0EU
M6 GKN G. Davies 6 Bayleys Close, Empingham, Oakham LE15 8PJ
MW6 GKQ R. Buchan-Terrey Godre'R Coed, Aberhosan, Machynlleth SY20 8RA
M6 GKT Dr M. Gunn Brynywawr, Talybont SY24 5HJ
MW6 GKU P. Daniel 71 Coed Isaf Road, Pontypridd CF37 1EN
M6 GKV D. Crook 72 Rushleigh Avenue, Cheshunt, Waltham Cross EN8 8PS
M6 GKW G. Wade 43 Green Park, Cambridge CB4 1SX
M6 GKX E. York Flat 23, Paul Stacey House, Coventry CV1 5GU
MI6 GLC G. Crabbe 39 Arran Avenue, Ballymena BT42 4AP
M6 GLD R. Broughton 3 The Row, Weeting, Brandon IP27 0QG
MM6 GLI G. Irvine 120A Shore Road, Innellan, Dunoon PA23 7SS
M6 GLJ S. Solomon 71 Ashby Road, Moira, Swadlincote DE12 6DN
MW6 GLK S. Bateman 26 Kenneth Treasure Court, Bethania Row, Cardiff CF3 5UD
M6 GLM S. Weeks Edithmead, Highbridge TA9 4HE
M6 GLN G. Brittleton 1 Littler Lane, Winsford CW7 2NE
MM6 GLO G. Burns 25 Dunskey Road, Kilmarnock KA3 6FJ
M6 GLP G. Parker 23 Woodrush Road, Purdis Farm, Ipswich IP3 8RB
M6 GLS D. Hendricks 60 Heath View, Leiston IP16 4JP
M6 GLU M. Morgan-Lucas 33 Stark Close, Diss IP22 4BY
MW6 GLV C. Tanner Pen Y Gogarth Llaneilian, Amlwch LL68 9NH
M6 GLW G. Williams 7 The Grove, Patchway, Bristol BS34 6PE
M6 GLX Dr C. Cowen Rosita, White Street Green, Sudbury CO10 5JN
M6 GLZ G. Pardoe 3 Bar Meadow, Shobdon, Leominster HR6 9BZ
M6 GMA G. Marsden 38 Sandhill Road, Rawmarsh, Rotherham S62 5NT
M6 GMC T. Smith 109 Ferriston, Banbury OX16 1XA
M6 GMD M. Davison 27 Ford Street, Consett DH8 7AE
M6 GMF G. Fildes 10 Windmill Gardens, St. Helens WA9 1EN
M6 GMH G. Muirhead 13 Berry Street, Skelmersdale WN8 8QZ
MW6 GMJ C. Jones 28 Ivor Street, Maesteg CF34 9AH
MI6 GMK G. Mckinley 26 Newry Street, Warrenpoint, Newry BT34 3JZ
M6 GMM M. Dickinson 16 Shearwater Avenue, Newcastle upon Tyne NE12 8PH
M6 GMO M. Boland 24 Hallam Close, Moulton, Northampton NN3 7LB
M6 GMP C. Crank 12 Boundary Lane North, Cuddington, Northwich CW8 2PL
M6 GMQ B. Healey 14 Orchard Close, Ferring, Worthing BN12 6QP
M6 GMR G. Mather Shawdene House, Donnington, Newbury RG14 3AJ
M6 GMS G. Moss 125 Lavender Avenue, Mitcham CR4 3RS
M6 GMT W. Pattison 30 Main Street, Flixton, Scarborough YO11 3UB
M6 GMU L. Lewis 653 Main Road, Dovercourt, Harwich CO12 4NF
M6 GMV M. Shepherd 38 Pryors Lane, Bognor Regis PO21 4LH
M6 GMY L. Allison 25 Saddington Road, Fleckney, Leicester LE8 8AX
MW6 GNA A. Jones 23 Pinecroft Avenue, Aberdare CF44 0HY
M6 GNC B. Chandler 1 Rambridge Farm Cottages, Weyhill, Andover SP11 0QF
M6 GND A. Holden 4 Gilberts Drive, East Dean, Eastbourne BN20 0DJ
MI6 GNF R. Simpson 3 Portadown Road, Tandragee, Craigavon BT62 2BB
M6 GNG C. Wilsher 17 Milton Road, Blacon, Chester CH1 5XE
M6 GNH G. Harris 100 Bennett Lane, Batley WF17 6DB
M6 GNJ G. Hance 23 Catalina Avenue, Chafford Hundred, Grays RM16 6RE
M6 GNM K. Markham Hampers Cottage, Hampers Lane, Pulborough RH20 3HZ
M6 GNN G. Norris Trade Winds, Preston Road, Preston PR4 0TT
M6 GNO A. Vanderahe 28 Percival Close, Norwich NR4 7EA
MI6 GNP M. Parke 39 Crannog Park, Strathfoyle, Londonderry BT47 6NF
M6 GNR P. Wollen 41 Bernadette Close, Exeter EX4 8DU
M6 GNS S. Hance 23 Catalina Avenue, Chafford Hundred, Grays RM16 6RE
M6 GNU P. Myszka 116 Ridyard Street, Little Hulton, Manchester M38 9NF
M6 GNV A. Stoll 6 Dovestone Gardens, Littleover, Derby DE23 4EJ
M6 GNW A. Land 12 Beverley Close Holton Le Clay, Grimsby DN36 5HG
M6 GNX J. Elliot 82 Hollinside Road, Sunderland SR4 8BG
M6 GNY A. Ali 42 Blease Close, Staverton, Trowbridge BA14 8WD
MI6 GOA J. Proctor 16 Lisnavaragh Road, Scarva, Craigavon BT63 6NX
M6 GOB I. Pashley 45 Highfield View Road Newbold, Chesterfield S41 7JZ
M6 GOC J. Adams 10 Leckford Road, Oxford OX2 6HY
M6 GOE P. Everton 110 Sandstone Road, Sheffield S9 1AG
M6 GOF M. Stitt 17 Evans Street, Prescot L34 6HU
M6 GOG M. Henman 47 London Rd Kirton, Boston PE201JJ
M6 GOI E. Walsh 12 Ock Meadow, Stanford In The Vale, Faringdon SN7 8LN
M6 GOL M. Goldthorpe 84 Park Lane Allerton Bywater, Castleford WF10 2AP
MM6 GON J. Loughren 16 Merlewood Road, Inverness IV2 4NL
M6 GOO J. Goolden 34 Grimsby Road, Louth LN11 0DY
M6 GOQ J. Walsh Flat 4, 14 Chantry Road, Bristol BS8 2QD
MM6 GOR G. Campbell 98 Netherton Road, East Kilbride, Glasgow G75 9LB
M6 GOS S. Forshaw 8 Stavesacre, Leigh WN7 3LD
MM6 GOW N. Gowans 1 Dunmuir Road, Castle Douglas DG7 1LG
MW6 GOY C. Rees 5 Dare Villas, Aberdare CF44 8AH
MW6 GOZ C. Gozzard Craig Dulas, Rhydyfoel Road, Llanddulas, Abergele LL22 8EG
M6 GPA M. Phillips 36 Hyde Heath Court, Crawley RH10 3UQ
M6 GPB G. Beacher 22 Trowbridge Gardens, Luton LU2 7JY
M6 GPC G. Coleman 15 Redwood Drive, Ormskirk L39 3NS
MM6 GPD L. Macdonald 193 Den Walk, Buckhaven, Leven KY8 1DJ
M6 GPE G. Davison Lyncroft House, Swan Lane, Edenbridge TN8 6AJ
MM6 GPF G. Fleming Tarbat View, Achavandra Muir, Dornoch IV25 3JB
M6 GPG G. Bates 230 Brook Street, Erith DA8 1DZ
M6 GPJ M. Stevens 4 Wellwood Close, Horsham RH13 6AL
M6 GPM G. Mccaffery 7 Cliffe Court, Sunderland SR6 9NT
M6 GPO P. Jones 2 Herons Court, Doncaster DN2 4GD
M6 GPP G. Powell 7 Donstan Road, Highbridge TA9 3LA
MI6 GPQ D. Boyd 11 Abbey Gardens, Belfast BT5 7HL
M6 GPS G. Stevens 17 Manston Close, Ernesettle, Plymouth PL5 2SN
M6 GPT R. Hampson 12 Oakhays, South Molton EX36 4DB
M6 GPU D. Mclean 38 New Burlington Road, Bridlington YO15 3HS
MI6 GPV G. Cassidy 90 Ardmeen Green, Downpatrick BT30 6JL
MI6 GPZ B. Cousins 58 Bannview Heights, Banbridge BT32 4NA
M6 GQA L. Warren 85 Greenwood Avenue, Blackpool FY1 6PR
M6 GQB S. Mccormack 56 Careys Road, Pury End, Towcester NN12 7NX
MM6 GQC B. Angus 6 Inshes Crescent, Inverness IV2 3SP
MW6 GQD B. Lee 46 Knowling Mead, Tenby SA70 8EB
M6 GQF D. Thomas 18 Howard Close, Cambridge CB5 8QU
M6 GQG E. Thresher 18 Sandy Lane, Preesall, Poulton-le-Fylde FY6 0EH
M6 GQH J. Fitzpatrick 56 Littlehaven Lane, Horsham RH12 4JB
MI6 GQI J. Steuerwald 50 Myrtlefield Park, Belfast BT9 6NF
M6 GQK E. Beckett 4 Princes Avenue, Ramsgate CT12 6DW
MM6 GQP P. Hunter 3 Promenade, Leven KY8 4HZ
M6 GQR J. Townsend 47 Cosgrove Avenue, Sutton-in-Ashfield NG17 3JY
M6 GQS S. Klee 4 Abbots Road, Pershore WR10 1LL
M6 GQT I. Alderman 107 Manton Drive, Luton LU2 7DL
M6 GQU A. Scott 37 Pikestone Close, Hayes UB4 9QT
M6 GQW H. Beavis 8 Hookfield, Harlow CM18 6QG
M6 GQX M. Martin 19 Elmsleigh Road, Farnborough GU14 0ET
M6 GQY L. Brearley Ash Tree Lodge, Snaith Road, Goole DN14 0AT
M6 GQZ P. Collier Flat 9, Henry House, Wyvil Road, London SW8 2TF
M6 GRC G. Clarke 6 Coverdale Road, Scunthorpe DN16 2RP
MD6 GRD C. Schofield Rockside, Dreemskerry Road, Maughold IM7 1BL Isle of Man
M6 GRH G. Hooper 16 Trentham Drive, Orpington BR5 2EP
M6 GRJ R. Davison Lyncroft House, Swan Lane, Edenbridge TN8 6AJ
M6 GRK G. Kenyon 2 Langdale Terrace, Stalybridge SK151EX
M6 GRN A. Young 48 Sussex Street, Cleethorpes DN35 7NP
M6 GRP G. Priestley 53 Millfield Gardens, Crowland, Peterborough PE6 0HA
M6 GRQ D. Mcclurg 40 Forfar Drive, Bletchley, Milton Keynes MK3 7LS
M6 GRR G. Rodwell 87 Park Avenue, Ruislip HA4 7UL
M6 GRT J. Grant 18 St. Agnes Place, Chichester PO19 7TN
M6 GRU J. Stewart 43 Balcombe Gardens, Horley RH6 9BY
M6 GRV C. Rawlin 5 Japonica Hill, Immingham DN40 1LT
M6 GRX S. Tomlinson 8 Levett Road, Stanford-le-Hope SS17 0BB
M6 GRZ G. Kober 1 Sawston, King's Lynn PE30 4XT
M6 GSC G. Casey 10 Windermere Road, Dukinfield SK16 4SJ
M6 GSD C. Riley 2 Shottisham Hall Cottages, Alderton Road, Shottisham, Woodbridge IP12 3EP
M6 GSF G. Ferns Oxlea House, Meerbrook, Leek ST13 8SL
MI6 GSG G. Gregg 30 Claremont Avenue, Moira, Craigavon BT67 0SS
M6 GSH P. Pritchard Flat 148, Nine Acre Court, Salford M5 3HU
M6 GSJ G. Sawyer 432 Rowood Drive, Solihull B92 9LQ
M6 GSK M. Silver 52 Park Crescent, Elstree, Borehamwood WD6 3PU
MW6 GSL A. Cassar 48 High Street, Abergwynfi, Port Talbot SA13 3YW

M6 GSO G. Wren 14 Overdale, Triangle, Sowerby Bridge HX6 3HZ
M6 GSP P. Lumb 22 Lydney Road, Liverpool L36 2LT
M6 GSQ J. Creese Matrice, Church Road, Billericay CM11 1RR
M6 GSR G. Rogers 69 Leechcroft Avenue, Swanley BR8 8AP
MW6 GSS G. Smith Ty Clyd, Llanfihangel-Nant-Bran, Brecon LD3 9NA
M6 GST G. Starling 4 Three Corner Drive, Norwich NR6 7HA
M6 GSX D. Tinkler 19 Askew Dale, Guisborough TS14 8JG
MM6 GSY N. Blampied 10 Lithe Lochan, Longmorn, Elgin IV30 8SA
M6 GSZ D. Gray 26 Highgate Road, Portsmouth PO35AS
M6 GTB G. Bolan 82 Calve Croft Road, Manchester M22 5FU
M6 GTC J. Tanner 7 Larch Crescent Eastwood, Nottingham NG16 3RB
M6 GTD D. Garratt 215 Coalpool Lane, Walsall WS3 1RF
M6 GTE G. Tomkins 22 Arundel Drive, Orpington BR6 9JG
M6 GTH G. Hofman Brookside, High Street, Stockbridge SO20 6EY
M6 GTK G. Tagg Tinkers Cottage, Nevendon Road, Wickford SS12 0QB
M6 GTO B. Thomas 13 Killerton Close, Paignton TQ3 3FT
M6 GTR M. Bailey 17 Sparrowhawk Way, Hartford, Huntingdon PE29 1XE
M6 GTT J. Owen 8 Highridge Crescent, Bristol BS13 8HN
M6 GTU R. Burton 1 Broad Street, Long Eaton, Nottingham NG10 1JH
MI6 GTY G. Mamijs 33 Glenmore Walk, Lisburn BT27 4RY
M6 GUA J. Hammond 8 Rowntree Way, Saffron Walden CB11 4DG
M6 GUB N. Emberson Lion House, Audley End, Saffron Walden CB11 4JB
M6 GUC R. Millen 21 Sunnymead, Tyler Hill, Canterbury CT2 9NW
MU6 GUE S. Tostevin Hillside, Le Francais, Vale GY3 5NL Guernsey
M6 GUF A. Cook 35 Park Road, Cheveley, Newmarket CB8 9DF
M6 GUH D. Hughes Flat 2, 13 Thorgam Court, Grimsby DN31 2EU
M6 GUJ S. Noller 3 Thor Road, Norwich NR7 0JS
M6 GUM A. Bell 36 Schneider Road, Barrow-in-Furness LA14 5DW
M6 GUO M. Wilson The Avenue Care Home, 23 Avenue Road, Malvern WR14 3AY
M6 GUS Phoenix Radio Group c/o G. Frisholm 99 Greenfell Mansions, Glaisher Street, London SE8 3EX
M6 GUT G. Taljaard Flat 140, 105 London Street, Reading RG1 4QD
M6 GUU E. Shaw Upper Floor Flat, 21 Norfolk Road, Littlehampton BN17 5PW
M6 GUV D. Connolly 2 Layton Close, Birchwood, Warrington WA3 6PT
M6 GUW D. Kemp 4 Wells Close, Hainford, Norwich NR10 3NB
M6 GUX D. Wittering 156 Langland Road, Netherfield, Milton Keynes MK6 4HX
M6 GUY G. Westbrook The Haven, St. Johns Road, Norwich NR12 9BE
M6 GVA P. Kelly 22 Manchester Place, Norwich NR2 2SH
M6 GVC G. Clayton The Forge , High Street, Moreton-in-Marsh GL56 0LL
M6 GVD M. Mcsweeney Barn House, Stone Quarry Road, Haywards Heath RH17 7LP
M6 GVF G. Vailati Facchini 8 Orchard Crescent, Edgware HA8 9PW
M6 GVI J. Day 120 Goring Road, Colchester CO4 0DB
M6 GVJ D. Gough 29 Belvedere Road, Biggin Hill, Westerham TN16 3HX
M6 GVL B. Alston 9 Central Avenue, Church Stretton SY6 6EE
M6 GVM A. Spurling 62 Swains Meadow, Church Stretton SY6 6HT
M6 GVN G. Tilling 1 Bellington Cottages, Worcester Road, Kidderminster DY10 4NE
M6 GVP G. Haynes 25 Ladbroke Road, Bishops Itchington, Southam CV47 2RA
MW6 GVR G. Roberts 14 Abercarn Fach, Cwmcarn, Newport NP11 7EP
M6 GVS G. Greaves 183 Wordsworth Avenue, Sheffield S5 8NE
M6 GVT L. Edwards 10 Marsh View, Beccles NR34 9RT
M6 GVU E. Hammond 14 Wannock Gardens, Polegate BN26 5PA
M6 GVW G. Weston 131 Ringwood Road, Eastbourne BN22 8TQ
M6 GVY R. Kendall 24 Scotland Road, Cambridge CB4 1QG
M6 GWB G. Bunting 31 Hardwick Avenue, Allestree, Derby DE22 2LN
M6 GWE G. Wells 1 Seagrove Way, Seaford BN25 3QY
M6 GWF G. Waterfall 18 Sandbed Lane, Belper DE56 0SH
M6 GWG G. Weston 94 Redhill Road, Northfield, Birmingham B31 3LA
M6 GWI N. Shears 1 The Old Rectory, Rectory Lane, Little Bookham, Leatherhead KT23 4DY
M6 GWK K. Jones 45 Liskeard Way, Freshbrook, Swindon SN5 8NL
MW6 GWL S. Gwillym 5 Alder Close New Inn, Pontypool NP4 0DF
MW6 GWM M. Martin 1 Y Gorlan, Bryn Street, Newtown SY16 2HN
M6 GWN G. Wynne 19 Monks Orchard, Nantwich CW5 5TX
MW6 GWO W. Jones 39 Heolddu Grove, Bargoed CF81 8UX
MW6 GWP G. Price 53 Tyddyn Mostyn Estate, Menai Bridge LL59 5DS
M6 GWQ M. Priest 35 Albert Road, Chaddesden, Derby DE21 6SJ
M6 GWS G. Salter 9 Spring Gardens, Malvern Link, Malvern WR14 1AP
M6 GWT G. Toomer 7 Rosewood Drive, Barnby Dun, Doncaster DN3 1BJ
M6 GWU T. Barrett 43 The Ridgeway, Meols, Wirral CH47 9RZ
MM6 GWW N. Robertson Ladyburn, Port William, Newton Stewart DG8 9QN
M6 GWZ K. Webb 52 Princes Avenue, Walsall WS1 2DH
M6 GXC T. Hales Flat 6, St. Georges Court, Cambridge CB1 7UP
M6 GXD J. Puddy 9 Prospect Close, Easter Compton, Bristol BS35 5SB
M6 GXF G. Fernando 1 Rosemary Avenue, West Molesey KT8 1QF
M6 GXH B. Gates 3 Highfield, Taunton TA1 5JE
MW6 GXI D. Burt 2 Cae Masarn, Pentre Halkyn, Holywell CH8 8JY
M6 GXJ D. Ashton 4 Middleburg House, Cheriton High Street, Folkestone CT19 4HP
M6 GXK P. Blagden 30 Charlecote Avenue, Tuffley, Gloucester GL4 0TH
M6 GXL A. Jones 1A Invicta Road, Folkestone CT19 6EY
M6 GXP H. Butcher 12 Bath Road, Willesborough, Ashford TN24 0BJ
M6 GXR S. Thresher 4 Huntersway, Culmstock, Cullompton EX15 3HJ
M6 GXS J. Allen 29 Wood Cottage Lane, Folkestone CT19 4QG
MW6 GXU S. Gordon 8 Maesteg, Cymau, Wrexham LL11 5EP
M6 GXV M. Butler Longuenesse, Dodwell Lane, Southampton SO31 1AD
M6 GXW A. Ashby 3 The Caravan, Heather Bank, Maryport CA15 6PB
M6 GXZ M. George 26 Yew Tree Road, Ormskirk L39 1NU
M6 GYA A. Bairstow 12 Danesfield Avenue, Waltham, Grimsby DN37 0QE
MW6 GYB L. Brown 13 Station Road Loughor, Swansea SA46TR
M6 GYC G. Codrai 27 Howard Avenue West Wittering, Chichester PO20 8EX
M6 GYE D. Voak 63 Green Lane, Crawley RH10 8JX
M6 GYF S. Robottom-Scott 73 St. Bernards Road, Solihull B92 7DF
M6 GYG G. Christison 9 Victoria Avenue, Market Harborough LE16 7BQ
M6 GYH M. Thompson 305 Highters Heath Lane, Birmingham B14 4NX
M6 GYI A. Hofstedt 16 Montana Road, London SW17 8SN
M6 GYK S. Woodfield 1 Kingsley Court, Church Road, Birmingham B25 8XS
MM6 GYL M. Glasper 1 Lindertis Cottages, Kirriemuir DD8 5NT
M6 GYM J. Martin 20 Hall Green Road, West Bromwich B71 3LA
MM6 GYP A. Dobie 201 High Street, Dalbeattie DG54DW
M6 GYQ K. Holloway Flat 3, Selwyn Court, Castle Street, Eccleshall, Stafford ST21 6DG
MM6 GYR G. Brass 114 Torbrex Rd Cumbernauld, Glasgow G67 2JS
M6 GYS I. Cook 93 Cathedral View, Houghton le Spring DH4 4HN
M6 GYU D. Perry Manor Garth, Wesley Road, Robin Hoods Bay, Whitby YO22 4RW
M6 GYV D. Harris Gatehouse 19, Skitfield Road, Dereham NR20 5QN
M6 GYY T. Keep Coombe Cottage, Coombe Lane, Cradley, Malvern WR13 5JF
M6 GYZ D. Murray 19 Andrew Allan Road, Rockwell Green, Wellington TA21 9DY
M6 GZA G. Watson 20 Windermere Drive, West Auckland, Bishop Auckland DL149LF
M6 GZB R. Appleby 20 Mcconnell Close, Aston Fields, Bromsgrove B60 3SD
MW6 GZC C. Ezard 59 Station Farm, Croesyceiliog, Cwmbran NP44 2JW
M6 GZE R. Adams 40 Lichfield Road, Gloucester GL4 3AL
MI6 GZF V. Kinney 49 Lanntara, Ballymena BT42 3BE
M6 GZG J. Causer 2 Kidd Croft, Tipton DY4 0AF
M6 GZI B. Jones 37 Sedgefield Close, Wirral CH46 9RW
M6 GZJ N. Ngan 6 Wynton Grove, Walton-on-Thames KT12 1LW
M6 GZK N. Heywood 38 Thurne Rise, Martham, Great Yarmouth NR29 4PU
M6 GZL C. Snowden 2 Haddon Road, Lowton WA3 2JQ
M6 GZN M. Armstrong 5 Aireside, Cononley, Keighley BD20 8LT
M6 GZO R. Bruce 19 Spindle Beams, Rochford SS4 1EH
M6 GZR G. Foster 248 Harbour Lane, Milnrow, Rochdale OL16 4EL
MM6 GZS G. Sinclair 33 Keptie Road, Arbroath DD11 3EF
M6 GZT T. Marshall 63A Newport Road, Ventnor PO38 1BD
M6 GZU C. Waters 45 Elmdale Road, Bedminster, Bristol BS3 3JF
M6 GZW C. Hughes 41 Rotherham Road, Dinnington, Sheffield S25 3RG
MW6 GZX L. Kurdi 8 Gwel Afon, Penparcau, Aberystwyth SY23 3PL
M6 GZY P. Garrett April Cottage, Castle Avenue, Blandford Forum DT11 0RY
MW6 GZZ G. Williams Flat 137, Rosser, Aberystwyth SY23 3LH
M6 HAC P. Selby 24 Juniper Close, Guildford GU1 1PA
MI6 HAD A. Mcmillen 55 Northwood Road, Belfast BT15 3QS
M6 HAE A. Watts 175 Ber Street, Norwich NR1 3HB
MI6 HAF A. Mckinley 36 Grangewood Drive, Londonderry BT47 5WN
M6 HAG S. Haigh 17 Glebe Street, Swadlincote DE11 9BW
MM6 HAH H. Halley 1 Grant Crescent, Renton, Dumbarton G82 4NH
M6 HAK R. Singer 19 Rosalind Avenue, Bebington, Wirral CH63 5JR
M6 HAM D. Trotter 48 Swindon Road, Sunderland SR3 4EE
MW6 HAR K. Harbour Glaslwyn, Cwmoernant, Carmarthen SA31 1EG
M6 HAS H. Searle 2 Tukes Avenue, Gosport PO13 0SE
M6 HAT Darwen ARC c/o R. Hatton 18 Tangier Road, Guildford GU1 2DF
M6 HAU A. Coghlan 1 Bell View Cross Houses, Shrewsbury SY5 6JR
MM6 HAV H. Venries 24 Lady Place, Livingston EH54 6TB
M6 HAY H. Harding 1 Saddleton Grove, Saddleton Road, Whitstable CT5 4LY
M6 HBB H. Buckley 10 Lower Hey Lane, Mossley, Ashton-under-Lyne OL5 9DE
M6 HBC J. Moore 118 Heneage Road, Grimsby DN32 9JQ
M6 HBD K. Mills 72 Sycamore Road, Ecclesfield, Sheffield S35 9YW
MM6 HBF J. Hawkins 113 Meadowpark Avenue, Bathgate EH48 2ST
M6 HBG H. Hutasuhut Hawkridge, Warden Road, London NW5 4SA
M6 HBH B. Wilson 38 Cotleigh Drive, Sheffield S12 4HU
MI6 HBI B. Huddleson 4 Knightsbridge Court, Bangor BT19 6SD
MW6 HBK A. Curry 58 Greenfields, St. Martins, Oswestry SY11 3AH
M6 HBN R. Tongs 9 Woodland Drive, Winterslow, Salisbury SP5 1SZ
M6 HBP H. Pascall 60 Weyland Road, Witnesham, Ipswich IP6 9ET
M6 HBQ I. Newton 16 Cross Close, Newquay TR7 3LB
MM6 HBR H. Ling Leadburnlea, Leadburn, West Linton EH46 7BE
M6 HBS J. Hobbs 82 Perry'S Lane, Wroughton, Swindon SN4 9AP
M6 HBT S. Hopkins 3B Tolfa House Wellington Terrace, Truro TR1 3JA
M6 HBU J. Mcdonald 222 Bristol Avenue, Farington, Leyland PR25 4QZ
M6 HBV J. Haynes 16 Mountsfield, Frome BA11 5AR
MM6 HBY J. Church 4 Weir Place, Perth, Ph1 3Gp, Perth PH1 3GP
M6 HCA C. Edwards Apartment, Maiori, Downs Lane, West Looe, Looe PL13 2HX
M6 HCB D. Mchugh 38 Quarryhill Road, Wath. upon Dearne S63 7TD
M6 HCD D. Humphries 100 Sunnyside Avenue, Stoke-on-Trent ST6 6EB
M6 HCE S. Ellis 9 Deanwood Close, Whiston, Prescot L35 3UX
M6 HCI L. Call 9 Hyperion Avenue, Polegate BN26 5HT
M6 HCJ S. Shone 22 Fenwick Road, Great Sutton, Ellesmere Port CH66 4UF
MM6 HCK C. Northcott 4/6 Castleview House, 2 Craigour Place, Edinburgh EH17 7RT
M6 HCO M. Moore 46 Scholes Park Road, Scarborough YO12 6QY
MI6 HCP S. Mawhinney 14 Beech Park, Portadown, Craigavon BT63 5ES
M6 HCR H. Rogers 2 Broom Close, Waterlooville PO7 8DP
MW6 HCS R. Hicks 14 Carn Celyn Beddau, Pontypridd CF38 2TF
M6 HCT J. Poulter 71H, Highstreet, Canvey Island SS8 7RD
M6 HCU P. Lockwood 80 Falmouth Road Le5 4Wh, Leicester LE5 4WH
M6 HCV D. Poulter 1 Deacon Drive, Laindon, Basildon SS15 5FY
M6 HCW C. Haynes 4 Thorn Close, Rugby CV21 1JN
MW6 HCY R. Porcher 9 Blenheim Close, Oswestry SY11 2UN
M6 HCZ C. Young 21A Union Crescent, Margate CT9 1NS
M6 HDA D. Harley 39 Tweedale Crescent, Madeley, Telford TF7 4EA
MW6 HDB H. Bancroft Stop And Call, Goodwick SA64 0EX
M6 HDD M. Hurst 48 Radcot Close, Woodley, Reading RG5 3BG
M6 HDE D. Edmondson 21 Hawthorne Close, Heathfield TN21 8HP
M6 HDG E. Majoch 66 Boughton Green Road, Northampton NN2 7SP
MI6 HDH G. Plunkett 18 Carnhill Place, Carrickfergus BT38 7RL
M6 HDI S. Ruddy 27 Grove Park Walk, Harrogate HG1 4BP
M6 HDK H. Karpuk Clarence Drive, Harrogate HG1 2QG
M6 HDM D. Money 2 Lodge Farm Cottage, Black Horse Road, Norwich NR10 5DJ
M6 HDO I. Johnson 35 Church Parade, Canvey Island SS8 9RQ
MW6 HDP M. Waldman 9 David Street, Cwmbwrla, Swansea SA5 8NX
M6 HDS D. Stapleton 179 Woodcock Road, Norwich NR3 3TQ
MW6 HDT D. Jenkins 12 Llys Yr Onnen, Dafen, Llanelli SA14 8PP
M6 HDU Dr S. Chipperfield 3 Clayton Avenue, Upminster RM14 2EZ
MW6 HDV J. Hodson 6 Heol Pentwyn, Tonyrefail, Porth CF39 8DF
M6 HDW D. Wilson Flat 6, Saxon House, Draymans Way, Alton GU34 1AY
M6 HDY D. Hardy Flat 10, Bridport House, Hillwood Road, Birmingham B31 1DN
MM6 HDZ H. De Zeeuw Eastfield Farm, Ballater AB35 5SH
MW6 HEA T. Heath 16 Beacons Park, Brecon LD3 9BR
M6 HEB S. Drakeley 1 Council Houses, Churchtown, Wadebridge PL27 7QA
MW6 HED M. Hedley Windberry Top Begelly, Kilgetty SA68 0XA
M6 HEE S. Marr 49 Gallows Hill, Ripon HG4 1RG
M6 HEF D. Robinson Height End Farm, Kirk Hill Road, Haslingden, Rossendale BB4 8TZ

M6 HEG P. Sellick 1, The Gallery, Northwick Park, Blockley GL56 9RJ
MW6 HEI R. Williams 88 Parc Pendre, Kidwelly SA17 4TE
M6 HEJ N. Lunden Flat 15 Western House, 8 Woodfield Place, London W9 2BJ
M6 HEK E. Wright 351 Market Street, Droylsden, Manchester M43 7EA
MM6 HEO L. Davis-Edmonds 6 Barlockhart Park, Glenluce, Newton Stewart DG8 0JQ
M6 HEP H. Purdy 13 St. Cuthbert Street, Worksop S80 2HN
MM6 HEQ R. Wilson 3 St. Peters Park, Stromness KW16 3EH
M6 HET I. Szabo 95B Westgate, Grantham NG31 6LE
M6 HEW W. Fearby 165 Brandsfarm Way, Telford TF3 2JJ
M6 HEX J. Ash 47 Stein Road, Emsworth PO10 8LB
MW6 HEY C. Hey 84 Trefelin, Aberdare CF44 8LF
MM6 HEZ W. Demczur 25 Maitland Court, Helensburgh G84 7EE
M6 HFA J. Wilson Flat 5, Blake House, London SE1 7DX
M6 HFF M. Collins The Haven, Kettleby Lane, Wrawby, Brigg DN20 8SW
M6 HFG H. Furniss 3 Byron Avenue, Chapeltown, Sheffield S35 1SQ
M6 HFH S. Harding Acres Hill, Jacobs Lane, Alhampton BA4 6PZ
M6 HFJ A. Cheung 197 St. Lukes Avenue, Ramsgate CT11 7HS
M6 HFK C. Boal Chindits, Wagg Lane, Probus, Truro TR2 4JX
M6 HFM H. Mcevoy 18 Brookfield Gardens, Wirral CH48 4EL
M6 HFN D. Kelly 6 Sandham Walk, Bolton BL3 6RA
M6 HFP J. Darley 159 Main Road, Hawkwell, Hockley SS5 4EL
M6 HFQ A. White 56 Seagrave Close, Oakwood, Derby DE21 2HZ
M6 HFR R. Heffer 36 Raven Avenue, Tibshelf, Alfreton DE55 5NR
M6 HFT University of Essex ARS (Ears). c/o D. Merridale The Granary, Falledge Lane, Upper Denby HD8 8YH
M6 HFU J. Russell Flat 1, Knapp Cottage, Roadwater, Watchet TA23 0QY
M6 HFV M. Johnson Charnwood, The Close, Ringwood BH24 2PE
M6 HFX T. Crawshaw 208 Ovenden Road, Halifax HX3 5QG
MW6 HFY N. James 8 Garth Lwyd, Caerphilly CF83 3QB
MM6 HFZ P. Mcnally 35 Crosshill Avenue, Glasgow G42 8BZ
M6 HGA L. Brookhouse 16 Clockmill Road, Walsall WS3 4AH
M6 HGD L. Brookhouse 16 Clockmill Road, Walsall WS3 4AH
M6 HGE J. Webber 14 Raleigh Street, Scarborough YO12 7JZ
M6 HGF A. Highfield 38 Brunswick Gardens, Garforth, Leeds LS25 1HF
M6 HGG S. Entwisle 30 Arden Mhor, Pinner HA5 2HR
M6 HGH H. Hamilton Flat B, 9 Cambridge Drive, London SE12 8AG
M6 HGJ M. Hadley 75 Glendower Avenue, Coventry CV5 8BD
M6 HGM B. Holgate 20 Milford Avenue, Bridlington YO16 7AU
M6 HGN J. Allen Flat 7, The Fairways, 35 The Esplanade, Knott End-On-Sea, Poulton-le-Fylde FY6 0AD
M6 HGO P. Niewiadomski Flat1 79A Dartmouth Road, London SE23 3HT
M6 HGR S. Yardley 22 Wedgwood Road, Clifton, Manchester M27 8RT
MI6 HGS M. Doogan 54 Birchdale Manor, Lurgan, Craigavon BT66 7SY
M6 HGU V. Keith Valentine Cottage, Frankton Road, Rugby CV23 9QT
MI6 HGV E. Rainey 22 Cherry Gardens, Ballymoney BT53 7AS
M6 HGW G. Whitton 40 Louville Avenue, Withernsea HU19 2PB
M6 HGX E. Keith Valentine Cottage, Frankton Road, Rugby CV23 9QT
M6 HGY A. Flintoft 71 Huntsman Lane, York YO41 1ET
MW6 HGZ R. Taylor 4 Castell Morgraig, Caerphilly CF83 3JH
M6 HHA G. Coldham 27 Welsby Road, Leyland PR25 1JA
MM6 HHB D. Burgess Quendale Farm, Quendale, Shetland ZE2 9JD
M6 HHC K. Townsend Flat 3, Drake House, Bexhill-on-Sea TN39 3TS
M6 HHF J. Sim 11 Haven Close Istead Rise, Gravesend DA13 9JR
M6 HHH J. Hoare Flat 3, 6 High Street, Watlington OX49 5PR
MM6 HHJ D. Jappy 21 Primrose Avenue, Grangemouth FK3 8YG
M6 HHL J. Wang 54 Mistle Thrush Drive, Cambridge CB24 1BS
M6 HHM H. Mccarthy 80 Vaughan Williams Way, Warley, Brentwood CM14 5WT
M6 HHO D. Przybylski 76 Turnberry, Skelmersdale WN8 8EQ
MM6 HHP R. Jenkins 163 Hurlford Road, Kilmarnock KA13QA
M6 HHQ B. Harvey Bennettshayes Barn, Awliscombe, Honiton EX14 3PY
M6 HHR R. Joyce 42 Long Close Station Road, Lower Stondon, Henlow SG16 6JS
MD6 HHT R. Jefferies The Old Police Station, Bay View Road, Port St. Mary IM9 5AW Isle of Man
M6 HHW C. Ogidih 89 West Road, Birmingham B43 5PG
M6 HHY C. Taylor Willow Court, Lincoln LN4 1AS
MM6 HIA A. Crichton 11 Baillie Court Sauchie, Alloa FK10 3FG
M6 HIB D. Bedworth 96 Balmoral Road, Stourbridge DY8 5JB
MM6 HIG H. Glennie 97 Smithfield Crescent, Blairgowrie PH10 6UE
M6 HIJ M. Willis 7 Belvawney Close, Chelmsford CM1 4YR
M6 HIK A. Gretton 67 Hawthorn Crescent, Arnold, Nottingham NG5 8BE
M6 HIL H. Penfold 15 Carmans Close, Loose, Maidstone ME15 0DR
M6 HIM Rssdale Rayt Gr c/o P. Green 15 Dickenson Road, Chesterfield S41 0RX
M6 HIP S. Admans 77 Beaumont Road, Birmingham B30 2EB
M6 HIT I. Petrie 88 Vicarage Road, Henley-on-Thames RG9 1JT
M6 HIU J. Leatherd 115 Rothesay Road, Blackburn BB1 2ER
MM6 HIZ D. Aitken 64 Brunton Street Cathcart, Glasgow G44 3NQ
M6 HJA K. Rouse 42 Berkley Close Highwoods, Colchester CO4 9RR
M6 HJB H. Barton 86 Forge Lane, Kingswinford DY6 0LG
M6 HJC H. Christie 15 Heilsburg Road, Canvey Island SS8 8HH
M6 HJD J. Hawley 11 Upper Green Way, Tingley, Wakefield WF3 1TA
M6 HJE M. Jessop 24 Woodcroft Avenue, Hull HU6 8LH
M6 HJF S. Rattley 2 Burnt Cottages Beanacre, Melksham SN12 7PT
M6 HJG A. Lawrence 6 Beaver Court, Ashford TN23 5QR
M6 HJH D. Money Flat 1, Lewin Court 24B Plumstead High Street Plumstead, London SE18 1SL
M6 HJI N. Crudgington Appledore Blackness Lane, Keston BR2 6HL
M6 HJK A. Scott 19 Estuary Drive, Felixstowe IP11 9TL
M6 HJM H. Mcneill 44 Anglesey Road, Wirral CH48 5EG
M6 HJQ G. Cairns Thomas 121 London Road, Bagshot GU19 5DH
M6 HJR D. Riman 22 Princess Road, Hinckley LE10 1EB
M6 HJT H. Hughes 27 The Holt, Hailsham BN27 3ND
M6 HJU C. Cairns Thomas 121 London Road, Bagshot GU19 5DH
M6 HJV J. Moore 65 Hamsey Green Gardens, Warlingham CR6 9RT
M6 HJX N. Fulcher 33 Water Lane, London SE14 5DN
M6 HJZ V. Lucock 34 Wentworth Drive, Ipswich IP8 3RX
M6 HKA J. Daniels 27 Hammerwater Drive, Warsop, Mansfield NG20 0DJ
M6 HKB D. Lock 20 Jasmine Close Trimley St. Martin, Felixstowe IP11 0UY
M6 HKC A. Chambers 5 Blandford Road, Shepton Mallet BA4 4FB
M6 HKI M. Davies 2 Ellins Terrace, Normanton WF6 1BL
M6 HKJ D. Killingley 17 Colbert Drive, Leicester LE3 2JB
MW6 HKL L. Chang 14 Greenfield Gardens, Pentrebach, Merthyr Tydfil CF48 4BQ
M6 HKN Dr D. Potts 90 Albemarle Road, Willesborough, Ashford TN24 0HN
M6 HKO S. Marden 65 Hedley Way, Hailsham BN27 3FZ
MI6 HKP S. Mccartney 35 Blackstaff Road Clough, Downpatrick BT30 8SR
M6 HKQ A. Lamont 89 Newlands, Whitfield, Dover CT16 3ND
M6 HKS D. Collier 7 Compass Close, Ashford TW15 1UT
M6 HKT B. Scott 3 Chaplin Close, Basildon SS15 4EJ
M6 HKU D. Holdbrook 9 Johns Terrace, Harold Park, Romford RM3 0AW
M6 HKV J. Robinson Old Laira Road, Plymouth PL3 6DH
M6 HKW H. Knowles 16 Farmcote Road, Coventry CV2 1SA
M6 HKX H. Khayer 24 Clyde Road, Stoke-on-Trent ST6 3DJ
M6 HKY A. Hickey 144 Gisburn Road, Barnoldswick BB18 5LQ
M6 HKZ J. Poole 61 Lower Vickers Street, Miles Platting, Manchester M40 7LX
M6 HLA A. Harvey 20 Fellowes Place, Plymouth PL1 5NB
MI6 HLC S. Savage 469 Old Belfast Road, Bangor BT19 1RQ
M6 HLE P. Bray 24 Eldon Terrace, Bristol BS3 4NZ
M6 HLF J. Taylor 41 Waters View, Yarwell Mill, Yarwell, Peterborough PE8 6EU
MW6 HLG H. Griffiths 5 Heol-Y-Sarn, Llantrisant CF72 8DA
M6 HLL Dr M. Beharrell 110 Scotforth Road, Lancaster LA1 4SQ
M6 HLP A. Wilson 39 Rochford Garden Way, Rochford SS4 1QH
MW6 HLQ L. Stevens 32 St Daniels Drive, Pembroke SA71 5QQ
M6 HLR L. Richardson 35 Vidgeon Avenue, Hoo, Rochester ME3 9DE
M6 HLS P. Hollis Flat 7, 20 Calais Hill, Leicester LE1 6FF
M6 HLT A. Tomkins 28 Newborough Close, Austrey, Atherstone CV9 3EX
MW6 HLU A. Garner 15 Midland Place, Llansamlet, Swansea SA7 9QU
M6 HLW C. Robinson 69 Sanger Avenue, Chessington KT9 1BY
MI6 HLY R. Lawrence 8 Wynford Park, Lisburn BT27 5HJ
MM6 HLZ H. Ross 16 Myreton Drive, Bannockburn, Stirling FK7 8PX
MM6 HMB H. Brown 7 Maple Avenue, Milton Of Campsie, Glasgow G66 8BB
MM6 HMC H. Campbell 10 Stewart Avenue, Linlithgow EH49 6DQ
M6 HME W. Mayall 17 Norrington Grove, Birmingham B31 5NY
M6 HMG M. Hughes 129 Jackson Road, Scunthorpe DN15 8JT
M6 HMK H. Melhuish 22 Mayflower Close, Glossop SK13 8UD
M6 HML K. Lees 24 Marks Road, Wokingham RG41 1NN
M6 HMM P. Bromley Broadwood Treovis Upton Cross, Liskeard PL14 5BQ
M6 HMQ J. Webb 49 Perth Avenue, Leicester LE3 6QQ
M6 HMR H. Pittard 21 The Oaklands, Church Eaton, Stafford ST20 0BA
M6 HMS M. Taylor 35 Valley Drive, Wilnecote, Tamworth B77 5FL
M6 HMU M. Gosi 49 Elms Drive, Marston, Oxford OX3 0NW
M6 HMV O. Wood 26 Parkfield Crescent, Kimpton, Hitchin SG4 8EQ
M6 HMW J. Walker 5 Preston Avenue, Alfreton DE55 7JX
M6 HMZ R. Dalziel 24 Horringford Road, Liverpool L19 3QX
M6 HNA M. Rosewell 54 Alder Drive, Chelmsford CM2 9EZ
M6 HNB H. Baird 35 St. Peters Road, Wolvercote, Oxford OX2 8AX
M6 HND A. Haylor 43 Church Road, Old Windsor SL4 2PH
M6 HNF C. Seymour 12 Silver Street Riccall, York YO19 6PB
M6 HNG N. Kerry 47 Harpley Dams Hillington, King's Lynn PE31 6DP
M6 HNH S. Kozlowski 28 Osney Crescent, Paignton TQ4 5EY
M6 HNI M. Lawson 131 Windermere Avenue, Ilkeston DE7 4EZ
M6 HNJ E. Potter 3 Thomson Court Chadwick Close, Crawley RH11 9LH
MW6 HNN D. Gibson 38 Newellhill, Tenby SA708EN
M6 HNO P. Byrne 18 St. Aidans Square, Bingley BD16 2BN
MM6 HNQ A. Macintyre 2 Memorial Square, Main Street, Castletown ?, Thurso KW14 8TU
M6 HNS H. Stanley 253 Brownley Road, Wythenshawe M22 9UX
M6 HNT M. Hunt 13 Pine Halt, Station Road, Cheltenham GL54 4JX
M6 HNV P. Sargeant 6 Meldon Way, Winlaton., Blaydon-on-Tyne. NE21 6HJ
MW6 HNW T. Fletcher 15A Les Maisonette Penryhn Road, Colwyn Bay LL29 8LG
M6 HNX A. Lorne 8 Campbell Close, Grantham NG31 8AW
M6 HNZ M. Bruce 28 Pheasants Way, Rickmansworth WD3 7ES
M6 HOF S. Baines 23 Gladstonbury Close, Belmont, Hereford HR2 7YL
M6 HOG G. Snape 3 Jasper Close, Barlaston, Stoke-on-Trent ST12 9BL
MW6 HOH P. Gostelow 6 Tree Field Caerau Farm, Llanidloes SY18 6LL
M6 HOI M. Simkins 11A Commercial Road, Southampton SO15 1GF
M6 HOK K. Hough 15 Moorside Road Endmoor, Kendal LA8 0EN
M6 HOM S. Mirjalili Mohanna Apartment 18B, White Croft Works, 69 Furnace Hill, Sheffield S3 7AH
MM6 HOO G. Freeburn 31 Courthill, Rosneath, Helensburgh G84 0RN
M6 HOP R. Hope 32 Winstanley Place, Rugeley WS15 2QB
M6 HOQ D. Hampson 29 Holywell Road Kilnhurst, Mexborough S64 5UQ
M6 HOS B. Mulholland 6 Burnside, Longhoughton, Alnwick NE66 3JQ
M6 HOT D. Hubbard 14 Parkfield Crescent, Kimpton, Hitchin SG4 8EQ
M6 HOU R. Houlton L11 (Ash 7) Woodlands Caravan Park, The Marshes Lane, Preston PR4 6JS
M6 HOV N. Fox Nik.Fox@Icloud.Com, Not Relevant PE100FG
M6 HOY J. Wilson 5 Queens Road Hoylake, Wirral CH47 2AG
MW6 HPC H. Cooper 3 Waunddu, Pontnewynydd, Pontypool NP4 6QZ
M6 HPD R. Hodson 99 Alcester Road, Hollywood, Birmingham B47 5NR
M6 HPF R. Stringer 9 Pershore Close, Walsall WS3 2UQ
M6 HPG H. Bell 7 Rosecomb Way, Haxby, York YO32 3ET
M6 HPH H. Heathfield Apartment 27, The Fitzgerald, 1 West Bar, Sheffield S3 8PQ
MW6 HPK R. Hopkins 132 Laurel Road Bassaleg, Newport NP10 8PT
M6 HPL C. Gorse Apartment 11 , 30 Stockton Road, Hartlepool TS25 1RY
MW6 HPN A. Hampson 30 The Crescent, Caerphilly CF83 2SW
M6 HPR I. Hooper 25 Honey Lane, Buntingford SG9 9BQ
M6 HPS H. Partridge 19 Dickens Drive, Melton Mowbray LE13 1HZ
M6 HPT S. Grubb 28 Bury Road, Newmarket CB8 7BT
M6 HPV C. Ward 325 Cedar Road, Nuneaton CV10 9DQ
M6 HPX W. Rainbow 69 Abbotsweld, Harlow CM18 6TG
M6 HQB J. Bristow 55 Haldon Close, Bristol BS3 5LR
MM6 HQC G. Fleming 169 Oldtown Road, Inverness, Scotland IV2 4QD
M6 HQD J. Hobbs 1 Hotground Cottage Branfield, Hertford SG14 2QG
MM6 HQE A. Mccormick 79 Kennedy Crescent, Tranent EH33 1DN
MM6 HQI W. Mcbain 56 Scotstoun Park, South Queensferry EH30 9PQ
MM6 HQK B. Mccabe 173 Marmion Road Cumbernauld, Glasgow G67 4AW
M6 HQL M. Dimmick 16 Bushell Way, Kirby Cross, Frinton-on-Sea CO13 0TW
M6 HQM R. Shulver 63 Hill Farm Way, Southwick, Brighton BN42 4YG

M6 HQN W. Hamlet 18 Bridle Lane, Alfreton DE55 1LG
M6 HQO R. Olive Lorien, The Ridge, Thatcham RG18 9HZ
M6 HQR P. Nolan 14 Woodlands Road, Stafford ST16 1QR
M6 HQW K. Kimura Cornerways, Tennyson Road, Yarmouth PO41 0PX
M6 HQZ H. Bamford 14 Calf Close, Haxby, York YO32 3NS
M6 HRC P. Humphreys 30 The Chestnuts, Hinstock, Market Drayton TF9 2SX
M6 HRD V. Greatwood 11 The Green, Long Preston, Long Preston, Skipton BD23 4PQ
M6 HRE S. Bligh-Wall 81 Warmingham Road, Leighton, Crewe CW1 4PS
MM6 HRF M. Reynolds 22 Fergus Place, Dyce, Aberdeen AB21 7DD
M6 HRG S. Probert 10 The Green, Church Lawton ST73ED
M6 HRI H. Hughes Cefn Glass, Clyro, Hereford HR3 5JT
M6 HRJ R. Huelin 15 Hill Chase, Walderslade, Chatham ME5 9HE
MW6 HRK G. Hughes 19 Gernant Braichmelyn Bethesda, Bangor LL57 3RE
M6 HRL H. Russell 4 Dearnsdale Close, Stafford ST16 1SD
M6 HRM H. Mcbrien Hamilton House, Hayes Lane, Wokingham RG41 4TA
M6 HRN B. Hurren 29 Chalk Lane, Ixworth, Bury St. Edmunds IP31 2JQ
M6 HRO J. Mills 24 Charles Street, Ryhill, Wakefield WF4 2BU
M6 HRQ S. Duncan 5 Spring Vale Bilton, Hull HU11 4DN
MW6 HRR H. Roberts 12 Rheidol Terrace Ceredigion, Aberystwyth SY23 1JU
M6 HRS H. Richardson 103 Marys Mead, Hazlemere, High Wycombe HP15 7DT
M6 HRT J. Krol 40 Hampton Gardens, Southend-on-Sea SS2 6RW
M6 HRV H. Harkishin 2 Kingfisher Close, Bournemouth BH6 5BB
M6 HRW H. Woolley 84 Bowthorpe Road, Norwich NR2 3TP
M6 HRX R. May 6 Gordon Court Well Street, Loose, Maidstone ME15 0QF
MM6 HRZ W. Mcewan 240 Turriff Brae, Glenrothes KY7 6UT
M6 HSA P. Sousa 11 Broom Crescent, Ipswich IP3 0EE
M6 HSB P. Henry 6 Greenwood, Bamber Bridge PR58JS
MM6 HSC H. Campbell 8B Hawthorn Place, Uphall, Broxburn EH52 5BX
M6 HSE H. Evans 26 Peartree Court, Welwyn Garden City AL7 3XN
M6 HSF F. Hayati Apartment 19, 29 Longleat Avenue, Birmingham B15 2DF
M6 HSH M. Williams 21 Elmbrook Close, Basildon SS14 2FH
M6 HSI J. Ostapiuk 33 Kent Terrace, Haswell DH62EL
MM6 HSK H. King 19 Gleneagles Way, Deans, Livingston EH54 8EW
MI6 HSL N. Davis 19 Toberhewny Hall, Lurgan BT66 8JZ
M6 HSP N. Hawkins Deganwy Hardwick Road, King's Lynn PE30 5BB
M6 HSQ R. Terry 31 Barnwood Road, Birmingham B32 2LY
M6 HSR H. Stewart-Roberts Cinderfield House, Cornwells Bank, Lewes BN8 4RH
MM6 HSS Prof. S. Skerratt 51 Vogrie Road, Gorebridge EH23 4HL
M6 HST M. Ghost 59 Kingswood Avenue, Park North, Swindon SN3 2RB
M6 HSX H. Sati 83 Edenfield Gardens, Worcester Park KT4 7DX
M6 HSY G. Kokinis 16 Wellesley Avenue, Beverley Road, Hull HU6 7LW
M6 HTB S. Crane 2 Wolsey Close, Ashton-In-Makerfield, Wigan WN4 8DL
M6 HTC A. Porter 35 St. Andrews Crescent, Hindley, Wigan WN2 3EQ
M6 HTF S. Griffiths 8 Alldicks Road, Hemel Hempstead HP3 9JJ
M6 HTG D. Quinney 8 Crabwood Road, Southampton SO16 9EZ
M6 HTI J. Trimmer 13 Cragside Place, Glenhaven 2156 Australia
M6 HTJ T. Hare 2 Primula Drive, Norwich NR4 7LZ
M6 HTK J. Ormston 15 Thackeray End, Aylesbury HP19 8JE
M6 HTM J. Clyne Ravenswood, Green Lane, Wisbech PE14 7BJ
M6 HTN A. Horton 51 Walsingham Gardens, Epsom KT19 0LS
MM6 HTS S. Spencer 55 Blackwell Court, Inverness IV2 7AR
MW6 HTT H. Thomas 2 Ffordd Donaldson, Copper Quarter, Swansea SA1 7FJ
M6 HTU T. Leatherbarrow 17 Egerton, Skelmersdale WN8 6AA
M6 HTW C. Brooks 12 Beeches Court, Thornton-Cleveleys FY5 4PZ
M6 HTX H. Banasiak 16 St Christophers Close, Bath BA2 6RG
M6 HTZ L. Shearson 23 Thumpers, Hemel Hempstead HP2 5SL
M6 HUC C. Lycett 2 Royce Avenue, Hucknall, Hucknall, Nottingham NG15 6FU
MM6 HUE J. Hughes 2 Forest Place, Townhill, Dunfermline KY12 0EP
M6 HUF Medway RS c/o S. Pennell 272 Coal Clough Lane, Burnley BB11 5BS
M6 HUH W. Lam Clarence Drive, Harrogate HG1 2QG
M6 HUI A. Windle 10 Longshaw Street, Blackburn BB2 4HS
M6 HUK A. Thomas 6 Coniston Avenue, Grimsby DN33 3EF
M6 HUL S. Lyon 10 Sycamore Close, Preston, Hull HU12 8TZ
M6 HUM D. Wood School Farm, Brock Road, Preston PR3 0XD
M6 HUN J. Collier 133 Woodstock Road, Moston, Manchester M40 0DG
M6 HUP M. Arif 171 Henley Road, Bedford MK40 4FZ
M6 HUQ R. Harwood 5 Cloakham Drive, Axminster EX13 5GT
M6 HUR J. Jefferies Millfield Cottage, Wild Meadow Bolnhurst Road, Colmworth, Bedford MK44 2LF
M6 HUS M. Hussain 10 Mercia Crescent, Stoke-on-Trent ST6 3JB
M6 HUV I. Barnes 35 Copley Road, Stanmore HA7 4PF
M6 HUW J. Mclean 24 Durham Drive, Oswaldtwistle BB5 3AT
M6 HUX M. Lisle 16 Collegiate Crescent, Sheffield S10 2BA
MW6 HUY D. Johnson 12 Bro Dedwydd Dunvant, Swansea SA2 7PR
M6 HUZ L. Hughes 32 Calder Road, Blackpool FY2 9TX
M6 HVA J. Wills 1 Aberdeen Road, Harrow HA3 7NF
M6 HVD L. Evans 8 Meadowbrook, Blackpool FY3 9UE
M6 HVE M. Simonsohn 5 Pitt Close, Blandford St. Mary, Blandford Forum DT11 9PS
M6 HVF S. Pearson 3 Berkeley Road, Shirley, Solihull B90 2HS
M6 HVH M. Rose 149 Claremont Road, Blackpool FY1 2QJ
M6 HVI A. Corbett 27 Larch Grove Blurton, Stoke-on-Trent ST32BN
M6 HVK K. Orchard 47 Trezaise Road Roche, St. Austell PL26 8HD
M6 HVL H. Ly 11 Nodes Drive, Stevenage SG2 8AL
M6 HVM P. Ashton 32 Sycamore Road, New Ollerton Nr Newark, Notts NG22 9PS
M6 HVO A. Jarvis 10 West Park, Wadebridge PL27 6AN
M6 HVP C. Harwood 5 Cloakham Drive, Axminster EX13 5GT
M6 HVR M. Murray 31 Feeny Street Sutton Manor, St. Helens WA9 4BJ
M6 HVS B. Smithers 4 Bidmead Court Kent Way, Surbiton KT6 7SX
M6 HVU T. Raymond 13 Riverside Wolsingham, Bishop Auckland DL13 3BP
M6 HVY E. Money 4 Cromes Place, Badersfield, Norwich NR10 5JT
M6 HWC H. Cheesman 49 Front Street, Chirton, North Shields NE29 7QN
M6 HWD J. Redgrave 24 Burnham Close Trimley St. Mary, Felixstowe IP11 0XG
M6 HWE A. Jepson Edmonton Road, Mansfield NG21 9AH
M6 HWH M. Milioto 7 Bennett Green, Colchester CO4 5ZR
M6 HWJ T. Armstrong 12 Mayhouse Road, Burgess Hill RH15 9RF
M6 HWL J. Dyson Fy8 3Tl, Lytham St. Annes FY8 3TL
M6 HWM D. Hatchman 14 Rudyard Close, Brighton BN2 6UA
M6 HWN J. Laws 47 Hampshire Place, Peterlee, SR8 2HE.
M6 HWO M. Wells 14, Werrington Grove, Peterborough PE46NT
M6 HWQ N. Williams 245 Central Drive, Bilston WV14 8JE
M6 HWT G. Scholey Ln9 6Jh, Horncastle LN9 6JH
MI6 HWV G. Osprey 24 The Hollies Carrickfergus Bt388Ha, Carrickfergus BT388HA
M6 HWW A. Graham 24 Primrose Bank, Wigton CA7 9JW
M6 HWX G. Mcritchie 21 St Marys Field, Colchester CO3 3BP
M6 HXA P. Barden 17 Chapel Fields, Charterhouse Road, Godalming GU7 2BS
M6 HXB D. Jones 91 Laburnum Road, High Wycombe HP12 3LP
M6 HXC H. Chawdhry Trinity College, Cambridge CB2 1TQ
M6 HXD M. Shelley 21 Ripley Close New Addington, Croydon CR0 0RP
M6 HXE A. Moss Winstons, Mayfield Lane Durgates, Wadhurst TN5 6DG
M6 HXF J. Phillips 10 Turners Avenue, Fleet GU51 1DX
M6 HXG P. Stokes 26 Ashford Road, Hastings TN34 2HA
M6 HXI R. Bowen 4 Crossley Gardens, Halifax HX1 5PU
M6 HXK R. Calvert 2 Coneyburrow Road, Tunbridge Wells TN2 3NA
M6 HXN R. Hammond 28 Birch Way, Hastings TN34 2JZ
M6 HXO G. Dyson 4 Davenport Avenue, Blackpool FY2 9EP
M6 HXR D. Goodchild Gravel Lane, Ringwood BH24 1XY
M6 HXT K. Ede 7 Corner Garth Ferring, Worthing BN12 5EL
M6 HXU A. Loader Rowan Tree House, Crowfield, Brackley NN13 5TW
M6 HXV H. Redington Galeholm, Whitecroft, Gosforth, Seascale CA20 1AY
M6 HXW B. Fitchett 1 Hilly Fields Mews, Parsonage Estate, Rogate, Petersfield GU31 5BF
M6 HXX A. Sayer 23 Toronto Street Lincoln, Lincoln LN2 5NN
MM6 HXY G. Towell 265 Stirling Street, Denny FK6 6QJ
M6 HYB G. Marfell Springfields Bungalow, Drybrrok GL17 9BW
M6 HYF R. Hubbard Southbroom School House, Estcourt Street, Devizes SN10 1LW
M6 HYH P. Thomas 5 Mallard Close Chipping Sodbury, Bristol BS37 6JA
M6 HYJ H. Jeram 6 Lavender Lane, Rowledge, Farnham GU10 4AX
M6 HYM A. Mears Waterways. Scotland Yard, Priors Leaze Lane, Hambrook, Chichester PO18 8RQ
M6 HYQ G. Oliver Bernard Road, Cromer NR27 9AW
M6 HYR D. Spencer 17 Rightup Lane, Wymondham NR18 9NB
MW6 HYS C. Lowes 3 Castle Close Creigiau, Cardiff CF15 9NJ
M6 HYT P. Carr 9 Hollydene Villas, Hythe, Southampton SO45 4HU
M6 HYW H. Wong Clarence Drive, Harrogate HG1 2QG
M6 HYX S. Short 16 Melrose Drive Fletton, Peterborough PE2 9DN
M6 HYY S. Greenaway 166 Marlborough Road, Swindon SN3 1LU
M6 HYZ K. Evennett 13 Elm Street, Dereham NR20 3FN
MM6 HZA S. Gibb 102 Whitehill Avenue, Musselburgh EH21 6PE
M6 HZD H. Donovan 2 All Saints, Weeting IP270QH
M6 HZF J. Franklin 31 Haining Gardens, Mytchett GU16 6BJ
M6 HZJ L. Gynn-Burton 1 Broad Street Long Eaton, Nottingham NG10 1JH
M6 HZL H. Smith 12 Lockgate East, Windmill Hill, Runcorn WA7 6LB
M6 HZM D. Walker 44 Albany Road Kilnhurst, Mexborough S64 5UG
MI6 HZN D. Henderson 16 Elliott Place, Enniskillen BT74 7HQ
MM6 HZO A. Artamonovs Tern Brae 15, Livingstone EH546UQ
MW6 HZP T. Tilbrook 17 Jubilee Close, Letterston, Haverfordwest SA62 5SW
M6 HZQ J. Platt 12 Tawny Grove Four Marks, Alton GU34 5DU
M6 HZT S. Rickard 66 Sinclair Drive, Basingstoke RG21 6AD
M6 HZU J. Howarth 19 Farnham Croft, Leeds LS14 2HR
MM6 HZW P. Majumdar 12 Richmond Avenue Clarkston, Glasgow G76 7JL
M6 HZX B. Bailey 215 Stretton Avenue, Coventry CV3 3HQ
M6 HZZ S. Conway 21 Milcote Avenue, Hove BN3 7EJ
MM6 IAB J. Joyce 7 Leitch Street, Greenock PA15 2HJ
M6 IAC Z. Cole 31 High Street, Kimpton, Hitchin SG4 8RA
M6 IAF S. Shambhu 34 Gascoigns Way, Patchway, Bristol BS34 5BY
MW6 IAG J. Richards 210A Pandy Road, Bedwas, Caerphilly CF83 8EP
MM6 IAI I. Brown 28 Garden Road, Cults, Aberdeen AB15 9RE
M6 IAJ I. Jones 21 Kennet Green, Worcester WR5 1JQ
M6 IAL I. Lambert 69 Anvil Crescent, Broadstone BH18 9DZ
M6 IAN I. Shires 19 Prince Charles Avenue, Sittingbourne ME10 4NA
M6 IAO I. Phillips 124 Brookwood Drive, Stoke-on-Trent ST3 6LP
M6 IAQ A. Instone 63 Larch Road, New Ollerton, Newark NG22 9SX
MM6 IAR A. Rees Mains Of Atherb, Maud, Peterhead AB42 4RD
M6 IAS I. Scott 21 Field Avenue, Shepshed, Loughborough LE12 9SH
M6 IAT D. James 53 Whittingham Road, Ilfracombe EX34 9LL
M6 IAV I. Avery 4 Southampton Drive, Liverpool L19 2HE
MM6 IAW I. Ferguson 5 Grahamsfield Court Kirkpatrick Fleming, Lockerbie DG7 3BD
M6 IAX I. Lawton 11 Goosewell Terrace, Plymstock, Plymouth PL9 9HW
MM6 IAY A. Maclennan 5 Baluachrach, Culbokie, Dingwall IV7 8FP
M6 IAZ D. Goldthorpe 6 West End Grove Haydock, St. Helens WA11 0AP
MM6 IBB I. Bainbridge 2 Courthill Road Cottage, Arbroath DD11 4UX
M6 IBC I. Barber 8 Newlands Close, Lowestoft NR33 7EY
MW6 IBD I. Daniel 35 New Road, Upper Brynamman, Ammanford SA18 1AF
M6 IBF I. Graham 49 Eagle Close, Leighton Buzzard LU7 4AT
M6 IBG I. Bruno-Gaston 83 Althorne Gardens, London E18 2DB
M6 IBH C. Tate 87 Overdale Road, Middlesbrough TS3 7NQ
M6 IBI A. Douglas 3 Beech Avenue, Bilsborrow, Preston PR3 0RH
M6 IBJ S. Moorby 11 Vespasian Gardens Rooksdown, Basingstoke RG24 9SH
M6 IBL Dr G. Eibl-Kaye 1 Main Road, Littleton, Winchester SO22 6PS
M6 IBO A. Lawrence 5 Heighton Close, Bexhill-on-Sea TN39 3UP
M6 IBP G. Stamp 17 Harlow Manor Park, Harrogate HG2 0EG
MM6 IBQ Dr J. Simpson 18 Corrour Road, Glasgow G43 2DX
MI6 IBR B. Rocks 4 Millview, Randalstown, Antrim BT41 3BA
M6 ICA C. Irons 11 Elm Grove, Moira, Swadlincote DE12 6HH
M6 ICB I. Bushnell 4 Upper Tail, Watford WD19 5DF
MI6 ICD I. Cairns 18 Molyneaux Avenue, Larne BT40 2TU
M6 ICH M. Pullen 2 Wyatts Lane, Little Cornard, Sudbury CO10 0NT
M6 ICJ A. Lorentsen 54 Kingsnorth Road, Gillingham ME8 6QY
M6 ICK M. Ginty 34 High Street, Branston, Lincoln LN4 1NB
M6 ICL M. Dominguez 52 St. Leonards Road, Amersham HP6 6DR
MW6 ICM A. Moody Perthiteg, Cwmhiraeth, Llandysul SA44 5XJ
M6 ICO J. Stevenson 18 Drakehouse Laneá, Sheffield S20 1FW
M6 ICP I. Pears 10 Pixley Dell, Consett DH8 7DB
M6 ICQ D. Banks 41 East Road, Rotherham S65 2UX
M6 ICR S. Barlow Apartment 34, Jet Centro, Sheffield S2 4AH
MW6 ICU A. Jones 8 Arles Road, Cardiff CF5 5AP

M6 ICX D. Lucock 34 Wentworth Drive, Ipswich Borough Council IP8 3RX
M6 ICZ O. Barber Homedale, St. Monicas Road Kingswood, Tadworth KT20 6ET
M6 IDB M. Parker 1 Ham Road, Wanborough, Swindon SN4 0DF
M6 IDC I. Cosham 54 Hawkins Crescent, Shoreham-by-Sea BN43 6TP
M6 IDD P. Wilcox 2 Merryhill Terrace, Belmont, Hereford HR2 9RT
M6 IDE A. Currie 31 Launceston Road, Bodmin SO50 6AY
M6 IDF I. Firth 124 Viking Road, Bridlington YO16 6TB
M6 IDG I. Garrard 33 Uplands Road, Hockley SS5 4DL
MI6 IDJ K. Mclaverty 123A Castle Road, Antrim BT41 4NG
M6 IDK I. King 7, Greenacres Avenue, Blythe Bridge, Stoke on Trent ST11 9HU
M6 IDM E. Ogbua 11 Coltness Crescent, London SE2 0UY
M6 IDN I. Norfolk Arwelfa, High Street, Uckfield TN22 3LP
M6 IDO Y. Wang 50 Sleaford Street, Cambridge CB1 2PU
M6 IDP I. Nelson 38 Warbro Road, Torquay TQ1 3PW
M6 IDR I. Reeve 36 Stone Pippin Orchard, Badsey, Evesham WR11 7AA
M6 IDS A. Pulman 3 Brunner Road, Billingham TS23 1HW
M6 IDU M. Brannon Leonard Cheshire Close, Salisbury SP47RN
M6 IDX J. Ledger 39Eascroft Drive, Sheffield S20 8JG
M6 IEA E. Aspden 4 Lilac Close, Newcastle ST5 7DH
M6 IEB I. Beales 6 Edge Well Rise, Sheffield S6 1FB
M6 IEC C. Williams Flat 21, Holly House Holmes Street, Burnley BB11 3BE
M6 IEI J. Lowe 26 Rectory Drive Yatton, Bristol BS49 4HF
M6 IEJ L. Thornton 11 Seabourne Ave Blackpool Fy4 1Eh, Blackpool FY4 1EH
M6 IEM J. Paine 11 Ferndale Park Fifield Road, Bray, Maidenhead SL6 2DZ
M6 IEO L. Metcalfe 40 St. Anns Court, Hartlepool TS24 7HY
M6 IEQ J. Skitt 38 Bodiam Avenue Tuffley, Gloucester GL4 0TJ
M6 IER A. Unwin 152 Epsom Road, Guildford GU1 2RP
M6 IEU R. Husher 194 Leybourne Ave, Bournemouth BH10 5NR
M6 IEW I. Williams 23 Symons Close, Blackwater, Truro TR4 8ER
M6 IEZ T. O'Gorman Spencer Avenue, Birkenhead CH42 2DN
M6 IFA J. Arnold 17 Larch Road Roby, Liverpool L369TY
M6 IFB M. Glen 6 Willow View, Catterick, Richmond DL10 7PD
M6 IFC M. Harvey May Tree Cottage, Kelvedon Road, Tiptree, Colchester CO5 0LJ
MW6 IFE L. Pickering 14 Bryn Wyndham Terrace, Cardiff CF42 5NG
M6 IFF J. Sharrad 52 Springwood Drive, Ashford TN23 3LQ
M6 IFG A. Kreissl 382 Oldfield Road, Altrincham WA14 4QT
M6 IFH I. Harrison 8 Jeffrey Avenue, Longridge, Preston PR3 3TH
M6 IFI I. Iremonger 2 Harbord Road, Cromer NR27 0BP
M6 IFN T. Sargeant 27 Digby Close Tilton On The Hill, Leicester LE7 9LL
M6 IFO D. Harris 35 Itchenor Road, Hayling Island PO11 9SN
M6 IFQ M. Murray Victor Court, Rainham RM13 8EL
M6 IFS G. Milner 9 Lilydene Avenue Grimethorpe, Barnsley S72 7AA
M6 IFT G. Saunders 140 Highbridge Road, Burnham-on-Sea TA8 1LW
M6 IFV P. Kearney 22 Kingston Drive, Cheltenham GL51 0UB
M6 IFW I. Warman 8 Burley Road, Bishop's Stortford CM23 3LR
M6 IFX S. Wade 2 Ealing Square, Beacon Lane, Cramlington NE23 8JB
M6 IFY M. Gratton The Static, Brandywine, Walton Rd., Wisbech PE14 7AF
M6 IFZ I. Huggett 104 Hinchcliffe Orton Goldhay, Peterborough PE2 5SS
M6 IGA E. O'Neill 13 Goodwood Close, Market Harborough LE168JF
MW6 IGC I. Curnock 62 Heol Y Banc, Bancffosfelen, Llanelli SA15 5DL
M6 IGH I. Garforth 63 Upper Perry Hill, Bristol BS3 1NJ
M6 IGJ I. Jackson 22 Greenwood Avenue, Congleton CW12 3HH
M6 IGK I. Wideman Silcoates Lane, Wrenthorpe, Wakefield WF2 0PD
M6 IGM G. Benford 34 Victoria Gardens, Colchester CO4 9YD
M6 IGP Dr J. Harmer 1 Wynford Rise, Leeds LS16 6HX
M6 IGQ S. Nelson 2 Boulsworth Road, North Shields NE29 9EN
M6 IGR D. Drummond 21 Beveland Road, Canvey Island SS8 7QU
M6 IGS N. Hawkins 17 Meddins Lane, Kinver, Stourbridge DY7 6BZ
M6 IGT A. Davies 6 Stonefield Close Shrivenham, Swindon SN6 8DY
MI6 IGV R. Freeburn 24Highfern Gardens, Belfast BT133RD
M6 IGW A. Wilson 38 Cotleigh Drive, Sheffield S12 4HU
M6 IGZ A. Hogg 17 Kedleston Close, Stretton, Burton-on-Trent DE13 0FN
M6 IHA C. Haworth Salisbury, Station Road Keyingham, Hull HU12 9SZ
M6 IHC I. Clement 24 Millais, Horsham RH13 6BS
M6 IHD J. Booth 27 Moorlands Scholes, Holmfirth HD9 1SW
M6 IHE K. Harcombe 29 Jurston Fields, Wellington TA21 9FX
M6 IHH I. Hutchinson 41 Eastbourne Avenue, Gosport PO12 4NU
M6 IHM A. Walker 39 Delves Road Killamarsh, Sheffield S21 1AW
M6 IHN N. Gooding 41 The Crescent, Wolverhampton WV6 8LA
M6 IHO B. Domigan 66 Nunts Lane, Coventry CV6 4HA
MM6 IHQ T. Rogers Marypark Farm, Marypark, Ballindalloch AB37 9BG
M6 IHR C. Witham 162 Stockbridge Lane, Liverpool L36 8EH
M6 IHS R. Hunter Poplars, March Road Guyhirn, Wisbech PE13 4DA
M6 IHT G. Lamming 8 Meadow Lane Newport, Brough HU15 2QN
M6 IHZ E. Pardoe 2 Bar Meadow, Shobdon HR6 9BZ
M6 IIB D. Faulkner 21 Chestnut Way, Bromyard HR7 4LG
M6 IIF D. Kerby-Collins Wr158Qn, Worcester WR158QN
M6 IIG R. Blair Springfield, Pewsey Road, Rushall, Rushall, Pewsey SN9 6EN
M6 IIJ I. Bailey 33 Govett Road, St. Helens WA9 5NQ
M6 IIL L. Furr 158 Eastern Avenue, Southend-on-Sea SS2 4AZ
MM6 IIO R. Treble 2 Baidland Meadow, Dalry KA24 5HP
MM6 IIP P. Wilson 2 Southerhouse Scousburgh, Dunrossness, Shetland ZE2 9JE
M6 IIQ P. Baillie 23 Brackens Drive Warley, Brentwood CM145UE
MW6 IIU R. Selby 7 West Hook Road Hook, Haverfordwest SA62 4LS
M6 IIW J. Kelly 25 Blunts Avenue Sipson, West Drayton UB7 0DR
M6 IIX P. Hill 83 Gladeside, Croydon CR0 7RW
M6 IIY A. Wood 85 Love Lane, Rayleigh SS6 7DX
M6 IIZ A. Billett 1 Tortoiseshell Drive, Attleborough NR17 1GU
M6 IJB I. Boddy 5 Boverton Avenue, Brockworth, Gloucester GL3 4ER
MW6 IJD N. Dimonaco 41 Brongwinau, Comins Coch, Aberystwyth SY23 3BQ
M6 IJG D. Mowbray 5 Heath Lane Leasingham, Sleaford NG34 8JF
M6 IJH I. Holdford 46 Hildreth Road, Prestwood, Great Missenden HP16 0LY
MW6 IJJ I. Johns Flat 16, Baker Street House James Street, Blaenavon, Pontypool NP4 9EH
M6 IJK C. Birch 5 Newport, Barton-upon-Humber DN18 5QJ
M6 IJL B. Damazer The Manse East Street, Crowland PE6 0EN
M6 IJM I. Morgan 30 Farm Road, Hutton, Weston-Super-Mare BS24 9RH
M6 IJO S. Deere 12 Leverton Road, Retford DN229HE
MM6 IJP I. Purkis Lochaber Croft, Tullynessle, Alford AB33 8QQ
M6 IJQ R. Bedford 29 Kent Road, Brookenby, Market Rasen LN8 6EW
M6 IJR M. Elkins 18 Scutts Close Lytchett Matravers, Poole BH16 6HB
M6 IJT C. Partington 55 Caldercroft, Elland HX5 9AY
M6 IJV K. Thwaites 65 Chessington Park Hill, Chessington KT9 2BJ
MW6 IJW P. Williams 52 Rhoslan, Tredegar NP22 4PF
M6 IJX D. Wilderspin 14 Tannery Court, North Street, Crewkerne TA18 7AY
M6 IJZ S. Cross 22 Park Avenue Washingborough, Lincoln LN4 1DB
M6 IKA I. Abraham 12 Graham Road, Halesowen B62 8LJ
M6 IKD M. Draper 160 Chanctonbury Road, Burgess Hill RH15 9HA
M6 IKF M. Blyth 9 High Grove, Ryton NE40 3JN
M6 IKH J. Jarvis The Wheel House, Ellerton Hill Ellerton Upon Swale, Richmond DL10 6AL
M6 IKI M. Anostalgia 136 Avenue Road Extension, Leicester LE2 3EH
M6 IKJ C. Smith 2 Burley Gardens, Street BA16 0SN
M6 IKL M. Mccallister 13 Darvel Avenue Ashton-In-Makerfield, Wigan WN4 0UA
M6 IKM M. Moffat 7 Lingfell Avenue, Cockermouth CA13 9BE
M6 IKN T. Caldwell 29 The Vale, Coventry CV3 1DW
MD6 IKR D. Cain Flat, Ballavagher, Main Road, Isle of Man IM4 4AR
MM6 IKS I. Stobo 25 Greenfield Circle, Elgin IV30 5NF
MM6 IKT Dr K. Brooks 192/6 Causewayside, Edinburgh EH9 1PN
M6 IKX D. Mielczarek 7 Bryant Close, Camberley GU16 8AD
M6 IKY M. Rawson Flat 9, West Court, West Street, Oxford OX2 0NP
M6 ILB R. Brocklehurst 12 Harriers Close, Christchurch BH23 4SL
M6 ILC C. Ng 45 Langdown Road, Southampton SO45 6EX
MI6 ILF I. Forsythe 45 Kensington Park, Portadown, Craigavon BT63 5PQ
M6 ILH I. Hobbs 115 Adams Way, Croydon CR06XR
M6 ILI N. Winfield 1 Southview, School Lane, Stoke Row, Henley-on-Thames RG9 5QX
M6 ILJ W. Fisher 15A Trevenner Lane, Marazion TR17 0BL
MW6 ILK T. Rathbone 13 North Avenue, Rhyl LL18 1HT
M6 ILM D. Sharpe 31 Malyons, Basildon SS13 1PJ
M6 ILN D. Bee 25 Blatcher Close, Minster On Sea, Sheerness ME12 3PG
M6 ILO M. Noblet 1 Lingdale Road, Wirral CH48 5DG
M6 ILP I. Patterson 63 Orchard Road, South Ockendon RM15 6HP
M6 ILR V. Seabright 208 Park Way, Rubery, Birmingham B45 9WA
M6 ILS J. Anthony 21 Belgrave Street, Denton, Manchester M34 3WP
M6 ILT P. Mccormick Fieldview, Crown East Lane, Lower Broadheath, Worcester WR2 6RH
M6 ILX M. Bennett 33 Charles Street, Redditch B97 5AA
M6 ILY D. Stuart 5 Woodlands Park, Shrewsbury SY26JN
MJ6 ILZ M. Thorpe Dolphin Cottage, Union Road, Grouville, Jersey JE3 9ER
M6 IMA I. Maley 6 Meadow View Close, Newport TF10 7NN
MM6 IMB M. Breimann 5 Leaside, Mossbank, Shetland ZE2 9TF
MM6 IMF I. Fairbairn 2 The Steadings, Slackend, Buckie AB56 5BS
M6 IMH I. Hickinbottom Clover Cottage, Snead, Montgomery SY15 6EB
M6 IMI A. Rayner 5 Buttermere Close, Lincoln LN6 0YD
MW6 IMM C. Milne 18 Chapelfield, Deganwy LL31 9BF
MM6 IMP C. Mcconnachie Flat C, 5 Whitefaulds Crescent, Maybole KA19 8AY
MI6 IMQ J. Crowe 36 A Ballynahatty Road, Belfast BT8 8LE
M6 IMR I. Rich 39 Wren Close, Heathfield TN21 8HG
M6 IMS M. Sims 5 Sandy Leaze, Bradford-on-Avon BA15 1LX
MW6 IMU M. Mcdonald Falcondale, Pleasant Valley, Stepaside, Wisemans Bridge, Narberth SA67 8NT
MI6 IMV A. Brashaw 3 The Straits, Lisbane, Comber, Belfast BT23 6AQ
M6 IMW I. Walker 24 Hawthorn Road, Norwich NR5 0LP
MI6 IMY P. Calvin 49 Gobhan Close Portadown, Craigavon BT63 5QZ
M6 IMZ I. Ross 48 Henry Drive, Leigh-on-Sea SS9 3QF
M6 INA G. Allen 13 Strathmore Avenue, Hull HU6 7HJ
MI6 INB M. Gibson 12 Drumcarn Gardens Portadown, Craigavon BT62 4DH
MM6 INC A. Mcmath 57 Hillhouse Avenue, Bathgate EH48 4BB
M6 IND P. Ind 30 Thompson Road, Stroud GL5 1SY
M6 ING D. Singh 28 Chadview Court, Chadwell Heath Lane, Romford RM6 4BF
M6 INI M. Smith Church Farm, Mucklestone, Market Drayton TF9 4DN
M6 INM M. Davies 91 Ameysford Road, Ferndown BH22 9QD
MM6 INN A. Macinnes 1 Brent Place, Glenrothes KY7 6TA
M6 INP J. Walton Walton Holt, Marsh Road Orby, Skegness PE24 5HZ
MM6 INS C. Ralph 37 Seaview Terrace, Edinburgh EH15 2HE
M6 INT M. Pomfret 5 Malvern Crescent, Ince, Wigan WN3 4QA
M6 INV M. Basford 4 Renoir Close, St. Ives PE27 3HF
M6 INW D. Inwood 2B Abbots Way, Northampton NN5 5DB
M6 INX J. Porter Flat 2, 14 Trafalgar Square, Scarborough YO12 7PY
MW6 IOA I. Dumitrescu The Rectory, Halkyn, Holywell CH8 8BU
M6 IOD S. Mowbray 5 Heath Lane Leasingham, Sleaford NG34 8JF
M6 IOG F. Cooper Needhams Farm House, Spittal Hill Road, Boston PE22 0PA
M6 IOI L. Emanuel 20 Wychwood Drive, Redditch B97 5NW
M6 IOL M. Scott 19 Saltburn Road, Sunderland SR3 4DJ
M6 ION L. Rimington 3 Amicombe, Wilnecote, Tamworth B77 4JJ
M6 IOO J. Berrio 29 Scalborough Close, Countesthorpe, Leicester LE8 5XH
M6 IOQ G. Rowland 52 Victoria Road, Mablethorpe LN12 2AJ
M6 IOR S. Engelmann 83C Downs Road, London E5 8DS
M6 IOS G. Baker 56 Chalklands, Bourne End SL8 5TJ
M6 IOW F. Alfrey 16 Walls Road, Bembridge PO35 5RA
M6 IOY R. Melbourne Barton Marina, Barton-under-Needwood DE13 8DZ
MW6 IOZ M. Simons 68 Harbour Village, Goodwick SA64 0DZ
M6 IPA R. Howitt 7 Badgers Close, Chelmsford CM2 8QB
MI6 IPB D. Neill 8 Castle Meadows Carrowdore, Newtownards BT22 2TZ
M6 IPE R. Fulcher 1 Edwards Close Hutton, Brentwood CM13 1BU
M6 IPF A. Allen 2 Robson Way, Blackpool FY3 7PP
M6 IPG N. Head-Jenner 5 Ponders Road, Colchester CO63LX
M6 IPH J. Dunn 39 Bramble Lane, Wye TN25 5AB
M6 IPI A. Cook 188 Vicarage Drive, Kendal LA9 5BS
M6 IPJ P. Allen 45 Under Knoll, Peasedown St. John, Bath BA2 8TY
M6 IPL I. Pilton Caleril Barn, Pool Foot Farm Haverthwaite, Ulverston LA12 8AA
MM6 IPP Lord I. Patterson 8 Jane Street, Dunoon PA23 7HX
M6 IPQ T. Baker 92 Conway Avenue, Derby DE723GR
M6 IPW A. Lee 1 Lower Stoke, Limpley Stoke, Bath BA2 7FU
M6 IPY P. Sayer 443 Sutton Road, Southend-on-Sea SS2 5PJ

M6 IQA P. Patmore 85 Frenchs Wells, Woking GU21 3AU
M6 IQC R. Clark 67 Seymour Street, Chorley PR6 0RR
M6 IQE D. Hopwood 64 Ridge Road, Stoke on Trent, Staffordshire ST6 5LP
MW6 IQF D. Jackson 10 Hafod Wen Johnstown, Wrexham LL14 2AT
M6 IQG J. Wooldridge 6 Heskin Road, Lydiate, Liverpool L31 0BS
M6 IQK C. Price 270A Chorley New Road, Horwich BL6 5NY
M6 IQL N. Green 44 Rushyford Drive, Chilton, Ferryhill DL17 0EQ
M6 IQM A. Sharman 8 Knowle Road, Biddulph, Stoke on Trent ST8 6LH
M6 IQN J. Holford 30 Meadow Avenue, Newcastle ST5 9AE
M6 IQO P. Taylor 32 Heliers Road, Liverpool L13 4DH
M6 IQT K. Carey 28 Honeysuckle Avenue Hailsham Bn274Fp , Hellingly, Hailsham BN27 4FP
M6 IQU A. Pym 58 Eastern Avenue, Chippenham SN15 3LW
M6 IQW D. Parker 51 Parsonage Road, Henfield BN5 9HZ
M6 IQX D. Vaughan 1 Boulnois Avenue, Poole BH14 9NX
M6 IQY H. Fisher 19 Mccabe Close, Bell Lane, Staplehurst, Tonbridge TN12 0BW
M6 IQZ D. Cassidy 30 Oakland Road, Botley, Southampton SO32 2SX
M6 IRB R. Bewick 357 Franklands Village, Haywards Heath RH16 3RP
M6 IRC I. Crowson 19 Burgoyne Road, Southsea PO5 2JJ
MI6 IRE Anglo Americn Rd c/o P. Mcaleer 24 Wansbeck Street, Belfast BT9 5FQ
M6 IRJ M. Impey 20 Chalton Road, Luton LU4 9ER
M6 IRK O. Kirton Woodland, Moretonhampstead, Newton Abbot TQ13 8SD
M6 IRL D. Lovejoy 9 West View Close, Middlezoy, Bridgwater TA7 0NP
M6 IRM M. Raine 91 Lulworth Avenue, Jarrow NE32 3SB
M6 IRP I. Pipe 9 Sherlock Hoy Close, Broseley, Telford TF125JB
M6 IRU S. Airey 22 Primrose Street, Lancaster LA1 3BN
M6 IRW I. Wilkes 7 Oriel Close, Dudley DY1 2UW
M6 IRX G. Harman 58 Laurence Avenue, Witham CM8 1JB
M6 IRZ W. Drozdz 29 Breedon Road, Birmingham B30 2HT
M6 ISA I. Stone 169 Booth Road, Wednesbury WS10 0EW
M6 ISB S. Brown 18 Goring Ave. Gorton., Manchester M18 8WW
M6 ISD S. Dodd 21 Sunningdale, Grantham NG31 9PF
M6 ISE M. Petre 52 Fremantle Road, Ilford IG6 2AZ
M6 ISG T. Holland 68 Church Street, Billericay CM11 2TS
M6 ISH R. Kinder 21 Oakdene, Chobham, Woking GU24 8PS
M6 ISJ S. Jones 5 Meadowlands, Kirton, Ipswich IP10 0PP
M6 ISK N. Butler 75 Rutland Street, Derby DE23 8PR
M6 ISN A. Oakey 53 Appledore Close, Margate CT9 3RG
M6 ISQ L. Boylan 30 Pembrey Way, Liverpool L259SN
M6 ISR I. Robinson 16, Stobarts Field New Ridley, Stocksfield NE437RL
M6 ISU M. Dunstan 5 Polgooth Close, Redruth TR15 1QL
M6 ISW J. Atkinson Magistratsvögen 55 H, Lgh 1220, Lund 226 44 Sweden
M6 ISX J. Rogerson 8 Bonser Crescent Huthwaite, Sutton-in-Ashfield NG17 2RE
MM6 ISY I. Phillips 16 Seton Court, Port Seton, Prestonpans EH32 0TU
M6 ISZ I. Townsend 263D Stapleton Road, Bristol BS5 0PQ
M6 ITD C. Knowles 10 The Drive, Southwick, Brighton BN42 4RR
M6 ITI C. Johnson Suite 204, 33 Queen Street, Wolverhampton WV13AP
M6 ITL J. White 4 Kings Road, New Milton BH25 5AY
M6 ITN R. Goodall 76 Beaconfield Road, Plymouth PL2 3LF
M6 ITP M. Baynes 92 Belgrave Drive, Hull HU4 6DW
M6 ITQ J. Swales 90 Earlswood Road, Dorridge, Solihull B93 8RN
M6 ITU P. James Flat 2, Marigold House, 2 Ironbridge Road, Twigworth, Gloucester GL2 9GS
M6 ITV I. Roberts 15 Broadcroft, Hemel Hempstead HP2 5YX
M6 ITX M. Reynolds 24 Burton Close, Corringham, Stanford-le-Hope SS17 7SB
M6 ITY A. Praveen 28 Long Deacon Road, London E4 6EG
M6 ITZ C. Mayer 81 Bohelland Road, Penryn TR10 8DY
MM6 IUE A. Young 2 Dunvegan Place, Ellon AB41 9TF
M6 IUF L. Barrett 2 Barton Rise Feniton, Honiton EX14 3HW
MD6 IUH K. Payne 2 Dreeym Balley Cubbon, Ballacubbon, Isle of Man IM9 4PR
M6 IUI D. Hendy 12 Rumsam Gardens, Barnstaple EX32 9EY
M6 IUK D. Grayson 79 Errington Avenue, Sheffield S2 2EA
M6 IUM R. Lincoln 7 John Carrs Terrace, Bristol BS8 1DW
MW6 IUN I. Jones 21 Albert Street, Maesteg CF34 0UF
MW6 IUP D. Flewin Huntingdon Way, Swansea SA2 9HN
MM6 IUR B. Findlay 39 Barrhill Court Kirkintilloch, Glasgow G66 3PL
M6 IVB C. Payne 11 The Paddocks Great Chart, Ashford TN23 3BE
M6 IVC I. Stephens 14 Vardon Close Kingston Hill, Stafford ST16 3YW
M6 IVE C. Cavalcante Pinheiro Filho Flat 7, Hollybush House, Hollybush Gardens, London E2 9QT
M6 IVG N. Simmons 28 Odo Road, Dover CT17 0DP
M6 IVI J. Knights 18 Kenilworth Gardens, Blackpool FY4 1JJ
MW6 IVK P. John 136 Birchgrove Road Birchgrove, Swansea SA7 9JT
M6 IVO I. Yovchev 11 Beverley Drive, Edgware HA8 5NQ
MM6 IVP S. Morrison 24 High Street, Fochabers IV32 7DX
M6 IVQ R. Cutting 19 Laburnum Close, Wisbech PE13 3RJ
M6 IVR I. Goodman 26 Vallansgate, Stevenage SG2 8PY
M6 IVT C. Drake 115, Woodlands Road, Gillingham ME72DX
MW6 IVW D. Morgan 28 Harbour Village, Goodwick SA64 0DY
M6 IWA I. Wright 1 Greaves Avenue, Old Dalby, Melton Mowbray LE14 3QE
M6 IWB I. Bunting 14 Mill Pightle, Aylsham, Norwich NR11 6LX
M6 IWD A. Sirrell 4 White Oaks North Pickenham, Swaffham PE37 8LB
M6 IWE G. Cousins 47 Nightingale Drive Taverham, Norwich NR8 6LA
M6 IWF I. Francis 34 Furlong Road, Bourne End SL85AA
M6 IWI D. Gilbertson 6 Lewens Lane, Wimborne BH21 1LE
MM6 IWL C. Somerville 112 Dickens Avenue, Glasgow G813EP
M6 IWO C. Hargrave 20 Gainsborough Road, Ashley Heath, Ringwood BH24 2HY
M6 IWP I. Perkins 49 Salters Lane, Tamworth B79 8BH
M6 IWX A. James 36 Cedar Close, Walton on the Naze CO14 8NJ
MW6 IXD J. Harrington 23 Heol Emrys, Fishguard SA65 9EE
M6 IXF M. Cox 17A Church End, Weston Colville, Cambridge CB21 5PE
M6 IXG P. Gretton 159 Birchfield Road Arnold, Nottingham NG5 8BP
MM6 IXH R. Stewart 31 Seatown, Lossiemouth IV 31 6 JJ
M6 IXI R. Forss Flat 1, 18 Park Lane, Bath BA1 2XH
M6 IXJ R. Reynolds 3 Canna Park Drive, Highampton, Beaworthy EX21 5AY
M6 IXM Dr D. Buckley 7 Clary Meadow, Northwich CW8 4XG
MM6 IXO K. Goggins Farm Road, Glasgow G81 6HH
M6 IXR D. Brotheridge 14 Nuthatch Drive, Torquay TQ2 7GF
M6 IXY A. Champkin 17 Blundell Place, Bedford MK42 9XB
MW6 IXZ C. Purviss 45 Heol Camlas Gwersyllt, Wrexham LL11 4HF
MM6 IYA D. Kean 24 Morar Drive, Clydebank G81 2YB
M6 IYC J. Mcintyre 10 Fern Drive Dudley, Cramlington NE23 7AF
M6 IYG G. Osborne Dunelm Cottage, High Street Castle Camps, Cambridge CB21 4SN
M6 IYJ M. Futcher 46 Houldey Road, Birmingham B313HJ
M6 IYL G. Bryant 15 The Clock Inn Park Lydeway., Devizes. SN10 3PP
M6 IYM P. Hutchinson 75 Windermere Avenue Orford, Warrington WA2 0NB
M6 IYN A. Bailey 52 Berkeley Road, Shirley, Solihull B90 2HT
M6 IYS E. Crafter 90 Connaught Avenueá, Grays RM162XT
M6 IYT M. Robino 61 Mill Hill Little Hulton, Manchester M38 9TN
M6 IYU S. Helm 10 St. Annes Avenue, Middlewich CW10 0AE
M6 IYV K. Smith London Road, Grays RM20 4AA
MW6 IYW P. Beech 9 Wesley Court, Pembroke Dock SA72 6NE
M6 IZA I. Brunt 62 Greenwood Drive, Watford WD25 0HX
M6 IZB A. Bowes 178 Saxton Road, Abingdon OX14 5HF
M6 IZE A. Hampson 12 Oakhays, South Molton EX36 4DB
M6 IZF C. Scouller 72 Brookmans Avenue Brookmans Park, Hatfield AL9 7QQ
MM6 IZH A. Sommerville Smeaton Farm, Dalkeith EH22 2NL
MD6 IZI I. Dorman 1 Sprucewood Rise, Foxdale, Douglas IM4 3JP Isle of Man
MM6 IZK I. Low 124 Sheephousehill, Fauldhouse, Bathgate EH47 9EL
M6 IZM J. Mccaffery 56 Churchill Avenue Bulford, Salisbury SP4 9HE
M6 IZN A. Mccaffery 57 St.Leonards Road, Newton Abbot TQ12 1JY
M6 IZO D. Drake 16 Princess Street, Blackpool FY1 5BZ
M6 IZP A. Jakstas 36 Boxridge Avenue, Purley CR8 3AQ
M6 IZQ W. Townsend 8 Beech Avenue Kirkham, Preston PR4 2UE
M6 IZT M. Scott 51 Devona Avenue, Blackpool FY4 4NU
M6 IZU B. Griggs 61 Langdale Road, Blackpool FY4 4RR
M6 IZW I. Whiteley 2 The Meade, Manchester M21 8FA
M6 IZY E. Coffey 133A Long Street, Atherstone CV9 1AD
M6 JAB J. Burrows 78A Coronation Road, Earl Shilton, Leicester LE9 7HJ
MI6 JAD S. Mullan 135 Ballyavelin Road, Limavady BT49 0QB
MM6 JAE J. Dalgety 5 East Court, Edinburgh EH16 4ED
M6 JAF D. Oliphant 29 Gannet Drive, Amble NE65 0FR
M6 JAG R. Price 92 Lincoln Road, Ingham, Lincoln LN1 2XF
M6 JAJ J. Smith 46 Mulberry Close, Goldthorpe, Rotherham S63 9LB
M6 JAK J. King 18 Ross Road, Wallington SM6 8QB
M6 JAL M. Leggett 7 Barley Way, Thetford IP24 1LG
M6 JAM J. Hunter 164 Grange Road, Newark NG24 4PP
MM6 JAN J. Mcclure Bridgefoot Croft, Udny, Ellon AB41 6RT
M6 JAQ J. Snell 32 Meadow Halt, Ogwell, Newton Abbot TQ12 6FA
M6 JAR J. Revell 63 Mountbatten Road, Bungay NR35 1PP
M6 JAS J. Wells 18 Roewood Road, Holbury, Southampton SO45 2JH
M6 JAU J. Goldsmith 20 Trinity Mews, Bury St. Edmunds IP33 3AT
M6 JAV M. Sharman 33 Bungalow Estate, Lady Lane, Coventry CV6 6BD
MM6 JAW J. White 63 Allershaw Tower, Wishaw ML2 0LP
M6 JAZ J. Cleeter 49 Hunters Field, Stanford In The Vale, Faringdon SN7 8LZ
M6 JBA J. Ashley Rowborough Farm Cottages, Brading, Sandown PO36 0BA
M6 JBB J. Berry 31 New Hall Way, Flockton, Wakefield WF4 4AX
M6 JBE J. Nicholson Flat 1 Kingsfield Court Orchard Vale, Kingswood Bristol BS159PB
M6 JBG J. Bethell 19 Kiln Cottages, The Brickfields, Stowmarket IP14 1RY
M6 JBI J. Innes Trelawny, Marine Drive, Bude EX23 0AH
M6 JBL B. Jedryka 31 Backhold Lane, Halifax HX3 9DR
M6 JBM J. Russell 81 Chapman Street, Loughborough LE11 1DD
MM6 JBN J. Breen 26 Maxwelton Road, Glasgow G33 1LR
M6 JBO J. Davies 66 Barnstaple Road, North Shields NE29 8QG
MW6 JBQ G. Smith 49 Coed Cae, Caerphilly CF83 1RU
M6 JBR J. Whitton 11 Dursley Road, Bristol BS11 9XB
M6 JBT J. Hawkins 24 Green Lane, Stourbridge DY9 7EW
M6 JBU J. Burnett Hobart House, 16 Church Lane, Reepham, Lincoln LN3 4DQ
M6 JBV M. Redmore 8 Hambleden Rise, Northampton NN4 8TT
M6 JBX J. Blower 8 The Laundry, Seifton, Ludlow SY8 2DH
MI6 JBZ J. Given 15 Middle Road, Lisburn BT27 6UU
M6 JCA J. Aubury 27 Gravel Walk, Tewkesbury GL20 5NH
M6 JCB C. Daniels 19 Gerrard Street, Rochdale OL11 2EB
M6 JCC J. Clare Kimberley, Boston Road, Boston PE20 3AP
M6 JCE J. Crewe 22 Myrtle Tree Crescent Sandbay, Weston-Super-Mare BS22 9UL
M6 JCF J. Faulkner Mount Pleasant, Elkstones, Buxton SK17 0LU
M6 JCG J. Anderson The Firs, Crapstone, Yelverton PL20 7PJ
MW6 JCH J. Hudson 50 Dan Y Cwarre, Kidwelli SA17 4JA
M6 JCJ J. Ward 6 Fairfield, Telegraph Hill, Redruth TR16 5AH
MM6 JCL J. Littlefair 30 Maple Crescent, Cambuslang, Glasgow G72 7NN
M6 JCN J. Barma 28 Briarfield Road, Timperley, Altrincham WA15 7DB
M6 JCP J. Preece 42 Henderson Road, Widnes WA8 7LR
M6 JCQ J. Stevens 51 Cheddington Road, Pitstone, Leighton Buzzard LU7 9AQ
M6 JCR J. Reynolds 15 Chestnut Mead, Oxford Road, Redhill RH1 1DR
M6 JCS J. Shettler 504 Leeds Road, Huddersfield HD2 1YW
M6 JCT J. Anderson 139 Cromwell Road, Rushden NN10 0EG
MM6 JCU J. Coubrough 41 Bridge Court, Alexandria G83 0BZ
MW6 JCV J. Baldwin 94 King Street Abertridwr, Caerphilly CF83 4BG
M6 JCW J. Wright 19 Halstead Close, Woodley, Reading RG5 4LD
M6 JCX J. Coxon 48 High Street, Great Easton, Market Harborough LE16 8ST
MM6 JCZ J. Mackenzie 40 Deanswood Park, Deans, Livingston EH54 8NX
M6 JDA J. Dale Corydon, Church Street, Sevenoaks TN14 7SW
M6 JDC J. Collier Benallack Farm, Grampound Road, Truro TR2 4BY
M6 JDD D. Riley 14 Mond Road, Widnes WA8 7NB
M6 JDF J. Firth 7 Manor Avenue, Derby DE23 6EB
M6 JDL J. Harding 181 South Coast Road, Peacehaven BN10 8NS
M6 JDM J. Briggs 47 Greenland Avenue, Derby DE22 4AQ
M6 JDQ J. Hole 17 Cromford Street, Sheffield S2 4BP
M6 JDU J. Clarke 160 Hall Lane Estate, Willington, Crook DL15 0PP
MW6 JDY J. Evans 311 Delffordd, Rhos, Swansea SA8 3ER
MM6 JEA J. Addison Lambhill Bungalow, St. Katherines, Inverurie AB51 8TS
M6 JED P. Carberry 203 Fitzwilliam Road, Rotherham S65 1NB
M6 JEF J. Marland 34 Barnfield, Stoke-on-Trent ST4 5JE
M6 JEG D. Stephens 37 Pimpernel Close, Southampton SO316TN
M6 JEH J. Harris 45 Sleigh Road, Sturry, Canterbury CT2 0HT

M6 JEJ J. Brennan 42 Maryland Road, Thornton Heath CR7 8DF
M6 JEK J. Durey 7 Staplers Close, Great Totham, Maldon CM9 8UN
MW6 JEL J. Christensen 29 Caerphilly Close, Rhiwderin, Newport NP10 8RF
M6 JEM J. Shaw 2A Priory Avenue, Petts Wood, Orpington BR5 1JF
M6 JEO J. Saunders 123 Medway Road, Ferndown BH22 8UR
M6 JEP D. Jepson 104 Norris Street, Warrington WA2 7RW
M6 JEQ J. Mcdermott 2 Denton Terrace, Castleford WF10 4LN
M6 JER J. Harley 170 Windsor Road, Hull HU5 4HH
M6 JET T. Edwards 17 The Green, Woodbastwick, Norwich NR13 6HH
M6 JEV J. Gourley 19 Walter Nash Road East, Kidderminster DY11 7EA
M6 JEY J. Johnson 49 Beach Priory Gardens, Southport PR8 2SA
M6 JEZ J. Mitchell 11 Brookside Drive, Oadby, Leicester LE2 4PB
M6 JFA J. Worsley 3 Sheephouse Road, Hemel Hempstead HP3 9LW
MM6 JFB J. Copland Janefield, Colvend, Dalbeattie DG5 4QN
M6 JFE D. Foley 1 Hill Rise Close, Harrogate HG2 0DQ
M6 JFF J. Carpenter 20 Barton Grove, Kedington, Haverhill CB9 7PT
M6 JFH J. Feltham 12 Penrith Way, Eastbourne BN23 8NS
MW6 JFI J. Power 27 Seaview Crescent, Goodwick SA64 0AZ
M6 JFJ J. Spratley 10 The Willows, Jarrow NE32 4QN
M6 JFK J. Knight 30 Ash Meadow Lea, Preston PR2 1RX
MW6 JFL J. Lewis 9 Llwyn Bedw, Cefn Pennar, Mountain Ash CF45 4DZ
M6 JFN S. Butler 41 Angus Close, Chessington KT9 2BN
MI6 JFO W. Forde 35 Torr Gardens, Larne BT40 2JH
M6 JFP J. Poole 82 Merchants Way, Canterbury CT2 8PN
M6 JFR A. Bass 7 Woodlands, Horbury, Wakefield WF4 5HH
M6 JFS R. Jones 32 Remington Drive, Sheffield S5 9AH
MM6 JFU J. Frew 57 Ford Avenue, Dreghorn, Irvine KA11 4BN
M6 JFV J. Hardingham 11 All Saints Close, Weybourne, Holt NR25 7HH
M6 JFW J. Leonard 51 Molyneux Drive, Bodicote, Banbury OX15 4AX
M6 JFY Dr C. Holmes Old Vicarage Farmhouse, Course Lane, Wigan WN8 7LA
MW6 JGC J. Sollis 15 Llanwonno Road, Mountain Ash CF45 3NB
M6 JGE J. Glover 12 Willow Street, London E4 7EG
M6 JGG J. Goodwin 6 Worrall Street, Congleton CW12 1DT
M6 JGH R. Hart Jays, South Street, Kington Magna, Gillingham SP8 5ET
M6 JGI J. Griffiths 83 Golborne Road Ashton-In-Makerfield, Wigan WN4 8XA
MI6 JGK G. Gardiner 60 Limestone Meadows, Moira, Craigavon BT67 0UT
M6 JGM J. Gould-Martin 203 Maple Crescent, Leigh WN7 5SW
M6 JGN J. Elsey 6 Lewis Avenue, Walthamstow, London E17 5BL
M6 JGP J. Paradi 168 Castle Road, Northolt UB5 4SG
M6 JGQ J. Chapman Boundary Farm, Garlic Street, Diss IP21 4RL
M6 JGR J. White 62 Dalebrook Road, Burton-on-Trent DE15 0AD
M6 JGS J. Seaton 88 Stanley Road, Cambridge CB5 8LB
M6 JGT J. Jeffryes 15 Shirley Road, St. Albans AL1 5ES
M6 JGU J. Borrett 84 Kingsway, Mapplewell, Barnsley S75 6EX
M6 JGV J. Chown 40 Kemps Green Road, Balsall Common, Coventry CV7 7QF
M6 JGY J. Diez 174 Humber Avenue, Coventry CV1 2AR
M6 JGZ J. Gilbert 101 Eastbrook Road, Lincoln LN6 7EW
M6 JHB J. Burdett 5 Winston Drive, Wainscott, Rochester ME2 4LJ
M6 JHC M. Clark 10 Priory Close, Sporle, King's Lynn PE32 2DU
MI6 JHE J. Evans 404 Foreglen Road, Dungiven, Londonderry BT47 4PN
M6 JHF P. Attwater 42 Danescourt Crescent, Sutton SM1 3EA
MM6 JHH J. Hutchinson 169 Glen Avenue, Largs KA30 8QQ
M6 JHI A. Randall 7 Blenheim Place Buckshaw Village Chorley Pr7 7Lt, Leyland PR7 7LT
MW6 JHJ J. Cook 40 Cemaes Crescent, Rumney, Cardiff CF3 1TA
M6 JHK J. Hele Kergozou De La Boessiere Lilac Cottage, The Street, Cheddar BS27 3TH
M6 JHL J. Lumm 25 Knowsley Way, Hildenborough, Tonbridge TN11 9LG
M6 JHN J. Curwen Corner Cottage, Tailors Green, Stowmarket IP14 4LL
M6 JHO D. Carey 7 Pelman House Pelman Way, Epsom KT19 8HH
M6 JHP J. Palmer 14 Linnet Close, Luton LU4 0XJ
M6 JHQ J. Rimmer 33 New Cut Lane, Southport PR8 3DW
M6 JHR J. Ross Foundry Cottage, Crowders Lane, Battle TN33 9LP
M6 JHS J. Harris 36 Northmoor Way, Wareham BH20 4SJ
M6 JHT J. Hill 45 Venus Street, Congresbury, Bristol BS49 5HA
M6 JHU B. Jones 64 Church Lane, Barwell, Leicester LE9 8DG
M6 JHV J. Hancox 10 Dunham Close, Newton-On-Trent, Lincoln LN1 2LH
M6 JHZ J. Hazeltine 21 Hassock Way, Wimblington, March PE15 0PJ
M6 JIC R. Price 17 Treleven Road, Bude EX23 8SA
M6 JID J. Darwin 14 Croftwood Grove, Whiston, Prescot L35 3UT
MI6 JIE J. Evans 404 Foreglen Road, Dungiven, Londonderry BT47 4PN
M6 JIF J. Miazek 2 Oak Grove, Armthorpe, Doncaster DN3 2DJ
M6 JIG C. Ember 35 Mattock Lane, Ealing, London W5 5BH
M6 JIH Rvd. J. Horsley 33 Amalfi Tower, Sunderland SR3 3AN
M6 JII J. Donovan 18 Ellesmere Street, Eccles, Manchester M30 0JN
M6 JIK B. Ferry 1 New Front Street, Anfield Plain., Stanley DH9 8JG
M6 JIN J. Ramachandran 1 Keplerlaan, Estec, Tec-Sws, Noordwijk 2201AZ Netherlands
M6 JIQ J. Ritson 38 Hawkshead Road, Burtonwood, Warrington WA5 4PW
M6 JIR J. Barrett 9 Hook Road, Goole DN14 5JB
M6 JIS M. Furby 14 Larch Avenue, Wickersley, Rotherham S66 2PQ
M6 JIT J. Topping 10 St. Pauls Road, Blackpool FY1 2NY
M6 JIW J. Barker Pearl Bungalow, Killerby Cliff, Cayton Bay, Scarborough YO11 3NR
M6 JIX J. Packer 20 Shipman Road, Market Weighton, Market Weighton YO43 3RB
M6 JIY D. Baker 22 Cleveland Road, Plymouth PL4 9DF
M6 JIZ J. Watson 82 Glendale Avenue, Washington NE37 2JS
MM6 JJB J. Barclay 24 Wellcroft Road, Hamilton ML3 9SG
M6 JJD J. Smyth 5 Lime Close, Lakenheath, Brandon IP27 9AJ
M6 JJE S. Kelly 7 Cedar Grove, Greetland, Halifax HX4 8HT
M6 JJF J. Fradley 9 Hagley Park Gardens, Rugeley WS15 2GY
M6 JJG J. Greenhalgh 7 Swynford Close, Kempsford GL7 4HN
M6 JJH J. Holbrook 24 Birks Holt Drive, Maltby, Rotherham S66 7JZ
M6 JJI J. Roberts 27 Pike Purse Lane, Richmond DL10 4PS
M6 JJK Letchworth Garden City ARC c/o C. Webb 16 Bexfield Road, Foulsham, Dereham NR20 5SB
M6 JJL J. Jackson 23 Hawkley Drive, Tadley RG26 3YH
M6 JJM J. Marriott 55 Ellis Avenue, Stevenage SG1 3SL
M6 JJN J. Nicholls 6 Marasca End, Holt Drive, Colchester CO2 0DL
MW6 JJO J. Owens 16 Blanche Street, Dowlais, Merthyr Tydfil CF48 3PE
MM6 JJQ J. Mccrae 50 Cuilmuir View, Croy, Glasgow G65 9HQ
M6 JJS J. Stewart 29 Cornwall Close, Leamore, Walsall WS3 2AR
M6 JJU K. Weston 10 Hollow Hill Road Ditchingham, Bungay NR35 2QZ
M6 JJV S. Virgo 41 Lynch Road, Berkeley GL13 9TE
MM6 JJW J. Wallace 323 High Street, Dalbeattie DG5 4DX
M6 JJX J. Adams 18 Vanguard Court, Sleaford NG34 7WL
MI6 JJZ J. Clark Apartment 16, Pipers Field, 16B Comber Road, Dundonald BT16 2AB
M6 JKA A. Kirby 61 The Chase, Harlow CM17 9JA
M6 JKB S. Lucas 31 Lilian Close, Norwich NR6 6RZ
M6 JKD J. Davies 135 Silvercourt Gardens, Brownhills, Walsall WS8 6EZ
M6 JKH J. Horry 20 Churchgate, Sutterton, Boston PE20 2NS
M6 JKM J. England 15 Woodward Avenue Radford Gardens, Hereford HR2 7FH
M6 JKQ R. Poulson 9 Scattergate Green, Appleby-in-Westmorland CA16 6SP
M6 JKR J. Krinks 29 Swaledale Avenue, Congleton CW12 2BY
M6 JKS J. Shaw 25 High Street, Gorleston, Great Yarmouth NR31 6RT
M6 JKT C. Baines Donative Farm, Warton, Tamworth B79 0JR
M6 JKW J. Gelder 36 Westcombe Court, Wyke, Bradford BD12 8PT
M6 JKX J. Kitto 306 Lewisham Road, London SE13 7PA
M6 JKY K. Young 48 Sussex Street, Cleethorpes DN35 7NP
M6 JLA J. Tayler 22 Wheatley Road, Leicester LE4 2HN
M6 JLD J. Drinkell 11 Valley Walk, Kettering NN16 0LY
M6 JLE J. Wiskow 15 Ferndale Close, Sandbach CW11 4HZ
M6 JLH J. Hughes 27 Mitchell Street, Stoke-on-Trent ST6 4EX
M6 JLL J. Caswell 10 Beech Close, Scole, Diss IP21 4EH
M6 JLM J. Mould 30 Northfields, Dunstable LU5 5AL
M6 JLQ M. Adams 10 St. Johns Road, Wallasey CH45 3LU
M6 JLR J. Ridley 73 The Markhams, New Ollerton, Newark NG22 9QY
M6 JLT J. Trunks 37 Carlton Street, Haworth, Keighley BD22 8JY
M6 JLU L. Adams 10 St. Johns Road, Wallasey CH45 3LU
M6 JLW J. Hart 420 Butts Road, Southampton SO19 1DD
MM6 JLX J. Leith 13 Chesterhall Avenue, Macmerry, Tranent EH33 1QJ
M6 JLY J. Lyall 11 Bowmont, Ellington, Morpeth NE61 5LT
M6 JLZ J. Beccles 15 Lambert Avenue, Shepshed, Loughborough LE12 9QH
MW6 JMB M. Bloore Halfway House, Hyfrydle Road, Talysarn, Caernarfon LL54 6HG
MI6 JMC J. Mccloskey 19 Kinnyglass Road, Coleraine BT51 3SN
MI6 JMD J. Mcdonald 70 Allen Park Dunamanagh, Strabane BT82 0PD
M6 JMF J. Fearn 16 Sandringham Road, Retford DN22 7QW
M6 JMG J. Mobbs 6 School View, Banbury OX16 4SD
MW6 JMH J. Hewitt 1 Highfield, Gloucester Road, Chepstow NP16 7DF
MM6 JMI J. Wilson 1 West Long Cottages, Livingston EH54 7AB
M6 JMJ M. Grimsley 80 Holborn Avenue, Coventry CV6 4FZ
M6 JMM J. Munton 14 Serpells Meadow, Polyphant PL157PR
M6 JMN J. Mason 9 Little Warton Road, Warton, Tamworth B79 0HR
M6 JMO C. Monahan 48 Church Road, Earley, Reading RG6 1HS
M6 JMP J. Pearse 23 Buckingham Drive, Knutsford WA16 8LH
M6 JMR J. Randall 18 Therty First Avenue, Kingston Up on Hull HU6 8DB
M6 JMX J. Matthews 50A South Farm Road, Worthing BN14 7AE
M6 JNA N. Watling 8 Badger Close, Solihull B90 4HR
MM6 JNB J. Beedie 21 Marywell Village, Arbroath DD11 5RH
M6 JNC J. Clay Willow Brook, Bairstow Lane, Sowerby Bridge HX6 2SY
M6 JNE J. Nobes 22 Mansfield Road, Edwinstowe, Mansfield NG21 9NJ
MM6 JNF J. Fyall 56 Whins Road, Alloa FK10 3RE
M6 JNH H. Blount 70 Harrisons Lane, Ringmer, Lewes BN8 5LJ
M6 JNI D. Storton 50 Cromwell Road, Blackpool FY1 2RG
M6 JNJ J. Dray Rose Cottage, The Street East Brabourne, Ashford TN25 5LR
M6 JNL S. Clarke Bugbrooke Marina, The Wharf, Bugbrooke, Northampton NN7 3QB
M6 JNM J. Nofrerias Mondejar 31 Emmendingen Avenue, Newark NG24 2FX
MM6 JNN J. Needham 154 Ravenswood Rise, Dedridge, Livingston EH54 6PQ
M6 JNO C. Pryke 50 Raglan Gardens, Watford WD19 4LL
MW6 JNP J. Powell 31 Laburnum Close Gurnos Estate, Merthyr Tydfil CF47 9SN
M6 JNQ J. Nicholas 14 Poplar Close, Frome BA11 2UH
M6 JNR Dr J. Reynolds 38 Spring Lane, Hockley Heath, Solihull B94 6QY
M6 JNS J. Shatford 31 Pinner Park Avenue, Harrow HA2 6LG
M6 JNV T. Wild 31 Ivel Road, Shefford SG17 5LB
M6 JNW J. Walker 232 Bideford Green, Leighton Buzzard LU7 2TS
M6 JNZ S. Jessup 6 Fifth Ave, Catterick DL9 4RJ
M6 JOA M. Hall Ward 2 Berth 15, Royal Hospital, Chelsea Royal Hospital Road, London SW3 4SR
M6 JOD J. O'Driscoll 48 St. Ives Road, Coventry CV2 5FZ
M6 JOE J. Bell 8 Firsleigh Park, Roche, St. Austell PL26 8JN
M6 JOG S. Mayor 12 Yealand Avenue, Heysham, Morecambe LA3 2LT
MM6 JOH J. Hutton 2 Watson Place, Dunfermline KY12 0DR
M6 JOI W. Pickles 31 Longfield Road, South Woodham Ferrers CM3 5JL
MM6 JOK J. Stewart 1 Barns Park, Dalgety Bay, Dunfermline KY11 9XX
M6 JON J. Oakley 13 Worsley Street, Warrington WA5 0NA
M6 JOO J. Fairhall 122 Thornhill Rise Portslade, Brighton BN41 2YL
M6 JOP S. Gowers High Elms Burnt House Lane, Dartford DA2 7SP
M6 JOR J. Oliver 16 Eastdale Road, Burgess Hill RH15 0NH
MI6 JOS J. Millar 3, Ahoghill BT42 1JN
M6 JOU J. Best 7 Lawns Court, Carr Gate, Wakefield WF2 0UT
M6 JOV M. Dale 50 St. Cuthberts Avenue, Colburn DL9 4NT
M6 JOW J. Whitmore 33 Wanton Lane Terrington St Clement, Kings Lynn PE34 4NR
MM6 JOX J. Greer 2/2 49 Strathcona Drive, Glasgow G13 1JH
MI6 JOY J. Ruddell 16 Beechfield Manor, Aghalee, Craigavon BT67 0GB
M6 JPC P. Crossley 4 Bartons Garth, Selby YO8 9RR
M6 JPD D. Bache 62 Whittingham Road, Halesowen B63 3TP
M6 JPF J. Franks 14 The Hamlet, Slades Hill, Templecombe BA8 0HJ
M6 JPH J. Higginson 187 Birmingham Road, Ansley Village, Warwickshire CV10 9PQ
M6 JPI J. Pinder 11 Andrews Close, Louth LN11 0BP
M6 JPJ J. Leeson 2 Hawthorn Road, Radstock BA3 3NW
M6 JPK J. Knight 58 Winchester Tower, Vauxhall Street, Norwich NR2 2SF
M6 JPL J. Lynch Beechway, Raddel Lane, Warrington WA4 4EE
MM6 JPM J. Moffat 6 Park Grove, Belhelvie, Aberdeen AB23 8YG
M6 JPO J. Owen 21 Marlborough Road, Luton LU3 1EF
M6 JPR J. Raine Braemar, Sandy Lane, Crawley RH10 4HS
M6 JPS J. Potts 103 Etherstone Street, Leigh WN7 4HY

M6 JPT J. Thompson 7 Hawthorne Terrace, Crosland Moor, Huddersfield HD4 5RP
M6 JPU S. Stretton 9 Kilton Road, Worksop S80 2EG
M6 JPV C. Bone 65 Walesmoor Avenue Kiveton Park, Sheffield S26 5RF
M6 JPW P. Whitehead 29 Cleveland, Bradville, Milton Keynes MK13 7AZ
M6 JPX J. Powell Temple Cottage, Monkey Island Lane, Maidenhead SL6 2ED
MM6 JQA M. Batho 25 Burghmuir Road, Perth PH1 1LU
M6 JQC K. Swift 1 Cedar Nook, Sheffield S26 5HL
M6 JQE J. Drea 25 Loxley Gardens, Burnley BB12 6PW
MM6 JQF A. Aitken Cherlor Mosscastle Road, Slamannan FK13EL
M6 JQG D. Start The Rise, Valley Lane Swaby, Alford LN13 0BH
M6 JQK C. Gilbert 3 Marston Close, Eastham CH629EA
M6 JQL N. Eyers Flat 9, Pennys Court, 450 New Road, Ferndown BH22 8EX
M6 JQQ C. Murphy 20 Bridge Close, Burgess Hill RH15 8PD
M6 JQS J. Wort 25 Renton Cottage, Ellery Grove, Lymington SO41 9DX
M6 JQU C. Hodson 64 Cradley, Widnes WA8 7PL
M6 JQW J. Wang Flat 31, 74 Arlington Avenue, London N1 7AY
M6 JQZ G. Hardman 12 Fernleigh Chorley New Road, Horwich, Bolton BL6 6HD
M6 JRB J. Brown 13 Waterlakes, Edenbridge TN8 5BX
M6 JRC J. Clay Riverside Mews, Nichols Yard, Sowerby Bridge HX6 2EE
M6 JRE J. Redfern Flat 11, Poplar Court, Poplar Street, Manchester M34 5EJ
M6 JRH J. Harrison 20 Somerton Avenue, Wilford, Nottingham NG11 7FD
MW6 JRI J. Richards 29 Kingsland Crescent, Barry CF63 4JQ
M6 JRJ J. Rhodes 13 Blake Hall Drive, Mirfield WF14 9NL
M6 JRK J. De Vantier 167 Ivy Road, Bolton BL1 6EF
M6 JRL J. Peacock 4 Luzley Cottages Luzley Road, Ashton-under-Lyne OL6 9AL
M6 JRO J. Oldman High Waters, Bentfield Green, Stansted CM24 8HX
M6 JRS J. Sadler 10 Spindle Warren, Havant PO9 2PU
M6 JRT N. Wing 39 Whittington Road, Hutton, Brentwood CM13 1JX
M6 JRW J. Wallis 20 Green Leys, West Bridgford, Nottingham NG2 7RX
M6 JSH J. Horner 36 Chadsfield Road, Rugeley WS15 2QP
M6 JSI G. Harvey 55 Trelawney Road Hainault, Ilford IG6 2NJ
M6 JSJ M. Burrows 1 Hedgemere, Taverham, Norwich NR8 6GG
M6 JSM J. Mcclure 25 Bluebell Avenue, Wigan WN6 8NS
MM6 JSN J. Drummond 1/1 2 Lethamhill Place, Glasgow G33 2SD
M6 JSP J. Parris 10, Wharfedale Grange, Ben Rhydding Drive, Ilkley, West Yorkshire LS29 8AR
MI6 JSQ J. Quinn 44 Gleanniseal, Dungannon BT70 3BE
M6 JSR I. Smith 2 George Street, Somercotes, Alfreton DE55 4JT
M6 JSS J. Stones Eads Astrium Ltd, Anchorage Road, Portsmouth PO3 5PU
M6 JST J. Tidmarsh 16 Birch Road, Wellington TA21 8EP
M6 JSU J. Neal 5 Shelley Close, Huntingdon PE29 1NF
M6 JSW J. Welsh 16 Colton Crescent, Dover CT16 2EP
M6 JSX Lord C. Burridge 43 Rackenford Road, Tiverton EX16 5AF
M6 JTA J. Allen 15 Wessington Drive, Hereford HR1 1AH
M6 JTC P. Askey The Maltings, Brewery Yard, Kettering NN14 3BT
M6 JTD D. Connor 145 Welsby Road, Leyland PR25 1JH
MM6 JTG J. Mcdowall 50 Dailly Road, Maybole KA19 7AU
M6 JTH J. Thomas 14 Lower Meadow, Cheshunt, Waltham Cross EN8 0QU
M6 JTI J. Smith 8 Westbourne Terrace, Thirsk YO7 1QD
M6 JTK D. Maitland 214 Springfield Road, Birmingham B13 9NE
M6 JTL M. Simons 123 Main Street, Little Harrowden, Wellingborough NN9 5BA
MW6 JTM J. Morgan The Lodge, Old Racecourse, Oswestry SY10 7PQ
M6 JTN M. Chivers 4 Hunters Lodge, Fareham PO15 5NF
M6 JTO J. O'Reilly 116 Coleridge Way, Crewe CW1 5LF
M6 JTP J. Payne 27 Newton Hall Gardens, Rochford SS4 3EP
M6 JTQ J. Cook 28 Wenny Estate, Chatteris PE16 6UX
M6 JTU J. Bridge 4 Knights Way, Camberley GU15 1EQ
M6 JTW J. Waring 12 Mary Street, Farnhill, Keighley BD20 9AU
M6 JUB A. Casson 7 Leebrook Court, Owlthorpe, Sheffield S20 6QJ
M6 JUC B. Marsh 21 Edward Road, Eynesbury, St. Neots PE19 2QF
MM6 JUE R. Cormack 2 Coghill Street, Wick KW1 4PN
M6 JUF J. Green 26 Foxhill Road Burton Joyce, Nottingham NG14 5DB
M6 JUG S. Jones 25 Kents Lane, Crewe CW1 4PX
M6 JUH M. Stanford-Taylor Holly Flat, School Lane, St. Johns, Crowborough TN6 1SE
M6 JUI Dr M. Ovenden 59 Cemetery Road Woodlands, Doncaster DN6 7RY
M6 JUJ W. Bruen 25 Carlton Avenue Upholland, Skelmersdale WN8 0AE
M6 JUK J. Rhodes 73 Keats Way, Hitchin SG4 0DP
M6 JUP A. Drewitt 12 Kinross Crescent, Loughborough LE11 4UQ
MW6 JUQ A. Morgan 4 Courtney Street Manselton, Swansea SA5 9NY
M6 JUT J. Thorne 53 Elfleda Road, Cambridge CB5 8NA
M6 JUU L. Hagan 7 Betjeman Mews, Southend-on-Sea SS2 5EJ
M6 JUW G. Rodriguez 32 Mount Pleasant, Prestwich, Manchester M25 2SD
M6 JUX A. Brown 17 Quail Ridge, Ford, Shrewsbury SY5 9LF
M6 JVB J. Cobb 32 Dellmont Road, Houghton Regis, Dunstable LU5 5HU
MI6 JVC J. Heyburn 20 Victoria Street, Armagh BT61 9DT
M6 JVD J. Boyd 125 High Street, Great Wakering, Southend-on-Sea SS3 0EB
M6 JVF W. Terry 11 Crescent Ave, Overhulton., Bolton BL51EN
M6 JVK J. Kinsey 1 Keystone Close, Goring-By-Sea, Worthing BN12 6GA
MI6 JVM O. Hart 18 Sandringham Court Portadown, Craigavon BT63 5BF
M6 JVO J. Simmonds Weir Cottage Lugwardine, Hereford HR1 4AS
MM6 JVQ S. Cummings 1 Inveresk Gardens, Dundee DD4 0XZ
M6 JVS C. James Ct93Nx, Margate CT92NX
M6 JVW M. Fogerty 25 Noel Street, Gainsborough DN21 2RY
M6 JVX G. Cope 171 Ockford Ridge, Godalming GU7 2NN
M6 JVY P. Smythe 15 Falcon Drive, Trowbridge BA14 7GE
M6 JVZ E. Walker St. Govans Ledbury Road, Ross-on-Wye HR9 7BG
M6 JWC J. Cater 5 Shady Grove, Hilton, Derby DE65 5FX
M6 JWD J. Wilson 35 Lawson Avenue, Jarrow NE32 5UF
M6 JWE J. Loveday 27 New Road, Chatteris PE16 6BJ
M6 JWF A. Marsh 140 Church Road, Redfield, Bristol BS5 9HN
M6 JWG J. Whitworth 31 Shirley Close, Chesterfield S40 4RJ
M6 JWI D. Milne 154 Cambridge Road, Ellesmere Port CH65 5BW
M6 JWO J. Whalley 28 Hatton Lane, Stretton, Warrington WA4 4NG
M6 JWP J. Wade 44 Newton Park Homes, Newton St. Faith, Norwich NR10 3LP
M6 JWQ J. Oldfield 2 Bailie Cross Cottages, Poole Road, Wimborne BH21 4AE
M6 JWT J. Thompson 19 Taverner Road, Boston PE21 8NL
M6 JWW J. Whiteside The Old Antique Shop, Bank Street Pulham Market, Diss IP21 4TG
M6 JWX J. Wade 105 Western Avenue, Woodley, Reading RG5 3BL
M6 JWY J. Willby 10 Sunbury Road, Birmingham B31 4LJ
M6 JXD J. Darragh 20 Templar Place, Hampton TW12 2NE
M6 JXE G. Spiers 2 Sponnes Road, Towcester NN12 6ED
M6 JXF D. Wells 96 Tennyson Avenue, Rugby CV22 6JF
MM6 JXH J. Hinchcliffe 27 Forest Glade, East Calder, Livingston EH53 0FQ
M6 JXI J. Jago 5 Horrell Court, Rushden NN10 9EG
M6 JXJ C. Connell 16 Woodman Drive, Bury BL9 5HQ
M6 JXK J. Kelly Flat 4, The Corner Place, 1 North Road, Harborne, Birmingham B17 9PA
M6 JXN J. Culshaw 13 Ravens Close, Knaphill, Woking GU21 2LD
M6 JXO P. Mcnulty 2 Belvedere Road Hanford, Stoke-on-Trent ST4 8RL
M6 JXY J. Yendole 15 Borgie Place, Weston-Super-Mare BS22 9HG
M6 JYB J. Brown 37 Calshot Avenue, Chafford Hundred, Grays RM16 6NS
M6 JYF A. Browning 1 Stamford Ave, Blackpool FY42BH
M6 JYI M. Piatkowski 152 Anchorway Road, Coventry CV3 6JG
M6 JYJ L. Richardson 28 Durham Way Wyton, Huntingdon PE28 2EQ
M6 JYL Y. Lai Harrogate Ladies' College, Clarence Drive, Harrogate HG1 2QG
M6 JYO S. Valsala 12 Montague Street, Rainham Essex RM13 8LW
M6 JYQ R. Cichocki 12 Crossland Crescent, Wolverhampton WV6 9JY
M6 JYR G. Hepworth 12 Fourlands Gardens, Bradford BD10 9SP
M6 JYS J. Pengilly 8 Willard Close, Eastbourne BN22 8SX
M6 JYT A. Lane 26 Astral Gardens Sutton-On-Hull, Hull HU7 4YS
M6 JYV K. Riddiough 47 Brooke Street Hoyland, Barnsley S74 9DP
M6 JYX N. Peppe 5 Cherford Road, Bournemouth BH11 8SU
M6 JYY R. Burton 152 Hough Road, Walsall WS2 9BQ
M6 JYZ J. Sanders 2 Manor Court Manor Grove, Mangotsfield, Bristol BS16 9LF
M6 JZC J. Clark 21 Bilton Way Crewe Cheshire , Cheshire CW2 8SN
M6 JZF D. Ward 10 Royston Court, Potton, Sandy SG19 2NJ
M6 JZH C. Garratt 309, Brookles Mead, Harlow CM19 4QD
M6 JZK J. Howlett 29 Little London, Heytesbury, Warminster BA12 0ES
M6 JZL J. Hope 4 Beaumont Way, Prudhoe NE42 6RA
M6 JZM K. Wood 89 Oriel Road, Portsmouth PO2 9EG
M6 JZN R. Rudelic Flat 26 Neptune House, 1 Neptune Way, Southampton SO14 3FN
M6 JZU P. Holmes 82 Moore Avenue, Norwich NR6 7LG
M6 JZV B. Shackleton 7 Erringden Street, Todmorden OL14 6AW
MM6 JZW S. Gillion 13 Aignish, Isle of Lewis HS20PB
M6 JZY J. Smale 30 Gillian Close, Aldershot GU12 4HU
MW6 JZZ J. Trahearn-O'Brien 148 Gladstone Road, Barry CF62 8ND
MW6 KAA K. Parry 80 Cripps Avenue, Cefn Golau, Tredegar NP22 3PB
MW6 KAB K. Boulter 16 Danygraig, Pontlottyn, Bargoed CF81 9RS
MW6 KAC M. Woodington 2 Tir Founder Fields, Aberdare CF44 0DT
M6 KAE J. Withall Humphreys Cottage, Fords Green, Uckfield TN22 3LJ
M6 KAG K. Goodyer 5A Station Road Bow Brickhill, Milton Keynes MK17 9JU
M6 KAH K. Holloway 6 Britons Lane Close, Beeston Regis, Sheringham NR26 8SH
M6 KAM K. Mcgill 25 Gladstone Street, Workington CA14 2XH
M6 KAN K. Sharpe 18 Dudhill Road, Rowley Regis B65 8HT
M6 KAO K. Gibbs 30 George Road, Water Orton, Birmingham B46 1PE
M6 KAQ K. Sidaway 19 Larkwhistle Walk, Havant PO9 4JA
M6 KAR K. Slotwinski 51 Moorland Gate, Heathfield, Newton Abbot TQ12 6TX
M6 KAS J. Gilhespy 106 Durham Drive, Jarrow NE32 4QY
M6 KAT C. Gibson 16A Hillside Road, Wool, Wareham BH20 6DY
MM6 KAU M. Krawczyk 19 Wishart Archway, Dundee DD1 2JA
M6 KAV K. Booth 28 Farndale Gardens, Lingdale, Saltburn-by-the-Sea TS12 3EW
M6 KAX J. Killman 19 Moorland Avenue, Walkeringham, Doncaster DN10 4LG
M6 KAZ K. Hamilton 171 Jackmans Place, Letchworth Garden City SG6 1RG
M6 KBA D. Reeve 5 Antelope Avenue, Grays RM16 6QT
MI6 KBB J. Elliott 12 Drumbeg Drive, Lisburn BT28 1NY
M6 KBC S. Williams Flat 4, The Crown Mews, 24 Station Road, Midsomer Norton, Radstock BA3 2FE
M6 KBE K. Bantock 22 Deepdale Drive, Consett DH8 7EH
M6 KBF B. Wiggins The Wigwam 13 Hastings Road, Bromsgrove B60 3NX
M6 KBG K. Glaysher 66 Talbot Road, Farnham GU9 8RR
M6 KBH K. Burt 2 Mousehole Close Dalton, Rotherham S65 4JF
M6 KBI M. Symmonds 24 Woodville Grove, Stockport SK5 7HU
M6 KBJ K. Jewell 25 Park View, Liskeard PL14 3EE
MW6 KBK B. Hopkins 28 Ivor Street, Maesteg CF34 9AH
M6 KBO R. Waterhouse Sunnymead, Beckley Road TN31 6JB
M6 KBS K. Burness 4 Fenwick Street, Boldon Colliery NE35 9HU
M6 KBT R. Mundy 9 Gable Court, Liverpool L11 7DS
M6 KBV K. Clarke 41 St. Hilda Street, Bridlington YO15 3EE
MD6 KBW K. Whittle 113 Ballaquark, Douglas, Isle of Man IM2 2EU
M6 KBX J. Greenwood 38 Baskerville Road, Sonning Common, Reading RG4 9LS
M6 KBY A. Davies 6 Gribble Road, Liverpool L10 7NF
M6 KCA K. Millward 133 Birchfield Way, Telford TF3 5HN
MM6 KCB S. Gallagher 29 Roslyn Drive Bargeddie, Baillieston, Glasgow G69 7QZ
M6 KCC K. Cole 16 Minster Road, Westgate-on-Sea CT8 8BP
M6 KCD K. Crosby 28 Exning Road, London E16 4NA
M6 KCE K. Bradford 3 Haven Close, Sutton-in-Ashfield NG17 2DG
M6 KCF City And Sticks ARG c/o L. Bee 25 Blatcher Close, Minster On Sea, Sheerness ME12 3PG
M6 KCG K. Allen 20 Rookery Road, Innsworth, Gloucester GL3 1AT
M6 KCI D. Orme 47 Poplar Avenue, Oldham OL8 3TZ
MW6 KCJ K. Uprichard 111 Fernhill, Mountain Ash CF45 3EF
M6 KCK L. Wood 2 The Bungalows, North Green, Woodbridge IP13 9NP
MM6 KCM K. Mair 92 Graham Street, Wishaw ML2 8HR
M6 KCP A. Ostatek 38 Coronation Way, Keighley BD22 6HF
MW6 KCQ M. Cook 40 Cemaes Crescent, Rumney, Cardiff CF3 1TA
MM6 KCR K. Riddick Davah, Port Road, Castle Douglas DG7 3JW
M6 KCS A. Stooke 128 Stamford Street, Grantham NG31 7BP
M6 KCT K. Coton 11 Pickers Way, Clacton-on-Sea CO15 5RU
M6 KCU C. Pavey 143 Queen Elizabeth Way, Colchester CO2 8LT
MM6 KCV C. Vines 49 Annandale Gardens, Glenrothes KY6 1JD
M6 KCY R. Fisher 14 Fraser Avenue, Caversham, Reading RG4 6RT
M6 KCZ R. Robinson 22 Riddings Court, Timperley, Altrincham WA15 6BG
MW6 KDA D. Henderson 1 Court Place, Tonypandy CF40 2RE
M6 KDB K. Beech 8 Gateford Drive, Worksop S81 7HL
M6 KDC D. Stewart 67 Little Moss Hey, Liverpool L28 5RJ

M6 KDD K. Daniels 64 Skelton Road, Norwich NR7 9UH
M6 KDF J. Nash 124 Tanhouse Avenue, Birmingham B43 5AG
M6 KDG C. Harman 58 Laurence Avenue, Witham CM8 1JB
M6 KDJ K. Hall 11 Hinton Villas, Hinton Charterhouse, Bath BA2 7SS
M6 KDK K. Khan 3 Marshall Close, Frimley, Camberley GU16 9NY
M6 KDM K. Middlehurst 7 Statham Drive, Lymm WA13 9NW
M6 KDO K. Porter 58 Panama Drive, Atherstone CV9 3HJ
M6 KDP R. Hutton 22A Victoria Road, Maldon CM9 5HF
M6 KDQ A. Bowles Golden Yews, Burnt Hill, Yattendon, Thatcham RG18 0XD
M6 KDR C. Pearcey 17 Peppercorn Close, Christchurch BH23 3BL
M6 KDV M. Pulling 21 Heathlands Avenue, West Parley, Ferndown BH22 8RW
M6 KDW K. Dillow 101 Martins Lane Hardingstone, Northampton NN4 6DJ
M6 KDZ I. Whiteley 29 Harvey Avenue, Wirral CH49 1RT
M6 KEA A. Wheeler 10 Handsacre Crescent, Rugeley WS15 4DQ
M6 KEC D. Preston 9 Daubney Street, Cleethorpes DN35 7BB
MW6 KED S. Kedward 9 Lawrence Avenue, Aberdare CF44 9EW
M6 KEE M. Keeler 15 Grove Park, Kingswinford DY6 9AD
M6 KEF C. Thompson 5 Oak Avenue, Charlton Kings, Cheltenham GL52 6JG
M6 KEG P. Selwood 27 Sharp Street, Warrington WA2 7AP
M6 KEH K. Dalby 52 Narborough Road South, Leicester LE3 2FN
MW6 KEL K. Gemmell 93 North Road, Ferndale CF43 4RG
M6 KEM K. Morris 1A Littlemoor, Queensbury, Bradford BD13 1DB
M6 KEP K. Applegarth 28 South View Gardens, Pontefract WF8 2HW
MW6 KEQ K. Mogford 49 Cefn Road, Rogerstone, Newport NP10 9AQ
MM6 KER K. Hamilton 25 Abbotsford Street, Falkirk FK2 7NH
M6 KES M. Shore 43 Daniel Fold, Rochdale OL12 7JU
M6 KET J. Daws 1157 Evesham Road, Astwood Bank, Redditch B96 6DY
M6 KEU Y. Law 59 Lake View, Edgware HA8 7SA
M6 KEV K. Taylor Bramley House, 277 Salisbury Road, Totton SO40 3LZ
M6 KFB Dr F. Kuttikkate 34 Shetland Crescent, Rochford SS4 3FJ
MM6 KFE K. Farnington 52 Glebe Park, Duns TD11 3EE
M6 KFF S. Coupe 47 Burn Street, Sutton-in-Ashfield NG17 4LL
M6 KFG J. Stokes 30 Anglers Reach, Grove Road, Surbiton KT6 4EX
MD6 KFH K. Gascoyne 22 Broogh Wyllin, Kirk Michael, Isle of Man IM6 1HU
MM6 KFJ A. Keogh 251 Main Street Plains, Airdrie ML6 7JH
M6 KFK K. Kozlowski Woking Homes / Flat 2, Oriental Road, Woking GU22 7BE
M6 KFP K. Prentice 24 Sulgrave Close, Liverpool L16 6AD
MW6 KFQ J. Turner 4 Imble Street, UK SA726QL
M6 KFS K. Fraser-Smith 70 Glen Eyre Road, Southampton SO16 3NL
M6 KFW K. Wells 5 Cook Avenue, Newport PO30 2LL
M6 KFX R. Stone 63 Sands Lane, Lowestoft NR32 3ER
M6 KGA K. Allgar 13 Deacon Avenue, Kempston, Bedford MK42 7DU
MW6 KGB W. Oliver Pwllmeyric, Chepstow NP16 6LE
M6 KGC K. Cossey 34 Pinewood Road, Hordle, Lymington SO41 0GP
M6 KGD M. Tabberer 29 Chase Vale, Chasetown, Burntwood WS7 3GD
M6 KGE K. Gibbins 11A Wood End, Banbury OX16 9ST
M6 KGF G. Flew 58 Streamside, Mangotsfield, Bristol BS16 9EA
M6 KGI D. Cooper Shirley Villa Morley Road North, Sheringham NR26 8JB
M6 KGK K. Goddard 49 Chalet Hill, Bordon GU35 0EF
M6 KGL M. Horwell 3 Vanity Close, Oulton, Stone ST15 8TZ
MM6 KGM K. Mitchell 92 Whiteside Court, Bathgate EH48 2TN
M6 KGN J. Badham Comwillgur House Ross Road, Longhope GL17 0LP
M6 KGO Dr B. Osborne 15 Ferguson Avenue, Surbiton KT5 8DS
M6 KGQ J. Loveland 6 Wren Drive Bradwell, Great Yarmouth NR31 8JW
M6 KGR K. Carter 50 Elliman Avenue Bottom Flat, Slough SL2 5BG
M6 KGS K. Stow 39 Biverfield Road, Prudhoe NE42 5ER
MI6 KGU D. Holloway 3 Iniscarn Close, Lisburn BT28 2BX
M6 KGV G. Carter 54 Wood Street, Taunton TA1 1UW
MM6 KHA K. Adrian 1C Oliphant Court, Paisley PA2 0DP
M6 KHB M. Heaton-Bentley 65 Brookfield Road, Thornton-Cleveleys FY5 4DR
M6 KHE K. Ballard 34 Orange Street South Wigston, Leicester LE184QB
M6 KHF R. Boden 44 High Street, Blackpool FY1 2BN
MM6 KHM K. Macrae 2 Netherhill Avenue, Glasgow G44 3XG
M6 KHP S. Smith 2 Burnsall Avenue, Blackpool FY3 7LQ
M6 KHQ C. Thompson 12 Otho Way, North Hykeham, Lincoln LN6 9ZD
M6 KHS K. Shaw 3 The Grange, Woolley Grange, Barnsley S75 5QP
M6 KIA A. Taylor 130A Hazelwood Avenue, Eastbourne BN22 0UX
M6 KIE K. Alejo-Blanco 23 Southern Road, Thame OX9 2EE
M6 KIF M. Chalkley 36 Cowper Road, Bournemouth BH9 2UJ
M6 KIG K. Greenfield 49 Railway Street, Northfleet, Gravesend DA11 9DU
M6 KII S. Hopper Coombe Road, Brighton BN24EA
M6 KIK W. Dover Windcrest, Fox Lane, Basingstoke RG23 7BB
M6 KIL B. Fryer 33 Flaxdown Gardens, Rugby CV23 0GX
M6 KIN M. Makin 34 Carlton Gardens, Farnworth, Bolton BL4 7TH
M6 KIP K. Davies 29 Corser Street, Stourbridge DY8 2DE
M6 KIR K. Halloway 41 Trenoweth Estate, North Country, Redruth TR16 4AQ
M6 KIT C. Ellis 1 Rugby Road, Stockton-on-Tees TS18 4AZ
M6 KIU S. Kato 38-9-6, Kamikizaki6-Chome, Urawa-Ku, Saitama-City 3300071 Japan
M6 KIV V. Ilic Vlade Djordjevica 41, Leskovac 16000 Serbia and Montenegro
MM6 KIW C. Ross 4 Weir Place, Greenock PA15 2JD
M6 KIX C. Hill 3 Beechmount Rise, Stafford ST17 4QR
M6 KJB K. Black 1 Woodcock Close, Haxby, York YO32 3NQ
M6 KJD K. Davison 1, Barnsley S71 2HS
M6 KJG K. Gallyot 18, High Pines, St. Georges Close, Christchurch BH23 4LN
M6 KJI M. Bidwell 3 Walsingham Place, London SW4 9RR
M6 KJK K. Kolesnik 15 Steer Road, Swanage BH19 2RU
M6 KJL N. Rasevic Bulevar Despota Stefana 7, Novi Sad 21000 Serbia and Montenegro
M6 KJM K. Mitchell 5 Finch Crescent, Linslade, Leighton Buzzard LU7 2PE
M6 KJP K. Parker Flat 3, Gleadless Court, Sheffield S2 3AE
M6 KJQ K. Freer 1 Masefield Flats, Masefield Road, Rotherham S63 6NQ
M6 KJR J. Cogman 6 Stockholm Way, Dereham NR19 1XF
MI6 KJW K. Watt 2A Drumalane Road, Newry BT35 8AP
MW6 KJX C. Hodson 6 Heol Pentwyn Tonyrefail, Porth CF39 8DF
M6 KJY K. Younger 29 Eagle Walk, Bury St. Edmunds IP32 6RJ
M6 KKA R. Pallister 15 High Row, Washington NE37 2LZ
M6 KKD K. Edmunds 61 Somers Road, Keresley End, Coventry CV7 8LE
M6 KKF F. Fitton 6 Leaford Close, Denton, Manchester M34 3QH
M6 KKG K. Komenan 113 Wightman Road, Finsbury Park, London N4 1RJ
M6 KKI R. Young 10 Hareholme Lane, Rossendale BB4 7JZ
M6 KKJ N. Dyer 79 Saltmarsh Drive, Bristol BS11 0NL
M6 KKL T. Cheng Harrogate Ladies' College, Clarence Drive, Harrogate HG1 2QG
M6 KKM D. Linnett Flat19. Kestrel House. Osprey Close. Sinfin. Derby., Derby DE24 3DD
MI6 KKN A. Conboy 16 Tullaghmurry Fold, St. Johns Close, Portstewart BT55 7DT
M6 KKO T. Stack 31A Chester Road South, Kidderminster DY10 1XJ
MI6 KKU S. Mc Court 17 Windermere Road, Lisburn BT28 2WY
M6 KKW N. Webb 27A Orchard Way, Bognor Regis PO22 9HJ
M6 KKY M. Douglas 26 Clumber Drive, Northampton NN3 3NX
M6 KLA D. Bourne 4 Market Street, Cheltenham GL50 3NH
M6 KLC K. Chappuis Priory Cottage, Priory Lane, Bridport DT6 3RW
M6 KLD T. Lovell 7 Victoria Wharf Victoria Street North, Grimsby DN31 1PQ
M6 KLF K. Francis 203 Colchester Road Lawford, Manningtree CO11 2BU
M6 KLI J. Heng 1 St. Giles Croft, Beverley HU17 8LA
M6 KLK J. Li Harrogate Ladies' College, Clarence Drive, Harrogate HG1 2QG
MW6 KLL K. Linahan 19 Heol Bryn Hebog, Merthyr Tydfil CF48 1HH
M6 KLM K. Mason 18 Goose Lane, Sutton, Norwich NR12 9SE
M6 KLN C. Dzundza 67 Sidegate Lane, Ipswich IP4 4HY
M6 KLR S. St George 37 Highfields, Bromsgrove B61 7DA
MM6 KLT C. Sharp 16 Wood Lane, Monifieth, Dundee DD5 4HS
M6 KLV P. Ayre 11 Paradise Road, Waltham Abbey EN9 1RL
M6 KLW L. Woods 193 Wimberley Street, Blackburn BB1 8HU
M6 KLX A. Swain 43 Stretton Road, Morton, Alfreton DE55 6GW
MM6 KLZ N. Stewart 35 Newbattle Gardens, Dalkeith EH22 3DR
M6 KMA K. Machen 18 Peveril Close, Whitefield, Manchester M45 6NR
M6 KMC S. Day 9 Lynton Street, , Derby DE22 3RW
M6 KMD K. Deans 31 Northcroft, Sandy SG19 1JJ
M6 KMF N. Powell 21 Hitchen, Merriott TA16 5QX
M6 KMG P. Matthew 24 Jubilee Close, Pamber Heath, Tadley RG26 3HP
M6 KMI N. Dobson 2 Hills Road, Breaston, Derby DE72 3DF
MW6 KMJ M. Beasley Ffynnon Wen, Bontnewydd, Aberystwyth SY23 4JJ
M6 KMK A. Platt 43 The Butts, Frome BA11 4AB
M6 KML K. Lavelle 12 Shalewood Court, Atherton, Manchester M46 0SN
M6 KMN K. Nilan 15 Broomhall Road Pendlebury, Swinton, Manchester M27 8XP
MM6 KMQ W. Donnelly 10 Sentry Knowe, Selkirk TD7 4BG
M6 KMR K. Riding 8 Kingston View, Barton-upon-Humber DN18 6DN
M6 KMS M. Harrison 70 Hope Avenue, Goldthorpe, Rotherham S63 9EA
MM6 KMU D. Cowe 1 Pentland Road Bonnyrigg Midlothian, Edinburgh EH192LG
M6 KMV A. Waller 155 Bridgemary Road, Gosport PO13 0UT
M6 KMW K. White 12 Lakefields, West Coker, Yeovil BA22 9BT
M6 KMX E. Rhodes 10 New Road Cottages, Selborne Road, Selborne, Alton GU34 3JA
M6 KMY K. Murray 5 Princes Crescent, Basingstoke RG22 6DP
M6 KMZ Dr K. Martinez Forest View, Forest Road, Salisbury SP5 2BP
M6 KNB M. Narayanankutty Karikkattu House , Chelamattom, Ernakulam 683550 India
M6 KNC J. Richardson Berwick Cottage, Bailes Lane, Guildford GU3 2AX
M6 KND D. Simmons 23 Fairey Street, Cofton Hackett, Birmingham B45 8GU
M6 KNG A. King 23 Tower Crescent, Lincoln LN2 5QF
MM6 KNJ J. Wilson 16 Big Brigs Way, Newtongrange EH22 4DG
M6 KNL G. Knowles 29 Shepherds Cote Drive, Hepscott Park, Stannington, Morpeth NE61 6FN
M6 KNM M. Hill 109 Kitchener Street, St. Helens WA10 4LU
MM6 KNO T. Knox 23 Hill Street, Alness IV17 0QL
M6 KNP M. Lewis 312 Twyford Avenue, Portsmouth PO2 8NT
M6 KNS K. Fletcher Beverley Hotel, 55 Old Brumby Street, Scunthorpe DN16 2AJ
M6 KNU R. Weightman 2 Bannister Grove, Winsford CW7 1RJ
M6 KNV M. Brown 99 Apprentice Drive, Colchester CO45SE
M6 KNW K. Woolsey 1 Park Farm Cottage, Westhorpe Road, Stowmarket IP14 4SP
M6 KNY R. Dunnaker 12 Dagger Lane, West Bromwich B71 4BA
M6 KNZ P. Webb 473 The Manor, Billing Garden Village, The Causeway, Great Billing, Northampton NN3 9EX
M6 KOA L. Lemmon 4 Honington Close, Wickford SS11 8XB
MI6 KOB K. Boyle 764 Springfield Road, Belfast BT12 7JD
M6 KOD A. Adams 27 George Street Stockton, Southam CV47 8JS
MM6 KOE D. Robinson 2 Brixwold Rise, Bonnyrigg EH19 3FG
M6 KOH D. Elcock 27 Harefield Road, Southampton SO17 3TG
MW6 KOI C. Young 13 Bryn Mawr Road, Holywell CH8 7AP
MM6 KOJ C. Smith 68 Craigmore Street, Dundee DD3 0EA
M6 KOM D. Slater 13 Longford Close, Rainham, Gillingham ME8 8EW
M6 KON S. Rouse 7 Cranbrook Road, Thurnby, Leicester LE7 9UA
MM6 KOS J. Kosarzecki 96 Colinton Mains Drive, Edinburgh EH13 9BL
M6 KOV M. Burton The Old Bakehouse, Lincoln Road, Goltho, Market Rasen LN8 5NF
M6 KOZ R. Koziolek 30 Lammas Beanhill, Milton Keynes MK6 4LA
M6 KPA H. Ilie Sos. Iancului Nr. 33 Bloc 105A Scara B Apt 63, Bucharest 21717 Romania
M6 KPC K. Cooper 12 Waverley Crescent, Brighton BN1 7BG
M6 KPD K. Dance 20 Harmers Hay Road, Hailsham BN27 1SU
M6 KPE A. Donnabella 124 Heron Way, Harwich CO12 3FF
MW6 KPF K. Foster Prince Of Wales House Short Bridge Street Sy18 6Ad, Llanidloes SY18 6AD
M6 KPG K. Gallery 89 Mavis Drive, Coppull, Chorley PR7 5AE
M6 KPI J. Barnes Landhill Farm, Halwill, Beaworthy EX21 5TX
M6 KPJ M. Barrett 57 Marlborough Avenue, Hornsea HU18 1UA
M6 KPK K. Killingback Flat 4 Shorland House, 6 Elm Grove Road, Dawlish EX7 0BZ
M6 KPL A. Lloyd 34 Diamond Way, Ellesmere SY120FH
MW6 KPN A. Thomas 10 Chapel Street, Gorseinon, Swansea SA4 4DT
M6 KPP S. Hunter 8 Clyde Street, Plymouth PL2 1QQ
M6 KPQ S. Brooks 68 Daws Heath Road, Benfleet SS7 2TA
M6 KPR C. Eales 7 Prima Cresent, Van Riebeeck Park 1619 South Africa
M6 KPT K. Thrower 19 Blyford Road, Lowestoft NR32 4PZ
MW6 KPV D. Heywood 6 Wentworth Close, Buckley CH7 2QX
M6 KPX D. Cooper Needhams Farm House, Spittal Hill Road, Boston PE22 0PA
M6 KPZ R. Ballard 11 Thurston, Skelmersdale WN8 8QU
M6 KQA J. Miller 21 Hedgerow Grove, Dunmow CM6 4AS
M6 KQC H. Klettke 13 Hastings Close, Wythall, Birmingham B47 6AW
M6 KQD J. Bovey 17 North Bank Road, Bingley BD16 1UH
M6 KQF J. Casey 81 Undercliffe Road, Bradford BD2 3BP

MM6 KQH S. Dunn 2 Glendaruel Avenue Bearsden, Glasgow G61 2PR
M6 KQJ J. Clarke 49 Brunel Close, Hartlepool TS24 0UF
M6 KQK P. Constantine 11 Lingwell Walk, Leeds West Yorkshire South Leeds LS10 4TH
MW6 KQL M. Castle 10 West Place, Gobowen, Oswestry SY11 3NR
M6 KQM A. Green 62 Briarwood Drive, Blackpool FY2 0EB
M6 KQP L. Barlow 51 Hare Hill Road, Littleborough OL15 9HE
M6 KQR A. Kinsella Apartment 28 241 Liverpool Road, Widnes WA8 7HL
M6 KQS R. O'Sullivan Woodhall Drive, Derby DE23 4RS
M6 KQW E. Almas 10, Rindal 6657 Norway
M6 KQZ P. Foster 95 Yew Tree Drive, Bristol BS15 4UF
M6 KRD D. Niggemann 35 Holm Court, Twycross Road, Godalming GU7 2QT
M6 KRF K. Furlong 22 Swaisland Road, Dartford DA1 3DE
M6 KRH K. Hall 130 Aylestone Lane, Wigston LE18 1BA
M6 KRI K. Irvin 21 Fremantle Crescent, Middlesbrough TS4 3HR
M6 KRK J. Birch 3 Partridge Way, High Wycombe HP13 5JX
M6 KRL P. Dudding 14 St. Levan Close, Marazion TR17 0BP
M6 KRM K. Mcleod 7 Priory Place, Sporle, King's Lynn PE32 2DT
MI6 KRP K. Pritchard 18 Ashbrooke, Donaghadee BT21 0EY
M6 KRQ M. Rotheram 19 Oteley Avenue, Wirral CH62 7DJ
M6 KRR G. Newton 20 East Avenue, Syston, Leicester LE7 2EH
M6 KRV K. Bott 294 Walthall Street, Crewe CW2 7LE
M6 KRW K. Wiles 24 Cromwell Way, Witham CM8 2ES
M6 KRX K. Rosema Apartment 801, 25 Goswell Road, London EC1M 7AJ
M6 KRZ K. Brazier Elgin House, Seaside Lane, Easington Colliery, Peterlee SR8 3JZ
M6 KSA C. Sibley 57 Palatine Road, Thornton-Cleveleys FY5 1EY
M6 KSC D. Clark 34 Magdalene Road, Owlsmoor, Sandhurst GU47 0UT
M6 KSD K. Darwin 38 Springbank Road, Gildersome, Leeds LS27 7DJ
M6 KSE A. Went 25\\\, Wayletts\, Laindon, Basildon SS16 6RN
M6 KSG K. Stevens 61A Main Road, Hoo, Rochester ME3 9AA
M6 KSH K. Haywood 126 Derby Street, Sheffield S2 3NF
M6 KSI A. Thompson 10 Belgrave Close, Hersham, Walton-on-Thames KT12 5PH
MM6 KSJ J. Blick Lower Adelaide House, 8 Mountstuart Road, Rothesay, Isle of Bute PA20 9DY
MW6 KSL S. Davies 5 Maldwyn Street, Cardiff CF11 9JR
M6 KSP S. Pegg 24 Fleetwood Close, Minster On Sea, Sheerness ME12 3LN
M6 KSS S. Evennett The Homestead, Pound Green Lane, Thetford IP25 7LS
MM6 KSU G. Smith 2 Croftfoot Place Gartcosh, Glasgow G69 8EG
M6 KSV S. Karpukhina 13 Rushmon Court, Barker Road, Chertsey KT16 9HP
MW6 KSW M. Hancock Halfway House, Hyfrydle Road, Talysarn, Caernarfon LL54 6HG
M6 KSX B. Elms-Lester Ferndale House Kerry'S Gate, Hereford HR2 0AH
M6 KSZ P. Morris Lindum, Dunholme Road Welton, Lincoln LN2 3RS
M6 KTA K. Tilly 14 Mcnally Place, Durham DH1 1JE
M6 KTC I. Mcewen 4 The Pantyles, Nightingale Lane, Sevenoaks TN14 6BX
M6 KTH K. Thompson 32 Westfield Harwell, Didcot OX11 0LG
M6 KTI K. Raynor 6 New Street, Castleford WF10 2RN
M6 KTJ K. Jones 7 Castlemans Cottages, Castlemans Lane, Hinton St. Mary, Sturminster Newton DT10 1LY
M6 KTK N. Edwards 42 Cambrai Avenue, Chichester PO19 7UY
MW6 KTM P. Jones 16 Severn View, Garndiffaith, Pontypool NP4 7SN
M6 KTN K. Turner 3 Beech Street, Sutton-in-Ashfield NG17 3FL
M6 KTO K. Oxford 5 Hazel Close, Rendlesham, Woodbridge IP12 2UR
M6 KTQ Dr R. Pediani Old School House Great Coxwell, Faringdon SN7 7NB
MW6 KTS I. Williams 5 Fron Goch, Llanberis, Caernarfon LL55 4LE
M6 KTT K. Thistlethwaite 140 Kingstown Road, Carlisle CA3 0AY
M6 KTV B. Davies 16 Pearmains Close, Orwell, Royston SG8 5QY
M6 KTW K. Wise 4 Cherrydown, Rayleigh SS6 9ND
M6 KTX K. Taylor 26 Elmbridge Road, Birmingham B44 8AB
MM6 KTY K. Heron 26 Lochancroft Lane, Wigtown, Newton Stewart DG8 9JA
M6 KTZ K. Tasker 16 Chopin Road, Basingstoke RG22 4JN
M6 KUC C. Elliott Vicarage Hill, Kingsteignton TQ12 3BA
MW6 KUD L. Ansell 114 Bowleaze, Greenmeadow, Cwmbran NP44 4LG
M6 KUE I. Vernon 7 Seaton Road Wick, Littlehampton BN17 7LG
M6 KUG A. Jacobs 41 Drake Ave, Sheerness ME123SA
M6 KUH S. Steinhoefel 222 Stretford Road, Urmston, Manchester M41 9NT
M6 KUI M. Diver 2 West Lodge Lane Sutton, Ely CB6 2NX
MI6 KUJ T. Calka 71 Willowfield Street, Belfast BT6 9AW
M6 KUK M. Decruz 124 Antrim Road Woodley, Reading RG5 3NY
M6 KUP A. Stain Apartment 5, 1 Bramble House, Bramble Road, Bridgwater TA52FT
M6 KUR P. Fletcher 14 Orchard Avenue, Aylesford ME20 7LY
M6 KUT A. Roberts 29 Manor Lane, Stourbridge DY8 3ER
M6 KUU J. Tobin 47C Crystal Palace Road East Dulwich, London SE22 9EX
M6 KUY J. Moore 9 Goldsmith Avenue, London RM7 0EX
MW6 KVA P. Barnes 55 Erw Werdd, Birchgrove, Swansea SA7 0HF
M6 KVB K. Bushell 4 Birch Grove, Harrogate HG1 4HR
M6 KVD J. Miller 5 Gilver Lane, Hanley Castle, Worcester WR8 0AT
MM6 KVE M. Hodgson 5 Jocks Loaning, Dumfries DG2 0NQ
M6 KVF K. Emery-Ford 48 Welham Grove, Retford DN22 6TS
M6 KVG D. Miles 133 Marston Lane, Nuneaton CV11 4RE
M6 KVH S. Hepple 20 Gower Road, Shaftesbury SP7 8RU
MM6 KVI G. Hepburn 33 Saxon Road, Glasgow G13 2YQ
M6 KVK G. Kirk 15 Underwood Avenue, Ash, Aldershot GU12 6PP
M6 KVL G. Dougherty 22 Mayplace Avenue, Dartford DA1 4PZ
M6 KVM Rvd. D. Palmer 14 Walcot Parade, Bath BA1 5NF
M6 KVN K. Sewell 12 Haylands Square, South Shields NE34 0JB
MW6 KVO K. Owen 27C Waen Fawr Estate, Holyhead LL65 1LT
M6 KVP J. Wilkins 3 Ganesfield Doulting, Shepton Mallet BA4 4QA
M6 KVQ A. Taylor 60 Wood Ride, Petts Wood, Orpington BR5 1PY
MW6 KVS M. Kveksas 4 Llys Coed Derw, Llantwit Fardre, Pontypridd CF38 2JB
M6 KVT R. Simpson 48 Weatherhill Road, Horley RH6 9LY
M6 KVU W. Cuddeford 5 Rosevalley Threemilestone, Truro TR3 6BH
M6 KVV G. Easton Cowpen Road, Blyth NE24 5TS
M6 KVW L. Sutherland 198 Norbury Crescent, London SW16 4JY
M6 KVX R. Timmons 36 Worrall Road High Green, Sheffield S35 3LP
MM6 KWA K. Mackenzie Graceland, Crosshill, Duns TD11 3UF
M6 KWC K. Chapman 227 Raglan Street, Lowestoft NR32 2LA
M6 KWG Dr W. Wightman 36 Holyoake Avenue, Woking GU21 4PW
M6 KWI K. Irwin 11 The Crofts, Silloth, Wigton CA7 4EU
M6 KWK D. Barnett 12 Craig Walk Alsager, Stoke-on-Trent ST7 2RJ
M6 KWL C. Henniker 1 Brook House Drive, Fairfield, Buxton SK17 7HW
M6 KWM D. Wright 203 Winn Street, Lincoln LN2 5EY
M6 KWP K. Pritchard 124 Milburn Road, Ashington NE63 0PQ
MM6 KWQ M. Brock 18 Birchfield Place, Dumfries DG1 4SD
M6 KWT J. Caulfield 2 Thornley Road, Tow Law, Bishop Auckland DL13 4ED
M6 KWU S. Penhaligan 7 Trembath Crescent, Newquay TR7 2DX
MM6 KWV I. Mcculloch 4 Lady Street, Brydekirk, Annan DG12 5LZ
M6 KWW B. Covill Walnut Tree Cottage, Holt Street, Dover CT15 4HX
M6 KWX K. Woodard 4 Pingo Road, Watton IP25 6ZB
M6 KWZ M. Cross 43 Queens Park, Wadebridge PL27 7PR
M6 KXA K. Adams 57 Haddenham Road, Leicester LE3 2BH
MM6 KXF D. White 31 Dunmuir Road, Castle Douglas DG7 1LQ
M6 KXH P. Hambly Mendip Road, Stoke St. Michael BA35JU
M6 KXI D. Cook 14 Filbridge Rise, Sturminster Newton DT10 1AA
MI6 KXM K. Mitchell 39 Lissize Avenue, Rathfriland, Newry BT34 5DE
M6 KXN T. Canning Whitegates, Mayfield Avenue New Haw, Addlestone KT15 3AG
M6 KXO G. Barnard 35 High Street, Hitchin SG4 8AJ
M6 KXP C. Wooldridge 26 Grieg Close, Basingstoke RG22 4DU
M6 KXQ B. Wogden 3 Gordonstoun Place, Blackburn BB2 2PT
M6 KXR K. Golloghly 7 Jenner Close, Bungay NR35 1QR
M6 KXS R. Rigden 36A Atherston, Bristol BS30 8YB
M6 KXU B. Wood 25 Chestnut Terrace, Lamerton PL19 8RL
M6 KXV A. Jordan 19 Arrow Lane, Newhaven BN9 0FG
M6 KXW A. Jordan 20 Normandy Close, Crawley RH107XP
M6 KXX B. Cook 11 Arbrook Lane, Esher KT10 9EG
M6 KYB K. Greenshields 3 Lovers Walk, Wells BA5 2QL
M6 KYC M. Cook 12 Davison Street, Lingdale, Saltburn-by-the-Sea TS12 3DX
M6 KYE A. Letchford 64 Medway Road, Sheerness ME12 1DR
M6 KYI G. Virostek 58 Hendon Street, Brighton BN2 0EG
M6 KYK J. Slim Troublesome Reach, Playford Road, Woodbridge IP13 6ND
MW6 KYN K. O'Brien Flat 8, 4 Market Street, Newport NP20 1FU
M6 KYO A. Noriega 363 Pracha Uthit Road, Don Mueang BANGKOK 10210 Thailand
M6 KYP P. Higgins 27 Middleton Avenue, Ross-on-Wye HR9 5BD
M6 KYS E. Panayiotou 13 Oakmere Close, Potters Bar EN6 5JQ
MM6 KYV W. Hannah 71 Herriot Avenue, Kilbirnie KA25 7JB
M6 KYX A. Tandler 6 Field View Cottages, Brimfield, Ludlow SY8 4LB
M6 KYZ G. Amos 9 Manor Road Sundridge, Sevenoaks TN14 6DL
M6 KZB C. Wilkinson 67 Middleton Park Grove, Leeds LS10 4BG
M6 KZC J. Neal 40 Channel View Road, Portland DT5 2AY
M6 KZE P. Orwin 108 Cordwell Avenue, Chesterfield S41 8BN
M6 KZF E. Cook 927 Manchester Road Linthwaite, Huddersfield HD7 5NE
M6 KZH L. Gostelow 7 Coronation Drive, Donhead St Mary, , Shaftesbury SP7 9NA
M6 KZI D. Ball 28 Bryony Gardens, Gillingham SP8 4TR
M6 KZK A. Whybrow 64 Church Road Sevington, Ashford TN24 0LF
M6 KZM K. Maddy 56 Coachwell Close, Telford TF3 2JB
M6 KZN J. Spence Officers Mess, Royal Air Force, Brize Norton, Carterton OX18 3LX
M6 KZQ M. Pott 29 Lilliebrooke Crescent, Maidenhead SL6 3XJ
M6 KZR D. Ellens 150 Lumley Avenue, South Shields NE34 7DJ
MI6 KZS K. Sullivan 149 Largy Road Ahoghill, Ballymena BT42 2RG
M6 KZU M. Macrae 91 Chosen Way, Hucclecote, Gloucester GL3 3BX
M6 KZW O. Dawkins 45 High Street, Astcote NN12 8NW
M6 KZX G. Burtenshaw The Blenheims, Keymer Road, Burgess Hill RH15 0BA
M6 KZZ P. Keenan 5 Downton Walk Tiptree, Colchester CO5 0DH
M6 LAA L. Atkins Mouse Hall, Low Row, Richmond DL11 6PY
M6 LAC L. Ainger 41 Gilbert Road, Camberley GU16 7RD
MM6 LAD J. Cattigan Lunan Home Farm Cottage, Lunan Bay, Arbroath DD11 5ST
M6 LAE L. Thompson 43 Manor Road, Horsham St. Faith, Norwich NR10 3LF
M6 LAF S. Dooley 1 Rosewood Close, Little Sutton, Ellesmere Port CH66 4AJ
M6 LAG L. Burgess 40 Sheridan Terrace, Hove BN3 5AF
MM6 LAH L. Hail 70 Nobleston Estate, Alexandria G83 9DB
M6 LAI L. Lewczenko 10 Saxon Court, Swaffham PE37 7TP
MM6 LAK S. Ramsay 6 Cross Road, Peebles EH45 8DH
M6 LAL L. Cowley 3 Park Villas, Keswick CA12 5LQ
M6 LAM M. Quemby 19 Oak Close, Coalville LE67 4JU
MM6 LAO W. Neish 29 Currievale Drive, Currie EH14 5RN
M6 LAP S. Potts 22 Valebridge Road, Burgess Hill RH15 0QY
M6 LAQ L. Briggs 5 Links Avenue, Cromer NR27 0EQ
M6 LAS L. Scambell 8 South Bank Road, East Cowes PO32 6JE
M6 LAV L. Hand 168 Barcroft Street, Cleethorpes DN35 7DX
M6 LAY S. Lay 7 Hunt Street, Swindon SN1 3HW
M6 LAZ G. Sinclair 23 Cummings Square, Wingate TS28 5JF
M6 LBC L. Nicholas 9 Lila Place, Swanley BR88JB
M6 LBE S. Blackburn 20 Seascale Close, Blackburn BB2 3TP
M6 LBI I. Hyde 3 Hibbert Avenue, Denton, Manchester M34 3NZ
M6 LBK I. Macdonald Broomhill Mill Lane, Worthing BN13 3DH
M6 LBL L. Bell 56 Boyd Road, Wallsend NE28 7SQ
M6 LBM L. Mason 86 The Street, Rockland St. Mary, Norwich NR14 7AH
M6 LBN P. Tolcher 15 Langstone Close, Torquay TQ1 3TX
M6 LBQ B. Li Harrogate Ladies' College, Clarence Drive, Harrogate HG1 2QG
M6 LBR S. Hayward 20 Abbey Square, Walsall WS3 2RJ
MM6 LBS W. Thomson 20 Greenfield Road, Glasgow G32 0LP
M6 LBT L. Thomas 66 Sturdee Avenue, Great Yarmouth NR30 4HL
M6 LBU L. Baldwin 24 Rockrose Way, Portsmouth PO6 4EZ
M6 LBV E. Price Little Acre Eardisley, Hereford HR3 6LX
M6 LBW E. White 16 Illingworth Way, Foxton, Cambridge CB22 6RY
M6 LBX A. Walker 27 Fielding Avenue, Poynton, Stockport SK12 1YX
M6 LBY A. Smith 69 Keele Road, Newcastle under Lyme ST52JT
M6 LCC P. Harris 16 Laxton Gardens, Baldock SG7 6DA
M6 LCF M. Macdonell 54 Cinque Foil, Peacehaven BN10 8DZ
M6 LCG E. Ayre 1 Spring Gardens, Broadmayne, Dorchester DT2 8PP
M6 LCH M. Balshaw 16 East Avenue, Heald Green, Cheadle SK8 3DL
M6 LCI L. O'Brien 172 Cheriton High Street, Folkestone CT19 4HN
M6 LCK L. Kerr 24, Lindisfarne Street, Carlisle CA1 2ND
MM6 LCL L. Clark 9 Aileymill Gardens, Greenock PA16 0QF
M6 LCP L. Bourn 4 Fell Wilson Street, Warsop, Mansfield NG20 0PT
M6 LCQ L. Man 115 Northdown Park Road, Margate CT9 3PX
MI6 LCR L. Robinson 32 Corrycroar Road, Pomeroy, Dungannon BT70 3DY
M6 LCT C. Lai Clarence Drive, Harrogate HG1 2QG

M6 LCU C. Keszei 2 Blackmore Hill Farm Cottages, Calvert Road, Buckingham MK18 2HA
M6 LCV A. Grant 15A Elmdon Road Marston Green, Birmingham B37 7BU
M6 LCW L. Weeks 11 Brock Close, Deepcut, Camberley GU16 6GA
MW6 LCX J. Hawkins 104 Ty Fry, Aberdare CF44 7PP
M6 LCY L. Heron 301 Marton Road, Middlesbrough TS4 2HG
M6 LDA M. Leggett 55 Colomb Road, Gorleston, Great Yarmouth NR31 8BU
M6 LDB L. Hoddinott 30 Deans Mead, Bristol BS11 0QX
M6 LDD E. Daniels 2 Garstons Close Titchfield, Fareham PO14 4EN
M6 LDE D. Simmons 8 Lower Grange, Huddersfield HD2 1RU
M6 LDF L. Ferguson 67 Knowlton Road, Poole BH17 9EE
M6 LDG R. Hazel 2 Lynwood Grove Hull, Hullá HU52BE
M6 LDH O. Lai Harrogate Ladies' College, Clarence Drive, Harrogate HG1 2QG
MI6 LDI P. Dorrian 12 Gortnamona Place, Belfast BT11 8PP
M6 LDJ L. Jepson 143 Walnut Avenue Weaverham, Northwich CW8 3DX
M6 LDK L. Kelley 51 Grasmere Street, Liverpool L5 6RH
M6 LDL J. Hatton 49 Buxton Street, Morecambe LA4 5SR
MW6 LDM L. Martin 2 Ael Y Glyn, Nant Road, Harlech LL46 2UJ
M6 LDQ L. Dobinson 20 Newholme Crescent, Evenwood, Bishop Auckland DL14 9RY
M6 LDR L. Roworth 27 Bury Road, Dagenham RM10 7XR
MW6 LDS T. Jones 13 Bond Street, Aberdare CF44 7HA
M6 LDU L. Dumbleton 9 Wareham Road, Rubery, Birmingham B45 0JS
M6 LDW T. Evans 70 Tremarle Home Park, North Roskear, Camborne TR14 0AR
M6 LDY A. Collins 15 North River Road, Great Yarmouth NR30 1JY
MW6 LDZ T. Clapp Crunns Farm. Coxhill, Narberth SA67 8EH
M6 LEA J. Bridgehouse 43 Age Croft, Oldham OL8 2HG
M6 LEC L. Collinson 26 Westway Avenue, Hull HU6 9SA
M6 LEE L. Davies 94 Trent Way Kearsley , , Bolton BL4 8PS
M6 LEF L. Faulkner Mount Pleasant, Elkstones, Buxton SK17 0LU
M6 LEG K. Foulger 89 Blaney Crescent, London E6 6BB
M6 LEH L. Hargreaves 32 Bank Road, Carrbrook, Stalybridge SK15 3JX
M6 LEJ C. Calvert 1 Moorsholme Avenue, Manchester M40 9BW
M6 LEQ L. Pinkney 18 Bridlington Road, Driffield YO25 5HZ
MD6 LET D. Holohan 37 Claughbane Drive, Ramsey, Isle of Man IM8 2BH
M6 LEU L. Chadwick 78 Blakemore, Telford TF3 1PT
MM6 LEW M. Strachan 62 Charleston Drive, Dundee DD2 2EZ
M6 LEX K. Rowland 9 Churchlands, North Bradley, Trowbridge BA14 0TD
M6 LEY C. Harris 4 Coronation Drive, , Leigh WN7 2UU
M6 LFB L. Bell 42 Ocean Road, Walney, Barrow-in-Furness LA14 3DX
M6 LFC D. Smith 5 Verbena Close, Beechwood, Runcorn WA7 3JA
M6 LFD L. Drew 7 Bronte Court, Tamworth B79 8DN
M6 LFE B. Lewin 68 Brackley Square, Woodford Green IG8 7LS
MW6 LFG J. Beavan 21 Llannerch Road West, Rhos On Sea, Colwyn Bay LL28 4AU
M6 LFJ M. Stevens The Glen, Sunnyside Avenue, Sheerness ME12 2RA
MI6 LFK L. Corr1 6 Lime Park Balnamore, Ballymoney BT537QG
M6 LFL L. Theobold 25 Aysgarth Road, Leicester LE4 0ST
M6 LFM A. Campbell 35 Goodwood Road, Gosport PO12 4HN
MM6 LFN C. Bolton 30 Brackenhill Drive, Hamilton ML3 8AY
M6 LFO P. Noble 30 Whitewater Rise, Dibden Purlieu, Southampton SO45 4BY
M6 LFQ K. Blatch 61 Linden Way, Haddenham, Ely CB6 3UG
M6 LFR L. Davison 58 Priestley Court, South Shields NE34 9NQ
MM6 LFS A. Miles 9 Buchanan Drive, Lenzie, Kirkintilloch, Glasgow G66 5HS
M6 LFT Dr M. Cianni 121 Springfield Park Avenue, Chelmsford CM2 6EW
MI6 LFU C. Mccartney 62 Lakeview, Crumlin BT29 4YA
M6 LFW M. O'Donovan Wyllsden House, Stroud GL52PA
M6 LFX J. Lamb 9A Matlock Road, Canvey Island SS8 0EW
M6 LFY J. Wells 27 Victoria Avenue, Camberley GU15 3HT
M6 LFZ E. Fish 16 Cartmel Place, Ashton-On-Ribble, Preston PR2 1TY
M6 LGB L. Brazier 165 Avon Road, Worcester WR4 9AH
M6 LGD L. Thomas Fair View, Close Hill, Redruth TR15 1EP
M6 LGF D. Riches 48 Turner Road, Ipswich IP3 0LX
M6 LGH J. Abraham 18 Ferneley Crescent, Melton Mowbray LE13 1RZ
M6 LGI L. Grover Room 4 1 Oak Villas Winchester Road Bishops Waltham, Southampton SO32 1BR
M6 LGJ L. Sparkes 40 Cambridge Road, Eastbourne BN22 7BT
M6 LGL M. Rusu 1 Turnbull Road, March PE15 9RX
M6 LGM L. Mann 14A Orchehill Avenue, Gerrards Cross SL9 8PX
M6 LGQ M. Lee 4 Cluny Court, Wavendon Gate, Milton Keynes MK7 7TT
MM6 LGS C. Sloan 7 Clova Street, Thornliebank, Glasgow G468NA
M6 LGT J. Leggett 78 Spruce Avenue, Ormesby Saint Margaret, Great Yarmouth NR29 3RQ
M6 LGV L. Vickers 24 Hearnes Meadow, Seer Green, Beaconsfield HP9 2YJ
MI6 LGX C. Sheppard 4 Fairview Drive, Whitehead, Carrickfergus BT38 9NT
M6 LHA A. Crawford 4 Trimpley Drive, Kidderminster DY11 5LB
MW6 LHB S. Hillman 8 Brigantine Grove, Duffryn, Newport NP10 8ET
M6 LHD K. Newbould 47 Old Barber, Harrogate HG1 3DF
M6 LHE L. Heppenstall 15 Gibraltar Road, Hemswell Cliff, Gainsborough DN21 5XJ
M6 LHF M. Pickering 30 Hotspur Avenue, Whitley Bay NE25 8RP
M6 LHG L. Hagan 7 Betjeman Mews, Southend-on-Sea SS2 5EJ
M6 LHH B. Cummings 8 Carshalton Way, Lower Earley, Reading RG6 4EP
M6 LHI Dr M. Depardieu 4 Belvedere Fff, Bath BA1 5ED
M6 LHJ L. Smith 17 Grove Street, Kirton Lindsey Lincs DN21 4BY
M6 LHL J. Ablett 33 Langley Road, Leeds LS13 1AX
M6 LHM L. Mason 2 Iris Close, Widnes WA8 4GA
M6 LHN L. Hulse 202A Shooters Hill Road, London SE3 8RP
M6 LHO A. Rowan 14 Craven Lea, Liverpool L12 0NF
M6 LHP L. Pass 14A Elm Avenue, Hucknall, Nottingham NG15 6GE
M6 LHQ L. Daniels 1 Church Road Cottages, Chart Road, Chart Sutton ME17 3RG
M6 LHR L. Rich 39 Wren Close, Heathfield TN21 8HG
M6 LHS H. Ren 50 Highwoods Drive, Marlow SL7 3PY
M6 LHV G. Mccarthy 14 Cosedge Crescent, Croydon CR0 4DN
M6 LHW H. Lau 113 Ruxley Lane, Epsom KT19 9EX
M6 LIB E. Bayliss 19 Rugby Road, Dunchurch, Rugby CV22 6PG
M6 LID J. Eades 128 Russells Hall Road, Dudley DY1 2NN
M6 LIE L. Gregory 14 Anderson Road, Hemswell Cliff, Gainsborough DN21 5XP
M6 LIH N. Hindl Hill House Haulage 67 Yarmouth Road, Ellingham, Bungay NR35 2PH
M6 LII A. James 45 Knockholt Road Cliftonville, Margate CT9 3HL
M6 LIJ J. Mason 14A Brooke Avenue, Margate CT9 5NG
M6 LIK T. Russell 38 Speedwell Close, Witham CM8 2XL
MM6 LIL L. Clark 17 Lewis Rise, Broomlands, Irvine KA11 1HH
M6 LIN L. Briggs 20 Broad Lane, Pelsall, Walsall WS4 1AP
M6 LIP S. Shakespeare 36 Maitland Road, Russells Hall Estate, Dudley DY1 2NU
M6 LIQ G. Pollard 22 Girton Avenue, Ashton in Makerfield WN4 9SA
M6 LIS M. Ramsbottom 1 Barn Gill Close, Blackburn BB2 3HU
MI6 LIT C. Brush 56 Larchwood, Banbridge BT32 3UT
M6 LIU D. Petrauskas Ateities 7, Jurbarkas LT74208 Lithuania
M6 LIV J. Hay 13 Windsor Road, Workington CA14 5BQ
MW6 LIW D. Evans 23 Wellington Street, Aberdare CF44 8EW
MW6 LIZ E. Martin 62 Llwyn Ynn, Talybont LL43 2AL
M6 LJA M. Duxbury 32 Radford Street, Darwen BB3 2PB
M6 LJB A. Billingham 6 Kemble Close, Lincoln LN6 0NR
M6 LJD L. Denham 92 Windermere Avenue, Southampton SO16 9GF
MM6 LJE L. Treble 2 Baidland Meadow, Dalry KA24 5HP
M6 LJF J. Bogdaniec 3 Cavalry Chase, Okehampton EX20 1GR
M6 LJG L. Goldsmith Hunters Cottage, 61 Fengate Drove, Weeting, Brandon IP27 0PW
M6 LJJ L. Jones Brimham Lodge Farm, Harrogate HG3 3HE
M6 LJK L. Kirkpatrick The Cleave, Nine Oaks Estate, Yelverton PL20 6ND
M6 LJM L. Marriott 94 Lyndhurst Road, Worthing BN11 2DW
MI6 LJO O. Conaghan 94 Curlyhill Road, Strabane BT82 8LS
M6 LJP L. Passam Glanceiro, Llandre, Bow Street SY24 5BS
M6 LJR C. Lonie Jr 41 De La Hay Avenue, Plymouth PL3 4HS
M6 LJS L. Smith 177 Waterloo Road, Stoke-on-Trent ST6 2ER
M6 LJT M. Rushton 17 Highbury Gardens, Ramsgate CT12 6QG
M6 LJU R. Gilliam 7 Hamble Street B, London SW6 2RT
M6 LJV S. Cilliers 61 Heathwood Gardens, London SE7 8ET
MI6 LJZ A. Porter 8 Ballyregan Avenue Dundonald, Belfast BT16 1JW
M6 LKA M. Wilkinson 3 Balsams Close, Hertford SG13 8BN
M6 LKD M. Dust 1A Elm Road, Erith DA8 2NN
M6 LKE L. Huddart 1 Rydal Court, Penrith CA11 8PN
M6 LKF L. Chapman 4 Russell Road, Clacton-on-Sea CO15 6BE
M6 LKI K. Lockstone Oceana, The Parade Pevensey Bay, Pevensey BN24 6LX
M6 LKK H. Freeman 95 Raleigh Road, Wirral CH46 2QY
M6 LKL G. Urban 33 High Meadow, Hathern, Loughborough LE12 5HW
M6 LKM L. Mcdonnell 108 Long Lane, Garston, Liverpool L19 6PQ
M6 LKW L. Whitby 45 Trent Road Shaw, Oldham OL2 7YG
M6 LKY A. Lee 27 Victoria Avenue, Camberley GU15 3HT
M6 LLB L. Brigham 42 Cayley Close, Clifton, York YO30 5PT
M6 LLC L. Clark 1 Brooklime Road, Liverpool L11 2YH
M6 LLD A. Clark 1 Brooklime Road, Liverpool L11 2YH
M6 LLE R. Clark 1 Brooklime Road, Liverpool L11 2YH
MI6 LLG S. Horner 10 Meadow Court, Bushmills BT57 8SD
M6 LLH A. Hoyte 43 Orchard Drive Mayland, Chelmsford CM3 6EP
MI6 LLI R. Hetherington 112 Screeby Road, Fivemiletown BT75 0LG
M6 LLK Dr D. Roberts Beggarwood House Ravensworth Park Estate, Gateshead NE11 0HQ
M6 LLL L. Foulkes 43 Mill Hayes Road, Stoke-on-Trent ST6 4JB
M6 LLM L. Meredith 131 Trimdon Avenue, Middlesbrough TS5 8RY
M6 LLN A. Allen 59 Bottels Road Warboys, Huntingdon PE282RZ
M6 LLO S. Hamilton 11 Jubilee Gardens, Telford TF3 2BR
MI6 LLS B. Kelly 153 Ardanlee, Ballynagard, Londonderry BT48 8RT
M6 LLW H. Liu Harrogate Ladies' College, Clarence Drive, Harrogate HG1 2QG
MI6 LLZ K. Dorman 25 Blackthorn Road, Newtownabbey BT37 0GH
M6 LMB L. Bate 4 Yoxall Road, Newborough, Burton-on-Trent DE13 8SU
M6 LMC L. Chatt Rosemary Cottage, Causeway End Road, Dunmow CM6 3LU
M6 LMG J. Porter 3 The Walks, Main Road, Woodbridge IP12 3DZ
MM6 LMH L. Mitchell Hynd Smithy House, Bruichladdich, Isle of Islay PA49 7UN
M6 LMI L. Mikolka Flat, 1 Scotney Court, Romney Marsh TN29 9JP
M6 LMJ L. Mathlin 29 Wagtail Drive, Stowmarket IP14 5GH
M6 LMU J. Jarvis 8 Wesleys Fold, Pinfold Street, Wednesbury WS10 8UN
M6 LMW A. Thomson Shire Jee Neevas, Cold Ash Hill, Thatcham RG18 9PH
MM6 LNB S. Anderson Middleton Mains, Gorebridge EH23 4RL
M6 LNC D. Lancaster Linkhill View, Frith Common, Eardiston, Tenbury Wells WR15 8JX
M6 LND L. Dawson 5 Harbour View, Roker, Sunderland SR6 0NL
M6 LNE C. Bruce Nunfield House, Bull Lane, Sittingbourne ME9 7SL
M6 LNI G. Peters Flat 6, Park Court, 46 North Park Road, Harrogate HG1 5AD
M6 LNM W. Walters 84 Green Close, Sturminster Newton DT101BL
MI6 LNP N. Scott 5 Sharonmore Parade, Newtownabbey BT36 6PR
M6 LNQ N. Sargent 21 St. Michaels Road, Claverdon, Warwick CV35 8NT
M6 LNR A. Evetts 21 Instone Road, Halesowen B63 4SA
M6 LNS D. Williams Apartment 54, 7 Tiltman Place, London N7 7EL
MW6 LNU R. Sullivan 58 Ynyscynon Road, Tonypandy CF40 2LN
M6 LNV R. Mayes Crescent Avenue, London RM17 6AZ
M6 LNX S. Wareham 8 Simon Road, Longlevens, Gloucester GL2 0ER
M6 LOB A. Brash 44 Broadway East, Chester CH2 2DP
M6 LOC L. Barker 88 Cecilia Road, London E8 2ET
MW6 LOD L. Broadhurst 4 Lilburne Drive, Newport NP19 0ET
M6 LOE N. Chaplin 5 Maxwell Street, Bury BL9 7QA
M6 LOF S. Mcilwaine 70 Autumn Drive, Sutton SM2 5BA
M6 LOG Dr I. Van Der Linde 77 Port Vale, Hertford SG14 3AF
M6 LOK B. Wolff 7 Church Terrace, Reading RG1 6AS
M6 LOL R. Cooper 11A Ambleside, Gamston, Nottingham NG2 6NA
MM6 LON L. Nicoll 15 Redford Walk, Edinburgh EH13 0AF
M6 LOQ P. Wright 25, Paynes Meadow, Whitminster, Gloucester U. K. GL2 7PS
M6 LOS J. Hawthorn 1 Tudor Close, Leigh-on-Sea SS9 5AR
MI6 LOT L. Treanor 1 Granemore Park, Keady, Armagh BT60 2GP
M6 LOW J. Lowenthal 133 Marshalswick Lane, St. Albans AL1 4UX
M6 LOZ L. Shaw 35 Richmond Street, Stoke-on-Trent ST4 7DZ
M6 LPB Verulam ARC c/o Lord N. Petit-Brown 6 Cope Avenue, Nantwich CW5 5JE
M6 LPD D. Lester 171 Glenavon Road, Birmingham B14 5BT
M6 LPF A. Kuba Flat 291-295, Jellicoe Court, Southampton SO16 3UJ
M6 LPH F. Quinn Halley Cottage Beech Grove Gardens, Carlisle CA30LR
M6 LPI N. Stone Flat, 97-99 Stoke Road, Gosport PO12 1LR
M6 LPK L. Kiddell 1 Sparham Hill, Sparham, Norwich NR9 5QT
M6 LPN J. Leach 9 Crown Point Drive, Ossett WF5 8RQ

M6 LPO R. Cowperthwaite 30 Glover Place, Bootle L20 4QR
M6 LPP P. Petersen 15 Kent Gardens, Birchington CT7 9RS
M6 LPS D. Clement 24 Millais, Horsham RH13 6BS
MM6 LPT J. Mckinnon 81 Willow Drive, Johnstone PA5 0DA
M6 LPV D. Churchill 39 East Street, Corfe Castle, Wareham BH20 5EE
M6 LPW L. Walker 7 Stroudley Close, Ashford TN240TY
M6 LPX S. Kembrey 101 Yew Tree Drive, Bristol BS15 4UF
M6 LPY S. Taylor 27 Anchorsholme Lane East, Thornton-Cleveleys FY5 3QH
MW6 LPZ D. Stevens 12 Firs Avenue Fairwater, Cardiff CF53TH
M6 LQA J. Strickler Friedlebenstra□e 48, Frankfurt/Main 60433 Germany
M6 LQC D. Tofield 6 Church Grove, Barnstaple EX32 9DJ
MM6 LQF J. Bennet 24/11 Greenpark, Edinburgh EH177TA
MW6 LQG S. Trump 168 Brynglas Avenue, Newport NP20 5LP
MW6 LQK M. Stanger 66 Heather Court, Ty Canol, Cwmbran NP44 6JR
M6 LQM B. Poulter 1 Heathyfields Road, Farnham GU9 0BN
M6 LQO J. Thompson 32 Coult Avenue, North Hykeham, Lincoln LN6 9RG
M6 LQP A. Clack 3 Darwin Close, Swindon SN3 3NF
M6 LQR M. Constantine 82 The Oval, Brough HU15 1DD
M6 LQT Dr P. Patureau 4 Belvedere Fff, Bath BA1 5ED
MM6 LQU S. Clayton 19 Howe Park, Edinburgh EH10 7HF
MW6 LQW M. Downward 17 Beeston Terrace, Wrexham LL139NN
M6 LQY I. Gilmore 19 Green Sward Lane, Redditch B98 0EN
M6 LRA L. Akred 25 Kitchener Street, Walney, Barrow-in-Furness LA14 3QW
M6 LRB D. Williamson 4 King Edward Road, Northampton NN1 5LU
M6 LRF R. Deller Four Winds Farm, Buckworth Road, Huntingdon PE28 4JX
M6 LRG M. Parker Ridgeways, Mill Common, Westhall, Halesworth IP19 8RQ
M6 LRH L. Hancock 106 Hoyle Street, Warrington WA5 0LW
MM6 LRK A. Lark 20 Lawfield, Coldingham, Eyemouth TD14 5PB
M6 LRL S. Vane 17 Knights Walk, Abridge, Romford RM4 1DR
MI6 LRN K. Bell 3 Alexandra Crescent, Larne BT40 1NE
MW6 LRO O. Thomas Garth Celyn, St. Davids Road, Aberystwyth SY23 1EU
M6 LRQ M. Bacon 69 Doverdale Close, Redditch B98 7SD
M6 LRR A. Nielsen 29 Greygarth Close Bransholme, Hull HU7 5AP
M6 LRU R. Walker 125 Devereux Road, West Bromwich B70 6RQ
M6 LRV H. Moore 52 Limefield Street, Accrington BB52AF
M6 LRW J. Welch 49 Walshs Manor, Stantonbury, Milton Keynes MK14 6BU
MM6 LRX G. Morrison 11 Goodman Place, Maddiston FK2 0NB
M6 LRZ S. Carr 24 Park Road, Blyth NE24 3DH
M6 LSA L. Allcock 26 Castleton Grove, Inkersall, Chesterfield S43 3HU
M6 LSB L. Catterall 14 Dunham Drive, Whittle-Le-Woods, Chorley PR6 7DN
M6 LSE C. Stoten 12 Boyd Avenue, Dereham NR19 1LU
M6 LSG L. Spong 2 Strathmore Drive, Charvil, Reading RG10 9QT
M6 LSH L. Shaw 47 Beechfields, Eccleston, Chorley PR7 5RF
M6 LSJ L. Sawkins 20 Nye Close, Cheddar BS27 3PB
M6 LSK C. Hare 25 Southend Place, Sheffield S2 5FQ
M6 LSL A. Wood 4 St. Andrews, The Common, Cranleigh GU6 8NX
M6 LSN P. Bamber 15 Grantley Avenue Kingswood Oak, Shrewsbury SY3 5LA
M6 LSO G. Coltman The Oaks Rushden, Buntingford SG9 0SN
M6 LSP L. Phillips 5 Barnes Green, Wirral CH63 9LU
M6 LST D. Lui Clarence Drive, Harrogate HG1 2QG
M6 LSU M. Pesendorfer 13 Blake Road, London N11 2AD
M6 LSV S. Ludziss 82 Trinity Avenue, Mildenhall, Bury St. Edmunds IP28 7LS
MW6 LSW D. Johns 35 Bronhaul, Talbot Green, Pontyclun CF72 8HW
MI6 LSY L. Stock 15 Mahon Drive, Portadown, Craigavon BT62 3JB
M6 LTB L. Burke Teviot, Malthouse Lane, Peasmarsh TN31 6TA
MM6 LTC T. Craig Cemetery Lodge, Lochmaben, Lockerbie DG11 1RL
M6 LTD P. Asher 124 Bath Street, Market Harborough LE16 9JL
M6 LTL L. Layland 3 Thirlmere Road, Golborne, Warrington WA3 3HH
M6 LTM L. Simons 123 Main Street, Little Harrowden, Wellingborough NN9 5BA
MW6 LTN M. Ashford 3 Candleston Place Bonymaen, City & County Of Swansea, Swansea SA1 7JB
M6 LTO L. Trend 140 Ardleigh, Basildon SS16 5RW
M6 LTQ W. Coates 36 The Crescent, Welwyn AL6 9JQ
MW6 LTR C. Rowe 21 Graig Terrace, Abercwmboi, Aberdare CF44 6AH
M6 LTS L. Stant 3 Uffa Fox Place, Cowes PO31 7NX
M6 LTU M. Brazinskas 25 Elswick Road, London SE13 7SP
M6 LTV N. Griffin 3 Gooseander House, Cirencester GL7 5FH
M6 LUA T. Goddard 217 Speedwell Road, Bristol BS5 7SP
M6 LUC D. Boden 249 Nottingham Road, Ilkeston DE7 5AT
M6 LUD K. Willson Ludpit Cottage, Ludpit Lane, Etchingham TN19 7DB
M6 LUG P. Schoenmaker 24 Greenheys Drive, London E18 2HB
M6 LUI G. Conboy 7 Bell Clough Road, Droylsden, Manchester M43 7NS
M6 LUJ S. Cottam 14 Barnard Close, Rednal, Birmingham B45 9SZ
M6 LUK L. Johnson 7 Southover Way, Hunston, Chichester PO20 1NY
M6 LUM L. Medza 16 Huntroyde Avenue, Bolton BL2 2ET
MI6 LUP M. Sullivan 149 Largy Road Ahoghill, Ballymena BT42 2RG
MW6 LUQ E. Williams 56 Heol Llansantffraid Sarn, Bridgend CF32 9NH
M6 LUR P. Hadley 20 Merrybrook, Evesham WR11 2QF
M6 LUT A. Lutley Springfield, Rookery Hill, Ashtead KT21 1HY
M6 LUW P. Steadman 82 Hereford Road, Shrewsbury SY3 7RA
M6 LUY G. Hinson The Leys, Brierley Hill DY5 3UJ
M6 LUZ G. Luscombe 28 St. Giles Gate, Doncaster DN5 8PQ
M6 LVA R. Scullion 41 Myrica Grove, Hoole, Chester CH2 3EW
M6 LVC I. Collins 19 Peel Street, Kidderminster DY11 6UG
MW6 LVD K. West 58 Claerwen Gelligaer, Hengoed CF82 8EX
M6 LVE C. Johnson 25 Pelham Street, Worksop S80 2TW
MW6 LVH E. Parker Gadlys, London Road Valley, Holyhead LL65 3DP
M6 LVJ I. Humphries 13 Malvern Close, Banbury OX16 9EL
M6 LVK K. Jones 7 Fazan Court, Wadhurst TN5 6BT
M6 LVN Dr J. Clark 14 Portobello Terrace Birtley, Chester le Street DH3 2JS
M6 LVR A. Humphreys 2 Harrop Place Ribbleton, Preston PR2 6TD
M6 LVS E. Pasqual Flat 203 2 South Ealing Road, London W5 4BY
MM6 LVV R. Holtom Old Post Office, Church Road, Laurencekirk AB30 1YS
M6 LVW L. Walker 17 Carr House Road, Halifax HX3 7QY
M6 LVX T. Von Bergmann 17 Seeley Crescent, Street BA16 0RN
M6 LWA L. Alderson 40 Hillside Crescent, Skipton BD23 2LE
MM6 LWB L. Bradley Amon Sul, Kiltarlity, Beauly IV4 7HT
M6 LWE A. Martin 34 Allpits Road, Calow, Chesterfield S44 5AT
MW6 LWF L. Fish Iddon Cottage, Bronygarth, Oswestry SY10 7NF
M6 LWG C. Rowland 20 Bell Hill Park Lindale, Grange-over-Sands LA11 6JZ
MI6 LWI A. Lewis 8 Liester Park, Ballyclare BT39 9RZ
M6 LWJ L. Webb Fern Bank, Wood Lea, Rotherham S66 8NN
M6 LWM C. Smithen 10 High Street, Temple Ewell, Dover CT16 3DU
M6 LWP A. Lawler 58 Bahram Road, Costessey, Norwich NR8 5EY
M6 LWQ G. Ford The Lodge, Home Farm Lane, Rimpton, Yeovil BA22 8AS
M6 LWR S. Ross 51 Claypiece Road, Bristol BS13 9DR
M6 LWS W. Sawyer 20 Park Terrace Willington, Crook DL15 0QL
M6 LWT M. Hennessey 3 Northgate Cottage, Falmer Road, Rottingdean, Brighton BN2 7DT
M6 LWV L. Brown 99 Apprentice Drive, Colchester CO4 5SE
M6 LWY G. Fenton 40 High Street, Easington Lane, Houghton le Spring DH5 0JN
M6 LWZ R. Marshall 17 Haywards Place, Easterton, Devizes SN10 4PP
M6 LXC Rvd. L. Clark 226 Philip Lane Tottenham, London N15 4HH
M6 LXE A. Elena 31 Larksfield Avenue, Bournemouth BH9 3LW
M6 LXH S. Li Clarence Drive, Harrogate HG1 2QG
M6 LXI E. Pestano 38 Third Avenue, Bexhill-on-Sea TN40 2PA
M6 LXK A. Weston 37-39 Clover Hill, Sunniside, Newcastle upon Tyne NE16 5PT
M6 LXM A. Brighton 67 Wilks Farm Drive, Sprowston, Norwich NR7 8RG
M6 LXN R. Froggatt 39 Spitfire Way, Hamble, Southampton SO31 4RT
M6 LXP M. Ling Flat 14, Rowan Court, London SW20 0BA
M6 LXQ Dr G. Turner 8 Scarborough Terrace, York YO30 7AW
M6 LXR L. Rhodes 91 Bedonwell Road, Bexleyheath DA7 5PS
M6 LXS M. Konrad 12 Dale View Silsden, Keighley BD20 0JP
M6 LXU D. Walton 45 Wells Hall Road Great Cornard, Sudbury CO10 0NH
M6 LXW C. Stickley 152 Fore Street, Pinner HA5 2NE
M6 LXX Dr A. Erlank 38 Elsley Road, London SW11 5LL
M6 LXY M. Oxley 49 Dalton Crescent, Shildon DL4 2LE
M6 LYA A. Polakovs 76 Sandringham Crescent, Leeds LS17 8DF
M6 LYB G. Turner 43 Harding Avenue, Eastbourne BN22 8PL
M6 LYC M. O'Connor 5 Kesbrook Drive Ashwood Park, Overseal DE126NS
M6 LYD R. Lyddall 102 Chapel Road, Brightlingsea, Colchester CO7 0HE
M6 LYF R. Downhill 40 Collingwood Close, Eastbourne BN23 6HW
M6 LYG D. Bache 81 Westgate, Driffield YO25 6TA
MM6 LYH R. Latimer 6 Crawford Avenue Rosemarkie, Fortrose IV10 8UX
M6 LYJ W. Bradley 67 Bury Road, Leamington Spa CV31 3JD
M6 LYN L. Groves 3 Hudson Davies Close, Pilley, Lymington SO41 5PA
M6 LYO O. Lyon Splinters, Nelson Park Road, Dover CT15 6HL
M6 LYP M. Jonusas 70 Methuen Street, Southampton SO14 6FR
M6 LYQ T. Mann 6 Kenley Close, Wickford SS11 8XL
M6 LYS A. Booth 27 Shenstone Road, Rotherham S65 2JR
M6 LYU J. Swanbrow 7 Manor Crescent, Rookley, Ventnor PO38 3NS
M6 LYY A. Allgood 39 Eastwood, Chatteris PE16 6RX
M6 LYZ S. Fearn 79 Maudlin Drive, Teignmouth TQ14 8SB
M6 LZA L. Jackson 90 Horne Street, Bury BL9 9HS
M6 LZC Zk CG c/o D. Cassidy 172 Lyde Road, Yeovil BA21 5PN
M6 LZD G. Kelley 31 Cherry Park, Brandon, Durham DH7 8TN
M6 LZF M. Clitheroe Green End, North Street, Castle Acre, King's Lynn PE32 2BA
M6 LZG T. Coldham 4 Carr Meadow Bamber Bridge, Preston PR5 8HS
M6 LZI J. Miller 9 St. Nicholas Road Tillingham, Chelmsford CM0 7SQ
MI6 LZL P. Robinson 119 Avenue Road, Lurgan, Craigavon BT66 7BD
M6 LZM M. Radulov 14 Grove Road, Chatham ME4 5HS
M6 LZN J. Marshall 6 Foster Walk, Sherburn in Elmet LS25 6EU
M6 LZP J. Lovelock Sea Spray, Lighthouse Road The Lizard, Helston TR12 7NU
M6 LZQ L. Hollingworth 43 Wingfield Road, Hull HU9 4PR
M6 LZS C. Dyson 5 Welton Park Welton, Daventry NN11 2JW
M6 LZT C. Plant 6 Talsarn Grove, Stoke-on-Trent ST4 8YL
MW6 LZU P. Matthews The Chateau, Wynnstay Hall Estate, Ruabon, Wrexham LL14 6LA
MW6 LZV K. Mccafferty 19 Quarr Road, Pontardawe SA8 4JD
MI6 LZW R. Lewis 8 Liester Park, Ballyclare BT39 9RZ
M6 LZX B. Siddle 7 Farebrother Street, Grimsby DN32 0NH
M6 LZY C. Hillcox 2 New Hall Drive, Sutton Coldfield B75 7UU
MW6 LZZ R. James 168 Brynglas Avenue, Newport NP20 5LP
M6 MAA M. Meadowcroft 210 Dickinson Close, Blackburn BB2 2LT
M6 MAD D. Binnall 21 Appletree Road, Featherstone, Pontefract WF7 5EA
M6 MAG C. Lavery 12 Jackson Close, Kesgrave, Ipswich IP5 2QL
M6 MAH M. Hyett 1 Darell Close, Quedgeley, Gloucester GL2 4YR
M6 MAJ M. Kealey 24 Ben Nevis Road, Birkenhead CH42 6QY
M6 MAK P. Mcgrath 24 Broadoak Drive, Lanchester, Durham DH7 0QA
M6 MAL M. Wallace 4 Edmund Street, Kettering NN16 0HU
M6 MAM M. Mcdougall 122 Lee Lane, Horwich, Bolton BL6 7AF
M6 MAO I. Connors 3 Wheatfield Way, Chelmsford CM1 2QZ
M6 MAP M. Peters 25 Windsor Court, Falmouth TR11 3DZ
M6 MAS S. Shailes 9 Ingham Street, Padiham, Burnley BB12 8DR
M6 MAT M. Pye 42 Milford Street, Colne BB8 9QH
M6 MAW M. Ansell-Wood Sanju, Old Lane, Nethertown, Bradford BD11 1LU
M6 MAX M. Trivett 36 Edward Street, Hartshorne, Swadlincote DE11 7HG
M6 MAY M. Buist 23 St. Chads Drive, Gravesend DA12 4EL
M6 MBB M. Bennett-Blacklock 46 Friern Road, London SE22 0AX
MW6 MBC S. Cook 114 Caerphilly Road, Bassaleg, Newport NP10 8LJ
M6 MBD M. Bowell 28 Jubilee Close, Byfield, Daventry NN11 6UZ
M6 MBF M. Bailey 34 Jephson Drive, Birmingham B26 2HW
M6 MBG C. Collins The Coppice, Old Coach Road, Sheffield S6 6HX
M6 MBH M. Creedy 25 Ryton Close, Redditch B98 0EW
M6 MBI M. Collis 35 Fishergate, Norwich NR3 1SE
M6 MBK J. Seaman 6 Gibbets, Hale Road, Thetford IP25 7QX
M6 MBO M. Siddall 10 Foston Drive, Chesterfield S40 4SJ
M6 MBP N. Challis 48 Brunsfield Close, Wirral CH46 6HE
M6 MBQ M. Ball 22 Wheatley Drive, Mirfield WF14 8NW
M6 MBR M. Burr 49 Knightsbridge Way, Hemel Hempstead HP2 5ES
M6 MBS M. Smith 78 New Croft, Weedon, Northampton NN7 4RL
M6 MBU M. Burnett 218 High Street, Clapham, Bedford MK41 6BS
M6 MBX M. Bell 68 Hereford Drive, Liverpool L30 1PR
M6 MBY M. Cox 120 Helmsdale, Bracknell RG12 0TB
MW6 MBZ M. Argyle 17 Heol Cae-Rhys, Cardiff CF14 6AN
MM6 MCA J. Mcardle 1 Queen Street, Hamilton ML3 9JR
M6 MCB M. Cooper 6 The Crescent, Cookley, Kidderminster DY10 3RY

M6 MCC S. Allaker 61 West Street, Winterton, Scunthorpe DN15 9QG
M6 MCE G. Mcewen 37 Malvern Way, Twyford, Reading RG10 9PY
MI6 MCF J. Macfarlane 5 Sunbeam Terrace, Lisnaskea, Lisnaskea BT92 0LL
M6 MCH M. Hill 10 The Moorings, Littlehampton BN17 6RG
M6 MCJ M. Coiley 25 Spring Garden Street, Queensbury, Bradford BD13 2AE
MI6 MCK S. Mckay 27 Rathbeg Crescent, Limavady BT49 0AT
M6 MCL M. Chaffey 46 Bartlett Way, Poole BH12 4FD
M6 MCM S. Mcmurtrie 5 Hill Road, Carshalton SM5 3RA
M6 MCO M. Denham1 209 Coppice Road, Arnold, Nottingham NG5 7HD
M6 MCP M. Pennington 88 Gillsmans Hill, St. Leonards-on-Sea TN38 0SL
M6 MCR K. Wills 24 Bitten Court, Northampton NN3 8HH
M6 MCS M. Statham Broad Oak Bungalow, Manston, Sturminster Newton DT10 1EZ
MM6 MCT J. Leitch 25 Lime Street, Grangemouth FK3 8LZ
M6 MCU M. Barker 18 Nickleby Road, Waterlooville PO8 0RH
M6 MCW M. Wilson 11A St. Julians Road, London NW6 7LA
MM6 MCX B. Bannister 28 Mansefield Crescent, Old Kilpatrick, Glasgow G60 5JJ
M6 MCY M. Attree 52 The Ridgeway, St. Albans AL4 9PS
M6 MCZ A. Davis Old Malt Kiln House, Barden, Leyburn DL8 5JS
M6 MDB T. Brown 69 Lawn Closes Alt, Oldham OL8 2HB
M6 MDC D. Smith 193 Brooke Road, Oakham LE15 6HQ
M6 MDG D. Green 45 Highthorn Road, Kilnhurst, Mexborough S64 5UP
M6 MDJ D. Jefferson 74 Cloisters Avenue, Barrow-in-Furness LA13 0BB
M6 MDL M. Luttrell 5 Byron Avenue, Radcliffe, Manchester M26 3QX
M6 MDM S. Clarke 27 Netherhouse Moor, Church Crookham, Fleet GU51 5TZ
M6 MDN M. Norman 28 Cumberland Close, Twickenham TW1 1RS
M6 MDR M. Riley 16 Dudley Avenue, Leicester LE5 2EE
M6 MDT M. Taylor 11 Holly Crescent, Sunnyside, Rotherham S66 3PL
M6 MDU M. Duchar 13 Thirlmere Avenue, Chester le Street DH2 3ED
M6 MDX M. Abraham 12 Graham Road, Halesowen B62 8LJ
M6 MDZ M. Smith 31 Atlantic Crescent, Sheffield S8 7FW
M6 MEB P. Shires 30 Philip Garth, Wakefield WF1 2LS
M6 MEC M. Carroll 11 Old Hall Court, Old Hall Street, Malpas SY14 8NE
M6 MED R. Muswell 7 Stoneyfields Gardens, Edgware HA8 9SP
M6 MEJ M. Bray 13 Rosebay Close, Hartlepool TS26 0ZL
M6 MEK A. Walters 32 Lincoln Road, Tuxford, Newark NG22 0HP
M6 MEL M. Lewis 73 Addenbrooke Street, Wednesbury WS108HJ
M6 MEN M. Lucas 8 Greenwood Close, Bury, Ramsey, Huntingdon PE26 2NZ
M6 MEO B. Clements 129 Gunners Road, Southend-on-Sea SS3 9SB
M6 MEP M. Lawton 20 Wharfedale Walk, Stoke-on-Trent ST3 2RS
M6 MEQ A. Riley 35 Ross Avenue, Wirral CH46 2SA
M6 MES J. Reeve 5 Antelope Avenue, Grays RM16 6QT
M6 MEU A. Sharif 6 Buckle Rise, Seaford BN25 2QN
M6 MEV M. Sanderson 20 East View, Castleford WF10 1PZ
MW6 MFB J. Murray 17 Bro Dawel, Bodedern, Holyhead LL65 3TB
M6 MFC C. Parkes 3 Greenham Close, Middlesbrough TS3 9NT
M6 MFD G. Mansfield 2 School Street, Syston, Leicester LE7 1HN
M6 MFF M. File 52 Helmdon Close, Ramsgate CT12 6TT
M6 MFG M. Gainza Stanhope, High Street, Saxmundham IP17 3EP
M6 MFJ M. Coleman 3 Tummon Road, Sheffield S2 5FD
M6 MFK A. Powell 37 Newnham Close, Mildenhall, Bury St. Edmunds IP28 7PD
M6 MFL M. Ross 31 Duke Street, Oswaldtwistle, Accrington BB5 3PN
M6 MFM M. Maheshwarappa R43 Room 2 International House, University Of Surrey, Guildford GU2 7JL
MW6 MFN A. Evans Maesyronnen, Sarnau, Llanymynech SY22 6QL
M6 MFO Lt. Col. M. Foster 21 The Bourtons, Newton Road, Totnes TQ9 6LS
MI6 MFR A. Morrow 769 Farranseer Park, Macosquin, Coleraine BT51 4NB
M6 MFS S. Finlayson 41 Low Catton Road, Stamford Bridge, York YO41 1DZ
M6 MFU K. Ledson 202 Brodick Drive, Bolton BL2 6UE
M6 MFZ M. Fitzgerald Flat 35, Winterton House, London E1 2QR
M6 MGA N. Valvona 63 Vale Road, Ash Vale, Aldershot GU12 5HR
MW6 MGC M. Carwardine Buttington Lodge, Sedbury, Chepstow NP16 7EX
M6 MGD L. Addison 25 Gladstone Street, Workington CA14 2XH
M6 MGF K. Holman 39 Trellech Court, Yeovil BA21 3TE
M6 MGG R. Wenlock 8 Dinchope Drive, Telford TF3 2ES
M6 MGH M. Hopewell 4 Cotes Crescent Bicton Heath, Shrewsbury SY3 5AS
M6 MGJ B. Hardy 10 Spring Farm Road, Burton-on-Trent DE15 9BN
MW6 MGM M. Margetts Central House, Llanfechain SY22 6UJ
M6 MGN N. Cook 210 Cemetery Road, Wath-Upon-Dearne, Rotherham S63 6HZ
M6 MGO M. Wenlock 3 Kennels Cottages, Hall Lane, Stone ST15 0RD
MD6 MGP A. Breen 1 Snugborough Close, Union Mills, Isle of Man IM4 4NZ
M6 MGR M. Reeks 33 Madresfield Village, Madresfield, Malvern WR13 5AA
M6 MGT J. Dickenson 2 Kirkleys Avenue North, Spondon, Derby DE21 7FX
M6 MGU M. Golding 11 Southwold Crescent, Broughton, Milton Keynes MK10 7BW
M6 MGV M. Pike 21 Watersmeet Close, Guildford GU4 7NQ
M6 MGW M. Walker 50 College Grove Road, Wakefield WF1 3RL
M6 MGX M. Gillard 66 West End Rd, Bradninch EX5 4QP
MW6 MGY M. Gray 15 The Circle, Two Locks NP44 7JP
MW6 MGZ M. Bannister 45 Queens Drive, Llantwit Fardre, Pontypridd CF38 2NT
M6 MHA M. Hersom Room 303, 95-98 Talbot Street, , Dublin DO1 WR94
M6 MHD M. Jodrell 2 Charlesworth Street, Crewe CW1 4DE
M6 MHE T. Harvey 12 Woodkirk Avenue, Tingley, Wakefield WF3 1JL
MI6 MHI A. Menzies 21 Woodview Park, Tandragee, Craigavon BT62 2DD
M6 MHJ M. Jackson 64 Main Road Moulton, Northwich CW9 8PB
MI6 MHK C. Moonie 21 Woodview Park, Tandragee, Craigavon BT62 2DD
M6 MHL M. Lacey 82 Bowerings Road, Bridgwater TA6 6HF
M6 MHM M. Hashim 1 Cheylesmore Drive, Frimley, Camberley GU16 9BL
MM6 MHN J. Mulhern 10 Fisher Court, Knockentiber, Kilmarnock KA2 0DS
M6 MHO M. Hossell 80 Murray Road, Sheffield S11 7GG
M6 MHQ M. Rea 15 Wensleydale Close, Royton, Oldham OL2 5TQ
M6 MHU M. Humphries 5 Coppice Mead, Stotfold, Hitchin SG5 4JX
M6 MHV M. Clarke 54 Stafford Grove, Shenley Church End, Milton Keynes MK5 6AZ
M6 MHW M. Hall 259 Lambourn Drive, Allestree, Derby DE22 2UR
M6 MHY M. Williams 37 Clarendon Road, Weston-Super-Mare BS23 3EE
M6 MIA M. Andrews 286 Huddersfield Road, Mirfield WF14 9PY
MI6 MIB P. Cobain 15 Drumfad Avenue, Millisle, Newtownards BT22 2GS
M6 MIC M. Taylor 24 Crowley Lane, Oldham OL4 2PN
M6 MID I. Shiradski 69 Masefield Avenue, Borehamwood WD6 2HG
M6 MIE M. Johnson 3 Bersted Mews Bersted Street, Bognor Regis PO22 9RR
M6 MIF M. Vaughan 2 Hendingham Close, Gloucester GL4 0XS
M6 MIG M. Greenwood Stone Fell Gate, Swallow Hill, Distington, Workington CA14 4PR
MI6 MIH J. Mercer 32 Templemore Avenue, Belfast BT5 4FT
M6 MII M. Bell 2 Fox Close, Dunton, Biggleswade SG18 8RF
M6 MIK M. Harris 27 Great Braitch Lane, Hatfield AL10 9FD
M6 MIL J. Milbourne 102 Bells Marsh Road, Gorleston, Great Yarmouth NR31 6PR
M6 MIN D. Mitchell Flat 2, Weavers Court, 51 Unwin Street, Sheffield S36 6EH
M6 MIO F. Miocinovic 14 Huxloe Rise, Northampton NN3 8YA
M6 MIP M. Payne 14 Linnell Road, Rugby CV21 4AN
MW6 MIQ V. Mckendley 9 Mary Street, Aberdare CF44 7NF
M6 MIR H. Mir 13-15 Wain Street, Stoke-on-Trent ST6 4ES
MM6 MIS J. Marsh 8 Hazelton Way, Broughty Ferry, Dundee DD5 3BT
M6 MIT M. Tarling 16 Cross Walk, Bristol BS14 0RX
M6 MIU J. Marsh 14 Eyam Road, Hazel Grove, Stockport SK7 6HP
M6 MIV B. Davies 60 Queensway, Blackburn BB2 4QT
M6 MIY P. Billingham 393 Landseer Road, Ipswich IP3 9LT
M6 MIZ M. Kwiatkowski-Zelazny 56 York Road, Hove BN3 1DL
M6 MJA M. Austwick 6 Worlaby Road, Grimsby DN33 3JY
MM6 MJC M. Clifford Bridgeton Castle, St. Cyrus, Montrose DD10 0DN
M6 MJD P. Smith 99 Racecourse Road, Chesterfield S41 8NW
M6 MJF M. Fysh 3 Jeffrey Close, Kings Lynn PE30 2HX
MM6 MJG M. Gilbert 3 Hoymansquoy, Stromness KW16 3DR
M6 MJI M. Greensmith 14 Fountain Road, Draycott-In-The-Clay, Ashbourne DE6 5HP
M6 MJL M. Lawrance 17 Wren Crescent, Scartho Top, Grimsby DN33 3RA
M6 MJM M. Mccormack Flat 13, 29 Stoneygate Road, Leicester LE2 2AE
M6 MJN M. Neale 41 Langford Road, Weston-Super-Mare BS23 3PQ
M6 MJP S. Parker 36 Eton Close, Lincoln LN6 0YF
MM6 MJR M. Robertson Tigh Jenny, Strath, Gairloch IV21 2BX
M6 MJS S. Shields 9 Berrington Drive, Newcastle upon Tyne NE5 4BG
M6 MJV D. Morley 9 Field Close, Thringstone, Coalville LE67 8PU
MM6 MJY M. Yarrow Lomond Villa, Downies Village, Aberdeen AB12 4QX
MW6 MJZ M. Shepley 16 Heulwen Close, Hope, Wrexham LL12 9PR
M6 MKB M. Buchanan 36 Church Lane, Manby, Louth LN11 8HL
M6 MKD D. Bell-Stephens 6 Woodland Road, Warminster BA12 8HJ
M6 MKE M. Gregory 65 Nursery Crescent North Anston, Sheffield S25 4BR
M6 MKH A. Freeth 121 Highfield Road, Burntwood WS7 9DA
M6 MKJ K. Juszczak 68 College Road, Sandy SG19 1RH
M6 MKK M. Kendall 53 Ellerker Rise, Willerby, Hull HU10 6EU
M6 MKM J. Mckie 59 Leaholme Terrace, Blackhall Colliery, Hartlepool TS27 4AB
M6 MKN N. Driscoll 42 Adelaide Square, Shoreham-by-Sea BN43 6LN
M6 MKO M. Kent 7 Lockyers Drive, Ferndown BH22 8AJ
M6 MKU D. Mcdonald 29 Highfield Crescent, Rayleigh SS6 8JP
M6 MKV M. Vardy 60 Hucklow Avenue, North Wingfield, Chesterfield S42 5PU
M6 MKW M. Wharton Sea Cadets, Riverside Road, Great Yarmouth NR31 6PX
M6 MKX S. Mcguckian 71 Heathfield Drive, Tyldesley, Manchester M29 8PJ
M6 MKY M. King Flat 6, Derwent Court, Solihull B92 7BU
M6 MLA M. Lovatt 3 Withington Close, Atherton, Manchester M46 0EZ
M6 MLE D. Pilkington 197 Saltings Road, Snodland ME6 5HP
M6 MLF J. Kiely 35 Chestnut Avenue, Todmorden OL14 5PH
M6 MLG P. Arnold 25 Arliston Drive, Woodville, Swadlincote DE11 8FS
M6 MLH M. Hoyland 3 Telford Street, Barrow-in-Furness LA14 2ER
M6 MLI R. Parker 58 Bryncastell, Bow Street SY24 5DF
M6 MLK M. Lake 64 Womersley Road, Norwich NR1 4QB
M6 MLL D. Neumunn 92 Miner Street, Walsall WS2 8QL
M6 MLM M. Milano 35 Orion Road, Rochester ME1 2UL
M6 MLN A. Mcdermid 49 Jubilee Street, Irthlingborough, Wellingborough NN9 5RL
M6 MLO M. Broyd 16 George Downing House, 4A Bickleigh Close, Plymouth PL6 5XJ
M6 MLP M. Le-Petit 3 Stanley Drive, Hatfield AL10 8XX
M6 MLQ M. Byard 1 Fieldside, Long Wittenham, Abingdon OX14 4QB
M6 MLR L. Rolt 40 Water Meadow Drive, Bradford BD13 4EX
MM6 MLT P. Connon 4 Highfield Court, Stonehaven AB39 2PL
M6 MLU P. Rath 60 Elstree Road, Bushey Heath, Hertfordshire WD23 4GL
M6 MLV D. Malcolm 66 Bracken Bank Grove, Keighley BD22 7AU
M6 MLX M. Pacitti-Lamb 41 Cowell Grove, Highfield, Rowlands Gill NE39 2JQ
M6 MLY M. Livesey 24 St. Marys Road, Bamber Bridge, Preston PR5 6TD
M6 MMB M. Parkes 2 Woodhouse Mount, Normanton WF6 1BN
M6 MMC Dr M. Mcintyre 18 Norlands Crescent, Chislehurst BR7 5RN
MM6 MMG D. Anderson Dail Darach, Monydrain Road, Lochgilphead PA31 8LG
M6 MMH M. Houghton 18 Leopold Way, Blackburn BB2 3UE
M6 MMI R. Hewson 10 Miriam Grove, Leigh WN7 3EX
MM6 MML J. Lucas 7 Rysland Avenue, Newton Mearns, Glasgow G77 6EA
M6 MMM M. Hunter 126 Turner Street, Stoke-on-Trent ST1 2NE
M6 MMN M. Newbury 2 Rowan Close, Clacton-on-Sea CO15 2DB
M6 MMP M. Chu Harrogate Ladies' College, Clarence Drive, Harrogate HG1 2QG
M6 MMQ M. Majhail 3 Poynders Hill, Hemel Hempstead HP2 4PQ
M6 MMR S. Wellsted 127 Goldthorn Hill, Wolverhampton WV2 4PS
M6 MMS M. Strange 101 Southbroom Road, Devizes SN10 1LY
MI6 MMT M. Torley 4 Yew Tree Park, Newry BT34 2QP
MW6 MMU G. Edwards 54 Old Street, Tonypandy CF40 2AF
M6 MMX A. Iggulden 78 Wrensfield Road, Stockton-on-Tees TS19 0BD
M6 MMY M. Barber 3 Baxter Road, Sunderland SR5 4LH
M6 MNC A. Cockburn 20 Hexham Avenue, Hebburn NE31 2HN
M6 MND D. Rogers 21 Belmont Park, Pensilva, Liskeard PL14 5QT
MM6 MNE C. Macnee 7 Church Street Chapelton, Strathaven ML10 6SD
M6 MNG N. Giuliano 13 Walton Drive, Littleover, Derby DE23 1GN
M6 MNH M. Hunt 37 Shortlands Avenue, Ongar CM5 0BL
M6 MNI M. Walton 38 Wingate Road, Grimsby DN37 9DU
M6 MNK P. Roberts 17 Cannon Hill, Prenton CH43 4XR
MI6 MNL J. Johnston 19 Killowen Grange, Lisburn BT28 3HQ
M6 MNO A. Hood 109 Trotters Field, Braintree CM7 3NW
M6 MNP M. Killoran 4 Victoria Road, Pudsey Leeds LS28 7SR
MM6 MNQ R. Boan 6 Philip Avenue, Newton Stewart DG8 6HF
M6 MNS T. Hodson 25 The Rise, Amersham HP7 9AG
M6 MNT S. Hamer Flat 5, 19 Frimley Road, Camberley GU15 3EN
M6 MNU J. Williams 1 Lower Meadow Drive, Congleton CW12 4UX
M6 MNV F. Brunt 74 Bardley Crescent, Tarbock Green, Prescot L35 1RJ
MM6 MNW D. Ryan 89 The Braes, Saltcoats KA21 5EP

M6 MNX M. Norfolk 185 Heath Road, Leighton Buzzard LU7 3AD
M6 MNZ M. Bartlett 34 Yarrow Drive, Birmingham B38 9QR
M6 MOB P. Hawes 6 Robert Street, Sunderland SR4 6EY
M6 MOC T. Forss Lower Conghurst Oast, Conghurst Lane, Cranbrook TN18 4RW
M6 MOD M. O'Driscoll 17 Petherton Gardens, Bristol BS14 9BT
M6 MOF C. Moffitt 5 Foxton Terrace, Horstead Avenue, Brigg DN20 8QR
M6 MOG N. Saville 4 Shannon Court, Downs Barn, Milton Keynes MK14 7PP
M6 MOH M. Hinds 10 Lustrells Close, Saltdean, Brighton BN2 8AS
M6 MOI S. Lewis 58 Ocean Close, Fareham PO15 6QP
M6 MOJ A. Mckie 5 Greenway Northenden, Manchester M22 4LW
M6 MOK M. Lowin 1A Burnside Avenue, Blackpool FY4 4AF
M6 MOM J. Clare Kimberley, Boston Road, Bicker, Boston PE20 3AP
M6 MON P. Montgomery 14 Saxon Crescent, Horsham RH12 2HU
M6 MOP D. Pomeroy 73 Pinewood Gardens, North Cove, Beccles NR34 7PG
M6 MOQ K. Hamilton 73 Leven Road, Stockton-on-Tees TS20 1DB
M6 MOS M. Steele 10 Green Lane, Houghton, Carlisle CA3 0NT
M6 MOU D. Dormer 69 Favell Drive, Furzton, Milton Keynes MK4 1AX
M6 MOV C. Moverly 17 Trefoil Road, Hailsham BN27 4FR
M6 MOW J. Horry 5 Donington Road, Bicker, Boston PE20 3EF
M6 MOX M. Cox 3 Cromwell Road, Hertford SG13 7DP
MM6 MOY P. Mcdonald 9 Partan Skelly Way, Cove Bay, Aberdeen AB12 3PH
M6 MOZ M. Meadowcroft 8 Lamlash Road, Blackburn BB1 2AS
M6 MPB M. Bridger 11, Beecham Close, Newcastle upon Tyne NE15 6LG
M6 MPD R. Yarrow 27 Staplers Road, Newport PO30 2DB
M6 MPE M. Evans 48 Paddock Lane, Aldridge, Walsall WS9 0BP
M6 MPF J. Dunn 3 Hobbs Way, Rustington, Littlehampton BN16 2QU
MI6 MPH J. Martin 22 Lisbane Road, Saintfield, Ballynahinch BT24 7BS
M6 MPK J. Garwood 4 Ryedale, Carlton Colville, Lowestoft NR33 8TB
M6 MPL G. Clark 65 Chyvelah Vale, Gloweth, Truro TR1 3YJ
M6 MPM M. Mccall 1 Jaunty Road, Sheffield S12 3DT
M6 MPO M. Woolger 25 Rookwood Park, Horsham RH12 1UB
M6 MPP M. Michalowski 39 Towan Avenue, Fishermead, Milton Keynes MK6 2DS
MW6 MPQ P. Harris 123 St. Georges Court, Tredegar NP22 3DD
M6 MPR P. Rodgers 25 Eagle Avenue, Barnsley S75 1FE
M6 MPS S. O'Riordan 46 Grange Road, London HA20LW
M6 MPT M. Thompson 35 Princes Avenue, Desborough, Kettering NN14 2RQ
M6 MPV M. Varley 50 Gorse Valley Road, Hasland, Chesterfield S41 0JP
M6 MPW M. Whotton 7 Heatherdale, Ibstock LE67 6JU
M6 MPX M. Phillips The Well House, Eastbury, Hungerford RG17 7JL
M6 MPY R. Greenwood 3, Whitescottages, Church Road, Kilndown, Cranbrook TN17 2SA
M6 MPZ M. Edmonds 20 Tomline Road, Ipswich IP3 8BZ
M6 MQB M. Bailey 1 Oriel Drive, Glastonbury BA6 9PA
M6 MQC M. Le Moine 115 Rothesay Road, Blackburn BB1 2ER
M6 MQD M. Oliver 14 Ash Road, Ashurst, Southampton SO40 7AT
M6 MQE M. Ayres 32 Kinterbury Close, Hartlepool TS25 1GQ
MI6 MQF D. Mulligan 10 Seaview, Ardglass, Downpatrick BT30 7SQ
M6 MQH A. Jovanovic 33 Seward Road, London W7 2JS
M6 MQJ M. Trathen Gweal An Mayn Cottage, Newtown, St. Martin, Helston TR12 6DP
M6 MQK D. Ferguson 39 Fouracres, Maghull, Liverpool L317BP
M6 MQL M. Boyle 64 Spencerfield Crescent, Middlesbrough TS3 9HD
M6 MQM M. Tozer 4 The Grange Dousland, Yelverton PL206NN
M6 MQN J. Milner 30 Rowena Drive, Thurcroft, Rotherham S66 9HT
MM6 MQO S. Doonan West Clanfin Farm Waterside, Kilmarnock KA3 6JQ
MM6 MQR Dr H. Erwood Lunna House, Lunna Vidlin, Vidlin ZE2 9QF
M6 MQT T. Mloduchowski Flat 4, Gwynne House, London E1 2AG
MW6 MQU R. Thomas 49 Viscount Evan Drive, Newport NP10 8HJ
MM6 MQV D. Robertson Grindhus, , Gonfirth, Voe ZE2 9PY
M6 MQW J. Evans Room 7, Mcnab Block, Royal Air Force, Digby, Lincoln LN4 3LH
MI6 MQX M. Elliott 17 Milebush Road, Dromore BT25 1RT
M6 MQY L. Andrews 72 Grange Road, Alresford SO24 9HF
M6 MRC M. Bridges 2 Bridgman Court, St John Moore, Hythe CT21 5EB
M6 MRD P. Sephton 11 Moss Avenue, Leigh WN7 2HH
M6 MRG H. Martin 27 Gordon Road, Fleetwood FY7 6UE
M6 MRH M. Hayward 8 Kiln Shaw, Langdon Hills, Basildon SS16 6LE
MI6 MRI F. Rafferty 96 Collinward Avenue, Newtownabbey BT36 6DZ
MI6 MRJ J. Martin 23 Winters Gardens, Omagh BT79 0DZ
M6 MRK M. Mckenna 54 Whickham Road, Hebburn NE31 1QU
M6 MRM M. Barnfather 2 Brazenhill Lane, Haughton, Stafford ST18 9HS
M6 MRN P. Fisher 26 Rydal Avenue, Barrow-in-Furness LA14 4NW
M6 MRP P. Boxx 19 Glenthorpe House, 54 West Stevenson Street, South Shields NE33 4DL
M6 MRR M. Rutter-Dacosta 144 Bellingdon Road, Chesham HP5 2HF
M6 MRS C. Davies 68 Wood Street, Castleford WF10 1LN
M6 MRT A. Blamires 2 Foldings Grove, Scholes, Cleckheaton BD19 6DQ
M6 MRU A. Owen 5 Kirkham Court, Goole DN14 6JU
M6 MRV R. Martin 21 Lonsdale Crescent, Dartford DA2 6LQ
M6 MRW M. Willison 8 Summervale Mews Wharf Lane, Ilminster TA19 0BA
M6 MRX R. Brough 36 Salstar Close, Aston, Birmingham B6 4PP
M6 MRY B. Murray 23 Tillotson Close, Crawley RH10 7WQ
MM6 MSA M. Saddler 12 Carnbee End, Edinburgh EH16 6GJ
M6 MSB M. Shopland 128 Whitewood Park, Liverpool L9 7LG
M6 MSC M. Colman 4 Northmead Drive, North Walsham NR28 0AU
MW6 MSE M. Roblin 6 Gethin Street, Briton Ferry, Neath SA11 2LU
M6 MSF M. Farnham 94 Rochester Way, Crowborough TN6 2DU
MI6 MSG G. Graham Flat J 86 Sunningdale Gardens, Belfast BT14 6SL
MM6 MSH M. Hatfull Flat 2 1 Northumberland Place Lane, Edinburgh EH3 6LD
M6 MSI M. Scott 25 Smithburn Road, Gateshead NE10 9DT
M6 MSJ C. Gibson 17 Clyde Court, Grantham NG31 7RB
MW6 MSN P. Evans Flat 1 Red Cow Annex, Lloyds Terrace, Adpar, Newcastle Emlyn SA38 9EH
M6 MSO S. Marsh 6 Sparsholt Road, Southampton SO19 9NJ
M6 MSP S. Palmer Sawrey Ground, Crosby, Maryport CA15 6SH
MI6 MSR M. Ruddy 34 Glenveagh Hilltown, Newry BT34 5US
M6 MSS M. Smith The Lawns Tylers Green, High Wycombe HP108BH
M6 MST M. Streeter Fairway, West Chiltington Road, Pulborough RH20 2EE
M6 MSU Denton Park Middle School RC c/o S. Hodder 19 Kingsclere, Huntington, York YO32 9SF
MI6 MSV M. Steele 134 Knock Road, Dervock, Ballymoney BT53 8AB
MW6 MSX M. Street The Lamb Inn, Hermon, Glogue SA36 0DS
M6 MSY M. Saysell Gordano Valley Riding Centre, Moor Lane, Bristol BS20 7RF
M6 MTA M. Ansari 37 Lizmans Court, Silkdale Close, Oxford OX4 2HF
M6 MTC D. Godfrey 25 Bosworth Crescent, Romford RM3 8JZ
M6 MTD P. Williams 4 Red Gables, Shap, Penrith CA10 3NL
MW6 MTE E. Thomas 1 Cambrian Gardens Y Drenewydd, Newtown SY16 2AW
MW6 MTG M. Blomfield 99 Mountain Road, Upper Brynamman, Ammanford SA18 1AN
M6 MTH M. Horton 8 Liptraps Lane, Tunbridge Wells TN2 3BS
M6 MTI M. Ilsley 38 Coleridge Road, Ottery St. Mary EX11 1TD
M6 MTJ M. Jones 18 Cleveleys Avenue, Heald Green, Cheadle SK8 3RH
M6 MTK A. Morris 4 Saunders Close, Uckfield TN22 2BX
M6 MTL D. Baldwin 19 Bramble Grove, Wigan WN5 9PR
M6 MTM M. Tyler-Moore 34 Guildford Road, Horsham RH121LS
M6 MTN M. Banks 37 Havelock Road, Southsea PO5 1RU
MI6 MTO T. Mehaffey 33 Lenaderg Road, Banbridge BT32 4PT
M6 MTQ B. Turnbull Brooklet Road, Heswall CH601UL
M6 MTR Prof. P. March 46 Christchurch Road, Tilbury RM18 8XP
M6 MTS M. Smith 5 Newland Avenue, Stafford ST16 1NL
M6 MTT P. Crosweller Flat 3, 18 Pelham Road, Seaford BN25 1ES
M6 MTU D. Atkins 32 Braybrook, Orton Goldhay, Peterborough PE2 5SH
M6 MTV P. Bailey 4 Roving Bridge Rise Clifton, Swinton, Manchester M27 8AF
MW6 MTW M. Ellis Heddwch Brithdir, Dolgellau LL40 2SF
M6 MTZ R. Lenicker 136 Nab Wood Drive, Shipley BD18 4EW
MM6 MUA M. Mulligan 18 Camperdown Road, Nairn IV12 5AR
M6 MUB M. Masters 49 St. Johns Avenue, Bridlington YO16 4ND
M6 MUD C. Brock 30 Cromer Road, Norwich NR6 6LZ
M6 MUF W. Burridge 20 Archer Close, Kingston upon Thames KT2 5NE
MM6 MUH C. Fenton 2 Rothiebrisbane Cottages Fyvie, Turriff AB53 8LE
M6 MUJ I. Fores 27 Southfield Lane, Whitwell, Worksop S80 4NS
M6 MUK M. King 7 Battismore Road, Morecambe LA4 4QG
MM6 MUO A. Flett Southview, Lein Road Kingston, Fochabers IV32 7NW
M6 MUP C. Meredith 6 North End, Shortstown, Bedford MK42 0XB
MM6 MUR G. Murray The Barn House, Springfield Farm, Carluke ML8 4QZ
M6 MUS A. Sutton 3 Grotes Buildings, London SE3 0QG
M6 MUT M. Northwood 34 Whitehead Drive, Wellesbourne, Warwick CV35 9PW
M6 MUY F. Boyce 277 Manor View, Par PL24 2EP
M6 MUZ M. Colpman 20 Rochford Road, Basingstoke RG21 7TQ
M6 MVA M. Johnson 143 Swan Lane, Wickford SS11 7DG
M6 MVB M. Bradley 13 Elizabeth Avenue, Bilston WV14 8EA
M6 MVD J. Dilworth 808 Liverpool Road, Southport PR8 3QF
M6 MVE W. Cartledge 20 Ariel Street, Newcastle NE639EZ
M6 MVF M. Findon 64 Fur Tree Grove, Birmingham B735UN
M6 MVH R. Barker 6 Trenoweth Cressent, Penzance TR184RY
M6 MVK J. Hart 57 Brentleigh Way, Stoke-on-Trent ST1 3GX
MM6 MVM V. Mcgowan 112 Oronsay Avenue, Port Glasgow PA14 6EF
M6 MVN S. Forshaw 22A Barley Hall Street, Heywood OL10 4DH
M6 MVT M. Bell 7 Shiregreen Lane, Sheffield S5 6AA
M6 MVV M. Mcarthur 25 Lingfield Road, Edenbridge TN8 5DS
M6 MVW M. Ward 13 Hyde Lane, Danbury CM3 4QT
MM6 MVX D. Wilson Rivendell Lodge, Glenkindie, Alford AB33 8RN
M6 MVZ D. Money Flat 1, Lewin Court 24B Plumstead High Street, London SE18 1SL
M6 MWA S. Hyland-Davis 34 Melody Close, Warden, Sheerness ME12 4PU
M6 MWB M. Bryant 284 Brantingham Road Chorlton Cum Hardy, Manchester M21 0QU
M6 MWC M. Curwen 40 Grange Street, Morecambe LA4 6BW
M6 MWD M. White 27 Winstone Close, Redditch B98 8JS
MM6 MWE L. Stoneham Kirkton Farm Balblair, Dingwall IV7 8LG
MM6 MWF M. Flynn 15 Riselaw Crescent, Edinburgh EH10 6HN
MW6 MWG J. Davies 1 South View, Pontycymer CF32 8LE
MW6 MWN W. Noble 2 Harriet Town, Troedyrhiw, Merthyr Tydfil CF48 4HJ
M6 MWP M. Poole 22 Padstow Gardens, Leeds LS10 4NQ
M6 MWQ C. Cromie 140 Whalley Road Wilpshire, Blackburn BB1 9LJ
MW6 MWS A. Sneddon 3 Marigold Close, Gurnos, Merthyr Tydfil CF47 9DA
M6 MWT M. Tolmie 28 Spencer Road, Rendlesham, Woodbridge IP12 2TJ
M6 MWV K. Beal 22 Harpers Road, Newhaven BN9 9RR
M6 MWW M. Watts 47 Westbury Crescent, Weston-Super-Mare BS23 4RF
M6 MWZ M. Way 9 Railton Avenue, Stoke on Trent ST34BN
M6 MXA A. O'Reilly 3B Summerleys, Edlesborough, Dunstable LU6 2HR
M6 MXB M. Boddy 26 Tulip Tree Road, Bridgwater TA6 4XD
M6 MXC M. Craven 78 Connaught Road, Brookwood, Woking GU24 0HF
M6 MXD Sqdn. Ldr. M. Dalziel 4 Meadow Close, St. Albans AL4 9TG
M6 MXF L. Watson 616 Beverley Drive, Stoke-on-Trent ST2 0RE
M6 MXH M. Head 9 Kinsbourne Grreen, Dunscroft, Doncaster DN7 4BL
MM6 MXK M. Rawlings Brownstone, 2 Springfield Terrace, Alness IV17 0SP
M6 MXM M. Meehan 14 Grosvenor Road, Walton, Liverpool L4 5RB
M6 MXO V. Adedeji 3 Royal Troon Mews, Wakefield WF1 4JL
M6 MXP D. Burrows 3 Novello St, Maltby Maltby, Rotherham S66 7QB
M6 MXR M. Russell Cowmans Cottage Spring Lane, Flintham, Newark NG23 5LB
MM6 MXU A. Mcbain 9/12 Tower Place, Edinburgh EH6 7BZ
MW6 MXV W. Cooper 1 Bedw Street, Maesteg CF34 0TF
M6 MXX C. Morrow 194 Cedar Road, Doncaster DN4 9ET
M6 MXY C. Simpson 2 Bradley Road, Haslington, Cheshire. Cw1 5Pw Haslington, Crewe CW1 5PW
MI6 MXZ M. Masterson 2 Pinley Drive, Banbridge BT32 3TZ
M6 MYB M. Bannister 110 Cotswold Crescent, Billingham TS23 2QB
M6 MYC M. Beddall 11 Sinodun Road, Wallingford OX10 8AD
M6 MYD A. Middleton 2 Moor View, Godshill, Ventnor PO38 3HW
MW6 MYE A. Twose 4 Bradford Street, Caerphilly CF83 1GA
M6 MYF L. Thompson 33 Dalton Crescent, Shildon Co Durham DL4 2LE
M6 MYG M. Young 72 Goddard Way, Saffron Walden CB10 2EB
M6 MYH S. Elliott 79 Somerton Road, Bolton BL2 6LN
M6 MYI O. Makhura H-1-6-S, The Maltings Haven Road, Colchester CO2 8FU
MM6 MYK M. Cheetham Plowvent, Muir Of Fowlis, Alford AB33 8NX
M6 MYL J. Swann 5 Lanark Close, Hazel Grove, Stockport SK7 4RU
M6 MYM M. Beniston 2 Railway View, Billington, Clitheroe BB7 9NJ
M6 MYN R. Eaton 31 Pinfold Lane Ruskington, Sleaford NG34 9EU
M6 MYR M. Moss 27 Dunn Side, Chelmsford CM1 1DL

UK Callsigns

M6 MYS A. Smart 7 Hinton Grove, Hyde SK14 5ST
M6 MYT B. Hilton 17 Bellwood, Westhoughton, Bolton BL5 2RT
M6 MYU J. Othen 149 Blumfield Crescent, Slough SL1 6NN
M6 MYV R. Flux 18 Ashdene Road, Ashurst SO40 7DP
MI6 MYW M. Mcwilliams 84 Syerla Road, Dungannon BT71 7ET
M6 MYY M. Ffrench 24 Monkwood Road, Chesterfield S41 8DG
M6 MZB M. Bauer Flat 21, 5 Queensland Road, London N7 7FE
M6 MZC G. Fox 23 The Driveway, Canvey Island SS8 0AB
M6 MZD B. Greenwood 13 Mayflower Street, Blackburn BB2 2RX
MM6 MZE C. Menzies 0/2 47 Wellshot Road, Glasgow G32 7XL
MM6 MZG K. Jones 6 Traill Street Castletown, Thurso KW14 8UG
M6 MZH M. Hubbard 14 Parkfield Crescent, Kimpton, Hitchin SG4 8EQ
M6 MZJ M. Juniper 2 Cranbourne Drive, Hoddesdon EN11 0QH
M6 MZL G. Johnson 22 Beechwood Close, Blythe Bridge, Stoke-on-Trent ST11 9RH
M6 MZN J. Forshaw 22A Barley Hall Street, Heywood OL10 4DH
M6 MZQ G. Peters Curlew Court, Guys Head Road, Sutton Bridge, Spalding PE12 9QQ
MI6 MZR M. Armstrong 2 Thralcot Link, Carrickfergus BT38 9RG
MW6 MZU T. Baldwin Rose Cottage, High Street, Pontypool NP4 6HE
M6 MZV A. Watson 67 Alamein Drive, Romiley, Stockport SK6 4JN
M6 MZX M. Pinter Bercsenyi Str. 47, Totkomlos 5940 Hungary
M6 MZY K. Phillips 12 Copland Avenue, Minster On Sea, Sheerness ME12 3PJ
M6 MZZ M. Driscoll 14B Pretoria Road Hedge End, Southampton SO30 0BS
MM6 NAA N. Mcdonald 8 Newton Place, Perth PH1 2QJ
MM6 NAB D. Bell 82 Campbell Avenue, Stevenston KA20 4BP
M6 NAC N. Carter 25 Breachfield, Burghclere, Newbury RG20 9HY
MM6 NAD N. Anderson The Cedars, Church Street, Keith AB55 4AR
MW6 NAG N. Berrall 41 Nantgarw Road, Caerphilly CF83 3FB
MM6 NAI A. Anderson 18 Selkirk Street, Wishaw ML2 8RA
M6 NAK N. Leech 59 Lakeside Court, Brierley Hill DY5 3RQ
M6 NAL N. Parry 125 Lawsons Road, Thornton-Cleveleys FY5 4PL
M6 NAM M. Cody 139 Vicarage Road, Watford WD18 0HA
M6 NAN Dr L. Alconcel Top Lock Cottage, Stoke Pound Lane, Bromsgrove B60 4LH
M6 NAO N. Griffiths 67 Warstones Drive, Wolverhampton WV4 4PF
M6 NAQ A. Cowley 7 Harwood Road, Gosport PO13 0TU
M6 NAS S. Nash Ashmead, Hemel Hempstead HP3 0BU
M6 NAT N. Jones 5 Montgomery Crescent, Quarry Bank, Brierley Hill DY5 2HB
M6 NAU N. Alders 14 Forest Rise, Crowborough TN6 2ES
M6 NAV J. Glicklich 86 Ainsdale Road, Bolton BL3 3ER
M6 NAW N. White 5 Badgers Walk, Burgess Hill RH15 0AE
MW6 NAX N. Jones 7 Dyffryn, Burry Port SA16 0TE
MW6 NAZ D. Tippett 30 Berw Road, Tonypandy CF40 2HD
M6 NBG J. Mccosh The Mill House, Moorlands Road, Merriott TA16 5NF
M6 NBH A. Barrett 2 Friars Close, Clacton-on-Sea CO15 4EU
M6 NBL Lady A. Mackendrick 65 Sandringham Court, Dorchester DT1 2BL
M6 NBN R. Anderson 26 Bowness Avenue, Warrington WA2 9NQ
M6 NBO B. Vickers 52 Edward Street, Grimsby DN32 9HJ
M6 NBP N. Williams 114 Essex Place Montague Street, Brighton BN2 1LL
M6 NBQ Dr P. Fabrega Flat 3, Post Office Court, Whitchurch SY13 1QT
M6 NBS N. Barker 17 Pippin Walk, Hardwick, Cambridge CB23 7QD
MW6 NBU N. Bruetsch The Firs, Penglais Road, Aberystwyth SY23 2EU
M6 NBV D. Whitelock 2 Shippards Road, Brighstone, Newport PO30 4BG
M6 NBW N. Warden 1 Forge House The Street, Earl Soham, Woodbridge IP13 7RT
M6 NBX N. Cohen 8 Henry Gepp Close, Adderbury, Banbury OX17 3FE
M6 NBY N. Thompson 10 Belgrave Close, Hersham, Walton-on-Thames KT12 5PH
MW6 NCA N. Curry 58 Greenfields, St. Martins, Oswestry SY11 3AH
M6 NCB N. Bettridge 37 Princess Avenue, Warsop, Mansfield NG20 0PY
MI6 NCC N. Corbett 10 Main Street, Rosslea, Enniskillen BT92 7PP
M6 NCD D. Senior 16 Cherry Tree Close, Billingshurst RH14 9NG
M6 NCE C. Mclennan 18 Loveridge Way, Eastleigh SO50 9PW
M6 NCF N. Froude 6 Park Road West, Chester CH4 8BG
MI6 NCG N. Griffin 327 Clonmeen, Drumgor, Craigavon BT65 4AT
M6 NCI Sqdn. Ldr. B. Dowley 120 Capel Street, Capel-Le-Ferne, Folkestone CT18 7HB
M6 NCK N. Taylor 212 Plantation Hill, Worksop S81 0HD
M6 NCL C. Braddock 22 Anncroft Road, Buxton SK17 6UA
M6 NCM N. Mcveagh 14 Chapel Close, Ravenstone, Coalville LE67 2JT
M6 NCO K. Tonge 25 Southcote Grove, Birmingham B38 8ED
MM6 NCP N. Pollard 30 Abbeyhill Crescent, Edinburgh EH8 8DZ
M6 NCR G. Paton 17 Blakeney Road, Stevenage SG1 2LH
M6 NCU C. Johnson 22 Carleton Close, Great Yeldham, Halstead CO9 4QJ
M6 NCV G. Killpack 20 Fisher Close, Banbury OX16 3ZW
M6 NCY J. Mooneapillay 354 Upper Elmers End Road, Beckenham BR3 3HG
M6 NDB N. Brown 9 Devonshire Avenue, Wigston LE18 4LP
M6 NDC N. Charlotte 26 Nettleton Avenue, Mirfield WF149AN
M6 NDE N. Evans 46 Furzehill Road, Plymouth PL4 7LA
M6 NDF D. Arnold 15 Chiltern Drive, Verwood BH31 6US
M6 NDG N. Graven 33 Sheldrake Road, Broadheath, Altrincham WA14 5LJ
M6 NDI A. Kaluarachchi 103 Bentinck Road, Newcastle upon Tyne NE4 6UX
M6 NDK C. Lucas 15 Higher Moor, Ruan Minor, Helston TR12 7JJ
M6 NDM N. Mason 11 Chapel Court, Brierley Hill DY5 2UT
M6 NDN K. Hutchens 7 Lapwing Close, Thurston, Bury St. Edmunds IP31 3PW
M6 NDO A. Ashmore 42 Holme Road, Chesterfield S41 7JF
M6 NDP N. Plunkett 11 Stoneleigh Gardens, Grappenhall, Warrington WA4 3LE
M6 NDR N. Reeve 4 Ash Grove, Swindon SN2 1RX
M6 NDT R. Thatcher 83 Westfield Drive, North Greetwell, Lincoln LN2 4RE
M6 NDY M. Twitchen 72 Finedon Road, Burton Latimer, Kettering NN15 5QB
M6 NEA N. Asher 17 Ashby Road, Cleethorpes DN35 9PF
M6 NEC P. Bean 14 St. Andrews Lane, Necton, Swaffham PE37 8HY
MM6 NED E. Brophy 44 Annieshill View, Plains, Airdrie ML6 7NT
M6 NEE J. Todd 20 Hexham Avenue, Hebburn NE31 2HN
M6 NEF N. Chapman 13 Clayton Grove, Bracknell RG12 2PT
M6 NEG A. Brown 19 Main Street Leconfield, Beverley HU17 7NQ
M6 NEH Dr N. Hoare 5 Kelsey Head, Port Solent, Portsmouth PO6 4TA
M6 NEI N. Yorke 30 Bramdene Avenue, Nuneaton CV10 0DH
M6 NEL N. Price 68 Powke Lane, Rowley Regis B65 0AG
M6 NEM E. Chance 33 Larkfield Avenue, Kirkby-In-Ashfield, Nottingham NG17 9FE
M6 NEO C. Radford 86 Flamstead Avenue Loscoe, Heanor DE75 7RP
M6 NEQ J. Clarke 24 Telford Court, East Howdon NE280JH
M6 NES N. Preval 63 Dudley Avenue, Leicester LE5 2EF
M6 NET D. Bridges 176 Coombe Valley Road, Dover CT17 0HE
M6 NEV N. Chambers 16A Hillside Road, Wool, Wareham BH20 6DY
M6 NEW W. Chesworth 28 Chapel Close, Gunnislake PL18 9JB
M6 NEY 350 DX c/o M. Bullions 33 Mounthurst Road, Bromley BR2 7QW
M6 NEZ N. Jones 46 Devon Street, Barrow-in-Furness LA13 9PX
M6 NFB M. Gregory 40 Withycombe Drive, Banbury OX16 0SR
M6 NFC A. Cockburn 52 Devon Road, Hebburn NE31 2DW
M6 NFD P. Rickman 2 Woodley Lane, Romsey SO51 7JN
M6 NFE G. Singleton 2 Rome Avenue, Burnley BB115LQ
MW6 NFG N. Terrell 82 Baglan Street, Treherbert, Treorchy CF42 5AR
MW6 NFH N. Holloman The Cloisters, Llanvihangel Crucorney, Abergavenny NP7 8DH
M6 NFI N. Mooney 60 Rhyddings Street, Oswaldtwistle., Accrington BB5 3EY
MW6 NFJ M. Woffindale 108 College Road, Oswestry SY11 2SB
M6 NFK N. Lee 2 The Green Ormesby, Great Yarmouth NR29 3JX
M6 NFL R. Wraith 53 Links View Road, Croydon CR0 8ND
MW6 NFN C. Davies Foelallt, North Road, Aberystwyth SY23 2EL
M6 NFP G. Robinson 12 Donnington Street, Grimsby DN32 9EN
M6 NFQ G. Brindle 25 Chedworth Drive, Witney OX28 5FS
M6 NFS J. Haigh 1 Smithy Site, Farnborough, Wantage OX12 8NS
M6 NFT N. Thomson 92 Sutton Road Hull, Hull HU67DT
M6 NFW P. Setter 199 Southbourne Grove, Westcliff-on-Sea SS0 0AN
M6 NFX N. Furneaux Hill View, Highridge Road, Bristol BS41 8JU
MW6 NFY R. Lannon 14 Fairwood Close, Cardiff CF5 3QP
M6 NFZ D. Creech 29 Lake Road, Poole BH15 4LE
M6 NGA S. Howell 7A Amelia Close, Probus, Truro TR2 4TS
M6 NGD N. Goode 46 Robert Road, Tipton DY4 9BJ
MW6 NGE G. Dixon 9 The Glen, Bryncethin, Bridgend CF32 9LX
M6 NGK G. Wheeler 3 Bank Cottages, Shalfleet PO30 4NQ
MI6 NGM A. Mckay 17 Thorn Hill Road, Banbridge BT32 3TL
M6 NGN G. Nelson 4 Garnet Field, Yateley GU46 6FN
M6 NGP A. Armstrong 9 Cumberland Drive, Mansfield NG19 6LS
M6 NGR N. Griffiths 85 Foljambe Road, Chesterfield S40 1NJ
M6 NGU R. Gwillym Flat 3, Ady House 4A Alexandra Road, Farnborough GU14 6DA
M6 NGW N. Lang 1 Peartree Court, Old Orchards, Lymington SO41 3TF
M6 NGX M. Royall 7 Stacklands, Welwyn Garden City AL8 6XW
M6 NGY R. Hawkes Littleworth Lane, Littleworth RH13 8JX
M6 NHA S. Ratcliff 27 Furlong Road, Manchester M22 1UD
MW6 NHC A. Taylor 74 Fidlas Avenue, Cardiff CF14 0NZ
M6 NHD D. Baker 34 Farnham Road, Durham DH1 5LA
M6 NHF A. Barford 11 Tuffnells Way, Harpenden AL5 3HJ
M6 NHG P. Spencer 146, Winchester Road, Wolverhampton WV10 6EZ
MI6 NHH E. Currid 41 Mill Road, Portstewart BT557PQ
M6 NHJ N. Jones 18 Cleveleys Avenue, Heald Green, Cheadle SK8 3RH
M6 NHK N. Bates 40 First Street, Bradley Bungalows, Consett DH8 6JT
MW6 NHL J. Keogh 1 Dyffryn lg Talley, Llandeilo SA19 7YP
MM6 NHM N. Morris 23 Sedgebank, Sedgebank, Livingston EH54 6HE
M6 NHN N. Collins 12 Fern Close, Eastbourne BN23 8AQ
M6 NHO D. Leach 35 Victoria Road, Alton GU34 2DG
M6 NHP N. Pettitt 2A The Oval, Bulford Road, Tidworth SP9 7SB
M6 NHS N. Dodge Crossways, Culford, Bury St. Edmunds IP28 6DT
M6 NHT M. Sheppard 113 Chesterfield Road North, Mansfield NG19 7JB
M6 NHU T. Butcher Flat 1, Whitstone Orchard Whitstone Road, Paignton TQ4 6EY
M6 NHW J. Foxall 2 Millers Walk Pelsall, Walsall WS3 4QS
M6 NHX N. Hurlock 9 Little Meadow, Exmouth EX8 4LU
M6 NHY M. Pittas 89 Seddon Road, Morden SM4 6ED
MM6 NIA N. Hague 11 Auchriny Circle, Bucksburn, Aberdeen AB21 9JJ
M6 NIB N. Bennett 44 Glenmoor Road, Buxton SK17 7DD
M6 NIC N. Bowker 16 Farncombe Close, Wivelsfield Green, Haywards Heath RH17 7RA
MI6 NID P. Moore 32 Kinnegar Rocks, Donaghadee BT21 0EZ
M6 NIE P. Martin 26 Kingfisher Close, Chatteris PE16 6TP
M6 NIK N. Armstrong 1 Sea View Terrace, Churchtown, Helston TR12 7BZ
M6 NIL J. Stacey 47 Station Road, Foulsham, Dereham NR20 5RD
MI6 NIM H. Mcallister Ballycairn Drive, Belfast BT8 8HG
M6 NIN J. Dewhirst Flat 12, Lewis Court, Tamworth B79 8BE
M6 NIQ N. Stokes 618A Thorne Road, Netheravon, Salisbury SP4 9QG
M6 NIS F. Nisar 19A Cromwell Road, Basingstoke RG21 5NR
MW6 NIT T. Dixon Troedyrhiw, Abercych, Boncath SA37 0EY
M6 NIU J. Coombs 1 Beeston Avenue, Northampton NN3 9UG
M6 NIV J. Snowden 30 St. Christophers Walk, Wakefield WF1 2UP
M6 NIX A. Wilkinson Central House, Main Road, Hull HU11 4DJ
M6 NJB N. Bennett 35 West Shepton, Shepton Mallet BA4 5UD
M6 NJD N. Dixon 39 Urswick Green, Barrow-in-Furness LA13 0BH
M6 NJE N. Spencer 47 Tyne Road, Oakham LE15 6SJ
M6 NJG N. Grey 131 Links Avenue, Hellesdon, Norwich NR6 5PQ
M6 NJH N. Hall 24C Oakleigh Court, Church Hill Road, Barnet EN4 8UX
M6 NJI N. Isherwood 41 Livingstone Road, Blackburn BB2 6NE
M6 NJK D. Hawes 5 Valleyside Road, Hastings TN35 5AD
MW6 NJM N. Morris Fairview, Trefonen, Oswestry SY10 9DP
M6 NJN K. Shaw Flat 40, Bamford House, Bamford Road, Walsall WS3 3SA
M6 NJO N. Owen 32 Westfield Road, Dudley DY2 8LE
M6 NJP N. Pipkin 46 Charles Avenue, Albrighton, Wolverhampton WV7 3LF
M6 NJS N. Sandy 5 High Ercal Avenue, Brierley Hill DY5 3QH
M6 NJT N. Jones 63 Bernwell Road, London E4 6HX
M6 NJV D. Villa 33 North Street, Tywardreath, Par PL24 2PW
M6 NJX Dr M. Nassau 4A London Road, Liphook GU30 7AN
M6 NKB N. Booth 10 Games Walk, Wythenshawe, Manchester M22 1SN
M6 NKC N. Carey 28 Tremayne Road, St. Austell PL25 4NE
M6 NKE Northern Hgts Rd c/o N. Rotherham 4 Spenser Road, Cheltenham GL51 7EA
M6 NKG J. Maddams 18 Roche Way, Wellingborough NN8 5YD
M6 NKH N. Hammond 453 Smorrall Lane, Bedworth CV12 0LD
M6 NKJ S. Teasdale 63 Copley, Bishop Auckland DL13 5LS
M6 NKK A. Hoad 189 The Diplocks, Hailsham BN27 3JZ
M6 NKM N. Morse 33 Tower Close, Bassingbourn, Royston SG8 5JX
M6 NKP N. Palin 21 Ford Lane, Crewe CW1 3EQ
M6 NKQ S. Gudgeon Animals In Need, Pine Tree Farm London Road, Little Irchester, Wellingborough NN8 2EH
M6 NKR R. Spearman 36 Beechwood Road, Northampton NN5 6JT

M6 NKY N. Brown 12 Forest Close, Newport PO30 5SF
MW6 NLA R. Williams Bardsville, Porthdafarch Road, Holyhead LL65 2LL
M6 NLB N. Burnet 27 Mackenzie Way, Tiverton EX16 4AW
M6 NLF D. Hydes 31 Ridgehill Avenue, Sheffield S12 2GL
MW6 NLG N. Gladding 19 Laleston Close, Nottage, Porthcawl CF36 3HW
M6 NLJ A. Stokes 34 Watercress Close, Hartlepool TS26 0QY
M6 NLK J. Cutter 34 Greengate Lane, Knaresborough HG5 9EL
M6 NLL G. Fores 120 Whiteleas Avenue North Wingfield, Chesterfield S42 5PW
M6 NLO P. Burt 56 Winslade Road, Sidmouth EX10 9EX
MM6 NLP M. Lawson 23 Kirkfield View, Livingston Village, Livingston EH54 7BP
M6 NLQ S. Elms-Lester Ferndale House Kerrys Gate, Hereford HR2 0AH
M6 NLR D. Miller 11 Church Road, Hoveton NR12 8UG
MM6 NLV J. Heywood 2 Broaddykes Drive Kingswells, Aberdeen AB15 8UE
M6 NLW S. Jones 30 Crown Fields Close, Newton-le-Willows WA12 0JW
M6 NLX P. Cooke 26 Welby Way, Coxhoe, Durham DH6 4BT
M6 NLZ J. Wainwright 8 Mount Drive Purdis Farm, Ipswich IP3 8UU
M6 NMD E. Martin 61 Uffington Avenue, Lincoln LN6 0AG
M6 NME E. Nudd 14 Birkbeck Way, Norwich NR7 0XZ
MW6 NMG M. Glenn 31 South Drive, Rhyl LL18 4SU
M6 NMH N. Hoddinott 30 Deans Mead, Bristol BS11 0QX
M6 NML K. Taylor 9 Whitcliffe Grange Richmond North Yorkshire, Richmond DL10 4ES
M6 NMN N. Norsworthy Pippins, Trusham, Newton Abbot TQ13 0NW
MW6 NMP S. Quinn 1334 Carmarthen Road Fforestfach, Swansea SA5 4BR
M6 NMQ B. Clark 46 Fraser Close, Laindon, Basildon SS15 6SU
M6 NMR K. Wright 6 Windsor Park, Dereham NR19 2SU
M6 NMS J. Blackwell 68 Scarborough Drive, Minster On Sea, Sheerness ME12 2NQ
M6 NMT M. Afolabi Clarence Drive, Harrogate HG1 2QG
M6 NMU E. Riddle 37B Stubbs Lane, Braintree CM7 3NR
M6 NMW C. Marcus 43 Townsend Square, Oxford OX4 4BB
M6 NMZ R. Rutland 53 Downs Avenue, Whitstable CT5 1RR
M6 NNA L. Wilkinson 20 Coniston Road, Chorley PR7 2JA
M6 NNB A. Hopper 7 Holmesdale Villas, Swallow Lane, Dorking RH5 4EY
M6 NNC N. Hanson 8 Oak Street, Skegby, Sutton-in-Ashfield NG17 3FF
M6 NND N. Davies 29 Burns Road, Congleton CW12 3EE
M6 NNE D. Hanwell 28 Chipperfield Road, Norwich NR7 9RR
M6 NNF N. Ferenc 2A Rosedene Ave, London SW16 2LT
M6 NNJ P. Johnson 25 Pelham Street, Worksop S80 2TW
M6 NNK P. Banks 110 Cherwell Drive, Walsall WS8 7LL
M6 NNL D. Connolly 5 Cotton Fold, Rochdale OL16 5HJ
M6 NNM I. Friend 12 Bladen Valley, Briantspuddle, Dorchester DT2 7HP
M6 NNO P. Livermore Cm31Pn, Chelmsford CM3 1PN
M6 NNP J. Pinto 7 Wright'S Way, Colchester CO6 4NS
MM6 NNS M. Juknevicius 13/3 Wester Hailes Park, Edinburgh EH14 3AE
M6 NNT D. Mallinson 7 Abbots Way, Yeovil BA21 3HX
M6 NNU D. Johnson 25 Pelham Street, Worksop S80 2TW
M6 NNV K. Jaroonsungnoen 10 Stainsbury Street, London E2 0NF
M6 NNW D. Salmon Old School House, Ford Lane Alresford, Colchester CO7 8AU
M6 NNY D. Richardson 10 Maple Drive, Penrith CA11 8TU
M6 NOA J. Laszlo 135 Willifield Way, London NW11 6XY
M6 NOC C. Singfield 35 Leamington Drive, Sutton-in-Ashfield NG17 5BA
M6 NOD A. Smith Flat 19, Crown Terrace, 10 High Street, Leamington Spa CV31 3AN
MI6 NOE G. Mccann 2 Mahon Close, Portadown, Craigavon BT62 3JF
MM6 NOF K. Leiper West Wing North Newseat Of Ardo Methlick, Aberdeen AB417HQ
M6 NOI M. Richards 13 Rosemary Gardens, Sudbury CO10 1WL
M6 NOJ C. Denman 16 Upper Oak Street, Windermere LA23 2LB
M6 NOK N. Kibble 5 Dunbar Drive, Thame OX9 3YD
M6 NOL M. Roys Flat 13, Gregory House, Lister Avenue, Rotherham S62 7JA
M6 NOM N. O'Mahony 23 Main Road, Broomfield, Chelmsford CM1 7BU
MW6 NON A. Rosser 25 Clos Tir-Y-Pwll, Newbridge, Newport NP11 5GE
M6 NOQ P. Turner 8 The Retreat, Ramsgate CT126ET
MM6 NOR N. Turnbull 16 Barntongate Terrace, Edinburgh EH4 8BA
M6 NOT Dr S. Leask 1 Collington Street, Beeston, Nottingham NG9 1FJ
M6 NOY D. Noyek 34 The Spinney, Sidcup DA14 5NF
M6 NPD N. Dagger 23 Vassall Road, Bristol BS16 2LH
M6 NPF N. Fairhurst 7 Heatherlands, Sunbury-on-Thames TW16 7QU
M6 NPL A. Lyman 12 Chicheley Street, Newport Pagnell MK16 9AR
M6 NPN R. Sandwell 2 Alder Meadows, Cullompton EX15 1TA
M6 NPO R. Chambers ?Tylers?, Fitzhead, Taunton TA43JN
M6 NPV A. Price 87 Derricke Road, Bristol BS14 8NH
MW6 NPW N. Pitt 33 Scotchwell View, Haverfordwest SA61 2RD
M6 NPX N. Paxman 128 Coggeshall Road, Braintree CM7 9ES
M6 NPZ J. Freeman 46 Wargrove Drive, Sandhurst GU470DU
M6 NQA J. Abrahart Coombe Park, Bickington, Newton Abbot TQ12 6NZ
M6 NQB D. Hickson 19 Emerson Avenue Stainforth, Doncaster DN7 5QL
M6 NQC R. Lane 2 Dickiemoor Lane, Plymouth PL5 3NU
MI6 NQD I. Taylor 26 Shandon Park, Newry BT34 1QD
M6 NQI P. Lounton 107 Browning Hill Coxhoe, Durham DH6 4SA
M6 NQJ A. Fraley 28 Riverside Park Colehouse Lane, Clevedon BS21 6TQ
M6 NQK T. Batchford 22 Roberts Court.., Chester Road Erdington Chester Road Erdington, Erdington B240BX
M6 NQL G. Norbury 3 Sherard Croft, Birmingham B36 0LS
M6 NQM P. Stevens 36 Cooksey Lane, Birmingham B44 9QN
M6 NQN S. Reason 49 Highfield Road, Pudsey LS28 7JW
M6 NQP K. Dibling 23C Woodend Ave, Liverpool L23 2TY
M6 NQQ P. Bridgwater 18 Angelica, Amington, Tamworth B77 3JZ
M6 NQR B. Somerville Roberts 21 Regency Way, Ponteland, Newcastle upon Tyne NE20 9AU
M6 NQT C. Gamble 60 Hazelbury Road, Bristol BS14 9ET
M6 NQU J. Brookhouse 10 Davenport Drive Castle Vale, Birmingham B35 7NT
M6 NQV T. Harris The Lodge Chenies, Rickmansworth WD3 6ER
M6 NQW J. Williams 7 Spinney Lane, Nuneaton CV10 9JA
M6 NRA R. Nagy 40 Oakhampton Road, London NW7 1NH
M6 NRB M. Beckett 4 Sandcross Close, Orrell, Wigan WN5 7AH
M6 NRC C. Willson 17 Knightons Way, Brixworth, Northampton NN6 9UE
MM6 NRE R. Hudson 128 Neilston Road 2/2, Paisley PA2 6EP
M6 NRF C. Halloway 149 Grenville Road, Plymouth PL4 9QD
M6 NRG N. Genge 21 Castle Mead, Washford, Watchet TA23 0PZ
M6 NRH N. Holmes 32 Spinney Close, Kidderminster DY11 6DQ
MD6 NRI N. Rice The Asters, 33 Upper Dukes Road, Douglas, Isle of Man IM2 4AT
M6 NRJ N. Johnson Belair, Western Road, Crediton EX17 3NB
MM6 NRK G. Kerr 43 Blairpark Avenue, Coatbridge ML5 2ES
M6 NRM N. Miles 28 Oakdale Road, Witney OX28 1BJ
M6 NRN D. Reason 3 Stanford Road, Thetford IP24 1FH
M6 NRO N. Rostant 14 Gas Street, Leamington Spa CV31 3BY
MM6 NRQ A. Yates 111 Oldmeldrum Road, Inverurie AB51 6BB
M6 NRR K. Bostock 6 Jubilee Road, Daventry NN11 9HB
M6 NRS J. Swain 35 Heygate Close, Baildon, Shipley BD17 6RT
MW6 NRT N. Tanner 3 Maes Y Tyra, Resolven, Neath SA11 4NN
M6 NRU D. Warden 9 Chauntry Place, Coventry CV1 1JR
M6 NRV I. Hall 119 Linneaus, Hull HU3 2QT
M6 NRW N. Waters 9 Shirley Road, Droitwich WR9 8NR
M6 NRX S. Cook Deganwy, Hardwick Road, King's Lynn PE30 5BB
M6 NSC C. Kemp Forest Edge, Deer Park, Blandford Forum DT11 0AY
M6 NSD C. Coppins 17 Balmoral Road, Gillingham ME7 4PY
M6 NSE H. Kennedy 11 Green Road, High Wycombe HP13 5BD
M6 NSG S. Green 11 Lavender Walk, Beverley HU17 8WE
M6 NSH B. Skelton 1 Summer Hill Road, Bexhill-on-Sea TN39 4LN
M6 NSI K. Maclean Ch48Nz, Chester CH48NZ
M6 NSJ N. Inglis 74 Runswick Avenue, Whitby YO21 3UE
M6 NSK T. Pennell 99 Westheath Avenue, Sunderland SR2 9LQ
MM6 NSM C. Morris 23 Sedgebank, Livingston EH54 6HE
M6 NSO N. Osborne 12 Spiller Road, Chickerell, Weymouth DT3 4AX
M6 NSQ A. Skelton 1 Summer Hill Road, Bexhill-on-Sea TN39 4LN
M6 NSR E. Parrish 89 Delamere Drive, Macclesfield SK10 2PS
MD6 NSS N. Smith 4 Cooil Farrane, Douglas IM2 1NX
M6 NST K. Theobald 21 Stirling Close, Rochester ME1 3AJ
M6 NSW S. Wilson 37 New Road, Minster On Sea, Sheerness ME12 3PU
MW6 NSY B. Adamson 16 Glan Garth, Wrexham LL12 7DS
M6 NSZ K. Pesztranszki Flat 5 72 Welldon Crescent, Harrow HA1 1QR
M6 NTB N. Turrell 30 Old Road, Longtown, Carlisle CA6 5TH
M6 NTJ N. Jones 27 Gatcombe Gardens, West End, , Southampton SO183NA
M6 NTL B. Porter 74 Whalley Road, Heywood OL103JG
M6 NTM N. Mcniece 17 Hempdyke Road, Scunthorpe DN15 8LA
M6 NTN N. Hazlehurst 4 Titchfield Close, Wolverhampton WV10 8UN
MI6 NTP N. Prentice 26 Claranagh Road, Claranagh, Enniskillen BT94 3FJ
M6 NTR N. Reeve 124 Greenhills Road, Eastwood, Nottingham NG16 3FR
M6 NTV Rvd. N. Wood 12 Spring Close, Verwood BH31 6LB
MW6 NTW N. Davies 99 Sandpiper Way, Duffryn, Newport NP10 8WY
M6 NTY D. Balderson 19 Valley Road, Wellingborough NN8 2PH
M6 NUD N. Cull 8 Eaton Road, Norwich NR4 6PY
M6 NUF D. Metcalfe 25 St. Abbs Walk, Hartlepool TS24 7NW
MD6 NUG Capt. D. Cheadle Bay House, College Green, Castletown, Isle of Man IM9 1BE Isle of Man
M6 NUJ S. Burgess 24 Read Avenue, Stafford ST16 3NP
M6 NUL T. Moye 33 Prince Charles Road, Colchester CO2 8NS
MM6 NUO Dr C. Brown 4 Damselfly View, Edinburgh EH17 8XH
MM6 NUP J. Mack 5 Glen Affric Court, Dumbarton G82 2BN
M6 NUQ A. Louko Gummeruksenkatu 3, Jyvõskylõ 40100 Finland
M6 NUR D. Harrod 44 Kenning Street, Clay Cross, Chesterfield S45 9LE
M6 NUV D. Wright 1 Prince Andrew Drive Dersingham, King's Lynn PE31 6JW
MW6 NUW L. Newton 45 Commercial Street, Risca, Newport NP11 6AW
MI6 NUX J. Denvir 14 Cleland Park Central, Bangor BT20 3EP
M6 NUY W. Day 137 Tuffley Lane, Gloucester GL40NZ
M6 NVA M. Phillips 2 Millstream Close, Goostrey, Crewe CW4 8JG
M6 NVB N. Brown 28 Broadway, Stockport SK2 5SN
M6 NVG R. Adiga 41 St. Pauls Road, Staines-upon-Thames TW18 3HQ
M6 NVJ C. Wellman 3 Church Street, Warwick CV344AB
MI6 NVM N. Mccann 3 Portadown Road, Tandragee, Craigavon BT62 2BB
M6 NVO R. Marshall 1 Hill Street, Upper Gornal, Dudley DY3 2DE
M6 NVR N. Ross 38 Jamaica Road, Malvern WR14 1TU
M6 NVT N. Tennant 22 The Lizard, Wymondham NR18 9BH
MM6 NVW V. Corcoran Talland Brae, Beley Bridge, St. Andrews KY16 8LX
MM6 NVY M. Fitzgerald 42 Rosetta Road, Peebles EH45 8HJ
M6 NVZ D. O Brien 5 Orson Leys, Rugby CV22 5RG
M6 NWA N. Wong Montefiore House, Wessex Lane, Southampton SO18 2NU
M6 NWB J. Benbow 44 Copthorne Park, Shrewsbury SY3 8TJ
M6 NWC N. Clark Denemead, Cromwell Road, Waltham Cross EN7 6AS
M6 NWF A. Herbert 42 Paulsgrove Orton Wistow, Peterborough PE2 6YE
MM6 NWH N. Holgate 12 Talisman Road, Glasgow G13 3QN
M6 NWI A. Howard 70 Victoria Road, Dukinfield SK16 4UN
MM6 NWK I. Smith 92 Malvina Place, Perth PH1 5FJ
M6 NWL M. Webster Summerfield Cottage, Walker Lane Wadsworth, Hebden Bridge HX7 8SJ
MW6 NWM N. Miles 16 Oakland Crescent, Cilfynydd, Pontypridd CF37 4HD
M6 NWN J. Smith 54 Sneinton Hermitage, Sneinton, Nottingham NG2 4BS
M6 NWO D. Etherington 20 Sandringham Drive, Leeds LS17 8DA
M6 NWP C. Jeary 2 Baker Close, North Walsham NR28 9JE
M6 NWQ D. Colborne 11 Martindale Court, Martindale Road, Weston-Super-Mare BS22 8QQ
M6 NWS S. Wright 139 Aberfield Drive, Leeds LS10 3QA
M6 NWT J. Turner 2 South Drive, Padiham, Burnley BB12 8SH
M6 NWV J. German 89 Sandham Grove Heswall, Wirral CH60 1XW
M6 NWW N. Fields 31 Oxford Drive, Wirral CH63 1JG
MM6 NWX C. Mcintyre 10 Castle Grove, Glasgow G814AW
M6 NWY S. Newhouse 28 Hillmorton Lane, Lilbourne, Rugby CV23 0SS
M6 NWZ E. Griffin 26 Pebworth Drive Hatton, Warwick CV35 7UD
M6 NXA A. Milner Smith 10 Clover Lane, Cricklade, Swindon SN6 6FJ
M6 NXE P. Holroyd 39 Ellerbeck, Tamworth B77 4PP
M6 NXF W. Rowland 5 Gwinear Downs, Leedstown, Hayle TR27 6DJ
M6 NXH D. Gunter 1 Clipstone Mews Bedford Road, Barton-Le-Clay, Bedford MK45 4FN
M6 NXL P. Mcgee 2 Fishers Close, Norwich NR5 0QH
M6 NXO W. Byers 37 Windermere Drive, Bletchley, Milton Keynes MK2 3NR
M6 NXP L. Antins 1 Green Lane Cottages, Penrith CA11 0HN
M6 NXQ S. Donovan 3 Speldhurst Court, Kents Hill, Milton Keynes MK7 6JA
M6 NXR M. Rollason 6 Shenton Road Barwell, Leicester LE9 8AR

UK Callsigns

MW6 NXS A. Williams Tyn Coed, Bodffordd, Llangefni LL77 7LJ
M6 NXT L. Stevens 55 Silverlands Road, St. Leonards-on-Sea TN37 7DF
MW6 NXV B. Williams The Grange Llanddewi, Llandrindod Wells LD1 6SF
M6 NXY M. Leach 64 Grove Street, Wantage OX12 7BG
M6 NXZ N. Dexter 49 Kennedy Road, Horsham RH13 5DB
M6 NYA N. Fill 3 St. Albans Crescent, Flat B, London N22 5NB
M6 NYB M. Matthews 43 Nightingale Drive, Weymouth DT3 5ST
MW6 NYE A. Minton 22 Heol Serth, Caerphilly CF83 2AN
M6 NYF D. Lamble 4 Laburnum Road, Chorley PR6 7BG
M6 NYG A. Oxtoby 29 Masons Road, Hemel Hempstead HP2 4QP
M6 NYJ A. Harvey 17 Kerry Drive Smalley, Ilkeston DE7 6ER
M6 NYM Prof. S. Braunstein Crossways, Roundhay Park Lane, Leeds LS17 8AR
M6 NYO J. Raybould 33 Lincoln Road, Dorrington, Lincoln LN4 3PT
M6 NYP J. Dyson 77 Grantham Road, Southport PR8 4LT
M6 NYR S. Hartley 30 Coltman Avenue, Beverley HU17 0EY
M6 NYT J. Woodcock 31 Moorway Lane Littleover, Derby DE23 2FR
M6 NYW A. Sim 273A James Recketts Ave Hull, Hull HU8 8LQ
M6 NYX S. Dyer 24 Fossil Road, London SE13 7DE
M6 NYY D. Paterson 48 Sissons Road, Leeds LS104JT
M6 NYZ N. Mcgroarty 8 Bakers Lane, Weldon, Corby NN17 3LR
M6 NZB N. Woodruffe 139 North Home Road, Cirencester GL7 1DY
M6 NZE D. Mills 8 Chestnut Avenue Bucknall, Woodhall Spa LN10 5DU
M6 NZF K. Surtees Flat 10 2 Beaufort Road, Bournemouth BH6 5AL
M6 NZH B. Moore 10 Shakespeare Drive, Leicester LE3 2SP
M6 NZI S. Morriss 10 Cherry Tree Avenue, London Colney AL2 1RU
M6 NZJ B. Clark 8 Langdale Close, Farnborough GU14 0LQ
M6 NZK G. Jenkin 24 Coronation Avenue, Camborne TR14 7PE
M6 NZL P. Payne 3 Queens Court, Woking GU22 7NE
M6 NZM C. Mccallion-Gow 70 Cheaton Close, Leominster HR6 8EW
M6 NZN C. Naylor 16 Boston Avenue, Norbreck Blackpool FY29BZ
MD6 NZO G. Wilby Byways, Glenlough Circle, Glen Vine, Isle of Man IM4 4AX
M6 NZR R. Sadowski 35 Fletcher Gardens, Bracknell RG42 1FJ
M6 NZS A. Jackson 13 Thorness Close Alvaston, Derby DE24 0UY
M6 NZT T. Reseigh 10 Higher Croft Parc, The Lizard, Helston TR12 7RL
M6 NZV R. Todd Lavender Cottage Hillcrest Road, Edenbridge TN8 6JS
M6 OAB E. Colbert 176 Carters Mead, Harlow CM17 9EU
M6 OAD T. Waters 1 Meadow Cottages, Church Lane, East Winch, King's Lynn PE32 1NQ
M6 OAE M. Snow 6 Evelyn Close, Doncaster DN2 6PA
M6 OAF I. Parbery 38 Moor End, Holyport Holyport, Maidenhead SL6 2YJ
MM6 OAG M. Scott 71 Craigie Road, Perth PH2 0BL
MM6 OAI S. Chacko 25 Wemyss Court, Rosyth, Dunfermline KY11 2LL
M6 OAL S. Chng Cambridge, CCB3 9BB
M6 OAN A. Neale 4 Elizabeth Road, Rothwell, Kettering NN14 6AJ
M6 OAO J. Wilkinson 40 Station Road, Kenilworth CV8 1JD
M6 OAP M. Deary 7 Newbold Avenue, Sunderland SR5 1LG
M6 OAQ D. Byng 1 Ambell Close, Rowley Regis B65 8PB
M6 OAS L. Rowlands 6 St. Michaels Avenue, Clevedon BS21 6LL
M6 OAT J. Bullock 62 Hawthorn Close, Halstead CO9 2SF
M6 OAV J. Mennell 16 Trafalgar Street West, Scarborough YO12 7AU
M6 OAW A. Willis 29 Bonington Drive, Three Elms Road, Hereford HR4 0RU
M6 OAX A. Edmonds 20 Tomline Road, Ipswich IP3 8BZ
MI6 OAZ N. Armstrong 1 Diamond Cottages, Ardmore Road, Crumlin BT29 4QU
M6 OBB B. Beard 8 Monks Close, Newcastle-under-Lyme ST5 3QU
M6 OBC Dr C. Bridges 23 Bramley Vale, Cranleigh GU6 7FY
M6 OBH M. Lachs 3 King Henrys Walk, Epping CM16 6FH
M6 OBJ J. Bookham 116 Clare Gardens, Petersfield GU31 4EU
M6 OBK D. Cull 6 Compass Way, Bromsgrove B60 3GP
M6 OBO B. Bentham 7 Maypole Crescent, Abram, Wigan WN2 5YL
M6 OBP P. Percival 2 St. Catherine Street, Ventnor PO38 1HG
M6 OBQ D. Gaskell 126 Higher Green Lane Astley, Tyldesley, Manchester M29 7JB
M6 OBR B. O'Brien 48 Wright Crescent, Bridlington YO16 4RG
M6 OBS S. Martin 77 Chatford Drive, Shrewsbury SY3 9PH
MM6 OBU K. Moore 26 Summerhill Place, Glasgow G15 7JA
M6 OBX B. Gilbert 34 Ramsey Road, St.Ives PE275RD
M6 OBY P. Cairns Thorverton, Exeter EX5 5NB
M6 OBZ P. Dimes 5 Meadowbrook, Oxted RH8 9LT
M6 OCB M. Wilsher Flat K, Wellington Court, Bedford MK40 2HY
MW6 OCC B. Werrell 26 Glynhafod Street, Cwmaman, Aberdare CF44 6LD
M6 OCD Lord J. Gilbody Winter Meadows, Puxton, Weston-Super-Mare BS24 6TH
M6 OCH C. Howe 21 Gotham Road, Birmingham B26 1LB
M6 OCI J. Bruce 17 Highcrest Grove Tyldesley, Manchester M29 8GH
M6 OCJ O. Jones 6 Pickworth, Stamford PE9 4DJ
M6 OCK A. Mock 60 Effra Road Wimbledon, London SW19 8PP
M6 OCM I. Johnson 9 Brook Road, Pontesbury, Shrewsbury SY5 0QZ
MM6 OCP H. Adkins 270 Cleeves Quadrant, Glasgow G53 6NR
M6 OCR D. Meakin 27 Spencer Road, Long Buckby, Northampton NN6 7YP
M6 OCS O. Smith 19 Tedder Road, Bournemouth BH11 8BT
MW6 OCT A. Davies 28 Y Fron, Cefneithin, Llanelli SA14 7DN
M6 OCU E. Joynson 15 Home Farm Lane, Bury St. Edmunds IP33 2QJ
M6 OCX D. Mccarthy-Stewart 4 Hurst Hill, Chatham ME5 9BX
M6 OCY L. Horne 1268 Evesham Road Astwood Bank, Redditch B96 6AX
M6 OCZ B. Salt 1 Chantry Close, Harrow HA3 9QZ
M6 ODA B. Naylor 45 Meadowgrass Gardens, Worsley, Manchester M28 1PS
M6 ODC G. Wellstead 14A Hardy Road, West Moors, Ferndown BH22 0EX
M6 ODD J. Saunders 9 Capitol Close, Bolton BL1 6LU
M6 ODE K. Hadley 5 The Mead, Clutton, Bristol BS39 5RF
M6 ODF L. Halloway 82 Northwall Road, Deal CT14 6PP
MI6 ODG D. Gallagher 33 Mulnafye Road, Omagh BT79 0PG
M6 ODH M. Vlachos 2 Cumbrian Gardens, London NW2 1EF
M6 ODK M. Nash Unit 1 Bull Lane Ind Est Acton Sudbury Suffolk Co10 0Bd, Sudbury CO100BD
M6 ODL M. Cavanagh 97 Denecroft Crescent, Uxbridge UB10 9HZ
M6 ODM J. Barrett 5 Oakapple Drive, Dereham NR19 2SR
MI6 ODN T. Pashler 29 Clogher Road, Ballymena BT43 6TD
M6 ODP Dr O. De Peyer Flat 5, Molasses House, Clove Hitch Quay, London SW11 3TN
M6 ODR M. Hornibrook 16 Hamilton Road, Ryde PO33 3QZ
M6 ODY P. Lane 5 Swan Court, Middle Watch, Swavesey, Cambridge CB24 4AG
MM6 OEA D. Henretty 52 Woodburn Avenue, Falkirk FK2 9YG
M6 OEB D. James 23 Hedingham Close Plympton, Plymouth PL7 2FJ
MM6 OEC J. Maclean Columbia 42B Coll, Isle of Lewis HS2 0LR
M6 OEF S. Dickens 73 Hengist Way Wallington, London SM69BP
M6 OEG O. Blagrove The Willows, Ford Heath, Shrewsbury SY5 9GZ
M6 OEH K. Hillier 8 Shuttle Close Rossington, Doncaster DN11 0FR
MM6 OEI M. Gordon 41 Jan Mayen Drive, Peterhead AB42 3PX
MM6 OEJ R. Sutherland 25 Burns Wynd, Maybole KA19 8FG
M6 OEM M. Faulkner 6 Stanley Avenue, Queenborough ME11 5DT
M6 OEN T. Cummings 45 Sutton Way, Shrewsbury SY2 6EE
M6 OER K. Sadlik 35 Tournament Road, Salisbury SP2 9LQ
M6 OES I. Jones 33 Cobham Avenue, Liverpool L9 3BP
M6 OET B. Mead 8 Wordsworth Road, Kettering NN16 9LB
M6 OEU S. Fraser Old Post Office, Mill Road Barton St. David, Somerton TA11 6DF
M6 OEV M. Appleton The Old School, Martin Street Baltonsborough, Glastonbury BA6 8QS
M6 OEW L. Hudspith 20 Low Wood Road, Denton, Manchester M34 2PD
M6 OEZ C. Chandler 20 Trumpeter Place, Dawlish EX7 0RN
MM6 OFA J. Tervit 94 Covington Road Thankerton, Biggar ML12 6NE
M6 OFD G. Round 128 Leicester Road, Shepshed, Loughborough LE12 9DH
M6 OFE H. Bond Flat 4, Athrington Court, First Avenue, Felpham, Bognor Regis PO22 7LB
M6 OFF C. Gibson 3 Conway Drive, Billinge, Wigan WN5 7LH
M6 OFG D. Walker 39 Spinkhill Avenue, Sheffield S13 8FA
M6 OFH R. Williams 16 Horseshoe Close, Cowes PO31 8PZ
M6 OFJ D. Walker 39 Spinkhill Avenue, Sheffield S13 8FA
M6 OFL R. Seville 29 Blakesley Lane, Portsmouth PO3 5UG
M6 OFM J. Joyce 2 Harold Road Cuxton, Rochester ME2 1EE
MI6 OFN C. May 20 Harryville Street, Drumgoon, Enniskillen BT94 4QX
M6 OFQ C. Leece 101 Wellstead Way, Hedge End, Southampton SO30 2BH
MI6 OFZ A. Canning 89 Drumachose Park, Limavady BT49 0NY
M6 OGA C. Ingram 17 Stackfield, Harlow CM202LA
M6 OGC A. Chapman 44 Malcolm Road, Hartlepool TS25 3QR
M6 OGE T. Cullis 98 Pendragon Road, Bromley BR1 5LH
MI6 OGF J. Mcclarence 10 Cloyfin Park, Coleraine BT52 2BL
M6 OGH G. Holmes 3 Honeysuckle Square, Wymondham NR18 0FH
M6 OGJ J. Smith 19 Primrose Terrace St. Michaels Street, Shrewsbury SY1 2EY
M6 OGK V. Sim 49 St. Julians Wells, Kirk Ella, Hull HU10 7AF
M6 OGO G. O'Gorman Potenza, Chapel Lane, Tewkesbury GL20 8HS
M6 OGP A. Bhogal 28 Leigham Drive, Isleworth TW7 5LU
M6 OGQ P. Kelly 32 Charlbury Close, , Maidstone ME16 8TE
M6 OGS G. Saunders Hastings Road, Battle TN330TA
MW6 OGT T. Cording 17 Maes Y Derwen, Llanrhaeadr Ym Mochnant, Oswestry SY10 0LE
MM6 OGU H. Knox Shalderha Holm, Orkney KW17 2SA
M6 OGX G. Green 90 Princes Way, Fleetwood FY78DX
MM6 OHE P. Gush 1 Gallanach Lochgair, Lochgilphead PA31 8SD
M6 OHF K. Charlton 14 Rubens Close, Aylesbury HP19 8SW
M6 OHL M. O'Connor 28 Cardigan Road, Southport PR8 4SF
M6 OHN Taff Vale ARC c/o J. Jones 3 Orchard Way, Oxted RH8 9DJ
M6 OHO R. Owen 41 Mount Road, Cosby LE91SX
M6 OHP M. Watson 59 Willowtree Avenue, Durham DH1 1EA
M6 OHQ C. Smith 15 Acorn Close, Wynyard TS22 5UZ
M6 OHS J. Hodgson Stokes 40 Homewood Road, St. Albans AL1 4BQ
M6 OHT J. Holley Lookers, Blenheim Road Littlestone, New Romney TN28 8PR
M6 OHU T. Allen 11 Church View Highworth, Swindon SN6 7ER
M6 OHZ N. Holden Plum Cottage, Avon Dassett, Southam CV47 2AP
M6 OIC G. Chinnappa 34 Barker Road, Chertsey KT16 9HX
MI6 OIM A. Mcguigan 34 Ardymagh Road, Ballyclare BT39 9TJ
MM6 OIR P. Riddiough 1 Cedar Road, Ayr KA7 3PE
MM6 OIT L. Elder 18 Stewart Gardens, Airdrie ML6 9AQ
M6 OIU J. Fox 23 The Driveway, Canvey Island SS8 0AB
M6 OIW C. Thurling 32 Birch Avenue, Chatteris PE16 6JJ
M6 OIY S. Davis 74 The Driveway, Canvey Island SS8 0AD
M6 OIZ A. Papiewski 42 Balmoral Avenue, Spalding PE11 2RU
M6 OJB O. Beck 11 Clarefield Drive, Maidenhead SL6 5DW
M6 OJC A. Bruton 29 Helyers Green, Wick, Littlehampton BN17 7HB
M6 OJD K. Winton 130 George V Avenue, Worthing BN11 5RX
M6 OJF J. Brooks 19 Cromer Street, Middlesbrough TS4 2DN
M6 OJI K. Schofield 18 Berrow Walk, Bristol BS3 5ES
MI6 OJK J. Kavanagh 11 Pinewood Crescent, Claudy BT47 4AB
M6 OJL L. Johnson 33 Yetholm Place, Newcastle upon Tyne NE5 4ED
M6 OJM J. Marriott 66 Latimer Road, Cropston, Leicester LE7 7GN
M6 OJO E. Mehmet 8 Hailsham Road, Tooting, London SW17 9EN
M6 OJP C. Mead 17 Nelson Crescent, Ramsgate CT11 9JF
M6 OJQ W. Tranter 10 Wellpark Walk Newton Heath, Manchester M40 1JE
M6 OJS A. Snow 8 Summerhill Grove, Enfield EN1 2HY
M6 OJT O. Trehearne Claywood House, East Mascalls Lane, Lindfield RH16 2QJ
M6 OJU M. Ross Whitewalls Off Golf Road, Mablethorpe LN12 1LP
M6 OJZ J. Mees 4 Heltwate Bretton, Peterborough PE3 8RL
M6 OKD J. Yates 12 Iris Gardens, Gillingham SP8 4QY
M6 OKH K. Humble 2 Woodford Close, Sunderland SR5 5SA
M6 OKI O. Kelland 16 Esher Place Avenue, Esher KT10 8PY
MW6 OKJ J. Orchard The Burrows, Spring Gardens, Whitland SA34 0HL
M6 OKK G. Cooper Holmfield, Chelmorton, Buxton SK17 9SG
M6 OKL A. Gates 46 Gloucester Place, Littlehampton BN17 7AL
M6 OKM T. Banham 16 Mallard Court, Oakham LE15 6RQ
M6 OKO S. Neale 13 Hilltop, St. Anns Chapel, Bigbury, Kingsbridge TQ7 4HG
M6 OKP M. Kitching 70 Harrington Street, Bourne PE10 9HB
M6 OKQ M. Broum 137 Culvers Avenue, Carshalton SM5 2BA
M6 OKR M. Cole 4 Park Manor, Britton Street, Gillingham ME7 5EX
MI6 OKS A. Hanna 21 Elms Park, Coleraine BT52 2QF
M6 OKY C. Bietz 37 Alan Avenue, Newton Flotman, Norwich NR15 1PY
M6 OKZ J. Moore Flat 4, 219 Holland Road, Holland On Sea, Clacton-on-Sea CO15 6NL
M6 OLB R. Carter 39 Ambleside Court Marine Parade East, Clacton-on-Sea CO15 6JL
M6 OLD G. Waters 7 Roeselare Close, Torpoint PL11 2LP
M6 OLE O. Ward 33 Seventh Avenue, Oldham OL8 3RY

M6 OLF M. Mills 17 Hornby Street, Plymouth PL2 1JD
M6 OLH P. Costall 3 Gaynsford Place, Little Canfield, Dunmow CM6 1WB
M6 OLI J. Barnett 16 Beryl Avenue, Blackburn BB1 9RR
MI6 OLJ J. Rankin 36 Glenbank Place, Belfast BT14 8AN
MW6 OLL O. Booth Oak Cottage, Knockin, Oswestry SY10 8HQ
M6 OLP K. Baker Flat 26, Redwing Court 58 High Street, Orpington BR6 0LE
M6 OLQ R. Gaskell Cottonwood California, Baldock SG7 6NU
M6 OLR S. Jenkinson 17A Britannia Road Burbage, Hinckley LE10 2HE
M6 OLS J. Myers 43 Boggart Hill Crescent, Leeds LS14 1LF
M6 OLT A. Hicks 15 West Road, Ruskington, Sleaford NG34 9AL
M6 OLU J. Bugden 57 Henley Meadows, Tenterden TN30 6EN
M6 OLW O. Wood 2 The Bungalows, North Green, Woodbridge IP13 9NP
M6 OLY D. Goodman 44 Roberts Road, Madeley, Telford TF7 5JJ
MI6 OMA A. O'Brien 15 Oldcastle Road, Newtownstewart, Omagh BT78 4HX
M6 OME M. Edge 15 Littlemoor Avenue, Kiveton Park, Sheffield S26 5NZ
M6 OMF M. Flower 1 Ridge Farm Cottages, Turnpike, Chilmark, Salisbury SP3 5AD
M6 OMG N. Patel 78 Wesley Close, South Harrow, Harrow HA2 0QE
M6 OMH M. Hawkridge 27 Northdale Road, Bradford BD9 4HG
MM6 OMJ M. Reid 16 Hayfield Road, Kirkcaldy KY2 5DG
M6 OMK L. Flis 15 Rifle Hill, Braintree CM7 1DG
M6 OML M. Lewis 1 Kingsmead Stretton, Burton on Trent DE13 0FQ
M6 OMP O. Powell 2 Little Bowden Manor Kettering Road, Market Harborough LE16 8AW
M6 OMR B. Withers 29 Yeoman Way, Trowbridge BA14 0QL
M6 OMS M. Stradling 25 Maple Court, Acacia Grove, New Malden KT3 3BX
MM6 OMT C. Ingram 22 The Square, Mintlaw, Peterhead AB42 5EH
M6 OMX M. Amos Willow Tree House, Deers Green, Clavering, Saffron Walden CB11 4PX
M6 OMZ S. Smith 8 Mill Field Close, South Kilworth, Leicester LE17 6FE
M6 OND C. Howard 75 Gordon Road, Herne Bay CT6 5QX
M6 ONH A. Brown 51 Towncroft, Chelmsford CM1 4JX
M6 ONJ T. Roberts 16 Malfort Road, London SE5 8DQ
M6 ONL P. Destoop 73, Oeselgemstraat, Wakken 8720 Belgium
M6 ONM M. Kulakowski Flat 4 Clement Mellish House East Stockwell St., Colchester CO1 1GJ
M6 ONN T. Newberry 2 Brodsworth Road, Peterbrough PE28XF
M6 ONO R. Cobern 21 Hopgarden Close, Lamberhurst, Tunbridge Wells, TN3 8DY
M6 ONP P. Maes Schoolstraat 111, Sint-Niklaas 9100 Belgium
M6 ONS J. Wood 3 Onslow Mews, Cranleigh GU6 8FD
M6 ONT M. Newman 274 Long Drive, Ruislip HA4 0HY
M6 ONU A. Kazmi 39 Herga Road, Harrow HA3 5AX
M6 ONW J. Gray 7 Ruskin Avenue, Melksham SN12 7NG
MW6 ONZ D. Cheeseman 61 Ffordd Y Millenium, Barry CF61 5BD
M6 OOA P. Ellis 14 Mountfield Gardens, Newcastle upon Tyne NE3 3DB
M6 OOB L. Atkinson 55 Warkworth Crescent, Seaham SR7 8JT
M6 OOC S. Kiely 35 Chestnut Avenue, Todmorden OL14 5PH
M6 OOD M. Holbrook-Bull 66 Wayman Road, Corfe Mullen, Wimborne BH21 3PN
M6 OOJ J. Edwards 23 Rectory Lane, Great Ellingham, Attleborough NR17 1LD
MM6 OOK A. Currie 2/2 44 Robertson Street, Greenock PA16 8QB
M6 OOL S. Latimer 40 Petersham Road, Long Eaton, Nottingham NG10 4DD
M6 OOM A. Shepherd 33 Meadow Way, Turton, Bolton BL7 0DE
M6 OOO P. France Flat 39, Netley House, Birmingham B32 2BT
MM6 OOP A. Brown 11A Bowfield Road, West Kilbride KA23 9LB
M6 OOQ M. Waite 108A Chantry Gardens, Southwick, Trowbridge BA14 9QS
M6 OOV G. Long Maple House. 13 Welfen Lane, Claypole, Newark NG23 5AL
M6 OOW W. Waterson 43 Highland Road, Bath BA2 1DY
M6 OOX M. Emans 212 Hilldene Avenue, Romford RM3 8DB
M6 OOZ J. Maclean 6 Matford Mews Matford Alphington, Exeterá EX2 8XP
M6 OPA G. Mourelatos 13 Stirling Road, London N22 5BL
M6 OPB S. O'Loughlin 1 Pine View, Headley Down, Bordon GU35 8AX
MW6 OPC O. Campbell 16 Whitcliffe Drive, Penarth CF64 5RY
MI6 OPD J. Reilly 220 Ardmore Road, Londonderry BT47 3TE
M6 OPE O. Evers 7 Rosebury Street, Hull HU3 6PQ
M6 OPF N. Prater 100 Pitfold Road, London SE12 9HY
M6 OPG A. Holmes Bolehall Manor Club Ltd Amington Road, Tamworth B77 3LH
M6 OPJ J. Viswambaran 103 Inglehurst Gardens, Ilford IG4 5HA
M6 OPL R. Topley 85 Stuart Road, Aylsham, Norwich NR11 6HW
M6 OPM O. Murphy 7 Charlotte Street, Leamington Spa CV31 3EB
M6 OPO O. Pochat Hame, Bealswood Road, Gunnislake PL18 9DA
M6 OPS B. Hillson Flat 2, Heatherton Park House, Heatherton Park, Taunton TA4 1EU
MI6 OPT A. Bailie 130 John Street, Newtownards BT23 4NA
M6 OQD L. Sproule Dragonpits North Perrott, Crewkerne TA18 7TH
M6 OQG G. Sampson Green Head Cottage, More Hall Lane Bolsterstone, Sheffield S36 3ST
M6 OQH C. Pendlebury 73 Pool Street, Wigan WN3 5BT
M6 OQL D. Cooper 129 Church Street Eastwood, Nottingham NG16 3HR
M6 OQN A. Tyrrell Flat 1, 61 Vicarage Road, Eastbourne BN20 8AH
M6 OQQ R. Webster 31 Beech Avenue, Alfreton DE55 7EW
M6 OQR S. Dolby 198 Cavendish Road, Worksop S80 2SH
M6 OQT J. Rymell 9 Carter Avenue Ruddington, Nottingham NG11 6NP
M6 OQX M. Nicholson 104 Harvesters Close Rainham, Gillingham ME8 8PA
M6 OQZ R. Carpenter 8 New Buildings Clee Hill, Ludlow SY8 3PN
M6 ORB P. Thomas 17 Vine Court, St. Pauls Road, Cheltenham GL50 4LL
M6 ORC A. Cross 12 Appleby Drive, Langdon Hills, Basildon SS16 6NU
M6 ORE E. Moore 33 Avon Drive, Congleton CW12 3RQ
MW6 ORF S. Leigh 27 Cae Clyd, Llandudno LL30 1BL
MJ6 ORG R. Spencer 1 Excelsior Villas, Le Mont Les Vaux, Jersey JE3 8LS
MW6 ORH O. Hopkin The Forge, Rock Road, , St. Athan CF62 4PG
M6 ORI D. Dart 164A Salterton Road, Exmouth EX8 2PA
MW6 ORJ K. Collins 27 Cae Clyd, Llandudno LL30 1BL
MM6 ORK M. Herridge The Hollies, Petticoat Lane, Orkney KW17 2RP
M6 ORM R. Turner 38 Cotman Road, Clacton-on-Sea CO16 8YB
M6 ORO O. Gascoigne 19 Shaftesbury Close, Nailsea, Bristol BS48 2QH
M6 ORT Rvd. G. Wellington 94 Arlington Road, London N14 5AT
MM6 ORU M. Drever 7 Abbotswell Crescent, Aberdeen AB12 5AQ
MW6 ORW O. Wilson 45 Dol Helyg, Penrhyncoch, Aberystwyth SY23 3GZ
M6 ORX D. Gibbeson 33 Treefield Walk, Barnstaple EX32 8PE
M6 OSC B. Wilson 119 Fountains Close, Washington NE38 7TQ
M6 OSF J. Clover 174 Sturton Street, Cambridge CB1 2QF
M6 OSI M. Kent 56 Hill Street, Hilperton, Trowbridge BA14 7RX
M6 OSJ S. Mcintosh 12 Cuthwine Place, Lechlade GL7 3EG
M6 OSL R. Kemish Damons Cottage, Romsey SO51 0GD
M6 OSO J. Coombe 21 Lesnewth, Par PL24 2DE
M6 OSP D. Park 29 Tresco Close, Blackburn BB2 4RT
M6 OST R. Frost 40 Greenwich Close, Downham Market PE38 9TZ
M6 OSU O. Summerfield 2 Walnut Close, Broughton Astley, Leicester LE9 6PY
M6 OSW J. Halford Flat 5, Lattice Court, 2 Leonora Walk, Campbell Park, Milton Keynes MK9 4BA
M6 OSX M. Rawson 71 Corn Mill Drive, Farnworth, Bolton BL4 9EN
M6 OSY D. Coppenhall 55 Vicarage Lane, Elworth, Sandbach CW11 3BU
M6 OTB D. Tate 14 Wordsworth Avenue, Hartlepool TS25 5NG
M6 OTH M. Riches Fransham Road Farm, Beeston, King's Lynn PE32 2LZ
M6 OTI J. Bryan 7 Riverside Apartments 43A Mowbray Street, Sheffield S3 8EP
M6 OTJ A. Packer 20 Shipman Road, , Market Weighton YO43 3RB
MW6 OTK J. Gilmour Llyswen, Crick, Caldicot NP26 5UW
M6 OTL A. Rawlins 7 Kiln Hill, Slaithwaite, Huddersfield HD7 5JS
M6 OTN T. Bluck 2 Pendas Meadow, Pinvin, Pershore WR102DQ
MI6 OTP T. Pearson 20 Lammy Walk, Omagh BT78 5JE
M6 OTR T. Rood The Beeches, Frodingham Road, Brandesburton, Driffield YO25 8QY
M6 OTS M. Clapham 12 Penkridge Grove Stechford, Birmingham B33 9JX
MW6 OTT M. Locke 18 Tan Y Capel Llanddaniel, Gaerwen LL60 6EA
M6 OTU R. Hallam 18 Trent Vale Walney, Barrow-in-Furness LA14 3NB
MI6 OTW P. Fallon 18 Church View, Killough, Downpatrick BT30 7RJ
M6 OTY N. Payne 221 Orchard Avenue, Bridport DT6 5RJ
M6 OUA A. Regmi 29 Birchwood Avenue Littleover, Derby DE23 1QA
M6 OUD A. Cattell 2 St. James Close, Ruscombe, Reading RG10 9LJ
M6 OUI B. Gimbert 49 Ratcliffe Road, Loughborough LE11 1LF
M6 OUL J. Nield 24 Bagot Street South Shore Blackpool, Blackpool FY1 6EZ
M6 OUP A. Wheatley 17 Cambridge Road, Southport PR9 9NQ
M6 OUR J. Miall 42 Garrard Road, Slough SL2 2QW
M6 OUS M. Mckean 107 Kings Grove, Cranfield, Bedford MK431AJ
MM6 OUU C. Mcconnochie 72 Duddingston Avenue, Kilwinning KA13 6RS
MW6 OUW B. Davies 8 Ferry Road, Kidwelly SA17 5BJ
M6 OUY I. Boffey 29 Elm Road Seaforth, Liverpool L21 1BJ
M6 OVB R. Gowers 43 Tungstone Way, Market Harborough LE16 9GA
M6 OVE G. Whale 1 Rokeby Gardens, Bradford BD10 0DN
MW6 OVF M. Rowan Flat 4, Albion Court, Wellington Street, Pembroke Dock SA72 6JR
M6 OVH S. Langton 44 Poppyfields Hesketh Bank, Preston PR4 6TJ
M6 OVI Dr O. Rominger 27 Paddock Close, Sixpenny Handley, Salisbury SP5 5NZ
M6 OVN G. Ayres 4 Bentley Way, Winterslow, Salisbury SP5 1PF
M6 OVR M. Barry 58 Bury Road, Radcliffe, Manchester M26 2UU
M6 OVS R. Yilmaz 2 Cumbrian Gardens, London NW2 1EF
M6 OVT J. Priddin 4, Priory Close Pen-Y-Ffordd Ch4 0Jb, Chester CH4 0JB
MW6 OVV D. Jonker 8 Angle Village Angle, Pembroke SA71 5AT
M6 OVW G. Sandell 1 St. Margaret Road, Ludlow SY8 1XN
M6 OVX M. Walton 358 Old Heath Road, Colchester CO2 8DD
M6 OVY H. Greenhill 39 Pook Lane East Lavant, Chichester PO18 0AH
M6 OWA J. Wright 10 Whalley Road, Heskin, Chorley PR7 5NY
M6 OWC S. Iles Bigbury Bay Holiday Park, Challaborough, Kingsbridge TQ7 4HS
MI6 OWD A. Kirkpatrick 2 James Street, Newtownards BT23 4DY
M6 OWG D. Burgfeld 20 Wilson Row, Crowthorne RG45 6WE
M6 OWI C. Pierce 42 Staplehurst Road, Reigate RH2 7PY
M6 OWJ D. Owen 1 Nightingale Road, Malvern WR14 2QA
MM6 OWL B. Ewart 94 Kirkness Street, Airdrie ML6 6ET
M6 OWM J. Thorpe 57 Bratch Lane, Wombourne, Wolverhampton WV5 8DL
M6 OWO A. Whitley 37 Park Avenue Saltney, Chester CH4 8TR
M6 OWP A. Stinton 22 Durham Road. Ronkswood, Worcester WR5 1NL
M6 OWS P. Blunden 64 Sussex Drive, Banbury OX16 1UN
M6 OWT R. Harwood 24 Firle Crescent, Lewes BN7 1QG
M6 OWU J. Woodcock 1 Main Street Brandesburton, Driffield YO258RL
MM6 OXA S. Waddington Wester Lathallan, Leven KY8 5QP
M6 OXB D. Whittaker Flat 1, Clement Court, Manchester M11 1EQ
M6 OXC I. Davison 35 Newhouse Avenue, Esh Winning, Durham DH7 9JH
M6 OXD S. Holt 65 Brown Royd Avenue, Huddersfield HD5 9QA
M6 OXF D. East 43 Conduit Hill Rise, Thame OX9 2EJ
M6 OXJ R. Low 4A Bellevue Road, Romford RM5 3AY
MI6 OXM C. O'Neill 53A Coole Road, Dungannon BT71 5DP
M6 OXS C. Gibson 20 Dordon Road Dordon, Tamworth B78 1QN
M6 OXT R. Ingram 17 Stackfield, Harlow CM20 2LA
M6 OXU G. Eaton 16 Holly Walk, Nuneaton CV11 6UU
MW6 OXV S. Davies 14 Elm Walk Mynydd Isa, Mold CH7 6XZ
M6 OXW D. Ellis 4 Vane Terrace, Darlington DL3 7AT
MM6 OXX A. Berry 8 Hill Street Striling Fk7 0Dh, Striling FK7 0DH
M6 OXY A. Oxborrow 24 Ickworth Crescent, Rushmere St. Andrew, Ipswich IP4 5PQ
M6 OXZ A. Santese 4 Vane Terrace, Darlington DL3 7AT
M6 OYA K. Peakman 32 Mob Lane, Walsall WS4 1BB
MI6 OYB P. Magee 11 Elm Corner, Dunmurry, Belfast BT17 9PZ
M6 OYC S. Boyce 15 Verbena Avenue Farnworth, Bolton BL4 0EN
M6 OYD M. Boyd 1 Harris Close, Brackley NN13 6NS
M6 OYE A. Hatton 270 Little Brays, Harlow CM18 6HD
M6 OYF G. Bailey 28 Station Road, Polesworth, Tamworth B78 1BQ
M6 OYH C. Staples 32 Browns Lane, Netherton, Bootle L30 5RW
MM6 OYJ B. Smith 88 Regent Court, Aberdeen AB24 1ZS
MW6 OYL A. Owen Bro Dawel, Brynrefail, Caernarfon LL55 3NR
M6 OYM M. James 50 Lambert Drive Acton, Sudbury CO10 0BZ
M6 OYN D. Applewhite 84 Eringden, Wilnecote, Tamworth B77 4DB
MW6 OYO T. Puxley 9 Wesley Court, Pembroke Dock SA72 6NE
M6 OYP M. Curran 32 Wedgewood Road, Lincoln LN6 3LU
M6 OYU L. Rowe 4 Allandale View, Lincoln LN1 3RD
M6 OYX S. Rees 24 Carlton Road, Helmsley YO625HA
M6 OYY N. Weiner 3 Bluebell Court Lower Mardyke Avenue, Rainham RM13 8GF
M6 OYZ S. Bushnall 7 Faraday Terrace Haswell, Durham DH6 2DT
MW6 OZB D. Gardner 1 Neath Court, Thornhill, Cwmbran NP44 5UH
MW6 OZF C. Lyle 41 Mount Pleasant Avenue, Llanrumney, Cardiff CF3 5SY
M6 OZG K. Randle 36 All Saints Road, Sittingbourne ME10 3PB
M6 OZI S. Carpenter 52 Mewstone Avenue, Wembury, Plymouth PL9 0JZ

M6 OZL A. Vickers 2 Overton Lane Hammerwich, Burntwood WS7 0LH
M6 OZM M. Osborne 6 Walnut Tree Way, Worthing BN13 3QQ
M6 OZN R. Jacobs 21 Queens Avenue, Ramsgate CT12 6DQ
MW6 OZO S. Hayward 22 Dewsland Street, Milford Haven SA73 2AU
M6 OZQ N. Silvester 34 The Burgage Eccleshall, Stafford ST21 6DR
M6 OZR I. Plimmer 120 Spitfire Road Castle Donington, Derby DE74 2AU
M6 OZS T. Falkingham 22 Almond Tree Avenue Carlton, Goole DN149QQ
M6 OZT C. Moore 2 Chapel Rise, Cross Hill, Filey YO14 0JA
M6 OZW E. Duffy 75 West Park, Selby YO84JN
M6 OZY K. Johnson 32 Redmire Close, Bransholme, Hull HU7 5AQ
M6 OZZ M. Crockford Centre Cottage Kelk, Kelk YO258HL
M6 PAA P. Sadler 52 Kent Avenue, Weston-Super-Mare BS24 7FH
M6 PAD Leek And District ARC c/o Dr P. Darlington 8 Uplands Road, Urmston, Manchester M41 6PU
M6 PAE P. Lobo-Kazinczi Flat 1, 52 Park Road, Hull HU5 2TA
M6 PAG W. Johnson 15 Sanders Road, Hemel Hempstead HP3 9UB
MW6 PAI P. Blackburne 16 College View, Connah'S Quay, Deeside CH5 4BY
M6 PAJ P. Moore 112 Westbury Lane, Bristol BS9 2PU
MW6 PAM G. Williams 99 Maes Llwyn, Amlwch LL68 9BG
M6 PAN P. Naylor 38 Piggott Grove, Stoke-on-Trent ST2 9BZ
M6 PAO P. Alley 21 Orchard Court, Thorney PE6 0QW
M6 PAS P. Simcox 67 Mervyn Road, Bilston WV14 8DB
MI6 PAT P. Coogan Flat B, 44 Ramoan Gardens, Belfast BT11 8LL
MM6 PAU M. Scott 28B Highfield Place, Birkhill, Dundee DD2 5PZ
M6 PAV P. Villette 5 Regent Close, Burton Latimer, Kettering NN15 5QS
M6 PAW P. Wallis 28 Munro Street, Stoke-on-Trent ST4 5HA
M6 PAX S. Pashley 45 Highfield View Road Newbold, Chesterfield S41 7JZ
M6 PAY D. Potter The Lines, Commonside, Boston PE22 9PR
MW6 PBF P. Martin 6 Herrick Place, Machen, Caerphilly CF83 8TA
MI6 PBI L. Thompson 28 Kirkdale, Newtownabbey BT36 5BX
M6 PBK P. Browne Ham Cottage, Hammingden Lane, Haywards Heath RH17 6SR
M6 PBL P. Parkin Hawksworth House, Main Street, Frolesworth, Lutterworth LE17 5EG
M6 PBM P. Mills 10 Laurel Estate, Cowes PO31 7HW
MW6 PBO Dr W. Dickson The Rowans, Pwllmeyric, Chepstow NP16 6LA
M6 PBP S. Bradley 90 Occaiso House, 90 Play House Square, Harlow ME20 1AP
M6 PBR M. Massie 3 East Mount, North Ferriby HU14 3BX
M6 PBT P. Burgess 61 Grosvenor Avenue, Torquay TQ2 7JX
M6 PBV J. Brown 26 Lynnes Close, Blidworth, Mansfield NG21 0TU
MI6 PBW P. White 46 Pine Cross, Dunmurry, Belfast BT17 9QY
M6 PBY I. Maltby 22 Fern Avenue, Doncaster DN5 9QX
MI6 PBZ P. Bell 3 Alexandra Crescent, Larne BT40 1NE
M6 PCB C. Vincent 81 Trethannas Gardens, Praze, Camborne TR14 0LL
M6 PCC P. Cumbers Azaiba, Fielden Road, Crowborough TN6 1TP
M6 PCD P. Cheek 55 Turpin Green Lane, Leyland PR253HA
M6 PCE C. Fryer 16 Elston Place, Aldershot GU12 4HY
M6 PCF P. Faulkner 32 Manvers Road, Beighton, Sheffield S20 1AY
MI6 PCJ P. James 14 Brookmount Crescent, Omagh BT78 5HG
M6 PCL P. Lambert 10 Cranmore, Brompton Road, Weston-Super-Mare BS24 9BU
M6 PCM P. Mason 1 Keepers Coombe, Bracknell RG12 0TN
M6 PCN P. Newth 3 Mulberry Court, Mulberry Street, Stratford-upon-Avon CV37 6RT
M6 PCP P. Couch 28 Polgover Way, St. Blazey, Par PL24 2DL
M6 PCQ P. Morris 14 Marina Road, Darlington DL3 0AL
M6 PCR P. Raine 29 Beech Gardens, Crawley Down, Crawley RH10 4JB
MW6 PCT S. Gau Disgwylfa, The Downs, Cardiff CF5 6SB
M6 PCU T. Prince 6 Hyperion Avenue, South Shields NE34 9AE
M6 PCV B. Eastman 49 Calshot Close, Calshot SO451BP
M6 PCW P. Cullen 57 Hall Close, Stafford ST17 4JJ
M6 PCX P. Coombes 2 Bissoe Cottages, Bissoe, Truro TR4 8SU
M6 PCZ P. Colyer 23 Florida Road, Torquay TQ1 1JY
MM6 PDA S. Haynes 86 Barrowfield Street, Coatbridge ML5 4BJ
M6 PDB P. De Basto 27 Cloudesley Square, London N1 0HN
M6 PDC P. Chell 174D South Road, Stourbridge DY8 3RN
M6 PDD A. Coleman 63 Newtown Green, Ashford TN24 0PL
MW6 PDE P. Devlin Brynteg, Fron Bache, Llangollen LL20 7BP
MW6 PDG P. Gough 92 Pendwyallt Road, Whitchurch, Cardiff CF14 7EH
M6 PDI M. Lidster 83 Stroma Gardens, Hailsham BN27 3AZ
M6 PDJ P. Davies 15 Carlyle Road, Wolverhampton WV10 8SL
M6 PDK P. Coote 251 Alfreton Road, Pye Bridge, Alfreton DE55 4PB
M6 PDL M. James 42 Doone Way, Ilfracombe EX34 8HS
M6 PDM P. Moffitt Wrawby Farm, Star Carr Lane, Brigg DN20 8SG
M6 PDN D. Ebbs 28 Betterton Court, Chapmangate, York YO42 2ET
M6 PDO A. Dingwall 48 Village Farm Caravan Site, Bilton Lane, Harrogate HG1 4DL
MW6 PDP D. James Coombe House, Coombe, Presteigne LD8 2HL
M6 PDQ M. Summers 22A Footners Lane, Burton, Christchurch BH23 7NT
M6 PDR P. Roberts 1 Winston Close, Eastleigh SO50 4NS
M6 PDS P. Stean 100 Faringford Road, London E15 4DF
M6 PDT C. Devlin 32 Kestrel Drive, Dalton-in-Furness LA15 8QA
MW6 PDU L. Davies 16 Pembrey Gardens, Pontllanfraith, Blackwood NP12 2LR
M6 PDV F. Farrer 16 High Street, Eagle, Lincoln LN6 9DH
M6 PDW P. Willis 21 Alton Close, Swindon SN2 5HF
MM6 PDX W. Davison 16 Millfore Court, Bourtreehill North, Irvine KA11 1LT
MI6 PDY P. Davis 35 Culross Drive, Dundonald, Belfast BT16 2SQ
M6 PDZ P. Bromfield 32 Mount Pleasant, Halesworth IP19 8JF
MM6 PEA A. Pemberton 111 Henderson Street, Bridge Of Allan, Stirling FK9 4HH
M6 PEB P. Beckwith 1 Whincroft Drive, Ferndown BH22 9LH
M6 PEC B. Clayton 1 Maude Crescent Sowerby Bridge, Halifax HX6 1LB
M6 PED J. Pedley 19 Ellis Drive, New Romney TN28 8XH
M6 PEG P. Boast 104 Whittier Road, Nottingham NG2 4AS
M6 PEI J. Brown The Chapel, Heath Green, Leighton Buzzard LU7 0AB
MI6 PEJ J. Beckett 20 Old Forge, Banbridge BT32 4AH
MW6 PEL P. Day 15-16 Troedrhiw-Trwyn, Pontypridd CF37 2SE
M6 PEM P. Metters 137 Nevinson Avenue, South Shields NE34 8NE
M6 PEN P. Kiley 178 Kingfisher Drive, Woodley, Reading RG5 3LQ
M6 PEO P. Owen 17 Jameston, Bracknell RG12 7WZ
M6 PEP S. Hill 35 Longs Way, Wokingham RG40 1QW
M6 PEQ P. Scarlett 75 Ilfracombe Road, Southend-on-Sea SS2 4PA
M6 PER P. Rimmington 28 Skipton Road, Swallownest, Sheffield S26 4NQ
M6 PEU G. Wale 21A Pole Barn Lane, Frinton-on-Sea CO13 9NH
M6 PEW P. Woodburn 21 The Row, Silverdale, Carnforth LA5 0UG
M6 PEX C. Rumball 4 Hasted Close, Bury St. Edmunds IP33 2UA
MW6 PEY A. Fey 28 Bryn Rhedyn, Caerphilly CF83 3BT
M6 PEZ C. Perrin 66 Central Avenue, Farnworth, Bolton BL4 0AU
M6 PFA D. Rees 16 Hopton Close, Hereford HR1 4DQ
M6 PFB P. Payne 5 Hurstwood Close, Bexhill-on-Sea TN40 2TA
MI6 PFD P. Dallas 12 Glendun Crescent, Coleraine BT52 1UJ
M6 PFG P. Roberts 1 Ballard Road, Wirral CH48 9XU
M6 PFH P. Holmes 18 Raleigh Avenue, Whiston, Prescot L35 3PL
MI6 PFI P. Sweeney 83 Rashee Park, Ballyclare BT399AS
M6 PFK P. Lewis 9 The Hill, Glapwell, Chesterfield S44 5LX
M6 PFL P. Flatt 8 Lingfield Crescent, Queensbury, Bradford BD13 2SA
M6 PFM T. Sarosi The Grange, Warwick Road, London W5 3XH
M6 PFO P. Noble 14 Park Street, Swallownest, Sheffield S26 4UP
M6 PFP B. Robinson 11 Wimbledon Drive, Stockport SK3 9RZ
M6 PFQ J. Beeley 47 Claremont Avenue, Hindley, Wigan WN2 4JG
M6 PFR P. Reeves 18 Ploughmans Lea, East Goscote, Leicester LE7 3ZR
MM6 PFT D. Barclay 45 Milton View, Gatehead, Kilmarnock KA2 0AY
M6 PFU P. Ogle 29 Patrick Street, Grimsby DN32 0JQ
M6 PFV F. Popa Flat 183, Edinburgh House, Edinburgh Gate, Harlow CM20 2TJ
M6 PFW P. Woodmass 29 Thirlmoor, Blackfell, Washington NE37 1HT
M6 PFY P. Knox 3 Woodside View, Benfleet SS7 4PB
M6 PFZ P. Catterall 117 Beech Hill Lane, Wigan WN6 8PJ
M6 PGA P. Burrows 31 Spencer Road, Lutterworth LE17 4PG
M6 PGB P. Boultwood 32 Makepiece Road, Bracknell RG42 2HJ
MW6 PGC G. Shepherd 48 Lasgarn View, Varteg, Pontypool NP4 7RZ
MW6 PGD P. David 5 The Bungalows, Oakfield Terrace, Bridgend CF32 7SP
M6 PGF P. Bagely 28 Tenacre Lane, Dudley DY3 1XQ
M6 PGH P. Hill 14 Drovers Way, Woodlands, Ivybridge PL21 9XA
MI6 PGI P. Brown 2 Legaterriff Road, Ballinderry Upper, Lisburn BT28 2EY
M6 PGL P. Lewis 154 Meadow Head, Sheffield S8 7UF
M6 PGN D. Martin 8 Garden City, Langport TA10 9ST
M6 PGO J. Barnes 6 Windsor Drive, Bredbury, Stockport SK6 2EH
M6 PGP P. Pearce 41 Tennyson Avenue, Boldon Colliery NE35 9EP
M6 PGQ P. Challans Flat 4, Sandringham Court, 2 Chandos Square, Broadstairs CT10 1QN
MM6 PGT M. Paget 40 Davaar Drive, Kilmarnock KA3 2JG
M6 PGU P. Hall 11 Middleton Court, Mansfield NG18 3RN
M6 PGW P. Gonczarow 25 Ribchester Avenue, Burnley BB10 4PD
M6 PGX B. Williams 8 Lindberg Way, Woodley, Reading RG5 4XE
M6 PGY P. Chapman 30 Roman Way Shrivenham, Swindon SN6 8FA
M6 PGZ P. Barkley 32 Tudor Road, Doncaster DN2 6EN
M6 PHA P. Biggs 4 Pendle Close, Southampton SO16 4QT
M6 PHC P. Conduit 16 Rectory Avenue, High Wycombe HP13 6HW
M6 PHE P. Swinyard 72 London Road, Wokingham RG40 1YE
M6 PHF S. Shaw Riverside, Calderbridge, Seascale CA20 1DN
MM6 PHG P. Groundwater Vollbekk, 14 Burnside, Kirkwall KW15 1TF
MI6 PHH P. Kinney 23 Glenariff Crescent, Ballymena BT43 6ET
M6 PHJ D. Collier 133 Woodstock Road, Moston, Manchester M40 0DG
M6 PHK P. Bentley The Vauce Farm, Langley-On-Tyne, Hexham NE47 5NA
M6 PHL P. Hughes 111 Wisbech Road, Littleport, Ely CB6 1JJ
M6 PHM P. Meerman 24 Horseshoe Crescent, Burghfield Common, Reading RG7 3XW
MM6 PHO C. Hunter 1 North Gate Lodge, Erines, Tarbert PA29 6YL
M6 PHP J. Dando 70 Lydgate Road, Southampton SO19 6NG
MI6 PHQ P. Wilkin 55 Clare Heights, Ballyclare BT39 9SB
M6 PHS P. Swannick 11 Derwent Road, Scunthorpe DN16 2PA
M6 PHT P. Hart 11 Sadlers Ride, West Molesey KT8 1SU
M6 PHU J. Steel 4 Station Terrace, Boroughbridge, York YO51 9BU
M6 PHV P. Lowe 38 Romany Road, Great Ayton, Middlesbrough TS9 6BX
M6 PHW P. Walsh 55 Fore Street, St. Marychurch, Torquay TQ1 4PU
M6 PHZ J. Smith 51 Myrtle Avenue, Peterborough PE1 4LR
M6 PIA M. Moriarty 30 Longmead, Abingdon OX14 1JQ
MM6 PIB M. Thiebaut 1/1 Block A, 2 Barrack Street, Hamilton ML3 0HZ
M6 PIC D. Berry 27 Harcourt Terrace, Headington, Oxford OX3 7QF
M6 PID P. Hodgson 19 Raymonds Drive, Benfleet SS7 3PL
MW6 PIH P. Humphreys 14 Woodside, Oswestry SY11 1EP
M6 PII S. Whitwell 12 Lambton Road, Stockton-on-Tees TS19 0ER
MM6 PIJ G. Wilson Flat 7 29 Second Avenue, Clydebank G813AB
M6 PIK D. Pike 46 Haymans Close, Cullompton EX15 1EH
M6 PIP P. Pritchard 12 Easton Crescent, Billingshurst RH14 9TU
MI6 PIR D. Hamilton 41 Moss Lane, Ballynahinch BT24 8FE
M6 PIT R. Peacock 24 Vicarage Estate, Wingate TS28 5BP
M6 PIU Dr A. Jackson 22 Albert Avenue, Nottingham NG8 5BE
M6 PIY B. Kay 17 Upper Abbots Royd Caravan Park Barkisland, Halifax HX4 0DE
M6 PJA P. Archer 31 Stoney Bank Drive, Kiveton Park, Sheffield S26 6SJ
M6 PJB P. Bacon 124 Aughton Road, Swallownest, Sheffield S26 4TH
M6 PJD P. Doyle 27 Brockenhurst Avenue, Havant PO9 4NS
M6 PJE P. Elmore 8 Gray Street, Elsecar, Barnsley S74 8JR
M6 PJH P. Higginson 187 Birmingham Road, Ansley Village, Warwickshire CV10 9PQ
M6 PJI P. Carne 1 Curlew Close, Letchworth Garden City SG6 4TG
MW6 PJJ P. Jones 23 Pinecroft Avenue, Aberdare CF44 0HY
M6 PJM P. Miller 46 Great Brooms Road, Tunbridge Wells TN4 9DH
MW6 PJN P. Noakes 8 Ellis Avenue, Old Colwyn, Colwyn Bay LL29 9LB
MW6 PJP P. Parsons 31 Clos Tan-Y-Fron, Bridgend CF31 2BY
MM6 PJR P. Russell 21 St. Andrews Drive, Law, Carluke ML8 5GB
M6 PJV K. Rutter 72 Astbury Drive, Barnton, Northwich CW8 4PX
M6 PJX S. Rodgers 22 Crabtree Lane, Sutton-On-Sea, Mablethorpe LN12 2RT
MM6 PJY K. Govan 30 Glenwell Avenue, Stranraer DG9 7BA
MM6 PKC I. Macnab 4/4 27 St. Andrews Crescent, Glasgow G41 5SD
M6 PKD C. Barker 52 Hazelmoor, Hebburn NE31 1DH
M6 PKL D. Cadet 2 Paddockside, Middleton, Ludlow SY8 3EB
M6 PKN P. Winterbottom 25A Celandine Rise, Swinton, Mexborough S64 8NZ
MM6 PKO G. Rothera Greystones, Watten, Wick KW1 5UG
M6 PKQ P. Robbins 2 Bramble Drive Claremont Park, Berrow, Burnham-on-Sea TA8 2NH
M6 PKR P. Karuppannan Rajan 1 Jacquard Close, Coventry CV3 5NG
M6 PKT P. Hall 112 Osmaston Park Road, Derby DE24 8EX
M6 PKV N. Beer North View, 30 Wood End Bluntisham, Huntingdon PE28 3LE

M6 PLB D. Smith 13A, Barnes Road, Skelmersdale WN8 8HN
MW6 PLC P. Carroll 22 Llanbad, Brynna, Pontyclun CF72 9QQ
M6 PLE P. Jenkins 4 Boulton Avenue, West Kirby, Wirral CH48 5HZ
M6 PLH P. Simmonds 25 Dimlington Bungalows, Easington, Hull HU12 0TH
M6 PLI P. Hartley 99 Cyprus Street, Stretford, Manchester M32 8BE
M6 PLK A. Jedryka 71 West Royd Drive, Shipley BD18 1HL
M6 PLO P. Thomas 77 Hawthorn Avenue, Lowestoft NR33 9BB
MW6 PLP P. Price 80 Maesglas, Pontyates, Llanelli SA15 5SH
M6 PLQ M. Kincaid 40 Sorrel Close, Uttoxeter ST14 8UP
M6 PLR P. Rasmussen 7 Portal Drive North, Upper Heyford, Bicester OX25 5TH
M6 PLS A. Hornby 54 Amberley Road, Horsham RH12 4LN
M6 PLV P. Le Vallois 14 London Row, Arlesey SG15 6RX
M6 PLZ L. Perry 23 Victors Crescent, Hutton, Brentwood CM13 2HZ
M6 PMA P. Ambrose 61 Greenleas Road, Wallasey CH45 8LR
M6 PMB P. Mcmullan-Bell 57 Melford Avenue, Barking IG11 9HS
M6 PMD P. Davies 68 Sidmouth Avenue, Stafford ST17 0HF
M6 PMF P. Fenney 8 Lynton Way, Bristolá BS16 1QP
M6 PMG P. Gorman 1 Hill Top, Stourbridge DY9 9BZ
M6 PMJ P. Mullen 14 Anderson Road, Hemswell Cliff, Gainsborough DN21 5XP
M6 PMK T. Karpasitis 4 East Crescent, Enfield EN1 1BS
MI6 PML P. Mclarnon 8C Greenview Way, Antrim BT41 4EG
MI6 PMM P. Maguire 139 Carrowshee Park, Drumhaw, Enniskillen BT92 0FS
M6 PMP P. Parker 3 Little Charlton, Basildon SS13 2EJ
M6 PMR B. Rubery 142 Chapel Street, Pensnett, Brierley Hill DY5 4EQ
M6 PMT P. Troth Beuna Vista, Hawford Wood, Droitwich WR9 0EZ
M6 PMY P. Imm 29 Naunton Crescent Leckhampton, Cheltenham GL53 7BD
M6 PMZ P. Mansfield 27 Popplechurch Drive, Swindon SN3 5DE
MI6 PNB P. O'Kane 31 Larchfield Gardens, Kilrea BT51 5SB
M6 PNC P. Griffiths 930 Cedar Point Cir, Rose Hill 67133 United States
MM6 PND C. Lucking 39 Overdale Street, Glasgow G42 9PZ
M6 PNG D. Golik Primrose Cottage, Newport Road Woodseaves, Stafford ST20 0NP
M6 PNH S. Nevins 4 Ubbanford Norham, Berwick-upon-Tweed TD15 2LA
M6 PNI M. Roe Flat 20, 150 London Road, , Guildford GU11UF
M6 PNJ P. Jackson 1 Osbern Road, Preston, Paignton TQ3 1HN
M6 PNK C. Taylor 212 Plantation Hill, Worksop S81 0HD
M6 PNL P. Neil 1 High Graham Street, Sacriston, Durham DH7 6LZ
M6 PNP A. Pepler Flat 2, 5 Fore Street, Westbury BA13 3AU
M6 PNR P. Nowak 38 Brook Road, Craven Arms SY7 9RF
M6 PNT A. Huby 38 Ferryhill Road, Irlam, Manchester M44 6DD
MM6 PNU I. Swanston 6 Roberts Grove, Galashiels TD1 2BJ
MW6 PNW P. White 13 Bridge Street, Maesteg CF34 9BJ
M6 PNX C. Peace 49 Rossefield Way, Leeds LS13 3RS
M6 PNY P. Sayles 11 Malton Close, Monkston, Milton Keynes MK10 9HR
MW6 PNZ C. Eckersley 14 Bro Nantfer, Gwaun Cae Gurwen, Ammanford SA18 1BR
M6 POA J. Lambert 205 Reading Road, Wokingham RG41 1LJ
M6 POB B. Campbell 9 Granams Croft, Liverpool L30 0PH
M6 POC Z. Schwingen 1 Elizabeth Lockhart Way, Braintree CM7 9RH
M6 POD D. Pochat Hame, Bealswood Road, Gunnislake PL18 9DA
M6 POE S. Barber 31 Aysgarth Avenue, Crewe CW1 4QE
MI6 POF A. Stewart 35 West Wind Terrace, Hillsborough BT26 6BS
M6 POG P. Gupta 36 Slimmons Drive, St. Albans AL4 9AP
MI6 POH P. O'Hare 15A Lisdoo Road, Clady, Strabane BT82 9RQ
M6 POP P. Barrow Beach View, Withernsea HU19 2DS
M6 POQ H. Thomas 5 Silver Drive, Frimley, Camberley GU16 9QN
M6 POU C. Moye 33 Prince Charles Road, Colchester CO2 8NS
MM6 POV B. Barclay 24 Hillcrest, Dalmellington KA67ST
M6 POW L. Patton 17 Hardwick Street, Weymouth DT4 7HT
M6 POY M. Moore 63 Homewater House Hulbert Road, Waterlooville PO7 7JY
M6 POZ P. Osborne 1, Ingate, Beccles, Great Yarmouth NR34 9RU
MW6 PPD J. Lawson 138 Fosse Road, Newport NP19 4TB
M6 PPE P. Wainwright 30 Bank View Earlsheaton, Dewsbury WF12 8HH
M6 PPJ P. Pearson 22 Norris Street, Darwen BB3 3DR
M6 PPK P. Kay 30 Broadway, Grange Park, St. Helens WA10 3RX
M6 PPL G. Smith 36 Palma Park Homes, Shelly Street, , Loughborough LE11 5LB
M6 PPN T. Penberthy 34 Aldrin Road, Exeter EX4 5DN
M6 PPP R. Parker 5 Hawksland Close, Gainsborough DN21 5FF
M6 PPQ C. Rainbow 13 Audley Road, Talke Pits, Stoke-on-Trent ST7 1UG
M6 PPS J. Eyre 13 Beech Close, Great Ayton TS96NQ
M6 PPT G. Owen 38 Trentham Drive, Bridlington YO16 6ES
M6 PPV J. Smith 48 London Road, Wymondham NR18 9BP
M6 PPY J. Dales 6 Woodfield Drive, Sawtry, Huntingdon PE28 5TZ
M6 PPZ P. Lloyd 8 Maydor Avenue, Saltney Ferry, Chester CH4 0AH
M6 PQA P. Casado Arias 24 Oldridge Road, London SW12 8PJ
M6 PQD P. Coxon 14 Barwick Street, Peterlee SR8 3SA
M6 PQF P. Ferreira Flat 48, Lambert Court, Bushey WD23 2HF
M6 PQI K. Meakin 8 Over Ross Street, Ross-on-Wye HR9 7AS
M6 PQL L. Jansen 45 Clive Road Failsworth, Manchester M35 0NN
M6 PQO N. Morris 19 Marshalswick Lane, St. Albans AL1 4UR
MW6 PQQ C. Kibblewhite 2 Maes Ceidio, Llannerch-Y-Medd LL71 7AE
M6 PQS R. Cowap 53 Cross Lane, Newton-le-Willows WA12 9QA
MW6 PQU E. Maher 25 Ffordd Cadfan, Bridgend CF31 2DP
M6 PQV B. Edwards 34 Alderley Digmoor, Skelmersdale WN89LZ
M6 PQW K. Matley Fleetwood, Fleetwood FY77HZ
M6 PRA A. Mcewen 4 The Pantyles, Nightingale Lane, Sevenoaks TN14 6BX
M6 PRB P. Bowes 72 Keswick Road, Worksop S81 7PS
M6 PRC L. Piercy 35 Claremont Road, Grimsby DN32 8NU
M6 PRE P. Evans 67 Grenville Street, Stokport SK3 9ER
M6 PRF J. Rowland-Stuart 86 Wiltshire House, Lavender Street, Brighton BN2 1LE
MM6 PRH P. Higgins South Fallaws Farm, Strathmartine, Dundee DD3 0PR
M6 PRM P. Munson 8 Longley Lane, Spondon, Derby DE21 7AT
M6 PRN R. Raine 110 Stirling Avenue, Jarrow NE32 4HS
M6 PRO A. Forrest 1 Errington Bungalows, Sacriston, Durham DH7 6NE
M6 PRP P. Punjabi 62 Cleveland Road, London W13 8AJ
M6 PRS P. Skidmore 36 Princes Drive, Harrow HA1 1XH
M6 PRV S. Bullock 25 Greenfield Street, Haslingden, Rossendale BB4 5TG
M6 PRW P. White 63 Garden Drive, Brampton, Barnsley S73 0TN
MM6 PRZ P. Misson 95 Alexandra Street, Devonside FK13 6JA
M6 PSC P. Croxford 1 Meteor Close, Bicester OX26 4YA
M6 PSG P. Griffith Long Field Barn, Clarkes Lane, Beccles NR34 8HR
M6 PSJ D. Williams 2 Tyning Road, Peasedown St. John, Bath BA2 8HT
M6 PSM P. Mansfield 15 Earl Avenue, New Waltham, Grimsby DN36 4NE
M6 PSN G. James 28 Redcar Road, Romford RM3 9PT
M6 PSO P. Szabo 93 Norham Avenue, Southampton SO16 6QB
M6 PSQ J. Anderson 18 Blackwater Grove Alderholt, Fordingbridge SP6 3AD
M6 PSR D. Rowe 18 Burman Road, Wath-Upon-Dearne, Rotherham S63 7ND
M6 PST P. Lester-Tugwell 71 Dunkellin Way, South Ockendon RM15 5ES
M6 PSV M. Beard 4A St. Aidans Grove, Liverpool L36 8JE
MM6 PSX A. Ross 29 East Banks, Wick KW1 5NL
M6 PSY M. O'Halloran 7 Waver Close, Corby NN18 8LL
M6 PSZ S. Macdonald Woodside Cottage, Horton Way, Verwood BH31 6JJ
MM6 PTE P. Davis 4 Daisy Park, Baltasound, Unst, Shetland ZE2 9EA
M6 PTF M. Haddleton 42 Grove Road, Pontefract WF8 2AB
M6 PTH P. Thomasson 13 Ringway, Neston CH64 3RS
M6 PTI S. Machon 19 Houfton Road, Mansfield NG18 2DG
M6 PTM P. Millington 125 Telford Way, High Wycombe HP13 5SZ
M6 PTO R. Morgan 17 Forbes Close, Glenfield, Leicester LE3 8LF
M6 PTP M. Taylor 84 Elgar Crescent, Brierley Hill DY5 4JJ
MM6 PTR P. Owen 4 Doonhill Wood, Newton Stewart DG8 6NU
M6 PTU Prof. P. Curtis Cotswold, Salisbury Road, Abbotts Ann, Andover SP11 7NX
M6 PTV C. Brough 40 Denby Grange, Harlow CM17 9PZ
M6 PTX P. Hutchins 25 The Paddock, Maidenhead SL6 6SD
M6 PTY P. Ryan 3 St. Quentin Close, Derby DE22 3JT
M6 PUA A. De Mora 36 West Park, Minehead TA24 8AN
M6 PUD J. Crudgington 29 Wild Flower Way, Ditchingham, Bungay NR35 2SF
M6 PUF R. Bath 226 Chaplin Road, Stoke-on-Trent ST34NP
M6 PUG A. Ward 29 Mainwaring Road, Wallasey CH44 9DN
M6 PUH Dr M. Foster 58 New Terrace, Staverton, Trowbridge BA14 6NY
M6 PUL D. Pullen 5 Weldon Close, Shotton Colliery, Durham DH6 2YJ
M6 PUN J. Schofield 21A Millgate, Thirsk YO7 1AA
M6 PUP S. Nichols 61B Norwich Common, Wymondham NR18 0SW
M6 PUQ R. Vincent 81 Trethannas Gardens, Praze, Camborne TR14 0LL
M6 PUS A. Russell 26 Diamond Ridge, Camberley GU15 4LD
MW6 PUT K. Duggan 97 Maesglas Grove, Newport NP20 3DN
M6 PUU M. Coleman 60 Church Rd, Fleet GU514LY
MI6 PUX N. Jenkinson 35 Scarvagh Heights, Scarva, Craigavon BT63 6LY
M6 PUY E. Barrett 54 The Parade, Greatstone, New Romney TN28 8SU
M6 PUZ B. Titmus 68 Hobart Road, Cambridge CB1 3PT
MJ6 PVB A. Bertram Roz-Den La Rue De La Guilleaumerie, Jersey JE2 7HQ
M6 PVC C. Constantine 257 Kings Road, Ashton-under-Lyne OL6 9EG
M6 PVI J. Griffiths Mill House, High Street, Waddingham, Gainsborough DN21 4SW
M6 PVL P. Lansdown 30 Bromley Road Kingsway Quedgeley, Gloucester GL2 2JB
M6 PVM P. Massey 11 Shakestone Close, Writtle, Chelmsford CM1 3HS
M6 PVN P. Nicholls 23 Bishops Gate, Birmingham B31 4AJ
M6 PVR P. Richards Hills & Hollows Barnwell Road Oundle, Peterborough PE8 5PB
M6 PVV B. Woods 193 Wimberley Street, Blackburn BB1 8HU
M6 PVW M. Lawrence Pheasant View, Church Lane Thwaite, Eye IP23 7EJ
MM6 PVY C. English 6 Torrance Court East Kilbride, Glasgow G75 0RU
M6 PVZ D. Till 157 Malkin Drive, Church Langley, Harlow CM17 9HL
M6 PWB P. Barrett 8 Durban Road, Portsmouth PO1 5RR
M6 PWE G. Corkett 28 Underwood, Hawkinge, Folkestone CT18 7NT
M6 PWG P. Green 3 Beech Court, Long Stratton, Norwich NR15 2WY
M6 PWH P. Whitworth 34 Rye Crescent, Danesmoor, Chesterfield S45 9HH
M6 PWK P. Martin 5 Shropshire Drive, Wilpshire, Blackburn. BB1 9NF
M6 PWL M. Mynn The Town House, Parsons Field, St. Mary'S, Hugh Town TR21 0JJ
MW6 PWM J. Pattinson 79 Waterloo Road, Penygroes, Llanelli SA14 7PN
MW6 PWO P. Oseland 6 Oaklands Close, Bridgend CF31 4SJ
MI6 PWR P. Warriner 25 St. Marys Road, Omagh BT79 7JX
MW6 PWS P. Spong 209 Neath Road, Briton Ferry, Neath SA11 2BJ
M6 PWT J. Girard 49 Beech Crescent, Hythe, Southampton SO45 3QF
M6 PXD P. Donaghy 67 Brockenhurst Way, Bicknacre, Chelmsford CM3 4XN
M6 PXF P. Forbes 55 The Henrys, Thatcham RG18 4LS
M6 PXG P. Guest 88 Windsor Drive, Wigginton, York YO32 2YE
M6 PXK C. Smith 83 Sea View Street, Cleethorpes DN35 8HY
M6 PXL C. Hembrow 5 Wetherby Road, Stoke-on-Trent ST4 8AZ
M6 PXP J. Maudsley Knight Stainforth Hall, Little Stainforth, Settle BD24 0DP
M6 PXT L. O'Connor Newton, Maldon Road, Witham CM8 1HP
M6 PXW P. Wright 4A Alma Street, Melbourne, Derby DE73 8GA
M6 PXY A. Schofield 24 Oldbrook, Bretton, Crowland PE3 8SH
M6 PYD D. Robinson 27 Abbeylea Drive Westhoughton, Bolton BL5 3ZD
M6 PYE Dr D. Pye 151 Smallbrook Lane, Leigh WN7 5PZ
M6 PYF A. Whitmore 12 Rutland Close, Congleton CW12 1LT
M6 PYG G. Fielding Chapel Court, Chapel Lane, Cradley, Malvern WR13 5HX
MW6 PYH G. Round 3 Pen Yr Hwylfa, Harlech LL46 2UW
MI6 PYL S. Begley 46 Lany Road Moira, Craigavon BT67 0NZ
M6 PYO P. Pritchard 1 Winceby Road, Wolverhampton WV6 7SY
MW6 PYP R. Hall 8 Adam Street, Abertillery NP13 1EX
M6 PYR P. Robinson 15 Cornelius Drive, Wirral CH61 9PY
M6 PYZ T. Brown 69 Lawn Closes Alt, Oldham OL82HB
M6 PZA R. Hawes 40 Nightingale Way, Thetford IP24 2YN
M6 PZB P. Brown 64 St. Johns Road, Swinton, Mexborough S64 8QW
M6 PZF P. Fairbourn 17 Perry'S Lane, Wroughton, Swindon SN4 9AX
MM6 PZG Dr J. Wills 2 Old Dalmore Gardens Auchendinny, Penicuik EH26 0RR
M6 PZM A. Aslam 101 The Oval, Guildford GU2 7TP
M6 PZO S. Mitchell 51 Chatsworth Avenue Tuffley, Gloucester GL4 0SH
MW6 PZP S. Owen 8 Old Tanymanod Terrace, Blaenau Ffestiniog LL41 4BU
M6 PZW M. Heenan 15 Woodacre Green Bardsey, Leeds LS17 9AB
M6 PZY C. Puzey 17 Iverdale Close, Iver SL0 9RJ
M6 PZZ L. Evans 1A South View, Fryston, Castleford WF10 2QF
M6 QFG F. Gardner 67 Woodside Road, Tunbridge Wells TN4 8PY
MI6 RAC R. Mcdaid 29 Linen Green, Sion Mills, Strabane BT82 9TL
MI6 RAD R. Todd 58 Kilrea Road, Portglenone, Ballymena BT44 8JB
M6 RAH R. Harriman 9 Millers Close, Rushden NN10 9RP
M6 RAL C. Jewell 3 Marsh Gate, Clee St. Margaret, Craven Arms SY7 9DU
MM6 RAM R. Law 4A Burnside Court, Dundee DD2 3AF
MU6 RAN R. Phibbs Aqeb, St. Martins GY4 6AD Guernsey
M6 RAO R. Rider 25 Kimber Close, Lancing BN15 8QD

MD6 RAQ R. Shooter 14 Cooyrt Shellagh, Ballasalla, Isle of Man IM9 2EU
M6 RAR R. Calleja 3 Stoutsfield Close, Yarnton, Kidlington OX5 1NX
MI6 RAS G. O'Neill 46 Ashgrove Road, Newtownabbey BT36 6LJ
MM6 RAV R. Nelson 8 South Street Cambus Fk102Pa, Stirling FK102PA
MW6 RAW A. Cotter 3 George Street, Aberdare CF44 6RY
M6 RAZ R. Booker 6 Kipling Road, Dursley GL11 4QB
M6 RBA R. Bristow 6 Blackness Road, Crowborough TN6 2LY
MM6 RBE J. Howison 54 Whiteside, Bathgate EH48 2RG
M6 RBF R. Fry 20 St. Andrews Road, Ellesmere Port CH65 5DG
M6 RBH R. Haynes 28 Ridgeway View, Montgomery SY15 6BF
MJ6 RBI R. Rumboll Windsor House, La Grande Route De St. Laurent, Jersey JE3 1NL
M6 RBK Dr B. Kalogerakis Inglewood, Madingley Road, Cambridge CB23 7PH
M6 RBO R. Axtell 74 Elmshott Lane, Slough SL1 5QZ
M6 RBP R. Lyttle 14 Rosehead Drive, County Antrim BT14 7BF Ireland
M6 RBQ R. Loseby 22 Stanley Gardens, Doncaster DN4 0JG
MM6 RBT R. Turpie 11 Ashkirk Place, Dundee DD4 0TN
M6 RBU R. Byard Wynbourne, Wynolls Hill Lane, Coleford GL16 8BP
M6 RBV R. Ford 7 Briar Court, Wickersley, Rotherham S66 1AF
M6 RBY R. Johnston-Stuart 35 Robbins Close, Bradley Stoke, Bristol BS32 8AS
MW6 RBZ R. Bishop 96 Heol Homfray, Cardiff CF5 5SB
M6 RCA R. Broadwith Keepers Cottage, Rookwith, Ripon HG4 4AY
M6 RCB R. Brignall 53 Grange Crescent, St. Michaels, Tenterden TN30 6DY
M6 RCD R. Coates 123 Mansfield Crescent, Armthorpe, Doncaster DN3 2AR
M6 RCE C. Mandville Garth Hill College, Bull Lane, Bracknell RG42 2AD
M6 RCF R. Goodier 56 Chariot Street, Manchester M11 1DP
M6 RCG P. Ballington 7 Links Close, Sinfin, Derby DE24 9PF
M6 RCI L. Rodgers 27 Arden Houses Normanby-By-Spital, Market Rasen LN8 2HE
MM6 RCK T. Mcgurk 18 Sheil Lane, East Calder, Livingston EH53 0FB
M6 RCN S. Francis 17 Garden Close, Rough Common, Canterbury CT2 9BP
MM6 RCO R. Nicoll 15 Redford Walk, Edinburgh EH13 0AF
M6 RCP C. Edwards 19 Bells Place, Coleford GL16 8BX
MI6 RCR R. Reid 13 Gelvin Grange, Londonderry BT47 2LD
M6 RCS R. Codrai 46 The Dale, Waterlooville PO7 5DE
MI6 RCV H. Ashe 74 Markville Portadown, Craigavon BT63 5SZ
M6 RCW R. Weaver 15 Sharps Field, Headcorn, Ashford TN27 9UF
M6 RCY C. Radley 7 Garforth Crescent, Droylsden, Manchester M43 7SW
MM6 RCZ R. Taylor Station Cottage, Main Street, Bathgate EH48 3BU
M6 RDA A. Hill 7 Knebworth Court, Congleton CW12 3SW
MW6 RDC R. Cole 14 Inner Loop Road, Beachley, Chepstow NP16 7HF
M6 RDD B. Barwell 48 Locke King Road, Weybridge KT13 0TB
M6 RDF I. Sharman 63 Duston Road, Northampton NN5 5AR
M6 RDI E. Vecenans 155 Upper Dale Road, Derby DE23 8BP
M6 RDK R. Starr 43 Newington Avenue, Southend-on-Sea SS2 4RD
MW6 RDL R. Green Hillside, Meinciau Road, Kidwelly SA17 4RA
MM6 RDM R. Matthews 5 Dollar Avenue, Falkirk FK2 7LD
M6 RDN R. Vickers 57 Cecil Road, Selly Park, Birmingham B29 7QQ
M6 RDP A. Toynton Flat 2, 7 Sea Lawn Terrace, Dawlish EX7 0AD
M6 RDQ R. Ellison 66 Coronation Road, Wingate TS28 5JW
MI6 RDR Rvd. R. Rowe 31 Main Street, Brookeborough BT94 4EZ
M6 RDS R. Sutton Yew Tree Farm, Paddol Green, Shrewsbury SY4 5QZ
MM6 RDT R. Tourish 8 Linnpark Gardens, Johnstone PA5 8LH
MM6 RDU R. Drummond 11 Firwood Drive, Bo'ness EH51 0NX
M6 RDV R. De Vries Corner Cottage, Hillcrest Close, Sturminster Newton DT10 2DL
M6 RDW I. Hartless 32 Long Acre, Mablethorpe LN12 1JF
M6 RDX A. Griffiths 61 Hawarden Road, Penyffordd, Chester CH4 0JD
M6 RDZ Wessex Ham c/o R. Sidwell 27 Gressingham Drive, Lancaster LA1 4RF
M6 REB R. Beardsley 10 Moreton Close, Church Crookham, Fleet GU52 8NS
M6 RED I. Sanderson 9 Haigh Street, Cleethorpes DN35 8QN
M6 REE R. Garvey 2 Link Lane, Oldham OL8 3AD
M6 REF C. Newsam 15 Devonshire Avenue North, New Whittington, Chesterfield S43 2DF
M6 REH R. Ashley 15 Wimbourne Drive, Gillingham ME8 9EN
M6 REI C. Reid 28 Albion Road, London N16 9PH
M6 REJ R. James 31 Tehidy Gardens, Camborne TR14 0ET
M6 REL R. Allen 44 Overing Avenue, Great Waldingfield, Sudbury CO10 0RJ
MI6 REM Dr R. Richey 10 Crest Road, Enniskillen BT74 6JJ
M6 REO D. Curry 2 Norham Avenue South, South Shields NE34 7LP
M6 REQ R. Saunders 128 Foxcroft Drive, Wimborne BH21 2LA
M6 RER R. Ridge Roskellan House, Maenlay, Helston TR12 7QR
M6 RES R. Strong 2 Dean Avenue, Thornbury, Bristol BS35 1JJ
MM6 REV T. Paterson Free Church Manse, Church Street, Golspie KW10 6TT
M6 REW Grimsby & Cleethorpes District Sas Radio Scouting Team c/o R. Wells 37 Water Meadows, Worksop S80 3DF
M6 REX R. Moldoveanu 17 Lynchford Road, Farnborough GU14 6AR
M6 REY T. Surrey 53 Central Avenue, North Shields NE29 7JB
M6 RFA Basingstoke ARC c/o R. Allen Milverton, Mill Road, Pulborough RH20 2PZ
M6 RFB R. Woodrow 190.Sedlescombe Road North, St. Leonnards on Sea TN377EN
M6 RFC R. Crockford 17 Tadcroft Walk, Calcot, Reading RG31 7JR
MI6 RFD A. Hudson 31 Knollwood, Seapatrick, Banbridge BT32 4PE
M6 RFE A. Braeman 59 Freshbrook Road, Lancing BN15 8DE
M6 RFF R. Hancock 125 Fairham Road, Stretton, Burton-on-Trent DE13 0BT
M6 RFG R. Morgan 14 Ash Road, Ashurst, Southampton SO40 7AT
M6 RFI G. Sullivan 9 Hine Close, Gillingham SP8 4GN
M6 RFJ D. Jacobs 7 Coppice Close Ravenstone, Coalville LE67 2NS
M6 RFK R. Brookes 5 Radstock Road, Stretford, Manchester M32 0AJ
M6 RFL R. Long 45 Dean Street, Low Fell, Gateshead NE9 5XL
M6 RFM R. Corney Lavender Cottage, Worlds End, Hambledon, Waterlooville PO7 4QU
M6 RFN S. Hinton 33 Park Street, Kidderminster DY11 6TP
M6 RFO A. Trohear 3 Frampton Crescent, Bristol BS16 4JA
M6 RFP M. Bruce 27 Blaenant, Emmer Green, Reading RG4 8PH
M6 RFQ R. Newton 38 Bedford Road, Denton, Northampton NN7 1DR
M6 RFR R. Wheddon 10 Penharget Close, Pensilva, Liskeard PL14 5SA
MW6 RFS R. Shipman 1 Lledfair Place, Heol Pentrerhedyn, Machynlleth SY20 8DL
M6 RFW P. Wilson 103 Thors Oak, Stanford-le-Hope SS17 7BZ
M6 RFZ R. Finch Garth Cottage North Cowton, Northallerton DL7 0HL
MW6 RGA R. Anderson 156 Cockett Road, Cockett, Swansea SA2 0FQ
M6 RGC D. Clavey 32 Apollo Close, Dunstable LU5 4AQ
M6 RGD R. Yates West Wing Manor House, 20 Smith Street, Elsworth, Cambridge CB23 4HY
M6 RGE W. Ridge 91 Ridgeway, Rotherham S65 3NL
M6 RGF R. Gott 21 Broughton Road, Crewe CW1 4NW
M6 RGI R. Griffiths 2 Old Market Close Acle, Norwich NR13 3EY
MW6 RGK T. Bourner 23 St. Michaels Road, Pembroke SA71 5JQ
MI6 RGM R. Murphy 40 Stoneypath, Londonderry BT47 2AF
M6 RGO R. Hunter 9 Gelt Burn, Didcot OX11 7TZ
M6 RGP R. Pearson 12 Mere View, Great Livermere, Bury St. Edmunds IP31 1JU
M6 RGQ R. Hobbs 2 Jenkyns Close Botley, Southampton SO30 2UQ
M6 RGR R. Reeves 20 Warren Close, Irchester, Wellingborough NN29 7HF
M6 RGS R. Sneddon 3 George Street, Dipton, Stanley DH9 9HD
MM6 RGT R. Thomson Inchully, Myreriggs Road, Blairgowrie PH13 9HS
M6 RGU R. Emery 115 Humberstone Road, Grimsby DN32 8DR
MW6 RGW R. Woodland Flat 14, Mays Court, Windsor Road, Neath SA11 1NG
M6 RGX W. Morrison-Bates 95 Beaulieu Close, Toothill, Swindon SN5 8AJ
M6 RGY P. Jay 113 Stanks Lane North, Leeds LS14 5AS
M6 RGZ R. Hughes 3 Wood Avens Close, Northampton NN4 9TX
M6 RHC R. Cook11 16 First Square, Stainforth, Doncaster DN7 5RH
MW6 RHD R. Hughes-Burton 6 Troed Y Garn, Llangybi, Pwllheli LL53 6DQ
MM6 RHI A. Hunsley 21 Dalkeith Street, Edinburgh EH15 2HP
M6 RHK R. Kirby 50 Harcourt Avenue, Harwich CO12 4NT
MI6 RHL A. Calvin 65 Tannaghmore Road, Markethill, Armagh BT60 1TW
MD6 RHN D. Heaton Rushen Vicarage Barracks Rd, Port St. Mary IM9 5LP Isle of Man
MM6 RHO R. Hendry 3 Claddyburn Terrace, Cairnryan, Stranraer DG9 8RD
MM6 RHQ R. Hepburn 33 Saxon Road, Glasgow G13 2YQ
MW6 RHR N. Oldham Arhosfa, Carmel, Llannerch-Y-Medd LL71 7DH
MM6 RHT Dr R. Harkness Fernwood, 4 Cassalands, Dumfries DG2 7NS
M6 RHU M. Ricketts 45 Jesmond Road, Grays RM16 2QS
M6 RHX A. Richardson 18 Warkworth Road, Durham DH1 5PB
MW6 RHY R. Cooper 3 Waunddu, Pontnewynydd, Pontypool NP4 6QZ
M6 RIA J. Aldersley 6 Cavendish Close, Kingswinford DY6 9PR
MM6 RIE E. Soames Viera View, Rousay, Orkney KW17 2PU
M6 RIH R. Hayes 32 Green Road, Newmarket CB8 9BA
M6 RIJ S. Blades 18 Hall Cliffe Road, Horbury, Wakefield WF4 6BX
M6 RIL R. Studeny 36 Maples Street, Nottingham NG7 6AD
M6 RIN A. Barker 3A Hillcrest Close, Castleford WF10 3QS
M6 RIO J. Roberts Worlds Wonder, Warehorne, Ashford TN26 2LU
M6 RIQ R. Gallagher 56 Chesterton Square, London W8 6PJ
MI6 RIR E. Stevenson 25 Woodlands Manor, Portadown, Craigavon BT62 4JP
M6 RIS R. Sowden Flat 8, Bramble Court Bramble Street, Derby DE1 1HW
M6 RIT R. Williams 25 Wolds Retreat, Brigg Road, Ln6 7Ru, Market Rasen LN7 6RU
M6 RIU R. May 8 Boscaswell Estate, Pendeen, Penzance TR19 7EU
M6 RIV C. Jenkin 64 Rivers Road, Yeovil BA21 5RJ
MW6 RIY A. Holland Four Winds, Nebo, Llanon SY23 5LF
M6 RIZ R. Harris 19 Nightingale Walk, Stevenage SG2 0QE
M6 RJB R. Beardmore 37 Fernhill Road, Solihull B92 7RU
M6 RJE R. Edwards 23 Queens Walk, Ruislip HA4 0LX
M6 RJF G. Ropinski 38 The Leys, Little Eaton, Derby DE21 5AR
M6 RJH R. Harrison 18-20 Hall Lane, Kirkburton, Huddersfield HD8 0QW
M6 RJJ R. Pettigrew 26 Middlewich Road, Northwich CW9 7AN
M6 RJK R. Killner Oak Cottage, Priors Byne Farm, Lock Lane, Horsham RH13 8EF
M6 RJL J. Rose 9 Denewood, New Barnet, Barnet EN5 1LX
M6 RJM R. Meehan 2A Pinfold Street, Macclesfield SK11 6HA
MM6 RJN N. Robertson 3 Wallach Brae, Dalbeattie DG5 4GY
M6 RJP R. Parsons Aintree, Hurn Bridge Road, Lincoln LN4 4XT
M6 RJQ R. Munir 29 Hamilton Drive, Guildford GU2 9PL
M6 RJR S. Slater 6 Hazel Dene, Hurstfarm Estate, Matlock DE4 3TG
M6 RJS J. Statham Oakwoods, School Lane Upper Basildon, Reading RG8 8LT
M6 RJU R. Uzzell 68 St.Michaels Road, Cirencester GL7 1ND
M6 RJW R. Wakelam 1 Cleeve Park Mews, Cleeve Park, Minehead TA24 6JH
MW6 RJX M. Burns 40 Matthysens Way, St. Mellons, Cardiff CF3 0PS
MW6 RJY J. Reason 158 Caerau Lane, Cardiff CF5 5JS
M6 RJZ R. Duthie 14 Kettles Close, Oakington, Cambridge CB24 3XA
M6 RKA M. Halliburton 7 Penrose Avenue West, Liverpool L14 6UT
M6 RKB K. Bache 49 Cheviot Way, Halesowen B63 1HD
M6 RKC M. Callow 4 The Firs, Canvey Island SS8 9TW
M6 RKD M. Davey Romea, Long Street, Great Ellingham, Attleborough NR17 1LW
M6 RKE R. Elger 3 Alexander Square, Eastleigh SO50 4BW
MW6 RKF W. Lewis 53 Leyshon Road, Gwaun Cae Gurwen, Ammanford SA18 1EN
M6 RKI R. Hogan 10 Forest Way, High Wycombe HP13 7JF
M6 RKK R. Dick 15 Havenwood, Arundel BN18 0AH
M6 RKL L. Hall 180 Moor Road, Chorley PR7 2NT
M6 RKM R. Mason 22 Coronation Close, Happisburgh, Norwich NR12 0RL
M6 RKO R. Oczerklewicz 4 Grisedale Place, Chorley PR7 2JW
M6 RKP Triple B.C.G c/o R. Page 4 Vale Cottages, Kirmington Vale DN38 6AF
M6 RKQ N. Monaghan 7 Hoyle Road, Wirral CH47 3AG
M6 RKR G. Hirst 94 Upper Brighton Road Sompting, Lancing BN15 0LB
M6 RKS M. Swanston 309 Main Road, Wharncliffe Side, Sheffield S35 0DQ
MM6 RKT J. Browne 33 Pilgrims Hill, Linlithgow EH49 7LN
M6 RKU A. Rickwood 10 Wardrop Road, Catterick Garrison DL9 3BW
M6 RKW M. Ware 20 Brentwood Avenue, Thornton-Cleveleys FY5 3QR
M6 RKX M. Hemming 11 Blackberry Way, Evesham WR11 2AH
MW6 RLA M. Lewis 7 Yew Tree Terrace, Cwmbran NP44 3NU
M6 RLB R. Brown 24 Malthouse Court, Wellington, Telford TF1 1QJ
M6 RLC R. Lee 28 Champion Way, Oxford OX4 4NS
MM6 RLD D. Beeton The Old School House, Sliddery KA27 8PB
MW6 RLE R. Lewis 5 Elizabeth Sparkes Close, Rogiet, Caldicot NP26 3UT
M6 RLG J. Berrisford 18 Trowels Ln, Derby DE22 3LS
MM6 RLL J. Young 26 Forsyth Avenue, Sanquhar DG4 6AT
M6 RLN D. Carroll 48 Foster Road, Trumpington, Cambridge CB2 9JR
M6 RLP R. Palmer Newhaven Stoneraise Durdar, Carlisle CA57AX
M6 RLQ R. Skingley Tregwin, The Commons, Helston TR12 7HZ
M6 RLR R. Hanson 664 Leeds Road, Huddersfield HD2 1UB
M6 RLT R. Trim 23 Coleman Road, Bournemouth BH11 8EQ
MI6 RLU F. Burgess 65 Sunnylands Avenue, Carrickfergus BT38 8JT
M6 RLV R. Oliver 3 Histon Road Cottenham, Cambridge CB24 8UF
M6 RLW R. Wood 7 Wishart Green, Old Farm Park, Milton Keynes MK7 8QB

M6 RLX R. Lang 31 Ewan Close, Barrow-in-Furness LA13 9HU
M6 RLZ K. Lewis 15 Gilbert Scott Way, Kidderminster DY10 2EZ
M6 RMA R. Armstrong 11 High Ditch Road, Fen Ditton, Cambridge CB5 8TE
M6 RMD R. Dockray 54 Kelsick Park, Seaton, Workington CA14 1PY
MI6 RME R. Mcdonald 6 Rose Park, Limavady BT49 0BF
M6 RMG R. Grout 5 Branton Close, Great Ouseburn, York YO26 9SF
M6 RMI R. Le Feuvre 14 Elm Drive, Brockworth, Gloucester GL3 4DH
M6 RMK R. Keith-Hill 32 Thornhill Way, Plymouth PL3 5NP
M6 RML J. Hodson 77 Waltham Road, Woodford Green IG8 8DW
M6 RMN T. Ahern 39 Essex Road, Romford RM7 8BE
M6 RMO C. Larner 98 Allandale, Hemel Hempstead HP2 5AT
MW6 RMR R. Russell 1 Horeb Cottages, Rhiw Road, Colwyn Bay LL29 7TL
M6 RMS R. Somerville 28 Yeathouse Road, Frizington CA26 3QJ
M6 RMU R. Mitchell 8 Prestwood Close, Benfleet SS7 3LD
MM6 RMV J. Vine Seaview, Half Of 4 Kilvaxter, Kilmuir IV51 9YR
M6 RMW M. Willmott 25 Clent Avenue, Kidderminster DY11 7EH
M6 RNA P. Bannon 73 London Road, Worcester WR5 2DU
M6 RNC R. Blow 131 Fairfax Avenue, Harrogate HG2 7RU
M6 RND B. Starling 4 Three Corner Drive, Norwich NR6 7HA
M6 RNE S. Wilson 2 Bydales Drive, Marske-By-The-Sea, Redcar TS11 7HJ
M6 RNF D. Meekins 30 Lorton Avenue, Workington CA14 3JF
M6 RNI R. Hillier 10 Buttermere Close, Folkestone CT19 5JH
M6 RNM A. Atkinson 2E Bagridge Road, Wolverhampton WV3 8HW
M6 RNN N. Renny 48 Tower Road Tividale, Oldbury B69 1NA
M6 RNO R. Noakes 3 Firtree Road, Hastings TN34 3TR
M6 RNQ R. Stevens 7 Dunbreck Grove, Sunderland SR4 7LL
M6 RNS J. Grint 9 Mountbatten Drive, Leverington PE13 5AF
M6 RNU M. Nutt 110 Birkinstyle Lane, Shirland, Alfreton DE55 6BT
M6 RNV Dr A. Green 14 Teversham Way, Sawston, Cambridge CB22 3DF
M6 RNZ R. Baldwin 98 Rosemary Avenue, Braintree CM7 2TA
M6 ROA R. Parrott 37 Wren Place, Gillingham SP8 4WE
M6 ROB R. Smith 21 Canal Road Crossflatts, Bingley BD16 2SR
M6 ROC C. Martin Kiln Close, Main Road, Lincoln LN4 4QH
M6 ROE P. Roe 36 Rutland Crescent, Harworth, Doncaster DN11 8HZ
M6 ROF A. Martin 81 Langstone Road, Dudley DY1 2NL
MM6 ROH R. Hutton 2 Watson Place, Dunfermline KY12 0DR
M6 ROI P. King 18 Paddock Close, Wantage OX12 7EQ
M6 ROJ R. Gudgeon 6 The Choakles, Wootton, Northampton NN4 6AP
M6 ROL R. Rollinson Beauty Bank Farm, Six Ashes, Bridgnorth WV15 6ER
M6 ROM C. Romocea 14 Foxgrove Path, Watford WD19 6YL
M6 ROP R. Pochat Hame, Bealswood Road, Gunnislake PL18 9DA
M6 ROU S. Courree 4-6 Alhambra Road, Southsea PO4 0RL
M6 ROV R. Van-Der-Wijst 6 Willow Street, Romford RM7 7LJ
M6 ROW C. Hartshorn 63 Mill Road, Maldon CM9 5HY
MW6 ROX S. Baker Sunville, Village Road, Mold CH7 6HT
M6 ROY R. Trent 62 Dean Street, Radcliffe, Manchester M26 3TZ
M6 RPC R. Cobb 57 Adams Drive, Willesborough, Ashford TN24 0FX
M6 RPE Dr A. Wallman Oakwood, 30 Elmsway, Bramhall, Stockport SK7 2AE
M6 RPH R. Hunt 7 Tinsley Close, Luton LU1 5QD
M6 RPL R. Lee 13 Earle Street, Barrow-in-Furness LA14 2PZ
MI6 RPM D. Williams 19 Ballymacruise Park, Millisle, Newtownards BT22 2NW
MM6 RPN P. Toner 1 Milton Mains Road, Clydebank G81 3NF
M6 RPO W. Donnelly 9 Old Laundry Mews, Laundry Lane, Ingleton, Carnforth LA6 3GH
M6 RPP J. Anderson 19A Buttermarket, Thame OX9 3EP
M6 RPQ R. Back 79 Aspin Close Wellington Home, Somerset TA21 9EG
M6 RPR M. Roper 13 St. Cuthbert Street, Worksop S80 2HN
MW6 RPS C. Hurley 14 Magazine Street, Maesteg CF34 0TG
M6 RPU A. Smith 20 Bishops Avenue, Worcester WR3 8XA
M6 RPV R. Connolly 29 Gayer Street, Coventry CV6 7EU
M6 RPW C. Allan-Mcwilliams 207 Rue De Bulliez, Lagnieu 1150 France
M6 RPX R. Price 1 Blakiston Close, Ashington, Pulborough RH20 3GL
MW6 RQC R. Chegwin 17 Cyncoed Crescent, Cardiff CF23 6SW
MM6 RQD R. Dean Lower Cross Head, Crieff PH7 3BT
M6 RQE R. Fakes 24 Fairways, Weyhill, Andover SP11 8DW
M6 RQH R. Hawkes The Old Oak Bungalow, Crossway Green, Stourport-on-Severn DY13 9SJ
MW6 RQM J. Morris 1 Minafon, Newtown SY16 1RH
M6 RQN J. Foster 23 High Street, Cumnor, Oxford OX2 9PE
M6 RRA R. Gibbons 18 Langdale Road, Ribbleton, Preston PR2 6AN
M6 RRC K. Robinson 103 Recreation Street, Mansfield NG18 2HP
M6 RRD R. Davies 76 Berry Park, Saltash PL12 6EN
M6 RRF T. Allman 46 Belmont Road, Rugby CV22 5NZ
M6 RRH S. Walters 87 Fairbourne Close, Bransholme, Hull HU7 5DH
M6 RRJ R. Jobber 79 Falcon Way, Ashford TN23 5UR
M6 RRK R. Rao 13 High Oaks, Enfield EN2 8JJ
M6 RRL K. Woodhams 83 Langdale Place, Newton Aycliffe DL5 7DY
M6 RRR R. Capon 49 Littlemoor Road, Illingworth, Halifax HX2 9EF
M6 RRV R. Oxley 17 Hardhurst Road, Alvaston, Derby DE24 0LF
MW6 RRW K. Hazlewood 22 Lon Cwm, Llandrindod Wells LD1 6BE
M6 RSB R. Blackman 32 Kingfisher Road, Sprowston NR7 8GX
M6 RSD R. Davies 8 The Brambles, Bar Hill, Cambridge CB23 8TA
M6 RSF A. Robson 270 Calder Road, Lincoln LN5 9TL
MI6 RSH M. Rush 122 Cullaville Road Crossmaglen, Newry BT35 9AQ
M6 RSI R. Best 14 Charles Street, Bugle, St. Austell PL26 8PS
M6 RSJ S. Rebisz 29 Rosecroft Drive, Nottingham NG5 6EH
M6 RSL B. Leeson 18 Georgeham Close, Wigston LE18 2HZ
M6 RSO R. Oliver 49 Albany Way, Skegness PE25 2NB
MM6 RSP R. Phillips 32 Parkview, Ayr KA7 4QF
M6 RSR S. Richtering 22 Darwin Drive, Driffield YO25 5PF
M6 RSS R. Brooks 52 Harebell Drive, Portslade, Brighton BN41 2UZ
M6 RST M. O'Donoghue 24 Whitelock Road, Abingdon OX14 1NZ
MD6 RSV N. Hazell 43 Allan Street, Douglas, Isle of Man IM1 3DP
M6 RSY R. Sayre 8 Lorne Road, Richmond TW10 6DS
M6 RTA S. Himsworth 17 Forber Road Saltersgill, Middlesbrough TS4 3HJ
M6 RTC R. Churchill 16 Mansfield Close, Worthing BN11 2QR
M6 RTD R. Wilson 18 Baldwin Road, Bewdley DY12 2BP
M6 RTE S. Warde 36 Cornwallis Road, London E17 6NN
M6 RTG I. Sharpe 4 Low Dowfold, Crook DL15 9AE
M6 RTI R. Todd 42 K D Tower, Cotterells, Hemel Hempstead HP1 1AS
M6 RTM R. Townsend 28 Steele Street Hoyland Common, Barnsley S74 0PS
M6 RTN M. Fletcher 8 Brigham Hill Mansion, Brigham Great Broughton, Brigham, Cockermouth CA13 0TL
MM6 RTO M. Smith 2 Richmond Court, Dundee DD2 1BF
M6 RTP R. Paster 8 Rachaels Lake View, Warfield, Bracknell RG42 3XU
M6 RTQ Dr R. Tempo 35 Warminster Road, Bath BA2 6XG
M6 RTR A. Gill 12 Andersons Court , Plymyard Avenue, Wirral CH62 6EF
M6 RTU A. Loukes 14 Batchwood View, St. Albans AL3 5TD
M6 RTX B. Purvis 23 Deans Gardens, St. Albans AL4 9LS
M6 RTZ A. Lewis 43 High Street, Colney Heath, St. Albans AL4 0NS
M6 RUB R. Beck 13 Chaseside Avenue, Twyford, Reading RG10 9BT
MI6 RUC W. Rafferty 44 Lestannon Avenue, Whitehead, Carrickfergus BT38 9NN
M6 RUH R. Plumley 9 Haynes Road, Kettering NN16 0NG
M6 RUI N. Ginns 45 Gapstile Close, Desborough, Kettering NN14 2TZ
M6 RUK M. Shockness 38 Park Grange Court, Sheffield S2 3SY
M6 RUM L. Rumbelow 71 Anchorage Lane, Doncaster DN5 8EB
M6 RUN M. Saunders 68 Heywood Road, Prestwich, Manchester M25 1FN
M6 RUP A. Tomlinson 10 Beech Street, Hollingwood, Chesterfield S43 2HN
M6 RUS R. Trickey 19 Eastfields, Folkestone CT19 5RU
MW6 RUT R. Williams 24 Heathfield Place, Cardiff CF14 3JZ
M6 RUZ R. Brierley 39 Hatfield Road, Alvaston, Derby DE24 0BU
M6 RVA V. Davies 49 Victoria Court, Birkdale PR8 2DN
M6 RVD R. Davies 4 Haven Villas, Ferry Road, Exeter EX3 0JW
M6 RVE A. Mitchell 121 Totteridge Lane, High Wycombe HP13 7PH
M6 RVG R. Kobela 32 Allen Road, Rushden NN10 0FT
M6 RVH R. Hughes 19 Pendine Crescent, North Hykeham, Lincoln LN6 8UW
M6 RVJ J. Roa Vicens 80 Queensway, London W2 3RL
M6 RVM R. Moore 22 Chase Close, Nuneaton CV11 6AJ
M6 RVN I. Smith 2 Frederick Street, Woodville, Swadlincote DE11 8BX
M6 RVR R. Varasani 34 Freemans Rd, Minster, Ramsgate CT12 4EL
M6 RWB W. Stobbs 19 Stonecliffe Bank, Leeds LS12 5BL
M6 RWE D. Woodhouse 18 Kirk Close, Ripley DE5 3RY
MM6 RWI R. Wilinski 1 Keil Gardens Benderloch, Oban PA37 1SY
M6 RWJ A. Nelson 38 Warbro Road, Torquay TQ1 3PW
M6 RWP R. Whalley 65 Stanley Street, Nelson Lancashire BB9 7ET
M6 RWS K. Saville Flat 4, Birchitt Court, 152 Bradway Road, Sheffield S17 4QX
M6 RWT R. Trueman 83 Morton Street, Middleton, Manchester M24 6AX
M6 RWV A. Durrant 22 Supple Close, Norwich NR1 4PP
M6 RWX C. Anderson 191 Waveney Road, Hull HU8 9NA
M6 RXA R. Arnold 20 South View, Frampton Cotterell, Bristol BS36 2HT
MI6 RXC T. Masterson 2 Pinley Drive, Banbridge BT32 3TZ
M6 RXE R. Ebdon 6 Bankside, Headington, Oxford OX3 8LT
M6 RXG M. Robson 270 Calder Road, Lincoln LN5 9TL
MI6 RXH R. Harvey 464 Drumbeg South, Craigavon BT65 5AQ
MM6 RXJ R. Johnstone 14 Hoover Driver, Cambuslang, , Glasgow G72 7EF
M6 RXL H. Leach 34 Forest Close, Crawley Down, Crawley RH10 4LU
M6 RXR R. Rich 54 Paddocks Way, Ferndown BH22 9FW
M6 RXS R. Sanders-Hewett 26 Verulam Road, Hitchin SG5 1QE
M6 RXT R. Thomas 43B Napier Road, Bromley BR2 9JA
M6 RXU C. Attrill 42 Purley Way, Clacton-on-Sea CO16 8YX
M6 RXX A. Tomczynski 66 Commercial Street, London E1 6LT
MW6 RXZ J. Davies 8 Coronation Road, Upper Brynamman, Ammanford SA18 1BB
M6 RYA R. Tarr 16 Stoneleigh Court, Frimley, Camberley GU16 8XH
MI6 RYC R. Carmichael 21 Stuart Park, Ballymoney BT53 7BE
MW6 RYD R. Davies 63 Maes Y Gwernen Road, Cwmrhydyceirw, Swansea SA6 6LL
MM6 RYG R. Young 22 Station Road Armadale, Bathgate EH48 3LN
M6 RYK G. Mackay 5 Furze Street, Carlisle CA1 2DL
M6 RYL R. Harris 183A Painswick Road, Gloucester GL4 4AG
M6 RYN R. Clough 8 Skeldyke Road, Kirton, Boston PE20 1LR
M6 RYR R. Brown 2 Westgate, Leominster HR68SA
M6 RYZ J. Fryd Keepers, Yeoviliton BA22 8EX
M6 RZA M. Box 273 Broad Lane, Birmingham B14 5AF
M6 RZD R. Dainton 1 The Woodlands, Stroud GL5 1QE
M6 RZE J. Stephenson Flat 12, St. Clares Court, St. Clares Avenue, Havant PO9 4JF
M6 RZI T. Hurst 22 Coronation Street, Mansfield NG18 2QL
MW6 RZL R. Higgins 44 Maeshyfryd Road, Holyhead LL65 2AL
M6 RZO D. Bardell 32 Bridle Road, Watton, Thetford IP25 6NA
M6 RZP R. Williams 10 Bramley Close, Twickenham TW2 7EU
M6 RZW R. Laye 11 Hillhead Rise, Falmouth TR11 5GZ
M6 RZX Gpt(Coventry)ARS c/o B. Forrest 32 Idonia Road, Wolverhampton WV6 7NQ
M6 RZY R. Howard 5 Wells Close, Woolston, Warrington WA1 4LH
M6 RZZ R. Reynolds 22A Beech Avenue, Shepton Mallet BA4 5XW
MW6 SAA S. Llewellyn 53 Cripps Avenue, Cefn Golau, Tredegar NP22 3PF
M6 SAC M. Clark 99 Cheylesmore Drive, Frimley, Camberley GU16 9BW
M6 SAD S. Abraham 12 Graham Road, Halesowen B62 8LJ
M6 SAE S. Everett 12 Broadlands, Netherfield, Milton Keynes MK6 4HL
M6 SAF S. Fisher 25 Tennyson Road, Diss IP22 4PY
M6 SAH S. Painter 47 Longmead Drive, Nottingham NG5 6DP
M6 SAI T. Beighton 97 Great North Road, Woodlands, Doncaster DN6 7NH
M6 SAL S. Lester 71 Ronelean Road, Surbiton KT6 7LL
M6 SAM M. Sampson 45 Carron Street, Stoke-on-Trent ST4 3DT
M6 SAN S. Weightman 131 Leeds Road, Birstall, Batley WF17 0JZ
M6 SAQ S. Wills 8 Amherst Road, Newcastle upon Tyne NE3 2QQ
M6 SAS A. Speight 112 Fern Avenue, Staveley, Chesterfield S43 3RA
M6 SAU C. Saunders 68 Heywood Road, Prestwich, Manchester M25 1FN
M6 SAV S. Morrell 20 Brocklebank Close, Bassingham, Lincoln LN5 9LJ
MW6 SAW P. Campigli 43 Waterloo Road, Penylan, Cardiff CF23 9BJ
MW6 SAX J. Elsmore 8 Clos Aberconway, Prestatyn LL19 9HU
M6 SAY S. York 1 The Cottage, Dogdyke Bank, Lincoln LN4 4JQ
M6 SBB S. Jeffery 79 Greenbank Road, Watford WD17 4FJ
M6 SBD S. Radford Littlehampton Marina, Ferry Road, Littlehampton BN17 5DS
M6 SBE S. Bennett 154 Humberstone Road, Grimsby DN32 8HR
M6 SBF S. Raine 91 Lulworth Avenue, Jarrow NE32 3SB
M6 SBH T. Thomas 269 Church Road, St. Annes, Lytham St. Annes FY8 3NP
M6 SBI S. Bassett 3 Lower Merryfield, Anchor Road, Radstock BA3 5PG
M6 SBJ S. Crombie 8 Hazel Close, Haverhill CB9 9LY
M6 SBK R. Colman-Whaley 37 Suters Drive, Taverham, Norwich NR8 6UU

M6 SBL S. Blaney 213 Albert Road, Poole BH12 2EZ
M6 SBM S. Brightmore 36 Laburnum Road, Langold, Worksop S81 9RR
M6 SBN S. Hanson 59 Weetwood Avenue Ne71 6Af, Wooler NE71 6AF
M6 SBO B. Mcglynn 22 Bracken Bank Way, Keighley BD22 7AB
M6 SBR S. Flowers 56 Pilkington Road Braunstone, Leicester LE3 1RA
M6 SBS S. Stiddard 21 Studland Park, Westbury BA13 3HQ
MW6 SBT A. Jones 10 Dan Y Bryn, Caerau, Maesteg CF34 0UW
M6 SBU S. De Chastelain 31 Campion Court, Northampton NN3 9BW
M6 SBV D. Hindle 18 Haig Street, Selby YO8 4BY
M6 SBW J. Stephenson 4 Carlow Drive West Sleekburn, Choppington NE62 5UT
M6 SBY S. Hollis 24 Bosley View, Congleton CW12 3TU
M6 SBZ D. Sibley 85 Brick Crescent, Stewartby MK43 9GG
M6 SCA S. Frampton 4 Sussex Gardens, Fleet GU51 2TL
M6 SCB S. Burnage 124 Mayfields, Spennymoor DL16 6TT
MI6 SCC S. Cole 16 Otterbank Road, Strathfoyle, Londonderry BT47 6YB
MW6 SCD S. Davies 28Y Fron, Llanelli SA14 7DN
M6 SCF S. Faulkner Mount Pleasant, Elkstones, Buxton SK17 0LU
MM6 SCG S. Greenland 0/1 22 Linden Street, Anniesland, Glasgow G13 1DQ
M6 SCH S. Hall 50 Charnock Wood Road, Sheffield S12 3HN
M6 SCI M. Pope Drove Lodge 39 The Drove Barroway Drove, Downham Market PE38 0AJ
M6 SCK J. Lynch 19 Riverside, Clitheroe BB7 2NP
M6 SCM S. Cumberland 11 Cleveleys Road, Blackburn BB2 3JS
M6 SCN R. Rothery 24 Maes Y Dafarn, Carno, Caersws SY17 5NG
M6 SCP S. Pettit 24 Bickington Lodge Estate, Bickington, Barnstaple EX31 2LH
M6 SCQ A. Chamberlain 35 Wexham Close, Luton LU3 3TU
M6 SCU D. Willetts Domus, Broyle Lane Ringmer, Lewes BN8 5PG
M6 SCV S. Chandler 7 Clinton Road, Redruth TR15 2LL
M6 SCW S. Whittaker 25 Cleveleys Road, Blackburn BB2 3JS
M6 SCX S. Carpenter Field View, Old Lyndhurst Road, Southampton SO40 2NL
M6 SDA O. Suda 52 Coltman Street, Hull HU3 2SG
MW6 SDE D. Ezard 59 Station Farm, Croesyceiliog, Cwmbran NP44 2JW
M6 SDF A. Hulok 3 Dickson Court, Sittingbourne ME10 3LG
M6 SDG S. Mallows 3 Village View, Chatham ME5 7TR
M6 SDH S. Hopkins 61 Green Avenue, Astley, Tyldesley, Manchester M29 7FF
M6 SDI S. Legg Woodview, Clay Lane, Clacton-on-Sea CO16 8HH
M6 SDL J. Layden 51 Markham Road, Langold, Worksop S81 9SH
M6 SDN A. Mcconkey 7 Darwin Road, Stevenage SG2 0DE
M6 SDP S. Porter Apartment 4, City Towers, 1 Watery Street, Sheffield S3 7ET
M6 SDQ S. Dean 39 Low Grange View, Leeds LS10 3DT
M6 SDS S. Brown 4 Bishop Fox Drive, Taunton TA1 3HQ
M6 SDT D. Grace 107 Bush Avenue, Little Stoke, Bristol BS34 8NG
M6 SDU S. Seddon 3 Kinsley Close, Ince, Wigan WN3 4PQ
MW6 SDV S. Price 27 Gilfach Road, Penygraig, Tonypandy CF40 1EN
M6 SDW R. Wilson 84 Sir Thomas Whites Road, Coventry CV5 8DR
M6 SDX S. Page Flat 25 The Riverfront , Eastern Esplanade, Canvey Island SS8 7DN
M6 SDZ S. Dimopoulos Flat 8, 31A High Street, Cricklade, Swindon SN6 6AB
M6 SEA S. Aspinall 33 Covertside Road, Scarisbrick, Southport PR8 5HB
M6 SEB S. Plowman 7 Birkdale Close, Cudworth, Barnsley S72 8EW
M6 SEC T. Summers 5 Fairclough Place, Adlington, Chorley PR7 4AN
M6 SEE A. Norton 9 The Common, West Tytherley, Salisbury SP5 1NS
MW6 SEF M. Williams 2 St. Andrews Road, Wenvoe, Cardiff CF5 6AF
M6 SEG S. Todd 24 Pentland Avenue, Redcar TS10 4HD
MM6 SEI D. Crane Otterburn, Dervaig, Isle of Mull PA75 6QL
M6 SEJ S. Jeronimo 46 Pretoria Road, Chertsey KT16 9AZ
M6 SEK A. Collins 5 Grove Close, Basingstoke RG21 3AS
M6 SEN A. Snelling 6 Aylsham Road, Buxton, Norwich NR10 5EX
M6 SEO A. Gardner 8 Azalea Road, Wick St. Lawrence, Weston-Super-Mare BS22 9TJ
MW6 SEP D. Jones 46 East Avenue, Caerphilly CF83 2SR
M6 SER S. Richards 5 Bloomfield Drive, Bracknell RG12 2JW
M6 SES S. Stoney 2 Fishers Mead, Dulverton TA22 9EN
M6 SEU M. Fairbairn 36 Avebury Place Eastfield-Lea, Cramlington NE23 2UR
M6 SEV S. Greaves The Grange Farmhouse, Leicester Forest East LE3 3GA
M6 SEY S. Borrell Rose Cottage, Colchester Main Road, Colchester CO7 8DD
MI6 SEZ Dr P. Wilson 2 Tweskard Lodge, Belfast BT4 2RH
MW6 SFB P. Latham 20 Kenyon Avenue, Wrexham LL11 2ST
M6 SFC P. Craig 4 Poolside, Burston, Stafford ST18 0DR
MM6 SFF S. Ferguson 17 Brown Street, Shotts ML7 5HW
MI6 SFH S. Hand 12 Church Street, Rosslea, Enniskillen BT92 7DD
M6 SFJ A. Morton 6 Chessar Ave, Chessar Ave Blakelaw, Newcastle NE5 3RE
MI6 SFK S. Mcelmurray 43 Main Street, Sixmilecross, Omagh BT79 9NH
M6 SFL D. Webber 11 Waghorn Road, Harrow HA3 9ET
M6 SFM S. South Ivy Cottage, Finkle Street Lane, Sheffield S35 7DH
M6 SFQ B. Cross 22 Park Avenue Washingborough, Lincolnshire LN 4 1 DB
MM6 SFR A. Marshall 6 Heron Way, Minnigaff, Newton Stewart DG8 6PZ
M6 SFS S. Spooner 51 Lewisham Court, Morley, Leeds LS27 8QB
M6 SFT S. Foster 16 Birchcroft Road, Ipswich IP1 6PA
M6 SFV S. Fry Piccadilly Farm, Aggs Hill, Cheltenham GL54 4ET
M6 SFW J. Walker 6 Wellington Terrace, Islip, Kettering NN14 3LJ
M6 SGA S. Gregory 9 Croftlands Road, Wythenshawe M22 9YE
M6 SGC S. Latter 1539 Great Cambridge Road, London EN1 4SY
M6 SGD S. Gibbs 35 St. Michaels, Houghton le Spring DH4 5NR
M6 SGE N. Pulev Zapaden Park, 79-B, Flat 4, Sofia 1373 Bulgaria
MM6 SGF S. Gray 9 Caledonian Crescent, Prestonpans EH32 9GF
M6 SGG S. Gillett 2 Chalkrow Cottages, , King's Lynn PE33 9BW
M6 SGH S. Holt 14 Fir Street, Cadishead, Manchester M44 5AU
M6 SGJ S. Powell Crosstrees, Main Road, Theberton, Leiston IP16 4RX
M6 SGL S. Gillard 1 Chevening Close, Stoke Gifford, Bristol BS34 8NJ
M6 SGM S. Roberts 17 East View Marsh, Huddersfield HD1 4NU
M6 SGN B. Frith 159 Milton Road, Grimsby DN33 1DN
MM6 SGO S. Green 4 Mid Avenue, Port Glasgow PA14 6PL
M6 SGP S. Phipps 53 Pioneer Avenue, Burton Latimer, Kettering NN15 5LJ
MM6 SGQ S. Gilruth 20 The Aspens, Carberry Crescent, Dundee DD4 0XJ
M6 SGS G. Street 105 Jeals Lane, Sandown PO36 9NS
M6 SGT A. Williams 11 Broadsmith Avenue, East Cowes PO32 6QW
M6 SGV M. Alcaino Pizani Flat 45, Brian Redhead Court, 123 Jackson Crescent, Manchester M15 5RR
M6 SGY S. Johnson 2 North Square, Edlington, Doncaster DN12 1ED
M6 SGZ B. Fitzsimmons 10 Windhill View, Wakefield WF1 4AB
M6 SHA M. Callaway 19 The Oval, Dudley DY1 2LN
MM6 SHB X. Bradshaw 55 Headwell Avenue, Dunfermline KY12 0JX
M6 SHC S. Corden 59 Brindles Field, Tonbridge TN9 2YR
M6 SHD G. Fisk 22 Church Street, Wangford, Beccles NR34 8RN
M6 SHF B. Thwaites 109 Emerson Crescent, Sheffield S5 7SW
M6 SHH R. Devos 76 North Parade, Sleaford NG348AW
M6 SHI S. Shillabeer 6 Meadow Court, Nomans Heath, Malpas SY14 8DU
MW6 SHJ S. Jones 56 Ffordd Offa, Rhosllanerchrugog, Wrexham LL14 2EY
M6 SHK A. Al-Shakarchi 17 Fairfax Place, London NW6 4EJ
M6 SHL S. Lalji 76 Abbotts Drive, Wembley HA0 3SG
MM6 SHM I. Mcewan 5 Creamery Row, Dunlop, Kilmarnock KA3 4DA
M6 SHP A. Sharp 30 Burbidge Close, Calcot, Reading RG31 7ZU
M6 SHQ J. Butcher 14 Park Road, Lowestoft NR32 1SW
M6 SHU S. Maton 41 Bemerton Gardens, Kirby Cross, Frinton-on-Sea CO13 0LQ
M6 SHV S. Roberts 10 Morningside Way, Liverpool L11 1BD
M6 SHW S. Wraith Old School House, West Knighton DT28PE
M6 SHZ S. Dainton 1 The Woodlands, Stroud GL5 1QE
MW6 SID S. Jones 25 The Crescent, Tredegar NP22 3HN
M6 SIE B. Brown 114 Woodhorn Road, Ashington NE63 9EN
M6 SIF S. Frost 4 Banister Way, Shipston-on-Stour CV36 4JU
M6 SIH S. Tomkins Flat 2, 18 Brandon Way, Birchington CT7 9XE
M6 SII S. Gray 15 The Green, Marsh Baldon, Oxford OX44 9LJ
M6 SIJ J. Simkins 37 St. Andrews Meadow, Harlow CM18 6BL
M6 SIK W. Henson 24 Grimshaw Close, North Road, London N6 4BH
MM6 SIM S. Beeson Muir Cottage, The Muirs, Huntly AB54 4GD
M6 SIP A. Davies 32 Kinross Road, Wallasey CH45 8LH
M6 SIR B. Greaves 2 Peel Street, Padiham, Burnley BB12 8RP
MM6 SIS G. Mccaughey 24A, Williamson Place, Johnstone PA5 9DW
M6 SIU S. Murdoch-Mckay 19 Beachy Road, Crawley RH11 9HN
MM6 SIV M. Sives 4 Fir Grove, Livingston EH54 5JP
M6 SIW S. Cartwright-Proctor 448 Tuttle Hill, Nuneaton CV10 0HR
M6 SIX B. North 54 Parklands, Mablethorpe LN12 1BY
M6 SIY S. Lilley 34 Rye Crescent, Danesmoor, Chesterfield S45 9HH
M6 SIZ W. Easdown 11 Mulcaster Avenue, Kidlington OX5 2HG
M6 SJA S. Bridges 65 Abbots Gate, Bury St. Edmunds IP33 2GB
M6 SJB S. Buckley 64 Wolseley Road, Rugeley WS15 2ES
M6 SJC S. Charles 29 Woolford Close, Winchester SO22 4DN
MI6 SJD S. Dallas 101 Coagh Road Stewartstown, Dungannon BT71 5JL
M6 SJE J. Hill 7 Knebworth Court, Congleton CW12 3SW
M6 SJF S. Fox 77 The Grove, London N13 5JS
M6 SJH S. Hayden 5 Blackbird Close, Thurston, Bury St. Edmunds IP31 3PF
M6 SJJ S. Jackson 54 Sefton Avenue, Poulton-le-Fylde FY6 8BL
MM6 SJK Dr D. Krauskopf Cocklehaa, Urafirth, Shetland ZE2 9RH
M6 SJN D. Lawson 56 River Bank East, Stakeford, Choppington NE62 5XA
M6 SJO S. Boniface 19 Toronto Drive, Smallfield, Horley RH6 9RB
M6 SJQ C. Gyngell 54 Association Walk, Rochester ME1 2XD
M6 SJR S. Ray 18 Crescent Way, Cholsey, Wallingford OX10 9NE
M6 SJU J. Searle 18 The Avenue, Bloxham, Banbury OX15 4QU
MI6 SJV D. Parkinson 16 Beechwood Gardens, Moira, Moira BT67 0LB
M6 SJW E. Watson 4 Glenluce Drive, Preston PR1 5TB
M6 SJX S. Bates 6 Foxdell, Northwood HA6 2BU
M6 SJZ S. Lampard 63 Broadmayne Road, Poole BH12 4EH
M6 SKF S. Froggett 1 The Paddocks, Pilsley, Chesterfield S45 8ET
M6 SKH S. Hemmings 26 Austin Drive, Banbury OX16 1DJ
M6 SKL S. Maqbool 69 Waltham Close, West Bridgford, Nottingham NG2 6LD
M6 SKN T. Watkins 72A St. Clements Road, Keynsham, Bristol BS31 1BA
M6 SKP S. Pryer 16 Wayside Avenue, Worthing BN13 3JU
M6 SKQ M. Berrisford 5 Branwell Drive, Haworth, Keighley BD22 8HG
MW6 SKR S. Rogers 30 Coed Celynen Drive, Abercarn, Newport NP11 5AU
M6 SKT T. Robinson 39 Cemetery Road, Laceby, Grimsby DN37 7ER
M6 SKU M. Baker 14 Thornberry Drive, Dudley DY1 2PL
M6 SKW S. Kneeshaw 39 Cherrytree Walk East Ardsley, Wakefield WF32HS
M6 SKX S. Key 29 Ellesmere Crescent, Brackley NN13 6BP
M6 SKY R. Coles Bay Cottage, St. Catherines Road, Ventnor PO38 2NE
M6 SKZ T. Barham 88 Rundells, Harlow CM18 7HD
M6 SLA S. Harvey 40 Littlemoor Lane, Newton, Alfreton DE55 5TY
M6 SLB S. Berry 9 Magnolia Street, Winnington, Northwich CW8 4EH
M6 SLC D. Morphew 8 Blinco Lane, George Green, , Slough, SL3 6RQ
M6 SLD S. Light 61 Eversfield Road, Horsham RH13 5JS
M6 SLE S. Lee 360 Ringley Road, Stoneclough, Manchester M26 1EP
M6 SLG S. Gash 22 Wood View, Rugeley WS15 1AT
M6 SLI S. Lisi 60 Middlecroft Road, Staveley, Chesterfield S43 3XH
M6 SLJ L. Jones 137 Breck Road, Poulton-le-Fylde FY6 7HJ
M6 SLK S. Lake 64 Womersley Road, Norwich NR1 4QB
M6 SLL L. Butler 42 Station Road, Stanbridge, Leighton Buzzard LU7 9JF
M6 SLR S. Perry 44 Ashcombe Road, Dorking RH4 1NA
M6 SLT S. Townsend 28 Steele Street, Hoyland, Barnsley S74 0PS
M6 SLU K. Cockburn 20 Hexham Avenue, Hebburn NE31 2HN
MM6 SLV K. Slaven Drymuir, Drymuir, Peterhead AB42 5RR
M6 SLZ S. Eloie 26 Halsbrook Road, London SE3 8QY
M6 SMA S. Hindmarsh 7 George Street Murton, Murton SR7 9BN
M6 SMB S. Bide 9 Greenway, Watchet TA23 0BP
M6 SMD S. Davies 95 Lathkill Drive, Ashbourne DE6 1TZ
M6 SME S. Elliott 74 Preston Avenue, Alfreton DE55 7JX
M6 SMF C. Smith 21 Mill House Drive, Cheltenham GL50 4RG
M6 SMG S. Hampshire Rose Cottage, Heath Road, Norwich NR12 0SU
M6 SMK S. Kendrick 103A Latimer Street, Liverpool L5 2RF
MM6 SML J. Macleod 388 Garrynamonie, Lochboisdale, Isle of South Uist HS8 5TX
MI6 SMM S. Maguire 139 Carrowshee Park, Drumhaw, Enniskillen BT92 0FS
MM6 SMN S. Nicoll 15 Redford Walk, Edinburgh EH13 0AF
M6 SMQ B. Mcgowan 200 Thomas Drive, Liverpool L14 3LE
M6 SMR G. Tudor 31 Church Road, Little Sandhurst, Sandhurst GU47 8HY
M6 SMX M. Feast 30 Peveril Drive, Riddings, Alfreton DE55 4AP
MM6 SMY S. Young 103 Feorlin Way, Garelochhead, Helensburgh G84 0EB
M6 SNC J. Sinclair 19 Ridge Way, Edenbridge TN8 6AU
M6 SNE S. Solanki 2 Churchill Mews, Newcastle upon Tyne NE6 1BH
M6 SNF S. Wall Flat 1, 41 Alexandra Road, Cleethorpes DN35 8LE

M6 SNG H. Pike 14 Milton Avenue, Barnet EN5 2EX
M6 SNH S. Hawkes 2 St Mary'S Close, Aston SG2 7EQ
M6 SNJ S. Briggs 3 Chapel Road, Southrepps, Norwich NR11 8UW
M6 SNN S. Ballard 77 Lanchester Road, Birmingham B38 9AG
M6 SNO X. Christofi 19 Kingsland Avenue, Northampton NN2 7PP
M6 SNP A. Alkhateb 51 Mendip Crescent, Bedford MK41 9ER
M6 SNQ S. Harrison 38 Alma Road, Bournemouth BH9 1AN
MM6 SNR S. Russell 3 Rankin Road, Wishaw ML2 8PG
M6 SNS S. Button 6 Farfield, Retford DN22 7TL
M6 SNU T. Lewis 653 Main Road, Harwich CO124NF
MD6 SNV A. Elliott Round Table House, Ronague, Castletown, Isle of Man IM9 4HJ
M6 SNW D. Snowden 10 Woodcroft, Wakefield WF2 7LS
MM6 SNY A. Dobie 7 Urr Terrace, Castle Douglas DG7 1BL
M6 SNZ S. Roshanmanesh 123 The Vale, Edgbaston, Birmingham B15 2RU
M6 SOC A. Wallace 17 Dennis Road, Liskeard PL14 3NS
M6 SOE P. Strickland 100 Spitfire Road, Castle Donington, Derby DE74 2AU
MW6 SOF D. Parry 45 Taff Court Thornhill, Cwmbran NP44 5UU
M6 SOG S. Jackson 50 Leicester Road, Sharnford, Hinckley LE10 3PR
M6 SOK J. Neary 3 Wordsworth Close, Torquay TQ2 6EA
M6 SON G. Stewart 35 Castle Crescent, Thornhill WF120EQ
M6 SOO R. Landragin 147 Taunton Road, Romford RM3 7PJ
MM6 SOR R. Ewing Kildonan House, Caerlaverock Farm, Crieff PH5 2BD
M6 SOT C. Scrivens 28 Bank Hall Road, Stoke-on-Trent ST6 7DL
M6 SOU A. Nur 92 Bramble Avenue, Conniburrow, Milton Keynes MK14 7AP
M6 SOV M. Mcdonald 55 Bournemouth Avenue, Middlesbrough TS3 0NN
M6 SOZ G. Knowles 31 Gibb Lane, Catshill, Bromsgrove B61 0JP
M6 SPA S. Etheridge 16 Crown Meadow Way, Newton St. Faith, Norwich NR10 3GW
M6 SPC J. Lyon 16 Well Court, St. Anns Close, Andover SP10 2FS
M6 SPG S. Coxon 31 Marden Road, Staplehurst, Tonbridge TN12 0NE
M6 SPH S. Harrison 26 Lambert Road, Lancaster LA1 2NA
M6 SPJ S. Johnston 67 Eversfield Road, Horsham RH13 5JS
M6 SPK S. Kay 32 Glossop Street, Derby DE24 8DU
M6 SPL S. Lycett 58 Hazel Grove, Hucknall, Nottingham NG15 6ED
M6 SPM S. Morris 23 De Courtenai Close, Bournemouth BH11 9PG
MW6 SPN S. Nelson 13 Llwyn Briscoe, Holyhead LL65 1HT
M6 SPP S. Fawcett Hollins Farm, Marske, Richmond DL11 7NH
M6 SPS S. Summers 450 Baddow Road, Chelmsford CM2 9RD
M6 SPT J. Gardner Silverdale, Vicarage Lane, Ormskirk L40 6HQ
M6 SPU Milton Keynes Raynet c/o G. Mason 21 High Street, Carisbrooke PO30 1NR
MI6 SPY J. Woods 39 Shetland Street, Antrim BT41 2TG
M6 SPZ S. Pendlebury 6 Normanby Close, Bewsey, Warrington WA5 0GJ
M6 SQB S. Stevens 40 Heath Road, Exeter EX2 5JX
M6 SQC S. Burton 2 West Batter Law Farm Cottages, Hawthorn, Seaham SR7 8RZ
M6 SQD P. Riding 160 Capel Road, London E7 0JT
M6 SQE S. Tudor 201 Cruddas Park, Westmorland Road, Newcastle upon Tyne NE4 7RG
M6 SQF Dr S. Pearce 15 Hillfield Court Road, Gloucester GL1 3QS
M6 SQI H. Blythe 14 The Green, South Creake, Fakenham NR21 9PD
M6 SQK C. Southey The Taverners, Crix Green, Felsted, Dunmow CM6 3JT
M6 SQL A. Burton 12 Munden Grove, Watford WD24 7EE
MI6 SQN A. Reid 40 Hillfoot Street, Belfast BT4 1PR
M6 SQO S. Macmurray 21 Dymoke Green, St. Albans AL4 9LX
M6 SQU B. Turner 35 Gosforth Lane, Watford WD19 7AY
M6 SQW S. Waite 61 King Edward Crescent, Leeds LS18 4BE
M6 SRA S. Antill Lodge Farm, Mansfield Road, Worksop S80 3DL
M6 SRB S. Robinson 2 Old Hall Crescent, Bentley, Doncaster DN5 0DW
M6 SRC S. Cash The Warren, Kent Hatch Road, Edenbridge TN8 6SX
M6 SRD R. Strong 82 Oakmount Road, Chandler'S Ford, Eastleigh SO53 2LL
M6 SRE S. Bradley 75 Weald Bridge Road North Weald, Epping CM16 6ES
M6 SRF J. Cranston 7 Cowen Gardens, Gateshead NE9 7TY
M6 SRG S. Gibson 66 Kinoulton Court, Grantham NG31 7XP
M6 SRI J. Bibby 12 James Street, Burton-on-Trent DE14 3SB
MM6 SRL S. Leeman 35 Foudland Court, Church Avenue, Insch AB52 6JZ
M6 SRN R. Dunn 15A Connaught Drive, Chapel St. Leonards, Skegness PE24 5YS
M6 SRO C. King 8A Barton Road, Bedford MK42 0NA
M6 SRS S. Tait 29 Hotspur Avenue, Bedlington NE22 5TD
M6 SRV S. Vickers 22 Thistle Green, Birmingham B38 9TT
M6 SRZ P. Willetts 193 Eaves Lane, Chorley PR6 0TR
M6 SSA S. Aldersley 245 St. Johns Road, Chesterfield S41 8PE
M6 SSB M. Saywell 8 Stanley Street North, Bristol BS3 3LU
M6 SSC S. Azzaro 5 Rye Hill Close, Bere Regis, Wareham BH20 7LU
M6 SSD S. Shephard 17 Grimsby Road, Laceby, Grimsby DN37 7DF
MJ6 SSF S. Foot 4 Aubin Place, Aubin Lane, Jersey JE2 7PP
M6 SSM S. Millard 44 South Bank Road, East Cowes PO32 6JD
M6 SSN B. Woods 28 Delph Drive, Burscough, Ormskirk L40 5BE
MM6 SSO S. Jones 14F High Street, Montrose DD10 8JL
M6 SSP S. Pashley 45 Highfield View Road Newbold, Chesterfield S41 7JZ
M6 SSR S. Rymel 3 Southview, Smalldale, Hope Valley S33 9JQ
M6 SSS Dr S. Rowe 9 Corfield Close, Finchampstead, Wokingham RG40 4PA
M6 SST S. Slapper 1 Standards Keep, Standards Road, Bridgwater TA7 0EZ
M6 SSV K. Niendorf 84 St. Marys Road, Stratford-upon-Avon CV37 6XQ
M6 SSW B. Hood 168 Shay Lane, Walton, Wakefield WF2 6NP
M6 SSX J. Prout 110 Dorothy Avenue North, Peacehaven BN10 8DP
M6 SSY N. Hancock 3 Market Street, Shipdham, Thetford IP25 7LY
M6 SSZ A. Jarre 8 Lingford Court Bishop Auckland County Durham Dl14 7Dz, Bishop Auckland DL14 7DZ
M6 STA S. Ancill 2 Lawrence House, Southampton SO456EA
M6 STC S. Cousins West Mill, Wareham Common, Wareham BH20 6AA
M6 STF S. Briant 1B College Road, Haywards Heath RH16 1QN
M6 STH A. Huckle 36 Weldbank Close, Beeston, Nottingham NG9 5FU
M6 STI B. Richards 58 Holm Lane, Oxton, Prenton CH43 2HS
M6 STJ S. Sheppard 34 Hardwicke Walk, Birmingham B14 5XX
MU6 STK C. Stockwell Fleurs Des Champs, La Colline Des Bas Courtils, St. Saviours GY7 9YQ Guernsey
MM6 STM S. Mcmillan 56 Mount Pleasant, Armadale, Bathgate EH48 3HB
M6 STP S. Phythian 24 Water Grove Road, Dukinfield SK16 5QS
M6 STQ S. Totham Cavaliers, 11 Dane Court Manor, School Road, Tilmanstone, Deal CT14 0JL
M6 STR M. Stroud 1 Sefton Court, Welwyn Garden City AL8 6WW
M6 STU S. Vzor 40 Henlow Road, Birmingham B14 5DS
M6 STV G. Wadsworth 21 Appletree Road, Featherstone, Pontefract WF7 5EA
M6 STY T. Fletcher 9 Lawn Avenue, Kimpton, Hitchin SG4 8QD
MM6 SUB S. Boyd 269 Cloch Caravans Cloch Road, Gourock PA19 1AZ
M6 SUC J. Murphy Wessex House, Drake Avenue, Staines-upon-Thames TW18 2AP
M6 SUD S. Griffiths 37 Stourton Close, Knowle, Solihull B93 9NP
M6 SUE S. Hadley 60 Chapel Street, Pensnett, Brierley Hill DY5 4EF
M6 SUF S. Fotheringham 8 Ivanhoe Court, Ulrica Drive, , Thurcroft S66 9QP
M6 SUH S. Halewood 12 Silver Street Riccall, York YO19 6PB
M6 SUI S. Brooker 4 Fleet End Close, Havant PO9 5ED
M6 SUJ S. Jones Marvin House, Ryhill Pits Lane, Cold Hiendley, Wakefield WF4 2DU
M6 SUL R. Sullivan 14 Colleton Drive, Twyford, Reading RG10 0AU
M6 SUM P. Smith 14 Highfield Crescent, Kettering NN15 6JS
M6 SUN R. Beaumont 61 Mitcham Road, Camberley GU154AR
M6 SUP D. Markey 7 Knightsway, Wakefield WF2 7EG
MM6 SUR R. Fair 6 Fairways, Stewarton, Kilmarnock KA3 5DA
M6 SUU S. Coombes 33 Clarence Park Road, Bournemouth BH7 6LF
M6 SUV D. Arnold The Chase, Rectory Road, Penzance TR19 6BB
M6 SUW S. Urwin-Wright 3 Beacon Hill, Newton Ferrers PL8 1DB
M6 SUX S. Harris 17 Swindale, Wilnecote, Tamworth B77 4LD
M6 SUY A. Fact 50 Imison Street, Liverpool L9 1EF
M6 SUZ S. Trigg 2 Langley Common Road, Barkham, Wokingham RG40 4TS
M6 SVF L. Garnett 6 Tremaine Close, Norwich NR6 5EL
M6 SVG J. Savage 44 Hastings Road, Maidstone ME15 7SP
M6 SVJ S. Carpenter 18 Warwick Drive, Bury St. Edmunds IP32 6TF
MW6 SVM P. Edwards Delfryn, Capel Dewi, Aberystwyth SY23 3HU
M6 SVN S. Jackson 18 Kentish Gardens, Tunbridge Wells TN2 5XU
M6 SVT L. Tayler 22 Wheatley Road, Leicester LE4 2HN
M6 SVU S. Logan Cedar House Reading Road North, Fleet GU51 4AQ
MW6 SVV S. Berrow 40 Priorsgate Oakdale, Blackwood NP12 0EL
M6 SVY D. Southernwood 24 Silver Gardens, Belton, Great Yarmouth NR31 9PD
MW6 SVZ J. Pauline 54 Laurel Road, Bassaleg, Newport NP10 8NY
M6 SWA Severn Valley Rd c/o S. Worger 6 Glendale Terrace, Mornington Road, Whitehill Bordon GU35 9AJ
MI6 SWB S. Beatty 132 Joanmount Gardens, Belfast BT14 6NZ
MM6 SWC S. Caldwell 45 Pappert, Alexandria G83 9LE
M6 SWD C. Rich Red Oak House, Summer Lane, Woodbridge IP12 2QA
M6 SWG G. Wicks 4 Bedford Street, Barnstaple EX32 8JR
M6 SWH S. Hughes 104 Thornley Road, Stoke-on-Trent ST6 7BA
M6 SWI A. Waller 64 Heaton Road, Billingham TS23 3GP
M6 SWK K. Michael 4 Newcroft Gardens, Christchurch BH23 2AS
M6 SWL P. Dekkers 21 Nodens Way, Lydney GL15 5NP
MW6 SWN S. Butler 33 Heol Penlan, Neath SA10 7LB
M6 SWO S. Woods 92 Rubens Avenue, South Shields NE34 8JT
M6 SWP S. Plume 1A Handside Lane, Welwyn Garden City AL8 6SE
M6 SWS I. Franklin 23 Ingle Drive, Ashby-de-la-Zouch LE65 2LW
MM6 SWT Z. Mckinnon 8 Rowanlea Avenue, Paisley PA2 0RP
M6 SXA N. Carter 16 Swinburne Close Sutton Heights, Telford TF7 4PZ
M6 SXB M. Comber 9 Blackwell Road, East Grinstead RH19 3HP
M6 SXC M. Revell 13 Mount Pleasant, Framlingham, Woodbridge IP13 9HQ
M6 SXD J. Brookes 177 Charnwood Close, Rubery, Birmingham B45 0JY
M6 SXG C. Champion 155 Walton Road, Walton on the Naze CO14 8NF
M6 SXI S. Robinson 47 Platt Hill Avenue, Bolton BL3 4JU
M6 SXM M. Cook 30 St. Peters Crescent, Selsey, Chichester PO20 0NA
M6 SXP S. Price 43 Charter Road, Weston-Super-Mare BS22 8LN
M6 SXT M. Dengate 15 Barn Close, Pease Pottage, Crawley RH11 9AN
M6 SXU K. Helm 90 Horne Street, Bury BL9 9HS
MM6 SXY L. Ritchie Braiklaw, Blackhills, Peterhead AB42 3LA
M6 SYB S. Baxby 197 Cemetery Road, Wath-Upon-Dearne, Rotherham S63 6LJ
M6 SYF H. Starling 23 Lathom Road, Manchester M20 4NX
M6 SYG S. Dibben 2 Taynton Covert, Birmingham B30 3QR
M6 SYH B. Hooper 55 Gildas Avenue, Birmingham B38 9HS
M6 SYI S. De Koster 4 Cambridge Way, Cullompton EX15 1GQ
M6 SYK P. Sykes 16 Hill Fold, South Elmsall, Pontefract WF9 2BZ
M6 SYW S. Ward Wellbeck, Wheel Road, Alpington, Norwich NR14 7NH
M6 SYX A. Davies 4 Capella Path, Hailsham BN27 2JY
MM6 SZK T. Truesdale 9 Cherry Lane, Banknock, Bonnybridge FK4 1JY
MW6 SZL D. Phillips 9 Baldwin Street, Newport NP20 2LT
M6 SZP W. Davies 61A Lightridge Road, Huddersfield HD2 2HF
MW6 SZW S. Sewell 53 Bryn Castell, Abergele LL22 8QA
M6 TAA T. Akay Caddebostan Mah. Plaj Yolu Sok. No25/15 Kad?K÷y, ?Stanbul (Asya) 34710 Turkey
M6 TAD G. Tew 66 St. Nicholas Estate, Baddesley Ensor, Atherstone CV9 2EZ
M6 TAF A. Cornelius 16 Crown House, North Street, Nailsea, Bristol BS48 4SX
M6 TAG D. Cutter 92 Hillcrest, Bar Hill, Cambridge CB23 8TQ
M6 TAH T. Hutchinson 134 Wingate Square, London SW4 0AN
M6 TAJ D. Jarvice 15 Meden Avenue, Warsop, Mansfield NG20 0PS
MW6 TAK C. Smith 7 Betws Avenue, Kinmel Bay, Rhyl LL18 5BN
M6 TAL T. Lewis 154 Meadow Head, Sheffield S8 7UF
M6 TAO P. Washbrook 1 Berry Hill Cottages, Berry Hill, Seaton EX12 3BD
M6 TAU A. West 33 Mundays Row, Waterlooville PO8 0HF
MI6 TAW T. Wilmot 4 Esdale Park, Bushmills BT57 8RB
MW6 TBC A. Doyle 54 Bro Syr Ifor, Tregarth, Bangor LL57 4AS
MW6 TBD A. Smith In2Change 3 Palmyra Place, Newport NP20 4EJ
M6 TBG A. Gravell 21 Wickridge Close, Stroud GL5 1ST
M6 TBH P. Bowman 26 Albany Hill, Tunbridge Wells TN2 3RX
M6 TBJ L. Perry 56 Lambrook Road, Taunton TA1 2AF
M6 TBK A. Green Flat 19, 19-25 Marine Parade East, Clacton on Sea CO15 1UX
MW6 TBL T. Lander Glopa Cottage, Old Racecourse, Oswestry SY10 7HP
M6 TBM T. Behan 48 Montrose Avenue, Datchet, Slough SL3 9NJ
M6 TBO M. Bradbury 104 Lilly Hall Road Maltby, Rotherham S66 8AT
M6 TBQ A. Trick 2 Newell Close, Aylesbury HP21 7FE
M6 TBR C. Herd 10 Amethyst Close, Rainworth, Mansfield NG21 0GH
M6 TBS D. Fletcher 2 Hillside Close, Heddington, Calne SN11 0PZ
M6 TBT A. Talbot 30 Irwell Road, Walney, Barrow-in-Furness LA14 3UZ
MW6 TBU M. Johns 151 Somerset Street, Abertillery NP13 1DR
M6 TBV R. Day 152 Swievelands Road Biggin Hill, Biggin Hill Westerham TN16 3QX

M6 TBX T. Burkinshaw 78 Binsted Road, Sheffield S5 8LL
M6 TCB P. Deluce 114 Townsfield Road, Westhoughton, Bolton BL5 2NT
M6 TCD T. Denny 20 School Lane, Surbiton KT6 7QH
M6 TCE R. Hannant 24 Tower Hill Park Costessey, Norwich NR8 5AT
M6 TCF I. Moore 128 Dalestorth Street, Sutton-in-Ashfield NG17 4EY
M6 TCG A. Gilberts 22 Granby Road, Buxton SK17 7TW
M6 TCH T. Hewlett Tuckers Cottage, Alfold Road, Cranleigh GU6 8NB
M6 TCI S. Friend 1 Orford Road, Tunstall, Woodbridge IP12 2JH
M6 TCK P. Pritchard 7 Warbler Road, Leighton Buzzard LU7 4DA
M6 TCL W. Hand 168 Barcroft Street, Cleethorpes DN35 7DX
MM6 TCM R. Fowler Windycroft, Whistlefield, Garelochhead G840EY
M6 TCN D. Woodhams 83 Langdale Place, Newton Aycliffe DL5 7DY
MI6 TCO J. Allen 3 Malwood Close, Belfast BT9 6QX
MM6 TCS T. Moffat 16 Drumellan Road, Ayr KA7 4XQ
M6 TCX T. Atherton 16 Steeple View, Ashton-On-Ribble, Preston PR2 2PX
M6 TCY A. Todd 5 Dom Pedro Cottages, Normanton WF6 1RS
M6 TCZ T. Clarke 329 Burton Road, Lincoln LN1 3XD
M6 TDA M. Aubrey 4 Broadleaf Close, Oakwood, Derby DE21 2DH
MW6 TDB L. Brookes Tyn Llidiart, Llanfairpwllgwyngyll LL61 6EQ
MW6 TDC D. Cowling Dorset Cottage, Alltyblacca, Llanybydder SA40 9SU
MW6 TDE P. Phillips Sedgemoor, Station Road, Kilgetty SA68 0XS
M6 TDF T. Peck 2 Primrose Lane, Miami Beach, Sutton on Sea LN12 2JZ
M6 TDI A. Jones 4 Park Close, Wymondham NR18 9BA
M6 TDJ J. Stringer 21 Ladywalk, Maple Cross, Rickmansworth WD3 9YZ
M6 TDK I. Radford 7 Eastmount Avenue, Hull HU8 9EW
MW6 TDL T. Lawrence 14 Railway Terrace, Caerau, Maesteg CF34 0UE
M6 TDM S. Brown 4 Dorado Gardens, Orpington BR6 7TD
M6 TDO S. Hogg 57 The Grange, Burton-on-Trent DE14 2EX
M6 TDP P. Thorley 57 Riverside Drive, Hambleton, Poulton-le-Fylde FY6 9EH
M6 TDR T. Robertson 6 Sandringham Court, Bircotes, Doncaster DN11 8QU
MD6 TDU K. Dodds 16 Second Avenue, Onchan, Onchan IM3 4LE Isle of Man
M6 TDV T. Mcnamara 19 Abbey Mews, Pontefract WF8 1TD
M6 TDY E. Ashford 56 Finch Close, Shepton Mallet BA4 5GL
M6 TDZ A. Thomas 5 Thornley Road, Wolverhampton WV11 2HR
M6 TEE P. Hardacre1 13 St. Johns Street, Bridlington YO16 7NL
M6 TEF R. Ashman 85A Fakenham Road, Great Ryburgh, Fakenham NR21 7AQ
M6 TEG Lapworth RC c/o T. Guy 16 Cogdeane Road, Poole BH17 9AS
M6 TEJ T. Jones Formula Cars, Wellington TA21 9HW
M6 TEK A. Sood Parima, Sewardstone Road, London E4 7RA
M6 TEM J. Turner 34 Vaughan Road, Stotfold, Hitchin SG5 4EH
M6 TEO M. Prentice Flat 3, 13 Meadow Road, Salisbury SP2 7BN
M6 TEP T. Coles 88C Dursley Road, Trowbridge Ba14 0Ns, Trowbridge BA14 0NS
MM6 TEQ K. Ruchomski 53A Barnton Avenue, Edinburgh EH4 6JJ
M6 TER Cambridgeshire Raynet c/o B. Lye 83C Lyngford Road, Taunton TA2 7EJ
MW6 TES A. Bailey 6 Trenos Gardens, Bryncae, Pontyclun CF72 9SZ
M6 TEV T. Howson 82 Bank End Avenue, Worsbrough, Barnsley S70 4QN
MM6 TEW T. Warr 407 Smerclate, Isle of South Uist HS8 5TU
M6 TEY A. Brown 3 Alston Road, New Hartley, Whitley Bay NE25 0ST
M6 TEZ T. Archer 1 Banks Road, Ashford TN234NR
MW6 TFB T. Blanchard 17 Heol Tyn-Y-Fron, Penparcau, Aberystwyth SY23 3RP
M6 TFD D. Lyes 2 Thelnetham Road, Blo Norton, Diss IP22 2JQ
M6 TFE T. Fletcher 3 Moorend Glade, Charlton Kings, Cheltenham GL53 9AT
M6 TFF P. Ratcliffe 61 Queens Avenue, Ilfracombe EX34 9LS
MI6 TFG T. Mckee 4 Earlford Heights, Newtownabbey BT36 5WZ
M6 TFJ T. Johnsen 5 Willow Lane, Billinghay, Lincoln LN4 4FN
M6 TFK T. Kightly Ferry Hill Farm, London Road, Chatteris PE16 6SG
MW6 TFL T. Fletcher 21 Sandpiper Road, Llanelli SA15 4SG
MI6 TFN R. Taylor 3 Austin Drive, Tandragee, Craigavon BT62 2AR
M6 TFO A. Theobold 25 Aysgarth Road, Leicester LE4 0ST
M6 TFP T. Chambers 104 Blue Hill Crescent, Leeds LS12 4PB
M6 TFQ D. Thomson 11 Uranus Road, Hemel Hempstead HP2 5QF
M6 TFT H. Salter 1 Rock Farm Cottages, Gibbs Hill, Maidstone ME18 5HT
M6 TFV D. Forshaw 14 Hope Carr Road, Leigh WN7 3ET
M6 TFW F. Worrall 297 Tamworth Road, Amington, Tamworth B77 3DG
MM6 TFY T. Yates 12 Fulmar Court, Newtonhill, Stonehaven AB39 3QG
M6 TFZ E. Walton 11 Parkfield Road, Northwich CW9 7AR
MM6 TGB T. Bell 22 Queens Road, Elderslie, Johnstone PA5 9LJ
M6 TGC B. Grainger 42 Madeira Avenue, Leigh-on-Sea SS9 3EB
MM6 TGD I. Currie 4 Greendyke Cottage, Falkirk FK2 8PP
MW6 TGF M. Malone 5 Brook Park Avenue, Prestatyn LL19 7HH
M6 TGG T. Sutton Yew Tree Farm, Paddol Green, Shrewsbury SY4 5QZ
M6 TGH A. Hardy 54 Trueway Drive Shepshed, Loughborough LE12 9HG
M6 TGI T. Hunt Mad Bess Cottage, Breakspear Road North, Uxbridge UB9 6LZ
M6 TGJ T. Woolvin 68 Jiggins Lane, Birmingham B32 3LA
M6 TGL C. Dolphin 24 Saughall Road, Wirral CH46 6DS
M6 TGM S. Thorpe 4 Moredon Road, Swindon SN25 3DQ
MM6 TGN A. Barclay 21 Netherlea, Scone, Perth PH2 6QA
M6 TGP R. Chambers 3 Westby Way, Poulton-le-Fylde FY6 8AD
M6 TGQ T. Webb 24A Baldmoor Lake Road, Birmingham B23 5QA
M6 TGR H. Mcphillips 160 Pasley Street, Plymouth PL2 1DT
M6 TGS A. Stubbs 20 Mossfields, Crewe CW1 4TD
M6 TGU M. Bostock 86 Beauvale Drive, Ilkest on DE7 8SJ
M6 TGV B. Mountford 189 Lloyd Street, Stockport SK4 1NH
M6 TGY T. Hopkins Horrabridge Stores, Commercial Road, Horrabridge, Yelverton PL20 7QB
M6 TGZ S. Richardson 12 Good Road, Poole BH12 3PJ
M6 THA K. Thacker 2 New Cottages, Selham, Petworth GU28 0PJ
M6 THB Dr T. Barratt Flat 41, The Panoramic, 30 Park Row, Bristol BS1 5LS
M6 THC M. Hellon 75 Brassey Street, Birkenhead CH41 8BZ
M6 THI K. Barker 10 Corn Mill Close, Rochdale OL12 9UW
M6 THJ T. Hunt 127 St. Michaels Road, Salisbury SP2 9EQ
M6 THO Dr A. Thornett Ty Bedwen, Birch Grove, Lichfield WS13 6EP
MJ6 THP T. Pallot Biltmore, La Grande Route De St. Laurent, Jersey JE3 1NH
M6 THQ C. Taylor 9 Cesson Close, Chipping Sodbury, Bristol BS37 6NJ
M6 THS T. Shevchenko 25 Kirby Road, Dartford DA2 6HE
MM6 THU A. Gray 28 Le Roux Drive, Oldmeldrum, Inverurie AB51 0PJ
M6 THV T. Hunt 14 Wenlock Road, South Shields NE34 9BA
MW6 THW T. Woodley 2 Parc Onen, Neath SA10 6AA
M6 THY H. Tang 14 Bridle Close, Borehamwood WD6 5QA
M6 TIA C. Lyne 4 Bridge Close, Catterick Garrison DL9 4PG
M6 TIC R. Wainwright 38 Stanway Close, Worcester WR4 9XL
M6 TID A. Gilham 38 Trojan Way , Waterlooville, Portsmouth P07-8AL
M6 TIF C. Townsend 64 Burnbridge Road Old Whittington, Chesterfield S41 9LR
M6 TIH T. Mackenzie 2 Newcastle Street, Carlisle CA2 5UH
M6 TII T. Ince 18 Holly Walk, Keynsham, Bristol BS31 2TU
MI6 TIJ C. Thompson 86 Huguenot Drive, Lisburn BT27 4YD
M6 TIO T. Allsopp 413 Boulton Lane, Derby DE24 9DL
MW6 TIQ R. Baker 20 Hawthorne Terrace, Aberdare CF44 7HE
MM6 TIR S. Taylor Pine Lodge, Easterton Of Auchleuchries, Hatton, Peterhead AB420TQ
M6 TIU D. Firth 5 Mowhay Gardens, Hatherleigh EX20 3FE
M6 TIW T. Heath 48 Brightholmlee Lane, Sheffield S35 0DD
M6 TIY T. Cooper 11 Warwick Road, Totton, Southampton SO40 3QP
M6 TJA T. Atkin 1 Wrangle Farm Green, Clevedon BS21 5DR
M6 TJB T. Bexon 51 Hookstone Drive, Harrogate HG2 8PR
M6 TJD T. Davis 11 Hinton Villas, Hinton Charterhouse, Bath BA2 7SS
MW6 TJE G. Hodges 79 Trefelin, Aberdare CF44 8LF
MI6 TJF T. Ferguson 3 Wheatfield Park, Ballybogy, Ballymoney BT53 6NT
MW6 TJG D. Watts 52 Ridgeway Avenue, Newport NP20 5AH
MW6 TJH R. Orchard The Burrows, Spring Gardens, Whitland SA34 0HL
M6 TJI T. Dyer 180 Seaton Lane, Hartlepool TS25 1HF
M6 TJJ T. Hempsall 16 Central Avenue, Warrington WA2 8AJ
M6 TJL T. Lake 64 Lytchett Drive, Broadstone BH18 9LB
M6 TJN C. Austen Fenbank House, Roman Bank, Holbeach Clough, Spalding PE12 8DH
M6 TJO R. Austen Fenbank House, Roman Bank, Holbeach Clough, Spalding P E12 8DH
M6 TJQ D. Roberts 61 Teign Bank Road, Hinckley LE10 0ED
MW6 TJS T. Skerritt 94 Turberville Street, Maesteg CF34 0LU
M6 TJU T. Idiculla 3 Stanhope Road, Slough SL1 6JR
M6 TJV T. Humphreys 93 Cornwall Crescent Yate, Bristol BS37 7RU
M6 TJX T. Dalby 52 Narborough Road South, Leicester LE3 2FN
M6 TJY F. Grimley 37 Reginald Road, Bexhill-on-Sea TN39 3PH
M6 TJZ T. Johnson 15 Tennyson Road, Creswell, Worksop S80 4DW
M6 TKA K. Thompson 184 Tickhill Road, Doncaster DN4 8QS
M6 TKC T. Corcoran 191 Queensway, Rochdale OL11 2NA
M6 TKE N. Crabb 1 Council Houses, Hall Lane, Crostwick, Norwich NR12 7BB
M6 TKI J. Lewis 93 Eastcliff, Portishead, Bristol BS20 7AD
M6 TKK T. Kosteletos 10 Church Lane, Southwick, Brighton BN42 4GD
M6 TKN J. Atkinson 8 Cragton Gardens, Blyth NE24 5PR
M6 TKP A. Pickavance 14 Loughrigg Avenue, St. Helens WA11 7AP
M6 TKR T. Berisford 18 Cambridge Avenue, Winsford CW7 2LL
M6 TKU S. Heys 15 Heathfield Road, Fleetwood FY7 7LY
MM6 TKV D. Hastings Flat 38, Ericht Court, Upper Mill Street, Blairgowrie PH10 6AE
M6 TKW T. Ward 1 Darrismere Villas, Edinburgh Street, Hull HU3 5AS
M6 TKX T. Keable 5 Redhills Way, Hetton-Le-Hole, Houghton le Spring DH5 0ES
M6 TLB R. Lawrence 20 Coronation Drive, Forest Town, Mansfield NG19 0AJ
M6 TLC T. Christopher 27 Brinkhill Crescent, Nottingham NG11 8GN
MW6 TLF T. Ford Clwt Joli, Llanfrothen, Penrhyndeudraeth LL48 6DU
MI6 TLG R. Greer 11 , Mullaghcarton Road, Lisburn BT28 2TE
MM6 TLI A. Pates 7/5 Sciennes, Edinburgh EH9 1NH
M6 TLK A. Kyte 42 Radford Street, Alvaston, Derby DE24 8NS
M6 TLL A. Lear 38 The Roundway, Claygate, Esher KT10 0DW
M6 TLM I. Williams 36 Telford Road, Tamworth B79 8EY
MW6 TLN N. Weightman 39 Brynamman Road, Lower Brynamman, Ammanford SA18 1TR
M6 TLP T. Porter 30 Woodbridge Close, Appleton, Warrington WA4 5RD
MW6 TLR T. Rogers 59 Park Place, Newport NP11 6BN
M6 TLS J. Olver 43 Heron Gardens, Portishead, Bristol BS20 7DH
M6 TLX M. Egan 4 Rutter Avenue, Warrington WA5 0HP
M6 TLY A. Thompson 25 Ardingly Road, Cuckfield, Haywards Heath RH17 5HD
M6 TMA R. Smith 3 Plane Tree Close Marple, Stockport SK6 7RJ
M6 TMC S. Oram 7 West Bank, Main Street, Old Weston, Huntingdon PE28 5LJ
M6 TMD P. Kirby 36 Durham Road, London E12 5AX
M6 TMF D. Fletcher 97 Wallace Crescent, Carshalton SM5 3SU
M6 TMG M. Galloway 2 Edendale Terrace, Horden, Peterlee SR8 4RD
M6 TMH T. Hoyle 60 Greenbank Crescent, Marple, Stockport SK6 7PB
M6 TMJ M. Jones 2 Home Park Road, Saltash PL12 6BH
M6 TMK T. Walker 11 Banburies Close, Bletchley, Milton Keynes MK3 6JP
M6 TML T. Longmore 3 Dairy Farm Cottages, Northlands Road, Gainsborough DN21 5DN
M6 TMM T. Marsland 34 Wakefield Road, Stalybridge SK15 1AJ
M6 TMO T. Williams Moor Farm, Moor Lane, Lincoln LN3 4EG
M6 TMP T. James 87A Beaconfield Road, Epping CM16 5AT
MW6 TMQ D. Mottram 35 Trem Arfon, Llanrwst LL26 0BP
MI6 TMR M. Graham 6 Sir Richard Wallace Drive, Lisburn BT28 3BA
M6 TMY M. Thackery 1 Bockings Grove, Clacton-on-Sea CO16 8DL
MI6 TMZ J. Allen 192 Joanmount Gardens, Belfast BT14 6PA
M6 TNA A. Golden 3 Ampleforth Road, Middlesbrough TS3 7PU
M6 TNB P. Bozikis 336 Higham Hill Road, London E17 5RG
M6 TNC J. Winstone 31 Setterfield Way, Rugeley WS15 1BJ
MI6 TND R. Dobson 2 Mourne View, Crossgar, Downpatrick BT30 9HW
M6 TNH L. Macdougall-Smith 160 Thingwall Park, Bristol BS16 2BU
MI6 TNI T. Nelson 25 Monaghan Road Annashanco Rosslea, Belfast BT927PT
M6 TNL C. Aston 21 Hawks Drive, Winshill, Burton-on-Trent DE15 0DL
MM6 TNO D. De Freitas 14 York Street, Clydebank G81 2PH
M6 TNR T. Robinson 39 Cemetery Road, Laceby, Grimsby DN37 7ER
M6 TNU J. Lambert 69 Anvil Crescent, Broadstone BH18 9DZ
M6 TNY A. Deeming 16 Marshall Road, Exhall, Coventry CV7 9BX
MI6 TNZ A. Mccullough 45 Eglantine Park, Hillsborough BT26 6HL
MI6 TOA T. Mcpolin 40 Olympia Drive, Belfast BT12 6NH
MI6 TOC T. Carvill 18 Hospital Road, Newry BT35 8PW
M6 TOD T. Sandham 96 South Road, Morecambe LA4 6JS
M6 TOE A. Smyth 2 North Edge, Leigh WN7 1HW
M6 TOF J. Morgan Glas Y Dorlan, Pontrhydfendigaid, Ystrad Meurig SY25 6EJ
M6 TOG P. Joyce 2 Harold Road, Cuxton, Rochester ME2 1EE

M6 TOK S. Halliday 8 Newby Farm Road, Scarborough YO12 6UN
M6 TOL C. Higgins 27 Chetwynd Road, Birmingham B8 2LB
M6 TON T. Ralph 54 Clacton Road, Portsmouth PO6 3QY
M6 TOT S. Dix 8 Beaumont Road, Longlevens, Gloucester GL2 0EJ
MW6 TOU W. Redbourn 20 Radnor Drive Tonteg, Pontypridd CF38 1LA
M6 TOV M. Murdoch 19 Beachy Road, Crawley RH11 9HN
M6 TOW A. Smith 206 Essex Square, Salisbury SP2 8HZ
M6 TOZ D. Torrance 30 St. Norbert Drive, Ilkeston DE7 4EH
M6 TPA A. Austin 4 Cornwall Avenue, Oldbury B68 0SW
MI6 TPC T. Crozier 6 Garden Of Eden, Carrickfergus BT38 7LS
M6 TPD T. Davies 7 Crescent Road, Warley, Brentwood CM14 5JR
M6 TPE A. Smith 14 Queenstock Lane, Buxted TN224AR
M6 TPG T. Gollins 17 John Street, Stafford ST16 3PJ
M6 TPI T. Kilfeather Flat 1, 57 Chalk Hill, Watford WD19 4DA
MW6 TPM T. Mckeown 13 George Street, Treherbert, Treorchy CF42 5AH
M6 TPO S. Bunworth 21 Slattsfield Close, Selsey, Chichester PO20 0EB
M6 TPR E. Butler Tanglewood, Elms Lane, Wolverhampton WV10 7JS
M6 TPS J. Bramley 13 Bolton Street, Stockport SK5 6BE
M6 TPT T. Taylor 35 Albany Way, Skegness PE25 2NB
M6 TPV P. Watson 6 Walnut Grove, Shepton Mallet BA4 4HX
M6 TPX A. Patrick 2 Beacon Grange, Malvern WR14 3EU
M6 TPY T. Pollett 10 Bridport Road, Poole BH12 4BS
M6 TQF K. Taylor 44 Main Street, Willoughby, Rugby CV23 8BH
M6 TQW S. Moss 27 Dunn Side, Chelmsford CM1 1DL
M6 TRC N. Kent Flat 1, Manor House, Redruth TR15 1AX
M6 TRD M. Read 133 Somersall Street, Mansfield NG19 6EL
MI6 TRF D. Foy 125 Drumbeg Tullygally, Craigavon BT65 5AE
M6 TRH D. Constable 7 Mill Field, Sutton, Ely CB6 2QB
M6 TRI D. Freeman 37 Bedale Road, Nottingham NG5 3GL
MW6 TRK T. Keane 19 Williton Road Llanrumney, Cardiff CF3 5QE
MW6 TRL P. Terrell 82 Baglan Street, Treherbert, Treorchy CF42 5AR
MM6 TRO S. Troscheit 20 James Street, St. Andrews KY16 8YA
M6 TRP T. Green Bentley Country Park Ltd, Flag Hill, Great Bentley, Colchester CO7 8RF
M6 TRS T. Steel Barrowfield House, Much Hadham SG10 6BD
M6 TRT J. Fry Deal Cottage, Ipswich Road, Long Stratton, Norwich NR15 2TF
M6 TRU G. Truman 3 Mulberry Road, North Anston, Sheffield S25 4BH
M6 TRV T. Meakin 33 The Markhams, New Ollerton, Newark NG22 9QX
M6 TRW T. Wheatley Three Chimneys, Ricket Lane Blidworth, Mansfield NG21 0NA
MM6 TRX A. Mcneill 13 Spinkhill, Laurieston, Falkirk FK2 9JR
M6 TSA T. Stamp 187 Knapmill Road, London SE6 3TF
MM6 TSC T. Couper 10 Sclandersburn Road, Denny FK6 5LP
MI6 TSH S. Hamilton 38B Doagh Road Kells, Ballymena BT42 3ND
M6 TSI S. Gilham 32 Whitby Road, Lytham St. Annes FY8 3HA
M6 TSJ S. Mcbain 13A St. Lukes Close, Cherry Willingham, Lincoln LN3 4LY
M6 TSM T. Keay 12 Quarryfields, Seahouses NE68 7TB
MM6 TSN R. Thomson Galloquhine Cottage, Auchenblae AB30 1TT
M6 TSP C. Power 23 Manor Road, Killamarsh, Sheffield S21 1BU
M6 TSR T. Scopes The Garden House, West Common Road, Keston BR2 6AJ
M6 TST the Jaggy Thistles c/o T. Tate 339 West Dyke Road, Redcar TS10 4PS
MD6 TSW D. Williamson 45 Bluebell Close, Peel IM5 1GH Isle of Man
M6 TSZ M. Wickens St. James Vicarage, The Parade, Dudley DY1 3JA
M6 TTB T. Bate 4 Yoxall Road, Newborough, Burton-on-Trent DE13 8SU
M6 TTE T. Battelle 207 Alfreton Road, Blackwell DE55 5JH
M6 TTF M. Lanham 3 Park Cottages, New Common, Bishop's Stortford CM22 7RT
M6 TTH T. Horsten Kastelsvej 4, 2.Tv, Copenhagen E 2100 Denmark
M6 TTK R. Rushton 65 Wilton Bank, Saltburn-by-the-Sea TS12 1PD
M6 TTM J. Aldwinckle-Day 45 Felicia Way, Grays RM16 4JF
M6 TTN M. Newham 3 Laurel Drive, Brockworth GL3 4GF
M6 TTO F. Armstrong 38 Dovecote Drive, Haydock, St. Helens WA11 0SD
M6 TTP M. Mcgowan 48 Alderley Road, Thelwall, Warrington WA4 2JA
M6 TTR S. Chuck 82 Redmayne Drive, Chelmsford CM2 9AG
MW6 TTS M. Johns 39 Tyla Coch, Llanharry, Pontyclun CF72 9LT
M6 TTT A. Allen 65 Hanbury Road, Dorridge, Solihull B93 8DN
M6 TTV D. Hargrove 24 Slade Road, Wolverhampton WV10 6QS
M6 TTW M. Berlyn 13 Hopping Jacks Lane, Danbury, Chelmsford CM3 4PN
M6 TTY T. Tong Lincoln House, Clarence Drive, Harrogate HG1 2QD
M6 TUD C. Barker 18 Nickleby Road, Waterlooville PO8 0RH
MM6 TUG A. Cairns Tigh Bruadair, Achmore, Strome Ferry IV53 8UX
M6 TUH R. Taylor Penhewer Park, Liskeard PL14 3LW
M6 TUK C. Yunnie Lighthouse, Park Lane, Totnes TQ9 7BD
MI6 TUM S. Tumilty 18 Cluain-Air, Newry BT34 1PW
M6 TUN A. Tunney 79 Scott Street, Burnley BB12 6NJ
M6 TUR D. Slee Turnpike, Brearton, Harrogate HG3 3BX
M6 TUT A. Mars 14 Woodyard Close, London NW5 4BX
MI6 TUV C. Bailie 26 Moatview Park, Dundonald, Belfast BT16 2BE
MI6 TUX E. Murray Apartment 4, 20 Global Crescent, Belfast BT6 8LN
M6 TVC L. Lovell 28 Park Lane, Blunham, Bedford MK44 3NJ
M6 TVH E. Grabham 37 Redstart Road, Chard TA20 1SD
M6 TVL T. Cannon 242 Hall Road, Norwich NR1 2PW
M6 TVM T. Melia 6 Burnley Close, Blackburn BB1 3HL
M6 TVN D. Royall 7 Stacklands, Welwyn Garden City AL86XW
M6 TVS N. Simmonds 3 Noneley Hall Barns, Noneley, Wem, Shrewsbury SY4 5SL
M6 TVW T. Hawkesford 24 Greenaway, Morchard Bishop, Crediton EX17 6PA
M6 TVX D. Mills 91 Harp Road Hanwell, London W7 1JQ
M6 TWA T. Alvey 2 Sunny View, Back Road, Saxmundham IP17 3NY
MM6 TWE S. Tweedie 3 Edgar Road, Westruther, Gordon TD3 6ND
MW6 TWH J. Hughes Brynawelon, Southgate, Aberystwyth SY23 1SE
M6 TWI J. Truman 9 Riddell Avenue, Langold, Worksop S81 9SS
M6 TWL T. Willis 8 Windsor Court, York YO31 7RY
M6 TWQ J. Attwood 13 John Winter Court, Euston Road, Great Yarmouth NR30 1DU
MM6 TWS A. Gray 73 Ash-Hill Drive, Aberdeen AB16 5YR
M6 TWV T. Lamb Marsh Green, Tn7 4Et, Colmans Hatch TN7 4ET
M6 TWY T. Wood 26 Parkfield Crescent, Kimpton, Hitchin SG4 8EQ
MM6 TWZ Dr T. Dodd 2 White Wisp Gardens, Dollar FK14 7BH
M6 TXG C. Whatmough 11 Blackchapel Drive, Rochdale OL16 4QU
M6 TXH J. Marriott-Levett 37 Christleton Drive, Ellesmere Port CH66 3NN
M6 TXI B. Cave 2 Beaufort Close, Newcastle upon Tyne NE5 3XL
M6 TXJ J. Hopkins 53 Sprules Road, London SE4 2NL
MM6 TXK Prof. T. Kerby 1 St. Mark'S Lane, Edinburgh EH15 2PX
M6 TXM A. Miller Tile Barn, Straight Mile, Bourton, Rugby CV23 9QQ
M6 TXN S. Moore 5795 Jefferson Commons Way, Apt. 203, Kalamazoo, Michigan 49009 United States
M6 TXP P. Cassells 49 Dodds Lane, Maghull, Liverpool L31 0BD
MI6 TXS R. Bates 45 Upper Ballyboley Road, Ballyclare BT39 9ST
M6 TXT M. Bookham 116 Clare Gardens, Petersfield GU31 4EU
M6 TXU A. Gillard 58 Queens Road, Thame OX9 3NQ
M6 TXX A. Gibbins Flat 120, Greenhill, London NW3 5TY
M6 TXZ R. Bolton 18 Slimbridge Road, Tuffley, Gloucester GL4 0NB
M6 TYB R. Thornett Ty Bedwen, Birch Grove, Lichfield WS13 6EP
M6 TYC T. Corcoran 50 Grange Road, Bracebridge Heath, Lincoln LN4 2PW
M6 TYE S. Connolly 82 Cheswood Drive, Minworth, Sutton Coldfield B76 1YE
MW6 TYG C. Morris 17 Percy Road, Wrexham LL13 7EA
M6 TYH T. Hill 14 Hunters Mead, Motcombe, Shaftesbury SP7 9QG
M6 TYK J. Goddard 217 Speedwell Road, Bristol BS5 7SP
M6 TYL T. Ward Flat 26, Bassett Court, Bassett Avenue, Southampton SO16 7DR
M6 TYN M. Tynan 57 Alpine Drive, Wardle, Rochdale OL12 9NY
MI6 TYR V. Brogan 7 Richmond Park, Omagh BT79 7SJ
M6 TYS S. Tyrrell 14 Town Orchard, Southoe, St. Neots PE19 5YJ
M6 TYT D. Redhead 113 Blythway, Welwyn Garden City AL7 1DL
M6 TYU B. Grzelak Flat 20, Cotterell Court, Southern Way, London SE10 0DW
M6 TYZ G. Hudson 74 Upperfield Drive, Felixstowe IP11 9LS
M6 TZA A. Zaulincy Adams 261 Nottingham Road, Derby DE21 6AP
M6 TZD N. Cripps 66 Forest Road, London E8 3BT
M6 TZE S. Johnston 62 Charlecote Park, Telford TF3 5HD
M6 TZI I. Tighe 4 Beach Walk, Broadstairs CT10 1FA
M6 TZO J. Morton 44 Silam Road, Stevenage SG1 1JJ
MI6 TZP C. Gault 7 Gardenmore Place, Larne BT40 1SE
M6 TZR T. Crouch 2 Park Farm Close, Horsham RH12 5EU
M6 TZU S. Moore 32 Broadgreen Close, Leyland PR25 2XA
MM6 TZX S. Mccorkell 28 Leven Road, Hamilton ML3 7WS
M6 TZY S. Crabb 1 Council Houses, Hall Lane, Crostwick, Norwich NR12 7BB
M6 TZZ P. Moore 24 Plough Road, Dormansland, Lingfield RH7 6PS
MI6 UAB A. Pritchard 16 Ballymaconnell Road South, Bangor BT19 6DQ
M6 UAD M. Kimber 2 Church Road, Brandon IP27 0EN
M6 UAE P. Potter 21 Ulverston Crescent, Lytham St. Annes FY8 3RZ
M6 UAF J. Glynn 106 Fairway Avenue, West Drayton UB7 7AP
M6 UAJ A. Ibbotson 323 Brincliffe Edge Road, Sheffield S11 9DE
M6 UAL A. Laidlaw 3 Litcham Close Litcham, King's Lynn PE32 2QX
M6 UAO N. Smith 2 Tibbett Close, Dunstable LU6 3TT
M6 UAP Dr S. Ando 84 New Hey Road, Cheadle SK8 2AQ
M6 UAR D. Randles 12 Wain Court. Rakeway, Saughall. Chester CH16BF
M6 UAS A. Cudworth 30 Compass Tower, Munnings Road, Norwich NR7 9TW
M6 UAW A. Wragg 14 Grizedale Avenue Sothall, Sheffield S20 2DL
M6 UAX A. Macdonald Woodside Cottage, Horton Way, Verwood BH31 6JJ
M6 UBC B. Clifford 7 Broke Court, Guildford GU4 7HQ
MI6 UBE L. Calderwood 43 Rathview Park, Mullybritt, Enniskillen BT94 5EW
M6 UBH R. Tolman 10 Woodcote Way, Abingdon OX14 5NE
M6 UBI C. Cotton 49 Cornwall Road, Portsmouth PO1 5AR
M6 UBM B. Mcdowell 10 Vineyard Lane, Kingswood, Wotton-under-Edge GL12 8SB
M6 UBN B. Shephard 13 Forest Street, Annesley Woodhouse, Kirkby In Ashfield, Nottingham NG17 9HE
M6 UBR M. Moore 3 St. Pauls Road Birkenshaw, Bradford BD11 2JY
M6 UBS N. Evetts 35 Wood End Road, Kempston, Bedford MK43 9BB
MI6 UBT T. Mcwilliams 221 Kings Road, Belfast BT5 7EH
M6 UCE S. Roderick 47 Cuckmans Drive., St. Albans AL2 3AY
M6 UCP D. Ali 20 Millwall Close Gorton Lancshire, Manchester M18 8LL
M6 UCR C. Rogers 1 Tregurthen Close, Camborne TR14 7EB
M6 UCS M. Coote 4 Hunters Oak, Watton, Thetford IP25 6HL
M6 UDA L. Symonds 10 Cowper Court, Willunga 5172 Australia
M6 UDB D. Hudson 34 Upton Gardens, Upton upon Severn WR8 0NU
M6 UDC D. Cole 39 Hillside Road, Southminster CM0 7AL
M6 UDM D. Meehan 47 Clinton Road, Shirley, Solihull B90 4RN
MI6 UDR R. Dunwoody 16 Dernalea Road, Milford, Armagh BT60 4DZ
M6 UDS G. Chapman 253 St. Pauls Road, Preston PR1 6NS
M6 UDX R. Berwick 10 Hall Lane, Wacton, Norwich NR15 2UH
M6 UDY D. Smout Sunrays, Warbage Lane, Bromsgrove B61 9BH
M6 UEA M. Robinson 58 Haworth Road, Cross Roads, Keighley BD22 9DL
M6 UEB J. Masters 31 Lower Beeches Road, Birmingham B31 5JB
M6 UED M. Tanji Apartment 63, Westside One, Birmingham B1 1LS
M6 UEE R. Goldsack 5 Parc Dellen, Croft Farm Park, Luxulyan, Bodmin PL30 5EW
M6 UEH Dr S. Smith 557, Riverside Island Marina, Isleham, , Ely, CB7 5SL.
MM6 UEN I. Macdonald 20 Newbigging Terrace, Auchtertool, Kirkcaldy KY2 5XL
M6 UES R. Hawes 40 Nightingale Way, Thetford IP24 2YN
M6 UET R. Taylor 68 Charter Road, Chippenham SN15 2RA
M6 UFC A. Hunt 49 Freame Close, Stroud GL68HG
M6 UFF T. George 188 Heathfield Avenue, Dover CT16 2PA
M6 UFO C. Sims 133 Canterbury Road, Hawkinge, Folkestone CT18 7BS
M6 UFR S. Sinderbury 40 Norfolk Gardens, Urmston, Manchester M41 8RE
M6 UGA G. Amos Willow Tree House, Deers Green, Clavering, Saffron Walden CB11 4PX
MI6 UGE G. English 232 Kernan Hill Manor, Portadown, Craigavon BT63 5WU
M6 UGF S. Peare 15 Clydesdale Gardens, Bognor Regis PO22 9BE
M6 UGN S. Nesling 64 Ruskin Avenue, Lincoln LN2 4BT
M6 UGP G. Palin 104 Nelson Street, Crewe CW2 7LN
M6 UGX M. Street 7 Salisbury Street, Sowerby Bridge HX6 1EE
M6 UHF J. Smith 24 Monk Road, Wallasey CH44 1AJ
M6 UHM C. Imm 29 Naunton Crescent, Cheltenham GL53 7BD
M6 UHN B. Marks 167 Linnet Drive, Chelmsford CM2 8AH
M6 UHT P. Lunn 179 Coventry Road, Nuneaton CV10 7BA
M6 UHU J. Willetts 102 Welch Road, Cheltenham GL51 0EG
M6 UHV S. King 29 Hall Road, Chatham ME5 8PL
MI6 UIM F. Hand 12 Church Street, Rosslea, Enniskillen BT92 7DD
M6 UIR R. Muir 4 Blandford Avenue, Worsley, Manchester M28 2JE
M6 UIT A. Rackett 151 Stoke Road, Gosport PO12 1SE
M6 UJD J. Dixey 108 Colchester Road West Bergholt, Colchester CO6 3JS

M6 UJK J. Keddy 63 High Street Kimpton, Hitchin SG4 8PU
M6 UJM R. Banks 3 Parkhayes, Woodbury Salterton, Exeter EX5 1QS
M6 UJR J. Risby 112 Stratton Heights, Cirencester GL7 2RL
M6 UKA G. Curtis 129A Salisbury Road, Blandford Forum DT11 7SW
M6 UKB J. Lightfoot 23 Nichols Street Desborough, Desborough NN142QU
M6 UKC P. Williams 53 Lilac Avenue, Cannock WS11 0AR
M6 UKF F. Hennigan 18 Friary Gardens, Newport Pagnell MK16 0ZJZ
M6 UKG J. English 1 Niton Cottage Pound Lane, Meonstoke, Southampton SO32 3NP
M6 UKH A. Shaw 9 Stamford Lane, Warmington, Peterborough PE8 6TW
M6 UKI U. Trnjakov 63 Welford Gardens, Abingdon OX14 2BH
M6 UKJ J. Pithers 77 Victoria Road, Laindon, Basildon SS15 6RA
M6 UKM M. Reeves 41 Hogarth Road, Whitwick, Coalville LE67 5GF
M6 UKT G. Eason Whitegates, Parsonage Road, Takeley, Bishop's Stortford CM22 6QX
M6 UKX G. Radulescu 41 Sherard Road, London SE9 6EX
M6 ULA J. Van Der Elsen 6 Kent Close, Churchdown, Gloucester GL3 2HQ
MM6 ULL K. Mackenzie Alderwood, Braes, Ullapool IV26 2TB
MW6 ULX C. Griffiths 21 Sandpiper Road, Sandy, , Llanelli SA15 4SG
M6 ULY B. Page 12 Hitchens Close, Hemel Hempstead HP1 2PP
M6 UMB M. Bramble 133 Fane Way, Maidenhead SL6 2TX
M6 UMD A. Slee Foal Cottage Mare Hill Common, Pulborough RH20 2DX
M6 UMI A. Smith 83 Lottbridge Drive, Eastbourne BN22 9PA
M6 UMM F. Colquhoun 2 Heslop Road, London SW12 8EG
M6 UMR U. Munir Flat 6, Horton House, Field Road, London W6 8HW
M6 UNA J. Taberner 32 Bell Lane, Sutton Manor, St. Helens WA9 4BD
MI6 UNC C. Harper 2 The Courtyard, Rathfriland, , Newry BT34 5PU
M6 UNI M. Knapton The Orchard, Rucklers Lane, Kings Langley WD4 9NA
M6 UNN D. Munn 36 Moor Lea, Braunton EX33 2PE
M6 UNS P. Unsworth 83 Newbold Avenue, Sunderland SR5 1LL
MW6 UNY L. Betts 12A Maesgwyn, Pontnewydd, Cwmbran NP44 1BQ
M6 UON J. Fernßndez Silva 45 Veals Mead, London CR4 3SB
M6 UPE R. Saddler 41 Clapham Close, Swindon SN2 2FL
MW6 UPH A. Williams 8 Old Tanymanod Terrace, Blaenau Ffestiniog LL41 4BU
M6 UPL M. Juuso 32 Hart Road, St. Albans AL1 1NF
M6 UPS A. Morehen 20 Castleton Grove, Inkersall, Chesterfield S43 3HU
M6 UPT P. Turner 59 Fairburn Avenue, Crewe CW2 7SY
M6 UPU A. Stirk 59 West Avenue, Lightcliffe, Halifax HX3 8TJ
M6 UPW A. Wood Berrybrook Northall Road, , Eaton Bray LU6 2DQ
M6 URA D. Kveksas 36, Purley CR8 3AQ
M6 URD J. Ratcliffe 63 Dickens Lane, Poynton, Stockport SK12 1NN
M6 URG S. Kitchen 16 Crown Avenue, Cudworth, Barnsley S72 8SE
M6 URH R. Roebuck 469 East Bank Road, Sheffield S22AE
MW6 URI M. Ouseley Brynsiriol Manorowen, Fishguard SA65 9PT
M6 URK M. Mcnamara 50 Portsdown Road, Portsmouth PO6 4QH
M6 URM A. Stansfield 59 Prankerds Road, Milborne Port, Sherborne DT9 5BX
MM6 URP S. Gray-Jones Flat C, 7 Nelson Street, Aberdeen AB24 5EP
MM6 URR M. Riddick Davah, Port Road, Castle Douglas DG7 3JW
M6 URS S. Banda 45 High View Avenue, Grays RM17 6RU
M6 URT B. Burnett 44 Westoe Drive, South Shields NE33 3EL
M6 URX M. Urban 59 Sterling Gardens, London SE14 6DU
M6 USA M. Dexter 70 Rutland Street, Derby DE23 8PR
MI6 USC W. Bradley 14 Ardmore Grange, Ballygowan, Newtownards BT23 5TZ
M6 USD J. Rufes Flat 1 & 3-8 12 Smyrna Road, London NW6 4LY
M6 USE D. Hardy 186 Sneyd Hill, Stoke-on-Trent ST6 1RA
M6 USG Dr B. Featherstone 1 Hall Terrace, Bodinnick, Fowey PL23 1LX
MW6 USK C. Moreton 20 Millbrook Court, Little Mill, Pontypool NP4 0HT
M6 USM J. Wright 32 Carlton Road, Nottingham NG10 3LF
M6 USO I. Usov Flat 12, 48 The Avenue, Southampton SO17 1XQ
M6 USP A. Brighton 38 Greenfield Crescent, Nailsea, Bristol BS48 1HR
MM6 USS E. Milligan 46 Arkaig Drive Crossford, Dunfermline KY12 8YW
M6 USV D. Soames 40 Woodland Drive, North Anston, Sheffield S25 4EP
M6 UTB E. Harrington 53 Kingscroft Road, Banstead SM7 3NA
M6 UTC A. Driscoll 82 Station Road, Langford, Biggleswade SG18 9PQ
M6 UTI J. Blay 31 Woodcote House, Queen Street, Hitchin SG4 9TL
M6 UTP C. Walker-Riley 1 Farmcote Court, Hemlington, Middlesbrough TS8 9LJ
M6 UTT C. Dyer 74 Godstone Road, Lingfield RH7 6BT
MI6 UTV T. Maclaine 172 Moylagh Road, Seskanore, Omagh BT78 2PN
M6 UTX P. Begg Green Moss, Yew Tree Road, Grange-over-Sands LA11 7AA
MM6 UTZ C. Mackie Hanover Court, 5/3 Slaeside, Balerno EH14 7HL
M6 UUA D. Sealey 6 Mizzen Road, Hull HU6 7AG
M6 UUE N. Busley Busley, South Drove, Spalding PE11 3BD
M6 UUH I. Ashmore 36 West End Avenue Royston, Barnsley S71 4LQ
M6 UUU A. Castell 35 Plantation Road, Chippenham SN14 0EX
M6 UVB B. Axcell 20 Sheerways, Faversham ME13 8TP
M6 UVD D. Smith 17 Stonepound Road, Hassocks BN6 8PP
M6 UVF E. Goodwin 55 Twickenham Road, Sunderland SR3 4JN
MI6 UVS Q. Church 51 Houston Park, Broughshane, Ballymena BT42 4LB
M6 UWK J. Hadley 75 Glendower Avenue, Coventry CV5 8BD
M6 UWS M. Sharman 3 Deben Crescent Haydonwick, Swindon SN25 3QB
M6 UWU A. Siabi 10 Ayloffs Walk, Hornchurch RM11 2RJ
M6 UXB C. Hurley 12 Grocott Road, Wednesbury WS10 8RQ
M6 UYF E. Martin 28 Mountford Close, Wellesbourne, Warwick CV35 9QQ
M6 UYP A. Tomlinson 45 Well Lane Willerby, Hull HU10 6HB
M6 UZH U. Hussain Cardwell Barn, Hostingley Lane, Dewsbury WF12 0QH
M6 UZK A. Sweet 3 Beechwood Grove, Blackpool FY2 0DZ
M6 UZT L. Howe 29 Abdy Avenue, Harwich CO12 4QP
M6 UZZ D. Willis 7 Shearman Avenue, Kimberworth Park, Rotherham S61 3AG
MJ6 VAA V. Atherton 4 Clos De La Mer, La Route De Noirmont, Jersey JE3 8AL
MM6 VAB V. Julius 100 Hogarth Drive, Cupar KY15 5YU
M6 VAG K. Lane 105 Robin Hood Lane, Chatham ME5 9NN
M6 VAH V. Downes 55 Ashfield Road Bromborough, Wirral CH62 7EE
MI6 VAI L. Hambleton 9 Springvale Road, Ballywalter, Newtownards BT22 2PE
M6 VAJ A. Coomber 3 Dolly Drove, Chard TA20 1PF
M6 VAK D. Evans 1 Copeland Street, Hyde SK14 4TD
M6 VAM E. Driscoll 42 Adelaide Square, Shoreham-by-Sea BN43 6LN
M6 VAN E. Tsang Harrogate Ladies' College, Clarence Drive, Harrogate HG1 2QG
M6 VAP A. Mason 13 Welton Gardens, Lincoln LN2 2AY
M6 VAR Dr T. Varoudis 179 Lymington Avenue, London N22 6JL
M6 VAV A. Vasarhelyi 1 Eldon Close, Langley Park, Durham DH7 9FR
M6 VAW D. Vassie Ashleigh, Filching, Polegate BN26 5QA
M6 VAY A. Vassie Teddards, Filching, Polegate BN26 5QA
M6 VAZ J. Bevan 18 Martin Road, Diss IP22 4HR
MI6 VBB D. Twaddle 19 Lisanduff Park, Portballintrae, Bushmills BT57 8RY
M6 VBD T. Hillier 16 Priory Walk, Leicester Forest East, Leicester LE3 3PP
MW6 VBE J. Jones 7 St. Mary Street, Trelewis, Treharris CF46 6AL
M6 VBF P. Staite Chestnut Farm, Eastville, Boston PE22 8LX
MI6 VBH V. Hazelton 12 The Elms Bush, Dungannon BT71 6UE
M6 VBJ J. Vanbesien Flat 8, Suffolk House, Chester CH1 3BZ
M6 VBN D. Melbourne 160 Markfield, Court Wood Lane, Croydon CR09DW
M6 VBP S. Vyner 20 Ryde Lands, Cranleigh GU6 7DD
M6 VBR J. Baughan Chestnut Farm, Eastville, Boston PE22 8LX
M6 VBS A. Cairns 202 The Ridgeway, St. Albans AL4 9XJ
M6 VBX P. Otterwell 50 Hythe Road, Staines-upon-Thames TW18 3EE
M6 VBZ D. Scott 13 Grey Close, Bredbury, Stockport SK6 1HA
M6 VCA V. Chady 12 Brent Place, Barnet EN5 2DP
M6 VCB D. Bonney 14 Jersey Close, Congleton CW12 3TW
M6 VCC Dr O. Blacklock 15 Kings Crescent, Lymington SO41 9GT
M6 VCH C. Wood 169 Bramerton Road, Nottingham NG8 4NB
M6 VCK V. Corcoran 50 Grange Road, Bracebridge Heath, Lincoln LN4 2PW
M6 VCM V. Millson Harrogate Ladies' College, Clarence Drive, Harrogate HG1 2QG
M6 VCN D. Williams 6 Homestone Gardens, Leicester LE5 2LJ
M6 VCP C. Pinder 70 Highfield Road, Beverley HU17 9QR
M6 VCS V. Newell 15 The Grove, Luton LU1 5PE
M6 VCU K. Beswick 29 Hops Lane, Halifax HX3 5FB
M6 VCV Dr C. Jenkins 52 Warden Abbey, Bedford MK41 0SN
M6 VDC N. Fairbairn 15 Hewitt Road, Dover CT16 1TH
MW6 VDE D. Evans 12 Caerfallwch, Rhosesmor, Mold CH7 6PN
M6 VDF J. Roser St Marys Terrace, Hastings TN343LR
M6 VDH V. Harrington 25 Victoria Street, Norwich NR1 3QX
M6 VDJ D. Jones 68 Dunelm Road Thornley, Durham DH6 3HW
M6 VDL D. Lodwig 15 Frithwood Park, Brownshill, Stroud GL6 8AB
MM6 VDP D. Pounder Birchbank, Lindean, Galashiels TD1 3PA
M6 VDR C. Johnson The Hollies, Belaugh Green Lane, Norwich NR12 7AJ
M6 VDT C. Stephens 8 New Molinnis, Bugle, St. Austell PL26 8QL
M6 VDX P. Martin 108 Headlands Grove, Swindon SN2 7HP
M6 VEC M. Broadhurst 23 Hucklow Avenue, North Wingfield, Chesterfield S42 5PX
M6 VED R. Dawson 6 Harleys Field, Abbeymead, Gloucester GL4 4RN
M6 VEG J. Waldron 55 Sheringham Road, Poole BH12 1NS
M6 VEL D. Rogers 6 Guildford Street, Plymouth PL4 8DS
M6 VEN A. Fisher 15 Washdyke Lane, Osgodby, Market Rasen LN8 3PB
M6 VEP E. Parker 3 Dowgate Close, Tonbridge TN9 2EH
M6 VET A. Williamson 4 Garden End, Melbourn, Royston SG8 6HD
M6 VFA A. Colcombe 217 Church Drive, Quedgeley, Gloucester GL2 4US
M6 VFC M. Chipperfield 5 Lullingstone Close, Hempstead, Gillingham ME7 3TS
M6 VFD H. Campbell 50 Northwood Road, Whitstable CT5 2ES
MM6 VFL B. Sonnet 12 Winton Circus, Saltcoats KA21 5DA
M6 VFN B. Bewick 88 Bentley Way, Weston Road, Norwich NR6 6TS
MW6 VGA J. Duffain 37 Mount Pleasant Street, Dowlais, Merthyr Tydfil CF48 3AF
M6 VGB A. Green 139 Clough Lane, Halifax HX2 8SN
M6 VGC N. Lupton 134 Ridgeway Road, Sheffield S12 2SZ
M6 VGE S. Coates 27 Primula Way, Chelmsford CM1 6QT
M6 VGR I. Brodie 38 Chatsworth Place, Stoke-on-Trent ST3 7DP
MM6 VGS R. Adams 1/R 61 Capelrig Street, Thornliebank, Glasgow G46 8LP
M6 VGU K. Haythornwhite 148 Nairne Street, Burnley BB11 4NP
M6 VGV G. Venugopalan 3 Southwater Close, London E14 7TE
M6 VHA V. Gudgeon 6 The Choakles, Wootton, Northampton NN4 6AP
M6 VHB H. Blayer 10 Riverside Close, Staines-upon-Thames TW18 2LW
M6 VHM L. Mooney Calde Cottage, Mannings Lane, Chester CH2 2PB
M6 VHZ J. Telfer 50 Agraria Road, Guildford GU2 4LF
M6 VIA Dr L. Kirkcaldy 19 Itchen Avenue, Bishopstoke, Eastleigh SO50 8JW
M6 VID D. Mason 94 Guessburn, Stocksfield NE43 7QR
M6 VIE S. Croston 12 Sefton Street, London SW15 1LZ
M6 VIG C. Harvey 1 New Pit Cottages, Bridge Place Road, Bath BA2 0PE
M6 VIO S. Morgan 8 Williams Drive, Braintree CM7 5QJ
M6 VIT A. Vitiello 8 Pegasus Road, Leighton Buzzard LU7 3NJ
M6 VIV V. Lee 56 Gordon Road, Fishersgate, Brighton BN41 1PT
M6 VIX W. Reynolds 4 Underwood Close, Stafford ST16 1TB
M6 VJG V. Gallagher 4 Kingsdown Way, Bromley BR2 7PT
MI6 VJR R. Russell 70 Allen Park, Dunamanagh, Strabane BT82 0PD
M6 VJX B. Walker 255 Packington Avenue, Birmingham B34 7RU
M6 VKA S. Rush 88 Mountview, Borden, Sittingbourne ME9 8JZ
M6 VKB Stewart Bryant Radio Station c/o N. Morehen 20 Castleton Grove Inkersall, Chesterfield S43 3HU
M6 VKE D. Lee 26 Church Lane, Chalgrove, Oxford OX44 7TA
M6 VKG M. Gibson 6 Harrison Road, Mansfield NG18 5RG
M6 VKH S. Hardy 55 Westwood Road, East Peckham, , Tonbridge TN12 5DB
M6 VKK R. Cresswell Meadow View, Hulver Road, Beccles NR34 7UW
M6 VKM K. Minett Rosedene, Honey Hill, Wimbotsham, King's Lynn PE34 3QD
M6 VKN R. Reilly 29 Burwell Close, Plymouth PL6 8QD
M6 VKW V. Williams Moor Farm, Moor Lane, Lincoln LN3 4EG
MM6 VKY V. Fleming Tarbat View, Achavandra Muir, Dornoch IV25 3JB
M6 VKZ G. Delaforce 29 Littlebridge Meadow Bridgerule, Holsworthy EX22 7DU
M6 VLB K. Earwicker Greenbank, Cripplestyle, Fordingbridge SP6 3DU
M6 VLD V. Mereuta 62 Sutherland Avenue, London W9 2QU
M6 VLE V. Sayer Observatory, Mount Bures Road Wakes Colne, Colchester CO6 2AS
MM6 VLG W. Guy 121 Stewarton Drive, Cambuslang, Glasgow G72 8DH
M6 VLH D. Smith 23 Old School House, Shotley Gate, Ipswich IP9 1QP
MW6 VLO B. Henley Rhewin Glas, Llandysul SA44 6PR
M6 VLQ D. Volkov 2 Robert Close, Potters Bar EN6 2DH
M6 VLR Rvd. R. Blunden 34 Stephenson Road Arborfield, Reading RG2 9NP
MW6 VLS S. Morris 21 Heritage Way Badgers Green, Llanymynech SY226LL
M6 VMA C. Skupski 57 Three Nooks, Bamber Bridge, Preston PR5 8EN
M6 VMF J. Jackson 80A Clarence Road, Leighton Buzzard LU7 3EL
M6 VMJ R. Atkinson 75 Talbot Road, Penistone, Sheffield S36 9ED
M6 VMK R. Whatling 41 Lawrence House Camperdown Street, Bexhill on Sea TN39 5EN

M6 VMO F. Deschacht 28 Barnfield, Slough SL1 5JW
M6 VMP I. Rotheram 60 Whitewood Park, Liverpool L9 7LG
M6 VMS M. Wickens Haven Lea, Queens Drive, Windermere LA23 2EL
M6 VMY V. Gunnoo 76 Katherine Drive, Dunstable LU5 4NU
M6 VNL M. Harris 5 Lynmore Close, Northampton NN4 9QU
MW6 VNP S. Beer 67 Killan Road, Dunvant, Swansea SA2 7TH
M6 VNT V. Sheppard Flat 8, Riverside Court Cambridge Road, Harlow CM20 2AD
M6 VNV D. Nicholls 12 Northview, Tufnell Park Road, London N7 0QB
MW6 VOC A. Parsons 21 Rectory Drive, St. Athan, Barry CF62 4PD
M6 VOD A. Motion 45 Bradgate Lane, Asfordby, Melton Mowbray LE14 3YB
MI6 VOF P. Mcfadden 35 West Wind Terrace, Hillsborough BT26 6BS
M6 VOG B. Rutkowski 18 Squirrel Close, Coventry CV2 1FP
M6 VOL S. Hill 86 Gilbert Road, Chichester PO19 3NL
M6 VOS S. Devos Applecross Cottage, Main Road, Newark NG23 7HR
MW6 VOW B. Moore Lower Bulford Farm, Bulford Road, Haverfordwest SA62 3ET
M6 VOX F. Riche 1 Lovelstaithe, Norwich NR1 1LW
M6 VOY M. Harrison 2 Broad Farm Cottages, North Street, Hailsham BN27 4DS
MI6 VOZ J. Kacprzyk 80 The Brambles, Magherafelt BT45 5RZ
MM6 VPF H. Phillips Maplebank, Leithen Road, Innerleithen EH446NJ
MW6 VPL S. Mccaffery 5 Troedyrhiw, Caerlan, Abercrave, Swansea SA9 1SX
MM6 VPM A. Cowan 32 Esk Valley Terrace, Dalkeith EH22 3FT
M6 VPS M. Partyka Ridgebourne Cottage, Ridgebourne Road, Kington HR5 3EG
MW6 VPX S. Chapman 2 Pensingrig, Rhoslan, Criccieth LL52 0NB
M6 VPZ T. Godfrey 43B Broadbury Road, Bristol BS4 1JT
M6 VQC C. Powell 52 Primley Road, Sidmouth EX10 9LF
M6 VQV T. Page 19 Lamorna Drive, Callington PL17 7QH
M6 VQX R. Olszewski 24A Nightingale Road, South Croydon CR2 8PT
M6 VRC A. Newbould 9 Laburnum Road, Rudheath, Northwich CW9 7JT
M6 VRD V. Bowen 4 Crossley Gardens, Halifax HX1 5PU
M6 VRE E. Vrentzos 12 Floyer Close, Queens Road, Richmond TW10 6HS
MM6 VRH C. Harvey Hillside, Sarclet, Wick KW1 5TU
M6 VRJ R. Jermy 15 Oak Tree Close Martham, Great Yarmouth NR29 4QN
M6 VRL R. Liszewski 79 Malvern Drive, Warmley, Bristol BS30 8XA
M6 VRO R. Sheen 29 Aldermoor Avenue, Southampton SO16 5GJ
M6 VRP I. Bell 164 The Broadway, Herne Bay CT6 8HY
M6 VRR K. Bird Mount View, Hill Top Road Ashover, Chesterfield S45 0BZ
M6 VRT A. Calvert 5 Pond Cottages Butts Pond, Sturminster Newton DT10 1BE
MI6 VRX D. Todd 2 Innishargie Gardens, Bangor BT19 1SN
M6 VSA A. Chubb 70 Goldcrest Road, Chipping Sodbury, Bristol BS37 6XQ
M6 VSB B. Rhodes 81 Witherston Way, Eltham, London SE9 3JL
M6 VSC V. Schultz 223 Long Down Avenue, Bristol BS16 1GE
M6 VSE Dr S. Vellaichamy 37 Briarswood, Chelmsford CM1 6UH
MM6 VSI S. Irvine 9 Pearce Grove, Edinburgh EH12 8SP
M6 VSJ S. Jermy 15 Oak Tree Close, Martham, Great Yarmouth NR29 4QN
M6 VSL S. Lawrance 69 Athelstan Gardens, Wickford SS11 7EF
MM6 VSM S. Mulligan 17 Crawfurd Gardens, Rutherglen, Glasgow G73 4JP
M6 VSS M. Petrea-Goanta 14 Oakdene Road, Watford WD24 6RW
M6 VST J. Baker 70 Vane Close, Norwich NR7 0US
MW6 VTA J. Roberts Hafan Llewelyn, Llanystumdwy, Criccieth LL52 0SS
M6 VTB L. Rudge 10 St. Johns Close, Aldingbourne, Chichester PO20 3TH
M6 VTC S. Dillon 33A Main Road, Cleeve, Bristol BS49 4NS
M6 VTG D. Coles 36 York Hill, Loughton IG10 1HT
MW6 VTP M. Davies 26 Ty Nant, Caerphilly CF83 2RA
MM6 VTS S. Davidson 6 Sutherland Walk, Mintlaw, Peterhead AB42 5GT
M6 VTT D. Bedford 55 Sycamore Avenue, Wickersley, Rotherham S66 2NS
M6 VTZ H. Hatchard 118 Marina, St. Leonards on-Sea TN380BN
M6 VUD A. Burr 3 Larch Close Attleborough, Attleborough NR 172HB
MM6 VUS C. Doherty 22 Castlelaw Street, Glasgow G32 0NF
MM6 VUV R. Fraser 72 Ferguson Drive, Denny FK6 5AG
M6 VVA A. Amos Willow Tree House, Deers Green, Clavering, Saffron Walden CB11 4PX
M6 VVB C. Harper 21 Holly Close Farnham Common, Slough SL2 3QT
M6 VVE R. Cook 62 High Hazel Road, Moorends, Doncaster DN8 4QN
M6 VVN V. Walker 291 Park Road, Blackpool FY1 6RR
M6 VVT M. Kenny 5 Fallowfield, Hazlemere, High Wycombe HP15 7RP
M6 VVX P. Buck Westlears Farm Chard Junction, Chard TA20 4LN
M6 VWA W. Vinnicombe 107 Coleridge Road, Cambridge CB1 3PN
M6 VWD J. Bartley 29 Cheltenham Road, Blackburn BB2 6HR
M6 VWE C. Brown 123 Godinton Road, Ashford TN23 1LN
MM6 VWH C. Robertson Flat 3, Redheugh House, Redheugh Court, Kilbirnie KA25 7JF
M6 VWN S. Mathew 183 West End, Costessey, Norwich NR8 5AW
M6 VWP E. Byrne 4 Centre Vale, Littleborough OL15 9EL
M6 VWR R. Hirons 28 Crossfell Road, Hemel Hempstead HP3 8RG
MM6 VWT W. Scott 11 The Marches Armadale, Bathgate EH48 2PG
M6 VWV J. Jackson 65 Berkeley Close, Abbots Langley WD5 0XD
M6 VWW A. Wilson 28 Langham Road, Bristol BS4 2LJ
M6 VWX G. Brownlow 21 Church Lane Tittleshall, King's Lynn PE32 2QD
MM6 VXB M. Al Saeed 9 Appin Place, Edinburgh EH14 1NJ
M6 VXI D. Holden 24 Penny Gate Close, Hindley, Wigan WN2 3DP
MM6 VXR A. Latto 8 Aspen Avenue, Glenrothes KY7 5TA
M6 VXT T. Searle 18 Witton Lane, Little Plumstead, Norwich NR13 5DL
M6 VYN M. Roberts 11 Oakleigh Road, Pinner HA5 4HB
M6 VZA Dr J. Lauxman Mccorkell Suite 77 City House, 131 Friargate, Preston PR1 2EF
M6 VZF E. Clarke 1 Farm Drive, Croydon CR0 8HX
M6 VZV M. O'Donovan 10 Rockfield Close, Teignmouth TQ14 8TS
MI6 VZW M. Mcpeake 47 Dunclug Gardens, Ballymena BT43 6NN
M6 WAA S. Myciunka 67 Templeton Drive, Fearnhead, Warrington WA2 0WR
MI6 WAB W. Mcdonald 14 Edenmore Park, Limavady BT49 0RG
M6 WAD P. Waddington 98 Harrogate Street, Barrow-in-Furness LA14 5LY
MI6 WAF W. Hawkes 12 Meadow Court, Newtownards BT23 8YE
MI6 WAG M. Barr 2 Willowvale Close, Islandmagee, Larne BT40 3SD
MM6 WAI W. Inglis 13 Princes Street, California, Falkirk FK1 2BX
M6 WAJ J. Watts 31 Ingledene Close, Havant PO9 1DG
MI6 WAM W. Mccormick Dernawilt Road, Crocknaboghill, Rosslea, Enniskillen BT92 7GG
MI6 WAN W. Nelson 44 Lacky Road, Tattynageeragh, Rosslea, Enniskillen BT92 7GA
M6 WAP W. Patterson 6 Lea Drive, Nantwich CW5 5JS
M6 WAQ J. Carn 12 Woodcock Hill Estate Harefiled Road, Rickmansworth WD3 1PQ
M6 WAS D. Simpson Flat 4, Victoria Mansions, East Street, Harwich CO12 3AS
M6 WAT P. Watson 10 Whitelands Crescent, Baildon, Shipley BD17 6NN
M6 WAV A. Snelson 6 Rayleigh Close, Braintree CM7 9TX
MM6 WAW M. Beaton 76 Fergus Avenue, Howden, Livingston EH54 6BG
M6 WAY B. Waymark 13 Beech Ave, Nottingham NG7 7LJ
M6 WBA J. Edridge 36 Conyngham Lane, Bridge, Canterbury CT4 5JX
M6 WBB J. Whitehead 12 Polkerris Road, Carharrack, Redruth TR16 5RJ
M6 WBE D. Buckley 99 Kings Drive, Bradwell, Great Yarmouth NR31 8TF
M6 WBF I. Westwood 1 Cook Avenue, Maltby, Rotherham S66 8QZ
M6 WBH D. Thompson 17 Sandpiper Close, Blyth NE24 3QN
MM6 WBJ W. Jackson 3 Annick Road, Dreghorn, Irvine KA11 4EY
M6 WBK D. Beck 13 Chipstead, Chalfont St. Peter, Gerrards Cross SL9 9JZ
M6 WBM A. Buckley 25 Queensway, Pilsley, Chesterfield S45 8EJ
MI6 WBN M. Parke 18 Brandywell Court, Londonderry BT48 9HL
M6 WBO W. Beacroft 3 Rose Farm Rise, Normanton WF6 2PL
MM6 WBP W. Ferguson 1D Macphail Drive, Kilmarnock KA3 7EJ
M6 WBR J. Webber 16 Marythorne Road, Bere Alston, Yelverton PL20 7BZ
MI6 WBT W. Turkington 8A Drummullan Road, Moneymore, Magherafelt BT45 7XS
MM6 WBU J. Mclelland 24 Hyslop Street, Airdrie ML6 0ES
M6 WBV W. Reeves 33 Pond Bank, Blisworth, Northampton NN7 3EL
M6 WBX D. Mills 79 Eastbourne Avenue, Gosport PO12 4NX
M6 WBZ K. Hunt 11 De Marnham Close, West Bromwich B70 6RJ
M6 WCA B. Neal 6 Canterbury Street, Chaddesden, Derby DE21 4LG
M6 WCB A. Williams 74 Broadfield Road, Bristol BS4 2UW
M6 WCC P. Homer 20 Sunningdale Road, Middlesbrough TS4 3HU
M6 WCE W. Steingold 36 Hailsham Road, Romford RM3 7SP
M6 WCF P. Wade 85 Penymynydd Road, Penyffordd, Chester CH4 0LF
M6 WCG C. Jones 90 Station Rod, Wem SY4 5BL
MW6 WCI W. Johns 5 Heol Pantgwyn, Llanharry, Pontyclun CF72 9HU
MM6 WCK D. Ross 29 East Banks, Wick KW1 5NL
M6 WCL O. Fallon 26 Central Avenue, Corfe Mullen, Wimborne BH21 3JD
M6 WCN D. Jones 85 Black Butts Lane, Walney, Barrow-in-Furness LA14 3JL
M6 WCR C. Bavister 10 Pheasant Grove, Wixams, Bedford MK42 6AH
MW6 WCT M. Cooper 3 Waunddu, Pontnewynydd, Pontypool NP4 6QZ
MW6 WCU D. Crowe 2 Severnside Cottages, Canal Road, Newtown SY16 2JN
M6 WCV W. Marshall 24 Union Road, Thorne, Doncaster DN8 5EL
M6 WCX W. Dix 21 Pine Vale Crescent, Bournemouth BH10 6BG
M6 WCY A. Whyman 2 Wilson Close, Thelwall, Warrington WA4 2ET
M6 WCZ D. Howden 46 Chestnut Avenue, Armthorpe, Doncaster DN3 2EP
MI6 WDB W. Bradshaw 206 Manor Street, Belfast BT14 6ED
M6 WDC M. Moseley 86 Sissons Terrace, Leeds LS10 4LH
MI6 WDD D. Milligan 30 Belgrano Ahoghill, Ballymena BT42 2QQ
M6 WDF D. Leyland 20 Newton Heath, Middlewich CW10 9HL
M6 WDG D. Goodwin 33 York Road Dunscroft, Doncaster DN7 4LZ
M6 WDH W. Hunt 128 Heath Road, Dudley DY2 0AU
M6 WDI K. Lynch Medindie, Woodside, Ryton NE40 4SY
M6 WDL W. Leake 61 Jubilee Road, Sutton-in-Ashfield NG17 2DD
MI6 WDM W. Mccormick 1 Mullagreenan Court, Rosslea, Enniskillen BT92 7PR
M6 WDN Dr A. Brito Da Silva 21 Grosvenor Gardens, Newcastle upon Tyne NE2 1HQ
M6 WDO T. Smith Chy Crowshensy, Clifton Road, Redruth TR15 3UD
M6 WDQ W. Carey 6 Gainsborough Road, Bexhill-on-Sea TN40 2UL
M6 WDR A. Wallace 46 Heathfield Road, Grantham NG31 7NH
MW6 WDX P. Dalton 89 Hillcrest, Brynna, Pontyclun CF72 9SL
M6 WDY C. Johnson 45 Gordon Road, Chelmsford CM2 9LN
MM6 WEB J. Webster 6 Livingston Crescent, Winchburgh, Broxburn EH52 6FX
MW6 WEE A. Morris Fairview, Trefonen, Oswestry SY10 9DP
M6 WEJ C. Buswell 34 Hatchmore Road, Denmead, Waterlooville PO7 6TF
M6 WEK K. Alabaster 16 Butlers Road, Horsham RH13 6AJ
M6 WEL D. Wells 40 Barnham Broom Road, Wymondham NR18 0DF
M6 WEN W. Jefferson Mk451Lf, Flitwick MK451LF
M6 WEO S. Kiel 32 Weavers Avenue, Frizington CA26 3AT
M6 WEP E. Raine Stable Cottage, St. Martins, Richmond DL10 4SJ
MI6 WEQ R. Skelton 16 Demiville Avenue, Lisburn BT27 5RE
MM6 WER J. Weir 36 Main Rd, Ferguslie, Paisley PA12QT
M6 WET D. Bilton 9 Ashby Street, Allenton, Derby DE24 8JR
M6 WEU C. Massimiani Flat 2 37 York Road, Guildford GU1 4DN
M6 WEV A. Weatherall The Old Telephone Exchange, The Street, Canterbury CT3 1ED
M6 WEW L. Wnekowski 22 Simonds Grove, Spencers Wood, Reading RG7 1BH
MI6 WEZ W. Todd 2C Knockwood Crescent, Belfast BT5 6GE
M6 WFA W. Durrant 19 Rydal Rd, Gosport PO12 4ES
MW6 WFB L. Bowman Chanrick, Penderyn Road, Aberdare CF44 9RU
M6 WFC P. Morgan 25 Junction Street, Dudley DY2 8XT
M6 WFE T. Mckenna 67 Raynsford Road, Great Whelnetham, Bury St. Edmunds IP30 0TN
MW6 WFF W. Webb 70 Beech Court, Bargoed CF81 8NS
M6 WFG S. Woodruff 172 Windsor Street, Wolverton, Milton Keynes MK12 5DR
M6 WFH M. Sharp Woodgate Farm, Livery Road, Salisbury SP5 1RJ
M6 WFI I. Warnecke 12 Caxton Road, Margate CT9 5NP
MW6 WFJ T. Penman 21 Maelog Road, Cardiff CF14 1HP
MD6 WFK R. Kissack 6 Falcon Cliff Court, Douglas, Isle of Man IM2 4AQ
M6 WFL C. Coles 2 Fern Square, Chickerell, Weymouth DT3 4NZ
M6 WFR R. Pickwoad 2 The Tannery, Barrowden, Oakham LE15 8EA
M6 WFV J. Stevens Parc-An-Lower Prospidnick, Helston TR13 0RY
M6 WFW W. Fletcher-Wells 46 London Road, Buxton SK17 9NU
M6 WFX C. Giddens 3 Neville Walk, Richmond DL10 5AG
M6 WFY K. Payne Eastern Esplanade, Broadstairs CT10 1DR
M6 WFZ M. Toomey Ravens Rock, 10A Castlebank, Northwich CW8 1BL
M6 WGB G. Birch 23 Hanson Street, Great Harwood, Blackburn BB6 7LP
M6 WGC T. Buck 6 Lynn Road, Terrington St. Clement, King's Lynn PE34 4JX
MM6 WGD R. Ward 35 Mill Street, Drummore, Stranraer DG9 9PS
MI6 WGE G. Orderly 30 Moatview Park, Dundonald, Belfast BT16 2BE
M6 WGF G. Harris 9 Penmeneth Trewennack, Helston TR13 0PU
M6 WGG I. Wagstaff 25 Bowbridge Gardens Bottesford, Nottingham NG130AZ
M6 WGH I. Mccorquodale 4 Wise Grove, Warwick CV34 5JW
M6 WGJ Sw London Rayne c/o W. Joyce 2 Palmers Cottage, Main Street, Oakham LE15 8DH
MI6 WGL W. Leonard 57 Old Coach Road Mullanavehy, Enniskillen BT92 2EW
MI6 WGM G. Mccusker 10 Birchdale, Lurgan, Craigavon BT66 7TR

M6 WGS C. Smart 35B Church Lane, Melksham SN12 7EF
MM6 WGT M. Gilchrist 17 Colinslee Crescent, Paisley PA2 6SD
M6 WGU G. Head 45 Beechnut Road, Kendal LA9 7FF
MI6 WGX J. Johnston 16 Fort Road, Bangor BT19 7BS
M6 WHD R. Searby 5 Bosvean Gardens, Truro TR1 3NQ
M6 WHF D. Cracknell 120 Woodhill, London SE18 5JL
MD6 WHG W. Hogg Medhamstead, Lhergydhoo, Isle of Man IM5 2AE
MM6 WHI R. Leslie 118 Western Avenue, Ellon AB41 9EU
M6 WHS D. Whitehouse 6 Larch Close, Heathfield TN21 8YW
M6 WHT K. Cullen 29 Colman Road, London E16 3JY
M6 WHU A. Hodgson 515 Ashingdon Road, Rochford SS4 3HE
MM6 WHW W. Woods 1 Nursery Lane, Kilmacolm PA13 4HP
MM6 WHX D. Strachan 30 Belhaven Park, Muirhead G69 9FB
M6 WHZ M. Underwood 39 Westbury Crescent, Oxford OX4 3SA
M6 WIB R. Wilson 171 Cooden Drive, Bexhill-on-Sea TN39 3AQ
M6 WIC Rvd. B. Jackson 46 Princess Road, Market Weighton, York YO43 3BR
M6 WID G. Holland 6 Moorfield Road, Widnes WA8 3JE
M6 WIF M. Landragin 101 Linden Gardens, Enfield EN1 4DY
M6 WII W. Smith 40 Westbury Rise, Harlow CM17 9NS
M6 WIK W. Trofimiuk 120A The Fairway, Northolt UB5 4SW
M6 WIL W. Outram 7 Bakewell Road, Baslow, Bakewell DE45 1RE
M6 WIM A. Forrester 14 Calder Close, Lytham St. Annes FY8 3NH
M6 WIR D. Daniels 60 Hawthorn Way, Northway, Tewkesbury GL20 8TQ
M6 WIS D. Wiskow 15 Ferndale Close, Sandbach CW11 4HZ
M6 WIT B. Witt 20 Foxglove Close, Newton Aycliffe DL5 4PF
M6 WIW C. Ring Acorn Cottage Prospect Place, Helston TR13 8RU
M6 WIX J. Unwin 39 Whinfell Drive, Normanby, Middlesbrough TS6 0BG
M6 WIZ S. Keen 13 Ivy Road, Kettering NN16 9TG
M6 WJA A. Woodhouse 14 Wrens Gardens, Wath-Upon-Dearne, Rotherham S63 7GD
M6 WJB W. Buss 83 Gorse Avenue, Chatham ME5 0UP
M6 WJF W. Foster 55 Drake Avenue Minster On Sea, Sheerness ME12 3SA
M6 WJJ K. Jones 98 Common View, Stedham, Midhurst GU29 0NU
MI6 WJK J. Kelso 32 Old Park Manor, Ballymena BT42 1RW
M6 WJL G. Hayers 87 Bradleigh Avenue, Grays RM17 5RH
M6 WJM J. Wheeler 14 Doomgate Asby, Appleby-in-Westmorland CA16 6RB
M6 WJN W. Naughton 14 Carisbrooke, Frimley, Camberley GU16 8XR
MM6 WJP W. Paterson 1 Burnside Terrace, Stranraer DG9 8HH
MM6 WJS S. Wilson 2 Kinnear Court, Guardbridge, St. Andrews KY16 0UE
MI6 WJW W. Wilson 9 Benbradagh Avenue, Limavady BT49 0AP
M6 WKB K. Brett 1 Tower Court, Peterborough PE2 9AT
MM6 WKC J. Cormack 2 Coghill Street, Wick KW1 4PN
M6 WKD R. Hammond Bye Road Cottage Pot Kiln Chase Co93Bh, Gestingthorpe CO93BH
MI6 WKE R. Wilkinson 11 Fairview Park, Dromore BT25 1PN
M6 WKF A. Rowland-Stuart 86 Wiltshire House, Lavender Street, Brighton BN2 1LE
M6 WKG A. Cole 104 Newport Road, Cowes PO31 7PS
M6 WKI M. Hawker Popes Corner Marina, Holt Fen, Little Thetford, Ely CB6 3HR
M6 WKL P. Houghton 6 Olivers Court, Calne SN11 0FL
MI6 WKN W. Gamble 11 Anderson Park, Limavady BT49 0RH
M6 WKR G. Walker 2 Cliffe Bank Cottages, Piercebridge DL2 3SX
M6 WKS A. Wilkins 124 Fullers Mead, Harlow CM17 9AU
M6 WKT G. Williams 18 Elmsleigh Road, Farnborough GU14 0ET
M6 WKW K. Davies 23 Egmanton Road, Meden Vale, Mansfield NG20 9QN
M6 WKY C. Wilkinson 40 Lumley Drive, Consett DH8 7DT
M6 WKZ A. Norden 10 School Lane, Watton At Stone, Hertford SG14 3SF
M6 WLA S. Fearnhead 1B Hampden Grove, Eccles, Manchester M30 0QU
MW6 WLB K. Clarke 13 Caerphilly Close, Dinas Powys CF64 4PZ
M6 WLC W. Cogdon 15 Strafford Avenue, Worsbrough, Barnsley S70 6SU
M6 WLD W. Daley 27 Rosebery Street, Manchester M14 4UR
M6 WLE E. Williams 24 Astbury Street, Congleton CW12 4EQ
M6 WLF K. Smith 3 Hawley Vale, Hawley Road, Dartford DA2 7RL
M6 WLG D. Collins-Cubitt 75 Anthony Drive, Norwich NR3 4EW
M6 WLR A. Coats 57 Mill Hill, Boulton Moor, Derby DE24 5AF
M6 WLS W. Le Serve 120 Cheam Road, Cheam, Sutton SM1 2EB
MW6 WLY L. Lane-Wells Gwaelod, Pool Quay, Welshpool SY21 9LH
M6 WMA M. Wall 227 Wayfield Road, Chatham ME5 0HJ
M6 WMF N. Riggall 41 Manor Drive, Waltham, Grimsby DN37 0NS
M6 WMG M. Carpenter The Retreat, High Lane Manaccan, Helston TR12 6HT
M6 WMH W. Clarke 1 Alvey Terrace, Nottingham NG7 3DF
M6 WMJ M. Watts 48 Tasburgh Street, Grimsby DN32 9LB
MM6 WMK M. Walker 1 Gilmerton Street, Glasgow G32 7SQ
MM6 WMM W. Moodie Barcloy Cottage, Kirkcudbright DG6 4XR
MI6 WMN W. Mcmullen 69 Lissize Avenue Rathfriland, Newry BT34 5DE
M6 WMP M. Weaver 16 Avocet Drive, Kidderminster DY10 4JT
M6 WMS M. Smith 78 Shiregate, Metheringham, Lincoln LN4 3DR
M6 WMV J. Hancock 162 Preston Road, Yeovil BA20 2EQ
M6 WMZ A. Law 2 The Bank, Somersham, Huntingdon PE28 3DJ
M6 WNB J. Hocking 26 Musket Road, Heathfield, Newton Abbot TQ12 6SB
M6 WNC C. Lindley 187 Alexandra Road, Sheffield S2 3EH
M6 WNE W. Gough 42 Merevale Avenue, Hinckley LE10 0PY
M6 WNF N. Waller Olive Cottage, 6 Church Road Chelmondiston, Ipswich IP9 1HS
M6 WNM W. Mccoo Ivy House, West Drove South, Walpole Highway, Wisbech PE14 7RA
M6 WNN W. Newey 3 Forrester Close, Flanderwell, Rotherham S66 2NL
M6 WNR N. Rapson 15 School Close, Bampton, Tiverton EX16 9NN
M6 WNS M. Tsun Dyson'S Farm, Long Row, Tibenham NR16 1PD
M6 WNY N. Yeates 149 Hallam Way West Hallam, Ilkeston DE7 6LP
M6 WNZ A. Argent-Wenz 6 Ashley Brake West Hill, Ottery St. Mary EX11 1TW
M6 WOB J. Blow 23 Leaders Way, Lutterworth LE17 4YS
MM6 WOC W. O'Rourke 6 Busby Place, Kilwinning KA13 7BA
MW6 WOD W. Davies Foelallt, North Road, Aberystwyth SY23 2EL
M6 WOE G. Needle 195 Kingsley Avenue, Kettering NN16 9ET
MI6 WOF A. Pulman 69 Islandarragh Road, Cape Castle, Ballycastle BT54 6HS
M6 WOH H. Pickett 13 Clifden Close Mullion, Helston TR12 7EQ
M6 WOI M. Jenkins Flat 2, 66 Union Street, Ryde PO33 2LG
M6 WOK M. Smith 7 Cherry Tree Close, Everton, Lymington SO41 0ZG
M6 WOL M. Walters 39 Portland Place, Coseley, Bilston WV14 9TB
M6 WOM M. Wilkes 1 Hawburn Close, Bristol BS4 2PB
M6 WOO J. Wooldridge 7 Heather Gardens, Belton, Great Yarmouth NR31 9PP
M6 WOS W. Briggs 88 Guild Avenue, Walsall WS3 1LQ
M6 WOT D. Bowers-Edgley 8 The Slopes, Lower Henley Road, Reading RG4 5LE
MW6 WOV V. Lyall 29 King Alfreds Road, Sedbury, Chepstow NP16 7AQ
M6 WPA W. Armsden 2 Rowan Close, Middlewich CW10 0TA
MW6 WPB W. Thomas 10 John Lewis Street Hakin, Milford Haven SA73 3HS
M6 WPI T. Hobson-Smith 15 Henconner Lane, Chapel Allerton, Leeds LS7 3NX
M6 WPN S. Comper 105 Westway, Copthorne, Crawley RH10 3QS
MI6 WPP J. Quigley 25 Rathfern Way, Newtownabbey BT36 6BX
MI6 WPT W. Turkington 230 Whitechurch Road, Ballywalter, Newtownards BT22 2LB
MI6 WPW W. Wilson 9 Benbradagh Avenue, Limavady BT49 0AP
M6 WRB W. Bellamy Flat 6, 26 Moremead Road Lewisham, London SE6 3LP
MW6 WRC W. Lockyer 100 Dan-Yr-Heol, Aberdare CF44 9ED
M6 WRD W. Davis 24 Strensall Road, Hull HU5 5TD
M6 WRE W. Rendell 71 Hardwick Bank Road, Northway, Tewkesbury GL20 8RP
M6 WRG M. Richardson 64 Henry Road, West Bridgford, Nottingham NG2 7ND
M6 WRH W. Hartley 30 Coltman Avenue, Beverley HU17 0EY
M6 WRJ W. Jones 11 Wykeham Place, Fareham PO16 0FA
M6 WRL S. Lacey 2 Purbeck Cottages, Acton, Swanage BH19 3LU
M6 WRN M. Wren-Hilton 28 Arwenack Avenue, Falmouth TR11 3JW
M6 WRO J. Wrobel Flat 27, Tenney House, Curzon Drive, Grays RM17 6SG
M6 WRR Dr C. Hall 6 Browning Road Church Crookham, Fleet GU52 0YJ
M6 WRS W. Sankey 14 Carter Grove, Hereford HR1 1NT
MI6 WRT R. Jameson 64 Blackhill Road, Dromore, Omagh BT78 3HL
M6 WRU W. Clough 19 Cardigan Road, Bedworth CV12 0LY
M6 WRV M. Wright 4 Woodlands Road, Headington, Oxford OX3 7RU
M6 WRW D. Hampton 6 St. Georges Lane South, Worcester WR1 1QZ
M6 WRX I. Scholey 36 Longroyd Avenue, Leeds LS11 5HA
M6 WRY L. Curtis 39 Mount Stewart Street, Seaham SR7 7NG
M6 WSB A. Mcallister-Bowditch 24 Morse Close, Chippenham SN15 3FY
M6 WSC D. Waterhouse 10 Falconers Drive, Battle TN33 0DT
MM6 WSG G. Greenwood 92 Porterfield, Comrie, Dunfermline KY12 9XG
M6 WSH Dr S. Halsey 15 Rectory Drive, Yatton, Bristol BS49 4HF
M6 WSM W. Gee 2 Milton Walk, Worksop S81 0DH
M6 WSO D. Mathison 74 Victoria Road, Tuebrook, Liverpool L13 8AW
MI6 WSP T. Mullan 35 Finn Park Rosslea, Enniskillen BT92 7LJ
M6 WSR R. Emery 15 Shaw Crescent, Hutton, Brentwood CM13 1JD
M6 WSS S. Roberto-Southall 121 Dunsheath, Telford TF3 2DA
M6 WST R. West 557 East Bank Road, Sheffield S2 2AG
M6 WSU S. Blackburn 36 Mardale Grove, Barrow-in-Furness LA13 9QG
M6 WSW D. Rolph 17 Moorlands Park Ashby Road, Sinope LE67 3BD
M6 WTD T. Dallas 74 Main Street, Asfordby, Melton Mowbray LE14 3SA
M6 WTE A. White 56 Raleigh Close, Churchdown, Gloucester GL3 1NT
M6 WTF S. Searles-Bryant 14 Canuden Road, Chelmsford CM1 2SX
M6 WTG W. Jones 8 Oakbrook Close, Ewyas Harold, Hereford HR2 0NX
M6 WTL A. Page 12 Hitchens Close, Hemel Hempstead HP1 2PP
M6 WTM D. Parsloe 13 Highview Park New Dover Road, Capel-Le-Ferne, Folkestone CT18 7GA
M6 WTP W. Pascoe 34 Jasper Road, London E16 3TR
M6 WTR S. Whittaker 115 Grange Close, Hoveton, Norwich NR12 8EB
M6 WTT W. Barnard 34 Forsyth Drive, Braintree CM7 1AR
M6 WTW T. White 9 Dashwood Court, Banbury OX16 5DF
M6 WTX S. Jackson 64 Main Road, Moulton, Northwich CW9 8PB
MI6 WTZ T. Saunderson 40 Hillfoot Street, Belfast BT4 1PR
M6 WUB A. Best 18 Chestnut Close Rendlesham, Woodbridge IP12 2UW
M6 WUG C. Humphries 44 Linksway, Folkestone CT19 5LS
M6 WUH C. Hopkins 24 Battle Road, Tewkesbury GL20 5TZ
MW6 WUK G. Reason 454 Cowbridge Road West, Cardiff CF5 5BZ
MM6 WUL W. Fulton 15 Staffa Avenue, Port Glasgow PA14 6DT
M6 WUM J. Griffiths 9 Cauldale Close, Middleton, Manchester M24 5SU
M6 WUN H. Cheung Clarence Drive, Harrogate HG1 2QG
M6 WVB V. Brett 1 Tower Court, Peterborough PE2 9AT
MW6 WVC T. Rowlands Caer Gog Farm, Bodffordd LL77 7BX
M6 WVE M. Huggett 12 West View Cottages Lewes Road, Lindfield, Haywards Heath RH16 2LJ
M6 WVG D. Millward 77A Meadowcroft, St. Albans AL1 1UG
M6 WVH P. Wood 26 Wamil Way, Mildenhall, Bury St. Edmunds IP28 7JU
M6 WVK K. Andrews-Mead 10 Alderlands Close Crowland, Peterborough PE6 0BS
M6 WVV S. Legg 4 Riverside Avenue, Fareham PO16 8TF
M6 WWB W. Buczkowski 54 Woodfield Heights, Wolverhampton WV6 8PT
M6 WWC C. West 112A Hawthorne Way Shelley, Huddersfield HD8 8PX
M6 WWD R. Howard 13 Top Common, East Runton, Cromer NR27 9PW
M6 WWE S. George 84 Kenilworth Crescent, Enfield EN1 3RG
M6 WWF M. Johnson 6 Swainson Road, Leicester LE4 9DQ
M6 WWI B. Kuttikkate 34 Shetland Crescent, Rochford SS4 3FJ
M6 WWJ J. Harrison 20 Kelcliffe Avenue, Guiseley, Leeds LS20 9EW
M6 WWK M. Wilks Flat 16, Overton Court, Cheltenham GL50 3BW
M6 WWM M. Barnard 47 Springfields, Dunmow CM6 1BP
M6 WWR I. Coleman 15 St. Andrews Close, Holme Hale, Thetford IP25 7EH
MM6 WWS W. Stevenson 28 Lightburn Road , Halfway, Cambuslang, Glasgow G72 8UE
M6 WWT A. Walls 7 Waveney Grove, York YO30 6EQ
MW6 WWY K. Macleod 11 Hillside Close, Goodwick SA64 0AX
MW6 WWZ A. Harris 4 Dol Elian, Old Colwyn, Colwyn Bay LL29 8YZ
M6 WXD B. Wild 1 Sunnymount, Midsomer Norton, Radstock BA3 2AS
M6 WXF J. Paddy Alansbur-W-Macbyrne 4 Rayner Street, Stockport SK1 4HX
M6 WXH D. Purnell 30 Conygre Green Timsbury, Bath BA2 0JU
M6 WXK P. Marsh 30 Mount Pleasant, Aylesford ME20 7BE
M6 WXN S. Harcourt 71 Ingleby Road, Long Eaton, Nottingham NG10 3DG
M6 WXP M. Howes 8 Oat Hill Drive, Northampton NN3 5AL
MM6 WXS A. Mccall Shielhill, Greengairs, Airdrie ML6 7TJ
M6 WXX D. Watts 3 Witney Road, Crawley RH10 6GJ
M6 WXY G. Hogan 58 Weeley Road, Little Clacton, Clacton-on-Sea CO16 9EN
M6 WYD G. Holt 6 Highfields, Holmfirth HD9 2PZ
M6 WYG D. Griffiths 5 New Grange Terrace, Pelton Fell, Chester le Street DH2 2PB
M6 WYM S. Koskinas 50 Amderley Drive, Norwich NR46HZ
M6 WYN D. Griffith The Old Stables, Briantspuddle Dairy, Dorchester DT2 7HT
M6 WYR P. Attwood 3 Stratford Grove, Evesham WR11 2SD
M6 WYT T. Webster 1 Fen Close, Newton, Alfreton DE55 5TD

M6 WYW M. Whatling 34 Cannon Park Road, Coventry CV4 7AY
M6 WYX N. Soane 8 Martello Mews, Martello Road, Seaford BN25 1JT
M6 WYY L. O'Neill Flat 6, Freshwater Court, Lee-on-the-Solent PO13 9BB
M6 WYZ A. Tanseli 157 Warwick Road, Rayleigh SS6 8SG
M6 WZB S. Burdis 10 Johnston Avenue, Hebburn NE31 2LJ
M6 WZF A. Downing 67 Mayfield Drive Caversham, Reading RG4 5JP
M6 WZK P. Goodhand 2A Moor Road, Sutton, Norwich NR12 9QN
M6 WZV T. Cole 36 Fairlee Road, Newport PO30 2EJ
MU6 WZY S. Kirkpatrick Ste Helene Manor, St. Andrew GY6 8XN Guernsey
M6 XAA J. Daniels 33 Park View, Crewkerne TA18 8HS
M6 XAB A. Clay 4 Craiglands Park, Ilkley LS29 8SX
MW6 XAC A. Oconnell 31 North Avenue, Tredegar NP22 3HF
M6 XAD J. Hyde 3 Carlton Road, Newport SK14 2PW
MW6 XAE J. Smith Llanstinan Fach, Letterston, Haverfordwest SA62 5XD
M6 XAF T. Chapman 12 Greenways Chilcompton, Radstock BA3 4HT
MW6 XAG S. Broderick 179 Malpas Road, Newport NP20 5PP
M6 XAI A. Rumsby 31 Howell Road, Drayton, Norwich NR8 6BU
M6 XAK S. Ferguson 80 London Road, Retford DN22 7DX
M6 XAL A. Darlington 92 Lancaster Road, St. Albans AL1 4ES
MI6 XAM M. Green 7 Liester Park, Ballyclare BT39 9RZ
M6 XAO A. Gray 18 St. Botolphs Green, Leominster HR6 8ER
M6 XAQ M. Dennis 40 Windsor Road, Linton, Swadlincote DE12 6PL
M6 XAS A. Smith 91 Norcliffe Road Bispham, Blackpool FY2 9EN
M6 XAT T. Whelan 243 Duke Street, Barrow-in-Furness LA14 1XU
M6 XAW A. Wardle 2 Deer Park Place, Sheffield S6 5ND
M6 XAX K. Poulton 21 East View, London E4 9JA
M6 XAY A. Laxton 7 St. Christophers Green, Broadstairs CT10 2SS
MI6 XBA T. Agnew 28 Knockdhu Park, Larne BT40 2EJ
M6 XBB J. Clarke 8 Waveney Heights, Brockdish, Diss IP21 4LD
MM6 XBD S. Boyd 1 St. Marks Lane, Edinburgh EH15 2PX
M6 XBJ D. Mccarthy 7 Shenley Close, Wirral CH63 7QU
MI6 XBL D. Mooney 12 Curragh Walk, Londonderry BT48 8HX
M6 XBM B. Johnson 6 Trevor Road, Swinton, Manchester M27 0YH
M6 XBO S. Wilson 7 Dalglish Drive, Blackburn BB2 4FU
M6 XBP B. Pettit 30A Station Road, Corton, Lowestoft NR32 5BE
M6 XBQ M. Doherty 151 College Croft, Eccles, Manchester M30 0AN
M6 XBR J. Breward 399 St. Margarets Road, Isleworth TW7 7BZ
M6 XBS B. Scott 8 Chelsea Close, Tilehurst, Reading RG30 6EP
M6 XBV C. Somerville Close House, Gretton, Cheltenham GL54 5EP
M6 XBW B. Woollett 23 Kinglake House, Southall UB24FZ
M6 XBX M. Wheeler 6 Severn Road, Melksham SN12 8BQ
M6 XCA M. Jurisic Flat 7, 27 Slade Way, Mitcham CR4 2GA
M6 XCC S. Derner 14 Elmhurst Drive, Huthwaite, Sutton-in-Ashfield NG17 2NP
M6 XCF C. Fish New House Farm, Peaton, Craven Arms SY7 9DW
M6 XCJ C. Jones 8 Henry Gepp Close Adderbury, Banbury OX17 3FE
M6 XCK C. Kellers 21 Aspen House West Terrace, Folkestone CT20 1TH
M6 XCL L. Robertson Clayhill Cottage Aldham, Ipswich IP76NN
M6 XCO R. Kempton Choices Folly, Marsh Road Gedney Drove End, Spalding PE12 9PJ
M6 XCP C. Poole 15 Devon Close, Macclesfield SK10 3HB
M6 XCZ M. Hartley 24 Burnham Avenue, Bognor Regis PO21 2JU
MW6 XDA M. Herbert 6 Yew Street Taffs Well, Cardiff CF15 7PT
M6 XDB Dr D. Buttle 29 Upthorpe Drive, Wantage OX12 7DF
M6 XDG Dr G. Welch Amazonas, Sandy Lane, Hightown, Liverpool L38 3RP
M6 XDI I. Doolan-Tanner 150 Milward Road, Hastings TN34 3RT
M6 XDJ S. Scott 13 Silver Close, Harrow HA36JT
M6 XDO D. Witt 20 Foxglove Close, Newton Aycliffe DL5 4PF
M6 XDR C. Darby Flat 4 146 Newark Road, Lincoln LN5 8QF
M6 XDT I. Hulme 10 Coney Green, Bicton Heath, Shrewsbury SY3 5AP
M6 XDW D. Tams 2 Jacksons Lane, Hazel Grove, Stockport SK7 6EL
M6 XDX S. James 124 Alcock Avenue, Mansfield NG18 2NF
M6 XDZ G. Parsons 2 The Close, East Grinstead RH191DQ
M6 XEE S. Morton 77 Hepworth Road, Stanton, Bury St. Edmunds IP31 2UA
M6 XEL L. Cook Luckfield, Ellis Road Boxted, Colchester CO4 5RN
MI6 XEM A. Mccusker 41 The Granary, Waringstown, Craigavon BT66 7TG
M6 XEP D. Heath 3 Elm Road, Congleton CW12 4PR
M6 XER S. Wood 3 Lion Lane, Haslemere GU27 1JF
M6 XEU L. Seddon 53 Reedswood Crescent, Cramlington NE23 6RW
M6 XEV N. Wright 10 Olden Mead, Letchworth Garden City SG6 2SP
MI6 XEX A. Rowan-Jenkins 143 Glenkeen Avenue, Greenisland, Carrickfergus BT38 8ST
M6 XEY J. Chadwick 21 Horridge Street 21, Bury BL8 1TN
M6 XFI F. Beckitt-Marshall 13 Colegrave Street, Lincoln LN5 8DR
MW6 XFM C. King Nyth Dedwydd, Llywernog, Aberystwyth SY23 3AB
MW6 XFU C. Jenkins Flat 5, The Lawns, Usk NP15 1BA
M6 XGB G. Bogg The Shires, Main Street, Pickering YO18 7PG
M6 XGC J. Beckitt-Marshall 13 Colegrave Street, Lincoln LN5 8DR
MW6 XGD G. Jukes 52 Beach Road, Pyle, Cardiff CF33 6AS
M6 XGF S. Wellborn 36 Lidgett Park Court, Leeds LS8 1ED
MM6 XGJ G. Jamieson 6 Maryville Park, Aberdeen AB15 6DU
MI6 XGN G. Houston 51 Rockfield Heights, Connor, Ballymena BT42 3LH
M6 XGS G. Stanley 95 Old Vicarage, Westhoughton, Bolton BL5 2EG
MM6 XGT G. Taylor 21 Glenesk Avenue, Montrose DD10 9AQ
M6 XGX J. Parry 12 Kerrysdale Close Sutton, St. Helens WA9 3WA
MI6 XGZ J. Downing 15 Knockburn Gardens, Lisburn BT28 2QL
M6 XHB B. Hobbs 2 Miller Court, Bedford MK42 9PB
M6 XHF H. Beer 12 Cross Street, Northam, Bideford EX39 1BS
M6 XHI C. Smith 17 Essex Street Wash Common, Newbury RG14 6QJ
MI6 XHL S. Kerr Bt71 6Sq, Dungannon BT71 6SQ
MM6 XHU R. Giles 304 Ellon Park Collydean, Glenrothes KY7 6UY
M6 XIE S. Reveley Neuadd Wen, Pontrhydygroes, Ystrad Meurig SY25 6DQ
M6 XIP P. Armstrong 10 Shirdley Avenue, Liverpool L32 7QG
M6 XIT D. Bond 53 Cotswold Grove, St. Helens WA9 2JD
M6 XJC J. Connett 81 Holwick Close, Consett DH8 7UJ
M6 XJD J. Cook 10A East Street, Newton Abbot TQ12 1AG
MW6 XJL J. Henley Rhewin Glas, Llandysul SA44 6PR
M6 XJM J. Mcfarlane 16 Lake View Houghton Regis, Dunstable LU5 5GJ
M6 XJO J. Ochalek 21 Fairfield Road, Buxton SK17 7DN
M6 XJP C. Dennis Hillsdene, Plex Lane, Ormskirk L39 7JY
M6 XJR J. Redhead 28 Sandfields, Frodsham WA6 6PT
M6 XJS D. Washington 13 Cheddar Waye, Hayes UB4 0DZ
M6 XJW J. Wraith 6 Ashfield Road, Chippenham SN15 1QQ
MM6 XKC C. Robertson 5 Broomlands Place, Irvine KA12 0DU
MI6 XKE W. Duncan 74 Bunderg Road, Douglas Bridge, Strabane BT82 8QQ
M6 XKG K. Abel 7 Foldgate View, Ludlow SY8 1NB
M6 XKJ K. Cook 4 Chelmscote Row, Upper Wardington OX17 1SS
M6 XKM M. Koster Stalworthy Manor Farm, Suton Lane, Suton, Wymondham NR18 9JG
M6 XKN K. Northrop 200 Nevells Road, Letchworth Garden City SG6 4TZ
M6 XKT K. Todman 12 Winscombe, Bracknell RG12 8UD
M6 XKW A. Summerfield Drovers Rest, Back Lane, Cross In Hand, Heathfield TN21 0QA
M6 XKY K. Cokayne 46A Roscoe Road, Irlam, Manchester M44 6AR
M6 XLC L. Cockburn 58 Priestley Court, South Shields NE34 9NQ
M6 XLE J. Walker 22 Lower Road, Malvern WR14 4BX
M6 XLF L. Fairs 140 Manchester Road, Chapel-En-Le-Frith, High Peak SK23 9TP
M6 XLG L. Graham Yews Avenue, Barnsley S70 4BW
M6 XLN C. Beckitt-Marshall 13 Colegrave Street, Lincoln LN5 8DR
M6 XLP L. Pawson 32 Cross Street, Upton, Pontefract WF9 1EU
M6 XLR N. Sayle 6 Glen View, Wigmore, Leominster HR6 9UU
M6 XLT M. Roach 31 Brumby Wood Lane, Scunthorpe DN17 1AA
M6 XLX J. Gascoigne 1 Mill Meadow, Aylesbury HP19 8GW
M6 XMA D. Jasper 38 Ingleway Avenue, Blackpool FY3 8JJ
M6 XMB B. Wale 23 Castleton Avenue, Bournemouth BH10 7HW
M6 XMC M. Callis 1 Webb Close, Letchworth Garden City SG6 2TY
M6 XME D. Evans 15 Cider Mill Court, Hereford HR2 6RY
MI6 XMG A. Mcgarvey 66A Scaddy Road, Downpatrick BT30 9BS
M6 XMH M. Hoult 43 Hutcliffe Wood Road, Sheffield S8 0EY
M6 XMJ M. Jermy 15 Oak Tree Close Martham, Great Yarmouth NR29 4QN
M6 XMN M. O'Brien Flat 9, Neilson Court, Manchester M23 1LE
MW6 XMO J. Morris 70 Forge Road, Port Talbot SA13 1PF
MM6 XMQ L. Anderson 6 St. Keiran Crescent, Stonehaven AB39 2GQ
M6 XMR M. Robinson 59 The Avenue Kennington, Oxford OX1 5PP
M6 XMS P. Kentish 17 Rose Drive, Walsall WS8 7EB
M6 XMX D. Lyon 1 Garnsgate Road, Long Sutton, Spalding PE12 9BT
M6 XNC R. Jeffery 126 Woodham Lane, New Haw, Addlestone KT15 3NQ
M6 XNG J. Blackwell 164 York Road Haxby, York YO32 3EL
M6 XNL N. Lamerton 51 High Street, Knaphill, Woking GU21 2PX
M6 XNO J. Mirfield 14 King Edward Avenue Horsforth, Leeds LS18 4BD
M6 XNU S. Grace-Bolton 26 Fairway Road, Blackpool FY4 4AZ
M6 XNX M. Powell 14 Moneta Rise, Leighton Buzzard LU7 9SN
MI6 XOD H. Mcdowell 51 Rockfield Heights, Connor, Ballymena BT42 3LH
M6 XOJ J. Gunson 66 Milward Crescent, Hastings TN34 3RU
M6 XOR M. Hauser 23 Prince William Way Sawston, Cambridge CB22 3SZ
MI6 XOS J. Mcgoldrick 45 Stewarts Road, Dromara, Dromore BT25 2AN
MI6 XOX S. Mccartan 13 Lecale Park, Downpatrick BT30 6ST
M6 XPB M. Partner 22 Moordale Avenue, Bracknell RG42 1RT
M6 XPC P. Cox 75 Rosecroft Gardens, Swadlincote DE11 9AF
M6 XPD P. Douglas 76 Woodside Avenue, Benfleet SS7 4NY
M6 XPH R. James 159 Orton Avenue, Sutton Coldfield B76 1JN
M6 XPK P. Davies 2 Lynfords Drive, Runwell, Wickford SS11 7PP
M6 XPN D. Rutter 42 Great Coates Road, Grimsby DN34 4ND
M6 XPT T. Parfitt 5 Sheridan Road, Frimley, Camberley GU16 7DU
M6 XPW P. Williams 3 Swift Close, Cottingham HU16 4DQ
M6 XQS P. Jarvis 50 Northcott, Bracknell RG12 7WR
M6 XQX S. Lowe 14 Windmill Rise, York YO26 4TX
MI6 XRC D. Mclaughlin 134 Castleroe Road, Coleraine BT51 3RW
M6 XRD R. Daniell Wits End, Barley Mow Lane, Woking GU21 2HY
M6 XRF R. Ford 13 Green Street, Hereford HR1 2QG
MM6 XRI R. Irvine 9 Pearce Grove, Edinburgh EH12 8SP
M6 XRL R. Lamerton 51 High Street Knaphill, Woking GU21 2PX
MW6 XRO A. Howard 4 Fern Rise Neyland, Milford Haven SA73 1RA
MM6 XRS C. Rankin 2 Thomson Green, Livingston EH54 8TA
M6 XRT Dr R. Tofts Elmcroft, Redhill Road, Ross-on-Wye HR9 5AU
M6 XSD C. Catlin 27 Main Street, Frizington CA26 3SA
M6 XSF S. Dockray 54 Kelsick Park, Seaton, Workington CA14 1PY
MW6 XSI N. Jones 30 Cardiff Road, Pwllheli LL53 5NU
M6 XSN S. Bolton The Conifers, Methwold Road Methwold Hythe, Thetford IP26 4QW
M6 XSO S. O'Brien 15 Fair Ave, Bedlington NE22 7BR
M6 XSR S. Mansfield The Old Piggery, Ham Lane, Compton Dundon, Somerton TA11 6PQ
M6 XSS A. Corless 4 Mayfield Road, Bentham, Lancaster LA2 7LP
M6 XST S. Harper 61 Elmhurst, Tadley RG263LF
M6 XSZ B. Staszewski 10 Priory Road, Stanford-le-Hope SS17 7EW
M6 XTA T. Windle 48 Sheridan Road, South Shields NE34 9JJ
M6 XTB T. Beale 28 Redhouse Way, Swindon SN25 2AZ
M6 XTD S. Brookes 71 Friends Road, Norwich NR5 8HW
M6 XTF S. Ganti 59 Fitzroy Road, Blackpool FY2 0RJ
M6 XTK C. Walsh 72 William Street, Totterdown, Bristol BS3 4TX
M6 XTL D. Baker 32 Richardson Crescent, Hethersett, Norwich NR9 3HS
M6 XTM T. Mercer Ralph Allen House, Railway Place, Bath BA1 1SR
M6 XTR M. Moggeridge 19 Sherwood Close, Fetcham, Leatherhead KT22 9QT
M6 XTU S. Baker 61 Victoria Road, Coleford GL16 8DS
M6 XUI J. Ribbands Dyson'S Farm, Long Row, Tibenham NR16 1PD
M6 XUM W. Webb 84 Bruce Street, Swindon SN2 2EN
M6 XUU P. Sanders Alresford Road, Wivenhoe CO7 9JX
M6 XUV S. Ellinor 53 Hillside, Banstead SM7 1HG
M6 XUX S. Andrews 9 Southernhay Crescent, Bristol BS8 4TT
M6 XVB J. Watson Croft Holm New Market Street, Buxton SK17 6LP
M6 XVC N. Booth 30 Kingfisher Drive, Cheltenham GL51 0WN
M6 XVJ C. Wallwork 8 Daneswood Close, Whitworth, Rochdale OL12 8UX
MJ6 XVL J. Bertram Roz-Den, La Rue De La Guilleaumerie St. Saviour, Jersey JE2 7HQ
M6 XVM E. Smith Flat 29 Heron Court, Emscote Drive, , Sutton Coldfield B73 5NF
M6 XVS S. Cooksey 22 Fitzgerald Place, Brierley Hill DY5 2SZ
MW6 XVT C. Williams 1 South View, Freeholdland Road, Pontypool NP4 8LL
M6 XVX M. Lawrence 16 Timson Close, Market Harborough LE16 7UU

M6 XWB P. Forrest 6 Scarisbrick Place, Liverpool L11 7DJ
M6 XWD N. Bishop 15 Katrine Road, Stourport-on-Severn DY13 8QB
M6 XWG P. Matthewson 30 Harvey Road, Rugeley WS15 4HF
M6 XWR C. Wrobel 33 Harsnett Road, Colchester CO1 2HS
M6 XWS R. Molli Boulock 3 Bray Lane, Telford TF3 5HH
M6 XXB B. Bentham 60 Chapel Lane, Burtonwood, Warrington WA5 4PQ
MW6 XXC D. Clark 137 Llanedeyrn Road, Penylan, Cardiff CF23 9DW
M6 XXD P. Jones 222 York Road, Shrewsbury SY13QE
M6 XXH C. Troughton 20 Oakleigh Road, Uxbridge UB10 9EL
M6 XXI B. Littlechild 85 Long Road, Canvey Island SS8 0JB
M6 XXS S. Harvey Penleigh, Cherry Cross Totnes Down Hill, Totnes TQ9 5EU
MM6 XXV J. Kerr 2 Burngrange Cottages, West Calder EH55 8EW
M6 XXZ T. O'Sullivan Onibury Hedsor Road, Bourne End SL8 5DH
M6 XYH D. Hatch 18 Victory Park Road, Addlestone KT15 2AX
M6 XYJ S. Jefferies 28 Curlew Road Abbeydale, Gloucester GL4 4TF
M6 XYL C. Harris 50 The Warren, Horsham St Faith, Norwich NR10 3JT
M6 XYY P. Liu International Hall, University Of London, London WC1N 1AS
M6 XZK E. Arthur 7 Magpie Close, Coulsdon CR5 1AT
M6 XZY A. Mclean 141 Crawford Avenue, Tyldesley, Manchester M29 8LS
M6 YAD G. Boam 36 Merlin Way, Woodville, Swadlincote DE11 7QU
MW6 YAE D. Bannister 45 Queens Drive, Llantwit Fardre, Pontypridd CF38 2NT
M6 YAF A. Garn 5 Bassett Street, Walsall WS2 9PZ
MW6 YAG A. Graham 2 Heol Undeb, Beddau, Pontypridd CF38 2LB
M6 YAH C. Bowden 12 Butts Green, Stoke-on-Trent ST2 8EH
M6 YAJ R. Harrison 268 Central Drive, Blackpool FY1 5JB
MI6 YAM J. Wilkinson 11 Fairview Park, Dromore BT25 1PN
M6 YAN Y. Watkins 1 St. Saviour Close, Colchester CO4 0PW
MM6 YAP C. Phillips 8 The Square, Newtongrange, Dalkeith EH22 4QD
M6 YAR R. Parker 4 Hall Yard Metheringham, Lincoln LN4 3BY
M6 YAS P. Savage 53 Ashby Road, Braunston, Daventry NN11 7HE
M6 YAT L. Jones 8 Oakbrook Close, Ewyas Harold, Hereford HR2 0NX
M6 YAW A. Nicholls 2 Park End, Forsbrook, Stoke-on-Trent ST11 9DR
M6 YAY C. Jack 18 Swan Close, Martlesham Heath, Ipswich IP5 3SD
M6 YBA M. Adams 18 Vanguard Court, Sleaford NG34 7WL
M6 YBB L. West 153 Chartist House, Mount Street, Hyde SK14 1RP
M6 YBC J. Bernard-Cooper 12 River Court, Green Lane, Durham DH1 3UA
M6 YBD R. Cook 8 Kingsdown Way New Marske, Redcar TS11 8JJ
MD6 YBE H. Jones 77 Royal Park, Ramsey IM8 3UH Isle of Man
MI6 YBH J. Mccourt 70A Sessiagh Scott Road Rock, Dungannon BT70 3JU
M6 YBK B. Watson 173 Churchfield Lane, Darton, Barnsley S75 5EA
M6 YBT M. Harrison 2 Rosemount Court, Holly Bank Road, York YO24 4EG
M6 YBV B. Young 11 Gainsborough Avenue, Washington NE38 7EF
MW6 YBZ P. Smith 29 Heol Cwarrel Clark, Caerphilly CF83 2NE
M6 YCA A. Young 53 Arnold Grove, Newcastle ST5 8LD
MM6 YCB J. Williamson Clunie Cottage, Tullibardine Road, Auchterarder PH3 1LX
M6 YCG C. Gouveia 86A Spareacre Lane, Eynsham, Witney OX29 4NP
M6 YCH C. Hohmann 13 Woodbine Terrace Bensham, Gateshead NE8 1RU
MM6 YCJ Prof. C. Jones 11B Ettrick Road, Edinburgh EH10 5BJ
M6 YCP C. Pearce 9 Swilgate Road, Tewkesbury GL20 5PQ
M6 YCR W. Jones 50 Bridge Place, Croydon CR0 2BB
M6 YCT D. Ashton-Hilton 14 Weetwood Road, Congresbury, Bristol BS49 5BN
M6 YDA K. Abel 7 Foldgate View, Ludlow SY8 1NB
M6 YDB D. Bower 4 Winsford Road, Sheffield S6 1HT
M6 YDC D. Cook 14 Alberta Avenue Selston, Nottingham NG16 6GN
M6 YDD R. Howson 37 Ghyllroyd Drive, Birkenshaw, Bradford BD11 2ET
M6 YDF S. Gains 14 Ainslie Street, Grimsby DN32 0LU
M6 YDG D. Lyall Royal Hospital Road, London SW3 4SR
M6 YDN P. Norman 11 Windmill Road, Irthlingborough, Wellingborough NN9 5RJ
MW6 YDP P. Warburton 71 Richards Terrace, Cardiff CF24 1RW
M6 YDV D. Powell 29 Kelston View, Bath BA2 1NW
M6 YEB M. Clayton 2 West Street, Wroxall, Ventnor PO38 3BU
M6 YEF S. Shore 8 Bromley Road, Colchester CO4 3JE
M6 YEG J. Fenton 4 Forest Hills, Newport PO30 5NG
M6 YEH A. Bate 16 East Avenue, Heald Green, Cheadle SK8 3DL
M6 YEL J. Callaghan 46 Highfields, Great Yeldham, Halstead CO9 4QQ
M6 YEO A. Yeo Bridge House, Tresillian, Truro TR2 4AU
M6 YEQ G. Browning 6 The Drive, Rickmansworth WD3 4EB
M6 YES S. Crabb 22 Mary Warner Road Ardleigh, Colchester CO7 7RP
M6 YEU M. Read 3613 Clary Ave, Fort Worth 76111 United States
M6 YEY T. Windass 58 Nicholas Gardens, High Wycombe HP13 6JG
MM6 YEZ A. Ewing Kildonan House, Caerlaverock Farm, Crieff PH5 2BD
M6 YFN A. Boubaker Citee Sonboula N5, Grombalia 8030 Tunisia
M6 YFT M. Hubbard Millstone, Highfield Road, Truro TR4 8DZ
M6 YFU M. Coats 3 Broadleaf Close, Exeter EX1 3XA
M6 YFX I. Cox 2 Welby Close Tadpole Garden Village, Swindon SN25 2RJ
M6 YFZ N. Mason 6 Maldern Avenue, Poulton-le-Fylde FY6 7TL
M6 YGD D. Young 35 Limber Hill, Cheltenham GL50 4RJ
M6 YGL G. Hansell 196 The Downs, Harlow CM20 3RH
M6 YGN G. Nicholls 65 Boleyn Way New Barnet, Barnet EN5 5LH
M6 YGR G. Robertson 12 Chester Close Ince, Wigan WN3 4JP
MM6 YGT G. Christie 10 Calcots Crescent, Elgin IV30 6GL
M6 YHM H. Marchington 30 Warwick Avenue Golcar, Huddersfield HD7 4BX
MM6 YHO D. Chisholm 111 Philips Wynd, Hamilton ML3 8PH
M6 YHQ P. May 1 Ballowall Terrace St. Just, Penzance TR19 7BG
MW6 YHR R. Boulton 41 St. Andrews Drive, Libanus Fields, Blackwood NP12 2ET
M6 YHS D. Stevens 67 New Road, East Hagbourne, Didcot OX11 9JX
M6 YHV C. Strange 18 Glenwood Court, Farnborough GU14 7TB
M6 YHW R. Lloyd 18 Newbury Grove Blurton, Stoke-on-Trent ST3 3DD
MW6 YIG M. Graham 60 Heol Seward Beddau, Pontypridd CF38 2SR
M6 YIK F. Nisbet 3 Fern Crescent, Seaham SR7 7UJ
M6 YIN D. Horton 27 Fishers Lane, Wirral CH61 9NT
M6 YJB J. Wilson 55 Chestnut Avenue Kirkby-In-Ashfield, Nottingham NG17 8BA
M6 YJD D. Sanders 1 Berkeley Grange, Carlisle CA2 7PN
M6 YJH J. Harman 1 Vernons Road, Newick BN84NF
M6 YJK J. Williams 56 Hillingdon Rise, Sevenoaks TN13 3SB
M6 YKC P. Purnell 89 St. Aubyns, Goldsithney, Penzance TR20 9LS
MW6 YKI K. Griffiths 11 Hatfield Meadow, Knighton LD7 1RY
MW6 YKS B. Roberts 6 Trem Y Moelwyn, Tanygrisiau, Blaenau Ffestiniog LL41 3SS
MM6 YLA W. Lawson 60 Inglis Avenue, Port Seton EH32 0AQ
M6 YLC C. Rawlinson Westfield Farm, Risden Lane, Cranbrook TN18 5DU
M6 YLD D. Braithwaite 300 Arnold Estate, Druid Street, London SE1 2XN
M6 YLE A. Kelly 62 Old Warren, Taverham, Norwich NR8 6GA
MI6 YLG G. Mccormick 24 Warren Park Drive, Lisburn BT28 1HF
M6 YLH H. Earnshaw 16 Estill Close, Cayton, Scarborough YO11 3TA
M6 YLI E. Fisher 1 Eaton Place, Kingswinford DY6 8JU
M6 YLJ R. Whitehead 29 Coulsons Road, Bristol BS14 0NN
MM6 YLO G. Davis 2 Virkie Cottages, Virkie, Shetland ZE3 9JS
M6 YLP R. Campbell-Black 10 Wren Close, Towcester NN12 6RD
M6 YLR D. Walker 290 Shannon Road, Hull HU8 9RY
M6 YLS S. Woolley Babbington House, Church Road, Crowborough TN6 3LG
MI6 YLT S. Mccormick 46 Lany Road, Moira, Craigavon BT67 0NZ
M6 YLW T. Stromberg Knowle Cottage, 108 Bath Road, Knowle, Bawdrip TA7 8PJ
M6 YLY F. Titeiu Hillary Road, 18, Kidsgrove ST74DN
M6 YMA A. Andrew 80 Hamble Drive, Abingdon OX14 3TE
M6 YMB L. Latimer 40 Petersham Road, Long Eaton NG10 4DD
M6 YME M. Smith Murreagh, Waterville V23 XY32 Ireland
M6 YMF K. Mbachu 46 Roseberry Gardens, Upminster RM14 1NW
M6 YMM T. Daniel 28 Polefield Circle Prestwich, Manchester M25 2WP
M6 YMN M. Lee 4 Addison Place, Bilston WV14 7BD
M6 YMR L. Humphreys Flat 38, Wimborne House, 2 Stokewood Road, Bournemouth BH3 7NP
M6 YMW M. Williamson 5 John F Kennedy Walk, Tipton DY4 0SF
M6 YMX M. Allen 55 The Meadows, Skegness PE25 2JA
M6 YMY K. Nelhams-Wright The Gatehouse, Frogmore Lane, Aylesbury HP18 9DZ
M6 YNA B. Wright 5 Ridge Road, Kingswinford DY6 9RB
MM6 YND A. Gillies 43 Oak Wynd, Cambuslang, Glasgow G72 7GS
MM6 YNG C. Young 0/2 27 Cardwell Road, Gourock PA19 1UW
M6 YNK S. Higton 25 Shady Grove Hilton, Derby DE65 5FX
M6 YNL O. Price 5 Redicliff Way Sw, Redcliff T0J Canada
M6 YNN A. Green 10 Coopers Hill, Eastbourne BN20 9HX
M6 YNT A. Ashby 5 New Street, Osbournby, Sleaford NG34 0DL
M6 YNY K. Edwards 352 Plessey Road, Blyth NE24 3RD
M6 YOB C. Burns 29 Foxglove Fold, Castleford WF10 5UJ
M6 YOG S. Lee 7 Lilac Avenue, Lower Quinton, Stratford upon Avon CV37 8US
M6 YON A. Parkhouse 3 St. Margarets Avenue, Ashford TW15 1DR
M6 YOO M. Akena 6 Concord Terrace, Coles Crescent, Harrow HA2 0HJ
M6 YOS L. Shackleton 1 Mayfield Road, Wooburn Green, High Wycombe HP10 0HG
M6 YOT C. Watson 51 Lodge Way, , Weymouth DT4 9UU
M6 YOU B. Roots 212 Hunt Road, Tonbridge TN10 4BJ
M6 YOW E. Murchie 52 Kingswood Road, Manchester M14 6SA
MM6 YOY J. Moir 41 Brisbane Terrace, East Kilbride, Glasgow G75 8DL
M6 YOZ S. Green 28 Courtney Drive, Sunderland SR3 1JR
M6 YPD D. Hawkes 62 Park Street, Wollaston, Wellingborough NN29 7RR
M6 YPE J. Pye 17 Langden Brook Mews, Morecambe LA3 3SN
M6 YPG G. Evans 19 Windsor Street, Thurnscoe S63 0HB
MM6 YPN G. Lailvaux 4 Oxenfoord Avenue, Pathhead EH37 5QD
M6 YPS P. Smith Meadows, Menthorpe Lane, North Duffield, Selby YO8 5RL
M6 YPW Dr P. Woodfin 14 Broomcroft Road, Bognor Regis PO22 7NJ
M6 YPX A. Gillard 4 Horton Avenue, Thame OX9 3NJ
MI6 YPY G. Kelly 50 Tullyrain Road, Beagh, Enniskillen BT94 2AS
M6 YPZ M. Meyer 22 Orchard Grove, Newton Abbot TQ12 1FZ
MM6 YQP A. Sharples 49 Ramsay Road, Banchory AB31 5TS
M6 YRB R. Braisby Flat 3 Rose House Victoria Road, Woking GU21 2AT
M6 YRC R. Curtis 3 Coniston Close Peterlee Co.Durham, Peterlee Co Durham SR85LW
M6 YRD S. Jones 19 Tipnall Road, Castle Donington, Derby DE74 2JY
M6 YRF Dr A. Saje 72 Bedworth Road, Bulkington, Bedworth CV12 9LL
M6 YRL R. Yerrell 88 Woodcock Road, Norwich NR3 3TD
MM6 YRO D. Kennedy 7 Burns Terrace, Cowie, Stirling FK7 7BS
M6 YRS J. Stallard 6 Richmond Crescent, Leominster HR6 8RX
M6 YRW R. Weber 11 Harewood Road, Calstock PL18 9QN
M6 YRY R. Lambert 7 Templeway, Lydney GL15 5HU
MW6 YSA B. Paterson 35 Ynys Yr Afon, Neath SA114BP
M6 YSB P. Bunting 29 Marion Avenue, Alverthorpe, Wakefield WF2 0BJ
M6 YSD S. Daniels 48 Gason Hill Road, Tidworth SP9 7JX
M6 YSF S. Fletcher Beverley Hotel, 55 Old Brumby Street, Scunthorpe DN16 2AJ
M6 YSK A. Bragg Buena Vista Low Moresby, Whitehaven CA28 6RR
M6 YSM M. Young 39 Hollins Lane, Sheffield S6 5GQ
MM6 YST A. Thomson 15 Waverley Road, Nairn IV12 4RH
M6 YSU J. Wright Paddock Views, Main Street, Leicester LE7 9YB
MI6 YSW D. Lappin 46 Grange Road, Kilmore, Armagh BT61 8NX
MJ6 YSY G. Antcliff Andylor Cottage, 6 Dicq Road, St. Saviour JE2 7PD Jersey
M6 YTB R. Thomas 9 Spa View Terrace, Sheffield S12 4HG
M6 YTC T. Campbell 3 Hillside Close, Helsby, Helsby WA69LB
MM6 YTD T. Davidson 141 Huron Avenue, Livingston EH54 6LQ
M6 YTF T. Fox 28 Parishes Mead, Stevenage, Stevenege SG2 9QD
M6 YTI H. Cropp 26 Coventry Street, Brighton BN1 5PQ
M6 YTL J. Watkins The Hundred, The Hundred Middleton On The Hill, Leominster HR6 0HZ
M6 YTM N. Allison 2 Gables Cottages Argos Hill, Rotherfield, Crowborough TN6 3QH
M6 YTS J. Stacey 12 Kimbridge Road, East Wittering, Chichester PO20 8PE
M6 YTU A. Coward 11 Oaklands, Ardingly, Haywards Heath RH17 6UE
M6 YTX H. Taylor 25 Northolme Avenue, Nottingham NG6 9AP
M6 YUG G. Harmath 101 Whitton Avenue East, Greenford UB6 0QE
MM6 YUI I. Page 2F1, 18 Milton Street, Edinburgh EH8 8HF
MM6 YUJ C. Duncan 36 Commore Drive, Glasgow G13 3TT
M6 YUK D. Barker 12 The Weavers, Denstone, Uttoxeter ST14 5DP
M6 YUL G. Szabo 6 Stanway Road, Gloucester GL4 4RE
M6 YUM T. Benson 3 Tey Road, Coggeshall, Colchester CO6 1SY
MM6 YUP B. Rankin 5 Glen Affric Court, Dumbarton G82 2BN
M6 YVN A. Lowe 17 Burnside, Telford TF3 1SS
MM6 YVO Y. Oldfield 1 Gallanach, Lochgair, Lochgilphead PA31 8SD
M6 YVR S. Preece 14 Bettespol Meadows, Redbourn, St. Albans AL3 7EW
M6 YVT P. Van Staveren 14 Fortune Green Road Flat 3, London NW6 1UE
M6 YWA M. Smith 51 Town Lane, Shepton Mallet BA4 5LX
M6 YWG R. Holland 431 Beverley Road, Newland, Hull HU5 1LX
M6 YWO M. Bruyneel 14 Riversmead, St. Neots PE19 1HA

M6 YWX D. Hartnell 3 Fairmead, Sidmouth EX10 9SU
M6 YXD D. Yeaman Flat 3, 62 Madeley Road, Ealing W5 2LU
M6 YXZ C. Cutting 5A Clifton Mansions, Clifton Road, Folkestone CT20 2EJ
MM6 YYB I. Nicolson 1 Gullane Place, Dundee DD2 3BF
M6 YYD A. Edge 79 Waterside Drive, Stoke-on-Trent ST3 3NU
M6 YYG D. Harwood 32 John Reid Road, South Shields NE34 9EB
M6 YYK S. Hoyle 20 Brandsby Grove, Huntington, York YO31 9HL
M6 YYL R. Ashton-Cox 10 Princes Avenue, , Benfleet SS7 3AZ
M6 YYM Viscount M. Thomas 13 Christchurch Gardens, Reading RG2 7AH
M6 YYT G. Finney 121 School Lane, Caverswall, Stoke-on-Trent ST11 9EN
M6 YYU J. Underwood 15 Fawsley Road, Northampton NN4 8NR
M6 YYY S. Christofi 19 Kingsland Avenue, Northampton NN2 7PP
MW6 YZF G. Jones 31 Liverpool Road, Buckley CH7 3LH
M6 YZH B. Yu 210 Ramsgate Road, Broadstairs CT10 2EW
M6 YZM G. Brown 134 Skipper Way, Lee-on-the-Solent PO13 8HD
M6 YZO C. Gregory 28 Rynal Street, Evesham WR11 4QA
M6 YZR D. King 170 Stepney Road, Scarborough YO12 5NH
M6 YZX S. Forber 32 Larch Avenue, Newton-le-Willows WA12 8JF
M6 ZAC M. Holmes 6 Wells Court, Saxilby, Lincoln LN1 2GY
M6 ZAD A. Ghafoor 20 Maureen Avenue, Manchester M8 5AR
M6 ZAF A. Barter 4 Blackberry Way Kingsteignton, Newton Abbot TQ12 3QX
M6 ZAH T. Hayden The Ferns, Yarmouth Road, North Walsham NR28 9LX
MW6 ZAN P. Smith 3, Digby Street, Barry CF63 4NP
M6 ZAQ Z. Gale 7 Somerset Road Droylsden, Manchester M43 7PF
M6 ZAS N. Zulkfli Harrogate Ladies' College, Clarence Drive, Harrogate HG1 2QG
M6 ZAU C. Sommers 11 Westlands Grove, Fareham PO16 9AA
M6 ZAV B. Appleby 1E Wilfred Owen Close, London SW19 8SW
M6 ZAW D. Packham 5 The Avenue, Yate, Bristol BS37 4PN
M6 ZAX M. Southgate 107 Englands Lane, Loughton IG10 2QL
M6 ZAY A. Carter 19 Burn Lane, Newton Aycliffe DL5 4HX
MM6 ZAZ G. Fyvie Bridgefoot Of Gaval, Mintlaw, Peterhead AB42 4HA
M6 ZBB Dr M. Palmer 116 Claverham Road, Yatton, Bristol BS49 4LE
M6 ZBD I. Painter 47 Longmead Drive, Nottingham NG5 6DP
MW6 ZBE S. Berrow 40 Priorsgate Oakdale, Blackwood NP12 0EL
MM6 ZBG C. Halcrow Hellia, Cunningsburgh, Shetland ZE2 9HG
M6 ZBL Z. Raybould 31 Meadow Road, Quinton, Birmingham B32 1AY
M6 ZBM C. Abbott 38 Foxcover, Linton Colliery, Morpeth NE61 5SR
M6 ZBQ A. Kerr 9 Martindale Way, Sawston, Cambridge CB22 3BT
MW6 ZBR B. Morgan 1 Y Dolydd, Aberdare CF44 8EX
M6 ZBS K. Florence 30 Lancaster Gardens, Ealing, London W13 9JY
M6 ZBT B. Tomlinson 7 Springwell Close, Crewe CW2 6TX
M6 ZBW B. Waddingham 11 Chandlers Court, Tidworth SP9 7FN
M6 ZCA C. Archer 31 Stoney Bank Drive Kiveton Park, Sheffield S26 6SJ
M6 ZCB C. Burdett 44 Emmett Carr Lane, Renishaw, Sheffield S21 3UL
MM6 ZCD C. Docherty 23 The Maltings, Haddington EH41 4EF
M6 ZCL C. Lynch 26 Warwick Street, Rochdale OL129SW
M6 ZCM C. May 18 Ravenscroft, Salisbury SP2 8DL
M6 ZCP C. Palawinna 3 Stirling Court Road, Burgess Hill RH15 0PS
M6 ZCR C. Rhodes 93 Southwark Close, Stevenage SG1 4PH
MW6 ZCT C. Tsoi Fairview, Llanbadarn Fawr, Aberystwyth SY23 3QU
M6 ZDA D. Harris 27 Ashley Road, Poole BH14 9BS
M6 ZDB D. Brownsea 47 Southill Road, Bournemouth BH9 1SH
M6 ZDC Dr D. Cooke Apartment 9, 27 Sheldon Square, London W2 6DW
M6 ZDF D. Forbes Flat 3, Arundel Court 1 Cherrywood Drive, London SW15 6DS
MM6 ZDG D. Gillies 43 Oak Wynd, Cambuslang, Glasgow G72 7GS
M6 ZDL A. Abson 117 Lysander Road, Rubery, Birmingham B45 0EN
MM6 ZDW D. Woodford Nordkette, Evnabrek, Levenwick, Shetland ZE2 9GY
MM6 ZDY D. Young Fada Fuireach, Erbusaig, Kyle IV40 8BB
M6 ZEA C. Scott 40 The Close, Skipton BD23 2BZ
M6 ZEB A. Hulme 6 Asmall Close, Ormskirk L39 3PX
MD6 ZEE C. Glaister Balleigh Villa, Jurby Road, Isle of Man IM8 3NZ
M6 ZEF J. Ferry 7, Pangbourne Road Thurnscoe, Rotherham S63 0LQ
M6 ZEJ M. Jeffrey 9 Stoney Lands, Plymouth PL125DF
M6 ZEK E. Short 2 Richmond Street, Kings Sutton, Banbury OX17 3RS
M6 ZEL S. King 105 Bluebell Street, Plymouth PL6 8EQ
M6 ZEN J. Charters-Reid Woodlands Farm, Flaxton, York YO60 7RJ
M6 ZEP S. Redford 78 Boyds Walk, Dukinfield SK16 4AU
M6 ZES G. Dimitrov Flat 6, Blenholme Court, Hampton TW12 2BL
M6 ZET A. Balmer 11 China Street, Darlington DL3 0EJ
M6 ZEW W. Woodhead 1 Hayle Road, Oldham OL1 4NP
M6 ZFE J. Mcinnes-Boylan 10 Faith Street, Leigh WN7 4TS
MM6 ZFG S. Street Tangaroa, Fairfield Gardens, Kilcreggan, Kilcreggan G84 0HS
M6 ZGR C. Panaitescu 131 Stafford Road, Croydon CR0 4NN
MM6 ZGS G. Sneddon 26 Linkwood Road, Airdrie ML6 6GP
M6 ZGY T. Talbot-Humphries 22 Vicar Street, Wednesbury WS10 9HF
M6 ZGZ N. Glazzard Snitton Gate Cottage, Snitton, Ludlow SY8 3JX
M6 ZIA Z. Sati 27 Hanover Road, London SW19 1EB
M6 ZIB S. Barker-Mawjee 23 Raleigh Road, London N8 0JB
M6 ZIP A. Bradley Flat 3, 56 Chase Green Avenue, Enfield EN2 8EN
M6 ZIY Dr J. Medland 34 Allderidge Avenue, Hull HU5 4EQ
MW6 ZIZ A. Morgan 2 Parc Cambria, Old Colwyn, Colwyn Bay LL29 9AJ
M6 ZJB J. Brownsea 47 Southill Road, Bournemouth BH9 1SH
M6 ZJH J. Hughes Herons Way, Munslow, Craven Arms SY7 9ET
M6 ZJP J. Povall Elsich Barn Farm, Seifton, Craven Arms SY7 9LF
M6 ZJT T. Heyes The Coach House, Mossley Hall, Congleton CW12 3LZ
M6 ZJW Z. Wilcoxen 5 Douglas Lane, Wraysbury, Staines-upon-Thames TW19 5NF
M6 ZKA D. Owings 11 Thingwall Road East, Thingwall, Wirral CH61 3UY
MD6 ZKK A. Kissack 6 Falcon Cliff Court, Douglas, Isle of Man IM2 4AQ
M6 ZLA Z. Lucas 31 Lilian Close, Norwich NR6 6RZ
M6 ZLC S. Rice 30 Oveton Way, Bookham, Leatherhead KT23 4ND
M6 ZLD P. Frost 26 Hollies Court, Britannia Road, Banbury OX16 5DR
M6 ZLL T. Wiggins 158 Prince Charles Avenue, Derby DE22 4LQ
M6 ZLM M. Swindells 35 Rivington Street, St. Helens WA10 4BL
M6 ZLN Z. Newell 15 The Grove, Luton LU1 5PE
M6 ZLP Z. Pepper 2 Greenways, Walton On The Hill, Tadworth KT20 7QE
M6 ZLR A. Zeller Flat 1, 57 Chalk Hill, Watford WD19 4DA
M6 ZMC M. Clayton11 64 Westwood Green, , Cookham SL6 9DE
M6 ZMI T. Kelly 2 Weaver House, Chester Road, Runcorn WA7 3EG
M6 ZMR M. Richards 33 Daleside, Dewsbury WF12 0PJ
M6 ZNF N. Farrington The Old Hall, Main Street, Elmley Castle, Pershore WR10 3HS
M6 ZNZ Dr G. Richardson Berwick Cottage, Bailes Lane, Guildford GU3 2AX
M6 ZOC M. Cozens 17 Ash Grove, Norwich NR3 4BE
MW6 ZOD C. Williams 1 Lawrence Terrace, Llanelli SA15 1SW
M6 ZOG S. Littlechild 539 Tyburn Road, Birmingham B24 9RX
MW6 ZOL R. Bowen 25 Maendu Terrace, Brecon LD3 9HH
M6 ZOM S. Corbishley 86 Boundary Lane, Congleton CW12 3JA
M6 ZOO N. Weale Flat 85, Redbridge Tower, Southampton SO16 9AW
M6 ZPB P. Bannon Rose House, 73 London Road, Worcester WR5 2DU
M6 ZPE M. Hills 19 Wilkes Road, Broadstairs CT10 2HL
M6 ZPH C. Ward 1 Granary Close, Codford, Warminster BA12 0PR
M6 ZPL R. Manning 21 Whitethorn Way, Oxford OX4 6ER
M6 ZPS Z. Svanda 5 Red River Country Park, Hullbridge, Hockley SS56EP
M6 ZPT A. Pattison-Turner 47 Jefferson Way, Bannerbrook Grange, Coventry CV4 9AN
M6 ZPY R. Pyner 1 Avon Court, 63 Shakespeare Road, Bedford MK40 2DS
M6 ZPZ P. Zhang Harrogate Ladies' College, Clarence Drive, Harrogate HG1 2QG
M6 ZRG G. Reynolds 43 Orchard Drive, Watford WD17 3DX
M6 ZRJ S. Rogers 28 Damian Way, Hassocks BN6 8BJ
M6 ZRL G. Lockett 15 Deepdale Street, Hetton-Le-Hole, Houghton le Spring DH5 0DQ
M6 ZRO M. Brough 40 Denby Grange, Harlow CM17 9PZ
M6 ZRT T. Cooper 9 Websters Close, Shepshed, Loughborough LE12 9AT
MM6 ZRX K. Scollay Nirvana, Orphir, Orkney KW17 2RB
M6 ZRZ D. Gill 60 Woodcock Court, Three Mile Cross RG7 1BZ
M6 ZSA S. Evans Le3 0Ea, Leicester LE3 0EA
M6 ZSB S. Bache 62 Whittingham Road, Halesowen B63 3TP
M6 ZSD S. Adams 31 Northway, Dudley DY3 3PH
M6 ZSE P. Holmes 53 Bishops Hull Road, Bishops Hull, Taunton TA1 5EP
M6 ZSH M. Gupta Flat 26, Bassett Court, Southampton SO16 7DR
M6 ZSK S. Kneller 12A Richard Street, Crewe CW1 3AF
M6 ZST S. Harris 61 Monks Park Avenue, Bristol BS7 0UA
M6 ZSV S. Trevor 8 Aldin Grange Terrace, Bearpark, Durham DH7 7AN
M6 ZTA H. Glendinning 9 Spinners Avenue, Wakefield WF1 3QD
M6 ZTB R. Janezko 10A Lens Road, Allestree, Derby DE22 2NB
M6 ZTC S. Broom 128 Springhill Road, Wolverhampton WV11 3AQ
M6 ZTD A. Maiden 79 Green End Road, Manchester M19 1LE
MI6 ZTM M. Meagher 42 Mourne View Park, C.Down, Newry BT35 6BZ
M6 ZTO K. Balls 7 Rowan Close, Holbeach, Spalding PE12 7BT
M6 ZTP E. Pritchard Lilliput, Doctors Commons Road, Berkhamsted HP4 3DR
M6 ZTS G. Harris Sunnyside Lodge, Mongeham Road, Deal CT14 8JW
M6 ZUB R. Landragin 101 Linden Gardens Enfield En1 4Dy United Kingd101, England EN1 4DY
M6 ZUF P. Norman Four Acres Bungalow Farm, Winwick Gated Road West Haddon, Northampton NN6 7BH
M6 ZUT M. Beyoglu Flat 22, Globe House, 30 Southall Street, Manchester M3 1LP
MM6 ZUY J. Woods 33A Dalrymple Street, Girvan KA26 9EU
M6 ZVD M. Ratcliffe 30 Newton Cross Road, Newton In Furness, Barrow-in-Furness LA13 0NB
M6 ZVL V. Lynch 16 Okehampton Crescent, Sale M33 5HR
M6 ZWB M. Wilson 10 Willow Lane, Langar, Nottingham NG13 9HL
M6 ZWP J. Scarcliffe 17 Fillingfir Drive, Leeds LS16 5EG
M6 ZWT A. Gow Flat 203, Viotti Heights Sandy Hill Road, London SE18 6PA
M6 ZXB P. Holloway Flat 26, Swan House, Romford RM7 8AZ
M6 ZXC T. Hunter 5 Regal Court, Dewsbury WF12 7DE
M6 ZXI J. Clark 54 Coleshill Place, Bradwell Common, Milton Keynes MK13 8DP
M6 ZXL C. Cox 47 Thurlstone Road, Penistone, Sheffield S36 9EF
MW6 ZXO R. Zlotnicki Gwynfryn, Goginan, Aberystwyth SY23 3PD
M6 ZXQ A. Cockle 32 Old Coach Road, Playing Place, Truro TR3 6ES
M6 ZXR D. Thomas 10 Priory Drove, Great Cressingham, Thetford IP25 6NJ
MI6 ZXT C. Mckinley 36 Grangewood Drive, Londonderry BT47 5WN
M6 ZXZ J. Richards 41 Mount Pleasure, Camborne TR14 7RR
M6 ZYE G. Winnett 24 Underleys, Beer, Seaton EX12 3LT
M6 ZYG C. Richardson Heathercroft, Kirkby Mills, Kirkbymoorside, York YO62 6NN
M6 ZYH Z. Hill 101 Abbeydale Road, Sheffield S7 1FE
M6 ZYK E. Matwiejczyk 3 Tippett Avenue, Swindon SN25 2GQ
MW6 ZYQ J. Moorhouse 33 Goylands Close, Llandrindod Wells LD1 5RB
M6 ZYX W. Coburn 42 Hinton Wood Avenue, Christchurch BH23 5AH
MM6 ZYZ R. Mcglynn 35 Barrie Quadrant, Clydebank G81 3EH
M6 ZZA J. Isles 128 Ditton Lane, Fen Ditton, Cambridge CB5 8SS
M6 ZZB S. Bird Milestone Cottage, Station Road Kimberley, Wymondham NR18 9HQ
M6 ZZC S. Alexander 13 Padgate, Thorpe End, Norwich NR13 5DG
M6 ZZD J. Nixon East View, Cloudside, Congleton CW12 3QG
M6 ZZE C. Hogan 26 Mowbray Avenue, St. Helens WA11 9JD
MM6 ZZG G. Runcie Smithy Cottage, Brotherton, Montrose DD10 0HW
M6 ZZK J. Gooch 38 Wards Crescent, Bodicote, Banbury OX15 4DY
M6 ZZQ S. Collins 5 Fernleigh Gardens, Stafford ST16 1HA
M6 ZZS A. Barrett 24 Burhill Way, St. Leonards-on-Sea TN38 0XP
M6 ZZV A. Siddall 12 Russell Gardens, Sipson, West Drayton UB7 0LS
M6 ZZY R. Soar 31 Sidlesham Close, Hayling Island PO11 9ST

M7

MI7 AAC A. Cairns 14 Otterbank Road, Strathfoyle, Londonderry BT47 6YB
MM7 AAE G. Mann Halcyon, Strachan, Banchory AB31 6NL
M7 AAH B. Lavery 113 Ena Crescent, Leigh WN75ET
M7 AAK E. Blackford 141 Hartside, Newcastle upon Tyne NE15 8BZ
M7 AAL D. Hollbrook 64 Stonycroft Albany, Washington NE37 1UN
M7 AAP G. Walton 141 Lunsford Lane Larkfield, Aylesford ME20 6HP
M7 AAQ S. Glover 20 Lilly Street, Bolton BL1 3AU
M7 AAT Dr D. Lester 87 Lime Road Normanby, Middlesbrough TS6 0BZ
M7 AAU R. Lucas 4 Midwinter Gardens, Swindon SN3 4NZ
M7 AAV I. Blofield 13 Wimborne Avenue, Ipswich IP3 8QW
M7 AAW P. Mabbott Flat 16, Tintagel Court, 1-3 Arthur Street, Hove BN3 5EY
M7 AAX O. Pollard Flat 1, Kent Mansions Brighton Road, Worthing BN11 3EH
M7 AAY S. Britt 2 West Terrace West End, Herstmonceux, Hailsham BN27 4NT
M7 AAZ P. Hutchinson Corby Steps, The Street, Woodbridge IP12 2QG
M7 ABA M. Voros Flat 4, 5 Spencer Road, Eastbourne BN21 4PB
M7 ABB P. Chadwick 1 Latrigg Crescent Middleton, Manchester M24 4LU

M7 ABC S. Dearne 26 Dilly Lane Barton On Sea, New Milton BH25 7DQ
M7 ABD A. O'Reilly 45 Commercial Street, Southampton SO18 6LY
M7 ABE A. Hyndman 7 Field View, Castleford WF10 5TU
M7 ABG W. Watson 54 Pannal Ash Grove, Harrogate HG2 0HZ
M7 ABM C. White Flat 2 55 Church Road, London SE19 2TE
M7 ABN R. Fletcher Riversdale Drive, Goole DN14 5LH
M7 ABP R. Miller 175 Warrington Road, Widnes WA8 0BA
M7 ABQ D. Pacey 35 Regent Court, Lord Street, Southport PR9 0QQ
M7 ABR A. Brook Inkerman House 113 Clovelly Road, Bideford EX39 3BY
M7 ABT A. Ryan Flat 8/91 Westward Road, Stroud GL5 4LF
MI7 ABU D. Clarke 117 Gregg Street, Lisburn BT27 5AW
MI7 ABV M. Robinson 31 Glenavon Crescent, Lurgan, Craigavon BT66 8JR
M7 ABW K. Nelson 185 Headlands, Fenstanton, Huntingdon PE28 9LP
M7 ABX R. Culley 6 The Hollies, Straight Road, Foxhall, Ipswich IP10 0FN
MM7 ABY I. Mcgowan Carbeth Road, Glasgow G62 7PT
M7 ACA J. Smith 44 Old Station Way, , Goldalming GU7 3HA
MI7 ACD J. Miskimmin 3 Orchard House, Mark Street, Newtownards BT23 4WS
M7 ACE V. Foster 8 Skirlington Grove, Hull HU9 3RH
M7 ACG M. Mann Woodlands, Coston, Norwich NR9 4DT
MW7 ACK B. Davies 10 Long Acre Court, Bishopston, Swansea SA3 3AY
MI7 ACL J. Mcpeake 47 Dunclug Gardens, Ballymena BT436NN
M7 ACM B. Chamberlin 349A Hungerdown Lane, Chippenham SN14 0JW
M7 ACN M. Wandby 44 Windrush Road, Hollywood, Birmingham B47 5QA
MM7 ACS C. Chroston 6 Queeness Road, Vidlin, Shetland ZE2 9UB
M7 ACV West Coast Rollers (Science And Engineering Club) c/o T. Abel 20 Falcon Fields, Maldon CM9 6YA
M7 ACW R. Gower 30 Westerham Road, Sittingbourne ME10 1XF
M7 ACX V. Oliveira 60 Hermine House Moselle Street, London N17 8DE
M7 ACY P. Young 11 St. Andrews Avenue, Washington NE37 1AH
M7 ADC A. Capon 1 Windermere Way, North Common BS305XN
M7 ADF N. Barrett 7 Tannersfield Shalford, Guildford GU4 8JW
MM7 ADI A. Mcmillan 13 Roxburgh, Greenock PA15 4PU
MW7 ADL C. Walker Perthi, Llaneilian, Amlwch LL689LY
M7 ADN A. Thomas Penberthy Road, Portreath TR164LU
M7 ADO S. Steppens 21 Keynes Close, London N2 9NE
MW7 ADT P. Hoath 8 Liverpool Terrace, Llithfaen, Pwllheli LL53 6NN
M7 ADU I. Jackson 40 Highgate, Cleethorpes DN35 8NT
M7 ADW R. Naden 10 Suffolk Close, Holland On Sea, Clacton-on-Sea CO15 5SQ
M7 ADX M. Lindup 87 Oakwood Avenue, West Mersea CO5 8BD
M7 AEA R. Reed 27 Howard Close, Braintree CM7 3DT
M7 AEB A. Winton 10 Old Parsonage Court Guithavon Street, Withem CM8 1XP
M7 AED J. Searl Studio 1 34-36 Crown Street, Reading UK RG1 2SE
M7 AEF K. Duncan 72 Elmswood Road, Tranmere CH42 7HR
MM7 AEI R. Kemp 14 Marshall Road, Luncarty, Perth PH1 3UT
M7 AEJ L. Marsh 11 Woodlands Road, Haywards Heath RH16 3JU
M7 AEL G. Mayell Flat 11, Eagle House, Goldsmiths, Grays RM17 6PX
MW7 AEN C. Trigg 23 Crystal Glen, Cardiff CF14 5QH
M7 AER K. Weston Flat 4 505 Grimsby Road, Cleethorpes DN35 8AN
M7 AES R. Carter-Sutherland 74 Legsby Avenue, Grimsby DN320NE
MM7 AEU B. Weir 22 Belvedere Road, Bathgate EH48 4AX
MI7 AEW G. Colgan 42 Loughview Village, Carrickfergus BT38 7PD
M7 AEX P. King 29 Newbury Road, Manchester SK83PA
M7 AEY A. Clarke Little Acre, Pound Lane, Hardwicke, Gloucester GL2 4RJ
M7 AEZ C. Bown 19 Victory Villas, Hatherop Road, Fairford GL7 4JU
M7 AFA L. Carmon 21 Wyncroft Road, Widnes WA88QE
M7 AFD T. Quinlan 29 Sekiton Road, Diss IP22 4PW
M7 AFE G. Ede 53 Mayfield Close, Bognor Regis PO21 3PS
M7 AFH S. Ali 13 Stonehill Drive, Rochdale OL12 7JN
M7 AFI N. Cheetham 39 Burns Avenue, Church Crookham, Fleet GU52 6BN
M7 AFK B. Fox Heath Gardens, Stone ST15 0AW
M7 AFL G. Clarke High Gables, Sandstone Close, Dudley DY3 2EQ
M7 AFO O. Phillips 54 Marshall Road, Cambridge CB1 7TY
MW7 AFP S. Cawsey 8 Hickman Road, Penarth CF64 2AJ
M7 AFR P. Gianrossi 2 Limetree Close, Cambridge CB1 8PF
M7 AFS S. Williamson 3 Mill Green, Warboys, Huntingdon PE28 2SA
M7 AFU R. Cardus 3 Washington Avenue Chaddesden, Derby DE216JS
M7 AFW G. Appleton Hawthorn Lodge, Ludborough Road, North Thoresby, Grimsby DN36 5RF
M7 AFZ C. Pike 47 Whitstone Rise, Shepton Mallet BA4 5QA
M7 AGB A. Boocock 25 Smallwood Road, Dewsbury WF12 7RU
M7 AGD D. Templar 6 Highfield Court, Highfield Avenue, Harwich CO12 4JG
MM7 AGE J. Groundwater Newhall, St. Margarets Hope KW17 2RW
M7 AGG A. Scotchmer 207 Cornelian Street, Blackburn BB1 9QN
M7 AGH A. Hill 24 Keelham Drive, Leeds LS19 6SG
M7 AGJ S. Eardley 20 Nash Peake Street Tunstall, Stoke on Trent ST6 5BT
MM7 AGL E. Barton Old Bakery Cottage, Main Road St. Cyrus, Montrose DD10 0BA
M7 AGN M. Quinlan-Sandy 173 Melfort Road, Thornton Heath CR7 7RU
M7 AGO J. Maxwell The Bungalow, Latton Bush Centre, Southern Way, Harlow CM18 7BL
M7 AGQ T. Tofts 5 Woodcroft, Harlow CM18 6XX
M7 AGS B. Gojkovic 42B Hogarth Road, Sw5opu Londonuk SW5OPU
M7 AGU S. Mcdermott 19 Walpole Street, Wolverhampton WV6 0AT
M7 AGV D. Hart 7 Penrose Road, Ferndown BH22 9JF
M7 AGX V. Fox 48 Grenham Avenue, Manchester M15 4HD
MM7 AGY K. Walsh 12B Humbie Holdings, Kirknewton EH27 8DS
M7 AHA R. Bishop 12A Goseley Avenue, Hartshorne, Swadlincote DE11 7EZ
M7 AHE L. Rawlings 9 Gladstone Road, Kingswood, Kingswood BS151SW
M7 AHG D. Sherlock London Road, Swanley BR8 7HA
M7 AHH J. Kent 28 Connaught Drive, Chapel St. Leonards, Skegness PE24 5YS
MW7 AHI R. King 29 Heol Y Waun, Seven Sisters, Neath SA10 9BL
MM7 AHK J. Martin 3F Kirkgate, Irvine KA12 0DF
MW7 AHM K. Burrows 25 Mill Road, Pontllanfraith, Blackwood NP12 2GE
M7 AHN F. Tombling 48 Venables Road, Guisborough TS14 6LQ
M7 AHT S. Horne Beaucroft, Keswick Road, Benfleet SS7 3HU
M7 AHU D. Newton 22 Cleves Road, Richmond TW10 7LD
M7 AIE R. Ward 45 Lumbards, Welwyn Garden City AL7 1PJ
MW7 AIF A. Hatch Flat B, 27 Woodland View, Abercarn, Newport NP11 4AP
MW7 AIH J. Harris 197 Henllys Way, Cwmbran NP44 7LB
M7 AII A. Ifrim 46 Northdown Road, Solihull B91 3NB
MM7 AIK Dr J. Emery-Barker 3 Main Street, Newmills, Dunfermline KY12 8SR
M7 AIM I. Sproson Flat 15, Olive Standring House, Todmorden Road, Littleborough OL15 9AH
MM7 AIP R. Stewart 14 Marshall Court, Queen Street, Dunoon PA23 8BA
M7 AIQ G. Hunt 21 Needham Street, Codnor, Ripley DE5 9RR
M7 AIR L. Sanduly 1 Archford Croft, Emerson Valley, Milton Keynes MK4 2EZ
M7 AIV V. Wheatley 22 Woodlands Avenue, Shelton Lock, Derby DE24 9FQ
M7 AIW W. Lancaster 39 Chester Grove, Seghill, Cramlington NE23 7TR
M7 AIZ A. Bailey 69 New Street, Baddesley Ensor CV9 2DN
M7 AJA A. Ashurst 66 Arcot Avenue, Nelson Village, Cramlington NE23 1EY
M7 AJB A. Burgess 5 Wilkie Road, Birchington CT7 9HE
MI7 AJD D. Hurst 85 Market Street, Ballycastle BT54 6DS
M7 AJF M. Soden 24 Barn Rise, Seaford BN25 3BY
M7 AJG A. Golding 44 Blendon Drive, Andover SP10 3NG
M7 AJH A. Holmes 23 Bowen Road, Darlington DL3 0TH
MM7 AJJ M. Mersinis Flat 30, 88 Albion Street, Glasgow G1 1NY
M7 AJK M. Collins Purcell Walk, Bristol BS4 1XT
M7 AJM A. Mauger 51 Pethertons, Halberton, Tiverton EX16 7AZ
M7 AJO C. Doggrell 16 Coronation Road, Frome BA11 2BJ
M7 AJP A. Pickrell 22 Fosseway, Lichfield WS14 0AD
MM7 AJQ J. Brummell 1 Moorpark Cottages, Kirkinner, Newton Stewart DG8 9BY
M7 AJT B. Starkey 109 Parc An Tansys, Pengegon, Camborne TR14 7PH
M7 AJU A. Taylor-Roberts 38 Old Coach Road, Bulford, Salisbury SP4 9DA
M7 AJX S. Evans 7 The Lampreys, Gloucester GL4 6QD
M7 AJZ O. Payne Holland Hall, Clydesdale Road, Exeter EX4 4SA
M7 AKA J. Harding 28 Berecroft, Harlow CM18 7SA
M7 AKC T. Farr 2 Westonia Court 797 Hertford Road, Enfield EN3 6UQ
M7 AKG M. Chapman 21 Connaught Street, Northampton NN1 3BP
M7 AKM S. Hey 3 Queens Square, Kirkby Lonsdale, Carnforth LA6 2AZ
M7 AKN W. Clarke 30 Swalecliffe Court Drive, Whitstable CT5 2LZ
M7 AKP J. Hockridge 7 Queen Street, Great Oakley, Harwich CO12 5AS
M7 AKS D. Jackson 22 Trent Road, Walsall WS3 4DQ
M7 AKU R. Mortlock 11 Thompson Avenue, Richmond TW9 4JP
M7 AKX F. Cairns 20 St. Davids Close, Maidenhead SL6 3BB
M7 AKZ A. Frago Psc 37 Box 3828, Bury St. Edmunds IP28 8NG
M7 ALC A. Bell 15 Kininvie Close, Redcar TS104JQ
M7 ALF A. Hawkes 9 Ferneley Avenue, Hinckley LE100FE
MM7 ALI A. Oakes 55 Northfield Meadows Longridge, Bathgate EH47 8SA
MM7 ALL G. Allathan 1A Irvine Road, Kilmarnock KA1 2JN
M7 ALN A. James 36 Westcote Close, Solihull B92 8PL
MM7 ALO A. Chambers 35 Echline Grove, South Queensferry EH30 9RU
M7 ALR A. Rudgley The New House, Plymouth Road, Buckfastleigh TQ11 0DB
M7 ALS A. Stanley 75.Birmingham Road, Kidderminster DY10 2SR
MW7 ALT D. Clarke 8 East Walk, Barry CF62 8DA
MM7 ALW S. Doyle 19 Cairns Gardens, Balerno EH14 7HJ
M7 AMA A. Martinavarro Twyford Court, University Campus, Guildford GU2 7JP
M7 AMB A. Bennett 112 Vicarage Crescent, Redditch B97 4RP
M7 AMC K. Conlon 136 Chart Downs, Dorking RH5 4DG
MJ7 AME P. Greenaway Plat Douet Road, St. Saviour S JE27PN Jersey
M7 AMJ M. Napiorkowski 74 Wigley Road, Feltham TW13 5HE
M7 AMK Dr T. Hoban Hillside Barn, London End, Priors Hardwick, Southam CV47 7SL
M7 AMM C. Quinn 3 Hunters Ride, Stafford ST17 9HU
M7 AMP R. Barker 33 Milton Brow, Weston-Super-Mare BS22 8DB
M7 AMQ A. Wilkinson 67 Middleton Park Grove, Leeds LS10 4BG
M7 AMR A. Rowland 5, Gwinear Downs, Leedstown, Hayle TR27 6DJ
M7 AMS A. Sonnex 49 Salisbury Road, Tonbridge TN10 4PD
M7 AMT R. Pickard 23 Blackberry Way, Midsomer Norton, Radstock BA3 2RN
M7 AMW P. Hind 19 Egmont Drive, Ringwood BH24 2BN
M7 AMX J. Wright 149 Wroslyn Road, Freeland, Witney OX29 8HR
M7 ANA A. Gimenez Flat 27, Lamb Court, 69 Narrow Street, London E14 8EJ
M7 ANB D. Smith 42 Channel View Road, Portland DT5 2AY
M7 AND A. Goncalves 21A Nashleigh Hill, Chesham HP5 3JQ
M7 ANE A. Simpson 18 Clavell Road, Liverpool L19 4TR
M7 ANG G. Gordon 1 Waverland Terrace, Gillingham SP8 4NT
M7 ANH J. Birkinshaw 57 Walnut Tree Avenue, Hereford HR2 7JU
M7 ANI D. Rockey 113 Bradwell Lane, Newcastle ST5 8QD
M7 ANJ N. Moorby 5 Nightingale Close Lower Tean, Stoke on Trent ST10 4LX
M7 ANK C. Tate 18 Spenser Walk, South Shields NE34 9NF
M7 ANM P. Ashworth 3 Hardman Close, Rossendale BB4 7DL
M7 ANN L. King 2 Ebenezer Cottages, Thorney Road, Eye, Peterborough PE6 7UB
M7 ANO P. Kilkenny 11 Ash Close, Rochdale OL12 9AN
M7 ANP S. Langton 44 Poppyfields, Hesketh Bank, Preston PR4 6TJ
M7 ANR A. Rawson 64 Dukes Mead, Fleet GU51 4HE
MW7 ANT A. Lockwood Church Road, Newport NP197EL
MM7 ANU S. Archer 34 Canal Street, Saltcoats KA21 5HZ
M7 ANV A. Young 22 Martin Close, Cambridge PE292WA
M7 ANX A. Matsell 1 Anglia Close, Quarrington, Sleaford NG34 8WX
M7 ANY N. Glass 79 Marsh Lane Farndon, Newark NG24 4TA
M7 AOC S. James 14 Dene Walk, West Parley, Ferndown BH22 8PQ
M7 AOD L. Weston 7 Chestnut Grove, Arnold, Nottingham NG5 8BD
M7 AOI M. Firth 12 Pendle Street West, Sabden, Clitheroe BB7 9EG
M7 AOL S. Wright 18 Garratts Road, Bushey WD24 4LA
M7 AOM R. Phillips 148 Northborough Road, Slough SL2 1TA
M7 AOO A. Watling 1 Dovendale Cottage Horncastle Road, Tathwell LN11 9SA
M7 AOR P. Rollason 68 Heathfield Lane West, Wednesbury WS10 8QP
MI7 AOU S. Sheridan 90 The Meadows Randalstown, Randalstown BT412JB
M7 AOV N. Austerfield Princes Avenue, Withernsea HU192JA
MM7 AOY A. Liddell 22 Tassie Court, Leargan, Leven KY8 5FL
M7 AOZ D. Hughes 14 Holts Lane, Clayton, Bradford BD14 6BL
MM7 APB J. Quinn 2 Manse Road, Shotts ML7 5EL
M7 APC R. Nicholson 27 Bishopton Way, Hexham NE46 2LR
M7 APF P. Richardson 91 Park Road, Blackpool FY1 4JE
M7 APH A. Harcourt Flat 6, 93 Westridge Road, Southampton SO17 2HJ
M7 APM G. Booton 69 The Street, Deal CT14 0AJ
M7 APN T. Gore 22 Stoppard Road, Burnham-on-Sea TA8 1QB
M7 APO L. Martin 51 Meadow Close, Stretton On Dunsmore, Rugby CV23 9NL
M7 APV C. West 11 Peterswood, Harlow CM18 7RJ

MM7 APW A. Ward 11 Ashgrove Place, Elgin IV30 1UJ
M7 APY D. Appleton Field Farm, Mumby Road, Hogsthorpe, Skegness PE24 5PD
M7 APZ B. Hender 11 Cambridge Road, Walton-on-Thames KT12 2DP
M7 AQC R. Pearce 5 Jubilee House, Chapel Street, Bicester OX26 6FE
M7 AQF R. Tew 61 Magna Road, Bournemouth BH11 9ND
M7 AQG K. Wallis-Gare 40 The Forge The Green, Upton, Norwich NR13 6AY
M7 AQI M. Pownall 43A Hughes Ave, Warrington WA29EW
M7 AQL M. Turnier Unit F, Winston Business Park, Churchill Way, Chapeltown, Sheffield S35 2PS
M7 AQM S. Cleaver 77 Main Street, Cockermouth CA13 9JS
M7 AQO G. Shaw 1 Platts, Lydlinch DT10 2HX
M7 AQP C. Watts 17 Hollycroft Road, Emneth, Wisbech PE14 8AY
M7 AQQ I. Dodds 54 Philip Road, Newark NG24 4PD
M7 AQR S. Salisbury 6 Ryan Close, Leyland PR25 2XW
M7 AQS M. Davys 9 Littlecombe Close, Kersfield Road, London SW15 3HR
M7 AQY M. Parker 113 Burringham Road, Ashby, Scunthorpe DN17 2DF
M7 AQZ T. O'Donnell 911A New Hey Road, Halifax HD3 3FH
M7 ARA M. Hicks 108 Northorpe, Thurlby, Bourne PE10 0HZ
M7 ARC M. Johnson 24 Stjames'S Park, Wakefield WF1 4EU
MW7 ARE J. Browne 4 Bro Walker, Ammanford SA18 3EF
M7 ARH C. Pilling 28 Waverley Rd, Exmouth EX8 3HJ
M7 ARL E. Hartopp 38 Lanchester Road, Middlesbrough TS6 7HG
M7 ARN M. Head 99 Bell Hill Road, St George, Bristol BS5 7LY
M7 ARO Lord K. Powell 57 Valletts Lane, Bolton BL1 6DW
M7 ARP A. Pomfret 40 Alwyne Grove, York YO30 5RT
M7 ARQ M. Feakins 29 Whitesfield Road Nailsea, Bristol BS48 2DY
M7 ART A. Fourie 5 Old Farm Road, Bexhill-on-Sea TN39 4DN
M7 ARU A. Bird 18 Welbeck Drive, Spalding PE11 1PD
M7 ARW D. White 15 Ropehaven Road, St. Austell PL25 4DU
M7 ARZ D. Trickett 22 Spring Croft, Rotherham S61 3RF
M7 ASA B. Chislett Seavale Road, Clevedon BS217QB
M7 ASB K. Strykowski Flat 2, Leyla House, 2 Dunn Street, London E8 2DB
MI7 ASC R. Barker 5 Castle Manor, Ballynure, Ballyclare BT39 9GW
M7 ASE M. Wilson 11 Rushdene Avenue, Barnet EN4 8EN
M7 ASF A. Axtell Flat 11, Owen House, 6 College Road, Ripon HG4 2AP
M7 ASH S. Ashfield 9 Penny Hill Close, Spalding PE11 2DE
M7 ASJ A. Reader 4 Hollicombe Close, Tilehurst, Reading RG30 4PA
M7 ASK J. Postlethwaite 9 Grange Street, Morecambe LA4 6BW
MW7 ASM P. Hodgkiss 35 Princess Avenue, Buckley CH72LN
MM7 ASN D. Macfarlane 8 Kingsmill Drive, Kennoway, Leven KY8 5LX
M7 ASO K. Hedges 28 Hill Park, Congresbury, Bristol BS49 5BT
MM7 ASP A. Aspinall 4 The Maples, Dundee DD40XQ
M7 ASU R. Crosby 85A Mendip Road, Portishead, Bristol BS20 6DF
M7 ASW A. Ward Westways, Rudston Road, Burton Agnes, Driffield YO25 4NE
MU7 ASX R. Best Les Mauxmarquis Farm Route De St Andre, St. Andrew GY6 8TU Guernsey
M7 ASY R. Samson 20 Chichester Park, Westbury BA133AN
M7 ASZ K. Burt 38 Church Lane, Moldgreen, Huddersfield HD5 9EB
M7 ATA A. Atack 601 Long Cross, Bristol BS110TX
M7 ATB Z. Berces 22 Breydon Walk, Crawley RH10 6RE
M7 ATC A. Hayler 19 Newgate Road, St. Leonards-on-Sea TN37 6SA
M7 ATD I. Leonard 10 Bouchers Mead, Chelmsford CM1 6PJ
M7 ATE C. Stubbs 114 High Street, Stoke on Trent ST71PY
MM7 ATF J. Shaver 1 Glendale Crescent, Ayr KA7 3SQ
M7 ATG M. Cowell 3 Selborne Mews, Blackburn BB2 2SQ
M7 ATH A. Higginbottom 2 Claremont Gardens, Waterlooville PO7 5LL
M7 ATI A. Toth 14 Runcorn Road, Sunderland SR5 5ET
M7 ATJ A. Jemmett 46 Barkway Road, Royston SG8 9EB
M7 ATL G. Des Jardins 9A Spofforth Hill, Wetherby LS22 6SF
M7 ATN A. Nelson 125 Brighton Road, Crawley RH10 6TE
M7 ATO Dr J. Hunt 10 Couzens Close, Chippenham SN15 1US
MW7 ATR T. Brown 12 George Lansbury Drive, Newport NP19 9DS
M7 ATT N. Watson 12 Gold Croft Gold Street, Barnsley S70 1TZ
M7 ATV J. Robinson 12 Rising Side, Barrow-in-Furness LA13 9ES
M7 ATW A. Wiseman 16 Hartford Close, Essex SS69DQ
M7 ATX D. Pyke 38 Mayflower Close, South Killingholme, Immingham DN40 3HF
M7 AUC S. Cooney 12 Angier Grove, Denton, Manchester M34 6HG
M7 AUD A. Valvona 26 Craigbank Court, Fareham PO14 1AQ
MW7 AUE M. Hallett 9 Chapel Close, Monmouth NP25 3NN
M7 AUF I. Edwards 9 Sanctuary Road, Holsworthy EX22 6DQ
MM7 AUH K. Stankiewicz 8 Victoria Place, Cullen AB56 4TU
M7 AUI S. Mauer Flat 18, Channing Court, Osborne Road, London W3 8SY
M7 AUL B. Hatwell 15 Northey Road, Bodmin PL31 1JE
M7 AUM J. Shaughnessy 37 Belton Road, London NW2 5QE
M7 AUO D. Tonks 54 Pageant Drive, Aqueduct, Telford TF4 3RF
M7 AUP J. Matthews 2 Farm Close, Bungay NR35 1JG
M7 AUQ C. Adams 44 Cranshaw Drive, Blackburn BB1 8RE
M7 AUR M. Gibbon 14 Hillside Way, West Lutton, Malton YO17 8TE
MW7 AUU M. Waller 9 Tre Rhoser, Newbourgh LL61 6TG
MW7 AUX A. Jenkins 13 Church Street, Bargoed CF81 8RN
M7 AUY D. Higton 24 Holly Avenue, Bradwell NR31 8NL
M7 AUZ G. Muir The King William Iv, Station Road, Wanstrow.., Shepton Mallet BA4 4SZ
M7 AVA I. Titchener 18 King Edgar Close, Ely CB6 1DP
M7 AVB M. Bareham 136 Coppins Road, Clacton on Sea CO153LA
M7 AVC K. Spowage 3 Arcadia Avenue, Mansfield NG20 8JS
M7 AVE S. Wright 115 Marlborough Road, Derby DE24 8DS
M7 AVF A. Hooper 13 Westend Avenue, Coppull PR7 5DB
M7 AVG A. Thompson-Dale 6 Emsley Avenue, Cudworth, Barnsley S72 8HU
M7 AVH S. Ackerley Pierremont Crescent, Darlington DL3 9PB
MM7 AVI B. Horan 82 Alloway Road, Lochside, Dumfries DG2 9LT
M7 AVL A. Hayter 50 Hamilton Road, Alford LN13 9AX
M7 AVN T. Ballam á28 Lannett Road, Gloucesterá GL1 5DE
M7 AVO D. Eccles 16 Pasture Close, Kelsall, Tarporley CW6 0PN
M7 AVP G. Prenas 58 Losinga Road, King's Lynn PE30 2DH
M7 AVQ J. West 61 Teign Bank Road, Hinckley LE10 0ED
M7 AVR D. Singleton 3 Swansbury Drive, Bournemouth BH80LB
M7 AVS P. Smith 431 New Street, Biddulph Moor, Stoke-on-Trent ST8 7NG
M7 AVT A. Velveris 32 Harrison Street, Carlisle CA2 4EP
M7 AVW M. Simpkins 67 Kirkdale, London SE26 4BL
M7 AWD A. Walkden 5 Ivy Close, Shaw, Oldham OL2 7TQ
M7 AWE S. Vijaysampath 19 Wallbrook Avenue, Macclesfield SK10 3GL
M7 AWF S. Allen Bar-Point, Carpalla, Foxhole, St. Austell PL26 7TY
M7 AWG A. Goldsmith 13 Skinners Lane, Galleywood, Chelmsford CM2 8RH
M7 AWH B. Knight 5 Lapwing, Tamworth B77 5NW
M7 AWJ I. Parkin 93 Lakeside, Isleham Marina, Fen Bank, Isleham, Ely CB7 5ZD
M7 AWK A. Lowe 91 Maidenway Road, Paignton TQ3 2AQ
MM7 AWL M. Love 17 Lindsay Road, East Kilbride, Glasgow G74 4HZ
M7 AWM S. Baker 76 Malvern Road, Bournemouth BH9 3AJ
M7 AWN J. Coles 4 Carloggas Close, St Mawgan. Newquay, Truro TR8 4HJ
MW7 AWO Dr R. Stoyle 37 Geiriol Road, Townhill, Swansea SA1 6QP
M7 AWQ A. Williams 13 Birch Close Colden Common, Colden Common SO21 1XE
M7 AWW B. Andrews 18 Parkside, Wilnecote, Tamworth B77 2JU
M7 AXA B. Thomas 33 Pelham Road, Southsea PO5 3DT
MM7 AXB A. Locke Cotswold 40 Craigton Av, Inverness IV3 8AZ
M7 AXC K. Davies 15 Sandpiper Drive, Weston-Super-Mare BS22 8UH
M7 AXD J. Warburton 82 Hampton Drive, Newport TF10 7RF
M7 AXE J. King 140 Valley Drive, Harrogate HG2 0JS
M7 AXF D. Entwisle 3 Ainley Street, Elland HX5 0AJ
M7 AXJ S. Brett 51 West End, Wirksworth DE44EG
M7 AXM D. White 11 Seymour Place, Canterbury CT1 3SF
M7 AXU R. Mills 6 Gould Close, Bristol BS13 0BJ
M7 AXV S. Kelly 12 Carmarthen Way, Rushden NN10 0TN
MM7 AXX M. Ronaldson 59B Glasgow Street, Ardrossan KA22 8EP
M7 AXY J. Cooper 42 St. Augustines Crescent, Chesterfield S40 2SD
M7 AYF M. Young 155 Arctic Road, Cowes PO31 7XS
MM7 AYG P. Payne 116 Main Street, West Kilbride KA23 9AR
M7 AYJ V. Sajanani University Of Warwick, Post Room, Gibbet Hill Road, 13 F Whitefields, Coventry CV4 7ES
M7 AYP T. Hobbs 10 Chervil Close, Chandler'S Ford, Eastleigh SO53 4JL
M7 AYR S. Hoare 47 Bright Street, Darlington DL1 4EY
M7 AYS A. Sullivan 155 Kings Road, Chelmsford CM12BA
M7 AYU P. Hasney 15 Mansfield Court, West Boldon, East Boldon NE36 0PL
M7 AYW R. Chandler 48 Middletons Road, Yaxley, Peterborough PE7 3NU
M7 AYZ S. Benniman 5 Round Hill Lane, Shrewsbury SY1 2NE
M7 AZA K. Brooks 1 Woodland Vale Cottages Wannock Road, Polegate BN265ED
M7 AZG D. Smith 27A Priory Road, Newbury RG14 7QS
M7 AZH P. Maxfield 1 Lynchet Lane, Worksop S817AN
M7 AZJ A. Clarke 48 East Street, Okehampton EX20 1AU
M7 AZK M. Wills 38 New Park House, New Park Road, Shrewsbury SY1 2RT
M7 AZN Dr A. Rennie 80 Bartons Drive, Yateley GU46 6DP
M7 AZP D. I?Onn 1 Bligh Road, Westhoughton, Bolton BL5 3TR
M7 AZQ A. Williams 11 The Fairway, Farnham GU9 9BB
M7 AZR R. White 7 Cornwell Close, Redditch B98 7TG
M7 AZS M. Hughes 16 Marlborough Road, Urmston, Manchester M41 5QG
M7 AZV W. Mcintosh 69A Park Hill, London SW4 9NS
M7 AZW J. Hartley 9 Westfield, Bradninch, Exeter EX5 4QU
M7 BAA T. Stockings 9 Queen Street, Driffield YO25 6QJ
M7 BAC N. Bacala Flat 3, 232A Seven Sisters Road, London N4 3NX
M7 BAE S. Stenning 18 Knights Mead, Chudleigh Knighton TQ13 0RE
M7 BAF P. Symon 8 Glebe Close, St. Columb TR9 6TA
M7 BAG R. Banks 1 Holly Meadows, Ashford TN23 3QR
M7 BAJ A. Carbonell 2 Wolvesmere Woolmer Green, Knebworth SG3 6JW
M7 BAK Dr A. Baker Victoria Cottage, Swan Lane, Aughton, Ormskirk L39 6SU
M7 BAN A. Wilson 4 Oxford Street, Doe Lea, Chesterfield S44 5PH
MW7 BAO B. Williams Bryn Celyn, Hendre Road, Conwy LL32 8RJ
M7 BAQ D. Wills 37 Hawden Road, Bournemouth BH118RP
M7 BAR D. Saunders 68 Heywood Road, Prestwich, Manchester M25 1FN
M7 BAS B. Rice 178 Hungerhill Road, Nottingham NG3 3LL
M7 BAV K. Bavin Browns Lane, Tamworth B798TA
M7 BAW B. Withington 20 Bond Way, Hednesford, Cannock WS12 4SN
M7 BAY B. Young 1 Bugle Place, Newton Abbot TQ12 1GZ
M7 BBE B. Eastland 4 Bergamot Close Manton, Marlborough SN8 4HT
MM7 BBG B. Gordon 125 Caird Street, Hamilton ML3 0AL
M7 BBI R. Rudling 18 Bournemouth Avenue, Gosport PO12 4NP
M7 BBK A. Hedgecock 4 Dean Garden Rise, High Wycombe HP11 1RE
M7 BBM A. Uren 12 Frewin Gardens, Plymouth PL6 6PN
M7 BBO E. Hanson 44 Dovehouse Road, Haverhill CB9 0BZ
M7 BBS S. Dean 154 Broad Lane, Walsall WS3 2TQ
M7 BBT J. Mcnaught Flat 2, 29 Prowse Place, London NW1 9PN
M7 BBV C. Butler Wv14 7Np, Wolverhampton WV14 7NP
M7 BBW G. Davison 29 Lansbury Way, Sunderland SR5 3DD
M7 BBY D. Lovick 328 Cranbrook Avenue, Hull HU6 9PH
M7 BBZ N. Granville Honeygarston Road, Bristol BS139LY
M7 BCA T. Pearson 4 Oakwood Drive, Barrow-in-Furness LA13 0UB
M7 BCE D. Davison 15 Primrose Drive, Bingley BD16 4QT
M7 BCH S. Bailey 39 Aintree Drive, Balby, Doncaster DN4 8TU
M7 BCI A. Leonard 109 Webb Crescent, Dawley, Telford TF4 3DX
MM7 BCL P. Siviter 32 Cairndhuna Terrace, Wick KW1 5BJ
MW7 BCM T. Jones Gowan Bank Aberarth, Aberaeron SA460LP
M7 BCN B. Nicholson 5 Arundel Road, Billingham TS23 2DJ.
M7 BCV B. Chadwick 44 Glendale Drive, Mellor, Blackburn BB2 7HD
M7 BCW S. Ferguson 266 Central Drive, Blackpool FY1 5JB
MW7 BCY E. Jones 34 Llys Charles, Towyn LL229NT
M7 BCZ D. Williams Little Benifold, Headley Hill Road, Headley, Bordon GU35 8DU
M7 BDB D. Bandiera 1 Broadhill Road, Kegworth, Derby DE74 2DQ
MM7 BDC P. Chant 22 Lemon Terrace, Leven KY8 4QQ
M7 BDE L. Kingaby 5 School Lane, Aldford, Chester CH3 6HZ
M7 BDF W. Sinnott 33 Mayfield Street, Atherton, Manchester M46 0AQ
M7 BDG D. Dheerasinghe 88 Buckmaster Avenue, Newcastle ST5 3AN
M7 BDO J. Middlemiss Kenmore, Marlpits Lane, Ninfield, Battle TN33 9LD
M7 BDP D. Taft 16 Briar Place, Eastbourne BN23 8DB
MW7 BDT R. Bowen 16 Lon Hywel, Whitland SA34 0BE
M7 BDV H. Woolston 32 Brodrick Road, Eastbourne BN22 9NR
MM7 BDW A. Mccubbin 1 Clark Place, Saltcoats KA21 6JU
M7 BDX S. Cooper 19 West End Road, Silsoe MK45 4DU

M7 BDZ B. Hickey 44 Nickelby Road, Chelmsford CM1 4UF
M7 BEE R. Yaxley 10 South Rise, North Walsham NR280EE
MW7 BEF M. Hopkins 95 Heol-Y-Parc, North Cornelly, Bridgend CF33 4LY
M7 BEG K. Greenwood 1 Eastfield Close, Clipstone Village, Mansfield NG21 9AZ
MM7 BEI R. Hoey 12 Chamberlain Street, St. Andrews KY16 8JF
MW7 BEK A. Kendrick Cilgwyn Isaf, Carmarthen SA33 6NE
M7 BEL M. Tielemans 2 Parr Close, Grange Park, Swindon SN5 6JY
M7 BEM S. Christie 124 Bickershaw Lane, Abram, Wigan WN2 5PP
M7 BEO M. Carter 32 Victoria Avenue, Brighouse HD6 1QT
M7 BEP D. Clarke 22 Hornsby Avenue, Harley-Goodacre, Worcester WR4 0PN
M7 BET M. Wu 1 Cookson Close, Newcastle upon Tyne NE4 5RY
M7 BEV P. Cheshire 43 Fremantle Avenue, St. Helens WA95SN
M7 BEX A. Sluijters 9 George Street, Ashford TN23 7AF
M7 BFA M. Ashford 11 Pond Close, Felixstowe IP112JW
M7 BFB J. Hunter 477 Tonge Moor Road, Bolton BL2 3BG
M7 BFG A. Rogers 60 Wellington Road, Bury BL9 9BQ
M7 BFJ P. Bragg 10 Wheelers Riseá, Brackley NN135ND
M7 BFK J. Marchant 129 Highbury Grove, Clapham, Bedford MK41 6DU
MM7 BFL L. Robertson 3 Greenhead Street, Dailly KA26 9SN
M7 BFN A. Threlfall 6 Brookside Lane, High Lane, Stockport SK6 8HL
MI7 BFP T. Smyth 139 Tullyreagh Road, Gorteen, Tempo, Enniskillen BT94 3PH
M7 BFQ J. Jenkins 28, Fitzgerald Avenue, Herne Bay CT6 8NA
M7 BFV S. Thornton Low Fold Barn, Orton, Penrith CA10 3RX
M7 BFX N. Daley 1 Lewis Road, Chipping Norton OX7 5JT
M7 BFY A. Evans 44 Wordsworth Ave Headless Cross, Redditch B97 5BH
M7 BFZ I. Cooper Fettlers Wharf Marina Ltd, Station Road, Rufford, Ormskirk L40 1TB
M7 BGB M. Wanless 3 Bromlow Hall Barns, Bromlow, Minsterley, Shrewsbury SY5 0DX
MW7 BGD B. Davies 3 Yr Hafan, St. Davids, Haverfordwest SA62 6RA
MI7 BGE J. Mccluskey 10 Arosa Parade, Belfast BT15 3JF
M7 BGF B. Grospe 8, Lincombe Road, Manchester M22 1GA
M7 BGG A. De Kock Cartwright Road, London RM9 6JL
M7 BGH G. Cockle 2 Beccles Road, Lowestoft NR33 8QX
M7 BGO P. Chinn 14 Gwelmeneth, Albion Road, Helston TR13 8JH
M7 BGR P. Gardner 7 Great Blakelands, Marston Moretaine, Bedford MK43 0WY
M7 BGU N. Rendell 95 Parkstone Road, Poole BH15 2NZ
M7 BGV T. Vines 14, Finningham Road, Old Newton, Stowmarket IP14 4EG
MM7 BGW J. Galbraith 16 Cameron Way, Prestonpans EH32 9FH
MW7 BGX P. Harding 6 Ridgeway View, Newport NP205AW
M7 BHC R. Mcmanus 125 Whitaker Road, Derby DE23 6AQ
M7 BHK C. Fox 209 Sulgrave Road, Washington NE37 3DE
M7 BHL S. Gillen 3 Shetland Road, Dronfield S18 1WB
M7 BHM R. Billington 77 Clacton Road St Osyth, Clacton on Sea CO168PD
M7 BHN P. Dawson 2 Harman Walk, Clacton on Sea CO16 8UN
M7 BHO J. Dobrucki 43 Priory Lane, Macclesfield SK10 3HJ
M7 BHQ P. Green 26 All Saints Close, Doddinghurst, Brentwood CM15 0NH
MM7 BHR D. Mcnelis 35 Parkinch, Erskine PA8 7HZ
M7 BHT S. Dickson 11 Benfield, Grasmere, Ambleside LA22 9RD
M7 BHU D. Cruz 16 Addison Close, Winchester SO22 4ED
MM7 BIC A. Knox 5 Blackdales Avenue, Largs KA30 8HU
M7 BID G. Russ 244 Goresbrook Road, Dagenham RM9 6XU
M7 BIE B. Whitfield 39 Moralee Close, Gateshead NE404QE
MW7 BIH D. Grewar 19 Holland Close, Rogerstone, Newport NP10 0AU
M7 BIL W. Hughes The Old Cider House, Well Farm Lane, Alford BA7 7PW
M7 BIO B. Leeson-Earle 7 Albany Court, Albany Road, Fleet GU51 3PX
M7 BIP J. Turner 12 Southfield Lane, Addingham, Ilkley LS29 0SS
M7 BIQ R. Tuthill 27 Woodbury House Woodbury Rise, Salisbury SP2 8FG
M7 BIS C. Lewis 55 Glynn Road West, Peacehaven BN10 7SL
MI7 BIU I. Kondrashenkov 44 Forge Manor, Magheralin, Craigavon BT67 0XP
MM7 BIW J. Campbell 113 Brent Field Circle, Ellon AB41 9DB
M7 BIX S. Eggleson 17 Fourth Ave, Raf Scampton, Lincoln LN1 2UP
MM7 BJA B. Anderson 22 Gladstone Gardens, Fettercairn, Laurencekirk AB30 1FR
MM7 BJC B. Corkindale 8 Rockland Park, Largs KA30 8HB
MM7 BJG P. Ling 5 Whinstone Place, Ratho, Newbridge EH28 8AD
M7 BJH B. Havers 81 Redwood House Cheviot Road Langley, Slough SL3 8UE
M7 BJI A. Edwards 15 Hindsford Bridge Mews, Atherton, Atherton M46 9QZ
M7 BJK B. Kemble 3 Red House Close, Chudleigh Knighton, Chudleigh, Newton Abbot TQ13 0RH
MD7 BJL B. Lerigo 24 Rheast Mooar Avenue, Ramsey, Isle of Man IM8 3LR
M7 BJN K. Smith 31 Greenfinch Road, Easington Lane, Houghton le Spring DH5 0GG
MW7 BJQ A. Jones 12 Uwch Y Nant, Mynydd Isa, Mold CH7 6YP
MM7 BJR J. Burton 6 Kilfinnan Lodges, Spean Bridge PH34 4EB
M7 BJS B. Slater Flaxman Rise, Greater Manchester OL1 4QB
M7 BJT P. Edwards 4 Stodart Road, London SE20 8ET
MM7 BJU B. Marshall 13 Alder Grove, Westquarter FK2 9SU
M7 BJV D. Shepherd 10 Luscombe Crescent, Paignton TQ3 3TW
MW7 BJW C. Weaver 2 Ty Cerrig, Llanddoged, Llanrwst LL26 0TY
M7 BJY A. Dawkins 12 Salisbury Road, Chatham ME4 5NW
MM7 BKA A. Stanley 22 Braid Mount, Edinburgh EH10 6JJ
M7 BKB A. Page 49 Chiltern Close, Shoreham-by-Sea BN43 6LE
M7 BKE A. Siani Flat 3, Windsor Lodge, High Street, Brighton BN2 1RP
M7 BKG T. Ward 83 Buxton Road, Congleton CW12 2DX
M7 BKH J. Gray 5 Oldfield Grove, London SE16 2NA
M7 BKI S. Manley The Cottage, Crow Ash Road, Berry Hill, Coleford GL16 7RB
M7 BKJ M. Mcmullen 11 Benlaw Grove, Felton, Morpeth NE65 9NG
M7 BKL J. Baxter 6 Clayton Mead, Godstone RH9 8NX
M7 BKM Dr S. Shales 47 Bradley Avenue, Winterbourne, Bristol BS36 1HX
M7 BKN J. Forrester 4 Hingley Street, Cradley Heath B64 5LA
M7 BKO P. Gregson 41 Newhouse Road, Blackpool FY4 4JJ
M7 BKP D. Scrivens Buzon De Correos En Los Patricios, Albox, Albox 4800 Spain
M7 BKS M. Gorrill 30 Higher Dunscar Egerton, Bolton BL7 9TF
M7 BKT P. Smith 3 Crown Cottages School Road, King's Lynn PE32 1UX
M7 BKU R. Lowe 31 Chapel Road, Habrough, Immingham DN40 3AE
M7 BKX M. Gannon 13 Holyoake Avenue, Woking GU21 4PW
M7 BKZ A. Prince 1A Greenacres, Crewe CW1 4JU
M7 BLC J. North 7 Kingsbridge Road, Newbury RG14 6DY
M7 BLD T. Bartholomew 6 Kilmaine Road, Harwich CO12 4UZ
M7 BLG V. Robinson Rectory Cottage Cuxham, Watlington OX49 5NQ
M7 BLI R. Thompson 1 Moat Farm Cottage Ashford Road Great Chart, Ashford TN23 3DH
M7 BLJ J. Utley 43 Wheatcrofts, Barnsley S70 6BZ
M7 BLL B. Louder Flat 1, 30 Mill Street, Bideford EX39 2JJ
M7 BLM A. Harris 14 All Saints Road, Bedworth CV12 0BL
M7 BLO B. Wright 60 The Ryde, Leigh-on-Sea SS9 4TN
MM7 BLP R. Mcdonald 135 Feorlin Way, Garelochhead, Helensburgh G84 0EB
M7 BLR J. Fitzsimmons 42 Handcross Road, Luton LU2 8JF
MW7 BLU D. Bluer 12 Llys Dewi Sant, Bangor LL57 2UJ
M7 BLX D. Cooke 2 Wiffens Loke, Hethersett, Norwich NR9 3RH
M7 BLY A. Woolhouse 70 Ditmas Avenue, Kempston, Bedford MK42 7DW
M7 BMC N. Rijckmans 1 Nurseries Avenue, Brundall, Norwich NR13 5NS
M7 BME R. Hale 19 Glen Mobile Home Park, Colden Common, Winchester SO21 1TE
M7 BMG R. Huntrod 64 Escallond Drive, Seaham SR7 8JZ
M7 BMH B. Hobbs 14 Durleigh Hill, Bridgwater TA5 2AG
M7 BMJ B. Josyfon 25 Norrice Lea, London N2 0RD
MM7 BMK K. Biegun 81 Highfield, New Pitsligo AB43 6PZ
M7 BMO L. Taylor 84 Victoria Way, Stafford ST17 0NX
M7 BMR B. Royles 18 Crown Avenue, Cudworth, Barnsley S72 8SE
MW7 BMU N. Parker 18 Woodland Terrace, Abercarn, Newport NP11 4SQ
MW7 BMW J. Angeles 3 Tolpath, Coed Eva, Cwmbran NP44 6UD
M7 BMX V. Cepraga 9 Kynon Close, Gosport PO12 4LW
MM7 BMY R. Fairman 52 Broomhill Crescent, Alexandria G83 9PW
MI7 BMZ K. Mcshane 40 Oar Building Annadale Crescent, Belfast BT7 3NB
M7 BNB D. Gerrard 268A Whelley, Wigan WN2 1DA
M7 BNE C. Lees Flat 8, Atlantic House, 57 Sandylands Promenade, Heysham, Morecambe LA3 1DW
M7 BNG M. Byng 177 Rosefield Road, Smethwick B67 6DY
M7 BNH D. Daly 9 Oxford Street Finedon Nn95Ez, Northants NN95EZ
M7 BNI R. Dovey Green Tiles, 138 Frinton Road, Holland-On-Sea, Clacton-on-Sea CO15 5PN
M7 BNK P. Amos Flat 24, Saville Court, Queen Street, Ravensthorpe, Dewsbury WF13 3BT
M7 BNL S. Jones 19 Runshaw Lane, Euxton, Chorley PR7 6AU
M7 BNN D. Gelkin Flat 5, Rock House, Station Hill, Chudleigh, Newton Abbot TQ13 0EE
M7 BNP N. Mason 4 Berry Hill Close, Mansfield NG18 4RS
M7 BNR K. Roast 11 Bracken Road, Keighley BD227DF
M7 BNS D. Croft Wn49Xa, Wigan WN4 9XA
MI7 BNV D. Thompson 45 Ardnabrocky Drumahoe, Londonderry BT47 3BF
M7 BNW G. Mustoe 171 Buckingham Road, Aylesbury HP19 9QF
M7 BNZ A. Sanders 24 Leeming Lane South, Mansfield Woodhouse, Mansfield NG19 9AB
M7 BOA A. Woodhead The Briggs, Ingbirchworth Road, Thurlstone, Sheffield S36 9QN
M7 BOC J. Woodford 58 Benedict Close, Romsey SO51 8PN
M7 BOD G. Stafford 25 Chilton Gardens, Houghton le Spring DH4 6LD
M7 BOE A. Beveridge Kernick Cottage, Sparry Bottom, Carharrack, Redruth TR16 5SH
M7 BOF K. Clegg 11 Birchwood Drive, Fulwood, Preston PR2 9UJ
M7 BOJ C. Eggleton 20 Arkwright Court, Leominster HR6 8NF
M7 BOK G. Boka B Flat, 41 Kimberley Gardens, London N4 1LB
M7 BON D. Harris 12A Clevelands Park, Northam, Bideford EX39 3QH
M7 BOP T. Gambrell 18 Moorland Park, Old Newton Road, Bovey Tracey TQ13 9DU
M7 BOQ C. Carr 52 Blundell Lane, Penwortham, Preston PR1 0AX
M7 BOS M. Fairchild Brooklyn Caravan Park 21 Almond Brow Gravel Lane Banks, Southport PR98BU
M7 BOU A. Clothier 57 Middle Road, Southampton SO19 8FT
M7 BOW D. Taylor 62 Floyds Lane, Walsall WS4 1LE
M7 BOZ S. Ghent 86 Tamworth Road, Long Eaton, Nottingham NG10 3NA
M7 BPA M. Pr³stel Flat 7, Cedars House, Cedars Road, Maidenhead SL6 1RY
M7 BPC B. Coley 17 Livingstone Road Handsworth, Birmingham B20 3LS
M7 BPD G. Glover 834 Moss Bank Way, Johnson Fold, Bolton BL1 5TB
M7 BPE W. Smith 69 Victoria Road Netley Abbey, Southampton SO315DQ
M7 BPH B. Heley 7 Ringwood Grove, Weston-Super-Mare BS23 2UA
M7 BPJ A. Odedra 8 Littlecote Grove, Peterborough PE4 6BJ
M7 BPN R. Wardley 218 Chester Road, Watford WD18 0LJ
M7 BPO T. Lacey 5 Pilgrims Way, Hastings TN34 2LF
MW7 BPQ S. Jha 19 Armoury Drive, Cardiff CF14 4NP
M7 BPS P. Ross 12 Haig Avenue, Great Sankey, Great Sankey Great Sankey, Warrington WA5 2TG
M7 BPW K. Bonfield 625A Burton Road, Midway, Swadlincote DE11 0DH
M7 BPX C. Esposito 1 Chestnut Avenue Wicken Green Village, Fakenham NR21 7QL
MW7 BPZ S. Hughes 7 Lower Row, Dowlais, Merthyr Tydfil CF48 3ND
M7 BQB Rvd. P. Callway 7 Heath Road, Langley, Maidstone ME17 3LH
M7 BQC P. Van-Embden 14 Maurice Cullen Close Shotton Colliery, Durham DH6 2FA
MM7 BQF D. Michael 24, Taylor Court, Aberlour AB389LA
MM7 BQG Sir R. Cherry 1/2 282 Royston Rd, Glasgow G212JB
M7 BQH A. Gambrell 7 St. Michaels Close, Shipton-Under-Wychwood, Chipping Norton OX7 6BE
M7 BQI T. Searchfield 2 Wares Field, Ridgewood, Uckfield TN22 5SG
M7 BQJ S. Porter 8 Brickwork Avenue, Liphook GU30 7WP
MW7 BQK T. Edge Llys Y Gwynt, Rhoscefnhir, Pentraeth LL75 8YU
M7 BQL A. Arcia Cosmopolitan Tower#400 Apt 2D, Panamß 0 Panama
MW7 BQM G. Edwards 9 Dalton Close, Merthyr Tydfil CF47 0TE
MM7 BQP C. Duncan 22 Dalgleish Avenue, Cumnock KA18 1QU
M7 BQX A. Toal 29 Highland Drive, Oakley, Basingstoke RG23 7LF
M7 BQY M. Smith 52 Hungerford Gardens, Bristol BS4 5HB
M7 BQZ G. Manzetti 98 Mold Crescent Banbury Ox160Ex, Banbury OX16 0EX
M7 BRA S. Bracegirdle Flat 28, West Fryerne, Parkside Road, Reading RG30 2BY
M7 BRC D. Dobbie 24 Harrow Road, Leighton Buzzard LU7 4UQ
MM7 BRG A. Brown 4 Damselfly View, Edinburgh EH17 8XH
M7 BRH B. Harwood 9 Wyndham Rd, Newstead, Stoke on Trent ST3 3LX
M7 BRI B. Freeman 11 Rachel Grove, Stoke-on-Trent ST4 3QX
M7 BRJ D. Withnall 35 Dalton Bank, Warrington WA1 3AH
M7 BRM J. Holden 47 Copse Hill, London SW20 0NJ
M7 BRO S. Finch 2 De Grey Road, Colchester CO4 5YE
MW7 BRQ J. Mann 11 Hilltop Close, Port Talbot SA12 8YH
M7 BRR M. Ding 130 Windermere Avenue, Warrington WA2 0NE
M7 BRX D. Reed 12 George Trollope Road, Watton, Thetford IP25 6AS
M7 BRY B. Bell 101 Victoria Road, Gateshead NE8 2SY
M7 BRZ D. Bishop 19 Cambrian Way, Winsford CW7 1QT

M7 BSF D. Westlake 11 Chapel St, Buckfastleigh TQ110AB
M7 BSH A. Gladosz 45 Bushey Ley, Welwyn Garden City AL7 3HB
M7 BSI A. Stockton 31 Jones Close, Brackley NN13 6JD
MW7 BSK M. Watts 2 Cysgod-Y-Cwm, St. Dogmaels, Cardigan SA43 3DS
M7 BSL E. Crobu 16 York Street, Flat 6, London W1U 6PS
M7 BSM B. Smith Bees Corner, Ideford, Chudleigh, Newton Abbot TQ13 0AZ
M7 BSN B. Shaw 9 Pheasant Close, Mulbarton, Norwich NR14 8BL
M7 BSS A. Jackson 6 Highbury Close, Westhoughton, Bolton BL5 2QU
M7 BSU S. Ash 10 Anchor Court 132 Bury Old Road, Manchester M85DR
M7 BSV W. Baron 10 Hembury Close, Middleton, Manchester M24 2SX
M7 BSX B. Shepard 3 The Leys, Oxhill CV35 0QX
M7 BSZ A. Watson 14 Linlithgow Close, Papworth Everard, Cambridge CB23 3RX
M7 BTD D. Sampson 116 South Mossley Hill Road, Liverpool L19 9BJ
MW7 BTF S. Nutting 9 Lugg View, Presteigne LD8 2DG
M7 BTG W. Trott 8 Dunlop Crescent, South Shields NE34 6QG
MI7 BTH R. Hannon 10 Malone Meadows, Belfast BT9 5BG
M7 BTL S. Parrott 39 Chichester Rd, Cleethorpes DN35 0HY
M7 BTO L. Tod 6 Middleham Close, Sandy SG19 1TU
M7 BTQ L. Jones Warren Farm Road, Birmingham B44 0QB
M7 BTS M. Evans Carlton Manor Caravan Park, Carlron Coleville, Chapel Road NR33 8BL
M7 BTU O. Schonrock 1 Powell Road, Poole BH14 8SG
M7 BTV G. Hodkinson 14 Mitton Close, Culcheth, Warrington WA3 4EU
M7 BTW B. Wyatt 2 Birchwood Gardens, Southampton SO30 2AR
MM7 BTX I. Morrison Coach House, Olivers Brae, Stornoway HS12SX
M7 BTY R. Chandler 56 Marsh Lane, Addlestone KT15 1UN
M7 BTZ W. Pearce 6A Havelock Street, Desborough, Kettering NN14 2LU
M7 BUB G. Veale 8 Duchy Cottages, Stoke Climsland, Callington PL17 8PA
M7 BUE I. Shepherd 61 Chatsworth Place Meir, Stoke on Trent ST37DP
M7 BUG A. Daimhin Simpson-Coyle 7 The Willows Bilsborrow, Preston PR3 0SG
M7 BUH G. Butler 4 Mercia Close Worksop, Worksop S81 0SS
M7 BUI D. Nixon 20 Oakdale Road, Retford DN22 7GX
M7 BUK A. Brown 17 Harley St, Coventry CV2 4EZ
M7 BUL N. Bull Eldoret, Castle Street Bampton, Tiverton EX16 9NS
M7 BUN D. Shingleton 6 Newsham Walk, Manchester M12 5QB
M7 BUR D. Burton 16 Cage Lane, Great Staughton, St. Neots PE19 5DB
M7 BUS G. Busacca Flat 6, Leigh Court, Tavistock Place, Bedford MK40 2XU
M7 BUV A. Harding 32 North Falls Road, Canvey Island SS87QG
M7 BUY K. Hodgson 411 Moss Bay Road, Workington CA14 5AB
M7 BVA I. Gourlay 1 Littlemoor Road, Mirfield WF149AL
M7 BVC A. Marvel Dean House Farm, Nordan, Leominster HR6 0AW
M7 BVF C. Morris 4 Wroxeter, Shrewsbury SY5 6PH
M7 BVH D. Clews 25 Lea Green Lane, Wythall, Birmingham B47 6HE
M7 BVI J. Lambert 7 Fawn Close, Huntington, Cannock WS12 4UP
M7 BVK P. Macfarlane 26 Whittingham Drive Ramsbottom, Bury BL0 9LZ
MM7 BVN R. Bevan 46 Oak Drive, Portlethen, Aberdeen AB12 4XF
M7 BVO D. Blencowe 25 Harrington Road Kelmarsh, Northamptonshire NN6 9LX
M7 BVQ D. Lamrhari 16 The Close, Potters Bar EN6 2HY
M7 BVS D. Szmytkowski Van Crombrugghe, Brussels 1150 Belgium
M7 BVT R. Morris 45 St. Kildas Road, Bath BA2 3QL
M7 BVV P. Pullinger 10 Cowley Drive Lancing, Lancing West Sussex BN15 8DJ
M7 BVZ B. Mockford 322 High Street, Eastleigh SO50 5ND
M7 BWE N. Hughes Hawes Bank Pow Lane, Shap CA10 3NB
M7 BWG B. Gentry 32 Abbey Meadow, Sible Hedingham, Halstead CO9 3QS
M7 BWH J. Lane 112 Honeycomb Valeá, Chard TA20 1GU
M7 BWJ W. Jacques 3 The Butts, Crudwell SN16 9HF
MM7 BWK M. Borthwick Shepherds Cottage, Sunnyside, Hawick TD9 9SS
M7 BWL B. Langford North Street House, North Street, Langport TA10 9RL
M7 BWM K. Brennan 14 West View, Silsden, Keighley BD20 9JY
MM7 BWN G. Scott 29 Kennedy Drive Dunure, Dunure KA7 4LR
M7 BWO R. Yarnall 279 Longfield, Falmouth TR11 4SS
M7 BWP A. Gornall 29 Clifton Street, Darwen BB3 0BR
MM7 BWT N. Mansfield Blencathra House, Township Road, Auckengill KW14XP
M7 BWW T. Smith Grange De Lings, Lincoln LN2 2LX
MM7 BWZ A. Marshall 46 Langton Road, Westquarter, Falkirk FK2 9SZ
M7 BXA R. Angus 48 Castle Avenue, Epsom KT17 2PH
M7 BXC P. Wilbraham 122 Prenton Hall Road, Prenton CH43 3BJ
M7 BXE P. Martin 46 Melody Road, Biggin Hill, Westerham TN16 3PH
M7 BXF G. Payne 508 Crownhill Road, Plymouth PL5 2QU
MM7 BXL B. Longmuir 12 Cloverfield Gardens, Bucksburn, Aberdeen AB21 9AY
M7 BXM K. Scott Flat 16, Quayside Court, The Quay, Harwich CO12 3HH
M7 BXO P. Wade 89, Top Fair Furlong, Redhouse Park, Milton Keynes MK14 5FQ
M7 BXT J. Stanley 57 Walterbush Road, Chipping Norton OX7 5DP
M7 BYC R. Bagnall 9 Anson Road, Stoke-on-Trent ST3 7AT
M7 BYD K. Potter 7 Leesfield Road, Meadowfield, Durham DH7 8NJ
M7 BYE G. Frost The Old Chestnut, 51 Knightcott Gardens, Banwell BS29 6HD
M7 BYF S. Austin 35 Flordon, Skelmersdale WN8 6PA
MM7 BYH W. Gordon 5 Balgarvie Road, Cupar KY15 4AH
MM7 BYK M. Weeks 109 East Princes Street, Helensburgh G84 7DN
M7 BYL M. Kimberley 45 Harbord Close, North Walsham NR28 0TA
M7 BYM G. Young 65 Boyton Road, Ipswich IP3 9PD
MM7 BYQ G. Nicholson 22Kilbowie Rd South Carbrain, Cumbernauld G67 2PX
M7 BYS M. Bosman Lindridge, Church Lane, Baschurch, Shrewsbury SY4 2ED
MM7 BYT Lord M. Bryant 46 Waverley Drive Glenrothes, Glenrothes KY6 2LU
M7 BYU M. Annan 43 Cleve Leaze Thornbury, Bristol BS35 2FW
MW7 BYV S. Richards 62 Glyn Bedw, Llanbradach, Caerphilly CF83 3PG
MW7 BYW P. Whitby 28 Bryn Dyrys, Bagillt CH6 6BX
M7 BZA J. Fagan 20 O'Sullivan Crescent, Blakbrook, St. Helens WA11 9RB
M7 BZB G. Woodham 5 Colne Rise Rowhedge, Colchester, Rowhedge CO5 7EH
M7 BZC T. Mather 40 Sandown Road, Haslingden, Rossendale BB4 6PL
M7 BZK J. Hammond 2 Blackmans Close, Newent GL18 1AL
M7 BZL N. Ormrod 13 Boscombe Ave, Peel Green Peel Green, Manchester M307DU
M7 BZO D. Neill Burrs Shoes, 7-9 Leys Avenue, Letchworth Garden City SG6 3EA
M7 BZP S. Tudge 11 Boar Croft, Coventry CV4 9SJ
MI7 BZR A. Bandeja 16, Craigavon Crescent, Dungannon BT71 7BD
M7 BZU A. Moore 178 Park Road, Bedworth CV12 8LA
M7 BZV K. Houghton 26 Heathfield Road, Grantham NG31 7NH
M7 BZZ A. Busby 24 Hall Close, Bourn, Cambridge CB23 2SW
M7 CAA D. Morland Greystone, Cliburn, Penrith CA10 3AL
M7 CAD G. Jones 7 Wheatlands Close, Cannock WS12 3XL
MW7 CAJ N. Hewelt 17 Saint Marys Road, Llandudno LL30 2UB
M7 CAK Dr L. Walker 12 Priory Gardens, Corbridge NE45 5HZ
M7 CAL J. Groombridge 57 Lower Road, Swanley BR8 7RY
MM7 CAN J. Cannon 31 Ellismuir Street, Coatbridge ML5 5BH
M7 CAQ D. Martin 179 Dartford Road, Dartford DA1 3EW
MI7 CAT C. Todd 75 Suffolk Road, Belfast BT11 9PU
M7 CAW C. Gray The Pigsty, Cleverton, Chippenham SN15 5BT
M7 CAZ M. Ball 16 King Street, Burntwood WS7 4QJ
MW7 CBF D. Holdcroft 7 Llys Mai, Buckley CH72GZ
M7 CBG A. Rizvi 41 Clarence Close, Manchester BL9 6HE
M7 CBH C. Heath 130 Holyfields, West Allotment, Newcastle upon Tyne NE27 0EY
M7 CBL A. Evans 45 Bonington Road, Mansfield NG19 6QH
M7 CBP D. Mcdonald 20 Montague Road, Bournemouth BH5 2EP
MW7 CBT P. Rogers 109 North Road, Pontywaun, Cross Keys, Newport NP11 7FS
M7 CBX C. Barber 32 Ashcroft Road, Ipswich IP1 6AB
M7 CBY N. Porter 37 Windermere Park, Lowestoft NR32 4UD
MM7 CCB C. Boal 20 Fairinsfell, Broxburn EH52 6AL
MM7 CCC B. Cook 2 Marlfield Farm Cottages, Kelso TD5 8ED
M7 CCF S. Greenan 1 Bramley Drive, Frome BA112EY
M7 CCG A. Jones 15 Wheatfields, Bradeley, Stoke-on-Trent ST6 7QD
M7 CCH C. Chowdhury-Hanscombe 35 Clavering Avenue, London SW13 8DX
M7 CCK G. Allen 22 Acrefield, Padiham, , Burnley BB12 8HN
M7 CCL C. Crawford 31 Marston Drive, Newbury RG14 2SQ
M7 CCM N. Pilling 12 Brooke Close, Baxenden BB5 2QX
M7 CCO B. Wyatt 20 Britannia Gardens, Westcliff-on-Sea SS0 8BN
M7 CCQ A. Woosnam 15 Saddlers Square, Northampton NN3 5AY
M7 CCR I. Read 8 The Moors Cressage, Shrewsbury SY5 6DA
M7 CCS C. Smith 14 Sinclair Road Shurdington, Cheltenham GL51 4SL
M7 CCU F. Richardson 7 Marton Grove, Grimsby DN33 1JF
M7 CCW C. Wright 11 Trinity Fields, Lower Beeding, Horsham RH13 6GH
M7 CCZ B. Crudgington 1 Sharmford Meadows, Barham, Ipswich IP6 0QY
M7 CDB P. Wood 11 Honey Hill Lees Oldham 0L4 5Dp Lancashire, Oldham OL4 5DP
M7 CDF C. Forrester 21 The Crescent, Northfleet, Gravesend DA11 7EB
M7 CDH Dr C. Haley Cold Norton Farm Ockham Lane Hatchford, Cobham KT11 1LW
M7 CDK I. Casbon 117A Pasture Street, Grimsby DN32 9EE
M7 CDO W. Harvey 17 Saunton Road, Braunton EX331HB
M7 CDQ J. Mastros 41 Radnor Street, Swindon SN1 3PR
M7 CDS C. Sole 2 Shepley Street, Manchester M359DY
M7 CDT C. Smith 17 Midwinter Avenue, Milton, Abingdon OX14 4XA
M7 CDU A. Zahid Begum 20 Hill Top, Sutton SM3 9JH
M7 CDW C. Woolf 127 Lavender Avenue Mitcham, Mitcham CR4 3RS
M7 CDX C. Dibben 30 Harty Road, Haydock WA11 0YY
M7 CEB C. Brennan 19 The Furrow, Littleport, Ely CB6 1GL
M7 CEF A. Lendrum 35 Hardfield Road, Alkrington, Middleton, Manchester M24 1JA
MM7 CEH C. Harris 27 Avon Road, Bathgate EH48 4AH
MI7 CEK J. Harvey 6 Montrose Street, Belfast BT5 4HY
M7 CEL C. Dench Thornby, Down St. Mary, Crediton EX17 6DU
MW7 CEO D. Hughes 6 Cambrian House, Old School Lane, Pontypridd CF37 2DD
M7 CEP S. Dunklin 2 High Street, Boxworth CB234LY
M7 CER C. Storer 90 Midland Road, Ellistown, Coalville LE67 1EH
MM7 CES D. Woods 3 Walnut Crescent, Johnstone PA5 9QJ
M7 CEV A. Peet Baystone Bank Farm, Whicham, Millom LA18 5LY
MW7 CEW C. Worka 9 Bertha Street, Pontypridd CF37 1TS
M7 CEY I. Rehacevs Flat 13, Rowan Court, 19 The Avenue, Beckenham BR3 5LH
M7 CEZ C. Neagu 26 Shaftesbury Avenue, Folkestone CT19 4NS
MM7 CFC S. Smith 10 Munro Street, Stenhousemuir, Larbert FK5 4QF
MM7 CFE G. Pow 29 Seafar Drive, Kelty KY4 0JX
M7 CFF D. Adelodun 128 Olney Road, London SE17 3HR
M7 CFG P. Luck Durban House, Merrymeet, Liskeard PL14 3LP
M7 CFL C. Locock 12 Park Road, Moseley, Birmingham B13 8AB
M7 CFR C. Ralston 11 Back Gillmoss Lane, Liverpool L11 0AY
MM7 CFU H. Mackenzie 31 Ulladale Crescent, Strathpeffer IV14 9AQ
M7 CFW D. Cadden 24 Lindenbrook Vale, Stafford ST17 4QN
MI7 CFX M. Douglas 7 Beechill Crescent Tandragee, Craigavon BT62 2BN
M7 CGD P. Miller 21 Golborne Road, Lowton, Warrington WA3 2DP
M7 CGH C. Heys 32 Wolseley Road, Sale M33 7AU
M7 CGI M. Gilliver 24 Fritchley Lane, Fritchley, Belper DE56 2FN
M7 CGK G. Steel 4 Doublegates Close, Ripon HG4 2TU
M7 CGM A. Davies 4 Erpingham Road, Poole BH12 1EX
M7 CGO M. Flint 22A High Street, Hinxton, Saffron Walden CB10 1QY
M7 CGQ K. Slater 4 Ridgewell Close, Lincolon LN63QQ
MM7 CGS C. Stephens Blairmains Farm, Harthill, Shotts ML7 5TJ
M7 CHA M. Charlesworth 18 Bushell Close, Leighton Buzzard LU7 4TQ
M7 CHB M. Bell 11 Greaves Road, Sheffield S5 9DB
M7 CHC D. Woodward 21 Northolt Avenue, Bishop's Stortford CM23 5DR
M7 CHD C. Copley 18 Pavement Lane, Halifax HX2 9JJ
M7 CHE R. Drage 4 Bruce'S Close Conington, Peterborough PE73QW
M7 CHF S. Brown 313 Milton Road, Cambridge CB4 1XQ
M7 CHG A. Salerno 1A Woodland Way, Marlow SL7 3LD
M7 CHH C. Haycock 45 Barden Road, Wakefield WF1 4HP
M7 CHI S. Speak 32 Noyes Avenue, Laxfield, Woodbridge IP13 8EB
M7 CHJ O. Burr 6 Curson Road, Tasburgh, Norwich NR15 1NH
M7 CHN I. Bott 22 Middlefield Road, Cossington, Leicester LE7 4UT
M7 CHP R. Chappell 24 Woodend Drive, Shipley BD18 2BW
M7 CHR C. Pascoe Treleven, Primrose Hill, Goldsithney, Penzance TR20 9JR
M7 CHS C. Schroth Flat 3, 15B Cavendish Road, Bournemouth BH1 1QX
MW7 CIE M. Thomas 88 Dafodil Court Ty Canol, Cwmbran NP44 6JF
MW7 CIH M. Knight 28 Ridgeway Hill, Newport NP20 5DG
M7 CII T. Bickerstaff 43 Riverway, Durrington, Salisbury SP48ES
M7 CIJ S. Mcdermott 48 Horseshoe Crescent, Bordon GU35 0DP
M7 CIL S. Perera 33 Lockington Croft, Halesowen B62 9BP
MU7 CIM R. Price Les Vaurioufs, St. Martins GY4 6TE Guernsey
M7 CIN K. Darker 19, The Horseshoe, Hemel Hempstead HP3 8QT
M7 CIO C. Green 36 Wren Gardens Alderholt, Fordingbridge SP63PJ
M7 CIS R. Scott 34 Moorfield Road, Birmingham B34 6QY
M7 CIT A. Jacobs 11 Office Road, Cinderford GL14 2HZ

M7 CIU J. Weber 9 Orpwood Way, Abingdon OX14 5PX
M7 CIX A. Williamson 15 Montgomery Road, Gosport PO13 0UZ
M7 CIZ P. Upton 27 Pavilion Way, Little Chalfont, Amersham HP6 6PZ
M7 CJB C. Braiden Flat 5, Lansdowne House, 12 Twickenham Close, Swindon SN3 3FQ
M7 CJD D. White Flat 12, Lansdowne House, Inverness Road, Gosport PO12 3HL
M7 CJE M. Hill 75 Orpen Road, Sholing, Southampton SO19 0EH
M7 CJG C. Gosling 34 Coldhams Crescent, Huntingdon PE29 1UG
M7 CJH S. Seddon 3 Riverside Wharf, Dartford DA1 5TN
M7 CJL C. Davey 1 Hillend Cottages, Eastnor HR8 1RF
M7 CJM J. Mccarthy Flat 2, Highview, 75 Eglinton Hill, London SE18 3PB
MM7 CJS C. Sutton 29 Hazel Dene, Methil, Leven KY8 2JL
M7 CJU C. Longster 57 Newtown Road The Offices, Hove BN3 7BA
M7 CJV D. Cooke 42 Branston Road, Uppingham, Oakham LE15 9RS
M7 CJX P. Jones 8 Normanton Close, Edwinstowe, Mansfield NG21 9PF
M7 CJY C. Yates 10 Grosvenor Avenue, Torquay TQ2 7LA
M7 CKA M. Angus 4 Irlam Road, Sale M33 2BH
M7 CKC J. Winsor 8 Sages Lea, Woodbury Salterton, Exeter EX5 1RA
M7 CKF B. Morrissey Fiboard House, 5 Oakleigh Gardens, London N20 9AB
M7 CKG L. Staples 13 Wedgewood Road, St. Austell PL25 3HY
M7 CKN G. Harris 18 Waggoners Way Bugbrooke, Northampton NN7 3QT
M7 CKR A. Shilling 18 Simmons Way, Okehampton EX2O 1PY
M7 CKT Rvd. S. Perry 8 Collingwood Avenue, March PE15 9EF
M7 CKV M. Sanger 45 Pier Plain, Gorleston, Great Yarmouth NR31 6PS
M7 CKW C. Wise 32 Commercial Street, Willington DL15 0AD
M7 CKX C. Whittaker Little Polmarth Farm Polmarth Carnminellis Nr Redruth Cornwall, Turo TR166NT
M7 CKY M. Logan 14 Long Ley, Langley Upper Green, Saffron Walden CB11 4RX
M7 CLC G. Walsh 42 St. Huberts Road, Great Harwood, Blackburn BB6 7AR
MW7 CLE P. Roberts 18 Maes Mawr, Llanrwst LL26 0HW
MM7 CLH C. Hornby 2 Galabank Gardens, Annan DG12 5FA
M7 CLI C. Davies 51 Holland Street, Crewe CW1 3TT
M7 CLJ C. Streader 56 Lowtown Street, Worksop S80 2JR
M7 CLK D. Rigby 4 Avalon Close Tottington, Bury BL8 3LW
M7 CLM C. Morse Redhouse, The Green, Dauntsey, Chippenham SN15 4JH
MM7 CLN G. Mcara 24 Balfour Street, Alloa FK10 1RU
M7 CLP W. Pryce 6 Banbury Close, Shrewsbury SY2 6TE
M7 CLQ R. Davies 32 Saddleback Road, , Swindon SN5 5RL
M7 CLU D. Pollard 6 Amble Road, Callington PL17 7QE
M7 CLW C. Wishart Flat 22, Lake View, Alcove Road, Bristol BS16 3AG
M7 CLX C. Greenhalgh 13 Barley Meadows, Abbeytown, Wigton CA7 4BF
MM7 CMB C. Mcconnell 4 Doune Road, Bragar HS2 9DF
MM7 CMC C. Mcintyre 70 Oldwood Place, Livingston EH54 6US
M7 CMD E. Hauser 23 Prince William Way Sawston, Cambridge CB22 3SZ
M7 CME A. Banwell Hazel Cottage, Chapel Lane, Hinton, Chippenham SN14 8HD
M7 CMI A. Castle 12 Churchfield House, Guessens Road, Welwyn Garden City AL8 6RJ
MW7 CMJ G. Hillman 5 Avondale Street, Abercynon, Abercynon CF454YU
M7 CML C. Cresswell 23 Wellfield Close, Newcastle upon Tyne NE15 9JL
MM7 CMM M. Duncan 72 Scalloway Road, Gartcosh, Glasgow G69 8LH
M7 CMS M. Cocciu 92 Dowdeswell Close, London SW15 5RL
M7 CMW C. Wardrop 5, Highstreet, Bidborough, Tunbridge Wells TN3 0UJ
MM7 CMY A. Mccall Smith 16 Napier Road, Edinburgh EH10 5AY
M7 CNA M. Higginbottom 6 Stuart Close, Strensall, York YO32 5ZP
M7 CNB J. Gulliver 32 The Row, Welford, Newbury RG20 8HS
M7 CND W. Johns 19 Kingston Close, Seaford BN25 4NF
M7 CNE A. Clarkson 9 St. Andrews Square, Stoke-on-Trent ST4 7GA
M7 CNI S. Arnull 1 Limetree Cottage, The Street, Erpingham, Norwich NR11 7QD
M7 CNJ T. Grear 8 Petworth Close, Frimley, Camberley GU16 8XS
M7 CNK K. Robinson 1 Berry Close, Earls Barton, Northampton NN6 0HU
M7 CNL M. Mcneil 58 Fairville Road, Stockton-on-Tees TS19 7NF
M7 CNM F. Sheane-Smith 9 Tagwell Grange, Droitwich Spa WR9 7FD
M7 CNN R. Sissons 57 Sedgehill Road, London SE6 3QR
MM7 CNQ H. Brown 6 Rupert Street 1/2, Glasgow G4 9AR
MM7 CNX L. Day 25 Meiklerig Court, Glasgow G53 5UY
M7 CNY A. Taylor 23 Park Lane, Southwick, Brighton BN42 4DL
M7 CNZ J. Greaves Petherton Preston New Road, Nr. Blackburn BB2 7PU
M7 COB P. Cobb 166 Milton Street, Southport PR97AP
M7 COD M. Ray 60 Albemarle Road, Gorleston, Great Yarmouth NR31 7AS
M7 COF P. Perkins 47 Ranulf Road, Flitch Green, Dunmow CM6 3GR
M7 COH S. O'Hanlon 113 Taunton Drive, Liverpool L108JN
M7 CON D. Wilkinson 8 Brecon Close, Newcastle upon Tyne NE54TD
M7 COP C. Speak 32 Noyes Avenue, Laxfield, Woodbridge IP13 8EB
M7 COQ J. Trajdos 43 Mount Road, Manchester M18 7BX
M7 COR D. Jones Ch658Hn, Ellesmere Port CH65 8HN
M7 COU M. Cartwright 7 Old Mill Close, Grove, Wantage OX12 7LD
M7 COV M. Higginson 190 St. Georges Road, Coventry CV1 2DF
M7 COX G. Cox 2 Poulteney Road, Stansted CM24 8ED
M7 COY J. Voak Flat 3, Dove Court, Packers Lane, Ramsgate CT11 8QA
M7 CPA C. Wilkins Alma Place, North Road, Barnstaple EX31 1PA
M7 CPC C. Clark 90 Castlebridge Gardens, Wolverhampton WV11 3NQ
M7 CPG D. Cooper 21 Staley Drive, Glapwell, Chesterfield S44 5QG
M7 CPH P. Hazelton 20 Greenways, Chelmsford CM1 4EF
M7 CPJ G. Pawson 9 Ivan Clarks Corner Abington, Cambridge CB216XR
M7 CPO R. Tullock 22A Tunstall Avenue, Billingham TS23 3SP
M7 CPQ R. Pink 75 Greenfinches, Hempstead, Gillingham ME7 3PW
M7 CPT J. Cressey 32 Shakespeare Road, Bredbury, Stockport SK6 2HS
M7 CPW A. Wells 8 Ward Street, Earls Barton, Northampton NN6 0JW
MM7 CPZ F. Dove 5 Cairnorchies, Mintlaw, Peterhead AB42 4LH
M7 CQA S. Jones 36 Woodbury, Lambourn, Hungerford RG17 7LT
M7 CQD D. Ransome Ashleigh, Coombe Lane, Shepton Mallet BA4 5UY
M7 CQE D. Harmer 98 King Georges Avenue, Coventry CV6 6FF
M7 CQF D. Owens Lucas Green Nurseries, Lucas Green, West End, Woking GU24 9LY
M7 CQJ D. Nicoll Dingle Lane Farm, Dingle Lane, Hilderstone, Stone ST15 8SG
M7 CQK R. Baber 38 Sinope Street, Gloucester GL1 4AR
M7 CQL Z. Gong Room 116, Walmsley Studio, 218 Saint John Street, London EC1V4AT
M7 CQM J. Law 58 Westfields, Zeals BA126PW
M7 CQS S. Peters 7 Greenfield Avenue, Chatburn BB7 4AJ
M7 CQU V. Rotaru 15 Leicester Drive, Glossop SK13 8SH
M7 CQX N. Horton Parkside Farm, Shripney Lane, Bognor Regis PO22 9NU
M7 CQY M. Blencowe 11 Gendalls Way, Launceston PL15 8SE
MM7 CQZ P. Addison 29 Clachnaharry Road, Inverness IV3 8RA
M7 CRA B. Tilley 46 Longway Avenue, Charlton Kings, Cheltenham GL53 9JJ
MM7 CRB C. Balmain 331 Dunecht Court, Glenrothes KY7 6UQ
M7 CRC C. Clark 48 Church Drove, Wisbbech PE148RH
M7 CRD C. Davis 47 Mendip Road, Weston-Super-Mare BS23 3HB
M7 CRF Dr C. Fletcher 7 Highfield Crescent, Baildon, Shipley BD17 5NR
M7 CRG C. Fryer 35 Palmer Avenue Abbeymead, Gloucester GL4 5BH
M7 CRK C. Hughes 120 Treningle View, Bodmin PL311PD
M7 CRM C. Marsh 75 Priorswood Road, Taunton TA2 7PT
MW7 CRP D. Crisp 5 Gyrnosfa, Lower Cwmtwrch, Swansea SA9 1DR
M7 CRQ L. Adams 38 St. Albans Road, Kingston upon Thames KT2 5HQ
M7 CRR M. Ketteringham 4 Westgate Street, Downham Market PE38 0PA
M7 CRS C. Oyitch 69 Melrose Road, Gainsborough DN21 2SA
M7 CRU L. Oconnor 7 Wentworth Way, St. Leonards-on-Sea TN38 0XG
M7 CRV Z. Mohammad 35 Fairlands Ave, London CR7 6HD
M7 CRW C. Smith 6 Sawley Avenue Lowton, Warrington WA32EW
MM7 CRY M. Napier-Holford 26 Craig Street Rosyth., Dunfermline KY112NG
M7 CSA C. Afonso 4 Adams Forge, Littleport, Ely CB6 1FA
MM7 CSC G. Porter Brae, Port William, Newton Stewart DG8 9RT
M7 CSF D. Hensman 31 Farmer Ward Road, Kenilworth CV8 2DJ
M7 CSI P. Johnson 135 Addison Road, Bilton, Rugby CV22 7HB
M7 CSK C. Kent 20 Cherrytree Grove, Spalding PE11 2NA
M7 CSL C. Lowe Keynsham Water Front, Stidham Lane, Keynsham, Bristol BS31 1GB
MI7 CSM Y. Kyle 2 Wattstown Crescent, Coleraine BT52 1SP
M7 CSN L. Beynon Flat 3, 2 Gloucester Mews, Weymouth DT4 7DA
MW7 CSO J. Healey 1 Park Crescent, Brynmawr, Ebbw Vale NP23 4HR
M7 CSP C. Pyne Blenheim Road, Bridgwater TA6 4HE
M7 CSR C. Reeves 36 Oakfield Avenue, Wrenbury, Nantwich CW5 8ER
M7 CSU R. Morley 7 Eliot Road, St. Austell PL25 4NL
M7 CSW P. Williams 14 Springfield Road, Flat 3, Ilfracombe EX34 9JW
M7 CSX R. Whiting Swanwick, Church Green, Churchfields, Wellington TA21 8SF
MW7 CTA A. Smedley 27 West Place, Gobowen, Oswestry SY11 3NR
M7 CTB C. Thorne 67 Devon Road, Cadishead, Manchester M44 5HB
M7 CTF Dr R. Kisiel 98 Iolanthe Drive, Exeter EX4 9EA
MM7 CTG T. Gallagher 18 Argyle Drive, Hamilton ML3 9EB
M7 CTI S. Stokes 2C The Street, Holywell Row, Bury St. Edmunds IP28 8LS
MI7 CTL G. Goodwin 29 Glenfield Road, Lurgan, Craigavon BT66 8ER
M7 CTM C. Hoult 11 Queen Street, Alnwick NE66 1RD
M7 CTN C. Cantral 38 Langtree Avenue, Old Whittington, Chesterfield S41 9HP
M7 CTP C. Parkyn 3 Walnut Road, Honiton EX14 2UG
M7 CTT M. Steeples 44 Trunch Road, Mundesley, Norwich NR11 8JX
M7 CTV C. Thompson 75A Park Lane, Bonehill, Tamworth B78 3HZ
M7 CTY P. Murton 7 Stony Close, Long Meadow, Worcester WR4 0JY
M7 CUA J. Woollett 3 Wishingtree Close, St. Leonards-on-Sea TN38 9JG
M7 CUD C. Dewhirst 40 Mountain Road Thornhill, Dewsbury WF120BW
M7 CUG D. Parsons 12 Heraldry Way, Exeter EX2 7RA
MM7 CUH C. Mitchell 25 Shoulderigg Road, Coalburn ML110EL
MW7 CUI T. Tansley 7 Edward Street, Oswestry SY112BL
M7 CUJ P. Kane The Snuggle Up, Sandhutton YO7 4RW
M7 CUN A. Newman 25 Pensford Court, Craydon Road, Bristol BS14 8EQ
MW7 CUO C. Evans 16 Oak Road, Llanharry, Pontyclun CF72 9HT
M7 CUP C. Hatton 54 St. Josephs Avenue, Birmingham B31 2XQ
M7 CUQ P. Budgen 9 Hornfair Road, London SE77BE
M7 CUR R. Jones St. Fillans, The Warren, East Horsley, Leatherhead KT24 5RH
M7 CUS A. Otton 21 Marshall Avenue, Bognor Regis PO21 2TJ
M7 CUU C. Lloyd 53 Gleneagles Road, Birmingham B26 2HT
M7 CUW P. Beckett 19 Greave Clough Close, Bacup OL13 9HS
M7 CUZ A. Sequeira 15 Bennetts Court, Garibaldi Street, London SE18 1DP
M7 CVC C. Chisholm Hill Top Kellah, Haltwhistle NE49 0JL
MI7 CVD D. Harrison Carnmoney Road, Belfast BT366HT
M7 CVH A. Pickering 4 Simpson Close, Chaple St. Leonards PE245JU
M7 CVJ A. Thornton 4 Riley Lane, Kirkburton, Huddersfield HD8 0RX
M7 CVM D. Wesley 14 Boylan Road, Coalville LE67 3JG
M7 CVQ I. Giles 4 Dainton Mews Fisher Street, Paignton TQ4 5UA
MW7 CVT C. Thorley Helston, The Mountain, Holyhead LL65 1YR
M7 CVW R. Nesbitt 7 Teal Court Juniper Close, St. Leonards-on -Sea TN38 9RW
M7 CVX J. Miller 62 Sheffield Road, Penistone, Sheffield S36 6HE
M7 CVZ R. Perry 24 Conrad Road, Stanford le Hope SS170AT
M7 CWA C. Walson 30 West Crescent, Duckmanton, Chesterfield S44 5HE
M7 CWB C. Bell 56 Derby Drive Moorside, Durham DH8 8DX
M7 CWD C. Dabrowski Overton, Salisbury Terrace, Mytchett, Camberley GU16 6DB
M7 CWK A. Wade 6 Elliott Grove, Brixham TQ5 8RT
M7 CWL K. Horigan 21 Churchtown, Gwinear, Hayle TR27 5JL
M7 CWM C. Mercer 31 Downside Avenue, Plymouth PL6 5SD
M7 CWN C. Woodall 73 Ridyard Street, Wigan WN5 9QD
M7 CWS I. Handley 110 Longslow Road, Market Drayton TF9 3BW
M7 CWT M. Crighton Flat 4 Charles Court 20 Avenue Road Erith Da8 3By, Erith DA8 3JY
M7 CWX C. Wright 151 Atkinson Road, Sunderland SR6 9AY
M7 CXB C. Beachell 19 Oldfield Avenue, Stannington, Sheffield S6 6DQ
M7 CXD J. Tunstall 46 The Hawthorns, West Kyo, Stanley DH9 8TX
M7 CXE W. Softley 15 Hare Law Gardens, Stanley DH9 8DE
M7 CXF S. Higginson 14 Hans Price Close, Weston-Super-Mare BS23 1NG
M7 CXJ D. Garside 14 Roche Avenue, York YO31 9BB
M7 CXL P. Needham 29 Sandhill Close, Bedford MK45 2JD
M7 CXP A. Jackson 21 Hibson Avenue Norden, Rochdale OL12 7RU
M7 CXQ D. Cumming 47 The Thicket, Fareham PO16 8QA
MM7 CXR M. Irons 70 Fairlie Street, Camelon, Falkirk FK1 4NL
M7 CXT G. Boxer 31 Barnfield Road Flat 2, Orpington BR5 3LP
M7 CXU R. Gilbert 51 Maudslay Road, London SE9 1LH
M7 CXV R. Jaya Chandran 14 Egerton Road, Reading RG2 8HQ
M7 CXW E. Mcgee 6 Longton Grove Road, Weston Super Mare BS23 1LT
M7 CXY S. Stray 9 Highfields Mews, Great Gonerby, Grantham NG31 8XA

M7 CXZ J. Turner 2 Dukewood Road, Clayton West, Huddersfield HD8 9HF
M7 CYA C. Yates 7 Plessey Terrace, Newcastle upon Tyne NE7 7DJ
M7 CYC M. Harrison Oldcastle Avenue, Newcastle under Lyme ST5 8HF
M7 CYE A. Done 3 Home Farm Cottages, Old Warden Park, Old Warden, Biggleswade SG18 9DU
M7 CYG P. Leftwich 43 Norman Drive, Old Sarum, Salisbury SP4 6FP
M7 CYH I. Sweet 70 Moor Park Drive, Ilkley LS290PT
M7 CYI R. Hawley Anni Healey Close, Woodbridge IP12 1GZ
MI7 CYL L. Miskimmin 3 Orchard House, Mark Street, Newtownards BT23 4WS
MW7 CYN C. Vugts Coed Coch, Llangammarch Wells LD4 4BS
M7 CYQ T. Considine Signal House Jacklyns Lane, Alresford SO24 9JJ
M7 CYR C. Dagga 12 Columbine Road, Rochester ME2 2XZ
M7 CYS S. Cole 8 Wayfarer Close, Warsash, Southampton SO31 9AU
M7 CYU A. Todd 16 Watling Close, Lincoln LN42BD
M7 CYV D. Bradbury 26 Sandbanks Way, Hailsham BN27 3LL
MI7 CYX T. Wojciechowski Flat 7, Easton House, 69-71 Cliftonville Road, Belfast BT14 6JP
M7 CYY L. Hawkins 11 Coronation Place, St. Budeaux, Plymouth PL5 1UP
M7 CZA R. Myrcha 100 Third Street, Horden, Peterlee SR8 4EH
M7 CZB R. Hyland 35 Brodrick Road, Eastbourne BN22 9NR
MM7 CZD M. Rostant 5 Baird Crescent, Alexandria G83 0TX
M7 CZF Y. Chadun 77 John Archer Way, London SW18 2TS
M7 CZM Dr D. Shephard 2 Charleville Mews, Isleworth TW7 7BW
M7 CZQ C. Ballard 3 Pope Court The Galleries Warley, Brentwood CM14 5FR
M7 CZR S. Cartwright Bs30 5Qu, Bristol BS 30 5QU
M7 CZU A. Durucz 7 Quayside, St. Marks Square, Lincoln LN5 7EX
M7 CZW E. Clothier 15 Trafalgar Road, Newport PO30 1QD
M7 CZZ T. Brown Bridge House, Redenhall Road, Harleston IP20 9QN
M7 DAB D. Beard 1 Bond Close, Leonard Stanley, Stonehouse GL10 3GQ
M7 DAC A. Capitan 41 Cunningham Drive, Runcorn WA7 4DL
M7 DAD M. Bland 10 Pennycross Square, Sunderland SR40HR
M7 DAE D. Elley 34 Nash Way, Coleford GL16 8RQ
M7 DAF R. Jones 34 Lytchett Way, Poole BH16 5LS
MM7 DAH D. Hancox 52 High Main Street, Dalmellington, Ayr KA6 7QN
M7 DAJ D. Jarman 62 Combe Drive, Dunstable LU6 2AE
M7 DAK N. Goult 5 Colbourne Close, Bransgore, Christchurch BH23 8BW
MM7 DAM D. Mcintosh Crunes Way, Greenock PA15 2WH
M7 DAN D. Savage Rufford, Barnes Lane Milford On Sea, Lymington SO41 0RR
M7 DAP D. Pope 176 Innsworth Lane, Gloucester GL3 1DX
M7 DAR D. Rockall 24 Alvington Road, Newport PO30 5AR
M7 DAY D. Allen 19 Brooklands Close, Uttoxeter ST148UH
M7 DAZ D. Walton 106 Cumberworth Lane, Lower Cumberworth, , Huddersfield HD8 8PG
M7 DBB C. Poole 128 Ansley Common, Nuneaton CV10 0QA
M7 DBD N. Dyer 80 Mosse Gardens, Chichester PO193PQ
MW7 DBF M. Bugby Derwen Deg, Fron Goch, Bala LL23 7NT
M7 DBG A. Lydiate 89 Berrylands Road, Wirral CH46 7TY
M7 DBH B. Taylor 43 Melbourne Way, Bush Hill Park, Enfield EN1 1XG
M7 DBJ A. Millington 8 Bede Road, Nuneaton CV10 8HP
M7 DBL O. Cable 116A London Road, Hadleigh SS7 2PG
M7 DBM D. Edwards 20 Keelson Pointe, Carl Street, Walsall WS2 7DB
M7 DBN N. Dinsdale 3 Pond Cottages, Faulkland, Radstock BA3 5XB
M7 DBP D. Pickles 27 Keelham Lane, Keighley BD20 6DE
MI7 DBR D. Bennett 15 Elgin Street, Belfast BT7 3AG
M7 DBX S. Ranger 137 Warren Avenue, Southampton SO16 6AF
M7 DBZ M. Webb 10 Sausthorpe Street, Lincoln LN5 7XW
M7 DCA D. Acton 31 Harrogate Terrace, , Murton SR7 9PQ
MM7 DCD D. Clark 9D West Court Clydebank, Glasgow G81 4PG
M7 DCK D. Sutton 18 Knights Hill Severn Stoke, Worcester WR8 9JD
M7 DCL P. Willingale 10 Therfield Walk Houghton Regis, Dunstable LU55QB
M7 DCM D. Clarke 34 Kimberlee Avenue, Cookley, Kidderminster DY10 3TN
M7 DCN D. Newman Common Mead Lane, Gillingham SP8 4RE
M7 DCP D. Potter 33 Wellington Grove, Doncaster DN5 9RJ
M7 DCR D. Ross Flat A, 174 Estcourt Road, London SW6 7HD
M7 DCT D. Thomas 8 Cedar Avenue, Weston-Super-Mare BS22 8HL
M7 DCV M. Dutton 10 Braemore Close Shaw Ol2 7Na, Oldham OL2 7NA
M7 DCX C. Dunn 27A Prospect Road, Longwood, Huddersfield HD3 4UY
M7 DDA D. Daniel 38 Netherthorpe Lane, Killamarsh, Sheffield S21 1DA
M7 DDC R. Williams 4 Bluebell Close, Barlborough, Barlborough S43 4WT
M7 DDE S. Kirby 4 Tittesworth Avenue, Leek ST13 6PS
M7 DDF J. Sivyour 16, Highlands Way., Salisbury SP52SZ
M7 DDK P. Cowling 11 Palmerston Road, Shanklin PO376AU
M7 DDM P. Hennings 12 Pendleway, Pendlebury, Swinton, Manchester M27 8QR
M7 DDN D. Reed Ivy Cottage, Foundry Yard, Ridsdale, Hexham NE48 2TG
MW7 DDP R. Mitchell The Hayloft, Plas Devon Court, Rossett Road, Commonwood, Holt, Wrexham LL13 9SY
M7 DDQ S. Brunsden 43 Porthcawl Drive, Washington NE37 2LT
M7 DDR D. Rushton 4 Avocet Drive Irlam, Salford M44 6PJ
M7 DDT A. Dryburgh 201 Watling Street, Grendon, Atherstone CV9 2PJ
MM7 DDV J. Mcintosh 1 Marchburn Court Northfield, Aberdeen AB16 7PQ
M7 DDW D. Wilde 12 Windhill Road Walker, Newcastle upon Tyne NE6 3TQ
MM7 DDX D. Cunningham 3 Dallerie, Crieff PH7 4JH
M7 DDY C. Westlotorn 84 Alverstone Road, Whippingham, East Cowes PO32 6NX
M7 DEA D. Sanders 37 Maple Avenue, Little Sutton, Ellesmere Port CH66 3QU
M7 DEE D. Edwards 2 Dingwall Road, London SW18 3AZ
M7 DEI D. Taylor 18 Moorside Place, Dewsbury WF136QJ
M7 DEJ P. Goodwin 60 Redburn Drive, Shipley BD18 3AZ
M7 DEK D. Howard 31 White Mullein Drive, Redlodge IP288XP
M7 DEL M. Sprague 11 The Orchards, Witcham, Ely CB6 2LR
M7 DEP D. Damodar 11, Tudor Court, Church Lane, , Rickmansworth WD3 8PX
M7 DER D. Jacobson 17 Robinia Walk Whitchurch, Bristol BS14 0SH
M7 DET D. Tipping 128 Bird Hill Road, Woodhouse Eaves, Loughborough LE12 8RR
M7 DEW D. Wade 2 Springwood Square, Back Spring Street, Huddersfield HD1 4AJ
M7 DEX D. Flood 506 Preston Old Road, Blackburn BB2 5LY
M7 DEY S. Kitchener 14A Lower Road, Malvern WR14 4BX
MM7 DFC D. Clark 106 Braes Avenue, Clydebank G81 1DP
M7 DFD D. Ford 9 The Crofts, Hatch Warren, Basingstoke RG22 4RE
MW7 DFE K. Brown 1 Heol Arthur Fear, Abertillery NP13 3JQ
M7 DFF P. Huber 15 Chelford Road, Bromley BR1 5QT
M7 DFI P. Robins 20 Saffron Close, Chineham, Basingstoke RG24 8XQ
M7 DFK C. Latham Hunter Road, Wigan WN5 0QD
M7 DFM D. Mccarron 22 Peel Close, Woodley, Reading RG5 4SR
M7 DFO J. Ball 11Lydieth Lea, Liverpool L278YL
M7 DFP M. Cavill 4 Corner Farm, Luke Lane, Brailsford, Ashbourne DE6 3BQ
MW7 DFX D. Fowler 5 Cotswold Way, Risca, Newport NP11 6QT
M7 DFY D. Ford 17 Adelaide Crescent, Burton-on-Trent DE15 0PA
M7 DGB U. Decruz Victoria Road, Lowestoft NR33 9LR
MM7 DGC D. Chappelle 10 Whitefield Terrace, Lennoxtown, Glasgow G66 7JT
M7 DGE L. Dauphin 22 Lincoln Place, Wigan WN50RA
M7 DGF D. Bailey 1 The Magpies, Maulden MK45 2EG
M7 DGG K. Mann 29, Wisteria Avenue, Branston Branston, Lincoln LN4 1QE
M7 DGH D. Henderson 75 Laburnum Road, Waterlooville PO7 7EW
MM7 DGK M. Greig 22 Rowanhill Close, Port Seton, Prestonpans EH32 0SY
MI7 DGL D. Gordon 48 Dunsy Way, Comber, Newtownards BT23 5DF
M7 DGN D. Rose 3 Iona Way, Urmston, Manchester M41 7EY
M7 DGP D. Potter 1 Upper Park Road, Colchester CO7 0JP
MM7 DGS T. Sneddon Bankend Cottage, The Wilderness, Airth, Falkirk FK2 8LN
M7 DGU S. Finn 12 Longfield Place, Maidstone ME15 9AJ
M7 DGW D. Warriner 1 St. Johns Avenue, North Hykeham, Lincoln LN6 8QR
M7 DGZ D. Godfrey 7 Austin Close, Dudley DY1 2ST
M7 DHA D. Cook 20 The Chine, London N10 3PY
MM7 DHB M. Kinnon 45D Blackthorn Place, Blairgowrie PH10 6FH
MM7 DHC D. Canavan 9 Ormelie Terrace, Edinburgh EH15 2EX
M7 DHD R. Dodge 4 Haunts Cottages Middle Chinnock, Crewkerne TA18 7PW
MM7 DHE N. Milne 83 Dalfarson Avenue, Dalmellington, Ayr KA6 7TX
M7 DHF R. Wildman 2 St. Leonards Way, Barnsley S71 5BS
MW7 DHG R. Hood Coed Bach, Morgan Street, Abercrave, Swansea SA9 1TS
M7 DHL D. Watson 30 Campbell Street, Tow Law, Bishop Auckland DL13 4DX
MW7 DHM R. Freeman Rock Cottage, Devizes SA3 1HD
M7 DHP C. Taylor Innismore, Old Road, Liskeard PL14 6DL
M7 DHT R. Groves 34 Clover Way, Paddock Wood TN12 6BQ
M7 DHZ S. Harriss 30 Chatsworth Place, Harrogate HG1 5HR
M7 DIF C. Walters 10 Grampian Crescent, Chesterfield S40 4QB
MJ7 DIJ D. Mashev Flat 23, 1875 Wesley Street, St. Helier, Jersey JE2 4DA
M7 DIL P. Bartlett 58 Ashdown Road, Chandler'S Ford, Eastleigh SO53 5QJ
M7 DIM G. Jackson 14 Southdown Crescent, Cheadle Hulme, Cheadle SK8 6EQ
M7 DIO K. Grzywaczewski Flat 2, Abbot House, Smythe Street, London E14 0HD
M7 DIP J. Grainger 42 Jackson Road, South Cerney, Cirencester GL7 6JB
M7 DIQ R. Duffy 27A Mortimer Close, Totton, Southampton SO40 2QE
M7 DIS Prof. R. Houlston 51 Adelaide Road, Surbiton KT6 4SR
M7 DIU B. Sancto 133 Hawthorne Avenue, Gillingham ME8 6YE
MM7 DIV D. Frazer 32 Loudoun Avenue, Kilmarnock KA1 3RZ
M7 DIW W. Lockley-Gardiner 193 Westbourne, Telford TF7 5QP
MW7 DIZ C. Cooper 26 Dunlin Court, Barry CF63 4JY
M7 DJA A. Deakin 69 Hammonds Ridge, Burgess Hill RH15 9QW
M7 DJB D. Bowles 3 Iris Road, Southampton SO16 3GU
M7 DJC D. Cleak 8 St Davids Road North, Lytham St. Annes FY8 2BL
M7 DJD D. De-Steunder 67 Mow Lane, Gillow Heath, Stoke-on-Trent ST8 6QB
M7 DJG D. Gorst 8 Devonshire Road, Padgate, Warrington WA1 3JS
M7 DJK D. Kearley 101 Ashdale, Bishop's Stortford CM23 4EB
M7 DJM D. Moseley 50 Chestnut Road, Walsall WS3 1BE
M7 DJN D. Norry 87 Waltondale, Telford TF7 5NJ
M7 DJP D. Peagram 1 Larke Rise, Southend-on-Sea SS2 6GQ
MW7 DJQ D. Cooper 1 Bedw Street, Caerau, Maesteg, Cardiff CF340TF
M7 DJS D. Spear 72 Hunters Road, Leyland PR25 5TT
MI7 DJT E. Morgan 12A Garden Street, Magherafelt BT45 5DD
M7 DJW D. Wyld 33 Pinewood Drive, Accrington BB5 6UG
M7 DJY M. Cowsill 23 Laithes Croft, Earlsheaton, Dewsbury WF12 8BN
MI7 DKB J. Shaw 14 Milewater Terrace, Onewtownabbey BT36 5UY
M7 DKC G. Bell 1 Bailey Close, Haverhill CB9 0LH
M7 DKE D. Kane 6 Bentry Road, Dagenham RM8 3PA
MU7 DKI J. Donaldson 62 Le Banquage Rue De Beaumont, Alderney GY9 3YP Guernsey
M7 DKJ J. Bailey 48 Blackmore Road, Shaftesbury SP7 8RD
M7 DKK D. Kirk 57 Sandringham Avenue, London SW20 8JY
M7 DKL T. Lee 6 Kerry Close, Upminster RM14 1JD
M7 DKM J. Lyth 170 Robin Hood Road, Coventry CV33AU
M7 DKS D. Kerr 19 Park Lane, Sutton Bonington, Loughborough LE12 5NQ
M7 DKU P. Piper 53 Garfield Avenue, Bournemouth BH1 4QT
MM7 DKX M. Jakob 11 Dunnet Place, Thurso KW14 8JE
MM7 DKY S. Drummond 45C Dunbeth Avenue, Coatbridge ML5 3JD
M7 DLB D. Barton 30 Mayfield Close, Bognor Regis PO213PL
MI7 DLD D. Donley 64 Glenbank Place, Belfast BT14 8AN
M7 DLK D. Swann Flat 4, 70 Hamstead Road, Handsworth B191DG
M7 DLO J. Miles 14 Westville Road, Bexhill on Sea TN393QB
M7 DLQ R. Clark 7 The Square, Chipping, Buntingford SG9 0PJ
M7 DLR B. Bleier Whipley Manor Cottage, Palmers Cross, Bramley, Guildford GU5 0LL
M7 DLS D. Salt Christmas Cottage, Ram Hill, Coalpit Heath, Bristol BS36 2TX
M7 DLV A. Yakimov 4 Erin Close, Bromley BR1 4NX
M7 DLZ S. Cockerton Hall Lodge, Wood Dalling Road, Wood Dalling, Norwich NR11 6SG
M7 DMA D. Andrijauskas 7 Hawthorn Grove, Enfield EN2 0DU
M7 DMB D. Buschbaum 42 Cordery Road, Leicester LE56DE
M7 DMF A. Fielding 4 Poot Hall, Rochdale OL120AS
M7 DMH R. Hennefer 4 Stirling Avenue, Ince, Wigan WN2 2JG
M7 DMI M. Tyson 93 Wroxall Drive Grantham, Grantham NG317EG
M7 DMJ D. Jones 53 Bradley Road, Stourbridge DY8 1UX
M7 DMK J. Davidson 78 Old Heath, Shrewsbury SY1 4SE
M7 DMP S. Buckley 13 Sherwell Road, Bristol BS4 4JX
M7 DMS D. Steel 36 Silver Street, South Petherton TA13 5AL
M7 DMV M. Chequer 14 Pasture Lane, Lazenby, Middlesbrough TS6 8EG
MM7 DMW D. White 9 Tweedsyde Park, Kelso TD5 7RF
M7 DMY D. Westwood 28 Weybridge Mead, Yateley GU46 7UY
MM7 DNE K. Thomson Inverluther, Laurencekirk AB30 1QT
MI7 DNG N. Foster 48 Beatrice Villas Bellaghy, Magherafelt BT45 8JA
M7 DNH D. Horne Alias, Lowthorpe, Southrey, Lincoln LN3 5TD

M7 DNM D. Simpson 12 Guildford Avenue, Midway, Swadlincote DE11 7LN
M7 DNN C. Norris Heather View, Forest Front, Hythe, Southampton SO45 3RJ
M7 DNO N. Wright 61 High Street, Arlesey SG15 6SW
MW7 DNQ T. Allen 53 Pendre Close, Brecon LD3 9EH
M7 DNW J. Armour Room 8, Beechwood Grove Care Home 42-44 East Dean Road, Eastbourne BN0 8EH
M7 DOD R. Dodd 82 Dickens Road, Coppull, Nr Chorley, Preston PR7 5BH
M7 DOG J. Warman 6 Down View Road, Denbury, Newton Abbot TQ12 6ER
M7 DOK D. Kelly 23A Baldwyns Road, Bexley DA5 2AB
M7 DON D. Buck 9 Ashwood Close, Mansfield Woodhouse NG19 9HD
M7 DOS S. Redmond 12 Hemlock Way, Manchester M9 7GR
M7 DOX A. Pinder 13 Hatfield Road, Bradford BD2 4QX
M7 DOZ S. Edwards 87 Garden Village, Micklefield L5254AD
M7 DPB D. Bristoll 138 Chatham Road, Northfield, Birmingham B31 2PL
M7 DPC D. Cardoso 11 Coppice Way, Aylesbury HP20 1XG
M7 DPD D. Street 18 Meadow Gardens, Bath BA1 3RY
M7 DPE P. Ashton 5 New Barns Avenue, Mitcham CR4 1LG
MW7 DPG T. Pigott 16A East Road, Tylorstown, Tylorstown CF43 3HF
M7 DPH D. Hozman 12 Bober Court, Holt Drive, Colchester CO2 0DR
MM7 DPL D. Lambert 76 Martin Crescent, Ballingry, Lochgelly KY5 8QA
MW7 DPO R. Davis 19 Ael-Y-Bryn, Caerphilly CF83 2QX
MM7 DPP S. Prime Torcroft Lodges, , Balnain, Drumnadrochit IV63 6TJ
M7 DPT M. Perescu 8 Marauder Road, Norwich NR66HD
M7 DPU M. Birkett 5 King Oswald Drive, Blaydon-on-Tyne NE21 4FD
M7 DPW D. Watmough 2 Alston Road, Wigan WN21AU
M7 DPX D. Palacionis 7 Swallow Road, Crawley RH11 7RF
M7 DPY D. Skinner Albany House, 5 Butler Road, Shrewsbury SY3 7AJ
M7 DPZ K. Harte 44 Mowrick Road, North Shields NE29 8JB
M7 DQC S. Tait 29 Hillside View Sherburn Village, Durham DH61DZ
M7 DQD S. Nelson-Smith Welbeck, Wickham Road, Fareham PO17 5BU
M7 DQF M. Johnson 2 Himley Close, Willenhall WV12 4LX
M7 DQG M. Mccarthy 1 Scargells Yard, High Street, March PE15 9LA
M7 DQI C. Male 4 Orton Lane, Twycross, Atherstone CV9 3HA
M7 DQJ M. Hine 5 Tamar Close, Spalding PE11 3GZ
M7 DQO D. Coward R F Welding Ltd, Jekils Bank, Holbeach St. Johns, Holbeach, Spalding PE12 8RQ
M7 DQQ C. Horne 54 Bridgnorth Road, Wollaston, Stourbridge DY8 3QG
M7 DQU D. Sharp 197 Preston Down Road, Preston, Paignton TQ3 1DL
MW7 DQV D. Rhodes 110 Swansea Road, Trebanos, Pontardawe, Swansea SA8 4BN
MW7 DQW N. Sparkes Oxford Lodge, 15 Leg Street, Oswestry SY11 2NL
M7 DQY S. Bone 20 Keld Avenue, Uckfield TN22 5BN
M7 DRB Dr B. Deane 2 Seymour Road, Preston PR2 2EU
M7 DRC D. Chalk Garretts Close, Southampton SO19 9RW
M7 DRE B. Owen 39 Bernard Avenue Appleton, Warrington WA4 3BA
MM7 DRG D. Gray 23 South End, Stromness KW16 3DJ
M7 DRH Dr J. Horlock 7 Laurel Drive, Prestbury, Cheltenham GL52 3DE
M7 DRJ D. Jones 20 Nightingale Road, Guisborough TS14 8HA
M7 DRL D. Lamb Unit 7, The Craft Workshops, Hutton-Le-Hole, York YO62 6UA
M7 DRR D. Robson 8 The Vale, Stockton-on-Tees TS19 0XL
M7 DRS Dr M. Sinclair 40 Grotto Road, South Shields NE34 7AH
M7 DRT R. Thomas 10 Red Salmon Road, Wixams, Bedford MK42 6ET
MM7 DRV S. Paton 31 Woodlands Place, Inverbervie DD100SL
M7 DRW A. Harcourt 12 Woodside Close, Finchampstead, Wokingham RG40 4EY
M7 DRY D. Young 1 Bugle Place, Newton Abbot TQ12 1GZ
MM7 DSA D. Sewell 51Main Street, Swinton, Duns, Berwickshire Main Street, Swinton, Duns, TD11 3JJ
M7 DSH D. Shaw 81 Chesterfield Road Tibshelf, Alfreton DE55 5NJ
M7 DSL D. Livingstone 4 Hatfield Road, Southport PR8 2PE
MW7 DSP D. Parfitt 16, Gladstone Street, Blaina NP13 3HJ
M7 DSQ D. Chalton 36 Coronation Road, Wingate TS28 5JN
M7 DSR D. Rudgley The New House, Plymouth Road, Buckfastleigh TQ11 0DB
M7 DSZ C. Terry 65 Intrepoid Close, Hartlepool TS25 1GF
M7 DTA T. Cocks Flat 1 1 Beaufort Road, Erdingington B23 7NB
M7 DTD D. Diiorio Aspley Close, Chesterfield S404HG
M7 DTF D. Fincham 2 Glebe Close, St. Columb Major TR9 6TA
M7 DTG D. Gryta 5 Sefton House, Tenbury Fold, Bradford BD4 0BD
MM7 DTL E. Blakeway 25 Allanton Grove, Wishaw ML2 7LL
M7 DTR B. Million 53 Chestnut Avenue Rossington, Doncaster DN11 0DF
M7 DTW D. Twaites 23 Church Street, Hadfield SK13 2AD
M7 DTZ P. Coulson 52 Milbank Terrace, Station Town TS285EF
M7 DUB D. Wood 3 Ripley Close, Wakefield WF3 2FG
M7 DUC D. Garland 18 Sandgate Road, Tipton DY4 0SX
M7 DUD I. Atanasov 12 Seymour Drive, Eaglescliffe, Stockton-on-Tees TS16 0LQ
MM7 DUF A. Duff 170 Saint Leonard Street, Lanark ML11 7DU
M7 DUH J. Edwards 23 Glossop Brook View, Glossop SK13 8BF
MW7 DUI M. Bennewith 1 Leisure Centre House, Chester Road West, Queensferry, Deeside CH5 1SA
M7 DUJ D. Mcclelland 4 Rectory Drive, Coppull, Chorley PR7 4QE
M7 DUM R. Walker 36, Lindrick Rd, New Marske TS11 8HT
M7 DUP S. Miller 39 Busticle Lane Sompting, Lancing BN15 0DJ
M7 DUR P. Stagg 70 Turin Court Roman Way Estate, Andover SP10 5LD
M7 DUT P. Ajayi 149 Albert Road, London N22 7AQ
M7 DUU A. Munson The Priory, Flowton, Ipswich IP8 4LH
M7 DVB J. Sandon 461 Archer Road, Pin Green, Stevenage SG1 5QP
M7 DVD D. Van Dijk 76 High Street, Tetsworth, Thame OX9 7AE
MM7 DVE D. Thompson 19 Tulloch Terrace, Perth PH1 2PF
M7 DVJ P. Calvert 9 Stanley Road, Keighley BD22 7DE
M7 DVK D. Kirk 26 Ingham Grove, Cramlington NE23 3LH
MM7 DVM B. Webb 14A Greendykes Road, Dundee DD4 7NA
M7 DVO M. Welsh Ch41 8Jf, Merserside CH41 8JF
M7 DVP D. Ward 6 Hillhead Bungalows Hillhead, Colyton EX24 6NN
M7 DVQ A. Vickers 16 Birdham Road, Brighton BN2 4RF
M7 DVS C. Lord 25 Thurso Close, Romford RM3 0YR
M7 DVT J. Harrison 20 Kelcliffe Avenue, Guiseley, Leeds LS20 9EW
M7 DVX R. Evmez Luscot, North Allington, Bridport DT6 5EQ
M7 DVY R. Werner 58 Waterside Lane, Colchester CO2 8HZ
M7 DWD A. Storr 4 Shipley Road, Southwater, Horsham RH13 9BD
M7 DWE D. Waller 4 Cymens Ora, Keynor Lane, Sidlesham, Chichester PO20 7NL
M7 DWF D. Fear 38 Axbridge Road, Bristol BS4 2RX
MM7 DWK D. Keay Parkhill, Cromwell Park, Almondbank, Perth PH1 3LW
M7 DWP D. Page Tamar Way, Slough SL3 8SY
MM7 DWQ W. Henderson 106 Gilberstoun, Edinburgh EH15 2QZ
M7 DWT A. Dykes 16 Paterson Drive, Stafford ST16 1WH
MM7 DWV L. Davidson 106 Fintry Drive, Dundee DD4 9HH
M7 DWW D. Woodall 9 Calmington Lane, Sandymoor, Runcorn WA7 1QE
M7 DX Newton Le Willows ARC c/o K. Horsfield 59 Queens Drive, Newton-le-Willows WA12 0LY
M7 DXF M. Broad 7 Long Ley, Harlow CM20 3NH
M7 DXH N. Mason 8 Barrowby Gate, Grantham NG31 7LT
M7 DXJ K. Coleman Nepturn Cottage, School Lane, North Newington, Banbury OX15 6AQ
MI7 DXL D. Lyons 16 William Street, Donacloney BT667LS
M7 DXN P. Brant Deben Hard, Lime Kiln Quay, Woodbridge IP12 1BD
M7 DXP C. Swift 10 Sankey Road, Haydock, St. Helens WA11 0DD
M7 DXQ R. Foulds 40 Broome Avenue, Swinton, Mexborough S64 8QQ
M7 DXS C. Bassett 2 Culverden Square, Tunbridge Wells TN4 9NS
M7 DXT A. Wickins Flat 8, Victoria Court, Milton Green, Langdon Hills, Basildon SS16 6GB
MI7 DXU X. Qin 9 Glenbrook Avenue, Belfast BT5 5JP
M7 DXX P. Kilby 1 Home Farm Barns, South End, Milton Bryan, Milton Keynes MK17 9HS
M7 DXZ J. Cheeran 37 Farnol Road, Dartford DA1 5NG
MM7 DYA C. Buchanan 4 Smiddy Place Letham, Forfar DD8 2SD
M7 DYB N. Abdallahi 66 West Street, Stratford-upon-Avon CV37 6DR
M7 DYG Dr M. Ali 4 Woolcombers Way, Bradford BD4 8JF
MW7 DYH D. Haden 1 Barham Road, Trecwn, Haverfordwest SA62 5XX
M7 DYI A. Giles Apartment 437, Holden Mill, Blackburn Road Blackburn Road, Bolton BL1 7QJ
M7 DYM S. Medhurst 50 Robson Drive, Hoo, Rochester ME3 9EA
M7 DYO G. Shaw 12 Aged Miners Homes, Boldon Colliery NE35 9JE
M7 DYP R. Forrest 5 Harmony Hill, Milnthorpe LA7 7QA
M7 DYR A. Loveday 50 Tenderah Road, Helston TR13 8NT
M7 DYS D. Sandland 54 Bishopdale Drive, Rainhill, Prescot L35 4QH
M7 DYU A. Styles 6 Morpeth Avenue, Darlington DL1 2QG
MD7 DYW S. St.John 8 Slieau Ree Apartments, Main Road, Union Mills, Isle of Man IM4 4ND
M7 DYZ B. Humble Linden Close, Dewsbury WF128PL
M7 DZC J. Baker 19 Littlebrook Lane, Bedford MK43 9RA
M7 DZE D. Simmons 23 Parker Road, Ashmore Park, Wednesfield, , Wolverhampton WV11 2HL
M7 DZF K. Phillips 89 Cochrane Road, Dudley DY2 0RU
M7 DZG D. Martin Fawcett House 34 Forton Road, Chard TA20 2HL
M7 DZI G. Hales Flat 19, The Cross, 101-107 Commercial Road, Poole BH14 0DL
M7 DZK D. Pyott 64 Carlton Avenue, Worksop S81 7JZ
M7 DZL D. Silsbury 35A Churchill Close, Wimborne BH21 4BH
M7 DZM D. Fox 183 Grange Road, Letchworth Garden City SG6 4LP
M7 DZO L. Pearson 45 Clementine Drive Mapperley, Nottingham NG3 5UX
M7 DZP R. Newbon Rose Cottage, Sowters Lane, Burton-On-The-Wolds, Loughborough LE12 5AL
M7 DZQ I. Coleby Engine House Cottage, Creech St. Michael, Taunton TA3 5RA
MM7 DZS K. Smillie 1 Blaven Court, Baillieston, Glasgow G69 7HY
M7 DZT S. Bradley Eden Cottage, Salters Lane, Shotton Colliery DH6 2PZ
M7 DZU S. Clements Church Farm, South Scarle NG237JH
M7 DZW H. Harclerode The Bungalow, Silverlace Green, Parham, Woodbridge IP13 9AD
MM7 EAB R. Boyd 9 Abbey Road, Scone PH2 6LWL
M7 EAC Dr R. Brown 46 Mandarin Way, Derby DE24 8YE
M7 EAE E. Barrett 4 Corinth Walk, Worsley, Manchester M28 3EB
M7 EAH C. Taylor Pond Farm, Sandy Lane, East Tuddenham, Dereham NR20 3JF
M7 EAK D. Smith 3 Littleton Crescent, Penkridge, Penkridge ST19 5BQ
M7 EAM E. Millett Church Road, Skeyton NR10 5AX
M7 EAN I. Crossley 81 Quinton Drive, Bradwell, Milton Keynes MK13 9HP
M7 EAO M. Lake 18 Keppel Road, Dagenham RM9 5LT
M7 EAP G. Watkins 10 Maple Drive, Newport PO30 5QP
MM7 EAR J. Eardley 8 Greenbraes Crescent, Gourdon, Montrose DD10 0NG
M7 EAS J. Chidley 6 Rhodes Avenue, London N22 7UT
M7 EBC A. Turner 602 Manchester Road, Sheffield S10 5PT
M7 EBI P. Watkins 10 Lea Lane, Cookley, Kidderminster DY10 3TA
M7 EBJ Z. Mcmillan 2 Burgess Close, Whitfield, Dover CT16 3NP
M7 EBL C. Edwards 57 Shawbrooke Road, Eltham SE96AL
M7 EBU A. Crombie 22 Withdean Court, London Road, Preston, Brighton BN1 6RN
M7 EBW L. Culshaw Barnsole Farm, Fleming Road Staple, Canterbury CT3 1LG
M7 EBY P. Norwell Russet, Litcham Road, Great Dunham, King's Lynn PE32 2LJ
M7 EBZ C. Parrott 15 Park Lane Snitterfield, Stratford upon Avon CV370LT
M7 ECF S. Phillipo 16 Leach Close, Chelmsford CM2 7DS
M7 ECG F. Ehlers North Lodge, Stanford Hall Estate, Stanford On Soar, Loughborough LE12 5QW
M7 ECM A. Warr 27 Blundell Place, Bedford MK42 9XB
M7 ECN S. Titley 20 Tower Road, Portishead BS20 8RE
M7 ECP Lord A. Aspinall 27 Swanswell Street, Coventry CV1 5FZ
M7 ECT E. Tattersall 17 Malt Kiln Way, Sandbach CW11 1JL
MM7 ECV A. Brand 6 Uraghag .Garenin, Stornoway HS29AJ
M7 ECW R. Holmes 21 Castle Hill Close, Eckington, Sheffield S21 4BJ
M7 ECY Lord D. Giles 22 Boston Close, Chaddesden, Derby DE21 6WB
M7 ECZ R. Jobson 25A Admirals Way, Thetford IP24 2LB
M7 EDE D. Edwards 21 Burton Old Road, Streethay, Lichfield WS13 8LJ
M7 EDK D. Kennett Twinstead Cottage, Sudbury CO10 7NA
M7 EDL E. Long 18 Prunus Road, Crewe CW1 4HB
MI7 EDO E. Mclaughlin 1 Stewarts Terrace, Derry BT487LH
M7 EDS E. Stewart Flat 1, 33 Robert Street, Harrogate HG1 1HP
M7 EDW E. Wright 2 Wulfrath Way, Ware SG12 0DN
M7 EDY S. Clarkson 14 Hambleton View, Tollerton, York YO61 1QW
M7 EDZ K. Edwards 24 Abbey Road, Halesowen B63 2HE
M7 EEC C. Johnson 40 Springfield Elstead, Godalming GU8 6EG
M7 EEE R. Vickery 17 Plain-An-Gwarry, Redruth TR15 1JB
M7 EEG M. Gee 97 Lincolns Mead, Lingfield RH76TA

M7 EEH J. Butters Overlands, Green Lane, Hyde Lea, Stafford ST18 9AY
M7 EEL L. Smith 10 Lee Road, Dewsbury WF13 3AX
M7 EEM V. Gibson 9 Dorchester Drive, Hartlepool TS24 9QY
MD7 EEO S. Williamson 45 Bluebell Close, Peel IM5 1GH
MM7 EEQ R. Hamlet 4 Broaddykes Avenue, Kingswells, Aberdeen AB15 8UH
M7 EEU W. Wren 9 Church Road Terrace, High Harrington, Workington CA14 5PL
M7 EEW N. Robson Roseland House, Pidney, Hazelbury Bryan, Sturminster Newton DT10 2EB
M7 EFA A. Corbett 3 Tretawn Gardens, Selsey, Chichester PO20 0DW
M7 EFB J. Armstrong Flat 2, 8 Thornfield Gardens, Tunbridge Wells TN2 4RZ
M7 EFC S. Walmsley 45 Johnston Avenue, Bootle L20 6HF
MW7 EFH A. Phillips Trefair, Tan Y Fron, Bylchau, Denbigh LL16 5NP
MM7 EFJ L. Yip 70 Kenley Road, Renfrew PA4 8YW
M7 EFM D. Compton Pin Lan, Far Vallens, Hadley, Telford TF1 5SE
M7 EFO A. Johnson 25 Eton Avenue, East Barnet, Barnet EN4 8TU
MM7 EFP E. Peat 1 Blackwood Drive New Cumnock, Cumnock KA18 4BW
M7 EFR M. Lane Kerity, Northchurch Common, Berkhamsted HP4 1LR
M7 EFT C. Towle 23 More Avenue, Aylesbury HP21 8JY
M7 EFU P. Jones Sussex Cottage, The Limes, Felbridge, East Grinstead RH19 2QY
M7 EFV N. Bonnar 5 Lavant Close, Waterlooville PO8 8BQ
M7 EFX G. Abson 62 New Hall Street, Burnley BB10 1PT
M7 EFZ D. Thomas 43 Buckland Road, Taunton TA2 8EW
M7 EGA N. Allsworth 6 Woodcroft, Oxford OX15NH
M7 EGB J. Smith 14 Alexander Avenue, Droitwich WR9 8NH
M7 EGE R. Jarvis Po Box 428 365 Waterloo Road Cheetham, Manchester M8 2FS
M7 EGJ A. Prime 68 Langley Drive, Crewe CW2 8LN
M7 EGK D. Frost 169 West Auckland Road, Darlington DL3 0SP
MM7 EGM E. Mackay 10 Blackpark Avenue, Invergordon IV18 0HY
M7 EGP O. Lazar 84 Bruce Street, Swindon SN2 2EN
MI7 EGS E. Paulikas 33 Clarefield, Dungannon BT71 6TQ
M7 EGT E. Travers 24 Mittell Court, Lydd, Romney Marsh TN29 9BJ
M7 EGW P. Farman 84, Hawthorn Road., Exeter EX2 6ED
M7 EGY J. Beer 92 Ashby Road, Hull HU4 7JT
M7 EHB M. Smith 30 Pitfold Road, London SE12 9HX
M7 EHC C. Cannon 7 Adelaide Close, Seaford BN25 2XB
M7 EHD I. Scriviner 32 Mountbatten Way, Plymouth PL9 9EJ
M7 EHG E. Hodgkiss 190 Ulverely Green Road, Solihull West Midlands B92 8AD
M7 EHL C. Talbot 24A Calder Road, Stourport-on-Severn DY13 8QD
M7 EHQ S. Al-Hajri 806 Witham Wharf Brayford Street, Lincoln LN5 7DL
MW7 EHV R. Scammells 11 Gower Green, Croesyceiliog, Cwmbran NP44 2QL
M7 EHW E. White 16 Sisley Way, Hinckley LE10 0GJ
MM7 EID D. Fitzgibbon 1 Walton Road, Kirkpatrick Durham, Castle Douglas DG7 3HG
MW7 EIJ R. Ramsay 29 Cardigan Close, Tonteg Tonteg, Pontypridd CF38 1LB
M7 EIK M. Rowlands 69 Greenvale Road, London SE91PB
M7 EIP D. Stanyer 37 Heron Crescent, Melton Mowbray LE13 1FT
M7 EIR I. Steele 4 Godwin Close, Colchester CO3 4BU
MM7 EIT D. Sipetan 10 North Square, Coatbridge ML5 2HB
M7 EIU M. Williams 70 Barton Drive, Barton On Sea, New Milton BH25 7JL
M7 EIW J. Clinton Flat 2, Watson Court, 12A The Crescent, Netley Abbey, Southampton SO31 5HP
M7 EJB P. Barnett 6 Flintsham Grove, Stoke-on-Trent ST1 5QS
MW7 EJC Dr E. Harries Ty Traeth, Caerwedros, Llandysul SA44 6BS
MW7 EJE E. Evans 10 Bryn Drive, Coedpoeth, Wrexham LL11 3LJ
M7 EJH L. Batten Y Felin Barn, Llawr-Y-Glyn, Caersws SY17 5RH
M7 EJI J. Earley White Dymes, Winford Road, Newchurch, Sandown PO36 0NE
MW7 EJJ E. Jenkins 41 Park Street, Bridgend CF31 4AX
M7 EJK R. Watkins 31 Talbot Road, Maidstone ME160HB
M7 EJM E. Mcmillan 12 Longendale Road, Standish, Wigan WN6 0UE
M7 EJN H. Williams 91 Sunholme Drive, Newcastle-upon-Tyne NE28 9YW
M7 EJP M. Zuber 36 Cashs Lane, Coventry CV1 4DS
M7 EJS E. Stammers 40 Tillingbourne Road, Shalford, Guildford GU4 8EY
MI7 EJV W. Moorhead 3 Ballyminstragh Road, Killinchy, Newtownards BT23 6PE
M7 EJW E. Weatherill Northend Cottage, North End Road, Yatton, Bristol BS49 4AS
M7 EJY J. Tarr 33 Easebourne Road, Dagenham RM8 2DW
M7 EKA G. Leach 43 Aldbanks, Dunstable LU6 1AH
M7 EKD S. Brown 56 Ashwood Avenue, Abram, Wigan WN2 5YE
M7 EKK G. Neagle 51 Kingston Lane, Southwick, Brighton BN42 4SJ
M7 EKL R. Slater 23 Lawn Road, Eastleigh SO50 4GT
MM7 EKP J. M Voinot 30 Argyll Street, Lochgilphead PA31 8NE
MM7 EKQ G. Gallagher 5 Mcnay Crescent, Saltcoats KA21 6AX
M7 EKR P. Clarke 19 Woodlands Close, Denby Dale, Huddersfield HD8 8RH
M7 EKS E. Schonrock 1 Powell Road, Poole BH14 8SG
M7 EKT N. Thorpe Dove Cottage, Kithurst Park, Storrington, Pulborough RH20 4JH
M7 EKU M. Butterworth 228 High Street, Leeds LS23 6AD
M7 EKV T. Cooper 14 Hawthorn Drive, Brackley NN13 6PA
MM7 EKZ S. Strachan 8 Pitforthie Place, Brechin DD9 7AX
M7 ELE E. Chauvelaine 292 Mount Pleasant, Redditch B97 4JL
M7 ELG K. Pollard 1 Ruskin Close, Selsey PO20 0TE
MW7 ELL P. Ellis Flat 4, 32 Lawson Road, Colwyn Bay LL29 8HE
M7 ELN A. Miller East Lyn, Lazonby, Penrith CA10 1BX
M7 ELO J. Myers 3 Cavendish Street, Leigh WN7 1SG
M7 ELS E. Sanders 1A Lychgate Drive, Waterlooville PO8 9QE
M7 ELU M. Morton 40 Mulgrave Road, London SE18 5TY
M7 ELZ E. Hase 23 Southdown Road, Thatcham RG19 3BF
M7 EMC J. Reed 3 Kings Croft, Ealand, Scunthorpe DN17 4GA
MM7 EMD P. Turner 7 Gartcows Drive, Falkirk FK1 5QQ
M7 EME J. Smethurst 57 Willoughby Close, Warrington WA5 9QP
M7 EMH C. Holmes Plot 4 Van 1 Hygrove Residential Park, Hygrove Lane, Minsterworth, Gloucester GL2 8LE
M7 EMI J. Isaacs 20 Vine Street, Worcester WR3 7DY
M7 EMJ E. Callow Orchardside, Lampley Road, Kingston Bridge, Clevedon BS21 6TY
MW7 EMK H. Parker Dolifor, Llanwrthwl, Llandrindod Wells LD1 6NU
M7 EMO M. Hall 6 Cunliffe Avenue, Newton-le-Willows WA12 0JX
M7 EMU D. Lambert-Barnett Fordlands Farm, Exeter EX6 7ST
M7 EMV M. Blain 63 Walmington Fold, London N12 7LD
M7 EMW D. Pearson 74 Arena Park, Exeter EX4 8RD
M7 EMY D. Emerton 15 Swallow Close, Stoke-on-Trent ST3 7FN
M7 ENB M. Glen 134 Spen Lane, Leeds LS6 3NA
M7 ENC J. Hoang 2 Chichester Avenue, Ruislip HA4 7EH
MM7 ENF D. Line 55 Burn Brae Westhill, Inverness IV2 5RH
M7 ENH I. Hodgkiss 190 Ulverley Green Road, Solihull B92 8AD
M7 ENI V. Curtis 52 Nairn Road, Lancaster LA1 5UY
M7 ENJ M. Faben 15 Monks Close, Doncaster DN7 4QL
MW7 ENN I. Wright Wernelli, Llannon, Llanelli SA14 6AP
MM7 ENQ S. Wright 67 Park Terrace, Broxburn EH52 6AP
M7 ENU M. Dadswell Avaeya Villa, Winchester Road, Waltham Chase, Southampton SO32 2LX
M7 ENV M. Savage 5 Mere Hall Barns, Mere Lane, Enville, Stourbridge DY7 5JL
M7 ENW Dr M. Heywood 17 Lavenham Close, Macclesfield SK10 2TS
M7 EOA M. Allen 8 Oakdene Road, Watford WD246RW
M7 EOB S. Tyler-Murphy 63 Sedlescombe Gardens, St. Leonards-On-Sea, St. Leonards on Sea TN38 0YT
M7 EOD A. Parker 51 Smithy Lane, Tingley, Wakefield WF3 1QH
M7 EOE T. Fowler 16 Walsingham Drive, Taverham, Norwich NR8 6FZ
M7 EOH E. Crisp-Mcneil Jacaranda, Gandish Road, East Bergholt CO7 6TP
M7 EOK Z. Downer 16 Windmill Lane Freshwater Isle Of Wight, Freshwater PO40 9DX
M7 EOL M. Williams 19 Cotebrook Drive, Upton, Chester CH2 1RA
M7 EOQ J. Kirk 108 Lodge Road, Heacham, King's Lynn PE31 7SZ
M7 EOS S. Allsopp 5 Grebe Road, Banbury OX16 9YZ
M7 EOT A. Highfield 54 Parkfield Street, Rowhedge, Colchester CO5 7EL
M7 EOU E. Pereira 15 Burgess Green Close, St. Annes Park, Bristol BS4 4DG
M7 EOV A. Beauchamp Nightingale Lane, Burgess Hill RH15 9JJ
M7 EOZ Z. Howling Sysonby Knoll Hotel Asfordby Road Melton Mowbray, Melton Mowbray LE13 0HP
MM7 EPD K. Macandie 33 Jordanhill Crescent, Glasgow G13 1UN
M7 EPF C. Miller Whitechapel Mission, 212 Whitechapel Road, London E1 1BJ
M7 EPG E. Gresty 7 Chester Road, Winsford CW7 2NG
MW7 EPH C. Howells 22 Hafod Arthen Estate, Brynithel, Abertillery NP13 2HX
M7 EPI T. Flude 32 Borough Street, Brighton BN1 3BG
M7 EPK N. Shave 52 Da Volls Court, Gorleston, Great Yarmouth NR31 6NH
M7 EPN M. Levett 16 Firs Avenue, Waterlooville PO8 8RS
M7 EPS B. Irving 25 Doniford Road, Williton, Taunton TA4 4SG
M7 EPT C. Duncombe 17 South Lynn Crescent, Bracknell RG12 7JU
M7 EPV S. Mawn 1 Blue Water Drive, Elborough, Weston-Super-Mare BS24 8PF
MM7 EPW A. Philip 10 Gordon Brown Place, Mallaig PH41 4RL
M7 EPY C. Ward 1 Castle View, Pontefract WF81EH
M7 EQC E. Coates 3 Glenview Close, Crawley RH10 8AS
MM7 EQF O. Nicol 34 Torvean Crescent, Kirkcaldy KY2 6FT
M7 EQG A. Ratowski 18B Hurstbrook Drive, Stretford, Manchester M32 9JQ
M7 EQI J. Mc Elhinney 231 Eastern Avenue, Sheffield S2 2GP
M7 EQM S. Cartlidge 45 Forest Road, Market Drayton TF9 3HX
M7 EQN C. Evans Lilly Cottage, Wood Enderby, Boston PE22 7PQ
M7 EQP K. Knight 72 Beaconsfield Villas, Brighton BN1 6HE
M7 EQQ M. Steed 38 Rivelands Road Swindon Village, Cheltenham GL51 9RF
M7 EQR F. Liput 8 Willow Way, Horsford, Norwich NR10 3GE
M7 EQS K. Whitman Flat 45, Regis Gate, North Street, Milton Regis, Sittingbourne ME10 2FA
MW7 EQT M. Rumball Crex, Tregynon, Newtown SY16 3PY
MW7 EQU D. Adams 1 Parkhall Caravan Park Penycwm, Haverfordwest SA626LS
MM7 EQX S. Wallace 28 Laurel Bank Terrace, Castle Douglas DG7 1BP
M7 EQZ A. Cartwright 154 Camp Hill Road, Nuneaton CV10 0JJ
M7 ERB E. Russell-Brown 2 Churn Hill, North Cerney, Cirencester GL7 7DN
M7 ERE C. Dodds 22 Cambridge Street, Wolverton, Milton Keynes MK12 5AJ
M7 ERF I. Thomson 23B Vicarage Gardens, Netheravon, Salisbury SP4 9RW
M7 ERG E. Earle 20 Queens Crescent, Gorleston, Great Yarmouth NR31 7NN
M7 ERI D. Conlan 33 Urmson Road, Wallasey CH45 7LE
M7 ERJ C. Whittenbury Little Orchard, The Common, Abberley, Worcester WR6 6AY
M7 ERK E. Keeley 10 Farringdon Close, Peterborough PE1 4RQ
MW7 ERP T. Clode 31 Heol Rhos, Caerphilly CF83 2BE
M7 ERQ M. Ward 81 Houldey Road, Birmingham B31 3HH
M7 ERS S. Stephens 65 Summerlin Drive, Milton Keynes, Buckinghamshire MK17 8GP
M7 ERT A. Eastwood 3 Bowes Nook, Bradford BD6 2BJ
M7 ERU A. Usher 3 Meadow View, Patrington HU12 0QG
M7 ERV M. Smith 34 Winnycroft Lane Matson, Gloucester GL4 6EJ
M7 ESB D. Wilson 7 Wisteria Grove, Birmingham B44 9AX
M7 ESC S. Christie 21 Lovat Lane, London EC3R 8EB
M7 ESE S. Ambrosio 8 The Pines, Long Lane, Chester CH2 2QF
M7 ESF J. Philps 14 Royston Gardens, Bexhill-on-Sea TN40 2PB
MM7 ESG A. Xenos 8 Hayburn Crescent Main Door, Glasgow G11 5AX
M7 ESH E. Sheremetyeva 6 Pixey Close, Yarnton, Kidlington OX5 1FY
M7 ESI A. Clark 62 New Road, London SE2 0QG
M7 ESJ D. Edmonds 5 Sykes Croft Emerson Valley, Milton Keynes MK4 2DX
M7 ESK S. Shields 28 Bank Field Westhoughton, Bolton BL5 2QG
M7 ESM H. Qureshi 33 St. Michaels Crescent, Luton LU3 1LZ
M7 ESQ B. Chapman 62 Spring Grove, Loughton IG10 4QE
M7 ESR N. Rajendran 1 Sharp?S Yard, London Road, Six Mile Bottom, Newmarket CB8 0UH
M7 ESS M. Taylor 168 Long Furrow, East Goscote, Leicester LE7 3SU
MM7 EST E. Thorpe 27 Castle Street, Dunbar EH42 1EX
M7 ESV R. Tadimalla 5A Canterbury Road, Croydon CR0 3PY
M7 ESW J. Roberts 34 Gloucester Road, Consett DH8 7LL
M7 ESX K. Polston 10 Marsh Farm Road, South Woodham Ferrers, Chelmsford CM3 5WP
M7 ESY S. Foster 84 Peveril Road, Tibshelf, Alfreton DE55 5LR
M7 ESZ E. Zaralli 53A Winchester Street, London SW1V 4NY
M7 ETB E. Ball 270 German Church Road, Mount Cotton 4165 Australia
M7 ETF D. Gill 38 Churnwood Road, Colchester CO4 3HG
M7 ETG S. Plant 32 Maple Avenue, Eccles, Manchester M30 8JU
M7 ETH E. Wiggins 32 Prestbury Road, Duston, Northampton NN5 6XP
MM7 ETI J. Stokroos 20 Ochlochy Park, Dunblane FK15 0DU
M7 ETK V. Edwards 2 Austin Road, Bromsgrove B603LZ
M7 ETL G. Turner 71 Caymer Road, Eastfield, Scarborough YO11 3HQ
M7 ETM J. Parish 33A The Broadway, Hornchurch RM12 4RN
M7 ETN R. Sheffield 7 Elysium Terrace, Northampton NN2 6EN
M7 ETQ T. Young 22, Martin Close Godmanchester, Huntingdon PE29 2WA
M7 ETS M. Parry 2 Conway Gardens, Falmouth TR11 4LN

MM7 ETU A. Mackellaig Glengarry, Marine Place, , Mallaig PH41 4RB
MW7 ETY A. Williams Dros-Y-Glyn, Bellfountain Road, Crickhowell NP8 1SN
M7 ETZ I. Maddock 54 First Avenue, Kidsgrove, Stoke-on-Trent ST7 1DW
M7 EUA S. Bowen Flat 6, Wyndley House, 9 Welshmans Hill, Sutton Coldfield B73 6RY
M7 EUC E. Cubukcu 27 Chapel Close, Rushden NN10 0FH
M7 EUF M. Szylko 9 Kimberley Road, Gillingham ME7 4NE
MW7 EUG D. Monks Bryn, Church Road, Talysarn, Caernarfon LL54 6HP
MW7 EUH G. Booth 6 Cerdin Avenue, Pontyclun, Rct CF72 9ER
M7 EUI A. Murphy Linnets, Prospect Road, Upton OX11 9HT
M7 EUM T. Dane 73 Firbeck Crescent, Langold, Worksop S81 9SA
M7 EUP M. Hardy 4 Acre Villas, Mytholmroyd, Hebden Bridge HX7 5DF
M7 EUS M. Allington Flat 16, Isambard Court, Damers Road, Dorchester DT1 1RQ
M7 EUW I. Mcdonnell 5 Mythop Place, Ashton-On-Ribble, Preston PR2 1LL
MW7 EUX R. Ayres 36 Ariel Reach, Newport NP20 2FP
MM7 EUY R. Allen 4 East Mains Of Dysart, Montrose DD10 9TH
M7 EVB K. Western 27 Main Street, Pymoor, Ely CB6 2ED
M7 EVF J. Best 40 Queens Avenue, Gedling, Nottingham NG4 4EJ
M7 EVG D. Robinson 20 Joys Bank, Holbeach St. Johns, Holbeach, Spalding PE12 8SD
M7 EVH P. Morgan Pen Y Maes Cottage, Brecon Road, Hay On Wye, Hereford HR3 5PP
MM7 EVK R. Stacey G1 23 Kelburn Street, Millport KA28 0DU
M7 EVM E. Merritt 35 Jocelyn Drive, Wells BA5 2ER
M7 EVP G. Hayes 126 Lincoln Road, Skegness PE25 2DN
M7 EVQ T. Gyalai 11 Victoria Road North, Southsea PO5 1PL
MM7 EVR A. Macleod 25 Woodlands Brae, Westhill, Inverness IV2 5JH
M7 EVZ Dr T. Lachlan-Cope 38 Cambridge Road Waterbeach, Cambridge CB25 9NJ
M7 EWB E. Beckett 55 Schoolfield, Preston PR58BH
MW7 EWF J. Butterfield Craig Y Nos, Llangammarch Wells LD4 4EW
MM7 EWH S. Sime 33 Inchkeith Crescent, Kirkcaldy KY1 1GL
M7 EWI S. Flay 20 Elmwood, Chippenham SN151AP
M7 EWM B. Stace 4, Langley Grove, , Bishop Auckland DL14 6UJ
M7 EWN F. Spencer 19Trafalgar Road Cirencester, Cirencester GL72EJ
MW7 EWP R. Stark 10 Stanley Park Avenue, Rhyl LL18 4SB
M7 EWQ I. Jackson 32 Cromwell Road, Poole BH12 2NS
M7 EWT A. Newton 4 Acre Close, Market Rasen LN8 3DL
MD7 EWU N. O'Dell 30 Magher Breek, Peel IM5 1XD
M7 EWY M. Leatherland 352 High Street, Chatham ME4 4NP
M7 EXB T. Ranger 24 Jasmine Way, Trowbridge BA14 7SW
M7 EXF R. Jacklin 43 Garrett Close, Kingsclere, Newbury RG20 5SD
M7 EXG Prof. L. Hall 5 Merlin Avenue, Nuneaton CV10 9JY
M7 EXI A. Edwards 23 Chalky Road Broadmayne, Dorchester DT2 8EJ
M7 EXJ M. Freeman 40 Springfield Road, Twickenham TW2 6LQ
M7 EXL A. Rimmer 9 Garth End, Huntington, York YO329QU
MM7 EXN T. Edgar 2 Whyte Place, Milnathort, Kinross KY13 9YL
M7 EXQ J. Townend St Annes Way, Birmingham B44 0HN
MW7 EXS S. Sayle 5 Lle Eirlys, Cwmbran NP44 1FE
M7 EXT C. Newman 3 Orchard Way, Northstowe, Cambridge CB24 1AG
M7 EXV P. Dobson Barton Cottage, Court Barton, Higher Rocombe, Stokinteignhead, Newton Abbot TQ12 4QL
M7 EYB S. Eyers 35 Milburn Road, Gillingham ME7 1PQ
MW7 EYC M. Cannaby 33 Victoria Road Bulwark, Chepstow NP165QW
MM7 EYO E. Young 31 Hilton Road, Milngavie, Glasgow G62 7DN
M7 EYP S. Wilkinson 16 Chaucer Street, Hull HU8 8NA
M7 EYX A. Twigg 19 Makepeace Avenue, Warwick CV34 5SB
M7 EYY S. Perks 8 Holcot Road, Coalway, Coleford GL16 7HH
M7 EZB N. Hall 11 Woodstock Road, Weston-Super-Mare BS22 8AH
M7 EZC M. Beckett 10 Elmfield Road, Huddersfield HD2 2XH
M7 EZG S. Bunting 15 Newbegin Close, Norwich NR14PU
M7 EZL A. Musgrave 179 High Street, Kimpton SG48QN
M7 EZO N. Finch Melbourne, Bridewell Street, Clare, Sudbury CO10 8QD
MW7 EZP M. O'Brien 37 Ty Glas Road Llanishen, Cardiff CF14 5EB
M7 EZQ S. Dewhurst 32 Humber Road Thornaby, Stockton-on-Tees TS17 8JB
M7 EZR J. Peat Flat 12, 1 Whyteleafe Hill, Whyteleafe CR3 0FA
M7 EZS R. Harris 5 Alric Drive, Brinsworth, Rotherham S60 5AJ
MW7 EZW V. Hawkins 2 Rumsey Drive, Neyland, Milford Haven SA73 1QQ
M7 EZX N. Harrison 14 Prior Dene, Darlington DL3 9EW
M7 EZZ I. Gray 14 The Drift, Barrowby, Grantham NG32 1DQ
M7 FAB F. Brazier 277 Feltham Hill Road, Ashford TW15 1LT
M7 FAF A. Davidson 2 Neil Street, Widnes WA86RH
MW7 FAI G. Waldron Glenafton, Pen Y Fan Road, Brecon LD3 8DB
M7 FAJ C. Radley 4 Sarum Road, Tadley RG26 4ES
M7 FAL M. Roberts Bowerbank Hall, Scotland Road, Penrith CA11 9HL
M7 FAM J. Jordaan 7 Springfields, Bugle, St. Austell PL26 8SJ
M7 FAN R. Fanyinka 9 Orchard Rise, Chard TA20 1FX
M7 FAO S. Hassan 24 Ranworth Road, London N9 0LN
M7 FAP D. Briton 77 Crabtree Avenue, Rossendale BB4 9TB
M7 FAV Dr S. Favell Lodge Farm, Moor Lane, Reepham, Lincoln LN3 4EE
M7 FAX N. Tarrant 87 Bideford Green, Linslade, Leighton Buzzard LU7 2TJ
M7 FBC J. Gittins 9 Beechfield, Grasscroft, Oldham OL4 4EN
M7 FBJ N. Wache 11 Horselees Road, Boughton-under-Blean ME13 9TG
M7 FBM C. Evans 7 Westfield View, Wakefield WF1 3RU
MM7 FBN J. Bolte 23 Dalblair Road, Ayr KA7 1UF
M7 FBO C. Hoy 27 Newminster Road, Morden SM4 6HJ
MM7 FBP F. Pate 55 Hamilton Crescent, Bishopton PA7 5JT
MM7 FBQ D. Mcknight 13 Eastfield Place Fauldhouse, West Lothian, Bathgate EH47 9BF
M7 FBR C. Wilson Collingwood Avenue, Tolworth KT59PT
M7 FBW D. Bevan Flat, The Roost, 30 Treverbyn Road, Stenalees, St. Austell PL26 8TJ
M7 FBX Lady A. Paul 38 Ploverly, Peterborough PE4 6HZ
M7 FBY D. Hampson 120 Boyds Walk, Dukinfield SK16 4AU
M7 FBZ R. Rose 9 Rowston Close, Gainsborough DN21 2SF
M7 FCA D. Grant 22 Alba Gardens, London NW11 9NR
M7 FCB C. Barrett 40 Hillside Road, Dover CT17 0JQ
M7 FCD T. Butterworth Flat 1. 15Treyew Road, Truro TR1 2BY
M7 FCG A. Rout 4 North Road, Banwell BS29 6AY
M7 FCH C. Hoyle 2 Westfield Street Chadderton, Oldham OL96RQ
MW7 FCN B. Hoskins 23A Llannon Road, Upper Tumble, Llanelli SA14 6BW
MM7 FCT J. Stone Eilean Fraoich North Connel, Oban PA37 1QX
MI7 FCW D. Gilroy 28 Sandyknowes Park, Newtownabbey BT36 5DE
M7 FCX M. Rossini 224 Homestead Way, New Addington, Croydon CR0 0DU
M7 FCY D. Horne Flat 12, The Banks, 287 Blackbird Road, Leicester LE4 0DF
M7 FDC P. Ainsworth 4 St. Johns Road, Wirral CH62 0BN
M7 FDD F. Done 3 Home Farm Cottages, Old Warden Park, Old Warden, Biggleswade SG18 9DU
MW7 FDI S. Turner 9 Baldwin Street, Newport NP20 2LT
M7 FDJ P. Freeman 53 Spillmans Road, Stroud GL5 3LS
MW7 FDL Dr R. Jones 2 Heathlands, Ystrad Mynach CF82 7AZ
MM7 FDM D. Harcus Flat 4, 26 Lumsden Square, Edinburgh EH17 8NN
M7 FDQ C. Taylor 80 Kingsway, Petts Wood, Orpington BR5 1PT
M7 FDR J. Nasson 25 Lamplugh Crescent, Bishopthorpe, York YO23 2SR
M7 FDS F. Da Silva Flat 72, Vesta House, 4 Liberty Bridge Road, London E20 1AN
M7 FDT C. Inman 11 Sylvan Way Gillingham Dorset, Gillingham SP8 4EQ
M7 FDW F. Woolley 10 Hazelmoor Fold, Elland HX5 0DR
M7 FDY G. Wheeler 1, The Buildings. The Street, , Stinchcombe GL11 6AW
M7 FDZ A. Flint 32 Trent Road, Shaw OL2 7YL
M7 FEA M. Eldicott 84 Frost Road, Bournemouth BH11 8HR
MW7 FEC J. Williams 12 Pantycelyn, Fishguard SA65 9EH
M7 FEH A. Stanley Penrith, Park Avenue, Madeley, Telford TF7 5AB
MM7 FEM J. Small 7 Gemmell Crescent, Ayr KA8 0JR
M7 FEO A. Ponting 16 Cavan Walk, Bristol BS4 1PN
M7 FEP A. Thickett 12 Waverley Road, Darlaston WS108ED
M7 FET D. Purslow 169 Hoo Road, Kidderminster DY10 1LP
M7 FEU G. Godfrey 11 Defrene Road, London SE26 4AB
M7 FEV S. Gee 7 Orton Way, Ashton-In-Makerfield, Wigan WN4 9NQ
MM7 FEW J. Frew 125 Minnoch Crescent, Maybole KA19 8DR
M7 FEX J. Longmuir 62 Ravenswood Avenue Rock Ferry, Wirral CH42 4NX
M7 FFL A. Beecroft 10 Marchant Close, Beverley HU17 9GE
M7 FFP J. Ross 5 Eastgate, Heckington NG349RB
M7 FFU N. Limb Ripston, Alford Road, Sutton-On-Sea, Mablethorpe LN12 2RL
MI7 FFV A. Neill 19 Huntingdale Lodge, Ballyclare BT39 9FB
M7 FFW A. Martin The Grange, Hereford HR29SD
M7 FFX A. Brind 77 Bridge Street, Runcorn WA7 1BE
M7 FGB F. Byrne 144 Meridian Road, Boston PE21 0NF
M7 FGC F. Chapman 77 Cedar House, Spelthorne Grove, Sunbury-on-Thames TW16 7DD
M7 FGD I. Wilson 20 Viburnum Close, Ashford TN23 3LB
MW7 FGG J. Waite 26 Glas Y Gors, Aberdare CF44 0BQ
M7 FGH D. Barwick 2 Sutton Close, Bury St. Edmunds IP327EP
MI7 FGI P. Gartland 79 Ballycowan Road, Ballymena BT42 3HS
M7 FGJ R. Mann Mill Pond View, Retford Road Blyth Notts, Worksop S81 8EY
MM7 FGL A. Hall 94 Muirside Avenue, Kirkintilloch, Glasgow G66 3PH
M7 FGM P. Jeffery Littlecroft, Stanton Road, Forest Hill, Oxford OX33 1DT
M7 FGQ L. Hardstaff 12 Daisy Dale, Boston PE21 6DR
M7 FGS P. Pendlebury 79 Pinfold Close, Westhoughton, Bolton BL5 2RN
M7 FGT F. Tozzi 120 Wellhouse Lane, Barnet EN5 3DP
MM7 FGU D. Price 72 St Michael Street, Dumfries DG1 2QF
M7 FGV D. Trigg 7 Severn Road, Ashton-In-Makerfield, Greater Manchester WN4 8UE
MW7 FGW R. Hook 27 Gwaun Ruperra Close, Llantrisant, Pontyclun CF72 8QR
M7 FGY W. Evenden 15 Knights Garden, Hailsham BN27 3JR
M7 FHC T. Liviu 6 Hales Park Close, Hemel Hempstead HP2 4TJ
M7 FHF G. Lesmoir-Gordon 48 Chichester Walk, Wimborne BH21 1SN
M7 FHG J. Akhtar 986 London Road, Alvaston, Derby DE24 8PY
M7 FHK F. Badcock 18A Main Street, Wardy Hill, Ely CB6 2DF
MI7 FHM F. Mcalister 24 Ballyhone Road Gleno, Larne BT403LW
M7 FHN J. Miles 55 Longbeech Park, Canterbury Road, Charing, Ashford TN27 0HA
M7 FHO M. Brabben 31 Tivoli Gardens 31, Grimsby DN32 7SE
M7 FHU Dr R. Munro 14 Homefield Gardens, Tadworth KT20 5HP
MM7 FHW A. Mcharg 11 Linfern Avenue East Kilmarnock, Kilmarnock KA1 3LL
MI7 FHY J. Wallace 207 Victoria Road, Bready, Strabane BT82 0EB
MD7 FHZ R. Annett 6 Carnane View, Ballakilley, , Port St. Mary IM9 5NR Isle of Man
M7 FIA S. Ezzard 65 Pettycot Crescent 65, Gosport PO13 0SJ
M7 FIB R. Orourke 33 David Street, Grimsby DN32 9NL
M7 FIE A. Taylor Old Rectory Cottage Faversham Road, Boughton Aluph TN25 4HS
MW7 FIJ J. Beards 16 Moorview Close Gendros, Swansea SA5 8BZ
M7 FIN P. Webber 17 Lilleshall Drive, Elstow, Bedford MK42 9FG
M7 FIP M. Tetlow 21 Alexander Road, Quorn, Loughborough LE12 8EQ
M7 FIQ L. Hibbs 60 Tremarle Home Park, North Roskear, Camborne TR14 0AT
M7 FIV S. Cockcroft 20 Essex Drive, Bury BL9 9JH
M7 FIW D. Parnham 79 Queen Street, Doncaster DN4 8AB
M7 FIZ M. Charnley 33 Cromdaleway Great Sankey, Warrington WA53NR
M7 FJC J. Ball Dumpford Park Farm, Trotton, Petersfield GU31 5JW
M7 FJE N. Larby 34B Portchester Road, Bournemouth BH8 8JY I
MM7 FJG D. Spooner Glachbeg, Allanglach Wood, North Kessock, North Kessock IV1 3XD
M7 FJL T. Price Flat D Rear Of Blackhall Yard Stricklandgate, Kendal LA9 4LU
M7 FJP P. Slade 2 The Coach House Burnby Lane, Pocklington YO42 2QE
M7 FJR A. Brown Highgate Pennymoor, Tiverton, Ex16 8Ls Tiverton, Tiverton EX16 8LS
M7 FJS F. Simpson 36 Lyndhurst Ave, Blackpool FY4 3AX
M7 FJT A. Thompson 7 Willetts Field, Muddles Green, Chiddingly, Lewes BN8 6HU
M7 FJU G. Manchester 2 B The Avenue, Eccleston Eccleston, St. Helens WA10 5NP
M7 FJY C. Fleming 18 Parkfield Crescent, Lea, Preston PR2 1QU
MW7 FKA S. Thomas Flat 2 Tai?R Grawen, Merthyr Tydfil CF478PP
M7 FKG N. Ryan 35 Pightle Close, Elmswell, Bury St. Edmunds IP30 9EJ
M7 FKJ J. Barnacle 4 Mariners Lea, Broadstairs CT10 2QB
M7 FKK M. Donnelly 4 Alstone Court, Choppington NE62 5BU
MM7 FKL R. Docherty 207 Greengairs Road, Greengairs, Airdrie ML6 7SZ
M7 FKO T. Dyer 6 Strathwell Crescent, Whitwell, Ventnor PO38 2QZ
M7 FKP M. Pinchbeck 52 Fellow Lands Way, Chellaston, Derby DE73 6SW
M7 FKQ M. Da Silva 7 Buttell Close, Grays RM17 6UN
M7 FKR A. Ferrer 26 Rochester Drive, Westcliff-on-Sea SS0 0NL
M7 FKT R. Stiff Hawthorn House, Flegg Green, Wereham, King's Lynn PE33 9BA
M7 FKX F. Bila-Nicola 48 Shepherds Pool, Evesham WR11 4JG
M7 FLA Lord R. Turner-Trotter 23 Nodes Drive, Stevenage SG2 8AL
M7 FLD S. Muller 426A St. John Street, London EC1V 4NJ
M7 FLE M. Appleton 88 Woolsington Road, North Shields NE29 8RS

M7 FLG F. Gardner 2 Austen Road, Guildford GU1 3NP
M7 FLJ J. Cross 18 Brickly Road, Luton LU4 9EU
MW7 FLK R. Sawle Westering, Stonewall Hill, Presteigne LD8 2HB
M7 FLL R. Collins 42B Bredfield Road, Woodbridge IP12 1JE
MM7 FLN A. Kennedy 40/2 Marionville Road, Edinburgh EH7 5UB
MW7 FLP D. O'Dea 12 Trewaun, Hirwaun, Aberdare CF44 9HN
M7 FLR C. Day 14 Windsor Drive, Ramsey Forty Foot, Ramsey, Huntingdon PE26 2XX
M7 FLY W. Mcguire 16 Piel View Grove, Barrow-in-Furness LA13 0EF
MM7 FLZ C. Coutts 1 Springfield Avenue, Duns TD11 3BF
M7 FMA F. Allen 109 Barston Road, Oldbury B680PU
M7 FMB F. Mclean-Brown 101 Mayfield Avenue, Southend-on-Sea SS2 6NR
MW7 FMD F. Mcdermott Maesnewydd Talsarn, Lampeter SA48 8RE
M7 FME D. Blowers Annexe, Garden Cottage, Trumpets Hill Road, Garden Cottage, Reigate RH2 8QY
M7 FMF S. May 146 Lodge Hill, Welling DA16 1BL
M7 FMG Dr F. Morales Gundin Corner Close Broadmead Road Send, Woking GU23 7AD
M7 FMI M. Dixon 15 Brandon Road, Braintree CM7 2NL
M7 FMK J. O'Riordan 341 Eastwood Road, Rayleigh SS6 7LH
M7 FMM C. Wood 17 Chestnut Close, Greetland, Halifax HX4 8HX
M7 FMO M. Difelice 62 Cranleigh, Wigan WN6 0EU
M7 FMR R. Clark Stone Ends, Hesket Newmarket, Wigton CA7 8JS
M7 FMT F. Tang 78A Church Lane, Mill End, Rickmansworth WD3 8HE
MW7 FMV R. Hardick 18, Pembroke Dock SA72 6PZ
M7 FMZ R. Vogelsinger 49 Butterton Drive, Manchester M18 8GZ
MU7 FNB Ncc Group RS c/o I. Collins Chalet Blanc, La Vallee, St. Anne GY9 3XA Guernsey
MW7 FNC M. Ormrod 22 Avalon Court, Tranch, Pontypool NP4 6AH
MM7 FNG E. Boardman 86 James Park, Burntisland KY3 9EP
M7 FNI R. Stuart 4 Red Lion Cottages, High Street, Culworth, Banbury OX17 2BD
M7 FNK T. Wilkes 95 Grove Lane, Coulsdon CR5 2QD
M7 FNL Licolnshire Scout RC c/o E. Thompson 7 Turnpike, Helston TR13 8LR
M7 FNP N. Sharp 5 Penrwyn Court, Eynesbury, , St. Neots PE192SU
MM7 FNS M. Kabala St.Molios Park 55, Lamlash KA278JQ
MI7 FNY M. Feeney Leamington Place, Lisburn BT27 4UL
M7 FOI R. Toplis 13 Aston House, Chesterfield S402FF
M7 FOJ N. Tsourdalakis 12 Widbrook Meadow, Trowbridge BA14 9SD
M7 FOK I. Wilkinson The Beeches, Whitchurch Road, Aston, Nantwich CW5 8DJ
MM7 FOM I. Stewart 7 Glebe Road, Bathgate EH48 1DG
MW7 FON G. Pritchard 3 Llwyn, Rhosgadfan, Caernarfon LL54 7HN
M7 FPA I. Butler 9 Sulgrave Close, Dudley DY1 3DE
M7 FPB M. Kaszewski 10A Robertson Road, New Alresford SO249LQ
M7 FPC D. Bevan 21 Saxon Square, Thame OX9 2FS
MM7 FPD J. Hill 07 Naismith Court, Grangemouth FK39BQ
MI7 FPG M. Mccarron 28 Bennett Street, Londonderry BT48 6SG
M7 FPH N. Marshall 19 Manor Road, Bude EX238PY
M7 FPJ M. Wilmshurst 92 Croxted Road, London SE21 8NP
M7 FPK P. Kelly Apartment 5, Lind House, 3 Ellerslie Drive, Malvern WR14 3LQ
M7 FPN A. Pearce 230 Farmers Close, Witney OX28 1LH
M7 FPO A. Jones-Turner 2 Meadows Place, Hadleigh, Ipswich IP7 5FD
M7 FPP J. Seyforth 2 Waterloo Cottages, Worthing Road, West Grinstead, Horsham RH13 8LG
M7 FPT T. Fletcher 5 Lower Buckfeild Cottages Barons Cross Road, Leominster , Herefordshire HR6 8RN
M7 FPW G. Tame 1 Weydale Grove Hindley, Wigan WN2 4NG
M7 FPZ D. Gill 7 Westway Hanging Heaton, Batley WF17 6DF
MM7 FQA R. Pease The Treehouse, Inverkirkaig IV274LR
M7 FQB G. Reeve 13 Fitzgilbert Road, Colchester CO2 7XB
M7 FQC A. Smith 42 Grants Avenue, Bournemouth BH1 4NS
M7 FQD S. Gore 240 Main Road West Winch, King's Lynn PE33 0NZ
MW7 FQF C. Shaw 86 West Park Drive, Porthcawl CF36 3RN
M7 FQG L. Holden 185 Darfield Road Cudworth, Barnsley S72 8SF
MW7 FQN S. Berry 2 Morfa Crescent, Tywyn LL36 9AU
M7 FQO H. Burden Fff 55 King Edward Avenue, Bournemouth BH9 1TZ
MM7 FQP W. Crossley 2 Bankhead Ave, Glasgow G13 3TD
MW7 FQQ C. De Winton Tymawr Farm, Llanfrynach, Brecon LD3 7BZ
M7 FQR B. Spencer 27 Longlands Rd, Sidcup DA157LU
M7 FQT A. Scott-Myers 1 Hyacinth Grove, Wirral CH46 1SW
M7 FQV Z. Baronas 51 Keats Way, West Drayton UB7 9DS
MM7 FQW L. Thom Flat 804D, 110 St. James Road, Glasgow G4 0PS
M7 FQY G. Iles Brooklyn Bath Road Hardwicke, Gloucester GL2 2RG
M7 FQZ D. Garner 1 Rowland Avenue, Field Street, Hull HU9 1HR
M7 FRD R. Haughton 5 Fairfield Way, Wesham, Preston PR4 3EP
M7 FRE P. Freeman Grove Orchard, Knapp Lane Coaley, Dursley GL11 5AR
MM7 FRF J. Pearson 25 Langside Avenue, Kilmarnock KA1 4SP
M7 FRG H. Glover 26 Mersey Walk, Birkenhead CH42 3UN
M7 FRI C. Brown 17A Langley Drive, Derby DE74 2DN
M7 FRK H. Simmonds Unit D5 Devon Business Park, Saunders Way, Cullompton EX15 1BS
M7 FRN G. White 153 Wenming No2 Road, Beihai 536000 China
M7 FRR F. Ruddick 12 Davids Avenue, Great Sankey, Warrington WA5 1LN
M7 FRS B. Horne 31 Grampian Way, Oulton, Lowestoft NR32 3EP
M7 FRT P. Smith 11 Brine Road, Nantwich CW5 7BA
M7 FRW L. Finlayson 44 School Lane, Ashurst Wood, East Grinstead RH19 3QP
MM7 FRZ F. Laidlaw 12 Limeview Avenue, Paisley PA2 8NB
M7 FSC A. Weavers 76 Collindale Avenue Erith, Erith DA8 1EE
M7 FSE T. Grigg 7Westfield Avenue, Norton Malton YO17 8DN
M7 FSJ J. Dale 67 Arun Road, Billingshurst RH14 9PE
M7 FSK T. Stock 73 Walsingham Road, Southend-on-Sea SS2 4AL
M7 FSL D. Coombe 1 Cliff Cottages Cliff Hill, Maidstone ME174NQ
MM7 FSR R. Hamilton Brodie 18 Torcraik Crescent, North Middleton EH23 4SU
M7 FSU B. Crump Apple Tree Cottage, Kingsland, Leominster HR6 9PY
M7 FSV C. Dabbs 4 Sutherland Close, Bexhill-on-Sea TN39 3QJ
MW7 FSY M. Selby Tyn Y Buarth Brithdir, Dolgellau LL40 2RS
M7 FTB S. Mythen Flat 4, Invicta House Millmead Road, Margate CT9 3RN
M7 FTD S. Lord 18 Caunsall Road, Kidderminster DY11 5 YB
M7 FTF A. Redshaw 155 Wainbody Avenue South, Coventry CV3 6BY
M7 FTG P. Ashall 25 Aughton Avenue, Aughton, Sheffield S26 3XB
MM7 FTH G. Hill 21 Affric Drive, Falkirk FK27UF
MW7 FTI C. Houghton 5 Bank Street, Southsea, Wrexham LL11 5PE
M7 FTK A. Mullett 13 Babbacombe Road, Coventry CV3 5PE
M7 FTL Dr S. Frankau 79 Boyne Road, London SE13 5AN
M7 FTN K. Close Flat 2, Grosvenor House, Grosvenor Street, Southsea PO5 4JQ
M7 FTQ K. Gonulkirmaz 14 Ashbourne Avenue, London NW11 0DR
M7 FTR P. Moseley 14 Ashbourne Gardens, Hindley, Wigan WN2 4JH
M7 FTS P. Johnson 12A Beechroyd, Pudsey LS28 8BH
M7 FTT I. Harrison 50 Derwent Drive, Tibshelf, Alfreton DE55 5LT
M7 FTV S. Colley 10 Hoads Wood Road, Hastings TN34 2BJ
M7 FTX J. Hole 40 Campkin Road, Wells BA5 2DG
M7 FTZ P. Sharpe 23 Sparrows Mead, Redhill RH12EJ
MW7 FUA J. Davies 27 Pen Y Pwll, Pontarddulais, Swansea SA4 8EH
MM7 FUE W. Hamilton 13 Kirkfield West, Livingston Village, Livingston EH54 7BD
M7 FUG M. Bishop-Saunders 6 Stradsett Close, Downham Market PE38 9NY
M7 FUH C. Chryssaphes 200 Cavendish Avenue, London W13 0JW
M7 FUL M. Sims 85 Critchill Road, Frome BA11 4HQ
M7 FUN D. Hills Birchglade, Southampton SO40 2GP
MM7 FUO K. Simpson 23 Deanshaugh Terrace, Elgin IV30 4EZ
MM7 FUR Q. Misell Flat 59 St Peter'S Studios, Aberdeen AB24 3HQ
M7 FUW T. Jones 14 Brutus Close, Dorchester DT1 2TJ
MW7 FVB F. Steele Eboracum, Mold Road, Buckley CH7 2NH
M7 FVF R. Metcalfe 24 Hall Carr Road, Rossendale BB4 6AW
M7 FVG S. Mak 6 Wootton Way, Cambridge CB3 9LX
MW7 FVH A. James 57 Moorland Crescent, Beddau, Pontypridd CF38 2DN
M7 FVK C. Fithian-Franks 9 Prospect Close, Camblesforth, Selby YO8 8HG
M7 FVM D. Goodwin 26 Ravenscroft Close, Middlewich CW10 9PX
M7 FVS P. Squires 3 Kenan Drive, Attleborough NR17 2RJ
M7 FVY S. Purchase 17 Lower Ream, Yeovil BA21 3SB
M7 FWA M. Hurson Hq Company Clive Barracks, Ternhill TF9 3QE
M7 FWB D. Lewis 23 Malvern Road, Southampton SO16 6PZ
M7 FWC V. Morrell 43 Willow Drive, Carterton OX181JU
M7 FWD A. Wilson 42 North Church St, Fleetwood FY7 6AX
M7 FWE A. Turner 11 Battens Way, Havant PO9 2DX
M7 FWF N. Darby 28 Denefields Court, Matlock DE4 3EY
M7 FWG N. Stevenson 2 Broom Green Road North Elmham, Dereham NR20 5JZ
M7 FWH F. West Harrington Lodge, Orton, Kettering NN14 1LN
M7 FWI K. Sanders Holly Heights, Ulverston LA12 0SU
MM7 FWK F. Kinloch 53 Belmont Street, Newtyle, Blairgowrie PH12 8UB
M7 FWR W. Rogerson 51 Grasmere Avenue, Ryde PO33 1UN
M7 FWT A. Pengelly 11 Insworke Crescent, Millbrook, Torpoint PL10 1EP
M7 FWU D. Eaton 11 Penny Piece Place, North Anston, Sheffield S25 4JZ
MW7 FWV S. Williams Harbour Masters Office, Penarth Portway, Penarth CF64 1TQ
M7 FWW A. Crocker 32 Godmanston Close, Poole BH17 8BU
MW7 FWX A. Cunningham 5 Delafield Road, Abergavenny NP7 7AW
M7 FWZ J. O'Connor 7 Peverill Close Attleborough, Nuneaton CV11 4TN
MW7 FXA M. Roberts 25 Maesderi, Hendy, Pontarddulais, Swansea SA40XG
MJ7 FXC F. Chesnay 3 Santa Rosa, La Ruette De Patier, St. Saviour, Jersey JE2 7LQ
M7 FXD S. Pallister 32 Greensnook Lane, Bacup OL13 9DQ
MW7 FXH K. Hawkins 34 Maes Derw, Llandudno Junction LL31 9AN
M7 FXI S. Gabriel 48 Glenthorne Road, Threemilestone, Truro TR3 6UA
M7 FXK B. Norfolk 4 Esmond Close, Rainham RM13 7EB
MW7 FXN J. Calway 2 Stanley Square, Stanleytown, Ferndale CF43 3ER
M7 FXP A. Mccaffrey 32 Dragonfly Way, Rhodesia, Worksop S80 3GW
M7 FXR J. Nelson Jasmine Cottage Holmans Moor Road, Towednack TR263AR
MW7 FXV I. Thomas 4 Narberth Court Caldy Close, Barry CF62 9DS
M7 FXW C. John Swallows Chapel Lane, Harlow CM17 9AJ
MW7 FXZ H. Swettenham Apartment, Casino Mansions Promenade, Llanfairfechan LL330DA
M7 FYB I. Schofield Enterprise Centre 6 David Street Basford, Nottingham NG6 0JU
M7 FYC P. Haigh 43 Acaster Drive, Garforth, Leeds LS25 2BH
M7 FYD D. Buckfield 25 Churchmead Close, Lavant, Chichester PO18 0AY
M7 FYF K. Fyfe 40 Hampden Road, Ashford TN23 6JL
M7 FYG K. Hajji 6 Spencer Close, London NW10 7DU
M7 FYJ G. Milner 48 Mildmay Street, Lincoln LN1 3HR
M7 FYL P. Smith 29 Fowlmere Road Foxton, Foxton, Cambridge CB226RT
M7 FYM B. Friedman 22 Rossiter Road, London SW12 9RU
MM7 FYP C. Carnduff 40 Glengarry Road, Perth PH2 0AQ
M7 FYQ N. Bostock 112 Whitegate Vale, Nottingham NG11 9NE
M7 FYR G. Cleary 2 Loanda Crescent, Newry BT35 8EZ Ireland
M7 FYU R. Cole 43 Barbury Crescent, Plymouth PL6 7EL
M7 FYV S. Mcdonald Stonefield, Swanton Road, West Peckham, Maidstone ME18 5JY
M7 FYZ M. Szucs 18 Avenue Road, Wath upon Dearne S63 7AJ
M7 FZA B. Mellor Flat 1, 138 Street Lane, Gildersome, Morley, Leeds LS27 7JB
M7 FZB D. Fell 2 Weddell Road, Crawley RH10 5BZ
M7 FZC J. Russell Flat 28, Clevedown, Barons Down Road, Lewes BN7 1EY
M7 FZE J. Brown 5 Church Road, Slapton, Leighton Buzzard LU7 9BX
M7 FZG D. Berridge 88 Greenacres Way, Newport TF10 7PJ
M7 FZM S. Allard 1 Whimberry Close, Salford M5 3WL
MM7 FZN T. Macartney Westside Farmhouse Pitcaple, Inverurie AB515HA
M7 FZO G. Cunningham 9 Kearney Drive Oulton, Lowestoft NR32 3FJ
MW7 FZW G. Watts C/O 136 Brynglas Hollybush Cwmbran Np44 7Ll, Cwmbran NP44 7LL
MM7 GAC G. Cattanach 14 Stewart Park Place, Aberdeen AB24 4GA
MM7 GAE L. Reke 21 Stafford Street Flat G, Top Floor, Aberdeen AB25 3UN
MW7 GAH G. Howard Sport Y Gwynt, Middle Road, Llandegfan (Nr Menai Bridge) LL595YD
MW7 GAN S. Ganley 1 Bron Wern Llanddulas, Abergele LL22 8JD
M7 GAP P. Grange 14 Westley Avenue, Whitley Bay NE26 4PA
M7 GAQ A. Lawson Parkinson 16 Baldings Cottages, Green Road, Wivelsfield Green, Haywards Heath RH17 7QD
MW7 GAR G. Williams 10 Strawberry Place Morriston, Swansea SA6 7AG
M7 GAT J. Reynolds 7A New Inn Lane, Trentham, Stoke-on-Trent ST4 8HA
M7 GAU D. Smithson 89 Evelyn Rd, Sheffield S105FG
M7 GAW T. Stephenson 10 Lower Heyshott, Petersfield GU31 4PZ
M7 GAX G. Rowntree 86 Clog Mill Gardens, Selby YO8 3EH
MU7 GBG G. De Putron Le Murier Cottage, Le Murier, St. Sampson, Guernsey GY2 4HQ
M7 GBH G. Hodgson 6 Gilbert Road, Sunderland SR4 9QU

MW7 GBJ G. Britten-Jones 101 Mill View Estate, Maesteg CF34 0DE
MM7 GBL G. Lennox 49 Sorbie Road, Ardrossan KA22 8AP
M7 GBN J. Gillespie Flat C, Swan House, 5 All Souls Place, London W1B 3DB
M7 GBO A. Woods 4 Sedgwick Close, Atherton M46 9EG
M7 GBP M. Finch 186 Hookfield, Harlow CM18 6QW
M7 GBQ H. Rowland 5, Gwinear Downs, Leedstown, Hayle TR27 6DJ
M7 GBR E. Wilson 51 Pincroft Wood, Longfield DA3 7HB
M7 GBS D. Hockley Green Farm, Mendlesham Green, Stowmarket IP14 5RE
M7 GBU K. Madden 44 Shutlock Lane, Birmingham B13 8NZ
MW7 GBW M. Williams 42 Parkside, Overton, , Wrexham LL13 0HA
MI7 GBX S. Golemboski-Byrne 79 Lackan Road, Ballyroney, Banbridge BT32 5HR
M7 GBZ P. Pearce Criggion Mw Radio Station, Back Lane, Criggion, Shrewsbury SY5 9BE
M7 GCA G. Allison 20 Joseph Johnson Road, Sandbach CW11 3TE
M7 GCD J. Swiestowski 54 Brentwood Ave, Coventry CV3 6FN
MW7 GCE R. Wells Dale House , Dale, Haverfordwest SA62 3QU
M7 GCH G. Chlapoutakis 10A Thorpes Avenue, Denby Dale HD8 8SP
M7 GCL T. Cox 74 Dean Way, Storrington, Pulborough RH20 4QS
M7 GCO K. Rutkowski 10 Fitzwarren Close, Chippenham SN15 3UF
M7 GCQ A. Smith 64 High Dane, Hitchin SG4 0BD
M7 GCR G. Roberts 10 Haslam Close, Ickenham UB10 8TJ
MM7 GCS G. Stuart 12 Todshaugh Gardens, Kirkliston EH29 9GE
M7 GCT P. Hughes 17 Worcester Park, Bath BA1 6QU
M7 GCU S. Norris 386 Liverpool Road Eccles, Manchester M30 8QD
MI7 GCV L. Moore 83 Cloyfin Road, Coleraine BT52 2NZ
M7 GCW D. Booth 19 Oak Avenue, Elloughton, Brough HU15 1LA
M7 GCX G. Churcher 4 The Crescent, Tonbridge TN9 1JH
M7 GCY G. Chedgy 67 Studland Park, Westbury BA13 3HN
MI7 GCZ S. Power 68 Quarterlands Road, Dunmurry BT17 9LN
M7 GDF D. Coombs-Farnsworth 27 Stork Lane, Rothwell NN14 6GE
M7 GDG G. Gibson 4 Swan Close, Swannington, Norwich NR9 5NL
MM7 GDI G. Irvine 5 Carlingnose Court, North Queensferry, Inverkeithing KY11 1EP
M7 GDQ D. Pardue 81 Beaumont Terrace, Newcastle upon Tyne NE5 5JQ
M7 GDT G. Tate 7 Chapel Court, Barnsley S71 5FA
M7 GDW G. Wills 15 Markham Avenue, Bournemouth BH10 7HL
M7 GDY A. Good 19 Milfields, Barton-upon-Humber DN185NA
M7 GEB J. Nel 12 Jodrell Close, Waterlooville PO8 9NH
M7 GEC J. Streeter 208 Old Church Road, Clevedon BS21 7UB
M7 GEF G. Barden 9 Nene Close, Wellingborough NN8 5WB
MW7 GEH G. Herd Cartref, Penisarwaun, Caernarfon LL55 3BS
MW7 GEL N. Shepherd Prospect, Newchapel, Llanidloes SY18 6JY
M7 GEN T. Gentry 8 Bowldown Cottages, Bowldown Road, Westonbirt GL8 8UD
M7 GEO G. Salt Christmas Cottage, Ram Hill, Bristol BS36 2TX
M7 GEP P. Griffiths 12 Plant Farm Crescent, Waterlooville PO7 3DB
M7 GER G. Tilley 24 Viador, Chester le Street DH3 3TP
M7 GET R. French South Cottage, Barns Green Farm, Chapel Road, Barns Green, Horsham RH13 0PR
M7 GEY H. Betenson 45 Doncaster Road, Eastleigh SO50 5QP
M7 GEZ G. Hever 15 Juniper Close, Reigate RH2 7NQ
M7 GFA A. Grime 13 Yarrow Gate, Chorley PR7 3AZ
M7 GFB G. Bingley 18 Seaton Close, Nuneaton CV11 6YX
M7 GFC C Roy Hill Club c/o B. Capewell 18 Westminster Road, Kidderminster DY11 6HG
M7 GFE M. Abba 18 Brownspring Drive, London SE9 3JX
M7 GFI M. Manuel 23 Kerr Place, Aylesbury HP21 7BB
M7 GFJ J. Lancashire 9 Moorfields, Colyton EX246PT
M7 GFP T. Humphreys The Farmhouse, Ham Middle Farm, Shepton Mallet BA44JA
M7 GFQ Lord R. Oneill 48 Leebank Road, Halesowen B63 1AE
M7 GFR N. Ramsey 16 Sunnycroft Close Bishops Cleeve, Cheltenham GL52 8AU
M7 GFU C. Macklin 21 C Kremlin Drive Old Swan, Liverpool L13 7BU
M7 GFW J. Neath 32 Richmond Avenue, Oldham OL9 8LG
M7 GFX S. Hallett 53 Hylstone Cresent Wednesfield, Wolverhampton WV11 3EZ
M7 GFY D. Ward 20 Norbury Road, Wolverhamplton WV109RL
MU7 GGA J. Clarke Top Flat The Wing La Croix, La Biloterie Road, St. Saviour GY7 9QX Guernsey
MM7 GGD G. Sneddon 34 Myreside Avenue, Kennoway KY8 5EN
M7 GGF G. Farmer 5 Atlas Road Earls Colne, Colchester CO62LU
M7 GGH Dr G. Howling Sysonby Knoll Hotel, Asfordby Road, Melton Mowbray LE13 0HP
M7 GGK G. Kerekes 10 Brookfield Road, Bury BL9 5JZ
MM7 GGL G. Ledgerwood 28 Cannerton Park, Milton Of Campsie, Glasgow G66 8HR
M7 GGM G. Mariner 6 Richmond Road, Manton, Worksop S80 2TP
M7 GGT J. Hall 93 Jepson Road, Sheffield S5 6AN
M7 GHB G. Brinklow 45 Gillsway, Northampton NN2 8HT
M7 GHE University of Exeter Amateur Radio c/o M. Ghent 76 Blue Hill Crescent, Wortley, Leeds LS12 4PB
MM7 GHF J. Hazle Grange Hill Farm, Kinghorn, Burntisland KY3 9YF
M7 GHH G. Hodgskins 2 Lime Tree Close, Hessett, Bury St. Edmunds IP30 9AY
M7 GHI G. Hewitt 5 Station Road, Dinnington, Sheffield S25 3RW
M7 GHJ K. Selvaratnam 15 Mark Close, Southall UB1 3QJ
M7 GHK G. Kerr 6 Cressages Close, Felsted, Dunmow CM6 3NW
M7 GHO S. Wheelhouse 30 Highfield Drive, Gildersome, Morley, Leeds LS27 7DW
M7 GHP G. Pinkstone Little Owl Barn, The Street, Helhoughton, Fakenham NR21 7BL
M7 GHS A. Jurd 54 Caesar Close, , Andover SP10 5JR
M7 GHW G. Heaton Crooke Village Marina, Crooke Road, Standish Lower Ground, Wigan WN6 8LR
M7 GHZ S. Rankin 2 Thompson Road, Broadbridge Heath, Horsham RH12 3TT
MW7 GIA G. Jacobsen 98 Laburnum Drive, Newport NP19 9AN
M7 GIF C. Arthur 25 Pump Hollow Lane, Mansfield NG18 3DU
M7 GIL S. Gilmartin 6 Newport Crescent, Leeds LS6 3BY
M7 GIM G. Marshall 23 Birch Grove, Felbridge, East Grinstead RH19 2TS
M7 GIN W. Pilling 12 Brooke Close, Accrington BB5 2QX
M7 GIP P. Malinowski 12 Lodge Road, Southampton SO14 6RN
M7 GIQ D. Balding 9 Arkwright Road, Tilbury RM18 8XL
MM7 GIR W. Girdwood 95 Parkhead Crescent, West Calder EH55 8BE
M7 GIS Dr D. Rodriguez 23 Thorpe Way, Cambridge CB5 8UJ
M7 GIU N. Dinham 95 The Westlands, Congleton CW12 4EH
M7 GIV H. Mosleh Adolf-Friedrich-Str.2, Oldenburg in Holstein 23758 Germany
M7 GIY G. Atkinson 6 Burnt Stones Dr Sheffield, Sheffield S10 5TT
M7 GJA G. Finnigan 33 Birchwood Drive, Ulverston LA12 9PN
M7 GJF G. Flatman Devenport Cresent, Manchester OL26JX
M7 GJG G. Graham Todcroft, Elsdon, Newcastle upon Tyne NE19 1AA
M7 GJJ G. Johnson 4 Delta Park Drive, Hesketh Bank, Preston PR4 6SE
M7 GJL G. Larner 58 Heather Close, , Carterton OX181TH
M7 GJP A. Bonwitt 60 Wellhouse Road, Beech, Alton GU34 4AG
MM7 GJQ J. Keenan Ingimster Farmhouse, Wick KW1 4TA
M7 GJT Lord G. Thompson 2 Beech Avenueravenshead, Nottingham NG15 9GN
M7 GKA H. Griffiths The Old Church Hall, Christmas Common, Watlington OX49 5HL
M7 GKB A. Hewitt 64 Ridgeway, Clowne, Chesterfield S43 4BD
MM7 GKD Rvd. J. Maclennan 87 Glenfruin Road, Blantyre, Glasgow G72 9RJ
M7 GKI S. Peachey 14 Morris Crescent, Thornley DH6 3DH
M7 GKL A. Gkelios 25 Newlands Avenue Norton, Stocton on Tees TS202PQ
M7 GKO G. Kirby The Chimneys, Woodsetts Road, Gildingwells, Worksop S81 8AU
M7 GKR D. Radulescu 8 White Street, Hull HU3 5PS
MW7 GKT G. Thomas 37 Broadstairs Road, Cardiff CF11 8DE
M7 GLB G. Briggs 37 Knutton Crescent, Sheffield S5 9NW
M7 GLE G. Porritt 23 Greenlands Road Pickering, Pickering YO18 8BQ
MW7 GLF N. Jones 11 Wood Green, Mold CH7 1UG
M7 GLK D. Hancock 7 Runswick Close Silksworth, Sunderland SR3 2YG
M7 GLL Dr R. Anbarasan 31 New Road, Hethersett, Norwich NR9 3HJ
M7 GLN G. Strong 11 Tamar Walk, Scunthorpe DN17 1UE
M7 GLO J. Elton 63 Bazeley Road, Gloucester GL4 6JE
M7 GLS G. Sangwell 34 Gatcombe Close, Calcot, Reading RG31 4XQ
M7 GLW G. Williams 329 Longford Lane Longford, Gloucester GL2 9ES
M7 GLY G. Freeman 20 Pitcairn Avenue, Eastbourne BN23 5BB
M7 GMA C. Clunn 38 Byfleet Avenue, Basingstoke RG247HR
M7 GMC G. Mcloughlin 12 Harwood Road, Norwich NR1 2NG
M7 GMD G. Doggett 68 Pilgrims Way, Harleston IP20 9QE
M7 GMF Dr G. Fahy 40 Mayfair Close, Great Sankey, Warrington WA5 3PL
M7 GMG A. Beckett 23 Lingcrest Close, Manchester M19 2WJ
M7 GMH G. Heath 2 Lower Drake Fold, Westhoughton, Bolton BL5 2RE
M7 GMJ I. Matthews Atrial, South Dock Marina, Rope Street, London SE16 7SZ
M7 GMK G. Madgwick 5 Plantation Drive, Walkford, Christchurch BH23 5SE
M7 GMN D. Jowle 6 Newton Ave, Sheffield S36 1EL
M7 GMP G. Preston 3 Thirlwall Avenue, Conisbrough, Doncaster DN12 3JZ
M7 GMR G. Mylonas 68 Lancaster Gate, Cambourne CB236AT
MM7 GMS G. Smith 37 Kilmorie Drive, Rutherglen, Glasgow G73 2ER
MM7 GMW G. Mcfarlane 18 Stanistone Road, Carluke ML8 4DY
M7 GNA G. Gray 9 Bevan Way, Market Drayton TF9 3US
M7 GNB G. Baguley 58 Chestnut Avenue, Great Notley, Braintree CM77 7YJ
M7 GNC G. Chalkin 18 Trinity Close, West Mersea, Colchester CO5 8RW
M7 GNH G. Harris 6 Penling Close, Cookham, Maidenhead SL6 9NF
MI7 GNI G. Baine 13 Cherrymount, Newtownabbey BT36 5NH
MW7 GNN S. Horan 26 Ffordd Ysgubor Goch, Caernarfon LL55 2RU
M7 GNR B. Ashcroft 7 Ash Crescent, Seaham SR7 7UE
MM7 GNX A. Cybulski 46 Waulkmill Avenue, Barrhead, Glasgow G78 1DD
M7 GOB J. Holden 262 St. Margarets Road, Twickenham TW1 1PR
M7 GOG D. Rolf 28 Titchener Way, Hook RG27 9GB
MW7 GOM G. O'Meara 30 West View, Chirk LL14 5HN
M7 GOO A. Goose 12 Brown Street, Rainham, Gillingham ME8 7JN
M7 GOT A. Croot 13 Bickford Road, Wolverhampton WV10 0NH
M7 GOU P. Gould 9 Crowden Way Thamesmead, London SE28 8HE
M7 GOW K. Laing 16 Cherrywood Drive Gonerby Hill Footá, Grantham NG31 8QL
M7 GOZ S. Gosbee Water Hall Farm, Wixoe, Stoke By Clare, Sudbury CO10 8UA
M7 GPH G. Hawes 59 Hartlands, Bedlington NE22 6JG
M7 GPI G. Pipkin 22 Orchard Way, Sutton SM1 3QQ
M7 GPJ G. Jones 39 Bark Road, Liverpool L21 7QN
M7 GPL G. Lomas 28 Moat Bank, Bretby, Burton-on-Trent DE15 0QJ
M7 GPM G. Manning 8 Devonshire Row, Princetown, Yelverton PL20 6QD
M7 GPP G. Phillips Labourn Fell Farm, Chopwell NE17 7AY
MI7 GPT P. Hegarty 26 Ashveagh Benburb, Dungannon BT717TS
M7 GPW C. Godfrey 3 Moor Park, Neston, Corsham SN13 9YJ
M7 GPX R. Haines 47 Upshire Road, Waltham Abbey EN9 3NZ
M7 GPY D. Twells Preston Lodge 3 Willoughby Court, Norwell, Newark NG23 6JJ
M7 GRD G. Downing Fort Road, Gosport PO12 2DT
MM7 GRG G. Manning Snipefield, Culsalmond, Insch AB526TU
M7 GRH I. Justice 4 Saxon Street, Droylsden, Manchester M43 7FR
M7 GRK J. Honeyands 154 Burnham Road, Hullbridge, Hockley SS5 6HJ
M7 GRM Dr G. Lioliou Flat 1 22 Grand Parade, Brighton BN2 9QB
M7 GRP G. Purser 1 Lichfield Gardens, Aldwick, Bognor Regis PO21 3RB
M7 GRR A. Moorhouse 66 Wicor Mill Lane, Fareham PO16 9EG
M7 GRS G. Smith 148 Firhill Road, London SE6 3SQ
M7 GSA K. Hemmings 90 Grovewood Avenue, Leigh-on-Sea SS9 5EG
M7 GSD A. Smith 30 Knutsford Road, Alderley Edge SK9 7SD
MW7 GSH G. Surendran 115 Western Avenue, Port Talbot SA12 7NB
MI7 GSI R. Nelson 61 Ballycrune Road, Hillsborough BT26 6NH
M7 GSP G. Patterson 3 Sunte Close, Haywards Heath RH16 1QT
M7 GSS A. Netting 33 Weatherdon Drive, Ivybridge PL21 0DD
M7 GST R. Budinger Kings Arms, Bexhill Road, Ninfield, Battle TN33 9JB
M7 GTA G. Ashcroft 8 Launceston Close, Winsford CW7 1LY
M7 GTC G. Bradley Greentree Cottage, Town End, Broadclyst, Exeter EX5 3HW
M7 GTD R. Dunnigan 21 Coldstone Drive Garswood, Ashton - in - Makerfield WN40RW
M7 GTE P. Bleasdale 12 Malvern Ave, Padiham BB127DT
M7 GTI S. Lifely The Farm, Moreton Jefferies, Hereford HR1 3QY
M7 GTJ G. Lawton 11 Highlands Drive, Stockport SK2 5HX
M7 GTL Dr I. Chessell Pondfield House, Pondfield, Great Dunmow CM6 1FX
M7 GTM G. Moulds Flat 88, Barkis House, Brownlow Close, Portsmouth PO1 4ES
MM7 GTN M. Gratton 103 Wylie Crescent, Cumnock KA18 1LT
MM7 GTS G. Gault Ka73Sb, Ayr KA73SB.
MW7 GTW G. Whitley 14 Erw Wen Road, Colwyn Bay LL29 7SD
M7 GTX R. Twyning 64 Hornbeam Way, , Kirkby-in-Ashfield NG17 8RL
M7 GTZ C. Aldous 342 Feltham Hill Road, Ashford TW151LW
M7 GUK C. Walter 5 Chyandour, Redruth TR15 3AB
M7 GUM B. Mcbride Apartment 377, Flint Building, 11 Stanley Street, Salford M3 5GL
M7 GVA A. Giles 137 Westhall Road, Warlingham CR6 9HJ
M7 GVM A. Miles The Cider House Weston Street East Chinnock, Yeovil BA22 9EJ

M7 GVP J. Hunt 18 Oakville Close, Worcester WR2 4XL
MW7 GVR P. Perrett 1 Llys Y Wennol Northop Hall, Mold CH7 6GE
M7 GWC G. Castledine 11 Cam Road, Cheltenham GL52 5QS
M7 GWD A. Dainty 6 St Nicholas Way Wygate Park Spalding, Spalding PE11 3GF
M7 GWF R. Lewis 46 Graydon Avenue, Chichester PO19 8RG
M7 GWH J. Housley Lesede Cottage, The Town, Carsington, Matlock DE4 4PX
M7 GWI G. Wynn 10 Mulberry Grove Bradwell, Great Yarmouth NR31 8QJ
M7 GWJ S. Joyner 34 Hycliffe Gardens, Chigwell IG7 5HJ
M7 GWK M. Kirton 6 Rowe Court, Dobwalls, Liskeard PL14 6AU
M7 GWM G. Goddard 13 Rosslyn Avenue, Coventry CV6 1GL
M7 GWW G. Worthington 190 Higher Road, Liverpool L26 1UW
M7 GXG M. Domingos 42 Stretton Road, Leicester LE3 6BJ
M7 GXS A. Royle 154 Baddow Hall Crescent, Chelmsford CM2 7BU
M7 GYT C. Clark 2B Hayling Rise High Salvington, Worthing BN13 3AL
M7 GYV S. Boothman 2 Rushey Close, Rawmarsh, Rotherham S62 7QP
M7 GYY D. Woods 37 Grantham Avenue, Grimsby DN33 2HQ
M7 GYZ M. Rothery 24 Maes Y Dafarn, Caersws SY17 5NG
M7 GZA J. Gresty 81 Old Hall Road, Sale M33 2HU
MI7 GZO J. Montgomery 27 Fir Park Broughshane, Ballymena BT42 4DH
MW7 GZR N. Davies 24 Tal Y Fan, Glan Conwy, Colwyn Bay LL28 5NG
M7 GZV P. Fell 1 Victoria Terrace, Church Walk, Melksham SN12 6NA
MI7 GZX P. Brennan 23 Ardchrois, Donaghmore, Dungannon BT70 3LB
MW7 HAD J. Hadley Rhianfryn, Caergelach, Llandegfan, Menai Bridge LL59 5UF
M7 HAF H. Fletcher Flat 4, Chase Court, High Street, Theale, Reading RG7 5AR
M7 HAL M. Iddon 40 Winchester Avenue, Waterloo, Liverpool L22 2AT
M7 HAM N. Parker 18 Sandown Road Bishops Cleeve, Cheltenham GL52 8BZ
M7 HAP A. Pursell 48 St. Stephens Avenue, Ashtead KT21 1PJ
M7 HAV Dr D. Argles 5 Harland Crescent, Southampton SO15 7QB
MW7 HAW A. Hawkins 5 Keats Road, Caldicot NP26 4LH
MW7 HAY M. Hay Helyg, Abernant, Carmarthen SA33 5RR
M7 HBA K. Anand 19 Tempest Road, Upper Cambourne, Cambridge CB23 6HW
M7 HBB T. Delsi 42 Westergate Avenue, Brooklands, Milton Keynes MK10 7LQ
M7 HBG H. Gruen Allfrey House, Herstmonceux, Hailsham BN27 4RS
M7 HBS B. Harris 10 Fish Street, Exeter EX2 7TR
M7 HBZ R. Allen 17 The Paddocks, Herne Bay CT66QX
M7 HCD D. Hall 16 Larkspur, Rugby CV23 0UW
MM7 HCM H. Mcintosh 43 Woodlands Avenue, Cults, Aberdeen AB15 9DE
M7 HCR H. Crump Apple Tree Cottage, Kingsland, Leominster HR6 9PY
M7 HCT A. Sykes 1 Stanhope Gardens, Boston PE21 0PJ
M7 HDC D. Horridge Flat 8, Blenheim Court, Marlborough Road, London N19 4HR
M7 HDJ J. Denton 7 Cedda Place, Sandbach CW11 3SW
M7 HDM Z. Harmath 2 Colyton Close, Wembley HA0 2HG
M7 HDR I. Hammond 88 Great Innings North, Watton At Stone SG14 3TD
M7 HDS D. Samson 20 Chichester Park, Westbury BA13 3AN
M7 HDX P. Matthews 6, Willow Road, Sunderland DH45QF
M7 HEF J. Kelly 8 Paddocks Close, Upper Lighthorne CV338AZ
M7 HEL S. Helyer 21 Staddons View, Bovey Tracey, Newton Abbot TQ13 9HN
MW7 HEN D. Edkins Penhower Newydd Caerhun, Bangor LL57 4DT
M7 HES Dr F. Sedgemore 12 The Brambles, Royston SG8 9NQ
MW7 HEW I. Hewitt Gwern Borter Cottage Rowen, Conwy LL32 8YL
M7 HEX K. King Flat 18, St. Peters House, Jacobs Wells Road, Bristol BS8 1DY
MM7 HFB Y. Sun 9, 39, Braid Square, Glasgow G4 9YQ
MM7 HFC S. Woodburn 6 Roman Place, Bellshill ML4 2AU
M7 HFD R. Beare 12 Granary Close, Rushwick, Worcester WR2 5QG
M7 HFX T. Clay Dene Farm, Dean Lane, Triangle, Sowerby Bridge HX6 3EA
M7 HFY S. Spencer 9 Racecourse Marina, Boroughbridge Road, Ripon HG4 1UG
M7 HGF H. Fernandes 118 Bushfield Drive, Redhill RH1 5BW
M7 HGK P. Morris 5 Pebble Court 129-131 Henver Road, Newquay TR7 3DT
M7 HGL I. Atkins 14 Ash Walk, Warminster BA12 8PY
M7 HGM J. Moran 12 Well Lane, Shaftesbury SP7 8LP
M7 HGN H. Nixon 24 Lingfield Road, Worcester Park KT4 8TG
M7 HGO P. Holpin 32 Friars Way, Bushey WD23 2BT
M7 HGV S. Jordan 74 Soane Gardens, South Shields NE34 8NN
M7 HHD G. Deans 4 Grovelands Road, Winchester SO225JU
M7 HHH H. Hughes 24 Broadley Green, Windlesham GU20 6AL
M7 HHI A. Arranz Garcia 196 Whitley Wood Lane, Reading RG2 8PR
M7 HHK H. Khoo Flat 2, 59 Lilac Road, Southampton SO16 3DA
M7 HHS D. Sutch 2 Elizabeth Cottages, Lutterworth Road, Shawell, Lutterworth LE17 6AE
M7 HHY M. Marsh 64 Hewitt Avenue, St. Helens WA10 4EW
M7 HHZ A. Smith 1 Henley Farm Cottages, Wantage Road, Great Shefford RG177DH
M7 HIA Dr D. Perrin 2A The Crescent, Hampton In Arden, Solihull B92 0BP
MM7 HIC Dr L. Campbell 21 Fordyce Way, Auchterarder PH3 1BE
M7 HIL G. Lawry 5 Trevithick Road, Pool, Redruth TR15 3NW
MM7 HIM A. Harker 29 Overmills Crescent, Ayr KA7 3LN
MM7 HIR M. Joglekar 21 Lovat Avenue, Bearsden, Glasgow G61 3LQ
M7 HJB H. Bingham 14 Slaters Close, Rushden NN10 0EE
MW7 HJE H. Evans 20 South Hook Road, Gelliswick, Milford Haven SA73 3RU
M7 HJH H. Hughes 346 Griffiths Drive, Wolverhampton WV11 2LB
MI7 HJK A. Mccabe Scaddin House 15 Myra Road, Downpatrick BT30 7JX
M7 HJM H. Mclean 3 Yarnton Close, Royton, Oldham OL2 6PF
M7 HJO J. Holroyd 31A Cross Street, Tenbury Wells WR15 8EF
M7 HJP H. Piper 14 Ashwood Close, Worthing BN11 2AF
M7 HJR H. Robinson 7 Higney Road, Hampton Vale, Peterborough PE7 8LZ
M7 HJS A. Stocks 82 Carr Street, Huddersfield HD3 4BQ
MI7 HJW J. Watson 74 Bann Meadows, Bann Road, Ballymoney BT53 7RN
M7 HKA J. Durkin 18 Woodlands Road, Harrow HA1 2RS
M7 HKP D. Cotton 53 Brookdale Road, Hartshorne, Swadlincote DE11 7HH
M7 HKY J. Hockey 5 Old Farm Road, Birmingham B33 9HH
MM7 HLD A. Mac An Iasgair 3 Poppy Court, Abbey Road, Scone, Perth PH2 6GD
M7 HLG G. Williams 40 Manion Avenue, Liverpool L31 4ED
M7 HLK S. Fisher 37 Lancaster Crescent, St. Eval, Wadebridge PL277TP
M7 HLN H. Fox 24 Northwood Street, Stapleford NG9 8GH
M7 HLR M. Hasler 15 Priory View, Cornworthy, Totnes TQ9 7HN
M7 HMB D. Williams 6 Swettenham Close Alsager, Stoke-on-Trent ST72XG
M7 HMC Dr H. Colman 16 Gleneagle Road, Plymouth PL3 5HJ
MI7 HMD D. Hargreaves 19 Laurel Park, Ahoghill, Ballymena BT42 1LN
M7 HME S. Swinton 9 Carnoustie, Bolton BL3 4TF
M7 HMO R. Jones 43 Walton Avenue, Penwortham, Preston PR1 0XR
M7 HMR W. Andes 141 Sunningdale Avenue, Hanworth, Feltham TW13 5JS
M7 HMS A. Draycott 56 Jacks Lane, Marchington, Uttoxeter ST14 8LW
MW7 HMT H. Todd 29 Maes Maelor , Penparcau, Aberystwyth SY23 1SZ
MI7 HNC C. Oliver 11 Lancastrian Court, Banbridge BT32 4QL
M7 HNK J. Henkins 69 Verdayne Gardens, Warlingham CR6 9RP
M7 HNU B. Gannamani 13 St. Anthony Road, Basingstoke RG24 9XP
M7 HOF J. Aldenhoff 5 Ellachie Gardens, Gosport PO12 2DS
M7 HOG S. Hogg 38 The Gables, Widdrington, Morpeth NE61 5RA
M7 HOM S. Mcgaughey 60 Newington Street, Hull HU35LX
M7 HOP Dr J. Hopcroft 18 Heron Road, Honiton EX14 2GL
MM7 HOS S. Hosseinzadeh 4/1, 3 Duke Wynd, Glasgow G4 0WX
M7 HOU G. Hoult 47 Woodborough Road, Mansfield NG19 6NN
M7 HOY J. Roberts 65 Westwoods Park, Bashley Cross Road, New Milton BH25 5TB
M7 HPS J. Sykes 79 Owler Park Road, Ilkley LS29 0BG
M7 HPW H. Whittle 22 Tinkers Field, Royal Wootton Bassett, Swindon SN4 8AE
MM7 HPX N. Taylor 9 Tigh Na Mara, Main Street, Gaza, Portmahomack, Tain IV20 1YS
MM7 HQI Q. Huang Flat 6, 14 Bothwell Street, Edinburgh EH7 5PS
MM7 HQS H. Matthews 88 Nevis Crescent, Alloa FK10 2BN
M7 HRL P. Sheldon 3 Westgate Terrace, Radnor Bridge Road, Folkestone CT201WT
MM7 HRR H. Rivett 22 Tailor Place, Aberdeen AB24 4RU
M7 HRS E. Harris 1 Pegasus Way, Haddenham, Aylesbury HP17 8SB
M7 HRT R. Hart 26 Bell Road, Norwich NR3 4RA
M7 HSF M. Deltrice 12 Back Street Horsham St Faith, Norwich NR103JP
M7 HSG G. Slater 38 Bradstock Road, Epsom KT17 2LH
M7 HSI D. Markland 1 Andromeda Way, Brackley NN13 6GU
M7 HSN M. Hodgkinson 116 Bidwell Hill, Houghton Regis, Dunstable LU5 5EP
M7 HTC H. Castree 31 Fairview Thickwood, Colerne, Chippenham SN14 8BS
M7 HTK H. Krishnan 17 Lugtrout Lane, Solihull B91 2SB
M7 HTL S. Pelc 44 Frithwood Crescent, Kents Hill, Milton Keynes MK7 6BQ
M7 HTR R. Hopkins - Esteris 1 Edendale Terrace, Horden, Peterlee SR8 4RD
M7 HTS J. Hogarth 23 Willow Road East, Darlington DL3 6PY
M7 HTU E. Lewis 49 Ridgebourne Road, Shrewsbury SY3 9AB
M7 HTZ R. Ives Crowhurst Road, London RH76DG
M7 HUB D. Hubbard 66 Parkfield Road, Rainham ME8 7SZ
M7 HUD D. Hudson 30 Sarmatian Fold, Ribchester, Preston PR3 3YG
MM7 HUE Dr Z. Yang 48 Foxglove Road, Newton Mearns, Glasgow G77 6FP
M7 HUG M. Townsend 35 Church Way, Hungerford RG17 0JP
M7 HUH J. Eaves 8 Silver Birch Close, Huntingdon PE29 7BW
M7 HUL A. Shakesby 14 Dawnay Road, Bilton, Hull HU11 4HB
M7 HUX D. Robertson Lingley, Cold Pool Lane, Badgeworth, Cheltenham GL51 4UP
M7 HVA H. Valentine 117 Great North Road, New Barnet, Barnet EN5 1AW
M7 HVH C. Sidaway 6 Rookery Chase Deepcar, , Sheffield S36 2NF
M7 HVL K. Howell 4 Wisteria Gardens Harlow Green, Crag Lane, Harrogate HG3 1FP
M7 HVM M. Perkins 53 Barn End Road, Warton, Tamworth B79 0JD
M7 HWB D. Maloney 21 Poplar Court, Northampton NN3 6SE
M7 HWH M. Street Ridge End Cottage Crawley Ridge, Camberley GU15 2AL
M7 HWK A. Hawkins Marellis, North Lane, Swaby, Alford LN13 0BD
M7 HWL D. Laker 85 The Gore, Basildon SS14 2DB
M7 HWO Dr H. Orridge Stonecross Cottage, Wadworth Hall Lane, Wadworth, Doncaster DN11 9BH
M7 HWT M. Barmby 1 Cardigan Road Bridlington, Bridlington YO15 3HG
MM7 HXI M. Moir 21 Primrosehill Gardens, Aberdeen AB24 4EQ
M7 HXO R. James 22 Moss Road, Watford WD25 0EN
M7 HYB M. Greenway 28 Purbrook, Tamworth B77 2NB
M7 HYE A. Dandy Longridge, Oakridge Lynch, Stroud GL6 7NZ
M7 HYL F. Stephenson 67 Geneva Avenue, Lincoln LN2 4EB
M7 HYM C. Mears Waterways, Scotland Yard, Priors Leaze Lane Hambrook, Chichester PO18 8RQ
M7 HYN H. James 59 Campbell Bannerman Way, Tividale, Oldbury B69 3NE
M7 HYP D. Cook 5 Pendennis Court, Red Lane, Bugle, St. Austell PL268QP
M7 HZY C. Barker 39 Jethro Street, Bolton BL2 2PU
MM7 IAB I. Black 30 Victoria Street, Dumbarton G82 1HP
M7 IAD I. Doutre 2 Oak Cottages Piddinghoe, Newhaven BN9 9AT
MM7 IAM I. Mort 5 Rockfield Road, Tobermory, Isle of Mull PA75 6PN
M7 IAP I. Pennington 12 Coral Place, Blackpool FY44PS
M7 IAZ I. Lowbridge 4802 East Ray Road #23-513, Phoenix 85044 United States
MM7 IBG I. Bagleybter 2F3, 20 Springvalley Gardens, Edinburgh EH10 4QE
MM7 IBS I. Burnett Green 10 Willow Place, Lanark ML11 9FY
M7 ICD J. Caine 16 Wheatfield Drive, Shifnal TF11 8HL
M7 ICG I. Gilbert 84 Carlyle Road, London W5 4BJ
M7 ICH C. Healey 37 Ennerdale Drive, Manchester M33 5NF
M7 ICK M. Percival 10 Binder Close, Higham Ferrers NN108PH
M7 ICL I. Larman 30 Kiln Drive, Tydd St Mary, Wisbech PE13 5RA
M7 ICP A. Moran 451 Haworth Road, Allerton, Bradford BD15 9LL
M7 ICQ D. Mason 25 Culm Valley Way, Uffculme, Cullompton EX15 3XZ
MM7 ICR I. Crosbie West Gates Avenue, Lochwinnoch PA12 4HG
M7 ICS I. Sherratt 52 Corran Close, Northampton NN5 7AL
M7 ICY S. Brace 2 Greenfields Cottages, Brockhampton Estate, Bringsty WR6 5TB
M7 IDD N. Iddison 25 Westfield Southway, Westfield, Sheffield S20 8EB
MW7 IDI I. Williams 72A Stryt Issa, Pen-Y-Cae, Wrexham LL14 2PN
M7 IDM I. Mills Garfield Station Road, Ganton YO12 4PB
M7 IDP I. Eaton 12 St. Christopher Road, Uxbridge UB8 3SG
M7 IDS I. Stanley 11 Reading Road, Burghfield Common RG7 3PY
M7 IDT I. Briley 12 Stirling Avenue, Waterlooville PO7 7NH
M7 IDY I. Duxbury 8 Eden Gardens, Longridge, Preston PR33WF
M7 IED M. Belch The Nook, Lower North Street, Cheddar BS27 3HA
M7 IEE S. Spray Flat 31 Devon House, Devon House Drive, Bovey Tracey TQ139HB
MM7 IEF A. Mclennan 10 Trefoil Place, Ayr KA73XG
MM7 IFC I. Collins 22 Ninian Quadrant, Glenrothes KY7 4HP
M7 IFR I. Rudge 24 Granville Road, Cradley Heath B64 7QH
MW7 IFZ I. Jones 27 Cae Bach Aur Estate, Bodffordd, Llangefni LL77 7JS
M7 IGC I. Costan 1 Willowherb Walk, Romford RM3 8JP
M7 IGG G. Gray 20 Philip Road, Staines-upon-Thames TW18 1PW
M7 IGN Prof. I. Neal 8 Rushey Gill, Brandon, Durham DH7 8BL
M7 IGT S. Mucklow Blithbury House, Blithbury Road, Rugeley WS15 3HR
M7 IGW I. Westland 123 Eton Hill Road Radcliffe, Manchester M26 2XQ
MM7 IIE H. Chen Iq Fountainbridge Room 118E 114 Dundee Street, Edinburgh EH11 1AD

MM7 IIH J. Jarvie Berryhill Farm, Tak-Ma-Doon Road, Kilsyth, Glasgow G65 0RY
M7 IIO T. Groves 34 Maple Way, Coulsdon CR53RN
M7 IIZ C. Bohoris 5 Fawcett Place, Crewe CW1 4UL
M7 IJB I. Bentley 34 Knapton Close, Chelmsford CM1 6UL
M7 IJC I. Chilley 24 Blackheath Road, Lowestoft NR33 7JG
MM7 IJK I. Killoh 49 May Place, Perth PH1 3BJ
M7 IJM I. Mcadam 43 Hightown Road, Ringwood BH24 1NQ
M7 IJP S. Holmes Plot 4 Van1 Hygrove Residential Park, Hygrove Lane, Ministerworth, Gloucester GL2 8LE
M7 IJT A. Purnell The Old Chapel Lydford, Okehampton EX20 4AW
M7 IKA J. Hawkes 142 Mercian Way, Slough SL1 5LE
M7 IKB I. Bourne The Granary, Pincott Farm, Nuthill, , Brockworth GL3 4RL
M7 IKC I. Colebourn 21 Well Close, Addingham, Ilkley LS29 0SH
MW7 IKH I. Hooper 43 Meadow Lane, Porthcawl CF365EY
M7 IKO B. Giraud 1 Carlton Terrace, Dewsbury WF12 9LD
M7 ILA A. Nicholson 2 Broom Close, Leyland PR25 5RQ
MM7 ILB Capt. B. Anderson 1 New Houses, East Fortune, North Berwick EH39 5JZ
M7 ILD R. Silva Coelho 114 Cannon Hill Lane, London SW20 9ET
MW7 ILL R. Bolger 49 St. Benedict Crescent Heath, Cardiff CF14 4DP
M7 ILM D. Roberts 14 North Street, Digby LN4 3LY
MM7 ILY S. Urquhart Sornbank, Bridgend, Isle of Islay PA44 7PQ
MM7 IME S. Boal 20 Fairinsfell, Broxburn EH526AL
M7 IMH I. Mahmood Loughborough University, B919111 Faraday Hall, Ashby Road, Loughborough LE11 3TY
MW7 IMJ I. Jones 4 Cwm Tecaf, Rhosybol, Amlwch LL68 9PU
MW7 IML P. Lowe 9 Felin Uchaf, Dolgellau LL40 1NS
MW7 IMO I. Wright The Barn, Penfeidr Farm, Treffgarne, Haverfordwest SA62 5PL
M7 INA M. Mitchell-Hardy Northcote Road, Norwich NR3 4QF
M7 INC G. Jones 158 Withington Lane, Aspull, Wigan WN2 1JE
M7 IND A. Tree 7 Blackthorne Road Biggin Hill, Westerham TN16 3SH
M7 INE J. Miller Manor Farm House, 16 Cause End Road, Bedford MK43 9DB
M7 ING R. Smith Guyatt Court, Burwell CB25 0DP
M7 INH T. Sorrell 20 Orchard Close, Saffron Walden CB11 4DQ
M7 INN A. Seal 33 Daleham Mews, London NW3 5DB
M7 INS S. Coppins 33 Honywood Road, Lenham, Maidstone ME17 2HH
M7 INT G. Farmer 288 Margate Road, Ramsgate CT12 6AJ
MM7 INV I. Hamilton 27 Cameron Avenue, Balloch, Inverness IV2 7JT
M7 INX A. Henderson Higginstown, Ballyshannon F94 ND35 Ireland
MM7 IOD J. Jarvis 39 Ladyacre Wynd, Irvine KA112FY
MW7 IOL I. Roberts Llwydiarth, 38 Tal Y Cae, Tregarth, Bangor LL57 4AE
M7 ION T. Heyes 11 Roughtor View Planet Park, Delabole PL33 9BX
M7 IOR I. Rutherford 12 Madox Brown End, College Town, Sandhurst GU47 0GJ
M7 IOS P. Lisewski 40D Gledstanes Road, London W14 9HU
MM7 IOW S. Mason 31 Clifton Road, Lossiemouth IV31 6DJ
M7 IPL I. Leyland 61 Stacey Avenue, Milton Keynes MK12 5DN
M7 IPM I. Melling 33 Brookfield Road, Market Harborough LE16 9DU
M7 IPY D. Martin 17 Fleet Road, Dartford DA2 6JE
M7 IRB I. Bell 11 Cherry Blossom Close, Northampton NN3 9DN
MM7 IRC N. Dean 2/4 Old Tolbooth Wynd, Edinburgh EH8 8EQ
MM7 IRG Dr M. Sutcliffe 11 Low Borland Way, Eaglesham, Glasgow G76 0BP
M7 IRJ J. Abel-Moir 6 Egypt Hill, Leckhampstead, Newbury RG20 8QF
M7 IRK I. Kocher 3 Eastbrook Close, Gosport PO12 3BP
MM7 IRM Dr I. Maoileoin Wester Blackfold, Abriachan, Inverness IV3 8LB
MM7 IRQ G. Stirling 45 Mains Ave, Invergordon IV18 0JT
M7 IRW R. Irwin 2 Sutton Close, Redditch B98 0JR
M7 ISB B. Isac 66 Dacre Gardens, Borehamwood WD6 2JW
M7 ISC I. Clifton 44 Clarence Road, Horsforth, Leeds LS18 4LB
MI7 ISF I. Forde 101 Princess Way, Portadown BT635EJ
M7 ISG I. Siomadis 15 Maidstone Road, Norwich NR1 1EA
M7 ISH I. Hasulyo 24 Chaucer Court, Guildford GU2 4DB
M7 ISM P. Lyba 6 Ambassadors Way, North Shields NE29 8ST
MM7 ITA C. Di Bona 294 Broadholm Street, Glasgow G22 6DN
M7 ITD D. Ward Flat 15 Barnes Court Whitley Mead, Stoke Gifford BS348XT
M7 ITI D. Foxcroft 8 Cowlard Close, Launceston PL15 7EQ
M7 ITR I. Robinson 220 Collis Street, Stourbridge DY8 4EQ
M7 IVA I. Lucas Orde House, Whitchurch, Ross-on-Wye HR9 6DQ
M7 IVN I. Peikov 41 Wellfield Road, London SW16 2BT
M7 IVY A. Gilbert 31 Elizabeth Road, Leamington Spa CV31 3LJ
M7 IWN I. Antoniw 128 Rosemary Rd Beighton, Sheffield S20 1DA
M7 IWT I. Thompson 26 Countrymans Way, Shepshed, Loughborough LE12 9RB
M7 IWW I. Watts The Walled Cottage 9 Holland Road, Frinton-on-Sea CO13 9DH
M7 IXD E. Turner. 80 Blackburn Road, Padiham, Burnley BB12 8JZ
M7 IXH S. Simon 4 Mina Close, Peterborough PE2 8TG
M7 IXI T. Hawksley Field Cottage, Stowey, Bishop Sutton, Bristol BS39 5TH
MW7 JAB J. Byast 10 Chapel Street, Amlwch Port, Amlwch LL68 9HT
M7 JAC J. Parkes 4 Round Saw Croft, Rubery, Rednal, Birmingham B45 9TT
M7 JAD J. Davis Wakefield, The Street, Tirley, Gloucester GL19 4ES
M7 JAG J. Stone 55 Austin Waye, Uxbridge UB8 2RQ
M7 JAJ J. Jobling 152 Derwent Way, York YO31 0RQ
M7 JAP P. Whiteley 16 Fustian Avenue, Heywood OL10 3FN
M7 JAQ J. Quinlan 45 Mill Road, Beccles NR349UT
M7 JAR J. Robinson 6 Oswin Place, Walsall WS3 1PU
M7 JAS J. Singh Lynx House, School Lane, , Hadlow Down TN22 4JE
M7 JAT J. Tesseyman 31 Tennyson Street, Guiseley, Leeds LS20 9LN
M7 JAV J. Singh 35 Eaton Rise, London W5 2HE
M7 JAW J. Withington 20 Bond Way, Hednesford, Cannock WS12 4SN
M7 JAY J. Buttery 11 Warton Lane, Austrey, Atherstone CV9 3EJ
M7 JAZ P. Pennington 10 Rosina Street, Manchester M11 1HX
M7 JBB B. Butler The Mount, Cobmoor Road, Kidsgrove, Stoke-on-Trent ST7 4DF
M7 JBE J. Etheridge 34 Westonzoyland Road, Bridgwater TA6 5BN
M7 JBL J. Booth 2 Fairfax Mews, London E161TY
M7 JBM J. Minterne 43 Lawn Lane, Little Downham, Ely CB6 2TS
M7 JBO J. Boslem 14 Morrins Close, Great Wakering SS3 0DY
M7 JBP P. Kinch 58 Harepath Road, Seaton EX12 2RX
M7 JBS J. Stone Flat 3, Conel Court, 66 Talbot Road, Bournemouth BH9 2EU
M7 JBT B. Thompson 301 Pleck Road, Walsall WS2 9HA
M7 JBW J. King 10 Holly Park View, Gateshead NE10 9NH
M7 JBY J. Youel 18 Park Avenue, Penistone, Sheffield S36 6DN
M7 JCA J. Carter 4 Pewter Court, Wilnecote, Tamworth B77 5FX
MI7 JCB J. Bradshaw 51 Albany Drive, Carrickfergus BT38 8BF
M7 JCC J. Clark 7 Moorland Avenue, Blackburn BB2 5EQ
M7 JCD J. Diss 18 Rutland Avenue, Thornton-Cleveleys FY5 2DU
M7 JCE J. Clark 4 Exeter Street, North Tawton EX20 2HB
M7 JCH J. Howarth 67A Crawford Ave, Manchester M298ET
M7 JCK J. Macnally 3 Sydenham Terrace, Covington Road, Westbourne, Emsworth PO10 8SZ
M7 JCL J. Campion 44 Church Street, Aldbrough, Hull HU11 4RN
M7 JCM J. Mcleod 11 Ryebank Close, Birmingham B30 1SN
M7 JCN R. Needs 33 St Andrews Place Melton, Woodbridge IP12 1PZ
M7 JCP J. Pablo 22 Nightingale Lane, London N8 7QU
M7 JCR J. Askew 41 Hodson Close, Skellingthorpe, Lincoln LN6 5XB
M7 JCS J. Sydney 17 Streamside, Tonbridge TN10 3PU
MW7 JCX J. Clare Ty Capel Tanrallt, Llanllyfni, Caernarfon LL54 6RR
M7 JDB J. De Beer 5 Kyte Close, Warminster BA12 8GE
M7 JDC J. Cumming Farley Court, 100 Homer Close, Gosport PO13 9TL
M7 JDH J. Hunter 53 Grove Road, Tiptree, Colchester CO5 0JJ
M7 JDI J. Elliott 22 Bath Crescent, Huntingdon PE28 2EH
MM7 JDL J. Lockhart 41 Westfield Ave, Cupar KY15 5AA
M7 JDM D. Moyse 2 Station Road, Princetown, Yelverton PL20 6QX
M7 JDP J. Powell The Gables, Whites Lane, Kessingland, Lowestoft NR33 7TF
M7 JDW J. Wilkinson 8 Hunters Point, Chinnor OX39 4TG
M7 JEE S. Mukherjee 28 Culverlands Close, Stanmore HA7 3AG
M7 JEF J. Sparrow 12 St. Huberts Close, Gerrards Cross SL9 7EN
M7 JEH J. Hartley 20A Park Homer Drive, Wimborne BH21 2SR
MW7 JEL J. Slade Flat 1, 29 Broad Street, Welshpool SY21 7RW
M7 JEN J. Swift 16 Lime Tree Avenue, Kiveton Park, Sheffield S26 5NY
M7 JEP D. Palmer 133 West Lane, Middlesbrough TS5 4EH
M7 JEX C. Lawton 15 Grenville View, Cotford St. Luke, Taunton TA4 1JH
MM7 JFA J. Faulds 123 East Main Street, Darvel KA17 0JG
MM7 JFB J. Burns 6 Bonnyvale Place, Bonnybridge FK4 1DG
M7 JFF F. Gutai Flat 1-082, Arthur Sanctuary House, Sandfield Road, Headington, Oxford OX3 7RH
M7 JFH J. Fisk 2 Garden Cottage, Hartsholme Park, Lincoln LN6 0EY
M7 JFJ J. Forte 109 Lower Galdeford, Ludlow, Shropshire, Sy8 1Ru, Ludlow SY81RU
M7 JFK J. King 99 Devizes Road, Salisbury SP27LQ
M7 JFL D. Searley 18 Plantagenet Crescent, Bournemouth BH11 9PJ
M7 JFR B. Roberts Oak House, Gorsley, Ross-on-Wye HR9 7SW
M7 JFX J. Fox 2 Rhodes Avenue, Rotherham S61 3LG
M7 JGB J. Davies 166 Victoria Avenue, Princes Avenue, Hull HU5 3DY
M7 JGC J. Cadman 1 Kittiwake Road, Chorley PR6 9BA
M7 JGD J. Green 42 Longfield Ave, Newbarn DA3 7LA
M7 JGH J. Harston 70 Camm Street, Sheffield S6 3TR
M7 JGP J. Pearce 53 Kettlewell Close, Woking GU21 4HY
M7 JGS J. Southan 39 Hardy Way, Enfield EN2 8NW
M7 JGW J. Weatherill Northend Cottage, North End Road, Yatton, Bristol BS49 4AS
M7 JGX J. Greenwood 45, Billington Gardens Hedge End, Southampton SO30 2AX
MM7 JGY G. Yang 48 Foxglove Road, Newton Mearns, Glasgow G77 6FP
M7 JGZ C. Millar 195 Worsley Road, Frimley, Camberley GU16 9BH
M7 JHA J. Ratcliffe 7 De Havilland Drive, Hazlemere HP15 7FP
M7 JHD J. Haswell 15 Oaklands, Swalwell, Newcastle upon Tyne NE16 3EJ
M7 JHE J. Headland 4 Spring Close, Histon, Cambridge CB24 9HT
M7 JHM C. Melton 25 Evergreen Way, Brayton, Selby YO8 9RD
MW7 JHN J. Hardman Rosedale, Abergele Road, Old Colwyn, Colwyn Bay LL29 8AS
M7 JHR J. Ratty 12 Challoners Close, East Molesey KT8 0DW
M7 JHT J. Turrell 24 Barnfield, East Allington Totnes Devon, Totnes TQ9 7QR
M7 JHW J. Wheaton Flat 4, 25 Greenside, Waterbeach, Cambridge CB25 9HW
MW7 JHX J. Humphreys 134 Mill Street, Tonyrefail, Porth CF39 8AF
M7 JHY J. Hurley 64 Carleton Road, Chorley PR6 8UB
M7 JIA R. Jia 48 Boston Av, Reading RG1 6JU
M7 JIB G. Ogram 42 Juniper Way, Bradley Stoke, Bristol BS32 0BR
M7 JIM J. Austin 25 Cotmore Gardens, Thame OX9 3LZ
MJ7 JIV A. Gouveia Flat 2 Ronceville St. Clements Road, St. Saviour JE2 7PX Jersey
M7 JIZ J. Cook 20 Waverley Avenue, Kidlington OX5 2NA
M7 JJC J. Claridge 31 Conifer Crest, Newbury RG14 6RS
MI7 JJE J. Mcerlain 40A Kilnacolpagh Road, Aughafatten Aughafatten, Ballymena BT42 4LN
M7 JJG J. Glass 8 Marlborough Close, Ramsbottom, Bury BL0 9YU
M7 JJI J. Iseton 23 Heddon Grove, Ingleby Barwick, Stockton-on-Tees TS17 0FT
M7 JJK J. Page 22 Coppersmith, Combs, Stowmarket IP14 2FD
M7 JJO D. Simpson 1 Bridge Cottage, Blackpool, Dartmouth TQ6 0RG
M7 JJT J. Turner 24 Edith Wood Close, Alvaston, Derby DE24 0HJ
MW7 JJY J. Young Rhos Owen, Llangristiolus, Bodorgan LL62 5RD
M7 JKA Dr J. Kinrade 112 Ullswater Road, Lancaster LA1 3PX
M7 JKC D. Langford 1 Peveril Street Allenton, Derby DE248DG
M7 JKD J. Dalton 166 Ennerdale Road, Cleator Moor CA25 5LF
M7 JKF J. Korzybski 24 Brabazon Close, Shortstown, Bedford MK42 0FF
M7 JKM J. Moore 19 Exe View, Exminster, Exeter EX6 8AL
M7 JKN M. Jackson Paragon Hotel 123 North Marine Road, Scarborough YO12 7HU
M7 JKO J. Castledine 9 Creswick Close, Chesterfield S40 3PX
M7 JLC J. Cole 68 Northfield Road, Gloucester GL4 6TX
M7 JLD J. Dickins 78 Gobions, Basildon SS16 5AY
M7 JLG S. Raneses Grospe 8 Lincombe Road, Manchester M22 1GA
MW7 JLM J. Simons 24 Church Terrace, Porth CF39 0ET
M7 JLW J. Walkey 1 Elm Avenue, Poulton-le-Fylde FY6 7SP
M7 JLX J. Laud 2F Manor Road, Atherstone CV9 1QJ
M7 JMA J. Mclachlan 3 Wentworth Road, Oxford OX2 7TG
M7 JMC J. Mcdonald 18 Ridgeway, Gateshead NE10 8DD
M7 JMD J. Davidson 4 Willows Avenue, Alfreton DE55 7ER
M7 JME J. Earle 24 Brook Vale, Charlton Kings, Cheltenham GL52 6JD
M7 JMF J. Marvel Dean House Farm, Nordan, Leominster HR6 0AW
M7 JMH M. Hill Buzon 169, Avd De La Condomina 53, Local 3, , Alicante 3540 Spain
M7 JMJ J. Jimenez Sevilla 14Bentley Close Bentley Close, Loughborough LE11 1SY
M7 JMO J. Partovi 58 Main Street, Burley in Wharfedale LS29 7DF
MM7 JMP C. Jump 15 The Lade, Bonhill, Alexandria G83 9JR
MM7 JMR J. Rennie 39 Birdland Avenue, Bonesss EH519LX
MI7 JMS J. Mcshane 14 Annahugh Road Loughgall, Armagh BT61 8PQ

MW7 JMW J. Whale 60 Thornhill Road, Cardiff CF14 6PF
M7 JMX J. Mullings 112 Vicarage Lane, Great Baddow, Chelmsford CM2 8JD
MW7 JNC J. Craig 32 The Uplands, Port Talbot SA13 2EW
MM7 JNE J. Wilding 9, Dalavich Taynuilt, Oban PA35 1HN
M7 JNG J. Green Belvoir Vale Cottage, Barrowby Stenwith, Grantham NG32 2HE
M7 JNH S. Johnston 33 Angus Crescent, North Shields NE29 6UF
MI7 JNJ J. Agg 70C Cloghoge Road, Tandragee, Craigavon BT62 2HB
MM7 JNK J. Park 6 Robsland Avenue, Ayr KA7 2RW
M7 JNQ A. North 37 Cromwell Road, Southend SS25NG
M7 JNR A. Primo 68 Menzies Avenue, Basildon SS15 6SY
M7 JNU J. Nuttall 73 Severn Drive, Walton-Le-Dale, Preston PR5 4TD
M7 JNY J. Curtis Juniper, Broadway Lane, Fladbury, Pershore WR10 2QF
M7 JOC J. O'Connell 23 Halstead Road, Gosfield CO9 1PG
M7 JOD J. Pollard 9 Challum Drive, Chadderton, Oldham OL9 0LY
MI7 JOE J. Hannon 75 Suffolk Road, Belfast BT11 9PU
MI7 JOJ G. Travers 88 Palmerston Road, Belfast BT4 1QD
M7 JOL J. Wainwright 83 Bronte Paths, Stevenage SG20PJ
M7 JOM J. Moreton 1 Andrew Mulligan Close, Stoke-on-Trent ST6 5XF
MW7 JON J. Loisz 46 Fairview Avenue, Risca, Newport NP11 6HW
M7 JOP A. Jopson 43 Lower Landedmans, Westhoughton, Bolton BL5 2QL
M7 JOS J. Stringer 2 Westend Marston Magna, Somerset BA22 8BW
M7 JOX K. Brailsford 4 Maple Drive Gedling, Nottingham NG44AF
M7 JOY J. Organ 58 Sycamore Close, Taunton TA1 2QJ
M7 JOZ A. Jozwik 68 Pipit Rise, Bedford MK41 7JT
M7 JPC J. Christiaens 6 Richard Hicks Drive, Scarning, Dereham NR19 2TN
M7 JPF J. Fulcher 2 Bagot Grove, Stoke-on-Trent ST1 6JD
M7 JPM J. Meredith Pearl House, 1 Quarry Close, Myddle SY4 3SB
M7 JPP J. Parish 60 New Drove, Wisbech PE13 2RZ
M7 JPS N. Mantle St36Qj, Stoke-on-Trent ST36QJ
M7 JPX J. Walters-Pennell Clopton Grange, Clopton, Woodbridge IP13 6QR
M7 JRA Dr J. Addis 28 Shaw Street Culcheth, Warrington WA3 5EX
M7 JRB J. Baines 34 Crompton Road, Stone ST15 8NL
M7 JRG J. Godbold 22 Lark Hill Rise, Winchester SO22 4LX
M7 JRH J. Howling Sysonby Knoll Hotel, Asfordby Road, Melton Mowbray LE13 0HP
M7 JRK J. Kermode 17A Vancouver Road, London SE23 2AG
M7 JRM J. Bailey 97 Cherry Tree Ave, Cowplain PO88AX
MM7 JRS J. Scott 81 Kinellar Drive, Glasgow G14 0EU
MM7 JRX J. Hutton 66/2 Craigmount Brae, Edinburgh EH12 8XF
M7 JRY A. Wilson 38 Saxons Heath, Long Wittenham OX14 4PU
M7 JSA J. Stephens 16 Northfield Road, Harborne, Birmingham B17 0SU
M7 JSC J. Scully 22, Lansdowne Terrace West North Shields, Newcastle upon Tyne NE29 0RW
M7 JSD J. Dale 123 Rosedale Gardens Belton, Great Yarmouth NR31 9PL
M7 JSF J. Fletcher 3 Moorend Glade Charlton Kings, Cheltenham GL53 9AT
MW7 JSG E. Phillips 60 Gittin Street, Oswestry SY111DS
M7 JSH J. Shears 7 Tinderley Grove, Huddersfield HD5 8PE
M7 JSI B. Wolfe 60 Lindley Street, Rotherham S65 1RT
M7 JSK J. King 1 Kennion Road, Wells BA5 2NP
M7 JSL A. Cramp 30 Park Drive, Hastings TN34 2PR
M7 JSP J. Pope 5 Cotswold Rd Little Sandhurst, Sandhurst GU47 8NA
M7 JSS W. Fisher 24 Lisle Road, Weston-Super-Mare BS22 7UA
M7 JST J. Talbot Hales Road, Wednesbury WS109BS
M7 JSW J. Spooner 82 Reedswood Lane, Walsall WS2 8QP
M7 JSY J. Yapp 6 Avondale, Dawley Bank, Telford TF4 2LW
M7 JTF A. Jones 86 Church Lane Gomersal, Cleckheaton BD19 4QL
MW7 JTH J. Turobin-Harrington Michaelmas Barn, Sawmills, Kerry, Newtown SY16 4LL
M7 JTI J. Isaac 103 Field Crescent, Shrewsbury SY1 4PG
M7 JTN T. Evans 69 Lincoln Street, Liverpool L19 8LF
MM7 JTR J. Robertson Pinewood Place, Edinburgh EH225JA
M7 JTU J. Tuohy 13 Ely Place, Woodford Green IG8 8AG
M7 JTZ J. Skinner 35 Holbein Close, Bedworth CV12 8TA
MW7 JUE J. Burford 13 Maes Ceidio, Llannerch-Y-Medd LL71 7AE
M7 JUK S. Jukes 147 Vicarage Road, West Bromwich B71 1AE
M7 JUL J. Brewer 1 Bentley Road, Forncett St. Peter, Norwich NR16 1LH
M7 JUS J. Doherty 35 Eastern Way, Letchworth Garden City SG6 4PE
MM7 JVA J. Howes 22 Criffel Drive, Lincluden, Dumfries DG2 0PE
M7 JVC L. Shaw 23 Church Street, Buckingham MK18 1BY
M7 JVH J. Henry 32 Kendricks Fold, Rainhill L35 9LX
MM7 JVR J. Rowlands 3 Earls Gate, Slackbuie, Inverness IV2 6FF
M7 JVS V. Jagannathan 19 Wallbrook Avenue, Macclesfield SK10 3GL
M7 JWE J. Emery Flat 3, 12 Buxton Road, Ashbourne DE6 1EX
M7 JWF J. Willis-Fisher Linden House, Portman Road Pimperne, Blandford Forum DT11 8UJ
M7 JWH J. Hunter 53 Grove Road, Tiptree, Colchester CO5 0JJ
MM7 JWK J. Keogh The Mound, Dornoch IV25 3JE
M7 JWL J. Limer 86 Elldawn Avenue Milton, Stoke-on-Trent ST68XE
M7 JWM Dr M. Moss 2 Crosby Avenue, Worsley, Manchester M28 3FQ
M7 JWP J. Plant Longacres, Station Road, Old Leake, Boston PE22 9RF
M7 JWR J. Wright-Roberts 19 Field Lane, Dursley GL116JF
M7 JWW J. Walker-Wilson Rest Harrow Southside, Scorton, Richmond DL10 6DN
M7 JWZ J. Marek 91 Dial House Road, Sheffield S6 4WU
M7 JXR J. Richmond 18 Ullswater Drive, Bradford BD6 2TE
M7 JXW J. Warrington 22 Knights Close, Buntingford SG9 9SE
M7 JXX J. Thompson 11 Foxbury Drive, Orpington BR66EJ
M7 JYE D. Basham 15 Arrendene Road, Haverhill CB9 9JQ
M7 JYP J. Capstick 186 Forest Lane, Harrogate HG2 7EE
MM7 JYS J. Steven 34 Queens Drive, Troon KA10 6SE
M7 JZN J. Nichols 4 Elmlee Close, Chislehurst BR7 5DU
M7 KAA K. Abrey 20 Shipfield, Norwich, Nr34dx NR34DX
M7 KAD N. Jaggs 15 St. Anthonys Road, Kettering NN15 5HT
M7 KAE K. Barnett 22 Highclere Road, Southampton SO16 7AW
M7 KAG K. Gillkerson Flat 14, Target Place, 489 Butts Road, Southampton SO19 1AD
M7 KAK K. Kariraman 17 Kipling Close, Galley Common, Nuneaton CV10 9SJ
MW7 KAP K. Gamwasam 01, Churton Drive, , Wrexham LL13 8RU
M7 KAS K. Fenton 40 High Street, Easington Lane, Houghton le Spring DH5 0JN
MI7 KAT K. Nesbitt 2 Church Road, Gracehill, Ballymena BT42 2NL
M7 KAU A. Mahal Kingsley Hall Stretton Road, Great Glen, Leicester LE8 9GP
M7 KAV P. Walmsley 24 Camton Road, Middleleaze, Swindon SN5 5TP
M7 KBO T. Menzies 4 Meadow Road, Muxton, Telford TF2 8JH
M7 KBW K. Barnes Lenbar, Mill Lane, Weston, Hitchin SG4 7AJ
MW7 KCA G. Roberts 31 Maes Y Dre Holywell Road, Caerwys Mold CH7 5AS
M7 KCB K. Beckett 29 Park Road, Camberley GU15 2SP
M7 KCD C. Daly 2 Withen Cottages, Withen Lane, Aylesbeare, Exeter EX5 2JQ
M7 KCE B. Toombs 21 Hawthorn Crescent, Shepton Mallet BA4 5XR
M7 KCK K. Kay Field House, Foxholes, Driffield YO25 3QF
M7 KCP K. Pugh 4 Salt Boxes, Pinvin, Pinvin WR102LB
M7 KCR K. Charkseliani 2, Raymond Crescent, Guildford GU2 7SX
M7 KCY C. Jones 4 Guestwick, Tonbridge TN10 4HU
MW7 KDA K. Smith 48 Ffordd Gryffydd, Llay, Wrexham LL12 0RT
M7 KDB K. Brough 161 Gloucester Road, Kidsgrove, Stoke-on-Trent ST7 1EH
M7 KDH K. Dhun 2 Sparvell Road Knaphill, Woking GU21 2RR
M7 KDR D. Brook 24 Botany Avenue, Bradford BD2 1EU
M7 KDU S. Callen 3 Pendleton Gardens, Blackfield, Southampton SO45 1DQ
M7 KDV P. Field 63 Hartford Road, Davenham, Northwich CW9 8JE
M7 KDZ K. Davenport 6 Wolfe Close, Walton, Chesterfield S40 2DF
M7 KEC J. Crawley 65 Wrentham Avenue, Herne Bay CT6 7UX
MW7 KEE S. Keeble Mynachlog, Tyn Y Gongl LL74 8SG
MM7 KEF K. Sterling 1 Glamis Road, Kinghorn, Burntisland KY3 9UR
MM7 KEH J. Hunter 22 Watson Place, Dunfermline KY12 0DR
M7 KEI K. Evans 10 Wharfedale Close, Blackburn BB2 5EY
MM7 KEL S. Kelly 15 Shinwell Avenue, Linnvale, Clydebank G81 2RA
M7 KEN K. Spilsbury 81 Leacroft Avenue, Wolverhampton WV10 9DB
M7 KEP P. Kenderes 43 Fairlight Road, London SW17 0JE
M7 KEU P. Humphreys 35 Doctors Hill, Stourbridge DY9 0YE
M7 KEV K. Allison 4 Ashwood Groveá, Sunderland SR53BU
M7 KEZ K. Buckwell 31 West Green, Stokesley TS95BE
M7 KFC P. Roppa 7 Kilowna Close, Charvil, Reading RG10 9QU
M7 KGA K. Gracey Gwendoline Close, Merseyside CH611DL
MM7 KGC R. Condie 48 Carleton Avenue, Glenrothes KY7 5AJ
M7 KGF J. Mussell Pool View Barn, Grafton, Hereford HR2 8BW
MM7 KGT K. Thomson 26 Canal Court, Linlithgow EH49 6LZ
M7 KGW K. Wheaton 3 Walnut Tree Crescent, Fenstanton PE289LE
MM7 KGX C. Thompson 5 Church Road, Duffus, Elgin IV30 5QQ
M7 KHA H. Khan 9 Prestwood Close, High Wycombe HP12 3DE
M7 KHB K. Bridge 22 Oaksfield, Darwen BB3 0AU
M7 KHL K. Hill Ash Cottage, 96A The Street, Carlton Colville, Lowestoft NR33 8JR
M7 KHS K. Summers 5 Arnison Close, Carlisle CA2 6RN
M7 KHX K. Heydon 34 Charles Road, Amble, Morpeth NE65 0SQ
M7 KIC M. Chrzanowski 4 Boulters Close, Slough SL1 9BQ
MW7 KID B. Ryan 8 Beechfield Close, Garnlydan, Ebbw Vale NP23 5EN
MM7 KIE R. Mackie 72 Rosemount Crescent, Carstairs ML118QW
M7 KIK B. Brightwell 3 Longedryve, Off Wavell Avenue, Colchester CO2 7HH
M7 KIM S. Kimber 8 Westfield View, Wakefield WF1 3RU
MI7 KIR C. Kirkwood 6F Ardcarn Green, Belfast BT5 7RR
M7 KIT A. Snell-Pym 50 Newton Avenue, Gloucester GL4 4NU
MW7 KJC K. Cox 61 Folly View, Penygarn, Pontypool NP4 8BU
M7 KJD J. Drew Slough Hall, Slough Lane, Little Cornard, Sudbury CO10 0NY
M7 KJF M. Farrar 8 Kingswood Chase, Trowbridge BA14 9GD
M7 KJG K. Green 9 Oakfield Road, Bishops Cleeve, Cheltenham GL52 8LA
M7 KJK K. Kolski 1 Buckingham Grove Scartho Top, Grimsby DN333RR
M7 KJL K. Lever 121, Greenwater Court, Mainway, Lancaster LA1 2AY
M7 KJM C. Mcmahon 42 St. Mildreds Avenue, Broadstairs CT10 2BX
MW7 KJN K. Nicholls 35 Partridge Road, Cardiff CF24 3QW
M7 KJP K. Price 5 Wheelers Lane, Redditch B97 6GT
M7 KJS K. Smith 30 Azalea Drive, Swanley BR8 8HZ
MM7 KJV K. VielhaberNether Littlefold, Crieff PH7 3NY
M7 KJW K. Webster 17 Cardinal Close, Tonbridge TN9 2EN
M7 KKB K. Bennett 24 The Quadrant, Uppingham, Oakham LE15 9QP
M7 KKG K. Gibson 44 Churchville, Micklefield, Leeds LS25 4AP
M7 KKM N. Gregory 8 Firman Close, , New Malden KT3 4BN
M7 KKO K. Oliver 50 Tewkes Road, Canvey Island SS8 8HG
MW7 KKW K. Kirkpatrick 151 Bridgend Road, Aberkenfig CF32 9AE
M7 KKY Y. Jo 103-503 Lg Apt., Jure2-Dong, Busan 46997 Korea, republic of
M7 KLA K. Adcock 37 Flansham Park, Bognor Regis PO22 6QH
M7 KLB K. Bragg Flat 5 126 Harrington Road, Workington CA14 2UW
MW7 KLC K. Evans 21 Orchard Street, Pontardawe, Swansea SA8 4ER
M7 KLF J. Rosewarn 47 Bond Street, Trowbridge BA14 0AS
M7 KLJ P. Jones 2 Whitley Place, Stoneley Park, Crewe CW1 4GH
M7 KLM N. Winchcombe 1 Talbot Villa, Talbot Street, Glossop SK13 7DG
M7 KLR S. Mccarthy 1 Fishmoor Drive, Blackburn BB2 3TJ
M7 KLT K. Louth 36 Copelea Cheswardine, Market Drayton TF9 2RX
M7 KLW S. Walker 2 Williamson Avenue, Stoke-on-Trent ST6 8AB
MW7 KMB K. Byast 10 Chapel Street, Amlwch Port, Amlwch LL68 9HT
M7 KMF K. Fell 211 Lower Way, Thatcham RG19 3TN
M7 KMO O. Glover 9 Lea Road, Grays RM16 4DD
M7 KMP G. Kemp 21 Chasten Hill, Letchworth Garden City SG6 4YN
MI7 KMQ K. Mchugh 31 Moyle Avenue, Ballycastle BT546NX
M7 KMS K. Singleton 2 Rome Avenue, Burnley BB115LQ
M7 KMV J. Varkevisser Stocks Road, Stocks Oast, Wittersham TN307EY
M7 KMW J. Nihill 14 Hereford Avenue, Clayton, Newcastle ST5 3ED
MM7 KNC C. Mackenzie 10 Strait Close, Strathaven ML10 6LN
MW7 KND N. Dewar 2 Woodside Cresent, Ebbw Vale NP23 6DD
MD7 KNE W. Kneale 2 The Crescent, Douglas, Isle of Man IM2 5ET
M7 KNG A. King 46 Purbeck View, Bovington, Wareham BH20 6PJ
M7 KNI R. Knight 10 Hargreaves Close, Basingstoke RG24 9SS
M7 KNO R. Massie 13 Noel Coward Gardens, Aldington, Ashford TN25 7EU
M7 KNR D. Tickner 26 Wye Road, Borough Green, Sevenoaks TN15 8DX
M7 KNV V. Kontham Flat 3, 10 Haydon Place, Guildford GU1 4LL
M7 KOF S. Kofler 29 Fownes Street, London SW11 2TJ
M7 KOL K. Dowling 8 Chathill Close, Morpeth NE61 2TH
M7 KOP S. Allen 26 Gloucester Avenue, Rayleigh, Essex SS68XR
M7 KOR D. Korzeniewicz 6 Barley Down Drive, Winchester SO22 4LS
M7 KOW B. Staniforth The Mill House, Chantry Mill, Chantry Lane, Storrington, Pulborough RH20 4AB
M7 KOY T. Headley 73 Tudor Green, Jaywick, Clacton-on-Sea CO15 2PB
M7 KPA K. Parkes 7 Primrose Avenue, Rotherham S60 5JX

M7 KPB K. Birkett 17 Torridon Crescent., Bradford BD62TY
M7 KPM J. Ward 66 Abbey Road, Astley, Tyldesley, Manchester M29 7WG
M7 KPS K. Snipe 5 Draycott Road, Chiseldon, Swindon SN4 0LT
MW7 KPV A. Mcgowan 5 Tollstone Way Grosmont, Abergavenny NP7 8ER
M7 KPW K. Watson 43 Rectory Place, Weyhill, Andover SP11 0PZ
M7 KRA K. Hiltz 35 Highview Road, London W13 0HA
M7 KRD K. Dodwell 29 Cleaveside Close, Queen Camel, Yeovil BA22 7NR
M7 KRE K. Mills 3 The Seasons, Summerway, Exeter EX4 8DQ
M7 KRH J. Burton 2A Empire Road, Salisbury SP2 9DF
MI7 KRJ K. Riddle 4 School Road, Newtownhamilton BT350DQ
M7 KRK C. Guri Comallonga 10 Collingwood Road, Sutton SM1 2RZ
M7 KRN K. Ayre St. Andrews Farm East, Woodbridge Lane, Bedchester, Shaftesbury SP7 0BF
MW7 KRP K. Poynter 19 Bryn Ebbw, Ebbw Vale NP23 5LU
M7 KRW K. Windle 2 Westerley Close, Shelley, Huddersfield HD8 8HL
M7 KSD K. Driver 91 Springfield Gardens, Ilkeston DE7 8JA
MW7 KSE T. Bancroft 4 Bryn Heli, Old Colwyn, Colwyn Bay LL29 9ER
M7 KSF A. Brooking 1 West Villas Cotford St Luke, Taunton TA4 1DF
MD7 KSL C. Larkham Monte Rosa , 7 Ballaughton Close, , Douglas IM2 1JE
M7 KST K. Turner 40 Walsingham Close, Eastbourne BN22 0UD
MW7 KSW K. Stephenson 6 Ffordd Brannan, Buckley CH7 3DE
M7 KSY L. Davey 5 The Caravan 53 Drump Road, Redruth TR151PR
M7 KTA K. Capon 17 Bonython Close, Mylor Bridge, Falmouth TR11 5NF
M7 KTB J. Sanderson 4 Elmgrove Close, Woking GU21 8XL
M7 KTL J. Boutros 15 Navigation House, Whiting Way, London SE16 7EG
MM7 KTO K. Todd 19 Coaledge, Cowdenbeath KY48HB
MW7 KTP P. Joyce 73 Glenwood Close, Coychurch, Bridgend CF35 5EU
M7 KTR K. Railton 102 Shieldfield House, Barker Street, Newcastle upon Tyne NE2 1BQ
M7 KTT P. Trent 3.Berkshire Close, Leigh-0n-Sea SS9 4RT
MM7 KTW R. Rae 1 Jura Drive, Tweedbank, Galashiels TD1 3ST
MI7 KTY R. Maguire 45 Dorchester Park Portadown, Craigavon BT62 3EB
M7 KVM S. Ahmed 445 Ashingdon Road, Rochford SS4 3EN
M7 KVS K. Slaney 69A Victoria Street, Littleport, Ely CB6 1NA
M7 KVZ R. Elles 8 Dilton Close, Trowbridge BA140FS
MM7 KWA K. Waitz C/O Waitz-Rainey, 16 Inverkeithing Road, Aberdour, Burntisland KY3 0RS
M7 KWB K. Buckley 44 Greenfields, Gnosall, Stafford ST20 0HR
M7 KWG W. Green 48 Springfield Rd Lofthouse, Wakefield WF33FN
M7 KWI K. Wilczynski 45 Caroline Street, Preston PR1 5UY
M7 KWS F. Stepien 11A Wirral Gardens, Wirral CH63 3BD
M7 KWT K. Tyler Ground Floor Flat 20 Brighton Road Newhaven, Newhaven BN9 9NB
M7 KXD K. Davis 38 South Lynn Crescent, Easthampstead, Bracknell RG127JY
MM7 KXI A. Stewart 5 Mosswater Wynd, Cumbernauld, Glasgow G68 9JU
M7 KXL K. Lee 8 Hillcrest, Barnsley S74 0BG
MW7 KYL K. Jones 24 Crown Street, Swansea SA6 8BD
M7 KZD O. Guest 36 Dace Road, Worcester WR5 3FD
M7 KZE J. Taylor 97 George Street, Cleethorpes DN35 8PL
M7 KZF C. Phillips 135 Misty Road, Harmony 28634 United States
M7 KZW K. Watson 24 Pine Close, Rishton, Blackburn BB1 4JX
M7 LAA A. Lester Flat 7 Wyre View 28 Queens Terrace, Fleetwood FY7 6BT
M7 LAE F. Khatir 12 Ashburn Garth, Hightown, Ringwood BH24 3DS
M7 LAG L. Gutierrez Flat 2, St. Albans, Elmwood Avenue, Feltham TW13 7AA
M7 LAH L. Horsfall 48 Orchard Drive, Grimsby DN32 7AW
MI7 LAJ J. O'Shaughnessy 17 Hood Court, Antrim BT41-4HW
M7 LAK L. Kroon Flat 102, The Courtyard, Circus Street, Brighton BN2 9AL
M7 LAN C. Langman 24 St. Marys Road Abbey Green, Nuneaton CV11 5AU
M7 LAO M. Thompson 11 Garston Road, Corby NN18 8NG
MM7 LAP P. Lapworth Meadow View, Borgue, Kirkcudbright DG6 4SH
M7 LAT L. Thompson 3 Leasway, Grays RM16 2HD
M7 LAW L. Wright 16 Sheffield Close, Potton SG19 2NY
M7 LBG L. Bingham 11 St Marys Av Barnetby, Brigg DN38 6HU
M7 LBI I. Brown 11 Snowdon Close, Warrington WA5 3HD
M7 LBM L. Munckton 20 Malvern Close, Bishops Waltham, Southampton SO32 1AY
M7 LBO Dr M. Coombes 49 Garendon Road, Shepshed, Loughborough LE12 9NU
MI7 LBW C. Corken 14 Teal Rocks, Newtownards BT23 8GL
M7 LBX M. Dickinson 16A Horsebrook Park, Calne SN11 8EY
M7 LBY C. Lazenby 33 Towning Close Deeping St. James, Peterborough PE68HS
M7 LCA M. Piatek 3 Chestnut Close, Wilmslow SK9 2NT
M7 LCB L. Baker 1B The Parade, Moss Road, Askern, Doncaster DN6 0LF
M7 LCD N. Omer Flat 12, Mallard Court, 1 Piper Close, London N7 8TQ
M7 LCF Prof. M. Atkinson 21 Dennington Crescent, Fryerns, Basildon SS14 2FF
M7 LCJ C. Ludlow M33 3An, Manchester M33 3AN
M7 LCL J. Phypers Cranborne Farm Bungalow, Blandford Road, Coombe Bissett, Salisbury SP5 4LF
M7 LCP L. Pickering 7 Ashmore Terrace, Sunderland SR2 7DE
M7 LCR L. Richardson 33 Dingley Road, Wednesbury WS10 9PU
M7 LCS T. Bayley 67 Shaftesbury Avenue, Feltham TW14 9LN
M7 LCW L. Wilkin 20 South Close, Halstead CO9 1NJ
M7 LCY A. Gee 14 Pipistrelle Drive, Onehouse IP14 1GS
M7 LDJ D. Loveridge 2 Abbots Mead, Cholsey OX10 9RJ
M7 LDK E. Priester Flat 5, 96 Osborne Road, Windsor SL43EN
M7 LDS L. Sawyer 85 , Beechwood Road, Sheffield S6 4LQ
M7 LDU G. Burridge 88 Eaton Road, Kempston, Bedford MK42 7RX
M7 LDV P. Copeland 15 Station Gardens, Woodford Halse, Daventry NN11 3PX
M7 LDW D. Whewell 41 Fordway Avenue, Blackpool FY3 8JL
M7 LDX G. Taylor 64 St. Elphins Close, Warrington WA1 2EH
M7 LED M. Clayton 16 Westwood Close, Scarborough YO11 2JB
MW7 LEJ L. James 103 Caer Wenallt, Cardiff CF14 7TJ
M7 LEL L. Lovell 28 Park Lane, Blunham MK44 3NJ
M7 LEM E. Lemaire 238 Inskip, Skelmersdale WN8 6JX
M7 LEN L. Wharram 8 Bourne Way, Sutton SM12EN
MM7 LEO L. Taylor 22 Briggies Wynd, Kintore, Inverurie AB51 0TX
M7 LEP L. Potter 201, Shadyside, Doncaster DN4 0HE
M7 LEW S. Freestone 47 Salisbury Street, Gainsborough DN21 2RS
M7 LEX J. Lewthwaite 12 Fullers Close, Melksham SN12 7BX
MM7 LEY R. Farley 16 Greenbank Road, Darvel KA17 0NN
M7 LEZ M. Jakes Crimond, Coles Lane, Kingskerswell TQ12 5BQ
M7 LFF A. Buck 1 Sandbeck Cottage, Stragglethorpe, Lincoln LN5 0QZ
M7 LFG M. Sykes 5 Greenfield Cottages, Barnsley S71 3LF
M7 LFI C. Sotiriou Flat B, 2 Thornton Road, Northampton NN2 6LS
M7 LGD A. Bent Lime Garth Sherburn Road, Durham DH1 2JR
M7 LGG L. Goodfellow 19 Quarry Bank, Newcastle under Lyme ST5 5AG
M7 LGT L. Gilbert 21 Maisemore, Bristol BS37 8UR
M7 LGV O. Vieira Leite Flat 39, Kedleston Court, Norbury Close, Allestree, Derby DE22 2QF
M7 LGW N. Westhead 33 Theobalds Way, Frimley, Camberley GU16 9RF
M7 LHR D. Langley 2 Fulwood Close, Hayes UB3 2NF
M7 LIE E. Wall 7 Fearnham Close, Leigh WN7 3LB
M7 LIK M. Atkinson 21 Arbroath Road, Sunderland SR3 3LA
M7 LIN L. Chadwick 19Regent Crescent, Manchester M350LR
M7 LIP Dr P. Lidstone Maddox, Bradworthy, Holsworthy EX22 7QY
M7 LIU H. Liu Flat 11, Albany Court, Palmer Street, London SW1H 0AA
M7 LIW M. Williams 59 Scrapsgate Road, Sheerness ME12 2EA
MM7 LJD L. Dickson 8 Hailes Park, Edinburgh EH13 0NG
M7 LJH L. Healy 31 Roach Road, Sheffield S11 8UA
M7 LJM L. Moyle 14 St. Lukes Close, Kettering NN15 5HD
MI7 LJR L. Roberts Clonduff Drive, Belfast BT6 9NT
M7 LKS A. Melchior 12 Midge Hall Drive, Rochdale OL11 4AX
M7 LKY M. Hackney 15 Station Drive, Winchester SO21 3FS
M7 LKZ A. Frydrych 17 Underhill Road, Hereford HR1 1SY
M7 LLB F. Cunliffe 33 Moor View Road, Sheffield S8 0HH
M7 LLE A. Miller 7 Alexandra Gardens, Shildon DL4 2EX
MW7 LLJ L. Jordan 74 Ladyhill Road, Newport NP199RZ
MM7 LLK D. Tanner 14 Kincaldrum Place, Dundee DD3 7HG
MM7 LLM J. Rae 4 Hillside Crescent, Langholm DG13 0EE
M7 LLN L. Minchella 30 Larksmead, Blandford Forum DT11 7LU
M7 LLR T. Gregory 50 Croft Street Lincoln England Ln25Az, Lincoln LN25AZ
M7 LLS M. Duckett 17 High Way, Lingwood, Norwich NR13 4BU
M7 LLY C. Salisbury 9 Oakville Road, Heysham, Morecambe LA3 2TB
M7 LMA L. Afonso Flat 1, 59 Croham Road, South Croydon CR2 7HF
M7 LMB R. Barwell 13 Pembroke Avenue, Orton Waterville, Peterborough PE2 5EY
M7 LMC L. Coupe 51 Leamington Drive, Sutton-in-Ashfield NG17 5BA
M7 LMF W. Turner Low Farm, Bishop Norton, Market Rasen LN8 2AJ
M7 LMG S. Crowther 4 Brigsteer Close, Clayton Le Moors, Accrington BB5 5GE
M7 LMK L. Mery-Kennedy 69 Alexandea Rd, Ipswich IP42RN
MM7 LMM L. Mullaney 8 School Road, Wellbank, Broughty Ferry, Dundee DD5 3PL
M7 LMN M. Horwich H Horwich Motor Engineers, Unit 2, Peel Green Trading Estate, Green Street, Eccles, Manchester M30 7HF
M7 LMP I. Jabegu 51 South Crescent, Blandford Camp, Blandford Forum DT11 8AJ
M7 LMS L. Scully 17 Fenwick Close, Woking GU21 3BY
M7 LNA N. Arnold 44 Small Drove Weston, Spalding PE12 6HS
M7 LND N. Legall 122 Cheddon Road, Taunton TA2 7DN
M7 LNG A. Laing 37 Crossways, South Croydon CR2 8JQ
M7 LNO T. Sidaway 6 Rookery Chase, Deepcar, Sheffield S36 2NF
M7 LOA L. Owen 49 Glycena Road, London SW11 5TP
M7 LOC Prof. J. Stochl 23 Hardy Close, Huntingdon PE29 1RR
MW7 LOI L. Illingworth 54 Swansea Road, Garden Village, Gorseinon, Swansea SA4 4HE
MM7 LOL A. Bennett 8 Buckstone Loan East, Edinburgh EH10 6XD
M7 LOT R. Coates 1 Audrey Road, Sheffield S13 8DQ
MW7 LOV M. Chacon-Dawson 57 Fair View, Blackwood NP12 3NR
M7 LOW M. Rook 116 Ollerton Road, Athersley North, Barnsley S71 3DL
MM7 LOX M. Macfarlane 9 Dreghorn Park, Colinton, Edinburgh EH13 9PH
M7 LOZ L. Woodward Oaklands, Ashford Lane, Hockley Heath, Solihull B94 6RH
M7 LPB A. Geldard Lunesdale, Swanton Road, Dereham NR20 4PS
M7 LPL P. Stanton 12 St. Lawrence Close, Liverpool L8 4XP
M7 LPN M. Lappin Willow Cottage, 16 Chapel Road, Old Newton, Stowmarket IP14 4PP
M7 LPP Dr A. Holt Ellon House, Church Road, Sutton, Norwich NR12 9SG
M7 LRB L. Bennett North Cottage Frilsham, Thatcham RG18 9UZ
M7 LRJ L. Jarrett 7 Valley View Road, Plymouth PL3 6QJ
M7 LRM R. Millington 34 Grosvenor Road, Rotherham S65 1QP
M7 LRP L. Paston 138A Forest View Road, Manor Park, London E12 5HX
M7 LRQ E. Rowe Allandale View Ermine West, Lincoln LN13RD
M7 LRT L. Rayment 8 Hartley Way, Taunton TA1 2LJ
MM7 LRV J. Sinclair El-Alto 13 Duthie Road, Gourock PA19 1XS
M7 LRY L. Young 5A Fernlea Road, Burnham-on-Crouch CM0 8EJ
M7 LSA S. Deakin 20 Riccat Lane, Stevenage SG1 3XY
M7 LSB P. Curtis 112 Goring Road, Colchester CO4 0DA
M7 LSE S. Evans 2A Manor Road, Merstham RH1 3LT
M7 LSL Rvd. L. Williams 36 Royd Court, Mirfield WF14 9DJ
M7 LSS P. Lu 34 Baker Street, Potters Bar, , London EN6 2EB
M7 LST L. Turnbull 56 Wyngates, Leighton Buzzard LU7 2LE
MW7 LTE L. Evans 32 Breakwater House, Cardiff CF11 0JQ
M7 LTH L. Hopkins 4 Deepfield Way, Coulsdon CR5 2SY
M7 LTM L. Marrable 6 Piccadilly Close, Northampton NN4 8RU
M7 LTR R. Latter Longreach, Chapel End, Broxted, Dunmow CM6 2BW
MW7 LTS C. Morris Hideaway Bettws Cedewain, Newtown, Powys SY16 3DS
MD7 LTT R. Talbot 43 Harcroft Meadow, New Castletown Road, Douglas IM2 1JT Isle of Man
M7 LTY G. Alty 20 Long Meadows, Everton, Doncaster DN10 5BL
MM7 LUB L. Balmforth 1 Laburnum Avenue, Beith KA15 1BQ
M7 LUF S. Luffingham 25 Taylor Close, Tonbridge TN9 2FE
M7 LUK L. Patalon 54A Castle Road, Bournemouth BH9 1PJ
M7 LUT A. Moss 2 Saxted Close, Luton LU2 9SQ
M7 LUV C. Harrod 99 Bergholt Road, Colchester CO4 5AF
M7 LVC S. Raby 55 Rainbird Road, Bishop's Stortford CM23 2ZR
M7 LVW J. Thurgood 28 Ploughmans Way, Rainham Rainham, Gillingham ME8 8LH
MM7 LWC L. Copland 67 Naughton Road, Wormit, Newport-on-Tay DD6 8NG
M7 LWF C. Bowen 103 Hilton Lane Great Wyrley, Walsall WS6 6DT
M7 LWI L. Wilde 23 Colliers Grove, Atherton M46 0GT
M7 LWP L. Pearson 136 Nottingham Road, Stapleford, Nottingham NG9 8AR
M7 LWR L. Russell 11 Hartbushes Station Town, Wingate TS28 5GA
MW7 LWX P. Lewis 9 Heol Aradur, Cardiff CF5 2RE
MM7 LWZ L. Macinally 3 Aglath, Stenness, Stromness KW16 3HA

M7 LXB R. Thompson Croft Michael Farm, Croft Mitchell, Troon, Camborne TR14 9JJ
M7 LXD L. Davis 9 High Street, Chapel-En-Le-Frith, High Peak SK23 0HD
MM7 LXI L. Hunter 9 Glenview, Dalmally PA33 1BE
M7 LXT D. Ingram 19A Parkville Road, London SW6 7DA
M7 LXY J. Grant 129 Eastern Avenue, Lichfield WS13 6RL
M7 LYA A. Pruszak 32 Whitefriars Road, Hereford HR2 7XF
M7 LYD L. Gill Sixhills Street, Grimsby DN32 9HT
M7 LYF Capt. J. Pearce Clematis Cottage, 52 Cheselbourne, Dorchester DT2 7NP
M7 LYM S. Owens 18 Harvester Way, Lymington SO41 8YD
M7 LYN L. Walsh 4 Musbury Crescent, Rossendale BB4 6AY
MM7 LYS R. Howell 338 Albert Drive, Glasgow G41 5HH
MM7 LYZ L. Johnston 61 Stafford Road, Greenock PA16 0TG
M7 LZR C. Simons Tarn Hows, Woburn Lane, Aspley Guise MK17 8JN
M7 LZT L. Telfer 7 Beech Street, Failsworth, Manchester M35 0BE
M7 MAA G. Sangiorgi Flat 3, Windsor Lodge, 84 High Street, Brighton BN2 1RP
MM7 MAB S. Kent 7 Craigie Place, Galston KA4 8AX
M7 MAE M. Parsons 16 Latimer Road, Teddington TW11 8QA
MW7 MAI C. O'Connor Tan Y Bryn Pant Y Dwr, Rhayader LD65LR
M7 MAM M. Matis Unit 15, Old Cement Worksá, Newhaven BN9 0HS
MD7 MAN Dr J. Daniels 24 King Orry Road Glen Vine, Isle of Man IM4 4ES
MW7 MAP M. Pluke 3 Maes Y Ficerdy, Rhosllanerchrugog, Wrexham LL14 2EJ
M7 MAS M. Szpytko 21 California Avenue, Scratby NR29 3PE
MW7 MAT M. Burnell Ty Talwyn Farm, Cefn Cribwr, Bridgend CF32 0BP
M7 MAW A. Whyatt 8 Acacia Avenue, Midway, Swadlincote DE11 0HE
MM7 MBB M. Burke 103 Seaforth Road, Falkirk FK2 7TQ
M7 MBC D. Cooper 38 Barnsdale Road, Leicester LE4 1AR
M7 MBE R. Baldwin 10 Queen Elizabeth Avenue, South Ferriby, Barton-upon-Humber DN18 6HJ
M7 MBI M. Bird 4 Saint Andrews Road Avonmouth, Bristol BS11 9EU
M7 MBK M. Booker 19 The Bentleys, Southend-on-Sea SS2 6UJ
M7 MBO M. Booth 59 Kenmore Road, Sale M33 4LG
M7 MBT T. Youp 123 Tanfields, Skelmersdale WN8 8NS
MM7 MCA C. Anderson 57 Fa'Side Avenue South, Wallyford, Musselburgh EH21 8AN
M7 MCB M. Crosby 40 Galtres Drive, Easingwold, York YO61 3DJ
MI7 MCC W. Mcclean 6 Alveston Drive Carryduff, Belfast BT8 8RL
MM7 MCD A. Mcdonald Kettlehills, Cupar KY15 7TW
M7 MCF M. Lambert 311 Manchester Road, Manchester M34 5GR
M7 MCG M. Godfrey 2 Park Street, Barnoldswick BB18 5BT
M7 MCI M. Inseal 63 Woodside Avenue, Benfleet SS7 4NX
M7 MCJ M. Johnson 27 Tyndall Walk, Birmingham B32 3UN
MI7 MCK A. Mckenzie 32 Abbot Gardens, Newtownards BT23 8UL
MM7 MCM A. Mcmillan 275 Saughs Drive Robroystone, Glasgow G33 1BN
MM7 MCN J. Mclaren 33 Foulford Street, Cowdenbeath KY4 9NB
M7 MCO M. Onassis 12 Gogle Close, Mattishall, Dereham NR20 3SY
M7 MCQ T. Mcquiggan 28 Stocks Park Drive, Horwich, Bolton BL6 6DD
M7 MCR T. Carroll 37 Mayswood Gardens, Dagenham RM10 8UU
MI7 MCS A. Davidson 8 Rashee Court, Ballyclare BT39 9SE
MM7 MCT F. Genolini 29 Primrosehill Drive, Aberdeen AB24 4ER
M7 MCU J. Whitworth 5 Hilltop Close, Shrewton SP3 4EB
M7 MCY M. Charity 36A Beeston Street, Northwich CW8 1ER
M7 MDA M. Andrijauskas 7 Hawthorn Grove, Enfield EN2 0DU
M7 MDB M. Burden 17 Water Meadow, Cullompton EX15 1QS
M7 MDD J. Dee 4 Marlborough Way, Kennington, Ashford TN24 9HH
M7 MDE A. Buck 49 Holtdale Garth, Leeds LS16 7SH
M7 MDH M. Haffenden 147 The Diplocks, Hailsham BN27 3JZ
M7 MDR M. Read 4 Farleigh Road, Shrivenham, Swindon SN6 8BD
M7 MDY A. Waters 21 Kings Road, Lee-on-the-Solent PO13 9NU
M7 MDZ P. Wilkinson 11, St Marys Row Aldeby, Beccles NR34 0AL
MD7 MEB M. Behrman 60 Scarlett Road, Castletown IM9 1PN Isle of Man
M7 MEC M. Crook 25 Race Hill, Launceston PL15 9BD
M7 MED C. Babb 35 James Crescent, Werrington, Stoke-on-Trent ST9 0DZ
MW7 MEJ E. Jones 8 Ty Newydd Court, High Street, Ruabon, Wrexham LL14 6BF
M7 MEK G. James 19 Fairway, Chatteris PE16 6SX
M7 MEO W. Beckett 20 St Fabians Close Newmarket, Suffolk CB8 0EJ
M7 MEP M. Price 31 Covert Mead Handcross, Haywards Heath RH17 6DL
M7 MET A. Morel Flat 67, Chadbrook Crest, Richmond Hill Road, Birmingham B15 3RN
M7 MEX A. Delchini 5 Hillside, Harefield, Uxbridge UB9 6AU
M7 MEZ Dr R. Mackie 31C Champion Hill, London SE5 8BS
M7 MFD M. Dougan Flat 7 Brooke Court 300 Kilburn Lane, London W10 4BW
M7 MFF M. Fryer 81 Warren Road, Twickenham TW2 7DJ
M7 MFH Dr M. Al-Naday 39 Quayside Drive, Colchester CO2 8GE
MM7 MFR M. Ritchie 97 Woodburn Crescent, Bonnybridge FK4 2DL
M7 MFV M. Frolov 135 Sapphire Road, Bishops Cleeve, Cheltenham GL52 7YT
MI7 MFW D. Hall 7 Kingsbury Gardens, Coleraine BT52 2JE
M7 MGD M. Drury 15 Averingcliffe Road, Bradford BD10 9HQ
M7 MGG M. Garth 77 Centenary Way, Chelmsford CM1 6AU
M7 MGH M. Harvey 4 Scilly Close, Ellesmere Port CH65 9JU
M7 MGL C. Mcglinchey 9 Spring Rise, Glossop SK13 6US
M7 MGM Y. Weng 20 Kendal Grove, Leeds LS3 1NS
M7 MGU M. Guidolin Flat 3, 3 Grimston Gardens, Folkestone CT20 2PT
M7 MGW M. Waterman 3 School Farm Cottages, Graveney Road, Faversham ME13 8UR
M7 MGX S. Swinton 4 Fallow Road, Newton Aycliffe DL5 4SU
M7 MHA M. Harper 67 Ludsden Grove, Thame OX9 3BY
M7 MHB M. Balyuzi 48 Cleveland Gardens, London NW2 1DY
MW7 MHC D. Roberts 25 Peibio Close, Holyhead LL65 2EG
MI7 MHG M. Hughes 96 White Rise, Dunmurry, Belfast BT17 0XD
MM7 MHJ M. Hancox 33 Caledonian Road Brechin, Brechin DD9 6BG
M7 MHM Capt. R. Massingham 6 Larch Crescent, Holt NR25 6TU
MW7 MHO Dr M. Hoptroff Pencae Bryn Rd Upper Brynamman, Ammanford SA181AR
M7 MHS M. Smith 32 Charity William Way, Stanton, Bury St. Edmunds IP31 2FB
M7 MIQ A. Gardner 16A Spring Lane, Shepshed LE12 9JE
M7 MIR G. Buaras 16 Potterswood Close, Bristol BS15 8LW
M7 MIT T. Haynes 30A London Road, Wymondham NR18 9JD
M7 MJB J. Brittain 66 High Street Swavesey, Cambridge CB24 4QU
M7 MJC M. Crees Flat 3 Fox House, Fox Lane North., Chertsey, KT16 9GY
M7 MJG M. Gardiner 6 Garston Grove, Hartlepool TS25 1HL
M7 MJH M. Hall 88 Kingscote Yate, Bristol BS378YE
MM7 MJK M. Keay Coomb Burn, Wamphray, Moffat DG10 9LZ
M7 MJO M. Jolly 7 Castle Road Colne Lancashire, Pendle BB8 7AR
M7 MJR M. Rigby 28A Lower Boston Rd Flat A, Hanwell W7 2NR
M7 MJV M. Viney 33 Denham Road, Burgess Hill RH15 9TE
M7 MJX M. Kerslake 48 Coresbrook Way, Knaphill GU21 2TP
M7 MKB M. Bradshaw 70 Brook Gardens, Emsworth PO10 7LB
MI7 MKD Rvd. M. Donald Station Road, Garvagh BT51 5LA
M7 MKJ M. Jarkiewicz 22 Lower Anchor Street, Chelmsford CM2 0AS
M7 MKK M. Tuck 12 Greenside Hill, Milton Keynes MK4 2DF
M7 MKO M. Conlin 11 James Street, Sunderland SR5 2DJ
MM7 MKV W. Macleod 26 Upper Barvas, Isle of Lewis HS2 0QX
MM7 MKX M. Gee Cruachan, Annfield Crescent, Kirkwall KW15 1NS
M7 MKY D. Curtis 62 Hartley Road, Birmingham B44 0RD
M7 MKZ Capt. B. Monks 7 Springwell Road, Bootle L20 6LU
M7 MLB L. Bennett 19 Campion Crescent, Cranbrook TN17 3QJ
M7 MLC M. Coombes 33 Woodside Road North Baddesley, Southampton SO52 9NB
M7 MLD M. Dudley 88 King Street, Dawley, Telford TF4 2AH
M7 MLE E. Groves 350 Middle Deal Road, Deal CT14 9SN
M7 MLK M. Kubiak 132 Evistone Gardens, Newcastle upon Tyne NE63RU
M7 MLL G. Genney 31 Columbus Way, Grimsby DN33 1RP
M7 MLM A. Stevens 5 Ifield Mill Close, Stone Cross BN245PF
M7 MLP M. Pullen 8 Athena Avenue, Waterlooville PO7 8AE
M7 MLU A. Janczura 63-65 Castle Streat Flat 7, Luton LU13AG
M7 MLY Dr E. Billinge 44 Spitfire Road, Sheffield S13 7AD
M7 MLZ D. Mullett 17 Green Crescent, Bucklesham, Ipswich IP10 0EA
MJ7 MMA M. Aubert Deansway La Grande Route De St. Martin, St. Saviour, Jersey JE2 7GR Jersey
MM7 MMB A. Hood 26 Annan Avenue, East Killbride G75 8XT
M7 MMC M. Mccarthy Flat 45, Beech, Beresford Road, Brighton BN2 5DD
MW7 MME E. Holmes 111 Llanfabon Drive, Trethomas, Caerphilly CF83 8GX
M7 MMK G. Cochrane Flat 11, Carlton Mansions, Sweyn Road, Cliftonville, Margate CT9 2DJ
M7 MML M. Larke 16 Linton Road, Tamerton Foliot, Plymouth PL5 4PG
M7 MMM P. Mocker Flat 24 The Gardens Clapton Common, London E5 9AZ
MM7 MMO S. O'Neill Flat 5-8 460 Sauchiehall Street, Glasgow G2 3JW
M7 MMP Q. Cope 102 Moreton Road, Buckingham MK18 1PW
M7 MMQ M. Qahwaji 22 Tesla Lane, Guiseley, Leeds LS20 9DS
M7 MMT M. Truman 3 Mulberry Road, North Anston, Rotherham S254BH
MI7 MMV A. Mercer 72A Trostan Avenue Avenue, Ballymena BT43 7BL
MM7 MMW M. Wylde 73 Glendale Crescent, Ayr KA7 3RZ
MI7 MNB M. Briggs 48 Bolton Road Loughgilly, Glenanne BT60 2DU
M7 MND M. Soles 70 Mitchell Road, Bedworth CV12 9HP
M7 MNJ J. Higgins Flat 7, 8 Hampton Park, Bristol BS6 6LP
MM7 MNY M. Cowie Larachmohr, Hawkhill, Keiss, Wick KW1 4XF
M7 MNZ A. Hall 16 Larkspur, Rugby CV23 0UW
MW7 MOC M. Angell Fronheulog, Cloth Hall Lane, Cefn Coed, Merthyr Tydfil CF48 2NT
M7 MOE M. Moen 37 Thistledown Road, Birmingham B34 7EG
M7 MOH H. Singh 19 Elmore Road, Peterborough PE3 9PS
M7 MOI S. Schonrock 1 Powell Road, Poole BH14 8SG
M7 MOK M. Swaby 4 Flamborough Close, Biggin Hill, Westerham TN16 3PB
M7 MOL H. Molyneux 16 Millbrook Close, Shaw, Oldham OL2 8QA
MW7 MON J. Byast Ty Arbennig Bull Bay Road, Amlwch LL68 9EA
M7 MOR M. Robson Induno, Hallgate, Moulton, Spalding PE12 6QG
M7 MOT T. Wilcox 7 Jacob Court, Billinge, Wigan WN5 7GE
M7 MOU P. Thomas 2 Darwin Close, Stamford PE9 1JL
M7 MOW M. Mower 83 Sipson Road, West Drayton UB7 9DH
M7 MOY M. Moyse 37 Kingfisher Road, Buckingham MK18 7EX
MW7 MOZ B. Morris 20, Poets Field Road, Barry CF62 9TY
M7 MPD M. Daryanani 20 Blossom Way, Uxbridge UB10 9LN
M7 MPL M. Loveridge 63 Guildford Road, Portsmouth PO1 5HU
MW7 MPM Essex Raynet c/o M. Moyse 10 Clifton Rise, Abergele LL22 7DN
M7 MPQ Dr A. Gair-Harris Osterley, White Lackington, Piddletrenthide, Dorchester DT2 7QU
M7 MPS M. Spikings 84 Northampton Road, Broughton, Kettering NN14 1NS
MI7 MQX M. Colhoun 102 Dowland Road, Limavady BT49 0HR
M7 MRA R. Ayres 11 Sunningdale Crescent, Bournemouth BH10 5LL
M7 MRB A. Bugler 2 Roman Barn, Worth Matravers, , Swanage BH19 3LZ
M7 MRJ M. Jones 4 Farmland Way, Hailsham BN27 1SP
MW7 MRK M. Morgan 30 Hafan Werdd, Mornington Meadows, Caerphilly CF83 3BU
M7 MRL M. Lee 2 Rupert Road Chaddesden, Derby DE214ND
M7 MRN R. Hunt 17 Robartes Road St Dennis, Saint Austell PL26 8DS
M7 MRO M. Roffe 27 Athena Close, Southend on Sea SS2 4GL
M7 MRP L. Pares 22 Hawthorn Road, Bourne PE10 9SN
M7 MRQ G. Newell 12 Firecrest Close, Wymondham NR18 9FA
M7 MRR M. Rogers 26 Boldrewood, Swindon SN3 6JP
M7 MRT M. Twyman Hazeldene, St. Marys Road, Great Bentley, Colchester CO7 8NN
M7 MRX A. Stokes 175 Exhall Close, Church Hill South, Redditch B98 9JA
M7 MRY J. Emery 52 Wordsworth Avenue, Sheffield S5 8NA
M7 MSA M. Simpson Elm Crescent, Peterborough PE67LE
M7 MSB M. Bridges 17 Widdowson Close, Didcot OX11 9GF
M7 MSD M. Dawson 26 Goodwin Close, Wellingborough NN8 4BS
M7 MSF P. Reynolds 1 Aspen Grove, Newcastle upon Tyne NE16 6QP
M7 MSJ S. Smith 14 Benbow Close, Malvern WR14 4JJ
MM7 MSL M. Craig 20B Duke Street, Coldstream TD12 4BW
M7 MSM M. Boisriveau-Mitchell 16 Pochard Close, Quedgeley, Gloucester GL2 4LL
M7 MSR M. Ramsay 28A Strathmore Drive, Charvil, Reading RG10 9QT
M7 MSV A. Varney 11 Redburn Drive, Birmingham B14 5XA
M7 MSW M. Wood 44 Harveyfields, Waltham Abbey EN9 1HN
MI7 MSX M. Silva Apartment 205, 2 William Street South, Belfast BT1 4FJ
MM7 MTB A. Campbell 10 Crathes Close, Glenrothes KY7 4SS
M7 MTE P. Polanyk 44 Lowther Hill Honor Oak, London SE23 1PY
M7 MTG M. Gale 3 - 4 Mill Rythe Lane, Hayling Island PO11 0QG
M7 MTH M. Thompson 49 Beechburn Park, Crook DL15 8NA
MD7 MTM D. Wilson 23 Snugborough Avenue, Union Mills, Braddan, Isle of Man IM4 4LT
M7 MTS M. Shepard 3 The Leys, Oxhill, Warwick CV35 0QX
MM7 MTU J. Wheatley Flat 1, 16 Chandler Crescent, Edinburgh EH6 7AL
M7 MTV M. Venters 18 Foxcovert Road, Werrington, Peterborough PE4 6RF

MI7 MTY M. Thompson 25 Ivy Mead Mews, Altnagelvin, Londonderry BT47 3FH
M7 MUB S. Ainsworth 15 Southfield Drive, Westhoughton Bolton, Manchester BL5 2PP
M7 MUF P. Dilly 4 Marsh Farm Cottages, Marsh Farm Lane, Alresford, Colchester CO7 8BQ
M7 MUJ B. Isaac 19 East Lodge, Fareham PO15 5LZ
M7 MUS D. Mcdean 2 Rigbys Row, Nantwich CW5 5RX
M7 MUT L. Lewin 83 Fulbourn Road, Cambridge CB1 9AJ
M7 MVB M. Boldero 1 Grasleigh Avenue, Allerton, Bradford BD15 9AR
M7 MVD D. Matthews 13 East Road, Egremont CA22 2ED
M7 MVF Rvd. P. Campbell 116 Kingsway Park, Urmston, Manchester M41 7FH
MM7 MVR D. Hood 70 Kirkfield Gardens, Renfrew PA4 8JE
M7 MVS A. Lammiman 177 Obelisk Rise, Northampton NN2 8TX
M7 MWG M. Gittins 4 Old Mill Close, East Knoyle, Salisbury SP3 6EX
MM7 MWL M. Lowson Old Leslie Farmhouse, Leslie, Insch AB52 6NS
M7 MWM M. Warwick 63 The Crescent, Irlam, Manchester M44 6FG
MW7 MWO R. Williams 12 Neville Place, Cardiff CF11 6EP
M7 MWT G. Griffiths 23 Kilnwick Court, Mill Lane, Northallerton DL7 8XS
M7 MWW M. Weekes Lorraine, Winchester Road, Hawkhurst, Cranbrook TN18 4DQ
M7 MWY M. Wilson 35 Allen Close, York YO10 3TS
MM7 MXB M. Butterworth 97A Dunearn Drive, Kirkcaldy KY2 6AL
M7 MXF M. Foxley 15 Ditchfield Lane, Finchampstead, Wokingham RG40 4HP
M7 MXM C. Ford 23 Seaton Orchard, Sparkwell PL75HX
M7 MXS M. Shahid 19 Kingswell Ride, Cuffley, Potters Bar EN6 4LH
M7 MXU C. Xu Studio 3.13 Block J, Birks Hall, New North Road, Exeter EX4 4GD
M7 MXY J. Crewe-Read The Branches, Mill Lane, Hartley Wespall, Hook RG27 0BQ
M7 MYA M. Joseph 53 Wilman Road, Tunbridge Wells TN4 9AL
M7 MYB M. Blicharz 58 Park Barn Drive, Guildford GU2 8ES
M7 MYC D. Mycroft 11 Paisley Walk, Church Gresley, Swadlincote DE11 9FF
M7 MYK M. Nicholas 1 Buckland Barton Cottages, Newton Abbot TQ12 4SA
M7 MYM M. Malik School Of Mechanical, Aerospace & Civil Engineering University Of Manchester Oxford Road, Manchester M13 9PL
MM7 MYS M. Szlama 27 Thriepland Wynd, Perth PH1 1RQ
M7 MYU K. Jenner 24 The Willows, Nailsea, Bristol BS48 1JQ
M7 MZA D. Robinson 20 Rowan Way, Rottingdean, Brighton BN2 7FP
M7 MZM R. Hopkins 26 High Firs Road, Romsey SO51 5PZ
M7 MZN M. Nabi 59 Sunninghill Avenue, Hove BN3 8JB
M7 MZO D. Wilson 3 Primrose Court, Moreton-in-Marsh GL56 0JG
M7 MZT J. Maertens 23 Cottingham Street, Goole DN14 5RR
M7 NAB M. Bradbury 7 Lulworth Avenue, Ipswich IP38RW
M7 NAC A. Campbell 21 Sherbrook Gardens, London N21 2NX
M7 NAG N. Nicholls 16 King Edwards Way, Edith Weston, Oakham LE15 8EZ
M7 NAI I. Goldsmith Foxwood Lane, York YO24 3LT
M7 NAJ N. Musgrave 14 Clarendon Green, Orpington BR5 2PA
M7 NAL A. Davies 5 Worths Way, Stratford-upon-Avon CV37 0RR
M7 NAM Dr N. Mirza 21 Knoll Road, London SW18 2DF
M7 NAN S. Morozov Haroldene, Towpath, Shepperton TW17 9LL
M7 NAS N. Stratford 108 Nottage Crescent, Braintree CM7 2TX
MW7 NAT N. Morris Llwyn-Y-Gorras, Castlemorris SA62 5ES
M7 NAV R. Matthewman 44 Huntwick Crescent Featherstone, Pontefract WF75JQ
M7 NAW C. Smith 3 Park Rd Sunderland, Sunderland SR2 8HR
M7 NAZ R. Messen 45 Church Lane North Bradley, Trowbridge BA14 0TE
M7 NBA C. Sims 52 Station Road, Royston SG8 0NP
MW7 NBC N. Conway 7, Heatherdale Close, Stansty Park. Gwersyllt LL11 4SZ
MI7 NBR P. Borland 27 Emerson Street, Londonderry BT47 6EL
M7 NBX N. Brown 22 Cherrytree Road, Bristol BS16 4EX
M7 NBZ S. Johnson 43 Rutherford Street Howdon, Wallsend NE28 0AY
M7 NCA N. Atkins 41 Church Lane, Bardsey, Leeds LS17 9DR
M7 NCB N. Brooks 105B Upper Woodcote Road, Caversham, Reading RG4 7JZ
M7 NCC A. Whiteley 3 Ruskin Court, Farnworth, Bolton BL4 9EQ
MM7 NCD N. Coulter 17 Gartinny, Coalsnaughton, Tillicoultry FK13 6LF
M7 NCM N. Moore 2 The Paddocks, Off The Croft, Longhoughton NE66 3DD
M7 NCN N. Nurse 3 Portnall Place, Cranfield MK430JW
M7 NCO N. Cooke 10 Westbury Road, Shrewsbury SY1 3HF
M7 NCR N. Roots 48 Grosvenor Road, , Stalbridge DT10 2PN
M7 NCV C. Nutu 37 Brading Road, London SW2 2AP
M7 NCW N. Worthington 190 Higher Road, Liverpool L26 1UW
MI7 NCZ N. Conway 30 Greenvale Manor, Antrim BT41 1SB
M7 NDA N. Ayre 58 Burford Avenue, Swindon SN3 1BN
M7 NDB J. Jones 227 George Lane, Bredbury, Stockport SK6 1DJ
M7 NDD N. Dsouza 75 Longstork Road, Rugby CV23 0GB
M7 NDH D. Hall 16 Larkspur, Rugby CV23 0UW
M7 NDI A. Roe 3A Fairfield Avenue, Rossendale BB4 9TG
M7 NDK D. Needham 6 Apley Drive, Wellington, Telford TF1 3DU
M7 NDL D. Callery 15 Cliff Road, Holland-On-Sea, Clacton-on-Sea CO15 5QQ
M7 NDM A. Mackey Kings Barrow, North Bovey, Newton Abbot TQ13 8RS
M7 NDS N. Spencer 11 Hitchcock Close, Shepperton TW17 0QT
M7 NDT A. Tunley 5 Camborne Close, Lower Earley, Reading RG6 4EN
M7 NEA N. Morris 31 Endeavour Place, Stourport-on-Severn DY13 9RL
M7 NED N. Hambly 172 Wareham Road Lytchett Matravers, Poole BH16 6DT
MW7 NEF S. Williams Dee View Cottage Pen Y Ball Hill, Holywell CH8 8SZ
M7 NEI P. Allen 58 Baltic Court, South Shields NE333NT
M7 NEJ J. Adeniji 28 Wigeon Path London, London SE28 0DS
M7 NEL D. Nelson Chapman Way, Hayward?S Heath RH164UL
M7 NEN N. Nowell 12 Townhead Court, Settle BD24 9HY
M7 NES N. Smith 17 Lady Margaret Gardens, Woodbridge IP124EZ
M7 NET D. Picker 12 Crown Close, Barnsley S70 4DB
M7 NEW A. Fleck 108 Hillsview Avenue, Newcastle upon Tyne NE3 3LA
M7 NEZ L. Ponton 3 Capon Hill, Brampton CA8 1QJ
M7 NFC A. Bulgakov 20 Torque Close, Basingstoke RG24 9YL
M7 NFL S. Bradley Leacroft Road, Penkridge ST19 5BX
M7 NFX N. Friend 4 Bartholomew Street, Dover CT16 2LH
M7 NFZ N. Fenn 7 Pottery Road, Tilehurst, Reading RG30 6BA
M7 NGC N. Connolly 32 St. Oswald Road, Bridlington YO16 7SD
M7 NGL N. Bennetts 1 The Moor, Falmouth TR11 3QA
MM7 NGZ N. Grainger 4 Hillhead, Inverfarigaig, Inverness IV2 6XS
M7 NHF N. Frost 5 George Street Elworth, Sandbach CW11 3BL
M7 NHG N. Groom 4 Burley Hill, Allestree, Derby DE22 2ET
M7 NHJ T. Sidebottom 256 Rolleston Road, Burton-on-Trent DE13 0AY
M7 NHS Dr R. De Silver 37 Hopeville Avenue, Broadstairs CT10 2TR
M7 NIB N. Beresford 2 Meadow View, Great Addington, Kettering NN14 4BN
MI7 NIH H. Drinkwater 26 Hillview Place, Holywood BT18 9DL
M7 NIK Dr N. Crisp Apartment 88, Advent House, 2 Isaac Way, Manchester M4 7EP
M7 NJD N. Dowe 47 Reachfields, Hythe CT21 6LS
M7 NJJ N. Foreman 10 Oak Road, Stilton, Peterborough PE7 3RB
M7 NJM N. Manning 9 Hawthorne Grove, Stockport SK6 2PJ
M7 NJR N. Roche 4 Kennedy Close, Rayleigh SS6 8UW
M7 NKA N. Aldrich 11 Town Mead, West Green, Crawley RH11 7EG
M7 NKE Dr J. Holmes Sexeys Farmhouse, Spring Street, Wool, Wareham BH20 6DB
M7 NKN N. Newens 5 Lillian Place, Gosport Road, Lower Farringdon, Alton GU34 3DH
M7 NKW P. Rowland 27 Parker Terrace, Ferryhill DL17 8JT
M7 NKZ N. Ellis 25 George Butler Close Laceby, Grimsby DN37 7WA
M7 NLA S. Fletcher 43 Castledale Grove, Sheffield S2 1NJ
MM7 NLH G. Frame 27 Oswald Court, Ayr KA8 8NL
MJ7 NLK N. King 15 Le Clos De Noirmont, La Route De Noirmont, St. Brelade, Jersey JE3 8AP
M7 NLP N. Pereira 55B North Street, Carshalton SM5 2HG
M7 NLS N. Andre Churchill College, Storey'S Way, Cambridge CB3 0DS
MM7 NLW N. Wilson 92 Inveroran Drive, Bearsden, Glasgow G61 2AT
MI7 NMA M. Dowie 16 Castle Meadow Park, Cloughey, Newtownards BT22 1GB
M7 NMC N. Cardoso 24 Northcroft Road, Gosport PO12 3DR
M7 NMD L. Lynch 10 Blackthorn Grove, Woburn Sands, Milton Keynes MK17 8PZ
M7 NML C. Dolakumbura 12 Kings Drive, Stoke Gifford, Bristol BS34 8RD
M7 NMM Dr H. Chen Flat 37, Castel Mill, Rodger Dudman Way, Oxford OX1 1AD
MM7 NMO C. Nimmo 6 Raith Avenue, Prestwick KA9 1DL
MW7 NMT N. Thomas 82 Fairhill, Fairwater, Cwmbran NP44 4RB
MM7 NNM N. Milne 27 Brechin Road, Arbroath DD11 1ST
MW7 NNR U. Miranovich Flat 1, 35 Queens Road, Aberystwyth SY23 2HN
M7 NNY W. Heaney 9 Copeland Avenue Egremont, Cumbria CA222QT
MW7 NOC P. Parkin Maengwyn, Esplanade, Penmaenmawr LL34 6LT
M7 NOH M. Vine 95 Cherry Tree Street, Elsecar, Barnsley S74 8DG
M7 NOL M. Nolan 7 Freston Gardens, Barnet EN4 9LX
M7 NOO J. Coakley 44 Haydock Close, Birmingham B36 8UN
M7 NOV J. Boase 28 Broad Lane, Rochdale OL16 4PG
M7 NOW N. Wadsley 5 Paget Close, Needham Market, Ipswich IP6 8XF
MM7 NOZ S. Wallace 8 Gullane Avenue, Dundee DD2 3BU
M7 NPA S. Walker 3 Lord Avenue, Bacup OL13 0RY
M7 NPC N. Cox 182 North Tenth Street, Milton Keynes MK9 3AY
M7 NPH N. Haigh High Farm Cottage, Blindcrake, Cockermouth CA13 0QP
MM7 NPT I. Stewart Douneside, Allanfearn, Inverness IV2 7HX
M7 NRC N. Cooper 3 Dalwood Gardens, Benfleet SS7 2NN
M7 NRH A. Ball 7 Primrose Avenue The Drive, Horley RH67JW
M7 NRJ N. Raj 52 Cranley Road, Hersham, Walton-on-Thames KT12 5BS
M7 NRL D. Norell 81 Station Road, Oakham LE15 6QT
M7 NRM N. May 37 Brightling Avenue, Hastings TN35 5EG
MM7 NRN T. Laurenson Flat G, 4 South Lodge Court, Ayr KA7 2TA
M7 NRO N. Olsson 60 Bebington Road, Birkenhead CH42 6PX
M7 NRR R. Levitt 4 Ashley Row, Aylesbury HP20 1HJ
M7 NRS N. Hills 8 Larch Crescent, Tonbridge TN10 3NN
M7 NRT M. Norton 56 Melrose Avenue Vicars Cross, Chester CH3 5JB
MI7 NRW R. Bestek 111 Cloughwater Road, Ballymena BT43 6SZ
M7 NSC N. Cooper Romer Cottage, Long Reach, Ockham, Woking GU23 6PF
M7 NSP M. Pinder 3 Hatherop, Cirencester GL73NA
M7 NSS V. De Souza Flat3 69-71, Gloucester Street, London W2 3DH
MI7 NSU Dr A. Fogarty 14 Brownstown Park, Portadown, Portadown BT62 3QJ
MM7 NSY N. Young 61 Park Terrace, Broxburn EH52 6AP
M7 NTD N. Tate 418 Broad Lane, Leeds LS13 3DF
M7 NTE D. Morgan 9 Florida Street, Sunderland SR4 6TE
M7 NTH N. Howells 1 Eunice Way, Newdale, Telford TF3 5FH
M7 NTP J. Reed 34 Showfields Road, Royal Tunbridge Wells TN2 5PW
M7 NTT S. Bishop 5 Mulberry Court, 266 Goring Road, Goring-By-Sea, Worthing BN12 4PF
M7 NUK L. Colley 31 Wharton Drive, North Walsham NR280UG
M7 NUT T. Gifford 53 Haydn Avenue, Purley CR8 4AJ
M7 NVC R. Price 253 Princes Road, Ellesmere Port CH65 8ES
M7 NVM D. Mills 31 Gorseway, Hatfield AL10 9GS
M7 NVY J. Brodie 10 Hillhead Rise, Falmouth TR11 5GZ
MI7 NWA A. Carlisle 21 Agnes Close, Belfast BT13 1DJ
M7 NWE O. Rose 30 Windsor Drive, Hertford SG14 2HU
MW7 NWJ Dr N. Jones 54 Glanrhyd, Coed Eva, Cwmbran NP44 6TY
M7 NWL R. Heaton 19 Douglas Drive, Ormskirk L39 1LJ
M7 NWT N. Trice 4 Atkins Crescent, Maldon CM9 6JB
M7 NWV N. Wright 12 Hartington Road, Oakhurst, Swindon SN25 2EF
M7 NWZ N. Willis 3 Harpur-Crewe Cottages, Alstonefield DE6 2GD
M7 NXC E. Walley 4 Langley Mill Close, Sutton Coldfield B75 7BY
M7 NXG N. Grundy 14 Townend Close, Asfordby, Melton Mowbray LE14 3TY
MW7 NYE A. Price 1 Graig Park Villas, Newport NP20 6GU
M7 NYH N. Hodson 163 Lincoln Road Branston, Lincoln LN4 1NS
MI7 NYK N. Montgomery 2 Cappagh Park, Portstewart BT55 7SZ
M7 NYY R. Coleman 41 Lawrence House, Camperdown Street, Bexhill-on-Sea TN39 5EN
M7 NZL M. Francis 11 Barton Avenue, Paignton TQ3 3JQ
MW7 NZP Z. Pugh 28 Hill View Road, Llandudno LL30 1SL
M7 OAD D. Drynski Flat 22, Nicholas Court, Corney Reach Way, Chiswick, London W4 2TS
MM7 OAI K. Blair 148A Barrangary Road, Bishopton PA7 5FR
M7 OAK D. Dalton 8 Moorland Crescent, Pudsey LS28 8EW
M7 OAN J. Robinson 96 White Lodge Park, Shawbury, Shrewsbury SY4 4NU
M7 OAZ P. Negros 14 Amersham Road, London SE14 6QE
M7 OBF M. Hamilton 45 Silkin Way, Newton Aycliffe DL5 4HE
M7 OBI D. Goatcher 5 Coopers Close, Burgess Hill RH158AN
M7 OBJ D. Evans 134 Somerton Road, Bolton BL2 6LW
M7 OBX A. Wilkinson 88 Main Street, Balderton, Newark NG24 3NU
M7 OBY G. Inman 11 Common Road, Sixpenny Handley, Salisbury SP5 5NJ
M7 OCB C. Burls 86 All Saints Road, Kings Heath, Birmingham B14 7LN
M7 OCJ O. Cox 110 Maidstone Road, Borough Green, Sevenoaks TN15 8HG
M7 OCM M. Thirkettle 320 Raglan Street, Lowestoft NR32 2LB

M7 OCN E. Robinson 2 Egerton Road, Plymouth PL4 9BR
M7 OCO K. O'Connell 63 Hazelton Road, Colchester CO4 3DS
MI7 OCP S. Spallen 65 Tirgarvil Road, Upperlands, Maghera BT46 5UW
M7 OCT S. Collins 15 Twiss Grove, Hythe CT21 5PA
MM7 OCX A. Mather 12 An Creagan Place, Port Charlotte PA48 7UF
M7 OCZ P. Morgan 41 Silver Street, Nailsea BS48 2AA
M7 ODC N. Southall 47 Chaffinch Drive, Kidderminster DY10 4SZ
M7 ODD S. Aynsley 24 Grange Avenue, Bedlington NE22 7EW
M7 ODG A. Cleveland The Flat, 2 Church Street, Wells-next-the-Sea NR23 1JA
M7 ODH D. Harris 381 Littleworth Road, Cannock WS12 1HY
M7 ODO C. Dawson 9 Telford Road, Sunderland SR3 4HZ
M7 ODT D. Timmins 1 Meryhurst Road, Wednesbury WS10 9BX
M7 ODY Dr M. Venables 50 Church View Close, Norwich NR7 8QA
M7 ODZ D. O'Dee 4 Mortomley Croft, Sheffield S35 3XS
M7 OEL I. Robinson 12 Osprey Road, Flitwick MK45 1RU
M7 OEN M. Nicholson 10 Beechfield Road, Cheadle Hulme, Cheadle SK87DS
M7 OFF M. Faulkner 23 Davenport Road, Sidcup DA14 4PN
M7 OFG S. Wilson 64 Deer Park Close, Sheffield S6 5NA
MM7 OFH A. Gajdos 55 Kirkfield West, Livingston Village, Livingston EH54 7BE
M7 OFM J. Joyce 2 Harold Road Cuxton, Rochester ME2 1EE
MW7 OFS A. Williams 49 Maesheli, Penparcau, Aberystwyth SY23 1TB
M7 OGB D. Fixter 68 Southwood, Coulby Newham, Middlesbrough TS8 0UF
M7 OGC S. Grey 40 Almery Drive, Carlisle CA2 4EX
M7 OGL O. Grein 6 Green Terrace, Mirfield WF14 9BG
M7 OHM S. Hagan 5 High Street, Brampton, Huntingdonshire PE28 4TG
M7 OJA O. Acton West Park Hospital, Edward Pease Way, Darlington DL2 2TS
M7 OJO C. Strikes American Church In London, 79A Tottenham Court Road, London W1T 4TD
M7 OJS J. Siddall 16 Alandale Avenue, Langwith Junction, Mansfield NG20 9RU
M7 OJV O. Verity 2 Mill Hill Cottage Masham, Ripon HG4 4BP
M7 OJW O. Williams 6 Millfield Drive, Bristol BS30 5NR
M7 OKG K. Gregson 14 Tarragon Drive, Stoke-on-Trent ST3 7YE
MM7 OKO J. Jasiewski 5 Katrine Road, Shotts ML7 4JA
M7 OKR K. Rowsell Crofts Edge, Loud Hill, Stanley DH9 8PL
M7 OKV P. Skermer 23 Lea Grove Bardney, Lincoln LN35XN
M7 OLA S. Coombe 66 Tungstone Way, Market Harborough LE16 9GG
MW7 OLF J. Wolfe 7 Cedar Close, Sealand, Deeside CH5 2RB
M7 OLM R. Holtom 79 Bowling Green Rd, Stourbridge DY8 3RZ
M7 OLO K. Elms-Lester Ferndale House, Kerrys Gate, Hereford HR2 0AH
M7 OLP A. Downey 1 Roslyn Avenue, Middlesbrough TS3 8QX
M7 OLY O. Botwright 16 Upper Whistler Walk, World'S End Estate, London SW10 0ER
M7 OMA T. Anable The Old Court House, South Street, Winterton, Scunthorpe DN15 9RP
M7 OMB P. Cummings 12 Odile Mews, Gilstead, Bingley BD16 3QL
M7 OMD C. Penny 9 Winifred Road, Dartford DA1 3BL
M7 OMG S. Orme 78 Castle Road, Northolt UB5 4SE
M7 OMM R. Reid Dart Cottage, Postbridge, Yelverton PL20 6TJ
M7 OMO P. Hickey 11 Border Brook Lane, Boothstown M281XJ
M7 OMP O. Phillips 1 Gate Foot Lane, Shepley, Huddersfield HD8 8AZ
M7 OMS C. Smith 2 Neville Road, Castle Bromwich, Birmingham B36 9HP
M7 OMW O. Williams Apartment 20, 104 Pensby Road, Heswall CH60 7RF
M7 OMY T. Sanduly 1 Archford Croft, Emerson Valley, Milton Keynes MK4 2EZ
M7 ONO M. Winkle 75 Primrose Ave Haslington, Crewe CW15NY
M7 ONR K. Mair 12 Cheltenham Crescent, Moreton, Wirral CH46 1PU
MM7 OOF B. Blyth 58 Spencerfield Road, Inverkeithing KY11 1PG
M7 OOI A. Wood Chicago House, The Bowjey Hill, Newlyn, Penzance TR18 5LW
M7 OOW J. Early 6 Goverseth Hill, Foxhole, St. Austell PL26 7UZ
M7 OPA E. Kane Talbot House, Queen Street, Hook Norton, Banbury OX15 5PJ
M7 OPB C. Mann Lunesdale, Swanton Road, Dereham , Norfolk, Dereham NR204PS
M7 OPH S. Richards 60 Brooklyn Avenue, Loughton IG10 1BN
M7 OPM P. Mcmillan 44 Wheatfield Grove, Benton NE12 8DP
M7 OPT N. Joicey Flat 2 2 Druid Street Hinckley, Hinckley LE10 1QH
M7 OPY P. Hope 22 Ayrshire Close, Salisbury SP2 9PF
M7 OQN T. Schonrock 1 Powell Road, Poole BH14 8SG
M7 ORC M. Minett 82 Lower Market Street, Penryn TR10 8BH
MM7 ORE J. Moore 43 Alexander Grove, Kilmardinny Grange, Bearsden G61 3EF
M7 ORG A. Angell Nightingales, Greys Green, Rotherfield Greys, Henley-on-Thames RG9 4QQ
MM7 ORR K. Orr 9 Wallace Avenue, Stevenston KA20 4BN
M7 OSD S. Mccallister 44 Chester Street, Accrington BB5 0SD
M7 OSF S. Fagg 65 Mountview, , Sittingbourne ME98JZ
M7 OSH O. Hirst 7 The Crescent Charles Street, Elland HX5 0HR
M7 OSM O. Morris Linden House Church End Weston Colville Cambridge Cb21 5Pe, Weston Coville CB215PE
M7 OSP J. Podlas 17 Hallett Close, Southampton SO18 2FD
M7 OSR S. Richardson 20 Lupin Way, Willand, Cullompton EX15 2SB
M7 OSS C. Donachie 22 Glasgow Street, Hull HU3 3PR
M7 OST A. Fuller 44 Carbeile Road, Torpoint PL11 2HR
M7 OSX S. Cartwright 132 Newry Park East, , Chester CH2 2BE
MW7 OTK O. Knowles 4 Church Street, Pen-Y-Cae, Wrexham LL14 2RL
M7 OTO P. Chamberlain 37 Curzon Street, Colne BB8 0HE
M7 OTZ J. Bithell 118 Hassam Parade, Newcastle ST5 9DN
M7 OUT K. Mason 214 Carfield, Skelmersdale WN8 9DW
M7 OVM J. Harley 33 Miranda Rd Archway, London N193RA
M7 OWA A. Oliver 49 Wharf Close, St. Georges, Telford TF2 9PX
M7 OWD V. Hodge 37 Monkswood Crescent, Tadley RG26 3UE
M7 OWJ O. Jackson 38 Bishop Close, Poole BH12 5HT
M7 OWL W. Burgess 241 King Avenue, Bootle L20 0BY
MM7 OWN O. Mcdermid 38 Steading Drive, Alexandria G83 9EB
MM7 OWO L. Tarvit 1/10 Chapel Lane, Leith, Edinburgh EH6 6ST
M7 OWT A. Hopkins 12 Woodway Horsforth, Leeds LS18 4HY
M7 OXO J. Bliszko 34 Hambledines, Redhouse Park, Milton Keynes MK14 5FS
M7 OXS O. Spearing 2 Goldcroft Road, Weymouth DT4 0DZ
M7 OXY D. Oxendale Flat 1, Quay House, Middle Wharf, Mevagissey, St. Austell PL26 6UP
M7 OYR S. Roberts 77 Lambwath Road, Hull HU8 0HB
M7 OZA A. Ozarski 86 Howards Way, Newton Abbot TQ124HX
M7 OZC O. Sloane-Hase 23 Southdown Road, Benham Hill, Thatcham RG19 3BF
M7 OZE O. Martin 22 Frankton Avenue, Hayward's Heath RH16 3QX
M7 PAA P. Rossi Flat 1, Claret Court, 125 Connersville Way, Croydon CR0 4FT
M7 PAC P. Cann 47 Drury Lane, Houghton Regis LU55ED
M7 PAG P. Goodier Macaulay Road, Hartlepool TS25 4NE
MW7 PAJ P. Jewell 27 Overdale Avenue, Mold CH7 6US
MW7 PAN S. Carter The Stables, Penllyn, Cowbridge CF71 7RQ
M7 PAR P. Pankhania 20 Dean Road, Hounslow TW3 2EZ
M7 PAU Dr J. Hamilton Flat 3, Tranmere House, Morgan Road, London N7 8NB
M7 PAY S. Payne 82 Lower Barnes Street, Clayton Le Moors, Accrington BB5 5SW
M7 PBB P. Boggis 69 Daffodil Way, Chelmsford CM1 6XE
M7 PBD P. Bowers-Davis 28 Muchelney Road, Morden SM4 6HU
M7 PBE J. Tan 7 The Drove, Brighton BN1 5NN
M7 PBN D. Newman Finsbury Ave, Blackpool FY16QN
M7 PBO A. Bogg 13, Emerald Grove, Hull HU35AE
M7 PBS P. Baskerville 4 Fosse Close, Swindon SN2 2BP
M7 PBT P. Townsend 80 Uplands Road, West Moors, Ferndown BH22 0BT
M7 PBW P. Winn Glen Lyon, Hoggs Drove, Marham, King's Lynn PE33 9JW
M7 PBX P. Barnett 18 Cloverdale, Northwich CW8 4UE
M7 PCA J. Forbes Sweetbriar, Southrepps Road, Antingham, North Walsham NR28 0NP
M7 PCB P. Cridland 49 North Drive, Thornton Cleveleys FY5 3AJ
M7 PCC P. Culmer 56 Main Road Washingborough, Lincoln LN4 1AU
M7 PCD P. Delaney 32 Netherfield View, Rotherham S65 3RB
M7 PCE P. Clark 8 Kyreside, Tenbury Wells WR15 8BX
M7 PCF P. Finch 6 Holmer Manor Close, Hereford HR4 9QZ
MW7 PCG P. Gorton Bradnant, Tylwch, Llanidloes SY18 6JZ
M7 PCL J. Perry 198 Totteridge Road, High Wycombe HP13 7LF
M7 PCQ N. Reid 14 Holly Lane, Margate CT9 3NA
M7 PCS P. Sanders 6 Primrose Hill, Warwick CV34 5HW
M7 PCW S. Cheetham Ward 21 Scarborough Close, Walsall WS2 9TN
M7 PCX P. Eley 172 New Bristol Road, Weston Super Mare BS226BG
M7 PDK P. Thomas 6 Middleton Road, Acomb, York YO24 3AS
M7 PDQ L. Pooler 56 Jacks Lane, Marchington, Uttoxeter ST14 8LW
M7 PDR P. Ryder Windrush Pudney Pie Lane Chalford, Stroud GL6 8FT
M7 PDS P. Steels 74 Conway Crescent, Carlton, Nottingham NG4 2PZ
M7 PDT M. Morris 111 Caunce Street, Blackpool FY13NG
MM7 PDW P. Whitbread Mains Of Whitehill, Banff AB453ER
M7 PDX S. Skorzewski 1K Penistone Road, London SW16 5LU
M7 PDZ R. Attan 45 Village Road, London N3 1TJ
M7 PEB P. Brennan 2 Ethel Road, Birmingham B17 0EL
M7 PEJ J. Puzhiveliparambil James 279 New Road, Dagenham RM10 9ND
M7 PEL D. Pelling 8 Fowler Close Earley, Reading RG6 7SS
M7 PEN J. Penn 8 Doulton Close, Harlow CM17 9RG
M7 PES P. Dale 14 Locksley Drive, Rotherham S66 9NU
M7 PEV D. Maxwell 28 High Street, Wilden MK44 2PB
M7 PEW A. Pewsey Pineholm, High Close, Bovey Tracey, Newton Abbot TQ13 9EX
M7 PFB P. Billington 10 East Crescent, Accrington BB5 5BS
M7 PFC P. Colman 24 Torr Lane, Plymouth PL3 5NY
M7 PFD Prof. P. Davis 83 South Street, Crowland PE6 0AH
MW7 PFF P. Francis 16 St. Maddocks Close, Brackla, Bridgend CF31 2BL
M7 PFK P. Kusiak 17 Kingsway Black, Camberley GU17 0JW
MD7 PFM M. Squires Eaglehurst, The Crescent, , Ramsey IM8 2JN Isle of Man
M7 PFT A. Richardson Valley Farm, Long Lane, Dunston Heath, Stafford ST18 9FB
M7 PFV M. Boyce 15 Gosport Grove, Stoke-on-Trent ST4 8FY
M7 PGA P. Greenough 26 Beech Grove, Liverpool L21 1BP
M7 PGE P. Glaze 28 Meddins Lane, Kinver DY7 6BY
M7 PGF P. Featherstone 122 Addiscombe Court Road, Croydon CR0 6TS
MM7 PGG P. Greig 22 Rowanhill Close, Port Seton, Prestonpans EH32 0SY
M7 PGH P. Hodgson 32 Main Road, Seaton, Workington CA14 1HS
M7 PGI P. Geary 1 Soar Lane, Sutton Bonington LE12 5PH
MM7 PGL M. Saunders 8D Springhill Road, Port Glasgow PA14 5QP
M7 PGM P. Millington 81 Kilpin Green, North Crawley, Newport Pagnell MK16 9LZ
M7 PGN A. Swenson Bankholme, Ashleigh Road, Arnside, Carnforth LA5 0HE
M7 PGP S. Hunt 12 Exmoor Drive, Leamington Spa CV32 7BB
M7 PGX P. Green 1 Otwell Close, Abingdon OX14 2QR
M7 PGY P. Gray 3 Herrington Court, Newton Aycliffe DL5 4RA
M7 PHA P. Alves 2 Sherwood Court High Road Leavesden, Watford WD25 7PA
M7 PHB P. Baines 11 Orchid Avenue, Farnworth, Bolton BL4 0ES
M7 PHE K. Munson Birchwood, Green Lane, North Kelsey, Market Rasen LN7 6FH
MW7 PHI P. Marks 61 Trefelin Crescent, Port Talbot SA13 1DX
MW7 PHY P. Williams 14 Penyfan Road, Llanelli SA15 1JP
M7 PHZ P. Hunt 21 Lawrence Drive, Cobham, Gravesend DA12 3BU
MM7 PIK J. Pike Mosside Farmhouse, Culsalmond, Insch AB52 6TU
M7 PIX T. Bacala Flat 3, 232 A, Seven Sisters Rd, London N4 3NX
M7 PJB P. Black 7 Birch View, Manchester OL129PZ
M7 PJD P. Dee 17 William Way, Lawford Manningtree CO11 2GE
M7 PJF P. Foster 40 South Lynn Crescent, Bracknell RG12 7JY
MW7 PJH P. Houghton 48, Woodlands Caravan Park, Tyn Y Morfa, Holywell CH8 9JN
M7 PJI Dr D. Mcconnell 2 Hancock Court, Norwich NR5 9NN
M7 PJK P. Keightley 2 Spring Meadow, Upton Bishop, Ross-on-Wye HR9 7SS
M7 PJM P. Mackwell Wistaria, Hawthorn Lane, Pickering YO18 7EA
M7 PJO P. Offord 1 Adare Close, Dunmow CM6 2GR
M7 PJS P. Lawrence 13 Woodcock Close, Haverhill CB9 0JP
M7 PJT P. Johnson 16 Peregrine Close, Hythe CT21 6QZ
M7 PJY P. Abraham 211B Hornby Road, Blackpool, Lancashire FY1 4JA
M7 PKA E. Fergus 98A West Vale, Neston CH64 0TL
M7 PKC P. Casey 29 Norton Avenue, Stockton-on-Tees TS20 2JJ
M7 PKK P. Kitching 83 Birchlands Avenue Wilsden, Bradford BD15 0HB
M7 PKP P. Paskiewicz 47 Great Gull Crescent, Northampton NN3 5AZ
MW7 PKT Dr K. Trimmis 142 Arran Street, Cardiff CF24 3HU
M7 PKW D. Woodward 178A Worcester Road, , Malvern WR14 2HL
M7 PLA P. Larcombe 55 Forest Drive, Weston-Super-Mare BS23 2UG
M7 PLB P. Brown 28 Farley Way, Stockport SK5 6JD
M7 PLD C. Abbott 1 Milton Road, Ellesmere Port CH65 5AT
MM7 PLG D. Johnson 8 Sandmartin Grove, Lenzie, Kirkintilloch, Glasgow G66 3WF
M7 PLH P. Hilton Shankly Cottage 161 Highgate, Jennings Yard, Kendal LA9 4EN
M7 PLI P. Newman Butlers Hill Farm Weights Lane, Redditch B97 6RQ

MW7 PLL J. Jillings Cane Garden, Dolau, Llandrindod Wells LD1 5TE
M7 PLN P. Neal 65 Van Diemans, Stanford in the Vale SN7 8HW
M7 PLO P. Osborne 28 Elizabeth Way, Rushden NN108JR
M7 PLP P. Dingsdale 51 Adelaide Avenue, Thatto Heath, St. Helens WA9 5RU
M7 PLS P. Shakespeare 12 Brickley Lane, Devizes SN10 3BQ
M7 PLT G. Lunt 9 Hereford Way, Aylesbury HP19 9GY
M7 PLW P. Walker 33 Bradshaw Street, Manchester M24 2AG
MW7 PLY D. Evans Bleddfa, Maes Meyrick, Heolgerrig, Merthyr Tydfil CF48 1RZ
MM7 PMB P. Barton 2/2 5 Park Quadrant, Glasgow G3 6BS
M7 PME P. Eachus 251 Forton Road, Gosport PO12 3HD
M7 PMF P. Fotherby Furndale Foxhill Court, Leeds LS16 5PL
M7 PMH P. Harston 70 Dairymans Walk, Burpham, Guildford GU4 7FF
M7 PMK P. Mckay 18 Dalewood Crescent, Elton, Chester CH2 4PR
M7 PML P. Long 206A Henver Road, Newquay TR7 3EH
M7 PMY Dr P. Ylioja 12 Nelson Crescent, Cambridge CB24 3GN
M7 PNG P. Garbett 827C Warwick Road, Birmingham B11 2EL
M7 PNO M. Troscianczyk 5 Haldane Court, Hull HU4 6ST
M7 PNR P. Whiterod 9 Brewhouse Lane, Soham, Ely CB7 5JD
M7 PNS N. Wheeler 6 Fishermens Court, Attleborough NR172QW
M7 PNY P. Yarrow 54 Wansbeck Road, Jarrow NE32 5SS
MJ7 POA L. Coates The Manse Clos De Carrel, St. Brelade JE3 8LJ Jersey
M7 POC P. O'Callaghan 220 Chapel House Road, Nelson BB90QR
M7 POE Lord J. Williams 7 Southrop Road Kingsway, Quedgeley, Gloucester GL2 2HN
M7 POF S. Carmichael 6 Shelley Gardens, Pelton Fell, Chester le Street DH2 2PN
M7 POG Dr J. Pogmore Sunnybanks, West End, Barlborough, Chesterfield S43 4HE
M7 POI R. Lanham 16 Pinewood Road, Ferndown BH22 9RW
MM7 POL P. Mccandlish 2 Dumbarton Road, Glasgow G11 6PB
M7 POV J. Symonds 13 Noel Coward Gardens, Aldington, Ashford TN25 7EU
M7 POW J. Pownall 75 Park Barn Drive, Guildford GU2 8ER
M7 PPF R. Mayo 8 Station Terrace, Kempley Road, Dymock GL18 2BD
MM7 PPH P. Hickman 76 Ettrick Terrace, Selkirk TD7 4JR
M7 PPL P. Latham 135 Ashgate Road, Chesterfield S40 4AN
M7 PPM P. Marks 4 Belmont Drive, Lymington SO41 3AE
M7 PPQ M. Deer 126 Samuel Jones Crescent Little Paxton, St. Neots PE196QY
M7 PPR P. Rush 8 Ashworth Street, Radcliffe, Manchester M26 2XU
MM7 PPW P. Wilkie 1 Kinnaber Road, Hillside, Montrose DD10 9HE
M7 PRA R. Park 19 Rosedale, Bedlington, Ne22 6ej NE22 6EJ
M7 PRE A. Hill 39 Powis Road, Ashton-On-Ribble, Preston PR2 1AD
M7 PRF B. Fulford 43 St. Peters Court, Whitby YO22 4JQ
M7 PRG P. Gill 102 Leighton Road, Sheffield S14 1SS
M7 PRH Dr P. Holford 52 Coldstream Close, Warrington WA20LL
M7 PRI P. Acharya 61 Rose Court Nursery Road Pinner, London HA5 2AR
M7 PRN D. Crisan Flat A, 9 Fortis Green Avenue, London N2 9LY
M7 PRW P. West Stour House, Church Street, West Stour, Gillingham SP8 5RL
M7 PRZ A. Prozapas 149 Osborne Road, Wisbech PE13 3JP
M7 PSB P. Browning 38 Moorsfield, Houghton le Spring DH4 5PF
M7 PSC Capt. P. Cunningham Flat 1, Torvesco House, Knowles Hill Road, Newton Abbot TQ12 2PW
MM7 PSG P. Anderson 76 Devonway, Clackmannan FK10 4LF
M7 PSI J. Stokes Orchard House, Old Smithy Close, Pattingham, Wolverhampton WV6 7AZ
M7 PSJ S. Jackson 8 Buttree Court, South Kirkby, Pontefract WF9 3NB
M7 PSM S. Miles 21A Bearwood Road, Barkham, Wokingham RG41 4TB
M7 PSS P. Sanderson 26 Ponsford Road, Minehead TA24 5DY
M7 PSV C. Karslake 14 Gilbert Close, Popley, , Basingstoke, RG24 9PA
M7 PSX A. Faulkner 13 Cherry Orchard, Woodchurch, Ashford TN26 3QX
MI7 PSY S. Redmond 79A Bridge Street, Portadown, Co Armagh, Belfast, Portadown BT63 5AA
M7 PSZ S. Marek 91 Dial House Road, Sheffield S6 4WU
M7 PTF T. Facemire 7 Crofton Avenue, London W4 3EW
M7 PTP C. Laycock 35A High Street, Henlow SG16 6AA
MW7 PTW P. Williams Gorn Newydd, The Gorn, Llanidloes SY18 6LA
M7 PTY P. Taylor 89 Sunningdale Gardens, Bognor Regis PO22 9LE
M7 PUF D. Williams 11 Friars Place, Littleport, Ely CB6 1LG
MI7 PUI K. Mc Erlean 33 Balmoral Avenue, Ballymena BT43 5ED
M7 PUK P. Wilson 90 Kenwood Crescent, Ingleby Barwick, Ingelby Barwick TS17 5BS
M7 PUL P. Parkinson 5 Gail Close, Failsworth, Manchester M35 0TG
M7 PUP R. Strutt Collingwood, 39 Henley Court, Ipswich IP1 3SD
M7 PUZ N. Titmus 68 Hobart Road, Cambridge CB1 3PT
M7 PVP R. Armshaw 11 Collingwood Crescent, Matlock DE4 3TB
M7 PVS P. Sloper 9 Rosoman Road, Southampton SO19 2PH
M7 PVW M. Vaites 19 Campion Drive, Sheffield S21 1TG
M7 PVX A. Mount 48 Cow Lane, Wakefield WF4 1BA
M7 PWB P. Bone 11 Fox Hill Drive, Stalybridge SK15 2RP
MW7 PWC P. Colwill 85 Davis Avenue, Bryncethin, Bridgend CF32 9JL
M7 PWD J. Savage 7 Elmsleigh Drive, Swadlincote DE110ET
MW7 PWH P. Hughes Awel Menai Penmaen Park, Llanfairfechan LL33 0RN
M7 PWK S. Padwick 37 The Fieldings, Southwater RH13 9LZ
M7 PWL R. Sparrow 97 Lower Street, Horning, Norwich NR12 8PF
M7 PWT G. Killeen 119 Sutton Road, Walsall WS5 3AG
M7 PWW P. Wilmore 1 Highfield Terrace Lower Bentham, Lancaster LA2 7EP
M7 PXJ P. Jobling 7 Caswell Close, Leicester LE4 2GH
M7 PXM P. Melichar 10 Thistlemead, Chislehurst BR7 5RF
M7 PYA P. Yates 7 Plessey Terrace, Newcastle upon Tyne NE7 7DJ
M7 PYE T. Johnson 7 Island Green, Stafford ST17 0QB
MW7 PYM P. Man 17A Cradock Street, Swansea SA1 3HE
MW7 PYS C. Sandever 22 Cae Melin, Little Mill, Pontypool NP4 0HX
M7 PZG P. Gahagan 28 Brow Hey, Bamber Bridge, Preston PR5 8DS
M7 RAB R. Bolger Little Annaside, Bootle LA19 5XL
M7 RAD R. Day Flat 22, Jefferies Lodge, 48-60 Footscray Road, London SE9 2SU
M7 RAE R. Evans 11 Swane Road Stockwood, Bristol BS14 8NQ
M7 RAJ R. Kumar 1 Stilton Close, Aylesbury HP19 8JH
M7 RAS R. Statham 47 Essington Way, Stoke-on-Trent ST6 5EE
MW7 RAT R. Couper 58 York Place, Newport NP20 4GD
M7 RAV A. Ravenscroft 80A Adelaide Street, Fleetwood FY7 6EE
M7 RAW R. Wallace Killerton Road, Bude EX238EN
M7 RAY R. Lovell Formby, Formby, Livepool L37 4BP
M7 RAZ A. Smith 55 Van Diemans Road, Thame OX9 2DH
M7 RBC R. Coomer 30 Torquay Road, Kingskerswell, Newton Abbot TQ12 5EZ
M7 RBE R. Beardmore 28 Broadway, Ilkeston DE7 8TD
MW7 RBO K. Ashman 21 Greenwood Avenue, Cwmbran NP44 5JF
M7 RBP R. Peica-Balosache 125 South East Road, Southampton SO19 8JS
MM7 RBT R. Telford 4 Cumming Avenue, Carluke ML8 4RL
M7 RBY A. Fox 8 Giles Road, Swindon SN25 1QD
M7 RCE J. Beniston 100 The Ridgeway, Croydon CR0 4AF
M7 RCG M. Hiscock 12 Pickwick Close, Basildon SS155SW
M7 RCH R. Lounton 107 Browning Hill, Durham DH6 4SA
M7 RCI A. Cottell 3 Honeylight View, Swindon SN25 4XS
M7 RCM J. Lingard 28 Pendine Crescent, North Hykeham, Lincoln LN6 8UR
M7 RCR R. Carter 5 Englefield Close, Crewe CW1 3YN
M7 RCS J. Mclean 3 Burgess Field, Woodlands, Wimborne BH21 8LQ
MM7 RCT T. Bray 2/1 West Grange Gardens, Edinburgh EH9 2RA
MM7 RCU R. Abringe 5 Suisnish Place, Broadford, Isle of Skye IV49 9BZ
MM7 RCV R. Campbell 3/1, 490 Tantallon Road, Glasgow G41 3HX
MM7 RCW R. Walker 8 Laverock Braes Road, Grandhome, Aberdeen AB22 9AE
MM7 RCX R. Caves West Lamberkin Farmhouse, Tibbermore, Perth PH1 1QA
MI7 RDA G. Mulholland 62 Castleward Road, Downpatrick BT30 7JT
M7 RDB R. Boag Sudeley Rd, Nuneaton CV10 7AJ
M7 RDC R. Cunningham 130 Station Road, Shimpling, Diss IP21 4UA
M7 RDH R. Holder 34 Looseleigh Lane, Plymouth PL6 5HQ
MI7 RDK D. Vanbossuyt 5 Longstone Way, Dundonald, Belfast BT16 2ED
M7 RDN A. Robinson 10A Denver Court Stapleford, Nottingham NG9 8LN
M7 RDS C. Dannemann Bracelands, The Headlands, Stroud GL5 5PS
M7 RDZ R. Diamond 51 Huddersfield Road, Diggle, Oldham OL3 5NT
M7 RED A. Rodionov 29 West Parkside, London SE10 0JT
M7 REL R. Lee 80 Bailey Drive Bootle, Bootle L20 6HB
M7 REO D. Cooper Seinna Bryne Lane, Padbury MK18 2AL
M7 REP S. Roberts 21 Hednesford Road, Walsall WS8 7LS
M7 RER R. Dixon Coach House, Chapel Lane, Ellel, Lancaster LA2 0PN
MW7 RES P. Leitch 57 Stephen Road, Prestatyn LL19 7EH
M7 RET P. Barrett 58 Friars Avenue, Northampton NN4 8PX
M7 REV P. Howard 4 Wayside, Telford TF7 5NF
M7 REX A. Palmer 7 Oak Street, Lechlade GL7 3AX
M7 REY J. Stanesby 21 Godwin Close, Wokingham RG41 2AH
M7 RFA M. Culley 99 Mill Lane, Stockport SK6 1QL
M7 RFD G. Davies 11, Liverpool L12 3HS
M7 RFF R. Fonseca 8 Bolton Drive, Reading RG29RD
M7 RFM R. Fox 8 Croome Close, Drakes Broughton, Pershore WR10 2BH
MM7 RFP M. Stephenson 20 Balmoral Road, Galashiels TD1 1JL
M7 RFU C. Foord 7 Highnam Crescent, Clee Hill, Ludlow SY8 3RE
MM7 RFW D. Wells 17. Henry Street, Kirriemuir DD85DL
M7 RGB R. Bamber 28 Market Street, Cheltenham GL50 3NH
M7 RGD R. Dukes 40 Eastern Road, Brightlingsea, Colchester CO7 0HU
M7 RGF R. Gallagher 8B Eden Close, Beverley HU177HE
M7 RGG R. Hearn 39 John Gaskell Court, Hensingham, Whitehaven CA28 8PH
M7 RGJ I. Harding 7 Hawthorne Close, River, Dover CT17 0NG
M7 RGK G. Clarke 1 Huntley Crescent, Sheringham NR26 8QQ
MM7 RGM R. Mcbride 7 Garnock Road, Stevenston KA20 3BA
MI7 RGN A. Fleming 7 Laurel Hill Gardens, Coleraine BT51 3GW
M7 RGP G. Peck 90 Chadwell Avenue, Cheshunt, Waltham Cross EN8 0ER
M7 RGR R. Robinson 4 Limetree Court, Taverham, Norwich NR8 6QY
M7 RGS R. Selby 1 Wakeling Close, Southwell NG25 0JF
M7 RHA R. Ashitaka 148 Coppice Lane, Basildon SS15 4JS
M7 RHB R. Butt Waterloo Cottage, Barrack Hill, Little Birch, Hereford HR2 8AX
MW7 RHC H. Mckirdy 7 Pritchard Court, Albany Road, Cardiff CF24 3RW
MW7 RHF P. Passmore 298 Dinas, Newtown SY16 1NW
M7 RHH Dr R. Hopper 44 Ditton Walk, Cambridge CB5 8QE
M7 RHI R. Harris 3 Bramble Close, Guildford GU3 3BQ
M7 RHM R. Ham 232 Moor Road, Chorley PR7 2NT
MW7 RHR R. Roberts 2 Trearddur Mews, Trearddur Bay Trearddur Bay, Holyhead LL65 2TT
MW7 RHS R. Jones 6 Treflan, Y Ffor, Pwllheli LL53 6UN
M7 RHW R. Haines-White 111 Glendower Crescent, Orpington BR6 0UP
M7 RIA A. Edwards 124 Ardrossan Gardens Worcester Park, Worcester Park KT47AY
MM7 RIK R. Maccormack 90 Wardlaw Crescent East Kilbride, Glasgow G75 0PY
M7 RIL J. Riley 67 Moss Bank Road, St. Helens WA117DE
MM7 RIN I. Gubaciova 38/25 South Clerk Street, Edinburgh EH8 9PS
M7 RIS R. Stone 374 Bourne Road, Spalding PE11 3LL
M7 RIZ M. Heywood-Macdonald 12 Burton Road, Kendal LA9 7JA
M7 RJB Sir R. Lindeman 26 Seattle Avenue, Blackpool FY2 0PW
M7 RJD R. Denby 168 Nottingham Road, Nuthall, Nottingham NG16 1AB
M7 RJG R. Gilbert 15 Marshfield Close Goonhavern, Truro TR4 9FN
M7 RJH R. Herbert 24 Green Acres Road, Birmingham B38 8NH
M7 RJI R. Ibbotson 33 St. Peters Avenue, Caversham, Reading RG4 7DH
M7 RJJ R. Jarrott 11 Stewkins, Stourbridge DY8 4YW
M7 RJL R. Land 31 Winifred Road Cale Green, Stockport SK2 6HF
M7 RJR R. Riley 117 Woodgrove Road, Burnley BB11 3EJ
M7 RJT I. Valasakis 16 Seaton Square, London NW7 1GB
M7 RKA R. Arnold 68 Springfield Drive, Calne SN11 0UG
M7 RKC M. Coull 20A Queen Street, Warley, Brentwood CM14 5JZ
M7 RKD R. King 14 Roper Avenue, Heanor DE75 7BZ
M7 RKE D. Roake 9 Falcondale Walk, Westbury On Trym, Bristol BS93JG
MM7 RKJ K. Roy 21 Irvine Place Stirling, Stirling FK8 1BZ
M7 RKL R. Kelleher 65Rugby Road, Leigh WN73HB
M7 RKN M. Corristine 16B Daventry Road, Bristol BS4 1DG
M7 RKS M. Stockdale 3 Manor Close, Sproatley, Hull HU11 4PY
M7 RKW M. Wallis 3 Lawson Court, Boldon Colliery NE35 9NH
M7 RKY M. Jordan-Reed Pump Corner, High Street Green, Sible Hedingham, Halstead CO9 3LG
M7 RKZ R. Keene 33 Jasmine Gardens, Bradwell, Great Yarmouth NR31 8HU
M7 RLA R. Adams N1 Creeksea Place Caravan Park, Ferry Road, Burnham on Crouch CM08PJ
MI7 RLC R. Calvert 29 Coachmans Way, Hillsborough BT26 6HQ
M7 RLD R. Lapinskas 38 Forest Road, Enfield EN3 6ST
M7 RLL R. Lester Finkle Street, Doncaster DN5 0RP
M7 RLP J. Patrick 57 Newick Road, Brighton BN1 9JL

M7 RLS R. Sharpe 23 Sparrowsmead, Redhill RH1 2EJ
MW7 RLT A. Noakes Westra Holt, Westra, Dinas Powys CF64 4HA
M7 RLW R. Walker Pilgrims, Gore Road, Burnham-on-Sea TA8 2HL
MW7 RMC Capt. R. Cooke 6 The Forestry, Trecastle, Brecon LD3 8YA
M7 RMH M. Hunt The Coach House, Church Lane, Gloucester GL19 4AD
M7 RMI R. Macefield 14 Marsh Drive, Kibworth Harcourt, Leicester LE8 0NT
M7 RML R. Little Phoenix Cottage, Frolesworth Lane, Claybrooke Magna, Lutterworth LE17 5AS
M7 RMP R. Pearson 14 Highfields, Lakenheath, Brandon IP27 9DZ
M7 RMU R. Underwood 32 Lowlands Close, Rectory Farm, Northampton NN3 5EP
M7 RMW R. Ambler 21 Whitley Spring Road, Ossett WF5 0QA
M7 RMZ S. Lilley 1 Ormonde Road, Chester CH2 2AH
MM7 RNB R. Bull Pinewood 6 Drumindorsair Beauly, Beauly IV4 7AH
M7 RNE R. Evans 364 Aldermans Green Road, Coventry CV2 1NN
M7 RNI P. Whitehouse 52 Hagley Road, Rugeley WS15 2AW
M7 RNL A. Scoggins 7 North Road, Southwold IP18 6BG
M7 RNO R. Fisher 6 Armitstead Road, Wheelock, Sandbach CW11 3LP
M7 RNS A. Atkinson 21 Dennington Crescent, Basildon SS14 2FF
M7 RNU I. Eliade 32 Meadow Sweet Road, Stratford-upon-Avon CV37 0TH
M7 RNX R. Nixon 32 Albany Road, Hornchurch RM12 4AF
M7 RNZ S. Hughes 23 Songbird Croft, Liverpool L144BF
M7 ROC C. Rockey 10 Doulton Drive Porthill, Newcastle under Lyme ST5 8SE
M7 ROI A. Ingham Rosemount, Church Whitfield Road, Whitfield, Dover CT16 3HZ
M7 ROO M. Ratcliffe 76 Churchill Road, Stone ST15 0DY
M7 ROT F. Sacca 78 St. Johns Avenue, London NW10 4EG
M7 ROV J. Mckeown 5 Sidney Avenue, Hesketh Bank, Preston PR4 6SU
M7 ROX E. Hauser 23 Prince William Way, Sawston, Cambridge CB22 3SZ
MW7 ROY D. Underwood 1 Bro Hafan, Cross Inn, Llandysul SA44 6NQ
M7 RPA R. Parker 21 Broadbridge Lane, Smallfield, Horley RH6 9RE
M7 RPF R. Fuller 27, Main Street, East Ayton, Scarborough YO13 9HJ
M7 RPI A. Wynne 7 Ruskin Close, Oxford OX2 9FU
M7 RPK R. Pike 44 Falkirk Close Bransholme, Hull HU7 5BX
M7 RPL R. Lyons 103A Oxney Road, Peterborough PE1 5NG
MI7 RPP M. Burns 1 Ashley Close, Armagh BT60 1EX
M7 RPR P. Shingler 4 Lawnswood Avenue, Burntwood WS74YD
M7 RPU R. Pugh 1 The Hemsleys, Pease Pottage, Crawley RH11 9BX
M7 RPY R. Young 23 Brewton Drive, Deeping St. James, Peterborough PE6 8GR
M7 RRG S. Brown 48 Austerby, Bourne PE10 9JG
M7 RRT B. Smith 53 Fareham Crescent 53, Wolverhampton WV44YN
MW7 RRW T. Williams Sea View House, Calon Fawr, Lon Masarn, Tycoch, Sketty, , Swansea SA2 9EX
M7 RSC J. Lawrence Flat 19, Badminton, 14 Manilla Crescent, Weston-Super-Mare BS23 2BP
M7 RSD J. Stannard Rose Cottage, Hereford HR69SY
M7 RSH R. Davies Lamb Cottage 3 Manor Barns, Snowshill, Broadway WR12 7JR
M7 RSJ S. Riley 51 Kenilworth Avenue, Reading RG30 3DL
M7 RSM P. Collins 7 Fitzmaurice Square, Calne SN11 8NL
M7 RSQ R. Snow 6 The Orchard, Leven, Beverley HU17 5QA
M7 RST R. Upton 139 Withens Lane, Wallasey CH45 7NF
M7 RTB J. Blundell 17 Griffin Close, Chester CH1 5TX
MI7 RTD R. Donaghy 35 Garvagh Road, Dungiven, Dungiven BT47 4LU
M7 RTE R. Logue 21 Hammers Lane, London NW7 4BY
M7 RTK R. Kelly 48 Strathearn Drive, Bristol BS10 6TJ
M7 RTO A. Rutson-Edwards 89 Bemerton Gardens, Kirby Cross, Frinton-on-Sea CO13 0LQ
M7 RTS P. Craske 1 Hythe Lane, Burwell, Cambridge CB25 0EH
M7 RTT A. Tait The Reddings, Dirty Lane, Beausale, Warwick CV35 7AQ
M7 RTU C. Rees 14 Elizabeth Court, Droitwich Spa WR9 8RU
M7 RTV M. Broadhead 271 Goodison Boulevard, Doncaster DN4 6TP
MW7 RTY M. Hayden 17 Haulfryn, Kenfig Hill, Bridgend CF33 6EJ
M7 RUE P. Brindley 70 Atherfield Road, Reigate RH2 7PS
M7 RUI R. Dos Santos Amorim 20 Urswick Road, Dagenham RM9 6EA
M7 RUK A. Johnson 7 Island Green, Stafford ST17 0QB
M7 RUS R. Downham 7 Cherry Garden Road, Maldon CM9 6ES
M7 RUT I. Rutlidge 42 Crispin Close, Locks Heath, Southampton SO31 6TD
M7 RUZ R. Cross 9 Mikado Road, Long Eaton NG103GN
MM7 RVC C. Cooper Beech Tree House, Bogriffie, Fintray, Aberdeen AB21 0YQ
M7 RVD W. Williams White Gates, Skinners Lane, Churchill, Winscombe BS25 5PW
M7 RVE R. Burgess 13 Knockwood Road, Tenterden TN30 7AP
M7 RVF D. Grear 8 Petworth Close, Frimley, Camberley GU16 8XS
M7 RVH C. Goodall 66 Coleshill Road, Birmingham B377HW
MM7 RVI F. Liddell 36 Ladeside, Newmilns KA16 9BE
MI7 RVJ R. Johnston 4 Fennel Drive, Antrim BT41 4FN
M7 RVL P. Moxhay 6 York Road, Torpoint PL11 2LG
M7 RVM A. Burn 23 Tyle Place, Old Windsor, Windsor SL4 2QR
MM7 RVN N. Fox 9 Springfield Terrace, Alness IV17 0SP
M7 RVT C. Patching 28 Chenies Drive, Basildon SS15 4AE
MJ7 RVV W. Stewart 9 Rue Verte Villas La Grande Route De Saint Laurent, St. Lawrence JE3 1NJ Jersey
M7 RWA R. Ambrose 217A Woodgrange Drive, Southend-on-Sea SS1 2SG
M7 RWE R. Evans 32 Bracken Lane, Higher Bebington CH63 2LZ
M7 RWF P. Baker 3 Carrion View, Gateford, Worksop S81 8UZ
M7 RWH R. Hart 15 West Street, Rottingdean BN2 7HP
M7 RWJ R. Jones 31 Mary Mead, Warfield, Bracknell RG42 3SZ
MI7 RWT R. Thompson 18 Parklands, Ballymena BT43 6FD
M7 RWW Capt. R. Wilkins 27 Northside Drive, Sutton Coldfield B74 3QQ
MI7 RXD R. Kearney 4 Flat P Clarendon Street, Derry BT487ES
MM7 RXH R. Haddow Pointhouse, King'S Cross, Isle of Arran KA27 8RG
M7 RXK R. Kilcommons Haven, Meadow View Drive, Ravenfield, Rotherham S65 4RJ
M7 RXM J. Jordan 8 Leven Road, Tamworth B77 2TX
M7 RXW R. Winkley 109 Furze Park Road, Bratton Fleming EX31 4TA
MI7 RYC R. Colhoun 118 Whitehill Park, Limavady BT49 0QG
M7 RYD D. Ryder 2 The Birches Canterbury Road, Herne Common CT6 7LE
MW7 RYK Dr A. Rykala 20 Mount Pleasant, Blaina, Abertillery NP13 3DD
M7 RYO G. Cougill 45 Grange Road, Fleetwood FY7 8DD
MI7 RYP G. English 15 Murrays Hollows, Ballyroney, Banbridge BT32 5ES
M7 RYV C. Ryves 21 Augustus Way, Chandler'S Ford Chandler'S Ford, Eastleigh, Eastleigh SO53 2BD
M7 RZA A. Keen 54 Leigh Road, Hale, Altrincham WA15 9BD
M7 RZE D. Rezaie 25 Coach Grove Road Green, Gateshead NE10 0EH
M7 RZO A. Brzosko 32 Lower Park Street, Cambridge CB5 8AR
M7 RZS R. Scott 57 Beaulieu Close, Southampton SO16 8ED
MM7 RZT J. Baugh-Clark 10 Deveron Street, Huntly AB54 8BY
M7 SAB S. Bodsworth 51 Berry Way, Andover SP10 3RZ
M7 SAC S. Carro 149 Ruston Rd, London SE18 5QY
MM7 SAH S. Hancox 45 Park Road, Brechin DD9 7AE
M7 SAK P. Phizackerley 19 Fir Court Avenue, Churchstoke , Montgomery SY15 6BA
M7 SAP S. Elton 55 Dugdell Close, Ferndown BH22 8BQ
M7 SAV S. Sedgwick Flat 2/A, St. Georges Court 44 Thorne Road, Doncaster DN1 2JA
M7 SAZ S. Ziya 32 Latchford Road, Wirral CH60 3RW
M7 SBA A. Shayes 4 Vicarage Road Deopham, Wymondham NR18 9DR
M7 SBB S. Swallow 16 Quarry Lane, Chesterfield S40 3AS
MW7 SBD S. Davies Llanbadarn Fynydd, Llandrindod Wells LD16YE
M7 SBM S. Brackstone 75 Mozart Close, Basingstoke RG22 4HZ
M7 SBN Dr S. Georgescu 42A Denbigh Road, London W13 8NH
M7 SBO S. Bickham 114 Parkway, Bridgwater TA6 4HT
M7 SBW S. Buchanan 66 Racemeadow Crescent, Netherton, Dudley DY2 0DX
M7 SBY C. Sobey 9 Waterhouse Lane, Chelmsford CM1 2TE
M7 SBZ S. Ball 23 High Heath Close, Birmingham B30 1HU
M7 SCB S. Buckley 341 Manchester Road, Northwich CW9 7NL
M7 SCC S. Cartwright 41 North Avenue, Stafford ST16 1NJ
M7 SCD S. Cooper 19 Old Chapel Road, Smethwick B67 6JA
M7 SCF S. Fadhley 36 Torrington Gardens, London N112AB
M7 SCG S. Groves 152 Rundells, Harlow CM18 7HF
M7 SCH S. Head 42 Bitterne Drive, Woking GU21 3JU
M7 SCK S. King 8 Stafford Road, Crawley RH117LJ
M7 SCN S. Nuttall 17 Redgate, Northwich CW8 4TQ
MM7 SCO D. Raeburn 168 Stonylee Road, North Carbrain, Cumbernauld G67 2LU
M7 SCP S. Palmer 173 Sorrel Bank, Linton Glade, Croydon CR0 9LZ
M7 SCR S. Ride 61 Pepper Street, Sutton-in-Ashfield NG17 5GD
M7 SCT G. Nightingale 21 Turners Close, Ongar CM5 9HH
M7 SCW S. Wormleighton 4 Selwyn Avenue, Wick, Littlehampton BN17 7NF
M7 SCY B. Tocknell 16 Calder Avenue, Walsall WS1 2BQ
M7 SDB S. Hartshorne 143 Winchester Drive Chelmsley Wood, Birmingham B37 5QL
M7 SDD S. Dawe Whiterocks House, St. Anns Chapel PL18 9HN
M7 SDE S. Evans 20 Edgbaston Way, , Doncaster DN12 1SQ
M7 SDI S. Unsworth Heath Terrace, Towcester NN128UP
MW7 SDJ S. Crabbe 58 Severn Crescent, Chepstow NP16 5EA
M7 SDX J. Moore 45 Northgate, Wiveliscombe, Taunton TA4 2LF
M7 SDZ S. Dunkerley 4 Ballantyne Grove, Bootle L20 0AD
M7 SEA S. Hodgson 35 Hamilton Drive, Guildford GU2 9PL
M7 SEG S. Green Meadow View, Ewleaze Farm, Tincleton, Dorchester DT2 8QR
M7 SEH S. Hodds 4 Alconbury Way, Middlesbrough TS3 9QW
M7 SEJ S. Adams 1 Byford Way, Winslow, Buckingham MK18 3RJ
M7 SEK R. Sek 6 Tanners Road, Cheltenham GL51 7LH
MW7 SEM S. Crickett 16 Llys Ambrose, Mold CH7 1GU
M7 SES S. Smith 133 Radcliffe New Road, Whitefield, Manchester M45 7RP
M7 SEW I. Hunt 17 Robartes Road, St. Austell PL26 8DS
M7 SEY B. Ramsey 2 Hamilton Walk, Great Yarmouth NR29 4TB
M7 SEZ A. Kaur 5 Hazel Drive, Leicester LE3 2JE
M7 SFB S. Brown 1 Boundary Cottage, Bures Road, Little Cornard, Sudbury CO10 0NN
M7 SFD S. Doust 109 Athol Mount, Halifax HX3 5RH
M7 SFY S. Hall 16 Larkspur, Rugby CV23 0UW
M7 SGA M. Riches 7 Leighton Square, Ipswich IP3 0LL
M7 SGB S. Gee 17 Newchurch Street, Rochdale OL11 2TA
M7 SGD P. Beard Flat 3, Laurel Court, 33 Cavendish Road, Bournemouth BH1 1QZ
M7 SGF S. Fowler 74 Osborne Road, Wisbech PE13 3JW
M7 SGM M. Stringer Castle Street, Keinton Mandeville TA11 6DX
M7 SGT R. Hunter 134 Casttoecoombe Drive London Wimbledon Sw196Rt, London SW19 6RT
MI7 SHB Quantum Amateur Radio & Technology Society c/o J. Crawford 50 Woodgreen Road, Shankbridge, Ballymena BT42 3DR
M7 SHF Rvd. T. Cundy Flat 1, Walnut House, 139-145 Tavistock St, Bedford MK40 2SB
M7 SHN S. Gannaway 8 Stubbington End, Evesham WR11 2SF
M7 SHO S. Blackwell The Forge, School Road, Thorney Hill, Bransgore BH23 8DS
M7 SHP R. Shepherd 19 Elford Avenue, Newcastle upon Tyne NE13 9AP
MM7 SHT J. Short 7 Macinnes Drive Newarthill, Motherwell ML1 5TY
M7 SHW S. Webster 22, St Michael'S Gardens, South Petherton TA13 5BD
M7 SHZ S. Baines 65 Milner St, York YO24 4NJ
M7 SIB S. Blight 5 Addison Street, Stoke-on-Trent ST1 6NY
M7 SID Dr P. Bunyan 228 Aldwick Road, Bognor Regis PO21 3QH
MW7 SIF S. French 2 Cambrian Road, Tywyn LL36 0AG
M7 SIJ S. Smart 26 Wycote Road, Gosport PO13 0TG
M7 SIN D. Murphy 29 Beverley Crescent, Northampton NN3 2PY
M7 SIS S. Issaias 35 Cowper Road, London N14 5RR
M7 SIW S. Wallace 6 Nuttall Gardens, Cranleigh GU67FQ
M7 SJB S. Bagnall 30 Bideford Avenue, Stafford ST17 0HB
MM7 SJF S. Ferguson 1 Rhindmuir Grove, Baillieston, Glasgow G69 6NE
M7 SJG S. Gardner 137 Allerburn Lea, Alnwick NE66 2QP
M7 SJH S. Hearne 8 Morgan?S Road, Calne SN11 0FH
M7 SJL S. Lovell 20 Courtenay Walk, Weston-Super-Mare BS22 7TQ
M7 SJM S. Martin 6 Cherry Tree Drive, Sedgefield, Stockton-on-Tees TS21 3DN
M7 SJT J. Thirlwell 6A Brook Street, Warminster BA12 8DN
M7 SJW S. Wilson 21 Earlham Grove, Weston-Super-Mare BS23 3JH
M7 SKA L. Steingold 20 Spenser Avenue, Exeter EX2 6BW
M7 SKB Capt. S. Bahulayan 23, Pennine Way, Farnborough GU14 9HT
MM7 SKE M. Chomentowski 24 Arbroath Lane, Aberdeen AB12 5BY
M7 SKG S. King 6 Wilne Road, Draycott, Derby DE72 3NG
M7 SKI H. Allen 8 Winchcombe Gardens, South Cerney, Cirencester GL7 5WJ
MM7 SKL M. Johnston 10 Montgomerie Drive, Skelmorlie PA17 5AG
M7 SKN C. Hodgskin 2 Roberta Walk, Bridgwater TA5 2GW
M7 SKO S. Ochot 6 Ufford Close, Harrow HA3 6PP
M7 SKP Capt. M. Williams 2 Addington Court, Madeira Road, Weston-Super-Mare BS23 2EY
M7 SKS K. Stevens 32 Christchurch Road Hucknall, Nottingham NG156SA

M7 SKT S. Turner 326 Middle Park Way, Havant PO9 5DS
M7 SKV L. Paplauskas 19 Watery Lane, Coventry CV6 2GE
M7 SKW S. Wickham 45 Nicholson Drive, Beccles NR34 9UX
M7 SLA S. Smith 103 Charles Road Quarry Bank Brierley Hill, Dudley DY5 1AE
M7 SLB S. Appleby 21 Osborne Park, Scarborough YO12 5QF
M7 SLD Capt. M. Bazzocchi 29 Turners Avenue, Fleet GU51 1DU
M7 SLE S. Griffin 54 Marine Drive, Hartlepool TS24 0DY
M7 SLG S. Graham 23 North Hill, Fareham PO16 7HN
M7 SLH S. Hodgkiss Estren, Station Road, Wendling, Dereham NR19 2NE
M7 SLI S. Lane 12 Ashenden Road, Guildford GU2 7UU
M7 SLK B. Slack 3 The Ridge, Broad Road, Braintree CM7 9RX
M7 SLO R. Mallinson 20 Moorside Court, Moorends, Doncaster DN8 4SL
M7 SLP M. Jones Flat, 2 The Anchor Centre, Bridge Street, Kingsbridge TQ7 1SB
M7 SLR S. Bennett 7 Holme Park, High Bentham LA2 7ND
M7 SLW S. Whitfield 20 Cherrytree Drive, , Langley Park DH7 9FX
MM7 SLY M. Sisley 3 Harley Terrace, Portlethen, Portlethen Village Portlethen, Aberdeen AB12 4NS
M7 SMC P. Harding 16 Wightman Road, London N4 1SQ
M7 SME R. Hauk 6 Hall Road, Stowmarket IP14 1TN
MM7 SMG S. Mcgill 39 Dudley Drive, Townhead, Coatbridge ML5 2PJ
M7 SMH S. Hammond Ellsworth, Thrigby Road, Filby, Great Yarmouth NR29 3HJ
MI7 SML R. Robinson 41 Baltylum Meadows, Portadown, Craigavon BT62 4AB
M7 SMN S. Trenchard 1 The Willows Worplesdon Road, Guildford GU3 3LL
M7 SMP S. Pressland 14 Conyerd Road Borough Green, Sevenoaks TN15 8RJ
M7 SMX S. Mansfield 545 Portswood Road, Southampton SO17 3SA
M7 SNB S. Brown 5 Overbrook Nursery, Green End, Landbeach, Cambridge CB25 9FD
MM7 SNP R. Cosh Lyndhurst, Kingston Road, Neilston, Glasgow G78 3DY
M7 SNR S. Nolan 24 West Avenue, Chelmsford CM1 2DE
M7 SNS S. Sampson 5 Healdwood Close, Castleford WF10 3AR
M7 SOE C. Haffenden 147 The Diplocks, Hailsham BN27 3JZ
M7 SON S. O'Neill 33 Norlands Park, Widnes WA8 5BH
M7 SOP P. Weaver 2 Almer Road, Poole BH15 4JR
M7 SOT D. Baddeley The Old Cottage, Sandy Lane, Baldwins Gate, Newcastle ST5 5DP
M7 SOX R. Keeys 5 Little Fields, High Street, Hadlow, Tonbridge TN11 0ED
M7 SOZ M. Bissell 10, Montgomery Road, Montgomery Road Montgomery Road, Andover SP11 6HB
M7 SPA M. Caddick 97 Hollemeadow Avenue, Walsall WS3 1JB
M7 SPC S. Curtis Flat 10, Maurice House, Southmill Road, Bishop's Stortford CM23 3DH
M7 SPE A. Campbell 26 Castle Grove, Horbury, Wakefield WF4 5DX
M7 SPH S. Harris Lackington Drove, Dorchester DT2 7QU
M7 SPJ S. Johnson 8 Foster Close, Seaford BN25 2JL
M7 SPM S. Manon 30 Glycena Road, London SW11 5TR
M7 SPO N. Booth Laburnum, Landcross, Bideford EX39 5JA
M7 SPP S. Parkes 19 Keats Road Harden, Walsall WS3 1DS
M7 SPS S. Polap 42 Woodland Drive, Leicester LE3 3EB
M7 SPW S. Worsley 2 Cassidy Close, Pennington Wharf, Leigh WN7 4GB
M7 SPX S. Page 20, Clarence Avenue, Bromley BR12DL
M7 SPY N. Ashton 33 Paxford Road, Wembley HA0 3RQ
M7 SRB B. Briggs Blackthorn Road, Northallerton DL7 8WB
M7 SRE S. Holroyd 31A Cross Street, Tenbury Wells WR15 8EF
M7 SRH S. Harvey 3 North Holmes Close, Horsham RH12 4HB
M7 SRP S. Robinson 7 Higney Road, Hampton Vale, Peterborough PE7 8LZ
MM7 SRR S. Rastocky Dudley Dr, Glasgow G129SB
M7 SRT T. Keloglou 52 Nutcroft Road, London SE15 1AF
M7 SRV A. Insarov Broadway Chambers, 20 Hammersmith Broadway, London W6 7AF
M7 SRW S. Wood 18 Eller Drive, West Winch, King's Lynn PE33 0NN
M7 SRZ S. Roberts 16 Corelli Road, Basingstoke RG22 4NB
M7 SSB J. Virgo 41 Lynch Road, Berkeley GL13 9TE
M7 SSC S. Compagno 26 Barnet Road Erdington Birmingham B23 6JI, Birmingham B23 6JL
M7 SSE S. Evans 78 High Brooms Road, Tunbridge Wells TN4 9BN
M7 SSG S. Clutton 14 Gorsey Lane Burtonwood, Warrington WA5 4HP
MI7 SSJ S. Spallen 65 Tirgarvil Road, Upperlands, Maghera BT46 5UW
M7 SSN S. Spraggon 71 Lane End Drive Knaphill, Woking GU21 2QG
M7 SSO S. Thompson 19 Tewkesbury Place Nether Street, Beeston, Nottingham NG9 2BA
M7 SSS S. Tierney 40 Gypsy Lane, Watford WD48PR
M7 SSW S. Webb Wa103Ht, St. Helens WA10 3HT
M7 SSY A. Parrott 24 The Bungalows, Savory Road, Wallsend NE28 7HX
MI7 SSZ S. Stewart 7 Orangefield Crescent, Armagh BT60 1DS
MM7 STF S. Ferguson Stewards House, Ardeer Golf Club, Stevenston KA20 4LB
M7 STH S. Hoyle 7 Horderns Road, Chapel-En-Le-Frith, High Peak SK23 9ST
MM7 STK T. Storkey 1 Hartington Gardens, Edinburgh EH10 4LD
MM7 STL S. Laing 57 Rosebery Avenue, South Queensferry EH30 9JQ
MM7 STM S. Mitchell 97 Barbieston Road, Auchinleck KA18 2ED
MM7 STS S. Sim 3 Deveron View, Glass, Huntly AB54 4XP
MM7 STT S. Thom 47 Muirhouse Green, Edinburgh EH4 4RB
M7 STU S. Dingle 7 Fieldside, Long Wittenham OX14 4QB
M7 STV S. Elliott Lower Ground Flat, 34 Maxwell Road, Fulham, London SW6 2HR
M7 SUE S. Braisby 4 Langmans Way, Woking GU21 3QY
M7 SUM S. Dossa 24 Warwick Drive, Cheshunt, Waltham Cross EN8 0BW
M7 SUN L. Bell 21 Great North Road, New Barnet, Barnet EN5 1EJ
M7 SUP N. Paul Enfield, Gunton Road, Wymondham NR18 0QP
MW7 SUS B. Williams 25 Park Avenue, Flint CH6 5DW
M7 SUZ P. Hiles 209 Oldbrook Boulevard, Oldbrook, Milton Keynes MK6 2QB
M7 SVA A. Mystikis 23 Barrier Point Road, London E16 2SB
M7 SVB C. Bolton 1A Salterns Terrace, Bideford EX39 4AG
M7 SVE S. Coles 5 Pinfold Road, Giltbrook, Nottingham NG16 2FT
MM7 SVI A. Robertson 1 Greenbank Street, Galashiels TD1 3BL
M7 SVK D. Durkot 15 Thornhill Road, Chorley PR60JB
M7 SVL L. Smith 2 Hope Cottages, Inworth Lane, Wakes Colne, Colchester CO6 2BE
M7 SVP Simon Langton Grammar School For Boys c/o V. Sterea 67 Lamberhead Road, Wigan WN5 9TU
MM7 SVR W. Mcglone 90 Nevis Park, Inverness IV3 8PP
M7 SVT S. Purcell 39 Broadfield Road, Stoke-on-Trent ST65PW
M7 SWA S. Allison 14 Thameside Court, Northmoor, Witney OX29 5BL
MM7 SWC S. Copland 91 Nimmo Avenue, Perth PH1 2PU
M7 SWD J. Marcelino 18 Blayney Row, Newcastle upon Tyne NE15 8QD
M7 SWE C. Sjostedt 14 New Road, Marlow SL7 3NG
M7 SWK M. Sedgwick 44 Cundall Road, Hartlepool TS26 8LG
MM7 SWM Dr S. Mackie Flat 115, 70 Kennishead Avenue, Thornliebank, Glasgow G46 8RS
MM7 SWO L. Mackay 16 Swordale, Isle of Lewis HS2 0BP
M7 SWR S. Prandoczky 8 South View, Kimblesworth, Chester le Street DH2 3QN
M7 SWS S. Walters-Smith 83 Chesterfield Road, Tibshelf, Alfreton DE55 5NJ
M7 SWT S. Tredwell 9 Ferry Road, Iwade, Sittingbourne ME9 8RG
M7 SWW W. Taylor 10 Miller Close, Redbourn, St. Albans AL3 7BG
M7 SWX S. Wilson 7 Boreham Road, Great Leighs, Chelmsford CM3 1NH
M7 SWY J. Yeo 10, Bradgate Close, Leicester LE79NP
M7 SXH A. Dawson 30 Dukes Road, Lindfield, Haywards Heath RH16 2JQ
MM7 SXM S. Canning 27 Culcreuch Avenue, Fintry, Glasgow G63 0YB
M7 SXZ Dr A. Paterson Flat 88, Abel Yard, Frys Lane, Bristol BS1 6ZN
M7 SYB D. Bunting 19 Hornscroft Park, Kingswood, Hull HU7 3GS
M7 SYD J. Sydenham 23 Lawrence Way, Bicester OX26 2FP
M7 SYE S. Stuart 18 Park Avenue, Castle Donington, Derby DE74 2JT
MW7 SYK D. Burns Old Tavern, Llangeinor, Bridgend CF32 8PE
MI7 SYL C. Waldie 47 Oakfield Road, Oakfield, Letterbreen, Enniskillen BT92 2GJ
M7 SYW S. Yem 8 Beechwood Avenue, Wallasey CH45 8NX
M7 SYZ S. Morfitt 10 Greenway, Ryde PO33 3SD
M7 SZD S. Dossa 83 Weston Road, Guildford GU2 8AS
M7 SZO Z. Szot 45 Ealing Park Gardens, London W5 4EX
M7 SZT A. Szot 90 Portland Avenue, New Malden KT3 6BA
M7 SZU M. Tarrant Holly Cottage Nanstallon, Bodmin PL30 5JZ
M7 TAA T. Aavola 25 Honiton Road, Southend-on-Sea SS1 2RT
M7 TAD D. Prior Holly House, Cross Street Drinkstone, Bury St. Edmunds IP30 9TP
MW7 TAF J. Watkins 6 High Street Abersychan, Pontypool NP4 7AB
M7 TAG B. Murtagh 4 Sandcroft Court, 76 Garlands Road, Redhill RH1 6GZ
MW7 TAH T. Hughes 1 Craig Y Don, Amlwch LL68 9DN
M7 TAS T. Jones 2 Woodcote, Hanham, Bristol BS15 8QS
M7 TAT M. Humphries Meadow View 40 Main Road, Cleeve, Bristol BS49 4NR
M7 TAW A. Walker 5 Bepton Down, Petersfield GU31 4PR
M7 TAY K. Taylor 56 Gibraltar Lane, Haughton Green Denton, , Manchester M34 7GG
M7 TAZ A. Allen 5 Davison Court, Longhorsley NE65 8LD
M7 TBA A. Greff 28 Merrylands Road, Great Bookham, Leatherhead KT23 3HW
M7 TBC R. Boland 10 Kenilworth Drive, Kidderminster DY10 1YD
M7 TBD D. Harvey 5 Tithe Barn Drive Bray, Maidenhead SL6 2DF
M7 TBF F. Grande Restrup Engvej 26, Aalborg 9000 Denmark
M7 TBI T. Baker 23 Ferrars Road, Tinsley, Sheffield S9 1RX
M7 TBM T. Mannion 31 Lockwell Road, Dagenham RM107RE
M7 TBO T. Wright 161 Golf Road, Mablethorpe LN12 1EZ
M7 TBR Dr D. Elliott Maynestone Road, Chinley SK23 6AQ
M7 TBZ B. Miller 1A Crowthorne Close, Cambridge CB1 9LZ
M7 TCA I. Daraban 10 Holly Blue Close, Little Paxton, Saint Neots PE19 6TD
M7 TCF J. Ashcroft 104 Deneside, Seghill, Cramlington NE23 7EU
M7 TCK R. Wicks 30 Ashridge Close, Rushden NN10 9HS
M7 TCM P. Twyman 2 Bronte Cottages, Blackmore End, Braintree CM7 4DG
M7 TCS J. Mutter 27 Snowdonia Way, Huntingdon PE29 6XP
M7 TCT M. Burrows 17 Jessie Road Bedhampton Havant, Portsmouth PO93TH
M7 TCV M. Vincent Clouds Hill, White Pit, Shillingstone, Blandford Forum DT11 0SZ
M7 TCW I. Sykes 31 The Park, Warrington WA5 2SG
M7 TDA A. Danks Flat 6, Cartwright House, Wolverhampton Road, Bloxwich, Walsall WS3 2HD
M7 TDJ S. Brooks 42 Brownlow Crescent, Melton Mowbray LE13 0QS
M7 TDK T. Duke 49 Stanley Webb Close, Sawston, Cambridge CB22 3FE
M7 TDL T. Large 4 Blackfirs Lane Somerford, Congleton CW12 4QQ
MW7 TDR J. Taylor 12 Badgers Mead, Brackla Bridgend, Bridgend CF31 2PZ
M7 TDS J. Rushton 7 Torpoint Walk, New Most On, , Manchester M400FZ
M7 TDT P. Elcombe 23 Merton Avenue, Hartley Longfield DA3 7EB
M7 TDV N. Dorling 51 Haygarth Close, Cirencester GL7 1WY
MI7 TDY J. Field 9 Broughton Park, Belfast BT6 0BD
M7 TEG T. Goddard 60 Lake Road, Bristol BS10 5JF
M7 TEJ J. Jewitt 148 Headley Way Headington, Oxford OX3 7SZ
M7 TEO T. Orzechowski 25 Lutterworth Road, Northampton NN1 5JR
M7 TER T. Hedger 1 Berry Terrace, Acton Square, Sudbury CO10 1HT
M7 TEU Dr S. Hill 36 The Woodlands, Market Harborough LE16 7BW
MW7 TEW M. Tew 2 Allen Street Caegarw, Mountain Ash CF45 4BD
M7 TEY A. Jordan 6 Conger Lane, Toddington, Toddington LU5 6BT
M7 TFA A. Brown 56 Ashwood Avenue, Wigan WN2 5YE
M7 TFD D. Johnson 5 Keats Walk, Hutton, Brentwood CM13 2RY
M7 TFG T. Gallagher 67 Denfield Crescent, Halifax HX3 5NQ
M7 TFN M. Rodgers 53 Shaw Avenue, Normanton WF6 2TS
M7 TFT C. Walker 14 Oaks Meade, Carterton OX18 1JX
M7 TFW W. Johnson Clock House, Main Road, Boreham, Chelmsford CM3 3JD
M7 TFX T. Forshaw 2A, Victoria Road, Fallowfield, Manchester M14 6AP
MI7 TFZ D. O'Hagan 45 Mountjoy Road, Coalisland BT715DH
M7 TGD D. Edwards 8 Carey Avenue Higher Bebington, Wirral Merseyside CH63 8LU
M7 TGE I. Cartwright 6 Overwood Place, Stoke-on-Trent ST6 6XD
M7 TGF T. Griffiths 6 Witney Lane, Edge, Malpas SY148JJ
M7 TGH A. Walker 196 High Street Worsbrough Dale, Barnsley S704SQ
M7 TGJ T. Jones 27 Chamberlain Grove, Fareham PO14 1HH
MW7 TGM M. Thomas 46 Westfield Way, Newport NP20 6EW
M7 TGR T. Greenbank 13 Three Stiles Road, Farnham GU9 7DE
M7 THA W. Thomas Southways, Butts Lane, Lumby, South Milford, Leeds LS25 5JA
M7 THF T. Fulbrook 2 Cob Place, Westbury BA13 3GS
M7 THL P. Luis 170 Bourton Way, Wellingborough NN8 2NU
M7 THO A. Swallow 16 Quarry Lane, Chesterfield S40 3AS
M7 THR J. Summers 263 Stroud Road, Gloucester GL1 5JZ
MM7 THU A. Ridley 12 Meadow Court, Thurso KW14 8DD
M7 THX L. Frohock Flat 3, Parkham Court, Shortlands Road, Bromley BR2 0JF
M7 THY E. Finlayson 4 Sycamore Way, Brixham TQ5 0DF
M7 TIA N. Truchla 15 Portchester Close, Peterborough PE2 8UP
M7 TIC J. Sanders 11 Butler Close, Plymouth PL6 6PL

M7 TIE B. Shearer 22 Linkscroft Avenue, Ashford TW15 2BE
M7 TIM Penzance RC c/o T. Eilec 5 Balata Way, Basingstoke RG24 9YP
M7 TIN D. Smith 82 Armstead Road, Beighton, Sheffield S20 1ET
M7 TJB T. Bowron 1, Broughton Avenue. Lu5 6Bq, Toddington, Nr. Dunstable, LU5. 6BQ
M7 TJC T. Tredwell-Coleman 22 The Quadrant, Hull HU6 8NX
M7 TJD T. Dix Willow Cottage 31 London Road, Woolmer Green SG3 6JE
M7 TJE J. Edwards The Granary Old Hall Farm School Road Tunstall, Woodbridge IP122JQ
M7 TJF T. Franklin Flat 57 The Maltings Clifton Road, Gravesend DA11 0AH
M7 TJM T. Mcgoun 64 Buttfield Lane, Howden, Goole DN14 7DS
M7 TJN T. Newton 4 Manor Close, Bradford Abbas, Sherborne DT9 6RN
M7 TJS T. Sherwin 58 Whaddon Chase, Aylesbury HP19 9QP
M7 TJT T. Treanor 2 Waltham Drive, Doncaster DN6 8NJ
M7 TJX T. Jinkerson 104 Foxcote, Finchampstead, Wokingham RG40 3PE
M7 TKB T. Bishop 12 Athelstan Road, Folkestone CT19 6EU
M7 TKD R. Mackay 7 Darwin Close, Lee-on-the-Solent PO13 8LS
M7 TKH T. Howell London Southend Airport Southend Airport, Southend-on-Sea SS2 6YF
M7 TKI T. King 83 Westbury Rise, Harlow CM17 9NT
M7 TKL T. Klimczak 3 Mariner Walk, Chorley PR6 9FF
M7 TKR P. Dulac 22 Colliery Green Drive, Little Neston, Neston CH64 0UA
M7 TKV L. Clinton 16A Melrose Avenue, London NW2 4JS
M7 TKW B. Whaley 14 Kings Way, Harrow HA1 1XU
M7 TKZ T. Lindsay 38 St. Vincents Close, Littlebourne, Canterbury CT3 1TZ
M7 TLA P. Barrows 5A Magdalen Road, Willoughby, Rugby CV23 8BJ
M7 TLC J. Phillips 1 Wath Cottages, Cundall, York YO61 2RL
M7 TLG J. Turner 3 Lulworth Close, Winsford CW7 1LZ
M7 TLJ R. Mcwilliam 3 Fountains Close, Riccall, York YO19 6QN
M7 TLM T. Miller 24 Waveney Residential Park Pound Road, Beccles NR34 9BJ
MW7 TLS M. Comerford 5 Oak Avenue, Penley LL13 0NW
MW7 TLX M. Blandford 33 Greenmeadow Drive, Parc Seymour, Penhow, Caldicot NP26 3AW
MW7 TMA T. Allen 32 Tir Morfa Road, Port Talbot SA12 7PF
M7 TMC C. Marris 16 Cedar Close, Louth LN11 0EH
M7 TMD M. Mcdowell 22 Paddock Close, Clapham, Bedford MK41 6BD
M7 TMH T. Hudson Mimosa Cottage, Ashley Gardens, Mayfield TN20 6DU
M7 TMM A. Mira Quinones 14, Langwood House, 63-81 High Street, Rickmansworth WD3 1EQ
MM7 TMO T. Oakes 55 Northfield Meadows Longridge, Bathgate EH47 8SA
M7 TMP M. Firestone 1 Reeds Close, Rossendale BB4 8ND
M7 TMR T. Reaney Odessa Marine, Little London, Newport PO30 5BS
M7 TMW F. Pereira Da Silva 19 Hatton Park Road, Wellingborough NN8 5BA
MM7 TMZ J. Burwood 8 Barbours Park, , Stewarton KA3 5HS
M7 TNM T. Feierabend 18 Richmond Close, Keynsham, Bristol BS31 2PP
M7 TNY T. Holden 132 Sutherland Street, Barrow in Furness LA14 2BJ
M7 TOG G. Tant 61 Western Way, Sandy SG19 1DU
MM7 TOH M. Cameron 27 Alwyn Avenue, Houston, Johnstone PA6 7LH
M7 TON A. Naylor 2 Byron Close Huyton Merseyside, Liverpool L360UH
M7 TOT S. Turner 4 Bridge Close, Shoeburyness, Southend-on-Sea SS3 9PE
M7 TOW A. Martin 17 Suffolk Road, Bexhill-on-Sea TN39 5BH
M7 TOY T. Littlebury 71 Hampden Drive, Kidlington OX5 2LT
M7 TPD T. Davis 37 Curzon Avenue, Northwich CW8 4YU
M7 TPO P. Davis 7 Rothbury Green, Cannock WS12 2TR
M7 TPT P. Taylor 20 Burlingham Avenue, Worcestershire WR113EE
M7 TPU A. Usher White Cottage, Bratoft, Skegness PE24 5DD
M7 TPW P. Watt 22 Schofield Road, Eccles, Manchester M30 7LG
M7 TQY K. Hughes 26 Spindle Mews, Manchester M4 6DB
M7 TQZ C. Trew 5 Gifford Close, Fareham PO15 6PJ
M7 TRA T. Alexander 6 Moselle Close, Farnborough GU14 9YB
M7 TRB A. Naylor 1 Harvington Close, Kidderminster DY11 5LP
M7 TRG R. Gill Peartree Cottage, Chapel Road, Necton, Swaffham PE37 8JA
M7 TRH T. Hawkins Deganwy, Hardwick Road, King's Lynn PE30 5BB
M7 TRM S. Amos 1 Philbye Mews, Cippenham, Slough SL1 5US
M7 TRV R. Woodward 8 Fulwell Grove, Birmingham B44 0EF
M7 TRW K. Dunleavy 22 Gloucester Court, Newcastle upon Tyne NE3 2XJ
MW7 TRY S. Stupple 7 Glan Preseli, Llanddewi Velfrey, Narberth SA67 7PG
M7 TSA P. Markiewicz 7 Parkfields, Abram, Wigan WN2 5XR
M7 TSB A. Stanford-Beale 2A Albert Road, Caversham, Reading RG4 7PE
M7 TSG A. Rice 23 Chilver Drive, Tong, Bradford BD4 0TS
M7 TSH T. Harrand 17 Highfield Cresent, Selby YO85HD
M7 TSO G. Smith 37 The Crescent, Bracebridge Heath, Lincoln LN4 2NP
M7 TST A. Neighbour 210 Priory Rd, Wellingborough NN82JU
M7 TSU T. Morgan 26 Farlers End, Nailsea BS48 4PG
M7 TSZ F. Holmes 2 Station Lane, Hartlepool TS25 1AX
MW7 TTA A. Gunn 63 Ffordd Tudur, Holyhead LL65 2DU
M7 TTB M. Britton Butterwick Low, Hales Street Tivetshall St. Margaret, Norwich NR15 2EE
M7 TTS A. Thorpe 8 Syke Avenue, Tingley, Wakefield WF3 1LU
MW7 TTV O. Buxton 25 Pen Y Bryn, Sychdyn, Mold CH7 6EE
M7 TUC M. Tucker 42 Martock Avenue, Southend on Sea SS00HH
M7 TUF A. Hawkins 111 Overbrook, Swindon SN3 6AT
M7 TUN J. Tunstall 2 The Cottage, Hames Hall, Gote Road, Cockermouth CA13 0NN
M7 TUS K. Eastwood 3 Bowes Nook, Bradford BD6 2BJ
M7 TUT D. Mcenroe Wayside Lodge, Mitford, Morpeth NE61 3PT
M7 TUX D. Parker 51 Ruskin Road, Congleton CW12 4EA
MI7 TVA R. Moore 34 Rathvarna Walk, Lisburn BT28 2UD
M7 TVD D. Santos 12 Aberford Road, Borehamwood WD6 1PL
MW7 TVH A. Holland 1 Llys Colwyn, Fairmount, Old Colwyn, Colwyn Bay LL29 9NF
M7 TVP A. Nawrocki 132 Lowestoft Drive, Slough SL16PE
M7 TVS L. Jenkins 40 Apollo Drive, Bordon GU35 0DZ
M7 TVT M. Burge 24 Oakdale Road, North Anston, Sheffield S25 4EY
M7 TWA S. Waterson 7 West Farm Road, Newcastle upon Tyne NE6 4JA
M7 TWC T. Stevens 6 West Street, Ilchester, Yeovil BA22 8NN
MW7 TWG J. Pritchard 193 Main Road Bryncoch, Neath SA10 7TT
M7 TWK T. Kerswill 11A Vernon Drive, Prestwich, Manchester M25 9RA
MM7 TWN J. Keymer 7 Pladda Wynd, Broomlands, Irvine KA11 1DW
M7 TWO J. Watts 3 Witney Road, Crawley RH10 6GJ
M7 TWW D. Hughes 10 Kingscroft Court, Northampton NN3 9BH
M7 TXA E. Rippingale-Combes 3 Leyfield, Albourne, Hassocks BN6 9DA
M7 TXB A. Blews 57 Highfield Grove, Stafford ST17 9RA
M7 TXC A. Collins 32 Upper Weybourne Lane, Farnham GU9 9DF
M7 TXD C. Crawford 45 Sunderland Avenue, Horden, Peterlee SR8 4BH
M7 TXM T. Lockyer 2 Decoy Road, Borough Fen, Peterborough PE6 7QE
M7 TXN D. Palmer 44 Milburn Street, Crook DL15 9DZ
M7 TXR B. Roberts 20 Mitchell Road, Enderby LE19 4NX
MM7 TXY A. Dunlop 12 Kendal Avenue 2/1, Glasgow G12 0DL
MW7 TYB J. Procter Awelfa, Ffordd Glan Mor, Talybont LL43 2AR
MW7 TYD R. Grimes Tyddyn Iolyn, Pentrefelin, Criccieth LL52 0RB
M7 TYG C. Sandbrook 24 Greensome Lane, Stafford ST16 1HE
M7 TYM T. Boniface 51 Harebeating Crescent, Hailsham BN27 1JL
MW7 TYR N. Griffiths 30 Offas Green, Norton, Presteigne LD82NX
M7 TYS P. Tysoe 51 Grendon Road Polesworth, Tamworth B78 1NX
M7 TZF C. Cooper 2 Brook View, Railway Street, Slingsby YO62 4AN
M7 TZG T. Grabiec 16 Jubilee Crescent, Clowne, Chesterfield S43 4ND
M7 TZJ C. Stagg Flat16, Westwood Court, Barnsley S70 2LT
M7 TZW A. Saxton Clifton, Nottingham NG11 9HB
M7 TZX P. Whiting 37 Aspen Grove School Aycliffe, Newton Aycliffe DL5 6GR
M7 UAM Capt. G. Story 7 Wheelock Close, Northwich CW9 8TQ
M7 UAP P. Richards 22 Waterpump Court, Northampton NN3 8US
M7 UCA P. Wheeldon 6 The Crescent, Brimington, Chesterfield S43 1AZ
M7 UCB C. Bowcutt 2 Cunnack Close, Helston TR13 8XQ
M7 UCD S. Cole 13 Boscobel Road North, St. Leonards-on-Sea TN38 0NY
M7 UCL T. Sun Orchard Heights, 31 Frogmore Street, Bristol BS1 5BY
M7 UDC G. Loader 10 Rhee Spring, Baldock SG7 6TD
M7 UDS U. Banik 5 Colwyn Road, Swinton, Manchester M27 0EU
M7 UDV G. Lee Princess Beatrice House, 192 Finborough Road, London SW10 9BA
M7 UFC J. Ross 9 Cromford Way, Glossop SK13 0JG
M7 UFF K. Simkin The Flat, Cinque Ports, 49 High Street, Seaford BN25 1PP
M7 UFL Dr S. Sadashivajois 1951 Chestnut St, Apt 208, Berkeley 94702 United States
M7 UGH S. Nicklen-Bundy 74 Ravenhill Way, Luton LU4 0YG
M7 UGL Dr P. Hall 120 Miswell Lane, Tring HP23 4EU
M7 UGZ D. Southall 3 Clive Mews, Saighton, Chester CH3 6FJ
M7 UHU C. Matthews Manor Farm Knedlington, Goole DN14 7EU
M7 UIG R. Horne 114 Waterpark Road, Prenton CH43 0RS
M7 UKD G. Lowe Stone Cottage, Esplanade Lane, Watchet TA23 0AH
M7 UKK P. Beard 14 Kingshill, Cirencester GL7 1DE
M7 UKL A. Lovelace 16 Brandwood Drive, Weston ST18 0GH
M7 UKN K. Mcdonald 11 Micklegate Murdishaw, Runcorn WA7 6HT
M7 UKP P. Csorba 59 Brightgreen Street, Stoke on Trent. ST35DG
MJ7 UKT T. Fillieul 47 Pomme D'Or Farm Estate, West Hill, , Sthelier JE2 3HD
M7 UKU M. Fowler 191 Rayne Road, Braintree CM7 2QE
M7 UKW A. Wearne 58 Mainstone Avenue, Princerock, Plymouth PL4 9NB
M7 UKX N. Radulescu 41 Sherard Road, London SE9 6EX
M7 UKZ C. Conghos 11A Rowans Way, Leavenheath, Colchester CO6 4UU
M7 ULV S. Price 41 Beech Drive., Ulverston LA129EX
MM7 UMU H. Wong Flat B12/4, 20 Clyde Place, Glasgow G58DA
M7 UNC A. Webb 1 Parklands Avenue, Bognor Regis PO21 2BA
MI7 UNI C. Mccrea 20 Kilmore Park, Kilmore, Armagh BT61 8NT
M7 UOK R. Dutton 3 Kilkenny Road, Guisborough TS14 7LE
M7 UPS A. Upstone 19 Calfe Fen Close, Soham, Ely CB7 5GD
M7 UPT B. Fellowes 35 Bolburn, Gateshead NE10 8XB
MM7 URF M. Mcgregor 17 Edindiach Road, Keith AB55 5JW
MW7 URL C. Cowling 14 Shelley Drive, Bridgend CF31 4QA
M7 USA A. Carter 36 Marriotts Close, Ramsey Mereside, Ramsey, Huntingdon PE26 2TX
M7 USB J. Hunter 30 Glebelands, Pulborough RH20 2JJ
M7 USE J. Everatt 50 Barnsley Road, Thorpe Hesley, Rotherham S61 2RR
M7 USL A. Lee 76 Sherborne Road, Stockport SK3 0SN
M7 USV C. Carter Bungalow 7, Higher Ingsdon Quarry, Liverton, Newton Abbot TQ12 6JA
M7 UTD P. Brice-Bullows 3 Ashcroft, Chard TA20 2JH
M7 UTG D. Balch 41 Crosscombe Drive, Bristol BS13 0DE
M7 UTM P. Richmond 34 Garfield Road, Scarborough YO12 7LJ
MM7 UTR M. Hanratty 41 Croftend Avenue, Glasgow G44 5PD
M7 UTT C. Little 28 Cecil Avenue, Warmsworth, Doncaster DN4 9QW
MM7 UVI H. Yang Flat3/2 7 Cooperage Place, Glasgow G3 8QP
M7 UWL R. Scott 157 Walton Back Lane, Walton, Chesterfield S42 7LT
M7 UXO J. Coleman 67 Church Meadows, Deal CT14 9QZ
M7 VAC T. Easton 62 Gunton Drive, Lowestoft NR32 4QB
M7 VAD A. Dsouza 13A Sellons Avenue, London NW10 4HJ
M7 VAH J. Hickman 17 Bramley Avenue, Needingworth, St. Ives PE27 4UD
M7 VAL Midland ARS c/o G. Hall 124 Borfard Road, Liverpool L250PR
M7 VAT P. Cartwright 43 Tarragon Drive Bispham, Blackpool FY20WL
M7 VAV A. Venn 42 Chester Court, Lomond Grove, London SE5 7HS
M7 VBD C. Bailey Bridgefield House, Double Rivers, Crowle, Scunthorpe DN17 4DD
M7 VBN M. Halasa 16, Beaconsfield Court, Leicester Road, Nuneaton CV11 6AE
M7 VBO M. Apps 15 Green Court, Bridge, Canterbury CT4 5LU
M7 VBR A. Riddle 19 Cottey Crescent, Exeter EX4 9DT
M7 VCA A. Lee 64 Arthur Street, Netherfield, Nottingham NG4 2HN
M7 VCC S. Cairney 42 Gorse Avenue, Stretford, Manchester M32 0UE
M7 VCF D. Peacock 41 Oxford Meadow, Sible Hedingham, Halstead CO9 3QW
M7 VCH I. Vernall 1 Owen Place, Bridge Street, Kington HR5 3DH
M7 VCK K. Vickers 11 Kendal Drive, Rainhill, Prescot L35 9JQ
M7 VCL V. Forsyth 4 Charmouth Walk, Manchester M22 5BF
MW7 VCM R. Poyser Glandwr, Snowdon Street, Porthmadog LL49 9DF
M7 VCT D. Cooper 58 Serpentine Road, Widley, Waterlooville PO7 5EF
M7 VCZ P. Simcox 29 Oakamoor Road, Cheadle, Stoke-on-Trent ST10 1BS
MD7 VDL R. Langstaff 17A Elm Drive, Onchan, Onchan IM3 4EH Isle of Man
M7 VDO C. Seal 65 Sandpits, Leominster HR6 8HT
M7 VDX I. Mclean 209C Silver Road, Norwich NR3 4TL
M7 VEG S. Goode 135 Parkside Drive, Watford WD17 3BA
M7 VEN G. Venn 14 School Road, Wychbold, Droitwich WR9 7PU
MM7 VES E. Smith 15 School Wynd, Quarrier's Village PA11 3NL
M7 VET M. Williams 6 Wellhouse Avenue, West Mersea, Colchester CO5 8GF
M7 VEX H. Grahame 4 Madresfield Court, Shenley WD7 9JR

MI7 VFR W. Mccandless 35 Whinsmoor Park, Broughshane, Ballymena BT42 4JG
M7 VGG G. Gribbin 96 Old Church Lane, Stanmore HA7 2RR
MM7 VGR V. Grauso 31 Curlew Brae, Livingston EH54 6UG
MW7 VHT J. Banfield 188 Penderry Road Penlan, Swansea SA5 7ER
MW7 VIB A. Jones 5 Harlech Road, Llandudno LL30 1RQ
M7 VIK A. Cretu 14 Ashton Gate, Flitwick MK45 1AG
M7 VIN V. Brindle 185 Brunshaw Road, Burnley BB10 4DL
M7 VIP G. Matthews 255 Coach Road Estate, Washington NE37 2EU
M7 VIV V. Roy 12 Salisbury Road, Penenden Heath, Maidstone ME14 2TX
M7 VJB J. Brownridge 7 Moorfield Avenue, Stalybridge SK15 2SP
M7 VJC J. Cooke The Old Cottage, Church Lane, Lower Moor, Pershore WR10 2PJ
M7 VJH V. Hayes 68 Billingsley Road, Birmingham B26 2EA
M7 VJM N. Marchant 18, Allington Crescent, Allington Road, , Newick BN84NT
M7 VJT V. Thayil 18 Bargrove Avenue, Hemel Hempstead HP1 1QP
M7 VKO E. White 46 Grenville Road, Aylesbury HP21 8EY
MM7 VKR V. Lang Darnaway, Castleton Place, Braemar, Ballater AB35 5ZQ
M7 VKT P. Hinchliffe 22 Shakespeare Ave, Millbrook, Stalybridge, , Stalybridge SK15 3HE
M7 VLA V. Oleinicenco 91A Brent Street, London NW4 2DY
MM7 VLC A. Cruickshank 58 Cavalry Park Drive, Edinburgh EH15 3QG
M7 VLD G. Baciu 7A Holmstall Avenue, Edgware HA8 5JQ
M7 VLG A. Dmitrik 29B George Street, Pocklington, York YO42 2DG
M7 VLQ Dr A. Volkov 2 Robert Close, Potters Bar EN6 2DH
M7 VMB S. Howard 57 Mill Lane, Huthwaite, Sutton-in-Ashfield NG17 2SJ
MM7 VMC V. Mccutcheon 30 Belhaven Park, Muirhead, Glasgow G69 9FB
M7 VMR V. Roberts Cobden Avenue, Southampton SO18 1FW
M7 VMS S. Spurr 12 Millom Way, Grimsby DN32 7EJ
M7 VNX A. Wright 32 Temple Grove, Leeds LS15 0HT
M7 VOD J. Naylor 16 Eaton Street, Prescot L34 6HD
M7 VOT A. Hopper 38 Mitchell Avenue, Thornaby TS17 9AF
M7 VOX D. Fincham 2 Glebe Close, St. Columb TR9 6TA
M7 VPM V. Mann Ss0 9Rj, Westcliff on Sea SS0 9RJ
M7 VPS P. Shepherd Belswains Lane, Hemel Hempstead HP3 9XE
M7 VQV R. Morales 31 Kings Mead, Ripon HG4 1EJ
M7 VRA K. Cope 9 Amber Heights, Ripley DE5 3SP
M7 VRB T. Wolfe 77 Moorend Crescent, Cheltenham GL53 0EW
M7 VRM V. Mccullagh 4 Sholebroke Mount, Leeds LS7 3HG
M7 VSD S. Pester 2 Coronation Drive, Penketh, Warrington WA5 2DD
M7 VSZ F. Zanchi Flat 2 4 Helios Road, London SM6 7BZ
MW7 VTD I. Denning 6 West Roedin, Coed Eva, Cwmbran NP44 7EB
M7 VTT T. Turner 39 High Street, Brampton PE284TG
M7 VUK F. De Meira Lins 90 Nora Street, South Shields NE34 0RB
M7 VUX D. Contractor 43 The Spinney, High Wycombe HP11 1QE
M7 VVB D. Elliott 21A The Broadway, Swindon SN25 3BN
M7 VVS S. Swingler 2, Kimmeridge, Wareham BH20 5PE
M7 VVV H. Horan 5 Hunters Ride, Stafford ST17 9HU
M7 VVZ R. Hall 7 St Saviours Road, Totland Bay PO390EZ
M7 VWD D. Bean 59 Moorsholm Drive, Nottingham NG8 2EF
MM7 VWG D. Morrison Osbourne Cottage, Benderloch, Oban PA37 1QP
M7 VWS S. Broomberg 31 Crundwell Road, Tunbridge Wells TN4 0LL
MM7 VWT G. Henderson 2 Monks Moss, Ladybank, Cupar KY15 7NN
M7 VWX P. Furmedge Whitegate House, Hoe Road, Bishops Waltham, Southampton SO32 1DU
M7 VXE S. Holden 21 Warren Crescent, East Preston, Littlehampton BN16 1BH
M7 VXU M. Paddock 22 The Lant, Shepshed, Loughborough LE12 9PD
M7 VXW J. Halsall 53 Grasmere Ave Ince, Wigan WN2 2NN
M7 VYN M. Taylor 20 King Richard Drive, Bournemouth BH11 9PE
MM7 WAB P. Scott 62 Kerse Terrace, Rankinston, Ayr KA6 7HG
M7 WAC S. Preston 105 Lodge Road Thackley, Bradford BD10 0RF
MW7 WAD W. Dabrowski 2 Underhill Crescent, Knighton LD7 1DG
M7 WAF E. Rogers Manor Farm Close, Westont Turville HP225SD
M7 WAI D. Waite The Haven, Kemsley Road, Tatsfield, Westerham TN16 2BH
M7 WAM I. Hamilton Grenfell House, Kirkby Stephen CA17 4HL
M7 WAW C. Coath 17 Ashville Road, Bristol BS3 2AP
M7 WAY W. Hind 6 Whinfield Avenue, Shotton Colliery, Co Durham DH6 2HE
M7 WBB W. Bevan 25 Lymbrook Close South Leigh, Witney OX29 6XL
M7 WBF A. Smith 3 Marion Avenue, Middlesbrough TS5 5JG
M7 WBN N. Nelson 40 Delph Mount, Leeds LS6 2HS
M7 WBT C. Elliott 51 Lower Street Quainton, Quainton HP22 4BL
M7 WBX E. Smart 81 Hunt Close, London W11 4JX
M7 WBY P. Mulder 8 Chapel Close, Little Gaddesden, Berkhamsted HP4 1QG
M7 WCC C. Watson 30 Moorfield Parade, Irlam, Manchester M44 6FY
M7 WCF W. Coombs-Farnsworth 6 Burditt Close, Rothwell NN14 6LD
MM7 WCT W. Forbes Tillathrowie, Gartly, Huntly AB54 4SB
M7 WCW A. Abeka 35 Goldbeaters Grove, Edgware HA8 0QE
M7 WDC J. Bell 9 Somerville Green, Leeds LS14 6AY
MW7 WDP D. Wheatley 10 Castle Street, Pennar, Pembroke Dock SA72 6RH
MW7 WDQ J. Thomas 112 Neath Road, Swansea SA1 2LG
MI7 WDT D. Tagg 23 Cloverhill Park Coleraine, Coleraine BT51 3RF
M7 WDW W. Wakeman Blunts Hall Cottage, Blunts Hall Road, Witham CM8 1LX
M7 WDX L. Baker 3 Oak Court, Peterborough PE7 3FS
MW7 WDY C. Wood 50 Heather Court, Cwmbran NP446JR
M7 WEA M. Speller 38 Brockley Crescent, Romford RM5 3JX
M7 WEB M. Webb 60 Waddens Brook Lane, Wednesfield, Wolverhampton WV11 3SF
M7 WEF E. Fernandes 8 Harestock Close, Winchester SO22 6NP
M7 WEL S. Weller 25A Orchard Way, East Grinstead RH19 1AY
M7 WEP G. Brearley 8 Albert Road Eston, Middlesbrough TS6 9QW
M7 WER W. Rees 69 Pewley Way, Guildford GU13PZ
M7 WET J. Smith 39 Kemp Road, North Walsham NR28 0FP
M7 WEW D. Thorpe 135 Oakfield, Woking GU21 3QU
M7 WEZ I. Bird 2 Church Street, Wiveliscombe, Taunton TA4 2LR
M7 WFD J. Kershaw 17 The Sycamores, Wakefield, Horbury WF4 5QG
M7 WFG W. Gola 128 Bournville Road, Weston-Super-Mare BS23 3RS
MM7 WFL J. Mcauley 55 Clapperhowe Road, Motherwell ML1 4BZ
M7 WFR G. Kirby East Road, West Mersea CO5 8EB
MM7 WGB W. Barclay Bressa Watersideá, Strathdon AB36 8XA
MM7 WGC Dr A. Woodwark 4 Park Place, Edinburgh EH64LB
M7 WGF D. Bullen 5 Westview, Long Sutton, Langport TA10 9LT
M7 WGM A. Wigham 15 Owens Way, Croxley Green, Rickmansworth WD3 3PS
M7 WGR R. Woodford 7 Steer Road, Swanage BH19 2RU
MI7 WGS W. Giles 11 Old Antrim Mews, Ballymena BT42 2SP
M7 WGX M. Chenery 27 Dunstan Road, Glastonbury BA6 8EE
M7 WHA Rvd. W. Hackman Kynance, Barden Road, Speldhurst, Tunbridge Wells TN3 0QB
M7 WHC Q. Wang 14, Marquis House, 45 Beadon Road, London W6 0BT
M7 WHI P. Whitehouse 2 Nans Rosen, Threemilestone, Truro TR3 6FW
MI7 WHK W. Mcdonnell 20 Lynda Gardens, Newtownabbey BT37 0NP
MM7 WIC C. Aitken Windybraes, Upper Gills, Canisbay, Caithness, Canisbay KW1 4YB
M7 WID J. Butler 112 Netherfield, Widnes WA8 8BZ
M7 WIM W. Mckechnie 49 Bedeswell Close, Hebburn NE31 2GB
M7 WIS W. Stone 63 Sands Lane, Lowestoft NR32 3ER
M7 WIT V. Vaznais 35 Lynwood Drive, London KT4 7AA
M7 WIV R. Laidler Swallow Cottage, Wiverton, Plympton, Plymouth PL7 5AA
M7 WIX S. Wix 43 Radburn Close, Harlow CM18 7EE
M7 WJA W. Auger Augernik Fruit Farm, Hopton Wafers, Kidderminster DY14 0HH
M7 WJB J. Blackall 2 Ryson Avenue, Blackpool FY4 4DN
M7 WJC C. Andrews 10 Hartford Road, Hartley Wintney, Hook RG27 8QW
MI7 WJD W. Burrowes 208 Ballynure Road, Ballyclare BT39 9AJ
M7 WJG W. Garvey 254 Bury Road, Tottington, Bury BL8 3DT
M7 WJH W. Hoar 46 Pendean Avenue, Liskeard PL14 6DA
MI7 WJL W. Little 87 Meadowvale Park, Limavady BT49 0RD
MM7 WJM M. Wigham 5 Milton Park, Auchtertool, Kirkcaldy KY2 5QX
M7 WJR Dr W. Rickford Rushcroft, Green End, Weston, Hitchin SG4 7AL
M7 WKD A. Brown 11 Newlands Close, Chandler'S Ford, Eastleigh SO53 4PD
MM7 WKH W. Hiscocks 63 High Street, Cockenzie EH32 0DG
MM7 WKK G. Hamilton 45 Argyle Square, Wick KW1 5AJ
M7 WKR G. Walker 15 Glamis Court, Woodstone Village, Houghton le Spring DH4 6TR
M7 WKS S. Wilkins Colroger Close, Mullion TR12 7DZ
M7 WLD D. Wilde 28 Horsley Crescent, Langley Mill, Nottingham NG16 4FX
M7 WLF R. Flores De Guirior 17 Keyhaven Close, Derby DE21 4SQ
M7 WLH W. Hughes 346 Griffiths Drive, Wednesfield, Wolverhampton WV11 2LB
M7 WLL C. Norman 24 Jasmine Grove, Paignton TQ3 3TH
M7 WLT S. Connor 2 Swedish Houses, Shalbourne, Marlborough SN8 3PX
M7 WLZ D. Walter 22 Romeo Way, Wellingborough NN8 3AZ
MM7 WMG W. Mcguigan 1 West Drip Farm Cottages, Stirling FK9 4UJ
M7 WMK M. Bare 4 Elenors Grove, Ryde PO33 4HE
M7 WML Rvd. M. Legg 30 Canonbury Road, Enfield EN1 3LW
M7 WMM T. Cross Piccotts End Farm 117 Piccotts End Road, Hemel Hempstead HP13AU
MM7 WMN P. Lawley 19 Linnet Brae, Livingston EH54 6UE
MJ7 WMQ R. Raynes Mon Plaisir Rue De Samares, St. Clement JE2 6LZ Jersey
MW7 WMS P. Williams 127 Penchwintan Road, Bangor LL57 2YG
M7 WMU C. Percival 70 Currock Road, Carlisle CA2 4BJ
M7 WMY A. Henslok 31 Pritchard Street, Wednesbury WS10 9EW
MI7 WNY P. Ffitch 1 Lisburn Road, Ballynahinch BT24 8BL
MW7 WOC O. Crump 26 Stanly Road, Wrexham LL14 1HH
MM7 WOK H. Cameron 1 Glenloch View, Achintore Road, Fort William PH33 6TZ
M7 WOL P. Wallis 24 Goldsdown Road, Enfield EN37QZ
M7 WOZ W. Walton 1 West Hall, Yeadon, Leeds LS19 7AJ
MM7 WPC J. Pearson 4 Falsidehill Farm Cottages, Kelso TD5 7TT
M7 WPE W. Edmondson 9 Mayfield Avenue, Clitheroe BB7 1LB
M7 WPH W. Hearn 16 The Cobbins, Burnham on Crouch, Essex CM0 8QL
M7 WPK P. King 107 Mackie Road, Filton, Bristol BS34 7NB
M7 WPL P. Conyers 112 Weston Park, Crouch End, London N8 9PN
MI7 WPX R. Arbuckle 1 Greenfield Park, Strathfoyle, Derry BT47 6XE
MM7 WPZ J. Lanteri Laura Langamull, Calgary, Tobermory, Isle of Mull PA75 6QY
MW7 WRG G. Williams 8 Fron Deg, Pantymwyn CH7 5EU
M7 WRI S. Wright 4 Cassino Road, Watchet TA23 0TX
M7 WRM W. Maudsley Knight Stainforth Hall Little Stainforth, Settle BD24 0DP
MW7 WRZ D. Rees 19 Waunfawr Road, Cross Keys, Newport NP11 7PG
MM7 WSI B. Inglis 51 Blackcraig Brae, Blantyre, Glasgow G72 0TZ
M7 WSJ D. Cooper 58 Serpentine Road, Widley, Waterlooville PO7 5EF
M7 WSM S. Reed 101 Milton Road, Weston Super Mare BS23 2UX
M7 WSP G. Lebond-Carroll 9 Hawthorn Way Carlton In Lindrick, Worksop S81 9HN
MM7 WSS W. Shaw Shaws Farm, Selkirk TD7 4PR
M7 WST C. West 12 St. Georges Road, Wallington SM6 0AS
M7 WSX G. Kinder 6 Ten Acres, Shaftesbury SP7 8PP
M7 WTB G. Whitbread 2 Victory Terrace, Redcar TS10 1QN
M7 WTD A. Shankar 3 Long Ridings Avenue, Hutton, Brentwood CM13 1DZ
M7 WTE A. Staple 5 East End, Kirmington, Ulceby DN39 6YS
M7 WTG W. Grimes 17 Handley Close, Stockport SK3 8NQ
M7 WTJ T. Wright 26 Stamford Rd, Carrington M31 4BA
MI7 WTS B. Rollins 2 Tarragon Park, Antrim BT41 4PF
MM7 WTX S. Thomson Woodbine Cottage, Mosstodloch, Fochabers IV32 7HZ
MI7 WTZ J. Jordan 2 Mill House, Crumlin BT29 4XN
M7 WUT Dr D. Hall 53 Elfleda Road, Cambridge CB5 8NA
M7 WVC W. Clancy 58 Beatrice Place, Hitchin SG5 4RZ
MW7 WVK A. Boyle 50 Goylands Close, Llandrindod Wells LD1 5RB
M7 WVM M. Lomas 33 Carnation Road, Oldham OL4 5QD
M7 WVQ J. Rogers 288 Wareham Street, Middleborough 23462906 United States
M7 WVT P. Blount 115 Essington Road, Willenhall WV12 5DT
MW7 WWF M. Ransom 16 Glyn Garfield Close, Neath SA11 2JR
M7 WWL A. Walker 50 Parkstone Crescent, Hellaby, Rotherham S66 8HD
MM7 WWM J. Hale Glenburn Rd, Drynoch, Lochgilphead PA308EU
M7 WWR R. Wall 3 Stag Drive, Huntington, Cannock WS12 4UJ
M7 WWT M. Howard Flat 10, Oakdale, 6 Westgate Road, Beckenham BR3 5DY
M7 WWX S. Hammond 5 Nursery Gardens, Sturminster Marshall, Wimborne BH21 4AX
M7 WXD R. Springall 7 The Spinney Grange Park, Northampton NN4 5BT
M7 WXS S. Thomas 2820 Winding Creek Rd, Prosper 75078 United States
M7 WYD R. Jenkinson 8 Newstead Avenue Off Newstead Street, Hull HU5 3NE
M7 WYK A. Charlish 271 Laceby Road, Grimsby DN34 5DU
MW7 WYN A. Williams 5 Brynymor, Three Crosses, Swansea SA4 3PE
MM7 WYZ W. Zaczyk 3/1 Saughton Gardens, Edinburgh EH12 5TF
M7 WZM R. Turner 2 Martins Mews, Haverhill CB9 7FU
M7 WZO D. Mulder 217 Cloes Lane, Clacton-on-Sea CO16 8AG
MW7 WZQ G. Roberts Ynys 3 Rhodfa Gwilym, , Four Mile Bridge LL65 2TX

M7 WZY W. Cooke 3 Torrington Close, Potton, Sandy SG19 2SD
M7 WZZ N. Langiano 1 Hopgarden Close, Lamberhurst, Tunbridge Wells TN3 8DY
MM7 XAC A. Campbell Margreig, Laghall Court, Kingholm Quay, Dumfries DG1 4SX
MI7 XAG A. Gregson 16 Dunkirk Road, Waringstown, Craigavon BT66 7SW
M7 XAM A. Moore 11 Glen View, Wigmore, Leominster HR6 9UU
M7 XAN A. Nasseri 4 Kenbury Close, Ickenham, Uxbridge UB10 8HU
M7 XAP A. Parham 2 Finch Gardens North Bersted, Bognor Regis PO229EQ
MW7 XAW R. Brindle Flat 1, Hillcrest, Blue Gardens, Mill Street, Aberystwyth SY23 1JE
M7 XBS B. Saunders 60 Mountford Close, Wellesbourne, Wellesbourne CV35 9QQ
M7 XBX Dr R. Bland 5 Seafield Road, Lytham St. Annes FY8 5PY
M7 XCB D. Bayley 13 St. Patrick'S Drive Poolfields, Newcastle-under-Lyme ST5 2NS
M7 XCJ C. James 13 Langley Green Road, Oldbury B69 4TG
M7 XCS C. Shepherdson 149 Scarborough Road, Norton, Malton YO17 8AD
M7 XDE D. Edwards Hoo Meavy, Holm Lane, Oakerthorpe, Alfreton DE55 7LJ
M7 XDS D. Soulsby 93 Stonecross Road, Hatfield AL10 0HW
M7 XFC M. Christy 565 Warrington Road, Abram, Wigan WN2 5XY
M7 XFY R. Martin 22 Bonnards Road Newton Longville, Milton Keynes MK17 0DS
M7 XGX S. Phillips 2 Watermill Drive, St. Leonards-on-Sea TN38 8WD
M7 XHC R. Williams 98B, The Manor, Billing Garden Village, The Causeway, Great Billing, Northampton NN3 9EX
M7 XHF C. Collins 27 Arnold Close, Castle Gresley, Swadlincote DE11 9HF
M7 XIA P. Mayne 56 Palmerston Street, Plymouth PL1 5LJ
M7 XIX P. Cammish 18 Hertford Close, Eastfield, Scarborough YO11 3HJ
MW7 XJO J. Mcneil Mosaic House, Beachley Road, Tutshill, Chepstow NP16 7EG
M7 XJP A. Griffiths 13 Hill Place, Knowlwood Road, Todmorden OL14 6PN
MW7 XJQ D. Smith Tyddyn Bach, Bethel, Caernarfon LL55 1YD
M7 XJS A. Shirley 7 Shirley Close, Malvern WR14 2NH
M7 XJW J. Graham 17 Valley Road, Wotton-under-Edge GL12 7NP
MI7 XKA K. Allen 106 Whitehill Park, Limavady BT49 0QG
M7 XLA S. Mayes 14 Prior Street, Hereford HR4 9LB
M7 XLD P. Trew 5 Gifford Close, Fareham PO15 6PJ
M7 XLF T. Evans 36A Swanmore Road, Ryde PO33 2TQ
M7 XLH T. Kidds Laurels Garth, York YO606SE
M7 XLR P. Ffitch 22 Beeching Close, Halwill Junction EX21 5XY
M7 XLX A. Freeman 122 Tandra, Beanhill, Milton Keynes MK6 4LL
M7 XMB P. Foster 100 Howe Road, Norton, Malton YO17 9BL
MW7 XMD J. Rossiter 20 Bennions Road, Wrexham LL13 7AW
M7 XMG M. Griffiths 23 Gawdy Close, Harleston IP20 9ET
M7 XML R. Stone 51 St Paul'S Mews, York YO24 4BR
M7 XNG M. Galbraith Oaklands, Didcot Road Harwell, Didcot OX11 0DP
M7 XNR J. Masterton Nobes Avenue, Gosport PO130HX
M7 XNY K. Blowing 194 Cheltenham Road East, Gloucester GL3 1AL
M7 XOM K. Mckeown 57 Stratton Avenue, Wallington SM6 9LJ
M7 XON H. Tranter Flat 6 Oak Apple Court 25 Acorn Road Catshill, Bromsgrove B61 0TR
MM7 XOO A. Doak 113 South Beach, Troon KA10 6EQ
M7 XPG P. Goldsmith 20 Lichfield Drive, Prestwich, Manchester M25 0HX
M7 XPL Dr A. George Stephen 11 Whitby Gardens, Queensbury, London NW9 9TU
M7 XPT L. Cowles 5 Chalvedon Square, Pitsea, Basildon SS13 3QX
M7 XPX J. Porter 45A Wealden Way, Bexhill-on-Sea TN39 4NZ
M7 XQS D. Dalik 26 Hanson Street, London W1W 6UH
MW7 XRE G. Phillips 109 Mount Pleasant Road, Ebbw Vale NP23 6JL
M7 XRN A. Craven 14 Deverel Road, Charlton Down, Dorchester DT2 9UD
M7 XRO J. Roberts 11 Vulcan Way, Castle Donington, Derby DE74 2UJ
MM7 XRR J. Caldwell 35, Clunie Drive, Larbert FK5 4UA
M7 XRS R. Spencer 41B Angerstein Close, Weeting, Brandon IP27 0RL
M7 XRW R. Williamson 67C Wilton Road, Salisbury SP2 7ER
M7 XSB S. Bishop 168 Mansfield Road, Clipstone Village, Mansfield NG21 9AE
M7 XSE S. Edwards 24 Philips Lane Great Sutton, Ellesmere Port CH66 4TP
M7 XST S. Rouse 20 Chepstow Road, Corby NN18 8QR
M7 XTB M. Machyna 12 Victoria Street, Grantham NG31 7BW
M7 XTP P. Carter 1 Bank Drive, Bradford BD6 1AH
M7 XTW P. Bryant 4 Neptune Drive, Hemel Hempstead HP2 5QQ
M7 XTX T. Jones 3 Langford Road, Liverpool L19 3RA
M7 XUN A. Southern 89 Purlewent Drive Weston, Bath BA1 4BD
M7 XUP S. Davison Shinwell House, Central Road, Central Road, Central Road Central Road, Dearham CA15 7HD
M7 XVO L. Nightingale 21 Turners Close, Ongar CM5 9HH
M7 XXS E. Savill 61 Lower Road, Orpington BR5 4AH
MM7 XYF J. Major 14 Younger Gardens, St. Andrews KY16 8AB
M7 XYX R. Franklin 7 Harvest Way, Thornbury, Bristol BS35 1AL
M7 XYZ I. Gould 2 Harps Avenue Minster, Sheerness ME12 3PF
MI7 XZD M. Alsallal 9 Grove Street, Lisburn BT27 4YQ
M7 XZW Z. Williamson 9 Vine Avenue, Cleckheaton BD19 3DW
M7 YAC A. Cross 8 Briar Close, Lowestoft NR324SU
M7 YAD A. Dumbleton 1 Kingsmead Park, Swinhope, Market Rasen LN8 6HS
M7 YAG P. Rogers 9 William Burt Close, Weston Turville, Aylesbury HP22 5QX
M7 YAM G. Smith 47 Percy Road, Carlisle CA2 6ER
M7 YAN J. Wojcik 3 Whitehall, Lidlington, Bedford MK43 0RS
M7 YAU A. Yau Pembury Road, Tunbridge Wells TN2 4ND
M7 YAZ R. Dabrowski Overton, Salisbury Terrace, Mytchett, Camberley GU16 6DB
M7 YBB R. Walker 25 Hedges Close, Hatfield AL10 0HZ
M7 YBC C. Walker 8 Foxhollows, Hatfield AL10 0HX
M7 YBZ I. Bradley 41 Beckford Avenue, Bracknell RG12 7ND
M7 YCA E. Cooper 20 Manor Close, Baston, Peterborough PE6 9PH
M7 YCM D. Gale 3 Westgarth Terrace, Darlington DL1 2LA
M7 YCS C. Sythes 2 Wentworth Crescent, Whitby YO21 1LQ
M7 YCV D. Wright 14 Alexandra Road Peel Green Eccles, Eccles M30 7HJ
MI7 YDB D. Boucher 52 Mourneview Park, Kilkeel BT34 4NB
MM7 YDG D. Greig 22 Rowanhill Close, Port Seton, Prestonpans EH32 0SY
M7 YDK M. Thorpe 135 Oakfield, Woking GU21 3QU
M7 YDX A. Crudele 5 Greatfield Road, Winchester SO22 6HN
MM7 YEA D. Yeaman 7 Brimmond Crescent, Westhill AB32 6RD
M7 YER M. Meyer 6 The Gallops Newbury Road, Chilton OX110PF
M7 YET S. Reed 69 Kingfisher Drive, Lydney GL155FX
M7 YEW P. Cannings 30 Graham Gardens, Luton LU3 1NQ
M7 YGI R. Skuse 17 Church Way, Gloucester GL4 4NW
M7 YGK G. Kane 10 Orchard Close, St. Mellion, Saltash PL12 6UQ
M7 YGM G. Meadows 36 Belbroughton Road Blakedown, Kidderminster DY10 3JG
M7 YGR G. Riccardi 25 Montefiore Road Flat 2, Hove BN3 1RD
M7 YHZ J. Hiltz 35 Highview Road, London W13 0HA
M7 YIF J. Allen 23 Amwell Court, Waltham Abbey EN9 3EA
M7 YIN M. Cavallaro 51 Mccreery Road, Sherborne DT9 4DT
M7 YJW G. Gillies 58 Marshlands, Dymchurch, Romney Marsh TN29 0PZ
M7 YKD N. Stevens 8 Moorfields, West Moor Lane, Raskelf, York YO61 3UZ
M7 YKE A. Wood 19 The Shepherdies, Ripon HG43HU
M7 YKG G. Hussey 2 The Drive, Weston Super Mare BS23 2SR
M7 YKK R. Pas 12 Monument Way, Bodmin PL31 1NZ
M7 YKM P. Court 105 Water Orton Lane, Minworth, Sutton Coldfield B76 9BD
M7 YKN E. Aykin 12 Isambard Place, London SE16 7DA
M7 YLB A. Doyle 2 The Greenways Paddock Wood, Tonbridge TN12 6LS
M7 YMA A. Thomas 20 Coronation Road, Ludgershall, Andover SP11 9NN
M7 YMC D. Ruscoe Wayside, Chester High Road, Neston CH64 7TT
M7 YME M. Embleton 23 Holloway Gardens, Plymouth PL9 9TS
M7 YNC R. Dalumpines Flat 2, 27 Cotswold Way, Worcester Park KT4 8HD
MM7 YNK N. Kirtley 10 Millcroft Road, Auldearn, Nairn IV12 5TW
M7 YOB M. Miller 1 Tonbridge Road, West Molesey KT82EL
M7 YOD M. Jowett Flat 1, 64 Locking Road, Weston-Super-Mare BS23 3DN
M7 YOE G. Yeomans 98 Well Street, Biddulph, Stoke-on-Trent ST8 6HY
M7 YOL A. Fresta Flat 2, Dobson House, John Williams Close, London SE14 5XF
M7 YON J. Ardern 206 Tressillian Road, London SE4 1XY
M7 YOR R. Bertie 14 St. Giles Street, Padiham, Burnley BB12 8HL
M7 YOY M. Cooper 2 May Tree Close, Waterthorpe, Sheffield S20 7JB
MI7 YOZ M. Kelly 19 Gowanvale Drive, Banbridge BT32 3GD
MM7 YRN R. Thomson Meadow Steading Tornaveen, Torphins, Banchory AB31 4PJ
M7 YRR D. Brown 336 Heath Road South, Birmingham B31 2BL
M7 YSE R. Moyse 22 Sugden Road Wandsworth, London SW11 5EF
M7 YSU D. Farrell 15 Woodcote Way, Bexhill-on-Sea TN39 4GP
M7 YTO S. Scorer Flat 28, Wedderburn Lodge, Wetherby Road, Harrogate HG2 7SQ
MM7 YTR M. Sandor 77 Montgomery Avenue, Coatbridge ML5 1QT
MM7 YUG G. Ward 10 Harmony Walk, Bonnyrigg EH19 3NU
M7 YVR R. Minarro Escalona 134 Leopold Road, Kensington, Liverpool L7 8SS
MW7 YXD D. Taylor 29 Walters Road, Neath SA11 2DP
M7 YXI R. Lee Bine Hill View, Westfield, Curry Rivel TA10 0HX
M7 YXZ A. Manczuk Unit 15 Old Cement Works The Hollow, South Heighton, Newhaven BN9 0HS
M7 YYF T. Bradley 23 Grafton Road, Gloucester GL2 0QP
M7 YYQ H. Lee 71 Easemore Road, Redditch B98 8EY
M7 YYT P. Boulding 16 Higher Croft Road, Lower Darwen, Darwen BB3 0QR
M7 YYZ N. Williams 3 Sandfield Crescent, Camborne TR14 7DX
MW7 YZF A. Parry 2 Moorfield Avenue, Clarbeston Road SA63 4UU
M7 YZP N. Doyle 32 Donald Road, Bristol BS13 7BU
M7 YZY M. Davies 12 Hopewell Road, Baldock SG7 5AA
M7 YZZ D. Green 84 Clarondale, Hull HU7 4AR

United Kingdom

'Details withheld'

The callsigns in this list are those of active licences for which the owner has requested details not be given.

2#0

2#0 AAD
2#0 AAE
2#0 AAG
2#0 AAH
2#0 AAL
2#0 AAM
2#0 AAP
2#0 AAR
2#0 AAS
2#0 ABW
2#0 ACO
2#0 ACS
2#0 ADD
2#0 ADE
2#0 ADK
2#0 ADM
2#0 ADO
2#0 AER
2#0 AFA
2#0 AFO
2#0 AGT
2#0 AHO
2#0 AHX
2#0 AIB
2#0 AIH
2#0 AIV
2#0 AJS
2#0 AJZ
2#0 AKA
2#0 AKG
2#0 AKT
2#0 ALC
2#0 ALE
2#0 ALI
2#0 ALK
2#0 AMO
2#0 AOO
2#0 APR
2#0 AQH
2#0 AQQ
2#0 AQR
2#0 AQX
2#0 ARE
2#0 ARI
2#0 ARO
2#0 ARW
2#0 ASJ
2#0 ATF
2#0 AUH
2#0 AWA
2#0 AWF
2#0 AWL
2#0 AWM
2#0 AWO
2#0 AWP
2#0 AWX
2#0 AXX
2#0 AYA
2#0 AYD
2#0 AYR
2#0 AZC
2#0 AZN
2#0 AZQ
2#0 AZS
2#0 AZT
2#0 AZV
2#0 BAE
2#0 BAS
2#0 BAW
2#0 BBB
2#0 BBC
2#0 BBD
2#0 BBP
2#0 BBV
2#0 BCA
2#0 BCI
2#0 BCN
2#0 BCP
2#0 BCU
2#0 BCV
2#0 BCY
2#0 BDA
2#0 BDC
2#0 BDF
2#0 BDG
2#0 BDL
2#0 BDX
2#0 BDY
2#0 BDZ
2#0 BEJ
2#0 BEK
2#0 BEN
2#0 BES
2#0 BEU
2#0 BEX
2#0 BEY
2#0 BEZ
2#0 BFB
2#0 BFH
2#0 BFI
2#0 BFK
2#0 BFL
2#0 BFP
2#0 BFR
2#0 BFU
2#0 BFX
2#0 BFY
2#0 BGA
2#0 BGB
2#0 BGF
2#0 BGG
2#0 BGK
2#0 BGR
2#0 BHE
2#0 BHI
2#0 BHK
2#0 BHL
2#0 BHM
2#0 BHO
2#0 BHR
2#0 BHV
2#0 BHX
2#0 BIA
2#0 BIF
2#0 BIJ
2#0 BIV
2#0 BIX
2#0 BIZ
2#0 BJC
2#0 BJG
2#0 BJJ
2#0 BJN
2#0 BJO
2#0 BKC
2#0 BKF
2#0 BKR
2#0 BKS
2#0 BKW
2#0 BLB
2#0 BLE
2#0 BLH
2#0 BLO
2#0 BLP
2#0 BLU
2#0 BLV
2#0 BMA
2#0 BMC
2#0 BML
2#0 BMV
2#0 BMX
2#0 BMZ
2#0 BNL
2#0 BNM
2#0 BNN
2#0 BNO
2#0 BNP
2#0 BNQ
2#0 BNR
2#0 BNS
2#0 BNU
2#0 BOA
2#0 BOE
2#0 BOO
2#0 BOQ
2#0 BOU
2#0 BOW
2#0 BPA
2#0 BPB
2#0 BPC
2#0 BPD
2#0 BPQ
2#0 BPZ
2#0 BQS
2#0 BRA
2#0 BRG
2#0 BRO
2#0 BRV
2#0 BSD
2#0 BSL
2#0 BSV
2#0 BSZ
2#0 BTH
2#0 BTL
2#0 BTY
2#0 BUB
2#0 BUG
2#0 BUL
2#0 BUR
2#0 BUS
2#0 BUW
2#0 BVA
2#0 BVC
2#0 BVM
2#0 BVP
2#0 BVX
2#0 BVY
2#0 BWB
2#0 BWC
2#0 BWE
2#0 BWF
2#0 BWG
2#0 BWM
2#0 BWR
2#0 BWS
2#0 BWW
2#0 BWZ
2#0 BXH
2#0 BXL
2#0 BXP
2#0 BXR
2#0 BYE
2#0 BYP
2#0 BYS
2#0 BYV
2#0 BYX
2#0 BYZ
2#0 BZD
2#0 BZK
2#0 BZR
2#0 BZY
2#0 CAB
2#0 CAC
2#0 CAH
2#0 CAI
2#0 CAM
2#0 CAM
2#0 CAN
2#0 CAY
2#0 CAZ
2#0 CBD
2#0 CBK
2#0 CBM
2#0 CBP
2#0 CBR
2#0 CBS
2#0 CBW
2#0 CBY
2#0 CCM
2#0 CCN
2#0 CCP
2#0 CCS
2#0 CCV
2#0 CCX
2#0 CDB
2#0 CDC
2#0 CDD
2#0 CDG
2#0 CDH
2#0 CDN
2#0 CDP
2#0 CDX
2#0 CEC
2#0 CEL
2#0 CET
2#0 CEW
2#0 CEZ
2#0 CFF
2#0 CFR
2#0 CFS
2#0 CFU
2#0 CGA
2#0 CGC
2#0 CGF
2#0 CGY
2#0 CHB
2#0 CHD
2#0 CHF
2#0 CHG
2#0 CHI
2#0 CHJ
2#0 CHM
2#0 CHO
2#0 CHP
2#0 CHS
2#0 CIC
2#0 CIL
2#0 CIP
2#0 CIW
2#0 CIY
2#0 CIZ
2#0 CJC
2#0 CJH
2#0 CJN
2#0 CJR
2#0 CJT
2#0 CJU
2#0 CJY
2#0 CKA
2#0 CKD
2#0 CKF
2#0 CKG
2#0 CKK
2#0 CKU
2#0 CKZ
2#0 CLA
2#0 CLC
2#0 CLG
2#0 CLK
2#0 CLM
2#0 CLO
2#0 CLQ
2#0 CLR
2#0 CMB
2#0 CMG
2#0 CMI
2#0 CML
2#0 CMM
2#0 CMR
2#0 CMS
2#0 CMU
2#0 CMV
2#0 CMW
2#0 CMX
2#0 CNF
2#0 CNJ
2#0 CNK
2#0 COO
2#0 COP
2#0 COR
2#0 COS
2#0 COU
2#0 CPA
2#0 CPI
2#0 CPJ
2#0 CPM
2#0 CPU
2#0 CPW
2#0 CQA
2#0 CQE
2#0 CQP
2#0 CQV
2#0 CQW
2#0 CQY
2#0 CRE
2#0 CRF
2#0 CRG
2#0 CRJ
2#0 CRK
2#0 CRL
2#0 CRP
2#0 CRT
2#0 CRZ
2#0 CSA
2#0 CSC
2#0 CSI
2#0 CSR
2#0 CSS
2#0 CSW
2#0 CTG
2#0 CTH
2#0 CTP
2#0 CTS
2#0 CTX
2#0 CUF
2#0 CUN
2#0 CUP
2#0 CUQ
2#0 CUR
2#0 CUZ
2#0 CVA
2#0 CVB
2#0 CVI
2#0 CVT
2#0 CWD
2#0 CWG
2#0 CWN
2#0 CWP
2#0 CWT
2#0 CWU
2#0 CWX
2#0 CWY
2#0 CXC
2#0 CXR
2#0 CXS
2#0 CXY
2#0 CXZ
2#0 CYA
2#0 CYB
2#0 CYD
2#0 CYF
2#0 CYH
2#0 CYJ
2#0 CYK
2#0 CYN
2#0 CYQ
2#0 CYV
2#0 CZE
2#0 CZH
2#0 CZO
2#0 CZQ
2#0 CZV
2#0 CZX
2#0 CZY
2#0 DAD
2#0 DAF
2#0 DAG
2#0 DAK
2#0 DAM
2#0 DAS
2#0 DAZ
2#0 DBC
2#0 DBD
2#0 DBE
2#0 DBJ
2#0 DBR
2#0 DBT
2#0 DBU
2#0 DBV
2#0 DBX
2#0 DCB
2#0 DCC
2#0 DCE
2#0 DCG
2#0 DCO
2#0 DCQ
2#0 DCU
2#0 DDE
2#0 DDT
2#0 DDV
2#0 DDY
2#0 DEB
2#0 DED
2#0 DEF
2#0 DEL
2#0 DET
2#0 DFK
2#0 DFR
2#0 DFU
2#0 DFW
2#0 DFY
2#0 DGE
2#0 DGF
2#0 DGN
2#0 DGO
2#0 DGP
2#0 DGX
2#0 DGY
2#0 DGZ
2#0 DHH
2#0 DHL
2#0 DHN
2#0 DHU
2#0 DIA
2#0 DIR
2#0 DIW
2#0 DIY
2#0 DIZ
2#0 DJB
2#0 DJD
2#0 DJE
2#0 DJG
2#0 DJN
2#0 DJO
2#0 DJP
2#0 DKC
2#0 DKH
2#0 DKJ
2#0 DKK
2#0 DKL
2#0 DKN
2#0 DLG
2#0 DLI
2#0 DLM
2#0 DLQ
2#0 DLS
2#0 DLX
2#0 DLY
2#0 DMF
2#0 DMH
2#0 DML
2#0 DMO
2#0 DMT
2#0 DNA
2#0 DNK
2#0 DNN
2#0 DNQ
2#0 DNT
2#0 DNY
2#0 DNZ
2#0 DOA
2#0 DOB
2#0 DOK
2#0 DOM
2#0 DOO
2#0 DOS
2#0 DOT
2#0 DOU
2#0 DPE
2#0 DPG
2#0 DPJ
2#0 DPK
2#0 DPP
2#0 DPQ
2#0 DPU
2#0 DPV
2#0 DQE
2#0 DQF
2#0 DQI
2#0 DQK
2#0 DQM
2#0 DQR
2#0 DQV
2#0 DQW
2#0 DR
2#0 DRH
2#0 DRL
2#0 DRM
2#0 DRP
2#0 DRR
2#0 DRS
2#0 DRU
2#0 DRX
2#0 DSA
2#0 DSD
2#0 DSF
2#0 DSM
2#0 DSR
2#0 DSX
2#0 DSZ
2#0 DTA
2#0 DTI
2#0 DTK
2#0 DTU
2#0 DUC
2#0 DUG
2#0 DUK
2#0 DUT
2#0 DUX
2#0 DUY
2#0 DVA
2#0 DVB
2#0 DVE
2#0 DVJ
2#0 DVL
2#0 DVQ
2#0 DVR
2#0 DVS
2#0 DVT
2#0 DVZ
2#0 DWH
2#0 DWI
2#0 DWN
2#0 DWO
2#0 DWQ
2#0 DWX
2#0 DWY
2#0 DXB
2#0 DXD
2#0 DXE
2#0 DXG
2#0 DXH
2#0 DXI
2#0 DXM
2#0 DXN
2#0 DXP
2#0 DXQ
2#0 DXR
2#0 DXS
2#0 DXT
2#0 DXV
2#0 DXX
2#0 DXY
2#0 DYA
2#0 DYC
2#0 DYE
2#0 DYF
2#0 DYH
2#0 DYO
2#0 DYR
2#0 DYS
2#0 DYT
2#0 DYW
2#0 DZB
2#0 DZD
2#0 DZG
2#0 DZH
2#0 DZI
2#0 DZL
2#0 DZP
2#0 DZU
2#0 DZW
2#0 EAB
2#0 EAE
2#0 EAG
2#0 EAH
2#0 EAJ
2#0 EAK
2#0 EAP
2#0 EAQ
2#0 EBC
2#0 EBE
2#0 EBM
2#0 EBT
2#0 EBY
2#0 ECA
2#0 ECB
2#0 ECE
2#0 ECF
2#0 ECH
2#0 ECL
2#0 ECM
2#0 ECN
2#0 ECR
2#0 ECS
2#0 ECT
2#0 ECX
2#0 EDB
2#0 EDJ
2#0 EDK
2#0 EDO
2#0 EDT
2#0 EDU
2#0 EDW
2#0 EDZ
2#0 EEA
2#0 EEC
2#0 EEE
2#0 EEG
2#0 EEH
2#0 EEL
2#0 EEN
2#0 EEP
2#0 EEY
2#0 EFF
2#0 EFL
2#0 EFM
2#0 EFQ
2#0 EFT
2#0 EGD
2#0 EGF
2#0 EGG
2#0 EGH
2#0 EGJ
2#0 EGU
2#0 EGV
2#0 EGW
2#0 EGX
2#0 EGY
2#0 EGZ
2#0 EHC
2#0 EHE
2#0 EHF
2#0 EHG
2#0 EHI
2#0 EHM
2#0 EHU
2#0 EHY
2#0 EIA
2#0 EIC
2#0 EIF
2#0 EII
2#0 EIJ
2#0 EIK
2#0 EIL
2#0 EIM
2#0 EIN
2#0 EIO
2#0 EIP
2#0 EIQ
2#0 EIR
2#0 EIS
2#0 EIT
2#0 EIV
2#0 EIY
2#0 EJB
2#0 EJE
2#0 EJF
2#0 EJG
2#0 EJH
2#0 EJI
2#0 EJS
2#0 EJX
2#0 EJY
2#0 EKF
2#0 EKG
2#0 EKL
2#0 EKQ
2#0 EKV
2#0 EKY
2#0 EKZ
2#0 ELF
2#0 ELG
2#0 ELH
2#0 ELJ
2#0 ELL
2#0 ELM
2#0 ELN
2#0 ELS
2#0 ELU
2#0 ELW
2#0 ELX
2#0 ELZ
2#0 EMA
2#0 EMC
2#0 EMH
2#0 EMJ
2#0 EMO
2#0 EMR
2#0 EMT
2#0 EMU
2#0 EMW
2#0 ENA
2#0 ENH
2#0 ENK
2#0 ENL
2#0 ENR
2#0 ENS
2#0 ENT
2#0 ENU
2#0 ENX
2#0 ENY
2#0 EOA
2#0 EOB
2#0 EOE
2#0 EOG
2#0 EOJ
2#0 EOO
2#0 EOQ
2#0 EOR
2#0 EOW
2#0 EOY
2#0 EPD
2#0 EPI
2#0 EPL
2#0 EPO
2#0 EPQ
2#0 EPR
2#0 EPS
2#0 EPV
2#0 EPX
2#0 EPY
2#0 EQA
2#0 EQF
2#0 EQM
2#0 EQO
2#0 EQV
2#0 EQX
2#0 EQY
2#0 ERC
2#0 ERH
2#0 ERI
2#0 ERW
2#0 ERY
2#0 ESD
2#0 ESE
2#0 ESF
2#0 ESG
2#0 ESH
2#0 ESK
2#0 ESM
2#0 ESN
2#0 ESQ
2#0 ESR
2#0 EST
2#0 ESV
2#0 ESZ
2#0 ETM
2#0 ETO
2#0 ETR
2#0 ETS
2#0 EUA
2#0 EUC
2#0 EUG
2#0 EUL
2#0 EUT
2#0 EUU
2#0 EUX
2#0 EVD
2#0 EVG
2#0 EVJ
2#0 EVL
2#0 EVN
2#0 EVO
2#0 EVT
2#0 EVY
2#0 EVZ
2#0 EWC
2#0 EWD
2#0 EWG
2#0 EWI
2#0 EWJ
2#0 EWK
2#0 EWN
2#0 EWO
2#0 EWR
2#0 EWU
2#0 EWV
2#0 EWW
2#0 EWZ
2#0 EXF
2#0 EXT
2#0 EYA
2#0 EYD
2#0 EYJ
2#0 EYL
2#0 EYM
2#0 EYN
2#0 EYO
2#0 EYR
2#0 EYT
2#0 EYV
2#0 EYW
2#0 EYZ
2#0 EZB
2#0 EZD
2#0 EZE
2#0 EZH
2#0 EZJ
2#0 EZK
2#0 EZM
2#0 EZN
2#0 EZP
2#0 EZQ
2#0 EZR
2#0 EZT
2#0 EZU
2#0 EZW
2#0 FAD
2#0 FAF
2#0 FAI
2#0 FAL
2#0 FAO
2#0 FAW
2#0 FAX
2#0 FAZ
2#0 FBB
2#0 FBG
2#0 FBK
2#0 FBP
2#0 FBT
2#0 FBV
2#0 FBW
2#0 FBX
2#0 FBZ
2#0 FCA
2#0 FCC
2#0 FCD
2#0 FCI
2#0 FCK
2#0 FCL
2#0 FCN
2#0 FCP
2#0 FCR
2#0 FCU
2#0 FCV
2#0 FCX
2#0 FCY
2#0 FDA
2#0 FDF
2#0 FDK
2#0 FDL
2#0 FDM
2#0 FDN
2#0 FDO
2#0 FDR
2#0 FDV
2#0 FDX
2#0 FDY
2#0 FEA
2#0 FED
2#0 FEF
2#0 FEG
2#0 FEH
2#0 FEI
2#0 FEK
2#0 FEL
2#0 FEN
2#0 FES
2#0 FET
2#0 FEW
2#0 FFA
2#0 FFD
2#0 FFE
2#0 FFG
2#0 FFQ
2#0 FFT
2#0 FFU
2#0 FFX
2#0 FFZ
2#0 FGB
2#0 FGC
2#0 FGD
2#0 FGE
2#0 FGG
2#0 FGJ
2#0 FGK
2#0 FGL
2#0 FGU
2#0 FGX
2#0 FGZ
2#0 FHF
2#0 FHI
2#0 FHL
2#0 FHO
2#0 FHP
2#0 FHS
2#0 FHT
2#0 FHU
2#0 FHW
2#0 FHY
2#0 FHZ
2#0 FID
2#0 FIE
2#0 FIG
2#0 FIH
2#0 FII
2#0 FIN
2#0 FIQ
2#0 FIS
2#0 FIZ
2#0 FJB
2#0 FJF
2#0 FJH
2#0 FJL
2#0 FJO
2#0 FJR
2#0 FKA
2#0 FKI
2#0 FKL
2#0 FKO
2#0 FKP
2#0 FKQ
2#0 FKR
2#0 FKX
2#0 FKZ
2#0 FLB
2#0 FLC
2#0 FLD
2#0 FLE
2#0 FLP
2#0 FLS
2#0 FLT
2#0 FLX
2#0 FLY
2#0 FLZ
2#0 FMC
2#0 FMD
2#0 FMJ
2#0 FMM
2#0 FMU
2#0 FMW
2#0 FMZ
2#0 FNC
2#0 FND
2#0 FNE
2#0 FNF
2#0 FNK
2#0 FNL
2#0 FNP
2#0 FNU
2#0 FNW
2#0 FNX
2#0 FNZ
2#0 FOA
2#0 FOB
2#0 FOJ
2#0 FOM
2#0 FON
2#0 FOQ
2#0 FOS
2#0 FOU
2#0 FOV
2#0 FOW
2#0 FOY
2#0 FPE
2#0 FPG
2#0 FPH
2#0 FPL
2#0 FPN
2#0 FPP
2#0 FPR
2#0 FPS
2#0 FPU
2#0 FPV
2#0 FPZ
2#0 FQB
2#0 FQD
2#0 FQF
2#0 FQH
2#0 FQK
2#0 FQL
2#0 FQN
2#0 FQO
2#0 FQQ
2#0 FQV
2#0 FQX
2#0 FQY
2#0 FQZ
2#0 FRE
2#0 FRH
2#0 FRI
2#0 FRM
2#0 FRN
2#0 FRT
2#0 FRW
2#0 FSJ
2#0 FSL
2#0 FSO
2#0 FSU
2#0 FSW
2#0 FSY
2#0 FSZ
2#0 FTE
2#0 FTG
2#0 FTJ
2#0 FTK
2#0 FTN
2#0 FTR
2#0 FTY
2#0 FUB
2#0 FUI
2#0 FUM
2#0 FUU
2#0 FUV
2#0 FUY
2#0 FVA
2#0 FVG
2#0 FVH
2#0 FVI
2#0 FVN
2#0 FVP
2#0 FVQ
2#0 FVU
2#0 FVW
2#0 FVY
2#0 FWA
2#0 FWE
2#0 FWF
2#0 FWG
2#0 FWH
2#0 FWI
2#0 FWJ
2#0 FWL
2#0 FWM
2#0 FWP
2#0 FWQ
2#0 FWU
2#0 FWV
2#0 FWX
2#0 FXI
2#0 FXN
2#0 FXQ
2#0 FXR
2#0 FXS
2#0 FXT
2#0 FXZ
2#0 FYB
2#0 FYH
2#0 FYI
2#0 FYJ
2#0 FYK
2#0 FYM
2#0 FYV
2#0 FYY
2#0 FZH
2#0 FZI
2#0 FZL
2#0 FZP
2#0 FZQ
2#0 FZR
2#0 FZV
2#0 FZX
2#0 GAB
2#0 GAD
2#0 GAE
2#0 GAG
2#0 GAI
2#0 GAT
2#0 GAZ
2#0 GBC
2#0 GBL
2#0 GBQ
2#0 GBR
2#0 GBX
2#0 GBY
2#0 GBZ
2#0 GCA
2#0 GCK
2#0 GCR
2#0 GCU
2#0 GCZ
2#0 GDC
2#0 GDD
2#0 GDI
2#0 GDJ
2#0 GDK
2#0 GDP
2#0 GDQ
2#0 GDR
2#0 GDS
2#0 GDU
2#0 GDX
2#0 GEA
2#0 GEH
2#0 GEI
2#0 GEO
2#0 GEP
2#0 GEQ
2#0 GEU
2#0 GEW
2#0 GEY
2#0 GFA
2#0 GFD
2#0 GFG
2#0 GFH
2#0 GFI
2#0 GFL
2#0 GFM
2#0 GFN
2#0 GFS
2#0 GGB
2#0 GGC
2#0 GGD
2#0 GGE
2#0 GGF
2#0 GGH
2#0 GGK
2#0 GGL
2#0 GGR
2#0 GGS
2#0 GGX
2#0 GHI
2#0 GHL
2#0 GHM
2#0 GHN
2#0 GHO
2#0 GHQ
2#0 GHU
2#0 GHW
2#0 GIB
2#0 GIC
2#0 GID
2#0 GIG
2#0 GII
2#0 GIJ
2#0 GIK
2#0 GIL
2#0 GIM
2#0 GIN
2#0 GIO
2#0 GIP
2#0 GIQ
2#0 GIR
2#0 GIU
2#0 GJA
2#0 GJC
2#0 GJD
2#0 GJF
2#0 GJH
2#0 GJI
2#0 GJL
2#0 GJM
2#0 GJU
2#0 GJV
2#0 GJW
2#0 GJX
2#0 GJY
2#0 GJZ
2#0 GKC
2#0 GKE
2#0 GKG
2#0 GKH
2#0 GKI
2#0 GKK
2#0 GKO
2#0 GKV
2#0 GKW
2#0 GKY
2#0 GLB
2#0 GLE
2#0 GLF
2#0 GLG
2#0 GLH
2#0 GLK
2#0 GLN
2#0 GLO
2#0 GLP
2#0 GLX
2#0 GLZ
2#0 GME
2#0 GMH
2#0 GMI
2#0 GMJ
2#0 GMK
2#0 GMP
2#0 GMQ
2#0 GMT
2#0 GMV
2#0 GMX
2#0 GMY
2#0 GNA
2#0 GNB
2#0 GND
2#0 GNE
2#0 GNF
2#0 GNK
2#0 GNM
2#0 GNR
2#0 GNT
2#0 GNX
2#0 GNY
2#0 GNZ
2#0 GOA
2#0 GOB
2#0 GOC
2#0 GOF
2#0 GOH
2#0 GOI
2#0 GOJ
2#0 GOT
2#0 GOU
2#0 GOV
2#0 GOX
2#0 GPM
2#0 GPN
2#0 GPO
2#0 GPP
2#0 GPW
2#0 GQB
2#0 GQG
2#0 GQJ
2#0 GQQ
2#0 GQV
2#0 GQX
2#0 GQY
2#0 GRB
2#0 GRC
2#0 GRD
2#0 GRG
2#0 GRJ
2#0 GRO
2#0 GRT
2#0 GRU
2#0 GRV
2#0 GSD
2#0 GSM
2#0 GSQ
2#0 GSU
2#0 GSV
2#0 GSX
2#0 GSY
2#0 GSZ
2#0 GTD
2#0 GTF
2#0 GTG
2#0 GTH
2#0 GTI
2#0 GTK
2#0 GTP
2#0 GTQ
2#0 GTS
2#0 GTV
2#0 GTW
2#0 GTY
2#0 GUB
2#0 GUD
2#0 GUE
2#0 GUK
2#0 GUO
2#0 GUP
2#0 GUQ
2#0 GUW
2#0 GUX
2#0 GVA
2#0 GVB
2#0 GVD
2#0 GVE
2#0 GVF
2#0 GVH
2#0 GVJ
2#0 GVM
2#0 GVQ
2#0 GVS
2#0 GVT
2#0 GVU
2#0 GVW
2#0 GVX
2#0 GVY
2#0 GWG
2#0 GWL
2#0 GWN
2#0 GWQ
2#0 GWW
2#0 GWY
2#0 GXA
2#0 GXJ
2#0 GXL
2#0 GXS
2#0 GXV
2#0 GYA
2#0 GYE
2#0 GYG
2#0 GYI
2#0 GYO
2#0 GYR
2#0 GYU
2#0 GZB
2#0 GZC
2#0 GZE
2#0 GZG
2#0 GZI
2#0 GZK
2#0 GZL
2#0 GZR
2#0 GZY
2#0 GZZ
2#0 HAD
2#0 HAI
2#0 HAO
2#0 HAQ
2#0 HAR
2#0 HAU
2#0 HAX
2#0 HAY
2#0 HBA
2#0 HBC
2#0 HBF
2#0 HBG
2#0 HBH
2#0 HBI
2#0 HBR
2#0 HBU
2#0 HBZ
2#0 HCA
2#0 HCB
2#0 HCD
2#0 HCE
2#0 HCH
2#0 HCI
2#0 HCJ
2#0 HCN
2#0 HCP
2#0 HCR
2#0 HCT
2#0 HCU
2#0 HDJ
2#0 HDK
2#0 HDO
2#0 HDR
2#0 HDT
2#0 HDV
2#0 HED
2#0 HEE
2#0 HEH
2#0 HEJ
2#0 HEK
2#0 HEL
2#0 HEN
2#0 HER
2#0 HEV
2#0 HEW
2#0 HFB
2#0 HFC
2#0 HFF
2#0 HFH
2#0 HFL
2#0 HFM
2#0 HFP
2#0 HFQ
2#0 HFR
2#0 HFU
2#0 HFV
2#0 HFW
2#0 HFX
2#0 HFY
2#0 HFZ
2#0 HGB
2#0 HGC
2#0 HGH
2#0 HGJ
2#0 HGL
2#0 HGS
2#0 HGV
2#0 HGY
2#0 HGZ
2#0 HHD
2#0 HHH
2#0 HHI
2#0 HHL
2#0 HHM
2#0 HHO
2#0 HHQ
2#0 HHR
2#0 HHS
2#0 HHV
2#0 HHY
2#0 HIA
2#0 HID
2#0 HIF
2#0 HIH
2#0 HII
2#0 HIM
2#0 HIN
2#0 HIU
2#0 HIX
2#0 HJA
2#0 HJD
2#0 HJF
2#0 HJH
2#0 HJL
2#0 HJM
2#0 HJO
2#0 HJQ
2#0 HJU
2#0 HJW
2#0 HJX
2#0 HKA
2#0 HKB
2#0 HKD
2#0 HKF
2#0 HKJ
2#0 HKM
2#0 HKO
2#0 HKQ
2#0 HKS
2#0 HKT
2#0 HKV
2#0 HKX
2#0 HLA
2#0 HLB
2#0 HLD
2#0 HLE
2#0 HLI
2#0 HLJ
2#0 HLL
2#0 HLN
2#0 HLQ
2#0 HLU
2#0 HLV
2#0 HLX
2#0 HLY
2#0 HLZ
2#0 HMB
2#0 HME
2#0 HMF
2#0 HMI
2#0 HMK
2#0 HMN
2#0 HMO
2#0 HMT
2#0 HMW
2#0 HMX
2#0 HMY
2#0 HMZ
2#0 HNA
2#0 HNB
2#0 HND
2#0 HNE
2#0 HNI
2#0 HNL
2#0 HNN
2#0 HNO
2#0 HNS
2#0 HNT
2#0 HNU
2#0 HNV
2#0 HNW
2#0 HNY
2#0 HOC
2#0 HOD
2#0 HOE
2#0 HOF
2#0 HOI
2#0 HOJ
2#0 HOP
2#0 HOZ
2#0 HPC
2#0 HPG
2#0 HPH
2#0 HPN
2#0 HPP
2#0 HPQ
2#0 HPS
2#0 HPT
2#0 HPU
2#0 HPV
2#0 HPW
2#0 HPX
2#0 HPY
2#0 HPZ
2#0 HQC
2#0 HQE
2#0 HQG
2#0 HQH
2#0 HQI
2#0 HQK
2#0 HQL
2#0 HQM
2#0 HQN
2#0 HQP
2#0 HQR
2#0 HQS
2#0 HQT
2#0 HQW
2#0 HQX
2#0 HQZ
2#0 HRA
2#0 HRF
2#0 HRK
2#0 HRQ
2#0 HRW
2#0 HRX
2#0 HRZ
2#0 HSA
2#0 HSD
2#0 HSF
2#0 HSH
2#0 HSI
2#0 HSJ
2#0 HSK
2#0 HSM
2#0 HSN
2#0 HSO
2#0 HSQ
2#0 HSR
2#0 HSS
2#0 HSX
2#0 HSZ
2#0 HTA
2#0 HTD
2#0 HTE
2#0 HTH
2#0 HTL
2#0 HTP
2#0 HTT
2#0 HTU
2#0 HTY
2#0 HTZ
2#0 HUA
2#0 HUC
2#0 HUE
2#0 HUF
2#0 HUI
2#0 HUP
2#0 HUQ
2#0 HUS
2#0 HUV
2#0 HUW
2#0 HVA
2#0 HVC
2#0 HVI
2#0 HVJ
2#0 HVK
2#0 HVN
2#0 HVO
2#0 HVP
2#0 HVR
2#0 HVX
2#0 HWA
2#0 HWB
2#0 HWF
2#0 HWG
2#0 HWH
2#0 HWL
2#0 HWT
2#0 HWU
2#0 HWV
2#0 HWX
2#0 HWZ
2#0 HXA
2#0 HXB
2#0 HXD
2#0 HXE
2#0 HXI
2#0 HXK
2#0 HXM
2#0 HXO
2#0 HXR
2#0 HXS
2#0 HXT
2#0 HXV
2#0 HXZ
2#0 HYA
2#0 HYB
2#0 HYL
2#0 HYN
2#0 HYQ
2#0 HYR
2#0 HYT
2#0 HYV
2#0 HYX
2#0 HZD
2#0 HZF
2#0 HZG
2#0 HZJ
2#0 HZM
2#0 HZN
2#0 HZT
2#0 HZV
2#0 HZX
2#0 HZY
2#0 HZZ
2#0 IAA
2#0 IAB
2#0 IAD
2#0 IAE
2#0 IAI

2#0 IAP
2#0 IAR
2#0 IAU
2#0 IAW
2#0 IAY
2#0 IBC
2#0 IBE
2#0 IBF
2#0 IBG
2#0 IBH
2#0 IBJ
2#0 IBK
2#0 IBL
2#0 IBO
2#0 IBQ
2#0 IBS
2#0 IBV
2#0 IBX
2#0 IBZ
2#0 ICA
2#0 ICG
2#0 ICJ
2#0 ICL
2#0 ICM
2#0 ICO
2#0 ICQ
2#0 ICS
2#0 ICV
2#0 ICZ
2#0 IDD
2#0 IDH
2#0 IDI
2#0 IDP
2#0 IDQ
2#0 IDS
2#0 IDU
2#0 IDV
2#0 IDW
2#0 IDZ
2#0 IEF
2#0 IEG
2#0 IEH
2#0 IEJ
2#0 IEN
2#0 IES
2#0 IEU
2#0 IEV
2#0 IEY
2#0 IEZ
2#0 IFD
2#0 IFE
2#0 IFH
2#0 IFI
2#0 IFJ
2#0 IFL
2#0 IFM
2#0 IFQ
2#0 IFR
2#0 IFX
2#0 IFY
2#0 IGA
2#0 IGD
2#0 IGF
2#0 IGG
2#0 IGI
2#0 IGK
2#0 IGS
2#0 IGU
2#0 IGV
2#0 IGZ
2#0 IHA
2#0 IHC
2#0 IHD
2#0 IHF
2#0 IHG
2#0 IHJ
2#0 IHK
2#0 IHP
2#0 IHQ
2#0 IHX
2#0 IHY
2#0 IHZ
2#0 IIA
2#0 IIB
2#0 IIC
2#0 IIE
2#0 IIG
2#0 IIK
2#0 IIL
2#0 IIN
2#0 IIO
2#0 IIS
2#0 IIW
2#0 IIX
2#0 IJA
2#0 IJC
2#0 IJE
2#0 IJF
2#0 IJI
2#0 IJJ
2#0 IJK
2#0 IJM
2#0 IJN
2#0 IJO
2#0 IJS
2#0 IJU
2#0 IKA
2#0 IKD
2#0 IKF
2#0 IKI
2#0 IKJ
2#0 IKK
2#0 IKL
2#0 IKP
2#0 IKR
2#0 IKY
2#0 IKZ
2#0 ILA
2#0 ILB
2#0 ILG
2#0 ILH
2#0 ILI
2#0 ILK
2#0 ILP
2#0 ILR
2#0 ILT
2#0 ILU
2#0 ILV
2#0 ILY
2#0 IMA
2#0 IME
2#0 IMF
2#0 IMI
2#0 IMK
2#0 IML
2#0 IMQ
2#0 IMR
2#0 IMU
2#0 IMY
2#0 INB
2#0 INF
2#0 ING
2#0 INH
2#0 INI
2#0 INL
2#0 INM
2#0 INN
2#0 INO
2#0 INQ
2#0 INR
2#0 INU
2#0 INV
2#0 INW
2#0 IOA
2#0 IOC
2#0 IOF
2#0 IOH
2#0 IOO
2#0 IOP
2#0 IOQ
2#0 IOR
2#0 IOT
2#0 IOV
2#0 IPD
2#0 IPE
2#0 IPH
2#0 IPJ
2#0 IPM
2#0 IPN
2#0 IPP
2#0 IPQ
2#0 IPR
2#0 IPS
2#0 IPU
2#0 IPV
2#0 IPZ
2#0 IQA
2#0 IQB
2#0 IQC
2#0 IQD
2#0 IQG
2#0 IQH
2#0 IQI
2#0 IQJ
2#0 IQS
2#0 IQV
2#0 IQZ
2#0 IRF
2#0 IRI
2#0 IRK
2#0 IRM
2#0 IRP
2#0 IRQ
2#0 IRR
2#0 IRS
2#0 IRU
2#0 IRV
2#0 IRW
2#0 IRY
2#0 ISC
2#0 ISD
2#0 ISG
2#0 ISH
2#0 ISI
2#0 ISJ
2#0 ISL
2#0 ISR
2#0 IST
2#0 ISV
2#0 ISW
2#0 ITA
2#0 ITB
2#0 ITD
2#0 ITE
2#0 ITK
2#0 ITL
2#0 ITP
2#0 ITT
2#0 ITW
2#0 ITZ
2#0 IUB
2#0 IUG
2#0 IUP
2#0 IUQ
2#0 IUS
2#0 IUT
2#0 IUU
2#0 IUV
2#0 IUX
2#0 IUY
2#0 IUZ
2#0 IVA
2#0 IVG
2#0 IVH
2#0 IVM
2#0 IVN
2#0 IVT
2#0 IVV
2#0 IWA
2#0 IWC
2#0 IWE
2#0 IWG
2#0 IWH
2#0 IWK
2#0 IWO
2#0 IWP
2#0 IWV
2#0 IWW
2#0 IWX
2#0 IXB
2#0 IXD
2#0 IXF
2#0 IXG
2#0 IXL
2#0 IXN
2#0 IXO
2#0 IXQ
2#0 IXR
2#0 IXY
2#0 IYB
2#0 IYD
2#0 IYE
2#0 IYF
2#0 IYJ
2#0 IYO
2#0 IYP
2#0 IYQ
2#0 IYR
2#0 IYT
2#0 IYW
2#0 IYX
2#0 IZC
2#0 IZD
2#0 IZE
2#0 IZL
2#0 IZM
2#0 IZO
2#0 IZQ
2#0 IZS
2#0 IZV
2#0 JAD
2#0 JAG
2#0 JAH
2#0 JAS
2#0 JAU
2#0 JAV
2#0 JBB
2#0 JBG
2#0 JBH
2#0 JBN
2#0 JBO
2#0 JBP
2#0 JBR
2#0 JBU
2#0 JBV
2#0 JCJ
2#0 JDG
2#0 JDJ
2#0 JDN
2#0 JDQ
2#0 JDT
2#0 JDU
2#0 JDZ
2#0 JEB
2#0 JEF
2#0 JEG
2#0 JEI
2#0 JEJ
2#0 JEL
2#0 JEO
2#0 JEP
2#0 JER
2#0 JEU
2#0 JEX
2#0 JFA
2#0 JFB
2#0 JFC
2#0 JFG
2#0 JFH
2#0 JFJ
2#0 JFM
2#0 JFQ
2#0 JFS
2#0 JFT
2#0 JFX
2#0 JFZ
2#0 JGA
2#0 JGC
2#0 JGR
2#0 JHB
2#0 JHJ
2#0 JHL
2#0 JHM
2#0 JHR
2#0 JHW
2#0 JHX
2#0 JHY
2#0 JII
2#0 JIK
2#0 JIN
2#0 JJC
2#0 JJD
2#0 JJE
2#0 JJL
2#0 JJM
2#0 JJO
2#0 JJT
2#0 JJX
2#0 JKA
2#0 JKC
2#0 JKE
2#0 JKJ
2#0 JKK
2#0 JKL
2#0 JKM
2#0 JKO
2#0 JKV
2#0 JKZ
2#0 JLA
2#0 JLC
2#0 JLG
2#0 JLH
2#0 JLL
2#0 JLN
2#0 JLO
2#0 JLP
2#0 JLY
2#0 JMA
2#0 JMG
2#0 JMM
2#0 JMN
2#0 JMP
2#0 JMQ
2#0 JMS
2#0 JMT
2#0 JMU
2#0 JMZ
2#0 JNC
2#0 JND
2#0 JNI
2#0 JNK
2#0 JNL
2#0 JNQ
2#0 JNS
2#0 JNT
2#0 JNX
2#0 JNZ
2#0 JOB
2#0 JOJ
2#0 JOO
2#0 JOT
2#0 JOZ
2#0 JPG
2#0 JPI
2#0 JPJ
2#0 JPK
2#0 JPN
2#0 JPV
2#0 JQB
2#0 JQQ
2#0 JQR
2#0 JQY
2#0 JQZ
2#0 JRC
2#0 JRF
2#0 JRG
2#0 JRH
2#0 JRM
2#0 JRO
2#0 JRX
2#0 JRY
2#0 JSA
2#0 JSC
2#0 JSL
2#0 JSP
2#0 JSW
2#0 JTC
2#0 JTD
2#0 JTF
2#0 JTI
2#0 JTL
2#0 JTP
2#0 JTR
2#0 JTU
2#0 JTZ
2#0 JUB
2#0 JUE
2#0 JUG
2#0 JUK
2#0 JUS
2#0 JUZ
2#0 JVB
2#0 JVK
2#0 JVL
2#0 JVW
2#0 JVZ
2#0 JWA
2#0 JWF
2#0 JWI
2#0 JWK
2#0 JWO
2#0 JWR
2#0 JWT
2#0 JWX
2#0 JWZ
2#0 JXE
2#0 JXH
2#0 JXI
2#0 JXJ
2#0 JXL
2#0 JXP
2#0 JXR
2#0 JXS
2#0 JXV
2#0 JXW
2#0 JYK
2#0 JYL
2#0 JYT
2#0 JYX
2#0 JYY
2#0 JZA
2#0 JZB
2#0 JZF
2#0 JZL
2#0 JZN
2#0 JZR
2#0 JZX
2#0 JZZ
2#0 KAA
2#0 KAD
2#0 KAE
2#0 KAH
2#0 KAI
2#0 KAJ
2#0 KAM
2#0 KAO
2#0 KAQ
2#0 KAR
2#0 KAV
2#0 KAW
2#0 KAY
2#0 KAZ
2#0 KBH
2#0 KBI
2#0 KBT
2#0 KBV
2#0 KBW
2#0 KBY
2#0 KCA
2#0 KCC
2#0 KCE
2#0 KCI
2#0 KCK
2#0 KCL
2#0 KCM
2#0 KCS
2#0 KDA
2#0 KDD
2#0 KDK
2#0 KDL
2#0 KDM
2#0 KDN
2#0 KDO
2#0 KDP
2#0 KDU
2#0 KDW
2#0 KDX
2#0 KDY
2#0 KEB
2#0 KEE
2#0 KEF
2#0 KEH
2#0 KEL
2#0 KEM
2#0 KER
2#0 KET
2#0 KEU
2#0 KEX
2#0 KFC
2#0 KFF
2#0 KFU
2#0 KGE
2#0 KGG
2#0 KGI
2#0 KGL
2#0 KGO
2#0 KGS
2#0 KGX
2#0 KHB
2#0 KHD
2#0 KHF
2#0 KHH
2#0 KHR
2#0 KHS
2#0 KHT
2#0 KHZ
2#0 KIB
2#0 KIC
2#0 KID
2#0 KIE
2#0 KIF
2#0 KIK
2#0 KIP
2#0 KIR
2#0 KIW
2#0 KIX
2#0 KJB
2#0 KJC
2#0 KJM
2#0 KJN
2#0 KJP
2#0 KJR
2#0 KJS
2#0 KJU
2#0 KJV
2#0 KJW
2#0 KKA
2#0 KKB
2#0 KKD
2#0 KKP
2#0 KKS
2#0 KKU
2#0 KKV
2#0 KKW
2#0 KKX
2#0 KLC
2#0 KLG
2#0 KLO
2#0 KLU
2#0 KLX
2#0 KLZ
2#0 KMC
2#0 KME
2#0 KMH
2#0 KMJ
2#0 KMK
2#0 KMM
2#0 KMO
2#0 KMR
2#0 KMV
2#0 KMW
2#0 KMX
2#0 KNG
2#0 KNJ
2#0 KNO
2#0 KNR
2#0 KNS
2#0 KNY
2#0 KNZ
2#0 KOG
2#0 KOJ
2#0 KOL
2#0 KON
2#0 KOO
2#0 KOT
2#0 KOW
2#0 KOY
2#0 KPG
2#0 KPJ
2#0 KPK
2#0 KPY
2#0 KRA
2#0 KRJ
2#0 KRL
2#0 KRM
2#0 KRO
2#0 KRV
2#0 KRW
2#0 KRZ
2#0 KSB
2#0 KSE
2#0 KSF
2#0 KSS
2#0 KTA
2#0 KTB
2#0 KTC
2#0 KTE
2#0 KTF
2#0 KTI
2#0 KTO
2#0 KTT
2#0 KTY
2#0 KTZ
2#0 KUB
2#0 KUD
2#0 KUI
2#0 KUL
2#0 KUP
2#0 KUW
2#0 KUX
2#0 KVD
2#0 KVG
2#0 KVJ
2#0 KVL
2#0 KVO
2#0 KVP
2#0 KVT
2#0 KVW
2#0 KVX
2#0 KWD
2#0 KWH
2#0 KWI
2#0 KWK
2#0 KWN
2#0 KWR
2#0 KWT
2#0 KWV
2#0 KWX
2#0 KXB
2#0 KXO
2#0 KXQ
2#0 KXS
2#0 KXT
2#0 KXY
2#0 KXZ
2#0 KYA
2#0 KYC
2#0 KYE
2#0 KYF
2#0 KYG
2#0 KYM
2#0 KYN
2#0 KYO
2#0 KYR
2#0 KYS
2#0 KZA
2#0 KZP
2#0 KZZ
2#0 LAA
2#0 LAF
2#0 LAG
2#0 LAJ
2#0 LAM
2#0 LAN
2#0 LAP
2#0 LAT
2#0 LAU
2#0 LAZ
2#0 LBC
2#0 LBD
2#0 LBF
2#0 LBO
2#0 LBT
2#0 LBX
2#0 LCB
2#0 LCF
2#0 LCG
2#0 LCL
2#0 LCO
2#0 LCS
2#0 LCT
2#0 LCX
2#0 LDA
2#0 LDB
2#0 LDK
2#0 LDO
2#0 LDS
2#0 LDT
2#0 LDU
2#0 LDW
2#0 LEC
2#0 LEH
2#0 LEJ
2#0 LEK
2#0 LER
2#0 LES
2#0 LFA
2#0 LFB
2#0 LFD
2#0 LFJ
2#0 LFO
2#0 LFP
2#0 LFV
2#0 LGA
2#0 LGC
2#0 LGD
2#0 LGK
2#0 LGM
2#0 LGN
2#0 LGP
2#0 LGX
2#0 LHB
2#0 LHG
2#0 LHM
2#0 LHP
2#0 LHT
2#0 LHW
2#0 LIA
2#0 LIB
2#0 LID
2#0 LIE
2#0 LIG
2#0 LIK
2#0 LIL
2#0 LIN
2#0 LIP
2#0 LIS
2#0 LIU
2#0 LJA
2#0 LJM
2#0 LJU
2#0 LJW
2#0 LJZ
2#0 LKA
2#0 LKF
2#0 LKI
2#0 LKL
2#0 LKN
2#0 LKP
2#0 LKQ
2#0 LKY
2#0 LLB
2#0 LLH
2#0 LLP
2#0 LLR
2#0 LLS
2#0 LMA
2#0 LMB
2#0 LMC
2#0 LMF
2#0 LMJ
2#0 LMN
2#0 LMX
2#0 LMZ
2#0 LNA
2#0 LND
2#0 LNE
2#0 LNL
2#0 LNS
2#0 LNT
2#0 LNY
2#0 LOB
2#0 LOD
2#0 LOJ
2#0 LOK
2#0 LOM
2#0 LOT
2#0 LOV
2#0 LOY
2#0 LPA
2#0 LPB
2#0 LPF
2#0 LPI
2#0 LPK
2#0 LPL
2#0 LPS
2#0 LPT
2#0 LPU
2#0 LQX
2#0 LRF
2#0 LRH
2#0 LRM
2#0 LRT
2#0 LRW
2#0 LRX
2#0 LRY
2#0 LSD
2#0 LSJ
2#0 LSM
2#0 LSN
2#0 LSP
2#0 LSQ
2#0 LSY
2#0 LTA
2#0 LTE
2#0 LTN
2#0 LTQ
2#0 LTY
2#0 LTZ
2#0 LUA
2#0 LUB
2#0 LUC
2#0 LUI
2#0 LUV
2#0 LUX
2#0 LUZ
2#0 LVA
2#0 LVI
2#0 LVO
2#0 LVX
2#0 LVZ
2#0 LWC
2#0 LWF
2#0 LWG
2#0 LWH
2#0 LWP
2#0 LWS
2#0 LWX
2#0 LXE
2#0 LXH
2#0 LXL
2#0 LXM
2#0 LXN
2#0 LYL
2#0 LYO
2#0 LYT
2#0 LYX
2#0 LZF
2#0 LZH
2#0 LZO
2#0 LZR
2#0 LZS
2#0 LZZ
2#0 MAG
2#0 MAI
2#0 MAK
2#0 MAM
2#0 MAR
2#0 MAT
2#0 MAU
2#0 MAW
2#0 MBC
2#0 MBF
2#0 MBH
2#0 MBL
2#0 MBP
2#0 MBU
2#0 MBY
2#0 MCE
2#0 MCO
2#0 MCU
2#0 MCX
2#0 MCZ
2#0 MDA
2#0 MDD
2#0 MDI
2#0 MDJ
2#0 MDL
2#0 MDM
2#0 MDQ
2#0 MDV
2#0 MDW
2#0 MDY
2#0 MEB
2#0 MEC
2#0 MEE
2#0 MEF
2#0 MEJ
2#0 MEM
2#0 MEN
2#0 MEP
2#0 MER
2#0 MEW
2#0 MFE
2#0 MFH
2#0 MFL
2#0 MFM
2#0 MFO
2#0 MFP
2#0 MFQ
2#0 MFV
2#0 MFW
2#0 MGB
2#0 MGD
2#0 MGF
2#0 MGG
2#0 MGH
2#0 MGK
2#0 MGN
2#0 MGO
2#0 MGS
2#0 MGZ
2#0 MHA
2#0 MHB
2#0 MHF
2#0 MHH
2#0 MHK
2#0 MHO
2#0 MHQ
2#0 MHR
2#0 MHS
2#0 MHW
2#0 MHX
2#0 MHY
2#0 MHZ
2#0 MIA
2#0 MIE
2#0 MII
2#0 MIK
2#0 MIN
2#0 MIP
2#0 MIQ
2#0 MIR
2#0 MIT
2#0 MJB
2#0 MJF
2#0 MJI
2#0 MJN
2#0 MJR
2#0 MJU
2#0 MJV
2#0 MK
2#0 MKD
2#0 MKL
2#0 MKM
2#0 MKN
2#0 MKO
2#0 MKP
2#0 MKQ
2#0 MKR
2#0 MKS
2#0 MKU
2#0 MKY
2#0 MLM
2#0 MLN
2#0 MLO
2#0 MLP
2#0 MLW
2#0 MLZ
2#0 MMA
2#0 MMK
2#0 MMN
2#0 MMR
2#0 MMS
2#0 MMV
2#0 MMW
2#0 MMY
2#0 MMZ
2#0 MNE
2#0 MNI
2#0 MNK
2#0 MNM
2#0 MNN
2#0 MNO
2#0 MNR
2#0 MNT
2#0 MNV
2#0 MOA
2#0 MOC
2#0 MOD
2#0 MOE
2#0 MOG
2#0 MOH
2#0 MOI
2#0 MOP
2#0 MOS
2#0 MOV
2#0 MOX
2#0 MPD
2#0 MPH
2#0 MPI
2#0 MPK
2#0 MPL
2#0 MPM
2#0 MPQ
2#0 MPU
2#0 MPW
2#0 MPZ
2#0 MQK
2#0 MQM
2#0 MQN
2#0 MRB
2#0 MRE
2#0 MRH
2#0 MRI
2#0 MRK
2#0 MRL
2#0 MRN
2#0 MRR
2#0 MRS
2#0 MRU
2#0 MRV
2#0 MRW
2#0 MRX
2#0 MSD
2#0 MSF
2#0 MSJ
2#0 MSK
2#0 MSP
2#0 MST
2#0 MSV
2#0 MSW
2#0 MSX
2#0 MSY
2#0 MTF
2#0 MTG
2#0 MTI
2#0 MTJ
2#0 MTK
2#0 MTP
2#0 MTU
2#0 MTV
2#0 MTW
2#0 MUC
2#0 MUF
2#0 MUG
2#0 MUK
2#0 MUL
2#0 MUO
2#0 MUP
2#0 MUX
2#0 MUY
2#0 MVA
2#0 MVC
2#0 MVF
2#0 MVG
2#0 MVK
2#0 MVR
2#0 MVY
2#0 MVZ
2#0 MWK
2#0 MWM
2#0 MWR
2#0 MWU
2#0 MXD
2#0 MXE
2#0 MXH
2#0 MXI
2#0 MXJ
2#0 MXK
2#0 MXS
2#0 MXV
2#0 MXX
2#0 MXY
2#0 MXZ
2#0 MYC
2#0 MYF
2#0 MYJ
2#0 MYM
2#0 MYN
2#0 MYO
2#0 MYP
2#0 MYR
2#0 MYY
2#0 MZA
2#0 MZF
2#0 MZG
2#0 MZH
2#0 MZI
2#0 MZJ
2#0 MZO
2#0 MZT
2#0 MZW
2#0 MZZ
2#0 NAC
2#0 NAH
2#0 NAK
2#0 NAL
2#0 NAO
2#0 NAV
2#0 NAW
2#0 NBA
2#0 NBL
2#0 NBP
2#0 NBS
2#0 NBT
2#0 NBW
2#0 NBY
2#0 NCH
2#0 NCL
2#0 NCP
2#0 NCW
2#0 NCX
2#0 NDB
2#0 NDC
2#0 NDD
2#0 NDM
2#0 NDS
2#0 NDX
2#0 NEA
2#0 NEB
2#0 NEE
2#0 NEG
2#0 NEK
2#0 NEP
2#0 NES
2#0 NET
2#0 NEW
2#0 NEX
2#0 NFA
2#0 NFG
2#0 NFJ
2#0 NFL
2#0 NFO
2#0 NFR
2#0 NFT
2#0 NFU
2#0 NFY
2#0 NGA
2#0 NGE
2#0 NGQ
2#0 NGS
2#0 NGU
2#0 NGV
2#0 NGX
2#0 NGY
2#0 NHD
2#0 NHT
2#0 NHX
2#0 NIC
2#0 NID
2#0 NIH
2#0 NII
2#0 NIK
2#0 NIM
2#0 NIN
2#0 NIQ
2#0 NIS
2#0 NIU
2#0 NIV
2#0 NIY
2#0 NJA
2#0 NJD
2#0 NJH
2#0 NJP
2#0 NJR
2#0 NKB
2#0 NKD
2#0 NKJ
2#0 NKL
2#0 NKV
2#0 NKX
2#0 NKY
2#0 NKZ
2#0 NLC
2#0 NLD
2#0 NLF
2#0 NLG
2#0 NLH
2#0 NLJ
2#0 NME
2#0 NMG
2#0 NMH
2#0 NMI
2#0 NMJ
2#0 NML
2#0 NMM
2#0 NMO
2#0 NMP
2#0 NMR
2#0 NMT
2#0 NNA
2#0 NNC
2#0 NNN
2#0 NNS
2#0 NOG
2#0 NOO
2#0 NPA
2#0 NPB
2#0 NPC
2#0 NPD
2#0 NPH
2#0 NPI
2#0 NPJ
2#0 NPK
2#0 NPQ
2#0 NPV
2#0 NQP
2#0 NQQ
2#0 NRC
2#0 NRD
2#0 NRM
2#0 NRO
2#0 NRP
2#0 NRY
2#0 NRZ
2#0 NSD
2#0 NSE
2#0 NSK
2#0 NSM
2#0 NSO
2#0 NST
2#0 NSU
2#0 NSV
2#0 NSX
2#0 NSZ
2#0 NTA
2#0 NTB
2#0 NTE
2#0 NTO
2#0 NTU
2#0 NTX
2#0 NUE
2#0 NUF
2#0 NUK
2#0 NUM
2#0 NUN
2#0 NUR
2#0 NVC
2#0 NVD
2#0 NVG
2#0 NVH
2#0 NVL
2#0 NVM
2#0 NVR
2#0 NVS
2#0 NVY
2#0 NWG
2#0 NWI
2#0 NWM
2#0 NXJ
2#0 NXK
2#0 NXN
2#0 NXR
2#0 NXS
2#0 NXT
2#0 NYH
2#0 NYQ
2#0 NYR
2#0 NYY
2#0 NZB
2#0 NZL
2#0 NZS
2#0 NZY
2#0 NZZ
2#0 OAD
2#0 OAF
2#0 OAL
2#0 OAN
2#0 OAT
2#0 OAV
2#0 OAW
2#0 OBA
2#0 OBJ
2#0 OBR
2#0 OBW
2#0 OBX
2#0 OBY
2#0 OCA
2#0 OCD
2#0 OCE
2#0 OCK
2#0 OCO
2#0 OCR
2#0 OCX
2#0 OCZ
2#0 ODC
2#0 ODD
2#0 ODE
2#0 ODG
2#0 ODH
2#0 ODY
2#0 OED
2#0 OEI
2#0 OEO
2#0 OER
2#0 OEW
2#0 OFC
2#0 OFG
2#0 OFK
2#0 OFN
2#0 OFP
2#0 OFR
2#0 OFS
2#0 OFV
2#0 OFX
2#0 OFZ
2#0 OGC
2#0 OGI
2#0 OGL
2#0 OGM
2#0 OGP
2#0 OGS
2#0 OGW
2#0 OHG
2#0 OHH
2#0 OHI
2#0 OHM
2#0 OHR
2#0 OHV
2#0 OIC
2#0 OID
2#0 OII
2#0 OJA
2#0 OJP
2#0 OJR
2#0 OJW
2#0 OJX
2#0 OKE
2#0 OKF
2#0 OKI
2#0 OKM
2#0 OKW
2#0 OLA
2#0 OLB
2#0 OLD
2#0 OLH
2#0 OLI
2#0 OLL
2#0 OLN
2#0 OLO
2#0 OLS
2#0 OLV
2#0 OLW
2#0 OLX
2#0 OMB
2#0 OMC
2#0 OMK
2#0 OML
2#0 OMM
2#0 OMO
2#0 OMP
2#0 OMY
2#0 OMZ
2#0 OND
2#0 ONF
2#0 ONK
2#0 ONN
2#0 ONO
2#0 ONR
2#0 ONX
2#0 ONZ
2#0 OOA
2#0 OOB
2#0 OOD
2#0 OOE
2#0 OOF
2#0 OOK
2#0 OOL
2#0 OOP
2#0 OOQ
2#0 OOS
2#0 OOZ
2#0 OPA
2#0 OPD
2#0 OPG
2#0 OPP
2#0 OPQ
2#0 OPR
2#0 OPV
2#0 OPW
2#0 OPZ
2#0 OQU
2#0 ORA
2#0 ORB
2#0 ORG
2#0 ORN
2#0 ORO
2#0 ORU
2#0 ORZ
2#0 OSB
2#0 OSL
2#0 OSO
2#0 OSR
2#0 OSW
2#0 OSZ
2#0 OTD
2#0 OTH
2#0 OTJ
2#0 OTK
2#0 OTL
2#0 OTO
2#0 OTQ
2#0 OTS
2#0 OTV
2#0 OUC
2#0 OUE
2#0 OUN
2#0 OUT
2#0 OUZ
2#0 OVA
2#0 OVE
2#0 OVG
2#0 OVK
2#0 OVO
2#0 OWI
2#0 OWN
2#0 OWW
2#0 OWX
2#0 OWZ
2#0 OXE
2#0 OXH
2#0 OXI
2#0 OXJ
2#0 OXM
2#0 OXN
2#0 OXS
2#0 OXW
2#0 OYF
2#0 OYI
2#0 OYO
2#0 OZC
2#0 OZP
2#0 OZT
2#0 OZV
2#0 OZZ
2#0 PAI
2#0 PAM
2#0 PAQ
2#0 PAR
2#0 PAW
2#0 PAZ
2#0 PBA
2#0 PBD
2#0 PBG
2#0 PBI
2#0 PBK
2#0 PBQ
2#0 PCA
2#0 PCH
2#0 PCJ
2#0 PCK
2#0 PCM
2#0 PCS
2#0 PCY
2#0 PDA
2#0 PDC
2#0 PDD
2#0 PDE
2#0 PDF
2#0 PDI
2#0 PDK
2#0 PDN
2#0 PDS
2#0 PDW
2#0 PDY
2#0 PEA
2#0 PEB
2#0 PEK
2#0 PEO
2#0 PER
2#0 PES
2#0 PEU
2#0 PFC
2#0 PFI
2#0 PFM
2#0 PFX
2#0 PGA
2#0 PGD
2#0 PGE
2#0 PGF
2#0 PGN
2#0 PGW
2#0 PGX
2#0 PGY
2#0 PGZ
2#0 PHD
2#0 PHH
2#0 PHI
2#0 PHJ
2#0 PHN
2#0 PHR
2#0 PHT
2#0 PHW
2#0 PHY
2#0 PIC
2#0 PIE
2#0 PIF
2#0 PII
2#0 PIR
2#0 PIV
2#0 PIZ
2#0 PJA
2#0 PJB
2#0 PJG
2#0 PJP
2#0 PJW
2#0 PJX
2#0 PJZ
2#0 PKC
2#0 PKD
2#0 PKE
2#0 PKG
2#0 PKO
2#0 PKP
2#0 PKT
2#0 PKW
2#0 PKX
2#0 PKY
2#0 PKZ
2#0 PLD
2#0 PLF
2#0 PLG
2#0 PLI
2#0 PLL
2#0 PLM
2#0 PLO
2#0 PLP
2#0 PLQ
2#0 PLW
2#0 PLZ
2#0 PMF
2#0 PMH
2#0 PMJ
2#0 PMK
2#0 PMQ
2#0 PMS
2#0 PMT
2#0 PMY
2#0 PNE
2#0 PNF
2#0 PNO
2#0 PNS
2#0 PNT
2#0 PNX
2#0 PNY
2#0 POA
2#0 POC
2#0 POF
2#0 POG
2#0 POI
2#0 POJ
2#0 POM
2#0 POP
2#0 POS
2#0 POV
2#0 POW
2#0 POY
2#0 PPC
2#0 PPE
2#0 PPG
2#0 PPN
2#0 PPP
2#0 PPQ
2#0 PPS
2#0 PPT
2#0 PPU
2#0 PPV
2#0 PPX
2#0 PQP
2#0 PQV
2#0 PRB
2#0 PRF
2#0 PRH
2#0 PRI
2#0 PRJ
2#0 PRK
2#0 PRN
2#0 PRP
2#0 PRT
2#0 PRU
2#0 PRW
2#0 PRX
2#0 PRY
2#0 PS
2#0 PSB
2#0 PSI
2#0 PSJ
2#0 PSL
2#0 PST
2#0 PSU
2#0 PSV
2#0 PSX
2#0 PSY
2#0 PTD
2#0 PTH
2#0 PTK
2#0 PTL
2#0 PTP
2#0 PTQ
2#0 PTT
2#0 PTU
2#0 PTV
2#0 PTW
2#0 PTX
2#0 PUC
2#0 PUD
2#0 PUI
2#0 PUK
2#0 PUP
2#0 PUR
2#0 PUX
2#0 PVC
2#0 PVG
2#0 PVO
2#0 PVR
2#0 PVT
2#0 PVX
2#0 PWA
2#0 PWE
2#0 PWH
2#0 PWN
2#0 PWS
2#0 PWW
2#0 PWX
2#0 PWZ
2#0 PXB
2#0 PXC
2#0 PXG
2#0 PXK
2#0 PXM
2#0 PXQ
2#0 PXT
2#0 PXX
2#0 PYB
2#0 PYE
2#0 PYG
2#0 PYO
2#0 PYT
2#0 PYU
2#0 PYW
2#0 PYX
2#0 PZA
2#0 PZD
2#0 PZL
2#0 PZO
2#0 PZR
2#0 PZS
2#0 PZW
2#0 PZZ
2#0 RAC
2#0 RAE
2#0 RAN
2#0 RAO
2#0 RAR
2#0 RAT
2#0 RAV
2#0 RAX
2#0 RAY
2#0 RAZ
2#0 RBB
2#0 RBD
2#0 RBE
2#0 RBF
2#0 RBJ
2#0 RBL
2#0 RBM
2#0 RBS
2#0 RCH
2#0 RCK
2#0 RCQ
2#0 RCS
2#0 RCY
2#0 RDB
2#0 RDM
2#0 RDY
2#0 RED
2#0 REE
2#0 REF
2#0 REI
2#0 REK
2#0 RET
2#0 REV
2#0 RFA
2#0 RFB
2#0 RFJ
2#0 RFQ
2#0 RFR
2#0 RFX
2#0 RFZ
2#0 RGE
2#0 RGF
2#0 RGG
2#0 RGI
2#0 RGP
2#0 RGQ
2#0 RGV
2#0 RGX
2#0 RHD
2#0 RHG
2#0 RHK
2#0 RHO
2#0 RHR
2#0 RHU
2#0 RHV
2#0 RHX
2#0 RHY
2#0 RHZ
2#0 RIB
2#0 RIC
2#0 RIF
2#0 RIG
2#0 RIJ
2#0 RIP
2#0 RIV
2#0 RIX
2#0 RJC
2#0 RJF
2#0 RJI
2#0 RJK
2#0 RJN
2#0 RJP
2#0 RJS
2#0 RJT
2#0 RKA
2#0 RKC
2#0 RKE
2#0 RKI
2#0 RKJ
2#0 RKL
2#0 RKN
2#0 RKT
2#0 RKW
2#0 RKY
2#0 RLD
2#0 RLF
2#0 RLI
2#0 RLL
2#0 RLM
2#0 RLZ
2#0 RMA
2#0 RME
2#0 RMG
2#0 RMU
2#0 RMX
2#0 RNC
2#0 RNG
2#0 RNL
2#0 RNN
2#0 RNY
2#0 ROA
2#0 ROB
2#0 ROF
2#0 ROH
2#0 ROL
2#0 ROT
2#0 ROW
2#0 ROZ
2#0 RPG
2#0 RPI
2#0 RPJ
2#0 RPK
2#0 RPS
2#0 RPT
2#0 RPX
2#0 RPZ
2#0 RQB
2#0 RQF
2#0 RQO
2#0 RQT
2#0 RRG
2#0 RRJ
2#0 RRM
2#0 RRN
2#0 RRR
2#0 RRS
2#0 RRU
2#0 RRV
2#0 RRW
2#0 RRZ
2#0 RSF
2#0 RSJ
2#0 RSK
2#0 RSL
2#0 RSO
2#0 RSP
2#0 RSQ
2#0 RSR
2#0 RSS
2#0 RST
2#0 RSU
2#0 RSW
2#0 RSY
2#0 RTB
2#0 RTR
2#0 RTS
2#0 RTT
2#0 RTV
2#0 RTX
2#0 RUB
2#0 RUE
2#0 RUF
2#0 RUK
2#0 RUL
2#0 RUM
2#0 RUN
2#0 RVC
2#0 RVE
2#0 RVL
2#0 RVN
2#0 RVO
2#0 RVP
2#0 RVR
2#0 RVS
2#0 RVX
2#0 RVY
2#0 RWD
2#0 RWV
2#0 RXB
2#0 RXD
2#0 RXF
2#0 RXI
2#0 RXJ
2#0 RXM
2#0 RXN
2#0 RXO
2#0 RXP
2#0 RXQ
2#0 RXS
2#0 RYA
2#0 RYC
2#0 RYE
2#0 RYK
2#0 RYM
2#0 RYO
2#0 RYT
2#0 RYU
2#0 RYW
2#0 RYZ
2#0 RZA
2#0 RZB
2#0 RZG
2#0 RZR
2#0 RZV
2#0 RZW
2#0 RZZ
2#0 SAB
2#0 SAC
2#0 SAE
2#0 SAJ
2#0 SAL
2#0 SAM
2#0 SAO
2#0 SAQ
2#0 SAV
2#0 SBA
2#0 SBE
2#0 SBG
2#0 SBI
2#0 SBQ

2#0 SBT
2#0 SBU
2#0 SBV
2#0 SBY
2#0 SCI
2#0 SCT
2#0 SCY
2#0 SDB
2#0 SDF
2#0 SDH
2#0 SDI
2#0 SDL
2#0 SDN
2#0 SDX
2#0 SEI
2#0 SEM
2#0 SEN
2#0 SEQ
2#0 SEV
2#0 SFI
2#0 SFK
2#0 SFL
2#0 SFM
2#0 SFO
2#0 SFR
2#0 SFU
2#0 SGA
2#0 SGC
2#0 SGE
2#0 SGF
2#0 SGJ
2#0 SGN
2#0 SGP
2#0 SGR
2#0 SGT
2#0 SGU
2#0 SGX
2#0 SGZ
2#0 SHD
2#0 SHE
2#0 SHL
2#0 SHQ
2#0 SHS
2#0 SHT
2#0 SIE
2#0 SIG
2#0 SII
2#0 SIN
2#0 SIP
2#0 SIR
2#0 SJA
2#0 SJC
2#0 SJL
2#0 SJW
2#0 SJY
2#0 SKB
2#0 SKC
2#0 SKD
2#0 SKF
2#0 SKJ
2#0 SKO
2#0 SKS
2#0 SKT
2#0 SKW
2#0 SKX
2#0 SLB
2#0 SLC
2#0 SLE
2#0 SLF
2#0 SLQ
2#0 SLX
2#0 SMC
2#0 SME
2#0 SMG
2#0 SMH
2#0 SMP
2#0 SMQ
2#0 SMR
2#0 SMU
2#0 SMW
2#0 SNH
2#0 SNI
2#0 SNN
2#0 SNO
2#0 SNQ
2#0 SNR
2#0 SNT
2#0 SNX
2#0 SNY
2#0 SOA
2#0 SOC
2#0 SOG
2#0 SOH
2#0 SOL
2#0 SOM
2#0 SON
2#0 SOO
2#0 SOU
2#0 SOY
2#0 SPC
2#0 SPI
2#0 SPK
2#0 SPO
2#0 SPQ
2#0 SPV
2#0 SQA
2#0 SQJ
2#0 SQM
2#0 SQP
2#0 SQR
2#0 SQW
2#0 SQZ
2#0 SRE
2#0 SRK
2#0 SRL
2#0 SRM
2#0 SRN
2#0 SRS
2#0 SRU
2#0 SRW
2#0 SSH
2#0 SSR
2#0 SSS
2#0 SSV
2#0 SSW
2#0 STC
2#0 STD
2#0 STE
2#0 STF
2#0 STG
2#0 STJ
2#0 STM
2#0 STO
2#0 STR
2#0 STU
2#0 STW
2#0 STY
2#0 SUG
2#0 SUJ
2#0 SUL
2#0 SUM
2#0 SUN
2#0 SUP
2#0 SUQ
2#0 SUR
2#0 SUV
2#0 SUY
2#0 SVD
2#0 SVE
2#0 SVJ
2#0 SVL
2#0 SVM
2#0 SVO
2#0 SVR
2#0 SVS
2#0 SVX
2#0 SVY
2#0 SWC
2#0 SWD
2#0 SWI
2#0 SWL
2#0 SWW
2#0 SWX
2#0 SWY
2#0 SXA
2#0 SXB
2#0 SXD
2#0 SXG
2#0 SXH
2#0 SXL
2#0 SXN
2#0 SXO
2#0 SXR
2#0 SXX
2#0 SYR
2#0 SYZ
2#0 SZA
2#0 SZD
2#0 SZE
2#0 SZN
2#0 SZO
2#0 SZP
2#0 SZS
2#0 SZW
2#0 SZX
2#0 SZY
2#0 TAP
2#0 TBB
2#0 TBC
2#0 TBG
2#0 TBK
2#0 TBM
2#0 TBN
2#0 TBO
2#0 TBP
2#0 TBT
2#0 TBU
2#0 TBY
2#0 TCE
2#0 TCH
2#0 TCS
2#0 TDA
2#0 TDB
2#0 TDG
2#0 TDK
2#0 TDM
2#0 TDQ
2#0 TDS
2#0 TDW
2#0 TDX
2#0 TDY
2#0 TDZ
2#0 TEA
2#0 TEB
2#0 TEC
2#0 TEL
2#0 TEM
2#0 TEO
2#0 TEQ
2#0 TER
2#0 TES
2#0 TET
2#0 TEV
2#0 TEX
2#0 TFC
2#0 TFG
2#0 TFJ
2#0 TFL
2#0 TFP
2#0 TFR
2#0 TFS
2#0 TFU
2#0 TFZ
2#0 TGE
2#0 TGH
2#0 TGP
2#0 TGR
2#0 TGS
2#0 TGW
2#0 TGX
2#0 TGY
2#0 TGZ
2#0 THA
2#0 THB
2#0 THC
2#0 THD
2#0 THH
2#0 THI
2#0 THK
2#0 THL
2#0 THM
2#0 THO
2#0 THV
2#0 THW
2#0 THX
2#0 THY
2#0 TIB
2#0 TIC
2#0 TIE
2#0 TIG
2#0 TII
2#0 TIM
2#0 TIP
2#0 TIS
2#0 TIX
2#0 TJA
2#0 TJB
2#0 TJH
2#0 TJL
2#0 TKA
2#0 TKB
2#0 TKC
2#0 TKH
2#0 TKI
2#0 TKK
2#0 TKL
2#0 TKM
2#0 TKN
2#0 TKO
2#0 TKP
2#0 TKR
2#0 TKT
2#0 TKW
2#0 TKZ
2#0 TLK
2#0 TLS
2#0 TLV
2#0 TLZ
2#0 TMG
2#0 TMI
2#0 TMJ
2#0 TMR
2#0 TMT
2#0 TMU
2#0 TMZ
2#0 TNA
2#0 TNF
2#0 TNG
2#0 TNI
2#0 TNJ
2#0 TNK
2#0 TNL
2#0 TNR
2#0 TNS
2#0 TNT
2#0 TNX
2#0 TNY
2#0 TOC
2#0 TOD
2#0 TOE
2#0 TOH
2#0 TOZ
2#0 TPB
2#0 TPJ
2#0 TPK
2#0 TPW
2#0 TPX
2#0 TPZ
2#0 TQM
2#0 TQR
2#0 TQS
2#0 TQW
2#0 TQX
2#0 TRB
2#0 TRC
2#0 TRF
2#0 TRG
2#0 TRN
2#0 TRQ
2#0 TRT
2#0 TRV
2#0 TSB
2#0 TSD
2#0 TSF
2#0 TSG
2#0 TSH
2#0 TSL
2#0 TSN
2#0 TSS
2#0 TST
2#0 TSW
2#0 TSZ
2#0 TTI
2#0 TTK
2#0 TTL
2#0 TTO
2#0 TTQ
2#0 TTR
2#0 TTV
2#0 TTX
2#0 TTZ
2#0 TUP
2#0 TUZ
2#0 TVA
2#0 TVC
2#0 TVE
2#0 TVI
2#0 TVK
2#0 TVO
2#0 TVP
2#0 TVV
2#0 TWA
2#0 TWB
2#0 TWC
2#0 TWE
2#0 TWF
2#0 TWG
2#0 TWH
2#0 TWM
2#0 TWN
2#0 TWR
2#0 TXF
2#0 TXN
2#0 TXO
2#0 TXS
2#0 TXT
2#0 TXX
2#0 TXZ
2#0 TYB
2#0 TYD
2#0 TYF
2#0 TYK
2#0 TYM
2#0 TYN
2#0 TYO
2#0 TYP
2#0 TYU
2#0 TYV
2#0 TZA
2#0 TZC
2#0 TZH
2#0 TZJ
2#0 TZN
2#0 TZP
2#0 TZT
2#0 TZX
2#0 UAF
2#0 UAG
2#0 UAI
2#0 UAJ
2#0 UBA
2#0 UBD
2#0 UBS
2#0 UBZ
2#0 UCK
2#0 UCW
2#0 UDG
2#0 UDK
2#0 UDL
2#0 UDP
2#0 UDV
2#0 UDX
2#0 UEM
2#0 UEX
2#0 UFJ
2#0 UFX
2#0 UGD
2#0 UGS
2#0 UGX
2#0 UHD
2#0 UHH
2#0 UHU
2#0 UHV
2#0 UHZ
2#0 UIO
2#0 UIR
2#0 UKC
2#0 UKE
2#0 UKF
2#0 UKH
2#0 UKI
2#0 UKL
2#0 UKO
2#0 UKP
2#0 UKQ
2#0 UKR
2#0 UKS
2#0 UKV
2#0 UKZ
2#0 ULA
2#0 ULF
2#0 ULL
2#0 UMB
2#0 UMC
2#0 UMF
2#0 UMI
2#0 UMM
2#0 UMS
2#0 UMT
2#0 UND
2#0 UNE
2#0 UNK
2#0 UNM
2#0 UNO
2#0 UNX
2#0 UOD
2#0 UOG
2#0 UOH
2#0 UOO
2#0 UOS
2#0 UPG
2#0 UPN
2#0 UPP
2#0 URL
2#0 URM
2#0 URN
2#0 URS
2#0 USE
2#0 USP
2#0 USQ
2#0 USR
2#0 UST
2#0 USZ
2#0 UTM
2#0 UTO
2#0 UTS
2#0 UTU
2#0 UUC
2#0 UUF
2#0 UUK
2#0 UUN
2#0 UUO
2#0 UUP
2#0 UUU
2#0 UVA
2#0 UVV
2#0 UWB
2#0 UWU
2#0 UXA
2#0 UXL
2#0 UXP
2#0 UYI
2#0 UYP
2#0 UYU
2#0 UZA
2#0 UZE
2#0 VAB
2#0 VAD
2#0 VAE
2#0 VAK
2#0 VAL
2#0 VAP
2#0 VAR
2#0 VAX
2#0 VAZ
2#0 VBC
2#0 VBD
2#0 VBE
2#0 VBG
2#0 VBO
2#0 VBQ
2#0 VBV
2#0 VCF
2#0 VCH
2#0 VCJ
2#0 VCN
2#0 VCO
2#0 VCT
2#0 VCX
2#0 VDF
2#0 VDK
2#0 VDR
2#0 VDS
2#0 VDU
2#0 VDX
2#0 VEB
2#0 VEC
2#0 VED
2#0 VEE
2#0 VEI
2#0 VEL
2#0 VEM
2#0 VEN
2#0 VEP
2#0 VER
2#0 VEV
2#0 VFM
2#0 VFS
2#0 VFZ
2#0 VGD
2#0 VGG
2#0 VGH
2#0 VGI
2#0 VGN
2#0 VGO
2#0 VGP
2#0 VGW
2#0 VGX
2#0 VHB
2#0 VIC
2#0 VIE
2#0 VII
2#0 VIK
2#0 VIM
2#0 VIP
2#0 VIR
2#0 VIX
2#0 VIY
2#0 VJC
2#0 VJM
2#0 VJT
2#0 VKC
2#0 VKD
2#0 VKF
2#0 VKH
2#0 VKJ
2#0 VKP
2#0 VKT
2#0 VKV
2#0 VKX
2#0 VLC
2#0 VLK
2#0 VLS
2#0 VLU
2#0 VLV
2#0 VMB
2#0 VME
2#0 VMF
2#0 VMI
2#0 VMK
2#0 VML
2#0 VMM
2#0 VMO
2#0 VMR
2#0 VMS
2#0 VMX
2#0 VMZ
2#0 VNA
2#0 VNB
2#0 VNC
2#0 VNG
2#0 VNU
2#0 VOE
2#0 VOK
2#0 VOL
2#0 VOM
2#0 VON
2#0 VOO
2#0 VOS
2#0 VOV
2#0 VOW
2#0 VOX
2#0 VPC
2#0 VPF
2#0 VPR
2#0 VPS
2#0 VPV
2#0 VPW
2#0 VPX
2#0 VPY
2#0 VPZ
2#0 VQD
2#0 VRF
2#0 VRG
2#0 VRH
2#0 VRJ
2#0 VRM
2#0 VRN
2#0 VRS
2#0 VRV
2#0 VSA
2#0 VSO
2#0 VSR
2#0 VSS
2#0 VST
2#0 VSV
2#0 VSX
2#0 VSY
2#0 VTD
2#0 VTE
2#0 VTI
2#0 VTL
2#0 VTM
2#0 VTN
2#0 VTX
2#0 VTY
2#0 VUE
2#0 VUF
2#0 VUX
2#0 VVD
2#0 VVG
2#0 VVL
2#0 VVN
2#0 VVQ
2#0 VVR
2#0 VVV
2#0 VVZ
2#0 VWA
2#0 VWB
2#0 VWC
2#0 VWI
2#0 VWP
2#0 VWR
2#0 VWS
2#0 VWV
2#0 VXC
2#0 VXN
2#0 VXP
2#0 VXV
2#0 VXY
2#0 VYD
2#0 VYG
2#0 VYL
2#0 VYM
2#0 VYV
2#0 VYZ
2#0 VZC
2#0 VZJ
2#0 VZK
2#0 VZM
2#0 VZT
2#0 VZV
2#0 VZW
2#0 VZZ
2#0 WAB
2#0 WAC
2#0 WAM
2#0 WAN
2#0 WAR
2#0 WAW
2#0 WAZ
2#0 WBA
2#0 WBC
2#0 WBK
2#0 WBM
2#0 WBN
2#0 WBP
2#0 WBY
2#0 WCA
2#0 WCD
2#0 WCF
2#0 WCH
2#0 WCK
2#0 WCS
2#0 WCT
2#0 WCW
2#0 WDC
2#0 WDF
2#0 WDJ
2#0 WDK
2#0 WDT
2#0 WDX
2#0 WDZ
2#0 WEA
2#0 WEH
2#0 WEM
2#0 WEN
2#0 WEP
2#0 WER
2#0 WEW
2#0 WEY
2#0 WFA
2#0 WFL
2#0 WFM
2#0 WFT
2#0 WFU
2#0 WFX
2#0 WFZ
2#0 WGA
2#0 WGH
2#0 WGK
2#0 WGR
2#0 WGW
2#0 WGX
2#0 WGY
2#0 WGZ
2#0 WHC
2#0 WHE
2#0 WHG
2#0 WHI
2#0 WHJ
2#0 WHL
2#0 WHM
2#0 WHN
2#0 WHQ
2#0 WHR
2#0 WHY
2#0 WIA
2#0 WIB
2#0 WII
2#0 WIK
2#0 WIM
2#0 WIN
2#0 WIT
2#0 WIW
2#0 WIY
2#0 WJG
2#0 WJK
2#0 WKA
2#0 WKC
2#0 WKD
2#0 WKF
2#0 WKH
2#0 WKI
2#0 WKO
2#0 WKQ
2#0 WKS
2#0 WLB
2#0 WLC
2#0 WLF
2#0 WLG
2#0 WLJ
2#0 WLL
2#0 WLR
2#0 WLW
2#0 WLZ
2#0 WMA
2#0 WMF
2#0 WMI
2#0 WMO
2#0 WMR
2#0 WMS
2#0 WMV
2#0 WMX
2#0 WNA
2#0 WNJ
2#0 WOE
2#0 WOJ
2#0 WOK
2#0 WOM
2#0 WON
2#0 WOO
2#0 WOT
2#0 WPA
2#0 WPG
2#0 WPL
2#0 WPM
2#0 WPO
2#0 WPR
2#0 WPW
2#0 WQE
2#0 WQZ
2#0 WRA
2#0 WRB
2#0 WRC
2#0 WRD
2#0 WRE
2#0 WRG
2#0 WRJ
2#0 WRN
2#0 WRP
2#0 WRW
2#0 WRZ
2#0 WSA
2#0 WSB
2#0 WSD
2#0 WSE
2#0 WSF
2#0 WSL
2#0 WTB
2#0 WTC
2#0 WTF
2#0 WTL
2#0 WTM
2#0 WTO
2#0 WTV
2#0 WTW
2#0 WTX
2#0 WUE
2#0 WUH
2#0 WUS
2#0 WUT
2#0 WUV
2#0 WVA
2#0 WVB
2#0 WVX
2#0 WVY
2#0 WVZ
2#0 WWW
2#0 WXC
2#0 WXV
2#0 WXX
2#0 WYB
2#0 WYD
2#0 WYI
2#0 WYK
2#0 WYL
2#0 WYP
2#0 WYR
2#0 WYY
2#0 WZC
2#0 WZD
2#0 WZG
2#0 WZL
2#0 WZO
2#0 WZP
2#0 WZV
2#0 WZW
2#0 XAB
2#0 XAC
2#0 XAD
2#0 XAF
2#0 XAJ
2#0 XAK
2#0 XAQ
2#0 XAT
2#0 XBF
2#0 XBH
2#0 XBI
2#0 XBJ
2#0 XBS
2#0 XCE
2#0 XCG
2#0 XCK
2#0 XCQ
2#0 XCW
2#0 XCX
2#0 XDB
2#0 XDE
2#0 XDJ
2#0 XDL
2#0 XDO
2#0 XDP
2#0 XDW
2#0 XEC
2#0 XED
2#0 XEQ
2#0 XEX
2#0 XFC
2#0 XFI
2#0 XFK
2#0 XFL
2#0 XFS
2#0 XFU
2#0 XGH
2#0 XGI
2#0 XGK
2#0 XGL
2#0 XGM
2#0 XGN
2#0 XGP
2#0 XGR
2#0 XHA
2#0 XHD
2#0 XHF
2#0 XHK
2#0 XHP
2#0 XHX
2#0 XID
2#0 XIE
2#0 XIH
2#0 XIM
2#0 XIO
2#0 XIX
2#0 XIZ
2#0 XJA
2#0 XJB
2#0 XJC
2#0 XJO
2#0 XJS
2#0 XKB
2#0 XKE
2#0 XKF
2#0 XKG
2#0 XKI
2#0 XKN
2#0 XKV
2#0 XKY
2#0 XKZ
2#0 XLA
2#0 XLC
2#0 XLD
2#0 XLF
2#0 XLL
2#0 XLV
2#0 XMA
2#0 XMB
2#0 XMD
2#0 XME
2#0 XMJ
2#0 XML
2#0 XMM
2#0 XMR
2#0 XMX
2#0 XMZ
2#0 XNA
2#0 XNB
2#0 XNG
2#0 XNI
2#0 XNP
2#0 XNR
2#0 XNV
2#0 XNX
2#0 XOB
2#0 XOF
2#0 XOI
2#0 XOM
2#0 XOO
2#0 XOS
2#0 XOW
2#0 XOX
2#0 XOY
2#0 XPB
2#0 XPE
2#0 XPL
2#0 XPN
2#0 XPO
2#0 XPR
2#0 XPW
2#0 XPX
2#0 XQB
2#0 XQS
2#0 XQT
2#0 XQU
2#0 XQZ
2#0 XRA
2#0 XRB
2#0 XRC
2#0 XRE
2#0 XRF
2#0 XRI
2#0 XRJ
2#0 XRK
2#0 XRP
2#0 XRT
2#0 XRW
2#0 XRY
2#0 XSE
2#0 XSH
2#0 XSI
2#0 XSM
2#0 XSQ
2#0 XSR
2#0 XSS
2#0 XSX
2#0 XTE
2#0 XTJ
2#0 XTN
2#0 XTO
2#0 XTW
2#0 XTX
2#0 XTY
2#0 XUB
2#0 XUE
2#0 XUP
2#0 XUT
2#0 XUX
2#0 XVE
2#0 XVS
2#0 XWB
2#0 XWF
2#0 XWH
2#0 XWM
2#0 XWP
2#0 XWT
2#0 XWV
2#0 XWZ
2#0 XXA
2#0 XXD
2#0 XXG
2#0 XXH
2#0 XXJ
2#0 XXL
2#0 XXN
2#0 XXR
2#0 XXS
2#0 XXV
2#0 XXZ
2#0 XYE
2#0 XYL
2#0 XYN
2#0 XYR
2#0 XZC
2#0 XZF
2#0 XZL
2#0 XZQ
2#0 XZT
2#0 XZY
2#0 YAA
2#0 YAK
2#0 YAN
2#0 YAT
2#0 YAU
2#0 YAZ
2#0 YBC
2#0 YBG
2#0 YBK
2#0 YBM
2#0 YBO
2#0 YBQ
2#0 YBT
2#0 YBV
2#0 YBW
2#0 YCB
2#0 YCC
2#0 YCF
2#0 YCK
2#0 YCL
2#0 YCM
2#0 YCP
2#0 YCS
2#0 YCT
2#0 YCU
2#0 YCX
2#0 YCZ
2#0 YDE
2#0 YDG
2#0 YDL
2#0 YDS
2#0 YDW
2#0 YDZ
2#0 YED
2#0 YEE
2#0 YER
2#0 YFA
2#0 YFL
2#0 YFM
2#0 YFO
2#0 YFT
2#0 YFY
2#0 YGA
2#0 YGG
2#0 YGN
2#0 YGW
2#0 YHB
2#0 YHQ
2#0 YIC
2#0 YJH
2#0 YJM
2#0 YKF
2#0 YKG
2#0 YKM
2#0 YKS
2#0 YLK
2#0 YLM
2#0 YLO
2#0 YLW
2#0 YLY
2#0 YMB
2#0 YMC
2#0 YMD
2#0 YMI
2#0 YML
2#0 YMN
2#0 YMO
2#0 YNE
2#0 YNG
2#0 YNM
2#0 YNV
2#0 YNW
2#0 YOB
2#0 YOC
2#0 YOG
2#0 YOR
2#0 YOS
2#0 YOT
2#0 YOU
2#0 YOW
2#0 YOX
2#0 YPA
2#0 YPC
2#0 YPD
2#0 YPL
2#0 YPM
2#0 YPO
2#0 YPT
2#0 YPZ
2#0 YQF
2#0 YQN
2#0 YRA
2#0 YRD
2#0 YRE
2#0 YRG
2#0 YRJ
2#0 YRK
2#0 YRN
2#0 YRO
2#0 YSD
2#0 YSH
2#0 YSW
2#0 YTK
2#0 YTO
2#0 YTS
2#0 YTV
2#0 YTX
2#0 YUB
2#0 YUC
2#0 YUE
2#0 YUK
2#0 YUM
2#0 YUS
2#0 YUV
2#0 YVE
2#0 YVF
2#0 YVN
2#0 YVO
2#0 YVS
2#0 YVV
2#0 YVX
2#0 YVZ
2#0 YWF
2#0 YWH
2#0 YWK
2#0 YXP
2#0 YXS
2#0 YYX
2#0 YZE
2#0 YZF
2#0 YZG
2#0 YZV
2#0 ZAD
2#0 ZAK
2#0 ZAN
2#0 ZAR
2#0 ZAT
2#0 ZAV
2#0 ZAW
2#0 ZAX
2#0 ZAY
2#0 ZBA
2#0 ZBG
2#0 ZBU
2#0 ZCL
2#0 ZCR
2#0 ZDF
2#0 ZDK
2#0 ZDY
2#0 ZEC
2#0 ZEM
2#0 ZEO
2#0 ZEY
2#0 ZFM
2#0 ZFS
2#0 ZFU
2#0 ZFX
2#0 ZGB
2#0 ZGN
2#0 ZIG
2#0 ZIL
2#0 ZIM
2#0 ZIX
2#0 ZKX
2#0 ZLG
2#0 ZLK
2#0 ZLR
2#0 ZLT
2#0 ZMX
2#0 ZNH
2#0 ZOG
2#0 ZON
2#0 ZPM
2#0 ZPS
2#0 ZQR
2#0 ZRA
2#0 ZRN
2#0 ZRO
2#0 ZSP
2#0 ZTP
2#0 ZTX
2#0 ZUA
2#0 ZUT
2#0 ZUX
2#0 ZWH
2#0 ZWZ
2#0 ZXA
2#0 ZXP
2#0 ZXV
2#0 ZXW
2#0 ZXX
2#0 ZYF
2#0 ZYQ
2#0 ZZI
2#0 ZZM
2#0 ZZR
2#0 ZZW
2#0 ZZX
2#1 AAH
2#1 ACX
2#1 ADB
2#1 ADW
2#1 AED
2#1 AFH
2#1 AFO
2#1 AFT
2#1 AJK
2#1 AJL
2#1 AJM
2#1 AKC
2#1 AKO
2#1 AKU
2#1 ALF
2#1 ALQ
2#1 ANP
2#1 AOQ
2#1 AOR
2#1 APT
2#1 AQG
2#1 AQO
2#1 ART
2#1 ARV
2#1 ATB
2#1 AVX
2#1 AWM
2#1 AWW
2#1 AWX
2#1 AXG
2#1 AXN
2#1 AXX
2#1 AYY
2#1 AZD
2#1 AZE
2#1 AZL
2#1 BBN
2#1 BCW
2#1 BDF
2#1 BDL
2#1 BDM
2#1 BDU
2#1 BDX
2#1 BDZ
2#1 BEI
2#1 BFD
2#1 BFO
2#1 BGC
2#1 BHS
2#1 BJF
2#1 BKA
2#1 BKU
2#1 BKW
2#1 BMD
2#1 BMM
2#1 BNA
2#1 BNB
2#1 BNF
2#1 BNP
2#1 BOF
2#1 BOS
2#1 BPC
2#1 BPJ
2#1 BRE
2#1 BRM
2#1 BRT
2#1 BRY
2#1 BTD
2#1 BTH
2#1 BTS
2#1 BTX
2#1 BUC
2#1 BUF
2#1 BVM
2#1 BVS
2#1 BVX
2#1 BWZ
2#1 BZF
2#1 BZU
2#1 CAD
2#1 CBC
2#1 CBL
2#1 CCB
2#1 CCM
2#1 CDJ
2#1 CDL
2#1 CEJ
2#1 CEL
2#1 CER
2#1 CES
2#1 CFA
2#1 CFU
2#1 CGT
2#1 CIC
2#1 CIH
2#1 CII
2#1 CIR
2#1 CJK
2#1 CKE
2#1 CLU
2#1 CMH
2#1 CMP
2#1 CMZ
2#1 CNT
2#1 CQG
2#1 CSN
2#1 CTB
2#1 CTV
2#1 CTZ
2#1 CUK
2#1 CUT
2#1 CUZ
2#1 CVG
2#1 CXB
2#1 CYH
2#1 CYR
2#1 CZI
2#1 CZY
2#1 DAA
2#1 DAG
2#1 DAH
2#1 DCF
2#1 DCH
2#1 DCJ
2#1 DCL
2#1 DCU
2#1 DDD
2#1 DDN
2#1 DDT
2#1 DDU
2#1 DEU
2#1 DEV
2#1 DEW
2#1 DFD
2#1 DFI
2#1 DFR
2#1 DHH
2#1 DHL
2#1 DHN
2#1 DIK
2#1 DIQ
2#1 DJM
2#1 DJN
2#1 DJR
2#1 DJZ
2#1 DKI
2#1 DKT
2#1 DLO
2#1 DMM
2#1 DMW
2#1 DNH
2#1 DNZ
2#1 DOX
2#1 DPJ
2#1 DPU
2#1 DQR
2#1 DQU
2#1 DSD
2#1 DSG
2#1 DSM
2#1 DSV
2#1 DTI
2#1 DTR
2#1 DTT
2#1 DUE
2#1 DUN
2#1 DUU
2#1 DVE
2#1 DVS
2#1 DWM
2#1 DWY
2#1 DWZ
2#1 DXY
2#1 DYR
2#1 DZF
2#1 DZR
2#1 DZY
2#1 EAR
2#1 EAS
2#1 EBK
2#1 EBO
2#1 EBS
2#1 ECQ
2#1 ECY
2#1 EDF
2#1 EER
2#1 EEU
2#1 EFH
2#1 EFZ
2#1 EGQ
2#1 EIA
2#1 EIL
2#1 EJI
2#1 EJN
2#1 EKU
2#1 EKW
2#1 ELH
2#1 ELM
2#1 ELR
2#1 ELY
2#1 EMM
2#1 EMS
2#1 EMT
2#1 ENF
2#1 EOW
2#1 EQH
2#1 EQM
2#1 EQY
2#1 ESU
2#1 ESX
2#1 ESY
2#1 ETP
2#1 EUB
2#1 EUG
2#1 EUT
2#1 EUW
2#1 EVB
2#1 EWF
2#1 EXB
2#1 EXC
2#1 EXG
2#1 EYI
2#1 EYO
2#1 EZF
2#1 FAV
2#1 FCN
2#1 FCV
2#1 FDZ
2#1 FEK
2#1 FEX
2#1 FFC
2#1 FGK
2#1 FGY
2#1 FHJ
2#1 FIY
2#1 FJU
2#1 FKA
2#1 FKE
2#1 FMS
2#1 FNP
2#1 FOJ
2#1 FOK
2#1 FOP
2#1 FPD
2#1 FPH
2#1 FPR
2#1 FQP
2#1 FQU
2#1 FTP
2#1 FTU
2#1 FTW
2#1 FUE
2#1 FUF
2#1 FUN
2#1 FUO
2#1 FUZ
2#1 FVC
2#1 FVG
2#1 FWL
2#1 FWR
2#1 FXO
2#1 FYA
2#1 FYE
2#1 FYI
2#1 FYJ
2#1 FZF
2#1 FZK
2#1 FZP
2#1 GAD
2#1 GAJ
2#1 GCT
2#1 GCX
2#1 GDC
2#1 GEF
2#1 GEH
2#1 GEN
2#1 GEO
2#1 GEV
2#1 GEW
2#1 GFD
2#1 GFH
2#1 GFL
2#1 GFY
2#1 GGW
2#1 GGX
2#1 GHW
2#1 GIA
2#1 GIB
2#1 GIP
2#1 GIR
2#1 GJN
2#1 GJO
2#1 GJW
2#1 GKA
2#1 GLA
2#1 GLO
2#1 GMD
2#1 GMH
2#1 GMS
2#1 GNW
2#1 GOG
2#1 GON
2#1 GOO
2#1 GOQ
2#1 GOS
2#1 GPF
2#1 GPH
2#1 GPR
2#1 GPT
2#1 GQA
2#1 GQK
2#1 GQL
2#1 GQM
2#1 GQX
2#1 GQZ
2#1 GRD
2#1 GRE
2#1 GRY
2#1 GSH
2#1 GSJ
2#1 GSK
2#1 GSN
2#1 GSV
2#1 GSY
2#1 GTU
2#1 GTZ
2#1 GUA
2#1 GUB
2#1 GUT
2#1 GUU
2#1 GWB
2#1 GWI
2#1 GXD
2#1 GXN
2#1 GXP
2#1 GXT
2#1 GYJ
2#1 GZB
2#1 GZE
2#1 GZU
2#1 GZW
2#1 HAA
2#1 HAB
2#1 HAF
2#1 HAK
2#1 HAT
2#1 HAV
2#1 HBI
2#1 HBV
2#1 HCJ
2#1 HCL
2#1 HDL
2#1 HDU
2#1 HDW
2#1 HEB
2#1 HEC
2#1 HEH
2#1 HEL
2#1 HEN
2#1 HIB
2#1 HIE
2#1 HIT
2#1 HJK
2#1 HJP
2#1 HJQ
2#1 HJT
2#1 HJY
2#1 HKO
2#1 HKQ
2#1 HKR
2#1 HKT
2#1 HLB
2#1 HLR
2#1 HMH
2#1 HMW
2#1 HNK
2#1 HOJ
2#1 HOL
2#1 HOM
2#1 HOW
2#1 HPE
2#1 HPN
2#1 HPX
2#1 HQB
2#1 HQM
2#1 HQS
2#1 HRA
2#1 HRC
2#1 HRT
2#1 HRZ
2#1 HSF
2#1 HSK
2#1 HSQ
2#1 HSU
2#1 HSW
2#1 HTG
2#1 HTW
2#1 HTX
2#1 HVG
2#1 HWD
2#1 HWE
2#1 HXG
2#1 HYA
2#1 HYG
2#1 HYK
2#1 HYP
2#1 HYT
2#1 HYU
2#1 HYV
2#1 HZA
2#1 HZE
2#1 HZH
2#1 HZW
2#1 IAK
2#1 IAV
2#1 IBG
2#1 IBQ
2#1 IBW
2#1 IBY
2#1 ICA
2#1 ICB
2#1 ICH
2#1 ICK
2#1 ICY
2#1 ICZ
2#1 IDA
2#1 IDD
2#1 IDJ
2#1 IDK
2#1 IDL
2#1 IDN
2#1 IDP
2#1 IDQ
2#1 IDT
2#1 IDU
2#1 IEC
2#1 IED
2#1 IEH
2#1 IEU
2#1 IEW
2#1 IEY
2#1 IFC
2#1 IFG
2#1 IFH
2#1 IFK
2#1 IFO
2#1 IFP
2#1 IFS
2#1 IGD
2#1 IGE
2#1 IGH
2#1 IGN
2#1 IGW
2#1 IHC
2#1 IHH
2#1 IHI
2#1 IHL
2#1 IHP
2#1 IIH
2#1 IIK
2#1 IIN
2#1 IIQ
2#1 IIS
2#1 IIY
2#1 IIZ
2#1 IJB
2#1 IJZ
2#1 IKC
2#1 IKD
2#1 IPA
2#1 IPS
2#1 IXG
2#1 JAJ
2#1 JAN
2#1 JAY
2#1 JBA
2#1 JBC
2#1 JDT
2#1 JFJ
2#1 JJE
2#1 JMK
2#1 JTL
2#1 KAC
2#1 KGB
2#1 KLM
2#1 KLO
2#1 LBH
2#1 LCY
2#1 LMW
2#1 LNJ
2#1 LOK
2#1 LOM
2#1 LOU
2#1 LTW
2#1 MCT
2#1 MDD
2#1 MDM
2#1 MEX
2#1 MHH
2#1 MJE
2#1 MJG
2#1 MMH
2#1 MNP
2#1 MOG
2#1 MPC
2#1 MPH
2#1 MTA
2#1 NAI
2#1 NBC
2#1 NBY
2#1 NOJ
2#1 NOP
2#1 NOW
2#1 NPR
2#1 NRJ
2#1 OAT
2#1 OCT
2#1 ODB
2#1 OKR
2#1 OOP
2#1 PAK
2#1 PAN
2#1 PAT
2#1 PEF
2#1 PJE
2#1 PJT
2#1 PKY
2#1 PRK
2#1 RAN
2#1 RBS
2#1 RCL
2#1 RCM
2#1 RDX
2#1 REK
2#1 RIC
2#1 RNY
2#1 ROY
2#1 RSX
2#1 RUS
2#1 RWH
2#1 SAM
2#1 SIM
2#1 SKJ
2#1 SLE
2#1 SPS
2#1 SRM
2#1 SSB
2#1 STB
2#1 SWW
2#1 TAF
2#1 TEC
2#1 TER
2#1 TFJ
2#1 THX
2#1 TJH
2#1 TMP
2#1 USC
2#1 VAL
2#1 VHS
2#1 VIN
2#1 VIX
2#1 WCB
2#1 WND
2#1 WOW
2#1 WSR
2#1 WTG
2#1 WWC
2#1 YAP
2#1 YOT
2#1 ZFA
2#2 CTR
2#2 DUZ
2#2 EGK
2#2 GYP
2#2 HON

G0

G0 AAR
G0 AAV
G0 AAW
G0 ABY
G0 ABZ
G0 ACE
G0 ADR
G0 AEI
G0 AFB
G0 AFD
G0 AGH
G0 AGM
G0 AGQ
G0 AHN
G0 AJC
G0 AJQ
G0 AKA
G0 AKI
G0 AKQ
G0 ALD
G0 ALV
G0 AMB
G0 ANC
G0 ANS
G0 ANV
G0 AOU
G0 APE
G0 APM
G0 APQ
G0 APW
G0 AQL
G0 AQP
G0 ARC
G0 ARW
G0 ASC
G0 ASE

Withheld

G0 ATR
G0 ATV
G0 AUO
G0 AUQ
G0 AVO
G0 AVR
G0 AVW
G0 AWD
G0 AWL
G0 AWV
G0 AXN
G0 AYE
G0 AYN
G0 AYR
G0 AYU
G0 AYZ
G0 AZS
G0 AZZ
G0 BBH
G0 BBS
G0 BBU
G0 BBZ
G0 BCK
G0 BDX
G0 BDY
G0 BEA
G0 BEH
G0 BEO
G0 BEQ
G0 BEU
G0 BFO
G0 BGC
G0 BGD
G0 BGN
G0 BGQ
G0 BGW
G0 BHD
G0 BHX
G0 BHY
G0 BIS
G0 BJF
G0 BJH
G0 BJO
G0 BJV
G0 BJW
G0 BJX
G0 BKO
G0 BLD
G0 BLN
G0 BMA
G0 BMB
G0 BMX
G0 BNC
G0 BND
G0 BNI
G0 BOB
G0 BPB
G0 BPI
G0 BQN
G0 BQT
G0 BRC
G0 BRP
G0 BRV
G0 BRY
G0 BSG
G0 BSR
G0 BSS
G0 BSU
G0 BTO
G0 BTP
G0 BUS
G0 BUY
G0 BVH
G0 BVX
G0 BVZ
G0 BWS
G0 BWT
G0 BXT
G0 BYI
G0 BZI
G0 BZK
G0 BZY
G0 BZZ
G0 CAA
G0 CAC
G0 CAF
G0 CAR
G0 CAT
G0 CAU
G0 CAW
G0 CBE
G0 CBH
G0 CCI
G0 CCK
G0 CCP
G0 CCR
G0 CCW
G0 CDD
G0 CDF
G0 CDX
G0 CED
G0 CET
G0 CFL
G0 CFR
G0 CFS
G0 CFV
G0 CFY
G0 CGL
G0 CHT
G0 CIB
G0 CID
G0 CIF
G0 CIQ
G0 CIU
G0 CKG
G0 CKP
G0 CKT
G0 CKW
G0 CLB
G0 CLK
G0 CLN
G0 CLP
G0 CME
G0 CMF
G0 CMQ
G0 CNE
G0 CNH
G0 CNN
G0 CNR
G0 CNS
G0 COD
G0 CPK
G0 CPW
G0 CRH
G0 CRI
G0 CRQ
G0 CTO
G0 CTX
G0 CUD
G0 CUK
G0 CUQ
G0 CUT
G0 CVL
G0 CVT
G0 CVZ
G0 CWM
G0 CWT
G0 CWV
G0 CXF
G0 CXM
G0 CXN
G0 CXP
G0 CXQ
G0 CXT
G0 CYA
G0 CYP
G0 CYT
G0 CZA
G0 CZB
G0 CZE
G0 CZP
G0 CZT
G0 DBH
G0 DCA
G0 DCB
G0 DCD
G0 DCK
G0 DCN
G0 DDD
G0 DDF
G0 DDX
G0 DFK
G0 DFP
G0 DGV
G0 DGX
G0 DHN
G0 DIN
G0 DJE
G0 DJH
G0 DJR
G0 DKL
G0 DLD
G0 DLO
G0 DLV
G0 DLY
G0 DMI
G0 DML
G0 DMO
G0 DNC
G0 DNE
G0 DNJ
G0 DNM
G0 DOI
G0 DON
G0 DOT
G0 DPB
G0 DPH
G0 DPZ
G0 DQL
G0 DQX
G0 DRC
G0 DTU
G0 DTX
G0 DUU
G0 DVF
G0 DVR
G0 DVV
G0 DVZ
G0 DWT
G0 DXA
G0 DXC
G0 DXD
G0 DXQ
G0 DXS
G0 DYB
G0 DYI
G0 DYZ
G0 EAC
G0 EAD
G0 EAK
G0 EAR
G0 EAZ
G0 EBB
G0 EBR
G0 EBU
G0 ECE
G0 ECO
G0 EED
G0 EES
G0 EFX
G0 EGX
G0 EGY
G0 EHF
G0 EHG
G0 EHH
G0 EHU
G0 EHY
G0 EIK
G0 EIO
G0 EIU
G0 EIX
G0 EJG
G0 EJW
G0 EKJ
G0 EKL
G0 EKP
G0 EKQ
G0 EKR
G0 ELE
G0 ELM
G0 ELS
G0 ELT
G0 EME
G0 EMG
G0 ENC
G0 ENE
G0 ENG
G0 ENH
G0 ENR
G0 ENS
G0 EOA
G0 EOB
G0 EOC
G0 EOQ
G0 EOU
G0 EPG
G0 EPH
G0 EPX
G0 EPZ
G0 EQA
G0 EQG
G0 EQP
G0 EQZ
G0 ERI
G0 ERM
G0 ESB
G0 ESC
G0 ESE
G0 ESV
G0 ETB
G0 ETC
G0 ETN
G0 ETO
G0 ETY
G0 EUB
G0 EUF
G0 EUH
G0 EUI
G0 EUT
G0 EWM
G0 EWN
G0 EWS
G0 EXF
G0 EXH
G0 EXJ
G0 EXX
G0 EXZ
G0 EYJ
G0 EZA
G0 EZN
G0 EZP
G0 FAF
G0 FAI
G0 FAQ
G0 FAY
G0 FBI
G0 FBR
G0 FBY
G0 FCD
G0 FCP
G0 FDM
G0 FDN
G0 FEA
G0 FEF
G0 FEX
G0 FFH
G0 FFR
G0 FHM
G0 FHR
G0 FHV
G0 FII
G0 FIK
G0 FIL
G0 FIM
G0 FIR
G0 FIZ
G0 FJI
G0 FJT
G0 FJV
G0 FJY
G0 FKH
G0 FKL
G0 FKM
G0 FLB
G0 FLC
G0 FLN
G0 FLO
G0 FMM
G0 FMY
G0 FMZ
G0 FNK
G0 FNL
G0 FNY
G0 FOA
G0 FOD
G0 FOJ
G0 FOQ
G0 FOX
G0 FOZ
G0 FPD
G0 FPV
G0 FQB
G0 FQN
G0 FQT
G0 FRE
G0 FRJ
G0 FRK
G0 FRQ
G0 FSO
G0 FTD
G0 FTM
G0 FUQ
G0 FVG
G0 FVV
G0 FVX
G0 FWK
G0 FWL
G0 FWO
G0 FWU
G0 FWV
G0 FWX
G0 FXU
G0 FXW
G0 FXZ
G0 FYG
G0 FYN
G0 FYQ
G0 FYS
G0 FYV
G0 FZP
G0 GAC
G0 GAD
G0 GBD
G0 GBO
G0 GBX
G0 GCB
G0 GCD
G0 GCH
G0 GCL
G0 GCP
G0 GCV
G0 GCW
G0 GCY
G0 GDA
G0 GDO
G0 GDP
G0 GDU
G0 GDW
G0 GDX
G0 GDY
G0 GDZ
G0 GEM
G0 GEY
G0 GFX
G0 GGF
G0 GGU
G0 GGX
G0 GHI
G0 GHQ
G0 GIC
G0 GIG
G0 GIP
G0 GIQ
G0 GIV
G0 GIW
G0 GIY
G0 GJK
G0 GJP
G0 GJU
G0 GJZ
G0 GKW
G0 GLM
G0 GLT
G0 GMF
G0 GMG
G0 GMR
G0 GMT
G0 GMU
G0 GNL
G0 GNM
G0 GNR
G0 GNX
G0 GOA
G0 GPZ
G0 GQF
G0 GQM
G0 GQU
G0 GQZ
G0 GRA
G0 GRI
G0 GRK
G0 GRN
G0 GRP
G0 GRT
G0 GRT
G0 GSD
G0 GTK
G0 GTM
G0 GTO
G0 GTR
G0 GTS
G0 GTU
G0 GUE
G0 GUH
G0 GUK
G0 GUP
G0 GVC
G0 GVI
G0 GVM
G0 GVQ
G0 GVY
G0 GWK
G0 GXL
G0 GYK
G0 GYZ
G0 GZS
G0 HAD
G0 HAF
G0 HAH
G0 HAS
G0 HCO
G0 HCS
G0 HCT
G0 HDA
G0 HDN
G0 HEG
G0 HEH
G0 HFH
G0 HFN
G0 HFW
G0 HGK
G0 HGX
G0 HHG
G0 HHS
G0 HIA
G0 HIO
G0 HIQ
G0 HIS
G0 HIX
G0 HJQ
G0 HKA
G0 HKP
G0 HKS
G0 HKU
G0 HLD
G0 HLM
G0 HLN
G0 HLO
G0 HLR
G0 HLT
G0 HLX
G0 HMG
G0 HMW
G0 HMY
G0 HMZ
G0 HNA
G0 HNB
G0 HNF
G0 HOK
G0 HOL
G0 HON
G0 HOO
G0 HOR
G0 HPO
G0 HPZ
G0 HQB
G0 HQD
G0 HQE
G0 HQJ
G0 HQW
G0 HRA
G0 HSO
G0 HSS
G0 HSZ
G0 HUO
G0 HUP
G0 HUS
G0 HUU
G0 HUZ
G0 HVL
G0 HVM
G0 HVQ
G0 HVU
G0 HVV
G0 HVW
G0 HWJ
G0 HWN
G0 HWW
G0 HXO
G0 HXT
G0 HXW
G0 HXX
G0 HXY
G0 HXZ
G0 HYF
G0 HYI
G0 HYY
G0 HZP
G0 HZR
G0 HZU
G0 HZV
G0 HZW
G0 HZX
G0 IAM
G0 IAN
G0 IAT
G0 IBB
G0 IBD
G0 IBP
G0 ICF
G0 ICI
G0 ICX
G0 ICZ
G0 IDA
G0 IEX
G0 IFG
G0 IFI
G0 IFM
G0 IFP
G0 IFZ
G0 IHB
G0 IIR
G0 IIS
G0 IIZ
G0 IJJ
G0 IJM
G0 IJO
G0 IKA
G0 IKG
G0 IKH
G0 IKS
G0 IKV
G0 ILF
G0 ILG
G0 ILS
G0 ILV
G0 IMN
G0 IMT
G0 INP
G0 IOX
G0 IPS
G0 IPU
G0 IPY
G0 IQX
G0 IRG
G0 IRN
G0 IRO
G0 IRV
G0 IRW
G0 ISD
G0 ISV
G0 ISW
G0 ITA
G0 ITB
G0 IUE
G0 IUQ
G0 IVA
G0 IVM
G0 IWA
G0 IWM
G0 IWT
G0 IXB
G0 IXE
G0 IXI
G0 IXP
G0 IYF
G0 IYZ
G0 IZA
G0 IZM
G0 IZU
G0 IZX
G0 JAY
G0 JBD
G0 JBK
G0 JBW
G0 JCM
G0 JDF
G0 JDI
G0 JDJ
G0 JDK
G0 JDP
G0 JDU
G0 JED
G0 JEI
G0 JFN
G0 JGP
G0 JGZ
G0 JHA
G0 JHC
G0 JHF
G0 JHV
G0 JHY
G0 JIP
G0 JIZ
G0 JJL
G0 JKB
G0 JKN
G0 JKQ
G0 JKV
G0 JKX
G0 JKZ
G0 JLO
G0 JLW
G0 JLY
G0 JMC
G0 JMP
G0 JMT
G0 JMV
G0 JMX
G0 JNH
G0 JNL
G0 JNN
G0 JNP
G0 JNS
G0 JOH
G0 JOO
G0 JOW
G0 JOY
G0 JPA
G0 JPN
G0 JPO
G0 JPS
G0 JQG
G0 JQJ
G0 JQL
G0 JQM
G0 JQN
G0 JQO
g0 jrk
G0 JRP
G0 JSB
G0 JSH
G0 JSS
G0 JSV
G0 JTB
G0 JUU
G0 JUW
G0 JUX
G0 JVA
G0 JVY
G0 JWB
G0 JWQ
G0 JXF
G0 JXM
G0 JYB
G0 JYM
G0 JYO
G0 JYP
G0 JYT
G0 JZG
G0 JZI
G0 JZP
G0 KAC
G0 KAD
G0 KAE
G0 KAF
G0 KAG
G0 KAI
G0 KAJ
G0 KAL
G0 KAO
G0 KAR
G0 KAW
G0 KBE
G0 KCQ
G0 KDJ
G0 KEH
G0 KEN
G0 KET
G0 KFA
G0 KFO
G0 KFX
G0 KGH
G0 KGK
G0 KGO
G0 KGV
G0 KHB
G0 KHC
G0 KHD
G0 KHG
G0 KHO
G0 KHV
G0 KHW
G0 KIJ
G0 KIS
G0 KIW
G0 KIZ
G0 KJV
G0 KJY
G0 KKB
G0 KKI
G0 KKJ
G0 KLI
G0 KLM
G0 KLR
G0 KLX
G0 KME
G0 KMX
G0 KNK
G0 KOA
G0 KOR
G0 KPK
G0 KPM
G0 KPR
G0 KQL
G0 KQZ
G0 KRA
G0 KSB
G0 KSE
G0 KST
G0 KTM
G0 KUM
G0 KUN
G0 KUO
G0 KVH
G0 KVL
G0 KVT
G0 KVY
G0 KVZ
G0 KWT
G0 KWV
G0 KXA
G0 KXE
G0 KXM
G0 KYC
G0 KZC
G0 KZZ
G0 LAH
G0 LAI
G0 LAT
G0 LBK
G0 LBW
G0 LCL
G0 LCQ
G0 LCW
G0 LDD
G0 LDM
G0 LDW
G0 LEG
G0 LET
G0 LFK
G0 LFU
G0 LFW
G0 LGD
G0 LGN
G0 LGS
G0 LGT
G0 LGU
G0 LHF
G0 LHK
G0 LHT
G0 LIF
G0 LIJ
G0 LIL
G0 LIX
G0 LJS
G0 LJT
G0 LJX
G0 LKM
G0 LKQ
G0 LKZ
G0 LLW
G0 LMF
G0 LMI
G0 LMM
G0 LMO
G0 LMT
G0 LNR
G0 LNY
G0 LOG
G0 LOQ
G0 LOU
G0 LOV
G0 LPO
G0 LPS
G0 LQP
G0 LQY
G0 LRD
G0 LRH
G0 LRX
G0 LRY
G0 LSB
G0 LSH
G0 LTG
G0 LTN
G0 LTU
G0 LTW
G0 LUA
G0 LUJ
G0 LUS
G0 LUZ
G0 LVB
G0 LVN
G0 LVQ
G0 LVV
G0 LVZ
G0 LWH
G0 LWK
G0 LWS
G0 LWX
G0 LWY
G0 LWZ
G0 LXA
G0 LXJ
G0 LXS
G0 LXZ
G0 LYL
G0 LYV
G0 LZB
G0 LZO
G0 MAB
G0 MAO
G0 MAV
G0 MBE
G0 MBT
G0 MCU
G0 MDA
G0 MDG
G0 MDI
G0 MDL
G0 MEG
G0 MEJ
G0 MEK
G0 MEL
G0 MEM
G0 MEP
G0 MES
G0 MEU
G0 MGA
G0 MGE
G0 MGF
G0 MGK
G0 MHR
G0 MIN
G0 MIQ
G0 MIY
G0 MJI
G0 MJM
G0 MJN
G0 MJQ
G0 MJS
G0 MJW
G0 MKB
G0 MKI
G0 MKJ
G0 MKM
G0 MKT
G0 MLD
G0 MLG
G0 MLS
G0 MMN
G0 MMR
G0 MNE
G0 MNM
G0 MOD
G0 MOP
G0 MPD
G0 MPG
G0 MPX
G0 MQA
G0 MQF
G0 MQG
G0 MQP
G0 MQS
G0 MRV
G0 MSL
G0 MSP
G0 MTR
G0 MTS
G0 MTX
G0 MTZ
G0 MUO
G0 MUT
G0 MUU
G0 MUV
G0 MUW
G0 MUY
G0 MWR
G0 MXF
G0 MXN
G0 MYW
G0 MYX
G0 MYY
G0 MZB
G0 MZS
G0 MZT
G0 MZV
G0 MZX
G0 NAB
G0 NAN
G0 NBF
G0 NBM
G0 NBO
G0 NCF
G0 NCM
G0 NCN
G0 NCR
G0 NDN
G0 NDT
G0 NEH
G0 NEW
G0 NEY
G0 NFQ
G0 NGL
G0 NGT
G0 NGU
G0 NGV
G0 NGX
G0 NHF
G0 NHN
G0 NHY
G0 NIH
G0 NII
G0 NIK
G0 NIM
G0 NIS
G0 NIV
G0 NJF
G0 NJM
G0 NJY
G0 NKE
G0 NKR
G0 NKV
G0 NLE
G0 NLR
G0 NLS
G0 NME
G0 NMK
G0 NML
G0 NMQ
G0 NMT
G0 NND
G0 NNY
G0 NOK
G0 NOS
G0 NOT
G0 NOY
G0 NPH
G0 NQZ
G0 NRG
G0 NRL
G0 NTE
G0 NTK
G0 NTM
G0 NTO
G0 NTP
G0 NTQ
G0 NTX
G0 NVF
G0 NVN
G0 NVP
G0 NVW
G0 NWB
G0 NWM
G0 NWX
G0 NYG
G0 NZO
G0 NZQ
G0 NZW
G0 OAH
G0 OAI
G0 OAK
G0 OAN
G0 OAQ
G0 OBS
G0 OCH
G0 OCI
G0 ODF
G0 ODL
G0 ODV
G0 OEF
G0 OEP
G0 OFC
G0 OFJ
G0 OGO
G0 OGT
G0 OGU
G0 OGW
G0 OHB
G0 OHE
G0 OHI
G0 OIL
G0 OJA
G0 OJD
G0 OJH
G0 OJI
G0 OJZ
G0 OKB
G0 OKR
G0 OKS
G0 OKW
G0 OLB
G0 OLH
G0 OLI
G0 OMG
G0 OMK
G0 OMP
G0 OMS
G0 OMV
G0 ONC
G0 OOG
G0 OOH
G0 OOM
G0 OOT
G0 OOW
G0 OOY
G0 OPD
G0 OPH
G0 OPR
G0 OPS
G0 OPW
G0 OQB
G0 OQC
G0 OQD
G0 OQJ
G0 OQM
G0 OQU
G0 OQV
G0 OQW
G0 ORA
G0 ORH
G0 ORI
G0 OSY
G0 OTA
G0 OTO
G0 OUF
G0 OUI
G0 OUQ
G0 OVI
G0 OVU
G0 OXI
G0 OXJ
G0 OXO
G0 OXQ
G0 OYB
G0 OZK
G0 OZU
G0 OZW
G0 OZX
G0 OZY
G0 OZZ
G0 PAT
G0 PCL
G0 PDL
G0 PDR
G0 PDU
G0 PFC
G0 PGE
G0 PHS
G0 PHU
G0 PIF
G0 PIG
G0 PIH
G0 PIM
G0 PIP
G0 PJK
G0 PJX
G0 PKH
G0 PKS
G0 PKU
G0 PLM
G0 PLN
G0 PLP
G0 PLV
G0 PMA
G0 PMH
G0 PMJ
G0 PMW
G0 PNY
G0 PNZ
G0 POH
G0 POI
G0 PPB
G0 PPF
G0 PPN
G0 PPO
G0 PPP
G0 PQE
G0 PQK
G0 PQQ
G0 PQT
G0 PRT
G0 PSL
G0 PSM
G0 PSN
G0 PSU
G0 PTJ
G0 PTS
G0 PUA
G0 PUB
G0 PUF
G0 PUH
G0 PUU
G0 PVK
G0 PWR
G0 PWZ
G0 PYB
G0 PYN
G0 PYT
G0 PYZ
G0 PZA
G0 PZE
G0 PZI
G0 RAB
G0 RAH
G0 RAP
G0 RAY
G0 RAZ
G0 RBE
G0 RBT
G0 RCC
G0 RCK
G0 RCM
G0 RCS
G0 RCZ
G0 RDI
G0 RDQ
G0 RDW
G0 REJ
G0 REM
G0 RES
G0 REW
G0 RFD
G0 RFO
G0 RGB
G0 RGF
G0 RGR
G0 RGT
G0 RGV
G0 RGY
G0 RHH
G0 RHT
G0 RIN
G0 RIO
G0 RIT
G0 RIV
G0 RIV
G0 RJP
G0 RKF
G0 RKL
G0 RLR
G0 RMA
G0 RME
G0 RMQ
G0 RMZ
G0 RNE
G0 RNG
G0 RNR
G0 RNT
G0 RNU
G0 ROG
G0 ROI
G0 ROJ
G0 ROK
G0 ROV
G0 RPH
G0 RPP
G0 RPR
G0 RQK
G0 RRA
G0 RSH
G0 RSN
G0 RSP
G0 RSQ
G0 RTD
G0 RTG
G0 RTK
G0 RTX
G0 RUE
G0 RUJ
G0 RUP
G0 RVA
G0 RVF
G0 RVG
G0 RVJ
G0 RVL
G0 RVN
G0 RVT
G0 RVU
G0 RVY
G0 RWB
G0 RWD
G0 RWF
G0 RWK
G0 RWR
G0 RXG
G0 RXH
G0 RXL
G0 RXN
G0 RXT
G0 RXW
G0 RXY
G0 RYC
G0 RYF
G0 RYH
G0 RYN
G0 RYY
G0 RZC
G0 RZE
G0 RZF
G0 RZJ
G0 RZR
G0 RZV
G0 SAA
G0 SAG
G0 SAM
G0 SAN
G0 SAP
G0 SAQ
G0 SAS
G0 SAW
G0 SBD
G0 SBE
G0 SBN
G0 SBS
G0 SBW
G0 SCF
G0 SDI
G0 SDN
G0 SDO
G0 SDP
G0 SDV
G0 SDY
G0 SEA
G0 SEM
G0 SEV
G0 SFF
G0 SFH
G0 SFL
G0 SGB
G0 SGN
G0 SGO
G0 SHE
G0 SHQ
G0 SHW
G0 SIC
G0 SID
G0 SIF
G0 SIP
G0 SIT
G0 SIZ
G0 SJI
G0 SJK
G0 SJY
G0 SKB
G0 SKE
G0 SKO
G0 SKP
G0 SKT
G0 SKU
G0 SLO
G0 SLV
G0 SMA
G0 SMB
G0 SMD
G0 SMG
G0 SMI
G0 SML
G0 SMX
G0 SNC
G0 SNE
G0 SNJ
G0 SOB
G0 SOB
G0 SOJ
G0 SOL
G0 SOM
G0 SOO
G0 SOP
G0 SOR
G0 SOT
G0 SOZ
G0 SP
G0 SPE
G0 SPG
G0 SPR
G0 SQZ
G0 SRC
G0 SRM
G0 SSB
G0 SSI
G0 SSO
G0 STE
G0 STG
G0 STN
G0 STQ
G0 STU
G0 STV
G0 SUG
G0 SUJ
G0 SUM
G0 SVG
G0 SVT
G0 SWG
G0 SWP
G0 SWQ
G0 SWR
G0 SWV
G0 SXB
G0 SXC
G0 SXF
G0 SXH
G0 SYA
G0 SYJ
G0 SYM
G0 SYO
G0 SYU
G0 SYW
G0 SZP
G0 SZR
G0 SZS
G0 TAP
G0 TAU
G0 TBA
G0 TBK
G0 TBN
G0 TBR
G0 TBT
G0 TBX
G0 TCD
G0 TCX
G0 TCZ
G0 TDB
G0 TDD
G0 TDF
G0 TDI
G0 TDY
G0 TEC
G0 TEQ
G0 TER
G0 TET
G0 TEY
G0 TEZ
G0 TFH
G0 TFS
G0 TFW
G0 TGC
G0 TGD
G0 TGI
G0 TGJ
G0 TGK
G0 TGT
G0 TGW
G0 TGY
G0 TGZ
G0 THG
G0 THM
G0 THZ
G0 TIK
G0 TIM
G0 TIQ
G0 TIY
G0 TJB
G0 TJZ
G0 TKD
G0 TKO
G0 TKP
G0 TKV
G0 TLB
G0 TLC
G0 TLD
G0 TLE
G0 TLG
G0 TLK
G0 TLM
G0 TLY
G0 TMB
G0 TMC
G0 TMD
G0 TMP
G0 TMX
G0 TMY
G0 TNI
G0 TNJ
G0 TNR
G0 TNT
G0 TNV
G0 TOG
G0 TOR
G0 TOU
G0 TOV
G0 TOW
G0 TPC
G0 TPS
G0 TPT
G0 TPU
G0 TPV
G0 TPX
G0 TPZ
G0 TQM
G0 TQN
G0 TQO
G0 TRF
G0 TRT
G0 TRX
G0 TRZ
G0 TSM
G0 TSP
G0 TSW
G0 TTA
G0 TTD
G0 TTJ
G0 TTU
G0 TTV
G0 TTX
G0 TUA
G0 TUB
G0 TUD
G0 TUH
G0 TUK
G0 TUY
G0 TVE
G0 TVF
G0 TVH
G0 TVI
G0 TVJ
G0 TVK
G0 TVP
G0 TVY
G0 TWN
G0 TWP
G0 TWW
G0 TXE
G0 TXI
G0 TYH
G0 TYX
G0 TZA
G0 TZE
G0 TZK
G0 TZN
G0 TZS
G0 UAE
G0 UAL
G0 UAM
G0 UAN
G0 UAT
G0 UAX
G0 UAZ
G0 UBC
G0 UBD
G0 UBE
G0 UBQ
G0 UBV
G0 UCR
G0 UDA
G0 UDC
G0 UDE
G0 UDF
G0 UDK
G0 UDM
G0 UDX
G0 UEI
G0 UEJ
G0 UEN
G0 UER
G0 UES
G0 UEX
G0 UFH
G0 UFK
G0 UGC
G0 UGE
G0 UGF
G0 UGO
G0 UGT
G0 UHA
G0 UHE
G0 UHY
G0 UIB
G0 UII
G0 UIK
G0 UIM
G0 UIN
G0 UIS
G0 UJA
G0 UJK
G0 UKB
G0 UKI
G0 UKN
G0 UKR
G0 ULE
G0 ULK
G0 ULN
G0 ULR
G0 ULV
G0 UMF
G0 UNJ
G0 UNO
G0 UNU
G0 UOA
G0 UOG
G0 UOH
G0 UON
G0 UOO
G0 UOR
G0 UPA
G0 UPC
G0 UPQ
G0 UPX
G0 UQN
G0 UQR
G0 UQX
G0 URE
G0 URG
G0 URJ
G0 URM
G0 URR
G0 URY
G0 USD
G0 USH
G0 USL
G0 USP
G0 USR
G0 USU
G0 USX
G0 UTL
G0 UTS
G0 UTY
G0 UUK
G0 UVB
G0 UVM
G0 UWG
G0 UWM
G0 UWY
G0 UXA
G0 UXC
G0 UXE
G0 UXM
G0 UYB
G0 UYL
G0 UYN
G0 UZC
G0 UZG
G0 UZJ
G0 VAK
G0 VAW
G0 VBA
G0 VBC
G0 VBI
G0 VBJ
G0 VBS
G0 VCA
G0 VCE
G0 VCG
G0 VCH
G0 VCL
G0 VCM
G0 VCP
G0 VCQ
G0 VCS
G0 VDG
G0 VDI
G0 VEA
G0 VFG
G0 VFI
G0 VFK
G0 VFR
G0 VGF
G0 VGQ
G0 VGS
G0 VGX
G0 VGZ
G0 VHB
G0 VHM
G0 VHS
G0 VHZ
G0 VIH
G0 VIN
G0 VIP
G0 VIR
G0 VIU
G0 VIW
G0 VIZ
G0 VJD
G0 VJG
G0 VJV
G0 VJW
G0 VKP
G0 VKQ
G0 VKT
G0 VLG
G0 VLH
G0 VLN
G0 VLO
G0 VMG
G0 VML
G0 VMX
G0 VNG
G0 VNP
G0 VNR
G0 VNS
G0 VNZ
G0 VOM
G0 VOW
G0 VOY
G0 VOZ
G0 VPF
G0 VPL
G0 VPM
G0 VPR
G0 VRD
G0 VRI
G0 VRN
G0 VRR
G0 VRY
G0 VSV
G0 VTH
G0 VTR
G0 VTU
G0 VTX
G0 VTZ
G0 VUF
G0 VUO
G0 VUV
G0 VUW
G0 VUZ
G0 VVE
G0 VVM
G0 VVN
G0 VVS
G0 VWC
G0 VWG
G0 VWL
G0 VWR
G0 VWY
G0 VXN
G0 VXO
G0 VYH
G0 VYO
G0 VYS
G0 VZG
G0 VZJ
G0 VZM
G0 VZW
G0 VZY
G0 WAF
G0 WAG
G0 WAI
G0 WAJ
G0 WAK
G0 WAP
G0 WAU
G0 WAV
G0 WBE
G0 WBJ
G0 WBX
G0 WBY
G0 WCA
G0 WCF
G0 WCG
G0 WCP
G0 WCW
G0 WCY
G0 WDA
G0 WDD
G0 WDM
G0 WEC
G0 WEE
G0 WEH
G0 WEJ
G0 WEN
G0 WFJ
G0 WFS
G0 WFW
G0 WFY
G0 WGC
G0 WGF
G0 WGK
G0 WGQ
G0 WHA
G0 WHE
G0 WHF
G0 WHG
G0 WHI
G0 WHJ
G0 WHM
G0 WHP
G0 WHS
G0 WHX
G0 WIU
G0 WIV
G0 WJE
G0 WJQ
G0 WKB
G0 WKC
G0 WKD
G0 WKG
G0 WKK
G0 WKR
G0 WKV
G0 WLB
G0 WLL
G0 WLO
G0 WLS
G0 WLW
G0 WMF
G0 WMM
G0 WNA
G0 WNC
G0 WNH
G0 WNO
G0 WNQ
G0 WNV
G0 WOD
G0 WOE
G0 WOK
G0 WOZ
G0 WPA
G0 WPB
G0 WPG
G0 WPJ
G0 WPN
G0 WPP
G0 WPR
G0 WQB
G0 WQI
G0 WQJ
G0 WQJ
G0 WQK
G0 WQL
G0 WQS
G0 WQT
G0 WQU
G0 WQV
G0 WRB
G0 WRF
G0 WRJ
G0 WRO
G0 WRP
G0 WSE
G0 WSG
G0 WSM
G0 WSO
G0 WSQ
G0 WST
G0 WSV
G0 WSW
G0 WTH
G0 WTT
G0 WTZ
G0 WUD
G0 WUE
G0 WUQ
G0 WUT
G0 WVK
G0 WVL
G0 WVU
G0 WVX
G0 WWG
G0 WWV
G0 WWW
G0 WXB
G0 WXI
G0 WXN
G0 WXS
G0 WXT
G0 WXU
G0 WYL
G0 WZI
G0 WZR
G0 XAX
G0 XBI
G0 XBJ
G0 XBU
G0 XEL
G0 XEM
G0 XGK
G0 XJB
G0 XRO
G0 XTV
G0 XXL
G0 XYS
G0 YAF
G0 YAK
G0 YAP
G0 YEW
G0 YLM
G0 YNM
G0 YOY
G0 ZAA
G0 ZEN
G0 ZEP
G0 ZGB
G0 ZIG

G1

G1 AAA
G1 AAE
G1 AAS
G1 ABR

G1 ABY
G1 ACH
G1 ADD
G1 ADT
G1 AED
G1 AEG
G1 AES
G1 AEV
G1 AGI
G1 AGT
G1 AHA
G1 AIH
G1 AIV
G1 AJI
G1 AJR
G1 AKN
G1 AKR
G1 AKT
G1 AKX
G1 ALT
G1 AMI
G1 AMV
G1 AMW
G1 ANT
G1 AOU
G1 AOY
G1 APV
G1 AQQ
G1 ARG
G1 ARI
G1 ARN
G1 ARV
G1 ASE
G1 AST
G1 ATI
G1 AUJ
G1 AUN
G1 AUQ
G1 AVI
G1 AVQ
G1 AYL
G1 BAD
G1 BAF
G1 BCL
G1 BCP
G1 BCR
G1 BCZ
G1 BDK
G1 BED
G1 BEI
G1 BEM
G1 BFQ
G1 BFU
G1 BFZ
G1 BGY
G1 BHM
G1 BIQ
G1 BIR
G1 BJC
G1 BJH
G1 BJJ
G1 BKH
G1 BLD
G1 BMF
G1 BND
G1 BNH
G1 BOE
G1 BOP
G1 BOZ
G1 BPH
G1 BPR
G1 BQC
G1 BRM
G1 BRV
G1 BSN
G1 BSP
G1 BSQ
G1 BSR
G1 BTU
G1 BUO
G1 BVI
G1 BWE
G1 BWT
G1 BXB
G1 BXG
G1 BXL
G1 BYY
G1 BZB
G1 BZG
G1 CAB
G1 CAE
G1 CAJ
G1 CAT
G1 CBD
G1 CCC
G1 CDV
G1 CEB
G1 CEH
G1 CFM
G1 CHB
G1 CIK
G1 CKL
G1 CLG
G1 CLH
G1 CLL
G1 CMB
G1 CMU
G1 COR
G1 COT
G1 CPP
G1 CPQ
G1 CQB
G1 CQL
G1 CQN
G1 CQO
G1 CQP
G1 CQV
G1 CSC
G1 CSL
G1 CSM
G1 CSU
G1 CUX
G1 CWC
G1 CWN
G1 CWO
G1 CXA
G1 CXD
G1 CXG
G1 CYB
G1 DBC
G1 DBD
G1 DBS
G1 DCG
G1 DDH
G1 DDY
G1 DEA
G1 DET
G1 DFL
G1 DHD
G1 DHE
G1 DIQ
G1 DIW
G1 DJA
G1 DJG
G1 DJT
G1 DKA
G1 DLC
G1 DLL
G1 DMQ
G1 DMX
G1 DOK
G1 DOS
G1 DOW
G1 DPF
G1 DQR
G1 DQX
G1 DRF
G1 DRP
G1 DRX
G1 DSC
G1 DST
G1 DTV
G1 DUE
G1 DUH
G1 DUM
G1 DVA
G1 DVW
G1 DVX
G1 DWG
G1 DWH
G1 DWN
G1 DX
G1 DXS
G1 DYD
G1 DYE
G1 DYX
G1 EAG
G1 EAY
G1 ECA
G1 ECL
G1 ECO
G1 ECZ
G1 EEL
G1 EEP
G1 EEQ
G1 EEU
G1 EFI
G1 EFY
G1 EGC
G1 EHN
G1 EHT
G1 EIQ
G1 EJE
G1 EJX
G1 EJY
G1 EKH
G1 EKW
G1 ELB
G1 ELG
G1 EMU
G1 ENI
G1 ENX
G1 EQT
G1 ERJ
G1 ERW
G1 ESV
G1 ETH
G1 EUZ
G1 EWA
G1 EWP
G1 EWQ
G1 EYO
G1 EZG
G1 EZV
G1 FBH
G1 FBK
G1 FCG
G1 FDB
G1 FDK
G1 FDY
G1 FEV
G1 FGB
G1 FGT
G1 FHB
G1 FIF
G1 FJQ
G1 FLQ
G1 FMD
G1 FML
G1 FNR
G1 FOX
G1 FPS
G1 FPU
G1 FQE
G1 FQR
G1 FQV
G1 FSH
G1 FTF
G1 FTO
G1 FTQ
G1 FUA
G1 FUD
G1 FXE
G1 FXJ
G1 FYG
G1 FYX
G1 GAB
G1 GBL
G1 GBO
G1 GBQ
G1 GCT
G1 GCU
G1 GCX
G1 GDS
G1 GFJ
G1 GGC
G1 GGH
G1 GHF
G1 GHM
G1 GIB
G1 GIQ
G1 GJC
G1 GJF
G1 GJM
G1 GKU
G1 GKY
G1 GLA
G1 GOE
G1 GPO
G1 GRD
G1 GRG
G1 GRP
G1 GRW
G1 GTI
G1 GTO
G1 GTV
G1 GUE
G1 GUT
G1 GVE
G1 GVI
G1 GWH
G1 GWL
G1 GWN
G1 GWV
G1 GXM
G1 GZY
G1 HAG
G1 HAK
G1 HAN
G1 HEH
G1 HFC
G1 HFD
G1 HHL
G1 HIZ
G1 HJH
G1 HJW
G1 HKE
G1 HLF
G1 HLW
G1 HMA
G1 HMO
G1 HMP
G1 HNI
G1 HOK
G1 HOS
G1 HPD
G1 HRE
G1 HRI
G1 HRW
G1 HSN
G1 HSS
G1 HUI
G1 HUL
G1 HUP
G1 HUW
G1 HWH
G1 HWN
G1 HXM
G1 HYD
G1 HZA
G1 HZW
G1 IAH
G1 IAN
G1 IAR
G1 IAT
G1 IBH
G1 ICO
G1 IDN
G1 IFL
G1 IGK
G1 IGZ
G1 IID
G1 IIE
G1 IIN
G1 IKO
G1 ILB
G1 ILN
G1 ILW
G1 IMH
G1 IMJ
G1 IML
G1 IMN
G1 IMQ
G1 INC
G1 INF
G1 IOJ
G1 IOS
G1 IOV
G1 IPM
G1 IQZ
G1 IRB
G1 IRP
G1 ISG
G1 ISO
G1 IUN
G1 IVE
G1 IWX
G1 IXY
G1 IYM
G1 IYU
G1 IZF
G1 IZQ
G1 JAJ
G1 JAK
G1 JAN
G1 JAR
G1 JAS
G1 JBD
G1 JDY
G1 JET
G1 JEV
G1 JFV
G1 JFW
G1 JFX
G1 JGQ
G1 JHI
G1 JHL
G1 JJJ
G1 JLE
G1 JME
G1 JMO
G1 JOB
G1 JPL
G1 JPV
G1 JQI
G1 JQJ
G1 JQW
G1 JRG
G1 JRH
G1 JRT
G1 JSA
G1 JSF
G1 JSH
G1 JSZ
G1 JTD
G1 JTG
G1 JTO
G1 JTY
G1 JUX
G1 JWZ
G1 JXI
G1 JZK
G1 JZQ
G1 KAJ
G1 KAN
G1 KAW
G1 KBU
G1 KBV
G1 KCE
G1 KDB
G1 KDC
G1 KDQ
G1 KDU
G1 KEA
G1 KEJ
G1 KEK
G1 KEX
G1 KFC
G1 KFD
G1 KGZ
G1 KHL
G1 KIK
G1 KMA
G1 KMO
G1 KNP
G1 KOO
G1 KPS
G1 KQJ
G1 KRI
G1 KTJ
G1 KTN
G1 KUL
G1 KVD
G1 KVM
G1 KVN
G1 KWJ
G1 KXF
G1 KXW
G1 KXY
G1 KYO
G1 KZI
G1 KZL
G1 LAJ
G1 LCX
G1 LCZ
G1 LDI
G1 LDM
G1 LFU
G1 LGA
G1 LGX
G1 LHQ
G1 LIE
G1 LIF
G1 LII
G1 LIQ
G1 LIS
G1 LIW
G1 LIY
G1 LIZ
G1 LJA
G1 LJB
G1 LJD
G1 LJR
G1 LJU
G1 LJY
G1 LLB
G1 LNT
G1 LNY
G1 LOA
G1 LOP
G1 LOR
G1 LOT
G1 LPW
G1 LQN
G1 LRO
G1 LRO
G1 LSJ
G1 LUS
G1 LUV
G1 LVN
G1 LWA
G1 LXI
G1 LXP
G1 LXR
G1 LXW
G1 MAK
G1 MAW
G1 MBT
G1 MCD
G1 MCP
G1 MCR
G1 MCX
G1 MDK
G1 MEZ
G1 MFE
G1 MFE
G1 MFF
G1 MFG
G1 MFN
G1 MHU
G1 MIB
G1 MIC
G1 MII
G1 MIK
G1 MIL
G1 MIR
G1 MIS
G1 MJD
G1 MKH
G1 MLK
G1 MLO
G1 MLR
G1 MME
G1 MMS
G1 MNP
G1 MPH
G1 MSE
G1 MSU
G1 MTC
G1 MTT
G1 MTY
G1 MUK
G1 MVL
G1 MWI
G1 MWP
G1 MXJ
G1 MXJ
G1 MXS
G1 NAF
G1 NAG
G1 NAR
G1 NBQ
G1 NCD
G1 NCI
G1 NCS
G1 NDA
G1 NDG
G1 NEJ
G1 NES
G1 NFF
G1 NFT
G1 NGH
G1 NHB
G1 NHH
G1 NHW
G1 NIB
G1 NID
G1 NIG
G1 NIS
G1 NKC
G1 NKU
G1 NLD
G1 NLI
G1 NLV
G1 NLX
G1 NMH
G1 NNP
G1 NOC
G1 NOD
G1 NOL
G1 NON
G1 NOX
G1 NPT
G1 NPV
G1 NQG
G1 NQP
G1 NQW
G1 NRL
G1 NRS
G1 NRT
G1 NRU
G1 NRV
G1 NSA
G1 NSU
G1 NTE
G1 NTH
G1 NTY
G1 NUC
G1 NUJ
G1 NUN
G1 NUR
G1 NUU
G1 NVQ
G1 NYN
G1 NZB
G1 NZM
G1 NZO
G1 NZT
G1 OAC
G1 OAH
G1 OAT
G1 OAV
G1 OAY
G1 OBD
G1 OBF
G1 OBY
G1 OCA
G1 OCJ
G1 OCN
G1 OEN
G1 OEX
G1 OFA
G1 OFZ
G1 OGU
G1 OHH
G1 OHN
G1 OHS
G1 OHT
G1 OHZ
G1 OIF
G1 OIN
G1 OIP
G1 OJJ
G1 OJP
G1 OKD
G1 OKT
G1 OLJ
G1 OMK
G1 OMO
G1 OMQ
G1 ONL
G1 ONN
G1 ONY
G1 OOE
G1 OOV
G1 OPE
G1 OPH
G1 OPK
G1 ORY
G1 OTR
G1 OUR
G1 OUT
G1 OVA
G1 OVR
G1 OWP
G1 OXD
G1 OXL
G1 OXX
G1 OYQ
G1 OYS
G1 OYW
G1 OYY
G1 OZF
G1 OZT
G1 OZU
G1 OZZ
G1 PAC
G1 PBO
G1 PCM
G1 PCS
G1 PCW
G1 PCX
G1 PDZ
G1 PEP
G1 PFP
G1 PFQ
G1 PGF
G1 PHL
G1 PHQ
G1 PHZ
G1 PIB
G1 PIC
G1 PIE
G1 PIR
G1 PJW
G1 PKX
G1 PKY
G1 PKZ
G1 PLB
G1 PLM
G1 PLT
G1 PLY
G1 PMN
G1 PMR
G1 PMS
G1 PNP
G1 POS
G1 PPA
G1 PPH
G1 PPP
G1 PRA
G1 PRB
G1 PRX
G1 PSP
G1 PSP
G1 PTH
G1 PTO
G1 PVO
G1 PVY
G1 PWD
G1 PWW
G1 PXH
G1 PXJ
G1 PYA
G1 PYT
G1 RAH
G1 RBK
G1 RBM
G1 RCD
G1 RCL
G1 RCP
G1 RCQ
G1 RCT
G1 REQ
G1 RFH
G1 RFO
G1 RHV
G1 RIJ
G1 RJR
G1 RKF
G1 RKO
G1 RLP
G1 RMH
G1 RNA
G1 ROI
G1 ROL
G1 RRD
G1 RRM
G1 RSB
G1 RSH
G1 RSL
G1 RSW
G1 RSZ
G1 RTG
G1 RTJ
G1 RTN
G1 RUN
G1 RUV
G1 RUX
G1 RUY
G1 RUZ
G1 RVD
G1 RVE
G1 RVU
G1 RVV
G1 RVX
G1 RWA
G1 RWB
G1 RWK
G1 RWS
G1 RWU
G1 RXC
G1 RXR
G1 RYO
G1 RZT
G1 RZV
G1 SAA
G1 SAO
G1 SAR
G1 SBB
G1 SBG
G1 SBO
G1 SBV
G1 SCP
G1 SCX
G1 SDC
G1 SDD
G1 SDT
G1 SE
G1 SEN
G1 SFI
G1 SFN
G1 SFO
G1 SFS
G1 SGC
G1 SGX
G1 SHC
G1 SHX
G1 SJF
G1 SKL
G1 SLN
G1 SLW
G1 SMA
G1 SMG
G1 SMX
G1 SNH
G1 SNJ
G1 SNL
G1 SNW
G1 SOA
G1 SOU
G1 SPC
G1 SPO
G1 SPQ
G1 SQB
G1 SQH
G1 SQY
G1 SRX
G1 SRZ
G1 SSO
G1 SSR
G1 SST
G1 STE
G1 SUI
G1 SUT
G1 SUV
G1 SWN
G1 SWY
G1 SXO
G1 SXV
G1 SXW
G1 SXZ
G1 SYG
G1 SYS
G1 TAO
G1 TAQ
G1 TBA
G1 TBC
G1 TBL
G1 TCM
G1 TDB
G1 TDG
G1 TDM
G1 TED
G1 TEJ
G1 TFI
G1 TFV
G1 TGC
G1 TGJ
G1 TGM
G1 TKO
G1 TLB
G1 TLF
G1 TLG
G1 TLQ
G1 TLY
G1 TMJ
G1 TMN
G1 TNG
G1 TNL
G1 TOA
G1 TOP
G1 TOU
G1 TOX
G1 TPA
G1 TQC
G1 TQP
G1 TQW
G1 TRB
G1 TRD
G1 TSE
G1 TSJ
G1 TSQ
G1 TSY
G1 TSZ
G1 TTM
G1 TTR
G1 TTW
G1 TTY
G1 TUK
G1 TUO
G1 TVG
G1 TVJ
G1 TVL
G1 TXC
G1 TXD
G1 TXY
G1 TYH
G1 TYV
G1 TYY
G1 TZC
G1 TZG
G1 TZJ
G1 UAI
G1 UAJ
G1 UAN
G1 UAS
G1 UAU
G1 UBA
G1 UBB
G1 UBW
G1 UCA
G1 UCP
G1 UCW
G1 UDA
G1 UDH
G1 UDK
G1 UDL
G1 UDO
G1 UEB
G1 UEG
G1 UFN
G1 UFQ
G1 UGM
G1 UGY
G1 UHD
G1 UHE
G1 UHI
G1 UHQ
G1 UHU
G1 UHY
G1 UHZ
G1 UIF
G1 UJB
G1 UJO
G1 UJU
G1 UJW
G1 UKE
G1 UKI
G1 UKQ
G1 UKR
G1 UKU
G1 UKX
G1 UKY
G1 ULN
G1 ULS
G1 UME
G1 UMH
G1 UML
G1 UMR
G1 UMV
G1 UMX
G1 UND
G1 UNF
G1 UNL
G1 UNP
G1 UNW
G1 UOJ
G1 UOM
G1 UOZ
G1 UPC
G1 UPH
G1 UPK
G1 UPR
G1 UPU
G1 UPY
G1 UQO
G1 URC
G1 USC
G1 USS
G1 UTA
G1 UTI
G1 UVQ
G1 UVY
G1 UWF
G1 UWG
G1 UWL
G1 UWR
G1 UWW
G1 UWY
G1 UXA
G1 UXB
G1 UXC
G1 UXG
G1 UXH
G1 UXK
G1 UXL
G1 UXP
G1 UXQ
G1 UYA
G1 UYE
G1 UYM
G1 UYP
G1 UZL
G1 VAC
G1 VAE
G1 VAF
G1 VBI
G1 VBM
G1 VBR
G1 VBW
G1 VBX
G1 VCA
G1 VCD
G1 VCN
G1 VCP
G1 VCT
G1 VDA
G1 VDB
G1 VDY
G1 VEB
G1 VEC
G1 VEQ
G1 VFD
G1 VFE
G1 VFG
G1 VFJ
G1 VFS
G1 VGE
G1 VGF
G1 VGT
G1 VGZ
G1 VHD
G1 VHG
G1 VHQ
G1 VIE
G1 VIK
G1 VIM
G1 VJW
G1 VKP
G1 VKZ
G1 VLB
G1 VLC
G1 VLM
G1 VLR
G1 VLV
G1 VLW
G1 VMB
G1 VMG
G1 VMH
G1 VMJ
G1 VML
G1 VMS
G1 VMT
G1 VMU
G1 VMW
G1 VMY
G1 VNA
G1 VOA
G1 VOK
G1 VPB
G1 VPK
G1 VPL
G1 VPP
G1 VPR
G1 VPV
G1 VPW
G1 VPX
G1 VQN
G1 VQO
G1 VQU
G1 VQW
G1 VRG
G1 VRH
G1 VSE
G1 VSL
G1 VSV
G1 VSW
G1 VTA
G1 VTD
G1 VTI
G1 VUE
G1 VUH
G1 VUL
G1 VUS
G1 VUT
G1 VUU
G1 VUX
G1 VVP
G1 VWV
G1 VWW
G1 VXJ
G1 VXK
G1 VXL
G1 VXQ
G1 VYK
G1 VYZ
G1 VZC
G1 VZM
G1 VZP
G1 WAT
G1 WCB
G1 WCL
G1 WCO
G1 WCR
G1 WDF
G1 WDM
G1 WDW
G1 WEA
G1 WEB
G1 WEU
G1 WEX
G1 WFK
G1 WFV
G1 WGH
G1 WGJ
G1 WGP
G1 WGV
G1 WHC
G1 WHH
G1 WHN
G1 WHP
G1 WIA
G1 WIH
G1 WIO
G1 WJZ
G1 WKP
G1 WKR
G1 WKV
G1 WLM
G1 WMD
G1 WMJ
G1 WNS
G1 WOB
G1 WOQ
G1 WOV
G1 WPD
G1 WRK
G1 WRM
G1 WTA
G1 WTF
G1 WUJ
G1 WUP
G1 WVF
G1 WVN
G1 WVO
G1 WWV
G1 WWX
G1 WXA
G1 WXB
G1 WXI
G1 WXR
G1 WXZ
G1 WZA
G1 WZB
G1 XAD
G1 XAE
G1 XBH
G1 XBU
G1 XCC
G1 XCS
G1 XCW
G1 XCZ
G1 XDA
G1 XDC
G1 XDH
G1 XDM
G1 XDT
G1 XEL
G1 XEM
G1 XFQ
G1 XGF
G1 XHW
G1 XIG
G1 XIJ
G1 XIP
G1 XIX
G1 XJP
G1 XJS
G1 XJV
G1 XKM
G1 XKR
G1 XKT
G1 XLC
G1 XLP
G1 XLX
G1 XMF
G1 XMK
G1 XMV
G1 XNS
G1 XOL
G1 XOX
G1 XPC
G1 XPR
G1 XQX
G1 XSF
G1 XSZ
G1 XTB
G1 XTP
G1 XWJ
G1 XWU
G1 XWX
G1 XWY
G1 XYL
G1 XYU
G1 XYX
G1 XZA
G1 XZL
G1 YAI
G1 YAR
G1 YBJ
G1 YBN
G1 YBQ
G1 YCB
G1 YCC
G1 YDH
G1 YDN
G1 YDT
G1 YEF
G1 YEX
G1 YEY
G1 YFF
G1 YFR
G1 YFU
G1 YFV
G1 YFW
G1 YFX
G1 YFY
G1 YGD
G1 YGJ
G1 YHR
G1 YIB
G1 YIF
G1 YIJ
G1 YIK
G1 YIO
G1 YJE
G1 YKV
G1 YLU
G1 YMI
G1 YMU
G1 YMW
G1 YMY
G1 YNE
G1 YNG
G1 YNI
G1 YNL
G1 YNW
G1 YNY
G1 YNZ
G1 YOK
G1 YOM
G1 YOR
G1 YPA
G1 YPB
G1 YPG
G1 YPQ
G1 YQH
G1 YQJ
G1 YQK
G1 YRV
G1 YRW
G1 YSR
G1 YST
G1 YSZ
G1 YUM
G1 YUO
G1 YUV
G1 YVF
G1 YVH
G1 YVR
G1 YVT
G1 YWC
G1 YWR
G1 YXF
G1 YXM
G1 YYC
G1 YYE
G1 YYN
G1 YYV
G1 YZB
G1 YZR
G1 YZS
G1 YZY
G1 ZAF
G1 ZAL
G1 ZAT
G1 ZCV
G1 ZCW
G1 ZDB
G1 ZDC
G1 ZDI
G1 ZDK
G1 ZDM
G1 ZEB
G1 ZEF
G1 ZEO
G1 ZEZ
G1 ZFO
G1 ZGC
G1 ZGL
G1 ZGL
G1 ZGU
G1 ZGX
G1 ZGZ
G1 ZHM
G1 ZHS
G1 ZHX
G1 ZIE
G1 ZIF
G1 ZIT
G1 ZIY
G1 ZJC
G1 ZJM
G1 ZJV
G1 ZKC
G1 ZLG
G1 ZLH
G1 ZLM
G1 ZLU
G1 ZMA
G1 ZMM
G1 ZMS
G1 ZMV
G1 ZNE
G1 ZNI
G1 ZNL
G1 ZOI
G1 ZOM
G1 ZOV
G1 ZQA
G1 ZQC
G1 ZQI
G1 ZQZ
G1 ZRF
G1 ZRN
G1 ZRZ
G1 ZSJ
G1 ZUG
G1 ZUK
G1 ZVB
G1 ZVL
G1 ZVN
G1 ZVQ
G1 ZWA
G1 ZWT
G1 ZXA
G1 ZXH
G1 ZXQ
G1 ZXT
G1 ZYB
G1 ZYC
G1 ZYH
G1 ZYX
G1 ZZX
G1 ZZY

G2

G2 AA
G2 AP
G2 AVV
G2 AXQ
G2 BC
G2 BDV
G2 BFC
G2 BGI
G2 CWO
G2 DA
G2 DPQ
G2 DS
G2 DVP
G2 DX
G2 FQS
G2 FZ
G2 GS
G2 GT
G2 HC
G2 HPG
G2 HW
G2 IV
G2 JP
G2 LO
G2 LV
G2 ML
G2 MT
G2 NM
G2 NV
G2 OT
G2 PLX
G2 PY
G2 RW
G2 SM
G2 TT
G2 TV
G2 UWX
G2 WW
G2 XYL
G2 ZE

G3

G3 ABC
G3 ADZ
G3 AGW
G3 AHD
G3 AHW
G3 AJY
G3 AQT
G3 ARC
G3 ARK
G3 ARO
G3 ARX
G3 ATA
G3 ATI
G3 AYA
G3 BA
G3 BEM
G3 BK
G3 BRK
G3 BRW
G3 BSA
G3 BSQ
G3 BXF
G3 BYV
G3 BZW
G3 CAR
G3 CDR
G3 CEN
G3 CJL
G3 CKO
G3 CNX
G3 CPS
G3 CQQ
G3 CSO
G3 CXX
G3 DAM
G3 DBI
G3 DDI
G3 DIH
G3 DJR
G3 DKJ
G3 DKN
G3 DLS
G3 DNE
G3 DRK
G3 DX
G3 DZE
G3 EAR
G3 EFB
G3 EFM
G3 EFU
G3 EGY
G3 EHG
G3 EIG
G3 EJS
G3 EKT
G3 EML
G3 EMZ
G3 ENG
G3 EPN
G3 ESP
G3 EWE
G3 EYB
G3 FAB
G3 FKB
G3 FRR
G3 FRW
G3 FSO
G3 FVO
G3 GFA
G3 GNF
G3 GNX
G3 GOA
G3 GRM
G3 GUL
G3 GVB
G3 GWI
G3 GZJ
G3 HAX
G3 HCM
G3 HCN
G3 HCU
G3 HDS
G3 HEL
G3 HER
G3 HGI
G3 HIJ
G3 HPP
G3 HRL
G3 HRZ
G3 HSR
G3 HUE
G3 HZX
G3 IGR
G3 II
G3 IIA
G3 IJN
G3 IMR
G3 INF
G3 ITK
G3 IVB
G3 IZV
G3 JBQ
G3 JCP
G3 JEV
G3 JFH
G3 JFL
G3 JHB
G3 JJM
G3 JMC
G3 JMT
G3 JPL
G3 JRT
G3 JVB
G3 JW
G3 JXB
G3 KCJ
G3 KER
G3 KFN
G3 KLB
G3 KMZ
G3 KNE
G3 KNM
G3 KOU
G3 KPP
G3 KRC
G3 KSG
G3 KSU
G3 KSY
G3 KUM
G3 KUS
G3 KWA
G3 KXU
G3 KYG
G3 LBL
G3 LCT
G3 LFZ
G3 LHE
G3 LHI
G3 LKZ
G3 LLL
G3 LMD
G3 LMK
G3 LMO
G3 LPP
G3 LSS
G3 LX
G3 LXI
G3 LYE
G3 LYK
G3 LZG
G3 MA
G3 MAL
G3 MAN
G3 MD
G3 MDB
G3 MDC
G3 MDX
G3 MEJ
G3 MGM
G3 MGP
G3 MGV
G3 MHW
G3 MII
G3 MJO
G3 MMH
G3 MMP
G3 MMR
G3 MPD
G3 MTM
G3 MUI
G3 MXG
G3 MZK
G3 NAC
G3 NAS
G3 NCM
G3 NCR
G3 NFO
G3 NGD
G3 NHS
G3 NKB
G3 NKR
G3 NLN
G3 NLZ
G3 NNF
G3 NNH
G3 NPI
G3 NQV
G3 NRB
G3 NRK
G3 NRT
G3 NSB
G3 NSM
G3 NTK
G3 NUE
G3 NUF
G3 NWD
G3 OBX
G3 ODR
G3 OFA
G3 OFL
G3 OGF
G3 OGY
G3 ORH
G3 OSB
G3 OSW
G3 OU
G3 OUF
G3 OUK
G3 OUR
G3 OVM
G3 OVQ
G3 OWF
G3 OWS
G3 OXQ
G3 OYE
G3 OYP
G3 OYS
G3 OZQ
G3 OZS
G3 PBA
G3 PBS
G3 PBU
G3 PCN
G3 PCQ
G3 PCR
G3 PD
G3 PDG
G3 PDK
G3 PGD
G3 PGV
G3 PHA
G3 PHR
G3 PHU
G3 PJD
G3 PJI
G3 PKM
G3 PKW
G3 PLH
G3 PLU
G3 PMM
G3 PNY
G3 PPG
G3 PQU
G3 PRN
G3 PTD
G3 PTU
G3 PUK
G3 PUV
G3 PVS
G3 PVW
G3 PWH
G3 PXQ
G3 PXT
G3 PYA
G3 PYU
G3 RAS
G3 RAT
G3 RAX
G3 RAY
G3 RBA
G3 RBG
G3 RBX
G3 RCU
G3 REA
G3 REM
G3 REN
G3 RFF
G3 RFL
G3 RFS
G3 RFT
G3 RHN
G3 RIE
G3 RJQ
G3 RLB
G3 RLU
G3 RME
G3 RNQ
G3 ROI
G3 RRX
G3 RSS
G3 RSY
G3 RTG
G3 RTU
G3 RUK
G3 RVU
G3 RWA
G3 RWP
G3 RXF
G3 RXL
G3 RXQ
G3 RZA
G3 SAZ
G3 SCU
G3 SDQ
G3 SEV
G3 SEW
G3 SIP
G3 SJZ
G3 SKC
G3 SKP
G3 SKY
G3 SLR
G3 SMP
G3 SMX
G3 SPB
G3 SPE
G3 SPX
G3 SPZ
G3 SQV
G3 SR
G3 SRA
G3 SRO
G3 SSC
G3 SSG
G3 SSO
G3 STR
G3 STS
G3 SUF
G3 SUY
G3 SVL
G3 SWF
G3 SWM
G3 SYF
G3 SZO
G3 SZT
G3 TAH
G3 TAM
G3 TAN
G3 TBA
G3 TCA
G3 TCE
G3 TCR
G3 TDJ
G3 TFT
G3 TGP
G3 TGW
G3 TIB
G3 TIC
G3 TIQ
G3 TKV
G3 TKW
G3 TLE
G3 TMG
G3 TMH
G3 TMZ
G3 TOE
G3 TQH
G3 TRF
G3 TSH
G3 TSK
G3 TTK
G3 TVA
G3 TVQ
G3 TWE
G3 TWZ
G3 TXA
G3 TYB
G3 TYY
G3 UBU
G3 UBW
G3 UCM
G3 UCV
G3 UDU
G3 UFK
G3 UFP
G3 UGG
G3 UGK
G3 UGY
G3 UIE
G3 UKU
G3 ULF
G3 UMK
G3 UMW
G3 UNC
G3 UNH
G3 UNR
G3 UPK
G3 USB
G3 USE
G3 UUW
G3 UWI
G3 UXB
G3 VAA
G3 VAG
G3 VBS
G3 VCU
G3 VDX
G3 VEQ
G3 VFP
G3 VFU
G3 VFV
G3 VGF
G3 VGG
G3 VGU
G3 VHB
G3 VHF
G3 VHG
G3 VHJ
G3 VHY
G3 VII
G3 VIV
G3 VLP
G3 VLT
G3 VMB
G3 VMH
G3 VMS
G3 VNC
G3 VOD
G3 VPD
G3 VRL
G3 VSS
G3 VUF
G3 VUP
G3 VUQ
G3 VW
G3 VWF
G3 VWL
G3 VWP
G3 VXI
G3 VXP
G3 VXX
G3 VZZ
G3 WA
G3 WCS
G3 WFB
G3 WFK
G3 WGC
G3 WGM
G3 WGQ
G3 WHL
G3 WHR
G3 WIC
G3 WIE
G3 WIR
G3 WKD
G3 WKH
G3 WKU
G3 WM
G3 WMM
G3 WMR
G3 WNU
G3 WOF
G3 WOX
G3 WPO
G3 WQO
G3 WQP
G3 WQT
G3 WRS
G3 WRU
G3 WRV
G3 WSN
G3 WTA
G3 WVY
G3 WVZ
G3 WW
G3 WWW
G3 WXL
G3 WXX
G3 WZD
G3 WZN
G3 XAD
G3 XAJ
G3 XCR
G3 XDA
G3 XDE
G3 XDH
G3 XDQ
G3 XDV
G3 XEB
G3 XEK
G3 XFA
G3 XFE
G3 XFV
G3 XFZ
G3 XGM
G3 XGQ
G3 XGS
G3 XHF
G3 XJF
G3 XJK
G3 XJO
G3 XKQ
G3 XKT
G3 XLF
G3 XNH
G3 XNT
G3 XPE
G3 XPO

Withheld

G3 XQX
G3 XRE
G3 XSA
G3 XSH
G3 XTX
G3 XUA
G3 XVF
G3 XWJ
G3 XWM
G3 XXT
G3 XYS
G3 XYX
G3 XZS
G3 XZT
G3 YAM
G3 YAP
G3 YAQ
G3 YAS
G3 YBD
G3 YBT
G3 YC
G3 YCA
G3 YCR
G3 YDL
G3 YEH
G3 YFJ
G3 YFN
G3 YGS
G3 YIP
G3 YIS
G3 YIV
G3 YJO
G3 YJR
G3 YL
G3 YLE
G3 YLG
G3 YLO
G3 YMP
G3 YMX
G3 YNH
G3 YNT
G3 YOK
G3 YPP
G3 YPQ
G3 YPX
G3 YQD
G3 YRE
G3 YRJ
G3 YRO
G3 YRW
G3 YRZ
G3 YSC
G3 YSH
G3 YTH
G3 YTK
G3 YUE
G3 YUV
G3 YVF
G3 YWQ
G3 YXX
G3 YXZ
G3 YYD
G3 YYH
G3 YYO
G3 YYT
G3 YZG
G3 YZJ
G3 YZY
G3 ZAF
G3 ZCC
G3 ZCU
G3 ZDD
G3 ZDR
G3 ZDV
G3 ZEH
G3 ZEV
G3 ZFJ
G3 ZHH
G3 ZHI
G3 ZHW
G3 ZHX
G3 ZKF
G3 ZLB
G3 ZLW
G3 ZMC
G3 ZMD
G3 ZMS
G3 ZMU
G3 ZMY
G3 ZOD
G3 ZOE
G3 ZPF
G3 ZPQ
G3 ZQX
G3 ZRK
G3 ZRV
G3 ZSE
G3 ZSK
G3 ZSP
G3 ZSV
G3 ZTO
G3 ZUD
G3 ZWI
G3 ZXH
G3 ZXL
G3 ZXP
G3 ZXX
G3 ZYN
G3 ZYS
G3 ZYW
G3 ZZO

G4

G4 AAO
G4 AAW
G4 ABD
G4 ABF
G4 ABP
G4 ABT
G4 ADN
G4 ADX
G4 AFG
G4 AHS
G4 AJC
G4 AJK
G4 AJM
G4 AKU
G4 AKZ
G4 ALG
G4 ALH
G4 ALI
G4 ANA
G4 ANH
G4 ANS
G4 AOG
G4 AON
G4 AOV
G4 APT
G4 APV
G4 AQU
G4 AQV
G4 AQY
G4 ARG
G4 ARP
G4 ASB
G4 ASE
G4 AST
G4 ATI
G4 ATX
G4 AVB
G4 AVD
G4 AVU
G4 AWC
G4 AWP
G4 AXH
G4 AXN
G4 AXP
G4 AYF
G4 AYJ
G4 AYM
G4 AYT
G4 AYV
G4 AZF
G4 AZR
G4 AZZ
G4 BAZ
G4 BBS
G4 BBZ
G4 BCC
G4 BCK
G4 BDE
G4 BDO
G4 BDP
G4 BDY
G4 BEE
G4 BEG
G4 BEI
G4 BEN
G4 BEW
G4 BFJ
G4 BFP
G4 BFY
G4 BFZ
G4 BGQ
G4 BGX
G4 BIP
G4 BJM
G4 BKC
G4 BKM
G4 BLF
G4 BLH
G4 BLI
G4 BLJ
G4 BLM
G4 BLN
G4 BLQ
G4 BLX
G4 BMF
G4 BMH
G4 BNU
G4 BOY
G4 BPQ
G4 BSJ
G4 BSU
G4 BSX
G4 BTH
G4 BTS
G4 BUD
G4 BVH
G4 BWD
G4 BWQ
G4 BXO
G4 BXP
G4 BYY
G4 BZD
G4 BZO
G4 CAO
G4 CAW
G4 CBR
G4 CCX
G4 CDE
G4 CDM
G4 CDT
G4 CEH
G4 CFF
G4 CFJ
G4 CGS
G4 CIG
G4 CIL
G4 CIN
G4 CIV
G4 CJC
G4 CJF
G4 CJS
G4 CJX
G4 CJZ
G4 CKN
G4 CKR
G4 CLK
G4 CLQ
G4 CLT
G4 CLU
G4 CND
G4 CNE
G4 CNF
G4 CNG
G4 COK
G4 COO
G4 CPK
G4 CPO
G4 CPR
G4 CQJ
G4 CQU
G4 CRF
G4 CRJ
G4 CRV
G4 CSN
G4 CSS
G4 CTG
G4 CTL
G4 CTP
G4 CTR
G4 CTS
G4 CUJ
G4 CUS
G4 CVY
G4 CXH
G4 CXM
G4 CYE
G4 CYQ
G4 CZE
G4 CZJ
G4 CZS
G4 DAA
G4 DAE
G4 DAR
G4 DBI
G4 DBL
G4 DBW
G4 DCQ
G4 DCR
G4 DCV
G4 DDR
G4 DDS
G4 DED
G4 DEE
G4 DGK
G4 DGU
G4 DHA
G4 DHJ
G4 DHO
G4 DHP
G4 DIF
G4 DIW
G4 DJW
G4 DKB
G4 DKF
G4 DKK
G4 DKO
G4 DLB
G4 DLX
G4 DMA
G4 DMD
G4 DME
G4 DMN
G4 DMU
G4 DNI
G4 DOI
G4 DPE
G4 DPN
G4 DPX
G4 DPY
G4 DQY
G4 DRK
G4 DRM
G4 DRY
G4 DSB
G4 DST
G4 DTD
G4 DTF
G4 DUC
G4 DUH
G4 DUK
G4 DUZ
G4 DVF
G4 DVP
G4 DVQ
G4 DVS
G4 DVW
G4 DWQ
G4 DWV
G4 DX
G4 DXA
G4 DXG
G4 DXX
G4 DYA
G4 DYK
G4 DYT
G4 DYW
G4 DZE
G4 DZF
G4 DZL
G4 DZR
G4 EAA
G4 EAR
G4 ECK
G4 ECM
G4 ECY
G4 EDE
G4 EDS
G4 EEI
G4 EEM
G4 EER
G4 EFT
G4 EFW
G4 EGH
G4 EGT
G4 EHA
G4 EHH
G4 EHZ
G4 EIB
G4 EIS
G4 EIT
G4 EJB
G4 EJN
G4 EKK
G4 ELD
G4 ELN
G4 EMM
G4 EMO
G4 EMR
G4 EOZ
G4 EPJ
G4 EQB
G4 EQH
G4 EQQ
G4 ETH
G4 ETQ
G4 ETV
G4 EUE
G4 EUM
G4 EUQ
G4 EWC
G4 EWN
G4 EWS
G4 EWX
G4 EXA
G4 EXQ
G4 EXT
G4 EXW
G4 EYL
G4 EYR
G4 EZR
G4 EZT
G4 EZV
G4 EZW
G4 EZZ
G4 FBA
G4 FBJ
G4 FBW
G4 FBX
G4 FCR
G4 FDB
G4 FDL
G4 FDO
G4 FFJ
G4 FHA
G4 FHI
G4 FID
G4 FIK
G4 FIS
G4 FIX
G4 FJQ
G4 FJR
G4 FKK
G4 FLH
G4 FLN
G4 FLQ
G4 FLU
G4 FMG
G4 FOC
G4 FOK
G4 FP
G4 FPH
G4 FPK
G4 FPS
G4 FRE
G4 FRU
G4 FSC
G4 FSJ
G4 FTC
G4 FTF
G4 FTJ
G4 FUB
G4 FUF
G4 FUH
G4 FUN
G4 FUT
G4 FVG
G4 FVN
G4 FWC
G4 FWG
G4 FWX
G4 FXB
G4 FXO
G4 FXQ
G4 FYA
G4 FYF
G4 FYX
G4 FZE
G4 GAD
G4 GAS
G4 GAU
G4 GAV
G4 GBG
G4 GCB
G4 GCD
G4 GCG
G4 GCM
G4 GCP
G4 GCW
G4 GCZ
G4 GEF
G4 GER
G4 GES
G4 GFK
G4 GFX
G4 GGS
G4 GHD
G4 GHJ
G4 GHP
G4 GHU
G4 GIF
G4 GII
G4 GJC
G4 GJH
G4 GJL
G4 GJN
G4 GJW
G4 GKA
G4 GLT
G4 GMA
G4 GME
G4 GMJ
G4 GMO
G4 GMP
G4 GMV
G4 GNC
G4 GNK
G4 GOC
G4 GOH
G4 GOQ
G4 GOY
G4 GPK
G4 GPU
G4 GPZ
G4 GQB
G4 GRI
G4 GRW
G4 GSQ
G4 GSV
G4 GTA
G4 GTF
G4 GTG
G4 GTI
G4 GTP
G4 GTR
G4 GTT
G4 GTW
G4 GTY
G4 GUF
G4 GUI
G4 GUJ
G4 GVB
G4 GVC
G4 GVD
G4 GVN
G4 GVX
G4 GWD
G4 GWO
G4 GYB
G4 GYO
G4 GZ
G4 GZJ
G4 HAD
G4 HAQ
G4 HBJ
G4 HCF
G4 HCL
G4 HCW
G4 HDK
G4 HEL
G4 HEQ
G4 HFB
G4 HFF
G4 HFT
G4 HGI
G4 HGQ
G4 HGU
G4 HHI
G4 HHT
G4 HHV
G4 HIP
G4 HJG
G4 HJM
G4 HKF
G4 HKK
G4 HKM
G4 HKT
G4 HLP
G4 HLR
G4 HLU
G4 HLU
G4 HMG
G4 HMI
G4 HMJ
G4 HNB
G4 HNH
G4 HNY
G4 HOV
G4 HPF
G4 HPG
G4 HPI
G4 HPL
G4 HPW
G4 HRK
G4 HRM
G4 HRO
G4 HRV
G4 HSO
G4 HSR
G4 HSU
G4 HSY
G4 HTP
G4 HUK
G4 HVE
G4 HVK
G4 HVX
G4 HXD
G4 HXT
G4 HYC
G4 HYX
G4 HZA
G4 HZB
G4 IAL
G4 IAZ
G4 IBK
G4 IBZ
G4 ICD
G4 ICT
G4 ICV
G4 IDK
G4 IDO
G4 IDY
G4 IDZ
G4 IEJ
G4 IEY
G4 IEY
G4 IFB
G4 IFC
G4 IFU
G4 IGO
G4 IGW
G4 IGZ
G4 IHH
G4 IHL
G4 IIG
G4 IIQ
G4 IIU
G4 IJC
G4 IJE
G4 IJL
G4 IKH
G4 IKK
G4 IKP
G4 IKR
G4 IKZ
G4 IMD
G4 IMN
G4 IMO
G4 IMZ
G4 INK
G4 INT
G4 IOX
G4 IPP
G4 IPQ
G4 IPW
G4 IQE
G4 IQI
G4 IQP
G4 IRN
G4 IRQ
G4 IRW
G4 IRX
G4 ISI
G4 ISP
G4 ITF
G4 IUB
G4 IUG
G4 IUR
G4 IUW
G4 IVH
G4 IVK
G4 IVS
G4 IVV
G4 IWB
G4 IXL
G4 IXN
G4 IYJ
G4 IYL
G4 IYU
G4 IZC
G4 JBC
G4 JBI
G4 JBX
G4 JCI
G4 JCN
G4 JCO
G4 JCP
G4 JDL
G4 JEC
G4 JEG
G4 JEN
G4 JEQ
G4 JET
G4 JFB
G4 JFK
G4 JFO
G4 JFT
G4 JGL
G4 JGP
G4 JGR
G4 JHC
G4 JHD
G4 JHM
G4 JIB
G4 JIN
G4 JIS
G4 JJI
G4 JKO
G4 JLB
G4 JLQ
G4 JLT
G4 JLU
G4 JMH
G4 JMJ
G4 JMN
G4 JMP
G4 JMV
G4 JNC
G4 JNN
G4 JNW
G4 JNY
G4 JOJ
G4 JOK
G4 JON
G4 JOP
G4 JPI
G4 JQE
G4 JQM
G4 JQP
G4 JQT
G4 JRC
G4 JRH
G4 JRL
G4 JRN
G4 JRO
G4 JRT
G4 JRV
G4 JSL
G4 JTJ
G4 JTY
G4 JUT
G4 JUY
G4 JVF
G4 JVG
G4 JWB
G4 JWG
G4 JWT
G4 JXL
G4 JXS
G4 JXX
G4 JXY
G4 JZJ
G4 KAG
G4 KAJ
G4 KAN
G4 KAO
G4 KAW
G4 KAY
G4 KBC
G4 KBD
G4 KBF
G4 KCS
G4 KCW
G4 KDG
G4 KDX
G4 KEU
G4 KEV
G4 KFE
G4 KFK
G4 KFO
G4 KHA
G4 KHV
G4 KIJ
G4 KIO
G4 KIV
G4 KJQ
G4 KJW
G4 KLG
G4 KLH
G4 KLL
G4 KLW
G4 KMC
G4 KMV
G4 KND
G4 KNK
G4 KNM
G4 KNP
G4 KOB
G4 KOH
G4 KOP
G4 KPB
G4 KPJ
G4 KPT
G4 KPW
G4 KQU
G4 KRB
G4 KRM
G4 KRO
G4 KRV
G4 KSD
G4 KSE
G4 KSW
G4 KTE
G4 KTS
G4 KUI
G4 KVA
G4 KWG
G4 KXF
G4 KXN
G4 KYA
G4 KYC
G4 KYD
G4 KYF
G4 KYN
G4 KYV
G4 KZS
G4 KZY
G4 LAC
G4 LAH
G4 LAX
G4 LBB
G4 LBI
G4 LCD
G4 LCZ
G4 LDC
G4 LDG
G4 LDY
G4 LDZ
G4 LEB
G4 LEF
G4 LEL
G4 LFA
G4 LFM
G4 LFU
G4 LGD
G4 LGE
G4 LGF
G4 LGK
G4 LGZ
G4 LHN
G4 LHU
G4 LHX
G4 LIB
G4 LIP
G4 LJJ
G4 LJZ
G4 LKI
G4 LKV
G4 LLX
G4 LMJ
G4 LMS
G4 LMZ
G4 LNA
G4 LND
G4 LNF
G4 LOU
G4 LPB
G4 LQY
G4 LRR
G4 LRS
G4 LRX
G4 LTF
G4 LTX
G4 LUG
G4 LUL
G4 LUZ
G4 LVM
G4 LWR
G4 LXE
G4 LXI
G4 LXK
G4 LXL
G4 LXQ
G4 LXS
G4 LXT
G4 LXZ
G4 LZG
G4 LZL
G4 LZM
G4 LZW
G4 MAD
G4 MAM
G4 MAP
G4 MAW
G4 MBN
G4 MBS
G4 MBU
G4 MCL
G4 MCN
G4 MDI
G4 MDP
G4 MDX
G4 MEC
G4 MEZ
G4 MFU
G4 MGL
G4 MHN
G4 MHU
G4 MIN
G4 MIU
G4 MIY
G4 MJH
G4 MJO
G4 MJS
G4 MJZ
G4 MKS
G4 MLA
G4 MLJ
G4 MLP
G4 MNO
G4 MOF
G4 MOM
G4 MOX
G4 MPD
G4 MPU
G4 MPV
G4 MPX
G4 MQ
G4 MQB
G4 MQE
G4 MR
G4 MRE
G4 MRG
G4 MRH
G4 MRM
G4 MRV
G4 MSF
G4 MSH
G4 MSO
G4 MTK
G4 MUF
G4 MUR
G4 MVC
G4 MVH
G4 MVK
G4 MVL
G4 MVN
G4 MVR
G4 MVU
G4 MVV
G4 MWC
G4 MWK
G4 MWM
G4 MXA
G4 MXQ
G4 MXR
G4 MYC
G4 MYI
G4 MZJ
G4 NAB
G4 NAD
G4 NAR
G4 NAZ
G4 NBZ
G4 NCE
G4 NCN
G4 NCT
G4 NDH
G4 NDR
G4 NDW
G4 NEC
G4 NEV
G4 NEX
G4 NFB
G4 NFC
G4 NFD
G4 NFJ
G4 NFU
G4 NGH
G4 NHJ
G4 NHS
G4 NIA
G4 NIS
G4 NIW
G4 NJE
G4 NJH
G4 NKL
G4 NKM
G4 NKT
G4 NKV
G4 NLP
G4 NLX
G4 NLY
G4 NMI
G4 NMX
G4 NNA
G4 NNC
G4 NNU
G4 NNW
G4 NOV
G4 NPJ
G4 NPX
G4 NQF
G4 NQK
G4 NQX
G4 NRL
G4 NSG
G4 NSI
G4 NSU
G4 NTB
G4 NTF
G4 NTM
G4 NTN
G4 NUL
G4 NVB
G4 NVF
G4 NVO
G4 NVR
G4 NVZ
G4 NWP
G4 NXH
G4 NXK
G4 NXM
G4 NYS
G4 NZF
G4 NZV
G4 OAC
G4 OAO
G4 OAO
G4 OAQ
G4 OAT
G4 OBI
G4 OBL
G4 OCO
G4 OCP
G4 ODC
G4 OEL
G4 OFY
G4 OGD
G4 OGE
G4 OGF
G4 OGI
G4 OGP
G4 OGT
G4 OHK
G4 OHN
G4 OIP
G4 OIX
G4 OIZ
G4 OJC
G4 OJO
G4 OKK
G4 OKT
G4 OLN
G4 OLT
G4 OLV
G4 OMA
G4 OME
G4 OMF
G4 OMH
G4 OMX
G4 ONE
G4 ONK
G4 ONL
G4 ONW
G4 OOP
G4 OOR
G4 OOW
G4 ORF
G4 ORM
G4 ORR
G4 OSW
G4 OTK
G4 OTQ
G4 OUK
G4 OUL
G4 OVK
G4 OVP
G4 OVU
G4 OWF
G4 OWI
G4 OWO
G4 OWW
G4 OXC
G4 OXF
G4 OXW
G4 OXY
G4 OYD
G4 OYF
G4 OYS
G4 OZH
G4 OZZ
G4 PAB
G4 PAG
G4 PAN
G4 PAO
G4 PAP
G4 PBL
G4 PBV
G4 PC
G4 PCA
G4 PCI
G4 PCS
G4 PDF
G4 PDL
G4 PDS
G4 PDV
G4 PEB
G4 PEM
G4 PF
G4 PFC
G4 PFD
G4 PFN
G4 PGI
G4 PGP
G4 PGS
G4 PHJ
G4 PHO
G4 PHU
G4 PHW
G4 PIO
G4 PIV
G4 PIX
G4 PJA
G4 PJC
G4 PJF
G4 PJG
G4 PLC
G4 PLN
G4 PLP
G4 PLT
G4 PM
G4 PMN
G4 PMQ
G4 PMX
G4 PNF
G4 PNJ
G4 PNW
G4 POA
G4 POQ
G4 PPM
G4 PQA
G4 PQL
G4 PQT
G4 PRV
G4 PSA
G4 PSB
G4 PSF
G4 PSN
G4 PTB
G4 PTI
G4 PTJ
G4 PTS
G4 PTX
G4 PUF
G4 PUJ
G4 PUR
G4 PVA
G4 PVB
G4 PWO
G4 PXI
G4 PXL
G4 PXO
G4 PXW
G4 PXZ
G4 PYM
G4 PYR
G4 PZE
G4 PZK
G4 RAD
G4 RAF
G4 RBI
G4 RBO
G4 RBV
G4 RCA
G4 RDK
G4 REA
G4 REL
G4 REQ
G4 REV
G4 RFM
G4 RFQ
G4 RFT
G4 RFX
G4 RFZ
G4 RGK
G4 RGV
G4 RHD
G4 RHN
G4 RHO
G4 RIC
G4 RII
G4 RIT
G4 RIW
G4 RKJ
G4 RKY
G4 RLQ
G4 RMI
G4 RMK
G4 RMM
G4 RMR
G4 RNI
G4 RNL
G4 RNS
G4 ROE
G4 ROO
G4 RPF
G4 RPX
G4 RQE
G4 RQY
G4 RRC
G4 RRZ
G4 RS
G4 RSA
G4 RSB
G4 RSE
G4 RSH
G4 RSQ
G4 RSR
G4 RSY
G4 RTB
G4 RUV
G4 RVB
G4 RVM
G4 RVR
G4 RWA
G4 RWC
G4 RWO
G4 RWU
G4 RXN
G4 RXT
G4 RYD
G4 RYF
G4 RYN
G4 RYU
G4 RYX
G4 RZG
G4 RZK
G4 RZS
G4 SAE
G4 SAP
G4 SAQ
G4 SBC
G4 SBL
G4 SBT
G4 SBX
G4 SCP
G4 SDG
G4 SDM
G4 SDN
G4 SDP
G4 SDR
G4 SDW
G4 SFI
G4 SFM
G4 SFU
G4 SFX
G4 SGB
G4 SGM
G4 SGO
G4 SGZ
G4 SHD
G4 SHP
G4 SHS
G4 SHZ
G4 SIO
G4 SIU
G4 SIV
G4 SJS
G4 SJX
G4 SJY
G4 SKH
G4 SKK
G4 SKN
G4 SLU
G4 SLX
G4 SMN
G4 SMY
G4 SNK
G4 SNS
G4 SNT
G4 SON
G4 SOX
G4 SPG
G4 SPK
G4 SPO
G4 SPX
G4 SQU
G4 SRA
G4 SRB
G4 SRG
G4 SRH
G4 SRK
G4 SRS
G4 SSG
G4 SSI
G4 SSK
G4 SST
G4 SSX
G4 STA
G4 STM
G4 STR
G4 SUH
G4 SUI
G4 SVF
G4 SVN
G4 SVW
G4 SWI
G4 SWL
G4 SWW
G4 SWX
G4 SXR
G4 SXS
G4 SYJ
G4 SZD
G4 SZM
G4 TAA
G4 TAI
G4 TAS
G4 TAX
G4 TBA
G4 TBD
G4 TBL
G4 TBW
G4 TBZ
G4 TCG
G4 TCJ
G4 TCN
G4 TCU
G4 TCX
G4 TCZ
G4 TDS
G4 TEI
G4 TET
G4 TEV
G4 TFE
G4 TFN
G4 TFR
G4 THD
G4 THE
G4 THJ
G4 THL
G4 THR
G4 TIN
G4 TIO
G4 TJB
G4 TJC
G4 TJE
G4 TJJ
G4 TJT
G4 TKA
G4 TKG
G4 TKI
G4 TKJ
G4 TKN
G4 TKR
G4 TKV
G4 TLP
G4 TLV
G4 TMT
G4 TMW
G4 TNI
G4 TNN
G4 TNT
G4 TNW
G4 TNX
G4 TOC
G4 TOJ
G4 TOK
G4 TPA
G4 TQE
G4 TQF
G4 TQG
G4 TQJ
G4 TQM
G4 TQM
G4 TRK
G4 TSH
G4 TSM
G4 TSR
G4 TTI
G4 TTU
G4 TTW
G4 TUE
G4 TUG
G4 TUS
G4 TUT
G4 TUY
G4 TUZ
G4 TVH
G4 TVK
G4 TVP
G4 TVR
G4 TWE
G4 TWI
G4 TWJ
G4 TXB
G4 TXP
G4 TXR
G4 TXS
G4 TXX
G4 TYB
G4 TYC
G4 TZJ
G4 TZP
G4 TZT
G4 UCD
G4 UCO
G4 UDC
G4 UDJ
G4 UDR
G4 UDT
G4 UEM
G4 UEQ
G4 UFI
G4 UHC
G4 UHD
G4 UHN
G4 UHP
G4 UIN
G4 UIX
G4 UJD
G4 UJG
G4 UKF
G4 UKP
G4 ULA
G4 ULF
G4 ULS
G4 UMX
G4 UNG
G4 UNP
G4 UNR
G4 UNT
G4 UNU
G4 UOQ
G4 UPE
G4 UPF
G4 UPZ
G4 UQB
G4 UQC
G4 URH
G4 URZ
G4 USG
G4 USI
G4 UTT
G4 UTW
G4 UVE
G4 UVP
G4 UWA
G4 UWD
G4 UWQ
G4 UXR
G4 UXU
G4 UYL
G4 UZX
G4 VAD
G4 VAJ
G4 VBT
G4 VBW
G4 VCD
G4 VCH
G4 VCK
G4 VCT
G4 VCY
G4 VDE
G4 VDU
G4 VDV
G4 VEK
G4 VFA
G4 VFM
G4 VFS
G4 VFT
G4 VGE
G4 VGH
G4 VHF
G4 VHU
G4 VIN
G4 VJG
G4 VJM
G4 VJU
G4 VJW
G4 VJX
G4 VJY
G4 VKB
G4 VLC
G4 VLM
G4 VLO
G4 VLR
G4 VLY
G4 VOM
G4 VOQ
G4 VOX
G4 VPO
G4 VPT
G4 VQG
G4 VQO
G4 VQQ
G4 VRQ
G4 VRR
G4 VSF
G4 VSZ
G4 VTF
G4 VTH
G4 VTK
G4 VTX
G4 VUB
G4 VUJ
G4 VUO
G4 VUT
G4 VUX
G4 VVG
G4 VVU
G4 VXC
G4 VXO
G4 VXQ
G4 VXY
G4 VYM
G4 VYS
G4 VYW
G4 VYX
G4 VZG
G4 VZN
G4 VZO
G4 VZV
G4 VZZ
G4 WAD
G4 WAU
G4 WBB
G4 WBC
G4 WBM
G4 WBR
G4 WCS
G4 WDL
G4 WDU
G4 WEA
G4 WEO
G4 WFQ
G4 WHM
G4 WHO
G4 WHR
G4 WHX
G4 WID
G4 WIF
G4 WIX
G4 WJI
G4 WJT
G4 WKA
G4 WKH
G4 WKJ
G4 WKX
G4 WKY
G4 WLC
G4 WMI
G4 WMM
G4 WMT
G4 WNC
G4 WNE
G4 WNM
G4 WNO
G4 WNT
G4 WOC
G4 WOF
G4 WOV
G4 WOY
G4 WPD
G4 WPN
G4 WPS
G4 WQG
G4 WQI
G4 WQR
G4 WQV
G4 WQX
G4 WRI
G4 WRT
G4 WRW
G4 WRY
G4 WRZ
G4 WSM
G4 WSQ
G4 WSZ
G4 WTF
G4 WTV
G4 WVE
G4 WVG
G4 WVU
G4 WVX
G4 WWD
G4 WWK
G4 WWS
G4 WXM
G4 WXN
G4 WXP
G4 WXZ
G4 WYJ
G4 WZE
G4 WZK
G4 XAK
G4 XBT
G4 XCA
G4 XCP
G4 XCS
G4 XDA
G4 XDD
G4 XDF
G4 XDZ
G4 XEC
G4 XEH
G4 XEM
G4 XEN
G4 XFB
G4 XFD
G4 XGB
G4 XGU
G4 XHJ
G4 XHS
G4 XIB
G4 XID
G4 XIF
G4 XIH
G4 XIV
G4 XIY
G4 XJA
G4 XJB
G4 XJC
G4 XJP
G4 XJW
G4 XKB
G4 XKX
G4 XKY
G4 XLS
G4 XMG
G4 XMW
G4 XNG
G4 XNH
G4 XNJ
G4 XOB
G4 XOO
G4 XOT
G4 XOU
G4 XOX
G4 XPC
G4 XPE
G4 XPL
G4 XPM
G4 XQL
G4 XQS
G4 XRI
G4 XRS
G4 XRZ
G4 XSD
G4 XSH
G4 XTN
G4 XUR
G4 XUT
G4 XWG
G4 XWJ
G4 XWV
G4 XYE
G4 XYT
G4 XZL
G4 XZU
G4 XZV
G4 YAD
G4 YAR
G4 YBC
G4 YBO
G4 YBU
G4 YCB
G4 YCC
G4 YCI
G4 YCK
G4 YCT
G4 YDK
G4 YDU
G4 YEA
G4 YED
G4 YEH
G4 YEN
G4 YEY
G4 YEZ
G4 YFE
G4 YFN
G4 YFP
G4 YFQ
G4 YHE
G4 YHF
G4 YHV
G4 YHW
G4 YIB
G4 YIJ
G4 YIL
G4 YJO
G4 YKC
G4 YKJ
G4 YKS
G4 YKU
G4 YLA
G4 YLB
G4 YLJ
G4 YLV
G4 YMK
G4 YMP
G4 YMV
G4 YMX
G4 YNA
G4 YOB
G4 YOE
G4 YOH
G4 YOM
G4 YOQ
G4 YOU
G4 YOW
G4 YPX
G4 YPY
G4 YQY
G4 YRB
G4 YRG
G4 YRK
G4 YRS
G4 YTW
G4 YUM
G4 YUP
G4 YUQ
G4 YWL
G4 YWW
G4 YWY
G4 YXG
G4 YXL
G4 YXT
G4 YY
G4 YYB
G4 YZI
G4 YZX
G4 ZAJ
G4 ZAQ
G4 ZAT
G4 ZAU
G4 ZAZ
G4 ZBA
G4 ZBR
G4 ZBV
G4 ZBY
G4 ZCZ
G4 ZDK
G4 ZED
G4 ZEK
G4 ZEO
G4 ZEP
G4 ZFL
G4 ZFZ
G4 ZGC
G4 ZGH
G4 ZIA
G4 ZIC
G4 ZIO
G4 ZJF
G4 ZJN
G4 ZJW
G4 ZKK
G4 ZKU
G4 ZLH
G4 ZLZ
G4 ZMF
G4 ZMG
G4 ZMI
G4 ZOM
G4 ZOW
G4 ZQP
G4 ZQZ
G4 ZQZ
G4 ZRE
G4 ZRL
G4 ZRP
G4 ZSF
G4 ZSN
G4 ZU
G4 ZUF
G4 ZUP
G4 ZUQ
G4 ZUR
G4 ZVC
G4 ZVE
G4 ZVY
G4 ZWE
G4 ZWH
G4 ZWL
G4 ZWP
G4 ZXE
G4 ZXX
G4 ZYG
G4 ZYJ
G4 ZYT
G4 ZYU
G4 ZZB
G4 ZZR

G5

G5 AAA
G5 AIB
G5 AJK
G5 AKB
G5 ALF
G5 ALG
G5 BAY
G5 BMH
G5 BSD
G5 CAZ
G5 CHC
G5 CL
G5 CLR
G5 CTH
G5 CWP
G5 CWP
G5 CWP
G5 DDX
G5 DEM
G5 DKT
G5 DOC
G5 DPM
G5 DSG
G5 DWS
G5 DX
G5 DXC
G5 EA
G5 ECO
G5 EEK
G5 EPC
G5 FLY
G5 FS
G5 GDP
G5 GEI
G5 GEM
G5 GOA
G5 GPD
G5 HIR
G5 HRS
G5 HRY
G5 HVH
G5 HWB
G5 IOM
G5 IPX
G5 JDA
G5 JET
G5 JFC
G5 JNX
G5 KCI
G5 KEN
G5 KNY
G5 KSO
G5 KV
G5 LB
G5 LBS
G5 LDR
G5 LEA
G5 LO
G5 LOW
G5 MAT
G5 MCL
G5 MDR
G5 MHS
G5 MUP
G5 NG
G5 NI
G5 ODB
G5 OOW
G5 OPW
G5 PEN
G5 PH
G5 PIR
G5 PJH
G5 PJR
G5 RAF
G5 RJG
G5 RMH
G5 RMP

G5 ROB
G5 ROC
G5 RPG
G5 RRY
G5 RSD
G5 SKY
G5 SRO
G5 STN
G5 STO
G5 TA
G5 TEE
G5 TGX
G5 TM
G5 UHF
G5 UHF
G5 UK
G5 UM
G5 VHF
G5 VK
G5 VKA
G5 VU
G5 WCB
G5 WCG
G5 WCW
G5 WZY
G5 XV
G5 XW
G5 XX
G5 YTT
G5 ZL
G5 ZX

G6

G6 AAD
G6 AAL
G6 AAQ
G6 AAW
G6 ABF
G6 ABQ
G6 ABR
G6 ABU
G6 ACA
G6 ACF
G6 ACT
G6 ACY
G6 ADK
G6 ADP
G6 ADW
G6 AEG
G6 AEN
G6 AFY
G6 AGC
G6 AHY
G6 AIC
G6 AJY
G6 AKE
G6 AKJ
G6 AKW
G6 ALA
G6 ALI
G6 ALM
G6 AMP
G6 AMU
G6 AMY
G6 ANF
G6 ANG
G6 ANM
G6 ANN
G6 AOU
G6 APR
G6 APT
G6 APU
G6 AQF
G6 AQK
G6 AQN
G6 AQP
G6 AQQ
G6 ARB
G6 ARN
G6 ARU
G6 ASB
G6 ASI
G6 ASP
G6 ASV
G6 ASW
G6 ATJ
G6 ATO
G6 AWC
G6 AWI
G6 AWK
G6 AWS
G6 AWT
G6 AWV
G6 AXF
G6 AXL
G6 AXM
G6 AXW
G6 AYK
G6 AZA
G6 AZF
G6 AZJ
G6 AZS
G6 AZU
G6 AZV
G6 BAI
G6 BAJ
G6 BAK
G6 BAR
G6 BBC
G6 BBE
G6 BBM
G6 BCC
G6 BCT
G6 BDE
G6 BDV
G6 BDX
G6 BDZ
G6 BEX
G6 BEZ
G6 BFD
G6 BFT
G6 BFW
G6 BGD
G6 BGT
G6 BGW
G6 BHK
G6 BHM
G6 BHW
G6 BIR
G6 BJM
G6 BJV
G6 BKH
G6 BLL
G6 BNB
G6 BOD
G6 BOJ
G6 BOL
G6 BPC
G6 BPM
G6 BQI
G6 BQJ
G6 BRE
G6 BRH
G6 BRL
G6 BRO
G6 BSE
G6 BSK
G6 BSU
G6 BTL
G6 BTM
G6 BTN
G6 BUB
G6 BVN
G6 BWR
G6 BWX
G6 BXA
G6 BZA
G6 BZB
G6 BZJ
G6 BZO
G6 BZX
G6 CAF
G6 CAO
G6 CAT
G6 CBR
G6 CBX
G6 CCA
G6 CCM
G6 CCM
G6 CGE
G6 CHH
G6 CHO
G6 CIW
G6 CIX
G6 CJF
G6 CJK
G6 CJM
G6 CJN
G6 CKT
G6 CLE
G6 CLI
G6 CLO
G6 CMG
G6 CMK
G6 COQ
G6 COW
G6 CPC
G6 CPI
G6 CPZ
G6 CQA
G6 CQD
G6 CQM
G6 CQO
G6 CRB
G6 CRL
G6 CRP
G6 CRV
G6 CSY
G6 CVC
G6 CVI
G6 CVL
G6 CW
G6 CWD
G6 CWL
G6 CWR
G6 CWS
G6 CXB
G6 CYR
G6 CYZ
G6 CZC
G6 CZF
G6 CZQ
G6 DAZ
G6 DBE
G6 DBT
G6 DCB
G6 DCB
G6 DCM
G6 DDE
G6 DDQ
G6 DDS
G6 DEH
G6 DEN
G6 DEQ
G6 DES
G6 DFD
G6 DFF
G6 DGT
G6 DHJ
G6 DHS
G6 DHY
G6 DIK
G6 DIV
G6 DIY
G6 DJA
G6 DJI
G6 DJT
G6 DKB
G6 DKC
G6 DKQ
G6 DLA
G6 DMA
G6 DMR
G6 DNM
G6 DNN
G6 DOM
G6 DOP
G6 DPC
G6 DPP
G6 DPR
G6 DPW
G6 DQM
G6 DQN
G6 DRR
G6 DSM
G6 DTB
G6 DTM
G6 DTX
G6 DUG
G6 DUS
G6 DUZ
G6 DUZ
G6 DVU
G6 DVZ
G6 DWI
G6 DWX
G6 DX
G6 DXH
G6 DXM
G6 DYA
G6 DYC
G6 DYD
G6 DYE
G6 DYJ
G6 DYX
G6 DZC
G6 DZD
G6 DZH
G6 DZM
G6 DZW
G6 EAL
G6 EAS
G6 EAV
G6 EBN
G6 ECF
G6 ECH
G6 EDL
G6 EEA
G6 EEY
G6 EFH
G6 EFN
G6 EFX
G6 EFY
G6 EGG
G6 EGM
G6 EGV
G6 EHB
G6 EHO
G6 EHU
G6 EIA
G6 EIG
G6 EIJ
G6 EIL
G6 EIW
G6 EJA
G6 EJG
G6 EJP
G6 EJV
G6 EJY
G6 EKG
G6 EKJ
G6 ELL
G6 ELO
G6 ELQ
G6 ELR
G6 ELY
G6 EMH
G6 EMT
G6 ENC
G6 ENG
G6 ENH
G6 EOC
G6 EOF
G6 EOL
G6 EOY
G6 EPQ
G6 EPW
G6 EQJ
G6 ERC
G6 ERV
G6 ESW
G6 ETE
G6 ETT
G6 EVD
G6 EVT
G6 EWV
G6 EXF
G6 EXO
G6 EXY
G6 EYB
G6 EYD
G6 EYX
G6 EYY
G6 EZS
G6 FAB
G6 FBI
G6 FCW
G6 FCX
G6 FDN
G6 FEO
G6 FEP
G6 FGK
G6 FGO
G6 FGP
G6 FGS
G6 FHW
G6 FID
G6 FIQ
G6 FIR
G6 FIU
G6 FIV
G6 FIY
G6 FJT
G6 FKQ
G6 FLX
G6 FMZ
G6 FNE
G6 FNF
G6 FNI
G6 FNL
G6 FNS
G6 FOM
G6 FOP
G6 FPQ
G6 FQJ
G6 FQW
G6 FRT
G6 FRU
G6 FSB
G6 FSH
G6 FSS
G6 FTN
G6 FTP
G6 FUB
G6 FUD
G6 FUW
G6 FVK
G6 FWA
G6 FWC
G6 FZU
G6 GAX
G6 GBH
G6 GCX
G6 GDG
G6 GDT
G6 GEA
G6 GEC
G6 GEI
G6 GEJ
G6 GEY
G6 GGE
G6 GGP
G6 GGU
G6 GIR
G6 GIY
G6 GJQ
G6 GJT
G6 GLP
G6 GMC
G6 GN
G6 GNN
G6 GOA
G6 GOU
G6 GPS
G6 GQP
G6 GRE
G6 GRG
G6 GRX
G6 GTD
G6 GTN
G6 GTR
G6 GUN
G6 GUP
G6 GUR
G6 GUX
G6 GVG
G6 GVK
G6 GWR
G6 GXH
G6 GYJ
G6 GZH
G6 HAC
G6 HAE
G6 HAG
G6 HAL
G6 HBK
G6 HBX
G6 HCL
G6 HCU
G6 HDP
G6 HEE
G6 HEW
G6 HFI
G6 HHB
G6 HHP
G6 HIL
G6 HIZ
G6 HJD
G6 HJO
G6 HJP
G6 HJR
G6 HJX
G6 HKQ
G6 HKR
G6 HLZ
G6 HMM
G6 HNF
G6 HNJ
G6 HNM
G6 HNU
G6 HOL
G6 HOM
G6 HOQ
G6 HOU
G6 HPC
G6 HPO
G6 HRN
G6 HSF
G6 HSM
G6 HSV
G6 HTL
G6 HUA
G6 HUH
G6 HUL
G6 HUU
G6 HVI
G6 HVP
G6 HXK
G6 HXV
G6 HYN
G6 HYU
G6 HZW
G6 IA
G6 IBC
G6 IBH
G6 IBI
G6 IBZ
G6 ICF
G6 ICJ
G6 ICP
G6 IDI
G6 IDP
G6 IDZ
G6 IEA
G6 IEJ
G6 IEZ
G6 IFD
G6 IFF
G6 IFX
G6 IFZ
G6 IGA
G6 IGB
G6 IHN
G6 IIG
G6 IJJ
G6 IKT
G6 IKW
G6 ILE
G6 IMF
G6 IMR
G6 IMT
G6 IMU
G6 IPU
G6 IQA
G6 IQL
G6 IRB
G6 IRS
G6 ISC
G6 ISD
G6 ISO
G6 ISQ
G6 ISW
G6 ITD
G6 ITF
G6 ITH
G6 ITK
G6 ITO
G6 IVT
G6 IWY
G6 IXS
G6 IZA
G6 IZC
G6 IZI
G6 IZN
G6 IZS
G6 JAD
G6 JDZ
G6 JEA
G6 JEG
G6 JEJ
G6 JEL
G6 JEQ
G6 JEV
G6 JFS
G6 JFT
G6 JFY
G6 JGD
G6 JGS
G6 JHA
G6 JHD
G6 JHE
G6 JHR
G6 JID
G6 JIE
G6 JIV
G6 JKG
G6 JKL
G6 JKQ
G6 JKT
G6 JLH
G6 JLM
G6 JME
G6 JMF
G6 JMS
G6 JON
G6 JOX
G6 JPA
G6 JPJ
G6 JQV
G6 JQW
G6 JR
G6 JRC
G6 JRV
G6 JSB
G6 JSK
G6 JSO
G6 JSV
G6 JTB
G6 JTN
G6 JUU
G6 JUZ
G6 JVD
G6 JVS
G6 JVV
G6 JVY
G6 JWE
G6 JWR
G6 JXE
G6 JXN
G6 JYC
G6 JYG
G6 JYQ
G6 JZZ
G6 KAR
G6 KAS
G6 KAX
G6 KBG
G6 KBJ
G6 KBR
G6 KDF
G6 KDQ
G6 KEC
G6 KFN
G6 KFW
G6 KFY
G6 KGI
G6 KGW
G6 KGZ
G6 KHK
G6 KIL
G6 KJR
G6 KJU
G6 KLS
G6 KLW
G6 KMA
G6 KME
G6 KMM
G6 KMT
G6 KMV
G6 KMY
G6 KNA
G6 KNF
G6 KNH
G6 KNX
G6 KOA
G6 KOI
G6 KOS
G6 KOY
G6 KOZ
G6 KPM
G6 KQI
G6 KQP
G6 KRD
G6 KRF
G6 KRL
G6 KSM
G6 KSN
G6 KUA
G6 KUC
G6 KUG
G6 KUS
G6 KUT
G6 KWJ
G6 KWM
G6 KWN
G6 KWO
G6 KXO
G6 KYD
G6 KYX
G6 KZD
G6 KZJ
G6 KZT
G6 KZX
G6 KZZ
G6 LAM
G6 LBA
G6 LBC
G6 LBD
G6 LCC
G6 LDI
G6 LDK
G6 LDL
G6 LDN
G6 LDV
G6 LDZ
G6 LFN
G6 LGF
G6 LGZ
G6 LHD
G6 LHJ
G6 LHO
G6 LHW
G6 LIQ
G6 LIS
G6 LJB
G6 LJJ
G6 LLB
G6 LLO
G6 LLX
G6 LMZ
G6 LOH
G6 LPF
G6 LPY
G6 LPZ
G6 LQU
G6 LQW
G6 LRE
G6 LRQ
G6 LTG
G6 LTI
G6 LTO
G6 LTS
G6 LTZ
G6 LUG
G6 LUH
G6 LUW
G6 LUZ
G6 LVW
G6 LWY
G6 LXK
G6 LXQ
G6 LXX
G6 LYY
G6 LZU
G6 MAF
G6 MCI
G6 MCJ
G6 MDD
G6 MDT
G6 MDU
G6 MEK
G6 MEN
G6 MFD
G6 MGB
G6 MGS
G6 MHH
G6 MIF
G6 MII
G6 MIT
G6 MIW
G6 MIY
G6 MJG
G6 MJN
G6 MJQ
G6 MLN
G6 MMC
G6 MNM
G6 MNO
G6 MNQ
G6 MOG
G6 MOY
G6 MPG
G6 MQW
G6 MRF
G6 MRI
G6 MRJ
G6 MSQ
G6 MSS
G6 MTZ
G6 MVC
G6 MVO
G6 MWJ
G6 MWP
G6 MWR
G6 MXJ
G6 MXR
G6 MXT
G6 MYW
G6 MZK
G6 MZS
G6 MZX
G6 NAT
G6 NCF
G6 NCN
G6 NCQ
G6 NCR
G6 NCU
G6 NDI
G6 NDR
G6 NDR
G6 NEB
G6 NEM
G6 NER
G6 NET
G6 NFF
G6 NGO
G6 NHR
G6 NIL
G6 NIS
G6 NJF
G6 NJI
G6 NJM
G6 NJP
G6 NKD
G6 NKG
G6 NKM
G6 NLL
G6 NLW
G6 NNN
G6 NNV
G6 NOS
G6 NOT
G6 NOZ
G6 NPL
G6 NPQ
G6 NPR
G6 NQF
G6 NQO
G6 NQP
G6 NRC
G6 NSB
G6 NSY
G6 NTI
G6 NTK
G6 NUA
G6 NUD
G6 NUM
G6 NUN
G6 NUT
G6 NVA
G6 NVT
G6 NWF
G6 NWO
G6 NWR
G6 NWV
G6 NXN
G6 NXR
G6 NZB
G6 NZP
G6 OAJ
G6 OAX
G6 OAY
G6 OBW
G6 OCD
G6 OCE
G6 OCL
G6 OCQ
G6 OCT
G6 OCW
G6 OCX
G6 OEY
G6 OGY
G6 OHB
G6 OI
G6 OIL
G6 OJB
G6 OJU
G6 OLS
G6 OMT
G6 OMY
G6 ONC
G6 ONN
G6 OOO
G6 OPB
G6 OQK
G6 ORA
G6 ORC
G6 ORN
G6 OTI
G6 OTJ
G6 OTM
G6 OTU
G6 OVL
G6 OVN
G6 OVQ
G6 OVR
G6 OVU
G6 OVY
G6 OVZ
G6 OXV
G6 OXW
G6 OYE
G6 OZV
G6 OZW
G6 PAF
G6 PAI
G6 PBN
G6 PBS
G6 PBT
G6 PCA
G6 PCD
G6 PCU
G6 PDH
G6 PDV
G6 PDW
G6 PER
G6 PFE
G6 PHA
G6 PHB
G6 PHD
G6 PHJ
G6 PHQ
G6 PJA
G6 PJT
G6 PJV
G6 PJZ
G6 PLN
G6 PLP
G6 PLV
G6 PMN
G6 PMY
G6 PNC
G6 PNH
G6 PNX
G6 POL
G6 POQ
G6 POY
G6 PPJ
G6 PPX
G6 PQW
G6 PRB
G6 PRT
G6 PRV
G6 PSI
G6 PTJ
G6 PTL
G6 PTY
G6 PUB
G6 PUC
G6 PUX
G6 PUX
G6 PVF
G6 PVQ
G6 PWO
G6 PWT
G6 PWY
G6 PXI
G6 PXK
G6 PXT
G6 PZP
G6 PZW
G6 RAJ
G6 RAY
G6 RBH
G6 RBK
G6 RCA
G6 RCP
G6 RDG
G6 RFG
G6 RFQ
G6 RHZ
G6 RIB
G6 RIK
G6 RIO
G6 RJT
G6 RKE
G6 RKP
G6 RMS
G6 RNE
G6 RNM
G6 ROB
G6 ROP
G6 RPM
G6 RPX
G6 RQO
G6 RRL
G6 RSC
G6 RSE
G6 RST
G6 RSY
G6 RTV
G6 RUK
G6 RUQ
G6 RVB
G6 RVW
G6 RWI
G6 RWT
G6 RWV
G6 RYC
G6 RZA
G6 SAF
G6 SAG
G6 SAS
G6 SBF
G6 SBR
G6 SCD
G6 SCR
G6 SDJ
G6 SDP
G6 SEA
G6 SEA
G6 SEM
G6 SET
G6 SFB
G6 SFN
G6 SFP
G6 SHB
G6 SIC
G6 SIY
G6 SJF
G6 SJH
G6 SJJ
G6 SJO
G6 SJY
G6 SKC
G6 SKL
G6 SKN
G6 SKO
G6 SLK
G6 SLM
G6 SLU
G6 SMZ
G6 SNJ
G6 SNL
G6 SNO
G6 SNR
G6 SPA
G6 SPC
G6 SPE
G6 SQC
G6 SQD
G6 SSF
G6 SSH
G6 SSX
G6 SUD
G6 SVB
G6 SVN
G6 SXW
G6 SYC
G6 SYT
G6 SZF
G6 SZK
G6 SZQ
G6 TAT
G6 TAU
G6 TBA
G6 TBT
G6 TCO
G6 TDC
G6 TDL
G6 TDP
G6 TDR
G6 TEC
G6 TFB
G6 TFC
G6 TFU
G6 TFY
G6 TGD
G6 TGO
G6 TGP
G6 TGV
G6 TGZ
G6 THB
G6 THR
G6 THX
G6 TIH
G6 TIM
G6 TIX
G6 TJS
G6 TKI
G6 TKK
G6 TKS
G6 TKX
G6 TLH
G6 TLR
G6 TME
G6 TMK
G6 TMM
G6 TMO
G6 TMW
G6 TNO
G6 TNZ
G6 TOB
G6 TOC
G6 TOZ
G6 TPK
G6 TPV
G6 TQO
G6 TQT
G6 TQU
G6 TRD
G6 TRE
G6 TRS
G6 TRT
G6 TSH
G6 TTL
G6 TTW
G6 TUH
G6 TUW
G6 TUZ
G6 TXW
G6 TXZ
G6 TYJ
G6 TYQ
G6 TYS
G6 TZB
G6 UAD
G6 UAU
G6 UAX
G6 UAY
G6 UBB
G6 UBD
G6 UBL
G6 UBM
G6 UBP
G6 UCJ
G6 UDE
G6 UDN
G6 UEB
G6 UEJ
G6 UEO
G6 UFJ
G6 UFT
G6 UGD
G6 UGL
G6 UIA
G6 UID
G6 UIP
G6 UIU
G6 UJB
G6 UJH
G6 UKB
G6 UKK
G6 UKV
G6 ULP
G6 ULX
G6 UNM
G6 UOD
G6 UQB
G6 UQC
G6 UQQ
G6 URA
G6 URB
G6 URE
G6 URY
G6 USH
G6 UT
G6 UTC
G6 UTD
G6 UTP
G6 UTQ
G6 UUM
G6 UVE
G6 UVH
G6 UWH
G6 UXD
G6 UXN
G6 UYG
G6 UZF
G6 UZH
G6 UZT
G6 VAH
G6 VAM
G6 VAM
G6 VCY
G6 VDH
G6 VDP
G6 VGW
G6 VHC
G6 VHF
G6 VIR
G6 VIW
G6 VIX
G6 VJE
G6 VLB
G6 VLK
G6 VME
G6 VND
G6 VNH
G6 VQU
G6 VQX
G6 VRD
G6 VRH
G6 VRT
G6 VTR
G6 VTU
G6 VUI
G6 VUL
G6 VVA
G6 VVC
G6 VVM
G6 VVT
G6 VVX
G6 VWC
G6 VXB
G6 VXE
G6 VXM
G6 VXX
G6 VYS
G6 VYT
G6 WBJ
G6 WBV
G6 WDE
G6 WDV
G6 WEX
G6 WEZ
G6 WFQ
G6 WGD
G6 WGK
G6 WGP
G6 WHI
G6 WHR
G6 WII
G6 WIM
G6 WJP
G6 WKW
G6 WLA
G6 WLL
G6 WLN
G6 WLZ
G6 WNA
G6 WNO
G6 WNW
G6 WOE
G6 WOL
G6 WOM
G6 WOV
G6 WOZ
G6 WQB
G6 WQU
G6 WRA
G6 WRB
G6 WRG
G6 WRI
G6 WRQ
G6 WRW
G6 WSI
G6 WTA
G6 WTX
G6 WVK
G6 WVX
G6 WWO
G6 WXC
G6 WYN
G6 WZH
G6 WZO
G6 XAC
G6 XAI
G6 XBC
G6 XBF
G6 XBW
G6 XBX
G6 XCF
G6 XCM
G6 XDP
G6 XDQ
G6 XEG
G6 XFD
G6 XFI
G6 XFP
G6 XFW
G6 XGL
G6 XHM
G6 XIP
G6 XIQ
G6 XJU
G6 XLU
G6 XLV
G6 XMI
G6 XMO
G6 XMY
G6 XNC
G6 XOI
G6 XPM
G6 XQW
G6 XRA
G6 XRQ
G6 XRX
G6 XSP
G6 XTL
G6 XTN
G6 XTO
G6 XTR
G6 XTW
G6 XUB
G6 XVN
G6 XVO
G6 XVT
G6 XVW
G6 XWE
G6 XWK
G6 XWN
G6 XWO
G6 XWT
G6 XWU
G6 XXW
G6 XYE
G6 XYT
G6 XYZ
G6 XZL
G6 XZV
G6 YAF
G6 YAL
G6 YBM
G6 YBY
G6 YCK
G6 YDD
G6 YGD
G6 YGG
G6 YHT
G6 YHU
G6 YHV
G6 YJN
G6 YKT
G6 YLJ
G6 YMP
G6 YMV
G6 YNZ
G6 YON
G6 YOS
G6 YOV
G6 YPU
G6 YQY
G6 YRK
G6 YRM
G6 YTC
G6 YTZ
G6 YUH
G6 YUI
G6 YVE
G6 YVX
G6 YWB
G6 YWC
G6 YWF
G6 YWX
G6 YYV
G6 YYX
G6 ZAB
G6 ZBB
G6 ZBI
G6 ZC
G6 ZDQ
G6 ZFJ
G6 ZFY
G6 ZGD
G6 ZGO
G6 ZIM
G6 ZIS
G6 ZKH
G6 ZKT
G6 ZLA
G6 ZLI
G6 ZMC
G6 ZMF
G6 ZMK
G6 ZMY
G6 ZMZ
G6 ZOD
G6 ZOO
G6 ZOV
G6 ZPD
G6 ZQE
G6 ZQQ
G6 ZQX
G6 ZQY
G6 ZSD
G6 ZSN
G6 ZSR
G6 ZTA
G6 ZTV
G6 ZVE
G6 ZWQ
G6 ZXN
G6 ZXU
G6 ZYD
G6 ZYF
G6 ZYG
G6 ZYP
G6 ZZP
G6 ZZX
G6 ZZZ

G7

G7 AAC
G7 AAF
G7 AAQ
G7 AAW
G7 ABG
G7 ACE
G7 ACU
G7 ACW
G7 AEB
G7 AED
G7 AEJ
G7 AEO
G7 AEP
G7 AER
G7 AEU
G7 AFK
G7 AFM
G7 AGZ
G7 AHE
G7 AHH
G7 AHJ
G7 AHQ
G7 AHS
G7 AHT
G7 AHY
G7 AIE
G7 AII
G7 AIJ
G7 AIS
G7 AIT
G7 AIV
G7 AJC
G7 AJH
G7 AJM
G7 AKF
G7 AKL
G7 ALG
G7 ALN
G7 ALW
G7 ALZ
G7 AME
G7 AMF
G7 AMG
G7 AMH
G7 AMJ
G7 AMK
G7 AMO
G7 AMU
G7 ANF
G7 ANJ
G7 ANZ
G7 AOO
G7 AOP
G7 AOR
G7 AOZ
G7 APC
G7 AQR
G7 AQU
G7 AQX
G7 AQY
G7 ARC
G7 ASJ
G7 AST
G7 AST
G7 ASX
G7 ATQ
G7 ATV
G7 AUG
G7 AVO
G7 AWN
G7 AXB
G7 AXC
G7 AXF
G7 AXG
G7 AXJ
G7 AXU
G7 AXV
G7 BAP
G7 BBF
G7 BBL
G7 BBP
G7 BBQ
G7 BCB
G7 BCE
G7 BCL
G7 BCS
G7 BDP
G7 BDQ
G7 BEA
G7 BED
G7 BEH
G7 BEX
G7 BFD
G7 BFK
G7 BFO
G7 BFP
G7 BGA
G7 BHB
G7 BHH
G7 BHN
G7 BHQ
G7 BIU
G7 BJA
G7 BJH
G7 BJJ
G7 BJQ
G7 BJU
G7 BJY
G7 BKK
G7 BKU
G7 BKV
G7 BKX
G7 BLR
G7 BMA
G7 BMS
G7 BNV
G7 BPE
G7 BPO
G7 BQC
G7 BRC
G7 BRH
G7 BRV
G7 BSV
G7 BSZ
G7 BTA
G7 BUG
G7 BUQ
G7 BUT
G7 BUZ
G7 BVK
G7 BVY
G7 BWP
G7 BZN
G7 BZW
G7 BZZ
G7 CAK
G7 CAN
G7 CAQ
G7 CAR
G7 CAW
G7 CAY
G7 CBG
G7 CBL
G7 CBM
G7 CCI
G7 CCZ
G7 CDA
G7 CDK
G7 CDZ
G7 CFJ
G7 CFO
G7 CFR
G7 CFU
G7 CFV
G7 CFY
G7 CGG
G7 CGK
G7 CGV
G7 CHO
G7 CHT
G7 CHX
G7 CHY
G7 CIB
G7 CIY
G7 CJY
G7 CKD
G7 CKF
G7 CKR
G7 CKU
G7 CKX
G7 CLP
G7 CLR
G7 CLZ
G7 CMJ
G7 CMK
G7 CMS
G7 CNF
G7 CNH
G7 CNQ
G7 COI
G7 COU
G7 COV
G7 COW
G7 CPB
G7 CPC
G7 CRO
G7 CRX
G7 CTA
G7 CTD
G7 CTQ
G7 CTX
G7 CUG
G7 CUH
G7 CUI
G7 CVH
G7 CVJ
G7 CWA
G7 CWH
G7 CWQ
G7 CWY
G7 CXP
G7 CXP
G7 CXR
G7 CYK
G7 CYO
G7 CYP
G7 CYX
G7 CZH
G7 CZO
G7 CZT
G7 CZZ
G7 DAA
G7 DAK
G7 DBA
G7 DBE
G7 DBJ
G7 DBK
G7 DBL
G7 DBM
G7 DBS
G7 DCD
G7 DCG
G7 DCO
G7 DDD
G7 DDY
G7 DEL
G7 DEM
G7 DEQ
G7 DER
G7 DET
G7 DFI
G7 DFM
G7 DFO
G7 DGH
G7 DGR
G7 DGU
G7 DGW
G7 DGZ
G7 DHL
G7 DIH
G7 DIQ
G7 DJP
G7 DJW
G7 DKE
G7 DKJ
G7 DKL
G7 DKM
G7 DKQ
G7 DKX
G7 DLC
G7 DLH
G7 DLM
G7 DLS
G7 DLV
G7 DLW
G7 DMD
G7 DMJ
G7 DMO
G7 DOH
G7 DOI
G7 DOM
G7 DOO
G7 DOT
G7 DPJ
G7 DPP
G7 DPY
G7 DQC
G7 DQD
G7 DQF
G7 DQG
G7 DQM
G7 DQR
G7 DQS
G7 DRI
G7 DRK
G7 DRS
G7 DSG
G7 DSI
G7 DSJ
G7 DSK
G7 DSX
G7 DUO
G7 DUP
G7 DUR
G7 DUW
G7 DVG
G7 DVY
G7 DWD
G7 DWK
G7 DXP
G7 DXS
G7 DXY
G7 DYA
G7 DYG
G7 DYH
G7 DYI
G7 DYO
G7 DYQ
G7 DZI
G7 DZS
G7 DZX
G7 EAG
G7 EAK
G7 EAP
G7 EBK
G7 EBP
G7 EBY
G7 ECK
G7 ECL
G7 ECM
G7 ECV
G7 EDC
G7 EDE
G7 EDM
G7 EDQ
G7 EDT
G7 EDU
G7 EEP
G7 EET
G7 EEY
G7 EFC
G7 EFE
G7 EFN
G7 EFO
G7 EGI
G7 EGK
G7 EGN
G7 EGZ
G7 EIM
G7 EIO
G7 EIT
G7 EIX
G7 EJG
G7 EKI
G7 EKY
G7 ELF
G7 ELK
G7 ELM
G7 ELR
G7 EMF
G7 EMK
G7 EML
G7 EOB
G7 EOL
G7 EOO
G7 EOQ
G7 EOR
G7 EOT
G7 EOX
G7 EOZ
G7 EPC
G7 EPG
G7 EQM
G7 EQU
G7 EQZ
G7 ERB
G7 ERM
G7 ERO
G7 ERU
G7 ERX
G7 ESB
G7 ESK
G7 ESP
G7 ESR
G7 ETA
G7 ETI
G7 ETN
G7 ETR
G7 ETT
G7 ETU
G7 ETW
G7 ETX
G7 ETZ
G7 EUA
G7 EUH
G7 EUM
G7 EUW
G7 EVA
G7 EVE
G7 EVW
G7 EWG
G7 EXN
G7 EYX
G7 EZG
G7 EZM
G7 EZP
G7 EZR
G7 EZT
G7 EZW
G7 EZY
G7 FAL
G7 FAN
G7 FAX
G7 FBD
G7 FBX
G7 FCT
G7 FCV
G7 FDN
G7 FDP
G7 FDR
G7 FDT
G7 FDU
G7 FDZ
G7 FEK
G7 FEW
G7 FEX
G7 FEZ
G7 FFA
G7 FFE
G7 FFL
G7 FGF
G7 FHC
G7 FHI
G7 FHJ
G7 FHK
G7 FHN
G7 FHW
G7 FHY
G7 FIF
G7 FIG
G7 FIT
G7 FJA
G7 FJI
G7 FKB
G7 FKE
G7 FKH
G7 FLO
G7 FLR
G7 FLT
G7 FLW
G7 FMF
G7 FMM
G7 FMN
G7 FMX
G7 FNA
G7 FNE
G7 FNS
G7 FOF
G7 FOJ
G7 FOK
G7 FPE
G7 FPL
G7 FQF
G7 FQG
G7 FQH
G7 FQW
G7 FRA
G7 FRB
G7 FRI
G7 FSI
G7 FTB
G7 FTL
G7 FTN
G7 FTP
G7 FTQ
G7 FTR
G7 FTY
G7 FUJ
G7 FVB
G7 FVD
G7 FVF
G7 FVG
G7 FVT
G7 FVV
G7 FWL
G7 FWO
G7 FWR
G7 FWT
G7 FWV
G7 FXM
G7 FXV
G7 FYO
G7 FYR
G7 FZV
G7 FZX
G7 GAD
G7 GAF
G7 GAM
G7 GAN
G7 GAS
G7 GAV
G7 GAX
G7 GBA
G7 GBC
G7 GCK
G7 GCO
G7 GCQ
G7 GCR
G7 GDH
G7 GDR
G7 GEQ
G7 GFS
G7 GFU
G7 GFW

G7 GGS
G7 GGU
G7 GGV
G7 GGW
G7 GGX
G7 GGZ
G7 GHA
G7 GHK
G7 GHS
G7 GHV
G7 GJF
G7 GJH
G7 GKO
G7 GKV
G7 GLE
G7 GLK
G7 GLN
G7 GLO
G7 GLT
G7 GMF
G7 GMG
G7 GMJ
G7 GMN
G7 GNB
G7 GNF
G7 GNG
G7 GNK
G7 GNL
G7 GNM
G7 GOC
G7 GOP
G7 GPR
G7 GQF
G7 GQP
G7 GRB
G7 GRK
G7 GSH
G7 GSI
G7 GSL
G7 GSN
G7 GSP
G7 GSQ
G7 GSW
G7 GTB
G7 GTI
G7 GTO
G7 GTV
G7 GUD
G7 GUE
G7 GUF
G7 GUJ
G7 GUN
G7 GVK
G7 GVV
G7 GVZ
G7 GWB
G7 GWM
G7 GWR
G7 GWV
G7 GXF
G7 GXG
G7 GXK
G7 GXY
G7 GYB
G7 GYP
G7 GZE
G7 GZG
G7 GZH
G7 HAA
G7 HAT
G7 HAV
G7 HAW
G7 HCD
G7 HCE
G7 HDA
G7 HDB
G7 HDE
G7 HDI
G7 HDJ
G7 HDO
G7 HDX
G7 HEL
G7 HEM
G7 HER
G7 HEU
G7 HEX
G7 HGP
G7 HIF
G7 HIP
G7 HJY
G7 HKD
G7 HKF
G7 HKK
G7 HKM
G7 HKP
G7 HKY
G7 HLF
G7 HLH
G7 HLJ
G7 HLL
G7 HLO
G7 HMC
G7 HMJ
G7 HNH
G7 HNH
G7 HNK
G7 HNN
G7 HNO
G7 HNQ
G7 HNX
G7 HOB
G7 HOD
G7 HOI
G7 HOV
G7 HOW
G7 HPB
G7 HPE
G7 HPG
G7 HPH
G7 HPK
G7 HPM
G7 HPP
G7 HPS
G7 HPW
G7 HQB
G7 HQH
G7 HRC
G7 HRQ
G7 HRU
G7 HRW
G7 HSE
G7 HSG
G7 HST
G7 HTP
G7 HTS
G7 HUD
G7 HUH
G7 HUR
G7 HUW
G7 HVN
G7 HVT
G7 HVW
G7 HWC
G7 HWD
G7 HWF
G7 HWJ
G7 HWK
G7 HWL
G7 HWT
G7 HWU
G7 HWW
G7 HXC
G7 HXJ
G7 HXM
G7 HXQ
G7 HXS
G7 HXV
G7 HXW
G7 HXX
G7 HYF
G7 HYK
G7 HZI
G7 IBI
G7 IBO
G7 IBY
G7 ICH
G7 ICQ
G7 ICS
G7 ICV
G7 ICW
G7 ICX
G7 ICY
G7 IDB
G7 IDD
G7 IDF
G7 IDJ
G7 IDL
G7 IEE
G7 IEI
G7 IFD
G7 IFF
G7 IGB
G7 IGC
G7 IGH
G7 IGZ
G7 IHA
G7 IHS
G7 III
G7 IIM
G7 IIU
G7 IJK
G7 IJM
G7 IJR
G7 IJS
G7 IJT
G7 IJU
G7 IKR
G7 IKT
G7 IKU
G7 IKW
G7 IKY
G7 ILK
G7 ILV
G7 ILZ
G7 ILZ
G7 IMF
G7 IND
G7 INM
G7 INN
G7 INO
G7 INP
G7 INS
G7 INV
G7 INX
G7 IOV
G7 IOW
G7 IOY
G7 IPB
G7 IPK
G7 IPK
G7 IPT
G7 IQJ
G7 IQK
G7 IQT
G7 IQW
G7 IRV
G7 ISO
G7 ISY
G7 ISZ
G7 ITD
G7 ITL
G7 IUJ
G7 IUK
G7 IUO
G7 IUU
G7 IUV
G7 IVC
G7 IVH
G7 IVJ
G7 IVP
G7 IVR
G7 IVY
G7 IWG
G7 IWS
G7 IXA
G7 IXJ
G7 IXL
G7 IXM
G7 IXO
G7 IXU
G7 IYB
G7 IYD
G7 IYR
G7 IYS
G7 IYT
G7 IYW
G7 IYZ
G7 IZD
G7 IZG
G7 IZH
G7 IZI
G7 IZQ
G7 IZR
G7 IZX
G7 IZZ
G7 JAB
G7 JAC
G7 JAD
G7 JAH
G7 JAK
G7 JBB
G7 JBE
G7 JBN
G7 JBQ
G7 JBT
G7 JCC
G7 JCJ
G7 JCP
G7 JDD
G7 JDE
G7 JDU
G7 JDV
G7 JDW
G7 JEW
G7 JEX
G7 JFF
G7 JFH
G7 JFS
G7 JGD
G7 JGL
G7 JHF
G7 JHJ
G7 JHL
G7 JHO
G7 JHP
G7 JHQ
G7 JIC
G7 JID
G7 JIQ
G7 JIW
G7 JJF
G7 JJI
G7 JJZ
G7 JKA
G7 JKB
G7 JKO
G7 JKV
G7 JKX
G7 JKZ
G7 JLI
G7 JLM
G7 JLR
G7 JLZ
G7 JMG
G7 JMI
G7 JND
G7 JNT
G7 JNY
G7 JOB
G7 JOG
G7 JOK
G7 JON
G7 JOP
G7 JOQ
G7 JOV
G7 JOX
G7 JOY
G7 JPD
G7 JPE
G7 JPH
G7 JPU
G7 JPW
G7 JPY
G7 JQE
G7 JQK
G7 JQL
G7 JRE
G7 JRV
G7 JRW
G7 JRZ
G7 JSF
G7 JSP
G7 JTQ
G7 JTS
G7 JTT
G7 JUE
G7 JUG
G7 JUS
G7 JUY
G7 JVL
G7 JVW
G7 JWB
G7 JWR
G7 JYD
G7 JYE
G7 JYF
G7 JYM
G7 JYR
G7 JYS
G7 JYU
G7 JZD
G7 JZE
G7 JZL
G7 JZO
G7 JZP
G7 JZR
G7 JZU
G7 JZW
G7 KAD
G7 KAH
G7 KAT
G7 KAW
G7 KBF
G7 KBL
G7 KBP
G7 KBV
G7 KBY
G7 KCD
G7 KCF
G7 KCK
G7 KCS
G7 KCW
G7 KCZ
G7 KDA
G7 KDC
G7 KDP
G7 KDZ
G7 KEH
G7 KEJ
G7 KES
G7 KET
G7 KEZ
G7 KFA
G7 KFC
G7 KFD
G7 KFG
G7 KFH
G7 KFI
G7 KFL
G7 KFT
G7 KFW
G7 KFX
G7 KFY
G7 KGB
G7 KGC
G7 KHN
G7 KIG
G7 KIM
G7 KJF
G7 KJG
G7 KKH
G7 KKJ
G7 KKL
G7 KKN
G7 KKP
G7 KKR
G7 KKS
G7 KKV
G7 KLA
G7 KLD
G7 KLI
G7 KMJ
G7 KMX
G7 KMZ
G7 KNE
G7 KNF
G7 KNI
G7 KNO
G7 KNP
G7 KNV
G7 KNZ
G7 KOA
G7 KOQ
G7 KOU
G7 KPB
G7 KPF
G7 KPN
G7 KPO
G7 KPP
G7 KPQ
G7 KPR
G7 KPS
G7 KPT
G7 KPU
G7 KPX
G7 KPY
G7 KPZ
G7 KQA
G7 KQC
G7 KQF
G7 KQG
G7 KQI
G7 KQK
G7 KQM
G7 KQO
G7 KQY
G7 KSI
G7 KSW
G7 KTE
G7 KTK
G7 KTL
G7 KTM
G7 KTN
G7 KTO
G7 KUF
G7 KUJ
G7 KUZ
G7 KVE
G7 KVK
G7 KVO
G7 KWC
G7 KWI
G7 KWX
G7 KWZ
G7 KXB
G7 KXD
G7 KXE
G7 KXM
G7 KXW
G7 KXX
G7 KXY
G7 KYB
G7 KYC
G7 KYM
G7 KYM
G7 KYN
G7 KZA
G7 KZF
G7 KZX
G7 KZZ
G7 LAE
G7 LAS
G7 LBG
G7 LBS
G7 LBY
G7 LCB
G7 LCC
G7 LCG
G7 LCL
G7 LCM
G7 LCU
G7 LCY
G7 LDE
G7 LDF
G7 LEC
G7 LEE
G7 LEU
G7 LFR
G7 LFW
G7 LGD
G7 LGK
G7 LHA
G7 LHB
G7 LHE
G7 LHF
G7 LHG
G7 LHM
G7 LIM
G7 LIN
G7 LIZ
G7 LJN
G7 LJV
G7 LJW
G7 LJY
G7 LKB
G7 LKN
G7 LLQ
G7 LLS
G7 LMC
G7 LMF
G7 LMJ
G7 LMK
G7 LMN
G7 LMP
G7 LMY
G7 LNC
G7 LOF
G7 LOH
G7 LOI
G7 LOM
G7 LOP
G7 LOR
G7 LOS
G7 LOU
G7 LQE
G7 LQH
G7 LQR
G7 LQY
G7 LRD
G7 LRE
G7 LRF
G7 LRJ
G7 LRK
G7 LRO
G7 LRQ
G7 LRR
G7 LRS
G7 LSH
G7 LSL
G7 LSR
G7 LSS
G7 LST
G7 LSV
G7 LSY
G7 LTQ
G7 LUC
G7 LUD
G7 LUS
G7 LUU
G7 LUV
G7 LVF
G7 LVO
G7 LVV
G7 LVW
G7 LVX
G7 LVY
G7 LVZ
G7 LWC
G7 LWI
G7 LWK
G7 LWM
G7 LWN
G7 LWV
G7 LXC
G7 LXL
G7 LXP
G7 LXT
G7 LXX
G7 LYP
G7 LYQ
G7 LYR
G7 LZI
G7 LZK
G7 LZW
G7 MAJ
G7 MAX
G7 MAZ
G7 MCI
G7 MCP
G7 MDH
G7 MDR
G7 MDU
G7 MEI
G7 MFK
G7 MGK
G7 MGL
G7 MGN
G7 MHK
G7 MHM
G7 MHN
G7 MIJ
G7 MIL
G7 MIW
G7 MIX
G7 MJA
G7 MJG
G7 MJK
G7 MJL
G7 MJN
G7 MJO
G7 MJW
G7 MJZ
G7 MKK
G7 MKO
G7 MKR
G7 MKS
G7 MLF
G7 MLZ
G7 MMA
G7 MMF
G7 MMM
G7 MMS
G7 MMT
G7 MNH
G7 MNM
G7 MNN
G7 MNY
G7 MOI
G7 MOT
G7 MPW
G7 MQB
G7 MQD
G7 MQL
G7 MQT
G7 MQV
G7 MRB
G7 MRC
G7 MRD
G7 MRG
G7 MRI
G7 MRK
G7 MRN
G7 MRP
G7 MRT
G7 MRV
G7 MRY
G7 MSD
G7 MTK
G7 MTX
G7 MUI
G7 MUL
G7 MUQ
G7 MVC
G7 MVP
G7 MVR
G7 MWF
G7 MWN
G7 MXB
G7 MXJ
G7 MYB
G7 MYL
G7 MYR
G7 MYZ
G7 MZH
G7 MZN
G7 MZO
G7 MZR
G7 MZT
G7 NAD
G7 NAG
G7 NAM
G7 NAQ
G7 NAS
G7 NAU
G7 NBA
G7 NBK
G7 NBN
G7 NBP
G7 NBW
G7 NCK
G7 NCN
G7 NCO
G7 NCT
G7 NDF
G7 NDG
G7 NDH
G7 NDJ
G7 NDL
G7 NDY
G7 NEI
G7 NEL
G7 NEO
G7 NFJ
G7 NFP
G7 NFS
G7 NFU
G7 NGG
G7 NGS
G7 NGV
G7 NGZ
G7 NHA
G7 NHG
G7 NIE
G7 NIG
G7 NIP
G7 NIS
G7 NJC
G7 NJF
G7 NJK
G7 NJO
G7 NKK
G7 NKL
G7 NKN
G7 NKP
G7 NKT
G7 NLB
G7 NLQ
G7 NLV
G7 NLX
G7 NMF
G7 NMG
G7 NMO
G7 NNE
G7 NNF
G7 NNG
G7 NNY
G7 NOA
G7 NOF
G7 NOJ
G7 NOO
G7 NOS
G7 NOT
G7 NOV
G7 NOX
G7 NOY
G7 NOZ
G7 NPK
G7 NPP
G7 NPW
G7 NQH
G7 NQI
G7 NQL
G7 NQN
G7 NQO
G7 NQS
G7 NRE
G7 NRH
G7 NRI
G7 NSA
G7 NSC
G7 NSR
G7 NSU
G7 NSY
G7 NTB
G7 NTC
G7 NTY
G7 NUK
G7 NVA
G7 NVE
G7 NVK
G7 NVO
G7 NVQ
G7 NWA
G7 NWB
G7 NWI
G7 NWK
G7 NYC
G7 NYK
G7 NZK
G7 NZT
G7 OAE
G7 OAK
G7 OAU
G7 OBH
G7 OBI
G7 OBK
G7 OBO
G7 OBU
G7 OBY
G7 OBZ
G7 OCD
G7 OCG
G7 OCM
G7 OCO
G7 OCS
G7 OCT
G7 ODA
G7 ODD
G7 ODE
G7 ODJ
G7 ODO
G7 ODQ
G7 ODX
G7 ODY
G7 OEL
G7 OEM
G7 OEP
G7 OEX
G7 OFC
G7 OFR
G7 OFX
G7 OGA
G7 OGG
G7 OGX
G7 OHA
G7 OHK
G7 OHT
G7 OII
G7 OIJ
G7 OIW
G7 OJH
G7 OJN
G7 OJR
G7 OJW
G7 OKD
G7 OKE
G7 OKU
G7 OKZ
G7 OLA
G7 OLD
G7 OLI
G7 OLP
G7 OLS
G7 OML
G7 OML
G7 OMT
G7 OMZ
G7 OND
G7 ONG
G7 ONK
G7 ONM
G7 ONS
G7 ONX
G7 ONY
G7 OOA
G7 OOD
G7 OOJ
G7 OOR
G7 OOW
G7 OOX
G7 OOZ
G7 OPC
G7 OPO
G7 OPZ
G7 OQH
G7 OQJ
G7 OQK
G7 OQN
G7 OQS
G7 OQV
G7 OQZ
G7 ORH
G7 ORI
G7 OSG
G7 OSL
G7 OTY
G7 OUL
G7 OUN
G7 OVA
G7 OVC
G7 OVV
G7 OWE
G7 OWH
G7 OXM
G7 OXX
G7 OYC
G7 OYO
G7 OZF
G7 OZO
G7 OZY
G7 PAB
G7 PAJ
G7 PAL
G7 PAS
G7 PAV
G7 PAZ
G7 PBX
G7 PCA
G7 PCJ
G7 PCL
G7 PCM
G7 PCN
G7 PCQ
G7 PDL
G7 PDP
G7 PDY
G7 PDZ
G7 PEA
G7 PEK
G7 PEM
G7 PEQ
G7 PET
G7 PEY
G7 PFB
G7 PFE
G7 PFN
G7 PFO
G7 PFU
G7 PFX
G7 PGB
G7 PGS
G7 PGZ
G7 PHN
G7 PHU
G7 PHZ
G7 PIA
G7 PID
G7 PIM
G7 PIQ
G7 PJC
G7 PJH
G7 PJH
G7 PJT
G7 PJV
G7 PKF
G7 PKO
G7 PKS
G7 PKU
G7 PKV
G7 PKW
G7 PKX
G7 PLF
G7 PLG
G7 PLN
G7 PLT
G7 PLX
G7 PLY
G7 PLZ
G7 PME
G7 PMJ
G7 PMO
G7 PMS
G7 PNQ
G7 PNR
G7 PNU
G7 POJ
G7 POL
G7 POU
G7 PPB
G7 PPE
G7 PPF
G7 PPM
G7 PPT
G7 PQF
G7 PQN
G7 PQR
G7 PQU
G7 PQV
G7 PRA
G7 PRH
G7 PRJ
G7 PRN
G7 PRR
G7 PRY
G7 PSA
G7 PSQ
G7 PSX
G7 PSY
G7 PTL
G7 PTQ
G7 PTU
G7 PUD
G7 PUQ
G7 PUV
G7 PVA
G7 PVJ
G7 PVR
G7 PVS
G7 PVV
G7 PWE
G7 PWH
G7 PXA
G7 PXB
G7 PXD
G7 PXE
G7 PXF
G7 PXK
G7 PXP
G7 PXT
G7 PYH
G7 PYI
G7 PZI
G7 PZP
G7 PZS
G7 PZV
G7 PZX
G7 RAC
G7 RAD
G7 RAP
G7 RAR
G7 RBG
G7 RBV
G7 RBW
G7 RC
G7 RCD
G7 RCI
G7 RCQ
G7 RDC
G7 RDI
G7 RDM
G7 RDO
G7 RDV
G7 RDX
G7 REB
G7 RES
G7 RFG
G7 RFI
G7 RFR
G7 RGB
G7 RGH
G7 RGQ
G7 RGW
G7 RHU
G7 RID
G7 RIK
G7 RIQ
G7 RIY
G7 RJA
G7 RJC
G7 RJL
G7 RJT
G7 RJU
G7 RJV
G7 RJZ
G7 RKI
G7 RKK
G7 RKM
G7 RKY
G7 RLI
G7 RMB
G7 RMH
G7 RMI
G7 RMN
G7 RMV
G7 RNE
G7 ROF
G7 ROS
G7 RPF
G7 RPG
G7 RPH
G7 RPO
G7 RPR
G7 RPS
G7 RPX
G7 RPZ
G7 RQA
G7 RQB
G7 RQE
G7 RQG
G7 RQJ
G7 RQM
G7 RQN
G7 RQS
G7 RQT
G7 RQU
G7 RQX
G7 RRF
G7 RRI
G7 RRP
G7 RRQ
G7 RRT
G7 RSD
G7 RSI
G7 RSO
G7 RST
G7 RSV
G7 RSZ
G7 RTK
G7 RUA
G7 RUG
G7 RUL
G7 RUM
G7 RUT
G7 RUV
G7 RVH
G7 RVJ
G7 RVL
G7 RVS
G7 RWA
G7 RWH
G7 RWI
G7 RWJ
G7 RWO
G7 RWQ
G7 RWU
G7 RWX
G7 RXG
G7 RXH
G7 RXP
G7 RXR
G7 RXV
G7 RXY
G7 RYE
G7 RYR
G7 RYV
G7 RZA
G7 RZD
G7 RZF
G7 RZM
G7 RZT
G7 RZU
G7 SAB
G7 SBH
G7 SBQ
G7 SBU
G7 SBV
G7 SBW
G7 SCC
G7 SCF
G7 SCY
G7 SDJ
G7 SDL
G7 SDL
G7 SDV
G7 SEH
G7 SEI
G7 SET
G7 SFC
G7 SFQ
G7 SFR
G7 SFV
G7 SFW
G7 SGB
G7 SGG
G7 SGS
G7 SHG
G7 SHW
G7 SHX
G7 SIC
G7 SII
G7 SIV
G7 SJP
G7 SJV
G7 SKE
G7 SKI
G7 SKJ
G7 SLD
G7 SLK
G7 SLM
G7 SLO
G7 SLS
G7 SLT
G7 SMA
G7 SMD
G7 SMK
G7 SMO
G7 SMY
G7 SOG
G7 SOI
G7 SOJ
G7 SOQ
G7 SPG
G7 SPI
G7 SPK
G7 SPR
G7 SPS
G7 SPT
G7 SPU
G7 SPV
G7 SPW
G7 SQC
G7 SQK
G7 SQL
G7 SQR
G7 SQS
G7 SQU
G7 SQX
G7 SRE
G7 SRG
G7 SSD
G7 SSE
G7 SSZ
G7 STA
G7 STB
G7 SUD
G7 SUH
G7 SUI
G7 SUP
G7 SUR
G7 SUZ
G7 SVB
G7 SVG
G7 SVK
G7 SVW
G7 SWC
G7 SWP
G7 SXA
G7 SXE
G7 SXH
G7 SXQ
G7 SXR
G7 SXX
G7 SYB
G7 SYK
G7 SYO
G7 SYP
G7 SYW
G7 SYZ
G7 SZD
G7 SZJ
G7 SZN
G7 SZR
G7 SZT
G7 TAC
G7 TAG
G7 TAH
G7 TAI
G7 TAJ
G7 TAO
G7 TAR
G7 TAS
G7 TBE
G7 TBH
G7 TBT
G7 TCI
G7 TCK
G7 TCL
G7 TCM
G7 TDO
G7 TDP
G7 TDW
G7 TDX
G7 TDY
G7 TDZ
G7 TED
G7 TEF
G7 TEM
G7 TER
G7 TFD
G7 TFH
G7 TFJ
G7 TFP
G7 TGE
G7 TGL
G7 TGO
G7 TGR
G7 TGV
G7 TGY
G7 THC
G7 THD
G7 THL
G7 THT
G7 THX
G7 TIA
G7 TIC
G7 TID
G7 TII
G7 TIS
G7 TIZ
G7 TJE
G7 TJK
G7 TJS
G7 TJX
G7 TJY
G7 TKE
G7 TKJ
G7 TKK
G7 TKL
G7 TLG
G7 TLN
G7 TLP
G7 TMG
G7 TMJ
G7 TMW
G7 TNA
G7 TNC
G7 TND
G7 TNF
G7 TNL
G7 TOC
G7 TOG
G7 TOP
G7 TOW
G7 TOX
G7 TPI
G7 TPL
G7 TPP
G7 TPR
G7 TPV
G7 TPX
G7 TQG
G7 TQJ
G7 TQS
G7 TQY
G7 TQZ
G7 TRE
G7 TRJ
G7 TRN
G7 TRW
G7 TRZ
G7 TSA
G7 TSL
G7 TSX
G7 TTR
G7 TTT
G7 TUA
G7 TUJ
G7 TUY
G7 TVB
G7 TVD
G7 TVH
G7 TVI
G7 TVP
G7 TVU
G7 TVX
G7 TWI
G7 TWP
G7 TWR
G7 TWX
G7 TWZ
G7 TXH
G7 TXJ
G7 TXM
G7 TXU
G7 TYA
G7 TYV
G7 TYY
G7 TZE
G7 TZH
G7 TZY
G7 UAA
G7 UAU
G7 UBF
G7 UBJ
G7 UBL
G7 UBT
G7 UBU
G7 UCC
G7 UCF
G7 UCX
G7 UDH
G7 UDP
G7 UDQ
G7 UDS
G7 UDT
G7 UEB
G7 UED
G7 UEF
G7 UEH
G7 UEM
G7 UEO
G7 UEQ
G7 UEY
G7 UFA
G7 UFG
G7 UFJ
G7 UFO
G7 UFP
G7 UFS
G7 UGE
G7 UGF
G7 UGG
G7 UGK
G7 UGN
G7 UGP
G7 UGS
G7 UGT
G7 UGV
G7 UGX
G7 UHF
G7 UHN
G7 UHV
G7 UIE
G7 UIL
G7 UIQ
G7 UJD
G7 UJG
G7 UJI
G7 UJM
G7 UJN
G7 UKG
G7 UKT
G7 ULE
G7 ULM
G7 ULT
G7 UMB
G7 UME
G7 UMP
G7 UMR
G7 UMU
G7 UMX
G7 UND
G7 UNE
G7 UNF
G7 UNH
G7 UNN
G7 UNQ
G7 UOV
G7 UOZ
G7 UPC
G7 UPE
G7 UQF
G7 UQK
G7 UQN
G7 UQP
G7 URA
G7 URB
G7 URD
G7 URN
G7 URU
G7 USP
G7 UTK
G7 UTL
G7 UTN
G7 UUV
G7 UUV
G7 UVC
G7 UVD
G7 UVG
G7 UVH
G7 UVT
G7 UWA
G7 UWF
G7 UWR
G7 UXG
G7 UXI
G7 UXV
G7 UXW
G7 UXX
G7 UYD
G7 UYH
G7 UZM
G7 UZP
G7 UZX
G7 VAJ
G7 VAK
G7 VAU
G7 VAV
G7 VAX
G7 VBB
G7 VBC
G7 VBP
G7 VBQ
G7 VBR
G7 VBS
G7 VBX
G7 VCC
G7 VCD
G7 VCQ
G7 VCX
G7 VDM
G7 VDR
G7 VEO
G7 VEQ
G7 VEZ
G7 VFD
G7 VFK
G7 VFW
G7 VFZ
G7 VGG
G7 VGP
G7 VGU
G7 VGW
G7 VGZ
G7 VHP
G7 VHV
G7 VIA
G7 VJC
G7 VJL
G7 VJN
G7 VJR
G7 VJX
G7 VKF
G7 VKK
G7 VKP
G7 VKQ
G7 VKU
G7 VKZ
G7 VLG
G7 VLN
G7 VLY
G7 VMC
G7 VMG
G7 VMI
G7 VMX
G7 VNB
G7 VNR
G7 VNU
G7 VNY
G7 VOF
G7 VOZ
G7 VPI
G7 VPK
G7 VPO
G7 VPZ
G7 VQD
G7 VQQ
G7 VQU
G7 VQV
G7 VRC
G7 VRD
G7 VRI
G7 VRT
G7 VSH
G7 VSQ
G7 VTF
G7 VTO
G7 VTU
G7 VTV
G7 VUB
G7 VUF
G7 VUR
G7 VUV
G7 VUZ
G7 VVE
G7 VVG
G7 VVU
G7 VWH
G7 VXV
G7 VXX
G7 VYC
G7 VYH
G7 VYJ
G7 VYL
G7 VZE
G7 VZJ
G7 VZN
G7 VZT
G7 VZW
G7 VZX
G7 WAJ
G7 WAM
G7 WAP
G7 WAT
G7 WBB
G7 WBI
G7 WBX
G7 WCA
G7 WCC
G7 WCI
G7 WCM
G7 WCO
G7 WDK
G7 WDR
G7 WDT
G7 WEF
G7 WEL
G7 WEX
G7 WFA
G7 WFY
G7 WGF
G7 WGJ
G7 WGK
G7 WHC
G7 WHF
G7 WHJ
G7 WHK
G7 WHN
G7 WIA
G7 WIB
G7 WIE
G7 WII
G7 WIJ
G7 WIK
G7 WIP
G7 WIR
G7 WIT
G7 WIU
G7 WIW
G7 WIZ
G7 WJI
G7 WKD
G7 WKE
G7 WKN
G7 WKX
G7 WKY
G7 WLI
G7 WLS
G7 WLW
G7 WLZ
G7 WMA
G7 WMC
G7 WOS
G7 WOT
G7 WRG
G7 WRO
G7 WST
G7 XAL
G7 XJS
G7 XTC
G7 YLS
G7 ZZZ

G8

G8 AAG
G8 ACE
G8 ACM
G8 ACZ
G8 ADH
G8 ADM
G8 AFC
G8 AFZ
G8 AIT
G8 AJF
G8 AJN
G8 ALB
G8 AMP
G8 AMV
G8 AMZ
G8 ANJ
G8 AOA
G8 AQX
G8 ARG
G8 ARP
G8 ARV
G8 ASZ
G8 ATH
G8 ATO
G8 ATU
G8 ATV
G8 AUC
G8 AUT
G8 AWO
G8 AXA
G8 AXK
G8 AXW
G8 AXX
G8 AXZ
G8 AZC
G8 AZI
G8 BBC
G8 BBI
G8 BBS
G8 BDA
G8 BDU
G8 BEG
G8 BFT
G8 BGG
G8 BHB
G8 BIJ
G8 BJG
G8 BJY
G8 BKS
G8 BMD
G8 BNL
G8 BOU
G8 BPR
G8 BPV
G8 BQR
G8 BTS
G8 BTT
G8 BTW
G8 BTW
G8 BUJ
G8 BUN
G8 BUR
G8 BVI
G8 BXU
G8 BZY
G8 CAP
G8 CAR
G8 CBL
G8 CCG
G8 CCR
G8 CDF
G8 CDJ
G8 CHB
G8 CHP
G8 CHQ
G8 CHW
G8 CIL
G8 CJS
G8 CJZ
G8 CKF
G8 CKT
G8 CLE
G8 CME
G8 CMJ
G8 CML
G8 CNE
G8 COE
G8 COJ
G8 COV
G8 CPB
G8 CPZ
G8 CQA
G8 CQK
G8 CQP
G8 CTZ
G8 CUV
G8 CVL
G8 CVY
G8 CWM
G8 CZV
G8 CZZ
G8 DAZ
G8 DBN
G8 DCZ
G8 DEN
G8 DGZ
G8 DID
G8 DJK
G8 DJM
G8 DJN
G8 DKC
G8 DLA
G8 DLW
G8 DMP
G8 DMS
G8 DNW
G8 DOE
G8 DOG
G8 DQQ
G8 DQX
G8 DRL
G8 DVG
G8 DVO
G8 DVR
G8 DWA
G8 DWN
G8 DXT
G8 DYF
G8 DYK
G8 DZP
G8 EBH
G8 EBJ
G8 EDE
G8 EDH
G8 EDJ
G8 EEZ
G8 EFP
G8 EFV
G8 EGT
G8 EHK
G8 EIS
G8 EJG
G8 EJN
G8 EKO
G8 ELA
G8 ELR
G8 ELS
G8 ELZ
G8 EPA
G8 EPM
G8 EPO
G8 EQA
G8 ERJ
G8 ERT
G8 ESB
G8 ESZ
G8 ETC
G8 ETQ
G8 EUD
G8 EUK
G8 EUQ
G8 EWG
G8 EWJ
G8 EWV
G8 EYG
G8 EYH
G8 EZO
G8 FAW
G8 FBC
G8 FBO
G8 FBZ
G8 FCW
G8 FDO
G8 FDS
G8 FEC
G8 FED
G8 FEG
G8 FEL
G8 FGD
G8 FGF
G8 FGK
G8 FHF
G8 FHM
G8 FHN
G8 FHP
G8 FIW
G8 FMQ
G8 FPN
G8 FQE
G8 FQW
G8 FRF
G8 FRL
G8 FSO
G8 FTH
G8 FTI
G8 FTO
G8 FUP
G8 FUQ
G8 FUR
G8 FVY
G8 FWF
G8 FXB
G8 FXD
G8 FYE
G8 FYH
G8 FZA
G8 FZC
G8 FZH
G8 GBW
G8 GCU
G8 GD
G8 GDF
G8 GDS
G8 GEC
G8 GFG
G8 GFH
G8 GFI
G8 GGG
G8 GGY
G8 GHD
G8 GHF
G8 GHX
G8 GIJ
G8 GIY
G8 GKA
G8 GKQ
G8 GKS
G8 GKU
G8 GLM
G8 GLO
G8 GLW
G8 GMC
G8 GMD
G8 GME
G8 GMT
G8 GNC
G8 GND
G8 GNL
G8 GNW
G8 GOA
G8 GOI
G8 GOW
G8 GPQ
G8 GPQ
G8 GQT

G8 GRA
G8 GSQ
G8 GSR
G8 GTC
G8 GUV
G8 GVK
G8 GWD
G8 GWL
G8 GWT
G8 GXP
G8 GYW
G8 GYZ
G8 HAJ
G8 HAK
G8 HAX
G8 HBA
G8 HBM
G8 HBS
G8 HCB
G8 HCO
G8 HCU
G8 HCY
G8 HED
G8 HEM
G8 HFI
G8 HFP
G8 HGT
G8 HHV
G8 HHX
G8 HHZ
G8 HIK
G8 HIM
G8 HIP
G8 HJT
G8 HLU
G8 HLW
G8 HNB
G8 HNQ
G8 HOD
G8 HOH
G8 HOM
G8 HON
G8 HOP
G8 HOV
G8 HQJ
G8 HQV
G8 HQY
G8 HQY
G8 HRH
G8 HRK
G8 HRR
G8 HRZ
G8 HSO
G8 HTH
G8 HTX
G8 HUA
G8 HUE
G8 HUI
G8 HUL
G8 HVR
G8 HVY
G8 HWG
G8 HWZ
G8 HXT
G8 HYU
G8 HZW
G8 IAD
G8 IAY
G8 IB
G8 ICJ
G8 IDX
G8 IEJ
G8 IFR
G8 IFW
G8 IGP
G8 IGS
G8 IGV
G8 IGY
G8 IIC
G8 IIV
G8 IJA
G8 IJF
G8 IJN
G8 IJZ
G8 IKL
G8 IKP
G8 ILF
G8 ILM
G8 ILS
G8 IMD
G8 IML
G8 IMN
G8 IMP
G8 IMR
G8 INE
G8 INI
G8 INN
G8 INP
G8 INU
G8 INV
G8 IQG
G8 IQL
G8 IQP
G8 ISX
G8 IUV
G8 IVC
G8 IVI
G8 IWL
G8 IWN
G8 IXA
G8 IXE
G8 IYP
G8 IZU
G8 IZZ
G8 JAO
G8 JAS
G8 JAU
G8 JAV
G8 JAZ
G8 JBE
G8 JC
G8 JCT
G8 JCW
G8 JEF
G8 JFJ
G8 JFW
G8 JGC
G8 JGE
G8 JGI
G8 JGO
G8 JIX
G8 JKP
G8 JKP
G8 JKQ
G8 JKR
G8 JLE
G8 JLZ
G8 JNF
G8 JNJ
G8 JNS
G8 JOD
G8 JOJ
G8 JOR
G8 JPZ
G8 JQK
G8 JTJ
G8 JTY
G8 JUF
G8 JUX
G8 JUZ
G8 JVD
G8 JXA
G8 JXJ
G8 JYJ
G8 JYV
G8 JZA
G8 JZE
G8 JZY
G8 KAI
G8 KAV
G8 KBC
G8 KBI
G8 KBO
G8 KBV
G8 KBZ
G8 KCG
G8 KCM
G8 KCQ
G8 KCS
G8 KCV
G8 KDG
G8 KDI
G8 KFA
G8 KFB
G8 KFW
G8 KGA
G8 KGC
G8 KHJ
G8 KHR
G8 KHS
G8 KHW
G8 KHZ
G8 KIO
G8 KIT
G8 KJC
G8 KJV
G8 KKD
G8 KKS
G8 KLD
G8 KLQ
G8 KLR
G8 KMH
G8 KNV
G8 KOE
G8 KPM
G8 KPS
G8 KQH
G8 KQW
G8 KRA
G8 KSP
G8 KTO
G8 KUE
G8 KUH
G8 KUJ
G8 KUW
G8 KVM
G8 KVP
G8 KWC
G8 KWI
G8 KWQ
G8 KWR
G8 KWX
G8 KXM
G8 KYT
G8 KYU
G8 KZU
G8 LAW
G8 LBD
G8 LBV
G8 LCO
G8 LDP
G8 LEC
G8 LGI
G8 LGX
G8 LIB
G8 LIL
G8 LIR
G8 LIT
G8 LIY
G8 LJC
G8 LKD
G8 LKM
G8 LLB
G8 LLK
G8 LMD
G8 LMN
G8 LMS
G8 LMX
G8 LNR
G8 LNT
G8 LOE
G8 LOL
G8 LOY
G8 LSV
G8 LVI
G8 LVK
G8 LVZ
G8 LWY
G8 LXI
G8 LXO
G8 LYS
G8 LYT
G8 LZI
G8 LZR
G8 LZU
G8 MAC
G8 MAG
G8 MAQ
G8 MAU
G8 MBI
G8 MCD
G8 MCG
G8 MCV
G8 MCZ
G8 MDA
G8 MEJ
G8 MES
G8 MHK
G8 MHR
G8 MJW
G8 MKK
G8 MKM
G8 MLQ
G8 MMD
G8 MMH
G8 MML
G8 MNK
G8 MNZ
G8 MNZ
G8 MOA
G8 MOU
G8 MPP
G8 MPV
G8 MQH
G8 MQN
G8 MRB
G8 MRJ
G8 MSA
G8 MSG
G8 MSR
G8 MSW
G8 MTM
G8 MTQ
G8 MUE
G8 MUQ
G8 MVF
G8 MVP
G8 MVV
G8 MWF
G8 MWK
G8 MWQ
G8 MXE
G8 MXZ
G8 MZG
G8 NAX
G8 NBD
G8 NBR
G8 NCF
G8 NCW
G8 NCX
G8 NDA
G8 NDL
G8 NEH
G8 NEZ
G8 NHN
G8 NHR
G8 NIC
G8 NJD
G8 NJJ
G8 NJW
G8 NKD
G8 NKV
G8 NKX
G8 NKY
G8 NLB
G8 NLY
G8 NMC
G8 NMI
G8 NML
G8 NMU
G8 NMW
G8 NNT
G8 NOC
G8 NPF
G8 NPV
G8 NQA
G8 NRY
G8 NSS
G8 NSU
G8 NT
G8 NTI
G8 NTU
G8 NUC
G8 NUG
G8 NUP
G8 NVF
G8 NVJ
G8 NVY
G8 NWR
G8 NXI
G8 NXL
G8 NXV
G8 NYS
G8 NYW
G8 NZE
G8 NZH
G8 NZM
G8 NZU
G8 OBV
G8 OCN
G8 ODF
G8 ODP
G8 OEB
G8 OEF
G8 OEL
G8 OEM
G8 OER
G8 OES
G8 OFJ
G8 OFK
G8 OFS
G8 OFT
G8 OGI
G8 OGJ
G8 OGO
G8 OHL
G8 OHR
G8 OIS
G8 OIV
G8 OLD
G8 OLT
G8 OMG
G8 OML
G8 OMU
G8 ONK
G8 ONX
G8 OPV
G8 OQF
G8 OQN
G8 OQV
G8 ORE
G8 ORZ
G8 ORZ
G8 OSA
G8 OSD
G8 OSE
G8 OSN
G8 OTE
G8 OTJ
G8 OUA
G8 OUQ
G8 OUU
G8 OUX
G8 OVM
G8 OVT
G8 OWB
G8 OWL
G8 OWO
G8 OWX
G8 OZG
G8 OZJ
G8 OZV
G8 OZW
G8 OZZ
G8 PAF
G8 PAF
G8 PBJ
G8 PCB
G8 PCF
G8 PEF
G8 PEV
G8 PFE
G8 PFG
G8 PFS
G8 PGD
G8 PGL
G8 PHN
G8 PHP
G8 PIB
G8 PID
G8 PIZ
G8 PJC
G8 PJH
G8 PKL
G8 PKY
G8 PLE
G8 PLL
G8 PLR
G8 PLW
G8 PLZ
G8 PME
G8 PMQ
G8 PMU
G8 PNC
G8 PND
G8 PNO
G8 PNX
G8 POC
G8 POW
G8 PPE
G8 PPS
G8 PRF
G8 PRI
G8 PRR
G8 PRU
G8 PSE
G8 PSP
G8 PTD
G8 PTJ
G8 PTP
G8 PTR
G8 PUO
G8 PUX
G8 PVB
G8 PVD
G8 PVI
G8 PVJ
G8 PVT
G8 PWA
G8 PWC
G8 PWY
G8 PXB
G8 PXL
G8 PY
G8 PYP
G8 PYT
G8 PYX
G8 PZJ
G8 PZL
G8 PZP
G8 PZR
G8 PZT
G8 QR
G8 RAE
G8 RAK
G8 RAL
G8 RCP
G8 RDE
G8 RDH
G8 RDO
G8 RDX
G8 REM
G8 REU
G8 REY
G8 RFN
G8 RGQ
G8 RHT
G8 RIA
G8 RIE
G8 RJA
G8 RJD
G8 RJU
G8 RJX
G8 RKC
G8 RLJ
G8 RLZ
G8 RMM
G8 ROD
G8 ROM
G8 ROZ
G8 RPH
G8 RPK
G8 RPV
G8 RQP
G8 RQY
G8 RTC
G8 RTW
G8 RTZ
G8 RUM
G8 RVC
G8 RVG
G8 RVX
G8 RWT
G8 RWT
G8 RWX
G8 RXK
G8 RXL
G8 RXM
G8 RXP
G8 RXU
G8 RYC
G8 RZA
G8 RZJ
G8 SAE
G8 SAJ
G8 SAM
G8 SBE
G8 SBF
G8 SCH
G8 SCK
G8 SCU
G8 SDC
G8 SDT
G8 SEJ
G8 SEZ
G8 SFR
G8 SFU
G8 SHU
G8 SIJ
G8 SIQ
G8 SJP
G8 SKU
G8 SLN
G8 SLX
G8 SNG
G8 SNH
G8 SOZ
G8 SQF
G8 SQM
G8 SRD
G8 STU
G8 SUP
G8 SUT
G8 SVC
G8 SVD
G8 SVE
G8 SVF
G8 SVL
G8 SVO
G8 SVU
G8 SWG
G8 SWM
G8 SWZ
G8 SXC
G8 SXN
G8 SYR
G8 SYU
G8 SYZ
G8 SZY
G8 TA
G8 TAM
G8 TBI
G8 TBK
G8 TCM
G8 TDB
G8 TDH
G8 TDL
G8 TDW
G8 TEE
G8 TET
G8 TFH
G8 TFI
G8 TFK
G8 TFO
G8 TFV
G8 TGR
G8 THS
G8 TIC
G8 TIJ
G8 TIM
G8 TIS
G8 TJQ
G8 TKR
G8 TKV
G8 TLQ
G8 TLX
G8 TLZ
G8 TMY
G8 TNC
G8 TOK
G8 TOU
G8 TPA
G8 TPT
G8 TPX
G8 TPY
G8 TQY
G8 TRB
G8 TSE
G8 TSJ
G8 TT
G8 TV
G8 TVH
G8 TVT
G8 TWH
G8 TWJ
G8 TWL
G8 TXM
G8 TXO
G8 TYI
G8 TYV
G8 TZT
G8 UAW
G8 UCB
G8 UCM
G8 UDH
G8 UDR
G8 UDR
G8 UEC
G8 UEM
G8 UEP
G8 UEQ
G8 UFD
G8 UFK
G8 UFN
G8 UFR
G8 UGB
G8 UGC
G8 UGC
G8 UGD
G8 UGM
G8 UGT
G8 UGZ
G8 UHN
G8 UHV
G8 UIB
G8 UJB
G8 UJP
G8 UKN
G8 UKT
G8 ULG
G8 ULO
G8 ULV
G8 UMX
G8 UNZ
G8 UOD
G8 UON
G8 UOQ
G8 UQI
G8 UQQ
G8 URC
G8 URE
G8 URP
G8 USX
G8 UUA
G8 UUG
G8 UUH
G8 UVC
G8 UVD
G8 UVE
G8 UVK
G8 UVR
G8 UVS
G8 UWF
G8 UWJ
G8 UWO
G8 UWU
G8 UWW
G8 UXE
G8 UXU
G8 UYA
G8 UYC
G8 UYN
G8 UZJ
G8 UZL
G8 UZO
G8 UZV
G8 UZZ
G8 VAU
G8 VAZ
G8 VBO
G8 VC
G8 VCB
G8 VCW
G8 VDO
G8 VEB
G8 VFH
G8 VFN
G8 VFS
G8 VGC
G8 VHC
G8 VIQ
G8 VIW
G8 VJ
G8 VJH
G8 VJS
G8 VKB
G8 VKD
G8 VKT
G8 VKU
G8 VKV
G8 VLQ
G8 VNA
G8 VNE
G8 VNU
G8 VNW
G8 VOC
G8 VOP
G8 VOV
G8 VPQ
G8 VPS
G8 VQM
G8 VRI
G8 VSK
G8 VSU
G8 VTF
G8 VTN
G8 VUR
G8 VUW
G8 VVK
G8 VVQ
G8 VWN
G8 VXG
G8 VXV
G8 VXX
G8 VYD
G8 VZA
G8 VZE
G8 WAA
G8 WBB
G8 WBH
G8 WBI
G8 WCK
G8 WED
G8 WEG
G8 WEN
G8 WHS
G8 WHZ
G8 WID
G8 WIG
G8 WIO
G8 WIO
G8 WJZ
G8 WKU
G8 WMQ
G8 WMZ
G8 WND
G8 WNX
G8 WOF
G8 WOW
G8 WPD
G8 WQB
G8 WQD
G8 WQM
G8 WQQ
G8 WSG
G8 WT
G8 WTR
G8 WTV
G8 WVW
G8 WWB
G8 WWS
G8 WWZ
G8 WY
G8 WYZ
G8 WZB
G8 XAC
G8 XAH
G8 XBA
G8 XBC
G8 XBI
G8 XBP
G8 XBW
G8 XBZ
G8 XCH
G8 XCK
G8 XDU
G8 XEV
G8 XFA
G8 XFP
G8 XFX
G8 XGA
G8 XGE
G8 XGI
G8 XGM
G8 XGP
G8 XGR
G8 XHE
G8 XHO
G8 XIJ
G8 XIT
G8 XIX
G8 XJR
G8 XKB
G8 XKR
G8 XKV
G8 XLM
G8 XLV
G8 XLY
G8 XMP
G8 XNI
G8 XNQ
G8 XNV
G8 XOH
G8 XPA
G8 XPF
G8 XPL
G8 XPV
G8 XQF
G8 XQI
G8 XQQ
G8 XTK
G8 XTX
G8 XUJ
G8 XUX
G8 XVA
G8 XVC
G8 XVD
G8 XVV
G8 XWG
G8 XWI
G8 XWJ
G8 XWK
G8 XXE
G8 XXF
G8 XXK
G8 XYY
G8 XZD
G8 XZO
G8 YAJ
G8 YBL
G8 YCR
G8 YCV
G8 YDU
G8 YFG
G8 YFT
G8 YHH
G8 YHM
G8 YHS
G8 YHW
G8 YIH
G8 YIJ
G8 YJM
G8 YKF
G8 YKW
G8 YLB
G8 YLU
G8 YLX
G8 YMV
G8 YNJ
G8 YNM
G8 YOJ
G8 YOV
G8 YPC
G8 YPE
G8 YPJ
G8 YQD
G8 YQK
G8 YQQ
G8 YQT
G8 YRI
G8 YRJ
G8 YRY
G8 YSK
G8 YSO
G8 YTC
G8 YTD
G8 YUQ
G8 YVR
G8 YXE
G8 YXL
G8 YXY
G8 YYH
G8 YZD
G8 YZH
G8 YZZ
G8 ZAZ
G8 ZBT
G8 ZBW
G8 ZBX
G8 ZCQ
G8 ZCV
G8 ZDF
G8 ZDW
G8 ZEC
G8 ZEG
G8 ZEZ
G8 ZGA
G8 ZGI
G8 ZGX
G8 ZHC
G8 ZHD
G8 ZHJ
G8 ZHL
G8 ZHP
G8 ZID
G8 ZJD
G8 ZJE
G8 ZJJ
G8 ZJP
G8 ZKK
G8 ZKV
G8 ZKY
G8 ZKZ
G8 ZLO
G8 ZMP
G8 ZNL
G8 ZNU
G8 ZNW
G8 ZOH
G8 ZPL
G8 ZPU
G8 ZQO
G8 ZQR
G8 ZRL
G8 ZRO
G8 ZSG
G8 ZSH
G8 ZSN
G8 ZTQ
G8 ZUD
G8 ZVE
G8 ZVJ
G8 ZWE
G8 ZWQ
G8 ZWS
G8 ZXW
G8 ZYO
G8 ZZE
G8 ZZF
G8 ZZZ

M0

M0 AAI
M0 AAL
M0 AAO
M0 AAT
M0 AAX
M0 AAY
M0 ABE
M0 ABL
M0 ABM
M0 ABR
M0 ABS
M0 ACD
M0 ACE
M0 ACF
M0 ACG
M0 ACH
M0 ACL
M0 ACQ
M0 ACS
M0 ADC
M0 ADI
M0 ADM
M0 ADP
M0 ADS
M0 ADU
M0 ADX
M0 AED
M0 AEG
M0 AEH
M0 AEI
M0 AEM
M0 AEY
M0 AFA
M0 AFE
M0 AFH
M0 AFL
M0 AFO
M0 AGC
M0 AGG
M0 AGK
M0 AGN
M0 AGX
M0 AHN
M0 AHP
M0 AHX
M0 AIA
M0 AII
M0 AIO
M0 AIR
M0 AIU
M0 AJL
M0 AJM
M0 AJN
M0 AJO
M0 AJW
M0 AJY
M0 AJZ
M0 AKB
M0 AKH
M0 AKO
M0 AKT
M0 AKW
M0 ALA
M0 ALC
M0 ALJ
M0 ALL
M0 ALW
M0 AMO
M0 ANG
M0 ANT
M0 AOP
M0 AOV
M0 APG
M0 APO
M0 APS
M0 APT
M0 APX
M0 AQC
M0 AQM
M0 AQS
M0 AQY
M0 ARF
M0 ARM
M0 ARP
M0 ARW
M0 ASA
M0 ASH
M0 ASP
M0 AST
M0 ASW
M0 ASZ
M0 ATE
M0 ATM
M0 AUE
M0 AUI
M0 AUQ
M0 AUV
M0 AVC
M0 AVD
M0 AVE
M0 AVG
M0 AWA
M0 AWF
M0 AWR
M0 AWS
M0 AWV
M0 AWW
M0 AXB
M0 AXF
M0 AXH
M0 AXK
M0 AXM
M0 AXP
M0 AXQ
M0 AXT
M0 AXY
M0 AYD
M0 AYJ
M0 AYP
M0 AYR
M0 AYW
M0 AZA
M0 AZH
M0 AZI
M0 AZM
M0 AZO
M0 AZQ
M0 BAN
M0 BAS
M0 BBA
M0 BBB
M0 BBJ
M0 BBP
M0 BBX
M0 BBY
M0 BCD
M0 BCY
M0 BDG
M0 BDP
M0 BEF
M0 BEG
M0 BEN
M0 BEW
M0 BEZ
M0 BEZ
M0 BFE
M0 BFF
M0 BFN
M0 BFO
M0 BFU
M0 BFY
M0 BGA
M0 BGC
M0 BGD
M0 BGL
M0 BGM
M0 BGP
M0 BGQ
M0 BGV
M0 BHB
M0 BHG
M0 BHL
M0 BHU
M0 BHY
M0 BIM
M0 BIR
M0 BIU
M0 BIW
M0 BJA
M0 BJH
M0 BJW
M0 BJY
M0 BKB
M0 BKQ
M0 BKR
M0 BKW
M0 BKZ
M0 BLB
M0 BLE
M0 BLK
M0 BLP
M0 BLQ
M0 BMH
M0 BMK
M0 BMO
M0 BNA
M0 BNE
M0 BNH
M0 BNI
M0 BNJ
M0 BNK
M0 BNY
M0 BOC
M0 BOG
M0 BOJ
M0 BOP
M0 BOT
M0 BOV
M0 BOZ
M0 BPH
M0 BPI
M0 BPJ
M0 BPK
M0 BPL
M0 BPO
M0 BQA
M0 BQP
M0 BQU
M0 BQV
M0 BQW
M0 BRC
M0 BRK
M0 BRV
M0 BSA
M0 BSR
M0 BTQ
M0 BUD
M0 BUJ
M0 BUL
M0 BUO
M0 BUQ
M0 BUS
M0 BUS
M0 BUS
M0 BUS
M0 BUU
M0 BUX
M0 BVL
M0 BWR
M0 BXH
M0 BXR
M0 BXW
M0 BYH
M0 BYQ
M0 BYW
M0 BZG
M0 BZT
M0 BZW
M0 CAA
M0 CAH
M0 CAL
M0 CAP
M0 CBC
M0 CBE
M0 CBH
M0 CBJ
M0 CBU
M0 CBV
M0 CBW
M0 CCE
M0 CCO
M0 CDD
M0 CDI
M0 CDP
M0 CDX
M0 CEE
M0 CEF
M0 CEL
M0 CEM
M0 CEN
M0 CEP
M0 CET
M0 CEV
M0 CFI
M0 CFM
M0 CFN
M0 CFO
M0 CFU
M0 CFV
M0 CFW
M0 CGC
M0 CGD
M0 CGI
M0 CGL
M0 CGM
M0 CGQ
M0 CHK
M0 CHW
M0 CHZ
M0 CIJ
M0 CIL
M0 CIN
M0 CIQ
M0 CIU
M0 CIV
M0 CJX
M0 CKQ
M0 CKR
M0 CLA
M0 CLC
M0 CLF
M0 CLX
M0 CMB
M0 CMD
M0 CMM
M0 CMR
M0 CMV
M0 CMX
M0 CMY
M0 CNQ
M0 COG
M0 COH
M0 COK
M0 COX
M0 COY
M0 CPA
M0 CPJ
M0 CPP
M0 CPR
M0 CPX
M0 CPY
M0 CQA
M0 CQE
M0 CQG
M0 CQU
M0 CQY
M0 CRX
M0 CSH
M0 CSI
M0 CSL
M0 CSM
M0 CSN
M0 CSS
M0 CSX
M0 CTW
M0 CUB
M0 CUV
M0 CUX
M0 CVL
M0 CVM
M0 CVX
M0 CWL
M0 CWM
M0 CWO
M0 CXC
M0 CXF
M0 CXG
M0 CXM
M0 CXN
M0 CXR
M0 CXT
M0 CXX
M0 CYF
M0 CYP
M0 CYQ
M0 CYS
M0 CYW
M0 CYZ
M0 CZI
M0 CZJ
M0 CZV
M0 CZW
M0 DAF
M0 DAJ
M0 DAO
M0 DAQ
M0 DAU
M0 DAV
M0 DBL
M0 DBN
M0 DBP
M0 DBQ
M0 DCJ
M0 DCK
M0 DCX
M0 DDD
M0 DDJ
M0 DDL
M0 DDM
M0 DDN
M0 DDP
M0 DDR
M0 DDS
M0 DED
M0 DEG
M0 DEM
M0 DFC
M0 DFG
M0 DFI
M0 DFK
M0 DFM
M0 DFS
M0 DFT
M0 DFU
M0 DFV
M0 DGD
M0 DGF
M0 DGG
M0 DGL
M0 DGP
M0 DGW
M0 DGY
M0 DGZ
M0 DHA
M0 DHC
M0 DHD
M0 DHG
M0 DHH
M0 DHJ
M0 DHL
M0 DHR
M0 DHS
M0 DHZ
M0 DIA
M0 DIF
M0 DII
M0 DIP
M0 DIR
M0 DIU
M0 DIY
M0 DIZ
M0 DJE
M0 DJJ
M0 DJK
M0 DJM
M0 DJN
M0 DJP
M0 DJS
M0 DJU
M0 DJV
M0 DJY
M0 DJZ
M0 DKA
M0 DKE
M0 DKH
M0 DKI
M0 DKK
M0 DKO
M0 DKQ
M0 DKW
M0 DKZ
M0 DLD
M0 DLJ
M0 DLK
M0 DLQ
M0 DLT
M0 DLV
M0 DLW
M0 DMC
M0 DMG
M0 DMH
M0 DML
M0 DMM
M0 DMP
M0 DMQ
M0 DMW
M0 DNA
M0 DNC
M0 DNE
M0 DNG
M0 DNI
M0 DNL
M0 DNQ
M0 DNS
M0 DNT
M0 DOE
M0 DOF
M0 DOG
M0 DOI
M0 DOJ
M0 DOQ
M0 DOU
M0 DOV
M0 DPA
M0 DPB
M0 DPC
M0 DPD
M0 DPE
M0 DPM
M0 DPN
M0 DPO
M0 DPP
M0 DPR
M0 DPU
M0 DPX
M0 DPZ
M0 DQE
M0 DQJ
M0 DQM
M0 DQR
M0 DQT
M0 DQV
M0 DQW
M0 DQX
M0 DQY
M0 DRC
M0 DRH
M0 DRP
M0 DRV
M0 DRW
M0 DRX
M0 DSB
M0 DSD
M0 DSH
M0 DSJ
M0 DSK
M0 DSQ
M0 DST
M0 DSU
M0 DSV
M0 DTC
M0 DTG
M0 DTM
M0 DTP
M0 DTT
M0 DTU
M0 DTV
M0 DTX
M0 DUC
M0 DUE
M0 DUF
M0 DUG
M0 DUH
M0 DUI
M0 DUJ
M0 DUK
M0 DUL
M0 DUN
M0 DUS
M0 DUW
M0 DUZ
M0 DVA
M0 DVC
M0 DVE
M0 DVO
M0 DVP
M0 DVS
M0 DVV
M0 DVY
M0 DWA
M0 DWJ
M0 DWN
M0 DWV
M0 DWY
M0 DXB
M0 DXE
M0 DXK
M0 DXL
M0 DXO
M0 DXU
M0 DXW
M0 DXY
M0 DYD
M0 DYE
M0 DYF
M0 DYI
M0 DYK
M0 DYL
M0 DYM
M0 DYN
M0 DZI
M0 DZJ
M0 DZP
M0 DZQ
M0 DZR
M0 DZS
M0 DZU
M0 DZY
M0 DZZ
M0 EAA
M0 EAG
M0 EAH
M0 EAJ
M0 EAP
M0 EAV
M0 EAW
M0 EBA
M0 EBB
M0 EBC
M0 EBE
M0 EBF
M0 EBH
M0 EBK
M0 EBL
M0 EBM
M0 EBS
M0 EBW
M0 EBZ
M0 ECA
M0 ECB
M0 ECD
M0 ECE
M0 ECG
M0 ECH
M0 ECI
M0 ECO
M0 ECS
M0 ECU
M0 EDB
M0 EDC
M0 EDD
M0 EDG
M0 EDH
M0 EDI
M0 EDJ
M0 EDK
M0 EDT
M0 EDV
M0 EDW
M0 EDY
M0 EDZ
M0 EEA
M0 EEC
M0 EED
M0 EEE
M0 EEF
M0 EEI
M0 EEJ
M0 EEY
M0 EGH
M0 EGM
M0 EIB
M0 EIG
M0 EIP
M0 EIQ
M0 EJN
M0 ELG
M0 ELI
M0 ELT
M0 EM
M0 EMS
M0 ENA
M0 ENG
M0 EOG
M0 EPV
M0 EPW
M0 ERA
M0 ESE
M0 ESG
M0 ETM
M0 ETO
M0 EUG
M0 EVT
M0 EWS
M0 EXV
M0 EYB
M0 EZR
M0 FAB
M0 FAQ
M0 FBI
M0 FBT
M0 FDA
M0 FDB
M0 FDE
M0 FDF
M0 FDO
M0 FDS
M0 FDY
M0 FET
M0 FEW
M0 FFF
M0 FGH
M0 FGM
M0 FHB
M0 FHN
M0 FIN
M0 FMA
M0 FMC
M0 FME
M0 FMF
M0 FNJ
M0 FNR
M0 FOS
M0 FOZ
M0 FPQ
M0 FRZ
M0 FSD
M0 FWD
M0 FWZ
M0 FXP
M0 FYF
M0 FZQ
M0 FZY
M0 GAA
M0 GAD
M0 GAF
M0 GAJ
M0 GAK
M0 GAM
M0 GAP
M0 GAU
M0 GAW
M0 GAZ
M0 GBP
M0 GBQ
M0 GBR
M0 GBS
M0 GBV
M0 GBX
M0 GBY
M0 GCE
M0 GCG
M0 GCJ
M0 GCK
M0 GCL
M0 GCM
M0 GCO
M0 GCP
M0 GCW
M0 GCZ
M0 GDA
M0 GDB
M0 GDE
M0 GDF
M0 GDK
M0 GDN
M0 GDO
M0 GDQ
M0 GDR
M0 GDW
M0 GDY
M0 GDZ
M0 GEE
M0 GEG
M0 GEH
M0 GEJ
M0 GER
M0 GET
M0 GEV
M0 GFB
M0 GFQ
M0 GFS
M0 GFT
M0 GFU
M0 GFV
M0 GFY
M0 GFZ
M0 GGA
M0 GGE
M0 GGI
M0 GGR
M0 GGS
M0 GGY
M0 GHB
M0 GHD
M0 GHF
M0 GHG
M0 GHH
M0 GHP
M0 GHQ
M0 GHS
M0 GHU
M0 GIC
M0 GIH
M0 GIK
M0 GIO
M0 GIR
M0 GIV
M0 GIX
M0 GJE
M0 GJF
M0 GJO
M0 GJQ
M0 GJR
M0 GJW
M0 GJY
M0 GJZ
M0 GKC
M0 GKH
M0 GKI
M0 GKM
M0 GKX
M0 GKY
M0 GLC
M0 GLD
M0 GLH
M0 GLJ
M0 GLM
M0 GLN
M0 GLO
M0 GLR
M0 GLW
M0 GLZ
M0 GMB
M0 GML
M0 GMM
M0 GMV
M0 GMX
M0 GMY
M0 GNE
M0 GNT
M0 GNV
M0 GNZ
M0 GOE
M0 GOJ
M0 GOR
M0 GPA
M0 GPI
M0 GPM
M0 GPN
M0 GPR
M0 GPS
M0 GPT
M0 GQA
M0 GQC
M0 GQH
M0 GQK
M0 GQL
M0 GQN
M0 GQQ
M0 GQY
M0 GRK
M0 GRL
M0 GRQ
M0 GRV
M0 GSA
M0 GSE
M0 GSF
M0 GSG
M0 GSJ
M0 GSM
M0 GSU
M0 GSY
M0 GTA
M0 GTC
M0 GTD
M0 GTF
M0 GTK
M0 GTS
M0 GTW
M0 GUB
M0 GUI
M0 GUP
M0 GUQ
M0 GUS
M0 GUT
M0 GUV
M0 GUY
M0 GVA
M0 GVD
M0 GVG
M0 GVH
M0 GVO
M0 GVR
M0 GVS
M0 GVU
M0 GVV
M0 GWG
M0 GWI
M0 GWJ
M0 GWP
M0 GWS
M0 GWW
M0 GWX
M0 GXA
M0 GXD
M0 GXF
M0 GXJ
M0 GXL
M0 GXP
M0 GXR
M0 GXS
M0 GXT
M0 GXX
M0 GXZ
M0 GYD
M0 GYE
M0 GYQ
M0 GYT
M0 GZG
M0 GZO
M0 GZP
M0 GZQ
M0 GZR
M0 GZS
M0 GZT
M0 GZU
M0 GZV
M0 GZY
M0 HAA
M0 HAE
M0 HAI
M0 HAJ
M0 HAK
M0 HAQ
M0 HAV
M0 HAX
M0 HAY
M0 HBA
M0 HBB
M0 HBD
M0 HBF
M0 HBI
M0 HBK
M0 HBP
M0 HBQ

M0 HBR
M0 HBS
M0 HBZ
M0 HCB
M0 HCD
M0 HCF
M0 HCH
M0 HCJ
M0 HCK
M0 HCL
M0 HCS
M0 HCU
M0 HDB
M0 HDD
M0 HDF
M0 HDH
M0 HDL
M0 HDO
M0 HDX
M0 HDY
M0 HEA
M0 HEC
M0 HEE
M0 HEH
M0 HEI
M0 HEK
M0 HEL
M0 HEO
M0 HEQ
M0 HER
M0 HES
M0 HEU
M0 HEV
M0 HEZ
M0 HFD
M0 HFJ
M0 HFK
M0 HFL
M0 HFM
M0 HFN
M0 HFP
M0 HFR
M0 HFT
M0 HGB
M0 HGC
M0 HGE
M0 HGH
M0 HGI
M0 HGL
M0 HGP
M0 HGQ
M0 HGR
M0 HGT
M0 HGU
M0 HGW
M0 HGX
M0 HGZ
M0 HHJ
M0 HHK
M0 HHL
M0 HHN
M0 HHO
M0 HHQ
M0 HHS
M0 HHY
M0 HHZ
M0 HIF
M0 HII
M0 HIK
M0 HIR
M0 HIS
M0 HIU
M0 HJA
M0 HJH
M0 HJM
M0 HJS
M0 HJT
M0 HJV
M0 HJX
M0 HJZ
M0 HKD
M0 HKN
M0 HKO
M0 HKQ
M0 HKY
M0 HKZ
M0 HLA
M0 HLF
M0 HLG
M0 HLK
M0 HLL
M0 HLO
M0 HLT
M0 HLY
M0 HMA
M0 HMC
M0 HMD
M0 HMG
M0 HMH
M0 HMK
M0 HML
M0 HMM
M0 HMN
M0 HMP
M0 HMQ
M0 HMW
M0 HNB
M0 HNP
M0 HNR
M0 HNS
M0 HNU
M0 HNV
M0 HNZ
M0 HOC
M0 HOE

M0 HOG
M0 HOH
M0 HOS
M0 HOW
M0 HOX
M0 HPA
M0 HPD
M0 HPI
M0 HPO
M0 HPY
M0 HQH
M0 HQK
M0 HQN
M0 HQS
M0 HQT
M0 HQV
M0 HQW
M0 HQX
M0 HRE
M0 HRF
M0 HRJ
M0 HRK
M0 HRR
M0 HRS
M0 HRU
M0 HRX
M0 HSA
M0 HSD
M0 HSE
M0 HSF
M0 HSK
M0 HSL
M0 HSM
M0 HSN
M0 HSP
M0 HST
M0 HSY
M0 HTD
M0 HTH
M0 HTM
M0 HTN
M0 HTP
M0 HTT
M0 HTZ
M0 HUB
M0 HUC
M0 HUD
M0 HUE
M0 HUK
M0 HUO
M0 HUP
M0 HUT
M0 HUX
M0 HVH
M0 HVJ
M0 HVT
M0 HVY
M0 HWA
M0 HWB
M0 HWE
M0 HWK
M0 HWX
M0 HWY
M0 HXD
M0 HXJ
M0 HXP
M0 HXQ
M0 HXR
M0 HXT
M0 HXU
M0 HXW
M0 HXY
M0 HYB
M0 HYF
M0 HYO
M0 HYR
M0 HYS
M0 HYT
M0 HYU
M0 HYV
M0 HYW
M0 HYY
M0 HYZ
M0 HZG
M0 HZH
M0 HZQ
M0 HZS
M0 HZZ
M0 IAB
M0 IAI
M0 IAN
M0 IAO
M0 IAP
M0 IAQ
M0 IAR
M0 IAU
M0 IAV
M0 IAW
M0 IAY
M0 IBA
M0 IBB
M0 IBF
M0 IBG
M0 IBP
M0 IBS
M0 IBU
M0 ICC
M0 ICD
M0 ICF
M0 ICH
M0 ICM
M0 ICN
M0 ICR
M0 ICV
M0 ICW

M0 ICX
M0 ICY
M0 IDA
M0 IDB
M0 IDD
M0 IDF
M0 IDH
M0 IDP
M0 IDQ
M0 IDS
M0 IDU
M0 IDV
M0 IDX
M0 IDZ
M0 IEC
M0 IEE
M0 IEF
M0 IEG
M0 IEI
M0 IEN
M0 IEU
M0 IEV
M0 IEX
M0 IFE
M0 IFR
M0 IGA
M0 IGD
M0 IGE
M0 IGI
M0 IGK
M0 IGN
M0 IGQ
M0 IGS
M0 IGV
M0 IGZ
M0 IHA
M0 IHD
M0 IHF
M0 IHG
M0 IHH
M0 IHI
M0 IHK
M0 IHO
M0 IHP
M0 IHQ
M0 IHV
M0 IHZ
M0 IIB
M0 IIC
M0 IID
M0 IIH
M0 IIJ
M0 IIK
M0 IIL
M0 IIN
M0 IIO
M0 IIP
M0 IIS
M0 IIT
M0 IIV
M0 IIW
M0 IIX
M0 IIY
M0 IJD
M0 IJH
M0 IJK
M0 IJN
M0 IJS
M0 IJV
M0 IJW
M0 IJX
M0 IJZ
M0 IKF
M0 IKL
M0 IKO
M0 IKP
M0 IKQ
M0 IKR
M0 IKU
M0 IKX
M0 IKY
M0 IKZ
M0 ILB
M0 ILD
M0 ILE
M0 ILL
M0 ILQ
M0 ILR
M0 ILS
M0 ILV
M0 ILW
M0 IMA
M0 IMB
M0 IMC
M0 IMG
M0 IMH
M0 IMK
M0 IMO
M0 IMX
M0 INA
M0 ING
M0 INJ
M0 INK
M0 INN
M0 INO
M0 INQ
M0 INR
M0 INV
M0 IOD
M0 IOF
M0 IOG
M0 IOH
M0 IOJ
M0 ION
M0 IOP

M0 IOR
M0 IOX
M0 IOY
M0 IPA
M0 IPC
M0 IPF
M0 IPG
M0 IPK
M0 IPM
M0 IPN
M0 IPT
M0 IPU
M0 IPW
M0 IPX
M0 IPY
M0 IQA
M0 IQB
M0 IQD
M0 IQE
M0 IQI
M0 IQJ
M0 IQP
M0 IQS
M0 IQU
M0 IQV
M0 IQW
M0 IRM
M0 IRN
M0 IRQ
M0 IRR
M0 IRW
M0 IRY
M0 ISA
M0 ISB
M0 ISH
M0 ISM
M0 ISO
M0 ISR
M0 ISS
M0 ISV
M0 ISX
M0 ITD
M0 ITF
M0 ITH
M0 ITJ
M0 ITK
M0 ITL
M0 ITM
M0 ITN
M0 ITP
M0 ITQ
M0 ITR
M0 ITU
M0 IUF
M0 IUH
M0 IUL
M0 IUO
M0 IUP
M0 IUS
M0 IUW
M0 IUX
M0 IUY
M0 IVA
M0 IVB
M0 IVG
M0 IVI
M0 IVP
M0 IVQ
M0 IVR
M0 IVT
M0 IVY
M0 IWD
M0 IWE
M0 IWF
M0 IWH
M0 IWK
M0 IWL
M0 IWR
M0 IWY
M0 IXA
M0 IXE
M0 IXI
M0 IXK
M0 IXN
M0 IXO
M0 IXS
M0 IYA
M0 IYB
M0 IYI
M0 IYL
M0 IYR
M0 IYW
M0 IYY
M0 IZB
M0 IZD
M0 IZE
M0 IZH
M0 IZL
M0 IZO
M0 IZT
M0 IZY
M0 JAA
M0 JAB
M0 JAH
M0 JAL
M0 JAS
M0 JAU
M0 JBE
M0 JBI
M0 JBJ
M0 JBL
M0 JBM
M0 JBN
M0 JBO
M0 JBP

M0 JBQ
M0 JBR
M0 JBX
M0 JCA
M0 JCF
M0 JCI
M0 JCJ
M0 JCN
M0 JCO
M0 JCU
M0 JCW
M0 JCX
M0 JDC
M0 JDH
M0 JDI
M0 JDK
M0 JDM
M0 JDN
M0 JDO
M0 JDR
M0 JED
M0 JEE
M0 JEN
M0 JEV
M0 JEY
M0 JFA
M0 JFF
M0 JFG
M0 JFN
M0 JFP
M0 JFQ
M0 JFS
M0 JFT
M0 JFV
M0 JFZ
M0 JGA
M0 JGD
M0 JGI
M0 JGJ
M0 JGL
M0 JGN
M0 JGO
M0 JGQ
M0 JGT
M0 JGV
M0 JGW
M0 JGX
M0 JGZ
M0 JHA
M0 JHE
M0 JHI
M0 JHJ
M0 JHK
M0 JHQ
M0 JHR
M0 JHS
M0 JHU
M0 JHX
M0 JHZ
M0 JIA
M0 JIF
M0 JIG
M0 JIH
M0 JII
M0 JIK
M0 JIN
M0 JIR
M0 JIT
M0 JIV
M0 JIW
M0 JIY
M0 JJF
M0 JJG
M0 JJL
M0 JJO
M0 JJS
M0 JJT
M0 JJY
M0 JKA
M0 JKC
M0 JKD
M0 JKK
M0 JKO
M0 JKR
M0 JKT
M0 JKZ
M0 JLA
M0 JLD
M0 JLG
M0 JLJ
M0 JLK
M0 JLL
M0 JLQ
M0 JLX
M0 JLZ
M0 JMA
M0 JMG
M0 JMM
M0 JMO
M0 JMR
M0 JMX
M0 JMZ
M0 JNA
M0 JNB
M0 JNC
M0 JND
M0 JNE
M0 JNG
M0 JNH
M0 JNI
M0 JNK
M0 JNN
M0 JNO
M0 JNR
M0 JNT

M0 JNW
M0 JNZ
M0 JOA
M0 JOE
M0 JOF
M0 JOG
M0 JOS
M0 JOT
M0 JOU
M0 JOV
M0 JOZ
M0 JPF
M0 JPH
M0 JPI
M0 JPJ
M0 JPQ
M0 JPU
M0 JPY
M0 JPZ
M0 JQA
M0 JQC
M0 JQE
M0 JQF
M0 JQH
M0 JQI
M0 JQM
M0 JQN
M0 JQQ
M0 JQR
M0 JQT
M0 JQU
M0 JQX
M0 JQZ
M0 JRB
M0 JRC
M0 JRD
M0 JRG
M0 JRK
M0 JRM
M0 JRS
M0 JRT
M0 JRV
M0 JRY
M0 JSB
M0 JSC
M0 JSF
M0 JSI
M0 JSM
M0 JSO
M0 JSQ
M0 JST
M0 JTA
M0 JTF
M0 JTI
M0 JTK
M0 JTO
M0 JTP
M0 JTR
M0 JTS
M0 JTV
M0 JTZ
M0 JUD
M0 JUE
M0 JUF
M0 JUG
M0 JUH
M0 JUM
M0 JUO
M0 JUR
M0 JUS
M0 JUW
M0 JUY
M0 JVA
M0 JVB
M0 JVD
M0 JVF
M0 JVI
M0 JVK
M0 JVN
M0 JVP
M0 JVS
M0 JWB
M0 JWF
M0 JWG
M0 JWK
M0 JWO
M0 JWS
M0 JXA
M0 JXB
M0 JXC
M0 JXH
M0 JXJ
M0 JXK
M0 JXL
M0 JXN
M0 JXO
M0 JXP
M0 JXQ
M0 JXT
M0 JXV
M0 JXW
M0 JXZ
M0 JYA
M0 JYD
M0 JYE
M0 JYH
M0 JYI
M0 JYL
M0 JYT
M0 JYU
M0 JYX
M0 JYY
M0 JZA
M0 JZC
M0 JZD

M0 JZH
M0 JZJ
M0 JZL
M0 JZQ
M0 JZX
M0 JZY
M0 KAA
M0 KAF
M0 KAH
M0 KAL
M0 KAQ
M0 KAS
M0 KAT
M0 KAV
M0 KAY
M0 KBE
M0 KBF
M0 KBI
M0 KBM
M0 KBO
M0 KBQ
M0 KBS
M0 KBV
M0 KCB
M0 KCG
M0 KCM
M0 KCN
M0 KCQ
M0 KCR
M0 KCT
M0 KCU
M0 KCX
M0 KDB
M0 KDD
M0 KDF
M0 KDI
M0 KDJ
M0 KDK
M0 KDN
M0 KDP
M0 KDQ
M0 KDS
M0 KEA
M0 KEC
M0 KEH
M0 KEI
M0 KEK
M0 KEM
M0 KER
M0 KEX
M0 KEZ
M0 KFE
M0 KFG
M0 KFM
M0 KFN
M0 KFQ
M0 KFS
M0 KFT
M0 KFX
M0 KFZ
M0 KGB
M0 KGC
M0 KGD
M0 KGH
M0 KGJ
M0 KGL
M0 KGN
M0 KGO
M0 KGR
M0 KGX
M0 KGZ
M0 KHF
M0 KHH
M0 KHI
M0 KHJ
M0 KHL
M0 KHP
M0 KHR
M0 KHT
M0 KHU
M0 KHV
M0 KIA
M0 KIE
M0 KIH
M0 KII
M0 KIK
M0 KIS
M0 KIT
M0 KIU
M0 KIX
M0 KJA
M0 KJB
M0 KJD
M0 KJF
M0 KJI
M0 KJL
M0 KJP
M0 KJR
M0 KJV
M0 KJW
M0 KJX
M0 KJY
M0 KKI
M0 KKP
M0 KKR
M0 KKS
M0 KKT
M0 KKU
M0 KKV
M0 KKX
M0 KKY
M0 KLC
M0 KLD
M0 KLF
M0 KLI

M0 KLO
M0 KLP
M0 KLQ
M0 KLU
M0 KLV
M0 KLY
M0 KLZ
M0 KMA
M0 KMD
M0 KME
M0 KMF
M0 KMM
M0 KMP
M0 KMV
M0 KMY
M0 KNA
M0 KNC
M0 KNG
M0 KNJ
M0 KNO
M0 KNR
M0 KNT
M0 KNU
M0 KNZ
M0 KOA
M0 KOJ
M0 KOL
M0 KOP
M0 KOS
M0 KOU
M0 KOX
M0 KPG
M0 KPH
M0 KPL
M0 KPN
M0 KPR
M0 KPS
M0 KPZ
M0 KQA
M0 KQE
M0 KQF
M0 KQH
M0 KQL
M0 KQM
M0 KQP
M0 KQT
M0 KRB
M0 KRF
M0 KRG
M0 KRI
M0 KRJ
M0 KRT
M0 KSD
M0 KSH
M0 KSJ
M0 KSM
M0 KSN
M0 KSP
M0 KSW
M0 KTA
M0 KTB
M0 KTC
M0 KTH
M0 KTI
M0 KTK
M0 KTM
M0 KTX
M0 KTZ
M0 KUB
M0 KUC
M0 KUN
M0 KUZ
M0 KVC
M0 KVG
M0 KVI
M0 KVL
M0 KVP
M0 KVW
M0 KVX
M0 KW
M0 KWD
M0 KWG
M0 KXO
M0 KXS
M0 KXT
M0 KXX
M0 KYB
M0 KYE
M0 KYG
M0 KYJ
M0 KYM
M0 KYO
M0 KYT
M0 KYW
M0 KZL
M0 KZO
M0 KZR
M0 KZX
M0 LAC
M0 LAJ
M0 LAK
M0 LAM
M0 LAN
M0 LAP
M0 LAX
M0 LBC
M0 LBT
M0 LBV
M0 LBW
M0 LCB
M0 LCD
M0 LCE
M0 LCF
M0 LCG
M0 LCN

M0 LDA
M0 LDB
M0 LDF
M0 LDI
M0 LDM
M0 LDN
M0 LDU
M0 LDW
M0 LEG
M0 LEJ
M0 LEU
M0 LFA
M0 LFD
M0 LFG
M0 LFO
M0 LFR
M0 LFV
M0 LFX
M0 LGH
M0 LGJ
M0 LGO
M0 LGV
M0 LHD
M0 LHH
M0 LHT
M0 LIH
M0 LIL
M0 LIN
M0 LIP
M0 LIT
M0 LJB
M0 LJE
M0 LJM
M0 LJO
M0 LJP
M0 LKA
M0 LKB
M0 LKC
M0 LKQ
M0 LKW
M0 LLB
M0 LLC
M0 LLF
M0 LLL
M0 LLP
M0 LLT
M0 LLZ
M0 LMD
M0 LMF
M0 LMK
M0 LMX
M0 LMZ
M0 LNA
M0 LNB
M0 LNG
M0 LNT
M0 LNZ
M0 LOK
M0 LOM
M0 LON
M0 LOR
M0 LOT
M0 LOV
M0 LOY
M0 LPD
M0 LPE
M0 LPH
M0 LRE
M0 LRF
M0 LRI
M0 LRQ
M0 LRT
M0 LRV
M0 LRX
M0 LSE
M0 LSK
M0 LSP
M0 LTC
M0 LTF
M0 LTH
M0 LTM
M0 LTQ
M0 LTV
M0 LTY
M0 LUG
M0 LUI
M0 LUX
M0 LVE
M0 LVV
M0 LWB
M0 LWG
M0 LWI
M0 LWO
M0 LWQ
M0 LXB
M0 LXM
M0 LXQ
M0 LXR
M0 LXT
M0 LXX
M0 LYC
M0 LYM
M0 LYT
M0 LYX
M0 LZA
M0 LZP
M0 LZR
M0 LZY
M0 MAD
M0 MAE
M0 MAK
M0 MAM
M0 MAS
M0 MAV
M0 MBF

M0 MBJ
M0 MBK
M0 MBN
M0 MBP
M0 MBQ
M0 MBU
M0 MBX
M0 MBY
M0 MCD
M0 MCF
M0 MCK
M0 MCN
M0 MCR
M0 MCU
M0 MCX
M0 MCZ
M0 MDA
M0 MDB
M0 MDD
M0 MDI
M0 MDK
M0 MDL
M0 MDM
M0 MDR
M0 MDS
M0 MDU
M0 MDW
M0 MDX
M0 MDY
M0 MEB
M0 MEC
M0 MEE
M0 MEM
M0 MEP
M0 MES
M0 MEU
M0 MEZ
M0 MFD
M0 MFF
M0 MFM
M0 MFO
M0 MFR
M0 MFZ
M0 MGC
M0 MGD
M0 MGG
M0 MGH
M0 MGN
M0 MGO
M0 MGR
M0 MGT
M0 MGV
M0 MGX
M0 MGY
M0 MGZ
M0 MHA
M0 MHB
M0 MHC
M0 MHS
M0 MIA
M0 MII
M0 MIK
M0 MIN
M0 MIO
M0 MIP
M0 MIS
M0 MIW
M0 MIX
M0 MJA
M0 MJC
M0 MJE
M0 MJJ
M0 MJL
M0 MJO
M0 MJP
M0 MJR
M0 MJU
M0 MKA
M0 MKB
M0 MKD
M0 MKF
M0 MKI
M0 MKK
M0 MKL
M0 MKM
M0 MKS
M0 MKT
M0 MKU
M0 MKW
M0 MKX
M0 MLC
M0 MLN
M0 MLP
M0 MLQ
M0 MLX
M0 MMA
M0 MMB
M0 MME
M0 MMH
M0 MMK
M0 MML
M0 MMM
M0 MMQ
M0 MMW
M0 MMY
M0 MMZ
M0 MNE
M0 MNF
M0 MNJ
M0 MNK
M0 MNM
M0 MNT
M0 MOA
M0 MOE
M0 MOF

M0 MOG
M0 MOM
M0 MOX
M0 MOZ
M0 MPC
M0 MPG
M0 MPH
M0 MPJ
M0 MPL
M0 MPR
M0 MPX
M0 MPZ
M0 MQB
M0 MQM
M0 MQS
M0 MQT
M0 MQV
M0 MRA
M0 MRD
M0 MRN
M0 MRR
M0 MRU
M0 MRZ
M0 MSC
M0 MSK
M0 MSL
M0 MST
M0 MSU
M0 MSW
M0 MTB
M0 MTE
M0 MTH
M0 MTL
M0 MTM
M0 MTP
M0 MTT
M0 MTU
M0 MTV
M0 MTY
M0 MTZ
M0 MUD
M0 MUF
M0 MUG
M0 MUJ
M0 MUS
M0 MUT
M0 MUX
M0 MVA
M0 MVC
M0 MVK
M0 MVR
M0 MVV
M0 MWC
M0 MWD
M0 MWF
M0 MWG
M0 MWH
M0 MWI
M0 MWM
M0 MWV
M0 MWZ
M0 MXD
M0 MXH
M0 MXJ
M0 MXL
M0 MXO
M0 MXR
M0 MXV
M0 MXW
M0 MXY
M0 MYO
M0 MYR
M0 MYT
M0 MYY
M0 MZB
M0 MZF
M0 MZG
M0 MZH
M0 MZL
M0 MZM
M0 MZR
M0 MZW
M0 MZZ
M0 NAC
M0 NAD
M0 NAH
M0 NAN
M0 NAT
M0 NAV
M0 NBB
M0 NBG
M0 NBR
M0 NBT
M0 NBW
M0 NBX
M0 NBY
M0 NCB
M0 NCF
M0 NCH
M0 NCP
M0 NCS
M0 NCT
M0 NCW
M0 NCY
M0 NDB
M0 NDN
M0 NDS
M0 NDW
M0 NEA
M0 NEB
M0 NED
M0 NEE
M0 NEI
M0 NEL
M0 NEN

M0 NES
M0 NEY
M0 NFA
M0 NFC
M0 NFE
M0 NFJ
M0 NFL
M0 NFP
M0 NGA
M0 NGD
M0 NGE
M0 NGG
M0 NGO
M0 NGT
M0 NHC
M0 NHG
M0 NHQ
M0 NHR
M0 NHS
M0 NIA
M0 NIB
M0 NIK
M0 NIM
M0 NIN
M0 NIQ
M0 NIR
M0 NIT
M0 NIU
M0 NIX
M0 NIY
M0 NJA
M0 NJB
M0 NJC
M0 NJH
M0 NJM
M0 NJN
M0 NKA
M0 NKC
M0 NKD
M0 NKG
M0 NKI
M0 NKK
M0 NKL
M0 NKM
M0 NKP
M0 NLB
M0 NLG
M0 NLM
M0 NLO
M0 NME
M0 NML
M0 NMP
M0 NMQ
M0 NMS
M0 NNG
M0 NNK
M0 NNN
M0 NNO
M0 NNQ
M0 NNR
M0 NOD
M0 NOP
M0 NOT
M0 NOX
M0 NPE
M0 NPK
M0 NPN
M0 NPP
M0 NPT
M0 NPW
M0 NQN
M0 NQP
M0 NRA
M0 NRD
M0 NRE
M0 NRF
M0 NRN
M0 NRO
M0 NRT
M0 NRX
M0 NRZ
M0 NSA
M0 NST
M0 NSU
M0 NSZ
M0 NTD
M0 NTE
M0 NTL
M0 NTM
M0 NTO
M0 NTP
M0 NTR
M0 NTS
M0 NTV
M0 NTW
M0 NTX
M0 NUE
M0 NUL
M0 NUM
M0 NUR
M0 NVA
M0 NVH
M0 NVS
M0 NWB
M0 NWF
M0 NWH
M0 NWK
M0 NWR
M0 NWX
M0 NXN
M0 NXS
M0 NXT
M0 NYE
M0 NYL
M0 NYW

M0 NZG
M0 NZH
M0 NZO
M0 NZX
M0 NZY
M0 NZZ
M0 OAF
M0 OAH
M0 OAK
M0 OAP
M0 OAS
M0 OAV
M0 OBI
M0 OBJ
M0 OBO
M0 OBS
M0 OCB
M0 OCD
M0 OCH
M0 OCN
M0 OCS
M0 ODC
M0 ODH
M0 ODN
M0 ODP
M0 ODV
M0 ODY
M0 OEE
M0 OEI
M0 OEM
M0 OEY
M0 OFC
M0 OFI
M0 OFP
M0 OFR
M0 OFV
M0 OFX
M0 OFY
M0 OGG
M0 OGY
M0 OHC
M0 OHM
M0 OII
M0 OIL
M0 OJD
M0 OJE
M0 OJM
M0 OJR
M0 OJY
M0 OKE
M0 OKI
M0 OKO
M0 OKR
M0 OKW
M0 OLA
M0 OLF
M0 OLI
M0 OLO
M0 OLP
M0 OLT
M0 OLY
M0 OME
M0 OML
M0 OMM
M0 OMN
M0 OMO
M0 OMR
M0 OMY
M0 ONA
M0 ONE
M0 ONM
M0 ONN
M0 ONP
M0 OOA
M0 OOB
M0 OOE
M0 OOJ
M0 OOM
M0 OON
M0 OOP
M0 OOV
M0 OOX
M0 OPA
M0 OPE
M0 OPI
M0 OPO
M0 OPP
M0 OPR
M0 OPT
M0 OPV
M0 OPW
M0 OPX
M0 OQU
M0 ORA
M0 ORB
M0 ORD
M0 ORG
M0 ORH
M0 ORO
M0 ORY
M0 OSA
M0 OSC
M0 OSG
M0 OSK
M0 OSS
M0 OTM
M0 OTP
M0 OTQ
M0 OTV
M0 OTW
M0 OTX
M0 OTY
M0 OUE
M0 OUF
M0 OUK

M0 OUZ
M0 OVA
M0 OVE
M0 OVG
M0 OVL
M0 OVR
M0 OVY
M0 OWE
M0 OWN
M0 OWW
M0 OWZ
M0 OXF
M0 OXH
M0 OXI
M0 OXN
M0 OXS
M0 OYF
M0 OYG
M0 OZD
M0 OZZ
M0 PAB
M0 PAD
M0 PAE
M0 PAH
M0 PAN
M0 PAP
M0 PAS
M0 PAT
M0 PAU
M0 PBA
M0 PBD
M0 PBK
M0 PBM
M0 PBQ
M0 PBS
M0 PCF
M0 PCI
M0 PCM
M0 PCP
M0 PDJ
M0 PDM
M0 PDT
M0 PEE
M0 PEK
M0 PEL
M0 PEY
M0 PEZ
M0 PFD
M0 PFM
M0 PFX
M0 PGA
M0 PGE
M0 PGF
M0 PGG
M0 PGR
M0 PGT
M0 PGZ
M0 PHC
M0 PHE
M0 PHG
M0 PHI
M0 PHS
M0 PHT
M0 PIH
M0 PIO
M0 PIU
M0 PIW
M0 PIZ
M0 PJB
M0 PJE
M0 PJH
M0 PJO
M0 PJW
M0 PKB
M0 PKD
M0 PKG
M0 PKM
M0 PKO
M0 PKP
M0 PKR
M0 PKT
M0 PKZ
M0 PLA
M0 PLD
M0 PLF
M0 PLK
M0 PLL
M0 PLM
M0 PLZ
M0 PMD
M0 PME
M0 PMK
M0 PMQ
M0 PMS
M0 PMT
M0 PMZ
M0 PNF
M0 PNG
M0 PNP
M0 PNX
M0 POJ
M0 POL
M0 POW
M0 PPD
M0 PPE
M0 PPK
M0 PPL
M0 PPQ
M0 PPT
M0 PRJ
M0 PRL
M0 PRU
M0 PRY
M0 PSH
M0 PSL

M0 PSP
M0 PSQ
M0 PSU
M0 PSX
M0 PTC
M0 PTD
M0 PTH
M0 PTK
M0 PTM
M0 PTP
M0 PTU
M0 PTV
M0 PTW
M0 PUB
M0 PUG
M0 PUN
M0 PUP
M0 PUX
M0 PUZ
M0 PVA
M0 PVB
M0 PVG
M0 PVM
M0 PVX
M0 PWA
M0 PWG
M0 PWH
M0 PWJ
M0 PWK
M0 PWN
M0 PWW
M0 PWX
M0 PWZ
M0 PXB
M0 PXG
M0 PXH
M0 PYM
M0 PYU
M0 PYZ
M0 PZB
M0 PZG
M0 PZL
M0 PZS
M0 PZT
M0 PZW
M0 QVE
M0 RAA
M0 RAH
M0 RAJ
M0 RAO
M0 RAQ
M0 RAV
M0 RAY
M0 RBS
M0 RBW
M0 RBZ
M0 RCB
M0 RCO
M0 RCQ
M0 RCW
M0 RDE
M0 RDF
M0 RDH
M0 RDJ
M0 RDM
M0 RDN
M0 RDU
M0 RDW
M0 REE
M0 REK
M0 REL
M0 REN
M0 REO
M0 RER
M0 RES
M0 RET
M0 REU
M0 REW
M0 REY
M0 RFC
M0 RFD
M0 RFF
M0 RFI
M0 RFL
M0 RFN
M0 RFP
M0 RFR
M0 RFT
M0 RFT
M0 RFX
M0 RGA
M0 RGG
M0 RGJ
M0 RGM
M0 RGT
M0 RGU
M0 RGZ
M0 RHF
M0 RHJ
M0 RHK
M0 RHN
M0 RHV
M0 RHZ
M0 RIE
M0 RIJ
M0 RIP
M0 RIQ
M0 RIT
M0 RIX
M0 RIZ
M0 RJA
M0 RJC
M0 RJD
M0 RJF
M0 RJG

M0 RJL
M0 RJP
M0 RJV
M0 RKC
M0 RKJ
M0 RKL
M0 RKP
M0 RKS
M0 RLA
M0 RLB
M0 RLE
M0 RLG
M0 RLH
M0 RLK
M0 RLR
M0 RLX
M0 RMA
M0 RMC
M0 RMR
M0 RMX
M0 RNA
M0 RNE
M0 RNJ
M0 RNN
M0 RNS
M0 RNT
M0 RNY
M0 ROA
M0 ROB
M0 ROD
M0 ROE
M0 ROF
M0 ROI
M0 ROK
M0 ROQ
M0 ROS
M0 ROT
M0 ROU
M0 ROZ
M0 RPA
M0 RPC
M0 RPH
M0 RPP
M0 RPQ
M0 RPS
M0 RPW
M0 RPX
M0 RPY
M0 RRA
M0 RRB
M0 RRD
M0 RRI
M0 RRK
M0 RRO
M0 RRQ
M0 RRS
M0 RRU
M0 RRV
M0 RRW
M0 RSB
M0 RSK
M0 RSL
M0 RSS
M0 RTB
M0 RTG
M0 RTI
M0 RTR
M0 RTS
M0 RTX
M0 RTZ
M0 RUD
M0 RUF
M0 RUN
M0 RVA
M0 RVF
M0 RVI
M0 RVM
M0 RVP
M0 RVR
M0 RVV
M0 RVW
M0 RVX
M0 RVZ
M0 RWX
M0 RWY
M0 RXJ
M0 RXL
M0 RXO
M0 RXP
M0 RXQ
M0 RXW
M0 RXY
M0 RYC
M0 RYL
M0 RYN
M0 RYP
M0 RYU
M0 RYW
M0 RYZ
M0 RZA
M0 RZK
M0 RZM
M0 RZW
M0 SAE
M0 SAF
M0 SAG
M0 SAN
M0 SAS
M0 SAT
M0 SAW
M0 SBE
M0 SBG
M0 SBI
M0 SBM
M0 SBS

M0 SBU
M0 SBV
M0 SBW
M0 SBX
M0 SCF
M0 SCH
M0 SCI
M0 SCJ
M0 SCK
M0 SCL
M0 SCM
M0 SCQ
M0 SCZ
M0 SDI
M0 SDJ
M0 SDL
M0 SDN
M0 SDV
M0 SDX
M0 SDZ
M0 SEE
M0 SEF
M0 SEG
M0 SEH
M0 SEI
M0 SEN
M0 SEQ
M0 SES
M0 SEU
M0 SFB
M0 SFU
M0 SFZ
M0 SGB
M0 SGC
M0 SGE
M0 SGI
M0 SGL
M0 SGM
M0 SGN
M0 SGO
M0 SGP
M0 SGR
M0 SHL
M0 SHW
M0 SHX
M0 SIB
M0 SIE
M0 SIG
M0 SIM
M0 SIS
M0 SIZ
M0 SJA
M0 SJB
M0 SJC
M0 SJF
M0 SJH
M0 SJI
M0 SJM
M0 SJP
M0 SJT
M0 SJY
M0 SKB
M0 SKE
M0 SKJ
M0 SKL
M0 SKS
M0 SKU
M0 SLA
M0 SLJ
M0 SLK
M0 SLM
M0 SLS
M0 SLT
M0 SLW
M0 SLX
M0 SMI
M0 SMK
M0 SMM
M0 SMO
M0 SMR
M0 SMU
M0 SMX
M0 SNC
M0 SNE
M0 SNF
M0 SNI
M0 SNO
M0 SNR
M0 SNW
M0 SNX
M0 SOB
M0 SOM
M0 SON
M0 SOY
M0 SPF
M0 SPG
M0 SPI
M0 SPO
M0 SPR
M0 SPU
M0 SPW
M0 SPY
M0 SQD
M0 SQJ
M0 SQL
M0 SQO
M0 SQZ
M0 SRC
M0 SRF
M0 SRH
M0 SRI
M0 SRQ
M0 SRS
M0 SRW
M0 SRY
M0 SSC
M0 SSI
M0 STB
M0 STD
M0 STE
M0 STG
M0 STH
M0 STP
M0 STQ
M0 STR
M0 STW
M0 STZ
M0 SUC
M0 SUE
M0 SUL
M0 SUT
M0 SUV
M0 SVG
M0 SVK
M0 SVL
M0 SVM
M0 SVU
M0 SWA
M0 SWG
M0 SWJ
M0 SWP
M0 SWS
M0 SWW
M0 SWX
M0 SWY
M0 SXC
M0 SXD
M0 SXF
M0 SXP
M0 SXR
M0 SXY
M0 SXZ
M0 SYC
M0 SYP
M0 SYR
M0 SZ
M0 SZA
M0 SZE
M0 SZM
M0 SZP
M0 SZT
M0 TA
M0 TAH
M0 TAI
M0 TAU
M0 TBF
M0 TBL
M0 TBM
M0 TBO
M0 TBP
M0 TBT
M0 TBU
M0 TBX
M0 TCG
M0 TCH
M0 TCW
M0 TCX
M0 TDA
M0 TDI
M0 TDN
M0 TDR
M0 TEC
M0 TED
M0 TEE
M0 TEH
M0 TEM
M0 TEO
M0 TEU
M0 TEV
M0 TEY
M0 TFD
M0 TFG
M0 TFR
M0 TFT
M0 TFW
M0 TGA
M0 TGH
M0 TGK
M0 TGO
M0 TGR
M0 TGZ
M0 THC
M0 THD
M0 THI
M0 THL
M0 THR
M0 THW
M0 THX
M0 TIB
M0 TIC
M0 TID
M0 TIG
M0 TII
M0 TIO
M0 TIS
M0 TJH
M0 TJK
M0 TJP
M0 TKK
M0 TKL
M0 TKO
M0 TLA
M0 TLB
M0 TLE
M0 TLI
M0 TLS
M0 TLT
M0 TLW
M0 TLZ
M0 TME
M0 TMK
M0 TML
M0 TMR
M0 TMU
M0 TMW
M0 TNA
M0 TND
M0 TNI
M0 TNK
M0 TNM
M0 TNN
M0 TNR
M0 TNW
M0 TNY
M0 TOA
M0 TOC
M0 TOE
M0 TOT
M0 TOX
M0 TOY
M0 TPM
M0 TPN
M0 TPO
M0 TPR
M0 TPS
M0 TPV
M0 TPX
M0 TPZ
M0 TQM
M0 TQR
M0 TRA
M0 TRD
M0 TRI
M0 TRJ
M0 TRQ
M0 TRR
M0 TSF
M0 TSG
M0 TSH
M0 TSK
M0 TSL
M0 TSP
M0 TSR
M0 TSS
M0 TST
M0 TSU
M0 TSV
M0 TSX
M0 TTA
M0 TTD
M0 TTJ
M0 TTM
M0 TTP
M0 TTQ
M0 TTS
M0 TTT
M0 TTV
M0 TTW
M0 TTZ
M0 TUC
M0 TUE
M0 TUG
M0 TUL
M0 TUS
M0 TVE
M0 TVI
M0 TVJ
M0 TVM
M0 TVO
M0 TVQ
M0 TVZ
M0 TWA
M0 TWB
M0 TWD
M0 TWE
M0 TWI
M0 TWP
M0 TWT
M0 TWZ
M0 TXE
M0 TXG
M0 TXI
M0 TXR
M0 TXT
M0 TXU
M0 TYA
M0 TYB
M0 TYL
M0 TYM
M0 TYY
M0 TZX
M0 UAB
M0 UAF
M0 UAL
M0 UAN
M0 UAW
M0 UAX
M0 UBA
M0 UBC
M0 UBS
M0 UBT
M0 UCB
M0 UCI
M0 UCQ
M0 UCU
M0 UCW
M0 UCX
M0 UDP
M0 UDR
M0 UEA
M0 UED
M0 UEM
M0 UFM
M0 UFO
M0 UFP
M0 UFX
M0 UGA
M0 UGB
M0 UGS
M0 UHF
M0 UHR
M0 UIA
M0 UIB
M0 UIO
M0 UIT
M0 UKB
M0 UKD
M0 UKE
M0 UKG
M0 UKH
M0 UKJ
M0 UKK
M0 UKL
M0 UKP
M0 UKR
M0 ULF
M0 ULT
M0 UMZ
M0 UNC
M0 UND
M0 UNG
M0 UNO
M0 UNX
M0 UOC
M0 UOS
M0 UPM
M0 URA
M0 URC
M0 URH
M0 URM
M0 URP
M0 USA
M0 USE
M0 USH
M0 USI
M0 USP
M0 USR
M0 USS
M0 USX
M0 UTC
M0 UTU
M0 UUF
M0 UVA
M0 UVK
M0 UVM
M0 UVR
M0 UVU
M0 UXD
M0 UXE
M0 UXH
M0 UXP
M0 UZE
M0 UZI
M0 VAE
M0 VAF
M0 VAK
M0 VAX
M0 VAZ
M0 VBA
M0 VBD
M0 VBH
M0 VBM
M0 VBP
M0 VBX
M0 VCD
M0 VCH
M0 VCK
M0 VCM
M0 VCO
M0 VCT
M0 VCW
M0 VCZ
M0 VDC
M0 VDD
M0 VDL
M0 VDP
M0 VDS
M0 VDT
M0 VDZ
M0 VEA
M0 VEB
M0 VED
M0 VEE
M0 VEI
M0 VEM
M0 VEN
M0 VEV
M0 VFA
M0 VFB
M0 VFT
M0 VFU
M0 VFW
M0 VFX
M0 VGB
M0 VGP
M0 VGW
M0 VHF
M0 VHP
M0 VHR
M0 VHS
M0 VIA
M0 VIC
M0 VID
M0 VIE
M0 VIM
M0 VIO
M0 VIP
M0 VIV
M0 VIX
M0 VJL
M0 VJM
M0 VJS
M0 VKA
M0 VKD
M0 VKH
M0 VKM
M0 VKO
M0 VKV
M0 VLE
M0 VLF
M0 VMA
M0 VME
M0 VMF
M0 VMK
M0 VML
M0 VMO
M0 VMP
M0 VMT
M0 VMZ
M0 VNA
M0 VNL
M0 VNP
M0 VOA
M0 VOB
M0 VOC
M0 VOD
M0 VOO
M0 VOR
M0 VOW
M0 VOX
M0 VPH
M0 VPN
M0 VPS
M0 VPT
M0 VPZ
M0 VRA
M0 VRC
M0 VRF
M0 VRH
M0 VRL
M0 VRM
M0 VRR
M0 VRX
M0 VRZ
M0 VSA
M0 VSG
M0 VSI
M0 VSN
M0 VSS
M0 VST
M0 VSV
M0 VTB
M0 VTF
M0 VTM
M0 VTO
M0 VTT
M0 VTX
M0 VTZ
M0 VUG
M0 VUP
M0 VVF
M0 VVJ
M0 VVK
M0 VVS
M0 VVZ
M0 VWA
M0 VWB
M0 VWG
M0 VWH
M0 VWV
M0 VWX
M0 VXA
M0 VXK
M0 VXN
M0 VXO
M0 VXR
M0 VXT
M0 VYD
M0 VZO
M0 VZZ
M0 WAL
M0 WAN
M0 WAW
M0 WBA
M0 WBH
M0 WBN
M0 WBT
M0 WBX
M0 WCC
M0 WCF
M0 WCN
M0 WCQ
M0 WCY
M0 WDC
M0 WDF
M0 WDM
M0 WDN
M0 WDR
M0 WDT
M0 WDV
M0 WDW
M0 WDX
M0 WEA
M0 WEF
M0 WEH
M0 WEZ
M0 WFD
M0 WGD
M0 WGG
M0 WGH
M0 WGN
M0 WGT
M0 WGX
M0 WGZ
M0 WHD
M0 WHH
M0 WHI
M0 WHU
M0 WHW
M0 WID
M0 WIX
M0 WJB
M0 WJD
M0 WJE
M0 WJI
M0 WJJ
M0 WJN
M0 WJO
M0 WKC
M0 WKD
M0 WKS
M0 WKW
M0 WLC
M0 WLM
M0 WLO
M0 WLR
M0 WLW
M0 WLZ
M0 WMC
M0 WMD
M0 WME
M0 WMJ
M0 WMM
M0 WNH
M0 WNN
M0 WNP
M0 WNU
M0 WOE
M0 WOK
M0 WOL
M0 WOO
M0 WOZ
M0 WPD
M0 WPG
M0 WPH
M0 WQD
M0 WQE
M0 WQX
M0 WRB
M0 WRC
M0 WRF
M0 WRH
M0 WRL
M0 WRM
M0 WRN
M0 WRT
M0 WRZ
M0 WSO
M0 WSS
M0 WTA
M0 WTB
M0 WTD
M0 WTF
M0 WTH
M0 WTM
M0 WTP
M0 WTR
M0 WTS
M0 WTT
M0 WTU
M0 WTV
M0 WUE
M0 WUF
M0 WUK
M0 WUT
M0 WVR
M0 WVV
M0 WVX
M0 WWA
M0 WWG
M0 WWJ
M0 WWS
M0 WWW
M0 WXB
M0 WXG
M0 WXT
M0 WZC
M0 WZD
M0 WZF
M0 WZW
M0 WZY
M0 XAA
M0 XAF
M0 XAG
M0 XAH
M0 XAI
M0 XAL
M0 XAN
M0 XAP
M0 XAX
M0 XAY
M0 XAZ
M0 XBC
M0 XBE
M0 XBF
M0 XBO
M0 XBP
M0 XCB
M0 XCD
M0 XCG
M0 XCK
M0 XCM
M0 XCV
M0 XCW
M0 XDB
M0 XDD
M0 XDG
M0 XDI
M0 XDM
M0 XDP
M0 XDR
M0 XDW
M0 XDY
M0 XEC
M0 XED
M0 XEN
M0 XEQ
M0 XET
M0 XFD
M0 XFJ
M0 XFM
M0 XFN
M0 XFR
M0 XFT
M0 XGA
M0 XGC
M0 XGD
M0 XGM
M0 XGP
M0 XHE
M0 XHF
M0 XHM
M0 XHO
M0 XHS
M0 XHT
M0 XIE
M0 XII
M0 XIM
M0 XIN
M0 XIO
M0 XIP
M0 XIT
M0 XIX
M0 XJA
M0 XJB
M0 XJC
M0 XJH
M0 XJR
M0 XKC
M0 XKF
M0 XKG
M0 XKI
M0 XKN
M0 XKR
M0 XKV
M0 XLD
M0 XLE
M0 XLF
M0 XLG
M0 XLO
M0 XLS
M0 XLZ
M0 XMA
M0 XME
M0 XMF
M0 XMM
M0 XMR
M0 XNA
M0 XNI
M0 XNP
M0 XOE
M0 XOF
M0 XPC
M0 XPO
M0 XPU
M0 XPX
M0 XQZ
M0 XRB
M0 XRD
M0 XRE
M0 XRG
M0 XRJ
M0 XRK
M0 XRO
M0 XRS
M0 XSI
M0 XSP
M0 XSR
M0 XSW
M0 XSX
M0 XTA
M0 XTC
M0 XTE
M0 XTF
M0 XTF
M0 XTH
M0 XTO
M0 XTR
M0 XTY
M0 XUE
M0 XUP
M0 XUT
M0 XVM
M0 XVR
M0 XVT
M0 XVW
M0 XWB
M0 XWE
M0 XWF
M0 XWN
M0 XWR
M0 XXA
M0 XXF
M0 XXH
M0 XXQ
M0 XXR
M0 XXS
M0 XXU
M0 XXY
M0 XXZ
M0 XYB
M0 XYD
M0 XYF
M0 XYG
M0 XYK
M0 XZA
M0 XZF
M0 XZI
M0 XZL
M0 XZT
M0 XZZ
M0 YAA
M0 YAF
M0 YAP
M0 YAS
M0 YAT
M0 YAX
M0 YAZ
M0 YBG
M0 YBM
M0 YBO
M0 YBX
M0 YCC
M0 YCL
M0 YCM
M0 YCP
M0 YCT
M0 YDA
M0 YDC
M0 YDD
M0 YDG
M0 YDR
M0 YDT
M0 YED
M0 YEH
M0 YEL
M0 YEW
M0 YFL
M0 YFM
M0 YFO
M0 YFR
M0 YGF
M0 YGH
M0 YGR
M0 YHB
M0 YHC
M0 YHF
M0 YHS
M0 YIK
M0 YIP
M0 YJB
M0 YJG
M0 YJR
M0 YKK
M0 YKW
M0 YMC
M0 YME
M0 YMK
M0 YMX
M0 YNE
M0 YNG
M0 YNW
M0 YOH
M0 YOK
M0 YOR
M0 YOU
M0 YOY
M0 YPA
M0 YPT
M0 YPZ
M0 YQV
M0 YRK
M0 YRS
M0 YRU
M0 YRX
M0 YSC
M0 YSL
M0 YSO
M0 YSU
M0 YSZ
M0 YTB
M0 YTE
M0 YTF
M0 YTH
M0 YTK
M0 YTL
M0 YTM
M0 YTS
M0 YTT
M0 YTX
M0 YTY
M0 YUI
M0 YUL
M0 YUP
M0 YVE
M0 YWG
M0 YXB
M0 YXD
M0 YYC
M0 YYO
M0 YYZ
M0 YZA
M0 YZC
M0 YZL
M0 YZX
M0 YZZ
M0 ZAD
M0 ZAG
M0 ZAH
M0 ZAO
M0 ZAT
M0 ZAX
M0 ZAZ
M0 ZBB
M0 ZBU
M0 ZCC
M0 ZCH
M0 ZCR
M0 ZDF
M0 ZDR
M0 ZDZ
M0 ZEP
M0 ZER
M0 ZES
M0 ZFA
M0 ZFJ
M0 ZFR
M0 ZFX
M0 ZGD
M0 ZGN
M0 ZGY
M0 ZHA
M0 ZHN
M0 ZHX
M0 ZIE
M0 ZIX
M0 ZLB
M0 ZLR
M0 ZLW
M0 ZMF
M0 ZNH
M0 ZNK
M0 ZOC
M0 ZOD
M0 ZOL
M0 ZOS
M0 ZPK
M0 ZQI
M0 ZQR
M0 ZRA
M0 ZRN
M0 ZRQ
M0 ZRT
M0 ZSD
M0 ZSP
M0 ZSU
M0 ZTP
M0 ZTX
M0 ZUN
M0 ZUU
M0 ZVX
M0 ZWE
M0 ZXA
M0 ZXP
M0 ZXT
M0 ZXZ
M0 ZZB
M0 ZZK
M0 ZZM
M0 ZZR
M0 ZZW
M0 ZZX
M0 ZZY

M1

M1 AAA
M1 AAD
M1 AAG
M1 AAK
M1 AAM
M1 AAO
M1 AAT
M1 ABD
M1 ABE
M1 ABK
M1 ABP
M1 ACI
M1 ACM
M1 ACP
M1 ACS
M1 ACU
M1 ACZ
M1 ADA
M1 ADC
M1 ADG
M1 ADL
M1 ADW
M1 AEC
M1 AEM
M1 AES
M1 AFD
M1 AFM
M1 AFN
M1 AFR
M1 AGB
M1 AGF
M1 AGJ
M1 AGN
M1 AHB
M1 AHE
M1 AHO
M1 AHS
M1 AHZ
M1 AIF
M1 AIS
M1 AIT
M1 AIU
M1 AJB
M1 AJJ
M1 AJK
M1 AJR
M1 AJV
M1 AKJ
M1 AKS
M1 ALI
M1 ALN
M1 ALY
M1 AMF
M1 AMH
M1 AMK
M1 AMN
M1 AMT
M1 AMY
M1 ANH
M1 ANV
M1 ANW
M1 ANX
M1 AOA
M1 AOC
M1 AOH
M1 APN
M1 APU
M1 APX
M1 AQF
M1 AQH
M1 AQK
M1 AQM
M1 AQO
M1 AQR
M1 AQV
M1 AQW
M1 ARA
M1 ARI
M1 ARN
M1 ARW
M1 ARY
M1 ARZ
M1 ASF
M1 ASO
M1 ASW
M1 ASY
M1 ATF
M1 ATH
M1 ATM
M1 ATT
M1 ATW
M1 ATZ
M1 AUD
M1 AUS
M1 AUU
M1 AVE
M1 AVG
M1 AVP
M1 AWG
M1 AWH
M1 AWZ
M1 AXA
M1 AXI
M1 AXL
M1 AXN
M1 AXO
M1 AXV
M1 AXY
M1 AYE
M1 AYI
M1 AYM
M1 AYX
M1 AYZ
M1 AZS
M1 AZT
M1 AZZ
M1 BAK
M1 BAT
M1 BAU
M1 BAW
M1 BAZ
M1 BBM
M1 BBN
M1 BBQ
M1 BBT
M1 BBV
M1 BCF
M1 BCJ
M1 BCV
M1 BCX
M1 BDB
M1 BDI
M1 BDK
M1 BDM
M1 BEA
M1 BEG
M1 BEK
M1 BEY
M1 BFH
M1 BFP
M1 BFZ
M1 BGB
M1 BGC
M1 BGQ
M1 BGU
M1 BGV
M1 BHD
M1 BHR
M1 BHS
M1 BIA
M1 BIJ
M1 BIN
M1 BIR
M1 BIZ
M1 BJH
M1 BJI
M1 BJL
M1 BJP
M1 BJR
M1 BKC
M1 BKD
M1 BKM
M1 BKU
M1 BLM
M1 BLV
M1 BMJ
M1 BMN
M1 BNE
M1 BNJ
M1 BNL
M1 BNM
M1 BNP
M1 BNW
M1 BNZ
M1 BOC
M1 BOI
M1 BPJ
M1 BPT
M1 BRH
M1 BRQ
M1 BRR
M1 BSA
M1 BSC
M1 BSZ
M1 BTF
M1 BTJ
M1 BTY
M1 BUD
M1 BUV
M1 BUZ
M1 BVA
M1 BWB
M1 BWO
M1 BWP
M1 BWT
M1 BWW
M1 BWY
M1 BXE
M1 BXG
M1 BXI
M1 BXT
M1 BXW
M1 BXY
M1 BYA
M1 BYB
M1 BYN
M1 BYO
M1 BYY
M1 BZM
M1 BZN
M1 BZV
M1 CAF
M1 CAK
M1 CAM
M1 CAN
M1 CBP
M1 CBQ
M1 CBW
M1 CCC
M1 CCO
M1 CCP
M1 CDE
M1 CDG
M1 CDH
M1 CDM
M1 CDU
M1 CDZ
M1 CEB
M1 CEC
M1 CEE
M1 CEH
M1 CEK
M1 CEL
M1 CEN
M1 CEU
M1 CEV
M1 CEX
M1 CFJ
M1 CFK
M1 CFL
M1 CFN
M1 CFV
M1 CGB
M1 CGE
M1 CGG
M1 CGH
M1 CGX
M1 CHG
M1 CHN
M1 CHO
M1 CHV
M1 CHW
M1 CIA
M1 CID
M1 CIW
M1 CIX
M1 CIY
M1 CJK
M1 CJP
M1 CJU
M1 CJY
M1 CKC
M1 CKG
M1 CKM
M1 CKP
M1 CLF
M1 CLK
M1 CLN
M1 CLQ
M1 CLV
M1 CMC
M1 CMP
M1 CMS
M1 CMV
M1 CNG
M1 CNO
M1 CNU
M1 CNW
M1 COI
M1 COK
M1 CPE
M1 CPH
M1 CPM
M1 CPO
M1 CPU
M1 CPV
M1 CPW
M1 CPX
M1 CQO
M1 CRB
M1 CRG
M1 CRN
M1 CRS
M1 CRV
M1 CRX
M1 CSB
M1 CSK
M1 CSO
M1 CSQ
M1 CSS
M1 CTA
M1 CTN
M1 CTP
M1 CUE
M1 CUI
M1 CUJ
M1 CUW
M1 CUZ
M1 CVA
M1 CVE
M1 CVJ
M1 CVR
M1 CVZ
M1 CWM
M1 CXB
M1 CXE
M1 CXF
M1 CXI
M1 CXJ
M1 CXX
M1 CYE
M1 CYI
M1 CYS
M1 CYV
M1 CYW
M1 CYY
M1 CZB
M1 CZC
M1 CZR
M1 CZT
M1 CZW
M1 DAJ
M1 DAT
M1 DAZ
M1 DBB
M1 DBE
M1 DBG
M1 DBS
M1 DBT
M1 DBZ
M1 DCG
M1 DCJ
M1 DCN
M1 DCO
M1 DCW
M1 DDD
M1 DDE
M1 DDK
M1 DDL
M1 DDN
M1 DDV
M1 DDX
M1 DEC
M1 DED
M1 DEI
M1 DEL
M1 DEN
M1 DEO
M1 DFG
M1 DFX
M1 DGA
M1 DGC
M1 DGD
M1 DGF
M1 DGH
M1 DGN
M1 DGO
M1 DGZ
M1 DHB
M1 DHD
M1 DHH
M1 DHL
M1 DHP
M1 DIC
M1 DID
M1 DIG
M1 DII
M1 DIP
M1 DJE
M1 DJF
M1 DJK
M1 DJR
M1 DJU
M1 DKC
M1 DKU
M1 DKV
M1 DLB
M1 DLC
M1 DLF
M1 DLJ
M1 DLL
M1 DLW
M1 DLY
M1 DMC
M1 DMJ
M1 DMQ
M1 DMS
M1 DMX
M1 DNF
M1 DNL
M1 DNO
M1 DNS
M1 DNV
M1 DOC
M1 DOJ
M1 DOP
M1 DOW
M1 DPB
M1 DPD
M1 DPM
M1 DPV
M1 DPZ
M1 DQA
M1 DQF
M1 DQJ
M1 DQY
M1 DRC
M1 DRN
M1 DRQ
M1 DRT
M1 DRX
M1 DSB
M1 DSL
M1 DSR
M1 DST
M1 DTC
M1 DTI
M1 DTJ
M1 DTX
M1 DUL
M1 DUM
M1 DUN
M1 DUQ
M1 DUX
M1 DUY
M1 DUZ
M1 DVM
M1 DWM
M1 DWO
M1 DWX
M1 DXF
M1 DXK
M1 DXN
M1 DXP
M1 DXZ
M1 DYP
M1 DYR
M1 DYX
M1 DZA
M1 DZI
M1 DZI
M1 DZK
M1 DZO
M1 DZU
M1 EAC
M1 EAR
M1 EAU
M1 EAX
M1 EBM
M1 EBN
M1 EBO
M1 ECE
M1 ECF
M1 ECK
M1 ECN
M1 ECP
M1 ECR
M1 ECY
M1 EDB
M1 EDG
M1 EDN
M1 EDT
M1 EDU
M1 EDX
M1 EEM
M1 EEU
M1 EEV
M1 EFI
M1 EFL
M1 EFQ
M1 EFY
M1 EGF
M1 EGH
M1 EGI
M1 EGJ
M1 EGO
M1 EHL
M1 EHM
M1 EHU
M1 EHW
M1 EHY
M1 EIB
M1 EIK
M1 EIV
M1 EJC
M1 EJH
M1 EJK
M1 EJU
M1 EKG
M1 EKI
M1 EKT
M1 EKV
M1 EKX
M1 EKZ
M1 ELC
M1 ELF
M1 ELJ
M1 ELK
M1 ELT
M1 ELY
M1 EME
M1 EMF
M1 EMH
M1 EMI
M1 EMJ
M1 EMT
M1 EMZ
M1 ENG
M1 ENH
M1 ENI
M1 ENO
M1 ENR
M1 ENU
M1 ENV
M1 ENW
M1 ENY
M1 EOA
M1 EOE
M1 EOG
M1 EOM
M1 EON
M1 EOX
M1 EPH
M1 EPL
M1 EPM
M1 EPO
M1 EPQ
M1 EPW
M1 EPY
M1 EPZ
M1 EQN
M1 EQP
M1 EQS
M1 ERG
M1 ERM
M1 ESA
M1 ESC
M1 ESG
M1 ESL
M1 ESR
M1 EST
M1 ESU
M1 ESW
M1 ETI
M1 EUA
M1 EUI
M1 EUS
M1 EUW
M1 EUZ
M1 EVA
M1 EVG
M1 EVK
M1 EVV
M1 EVY
M1 EWH
M1 EWK
M1 EWZ
M1 EXA
M1 EXE
M1 EXL
M1 EXP
M1 EXR
M1 EXX
M1 EXZ
M1 EYB
M1 EYE
M1 EYK
M1 EYW
M1 EYX
M1 EYY
M1 EZA
M1 EZF
M1 EZS
M1 EZV
M1 EZW
M1 EZY
M1 FAM
M1 FAU
M1 FAV
M1 FBC
M1 FBE
M1 FBG
M1 FBH
M1 FBJ
M1 FBR
M1 FBT
M1 FBU
M1 FBV
M1 FBX
M1 FBY
M1 FBZ
M1 FCD
M1 FCP
M1 FDB
M1 FDI
M1 FDJ
M1 FDQ
M1 FDS
M1 FDT
M1 FDU
M1 FDZ
M1 FEB
M1 FEF
M1 FEV
M1 FFK
M1 FFQ
M1 FFZ
M1 FGA
M1 FGE
M1 FGF
M1 FGG
M1 FGI
M1 FGK
M1 FGL
M1 FGN
M1 FGQ
M1 FGR
M1 FGS
M1 FGU
M1 FGY
M1 FGZ
M1 FHD
M1 FHG
M1 FHI
M1 FHK
M1 FHM
M1 FHU
M1 FID
M1 FIF
M1 FIH
M1 FIJ
M1 FIM
M1 FIN
M1 FIO
M1 FIQ
M1 FIU
M1 FIW
M1 FJE
M1 FJI
M1 FJN
M1 FJO
M1 FJU
M1 FJV
M1 FMJ
M1 FMW
M1 FRH
M1 FTA
M1 FXY
M1 FZL
M1 GAC
M1 GAZ
M1 GGY
M1 GHO
M1 GLA
M1 GLO
M1 GMJ
M1 GMW
M1 GNB
M1 GPD
M1 GPO
M1 GRY
M1 GSE
M1 GSX
M1 GTR
M1 HAC
M1 HAT
M1 HAX
M1 HCO
M1 HDD
M1 HIS
M1 HOG
M1 HRQ
M1 HSB
M1 HTG
M1 HUD
M1 HUL
M1 HVA
M1 HVD
M1 HWH
M1 HZJ
M1 IJJ
M1 ISR
M1 IVA
M1 IWB
M1 JCM
M1 JDK
M1 JED
M1 JEN
M1 JJK
M1 JKL
M1 JMH
M1 JMR
M1 JMW
M1 JNC
M1 JPF
M1 JPM
M1 JRC
M1 JRG
M1 JRV
M1 JSC
M1 JUL
M1 JUN
M1 JWT
M1 KAB
M1 KAC
M1 KAH
M1 KCS
M1 KDA
M1 KE
M1 KEN
M1 KET
M1 KFM
M1 KGB
M1 KGC
M1 KLS
M1 KTM
M1 KTO
M1 LCR
M1 LEE
M1 LIV
M1 LKY
M1 LNB
M1 MAR
M1 MAS
M1 MAX
M1 MBZ
M1 MDH
M1 MHS
M1 MHZ
M1 MIS
M1 MJT
M1 MOW
M1 MOZ
M1 MPR
M1 MPW
M1 MSD
M1 MSG
M1 MWD
M1 MZX
M1 NCD
M1 NCP
M1 NCR
M1 NDF
M1 NED
M1 NIC
M1 NIX
M1 NJJ
M1 NRT
M1 NSP
M1 NTP
M1 NWT
M1 OAK
M1 OEY
M1 OHM
M1 OMW
M1 OOO
M1 PDJ
M1 PEP
M1 PGC
M1 PHS
M1 PJC
M1 PLA
M1 PMO
M1 POM
M1 POP
M1 PPP
M1 PRO
M1 PRV
M1 PSH
M1 PSY
M1 PUB
M1 PUP
M1 PVS
M1 RCA
M1 RDB
M1 REW
M1 RGN
M1 RGZ
M1 RIP
M1 RKH
M1 RMB
M1 ROK
M1 RON
M1 RPM
M1 RPW
M1 RPY
M1 RWN
M1 SAG
M1 SBH
M1 SCH
M1 SDH
M1 SJE
M1 SJM
M1 SJP
M1 SMH
M1 SMJ
M1 SPH
M1 SSY
M1 STA
M1 STE
M1 STX
M1 TAY
M1 TCH
M1 TDM
M1 TEX
M1 TGQ
M1 TIG
M1 TLD
M1 TLK
M1 TNT
M1 TNY
M1 TRV
M1 TTT
M1 TTY
M1 TUT
M1 TWO
M1 TXI
M1 TZR
M1 USA
M1 VAW
M1 VCG
M1 VFO
M1 VFR
M1 VKJ
M1 VOX
M1 VSA
M1 VTR
M1 VWB
M1 WCJ
M1 WEB
M1 WML
M1 WPB
M1 WRG
M1 XCQ
M1 XDX
M1 XED
M1 XER
M1 XJC
M1 XMP
M1 XOR
M1 XTN
M1 XUP
M1 XWD
M1 XXX
M1 YAP
M1 YJG
M1 ZRP

M3

M3 ACK
M3 ACV
M3 ADF
M3 ADV

Withheld

Withheld

M3 AEX
M3 AFV
M3 AFW
M3 AFZ
M3 AGC
M3 AGJ
M3 AGO
M3 AGQ
M3 AGS
M3 AHK
M3 AHM
M3 AHT
M3 AIX
M3 AJB
M3 AJG
M3 AKD
M3 ALP
M3 ALR
M3 ALT
M3 AMG
M3 AMN
M3 AMY
M3 ANN
M3 ANX
M3 ANY
M3 AOB
M3 AOI
M3 APF
M3 APK
M3 APL
M3 APM
M3 APP
M3 APR
M3 ARC
M3 ARV
M3 ARX
M3 ASQ
M3 ASU
M3 ATA
M3 ATH
M3 ATK
M3 ATM
M3 ATX
M3 AUD
M3 AUE
M3 AUH
M3 AUR
M3 AUV
M3 AUW
M3 AVB
M3 AVC
M3 AVG
M3 AVN
M3 AVP
M3 AVT
M3 AWA
M3 AWF
M3 AWJ
M3 AWL
M3 AWV
M3 AXJ
M3 AYG
M3 AYK
M3 AYR
M3 AZC
M3 AZD
M3 AZW
M3 BAF
M3 BAP
M3 BAV
M3 BAX
M3 BBE
M3 BBJ
M3 BBM
M3 BBT
M3 BBV
M3 BBW
M3 BCN
M3 BCO
M3 BCT
M3 BDE
M3 BDF
M3 BDI
M3 BDL
M3 BDM
M3 BDO
M3 BDS
M3 BEH
M3 BEI
M3 BEO
M3 BEY
M3 BFA
M3 BFS
M3 BFW
M3 BGO
M3 BHN
M3 BHV
M3 BHX
M3 BIJ
M3 BIL
M3 BIM
M3 BIN
M3 BJC
M3 BJD
M3 BJI
M3 BJU
M3 BJX
M3 BKF
M3 BKL
M3 BKR
M3 BKZ
M3 BLL
M3 BLS
M3 BLV
M3 BLW
M3 BMD
M3 BMY
M3 BNH
M3 BNI
M3 BNO
M3 BNP
M3 BNT
M3 BOM
M3 BOX
M3 BOY
M3 BPB
M3 BPE
M3 BPK
M3 BQA
M3 BQR
M3 BQU
M3 BQV
M3 BQW
M3 BRP
M3 BSA
M3 BSL
M3 BSQ
M3 BSX
M3 BTC
M3 BTD
M3 BTE
M3 BTV
M3 BTY
M3 BUB
M3 BUG
M3 BUK
M3 BUO
M3 BVC
M3 BVN
M3 BVZ
M3 BWM
M3 BWO
M3 BXA
M3 BXI
M3 BXL
M3 BYG
M3 BYH
M3 BYY
M3 BYZ
M3 BZG
M3 BZO
M3 CBD
M3 CBI
M3 CCR
M3 CDC
M3 CDD
M3 CDF
M3 CDH
M3 CDS
M3 CDZ
M3 CEC
M3 CED
M3 CEE
M3 CEF
M3 CEH
M3 CEI
M3 CEK
M3 CEY
M3 CFC
M3 CFE
M3 CFH
M3 CFK
M3 CFL
M3 CFN
M3 CFS
M3 CFW
M3 CFY
M3 CGB
M3 CGV
M3 CGW
M3 CGY
M3 CHA
M3 CHC
M3 CHD
M3 CHH
M3 CHO
M3 CIB
M3 CIN
M3 CJQ
M3 CJV
M3 CJW
M3 CKB
M3 CKG
M3 CKS
M3 CKV
M3 CLF
M3 CLN
M3 CLV
M3 CLZ
M3 CMD
M3 CMT
M3 CNU
M3 CNX
M3 COH
M3 COO
M3 COR
M3 COS
M3 COX
M3 COY
M3 CPA
M3 CPB
M3 CPJ
M3 CPP
M3 CPT
M3 CPU
M3 CPV
M3 CQE
M3 CQG
M3 CQI
M3 CQJ
M3 CQK
M3 CQL
M3 CQM
M3 CQZ
M3 CRB
M3 CRE
M3 CRG
M3 CRJ
M3 CRU
M3 CSP
M3 CSW
M3 CTF
M3 CTG
M3 CTH
M3 CTJ
M3 CTL
M3 CTW
M3 CUD
M3 CUG
M3 CUI
M3 CUO
M3 CUP
M3 CUQ
M3 CUS
M3 CVF
M3 CVG
M3 CVI
M3 CVT
M3 CWB
M3 CWD
M3 CWM
M3 CWR
M3 CWS
M3 CWT
M3 CWU
M3 CWW
M3 CXA
M3 CXL
M3 CXS
M3 CXW
M3 CYP
M3 CYR
M3 CYY
M3 CZC
M3 CZD
M3 CZH
M3 CZO
M3 CZZ
M3 DAC
M3 DAJ
M3 DAT
M3 DAY
M3 DBD
M3 DBI
M3 DBM
M3 DBO
M3 DBQ
M3 DCA
M3 DCB
M3 DCG
M3 DCO
M3 DCQ
M3 DCT
M3 DDL
M3 DDN
M3 DDO
M3 DDP
M3 DDV
M3 DEG
M3 DEK
M3 DEO
M3 DEQ
M3 DEU
M3 DFE
M3 DFI
M3 DGC
M3 DGE
M3 DHC
M3 DHL
M3 DHP
M3 DID
M3 DIF
M3 DII
M3 DJS
M3 DJT
M3 DKD
M3 DKF
M3 DKQ
M3 DKR
M3 DKX
M3 DKY
M3 DLA
M3 DLL
M3 DLS
M3 DLV
M3 DLY
M3 DMB
M3 DMC
M3 DMH
M3 DMV
M3 DNK
M3 DNL
M3 DNZ
M3 DOE
M3 DOI
M3 DOQ
M3 DOV
M3 DPD
M3 DPE
M3 DPL
M3 DPO
M3 DPR
M3 DPU
M3 DPW
M3 DQC
M3 DQF
M3 DQO
M3 DQT
M3 DRL
M3 DRP
M3 DRQ
M3 DRT
M3 DSN
M3 DSO
M3 DTF
M3 DTS
M3 DTW
M3 DUA
M3 DUF
M3 DUG
M3 DUV
M3 DUW
M3 DUX
M3 DUY
M3 DVE
M3 DVK
M3 DVS
M3 DVX
M3 DWJ
M3 DWN
M3 DWS
M3 DWT
M3 DWU
M3 DWY
M3 DXB
M3 DXC
M3 DXH
M3 DXJ
M3 DXP
M3 DYD
M3 DYG
M3 DYH
M3 DYI
M3 DYJ
M3 DYP
M3 DYS
M3 DYX
M3 DZJ
M3 DZY
M3 EAC
M3 EAG
M3 EAJ
M3 EAM
M3 EAO
M3 EAV
M3 EBN
M3 ECA
M3 ECH
M3 ECL
M3 ECM
M3 ECT
M3 ECX
M3 EDB
M3 EDE
M3 EDF
M3 EDM
M3 EEA
M3 EEB
M3 EEY
M3 EFE
M3 EFF
M3 EFM
M3 EFY
M3 EFZ
M3 EGN
M3 EGO
M3 EGP
M3 EHE
M3 EHN
M3 EHO
M3 EHQ
M3 EHW
M3 EHX
M3 EIE
M3 EII
M3 EIR
M3 EIS
M3 EJA
M3 EJG
M3 EJT
M3 EJZ
M3 EKB
M3 EKQ
M3 EKX
M3 ELE
M3 ELH
M3 ELJ
M3 ELU
M3 EMF
M3 EMH
M3 EMP
M3 EMT
M3 EMY
M3 ENB
M3 END
M3 ENQ
M3 ENT
M3 ENW
M3 ENX
M3 EOA
M3 EOG
M3 EOK
M3 EPA
M3 EPB
M3 EPE
M3 EPN
M3 EPP
M3 EPS
M3 EPY
M3 EPZ
M3 EQC
M3 EQI
M3 EQJ
M3 EQN
M3 EQR
M3 EQV
M3 EQX
M3 EQZ
M3 ERG
M3 ERK
M3 ERL
M3 ERW
M3 ESC
M3 ESR
M3 ETD
M3 ETP
M3 ETY
M3 EUC
M3 EUO
M3 EUT
M3 EVL
M3 EVZ
M3 EWC
M3 EWJ
M3 EWL
M3 EXE
M3 EXO
M3 EXQ
M3 EXU
M3 EYE
M3 EYG
M3 EYU
M3 EYV
M3 EZA
M3 EZD
M3 EZT
M3 EZW
M3 FAD
M3 FAQ
M3 FAV
M3 FBE
M3 FBH
M3 FBT
M3 FBZ
M3 FCC
M3 FCD
M3 FCE
M3 FCH
M3 FCI
M3 FCJ
M3 FDD
M3 FDE
M3 FDI
M3 FDK
M3 FDP
M3 FDT
M3 FDW
M3 FDY
M3 FEE
M3 FEH
M3 FEP
M3 FER
M3 FEV
M3 FEZ
M3 FFG
M3 FFW
M3 FGD
M3 FGE
M3 FGJ
M3 FHN
M3 FID
M3 FIF
M3 FIN
M3 FIR
M3 FIV
M3 FJU
M3 FKA
M3 FKF
M3 FKH
M3 FKJ
M3 FKT
M3 FKX
M3 FLF
M3 FLH
M3 FLQ
M3 FLW
M3 FLY
M3 FMA
M3 FMG
M3 FMJ
M3 FML
M3 FMO
M3 FMT
M3 FMX
M3 FMZ
M3 FNF
M3 FNK
M3 FNP
M3 FNU
M3 FNW
M3 FOB
M3 FOI
M3 FON
M3 FOT
M3 FOU
M3 FPP
M3 FPX
M3 FPY
M3 FQB
M3 FQE
M3 FQF
M3 FQH
M3 FQK
M3 FQL
M3 FQP
M3 FQV
M3 FQW
M3 FRA
M3 FRF
M3 FRI
M3 FRW
M3 FSA
M3 FSJ
M3 FSM
M3 FSO
M3 FST
M3 FTL
M3 FTO
M3 FTS
M3 FUE
M3 FUL
M3 FUU
M3 FVF
M3 FVI
M3 FVK
M3 FVQ
M3 FVT
M3 FVU
M3 FVV
M3 FWB
M3 FWG
M3 FWQ
M3 FXH
M3 FXN
M3 FXR
M3 FXS
M3 FXV
M3 FYD
M3 FYH
M3 FYO
M3 FYP
M3 FYT
M3 FYU
M3 FYX
M3 FZG
M3 FZH
M3 FZK
M3 FZN
M3 FZP
M3 FZQ
M3 FZU
M3 FZW
M3 FZY
M3 GAI
M3 GAJ
M3 GAR
M3 GAS
M3 GAU
M3 GBG
M3 GBL
M3 GBQ
M3 GBU
M3 GBV
M3 GCA
M3 GCE
M3 GCG
M3 GCL
M3 GCO
M3 GCQ
M3 GCU
M3 GCV
M3 GDA
M3 GDB
M3 GDJ
M3 GDP
M3 GDQ
M3 GDS
M3 GDW
M3 GDZ
M3 GED
M3 GEG
M3 GEO
M3 GEV
M3 GFC
M3 GFK
M3 GFN
M3 GFU
M3 GFV
M3 GFX
M3 GFY
M3 GGC
M3 GGG
M3 GGI
M3 GGK
M3 GGL
M3 GGP
M3 GGT
M3 GGU
M3 GGX
M3 GGY
M3 GGZ
M3 GHP
M3 GHQ
M3 GHV
M3 GIA
M3 GIB
M3 GIN
M3 GIO
M3 GIP
M3 GIU
M3 GJA
M3 GJF
M3 GJJ
M3 GJL
M3 GJP
M3 GJS
M3 GJZ
M3 GKC
M3 GKF
M3 GKP
M3 GKQ
M3 GKS
M3 GKV
M3 GLB
M3 GLE
M3 GLF
M3 GLG
M3 GLU
M3 GLW
M3 GLX
M3 GME
M3 GMH
M3 GMN
M3 GMO
M3 GMQ
M3 GMZ
M3 GNC
M3 GNE
M3 GNH
M3 GNL
M3 GNO
M3 GNQ
M3 GNS
M3 GNV
M3 GNW
M3 GNX
M3 GOC
M3 GOF
M3 GOG
M3 GON
M3 GOO
M3 GOP
M3 GPF
M3 GPO
M3 GPT
M3 GPU
M3 GPZ
M3 GQF
M3 GQG
M3 GQH
M3 GQQ
M3 GQU
M3 GRD
M3 GRE
M3 GRQ
M3 GRU
M3 GRW
M3 GRX
M3 GRZ
M3 GSA
M3 GSN
M3 GSS
M3 GSV
M3 GTI
M3 GTS
M3 GUA
M3 GUE
M3 GUK
M3 GUL
M3 GUP
M3 GUT
M3 GUW
M3 GVI
M3 GVP
M3 GVR
M3 GVX
M3 GVY
M3 GVZ
M3 GWE
M3 GWL
M3 GWM
M3 GWO
M3 GWX
M3 GWY
M3 GXC
M3 GXF
M3 GXS
M3 GXT
M3 GXV
M3 GXW
M3 GYG
M3 GYS
M3 GYX
M3 GZA
M3 GZB
M3 GZH
M3 HAA
M3 HAH
M3 HAO
M3 HAR
M3 HAY
M3 HBA
M3 HBI
M3 HBO
M3 HBZ
M3 HCZ
M3 HDB
M3 HDO
M3 HDR
M3 HDY
M3 HEF
M3 HEN
M3 HEP
M3 HES
M3 HEW
M3 HEZ
M3 HFB
M3 HFK
M3 HFM
M3 HFV
M3 HFY
M3 HGB
M3 HGD
M3 HGF
M3 HGS
M3 HGY
M3 HHJ
M3 HHT
M3 HHW
M3 HHY
M3 HIJ
M3 HIY
M3 HJD
M3 HJH
M3 HJL
M3 HJN
M3 HJR
M3 HKG
M3 HKI
M3 HKK
M3 HKO
M3 HKS
M3 HKX
M3 HLO
M3 HMX
M3 HNC
M3 HNH
M3 HNI
M3 HNY
M3 HNZ
M3 HOB
M3 HOF
M3 HOI
M3 HOJ
M3 HOO
M3 HOS
M3 HOZ
M3 HPF
M3 HPG
M3 HPJ
M3 HPL
M3 HPP
M3 HPR
M3 HPS
M3 HPX
M3 HQG
M3 HQI
M3 HQO
M3 HRC
M3 HRI
M3 HRP
M3 HRQ
M3 HRR
M3 HSA
M3 HSB
M3 HSN
M3 HSP
M3 HSY
M3 HTB
M3 HTC
M3 HTD
M3 HTT
M3 HTU
M3 HTV
M3 HTZ
M3 HUA
M3 HUC
M3 HUF
M3 HUH
M3 HUJ
M3 HVB
M3 HVF
M3 HVG
M3 HVI
M3 HVR
M3 HVT
M3 HWB
M3 HWE
M3 HWF
M3 HWG
M3 HWH
M3 HWK
M3 HWT
M3 HWZ
M3 HXA
M3 HXE
M3 HXI
M3 HXP
M3 HXT
M3 HXX
M3 HYA
M3 HYC
M3 HYK
M3 HYM
M3 HYW
M3 HYZ
M3 HZF
M3 HZG
M3 HZV
M3 HZY
M3 IAB
M3 IAJ
M3 IAN
M3 IAS
M3 IAT
M3 IBB
M3 IBD
M3 IBF
M3 IBH
M3 IBL
M3 IBN
M3 IBO
M3 IBW
M3 IBX
M3 ICJ
M3 ICX
M3 ICY
M3 ICZ
M3 IDJ
M3 IDL
M3 IDP
M3 IDS
M3 IDU
M3 IEO
M3 IES
M3 IEY
M3 IFC
M3 IFH
M3 IFX
M3 IGI
M3 IGK
M3 IGL
M3 IGP
M3 IGW
M3 IHE
M3 IHG
M3 IHH
M3 IHT
M3 IIB
M3 IIE
M3 IIF
M3 IIQ
M3 IIS
M3 IIY
M3 IJB
M3 IJH
M3 IKA
M3 IKF
M3 IKL
M3 IKX
M3 IKY
M3 ILE
M3 ILF
M3 ILI
M3 ILL
M3 ILO
M3 ILP
M3 ILX
M3 IMN
M3 IMT
M3 IMU
M3 IMW
M3 INB
M3 INI
M3 INK
M3 INV
M3 INX
M3 INZ
M3 IOB
M3 IOE
M3 IOP
M3 IOY
M3 IPB
M3 IPI
M3 IPN
M3 IPU
M3 IQE
M3 IQH
M3 IQK
M3 IQL
M3 IQR
M3 IRD
M3 IRI
M3 IRK
M3 IRL
M3 IRN
M3 ISF
M3 ISV
M3 ITE
M3 ITF
M3 ITJ
M3 ITP
M3 ITQ
M3 ITR
M3 IUB
M3 IUG
M3 IUL
M3 IUQ
M3 IUR
M3 IUY
M3 IVE
M3 IVH
M3 IVJ
M3 IVL
M3 IVP
M3 IWH
M3 IWS
M3 IWY
M3 IXI
M3 IXS
M3 IXV
M3 IXW
M3 IYC
M3 IYD
M3 IYF
M3 IYJ
M3 IYL
M3 IYN
M3 IYS
M3 IYT
M3 IYU
M3 IYZ
M3 IZC
M3 IZR
M3 IZS
M3 IZX
M3 JAH
M3 JAJ
M3 JAN
M3 JAT
M3 JBA
M3 JBC
M3 JBD
M3 JBL
M3 JBN
M3 JBO
M3 JBU
M3 JBX
M3 JCJ
M3 JCM
M3 JDC
M3 JDD
M3 JDE
M3 JDK
M3 JDV
M3 JDY
M3 JDZ
M3 JEG
M3 JEL
M3 JEO
M3 JEU
M3 JEV
M3 JEX
M3 JEZ
M3 JFG
M3 JFI
M3 JFJ
M3 JFM
M3 JFO
M3 JFU
M3 JGD
M3 JHD
M3 JHF
M3 JHI
M3 JHK
M3 JHO
M3 JHQ
M3 JHU
M3 JIB
M3 JIE
M3 JIO
M3 JIY
M3 JJA
M3 JJF
M3 JJK
M3 JJO
M3 JJR
M3 JJX
M3 JKC
M3 JKF
M3 JKO
M3 JKQ
M3 JKV
M3 JLC
M3 JLG
M3 JLN
M3 JLO
M3 JLP
M3 JLU
M3 JLY
M3 JLZ
M3 JMB
M3 JME
M3 JMF
M3 JMH
M3 JMN
M3 JNF
M3 JNO
M3 JNS
M3 JOB
M3 JOG
M3 JOK
M3 JOX
M3 JOZ
M3 JPE
M3 JPR
M3 JPT
M3 JQA
M3 JQE
M3 JQL
M3 JQO
M3 JQP
M3 JQQ
M3 JQZ
M3 JRB
M3 JRP
M3 JSC
M3 JSP
M3 JSR
M3 JSU
M3 JSW
M3 JSX
M3 JSY
M3 JTD
M3 JTG
M3 JTI
M3 JTL
M3 JTP
M3 JTQ
M3 JTR
M3 JTW
M3 JTY
M3 JUA
M3 JUG
M3 JUS
M3 JUT
M3 JUU
M3 JVE
M3 JVL
M3 JVN
M3 JVO
M3 JVU
M3 JVZ
M3 JWH
M3 JWI
M3 JWL
M3 JWO
M3 JWU
M3 JWX
M3 JWY
M3 JXB
M3 JXT
M3 JYJ
M3 JYQ
M3 JYU
M3 JZB
M3 JZH
M3 JZQ
M3 JZS
M3 JZU
M3 JZZ
M3 KAA
M3 KAO
M3 KAT
M3 KAZ
M3 KBA
M3 KBI
M3 KBK
M3 KBO
M3 KBU
M3 KBV
M3 KBW
M3 KCK
M3 KCN
M3 KCW
M3 KDB
M3 KDD
M3 KDT
M3 KEP
M3 KEX
M3 KFA
M3 KFD
M3 KFF
M3 KFX
M3 KFY
M3 KFZ
M3 KGA
M3 KGC
M3 KGH
M3 KGM
M3 KGR
M3 KGX
M3 KHD
M3 KHN
M3 KHO
M3 KHS
M3 KHV
M3 KIC
M3 KIH
M3 KII
M3 KIJ
M3 KIK
M3 KIV
M3 KIW
M3 KIX
M3 KJI
M3 KJN
M3 KJO
M3 KJX
M3 KJZ
M3 KKJ
M3 KKR
M3 KKV
M3 KLA
M3 KLD
M3 KLF
M3 KLG
M3 KLI
M3 KLQ
M3 KLR
M3 KLW
M3 KLX
M3 KMG
M3 KMI
M3 KMJ
M3 KMM
M3 KMV
M3 KNG
M3 KNI
M3 KNQ
M3 KNX
M3 KOD
M3 KOE
M3 KOH
M3 KOL
M3 KOO
M3 KOQ
M3 KOV
M3 KPA
M3 KPH
M3 KPN
M3 KPR
M3 KPS
M3 KPV
M3 KPX
M3 KPY
M3 KQA
M3 KQE
M3 KQH
M3 KQJ
M3 KQQ
M3 KQZ
M3 KRG
M3 KRJ
M3 KRU
M3 KRV
M3 KRW
M3 KSB
M3 KSF
M3 KSL
M3 KSR
M3 KST
M3 KSU
M3 KSY
M3 KSZ
M3 KTF
M3 KTN
M3 KTQ
M3 KTU
M3 KUA
M3 KUB
M3 KUF
M3 KUL
M3 KUR
M3 KUT
M3 KUW
M3 KVF
M3 KVS
M3 KWA
M3 KWK
M3 KWN
M3 KWO
M3 KWQ
M3 KWW
M3 KXN
M3 KXO
M3 KXQ
M3 KXW
M3 KXX
M3 KYF
M3 KYN
M3 KYS
M3 KYT
M3 KYU
M3 KYY
M3 KZG
M3 KZL
M3 KZM
M3 KZN
M3 KZO
M3 KZU
M3 LAA
M3 LAC
M3 LAE
M3 LAH
M3 LAI
M3 LAL
M3 LAM
M3 LAR
M3 LBF
M3 LBH
M3 LBI
M3 LBU
M3 LCA
M3 LCB
M3 LCK
M3 LCM
M3 LCQ
M3 LCV
M3 LCX
M3 LDG
M3 LDN
M3 LDP
M3 LDV
M3 LDZ
M3 LEA
M3 LEM
M3 LEQ
M3 LEU
M3 LEY
M3 LFA
M3 LFD
M3 LFF
M3 LFJ
M3 LFK
M3 LFM
M3 LFS
M3 LFT
M3 LFW
M3 LFY
M3 LGA
M3 LGG
M3 LGQ
M3 LGT
M3 LGV
M3 LHD
M3 LHH
M3 LHJ
M3 LHL
M3 LHS
M3 LHT
M3 LHY
M3 LIA
M3 LIF
M3 LII
M3 LIJ
M3 LIQ
M3 LIR
M3 LJD
M3 LJN
M3 LJO
M3 LJU
M3 LJV
M3 LKA
M3 LKG
M3 LKI
M3 LKK
M3 LKL
M3 LKX
M3 LKZ
M3 LLE
M3 LLI
M3 LLJ
M3 LLO
M3 LLR
M3 LLY
M3 LMO
M3 LMP
M3 LMT
M3 LND
M3 LNK
M3 LNS
M3 LNV
M3 LOD
M3 LOH
M3 LOJ
M3 LOM
M3 LON
M3 LOV
M3 LOW
M3 LPD
M3 LPO
M3 LPV
M3 LPX
M3 LPY
M3 LQG
M3 LQH
M3 LQM
M3 LQQ
M3 LQR
M3 LQS
M3 LQT
M3 LRA
M3 LRB
M3 LRD
M3 LRL
M3 LSD
M3 LSJ
M3 LSM
M3 LSN
M3 LSP
M3 LSQ
M3 LSY
M3 LSZ
M3 LTC
M3 LTD
M3 LTF
M3 LTJ
M3 LTK
M3 LTM
M3 LTN
M3 LTO
M3 LTQ
M3 LTZ
M3 LUA
M3 LUB
M3 LUE
M3 LUF
M3 LUH
M3 LUI
M3 LUJ
M3 LUM
M3 LUQ
M3 LUR
M3 LUT
M3 LVE
M3 LVG
M3 LVH
M3 LVL
M3 LVO
M3 LVQ
M3 LVS
M3 LVU
M3 LVV
M3 LVW
M3 LWA
M3 LWB
M3 LWD
M3 LWE
M3 LWH
M3 LWI
M3 LWL
M3 LWM
M3 LWN
M3 LWS
M3 LWY
M3 LXC
M3 LXG
M3 LXM
M3 LXO
M3 LXQ
M3 LXX
M3 LXY
M3 LYB
M3 LYD
M3 LYF
M3 LYJ
M3 LYK
M3 LYO
M3 LYT
M3 LYW
M3 LYY
M3 LZB
M3 LZE
M3 LZG
M3 LZI
M3 LZJ
M3 LZM
M3 LZN
M3 LZP
M3 LZQ
M3 LZW
M3 LZX
M3 LZZ
M3 MAF
M3 MAH
M3 MAI
M3 MAL
M3 MAQ
M3 MAW
M3 MAX
M3 MBN
M3 MBP
M3 MBW
M3 MCH
M3 MCJ
M3 MCL
M3 MCO
M3 MCP
M3 MCZ
M3 MDL
M3 MDQ
M3 MDT
M3 MDZ
M3 MEQ
M3 MEX
M3 MFI
M3 MFM
M3 MFQ
M3 MFW
M3 MGG
M3 MGM
M3 MGV
M3 MGX
M3 MHB
M3 MHJ
M3 MHO
M3 MHT
M3 MIB
M3 MIL
M3 MIY
M3 MJB
M3 MJC
M3 MJP
M3 MJQ
M3 MJT
M3 MJU
M3 MJW
M3 MKL
M3 MKR
M3 MKU
M3 MKX
M3 MLC
M3 MLD
M3 MLE
M3 MLN
M3 MLO
M3 MLS
M3 MLU
M3 MLY
M3 MMH
M3 MMR
M3 MMT
M3 MMX
M3 MNA
M3 MNE
M3 MNF
M3 MNH
M3 MNM
M3 MNN
M3 MNO
M3 MNP
M3 MNZ
M3 MOA
M3 MOI
M3 MOM
M3 MOS
M3 MOU
M3 MPD
M3 MPF
M3 MPG
M3 MPN
M3 MPQ
M3 MPU
M3 MPV
M3 MPX
M3 MPY
M3 MQD
M3 MQE
M3 MQF
M3 MQG
M3 MQK
M3 MQN
M3 MQQ
M3 MQS
M3 MQU
M3 MQW
M3 MQZ
M3 MRP
M3 MRR
M3 MRT
M3 MSE
M3 MSI
M3 MSM
M3 MSS
M3 MTA
M3 MTG
M3 MTH
M3 MTJ
M3 MTN
M3 MTV
M3 MTY
M3 MUC
M3 MUH
M3 MUJ
M3 MUM
M3 MUR
M3 MUV
M3 MUY
M3 MUZ
M3 MVA
M3 MVC
M3 MVD
M3 MVF
M3 MVG
M3 MVH
M3 MVQ
M3 MVS
M3 MVZ
M3 MWF
M3 MWK
M3 MWL
M3 MWP
M3 MWU
M3 MWY
M3 MXB
M3 MXD
M3 MXI
M3 MXS
M3 MXT
M3 MXY
M3 MYA
M3 MYB
M3 MYJ
M3 MYL
M3 MYN
M3 MYO
M3 MYP
M3 MYR
M3 MYU
M3 MYV
M3 MYX
M3 MYY
M3 MZH
M3 MZI
M3 MZL
M3 MZQ
M3 MZS
M3 MZU
M3 MZY
M3 NAA
M3 NAM
M3 NBA
M3 NBC
M3 NBF
M3 NBM
M3 NBO
M3 NBR
M3 NBT
M3 NBW
M3 NCF
M3 NCR
M3 NCU
M3 NCV
M3 NCW
M3 NCX
M3 NCY
M3 NCZ
M3 NDD
M3 NDH
M3 NDP
M3 NDV
M3 NDY
M3 NEJ
M3 NEK
M3 NEU
M3 NEZ
M3 NFC
M3 NFD
M3 NFI
M3 NGI
M3 NGL
M3 NGQ
M3 NGT
M3 NGW
M3 NHF
M3 NHJ
M3 NHL
M3 NHO
M3 NHQ
M3 NHT
M3 NHX
M3 NIB
M3 NIH
M3 NIL
M3 NIM
M3 NIO
M3 NIU
M3 NIY
M3 NJG
M3 NJL
M3 NJT
M3 NJZ
M3 NKA
M3 NKD
M3 NKH
M3 NKI
M3 NKJ
M3 NKM
M3 NKS
M3 NKT
M3 NKV
M3 NKY
M3 NLB
M3 NLC
M3 NLL
M3 NLO
M3 NLS
M3 NLU
M3 NLV
M3 NMA
M3 NMC
M3 NME
M3 NMF
M3 NMH
M3 NMN
M3 NMO
M3 NMT
M3 NMW
M3 NMY
M3 NMZ
M3 NNB
M3 NNC
M3 NND
M3 NNF
M3 NNK
M3 NNL
M3 NNR
M3 NNX
M3 NOC
M3 NOG
M3 NOK
M3 NOT
M3 NOU
M3 NPJ
M3 NPL
M3 NPP
M3 NPQ
M3 NPV
M3 NPY
M3 NQC
M3 NQD
M3 NQF
M3 NQG
M3 NQJ
M3 NQP
M3 NQQ
M3 NQV
M3 NQW
M3 NQX
M3 NQZ
M3 NRD
M3 NRO
M3 NRS
M3 NRU
M3 NRZ
M3 NSE
M3 NSW
M3 NSY
M3 NSZ
M3 NTF
M3 NTM
M3 NTO
M3 NTS
M3 NTU
M3 NUA
M3 NUD
M3 NUF
M3 NUJ
M3 NUK
M3 NUR
M3 NUS
M3 NUZ
M3 NVB
M3 NVI
M3 NVM
M3 NVT
M3 NVU
M3 NVZ
M3 NWB
M3 NWG
M3 NWI
M3 NWM
M3 NWN
M3 NXB
M3 NXG
M3 NXI
M3 NXL
M3 NXM
M3 NXN
M3 NXP
M3 NXU
M3 NXV
M3 NXW
M3 NXX
M3 NYE
M3 NYJ
M3 NYL
M3 NYO
M3 NYR
M3 NYT
M3 NYU
M3 NYV
M3 NZB
M3 NZC
M3 NZD
M3 NZF
M3 NZL
M3 NZP
M3 NZQ
M3 NZS
M3 OAD
M3 OAF
M3 OAN
M3 OAO
M3 OAU
M3 OAV
M3 OBE
M3 OBH
M3 OBL
M3 OBP
M3 OBY
M3 OCB
M3 OCD
M3 OCF
M3 OCH
M3 OCI
M3 OCM
M3 OCT
M3 OCU
M3 ODD
M3 ODF
M3 ODG
M3 ODI
M3 ODJ
M3 ODN
M3 ODQ
M3 ODS
M3 ODU

M3 ODZ
M3 OEA
M3 OEK
M3 OEP
M3 OET
M3 OEU
M3 OEW
M3 OEX
M3 OEZ
M3 OFG
M3 OFK
M3 OFL
M3 OFM
M3 OFP
M3 OFQ
M3 OFT
M3 OFY
M3 OFZ
M3 OGA
M3 OGE
M3 OGF
M3 OGH
M3 OGI
M3 OGJ
M3 OGO
M3 OGP
M3 OGQ
M3 OGW
M3 OGX
M3 OGZ
M3 OHA
M3 OHH
M3 OHM
M3 OHT
M3 OHU
M3 OHV
M3 OHW
M3 OIA
M3 OID
M3 OIE
M3 OIF
M3 OIG
M3 OIJ
M3 OIO
M3 OIP
M3 OIQ
M3 OIR
M3 OIT
M3 OIU
M3 OIV
M3 OIW
M3 OJA
M3 OJB
M3 OJH
M3 OJI
M3 OJM
M3 OJQ
M3 OJT
M3 OKD
M3 OKK
M3 OKL
M3 OKM
M3 OKN
M3 OKQ
M3 OKV
M3 OKW
M3 OLB
M3 OLG
M3 OLH
M3 OLJ
M3 OLL
M3 OLP
M3 OLR
M3 OLS
M3 OMB
M3 OMJ
M3 OMK
M3 OML
M3 OMM
M3 OMN
M3 OMO
M3 OMP
M3 OMQ
M3 OMR
M3 OMV
M3 OMW
M3 ONA
M3 ONJ
M3 ONL
M3 ONQ
M3 ONR
M3 ONS
M3 ONT
M3 ONU
M3 ONW
M3 ONZ
M3 OOB
M3 OOJ
M3 OOK
M3 OON
M3 OOO
M3 OOS
M3 OOV
M3 OOW
M3 OPE
M3 OPF
M3 OPH
M3 OPJ
M3 OPQ
M3 OPR
M3 OPY
M3 OQA
M3 OQE
M3 OQF
M3 OQM
M3 OQO
M3 OQP
M3 OQU
M3 OQW
M3 OQX
M3 OQY
M3 ORA
M3 ORF
M3 ORH
M3 ORI
M3 ORJ
M3 ORW
M3 ORX
M3 OSD
M3 OSH
M3 OSL
M3 OSM
M3 OSR
M3 OSX
M3 OTC
M3 OTD
M3 OTH
M3 OTK
M3 OTL
M3 OTN
M3 OTO
M3 OTX
M3 OTY
M3 OUA
M3 OUB
M3 OUD
M3 OUE
M3 OUJ
M3 OUK
M3 OUM
M3 OUO
M3 OUP
M3 OUR
M3 OUW
M3 OUX
M3 OUZ
M3 OVB
M3 OVH
M3 OVI
M3 OVL
M3 OVN
M3 OVP
M3 OVQ
M3 OVS
M3 OVU
M3 OVW
M3 OVY
M3 OWA
M3 OWC
M3 OWD
M3 OWJ
M3 OWK
M3 OWL
M3 OWM
M3 OWR
M3 OWT
M3 OWV
M3 OWW
M3 OWX
M3 OWY
M3 OXA
M3 OXF
M3 OXG
M3 OXI
M3 OXJ
M3 OXL
M3 OXM
M3 OXS
M3 OXT
M3 OXU
M3 OXW
M3 OXX
M3 OYA
M3 OYF
M3 OYG
M3 OYH
M3 OYK
M3 OYM
M3 OYO
M3 OYT
M3 OYV
M3 OYX
M3 OYY
M3 OZA
M3 OZD
M3 OZL
M3 OZQ
M3 OZR
M3 OZS
M3 OZV
M3 OZX
M3 OZZ
M3 PAA
M3 PAB
M3 PAJ
M3 PAK
M3 PAO
M3 PAQ
M3 PBD
M3 PBF
M3 PBH
M3 PBJ
M3 PBM
M3 PBZ
M3 PCA
M3 PCB
M3 PCM
M3 PCU
M3 PCZ
M3 PDA
M3 PDI
M3 PDS
M3 PEG
M3 PEI
M3 PEJ
M3 PEK
M3 PEO
M3 PEP
M3 PEW
M3 PEX
M3 PFD
M3 PFG
M3 PFI
M3 PFO
M3 PFP
M3 PFQ
M3 PFS
M3 PFT
M3 PFV
M3 PFW
M3 PFZ
M3 PGJ
M3 PGM
M3 PGQ
M3 PGR
M3 PGV
M3 PGW
M3 PGX
M3 PGZ
M3 PHB
M3 PHI
M3 PHJ
M3 PHK
M3 PHN
M3 PHV
M3 PIF
M3 PII
M3 PIJ
M3 PIR
M3 PJA
M3 PJB
M3 PJD
M3 PJY
M3 PKA
M3 PKD
M3 PKF
M3 PKG
M3 PKJ
M3 PKK
M3 PKN
M3 PKU
M3 PKV
M3 PKX
M3 PKY
M3 PLC
M3 PLF
M3 PLG
M3 PLK
M3 PLM
M3 PLQ
M3 PLR
M3 PLY
M3 PMA
M3 PMC
M3 PMD
M3 PMG
M3 PMJ
M3 PMP
M3 PMQ
M3 PMV
M3 PND
M3 PNJ
M3 PNM
M3 PNP
M3 PNS
M3 PNU
M3 PNX
M3 POA
M3 POG
M3 POS
M3 POU
M3 POY
M3 POZ
M3 PPF
M3 PPJ
M3 PPP
M3 PPS
M3 PPT
M3 PPV
M3 PQA
M3 PQC
M3 PQD
M3 PQH
M3 PQR
M3 PQW
M3 PQX
M3 PRC
M3 PRF
M3 PRI
M3 PRO
M3 PRQ
M3 PRT
M3 PRW
M3 PRX
M3 PSK
M3 PSP
M3 PSQ
M3 PST
M3 PSX
M3 PTJ
M3 PTK
M3 PTS
M3 PTW
M3 PTY
M3 PUG
M3 PUK
M3 PUM
M3 PUS
M3 PUX
M3 PUY
M3 PVA
M3 PVD
M3 PVH
M3 PVJ
M3 PVK
M3 PVL
M3 PVM
M3 PVN
M3 PVO
M3 PVS
M3 PVW
M3 PWA
M3 PWB
M3 PWF
M3 PWG
M3 PWI
M3 PWJ
M3 PWQ
M3 PWR
M3 PWU
M3 PWV
M3 PWX
M3 PXD
M3 PXH
M3 PXI
M3 PXJ
M3 PXS
M3 PYA
M3 PYF
M3 PYH
M3 PYN
M3 PYP
M3 PYU
M3 PYZ
M3 PZA
M3 PZD
M3 PZG
M3 PZH
M3 PZI
M3 PZM
M3 PZP
M3 PZT
M3 PZU
M3 RAO
M3 RAP
M3 RAX
M3 RAY
M3 RBV
M3 RBZ
M3 RCA
M3 RCF
M3 RCK
M3 RCN
M3 RCP
M3 RCU
M3 RDE
M3 RDG
M3 RDI
M3 RDN
M3 RDT
M3 RDZ
M3 REE
M3 REF
M3 RER
M3 RES
M3 RFP
M3 RFT
M3 RFU
M3 RFY
M3 RFZ
M3 RGI
M3 RGM
M3 RGQ
M3 RGT
M3 RGX
M3 RGY
M3 RHE
M3 RHM
M3 RHU
M3 RHV
M3 RHX
M3 RHY
M3 RIB
M3 RIH
M3 RIN
M3 RIR
M3 RIY
M3 RJA
M3 RJD
M3 RJQ
M3 RJR
M3 RJT
M3 RJV
M3 RJW
M3 RKA
M3 RKM
M3 RKP
M3 RKU
M3 RKX
M3 RLI
M3 RLL
M3 RLQ
M3 RLR
M3 RLU
M3 RLY
M3 RLZ
M3 RMC
M3 RME
M3 RMK
M3 RMM
M3 RMO
M3 RMR
M3 RNB
M3 RND
M3 RNE
M3 RNF
M3 RNI
M3 RNJ
M3 RNL
M3 RNP
M3 RNR
M3 RNV
M3 ROE
M3 ROH
M3 ROJ
M3 ROT
M3 RPC
M3 RPG
M3 RPL
M3 RPM
M3 RPN
M3 RPU
M3 RPY
M3 RQF
M3 RQL
M3 RQS
M3 RQT
M3 RQU
M3 RQZ
M3 RRD
M3 RRK
M3 RRL
M3 RRO
M3 RRS
M3 RRX
M3 RSK
M3 RSL
M3 RSP
M3 RSQ
M3 RSW
M3 RTF
M3 RTG
M3 RTX
M3 RUA
M3 RUD
M3 RUJ
M3 RUN
M3 RUU
M3 RUX
M3 RVA
M3 RVF
M3 RVG
M3 RVL
M3 RVU
M3 RVW
M3 RVY
M3 RVZ
M3 RWE
M3 RWG
M3 RWL
M3 RWT
M3 RWU
M3 RWY
M3 RXC
M3 RXI
M3 RXR
M3 RXS
M3 RXV
M3 RXX
M3 RYH
M3 RYK
M3 RYL
M3 RYM
M3 RYQ
M3 RYV
M3 RYX
M3 RZH
M3 RZK
M3 RZR
M3 RZS
M3 SBG
M3 SBH
M3 SBI
M3 SBK
M3 SBL
M3 SBU
M3 SBV
M3 SBZ
M3 SCE
M3 SCK
M3 SCS
M3 SCT
M3 SCY
M3 SCZ
M3 SDM
M3 SDO
M3 SDT
M3 SEH
M3 SEP
M3 SEQ
M3 SER
M3 SEW
M3 SFB
M3 SFE
M3 SFF
M3 SFH
M3 SFI
M3 SFL
M3 SFO
M3 SFQ
M3 SFR
M3 SFU
M3 SFV
M3 SGC
M3 SGN
M3 SGO
M3 SGY
M3 SHC
M3 SHD
M3 SHE
M3 SHH
M3 SHM
M3 SIA
M3 SIF
M3 SIH
M3 SIJ
M3 SIU
M3 SIV
M3 SJC
M3 SJG
M3 SJZ
M3 SKE
M3 SKG
M3 SKO
M3 SKS
M3 SLE
M3 SLR
M3 SLU
M3 SLV
M3 SMO
M3 SMS
M3 SMU
M3 SNC
M3 SND
M3 SNM
M3 SNP
M3 SNT
M3 SNU
M3 SNV
M3 SOH
M3 SOI
M3 SOJ
M3 SOK
M3 SOM
M3 SOO
M3 SOR
M3 SOU
M3 SOX
M3 SPB
M3 SPD
M3 SPF
M3 SPM
M3 SPO
M3 SPV
M3 SPW
M3 SPX
M3 SQB
M3 SQC
M3 SQD
M3 SQF
M3 SQR
M3 SRE
M3 SRN
M3 SRX
M3 SRZ
M3 SSE
M3 SSN
M3 SSP
M3 SST
M3 SSY
M3 STB
M3 STE
M3 STF
M3 STG
M3 STM
M3 STO
M3 STX
M3 STZ
M3 SUA
M3 SUH
M3 SUN
M3 SUQ
M3 SUZ
M3 SVA
M3 SVE
M3 SVG
M3 SVI
M3 SVQ
M3 SVR
M3 SVS
M3 SVV
M3 SVX
M3 SWJ
M3 SWL
M3 SWN
M3 SWR
M3 SWX
M3 SXA
M3 SXD
M3 SXG
M3 SXN
M3 SXO
M3 SYA
M3 SYE
M3 SYG
M3 SYK
M3 SZH
M3 SZJ
M3 SZL
M3 SZN
M3 SZP
M3 SZR
M3 SZU
M3 SZV
M3 SZW
M3 SZX
M3 SZZ
M3 TAI
M3 TAJ
M3 TAO
M3 TAQ
M3 TBB
M3 TBD
M3 TBE
M3 TBI
M3 TBN
M3 TBR
M3 TBY
M3 TCA
M3 TCH
M3 TCL
M3 TCM
M3 TCO
M3 TCV
M3 TCZ
M3 TDA
M3 TDF
M3 TDI
M3 TDN
M3 TDQ
M3 TDV
M3 TDW
M3 TDZ
M3 TEA
M3 TEC
M3 TEJ
M3 TEO
M3 TER
M3 TEU
M3 TEZ
M3 TFQ
M3 TFU
M3 TGF
M3 TGG
M3 TGH
M3 TGN
M3 TGS
M3 TGX
M3 TGY
M3 THG
M3 THL
M3 THT
M3 THU
M3 THZ
M3 TIH
M3 TIO
M3 TIP
M3 TIR
M3 TIW
M3 TJA
M3 TJN
M3 TJP
M3 TJV
M3 TJY
M3 TKA
M3 TKB
M3 TKC
M3 TKF
M3 TKK
M3 TKL
M3 TKY
M3 TKZ
M3 TLA
M3 TLE
M3 TLK
M3 TMD
M3 TMH
M3 TMI
M3 TMS
M3 TMV
M3 TNI
M3 TNR
M3 TNU
M3 TOH
M3 TOQ
M3 TOU
M3 TOX
M3 TPC
M3 TPL
M3 TPP
M3 TPX
M3 TQC
M3 TQE
M3 TQL
M3 TQQ
M3 TQR
M3 TQW
M3 TQZ
M3 TRD
M3 TRN
M3 TRQ
M3 TRS
M3 TRU
M3 TSC
M3 TSK
M3 TSL
M3 TSM
M3 TSS
M3 TSU
M3 TTG
M3 TTJ
M3 TTQ
M3 TTV
M3 TTW
M3 TUE
M3 TUK
M3 TUT
M3 TUV
M3 TVE
M3 TVS
M3 TVU
M3 TVX
M3 TWJ
M3 TWO
M3 TWU
M3 TWX
M3 TXB
M3 TXD
M3 TXE
M3 TXN
M3 TXO
M3 TXX
M3 TXZ
M3 TYD
M3 TYF
M3 TYN
M3 TYY
M3 TZA
M3 TZH
M3 TZJ
M3 TZL
M3 TZM
M3 TZT
M3 TZV
M3 TZX
M3 TZZ
M3 UAB
M3 UAD
M3 UAF
M3 UAU
M3 UBA
M3 UBI
M3 UBJ
M3 UBM
M3 UBO
M3 UBP
M3 UBV
M3 UBW
M3 UBZ
M3 UCB
M3 UCG
M3 UCK
M3 UCM
M3 UCQ
M3 UCT
M3 UCW
M3 UDE
M3 UDG
M3 UDH
M3 UDM
M3 UDP
M3 UEC
M3 UEM
M3 UEO
M3 UES
M3 UFB
M3 UFM
M3 UFN
M3 UFO
M3 UFP
M3 UFR
M3 UFV
M3 UGN
M3 UGP
M3 UHA
M3 UHD
M3 UHE
M3 UHM
M3 UHP
M3 UHR
M3 UHU
M3 UHZ
M3 UIH
M3 UIZ
M3 UJG
M3 UJI
M3 UJK
M3 UJM
M3 UJU
M3 UJW
M3 UKA
M3 UKE
M3 UKI
M3 UKL
M3 UKZ
M3 ULA
M3 ULF
M3 ULG
M3 ULH
M3 ULK
M3 ULR
M3 ULV
M3 UME
M3 UMI
M3 UMK
M3 UMO
M3 UMP
M3 UMS
M3 UMT
M3 UNA
M3 UNE
M3 UNM
M3 UNS
M3 UNV
M3 UNX
M3 UOA
M3 UOG
M3 UOH
M3 UOI
M3 UOQ
M3 UOT
M3 UOU
M3 UPC
M3 UPD
M3 UPE
M3 UPH
M3 UPM
M3 UPR
M3 UPU
M3 UPW
M3 UQS
M3 UQU
M3 UQV
M3 UQX
M3 URC
M3 URI
M3 URJ
M3 URM
M3 URP
M3 URY
M3 USE
M3 USG
M3 USI
M3 USL
M3 USO
M3 USQ
M3 USY
M3 UTA
M3 UTC
M3 UTD
M3 UTF
M3 UTG
M3 UTL
M3 UTW
M3 UUB
M3 UUC
M3 UUI
M3 UUM
M3 UUQ
M3 UUU
M3 UVE
M3 UVI
M3 UVL
M3 UVR
M3 UVY
M3 UVZ
M3 UWC
M3 UWG
M3 UWK
M3 UWQ
M3 UXA
M3 UXD
M3 UXJ
M3 UXP
M3 UXT
M3 UXV
M3 UXW
M3 UYB
M3 UYN
M3 UYR
M3 UZC
M3 UZD
M3 UZG
M3 UZR
M3 UZS
M3 UZY
M3 VAI
M3 VAL
M3 VAN
M3 VAO
M3 VAP
M3 VAZ
M3 VBD
M3 VBK
M3 VBO
M3 VBS
M3 VBX
M3 VCH
M3 VCL
M3 VCU
M3 VDJ
M3 VDK
M3 VDR
M3 VDS
M3 VEB
M3 VED
M3 VEE
M3 VEI
M3 VEO
M3 VEU
M3 VEV
M3 VFA
M3 VFD
M3 VFG
M3 VFH
M3 VFT
M3 VGC
M3 VGI
M3 VGO
M3 VGQ
M3 VGR
M3 VGU
M3 VGY
M3 VID
M3 VIE
M3 VIF
M3 VIK
M3 VIT
M3 VIX
M3 VIY
M3 VJQ
M3 VJR
M3 VJY
M3 VJZ
M3 VKC
M3 VKE
M3 VKH
M3 VKI
M3 VKK
M3 VKW
M3 VKZ
M3 VLC
M3 VLK
M3 VLR
M3 VLW
M3 VLY
M3 VLZ
M3 VMB
M3 VMG
M3 VMK
M3 VMO
M3 VMX
M3 VNA
M3 VNB
M3 VND
M3 VNE
M3 VNY
M3 VOD
M3 VOE
M3 VOF
M3 VOK
M3 VOM
M3 VOO
M3 VOV
M3 VPC
M3 VPF
M3 VPI
M3 VPM
M3 VPU
M3 VPZ
M3 VQB
M3 VQO
M3 VQT
M3 VQU
M3 VQW
M3 VRE
M3 VRG
M3 VRH
M3 VRJ
M3 VRS
M3 VRW
M3 VSA
M3 VSD
M3 VSM
M3 VSP
M3 VSR
M3 VSV
M3 VSY
M3 VTD
M3 VTI
M3 VTM
M3 VTY
M3 VTZ
M3 VUB
M3 VUE
M3 VUF
M3 VUG
M3 VUM
M3 VVG
M3 VVM
M3 VVP
M3 VWA
M3 VWB
M3 VWI
M3 VWL
M3 VWM
M3 VWN
M3 VWS
M3 VWT
M3 VWU
M3 VWX
M3 VWZ
M3 VXD
M3 VXQ
M3 VXT
M3 VXU
M3 VXW
M3 VXZ
M3 VYG
M3 VYI
M3 VYL
M3 VYW
M3 VYZ
M3 VZA
M3 VZB
M3 VZE
M3 VZG
M3 VZK
M3 VZM
M3 VZR
M3 VZT
M3 VZW
M3 VZY
M3 WAA
M3 WAG
M3 WAH
M3 WAI
M3 WAO
M3 WAQ
M3 WAU
M3 WAZ
M3 WBW
M3 WBX
M3 WCC
M3 WCD
M3 WCF
M3 WCH
M3 WCJ
M3 WCK
M3 WCL
M3 WCT
M3 WCU
M3 WDD
M3 WDG
M3 WDM
M3 WDQ
M3 WDS
M3 WDW
M3 WDX
M3 WEC
M3 WED
M3 WEK
M3 WEL
M3 WEN
M3 WEO
M3 WER
M3 WEX
M3 WFD
M3 WFI
M3 WFV
M3 WFW
M3 WFX
M3 WFZ
M3 WGF
M3 WGG
M3 WGH
M3 WGN
M3 WGQ
M3 WGT
M3 WGX
M3 WHD
M3 WHE
M3 WHF
M3 WHM
M3 WHN
M3 WHO
M3 WIB
M3 WIE
M3 WIF
M3 WIS
M3 WIY
M3 WJB
M3 WJC
M3 WJH
M3 WJJ
M3 WJQ
M3 WJS
M3 WJX
M3 WKA
M3 WKD
M3 WKG
M3 WKS
M3 WKT
M3 WKU
M3 WLC
M3 WLE
M3 WLF
M3 WLH
M3 WLJ
M3 WLM
M3 WLN
M3 WLP
M3 WLQ
M3 WLR
M3 WLT
M3 WMA
M3 WMF
M3 WMN
M3 WMT
M3 WMZ
M3 WND
M3 WNL
M3 WNO
M3 WNU
M3 WNW
M3 WNY
M3 WOE
M3 WOH
M3 WOJ
M3 WOM
M3 WON
M3 WOO
M3 WOU
M3 WPD
M3 WPR
M3 WPZ
M3 WQD
M3 WQH
M3 WQJ
M3 WQK
M3 WQP
M3 WQQ
M3 WQU
M3 WQZ
M3 WSA
M3 WSD
M3 WSG
M3 WST
M3 WSU
M3 WSX
M3 WSY
M3 WTJ
M3 WTQ
M3 WTX
M3 WTZ
M3 WUC
M3 WUF
M3 WUT
M3 WUY
M3 WVA
M3 WVE
M3 WVM
M3 WVS
M3 WVU
M3 WWB
M3 WWE
M3 WWF
M3 WWI
M3 WWL
M3 WWT
M3 WWX
M3 WXK
M3 WXM
M3 WXO
M3 WXQ
M3 WXV
M3 WXY
M3 WYD
M3 WYE
M3 WYH
M3 WYL
M3 WYP
M3 WYU
M3 WYX
M3 WZC
M3 WZD
M3 WZI
M3 WZK
M3 WZO
M3 WZQ
M3 WZU
M3 WZX
M3 XAA
M3 XAD
M3 XAL
M3 XAP
M3 XAQ
M3 XBA
M3 XBD
M3 XBI
M3 XBJ
M3 XBP
M3 XBQ
M3 XBU
M3 XBV
M3 XBW
M3 XBX
M3 XCC
M3 XCD
M3 XCG
M3 XCK
M3 XCL
M3 XCM
M3 XCO
M3 XCQ
M3 XCZ
M3 XDA
M3 XDC
M3 XDN
M3 XDO
M3 XDR
M3 XDY
M3 XEC
M3 XED
M3 XEE
M3 XEH
M3 XEM
M3 XEV
M3 XEW
M3 XEZ
M3 XFE
M3 XFF
M3 XFG
M3 XFJ
M3 XFL
M3 XFQ
M3 XFU
M3 XFV
M3 XFW
M3 XFY
M3 XGE
M3 XGF
M3 XGJ
M3 XGK
M3 XGM
M3 XGO
M3 XHA
M3 XHD
M3 XHI
M3 XHJ
M3 XHP
M3 XHR
M3 XHS
M3 XHV
M3 XIB
M3 XII
M3 XIR
M3 XIX
M3 XJD
M3 XJN
M3 XJR
M3 XJU
M3 XJV
M3 XJY
M3 XKB
M3 XKC
M3 XKK
M3 XKQ
M3 XKR
M3 XKS
M3 XKT
M3 XKV
M3 XKW
M3 XKZ
M3 XLA
M3 XLD
M3 XLF
M3 XLI
M3 XLN
M3 XLY
M3 XMF
M3 XML
M3 XMM
M3 XMR
M3 XMV
M3 XMX
M3 XND
M3 XNG
M3 XNI
M3 XNJ
M3 XNS
M3 XNY
M3 XNZ
M3 XOB
M3 XOF
M3 XOM
M3 XOO
M3 XOP
M3 XPA
M3 XPB
M3 XPC
M3 XPE
M3 XPO
M3 XPQ
M3 XPV
M3 XPZ
M3 XQA
M3 XQD
M3 XQR
M3 XQU
M3 XRA
M3 XRB
M3 XRF
M3 XRL
M3 XRR
M3 XRU
M3 XSB
M3 XSH
M3 XSL
M3 XSO
M3 XSQ
M3 XSS
M3 XSW
M3 XTB
M3 XTH
M3 XTO
M3 XTQ
M3 XTW
M3 XTX
M3 XUB
M3 XUK
M3 XUL
M3 XUM
M3 XUQ
M3 XUZ
M3 XVA
M3 XVE
M3 XVN
M3 XVV
M3 XWA
M3 XWD
M3 XWH
M3 XWJ
M3 XWL
M3 XWT
M3 XXC
M3 XXD
M3 XXF
M3 XXH
M3 XXN
M3 XYD
M3 XYF
M3 XYG
M3 XYL
M3 XYQ
M3 XYR
M3 XYY
M3 XZI
M3 XZL
M3 XZM
M3 XZO
M3 XZQ
M3 XZU
M3 XZZ
M3 YAF
M3 YAH
M3 YAP
M3 YAR
M3 YAU
M3 YAY
M3 YAZ
M3 YBE
M3 YBH
M3 YBO
M3 YBP
M3 YBS
M3 YBV
M3 YBY
M3 YBZ
M3 YCA
M3 YCC
M3 YCF
M3 YCH
M3 YCP
M3 YCY
M3 YDC
M3 YDE
M3 YDG
M3 YDO
M3 YDR
M3 YEB
M3 YED
M3 YEF
M3 YEH
M3 YEI
M3 YEO
M3 YEV
M3 YEX
M3 YEY
M3 YFC
M3 YFK
M3 YFO
M3 YFP
M3 YFU
M3 YGE
M3 YGH
M3 YGN
M3 YGX
M3 YHB
M3 YHE
M3 YHI
M3 YHS
M3 YIA
M3 YIB
M3 YIN
M3 YIP
M3 YIW
M3 YJC
M3 YJO
M3 YJQ
M3 YJX
M3 YKB
M3 YKD
M3 YKG
M3 YKJ
M3 YKQ
M3 YKU
M3 YKX
M3 YKY
M3 YLC
M3 YLF
M3 YLI
M3 YLY
M3 YMO
M3 YMP
M3 YMR
M3 YMT
M3 YMZ
M3 YNF
M3 YNR
M3 YNU
M3 YNV
M3 YNW
M3 YOA
M3 YOD
M3 YOJ
M3 YON
M3 YPC
M3 YPL
M3 YPM
M3 YPO
M3 YPT
M3 YPV
M3 YPZ
M3 YQB
M3 YQE
M3 YQF
M3 YQI
M3 YQK
M3 YQN
M3 YQO
M3 YQS
M3 YQV
M3 YQY
M3 YRD
M3 YRG
M3 YRI
M3 YRK
M3 YRL
M3 YRN
M3 YRQ
M3 YRU
M3 YRY
M3 YSB
M3 YSG
M3 YSH
M3 YSI
M3 YSK
M3 YSO
M3 YSP
M3 YSR
M3 YST
M3 YSX
M3 YTD
M3 YTJ
M3 YTK
M3 YTO
M3 YTT
M3 YTY
M3 YUG
M3 YUJ
M3 YUL
M3 YUO
M3 YUZ
M3 YVH
M3 YVI
M3 YVL
M3 YVM
M3 YVO
M3 YVP
M3 YVQ
M3 YVX
M3 YWA
M3 YWB
M3 YWN
M3 YWQ
M3 YWR
M3 YWS
M3 YWV
M3 YWX
M3 YWY
M3 YWZ
M3 YXG
M3 YXI
M3 YXO
M3 YXY
M3 YYA
M3 YYC
M3 YYL
M3 YYP
M3 YYY
M3 YZD
M3 YZF
M3 YZG
M3 YZL
M3 YZT
M3 YZY
M3 ZAB
M3 ZAG
M3 ZAH
M3 ZAJ
M3 ZAX
M3 ZBA
M3 ZBE
M3 ZBK
M3 ZBN
M3 ZBP
M3 ZBU
M3 ZBY
M3 ZCK
M3 ZCL
M3 ZCP
M3 ZCV
M3 ZCZ
M3 ZDO
M3 ZDP
M3 ZDY
M3 ZDZ
M3 ZEA
M3 ZEF
M3 ZEG
M3 ZEK
M3 ZEP
M3 ZEU
M3 ZEZ
M3 ZFA
M3 ZFE
M3 ZFM
M3 ZFP
M3 ZFQ
M3 ZFT
M3 ZFU
M3 ZFX
M3 ZGB
M3 ZGQ
M3 ZGV
M3 ZGY
M3 ZGZ
M3 ZHA
M3 ZHK
M3 ZHN
M3 ZHO
M3 ZHR
M3 ZHT
M3 ZIB
M3 ZIC
M3 ZIG
M3 ZIJ
M3 ZIP
M3 ZIQ
M3 ZIW
M3 ZJC
M3 ZJI
M3 ZJR
M3 ZJT
M3 ZJU
M3 ZJW
M3 ZJZ
M3 ZKB
M3 ZKC
M3 ZKG
M3 ZKH
M3 ZKK
M3 ZKM
M3 ZKN
M3 ZKU
M3 ZKV
M3 ZLB
M3 ZLC
M3 ZLG
M3 ZLH
M3 ZLM
M3 ZLN
M3 ZLQ
M3 ZLT
M3 ZMA
M3 ZMF
M3 ZMG
M3 ZMH
M3 ZMK
M3 ZND
M3 ZNE
M3 ZNH
M3 ZNI
M3 ZOF
M3 ZOS
M3 ZOT
M3 ZOW
M3 ZOX
M3 ZPA
M3 ZPF
M3 ZPH
M3 ZPI
M3 ZPK
M3 ZPL
M3 ZPN
M3 ZPP
M3 ZPQ
M3 ZPS
M3 ZPU
M3 ZPV
M3 ZPX
M3 ZQD
M3 ZQE
M3 ZQH
M3 ZQI
M3 ZQQ
M3 ZQR
M3 ZQT
M3 ZQY
M3 ZQZ
M3 ZRD
M3 ZRI
M3 ZRJ
M3 ZRN
M3 ZRT
M3 ZRU
M3 ZSN
M3 ZSP
M3 ZSR
M3 ZSS
M3 ZSX
M3 ZSZ
M3 ZTA
M3 ZTE
M3 ZTG
M3 ZTO
M3 ZTQ
M3 ZTY
M3 ZUE
M3 ZUH
M3 ZUQ
M3 ZUX
M3 ZVM
M3 ZVO
M3 ZVP
M3 ZVQ
M3 ZVZ
M3 ZWV
M3 ZWZ
M3 ZXB
M3 ZXJ
M3 ZXM
M3 ZXP
M3 ZXS
M3 ZXV
M3 ZXW
M3 ZYA
M3 ZYB
M3 ZYC
M3 ZYD
M3 ZYH
M3 ZYJ
M3 ZYL
M3 ZYN
M3 ZZB
M3 ZZC
M3 ZZZ

M5

M5 AAD
M5 AAE
M5 AAO
M5 AAS
M5 ABP
M5 ABV
M5 ACM
M5 ACU
M5 ACZ
M5 ADU
M5 ADX
M5 ADZ
M5 AED
M5 AEJ
M5 AEK
M5 AFD
M5 AFK
M5 AFO
M5 AFU
M5 AGC
M5 AGH
M5 AGT
M5 AHQ
M5 AIF
M5 AIL
M5 AIM
M5 AIR
M5 AJF
M5 AJT
M5 AJV
M5 AKA
M5 ALI
M5 AML
M5 ASF
M5 AX
M5 BEN
M5 BRY
M5 CDS
M5 CJW
M5 CSM
M5 DBH
M5 DEN
M5 DJS
M5 DOS
M5 DWI
M5 EEK
M5 EPA
M5 ESE
M5 EVT
M5 FDB
M5 FET
M5 FLY
M5 FUN
M5 GB
M5 GSK
M5 GTI
M5 GUM
M5 IA
M5 ICB

M5 IEP
M5 IIT
M5 IJH
M5 IPX
M5 IXG
M5 JAB
M5 JAN
M5 JDW
M5 JDZ
M5 JFS
M5 JSW
M5 JTS
M5 KAW
M5 KEV
M5 KR
M5 KRG
M5 LAM
M5 LLT
M5 LOE
M5 MAN
M5 MAT
M5 MD
M5 MDC
M5 MKW
M5 MOW
M5 MPC
M5 MSX
M5 NCW
M5 OFV
M5 PD
M5 PMJ
M5 RAO
M5 RIO
M5 RJL
M5 RS
M5 SPY
M5 ST
M5 STC
M5 TAW
M5 TGW
M5 TKA
M5 UF
M5 UFO
M5 UK
M5 URR
M5 UTC
M5 WMF
M5 WOB
M5 XT
M5 YRG
M5 ZZR

M6

M6 AAB
M6 AAI
M6 AAN
M6 AAS
M6 AAT
M6 ABA
M6 ABC
M6 ABH
M6 ABI
M6 ABJ
M6 ABL
M6 ABU
M6 ABW
M6 ABY
M6 ACG
M6 ACK
M6 ACL
M6 ACN
M6 ACO
M6 ACT
M6 ACU
M6 ADA
M6 ADH
M6 ADK
M6 ADN
M6 ADO
M6 ADP
M6 ADR
M6 ADW
M6 AEA
M6 AEH
M6 AEO
M6 AEP
M6 AEV
M6 AFD
M6 AFE
M6 AFJ
M6 AFO
M6 AFP
M6 AFR
M6 AFT
M6 AFV
M6 AFZ
M6 AGB
M6 AGC
M6 AGE
M6 AGM
M6 AGN
M6 AGP
M6 AGU
M6 AGW
M6 AGX
M6 AHC
M6 AHI
M6 AHM
M6 AIG
M6 AIX
M6 AIY
M6 AJE
M6 AJG
M6 AJK
M6 AJR
M6 AJS
M6 AJT
M6 AJU
M6 AJV
M6 AJW
M6 AJY
M6 AJZ
M6 AKB
M6 AKL
M6 AKM
M6 AKP
M6 AKR
M6 AKS
M6 AKU
M6 AKX
M6 ALA
M6 ALF
M6 ALN
M6 ALS
M6 ALV
M6 AMB
M6 AMH
M6 AMP
M6 AMQ
M6 AMS
M6 AMU
M6 AMY
M6 AMZ
M6 ANA
M6 ANC
M6 ANF
M6 ANG
M6 ANH
M6 ANJ
M6 ANK
M6 ANL
M6 ANQ
M6 ANU
M6 ANW
M6 ANZ
M6 AOA
M6 AOC
M6 AOI
M6 AOM
M6 AOO
M6 AOS
M6 APE
M6 APF
M6 APH
M6 APJ
M6 APK
M6 APL
M6 APN
M6 APP
M6 APQ
M6 APT
M6 APV
M6 APX
M6 APY
M6 APZ
M6 AQB
M6 AQC
M6 AQF
M6 AQQ
M6 AQS
M6 AQX
M6 AQZ
M6 ARA
M6 ARE
M6 ARG
M6 ARZ
M6 ASA
M6 ASB
M6 ASG
M6 ASR
M6 ATE
M6 ATN
M6 ATO
M6 ATT
M6 ATZ
M6 AUK
M6 AUN
M6 AUR
M6 AUS
M6 AUV
M6 AUW
M6 AUY
M6 AVB
M6 AVC
M6 AVI
M6 AVN
M6 AVP
M6 AWF
M6 AWH
M6 AWK
M6 AWM
M6 AWT
M6 AWX
M6 AXE
M6 AXK
M6 AXN
M6 AXO
M6 AXQ
M6 AXV
M6 AXY
M6 AXZ
M6 AYB
M6 AYF
M6 AYO
M6 AYT
M6 AYX
M6 AZB
M6 AZI
M6 AZJ
M6 AZL
M6 AZW
M6 AZZ
M6 BAE
M6 BAF
M6 BAL
M6 BAO
M6 BAP
M6 BAT
M6 BBI
M6 BBN
M6 BBV
M6 BBW
M6 BBY
M6 BBZ
M6 BCA
M6 BCD
M6 BCE
M6 BCI
M6 BCL
M6 BCM
M6 BCO
M6 BCP
M6 BCR
M6 BCS
M6 BCT
M6 BCX
M6 BDB
M6 BDC
M6 BDE
M6 BDF
M6 BDH
M6 BDK
M6 BDP
M6 BDT
M6 BDW
M6 BDY
M6 BDZ
M6 BEA
M6 BEO
M6 BEP
M6 BER
M6 BET
M6 BEV
M6 BEW
M6 BEY
M6 BFA
M6 BFD
M6 BFE
M6 BFT
M6 BFW
M6 BFY
M6 BGN
M6 BGQ
M6 BGR
M6 BGU
M6 BGX
M6 BGY
M6 BHG
M6 BHL
M6 BHR
M6 BHV
M6 BHW
M6 BHX
M6 BIC
M6 BIK
M6 BIL
M6 BIW
M6 BIZ
M6 BJD
M6 BJH
M6 BJR
M6 BJS
M6 BJU
M6 BJV
M6 BKB
M6 BKG
M6 BKW
M6 BKX
M6 BKY
M6 BKZ
M6 BLD
M6 BLJ
M6 BLL
M6 BLR
M6 BLU
M6 BMA
M6 BMD
M6 BMG
M6 BMH
M6 BML
M6 BMS
M6 BMU
M6 BMX
M6 BND
M6 BNE
M6 BNI
M6 BNR
M6 BNZ
M6 BOH
M6 BON
M6 BOO
M6 BOP
M6 BOU
M6 BPF
M6 BPN
M6 BPP
M6 BPT
M6 BPZ
M6 BQI
M6 BQT
M6 BQU
M6 BQX
M6 BQY
M6 BRA
M6 BRC
M6 BRD
M6 BRE
M6 BRF
M6 BRG
M6 BRM
M6 BRW
M6 BRX
M6 BSD
M6 BSE
M6 BSH
M6 BSJ
M6 BSM
M6 BSN
M6 BTD
M6 BTI
M6 BTJ
M6 BTL
M6 BTO
M6 BTU
M6 BTV
M6 BTW
M6 BTX
M6 BTZ
M6 BUK
M6 BUN
M6 BUO
M6 BUP
M6 BUQ
M6 BUT
M6 BUV
M6 BUZ
M6 BVA
M6 BVB
M6 BVF
M6 BVH
M6 BVL
M6 BVO
M6 BVT
M6 BVY
M6 BWA
M6 BWD
M6 BWI
M6 BWM
M6 BWS
M6 BWT
M6 BWX
M6 BXC
M6 BXE
M6 BXG
M6 BYB
M6 BYC
M6 BYG
M6 BYI
M6 BYW
M6 BZJ
M6 BZL
M6 BZO
M6 BZR
M6 BZS
M6 BZZ
M6 CAB
M6 CAC
M6 CAL
M6 CAM
M6 CAO
M6 CAX
M6 CBB
M6 CBK
M6 CBR
M6 CBS
M6 CBT
M6 CBV
M6 CBX
M6 CBZ
M6 CCC
M6 CCD
M6 CCK
M6 CCL
M6 CCP
M6 CCT
M6 CCZ
M6 CDA
M6 CDF
M6 CDI
M6 CDM
M6 CDP
M6 CDS
M6 CDX
M6 CDY
M6 CEC
M6 CEK
M6 CER
M6 CEV
M6 CFB
M6 CFC
M6 CFF
M6 CFG
M6 CFK
M6 CFM
M6 CFP
M6 CGA
M6 CGG
M6 CGI
M6 CGN
M6 CGP
M6 CGR
M6 CGU
M6 CHA
M6 CHE
M6 CHK
M6 CHO
M6 CHR
M6 CIB
M6 CIC
M6 CID
M6 CIF
M6 CIH
M6 CII
M6 CIL
M6 CIN
M6 CIO
M6 CIT
M6 CIV
M6 CJB
M6 CJO
M6 CKA
M6 CKB
M6 CKG
M6 CKH
M6 CKK
M6 CKN
M6 CKS
M6 CKW
M6 CKX
M6 CLD
M6 CLE
M6 CLH
M6 CLN
M6 CLR
M6 CLS
M6 CLU
M6 CLY
M6 CMA
M6 CMD
M6 CME
M6 CMH
M6 CMJ
M6 CMN
M6 CMP
M6 CMR
M6 CMS
M6 CMV
M6 CMX
M6 CNB
M6 CNE
M6 CNF
M6 CNH
M6 CNJ
M6 CNM
M6 CNU
M6 CNW
M6 COA
M6 COG
M6 COJ
M6 COO
M6 COQ
M6 COR
M6 COS
M6 COT
M6 COZ
M6 CPD
M6 CPF
M6 CPG
M6 CPI
M6 CPL
M6 CPY
M6 CQA
M6 CQC
M6 CQG
M6 CQI
M6 CQK
M6 CQM
M6 CQQ
M6 CQT
M6 CQW
M6 CQZ
M6 CRB
M6 CRD
M6 CRK
M6 CRL
M6 CRM
M6 CRN
M6 CRS
M6 CRV
M6 CRY
M6 CSB
M6 CSH
M6 CSQ
M6 CST
M6 CSV
M6 CSY
M6 CTF
M6 CTM
M6 CTP
M6 CTR
M6 CTU
M6 CUB
M6 CUG
M6 CUH
M6 CUI
M6 CUK
M6 CUO
M6 CUT
M6 CUY
M6 CVA
M6 CVD
M6 CVG
M6 CVI
M6 CVL
M6 CVR
M6 CVS
M6 CVT
M6 CVX
M6 CVZ
M6 CWE
M6 CWM
M6 CWS
M6 CWT
M6 CWY
M6 CXE
M6 CXG
M6 CXK
M6 CXL
M6 CXR
M6 CXS
M6 CXT
M6 CXU
M6 CXX
M6 CXY
M6 CYH
M6 CYI
M6 CYJ
M6 CYK
M6 CYZ
M6 CZE
M6 CZG
M6 CZK
M6 CZM
M6 CZP
M6 CZQ
M6 CZT
M6 CZX
M6 CZY
M6 CZZ
M6 DAA
M6 DAH
M6 DAJ
M6 DAV
M6 DBC
M6 DBM
M6 DBO
M6 DBR
M6 DBY
M6 DBZ
M6 DCC
M6 DCF
M6 DCG
M6 DCV
M6 DCX
M6 DDA
M6 DDE
M6 DDJ
M6 DDK
M6 DEB
M6 DEN
M6 DEQ
M6 DET
M6 DEU
M6 DEW
M6 DEX
M6 DFB
M6 DFF
M6 DFN
M6 DFO
M6 DFR
M6 DFV
M6 DGB
M6 DGE
M6 DGF
M6 DGL
M6 DGT
M6 DGV
M6 DGX
M6 DHA
M6 DHC
M6 DHD
M6 DHE
M6 DHH
M6 DHJ
M6 DHL
M6 DHM
M6 DHN
M6 DHO
M6 DHP
M6 DHQ
M6 DHR
M6 DHY
M6 DIA
M6 DID
M6 DIN
M6 DIP
M6 DIR
M6 DIS
M6 DIW
M6 DIX
M6 DIY
M6 DJB
M6 DJD
M6 DJE
M6 DJF
M6 DJH
M6 DJM
M6 DJO
M6 DJP
M6 DJR
M6 DJU
M6 DJV
M6 DJY
M6 DKB
M6 DKD
M6 DKE
M6 DKG
M6 DKH
M6 DKJ
M6 DKP
M6 DKQ
M6 DKV
M6 DKZ
M6 DLE
M6 DLF
M6 DLG
M6 DLK
M6 DLN
M6 DLS
M6 DLZ
M6 DMI
M6 DMJ
M6 DML
M6 DMM
M6 DMR
M6 DMT
M6 DMU
M6 DMV
M6 DMX
M6 DMZ
M6 DNA
M6 DNE
M6 DNH
M6 DNK
M6 DNT
M6 DNU
M6 DNW
M6 DNY
M6 DOE
M6 DOH
M6 DOI
M6 DOK
M6 DOL
M6 DOO
M6 DOP
M6 DOQ
M6 DOS
M6 DOT
M6 DOU
M6 DOV
M6 DOY
M6 DPB
M6 DPD
M6 DPE
M6 DPK
M6 DPM
M6 DPS
M6 DPT
M6 DQA
M6 DQE
M6 DQF
M6 DQI
M6 DQM
M6 DQX
M6 DRA
M6 DRB
M6 DRC
M6 DRD
M6 DRE
M6 DRL
M6 DRN
M6 DRQ
M6 DRS
M6 DRT
M6 DRX
M6 DRY
M6 DSA
M6 DSB
M6 DSF
M6 DSG
M6 DSI
M6 DSK
M6 DSL
M6 DSM
M6 DSX
M6 DTB
M6 DTI
M6 DTK
M6 DTW
M6 DTZ
M6 DUA
M6 DUC
M6 DUQ
M6 DUS
M6 DUU
M6 DVB
M6 DVD
M6 DVG
M6 DVI
M6 DVJ
M6 DVK
M6 DVL
M6 DVX
M6 DWB
M6 DWD
M6 DWH
M6 DWL
M6 DWN
M6 DWQ
M6 DWR
M6 DWU
M6 DWX
M6 DXB
M6 DXD
M6 DXE
M6 DXM
M6 DXP
M6 DXR
M6 DXT
M6 DXX
M6 DXY
M6 DYE
M6 DYJ
M6 DYK
M6 DYQ
M6 DYS
M6 DYT
M6 DYZ
M6 DZD
M6 DZE
M6 DZG
M6 DZI
M6 DZJ
M6 DZK
M6 DZL
M6 DZM
M6 DZO
M6 DZU
M6 DZW
M6 EAM
M6 EAR
M6 EAU
M6 EAV
M6 EAY
M6 EBF
M6 EBH
M6 EBK
M6 EBM
M6 EBT
M6 EBV
M6 ECA
M6 ECF
M6 ECG
M6 ECI
M6 ECK
M6 ECL
M6 ECN
M6 ECQ
M6 ECS
M6 ECU
M6 ECY
M6 EDE
M6 EDF
M6 EDT
M6 EDU
M6 EDX
M6 EEE
M6 EEI
M6 EES
M6 EEV
M6 EEY
M6 EFA
M6 EFN
M6 EFO
M6 EFQ
M6 EFT
M6 EFZ
M6 EGI
M6 EGQ
M6 EGR
M6 EGY
M6 EHG
M6 EHJ
M6 EHN
M6 EHS
M6 EHT
M6 EIP
M6 EIQ
M6 EIT
M6 EIV
M6 EJD
M6 EJE
M6 EJH
M6 EJN
M6 EJQ
M6 EJZ
M6 EKG
M6 EKH
M6 EKJ
M6 EKR
M6 EKU
M6 EKZ
M6 ELA
M6 ELG
M6 ELK
M6 ELM
M6 ELO
M6 ELP
M6 ELT
M6 EMB
M6 EMD
M6 EMN
M6 EMR
M6 EMU
M6 ENG
M6 ENO
M6 ENT
M6 EOD
M6 EOF
M6 EOG
M6 EOH
M6 EOI
M6 EOJ
M6 EOL
M6 EOO
M6 EPB
M6 EPC
M6 EPG
M6 EPI
M6 EPK
M6 EPP
M6 EPR
M6 EPU
M6 EPY
M6 EQP
M6 EQU
M6 EQX
M6 ERB
M6 ERE
M6 ERG
M6 ERJ
M6 ERK
M6 ERM
M6 ERV
M6 ERX
M6 ERY
M6 ESB
M6 ESC
M6 ESD
M6 ESE
M6 ESF
M6 ESG
M6 ESI
M6 ESK
M6 ESM
M6 ESN
M6 ESP
M6 ESQ
M6 ESU
M6 ESX
M6 ETB
M6 ETF
M6 ETI
M6 ETK
M6 ETM
M6 ETO
M6 ETQ
M6 ETT
M6 ETV
M6 ETX
M6 EUC
M6 EUD
M6 EUF
M6 EUG
M6 EUJ
M6 EUM
M6 EUN
M6 EUQ
M6 EUR
M6 EUS
M6 EUT
M6 EVF
M6 EVG
M6 EVJ
M6 EVN
M6 EVP
M6 EVQ
M6 EVT
M6 EVV
M6 EWB
M6 EWE
M6 EWF
M6 EWG
M6 EWM
M6 EWS
M6 EXM
M6 EYC
M6 EYE
M6 EYN
M6 EZB
M6 EZD
M6 EZH
M6 EZI
M6 EZJ
M6 EZK
M6 EZL
M6 EZQ
M6 EZU
M6 EZW
M6 FAA
M6 FAD
M6 FAI
M6 FAL
M6 FAO
M6 FAP
M6 FAQ
M6 FAR
M6 FAT
M6 FAV
M6 FAZ
M6 FBC
M6 FBG
M6 FBI
M6 FBL
M6 FBX
M6 FCK
M6 FCL
M6 FCO
M6 FCP
M6 FCU
M6 FCY
M6 FDA
M6 FDB
M6 FDC
M6 FDE
M6 FDJ
M6 FDM
M6 FDO
M6 FDP
M6 FDQ
M6 FEG
M6 FES
M6 FET
M6 FEV
M6 FEW
M6 FEY
M6 FFA
M6 FFC
M6 FFG
M6 FFH
M6 FFL
M6 FFX
M6 FGB
M6 FGD
M6 FGF
M6 FGL
M6 FGO
M6 FGP
M6 FGT
M6 FGU
M6 FGV
M6 FGX
M6 FHC
M6 FHL
M6 FHT
M6 FHU
M6 FHW
M6 FHZ
M6 FIE
M6 FIF
M6 FIH
M6 FIM
M6 FIP
M6 FIQ
M6 FIS
M6 FJB
M6 FJG
M6 FJK
M6 FJP
M6 FJR
M6 FJY
M6 FKK
M6 FKL
M6 FKM
M6 FKT
M6 FKX
M6 FLD
M6 FLI
M6 FLM
M6 FLP
M6 FLT
M6 FLV
M6 FMA
M6 FMB
M6 FMD
M6 FMH
M6 FMK
M6 FMQ
M6 FMR
M6 FMX
M6 FNC
M6 FND
M6 FNH
M6 FNI
M6 FNJ
M6 FNN
M6 FNS
M6 FNW
M6 FOB
M6 FOH
M6 FOI
M6 FOM
M6 FON
M6 FOP
M6 FOQ
M6 FOS
M6 FOU
M6 FOV
M6 FOW
M6 FPA
M6 FPB
M6 FPR
M6 FPS
M6 FPT
M6 FPZ
M6 FQB
M6 FQI
M6 FQQ
M6 FQS
M6 FQT
M6 FQV
M6 FRA
M6 FRB
M6 FRH
M6 FRL
M6 FRM
M6 FRU
M6 FRX
M6 FSF
M6 FSH
M6 FSK
M6 FSL
M6 FSS
M6 FSX
M6 FSY
M6 FTB
M6 FTD
M6 FTE
M6 FTH
M6 FTJ
M6 FTL
M6 FTR
M6 FTS
M6 FUB
M6 FUG
M6 FUI
M6 FUP
M6 FUS
M6 FUV
M6 FUZ
M6 FVF
M6 FVG
M6 FVI
M6 FVQ
M6 FVU
M6 FVX
M6 FWA
M6 FWF
M6 FWH
M6 FWI
M6 FWJ
M6 FWK
M6 FWN
M6 FWR
M6 FWU
M6 FXC
M6 FXF
M6 FXH
M6 FXK
M6 FXN
M6 FXR
M6 FXS
M6 FXT
M6 FXV
M6 FYC
M6 FYI
M6 FYO
M6 FYS
M6 FYU
M6 FYX
M6 FYZ
M6 FZA
M6 FZH
M6 FZI
M6 FZO
M6 FZP
M6 FZU
M6 GAJ
M6 GAY
M6 GAZ
M6 GBT
M6 GCA
M6 GCF
M6 GCG
M6 GCH
M6 GCR
M6 GCU
M6 GCZ
M6 GDF
M6 GDG
M6 GDK
M6 GDM
M6 GDR
M6 GDU
M6 GDW
M6 GDX
M6 GEB
M6 GEC
M6 GED
M6 GEH
M6 GEI
M6 GEM
M6 GEN
M6 GEO
M6 GER
M6 GES
M6 GEZ
M6 GFB
M6 GFC
M6 GFD
M6 GFG
M6 GFI
M6 GFJ
M6 GFL
M6 GFT
M6 GFU
M6 GFX
M6 GGA
M6 GGB
M6 GGC
M6 GGK
M6 GGP
M6 GGR
M6 GGU
M6 GGY
M6 GHE
M6 GHG
M6 GHI
M6 GHK
M6 GHO
M6 GHS
M6 GHU
M6 GHZ
M6 GIE
M6 GIG
M6 GIJ
M6 GIK
M6 GIO
M6 GIP
M6 GIR
M6 GJF
M6 GJK
M6 GJR
M6 GJU
M6 GJW
M6 GKA
M6 GKD
M6 GKJ
M6 GKL
M6 GKO
M6 GKP
M6 GKR
M6 GKY
M6 GLB
M6 GLE
M6 GLF
M6 GLG
M6 GLL
M6 GLQ
M6 GLR
M6 GLT
M6 GLY
M6 GMB
M6 GME
M6 GMG
M6 GML
M6 GMN
M6 GMW
M6 GMX
M6 GMZ
M6 GNB
M6 GNE
M6 GNI
M6 GNK
M6 GNL
M6 GNQ
M6 GNT
M6 GOH
M6 GOJ
M6 GOK
M6 GOP
M6 GOT
M6 GOV
M6 GPH
M6 GPN
M6 GPW
M6 GPY
M6 GQE
M6 GQL
M6 GQM
M6 GQN
M6 GQO
M6 GQQ
M6 GRE
M6 GRF
M6 GRG
M6 GRI
M6 GRL
M6 GRM
M6 GRO
M6 GRS
M6 GRW
M6 GSA
M6 GSB
M6 GSE
M6 GSI
M6 GSM
M6 GSN
M6 GSU
M6 GSV
M6 GTA
M6 GTG
M6 GTJ
M6 GTL
M6 GTN
M6 GTP
M6 GTV
M6 GTX
M6 GTZ
M6 GUD
M6 GUG
M6 GUI
M6 GUK
M6 GUL
M6 GUN
M6 GUP
M6 GUQ
M6 GUR
M6 GUZ
M6 GVB
M6 GVE
M6 GVG
M6 GVH
M6 GVK
M6 GVO
M6 GVQ
M6 GVX
M6 GVZ
M6 GWC
M6 GWR
M6 GWV
M6 GWY
M6 GXB
M6 GXE
M6 GXG
M6 GXM
M6 GXN
M6 GXQ
M6 GXT
M6 GXX
M6 GXY
M6 GYD
M6 GYJ
M6 GYN
M6 GYO
M6 GYT
M6 GYW
M6 GYX
M6 GZC
M6 GZH
M6 GZM
M6 GZP
M6 GZQ
M6 GZV
M6 HAA
M6 HAB
M6 HAI
M6 HAJ
M6 HAL
M6 HAN
M6 HAO
M6 HAP
M6 HAQ
M6 HAW
M6 HAX
M6 HAZ
M6 HBA
M6 HBL
M6 HBO
M6 HBW
M6 HBX
M6 HBZ
M6 HCC
M6 HCF
M6 HCH
M6 HCN
M6 HCX
M6 HDF
M6 HDJ
M6 HDL
M6 HDQ
M6 HDR
M6 HEC
M6 HEL
M6 HEM
M6 HES
M6 HEU
M6 HEV
M6 HFD
M6 HFI
M6 HFL
M6 HFS
M6 HFW
M6 HGB
M6 HGC
M6 HGK
M6 HGL
M6 HGP
M6 HGT
M6 HHD
M6 HHI
M6 HHK
M6 HHS
M6 HHU
M6 HHV
M6 HHZ
M6 HIC
M6 HIE
M6 HIF
M6 HIH
M6 HIN
M6 HIO
M6 HIQ
M6 HIR
M6 HIS
M6 HIW
M6 HIX
M6 HIY
M6 HJJ
M6 HJL
M6 HJN
M6 HJO
M6 HJP
M6 HJW
M6 HJY
M6 HKD
M6 HKE
M6 HKF
M6 HKG
M6 HKH
M6 HKK
M6 HKM
M6 HKR
M6 HLB
M6 HLD
M6 HLH
M6 HLI
M6 HLJ
M6 HLK
M6 HLM
M6 HLN
M6 HLO
M6 HLV
M6 HLX
M6 HMA
M6 HMF
M6 HMH
M6 HMI
M6 HMJ
M6 HMN
M6 HMO
M6 HMT
M6 HMX
M6 HMY
M6 HNE
M6 HNK
M6 HNL
M6 HNM
M6 HNP
M6 HNR
M6 HNU
M6 HNY
M6 HOB
M6 HOC
M6 HOD
M6 HOE
M6 HOJ
M6 HOL
M6 HOW
M6 HOX
M6 HPA
M6 HPB
M6 HPE
M6 HPI
M6 HPJ
M6 HPM
M6 HPO
M6 HPP
M6 HPQ
M6 HPU
M6 HPW
M6 HPY
M6 HQA
M6 HQF
M6 HQG
M6 HQH
M6 HQJ
M6 HQP
M6 HQQ
M6 HQS
M6 HQT
M6 HQU
M6 HQV
M6 HQY
M6 HRA
M6 HRB
M6 HRH
M6 HRP
M6 HRU
M6 HRY
M6 HSD
M6 HSG
M6 HSJ
M6 HSN
M6 HSO
M6 HSU
M6 HSW
M6 HSZ
M6 HTA
M6 HTD
M6 HTE
M6 HTL
M6 HTO
M6 HTP
M6 HTR
M6 HTV
M6 HTY
M6 HUA
M6 HUB
M6 HUD
M6 HUG
M6 HUJ
M6 HUO
M6 HUT
M6 HVB
M6 HVC
M6 HVG
M6 HVJ
M6 HVN
M6 HVQ
M6 HVT
M6 HVV
M6 HVW
M6 HVX
M6 HVZ
M6 HWA
M6 HWB
M6 HWF
M6 HWG
M6 HWK
M6 HWP
M6 HWS
M6 HWU
M6 HWY
M6 HWZ
M6 HXH
M6 HXJ
M6 HXL
M6 HXM
M6 HXP
M6 HXQ
M6 HXS
M6 HXZ
M6 HYA
M6 HYC
M6 HYD
M6 HYE
M6 HYG
M6 HYI
M6 HYL
M6 HYN
M6 HYO
M6 HYP
M6 HYU
M6 HYV
M6 HZB
M6 HZC
M6 HZG
M6 HZH
M6 HZK
M6 HZS
M6 HZV
M6 HZY
M6 IAD
M6 IAE
M6 IAM
M6 IAP
M6 IAU
M6 IBA
M6 IBE
M6 IBK
M6 IBM
M6 IBS
M6 IBT
M6 IBU
M6 IBV
M6 IBW
M6 IBX
M6 IBY
M6 IBZ
M6 ICC
M6 ICG
M6 ICI
M6 ICN
M6 ICS
M6 ICT
M6 ICV
M6 ICW
M6 ICY
M6 IDA
M6 IDH
M6 IDL
M6 IDT
M6 IDW
M6 IDY
M6 IED
M6 IEE
M6 IEG
M6 IEK
M6 IEL
M6 IEN
M6 IEP
M6 IES
M6 IET
M6 IEX
M6 IEY
M6 IFD
M6 IFJ
M6 IFK
M6 IFL
M6 IFM
M6 IFP
M6 IFR
M6 IFU
M6 IGB
M6 IGD
M6 IGE
M6 IGF
M6 IGG
M6 IGI
M6 IGL
M6 IGN
M6 IGO
M6 IGU
M6 IGX
M6 IGY
M6 IHB
M6 IHF
M6 IHG
M6 IHI
M6 IHJ
M6 IHK
M6 IHU
M6 IHV
M6 IHW
M6 IHX
M6 IHY
M6 IIA
M6 IIC
M6 IID
M6 IIE
M6 IIH
M6 III
M6 IIK
M6 IIM
M6 IIN
M6 IIS
M6 IIT
M6 IIV
M6 IJA
M6 IJC
M6 IJE
M6 IJF
M6 IJI
M6 IJS
M6 IJU
M6 IJY
M6 IKB
M6 IKC
M6 IKE
M6 IKG
M6 IKK
M6 IKO
M6 IKP
M6 IKQ
M6 IKU
M6 IKV
M6 IKW
M6 IKZ
M6 ILA
M6 ILD
M6 ILE
M6 ILG
M6 ILL
M6 ILU
M6 ILV
M6 ILW
M6 IMD
M6 IME
M6 IMG
M6 IMK
M6 IMN
M6 IMO
M6 IMT
M6 IMX
M6 INE
M6 INF
M6 INH
M6 INJ
M6 INK
M6 INL
M6 INO
M6 INQ
M6 INR
M6 INY
M6 INZ
M6 IOB
M6 IOC
M6 IOE
M6 IOJ
M6 IOK
M6 IOM
M6 IOP
M6 IOT
M6 IOU
M6 IOX
M6 IPC
M6 IPD
M6 IPN
M6 IPR
M6 IPS
M6 IPT
M6 IPU
M6 IPV
M6 IPX
M6 IQB
M6 IQD
M6 IQI
M6 IQJ
M6 IQP
M6 IQQ
M6 IQR
M6 IQS
M6 IQV
M6 IRD
M6 IRF
M6 IRG
M6 IRH
M6 IRN
M6 IRO
M6 IRQ
M6 IRR
M6 IRS
M6 IRT
M6 IRV
M6 IRY
M6 ISC
M6 ISL
M6 ISO
M6 ISP
M6 ISS
M6 IST
M6 ISV
M6 ITA
M6 ITB
M6 ITC
M6 ITE
M6 ITF
M6 ITG
M6 ITH
M6 ITJ
M6 ITK
M6 ITO
M6 ITS
M6 ITT
M6 IUA
M6 IUB
M6 IUC
M6 IUG
M6 IUJ
M6 IUL
M6 IUO
M6 IUQ
M6 IUS
M6 IUT
M6 IUU
M6 IUV
M6 IUW
M6 IUX
M6 IUY
M6 IUZ
M6 IVA
M6 IVD
M6 IVF
M6 IVH
M6 IVM
M6 IVS
M6 IVU
M6 IVV
M6 IVX
M6 IVY
M6 IVZ
M6 IWC
M6 IWG
M6 IWH
M6 IWJ
M6 IWK
M6 IWN
M6 IWR
M6 IWS
M6 IWT
M6 IWU
M6 IWV
M6 IWW
M6 IWY
M6 IWZ
M6 IXA
M6 IXB
M6 IXC
M6 IXE
M6 IXK
M6 IXL
M6 IXN
M6 IXQ
M6 IXS
M6 IXT
M6 IXU
M6 IXV
M6 IXW
M6 IXX
M6 IYB
M6 IYD
M6 IYE
M6 IYF
M6 IYI
M6 IYK
M6 IYO
M6 IYP
M6 IYQ
M6 IYR
M6 IYX
M6 IZC
M6 IZD
M6 IZG
M6 IZJ
M6 IZL
M6 IZR
M6 IZS
M6 IZV

M6 IZX
M6 IZZ
M6 JAA
M6 JAC
M6 JAH
M6 JAI
M6 JAO
M6 JAT
M6 JAX
M6 JAY
M6 JBD
M6 JBF
M6 JBJ
M6 JBK
M6 JBP
M6 JBS
M6 JBW
M6 JBY
M6 JCI
M6 JCK
M6 JCO
M6 JCY
M6 JDB
M6 JDE
M6 JDG
M6 JDH
M6 JDI
M6 JDJ
M6 JDO
M6 JDP
M6 JDR
M6 JDS
M6 JDT
M6 JDW
M6 JDZ
M6 JEE
M6 JEI
M6 JEN
M6 JER
M6 JES
M6 JEU
M6 JEX
M6 JFC
M6 JFD
M6 JFM
M6 JFQ
M6 JFT
M6 JFX
M6 JFZ
M6 JGA
M6 JGB
M6 JGD
M6 JGF
M6 JGJ
M6 JGL
M6 JGO
M6 JGW
M6 JGX
M6 JHA
M6 JHM
M6 JHW
M6 JHX
M6 JHY
M6 JIA
M6 JIB
M6 JIJ
M6 JIL
M6 JIM
M6 JIO
M6 JIP
M6 JIU
M6 JIV
M6 JJA
M6 JJC
M6 JJP
M6 JJR
M6 JJT
M6 JJY
M6 JKC
M6 JKE
M6 JKF
M6 JKG
M6 JKI
M6 JKJ
M6 JKK
M6 JKL
M6 JKN
M6 JKO
M6 JKP
M6 JKU
M6 JKZ
M6 JLC
M6 JLF
M6 JLG
M6 JLI
M6 JLK
M6 JLN
M6 JLO
M6 JLP
M6 JLS
M6 JLV
M6 JMA
M6 JME
M6 JMK
M6 JMQ
M6 JMS
M6 JMT
M6 JMU
M6 JMV
M6 JMY
M6 JMZ
M6 JND
M6 JNG
M6 JNK
M6 JNT
M6 JNX
M6 JNY
M6 JOB
M6 JOC
M6 JOF
M6 JOJ
M6 JOL
M6 JOM
M6 JOQ
M6 JOT
M6 JOZ
M6 JPA
M6 JPE
M6 JPG
M6 JPN
M6 JPQ
M6 JPY
M6 JQB
M6 JQD
M6 JQH
M6 JQI
M6 JQJ
M6 JQM
M6 JQN
M6 JQP
M6 JQR
M6 JQT
M6 JQX
M6 JQY
M6 JRA
M6 JRD
M6 JRG
M6 JRM
M6 JRP
M6 JRQ
M6 JRR
M6 JRU
M6 JRV
M6 JRX
M6 JRY
M6 JRZ
M6 JSA
M6 JSB
M6 JSC
M6 JSD
M6 JSE
M6 JSF
M6 JSG
M6 JSK
M6 JSL
M6 JSO
M6 JSV
M6 JSY
M6 JSZ
M6 JTB
M6 JTF
M6 JTJ
M6 JTR
M6 JTS
M6 JTT
M6 JTX
M6 JTY
M6 JTZ
M6 JUA
M6 JUD
M6 JUL
M6 JUM
M6 JUN
M6 JUO
M6 JUR
M6 JUS
M6 JUY
M6 JVA
M6 JVE
M6 JVH
M6 JVI
M6 JVJ
M6 JVL
M6 JVN
M6 JVP
M6 JVR
M6 JVT
M6 JWA
M6 JWB
M6 JWJ
M6 JWK
M6 JWM
M6 JWN
M6 JWR
M6 JWS
M6 JWV
M6 JWZ
M6 JXA
M6 JXB
M6 JXC
M6 JXG
M6 JXL
M6 JXM
M6 JXP
M6 JXQ
M6 JXR
M6 JXS
M6 JXT
M6 JXU
M6 JXV
M6 JXW
M6 JXZ
M6 JYA
M6 JYC
M6 JYG
M6 JYH
M6 JYK
M6 JYM
M6 JYN
M6 JYP
M6 JYU
M6 JYW
M6 JZA
M6 JZD
M6 JZE
M6 JZG
M6 JZJ
M6 JZO
M6 JZP
M6 JZQ
M6 JZR
M6 JZS
M6 JZX
M6 KAD
M6 KAI
M6 KAJ
M6 KAK
M6 KAL
M6 KAP
M6 KAW
M6 KAY
M6 KBL
M6 KBM
M6 KBP
M6 KBQ
M6 KBR
M6 KBU
M6 KBZ
M6 KCH
M6 KCN
M6 KCW
M6 KCX
M6 KDE
M6 KDH
M6 KDI
M6 KDL
M6 KDS
M6 KDT
M6 KDU
M6 KDY
M6 KEB
M6 KEI
M6 KEJ
M6 KEK
M6 KEN
M6 KEO
M6 KEW
M6 KEX
M6 KEY
M6 KEZ
M6 KFA
M6 KFD
M6 KFI
M6 KFL
M6 KFN
M6 KFR
M6 KFT
M6 KFU
M6 KFV
M6 KFY
M6 KFZ
M6 KGG
M6 KGH
M6 KGJ
M6 KGP
M6 KGT
M6 KGW
M6 KGY
M6 KGZ
M6 KHC
M6 KHD
M6 KHH
M6 KHI
M6 KHJ
M6 KHK
M6 KHL
M6 KHN
M6 KHO
M6 KHR
M6 KHT
M6 KHU
M6 KHV
M6 KHW
M6 KHY
M6 KHZ
M6 KIB
M6 KIC
M6 KID
M6 KIH
M6 KIJ
M6 KIM
M6 KIO
M6 KIQ
M6 KIS
M6 KIY
M6 KIZ
M6 KJC
M6 KJE
M6 KJH
M6 KJN
M6 KJO
M6 KJS
M6 KJT
M6 KJU
M6 KJV
M6 KKC
M6 KKE
M6 KKH
M6 KKP
M6 KKQ
M6 KKR
M6 KKS
M6 KKT
M6 KKV
M6 KKX
M6 KKZ
M6 KLB
M6 KLG
M6 KLJ
M6 KLO
M6 KLQ
M6 KLS
M6 KLY
M6 KMB
M6 KME
M6 KMH
M6 KMM
M6 KMO
M6 KMT
M6 KNA
M6 KNE
M6 KNH
M6 KNK
M6 KNN
M6 KNQ
M6 KNX
M6 KOF
M6 KOG
M6 KOL
M6 KOO
M6 KOP
M6 KOQ
M6 KOR
M6 KOT
M6 KOU
M6 KOW
M6 KOX
M6 KOY
M6 KPB
M6 KPH
M6 KPS
M6 KPW
M6 KPY
M6 KQB
M6 KQE
M6 KQG
M6 KQI
M6 KQN
M6 KQO
M6 KQQ
M6 KQT
M6 KQU
M6 KQV
M6 KQX
M6 KQY
M6 KRA
M6 KRB
M6 KRC
M6 KRG
M6 KRJ
M6 KRN
M6 KRO
M6 KRS
M6 KRT
M6 KRU
M6 KRY
M6 KSB
M6 KSK
M6 KSN
M6 KSO
M6 KSQ
M6 KSR
M6 KSY
M6 KTB
M6 KTD
M6 KTE
M6 KTF
M6 KTG
M6 KTL
M6 KTP
M6 KTR
M6 KTU
M6 KUA
M6 KUB
M6 KUF
M6 KUL
M6 KUM
M6 KUN
M6 KUO
M6 KUS
M6 KUW
M6 KUX
M6 KUZ
M6 KVC
M6 KVJ
M6 KVR
M6 KVY
M6 KVZ
M6 KWB
M6 KWD
M6 KWE
M6 KWF
M6 KWJ
M6 KWO
M6 KWR
M6 KWS
M6 KWY
M6 KXB
M6 KXC
M6 KXD
M6 KXG
M6 KXJ
M6 KXK
M6 KXL
M6 KXT
M6 KXY
M6 KXZ
M6 KYA
M6 KYD
M6 KYF
M6 KYG
M6 KYH
M6 KYJ
M6 KYM
M6 KYQ
M6 KYR
M6 KYT
M6 KYU
M6 KYW
M6 KYY
M6 KZA
M6 KZD
M6 KZG
M6 KZJ
M6 KZO
M6 KZP
M6 KZT
M6 KZV
M6 KZY
M6 LAJ
M6 LAN
M6 LAT
M6 LAU
M6 LAW
M6 LAX
M6 LBA
M6 LBB
M6 LBF
M6 LBG
M6 LBH
M6 LBJ
M6 LBO
M6 LBZ
M6 LCA
M6 LCB
M6 LCE
M6 LCJ
M6 LCO
M6 LCS
M6 LDC
M6 LDN
M6 LDO
M6 LDP
M6 LDT
M6 LDX
M6 LED
M6 LEI
M6 LEL
M6 LEM
M6 LEN
M6 LEO
M6 LER
M6 LES
M6 LEZ
M6 LFA
M6 LFF
M6 LFP
M6 LFV
M6 LGE
M6 LGG
M6 LGK
M6 LGN
M6 LGO
M6 LGP
M6 LGU
M6 LGW
M6 LGY
M6 LGZ
M6 LHC
M6 LHK
M6 LHT
M6 LHU
M6 LHX
M6 LHY
M6 LHZ
M6 LIA
M6 LIF
M6 LIG
M6 LIR
M6 LIX
M6 LIY
M6 LJC
M6 LJH
M6 LJI
M6 LJL
M6 LJN
M6 LJQ
M6 LJW
M6 LJX
M6 LJY
M6 LKB
M6 LKC
M6 LKG
M6 LKH
M6 LKJ
M6 LKN
M6 LKO
M6 LKP
M6 LKQ
M6 LKR
M6 LKT
M6 LKU
M6 LKX
M6 LKZ
M6 LLA
M6 LLF
M6 LLJ
M6 LLP
M6 LLQ
M6 LLR
M6 LLV
M6 LLX
M6 LMA
M6 LMD
M6 LME
M6 LMF
M6 LML
M6 LMM
M6 LMN
M6 LMO
M6 LMP
M6 LMR
M6 LMS
M6 LMT
M6 LMV
M6 LMX
M6 LMY
M6 LNA
M6 LNF
M6 LNH
M6 LNJ
M6 LNK
M6 LNL
M6 LNN
M6 LNO
M6 LNR
M6 LNT
M6 LNW
M6 LNY
M6 LNZ
M6 LOA
M6 LOH
M6 LOI
M6 LOJ
M6 LOM
M6 LOP
M6 LOR
M6 LOU
M6 LOV
M6 LOX
M6 LOY
M6 LPA
M6 LPE
M6 LPJ
M6 LPL
M6 LPM
M6 LPQ
M6 LPU
M6 LQD
M6 LQE
M6 LQH
M6 LQJ
M6 LQL
M6 LQQ
M6 LQS
M6 LQV
M6 LQZ
M6 LRC
M6 LRD
M6 LRE
M6 LRJ
M6 LRM
M6 LRP
M6 LRS
M6 LRT
M6 LRY
M6 LSC
M6 LSD
M6 LSF
M6 LSI
M6 LSM
M6 LSQ
M6 LSR
M6 LSS
M6 LSX
M6 LSZ
M6 LTA
M6 LTE
M6 LTF
M6 LTG
M6 LTH
M6 LTI
M6 LTP
M6 LTT
M6 LTW
M6 LTX
M6 LTY
M6 LTZ
M6 LUB
M6 LUE
M6 LUF
M6 LUL
M6 LUN
M6 LUO
M6 LUS
M6 LUU
M6 LUV
M6 LUX
M6 LVB
M6 LVF
M6 LVG
M6 LVI
M6 LVL
M6 LVO
M6 LVP
M6 LVQ
M6 LVT
M6 LVU
M6 LVY
M6 LVZ
M6 LWC
M6 LWD
M6 LWH
M6 LWK
M6 LWL
M6 LWN
M6 LWO
M6 LWU
M6 LWW
M6 LWX
M6 LXA
M6 LXB
M6 LXF
M6 LXG
M6 LXJ
M6 LXL
M6 LXO
M6 LXT
M6 LXV
M6 LXZ
M6 LYE
M6 LYI
M6 LYK
M6 LYL
M6 LYM
M6 LYR
M6 LYT
M6 LYV
M6 LYW
M6 LYX
M6 LZB
M6 LZE
M6 LZH
M6 LZJ
M6 LZO
M6 LZR
M6 MAC
M6 MAE
M6 MAI
M6 MAN
M6 MAQ
M6 MAR
M6 MAU
M6 MAV
M6 MAZ
M6 MBA
M6 MBJ
M6 MBL
M6 MBM
M6 MBT
M6 MBW
M6 MCD
M6 MCG
M6 MCI
M6 MCN
M6 MCV
M6 MDA
M6 MDD
M6 MDE
M6 MDF
M6 MDI
M6 MDK
M6 MDP
M6 MDQ
M6 MDS
M6 MDV
M6 MDW
M6 MDY
M6 MEE
M6 MEF
M6 MEH
M6 MEI
M6 MEM
M6 MER
M6 MET
M6 MEW
M6 MEX
M6 MEY
M6 MEZ
M6 MFH
M6 MFP
M6 MFQ
M6 MFT
M6 MFV
M6 MFW
M6 MFX
M6 MFY
M6 MGB
M6 MGE
M6 MGI
M6 MGK
M6 MGL
M6 MGQ
M6 MGS
M6 MHB
M6 MHC
M6 MHG
M6 MHP
M6 MHR
M6 MHS
M6 MHX
M6 MHZ
M6 MIJ
M6 MIW
M6 MIX
M6 MJB
M6 MJE
M6 MJH
M6 MJJ
M6 MJK
M6 MJQ
M6 MJU
M6 MJW
M6 MJX
M6 MKA
M6 MKC
M6 MKF
M6 MKG
M6 MKI
M6 MKL
M6 MKP
M6 MKQ
M6 MKR
M6 MKS
M6 MKT
M6 MKZ
M6 MLB
M6 MLC
M6 MLD
M6 MLJ
M6 MLS
M6 MLW
M6 MLZ
M6 MMA
M6 MMD
M6 MME
M6 MMF
M6 MMJ
M6 MMK
M6 MMO
M6 MMV
M6 MMW
M6 MMZ
M6 MNA
M6 MNB
M6 MNF
M6 MNJ
M6 MNM
M6 MNN
M6 MNY
M6 MOA
M6 MOE
M6 MOL
M6 MOR
M6 MOT
M6 MPA
M6 MPC
M6 MPG
M6 MPJ
M6 MPN
M6 MPU
M6 MQA
M6 MQG
M6 MQI
M6 MQQ
M6 MQZ
M6 MRA
M6 MRB
M6 MRE
M6 MRF
M6 MRL
M6 MRO
M6 MRQ
M6 MRZ
M6 MSK
M6 MSL
M6 MSQ
M6 MSW
M6 MSZ
M6 MTB
M6 MTF
M6 MTP
M6 MTX
M6 MTY
M6 MUC
M6 MUE
M6 MUG
M6 MUI
M6 MUM
M6 MUN
M6 MUQ
M6 MUV
M6 MUW
M6 MUX
M6 MVC
M6 MVG
M6 MVJ
M6 MVO
M6 MVP
M6 MVR
M6 MVS
M6 MVU
M6 MWH
M6 MWK
M6 MWL
M6 MWM
M6 MWR
M6 MWU
M6 MWX
M6 MWY
M6 MXG
M6 MXI
M6 MXJ
M6 MXL
M6 MXN
M6 MXQ
M6 MXS
M6 MXT
M6 MXW
M6 MYA
M6 MYJ
M6 MYP
M6 MYQ
M6 MYX
M6 MYZ
M6 MZA
M6 MZF
M6 MZI
M6 MZK
M6 MZM
M6 MZO
M6 MZP
M6 MZS
M6 MZT
M6 MZW
M6 NAE
M6 NAF
M6 NAH
M6 NAR
M6 NBA
M6 NBB
M6 NBC
M6 NBD
M6 NBE
M6 NBF
M6 NBJ
M6 NBK
M6 NBT
M6 NBZ
M6 NCJ
M6 NCN
M6 NCQ
M6 NCT
M6 NCW
M6 NCX
M6 NDA
M6 NDD
M6 NDH
M6 NDJ
M6 NDL
M6 NDQ
M6 NDS
M6 NDU
M6 NDV
M6 NDW
M6 NDX
M6 NEB
M6 NEJ
M6 NEK
M6 NEN
M6 NEP
M6 NER
M6 NEX
M6 NFA
M6 NFF
M6 NFO
M6 NFR
M6 NFU
M6 NFV
M6 NGB
M6 NGC
M6 NGG
M6 NGH
M6 NGJ
M6 NGL
M6 NGO
M6 NGQ
M6 NGS
M6 NGT
M6 NGV
M6 NGZ
M6 NHB
M6 NHE
M6 NHI
M6 NHQ
M6 NHZ
M6 NIF
M6 NIH
M6 NII
M6 NIO
M6 NIR
M6 NIW
M6 NIY
M6 NIZ
M6 NJA
M6 NJC
M6 NJF
M6 NJJ
M6 NJL
M6 NJQ
M6 NJR
M6 NJU
M6 NJW
M6 NJY
M6 NKD
M6 NKF
M6 NKL
M6 NKN
M6 NKO
M6 NKS
M6 NKT
M6 NKU
M6 NKV
M6 NKW
M6 NKX
M6 NKZ
M6 NLC
M6 NLD
M6 NLE
M6 NLH
M6 NLI
M6 NLM
M6 NLN
M6 NLT
M6 NLU
M6 NLY
M6 NMA
M6 NMB
M6 NMC
M6 NMF
M6 NMI
M6 NMJ
M6 NMK
M6 NMM
M6 NMO
M6 NMV
M6 NMX
M6 NMY
M6 NNG
M6 NNH
M6 NNI
M6 NNN
M6 NNQ
M6 NNR
M6 NNZ
M6 NOG
M6 NOO
M6 NOP
M6 NOU
M6 NOV
M6 NOW
M6 NOZ
M6 NPA
M6 NPB
M6 NPC
M6 NPG
M6 NPH
M6 NPI
M6 NPJ
M6 NPK
M6 NPM
M6 NPP
M6 NPQ
M6 NPR
M6 NPT
M6 NPU
M6 NPY
M6 NQE
M6 NQF
M6 NQG
M6 NQH
M6 NQO
M6 NQS
M6 NQX
M6 NQY
M6 NQZ
M6 NRD
M6 NRL
M6 NRY
M6 NRZ
M6 NSA
M6 NSB
M6 NSF
M6 NSL
M6 NSN
M6 NSP
M6 NSU
M6 NSV
M6 NSX
M6 NTA
M6 NTD
M6 NTE
M6 NTF
M6 NTG
M6 NTH
M6 NTI
M6 NTO
M6 NTQ
M6 NTS
M6 NTU
M6 NTX
M6 NTZ
M6 NUA
M6 NUB
M6 NUE
M6 NUH
M6 NUI
M6 NUK
M6 NUN
M6 NUS
M6 NUT
M6 NUU
M6 NUZ
M6 NVC
M6 NVD
M6 NVE
M6 NVF
M6 NVH
M6 NVI
M6 NVK
M6 NVL
M6 NVN
M6 NVP
M6 NVQ
M6 NVU
M6 NVV
M6 NVX
M6 NWD
M6 NWE
M6 NWG
M6 NWJ
M6 NWR
M6 NWU
M6 NXB
M6 NXC
M6 NXD
M6 NXG
M6 NXI
M6 NXJ
M6 NXK
M6 NXM
M6 NXN
M6 NXU
M6 NXW
M6 NXX
M6 NYC
M6 NYH
M6 NYI
M6 NYK
M6 NYN
M6 NYQ
M6 NYS
M6 NYU
M6 NYV
M6 NZA
M6 NZC
M6 NZD
M6 NZG
M6 NZP
M6 NZU
M6 NZW
M6 NZX
M6 NZZ
M6 OAA
M6 OAC
M6 OAK
M6 OAM
M6 OAY
M6 OBA
M6 OBD
M6 OBE
M6 OBF
M6 OBG
M6 OBI
M6 OBL
M6 OBM
M6 OBN
M6 OBT
M6 OBV
M6 OBW
M6 OCA
M6 OCE
M6 OCF
M6 OCG
M6 OCL
M6 OCN
M6 OCO
M6 OCQ
M6 ODI
M6 ODJ
M6 ODQ
M6 ODT
M6 ODU
M6 ODV
M6 ODW
M6 ODX
M6 ODZ
M6 OED
M6 OEE
M6 OEL
M6 OEO
M6 OEQ
M6 OEX
M6 OEY
M6 OFB
M6 OFC
M6 OFI
M6 OFO
M6 OFP
M6 OFR
M6 OFS
M6 OFT
M6 OFU
M6 OFV
M6 OFW
M6 OFX
M6 OFY
M6 OGB
M6 OGD
M6 OGG
M6 OGI
M6 OGL
M6 OGM
M6 OGN
M6 OGR
M6 OGV
M6 OGW
M6 OGY
M6 OGZ
M6 OHA
M6 OHB
M6 OHC
M6 OHD
M6 OHG
M6 OHH
M6 OHI
M6 OHJ
M6 OHK
M6 OHM
M6 OHR
M6 OHV
M6 OHW
M6 OHX
M6 OIA
M6 OIB
M6 OID
M6 OIE
M6 OIF
M6 OIG
M6 OII
M6 OIJ
M6 OIK
M6 OIL
M6 OIN
M6 OIO
M6 OIP
M6 OIQ
M6 OIS
M6 OIX
M6 OJA
M6 OJG
M6 OJH
M6 OJJ
M6 OJR
M6 OJV
M6 OJW
M6 OJX
M6 OJY
M6 OKA
M6 OKE
M6 OKF
M6 OKG
M6 OKN
M6 OKT
M6 OKU
M6 OKV
M6 OKW
M6 OKX
M6 OLA
M6 OLC
M6 OLK
M6 OLM
M6 OLN
M6 OLO
M6 OLV
M6 OLZ
M6 OMB
M6 OMC
M6 OMD
M6 OMI
M6 OMM
M6 OMN
M6 OMO
M6 OMQ
M6 OMU
M6 OMV
M6 OMW
M6 OMY
M6 ONA
M6 ONB
M6 ONC
M6 ONE
M6 ONF
M6 ONG
M6 ONI
M6 ONK
M6 ONQ
M6 ONR
M6 ONV
M6 ONX
M6 ONY
M6 OOE
M6 OOF
M6 OOH
M6 OOI
M6 OON
M6 OOR
M6 OOS
M6 OOT
M6 OOU
M6 OPH
M6 OPI
M6 OPK
M6 OPN
M6 OPP
M6 OPQ
M6 OPR
M6 OPU
M6 OPV
M6 OPW
M6 OPX
M6 OPY
M6 OPZ
M6 OQA
M6 OQC
M6 OQE
M6 OQF
M6 OQI
M6 OQJ
M6 OQK
M6 OQM
M6 OQO
M6 OQP
M6 OQS
M6 OQU
M6 OQV
M6 OQW
M6 OQY
M6 ORA
M6 ORD
M6 ORL
M6 ORP
M6 ORQ
M6 ORR
M6 ORS
M6 ORV
M6 ORY
M6 ORZ
M6 OSA
M6 OSB
M6 OSD
M6 OSE
M6 OSG
M6 OSH
M6 OSK
M6 OSQ
M6 OSR
M6 OSS
M6 OSV
M6 OSZ
M6 OTA
M6 OTC
M6 OTD
M6 OTE
M6 OTF
M6 OTO
M6 OTQ
M6 OTV
M6 OTZ
M6 OUB
M6 OUC
M6 OUE
M6 OUG
M6 OUH
M6 OUJ
M6 OUK
M6 OUM
M6 OUO
M6 OUQ
M6 OUT
M6 OUV
M6 OUX
M6 OUZ
M6 OVA
M6 OVC
M6 OVD
M6 OVG
M6 OVJ
M6 OVK
M6 OVL
M6 OVM
M6 OVO
M6 OVP
M6 OVQ
M6 OVU
M6 OVZ
M6 OWB
M6 OWE
M6 OWF
M6 OWH
M6 OWK
M6 OWN
M6 OWQ
M6 OWR
M6 OWV
M6 OWW
M6 OWX
M6 OWY
M6 OWZ
M6 OXG
M6 OXH
M6 OXI
M6 OXK
M6 OXL
M6 OXN
M6 OXO
M6 OXP
M6 OXQ
M6 OYG
M6 OYI
M6 OYK
M6 OYQ
M6 OYR
M6 OYS
M6 OYT
M6 OYV
M6 OYW
M6 OZA
M6 OZC
M6 OZD
M6 OZE
M6 OZH
M6 OZJ
M6 OZK
M6 OZP
M6 OZU
M6 OZV
M6 PAB
M6 PAH
M6 PAK
M6 PAL
M6 PAQ
M6 PAR
M6 PBA
M6 PBB
M6 PBC
M6 PBG
M6 PBH
M6 PBQ
M6 PBS
M6 PBX
M6 PCA
M6 PCG
M6 PCH
M6 PCI
M6 PCK
M6 PCO
M6 PCS
M6 PCY
M6 PDF
M6 PDH
M6 PEE
M6 PEH
M6 PEK
M6 PEV
M6 PFC
M6 PFE
M6 PFF
M6 PFS
M6 PFX
M6 PGE
M6 PGG
M6 PGJ
M6 PGK
M6 PGM
M6 PGR
M6 PGS
M6 PGV
M6 PHB
M6 PHD
M6 PHI
M6 PHN
M6 PHR
M6 PHY
M6 PIE
M6 PIF
M6 PIO
M6 PIQ
M6 PIV
M6 PIW
M6 PIX
M6 PIZ
M6 PJC
M6 PJF
M6 PJG
M6 PJL
M6 PJO
M6 PJQ
M6 PJU
M6 PJW
M6 PJZ
M6 PKA
M6 PKB
M6 PKE
M6 PKF
M6 PKG
M6 PKI
M6 PKJ
M6 PKM
M6 PKP
M6 PKS
M6 PKW
M6 PKX
M6 PKY
M6 PKZ
M6 PLA
M6 PLD
M6 PLF
M6 PLG
M6 PLJ
M6 PLL
M6 PLM
M6 PLU
M6 PLX
M6 PMC
M6 PME
M6 PMI
M6 PMN
M6 PMO
M6 PMQ
M6 PMS
M6 PMU
M6 PMV
M6 PMX
M6 PNA
M6 PNE
M6 PNF
M6 PNM
M6 PNO
M6 PNQ
M6 PNS
M6 PNV
M6 POI
M6 POJ
M6 POM
M6 POS
M6 PPA
M6 PPB
M6 PPC
M6 PPF
M6 PPG
M6 PPI
M6 PPM
M6 PPO
M6 PPR
M6 PPU
M6 PPW
M6 PPX
M6 PQB
M6 PQC
M6 PQE
M6 PQG
M6 PQJ
M6 PQN
M6 PQR
M6 PQT
M6 PQX
M6 PQY
M6 PQZ
M6 PRD
M6 PRG
M6 PRI
M6 PRJ
M6 PRK
M6 PRL
M6 PRQ
M6 PRR
M6 PRX
M6 PRY
M6 PSA
M6 PSB
M6 PSD
M6 PSE
M6 PSF
M6 PSH
M6 PSI
M6 PSK
M6 PSL
M6 PSP
M6 PSS
M6 PSU
M6 PSW
M6 PTA
M6 PTB
M6 PTC
M6 PTD
M6 PTJ
M6 PTK
M6 PTL
M6 PTN
M6 PTQ
M6 PTS
M6 PTT
M6 PTW
M6 PTZ
M6 PUB
M6 PUC
M6 PUE
M6 PUK
M6 PUR
M6 PVA
M6 PVG
M6 PVP
M6 PVX
M6 PWA
M6 PWI
M6 PWJ
M6 PWW
M6 PWX
M6 PWY
M6 PWZ
M6 PXB
M6 PXQ
M6 PXS
M6 PYA
M6 PYB
M6 PYK
M6 PYM
M6 PYS
M6 PYT
M6 PYU
M6 PYX
M6 PZD
M6 PZL
M6 PZT
M6 PZX
M6 QCB
M6 RAA
M6 RAB
M6 RAE
M6 RAF
M6 RAG
M6 RAI
M6 RAJ
M6 RAK
M6 RAP
M6 RAT
M6 RAU
M6 RAY
M6 RBC
M6 RBD
M6 RBG
M6 RBJ
M6 RBL
M6 RBM
M6 RBR
M6 RBS
M6 RBX
M6 RCH
M6 RCM
M6 RCQ
M6 RCT
M6 RCX
M6 RDB
M6 RDE
M6 RDG
M6 RDH
M6 RDJ
M6 RDO
M6 RDY
M6 REA
M6 REC
M6 REK
M6 REN
M6 REP
M6 RET
M6 REU
M6 REZ
M6 RFT
M6 RFU
M6 RFX
M6 RFY
M6 RGB
M6 RGG
M6 RGJ
M6 RGL
M6 RGV
M6 RHE
M6 RHG
M6 RHH
M6 RHJ
M6 RHM
M6 RHP
M6 RHS
M6 RHW
M6 RHZ
M6 RIC
M6 RIF
M6 RIG
M6 RII
M6 RIP
M6 RIW
M6 RIX
M6 RJ
M6 RJA
M6 RJC
M6 RJD
M6 RJG
M6 RJI
M6 RJT
M6 RJV
M6 RKG
M6 RKH
M6 RKJ
M6 RKN
M6 RKV
M6 RKY
M6 RKZ
M6 RLF
M6 RLH
M6 RLI
M6 RLJ
M6 RLK
M6 RLM
M6 RLO
M6 RLS
M6 RLY
M6 RMB
M6 RMC
M6 RMH
M6 RMM
M6 RMP
M6 RMQ
M6 RMY
M6 RMZ
M6 RNB
M6 RNG
M6 RNH
M6 RNJ
M6 RNL
M6 RNR
M6 RNW
M6 RNY
M6 ROK
M6 ROO
M6 ROQ
M6 ROR
M6 ROS
M6 ROT
M6 ROZ
M6 RPA
M6 RPG
M6 RPI
M6 RPK
M6 RPY
M6 RQB
M6 RQI
M6 RQL
M6 RRB
M6 RRE
M6 RRI
M6 RRM
M6 RRO
M6 RRP
M6 RRS
M6 RRT
M6 RRU
M6 RRX
M6 RRY
M6 RRZ
M6 RSC
M6 RSG
M6 RSK
M6 RSM
M6 RSN
M6 RSQ
M6 RSW
M6 RSX
M6 RSZ
M6 RTB
M6 RTH
M6 RTJ
M6 RTK
M6 RTL
M6 RTS
M6 RTV
M6 RTW
M6 RTY
M6 RUA
M6 RUD
M6 RUE
M6 RUF
M6 RUU
M6 RVB
M6 RVC
M6 RVF
M6 RVI
M6 RVK
M6 RVL
M6 RVP
M6 RVT
M6 RVV
M6 RVY
M6 RWC
M6 RWD
M6 RWH
M6 RWN
M6 RWQ
M6 RWR
M6 RWW
M6 RXD
M6 RXF
M6 RXI
M6 RXM
M6 RXN
M6 RXO
M6 RXP
M6 RYB
M6 RYE
M6 RYO
M6 RYS
M6 RYT
M6 RYW
M6 RZB
M6 RZJ
M6 RZM
M6 RZR
M6 RZS
M6 RZT
M6 SAB
M6 SAJ
M6 SAO
M6 SAP
M6 SAR
M6 SAT
M6 SBA
M6 SBC
M6 SBQ
M6 SBX
M6 SCE

M6 SCJ
M6 SCL
M6 SCO
M6 SCR
M6 SCS
M6 SCT
M6 SCY
M6 SDB
M6 SDC
M6 SDD
M6 SDM
M6 SDO
M6 SDY
M6 SED
M6 SEH
M6 SEL
M6 SEM
M6 SEQ
M6 SET
M6 SEW
M6 SEX
M6 SFA
M6 SFD
M6 SFE
M6 SFG
M6 SFI
M6 SFN
M6 SFO
M6 SFU
M6 SFX
M6 SFY
M6 SGI
M6 SGK
M6 SGR
M6 SGU
M6 SGW
M6 SGX
M6 SHE
M6 SHN
M6 SHO
M6 SHR
M6 SHS
M6 SHT
M6 SHX
M6 SHY
M6 SIB
M6 SIG
M6 SIN
M6 SIO
M6 SIQ
M6 SJI
M6 SJL
M6 SJM
M6 SJP
M6 SJS
M6 SJT
M6 SJY
M6 SKA
M6 SKB
M6 SKC
M6 SKD
M6 SKE
M6 SKG
M6 SKI
M6 SKJ
M6 SKK
M6 SKO
M6 SKS
M6 SLF
M6 SLH
M6 SLQ
M6 SLS
M6 SLW
M6 SLX
M6 SLY
M6 SMC
M6 SMJ
M6 SMO
M6 SMP
M6 SMS
M6 SMT
M6 SMU
M6 SMV
M6 SMW
M6 SMZ
M6 SNA
M6 SNB
M6 SND
M6 SNI
M6 SNK
M6 SNL
M6 SNM
M6 SNT
M6 SNX
M6 SOA
M6 SOB
M6 SOH
M6 SOJ
M6 SOL
M6 SOM
M6 SOP
M6 SOQ
M6 SOY
M6 SPB
M6 SPD
M6 SPE
M6 SPF
M6 SPI
M6 SPO
M6 SPQ
M6 SPR
M6 SPV
M6 SPW
M6 SPX
M6 SQA
M6 SQJ
M6 SQM
M6 SQR
M6 SQS
M6 SQT
M6 SQY
M6 SQZ
M6 SRH
M6 SRJ
M6 SRK
M6 SRP
M6 SRT
M6 SRU
M6 SRW
M6 SRX
M6 SRY
M6 SSE
M6 SSG
M6 SSH
M6 SSJ
M6 SSK
M6 SSL
M6 SSQ
M6 STB
M6 STE
M6 STG
M6 STL
M6 STO
M6 STS
M6 STT
M6 STW
M6 STX
M6 SUA
M6 SUG
M6 SUK
M6 SUS
M6 SVA
M6 SVB
M6 SVC
M6 SVD
M6 SVE
M6 SVK
M6 SVL
M6 SVO
M6 SVR
M6 SVS
M6 SVW
M6 SVX
M6 SWE
M6 SWM
M6 SWR
M6 SWW
M6 SWX
M6 SWY
M6 SWZ
M6 SXE
M6 SXH
M6 SXJ
M6 SXL
M6 SXR
M6 SXS
M6 SXW
M6 SXX
M6 SYA
M6 SYD
M6 SYL
M6 SYM
M6 SYN
M6 SYP
M6 SYR
M6 SYS
M6 SYU
M6 SYY
M6 SYZ
M6 SZE
M6 SZF
M6 SZR
M6 SZS
M6 SZT
M6 SZY
M6 TAB
M6 TAC
M6 TAE
M6 TAI
M6 TAM
M6 TAN
M6 TAP
M6 TAQ
M6 TAR
M6 TAS
M6 TAT
M6 TAV
M6 TAZ
M6 TBB
M6 TBE
M6 TBI
M6 TBP
M6 TBW
M6 TBZ
M6 TCC
M6 TCJ
M6 TCP
M6 TCQ
M6 TCR
M6 TCT
M6 TCU
M6 TCW
M6 TDG
M6 TDH
M6 TDN
M6 TDQ
M6 TDS
M6 TDT
M6 TDW
M6 TDX
M6 TEA
M6 TEB
M6 TEC
M6 TED
M6 TEH
M6 TEI
M6 TET
M6 TEX
M6 TFA
M6 TFC
M6 TFI
M6 TFM
M6 TFR
M6 TFS
M6 TFU
M6 TFX
M6 TGA
M6 TGE
M6 TGW
M6 TGX
M6 THD
M6 THE
M6 THF
M6 THG
M6 THH
M6 THK
M6 THL
M6 THM
M6 THN
M6 THR
M6 THT
M6 THX
M6 THZ
M6 TIB
M6 TIG
M6 TIK
M6 TIM
M6 TIN
M6 TIP
M6 TIS
M6 TIV
M6 TIX
M6 TIZ
M6 TJC
M6 TJK
M6 TJM
M6 TJR
M6 TJT
M6 TJW
M6 TKB
M6 TKD
M6 TKF
M6 TKH
M6 TKJ
M6 TKL
M6 TKM
M6 TKO
M6 TKT
M6 TLA
M6 TLD
M6 TLE
M6 TLH
M6 TLJ
M6 TLT
M6 TLU
M6 TLV
M6 TLW
M6 TLZ
M6 TMB
M6 TME
M6 TMI
M6 TMN
M6 TMT
M6 TMV
M6 TMW
M6 TMX
M6 TNE
M6 TNF
M6 TNG
M6 TNK
M6 TNP
M6 TNS
M6 TNW
M6 TNX
M6 TOB
M6 TOH
M6 TOJ
M6 TOP
M6 TOR
M6 TOX
M6 TOY
M6 TPF
M6 TPJ
M6 TPK
M6 TPL
M6 TPN
M6 TPP
M6 TPW
M6 TPZ
M6 TQB
M6 TQM
M6 TQP
M6 TQY
M6 TRA
M6 TRB
M6 TRG
M6 TRJ
M6 TRM
M6 TRN
M6 TRY
M6 TSB
M6 TSD
M6 TSE
M6 TSF
M6 TSG
M6 TSL
M6 TSO
M6 TSQ
M6 TSS
M6 TTA
M6 TTC
M6 TTD
M6 TTG
M6 TTI
M6 TTJ
M6 TTQ
M6 TUB
M6 TUC
M6 TUE
M6 TUI
M6 TUP
M6 TUS
M6 TUZ
M6 TVA
M6 TVB
M6 TVE
M6 TVG
M6 TVI
M6 TVO
M6 TVR
M6 TVT
M6 TVV
M6 TVZ
M6 TWB
M6 TWC
M6 TWD
M6 TWF
M6 TWG
M6 TWO
M6 TWP
M6 TWR
M6 TWT
M6 TWX
M6 TXB
M6 TXC
M6 TXL
M6 TXO
M6 TXW
M6 TYD
M6 TYF
M6 TYM
M6 TYO
M6 TYP
M6 TYW
M6 TYX
M6 TYY
M6 TZF
M6 TZG
M6 TZH
M6 TZL
M6 TZN
M6 TZT
M6 TZW
M6 UAA
M6 UAC
M6 UAH
M6 UAM
M6 UAQ
M6 UAT
M6 UAV
M6 UBA
M6 UBJ
M6 UBU
M6 UCC
M6 UCH
M6 UCK
M6 UCO
M6 UCW
M6 UCY
M6 UDD
M6 UDE
M6 UDF
M6 UDG
M6 UDH
M6 UDI
M6 UDK
M6 UDP
M6 UDQ
M6 UDW
M6 UDZ
M6 UEC
M6 UEX
M6 UFI
M6 UGB
M6 UGD
M6 UGG
M6 UGT
M6 UHD
M6 UHH
M6 UHK
M6 UHZ
M6 UIK
M6 UIN
M6 UIX
M6 UJB
M6 UJC
M6 UJI
M6 UKD
M6 UKE
M6 UKK
M6 UKL
M6 UKO
M6 UKP
M6 UKR
M6 UKS
M6 UKV
M6 UKW
M6 UKY
M6 UKZ
M6 ULF
M6 ULN
M6 ULP
M6 ULT
M6 ULU
M6 UMA
M6 UME
M6 UMF
M6 UMH
M6 UML
M6 UMP
M6 UNE
M6 UNG
M6 UNK
M6 UNO
M6 UNX
M6 UOJ
M6 UOK
M6 UOO
M6 UOS
M6 UPD
M6 UPP
M6 URB
M6 URC
M6 URF
M6 URL
M6 URZ
M6 USB
M6 USH
M6 USL
M6 USR
M6 UST
M6 UTA
M6 UTD
M6 UTE
M6 UTH
M6 UTM
M6 UTR
M6 UTS
M6 UTU
M6 UUB
M6 UUC
M6 UUF
M6 UUG
M6 UUL
M6 UUN
M6 UVE
M6 UVL
M6 UVR
M6 UWE
M6 UWF
M6 UWM
M6 UXH
M6 UXL
M6 UXO
M6 UXP
M6 UXY
M6 UYT
M6 UZI
M6 VAC
M6 VAD
M6 VAF
M6 VAO
M6 VAS
M6 VBA
M6 VBC
M6 VBG
M6 VBM
M6 VBY
M6 VCG
M6 VCJ
M6 VCO
M6 VCR
M6 VCT
M6 VCW
M6 VCX
M6 VCZ
M6 VDM
M6 VDN
M6 VDQ
M6 VDS
M6 VEE
M6 VEI
M6 VEK
M6 VEM
M6 VER
M6 VES
M6 VEU
M6 VEX
M6 VEY
M6 VEZ
M6 VFM
M6 VFO
M6 VFR
M6 VFT
M6 VFX
M6 VGF
M6 VGG
M6 VGH
M6 VGL
M6 VGM
M6 VGT
M6 VGW
M6 VGX
M6 VHC
M6 VHD
M6 VHF
M6 VHG
M6 VHP
M6 VHR
M6 VHS
M6 VII
M6 VIK
M6 VIM
M6 VIN
M6 VIP
M6 VIR
M6 VIS
M6 VJB
M6 VJK
M6 VJM
M6 VJP
M6 VJS
M6 VJT
M6 VKC
M6 VKD
M6 VKL
M6 VKO
M6 VKP
M6 VKR
M6 VKV
M6 VLC
M6 VLF
M6 VLJ
M6 VLK
M6 VLL
M6 VLM
M6 VMB
M6 VME
M6 VML
M6 VMM
M6 VMR
M6 VMW
M6 VMZ
M6 VNA
M6 VNB
M6 VNC
M6 VNE
M6 VNO
M6 VNR
M6 VNS
M6 VNY
M6 VOB
M6 VOE
M6 VOJ
M6 VOK
M6 VON
M6 VOO
M6 VOP
M6 VOV
M6 VPC
M6 VPN
M6 VPR
M6 VPT
M6 VPW
M6 VRA
M6 VRB
M6 VRS
M6 VRV
M6 VSG
M6 VSP
M6 VSY
M6 VTD
M6 VTF
M6 VTL
M6 VTR
M6 VTV
M6 VTW
M6 VTX
M6 VTY
M6 VUB
M6 VUP
M6 VVC
M6 VVD
M6 VVK
M6 VVM
M6 VVO
M6 VVP
M6 VVV
M6 VVW
M6 VWG
M6 VWY
M6 VXC
M6 VXD
M6 VXH
M6 VXL
M6 VXN
M6 VXO
M6 VXX
M6 VYD
M6 VYK
M6 VYL
M6 VYZ
M6 VZG
M6 VZM
M6 VZN
M6 VZO
M6 VZT
M6 VZY
M6 WAC
M6 WAE
M6 WAH
M6 WAK
M6 WAL
M6 WAO
M6 WAR
M6 WAU
M6 WAZ
M6 WBC
M6 WBD
M6 WBG
M6 WBL
M6 WCD
M6 WCH
M6 WCJ
M6 WCM
M6 WCO
M6 WCS
M6 WDA
M6 WDE
M6 WDJ
M6 WDK
M6 WDP
M6 WDT
M6 WDV
M6 WDZ
M6 WEA
M6 WEC
M6 WEF
M6 WEG
M6 WEH
M6 WEI
M6 WEY
M6 WFD
M6 WFM
M6 WFN
M6 WFS
M6 WGA
M6 WGP
M6 WGW
M6 WGY
M6 WHB
M6 WHC
M6 WHE
M6 WHH
M6 WHK
M6 WHL
M6 WHM
M6 WHN
M6 WHO
M6 WHQ
M6 WHR
M6 WHV
M6 WHY
M6 WIA
M6 WIQ
M6 WJD
M6 WJG
M6 WJH
M6 WJO
M6 WJR
M6 WJT
M6 WJV
M6 WJX
M6 WKA
M6 WKM
M6 WLJ
M6 WLK
M6 WLL
M6 WLM
M6 WLT
M6 WLZ
M6 WMC
M6 WMD
M6 WMI
M6 WMO
M6 WMQ
M6 WMR
M6 WMT
M6 WMU
M6 WMX
M6 WMY
M6 WNA
M6 WNT
M6 WNX
M6 WON
M6 WOZ
M6 WPF
M6 WPJ
M6 WPK
M6 WPL
M6 WPM
M6 WPX
M6 WPZ
M6 WQI
M6 WQP
M6 WQW
M6 WRA
M6 WRF
M6 WRI
M6 WRK
M6 WRP
M6 WRZ
M6 WSA
M6 WSD
M6 WSE
M6 WSF
M6 WSL
M6 WSN
M6 WSX
M6 WSZ
M6 WTA
M6 WTB
M6 WTH
M6 WTI
M6 WTK
M6 WTN
M6 WTS
M6 WTV
M6 WTY
M6 WUF
M6 WUS
M6 WUT
M6 WUV
M6 WUZ
M6 WVD
M6 WVI
M6 WVL
M6 WVM
M6 WVU
M6 WVW
M6 WVZ
M6 WWA
M6 WWP
M6 WWW
M6 WWX
M6 WXA
M6 WXC
M6 WXG
M6 WXM
M6 WXW
M6 WXZ
M6 WYE
M6 WYK
M6 WYL
M6 WYP
M6 WYS
M6 WZA
M6 WZG
M6 WZH
M6 WZL
M6 WZP
M6 WZS
M6 WZT
M6 WZZ
M6 XAH
M6 XAJ
M6 XAN
M6 XAZ
M6 XBE
M6 XBF
M6 XBG
M6 XBH
M6 XBK
M6 XBN
M6 XBT
M6 XCB
M6 XCD
M6 XCE
M6 XCG
M6 XCM
M6 XCS
M6 XCT
M6 XCV
M6 XCX
M6 XDD
M6 XDE
M6 XDF
M6 XDH
M6 XDK
M6 XDL
M6 XDM
M6 XDN
M6 XDP
M6 XDS
M6 XDV
M6 XDY
M6 XEB
M6 XED
M6 XEG
M6 XEH
M6 XEN
M6 XEO
M6 XEQ
M6 XES
M6 XET
M6 XEW
M6 XFC
M6 XFD
M6 XFF
M6 XFK
M6 XFV
M6 XFZ
M6 XGE
M6 XGK
M6 XGM
M6 XGP
M6 XGR
M6 XHA
M6 XHK
M6 XHT
M6 XHV
M6 XHX
M6 XII
M6 XIJ
M6 XIM
M6 XIO
M6 XIS
M6 XIV
M6 XJA
M6 XJB
M6 XJF
M6 XJH
M6 XJK
M6 XJT
M6 XJX
M6 XKA
M6 XKB
M6 XKD
M6 XKP
M6 XKR
M6 XKX
M6 XLA
M6 XLB
M6 XLD
M6 XLL
M6 XLM
M6 XLV
M6 XLW
M6 XLY
M6 XLZ
M6 XMD
M6 XMK
M6 XML
M6 XMM
M6 XMP
M6 XMT
M6 XNI
M6 XOC
M6 XOF
M6 XOL
M6 XOM
M6 XOO
M6 XOP
M6 XOV
M6 XOZ
M6 XPA
M6 XPE
M6 XPG
M6 XPM
M6 XPR
M6 XPS
M6 XPY
M6 XQD
M6 XQE
M6 XRA
M6 XRB
M6 XRE
M6 XRG
M6 XRH
M6 XRJ
M6 XRK
M6 XRM
M6 XRW
M6 XRX
M6 XRY
M6 XSB
M6 XSG
M6 XSJ
M6 XSM
M6 XSP
M6 XSQ
M6 XSW
M6 XSX
M6 XSY
M6 XTC
M6 XTE
M6 XTG
M6 XTH
M6 XTN
M6 XTP
M6 XTS
M6 XTV
M6 XTW
M6 XTX
M6 XTY
M6 XUD
M6 XUE
M6 XUK
M6 XUL
M6 XUP
M6 XVE
M6 XVN
M6 XVZ
M6 XWF
M6 XWH
M6 XWM
M6 XWP
M6 XWT
M6 XWX
M6 XWZ
M6 XXA
M6 XXE
M6 XXF
M6 XXG
M6 XXK
M6 XXO
M6 XXP
M6 XXR
M6 XXX
M6 XXY
M6 XYA
M6 XYE
M6 XYM
M6 XYO
M6 XYS
M6 XYT
M6 XYX
M6 XYZ
M6 XZA
M6 XZQ
M6 XZV
M6 XZX
M6 XZZ
M6 YAA
M6 YAB
M6 YAC
M6 YAI
M6 YAK
M6 YAL
M6 YAQ
M6 YAU
M6 YAX
M6 YAZ
M6 YBF
M6 YBG
M6 YBM
M6 YBP
M6 YBU
M6 YBW
M6 YBX
M6 YCC
M6 YCE
M6 YCL
M6 YCN
M6 YCQ
M6 YCS
M6 YCZ
M6 YDM
M6 YDS
M6 YDT
M6 YDX
M6 YDZ
M6 YEA
M6 YEC
M6 YED
M6 YEE
M6 YEM
M6 YEN
M6 YEP
M6 YER
M6 YEW
M6 YFC
M6 YFJ
M6 YFL
M6 YFS
M6 YFY
M6 YGC
M6 YGG
M6 YGH
M6 YGI
M6 YGK
M6 YGM
M6 YGW
M6 YGZ
M6 YHA
M6 YHF
M6 YHT
M6 YHZ
M6 YIF
M6 YIP
M6 YJC
M6 YJE
M6 YJM
M6 YJP
M6 YJS
M6 YKB
M6 YKK
M6 YKL
M6 YKV
M6 YKZ
M6 YLB
M6 YLF
M6 YLL
M6 YLM
M6 YLX
M6 YLZ
M6 YMC
M6 YMD
M6 YMG
M6 YMJ
M6 YMK
M6 YML
M6 YMO
M6 YMZ
M6 YNE
M6 YNF
M6 YOA
M6 YOK
M6 YOL
M6 YOM
M6 YOP
M6 YOR
M6 YOX
M6 YPC
M6 YPF
M6 YPH
M6 YPK
M6 YPL
M6 YPM
M6 YPT
M6 YQB
M6 YQY
M6 YRA
M6 YRG
M6 YRK
M6 YRM
M6 YRP
M6 YRT
M6 YRU
M6 YSL
M6 YSO
M6 YSZ
M6 YTE
M6 YTH
M6 YTK
M6 YTN
M6 YTY
M6 YUC
M6 YUR
M6 YVM
M6 YVX
M6 YWF
M6 YWK
M6 YWY
M6 YWZ
M6 YXA
M6 YXS
M6 YXX
M6 YXY
M6 YYA
M6 YYC
M6 YYP
M6 YYZ
M6 YZA
M6 YZG
M6 YZP
M6 YZV
M6 YZW
M6 YZZ
M6 ZAE
M6 ZAG
M6 ZAI
M6 ZAJ
M6 ZAM
M6 ZAO
M6 ZAP
M6 ZAT
M6 ZBC
M6 ZBF
M6 ZBI
M6 ZBK
M6 ZBO
M6 ZBP
M6 ZCC
M6 ZCJ
M6 ZDD
M6 ZDK
M6 ZDM
M6 ZDR
M6 ZDX
M6 ZEC
M6 ZED
M6 ZEY
M6 ZEZ
M6 ZFS
M6 ZFX
M6 ZGP
M6 ZHC
M6 ZIG
M6 ZIM
M6 ZIN
M6 ZIX
M6 ZJC
M6 ZJO
M6 ZJS
M6 ZKC
M6 ZKG
M6 ZKL
M6 ZKM
M6 ZKO
M6 ZKX
M6 ZKZ
M6 ZLG
M6 ZMD
M6 ZNQ
M6 ZOA
M6 ZOE
M6 ZON
M6 ZOT
M6 ZOZ
M6 ZPC
M6 ZPG
M6 ZPM
M6 ZQQ
M6 ZRA
M6 ZRB
M6 ZSC
M6 ZSR
M6 ZSU
M6 ZSY
M6 ZTE
M6 ZTT
M6 ZTZ
M6 ZUJ
M6 ZUK
M6 ZUM
M6 ZUU
M6 ZVV
M6 ZWD
M6 ZXA
M6 ZXJ
M6 ZXY
M6 ZZI
M6 ZZO
M6 ZZR
M6 ZZT
M6 ZZW
M6 ZZZ

M7

M7 AAD
M7 AAF
M7 AAG
M7 AAI
M7 AAJ
M7 AAM
M7 AAN
M7 AAO
M7 AAR
M7 ABF
M7 ABH
M7 ABI
M7 ABJ
M7 ABK
M7 ABL
M7 ABO
M7 ABS
M7 ACB
M7 ACC
M7 ACF
M7 ACI
M7 ACJ
M7 ACO
M7 ACP
M7 ACQ
M7 ACR
M7 ACT
M7 ACU
M7 ACZ
M7 ADB
M7 ADD
M7 ADE
M7 ADG
M7 ADH
M7 ADJ
M7 ADK
M7 ADM
M7 ADP
M7 ADQ
M7 ADR
M7 ADV
M7 ADY
M7 AEC
M7 AEE
M7 AEG
M7 AEH
M7 AEK
M7 AEM
M7 AEO
M7 AEP
M7 AEQ
M7 AET
M7 AEV
M7 AFB
M7 AFC
M7 AFF
M7 AFG
M7 AFJ
M7 AFM
M7 AFN
M7 AFQ
M7 AFT
M7 AFV
M7 AFX
M7 AFY
M7 AGA
M7 AGC
M7 AGF
M7 AGI
M7 AGM
M7 AGP
M7 AGR
M7 AGT
M7 AGW
M7 AGZ
M7 AHB
M7 AHC
M7 AHF
M7 AHJ
M7 AHL
M7 AHO
M7 AHP
M7 AHQ
M7 AHS
M7 AHV
M7 AHW
M7 AHX
M7 AHY
M7 AHZ
M7 AIA
M7 AIB
M7 AIC
M7 AIJ
M7 AIL
M7 AIN
M7 AIO
M7 AIS
M7 AIT
M7 AIX
M7 AIY
M7 AJC
M7 AJE
M7 AJI
M7 AJL
M7 AJS
M7 AJV
M7 AJW
M7 AJY
M7 AKB
M7 AKD
M7 AKE
M7 AKF
M7 AKH
M7 AKI
M7 AKJ
M7 AKK
M7 AKL
M7 AKO
M7 AKQ
M7 AKR
M7 AKT
M7 AKV
M7 AKW
M7 ALA
M7 ALB
M7 ALD
M7 ALE
M7 ALG
M7 ALH
M7 ALJ
M7 ALK
M7 ALP
M7 ALQ
M7 ALU
M7 ALV
M7 ALX
M7 ALY
M7 ALZ
M7 AMD
M7 AMF
M7 AMG
M7 AMI
M7 AML
M7 AMN
M7 AMO
M7 AMU
M7 AMV
M7 AMY
M7 AMZ
M7 ANC
M7 ANF
M7 ANL
M7 ANQ
M7 ANS
M7 ANW
M7 ANZ
M7 AOA
M7 AOB
M7 AOE
M7 AOF
M7 AOG
M7 AOJ
M7 AOK
M7 AON
M7 AOP
M7 AOQ
M7 AOS
M7 AOT
M7 AOW
M7 AOX
M7 APA
M7 APD
M7 APE
M7 APG
M7 API
M7 APK
M7 APL
M7 APP
M7 APQ
M7 APR
M7 APS
M7 APT
M7 APU
M7 APX
M7 AQA
M7 AQB
M7 AQD
M7 AQE
M7 AQH
M7 AQJ
M7 AQK
M7 AQN
M7 AQT
M7 AQU
M7 AQV
M7 AQW
M7 AQX
M7 ARB
M7 ARD
M7 ARF
M7 ARG
M7 ARI
M7 ARJ
M7 ARK
M7 ARM
M7 ARR
M7 ARS
M7 ARV
M7 ARX
M7 ARY
M7 ASD
M7 ASG
M7 ASI
M7 ASL
M7 ASQ
M7 ASR
M7 ASV
M7 ATK
M7 ATM
M7 ATP
M7 ATQ
M7 ATS
M7 ATU
M7 ATY
M7 ATZ
M7 AUA
M7 AUG
M7 AUJ
M7 AUK
M7 AUN
M7 AUS
M7 AUV
M7 AUW
M7 AVD
M7 AVJ
M7 AVK
M7 AVM
M7 AVU
M7 AVV
M7 AVX
M7 AVY
M7 AVZ
M7 AWA
M7 AWB
M7 AWC
M7 AWI
M7 AWP
M7 AWR
M7 AWS
M7 AWT
M7 AWU
M7 AWV
M7 AWX
M7 AWY
M7 AWZ
M7 AXG
M7 AXH
M7 AXI
M7 AXK
M7 AXL
M7 AXN
M7 AXO
M7 AXP
M7 AXQ
M7 AXR
M7 AXS
M7 AXT
M7 AXW
M7 AXZ
M7 AYA
M7 AYC
M7 AYD
M7 AYE
M7 AYH
M7 AYI
M7 AYK
M7 AYL
M7 AYM
M7 AYN
M7 AYO
M7 AYQ
M7 AYT
M7 AYX
M7 AYY
M7 AZB
M7 AZC
M7 AZD
M7 AZE
M7 AZF
M7 AZI
M7 AZL
M7 AZM
M7 AZT
M7 AZU
M7 AZX
M7 AZY
M7 AZZ
M7 BAB
M7 BAD
M7 BAH
M7 BAI
M7 BAM
M7 BAP
M7 BAT
M7 BAU
M7 BAX
M7 BAZ
M7 BBA
M7 BBB
M7 BBC
M7 BBF
M7 BBH
M7 BBL
M7 BBP
M7 BBQ
M7 BBR
M7 BBU
M7 BBX
M7 BCB
M7 BCC
M7 BCD
M7 BCF
M7 BCG
M7 BCJ
M7 BCK
M7 BCO
M7 BCQ
M7 BCR
M7 BCS
M7 BCT
M7 BCU
M7 BCX
M7 BDA
M7 BDD
M7 BDH
M7 BDI
M7 BDJ
M7 BDK
M7 BDK
M7 BDL
M7 BDM
M7 BDN
M7 BDR
M7 BDS
M7 BDU
M7 BDY
M7 BEA
M7 BEB
M7 BEC
M7 BED
M7 BEH
M7 BEJ
M7 BEN
M7 BEQ
M7 BER
M7 BES
M7 BEU
M7 BEW
M7 BEY
M7 BEZ
M7 BFC
M7 BFD
M7 BFE
M7 BFF
M7 BFH
M7 BFI
M7 BFM
M7 BFO
M7 BFR
M7 BFS
M7 BFT
M7 BFU
M7 BFW
M7 BGA
M7 BGC
M7 BGI
M7 BGJ
M7 BGK
M7 BGL
M7 BGM
M7 BGN
M7 BGP
M7 BGQ
M7 BGS
M7 BGT
M7 BGY
M7 BGZ
M7 BHA
M7 BHB
M7 BHD
M7 BHE
M7 BHF
M7 BHG
M7 BHH
M7 BHI
M7 BHJ
M7 BHP
M7 BHS
M7 BHV
M7 BHW
M7 BHX
M7 BHY
M7 BIA
M7 BIB
M7 BIF
M7 BII
M7 BIJ
M7 BIK
M7 BIM
M7 BIN
M7 BIR
M7 BIV
M7 BIY
M7 BIZ
M7 BJB
M7 BJD
M7 BJE
M7 BJF
M7 BJM
M7 BJO
M7 BJP
M7 BJX
M7 BJZ
M7 BKC
M7 BKD
M7 BKF
M7 BKK
M7 BKQ
M7 BKR
M7 BKV
M7 BKW
M7 BKY
M7 BLA
M7 BLB
M7 BLE
M7 BLF
M7 BLH
M7 BLK
M7 BLN
M7 BLQ
M7 BLS
M7 BLT
M7 BLV
M7 BLW
M7 BLZ
M7 BMA
M7 BMB
M7 BMI
M7 BML
M7 BMM
M7 BMN
M7 BMP
M7 BMQ
M7 BMS
M7 BMT
M7 BMV
M7 BNA
M7 BNC
M7 BNF
M7 BNM
M7 BNO
M7 BNQ
M7 BNT
M7 BNU
M7 BNY
M7 BOB
M7 BOH
M7 BOI
M7 BOR
M7 BOT
M7 BOV
M7 BOX
M7 BOY
M7 BPF
M7 BPG
M7 BPK
M7 BPL
M7 BPM
M7 BPP
M7 BPR
M7 BPT
M7 BPU
M7 BPV
M7 BPY
M7 BQA
M7 BQD
M7 BQE
M7 BQN
M7 BQO
M7 BQR
M7 BQS
M7 BQT
M7 BQV
M7 BQW
M7 BRB
M7 BRD
M7 BRE
M7 BRF
M7 BRK
M7 BRL
M7 BRN
M7 BRP
M7 BRT
M7 BRU
M7 BRV
M7 BRW
M7 BSA
M7 BSB
M7 BSC
M7 BSD
M7 BSE
M7 BSG
M7 BSJ
M7 BSO
M7 BSP
M7 BSQ
M7 BST
M7 BSW
M7 BSY
M7 BTA
M7 BTB
M7 BTC
M7 BTI
M7 BTJ
M7 BTK
M7 BTM
M7 BTN
M7 BTP
M7 BTR
M7 BTT
M7 BUA
M7 BUC
M7 BUD
M7 BUF
M7 BUJ
M7 BUO
M7 BUP
M7 BUQ
M7 BUT
M7 BUU
M7 BUW
M7 BUX
M7 BUZ
M7 BVB
M7 BVD
M7 BVE
M7 BVG
M7 BVL
M7 BVM
M7 BVP
M7 BVR
M7 BVU
M7 BVW
M7 BVX
M7 BVY
M7 BWA
M7 BWC
M7 BWD
M7 BWF
M7 BWI
M7 BWQ
M7 BWR
M7 BWS
M7 BWU
M7 BWV
M7 BWX
M7 BWY
M7 BXB
M7 BXD
M7 BXG
M7 BXH
M7 BXI
M7 BXJ
M7 BXK
M7 BXP
M7 BXQ
M7 BXR
M7 BXS
M7 BXU
M7 BXV
M7 BXW
M7 BXX
M7 BXY
M7 BXZ
M7 BYA
M7 BYB
M7 BYG
M7 BYI
M7 BYJ
M7 BYN
M7 BYO
M7 BYP
M7 BYR
M7 BYX
M7 BYY
M7 BYZ
M7 BZD
M7 BZE
M7 BZF
M7 BZG
M7 BZH
M7 BZI
M7 BZJ
M7 BZM
M7 BZN
M7 BZQ
M7 BZS
M7 BZT
M7 BZW
M7 BZX
M7 BZY
M7 CAB
M7 CAC
M7 CAE
M7 CAF
M7 CAG
M7 CAH
M7 CAI
M7 CAM

Withheld

M7 CAO
M7 CAP
M7 CAR
M7 CAS
M7 CAU
M7 CAV
M7 CAX
M7 CAY
M7 CBA
M7 CBB
M7 CBC
M7 CBD
M7 CBE
M7 CBI
M7 CBJ
M7 CBK
M7 CBM
M7 CBN
M7 CBO
M7 CBQ
M7 CBR
M7 CBS
M7 CBU
M7 CBV
M7 CBW
M7 CBZ
M7 CCA
M7 CCD
M7 CCE
M7 CCI
M7 CCN
M7 CCP
M7 CCT
M7 CCV
M7 CCX
M7 CCY
M7 CDC
M7 CDD
M7 CDE
M7 CDG
M7 CDI
M7 CDJ
M7 CDL
M7 CDM
M7 CDN
M7 CDR
M7 CDV
M7 CDY
M7 CDZ
M7 CEA
M7 CEC
M7 CED
M7 CEE
M7 CEI
M7 CEJ
M7 CEN
M7 CEQ
M7 CET
M7 CEU
M7 CEX
M7 CFA
M7 CFB
M7 CFD
M7 CFH
M7 CFI
M7 CFJ
M7 CFK
M7 CFN
M7 CFO
M7 CFP
M7 CFQ
M7 CFS
M7 CFT
M7 CFV
M7 CFY
M7 CFZ
M7 CGA
M7 CGB
M7 CGC
M7 CGE
M7 CGG
M7 CGJ
M7 CGL
M7 CGN
M7 CGP
M7 CGR
M7 CGT
M7 CGU
M7 CGV
M7 CGW
M7 CGX
M7 CGY
M7 CGZ
M7 CHK
M7 CHL
M7 CHM
M7 CHO
M7 CHQ
M7 CHT
M7 CHU
M7 CHV
M7 CHW
M7 CHX
M7 CHY
M7 CHZ
M7 CIA
M7 CIB
M7 CIC
M7 CID
M7 CIF
M7 CIG
M7 CIK
M7 CIP
M7 CIQ
M7 CIR
M7 CIV
M7 CIW
M7 CIY
M7 CJF
M7 CJI
M7 CJJ
M7 CJK
M7 CJN
M7 CJO
M7 CJP
M7 CJQ
M7 CJR
M7 CJT
M7 CJW
M7 CJZ
M7 CKB
M7 CKD
M7 CKE
M7 CKH
M7 CKI
M7 CKJ
M7 CKK
M7 CKL
M7 CKM
M7 CKO
M7 CKP
M7 CKQ
M7 CKS
M7 CKU
M7 CKZ
M7 CLA
M7 CLB
M7 CLD
M7 CLF
M7 CLG
M7 CLL
M7 CLO
M7 CLR
M7 CLS
M7 CLT
M7 CLV
M7 CLY
M7 CLZ
M7 CMA
M7 CMF
M7 CMG
M7 CMH
M7 CMK
M7 CMN
M7 CMO
M7 CMP
M7 CMQ
M7 CMR
M7 CMT
M7 CMU
M7 CMV
M7 CMX
M7 CMZ
M7 CNC
M7 CNF
M7 CNG
M7 CNH
M7 CNO
M7 CNP
M7 CNR
M7 CNS
M7 CNU
M7 CNV
M7 CNW
M7 COA
M7 COE
M7 COG
M7 COI
M7 COJ
M7 COM
M7 COO
M7 COS
M7 COT
M7 COZ
M7 CPB
M7 CPD
M7 CPE
M7 CPF
M7 CPK
M7 CPL
M7 CPM
M7 CPN
M7 CPP
M7 CPR
M7 CPS
M7 CPU
M7 CPV
M7 CPX
M7 CPY
M7 CQB
M7 CQC
M7 CQG
M7 CQH
M7 CQI
M7 CQN
M7 CQO
M7 CQP
M7 CQQ
M7 CQR
M7 CQT
M7 CQV
M7 CQW
M7 CRE
M7 CRH
M7 CRI
M7 CRJ
M7 CRL
M7 CRN
M7 CRO
M7 CRT
M7 CRX
M7 CSB
M7 CSD
M7 CSE
M7 CSG
M7 CSH
M7 CSJ
M7 CSQ
M7 CSS
M7 CST
M7 CSV
M7 CSY
M7 CSZ
M7 CTC
M7 CTD
M7 CTE
M7 CTH
M7 CTJ
M7 CTK
M7 CTO
M7 CTQ
M7 CTR
M7 CTU
M7 CTW
M7 CTZ
M7 CUB
M7 CUC
M7 CUE
M7 CUF
M7 CUK
M7 CUL
M7 CUT
M7 CUV
M7 CUX
M7 CUY
M7 CVA
M7 CVB
M7 CVE
M7 CVF
M7 CVG
M7 CVI
M7 CVK
M7 CVL
M7 CVN
M7 CVO
M7 CVP
M7 CVR
M7 CVS
M7 CVU
M7 CVV
M7 CVY
M7 CWC
M7 CWE
M7 CWF
M7 CWG
M7 CWH
M7 CWI
M7 CWJ
M7 CWO
M7 CWP
M7 CWQ
M7 CWR
M7 CWU
M7 CWV
M7 CWW
M7 CWY
M7 CWZ
M7 CXA
M7 CXC
M7 CXG
M7 CXH
M7 CXI
M7 CXK
M7 CXN
M7 CXO
M7 CXS
M7 CXX
M7 CYB
M7 CYD
M7 CYF
M7 CYJ
M7 CYK
M7 CYM
M7 CYO
M7 CYP
M7 CYT
M7 CYW
M7 CYZ
M7 CZC
M7 CZE
M7 CZG
M7 CZH
M7 CZI
M7 CZJ
M7 CZK
M7 CZL
M7 CZN
M7 CZO
M7 CZP
M7 CZS
M7 CZT
M7 CZV
M7 CZX
M7 CZY
M7 DAA
M7 DAG
M7 DAI
M7 DAL
M7 DAO
M7 DAQ
M7 DAS
M7 DAU
M7 DAW
M7 DAX
M7 DBA
M7 DBC
M7 DBE
M7 DBI
M7 DBK
M7 DBO
M7 DBQ
M7 DBS
M7 DBT
M7 DBU
M7 DBV
M7 DBW
M7 DBY
M7 DCB
M7 DCC
M7 DCE
M7 DCF
M7 DCG
M7 DCI
M7 DCJ
M7 DCO
M7 DCQ
M7 DCS
M7 DCU
M7 DCW
M7 DCY
M7 DCZ
M7 DDB
M7 DDD
M7 DDG
M7 DDH
M7 DDI
M7 DDJ
M7 DDL
M7 DDO
M7 DDS
M7 DDU
M7 DDZ
M7 DEB
M7 DEC
M7 DED
M7 DEF
M7 DEG
M7 DEH
M7 DEM
M7 DEN
M7 DEO
M7 DEQ
M7 DES
M7 DEU
M7 DEV
M7 DEZ
M7 DFA
M7 DFB
M7 DFG
M7 DFH
M7 DFJ
M7 DFL
M7 DFN
M7 DFQ
M7 DFR
M7 DFS
M7 DFT
M7 DFU
M7 DFV
M7 DFW
M7 DFZ
M7 DGA
M7 DGD
M7 DGI
M7 DGJ
M7 DGM
M7 DGO
M7 DGR
M7 DGT
M7 DGV
M7 DGX
M7 DGY
M7 DHH
M7 DHI
M7 DHJ
M7 DHK
M7 DHN
M7 DHO
M7 DHQ
M7 DHR
M7 DHS
M7 DHU
M7 DHV
M7 DHW
M7 DHX
M7 DIA
M7 DIB
M7 DID
M7 DIG
M7 DIH
M7 DII
M7 DIN
M7 DIR
M7 DIX
M7 DIY
M7 DJF
M7 DJH
M7 DJI
M7 DJJ
M7 DJO
M7 DJR
M7 DJU
M7 DJV
M7 DJX
M7 DJZ
M7 DKA
M7 DKD
M7 DKF
M7 DKG
M7 DKN
M7 DKO
M7 DKP
M7 DKQ
M7 DKR
M7 DKV
M7 DKW
M7 DKZ
M7 DLA
M7 DLC
M7 DLE
M7 DLF
M7 DLG
M7 DLH
M7 DLI
M7 DLL
M7 DLM
M7 DLN
M7 DLP
M7 DLT
M7 DLU
M7 DLW
M7 DLY
M7 DMC
M7 DME
M7 DMG
M7 DML
M7 DMM
M7 DMN
M7 DMO
M7 DMQ
M7 DMR
M7 DMT
M7 DMU
M7 DMX
M7 DMZ
M7 DNA
M7 DNB
M7 DND
M7 DNF
M7 DNI
M7 DNJ
M7 DNK
M7 DNL
M7 DNP
M7 DNR
M7 DNS
M7 DNT
M7 DNU
M7 DNV
M7 DNX
M7 DNY
M7 DNZ
M7 DOA
M7 DOB
M7 DOC
M7 DOE
M7 DOF
M7 DOH
M7 DOI
M7 DOJ
M7 DOL
M7 DOM
M7 DOO
M7 DOP
M7 DOQ
M7 DOT
M7 DOU
M7 DOV
M7 DOW
M7 DOY
M7 DPA
M7 DPF
M7 DPI
M7 DPJ
M7 DPK
M7 DPM
M7 DPN
M7 DPQ
M7 DPR
M7 DPS
M7 DPV
M7 DQA
M7 DQB
M7 DQE
M7 DQH
M7 DQK
M7 DQL
M7 DQM
M7 DQN
M7 DQP
M7 DQS
M7 DQT
M7 DQX
M7 DQZ
M7 DRA
M7 DRD
M7 DRF
M7 DRI
M7 DRK
M7 DRM
M7 DRN
M7 DRO
M7 DRP
M7 DRQ
M7 DRU
M7 DRX
M7 DRZ
M7 DSB
M7 DSD
M7 DSE
M7 DSF
M7 DSG
M7 DSI
M7 DSJ
M7 DSK
M7 DSM
M7 DSN
M7 DSO
M7 DSS
M7 DST
M7 DSU
M7 DSW
M7 DSX
M7 DSY
M7 DTB
M7 DTC
M7 DTE
M7 DTJ
M7 DTK
M7 DTM
M7 DTN
M7 DTO
M7 DTP
M7 DTQ
M7 DTS
M7 DTT
M7 DTU
M7 DTV
M7 DTX
M7 DTY
M7 DUA
M7 DUE
M7 DUG
M7 DUK
M7 DUL
M7 DUN
M7 DUO
M7 DUQ
M7 DUS
M7 DUV
M7 DUW
M7 DUX
M7 DUY
M7 DUZ
M7 DVA
M7 DVC
M7 DVF
M7 DVG
M7 DVH
M7 DVI
M7 DVL
M7 DVN
M7 DVR
M7 DVU
M7 DVV
M7 DVW
M7 DVZ
M7 DWB
M7 DWC
M7 DWG
M7 DWH
M7 DWI
M7 DWJ
M7 DWL
M7 DWM
M7 DWN
M7 DWO
M7 DWR
M7 DWS
M7 DWU
M7 DWX
M7 DWY
M7 DWZ
M7 DXA
M7 DXB
M7 DXD
M7 DXE
M7 DXG
M7 DXK
M7 DXM
M7 DXO
M7 DXR
M7 DXV
M7 DXW
M7 DXY
M7 DYC
M7 DYD
M7 DYE
M7 DYF
M7 DYJ
M7 DYK
M7 DYL
M7 DYN
M7 DYQ
M7 DYT
M7 DYV
M7 DYX
M7 DZA
M7 DZB
M7 DZD
M7 DZH
M7 DZJ
M7 DZN
M7 DZR
M7 DZX
M7 DZY
M7 DZZ
M7 EAA
M7 EAD
M7 EAF
M7 EAG
M7 EAI
M7 EAJ
M7 EAL
M7 EAQ
M7 EAT
M7 EAU
M7 EAV
M7 EAW
M7 EAX
M7 EAY
M7 EAZ
M7 EBB
M7 EBD
M7 EBE
M7 EBF
M7 EBG
M7 EBH
M7 EBK
M7 EBM
M7 EBN
M7 EBQ
M7 EBR
M7 EBS
M7 EBT
M7 EBV
M7 EBX
M7 ECA
M7 ECB
M7 ECC
M7 ECD
M7 ECE
M7 ECH
M7 ECI
M7 ECJ
M7 ECK
M7 ECL
M7 ECO
M7 ECQ
M7 ECR
M7 ECS
M7 ECU
M7 ECX
M7 EDA
M7 EDB
M7 EDC
M7 EDF
M7 EDG
M7 EDH
M7 EDI
M7 EDJ
M7 EDN
M7 EDP
M7 EDQ
M7 EDR
M7 EDT
M7 EDU
M7 EDV
M7 EDX
M7 EEA
M7 EEB
M7 EED
M7 EEF
M7 EEI
M7 EEJ
M7 EEK
M7 EEN
M7 EEP
M7 EER
M7 EES
M7 EET
M7 EEV
M7 EEX
M7 EEY
M7 EEZ
M7 EFD
M7 EFE
M7 EFF
M7 EFG
M7 EFI
M7 EFK
M7 EFL
M7 EFN
M7 EFQ
M7 EFW
M7 EFY
M7 EGC
M7 EGD
M7 EGF
M7 EGG
M7 EGH
M7 EGI
M7 EGL
M7 EGN
M7 EGO
M7 EGQ
M7 EGR
M7 EGU
M7 EGV
M7 EGX
M7 EGZ
M7 EHA
M7 EHE
M7 EHF
M7 EHH
M7 EHI
M7 EHJ
M7 EHK
M7 EHM
M7 EHN
M7 EHO
M7 EHP
M7 EHR
M7 EHS
M7 EHT
M7 EHU
M7 EHX
M7 EHY
M7 EHZ
M7 EIA
M7 EIB
M7 EIC
M7 EIE
M7 EIF
M7 EIG
M7 EIH
M7 EII
M7 EIL
M7 EIM
M7 EIN
M7 EIQ
M7 EIS
M7 EIV
M7 EIX
M7 EIY
M7 EIZ
M7 EJA
M7 EJD
M7 EJF
M7 EJG
M7 EJL
M7 EJO
M7 EJQ
M7 EJR
M7 EJT
M7 EJU
M7 EJX
M7 EJZ
M7 EKB
M7 EKC
M7 EKE
M7 EKF
M7 EKG
M7 EKH
M7 EKI
M7 EKJ
M7 EKM
M7 EKN
M7 EKO
M7 EKW
M7 EKX
M7 EKY
M7 ELA
M7 ELB
M7 ELC
M7 ELD
M7 ELF
M7 ELH
M7 ELI
M7 ELJ
M7 ELK
M7 ELM
M7 ELP
M7 ELQ
M7 ELR
M7 ELT
M7 ELV
M7 ELW
M7 ELX
M7 ELY
M7 EMA
M7 EMB
M7 EMF
M7 EMG
M7 EML
M7 EMM
M7 EMN
M7 EMP
M7 EMQ
M7 EMS
M7 EMT
M7 EMX
M7 EMZ
M7 ENA
M7 END
M7 ENE
M7 ENG
M7 ENK
M7 ENL
M7 ENM
M7 ENO
M7 ENP
M7 ENR
M7 ENS
M7 ENT
M7 ENX
M7 ENY
M7 ENZ
M7 EOC
M7 EOF
M7 EOG
M7 EOI
M7 EOJ
M7 EOM
M7 EON
M7 EOO
M7 EOP
M7 EOR
M7 EOW
M7 EOX
M7 EOY
M7 EPA
M7 EPB
M7 EPC
M7 EPE
M7 EPJ
M7 EPL
M7 EPO
M7 EPP
M7 EPQ
M7 EPR
M7 EPU
M7 EPX
M7 EQA
M7 EQB
M7 EQD
M7 EQE
M7 EQH
M7 EQJ
M7 EQK
M7 EQL
M7 EQO
M7 EQV
M7 EQW
M7 EQY
M7 ERA
M7 ERC
M7 ERD
M7 ERH
M7 ERL
M7 ERM
M7 ERN
M7 ERO
M7 ERR
M7 ERW
M7 ERX
M7 ERY
M7 ERZ
M7 ESA
M7 ESD
M7 ESL
M7 ESN
M7 ESO
M7 ESP
M7 ESU
M7 ETC
M7 ETD
M7 ETE
M7 ETJ
M7 ETO
M7 ETP
M7 ETR
M7 ETT
M7 ETV
M7 ETW
M7 ETX
M7 EUC
M7 EUD
M7 EUE
M7 EUJ
M7 EUK
M7 EUL
M7 EUN
M7 EUO
M7 EUQ
M7 EUR
M7 EUT
M7 EUU
M7 EUV
M7 EUZ
M7 EVC
M7 EVD
M7 EVE
M7 EVI
M7 EVJ
M7 EVL
M7 EVN
M7 EVO
M7 EVS
M7 EVT
M7 EVU
M7 EVV
M7 EVW
M7 EVX
M7 EVY
M7 EWA
M7 EWC
M7 EWD
M7 EWE
M7 EWG
M7 EWJ
M7 EWK
M7 EWL
M7 EWO
M7 EWR
M7 EWS
M7 EWV
M7 EWW
M7 EWX
M7 EWZ
M7 EXA
M7 EXC
M7 EXD
M7 EXE
M7 EXH
M7 EXK
M7 EXM
M7 EXO
M7 EXP
M7 EXR
M7 EXU
M7 EXW
M7 EXX
M7 EXY
M7 EXZ
M7 EYA
M7 EYD
M7 EYE
M7 EYF
M7 EYG
M7 EYH
M7 EYI
M7 EYJ
M7 EYK
M7 EYL
M7 EYM
M7 EYN
M7 EYQ
M7 EYR
M7 EYS
M7 EYT
M7 EYU
M7 EYV
M7 EYW
M7 EYZ
M7 EZA
M7 EZD
M7 EZE
M7 EZF
M7 EZH
M7 EZI
M7 EZJ
M7 EZK
M7 EZM
M7 EZN
M7 EZT
M7 EZU
M7 EZV
M7 EZY
M7 FAA
M7 FAC
M7 FAD
M7 FAE
M7 FAH
M7 FAK
M7 FAQ
M7 FAR
M7 FAS
M7 FAT
M7 FAU
M7 FAW
M7 FAY
M7 FAZ
M7 FBA
M7 FBB
M7 FBD
M7 FBE
M7 FBF
M7 FBG
M7 FBH
M7 FBI
M7 FBK
M7 FBL
M7 FBS
M7 FBT
M7 FBU
M7 FBV
M7 FCC
M7 FCE
M7 FCF
M7 FCI
M7 FCJ
M7 FCK
M7 FCL
M7 FCM
M7 FCO
M7 FCP
M7 FCQ
M7 FCR
M7 FCS
M7 FCU
M7 FCV
M7 FCZ
M7 FDA
M7 FDB
M7 FDE
M7 FDF
M7 FDH
M7 FDK
M7 FDN
M7 FDO
M7 FDP
M7 FDU
M7 FDV
M7 FDX
M7 FEB
M7 FED
M7 FEE
M7 FEF
M7 FEG
M7 FEI
M7 FEJ
M7 FEK
M7 FEL
M7 FEN
M7 FEQ
M7 FER
M7 FES
M7 FEY
M7 FEZ
M7 FFA
M7 FFB
M7 FFC
M7 FFD
M7 FFE
M7 FFF
M7 FFG
M7 FFH
M7 FFI
M7 FFJ
M7 FFK
M7 FFM
M7 FFN
M7 FFO
M7 FFQ
M7 FFR
M7 FFS
M7 FFY
M7 FFZ
M7 FGA
M7 FGE
M7 FGF
M7 FGK
M7 FGN
M7 FGO
M7 FGP
M7 FGR
M7 FGX
M7 FGZ
M7 FHA
M7 FHB
M7 FHD
M7 FHE
M7 FHH
M7 FHI
M7 FHJ
M7 FHP
M7 FHQ
M7 FHR
M7 FHS
M7 FHT
M7 FHV
M7 FHX
M7 FID
M7 FIF
M7 FIG
M7 FIH
M7 FII
M7 FIK
M7 FIL
M7 FIM
M7 FIO
M7 FIR
M7 FIS
M7 FIT
M7 FIU
M7 FIX
M7 FIY
M7 FJA
M7 FJB
M7 FJD
M7 FJF
M7 FJH
M7 FJI
M7 FJJ
M7 FJK
M7 FJM
M7 FJN
M7 FJO
M7 FJQ
M7 FJV
M7 FJW
M7 FJX
M7 FJZ
M7 FKB
M7 FKC
M7 FKD
M7 FKF
M7 FKH
M7 FKI
M7 FKM
M7 FKN
M7 FKO
M7 FKS
M7 FKU
M7 FKV
M7 FKW
M7 FKY
M7 FKZ
M7 FLB
M7 FLC
M7 FLF
M7 FLH
M7 FLI
M7 FLM
M7 FLO
M7 FLQ
M7 FLS
M7 FLT
M7 FLU
M7 FLV
M7 FLW
M7 FLX
M7 FMC
M7 FMH
M7 FMJ
M7 FML
M7 FMN
M7 FMP
M7 FMQ
M7 FMS
M7 FMU
M7 FMW
M7 FMY
M7 FNA
M7 FND
M7 FNE
M7 FNF
M7 FNH
M7 FNM
M7 FNN
M7 FNQ
M7 FNR
M7 FNT
M7 FNU
M7 FNV
M7 FNW
M7 FNX
M7 FNZ
M7 FOA
M7 FOB
M7 FOD
M7 FOE
M7 FOG
M7 FOH
M7 FOL
M7 FOQ
M7 FOR
M7 FOS
M7 FOT
M7 FOU
M7 FOV
M7 FOW
M7 FOX
M7 FOY
M7 FOZ
M7 FPE
M7 FPF
M7 FPI
M7 FPL
M7 FPM
M7 FPQ
M7 FPR
M7 FPS
M7 FPU
M7 FPV
M7 FPX
M7 FPY
M7 FQE
M7 FQH
M7 FQI
M7 FQJ
M7 FQK
M7 FQL
M7 FQM
M7 FQS
M7 FQU
M7 FQX
M7 FRA
M7 FRB
M7 FRC
M7 FRH
M7 FRJ
M7 FRL
M7 FRM
M7 FRO
M7 FRP
M7 FRQ
M7 FRU
M7 FRV
M7 FRX
M7 FRY
M7 FSA
M7 FSB
M7 FSD
M7 FSF
M7 FSG
M7 FSH
M7 FSI
M7 FSM
M7 FSN
M7 FSO
M7 FSP
M7 FSQ
M7 FSS
M7 FST
M7 FSW
M7 FSX
M7 FSZ
M7 FTA
M7 FTC
M7 FTE
M7 FTJ
M7 FTM
M7 FTO
M7 FTP
M7 FTU
M7 FTW
M7 FTY
M7 FUB
M7 FUD
M7 FUF
M7 FUI
M7 FUJ
M7 FUM
M7 FUP
M7 FUQ
M7 FUS
M7 FUT
M7 FUU
M7 FUV
M7 FUY
M7 FUZ
M7 FVA
M7 FVC
M7 FVD
M7 FVE
M7 FVI
M7 FVJ
M7 FVL
M7 FVN
M7 FVO
M7 FVP
M7 FVQ
M7 FVR
M7 FVT
M7 FVU
M7 FVX
M7 FVZ
M7 FWJ
M7 FWL
M7 FWM
M7 FWN
M7 FWO
M7 FWP
M7 FWQ
M7 FWS
M7 FWY
M7 FXB
M7 FXE
M7 FXF
M7 FXG
M7 FXJ
M7 FXL
M7 FXM
M7 FXO
M7 FXQ
M7 FXS
M7 FXT
M7 FXU
M7 FXX
M7 FXY
M7 FYA
M7 FYE
M7 FYH
M7 FYI
M7 FYK
M7 FYN
M7 FYO
M7 FYS
M7 FYT
M7 FYW
M7 FYX
M7 FYY
M7 FZD
M7 FZF
M7 FZH
M7 FZI
M7 FZJ
M7 FZK
M7 FZL
M7 FZP
M7 FZQ
M7 FZR
M7 FZS
M7 FZT
M7 FZU
M7 FZV
M7 FZX
M7 FZY
M7 FZZ
M7 GAA
M7 GAB
M7 GAD
M7 GAF
M7 GAG
M7 GAI
M7 GAJ
M7 GAK
M7 GAL
M7 GAM
M7 GAO
M7 GAS
M7 GAV
M7 GAZ
M7 GBA
M7 GBB
M7 GBC
M7 GBD
M7 GBE
M7 GBF
M7 GBI
M7 GBK
M7 GBM
M7 GBT
M7 GBV
M7 GBY
M7 GCB
M7 GCC
M7 GCF
M7 GCG
M7 GCI
M7 GCJ
M7 GCK
M7 GCM
M7 GCN
M7 GCP
M7 GDA
M7 GDB
M7 GDC
M7 GDD
M7 GDE
M7 GDH
M7 GDJ
M7 GDK
M7 GDL
M7 GDM
M7 GDN
M7 GDO
M7 GDP
M7 GDR
M7 GDU
M7 GDV
M7 GDX
M7 GDZ
M7 GEA
M7 GED
M7 GEE
M7 GEI
M7 GEK
M7 GEM
M7 GEQ
M7 GES
M7 GEU
M7 GEV
M7 GEW
M7 GEX
M7 GFD
M7 GFF
M7 GFG
M7 GFH
M7 GFK
M7 GFL
M7 GFM
M7 GFN
M7 GFO
M7 GFS
M7 GFT
M7 GFV
M7 GFZ
M7 GGB
M7 GGC
M7 GGE
M7 GGG
M7 GGI
M7 GGJ
M7 GGN
M7 GGO
M7 GGQ
M7 GGR
M7 GGW
M7 GGX
M7 GGY
M7 GGZ
M7 GHG
M7 GHL
M7 GHM
M7 GHQ
M7 GHR
M7 GHT
M7 GHV
M7 GIC
M7 GIG
M7 GIK
M7 GIO
M7 GIW
M7 GIX
M7 GIZ
M7 GJB
M7 GJC
M7 GJH
M7 GJM
M7 GJR
M7 GJS
M7 GJW
M7 GJZ
M7 GKC
M7 GKH
M7 GKK
M7 GKM
M7 GKN
M7 GKP
M7 GKS
M7 GKU
M7 GLA
M7 GLC
M7 GLD
M7 GLG
M7 GLH
M7 GLJ
M7 GLM
M7 GLP
M7 GLR
M7 GLT
M7 GLX
M7 GLZ
M7 GMB
M7 GME
M7 GMI
M7 GML
M7 GMM
M7 GMO
M7 GMT
M7 GMX
M7 GMZ
M7 GND
M7 GNG
M7 GNO
M7 GNS
M7 GNT
M7 GNU
M7 GNZ
M7 GOL
M7 GON
M7 GOR
M7 GOS
M7 GOV
M7 GPA
M7 GPB
M7 GPC
M7 GPD
M7 GPE
M7 GPF
M7 GPK
M7 GPO
M7 GPR
M7 GPS
M7 GPU
M7 GPV
M7 GPZ
M7 GRA
M7 GRB
M7 GRC
M7 GRI
M7 GRJ
M7 GRL
M7 GRN
M7 GRO
M7 GRT
M7 GRW
M7 GRX
M7 GRY
M7 GRZ
M7 GSB
M7 GSC
M7 GSE
M7 GSF
M7 GSK
M7 GSL
M7 GSM
M7 GSN
M7 GSO
M7 GSQ
M7 GSR
M7 GSW
M7 GSX
M7 GSY
M7 GTB
M7 GTF
M7 GTG
M7 GTH
M7 GTO
M7 GTP
M7 GTR
M7 GTT
M7 GTV
M7 GTY
M7 GUB
M7 GUE
M7 GUG
M7 GUI
M7 GUN
M7 GUP
M7 GUR
M7 GUS
M7 GUT
M7 GUU
M7 GUV
M7 GUY
M7 GUZ
M7 GVB
M7 GVE
M7 GVS
M7 GVZ
M7 GWA
M7 GWB
M7 GWN
M7 GWO
M7 GWP
M7 GWR
M7 GWS
M7 GWX
M7 GWY
M7 GWZ
M7 GXA
M7 GXB
M7 GXL
M7 GXN
M7 GXT
M7 GYB
M7 GYF
M7 GYH
M7 GYL
M7 GYM
M7 GYN
M7 GYP
M7 GZM
M7 GZT
M7 GZZ
M7 HAA
M7 HAB
M7 HAC
M7 HAH
M7 HAI
M7 HAK
M7 HAN
M7 HAQ
M7 HAR
M7 HAS
M7 HAT
M7 HAX
M7 HAZ
M7 HBC
M7 HBD
M7 HBF
M7 HBK
M7 HBL
M7 HBR
M7 HBT
M7 HBX
M7 HBY
M7 HCA
M7 HCB
M7 HCC
M7 HCF
M7 HCG
M7 HCI
M7 HCJ
M7 HCK
M7 HCL
M7 HCN
M7 HCW
M7 HCY
M7 HDB
M7 HDD
M7 HDG
M7 HDI
M7 HDL
M7 HDP
M7 HDZ
M7 HEB
M7 HEC
M7 HEG
M7 HEJ
M7 HEM
M7 HEY
M7 HEZ
M7 HFE
M7 HFW
M7 HGA
M7 HGG
M7 HGI
M7 HGS
M7 HGY
M7 HHA
M7 HHC
M7 HIG
M7 HIJ
M7 HIN
M7 HIP
M7 HIT
M7 HIX
M7 HJA
M7 HJD
M7 HJF
M7 HJL
M7 HJT
M7 HKE
M7 HKG
M7 HKL
M7 HKN
M7 HKR
M7 HKS
M7 HKW
M7 HLH
M7 HLI
M7 HLL
M7 HLM
M7 HLO
M7 HLP
M7 HLS
M7 HLT
M7 HLV
M7 HLW
M7 HLY
M7 HMF
M7 HMG
M7 HMJ
M7 HML
M7 HMM
M7 HMP
M7 HMV
M7 HNA
M7 HNM
M7 HNR
M7 HNT
M7 HNY
M7 HOB
M7 HOC
M7 HOE
M7 HOK
M7 HOL
M7 HOO
M7 HOT
M7 HOW
M7 HPG
M7 HPL
M7 HPM
M7 HPR
M7 HQA
M7 HRA
M7 HRB
M7 HRC
M7 HRD
M7 HRE
M7 HRH
M7 HRJ
M7 HRO
M7 HRV
M7 HRX
M7 HRY
M7 HRZ
M7 HSA
M7 HSB
M7 HSE
M7 HSK
M7 HSL
M7 HSQ
M7 HST
M7 HSW
M7 HSZ
M7 HTB
M7 HTD
M7 HTG
M7 HTM
M7 HTO
M7 HTP
M7 HTT
M7 HTW
M7 HTX
M7 HTY
M7 HUF
M7 HUK
M7 HUM
M7 HUN
M7 HUT
M7 HUW
M7 HUZ
M7 HVC
M7 HVR
M7 HVY
M7 HWA
M7 HWD
M7 HWF
M7 HWG
M7 HWM
M7 HWY
M7 HXL
M7 HXM
M7 HXR
M7 HXS
M7 HXX
M7 HYS

M7 HZA
M7 HZS
M7 HZZ
M7 IAC
M7 IAH
M7 IAI
M7 IAN
M7 IAO
M7 IAS
M7 IAT
M7 IBB
M7 IBC
M7 IBJ
M7 IBL
M7 IBM
M7 IBW
M7 ICA
M7 ICB
M7 ICC
M7 ICE
M7 ICJ
M7 ICM
M7 ICO
M7 ICT
M7 ICW
M7 ICX
M7 ICZ
M7 IDA
M7 IDB
M7 IDC
M7 IDE
M7 IDH
M7 IDK
M7 IDX
M7 IEG
M7 IEI
M7 IEL
M7 IFA
M7 IFB
M7 IFE
M7 IFS
M7 IGI
M7 IGK
M7 IGR
M7 IGX
M7 IHB
M7 IHH
M7 IHR
M7 IHS
M7 III
M7 IIM
M7 IJD
M7 IJF
M7 IJH
M7 IJL
M7 IJO
M7 IJV
M7 IKD
M7 IKE
M7 IKL
M7 IKZ
M7 ILE
M7 ILF
M7 ILI
M7 ILK
M7 ILN
M7 ILP
M7 ILS
M7 ILW
M7 ILZ
M7 IMA
M7 IMD
M7 IMF
M7 IMG
M7 IMM
M7 IMP
M7 IMR
M7 IMS
M7 IMT
M7 IMX
M7 INF
M7 INI
M7 INK
M7 INW
M7 IOB
M7 IOI
M7 IOM
M7 IOO
M7 IOP
M7 IOT
M7 IOU
M7 IPA
M7 IPC
M7 IPD
M7 IPH
M7 IPN
M7 IPO
M7 IPS
M7 IPT
M7 IPX
M7 IQQ
M7 IQT
M7 IRE
M7 IRF
M7 IRH
M7 IRL
M7 IRN
M7 IRV
M7 ISI
M7 ISK
M7 ISO
M7 ISP
M7 ISS
M7 IST
M7 ISW

M7 ITB
M7 ITC
M7 ITG
M7 ITH
M7 ITM
M7 ITO
M7 ITS
M7 ITT
M7 ITV
M7 ITX
M7 ITY
M7 IVL
M7 IVP
M7 IWH
M7 IXG
M7 IXN
M7 IXX
M7 IZD
M7 IZI
M7 IZY
M7 IZZ
M7 JAA
M7 JAF
M7 JAH
M7 JAI
M7 JAK
M7 JAL
M7 JAM
M7 JAN
M7 JAO
M7 JAU
M7 JAX
M7 JBA
M7 JBC
M7 JBD
M7 JBF
M7 JBG
M7 JBN
M7 JBR
M7 JBX
M7 JBZ
M7 JCF
M7 JCI
M7 JCJ
M7 JCO
M7 JCT
M7 JCU
M7 JCW
M7 JCY
M7 JDA
M7 JDD
M7 JDE
M7 JDF
M7 JDG
M7 JDJ
M7 JDK
M7 JDN
M7 JDO
M7 JDR
M7 JDS
M7 JDT
M7 JDV
M7 JDZ
M7 JEA
M7 JEB
M7 JEC
M7 JED
M7 JEG
M7 JEK
M7 JEM
M7 JER
M7 JES
M7 JET
M7 JEY
M7 JEZ
M7 JFC
M7 JFM
M7 JFS
M7 JFT
M7 JGA
M7 JGG
M7 JGJ
M7 JGK
M7 JGL
M7 JGT
M7 JHB
M7 JHC
M7 JHG
M7 JHH
M7 JHP
M7 JHS
M7 JHZ
M7 JIL
M7 JIN
M7 JIP
M7 JIT
M7 JJA
M7 JJB
M7 JJD
M7 JJF
M7 JJH
M7 JJM
M7 JJP
M7 JJR
M7 JJS
M7 JJW
M7 JKB
M7 JKI
M7 JKK
M7 JKL
M7 JKP
M7 JKR
M7 JKS
M7 JKT
M7 JKV

M7 JKX
M7 JLA
M7 JLB
M7 JLE
M7 JLH
M7 JLN
M7 JLO
M7 JLP
M7 JLR
M7 JLS
M7 JLT
M7 JLY
M7 JLZ
M7 JMB
M7 JMG
M7 JMI
M7 JMK
M7 JML
M7 JMM
M7 JMN
M7 JMQ
M7 JMT
M7 JMU
M7 JMV
M7 JMY
M7 JMZ
M7 JNA
M7 JND
M7 JNL
M7 JNM
M7 JNP
M7 JNS
M7 JNT
M7 JNX
M7 JNZ
M7 JOA
M7 JOB
M7 JOG
M7 JOH
M7 JOK
M7 JOO
M7 JOR
M7 JOT
M7 JOW
M7 JPA
M7 JPB
M7 JPD
M7 JPE
M7 JPG
M7 JPH
M7 JPK
M7 JPL
M7 JPN
M7 JPO
M7 JPQ
M7 JPT
M7 JPW
M7 JPZ
M7 JQI
M7 JQN
M7 JQR
M7 JRC
M7 JRD
M7 JRE
M7 JRF
M7 JRJ
M7 JRL
M7 JRN
M7 JRO
M7 JRP
M7 JRR
M7 JRT
M7 JRV
M7 JRW
M7 JRZ
M7 JSB
M7 JSE
M7 JSM
M7 JSN
M7 JSO
M7 JSR
M7 JSZ
M7 JTA
M7 JTB
M7 JTC
M7 JTD
M7 JTK
M7 JTL
M7 JTM
M7 JTO
M7 JTP
M7 JTQ
M7 JTS
M7 JTT
M7 JTW
M7 JTX
M7 JTY
M7 JUD
M7 JUG
M7 JVB
M7 JVE
M7 JVF
M7 JVL
M7 JVM
M7 JWA
M7 JWB
M7 JWD
M7 JWG
M7 JWN
M7 JWS
M7 JWT
M7 JWX
M7 JWY
M7 JXC
M7 JXK

M7 JXM
M7 JXN
M7 JXP
M7 JXS
M7 JXT
M7 JXZ
M7 JYD
M7 JYM
M7 JYN
M7 KAB
M7 KAH
M7 KAI
M7 KAJ
M7 KAL
M7 KAM
M7 KAN
M7 KAO
M7 KAQ
M7 KAR
M7 KAX
M7 KAY
M7 KAZ
M7 KBE
M7 KBL
M7 KBM
M7 KBR
M7 KBS
M7 KBT
M7 KBX
M7 KBZ
M7 KCC
M7 KCI
M7 KCL
M7 KCM
M7 KCN
M7 KCS
M7 KCT
M7 KCW
M7 KCX
M7 KDC
M7 KDE
M7 KDF
M7 KDK
M7 KDL
M7 KDM
M7 KDS
M7 KDW
M7 KDX
M7 KEB
M7 KED
M7 KEG
M7 KEK
M7 KEM
M7 KER
M7 KES
M7 KET
M7 KEX
M7 KEY
M7 KFE
M7 KFH
M7 KFM
M7 KFO
M7 KFX
M7 KGB
M7 KGE
M7 KGI
M7 KGL
M7 KGR
M7 KGS
M7 KHI
M7 KHK
M7 KHN
M7 KHR
M7 KHT
M7 KHZ
M7 KIA
M7 KIB
M7 KIG
M7 KIL
M7 KIN
M7 KIP
M7 KIS
M7 KIX
M7 KIZ
M7 KJA
M7 KJB
M7 KJE
M7 KJH
M7 KJI
M7 KJO
M7 KJR
M7 KJX
M7 KKA
M7 KKC
M7 KKE
M7 KKH
M7 KKI
M7 KKJ
M7 KKP
M7 KLE
M7 KLH
M7 KLI
M7 KLL
M7 KLN
M7 KLO
M7 KLS
M7 KLU
M7 KLV
M7 KLX
M7 KLY
M7 KLZ
M7 KMA
M7 KMC
M7 KMD
M7 KME

M7 KMH
M7 KMK
M7 KML
M7 KMM
M7 KMR
M7 KMT
M7 KMZ
M7 KNB
M7 KNJ
M7 KNS
M7 KNT
M7 KNZ
M7 KOA
M7 KOB
M7 KOE
M7 KOI
M7 KOJ
M7 KOM
M7 KON
M7 KOO
M7 KOQ
M7 KOS
M7 KOT
M7 KPD
M7 KPF
M7 KPH
M7 KPI
M7 KPK
M7 KPO
M7 KPP
M7 KPR
M7 KPT
M7 KPX
M7 KPZ
M7 KRB
M7 KRC
M7 KRG
M7 KRI
M7 KRM
M7 KRO
M7 KRS
M7 KRT
M7 KSA
M7 KSC
M7 KSG
M7 KSH
M7 KSI
M7 KSK
M7 KSP
M7 KSR
M7 KSS
M7 KTC
M7 KTE
M7 KTF
M7 KTG
M7 KTH
M7 KTI
M7 KTM
M7 KTN
M7 KTU
M7 KTX
M7 KTZ
M7 KUC
M7 KUH
M7 KUK
M7 KUL
M7 KVA
M7 KVH
M7 KVJ
M7 KVK
M7 KVN
M7 KVP
M7 KVT
M7 KVX
M7 KWD
M7 KWH
M7 KWK
M7 KWM
M7 KWP
M7 KWW
M7 KWZ
M7 KXA
M7 KXP
M7 KXS
M7 KYE
M7 KYN
M7 KYO
M7 KZB
M7 KZI
M7 KZN
M7 KZR
M7 KZZ
M7 LAB
M7 LAC
M7 LAD
M7 LAF
M7 LAL
M7 LAR
M7 LAS
M7 LAU
M7 LAV
M7 LAX
M7 LAY
M7 LAZ
M7 LBK
M7 LBL
M7 LBR
M7 LBS
M7 LCC
M7 LCE
M7 LCG
M7 LCH
M7 LCK
M7 LCM
M7 LCN

M7 LCO
M7 LCT
M7 LCX
M7 LDA
M7 LDB
M7 LDC
M7 LDF
M7 LDG
M7 LDH
M7 LDL
M7 LDN
M7 LDP
M7 LDQ
M7 LDR
M7 LDT
M7 LDY
M7 LDZ
M7 LEA
M7 LEB
M7 LEC
M7 LEE
M7 LEG
M7 LEI
M7 LEK
M7 LER
M7 LES
M7 LET
M7 LEU
M7 LEV
M7 LFB
M7 LFC
M7 LFM
M7 LFO
M7 LFR
M7 LFS
M7 LFT
M7 LFX
M7 LGA
M7 LGH
M7 LGJ
M7 LGL
M7 LGM
M7 LGN
M7 LGO
M7 LGR
M7 LGS
M7 LGX
M7 LGZ
M7 LHG
M7 LHJ
M7 LHL
M7 LHM
M7 LHN
M7 LHO
M7 LHS
M7 LHT
M7 LHU
M7 LHX
M7 LIA
M7 LIG
M7 LIJ
M7 LIL
M7 LIS
M7 LIT
M7 LIV
M7 LIZ
M7 LJB
M7 LJC
M7 LJF
M7 LJG
M7 LJN
M7 LJO
M7 LJP
M7 LJS
M7 LJW
M7 LKA
M7 LKB
M7 LKC
M7 LKD
M7 LKG
M7 LKL
M7 LKM
M7 LKR
M7 LLA
M7 LLC
M7 LLF
M7 LLG
M7 LLH
M7 LLI
M7 LLL
M7 LLO
M7 LLT
M7 LMD
M7 LMH
M7 LML
M7 LMO
M7 LMR
M7 LMT
M7 LMW
M7 LMX
M7 LMY
M7 LMZ
M7 LNC
M7 LNE
M7 LNR
M7 LNS
M7 LNX
M7 LNZ
M7 LOB
M7 LOD
M7 LOG
M7 LOK
M7 LOM
M7 LON
M7 LOP

M7 LOR
M7 LOS
M7 LOU
M7 LOY
M7 LPD
M7 LPE
M7 LPG
M7 LPH
M7 LPM
M7 LPO
M7 LPS
M7 LPW
M7 LPY
M7 LQE
M7 LRA
M7 LRD
M7 LRF
M7 LRG
M7 LRH
M7 LRI
M7 LRN
M7 LRO
M7 LRS
M7 LRW
M7 LRZ
M7 LSC
M7 LSD
M7 LSH
M7 LSI
M7 LSK
M7 LSM
M7 LSO
M7 LSP
M7 LSR
M7 LSW
M7 LSY
M7 LTB
M7 LTC
M7 LTD
M7 LTL
M7 LTN
M7 LTQ
M7 LTU
M7 LTW
M7 LTX
M7 LTZ
M7 LUC
M7 LUD
M7 LUM
M7 LUN
M7 LUX
M7 LVE
M7 LVN
M7 LVR
M7 LVS
M7 LWA
M7 LWB
M7 LWE
M7 LWG
M7 LWK
M7 LWN
M7 LWO
M7 LWS
M7 LXF
M7 LXH
M7 LXK
M7 LXL
M7 LXP
M7 LXX
M7 LYK
M7 LYO
M7 LZA
M7 LZB
M7 LZE
M7 LZW
M7 LZZ
M7 MAC
M7 MAF
M7 MAH
M7 MAJ
M7 MAK
M7 MAL
M7 MAO
M7 MAQ
M7 MAR
M7 MAU
M7 MAV
M7 MAX
M7 MAY
M7 MAZ
M7 MBA
M7 MBD
M7 MBF
M7 MBG
M7 MBH
M7 MBL
M7 MBM
M7 MBN
M7 MBR
M7 MBS
M7 MBU
M7 MBW
M7 MBX
M7 MBY
M7 MBZ
M7 MCE
M7 MCH
M7 MCL
M7 MCP
M7 MCV
M7 MCW
M7 MCX
M7 MCZ
M7 MDF
M7 MDG

M7 MDK
M7 MDL
M7 MDM
M7 MDO
M7 MDP
M7 MDS
M7 MDT
M7 MDV
M7 MDW
M7 MDX
M7 MEA
M7 MEE
M7 MEF
M7 MEG
M7 MEH
M7 MEI
M7 MEL
M7 MEM
M7 MEN
M7 MER
M7 MES
M7 MEU
M7 MEW
M7 MEY
M7 MFA
M7 MFC
M7 MFG
M7 MFJ
M7 MFP
M7 MFS
M7 MFT
M7 MGA
M7 MGC
M7 MGE
M7 MGI
M7 MGJ
M7 MGK
M7 MGN
M7 MGO
M7 MGP
M7 MGR
M7 MGT
M7 MGV
M7 MGY
M7 MHD
M7 MHL
M7 MHN
M7 MHT
M7 MHW
M7 MHX
M7 MHY
M7 MHZ
M7 MIA
M7 MIB
M7 MIC
M7 MID
M7 MIF
M7 MIG
M7 MIH
M7 MIJ
M7 MIK
M7 MIL
M7 MIM
M7 MIO
M7 MIP
M7 MIS
M7 MIW
M7 MIX
M7 MIZ
M7 MJA
M7 MJD
M7 MJE
M7 MJF
M7 MJJ
M7 MJM
M7 MJN
M7 MJP
M7 MJS
M7 MJT
M7 MJW
M7 MJY
M7 MJZ
M7 MKA
M7 MKE
M7 MKF
M7 MKL
M7 MKM
M7 MKN
M7 MKP
M7 MKR
M7 MKS
M7 MKT
M7 MKU
M7 MLA
M7 MLG
M7 MLH
M7 MLJ
M7 MLN
M7 MLO
M7 MLR
M7 MLS
M7 MLT
M7 MLW
M7 MLX
M7 MMD
M7 MMG
M7 MMH
M7 MMI
M7 MMJ
M7 MMR
M7 MMS
M7 MMX
M7 MMY
M7 MMZ
M7 MNA

M7 MNE
M7 MNG
M7 MNK
M7 MNL
M7 MNM
M7 MNO
M7 MNP
M7 MNS
M7 MNT
M7 MNW
M7 MNX
M7 MOA
M7 MOB
M7 MOD
M7 MOF
M7 MOG
M7 MOJ
M7 MOM
M7 MOP
M7 MOS
M7 MOV
M7 MOX
M7 MPA
M7 MPB
M7 MPC
M7 MPE
M7 MPF
M7 MPG
M7 MPH
M7 MPK
M7 MPP
M7 MPR
M7 MPT
M7 MPW
M7 MPX
M7 MPZ
M7 MQS
M7 MQT
M7 MRC
M7 MRD
M7 MRE
M7 MRF
M7 MRH
M7 MRM
M7 MRS
M7 MRU
M7 MRV
M7 MRW
M7 MRZ
M7 MSC
M7 MSN
M7 MSO
M7 MSP
M7 MSS
M7 MST
M7 MSY
M7 MSZ
M7 MTA
M7 MTC
M7 MTD
M7 MTF
M7 MTI
M7 MTJ
M7 MTK
M7 MTL
M7 MTN
M7 MTO
M7 MTP
M7 MTR
M7 MTT
M7 MTX
M7 MTZ
M7 MUC
M7 MUD
M7 MUG
M7 MUH
M7 MUK
M7 MUM
M7 MUN
M7 MUP
M7 MUR
M7 MUX
M7 MUZ
M7 MVC
M7 MVH
M7 MVK
M7 MVN
M7 MVP
M7 MWA
M7 MWB
M7 MWC
M7 MWH
M7 MWI
M7 MWJ
M7 MWK
M7 MWP
M7 MWR
M7 MWS
M7 MWV
M7 MWZ
M7 MXA
M7 MXC
M7 MXD
M7 MXE
M7 MXG
M7 MXJ
M7 MXL
M7 MXP
M7 MXR
M7 MXT
M7 MXX
M7 MXZ
M7 MYD
M7 MYN
M7 MYR

M7 MYT
M7 MYX
M7 MYY
M7 MZB
M7 MZC
M7 MZD
M7 MZK
M7 MZU
M7 MZX
M7 MZZ
M7 NAA
M7 NAD
M7 NAF
M7 NAK
M7 NAP
M7 NAR
M7 NAU
M7 NAY
M7 NBB
M7 NBD
M7 NBG
M7 NBI
M7 NBL
M7 NBQ
M7 NBS
M7 NBY
M7 NCE
M7 NCF
M7 NCG
M7 NCH
M7 NCJ
M7 NCK
M7 NCL
M7 NCT
M7 NCX
M7 NDC
M7 NDP
M7 NDR
M7 NDW
M7 NDX
M7 NDY
M7 NDZ
M7 NEB
M7 NEC
M7 NEE
M7 NEG
M7 NEO
M7 NER
M7 NEU
M7 NEV
M7 NEX
M7 NEY
M7 NFD
M7 NFG
M7 NFI
M7 NFJ
M7 NFM
M7 NFO
M7 NFR
M7 NFT
M7 NGA
M7 NGB
M7 NGD
M7 NGF
M7 NGM
M7 NGO
M7 NGR
M7 NGY
M7 NHC
M7 NHK
M7 NHL
M7 NHM
M7 NHT
M7 NIA
M7 NIC
M7 NIF
M7 NII
M7 NIN
M7 NIO
M7 NIS
M7 NIU
M7 NIX
M7 NJA
M7 NJB
M7 NJC
M7 NJE
M7 NJH
M7 NJL
M7 NJN
M7 NJP
M7 NJS
M7 NJT
M7 NJW
M7 NKD
M7 NKI
M7 NKJ
M7 NKO
M7 NKT
M7 NKX
M7 NKY
M7 NLB
M7 NLD
M7 NLM
M7 NLX
M7 NME
M7 NMR
M7 NMS
M7 NNN
M7 NOA
M7 NOD
M7 NOG
M7 NOK
M7 NOM
M7 NOP
M7 NOR

M7 NOS
M7 NOT
M7 NOX
M7 NPB
M7 NPF
M7 NPL
M7 NPN
M7 NPR
M7 NPS
M7 NPY
M7 NQL
M7 NRA
M7 NRB
M7 NRD
M7 NRF
M7 NRG
M7 NRP
M7 NSA
M7 NSD
M7 NSG
M7 NSH
M7 NSK
M7 NSM
M7 NSN
M7 NSR
M7 NSW
M7 NTA
M7 NTB
M7 NTS
M7 NTX
M7 NTY
M7 NUB
M7 NUC
M7 NUG
M7 NUL
M7 NUM
M7 NUS
M7 NUX
M7 NVA
M7 NVI
M7 NWB
M7 NWC
M7 NWG
M7 NWH
M7 NWO
M7 NWR
M7 NWW
M7 NXF
M7 NXT
M7 NXX
M7 NYA
M7 NYP
M7 NYT
M7 NYX
M7 NZA
M7 NZR
M7 NZZ
M7 OAB
M7 OAC
M7 OAF
M7 OAH
M7 OAL
M7 OAM
M7 OAP
M7 OAR
M7 OAT
M7 OAX
M7 OBE
M7 OBG
M7 OBK
M7 OBN
M7 OBO
M7 OBS
M7 OCA
M7 OCC
M7 OCD
M7 OCE
M7 OCG
M7 OCH
M7 OCS
M7 OCW
M7 ODB
M7 ODE
M7 ODN
M7 ODP
M7 ODR
M7 ODX
M7 OED
M7 OEM
M7 OES
M7 OET
M7 OEW
M7 OEY
M7 OFB
M7 OFC
M7 OFO
M7 OGA
M7 OGD
M7 OGG
M7 OGT
M7 OGV
M7 OGY
M7 OGY
M7 OHC
M7 OHO
M7 OHS
M7 OHY
M7 OIL
M7 OIO
M7 OIR
M7 OJB
M7 OJC
M7 OJG
M7 OJH
M7 OJK

M7 OJT
M7 OKA
M7 OKE
M7 OKI
M7 OKL
M7 OKT
M7 OKW
M7 OKY
M7 OLC
M7 OLD
M7 OLE
M7 OLG
M7 OLH
M7 OLI
M7 OLJ
M7 OLK
M7 OLL
M7 OLN
M7 OLS
M7 OLT
M7 OLW
M7 OLX
M7 OLZ
M7 OMC
M7 OML
M7 OMR
M7 OMT
M7 OMZ
M7 OND
M7 ONE
M7 ONG
M7 ONI
M7 ONK
M7 ONS
M7 ONX
M7 ONY
M7 OOB
M7 OOH
M7 OOK
M7 OOL
M7 OOM
M7 OON
M7 OOP
M7 OOR
M7 OOS
M7 OOT
M7 OOX
M7 OOZ
M7 OPC
M7 OPD
M7 OPG
M7 OPK
M7 OPL
M7 OPN
M7 OPO
M7 OPR
M7 OPS
M7 ORA
M7 ORB
M7 ORD
M7 ORF
M7 ORH
M7 ORK
M7 ORM
M7 ORN
M7 ORS
M7 ORT
M7 ORV
M7 ORX
M7 ORY
M7 ORZ
M7 OSA
M7 OSC
M7 OSI
M7 OSJ
M7 OSK
M7 OSL
M7 OSO
M7 OSU
M7 OTA
M7 OTC
M7 OTG
M7 OTH
M7 OTI
M7 OTJ
M7 OTP
M7 OTR
M7 OTS
M7 OTT
M7 OTW
M7 OUV
M7 OUZ
M7 OVE
M7 OVR
M7 OWC
M7 OWE
M7 OWQ
M7 OWW
M7 OXF
M7 OXG
M7 OXN
M7 OXT
M7 OXW
M7 OYD
M7 OYZ
M7 OZD
M7 OZI
M7 OZN
M7 OZO
M7 OZX
M7 OZY
M7 OZZ
M7 PAB
M7 PAD
M7 PAE

M7 PAF
M7 PAH
M7 PAK
M7 PAL
M7 PAM
M7 PAP
M7 PAQ
M7 PAS
M7 PAT
M7 PAW
M7 PAX
M7 PAZ
M7 PBA
M7 PBC
M7 PBK
M7 PBM
M7 PBP
M7 PBR
M7 PBU
M7 PBZ
M7 PCH
M7 PCJ
M7 PCK
M7 PCM
M7 PCO
M7 PCR
M7 PCT
M7 PCU
M7 PCZ
M7 PDB
M7 PDC
M7 PDD
M7 PDF
M7 PDG
M7 PDH
M7 PDI
M7 PDJ
M7 PDL
M7 PDM
M7 PDP
M7 PDY
M7 PEA
M7 PEC
M7 PED
M7 PEE
M7 PEF
M7 PEG
M7 PEK
M7 PEM
M7 PEP
M7 PER
M7 PET
M7 PEZ
M7 PFE
M7 PFL
M7 PFO
M7 PGB
M7 PGC
M7 PGD
M7 PGK
M7 PGS
M7 PGT
M7 PGV
M7 PGW
M7 PGZ
M7 PHD
M7 PHG
M7 PHK
M7 PHL
M7 PHM
M7 PHN
M7 PHO
M7 PHP
M7 PHS
M7 PHW
M7 PHX
M7 PIA
M7 PIC
M7 PIE
M7 PIL
M7 PIP
M7 PIR
M7 PIT
M7 PJA
M7 PJC
M7 PJE
M7 PJG
M7 PJL
M7 PJP
M7 PJR
M7 PJW
M7 PJX
M7 PKB
M7 PKD
M7 PKE
M7 PKJ
M7 PKL
M7 PKM
M7 PKR
M7 PKS
M7 PKY
M7 PLC
M7 PLE
M7 PLF
M7 PLJ
M7 PLM
M7 PLR
M7 PLU
M7 PLX
M7 PLZ
M7 PMA
M7 PMC
M7 PMD
M7 PMG
M7 PMI

M7 PMJ
M7 PMM
M7 PMN
M7 PMO
M7 PMP
M7 PMQ
M7 PMR
M7 PMS
M7 PMV
M7 PMX
M7 PMZ
M7 PNB
M7 PNF
M7 PNH
M7 PNI
M7 PNK
M7 PNP
M7 PNW
M7 PNX
M7 POB
M7 POD
M7 POH
M7 POM
M7 POP
M7 POY
M7 PPC
M7 PPK
M7 PPP
M7 PPS
M7 PPT
M7 PPY
M7 PQR
M7 PRB
M7 PRC
M7 PRD
M7 PRJ
M7 PRK
M7 PRL
M7 PRM
M7 PRO
M7 PRP
M7 PRR
M7 PRS
M7 PRT
M7 PRU
M7 PRX
M7 PRY
M7 PSA
M7 PSD
M7 PSH
M7 PSK
M7 PSL
M7 PSO
M7 PSP
M7 PSR
M7 PST
M7 PSW
M7 PTA
M7 PTB
M7 PTC
M7 PTD
M7 PTE
M7 PTG
M7 PTH
M7 PTK
M7 PTL
M7 PTM
M7 PTN
M7 PTO
M7 PTR
M7 PTS
M7 PTT
M7 PTX
M7 PTZ
M7 PUB
M7 PUD
M7 PUG
M7 PUH
M7 PUN
M7 PUR
M7 PVB
M7 PVD
M7 PVL
M7 PVM
M7 PVT
M7 PWE
M7 PWG
M7 PWJ
M7 PWM
M7 PWN
M7 PWO
M7 PWP
M7 PWR
M7 PWS
M7 PWZ
M7 PXD
M7 PXE
M7 PXK
M7 PXP
M7 PXR
M7 PXT
M7 PXX
M7 PYB
M7 PYF
M7 PYN
M7 PYO
M7 PYX
M7 PZK
M7 PZR
M7 PZT
M7 PZW
M7 RAA
M7 RAC
M7 RAG
M7 RAH

M7 RAK
M7 RAL
M7 RAM
M7 RAN
M7 RAP
M7 RAQ
M7 RAR
M7 RBA
M7 RBB
M7 RBD
M7 RBF
M7 RBH
M7 RBJ
M7 RBK
M7 RBM
M7 RBN
M7 RBS
M7 RBX
M7 RBZ
M7 RCA
M7 RCB
M7 RCD
M7 RCF
M7 RCJ
M7 RCK
M7 RCL
M7 RCN
M7 RCO
M7 RCY
M7 RDE
M7 RDF
M7 RDG
M7 RDJ
M7 RDL
M7 RDM
M7 RDO
M7 RDP
M7 RDR
M7 RDT
M7 RDV
M7 RDW
M7 RDX
M7 RDY
M7 REA
M7 REB
M7 REC
M7 REE
M7 REG
M7 REH
M7 REI
M7 REJ
M7 REM
M7 REN
M7 REW
M7 RFB
M7 RFC
M7 RFJ
M7 RFK
M7 RFN
M7 RFQ
M7 RFS
M7 RFT
M7 RFX
M7 RGA
M7 RGC
M7 RGL
M7 RGW
M7 RGX
M7 RHE
M7 RHK
M7 RHL
M7 RHO
M7 RHT
M7 RHU
M7 RIB
M7 RIC
M7 RIF
M7 RIG
M7 RIO
M7 RIP
M7 RIR
M7 RIX
M7 RJA
M7 RJE
M7 RJF
M7 RJK
M7 RJM
M7 RJO
M7 RJP
M7 RJS
M7 RJV
M7 RJW
M7 RJX
M7 RJY
M7 RJZ
M7 RKB
M7 RKF
M7 RKG
M7 RKI
M7 RKK
M7 RKM
M7 RKO
M7 RKP
M7 RKR
M7 RKT
M7 RKX
M7 RLB
M7 RLH
M7 RLK
M7 RLM
M7 RLO
M7 RLR
M7 RLX
M7 RLY
M7 RMA

M7 RMB
M7 RMD
M7 RME
M7 RMF
M7 RMG
M7 RMJ
M7 RMK
M7 RMM
M7 RMO
M7 RMR
M7 RMS
M7 RMT
M7 RMX
M7 RMY
M7 RNA
M7 RND
M7 RNF
M7 RNG
M7 RNJ
M7 RNM
M7 RNN
M7 RNR
M7 RNT
M7 RNW
M7 ROA
M7 ROB
M7 ROE
M7 ROF
M7 ROJ
M7 ROK
M7 ROL
M7 ROM
M7 ROP
M7 ROR
M7 ROS
M7 ROU
M7 ROW
M7 ROZ
M7 RPB
M7 RPC
M7 RPE
M7 RPG
M7 RPH
M7 RPJ
M7 RPM
M7 RPN
M7 RPS
M7 RPW
M7 RPX
M7 RPZ
M7 RRA
M7 RRB
M7 RRC
M7 RRD
M7 RRE
M7 RRK
M7 RRL
M7 RRM
M7 RRP
M7 RRR
M7 RRS
M7 RRY
M7 RSA
M7 RSB
M7 RSF
M7 RSG
M7 RSI
M7 RSK
M7 RSL
M7 RSN
M7 RSP
M7 RSR
M7 RSS
M7 RSU
M7 RSV
M7 RSW
M7 RSX
M7 RSZ
M7 RTA
M7 RTC
M7 RTF
M7 RTG
M7 RTH
M7 RTI
M7 RTJ
M7 RTL
M7 RTM
M7 RTN
M7 RTP
M7 RTR
M7 RTW
M7 RTX
M7 RTZ
M7 RUB
M7 RUC
M7 RUD
M7 RUF
M7 RUG
M7 RUM
M7 RUV
M7 RUX
M7 RVB
M7 RVK
M7 RVP
M7 RVR
M7 RVS
M7 RVX
M7 RVY
M7 RWB
M7 RWC
M7 RWD
M7 RWG
M7 RWI
M7 RWK
M7 RWL

Withheld

M7 RWM
M7 RWP
M7 RWR
M7 RWS
M7 RWX
M7 RWZ
M7 RXA
M7 RXB
M7 RXT
M7 RXX
M7 RXY
M7 RYA
M7 RYB
M7 RYE
M7 RYG
M7 RYM
M7 RYN
M7 RYU
M7 RZL
M7 RZN
M7 RZZ
M7 SAA
M7 SAD
M7 SAE
M7 SAF
M7 SAG
M7 SAI
M7 SAJ
M7 SAL
M7 SAM
M7 SAN
M7 SAO
M7 SAS
M7 SAT
M7 SAU
M7 SAW
M7 SAX
M7 SAY
M7 SBC
M7 SBE
M7 SBF
M7 SBG
M7 SBH
M7 SBJ
M7 SBL
M7 SBR
M7 SBS
M7 SBT
M7 SBX
M7 SCA
M7 SCE
M7 SCI
M7 SCJ
M7 SCL
M7 SCM
M7 SCU
M7 SCZ
M7 SDA
M7 SDC
M7 SDH
M7 SDK
M7 SDL
M7 SDM
M7 SDN
M7 SDO
M7 SDP
M7 SDR
M7 SDS
M7 SDT
M7 SDU
M7 SDV
M7 SDW
M7 SDY
M7 SEB
M7 SEC
M7 SED
M7 SEE
M7 SEI
M7 SEL
M7 SEN
M7 SEO
M7 SEP
M7 SER
M7 SET
M7 SFA
M7 SFC
M7 SFK
M7 SFL
M7 SFM
M7 SFN
M7 SFO
M7 SFP
M7 SFR
M7 SFS
M7 SFT
M7 SFV
M7 SFX
M7 SGC
M7 SGE
M7 SGH
M7 SGK
M7 SGL
M7 SGN
M7 SGP
M7 SGR
M7 SGS
M7 SGU
M7 SGW
M7 SGX
M7 SGZ
M7 SHA
M7 SHD
M7 SHE
M7 SHG
M7 SHH
M7 SHI
M7 SHK
M7 SHL
M7 SHS
M7 SHV
M7 SHX
M7 SHY
M7 SIA
M7 SIE
M7 SIL
M7 SIM
M7 SIO
M7 SIP
M7 SIR
M7 SIT
M7 SIU
M7 SIX
M7 SJA
M7 SJD
M7 SJE
M7 SJI
M7 SJJ
M7 SJK
M7 SJO
M7 SJP
M7 SJQ
M7 SJR
M7 SJS
M7 SKC
M7 SKD
M7 SKK
M7 SKM
M7 SKR
M7 SKX
M7 SKY
M7 SLF
M7 SLJ
M7 SLL
M7 SLM
M7 SLQ
M7 SLS
M7 SLX
M7 SMA
M7 SMB
M7 SMD
M7 SMF
M7 SMI
M7 SMJ
M7 SMK
M7 SMM
M7 SMO
M7 SMR
M7 SMS
M7 SMT
M7 SMU
M7 SMW
M7 SMY
M7 SNA
M7 SNC
M7 SNE
M7 SNF
M7 SNG
M7 SNK
M7 SNM
M7 SNO
M7 SNT
M7 SNW
M7 SNY
M7 SOA
M7 SOB
M7 SOF
M7 SOG
M7 SOK
M7 SOL
M7 SOM
M7 SOO
M7 SOR
M7 SOU
M7 SOV
M7 SOY
M7 SPB
M7 SPD
M7 SPF
M7 SPG
M7 SPI
M7 SPK
M7 SPN
M7 SPR
M7 SPT
M7 SPV
M7 SQD
M7 SQF
M7 SQI
M7 SQL
M7 SQN
M7 SQQ
M7 SQT
M7 SRA
M7 SRC
M7 SRD
M7 SRF
M7 SRI
M7 SRJ
M7 SRK
M7 SRL
M7 SRM
M7 SRO
M7 SRQ
M7 SRS
M7 SRX
M7 SRY
M7 SSA
M7 SSD
M7 SSH
M7 SSK
M7 SSL
M7 SSP
M7 SSR
M7 SST
M7 STA
M7 STB
M7 STC
M7 STD
M7 STE
M7 STG
M7 STI
M7 STJ
M7 STN
M7 STO
M7 STP
M7 STR
M7 STW
M7 STX
M7 STY
M7 STZ
M7 SUB
M7 SUC
M7 SUD
M7 SUI
M7 SUK
M7 SUR
M7 SUV
M7 SUX
M7 SVF
M7 SVG
M7 SVH
M7 SVN
M7 SVS
M7 SVV
M7 SWF
M7 SWG
M7 SWH
M7 SWI
M7 SWJ
M7 SWL
M7 SWN
M7 SWP
M7 SWZ
M7 SXA
M7 SXB
M7 SXC
M7 SXG
M7 SXI
M7 SXL
M7 SXP
M7 SXX
M7 SXY
M7 SYG
M7 SYM
M7 SYN
M7 SYR
M7 SZA
M7 SZL
M7 SZM
M7 SZP
M7 SZR
M7 SZS
M7 SZY
M7 SZZ
M7 TAB
M7 TAC
M7 TAE
M7 TAJ
M7 TAK
M7 TAL
M7 TAM
M7 TAN
M7 TAO
M7 TAP
M7 TAR
M7 TAU
M7 TAV
M7 TAX
M7 TBB
M7 TBE
M7 TBG
M7 TBH
M7 TBJ
M7 TBL
M7 TBN
M7 TBS
M7 TBT
M7 TBU
M7 TBY
M7 TCB
M7 TCC
M7 TCD
M7 TCE
M7 TCG
M7 TCH
M7 TCI
M7 TCJ
M7 TCO
M7 TCP
M7 TCR
M7 TCX
M7 TCY
M7 TCZ
M7 TDB
M7 TDC
M7 TDD
M7 TDF
M7 TDG
M7 TDH
M7 TDI
M7 TDM
M7 TDN
M7 TDO
M7 TDP
M7 TDW
M7 TDX
M7 TDZ
M7 TEA
M7 TEC
M7 TED
M7 TEE
M7 TEH
M7 TEK
M7 TEL
M7 TEM
M7 TEN
M7 TES
M7 TET
M7 TEV
M7 TEX
M7 TEZ
M7 TFF
M7 TFH
M7 TFJ
M7 TFL
M7 TFO
M7 TGB
M7 TGC
M7 TGI
M7 TGL
M7 TGP
M7 TGS
M7 TGV
M7 TGW
M7 TGZ
M7 THC
M7 THD
M7 THE
M7 THH
M7 THK
M7 THM
M7 THS
M7 THT
M7 THW
M7 THZ
M7 TIF
M7 TIG
M7 TIL
M7 TIO
M7 TIP
M7 TIR
M7 TIS
M7 TJA
M7 TJG
M7 TJH
M7 TJJ
M7 TJK
M7 TJL
M7 TJP
M7 TJR
M7 TJW
M7 TKA
M7 TKE
M7 TKF
M7 TKJ
M7 TKM
M7 TKS
M7 TLB
M7 TLD
M7 TLF
M7 TLH
M7 TLK
M7 TLL
M7 TLN
M7 TLP
M7 TLR
M7 TLT
M7 TLW
M7 TLY
M7 TLZ
M7 TMB
M7 TME
M7 TMG
M7 TMI
M7 TMJ
M7 TMK
M7 TML
M7 TMN
M7 TMQ
M7 TMS
M7 TMT
M7 TMU
M7 TMY
M7 TNA
M7 TNB
M7 TNC
M7 TND
M7 TNE
M7 TNG
M7 TNI
M7 TNK
M7 TNN
M7 TNO
M7 TNQ
M7 TNS
M7 TNT
M7 TNX
M7 TNZ
M7 TOA
M7 TOB
M7 TOC
M7 TOD
M7 TOE
M7 TOF
M7 TOJ
M7 TOK
M7 TOL
M7 TOM
M7 TOP
M7 TOR
M7 TOZ
M7 TPA
M7 TPB
M7 TPC
M7 TPE
M7 TPG
M7 TPH
M7 TPK
M7 TPL
M7 TPM
M7 TPP
M7 TPR
M7 TPS
M7 TPX
M7 TQJ
M7 TRC
M7 TRD
M7 TRE
M7 TRF
M7 TRI
M7 TRJ
M7 TRK
M7 TRL
M7 TRN
M7 TRO
M7 TRP
M7 TRR
M7 TRS
M7 TRT
M7 TRU
M7 TRX
M7 TSC
M7 TSD
M7 TSE
M7 TSF
M7 TSJ
M7 TSK
M7 TSL
M7 TSM
M7 TSN
M7 TSR
M7 TSV
M7 TSW
M7 TSX
M7 TTC
M7 TTE
M7 TTF
M7 TTG
M7 TTI
M7 TTJ
M7 TTL
M7 TTM
M7 TTN
M7 TTP
M7 TTR
M7 TTW
M7 TTX
M7 TTY
M7 TTZ
M7 TUB
M7 TUD
M7 TUG
M7 TUI
M7 TUK
M7 TUR
M7 TUV
M7 TVB
M7 TVC
M7 TVG
M7 TVK
M7 TVM
M7 TVN
M7 TVR
M7 TVZ
M7 TWB
M7 TWE
M7 TWH
M7 TWI
M7 TWL
M7 TWM
M7 TWP
M7 TWR
M7 TWS
M7 TWT
M7 TWX
M7 TWY
M7 TWZ
M7 TXK
M7 TXL
M7 TXP
M7 TXV
M7 TXX
M7 TXZ
M7 TYF
M7 TYK
M7 TYN
M7 TYO
M7 TYT
M7 TYY
M7 TZB
M7 TZM
M7 TZN
M7 TZR
M7 TZT
M7 TZZ
M7 UAA
M7 UAC
M7 UAE
M7 UAN
M7 UAS
M7 UAT
M7 UAU
M7 UAV
M7 UAX
M7 UCF
M7 UCH
M7 UDM
M7 UDP
M7 UDR
M7 UDT
M7 UDX
M7 UEI
M7 UEL
M7 UFO
M7 UFX
M7 UHC
M7 UHF
M7 UHK
M7 UHW
M7 UID
M7 UIM
M7 UKA
M7 UKC
M7 UKE
M7 UKG
M7 UKJ
M7 UKR
M7 ULF
M7 ULI
M7 ULL
M7 ULS
M7 UMA
M7 UMM
M7 UMP
M7 UNE
M7 UNK
M7 UNO
M7 UNR
M7 UNX
M7 UPC
M7 UPH
M7 UPP
M7 URC
M7 URG
M7 URM
M7 URR
M7 URW
M7 USK
M7 USS
M7 UTB
M7 UTC
M7 UTE
M7 UTH
M7 UTL
M7 UTN
M7 UTO
M7 UTS
M7 UTW
M7 UTX
M7 UVA
M7 UVW
M7 UWU
M7 UXA
M7 UXB
M7 UXG
M7 UXT
M7 UYB
M7 UYK
M7 UZI
M7 UZT
M7 VAA
M7 VAG
M7 VAI
M7 VAM
M7 VAN
M7 VAO
M7 VAP
M7 VAR
M7 VAS
M7 VAU
M7 VAX
M7 VAZ
M7 VBC
M7 VCB
M7 VCD
M7 VCG
M7 VCO
M7 VCR
M7 VCW
M7 VDK
M7 VDM
M7 VDR
M7 VEC
M7 VED
M7 VEE
M7 VEF
M7 VEJ
M7 VER
M7 VEV
M7 VFA
M7 VFD
M7 VFL
M7 VFO
M7 VFS
M7 VGA
M7 VGC
M7 VGD
M7 VGN
M7 VGO
M7 VGT
M7 VGX
M7 VHD
M7 VHF
M7 VHR
M7 VHS
M7 VIA
M7 VIC
M7 VID
M7 VIE
M7 VII
M7 VIM
M7 VIR
M7 VIS
M7 VIX
M7 VJE
M7 VJS
M7 VJW
M7 VKA
M7 VKG
M7 VKK
M7 VKN
M7 VKV
M7 VLB
M7 VLF
M7 VLM
M7 VLT
M7 VLX
M7 VMD
M7 VMG
M7 VMI
M7 VMM
M7 VMT
M7 VMX
M7 VNA
M7 VNC
M7 VND
M7 VNE
M7 VNG
M7 VNM
M7 VNO
M7 VNS
M7 VNU
M7 VOB
M7 VOE
M7 VOH
M7 VOK
M7 VOL
M7 VOM
M7 VON
M7 VOO
M7 VOR
M7 VOS
M7 VOZ
M7 VPB
M7 VPC
M7 VPE
M7 VPG
M7 VPH
M7 VPI
M7 VPN
M7 VPR
M7 VPT
M7 VQQ
M7 VQW
M7 VRE
M7 VRF
M7 VRJ
M7 VRK
M7 VRN
M7 VRO
M7 VRS
M7 VRT
M7 VSI
M7 VSM
M7 VST
M7 VSW
M7 VTC
M7 VTM
M7 VTR
M7 VTX
M7 VUG
M7 VUP
M7 VUU
M7 VVH
M7 VVR
M7 VVT
M7 VVW
M7 VWB
M7 VWC
M7 VWN
M7 VWV
M7 VXJ
M7 VXR
M7 VXT
M7 VXX
M7 VXZ
M7 VYZ
M7 VZZ
M7 WAE
M7 WAG
M7 WAH
M7 WAK
M7 WAL
M7 WAN
M7 WAP
M7 WAR
M7 WAS
M7 WAT
M7 WAV
M7 WAX
M7 WAZ
M7 WBC
M7 WBD
M7 WBE
M7 WBJ
M7 WBR
M7 WBS
M7 WBU
M7 WBV
M7 WCB
M7 WCD
M7 WCJ
M7 WCL
M7 WCM
M7 WCS
M7 WDB
M7 WDE
M7 WDG
M7 WDH
M7 WDI
M7 WDK
M7 WDM
M7 WDR
M7 WDS
M7 WEC
M7 WEE
M7 WEN
M7 WEO
M7 WES
M7 WEX
M7 WEY
M7 WFC
M7 WFK
M7 WFM
M7 WFS
M7 WFT
M7 WGA
M7 WGD
M7 WGG
M7 WGJ
M7 WGK
M7 WGL
M7 WGP
M7 WGW
M7 WHB
M7 WHE
M7 WHL
M7 WHM
M7 WHN
M7 WHO
M7 WHP
M7 WHT
M7 WHU
M7 WHW
M7 WHX
M7 WHY
M7 WIB
M7 WII
M7 WIL
M7 WIN
M7 WIO
M7 WIZ
M7 WJF
M7 WJI
M7 WJN
M7 WJP
M7 WJQ
M7 WJS
M7 WJT
M7 WJW
M7 WJX
M7 WKA
M7 WKC
M7 WKM
M7 WKN
M7 WKO
M7 WKP
M7 WKX
M7 WKY
M7 WKZ
M7 WLA
M7 WLC
M7 WLE
M7 WLG
M7 WLK
M7 WLM
M7 WLN
M7 WLS
M7 WMA
M7 WMC
M7 WMD
M7 WMH
M7 WMR
M7 WMT
M7 WMZ
M7 WND
M7 WNF
M7 WOA
M7 WOB
M7 WOD
M7 WOF
M7 WOM
M7 WON
M7 WOO
M7 WOT
M7 WOW
M7 WPA
M7 WPD
M7 WPJ
M7 WPR
M7 WPS
M7 WPT
M7 WPW
M7 WQE
M7 WQW
M7 WRA
M7 WRC
M7 WRD
M7 WRH
M7 WRJ
M7 WRK
M7 WRL
M7 WRS
M7 WRT
M7 WRX
M7 WRY
M7 WSC
M7 WSD
M7 WSE
M7 WSL
M7 WSN
M7 WSR
M7 WSW
M7 WSY
M7 WSZ
M7 WTA
M7 WTC
M7 WTF
M7 WTH
M7 WTL
M7 WTM
M7 WTN
M7 WTP
M7 WTR
M7 WTT
M7 WUB
M7 WUD
M7 WUF
M7 WUH
M7 WUJ
M7 WUK
M7 WVF
M7 WVX
M7 WWA
M7 WWB
M7 WWC
M7 WWD
M7 WWI
M7 WWP
M7 WWW
M7 WWZ
M7 WXM
M7 WXT
M7 WXW
M7 WXY
M7 WXZ
M7 WYA
M7 WYB
M7 WYC
M7 WYE
M7 WYO
M7 WYS
M7 WZA
M7 WZD
M7 WZE
M7 WZF
M7 WZL
M7 WZP
M7 WZT
M7 WZX
M7 XAB
M7 XAD
M7 XAF
M7 XAJ
M7 XAK
M7 XAL
M7 XAR
M7 XAT
M7 XAV
M7 XAX
M7 XAZ
M7 XBD
M7 XBJ
M7 XBM
M7 XBP
M7 XBR
M7 XBT
M7 XBW
M7 XCD
M7 XCG
M7 XCH
M7 XCQ
M7 XCR
M7 XCT
M7 XCV
M7 XCW
M7 XCX
M7 XCZ
M7 XDA
M7 XDB
M7 XDC
M7 XDD
M7 XDG
M7 XDH
M7 XDJ
M7 XDL
M7 XDM
M7 XDN
M7 XDR
M7 XDT
M7 XDX
M7 XDY
M7 XDZ
M7 XEA
M7 XED
M7 XEH
M7 XEL
M7 XEM
M7 XEN
M7 XER
M7 XEZ
M7 XFD
M7 XFH
M7 XFM
M7 XFR
M7 XFT
M7 XFX
M7 XGA
M7 XGB
M7 XGC
M7 XGP
M7 XGR
M7 XGW
M7 XHL
M7 XIM
M7 XIN
M7 XIO
M7 XIV
M7 XJB
M7 XJD
M7 XJE
M7 XJF
M7 XJG
M7 XJH
M7 XJI
M7 XJM
M7 XJN
M7 XJR
M7 XJT
M7 XKL
M7 XKP
M7 XKX
M7 XLB
M7 XLS
M7 XLT
M7 XLV
M7 XLW
M7 XLY
M7 XLZ
M7 XMA
M7 XMC
M7 XMJ
M7 XMM
M7 XMO
M7 XMP
M7 XMR
M7 XMS
M7 XMT
M7 XMW
M7 XMX
M7 XMY
M7 XNF
M7 XNS
M7 XNW
M7 XOC
M7 XOF
M7 XOI
M7 XOR
M7 XOX
M7 XOZ
M7 XPC
M7 XPP
M7 XPS
M7 XPZ
M7 XQU
M7 XRB
M7 XRC
M7 XRF
M7 XRM
M7 XRP
M7 XRT
M7 XRV
M7 XRY
M7 XSD
M7 XSM
M7 XSS
M7 XSW
M7 XTA
M7 XTC
M7 XTD
M7 XTF
M7 XTH
M7 XTL
M7 XTN
M7 XTR
M7 XTS
M7 XTT
M7 XUK
M7 XVF
M7 XVK
M7 XVT
M7 XVX
M7 XVZ
M7 XWK
M7 XWL
M7 XWT
M7 XWX
M7 XWZ
M7 XXC
M7 XXD
M7 XXL
M7 XXO
M7 XXT
M7 XXW
M7 XXY
M7 XXZ
M7 XYB
M7 XYG
M7 XYL
M7 XYM
M7 XZT
M7 XZX
M7 YAJ
M7 YAK
M7 YAL
M7 YAP
M7 YAR
M7 YAS
M7 YAT
M7 YAY
M7 YBA
M7 YBE
M7 YBG
M7 YBN
M7 YBS
M7 YBT
M7 YBV
M7 YCB
M7 YCW
M7 YDC
M7 YDE
M7 YDF
M7 YDT
M7 YDU
M7 YDZ
M7 YED
M7 YEE
M7 YEH
M7 YEP
M7 YES
M7 YEY
M7 YFB
M7 YFI
M7 YFR
M7 YFY
M7 YGG
M7 YGH
M7 YGN
M7 YGX
M7 YHF
M7 YHI
M7 YIO
M7 YIY
M7 YJC
M7 YKL
M7 YKS
M7 YKX
M7 YLA
M7 YLE
M7 YMO
M7 YMT
M7 YMY
M7 YNE
M7 YOG
M7 YOO
M7 YOT
M7 YOU
M7 YPE
M7 YRC
M7 YRK
M7 YRU
M7 YSC
M7 YSD
M7 YSL
M7 YST
M7 YSW
M7 YSX
M7 YSZ
M7 YTH
M7 YTI
M7 YTT
M7 YTX
M7 YUK
M7 YUM
M7 YUU
M7 YVZ
M7 YWB
M7 YWH
M7 YXC
M7 YXY
M7 YYL
M7 YYX
M7 YYY
M7 YZC
M7 YZD
M7 YZE
M7 YZV
M7 YZX

Special Contest Calls

Call	Holder
G0A	GW4SKA
G0B	M0BUL
G0C	G0CER
G0H	G0HEU
G0V	GM0OQV
G0W	G0VDZ
G0Z	G1RVD
G1A	M0NKR
G1B	G1PPA
G1C	GM1BSG
G1D	M0AQM
G1E	G1TPA
G1F	M1AOB
G1G	G4KIV
G1J	MM0BQI
G1K	M0RTQ
G1M	G0PZE
G1N	G0URR
G1P	M0IEP
G1R	GM5RDX
G1T	GM0ULK
G1V	G4CZP
G1W	M0HAO
G1X	GM0HBF
G2A	GJ3IT
G2B	G8VVY
G2C	M0VCB
G2D	M0HRF
G2E	M0ORD
G2F	M0URL
G2G	MM0DFV
G2K	M0ICR
G2L	G3SVJ
G2M	G0HDB
G2N	G4ARN
G2O	G0UVX
G2P	G0JCC
G2R	GW4BVE
G2S	GM3ZDH
G2T	MM0CPS
G2U	G0UGO
G2V	GM3WOJ
G2W	G4DBW
G2X	G0DCK
G2Y	MM0DXH
G2Z	G8JYV
G3A	MM0JOM
G3B	G2DPQ
G3C	GM0WED
G3D	G4PDS
G3F	G4AFF
G3G	G3XLG
G3J	G4RMV
G3L	G3LHJ
G3M	G5LK
G3N	G3OTK
G3P	G3WPH
G3Q	G3RXQ
G3R	G3CKR
G3S	GM3SEK
G3T	G3VGZ
G3U	G3UJE
G3V	G3VER
G3W	GM4RIV
G3X	GM3POI
G3Y	G3YBY
G3Z	G3ZME
G4A	G4TSH
G4B	MM0MUN
G4C	G1FCW
G4E	G4ENZ
G4G	M0RRG
G4H	GI4JTF
G4J	GW0ETF
G4K	G4OED
G4L	G4LDL
G4M	GM4UBJ
G4N	G4ZVB
G4O	GM0IIO
G4P	G3YPP
G4Q	G3PRI
G4R	G4RFR
G4S	G3TXF
G4T	MI0SMK
G4U	G4SGX
G4V	MM0CCC
G4W	GW4EVX
G4X	GM4WZG
G4Y	G0CCT
G4Z	GM4ZUK
G5A	GM8VL
G5B	G4SIV
G5C	G4OGB
G5D	M0TTG
G5F	GD4RFZ
G5G	MM0TGH
G5H	G8TRF
G5I	GI4DOH
G5K	G0BNR
G5L	GW4ZAR
G5M	MS0TJT
G5N	G0NWM
G5O	G6UQ
G5P	GW0EGH
G5Q	G3SVL
G5R	GW3YDX
G5T	G3XSD
G5U	G3RXP
G5V	G3KAF
G5W	G3BJ
G5X	GM4YXI
G6A	G3VDB
G6C	M0ITR
G6K	G0EAK
G6M	G4BYG
G6N	G0GDU
G6T	G4MKP
G6X	M0KLO
G7A	GM0ADX
G7C	M5ARC
G7D	G3XTZ
G7G	M0XAR
G7H	G7SYW
G7L	G7LRQ
G7M	M0SDC
G7N	GW4MVA
G7R	GM0NAI
G7T	M0VSE
G7V	GM2MP
G7W	GC0VPR
G7X	G0MCV
G8A	G3XSV
G8B	G3UBX
G8C	M0WLY
G8D	G3SJJ
G8K	GW4BRS
G8L	G8LZE
G8N	G8LED
G8P	G4LIP
G8S	G4IDF
G8T	M0BAA
G8W	M0VCT
G8X	G4FJK
G9A	GM4FDM
G9C	MM0GHM
G9D	G6NHU
G9F	G4BVY
G9J	GW0GEI
G9P	M0NCG
G9R	MS0EPC
G9T	GW4WXM
G9V	M0VSQ
G9W	M0DXR
G9X	M1LCR
G9Y	M0YHC
G9Z	GW1YQM
M0A	G8APB
M0B	G4KZD
M0C	MM0CEZ
M0H	MI0KOA
M0I	MI0RRE
M0K	G8FMC
M0M	MN0NID
M0N	M0NVS
M0P	M0RYB
M0Q	M0DSL
M0T	GM3WUX
M0X	M0RTI
M1A	MI0ULK
M1B	G1YBB
M1C	G1XOW
M1E	MM0GOR
M1F	M0NVK
M1G	G0UWS
M1K	M1ABK
M1L	G0LGS
M1M	MI0LLG
M1N	M1DST
M1R	MM0ZBH
M1T	M0KYB
M1U	M0UTD
M1V	M1VPN
M1W	M0ICK
M1X	G0CKP
M1Z	MS0KPV
M2A	G3SDC
M2C	MD0CCE
M2D	G4NVR
M2E	G0RPM
M2F	G0OOG
M2G	G4RCG
M2I	G0FRE
M2J	G4NBS
M2K	GU3HFN
M2L	M0BJL
M2N	MM0GPZ
M2O	G1GEY
M2P	G1ZJP
M2R	GM0GNK
M2S	G0MGM
M2T	GM0LIR
M2U	M0DHP
M2W	G0MIN
M2Y	MW0YVK
M3A	M0UKR
M3C	G4GFI
M3D	G3XTT
M3E	G4CWH
M3F	G3WZD
M3I	GM5TDX
M3M	G3PLE
M3N	GM4SID
M3O	M0SDB
M3P	G3PIA
M3R	G3RTU
M3S	G0MFR
M3T	MM0LCG
M3V	G0ORY
M3W	M0HDG
M3X	M0IHT
M4A	G6UW
M4C	G0FCT
M4D	G8DYT
M4I	GI4SJQ
M4J	G0DVJ
M4K	GD0EMG
M4M	M0PNN
M4N	G4IZZ
M4R	GD4XUM
M4T	M0BEW
M4U	G0RGH
M4W	G8SRC
M4X	G3SZU
M4Y	M5ESE
M5A	G0AAA
M5B	M0ZIP
M5C	G3CO
M5D	G4WQI
M5E	GU4YOX
M5G	M0ROA
M5I	GI0RQK
M5K	MI0SLE
M5L	M5LMG
M5M	G4BRK
M5N	G0GJV
M5O	G3LET
M5P	M5BIR
M5R	MW0EDX
M5S	G4IRN
M5T	M0RIU
M5W	M0HMJ
M5X	G3RLE
M5Y	G3UES
M5Z	M0CFW
M6A	G4KZY
M6C	M0HFC
M6M	GW4BVJ
M6N	M0NPK
M6O	G3WGN
M6T	G4MRS
M6W	G3WW
M7A	G4HVC
M7C	M5RIC
M7K	M0SDV
M7N	G3RWF
M7O	M0VKY
M7Q	G4PIQ
M7R	G0TPH
M7T	G3YYD
M7V	M0VAA
M7W	G3TBK
M7X	G0TSM
M7Z	G4BWP
M8A	M0HDF
M8C	G3RCV
M8K	M0HMO
M8M	G0JJG
M8P	MW0OMB
M8R	GW4SHF
M8T	MM0CWJ
M8Z	GM7VSB
M9A	G3ZVW
M9B	M0LKW
M9C	G0TKZ
M9I	GM0OPS
M9K	M0SIY
M9M	G4BEE
M9N	G7WHI
M9T	M0MCG
M9W	GW0KRL
M9X	M0PGX

Permanent Special Event Callsigns

GB0 MWM	47 Oakfield Rd, Kidderminster DY11 6PL
GB0 RSM	Raf Snaith Museum Long Lane Pollington, Goole DN14 0DF
GB0 SMA	Stow Maries Aerodrome, Hackmans Lane, Chelmsford CM3 6RN
GB0 WMM	Waterbeach Military Heritage Museum, Building No.3, Waterbeach Barracks, Cambridge CB25 9PA
GB1 AA	30 Belvoir Road, Widnes WA8 6HR
GB1 FBS	Stubb Lane , Brede TN31 6EH
GB1 NHS	42 Trafalgar Way Lichfield, Lichfield WS14 9FD
GB1 ROC	Northern Ireland Secret Bunker, Derrylettiff Road, Portadown BT62 1QU
GB1 TEN	Bolehall Manor Club, Tamworth B77 3LH
GB2 CPM	Amberley Museum, New Barn Road, Amberley BN18 9LT
GB2 CWP	East Kirkby Airfield, Spilsby PE23 4DE
GB2 EVR	Warcop Station, Warcop CA16 6PR
GB2 GM	The Marconi Centre, Poldhu TR12 7JB
GB2 HAM	Harrington Aviation Museum, Harrington NN6 9PF
GB2 HAM	Harrington Aviation Museum, Off Lamport Road, Harrington NN6 9PF
GB2 MOP	Museum Of Power Castell Pridd , Tan-Y-Groes SA43 2JS
GB2 NTM	Norfolk Tank Museum, Station Road, Norwich NR16 1HZ
GB2 SJ	Coast Rd, Sunderland NE34 9JJ
GB2 SPY	Rosemount, Church Whitfield Road, Dover CT16 3HZ
GB4 GCT	Greenham Common Control Tower, Burys Bank Road, Thatcham RG19 8BZ
GB4 LD	The Old Marcon Hut, Lloyds Lane, Lizard TR12 7AP
GB4 SMH	Signals Museum, Building 104, Raf Henlow, Henlow SG16 6DN
GB4 VLB	Brigade Watch House Museum, Tynemouth NE30 4DD
GB6 SL	12 Tramayne Avenue Brough, Hull HU15 1BL
GB8 LY	23 Worthington Place, Leigh WN7 2EJ
GB100 RSM	Blandford Camp, Blandford DT11 8RH
GB150 GM	Crofthandy Village Hall, Crofthandy TR16 5JQ
GBGR2 HQ	Moscar Cross House, Sheffield S6 6GL

Postcode Index

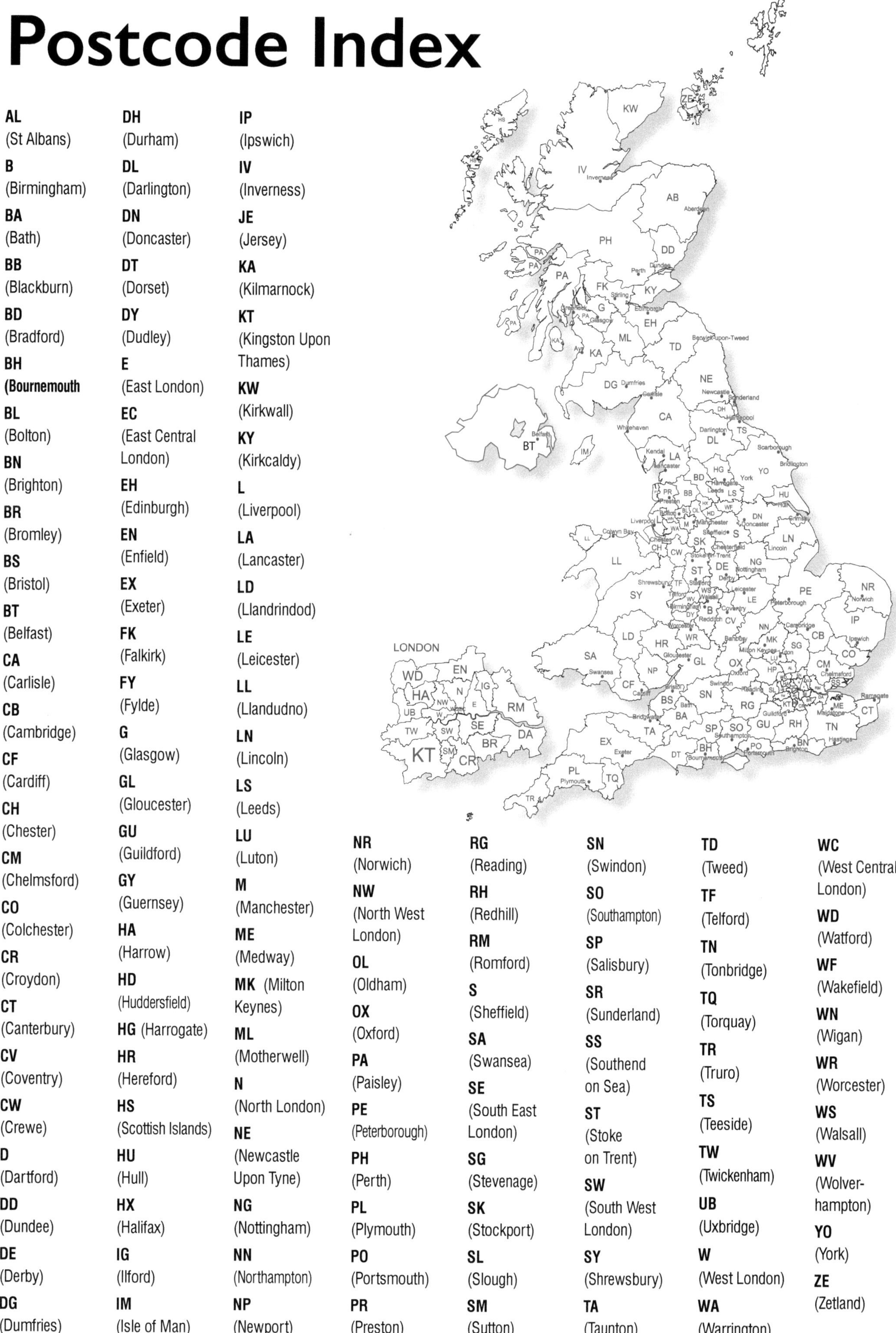

AL
(St Albans)

B
(Birmingham)

BA
(Bath)

BB
(Blackburn)

BD
(Bradford)

BH
(Bournemouth

BL
(Bolton)

BN
(Brighton)

BR
(Bromley)

BS
(Bristol)

BT
(Belfast)

CA
(Carlisle)

CB
(Cambridge)

CF
(Cardiff)

CH
(Chester)

CM
(Chelmsford)

CO
(Colchester)

CR
(Croydon)

CT
(Canterbury)

CV
(Coventry)

CW
(Crewe)

D
(Dartford)

DD
(Dundee)

DE
(Derby)

DG
(Dumfries)

DH
(Durham)

DL
(Darlington)

DN
(Doncaster)

DT
(Dorset)

DY
(Dudley)

E
(East London)

EC
(East Central London)

EH
(Edinburgh)

EN
(Enfield)

EX
(Exeter)

FK
(Falkirk)

FY
(Fylde)

G
(Glasgow)

GL
(Gloucester)

GU
(Guildford)

GY
(Guernsey)

HA
(Harrow)

HD
(Huddersfield)

HG (Harrogate)

HR
(Hereford)

HS
(Scottish Islands)

HU
(Hull)

HX
(Halifax)

IG
(Ilford)

IM
(Isle of Man)

IP
(Ipswich)

IV
(Inverness)

JE
(Jersey)

KA
(Kilmarnock)

KT
(Kingston Upon Thames)

KW
(Kirkwall)

KY
(Kirkcaldy)

L
(Liverpool)

LA
(Lancaster)

LD
(Llandrindod)

LE
(Leicester)

LL
(Llandudno)

LN
(Lincoln)

LS
(Leeds)

LU
(Luton)

M
(Manchester)

ME
(Medway)

MK (Milton Keynes)

ML
(Motherwell)

N
(North London)

NE
(Newcastle Upon Tyne)

NG
(Nottingham)

NN
(Northampton)

NP
(Newport)

NR
(Norwich)

NW
(North West London)

OL
(Oldham)

OX
(Oxford)

PA
(Paisley)

PE
(Peterborough)

PH
(Perth)

PL
(Plymouth)

PO
(Portsmouth)

PR
(Preston)

RG
(Reading)

RH
(Redhill)

RM
(Romford)

S
(Sheffield)

SA
(Swansea)

SE
(South East London)

SG
(Stevenage)

SK
(Stockport)

SL
(Slough)

SM
(Sutton)

SN
(Swindon)

SO
(Southampton)

SP
(Salisbury)

SR
(Sunderland)

SS
(Southend on Sea)

ST
(Stoke on Trent)

SW
(South West London)

SY
(Shrewsbury)

TA
(Taunton)

TD
(Tweed)

TF
(Telford)

TN
(Tonbridge)

TQ
(Torquay)

TR
(Truro)

TS
(Teeside)

TW
(Twickenham)

UB
(Uxbridge)

W
(West London)

WA
(Warrington)

WC
(West Central London)

WD
(Watford)

WF
(Wakefield)

WN
(Wigan)

WR
(Worcester)

WS
(Walsall)

WV
(Wolver-hampton)

YO
(York)

ZE
(Zetland)

Postcode

AB10 6ED 2M0EMM
AB10 6ED MM0MUN
AB10 6HP GM4BAP
AB10 6JD GM4HTU
AB10 6JU GM4VRE
AB10 6QH GM0NRT
AB10 6RA GM1MCN
AB10 6SB GM6GFQ
AB10 7BS MM3OMI
AB10 7JE GM3KJE
AB11 6XL MM0GAL
AB11 7WD 2M0RND
AB11 7WD MM0ROV
AB11 7WD MM3MVY
AB11 7WD MM3YBG
AB11 7WF GM7VZV
AB11 7WF MM0AOF
AB11 8FX 2M0AKS
AB11 8FX MM0GDG
AB11 8FX MM3JJC
AB12 3DF MM0KRZ
AB12 3DZ 2M1IIW
AB12 3JJ MM1ABA
AB12 3NG GM1THS
AB12 3PH MM6MOY
AB12 3RL GM3NUU
AB12 3RN GM8FFX
AB12 3SH GM7BYB
AB12 4NS MM7SLY
AB12 4NY GM4RAZ
AB12 4NY GM4RGS
AB12 4QA MM0AOY
AB12 4QX 2M0MJY
AB12 4QX MM0MJY
AB12 4QX MM6MJY
AB12 4TF GM3ZEU
AB12 4XF MM7BVN
AB12 4XL GM8GDN
AB12 4XT 2M0DWC
AB12 4XT MM3YXN
AB12 4XT MM3ZGK
AB12 5AB GM0DBW
AB12 5AQ 2M0FPI
AB12 5AQ MM6ORU
AB12 5BY MM7SKE
AB12 5DQ GM4UWN
AB12 5LJ MM0DBR
AB12 5QT 2M1HRS
AB12 5XT GM0VFY
AB13 0ER GM0OPX
AB13 0JB GM3TLA
AB14 0AB GM3CIO
AB14 0AB GM4EMX
AB14 0LN GM0FRT
AB14 0LN GM4ZUK
AB14 0NX GM4TEF
AB14 0TU GM0CQV
AB15 4BP GM1MYF
AB15 4UE GM1KBZ
AB15 4UH GM8ADK
AB15 5WJ MM0KSS
AB15 6AN GM3OUU
AB15 6BH 2M1HTR
AB15 6DU 2M0GEJ
AB15 6DU MM0TGG
AB15 6DU MM6XGJ
AB15 7QA GM4EKC
AB15 7SX GM3VEY
AB15 7XP GM7RNJ
AB15 8LQ GM3WIJ
AB15 8LQ GM4GTV
AB15 8LT GM3HGA
AB15 8SF GM3BSQ
AB15 8SF GM4GVK
AB15 8UE 2M0JUH
AB15 8UE MM6NLV
AB15 8UH MM7EEQ
AB15 9AR GM4NVI
AB15 9BU 2M0RVM
AB15 9BU GM5ALX
AB15 9DE MM7HCM
AB15 9JX GM4IXH
AB15 9PL GM8MFZ
AB15 9QF GM3VAP
AB15 9RE MM6IAI
AB15 9RH GM4EGX
AB16 5QG MM3RBJ
AB16 5RP 2M0RYL
AB16 5RP 2M0SYL
AB16 5RP MM0BCR
AB16 5RR MM6AIK
AB16 5SB MM3KQI
AB16 5SN 2M0NSA
AB16 5SN 2M0STB
AB16 5SN MM0ROR
AB16 5UY MM0HZO
AB16 5YR MM6TWS
AB16 6FN GM6JBF
AB16 6FU GM0MYQ
AB16 6NX GM7KBK
AB16 6PD GM8MHU
AB16 6XY MM6CIA
AB16 7AX MM6EUA
AB16 7DQ 2M0FYG
AB16 7PQ MM7DDV
AB21 0JZ GM4NNK
AB21 0LH MM0ALY
AB21 0NG GM8BZP
AB21 0NS MM6AGL
AB21 0QF 2M0MRO
AB21 0RJ GM8GHV
AB21 0RS GM1BNP
AB21 0TW GM0IPV
AB21 0TW GM4SID
AB21 0TY MM1FDF
AB21 0TZ GM6GMZ
AB21 0YQ MM7RVC
AB21 7DD MM6HRF
AB21 7FN GM3WSR
AB21 9AY MM7BXL
AB21 9HJ MM0CZH
AB21 9HS GM0PTY
AB21 9JJ 2M0NIA
AB21 9JJ MM0CUG
AB21 9JJ MM6NIA
AB21 9PQ MM0DOT
AB21 9QS 2M0BPV
AB21 9QS MM3WMQ
AB21 9QX GM7NHS
AB21 9RH GM0SZA
AB21 9SE MM1GAS
AB214NR MM3SWW
AB22 8LJ GM7MWL
AB22 8RW GM7MMI
AB22 8TP GM1AUZ
AB22 8WF 2M0HIB
AB22 8WF MM3ZRZ
AB22 8WG GM0MCJ
AB22 8WY GM0JOV
AB22 8XB GM6MJY
AB22 8YJ 2M0OQR
AB22 8YJ MM0OQR
AB22 8YJ MM3OQR
AB22 8ZB MM5ISS
AB22 8ZH 2M0DHI
AB22 8ZH MM6DHI
AB22 9AE MM7RCW
AB23 8BD GM8BNH
AB23 8EH GM1FSU
AB23 8PL GM4PMT
AB23 8PW 2M0RMH
AB23 8PW MM3NXY
AB23 8QD GM4YWV
AB23 8QN GM1LKD
AB23 8UT 2M1ENI
AB23 8WB 2M0PKA
AB23 8XY GM7SPB
AB23 8YG MM6JPM
AB24 1WT GM4BFX
AB24 1ZS MM6OYJ
AB24 2RP GM0AUL
AB24 2RP MM6BNO
AB24 2XS MM6BNN
AB24 3HQ MM7FUR
AB24 3JL GM8BSU
AB24 4EQ MM7HXI
AB24 4ER MM7MCT
AB24 4GA MM7GAC
AB24 4HX GM4CAU
AB24 4NJ GM6SDV
AB24 4NT GM7RXL
AB24 4RU MM7HRR
AB24 5BE MM0TJR
AB24 5EP 2M0URP
AB24 5EP GM6URP
AB24 5EP MM6URP
AB25 1DQ GM4HQF
AB25 1LW MM5SHA
AB25 2JX 2M1CUS
AB25 2QB GM1RDG
AB25 2YF MM0DBF
AB25 3UL MM6AOD
AB25 3UN MM7GAE
AB25 3XX MM6CJY
AB3 1NX GM7IHR
AB30 1FR MM7BJA
AB30 1QT MM7DNE
AB30 1TT MM6TSN
AB30 1YS MM6LVV
AB31 4AF GM4JLZ
AB31 4EN GM6EUC
AB31 4HG GM0TCU
AB31 4JU MM0IWS
AB31 4LS GM8GJI
AB31 4NB GM6UHC
AB31 4PJ 2M0OYS
AB31 4PJ MM7YRN
AB31 4QA MM3XAF
AB31 4RY GM0SXQ
AB31 5AU GM4MFB
AB31 5HA MM0ACR
AB31 5HA MM0ACT
AB31 5TE MM0BLW
AB31 5TS MM6YQP
AB31 5UY 2M0DJT
AB31 5XA GM0PKQ
AB31 5XW MM0DRT
AB31 5YG 2M0CXI
AB31 5YG MM0HVW
AB31 5ZF GM7BDD
AB31 6BL GM0AZV
AB31 6DT GM0RAO
AB31 6NL MM0IYD
AB31 6NL MM7AAE
AB32 6RD 2M0FTI
AB32 6RD MM0KKC
AB32 6RD MM7YEA
AB32 6TH 2M0YMA
AB32 6WS GM8BSQ
AB32 6WW GM1XEA
AB32 6XH MM0JOM
AB32 6XY GM0GAT
AB32 6YE MM3HLG
AB32 7EQ GM4NHI
AB33 8BH GM8ZFW
AB33 8ER 2M0PTE
AB33 8ER MM3UDI
AB33 8HQ GM0MHD
AB33 8HQ MM1MHD
AB33 8NX MM6MYK
AB33 8QA MM6ELU
AB33 8QQ MM6IJP
AB33 8QW MM5CFA
AB33 8RN 2M0MVX
AB33 8RN MM0MVX
AB33 8RN MM6MVX
AB33 8SH MM1DZW
AB33 8SH MM3RGH
AB33 8UB GM7IEU
AB34 4TA GM8AT
AB34 4UH MM3VNU
AB34 5JF GM4JXP
AB34 5JZ GM0KDP
AB34 5PQ 2M0EFI
AB34 5PQ MM0EFI
AB35 5QH GM4RLV
AB35 5SF GM8FVN
AB35 5SH MM6HDZ
AB35 5UT GM0DZW
AB35 5ZQ MM6CHN
AB35 5ZQ MM7VKR
AB36 8UJ GM3ALZ
AB36 8UJ MM0ODL
AB36 8XA MM7WGB
AB37 9BG MM6IHQ
AB37 9ET MM3WFU
AB37 9HW MM3WQO
AB38 7BD MM3YDK
AB38 7QZ GM0OSJ
AB38 9NU GM1TGY
AB38 9SQ MM3XMT
AB389LA MM7BQF
AB39 2AD GM1JPJ
AB39 2BZ 2M0DRY
AB39 2BZ MM6BRV
AB39 2EG GM4PVQ
AB39 2GF GM4HVS
AB39 2GQ 2M0XMQ
AB39 2GQ MM0XMQ
AB39 2GQ MM6XMQ
AB39 2JA MM0CFE
AB39 2LU GM4VQY
AB39 2PL 2M0YZT
AB39 2PL MM6MLT
AB39 2UA GM1JHM
AB39 3PF GM4KOI
AB39 3PF GM6OSZ
AB39 3QG MM6TFY
AB39 3SY MM0ALM
AB39 3UL 2M0MGM
AB39 3UL MM3RLG
AB39 3XW MM0GQF
AB4 5SE GM1VSR
AB41 6BJ GM4MBG
AB41 6RT MM6JAN
AB41 7DF GM4LTL
AB41 7DH GM7LDU
AB41 7DU GM7KRQ
AB41 7HS GM4IBI
AB41 7JY GM1WKH
AB41 7NZ MM3OZW
AB41 7PH MM6DXJ
AB41 8BA GM0PKF
AB41 8BH GM4FVS
AB41 8QW GM4GNR
AB41 8TF MM5AJN
AB41 8UJ GM7LAC
AB41 8UJ MM6BOS
AB41 8UW 2M0DDS
AB41 8UW MM0TDS
AB41 8UW MM3DDS
AB41 8YH GM2MP
AB41 8YH GM4YXI
AB41 9DB 2M0IAQ
AB41 9DB MM7BIW
AB41 9EU MM6WHI
AB41 9HF GM0JEF
AB41 9JB GM6AOR
AB41 9JF GM3UAG
AB41 9LW GM4OBD
AB41 9NF GM0MIS
AB41 9TF 2M0IUE
AB41 9TF MM6IUE
AB417HQ MM6NOF
AB42 0NG GM7LJE
AB42 0NY GM1TBW
AB42 0PP GM4PXB
AB42 0PZ MM3TAV
AB42 0RE MM1BMK
AB42 0TB GM8AGM
AB42 0TB GM8ZKU
AB42 0TQ MM6FYR
AB42 1HB MM0CJF
AB42 1HL GM1JNS
AB42 1NX GM4UFD
AB42 1RD GM6KAM
AB42 2UF GM7FYB
AB42 3AT MM3PDM
AB42 3AY GM1GCB
AB42 3DD 2M0CBE
AB42 3DD MM6CBE
AB42 3DN 2M0BVN
AB42 3DN MM0IHE
AB42 3DN MM3XIZ
AB42 3LA MM6SXY
AB42 3PX MM6OEI
AB42 3QH GM5JDG
AB42 4HA MM6ZAZ
AB42 4HX GM7NNS
AB42 4JU GM1XIN
AB42 4JU GM6FDQ
AB42 4LH MM0MPW
AB42 4LH MM7CPZ
AB42 4NL GM4EHP
AB42 4QG MM1EYZ
AB42 4RD MM6IAR
AB42 4XQ MM0DSP
AB42 5AY GM7OJJ
AB42 5AY GM7OTT
AB42 5BL MM0XTW
AB42 5DE GM3ZMA
AB42 5DG MM3WKF
AB42 5EH MM6OMT
AB42 5EJ GM6WTT
AB42 5ES GM0RSI
AB42 5GT MM6VTS
AB42 5HR GM1KUI
AB42 5LR GM0HAN
AB42 5RR MM6SLV
AB42 5WE MM0AOQ
AB420TQ MM6TIR
AB423HZ GM0DYU
AB43 6NE GM1KZG
AB43 6NN MM0CAE
AB43 6NQ GM7OGS
AB43 6PZ 2M0OKO
AB43 6PZ MM7BMK
AB43 6QN MM0VOK
AB43 6SY MM3FYN
AB43 6TJ MM0DVZ
AB43 7ED MM0JGP
AB43 7JS GM4PMH
AB43 7JT MM3LNT
AB43 7JT MM3LOF
AB43 7NW MM1DAK
AB43 8RU GM7FQZ
AB43 8WA 2M1VXB
AB43 9DH MM1CAC
AB43 9NL GM1INS
AB436RX GM6JEP
AB44 1RP GM3UBJ
AB44 1UL 2M0KIA
AB45 1DB GM1HRY
AB45 1DB GM4VVY
AB45 1DZ GM4TOE
AB45 1HS MM0HKU
AB45 2BQ GM0FHD
AB45 2BQ GM0JLJ
AB45 2JR GM1CCI
AB45 2JT MM0BUH
AB45 2PJ GM4PSJ
AB45 3BR GM6JOA
AB45 3UD GM1ROX
AB451FB GM4LFE
AB453ER MM7PDW
AB5 2BJ GM1VBD
AB51 0AB GM4ZEX
AB51 0ES GM4DZM
AB51 0JT GM0APN
AB51 0PJ MM6THU
AB51 0SN GM0PKX
AB51 0TF GM7NQP
AB51 0TX MM7LEO
AB51 0XA GM0TGE
AB51 0XA MM0DTL
AB51 0XX MM6ACM
AB51 3WJ GM0TFQ
AB51 4FL GM0VGI
AB51 4RQ GM4DIN
AB51 4RY GM7NUQ
AB51 5HE GM7UPD
AB51 5LY GM4YAU
AB51 5QT GM3XOQ
AB51 5QZ GM0FIQ
AB51 5RH GM1FSZ
AB51 6BB MM6NRQ
AB51 7HH MM3MXN
AB51 7QP 2M0GMB
AB51 8TQ GM1MKC
AB51 8TS MM6JEA
AB515HA MM7FZN
AB52 6HX MM5FWD
AB52 6JY GM1KHU
AB52 6JY MM6FYW
AB52 6JZ MM6SRL
AB52 6LH GM8UPC
AB52 6NS MM7MWL
AB52 6PD MM0TXO
AB52 6TU 2M0IVP
AB52 6TU MM0WJP
AB52 6TU MM7PIK
AB52 6UN MM3ERP
AB52 6ZD GM0UYS
AB526TU MM7GRG
AB53 4GN GM3OXX
AB53 4GS MM0UDI
AB53 4LX MM0ECV
AB53 4UD GM0EMQ
AB53 5QP GM4OCA
AB53 5QP GM6KMK
AB53 5RN GM0EQS
AB53 5TQ GM7CPJ
AB53 5WH 2M0TRA
AB53 5WH GM7SZA
AB53 5WH MM3OGU
AB53 5YH 2M0FTH
AB53 5YH MM3SZM
AB53 6SL GM4TRS
AB53 6TE GM0NGJ
AB53 6UP MM6ACI
AB53 8AE GM4ZDT
AB53 8ED GM1LXA
AB53 8HD GM0BNQ
AB53 8LE MM6MUH
AB54 4GD MM3VSX
AB54 4GD MM6FVD
AB54 4GD MM6SIM
AB54 4JB GM8MNM
AB54 4NN MM3VSU
AB54 4NN MM6EQM
AB54 4PF 2M0IOK
AB54 4SB MM7WCT
AB54 4XP MM7STS
AB54 5NY GM0WPU
AB54 5NY GM8JWQ
AB54 6AT GM6KDB
AB54 6HA GM1XLH
AB54 7LQ MM3TPF
AB54 7LR GM7ESM
AB54 7LR MM3IOM
AB54 7SY GM8REG
AB54 7XR GM4NXT
AB54 8BY MM7RZT
AB55 4AR MM6NAD
AB55 4AZ MM0SMD
AB55 4AZ MM3JIN
AB55 4EF GM7TWM
AB55 5AG GM0EIT
AB55 5AP GM0LVK
AB55 5AP GM1MRS
AB55 5BX GM1LOZ
AB55 5EB MM1DVC
AB55 5ED MM1DTN
AB55 5ED MM3TQI
AB55 5EG GM7DXT
AB55 5EH GM4FGL
AB55 5EQ MM0DVB
AB55 5FX MM3AWC
AB55 5JW MM7URF
AB55 6LP GM1HNZ
AB55 6LQ MM0IEL
AB55 6QU GM4WJA
AB55 6UR GM8PSV
AB56 1DD GM3UKG
AB56 4LD 2M0HJS
AB56 4LD MM3UDQ
AB56 4NB MM3TLQ
AB56 4NR GM4UWX
AB56 4QA GM4OEZ
AB56 4SD GM8JCF
AB56 4TU MM7AUH
AB56 5AL GM1KWG
AB56 5BS MM6IMF
AB56 5BW GM0ARY
AB56 5EP GM3KHH
AB56 5HB GM1OXB
AB56 5HP 2M0FSP
AB56 5HP MM6FSP
AB56 5YD GM4YGS
AL1 1DW 2E1BDB
AL1 1NF M6UPL
AL1 1UG 2E0WVG
AL1 1UG G8AMC
AL1 1UG M0TPT
AL1 1UG M6WVG
AL1 2PG G6BDW
AL1 2QS G6CMD
AL1 3AG G8VMP
AL1 4BQ M6OHS
AL1 4ES M6XAL
AL1 4JZ M6CYP
AL1 4PZ G3XXF
AL1 4RD G3YCY
AL1 4SN M6GJQ
AL1 4TT G0MVY
AL1 4UR M6PQO
AL1 4UX M6LOW
AL1 4XZ G0LQU
AL1 5AE M6ADI
AL1 5BX 2E0MRM
AL1 5ES M6JGT
AL1 5EX G4BIX
AL1 5LF G7ARF
AL1 5NJ G4DDV
AL1 5NS M0GTJ
AL1 5QB M0CZX
AL10 0HW M7XDS
AL10 0HX M7YBC
AL10 0HZ M7YBB
AL10 0UB G1ROH
AL10 8DQ G4RMD
AL10 8HE G6INV
AL10 8QF G1IHE
AL10 8XX 2E0FWS
AL10 8XX M6MLP
AL10 9FD M6MIK
AL10 9GS 2E0MVN
AL10 9GS M7NVM
AL10 9GW M0TMS
AL10 9LE G3VIX
AL10 9NT M1JES
AL10 9PB G4SNI
AL10 9PE 2E0BXF
AL10 9RW G3NQN
AL10 9SB G6UBH
AL10 9WN G4KDS
AL2 1HL M1FCX
AL2 1RU M6NZI
AL2 2AH G1ARL
AL2 2AR G8GYK
AL2 2QL G8UCP
AL2 2RD G6URU
AL2 3AY 2E0UCE
AL2 3AY M6UCE
AL2 3DD G7GJN
AL2 3ND G0BEE
AL2 3PU G1WVD
AL2 3QA G8NGR
AL2 3SE G7SDG
AL2 3SJ G0HTL
AL2 3SJ G0KVA
AL2 3SR G1IHS
AL2 3ST G6CKW
AL2 3TD G4LMN
AL2 3TD G8KGG
AL3 4HH G3GEX
AL3 4JZ G4WSL
AL3 4NJ G4HVG
AL3 4TL G6PWL
AL3 5RE G3PZF
AL3 5TD 2E0RTY
AL3 5TD M0JZG
AL3 5TD M6RTU
AL3 5TU 2E1HHE
AL3 5TX 2E0YXX
AL3 5UB 2E0FYZ
AL3 6HP G8SGF
AL3 6LR 2E0BIY
AL3 6LR M3KIZ
AL3 7BG 2E0IZF
AL3 7BG M7SWW
AL3 7EW 2E0CIM
AL3 7EW M6YVR
AL3 7HD 2E1EWN
AL3 7JB G0EHO
AL3 7NL G4YSJ
AL3 7PF G6USO
AL3 8EF G7MSG
AL3 8HW M3MNY
AL3 8JB G1NTJ
AL3 8JN G3LFV
AL4 0DH G0LZW
AL4 0DH G8BNR
AL4 0DR 2E1IDI
AL4 0EN G8UPX
AL4 0EX 2E0WAP
AL4 0EX M6APW
AL4 0EY M6GBF
AL4 0GE G6PFP
AL4 0JG G4HHJ
AL4 0NS M6RTZ
AL4 0QA G6DYM
AL4 0QR G0MDR
AL4 0QS 2E0KHA
AL4 0QS M3KHA
AL4 0XA G4XJS
AL4 8JD G4WNP
AL4 8NZ M6GDB
AL4 8PE G4BOU
AL4 8PR G0HGO
AL4 8RY G6AHE
AL4 8TP G0XBC
AL4 8TP M0ABY
AL4 9AF G4ZRA
AL4 9AP M6POG
AL4 9DW G4IXY
AL4 9EH G6PWS
AL4 9HD M6APR
AL4 9JU 2E0KIV
AL4 9JU G4BEO
AL4 9JX G4AUE
AL4 9LS M6RTX
AL4 9LX M6SQO
AL4 9NZ G4CZA
AL4 9PS G7IFJ
AL4 9PS M6MCY
AL4 9PW G7PKH
AL4 9QH M6BVS
AL4 9QN M6ACD
AL4 9SJ G3XYH
AL4 9TG G4DJX
AL4 9TG M0SCY
AL4 9TG M6MXD
AL4 9XB G4ZES
AL4 9XJ M6VBS
AL4 9XQ 2E0IOZ
AL4 9XQ M0IZA
AL5 1EF 2E1BZB
AL5 1EZ G7HDR
AL5 1HD G4DOC
AL5 1JQ G3UHN
AL5 1LL G1XEH
AL5 1RF 2E0IHB
AL5 1RF M0RBK
AL5 1RF M6AKY
AL5 1RP G8YNB
AL5 1SR G8CLY
AL5 1TD G8KEK
AL5 2AL G0DFF
AL5 2HJ M0AYS
AL5 2QJ G4GEZ
AL5 3HJ M6NHF
AL5 3LJ 2E0JSG
AL5 3NX G4OGZ
AL5 4BT G1ZYJ
AL5 4DB G0CPN
AL5 4DB G8ASP
AL5 4HN G4KCN
AL5 4LP G3YWA
AL5 4TE M5WIZ
AL5 5BT G4OAV
AL5 5DW G3MSW
AL5 5DW G8XOB
AL5 5HR G4RIK
AL5 5LY M0SUU
AL5 5QS G1ZGF
AL5 5QU M0MLT
AL5 5ST M1CBV
AL5 5SU G0RVH
AL5 5SU G3SDG
AL6 0DB G1MMZ
AL6 0DH G0URC
AL6 0DZ G0NYY
AL6 0EN G3AQF
AL6 0RU G8TRU
AL6 0RW G3KOX
AL6 9BB G6KJA
AL6 9JQ M6LTQ
AL6 9NW G3XDM
AL6 9PX G0WTF
AL6 9SZ G0IPJ
AL6 9TF G4SZI
AL7 1DL M6TYT
AL7 1DU G0IIG
AL7 1DW 2E0XYT
AL7 1NL G7BDS
AL7 1PB 2E1PDM
AL7 1PD 2E0JMJ
AL7 1PJ M7AIE
AL7 1QP G0KLU
AL7 1SD G0WAT
AL7 2BW G6BUP
AL7 2JS M1ACE
AL7 2QJ G5KW
AL7 2QJ G8FXM
AL7 2QJ M3FXM
AL7 3FA 2E0YUN
AL7 3FA M0LSY
AL7 3HB M7BSH
AL7 3LR G0NJS
AL7 3RJ G8TVW
AL7 3RL G0MKN
AL7 3TD G1PGJ
AL7 3TE G3TNE
AL7 3TW G0HVX
AL7 3XN M6HSE
AL7 4AQ G0JJM
AL7 4HB M0KQW
AL7 4JT G3HLN
AL7 4LR M0HSQ
AL7 4TG M1FEK
AL72HF G6POE
AL8 6HT G4JVA
AL8 6RF G4LWV
AL8 6RH G1LLW
AL8 6RJ M7CMI
AL8 6SE M6SWP
AL8 6TA G0CYC
AL8 6WW 2E0HOG
AL8 6WW M6STR
AL8 6XW M6NGX
AL8 7AW M6GCV
AL8 7BA M1PWT
AL8 7BG G4MDC
AL8 7EE M0GJK
AL8 7EN G0VDT
AL8 7HX M0JLW
AL8 7HZ G4POB
AL8 7LF G6CXY
AL8 7LH G4DRI
AL8 7NN G8TVZ
AL8 7SG G4TIM
AL8 7SN G6FVF
AL8 7ST G8CYX
AL86XW M6TVN
AL9 7AY M0BMU
AL9 7QQ 2E0IHT
AL9 7QQ M6IZF
AL95HQ G7CMB
B1 1LS M6UED
B1 1UJ G3OTY
B1 1UJ G4ACF
B10 0LE 2E0XTV
B10 0LE M3XTV
B10 9TA G7LOZ
B10 9TB G6HLR
B11 2EL M7PNG
B11 3QJ G8EMX
B11 4NJ G0IVD
B12 9ER G0NHP
B12 9JQ G1YIL
B13 0BA G6DSD
B13 0BH G7UGC
B13 0BH G8ENA
B13 0BH G8GKR
B13 0DL G4KZV
B13 0DL G4OHM
B13 0EJ G4PUD
B13 0JL M3JPP
B13 0NR G6DRN
B13 0PR G4GIG
B13 0RQ G4BJS
B13 8AB M7CFL
B13 8JZ G0EBW
B13 8NZ M0AEJ
B13 8NZ M7GBU
B13 9AD M6AKA
B13 9EN G4DFO
B13 9NE M6JTK
B13 9QR 2E0BTU
B13 9QR M3YGR
B13 9TY G3YHF
B13 9YP M0JMT
B14 4DS G4ETZ
B14 4LS G1POR
B14 4NX M6GYH
B14 4TE G8MYK
B14 4TG G1LZH
B14 4TJ 2E0MSE
B14 4TJ M0MSE
B14 4TW M3LTH
B14 5AF M6RZA
B14 5BT 2E0LPD
B14 5BT M0WYH
B14 5BT M6LPD
B14 5DS 2E0NYC
B14 5DS M0NYP
B14 5DS M6STU
B14 5EF G3PQP
B14 5HD G4MTG
B14 5HD G8TXL
B14 5LX M1CPC
B14 5XA M7MSV
B14 5XX 2E0SDD
B14 5XX M6STJ
B14 6AD G8YQC
B14 6DE G2BBC
B14 6DE G3YXM
B14 6DE M3YXM
B14 6HH G4ZVS
B14 6TN G1FDD
B14 7AS G4JGH
B14 7DB M0DGQ
B14 7DB M3ZCR
B14 7LN 2E0IKN
B14 7LN M0KNP
B14 7LN M7OCB
B15 2AA G7ARP
B15 2DF M6HSF
B15 2RU M6SNZ
B15 2TT G7AFQ
B15 2XA G3UOA
B15 2XA G8CQH
B15 3JB G3NMW
B15 3RL G4NMC
B15 3RN M7MET
B16 0EF G1JDE
B16 0EF G6AUO
B16 8HU M0LTL
B16 8UQ G6ZYZ
B16 9EY G1XFL
B17 0AQ G6DMQ
B17 0AT G3SJH
B17 0EL 2E0GXP
B17 0EL M7PEB
B17 0NW G0KOI
B17 0QT G4NZY
B17 0SU M7JSA
B17 8BN G4HWN
B17 8PY G4KXV
B17 8QH 2E1BHB
B17 8QH G4XDM
B17 9EB M0BZY
B17 9JU G4LQF
B17 9JX M6EFM
B17 9LE G4YDO
B17 9PA 2E0EXV
B17 9PA M6JXK
B17 9RE M5PIP
B17 9TB G6PGO
B18 4NE G1UBT
B18 6JQ 2E0HRS
B18 6JQ M3NBH
B19 1EH G6KEX
B191DG M7DLK
B20 1EQ 2E1TBW
B20 1EQ M3LTV
B20 2AQ G6EEB
B20 2AS M3OCQ
B20 2HN G6KHN
B20 2NX G7WKG
B20 3LS 2E0GJP
B20 3LS M0JNF
B20 3LS M7BPC
B21 9NU G1AJK
B21 9QD M3EWV
B23 5DF 2E0FOX
B23 5LL M3UGB
B23 5QA M6TGQ
B23 5RG G7RYA
B23 5XE G8LIX
B23 5XR G8NZO
B23 6BT G0RAR
B23 6DX G8NDT
B23 6JL M7SSC
B23 6NN G6IHW
B23 6PJ M3PKQ
B23 7JJ 2E0LPJ
B23 7JJ M3XIP
B23 7JR G0VBQ
B23 7LG G8ROU
B23 7LT G6JAC
B23 7NB M7DTA
B23 7PB G0LNE
B23 7XL G7EKJ
B24 0SN M0KRW
B24 0TQ 2E0GDM
B24 0TQ G4JDC
B24 8JP 2E0GOG
B24 8LX G4ZCR
B24 9BY 2E0XEB
B24 9BY M0DID
B24 9NB M0YJL
B24 9NE M6CKT
B24 9NF G8YJT
B24 9NF M0YJT
B24 9RJ M3UTV
B24 9RX M6ZOG
B240BX M6NQK
B25 8EJ 2E0DJV
B25 8EJ M3ZQN
B25 8JF G8AFI
B25 8LH 2E0IZJ
B25 8LQ G8XXZ
B25 8NE M3TBF
B25 8NZ G7PSC
B25 8TJ M6AUO
B25 8XS M6GYK
B26 1DY G3NXC
B26 1LB M6OCH
B26 1PR G3KYE
B26 1TT G6FGW
B26 2AF G3JTT
B26 2AW G1VSD
B26 2EA 2E0FMV
B26 2EA M0KBG
B26 2EA M7VJH
B26 2HT G6KHM
B26 2HT M7CUU
B26 2HW 2E0MCH
B26 2HW M6MBF
B26 2NY M6DFG
B26 2PP M0GNP
B26 2PP M0SAL
B26 3BX M0WSN
B26 3EL G7HNR
B26 3HY G8DHI
B26 3LA 2E0EUN
B26 3LA M0TAX
B26 3LS G8KGS
B26 3PW G4NPG
B26 3ST G0LBQ
B27 6BB G100W
B27 7LA M6CYY
B27 7PB 2E0DHZ
B27 7PB M3UNH
B27 7RU G6HFZ
B28 0DQ M6CUP
B28 0HH G1BLK
B28 0HR M6EQS
B28 0LA G8ADD
B28 0QX G6KVR
B28 0TE G7IJI
B28 0TW G3RGD
B28 0TZ M6EXQ
B28 8PF 2E0TBR
B28 8PF M6AUL
B28 9EQ G3KLD
B28 9NT G1SWU
B29 4AG G8GDZ
B29 4AH G3ZKQ
B29 4LT 2E1GYZ
B29 4LT M1EBK
B29 4NH G4KLA
B29 5DG M0PTG
B29 5DR M3ONM
B29 5LE G1LWH
B29 5NX G0HDF
B29 5PU G0ETJ
B29 6NH M0JZT
B29 6PX M0BCN
B29 7QJ G4RCZ
B29 7QQ M6RDN
B29 7SG G4NCY
B30 1DR G4NPA
B30 1DR G4NPB
B30 1DU G8LKW
B30 1HU M7SBZ
B30 1LY G8ORR
B30 1NG G0BWP
B30 1PU 2E0GAH
B30 1QN M3SIW
B30 1RF M1EAN
B30 1SN M7JCM
B30 1TH G8DOY
B30 2AX G6WPD
B30 2BA G4YUI
B30 2DY G0UYT
B30 2EB M6HIP
B30 2HT M6IRZ
B30 2LS M3VCB
B30 2SH G4OMP
B30 3BZ G1EAX
B30 3LG M3YIY
B30 3NG 2E0GLM
B30 3QJ G7PIR
B30 3QN G0MQC
B30 3QR M6SYG
B30 3RP M6BJN
B31 1AL G6VGT
B31 1DJ M6DJW
B31 1DN 2E0HBD
B31 1DN M6HDY
B31 1DX G4RPV
B31 1NE G7DNX
B31 1UQ 2E0DOQ
B31 1XH 2E0AZL
B31 1XH M3HTG
B31 2AX G8XIZ
B31 2BJ G8ZXY
B31 2BL M7YRR
B31 2EB G1FVC
B31 2EE M0RTE
B31 2EJ G4YTO
B31 2FG G0FPN
B31 2FS G8VBE
B31 2FT G8AHE
B31 2HD G6IRJ
B31 2JG G3XVW
B31 2LY G1JRF
B31 2PL M7DPB
B31 2PP G8VNL
B31 2QB G8KAE
B31 2QY M0DTR
B31 2QZ G2AZM
B31 2RP M0VAT
B31 2XQ M7CUP
B31 3GG G7BZM
B31 3HH M7ERQ
B31 3HU 2E0PGT
B31 3LA M6GWG
B31 3NL M6EBP
B31 3SY M3XHM
B31 3XE M6FBO
B31 4AJ 2E0PVN
B31 4AJ M0PVN
B31 4AJ M6PVN
B31 4AU G7FUW
B31 4BS M0BAZ
B31 4LJ 2E0JWY
B31 4LJ M0WBY
B31 4LJ M6JWY
B31 4LN M0ALH
B31 4PA M3XOQ
B31 4QN G1JON
B31 4QN G3OHM
B31 4QN G6BV
B31 4QN G8SH
B31 4SS G6TDG
B31 4SU 2E0INC
B31 4SU M0LWT
B31 4SU M6ERA
B31 4TA M3UTM
B31 4TQ G8MIT
B31 5AU 2E1BHF
B31 5DP M0BSP
B31 5HH 2E0UGO
B31 5HH M3UGO
B31 5HQ M0PGD
B31 5JB M6UEB
B31 5NT M3HNE
B31 5NY 2E0REN
B31 5NY M6HME
B31 5RD G6KAE
B313HJ M6IYJ
B32 1AY M6ZBL
B32 1DR G4SWA
B32 1EG G4XUA
B32 1EL G4CGU
B32 1HE G4XEJ
B32 1JA 2E0KEG
B32 1LB G1MAR
B32 1LB G3MAR
B32 1LB G8BHE
B32 1LB G8OHM
B32 1PG 2E1AIT
B32 1PG G0PWQ
B32 1QT G0LGO
B32 2AA G7TPB
B32 2BA G4JCZ
B32 2BS 2E0PPH
B32 2BT M6OOO
B32 2LY M6HSQ
B32 2NW G0PPX
B32 2SB G1NHX
B32 2TT G4ZTM
B32 2UX G4CKK
B32 3LA M6TGJ
B32 3NL M6EYH
B32 3TA G1PKG
B32 3UN 2E0ROX
B32 3UN M7MCJ
B32 4LW 2E0DLF
B32 4LW M6EPF
B33 0LH G4KII
B33 0NR G4SAS
B33 0RL G7DQQ
B33 0YL G6BNO
B33 0YP G4FGF
B33 8AB M0DZO

B33 8JP 2E0MJX
B33 8SL 2E0UOK
B33 8SL M3UOK
B33 8ST G6LKZ
B33 9BU G7FPZ
B33 9DL G8KCB
B33 9HH M7HKY
B33 9JX M6OTS
B33 9NZ G6SFW
B34 6JL G4LTT
B34 6PZ G6NOW
B34 6QY 2E0HMP
B34 6QY M0JVL
B34 6QY M7CIS
B34 7AY 2E1SJG
B34 7BU M0BEM
B34 7EG M7MOE
B34 7LP G4ZPI
B34 7QX G0EOS
B34 7RS G4LAJ
B34 7RU 2E0VJX
B34 7RU M0VJX
B34 7RU M6VJX
B35 6PZ M0LEE
B35 7NT M6NQU
B35 7PD G8TBW
B35 7PF M3VVH
B36 0AD G6WYQ
B36 0BX G6VMR
B36 0EH M1NPH
B36 0HH M3VVA
B36 0JT 2E0DEZ
B36 0LG G3OKH
B36 0LS 2E0GTN
B36 0LS M0XGN
B36 0LS M6NQL
B36 0PB G4OOX
B36 0UH G7UEJ
B36 8HA G0OTF
B36 8QG G3NQA
B36 8UN M7NOO
B36 9DY M3VVB
B36 9DY M3ZAU
B36 9HP M7OMS
B36 9HX G4SGA
B36 9JD G4IMB
B36 9JD G8PXU
B36 9JD M3SZY
B36 9LL G6VXL
B36 9LL G6YCI
B36 9LL G7OZE
B36 9LL M3OZE
B36 9SN G3DID
B36 9ST G4LRN
B36 9TS M3JYA
B36 9TW G1GFW
B36 9TY G4XPV
B37 5HS M3CIP
B37 5HX G0FCQ
B37 5QL M7SDB
B37 6DL G4SPY
B37 6JQ G0WFX
B37 6SB G6APQ
B37 7BU M0CDQ
B37 7BU M6LCV
B37 7EE G4HPT
B37 7HS G4TDF
B37 7PA M3XJP
B37 7QT M6FMI
B37 7RD G1PRF
B37 7RW M0ZDO
B377HW M7RVH
B38 0AB G1BUJ
B38 0AL M0BTO
B38 0DN G3YKO
B38 0EP G0MTN
B38 0EP G0WRC
B38 0EP G7WAC
B38 8AJ G1UOD
B38 8DA G8NLK
B38 8DT G4AEG
B38 8ED 2E0NCO
B38 8ED M6NCO
B38 8LB G0MKU
B38 8LS G1XPD
B38 8NH M7RJH
B38 8PH G4KRT
B38 8PW G3TGL
B38 8PW G8XUN
B38 8TH G0AHC
B38 9AG M6SNN
B38 9HS M6SYH
B38 9LA G0BOT
B38 9PA G0WYT
B38 9QR M6MNZ
B38 9QT G1MZM
B38 9RH M3UZV
B38 9RY G0OKI
B38 9TT 2E0SRV
B38 9TT M0SSV
B38 9TT M6SRV
B38 9UW M3VXB
B42 1EU G7TVL
B42 1HF G0FBQ
B42 1LP G1HUM
B42 1LY G4WSI
B42 1PL G0DOG
B42 1PZ G6UGA
B42 1RY G1EDT
B42 2BH M0MLS
B42 2BX G1MOV
B42 2DT G4NTV
B42 2EA M3EYK
B42 2EH M3PHX
B42 2HF M0WPS
B42 2HJ G0GPF
B42 2HT G8ASW
B42 2JW G8SFQ
B42 2LX M0GYA
B42 2NU G1LTE
B42 2PQ G6NHW
B42 2QB G6UGZ
B42 2QJ G4HOM
B42 2RL G6YQW
B42 2SQ G4NBW
B43 5AG M6KDF
B43 5EH M0ANK
B43 5HH 2E1ILM
B43 5JR G7IKG
B43 5LN M3OHX
B43 5ND 2E0XAW
B43 5ND G4YQD
B43 5ND M6ARW
B43 5PG 2E0EHX
B43 5PG M6GGX
B43 5PG M6HHW
B43 6BB G3FIA
B43 6HX G0TEM
B43 6NA M3EMO
B43 6QE G6YSB
B43 7HG G6LPB
B43 7HX 2E0NVP
B43 7HX M3NVP
B43 7JW M3NVR
B43 7PG G4PFK
B43 7PQ 2E0CVU
B43 7PQ M0TVU
B44 0AL G6WOI
B44 0AY 2E0VMV
B44 0AY M0VMV
B44 0AY M3VMV
B44 0EF M7TRV
B44 0HN M7EXQ
B44 0LB G1BJK
B44 0LF G0FOC
B44 0NF M3WMU
B44 0QB M7BTQ
B44 0QS 2E0OAM
B44 0RD M7MKY
B44 0SN M0WET
B44 8AB M6KTX
B44 8ER G7CWO
B44 8JB G4DIP
B44 8LQ G6WIG
B44 8RG 2E0ELD
B44 8RG M3ELD
B44 8RL G0WMU
B44 8RS G7OPD
B44 8SW 2E0GOM
B44 8SW G8CHA
B44 9AX M7ESB
B44 9BY G4RZD
B44 9DB G1XKB
B44 9NY G8TZW
B44 9QH G7ORT
B44 9QN M6NQM
B44 9RP G6NZW
B44 9SP G3URV
B44 9SS G4BTK
B440NG M3INC
B45 0DA 2E1IAT
B45 0DD 2E0GCB
B45 0DD M3GKH
B45 0DD M3HIO
B45 0EN 2E0SVZ
B45 0EN M6ZDL
B45 0JB G8VXY
B45 0JS M6LDU
B45 0JY G0HPG
B45 0JY G0HPH
B45 0JY M6BNP
B45 0JY M6SXD
B45 0NB G6ZSG
B45 8EH G0NJT
B45 8GU 2E0AAI
B45 8GU M6KND
B45 8HP G4TSB
B45 8NG M0VGG
B45 8NZ G7UMY
B45 8QU 2E0VUK
B45 8QU M3VUK
B45 8SJ M3JYE
B45 8TQ 2E1ANG
B45 8TQ 2E1ANH
B45 9DA G1XKY
B45 9EU 2E0AQE
B45 9EU M0JTB
B45 9HW G6YXW
B45 9HY G7RLV
B45 9LR G4YTJ
B45 9LW M0BYJ
B45 9RU M3KCQ
B45 9SZ 2E0LUJ
B45 9SZ G0HVN
B45 9SZ G3IUB
B45 9SZ G8IUB
B45 9SZ M6LUJ
B45 9TT 2E0FXJ
B45 9TT M6FXJ
B45 9TT M7JAC
B45 9WA M6ILR
B45 9XB G8IFT
B45 9XB M6GKG
B46 1EP G4OPE
B46 1HL G6NTY
B46 1NP 2E0TJE
B46 1PD G3UPA
B46 1PE 2E0KHG
B46 1PE M6KAO
B46 1RU G8MWE
B46 1SA 2E0MEY
B46 1SA M6CKR
B46 1SN G0CKE
B46 1SN G3ZUM
B46 1TW G6VNC
B46 1UF G1KAT
B46 2NH 2E0YOM
B46 2NH M0YOM
B46 2NH M3YOM
B46 3EH G0BUV
B46 3LZ G4YQD
B46 3NE G4FTY
B47 5EN G3OIC
B47 5HJ G8TWZ
B47 5HY G1WAC
B47 5HY G4VPD
B47 5HY G4WAC
B47 5LN M6DCW
B47 5NR 2E0HPD
B47 5NR M0IFT
B47 5NR M6HPD
B47 5PX G6JYO
B47 5QA 2E0YZW
B47 5QA M7ACN
B47 5QE G0NES
B47 6AW 2E0HKL
B47 6AW M6KQC
B47 6HE M7BVH
B47 6HP G4OJL
B47 6LX G0ICJ
B48 7NE G1PHN
B48 7PL G3UYX
B48 7TB G0BNZ
B48 7TB G4SVL
B49 5DD G1YFD
B49 5DD G4GYI
B49 5EG 2E0BOT
B49 5EG M0OJG
B49 5HA G0UGY
B49 5HY G6MOZ
B49 5LJ G0UMV
B49 5LJ G3UMV
B49 5LJ G4UMV
B49 5LJ G6LRT
B49 6AP M1EZD
B49 6DR 2E1DLA
B49 6DR G1IKL
B49 6HQ G3USA
B49 6LF G3EVT
B49 6LF G4ACZ
B49 6LX G1WLU
B49 6LX G1ZUU
B49 6PG G0RGW
B49 6PG G8EXF
B49 6QY G6BAY
B5 7NE G6DWS
B50 4AN G4OHJ
B50 4AN M3MEB
B50 4AP G0VBZ
B50 4AR G3VRF
B50 4HQ G4XVO
B50 4NP 2E1MGB
B50 4NP M0XLX
B50 4NX M0BHM
B50 4PH G7BRF
B6 4PP M6MRX
B6 5HW 2E0DHY
B60 1AD G1BHO
B60 1AL G1BIU
B60 1AW G1GDR
B60 1AZ G4LVK
B60 1AZ G6VGG
B60 1BN M0FRA
B60 1BP G1FPY
B60 1DG G8DEC
B60 1DW G0MET
B60 1DW G8VFP
B60 1DY G4OJS
B60 1DY G4WZA
B60 1DZ G6NQB
B60 1HE G0KHK
B60 1HN G4BBU
B60 1HN M3RXP
B60 1HW G4AHK
B60 1PJ M0WOD
B60 2DB 2E1GNE
B60 2EB G0BLT
B60 2EB G0HAW
B60 3AY G0EVY
B60 3EP G4UIW
B60 3GP 2E0OBK
B60 3GP M0OBK
B60 3GP M6OBK
B60 3HB G0BIR
B60 3HB G0EHQ
B60 3NX 2E0KBF
B60 3NX M6KBF
B60 3SD M6GZB
B60 4EB G1TCK
B60 4EB M3CBC
B60 4LH M6NAN
B60 4LS G6OPY
B60 4NF G4FNQ
B603LZ M7ETK
B61 0DX G6OIF
B61 0EL G4AHO
B61 0ER G6EET
B61 0ER M1CLZ
B61 0JP M6SOZ
B61 0JT G4LRL
B61 0LQ G0AQF
B61 0LU G4NZK
B61 0LU G6LPS
B61 0PA G8XUW
B61 0TR 2E0XON
B61 0TR M7XON
B61 0TT M3KGO
B61 7BE G1IEC
B61 7DA M6KLR
B61 7EB G4CQS
B61 7EB G8GEB
B61 7JG G0BGA
B61 7JT M1JSS
B61 7PR M0LLS
B61 8HY M3SKU
B61 8NB 2E0MJT
B61 8NQ G0SUQ
B61 8PE G4OTC
B61 8PN M3UHW
B61 8UA G4PKW
B61 9BH 2E0MIS
B61 9BH M0LIJ
B61 9BH M6UDY
B61 9JN M3XNE
B61 9JT G4CAF
B61 9JW M3AXT
B61 9LH M3OJS
B62 0HU G1WTH
B62 0HW 2E0UAE
B62 0HW M3ULS
B62 0JL M0JFW
B62 0LN G4OHB
B62 0NE G0EVM
B62 8EU G7IIS
B62 8EZ M0VCR
B62 8JA G7AXW
B62 8JS G2FXZ
B62 8JS G6GQJ
B62 8LJ M3SAR
B62 8LJ M6IKA
B62 8LJ M6MDX
B62 8LJ M6SAD
B62 8LR 2E0KPP
B62 8LU M6EFJ
B62 8QE M0CWX
B62 8SH G6IJQ
B62 8TH G1IAL
B62 9AW G3VBV
B62 9BP M7CIL
B62 9DX 2E0DSS
B62 9DX M6DSS
B62 9HJ M3CPY
B62 9NQ G4VPZ
B62 9NR G0AUK
B62 9NR G4DPZ
B62 9QJ 2E0MKB
B62 9QJ M0FCW
B62 9QJ M0KMW
B62 9QW G4JSV
B62 9RF G4PVP
B62 9SX G6EVW
B62 9TF 2E0SLJ
B629LP M0BRE
B63 1AE M7GFQ
B63 1BB G1FET
B63 1BZ G4OQX
B63 1DU G8TXA
B63 1EQ 2E1WIN
B63 1EQ G1PPZ
B63 1HD M6RKB
B63 1JQ G4LWF
B63 1JQ G4MEB
B63 1JY G4AYK
B63 1JY G4NQW
B63 2AY G7VJT
B63 2DW G6XYO
B63 2HE 2E0XSZ
B63 2HE M7EDZ
B63 2JA G7FIJ
B63 2JJ G0GUG
B63 2PP G4EHR
B63 2PR M3XMJ
B63 2PY G1VQV
B63 2SY G8TIA
B63 2SZ G8ZPD
B63 2TB G6LDA
B63 2UY G1KII
B63 2XH M0YIM
B63 3DN G8XSA
B63 3JX G4UMY
B63 3QR G4FAV
B63 3QR M3EFL
B63 3RP 2E0RWC
B63 3RZ M3FFA
B63 3RZ M3FFE
B63 3ST M6EUX
B63 3TJ G1BEG
B63 3TP 2E0CUY
B63 3TP 2E0JTM
B63 3TP G7JMZ
B63 3TP M6EDA
B63 3TP M6JPD
B63 3TP M6ZSB
B63 3TZ M6ARK
B63 4BX M6EMS
B63 4HD M0CTK
B63 4HG G6ALW
B63 4HQ G1VOB
B63 4HQ G3DTU
B63 4PB G0NLA
B63 4QG G1BRF
B63 4RN M1BZG
B63 4SA M6LNR
B63 4TJ M1EVF
B632BD G7VIE
B64 5BG G8MUV
B64 5EX G7TUV
B64 5LA M7BKN
B64 6DU G0PPJ
B64 6EA M0EHS
B64 6HP 2E0OFT
B64 6QX G0BHR
B64 6RB G4JFF
B64 7EZ G3VPX
B64 7EZ G7BRA
B64 7HH 2E0FGT
B64 7HJ G6NNO
B64 7JS G4HPB
B64 7PN G4YAH
B64 7QH M7IFR
B65 0AG M6NEL
B65 0HF G6LOR
B65 0NP G0WYA
B65 0QE M1FFG
B65 0QE M3FXU
B65 0RR 2E0JAM
B65 0RR 2E0LEE
B65 0RR 2E1HXM
B65 0RR 2E1HXN
B65 8DT G4YIV
B65 8DW M3VBI
B65 8DW M3VRP
B65 8HT 2E0KAS
B65 8HT M0KAU
B65 8HT M6KAN
B65 8NX M0GSX
B65 8PB M6OAQ
B65 9BQ G1OOZ
B65 9DZ G0PAI
B65 9DZ M3XNK
B65 9HZ G0OYF
B65 9JN G0KJG
B65 9JX 2E0HAZ
B65 9JX M3GHH
B65 9LG G0MJT
B65 9NJ M0IMM
B65 9NT G3TRG
B65 9NT G8TRG
B65 9QN M3FGU
B65 9SD G4FJJ
B67 5AY G1JVM
B67 5AY M0JVM
B67 5DH G1UCO
B67 5PD G7UNB
B67 6AD M6DKK
B67 6DY M7BNG
B67 6HU M0BBW
B67 6JA 2E0IRH
B67 6JA M7SCD
B67 6LA G1UUZ
B67 6PR G1FFO
B67 6QS M3JJH
B67 6QX G0TVR
B67 6QX M3DVU
B67 7BX G4KVC
B68 0BJ M0MEL
B68 0NA G8DEM
B68 0NU G8PTF
B68 0PU 2E0PYA
B68 0PU M0PYA
B68 0SW 2E0XYX
B68 0SW M6TPA
B68 8AQ G4SFG
B68 8BE M1AED
B68 8BP M3LXV
B68 8HY M3EFQ
B68 8LT G4OJJ
B68 8LT M3TOF
B68 8NG G0UFJ
B68 8PP M1WEH
B68 8PR G4OYT
B68 9DP G1XFO
B68 9DP G6IQY
B68 9DU M3XVZ
B68 9ES G8MKE
B68 9LQ G1IOU
B68 9LU G4VRX
B68 9PW G8RAO
B68 9TB G7RTQ
B68 9UL G0WXA
B680PU M7FMA
B69 1AJ G7IYY
B69 1BA 2E0TAM
B69 1BA G0NJZ
B69 1BA G7IYX
B69 1JU 2E0TAT
B69 1NA M6RNN
B69 1NP G4JVH
B69 1NP G8FCO
B69 1NT G3NZS
B69 1PW G4YFT
B69 1QA G1WHY
B69 1SE G6EOR
B69 1SW 2E0SOX
B69 1TP G8SPU
B69 1UD G8TFB
B69 3NE M7HYN
B69 4QA G7LOE
B69 4TG M7XCJ
B69 6RN G3VJX
B70 0SL G4DVM
B70 6RJ 2E0WBZ
B70 6RJ M0WBZ
B70 6RJ M6WBZ
B70 6RQ 2E1LJW
B70 6RQ M6LRU
B70 7AZ G7POG
B70 7AZ M3MHM
B70 7LS G1TBK
B70 8JY 2E0KVE
B70 8JY M3GUM
B70 8LD G0BZP
B70 8PL M3LMD
B70 8QR M3LLC
B70 9ES 2E0ZGL
B70 9ES M0VGL
B70 9ES M3NUE
B71 1AE M7JUK
B71 1DQ M6BSI
B71 1DX M3DIU
B71 1NJ G0WCK
B71 1RU G4PMW
B71 1RU G6FPN
B71 2AA 2E0SJS
B71 2AA M3MFS
B71 2DY G7RCP
B71 2DY M3APE
B71 2QJ 2E0BMU
B71 2QW 2E0RWN
B71 2QW G4LWN
B71 2QW G8UID
B71 2QW M0RWN
B71 2QW M3RWN
B71 3BT G6VFO
B71 3DA 2E0DEM
B71 3DA M6EED
B71 3EE M1DMT
B71 3HX G6MPT
B71 3LA 2E0LIT
B71 3LA 2E0XAV
B71 3LA M0PPY
B71 3LA M6GYM
B71 3LX G0RCX
B71 3NE 2E1IHE
B71 4BA 2E0LET
B71 4BA M6KNY
B71 4HN 2E0BCD
B71 4LQ M3OFU
B71 4LR M1GAP
B71 4LR M3TYZ
B72 1AG G1AAL
B72 1HB G7ETS
B72 1HP G3PLP
B72 1JP G0UQJ
B72 1JU G1NFN
B72 1LG G5GWH
B72 1YE G4TLR
B72 1YF G0VXK
B72 1YZ G8NFD
B73 5AR G7FBY
B73 5EA G4TYR
B73 5EH M3FSU
B73 5EL G8EZB
B73 5JY G1KYK
B73 5LD M3TGE
B73 5LF G0VZL
B73 5LT G0GEP
B73 5LT G1LTG
B73 5NF M6XVM
B73 5SP G7EKW
B73 6NZ G4OFN
B73 6NZ G7ORV
B73 6PG G8AMD
B73 6QA G4LBT
B73 6QR G4NBI
B73 6RY 2E0IOJ
B73 6RY M0KJZ
B73 6RY M7EUA
B73 6UQ G1NPA
B735UN M6MVF
B74 2AE M3LIY
B74 2BS G6HOC
B74 2BU M3IZN
B74 2DA 2E1EQE
B74 2DG M0CUD
B74 2EA G7JJW
B74 2JE G8KFF
B74 2LA G4DDD
B74 2PS G3MCB
B74 2QA G3JZF
B74 2TB G4ABW
B74 2UH G6AYI
B74 3JU G6VPL
B74 3LR G6RBO
B74 3NP G1MGZ
B74 3NP G1OBA
B74 3PG G4IWF
B74 3PQ G7OBC
B74 3QE G0PJG
B74 3QE M0AZK
B74 3QP G1EYT
B74 3QQ M7RWW
B74 3TU G1XDV
B74 3UD 2E0XCN
B74 3UD M3XCN
B74 4AN G0MBI
B74 4BL M3YOH
B74 4DQ G1AZE
B74 4DX G7NJW
B74 4DY G6REY
B74 4HL G8EFU
B74 4HL M0DBM
B74 4JS G3WWL
B74 4LQ 2E0CLI
B74 4SG G8XMH
B74 4UG 2E0MAJ
B74 4UG M0MAJ
B74 4UG M3OFJ
B74 4XG G8OHS
B74 4XR G1JXX
B74 4YD G1GFA
B74 4YD G7SER
B75 5EJ G7PZQ
B75 5EX G7LHK
B75 5LD G1NSG
B75 5LH G0KLK
B75 5PQ G6VIY
B75 5QA M0AZE
B75 5TJ G7VBJ
B75 6AU G0EVH
B75 6AX M3FQX
B75 6DB G6HNS
B75 6DH G7UCG
B75 6DW G3AVE
B75 6EN 2E0GTL
B75 6EN G1AZZ
B75 6EN G3MYC
B75 6EN M3PNF
B75 6SN G8TBB
B75 6TF G5RGS
B75 7AA G6AGO
B75 7AA G6DGR
B75 7BL G6BBR
B75 7BY M7NXC
B75 7TH G6HSR
B75 7UU M6LZY
B756AA M0BUC
B76 1EA G7IXH
B76 1FN M3WAP
B76 1HU G8VUU
B76 1HY 2E0ROJ
B76 1HY M3HOD
B76 1JN 2E0FDE
B76 1JN M6XPH
B76 1JQ G1BUQ
B76 1JR G1XKN
B76 1LQ G7PHF
B76 1LZ G8IOS
B76 1PJ G6KPX
B76 1QZ G8ERN
B76 1XR G6UED
B76 1YD G0HID
B76 1YE 2E0CCB
B76 1YE M6TYE
B76 1YR 2E1DSU
B76 2PT G3VNY
B76 2SY G1FQX
B76 9BD M7YKM
B77 1AB 2E0HZS
B77 1BT G0WKI
B77 1BY G6EOO
B77 1DF M6BTN
B77 1EG M0RQQ
B77 1HB G4GYA
B77 1JD G4MFN
B77 1NY 2E0CCC
B77 1NY M3FEG
B77 1PE G1MGN
B77 1QR G4ORW
B77 1QT G6RZY
B77 2HQ G3YUX
B77 2JA G0UYP
B77 2JU M7AWW
B77 2LD G0GUD
B77 2NA G0PFQ
B77 2NB M7HYB
B77 2QD G6SRU
B77 2RS M3JWN
B77 2RY G8SCG
B77 2TX M7RXM
B77 3BT G4LZV
B77 3BT G4NWM
B77 3BT G4NWN
B77 3BT G8UXX
B77 3DG M6TFW
B77 3JH G4NWO
B77 3JW M5ACJ
B77 3JZ 2E0KPB
B77 3JZ G8RFL
B77 3JZ M6NQQ
B77 3LA G6LPX
B77 3LH M1FEQ
B77 3LH M6OPG
B77 3NB M3FLB
B77 3PH G8MCT
B77 3PP M1FER
B77 3PR M6AQA
B77 3QG M1CVH
B77 4AA G4INA
B77 4AB 2E0DPS
B77 4AB M6DSZ
B77 4AQ G0SKK
B77 4BZ G0LHR
B77 4DB M6OYN
B77 4DF M6APU
B77 4DL G6NHG
B77 4EJ G0KFF
B77 4EJ G4ZPJ
B77 4EN G3DAT
B77 4EU G0LTR
B77 4EZ M3FLZ
B77 4HT G0IKQ
B77 4JJ 2E0LCR
B77 4JJ M6ION
B77 4JL G0TRB
B77 4JW M3ASN
B77 4LD M6SUX
B77 4LU M3RLT
B77 4LX 2E0BME
B77 4LX M3UWV
B77 4NA G1UZW
B77 4NA M3UZW
B77 4NA M3YII
B77 4PP M6NXE
B77 5EY G0FXL
B77 5FL M6HMS
B77 5FX M7JCA
B77 5GG M6DSO
B77 5JD 2E1FDY
B77 5JE 2E0ZAP
B77 5JE M3RJB
B77 5NW M7AWH
B77 5PL G1GAW
B77 5PQ G8VPH
B77 5QE M6GCJ
B77 5QF G6MOD
B77 5TD G6FBH
B78 1BQ M6OYF
B78 1DE G7ODM
B78 1EX 2E0EHB
B78 1EX M0NPL
B78 1EX M6EHB
B78 1JY 2E0RPU
B78 1JY M3ZQG
B78 1LS 2E0LLE
B78 1LS M0XVK
B78 1LS M6CKO
B78 1LW M6AMG
B78 1NJ G8NAP
B78 1NU G6DYR
B78 1NX M7TYS
B78 1QN M6OXS
B78 1QZ G7UKF
B78 1SY G2BZR
B78 2DR G4BKA
B78 2EP G4NRY
B78 2ER 2E0JBM
B78 2ER M1DNE
B78 2ER M3SWV
B78 2ER M6CSM
B78 2ET M6FAJ
B78 2JR M3OHJ
B78 2JU G0FEO
B78 2LA G0GEF
B78 2LG 2E0GEV
B78 2LG M3YFH
B78 2LH G4ORY
B78 2LW G6SJD
B78 2LX G6VWI
B78 3DE G3LGW
B78 3HS M6DEJ
B78 3HW G8OSX
B78 3HZ M7CTV
B78 3NW M6CRJ
B78 3QD G0NXY
B78 3RA G4NRX
B78 3SE G7HJG
B78 3SW G7UHG
B78 3SZ M1APQ
B78 3TJ G4ICI
B78 3YA G4SBS
B79 0DJ G6VBD
B79 0HR M6JMN
B79 0JD M7HVM
B79 0JR M6JKT
B79 0LD M1EDF
B79 7BE G0BFC
B79 7BQ G0FGK
B79 7EW G0FZM
B79 7RA M6FAH
B79 7SQ M6EEP
B79 8BE M6NIN
B79 8BH M6IWP
B79 8BZ 2E1SHE
B79 8DE G8KRV
B79 8DN M6LFD
B79 8EJ G8NCK
B79 8EY 2E0TLM
B79 8EY 2E1JMW
B79 8EY M0TLM
B79 8EY M6TLM
B79 8HZ 2E0JRW
B79 8JB G6DIF
B79 8JB M3SQU
B79 8LB G8TRS
B79 8NB M3SQV
B79 8NF 2E0FAJ
B79 8PE M0TSD
B79 8QB G7WFK
B79 8UA M3SUY
B79 9BP G3ENO
B79 9JA 2E0CLP
B79 9JA M0RGH
B79 9JA M6BHB
B79 9JG G8RVZ
B79 9JJ M6GJP
B798TA M7BAV
B8 2AU G3RSC
B8 2AU M1CQN
B8 2AU M3ASZ
B8 2AU M3BJV
B8 2AU M3LMA
B8 2AU M3SVN
B8 2EA G7GGT
B8 2LB M6TOL
B8 2PD G0NFR
B8 3LL M3AOM
B8 3LL M3REZ
B80 7HD G1YFA
B80 7HN G4YJC
B80 7JJ G7APL
B80 7LX G4LMF
B80 7PG G4STE
B80 7QL M3VDL
B80 7RD G6FDO
B80 7RR G3CON
B80 7SH G0WDU
B807BZ M3RKE
B9 5NG G7LTG
B9 5RY G7LXV
B90 1BS M6CGX
B90 1DS G7DDN
B90 1DS M6FAB
B90 1LF G4AQJ
B90 1RW G7JHX
B90 2BB G3XBY
B90 2BQ G7GFP
B90 2DR G7WBJ
B90 2EJ G1MJO
B90 2HB G4ZVZ
B90 2HS 2E0HVF
B90 2HS M6HVF
B90 2HT 2E0EPA
B90 2HT M0IQM
B90 2HT M6IYN
B90 2HY G4RTI
B90 2LN G7IMQ
B90 2PR G7IMR
B90 2PU M3UKX
B90 2QF M0GWM
B90 2QW G6AQW
B90 3DF G0IHU
B90 3DQ M3IBS
B90 3HX G4KOR
B90 3JE M1DJG
B90 3JE M3DJG
B90 3JF M3FAE
B90 3JR G8ZUL
B90 3JZ G6FIO
B90 3LG G4TBJ
B90 3LJ 2E0DBL
B90 3LJ M0PYT
B90 3LJ M6GAL
B90 3PL G8XQA
B90 3RE G6IHB
B90 3SA G3MRZ
B90 4BU G6BWT
B90 4BU M3DOX
B90 4BX G1ZQE
B90 4DX G4LCH
B90 4HR M6JNA
B90 4JT G4LPE
B90 4PH G4BQW
B90 4PH M6AXW
B90 4PN G6HNR
B90 4RN 2E0UDM
B90 4RN M6UDM
B90 4RU G6UUR
B90 4UU G4HOR
B90 4YB G0NFZ
B902TA M0DOH
B91 1BS G6VUJ
B91 1DD G4MPG
B91 1DQ G7KZV
B91 1DQ G7MAB
B91 1DX G4AMI
B91 1DY M3ZAR
B91 1LN 2E1GYC
B91 1LN 2E1GYD
B91 1LN G4WMH
B91 1PR G4LOE
B91 1QB G1FMW
B91 1TJ G3LUA
B91 1TQ G8GBM
B91 1TZ M0AEC
B91 2EG G3DEJ
B91 2EH G4EET
B91 2JD G0DJF
B91 2PL G6JEB
B91 2SB M7HTK
B91 3GA G1RBX
B91 3HH G0NOB
B91 3JY G1BHB
B91 3LL G6KMQ
B91 3LL M3YWD
B91 3NB 2E0XZM
B91 3NB M7AII
B91 3ND G4KSG
B91 3PW G1ZLC
B91 3UD G3WZI
B91 3XR G4NYG
B92 0BP M7HIA
B92 0BS G8SAN
B92 7BU M6MKY
B92 7DF 2E0ZSR
B92 7DF M0SRZ
B92 7DF M6GYF
B92 7ES G8GWR
B92 7EY G6BJG
B92 7HB G6DFH
B92 7HE G1STK
B92 7HE G6MCY
B92 7HH G4FJB
B92 7JA G8YMU
B92 7JB G0HLW
B92 7JH G4VMO
B92 7NU G4KQV
B92 7PN M1LCL
B92 7PN M3CTO
B92 7RU M6RJB
B92 8AD 2E0YZV
B92 8AD G4ENH
B92 8AD M7EHG
B92 8AD M7ENH
B92 8AL G6EDF
B92 8DP G7LED
B92 8DX G4CEX
B92 8EE G3GEI
B92 8EE G4BBT
B92 8HQ 2E0AXB
B92 8NB G4MVB
B92 8NN G4PCE
B92 8PL M0PPX
B92 8PU M7ALN
B92 8PU 2E1GXI
B92 8QA M1CXA
B92 8QS G8ULQ
B92 9BJ G1ASU
B92 9DB G1FUG
B92 9DQ G0PHR
B92 9EY M0WAO
B92 9HH G7OJO
B92 9LQ 2E0SSL
B92 9LQ M6GSJ
B92 9ND G4KWO
B92 9NW G7JVF
B92 9PQ G6VXN
B92 9PT G7GQX
B92 9QH G4EIG
B93 0BZ G4AOJ
B93 0HX G0IZQ
B93 0PT G3OIF
B93 8AW M0LYQ
B93 8DN G4RWG
B93 8DN M6TTT
B93 8NN G7FFS
B93 8QP G3UFQ
B93 8RA G8IK
B93 8RN 2E0ITQ
B93 8RN G1DWU
B93 8RN M0JSL
B93 8RN M6ITQ
B93 8SY M0XXT
B93 9AW G6VKS
B93 9EQ G3KEK
B93 9JL G4MAU
B93 9LA G3TZM
B93 9LG M0RXV
B93 9LQ G1JYR
B93 9LQ G3VPE
B93 9NP 2E0THE
B93 9NP M0SUD
B93 9NP M6SUD
B93 9PA G1VIW
B93 9PP G4CVM
B94 5AP G3GBS
B94 5EB G8NOF
B94 5LP G7OKF
B94 5LP M3JKZ
B94 5RZ G6HSD
B94 5SD G4NSB
B94 6LN M0HRT
B94 6PA 2E0DTV
B94 6PA M0REM
B94 6QH M3KKO
B94 6QY 2E0RNJ
B94 6QY M0RJH
B94 6QY M6JNR
B94 6RH 2E0LQZ
B94 6RH M7LOZ
B95 5BA G0EPL
B95 5HU G8JMY
B95 5LR G0CRB
B95 5NN G4PIP
B95 6AB 2E1FRC
B95 6BH G3UOC
B95 6BH G4CGR
B95 6HJ G2FHF
B95 6JA G3XTI
B96 6AF G7IRL
B96 6AX 2E0LHE

B96 6AX M6OCY
B96 6DY 2E0XET
B96 6DY M6KET
B96 6EB G7VTR
B96 6EB M0CUS
B96 6ED G1DCY
B96 6ED G3RZI
B96 6LT G6CUQ
B96 6NG 2E0EGP
B96 6NG M6EGP
B97 4JL 2E0MYD
B97 4JL G0MYD
B97 4JL G0TOX
B97 4JL M7ELE
B97 4LX G3WF
B97 4NE G1JJA
B97 4NP G3KWK
B97 4PL G1RFI
B97 4PN G8MGK
B97 4RL G8DSM
B97 4RP 2E0GUF
B97 4RP G8PYU
B97 4RP M7AMB
B97 4SP M0JEM
B97 5AA 2E0ILX
B97 5AA G2FXJ
B97 5AA M6ILX
B97 5AS G6EAY
B97 5AY G3TNI
B97 5BH M7BFY
B97 5DA G7UQS
B97 5DF G4ZWR
B97 5EP G4SGV
B97 5FP G4NTG
B97 5LX G0TPG
B97 5NG G0ORE
B97 5NG G8XUB
B97 5NW 2E0LRV
B97 5NW G0RMG
B97 5NW M0RGP
B97 5NW M6IOI
B97 5PZ G1XVY
B97 5QT G1TQU
B97 5RX G0EYO
B97 5RX G6SL
B97 5TB G4NRP
B97 5UA G3OXL
B97 5UN G4SRV
B97 5WB G4NYZ
B97 5WG G8RCE
B97 5XS G0NRF
B97 6GT M7KJP
B97 6JH M3PSF
B97 6LP M3JFS
B97 6NJ G2HHH
B97 6NS G0TNH
B97 6PG M1ERO
B97 6PH G1AJV
B97 6RQ 2E0WNP
B97 6RQ M7PLI
B97 6SE G3XHY
B97 6TS G0CLM
B97 6UF G7RBQ
B975YR G4TJS
B98 0AG G7GFM
B98 0AS G4MUV
B98 0BH G4SWR
B98 0BJ 2E1FPM
B98 0BJ M3DVN
B98 0BJ M3FPM
B98 0BJ M3HLN
B98 0BJ M5DRW
B98 0BX G1AVW
B98 0EN 2E0LQY
B98 0EN M0LQY
B98 0EN M6LQY
B98 0EW M6MBH
B98 0EY 2E0PWM
B98 0EY M3PWM
B98 0HH M3TXQ
B98 0HJ G3CKE
B98 0HJ G6WPE
B98 0JQ G0PLA
B98 0JQ G7HSL
B98 0JR M7IRW
B98 0LA M3VNM
B98 0NA G1DOA
B98 0NF G1RAG
B98 0NF M5AFV
B98 0PJ G0CAX
B98 0PP 2E0ULC
B98 0QT G7AZH
B98 0SB G4DFC
B98 0TQ G6DVP
B98 7NH G1DHM
B98 7NR G0NVV
B98 7PD G6FVL
B98 7PL M6AGA
B98 7PW G3TBW
B98 7QA G3WJN
B98 7QF 2E0ETH
B98 7QF M0NAK
B98 7QF M6BNQ
B98 7QW G4UCF
B98 7RF M3EFX
B98 7SD M6LRQ
B98 7SZ G6OUX
B98 7TG M7AZR
B98 7UT G4UDK
B98 7XE G0UQE
B98 7XT G1MZT
B98 7YD G6HCW
B98 7YL G4LDB
B98 7YR G3SAH
B98 7YZ M0AMF
B98 8EY M7YYQ
B98 8HT 2E0HOL
B98 8HT 2E0MNC
B98 8JS M6MWD
B98 8PL G0KWQ
B98 8QL G0LFY
B98 8QL G0LFZ
B98 8RL G7KMW
B98 8RW G6VUN
B98 8SQ G4KNX
B98 9JA M7MRX
B98 9JH G6ERJ
B98 9JX G4TRI
B98 9JX G8EOJ
B98 9LE M6DWY
BA1 1AR G6FPC
BA1 1SR M6XTM
BA1 2BL G4YNM
BA1 2QL G8OTA
BA1 2UU G0EJR
BA1 2XH 2E0IXI
BA1 2XH M3XUR
BA1 2XH M6IXI
BA1 3HH M0RCR
BA1 3LN G8DWW
BA1 3RB G6MZW
BA1 3RY M0PIZ
BA1 3RY M7DPD
BA1 4BD 2E0XUN
BA1 4BD M7XUN
BA1 4DZ G3UMM
BA1 4NR G1BCB
BA1 5ED 2E0TGA
BA1 5ED M6LHI
BA1 5ED M6LQT
BA1 5JU M3TCG
BA1 5NF 2E0EOL
BA1 5NF M6KVM
BA1 5SW G6UTK
BA1 5SY G1ZUC
BA1 5TB M0VNR
BA1 5TB M6EUL
BA1 5TW G6UZG
BA1 6EF M3GZP
BA1 6JR G1MDC
BA1 6NA G8DRK
BA1 6NP G7MZY
BA1 6QN G7VSJ
BA1 6QU 2E1XGX
BA1 6QU G5XGX
BA1 6QU M7GCT
BA1 6QX M0LAZ
BA1 7BA G0WZY
BA1 7EE G3PWX
BA1 7NX 2E1DAR
BA1 7SB G1OPW
BA1 7TJ G3RVX
BA1 7TT G4JQB
BA1 8AD G0WAW
BA1 8AD G7STQ
BA10 0BS G3KZR
BA10 0HR G0WRL
BA10 0JD G7VNC
BA10 0RJ G4WTX
BA11 1AQ G3XBW
BA11 1RR M3IKD
BA11 2BD G1JPK
BA11 2BJ M7AJO
BA11 2QD M0IBW
BA11 2TN G4OWH
BA11 2UH M6JNQ
BA11 3BL G0JDM
BA11 3DP G8VGI
BA11 3DY M3BPG
BA11 3FE 2E0BHQ
BA11 3FE M3SXZ
BA11 3LR M0EDA
BA11 3LR M3LPF
BA11 4AA M0RKM
BA11 4AB M6KMK
BA11 4HQ M7FUL
BA11 4JA G4XAG
BA11 4JA G4XAH
BA11 4JB M0HBH
BA11 4NR G7VQX
BA11 5AR 2E0HBV
BA11 5AR M0JCP
BA11 5AR M6HQX
BA11 6PR G1JAL
BA112EY M7CCF
BA12 0AE G7AZV
BA12 0ES 2E0JZK
BA12 0ES M0JZK
BA12 0ES M6JZK
BA12 0JW G0BGI
BA12 0PR 2E0FDH
BA12 0PR M6ZPH
BA12 0PW G7FHA
BA12 0RN G0AYD
BA12 6BP G0WSJ
BA12 6HH G3MAZ
BA12 6JX G4SSP
BA12 6LR G4KDK
BA12 7AG G7VHC
BA12 7AP G3NFJ
BA12 7BB M6FRP
BA12 7HE M0RSG
BA12 7PA G1ACY
BA12 8BU M3HNL
BA12 8DN 2E0HGD
BA12 8DN M0JUA
BA12 8DN M7SJT
BA12 8EB G0GGG
BA12 8EZ G4ILF
BA12 8GE M7JDB
BA12 8HJ M6MKD
BA12 8JR M0BYL
BA12 8JR M0HZT
BA12 8LY G7COA
BA12 8PY M7HGL
BA12 8TB G6MTV
BA12 8TF G4YMG
BA12 9BY G7HVL
BA12 9DU M0SJG
BA12 9EF M1SRP
BA12 9HX G4NKP
BA12 9PN G3MHV
BA12 9PN G4WHV
BA12 9PN M0ZYF
BA12 9PT 2E0PYC
BA12 9PT M0XAW
BA12 9PT M6BNG
BA126PW 2E0IEQ
BA126PW M7CQM
BA13 2GR 2E0TMX
BA13 3AE 2E0DNC
BA13 3AE M0SFT
BA13 3AE M6DNC
BA13 3AN M7HDS
BA13 3AQ M1BPW
BA13 3AU M6PNP
BA13 3ES G4IQZ
BA13 3GS M0WZT
BA13 3GS M3WZT
BA13 3GS M7THF
BA13 3HL 2E0JWJ
BA13 3HL M3JWJ
BA13 3HN M7GCY
BA13 3HP G0VFS
BA13 3HQ G4YSE
BA13 3HQ M6SBS
BA13 3HW G3ZQC
BA13 3JW G7LND
BA13 3LQ M0ADW
BA13 3QL M5EAY
BA13 3RW G6ZDE
BA13 3SH G1WFO
BA13 3UE G6ESJ
BA13 3UH G0GHW
BA13 3UH G8CPF
BA13 3UW G7PVY
BA13 3XG M0XBI
BA13 4AT G4YXS
BA13 4BH G3WZH
BA13 4EA G0PVN
BA13 4LG G4JQN
BA13 4NX G4JQX
BA13 4NY G4CLC
BA13 4QQ G0MZJ
BA13 4TH G0JYL
BA133AN M7ASY
BA133XQ G8FAS
BA14 0AS M7KLF
BA14 0HG G0RKB
BA14 0HG G0TFX
BA14 0HS M1WAZ
BA14 0LH G4KHK
BA14 0LH G6CKL
BA14 0LJ G7FXY
BA14 0NS 2E0BKU
BA14 0NS G0BKU
BA14 0NS M6TEP
BA14 0QL 2E0OMR
BA14 0QL M6OMR
BA14 0QP M3DHV
BA14 0QP M3DHW
BA14 0QS G7OIB
BA14 0RE G7NGI
BA14 0RX G7KNU
BA14 0TD G1UGV
BA14 0TD M6LEX
BA14 0TE 2E0XLI
BA14 0TE M0WIZ
BA14 0TE M7NAZ
BA14 0TX G0KCZ
BA14 0UJ G6PAE
BA14 0UX G6HQX
BA14 6EH M1CKJ
BA14 6EH M3ACA
BA14 6EZ G0HEL
BA14 6JG G6ZRL
BA14 6JG G7RAF
BA14 6JQ G0LJG
BA14 6JZ G3VMZ
BA14 6NA G4UJJ
BA14 6NY 2E0PUH
BA14 6NY M0PUH
BA14 6NY M6PUH
BA14 6QP G7JQW
BA14 6SA G4SPE
BA14 7GE 2E0JVY
BA14 7GE M0JVY
BA14 7GE M6JVY
BA14 7LE G5BBL
BA14 7PE G8GUA
BA14 7PG 2E0BUA
BA14 7PG G1HFY
BA14 7PG M3PVU
BA14 7PH G4GFJ
BA14 7PR G0VYU
BA14 7RS G3AZW
BA14 7RX M6OSI
BA14 7SW M7EXB
BA14 7TX G6MQH
BA14 7UN G0KHQ
BA14 8QP G1TST
BA14 8RY G0GKH
BA14 8RY G2BQY
BA14 8RY M1BQY
BA14 8WD 2E0DUQ
BA14 8WD M6GNY
BA14 9BQ G0WPL
BA14 9DA G6POW
BA14 9GD M7KJF
BA14 9GG M6ALE
BA14 9HH G0BQG
BA14 9HL G0BNG
BA14 9HS M6DNO
BA14 9JZ M3WAY
BA14 9LQ G3PYF
BA14 9PH G7PQW
BA14 9PW G0HFX
BA14 9PW G8BYI
BA14 9QS 2E0FPF
BA14 9QS M6OOQ
BA14 9RB G0TOE
BA14 9RB G6YDP
BA14 9SD M7FOJ
BA14 9TB G0EUR
BA14 9TP M3FMI
BA140FS M7KVZ
BA15 1AX G0IAK
BA15 1HZ G7KBD
BA15 1JF G4PLF
BA15 1LL G8MGQ
BA15 1LX 2E0MSI
BA15 1LX M0IMS
BA15 1LX M6IMS
BA15 1RJ G0OFT
BA15 1SE G4VVZ
BA15 1SE G4ZAP
BA15 1SJ M3OKE
BA15 1TB M3VAT
BA15 1TJ G3BBX
BA15 1UD G7SPM
BA15 2BH G1LCN
BA15 2HL M0AXW
BA15 2SB G3YIQ
BA152RZ G0JYF
BA16 0BY G4PLY
BA16 0HX G7SDD
BA16 0HX M0BRH
BA16 0HY G4JGW
BA16 0HY G8SCO
BA16 0RE G4XRM
BA16 0RL G4LWQ
BA16 0RN M6LVX
BA16 0SA G1ZTM
BA16 0SN 2E0WRT
BA16 0SN M0VUW
BA16 0SN M6IKJ
BA16 0TE G1FGK
BA16 0TE G4XLY
BA16 9PE G7NCE
BA16 9PF G0BNF
BA16 9QN G6YZF
BA16 9QX 2E0RWW
BA16 9QX M6FCT
BA16 9RJ G1OOB
BA2 0AP G3WEY
BA2 0DH G0LTE
BA2 0DH G8LJY
BA2 0DZ G4BHP
BA2 0EA G4OMG
BA2 0HB M1LRX
BA2 0HB M3LRX
BA2 0JU M6WXH
BA2 0PE M6VIG
BA2 1AE G4YTN
BA2 1DY M6FZF
BA2 1DY M6OOW
BA2 1HS G6VIF
BA2 1NW G4YCE
BA2 1NW G7ANB
BA2 1NW M6YDV
BA2 1PY 2E0CXK
BA2 2DL 2E0ICP
BA2 2DL M0ICP
BA2 2DL M3XKP
BA2 2LZ G0KTN
BA2 2NG M1ZZY
BA2 2NG M3BFY
BA2 2PG G8CJT
BA2 2PS 2E0PBL
BA2 2QB G6ZKU
BA2 2TB 2E0TFO
BA2 2TB M0TFO
BA2 2TB M3TFO
BA2 2UD G1EEZ
BA2 3BS G0FUW
BA2 3BS M3MOB
BA2 3DQ G8TYD
BA2 3JL G8XZB
BA2 3JW 2E0PPD
BA2 3NN M0JFJ
BA2 3PT G5RET
BA2 3QL 2E0HGQ
BA2 3QL M0VIS
BA2 3QL M7BVT
BA2 4DH M6FWZ
BA2 4HS G3NAW
BA2 4LP G0CQC
BA2 4RJ G4BBD
BA2 5AL G7NFN
BA2 5FQ 2E0IFG
BA2 5JE G8FRI
BA2 5NF G4NDT
BA2 5NF G4NDU
BA2 5PU 2E0MIZ
BA2 5PU M3VWR
BA2 6AL G4GON
BA2 6DE 2E1CZO
BA2 6DE G3VTO
BA2 6DE M5SUE
BA2 6DF G3VWC
BA2 6PJ G7AYL
BA2 6RG 2E0HTB
BA2 6RG M0HTB
BA2 6RG M6HTX
BA2 6UE G6MBF
BA2 6XG 2E0RTK
BA2 6XG M0RTK
BA2 6XG M6RTQ
BA2 7AF G4NBG
BA2 7AY M0HFO
BA2 7BA G4FEA
BA2 7FU M0OLS
BA2 7FU M6IPW
BA2 7GR G3TTJ
BA2 7SS M6KDJ
BA2 7SS M6TJD
BA2 7TJ G3TXF
BA2 8AF G0LIB
BA2 8EF G1WFU
BA2 8EQ M0WYB
BA2 8HT M0UAS
BA2 8HT M6PSJ
BA2 8JN M3ZJE
BA2 8SA G4PVX
BA2 8TG M6FIZ
BA2 8TY M6IPJ
BA2 9AF M6DBW
BA2 9DZ M1TAP
BA20 2AZ G7SFY
BA20 2BD G4JBG
BA20 2BD G4JBH
BA20 2BD G8TCQ
BA20 2BD M3VBH
BA20 2DB G8AFN
BA20 2ED 2E0JFW
BA20 2ED M3PIW
BA20 2EH G1FZL
BA20 2EQ M6WMV
BA20 2NR 2E0MJM
BA20 2NR M3OWO
BA20 2PD G4GNV
BA21 3AH G0UMS
BA21 3BE M0BAO
BA21 3BT G7LNJ
BA21 3BT M3LNJ
BA21 3HX 2E0EVU
BA21 3HX M6NNT
BA21 3JB G7AIB
BA21 3LF M6FMT
BA21 3QP M0DZT
BA21 3SB M7FVY
BA21 3TE 2E0KNH
BA21 3TE M0KNH
BA21 3TE M6MGF
BA21 3TW G7MSK
BA21 3UA G7KBE
BA21 4AW G7GQJ
BA21 4BA G8WBT
BA21 4DD G4EVI
BA21 4EX M3RKZ
BA21 4HF G7EAQ
BA21 4JF G3CMH
BA21 4JF M0WOB
BA21 4NN G3KCV
BA21 4PG G4KKG
BA21 4QD M3AIG
BA21 4RA M3VQF
BA21 4RJ M0BHO
BA21 5FP M3SJW
BA21 5FQ M0SCA
BA21 5HA G6FHR
BA21 5JB G0AIL
BA21 5JE G3QPA
BA21 5NY G7AIC
BA21 5PN 2E0XIG
BA21 5PN M0VLX
BA21 5PN M6LZC
BA21 5RJ M6RIV
BA21 5SH G3ZLQ
BA21 5SU G6IUQ
BA21 5XA G3OBL
BA21 5XQ G6LLP
BA213FJ G8VCN
BA22 7AP G8UED
BA22 7NQ M1FFO
BA22 7NR M7KRD
BA22 7QZ G8EGG
BA22 8AS 2E0LWQ
BA22 8AS M6LWQ
BA22 8BW G0LNI
BA22 8BW M7JOS
BA22 8EX M6RYZ
BA22 8HF G4SMD
BA22 8JY 2E0FFW
BA22 8NN M7TWC
BA22 8NS G1XNK
BA22 8NY G0WRK
BA22 8RB G7HRZ
BA22 8RB M3ARS
BA22 8UR G3YPL
BA22 9BT M3OQJ
BA22 9BT M6KMW
BA22 9EJ M7GVM
BA22 9EN G3AST
BA22 9HF 2E0BHH
BA22 9HF M0KRP
BA22 9LF G0EON
BA22 9LY G4BMO
BA22 9QW G1GAN
BA22 9RR G2AMG
BA228RB M0DZH
BA25BN M0HLV
BA25DN G8HKK
BA3 2AS 2E0WXD
BA3 2AS M0WLD
BA3 2AS M6WXD
BA3 2AX G8HKP
BA3 2FE M6KBC
BA3 2FG G7IRF
BA3 2JH G4KVI
BA3 2JH M3FED
BA3 2NB G7IYA
BA3 2PR G1IHI
BA3 2RA M3YOB
BA3 2RN M7AMT
BA3 2RZ G6VJP
BA3 2SD 2E0DFG
BA3 2SD M0HWP
BA3 2SD M6EQK
BA3 2SL 2E1DOZ
BA3 3JZ M3XSA
BA3 3NW 2E0EFU
BA3 3NW M6JPJ
BA3 4AN G1ORL
BA3 4BB G6EYI
BA3 4BR G3RHU
BA3 4BR G4PSP
BA3 4BZ G0TJP
BA3 4EX G0HKB
BA3 4GT G4VVP
BA3 4HA M6BEB
BA3 4HG G7SSA
BA3 4HT 2E0CSE
BA3 4HT M6XAF
BA3 4JL G4MKE
BA3 4QH G1ORN
BA3 4QH G8WKK
BA3 4QH G8WKL
BA3 4RW M0SKV
BA3 4SS G6KPD
BA3 5AL 2E1FPV
BA3 5ED G7LKI
BA3 5FP M6EPO
BA3 5PG M0SBY
BA3 5PG M6SBI
BA3 5PJ G7VCY
BA3 5PP G7KEP
BA3 5PT G0HIC
BA3 5XB M0PDN
BA3 5XB M7DBN
BA3 5XU M3VQG
BA34SF M3WNC
BA35JU M6KXH
BA4 4AZ M1ANL
BA4 4FB 2E0HKC
BA4 4FB M6HKC
BA4 4FD G4JJP
BA4 4FD M0WSX
BA4 4HX M6TPV
BA4 4LS M6AXR
BA4 4PS 2E0TUW
BA4 4PS M0TUW
BA4 4QA M6KVP
BA4 4QN G8EDQ
BA4 4SN G0WKM
BA4 4SZ M7AUZ
BA4 4TL G4YJH
BA4 5DH M1ERH
BA4 5GL M6DGA
BA4 5GL M6TDY
BA4 5JW M3BVK
BA4 5JW M3CGS
BA4 5JW M3FES
BA4 5LE M0PRF
BA4 5LG G7FPW
BA4 5LX M6YWA
BA4 5PX G4YJX
BA4 5QA M7AFZ
BA4 5UD 2E0FGQ
BA4 5UD M6NJB
BA4 5UY M7CQD
BA4 5XR G8KKA
BA4 5XR M7KCE
BA4 5XW M6RZZ
BA4 5YG G6RRY
BA4 6BB G7ORK
BA4 6LZ G8DMN
BA4 6NE M0IJT
BA4 6NG G4STH
BA4 6NG G4YLO
BA4 6NG G6SIG
BA4 6PN M3ULZ
BA4 6PZ 2E0PPF
BA4 6PZ M0PPF
BA4 6PZ M6HFH
BA44JA M7GFP
BA5 1AH M5ENM
BA5 1DG G3OJL
BA5 1JA M1FJL
BA5 1PD G3XTT
BA5 1SA G2BJK
BA5 1UD G1WQU
BA5 2DG M7FTX
BA5 2EN G4KQQ
BA5 2ER M7EVM
BA5 2FE M3MYM
BA5 2FF G6SIM
BA5 2FN G3IJU
BA5 2GA M6BJK
BA5 2HZ G4ZDR
BA5 2JU G3ZJF
BA5 2NP M7JSK
BA5 2QL G4FSU
BA5 2QL M6KYB
BA5 2XG G3IUZ
BA5 3AL G2RE
BA5 3AL M6EVE
BA5 3ED G4SFS
BA5 3FG G0AQA
BA5 3HY G4XWE
BA5 3QN 2E0JNM
BA52DG G8BFV
BA6 8AW G6UVO
BA6 8AW M0IOA
BA6 8EE M7WGX
BA6 8EJ G0MYH
BA6 8QS M6OEV
BA6 8RG M0BYZ
BA6 8UB G4FOB
BA6 9AN G0GJX
BA6 9ER G8HVZ
BA6 9JW 2E0FNT
BA6 9PA M6MQB
BA6 9PH G6HIQ
BA6 9PH G7SWB
BA6 9PH M0MSA
BA6 9PQ M6DBP
BA6 9PU G7LVN
BA6 9SH M6AWB
BA6 9SP G8JDN
BA6 9TT 2E0WWF
BA7 7EH M1FFP
BA7 7FE G1YWY
BA7 7HE G0ANP
BA7 7JY G6ZJK
BA7 7LA G7PBT
BA7 7LA G8YEO
BA7 7LB M3UED
BA7 7LT G7VCJ
BA7 7PW M7BIL
BA8 0BP G0ENW
BA8 0DN G6NDA
BA8 0ED M0GOT
BA8 0HJ 2E0RNO
BA8 0HJ M0VIT
BA8 0HJ M6JPF
BA8 0HR M1EYT
BA8 0JG G4KHU
BA8 0JP G4WJW
BA8 0TH G3BCE
BA9 8AL G6UZR
BA9 8LY M0HQP
BA9 9BL G6BOP
BA9 9BS G8GJA
BA9 9BZ G4YXX
BA9 9EJ M0FZR
BA9 9HH M5OMO
BA9 9LS G4BIN
BA9 9LT G7SSG
BA9 9PB G3TZT
BA9 9RB G6IRX
BA9 9SB G6RCT
BB1 2AS 2E0MOZ
BB1 2AS M6MOZ
BB1 2DR G3YWH
BB1 2ER 2E0MQC
BB1 2ER M6HIU
BB1 2ER M6MQC
BB1 2HB G0PXH
BB1 2HY G4XHZ
BB1 2JQ G0DTI
BB1 2NN G4FSD
BB1 3HL M6TVM
BB1 3LP G1DJQ
BB1 4BH G7WJC
BB1 4EE M3TIQ
BB1 4ES G6SUV
BB1 4JX G7WBZ
BB1 4JX M7KZW
BB1 4ND G7IJL
BB1 5HQ G4CDR
BB1 6LF 2E0DIL
BB1 6LF M0YKB
BB1 6LF M6DIL
BB1 7EX G4BJJ
BB1 7EX G8EMB
BB1 8HU 2E0KLW
BB1 8HU M6KLW
BB1 8HU M6PVV
BB1 8LP G1BBC
BB1 8NS 2E0BPP
BB1 8QZ M0SKA
BB1 8RE M7AUQ
BB1 9DP G3SQO
BB1 9LJ 2E0EUJ
BB1 9LJ M0IQY
BB1 9LJ M6MWQ
BB1 9NF 2E0PME
BB1 9NF M0NWI
BB1 9NF M6PWK
BB1 9NH M3JRI
BB1 9NS 2E0JCM
BB1 9NS M6AEK
BB1 9PE G8LTC
BB1 9PW G0GSN
BB1 9QN M7AGG
BB1 9QT G7MOB
BB1 9QY G0HTD
BB1 9RR M6OLI
BB1 9SA G0VOF
BB10 1BA G0KMK
BB10 1EU G4EOX
BB10 1HU 2E1FUH
BB10 1HZ M3LIU
BB10 1PT M7EFX
BB10 2HB G8YYX
BB10 2JT G0ECG
BB10 2LG M3OAT
BB10 2NP M3CFJ
BB10 2QP G8FLS
BB10 3AG G4YMQ
BB10 3DS G3KJY
BB10 3JF G6OHK
BB10 3JF M3JAB
BB10 3JG G8ZQF
BB10 3NG G4OKZ
BB10 3NU G6SYI
BB10 3PS M0CQB
BB10 3QH G3XAB
BB10 4AJ G6PPY
BB10 4BH 2E0FHR
BB10 4DL 2E0VNY
BB10 4DL M0VNY
BB10 4DL M7VIN
BB10 4HX 2E0LFI
BB10 4HX M0WBM
BB10 4JA G3MBU
BB10 4LA G8BEK
BB10 4LA M0BEK
BB10 4LB G0MLL
BB10 4NR M6CXW
BB10 4PD 2E0GON
BB10 4PD M6PGW
BB10 4QH G0OCK
BB10 4QH G0OCL
BB10 4SZ G6NGR
BB10 4TE G4JSK
BB11 1UG G8IUQ
BB11 3BE M6IEC
BB11 3EJ M7RJR
BB11 3NX M3HLV
BB11 3PR M0AIS
BB11 4BJ M3RFR
BB11 4BP G8XCE
BB11 4DN G0UAA
BB11 4LH 2E0YFR
BB11 4LH M3YFR
BB11 4NP M6VGU
BB11 4QG M3PFK
BB11 5BS M6HUF
BB11 5EA 2E0CSG
BB11 5HP G7BRS
BB11 5HX M3KVL
BB11 5QL M0CHJ
BB11 5RB G4OPN
BB115LQ 2E0EVQ
BB115LQ M0OGZ
BB115LQ M6NFE
BB115LQ M7KMS
BB12 0DF M3HLA
BB12 0DX G4GOU
BB12 0ED 2E0IOG
BB12 0EF G0DZC
BB12 0EF G1ZBP
BB12 0JG G0MUH
BB12 0JJ G6YRC
BB12 6AA G8IUQ
BB12 6DT 2E0CKT
BB12 6ET 2E1CTU
BB12 6JT G6PRA
BB12 6LZ M0ISU
BB12 6NG G6RIM
BB12 6NJ 2E0TUN
BB12 6NJ G6IFV
BB12 6NJ M6TUN
BB12 6NQ G3YGC
BB12 6NZ G7DMS
BB12 6PW M6JQE
BB12 6QE 2E0DWK
BB12 7AU 2E0BLZ
BB12 7AU M3SZS
BB12 7DB G4RQA
BB12 7HT G3ROS
BB12 7HY G0DLT
BB12 7QG G3XAC
BB12 7QH G4NYL
BB12 8DR 2E0SBD
BB12 8DR 2E1CXF
BB12 8DR M6MAS
BB12 8HL M7YOR
BB12 8HN M7CCK
BB12 8JB M0MMX
BB12 8JZ M7IXD
BB12 8NP G1JCW
BB12 8RP M6SRH
BB12 8SH M0NWT
BB12 8SH M6NWT
BB12 9LW M0IEY
BB12 9QA G0RTU
BB127DT 2E0PRV
BB127DT M7GTE
BB14NP G6SZS
BB18 5BS G1BDY
BB18 5BT M7MCG
BB18 5JB G4GOZ
BB18 5LB M1DHA
BB18 5LP 2E0BYM
BB18 5LQ 2E0CKS
BB18 5LQ M6HKY
BB18 5NH G7WAW
BB18 5NW G4LWG
BB18 5PD G7SNQ
BB18 5PD G7WEK
BB18 5PR M6EYF
BB18 6DD G6FEQ
BB18 6LX 2E0ZTM
BB18 6NA M3VPQ
BB18 6NA M3VQD
BB18 6WA G1VAL
BB2 2HL M1FJF
BB2 2LT M6MAA
BB2 2NQ G1WYB
BB2 2PT M6KXQ
BB2 2RX 2E0MZD
BB2 2RX M0MZD
BB2 2RX M6MZD
BB2 2SQ M7ATG
BB2 2TX G0TPE
BB2 3EH G4HYT
BB2 3HU M6FFD
BB2 3HU M6LIS
BB2 3JS 2E0SCW
BB2 3JS M0WSW
BB2 3JS M6SCM
BB2 3JS M6SCW
BB2 3JZ 2E0JQI
BB2 3LQ M3WOX
BB2 3NZ G1BWI
BB2 3ST M0CAS
BB2 3TJ M7KLR
BB2 3TP M6LBE
BB2 3UE M6MMH
BB2 4AU 2E0MPO
BB2 4AU M3NJO
BB2 4EU 2E0MFC
BB2 4EU M3JQM
BB2 4FT 2E0BGO
BB2 4FT M3OFB
BB2 4FU M6XBO
BB2 4HS M6HUI
BB2 4LX M0BXM
BB2 4NG G6ZAL
BB2 4QJ G4UCC
BB2 4QT 2E0MIV
BB2 4QT M6MIV
BB2 4RQ M0MCW
BB2 4RT 2E0FSD
BB2 4RT M6OSP
BB2 4TD G7PZT
BB2 4TY 2E0TRH
BB2 4TY M6ARH
BB2 5AH M0BHP
BB2 5DT G7VIY
BB2 5DU G6OWI
BB2 5EJ G6ZKZ
BB2 5EQ M3ZWF
BB2 5EQ M7JCC
BB2 5ER G7DEC
BB2 5EY M7KEI
BB2 5LE G3UUM
BB2 5LQ M0YCX
BB2 5LY 2E0FNH
BB2 5LY M0JFH
BB2 5LY M7DEX
BB2 5NN G3OZC
BB2 6ET G7VNK
BB2 6HR M6VWD
BB2 6LW G4TMY
BB2 6NE 2E0NJI
BB2 6NE G0ELR
BB2 6NE M0NJI
BB2 6NE M6NJI
BB2 7DS G3JXC
BB2 7HA G6ILH
BB2 7HD 2E0XUK
BB2 7HD M7BCV
BB2 7PA G4HUQ
BB2 7PN G1JHP
BB2 7PU M7CNZ
BB2 7QS G8CFD
BB25EJ 2E0BFN
BB25EJ M3JIT
BB3 0AJ G6UXU
BB3 0AQ 2E1HXP
BB3 0AQ G3PJV
BB3 0AU M0KHB
BB3 0AU M3KHB
BB3 0AY G1VLS
BB3 0BR M7BWP
BB3 0EH G8WZW
BB3 0HW G8YLM
BB3 0HZ G4JBY
BB3 0JB G6WGA
BB3 0JW G4IAT
BB3 0LU G6ETP
BB3 0QR 2E0YYM
BB3 0QR M7YYT
BB3 0QT G1ZEX
BB3 0QT G1ZFD
BB3 0QY G7DNM
BB3 0RG G1FDN
BB3 1EF M1EXS
BB3 1LQ G1JBE
BB3 1NP M3OYR
BB3 1NS G3KWO
BB3 2BS G4JGF
BB3 2DT G1BKZ
BB3 2HP G6CIE
BB3 2JH G6FKR
BB3 2LW G6RXP
BB3 2PB M6LJA
BB3 2SA G7VZS
BB3 2SF G1XNX
BB3 2SQ G0NPJ
BB3 2SQ G1ECC
BB3 2SQ G4PSE
BB3 2SS G4IBS
BB3 3AG G0BPQ
BB3 3AG M0DOC
BB3 3DR M0PJP
BB3 3DR M6PPJ
BB3 3EB 2E0FRC
BB3 3EB M0TAQ
BB3 3EB M6FRC
BB3 3GZ G0KXD
BB3 3JH G3YTI
BB3 3QA G8TJG
BB4 4AJ G4LPO
BB4 4AN G0OCW
BB4 4DZ G7JXD
BB4 4EA M3HZE
BB4 4EE G4NGV
BB4 4FN G7BXL
BB4 4HN M6FBR
BB4 4PB G0KIM
BB4 5BQ G0LRR
BB4 5BQ G4TWG
BB4 5EF M3ILV
BB4 5NA G1BPS
BB4 5NG G0PGW
BB4 5TE G8FGB
BB4 5TG M6PRV
BB4 6AW M7FVF
BB4 6AY 2E0GQE
BB4 6AY G7IAW
BB4 6AY M7LYN
BB4 6BE G4ZLJ
BB4 6DS G8JCN
BB4 6EE G3XWB
BB4 6PL M7BZC
BB4 6QN G7WGO
BB4 6RX G4UUE
BB4 6TH G6GLT
BB4 7DL M7ANM
BB4 7HN G0RFY
BB4 7HN G6MBV
BB4 7JA G0MEX
BB4 7JA M0AKS
BB4 7JZ M6KKI
BB4 7TH G7CZL
BB4 7TT G6VZZ
BB4 8AG M3HJG
BB4 8JG M3FHK
BB4 8LY 2E0ETA
BB4 8LY 2E0WIE
BB4 8LY M0GLI
BB4 8LY M3RYO
BB4 8ND M7TMP
BB4 8PY G4VVK
BB4 8QH G0VYX
BB4 8QL G0ZAP
BB4 8TZ 2E0HEF
BB4 8TZ M0LMN
BB4 8TZ M6HEF
BB4 8UW G0FCA
BB4 9BT 2E0PLA
BB4 9HG G6TKB
BB4 9JE G8GNO
BB4 9PX G0FQF
BB4 9SD G4MKT
BB4 9TB M7FAP
BB4 9TG G5GRG
BB4 9TG M0IXW
BB4 9TG M7NDI
BB4 9TQ G4LNE
BB4 9TR G6OIN
BB5 0FN G0JSA
BB5 0HQ G0SVP
BB5 0JJ M0BZS
BB5 0NA 2E0BBL

BB5 0NA M3HJV
BB5 0ND M3GMY
BB5 0NT G7MYJ
BB5 0SD M7OSD
BB5 0SQ G3ILE
BB5 1SL G4ZMB
BB5 2AF M0MMD
BB5 2AF M6DQT
BB5 2AS G6MKQ
BB5 2JD 2E0PCV
BB5 2JD M3LYV
BB5 2NF G1OPV
BB5 2NY G1ZED
BB5 2NY M3UOL
BB5 2PA G1BQQ
BB5 2PF G4DPU
BB5 2PR G6EYS
BB5 2QP G6VJM
BB5 2QX 2E0VIB
BB5 2QX 2E0XIA
BB5 2QX M7CCM
BB5 2QX M7GIN
BB5 2TQ G7PQL
BB5 3AT G0BMH
BB5 3AT M0HFI
BB5 3AT M0HIP
BB5 3AT M6HUW
BB5 3AU 2E0LGV
BB5 3BL M0CZU
BB5 3EY 2E0NFI
BB5 3EY G4ELR
BB5 3EY M0NFI
BB5 3EY M6NFI
BB5 3LH G4GQP
BB5 3PN M6MFL
BB5 3TA G0VGN
BB5 4AL G7DHD
BB5 4AR 2E0LLX
BB5 4AR G0UNK
BB5 4AR M3LLX
BB5 4AR M3UNK
BB5 4BZ M0JMY
BB5 4DX G4GHK
BB5 4HL G0SVH
BB5 4HL M3LPE
BB5 4JQ G4GQV
BB5 4JR G7ILY
BB5 4LY M1CEM
BB5 4PH G7WJK
BB5 5AU G2CJK
BB5 5BS M7PFB
BB5 5DH G0GBU
BB5 5GE 2E0UKU
BB5 5GE M7LMG
BB5 5GG G1CTQ
BB5 5GH G3RFN
BB5 5NQ G4FQW
BB5 5QD G3KEG
BB5 5QH G6ZFO
BB5 5QJ G6IXP
BB5 5RB G0LQO
BB5 5RB M1CVX
BB5 5RU G8JIE
BB5 5RX G3PUO
BB5 5SW M7PAY
BB5 5XF M0AFC
BB5 5XG G6LVM
BB5 5XP G4FRF
BB5 6BD G4BYL
BB5 6BJ G1BQI
BB5 6BS G4GLW
BB5 6JG G1VVM
BB5 6JQ G0CYR
BB5 6NB M3BLO
BB5 6PL G0TZY
BB5 6QU M6CND
BB5 6TD G8RBV
BB5 6UG M7DJW
BB52AF 2E0HTR
BB52AF M0WON
BB52AF M6LRV
BB55LA G4XEI
BB6 7AR M7CLC
BB6 7AS G0PXI
BB6 7HU G8UQY
BB6 7JE G6PXX
BB6 7JS G0TPP
BB6 7JS G8MEC
BB6 7JU G8VJO
BB6 7LP 2E0WBI
BB6 7LP M6WGB
BB6 7NF M0CRH
BB6 7NL G6LXU
BB6 7NL M0CKO
BB6 7PH G3SXC
BB6 7QF G6ENQ
BB6 7ST M6EIA
BB6 7UQ G8CZG
BB6 8BQ G6HSG
BB6 8DD M0YFT
BB6 8ET G4VEY
BB6 8HB G3DMO
BB67DJ G7DSV
BB7 1DG M0BXC
BB7 1LA M1DGL
BB7 1LB M7WPE
BB7 1ND G0AZH
BB7 1PD G7RDP
BB7 2LS G8XSU

BB7 2NP M6SCK
BB7 2NX M0BXF
BB7 2QD G4IJD
BB7 2QH M1BOZ
BB7 2QN G0RKP
BB7 2QW G4LKZ
BB7 3DA G0AZG
BB7 3HJ G4UEA
BB7 3JD G4UMB
BB7 3JW 2E0EAW
BB7 3JW M6AZH
BB7 3LB G4WJG
BB7 4AJ M7CQS
BB7 4ES G4LDD
BB7 4RR M0DFH
BB7 4RS G2FLW
BB7 9BJ G3YFG
BB7 9EG M7AOI
BB7 9JN G4XSG
BB7 9JW M1CLX
BB7 9NJ M6MYM
BB7 9TJ 2E0EXY
BB7 9XR G8IEZ
BB8 0HE M7OTO
BB8 0HN G0RFQ
BB8 0ND G6PDM
BB8 0TX G6BXS
BB8 7AA M3TBP
BB8 7AR G4UCU
BB8 7AR M7MJO
BB8 7HW G1WAP
BB8 7JY G3ZCX
BB8 7NH G4HXL
BB8 8BF G0JMR
BB8 8BW G6HMN
BB8 8DP G0OKV
BB8 8DT G3KLN
BB8 8JX G4DUI
BB8 8NR G4JMO
BB8 8QS G0IQM
BB8 8TB G0HBN
BB8 9AA M1EAZ
BB8 9AZ M0FIL
BB8 9DF G6DFV
BB8 9ED 2E0WLN
BB8 9ED M6EUP
BB8 9PE G0UGM
BB8 9QH M6MAT
BB8 9QJ G4RTS
BB8 9QR G0VQJ
BB8 9RS G7VZM
BB88JD 2E0MMP
BB9 0AP G0TFK
BB9 0EG G4ZFD
BB9 0EZ G0BQC
BB9 0HZ G4ZLU
BB9 0LE G0AQZ
BB9 0QD G0BPR
BB9 0RP G4MJX
BB9 0ST G4MLB
BB9 0TW M1BSV
BB9 0YH G4KFF
BB9 5BN M0ETY
BB9 5DR 2E0DLZ
BB9 5DR 2E0ZLD
BB9 5DR G0ECB
BB9 5DR M0WOW
BB9 5DR M0ZOE
BB9 5DR M3LHF
BB9 5DR M3LHI
BB9 5DR M3OFC
BB9 5DR M3OFD
BB9 5DR M3VSB
BB9 5DT G4HCC
BB9 5DT M0YRC
BB9 5RS G4MYU
BB9 5RX M6AZG
BB9 6BL G1LQE
BB9 6BQ G0LLL
BB9 6BT G3YGD
BB9 6EX G6JAL
BB9 6HE 2E0PIO
BB9 6HE M0JPN
BB9 6HE M3PIO
BB9 6LD G4WZM
BB9 6LZ G4SEG
BB9 7BD G4TSV
BB9 7BD G4UUA
BB9 7ET 2E0RWB
BB9 7ET M6RWP
BB9 7RA G0ABV
BB9 8AB G2IF
BB9 8DF G7PZU
BB9 8EE G6RUY
BB9 8SD G0JAF
BB9 8SQ M3NSJ
BB9 9HR G0DFO
BB9 9JL G4UNB
BB9 9LL M6ENS
BB9 9LN G3TRK
BB9 9RH 2E0VEK
BB9 9RH M3VYD
BB9 9RR G4KTR
BB90QR M7POC
BD10 0DN M6OVE
BD10 0LP G4ONZ
BD10 0NX M0AMM
BD10 0RF M7WAC
BD10 0RJ G4RSF

BD10 0UR G7ELE
BD10 8PU G1CHQ
BD10 8RZ M3UAV
BD10 9HQ M7MGD
BD10 9JW G0LJM
BD10 9SP 2E0FYR
BD10 9SP M6JYR
BD11 1ET 2E1GVD
BD11 1HE G4TCT
BD11 1HR G4YJM
BD11 1JL G3YXH
BD11 1LU M6MAW
BD11 2EE G8CHN
BD11 2EF M3LHW
BD11 2ET M6YDD
BD11 2JE M0BCE
BD11 2JY M6UBR
BD11 2NN G0FOI
BD11 2NN G3SVC
BD11 2PG G4UNJ
BD12 0JQ G6BIU
BD12 0UX G0WJC
BD12 8DN G0BVQ
BD12 8PT M6JKW
BD12 9DA 2E0YZQ
BD12 9DA M0XAK
BD12 9DA M6ALK
BD12 9HL 2E0OIR
BD12 9HL M0IIQ
BD12 9LS G7HSS
BD12 9NR G0LGB
BD13 1DB M6KEM
BD13 1LD 2E0WBQ
BD13 1LD M6EHU
BD13 1LF G6DDN
BD13 1LP G4XGN
BD13 1NE G7NEC
BD13 1PL G4NTA
BD13 2AE M6MCJ
BD13 2BE M6EJX
BD13 2FF M0AAM
BD13 2HQ 2E0FKU
BD13 2HQ M3AIE
BD13 2JN G0KVM
BD13 2LJ G4UKW
BD13 2QA M3JNR
BD13 2SA M6PFL
BD13 3BE G0SNV
BD13 3BE G7DFC
BD13 3BG M0NAP
BD13 3DQ G8FJR
BD13 3PQ G4JHS
BD13 4DX G0JBC
BD13 4EX M6MLR
BD13 4EY G0IBS
BD13 4LZ G4WFC
BD13 5BA G7HLV
BD133AT M6EHA
BD14 6BL 2E0GQC
BD14 6BL M0AIY
BD14 6BL M7AOZ
BD14 6JZ G3HMV
BD14 6PJ G1IEP
BD14 6PJ G1UHB
BD14 6QE M1EKK
BD14 6RY G6CMN
BD15 0BE G3UCK
BD15 0HB G7HEN
BD15 0HB M7PKK
BD15 0HH G4EZX
BD15 7AU G3NYR
BD15 7AU G4TIV
BD15 7PR G1OLQ
BD15 7QZ G1FKM
BD15 9AR 2E0MQB
BD15 9AR M7MVB
BD15 9BD G0IFT
BD15 9BD G4ZFT
BD15 9BT G6OSJ
BD15 9LD 2E0KEI
BD15 9LL M7ICP
BD15 9LP G7EVR
BD16 1AD 2E0PJH
BD16 1AD M0IRK
BD16 1BD M0EAD
BD16 1DA G8ESK
BD16 1HW G4MOY
BD16 1LN G3UOI
BD16 1LZ G0CEF
BD16 1NS G1CHY
BD16 1PU M3HHC
BD16 1QB M0CRD
BD16 1QD G7CKG
BD16 1RB G4SMK
BD16 1UF G0MPP
BD16 1UH 2E0FOP
BD16 1UH M0JQB
BD16 1UH M6KQD
BD16 2BN M6HNO
BD16 2EY G0MDO
BD16 2SR 2E0ERT
BD16 2SR M0XOM
BD16 2SR M6ROB
BD16 3AY G3ZXZ
BD16 3BX G3RXS
BD16 3DF G6KJT
BD16 3DH G4YCP
BD16 3LG G8UPK
BD16 3PL G3NN

BD16 3QL M7OMB
BD16 3QN 2E0PMM
BD16 4ED G8NTZ
BD16 4EE G8PZF
BD16 4EJ M0HVX
BD16 4HL G4BLL
BD16 4LB G7UTR
BD16 4QD G4YTI
BD16 4QT M7BCE
BD16 4RN G0HUK
BD17 5DW G8FTX
BD17 5NR 2E0FXM
BD17 5NR G0RJC
BD17 5NR G7HKU
BD17 5NR M0JIX
BD17 5NR M7CRF
BD17 5RL M6DER
BD17 5RS G0YKK
BD17 5TJ M1RAL
BD17 6DR G0FVO
BD17 6NN 2E0PAK
BD17 6NN M0PKW
BD17 6NN M6WAT
BD17 6RT M6NRS
BD17 7PQ 2E0HTS
BD17 7PQ M0YKS
BD17 7PQ M3TLL
BD18 1EH 2E0CDR
BD18 1EH M3RZE
BD18 1HL 2E0PLK
BD18 1HL M6PLK
BD18 2BW 2E0FKC
BD18 2BW M7CHP
BD18 2EY G1XJO
BD18 2JB 2E1ESD
BD18 2JB G7KHE
BD18 2LT G0OEJ
BD18 2NT G0GME
BD18 2NT M3VFU
BD18 3AZ M7DEJ
BD18 3JB G1LZF
BD18 3LW M0IZV
BD18 4AR G8ZMG
BD18 4AW G0VJB
BD18 4DY G8ORM
BD18 4EJ G0PCM
BD18 4EW M6MTZ
BD18 4JZ M0TAN
BD19 3BY G0DPX
BD19 3DG G1SKE
BD19 3DW M7XZW
BD19 3EJ 2E0ZDW
BD19 3EJ M0WIT
BD19 3EJ M6DFW
BD19 3PX M3ZPT
BD19 4JT G4OTL
BD19 4LG G1GQB
BD19 4NX M3TAN
BD19 4QL M7JTF
BD19 4RU G0MZZ
BD19 4SB G3WYP
BD19 5AY G4VRW
BD19 6DQ 2E0MBQ
BD19 6DQ M6MRT
BD19 6DU M0WRE
BD19 6JH G4OSO
BD19 6JH G4OSP
BD19 6LJ G3RGN
BD2 1BQ G8AQH
BD2 1EU M7KDR
BD2 1JP G4XYR
BD2 3BP M6KQF
BD2 3HD G0PVB
BD2 3RB G1OCS
BD2 4DP 2E0VPI
BD2 4DP M0VPI
BD2 4HX G8MVD
BD2 4HY G4TFD
BD2 4QX M7DOX
BD2 4SA G6PAR
BD2 4SG G1FVS
BD20 0JP M6LXS
BD20 0LD M3HOU
BD20 0LQ M0KWS
BD20 0LQ M0PLN
BD20 0LQ M6DXA
BD20 0ND G6IPB
BD20 0QG 2E0HZH
BD20 0QG M3VGH
BD20 0QQ G0JFC
BD20 5AN G3WZZ
BD20 5DB G0HOT
BD20 5HN M3YBB
BD20 5JT G7BUR
BD20 5NW G4AFS
BD20 5TD M1EKD
BD20 5TD M5LAR
BD20 5TL G0SUI
BD20 6DE M7DBP
BD20 6HA G7URG
BD20 6HP G4YXB
BD20 6SP 2E0DVI
BD20 6SP M0JTM
BD20 6SP M6GGI
BD20 6SZ G1OTA
BD20 6SZ G7KDH
BD20 7BH M0PWF
BD20 7DN G4CPA
BD20 7EU M0AEZ

BD20 7RW G0STK
BD20 8LT M6GZN
BD20 9AU 2E0JCB
BD20 9AU M6DWF
BD20 9AU M6JTW
BD20 9JY M7BWM
BD20 9LL G0LVT
BD205HN G0KLN
BD21 1HY G3VDK
BD21 2QJ G0OPT
BD21 2QJ G1ONJ
BD21 2RX M0DIT
BD21 4NP G8VPX
BD21 4TA 2E1HFV
BD21 4TD M0CEX
BD21 4TD M1DSV
BD21 4TF G0AJF
BD21 4TL G0BWY
BD21 4YG G0MJB
BD21 4YG G7BQU
BD21 5BD M0CAN
BD22 0AP G7SFJ
BD22 0HA G3TFF
BD22 0JQ M0MSS
BD22 0NU M0HSS
BD22 0QY G4AEE
BD22 6DD G0MEA
BD22 6ET 2E0RIZ
BD22 6EU M6AXH
BD22 6FF G4IUP
BD22 6HF M6KCP
BD22 7AB 2E0CQN
BD22 7AB M6SBO
BD22 7AU M6MLV
BD22 7BD 2E0CSD
BD22 7BP G1AVA
BD22 7DE M7DVJ
BD22 7DH 2E1FON
BD22 7DH M1BVX
BD22 7DN G0MMC
BD22 7DW M1BZD
BD22 7EX G4TFT
BD22 7NQ G0VXS
BD22 7PD G0GVE
BD22 7QS G3UBD
BD22 7RH G0BZH
BD22 7SH G4XNF
BD22 8BJ G0RLY
BD22 8BJ G0TSJ
BD22 8BY G6BAD
BD22 8HG 2E0MFI
BD22 8HG M6SKQ
BD22 8JY M6JLT
BD22 8PL G1NQU
BD22 8PR 2E0DOG
BD22 8PR M0UEZ
BD22 9DL M6UEA
BD22 9LE G1SRA
BD22 9LE G7KRG
BD22 9SS G3OPW
BD227DF M7BNR
BD23 1BB G0XDL
BD23 1BP 2E0VRX
BD23 1BP G0RAX
BD23 1BP M6BEQ
BD23 1RX M0IYF
BD23 1TL 2E0RCI
BD23 2BE 2E0BWN
BD23 2BE M3EHH
BD23 2BZ M6ZEA
BD23 2ET 2E0PHL
BD23 2ET M3NYF
BD23 2ET M3SKN
BD23 2LE M3CWC
BD23 2LE M6LWA
BD23 2PY 2E0JBC
BD23 2PY M0YBC
BD23 2PY M3CSZ
BD23 2QE M0NTT
BD23 2RR G0UCD
BD23 2SP M3SBY
BD23 3DW G8LKQ
BD23 3DW M0DRF
BD23 3DW M0YCG
BD23 3NT G7COD
BD23 3RY M0XLT
BD23 3SE G7CYN
BD23 3TH 2E0XLG
BD23 3TP G1LNQ
BD23 4HJ 2E0PIW
BD23 4HJ M0FTL
BD23 4HJ M3HLD
BD23 4HX G0EPY
BD23 4JE G6LYE
BD23 4JW G6RIY
BD23 4LB G6LFC
BD23 4PQ 2E0HRD
BD23 4PQ M0VHG
BD23 4PQ M6HRD
BD23 5EN G6LHA
BD23 5EZ M1DEY
BD23 5JA G8UHL
BD23 6SD G4YQA
BD24 0AG G4GLC
BD24 0AH 2E0PFL
BD24 0AH M6DLQ
BD24 0DP 2E0PXP
BD24 0DP M0PXP
BD24 0DP M6PXP

BD24 0DP M7WRM
BD24 0HG G4HOK
BD24 0LH G4ZVD
BD24 9DA 2E0RSM
BD24 9HL G7EKH
BD24 9HY M7NEN
BD24 9JP M5CMO
BD24 9QR M0ANP
BD3 7BU M3IAO
BD3 7BY G7FDW
BD3 9JT G0GNU
BD3 9NU G0JTP
BD4 0BD G1DTG
BD4 0BD M7DTG
BD4 0JJ 2E1HZM
BD4 0QH G4HAG
BD4 0SJ G1ANA
BD4 0TS M7TSG
BD4 6BY G0BZV
BD4 6JY M3TZS
BD4 6PJ M0CXL
BD4 7JD M1DYK
BD4 7JJ 2E1GDD
BD4 7JJ M3DXI
BD4 7JT 2E1GVC
BD4 8EN M0LMO
BD4 8JF M7DYG
BD4 8PB 2E0RAS
BD4 8PB M3RSX
BD4 9JH G1XGM
BD4 9JJ 2E0NOK
BD4 9LX G4KZW
BD5 8NX G4JRW
BD5 9AN M3XWN
BD5 9HB G0ITU
BD6 1AH M7XTP
BD6 1EJ M0WIG
BD6 1ET 2E0JPA
BD6 1ET G6FDG
BD6 1ET M0JPA
BD6 1ET M3JPW
BD6 1HL M6BRS
BD6 1PQ G0FUY
BD6 1QU G0IQH
BD6 1RP G4RFO
BD6 1RP M0AMB
BD6 1UL M6FKP
BD6 1UU G4YOR
BD6 2BJ M7ERT
BD6 2BJ M7TUS
BD6 2BT G6RFL
BD6 2TE M7JXR
BD6 3DJ G4UNH
BD6 3JQ G0VVK
BD6 3SW G7JZM
BD6 3SW M3XKY
BD6 3XE G0BTA
BD6 3YN M3AHL
BD62TY M7KPB
BD7 2LX G7RTA
BD7 4AS M3RQO
BD7 4BG 2E1HEM
BD7 4DB G6CGI
BD8 0AA G0OWI
BD8 0EN M0GEN
BD8 0EN M3CAP
BD8 0JZ 2E0GYW
BD8 7SZ G4VFU
BD8 9EX G7MMJ
BD9 4AX G8ZMQ
BD9 4ES M3FTU
BD9 4HG M6OMH
BD9 5PA M1DJI
BD9 6DQ G0TVM
BD9 6EZ G7PHW
BD9 6EZ G7PHY
BD9 6NP G1KWK
BD9 6PU M3XJE
BF19DB G7GGF
BFPO59 G1VJE
BH1 1QX 2E0GQK
BH1 1QX M0VKP
BH1 1QX M7CHS
BH1 1QZ M7SGD
BH1 2PG G3PUZ
BH1 3DX G4LXY
BH1 3SS G4UGV
BH1 3TE G3XFD
BH1 4NS M7FQC
BH1 4PH G0OFE
BH1 4QT M7DKU
BH10 4BB 2E0UXV
BH10 4EE G1INB
BH10 4EY G7IWW
BH10 4HP G3WZP
BH10 4HP M0LWZ
BH10 5AW G4AZT
BH10 5EP G1OQM
BH10 5JT G8CYT
BH10 5JT M0GHD
BH10 5LF G8UAD
BH10 5LG G4OEB
BH10 5LL M7MRA
BH10 5NR M6IEU
BH10 6BG 2E0WCX
BH10 6BG M6WCX
BH10 7AA G1XES
BH10 7EF G4ENM
BH10 7EU G3ZWD

BH10 7HL M7GDW
BH10 7HW M6XMB
BH10 7JL G4MFS
BH11 8AE M0KHA
BH11 8AW G8SLE
BH11 8BN M3MYZ
BH11 8BT M6OCS
BH11 8EE M0VCA
BH11 8EQ 2E0RAI
BH11 8EQ M6RLT
BH11 8HR M7FEA
BH11 8NN G1WAW
BH11 8PS G4DKM
BH11 8RB 2E0KSC
BH11 8RB M6CJW
BH11 8RH G3XBZ
BH11 8RH G4PRS
BH11 8RH M3BDC
BH11 8SL M3ZWN
BH11 8SU 2E0EQD
BH11 8SU M6JYX
BH11 9EQ G7TKG
BH11 9EU M0AUY
BH11 9HJ M3ZKE
BH11 9HN G7VJI
BH11 9JD G7VJJ
BH11 9LJ G4XOH
BH11 9ND 2E0WMT
BH11 9ND M7AQF
BH11 9PE M7VYN
BH11 9PG 2E0SPM
BH11 9PG M0SXM
BH11 9PG M6SPM
BH11 9PJ M7JFL
BH11 9SJ G7JNS
BH118RP 2E0BAQ
BH118RP M7BAQ
BH119BP G1ERS
BH12 1DU M3XCH
BH12 1EX 2E0JNR
BH12 1EX M7CGM
BH12 1NS G1WND
BH12 1NS M6VEG
BH12 1QE G0UCX
BH12 2EZ M6SBL
BH12 2HE G0ROZ
BH12 2HL M3JER
BH12 2HT 2E0TMN
BH12 2HT M3HCE
BH12 2HT M3KMN
BH12 2JF G4FDS
BH12 2JQ M0MMC
BH12 2LQ M3SUK
BH12 2NS M7EWQ
BH12 3AQ M6ANV
BH12 3AS M6GFS
BH12 3AW M0NMO
BH12 3BT M6GBH
BH12 3DA G4EOE
BH12 3HB G0CRY
BH12 3HE G6AHC
BH12 3HF M3KCA
BH12 3JW G7CAA
BH12 3LP G0GZN
BH12 3LP G0ISO
BH12 3ND G8HCC
BH12 3PJ M6TGZ
BH12 4BD G1UIO
BH12 4BS 2E0TPY
BH12 4BS M0TPY
BH12 4BS M6TPY
BH12 4DH G0AEP
BH12 4DY 2E1BVY
BH12 4EH M6SJZ
BH12 4FD M6MCL
BH12 4JH G0LOU
BH12 4JH M6EKB
BH12 4JQ G4JQW
BH12 4LD M0GUH
BH12 4LT G0ISC
BH12 4PJ M0BTJ
BH12 5AY M3TFW
BH12 5ET M3OEF
BH12 5HT M7OWQ
BH121JA 2E0WEG
BH13 6AU M0GWU
BH13 6EF G0MJO
BH13 6EN 2E0MIL
BH13 6EN M0IHM
BH13 7BE G4PVY
BH14 0DB G6EJH
BH14 0DL M7DZI
BH14 0LL G6CGQ
BH14 0PA M7NJH
BH14 0PF G8YYA
BH14 0PJ G3RRP
BH14 0PP G3WZP
BH14 0QS G3WJJ
BH14 0RZ G0RPA
BH14 8AD G6ABA
BH14 8DL M0PTR
BH14 8EG G3RGJ
BH14 8EY G8GAT
BH14 8NW G8IWE
BH14 8QY G3LUH
BH14 8SD M0BVI
BH14 8SG M3RLX
BH14 8SG M7EKS
BH14 8SG M7MOI

BH14 8SG M7OQN
BH14 8UE G0KKL
BH14 9AU G1NNR
BH14 9BS 2E0MVT
BH14 9BS M6ZDA
BH14 9EL G1TEX
BH14 9HU M6BLC
BH14 9JH 2E0DUD
BH14 9JH M3YVT
BH14 9LW G4UWS
BH14 9NX M0ILX
BH14 9NX M6IQX
BH14 9QP G0KCC
BH15 1EH G3YEK
BH15 1JY 2E0PWP
BH15 1JY G8FDZ
BH15 1JY M0PAA
BH15 1JY M3UDN
BH15 1TT G6OFM
BH15 1TU G4UTG
BH15 1UA M0LEZ
BH15 1UP M0GPV
BH15 1UR 2E0DMY
BH15 1UR M6EYL
BH15 1XL G7FKS
BH15 2BZ G1YHE
BH15 2DB G0CBP
BH15 2DW G0PRS
BH15 2DW G4WCK
BH15 2DW G6OAI
BH15 2EF M6CDL
BH15 2ES G4CQW
BH15 2ES M1EZG
BH15 2EX G0JII
BH15 2EX G0OCQ
BH15 2HQ G6NIO
BH15 2JQ G3WCU
BH15 2NZ M7BGU
BH15 2QG 2E0GVP
BH15 3AG M0IDL
BH15 3AQ G3NIL
BH15 3EE G6TEL
BH15 3ET G3HUO
BH15 3ET G3UEE
BH15 3HB M0RWB
BH15 3HB M1DYS
BH15 3NE G3FTK
BH15 3NR G3JTK
BH15 3PH G0WTG
BH15 3PU M0PXR
BH15 3QU G0IBJ
BH15 3QZ G3RZV
BH15 4DQ G3PLR
BH15 4DQ G8EPK
BH15 4DX M3YYO
BH15 4HP G1UEQ
BH15 4HP G4CVX
BH15 4JD G1FPZ
BH15 4JR M7SOP
BH15 4JS G0JJI
BH15 4JX M6BUW
BH15 4JZ G7TZB
BH15 4JZ M3TZB
BH15 4LE 2E0DJW
BH15 4LE M6NFZ
BH15 4LT G0KQH
BH16 5BZ G0UPG
BH16 5ED G8MCW
BH16 5EJ M0ATB
BH16 5EQ G1TQN
BH16 5HE G4NFT
BH16 5LA G2HKQ
BH16 5LS M7DAF
BH16 5LY M6EKE
BH16 5PB G0JBZ
BH16 5PP G3VOB
BH16 5QT G6AKG
BH16 5RA G0IWZ
BH16 5RT G6DUN
BH16 5RX G7MUT
BH16 6AP G0ICG
BH16 6DP G0FCV
BH16 6DP M3WHQ
BH16 6DT M3EZB
BH16 6DT M7NED
BH16 6EQ G0FRR
BH16 6EQ G3PYA
BH16 6EQ G6SFR
BH16 6EQ M1DKY
BH16 6ET G3PKL
BH16 6HB M6IJR
BH16 6HJ G1JUI
BH16 6JH G1YHI
BH16 6JS G0FAJ
BH16 6NB G4MPQ
BH16 6NB G7BIK
BH17 7AH G3ZPR
BH17 7AX 2E0CHN
BH17 7DW 2E1HBJ
BH17 7DW G7TEZ
BH17 7DW M3GCM
BH17 7EH M0AIJ
BH17 7EU G3MEC
BH17 7HA 2E1DBS
BH17 7SB G6MXL
BH17 7UP G0FVH
BH17 7UP M3RLX
BH17 7XT 2E1FDK
BH17 7YJ 2E0LGR

BH17 8BP G7SVI
BH17 8BU 2E0EXO
BH17 8BU G1ZSE
BH17 8BU M6EXO
BH17 8BU M7FWW
BH17 8DB G1HBF
BH17 8QW G4UKR
BH17 8SQ G7MZX
BH17 8SR 2E0AOL
BH17 8SR M0AKY
BH17 8SU G0RVI
BH17 9AH M1EEQ
BH17 9AS 2E0TCF
BH17 9AS M0TGY
BH17 9AS M6TEG
BH17 9BF G4IBM
BH17 9BF G8RSE
BH17 9BG G7OAX
BH17 9DU M5PDL
BH17 9EE M6LDF
BH17 9WE G7GNU
BH18 8AE M0HZU
BH18 8DY 2E0UEL
BH18 8DY M0HTY
BH18 8DY M6DWV
BH18 8HZ M3WBM
BH18 8JS 2E0AMW
BH18 8JS M0CJC
BH18 8LN M3FSB
BH18 8LN M3FSD
BH18 9AE G4BYO
BH18 9AJ G1YHJ
BH18 9AX G4SZO
BH18 9BD G3PSV
BH18 9BF G8HNA
BH18 9DZ G8TNU
BH18 9DZ M5ADL
BH18 9DZ M6IAL
BH18 9DZ M6TNU
BH18 9ED G8JMB
BH18 9EH G7EYA
BH18 9HQ G4BKE
BH18 9HY G6PIM
BH18 9JG G3SGX
BH18 9JG M0BHH
BH18 9LB M6TJL
BH18 9ND G0WKH
BH18 9NR G0BRQ
BH18 9QZ 2E1DFZ
BH19 1HY G1HEJ
BH19 1HY G4ZPB
BH19 1JT 2E0HVU
BH19 1LG G7FXO
BH19 1NF G6CAC
BH19 1PQ M3SGZ
BH19 1QF G8CLJ
BH19 1QU G7OVM
BH19 2DE G3MBM
BH19 2DL M0HRZ
BH19 2ET G3NR
BH19 2LB G4ZYY
BH19 2PX G4CEK
BH19 2PZ G7OYX
BH19 2RB G0WZK
BH19 2RG M1DCH
BH19 2RU 2E0KJK
BH19 2RU 2E0WRO
BH19 2RU M0KJK
BH19 2RU M6KJK
BH19 2RU M7WGR
BH19 2SL G0TOT
BH19 3LU 2E0WDR
BH19 3LU 2E0WVL
BH19 3LU M0WWU
BH19 3LU M6WRL
BH19 3LX G6BOF
BH19 3LZ M7MRB
BH2 5JT G8CJH
BH2 5NR G4FNI
BH2 5RA M3XZR
BH2 5RA M7XZS
BH20 4AQ G0WFE
BH20 4DR G7VXQ
BH20 4EL G4GHA
BH20 4HY G0BAO
BH20 4JA G3TIG
BH20 4QX G0TGP
BH20 4SD M3KQP
BH20 4SG G1HXR
BH20 4SJ M6JHS
BH20 5BB M0IXC
BH20 5EE M6LPV
BH20 5EL G8LUZ
BH20 5EN G6UBC
BH20 5ES G0MWV
BH20 5PE M7VVS
BH20 5RN G1LZZ
BH20 5RQ G6HCH
BH20 5RY G1EQJ
BH20 6AA M6STC
BH20 6DB M7NKE
BH20 6DY M6KAT
BH20 6DY M6NEV
BH20 6EQ G7MND
BH20 6HB G7EWY
BH20 6JZ M3MUF
BH20 6NN M0HXA
BH20 6PJ M7KNG
BH20 7BD 2E0XAM

BH20 7BD M0XAM
BH20 7BD M3XAM
BH20 7BX G1AAH
BH20 7DH G1PJB
BH20 7JF G7BRU
BH20 7LU M6BEE
BH20 7LU M6SSC
BH20 7PA G6XEL
BH20 7PE 2E0JPD
BH20 7PE 2E0VGC
BH20 7PE M3ZNP
BH20 7PE M6DKT
BH20 7QA 2E0CTE
BH20 7QA M6CCE
BH21 1HZ G7RMG
BH21 1HZ M0RMG
BH21 1HZ M3RMG
BH21 1JJ G3VQR
BH21 1LE 2E0IWI
BH21 1LE M0IWI
BH21 1LE M6IWI
BH21 1PJ M0JLE
BH21 1PJ M0TAK
BH21 1PL G4RLM
BH21 1QJ G4FUP
BH21 1QT G6NZN
BH21 1RG G6EOK
BH21 1RX M1EBH
BH21 1SL G8CEZ
BH21 1SN M7FHF
BH21 1SR G6JKK
BH21 1SX M0CGT
BH21 1TB G6SHS
BH21 1TE 2E0JWS
BH21 1TP G0GOX
BH21 1TU G4WEY
BH21 1TZ M3ONK
BH21 1UQ G8MXW
BH21 2AA M1ETT
BH21 2AU G0EEF
BH21 2BL M0MJS
BH21 2DH G0UPS
BH21 2EA G0MTV
BH21 2GF G8MAW
BH21 2HR G4NHE
BH21 2JB G7MQU
BH21 2JU G0DVE
BH21 2JZ G4RFV
BH21 2LA M1BAI
BH21 2LA M6REQ
BH21 2LD G3BHM
BH21 2LE G0MDK
BH21 2NG G8YLR
BH21 2NZ G0WAL
BH21 2PQ G3ZXW
BH21 2QB 2E0AKK
BH21 2QB M0GJV
BH21 2QB M3GTH
BH21 2SF G0SKN
BH21 2SR M7JEH
BH21 2ST G8CHI
BH21 2UW G0UGS
BH21 3AF M6CVF
BH21 3BG G3RJH
BH21 3BJ G4VZT
BH21 3DS G1VIP
BH21 3DS G8UCY
BH21 3EZ G0API
BH21 3EZ G3ZTY
BH21 3EZ G6NLC
BH21 3EZ G7MHO
BH21 3EZ G7PRO
BH21 3HL G0PRH
BH21 3HW M0EYT
BH21 3JD 2E0WCL
BH21 3JD M0WTL
BH21 3JD M6WCL
BH21 3LG G0FKY
BH21 3LW M0FBB
BH21 3LZ G1YHV
BH21 3NH G7AUF
BH21 3NN 2E0BNA
BH21 3PN 2E0ZXR
BH21 3PN M6MOO
BH21 3PU G8NVC
BH21 3PZ M1EZL
BH21 3QG M3YZP
BH21 3SB G0ODP
BH21 3SL G8KNS
BH21 3SP G0KFM
BH21 3SU G3NUN
BH21 3TB M1PMR
BH21 3TF G7AGB
BH21 3TL M0MRP
BH21 3TW M0GUJ
BH21 3UA G8CRV
BH21 3UW 2E0CMT
BH21 4AE M6JWQ
BH21 4AX M7WWX
BH21 4BH G4WUK
BH21 4BH M7DZL
BH21 4BH M7HYI
BH21 4DS M6FRV
BH21 4HD G8ZEK
BH21 5AF G6CHX
BH21 5BH M3RLB
BH21 5ET G4GXZ
BH21 5RD G7MYI
BH21 5RQ G7KJI
BH21 6RR M3XQM

Postcode

BH21 6SG G4MHF
BH21 7AN G3OAF
BH21 7AU G1AJH
BH21 7AZ G7WBU
BH21 7BB G1MKJ
BH21 7EZ G4MSA
BH21 7JZ M0MAI
BH21 8LQ M7RCS
BH211XQ G4YGM
BH22 0BQ G4PIJ
BH22 0BT M7PBT
BH22 0BW G1GRB
BH22 0BZ G1JBM
BH22 0DW G8TEO
BH22 0EE G0KLQ
BH22 0EX M6ODC
BH22 0EY 2E1FSZ
BH22 0EZ G3YQO
BH22 0HN G4FOL
BH22 0LG G4VLP
BH22 0ND G1ZXD
BH22 8AA G1BBI
BH22 8AJ M6MKO
BH22 8AL G0UJI
BH22 8AT G1LDJ
BH22 8BA G4VHI
BH22 8BQ 2E0GXM
BH22 8BQ M7SAP
BH22 8BY G1BRS
BH22 8BY G2BRS
BH22 8BY G3CPN
BH22 8DZ G8UJS
BH22 8EX M6JQL
BH22 8PA 2E1FSY
BH22 8PQ M7AOC
BH22 8PS M0WPR
BH22 8QN G8DJL
BH22 8QT M6FCD
BH22 8QW G8TSZ
BH22 8RG 2E1JME
BH22 8RS G0AEL
BH22 8RW M6KDV
BH22 8RZ 2E0ADN
BH22 8SB G3ZCL
BH22 8SF M0EWG
BH22 8SF M1EWF
BH22 8SF M5BMW
BH22 8UR 2E0DQX
BH22 8UR M6JEO
BH22 8XA M5ACX
BH22 8XN M0DOS
BH22 9EQ G8YHF
BH22 9ES G4LJN
BH22 9FW 2E0RXR
BH22 9FW M6RXR
BH22 9HN G1RSE
BH22 9HP G4ULQ
BH22 9HW G4CFW
BH22 9JE G3WPB
BH22 9JF 2E0XOE
BH22 9JF M0WEO
BH22 9JF M7AGV
BH22 9JH G0NLM
BH22 9JP G8DQE
BH22 9JQ G0CDY
BH22 9LA G4PAI
BH22 9LH M6PEB
BH22 9LJ G3WEG
BH22 9LY M3PIL
BH22 9NB G8XKD
BH22 9PD G0ENA
BH22 9QB M3ORN
BH22 9QD M6INM
BH22 9QE G3EBP
BH22 9QE M0DOY
BH22 9QQ G1DNK
BH22 9RW M7POI
BH22 9SD G4ACJ
BH22 9SG G4XEE
BH22 9SG G6LQG
BH22 9UB G1MMT
BH22 9UX M0VWP
BH220BU M0CPU
BH23 1FA G0JIL
BH23 1JU 2E0IYY
BH23 1QL 2E1CDS
BH23 1QL G0RCN
BH23 1QT G4AWA
BH23 1QX 2E0XAZ
BH23 2AF G8ZPW
BH23 2AG G4HEC
BH23 2AJ 2E0GFE
BH23 2AJ M6GFE
BH23 2AS 2E0YRU
BH23 2AS M6SWK
BH23 2AU 2E1CEQ
BH23 2DF G4XWT
BH23 2EH M3VRM
BH23 2HJ M0CDY
BH23 2LL 2E1FTI
BH23 2NG M3LAQ
BH23 2NP M1EGM
BH23 2NW G6MUQ
BH23 2NW M3TJO
BH23 2RP G3YPT
BH23 2SP G3UXR
BH23 2TW G6MYT
BH23 3BJ G4SEU
BH23 3BJ G6LUK
BH23 3BL 2E0KFR
BH23 3BL M6KDR
BH23 3EW G6SPB
BH23 3JA G0AHK
BH23 3JA G0MUD
BH23 3JA G7MUD
BH23 3NB 2E0CSF
BH23 3NB M3CRL
BH23 3NB M3CSF
BH23 3NS 2E1DHX
BH23 3QN G4MUU
BH23 3QQ G4SNR
BH23 4BP G6FAH
BH23 4DD G0NIQ
BH23 4ED G0TVS
BH23 4HB G1SLO
BH23 4HB G6MEH
BH23 4LN 2E0KJG
BH23 4LN M6KJG
BH23 4QQ 2E0CXM
BH23 4SL M6ILB
BH23 5AH 2E0CPS
BH23 5AH M0YAH
BH23 5AH M6ZYX
BH23 5BA G8GZN
BH23 5DB G7UWZ
BH23 5DF G3NDS
BH23 5DX G1HQG
BH23 5HD 2E0DBB
BH23 5HD M6DAG
BH23 5PN G3YNC
BH23 5QD M3EBA
BH23 5RF M0MRW
BH23 5SE M7GMK
BH23 5SG M6BCU
BH23 6BE G0IXT
BH23 7BE G1LLA
BH23 7LD G0IKN
BH23 7LE M0DEP
BH23 7NJ G4TKP
BH23 7NT M6PDQ
BH23 7NW G4FND
BH23 8BS G8YBZ
BH23 8BT G4OXK
BH23 8BW 2E0LYZ
BH23 8BW M7DAK
BH23 8DS M7SHO
BH23 8DU G3PIY
BH23 8DU G3SGL
BH23 8FH G8EII
BH23 8HG G1IOP
BH23 8HG G8JEM
BH23 8NA G0HJL
BH238BU G1BLO
BH24 1LL 2E0EIE
BH24 1LL M3FSS
BH24 1LS G1RRR
BH24 1LS G1SVJ
BH24 1NE G6SUC
BH24 1NQ M7IJM
BH24 1NX M3ZER
BH24 1PH G4ZLI
BH24 1PQ G4ZPW
BH24 1PU G8LHZ
BH24 1PX G8BDF
BH24 1QU G3OMY
BH24 1QX G6PBO
BH24 1UY M1EHF
BH24 1UY M3ANH
BH24 1XD G6PLU
BH24 1XL G1LOU
BH24 1XY 2E0EID
BH24 1XY M6HXR
BH24 2BH 2E1HGE
BH24 2BL G1MTA
BH24 2BN M7AMW
BH24 2BT M6DTY
BH24 2DU 2E0YKX
BH24 2HH G6EDU
BH24 2HH G6IMH
BH24 2HS 2E0DWS
BH24 2HS M0WCS
BH24 2HS M6BUI
BH24 2HY M6IWO
BH24 2JA G0SQH
BH24 2JP G0FUR
BH24 2JP G1VBA
BH24 2PE M6HFV
BH24 2PF G8SXJ
BH24 2PH G3YNJ
BH24 2QJ G3NWL
BH24 2QS M6DOB
BH24 2QW M5AGG
BH24 2QZ 2E0OEV
BH24 2QZ M3OEV
BH24 3BE M0HVO
BH24 3BP 2E0RFG
BH24 3BP M3RFG
BH24 3DS M7LAE
BH24 3DT G3ZCI
BH24 3DX 2E0IKC
BH24 3ER G1OCH
BH24 3HT G0MYL
BH24 3NQ G1WSN
BH24 4EL G0DBI
BH25 5AE M3WRN
BH25 5AF G4FDX
BH25 5AY M6ITL
BH25 5BQ M0JRE
BH25 5ED M0AYC
BH25 5EJ M0EVI
BH25 5GX G4ZKT
BH25 5JF M0DBX
BH25 5JG G7EPE
BH25 5JP 2E1GQN
BH25 5JP 2E1HAW
BH25 5JP 2E1HFA
BH25 5JP G1ZEC
BH25 5JP G1ZVC
BH25 5JP G8CON
BH25 5JP M3NBB
BH25 5NA G0RIX
BH25 5NB G0TMZ
BH25 5NL M6CFY
BH25 5NQ 2E1IHW
BH25 5NQ G4MVP
BH25 5NS M0IVZ
BH25 5PE G1HHO
BH25 5PW G0HOQ
BH25 5SD 2E1DUW
BH25 5SD G4DUW
BH25 5TB M7HOY
BH25 5XE M3RLH
BH25 5XP G1IAV
BH25 6AB G3DCO
BH25 6ES G4ODM
BH25 6ES G6JDP
BH25 6EY G4HFQ
BH25 6NW 2E0SEW
BH25 6NW 2E0VDE
BH25 6NW M0NFR
BH25 6NW M0RBF
BH25 6NW M3VDE
BH25 6NW M3YGF
BH25 6NZ G6LVC
BH25 6NZ G8WQW
BH25 6QE G1UWV
BH25 6SL G3UEZ
BH25 6TA G4XYU
BH25 7AJ 2E0TFE
BH25 7AJ M0ORC
BH25 7AJ M3TFE
BH25 7AW G7ILD
BH25 7DF M0ORE
BH25 7DQ 2E0GMF
BH25 7DQ M3IZQ
BH25 7DQ M7ABC
BH25 7HN G5HY
BH25 7HR 2E1IGY
BH25 7HR G8BKE
BH25 7JL M7EIU
BH25 7JT M1HHL
BH25 7LU G4BJB
BH25 7LU G7MER
BH25 7NX M0ABI
BH3 7AF G7HEY
BH3 7AF M3TIE
BH3 7AG M0MEY
BH3 7DG G3YUZ
BH3 7DY M1GPE
BH3 7NP 2E0YMR
BH3 7NP M6YMR
BH31 6BL G8YZL
BH31 6BL M0SPC
BH31 6BS G4GTH
BH31 6DX M6DYY
BH31 6EY G0SWF
BH31 6HX G8NAG
BH31 6JA M6FFM
BH31 6JJ 2E0EHV
BH31 6JJ 2E0FQD
BH31 6JJ G4WHY
BH31 6JJ M0PSZ
BH31 6JJ M0VLT
BH31 6JJ M6PSZ
BH31 6JJ M6UAX
BH31 6JP 2E1HVM
BH31 6LB 2E0NTV
BH31 6LB M6NTV
BH31 6LH G0WXZ
BH31 6LQ G4TMF
BH31 6QB G0VPJ
BH31 6US M6NDF
BH31 6XA G6SRY
BH31 7LE G3XYE
BH31 7PG G0DEJ
BH31 7PW G4HTW
BH4 8AB G8PKG
BH4 8AL M0DZT
BH4 8AL M3DUL
BH4 8AT G1HDO
BH4 8BX G7FPR
BH5 1LT M1BFG
BH5 1LY G8PHK
BH5 1NF 2E0IUO
BH5 2DJ G1PFY
BH5 2EP M7CBP
BH5 2HT M6FCN
BH6 3DU G3WAL
BH6 3HJ G8ARA
BH6 3JR M0TRF
BH6 3LU G7SUM
BH6 3ND M0MPS
BH6 3NN G4YNO
BH6 3PZ M3ZNZ
BH6 4AE G6EZM
BH6 4DT G4YRY
BH6 4DX G4RGP
BH6 4DX M3GPP
BH6 4HQ G4WLP
BH6 4HU G3IQX
BH6 4NB G8CRZ
BH6 5AL M6NZF
BH6 5BB M6HRV
BH6 5JL G8GLY
BH6 5LD G6INW
BH6 5NW G4JYH
BH6 5PY G8ASX
BH7 6LF 2E0SUZ
BH7 6LF G4ERV
BH7 6LF M0SUZ
BH7 6LF M6SUU
BH7 7AA G4ELC
BH7 7AS G4BNO
BH7 7BD G0OPI
BH7 7BD G4JDW
BH8 0BT 2E0RIS
BH8 0BT M0JAX
BH8 0BT M6BCY
BH8 0DS M1EGL
BH8 0JP 2E0DFA
BH8 0JQ 2E1ICM
BH8 0NL G4STB
BH8 8JY I M7FLE
BH8 8NP G8FXL
BH8 8PS 2E0BKE
BH8 8PS M0KUP
BH8 8PS M3UHX
BH8 8SF M5BUF
BH8 8SR G1BNG
BH8 8SR G4WFZ
BH8 8ST M3MNR
BH8 8UA M1ERU
BH8 9AE G8OWZ
BH8 9HW G0EJO
BH8 9HY G0EGR
BH8 9PJ G7HLG
BH8 9SG G8MQT
BH8 9UD G3IWV
BH80LB M7AVR
BH9 1AN M6SNQ
BH9 1DB G4KLB
BH9 1DB G8SWO
BH9 1LH G7ADH
BH9 1LR M1ELM
BH9 1PJ M7LUK
BH9 1SH 2E0ZDB
BH9 1SH M0ZDB
BH9 1SH M6ZDB
BH9 1SH M6ZJB
BH9 1SJ M6BPV
BH9 1TX G8RAJ
BH9 1UQ M7FQO
BH9 2BZ G4HBD
BH9 2EU M7JBS
BH9 2JE G3JAU
BH9 2ND G3AWP
BH9 2QJ G0UAP
BH9 2UD G6ZAX
BH9 2UJ M0KIF
BH9 2UJ M6KIF
BH9 3AJ M7AWM
BH9 3BB M3HDV
BH9 3BX M6FCI
BH9 3EH M3VUN
BH9 3LP G7VIK
BH9 3LW G1HRH
BH9 3LW M6LXE
BH9 3NS G0URB
BH9 3PZ M1PKB
BH9 3PZ M3BVA
BH9 3SB G3BHA
BL0 0DP G4PAS
BL0 0LD G4YVO
BL0 0LD G6ZVO
BL0 0QA 2E1EPD
BL0 0QA 2E1EPE
BL0 0RY G4PVS
BL0 9BR M3IMR
BL0 9DF 2E0CCD
BL0 9ED M0YYA
BL0 9JX G7CER
BL0 9LZ M7BVK
BL0 9ND G3XPZ
BL0 9ND G4SVA
BL0 9QG G6SHD
BL0 9RE G6JJF
BL0 9SB 2E0IFP
BL0 9UF G0UKM
BL0 9UN G7BYS
BL0 9UT G4APJ
BL0 9XE G6CHC
BL0 9YN G0JOG
BL0 9YP 2E0YLR
BL0 9YU M7JJG
BL1 1PJ G3MTL
BL1 2AE 2E0NNX
BL1 2JP G0WWH
BL1 2XP G1ITV
BL1 3AU M7AAQ
BL1 3DJ 2E0FNB
BL1 3EZ G8AJT
BL1 3LD G0IZJ
BL1 3PE G6VFB
BL1 3RH M3YWG
BL1 4HW G7ROM
BL1 4LZ G6TWX
BL1 4NJ G6YEA
BL1 4PA G0VUX
BL1 4RQ G0BWC
BL1 4RQ G6GVI
BL1 4RQ G6MRY
BL1 4SA G4NTC
BL1 4UA G0FRL
BL1 5FJ M0HZX
BL1 5RG 2E1HLA
BL1 5RG M0BGU
BL1 5TB M7BPD
BL1 5UN 2E1IAB
BL1 6AA G0PVP
BL1 6AH M3ZHP
BL1 6AJ M6BRP
BL1 6BL G1KQZ
BL1 6DW M7ARO
BL1 6EF M6JRK
BL1 6HZ M0CSP
BL1 6LU M6ODD
BL1 6NT M3TEY
BL1 6RR G7KON
BL1 7QJ M7DYI
BL1 7RJ G6ZBV
BL1 7RJ M3ZBV
BL1 8PA G4RWS
BL1 8RW G1PSL
BL1 8SD G4YNK
BL1 8SF M1AWC
BL2 1AD G7DAZ
BL2 1PA G1OUG
BL2 2DE 2E1EKQ
BL2 2ET M6LUM
BL2 2PU M7HZY
BL2 2SS M3DPI
BL2 2TA G0JFE
BL2 2TA G0TCY
BL2 2TA G1IOO
BL2 2UR G1JMV
BL2 3AY G1YQI
BL2 3AY G4XPU
BL2 3BG M7BFB
BL2 3BW M3UNT
BL2 3JJ M0ZVF
BL2 3NG G4CMH
BL2 3NS G0HYG
BL2 3QN G6MVR
BL2 3QT 2E0RIT
BL2 4AQ G1RRE
BL2 4DU G1ZWQ
BL2 4ET G0EWV
BL2 4LN G1CCD
BL2 4LZ 2E0MLK
BL2 4LZ M0MLK
BL2 4LZ M6FYL
BL2 4NU 2E0JAC
BL2 4NU M0JBC
BL2 5ED 2E1GNZ
BL2 5LY G0GRX
BL2 5LY G4MUQ
BL2 5NL G7HVO
BL2 5NS G4TZG
BL2 5QU G6ERK
BL2 5RJ G0KEX
BL2 6BJ G1VFW
BL2 6EU G1EVR
BL2 6JJ G3TVT
BL2 6LN 2E0MYX
BL2 6LN M6MYH
BL2 6LQ M3SVT
BL2 6LT G4AQB
BL2 6LW M7OBJ
BL2 6NX G8YOY
BL2 6TU M6BNB
BL2 6UD M3OPN
BL2 6UE M6MFU
BL2 6US G3ZIK
BL26HB M3ZPJ
BL3 1BA 2E0BVJ
BL3 1BA M3VPY
BL3 1BA M6MBA
BL3 1JU G4YKB
BL3 1NX G1BIF
BL3 1PN G6VPH
BL3 1PW G1EDM
BL3 1QE M3KKZ
BL3 1RG 2E0DAH
BL3 1RG 2E0RIN
BL3 1RG M0LSW
BL3 1RG M0TEG
BL3 1RG M3PYO
BL3 1RG M3YRH
BL3 1RN G8ROS
BL3 1TE G8RGN
BL3 1TR M3YTA
BL3 1UB G7GLS
BL3 1UD G6CTH
BL3 3AU G7MGC
BL3 3ER 2E0IRN
BL3 3ER M6NAV
BL3 3JD G1YYH
BL3 3JH G6KUJ
BL3 3LG M1FJJ
BL3 3NU M0IZG
BL3 3RB 2E0NSG
BL3 4BR M3GPC
BL3 4BW G0RIP
BL3 4JU M6SXI
BL3 4LF G4TNA
BL3 4PH 2E0JSR
BL3 4PP G4DFP
BL3 4QR G7RZW
BL3 4QZ G1AEQ
BL3 4QZ G1ONE
BL3 4QZ M3DEE
BL3 4TF M7HME
BL3 5HX G4PMV
BL3 5NU G0KAB
BL3 5PJ 2E0CAV
BL3 5PJ M0TJV
BL3 5PQ G7SKV
BL3 6QP M6EGJ
BL3 6RA M6HFN
BL3 6SJ M1FAF
BL3 6YH 2E0SIA
BL3 6YH M6ARL
BL33DT M0IWN
BL4 0AU M6PEZ
BL4 0AY M1TOM
BL4 0DS G6KQN
BL4 0EN M6OYC
BL4 0ES M7PHB
BL4 0HQ G1RWX
BL4 0RG 2E1AMW
BL4 7DF M3KVR
BL4 7DQ G0BMS
BL4 7DQ M3LUP
BL4 7HS M0JDE
BL4 7QD G3BRS
BL4 7QD M0OWS
BL4 7TG M0CPT
BL4 7TH M6KIN
BL4 8DW G1FGE
BL4 8NR G1YCK
BL4 8NT G6QA
BL4 8PS M6LEE
BL4 9EN M6OSX
BL4 9EQ M7NCC
BL4 9HT G3XUM
BL4 9LX G0KHJ
BL4 9NE G6SNZ
BL4 9QW G0TBW
BL40QH 2E1HDF
BL5 1BA G1JZX
BL5 1BL 2E0SXP
BL5 1BL M3SXP
BL5 1DL M0YOT
BL5 2AF M3BBC
BL5 2AF M3LMH
BL5 2EG 2E0XGS
BL5 2EG M0XGS
BL5 2EG M6XGS
BL5 2EU G7BKL
BL5 2GR G6OBE
BL5 2HR G2ANC
BL5 2LE G0AIM
BL5 2LY G6GVR
BL5 2NT M6TCB
BL5 2PJ G7CGN
BL5 2PJ M0SPH
BL5 2PP M7MUB
BL5 2QG M0KSE
BL5 2QG M7ESK
BL5 2QL G7GKC
BL5 2QL M7JOP
BL5 2QU M7BSS
BL5 2RE 2E0HGM
BL5 2RE M7GMH
BL5 2RN M7FGS
BL5 2RT M6MYT
BL5 2SD G6UVS
BL5 2SL M6ABV
BL5 3AX 2E1CKH
BL5 3DN G6GYC
BL5 3DP G0LBE
BL5 3HQ G4RIE
BL5 3TR M7AZP
BL5 3UZ G8IZR
BL5 3YB G4EGG
BL5 3ZD 2E0JBQ
BL5 3ZD M6PYD
BL51EN 2E0FMI
BL51EN M6JVF
BL6 4AZ G4PDD
BL6 4BB M0WHB
BL6 4BB M3BXC
BL6 4FA G4IAD
BL6 4HX G0MXH
BL6 4JF G1AQP
BL6 4LG G8WLL
BL6 4LR 2E0MEO
BL6 4PJ G3JNM
BL6 4RQ G1VHW
BL6 5BG G0MRL
BL6 5BG M3LBQ
BL6 5GZ 2E0HLG
BL6 5NY M6IQK
BL6 5QX 2E0EBJ
BL6 5RH G7CIQ
BL6 5TE G1VON
BL6 5TR 2E0DDN
BL6 5TR M6DNQ
BL6 5UG G7UXE
BL6 5UG G4WBO
BL6 6DD M7MCQ
BL6 6DJ 2E0HHT
BL6 6DJ M6CRR
BL6 6HD 2E0GXC
BL6 6HD M0MHW
BL6 6HD M6JQZ
BL6 6HN 2E1BKT
BL6 6JE G0UXF
BL6 6NR G6YQI
BL6 6PZ G7GFK
BL6 6PZ G7LWT
BL6 6QG G3KMS
BL6 7AF M6MAM
BL6 7BE G0WBT
BL6 7ED G0KGI
BL6 7HZ G7ETK
BL6 7JU G4FSN
BL6 7LF G6CLX
BL65SZ G1EFP
BL7 0DE M0XAS
BL7 0DE M6OOM
BL7 0HS G0RGO
BL7 0HS G0TWD
BL7 0HS G7PSU
BL7 0PA G3HCZ
BL7 0PW G4XTG
BL7 9BN G1HSG
BL7 9DX G8UQV
BL7 9LL G6MAJ
BL7 9RR M0HDQ
BL7 9SP 2E0LXF
BL7 9SP M0PSY
BL7 9SP M3NPO
BL7 9TF 2E0HBP
BL7 9TF M7BKS
BL8 1BJ G0HUQ
BL8 1DW G6DEG
BL8 1JA M0IWV
BL8 1JA M0IWW
BL8 1JB G0TGR
BL8 1RT G7SPL
BL8 1TN M6XEY
BL8 1UE G1RAO
BL8 1XB G0ROC
BL8 1XX G6BIT
BL8 2BQ G0BZU
BL8 2HD G4GSY
BL8 2HF M1BNH
BL8 2QG M3IXO
BL8 2RE G7NLP
BL8 2TS G6DGV
BL8 2UL 2E0DQJ
BL8 2UL M6EVH
BL8 3AU M3TIF
BL8 3BL G0VYP
BL8 3DB G6AMX
BL8 3DB G6PZE
BL8 3DB G8AEN
BL8 3DB G8EZU
BL8 3DT 2E0FPW
BL8 3DT M0WJG
BL8 3DT M7WJG
BL8 3DY 2E0DNB
BL8 3DY M3DNB
BL8 3DY M3DNC
BL8 3JA G0WVY
BL8 3JE G3DQQ
BL8 3LW M7CLK
BL8 4BG G3BQT
BL8 4EP M0BSB
BL9 5BT G0KAY
BL9 5DL G1POX
BL9 5DL G4KLT
BL9 5DQ G0GPH
BL9 5HQ M6JXJ
BL9 5JG G3UGC
BL9 5JZ M7GGK
BL9 6HE M7CBG
BL9 6JH 2E0PDG
BL9 6JH M3PDG
BL9 6NE M3HGH
BL9 6PP G4YYD
BL9 6RN G1YUS
BL9 6RN G4UYJ
BL9 7BU G4KQZ
BL9 7EL G8AJZ
BL9 7HB M6GJX
BL9 7QA 2E0LOE
BL9 7QA G1AKE
BL9 7QA M6LOE
BL9 7SG G7JUL
BL9 8DN G3TNQ
BL9 8HN G6MHO
BL9 8PD G1ADB
BL9 8PD G7DAL
BL9 9BQ M7BFG
BL9 9DQ G0LVX
BL9 9ET M6DSV
BL9 9EY M1EKU
BL9 9EY M3EKU
BL9 9HS M3SXU
BL9 9HS M6LZA
BL9 9HS M6SXU
BL9 9HZ G7TYO
BL9 9JH M7FIV
BL9 9PD M3OAZ
BL9 9QB 2E0UTD
BL9 9QB M3MNO
BL9 9QE M1AGA
BL9 9SD M0LZH
BL9 9TA G4XHT
BL9 9UD G0RAN
BN0 8EH M7DNW
BN1 3BG M7EPI
BN1 3LS G6LFJ
BN1 3PS G3XCT
BN1 3RU G6YRJ
BN1 4AB G0CIT
BN1 5AG G8KTG
BN1 5EL G3TDL
BN1 5EP G4CFB
BN1 5FA G0CJZ
BN1 5FN G4FNL
BN1 5GB G3UWZ
BN1 5HH G3CUY
BN1 5NH G6CXO
BN1 5NN M7PBE
BN1 5PQ M6YTI
BN1 6EB G3WBK
BN1 6FB G4HWF
BN1 6HE M7EQP
BN1 6RN M7EBU
BN1 6RZ 2E0WMY
BN1 6WG G7KIL
BN1 7BG M6KPC
BN1 7EG G3GZT
BN1 7EJ G6LMB
BN1 7GH G6JNW
BN1 7HP G3EWT
BN1 7HX G7UDE
BN1 8AH G0SFV
BN1 8HA G7FYT
BN1 8HF G8GEZ
BN1 8LF M6FMJ
BN1 8LP M3HXG
BN1 8NE G3WXG
BN1 8NL G2FSR
BN1 8NL G8VYP
BN1 8NL M1CTK
BN1 8NP G8VIC
BN1 8QW G8TTP
BN1 8RE G1GGB
BN1 8RH G1OYG
BN1 8SH G4WCP
BN1 8TQ G4ZWD
BN1 8UE M6FXI
BN1 8WE G3XBN
BN1 8XQ G0CKA
BN1 8YG G1GDJ
BN1 9AD M3TJT
BN1 9EB M6EWC
BN1 9JL M7RLP
BN10 7EF G0MIB
BN10 7LA M3RTR
BN10 7LS M0HEP
BN10 7NS 2E0JAW
BN10 7PS G0RNS
BN10 7RL M3UZN
BN10 7RS G4ASX
BN10 7SL M7BIS
BN10 8AB M3CEZ
BN10 8AW G7UBB
BN10 8BA G7FJC
BN10 8DP M6SSX
BN10 8DS G4XLM
BN10 8DZ 2E0XCO
BN10 8DZ M0XCO
BN10 8DZ M6XCO
BN10 8HX G8KHH
BN10 8HX M5KHH
BN10 8JB G8KFO
BN10 8JT G0TEL
BN10 8NS G7RUQ
BN10 8NS M6JDL
BN10 8PJ G8ZTG
BN10 8RP G4PNP
BN10 8SA M0AZT
BN10 8TE M6CWK
BN10 8TH M3ZZA
BN10 8TJ G0AWG
BN11 1LB G6GPF
BN11 1RD M6CJU
BN11 1UG 2E0WGB
BN11 1UG M0WGB
BN11 1UG M3ORE
BN11 2AF M7HJP
BN11 2DW 2E0LJK
BN11 2DW G4FFE
BN11 2DW M0LJK
BN11 2DW M6LJM
BN11 2JG G7LLD
BN11 2JG M3ZNM
BN11 2LT G4DYI
BN11 2PH M0GMT
BN11 2PH M1RAD
BN11 2PN M0JKW
BN11 2PN M6ARQ
BN11 2QE M0GEK
BN11 2QR M6RTC
BN11 2SS M3ZAM
BN11 3EH M7AAX
BN11 3LE M6FWS
BN11 3NE M6DTX
BN11 3RB G7LJY
BN11 3RU G8RKG
BN11 4JB G0HKN
BN11 5EF G3SPN
BN11 5HB G6JEY
BN11 5LP G0UZD
BN11 5QS G8RJB
BN11 5RX 2E0JD
BN11 5RX M0MCL
BN11 5RX M6OJD
BN12 4BH 2E1CPP
BN12 4BH M0CQQ
BN12 4EN G8MAD
BN12 4HG G3YZT
BN12 4LD G3SZM
BN12 4LH G4TWK
BN12 4PF 2E0KYT
BN12 4PF M7NTT
BN12 4PN G8ZNK
BN12 4PU M3RIP
BN12 4QA G4TRP
BN12 4TD M1KSB
BN12 4TQ 2E0VBY
BN12 4TQ M0VBY
BN12 4TQ M3VBY
BN12 4TU 2E0KZM
BN12 4TU M0KZM
BN12 4UB 2E0SRB
BN12 4UB M0HBJ
BN12 5BQ G8RHU
BN12 5BU G6IEQ
BN12 5DG G4UNE
BN12 5DJ G1DKX
BN12 5EL M6HXT
BN12 5JF G7RQD
BN12 5JF M0REG
BN12 5JG M6CCI
BN12 5JW M6CCJ
BN12 5PL G4GUO
BN12 5QA G4KHM
BN12 5RD G0PBV
BN12 6AB G6JVT
BN12 6DY 2E0CHK
BN12 6EY G4UWM
BN12 6GA M6JVK
BN12 6HR G4FZF
BN12 6JE M6FQN
BN12 6LA G4GOT
BN12 6LH 2E0GPX
BN12 6LH M6DCU
BN12 6QA G3HPB
BN12 6QP 2E0DYU
BN12 6QP M6GMQ
BN12 6QS G8MMF
BN13 1AE G4HNU
BN13 1BH M3NHU
BN13 1BH M3NHV
BN13 1DA G1OLE
BN13 1DA G3TGS
BN13 1DA G3YSW
BN13 1DA M3YSW
BN13 1DG G0MVP
BN13 1DX 2E0EKB
BN13 1DX M0EKB
BN13 1DZ G4CZX
BN13 1ET M0HVD
BN13 1JS G6HXW
BN13 1LQ G1ZZC
BN13 1LX G3SKI
BN13 1PQ G6BWP
BN13 1QX M6BMO
BN13 1RE M3DZQ
BN13 2AU G1YVI
BN13 2BH G7GOA
BN13 2DH G0WSS
BN13 2DH G1KIZ
BN13 2DH G4LKW
BN13 2EN 2E0TGQ
BN13 2EN M0MEW
BN13 2LP M3GXI
BN13 2NT G3UEQ
BN13 2PW G4BAQ
BN13 2QX G4OHC
BN13 2QY G3YHM
BN13 2SU G7EWA
BN13 2TT 2E0LDY
BN13 3AG G0JXX
BN13 3AG G4GFN
BN13 3AG G8GRQ
BN13 3AL M7GYT
BN13 3AT G3XIA
BN13 3BH G3SXE
BN13 3BH G4DXO
BN13 3BU M0STC
BN13 3BZ G0SBB
BN13 3DG G4ALZ
BN13 3DH 2E0IAL
BN13 3DH M0IAD
BN13 3DH M6LBK
BN13 3DS G7EQR
BN13 3HB G6FCS
BN13 3HG G4RRU
BN13 3HH G6FJL
BN13 3HZ M6EEA
BN13 3JD G8YGK
BN13 3JG G4XHE
BN13 3JU 2E0XBT
BN13 3JU M6SKP
BN13 3LG G4FPM
BN13 3PG G7NUE
BN13 3QG M3UJS
BN13 3QQ M6OZM
BN14 0EY 2E1AMB
BN14 0EZ G7FCU
BN14 0HR G7GSD
BN14 7 EG G4SVQ
BN14 7AE M6JMX
BN14 7BL M0AIB
BN14 7DB M6FMN
BN14 7EE G0ELN
BN14 7EG 2E0DSK
BN14 7EG M6DCQ
BN14 7HE G4NSJ
BN14 7HH G8MLH
BN14 7HW G8ZWC
BN14 7NE G7JRM
BN14 7PE G1KKH
BN14 7PY G6MBL
BN14 7QU 2E1EOO
BN14 7QU M3YEA
BN14 7QW 2E1HEG
BN14 7QW G8DHE
BN14 7QY M0PCR
BN14 7SB G0SWH
BN14 8AH G7PIK
BN14 8AZ G6MMJ
BN14 8DG G3UQD
BN14 8EL G6KHG
BN14 8EN G6PLL
BN14 8ET G6GAW
BN14 9AU G2ALM
BN14 9BB M3VEM
BN14 9DG M3CWH
BN14 9EA M3ZUJ
BN14 9JE G8RDK
BN14 9JT M0GHO
BN14 9NJ G3VXJ
BN14 9RJ G3UDH
BN140AH G3OUA
BN15 0AE G4GPW
BN15 0DJ G0WAM
BN15 0DJ M7DUP
BN15 0DX G4JEI
BN15 0HU G1EOM
BN15 0LB 2E0RKR
BN15 0LB M0RKR
BN15 0LB M6RKR
BN15 0LU 2E0EOI
BN15 0LU M3YFY
BN15 0ND G7FGZ
BN15 0NG G4ZZS
BN15 0NN G8XEU
BN15 0NZ G4OEH
BN15 8AS G3JSU
BN15 8DE G1UGB
BN15 8DE M1DNZ
BN15 8DE M6RFE
BN15 8DJ M7BVV
BN15 8EN G3IUE
BN15 8EQ M6BJI
BN15 8HX G6ENT
BN15 8JN G1OCL
BN15 8LF 2E0FVV
BN15 8LF M0WTO
BN15 8LF M6FVV
BN15 8LW G4JEY
BN15 8LZ G1BGC
BN15 8LZ G6BWN
BN15 8NY G0WMG
BN15 8PA G4RTX
BN15 8QD 2E0LNG
BN15 8QD M6RAO
BN15 8RL G0TLU
BN15 8RL G4GQR
BN15 8RL M0AAN
BN15 9BS G7NVI
BN15 9DD G1JPI
BN15 9PZ G3UFS
BN15 9QD G1DLB
BN15 9QL M0THT
BN15 9QL M3CCF
BN15 9RQ G4FIG
BN15 9UF M0DLE
BN159SY G8XRX
BN16 1AJ G4HNX
BN16 1AJ M1BXO
BN16 1BH M7VXE
BN16 1DG M6CFZ
BN16 1DT G7KAV
BN16 1DU G8JVE
BN16 1HB G6MIC
BN16 1HE G7IWZ
BN16 1HF G0LOF
BN16 1JS M0KEL
BN16 1QS G3WMU
BN16 1QS M0MNG
BN16 1QS M3DNX
BN16 1RD M0JGK
BN16 1SQ M0JIQ
BN16 2AB G8FNH
BN16 2EF G4TMG
BN16 2HH G3ZBG
BN16 2NY G0BHK
BN16 2QU M0RUX
BN16 2QU M6MPF
BN16 2RN G6CML
BN16 2RZ G4IWU
BN16 2UB M3RBQ
BN16 3ET G0GMC
BN16 3HW M0HJF
BN16 3JP M0RJK
BN16 3NP G8XDV
BN16 3PA M1BSP
BN16 3QG G8NXG
BN16 3RU G4KOU
BN16 3RZ G4EJG
BN16 3TS G1LTI

BN16 4AF G8MFO
BN16 4FH M1DPY
BN16 4HE G4RVE
BN16 4HE G7AIF
BN16 4HF G8VCH
BN163JQ M0GVB
BN163QU G6WHH
BN17 5BX M0IMP
BN17 5BX M3PGI
BN17 5DB M3RPQ
BN17 5DP G1WXC
BN17 5DS M6SBD
BN17 5DW G6TXY
BN17 5EY G0AMP
BN17 5HA G4XWD
BN17 5HE G7HNF
BN17 5HN M0FRH
BN17 5NU 2E0VKZ
BN17 5NU M0TVV
BN17 5PW 2E0GKL
BN17 5PW M6GUU
BN17 5QQ 2E0BDQ
BN17 5QQ M0WJT
BN17 5QQ M3KJE
BN17 6AS G7KEA
BN17 6AU 2E0KCB
BN17 6AU M3TKO
BN17 6BG G1AFK
BN17 6JN G0NGP
BN17 6LT G0MOU
BN17 6ND G4ZKE
BN17 6NN G1VYM
BN17 6PA G4ZFV
BN17 6QB G0XAG
BN17 6QD M0RDV
BN17 6QX G4RBC
BN17 6RG M6MCH
BN17 6RY G8DOW
BN17 6UL 2E0BHP
BN17 6UL M0GJS
BN17 6UP G3ZTZ
BN17 6UT M3CZJ
BN17 7AL 2E0OKL
BN17 7AL M0OKL
BN17 7AL M6OKL
BN17 7AT G0SPZ
BN17 7BW M3EPQ
BN17 7DF G1MOW
BN17 7HB 2E0TCB
BN17 7HB M6OJC
BN17 7HN 2E0CQZ
BN17 7HN M0JSR
BN17 7HN M6BXK
BN17 7HY M6DXI
BN17 7JS G3GGN
BN17 7LG 2E0KUE
BN17 7LG G5KUE
BN17 7LG M6KUE
BN17 7ND 2E0RCN
BN17 7NE M1BSO
BN17 7NE M3UJX
BN17 7NF M7SCW
BN17FB G7UCO
BN18 0AH 2E0WWD
BN18 0AH M6RKK
BN18 0DH G4YCS
BN18 0DJ G0JZU
BN18 0EG 2E0RKO
BN18 0FD 2E0MKJ
BN18 0JA G8DIU
BN18 0JE G4TSQ
BN18 0JE G4WYL
BN18 0JH G0LPF
BN18 0PA G8KFJ
BN18 0PW G7JXB
BN18 0RP G0RBQ
BN18 0SD G6FKY
BN18 0SX 2E0DEG
BN18 0SX M0MRI
BN18 0SX M6ELD
BN18 0TG M0CKM
BN180GB G0DLP
BN2 0EG M6KYI
BN2 0FZ M6GGL
BN2 0HP M1ATU
BN2 0JH G4SLW
BN2 1HE 2E0CKX
BN2 1JZ G7WBO
BN2 1LE 2E0CWW
BN2 1LE M6PRF
BN2 1LE M6WKF
BN2 1LL M6NBP
BN2 1RP M7BKE
BN2 1RP M7MAA
BN2 2EX G4AGQ
BN2 2EX G8AAC
BN2 3BG G7MRH
BN2 3DA G6NCE
BN2 3EQ G8ROF
BN2 3RH G1EXG
BN2 4LT G1KKF
BN2 4LU G4YLW
BN2 4RF M7DVQ
BN2 4TD G4YLG
BN2 5DA G0VAJ
BN2 5DD M7MMC
BN2 5GH G1BCG
BN2 5JA M3TGZ
BN2 5JS G8JFT
BN2 5PD M0IVW
BN2 5PU G0JKP
BN2 5TD M0DKD
BN2 5UP G3UON
BN2 6BE G6JNV
BN2 6DJ G1SBZ
BN2 6LH G8VTZ
BN2 6PF G4NQC
BN2 6RG G0ODI
BN2 6RH M3KXZ
BN2 6SF M0OTA
BN2 6SF M6GIH
BN2 6UA M6HWM
BN2 6UE 2E0ABD
BN2 6UE G7PTZ
BN2 7DP G3ITF
BN2 7DT 2E0SPG
BN2 7DT M0MBV
BN2 7DT M6LWT
BN2 7FD G3XDK
BN2 7FD G4AQG
BN2 7FP M7MZA
BN2 7GG G8ZRM
BN2 7HA G4GDP
BN2 7HH G1RSC
BN2 7HH M0ATQ
BN2 7HP M7RWH
BN2 8AF G2CHI
BN2 8AS M6MOH
BN2 8DH G8LGS
BN2 8DJ G0ENJ
BN2 8EJ G4CBZ
BN2 8EZ G7NKU
BN2 8EZ M3NKU
BN2 8FQ G0ECW
BN2 8FR M0EBI
BN2 8PF M3DVB
BN2 8PH G6CIT
BN2 9AL M7LAK
BN2 9QB M7GRM
BN2 9YE G4MFR
BN20 0AS G4CUG
BN20 0DJ M6GND
BN20 0EU G8GEA
BN20 7EU G6TIQ
BN20 7HD G4DRV
BN20 7HD M0AEK
BN20 7HS G8CRC
BN20 7QE G4NYA
BN20 7TT M0NZL
BN20 7TZ 2E1OLI
BN20 7XR G0UGD
BN20 8AH 2E0TIZ
BN20 8AH M0TIZ
BN20 8AH M6OQN
BN20 8AU M0JAO
BN20 8AU M3XTE
BN20 8DQ G3YXW
BN20 8HY G4MHK
BN20 8LP G4YJW
BN20 8LP G7ULN
BN20 8NT M6BVP
BN20 8SD G1KSN
BN20 8SY 2E0IGO
BN20 8UG G1AKV
BN20 8XD G0USA
BN20 8XD G7ALC
BN20 9DJ 2E1HKY
BN20 9DJ M3DVG
BN20 9DL M3NIC
BN20 9DP 2E1LGV
BN20 9DY M1EHJ
BN20 9HX M6YNN
BN20 9JA G1RMN
BN20 9NJ G0EDS
BN20 9NJ G1KAR
BN20 9NJ G3WQK
BN20 9NJ G3ZQB
BN20 9NS G1UUP
BN20 9PP G7VAE
BN20 9QD G3ZHZ
BN20 9QJ G1FNF
BN20 9QJ M0WNT
BN20 9RD G4UPY
BN20 9RJ G7GSC
BN20 9RQ 2E0WFC
BN20 9RQ M0WTC
BN20 9RQ M3WFC
BN20 9SQ 2E0JXU
BN20 9SQ M0LRR
BN20 9SQ M6EYO
BN21 1HB G3MHF
BN21 1LW G7NIZ
BN21 1PS M0RMT
BN21 1SH M3SPP
BN21 1UD G7GMB
BN21 1UJ M3OYU
BN21 2AH 2E0VHC
BN21 2AH M3WUK
BN21 2HR G4WLV
BN21 2LQ G1SSS
BN21 2LQ G8LQZ
BN21 2LQ M3BBA
BN21 2SD G0GZB
BN21 2TL M0LDH
BN21 2UR G1VGK
BN21 3JG M0IJC
BN21 3QD G0RSG
BN21 3QD M0THM
BN21 4HR G6KJM
BN21 4PB M7ABA
BN22 0AB G8NVB
BN22 0AJ G0JPY
BN22 0PZ M0DJQ
BN22 0TT G4YGJ
BN22 0UD M7KST
BN22 0UT G7DWI
BN22 0UX 2E0DIG
BN22 0UX M6KIA
BN22 0UZ G1MNX
BN22 0XH G4NBQ
BN22 0XH G7UWG
BN22 7BT M6LGJ
BN22 7JF M6CHG
BN22 7JG G8DXU
BN22 7JL M3EVM
BN22 8EH G0EBF
BN22 8EH G7PGH
BN22 8LW 2E1PJJ
BN22 8LW M3MJJ
BN22 8PL M6LYB
BN22 8RS G7MJI
BN22 8SX M6JYS
BN22 8TQ 2E0DMV
BN22 8TQ M6GVW
BN22 8UH G8HGM
BN22 8UQ G4SHM
BN22 9EJ G0MUC
BN22 9EZ G6RGA
BN22 9EZ M3IJE
BN22 9HP G0DAX
BN22 9HQ 2E0JRP
BN22 9JL 2E1IKB
BN22 9NR M7BDV
BN22 9NR M7CZB
BN22 9PA M6UMI
BN22 9QG G4RUL
BN22 9RH G7JVO
BN22 9RN M1DRB
BN22 9RN M1HLG
BN22 9RR M3ENE
BN22 9RR M3ENF
BN22 9SF 2E0SPU
BN22 9SF G0UAI
BN22 9SF M3SPU
BN227LL 2E0MKI
BN227LL M0IKM
BN227LL M3IKM
BN227SZ M3YHR
BN23 5AJ M3HIP
BN23 5BB M7GLY
BN23 5BN G1SGR
BN23 5BN M3NJM
BN23 5PG G4XXM
BN23 5PL G3IAZ
BN23 5RA G0GZP
BN23 5TH G0FIP
BN23 5TH G0LTD
BN23 5UB 2E0WGO
BN23 5UB M0WGO
BN23 5UB M3WGO
BN23 5UH G8CCO
BN23 6AB G0OZG
BN23 6AF G4EKS
BN23 6AL G6GVU
BN23 6HQ M3UDS
BN23 6HW M3OWF
BN23 6HW M6LYF
BN23 6JP G7TSP
BN23 6LN G0EYE
BN23 6LN G6UNU
BN23 6TQ G7PMY
BN23 7AQ M3PWZ
BN23 7BE G6GVL
BN23 7BH G1LEO
BN23 7BT M0DUP
BN23 7DS M0CHO
BN23 7NY G1FSF
BN23 7PF G4PRJ
BN23 7PF G8XCY
BN23 7QP G1FHY
BN23 7QT G4TWP
BN23 7QT G4ZQS
BN23 7RW M6DPL
BN23 7TP G8ANT
BN23 8AL M1APT
BN23 8AN M3LHU
BN23 8AQ M6NHN
BN23 8BR M3GZQ
BN23 8BX M0PHL
BN23 8DB M7BDP
BN23 8DX G0IGU
BN23 8FB G0GFP
BN23 8JD M3GUQ
BN23 8JQ 2E0SCR
BN23 8JQ M6CJP
BN23 8NS M6JFH
BN24 5AW M3TZE
BN24 5DT M0RFW
BN24 5DY M1ABM
BN24 5HW M0CUY
BN24 5HW M0RJO
BN24 5NL M3RVM
BN24 5NN M6COH
BN24 5QX G4PGA
BN24 6BE 2E0DWW
BN24 6DE G1NEZ
BN24 6DE G6GVS
BN24 6LX M0KIL
BN24 6LX M6LKI
BN24 6RT M6DQH
BN24 6RW G0RPJ
BN24 6SL G0HVO
BN245LB 2E0BRL
BN245LB M3UYO
BN245PF 2E0LML
BN245PF M7MLM
BN24EA M6KII
BN25 1ES M6MTT
BN25 1JT M6WYX
BN25 1PP 2E0PUF
BN25 1PP M0ABU
BN25 1PP M7UFF
BN25 1QF G8DWX
BN25 1SP G6FYA
BN25 1SW G3DQT
BN25 2EB G1OIO
BN25 2EW G3JKF
BN25 2EW G6GAF
BN25 2HA G0GJH
BN25 2HZ G8WHD
BN25 2JL M7SPJ
BN25 2JQ 2E0WCO
BN25 2JZ G4BMK
BN25 2NE G4MGN
BN25 2NZ G6OHR
BN25 2NZ G6VDD
BN25 2QN 2E0MEU
BN25 2QN G6MAR
BN25 2QN M6MEU
BN25 2RT M0RKF
BN25 2RU G0KDD
BN25 2RU M0ABT
BN25 2TX G1VAY
BN25 2UL G3AGF
BN25 2UN 2E1IIJ
BN25 2UN M3GLC
BN25 2XB M7EHC
BN25 2XG 2E0FJK
BN25 2XG M0USC
BN25 2XW G7AFZ
BN25 3AD G3KXF
BN25 3AX G3ZUN
BN25 3BY M7AJF
BN25 3EA M0KPX
BN25 3EZ G8XXJ
BN25 3HH 2E0BXE
BN25 3HH 2E0RBN
BN25 3HW G4BQA
BN25 3JB 2E0PQR
BN25 3JF 2E1MPN
BN25 3JR G6FLH
BN25 3LB M1EFP
BN25 3QY M6GWE
BN25 3ST M3TPY
BN25 3ST M3WCO
BN25 3TN G0IOF
BN25 3UE 2E0DMX
BN25 4DU G8FGN
BN25 4LZ G1BAB
BN25 4NF M7CND
BN25 4NU G8KUV
BN25 4PD 2E0MEZ
BN25 4PD M6DZP
BN25 4QL G7TOI
BN25 4QL M3PJG
BN25 4QR G1KVR
BN25 4QT 2E1JAC
BN25 4QT 2E1MJM
BN25 4QT M3JRM
BN25 4QT M3MJM
BN252RT M0BEL
BN26 5HT M6HCI
BN26 5NS G3CMU
BN26 5PA M6GVU
BN26 5QA M6VAW
BN26 5QA M6VAY
BN26 6DA G4BJP
BN26 6DQ M3VEJ
BN26 6LU G6NIX
BN26 6NE G0UOI
BN265ED M7AZA
BN27 1DS 2E0MED
BN27 1DS M0EBQ
BN27 1JL M7TYM
BN27 1NA M1BSY
BN27 1PJ G4ZKQ
BN27 1RS 2E0BVZ
BN27 1RS M0GUR
BN27 1RS M3YGV
BN27 1SP 2E0KFX
BN27 1SP M7MRJ
BN27 1SR G4RWM
BN27 1SU M6KPD
BN27 1TU G6YFF
BN27 1TW G6JTK
BN27 1UN G0UZY
BN27 2BT G8JLD
BN27 2BU G8NFZ
BN27 2JY 2E0CVF
BN27 2JY M0NAZ
BN27 2JY M6SYX
BN27 2LX G1GKD
BN27 2PE M3YNE
BN27 3AH G1RTW
BN27 3AQ M6EHY
BN27 3AT M3TEG
BN27 3AZ M6FJM
BN27 3AZ M6PDI
BN27 3BG M6CQX
BN27 3BZ G3JYG
BN27 3DJ G4BLS
BN27 3EG 2E0ERF
BN27 3EG M6BUJ
BN27 3FU G7JWX
BN27 3FZ M6HKO
BN27 3GP 2E0SEB
BN27 3GP M0STT
BN27 3HU G0RFF
BN27 3JR M7FGY
BN27 3JZ M6NKK
BN27 3JZ M7MDH
BN27 3JZ M7SOE
BN27 3LJ G8MAF
BN27 3LL M7CYV
BN27 3LS G1ATL
BN27 3ND 2E0BXG
BN27 3ND M0HJL
BN27 3ND M6HJT
BN27 3NP G4KAR
BN27 3NP G8TNH
BN27 3NX M1BMR
BN27 3QG 2E1GPE
BN27 3RA M3YDA
BN27 3TL G4LYU
BN27 3TP M0AWY
BN27 4BG G3RXO
BN27 4DS M6VOY
BN27 4FP M6IQT
BN27 4FR M6MOV
BN27 4JL G0CRD
BN27 4LN G8URG
BN27 4LN M0CTC
BN27 4NH 2E1BUR
BN27 4NT M7AAY
BN27 4PP 2E0UVK
BN27 4PP M3UWF
BN27 4RA 2E1GTT
BN27 4RS 2E0HBX
BN27 4RS M0TXH
BN27 4RS M7HBG
BN27 4TX G4HEJ
BN27 4UT 2E0ACA
BN27 4UT G3SGR
BN3 1DL M6MIZ
BN3 1RD M7YGR
BN3 1RL M1BIL
BN3 1RN G3LAZ
BN3 2BP G0CIM
BN3 2FF M0DTB
BN3 2LH G6ONI
BN3 2RT M0EEK
BN3 3RE G3LRU
BN3 3WJ G4GUX
BN3 3WW M6CZC
BN3 5AF M6LAG
BN3 5BE G3VBE
BN3 5DF G3MOL
BN3 5EY M7AAW
BN3 5HP G7WHX
BN3 5HP M5ABJ
BN3 5ND G3ECM
BN3 5NY 2E0BHS
BN3 5NY M3XKF
BN3 5SG G1PIN
BN3 5SQ G6JAY
BN3 6FZ M6ETL
BN3 6HP G1VUP
BN3 6HP G3WOR
BN3 6NT G3HZT
BN3 6PJ G4DGW
BN3 6PL G3VLC
BN3 6UJ M6CJK
BN3 6UQ G1NKN
BN3 6WQ G0IGA
BN3 7AQ G0UCF
BN3 7BA M7CJU
BN3 7EJ M6HZZ
BN3 7FW G7MGG
BN3 7GX G1GSK
BN3 7NS G4KIL
BN3 7QB M0LHB
BN3 7QX G3LCF
BN3 8BA G4ORP
BN3 8EE G1JYH
BN3 8EN G8BZL
BN3 8EQ G1KGL
BN3 8FE G0FUI
BN3 8JB M7MZN
BN3 8LQ G6FFH
BN3 8PP G8IQX
BN41 1GG G7UTH
BN41 1PT M6VIV
BN41 1SA M6CWV
BN41 1SJ G1FZS
BN41 1SJ M3LPN
BN41 1SW 2E0MRZ
BN41 2DN G4XKF
BN41 2DN M6FTX
BN41 2HN G1GID
BN41 2LS G7OBD
BN41 2PJ G0SJG
BN41 2PN M5AJZ
BN41 2RD G0NDF
BN41 2UZ M6RSS
BN41 2WU M3OHN
BN41 2YB G1ELE
BN41 2YE G7NLZ
BN41 2YF G0XAN
BN41 2YF M3TII
BN41 2YF M3TIJ
BN41 2YG G0SLW
BN41 2YL M6JOO
BN42 4AT M6BBQ
BN42 4DL M7CNY
BN42 4GD M6TKK
BN42 4LA G4YRT
BN42 4NH M3HXF
BN42 4QS M3XEX
BN42 4RR M6ITD
BN42 4SJ M7EKK
BN42 4YG 2E0MME
BN42 4YG M6HQM
BN424NE G0DWZ
BN43 5AA M1IRB
BN43 5AN G4ZHK
BN43 5AY M3GQB
BN43 5ES G8AIP
BN43 5HL G4OCZ
BN43 5JH M3ITK
BN43 5LG G7VML
BN43 5LN G0FJJ
BN43 5LW G1RDU
BN43 5NE G6YIK
BN43 5NN M3RQR
BN43 5NQ G4EUK
BN43 5TH G0FIG
BN43 5WY G4JBA
BN43 6AA G4XUZ
BN43 6BH G4GNX
BN43 6GH G7GMD
BN43 6GP G0FZB
BN43 6GR G0AQH
BN43 6HF G2DPY
BN43 6HY G3LLD
BN43 6LE M7BKB
BN43 6LN G3MZV
BN43 6LN M6MKN
BN43 6LN M6VAM
BN43 6NN G0SIU
BN43 6PF G6WDR
BN43 6PJ 2E0CCA
BN43 6RA G0SON
BN43 6TP 2E0IDC
BN43 6TP G3XOI
BN43 6TP G6ENS
BN43 6TP G7GFX
BN43 6TP M6IDC
BN43 6YB G0IFQ
BN44 3AG M1PKW
BN44 3FJ M5NEV
BN44 3FP G3REP
BN44 3GD G4GZH
BN44 3GF G0VLC
BN44 3HF M0FVV
BN44 3LH M0MBH
BN44 3LN G6AIK
BN44 3LR G0BAF
BN44 3RL G0VQD
BN44 3RQ G1HWY
BN44 3SN M6FLR
BN44 3TB G4RDH
BN44 3TF G4AND
BN44 3WE G3EUE
BN44 3WH G0CPR
BN44 3WH G0NRX
BN44 3WH G1WOR
BN44 3WH G4UDU
BN44 3WH M3ULO
BN44 3WH M3UPO
BN44 3WJ G0UQY
BN44 3WJ G7MGT
BN44 3WJ M3UQY
BN5 9DG G8RDG
BN5 9EZ G4GLO
BN5 9FR M6EWT
BN5 9HZ 2E0IQW
BN5 9HZ M0LDP
BN5 9HZ M6IQW
BN5 9SB G8XNB
BN5 9UX G3WMY
BN5 9YU G4XFV
BN6 8BB G4VTD
BN6 8BJ G8YKV
BN6 8BJ M6ZRJ
BN6 8DD G3XTH
BN6 8HR G6MJW
BN6 8HR G6YPY
BN6 8HZ G1JNX
BN6 8JL G4HS
BN6 8NB G0DKY
BN6 8NS G7VNG
BN6 8NU G8XYR
BN6 8PD G0GZE
BN6 8PD G4LYX
BN6 8PD M0NJD
BN6 8PE G4GDT
BN6 8PE M6UVD
BN6 8SB G4CYZ
BN6 9BZ G3SGF
BN6 9DA M7TXA
BN6 9RT 2E0LBW
BN6 9RT M5REW
BN6 9RZ G3NYX
BN6 9TR M3NIF
BN7 1BT M0GKP
BN7 1EN G0VPX
BN7 1EY M7FZC
BN7 1HT G7IIH
BN7 1HY M1ALO
BN7 1LT 2E1CCI
BN7 1LT G4XBG
BN7 1NP G3IIO
BN7 1QG G4OWT
BN7 1QG G6CWH
BN7 1QG M6OWT
BN7 1SP G8OYM
BN7 1SP M6AIT
BN7 1TY G4FIH
BN7 1XU G8UND
BN7 2BE G6RXK
BN7 2BU G4UNX
BN7 2DL G1WSE
BN7 2DL M1CNS
BN7 2EJ G4TBM
BN7 2EN G7MOW
BN7 2HY G4PZU
BN7 2SH G6GOS
BN7 2TX G1MXC
BN7 2TX G7PUW
BN7 2UB G4YGE
BN7 3JL G8BYC
BN7 3PB M0YDM
BN8 4DS G1RKR
BN8 4GD 2E1BWT
BN8 4HP G7JKW
BN8 4NA G6DGK
BN8 4NA M0SXE
BN8 4PG 2E1JBJ
BN8 4PG M3BER
BN8 4PG M3BJB
BN8 4PG M3IMB
BN8 4PS G2FQZ
BN8 4RH M6HSR
BN8 5HX G3YXO
BN8 5LJ M6JNH
BN8 5PA M3IQG
BN8 5PG M6SCU
BN8 6DR G1EJG
BN8 6DR 2E1HQA
BN8 6DR M3BOB
BN8 6DR M3HAT
BN8 6EQ M6CNZ
BN8 6HU M7FJT
BN8 6HW G4SVB
BN8 6LS G4MJC
BN8 6PS G3EKJ
BN84NF M6YJH
BN84NT M7VJM
BN9 0FG 2E0KXV
BN9 0FG M6KXV
BN9 0HS M7MAM
BN9 0HS M7YXZ
BN9 0JU G8ZVX
BN9 0ND G4KZX
BN9 0PS G8NQK
BN9 0QU G0CCX
BN9 0RD G4YRA
BN9 9AH G1AAR
BN9 9AT M7IAD
BN9 9DJ G8KAS
BN9 9DY M6ACB
BN9 9ER 2E0CAE
BN9 9ER M3XGU
BN9 9EZ M1UKC
BN9 9LJ G6FYR
BN9 9NB M7KWT
BN9 9NH G0TJH
BN9 9NR M3VQI
BN9 9QB G7WHU
BN9 9RR M6MWV
BN9 9SP M1DNC
BR1 2AT G4XGP
BR1 2NF G0JBP
BR1 2PR M3EFW
BR1 2UD 2E0JXB
BR1 3BL G4NCS
BR1 3PU G4FXR
BR1 4DB G4OIM
BR1 4LP G8OTG
BR1 4NX M7DLV
BR1 4QY G1GYF
BR1 5AZ G3YQN
BR1 5BT G4AHT
BR1 5EA 2E0RTM
BR1 5EG G7CRK
BR1 5EW G0UBA
BR1 5LH M6OGE
BR1 5LR M3RHB
BR1 5LR M6BHZ
BR1 5NA G0WYG
BR1 5NA G1WYG
BR1 5QT M3LBP
BR1 5QT M7DFF
BR12DL M7SPX
BR2 0DN M0HOV
BR2 0EE G3IUJ
BR2 0JF M7THX
BR2 0LS G1ROK
BR2 0RW G3IZA
BR2 0TN 2E0PAF
BR2 6AJ M6TSR
BR2 6BF G0REE
BR2 6BF G7HCC
BR2 6DG G6YRY
BR2 6HL 2E0EHQ
BR2 6HL M0ITG
BR2 6HL M6HJI
BR2 7AL M3UMB
BR2 7DY G3XDL
BR2 7DY G8ZYH
BR2 7EH M3FLV
BR2 7HE G3TSC
BR2 7HE G4XAT
BR2 7HE M3LSU
BR2 7HE M3TVC
BR2 7HU G4TYT
BR2 7HX G1OJO
BR2 7JA G4WJQ
BR2 7JS G4IPF
BR2 7LX G8SXU
BR2 7PT M6VJG
BR2 7QJ G8BVQ
BR2 7QN G3OQD
BR2 7QT G3NGK
BR2 7QT G8BUV
BR2 7QW M6NEY
BR2 8EY G4MSK
BR2 8FG G3WOE
BR2 8JB G3WIM
BR2 8LT G1BYS
BR2 8PF G1ZDR
BR2 8PP G3ZJX
BR2 8QQ G8ITB
BR2 8QQ M0XBY
BR2 9EN G0ELU
BR2 9JA M6RXT
BR2 9NW G1ZQN
BR2 9NW M0HTJ
BR2 9NW M0TZO
BR2 9UL M0HUN
BR3 1LE G7RUS
BR3 1NB M3UNP
BR3 1NN M6AWG
BR3 1NY G4HZX
BR3 3AY G7TOU
BR3 3BJ G7HNG
BR3 3HG 2E0NCY
BR3 3HG M6NCY
BR3 3JT G1HIG
BR3 3RL G7KWO
BR3 3SB G7CXO
BR3 4AE G7HYM
BR3 4JJ G4VJT
BR3 4JT M6DOM
BR3 4PZ M6FZC
BR3 4RU G8NKM
BR3 4SS G3SZR
BR3 4XS G4NPD
BR3 5AA G3EFS
BR3 5DY 2E0FPX
BR3 5DY M7WWT
BR3 5LH M7CEY
BR3 6LJ G4RTV
BR3 6LX M0ZVB
BR3 6SN 2E0JFL
BR3 6SN M0JKM
BR3 6SN M6BXL
BR4 0HA G8VZS
BR4 0HB M1ENE
BR4 0QH G1PGS
BR4 0SG G8OMQ
BR4 0TA G4IKL
BR4 9AH G3ZQF
BR4 9DZ M6GJA
BR4 9JA M1LES
BR4 9JA M1PAM
BR5 1AB G6MYZ
BR5 1AL 2E0BBG
BR5 1AL M3IXU
BR5 1BZ G8IVB
BR5 1EZ G3NAT
BR5 1EZ G4WGZ
BR5 1EZ M0UYR
BR5 1JF G8YTZ
BR5 1JF M6ELY
BR5 1JF M6JEM
BR5 1NF G7URM
BR5 1PT M7FDQ
BR5 1PY 2E0KVV
BR5 1PY M0KVB
BR5 1PY M6KVQ
BR5 2DL G8KPY
BR5 2EH G1FAA
BR5 2EP M6GRH
BR5 2JF G1BPD
BR5 2JR G7RUY
BR5 2JX G7KWM
BR5 2LZ G6LGR
BR5 2NT G7RFM
BR5 2NY G1UPT
BR5 2PA M7NAJ
BR5 2SB G7CEW
BR5 2SH M0LRH
BR5 2TF G0RRL
BR5 3AN G7BIP
BR5 3AT G0OMH
BR5 3AT M6AIZ
BR5 3LF G4DOE
BR5 3LG G6DEV
BR5 3LP M7CXT
BR5 3LQ M3TKT
BR5 3LY G7FRH
BR5 4AH M7XXS
BR5 4EQ G6XYD
BR5 4JP G1EYS
BR5 4JZ G7SAT
BR5 4LA G1VOJ
BR5 4LU G3BVA
BR5 4NS G0LXF
BR5 4PF G0JPM
BR5 4PF M1AHJ
BR5 4PN G4KMF
BR6 0AQ G7PJD
BR6 0BH G1VII
BR6 0BT G7HOL
BR6 0EJ G0TZR
BR6 0EQ G4NMT
BR6 0ER G8LSC
BR6 0LE M6OLP
BR6 0QE G1PJI
BR6 0QJ G0VPC
BR6 0TD G8LSI
BR6 0UP 2E0RHW
BR6 0UP M7RHW
BR6 6BE G7DRW
BR6 6HW M1DPO
BR6 6JF M3XFB
BR6 6JU G3SIU
BR6 7QG M0GZI
BR6 7QJ M3XEF
BR6 7QN M0HHE
BR6 7RT G7RGJ
BR6 7RT M0HAL
BR6 7SD G3VFD
BR6 7TD 2E0CGG
BR6 7TD M0REZ
BR6 7TD M6TDM
BR6 8BL G8KZJ
BR6 8DJ 2E0BWI
BR6 8DJ M0YRG
BR6 8HJ G3ZOH
BR6 8HP G3MCA
BR6 8JB G6GKL
BR6 8JP G4TQO
BR6 9AA 2E0AYQ
BR6 9BS G0BZX
BR6 9BZ G8HKN
BR6 9EW M0HRY
BR6 9JG M6GTE
BR6 9LE M3HVE
BR6 9LL 2E0PAT
BR6 9LL M6EWD
BR6 9LY G1JTX
BR6 9NF G8IKA
BR6 9PN G1EOH
BR6 9PN G6XVQ
BR6 9RS G7AQK
BR6 9RS M3TYX
BR66EJ 2E0JTT
BR66EJ M7JXX
BR7 5DU M0JZN
BR7 5DU M7JZN
BR7 5RF M7PXM
BR7 5RN M6MMC
BR7 6AG G7CXU
BR7 6JD G7ULL
BR7 6JR 2E0SPW
BR7 6JR M0GXN
BR7 6JR M6AEI
BR7 6LA G0GZV
BR7 6QT G6DNL
BR7 6SJ M1FAA
BR8 7HA M7AHG
BR8 7JA G1PUO
BR8 7JY G4IKQ
BR8 7LS M1LYE
BR8 7QT M3GML
BR8 7QZ M3KGG
BR8 7QZ M3KSG
BR8 7RJ G7LIT
BR8 7RY M7CAL
BR8 7SB G3IKQ
BR8 7SE G6SYA
BR8 7TW G7AKM
BR8 7UR G7PEB
BR8 8AP M6GSR
BR8 8DD 2E0SSI
BR8 8DQ 2E1DQQ
BR8 8DQ M3DQQ
BR8 8HZ 2E0FVB
BR8 8HZ M7KJS
BR8 8JL M0RAW
BR8 8LP G8VJG
BR8 8NZ G4AGC
BR8 8TN M1AEP
BR87JJ G0WBO
BR88JB M6LBC
BS 30 5QU M7CZR
BS1 4AU M0HKH
BS1 5BY M7UCL
BS1 5JR G4FAZ
BS1 5LS M0THB
BS1 5LS M6THB
BS1 6EH M3NWQ
BS1 6EH M3NYG
BS1 6ZN M7SXZ
BS10 5JF M7TEG
BS10 5SR M3HXQ
BS10 6BP G0PID
BS10 6JN G0JKU
BS10 6TJ M7RTK
BS10 7PH G0HNO
BS10 7TE G4SQQ
BS10 7XE G1EIR
BS11 0AG M6EVR
BS11 0EF G4HHL
BS11 0NL M6KKJ
BS11 0QX 2E0LHD
BS11 0QX M6LDB
BS11 0QX M6NMH
BS11 9AD M0KGA
BS11 9EU M7MBI
BS11 9JD M0HXI
BS11 9RT G0OKL
BS11 9RY M0GMW
BS11 9SR G1PBX
BS11 9SR G8OTH
BS11 9TX G3YHV
BS11 9TX G4AHG
BS11 9TX M3YHV
BS11 9UF G0XAK
BS11 9UP G3LWF
BS11 9XB 2E0JCC
BS11 9XB G1RWT
BS11 9XB M6JBR
BS110TX M7ATA
BS12 3LQ G4KXO
BS13 0BB 2E0SZH
BS13 0BB M3AMB
BS13 0BJ M7AXU
BS13 0DE M7UTG
BS13 0HS M0WLF
BS13 0JG G8VUM
BS13 0NL M0PCH
BS13 0QA G7VOK
BS13 7BU M7YZP
BS13 7BW G0RCU
BS13 7DB G3XED
BS13 7DF G3WCL
BS13 7LY G6FMN
BS13 7SA G7NQJ
BS13 8AQ G6SDW
BS13 8DB M0AKF
BS13 8EF 2E0WSW
BS13 8EF M3VIU
BS13 8EF M3WZS
BS13 8HN 2E0VTT
BS13 8HN M0JFO
BS13 8HN M6GTT
BS13 8HQ G1AVB
BS13 8HQ M3RWD
BS13 8HU G8GRS
BS13 8PY M1KEV
BS13 8SA M3NRJ
BS13 9DR M6LWR
BS13 9HS M1BOB
BS13 9RB G0CEM
BS13 9RL M3HBP
BS139LY M7BBZ
BS14 0DS G4OJU
BS14 0EF 2E0LRL
BS14 0EG G4YZR
BS14 0EH M0DIL
BS14 0EH M3SWS
BS14 0EJ 2E0WIL
BS14 0EJ M3WPW
BS14 0HH G4JFX
BS14 0HS G3IOI
BS14 0JF G6SPA
BS14 0NN 2E0JIL
BS14 0NN G3YLJ
BS14 0NN M0DXV
BS14 0NN M3YLJ
BS14 0NN M6BEX
BS14 0NN M6YLJ
BS14 0PB G4EPH
BS14 0PP G0DRX
BS14 0PZ G4WUB
BS14 0QG G0CCA
BS14 0RX M6MIT
BS14 0SH M7DER
BS14 8DY G4RAB
BS14 8EQ M7CUN
BS14 8LQ G0JLI
BS14 8NH 2E0FNV
BS14 8NH M6NPV
BS14 8NQ 2E0UGE
BS14 8NQ M0UGE
BS14 8NQ M7RAE
BS14 8PZ G0KDS
BS14 8SS G7PVG
BS14 8SZ G1SLU
BS14 8TT M0CRO
BS14 9BD 2E0HCC
BS14 9BT M6MOD
BS14 9ED G4XML
BS14 9ET M6NQT
BS14 9LW G4NFS
BS14 9ND M3CFU
BS14 9NT 2E0EZZ
BS14 9SF M3VRN
BS14 9TB M6FZM
BS15 1DE M0ORN
BS15 1EP G1HPZ
BS15 1HF G4CQI
BS15 1JR M0PHN
BS15 1TA G0VBK
BS15 1UW M0EAE
BS15 1XB G0TDV

BS15 3BY G3KZC
BS15 3JR G7MWJ
BS15 3JX G8DBP
BS15 3JZ G7IRP
BS15 3PW 2E0BYQ
BS15 3PW M0MMO
BS15 3PW M6BGT
BS15 3RA M1MVX
BS15 3TB G0JYX
BS15 3TJ G4EQP
BS15 4HN G6NQM
BS15 4HT G7BME
BS15 4JA G7KCC
BS15 4LT G7IPH
BS15 4PA G6XEB
BS15 4QB M3VAE
BS15 4QJ G4UBI
BS15 4RT G1FUJ
BS15 4UF 2E0EYU
BS15 4UF M0UTK
BS15 4UF M6KQZ
BS15 4UF M6LPX
BS15 8AA G8DQD
BS15 8DQ G4OPO
BS15 8ES G4BVK
BS15 8LW M7MIR
BS15 8NT G4WOD
BS15 8NX G4FJH
BS15 8NZ G1DFM
BS15 8NZ G1ODD
BS15 8QR M1BEP
BS15 8QS M7TAS
BS15 9NN G4DCX
BS15 9PU M1MAD
BS15 9QJ G1XYR
BS15 9QP G1IHL
BS15 9SH M0VVM
BS15 9UE G8ZEX
BS15 9ZE G0CFM
BS151SW M7AHE
BS159PB M6JBE
BS16 1DQ G3EWF
BS16 1FB G3TEX
BS16 1FD M0DEY
BS16 1GE M6VSC
BS16 1QP 2E0BZU
BS16 1QP M6PMF
BS16 1WL G3UHK
BS16 2AH M5AJO
BS16 2BU M6TNH
BS16 2LH M6NPD
BS16 2PN M0PLO
BS16 2SW G4CXW
BS16 2UB G3RUJ
BS16 2UD G4TVD
BS16 3AG M7CLW
BS16 3DR G1IXE
BS16 3DR G1IXF
BS16 3NG G1FWF
BS16 3QY G7NZZ
BS16 3TL M0KBB
BS16 4EX M7NBX
BS16 4JA M6RFO
BS16 4JD G7UUC
BS16 4PQ M0HDJ
BS16 4QE G4OPQ
BS16 4SQ M1BGK
BS16 5AU M6CJN
BS16 5BL G0PDV
BS16 5DE G7BLT
BS16 5EH 2E0FEP
BS16 5LE G3IZM
BS16 5LG G6YB
BS16 5QS G4DEU
BS16 5RU G4FHN
BS16 5RU G8LMC
BS16 5TN G4UGU
BS16 5UP G1LBM
BS16 6EG 2E0BLW
BS16 6EG M0LHS
BS16 6EG M3VHI
BS16 6HN G6CWF
BS16 6JG G4OEQ
BS16 6PN M6GFM
BS16 7BP G2BAR
BS16 7BP G4VKO
BS16 7DN G0FMJ
BS16 7EB 2E0TBS
BS16 7EB M6CXB
BS16 7HB G4JQK
BS16 7JL G6AWZ
BS16 9AZ G0KGL
BS16 9BQ G8HSR
BS16 9BZ G0XAF
BS16 9DR G0NQJ
BS16 9DW M0MAT
BS16 9EA M6KGF
BS16 9EY 2E0DSB
BS16 9EY M6DFS
BS16 9HN G4RXF
BS16 9LB 2E0BFF
BS16 9LF M6JYZ
BS16 9NH G7CWN
BS16 9QF G7DSQ
BS162UF G8JUT
BS17 1RH G4YCD
BS2 9TB G1ODE
BS2 9UB G3TVV
BS20 0DE G7IEB
BS20 0DR G6PHM
BS20 0EQ G0SNP
BS20 0EY M0BJS
BS20 0JR G4VYK
BS20 0JT 2E0DQQ
BS20 0JT M3GNM
BS20 0JX G4XDP
BS20 0JX G4YSZ
BS20 0LF G6TSE
BS20 0QB G4WRQ
BS20 6BB M6BGW
BS20 6DF M7ASU
BS20 6JQ G8YWQ
BS20 6JR G7AQF
BS20 6JX G4DRZ
BS20 6LA G4KMB
BS20 6LD G8EZR
BS20 6LU G4YNV
BS20 6PF G4UGT
BS20 6QY G4USQ
BS20 6RG G7UTS
BS20 6SR G0RWI
BS20 6YT G6EQZ
BS20 6YT M0DRD
BS20 7AD M6TKI
BS20 7BN M0JXR
BS20 7DH M6TLS
BS20 7DY G4EJH
BS20 7FG G0XZT
BS20 7FW G1UPP
BS20 7FW G4CDU
BS20 7HH M0RBH
BS20 7NY G1SWX
BS20 7NY M3CPH
BS20 7RF M6MSY
BS20 7TQ G6ETL
BS20 8AX G4NXI
BS20 8DD G0FKJ
BS20 8JB G6TNP
BS20 8JQ G4KPM
BS20 8JQ G6ZPV
BS20 8LG G4NVV
BS20 8LG G6ZOE
BS20 8RE M7ECN
BS20 8RQ G4VEH
BS20 8RW G1SEW
BS21 5AQ G0ELO
BS21 5DR M6TJA
BS21 5HB G4ONS
BS21 5HN G7ENR
BS21 6AY G6IQF
BS21 6EH M3YMX
BS21 6HR M1EPX
BS21 6JE 2E0CFY
BS21 6JJ G1CZW
BS21 6JU G1DAX
BS21 6LE G4IWQ
BS21 6LL 2E0LXR
BS21 6LL M6OAS
BS21 6LQ G0IMA
BS21 6LQ G1YDJ
BS21 6QS G4OCF
BS21 6RD G0AZE
BS21 6RG G4NAQ
BS21 6SN 2E0LZT
BS21 6SN M3LZT
BS21 6TQ 2E0KKG
BS21 6TQ M6NQJ
BS21 6TY M7EMJ
BS21 6UQ 2E0IVO
BS21 6UQ M0IVO
BS21 6UQ M3IVO
BS21 7AJ G4ULP
BS21 7DY G0AKS
BS21 7DY G1EFS
BS21 7LU G3WBA
BS21 7QZ G8SPC
BS21 7RL G4BBL
BS21 7RL G8CUW
BS21 7TN G6BGY
BS21 7UB M7GEC
BS21 7UP G6AEC
BS21 7US G7RKT
BS21 7YJ M3GYA
BS21 7YJ M5AFH
BS21 7YN G8OUG
BS217QB M7AVC
BS22 6DJ G1FWZ
BS22 6DQ G6AMV
BS22 6EN G8AVK
BS22 6NX M0RPZ
BS22 6NY 2E0HOQ
BS22 6NY M3HOQ
BS22 6NY M3UEQ
BS22 6PW 2E0JQF
BS22 6PW M3JQF
BS22 6RA 2E0DAR
BS22 6RA M0RGO
BS22 6RA M3PWO
BS22 6RL G0ATD
BS22 6XP G8PRP
BS22 7DR M3VUI
BS22 7FA G1OOJ
BS22 7FW G6OPD
BS22 7HP G4ANZ
BS22 7LU M1EYL
BS22 7PA M1NEW
BS22 7PF G8VZI
BS22 7PG M1EZB
BS22 7RE 2E0AFT
BS22 7SF G1MHP
BS22 7TJ G1PEI
BS22 7TQ 2E0GXQ
BS22 7TQ M0SLD
BS22 7TQ M7SJL
BS22 7TS 2E0WSM
BS22 7UA M7JSS
BS22 7YB G0IHE
BS22 8AD M0ONS
BS22 8AH M7EZB
BS22 8DB M7AMP
BS22 8DD 2E0UWI
BS22 8DD G7UWI
BS22 8DD M3UWI
BS22 8DD M3UWJ
BS22 8DD M6ABZ
BS22 8HL 2E0PHE
BS22 8HL M0OOR
BS22 8HL M7DCT
BS22 8JX G7HYS
BS22 8LN M6SXP
BS22 8PS 2E0BAP
BS22 8PT G4YQG
BS22 8QH M1BQD
BS22 8QN G0SVA
BS22 8QQ M6NWQ
BS22 8SE M1JMB
BS22 8UH M7AXC
BS22 8UY G8GYV
BS22 8XR 2E0AUV
BS22 8XR 2E1HTM
BS22 8XS M3KUS
BS22 9AL G4WAZ
BS22 9AQ G7OPJ
BS22 9AY G0WMW
BS22 9BD G1XRO
BS22 9BD G4JWV
BS22 9DZ M3NKB
BS22 9HA G0LOJ
BS22 9HG M6JXY
BS22 9HT G4ZUX
BS22 9HT M1BQC
BS22 9HT M1BQE
BS22 9HU G3PLT
BS22 9LG M0FXB
BS22 9LW G4DUQ
BS22 9QR G1WOV
BS22 9QS G0ADW
BS22 9SP G3GMC
BS22 9TJ M6SEO
BS22 9UL 2E0JCO
BS22 9UL M0JCE
BS22 9UL M6JCE
BS22 9UU M0CAZ
BS226BG M7PCX
BS23 1LT M7CXW
BS23 1NG M7CXF
BS23 1RQ G3PLJ
BS23 2BH G3LJD
BS23 2BP 2E0RSC
BS23 2BP M7RSC
BS23 2EY M7SKP
BS23 2HE G3RNX
BS23 2JR G3TJE
BS23 2SR M7YKG
BS23 2SY G6LQI
BS23 2UA G4UPR
BS23 2UA M7BPH
BS23 2UG 2E0VPA
BS23 2UG M7PLA
BS23 2US M3HPZ
BS23 2UX 2E0WXU
BS23 2UX M7WSM
BS23 3BX G0TCH
BS23 3DE G1CMZ
BS23 3DF G4CXQ
BS23 3DF G8WSM
BS23 3DN M7YOD
BS23 3EE M6MHY
BS23 3HB M7CRD
BS23 3JH M7SJW
BS23 3LY M3NFJ
BS23 3PQ M6MJN
BS23 3RR G4RNZ
BS23 3RS M7WFG
BS23 3RT M3NVC
BS23 3UY G3JLK
BS23 3XQ G4CDI
BS23 3XX M3LJX
BS23 4BG G0VJM
BS23 4DH 2E1KIP
BS23 4HR M6EDH
BS23 4HU G6YEK
BS23 4JR G0PZB
BS23 4JY G8JNO
BS23 4JZ G0VAZ
BS23 4LQ M6CTZ
BS23 4QQ G1VSX
BS23 4QS G4NMV
BS23 4RF M6MWW
BS23 4TU G7LXA
BS23 4TU M0VLA
BS23 4YH G0IYE
BS234QZ M3HBS
BS24 0AB G0TAT
BS24 0AN M3TQX
BS24 6TH M6OCD
BS24 7AS G7BRX
BS24 7DZ M3FIX
BS24 7EF G7USJ
BS24 7FH M6PAA
BS24 7GT G0WWE
BS24 7HS G8JMK
BS24 7SB M1EPN
BS24 7TJ 2E0ZDJ
BS24 8AD G6FEX
BS24 8AG M3PLV
BS24 8BH G4HJF
BS24 8DS G3LAA
BS24 8PA G8WAL
BS24 8PF M7EPV
BS24 8RZ G6XNQ
BS24 9BU M6PCL
BS24 9DY G7UUK
BS24 9EB M3UPP
BS24 9EH G0BMT
BS24 9JF G6EQI
BS24 9JG G0BOR
BS24 9JW G4PWP
BS24 9LH G3WXH
BS24 9LW G3RUD
BS24 9NJ M1BDO
BS24 9QX G0DKM
BS24 9RH M0KYX
BS24 9RH M6IJM
BS25 1AL G1HHQ
BS25 1AR 2E1EMH
BS25 1AT G7SMH
BS25 1HB G3GTA
BS25 1HD G7PRI
BS25 1HL G0CHJ
BS25 1HQ G8TTX
BS25 1JX 2E0GHA
BS25 1JX M0HAG
BS25 1JX M3ZWJ
BS25 1NB G1CGJ
BS25 1NH G3YOL
BS25 1SA G3RXG
BS25 1TG G4XYH
BS25 1TR G4EYZ
BS25 1UE G7DRO
BS25 5PD G7NJX
BS25 5PE M0GAC
BS25 5PW M7RVD
BS26 2AX G4MCE
BS26 2EH G3XLX
BS26 2QW G4YKG
BS27 3DH M1DIR
BS27 3HA M7IED
BS27 3HS G8OGP
BS27 3HZ G0VSS
BS27 3JH 2E0VHA
BS27 3JH M3VHA
BS27 3NY G6OSR
BS27 3PB M6LSJ
BS27 3TH M6CHD
BS27 3TH M6JHK
BS27 3UB G0VVR
BS27 3XN G4KAE
BS28 4BZ G3RBJ
BS28 4HL G4FXM
BS28 4SW G0HVB
BS29 6AY M7FCG
BS29 6AZ G3KVR
BS29 6DG M6KGL
BS29 6DG M6BJP
BS29 6HD 2E0GFJ
BS29 6HD M7BYE
BS29 6JD 2E1ADP
BS3 1NJ M6IGH
BS3 1XB G0VJN
BS3 2AP M7WAW
BS3 2BP G0FXI
BS3 3DY G7PXR
BS3 3EA G4XED
BS3 3EA G8BDZ
BS3 3JF M6GZU
BS3 3LU M6SSB
BS3 3PW G1HYQ
BS3 4LG M0SYJ
BS3 4NZ 2E0EKM
BS3 4NZ M6HLE
BS3 4TX M6XTK
BS3 5BT G8SUV
BS3 5EG G4EGR
BS3 5ES G4OJI
BS3 5ES M6OJI
BS3 5ET 2E0DDA
BS3 5LN G3IUO
BS3 5LR M6HQB
BS30 5JH M3OPG
BS30 5NR 2E0IHR
BS30 5NR M7OJW
BS30 5PN G8PQA
BS30 5PP G4VRP
BS30 5PP G6AUR
BS30 5PW G7VVO
BS30 5WJ G1AIG
BS30 6DY G4SVS
BS30 6EZ G8VZB
BS30 6JA G0VDJ
BS30 6RH G0JJS
BS30 7BS G1ABA
BS30 8EJ G4MQF
BS30 8UT G4MCQ
BS30 8UT G8KGE
BS30 8XA M6VRL
BS30 8YB 2E0RBG
BS30 8YB M6KXS
BS30 9DU G4YOE
BS30 9PX G8ZFL
BS30 9QQ G7IYK
BS30 9UE G7BTZ
BS30 9UE G4ULV
BS30 9XB G4SNU
BS30 9XB G7JVK
BS30 9YQ G4EXZ
BS305XN 2E0TPL
BS305XN M0TPL
BS305XN M7ADC
BS31 1AS 2E0DUH
BS31 1AS M6FKG
BS31 1BA M6SKN
BS31 1DB G1PCA
BS31 1GB M7CSL
BS31 1JW G4WIZ
BS31 1JX 2E1GOZ
BS31 1QE G0JZH
BS31 1QF M0HBT
BS31 1QF M6BDQ
BS31 1QG G0DPW
BS31 1QW G7CBI
BS31 1XD 2E1CCG
BS31 1XD G0DSB
BS31 1XG G3XAW
BS31 2EQ G7PHE
BS31 2LJ M6GDP
BS31 2NL M3PKE
BS31 2NL M3PKL
BS31 2PP M7TNM
BS31 2TR M6DYA
BS31 2TU M6TII
BS31 3BG G3TKF
BS31 3BG G4HTV
BS31 3BZ 2E0CTK
BS31 3BZ M6CDT
BS31 3DU G7TFU
BS31 3DX G8VPG
BS31 3LA G6AYY
BS32 0AP G6TJZ
BS32 0BR M7JIB
BS32 0DA G7IYM
BS32 0DW G1YHN
BS32 0DZ G4GSA
BS32 4HH 2E1CYP
BS32 4LQ G4OST
BS32 8AF M0ZLI
BS32 8AS G4FUA
BS32 8AS M0GBH
BS32 8AS M6RBY
BS32 8AU M0JOB
BS32 8BB G4BOL
BS32 8BP G0RVM
BS32 8DP M0SUG
BS32 9AR G0RAT
BS34 5AU G0HTS
BS34 5BG 2E0CNV
BS34 5BY M6IAF
BS34 5HH G0UMP
BS34 5HX G4IYE
BS34 5LF M6DJJ
BS34 5NP G0NFH
BS34 5NP G1DOX
BS34 5PW G3ZZU
BS34 5PY G6ZRS
BS34 5RN G0ECM
BS34 5SA G6RUP
BS34 6EB G4DVV
BS34 6EF G1YXA
BS34 6PE M6GLW
BS34 7LJ G7MNO
BS34 7LU M3WLO
BS34 7NB M7WPK
BS34 7RD G4YQQ
BS34 8GD G6TVJ
BS34 8GD G7RHT
BS34 8GD G8EKZ
BS34 8NG G3VM
BS34 8NG M6SDT
BS34 8NJ 2E0CCL
BS34 8NJ M0GYP
BS34 8NJ M6SGL
BS34 8PJ G0SXU
BS34 8QH G8IMB
BS34 8RD M7NML
BS34 8RZ G4LAW
BS34 8UD G4AEL
BS34 8XA 2E0PGS
BS34 8XA G8YMM
BS34 8XA M0NBC
BS34 8XA M3PGS
BS34 8XN G4CJV
BS348XT M7ITD
BS35 1AL M7XYX
BS35 1AY G1USV
BS35 1DP G4YZD
BS35 1HB G7FCO
BS35 1HX M0HFH
BS35 1JF G3XNN
BS35 1JH G4AGH
BS35 1JJ 2E0RES
BS35 1JJ M6RES
BS35 1SR G8AZT
BS35 1SX M3EQQ
BS35 1TB G0JPU
BS35 1UE 2E0DFV
BS35 1UL G4EDQ
BS35 2DN G8MXD
BS35 2EJ G7NVZ
BS35 2FH G6FLE
BS35 2FW M7BYU
BS35 2JE G4ZOG
BS35 2LL M6CPA
BS35 2LX G1PXW
BS35 2QX G0RYM
BS35 2YD G6OLJ
BS35 2YE G3XIY
BS35 3JG G0WMB
BS35 3JL G1FND
BS35 3LQ G4UGO
BS35 3LZ G4GMW
BS35 3NJ G0WOI
BS35 3NJ G4RAE
BS35 3PR G3HTJ
BS35 3RR G0HKC
BS35 3RR G6RAZ
BS35 3SB G6YNL
BS35 3TA G0NBP
BS35 3TL G6FFL
BS35 4AQ G0EBZ
BS35 4DX G0MGC
BS35 4DZ G4YHG
BS35 4HA M6ATI
BS35 4HJ G0CYD
BS35 4LZ M0LDG
BS35 4LZ M5JON
BS35 4PF G0WRN
BS35 5RE G3ZUT
BS35 5SB M6GXD
BS36 1BY G4PHK
BS36 1EP G3FYX
BS36 1HQ 2E0FKE
BS36 1HQ M0IYQ
BS36 1HQ M6EZX
BS36 1HX M7BKM
BS36 1NA G3GBD
BS36 2AT M3PSO
BS36 2EN G0LXC
BS36 2FD G4DEM
BS36 2HL G0NQG
BS36 2HT M6RXA
BS36 2NB G4FKA
BS36 2NQ G4RKG
BS36 2SB G4RWB
BS36 2TX M7DLS
BS36 2TX M7GEO
BS37 4DJ M6EIY
BS37 4EG G0IUD
BS37 4EY G3YAD
BS37 4GB G4FBK
BS37 4JX G7IBF
BS37 4LL 2E0JEZ
BS37 4LL M0JEZ
BS37 4LL M3YSQ
BS37 4LR G4IGU
BS37 4LS M3ZGM
BS37 4PF G6FDR
BS37 4PN M6ZAW
BS37 5DY M0TPW
BS37 5EX G1WVM
BS37 5PJ G6VEJ
BS37 5UR M0SVR
BS37 6HE G0LXL
BS37 6JA M6HYH
BS37 6JB 2E0PSM
BS37 6JB G4ABC
BS37 6JB M0PSM
BS37 6JB M6AVL
BS37 6JF G4EIJ
BS37 6JU G4FPI
BS37 6LA G1ZFF
BS37 6LD M6FUA
BS37 6NJ M6THQ
BS37 6XA G0JYN
BS37 6XB G7BYN
BS37 6XF M0EEP
BS37 6XF M1EEP
BS37 6XJ G1JOR
BS37 6XQ M6VSA
BS37 7AH M6DTT
BS37 7LL G7TZO
BS37 7NA G7GNS
BS37 7NA M3BUU
BS37 7RU 2E0TJV
BS37 7RU G7SGR
BS37 7RU M0TJX
BS37 7RU M6TJV
BS37 8SA G7CJD
BS37 8TG G0JMD
BS37 8TG G1YEH
BS37 8UA G0JTR
BS37 8UR M7LGT
BS37 8YW G6HKZ
BS37 9XN M0XXX
BS37 9XN M0YYY
BS375TF G0JZF
BS377SJ M0GTT
BS378YE M7MJH
BS39 4BH G4SND
BS39 4NT G1ZFG
BS39 5ED M3NEE
BS39 5PB G1FZV
BS39 5RF M6ODE
BS39 5RJ G0BLB
BS39 5SA G4OTJ
BS39 5TH 2E0OOI
BS39 5TH M7IXI
BS39 5UP G0KMV
BS39 6TW 2E0VHV
BS39 6TW M3VHV
BS39 6UD M5AKY
BS39 7LU G0RWT
BS39 7LW M1ALR
BS39 7NL M3LDS
BS39 7PN G1OWD
BS39 7PX G1DOJ
BS39 7PX G4LAF
BS39 7QB G7AEQ
BS39 7QS G4BAD
BS39 7QS G4PDC
BS39 7RP M0ALZ
BS39 7SE 2E0DMA
BS39 7XA G8OEU
BS39 7YX G1GFD
BS4 1BN G4YTH
BS4 1DG M7RKN
BS4 1JT M6VPZ
BS4 1PL G7LPP
BS4 1PN M7FEO
BS4 1QE M0RCE
BS4 1RN G4RZY
BS4 1SL 2E0WXT
BS4 1UQ G7BYK
BS4 1XT M7AJK
BS4 2DL G4KUQ
BS4 2DL G8SHR
BS4 2DX G6HMV
BS4 2LA G7WFH
BS4 2LJ 2E0MXR
BS4 2LJ M6VWW
BS4 2PB M6WOM
BS4 2QP G4KVT
BS4 2RX M7DWF
BS4 2UP G0EXU
BS4 2UW 2E0WCB
BS4 2UW G3ATI
BS4 2UW M6WCB
BS4 2XN M3KPZ
BS4 3BY M0IPH
BS4 3EY G0CCU
BS4 3JD M5AXA
BS4 3JF G6LPG
BS4 3QP G4GHR
BS4 4AF G1XXE
BS4 4BN 2E0JPH
BS4 4DG M7EOU
BS4 4HN G4XCB
BS4 4JL G3XOD
BS4 4JT G6VPW
BS4 4JX M7DMP
BS4 4LP G1XZV
BS4 4NX G7AES
BS4 4QX M1ACQ
BS4 4RA G4GTD
BS4 5DJ G0LTB
BS4 5DZ M6FWL
BS4 5ES G0NVJ
BS4 5HB M7BQY
BS4 5HQ M6EAN
BS4 5JA G0SYI
BS40 5EB G3XSV
BS40 5EG 2E0DPH
BS40 5HD M6DCP
BS40 5LD 2E1DTN
BS40 5QG M0FCP
BS40 6AD G4MKX
BS40 6AP G1KTY
BS40 6BJ G4AXX
BS40 6BJ M0CAM
BS40 6BZ G0PCQ
BS40 6HF G3XMC
BS40 6JE G3SDH
BS40 7TL G3ZMH
BS40 8BG M0TTI
BS40 8SS G8ZOO
BS41 8JA M0KEE
BS41 8JU M6NFX
BS41 9AQ G3KOS
BS41 9AZ 2E0EKT
BS41 9FE G7HMQ
BS41 9NA 2E0AKJ
BS41 9NA M3OQL
BS41 9NF G3YQV
BS48 1BB G8XXG
BS48 1HR M6USP
BS48 1JD G0CCB
BS48 1JL G7PVL
BS48 1JQ G4WAW
BS48 1JQ G7KNA
BS48 1JQ M7MYU
BS48 1LT 2E0GWF
BS48 1PS G8DEX
BS48 1QW G8KIJ
BS48 2AA M7OCZ
BS48 2AG G6ATS
BS48 2AL 2E0RUU
BS48 2BG M1EJD
BS48 2BH G0IFF
BS48 2BN M0HCT
BS48 2DS G6WLX
BS48 2DY 2E0GKF
BS48 2DY M7ARQ
BS48 2DZ G1ODB
BS48 2EE 2E1IHO
BS48 2EE M1FFN
BS48 2QH M6ORO
BS48 2UR G0CFA
BS48 2XD G0LHD
BS48 3BG G3ATX
BS48 3PB G3PRH
BS48 3RX G4DEQ
BS48 4PG G5OLD
BS48 4PG M7TSU
BS48 4RA G0GHM
BS48 4RT G0CEN
BS48 4SX 2E0CFN
BS48 4SX M6TAE
BS48 4YD 2E0DBW
BS48 4YD G8NQO
BS48 4YD M3NQO
BS48 4YD M3ZDW
BS48 4YH G4LSX
BS48 4YH G8HVT
BS49 4AS G0LCX
BS49 4AS M7EJW
BS49 4AS M7LGW
BS49 4DA G4DPH
BS49 4DF G0JLF
BS49 4EB G4FZV
BS49 4EB G6GVH
BS49 4ER G6DMC
BS49 4HF M6IEI
BS49 4HF M6WSH
BS49 4JU M6AEZ
BS49 4LE 2E0ZBB
BS49 4LE M6ZBB
BS49 4LN G6ANJ
BS49 4LS G4BWR
BS49 4NR M7TAT
BS49 4NS M6VTC
BS49 4RB M0SWH
BS49 4RG G4HBT
BS49 4RG G6BZW
BS49 5BN 2E0YCD
BS49 5BN M3FPZ
BS49 5BN M6YCT
BS49 5BT 2E0FYO
BS49 5BT M0JVE
BS49 5BT M7ASO
BS49 5ES G4DYM
BS49 5ES G4FDK
BS49 5EX M6DWD
BS49 5EX M3DWD
BS49 5EY G4KWL
BS49 5EY G4RSC
BS49 5HA 2E0JHT
BS49 5HA M0XOX
BS49 5HA M6JHT
BS49 5HB G0ALI
BS49 5HB G4TCI
BS49 5HL G4REH
BS49 5HQ G3SWH
BS5 0DL G4BWO
BS5 0PQ M6ISZ
BS5 0SE G1JOO
BS5 6RJ G4TPV
BS5 6SY G8BIR
BS5 6TN M0SVA
BS5 7BQ G4GGE
BS5 7EJ G0SCK
BS5 7LY M7ARN
BS5 7NE G0GJN
BS5 7QT 2E0GHE
BS5 7QZ M3WMC
BS5 7RG M0OAJ
BS5 7RL 2E0KAC
BS5 7RL M3KAC
BS5 7SP 2E0TUK
BS5 7SP M6LUA
BS5 7SP M6TYK
BS5 8DX G3ZKI
BS5 8DX G4JPS
BS5 8DX G8XAA
BS5 8HF G7GLQ
BS5 8JU 2E0LJT
BS5 8JU M0LJT
BS5 8JU M6CQJ
BS5 8LN G4ZBL
BS5 8RH G4EIA
BS5 8ST G8CKK
BS5 8ST M0BUV
BS5 8ST M6EZG
BS5 8SZ G0JLE
BS5 8TA G6YFG
BS5 8TW G0KWF
BS5 9HN M3UIC
BS5 9HN M6JWF
BS5 9HW M0IIU
BS5 9JT M3MCU
BS5 9QN G4TAH
BS6 5AH 2E0PFB
BS6 5AH M0HWQ
BS6 5BQ M1BBR
BS6 5DQ G8WVU
BS6 5EG M1KPZ
BS6 5HX G8NNU
BS6 6BE G3ORV
BS6 6DU G1YOY
BS6 6LP M7MNJ
BS6 6PD G4FTN
BS6 7LG G4LOX
BS6 7SU G1UGO
BS6 7XS G8UXB
BS6 7YW G4DHK
BS6 7YW G8HNS
BS66NS G4FVX
BS7 0HQ G8OQG
BS7 0HS G0NZT
BS7 0RG G0NZU
BS7 0RH G6BZG
BS7 0RH G8ZRN
BS7 0RP G8VYT
BS7 0RT G3XPJ
BS7 0SA G7UHS
BS7 0SG G7ISR
BS7 0SR G7DRU
BS7 0TT G4CSE
BS7 0UA 2E0ZST
BS7 0UA M6ZST
BS7 0UH G3FPY
BS7 0UQ G3IUV
BS7 0US G4CGF
BS7 8EX G3MGS
BS7 8JJ G8MHD
BS7 8LT G7AGI
BS7 8LU G0KKC
BS7 8QG G4YQH
BS7 9DW M0HCV
BS7 9JW G6RNZ
BS7 9ST G4ROX
BS7 9XS 2E0RBK
BS7 9YW G7PKJ
BS8 1DW M6IUM
BS8 1DY M7HEX
BS8 2HF G6GGN
BS8 2QD M6GOQ
BS8 3EG G7RLQ
BS8 3PE G6DUC
BS8 4DL G8OOQ
BS8 4TT M6XUX
BS9 1DR G4NYK
BS9 1NG G8HJD
BS9 1NP G3YER
BS9 1PL M1PUW
BS9 1QP G4FRO
BS9 1SN G0SDW
BS9 2BA 2E0KMI
BS9 2BA M0KMI
BS9 2BW G3TCO
BS9 2JF G0OBT
BS9 2LU G4WAD
BS9 2PU M6PAJ
BS9 2QP G0KJM
BS9 2QQ G8XIM
BS9 2QR G4ZBQ
BS9 2QT G0OER
BS9 2QU G8PSO
BS9 2RZ G3XSJ
BS9 3DQ G3XOB
BS9 3RN M0AWH
BS9 3SX G8GRD
BS9 3TU M0HVR
BS9 3UU G4HCB
BS9 3UW G3OWX
BS9 4AH G8KGH
BS9 4BU G3SJI
BS9 4DW G4NBN
BS9 4EL G2BQP
BS9 4QW G8FNR
BS9 4RH G0CJG
BS9 4RS G4KSR
BS9 4TB G3TZA
BS9 4TF G4OEP
BS93JG 2E0FJX
BS93JG M0RKE
BS93JG M7RKE
BS99 5LG G0TEI
BS99 5LG G1IBO
BT1 4FJ MI7MSX
BT10 0AN GI0JRI
BT10 0AS GI6JOP
BT10 0JQ GI0JRD
BT10 0QE MI0GPB
BT11 8BP GI0DPV
BT11 8LL MI6PAT
BT11 8LP GI7OMY
BT11 8PP MI6LDI
BT11 9ED 2I0TJM
BT11 9ED MI0TJM
BT11 9ED MI0TJV
BT11 9GF GI3UIH
BT11 9LU MI6CVO
BT11 9PU 2I0JHH
BT11 9PU MI7CAT
BT11 9PU MI7JOE
BT119ER MI1DAW
BT119ER MI5DAW
BT12 4NH 2I0GCC
BT12 4NH MI6GAQ
BT12 4QB GI7PJU
BT12 5NR MI1RDR
BT12 5NR MI3BRX
BT12 6GF GI0VWU
BT12 6JS MI0SAM
BT12 6JT MI6AHR
BT12 6NH MI6TOA
BT12 7JD 2I0KBS
BT12 7JD MI6KOB
BT12 7LG GI0XBH
BT12 7LZ MI0KYD
BT13 1DJ MI7NWA
BT13 2DR GI4XFN
BT13 2SB GI4GOL
BT13 3DZ GI0CDM
BT13 3LG GI4CFQ
BT13 3LR GI4PXM
BT13 3LR MI0MFI
BT13 3LR MI6EAI
BT13 3PQ MI6DCH
BT13 3PS MI3RLA
BT13 3XN GI0ZAK
BT133RD MI6IGV
BT14 6BT MI0HPE
BT14 6ED MI6WDB
BT14 6ES MI0ILJ
BT14 6ES MI6DND
BT14 6JP MI7CYX
BT14 6NZ MI6SWB
BT14 6PA 2I0CVR
BT14 6PA MI6TMZ
BT14 6RZ GI4ZOI
BT14 6SL MI6MSG
BT14 6TE GI4IKF
BT14 7GD GI7TPO
BT14 7NF MI3MPL
BT14 8AN MI6OLJ
BT14 8AN MI7DLD
BT14 8FP MI0BDZ
BT14 8HD GI4EIZ
BT14 8JX GI4CUV
BT14 8JY GI7MBP
BT14 8PP MI3GEI
BT14 8RE 2I0ZFZ
BT14 8RE MI6FNZ
BT146BE MI6FRQ
BT15 2BP MI0ISY
BT15 3FY GI0RWO
BT15 3JF MI7BGE
BT15 3LA MI5HNA
BT15 3NP MI0PCJ
BT15 3QS MI6HAD
BT15 4EP GI1KDS
BT15 4GR GI4MNN
BT15 4JU GI7IFW
BT15 5AJ 2I0EBB
BT15 5AJ MI6EDP
BT15 5GL GI3RQU
BT16 1JD 2I0DHR
BT16 1JD MI3DHR
BT16 1JW 2I0VEH
BT16 1JW MI6LJZ
BT16 1SL MI5AMO
BT16 1UU GI4SAM
BT16 1XD GI4NBO
BT16 1XU GI8SJS
BT16 2AB MI6JJZ
BT16 2BB GI4OCL
BT16 2BE 2I0TUV
BT16 2BE GI4TUV
BT16 2BE MI6TUV
BT16 2BE MI6WGE
BT16 2ED MI7RDK
BT16 2HB MI3MDV
BT16 2NT GI7KHR
BT16 2PQ MI3CST
BT16 2SQ MI6PDY
BT17 0AF GI1JQP
BT17 0JG GI4CSO
BT17 0NZ MI3OIB
BT17 0NZ MI6BFJ
BT17 0RZ GI1TYX
BT17 0XD MI7MHG
BT17 9LN MI7GCZ
BT17 9PY GI4KCO
BT17 9PZ MI6OYB
BT17 9QY 2I0DHC
BT17 9QY MI6PBW
BT18 0DY GI3GTR
BT18 0HG GI0TDP
BT18 0HH GI4JTF
BT18 0NB GI5SBZ
BT18 0PJ GI4ZTU
BT18 0PL GI3USK
BT18 9DL MI7NIH
BT18 9EL GI3WUO
BT18 9EU GI0USQ
BT18 9JQ MI0RJW
BT18 9NB GI8YJV
BT18 9NX 2I0VOQ
BT19 1AA GI4MUE
BT19 1BJ MI3SEV
BT19 1DQ GI6ATD
BT19 1EQ GI4POC
BT19 1FG MI3CCA
BT19 1GH MI3LXZ
BT19 1HD MI0WHX
BT19 1HG MI3LXN
BT19 1HQ MI0CBX
BT19 1HQ MI3OHF
BT19 1JA MI3XOI
BT19 1LB GI6UFU
BT19 1NP MI3XIU
BT19 1NS MI0NYC
BT19 1NT GI4OCK
BT19 1RQ 2I0DUR
BT19 1RQ MI0OBR
BT19 1RQ MI6DUR
BT19 1RQ MI6HLC
BT19 1SN MI6VRX
BT19 1SR MI1EVD
BT19 1YN MI3JSH

BT19 1YU GI3XRQ
BT19 1YU GI4DOH
BT19 1YU MI3FSW
BT19 1ZQ MI3FBX
BT19 6AE GI3TZB
BT19 6AF GI4TPY
BT19 6AF GI7DIT
BT19 6AR GI3OTU
BT19 6AY GI1VPA
BT19 6AY GI4NSS
BT19 6BA GI3MBB
BT19 6DG GI6BDN
BT19 6DQ 2I0UAD
BT19 6DQ MI6UAB
BT19 6DT GI7VIW
BT19 6DU GI3YMY
BT19 6EB MI6CBG
BT19 6FN GI3MMG
BT19 6HW GI4TCR
BT19 6HY GI6KJC
BT19 6JF GI6JGB
BT19 6LX MI3SRG
BT19 6LZ MI3PPI
BT19 6NJ 2I0POD
BT19 6NR GI7FOD
BT19 6NX GI1SZC
BT19 6SD GI3VAF
BT19 6SD MI6HBI
BT19 6XG GI6BNI
BT19 6ZH MI3FSR
BT19 6ZW GI6PLO
BT19 7BS MI6WGX
BT19 7EY GI4JNS
BT19 7FE 2I0DTE
BT19 7FE MI0OBC
BT19 7FE MI6DTE
BT19 7GB GI7GXZ
BT19 7GY MI0ILH
BT19 7HR GI4MRZ
BT19 7QB GI3WFP
BT19 7RB GI4WYE
BT20 3BF MI0RSO
BT20 3DN GI6IHM
BT20 3EP GI4LZS
BT20 3EP MI6NUX
BT20 3ER GI4JJF
BT20 3HA GI0HSB
BT20 3JD GI0POB
BT20 3PP GI3TZX
BT20 3PU GI8JPF
BT20 3TP GI7JEB
BT20 4DF GI0WPV
BT20 4HS GI6IVJ
BT20 4JD GI0FZT
BT20 4NN MI3BLN
BT20 4NP GI4OPH
BT20 4NS MI6EAC
BT20 4PE GI0HHV
BT20 4PP 2I0LNZ
BT20 4PT GI4SPU
BT20 4PT GI6SBW
BT20 4PX GI7ISX
BT20 4PX MI3ISX
BT20 4QB GI6FZI
BT20 4RS GI4TMB
BT20 4TG GI8YJD
BT20 4TX GI4MRN
BT20 4US GI0BEY
BT20 4UX GI7TVV
BT20 5LT MI3CBJ
BT20 5NT MI3FEO
BT20 5NU GI8ZHW
BT20 5PN GI3UBA
BT20 5RF GI4FLG
BT20 5RQ GI4TPI
BT21 0AX MI3CBL
BT21 0BN GI1SYM
BT21 0BQ GI7FNP
BT21 0DR 2I1HNZ
BT21 0ES GI8VTK
BT21 0EY MI6KRP
BT21 0EZ 2I0ETW
BT21 0EZ GI0SSA
BT21 0EZ MI0HWG
BT21 0EZ MI1EIH
BT21 0EZ MI3BPS
BT21 0EZ MI3CQO
BT21 0EZ MI6NID
BT21 0GA GI4BXB
BT21 0HL GI1WGK
BT21 0HZ MI6CUZ
BT21 0LN GI3UPG
BT21 0NR MI3VFJ
BT21 0NS GI7PIZ
BT21 0PU MI3FVW
BT21 0PY GI3TJM
BT21 0QR GI0BCP
BT21 0SH GI4PQV
BT21 0SH GI6IOU
BT210FB GI4OVN
BT22 1AF MI0AEX
BT22 1AF MI0JSJ
BT22 1AJ MI0BSU
BT22 1AU MI3MJI
BT22 1BW GI0PNP
BT22 1DZ MI0SRM
BT22 1EL GI0AQD
BT22 1GB MI7NMA
BT22 1HP GI0VTS
BT22 1HP GI4XSF
BT22 1JX GI0UAG
BT22 1LL GI4TTL
BT22 1ND GI4NAE
BT22 1NE GI6IES
BT22 1NQ 2I0GTO
BT22 1NQ GI3UZJ
BT22 1NQ MI3GTO
BT22 1QT MI0GLG
BT22 1QT MI3VFZ
BT22 1RB GI4YPR
BT22 2BG GI7NMK
BT22 2BN GI3OBO
BT22 2GS MI6MIB
BT22 2HW MI3CSS
BT22 2HY GI4PBS
BT22 2HY GI4PBT
BT22 2JQ MI0RPT
BT22 2LA GI4HCX
BT22 2LB MI6WPT
BT22 2LE MI0TMP
BT22 2NG MI0UST
BT22 2NQ MI0WST
BT22 2NQ MI3PYX
BT22 2NW 2I0DTW
BT22 2NW MI3VEQ
BT22 2NW MI6RPM
BT22 2PE MI6VAI
BT22 2QF GI4PGN
BT22 2TH GI4NKK
BT22 2TR GI4EQN
BT22 2TZ 2I0FPB
BT22 2TZ GI7VCR
BT22 2TZ MI0IGL
BT22 2TZ MI6IPB
BT23 3BN GI4AFH
BT23 4AN MI0JPC
BT23 4BN 2I0BXJ
BT23 4BN MI3LZF
BT23 4DY MI6OWD
BT23 4LY GI7ALH
BT23 4NA MI6OPT
BT23 4ND MI3WJO
BT23 4NT MI3ZMJ
BT23 4PD MI0RSN
BT23 4SF GI7KTU
BT23 4SQ GI4MCW
BT23 4TB GI1WLJ
BT23 4TN MI6BAI
BT23 4TP GI4JTS
BT23 4TQ GI4PGH
BT23 4UW GI0PFL
BT23 4WS 2I0FIP
BT23 4WS MI7ACD
BT23 4WS MI7CYL
BT23 5DF MI7DGL
BT23 5EW GI6EGE
BT23 5HA GI7FGQ
BT23 5HE GI0SMU
BT23 5HR GI4GST
BT23 5JJ GI4TSK
BT23 5LD MI1FAR
BT23 5LN GI6CAG
BT23 5LP GI4MBM
BT23 5LT GI6VLY
BT23 5QP GI3NYJ
BT23 5RJ GI4OSG
BT23 5RN GI0EEO
BT23 5TQ GI4OWB
BT23 5TZ 2I0WWB
BT23 5TZ GI5TKA
BT23 5TZ MI0WWB
BT23 5TZ MI6USC
BT23 5XA MI3SXI
BT23 5YX GI4IYO
BT23 6AQ MI6IMV
BT23 6BB GI4MQA
BT23 6DE MI3VFF
BT23 6EN GI3XZM
BT23 6ES GI3VHM
BT23 6PB MI1RGL
BT23 6PE MI7EJV
BT23 6RZ MI3ZMP
BT23 7AR GI0LTT
BT23 7BN 2I0LJQ
BT23 7BN MI3LJQ
BT23 7BW MI0ABD
BT23 7BZ GI0OHT
BT23 7ED MI6CMQ
BT23 7GQ GI6JMD
BT23 7QP GI4OYI
BT23 7RE GI0WYK
BT23 8GL MI7LBW
BT23 8NN MI0MGJ
BT23 8QS GI3AMY
BT23 8QT GI0HHZ
BT23 8RF MI3PMR
BT23 8RS MI0BVG
BT23 8RT GI0DUP
BT23 8SN MI5ALO
BT23 8TE GI3WHA
BT23 8UA MI3LVZ
BT23 8UL 2I0FZO
BT23 8UL MI7MCK
BT23 8XD MI5CFM
BT23 8YE 2I0WMH
BT23 8YE MI6WAF
BT24 7BE GI4BJK
BT24 7BS MI6MPH
BT24 7DP MI3EZF
BT24 7FQ GI0VIF
BT24 7HU MI3NSF
BT24 7PR 2I0LXS
BT24 7PR MI6CVW
BT24 8BL 2I0WKY
BT24 8BL MI7WNY
BT24 8FE MI6PIR
BT24 8GA GI4JOR
BT24 8HU GI8TME
BT24 8HW GI4MUN
BT24 8LB GI4TUJ
BT24 8LF GI4XLB
BT24 8LF GI6ATZ
BT24 8PT GI0WJI
BT24 8QQ GI4AXV
BT24 8QY GI6DCX
BT24 8QZ GI7MDP
BT24 8QZ MI3NJU
BT24 8YS GI4SZP
BT25 1BD 2I0YMF
BT25 1BD MI0YMF
BT25 1BD MI3YMF
BT25 1DD MI0BAT
BT25 1LL 2I0GIF
BT25 1LL MI6GIF
BT25 1NP MI0SRR
BT25 1NP MI3AJK
BT25 1PN 2I0WKE
BT25 1PN MI6WKE
BT25 1PN MI6YAM
BT25 1RT MI6MQX
BT25 2AF MI1CUS
BT25 2AN 2I0EIG
BT25 2AN 2I0ESA
BT25 2AN MI6EQA
BT25 2AN MI6XOS
BT25 2EQ GI0UXD
BT25 2HN GI6SFO
BT25 2HY GI7VXC
BT25 2NT GI0TSS
BT25 2PN MI0BTK
BT26 6BH GI4KSO
BT26 6BL MI0DGX
BT26 6BS 2I0VOF
BT26 6BS MI6POF
BT26 6BS MI6VOF
BT26 6BX GI4OZJ
BT26 6DJ 2I0LPG
BT26 6EG MI3VPO
BT26 6ES GI3VPV
BT26 6HL MI6TNZ
BT26 6HQ MI1BRS
BT26 6HQ MI7RLC
BT26 6HU GI7NFB
BT26 6LJ GI4TCS
BT26 6NH MI7GSI
BT26 6NS GI6UUT
BT26 6PW GI4RXM
BT27 4BH GI7TFK
BT27 4BS GI4NFH
BT27 4DA GI4ERM
BT27 4DA GI8KEP
BT27 4EF GI6ETQ
BT27 4EW GI6CMA
BT27 4JA GI6WFX
BT27 4PL 2I0SHM
BT27 4QA GI4NLQ
BT27 4QX GI4NKY
BT27 4RY MI6GTY
BT27 4UL MI7FNY
BT27 4YD MI6TIJ
BT27 4YQ 2I0FQW
BT27 4YQ MI7XZD
BT27 5AW MI7ABU
BT27 5BF GI4AHP
BT27 5BT GI6FOR
BT27 5BY 2I0GKB
BT27 5BY MI6GKB
BT27 5BZ MI0CZF
BT27 5DA 2I0EIU
BT27 5DA MI6GGF
BT27 5DB GI4XIR
BT27 5HJ MI6HLY
BT27 5LF MI0OPM
BT27 5LF MI1OPM
BT27 5LF MI3TEM
BT27 5LQ GI7PBQ
BT27 5LS GI1GKI
BT27 5LW MI1FRM
BT27 5PD GI0LWO
BT27 5PD GI8TSI
BT27 5RE 2I0EUV
BT27 5RE MI6WEQ
BT27 5RF 2I1AXH
BT27 6UU 2I0ITY
BT27 6UU MI0TUB
BT27 6UU MI6JBZ
BT27 6YW MI1ERL
BT276TS MI3GSW
BT28 1EX GI0RDJ
BT28 1HE GI0GDF
BT28 1HF 2I0GYL
BT28 1HF GI7PWQ
BT28 1HF MI6YLG
BT28 1LD GI0RYK
BT28 1NY 2I0KBB
BT28 1NY MI6KBB
BT28 1QD MI0JLC
BT28 1QD MI3OHG
BT28 1SQ GI0UVD
BT28 1TY GI0OKM
BT28 1YJ GI0PGC
BT28 2BX MI6KGU
BT28 2DN GI0TJV
BT28 2DR GI3LQY
BT28 2DW MI0BME
BT28 2EY MI6PGI
BT28 2HU MI3SYF
BT28 2HX MI3MOT
BT28 2JP GI0MKK
BT28 2LH GI8DGB
BT28 2PL MI0MSB
BT28 2QF MI0MOD
BT28 2QF MI0UGP
BT28 2QL MI6XGZ
BT28 2TE MI0TGL
BT28 2TE MI0TLG
BT28 2TE MI6TLG
BT28 2UD MI7TVA
BT28 2UN MI3NSR
BT28 2WY MI6KKU
BT28 2XU GI4PES
BT28 3BA MI6TMR
BT28 3BT GI4DNW
BT28 3DG GI7ROB
BT28 3DS GI4XTC
BT28 3HD MI3MRG
BT28 3HG GI4LKG
BT28 3HN 2I0PRL
BT28 3HN MI6FIU
BT28 3HQ MI6MNL
BT28 3JH GI4MEQ
BT28 3LL GI8YWE
BT28 3LP GI6GNA
BT28 3PD GI3ECQ
BT28 3QB GI4RKC
BT28 3QX MI0ASV
BT28 3QY MI0ABN
BT28 3QY MI3EOD
BT28 3RE GI4SNA
BT28 3RR GI0DVU
BT28 3TQ GI1YSG
BT28 3TR GI7GSB
BT28 3UT GI8UUN
BT28 3YB MI3TBL
BT29 4FX MI0IXH
BT29 4JL 2I0EQC
BT29 4JL 2I0EQR
BT29 4JL 2I0EQS
BT29 4JL GI4WTT
BT29 4JL MI3EQS
BT29 4JL MI3WTT
BT29 4JQ GI0AIJ
BT29 4JW MI5AJH
BT29 4QU 2I0OAZ
BT29 4QU MI0OAZ
BT29 4RH GI7IRJ
BT29 4SF GI7ULG
BT29 4SG GI7UBY
BT29 4WT GI7PJF
BT29 4XN MI7WTZ
BT29 4YA MI1DJW
BT29 4YA MI6LFU
BT30 6JL MI6GPV
BT30 6NS GI1XTK
BT30 6PZ GI4NJQ
BT30 6ST MI6XOX
BT30 7AZ MI3LXE
BT30 7DA GI8UIU
BT30 7JT MI7RDA
BT30 7JX 2I0ARM
BT30 7JX MI7HJK
BT30 7LY GI4RMA
BT30 7NU MI5AFL
BT30 7RJ 2I0OTW
BT30 7RJ MI6OTW
BT30 7SA GI4HYM
BT30 7SQ 2I0NEJ
BT30 7SQ MI6MQF
BT30 8SH GI3KME
BT30 8SH GI8PMA
BT30 8SR MI6HKP
BT30 9BS 2I0TXM
BT30 9BS MI0HNQ
BT30 9BS MI0TXM
BT30 9BS MI6XMG
BT30 9BU GI4MMJ
BT30 9BW MI3AIN
BT30 9EH GI4MAC
BT30 9EW GI4MHD
BT30 9GB GI3TAC
BT30 9HW MI6TND
BT30 9PD GI3HNM
BT30 9PX GI1RXM
BT31 9SJ GI0LAM
BT32 3AW MI0DNM
BT32 3BQ MI0MAP
BT32 3GD MI7YOZ
BT32 3QX GI7INR
BT32 3RD GI0USS
BT32 3TL 2I0NGM
BT32 3TL MI6NGM
BT32 3TZ MI6MXZ
BT32 3TZ MI6RXC
BT32 3UT MI6LIT
BT32 3YA GI0WZW
BT32 4AH MI6PEJ
BT32 4HF GI0HXH
BT32 4JL 2I0BSH
BT32 4JL MI0TFK
BT32 4LF GI3WEM
BT32 4NA MI6GPZ
BT32 4NU MI6EFD
BT32 4PE MI6RFD
BT32 4PT MI6MTO
BT32 4PY GI4GUH
BT32 4PZ 2I0SXM
BT32 4PZ MI3SXM
BT32 4QL MI7HNC
BT32 4RA GI8RQI
BT32 4RE GI0UQK
BT32 5ES 2I0GQR
BT32 5ES MI0SNY
BT32 5ES MI7RYP
BT32 5HR MI7GBX
BT32 5JF MI0CGV
BT32 5NN GI4GPC
BT32 5RD GI1JXE
BT323GD 2I0PBM
BT323GD MI3GJI
BT33 0AR GI6DWZ
BT33 0DT MI6AGV
BT33 0FA GI1VKJ
BT33 0HW MI1FCB
BT33 0NQ GI3FJX
BT34 1HL MI3PMW
BT34 1JG 2I0BIR
BT34 1JG MI3RIF
BT34 1JW MI0PYN
BT34 1NZ GI0LRZ
BT34 1PW MI6TUM
BT34 1QD MI6NQD
BT34 2BQ MI0AQX
BT34 2BQ MI0BES
BT34 2JB MI5AHG
BT34 2NA GI6JJR
BT34 2ND GI0THR
BT34 2NY GI0WAH
BT34 2PG GI0UYY
BT34 2PG GI1YEA
BT34 2PJ GI8ZFZ
BT34 2QP 2I0MMT
BT34 2QP MI6MMT
BT34 3AN MI6AOR
BT34 3BJ GI6RBD
BT34 3BL GI1WFP
BT34 3DX GI8YJF
BT34 3EP MI0ZKX
BT34 3HL GI4MBQ
BT34 3JZ MI6GMK
BT34 3NL MI3EOH
BT34 3PG GI0VVC
BT34 3SA GI1RAA
BT34 3SR MI0DDW
BT34 4AQ GI4BBE
BT34 4DA GI7IVX
BT34 4FT 2I0TXB
BT34 4JJ GI6EBX
BT34 4LP GI4XFY
BT34 4NB MI7YDB
BT34 4NU GI3HJH
BT34 4XW MI3SXQ
BT34 5DE 2I0KXM
BT34 5DE 2I0WMN
BT34 5DE MI6KXM
BT34 5DE MI6WMN
BT34 5EL 2I1ALE
BT34 5EL GI0UTE
BT34 5LS 2I0SEC
BT34 5LS MI0NLY
BT34 5LS MI6EEC
BT34 5PU MI6UNC
BT34 5TJ GI4WAH
BT34 5US MI6MSR
BT35 0JG GI4DQT
BT35 0PJ GI0VGL
BT35 6BZ 2I0ZXM
BT35 6BZ MI6ZTM
BT35 6DD 2I0DBK
BT35 6DS MI3RIV
BT35 6EH GI6FXY
BT35 6NA GI4OVE
BT35 6NS 2I0FEX
BT35 6NS 2I0JAP
BT35 6NS 2I0RRE
BT35 6NS MI0KAG
BT35 6NS MI0RRE
BT35 6NS MI3FEX
BT35 6NS MI3RRE
BT35 6NS MI3WBU
BT35 6SD 2I0BYL
BT35 6SD 2I0FUT
BT35 6SD MI0GQG
BT35 6SD MI0GQI
BT35 6SD MI3GMI
BT35 6SD MI3LXJ
BT35 6TW 2I0MBI
BT35 6TW MI3SYI
BT35 7FD MI0YCK
BT35 7GA MI6CQS
BT35 7HD GI4SZW
BT35 7HD MI0JWY
BT35 8AP MI6KJW
BT35 8PP MI6DVN
BT35 8PW GI1ANG
BT35 8PW MI6TOC
BT35 8XA MI3CQX
BT35 9AQ 2I0EIB
BT35 9AQ MI6RSH
BT35 9RR MI3CQR
BT35 9RR MI3NCC
BT35 9TX GI0CTI
BT350DQ MI7KRJ
BT36 4QT GI4OXO
BT36 4TP GI4BUJ
BT36 4WL GI4BTG
BT36 4WQ GI4SQL
BT36 5BX MI6PBI
BT36 5DE MI7FCW
BT36 5FG MI1AUI
BT36 5GD GI7AQO
BT36 5GZ GI4RXX
BT36 5JU GI7JYK
BT36 5JU GI8UCS
BT36 5JU MI5JYK
BT36 5JZ GI1WYZ
BT36 5JZ GI4RNP
BT36 5JZ MI0JQS
BT36 5LA 2I0BID
BT36 5LA MI0GTI
BT36 5LA MI3OEQ
BT36 5NH MI7GNI
BT36 5ST GI4KEQ
BT36 5ST MI1ASN
BT36 5UF GI5CEO
BT36 5UY MI7DKB
BT36 5WQ MI3LPH
BT36 5WR GI4JIW
BT36 5WZ 2I0TCJ
BT36 5WZ MI0TBV
BT36 5WZ MI6TFG
BT36 6BA GI4OTG
BT36 6BD 2I0EPC
BT36 6BX MI6WPP
BT36 6DZ MI6HMI
BT36 6LE MI3DNN
BT36 6LJ 2I0CLS
BT36 6LJ MI6VRS
BT36 6LS GI4KSH
BT36 6PA MI0NAJ
BT36 6PR MI6LNP
BT36 6QQ GI0BEB
BT36 6SP GI8RPT
BT36 6TZ GI8LIH
BT36 6UA MI1BOE
BT36 6UN GI4JWW
BT36 7DG MI3CXM
BT36 7DR MI1EZZ
BT36 7DR MI3STW
BT36 7HA GI0USW
BT36 7SU GI4GID
BT36 7TG 2I0TUI
BT36 7TG MI0HRO
BT36 7YP GI0JFF
BT366HT MI7CVD
BT37 0AZ GI4MCH
BT37 0EL GI6KVS
BT37 0GH 2I0KFD
BT37 0GH MI0KKD
BT37 0GH MI6LLZ
BT37 0JN GI0BDU
BT37 0LH MI0XLK
BT37 0NL MI0OIM
BT37 0NP MI7WHK
BT37 0RR GI0IQA
BT37 0TD GI0HHE
BT37 0UL GI4KBW
BT37 0XH GI4SFE
BT37 0XY GI4VWC
BT37 0ZY 2I0LOR
BT37 0ZY MI6BAJ
BT37 9JZ MI3FPN
BT37 9PD GI4RYL
BT37 9SH 2I0SHB
BT37 9SH GI6ANC
BT37 9SQ MI0TBE
BT379QZ 2I0RWG
BT379QZ MI0RGX
BT379QZ MI6BJG
BT38 7BL 2I0CYW
BT38 7BL MI3ZCY
BT38 7BY GI8UMV
BT38 7DJ MI3CGT
BT38 7DJ MI3CGU
BT38 7EH GI0USC
BT38 7EP MI3KIL
BT38 7HN MI0PQR
BT38 7HN MI3JTB
BT38 7HQ MI3WMK
BT38 7JT GI8LCJ
BT38 7LD MI0KMJ
BT38 7LL MI0PCW
BT38 7LS 2I0TPC
BT38 7LS MI6TPC
BT38 7LZ 2I1EXU
BT38 7NG GI3YRL
BT38 7NG GI4SBA
BT38 7PD 2I0FKM
BT38 7PD MI7AEW
BT38 7QD GI4DAV
BT38 7RB GI6ROI
BT38 7RL MI6HDH
BT38 7RT MI3FCK
BT38 7RU MI0ZSC
BT38 7UE MI0PJL
BT38 7XU 2I0TLT
BT38 7XU MI0TIP
BT38 8BF GI7VGR
BT38 8BF MI0SYN
BT38 8BF MI7JCB
BT38 8BQ 2I0HWA
BT38 8BQ MI0AFT
BT38 8BQ MI0LBS
BT38 8BS GI7GJX
BT38 8BY GI0IBC
BT38 8BY GI4GVS
BT38 8DP MI3VQH
BT38 8EW 2I0HRV
BT38 8EW MI0HRV
BT38 8EW MI6AOX
BT38 8FB GI4RVF
BT38 8FG GI4OYG
BT38 8GP MI3FBW
BT38 8GQ 2I0SMD
BT38 8HZ GI4GCN
BT38 8JT MI6RLU
BT38 8NE 2I0IYH
BT38 8NE MI3IYH
BT38 8NN 2I0CEI
BT38 8NN MI0HHV
BT38 8NN MI3YVB
BT38 8PS GI0XYZ
BT38 8SN GI0IJB
BT38 8ST MI6XKT
BT38 8SY GI6EJW
BT38 8TX GI0RUC
BT38 9AG 2I0UCS
BT38 9AG MI0GJN
BT38 9AG MI3UCS
BT38 9AP GI0DFD
BT38 9BB GI4FUE
BT38 9DL GI8SKN
BT38 9DL MI6EXU
BT38 9DN GI8KYI
BT38 9EA GI6TFF
BT38 9EG GI8WBZ
BT38 9EG MI0AWL
BT38 9GZ GI0PCU
BT38 9HE GI8KFG
BT38 9JD GI7NOW
BT38 9LF GI4IZF
BT38 9ND GI6GRV
BT38 9NN MI6RUX
BT38 9NT MI6LGX
BT38 9NZ GI4XHO
BT38 9RG MI6MZR
BT38 9RL GI1TFC
BT38 9SU MI0JPL
BT38 9XA GI7JEM
BT388HA MI6HWV
BT39 0BW GI0SRL
BT39 0BW GI3PDN
BT39 0FA GI4BWM
BT39 0HZ GI3ZVZ
BT39 0JP GI4KQA
BT39 0PH GI4SYZ
BT39 0PZ 2I0BAD
BT39 0QB GI4PRH
BT39 0QB GI6PAZ
BT39 0SE 2I0FMN
BT39 0SE GI4OSF
BT39 0SQ 2I0BAC
BT39 0SQ MI3IIH
BT39 0SQ MI3IIL
BT39 0TN GI3XDD
BT39 0TQ GI0LGV
BT39 9AJ 2I0SGK
BT39 9AJ MI7WJD
BT39 9DD MI3FXE
BT39 9FB MI7FFV
BT39 9FL 2I0GOK
BT39 9GN GI7FCM
BT39 9GW MI7ASC
BT39 9HE GI1CKU
BT39 9HT GI7GUT
BT39 9HZ GI4TAJ
BT39 9HZ GI4XGO
BT39 9HZ MI6ALL
BT39 9JS GI7DBZ
BT39 9NA MI0LGD
BT39 9NA MI6GDD
BT39 9PS GI7GVI
BT39 9QW GI4JYJ
BT39 9QW GI6UUC
BT39 9RZ 2I0LXW
BT39 9RZ MI3LXW
BT39 9RZ MI6LWI
BT39 9RZ MI6LZW
BT39 9RZ MI6XAM
BT39 9SB MI6PHQ
BT39 9SE 2I0KUN
BT39 9SE MI0HWE
BT39 9SE MI7MCS
BT39 9ST MI6TXS
BT39 9TJ 2I0INA
BT39 9TJ MI0IRX
BT39 9TJ MI6OIM
BT39 9TS GI8VKA
BT39 9WE GI4VKA
BT399AS MI6PFI
BT4 1LJ MI3IAI
BT4 1NA GI7PUG
BT4 1ND GI0RBC
BT4 1PR MI6SQN
BT4 1PR MI6WTZ
BT4 1QD 2I0JIJ
BT4 1QD MI0XOJ
BT4 1QD MI7JOJ
BT4 1QT GI7RAM
BT4 1QU GI4IOO
BT4 1RJ GI7HVC
BT4 2BH GI2BX
BT4 2BH GI4NKB
BT4 2BH GI6DEY
BT4 2BL GI4JLF
BT4 2BY GI3TNK
BT4 2DX GI4LGP
BT4 2EH GI4GOS
BT4 2HH GI7RAH
BT4 2HS GI4FZD
BT4 2HT GI6FEN
BT4 2JZ GI4ILZ
BT4 2PA GI4WRJ
BT4 2RB 2I0OHE
BT4 2RB MI3OHE
BT4 2RH MI6SEZ
BT4 3DE GI0VAB
BT4 3DJ GI6DRK
BT40 1BD GI1TRZ
BT40 1BD MI0CNI
BT40 1EB GI8WHP
BT40 1EN GI8NLF
BT40 1ET 2I0GFO
BT40 1ET MI6GFO
BT40 1EW MI3BKA
BT40 1NE 2I0LRN
BT40 1NE 2I0PBZ
BT40 1NE MI6LRN
BT40 1NE MI6PBZ
BT40 1QL GI6EWO
BT40 1SE 2I0RPM
BT40 1SE MI3TUZ
BT40 1SE MI6TZP
BT40 1UB GI6VCG
BT40 1UL GI7DZE
BT40 2DF GI4WMV
BT40 2EG GI4MVQ
BT40 2EJ MI6XBA
BT40 2ER MI3CIV
BT40 2ER MI3CIW
BT40 2JE MI3CIZ
BT40 2JH 2I0JFO
BT40 2JH MI6GHA
BT40 2JH MI6JFO
BT40 2JR MI0BOU
BT40 2PA GI8WFA
BT40 2PH GI4MAJ
BT40 2QA GI6DNI
BT40 2TU 2I0EPG
BT40 2TU MI6ICD
BT40 2WF 2I0TAA
BT40 2WF MI3VHW
BT40 3DH 2I0WNC
BT40 3DU MI3SRL
BT40 3FF GI7USA
BT40 3HL GI7MDK
BT40 3JG GI4UPC
BT40 3NF 2I0XDR
BT40 3NT MI3UMC
BT40 3SD GI1CET
BT40 3SD MI6WAG
BT40 3SJ MI0IZP
BT40 3SQ GI4RVT
BT40 3TT GI4MTZ
BT40 3TX GI0HWO
BT40 3TX GI7WCS
BT40 3TX GI8IZB
BT40 3TX MI3MJK
BT40 3UG GI0TWX
BT40 3UG GI0XAC
BT403LW MI7FHM
BT41 1AY MI3MRF
BT41 1HH GI3RNO
BT41 1HP MI0GFE
BT41 1LN 2I0RZT
BT41 1LN MI0GIJ
BT41 1LN MI3RZT
BT41 1SB 2I0HFM
BT41 1SB MI6FHG
BT41 1SB MI7NCZ
BT41 2DP MI3TUS
BT41 2DR GI0MQN
BT41 2EU GI6OQL
BT41 2EY GI1EOS
BT41 2HJ GI3YDM
BT41 2JD MI6CAD
BT41 2JH GI0NNK
BT41 2PN MI0DFH
BT41 2QT GI4PID
BT41 2RF MI6DUP
BT41 2TG MI6SPY
BT41 2TR MI1DPL
BT41 2TS GI7IPO
BT41 3BA MI6IBR
BT41 3BJ GI4VJZ
BT41 3DX GI8TWB
BT41 3FL GI0PML
BT41 3NH 2I0LPO
BT41 3NH MI0LPO
BT41 3NX GI0NQC
BT41 3NX GI0WVP
BT41 3RT GI4DCC
BT41 4EG MI6PML
BT41 4FG MI0HHU
BT41 4FG MI6BZC
BT41 4FN MI7RVJ
BT41 4HD GI4KUM
BT41 4HD MI3NOH
BT41 4JQ GI4IHY
BT41 4NG 2I0IDJ
BT41 4NG MI6IDJ
BT41 4NP GI0VLE
BT41 4NS MI6FNO
BT41 4PF MI7WTS
BT41 4QQ GI4DWZ
BT41 4SB GI4FUM
BT41 4SB GI4SIW
BT41 4SB GI8MIV
BT412JB MI7AOU
BT41-4HW MI7LAJ
BT42 1AX GI4SFZ
BT42 1AX MI3FHM
BT42 1DE GI3FFF
BT42 1DE GI4HCN
BT42 1GW GI1XIB
BT42 1JA GI0JEV
BT42 1JL MI3AVJ
BT42 1JN 2I0JOS
BT42 1JN MI6JOS
BT42 1JW MI0CUN
BT42 1LN MI7HMD
BT42 1PU GI6KBX
BT42 1RW 2I0EOS
BT42 1RW MI6WJK
BT42 1RX MI3PQM
BT42 1SB GI8MOV
BT42 2AU GI0AYG
BT42 2BJ GI4OYL
BT42 2DG GI4TOR
BT42 2DQ GI4OGQ
BT42 2LQ MI3PZV
BT42 2LR MI6AUP
BT42 2NL GI1XPV
BT42 2NL MI7KAT
BT42 2PX MI0WWF
BT42 2QH GI0THO
BT42 2QQ 2I0WDD
BT42 2QQ MI6WDD
BT42 2RG MI6KZS
BT42 2RG MI6LUP
BT42 2RJ MI1DEZ
BT42 2RP GI4KUZ
BT42 2SP MI7WGS
BT42 3BE MI6GZF
BT42 3DF MI3DSM
BT42 3DR MI7SHB
BT42 3GH 2I0RBV
BT42 3HS MI7FGI
BT42 3JE 2I0RGT
BT42 3LD GI0OHG
BT42 3LH MI6XGN
BT42 3LH MI6XOD
BT42 3ND MI6TSH
BT42 4AP 2I0GLC
BT42 4AP MI6GLC
BT42 4DH MI3LNC
BT42 4DH MI7GZO
BT42 4JG MI7VFR
BT42 4LB MI6UVS
BT42 4LN MI7JJE
BT42 4NJ MI3LRR
BT43 5ED MI7PUI
BT43 5HE GI4ESI
BT43 5NP GI0RBO
BT43 5PR GI3XDX
BT43 5PY GI0ITJ
BT43 6DT 2I1JMC
BT43 6DT MI0JML
BT43 6DT MI3JMC
BT43 6ET MI6PHH
BT43 6FD MI7RWT
BT43 6JG MI3OHP
BT43 6NF GI6OJC
BT43 6NN MI0GVC
BT43 6NN MI6VZW
BT43 6PB GI4NNM
BT43 6QE 2I0IPB
BT43 6QE MI3NUW
BT43 6SX MI0IOU
BT43 6SZ 2I0GQA
BT43 6SZ MI0REI
BT43 6SZ MI7NRW
BT43 6TA GI0LMR
BT43 6TD MI6ODN
BT43 6TL GI1KHF
BT43 7AP GI1GME
BT43 7BL MI7MMV
BT436NN MI7ACL
BT44 0AN GI7EBM
BT44 0NS GI4PCQ
BT44 0QZ GI1LGM
BT44 8AD GI4XFX
BT44 8AR GI6CXD
BT44 8EF GI1BSJ
BT44 8EF GI6VCL
BT44 8EF GI7UPQ
BT44 8HH GI4RXS
BT44 8JB 2I0RVT
BT44 8JB MI6RAD
BT44 8QH MI3DZD
BT44 9BN GI0WYB
BT44 9DL GI4SZU
BT44 9DT 2I0RHQ
BT44 9HZ MI0PJS
BT44 9JJ MI3XUC
BT44 9NA MI3SIL
BT44 9PE MI6FHJ
BT44 9QA MI0BBF
BT44 9RH MI3UHL
BT448NZ 2I0WAI
BT448NZ MI3UKW
BT45 5DD MI7DJT
BT45 5JF GI0EUG
BT45 5LY GI7CTW
BT45 5QA GI0SRP
BT45 5QG GI4NGP
BT45 5RP MI0UTY
BT45 5RZ MI6VOZ
BT45 6DN MI0GRN
BT45 6DN MI3XEY
BT45 6DS MI3UTY
BT45 6EX GI7MWA
BT45 6HW GI1BZT
BT45 6JU MI0GOZ
BT45 6LY MI3LQN
BT45 6ND MI3UFD
BT45 6NH GI4WNH
BT45 6PF MI0JPD
BT45 6PU GI0PJH
BT45 6PY GI4EQA
BT45 7DT 2I0CKB
BT45 7DT 2I0GQZ
BT45 7LY GI4BGB
BT45 7NU GI3TIJ
BT45 7QF GI4NFW
BT45 7TW GI7JKM
BT45 7XS 2I0WBF
BT45 7XS MI6WBT
BT45 8AL GI3RXV
BT45 8BZ GI4BDR
BT45 8HJ MI3FQD
BT45 8JA MI7DNG
BT45 8JE MI0LBA
BT45 8JE MI0TLF
BT45 8JE MI3EAQ
BT45 8NQ GI6OCC
BT45 8NQ MI3OYP
BT45 8PJ MI3OLM
BT457HT 2I0PND
BT46 5JR GI3ZTL
BT46 5NX MI3EYB
BT46 5SH 2I0ZCM
BT46 5SH MI3BCM
BT46 5TU MI3MIE
BT46 5UW MI7OCP
BT46 5UW MI7SSJ
BT47 2AF 2I0RGM
BT47 2AF MI0RYM
BT47 2AF MI6RGM
BT47 2BY GI7FJY
BT47 2ES GI8AFS
BT47 2HA MI3TXI
BT47 2HW GI4YWT
BT47 2HW MI0KQU
BT47 2HW MI3YWT
BT47 2LD MI0SDR
BT47 2LD MI3TPR
BT47 2LD MI6RCR
BT47 2NL GI3FTT
BT47 2PL 2I0HKW
BT47 2QR MI3RNN
BT47 2RD GI6AUI
BT47 2RD MI0AKU
BT47 2RD MI5TCC
BT47 2RY GI0AYB
BT47 2RY MI3NYB
BT47 2RY MI3XEB
BT47 2RY MI3XYB
BT47 2SL MI3LGL
BT47 3BF MI7THE
BT47 3DD 2I0MRY
BT47 3FH MI7MTY
BT47 3NY MI6FAU
BT47 3PR GI6BVQ
BT47 3PW GI0LDI
BT47 3PW MI0NWG
BT47 3PW MI3MFD
BT47 3QS MI5ROY
BT47 3SF MI3STY
BT47 3TE MI3CCN
BT47 3TE MI3GER
BT47 3TE MI3PPD
BT47 3TE MI6OPD
BT47 3TR MI3FGK
BT47 3YE 2I0TDL
BT47 3YE MI6HXB
BT47 4AB 2I0OJK
BT47 4AB MI6OJK
BT47 4AD MI3OZT
BT47 4AQ GI4AHD
BT47 4AQ GI8RPP
BT47 4AS MI3TFF
BT47 4FF MI3KKP
BT47 4GA 2I0VTZ
BT47 4HQ MI3EGJ
BT47 4JN MI3RXF
BT47 4LU MI7RTD
BT47 4PJ GI0EFW
BT47 4PN 2I0JIE
BT47 4PN GI0AZA
BT47 4PN GI0AZB

BT47 4PN	GI4MFT
BT47 4PN	GI5ALP
BT47 4PN	MI0OPC
BT47 4PN	MI6JHE
BT47 4PN	MI6JIE
BT47 4PP	2I0DMC
BT47 4PP	MI3RUV
BT47 4ST	MI6CEJ
BT47 4TJ	GI4HDJ
BT47 4TT	2I0EGN
BT47 5HE	GI0RJO
BT47 5JP	GI0AWK
BT47 5PS	MI3TJM
BT47 5QY	GI0IOT
BT47 5SZ	GI8ZDB
BT47 5WN	MI6HAF
BT47 5WN	MI6ZXT
BT47 5XS	GI4OWA
BT47 5YD	GI4FHB
BT47 6DU	MI0UNA
BT47 6DU	MI0WJM
BT47 6DU	MI1JOE
BT47 6DU	MI1UNA
BT47 6DU	MI3CEM
BT47 6DU	MI3SVM
BT47 6DU	MI3XRT
BT47 6EL	MI7NBR
BT47 6FE	GI0TSA
BT47 6FE	MI3TWM
BT47 6HY	MI6CGQ
BT47 6NF	MI6GNP
BT47 6SE	GI0WYO
BT47 6UG	GI0IUP
BT47 6UW	MI3IFO
BT47 6XA	MI3DBB
BT47 6XE	MI7WPX
BT47 6XL	MI3CXD
BT47 6YB	MI6SCC
BT47 6YB	MI7AAC
BT47 6YU	MI3WQC
BT48 0AD	GI4MJD
BT48 0AD	GI8YBU
BT48 0AD	MI0DVH
BT48 0AU	GI3TJJ
BT48 0BY	GI0NWN
BT48 0JX	MI3DKN
BT48 0QA	GI4ZLD
BT48 0RL	MI3CGA
BT48 0RS	GI0HDO
BT48 6SG	MI7FPG
BT48 7ER	GI0PXS
BT48 7JN	GI4DGI
BT48 7PN	GI4ZCH
BT48 7UA	MI3GJG
BT48 8AQ	GI0GGY
BT48 8BA	2I0HWW
BT48 8BA	MI0ITS
BT48 8BA	MI3PCF
BT48 8EF	GI0LRB
BT48 8GQ	GI3TME
BT48 8HN	MI3IYP
BT48 8HX	2I0NAT
BT48 8HX	MI3WFT
BT48 8HX	MI3WVL
BT48 8HX	MI6XBL
BT48 8JD	GI0SFT
BT48 8JD	GI7KVR
BT48 8JW	GI0TQD
BT48 8JW	GI1BEU
BT48 8JW	GI4EXI
BT48 8NT	GI7UPU
BT48 8PF	GI4EPK
BT48 8RT	MI6LLS
BT48 8TH	GI0DSG
BT48 8UJ	MI6BLW
BT48 9DU	GI0OZQ
BT48 9HL	MI6WBN
BT48 9LA	GI1PVE
BT487ES	MI7RXD
BT487LH	MI7EDO
BT49 0AE	GI4EBS
BT49 0AP	MI6WJW
BT49 0AP	MI6WPW
BT49 0AS	MI3VTJ
BT49 0AT	MI6MCK
BT49 0BB	MI3YXX
BT49 0BD	GI0RBS
BT49 0BF	2I0NWO
BT49 0BF	MI0NWO
BT49 0BF	MI3NWO
BT49 0BF	MI6EJK
BT49 0BF	MI6RME
BT49 0BH	MI0GKL
BT49 0BH	MI0JBT
BT49 0DQ	MI3JTM
BT49 0EQ	GI4OKU
BT49 0HR	MI7MQX
BT49 0HS	MI3XWW
BT49 0HS	MI3YYT
BT49 0NF	2I0KEW
BT49 0NF	MI3IHZ
BT49 0NT	GI4PMP
BT49 0NY	GI4MAZ
BT49 0NY	MI6OFZ
BT49 0QB	MI6JAD
BT49 0QF	GI0BZM
BT49 0QF	GI3KVD
BT49 0QF	MI3UIW
BT49 0QG	MI7RYC
BT49 0QG	MI7XKA
BT49 0RD	MI7WJL
BT49 0RG	2I0WBD
BT49 0RG	MI6WAB
BT49 0RH	MI6WKN
BT49 0RW	MI3MBM
BT49 0SF	GI3VAW
BT49 0SH	MI0DWD
BT49 0SH	MI3BIE
BT49 0UF	2I0BHT
BT49 0UF	MI0GGB
BT49 0UF	MI3IHY
BT49 9AD	GI4YDP
BT49 9BQ	GI7JRG
BT49 9BS	GI6VWS
BT49 9BS	MI3INS
BT49 9BS	MI3JDQ
BT49 9EN	MI3XGR
BT49 9EW	MI3RYJ
BT49 9HQ	GI4GNT
BT49 9HT	2I0BTT
BT49 9LR	GI4SIZ
BT49 9LY	GI4ODT
BT49 0AT	MI3UIV
BT5 4FE	2I0WSH
BT5 4FT	2I0HRM
BT5 4FT	MI6MIH
BT5 4HY	MI7CEK
BT5 4NT	GI7RCH
BT5 5HS	GI6BDI
BT5 5JP	MI7DXU
BT5 5LB	2I0HBO
BT5 5LB	MI0HZD
BT5 5LB	MI6EMI
BT5 5LT	2I0CGZ
BT5 5LT	MI6AOZ
BT5 5NT	GI8YTH
BT5 6AD	GI0UTV
BT5 6AG	GI7CMC
BT5 6AG	MI1BRA
BT5 6AL	GI7IMU
BT5 6BT	GI4FNU
BT5 6DD	GI7UQW
BT5 6ED	MI1DQB
BT5 6FN	GI7URC
BT5 6GE	MI6WEZ
BT5 6NG	GI0BRO
BT5 6NG	GI0UEG
BT5 6PU	GI0WCE
BT5 7DH	GI4CBG
BT5 7EH	MI6UBT
BT5 7EQ	GI3MMF
BT5 7EQ	GI6YM
BT5 7EY	GI0OUM
BT5 7EZ	GI4SSF
BT5 7GX	GI6HKE
BT5 7HL	2I0GPQ
BT5 7HL	GI8YYM
BT5 7HL	MI0RUR
BT5 7HL	MI6GPQ
BT5 7HW	GI4ZOS
BT5 7JB	GI7KMC
BT5 7JP	GI7MDJ
BT5 7LY	MI0ALS
BT5 7NR	GI0BDZ
BT5 7PS	GI4MYT
BT5 7RL	MI3SXR
BT5 7RR	MI7KIR
BT51 3GW	MI7RGN
BT51 3QZ	GI4OHH
BT51 3RA	MI0RTY
BT51 3RD	GI8LTB
BT51 3RE	GI4GPA
BT51 3RF	MI7WDT
BT51 3RW	MI6XRC
BT51 3RZ	GI4VIZ
BT51 3SN	MI6JMC
BT51 3TN	2I0SEK
BT51 3TN	MI0PLC
BT51 3TN	MI3SEK
BT51 4AR	MI1BNO
BT51 4BD	GI7LCQ
BT51 4DN	MI0PMX
BT51 4LZ	GI7SOB
BT51 4NB	2I0GCN
BT51 4NB	MI0ULK
BT51 4NB	MI6MFR
BT51 4NE	GI1LBI
BT51 4NG	MI3VXI
BT51 4RA	GI4ZAH
BT51 4SD	GI3PLL
BT51 4SZ	MI3CON
BT51 4TS	GI4HVI
BT51 4US	GI8OLH
BT51 5BJ	MI1CSA
BT51 5JP	MI0BYR
BT51 5LA	2I0LOP
BT51 5LA	MI0JQY
BT51 5LA	MI7MKD
BT51 5RZ	MI0HYQ
BT51 5RZ	MI0SLE
BT51 5SB	MI0DMT
BT51 5SB	MI0DNB
BT51 5SB	MI0MSM
BT51 5SB	MI6PNB
BT51 5TA	GI6JXG
BT51 5YR	GI6GBK
BT51 5YR	MI0AAZ
BT52 1EN	2I0TAN
BT52 1EN	MI0TBD
BT52 1EN	MI3ZYU
BT52 1EW	GI4JFP
BT52 1JR	GI7TEB
BT52 1JW	GI6IBL
BT52 1NG	GI1RXL
BT52 1RG	GI1JHQ
BT52 1SP	2I0BKI
BT52 1SP	MI0WGW
BT52 1SP	MI3UHI
BT52 1SP	MI7CSM
BT52 1TL	GI4WWF
BT52 1TU	MI3GFA
BT52 1TW	GI6EIR
BT52 1TY	MI3BRJ
BT52 1UJ	2I0PAC
BT52 1UJ	MI3UIA
BT52 1UJ	MI6PFD
BT52 1WN	MI3BYJ
BT52 2BL	2I0UDR
BT52 2BL	MI3VCI
BT52 2BL	MI6OGF
BT52 2DY	MI6BEF
BT52 2ES	MI0RJN
BT52 2EU	MI0HMY
BT52 2EX	MI3NWU
BT52 2HD	MI0HPX
BT52 2HW	MI0JZZ
BT52 2JE	MI7MFW
BT52 2JL	MI0WJC
BT52 2ND	2I0ETB
BT52 2ND	MI6ETE
BT52 2NY	GI7TTO
BT52 2NZ	MI7GCV
BT52 2QA	MI3CKF
BT52 2QE	2I0IMO
BT52 2QE	MI3IMO
BT52 2QF	MI6CWC
BT52 2QF	MI6OKS
BT521WN	MI0CIB
BT53 6BZ	MI0JAY
BT53 6EQ	MI0ENR
BT53 6ES	2I0BPO
BT53 6ES	MI0JAR
BT53 6ES	MI3FPE
BT53 6EX	MI0JPO
BT53 6JG	MI3MMC
BT53 6NT	2I0TJF
BT53 6NT	MI6TJF
BT53 6PT	MI6GAU
BT53 6QB	GI6GLE
BT53 6QF	GI3OYG
BT53 7AH	GI6GAQ
BT53 7AQ	MI3FOJ
BT53 7AS	2I0EAR
BT53 7AS	MI6HGV
BT53 7BB	MI3DCM
BT53 7BE	MI0MCC
BT53 7BE	MI6RYC
BT53 7BX	MI0VAC
BT53 7DE	MI0GHI
BT53 7QL	GI8DHW
BT53 7QS	MI0OBE
BT53 7RN	MI7HJW
BT53 8AB	MI6MSV
BT53 8DL	MI3LMR
BT53 8JT	GI4JRA
BT53 8NL	MI0JTE
BT53 8NL	MI3CQB
BT53 8NL	MI3TRR
BT53 8NY	MI6GGN
BT53 8PS	MI1WWG
BT53 8QQ	MI0MCB
BT53 8YH	2I1IKL
BT537QG	MI6LFK
BT54 6AN	MI3YJE
BT54 6BW	MI3BUT
BT54 6BW	MI3BYQ
BT54 6DQ	2I1SWD
BT54 6DQ	MI0CRQ
BT54 6DQ	MI1DOG
BT54 6DQ	MI3DMM
BT54 6DQ	MI3SWD
BT54 6DS	GI8WIU
BT54 6DS	MI7AJD
BT54 6DZ	2I0RMK
BT54 6DZ	2I0TJK
BT54 6DZ	MI0JBK
BT54 6DZ	MI0RMK
BT54 6DZ	MI3TJK
BT54 6HS	MI6WOF
BT54 6JH	MI3WES
BT54 6JL	GI0JQQ
BT54 6LE	MI3WDO
BT54 6LP	MI3POB
BT54 6PF	MI0MRV
BT54 6PF	MI3DDK
BT54 6PF	MI3TXT
BT54 6QZ	MI0CRR
BT54 6QZ	MI0MRG
BT54 6RT	MI1BLZ
BT546NX	MI7KMQ
BT55 7AU	2I0PHA
BT55 7DL	2I0ELR
BT55 7DL	MI3CDA
BT55 7DT	MI6KKN
BT55 7EG	GI7JLD
BT55 7HA	MI0KAM
BT55 7HS	2I0IOI
BT55 7HS	MI0OUT
BT55 7HS	MI3CAB
BT55 7SZ	MI7NYK
BT557PQ	MI6NHH
BT56 8AS	GI4OYM
BT56 8GP	GI4CZO
BT56 8HN	GI7TMQ
BT56 8JN	GI4OQY
BT56 8NJ	GI7SLN
BT56 8QA	GI8PGJ
BT56 8SP	GI0OTC
BT56 8ST	MI3DOD
BT57 8AE	MI0GBU
BT57 8AE	MI0RUC
BT57 8AE	MI3NRB
BT57 8RB	MI6TAW
BT57 8RY	MI6VBB
BT57 8SD	MI0LLG
BT57 8SD	MI6LLG
BT57 8UX	GI0ISQ
BT57 8YX	GI8AIR
BT6 0AD	MI3PUH
BT6 0BD	MI7TDY
BT6 0DG	GI4IBV
BT6 0DZ	GI7KEC
BT6 0ER	GI0DHW
BT6 0FN	MI0AAW
BT6 0NH	GI4FGH
BT6 0NR	GI3YDH
BT6 8BH	GI4RCK
BT6 8LN	MI6TUX
BT6 8NL	GI3LEG
BT6 9AW	2I0KUJ
BT6 9AW	MI0KUJ
BT6 9AW	MI6KUJ
BT6 9DS	2I0IZI
BT6 9DS	MI6AHO
BT6 9GJ	GI3UEX
BT6 9JF	GI4LZR
BT6 9NT	MI7LJR
BT6 9NU	MI3CID
BT6 9PJ	GI6NTP
BT6 9RB	MI5WJB
BT6 9RH	2I0OTC
BT6 9RH	MI0OTC
BT6 9RH	MI6GIN
BT6 9RP	GI0URN
BT6 9RP	MI0NOR
BT6 9RX	GI4GOV
BT6 9SE	GI3XEQ
BT6 9SF	GI4MDD
BT6 9SF	GI7JAM
BT6 9SR	MI3KDR
BT60 1DS	GI4XFE
BT60 1DS	MI7SSZ
BT60 1EX	MI7RPP
BT60 1HP	GI4RYP
BT60 1JU	MI1EQI
BT60 1LP	GI4SRQ
BT60 1LQ	2I0NYL
BT60 1LQ	GI3NYL
BT60 1LQ	MI3WLW
BT60 1LZ	MI0DBK
BT60 1NT	GI8OCR
BT60 1QR	GI6TBC
BT60 1TW	GI7WLA
BT60 1TW	MI6RHL
BT60 2BH	2I0FPT
BT60 2BN	GI0DWN
BT60 2DU	MI7MNB
BT60 2GP	MI3ZJN
BT60 2GP	MI6LOT
BT60 2JF	GI0MSI
BT60 2JQ	MI3RST
BT60 2NA	2I0GWA
BT60 2NA	GI0KOW
BT60 2UP	MI3WBL
BT60 3AA	GI0MSH
BT60 3JS	2I0CBV
BT60 3JS	GI0NWG
BT60 3JS	MI3YTH
BT60 3LJ	MI3LWU
BT60 3TS	GI0VHG
BT60 4AS	GI6NNP
BT60 4BL	GI0GPG
BT60 4BU	GI1VMF
BT60 4BU	MI3LFE
BT60 4DZ	2I0RHN
BT60 4DZ	MI6UDR
BT60 4NZ	2I0KPA
BT60 4NZ	MI0KPA
BT60 4NZ	MI0WHG
BT61 7JB	GI8RNG
BT61 7JD	GI0NYI
BT61 7PE	MI6DOD
BT61 7QU	GI0LTF
BT61 7SA	GI0MSG
BT61 7SA	GI7IEZ
BT61 8AH	GI4SOY
BT61 8BU	MI0CLP
BT61 8JD	GI6NFK
BT61 8ND	MI0GRG
BT61 8NP	GI0MTE
BT61 8NT	MI7UNI
BT61 8NX	MI6YSW
BT61 8PQ	MI1PDN
BT61 8QN	GI3WTG
BT61 8RW	GI4SYM
BT61 9BB	GI0MSK
BT61 9DT	MI6JVC
BT61 9HA	GI7DWF
BT61 9HA	MI3BSN
BT61 9HJ	GI8RLG
BT61 9JX	GI4FFL
BT61 9LD	GI0ADD
BT61 9LD	GI8RLE
BT61 9LT	GI7GRY
BT61 9RW	2I0NTH
BT61 9RW	MI6BZI
BT62 1EH	GI0KUH
BT62 1EW	MI3IRV
BT62 1EY	GI0RYU
BT62 1JN	GI7FCP
BT62 1JX	GI0AIQ
BT62 1RN	GI0UJG
BT62 2AR	MI6TFN
BT62 2BB	MI6GNF
BT62 2BB	MI6NVM
BT62 2BH	GI1RSR
BT62 2BL	MI0SAP
BT62 2BL	MI3SEO
BT62 2BN	MI7CFX
BT62 2DD	2I0PRM
BT62 2DD	MI0IZN
BT62 2DD	MI0PRM
BT62 2DD	MI3WWJ
BT62 2DD	MI6MHI
BT62 2DD	MI6MHK
BT62 2EJ	MI0IRZ
BT62 2HB	MI7JNJ
BT62 2NE	MI3VWH
BT62 2NF	MI6ECC
BT62 2NJ	GI3ONF
BT62 2NJ	GI4IVI
BT62 3BN	2I0EJT
BT62 3BN	MI6AJA
BT62 3EB	MI7KTY
BT62 3HH	MI3IFI
BT62 3HY	MI6CAV
BT62 3JB	2I0UCY
BT62 3JB	MI6LSY
BT62 3JF	MI6NOE
BT62 3QH	GI4SJQ
BT62 3QJ	MI7NSU
BT62 3QU	GI8TAX
BT62 3QX	GI0DQJ
BT62 3RN	GI4BQI
BT62 3TD	2I0SAI
BT62 3TD	2I0WAH
BT62 3TD	MI0NID
BT62 3TD	MI0SAI
BT62 3TD	MI3XUS
BT62 3TD	MI3YBI
BT62 3TF	2I0EBS
BT62 3TF	MI6EBS
BT62 4AB	MI7SML
BT62 4DH	MI6INB
BT62 4HL	MI0TRC
BT62 4HL	MI3DFR
BT62 4HP	GI4MDO
BT62 4HX	2I0NIO
BT62 4HX	MI6FWD
BT62 4JP	2I0RIR
BT62 4JP	MI6RIR
BT622BL	2I0SGM
BT63 5AA	MI7PSY
BT63 5AR	2I0JPP
BT63 5AR	GI8XSB
BT63 5AR	MI0UYD
BT63 5BF	MI0INL
BT63 5BF	MI6JVM
BT63 5DQ	GI1ACN
BT63 5ES	MI6HCP
BT63 5FY	GI1PXX
BT63 5HR	2I0TME
BT63 5LT	GI6FTM
BT63 5PF	GI0NOX
BT63 5PQ	MI6ILF
BT63 5QZ	MI6IMY
BT63 5RJ	MI3KFI
BT63 5RS	GI7TGJ
BT63 5RX	MI3ISC
BT63 5SW	GI4YCZ
BT63 5SZ	2I0MFJ
BT63 5SZ	MI6RCV
BT63 5TB	MI6ENR
BT63 5TL	GI6EEH
BT63 5UU	2I0TJR
BT63 5UU	MI3TJR
BT63 5WU	MI6UGE
BT63 5XT	MI3IRY
BT63 5YD	GI0PVG
BT63 5YD	GI6FHD
BT63 5YD	GI6NAQ
BT63 5YD	GI7GHC
BT63 5YH	GI4MXW
BT63 5YJ	GI8JOA
BT63 6AT	MI1AYL
BT63 6BH	GI3WWY
BT63 6FA	GI6MTL
BT63 6LY	2I0SFA
BT63 6LY	MI6PUX
BT63 6NX	MI6GOA
BT635EJ	MI7ISF
BT64 3AN	MI3PDN
BT65 4AB	GI6GAG
BT65 4AL	GI0STS
BT65 4AT	MI0NWA
BT65 4AT	MI6NCG
BT65 5AE	MI6TRF
BT65 5AQ	MI6RXH
BT65 5DH	GI0SZH
BT65 5JF	MI3LZA
BT66 6LD	GI0MHB
BT66 6PS	MI0TBN
BT66 6QW	GI6FQT
BT66 7AN	2I0TRM
BT66 7AN	MI0LRC
BT66 7AN	MI3RYD
BT66 7BD	2I0HEZ
BT66 7BD	MI6LZL
BT66 7EB	GI3POS
BT66 7HD	GI8GZM
BT66 7HH	GI4BDL
BT66 7LE	MI1FIS
BT66 7LP	2I0RGD
BT66 7LP	MI6BGD
BT66 7NE	GI4GEL
BT66 7NE	GI8LFY
BT66 7PD	MI3RBM
BT66 7PP	MI0BPB
BT66 7PQ	GI4YRP
BT66 7QY	GI0OHU
BT66 7RU	GI3WEL
BT66 7SW	MI7XAG
BT66 7SY	2I0OUI
BT66 7SY	MI6HGS
BT66 7TG	2I0GLY
BT66 7TG	MI6XEM
BT66 7TR	2I0WGM
BT66 7TR	MI0WGM
BT66 7TR	MI6WGM
BT66 7TY	MI0MEV
BT66 7UA	GI4ISR
BT66 7UT	2I0ROC
BT66 7UT	MI6DED
BT66 8DF	2I0SNG
BT66 8DF	MI0SNG
BT66 8EE	2I0FVX
BT66 8ER	MI7CTL
BT66 8JP	MI0EDF
BT66 8JZ	2I0HSL
BT66 8JZ	MI6HSL
BT66 8LE	GI3SCM
BT66 8LW	GI7OOM
BT66 8QT	MI3WQT
BT66 8RH	GI0OUZ
BT667LS	MI7NTE
BT67 0AU	2I0FNN
BT67 0AU	MI6FIK
BT67 0ED	MI0BMM
BT67 0FA	MI0SMV
BT67 0FQ	MI0GCV
BT67 0FQ	MI3OFX
BT67 0GB	GI4MWA
BT67 0GB	MI6AJN
BT67 0GB	MI6JOY
BT67 0GJ	2I0YDF
BT67 0GJ	MI0YAM
BT67 0GJ	MI3YDF
BT67 0HP	GI4AIO
BT67 0HW	MI6GBC
BT67 0JB	GI4MNF
BT67 0JZ	2I0FBY
BT67 0LB	2I0SJV
BT67 0LB	GI0OEH
BT67 0LB	MI6SJV
BT67 0LN	MI3UGI
BT67 0NL	GI1FSJ
BT67 0NZ	2I0YLT
BT67 0NZ	MI0YLT
BT67 0NZ	MI6PYL
BT67 0NZ	MI6YLT
BT67 0QT	GI0VGV
BT67 0RN	MI5HIL
BT67 0SS	2I0GSG
BT67 0SS	MI6GSG
BT67 0SZ	GI6IRL
BT67 0TU	2I0MOU
BT67 0TU	MI3GAM
BT67 0TU	MI5BUG
BT67 0UT	2I0DKQ
BT67 0UT	MI6JGK
BT67 0XP	2I0IGX
BT67 0XP	MI0LTX
BT67 0XP	MI7BIU
BT67 9BS	MI3PJM
BT67 9HN	MI5LYN
BT67 9JN	GI4ELQ
BT7 1BL	MI3BCR
BT7 2GF	GI8HUD
BT7 2GJ	MI0BOK
BT7 3AG	MI7DBR
BT7 3HA	GI6KDN
BT7 3NB	MI7BMZ
BT70 1SZ	GI3OZW
BT70 1TH	GI4XKI
BT70 1UH	GI0RDM
BT70 2EL	GI4VCZ
BT70 3BE	2I0JSQ
BT70 3BE	MI6JSQ
BT70 3BU	GI7FCW
BT70 3DY	GI0ZER
BT70 3DY	GI7TDA
BT70 3DY	MI6LCR
BT70 3JU	2I0YBH
BT70 3JU	MI0YBH
BT70 3JU	MI6YBH
BT70 3JY	GI4XAA
BT70 3LB	2I0GZX
BT70 3LB	MI0KBL
BT70 3LB	MI7GZX
BT70 3LU	GI4LDN
BT70 3PU	GI0STM
BT71 4AA	MI6FQY
BT71 4AG	GI7NEB
BT71 4AJ	GI7RTB
BT71 5BZ	GI0OJL
BT71 5DP	MI6OXM
BT71 5GA	MI0CAC
BT71 5JL	2I0SMY
BT71 5JL	MI0SMY
BT71 5JL	MI6SJD
BT71 5LY	GI0JHS
BT71 6DD	GI0URI
BT71 6DR	GI3NFM
BT71 6HX	MI6EQD
BT71 6PJ	GI4CSP
BT71 6PW	2I0EXP
BT71 6PW	MI6EXP
BT71 6QG	GI3NPP
BT71 6QN	GI3OQR
BT71 6SQ	MI6XHL
BT71 6TN	GI0IVJ
BT71 6TQ	2I1WBC
BT71 6TQ	MI7EGS
BT71 6UE	2I0VBH
BT71 6UE	MI6VBH
BT71 7AY	MI3IOH
BT71 7BD	2I0IIQ
BT71 7BD	MI7BZR
BT71 7BH	GI0KPF
BT71 7BN	MI6EQC
BT71 7ER	GI0KVQ
BT71 7ER	MI0AVI
BT71 7ET	MI6MYW
BT71 7HT	GI0VIB
BT71 7JN	GI8ITD
BT71 7NH	2I0BIQ
BT71 7NY	2I0WBU
BT71 7PY	MI0CXE
BT71 7QF	GI4TED
BT71 7QF	GI4XJD
BT71 7SQ	GI4SLQ
BT715DH	MI7TFZ
BT717TS	MI7GPT
BT74 4DY	2I0NIE
BT74 4DY	MI3NIE
BT74 4RB	MI3TMN
BT74 5ND	GI7NET
BT74 5NQ	GI4CZW
BT74 6AZ	MI6FEB
BT74 6BQ	MI3GHW
BT74 6EP	GI0BQX
BT74 6ET	MI6NTT
BT74 6EU	GI0BFD
BT74 6HD	MI1FHE
BT74 6JJ	MI6REM
BT74 6JP	2I0EAS
BT74 6JP	MI3YUT
BT74 7HQ	MI6HZN
BT74 7JN	GI4UHA
BT744JZ	GI7UIP
BT75 0LG	2I0EPN
BT75 0LG	MI0LLM
BT75 0LG	MI6LLI
BT76 0HJ	GI0EWE
BT78 1ES	MI6AUU
BT78 1QS	2I0CKN
BT78 1QS	MI3MWA
BT78 1QZ	GI4OHW
BT78 2DD	MI6BJZ
BT78 2DD	MI6BKD
BT78 2DL	GI0LXN
BT78 2EH	GI0KDH
BT78 2PN	GI7SZV
BT78 2PN	MI6UTV
BT78 2PP	GI7THY
BT78 3AW	2I0CFW
BT78 3AW	MI3DWQ
BT78 3HL	MI6WRT
BT78 3LR	GI0OVK
BT78 4HX	GI1CAI
BT78 4HX	MI6OMA
BT78 4JT	GI4KJC
BT78 4JZ	GI0HVJ
BT78 4LG	GI0TMS
BT78 5AL	MI1FCQ
BT78 5BB	MI3KRL
BT78 5HG	MI6PCJ
BT78 5JE	MI6OTP
BT79 0DW	2I0JXO
BT79 0DW	MI3JXO
BT79 0DZ	2I0OMA
BT79 0DZ	MI0UFT
BT79 0DZ	MI6MRJ
BT79 0EB	GI4RSI
BT79 0HF	GI3NVW
BT79 0HX	MI1CTQ
BT79 0PG	MI6ODG
BT79 0PH	GI0GQG
BT79 0PL	GI4UDI
BT79 0RZ	GI6EGJ
BT79 0WA	MI6BVN
BT79 0XS	GI3XCZ
BT79 7GL	2I0LDC
BT79 7GL	MI6BII
BT79 7GL	MI6ECV
BT79 7GL	MI6EJT
BT79 7JJ	GI4LIF
BT79 7JX	MI6PWR
BT79 7LA	GI7UCS
BT79 7LB	MI6EDK
BT79 7PQ	GI0EJT
BT79 7PQ	GI1RIB
BT79 7RA	2I0DJM
BT79 7RA	GI4SXV
BT79 7RA	MI3DJM
BT79 7SJ	GI7AUY
BT79 7SJ	GI7FHZ
BT79 7SJ	MI6TYR
BT79 7TJ	GI0EJU
BT79 7WJ	GI3YBZ
BT79 7WJ	GI8BNC
BT79 7XG	2I0MSO
BT79 7XG	MI0IFG
BT79 7XG	MI0MSO
BT79 9AY	MI3DLO
BT79 9NH	MI6SFK
BT79 9PH	MI1CCU
BT79 9PH	MI6CCU
BT79 9QA	GI4IWP
BT8 6HU	GI0BFA
BT8 6JX	MI5EEM
BT8 6LF	2I0JXA
BT8 6LF	MI3JXA
BT8 6RG	GI3YMT
BT8 6SB	GI4KIX
BT8 6WL	GI4JIC
BT8 7DB	GI4WME
BT8 7RJ	GI1VAZ
BT8 8BQ	GI4TAP
BT8 8HG	MI6NIM
BT8 8HQ	GI4TAV
BT8 8JZ	GI4TGR
BT8 8LE	MI6IMQ
BT8 8LX	GI8PDK
BT8 8NL	MI0BDX
BT8 8RL	2I0WJM
BT8 8RL	MI7MCC
BT8 8TG	GI4SIP
BT80 0JL	MI6CAK
BT80 8AS	MI0RJY
BT80 8BW	GI8XSY
BT80 8EZ	MI0LNL
BT80 8JH	MI0AHH
BT80 8LJ	GI7BET
BT80 8PL	GI3SOO
BT80 8PL	GI8YGG
BT80 8PW	MI0AHI
BT80 8PW	MI0BML
BT80 8PZ	2I0HGI
BT80 8QE	GI6JRY
BT80 8QJ	GI0JHR
BT80 8XW	GI8JRE
BT80 9LB	GI6WHZ
BT80 9RN	MI0AIH
BT808RS	GI6NDM
BT81 7DF	MI6DAY
BT81 7DR	MI3SLT
BT81 7NA	MI3NVX
BT81 7NW	MI3JQD
BT81 7QF	2I0NDJ
BT81 7QF	MI0NDK
BT81 7QY	MI0JAT
BT81 7RF	MI0HRG
BT81 7RF	MI1CCT
BT81 7TJ	MI3CCT
BT81 7TJ	MI3FCA
BT82 0AE	GI0VJE
BT82 0AE	MI3BXQ
BT82 0AE	MI3JVJ
BT82 0AE	MI3JVV
BT82 0AE	MI5ALJ
BT82 0AJ	MI3EZK
BT82 0BD	GI4DXK
BT82 0DN	MI3RXU
BT82 0DP	GI3CFH
BT82 0DP	GI4OUN
BT82 0DP	MI3OUN
BT82 0DW	MI0RIB
BT82 0DW	MI3ASH
BT82 0DW	MI3BEG
BT82 0DZ	MI3CGZ
BT82 0DZ	MI3FPB
BT82 0DZ	MI3IGO
BT82 0EB	MI7FHY
BT82 0GZ	GI0EWP
BT82 0HN	MI3NEN
BT82 0HN	MI3TLV
BT82 0LZ	2I0BSA
BT82 0LZ	2I0FLO
BT82 0LZ	MI3ATT
BT82 0PD	MI6EAF
BT82 0PD	MI6JMD
BT82 0PD	MI6VJR
BT82 0PP	GI4XGQ
BT82 0QQ	MI3RIL
BT82 0QQ	MI5RJS
BT82 8EL	GI7THH
BT82 8LD	GI4VQK
BT82 8LF	MI3NMG
BT82 8LS	MI0HOZ
BT82 8LS	MI6LJO
BT82 8NW	GI4VKS
BT82 8QQ	MI6XKE
BT82 9AJ	2I0VFO
BT82 9AJ	MI0GTM
BT82 9AJ	MI3OZK
BT82 9DT	2I0IMB
BT82 9DT	MI3KMB
BT82 9FZ	GI5JEB
BT82 9LA	MI1AWM
BT82 9PN	MI1DRP
BT82 9PN	MI3AGR
BT82 9QL	GI4DYE
BT82 9QS	MI3PDL
BT82 9RQ	MI6POH
BT82 9TL	MI6RAC
BT820DB	GI4OUP
BT9 5AA	GI3ZCK
BT9 5BG	MI7BTH
BT9 5EL	MI0EFM
BT9 5FQ	MI6IRE
BT9 5GZ	MI3WDI
BT9 5PG	GI1MJJ
BT9 5QL	GI4OMK
BT9 5QL	GI4VIP
BT9 5QP	GI4WVN
BT9 5QP	GI4XFS
BT9 6HA	GI3VYY
BT9 6JX	MI3GWQ
BT9 6LJ	GI8SKR
BT9 6NF	MI6GQI
BT9 6PJ	GI7JUH
BT9 6QR	GI3SG
BT9 6QX	2I0NGK
BT9 6QX	MI6TCO
BT9 6TJ	GI3PSQ
BT9 6UH	GI3VCI
BT9 7AX	GI4ISH
BT9 7JD	GI4DOM
BT92 0EQ	MI6AFI
BT92 0EQ	MI6AFM
BT92 0FS	MI1RAY
BT92 0FS	MI6PMM
BT92 0FS	MI6SMM
BT92 0FW	MI3LDO
BT92 0LL	2I0PPW
BT92 0LL	MI0PPW
BT92 0LL	MI6MCF
BT92 2DL	GI0LEC
BT92 2DL	GI6ZIR
BT92 2EW	2I0WGL
BT92 2EW	MI0WGL
BT92 2EW	MI6WGL
BT92 2GJ	MI7SYL
BT92 3BU	GI0CAH
BT92 7DD	2I0WMC
BT92 7DD	MI6BMC
BT92 7DD	MI6SFH
BT92 7DD	MI6UIM
BT92 7FN	MI6GDN
BT92 7GA	2I0XAN
BT92 7GA	MI6WAN
BT92 7GG	MI6WAM
BT92 7HP	MI6COM
BT92 7LJ	MI6WSP
BT92 7PP	MI6NCC
BT92 7PR	MI6WDM
BT92 7QY	GI7JGT
BT92 8AE	MI6CAY
BT927PT	2I0RVH
BT927PT	MI0RVH
BT927PT	MI6TNI
BT93 0FB	2I0JCH
BT93 0FB	GI7SBF
BT93 0FB	MI6EAS
BT93 0FQ	GI1OMD
BT93 1UH	GI4SZQ
BT93 3EZ	MI5IMB
BT93 3FT	GI0PTQ
BT93 5BR	MI0BHX
BT93 5FB	GI1INH
BT93 6FR	2I1GIH
BT93 7DN	GI6PYP
BT93 8DF	GI3WBR
BT94 1HQ	MI3JXG
BT94 1RN	MI5MTC
BT94 2AS	MI6YPY
BT94 2JA	GI1RBI
BT94 3DB	2I0GHY
BT94 3DB	MI0IIG
BT94 3DB	MI3GHY
BT94 3EH	GI8WJN
BT94 3EQ	2I0EKN
BT94 3FJ	2I0NTP
BT94 3FJ	MI0NPR
BT94 3FJ	MI6NTP
BT94 3PH	2I0HGF
BT94 3PH	MI7BFP
BT94 4EZ	2I0RDR
BT94 4EZ	MI6RDR
BT94 4QX	MI6OFN
BT94 5BU	2I0GFR
BT94 5BX	GI6RMO
BT94 5EW	2I0UBE
BT94 5EW	GI4VHO
BT94 5EW	MI0UBE
BT94 5EW	MI6UBE
BT94 5FB	GI6JPO
BT94 5FB	MI0RCF
BT94 5FL	MI0DWE
BT94 5HJ	GI6UFO
BT97HB	2I0IOE
CA1 1LX	G4RQW
CA1 2DL	M6RYK
CA1 2ND	M6LCK
CA1 2QE	M6ELH
CA1 2QJ	G3XWA
CA1 2SZ	G3WTO
CA1 2TA	M0SHD
CA1 2WF	M3OLQ
CA1 3AS	M3TUC
CA1 3BP	M0YDP
CA1 3HF	G4KVQ
CA1 3HT	G1COY
CA1 3JQ	M3ZYF
CA1 3QB	G1FBE
CA10 1AJ	G7UNZ
CA10 1AT	M3OKP
CA10 1BA	G1AEI
CA10 1BX	2E0AND
CA10 1BX	M7ELN
CA10 1DT	G4WKG
CA10 1PD	G0IYQ
CA10 1QF	G4JHV
CA10 1QR	G4EXD
CA10 1QX	G4GXO
CA10 1TA	G4XET
CA10 2AW	G4UWG
CA10 2LL	G0VRZ
CA10 3AL	G0TNF
CA10 3AL	M5FAB
CA10 3AL	M7CAA
CA10 3HD	G3WGV
CA10 3JE	G7VVX
CA10 3LT	M3ITH
CA10 3NB	M7BWE
CA10 3NL	M6MTD
CA10 3PF	G0VMP
CA10 3PF	G7DEG
CA10 3RX	M7BFV
CA10 3RZ	G8OZT
CA10 3TL	G6UPM
CA10 3TW	G0NAS
CA10 3UE	G3LRA
CA11 0AY	2E0FQI
CA11 0DT	G1XTD
CA11 0HN	M6NXP
CA11 0LS	M1XCG
CA11 0XB	G0AUG
CA11 0XB	G0AUH
CA11 7JY	G0MDV
CA11 7JY	G1FNP
CA11 7QT	M1EUM
CA11 7UW	G0TDM
CA11 7UW	G7GQL
CA11 7UW	G7GQM
CA11 8AL	G0MTD
CA11 8AX	G0OFR
CA11 8BJ	G3YEP
CA11 8EH	G4GLQ
CA11 8PN	M6LKE
CA11 8SD	G1AQV
CA11 8TU	M6NWY
CA11 8TW	G4GQZ
CA11 8UF	G4FUI
CA11 8UF	G8LBT
CA11 9HL	M7FAL
CA11 9HN	G1VKG
CA11 9NA	M5TNT
CA11 9PD	M6FWW
CA11 9PF	M3WBR
CA11 9SJ	M6CTO
CA11 9SS	G6XWD
CA11 9TR	G4YLI
CA12 4AP	G3ZED
CA12 4AZ	M0AOT
CA12 4HS	G0UQC
CA12 4NN	G4ERL
CA12 5LQ	M6LAL
CA12 5SA	G0END
CA13 0AR	2E1APW
CA13 0LA	M0GCU
CA13 0NN	M7TUN
CA13 0QP	M7NPH
CA13 0QR	G1GWE
CA13 0TJ	M0AYB
CA13 0TL	M6RTN
CA13 0UL	M6BUU
CA13 9BE	2E0IKM
CA13 9BE	M6IKM
CA13 9EQ	G7OHM
CA13 9JD	2E0FLT
CA13 9JS	M7AQM
CA13 9UQ	G0JSO
CA13 9UQ	M0WLA
CA14 1AA	G3WCM
CA14 1DT	2E0OMT
CA14 1DT	M3OMT
CA14 1DW	M0XAT
CA14 1EE	G1OAE
CA14 1HE	G3RRI
CA14 1HS	2E0KKT
CA14 1HS	M7PGH
CA14 1LL	G0HCE
CA14 1LP	G3WJH
CA14 1PY	G0EMM
CA14 1PY	M6RMD
CA14 1PY	M6XSF

CA14 1XJ 2E0MAX
CA14 1XJ G4ZFX
CA14 1XJ M0WCR
CA14 2PT 2E0IFF
CA14 2PT G8RZ
CA14 2PT M3IFF
CA14 2QY 2E0DNW
CA14 2QY M6CVP
CA14 2TP G6CSL
CA14 2UG G0OMB
CA14 2UW M7KLB
CA14 2XH 2E0LLC
CA14 2XH M6KAM
CA14 2XH M6MGD
CA14 2XZ M3HWP
CA14 3EH 2E0BEG
CA14 3EH M0XMD
CA14 3EH M3LCI
CA14 3EY G7NMB
CA14 3HR G6TAK
CA14 3JF M6RNF
CA14 3PE M1RIC
CA14 4HY G1LEH
CA14 4JA M3LDL
CA14 4NU G1GTA
CA14 4NX G4CJP
CA14 4PR M6MIG
CA14 4PU G6PTF
CA14 5AB M7BUY
CA14 5AF G1LZL
CA14 5BQ M6LIV
CA14 5HP G0DPE
CA14 5JA G1AQI
CA14 5LB 2E0YHI
CA14 5LZ M3SDN
CA14 5PL M7EEU
CA14 5QR 2E0NWT
CA14 5RE G7KLR
CA14 5UJ G1FYE
CA15 6BQ G0NNN
CA15 6BQ G4HLZ
CA15 6HR M3XVK
CA15 6PB M6GXW
CA15 6SH M6MSP
CA15 7BU G0TII
CA15 7BU M0BCV
CA15 7DB M3UVA
CA15 7DX 2E0GSR
CA15 7DX M6EZA
CA15 7EG G0MWE
CA15 7ER M1EBC
CA15 7ER M1EBD
CA15 7HD M7XUP
CA15 7JU M3SDK
CA15 7JW M1CIE
CA15 7QJ 2E1BMB
CA15 7QN G0ORM
CA15 7RG M0GDH
CA15 7SS G0MTQ
CA15 7SS G7DIS
CA15 8RX 2E0ENP
CA15 8RX G2KS
CA15 8RX M6ENP
CA15 8ST M0VLA
CA157LW 2E0IKX
CA157LW M0KNQ
CA16 6BD G0KDB
CA16 6BD G0VGJ
CA16 6DA G7ITT
CA16 6JS G1ZBW
CA16 6JZ 2E0AOK
CA16 6JZ 2E1EIO
CA16 6JZ G4XTA
CA16 6PX G0ANT
CA16 6PX G8TXJ
CA16 6PX M5TXJ
CA16 6QE G0OPM
CA16 6RB M6WJM
CA16 6RD G0NYQ
CA16 6RS G4PZN
CA16 6SP M0JKQ
CA16 6SP M3ZOP
CA16 6SP M6JKQ
CA16 6TA G7THI
CA16 6TU G7NRG
CA16 6XU G1AOZ
CA17 4AU G4TUA
CA17 4BZ G8OCA
CA17 4EX G7BNB
CA17 4HL M7WAM
CA17 4HR G4SPR
CA174SN 2E0TIN
CA174SN M0XXL
CA174SN M3ZFW
CA18 1RG 2E0GTM
CA18 1RG M6CWA
CA18 1SL 2E0DAL
CA18 1SL M3ZCX
CA18 1SW G1KGU
CA19 1UD M1PAF
CA19 1UL 2E0NEN
CA19 1XH G3PNU
CA2 4BJ M7WMU
CA2 4EP M7AVT
CA2 4EX 2E0BKV
CA2 4EX M3OZI
CA2 4EX M7OGC
CA2 4LD M1LSD
CA2 4PZ G0NUD
CA2 5DZ M0AOH
CA2 5UH 2E0TMS
CA2 5UH M6TIH
CA2 6AG M0ADG
CA2 6DD G4HZP
CA2 6ER 2E0UKY
CA2 6ER M0XOD
CA2 6ER M7YAM
CA2 6JJ G8CAU
CA2 6LZ G4FQM
CA2 6LZ G8KIZ
CA2 6RN M7KHS
CA2 6SQ 2E0TLF
CA2 7DA M6COB
CA2 7DW M0PTT
CA2 7LD M3VZQ
CA2 7LU G4ARS
CA2 7LU G6LSO
CA2 7LU M0SOL
CA2 7PN M6YJD
CA2 7RD G0JXO
CA2 7TB M3IRH
CA2 7XD G8GUH
CA2 7XY M3BSF
CA20 1AY 2E0EHA
CA20 1AY G0SVK
CA20 1AY M6HXV
CA20 1DN M6PHF
CA20 1EF M3IUX
CA20 1LB M3MLI
CA20 1PZ G1HZJ
CA21 2YY G1MPG
CA22 2AT 2E0CBB
CA22 2DN M0KHZ
CA22 2DT M0HCI
CA22 2ED M7MVD
CA22 2EH G4VFL
CA22 2NA G1GWF
CA22 2NA G3BJD
CA22 2NN 2E0OZM
CA22 2PW 2E0XUH
CA22 2PW M0XUH
CA22 2PW M3XUH
CA22 2QQ G1HDX
CA22 2RL M0EJB
CA22 2SG M5GVY
CA22 2TP 2E0DNX
CA222QT M7NNY
CA23 3DY G1LMW
CA23 3DY M0LMS
CA24 3JL G3YSD
CA25 5EU G4RVH
CA25 5LF M7JKD
CA25 5LR M0ALR
CA26 3AT 2E0WEO
CA26 3AT M0LVL
CA26 3AT M6WEO
CA26 3QJ M6RMS
CA26 3QS G0RZI
CA26 3RU M0ELC
CA26 3SA 2E0XSD
CA26 3SA M0XSD
CA26 3SA M6EPW
CA26 3SA M6XSD
CA26 3TA G4CIC
CA26 3XN M1DZR
CA27 0AN G4DSI
CA27 0AR G7KSE
CA27 0AR M0LCW
CA27 0AS G0ESI
CA27 0EG G3WIN
CA27 0EG G4PEW
CA28 6AQ M3USF
CA28 6EF M0SHP
CA28 6JZ G6WLP
CA28 6LS 2E0XGO
CA28 6PE M6FYM
CA28 6RR 2E0YSK
CA28 6RR M6YSK
CA28 6SG M0FWM
CA28 6SW G7RKV
CA28 6SX G8IMM
CA28 6XL 2E0MIX
CA28 6XL M6DEG
CA28 7AW G7SGO
CA28 7XG G4VIA
CA28 8EP G0LJB
CA28 8HS G0HYP
CA28 8JP G7MRL
CA28 8JP M0BEE
CA28 8JP M0CRM
CA28 8JP M7CLE
CA28 8PH M7RGG
CA28 8PX M3YRR
CA28 8PZ G0DIG
CA28 8XJ G4RHC
CA28 8YG M0SKY
CA28 9AH G4FBC
CA28 9LE G4YUZ
CA28 9PB M3AQP
CA28 9PD G4MTR
CA28 9PD G8ORO
CA28 9TG G0CZU
CA28 9UG 2E0BEE
CA3 0AW G0KEE
CA3 0AY M6KTT
CA3 0DB M1ETM
CA3 0LF G0TFC
CA3 0NT M6MOS
CA3 0NU G1FVA
CA3 9RE M1ERP
CA3 9RE M3MIR
CA3 9SN G0DHI
CA30LR 2E0FDW
CA30LR M6LPH
CA4 0JF G7SZZ
CA4 0RP M3BWT
CA4 8BD G8UTQ
CA4 8ER M3OLF
CA4 8FG G7DXB
CA4 8LD G4BUI
CA4 8PL G4ISS
CA4 9PX G7JKD
CA4 9QR G4IIB
CA4 9QY M3FZV
CA4 9QY M3ZTR
CA4 9RR 2E0TXG
CA4 9RR M3TXA
CA4 9TP M3WGV
CA5 6BE G1OFY
CA5 6EN G6PPD
CA5 6NS G0NVA
CA5 6NS G7EKM
CA5 6NS M3GMG
CA5 6PB M3OLN
CA5 6QW G4SCE
CA5 7AX 2E0MXP
CA5 7AX M6DLP
CA5 7DP G8KAP
CA5 7NS M3DLJ
CA5 7NW G7IVG
CA5 7NW M3VYT
CA5 7QH G8GJW
CA57AX M6RLP
CA6 4EA G0WFH
CA6 4LY G6WAM
CA6 4NN G8ITX
CA6 4PZ G0OYI
CA6 5SS G4LIL
CA6 5TH M6NTB
CA6 5TS M1JWR
CA6 5UJ M1JON
CA6 5UJ M3CFM
CA6 5UJ M3JDG
CA6 5UJ M3MDK
CA6 5XX M3FBJ
CA6 5XX M3HFX
CA6 6PA M3WLD
CA64QG 2E0IBT
CA64QG M3IBT
CA7 0BZ G3KEL
CA7 0DF 2E0COB
CA7 0DF M3OSC
CA7 0DF M3OSF
CA7 1JF G4NEH
CA7 1JF G8DVW
CA7 1JG M1ACO
CA7 2JF G1GDB
CA7 3AH M0IHX
CA7 3DJ M0BOA
CA7 3DJ M0DXT
CA7 3QX G0ZMH
CA7 3QX M0AMX
CA7 3RJ G7ATW
CA7 4BF M7CLX
CA7 4EU M6KWI
CA7 4JJ G0TNL
CA7 4QH G7KWQ
CA7 5AN G0TRW
CA7 5AS M0PMR
CA7 5AS M3AQN
CA7 8AU G0HNQ
CA7 8JS M7FMR
CA7 8PF 2E0BKP
CA7 8PF 2E1BKP
CA7 8PF G0ORO
CA7 9DR G8DNH
CA7 9HQ M3JMY
CA7 9JL 2E0PGR
CA7 9JL M0PEG
CA7 9JW M6HWW
CA7 9QW M3TEE
CA7 9RE M3UIK
CA8 1EN G3VUS
CA8 1EX G4LCE
CA8 1LE G6KXD
CA8 1QJ M7NEZ
CA8 1RZ M3TFM
CA8 1SR G0JVU
CA8 1SR G7ANH
CA8 1TT 2E0ACE
CA8 1TY M3XUO
CA8 2BB G4IIY
CA8 2NJ G3OHL
CA8 7AL G4RHC
CA8 9AA G8IBK
CA8 9DH G4LQM
CA8 9HF M6BAX
CA8 9JA G7NZR
CA81BN M6BJW
CA9 3DY G4ZXF
CA9 3LH G1RLT
CA9 3LZ G0NXS
CB1 1DP 2E0BPE
CB1 1DP M0GJU
CB1 1DP M3XQZ
CB1 1LE M0DVK
CB1 2DW 2E0BOV
CB1 2EW 2E0XZW
CB1 2EW M0XZW
CB1 2FJ M1DUO
CB1 2LG G6TIU
CB1 2LL 2E1DZM
CB1 2LP G3ROP
CB1 2PU 2E0IDO
CB1 2PU M6IDO
CB1 2QF M6OSF
CB1 3AJ M3KWR
CB1 3EF G7PSF
CB1 3JQ M6FUH
CB1 3LR M0GXM
CB1 3LW 2E0EFW
CB1 3PN M6VWA
CB1 3PT 2E0CDV
CB1 3PT 2E0OIK
CB1 3PT 2E0PUZ
CB1 3PT M0TFB
CB1 3PT M6PUZ
CB1 3PT M7PUZ
CB1 3PW G0IJZ
CB1 3QL M0GLT
CB1 3RS M6EIS
CB1 3RY G4EZN
CB1 3RY G6UW
CB1 3TB M0NMM
CB1 3UB G0SFA
CB1 3UB G1OKI
CB1 3UF G1KUN
CB1 6HS G0DEU
CB1 7TY 2E0FLQ
CB1 7TY M0JEI
CB1 7TY M7AFO
CB1 7UP M6GXC
CB1 7UR G0GVZ
CB1 7UR G3INR
CB1 8PF M7AFR
CB1 8QR G6ZEY
CB1 8RG G8OUS
CB1 8RX G0XAS
CB1 8SG G4EAG
CB1 8SH G4GLI
CB1 9AJ M7MUT
CB1 9AL M3UBS
CB1 9AZ G6EUI
CB1 9GF 2E0CFG
CB1 9GF M6AOT
CB1 9GH 2E0BLI
CB1 9GH M0SKI
CB1 9GH M3XSK
CB1 9JA G0PNQ
CB1 9LN G3UPJ
CB1 9LZ M7TBZ
CB1 9YL G7VCE
CB10 1LN G0TCI
CB10 1NR M3IAQ
CB10 1QD M0ECC
CB10 1QY G8ULM
CB10 1QY M7CGO
CB10 2AH G1WSF
CB10 2AJ G4YHN
CB10 2AP G4UHR
CB10 2EB 2E0MYG
CB10 2EB G6YBH
CB10 2EB M0MYG
CB10 2EB M6MYG
CB10 2HG G8SZR
CB10 2RW G4FBY
CB11 3BJ G8XLB
CB11 3DN G0GWM
CB11 3ED G0ICB
CB11 3ER G0JML
CB11 3HD G0IMG
CB11 3HU 2E0PKM
CB11 3PU G4FON
CB11 3RU G3WMS
CB11 3SL G7PNP
CB11 3SP G4HSN
CB11 4AU M1EMO
CB11 4BS M1CKQ
CB11 4DE G0CLJ
CB11 4DG 2E0GUA
CB11 4DG M1JDH
CB11 4DG M6GUA
CB11 4DJ 2E1HLH
CB11 4DJ G0HEM
CB11 4DJ G1LTL
CB11 4DQ M7INH
CB11 4DW 2E0NOZ
CB11 4DW M0NOZ
CB11 4DW M3TPN
CB11 4JB M6GUB
CB11 4JU G3YZO
CB11 4LX G0OJR
CB11 4PX 2E0GGA
CB11 4PX 2E0MXA
CB11 4PX 2E0VVA
CB11 4PX M0ONL
CB11 4PX M0VVA
CB11 4PX M6OMX
CB11 4PX M6UGA
CB11 4PX M6VVA
CB11 4RX M7CKY
CB11 4SJ G3TXC
CB11 4SJ G6FEM
CB11 4TH G8EBX
CB11 4UU G3NOX
CB11 4XG G3XHZ
CB2 0QZ G7PLE
CB2 1RL 2E0ION
CB2 1RL M0POE
CB2 1SS M0XUB
CB2 1ST M6BXU
CB2 1TA M3HZC
CB2 1TJ M0FFX
CB2 1TP M0ERY
CB2 1TQ 2E0TRY
CB2 1TQ M0WCK
CB2 1TQ M6HXC
CB2 3BU 2E0FJG
CB2 3BU M0JDY
CB2 3BU M0KMN
CB2 3BU M3VUS
CB2 3HU M0RSW
CB2 3JU M0TMN
CB2 3JU M3TNM
CB2 6SP M5AKW
CB2 9AQ M3VFC
CB2 9BG M0IPQ
CB2 9HR G3NHB
CB2 9HS G1YPU
CB2 9JR 2E0BUV
CB2 9JR M6RLN
CB2 9JW 2E0CFL
CB2 9NJ 2E0BCK
CB2 9NJ M0EAQ
CB21 4AD G4PDI
CB21 4EG G6BOX
CB21 4EG M0BOX
CB21 4EU M0RSP
CB21 4HS G7KYI
CB21 4JQ 2E1HLF
CB21 4JQ G7SRK
CB21 4JT G1VCU
CB21 4LY G1CQA
CB21 4RY G0CHC
CB21 4SN M6IYG
CB21 4XB G7RFY
CB21 5BG M6FQO
CB21 5BT G4YJK
CB21 5DT G8RYL
CB21 5EY G4VTU
CB21 5JY G4HJW
CB21 5LD G3UDI
CB21 5LR G3JQS
CB21 5LR G4FZM
CB21 5LU M0RBU
CB21 5LY G4FTW
CB21 5PE G3ZUK
CB21 5PE M0IJJ
CB21 5PE M6IXF
CB21 6BU G4MZN
CB215PE M7OSM
CB216XR M7CPJ
CB22 3BN G1GFC
CB22 3BT 2E0WWA
CB22 3BT M6ZBQ
CB22 3DF M6RNV
CB22 3DH G4WEZ
CB22 3DU G7KRT
CB22 3DU G8TMV
CB22 3DW M1AQP
CB22 3DW M1CKO
CB22 3EA G8LVU
CB22 3FE M7TDK
CB22 3HJ G6PUO
CB22 3LP M1DBK
CB22 3LP M3NQS
CB22 3SP G6ASH
CB22 3SZ 2E0XOR
CB22 3SZ M0XOR
CB22 3SZ M6XOR
CB22 3SZ M7CMD
CB22 3SZ M7ROX
CB22 3UR M6BSL
CB22 4LX G6BBN
CB22 4NE G4LCE
CB22 4NG G0PGQ
CB22 4QN 2E0BKQ
CB22 4QN M0ZDG
CB22 4QN M3UYC
CB22 4RR M0ZZE
CB22 4RT G6WJX
CB22 5AE G8MEI
CB22 5AW G0ETP
CB22 5BP M1KTA
CB22 5BP M3KTA
CB22 5BX G3WLD
CB22 5BX M0LAT
CB22 5EG G3UUT
CB22 5EG G5PI
CB22 5JJ G6WKQ
CB22 5JW G3TVM
CB22 5JZ M0GCX
CB22 5LG G1IMM
CB22 5LW M6FCG
CB22 6RT G6UFL
CB22 6RY M6LBW
CB22 6SP M0DCV
CB22 7QF M1HJE
CB22 7RP M3IQN
CB22 7RZ G8SHC
CB223HZ G3WFF
CB225JF G7UUT
CB226RT M7FYL
CB23 1LS G6KWA
CB23 2NG G6FSU
CB23 2NL G4VHK
CB23 2RE G4VYG
CB23 2RE M3CMX
CB23 2RY G7SPZ
CB23 2SJ G7ASZ
CB23 2SW M7BZZ
CB23 2TH G4IPM
CB23 2TU G0BOM
CB23 3PP 2E0DSW
CB23 3PP M0MRL
CB23 3PP M6DSW
CB23 3RX M7BSZ
CB23 3XP 2E0GGW
CB23 3XP M0HHB
CB23 4HY M6RGD
CB23 4JQ G4UGQ
CB23 4JQ G4VWS
CB23 5ED G0ULA
CB23 5EF M1AQR
CB23 5FN M6CWJ
CB23 5HY G0RQN
CB23 6FN M3WCX
CB23 6FP G0REU
CB23 6GU M0LCM
CB23 6HW 2E0VUL
CB23 6HW M7HBA
CB23 7AE G8MEX
CB23 7AZ G8GEV
CB23 7BU G4DGL
CB23 7DY G3PJT
CB23 7PG M0YCS
CB23 7PH 2E0VBK
CB23 7PH M6RBK
CB23 7PS G3RUH
CB23 7PW G3XZP
CB23 7QD 2E0NHB
CB23 7QD M0HZR
CB23 7QD M6NBS
CB23 7QL G4JZQ
CB23 7RY G8EYP
CB23 7UZ G7GCW
CB23 7XA 2E0BKD
CB23 7XA M3FHO
CB23 7XJ G4NBS
CB23 7XJ G4NXO
CB23 7XL G3YMW
CB23 7XR G6HFS
CB23 7XR G6NIW
CB23 8BT G4AKD
CB23 8BW 2E0JJA
CB23 8BW 2E0JOA
CB23 8BW M6CVB
CB23 8BX G7BIX
CB23 8EP 2E1AEJ
CB23 8EP G0NJJ
CB23 8EQ G3LOD
CB23 8HH G6KIB
CB23 8SA M0JMQ
CB23 8SF G0KRB
CB23 8SJ G8RPI
CB23 8SU G6VTN
CB23 8SZ G6TUS
CB23 8TA M6RSD
CB23 8TD G6HSW
CB23 8TQ M6TAG
CB23 8TR G1ZCC
CB234LY M7CEP
CB236AT 2E0GTJ
CB236AT M0SVP
CB236AT M7GMR
CB24 1AG M7EXT
CB24 1BS M6HHL
CB24 3AY G8NPR
CB24 3EH G7LBL
CB24 3GN M7PMY
CB24 3XA 2E0RJZ
CB24 3XA M6RJZ
CB24 4AG 2E0OTY
CB24 4AG M6ODY
CB24 4QU M7MJB
CB24 4RT G7CPQ
CB24 4RT G8JDQ
CB24 4SF G7NBL
CB24 4SR G3XIP
CB24 4TR 2E0BPX
CB24 4TR M0GUM
CB24 4TR M3VGZ
CB24 5HT 2E0DWV
CB24 5HT M0IMZ
CB24 5HT M6FVZ
CB24 5HY G1JZN
CB24 5HY G1XAA
CB24 5JU G4NEY
CB24 5LS G6XRY
CB24 5NB G0DAH
CB24 5NJ G8GJU
CB24 5PZ G4BHJ
CB24 5UR G1JKF
CB24 5UU G8FTE
CB24 5UX G8IWQ
CB24 6AD G7DJA
CB24 6BS G0OIR
CB24 6DA G1NWM
CB24 6DG G8CRB
CB24 6DT G4LNY
CB24 6EB G0WZD
CB24 6ET M0HKW
CB24 8AW G0MBD
CB24 8QG G1KIJ
CB24 8SA G7WFV
CB24 8SD G0EUZ
CB24 8TB M0BLF
CB24 8TG G1DEQ
CB24 8TG G3XXX
CB24 8TR M0VFC
CB24 8TS M1EMG
CB24 8TX G3XPW
CB24 8UF 2E0RLV
CB24 8UF M0RLV
CB24 8UF M6RLV
CB24 8UJ M6ESO
CB24 8UL G2CR
CB24 8UL M0VMC
CB24 9HT M7JHE
CB24 9JL G8XPQ
CB24 9JU G0THI
CB24 9PE G7KDG
CB24 9PP G8HSS
CB24 9PP M0AQF
CB25 0BA G4PZW
CB25 0BB M3MKZ
CB25 0DP M7ING
CB25 0DT G8NEO
CB25 0DT M0CNP
CB25 0DW M3HWW
CB25 0EH M7RTS
CB25 0EL G8DQF
CB25 0ER M0DRK
CB25 0ES G7SCO
CB25 0ES M1BIB
CB25 0HD G3TFX
CB25 0HL G3XBM
CB25 0HT G6ALB
CB25 0HT G8BXA
CB25 0JU G4FVA
CB25 0JX G0SBK
CB25 9BQ G3PTQ
CB25 9BS G6KSV
CB25 9DW 2E0HQW
CB25 9FD M7SNB
CB25 9HB G7BNM
CB25 9HW G4VME
CB25 9HW M7JHW
CB25 9JW 2E0SDY
CB25 9JW M0SND
CB25 9NB G8XLE
CB25 9NE G8ZRD
CB25 9NJ 2E0IWL
CB25 9NJ M7EVZ
CB25 9PX G4BAO
CB3 0BN 2E0TJG
CB3 0BN M0HUH
CB3 0BN M0JSH
CB3 0DG M0HFQ
CB3 0DG M6BHK
CB3 0DS 2E0IXU
CB3 0DS 2E0NLS
CB3 0DS M0XOS
CB3 0DS M7NLS
CB3 0GP G0VDE
CB3 0HX G4XIL
CB3 0LT M1MAJ
CB3 0LT M3ZCB
CB3 0NP G1DAZ
CB3 0NS 2E1FLN
CB3 0PE M0HOT
CB3 0PP G0ECL
CB3 6GA G3HZP
CB3 9BB 2E0YIP
CB3 9BB M0LAH
CB3 9BB M6OAL
CB3 9DY G8KDO
CB3 9LQ G3MRX
CB3 9LX M7FVG
CB4 1DG G3VGX
CB4 1ET G6AIG
CB4 1FD G0RDB
CB4 1LW G4DNG
CB4 1NP G6FKS
CB4 1NX M0ZCJ
CB4 1QG M6GVY
CB4 1SQ G3ZAY
CB4 1SX M6GKW
CB4 1TT G3RQS
CB4 1TZ G1DVD
CB4 1TZ G8CTX
CB4 1XG G3URX
CB4 1XQ M7CHF
CB4 1YZ G8NNA
CB4 2AJ G3OWB
CB4 2AS G7AZJ
CB4 2AU G1TPV
CB4 2BJ G8UDV
CB4 2DB G3RSE
CB4 2DB G8UDG
CB4 2HH M0HOJ
CB4 2HT G1VJG
CB4 2LD G7KHL
CB4 2NS M6EBG
CB4 2QR M5CHH
CB4 2RA G7DWU
CB4 2SH G7NZY
CB4 2UF G4NUA
CB4 2UH G8XYS
CB4 2UN G1AQX
CB4 3BW G4OSX
CB4 3ND G3HRE
CB4 3PA G4AJW
CB4 3PA G8KNN
CB5 0JF G1YAS
CB5 8AR M0RZO
CB5 8AR M7RZO
CB5 8LB 2E0JGS
CB5 8LB M0JGS
CB5 8LB M0WBJ
CB5 8LB M6JGS
CB5 8NA M6JUT
CB5 8NA M7WUT
CB5 8QE M0RHC
CB5 8QE M7RHH
CB5 8QU 2E0DVY
CB5 8QU M6FVM
CB5 8QU M6GQF
CB5 8SE G7SWW
CB5 8SS 2E0IHI
CB5 8SS M0RTW
CB5 8SS M6FHI
CB5 8SS M6ZZA
CB5 8TE M6RMA
CB5 8UB M6AWR
CB5 8UJ 2E0IFN
CB5 8UJ M0KGE
CB5 8UJ M7GIS
CB5 8US M6CFX
CB5 8XJ 2E0FPQ
CB5 8XJ M6FPQ
CB58DL M1CWA
CB6 1AB G0KFQ
CB6 1AY G0HCP
CB6 1BS G7VGH
CB6 1DA M3UPB
CB6 1DP 2E0XCF
CB6 1DP G8GLB
CB6 1DP M0XCF
CB6 1DP M7AVA
CB6 1DZ G6YTO
CB6 1FA M7CSA
CB6 1FH G7HMI
CB6 1GL 2E0LTP
CB6 1GL M0OEB
CB6 1GL M7CEB
CB6 1HU G3KKC
CB6 1JD M1BFY
CB6 1JJ 2E0DPH
CB6 1JJ M6PHL
CB6 1JS G6YQJ
CB6 1JW G0MGI
CB6 1LG M7PUF
CB6 1NA M7KVS
CB6 1RJ 2E0TTN
CB6 1RJ M6DXL
CB6 1RQ G7BNL
CB6 1SB G4HWJ
CB6 1TL 2E0RDF
CB6 1TL M0MTW
CB6 1TL M6DFK
CB6 1TP G4BUH
CB6 2AQ G6OIB
CB6 2DF M7FHK
CB6 2ED M7EVB
CB6 2EQ G1HMT
CB6 2HH G1DUI
CB6 2HS G1PEK
CB6 2LR 2E0FHA
CB6 2LR M7DEL
CB6 2NW G4FIT
CB6 2NX M6KUI
CB6 2PD G0PPL
CB6 2QB M0TVT
CB6 2QB M6TRH
CB6 2QF M0DBJ
CB6 2RL 2E0DNF
CB6 2RL G0PZJ
CB6 2RL M0ZEB
CB6 2RL M6DNF
CB6 2RX G3IUY
CB6 2ST G4WMZ
CB6 2TQ M1ZEM
CB6 2TS M7JBM
CB6 2UH M0BWH
CB6 3AL G0ICE
CB6 3AL G1XCY
CB6 3DA G4GXY
CB6 3DL G3MXA
CB6 3DN G1CQR
CB6 3HR M0RKA
CB6 3HR M6WKI
CB6 3JT G8OHP
CB6 3PP G4GND
CB6 3PW G0PPI
CB6 3PW G4KNS
CB6 3PW G7IEY
CB6 3SA G7VNP
CB6 3ST G4BEL
CB6 3UE 2E1NPH
CB6 3UE G0PYE
CB6 3UE G4NPH
CB6 3UE G8NPH
CB6 3UE M3NPH
CB6 3UG M6LFQ
CB6 3XE G1VLD
CB61DJ G0OKK
CB7 4BU G1UAF
CB7 4HU G6IOB
CB7 4PQ G6YPD
CB7 4SS M0CGR
CB7 4TH G4KPX
CB7 4UN G0LQV
CB7 4YJ G1AAQ
CB7 5AL M0TIX
CB7 5AL M3DCS
CB7 5AT G3UCL
CB7 5AT M0SNB
CB7 5AT M1GEO
CB7 5BL G8LWQ
CB7 5DH G6BUU
CB7 5DN M0KKZ
CB7 5DS G7BRP
CB7 5EL G8PAG
CB7 5EX 2E0VAS
CB7 5EX M0VAS
CB7 5EX M3VAS
CB7 5FN 2E0GXB
CB7 5GD M7UPS
CB7 5HJ M3YEJ
CB7 5HL 2E0FBE
CB7 5HL M6FTF
CB7 5JD 2E0RPN
CB7 5JD G8DIY
CB7 5JD M6DZV
CB7 5JD M7PNR
CB7 5NR G4UJV
CB7 5NW 2E0LWA
CB7 5QX G8WAV
CB7 5RU G0PIK
CB7 5SH G0TLQ
CB7 5SJ M6GEQ
CB7 5SL. 2E0UEH
CB7 5SL. M0UEH
CB7 5SL. M6UEH
CB7 5TN G4ERO
CB7 5TW G3FCM
CB7 5UU G0EIM
CB7 5XE G7FMV
CB7 5XH M0GUO
CB7 5XW M6BSF
CB7 5ZD M7AWJ
CB76AZ M6FVS
CB8 0BY 2E0BLR
CB8 0DJ G6GQI
CB8 0EJ M7MEO
CB8 0QD 2E0NEV
CB8 0SE G3LUN
CB8 0SE G8IDL
CB8 0SQ G1ACB
CB8 0UH M7ESR
CB8 7AZ G8WUR
CB8 7BQ G0SPK
CB8 7BT M6HPT
CB8 7DU G6JJK
CB8 7RT M3VSF
CB8 7SD G6BBK
CB8 7SD G8BBK
CB8 7YP G0ELK
CB8 7YP G1NHG
CB8 8AD G7JAX
CB8 8DT G4DSN
CB8 8HU G7VGN
CB8 8RL M6AWY
CB8 8SW M3YDY
CB8 8YH 2E0TUG
CB8 8YH M6EZN
CB8 9BA M6RIH
CB8 9DF M6GUF
CB8 9DG G3NUL
CB8 9DG G3SZY
CB8 9DQ G0ADZ
CB8 9NB G0FRU
CB8 9NR G4IQR
CB8 9PD M1BJE
CB8 9RX G4HVV
CB8 9RX G8HVV
CB8 9RY G6JDO
CB8 9SA G0HUG
CB8 9SA G0VQH
CB9 0AP G7MVU
CB9 0AU 2E0CMQ
CB9 0AU M1DQU
CB9 0AU M6BCQ
CB9 0BZ G0OJB
CB9 0BZ M7BBO
CB9 0EQ G8LHD
CB9 0JP M7PJS
CB9 0JR G7EQG
CB9 0LH M7DKC
CB9 0PH M3DMG
CB9 7DL 2E0ELA
CB9 7ED G4PQY
CB9 7EE G7LVM
CB9 7EW G7MQC
CB9 7FU G1WZM
CB9 7FU M7WZM
CB9 7GH M0JDD
CB9 7GH M1RTT
CB9 7NA G6MQN
CB9 7NN G4KVU
CB9 7PT M6JFF
CB9 7RF M3ZKI
CB9 7SG M6EZT
CB9 7ST G8FFF
CB9 7WG G8UBX
CB9 7XW G1DNO
CB9 8EH M6BEJ
CB9 8LX M5PYE
CB9 9BE M3ZCA
CB9 9DD M3XCA
CB9 9JQ M7JYE
CB9 9LE G4LYF
CB9 9LY M6SBJ
CB9 9ND 2E0PCI
CF1 8BZ GW3KXX
CF11 0AP MW1CQP
CF11 0JQ MW7LTE
CF11 6EP MW7MWO
CF11 7BG GW8WEY
CF11 7HF MW3ZYQ
CF11 8AJ 2W0HRL
CF11 8AJ MW3LLV
CF11 8DE MW7GKT
CF11 8HQ GW4ISJ
CF11 9JR 2W0TKS
CF11 9JR MW0KST
CF11 9JR MW6KSL
CF11 9NQ MW0CCK
CF14 0NZ MW6NHC
CF14 0RF GW8VRS
CF14 0RW GW3KNR
CF14 1EH MW6CJS
CF14 1EH MW6ESW
CF14 1HP MW6WFJ
CF14 1NQ GW4LFW
CF14 1UE GW4VGB
CF14 1UH GW6RCK
CF14 1UN GW4TJQ
CF14 2BR MW3RVP
CF14 2EH GW4NHB
CF14 2FU GW6LSL
CF14 2LR GW7ASL
CF14 2RY 2W0PCN
CF14 2TE GW4NHH
CF14 3DU GW0KRQ
CF14 3DU GW1WZI
CF14 3JZ MW7EJX
CF14 3PL MW3FPF
CF14 3QA GW8NCN
CF14 4AQ 2W1CYC
CF14 4DJ GW7EMV
CF14 4DP MW7ILL
CF14 4HN MW3XIJ
CF14 4HS GW4GZX
CF14 4JQ GW8LKX
CF14 4NP MW7BPQ
CF14 4NS GW0GDI
CF14 4SB GW1AWH
CF14 4SP GW8ARC
CF14 5EB MW7EZP
CF14 5HJ GW4JOG
CF14 5JW MW3HZB
CF14 5LR MW3WWO
CF14 5PW MW3DEM
CF14 5QA GW6UAS
CF14 5QH MW7AEN
CF14 6AE GW4LXO
CF14 6AN MW6MBZ
CF14 6BJ GW1BCI
CF14 6DF GW6DFX
CF14 6JZ GW0BBC
CF14 6JZ GW3RWX
CF14 6NS MW6FKC
CF14 6PE GW0WMT
CF14 6PF MW7JMW
CF14 6QH MW3PJU
CF14 6RL GW6ZHM
CF14 6SS GW6ITB
CF14 6SW GW3WLN
CF14 7FH 2W0CDZ
CF14 7EH GW3UZS
CF14 7EH MW6PDG
CF14 7TJ MW7LEJ
CF14 7TP GW0TPL
CF14 9AF GW4LWL
CF14 9AH GW7EYP
CF14 9LA GW3TKH
CF15 7PT MW6XDA
CF15 7SE MW3ZKX
CF15 8AL GW8EHO
CF15 8DD GW6MNC
CF15 8FE GW6VTZ
CF15 8RA GW8FNO
CF15 8RE GW0HCB
CF15 9JP GW1CUQ
CF15 9NJ MW6HYS
CF15 9NN MW0ZAP
CF15 9NZ GW7HOM
CF15 9PB GW3TOB
CF15 9QQ 2W0JYI
CF15 9QQ MW0WSH
CF15 9QQ MW6EDQ
CF15 9QR GW3VOL
CF15 9SJ GW4LFV
CF15 9SJ GW6IOA
CF15 9TJ GW8YYF
CF2 6JN GW3UMD
CF23 5BY GW8EHQ
CF23 5EZ GW0WLI
CF23 5HE MW6EPX
CF23 5NN GW4ZAW
CF23 5SB MW3PGN
CF23 6EJ GW6MHV
CF23 6EQ GW3XXB
CF23 6HN GW4JJV
CF23 6HN GW4JJW
CF23 6LF GW6AGS
CF23 6QU GW0MLN
CF23 6SW 2W0RQC
CF23 6SW GW0RQC

CF23 6SW MW6RQC
CF23 6SZ 2W0EUO
CF23 6SZ MW6EUO
CF23 7DA GW3MKT
CF23 7HN GW0VMD
CF23 7HQ GW4KOE
CF23 8XJ GW0HYU
CF23 9AZ GW7JDX
CF23 9BJ MW6SAW
CF23 9DW 2W0FCF
CF23 9DW MW6XXC
CF23 9HX MW0AQT
CF23 9NR 2W0YIF
CF23 9NR GW6CUR
CF23 9NR GW8NP
CF24 0HS GW0DGJ
CF24 1RW GW7HOC
CF24 1RW MW3YKL
CF24 1RW MW5HOC
CF24 1RW MW6YDP
CF24 2EL GW1SMT
CF24 2NW GW7EMO
CF24 3HU MW7PKT
CF24 3QW 2W0KXN
CF24 3QW MW0KXN
CF24 3QW MW7KJN
CF24 3RW MW7RHC
CF24 4BZ GW8AHB
CF24 4LS MW3KHH
CF24 4RU GW1XZI
CF24 4RU GW4SGR
CF24 4RU GW7BTC
CF3 0AQ GW6YML
CF3 0BY GW7KDU
CF3 0DF MW6FPL
CF3 0DQ GW0FVC
CF3 0DU MW6ATC
CF3 0NB 2W0OCF
CF3 0PS MW6RJX
CF3 1RE MW3IHD
CF3 1TA MW6JHJ
CF3 1TA MW6KCQ
CF3 2TX MW6ETY
CF3 2WL MW3ASG
CF3 3AG MW3UFH
CF3 3HQ GW4IUN
CF3 3LX 2W0BJE
CF3 3LX MW3RXK
CF3 3LZ GW0POA
CF3 5QE MW6TRK
CF3 5SY MW6OZF
CF3 5UD MW6GLK
CF3 6XW GW4BNC
CF3 7LL GW4OES
CF31 1HD 2W1GKJ
CF31 1PF GW0SXZ
CF31 1PQ GW0PYU
CF31 1QN MW6DHK
CF31 1RL 2W0ZAA
CF31 1RL MW3USX
CF31 2AT MW3KHY
CF31 2BL MW7PFF
CF31 2BY MW6PJP
CF31 2DP 2W0PQU
CF31 2DP 2W0YAB
CF31 2DP MW3YDS
CF31 2DP MW6PQU
CF31 2EA GW0RVR
CF31 2ED GW0FJP
CF31 2HL GW8PJE
CF31 2JA MW3THI
CF31 2JG MW3VEL
CF31 2JR MW0AXA
CF31 2LA GW1ERA
CF31 2LJ GW6HMJ
CF31 2ND GW4MTE
CF31 2ND GW8XJC
CF31 2NE GW0NKG
CF31 2NG GW3ZTH
CF31 2PF GW1ZNC
CF31 2PR MW0LCH
CF31 2PZ MW7TDR
CF31 2QF MW6AHT
CF31 2SS GW8KBK
CF31 3BA GW7RRS
CF31 3DA MW0GRZ
CF31 3LN GW4SUD
CF31 3LN GW8ZIL
CF31 4AX 2W0JKN
CF31 4AX GW7RQV
CF31 4AX MW7EJJ
CF31 4BB GW0SZU
CF31 4EY GW3SRF
CF31 4HD GW0SIS
CF31 4HD GW4NBY
CF31 4LU MW3OEJ
CF31 4NR GW1KQV
CF31 4PY MW0AEL
CF31 4QA 2W0HQA
CF31 4QA MW0SGU
CF31 4QA MW7URL
CF31 4QS GW4LFO
CF31 4QS MW0MPD
CF31 4QX GW1FKY
CF31 4QY 2W0GNG
CF31 4RN MW0FDG
CF31 4SJ 2W0PWO
CF31 4SJ MW0GYV
CF31 4SJ MW6PWO
CF31 4SR GW6MAB
CF31 4TG MW0VCV
CF31 5BA GW6VET
CF31 5BT GW7DVJ
CF32 0BJ GW0MNP
CF32 0BP 2W0SYT
CF32 0BP MW7MAT
CF32 0EW 2W0SNW
CF32 0EW MW3SNW
CF32 0HL GW0TTF
CF32 0LH GW1XJJ
CF32 0PJ GW4ZEA
CF32 0SL GW4SKP
CF32 7BB 2W0PEH
CF32 7BB MW0PEH
CF32 7BB MW3PEH
CF32 7NB MW3OEC
CF32 7SP MW6PGD
CF32 7SY GW1CXT
CF32 8AH GW7VBE
CF32 8BL MW1CFE
CF32 8HY GW8BWX
CF32 8LE MW6MWG
CF32 8PE MW7SYK
CF32 8SA MW0DCQ
CF32 8SD GW0GQC
CF32 8SE MW3BBQ
CF32 8UU MW3SLL
CF32 8YB GW1OUP
CF32 8YB GW1YHA
CF32 8YB MW3DIL
CF32 9AE MW7KKW
CF32 9AF GW4VKG
CF32 9AQ GW7TNS
CF32 9BJ MW3LEW
CF32 9DH GW3SSK
CF32 9DH MW3MRL
CF32 9JL MW7PWC
CF32 9LU GW0ACH
CF32 9LX MW6NGE
CF32 9NH MW6LUQ
CF32 9RJ MW0WNL
CF32 9SB GW0KAX
CF32 9SB GW4WVO
CF32 9SX MW6CLP
CF32 9UE MW3BOC
CF32 9UE MW3UBY
CF32 9UG GW4THK
CF32 9UR 2W0WCQ
CF32 9YH GW8CNS
CF32 9YN GW4SKA
CF33 4LT MW0COE
CF33 4LT MW0COF
CF33 4LY MW7BEF
CF33 4RQ GW4IFE
CF33 6AJ GW7PEX
CF33 6AP MW0AGE
CF33 6AS MW6XGD
CF33 6EJ MW7RTY
CF33 6EU GW0RTP
CF33 6HT GW6FNB
CF33 6HY MW0PDR
CF33 6HY MW0PRC
CF34 0AB 2W0EJP
CF34 0AB GW0CYK
CF34 0BB GW3WRE
CF34 0BP MW3LSG
CF34 0BT MW6BPB
CF34 0DE MW0WGR
CF34 0DE MW1WYN
CF34 0DE MW7GBJ
CF34 0DP GW6OMV
CF34 0HT 2W1AID
CF34 0HY MW3MVT
CF34 0LU MW6TJS
CF34 0NG GW7DUI
CF34 0NT GW0TOM
CF34 0NT GW0VMS
CF34 0RU MW0COD
CF34 0SW 2W0JAI
CF34 0TF 2W0ETZ
CF34 0TF MW6MXV
CF34 0TG 2W0ENV
CF34 0TG MW6RPS
CF34 0UE MW6TDL
CF34 0UF 2W0DRK
CF34 0UF 2W0IUN
CF34 0UF MW0HXX
CF34 0UF MW0IUN
CF34 0UF MW6DRM
CF34 0UF MW6IUN
CF34 0UR GW0IRP
CF34 0UW 2W0LFY
CF34 0UW 2W0LLY
CF34 0UW MW6SBT
CF34 0YG MW0GWT
CF34 0YP GW4HCA
CF34 9AH MW6GMJ
CF34 9AH MW6KBK
CF34 9BJ MW6PNW
CF34 9HP GW3XJC
CF34 9JL 2W0FAR
CF34 9JL MW6BAU
CF34 9JL MW6FAN
CF34 9ST MW0AWO
CF34 9TB 2W0DMG
CF34 9TB MW3GPG
CF340TB MW3NCS
CF340TF MW7DJQ
CF35 5BH GW4ZXG
CF35 5EG MW1MFY
CF35 5EU MW7KTP
CF35 5HX MW1BUN
CF35 5NA GW6ZGY
CF35 5NA MW0WIL
CF35 5NF MW0JAN
CF35 5QA GW6CZE
CF35 5RH GW7NJT
CF35 6ED GW8HWL
CF35 6FA GW8TGS
CF35 6HE MW0SWR
CF35 6JR GW0KFL
CF35 6PD GW3FSP
CF35 6SD GW3RVG
CF35 6TF MW0JLN
CF35 6TY GW1YBF
CF35 6TY MW0PSG
CF35 6YA 2W1FUD
CF35 6YD MW3NXD
CF35 6YD MW3WQE
CF35 6YF MW5MWR
CF35 6YG 2W0DJS
CF35 6YG MW3NUX
CF36 3HR GW0BAH
CF36 3HW MW6NLG
CF36 3PR GW7SAQ
CF36 3QH GW0VST
CF36 3RN MW7FQF
CF36 3SW GW6JFV
CF36 3UN GW7NUL
CF36 3YS GW0BZA
CF36 5AJ GW7DTB
CF36 5AN GW3RIY
CF36 5AU GW4BKG
CF36 5HN GW1MNT
CF36 5LP GW4ZVO
CF36 5SG GW3ARS
CF363EP GW4BCF
CF363EY GW4VWY
CF365EY MW7IKH
CF37 1EN MW6GKU
CF37 1HT GW0ONU
CF37 1HY GW0SQY
CF37 1NH GW8HF
CF37 1NW GW0RQP
CF37 1SS GW1FKL
CF37 1TS 2W0FMR
CF37 1TS MW7CEW
CF37 2DA GW4ZVV
CF37 2DD MW7CEO
CF37 2HF GW3YBN
CF37 2LG 2W0BAO
CF37 2SE 2W0PCD
CF37 2SE MW0HKA
CF37 2SE MW6PEL
CF37 3BW GW8LRO
CF37 3DF GW0SLC
CF37 3EU GW0JHH
CF37 4AN 2W0KKR
CF37 4DP MW0ATG
CF37 4HD MW6NWM
CF37 4NR MW3HBF
CF37 5DA MW0MWL
CF37 5ER MW6ARX
CF37 5HH 2W0LMM
CF38 1AU GW0MOQ
CF38 1BW GW1KQY
CF38 1BY GW4CNL
CF38 1DW MW3ZSO
CF38 1LA MW6TOU
CF38 1LB MW7EIJ
CF38 1LF 2W0BMR
CF38 1LF MW3PBV
CF38 1LY 2W0CBJ
CF38 1RP GW0WLN
CF38 1RY GW8RLI
CF38 1UE GW3ZXI
CF38 1UF GW4NPC
CF38 2AL MW3CHZ
CF38 2DN MW7FVH
CF38 2EN 2W0KPH
CF38 2EN MW6CDV
CF38 2JB 2W0VBT
CF38 2JB MW0VBT
CF38 2JB MW6KVS
CF38 2LB 2W0OAG
CF38 2LB MW0YAG
CF38 2LB MW6YAG
CF38 2LN 2W0CWL
CF38 2LN MW0HUY
CF38 2LN MW6CDZ
CF38 2LN MW6CWL
CF38 2NH GW4PAF
CF38 2NT 2W0RLB
CF38 2NT MW6MGZ
CF38 2NT MW6YAE
CF38 2SR 2W0YMG
CF38 2SR MW6YIG
CF38 2TF 2W0MEX
CF38 2TF MW6HCS
CF39 0EG MW3ZSM
CF39 0ET MW0DVM
CF39 0ET MW7JLM
CF39 8AF MW7JHX
CF39 8AL GW4RKZ
CF39 8AL GW6MIH
CF39 8DF MW6HDV
CF39 8DF MW6KJX
CF39 8DU GW7CSK
CF39 8PL GW0UHJ
CF39 8PL GW7PIN
CF39 8PP GW0WLQ
CF39 8TW 2W0GER
CF39 8UG 2W0BKA
CF39 8UG MW0HAB
CF39 9EL MW6BUH
CF39 9HT MW0GWL
CF39 9TE 2W0YMS
CF4 5SR GW8NNF
CF4 6HD GW0NIY
CF4 6HY GW1YKT
CF4 7PD GW4WQC
CF40 1BW MW6BZP
CF40 1BY GW7RKC
CF40 1DQ GW1BDF
CF40 1DQ GW1BDG
CF40 1DW MW0SBJ
CF40 1EN MW6SDV
CF40 1EQ MW0SKD
CF40 1EU 2W0BED
CF40 1HF MW0SWB
CF40 1HP 2W0DLE
CF40 1HR GW2FOF
CF40 1HR GW4BUZ
CF40 1JD MW0MWA
CF40 1JT GW0KQU
CF40 1LX GW6ZYI
CF40 1LX MW0WRQ
CF40 1NG MW6FIN
CF40 1PJ MW0PRP
CF40 1QY GW4YCO
CF40 1TA 2W0SCL
CF40 1TA MW3ZDQ
CF40 2AF MW6MMU
CF40 2AH MW6CGH
CF40 2DZ MW3FVH
CF40 2HD MW6NAZ
CF40 2HJ MW6AEX
CF40 2HN GW4SYO
CF40 2JD GW0LYF
CF40 2LN MW6LNU
CF40 2NX 2W1FVH
CF40 2RE MW6KDA
CF40 2UN MW3MKN
CF41 7HF GW3XHG
CF41 7JT MW1DTT
CF41 7NJ GW0DLA
CF41 7PW MW6CUA
CF41 7PW MW6FBA
CF41 7TW GW4FQV
CF41 7UD GW1NZF
CF41 7UE MW1COB
CF41 7UG GW4WBS
CF42 5AH GW1JNH
CF42 5AH MW6TPM
CF42 5AR 2W0KSA
CF42 5AR MW6DJT
CF42 5AR MW6NFG
CF42 5AR MW6TRL
CF42 5DR GW4VAZ
CF42 5DT 2W0YAE
CF42 5DT MW0YAE
CF42 5DT MW3YAE
CF42 5HL MW3TWQ
CF42 5NG 2W0EQQ
CF42 5NG MW6FBG
CF42 6AW GW7CYT
CF42 6DF GW0NDZ
CF42 6DT 2W0VDW
CF42 6DT 2W0VEH
CF42 6DT MW3VEH
CF42 6ED MW0TJD
CF42 6RL GW1VAW
CF42 6RL MW6AOY
CF42 6SN 2W0PHP
CF42 6SN MW3XZP
CF42 6TF GW8UAP
CF43 3BD MW0DCM
CF43 3ER MW7FXN
CF43 3HF MW7DPG
CF43 3LF GW7VSO
CF43 3LJ MW1BTM
CF43 3NW 2W0CJG
CF43 3NW MW3YHW
CF43 3RN MW0DRU
CF43 4EU GW3SUH
CF43 4HE 2W0VKA
CF43 4HE MW3IPK
CF43 4HE MW3VKA
CF43 4PB MW0HVL
CF43 4RG MW6KEL
CF44 0BQ MW7FGG
CF44 0DT 2W0TMB
CF44 0DT MW6KAC
CF44 0EJ 2W0XMG
CF44 0EJ MW0XMG
CF44 0EJ MW3USS
CF44 0EJ MW3XMG
CF44 0ER GW0HCS
CF44 0HR GW0UKC
CF44 0HY 2W0PJJ
CF44 0HY MW0HPH
CF44 0HY MW0PJJ
CF44 0HY MW6GNA
CF44 0HY MW6PJJ
CF44 0PB GW3SFC
CF44 0PF GW4UAJ
CF44 0PW MW0KAX
CF44 0PY GW4SUN
CF44 0TP MW6AUX
CF44 0TU GW6MKR
CF44 6AH MW6LTR
CF44 6DD GW0OUH
CF44 6EN 2W0DTR
CF44 6LD GW3ASW
CF44 6LD GW4RUX
CF44 6LD MW6OCC
CF44 6NG 2W0XTP
CF44 6NG MW0TMI
CF44 6NL 2W0KOP
CF44 6RY MW6ALC
CF44 6RY MW6RAW
CF44 6SH 2W1HFZ
CF44 6UW 2W0LJD
CF44 6UW MW6EFV
CF44 6YG MW1AND
CF44 7HA MW6LDS
CF44 7HE MW6TIQ
CF44 7NF MW6MIQ
CF44 7PP MW6LCX
CF44 8AH MW6GOY
CF44 8BW GW6YCT
CF44 8EW MW6BQA
CF44 8EW MW6LIW
CF44 8EX MW6ZBR
CF44 8HA GW4QPW
CF44 8LA GW4ZUA
CF44 8LA MW6DRV
CF44 8LF 2W0PPL
CF44 8LF MW6AZP
CF44 8LF MW6HEY
CF44 8LF MW6TJE
CF44 8TE GW4SXA
CF44 8TH MW0BWM
CF44 8TL MW3EQE
CF44 8UB GW0KZE
CF44 8YB GW4KCY
CF44 9ED MW6WRC
CF44 9EW 2W0KED
CF44 9EW MW0JZM
CF44 9EW MW6KED
CF44 9HN MW7FLP
CF44 9JU MW3FJW
CF44 9LY GW0UHX
CF44 9LY GW1SVV
CF44 9ND GW4VEI
CF44 9NG MW0DAR
CF44 9RU 2W0WFB
CF44 9RU MW0WFB
CF44 9RU MW6WFB
CF44 9SA MW6BHD
CF44 9TX 2W0DSO
CF44 9TX MW6CQL
CF44 9YJ 2W0XBC
CF45 3AB GW0WXO
CF45 3BS GW0UHO
CF45 3BS MW3VGJ
CF45 3BY MW6BVG
CF45 3EF MW6KCJ
CF45 3NB MW6JGC
CF45 3ND GW4LWD
CF45 3NU 2W0SKG
CF45 3YA GW0VQZ
CF45 3YW GW4ZUJ
CF45 3YW MW6ENY
CF45 4BD MW7TEW
CF45 4BG 2W0RAD
CF45 4BG MW0PIC
CF45 4BG MW3ZLX
CF45 4BL 2W0IWM
CF45 4DZ MW6JFL
CF45 4HH GW7HVA
CF454YU MW7CMJ
CF46 5HG GW0JTJ
CF46 5LA GW7SXN
CF46 5LB GW0OPP
CF46 5LN GW7HDS
CF46 5RL 2W0BGS
CF46 5RL MW3ONG
CF46 6AB GW0PSV
CF46 6AL MW6VBE
CF46 6AW GW0TOI
CF46 6HJ MW3NKG
CF46 6HQ 2W0APP
CF46 6HQ MW3XVB
CF46 6JF GW4ZCM
CF46 6LF MW3MOJ
CF46 6NS GW0OUV
CF46 6NT GW6OVD
CF46 6RY MW6EGH
CF46 6SD MW6DAI
CF47 0NB MW0DYS
CF47 0TA MW0KDL
CF47 0TB GW0UZX
CF47 0TE MW7BQM
CF47 0TG MW6CIY
CF47 0UX GW4PWZ
CF47 0YU 2W1SRB
CF47 8HJ GW3RYR
CF47 8LU GW1MTH
CF47 8NJ GW0VMZ
CF47 8RH MW3YCL
CF47 9DA 2W0MJA
CF47 9DA MW0DOZ
CF47 9DA MW0EQL
CF47 9DA MW0MUM
CF47 9DA MW6MWS
CF47 9HN MW0UKX
CF47 9SN MW6JNP
CF47 9YP GW3PYD
CF478PP MW7FKA
CF48 1EE GW1WRV
CF48 1EN GW0UXJ
CF48 1EN MW0IBI
CF48 1HH MW6KLL
CF48 1HW GW6AYR
CF48 1JH 2W0DTM
CF48 1JH MW0KIJ
CF48 1JS GW4GJA
CF48 1JU 2W1CLC
CF48 1LR 2W0BPJ
CF48 1LR MW3WFF
CF48 1LU MW3BOP
CF48 1PH GW4CGZ
CF48 1RZ 2W0HHN
CF48 1RZ MW7PLY
CF48 1TP GW6VRN
CF48 1YY GW0KIG
CF48 1YY MW3IXZ
CF48 2BB GW3LNR
CF48 2DG MW3TPJ
CF48 2NT MW7MOC
CF48 2SA MW3AFX
CF48 3AF 2W0GFX
CF48 3AF MW6VGA
CF48 3HE GW0BYZ
CF48 3ND MW7BPZ
CF48 3NY 2W0BOK
CF48 3NY GW3YUC
CF48 3PE MW3DTO
CF48 3PE MW6JKP
CF48 3PU GW0JTE
CF48 4BQ MW6HKL
CF48 4EE GW3RDB
CF48 4EE GW4ZRW
CF48 4HD GW4BIS
CF48 4HJ MW6MWN
CF48 4NG GW6RUE
CF48 4RN GW0WUL
CF5 1DH 2W0FCM
CF5 1GH GW1VRR
CF5 1GH GW1YXR
CF5 1HL MW3MHG
CF5 1JP MW0WZX
CF5 1QS MW1ABZ
CF5 1RF GW8HWS
CF5 2AU GW8REV
CF5 2DH 2W0BTP
CF5 2JF 2W0USN
CF5 2JN GW2DHM
CF5 2JR MW0CCL
CF5 2QP GW8RAS
CF5 2QS GW8THM
CF5 2QT GW3WCV
CF5 2RE MW7LWX
CF5 3EW 2W0UZO
CF5 3EW MW0UZO
CF5 3EW MW3UZO
CF5 3QP MW6NFY
CF5 3SS GW6BHH
CF5 4AP GW4KQ
CF5 4FG GW8VFF
CF5 4FH MW0DYZ
CF5 4HJ MW3VNZ
CF5 4PG MW3HAS
CF5 4PH GW6RDV
CF5 4PR 2W0LEN
CF5 4PS 2W0FFN
CF5 4QW MW3KCL
CF5 4QY MW6AYA
CF5 5AH MW6CTE
CF5 5AP MW6ICU
CF5 5AQ GW4FQA
CF5 5BZ 2W0GJR
CF5 5BZ MW0IHW
CF5 5BZ MW6WUK
CF5 5DD GW7ESF
CF5 5JS 2W0PEG
CF5 5JS MW6RJY
CF5 5NW MW3LGS
CF5 5PB 2W0YFC
CF5 5PL MW6BSR
CF5 5SB MW3HSZ
CF5 5SB MW6AWV
CF5 5SB MW6RBZ
CF5 6AF MW6SEF
CF5 6BQ 2W0SHX
CF5 6BQ GW1CHS
CF5 6BQ GW5NIC
CF5 6BQ MW6EEJ
CF5 6DA GW0WDV
CF5 6EP GW1OZW
CF5 6LQ GW3RIH
CF5 6SB 2W0PCT
CF5 6SB MW0PCT
CF5 6SB MW6PCT
CF5 6SS GW4CSY
CF5 6TN GW4JCK
CF53TH MW6LPZ
CF54LN 2W0ZBC
CF61 1GX GW8KEV
CF61 1GX GW8MFQ
CF61 1TA GW4RER
CF61 1TA GW6GSR
CF61 1YA 2W0DUN
CF61 1YD GW4LFF
CF61 2GS GW4MOG
CF61 2SG 2W0IDT
CF61 2SG GW7HFZ
CF61 2SG MW3GLN
CF61 2UF GW3NCT
CF61 2UG GW0CNJ
CF61 3XR MW3FYA
CF61 5BD MW6ONZ
CF62 3DS GW3YGH
CF62 3DZ GW4ZCL
CF62 3EA GW6ZMN
CF62 3FT GW8YTO
CF62 3HQ GW0LTC
CF62 3LB MW6AQJ
CF62 3LQ GW6OGD
CF62 4JD MW6ADS
CF62 4PD 2W0VOC
CF62 4PD MW6VOC
CF62 4PG 2W0ORH
CF62 4PG MW6ORH
CF62 5AB 2W0ODS
CF62 5AB MW6FNV
CF62 5BD GW1ABX
CF62 5US GW0TSL
CF62 6PB MW6COY
CF62 6RA GW1JVB
CF62 6RA GW1XFB
CF62 6RA GW8LJJ
CF62 7HX GW1VRW
CF62 7TG GW1ZLL
CF62 8BJ GW1XUD
CF62 8BL GW6CNS
CF62 8BR GW8KZA
CF62 8BU MW3DJV
CF62 8BU MW3DJZ
CF62 8BY GW0JCB
CF62 8DA MW7ALT
CF62 8HF MW3JDA
CF62 8ND GW4BRS
CF62 8ND GW6BRC
CF62 8ND MW0VRQ
CF62 8ND MW6JZZ
CF62 9DE MW6BVR
CF62 9DF 2W1SDR
CF62 9DR GW7PFK
CF62 9DS MW7FXV
CF62 9DW 2W0NCA
CF62 9HD GW0SQT
CF62 9HJ GW3WSU
CF62 9HJ GW3XNG
CF62 9SW GW3CBA
CF62 9TE MW6FYY
CF62 9TG MW0RUH
CF62 9TY MW7MOZ
CF63 1DW GW7AUQ
CF63 1FP MW0CWF
CF63 1FP MW1COY
CF63 1QJ GW7HQL
CF63 1RQ GW0NXW
CF63 2FE MW3WRH
CF63 3QD GW4ANK
CF63 4EF GW6ATT
CF63 4EF GW7VST
CF63 4JQ MW6JRI
CF63 4JY MW7DIZ
CF63 4NP MW6ZAN
CF63 4PQ GW0GRQ
CF64 1DD GW3VKL
CF64 1DD MW0DHF
CF64 1TN GW3MOV
CF64 1TQ GW4LDA
CF64 1TQ MW7FWV
CF64 1TR GW1LKG
CF64 1WW GW1LQE
CF64 2AJ 2W0VOG
CF64 2AJ GW5VOG
CF64 2AJ MW7AFP
CF64 2EZ MW3WXN
CF64 2JS MW3LZX
CF64 2PG GW0GHF
CF64 2QZ GW3WBU
CF64 2RX GW0PZZ
CF64 2SA GW0OTY
CF64 2TZ GW4UGI
CF64 3BE GW3UWL
CF64 3DD GW3SPA
CF64 3DH MW1EAA
CF64 3DH MW3RRW
CF64 3EN GW3FZV
CF64 3HF GW3VNZ
CF64 3HH GW1MBV
CF64 3HY MW0HAT
CF64 3JD GW4IUL
CF64 3PP MW3NHC
CF64 3QX 2W1YEG
CF64 3QX MW3YEG
CF64 3RB GW3PYX
CF64 4HA 2W0INY
CF64 4HA MW0RSR
CF64 4HA MW7RLT
CF64 4LR GW3SFQ
CF64 4PS GW1BAI
CF64 4PZ MW6WLB
CF64 5BW GW1DQV
CF64 5GA GW4EUA
CF64 5GA GW8JWL
CF64 5JW 2W0ICD
CF64 5QD MW0GBW
CF64 5QE MW6FTC
CF64 5RY 2W0OPC
CF64 5RY MW6OPC
CF64 5SX MW3UDO
CF64 5TQ GW4BEQ
CF64 5UG GW0CKK
CF71 7EX GW4MOZ
CF71 7LE GW7KQN
CF71 7LX GW0ANA
CF71 7PB 2W1ETN
CF71 7QR GW7NJQ
CF71 7RQ MW7PAN
CF72 8BG 2W0YYP
CF72 8DA GW0DIV
CF72 8DA MW6HLG
CF72 8DY 2W0CGM
CF72 8ED 2W0ITH
CF72 8EG 2W0BMO
CF72 8EG MW0OMB
CF72 8EG MW3RXH
CF72 8HQ GW4SML
CF72 8HR GW4VMT
CF72 8HW MW6LSW
CF72 8QR MW7FGW
CF72 9BP 2W0HPM
CF72 9DB GW4HDZ
CF72 9ER GW1DPV
CF72 9ER MW7EUH
CF72 9FZ GW6PFK
CF72 9HT MW7CUO
CF72 9HU MW6WCI
CF72 9JD 2W1ACM
CF72 9LT 2W0RTJ
CF72 9LT MW6TTS
CF72 9ND GW7VSF
CF72 9QQ MW6PLC
CF72 9SB MW6DYL
CF72 9SJ GW0VWD
CF72 9SL MW6WDX
CF72 9SZ MW6TES
CF72 9TF GW7VOO
CF72 9TX GW4DVB
CF729LQ MW0PVW
CF8 4AU GW4ZBN
CF8 4DL GW1SRB
CF8 8NA 2W0IVZ
CF81 8JS GW0LBI
CF81 8JX 2W1EKR
CF81 8NB GW0WGW
CF81 8NS MW6WFF
CF81 8NT 2W0WMB
CF81 8NW MW1AFW
CF81 8PP 2W0EGK
CF81 8PP MW3EGZ
CF81 8QJ MW3PKC
CF81 8RN MW7AUX
CF81 8TG MW0TAF
CF81 8US MW3FFL
CF81 8UU 2W0HAK
CF81 8UX MW6GWO
CF81 9AB GW8GOC
CF81 9DZ MW3IWC
CF81 9LQ 2W1JOL
CF81 9PT MW3XNW
CF81 9RS MW6KAB
CF82 7AZ MW7FDL
CF82 7DB 2W0UAA
CF82 7DB MW0MAU
CF82 7HF MW3ZTH
CF82 7JB MW0ATK
CF82 7JP MW0ICE
CF82 7JP MW3CCE
CF82 7NP GW4TPG
CF82 7QQ 2W1EUR
CF82 7RH GW0PCJ
CF82 7RZ GW6JSJ
CF82 8BR MW1FEU
CF82 8EG MW3NQH
CF82 8ET GW0TKX
CF82 8EW MW0IQZ
CF82 8EW MW6EOP
CF82 8EX MW6LVD
CF82 8FN MW0XMI
CF82 8LA GW4XMU
CF828EW 2W0KMU
CF828EW MW3KMU
CF83 1AL 2W0BVS
CF83 1AN MW6CRU
CF83 1GA MW6MYE
CF83 1JW MW6EXV
CF83 1LA MW3ZTB
CF83 1NQ MW6FNT
CF83 1RF 2W0RDD
CF83 1RF MW0NNX
CF83 1RF MW3DQB
CF83 1RU MW6JBQ
CF83 1SP GW4HZM
CF83 1SQ GW6IPR
CF83 1SZ GW0IRC
CF83 2AN MW6NYE
CF83 2BE MW7ERP
CF83 2JL 2W0BUQ
CF83 2JU MW0MEX
CF83 2LA 2W0PCE
CF83 2LA MW0WTK
CF83 2LL MW3USP
CF83 2NE 2W0CYY
CF83 2NE 2W0YBZ
CF83 2NE MW0YBZ
CF83 2NE MW6CVC
CF83 2NE MW6YBZ
CF83 2NJ 2W0RFT
CF83 2NJ MW3IQY
CF83 2NR GW7KKM
CF83 2NZ GW0MXG
CF83 2QX 2W0HZC
CF83 2QX MW7DPO
CF83 2RA 2W0VTP
CF83 2RA MW3HAQ
CF83 2RA MW6VTP
CF83 2RX GW8UAM
CF83 2SR MW6SEP
CF83 2SW MW6HPN
CF83 2TA 2W0LKD
CF83 2TX GW0OSB
CF83 2TX GW1MAX
CF83 2UN MW3TJG
CF83 2UW GW4HDF
CF83 3BT 2W0FEY
CF83 3BT MW6PEY
CF83 3BU 2W1MRK
CF83 3BU GW6WTK
CF83 3BU MW7MRK
CF83 3FB 2W0MSL
CF83 3FB MW6NAG
CF83 3FT GW0KWO
CF83 3HF GW4VEB
CF83 3JH MW6HGZ
CF83 3PG GW4SRR
CF83 3PG MW7BYV
CF83 3QB MW6HFY
CF83 3RT GW6DOC
CF83 4BG 2W0JCV
CF83 4BG MW6JCV
CF83 4BN GW4SLI
CF83 4DD MW3WLB
CF83 4DH MW6ECR
CF83 4EQ GW0CKL
CF83 4GG GW0PUP
CF83 4HJ 2W0BGQ
CF83 4HJ MW3IUS
CF83 8EP MW6IAG
CF83 8FY GW0MTI
CF83 8GX MW7MME
CF83 8TA MW6PBF
CF83 8TT 2W0NQE
CF83 8TT MW0GEI
CF83 8TT MW3NQE
CH1 2DW M0IGU
CH1 2NW G3XIR
CH1 3BZ M6VBJ
CH1 3HG G8WLD
CH1 3ND 2E0BSR
CH1 4AN G3SES
CH1 4BE 2E0IFW
CH1 4BE M0JXE
CH1 4BE M3JXE
CH1 5AF G6LRU
CH1 5AZ G0ULM
CH1 5DP 2E1CRA
CH1 5DZ M6BNK
CH1 5JQ G1URW
CH1 5NW G7VPQ
CH1 5PP G7JGS
CH1 5QX M1EQD
CH1 5RU M3LIN
CH1 5SW M6DOA
CH1 5SY M1DAP
CH1 5TX M7RTB
CH1 5XE M6GNG
CH1 6AQ G0JCG
CH1 6BF M0IVM
CH1 6JS G4SMM
CH1 6LU G4WXO
CH1 6QJ G0CGD
CH16BF 2E0DDR
CH16BF M0AUT
CH16BF M6UAR
CH2 1BB 2E0UPT
CH2 1BB M3UPT
CH2 1BS G6SJV
CH2 1DG G8GWX
CH2 1HF G6VGV
CH2 1HT G8BMH
CH2 1JG M0NUG
CH2 1LX G0NLQ
CH2 1LX G7HGB
CH2 1NF G3ZVH
CH2 1NF G7GFC
CH2 1NW M0AUA
CH2 1PF G0HEJ
CH2 1RA 2E0IVE
CH2 1RA G1BCW
CH2 1RA M7EOL
CH2 1RD G8RRS
CH2 2AH 2E0IIV
CH2 2AH M7RMZ
CH2 2AQ G3KJS
CH2 2BE M7OSX
CH2 2BX G0OIV
CH2 2DH M3UZA
CH2 2DP M6LOB
CH2 2HL M3IHA
CH2 2LA G0ROY
CH2 2PB M6VHM
CH2 2PL G0NMD
CH2 2QF 2E0SYV
CH2 2QF M7ESE
CH2 3EW 2E0GPE
CH2 3EW M0PEA
CH2 3EW M6GHP
CH2 3EW M6LVA
CH2 3HT M3OME
CH2 3JT M0CVP
CH2 3NL G0MTA
CH2 3RE G6SKP
CH2 4DQ M1RWB
CH2 4PR M7PMK
CH2 4QJ G7IDE
CH2 4RP M3HWV
CH3 5DY G1YNJ
CH3 5HA G1LML
CH3 5HD M3OAS
CH3 5HE 2E0WVD
CH3 5HE M3WVD
CH3 5HE M3WVF
CH3 5HE M3WVT
CH3 5HF M0PWS
CH3 5JB G7HEK
CH3 5JB M7NRT
CH3 5JF 2E0OBZ
CH3 5JF M3OBZ
CH3 5LA G8KKN
CH3 5LE G1XPW
CH3 5LU M3JAL
CH3 5LY G7BQY
CH3 5PT G8ZRE
CH3 5RR G4JMF
CH3 5RR G8GIZ
CH3 5RR M3XSD
CH3 5SN G8XDY
CH3 5XN G1DIA
CH3 6BL 2E0HQY
CH3 6BL G7REV
CH3 6FJ M7UGZ
CH3 6HZ M7BDE
CH3 6NE G0VAH
CH3 6QX M3CZX
CH3 7EJ G8KWP
CH3 7EL G6FMS
CH3 7NT G0DUB
CH3 7NT G0RCW
CH3 7NT G4NRC
CH3 7NT M0LSA
CH3 7QD G6FDK
CH3 7QD M3DVA
CH3 7QQ 2E0IBA
CH3 7QQ M3OCS
CH3 8EQ G6VUX
CH3 8LP G7NEH
CH3 8LR G0DYL
CH3 8LR G6MYL
CH3 9BE G7EEN
CH35LY G1CZU
CH4 0AH 2E0LUD
CH4 0AH M6PPZ
CH4 0AL M0CSO
CH4 0AQ G6FKB
CH4 0FT M6AQT
CH4 0FY M6FIB
CH4 0JB M6OVT
CH4 0JD M6RDX
CH4 0LF M6WCF
CH4 0NH G0EDC
CH4 0NZ 2E0BTA
CH4 0NZ 2E0BTD
CH4 0NZ M3LXB
CH4 0NZ M3LXK
CH4 0PT G8VNN
CH4 0PU M3LBC
CH4 0QE M6AKF
CH4 0QT G6OKC
CH4 0RJ M3KXV
CH4 0RL G7TZX
CH4 0RN G1LDY
CH4 0SG G7HXF
CH4 0SQ G8ZBC
CH4 0SZ G0ISE
CH4 7LE G6XKJ
CH4 7LU G8VAD
CH4 7LZ G1LED
CH4 7NF M1BZI
CH4 8BB G7KLN
CH4 8BG M6NCF
CH4 8BY M3IUV
CH4 8DE M1FJP
CH4 8DH 2E0JWE
CH4 8DH M3YYW
CH4 8FB 2E0CKL
CH4 8FB M0UFA
CH4 8FB M6BHF
CH4 8JG G8SLP
CH4 8LB G4UXD
CH4 8PA G8DOF
CH4 8PJ 2E0SEY
CH4 8SJ 2E0TWI
CH4 8SS M3JUY
CH4 8TR M6OWO
CH4 9DG G4YNS
CH4 9NG G0CTH
CH4 9NG G1SVN
CH4 9NN G0UVT
CH4 9NN G0UWA
CH4 9NU G6YCW
CH41 0AX M1AAS
CH41 3LF 2E0LGZ

CH41 4FF G4NCI
CH41 5LD G7EED
CH41 8BZ M6THC
CH41 8HJ 2E0MTH
CH41 8HJ M5TT
CH41 8HJ M6FTT
CH41 8JF M7DVO
CH416JA 2E0DEQ
CH42 0LB G8UWG
CH42 2BU G6RYM
CH42 2DN M6IEZ
CH42 3UN M7FRG
CH42 3XQ G6CHD
CH42 3YE 2E0OQH
CH42 3YE M3OQH
CH42 3YH 2E1SOX
CH42 3YH M3SHJ
CH42 4NX M7FEX
CH42 4PD G3OTW
CH42 4QX M0BZZ
CH42 5LH M5HFJ
CH42 5LR M3ELV
CH42 6PX 2E0IYA
CH42 6PX M7NRO
CH42 6QY 2E0YJY
CH42 6QY G4EWJ
CH42 6QY M6MAJ
CH42 6RG 2E0OOY
CH42 7HR M7AEF
CH42 7JA G4WVC
CH42 8JZ M0HGG
CH42 8QA G6JZE
CH42 9LJ G0KXL
CH42 9NZ G6RVH
CH42 9NZ M3RGJ
CH42 9PH G7CQZ
CH42 9QA 2E1PGB
CH42 9QA M3PGB
CH43 0RS M7UIG
CH43 0SZ 2E0IYZ
CH43 0TT G6HGI
CH43 0XB G1MHF
CH43 2 HE 2E1GMA
CH43 2HE G0PZO
CH43 2HS 2E0WED
CH43 2HS G6XHF
CH43 2HS M6CUL
CH43 2HS M6STI
CH43 2JL M3LXH
CH43 2LZ G0LZI
CH43 2NQ M1ARH
CH43 3AS 2E0GLR
CH43 3BB M3XCB
CH43 3BJ M7BXC
CH43 3DW 2E0ENZ
CH43 4UL G6VDX
CH43 4XR 2E0MFN
CH43 4XR M6MNK
CH43 5RP M0GNA
CH43 6TA G4ASM
CH43 6UT G0UCP
CH43 6UY G0MHF
CH43 7NX G4MUP
CH43 7SE G0WMY
CH43 7SQ G6NFB
CH43 7YQ G6EWK
CH43 8SU G8FPU
CH43 9YT M3LGX
CH43 9YT M3LGY
CH44 0AQ G1RCI
CH44 0BQ G4NXG
CH44 0BQ G6MSY
CH44 0BQ M0AJB
CH44 0BQ M3PSB
CH44 0BQ M3YJB
CH44 0EB 2E1GXL
CH44 1AJ M6UHF
CH44 1AT M0AXJ
CH44 2EY G1BYI
CH44 3AF G1PMJ
CH44 3DJ M6AOQ
CH44 3DZ G8NNS
CH44 3DZ G8TFW
CH44 3ED G0JQK
CH44 4DL G0DPO
CH44 5RQ G4ZCA
CH44 5RQ G7KOS
CH44 5UH 2E0AAN
CH44 5UH G8AA
CH44 5XB 2E0DHQ
CH44 5XB M6EKW
CH44 6PZ G7VQM
CH44 6QA M1MPB
CH44 7BJ 2E1TAB
CH44 7BJ M3ATB
CH44 8AB G7OKR
CH44 9AA G6LWC
CH44 9BR G4PGO
CH44 9DN 2E0WAE
CH44 9DN M6PUG
CH44 9DX G4KVP
CH44 9DZ G1NXB
CH45 0LG M6AAC
CH45 1HA G0NBD
CH45 1HD G6ABO
CH45 1JG G0WMQ
CH45 1JT G1JEZ
CH45 2LZ G4KRF
CH45 2NG M0NBJ
CH45 2NN M6EXS
CH45 2PE G8STF
CH45 3HA G6RUU
CH45 3HJ G0GQW
CH45 3LU M6JLQ
CH45 3LU M6JLU
CH45 4PN M6BTM
CH45 5AY G4ENK
CH45 5EJ 2E0PSO
CH45 5HB G0AWW
CH45 5HN G1FEP
CH45 6TD G3SEJ
CH45 6TD G8STD
CH45 6TE G3XCD
CH45 6TF M0ATZ
CH45 7LE M7ERI
CH45 7LQ 2E0PJD
CH45 7LQ M3IRU
CH45 7NF M7RST
CH45 7NZ G1PDA
CH45 7NZ G3VEB
CH45 7QF G8TRY
CH45 7QF G8WDC
CH45 7QG G4VQS
CH45 7QQ G4KOY
CH45 8JN M3DBX
CH45 8LH M6SIP
CH45 8LR G0PIT
CH45 8LR M6PMA
CH45 8NX 2E0SYA
CH45 8NX G8JUV
CH45 8NX G8UUC
CH45 8NX M7SYW
CH45 9JA G1EYJ
CH45 9LJ G6IIN
CH45 9LW G6IHD
CH46 0QT G3PEZ
CH46 0RS M6DXZ
CH46 0SN G6UFI
CH46 0TP 2E0RAA
CH46 0TU G3LCI
CH46 1PU M7ONR
CH46 1QH G0PTE
CH46 1QZ G7NHR
CH46 1RQ M3PUL
CH46 1RW 2E0NLP
CH46 1RW M6ATS
CH46 1RX M3HQP
CH46 1RX M3PAP
CH46 1SW M7FQT
CH46 2QR G6HBF
CH46 2QY M6LKK
CH46 2QZ G0WUO
CH46 2SA 2E0WBL
CH46 2SA M6MEQ
CH46 2SB G0PVY
CH46 3RR M6BXW
CH46 3RX 2E0AZK
CH46 3SG 2E1HOU
CH46 3SL M0SMC
CH46 5NE M6GAW
CH46 6AX G3YHB
CH46 6BJ G6USU
CH46 6DH G4WUA
CH46 6DH M0MTC
CH46 6DS M6TGL
CH46 6EL 2E0DLJ
CH46 6EN 2E0BDO
CH46 6EN M3KZW
CH46 6ET M1ELR
CH46 6FL G4VWL
CH46 6HB G4UJP
CH46 6HB G6WWS
CH46 6HE 2E0MTX
CH46 6HE M0WBG
CH46 6HE M6MBP
CH46 6HQ G3XJZ
CH46 7SJ G0CLV
CH46 7TY M7DBG
CH46 7UP G7FND
CH46 7UT G0DNQ
CH46 8TT M0SPD
CH46 8TX G0EQE
CH46 9PF G0LHN
CH46 9PS 2E1RSB
CH46 9RW M6GZI
CH46 9RY M3VYK
CH47 0LB G3OKA
CH47 0LQ G4ELA
CH47 2AE 2E0LGH
CH47 2AE M3LGH
CH47 2AG 2E0YOZ
CH47 2AG M6HOY
CH47 3AG M6RKQ
CH47 3AJ M3FLL
CH47 3BZ G4FPB
CH47 3DD G0KQY
CH47 3DD G8THZ
CH47 5AW G3VDS
CH47 7DA G1ZZA
CH47 9RW 2E0CDM
CH47 9RW M3VZC
CH47 9RZ M6GWU
CH48 3HE M0BFV
CH48 4EL 2E0ECJ
CH48 4EL M6HFM
CH48 5DG 2E0ILO
CH48 5DG M0ILO
CH48 5DG M6ILO
CH48 5EG M6HJM
CH48 5HQ G8ODK
CH48 5HZ M6PLE
CH48 6DA G0VAX
CH48 6DJ G1UHO
CH48 6ET M3IYO
CH48 7EU M0SJK
CH48 7EX G3OVE
CH48 7HR G8LEM
CH48 9UF G7NIR
CH48 9UL G0SSG
CH48 9UX G0GSF
CH48 9XP 2E0ETD
CH48 9XP M6ETD
CH48 9XS M5AOI
CH48 9XU 2E0PCU
CH48 9XU M6PFG
CH486EJ M1AOR
CH48NZ M6NSI
CH49 0TD G4BKF
CH49 1RT 2E0TFM
CH49 1RT M6KDZ
CH49 1SS G6PRE
CH49 2RQ G6KBQ
CH49 2RZ G8OBT
CH49 3AG G3YSM
CH49 3AW G3MAJ
CH49 3NG G0MQJ
CH49 3QH G0LPU
CH49 4GD G8WWD
CH49 4GS G8ZJH
CH49 4GY G3VUY
CH49 4QP G0TYN
CH49 4QP M3LDH
CH49 6JD G7MIZ
CH49 6LA G3ZIC
CH49 6PN 2E1JEF
CH49 6PW G8CVF
CH49 7LH M3TDM
CH49 7NJ 2E0BJW
CH49 7NT M6DYG
CH49 8EE G7GRU
CH49 9AW G6WCW
CH5 1EH GW7UVO
CH5 1EZ GW7GKX
CH5 1HH GW4KDI
CH5 1NU GW1ATZ
CH5 1NU GW3ATZ
CH5 1SA MW7DUI
CH5 1XE MW0SZR
CH5 1XU MW3JIJ
CH5 2BR MW3RPX
CH5 2DP GW7SBJ
CH5 2FX GW4MVA
CH5 2RB MW7OLF
CH5 3AL GW0PBJ
CH5 3BF GW8UEK
CH5 3BF MW3UEK
CH5 3EX GW0TCL
CH5 3HS GW4EYO
CH5 3JG GW4SII
CH5 3RW GW0DRS
CH5 3RW GW0HRG
CH5 4AL GW1RVC
CH5 4AU MW3XZV
CH5 4BY MW6PAI
CH5 4GG GW6POO
CH5 4HN 2W0DPI
CH5 4HN MW6AFK
CH5 4HW GW4ZAG
CH5 4JP GW0DSP
CH5 4JP GW1LFX
CH5 4LG GW3KNZ
CH5 4LG MW3JLK
CH5 4LG MW3MRK
CH5 4PA GW1LEL
CH5 4QA MW3FLI
CH5 4RE GW4RQS
CH5 4RE GW4VBM
CH5 4SH GW6NVJ
CH5 4SN GW7AAU
CH5 4SN GW7AAV
CH5 4TF MW3SUF
CH5 4TN GW0UEO
CH5 4WP GW0SXE
CH52HX GW1GBH
CH6 5AN GW0UGQ
CH6 5DW MW7SUS
CH6 5HU GW0UIP
CH6 5HU MW0GPP
CH6 5JP GW6LMI
CH6 5JU MW1CRE
CH6 5LN MW1LSG
CH6 5LP 2W0CCG
CH6 5LT MW1RES
CH6 5QA GW4ZAR
CH6 5QB GW4VFE
CH6 5RQ MW0GRJ
CH6 5RX GW4RYJ
CH6 5SB GW0GZR
CH6 5SB GW7FZW
CH6 5YE GW7CMM
CH6 6BX MW7BYW
CH6 6JE GW0POG
CH6 6JS 2W1RSS
CH6 6JS MW0RKD
CH6 6LS GW7CCR
CH60 0BP G3WIO
CH60 1UH G6LKG
CH60 1XW M6NWV
CH60 1YD G6AGP
CH60 2SG G0MQR
CH60 2TT G7SFM
CH60 2TT M0CKV
CH60 3RW 2E0SSZ
CH60 3RW M7SAZ
CH60 5RY G4EIC
CH60 5RZ M3YKZ
CH60 6RD M3REX
CH60 6TE G7FVR
CH60 7RA G3UVR
CH60 7RA G4MGR
CH60 7RF M7OMW
CH60 9JT G3RLA
CH601UL M6MTQ
CH61 0HD G4DBE
CH61 3UY 2E0EFC
CH61 3UY M6ZKA
CH61 3XH G7AAI
CH61 3XS G0WAB
CH61 4UA G1RIX
CH61 5UG G6KIV
CH61 6UP G1KJX
CH61 6UZ G4NOY
CH61 6XT G8DTS
CH61 6YH G8HLJ
CH61 7UL G7IIF
CH61 8SU G4ILA
CH61 8SX G8REQ
CH61 9NT M6YIN
CH61 9PR M0CNN
CH61 9PY 2E0PYR
CH61 9PY M6PYR
CH61 9QA 2E0XGA
CH61 9QA M3XGA
CH611DL 2E0IKS
CH611DL 2E0KGA
CH611DL M7KGA
CH62 0BD G3ODC
CH62 0BN M7FDC
CH62 1BJ M0BQH
CH62 1DR M3IRJ
CH62 1DY M3YWJ
CH62 3LH M0HVS
CH62 5DB M3TFI
CH62 5DB M3TXU
CH62 6AN G6HKY
CH62 6AW G7IGU
CH62 6BE G6PYL
CH62 6BE M3FMV
CH62 6BE M3ODC
CH62 6BR G8WHR
CH62 6BR G8WSR
CH62 6BU G0GBN
CH62 6DT G8IWJ
CH62 6DX G8LAU
CH62 6EA M0WAD
CH62 6EF M6RTR
CH62 7BA 2E0HLP
CH62 7BA M3HLP
CH62 7DJ M6KRQ
CH62 7EE 2E0DUP
CH62 7EE M0VWD
CH62 7EE M6VAH
CH62 7EX G3NWR
CH62 7EX G4YWD
CH62 7JY G3NNV
CH62 8BB G6AVY
CH62 8BN M6GHM
CH62 8BR G3YGL
CH62 8BS G0VEO
CH62 8DG G1WAS
CH62 8DL G4UCE
CH62 8EB G1NPN
CH62 8EQ G3TRR
CH62 8EX G0PZD
CH62 9AA G3AVL
CH62 9AZ 2E0ICE
CH62 9DF G6GFG
CH629EA M6JQK
CH63 0JA G6TPE
CH63 0JJ G8TSG
CH63 0JW G7NHD
CH63 0LA G8EYA
CH63 1JG M6NWW
CH63 1JS G7TIY
CH63 2JU G6NOI
CH63 2LW M6CXH
CH63 2LZ 2E0ARC
CH63 2LZ M7RWE
CH63 3BD 2E0HZE
CH63 3BD M0KSX
CH63 3BD M7KWS
CH63 5JR G4NSA
CH63 5JR M6HAK
CH63 5PA G0OXA
CH63 6HN M3PRU
CH63 7LR 2E0OWL
CH63 7QU M6XBJ
CH63 8LP G5YSS
CH63 8LU M7TGD
CH63 9AH G6BWO
CH63 9FP G8DDR
CH63 9FW G8DRB
CH63 9LP G0PXO
CH63 9LU M6LSP
CH63 9NG M0HWO
CH63 9YJ M1BZZ
CH64 0SG G6SLN
CH64 0TL M7PKA
CH64 0UA M7TKR
CH64 0UY G8VSI
CH64 0UZ G4EEQ
CH64 0XP G1NUH
CH64 1TN G4JZR
CH64 2XG M0PER
CH64 2XQ G8OKZ
CH64 3RS M6PTH
CH64 3TJ 2E0IGQ
CH64 4AN G6HQ
CH64 4AT G4HGL
CH64 4BJ G3NTI
CH64 5TJ 2E0CPB
CH64 5TJ M3YPB
CH64 6QB G0KQS
CH64 6RB 2E0MLD
CH64 6RB G6FTL
CH64 6SF G4ELK
CH64 7TR G6ADO
CH64 7TR G8MMM
CH64 7TR M3JLI
CH64 7TT M7YMC
CH64 9SW G7DEY
CH65 2BD 2E0GWI
CH65 2BH G1AOF
CH65 2BE M0COC
CH65 5AD M1SUM
CH65 5AT M7PLD
CH65 5BW M6JWI
CH65 5DG M6FEK
CH65 5DG M6RBF
CH65 5DX G0HFE
CH65 6PB G7JKH
CH65 6RW 2E0RFF
CH65 6RW G0RGG
CH65 6RW M6EBC
CH65 6SB 2E0BJX
CH65 6SB G3YCO
CH65 6SB M0GPG
CH65 6SB M3UQJ
CH65 6SE G4JTK
CH65 6TD G6ALN
CH65 7AZ M3LCL
CH65 7BW G0WVW
CH65 7DX M6BLV
CH65 7DZ 2E1LES
CH65 8EP 2E0IHO
CH65 8EP M3IHO
CH65 8ES M7NVC
CH65 8HN M7COR
CH65 9DZ G0PJY
CH65 9EN G0CTR
CH65 9EY G8RVY
CH65 9JU M7MGH
CH66 1JF G1PRL
CH66 1JG M3UHB
CH66 1JJ G3VZM
CH66 1JW G0JZJ
CH66 1JX M3AIS
CH66 1NY M3UQZ
CH66 1QT G5MUN
CH66 1RL M0AMS
CH66 1RW G0AKU
CH66 1TY G4STZ
CH66 2BG M0BAU
CH66 2BJ G4DBG
CH66 2BL G4YBJ
CH66 2GU G8XPB
CH66 2GX G6IIM
CH66 2SL G1GWS
CH66 2SX G4ZKG
CH66 2TD 2E0FWZ
CH66 2US G0IEQ
CH66 2WL M3LYU
CH66 2YJ G3CSA
CH66 2YJ G4WSE
CH66 3LL G0COJ
CH66 3NN M6TXH
CH66 3PF G7GQW
CH66 3PH G4UXL
CH66 3QU M7DEA
CH66 3SQ G3TXH
CH66 3TA 2E0IQM
CH66 4AJ M6LAF
CH66 4NA G7NHV
CH66 4PD G0RCY
CH66 4QY M1ZXZ
CH66 4SG G5CH
CH66 4SG G8IPT
CH66 4TP M7XSE
CH66 4UF M6HCJ
CH66 5NB M3YQH
CH66 5PB M1AOX
CH66DP G6WGY
CH7 1GU MW7SEM
CH7 1HH 2W0GDA
CH7 1HH MW0WML
CH7 1HH MW6AQU
CH7 1LD GW0TCV
CH7 1SU GW7MVG
CH7 1TH MW3RAA
CH7 1UG 2W0IQY
CH7 1UG MW7GLF
CH7 2AG GW3UOO
CH7 2AG GW4HER
CH7 2AX GW7MHB
CH7 2BS 2W1HUX
CH7 2BS GW3RIB
CH7 2EQ GW0EGQ
CH7 2HQ GW7AIY
CH7 2JA GW0GXQ
CH7 2JL GW0FHL
CH7 2JS MW6BPH
CH7 2LJ 2W0NAD
CH7 2LJ MW0JRX
CH7 2NH MW7FVB
CH7 2PA GW3HDF
CH7 2PB 2W0IBM
CH7 2QP MW0HLW
CH7 2QX MW6KPV
CH7 3BL GW3UVA
CH7 3BR GW7BOY
CH7 3DE MW7KSW
CH7 3DU MW0IEH
CH7 3EL GW1OTI
CH7 3ER 2W0VMC
CH7 3ER MW3NQK
CH7 3ET GW4GTE
CH7 3HN GW4HBZ
CH7 3JU MW6BNH
CH7 3LH 2W0RSV
CH7 3LH MW0RSV
CH7 3LH MW6YZF
CH7 3NB MW0RHT
CH7 3NG GW0PJA
CH7 3QA GW0OHJ
CH7 3QF MW6FEF
CH7 4NE GW0MDQ
CH7 4NE GW3YHR
CH7 4NS MW3JNX
CH7 4PY GW4GDM
CH7 4QD GW1PKW
CH7 4SB GW7APM
CH7 4SN GW6UTF
CH7 4SS GW3TMP
CH7 4TU GW4SDO
CH7 5AE GW1FWE
CH7 5AS MW7KCA
CH7 5BH GW4XUM
CH7 5DZ GW8EQI
CH7 5EL GW7VFJ
CH7 5EU MW7WRG
CH7 5HY MW0ISC
CH7 5NU GW0LNO
CH7 5RF GW8NZN
CH7 5UY MW0JFX
CH7 6DU GW6RNV
CH7 6ED GW7NFM
CH7 6EE 2W0CCK
CH7 6EE MW0GYF
CH7 6EE MW0TTK
CH7 6EE MW3MTB
CH7 6EE MW7TTV
CH7 6EF GW4EVX
CH7 6GE MW7GVR
CH7 6HT MW6ROX
CH7 6JF GW3FPH
CH7 6PN MW6VDE
CH7 6RT 2W0VAC
CH7 6RT MW0ZAQ
CH7 6RT MW3YCR
CH7 6RW GW8ICT
CH7 6SL GW3IVK
CH7 6SW GW4BTW
CH7 6TD GW4NEI
CH7 6TF GW0PZU
CH7 6TR MW0CVW
CH7 6TU MW3KML
CH7 6US GW6FKP
CH7 6US MW7PAJ
CH7 6UZ GW3LWU
CH7 6WD MW0RPE
CH7 6XZ MW6OXV
CH7 6YP GW3KJN
CH7 6YP MW7BJQ
CH7 6YU GW4GSG
CH72GZ MW7CBF
CH72LN MW7ASM
CH75BE MW0HCC
CH8 7AP MW6KOI
CH8 7AU MW3UDA
CH8 7DF GW0PFZ
CH8 7DR GW0HKQ
CH8 7EQ 2W0EUY
CH8 7PJ GW1CGD
CH8 7QR GW4MOK
CH8 7SJ GW4TIZ
CH8 7UG GW6WQJ
CH8 7UJ GW6VZB
CH8 7UR GW7KIS
CH8 7XG GW4VUH
CH8 8AX GW0UDJ
CH8 8BU MW6IOA
CH8 8DL GW4ZVR
CH8 8DU MW3PZC
CH8 8ES GW0FEU
CH8 8HA MW0CIH
CH8 8HE GW1PJL
CH8 8JY 2W0GXI
CH8 8JY MW0XRT
CH8 8JY MW6GXI
CH8 8NG MW3CBX
CH8 8NJ 2W0OFY
CH8 8SZ MW7NEF
CH8 9BZ MW0LCK
CH8 9HJ MW0ATT
CH8 9HY GW6RCX
CH8 9JB MW1ARM
CH8 9JB MW3ARM
CH8 9JN 2W0FXF
CH8 9JN MW7PJH
CH8 9JQ GW7KBI
CH8 9NZ MW3NGN
CH8 9NZ MW3NGP
CH8 9PQ MW3LBX
CH88ES MW0CRA
CM0 7AH 2E1AEU
CM0 7AL 2E0XDM
CM0 7AL M6UDC
CM0 7BA M3TCR
CM0 7HJ 2E0CUS
CM0 7NF G4JDH
CM0 7QE G6VQV
CM0 7QT G7JDR
CM0 7QT M0RBI
CM0 7RD 2E1EDV
CM0 7RD 2E1EDW
CM0 7RY M6EOV
CM0 7SQ 2E0EYG
CM0 7SQ M6LZI
CM0 7TH G0ATK
CM0 8BT G1IGN
CM0 8EH M0CZC
CM0 8EJ M7LRY
CM0 8EY G4POP
CM0 8EY G6CNQ
CM0 8PZ G6EUW
CM0 8QL M7WPH
CM0 8RB G1LUX
CM08PJ M7RLA
CM1 1DL 2E0NKR
CM1 1DL M6MYR
CM1 1DL M6TQW
CM1 1QB G7KOI
CM1 2AR G8UVY
CM1 2AT 2E0TXL
CM1 2AT M0TXL
CM1 2AT M6FIL
CM1 2BW 2E1GTQ
CM1 2DE M7SNR
CM1 2DZ M1EZE
CM1 2JA G3XRC
CM1 2JA G4YTG
CM1 2NH 2E1HRB
CM1 2NJ G7RFT
CM1 2NQ G0NVM
CM1 2PT G3SLT
CM1 2QZ 2E0CBX
CM1 2QZ M6MAO
CM1 2RR G0NAX
CM1 2SE G7GBZ
CM1 2SS G3NWG
CM1 2SX M6WTF
CM1 2SY 2E0FSX
CM1 2SY M6FNR
CM1 2TE M7SBY
CM1 3EA G4IMS
CM1 3GW M3ZJD
CM1 3HS M6PVM
CM1 3JB M0HBY
CM1 3JZ G4RAP
CM1 3NU G8GWB
CM1 4DA M3DIM
CM1 4DG 2E0JCS
CM1 4DG G1CSR
CM1 4DG G3CSR
CM1 4DG G8GFF
CM1 4DG M3AAY
CM1 4DG M3JCS
CM1 4DN G6ZVV
CM1 4DU G4TRF
CM1 4EF G3VCG
CM1 4EF M7CPH
CM1 4EJ G8AEU
CM1 4HJ G1NZD
CM1 4JQ 2E0KFM
CM1 4JQ M0KVR
CM1 4JQ M3GTV
CM1 4JX 2E0ONH
CM1 4JX M0ONH
CM1 4JX M6ONH
CM1 4JY G4FKH
CM1 4JY G4TOO
CM1 4NW M0ROO
CM1 4PG M0CSV
CM1 4SJ G6ZWC
CM1 4UF M7BDZ
CM1 4UN G4HSK
CM1 4UQ G6DJS
CM1 4XU G8SGM
CM1 4XZ M0BQC
CM1 4YA G4ADG
CM1 4YG M3YNM
CM1 4YR M6HIJ
CM1 6AU M7MGG
CM1 6BF 2E0KIL
CM1 6BF M3HNK
CM1 6BU G0URK
CM1 6FD G0LLB
CM1 6HY G4JJH
CM1 6JF 2E0RMI
CM1 6JF M3RMI
CM1 6JX G4DJC
CM1 6PJ M7ATD
CM1 6QR G3VSI
CM1 6QT M6VGE
CM1 6TX G7RCL
CM1 6UH 2E0VSE
CM1 6UH M6VSE
CM1 6UJ M3NOJ
CM1 6UJ M6ERR
CM1 6UL M7IJB
CM1 6UW G8AAE
CM1 6UX G1EFL
CM1 6UX M0VAM
CM1 6UX M3VAM
CM1 6XE M7PBB
CM1 6XY 2E0HDW
CM1 6YP G8PFL
CM1 7AN G8CDB
CM1 7BU M6NOM
CM1 7PG G8ETN
CM1 7RD M3GUO
CM1 7RX 2E0BWU
CM1 7RX M3OJN
CM1 7RY G6GOV
CM11 1AU G4EVZ
CM11 1ET G0ORT
CM11 1ET G4HTG
CM11 1HB G0NGD
CM11 1LH G0PXA
CM11 1QN G8CCN
CM11 1RR M6GSQ
CM11 2LL G1DHB
CM11 2NX G0TST
CM11 2NZ M0BZC
CM11 2PD G8URI
CM11 2QN G3UTC
CM11 2QN G4SIS
CM11 2RQ G8HGN
CM11 2RY G6KVY
CM11 2TS M6ISG
CM11 2XA G3MVV
CM11 2XE G4XTS
CM11 2YX M6EZP
CM12 0HH G3TTB
CM12 0JF G7PMI
CM12 0PF G8GZV
CM12 0PY M3MQB
CM12 0SU G1UKA
CM12 0SU M3CAD
CM12 0UD M6FVR
CM12 0YE M0DKS
CM12 9DS M0HSX
CM12 9JN G1DSG
CM12 9LX G8DQN
CM12 9NT M1EMU
CM12 9SN G5PMJ
CM129LH M3VUC
CM12BA M7AYS
CM12PR G4BNE
CM13 1BT G3JWI
CM13 1BU 2E0EKC
CM13 1BU M0IPE
CM13 1BU M6IPE
CM13 1DZ M7WTD
CM13 1JD M6WSR
CM13 1JX 2E0JRT
CM13 1JX M6JRT
CM13 1JZ M0TIU
CM13 1LW G6YUY
CM13 1RH G0UBL
CM13 1SJ G0PNO
CM13 2AG G7RCW
CM13 2HZ G8NYD
CM13 2HZ M6PLZ
CM13 2LA G7HRL
CM13 2NF G8CUB
CM13 2RY M7TFD
CM13 2UF G4WYI
CM13 3AL G0FMU
CM13 3DD M0OGS
CM13 3DZ G0LKY
CM13 3RJ M6DII
CM13 3TR 2E0RKB
CM14 4HE M0TKU
CM14 4UT G7HFL
CM14 5AZ G8APZ
CM14 5AZ G8PUR
CM14 5DB M0BAW
CM14 5DG G4XFZ
CM14 5FR M7CZQ
CM14 5JR 2E0DTD
CM14 5JR M6TPD
CM14 5JZ M7RKC
CM14 5NS M0GNC
CM14 5UE G6PZF
CM14 5WT M6HHM
CM145UE M6IIQ
CM14DU G1IBP
CM15 0BQ G4SYR
CM15 0LX G6TNJ
CM15 0NB G8DXV
CM15 0NH M7BHQ
CM15 0PP 2E0XDC
CM15 0PP M0XDC
CM15 0PP M3NUU
CM15 0QX G8SPP
CM15 8BW G3ZRH
CM15 8JL G4NGS
CM15 8PF G4YVW
CM15 8SA G8DWP
CM15 9ND M0MPY
CM15 9ND M0ZNP
CM15 9QA M3NRW
CM15 9QS G4BLD
CM15 9SG 2E0KWG
CM16 5AT M6TMP
CM16 5DL G4LGY
CM16 6BP M3WEA
CM16 6EF M5AGV
CM16 6ES M6SRE
CM16 6EW G0PYV
CM16 6FH M6OBH
CM16 6HA G6OIX
CM16 6JA G7CWX
CM16 6JQ M3EYY
CM16 6JQ M3JXX
CM16 6LF G4WWY
CM16 6PJ M3WYT
CM16 6RR M6ARR
CM16 7BB G1WXS
CM16 7ET G8DZH
CM16 7HT M3GKJ
CM16 7JX G4ACL
CM17 0EX M3MER
CM17 0HD G3UUY
CM17 0HN G8CUA
CM17 0HQ G8MHO
CM17 0JY G0BXL
CM17 0LH G3UEG
CM17 0LQ G0CPU
CM17 0QR G8DNP
CM17 0SB G7ULS
CM17 9AJ M7FXW
CM17 9AU G6DMF
CM17 9AU M6DMF
CM17 9AU M6WKS
CM17 9EU M0OAB
CM17 9EU M6OAB
CM17 9HL M6PVZ
CM17 9JA M6JKA
CM17 9NL 2E1HJW
CM17 9NS M6WII
CM17 9NT M7TKI
CM17 9PL G0AWY
CM17 9PL G6MZV
CM17 9PR G4XUQ
CM17 9PZ M6PTV
CM17 9PZ M6ZRO
CM17 9QG G0KNJ
CM17 9RG M7PEN
CM18 6BL 2E0SIJ
CM18 6BL G6BUT
CM18 6BL G7OBS
CM18 6BL M3OBS
CM18 6BL M6SIJ
CM18 6EB G7IZW
CM18 6ES G0HDZ
CM18 6EY G3RYK
CM18 6HD M6OYE
CM18 6HN M6BUY
CM18 6QE M0JPG
CM18 6QE M3ZBX
CM18 6QG M6GQW
CM18 6QN G8VQE
CM18 6QT M6FKI
CM18 6QW M7GBP
CM18 6QY 2E0AUU
CM18 6ST G8POE
CM18 6TG M6HPX
CM18 6XL 2E0TCI
CM18 6XL M6EJA
CM18 6XX M7AGQ
CM18 7BL M7AGO
CM18 7BY G3WRO
CM18 7BY M3ZSL
CM18 7EE M7WIX
CM18 7HD M6SKZ
CM18 7HF M7SCG
CM18 7PG G4JQU
CM18 7QD G0HRR
CM18 7QD G1FRL
CM18 7QZ 2E0VAO
CM18 7QZ G3ATC
CM18 7RE 2E0YND
CM18 7RE M3YND
CM18 7RF G1OSI
CM18 7RJ 2E0VLT
CM18 7RJ M0PHO
CM18 7RJ M3VLT
CM18 7RJ M7APV
CM18 7SA M7AKA
CM18 7SY G1FIM
CM19 4DB M3GOV
CM19 4DQ M6BUE
CM19 4HE M6FXL
CM19 4NJ G1SFU
CM19 4NW G3GWE
CM19 4QD M6JZH
CM19 4RD G7RFZ
CM19 4RR G0AGO
CM19 5LJ G0MGU
CM19 5NU G6GXO
CM19 5NZ G7AZA
CM19 5RN G0REF
CM19 5RR G1TZS
CM19 5SA G6TPO
CM2 0AS M7MKJ
CM2 0ES G1JIJ
CM2 0FJ G8NWL
CM2 0RX 2E0DNU
CM2 0RX M6DNV
CM2 0SA G8MYJ
CM2 0TY G0MWT
CM2 0TY G1FCW
CM2 0TY G4PVM
CM2 0TY G8CW
CM2 0TY M5BAE
CM2 0UA G6BZQ
CM2 6AP 2E1GQB
CM2 6AQ G4MDB
CM2 6AQ M3HPY
CM2 6AQ M3ZAN
CM2 6AZ G8DET
CM2 6BA G7RVY
CM2 6DH G6CNK
CM2 6DJ G1AJQ
CM2 6DN M0MVO
CM2 6DX M1SNM
CM2 6EB G1INA
CM2 6EW 2E0SES
CM2 6EW M0OER
CM2 6EW M6LFT
CM2 6HA G4JDS
CM2 6QE G6WFM
CM2 6QL G4KGL
CM2 6QN G4LAE
CM2 6QW G8UDD
CM2 6TN 2E0NNZ
CM2 6UH G4LQE
CM2 6XN M3LOT
CM2 7AU G6JYB
CM2 7AX M3IWO
CM2 7BU M7GXS
CM2 7DD G4OLU
CM2 7DS M7ECF
CM2 7JL G1NLZ
CM2 7JN G3PMW
CM2 7LP G0KOM
CM2 7LP G2AD
CM2 7LT G8BOS
CM2 7PP 2E0IWT
CM2 7PP M3IWT
CM2 7SF G3JOX
CM2 7TD G3FMO
CM2 8AH 2E0WHB
CM2 8AH M6UHN
CM2 8AL 2E0LZI
CM2 8AL M0LZI
CM2 8AL M6DNR
CM2 8HY G7KLV
CM2 8JD M7JMX
CM2 8NF G0RLI
CM2 8NT G3YDY
CM2 8NT G8BGV
CM2 8NY G0BDS
CM2 8PD G8GNZ
CM2 8PT G0KOU
CM2 8QB M6IPA
CM2 8QP G3WGE
CM2 8RH M7AWG
CM2 8RQ M6ERF
CM2 8RR M3OQI
CM2 8UR 2E0YYA
CM2 8UR M0MYM
CM2 8XJ G3VUO
CM2 8YY G6FJE
CM2 8YY G6ZNJ
CM2 9AG M6TTR
CM2 9BD 2E1RMD
CM2 9BD M3OGM
CM2 9BZ G8VU
CM2 9DD G3PEM
CM2 9DX G1MVI
CM2 9DZ G3UCD
CM2 9DZ G4BMW
CM2 9EZ M6HNA
CM2 9GJ 2E1GHE
CM2 9HA G0POK
CM2 9LF G1NNN
CM2 9LN 2E0WDY
CM2 9LN M6WDY
CM2 9LW G0SQP
CM2 9NY 2E1IDH
CM2 9NY M3IDH
CM2 9PW G0SXK
CM2 9RD M6SPS
CM2 9RE G3PBT
CM2 9SN G0NEM
CM2 9SN G8VYK
CM2 9XQ G6ZGI
CM20 1SW M0SLC
CM20 2AD 2E0VNT
CM20 2AD M6VNT
CM20 2JG M0TJB
CM20 2LA 2E0OXT
CM20 2LA M6OXT
CM20 2PA M3WMV
CM20 2PZ G4GGX
CM20 2TJ M6PFV
CM20 3BW M3APA
CM20 3DP G8CHO
CM20 3DW 2E0JOP
CM20 3DW M3ZLJ
CM20 3DY G4MQG
CM20 3EF G6UXX
CM20 3EY G7UFF
CM20 3EY M0NRY
CM20 3EY M3LEN
CM20 3HB G4YBN
CM20 3HL M6FQX

CM20 3JP M0XID
CM20 3LE G8XGO
CM20 3LY 2E0MBD
CM20 3LY M3NBD
CM20 3NH M7DXF
CM20 3RH M6YGL
CM202LA 2E0OGA
CM202LA M6OGA
CM21 0AU G4PGB
CM21 0BT G8WUG
CM21 0DZ G1XRT
CM21 0ER G4SNL
CM21 0HX G4ERS
CM21 9ED G7NBQ
CM21 9HG G8FUL
CM21 9NF G0UEU
CM21 9NF G4DMI
CM21 9NN G3NWX
CM21 9NT G3WOO
CM22 6BG M0KIB
CM22 6JP G1RBO
CM22 6LG G6IWT
CM22 6NT M3HKV
CM22 6QX 2E1UKT
CM22 6QX M6UKT
CM22 6SW G4PFJ
CM22 7AJ G7VAB
CM22 7AS G8BPH
CM22 7FF M6DJQ
CM22 7LT G4PJD
CM22 7NB G6ART
CM22 7PA M1SGA
CM22 7RT M6TTF
CM22 7SJ G4WWP
CM22 7SQ G7VRX
CM22 7TP G7VJE
CM226UT 2E0WWK
CM226UT M6BLB
CM23 1BD M0DXZ
CM23 1HX G6ECS
CM23 2BL G3WYD
CM23 2BQ G0CGH
CM23 2HU M0GCA
CM23 2HX G4CWH
CM23 2ZR M7LVC
CM23 3DH M7SPC
CM23 3EU G7CQX
CM23 3EX G3KCG
CM23 3EX M0AQO
CM23 3JN G7TET
CM23 3LR M6IFW
CM23 3NG M0AZR
CM23 3NP G4SUA
CM23 3NQ G4WPO
CM23 3PU 2E1WGB
CM23 3PU M5REG
CM23 3PX G1NRK
CM23 3PX G1OSH
CM23 3QH M3XBC
CM23 3QN M6BUA
CM23 3RJ G3PRI
CM23 3YW 2E0WAG
CM23 4AD G0PQF
CM23 4AD G1ARU
CM23 4AD G5ZG
CM23 4BL G4SNJ
CM23 4DU G0PLS
CM23 4EB M7DJK
CM23 4EY G0GZU
CM23 4HU G0XAD
CM23 4JL G3SUS
CM23 4JP G4OBV
CM23 4LL G6LIB
CM23 4LL G8THH
CM23 4PD G6YMU
CM23 5AP G3ZXF
CM23 5DE G8ZPH
CM23 5DR M7CHC
CM23 5DS G4HXH
CM23 5HY G4GIS
CM23 5JJ G7WJV
CM23 5LP M0ITY
CM23 5LS G4LSE
CM23 5NU G8SEY
CM23 5QH G0ODE
CM24 8ED M7COX
CM24 8HX 2E0EAL
CM24 8HX M0GZM
CM24 8HX M6JRO
CM24 8LR G3PEW
CM3 1AS G6ANO
CM3 1DA G1UZC
CM3 1DA M1BWN
CM3 1ND 2E0GZD
CM3 1NH M7SWX
CM3 1NP G8LVW
CM3 1NR G6YJH
CM3 1PN M6NNO
CM3 1QS G3WSV
CM3 1RB G6OTE
CM3 1RT G1GKN
CM3 1RT G6EEE
CM3 2AY G4VVQ
CM3 2DF G8RSX
CM3 2JA M3AAS
CM3 2JE 2E0NOD
CM3 2JE M3NOD
CM3 2LB G8LYV
CM3 2LJ G4OAU
CM3 2LP 2E0DEO
CM3 2LU M0VRS
CM3 2QS G4KTX
CM3 3AD G6CMS
CM3 3BY G4DBM
CM3 3EF M3MFX
CM3 3JD 2E0WDW
CM3 3JD M7TFW
CM3 3JJ 2E0BOB
CM3 4DP G3JCM
CM3 4HT G7KSQ
CM3 4HT M6BDL
CM3 4LS G3PJC
CM3 4LS G4EOG
CM3 4PG M0DTA
CM3 4PH G7UVP
CM3 4PN M6TTW
CM3 4PS G0TRM
CM3 4QT M6MVW
CM3 4RB M6BAK
CM3 4RP G4TNB
CM3 4SD G8CYK
CM3 4XL M3DRH
CM3 4XN 2E0PXD
CM3 4XN M0PXD
CM3 4XN M6PXD
CM3 5FP M0DVD
CM3 5FS G7HBV
CM3 5FS M3EQY
CM3 5GH G4GYJ
CM3 5GY G3STP
CM3 5JF G4DUJ
CM3 5JJ M0AMZ
CM3 5JL 2E0JOI
CM3 5JL M0XJL
CM3 5JL M6JOI
CM3 5NJ G0AUR
CM3 5NL 2E0RFH
CM3 5NL M3RFH
CM3 5NP 2E0MYZ
CM3 5NY G7VEE
CM3 5PF G7NBR
CM3 5PQ G3XMB
CM3 5SB G0BCW
CM3 5TS G0VGC
CM3 5TS G7MOY
CM3 5TY G1YJF
CM3 5WH G8NAM
CM3 5WP 2E0FEQ
CM3 5WP M0XYM
CM3 5WP M7ESX
CM3 5YG G6KWY
CM3 5ZT G7BHE
CM3 5ZU 2E0JPU
CM3 5ZU M0GGX
CM3 5ZU M3JPU
CM3 6AQ G4TWL
CM3 6BY G0IJN
CM3 6BY G0UTT
CM3 6DD G0LWL
CM3 6DF G4THV
CM3 6DP G7EHS
CM3 6DQ G6TXV
CM3 6DT G8WRB
CM3 6DU G3ZES
CM3 6EJ 2E0GRA
CM3 6EP 2E0IWZ
CM3 6EP 2E0THT
CM3 6EP M0IWZ
CM3 6EP M0UHH
CM3 6EP M3IWZ
CM3 6EP M6LLH
CM3 6EU G1SNU
CM3 6EU G7LIK
CM3 6JE G3HMQ
CM3 6LF G6AHD
CM3 6NF G4BUP
CM3 6NZ 2E0DJQ
CM3 6PZ G0MLO
CM3 6RX 2E0DVH
CM3 6TP G7TAV
CM3 7AH M1GUS
CM3 7BD 2E0FJJ
CM3 8AP M6GFF
CM3 8EL G3XYI
CM3 8EW G6BCL
CM3 8HJ G8CRX
CM3 8UT M3USH
CM3 8XA G4OGL
CM3 8XA G8EGE
CM4 0AL G3FJO
CM4 0HA G8YQO
CM4 0RS G8GIG
CM4 9DT G8LAB
CM4 9QA G7WKP
CM5 0BL M6MNH
CM5 0ES M6BUC
CM5 0QG G0KKH
CM5 9BG G4KVR
CM5 9HH 2E0WKU
CM5 9HH G4JLL
CM5 9HH M7XVO
CM5 9HW M3YIF
CM5 9QN G6NAX
CM5 9QW G4TOX
CM6 1AJ G7JXL
CM6 1BP M6WWM
CM6 1BY G7DPR
CM6 1BY M6DWO
CM6 1FX M7GTL
CM6 1WB 2E0RHA
CM6 1WB M0RXA
CM6 1WB M6OLH
CM6 1WU M6DVS
CM6 1XQ M6BCW
CM6 2AP M5BIL
CM6 2BL M0DXR
CM6 2BS M1SUE
CM6 2BW M7LTR
CM6 2DL G0PTG
CM6 2GR 2E0FGO
CM6 2GR M0PXO
CM6 2GR M7PJO
CM6 2JR 2E0MAA
CM6 2JR M0MAF
CM6 2JR M3RXQ
CM6 2PD M6DVY
CM6 2PU G1ILG
CM6 2PU M3ILG
CM6 2QZ 2E0XGY
CM6 2QZ G8GSY
CM6 2SQ M3ZIF
CM6 3AW G6WYH
CM6 3AX G7TKI
CM6 3BP M0DUT
CM6 3GR 2E0IWR
CM6 3GR M7COF
CM6 3JT M6SQK
CM6 3LU M6LMC
CM6 3NW G5TCP
CM6 3NW M7GHK
CM6 3PJ G8HUT
CM6 4AS M6KQA
CM7 1AB M3TPU
CM7 1AR M6WTT
CM7 1DG M6OMK
CM7 1DL M3RTI
CM7 1DR G1JSK
CM7 1EB G4TVX
CM7 1TD G3YVK
CM7 1XF G6HSC
CM7 2LR 2E0PAB
CM7 2LR M3XOL
CM7 2NL M7FMI
CM7 2QE M7UKU
CM7 2RZ G6XCU
CM7 2TA M6RNZ
CM7 2TX M7NAS
CM7 2WB G6OIY
CM7 3DT M7AEA
CM7 3JR M3TXH
CM7 3JX 2E0XRS
CM7 3JX M3YFW
CM7 3LG G4ICP
CM7 3NR 2E0JSE
CM7 3NR M0JBY
CM7 3NR M6NMU
CM7 3NW M6MNO
CM7 3PE G3TGB
CM7 3RN M0KJH
CM7 4BY G8PLO
CM7 4DG M7TCM
CM7 4JZ G4ONH
CM7 4LE G4YQL
CM7 5EG G3WVR
CM7 5FS 2E1IAI
CM7 5HN G4YWN
CM7 5HU G6DIO
CM7 5NB G0HWQ
CM7 5PY G0EMK
CM7 5PY G3XG
CM7 5PY M0CLO
CM7 5QJ M6VIO
CM7 5SD 2E0TKY
CM7 5SH G0OSI
CM7 5UE G7IRK
CM7 5UQ G1XWN
CM7 9DZ 2E0XLS
CM7 9DZ M3OIL
CM7 9ES 2E0NPX
CM7 9ES M6NPX
CM7 9LL G0DEC
CM7 9LL G1YYY
CM7 9LL G8XVO
CM7 9LR G3ITL
CM7 9NF G0DCR
CM7 9NJ G1WRH
CM7 9RH M6POC
CM7 9RX M7SLK
CM7 9TX 2E0WAV
CM7 9TX M0WAV
CM7 9TX M6WAV
CM7 9UG 2E1HYX
CM7 9UJ G7PCG
CM7 9UR G1TZZ
CM75LW G8DJO
CM75LW M0IAE
CM77 6BY G4WXT
CM77 6DE G7SLJ
CM77 7JW G6EJF
CM77 7PX G8ZIW
CM77 7UA G6IOW
CM77 7YJ M7GNB
CM77 8BP G1GNQ
CM77 8BP M0VZA
CM77 8BP M0XG
CM77 8HN G7HHN
CM8 1HG 2E1IGJ
CM8 1HP M6PXT
CM8 1HR G6CSK
CM8 1JB 2E0IRX
CM8 1JB 2E0KDG
CM8 1JB M6IRX
CM8 1JB M6KDG
CM8 1JZ G3MMA
CM8 1LX M7WDW
CM8 1QJ G0FWA
CM8 1TJ M3RYT
CM8 1XP 2E0FJC
CM8 1XP M7AEB
CM8 1XY G8NGM
CM8 2AQ M0IYZ
CM8 2ES M6KRW
CM8 2LH G6GXG
CM8 2LJ G8WSC
CM8 2LL G0DJK
CM8 2NP 2E0NRH
CM8 2NP M0NRH
CM8 2NP M3ILM
CM8 2NP M3TSA
CM8 2PA M6ENM
CM8 2PE G1SDN
CM8 2PT 2E0XIS
CM8 2SZ G1NNB
CM8 2SZ G7BIQ
CM8 2SZ G7LEY
CM8 2TY 2E0CSL
CM8 2TY M0GGL
CM8 2TY M3JYP
CM8 2UQ G7BND
CM8 2UQ M3BAW
CM8 2XG G3KXI
CM8 2XL M6LIK
CM8 2XQ G8CMO
CM8 3JR G1UZD
CM8 3LN G3LVL
CM8 3LT G7NAI
CM8 3LZ G3VYD
CM8 3NR G6AUD
CM8 3NS G3XAX
CM8 3NZ M0PWB
CM8 3PH G3XVV
CM8 3QY 2E1IIV
CM8 3QY M0UTH
CM8 3RZ 2E0RNS
CM8 3RZ M3RNS
CM8 3SN M0TCT
CM8 3SP 2E1HGG
CM8 3SP G4KQE
CM8 3SP G8WOZ
CM8 3SP G8WQZ
CM8 3SY G8UUO
CM9 4BL G6RKF
CM9 4BW M0RAX
CM9 4LF M1OCN
CM9 4PB G1WFA
CM9 4RQ G0AAN
CM9 4SE 2E0WAA
CM9 4SE M0WAG
CM9 4SE M3WPI
CM9 4TY G1PMU
CM9 4US G1YKZ
CM9 4YN G0MBY
CM9 4YN G8DHV
CM9 4YX G8JLM
CM9 5DP G3VNP
CM9 5EE G1BQG
CM9 5HF 2E0KCP
CM9 5HF M0KCP
CM9 5HF M6KDP
CM9 5HY M6ROW
CM9 5JJ G8GLP
CM9 5JQ M6BZW
CM9 6BE 2E0OVF
CM9 6BE M0SHQ
CM9 6BE M6AEE
CM9 6DJ G4HRB
CM9 6ES M7RUS
CM9 6EW G1NMN
CM9 6EW G4MOV
CM9 6JB M7NWT
CM9 6JF G7RFC
CM9 6JF G7SRC
CM9 6JF G8PYD
CM9 6JF M1BDR
CM9 6JH G6LJR
CM9 6LR G6PCP
CM9 6QP G3UYC
CM9 6UH G4MNT
CM9 6UQ G8WTM
CM9 6YA M7ACV
CM9 6YX G7PIX
CM9 8AY M1GFE
CM9 8AZ G8UVG
CM9 8BW G0UHD
CM9 8BY M1DOZ
CM9 8EH M3YWO
CM9 8HA M6EAP
CM9 8LL G0VDV
CM9 8NE G8CCL
CM9 8NE M3CCY
CM9 8PB G4EMB
CM9 8PB G6OXQ
CM9 8PJ G0VDP
CM9 8PR G3ZOX
CM9 8QA G0ASN
CM9 8RJ G6CYE
CM9 8SB G3GLL
CM9 8SR G4IIH
CM9 8UA G1FXM
CM9 8UD G1FWY
CM9 8UF G6IFH
CM9 8UN 2E1DET
CM9 8UN M3DVP
CM9 8UN M6JEK
CM9 8XB G0IBN
CO1 1GJ 2E0ONM
CO1 1GJ M6ONM
CO1 1PR G0VNY
CO1 1UR 2E0BYF
CO1 1UR M3BYF
CO1 2GQ 2E1FRI
CO1 2GQ M1CUY
CO1 2HS 2E0ETY
CO1 2HS M0IQR
CO1 2HS M6XWR
CO1 2HU G7LYH
CO1 2JJ M6AWP
CO1 2LF 2E0FTQ
CO1 2LF M6FTQ
CO1 2SN M0LCX
CO1 2UT M1FJD
CO1 2WA M0WNF
CO10 0BZ M6OYM
CO10 0DJ G0PAO
CO10 0DX M3LEL
CO10 0EP G0VAS
CO10 0FH G4LSK
CO10 0HH G1DEU
CO10 0JN G4GGC
CO10 0LJ M6DNX
CO10 0NH M6LXU
CO10 0NN M7SFB
CO10 0NS M0ICH
CO10 0NY M7KJD
CO10 0QP G8AAR
CO10 0RJ 2E0REL
CO10 0RJ M6REL
CO10 0RN G6RKG
CO10 0YE 2E0BYC
CO10 0YF G7SRA
CO10 0YF G8LTY
CO10 0YU G4JAA
CO10 1HT 2E0ORX
CO10 1HT M7TER
CO10 1JB G6DKE
CO10 1PJ G7HMF
CO10 1QX G7MLK
CO10 1WL M6NOI
CO10 1YU G3PZU
CO10 2BU G7NZV
CO10 2PP G7UTC
CO10 2QH M0JMV
CO10 2TP G1IUW
CO10 2TU M3TFP
CO10 5JA G1YJL
CO10 5JN 2E0TBZ
CO10 5JN M0TBZ
CO10 5JN M6GLX
CO10 6JB G1PMF
CO10 7AR 2E0CNE
CO10 7AR M0HGS
CO10 7AR M6AWL
CO10 7EZ G6VSM
CO10 7JX G4DHU
CO10 7LS G1EUF
CO10 7NA M7EDK
CO10 7PH G4OBB
CO10 7PN M0COJ
CO10 7PS M0AUR
CO10 7PS M0CEQ
CO10 7QZ M1COQ
CO10 7RL G3YAI
CO10 7RS G4SWQ
CO10 7RW G3WRD
CO10 7SJ G0PGS
CO10 7SJ G7KSH
CO10 8AS G7MTQ
CO10 8AS M3MTQ
CO10 8DA G4IMM
CO10 8HA G4HGH
CO10 8HA G1BWX
CO10 8JN G4HTZ
CO10 8JP G6FCL
CO10 8JS G8GRL
CO10 8LQ 2E1MJH
CO10 8LQ 2E1ORT
CO10 8LQ M0MJH
CO10 8LQ M3MJH
CO10 8NN G8CDG
CO10 8PD G6UQZ
CO10 8QD M7EZO
CO10 8UA M7GOZ
CO10 9LF G6CHJ
CO10 9QD G1TWY
CO10 9RU G0SNU
CO10 9SD M3NIT
CO100BD M6ODK
CO11 1DL M0CGE
CO11 1NS G1XMH
CO11 1RN G1NNA
CO11 2AL G0NXH
CO11 2AL G4JVM
CO11 2BU 2E0KLF
CO11 2BU M0JVC
CO11 2BU M6KLF
CO11 2DT G7EPM
CO11 2GE M7PJD
CO11 2HE M0BGE
CO11 2HE M6EWJ
CO11 2HX G3YAJ
CO11 2HZ 2E1RFS
CO11 2LH M0AFX
CO11 2QP M1DVO
CO11 2RB G7VFC
CO11 2SU M3JIG
CO112SP G6UWK
CO12 3AS M6WAS
CO12 3BD G4WHK
CO12 3DS M3DZT
CO12 3FF M6KPE
CO12 3HH M7BXM
CO12 3HS G0OEY
CO12 3JA G1PJJ
CO12 3JZ M0NWW
CO12 3NR 2E0TUD
CO12 3PL G8XAX
CO12 3PP M0DSY
CO12 3PP M6DSY
CO12 4BU M6AZX
CO12 4BW G3UJB
CO12 4DW M0RIC
CO12 4DW M0TYG
CO12 4EQ G1VCZ
CO12 4HR G7HMN
CO12 4JG M7AGD
CO12 4JU M6BLO
CO12 4LA M0BVQ
CO12 4NF 2E0NRN
CO12 4NF G4NDR
CO12 4NF M6GMU
CO12 4NT M6RHK
CO12 4QP M6UZT
CO12 4QU 2E0GYY
CO12 4QU G7OWB
CO12 4TS G0STW
CO12 4UZ 2E0HEI
CO12 4UZ M7BLD
CO12 4XN G3VTT
CO12 4XN M0GVP
CO12 5AS M7AKP
CO12 5BQ G0RDX
CO12 5BS G2ABR
CO12 5DH G4JIG
CO12 5EE G4YJQ
CO12 5EJ G3YYZ
CO12 5HE G4GBY
CO12 5JF G4EYE
CO12 5NE G8PAI
CO12 5NN G4TZM
CO124NF M6SNU
CO13 0AB 2E0UPM
CO13 0AN G6CDU
CO13 0AQ G0KCD
CO13 0BH G0NLG
CO13 0ET G1WJR
CO13 0EW M0LWM
CO13 0JL G0MJA
CO13 0LQ 2E0XLH
CO13 0LQ G6NHU
CO13 0LQ M6DHU
CO13 0LQ M6SHU
CO13 0LQ M7RTO
CO13 0QH G3OMB
CO13 0SF G4OAX
CO13 0SW 2E0MGP
CO13 0SW M0MGP
CO13 0SW M3NGC
CO13 0TG G4ZDY
CO13 0TH M3LIX
CO13 0TQ M6FCV
CO13 0TW M6HQL
CO13 9DG G8FRH
CO13 9DH M7IWW
CO13 9DW G4PBD
CO13 9HR G3OJS
CO13 9JE G6WHT
CO13 9JZ 2E0NTR
CO13 9LY G0HPM
CO13 9NH M6PEU
CO14 8BL G0WWR
CO14 8BU G0GYY
CO14 8DF G8HSI
CO14 8EW G0OFF
CO14 8LE G6GJV
CO14 8LT M6ALY
CO14 8NF 2E0FYA
CO14 8NF M0LTG
CO14 8NF M3MGP
CO14 8NF M6SXG
CO14 8NJ M6IWX
CO14 8PE 2E0DPX
CO14 8PX G3VLD
CO14 8RE G1ISS
CO14 8RG G3CO
CO14 8RG G6XOU
CO14 8RG M1COL
CO14 8RG M3LFO
CO14 8RL G3TUU
CO14 8RL G7BWF
CO14 8RR 2E1GXH
CO14 8SP M3ZLY
CO14 8SR G4MVX
CO14 8SS G0KKT
CO14 8SS G0OIF
CO14 8UD G1HZD
CO14 8UE 2E1FPU
CO15 1EJ G4AQZ
CO15 1HA G6DLZ
CO15 1HB 2E1HGU
CO15 1HB 2E1HLP
CO15 1HB G7PMU
CO15 1NQ 2E1HHB
CO15 1TX 2E1HAQ
CO15 1TX M0GWE
CO15 1TX M3KEV
CO15 1UW G6EAR
CO15 1UX M6TBK
CO15 2DA G1JOA
CO15 2DB M6MMN
CO15 2DD G6AGN
CO15 2JN G6WIL
CO15 2JW M6EVD
CO15 2NN G0KFV
CO15 2NU 2E1IGU
CO15 2PA G6JIR
CO15 2PB 2E0ANM
CO15 2PB M7KOY
CO15 2PE G4PQU
CO15 2QB 2E0DKM
CO15 2QB M6DKM
CO15 2QT G8DMT
CO15 2RW M0HTF
CO15 3DW 2E0IKH
CO15 3HS G6CYA
CO15 3JR M0CHL
CO15 3LA G1YJJ
CO15 3LA M0RBC
CO15 3LA M3ZSC
CO15 3LZ G1XZX
CO15 3NF G7AJG
CO15 3QX G0UHU
CO15 3SB M0DEX
CO15 3SE G6XJC
CO15 3SR G0DGH
CO15 4EU 2E0VBJ
CO15 4EU M6NBH
CO15 4JE G3ZEZ
CO15 4LH G8ZZL
CO15 4ND G0LQT
CO15 4NN G7UHX
CO15 4PA G4AEB
CO15 4PJ G8TOI
CO15 4RJ G4ZXS
CO15 4RJ G8ATL
CO15 4RU G0VAL
CO15 4RZ G0HWC
CO15 5DG 2E0FBQ
CO15 5DG M6EML
CO15 5DJ G3LZR
CO15 5JF 2E0COI
CO15 5JF M6BEZ
CO15 5LA G4ZJK
CO15 5NA G0SMJ
CO15 5PN M7BNI
CO15 5QQ M7NDL
CO15 5RH 2E1HIO
CO15 5RU M6KCT
CO15 5SQ 2E0TNO
CO15 5SQ G0NAD
CO15 5SQ M7ADW
CO15 5SR G7KME
CO15 5XH M3TLO
CO15 5XY M0DNO
CO15 6BE M6LKF
CO15 6DP G6CTY
CO15 6EN M0IWQ
CO15 6JL M6OLB
CO15 6LN G4TTQ
CO15 6NL 2E0XTA
CO15 6NL M6OKZ
CO15 6NX G7MAV
CO153LA M7AVB
CO16 0AX G0UBJ
CO16 0BU M0PJK
CO16 0EG 2E0CWZ
CO16 0EG M0KEB
CO16 0EG M6FGG
CO16 0EP G4CYF
CO16 0HT G1JCP
CO16 0NT G8JRI
CO16 7HF G0MBA
CO16 7HF G0PKT
CO16 7HG G8LBR
CO16 7HQ G4OJH
CO16 7HQ G8JPJ
CO16 8AG M7WZO
CO16 8BN G4RUJ
CO16 8DB G4WHZ
CO16 8DG G8LID
CO16 8DL M6TMY
CO16 8EP G1FWR
CO16 8FB G7RIA
CO16 8FE G7HJK
CO16 8FQ G0DZZ
CO16 8FQ M0JXY
CO16 8HB G0HCY
CO16 8HH M6SDI
CO16 8PH G3CRC
CO16 8PH M1EAK
CO16 8QE G1NQB
CO16 8RQ G0CKD
CO16 8RQ G0RIU
CO16 8TH M0CNL
CO16 8UA G6XHI
CO16 8UN M7BHN
CO16 8US G4FJT
CO16 8YB 2E0NNG
CO16 8YB M6ORM
CO16 8YD G6TNQ
CO16 8YE G0GEQ
CO16 8YU G4RJQ
CO16 8YX M6RXU
CO16 9AN G3VMP
CO16 9DH G6LTD
CO16 9EN M6WXY
CO16 9ES G3PQM
CO16 9NE G8CVP
CO16 9PS G0FIW
CO16 9PS G1OGY
CO16 9QP G0EBI
CO16 9RS G7BVZ
CO167DX 2E1PAW
CO168BX M0CHS
CO168PD M7BHM
CO169EN G1IPU
CO2 0DL 2E0JJN
CO2 0DL M0JJN
CO2 0DL M6JJN
CO2 0DR M7DPH
CO2 0JE G4YJN
CO2 0NQ G4YOF
CO2 7EN G6DNA
CO2 7HH M7KIK
CO2 7LD G7WLC
CO2 7LQ G4NNN
CO2 7PE M3WWR
CO2 7UG G7EQX
CO2 7UX G6FBB
CO2 7XB M7FQB
CO2 8AR G6XWY
CO2 8BH G1XAJ
CO2 8BP G7PST
CO2 8DD 2E0OVX
CO2 8DD M6OVX
CO2 8EE 2E0KFG
CO2 8EG M3EUM
CO2 8FU M6MYI
CO2 8GE M7MFH
CO2 8GY M3RPA
CO2 8HZ M7DVY
CO2 8LT M6KCU
CO2 8NS G7TAT
CO2 8NS M6NUL
CO2 8NS M6POU
CO2 8NZ G8UBU
CO2 8PA G7UFT
CO2 8QF G7RWF
CO2 8QY 2E0LSI
CO2 8QY M0LSI
CO2 8QY M6EPZ
CO2 8SJ G4DIS
CO2 8UB G7ELH
CO2 8UD G6FLW
CO2 8UD M0VLL
CO2 8UJ M6EBW
CO2 8UP G8RFC
CO2 8WW G6RVP
CO2 9DT M0KTR
CO2 9DZ M0KAP
CO2 9HN M3WMO
CO2 9JR M1CZY
CO2 9JR M3CZY
CO2 9NE M6ASK
CO3 0HP G4CTA
CO3 0HP G6WYL
CO3 0HY G8VJR
CO3 0NP G7CGC
CO3 0RZ M1BPY
CO3 0YA M0BCK
CO3 0YJ G0NHB
CO3 3BP M6HWX
CO3 3HE M0LFS
CO3 3HW G4UTJ
CO3 3QE M6ALU
CO3 3RS G0KOY
CO3 4BU M7EIR
CO3 4JJ G0WZV
CO3 4JP G4JKC
CO3 4NQ G6DFZ
CO3 4NU M6SCA
CO3 4PS 2E0JBK
CO3 4PS M3JBK
CO3 4QN G4SOB
CO3 9AQ G6RDD
CO3 9DP G0HHC
CO3 9TR M1XXT
CO4 0AJ G0OWK
CO4 0DA M7LSB
CO4 0DB M6GVI
CO4 0EA G6IIF
CO4 0HX G3MQR
CO4 0LD 2E0BMG
CO4 0LD M3LFQ
CO4 0LP G4OYH
CO4 0PJ G4AUG
CO4 0PW 2E0DPW
CO4 0PW M0YNK
CO4 0PW M6YAN
CO4 0QJ G6AQI
CO4 3AS G8SGB
CO4 3DS 2E0KFV
CO4 3DS M0KFV
CO4 3DS M7OCO
CO4 3EY G4HKC
CO4 3FD G4HKB
CO4 3FE G0VYC
CO4 3FN G4RKB
CO4 3HB G4JHP
CO4 3HG M7ETF
CO4 3JA M0ASG
CO4 3JE M6YEF
CO4 3JH G7NID
CO4 3JP G0PFM
CO4 3JR M1AMI
CO4 3JR M3DFC
CO4 3LX G7EHY
CO4 3NF G0VAE
CO4 3NF M0DEQ
CO4 3NF M3ZZN
CO4 3NL G1DTE
CO4 3SQ M0UOE
CO4 3SQ M3PQI
CO4 3SQ M3TFA
CO4 3SZ M3SCH
CO4 3UT 2E0CEA
CO4 3UT M6CEA
CO4 3XN M0CVH
CO4 3YA G6CLA
CO4 3YH G0SLB
CO4 3YP G3ZOL
CO4 5AF M7LUV
CO4 5BD M3KMS
CO4 5DU 2E0TFH
CO4 5FY G0ULY
CO4 5JT G8MKN
CO4 5JX 2E1DXB
CO4 5JX M3AJD
CO4 5LQ 2E1EAG
CO4 5PE G0AHE
CO4 5PY G8WBY
CO4 5RD G6KTR
CO4 5RN M6XEL
CO4 5RX M0TLR
CO4 5RX M3UGT
CO4 5SE M6LWV
CO4 5YA M3YAL
CO4 5YE M7BRO
CO4 5ZR M6HWH
CO4 9EA G4ZOR
CO4 9RE M3JBE
CO4 9RJ M1BQZ
CO4 9RR M6HJA
CO4 9SL 2E1EWK
CO4 9ST 2E0GMM
CO4 9ST M3OEG
CO4 9WQ G7BIV
CO4 9YD 2E0IGM
CO4 9YD M6IGM
CO4 9YU G1GWO
CO45SE 2E0EMZ
CO45SE M0ITB
CO45SE M6KNV
CO5 0AA G6XIF
CO5 0AE G6KVI
CO5 0AY G4TFI
CO5 0BN G7JMW
CO5 0DH M6KZZ
CO5 0DT G4KTB
CO5 0EF G4TFP
CO5 0EF G7NMI
CO5 0EN M0JJH
CO5 0HJ M1ASV
CO5 0HN M0HIG
CO5 0JG M6CHP
CO5 0JG M6FSC
CO5 0JJ 2E0VVB
CO5 0JJ M7JDH
CO5 0JJ M7JWH
CO5 0JU G4ZZL
CO5 0LJ 2E0MNH
CO5 0LJ M0JBV
CO5 0LJ M6IFC
CO5 0LQ M0ECL
CO5 0LR G0IZK
CO5 0QD G3GML
CO5 0QT G6DBY
CO5 0RP G6YWU
CO5 0SL M0XDA
CO5 5AD 2E0NOW
CO5 5AD M0HKG
CO5 5AD M3HHB
CO5 7AS G4PZX
CO5 7EH M7BZB
CO5 7EL M7EOT
CO5 7ET G6KNM
CO5 7HF G3KHK
CO5 7LB G4YK
CO5 7LJ G4PBR
CO5 7PT G7UQQ
CO5 8BD M7ADX
CO5 8DR G8SOI
CO5 8EB 2E0WFR
CO5 8EB M1RJS
CO5 8EB M7WFR
CO5 8GF 2E0DPV
CO5 8GF M7VET
CO5 8GG M5RHG
CO5 8GJ G8IFN
CO5 8JU M3BSH
CO5 8LN G0JDE
CO5 8PX M1EVN
CO5 8QP G0APP
CO5 8RW 2E0FIO
CO5 8RW M7GNC
CO5 8RY G6OOT
CO5 8SL M1CAY
CO5 8SS M0PUC
CO5 9AG G8ADZ
CO5 9AH G8AFQ
CO5 9TD G3NXK
CO5 9TH G4PYG
CO5 9TU M3VXK
CO6 1BP G0NMB
CO6 1DB G3YEC
CO6 1NY G1GMM
CO6 1RQ G1NPX
CO6 1SG G0WJD
CO6 1SY M6YUM
CO6 1UL G7VGC
CO6 1XG G1IUD
CO6 1XG G6LJF
CO6 1XG G8IUD
CO6 1YN 2E0JPR
CO6 1YN M0JAI
CO6 1YN M6ETU
CO6 2AS M6VLE
CO6 2BE M7SVL
CO6 2DZ G8LHF
CO6 2NH 2E0VAN
CO6 2NH M0YEP
CO6 2NH M3VXC
CO6 2PF G7LNU
CO6 3DB G0VHF
CO6 3DB G4ZTR
CO6 3DB M1CRO
CO6 3DX G8VAF
CO6 3EP G8YMZ
CO6 3HY 2E0FBA
CO6 3JS M6UJD
CO6 3LX 2E0CGV
CO6 3LX M0HBV
CO6 3LX M6ALT
CO6 3NJ 2E0WDM
CO6 3NJ M3VZV
CO6 3RY 2E0ORP
CO6 4BJ G6XMM
CO6 4EH 2E0ALB
CO6 4EH M3XMA
CO6 4LT G4DKX
CO6 4NS 2E0NNP
CO6 4NS M0IUI
CO6 4NS M6NNP
CO6 4PQ G7DTT
CO6 4QH G3LST
CO6 4UN G1AWK
CO6 4UU 2E0TLW
CO6 4UU M0WCE
CO6 4UU M7UKZ
CO6 5AG M0EJG
CO6 5AR G0IBZ
CO62LU M7GGF
CO63LX 2E0IPG
CO63LX M6IPG
CO7 0AQ 2E0KKC
CO7 0DU G6AEB
CO7 0HE 2E0DPZ
CO7 0HE M6LYD
CO7 0HU M7RGD
CO7 0JP M7DGP
CO7 0LA G4IZX
CO7 0LB G4CIA
CO7 0NA G1XUU
CO7 0NZ G6PMD
CO7 0PE G3ELS
CO7 0PR G3MGW
CO7 0QR G4JAC
CO7 0RH G7KRO
CO7 0RP G7TBU
CO7 0RS G0SOX
CO7 0RU G4EUW
CO7 0SJ G1BFF
CO7 0SJ M0UEI
CO7 0ST G8DRE
CO7 6ES G4PZL
CO7 6QY G1VZT
CO7 6RF G0VEI
CO7 6SJ G8CJD
CO7 6TP G1CHN
CO7 6TP M7EOH
CO7 6TX G6VJK
CO7 7AS M0PDE
CO7 7AS M0PDF
CO7 7EG G7USX
CO7 7GA G3YVW
CO7 7PQ 2E0HCO
CO7 7PQ M0OKD
CO7 7RP 2E0HUM
CO7 7RP M6YES
CO7 7RY G8GML
CO7 7SD 2E0BZM
CO7 7SD M3ZMQ
CO7 7SY G6IGU
CO7 8AU M6NNW
CO7 8BQ M7MUF
CO7 8DD 2E0DDU
CO7 8DD M0XLB
CO7 8DD M6SEY
CO7 8DH G0GYI
CO7 8HS 2E0WMG
CO7 8HS G8EWC

CO7 8JA G4AZD
CO7 8JB G4URA
CO7 8JB G1TWH
CO7 8JH M3DEB
CO7 8LH M6BAR
CO7 8LJ G4ZKS
CO7 8NN G0PJZ
CO7 8NN M7MRT
CO7 8RF M6TRP
CO7 9EH G0BEP
CO7 9JX 2E0WIV
CO7 9JX M6XUU
CO7 9LD G0GGM
CO7 9LG 2E0AIK
CO7 9LG G4SDI
CO7 9LS 2E0PXY
CO7 9LS 2E0PXZ
CO7 9LS M0PXZ
CO7 9LS M3PXY
CO7 9LS M3PXZ
CO7 9NH G4FTP
CO7 9QH M1BQF
CO7 9QQ G4NXR
CO7 9QZ G7TWA
CO7 9RP M3XGB
CO7 9SD M1DYW
CO8 5BN G6WPJ
CO9 1AS G1ZLD
CO9 1BJ M1CJF
CO9 1ED G6LMC
CO9 1EH G7UUA
CO9 1EH G7UUD
CO9 1JU G4YAX
CO9 1JU G6NEK
CO9 1LB G0GWN
CO9 1NJ M7LCW
CO9 1NY G4FJC
CO9 1PD G4BCV
CO9 1PD G7ORE
CO9 1PD M0NAS
CO9 1PD M3JDS
CO9 1PG 2E0JTO
CO9 1PG M0JOC
CO9 1PG M7JOC
CO9 1PY M3PSI
CO9 1QG G4LBJ
CO9 1TD G1CCW
CO9 1UB G6OXJ
CO9 1UG G4TEB
CO9 1YB M3IRX
CO9 2HF G0GII
CO9 2HJ M3JEP
CO9 2NW G7UVV
CO9 2SF M6OAT
CO9 2TA G6LHG
CO9 2TB G0CQI
CO9 2TB G1GRZ
CO9 2UA 2E0MKW
CO9 3AZ G0LZY
CO9 3BZ G4ZUL
CO9 3DP 2E0IXV
CO9 3HX G0GZF
CO9 3LG 2E0SLG
CO9 3LG M7RKY
CO9 3NX G1OFX
CO9 3QN G4WVH
CO9 3QS 2E0GFZ
CO9 3QS M7BWG
CO9 3QW 2E0FIW
CO9 3QW M0OBM
CO9 3QW M7VCF
CO9 3RN M5AJB
CO9 4AB 2E0AAF
CO9 4AB 2E0FLR
CO9 4AB M0JNQ
CO9 4AB M0VJR
CO9 4AB M6DTR
CO9 4JG G3PGN
CO9 4LN G6WHY
CO9 4LX G8JVV
CO9 4QJ M6NCU
CO9 4QQ M6GEY
CO9 4QQ M6YEL
CO93BH 2E0COM
CO93BH M6WKD
CR0 0AA G0DDT
CR0 0BL G7VTH
CR0 0DN M3XNR
CR0 0DU M7FCX
CR0 0JE G0VQM
CR0 0NU G0HSH
CR0 0PF M0RXZ
CR0 0RP M6HXD
CR0 1JS G7SFA
CR0 1XL G1ALR
CR0 2BB 2E0WBO
CR0 2BB M6YCR
CR0 2HX G1OIS
CR0 2LP G4NHA
CR0 2LW 2E0DIS
CR0 2LW M6EEW
CR0 2PF 2E0BPU
CR0 2PF M3FUB
CR0 3AD G4MZK
CR0 3JF G0GFY
CR0 3NF G1IEY
CR0 3PY M7ESV
CR0 3QP G8GYM
CR0 3SW G6FGY
CR0 4AF M7RCE
CR0 4DN M6LHV
CR0 4EG G4DAF
CR0 4FT M7PAA
CR0 4HG G7LTP
CR0 4JD M6BGG
CR0 4NN 2E0IPL
CR0 4NN M0UPL
CR0 4NN M6ZGR
CR0 4PU G4FFX
CR0 5BA G3WBN
CR0 5BP G8TBL
CR0 5HX G4STD
CR0 5PS G6DAY
CR0 5QA M0CCF
CR0 5QA M0VOG
CR0 5QA M1ACF
CR0 5QA M1CCF
CR0 5QA M3ACF
CR0 5ST G4KQO
CR0 5TB G3SCB
CR0 6AP M3WXH
CR0 6AQ M6CGF
CR0 6DE G4DPO
CR0 6TS M7PGF
CR0 7EB G4AVV
CR0 7HY G4RWW
CR0 7PR G3ZIO
CR0 7QP M0CVZ
CR0 7RW 2E0EQU
CR0 7RW M6IIX
CR0 7SH M3WRO
CR0 8HJ M3OSA
CR0 8HX M6VZF
CR0 8LG G7ODG
CR0 8ND M6NFL
CR0 8PN G3RMN
CR0 8SB G3ALG
CR0 8YQ M3YRV
CR0 9DG G7MSF
CR0 9HE G4UIO
CR0 9LN G4FUU
CR0 9LZ M7SCP
CR06SW G7NQW
CR06XR M6ILH
CR09DW 2E0IDM
CR09DW M6VBN
CR2 0BL M3YDJ
CR2 0DZ G0HWY
CR2 0EF G3WMT
CR2 0LB G4SXY
CR2 0RB M0CGF
CR2 6EE G0VVX
CR2 6EE G3ZRR
CR2 6EE G6CAU
CR2 6EJ G6CID
CR2 6NE G1WFG
CR2 7EF G0DJT
CR2 7ER G8MNY
CR2 7GE G3TCZ
CR2 7HF M7LMA
CR2 7HH G3MCX
CR2 7JJ G4DAC
CR2 7RE G4GFC
CR2 8AR M6CNX
CR2 8DW 2E0DCZ
CR2 8DW M6DCZ
CR2 8HR 2E1DRV
CR2 8JQ M7LNG
CR2 8PH G4LZE
CR2 8PT M6VQX
CR2 8RA G3ZYZ
CR2 8SL M3YGO
CR2 8SL M3ZGO
CR2 9DU M1ZAR
CR2 9ES G3KKZ
CR2 9JR G0TCE
CR2 9JY G8DNL
CR2 9LN G8IYS
CR3 0AJ G7VFX
CR3 0EP G4BWG
CR3 0FA M7EZR
CR3 5DL G1RCE
CR3 5EL G0SCR
CR3 5EL G4APL
CR3 5EL G7BSF
CR3 5EL M3BYL
CR3 5JN G6LTK
CR3 5LJ G4LXR
CR3 5QH G8LMI
CR3 5RB G7BWE
CR3 5RT G0SYR
CR3 5RT G8DTQ
CR3 5SG 2E0FSC
CR3 5SH M3SOF
CR3 5TG G4ECS
CR3 5XN G7APO
CR3 5ZU G7ONI
CR3 6AD G8PRK
CR3 6BA G3ODX
CR3 6DQ G6CDW
CR3 6HN G4DTC
CR3 6NJ M3YTG
CR3 6QX G0TNM
CR3 7EH G8HDP
CR36SA G7NGB
CR4 1JF M3WVG
CR4 1LF G0WCR
CR4 1LG M7DPE
CR4 1NY G7JMQ
CR4 1XJ M3MZP
CR4 1XN G1IZN
CR4 2GA M6XCA
CR4 2LF G0LUL
CR4 2LT M3URZ
CR4 3DW G7OZJ
CR4 3DZ G0VTM
CR4 3JS G4RBH
CR4 3LL 2E1GOE
CR4 3LW G1KGO
CR4 3RQ G7OLH
CR4 3RS 2E0YGS
CR4 3RS M0IYN
CR4 3RS M6GMS
CR4 3RS M7CDW
CR4 3SB M6UON
CR4 4HD G3LCH
CR4 4LZ G1FOF
CR5 1AT M6XZK
CR5 1BB G3MZA
CR5 1DF G8EIN
CR5 1HR G0KZT
CR5 1HR G4FUR
CR5 1HR M1FUR
CR5 1JS G1KGA
CR5 1NF M1DOR
CR5 1NL G8EDN
CR5 1QP G3CQU
CR5 1QS G0GZM
CR5 1RF G1TLW
CR5 2BL G4CJR
CR5 2EG G7OKY
CR5 2EJ G4FVL
CR5 2JF G8GAR
CR5 2LD G8RWG
CR5 2LF G3ZPB
CR5 2QD M7FNK
CR5 2SY M7LTH
CR5 3BP G0KUE
CR5 3DD G0FUH
CR5 3DE G2LW
CR5 3DE G3OOU
CR5 3DE M7SCP
CR5 3PH G6UVB
CR5 3SJ M6EBE
CR5 3SX G3GWC
CR53RN M7IIO
CR6 9EP M0UGM
CR6 9EP M0USM
CR6 9HJ M7GVA
CR6 9HZ G3WZK
CR6 9LB G4LLM
CR6 9LP G8JAC
CR6 9RP M7HNK
CR6 9RT 2E0HJV
CR6 9RT M6HJV
CR6 9TD 2E0FJA
CR6 9TD M0IDM
CR6 9TD M6FJA
CR7 6BX G1POM
CR7 6EB M0HNC
CR7 6HD M7CRV
CR7 7AF G3UFY
CR7 7HQ G4KRD
CR7 7JE G3JAL
CR7 7NP G1HER
CR7 7NQ G7HLW
CR7 7RU M7AGN
CR7 8DF G0IOO
CR7 8DF M6JEJ
CR7 8NY G4REK
CR7 8RP G1OTN
CR8 1JA G7JAQ
CR8 1JB G4BUK
CR8 1JG G0UCT
CR8 1JQ G7PWV
CR8 2DY 2E0BCF
CR8 2DY M0GNM
CR8 2DY M3IHV
CR8 2HQ G8FOT
CR8 2LR G6GFJ
CR8 3AQ 2E0LTU
CR8 3AQ M0PZD
CR8 3AQ M6IZP
CR8 3AQ M6URA
CR8 3PE G0UQO
CR8 4AJ M7NUT
CR8 4HH G3TWJ
CR8 4NB G8KWV
CR8 5DG M0CUP
CR8 5DG M3IDY
CR8 5JJ G3ZXV
CT1 1NR M3KRE
CT1 1PZ 2E1EHM
CT1 1QG G7RBB
CT1 1SJ M3AJA
CT1 1TS G6EGU
CT1 1TS G6YZU
CT1 1WX M0HWJ
CT1 1XH M0IBX
CT1 1YG 2E0WPJ
CT1 1YG M0WPJ
CT1 2AA G1AUI
CT1 3DH G4KGY
CT1 3JL M0ELS
CT1 3LD G3MMJ
CT1 3SF M7AXM
CT1 3UF M0PAM
CT1 3UP 2E0DBP
CT1 3UP M3AYQ
CT10 1DR 2E0RDP
CT10 1DR 2E0WFY
CT10 1DR G8FME
CT10 1DR M6WFY
CT10 1FA M6TZI
CT10 1HN M3TDP
CT10 1HR M0ATS
CT10 1PG 2E1EKM
CT10 1QN 2E0PGC
CT10 1QN M5PGC
CT10 1QN M6PGQ
CT10 1RP G0LGW
CT10 1SR G6SFC
CT10 1TL G4GAT
CT10 2BX M7KJM
CT10 2DT M0ASC
CT10 2EW M6YZH
CT10 2HA M0AQA
CT10 2HG G8DZN
CT10 2HL M6ZPE
CT10 2HY M3UVJ
CT10 2HY M6DWM
CT10 2JG G4AQE
CT10 2JL G8XJE
CT10 2ND G7FNU
CT10 2NG G0LEU
CT10 2PE 2E1HPT
CT10 2PL M0TFC
CT10 2QB M7FKJ
CT10 2SD G0RJJ
CT10 2SS M6XAY
CT10 2TR M7NHS
CT10 2TX G0KCA
CT10 2XN G7OHO
CT10 2XN M0DLI
CT10 2XN M0KNI
CT10 2XU 2E0NJJ
CT10 2XU M0GIG
CT10 2XU M3NJJ
CT10 3AA G1VID
CT10 3AH G4PTE
CT10 3AZ G0VUT
CT10 3DE M3JKJ
CT10 3DE M3JKP
CT10 3DL M3MQP
CT10 3DR G1KMJ
CT10 3ES G4GAP
CT10 3EY G0GUW
CT10 3HN G8WQT
CT10 3NE G0RXU
CT10 3QP G4VBI
CT10 3SD G4VPI
CT102XN 2E1CPF
CT11 0BH M3UCF
CT11 0DF M6DJA
CT11 0HT M3WOD
CT11 0JJ M3XJZ
CT11 0LL G3LHS
CT11 0LQ M1CCL
CT11 0ND G7MGZ
CT11 0PB G7FMI
CT11 0PF 2E0CVJ
CT11 0PF M0ZRF
CT11 0PL G7SYE
CT11 0PX G0UAK
CT11 0RN G1YZT
CT11 0RR G7DNQ
CT11 0RY G0AHA
CT11 7AS M0IJA
CT11 7DP M6FFU
CT11 7EF G6ENA
CT11 7HS M6HFJ
CT11 7HT G4ADS
CT11 7LP 2E0XDY
CT11 7NW G7ORS
CT11 8DF G6OJN
CT11 8JP M3TWY
CT11 8JY G6MWQ
CT11 8QA 2E1GDO
CT11 8QA M7COY
CT11 9DE G3TVD
CT11 9JF M6OJP
CT11 9LP G1ZFB
CT11 9LP M6AVQ
CT11 9LU G0JIF
CT11 9PB G4HAK
CT11 9QZ G4BXI
CT11 9RT M0IKS
CT110ND G0MBL
CT12 4AG 2E1HNF
CT12 4BE G6UXG
CT12 4EL M3UGH
CT12 4EL M6RVR
CT12 4EP 2E0MRJ
CT12 4EW G1UFJ
CT12 4HR G6OKB
CT12 5AW G0LFM
CT12 5ED G3YCV
CT12 5ED M1DEJ
CT12 5LD G0WWP
CT12 5LX G7IFU
CT12 6AJ M7INT
CT12 6DD 2E0IAJ
CT12 6DD M3INL
CT12 6DE G0TFB
CT12 6DQ M6OZN
CT12 6DW M6GQK
CT12 6DX 2E1HPS
CT12 6DX G7ITS
CT12 6EZ G3ZBF
CT12 6JQ G7SFD
CT12 6NX M0DIJ
CT12 6NX M3EOX
CT12 6NZ G1ODK
CT12 6QG M6LJT
CT12 6RY M6EAJ
CT12 6SW G8UHJ
CT12 6TL M0CAG
CT12 6TT M6MFF
CT12 6UG M3TNH
CT126ET 2E0OOW
CT126ET M6NOQ
CT13 0AQ 2E0PBY
CT13 0AQ 2E1ABW
CT13 0AQ 2E1ATH
CT13 0AS 2E1GTF
CT13 0BE M6GJY
CT13 0BG G7BPZ
CT13 0BG G7BQA
CT13 0DP G7VQJ
CT13 0DW M3SOQ
CT13 0EU G4RXG
CT13 0JZ G1EHK
CT13 0PE G0JBA
CT13 9AS G4KPF
CT13 9JB 2E0ALH
CT13 9JB M3IYY
CT13 9JB M3KRM
CT13 9JB M3VKJ
CT13 9JE G4JYU
CT13 9JE G8MLB
CT13 9JF G0BVA
CT13 9JF G1PJR
CT13 9NY G0MAZ
CT13 9NY G7BVH
CT13 9QA 2E0BZV
CT13 9QA M0GUG
CT13 9QA M3ZNT
CT13NN 2E0WJF
CT14 0AJ 2E0GYJ
CT14 0AJ M0JSK
CT14 0AJ M7APM
CT14 0BT G0LGK
CT14 0JH G4WOS
CT14 0JL 2E0EWF
CT14 0JL M6STQ
CT14 0LA G4SUS
CT14 6PP 2E0LJH
CT14 6PP M0PKH
CT14 6PP M6CEH
CT14 6PP M6ODF
CT14 7BL M3BWZ
CT14 7DA 2E0VWW
CT14 7DA M3VWW
CT14 7EZ G0RDN
CT14 7PX M3OZC
CT14 7QB 2E0AAK
CT14 7QB G0OXX
CT14 7SY M1AJU
CT14 8BT G8YXR
CT14 8DD G4YWA
CT14 8EB G0DQI
CT14 8JW G1IUA
CT14 8JW G7MSC
CT14 8JW M6ZTS
CT14 9AA M3WGZ
CT14 9AT 2E1ACK
CT14 9DQ G3MZI
CT14 9EE G1LLU
CT14 9EF 2E0JEN
CT14 9EF G8SOU
CT14 9EF M0EEH
CT14 9EW G3FBU
CT14 9FB G0EFA
CT14 9HW M6DWI
CT14 9JF G3VIR
CT14 9JF M3VIR
CT14 9JR M1DNA
CT14 9LL G4FJF
CT14 9LS G7HIX
CT14 9NB G0DUK
CT14 9NJ G0SRR
CT14 9NL M0ARX
CT14 9NP G4RXH
CT14 9QN M0SOE
CT14 9QZ 2E0UWO
CT14 9QZ M7UXO
CT14 9RG M1DEG
CT14 9RG M3AJU
CT14 9SN M0VRI
CT14 9SN M7MLE
CT14 9TW G7MSS
CT148ER M3ZIZ
CT15 4AW 2E1DFB
CT15 4AW G7SGH
CT15 4BT G6IOM
CT15 4HE M0HHI
CT15 4HH M0WPL
CT15 4HH G6BNW
CT15 4HX M6KWW
CT15 4HZ M3OVC
CT15 5BY G7FCL
CT15 5JD G8SMZ
CT15 5JW G6HIG
CT15 6AH G8YMD
CT15 6AH M1BKI
CT15 6BS M3VFM
CT15 6DD G7SXJ
CT15 6DL 2E1BTK
CT15 6ED G6PSQ
CT15 6EJ G4OPR
CT15 6HL 2E1BTG
CT15 6HL M6LYO
CT15 6JL G6WPK
CT15 7ES G4OJG
CT15 7HF G8MBV
CT15 7HJ G8UMB
CT15 7HR G3ZHU
CT15 7JN G4MIX
CT15 7JS G4AWW
CT15 7LJ G0AXD
CT15 7LP G4VRB
CT16 1EX 2E0CUV
CT16 1LH G0SET
CT16 1PX G4DDB
CT16 1TH 2E0VDC
CT16 1TH M0KID
CT16 1TH M6VDC
CT16 2AR G1PQX
CT16 2AX G4BBH
CT16 2AX M6JSW
CT16 2JW 2E1ABN
CT16 2LH 2E0NFX
CT16 2LH M7NFX
CT16 2NQ 2E1NFD
CT16 2PA M6UFF
CT16 2QP G6APW
CT16 2QW G7MBU
CT16 2SG 2E0LYD
CT16 2SG M0IML
CT16 2SG M6BDV
CT16 3BA M6FEN
CT16 3DU M6LWM
CT16 3EE M1EJO
CT16 3EJ G4HHX
CT16 3HA G4RLX
CT16 3HZ G0ROO
CT16 3HZ G3ROO
CT16 3HZ M7ROI
CT16 3LQ G4UHT
CT16 3LT G7JOW
CT16 3ND 2E0EHR
CT16 3ND G6HKQ
CT16 3NG M3WZP
CT16 3NP G4XDW
CT16 3NP M7EBJ
CT17 0BD 2E0BEH
CT17 0BD M0GGO
CT17 0BD M3JOM
CT17 0BS G8IYN
CT17 0BS M6GXL
CT17 0DP M6IVG
CT17 0DY M6BAQ
CT17 0ER G4WIY
CT17 0FP G6JHT
CT17 0HE M6NET
CT17 0JQ M7FCB
CT17 0NG 2E0FHK
CT17 0NG M7RGJ
CT17 0NP 2E1ABY
CT17 0NT G8YNG
CT17 0NX G0ADK
CT17 0PR G7BRM
CT17 0PS G3YMD
CT17 0PS G6USA
CT17 0RX M1CJB
CT17 9ES M3IWX
CT17 9JZ G6TRM
CT17 9LL 2E0CPL
CT17 9LL M0UGR
CT17 9LL M3UGR
CT17 9LQ G4FXE
CT17 9PN G4SMX
CT17 9QF M6ELZ
CT17 9QQ 2E0BHJ
CT17 9QQ 2E0OIN
CT17 9QQ G8WLY
CT17 9QQ M3KYH
CT17 9QQ M3OIN
CT17 9RB M0IER
CT17 9RQ G0CCG
CT17 9SZ G6AIC
CT18 7AP G4IMP
CT18 7AP G4RJZ
CT18 7AP M3PNA
CT18 7BS 2E0UGF
CT18 7BS M0WHO
CT18 7BS M3UGF
CT18 7BS M6UFO
CT18 7DS G6RRS
CT18 7DZ G4TGP
CT18 7EQ M0KBC
CT18 7FA M6AKH
CT18 7FU G0BPS
CT18 7GA M6WTM
CT18 7GZ G4MDZ
CT18 7HB 2E0NCI
CT18 7HB M6NCI
CT18 7JS G0IXV
CT18 7LB G3OJZ
CT18 7LH 2E1DYT
CT18 7LH M1CVF
CT18 7LQ G0VHL
CT18 7LS 2E0DTS
CT18 7LW 2E0WUN
CT18 7LW M3RVQ
CT18 7LZ M0DRN
CT18 7NT M6PWE
CT18 7QL G4DUE
CT18 7TG M3FAK
CT18 8BY G8KSD
CT18 8BY M0CFH
CT18 8DF 2E0JKG
CT18 8DF M0JKG
CT18 8DF M3XJG
CT18 8DS G0VAI
CT18 8LX M3PMN
CT19 4AF G1RQI
CT19 4BX M3RWI
CT19 4EQ G0NOH
CT19 4HN M6LCI
CT19 4HP M6GXJ
CT19 4JA 2E0DJF
CT19 4JA M3PQB
CT19 4JA M3ZEJ
CT19 4JT 2E0MTR
CT19 4JW 2E0FLV
CT19 4JW M6AZQ
CT19 4NS M7CEZ
CT19 4QG M6GXS
CT19 4QH G0VWB
CT19 4RL 2E0DLR
CT19 5AT G8YXQ
CT19 5BZ M1CMN
CT19 5DD M3HOM
CT19 5EL M3JBF
CT19 5HG G4IVL
CT19 5HG G7IWV
CT19 5JF G6GXE
CT19 5JH M6RNI
CT19 5LS 2E0NEC
CT19 5LS M6WUG
CT19 5NA G8SUQ
CT19 5NB G4EQJ
CT19 5NE 2E0VYW
CT19 5NE M0VYW
CT19 5NR M1DFB
CT19 5NR M3DFB
CT19 5PW M1AZO
CT19 5RA M3MSN
CT19 5RU M6RUS
CT19 5SH G8YNH
CT19 5SH M3BNU
CT19 5TA M3IZP
CT19 5UB G8EKD
CT19 6BA G3XVY
CT19 6BP M3OZP
CT19 6BS G8MFV
CT19 6EU M7TKB
CT19 6EY M6GXL
CT19 6NE 2E0ALL
CT19 6NN G4IDX
CT19 6PR G4MHS
CT19 6QP G4FVI
CT2 0HR G8WMK
CT2 0HR M3CSR
CT2 0HT 2E0EKR
CT2 0HT M0GRB
CT2 0HT M3EKR
CT2 0HT M6JEH
CT2 0HX G1LHL
CT2 0JZ G7NIN
CT2 0LA M1DFM
CT2 0LA M3DFM
CT2 0LA M3TTA
CT2 0LL M6DYV
CT2 0NT M0KTY
CT2 0QB G6ZEQ
CT2 0QH G3FMU
CT2 7BX 2E0MNY
CT2 7BX M0RDP
CT2 7BX M6FQU
CT2 7HH 2E0DFQ
CT2 7JZ G3HTT
CT2 7TA M3DKO
CT2 7TF M0DCO
CT2 8BA M0KKB
CT2 8DR G6WDS
CT2 8PN M6JFP
CT2 8PQ G8UIV
CT2 9BL G4PCN
CT2 9BP M0SGF
CT2 9BP M6RCN
CT2 9BU M3WTU
CT2 9DB 2E0IMJ
CT2 9DB M0YMJ
CT2 9DB M3IMJ
CT2 9DL G8BRD
CT2 9JX 2E0XXK
CT2 9JX G2DLX
CT2 9JX M0XXK
CT2 9JX M3XXK
CT2 9NW M6GUC
CT20 1DA G6ZNW
CT20 1DF G6WGM
CT20 1DL M3FSC
CT20 1HY G7HIY
CT20 1NN 2E1BYH
CT20 1TH M6XCK
CT20 2EJ 2E0LCK
CT20 2EJ M6YXZ
CT20 2PQ G8IWX
CT20 2PT 2E0MGU
CT20 2PT M7MGU
CT20 2RP M6ECZ
CT20 2TY G0SLJ
CT20 2TY G2FA
CT20 3BE M0DRO
CT20 3EJ M6ASH
CT20 3JY M6BOK
CT20 3LA M3BJL
CT20 3LH 2E0RDE
CT20 3QJ 2E1GTI
CT20 3SA 2E0AYY
CT20 3SA G4EGQ
CT20 3SA M0RNP
CT20 3TA G3XHW
CT20 3TL M0GCB
CT201WT M7HRL
CT21 4EA G7SZO
CT21 4JP G3LWD
CT21 4JQ G3ZHT
CT21 4SE G4SSZ
CT21 5EB M3VPD
CT21 5EB M6MRC
CT21 5PA M7OCT
CT21 5RY M3PLI
CT21 6AG G3VYW
CT21 6DG 2E0JXT
CT21 6LS M7NJD
CT21 6QZ M7PJT
CT3 1ED 2E0TWA
CT3 1ED M6WEV
CT3 1JX G3JIX
CT3 1LG M7EBW
CT3 1LL G7MZA
CT3 1LN G3TAJ
CT3 1LY G4FLR
CT3 1RQ M6FHQ
CT3 1SY G3KFG
CT3 1TZ G3KTZ
CT3 1TZ M7TKZ
CT3 1UA G3RWF
CT3 1UH G4KJS
CT3 1XW 2E0XFD
CT3 2LP G1MXM
CT3 2NH G8DBU
CT3 3AQ G6EPL
CT3 3HZ G6WSF
CT3 4DB G4NEE
CT3 4HL 2E1JEH
CT3 4JN G7VQW
CT3 4LN G4PLS
CT34DS G0TXA
CT4 5EX G4ZHN
CT4 5JX M6WBA
CT4 5LU M7VBO
CT4 5PN G6GDA
CT4 6DN G3MLO
CT4 6HX G4AZG
CT4 6JB G6GES
CT4 6JQ G7LNB
CT4 6NP G4GWI
CT4 6RT G4BQS
CT4 7AH G4ZIF
CT4 7ND G0VRW
CT4 7ND G7MPV
CT4 7NN G3XAQ
CT5 1EL G0IAA
CT5 1JQ M3PPU
CT5 1JQ M3RZY
CT5 1JZ 2E0GTB
CT5 1NS G4SIL
CT5 1QF G4RIS
CT5 1QF M3BQN
CT5 1RR 2E0RUT
CT5 1RR M6NMZ
CT5 2DH G7EOE
CT5 2DS G8BNK
CT5 2DY G8WIR
CT5 2ES 2E0VTF
CT5 2ES M0IUG
CT5 2ES M6VFD
CT5 2LA 2E0TDO
CT5 2LA M0ZTD
CT5 2LA M6CWX
CT5 2LB G8NXQ
CT5 2LE G4LTS
CT5 2LE M6BIE
CT5 2LZ M7AKN
CT5 2NW G4PYA
CT5 2PH G0FAE
CT5 3AR M0KFJ
CT5 3EJ G3KXB
CT5 3JZ G0IFS
CT5 3PE G4ELP
CT5 3PE G4ICM
CT5 3PJ G0AUW
CT5 3QD G4SEJ
CT5 3RF G8YMN
CT5 4BZ G0BEX
CT5 4DN G4GRJ
CT5 4DT G4WZZ
CT5 4DT G4XAB
CT5 4EL G0PPY
CT5 4EQ M1DHI
CT5 4HX G4CZU
CT5 4LA G1PVA
CT5 4LA M3MRS
CT5 4LQ 2E0BEI
CT5 4LY G7UWW
CT5 4LY M6HAY
CT5 4NL G8BUI
CT5 4ST G0LJD
CT5 4ST G8XAJ
CT5 4ST M0SKG
CT5 4TE M3PNB
CT5 4TH G0NBB
CT5 4TH G1PRH
CT6 5QX 2E0TUT
CT6 5QX G8NEL
CT6 5QX M6OND
CT6 6BH G1KQE
CT6 6DX G7EIA
CT6 6GZ G8ELP
CT6 6HG G0ANK
CT6 6HU G4ZIH
CT6 6JA G4WZQ
CT6 6JS M6AFC
CT6 6NT G0ILO
CT6 6PP G8PPQ
CT6 6QZ G1FNN
CT6 6RE 2E0AIT
CT6 6RF G0HVC
CT6 6RF M1DHY
CT6 6SB G8ESW
CT6 6SE G2FMW
CT6 6SR G4SUK
CT6 6SS M3BBY
CT6 6UQ M3UCJ
CT6 7DW G4EVD
CT6 7EQ G8FEZ
CT6 7EW G1TQH
CT6 7HB G0ALJ
CT6 7HD G1AGW
CT6 7HG G6CVD
CT6 7LE G1EDK
CT6 7LE M7RYD
CT6 7NA M3BOV
CT6 7QA M3CUH
CT6 7RS G0IHI
CT6 7RS G1VNM
CT6 7RS G4MKI
CT6 7TA M6BNY
CT6 7TB M3PSE
CT6 7UW G0ETI
CT6 7UX M7KEC
CT6 7XB M3SEY
CT6 7XF 2E0MDC
CT6 7XF G1DSZ
CT6 7XF M3YMG
CT6 7XG G6VRI
CT6 8AD G8KOL
CT6 8AE G4XOE
CT6 8AN G7MIF
CT6 8HG G4URD
CT6 8HX G4VMZ
CT6 8HX G4ZZK
CT6 8HY 2E0VRP
CT6 8HY M0VRP
CT6 8HY M6VRP
CT6 8JA M0HWM
CT6 8JS G6PKS
CT6 8JS M3KTV
CT6 8LN G6RMA
CT6 8LP 2E0ATZ
CT6 8LS G0LAA
CT6 8LT G0HMK
CT6 8LZ G7LFQ
CT6 8NA M7BFQ
CT6 8QW G8KDU
CT6 8QW M6CHZ
CT6 8RX G1DKY
CT6 8SD G4RKV
CT6 8TU G0NFG
CT6 8UQ G0EAG
CT65PZ M5GTA
CT66QX M7HBZ
CT7 0BT M0PJG
CT7 0DX G8NAV
CT7 0EL G0ANW
CT7 9AJ M6AYS
CT7 9AS G1UEA
CT7 9AZ G0INT
CT7 9DF G0CIX
CT7 9ED G1JEH
CT7 9HE 2E1AJB
CT7 9HE M7AJB
CT7 9NA 2E1EHY
CT7 9QD G3OND
CT7 9QH G0CTQ
CT7 9QN M1TCI
CT7 9RS 2E0EPP
CT7 9RS M6LPP
CT7 9SR G3VID
CT7 9UQ G0MUZ
CT7 9XE M6SIH
CT79TY M3TGJ
CT8 8AN G4SBD
CT8 8AN G4SEK
CT8 8AN G8NZA
CT8 8AP G1NLQ
CT8 8AP G7MEZ
CT8 8AR 2E1DDZ
CT8 8BP M6KCC
CT8 8BT G4KPV
CT8 8BX 2E1CQP
CT8 8BX G0TBU
CT8 8HR 2E0CNU
CT8 8HR M6BOA
CT8 8LW G3PNT
CT8 8RJ G0CEY
CT88AH G6KKW
CT9 1FE G0WVA
CT9 1HU G0GUE
CT9 1NS 2E0EMN
CT9 1NS M6HCZ
CT9 1TR G4YGB
CT9 2DJ M7MMK
CT9 2EJ M0AJC
CT9 2EN G3YEQ
CT9 2EN M0MVS
CT9 2NH M6EBL
CT9 2NS G0SJP
CT9 2PS G0GNQ
CT9 2SL M3CAE
CT9 2SW G0HRS
CT9 2SW G0PDZ
CT9 2SW G7HTI
CT9 2SW M5IC
CT9 2TD M6DGJ
CT9 3BH 2E0JFK
CT9 3DU G3YWF
CT9 3EF 2E1COV
CT9 3ES G8WMW
CT9 3HD G3SVJ
CT9 3HD G8ATD
CT9 3HJ 2E0XPY
CT9 3HL G0LGE
CT9 3HL M6LII
CT9 3JB G0DFI
CT9 3JB G1DUJ
CT9 3NA M7PCQ
CT9 3NX 2E0PZN
CT9 3PT M0RST
CT9 3PT M1CBT
CT9 3PX G6KNU
CT9 3PX M6LCQ
CT9 3RG M6ISN
CT9 3RN M7FTB
CT9 3RX G0CHN
CT9 3SL 2E1PDQ
CT9 3XJ G0NEP
CT9 3XT 2E0EMB
CT9 3XT M6AKJ
CT9 3YA M1BBS
CT9 4DH G3YUH
CT9 4HA G3SHX
CT9 4LL G6ILU
CT9 4NE G7NOR
CT9 5DT G7EYE
CT9 5EH G3ZZZ
CT9 5EH G8GHH
CT9 5HT G3TRX
CT9 5JA G4BSW
CT9 5JZ G0TBO
CT9 5LN G1AOC
CT9 5NE 2E1HQH
CT9 5NG 2E1HUC
CT9 5NG M6LIJ
CT9 5NP 2E0DUE
CT9 5NP M6WFI
CT9 5NT 2E1DNC
CT9 5PA G7MKF
CT9 5PF 2E0ZDE
CT9 5PF M0ZDE
CT9 5PF M6DES
CT9 5PN M6FZB
CT9 5QA G8DHJ
CT92NX M6JVS
CV1 1FZ G1XLL
CV1 1JR M6NRU
CV1 2AA G6AJC
CV1 2AL M6FKW
CV1 2AR M6JGY
CV1 2DF M7COV
CV1 4DS M7EJP
CV1 4EB M3YVZ
CV1 4HL M1DCV
CV1 5EA G6XMT
CV1 5FZ M7ECP
CV1 5GU M6GKX
CV10 0BA G1JWY
CV10 0BA G1WRN
CV10 0BX G4UQU
CV10 0DF G0MDN
CV10 0DH 2E0NEI
CV10 0DH M0NKE
CV10 0DH M6NEI
CV10 0DW G0VZO
CV10 0DW G3VDU
CV10 0DW G7SKL
CV10 0DZ G4NCV
CV10 0HP G8VHI
CV10 0HR G1VVL
CV10 0HR M6SIW
CV10 0JJ M7EQZ
CV10 0LB G0FBG
CV10 0LY M3TID
CV10 0NL G7ANQ
CV10 0QA M7DBB
CV10 0SL G0LLP
CV10 0SN G4KQL
CV10 0TD G8SYE
CV10 7AJ M7RDB
CV10 7AW G1FTH
CV10 7BA M6UHT
CV10 7DE G3ZSQ
CV10 7EE G4ZUE
CV10 7ES G4AEH
CV10 7ES G7BBC
CV10 7ET G3ZOY
CV10 7HG G7USI

CV10 7HP G4YOZ
CV10 7JX M6AIU
CV10 7PW G3YTW
CV10 7SW G0NZI
CV10 7SW M3BCW
CV10 8BB 2E0FQG
CV10 8BD G6DDA
CV10 8DL M3VCO
CV10 8EG G4BEH
CV10 8HL M3CBW
CV10 8HL M3CCB
CV10 8HP M7DBJ
CV10 8JQ G8ITJ
CV10 8NH G4NHF
CV10 8NU G4BBQ
CV10 8NW M6FSJ
CV10 8PA G0REO
CV10 8PT G7FSC
CV10 8PT M6CFI
CV10 8QA M6DZY
CV10 9AR 2E0NAG
CV10 9AR M3NWY
CV10 9DQ M6HPV
CV10 9DY G0SZG
CV10 9DY M3ETQ
CV10 9DY M3EYR
CV10 9HU 2E0CJF
CV10 9HU M6AHU
CV10 9HY G0PYL
CV10 9JA M6NQW
CV10 9JH G7JXF
CV10 9JY M7EXG
CV10 9JZ G1VOR
CV10 9PB G6OQV
CV10 9PQ M6JPH
CV10 9PQ M6PJH
CV10 9QJ G8OMB
CV10 9QJ M0NDA
CV10 9SD G1FLW
CV10 9SG G7CEN
CV10 9SJ 2E0KNK
CV10 9SJ M0KNK
CV10 9SJ M7KAK
CV107NH 2E0DRN
CV11 4LH G4SHY
CV11 4RE M6KVG
CV11 4TN M7FWZ
CV11 4XG G7LSD
CV11 4XQ G4APS
CV11 5AT G1VDP
CV11 5AT G3VLG
CV11 5AU M7LAN
CV11 5HP G8AWI
CV11 5LU G1EXV
CV11 5PD G6MTG
CV11 5RR M0GNB
CV11 5UB G1JVL
CV11 6AE M7VBN
CV11 6AJ M6RVM
CV11 6BQ G4LMK
CV11 6DW M6EMF
CV11 6DZ G4NQZ
CV11 6EE G0SBC
CV11 6EZ G8GXN
CV11 6FF G6SFY
CV11 6GA G4ASQ
CV11 6HN G3SLK
CV11 6JA G4HXC
CV11 6JR G3YTT
CV11 6UU 2E0GXG
CV11 6UU G0NCQ
CV11 6UU M6OXU
CV11 6YX M7GFB
CV12 0AX G0DCZ
CV12 0AZ G0OBE
CV12 0BL M7BLM
CV12 0DE G0JUN
CV12 0EL G8GMU
CV12 0JH M0HSJ
CV12 0LD M6NKH
CV12 0LY G0LDP
CV12 0LY M6WRU
CV12 0NS M3HBM
CV12 8AL M3HSH
CV12 8AR M0WWX
CV12 8BE G0IYW
CV12 8BQ M3ZNV
CV12 8DA G1LBH
CV12 8DG G1OJD
CV12 8DU M1BNI
CV12 8JG M3JOS
CV12 8LA 2E0HUY
CV12 8LA M7BZU
CV12 8QU M6ENZ
CV12 8TA M7JTZ
CV12 9AG G1VNB
CV12 9BX G0BVS
CV12 9DZ 2E0BGL
CV12 9EL M1JAK
CV12 9HP M7MND
CV12 9JB G0BIN
CV12 9LL 2E0YRF
CV12 9LL M0SLO
CV12 9LL M6YRF
CV12 9ND G3YQZ
CV12 9PR M3SYH
CV12 9PS G1ORG
CV12 9SG G6AD
CV13 0AU M3MFU
CV13 0HQ G6CLC
CV13 0NT G1LQP
CV13 0QH G3RHZ
CV13 0QH G4IQF
CV13 6BB G1SWI
CV13 6EP M0CJS
CV13 6HN G0OEM
CV2 1AF G0BJA
CV2 1DL G1HOP
CV2 1EF G1NAT
CV2 1FP M6VOG
CV2 1NN 2E0OGB
CV2 1NN M6DTL
CV2 1NN M7RNE
CV2 1PP M3XXU
CV2 1SA G1AVF
CV2 1SA M6HKW
CV2 2JD G6FXE
CV2 2JG G4HRY
CV2 2JS G4WKB
CV2 2JS G7DBO
CV2 2NE G6FYL
CV2 2QF G7PEN
CV2 3BH G1BNX
CV2 3DX G7JSQ
CV2 3NG G8JSC
CV2 4EU G0MTP
CV2 4EZ M7BUK
CV2 4GL G0UGX
CV2 4HT M0SQC
CV2 4HT M5BIR
CV2 4HT M6BIR
CV2 4JW G8MIE
CV2 4QY M0HWI
CV2 5BH G2ASF
CV2 5BH G7ASF
CV2 5BH G8SEQ
CV2 5BH M3MKV
CV2 5BT M3TSE
CV2 5EH G4GEE
CV2 5FZ 2E0FNG
CV2 5FZ M0ZAI
CV2 5FZ M6FNG
CV2 5FZ M6JOD
CV2 5GJ G6WVM
CV2 5GL G8TGD
CV2 5JW G8ZJO
CV2 5LG M0COV
CV2 5NQ G1OPA
CV2 5NU G4ZXN
CV21 1DE M6EEN
CV21 1FB M3JUO
CV21 1HW G3OQO
CV21 1HW M0RTM
CV21 1JB G6GND
CV21 1JN 2E0HCW
CV21 1JN M0WCH
CV21 1JN M6HCW
CV21 1NJ G1VAJ
CV21 1PG G0ZMC
CV21 1PG G7TMC
CV21 1PZ 2E0PPK
CV21 1PZ M0GWK
CV21 1PZ M3PPK
CV21 2QU G1FLI
CV21 2SZ 2E0SND
CV21 2SZ G0REQ
CV21 2SZ M3REQ
CV21 2TE G1ICA
CV21 3AB M0BBH
CV21 3BD G7BNI
CV21 3LH G8VTY
CV21 3NQ 2E1LOZ
CV21 3QH G3IKL
CV21 3RR M3XMP
CV21 3SZ G0COY
CV21 3TS G0FIN
CV21 4AN M6MIP
CV21 4BP G8EYY
CV21 4EF M0DEF
CV21 4HG G7KRE
CV21 4HJ M0HVV
CV21 4JY G7NFO
CV21 4LT G7SRL
CV21 4LT M0ASD
CV22 5BG G0UFW
CV22 5ET G7APD
CV22 5ET G8LYB
CV22 5HN G0DLB
CV22 5HN G0RLV
CV22 5JN M0CNE
CV22 5NY M3SWK
CV22 5NZ 2E0MTA
CV22 5NZ M6RRF
CV22 5QJ G8AQN
CV22 5RG 2E0EXA
CV22 5RG G4GZS
CV22 5RG M6NVZ
CV22 5RW G8FFZ
CV22 5RW M3AEA
CV22 6BG M3VIG
CV22 6BG M3XRP
CV22 6DD M1CIZ
CV22 6EH G1BIM
CV22 6EU M0LMQ
CV22 6HB G3ODB
CV22 6JF 2E0JXF
CV22 6JF M0WCZ
CV22 6JF M6JXF
CV22 6LG G4ACY
CV22 6NS M0ANH
CV22 6PG M6LIB
CV22 6PX G8IHF
CV22 7AP G6CYT
CV22 7BZ G8DLX
CV22 7EG G7RGO
CV22 7EW G4SSW
CV22 7GN M6EIE
CV22 7HB M7CSI
CV22 7PT G7GAB
CV22 7PT G7LIH
CV22 7RE G4AUB
CV22 7RG G8BLK
CV22 7RG M1AXE
CV22 7TN G1EBW
CV22 7TT G7JBW
CV23 0DB G1BIN
CV23 0DE G1UUV
CV23 0EE G3TQF
CV23 0GB M7NDD
CV23 0GX M6KIL
CV23 0JA G4JSX
CV23 0NN G4CFG
CV23 0NZ G0GZL
CV23 0PE G0GSA
CV23 0PH G0TZH
CV23 0PH G4AZH
CV23 0SP G7DDF
CV23 0SS 2E0NWY
CV23 0SS M0NWY
CV23 0SS M6NWY
CV23 0TS M0CNH
CV23 0UH G4IXT
CV23 0UW M7HCD
CV23 0UW M7MNZ
CV23 0UW M7NDH
CV23 0UW M7SFY
CV23 1BS 2E0HAB
CV23 1BS M0HOU
CV23 8BH 2E0HOV
CV23 8BH M0KNV
CV23 8BH M6TQF
CV23 8BJ 2E0VEZ
CV23 8BJ M0WDK
CV23 8BJ M7TLA
CV23 8DN G3ZJK
CV23 8HF M0GVW
CV23 8TR G3TYP
CV23 8UF G6IQM
CV23 8XQ G4VHW
CV23 9NL M7APO
CV23 9QA G8CDC
CV23 9QQ 2E0KWA
CV23 9QQ M6TXM
CV23 9QT M6HGU
CV23 9QT M6HGX
CV230LB 2E0CJW
CV230LB M6CJT
CV238UB G8AWY
CV3 1AN M0BVV
CV3 1AU G0DKR
CV3 1BB M0WFI
CV3 1DW M6IKN
CV3 1EQ G0RDU
CV3 1GG M6DZF
CV3 1GW G8MJX
CV3 1LY M6DIU
CV3 1NZ G1HGD
CV3 1QU 2E0VRZ
CV3 2AJ G4GDY
CV3 2AX G8YKY
CV3 2HA G3HLI
CV3 2HF G8OHG
CV3 2LA M3JMI
CV3 2LZ 2E0PDP
CV3 2LZ M3PDP
CV3 2NQ G4ZDG
CV3 2NQ M6FKA
CV3 2NW G1COV
CV3 2NW G8HPV
CV3 2NW M0HPV
CV3 2PE G4ZCP
CV3 2QG G0ZIL
CV3 3EQ G7MNT
CV3 3HH G1AMS
CV3 3HH M3WCQ
CV3 3HQ M6HZX
CV3 4BQ M1ECM
CV3 4FS G7LXY
CV3 5AG G0JXJ
CV3 5AU G4ROA
CV3 5DA M0IRT
CV3 5DS G6CTC
CV3 5DS G8ISJ
CV3 5GZ G1KUQ
CV3 5HH M0BBQ
CV3 5JT G3LMQ
CV3 5LF G3TZG
CV3 5NG M6PKR
CV3 5PE M7FTK
CV3 5PJ 2E0BLX
CV3 5PJ M0GKO
CV3 5PJ M3RVE
CV3 6BT G1EOK
CV3 6BT M3BMQ
CV3 6BT M3XGV
CV3 6BY M7FTF
CV3 6EB G7RWY
CV3 6EW M0BVW
CV3 6FN M7GCD
CV3 6JB M0AZY
CV3 6JG M6JYI
CV3 6JH G6RZS
CV3 6JT G7VCZ
CV3 6NY M5ZAP
CV3 6RD G4BBM
CV31 1DQ G4DXN
CV31 1GE M0JJR
CV31 1LA M5ALS
CV31 1RJ G3XKE
CV31 1WA 2E0CWC
CV31 1WA M0SER
CV31 1WA M3ZCW
CV31 2JX 2E0HFK
CV31 2JX M6EFW
CV31 2JZ G0GIL
CV31 2PB G0RTI
CV31 2RB G4GEP
CV31 2TQ G3SPL
CV31 3AN M6NOD
CV31 3BD M1ERJ
CV31 3BY 2E0TRI
CV31 3BY M6NRO
CV31 3EB M6OPM
CV31 3JD M6LYJ
CV31 3JE G0KWC
CV31 3LJ 2E0JTG
CV31 3LJ M7IVY
CV32 5AJ G4TYA
CV32 5JG G4PKT
CV32 5LU G6ISD
CV32 5NU G0KPH
CV32 6BA G8UIO
CV32 6BA G8WRG
CV32 6ES M6AVR
CV32 6HE G0BNE
CV32 6HY 2E1FQY
CV32 6LR G7DIW
CV32 6PB G3ZCG
CV32 6PW G3YJA
CV32 6QG G0LXG
CV32 7BB M7PGP
CV32 7DN G0FEH
CV32 7DY M1AUK
CV32 7EY G4CYG
CV32 7HT G6GSI
CV32 7JD G8AIM
CV32 7JD M3JHT
CV32 7JT G3HTB
CV32 7RX G0HCR
CV32 7SN G0DWJ
CV32 7TF G6NMA
CV33 9AQ G6VQW
CV33 9HG G1EAN
CV33 9LN G3YJN
CV33 9NP G6ESK
CV33 9PX M1CZL
CV33 9QL G1PVR
CV33 9RN G1APQ
CV33 9RZ G0ASX
CV33 9TE G4USP
CV33 9TE M3EBK
CV33 9TH G1FEO
CV338AZ M7HEF
CV33AU M7DKM
CV34 4SS G6BVR
CV34 5BX G0FGC
CV34 5FE G8GEE
CV34 5HW 2E0NEM
CV34 5HW M0NEM
CV34 5HW M7PCS
CV34 5HX G6VGZ
CV34 5HY G4DZY
CV34 5JW 2E0JGM
CV34 5JW M6WGH
CV34 5NJ M3NYM
CV34 5NL G7RMW
CV34 5NL G8XDL
CV34 5PB G1FIP
CV34 5PU G0OYR
CV34 5RQ G4TIF
CV34 5SB M7EYX
CV34 5SN G6YPK
CV34 5TG G1HRL
CV34 5TS G4JFS
CV34 5UJ G6YLB
CV34 5UL M0GWB
CV34 5XD G0GLU
CV34 5XD G1WXF
CV34 5YQ M0LCA
CV34 6FQ G1ABM
CV34 6JD G4RCR
CV34 6JQ G3UDN
CV34 6JQ G6WAR
CV34 6JQ M0OAE
CV34 6PF G3VRW
CV34 6QA G3TTC
CV34 6QJ G0GRM
CV34 6QX M3RYZ
CV34 6QY G3PXU
CV34 6XB G1XWM
CV344AB M6NVJ
CV345FE G3GYQ
CV35 0HH M0AYA
CV35 0QX M7BSX
CV35 0QX M7MTS
CV35 0RE G1YYP
CV35 0SB G6NGF
CV35 0SS G8GDC
CV35 0SS M0BXA
CV35 0UE M0BBT
CV35 7AQ G5ART
CV35 7AQ M7RTT
CV35 7HF G1WUM
CV35 7HF M3WUM
CV35 7TL G8NTQ
CV35 7TQ G1OUY
CV35 7UA G0OMN
CV35 7UD M6NWZ
CV35 8ED G1UTJ
CV35 8JX M0CLM
CV35 8NT 2E0RTU
CV35 8NT M6LNQ
CV35 8QP M0XTD
CV35 8TY G6XCC
CV35 9HQ G6HOB
CV35 9LY G6VGH
CV35 9NZ G3XMQ
CV35 9PW M6MUT
CV35 9QQ 2E0MVS
CV35 9QQ M6UYF
CV35 9QQ M7XBS
CV359AR 2E0IMG
CV36 4DJ M1DCX
CV36 4FD G0CHO
CV36 4FD G0SOA
CV36 4FD G1SWR
CV36 4HG 2E0HSC
CV36 4HS G1RLD
CV36 4JU 2E0SIF
CV36 4JU M6SIF
CV36 4NR G1AWJ
CV36 4PG G0CXJ
CV36 5DA G8HGG
CV36 5LN G0UPY
CV37 0AP G1HPB
CV37 0DN G8JXP
CV37 0DZ G3YED
CV37 0JD M6BAH
CV37 0PP G4RYM
CV37 0PP G6MMD
CV37 0RR M7NAL
CV37 0TH 2E0YOI
CV37 0TH G6JNY
CV37 0TH M7RNU
CV37 0TH G1UNN
CV37 6DD G0TVC
CV37 6DR M7DYB
CV37 6RT 2E0WTH
CV37 6RT M6PCN
CV37 6ST G6SGE
CV37 6TD G0REP
CV37 6TF G0JJY
CV37 6XG G0EDT
CV37 6XQ M6SSV
CV37 7DD G0CCJ
CV37 7DW M0LGW
CV37 7EX 2E0BAV
CV37 7EX M3YOE
CV37 7JF G4IQW
CV37 7LW G8RIM
CV37 7PL G1EIX
CV37 8EU M3XWC
CV37 8HJ M0HGA
CV37 8RA 2E1DMH
CV37 8US G3GXG
CV37 8US M6YOG
CV37 8XT 2E1DFS
CV37 8XT M0VKB
CV37 8XY G3XGV
CV37 9ED 2E0NDZ
CV37 9ED M3NDZ
CV37 9FZ G4YGY
CV37 9JN G0JUQ
CV37 9QL G0BKA
CV37 9QL G0BKB
CV37 9QW G8OCS
CV37 9SR 2E0VAV
CV37 9ST 2E1FCC
CV37 9ST 2E1FCD
CV37 9ST G8RDT
CV37 9XN 2E0BZA
CV37 9XN M3MXV
CV37 9XR G0WKT
CV370LT M7EBZ
CV377PL 2E0EKW
CV377PL M6AGT
CV4 7AY M6WYW
CV4 7DU G1KCR
CV4 7ED G6GLZ
CV4 7ES M7AYJ
CV4 8HF G0KNX
CV4 9AA G0OTT
CV4 9AN M6ZPT
CV4 9AU G6UCT
CV4 9AZ G8BZT
CV4 9BW G6PSO
CV4 9EX M0URX
CV4 9EZ G4VCX
CV4 9JH G1YFI
CV4 9LA G6WLM
CV4 9NR G1IUL
CV4 9SJ M7BZP
CV4 9TD G6JPR
CV47 0ES M3BOU
CV47 0HW G8OKN
CV47 1AS G1VBB
CV47 1EX G1XLG
CV47 1GH M3LQJ
CV47 1NF G3TFA
CV47 1PH G0AFZ
CV47 2AP 2E0OHN
CV47 2AP M0NPH
CV47 2AP M6OHZ
CV47 2AT G0CDO
CV47 2AT G0SDX
CV47 2BN G3ZST
CV47 2FJ G4EWW
CV47 2QS G0GNF
CV47 2RA M6GVP
CV47 2XY G1GAS
CV47 2XY G4YIE
CV47 7SL 2E0GHC
CV47 7SL M7AMK
CV47 8JS 2E0KOD
CV47 8JS M6KOD
CV47 8LG M0ZOM
CV47 8LG M6FZY
CV47 8LX G3SNO
CV47 8NE G1KCS
CV47 8NN G3OJI
CV47 8NN G6EVC
CV47AL M3WRK
CV5 6AR 2E0CBG
CV5 6AY G3VWD
CV5 6BB G4RUN
CV5 6DJ G3PUR
CV5 6DR G3WCQ
CV5 6FF G0OFF
CV5 6GH 2E1IAX
CV5 6LF G1MSA
CV5 6QH M6CVJ
CV5 7AD G8BHY
CV5 7AH G6MQP
CV5 7BY G8WYI
CV5 7DA M6BHT
CV5 7EH G3MAU
CV5 7FR G1KFQ
CV5 7GP G1PPB
CV5 7HP G7DTK
CV5 7JU G3NAP
CV5 7LJ G3UKD
CV5 7ND 2E0UAC
CV5 7NE G8AHR
CV5 7NR G4HOC
CV5 7QE G4MWP
CV5 7QE G4UVW
CV5 8BD 2E0UWK
CV5 8BD M6HGJ
CV5 8BD M6UWK
CV5 8DA M0XKW
CV5 8DE M1EQV
CV5 8DR 2E0RRC
CV5 8DR M0DTJ
CV5 8DR M6SDW
CV5 8FL G4DRU
CV5 9AS G0ENY
CV5 9EA M0NOI
CV5 9EN G8ALS
CV5 9HL G3JYS
CV5 9JT G4BCG
CV5 9JT G8GET
CV5 9JY G3NAY
CV5 9LN G4STI
CV5 9LQ G8RHM
CV5 9NA G4JDO
CV5 9NS G4KUE
CV6 1BE G7LAW
CV6 1BY G0EWZ
CV6 1GL 2E0LXC
CV6 1GL M7GWM
CV6 1LF G8KTX
CV6 1LH G1UUO
CV6 1NW G6KHA
CV6 1PW G7TFZ
CV6 2GE M7SKV
CV6 2GQ G1UWQ
CV6 2HB G8LGT
CV6 2NN G4KZU
CV6 2PG M3ORU
CV6 3GD M0HIN
CV6 3GD M0TAV
CV6 3GF G1ZNK
CV6 3GJ M3ECW
CV6 3GR G6IHO
CV6 3GS G7IZV
CV6 4EY G4LTK
CV6 4FH G3CWH
CV6 4FZ M6JMJ
CV6 4GZ G1NML
CV6 4HA M6IHO
CV6 4HL M3VYN
CV6 6BD M6JAV
CV6 6FF 2E0UPK
CV6 6FF G0UPK
CV6 6FF M7CQE
CV6 6GY G4OQV
CV6 6JZ G6ZTF
CV6 6NB M3JSF
CV6 7EU 2E0TSP
CV6 7EU M6RPV
CV6 7JJ M6EXH
CV6 7LB M3LVP
CV7 7AB G4LMV
CV7 7FX G4IEV
CV7 7FX G6SCM
CV7 7GG G6OES
CV7 7LU G0EPV
CV7 7LW G8XUU
CV7 7NB 2E0TUX
CV7 7NB M0YMM
CV7 7NB M3POV
CV7 7NQ 2E1ACG
CV7 7NQ G4KUR
CV7 7NQ G7UBX
CV7 7QF G3UKM
CV7 7QF M6JGV
CV7 7QL M0CAR
CV7 7QN G1UUS
CV7 7QW G4JYE
CV7 7QW M3HYF
CV7 8AQ M3UYF
CV7 8AU G4IAG
CV7 8BF G4RWH
CV7 8BY M6DAD
CV7 8ER G3SZE
CV7 8HL M0BVX
CV7 8HL M0BVZ
CV7 8HY G6GYN
CV7 8JJ G3SPY
CV7 8JJ G3UVW
CV7 8JJ G3ZFR
CV7 8JX M3CVO
CV7 8LB M0NIE
CV7 8LE M6KKD
CV7 8PR G1UUK
CV7 8PR G6KXN
CV7 9AQ M0BVU
CV7 9AU G6VJR
CV7 9AU G7AMQ
CV7 9BX M6TNY
CV7 9FH M0DVG
CV7 9LR G1BXQ
CV7 9LR G0AIZ
CV7 9LT G1OPJ
CV7 9NJ G4WXK
CV7 9NQ M0OCV
CV8 1AH G7RYO
CV8 1BE G1UUT
CV8 1DG G6XVZ
CV8 1DJ M0AZJ
CV8 1EZ G0KPG
CV8 1FL G4GJY
CV8 1GH M0LDY
CV8 1GL M3UKV
CV8 1HX G6GQF
CV8 1JD M6OAO
CV8 1JW G3VTL
CV8 1RB G8CYA
CV8 2AF M0BNC
CV8 2BB M0JDB
CV8 2BE G6XRI
CV8 2BX G4YMT
CV8 2DJ M7CSF
CV8 2DU G0DKV
CV8 2EE G0JMW
CV8 2JZ M6ENC
CV8 2PP G8YHO
CV8 2XH G7JYY
CV8 3ET G6WTD
CV8 3FR G0HOV
CV8 3GZ G4JAU
CV8 3HH M0JCT
CV8 3JF G3JZL
CV9 1AD M6IZY
CV9 1ET M0WHR
CV9 1ET M3WHR
CV9 1HN G1SIM
CV9 1HP 2E0EKA
CV9 1HP G8LQF
CV9 1JY M0VEX
CV9 1NZ G4IWA
CV9 1PS G0LKA
CV9 1QJ M7JLX
CV9 2AA G7TYJ
CV9 2DA G4XDE
CV9 2DN M7AIZ
CV9 2ET G4YNX
CV9 2EZ M1CNL
CV9 2EZ M6TAD
CV9 2PJ M7DDT
CV9 2PL G8SEI
CV9 2RP G6SGY
CV9 3BP M3GTQ
CV9 3EH 2E0NDW
CV9 3EJ G0VZK
CV9 3EJ M7JAY
CV9 3EX G0HUW
CV9 3EX M6HLT
CV9 3HA M7DQI
CV9 3HJ M6KDO
CV9 3LT G4KMX
CV9 3PJ G8NDB
CV9 3RE G2FXQ
CV91NX G6MUW
CV91QA G0KKV
CW1 3AF 2E0ZSK
CW1 3AF M6ZSK
CW1 3AX G8DTT
CW1 3AX M6FUJ
CW1 3EG M0MTF
CW1 3EQ 2E0NKP
CW1 3EQ M6NKP
CW1 3JN 2E1EHP
CW1 3JN M3EHP
CW1 3LE 2E0BDP
CW1 3NW 2E0FQC
CW1 3NW M6FQC
CW1 3PJ 2E0AZZ
CW1 3PJ 2E1UJE
CW1 3PJ M5WGD
CW1 3QD G7PEH
CW1 3SG 2E0KNC
CW1 3TT M7CLI
CW1 3XJ G6WZY
CW1 3XN M0MBG
CW1 3XN M6BCV
CW1 3YN M7RCR
CW1 4AL 2E0RMS
CW1 4AL 2E1RMS
CW1 4AL M0TDC
CW1 4AL M3AHS
CW1 4AL M3RMS
CW1 4AU M3KQW
CW1 4BS G7BMT
CW1 4DA 2E1RAO
CW1 4DE M3DMJ
CW1 4DE M6MHD
CW1 4DF 2E0NPP
CW1 4DF M3ZJF
CW1 4DF M3ZMV
CW1 4DU M3USB
CW1 4EL 2E1FIV
CW1 4ES G1DBI
CW1 4GH 2E0HBK
CW1 4GH M7KLJ
CW1 4HB M7EDL
CW1 4HL M3ISG
CW1 4HZ 2E0HST
CW1 4HZ M0KDA
CW1 4HZ M0WFX
CW1 4JU M7BKZ
CW1 4JZ M3VFL
CW1 4NW M6RGF
CW1 4PS M6HRE
CW1 4PX M6JUG
CW1 4QE 2E0FTV
CW1 4QE M6FTV
CW1 4QE M6POE
CW1 4RA G4DBX
CW1 4RP M0UTD
CW1 4RS 2E0CKM
CW1 4RS M6AVU
CW1 4TD M6TGS
CW1 4TT G6INM
CW1 4TX M6CSI
CW1 4TY G0HIZ
CW1 4UL M7IIZ
CW1 5LF 2E0FFM
CW1 5LF G7UUG
CW1 5LF M6JTO
CW1 5PB 2E0DCP
CW1 5PB M6DEO
CW1 5PW M6MXY
CW1 5PX 2E0CPE
CW1 5PX 2E0ECO
CW1 5PX M0HPW
CW1 5PX M0MRH
CW1 5PX M6EXC
CW1 5QE G6OBT
CW1 5RA G0CGQ
CW1 5SY G3ZFC
CW1 5YF G0AVE
CW1 5YQ M0CJK
CW1 6ES M3HQS
CW1 6HD G4DQG
CW1 6HN 2E0RAL
CW10 0AE 2E0SHK
CW10 0AE M0IUV
CW10 0AE M6IYU
CW10 0DG M3LVY
CW10 0DL M0BMR
CW10 0EA M6DMA
CW10 0HR G3MCC
CW10 0JS G4CFK
CW10 0PF 2E0CDQ
CW10 0PF M0WJH
CW10 0PF M6AFY
CW10 0PJ G4ZKR
CW10 0TA M6WPA
CW10 9AU 2E1GDA
CW10 9BY G0NTG
CW10 9HG G0GKN
CW10 9HL 2E0DWF
CW10 9HL M6WDF
CW10 9NW M3KKQ
CW10 9PX M7FVM
CW10 9QY M1BUR
CW11 1BL M3NNZ
CW11 1BN M0CJI
CW11 1BS G6MPN
CW11 1BZ G6ZGB
CW11 1EB G0BRZ
CW11 1EP G6RXD
CW11 1FW G4RXB
CW11 1FY G4GVE
CW11 1GF G7JHE
CW11 1JL 2E0NRL
CW11 1JL M7ECT
CW11 1RS M3TAE
CW11 1RW 2E0ZAC
CW11 1RW M0BZI
CW11 1SG G7OOI
CW11 1SQ G4MXE
CW11 1WN M3LIV
CW11 1XP 2E0FUH
CW11 1XP M0EEG
CW11 1XP M3FUH
CW11 1XP M3NBX
CW11 1XU M3SIS
CW11 3AR M6DBQ
CW11 3AS G0JNE
CW11 3BL 2E0CXN
CW11 3BL 2E0NHF
CW11 3BL M6DVA
CW11 3BL M7NHF
CW11 3BU 2E0OSY
CW11 3BU G4OHP
CW11 3BU G6IIU
CW11 3BU M6OSY
CW11 3FZ G1XRJ
CW11 3JF G7HOA
CW11 3JF M0OBW
CW11 3JF M1CNY
CW11 3LP M7RNO
CW11 3QL G4UCT
CW11 3RD G6BMG
CW11 3SW M7HDJ
CW11 3TA G7CTE
CW11 3TE M7GCA
CW11 4BP M0JEH
CW11 4HL G4VIF
CW11 4HS G7SHI
CW11 4HZ 2E0WIS
CW11 4HZ M6JLE
CW11 4HZ M6WIS
CW11 4PQ G6VPJ
CW11 4RE G4PMY
CW11 4RY G3UCW
CW11 4RZ G1GRN
CW11 4SP G3HUD
CW11 4SP G3JOE
CW11 4TN G6UKN
CW12 1AG 2E0GRS
CW12 1AG M3FIH
CW12 1DT M6JGG
CW12 1FS G3TVX
CW12 1HE G3WNW
CW12 1LT M6PYF
CW12 1NB G6BDA
CW12 1NU G0REB
CW12 1NU G4LTZ
CW12 1NY G7PZF
CW12 1QW M6CLX
CW12 1SD G7VTQ
CW12 1SE G4JYK
CW12 1SE G4NVN
CW12 1SH G6DZX
CW12 2BH 2E0FGI
CW12 2BH M6EGB
CW12 2BS 2E0SOT
CW12 2BS M3SOT
CW12 2BY M6JKR
CW12 2DX G6TDJ
CW12 2DX M7BKG
CW12 2EP M3KXE
CW12 2EQ M0OWO
CW12 2HN G3WRK
CW12 2HQ G7HIO
CW12 2JJ G7RMJ
CW12 2JU M3JCH
CW12 3DB M0AVQ
CW12 3DE G4GMZ
CW12 3DZ M3KWZ
CW12 3EE M6NND
CW12 3EP 2E1CPQ
CW12 3FF M6GCD
CW12 3HH M6IGJ
CW12 3HQ M6DEI
CW12 3HS G4UJO
CW12 3JA M6ZOM
CW12 3JJ M6BSO
CW12 3JP G7NCD
CW12 3JY G8JPU
CW12 3JY M3ZBL
CW12 3LD M0TCL
CW12 3LL G6VMV
CW12 3LZ M6ZJT
CW12 3PL G1LCE
CW12 3PL G4FSH
CW12 3QG G0DOC
CW12 3QG M6ZZD
CW12 3RB G3ZAU
CW12 3RQ 2E0ICY
CW12 3RQ M0RZY
CW12 3RQ M6ORE
CW12 3RX M6AMX
CW12 3SW M6RDA
CW12 3SW M6SJE
CW12 3TD 2E1CAT
CW12 3TD G8WWF
CW12 3TU M6SBY
CW12 3TW M1EER
CW12 3TW M6VCB
CW12 3TY G8RXY
CW12 4BS 2E0CJD
CW12 4BS M3SQT
CW12 4DY G4WTZ
CW12 4EA 2E0TXE
CW12 4EA M7TUX
CW12 4EH M7GIU
CW12 4EQ M6WLE
CW12 4FR G4HMX
CW12 4FR M6EAL
CW12 4HL G8UVZ
CW12 4JL M0VVG
CW12 4JL M1LOL
CW12 4JL M1REK
CW12 4JN M0GVE
CW12 4JR M3ZQB
CW12 4NE G0UXO
CW12 4NN G6ORT
CW12 4PH G0CQY
CW12 4PR M6XEP
CW12 4PS M3AVZ
CW12 4PY G7CPN
CW12 4QH G4IWV
CW12 4QQ M7TDL
CW12 4QR M6EHC
CW12 4QU G6XYV
CW12 4QX G6MAT
CW12 4RF G1VIY
CW12 4RR G6LAO
CW12 4RS G6WXI
CW12 4SN M3HXM
CW12 4TQ G4BOH
CW12 4UX M6MNU
CW13TL 2E0NNH
CW13TL M0NCC
CW13TL M0NNH
CW15NY M7ONO
CW15PN M3MVK
CW2 5AW G0NWF
CW2 5AW G4BZI
CW2 5AZ 2E1GDB
CW2 5AZ G8SIG
CW2 5AZ M3MAR
CW2 5AZ M3ZTT
CW2 5BJ G0TMK
CW2 5BY 2E1EGV
CW2 5EZ 2E0PWJ
CW2 5EZ M6EFP
CW2 5GP G4VXX
CW2 5GS G0FAU
CW2 5LX G0GRO
CW2 6HN 2E0ZHG
CW2 6HN M3ZHG
CW2 6JB G8DHQ
CW2 6RD G0DRM
CW2 6SQ M0XPA
CW2 6TG G1LSN
CW2 6TJ G6IGW
CW2 6TX M0VFR
CW2 6TX M0YVX
CW2 6TX M6ZBT
CW2 6XP G3NKC
CW2 7DP M3BMI
CW2 7HD M3KKF
CW2 7JY M3KPF
CW2 7LE M6KRV
CW2 7LN 2E0UGP
CW2 7LN M6UGP
CW2 7LQ G0DRN
CW2 7NY 2E0DYB
CW2 7NY M0SWC
CW2 7NY M6DYB
CW2 7SY M6UPT
CW2 8AP G0MMH
CW2 8AP G4RRM
CW2 8AP G6TW
CW2 8BD M1LEO
CW2 8BD M3AGB
CW2 8BN G4YFO
CW2 8BT G7IFI
CW2 8DS G1GRM
CW2 8EX G7GZZ
CW2 8HG G4MAG
CW2 8HJ G4LVR
CW2 8HP G0SIE
CW2 8HS G3ZBM
CW2 8LN 2E0IHL
CW2 8LN M7EGJ
CW2 8NA G8ZSK
CW2 8NH 2E0JDD
CW2 8NH M3LZK
CW2 8NS M3EUU
CW2 8PB G7PWU
CW2 8PL M3NVF
CW2 8QQ M3SDH
CW2 8RE M0RCT
CW2 8RL 2E0MCG
CW2 8RS G7OMN
CW2 8SH G6LVT
CW2 8SN 2E0JZC
CW2 8SN M6JZC
CW2 8SZ G8BTD
CW2 8UE G0TWH
CW3 0AH 2E0KZL
CW3 0AH M3LYP
CW3 0BA G3YMU
CW3 0BE 2E0COX
CW3 0BE M3MIO
CW3 0JB G0NBH
CW3 0JB G1XHR
CW3 9BS M3KEF
CW3 9DN G8FNA
CW3 9EG G4YPQ
CW3 9EL G3JJR
CW3 9EU G1VTN
CW3 9HX M0HAD

CW3 9HX M1KMT
CW3 9SZ M3YHQ
CW3 9SZ M3ZHY
CW4 7AS G3GMM
CW4 7BT G0CTP
CW4 7DR M3IFE
CW4 7HA 2E0MHT
CW4 7HA M0MHT
CW4 7LA G1DBL
CW4 8AS M3HYV
CW4 8DT G0CSY
CW4 8HX M0SUN
CW4 8JB G4NVA
CW4 8JG M0ESR
CW4 8JG M0ZED
CW4 8JG M6DNM
CW4 8JG M6NVA
CW4 8JU G4ZVA
CW4 8LJ M0YIJ
CW4 8LL 2E0OCG
CW4 8PG M3JWZ
CW4 8PP 2E0SKA
CW4 8PP M3BMV
CW5 5JE M6LPB
CW5 5JS M6WAP
CW5 5NW G0EEI
CW5 5QJ M0ELO
CW5 5RX M7MUS
CW5 5TX M6GWN
CW5 6AL G3KMV
CW5 6AL G8UHT
CW5 6DX G4XNR
CW5 6ED G0RBA
CW5 6HJ G4PCR
CW5 6HL G0FUZ
CW5 6JJ G6SNI
CW5 6NL G0RBW
CW5 6QF G0VJR
CW5 6SQ M0JEG
CW5 7BA M7FRT
CW5 7BB G3HEH
CW5 7BJ G8AVO
CW5 7EJ G3PFE
CW5 7ER G0OEW
CW5 7FL G4SHC
CW5 7GU 2E0JCD
CW5 7GU M0JCD
CW5 7HY G4DBD
CW5 7JA G0CZD
CW5 7NN G1BDQ
CW5 7NX G4MHX
CW5 7PN G4PUM
CW5 7QD G4PNC
CW5 8AQ M3FJN
CW5 8AQ M3FJR
CW5 8AQ M3HAK
CW5 8BG G6PGG
CW5 8DJ M7FOK
CW5 8ER M7CSR
CW5 8JD G0ZPV
CW6 0BA 2E0RJM
CW6 0BA M0RJM
CW6 0BA M3RZV
CW6 0DA G0TQJ
CW6 0GJ G8NXY
CW6 0HJ G3TRL
CW6 0NL G4VPC
CW6 0PN M7AVO
CW6 0PU G3NHP
CW6 0QR M3SHK
CW6 0SD G7VOI
CW6 0SX G3VUN
CW6 9AJ M0SOA
CW6 9BP M1EIU
CW6 9JT G4KZO
CW6 9QX G4XMX
CW6 9RT M6AYI
CW6 9TF G0MMT
CW7 1DZ M3RVX
CW7 1EU G1ILH
CW7 1FE M3JIC
CW7 1HA M0RDR
CW7 1LU G0GZI
CW7 1LU G6DSA
CW7 1LU G7DSA
CW7 1LU M3NLJ
CW7 1LY 2E0FZE
CW7 1LY M7GTA
CW7 1LZ M7TLG
CW7 1NE G7RYN
CW7 1NE M1ATI
CW7 1NJ M6CAG
CW7 1QL M1ARS
CW7 1QT M7BRZ
CW7 1QZ M3HGP
CW7 1RE G4TEZ
CW7 1RJ 2E0KNU
CW7 1RJ M6KNU
CW7 1SW G4JVX
CW7 2BJ G0SPH
CW7 2ED 2E1HEV
CW7 2ED G7GRO
CW7 2LJ M1DJP
CW7 2LL M6TKR
CW7 2NE M6GLN
CW7 2NG M7EPG
CW7 2QD G0EOL
CW7 2SE G0FOU
CW7 2SE M0TGS

CW7 2UE M3XFD
CW7 2UW G6WEL
CW7 2YR 2E0PXW
CW7 2YR M3PXW
CW7 3DE G7KFZ
CW7 3EN G0ADU
CW7 3EN G6WZZ
CW7 3JB 2E0RJD
CW7 3JB M0JFD
CW7 3JB M3LKD
CW7 3JU M1BXM
CW7 3LE G1DBR
CW7 3NG G8UYB
CW7 4DP M3YUA
CW8 1BL M6WFZ
CW8 1BW G0VOK
CW8 1ER M7MCY
CW8 1HX 2E0HGE
CW8 1HX M3HGE
CW8 1JR M3YET
CW8 1LA 2E0SDK
CW8 1LA M3KDK
CW8 1PL G6HXU
CW8 1PY G3ZTT
CW8 1PY G4XUV
CW8 1RD G1GYJ
CW8 2BQ G0JIT
CW8 2JB G3YWU
CW8 2LW G4MIH
CW8 2LZ G4XDG
CW8 2PB G1YJR
CW8 2PL M6GMP
CW8 2PP G6AFG
CW8 2XJ G4JYP
CW8 2XN G6DQO
CW8 3DX 2E0LDJ
CW8 3DX M0NEX
CW8 3DX M1ALA
CW8 3DX M6LDJ
CW8 3EZ M0XEY
CW8 3HD G4YVI
CW8 3HH G1GOP
CW8 3JD G7VDQ
CW8 3LP M3KJB
CW8 3PN G4KCZ
CW8 3PN G6NBP
CW8 3PT G4IAB
CW8 3RH G6LCS
CW8 4BA G0DQQ
CW8 4DF G0HXD
CW8 4EH M6SLB
CW8 4GS M0WKG
CW8 4GS M6CSS
CW8 4HR G6GAK
CW8 4JU G0WOA
CW8 4JX G4ROK
CW8 4PU G0BKH
CW8 4PU G1MCG
CW8 4PX M6PJV
CW8 4TQ M0SCN
CW8 4TQ M7SCN
CW8 4UE M7PBX
CW8 4XA G8NCS
CW8 4XG 2E0IXM
CW8 4XG M6IXM
CW8 4YU M7TPD
CW81SH M6ENX
CW9 5JL 2E0DHJ
CW9 5JL M6EXB
CW9 5LJ G0DND
CW9 5LJ M3CRV
CW9 5NT 2E0RNI
CW9 5NT M3CUU
CW9 5PS G1LDC
CW9 5QR G4OJF
CW9 6DA G8BJA
CW9 6DR G4XQB
CW9 6DS G1MVE
CW9 6EB G1URR
CW9 6ED M3GYH
CW9 6EZ G6LDM
CW9 6EZ M3OUQ
CW9 6EZ M3OVG
CW9 6HJ G0THJ
CW9 6LN 2E0XOJ
CW9 6LN M3XOJ
CW9 6PP G8XMZ
CW9 6PX G8VNX
CW9 7AN M6RJJ
CW9 7AR 2E0DHV
CW9 7AR M6EQO
CW9 7AR M6TFZ
CW9 7AS M3ISH
CW9 7BF M3GHG
CW9 7EP M3LZR
CW9 7JA G0OGJ
CW9 7JB 2E1GYG
CW9 7JB G1GOQ
CW9 7JQ M3XRQ
CW9 7JT 2E0VRC
CW9 7JT M6VRC
CW9 7LD M3CGI
CW9 7NL M7SCB
CW9 7QD M3XQV
CW9 7QD M3XQY
CW9 7QH 2E0HGX
CW9 7QH G4PKX
CW9 7QH M3HGX
CW9 7QJ M0CRP

CW9 7RG G0WPO
CW9 7TY M6EWU
CW9 8AR G6LJX
CW9 8BN M3NDJ
CW9 8GF G7KTH
CW9 8GG G6ZTT
CW9 8GG G7LQD
CW9 8GP G4MUA
CW9 8JE 2E0OJN
CW9 8JE M0RWT
CW9 8JE M7KDV
CW9 8PB 2E0STX
CW9 8PB M0WTX
CW9 8PB M0XSJ
CW9 8PB M6EAX
CW9 8PB M6MHJ
CW9 8PB M6WTX
CW9 8QA G0LBO
CW9 8QQ G4CAX
CW9 8RQ M0FCT
CW9 8TQ M7UAM
CW95PX M3KKN
DA1 1ND G4ZHX
DA1 1QT M3YJU
DA1 1TR G6UML
DA1 1YY M3ALB
DA1 2QN M1CXK
DA1 2RZ G6MVN
DA1 2TX G8PHJ
DA1 3AJ M1CYN
DA1 3AN G3ZPS
DA1 3AN M3RZX
DA1 3AP M0JVT
DA1 3BA G7KAO
DA1 3BL M7OMD
DA1 3BP G0RJL
DA1 3DE M6CAF
DA1 3DE M6GAF
DA1 3DE M6KRF
DA1 3EE M6BPC
DA1 3EW M7CAQ
DA1 3JU G4XKZ
DA1 3JX M0TGV
DA1 3NX G1RCV
DA1 3NX G3RCV
DA1 4PZ M6KVL
DA1 4RY G8ZHR
DA1 4SN G7WLL
DA1 4SU G4APB
DA1 5HT G6CMB
DA1 5JW G6TSC
DA1 5JW G8XJB
DA1 5NF M3YZN
DA1 5NG 2E0GLT
DA1 5NG M0HPF
DA1 5NG M7DXZ
DA1 5TN M7CJH
DA10 0LP G0KDV
DA10 0LT G0NZR
DA11 0AH M7TJF
DA11 0PH G4WTE
DA11 7AQ G7MIE
DA11 7BN G8JAD
DA11 7BN M1ENQ
DA11 7BN M3JUL
DA11 7EB G8TRF
DA11 7EB M0JKX
DA11 7EB M0JUQ
DA11 7EB M7CDF
DA11 7EZ G0DWS
DA11 7LB G8IEA
DA11 7LG G0RNP
DA11 7NY M3UUZ
DA11 7QG G3NPS
DA11 7QG G8LDV
DA11 8NH M1DUV
DA11 8NQ G4ALD
DA11 9DU M1CWG
DA11 9DU M6KIG
DA11 9DX G1YXY
DA11 9FF M0HZF
DA12 2LP G7EOC
DA12 3BU M7PHZ
DA12 4AR M3TJI
DA12 4BJ G3DCV
DA12 4EL 2E0OOM
DA12 4EL M6MAY
DA12 4HD G7CJS
DA12 4HJ G3RE
DA12 4HJ G4IYK
DA12 4NA G4BBJ
DA12 4NJ G8XIR
DA12 5BD G1PHJ
DA12 5BD G1PNL
DA12 5DN G1OFL
DA13 0BW G3KBH
DA13 0EA G0OAT
DA13 0LS G3BAC
DA13 0QA G0RJN
DA13 0SQ G4YGU
DA13 0TQ M6DDF
DA13 0TX 2E0RCL
DA13 0TX M0RBX
DA13 0TX M6DBI
DA13 0UD G0AFH
DA13 0UD G0FBB
DA13 0XB G1FCN
DA13 9DS G3PBF
DA13 9EJ G0SSN

DA13 9EJ G6CBY
DA13 9JR 2E0EDA
DA13 9JR M6HHF
DA14 4BE G8MCA
DA14 4ES G3XEF
DA14 4ET G4YMF
DA14 4LJ G7EDA
DA14 4PN M7OFF
DA14 4PS G6INU
DA14 4QU G0BAX
DA14 5LT G8PJF
DA14 5NF 2E0YEK
DA14 5NF G4FAA
DA14 5NF M0NOY
DA14 5NF M6NOY
DA14 5NG G6GTC
DA14 6JQ G3BNE
DA14 6JT G8XIN
DA14 6SG M1BXU
DA15 7NL G4ILH
DA15 8AR M0KUK
DA15 8AT 2E0SDW
DA15 8AT M0SJW
DA15 8AT M3NHI
DA15 8ER G1TXO
DA15 8JN G8JTG
DA15 8QL G1HBR
DA15 8QY G4NHP
DA15 8RX M0DDV
DA15 8SZ G8KAM
DA15 8TA G1PRZ
DA15 9DZ G8OPA
DA157LU M7FQR
DA158RD 2E0ERX
DA16 1BL M7FMF
DA16 1DE M1TAD
DA16 1EJ G0KPZ
DA16 2AW G8SWC
DA16 2BN G6NRH
DA16 2BP G0KTS
DA16 2HJ G2DZH
DA16 2HX G8GKC
DA16 2LH G1OAZ
DA16 2QD G0GKI
DA16 2RU G6ODW
DA16 3AW G0LZF
DA17 5BB M0KBX
DA17 5BG G8MIF
DA17 5BT G7UHW
DA17 5EW G3CUR
DA17 6HB G4WJH
DA17 6JE M0PBR
DA17 6LP G0VWF
DA2 6BN G7FIA
DA2 6DN G1JNQ
DA2 6HE M6THS
DA2 6HZ G1ZAW
DA2 6JE M7IPY
DA2 6JS G4OFU
DA2 6JX G6ZAY
DA2 6LQ M6MRV
DA2 6NB G4CVC
DA2 7ES G4IQQ
DA2 7LP G0HRD
DA2 7NW G0DVL
DA2 7PB G0KUU
DA2 7QQ G4LGU
DA2 7RL M6CMO
DA2 7RL M6WLF
DA2 7SP G3XVC
DA2 7SP M6JOP
DA2 7WB G1EWE
DA2 8BH G1UXZ
DA2 8BH G7JYL
DA2 8BZ M0DPS
DA26HZ G7PXM
DA3 7EB M7TDT
DA3 7HB M7GBR
DA3 7HD G3FGP
DA3 7LA 2E1JGD
DA3 7LA M7JGD
DA3 7NS G6XND
DA3 8DD G1FJJ
DA3 8EU G4NRV
DA3 8LG G8RW
DA3 8LG M0XPS
DA3 8LN 2E0RBI
DA3 8LN M0RJT
DA3 8LN M3RBI
DA4 0BE G7FUV
DA4 0BE M0UAT
DA4 0BE M3YBJ
DA4 0HA G0WMC
DA4 0HA M1ABU
DA4 9AR M1CGP
DA4 9DQ 2E0WAF
DA4 9DQ G4BWV
DA4 9DQ M3WPK
DA4 9EX G6YLD
DA4 9HS M3YOU
DA5 1LX G6IRP
DA5 1RJ G6TYE
DA5 2AB M7DOK
DA5 2ER G3XEW
DA5 2ES G4CW
DA5 2HN M3VSO
DA5 3AH G0FDZ
DA5 3BT G6YZB
DA6 7LA M1DYD

DA6 7NY G6AVP
DA6 7PA G4MB
DA6 8HU G4RPP
DA6 8JS G1MAV
DA7 4 G4DDP
DA7 4 G8BXC
DA7 4 M0THF
DA7 4AJ M6ASM
DA7 4JL G4XKV
DA7 4JZ G4KOW
DA7 4LY M3JPG
DA7 4PG G6CUE
DA7 4PQ 2E0WUF
DA7 4QD M6BZG
DA7 4RL G0FAS
DA7 4ST G8CXI
DA7 4UE M1CZA
DA7 4UX G7MJJ
DA7 5BT G7UUB
DA7 5BT M3FTI
DA7 5DG G6ENO
DA7 5DZ G3KHR
DA7 5HG M3USW
DA7 5JU G0PXT
DA7 5JU M3KVJ
DA7 5NP G8JZT
DA7 5NP G8TNK
DA7 5PS 2E0LRR
DA7 5PS M6LXR
DA7 5SL G7GPI
DA7 6AF G4DUM
DA7 6NL G0UAO
DA7 6QU G4GZN
DA7 6SG M6BLT
DA8 1DZ 2E0GPG
DA8 1DZ M0YUG
DA8 1DZ M6GPG
DA8 1EE M7FSC
DA8 1JQ M3JLV
DA8 1LU G4LSU
DA8 1NN G4MHJ
DA8 1NN M1AJT
DA8 1PZ M0JHP
DA8 2AQ M1CXN
DA8 2DR 2E0XUZ
DA8 2JD G4EGU
DA8 2JD M3VNX
DA8 2JG 2E0WTZ
DA8 2JG M1CBK
DA8 2NN M6LKD
DA8 3BA M0LHA
DA8 3BL G3NRZ
DA8 3JY M7CWT
DA8 3LZ G6NNU
DA8 3NG 2E0JET
DA8 3QT M0SHI
DA8 3SN G8YLC
DA9 9PG G8MHI
DD1 1JZ GM5LUK
DD1 2AP MM0GDI
DD1 2JA 2M0KAU
DD1 2JA MM6KAU
DD1 3AW GM7TTU
DD1 4LY GM0TOF
DD10 0BA MM7AGL
DD10 0DN MM6MJC
DD10 0HW MM6ZZG
DD10 0HX GM4PKJ
DD10 0HX GM8ZEQ
DD10 0NG MM7EAR
DD10 0PH 2M0BXN
DD10 0RY MM3NWF
DD10 0SB GM4TXN
DD10 0SL GM0ENQ
DD10 0SR GM1MLS
DD10 0SW GM3YAO
DD10 0TG GM1TDU
DD10 0TT GM0WRV
DD10 8AZ MM1COS
DD10 8JL MM6SSO
DD10 8TW MM3DFG
DD10 9AA GM3COQ
DD10 9AQ MM6XGT
DD10 9BX MM3PFA
DD10 9DD GM5CGA
DD10 9EJ GM0ARH
DD10 9HE MM7PPW
DD10 9RR MM0BIX
DD10 9TH MM7EUY
DD100SL MM7DRV
DD11 1PX MM0CXA
DD11 1ST MM7NNM
DD11 2LZ GM4SXJ
DD11 2NX GM0MIW
DD11 2NX GM4YWU
DD11 2RJ MM3KDN
DD11 3EF 2M0GXZ
DD11 3EF MM0KGS
DD11 3EF MM6GZS
DD11 4BH MM3NJV
DD11 4EZ 2M0DOI
DD11 4EZ MM6GFG
DD11 4HT 2M0ULD
DD11 4RA GM4YWS
DD11 4RH GM0WNS
DD11 4SR GM1XHZ
DD11 4SX GM0SHD
DD11 4SX GM3GBZ
DD11 4UX MM6IBB

DD11 5FG MM0HRI
DD11 5RH MM6CWB
DD11 5RH MM6JNB
DD11 5SS GM1MON
DD11 5ST 2M0MAV
DD11 5ST MM0LBX
DD11 5ST MM6LAD
DD11 5SY GM3OBG
DD2 1BF 2M0RTO
DD2 1BF MM6RTO
DD2 1JT MM0DNH
DD2 1PJ GM8JVZ
DD2 1QN MM6DBJ
DD2 2BU GM8UMN
DD2 2BU MM3XGS
DD2 2EZ 2M0LEW
DD2 2EZ MM0LWS
DD2 2EZ MM6LEW
DD2 2FB 2M0MFK
DD2 2NH MM6AKQ
DD2 2PW GM3OQI
DD2 2PW MM0AKX
DD2 3AF MM6RAM
DD2 3BF 2M0NYK
DD2 3BF MM6YYB
DD2 3BU MM7NOZ
DD2 3JR GM3WPA
DD2 4EG GM3WJE
DD2 4HP MM6BXH
DD2 4TT 2M0XDS
DD2 4UA GM4JCM
DD2 4XB GM4VXM
DD2 5AH 2M0SEF
DD2 5DL GM0KKE
DD2 5PX GM4LUD
DD2 5PZ 2M0INE
DD2 5PZ MM6PAU
DD2 5QJ GM1CMF
DD2 5QN GM0TAY
DD2 5QN GM4FLP
DD2 5RY GM8LON
DD3 0BN GM4ZFS
DD3 0EA 2M1KOJ
DD3 0EA MM6KOJ
DD3 0JR MM3NPG
DD3 0LT MM6AAR
DD3 0PH GM4API
DD3 0PR MM6PRH
DD3 0QN 2M0CVK
DD3 6AQ MM0DTW
DD3 6DD GM3ZXB
DD3 7HG MM7LLK
DD3 8AF 2M0XZX
DD3 8EJ GM3ZTP
DD3 8PY GM5BDW
DD3 9LH 2M0MGY
DD3 9PL GM0ROU
DD3 9RG GM0FSW
DD4 0AD GM1KCH
DD4 0NR GM4AAF
DD4 0NR MM0DXD
DD4 0NR MM3JFW
DD4 0PP GM7ONJ
DD4 0PW GM0GYQ
DD4 0QU GM0NLU
DD4 0TN 2M0RTA
DD4 0TN MM0RTT
DD4 0TN MM6RBT
DD4 0UJ GM7OWU
DD4 0XJ MM6SGQ
DD4 0XZ MM6JVQ
DD4 7EL GM0PIV
DD4 7NA MM7DVM
DD4 8AP MM6GID
DD4 8FE MM3PXG
DD4 8RP MM6DDX
DD4 9EX GM1VWA
DD4 9HH MM7DWV
DD4 9HQ GM0DNH
DD4 9LP GM0ISA
DD4 9ND MM0TMG
DD40XQ MM7ASP
DD5 1QT GM3NHQ
DD5 1QW GM1JTK
DD5 1QW GM7GOE
DD5 2EU GM4CAB
DD5 2RE GM1MUY
DD5 2RJ GM6BML
DD5 3AT GM0URU
DD5 3BN GM0RKU
DD5 3BT 2M0DOL
DD5 3BT MM6MIS
DD5 3DQ MM3SYU
DD5 3EF GM0CQL
DD5 3JG GM4JPZ
DD5 3LE GM3YVX
DD5 3PD GM8OEG
DD5 3PL 2M0LMY
DD5 3PL MM7LMM
DD5 3QN GM4UGF
DD5 3RE MM3NLH
DD5 3WN MM3HKE
DD5 4AA GM7DPI
DD5 4HG GM7AUX
DD5 4HS MM6KLT
DD5 4HT GM4MUZ
DD5 4LD MM1CFC
DD5 4SH 2M0JEY
DD5 4SW GM0VOL

DD5 4TS MM0DRA
DD51DY GM3LIW
DD6 8AP GM0TGG
DD6 8DT GM4FSB
DD6 8HL GM3MOR
DD6 8HL GM4RXW
DD6 8JH MM0BTD
DD6 8LF GM0CNW
DD6 8ND GM8RPE
DD6 8NG MM7LWC
DD6 9DT 2M0CLN
DD6 9EX MM3ZUP
DD6 9JD 2M1IBX
DD6 9LG GM4RDI
DD6 9NE MM3FZI
DD6 9NX GM4OLH
DD7 6DQ MM1ELE
DD7 6EX GM4FEI
DD7 6HT MM1EYD
DD7 6HW GM6RAK
DD7 6JY 2M0NGO
DD7 6JY MM3XLO
DD7 7BR MM0GGD
DD7 7BR MM1DSD
DD7 7PJ GM3XJE
DD7 7QF MM0EQE
DD7 7QF MM1EQE
DD7 7QQ MM3XFP
DD7 7SQ GM4JEJ
DD7 7TB GM0EFT
DD8 1AD GM3KEZ
DD8 1DQ GM1JWJ
DD8 1JR GM0DGK
DD8 1UE MM6BHS
DD8 1UF GM3ZBR
DD8 1UF GM4GOW
DD8 1UN GM0BTK
DD8 2AT 2M0YEQ
DD8 2AT MM0YEQ
DD8 2AT MM3YEQ
DD8 2PQ 2M0EDV
DD8 2PQ GM8RSC
DD8 2PQ MM3IZO
DD8 2SD MM7DYA
DD8 2SP GM0TWB
DD8 2UZ GM4RWE
DD8 3AG GM4XKP
DD8 3ES GM4WZY
DD8 3EY GM6AAJ
DD8 3EY MM0BSX
DD8 3TU GM1TCN
DD8 5AB MM3KVV
DD8 5ET MM3KVX
DD8 5LD GM4YHS
DD8 5NT MM6GYL
DD8 5NX MM0JNL
DD85DL MM7RFW
DD9 2PJ MM3HAF
DD9 6BG MM7MHJ
DD9 6JD GM4LFL
DD9 6RQ GM0JZV
DD9 6SD GM7DAP
DD9 6UH GM3KC
DD9 6UH MM0ERK
DD9 7AE 2M0SYH
DD9 7AE MM7SAH
DD9 7AX MM7EKZ
DD9 7QE MM0BDA
DD9 7SZ GM6RGD
DD9 7UX GM4YRE
DE1 1HW M6RIS
DE1 3RW M1CWY
DE11 0BH G0VVF
DE11 0DH M7BPW
DE11 0DS G1PKR
DE11 0EB G3OMT
DE11 0HE M7MAW
DE11 0JR G4WPE
DE11 0LY M0APK
DE11 0LZ G6DCS
DE11 0NB G1PKV
DE11 0RX M1FBF
DE11 0SL M6CGM
DE11 0SP 2E1FLD
DE11 0SR M3THN
DE11 0TG G7EJO
DE11 0UU 2E0KPO
DE11 0UU 2E0NBR
DE11 0UU G8UZQ
DE11 0UU M0KPO
DE11 0UU M0SCW
DE11 0UU M3KPO
DE11 7EZ 2E0FNO
DE11 7EZ M0JEO
DE11 7EZ M0URJ
DE11 7EZ M7AHA
DE11 7HG M6MAX
DE11 7HH M7HKP
DE11 7LN M7DNM
DE11 7QU G6NZL
DE11 7QU M6YAD
DE11 8AA 2E0EWL
DE11 8AA G4RJO
DE11 8BD G8VBC
DE11 8BG G0WHC
DE11 8BX M6RVN
DE11 8FS M6MLG
DE11 8HD 2E0XPK
DE11 8HD M3XPK

DE11 8LX G1ZMW
DE11 9AF M6XPC
DE11 9AQ M6EZF
DE11 9AT G7AYB
DE11 9BW 2E0FGM
DE11 9BW M0OSI
DE11 9BW M6HAG
DE11 9DZ G1OWK
DE11 9FB G7JHW
DE11 9FB M0GND
DE11 9FB M3JQY
DE11 9FF 2E0MYA
DE11 9FF G4UWA
DE11 9FF M7MYC
DE11 9FH G1SNQ
DE11 9FH G4ZDE
DE11 9HD G8FVC
DE11 9HF 2E0THF
DE11 9HF M7XHF
DE11 9PH M1BFV
DE11 9PJ 2E0BGV
DE11 9PJ M3FKV
DE11 9QN M6GHV
DE11 9QU M6DYP
DE11 9SG M0TRZ
DE110ET M7PWD
DE110RX G4EWK
DE12 6BL G4AYS
DE12 6DN M6GLJ
DE12 6EX 2E0DAQ
DE12 6EX M3VWD
DE12 6HH M6ICA
DE12 6HQ G1YBM
DE12 6JJ M3KBY
DE12 6JJ M3KRB
DE12 6JL M3XWP
DE12 6LU G4CRT
DE12 6PL M6XAQ
DE12 6PZ M3UDF
DE12 7AS G1OAX
DE12 7BD G1MVQ
DE12 7EG G6JVT
DE12 7JG M3XUT
DE12 7JH M0SDP
DE12 7JZ G7MGX
DE12 7JZ M0PCA
DE12 7JZ M1LYN
DE12 7NA G7KVZ
DE12 7PW G6MYO
DE12 7QG G7EIK
DE12 7QU G0HYR
DE12 8BJ G4YBP
DE12 8DL M0APD
DE12 8ES G1VQH
DE12 8JJ M0NXD
DE12 8JW G7EHU
DE12 8LB G0OES
DE12 8NJ G4HIW
DE126NS 2E0EZO
DE126NS M6LYC
DE128HG G6DIC
DE13 0AD G7JEJ
DE13 0AL G6NSQ
DE13 0AY M7NHJ
DE13 0BT M3BJE
DE13 0BT M6RFF
DE13 0DQ G4YFZ
DE13 0EE G4GKZ
DE13 0FD M6BSW
DE13 0FN M6IGZ
DE13 0FQ 2E0EZG
DE13 0FQ M0XMX
DE13 0FQ M6OML
DE13 0JD G4RAC
DE13 0JD G8VBW
DE13 0JH G5RJH
DE13 0JP G1ATA
DE13 0LD G1OYZ
DE13 0LD G1OZD
DE13 0LR M1CBO
DE13 0ND M1CDT
DE13 0NN M3GPM
DE13 0NT M3RCW
DE13 0NU G4XKL
DE13 0NY G4IPV
DE13 0RD M3FKN
DE13 0RD M3IEW
DE13 0RF G4HBY
DE13 0SQ M3HRV
DE13 0UU G4VPF
DE13 0WJ G1PRW
DE13 7AJ G3CRH
DE13 7EB G6ZVR
DE13 7EZ G1DQU
DE13 7EZ G4UJW
DE13 8AB G7VSM
DE13 8DS G7KIF
DE13 8DZ M6IOY
DE13 8JR G4RWD
DE13 8NT G4CHI
DE13 8PT G4ICZ
DE13 8SU 2E0TTB
DE13 8SU M6LMB
DE13 8SU M6TTB
DE13 9AA G6EIH
DE13 9AB G8VBA
DE13 9AR G0JLU
DE13 9AR G4LIR
DE13 9BJ G0FNH

DE13 9BJ G1OYU
DE13 9DE G4SXE
DE13 9DS G4KBP
DE13 9DY 2E1IIL
DE13 9EG G0SHO
DE13 9JR G0AOD
DE13 9JR M3RHK
DE13 9LE G3BZB
DE13 9NH G8OZP
DE13 9QD G0JEE
DE13 9RT G3TOY
DE14 1BS G6TJK
DE14 1BW M3WJY
DE14 1QS M6FDY
DE14 2EX 2E0SMX
DE14 2EX M6TDO
DE14 2FL G8OWA
DE14 2JS G4TBQ
DE14 2NB G8WSY
DE14 2NP G1NBO
DE14 2SN M3KQR
DE14 2SW M3VLH
DE14 3BY M0HFA
DE14 3DA G6FLY
DE14 3DY M3AUF
DE14 3DZ G8HZI
DE14 3FL G0HFL
DE14 3FQ 2E0GSB
DE14 3FQ M3VDF
DE14 3GJ M3OTS
DE14 3JE 2E0CGU
DE14 3JE M3HVA
DE14 3JF G6AZL
DE14 3LR G0FSF
DE14 3SB 2E0JBY
DE14 3SB M0ASW
DE14 3TX 2E0KLR
DE15 0AB G4DIH
DE15 0AD M6JGR
DE15 0AR G6OVA
DE15 0BA G7RUC
DE15 0BW 2E0CGJ
DE15 0BW M0KAJ
DE15 0BY G0VUC
DE15 0DL M6TNL
DE15 0EJ G0IDD
DE15 0PA M7DFY
DE15 0PT G0CQS
DE15 0QJ G4PGJ
DE15 0QJ M7GPL
DE15 0QW G0TZM
DE15 0RY G8EBM
DE15 0SL G4ELG
DE15 0TT G6JPQ
DE15 9BJ M0AGS
DE15 9BN 2E0ZAI
DE15 9BN M0ISZ
DE15 9BN M6MGJ
DE15 9EB G0HJR
DE15 9EN M1EWB
DE15 9EY G4PGX
DE15 9FR G6BRY
DE15 9LS 2E0TOX
DE15 9QN G3NFC
DE15 9QN G8EKG
DE15 9QW M3XXG
DE15 9SE G8UUR
DE21 2DH M6TDA
DE21 2HY 2E0ZCG
DE21 2HY M1BUQ
DE21 2HY M3ZCG
DE21 2HZ M6HFQ
DE21 2LH M1CFZ
DE21 2UG 2E0RCB
DE21 2UG M3LFP
DE21 4DX M0NMD
DE21 4GA 2E0TGV
DE21 4GB M3DGN
DE21 4HQ G3HCS
DE21 4HQ G8HCS
DE21 4JL G1VQB
DE21 4JY M6FFI
DE21 4LG 2E0TCU
DE21 4LG M6WCA
DE21 4LY 2E0JNH
DE21 4LY M6DDY
DE21 4RF G6VFA
DE21 4RL 2E0OES
DE21 4RL M0TRY
DE21 4RL M3XUG
DE21 4SQ M7WLF
DE21 5AP G4AOA
DE21 5AR 2E1ICU
DE21 5AR G4IBW
DE21 5AR M6RJF
DE21 5AX G4FQZ
DE21 5BN G4RAR
DE21 5DY M0IMF
DE21 6AP M6TZA
DE21 6DZ M3JVW
DE21 6FW M0CNY
DE21 6FX G4TBK
DE21 6FZ M1CWV
DE21 6LW M6GDL
DE21 6NX G1NZN
DE21 6NX G7AFW
DE21 6PH G1DHY
DE21 6PL G1UJX
DE21 6QB G6CWW

DE21 6SH G4EYN
DE21 6SJ 2E0EBQ
DE21 6SJ M0IEQ
DE21 6SJ M6GWQ
DE21 6TR G1XFE
DE21 6TZ 2E0BFA
DE21 6TZ M3MVV
DE21 6WB M7ECY
DE21 6WX G8BFA
DE21 7AT 2E0PET
DE21 7AT M0KEF
DE21 7AT M6PRM
DE21 7DS M5GAC
DE21 7DZ G4UMT
DE21 7EN G3OCA
DE21 7EN G3ZBI
DE21 7ES G3JFT
DE21 7FX M6MGT
DE21 7GL G1UYZ
DE21 7GT M0GPT
DE21 7JW G6WYD
DE21 7JZ G3OXN
DE21 7LR 2E1GOC
DE21 7LR G0IDL
DE21 7LX G6JPS
DE21 7QA G6FVJ
DE21 7RG 2E0SYD
DE21 7RG M3IIP
DE21 7TP M3NTZ
DE214ND 2E0MXL
DE214ND M7MRL
DE216JS M7AFU
DE22 1EQ G4OMT
DE22 1EY G8RBK
DE22 1GE M3AWQ
DE22 2BJ G3SZJ
DE22 2BJ G4EYM
DE22 2EH G6NLE
DE22 2ET M7NHG
DE22 2GN 2E1DIA
DE22 2GQ M1EAG
DE22 2GR M0ADL
DE22 2GW G6DNH
DE22 2HA 2E1FBS
DE22 2HF G1BEJ
DE22 2HF G4FPY
DE22 2HG G0RII
DE22 2HH M3WMS
DE22 2JR G5GIH
DE22 2LN 2E0GWB
DE22 2LN G3YOO
DE22 2LN M0HLP
DE22 2LN M6BJM
DE22 2LN M6GWB
DE22 2NA G6WXJ
DE22 2NB M0PUR
DE22 2NB M6ZTB
DE22 2QF M7LGV
DE22 2RE G0GDL
DE22 2SN M3ULN
DE22 2TA M6FHX
DE22 2UR 2E0WCE
DE22 2UR M6MHW
DE22 2XP G4LOF
DE22 3AS G0VVA
DE22 3EF G6EBL
DE22 3GL G7ODZ
DE22 3JP G4CPQ
DE22 3JT M6PTY
DE22 3LS 2E0RLG
DE22 3LS M6RLG
DE22 3RW M6KMC
DE22 3UE G4CZH
DE22 4AQ M6JDM
DE22 4BW G3SMV
DE22 4FL G7RWW
DE22 4HN G3PWS
DE22 4JU 2E1ADJ
DE22 4JX 2E1EDD
DE22 4LQ M6ZLL
DE23 1BZ G4CMZ
DE23 1DN 2E0KDH
DE23 1GN 2E0MNG
DE23 1GN M6MNG
DE23 1PR M3RCI
DE23 1QA M6OUA
DE23 1RH G6XOR
DE23 1WN G0IOZ
DE23 1XF 2E1GDF
DE23 2FR 2E0NYT
DE23 2FR M6NYT
DE23 2UB G4EJE
DE23 3RL G4COR
DE23 3XU M1MPK
DE23 4BJ M3JBZ
DE23 4EJ M6GNV
DE23 4RS M6KQS
DE23 6AQ 2E0NPF
DE23 6AQ M7BHC
DE23 6BL G7SJS
DE23 6BW 2E1DET
DE23 6DA G7GXE
DE23 6DL G3URU
DE23 6EB M6JDF
DE23 6EN G4BGD
DE23 6FB 2E0OJB
DE23 6FB M3NZK
DE23 6FT 2E0KRS
DE23 6GA M0HLS

DE23 6GA M6BFO
DE23 6HB G0WQY
DE23 6HB G8MXT
DE23 6JD G8TNE
DE23 6JD M3JBB
DE23 6JD M3TNE
DE23 6JY M0DVQ
DE23 6NF M0OTS
DE23 6NF M5EHG
DE23 6PF G1VAB
DE23 6PS 2E1CZJ
DE23 6TE M0GPK
DE23 6TG G3RLO
DE23 6TG G6XGF
DE23 6XL G3FUJ
DE23 8AX 2E0DLP
DE23 8AX M0OAC
DE23 8BP 2E0GCG
DE23 8BP M6GBE
DE23 8BP M6RDI
DE23 8LH M6EPJ
DE23 8PR 2E0SIK
DE23 8PR M6ISK
DE23 8PR M6USA
DE24 0AQ G8YCK
DE24 0AQ M0JBI
DE24 0BP G7LPV
DE24 0BU 2E0RUZ
DE24 0BU M0RUZ
DE24 0BU M6RUZ
DE24 0DJ G0PRY
DE24 0ER G0PHC
DE24 0FJ G6SOX
DE24 0FT G0DKX
DE24 0GH G1HEU
DE24 0GQ G4FAE
DE24 0HJ M7JJT
DE24 0HZ G6KUI
DE24 0JU G4YVV
DE24 0LF 2E0VRR
DE24 0LF M6RRV
DE24 0LH G0BIE
DE24 0LP G7SZG
DE24 0LQ G3ERD
DE24 0LQ G4AKE
DE24 0PX G8BFC
DE24 0SA G3KWY
DE24 0TA G4LNQ
DE24 0UQ 2E1HVZ
DE24 0UY M1EDL
DE24 0UY M6NZS
DE24 3AA G3WSM
DE24 3BH 2E0WHO
DE24 3BH G0WAC
DE24 3BS G1LKK
DE24 3DD M6KKM
DE24 3DT M0LKN
DE24 3DZ G0IWF
DE24 3EA G4MVD
DE24 3HE G0CPO
DE24 3JA G4PTZ
DE24 5AF 2E0BNH
DE24 5AF 2E0WBX
DE24 5AF M0XFX
DE24 5AF M3SKV
DE24 5AF M6WLR
DE24 8AZ G7BJB
DE24 8BD G4PDE
DE24 8DA M3PRS
DE24 8DS M7AVE
DE24 8DU M6SPK
DE24 8EX M6PKT
DE24 8GT G0RUR
DE24 8JR M6WET
DE24 8NF G0TJT
DE24 8NP G6UGT
DE24 8NS M6TLK
DE24 8PY M7FHG
DE24 8RN G4SNN
DE24 8WD G7VNO
DE24 8YE M7EAC
DE24 9BT G1IWT
DE24 9DL M6TIO
DE24 9FQ 2E0VEW
DE24 9FQ M7AIV
DE24 9GY 2E0FOR
DE24 9GY M6FOR
DE24 9LZ M1AJQ
DE24 9NG G6CXI
DE24 9PD G4YYI
DE24 9PF 2E0RTE
DE24 9PF M6RCG
DE24 9PT G6XEN
DE248AT 2E0JWW
DE248AT M3NVG
DE248DG M7JKC
DE24SD 2E0FMO
DE3 0DF G4HJL
DE3 0DU M3JIK
DE3 0ED M0WSB
DE3 0EE G4YIZ
DE3 0PA G7BHY
DE3 0PH M3LFG
DE3 0QQ G4ZHD
DE3 0QT G4MBH
DE3 0RD G0MSS
DE3 0RD G7SVQ
DE3 0RG M1FAJ
DE3 0RR G3TOV
DE3 0RS M3JKG
DE3 0TB M3JXI
DE3 4LQ G7BJC
DE3 9AH G8GBU
DE3 9AQ G6MWS
DE3 9BD M1MTV
DE3 9FJ G4KRW
DE3 9FL G7WGP
DE3 9FP G6WZE
DE3 9FT G8YNK
DE3 9FY G1UZS
DE3 9GH G7GJI
DE3 9GH G8QZ
DE3 9GH M3DPS
DE3 9GQ G0DMK
DE3 9GT G0CRN
DE3 9HB G4KOJ
DE3 9HZ G7GEX
DE3 9JT G1KKD
DE3 9JW G2DAN
DE3 9LN G4BKO
DE3 9LU M3JZE
DE4 2AH M3FUV
DE4 2GG G7VWN
DE4 2GL G0KKD
DE4 2HF G3RBP
DE4 2HF M5AA
DE4 2HP G0OTJ
DE4 2JJ G1CJC
DE4 2JW G4XTK
DE4 2PW G6HWR
DE4 3BA 2E0VKS
DE4 3BA M0VKS
DE4 3BA M3VKS
DE4 3BT G4ZEY
DE4 3EU G7NBJ
DE4 3EW 2E1AWZ
DE4 3EY M7FWF
DE4 3GP G7NLJ
DE4 3JW G3YGB
DE4 3JW G8GGR
DE4 3LL G0OWC
DE4 3QU G0FSB
DE4 3SD G3RLX
DE4 3SN G1TCH
DE4 3TB M7PVP
DE4 3TE G0BJD
DE4 3TG M6RJR
DE4 4AD G4UIQ
DE4 4AD G6NMU
DE4 4EX M0CVS
DE4 4FR G1SGP
DE4 4PG G0MMO
DE4 4PX 2E0HSY
DE4 4PX G5JON
DE4 4PX M7GWH
DE4 5AR M6BOR
DE4 5BH G8DLP
DE4 5DG G4HTW
DE4 5DJ G0FVU
DE4 5DJ G6JNZ
DE4 5DJ M3NVA
DE4 5DJ M3WTC
DE4 5EG G3VLF
DE4 5FF G8RNU
DE4 5FS G0ACZ
DE4 5GT G7EJK
DE4 5LT M3OGD
DE42DT 2E0GHJ
DE42DT M0JKS
DE44EG 2E0VMG
DE44EG M7AXJ
DE45 1BH G4JIK
DE45 1DD G6EAZ
DE45 1FG M0PKV
DE45 1JB G4GWX
DE45 1NJ G3VOT
DE45 1QQ G3SPV
DE45 1QS G7ABR
DE45 1QS M3DXD
DE45 1RE M6WIL
DE45 1TP G6CSC
DE5 3FG G0WWO
DE5 3FL G7GZU
DE5 3HD G1YPT
DE5 3JL 2E1HFN
DE5 3LJ G0NYM
DE5 3PJ 2E0GRM
DE5 3PY G4GGL
DE5 3RE G3NJX
DE5 3RR G3ZYC
DE5 3RR G8ZYC
DE5 3RY M6RWE
DE5 3SP 2E0VRA
DE5 3SP G5VRA
DE5 3SP M7VRA
DE5 3TR 2E1EOK
DE5 3TR G8ERV
DE5 8JG G0NNU
DE5 8JG G7OWZ
DE5 8PQ G4DXY
DE5 8PQ G6TFL
DE5 8RF 2E1AZA
DE5 9QN G8RNG
DE5 9RB G8ZIY
DE5 9RN M0XTX
DE5 9RR M7AIQ
DE5 9SP G3YQL
DE55 1LG M6HQN
DE55 1LN G6KPW
DE55 1RL 2E0IPA
DE55 2AJ G7DGF
DE55 2AJ M0GNU
DE55 2AS G1UAZ
DE55 2BJ G0DGQ
DE55 2BR M5RST
DE55 2DN G4CHM
DE55 2EJ G1BFK
DE55 2EP M0DAG
DE55 2ER M0HXO
DE55 2HE 2E0CND
DE55 2HE M3CND
DE55 2HE M6BYS
DE55 2HS G8VSN
DE55 2JD G0VBX
DE55 2JD G7SJD
DE55 2JD M3IQM
DE55 3AP G6TRQ
DE55 3AW M3DAF
DE55 3NW 2E0TSV
DE55 3NW M3FYM
DE55 4AA M6EPQ
DE55 4AF G4KBI
DE55 4AG G1NWH
DE55 4AG G7ANO
DE55 4AP 2E0MFS
DE55 4AP M0IEM
DE55 4AP M6SMX
DE55 4BL G1OGC
DE55 4BP G3JWQ
DE55 4EX M3TFK
DE55 4JT M6JSR
DE55 4LA 2E1DBQ
DE55 4LA M0SBA
DE55 4LH M1GHT
DE55 4LT G4OHV
DE55 4NF G6TLA
DE55 4PB M6PDK
DE55 5AJ M0TRP
DE55 5HD G4AMF
DE55 5HL M3HUN
DE55 5HT G7BHW
DE55 5HU M3EAT
DE55 5JH M6TTE
DE55 5JJ M3EUW
DE55 5JL G6LSD
DE55 5JL M5SRE
DE55 5JN G0OKD
DE55 5JN M1DDI
DE55 5LL G4VNG
DE55 5LQ G4PCL
DE55 5LR M7ESY
DE55 5LT M3VLO
DE55 5LT M7FTT
DE55 5LU G0DXT
DE55 5NA G4THI
DE55 5NJ 2E0FUO
DE55 5NJ 2E1SWS
DE55 5NJ G1SWS
DE55 5NJ M0XSH
DE55 5NJ M7DSH
DE55 5NJ M7SWS
DE55 5NR M6HFR
DE55 5QH G4EKD
DE55 5TD 2E0WYT
DE55 5TD M0WYT
DE55 5TD M6WYT
DE55 5TT G6ZDB
DE55 5TY 2E0SBK
DE55 5TY G6AZE
DE55 5TY M0BAY
DE55 5TY M0BMY
DE55 5TY M6SLA
DE55 6BT 2E0RNU
DE55 6BT M0RNU
DE55 6BT M6RNU
DE55 6DS G4UUQ
DE55 6DX M0SMG
DE55 6EH G3LGK
DE55 6GW M6KLX
DE55 6JG G1AII
DE55 6JW G4DTP
DE55 6LD M3ZHW
DE55 6LH G7GLL
DE55 7DG G4NID
DE55 7DG G4UTN
DE55 7ER G1XJM
DE55 7ER M0YJD
DE55 7ER M7JMD
DE55 7EW M6OQQ
DE55 7FL M3TXV
DE55 7HT G7SMZ
DE55 7HX G0GHD
DE55 7HZ G3GRL
DE55 7JN G4LSV
DE55 7JX 2E0SBX
DE55 7JX M6HMW
DE55 7JX M6SME
DE55 7JY G7TYP
DE55 7LJ M7XDE
DE55 7LP G4EWZ
DE55 7NH 2E0CEM
DE55 7NH M0ILM
DE55 7PL G3TIX
DE56 0EA G1RYQ
DE56 0EU M6GKM
DE56 0HD M0WMX
DE56 0HL M3FXQ
DE56 0HN G0ORC
DE56 0HN M0KBZ
DE56 0NS G4JTX
DE56 0PF 2E0CTM
DE56 0PF G7EMZ
DE56 0PF M3UJY
DE56 0PG G4ANV
DE56 0PQ G0RLJ
DE56 0PY G8IHA
DE56 0QH G7HHZ
DE56 0QR G4RVL
DE56 0SH M6GWF
DE56 0TG G4NXL
DE56 0UB G6SKK
DE56 1AP G1BZE
DE56 1BX G8VYO
DE56 1EE G0MGX
DE56 1EJ G3ZYD
DE56 1FQ M0TVX
DE56 1GH 2E0YPG
DE56 1JG G4VSI
DE56 1LX 2E1HES
DE56 1NE G0ROD
DE56 1PB G8GJC
DE56 1PD G0FRY
DE56 1RE M6FED
DE56 1UF G7NAE
DE56 1UT M0XXI
DE56 1UT M1EAI
DE56 2AL G3PDD
DE56 2BA G4UBR
DE56 2FN M7CGI
DE56 2GR G0UOV
DE56 2GR G8GTD
DE56 2GS G0JNK
DE56 2HP G6JFE
DE56 2LH 2E0FGY
DE56 2LH M6FGY
DE56 2TR M0NOV
DE56 2UW G0AEU
DE56 4BH G3SNR
DE56 4DF G4RVU
DE56 4DP G4DJP
DE56 4DR G3ROD
DE56 4FJ G7UCB
DE56 4FX G6TQC
DE6 1AT G4HIJ
DE6 1BB G8PKM
DE6 1BJ G1NDL
DE6 1BR G3TVU
DE6 1DF G1MUT
DE6 1EX 2E0HWE
DE6 1EX G0DLS
DE6 1EX M7JWE
DE6 1LZ G0BPZ
DE6 1QB G6YWV
DE6 1TZ M6SMD
DE6 1UD G7NZO
DE6 2AS G0LAU
DE6 2EH G3KAG
DE6 2FB G0DJQ
DE6 2GD M7NWZ
DE6 2GY G8NOP
DE6 2LP G0CQP
DE6 2LP G7KRM
DE6 3AG M3IJF
DE6 3AJ G3OQT
DE6 3BQ M7INT
DE6 3EF G4DKV
DE6 3FB G0CXD
DE6 3FE G6RBH
DE6 4JS G3UBS
DE6 4LP M0BAJ
DE6 4NG G7DDR
DE6 4NJ G7EUT
DE6 5BE M3MQI
DE6 5HP 2E0MGA
DE6 5HP M6MJI
DE6 5JZ G6YWN
DE6 5NY G3LKV
DE6 6NB G4MIT
DE65 5DL G1PUU
DE65 5ER G4PYI
DE65 5FD 2E1BUM
DE65 5FX 2E0JWC
DE65 5FX G0WOC
DE65 5FX G7RIS
DE65 5FX M0KCA
DE65 5FX M6JWC
DE65 5FX M6YNK
DE65 5FY G1JJT
DE65 5HP G7BJE
DE65 5JG M3LFV
DE65 5LP G4RJA
DE65 5QQ G1TKY
DE65 6DW G7ILL
DE65 6EB G0HLI
DE65 6ES G6CZD
DE65 6FX M3XHN
DE65 6GP G0MRY
DE65 6GR G0DMS
DE65 6GT G8VAN
DE65 6GT G8WSQ
DE65 6HD G4PWY
DE65 6HL G0SUT
DE65 6HT 2E0DWE
DE65 6HT M6GJD
DE65 6HY G3ZIF
DE7 4DA G1ZHZ
DE7 4DF M0AIT
DE7 4DL G6CDT
DE7 4EH 2E0GBU
DE7 4EH M6TOZ
DE7 4EZ M6HNI
DE7 4HD M3IZH
DE7 4HD M3IZM
DE7 4HQ M0CQF
DE7 4JY G3XRD
DE7 4LE 2E1GOK
DE7 4NE M0BCJ
DE7 4PW G0ITL
DE7 5AP 2E0CTC
DE7 5AT G0LUI
DE7 5AT M6LUC
DE7 5DW G6IQP
DE7 5EF G3IFX
DE7 5EX G4XCK
DE7 5HR G4AIB
DE7 5HT G8NBI
DE7 5NZ M0BCC
DE7 5RB M0BQT
DE7 6AU G4UFX
DE7 6AW G3RKZ
DE7 6AX 2E0VGK
DE7 6AX M3VGK
DE7 6DD G1GNP
DE7 6DE G8DKV
DE7 6EG G4NOB
DE7 6ER M6NYJ
DE7 6EZ G3NNN
DE7 6FR G4LPF
DE7 6GB 2E1GMV
DE7 6GB G4DAM
DE7 6GB G4EDD
DE7 6GU G8KSW
DE7 6HW G8RJF
DE7 6LP M6WNY
DE7 6LW 2E0TDT
DE7 6LW M6GKK
DE7 6PE M3ZYV
DE7 6XB G7UUP
DE7 8AL 2E1BFP
DE7 8AW G8WRY
DE7 8EY G1DMH
DE7 8JA M7KSD
DE7 8NL G0JVB
DE7 8PW 2E0FSH
DE7 8PX G6KBC
DE7 8SJ 2E0DTF
DE7 8SJ M0WCA
DE7 8SJ M6TGU
DE7 8TD 2E0HGA
DE7 8TD M0RLF
DE7 8TD M7RBE
DE7 8WB G0KBN
DE7 9HE G1YEV
DE7 9HJ M0IVL
DE7 9HJ M3AQK
DE7 9HJ M6EPD
DE7 9JW G3ZDK
DE7 9LF M1DYC
DE72 2AU G0JKC
DE72 2AU G0NHR
DE72 2BA G0PBM
DE72 2BJ G4NYB
DE72 2DF G4RBZ
DE72 2DQ 2E0GEN
DE72 2DQ M0TDY
DE72 2DQ M6CSX
DE72 2GE M6CDD
DE72 2GL 2E0AVW
DE72 2GL M0LJC
DE72 2GS G3ZOW
DE72 3AQ 2E0RRF
DE72 3DF M6KMI
DE72 3DZ G4EHX
DE72 3EE G4EAX
DE72 3GN G8KEA
DE72 3HR G7TLL
DE72 3JJ M0SDS
DE72 3LN G8YZC
DE72 3NG M7SKG
DE72 3NP G0DPQ
DE72 3PN G4OQL
DE72 3PN G6CHI
DE72 3RP G1DGY
DE72 3TE G1SGZ
DE72 3TF G1PWH
DE723GR 2E0EIZ
DE723GR M6IPQ
DE73 1AG M3IHU
DE73 1BW G3NYZ
DE73 5AS 2E1BHC
DE73 5BX M1BGT
DE73 5SJ G1SPT
DE73 5SU G7CYF
DE73 6PB 2E0PFT
DE73 6PB M0PFT
DE73 6PF G7DZR
DE73 6RD G4LPZ
DE73 6RP G4XYD
DE73 6SB M0CAT
DE73 6SW M7FKP
DE73 6XH G1PER
DE73 7AA M0JVR
DE73 7GY G0JCK
DE73 7HD M3IIN
DE73 7JJ G3XER
DE73 8AG G8DBO
DE73 8BD G8MZA
DE73 8BX G8NFM
DE73 8EB 2E0GRL
DE73 8FG M3ZZL
DE73 8GA 2E0WRI
DE73 8GA M6PXW
DE73 8JH M3RBT
DE74 2AU 2E0CPX
DE74 2AU M6OZR
DE74 2AU M6SOE
DE74 2DN 2E0IYS
DE74 2DN G0SOU
DE74 2DN G4IRH
DE74 2DN M7FRI
DE74 2DQ M7BDB
DE74 2DS G0WUG
DE74 2DT G0RDY
DE74 2EJ G4TCA
DE74 2GQ G3JRS
DE74 2JG 2E0MPJ
DE74 2JG M3WQX
DE74 2JT M7SYE
DE74 2JY M6YRD
DE74 2LF G0EKK
DE74 2NG 2E1CPI
DE74 2NG G8BAG
DE74 2PG 2E0RJU
DE74 2PJ G1ZEK
DE74 2PQ G3KTP
DE74 2QQ G8PTW
DE74 2RX G4UVG
DE74 2SR G7PNG
DE74 2UJ M7XRO
DE74 2XA G0VNQ
DE74 2XB G1LCR
DE74 2XB G1SPA
DE75 7AN G7LGY
DE75 7BN 2E1FAT
DE75 7BN G6XTD
DE75 7BZ 2E0KGY
DE75 7BZ M7RKD
DE75 7FW G7MLT
DE75 7GQ G4DMF
DE75 7HB G1IQG
DE75 7HF G7JJC
DE75 7HG G3LZC
DE75 7JX M1EUR
DE75 7LY G8IOW
DE75 7NE M6BIA
DE75 7NJ G4GBC
DE75 7NJ M3MBI
DE75 7PQ G0FEZ
DE75 7PQ G4TYY
DE75 7PQ G4ZNI
DE75 7PZ G1GKK
DE75 7QB G8UCC
DE75 7RL M3UFX
DE75 7RP M6NEO
DE75 7TJ G1SEA
DE75 7UN 2E0CID
DE75 7UN M6BCZ
DG1 1BT GM1OXQ
DG1 1PP GM4JKB
DG1 1QN GM1JVU
DG1 1RB MM5AKZ
DG1 1RT GM7TUD
DG1 1RZ MM6JWP
DG1 1UZ 2M0RDG
DG1 1UZ MM3RSR
DG1 2LU GM0PFH
DG1 2QF MM7FGU
DG1 3HE GM4JR
DG1 3HE GM6EPU
DG1 3HE GM7WJP
DG1 3LH MM6BXF
DG1 3RJ GM3OXK
DG1 3SL MM6BNT
DG1 4BU GM4TNJ
DG1 4DN GM0WIZ
DG1 4DW GM4RJF
DG1 4HU GM8VBX
DG1 4LU MM3OYW
DG1 4RP MM0ZEQ
DG1 4SD MM6KWQ
DG1 4SX MM7XAC
DG10 9DY GM1KBJ
DG10 9JU GM3OFT
DG10 9LZ 2M0MZK
DG10 9LZ MM7MJK
DG11 1JL 2M0NEO
DG11 1JL MM0NEO
DG11 1JL MM3VRX
DG11 1NF MM3MID
DG11 1RL 2M0NJS
DG11 1RL MM0NJS
DG11 1RL MM6LTC
DG11 1SA GM8UPI
DG11 2BA MM0WHA
DG11 2QL MM3FRU
DG11 2QP MM0KQG
DG11 3AE GM3UHT
DG11 3AT GM0WMH
DG11 3BA GM1VLA
DG11 3DU GM0CBC
DG11 3JH GM0BVG
DG12 5FA MM7CLA
DG12 5LN 2M0ETJ
DG12 5LZ MM6KWV
DG12 5QL GM3YLU
DG12 6BU GM6LYJ
DG12 6HX GM4WHA
DG12 6HX GM6SMW
DG12 6ND GM0GBV
DG12 6QX GM6TVR
DG12 6UH MM0JZI
DG13 0AT GM7IHH
DG13 0AX GM4BDJ
DG13 0BG GM8ETJ
DG13 0EE 2M0GQS
DG13 0EE MM0JYR
DG13 0EE MM7LLM
DG13 0HJ GM6NJL
DG13 0JW GM1GES
DG13 0JW GM1PGP
DG13 0PS MM6AEB
DG14 0RZ GM6LJE
DG14 0UY GM1CUC
DG16 5JS GM3MRV
DG2 0AJ GM4RPO
DG2 0NQ MM6KVE
DG2 0PE 2M0HDL
DG2 0PE MM7JVA
DG2 0QX GM0RJG
DG2 0QX GM1EOA
DG2 0UG GM4FZH
DG2 7AS 2M0EBW
DG2 7AS MM3RBF
DG2 7DH GM1SVQ
DG2 7DL GM8RQW
DG2 7LW GM6PFJ
DG2 7NS 2M0RHT
DG2 7NS MM6RHT
DG2 8DA MM6HSE
DG2 8LN GM8XKW
DG2 8NR GM0OHD
DG2 9BP GM4XUJ
DG2 9DZ MM3ELT
DG2 9LT MM7AVI
DG2 9NN MM3MFR
DG2 9NR GM0AVB
DG2 9NR GM1PKB
DG3 4HD GM1YRD
DG3 5HR GM3XNC
DG3 5HR GM7ENM
DG4 6AT MM6RLL
DG4 6JZ MM0CIK
DG4 6LD GM4NTL
DG4 6LD GM8ZXQ
DG4 6SN GM7AWK
DG5 4DX 2M0HOS
DG5 4DX MM6JJW
DG5 4EB GM6VVG
DG5 4EE GM4BRB
DG5 4EN GM3LJR
DG5 4FA MM1BHO
DG5 4FA MM3PBQ
DG5 4GY MM6RJN
DG5 4QN MM6JFB
DG54DW MM6GYP
DG6 4HL GM4DSO
DG6 4LH MM3ZJL
DG6 4PE GM4KOO
DG6 4QD MM3SLD
DG6 4RN GM6RXQ
DG6 4SH MM7LAP
DG6 4TT MM3UJQ
DG6 4XR MM6WMM
DG6 4XT MM0KBT
DG7 1AH 2M0WUL
DG7 1BL MM6SNY
DG7 1BP 2M0MNB
DG7 1BP MM7EQX
DG7 1EG MM6ACV
DG7 1EW GM0NCA
DG7 1GB GM6ZLY
DG7 1HS MM6ACY
DG7 1HX GM1BZR
DG7 1HX GM4VIS
DG7 1LG MM6GOW
DG7 1LQ MM6KXF
DG7 1LW GM4TPR
DG7 1LW GM6WNX
DG7 1LW MM6FNL
DG7 1TN MM6DPV
DG7 2EA MM0GOF
DG7 2JE MM6DMC
DG7 2JJ GM4DON
DG7 2NG MM3WDT
DG7 3BD 2M0ETL
DG7 3BD MM6IAW
DG7 3BF GM3JKS
DG7 3HG MM7EID
DG7 3JW MM6KCR
DG7 3JW MM6URR
DG7 3LD MM6DYH
DG7 3UJ GM7CTQ
DG7 3UR MM6DPQ
DG7 3XJ MM0YOS
DG72PW GM8GQS
DG8 0JQ 2M0HEO
DG8 0JQ MM6HEO
DG8 0PN GM6FXZ
DG8 6AB GM6AQL
DG8 6BH GM3MZX
DG8 6HF MM6MNQ
DG8 6NU MM6PTR
DG8 6PZ MM6SFR
DG8 7AE GM3ZYE
DG8 7AZ MM6DGZ
DG8 7HU 2M0TKE
DG8 7HU MM3TKE
DG8 7JB GM4SNW
DG8 8BA GM4DUX
DG8 8DQ 2M0NSW
DG8 8DQ GM3SEK
DG8 8DQ GM5RP
DG8 8DQ MM0WNW
DG8 8LD GM0TFF
DG8 8LD GM3AUE
DG8 8NZ GM0ECU
DG8 8NZ GM0IDJ
DG8 8PP GM6FSG
DG8 9AH GM4GDF
DG8 9AQ GM4UQK
DG8 9BY MM7AJQ
DG8 9EE GM8AVM
DG8 9EE MM5WIG
DG8 9HX 2M0GKD
DG8 9JA MM6KTY
DG8 9JG GM4XAW
DG8 9LJ GM7NFF
DG8 9LL MM6FKB
DG8 9QN 2M0NIX
DG8 9QN MM0WWH
DG8 9QN MM6GWW
DG8 9RT MM7CSC
DG9 0AU GM6WAZ
DG9 0DX GM0AGV
DG9 0DX GM0HPL
DG9 0LJ GM0UWV
DG9 0LN GM0AJT
DG9 0LW GM0VYB
DG9 0RY GM4ZIL
DG9 7BA 2M0FBU
DG9 7BA MM6PJY
DG9 7BX GM1FMX
DG9 7HP GM4YTD
DG9 7LL MM3NWZ
DG9 7TA GM0AQL
DG9 8BQ GM4LPT
DG9 8DB GM3TGG
DG9 8ET MM3MZO
DG9 8EY GM6KXP
DG9 8HH 2M0WJP
DG9 8HH MM6WJP
DG9 8LX 2M0BOY
DG9 8RD MM6DGY
DG9 8RD MM6RHO
DG9 8TF GM8IWR
DG9 9AW 2M0WML
DG9 9AW MM6FJV
DG9 9BZ GM0HPK
DG9 9BZ GM4RIV
DG9 9BZ MM3MCG
DG9 9EX GM1ZIV
DG9 9HF MM6DPO
DG9 9NE GM4DLG
DG9 9PS MM6WGD
DG9 9PT GM1HZL
DG9 9QE MM6EDB
DG97HR GM0CXY
DH1 1EA GM0VHU
DH1 1EB G7VIL
DH1 1HA G0PMX
DH1 1HW 2E0BYO
DH1 1HW M3YWU
DH1 1JE M6KTA
DH1 1LT G1GXW
DH1 2JR 2E0FYT
DH1 2JR 2E0OGZ
DH1 2JR M7LGD
DH1 2JU G7BYE
DH1 3TY M0RCL
DH1 3UA M6YBC
DH1 4DS 2E0WTQ
DH1 4DS M0HNN
DH1 4DS M6CPQ
DH1 4EN G6RMJ
DH1 4QJ M0BXP
DH1 5DQ G4EBN
DH1 5DR G1SLP
DH1 5EH G7DGD
DH1 5GL G8PSS
DH1 5HP G7VJH
DH1 5HZ M0CKC
DH1 5JN G0DGB
DH1 5JN G0DRJ
DH1 5LA M6NHD
DH1 5LN M6EUB
DH1 5NL G3JQL
DH1 5PB M6RHX
DH1 5PN M0HAM
DH1 5PU G4PFE
DH1 5RJ G4LOM
DH1 5ZG G0VLF
DH1 5ZG G4EUZ
DH11JA M0DIN
DH11JA M3ADB
DH2 1JJ G3ZIJ
DH2 1NR G1NCL
DH2 1QH G4OCQ
DH2 1QY M1ELW
DH2 1RS M6BRL
DH2 1TA 2E0DEK
DH2 1TA M6ATP
DH2 1UF G4PSS
DH2 2BY 2E0DGQ
DH2 2BY M6EOC
DH2 2LD G7TFX
DH2 2LQ 2E0KMT
DH2 2LQ M0KMX
DH2 2LX G4ODE
DH2 2PB M6WYG
DH2 2PN M7POF
DH2 2SD G7OES
DH2 3DD G8YWK
DH2 3EA G6LCL
DH2 3ED 2E0LOW
DH2 3ED M6MDU
DH2 3EU G8APW
DH2 3HY G7DPZ
DH2 3JS G4MJA
DH2 3LX M3NJD
DH2 3QN M7SWR
DH3 1AR G4RXQ
DH3 1HP G1JGE
DH3 2EY G0UEC
DH3 2JG G0CGW
DH3 2JL M0CBP
DH3 2JS M6LVN
DH3 2NW M0CZO
DH3 2PT M0CSF
DH3 3ED M0BPM
DH3 3PR G4OSK
DH3 3RY M3GQP
DH3 3TP M7GER
DH3 4HU G1REO
DH3 4LU G6GLR
DH3 4LU G8FBQ
DH4 4BW M0FCY
DH4 4BW M0HRA
DH4 4EF M1ADK
DH4 4FB G8KLT
DH4 4HN M6CLO
DH4 4HN M6GYS
DH4 4HU 2E0BMW
DH4 4HU M3OVO
DH4 4PE M3KFD
DH4 4XR 2E0GRI
DH4 4XR M0IAS
DH4 4XR M3XNC
DH4 5BB G2DWC
DH4 5BB M0EXM
DH4 5ND M3JEE
DH4 5NR M6SGD
DH4 5PF M7PSB
DH4 6EA G0AAU
DH4 6GG G0FRZ
DH4 6JG G0FDA
DH4 6JG G4VOK
DH4 6JG G6KVA
DH4 6LA M3IVI
DH4 6LD M7BOD
DH4 6PA G3NMD
DH4 6PA M0RZE
DH4 6TR M7WKR
DH4 7FB 2E0ODB
DH4 7FB M0ZXW
DH4 7HP 2E0JDH
DH4 7HP M0GDV
DH4 7HP M0HJD
DH4 7LP G0FBL
DH4 7RD G0OLO
DH4 7RZ M1FBI
DH4 7TD G0EHV
DH45JZ 2E0EYI
DH45QF M7HDX
DH5 0BA M0AHT
DH5 0BH M0ETP
DH5 0BH M0MPX
DH5 0DB 2E0AAC
DH5 0DQ M6ZRL
DH5 0EF G6VMF
DH5 0ES M6TKX
DH5 0GG M7BJN
DH5 0JN 2E0HIS
DH5 0JN M0ODZ
DH5 0JN M6LWY
DH5 0JN M7KAS
DH5 0NR M0EGA
DH5 0NW M3BFG
DH5 0SE G6BIM
DH5 8AG M0GNG
DH5 8DP 2E0HNJ
DH5 8EQ M3VQC
DH5 8EX G4WJV
DH5 8GB M0RNR
DH5 8HN M3JMQ
DH5 8HR G0CKI
DH5 9LB 2E0IAS
DH5 9LJ M3KUV
DH5 9LX G7GJV
DH5 9QX 2E0DLA
DH5 9QX M6DLB
DH5 9RD M6FMS
DH5 9RR G4HBR
DH5 9RW G8VLS
DH6 1LB 2E0MBB
DH6 1LS G6WML
DH6 1PA G4SYD
DH6 1QB G6CQH
DH6 1QN G4KOT
DH6 1RH G4RHL
DH6 1RJ G0WPM
DH6 2DT M6OYZ
DH6 2FA M7BQC
DH6 2HE 2E0WWY
DH6 2HE M7WAY
DH6 2NF G0WIC
DH6 2PZ M7DZT
DH6 2QD M0IOC
DH6 2RG M6BLK
DH6 2TR G0PXQ
DH6 2YJ 2E0PUL
DH6 2YJ M6PUL
DH6 3DB M3ZAL
DH6 3DH M7GKI
DH6 3HW M6VDJ
DH6 4BE 2E0NMK
DH6 4BT 2E0NLX
DH6 4BT M6NLX
DH6 4DG G7DIG
DH6 4DN G7MVX
DH6 4NN G4WUI
DH6 4NP G0DUG
DH6 4QW M6CAT
DH6 4SA 2E0PIU
DH6 4SA M0OPL
DH6 4SA M6NQI
DH6 4SA M7RCH
DH6 5ES G6LLD
DH6 5QA G7NFG
DH6 5QB G0SLP
DH61DZ M7DQC
DH61LB M3GBB
DH62EL 2E0DZQ
DH62EL M6HSI
DH7 0BA G7UXQ
DH7 0BQ G6CQC
DH7 0HL G0OGB
DH7 0QA 2E0MPG
DH7 0QA M6MAK
DH7 0QD G8AGR
DH7 6LZ M6PNL
DH7 6NE 2E0WET
DH7 6NE M6PRO
DH7 6SG G8FWY
DH7 6SQ G1WSZ
DH7 6TF G0CUO
DH7 6TW G3SNT
DH7 7AN M6ZSV
DH7 7DP M3ZGE
DH7 7FD G1RVT
DH7 7NN G7MBH
DH7 7RL M0BAR
DH7 8BL 2E0INX
DH7 8BL M0KEO
DH7 8BL M7IGN
DH7 8NJ M7BYD
DH7 8PQ G7UKA
DH7 8RY M0DOW
DH7 8RY M3DPX
DH7 8SQ G6YAR
DH7 8TG G1WRF
DH7 8TN 2E0LZD
DH7 8TN M6LZD
DH7 9AG M6FSZ
DH7 9AU G0MNH
DH7 9FR 2E0JVA
DH7 9FR M6VAV
DH7 9FX M7SLW
DH7 9JH 2E0FEJ
DH7 9JH M6OXC
DH7 9LY G7TLR
DH7 9NF M6ESH
DH7 9TB M0NIW
DH7 9UW G6XCO
DH7 9YA G6PTT
DH8 0RF G0XVC
DH8 5JF G1EJQ
DH8 5LS G0JRT
DH8 5LS G4PFQ
DH8 5PA G0OHA
DH8 6EZ M0HTV
DH8 6HZ 2E1EPA
DH8 6JG M3UJN
DH8 6JT M6NHK
DH8 6JZ 2E0GBB
DH8 6JZ M6AQO
DH8 6TL G0GKK
DH8 7AE G0SBU
DH8 7AE M6GMD
DH8 7DB M6ICP
DH8 7DT M6WKY
DH8 7DY G3UTS
DH8 7EH M0MFL
DH8 7EH M6KBE
DH8 7HS 2E0IGL
DH8 7JB 2E0CIN
DH8 7JB M6GAM
DH8 7JE M0WDJ
DH8 7LL M7ESW
DH8 7LX M6FOX
DH8 7QT G1NVN
DH8 7SB G8IAJ
DH8 7SJ G3EGF
DH8 7UJ M6XJC
DH8 8DX M7CWB
DH8 8TU G6WDM
DH8 9AP G3KMG
DH8 9JU G1BET
DH8 9JU M3BET
DH8 9RF G8KZN
DH88TQ 2E0FTT

DH9 0QD G1NTI
DH9 0QP G1HYG
DH9 0QR G7LQK
DH9 6BE G1HEX
DH9 6ET 2E0DYD
DH9 6ET M6EKI
DH9 6HX 2E0GBN
DH9 6HX M6GBN
DH9 6JD M3TSO
DH9 6JU M0BZX
DH9 6XE G7OVB
DH9 7AJ G6FLK
DH9 7HP G4APP
DH9 7JW M3EAS
DH9 7PH M5ABT
DH9 7TR G1AIF
DH9 7TZ M0AYI
DH9 8DE M7CXE
DH9 8DG G1KRX
DH9 8DG G6UEV
DH9 8EQ G7GJU
DH9 8JG M6JIK
DH9 8JW M1FIV
DH9 8PL M7OKR
DH9 8QY G1OGH
DH9 8TX M7CXD
DH9 8UH G4MHA
DH9 9HB G0KYL
DH9 9HD M6RGS
DH9 9LZ G0LEI
DH9 9PA G0IVX
DH9 9PT G0VII
DH9 9PT M3KQB
DH9 9UH M0CLD
DL1 1EQ 2E0AES
DL1 1HG G8BWH
DL1 1JG M6AAJ
DL1 2DX G0FGG
DL1 2HY G0LUY
DL1 2LA M7YCM
DL1 2QF G7CIU
DL1 2QG M7DYU
DL1 2TA G6VPV
DL1 2TU G4NYJ
DL1 3HH G6SPQ
DL1 3ND M0WLY
DL1 3ND M6DLY
DL1 4EY M7AYR
DL1 4LG G8OWS
DL1 4XG G4KEL
DL1 5BJ 2E0NYM
DL1 5BJ M3CRD
DL1 5DF G0TJC
DL1 5DF G1LEN
DL1 5DX G4AUN
DL1 5EA G6XCD
DL1 5LH G0JHD
DL1 5LX 2E0LRP
DL1 5LX M3HUG
DL10 4ES 2E0UXM
DL10 4ES M6NML
DL10 4ET M6ABR
DL10 4PQ G3XHB
DL10 4PQ G8YPN
DL10 4PS M6JJI
DL10 4QY M0OJC
DL10 4SJ 2E0YEA
DL10 4SJ M1SDE
DL10 4SJ M6WEP
DL10 5AG M6WFX
DL10 5BJ 2E0BNC
DL10 5BJ M3WUS
DL10 5DB G3EKL
DL10 5DB G3RLV
DL10 5DJ G0MQV
DL10 5PE M0BYU
DL10 5QD M0DBB
DL10 6AL M6IKH
DL10 6DN 2E0JTS
DL10 6DN G2MN
DL10 6DN M7JWW
DL10 6EE M0JWR
DL10 6EE M5JWR
DL10 6SB 2E1GUN
DL10 6SB G4FZN
DL10 6SB G8HQW
DL10 6SB M0RRG
DL10 6SB M1ECH
DL10 7AE G8ZNB
DL10 7AG G4VGY
DL10 7BG M3NPZ
DL10 7BQ G0RYS
DL10 7DL G4DBY
DL10 7DX M3VON
DL10 7JE 2E0WGC
DL10 7JE M0WGC
DL10 7JE M6CGW
DL10 7JP 2E0LXA
DL10 7JP G7LLY
DL10 7JP M6AXL
DL10 7LQ G6XJJ
DL10 7PB G6CUK
DL10 7PD 2E0MXG
DL10 7PD M6IFB
DL10 7QD G6RNR
DL10 7RT 2E0BEA
DL10 7SP M1EAB
DL10 7SS M3UBU
DL11 6HH G0JTA
DL11 6PY M6LAA
DL11 6QX G8TOQ
DL11 6TE G1FCX
DL11 7AJ M3CDI
DL11 7NH M6SPP
DL11 7RD G3XWV
DL117JH G6IET
DL12 0AH 2E0BWP
DL12 0AH M6DLM
DL12 0JD M6EVM
DL12 0QU G0ADO
DL12 0RP G4IPB
DL12 0UU G4AFE
DL12 8EB G0NRK
DL12 8LA G4OUB
DL12 8LF 2E0BFM
DL12 8LQ 2E0BPF
DL12 9BZ G7UGY
DL12 9PQ G8LIE
DL13 1JD G6FZC
DL13 2AY G4VZS
DL13 2NB G7GLP
DL13 2QR G4DWM
DL13 2XY G6OTL
DL13 3BP 2E0FTD
DL13 3BP M6HVU
DL13 4DS 2E0AIM
DL13 4DU G4TMI
DL13 4DX M7DHL
DL13 4ED M6KWT
DL13 4EE M6CZW
DL13 4JS G4OJY
DL13 5LS M6NKJ
DL13 5NF G0LLC
DL13 5RH G4KUX
DL14 0DH G0OCB
DL14 0DH G1XNI
DL14 0DH M6AMD
DL14 0RN M3WIT
DL14 0RP 2E1EYS
DL14 0RX G0EQD
DL14 0TG G1ASR
DL14 0TP 2E0MRQ
DL14 0TP M3ZJG
DL14 6AR M3RLO
DL14 6ET M3UOM
DL14 6LW G7RGR
DL14 6NT M5CJH
DL14 6TU M0BLN
DL14 6UB M3BAH
DL14 6UB M3EMA
DL14 6UJ M7EWM
DL14 7DS G0KIC
DL14 7DZ M6SSZ
DL14 7GH M3FSV
DL14 7LZ M0ESZ
DL14 9EJ G4LRG
DL14 9LG M0ACV
DL14 9LQ G1YNO
DL14 9PY G0JQP
DL14 9PY M3UOB
DL14 9QH M0DQD
DL14 9RY 2E0LDQ
DL14 9RY M6LDQ
DL149LF 2E0GWE
DL149LF M6GZA
DL15 0AD 2E0WKM
DL15 0AD M0WCW
DL15 0AD M7CKW
DL15 0BJ 2E0BJA
DL15 0BJ M0BPC
DL15 0BJ M3NAR
DL15 0DA G4STP
DL15 0DS M3XWM
DL15 0HN G6LQR
DL15 0LQ 2E0CEG
DL15 0LQ G1YUB
DL15 0LQ M3ALX
DL15 0NS M0MEG
DL15 0PP 2E0CWQ
DL15 0PP M6JDU
DL15 0QL 2E0LIW
DL15 0QL M6LWS
DL15 8AG G7VNM
DL15 8NA M7MTH
DL15 8NN G4FZS
DL15 8QU G4PSI
DL15 9AE 2E0RTG
DL15 9AE M0RRF
DL15 9AE M6RTG
DL15 9DB G7OCK
DL15 9DY G7PTT
DL15 9DZ 2E0YDP
DL15 9DZ M7TXN
DL15 9GH M0LYI
DL15 9LH M0KOE
DL15 9QX G0HCD
DL15 9SE G4VIO
DL15 9SF M1GSM
DL15 9UT G7BPN
DL16 6DW G7ESY
DL16 6ER M3AHR
DL16 6LY G0TSR
DL16 6NB G0RNY
DL16 6QU M6DTJ
DL16 6RN 2E0XVF
DL16 6RN G0KNN
DL16 6RN M0XVF
DL16 6RN M3XVF
DL16 6TJ M0PJM
DL16 6TT M6SCB
DL16 6TZ G7MKQ
DL16 7DZ G8NSX
DL16 7HF G7VYZ
DL16 7HN G1LPS
DL16 7HU G7KJR
DL16 7HZ G0JEC
DL16 7QS G0BVO
DL16 7TA M3AHO
DL16 7TU 2E0HDF
DL16 7UG G0PWK
DL166HP G7ESX
DL17 0EQ 2E0NGL
DL17 0EQ M0NGN
DL17 0EQ M6IQL
DL17 0RW G0BNY
DL17 0RW G4CAY
DL17 8DA G4MQV
DL17 8DA G6CII
DL17 8DA G8UYM
DL17 8DL G4JIX
DL17 8JT M7NKW
DL17 8NG G4XHP
DL17 9HE G1LGQ
DL17 9HE 2E0SLN
DL2 1AU G0AIH
DL2 1HF G3SGY
DL2 1JG 2E0UCH
DL2 1JG 2E0ZOT
DL2 1JG M0VLI
DL2 1JG M0XEE
DL2 1JG M3VFS
DL2 1JG M3VLI
DL2 1JG M3ZVA
DL2 2AH 2E0VVF
DL2 2AL 2E1DPK
DL2 2GL G0SBP
DL2 2LN M3KRX
DL2 2PX G4NLL
DL2 2SL G8MTV
DL2 2SQ M1SAZ
DL2 2TS M7OJA
DL2 3JJ G4MIJ
DL2 3NS 2E1HHL
DL2 3NS M1DYJ
DL2 3RY G3GJJ
DL2 3SX M0OAT
DL2 3SX M6WKR
DL3 0AH G3UHJ
DL3 0AL 2E0PCQ
DL3 0AL M6PCQ
DL3 0EJ M6ZET
DL3 0HX G8HZS
DL3 0JQ G1LEC
DL3 0JY G2HMK
DL3 0NW G4YFS
DL3 0SP 2E0IVX
DL3 0SP M7EGK
DL3 0TH 2E0FIK
DL3 0TH M7AJH
DL3 6ES G7TCD
DL3 6HN G4JIR
DL3 6HT M1CUB
DL3 6PY M7HTS
DL3 7AT M0OXW
DL3 7AT M6OXW
DL3 7AT M6OXZ
DL3 7HP G1DMN
DL3 7SJ G1BIA
DL3 8BH G4OXU
DL3 8BH G8YBO
DL3 8HU G6KMG
DL3 8HY G4FVP
DL3 8HY M0NFD
DL3 8LD G0ARZ
DL3 8ND M3ONH
DL3 8RP M6GHL
DL3 9AT 2E0GBA
DL3 9AT M3XNM
DL3 9DT G6BCG
DL3 9EW M7EZX
DL3 9PB G3CDM
DL3 9PB M7AVH
DL3 9PF G0MSZ
DL3 9SA G0VQG
DL3 9SR M3MCA
DL3 9XS G8TJR
DL4 1AP G7BPR
DL4 1BH G3LUC
DL4 1EB G0RYW
DL4 1EB M0SDE
DL4 1QA G1DFN
DL4 2AS M0MYE
DL4 2AZ M3NFU
DL4 2DN 2E0SUX
DL4 2DN M0SGA
DL4 2EX M7LLE
DL4 2LE 2E0XLY
DL4 2LE M0XLY
DL4 2LE M6LXY
DL4 2LE M6MYF
DL5 4HE M7OBF
DL5 4HX M6ZAY
DL5 4PF M6WIT
DL5 4PF M6XDO
DL5 4RA M7PGY
DL5 4SU 2E0VNE
DL5 4SU M7MGX
DL5 5EQ G0IDS
DL5 5ET 2E0DLL
DL5 5LZ G6VLL
DL5 6GR M7TZX
DL5 6RF G3OAL
DL5 6TA G1ZEU
DL5 7BD M0PST
DL5 7BQ M0WRI
DL5 7DR. G1AZC
DL5 7DX G8LCP
DL5 7DX M0BUG
DL5 7DY 2E0RRL
DL5 7DY 2E0WPD
DL5 7DY M0RRL
DL5 7DY M0UDL
DL5 7DY M6RRL
DL5 7DY M6TCN
DL5 7LN G3FBT
DL5 7NP M3VRY
DL5 7PS G4GMB
DL6 1DZ G0GBG
DL6 1EE G8FLV
DL6 1HP G1XRE
DL6 1QQ G0LEL
DL6 1QT G8IHT
DL6 1RQ G0GCK
DL6 1SF G1RQK
DL6 1SJ G7NKJ
DL6 2AA G7HHK
DL6 2BD G1JER
DL6 2BE G3MAE
DL6 2LB 2E0FBC
DL6 2LB M6FOY
DL6 2RD G0WWF
DL6 2RD G2PB
DL6 2SN G1HLV
DL6 3QA M0BBK
DL7 0HL 2E0DNH
DL7 0HL M6RFZ
DL7 8AD M3XYJ
DL7 8BN 2E1CQQ
DL7 8BN G7COC
DL7 8JF M0MFT
DL7 8WB M7SRB
DL7 8XS M7MWT
DL7 9RA G8CEE
DL8 1SX G7HSN
DL8 2DR G3GEJ
DL8 2HT G1WWH
DL8 2HT G3WBG
DL8 2JE G4AFU
DL8 2QF G4NUY
DL8 3HN G8AMJ
DL8 3RH G3ZMO
DL8 4LU M3LOA
DL8 4NA G1BLV
DL8 4QN 2E1AZW
DL8 4QN G0OLR
DL8 5BH G4HVF
DL8 5EL M1FGH
DL8 5HU G0BAQ
DL8 5HU G7SLP
DL8 5JQ G3XPQ
DL8 5JS 2E0CQX
DL8 5JS M6MCZ
DL8 5QN G8LNQ
DL8 5QN M0KXD
DL9 3BW M6RKU
DL9 3NJ 2E1COG
DL9 3NJ G1YQY
DL9 3NP M6FGM
DL9 3RA G7WBW
DL9 4HX M0DIG
DL9 4NT M6JOV
DL9 4PG 2E0TCN
DL9 4PG M0TCN
DL9 4PG M6TIA
DL9 4RJ M6JNZ
DN1 2JA 2E0MLC
DN1 2JA M7SAV
DN1 2NP G0UQU
DN1 2PZ G4OEM
DN1 2QU G1PBY
DN10 4BE G4OZN
DN10 4BU G0OKZ
DN10 4DP G8EGL
DN10 4EF M1FGM
DN10 4HT M1WDK
DN10 4LG 2E0CTO
DN10 4LG M6KAX
DN10 4NG G0EJD
DN10 5BL M7LTY
DN10 5BS G6GWP
DN10 6HE G8JFC
DN10 6NW G3RCW
DN10 6NW M0LLW
DN10 6QF M1CTB
DN10 6RT M3ZFS
DN10 6SW 2E0XAH
DN11 0AA G0PXD
DN11 0BT M0BZQ
DN11 0DF M7DTR
DN11 0FR G4ZLP
DN11 0FR M6OEH
DN11 0JQ M0YDB
DN11 0JP 2E0DMJ
DN11 0JP 2E0KPI
DN11 0NQ G0PXX
DN11 0NQ G6USL
DN11 0PU M3EGM
DN11 0QF M0CTN
DN11 0QF M3CTN
DN11 0QX G0DYE
DN11 0RW M3EYW
DN11 0RW M3EYX
DN11 0RZ M3BZQ
DN11 0SB 2E1IOS
DN11 0SB 2E1RAD
DN11 0SH M3WXB
DN11 0TS M3ZLS
DN11 0TU G7JDH
DN11 0UP G6DIE
DN11 0UR G8HIQ
DN11 8HP G4ZVP
DN11 8HT G0FVD
DN11 8HZ M6ROE
DN11 8LL G4BZG
DN11 8LU M3IOK
DN11 8PF G1BAL
DN11 8PF M0DRM
DN11 8PF M0MHY
DN11 8PW 2E0TCQ
DN11 8QP M6FYD
DN11 8QU M6TDR
DN11 8SN M6ENE
DN11 9BH 2E0HWO
DN11 9BH M0YZT
DN11 9BH M7HWO
DN11 9BW G0TTL
DN11 9ET G6HMS
DN11 9EW 2E0DFT
DN11 9PW M0BOI
DN11 9TE G7BGV
DN11 9UE M6BGP
DN11 9UH G7EDF
DN11 9UL M3XPN
DN12 1BD 2E0DKS
DN12 1BD 2E0IKB
DN12 1ED 2E0SEJ
DN12 1ED M0SBK
DN12 1ED M6SGY
DN12 1NQ M3SMD
DN12 1NW 2E0OXO
DN12 1NW M3KAK
DN12 1SQ M7SDE
DN12 2JG 2E0XPM
DN12 2JG M0XPM
DN12 2JW G3NXZ
DN12 3DF G3WBG
DN12 3DG M6EMJ
DN12 3JZ M7GMP
DN12 3LB 2E0RDX
DN12 4EN M0TOR
DN12 4LR M3LTG
DN14 0AT 2E0DXZ
DN14 0AT M6GQY
DN14 0BU M0ROW
DN14 0JX G4FPO
DN14 0LN G4III
DN14 0NW G6LUE
DN14 0PD G8DML
DN14 0PG M0JKP
DN14 0QY G4AFZ
DN14 0RD G8FCT
DN14 5HQ M3HVU
DN14 5JB M6JIR
DN14 5LH M7ABN
DN14 5NY G4GNP
DN14 5RR M7MZT
DN14 5XF G8FWC
DN14 6DH G0PQX
DN14 6DY M3BRW
DN14 6EB M6BPX
DN14 6EL M3NVE
DN14 6HJ 2E0EWX
DN14 6HJ M0JDU
DN14 6JU M6MRU
DN14 6JX G3LYZ
DN14 6LG M5RHS
DN14 6NA G0GLZ
DN14 6QW G8VHL
DN14 6SH 2E0CAD
DN14 6XN G4GRP
DN14 7DS 2E0TUM
DN14 7DS M7TJM
DN14 7EN G3ZGT
DN14 7EU M7UHU
DN14 7FD G4NLG
DN14 7HD G0KHZ
DN14 7HE G4DBN
DN14 7JL G1RFC
DN14 7LZ G6COE
DN14 7NT M6BIB
DN14 7TP M3KQS
DN14 7YA 2E0UIP
DN14 8DE 2E0XAG
DN14 8DE M3WQF
DN14 8DW M6AFH
DN14 8ER M1AKQ
DN14 8ET G0FRX
DN14 8ET M0UVZ
DN14 8EX G4BDX
DN14 8HL G4LKD
DN14 8HL M0NTG
DN14 8RP G6PXJ
DN14 8RW G7UJT
DN14 8RW M3UJT
DN14 9JG G7OYD
DN14 9JH M3SRI
DN14 9NG M3GUG
DN14 9NG M3PKH
DN14 9QR G0UYF
DN149QQ M6OZS
DN15 0AD G7MTJ
DN15 0BE M3OZB
DN15 6SN M6FJW
DN15 7AT G0RMD
DN15 7BT G0OQX
DN15 7EG G4KWW
DN15 7EH G1HKM
DN15 7EN 2E0XNL
DN15 7EN M6GFH
DN15 7EQ M0ADH
DN15 7HT 2E0PFY
DN15 7HT M6FXU
DN15 7PJ G0DLL
DN15 7PJ M6BYQ
DN15 8AB G3KFB
DN15 8AD 2E1HHJ
DN15 8AU G0EFY
DN15 8BP G0RKQ
DN15 8JT M6HMG
DN15 8LA M6NTM
DN15 8NS 2E1FVS
DN15 8NS G0GHK
DN15 8NS G7HAH
DN15 8NS M0HOM
DN15 8PE M0PDC
DN15 8UG G6JVO
DN15 9AG G0DVG
DN15 9DH G4IPE
DN15 9EW G4CDC
DN15 9EX M3IPM
DN15 9HH G0JRR
DN15 9NQ G1RUE
DN15 9NQ M0GBB
DN15 9NR G4YZH
DN15 9NW 2E0AWT
DN15 9NW M0GIQ
DN15 9QG M6MCC
DN15 9QP M6KYS
DN15 9QY G8TLL
DN15 9RP M7OMA
DN15 9RT G6CMX
DN15 9RU 2E0FTW
DN15 9RU M1DOS
DN158LL M3VTP
DN16 1EB G4JJY
DN16 1EY G0GWY
DN16 1HT 2E1MIN
DN16 1HT M1FFM
DN16 1NA G0OKF
DN16 1QH 2E0FQT
DN16 1RW 2E0MDN
DN16 2AJ 2E0LYF
DN16 2AJ M0LYF
DN16 2AJ M6KNS
DN16 2AJ M6YSF
DN16 2BE G4JRY
DN16 2ES G4HFZ
DN16 2LQ G1YNQ
DN16 2LR 2E0CQD
DN16 2PA M6PHS
DN16 2RP M6GRC
DN16 3DE G4MGD
DN16 3EN G8GIH
DN16 3LQ G0TNS
DN16 3PH G0PQY
DN16 3SA G8LIP
DN16 3SW G4NJA
DN16 3TT 2E0GYK
DN16 3TT M0XRV
DN17 1AA M6XLT
DN17 1AF G4CBY
DN17 1AS G0RUH
DN17 1DU G1ZRT
DN17 1SA 2E0DGC
DN17 1SA G0UTM
DN17 1SA M0HDV
DN17 1SA M6RCD
DN17 1SW 2E1CZB
DN17 1TS 2E1GXY
DN17 1TS M3BAT
DN17 1UE M7GLN
DN17 1YB 2E0CZA
DN17 1YB M6DQR
DN17 1YL 2E1HHA
DN17 2BD G8YVC
DN17 2DF M7AQY
DN17 2EQ G3KNU
DN17 2LW 2E0RNP
DN17 2NQ G7LAL
DN17 2QY G0JRB
DN17 2TP G0ECS
DN17 3HZ G7URL
DN17 3JR G8JET
DN17 3PB M0ACB
DN17 3RA G6PAJ
DN17 3SB G0JEU
DN17 3SB G4GKQ
DN17 3SZ G0PAB
DN17 3TT G4EQD
DN17 3TU G4SVW
DN17 4AT G6DBC
DN17 4DD M7VBD
DN17 4DG G4FZR
DN17 4ET G8MOF
DN17 4GA M7EMC
DN17 4HU G8ZGY
DN17 4LH G4LSL
DN17 4NB M3FWU
DN17 4PP G4VOV
DN17 4PQ M1LMO
DN17 4RZ G0NVD
DN18 5BS 2E0GDN
DN18 5BS M0IIE
DN18 5DP G1ZQG
DN18 5DZ G3RHQ
DN18 5HH G0HMD
DN18 5LE M0AEP
DN18 5LN 2E0RWX
DN18 5NH G4IWR
DN18 5QA G0PHP
DN18 5QJ M6IJK
DN18 5TD G8RQH
DN18 5TW M3USJ
DN18 6AE G4TGE
DN18 6DN M6KMR
DN18 6HJ M7MBE
DN18 6JS M0IUZ
DN185NA M7GDY
DN19 7AU G0KAT
DN19 7AX G8FWE
DN19 7EE M1JPK
DN19 7EE M1TAT
DN19 7EY G1CUM
DN19 7HG G4CFD
DN19 7QB G7PXX
DN19 7QG M3BEE
DN19 7QG M3PWS
DN19 7TL 2E0MPR
DN19 7TL M3JZI
DN2 4BB M1BUX
DN2 4DA M3KJY
DN2 4GD M6GPO
DN2 4JS M3LKY
DN2 4LF M3WPV
DN2 4QA G0IDZ
DN2 4RF M6DJL
DN2 4RJ 2E0PBP
DN2 4RJ M6DPJ
DN2 5HN G4PJL
DN2 5RF G3YQE
DN2 5TR 2E0IER
DN2 6AN G4LKX
DN2 6DT G4ZGM
DN2 6EN M6PGZ
DN2 6HF G0VID
DN2 6HF G7GJO
DN2 6HF M0BDL
DN2 6JQ M0JIC
DN2 6LG M1CHM
DN2 6LN 2E0ECG
DN2 6PA M6OAE
DN20 0AU M3UPN
DN20 0BB G3MHT
DN20 0NN G8YAU
DN20 0PP G0AOJ
DN20 0SE 2E0SXF
DN20 8PW M0ICJ
DN20 8QR M6MOF
DN20 8SG M6PDM
DN20 8SH G0CMK
DN20 8SW M6HFF
DN20 9BE G3MPW
DN20 9DG G4GNP
DN20 9FN M3YUN
DN20 9HU G8YYW
DN20 9JG G6GIO
DN20 9JG G6NIE
DN208RR 2E0NXM
DN21 1BQ M3VYF
DN21 1BT G3SCJ
DN21 1DD G1DKF
DN21 1DH G7AVU
DN21 1NA G7OLF
DN21 1PA G7GMR
DN21 1SS 2E0RKS
DN21 1SS M0ZRS
DN21 1SS M6NYB
DN21 1TT G8VRN
DN21 1WB G7MDM
DN21 1XQ M3LAJ
DN21 2HQ G6FOW
DN21 2JA G3YPS
DN21 2JZ G3UHS
DN21 2RS 2E1LEW
DN21 2RS M7LEW
DN21 2RY M6JVW
DN21 2SA M7CRS
DN21 2SF M7FBZ
DN21 2TH M3TNL
DN21 2TU G6YVS
DN21 3DL 2E0FAM
DN21 3DL G1CIM
DN21 3DL M0CGS
DN21 3DL M6FLC
DN21 3HY M3CVD
DN21 3JY G4XHC
DN21 3RR G3ZOT
DN21 3RU G7WGY
DN21 3RU M3WGY
DN21 3SL G7JQT
DN21 3SR G3TMD
DN21 3TZ G0PZW
DN21 4AF G4GZA
DN21 4BA 2E0BCX
DN21 4BA M0JAE
DN21 4BB G4SXZ
DN21 4BY M6LHJ
DN21 4DF M7GFW
DN21 4ER G4MBK
DN21 4NE 2E1HXY
DN21 4NE 2E1HXZ
DN21 4NE G4CCH
DN21 4NF G0MPK
DN21 4SW G6PHC
DN21 4SW M6PVI
DN21 5BZ G3RAU
DN21 5DN 2E0LQR
DN21 5DN M0LQR
DN21 5DN M6TML
DN21 5FF M6PPP
DN21 5LB G8SFF
DN21 5LY G7IYQ
DN21 5NL M0AYF
DN21 5PT G6WVO
DN21 5TQ G3SET
DN21 5UT G0SQS
DN21 5UT G7JVC
DN21 5XJ M6LHE
DN21 5XP 2E0CJK
DN21 5XP M0PMJ
DN21 5XP M6LIE
DN21 5XP M6PMJ
DN22 0AS G7SRI
DN22 0BY G6NRL
DN22 0FL G7KYL
DN22 0HU M3BAL
DN22 0JX G0TGB
DN22 0LG G4SOI
DN22 0LN G1EWH
DN22 0QQ G0RAF
DN22 0QQ G3VCA
DN22 6LR 2E1FRW
DN22 6NH G0MYA
DN22 6NW G4XTU
DN22 6QH G0CEB
DN22 6QU 2E0LDV
DN22 6QU M3GUQ
DN22 6SF G4YWX
DN22 6SQ M0PAL
DN22 6TS 2E0KVF
DN22 6TS M0KVF
DN22 6TS M6KVF
DN22 6UB G0HLU
DN22 6UF G3DXZ
DN22 7AD 2E1HYI
DN22 7BT G6GKK
DN22 7DX M6XAK
DN22 7GX M7BUI
DN22 7LJ G3NTD
DN22 7NX G4ZDN
DN22 7QW M6JMF
DN22 7TL 2E0SNS
DN22 7TL M6SNS
DN22 7UW G4WBH
DN22 8AJ G0IAS
DN22 8HQ G7KMM
DN22 8NE 2E0ZCB
DN22 8NE M0YCB
DN22 8NL G4VEO
DN22 8PQ G0NPG
DN22 8RS G4AWU
DN22 9NQ G0VUL
DN229HE M6IJO
DN3 1AN G1MCT
DN3 1AW M3XYN
DN3 1BJ M6GWT
DN3 1BS G4NZX
DN3 1DP G0UZW
DN3 1DP G7RXE
DN3 1DP M0CJZ
DN3 1DS G3UWT
DN3 1JU G4KKJ
DN3 1LY G0EZY
DN3 1LY G8CDV
DN3 1NY M6DMO
DN3 2AR M6RCD
DN3 2AZ 2E0PKB
DN3 2AZ G7MEX
DN3 2AZ M0BOH
DN3 2AZ M0EVE
DN3 2AZ M0MEO
DN3 2AZ M1DAH
DN3 2AZ M3DYF
DN3 2AZ M3HJB
DN3 2AZ M3PLB
DN3 2AZ M3WQI
DN3 2DE G1IND
DN3 2DE M3IND
DN3 2DJ M6AMC
DN3 2DJ M6JIF
DN3 2EP M6WCZ
DN3 2ES M3MSC
DN3 2ET G6XVH
DN3 2HE M3YJG
DN3 2HN G4PNT
DN3 2JF G7GEI
DN3 2PE G4MDE
DN3 3AJ G7JTF
DN3 3EH G7BGZ
DN3 3HE G7GZK
DN3 3HE G8XKI
DN3 3HG M3PTV
DN3 3HS G0GOH
DN3 3PB G0WKA
DN31 1PQ M0ZLF
DN31 1PQ M6KLD
DN31 2DN G6NDH
DN31 2EU 2E0GUH
DN31 2EU M6GUH
DN31 2NB G1WSC
DN31 2PW M3YAV
DN32 0HN 2E0OZW
DN32 0HN M3HVV
DN32 0JQ M6PFU
DN32 0LU M6YDF
DN32 0NG G8BMZ
DN32 0NH M6EJP
DN32 0NH M6LZX
DN32 0NQ M3WJA
DN32 7AW M7LAH
DN32 7EJ M7VMS
DN32 7JE G1HCI
DN32 7JE G1HCJ
DN32 7NF 2E1HHK
DN32 7NF G4TYL
DN32 7NJ M3DKW
DN32 7PL G0CJV
DN32 7SB G7HJR
DN32 7SE M7FHO
DN32 8DR M6RGU
DN32 8DZ G7VKY
DN32 8HR M6CBP
DN32 8HR M6RHS
DN32 8NU M6PRC
DN32 8NX M0DDY
DN32 8QS G7LIW
DN32 9EE M7CDK
DN32 9EN M6NFP
DN32 9HJ M6NBO
DN32 9HT M7LYD
DN32 9JQ M6HBC
DN32 9LB M6WMJ
DN32 9NL 2E0GYX
DN32 9NL M0XRF
DN32 9NL M7FIB
DN32 9NP M6GRV
DN32 9NU M3XPU
DN32 9NW G1CWJ
DN32 9PQ M1BYQ
DN32 9PY G1KBL
DN32 9QG M0DMY
DN32 9SW M6ARB
DN320NE M7AES
DN328BJ M3TYW
DN33 1BG G4HZF
DN33 1DE G3DAE
DN33 1DN M6SGN
DN33 1EZ G8XXI
DN33 1JB M0AJT
DN33 1JF M7CCU
DN33 1LU 2E0KQV
DN33 1NL M3UIP
DN33 1RP M7MLL
DN33 1SB M0SDG
DN33 2BB G0KUD
DN33 2BY 2E0UVO
DN33 2BY M3UVO
DN33 2EA G4NPS
DN33 2HF G3DSZ
DN33 2HQ M7GYY
DN33 2JS G1HZN
DN33 2LG G6EXZ
DN33 2NW G1FVE
DN33 2PL G0CGZ
DN33 2PL G0GIE
DN33 3BL G0CSV
DN33 3BS G1BKJ
DN33 3BW G4VIM
DN33 3EE G3JPH
DN33 3EF M6HUK
DN33 3JP G0JNT
DN33 3JT G3SWU
DN33 3JY M6MJA
DN33 3NG G4MSY
DN33 3RA M6MJL
DN333RR M7KJK
DN34 4AD 2E1FHZ
DN34 4DP 2E0ZQ
DN34 4ND M6XPN
DN34 4NJ M6AVY
DN34 4PN G0IGC
DN34 4PN G0IGH
DN34 4PW M0FIS
DN34 4QH G0HXL
DN34 4QN G0RIR
DN34 4SG M5ACD
DN34 4TY G8RHN
DN34 5DB G0RUS
DN34 5DU M7WYK
DN34 5JP 2E0GCI
DN34 5JP M0KIN
DN34 5JP M3LAP
DN34 5LX M1NCC
DN34 5PE G6RIQ
DN34 5RE G7EOG
DN34 5RE M3LEO
DN34 5TG 2E1HIL
DN34 5UP 2E1GCB
DN34 5UQ G4JMY
DN34 5UQ G4VUP
DN35 0HY M7BTL
DN35 0JQ G8ULH
DN35 0NN M1BYI
DN35 0NX G1GHU
DN35 0SE M0CDU
DN35 0SF G4YSO
DN35 7BB M6KEC
DN35 7DX G0LFA
DN35 7DX M6LAV
DN35 7DX M6TCL
DN35 7JJ 2E0YQT
DN35 7JJ M3YQT
DN35 7NG G4YVJ
DN35 7NP 2E0GRN
DN35 7NP 2E0JKY
DN35 7NP M6FFS
DN35 7NP M6GRN
DN35 7NP M6JKY
DN35 7SD G4ZVX
DN35 7UR M0KWK
DN35 8AN M7AER
DN35 8EA M0BZU
DN35 8HY M6PXK
DN35 8HZ G0MNI
DN35 8HZ G6MHY
DN35 8JA G8VSX
DN35 8LE M6SNF
DN35 8NT M7ADU
DN35 8PD G0GSY
DN35 8PL G6VJA
DN35 8PL M7KZE
DN35 8PR G7FFW
DN35 8QN G0PBR
DN35 8QN M6RED
DN35 9DE G4SRF
DN35 9HF G4ZRV
DN35 9JR M0SMT
DN35 9PF M6NEA
DN35 9PP G7CHB
DN35 9PP G7PSV
DN35 9PP M3RHP
DN36 4DE G1INK
DN36 4DF G8ZLF
DN36 4DS G0BBT
DN36 4EA G7DEU
DN36 4HQ M6CVK
DN36 4JA G0KFD
DN36 4LJ G1JZL
DN36 4NE 2E0PBJ
DN36 4NE M6PSM
DN36 4NS 2E0WNW
DN36 4NS M6DQK
DN36 4QY G8ECI
DN36 4RB M0ANU
DN36 4ST G6CCB
DN36 4TT G7IMZ
DN36 4UF G3ZSF
DN36 4WH M3UII
DN36 5AD G4HOF
DN36 5AQ G3VIP
DN36 5BG 2E1CAW
DN36 5BG G0NQW
DN36 5BH 2E0CTZ
DN36 5BH M6FBB
DN36 5BQ G4PYD
DN36 5DS G4GAB
DN36 5HG 2E0GNW
DN36 5HG M6GNW
DN36 5JE G4OII
DN36 5LP M0DER
DN36 5NF M1CYP
DN36 5NJ G4SEP
DN36 5PL G4GOJ
DN36 5QS G4LOY
DN36 5RF M7AFW
DN36 5SQ M6FJU
DN36 5TP G1IZB
DN37 0EE M1CDQ
DN37 0JR G3TVC
DN37 0NB G7NCV
DN37 0NS M6WMF
DN37 0QE 2E0VBM
DN37 0QE M6GYA
DN37 0UA G4WOH
DN37 0UZ G5ATT
DN37 0YJ G0TWE
DN37 7BA G0IIQ
DN37 7DB G0ATW
DN37 7DF M6DCS
DN37 7DF M6SSD
DN37 7ER M6SKT
DN37 7ER M6TNR
DN37 7WA M7NKZ
DN37 8LB G0DQD
DN37 9DU M6MNI
DN37 9EE M0COI
DN37 9HA G7PKP
DN37 9QJ G4EBK
DN37 9QL G3RGC
DN37 9QZ 2E0SFT
DN37 9RH G1HDT
DN37 9RN G8RIW
DN37 9RX G4DXB
DN37 9RZ G1BRB
DN37 9SQ M1RMO
DN38 6AF M6RKP
DN38 6AP G8HQI
DN38 6AU G7TRG
DN38 6BD G4FHF
DN38 6HF G7IJC
DN38 6HJ M6FOK
DN38 6HU M7LBG

Postcode

DN39 6RB M0PXS
DN39 6SG G0EFZ
DN39 6TB G1WSA
DN39 6UQ G0FEQ
DN39 6YS M7WTE
DN4 0EP 2E0SOF
DN4 0EP M3JZT
DN4 0HE M7LEP
DN4 0JG M6RBQ
DN4 0UD 2E0NDH
DN4 5EE G8XTU
DN4 5EG G6RQJ
DN4 5EQ 2E1BGN
DN4 5EY M1EUF
DN4 5QF G0PXE
DN4 6AL M3OYZ
DN4 6AU G0RFV
DN4 6DQ G3NYB
DN4 6DX G3AAF
DN4 6DX G7PUL
DN4 6EZ 2E0IMC
DN4 6EZ G4FYE
DN4 6HE G1HZR
DN4 6HX G6ZKS
DN4 6LF G0LOL
DN4 6PD 2E1DBZ
DN4 6PD M3DBZ
DN4 6QB G0GTI
DN4 6RE M3CMP
DN4 6SQ G4OEK
DN4 6TB M3WSI
DN4 6TP M7RTV
DN4 6UR M3KFH
DN4 6UT G7KEK
DN4 7AX G4WTR
DN4 7EG G7CDI
DN4 7HU G3VHZ
DN4 7JH G7KMO
DN4 7JQ G7TMO
DN4 7JY G0NMJ
DN4 7JY G4DMH
DN4 7RB M0MCH
DN4 8AB M7FIW
DN4 8QS M6TKA
DN4 8SX G8RFZ
DN4 8TN G6MWD
DN4 8TN M5ADA
DN4 8TU 2E0TTP
DN4 8TU M7BCH
DN4 9AJ G3OFP
DN4 9AR G3TTU
DN4 9DA M0KAI
DN4 9DB M0AGJ
DN4 9DQ G0SSC
DN4 9DT G6DJX
DN4 9EL M3KEW
DN4 9ET M6MXX
DN4 9LA G0LJV
DN4 9QR G0LHM
DN4 9QW 2E0GZJ
DN4 9QW M7UTT
DN40 1DW G4YAM
DN40 1EE 2E1HWU
DN40 1HH G0NBC
DN40 1LT 2E0CMZ
DN40 1LT M0PEM
DN40 1LT M6GRV
DN40 1NR G2HLB
DN40 1PT G0KTX
DN40 1PW G0DOB
DN40 1RB G7PLS
DN40 2AS G7GOV
DN40 2AZ G8XFY
DN40 2DQ G0HDS
DN40 2EQ G7BRZ
DN40 2JH G3VCX
DN40 2SG G1RSK
DN40 3AE M7BKU
DN40 3AX M0PLY
DN40 3DA G4YTQ
DN40 3HF M7ATX
DN40 3PS M3KYV
DN40 3PT G3YKK
DN402EU G0MAF
DN41 7JB G1STW
DN41 7RD G4YHP
DN41 7RE G1LWL
DN41 7SR G7RLK
DN41 8AN G8FKF
DN41 8ED G1OPD
DN41 8EL M0MAL
DN41 8ER G6PEH
DN41 8HE G4KAL
DN41 8HG G6SXN
DN41 8JB G1EXU
DN5 0DW M6SRB
DN5 0ED G0RLH
DN5 0EG M0BDU
DN5 0LH G7VYY
DN5 0NT G0IBG
DN5 0PQ M3GFO
DN5 0RP 2E0RXL
DN5 0RP M7RLL
DN5 7BL G7ANA
DN5 7EN G4OVS
DN5 7LH G3VLL
DN5 7LW G0UQQ
DN5 7QF G8LHT
DN5 7RE G7ONV
DN5 7RY G1SCO
DN5 7SD 2E0HNF
DN5 7TE G0GNP
DN5 8EB M6RUM
DN5 8EH G8JJR
DN5 8ER G6NBF
DN5 8NQ G0DQB
DN5 8PQ M6LUZ
DN5 8QN G4ANP
DN5 8QP G3IGU
DN5 8QW G0HLJ
DN5 8RR G6GNO
DN5 9DY M3KKA
DN5 9HD G7ACO
DN5 9QX M6PBY
DN5 9RJ M7DCP
DN5 9ST G4PCD
DN50RA M6BMW
DN6 0AL G4YTR
DN6 0AR G6GYV
DN6 0EZ G0TSH
DN6 0EZ G7AYO
DN6 0LF 2E0XLB
DN6 0LF M7LCB
DN6 0LU 2E0BSJ
DN6 0LU M3LCS
DN6 7AX G7CTG
DN6 7BT M0RWA
DN6 7EA G4JII
DN6 7EE G0GBQ
DN6 7JL M3LQB
DN6 7JQ M3UGZ
DN6 7JQ M3YEZ
DN6 7JZ G6HWI
DN6 7JZ G6NSZ
DN6 7LX M3WXP
DN6 7NH M6SAI
DN6 7QQ 2E0AWZ
DN6 7QQ 2E0AXA
DN6 7QQ M6DGG
DN6 7RY M6JUI
DN6 7TT M0FCI
DN6 8EB G3UWR
DN6 8EH G1JTZ
DN6 8JE G1ILF
DN6 8JL M0CCD
DN6 8JQ G4MRU
DN6 8NJ M7TJT
DN6 8PD G0AHV
DN6 8RJ G8WCX
DN6 9BY G3WZW
DN6 9HJ G7UBO
DN6 9HX G7JWY
DN6 9HX G7VMO
DN6 9JX G0DQM
DN6 9PY 2E0MEH
DN6 9PY M0TCB
DN6 9PY M6AXC
DN67NW M3JVK
DN69JU 2E0UBT
DN69JU M3UBT
DN7 4AH M0DLY
DN7 4AH M3GCN
DN7 4AH M3SJK
DN7 4AS M1EJX
DN7 4BL M6MXH
DN7 4BS G3VZE
DN7 4BU G1HOU
DN7 4EJ M1ALU
DN7 4HB G0TJR
DN7 4HB M3NVS
DN7 4HS 2E1SKA
DN7 4LZ M6WDG
DN7 4QL 2E0IZT
DN7 4QL M7ENJ
DN7 5BS G8TVU
DN7 5PE G4VVE
DN7 5PT G6OSK
DN7 5PU M0ASJ
DN7 5QL M6NQB
DN7 5RH 2E0RHC
DN7 5RH M6RHC
DN7 5TQ G4ALF
DN7 6AB G3OZD
DN7 6AH G0LCT
DN7 6AT M0THZ
DN7 6BD M6DGP
DN7 6DA G4RJS
DN7 6EH M6BTP
DN7 6ER G8PRH
DN7 6JX M3IKJ
DN7 6QQ G7BPF
DN7 6RY M0YRF
DN8 4NG M3RRV
DN8 4NU G0BKD
DN8 4QN M6VVE
DN8 4QT G0LNT
DN8 4SF M1EMR
DN8 4SL 2E0WUK
DN8 4SL M7SLO
DN8 4ST 2E0VVE
DN8 5BS G4HZN
DN8 5EA G7MGQ
DN8 5EL M6WCV
DN8 5JG G3KPU
DN8 5QS M0VYC
DN8 5QU M0GIW
DN8 5TL G4SZX
DN8 5YW G7MZW
DN84ST 2E1GLT
DN9 1DP G4WJE
DN9 1DY M6GCW
DN9 1HA G7ABT
DN9 1JL G4HOY
DN9 1JL M0XRC
DN9 1LB 2E0SWF
DN9 1LB G1IAB
DN9 1LB M3SWF
DN9 1LJ G7IMD
DN9 1LR G4GZC
DN9 1LT 2E0CMY
DN9 1LT M6FAS
DN9 1MB G4NEG
DN9 1NA G3OS
DN9 1NB G4YWJ
DN9 1NW 2E0EDI
DN9 1PD 2E0EZX
DN9 1PD M3EZX
DN9 1PG G0OPA
DN9 1RG G4ZWQ
DN9 1RG G6ONE
DN9 1RT G8EVI
DN9 1RT M0DMZ
DN9 1RT M0NLR
DN9 1SD G8NDK
DN9 2DG G4YUK
DN9 2FB G4OGB
DN9 2HL 2E0LYK
DN9 2HP M3LQX
DN9 2HX G8BPS
DN9 2JQ G0UQV
DN9 2JX 2E1PMT
DN9 2JX G0TTR
DN9 2JX M1IFT
DN9 2LH G6BWA
DN9 2LR G8BAQ
DN9 2NX G4XRO
DN9 2PE 2E0CRS
DN9 3AE 2E0GWP
DN9 3AE M3RCC
DN9 3AJ G4RHY
DN9 3AJ M3RYN
DN9 3BA M0CAX
DN9 3ED 2E0MRG
DN9 3EG M3XPW
DN9 3JR G4RHZ
DN9 3PE G0RON
DN9 3PQ G0TUU
DN91LL G4NSM
DO1 WR94 2E0CBT
DO1 WR94 M6MHA
DT1 1LH M0AFR
DT1 1LL G1DZZ
DT1 1NZ G1GMV
DT1 1RQ M7EUS
DT1 2BA G2DGB
DT1 2BL M6NBL
DT1 2BS M0EJW
DT1 2BY G6KXW
DT1 2DY M6DDB
DT1 2EF G4CFY
DT1 2EL M0MZX
DT1 2JJ 2E0VWK
DT1 2JJ M0VWK
DT1 2JJ M3VWK
DT1 2LZ G3YUD
DT1 2NR G6EAS
DT1 2PE G7JIM
DT1 2PE G8SDS
DT1 2PF G4ZPO
DT1 2PS G3XIG
DT1 2SB G1RUL
DT1 2TJ M7FUW
DT1 3RJ G6PBG
DT1 3RR M1CSG
DT1 3SU M1ALT
DT1 3WP G4EDK
DT10 1AA M6KXI
DT10 1AR M0GID
DT10 1BE 2E0VXT
DT10 1BE M0XAJ
DT10 1BE M6VRT
DT10 1BN M0OEK
DT10 1DD G4JBL
DT10 1EW M6FKJ
DT10 1EZ 2E0MSA
DT10 1EZ M6MCS
DT10 1HQ G4CEL
DT10 1HQ G4LRV
DT10 1LQ G8MNO
DT10 1LY M6KTJ
DT10 1PS G1GLD
DT10 2BE G3RTD
DT10 2DL 2E0SDP
DT10 2DL M6RDV
DT10 2DS G7HMU
DT10 2DT M3PAI
DT10 2EB M7EEW
DT10 2EU 2E0IZG
DT10 2HF G7GPJ
DT10 2HX M7AQO
DT10 2LZ G0UXZ
DT10 2PN M7NCR
DT101BL M6LNM
DT11 0AA G3ZDQ
DT11 0AY 2E0RHS
DT11 0AY G4RFR
DT11 0AY G7RDT
DT11 0AY M0LXS
DT11 0AY M0ODE
DT11 0AY M0RHS
DT11 0AY M0VYB
DT11 0AY M6NSC
DT11 0DQ G6XAG
DT11 0EF G8RGU
DT11 0HU G0NXQ
DT11 0JG G3VOO
DT11 0PP 2E0DKR
DT11 0PP M6DKR
DT11 0RY M6GZY
DT11 0SN M3GNY
DT11 0SS G6DAI
DT11 0SZ M7TCV
DT11 7DE G3KWN
DT11 7DL G1XCK
DT11 7GF 2E0NUQ
DT11 7GF M3NUQ
DT11 7HH G4GKX
DT11 7LU M7LLN
DT11 7LW G3TPH
DT11 7LX 2E1EJC
DT11 7LZ G7PCE
DT11 7PA G7JTB
DT11 7RT M1LXM
DT11 7RU G3XGY
DT11 7SW M6UKA
DT11 7UA G0UZL
DT11 7XQ G3PCW
DT11 7XQ M6ECX
DT11 7XU M0WAM
DT11 8AJ 2E0MHJ
DT11 8AJ M0MHJ
DT11 8AJ M7LMP
DT11 8BP M6FKH
DT11 8JD 2E1ZPR
DT11 8JQ G1RRW
DT11 8JY G4TEN
DT11 8RH G3RGE
DT11 8TA G7NUN
DT11 8UJ M7JWF
DT11 9BU G4TRM
DT11 9BY G8LSS
DT11 9DW G1YHG
DT11 9HG G6JSN
DT11 9NB G3RVS
DT11 9NN G8BXQ
DT11 9NP G6AOF
DT11 9PH G4ZLX
DT11 9PS 2E0HVE
DT11 9PS M6HVE
DT11 9QP G6PJD
DT12FE M6EKF
DT2 0AQ G8DJW
DT2 0AQ M3DOR
DT2 0BP G0OJP
DT2 0ER G0LOH
DT2 7HA G8GYL
DT2 7HP G8BAZ
DT2 7HP M6NNM
DT2 7HT M6WYN
DT2 7HX M1NAD
DT2 7NP 2E0LOA
DT2 7NP G8SEA
DT2 7NP M7LYF
DT2 7QQ G8XBY
DT2 7QS M3OPW
DT2 7QU 2E1MPQ
DT2 7QU 2E1SPH
DT2 7QU M7MPQ
DT2 7QU M7SPH
DT2 8AE G3KKJ
DT2 8BG G4DSS
DT2 8BG G4JQL
DT2 8BG M3WEZ
DT2 8BQ G1HYX
DT2 8EB G4PQX
DT2 8EG G8JUG
DT2 8EJ M7EXI
DT2 8EW M1EZX
DT2 8HS M0RRR
DT2 8JW G1WPG
DT2 8PE G0HVA
DT2 8PP 2E0SDZ
DT2 8PP G0MFR
DT2 8PP G4TFF
DT2 8PP M0ZGT
DT2 8PP M6LCG
DT2 8QJ G0FIT
DT2 8QR 2E1STU
DT2 8QR G5STU
DT2 8QR M7STU
DT2 8TS G1RPT
DT2 8TX G7TNO
DT2 9DD G4AXU
DT2 9DS M6BPO
DT2 9ES G3IVC
DT2 9HZ 2E1HSA
DT2 9HZ G0PGT
DT2 9HZ M3TEN
DT2 9JN G0KYH
DT2 9JT G8DTE
DT2 9JU G0DUD
DT2 9LT G1LUC
DT2 9QB G0TZO
DT2 9QB M0HUV
DT2 9QX G3ZGN
DT2 9RE M3FRS
DT2 9RG G4GUV
DT2 9RY G4JQS
DT2 9TW M0JOL
DT2 9UD 2E0FZT
DT2 9UD M7XRN
DT28PE M6SHW
DT3 4AX M6NSO
DT3 4BA 2E1LEC
DT3 4BA 2E1NRQ
DT3 4BA M3PSC
DT3 4BB G3SBT
DT3 4GU G0ROT
DT3 4HG G3VPF
DT3 4JP G7LPN
DT3 4JW M3YAD
DT3 4LD G2EC
DT3 4LD G4PTU
DT3 4LD M6DUK
DT3 4NZ M0XFL
DT3 4NZ M6WFL
DT3 5BG G3ZGP
DT3 5BP G3RAF
DT3 5BP G6RAF
DT3 5BP M0MJG
DT3 5HE G3OWE
DT3 5HF 2E0SEA
DT3 5HF M0ESU
DT3 5JS M0MCE
DT3 5LF G4XSC
DT3 5NG G3YWW
DT3 5PB G4RSL
DT3 5PF G3SDO
DT3 5QA G3RZG
DT3 5RP M6GKF
DT3 5RW 2E0VDD
DT3 5SA M0ACC
DT3 5ST M6NYB
DT3 6AA M0XDL
DT3 6AA M6DLO
DT3 6AS G4TRW
DT3 6AS G6ZMU
DT3 6BG G0LQI
DT3 6DS G0NEV
DT3 6HE G4XRR
DT3 6JW G1KDO
DT3 6LE G0FBS
DT3 6LF G0BMQ
DT3 6LF G3JRL
DT3 6LZ G6GYG
DT3 6NE G3EGV
DT3 6NG M6GFQ
DT3 6NL G2CO
DT3 6PD G3XCY
DT3 6PT G3LAG
DT3 6PT G6LBJ
DT3 6PT M0BQO
DT3 6QL G1HTO
DT3 6RB M3UYL
DT3 6RD G0SEC
DT3 6RE 2E0GNJ
DT3 6RH G6XSK
DT3 6RP G3PGK
DT3 6SG M3NMJ
DT4 0AS 2E1DNX
DT4 0AS G0RWL
DT4 0BE G0VLJ
DT4 0DS G1YHB
DT4 0DS M7OXS
DT4 0EA G1TRI
DT4 0ET G4ZXP
DT4 0EZ 2E1OZY
DT4 0EZ M0WIK
DT4 0EZ M3OZY
DT4 0FE G0EVW
DT4 0FE G3SDS
DT4 0FE G8WQ
DT4 0JX G0ROX
DT4 0LN G4OWY
DT4 0LN G6NLW
DT4 0NH 2E0IJP
DT4 0NJ G0ECX
DT4 0PW M0DUK
DT4 0QF M3LUU
DT4 0QY G1CAN
DT4 0SA 2E1TNE
DT4 0UE G6BSP
DT4 7DA M7CSN
DT4 7HQ M3WJU
DT4 7HT M6POW
DT4 7JH G3YIC
DT4 7LX M3ZUZ
DT4 7PS G4SDJ
DT4 8RF G3UHU
DT4 8SG G0LVF
DT4 8SQ 2E0TPH
DT4 8SQ M3TPH
DT4 8SQ M3YTQ
DT4 8TX M3KIO
DT4 9AL G4FYT
DT4 9AL M3MHZ
DT4 9AU M0SGV
DT4 9BH G0HOS
DT4 9EE 2E0PBS
DT4 9EZ 2E1HFS
DT4 9HJ G7OLW
DT4 9JU 2E1HSJ
DT4 9LE G8ZVS
DT4 9LY G7WAQ
DT4 9NB M5RIC
DT4 9PA 2E0HIO
DT4 9RN G3VYX
DT4 9RR G7DOW
DT4 9RX M3XPI
DT4 9SG G6LNF
DT4 9UE G1BBT
DT4 9UU M6YOT
DT48DN M6CBO
DT5 1AS 2E1RBH
DT5 1AS G0RYL
DT5 1BD G4WWH
DT5 1JP 2E0RJB
DT5 1JP M3KKS
DT5 1NH G0PNG
DT5 2AA 2E1DSX
DT5 2AA G4DOL
DT5 2AA G4ZIY
DT5 2AB G0BZW
DT5 2AY 2E0ERB
DT5 2AY G0ACQ
DT5 2AY M6KZC
DT5 2AY M7ANB
DT5 2DE G0CAE
DT5 2EA G4RAK
DT5 2EE G4CFZ
DT5 2HJ 2E1DQZ
DT5 2JG G7NIU
DT5 2JX M3VWP
DT5 2JZ 2E1NRL
DT5 2JZ G7EIS
DT6 3EB G7KLZ
DT6 3PW G0JUV
DT6 3RW M6KLC
DT6 3UB G4PPZ
DT6 4AN G3PXH
DT6 4DW G8JCC
DT6 4EH M0CYG
DT6 4ES G0EZJ
DT6 4ET G4MZL
DT6 4HX G4JFG
DT6 4NQ G4XMZ
DT6 4NU M6CIS
DT6 4NY G6LEU
DT6 4NY M5ABC
DT6 4QL M0AXC
DT6 4QN G8IUN
DT6 4RG G4JWA
DT6 4RY M3BPF
DT6 5EQ M7DVX
DT6 5HX G3MTP
DT6 5LS G6RMV
DT6 5QY G4UHU
DT6 5RF G4MZY
DT6 5RF G7BYG
DT6 5RJ M6OTY
DT6 6AH M0IJL
DT6 6AL M3FRE
DT6 6DF G0HET
DT6 6DF G1EFX
DT6 6DU G3ZUE
DT6 6DY G4HNG
DT6 6JE G6VSE
DT6 6LR M0DRB
DT6 6PE G8ZTN
DT6 6SA G1EFG
DT7 3QY G8FIF
DT7 3SL G1GTM
DT7 3SX G0AYY
DT7 3UR G1ENA
DT8 3BQ G1SCN
DT8 3ES G6BTP
DT8 3HD M6BRZ
DT8 3HU G8JBQ
DT8 3JT G8LOJ
DT8 3PJ G1XHA
DT8 3RF G4BQN
DT9 3EQ G0WDG
DT9 3RT G7AUU
DT9 3RT M3AHJ
DT9 4AT G3OPJ
DT9 4DT M7YIN
DT9 4NH M0MRB
DT9 4PZ M1EZT
DT9 4SW G4ASY
DT9 4SZ M1BUP
DT9 5BD M0ARO
DT9 5BX M6URM
DT9 5DD G8DXO
DT9 5ED G7UPP
DT9 5FD G4INX
DT9 5PD G8CIG
DT9 6AF M3DIB
DT9 6BU G4GHI
DT9 6DN G7BYV
DT9 6EJ M1FFA
DT9 6HL G3UCT
DT9 6HX M6DGM
DT9 6NQ G8AFA
DT9 6RG G2UH
DT9 6RN 2E0TJN
DT9 6RN M0TJN
DT9 6RN M7TJN
DY1 1SL G8ZZT
DY1 2AZ G0NXD
DY1 2EF M0MGS
DY1 2ET 2E0NZD
DY1 2EU M3LWG
DY1 2GG G0OIY
DY1 2LN M6SHA
DY1 2NL M6ROF
DY1 2NN M6LID
DY1 2NU M6LIP
DY1 2NZ 2E0LSB
DY1 2NZ M3OLU
DY1 2PD 2E0SWS
DY1 2PD M6YOT
DY1 2PL M0EAF
DY1 2PL M6SKU
DY1 2QZ G1BTV
DY1 2RT G7EDZ
DY1 2RX M6EVI
DY1 2SL G6VJC
DY1 2SN G4XME
DY1 2SN G6PPA
DY1 2ST M7DGZ
DY1 2UN G6ZSU
DY1 2UW M6IRW
DY1 3DE M7FPA
DY1 3DF G6IRG
DY1 3ED G7OLC
DY1 3JA M6TSZ
DY1 3JU M3UWU
DY1 3LE G0OWU
DY1 3RE M3XOV
DY1 4DZ G4ZLK
DY1 4LU G4OUH
DY1 4LU G4OUI
DY1 4NE 2E0COJ
DY1 4NE M0KCC
DY1 4NE M6BXO
DY1 4NP G7UHY
DY10 1LP M7FET
DY10 1LQ M3NGK
DY10 1LR G0NFO
DY10 1LR G1ZNV
DY10 1SE 2E0DUB
DY10 1SE M6DUB
DY10 1SS M0JMP
DY10 1XJ 2E0KKO
DY10 1XJ M0KKO
DY10 1XJ M1DTO
DY10 1XJ M6KKO
DY10 1YD 2E0XBB
DY10 1YD M0XBB
DY10 1YD M7TBC
DY10 1YH G7ABZ
DY10 2BZ G0MBG
DY10 2EZ 2E0RLX
DY10 2EZ M6RLZ
DY10 2HB G0MJX
DY10 2HZ G1EBB
DY10 2QD M0UCK
DY10 2RA 2E0ZVR
DY10 2RA M0ZVR
DY10 2RH G7EZE
DY10 2RY M3XNV
DY10 2SR M7ALS
DY10 2ST G0ISG
DY10 2SX M1DCY
DY10 2TH G1XJZ
DY10 2TP G4XCX
DY10 2UN G6BAM
DY10 2UT M0ORR
DY10 2XT G3SZG
DY10 2YB G0MJY
DY10 2YB G0WVT
DY10 3AG 2E1DLR
DY10 3AP G8WOX
DY10 3AQ G4TXV
DY10 3BH G4OBC
DY10 3DT 2E0DGG
DY10 3DT M0MCO
DY10 3DT M3TIC
DY10 3EG M0TTX
DY10 3EY M6GEU
DY10 3JG G0RQO
DY10 3JG M7YGM
DY10 3JR G0PEU
DY10 3LH G7ILG
DY10 3ND 2E1IFW
DY10 3ND M0MRB
DY10 3ND M0NYX
DY10 3ND M1EJG
DY10 3ND M3ZOG
DY10 3QR G8OXG
DY10 3QS G0EYW
DY10 3QS G4AFY
DY10 3QZ G6DQZ
DY10 3RY M6MCB
DY10 3TA M7MEL
DY10 3TL G8TPM
DY10 3TN M7DCM
DY10 3TU G1HOL
DY10 3UA M6DTD
DY10 3UB G0UQB
DY10 3XL G4OIK
DY10 3XP M3WVJ
DY10 3XS M6BXP
DY10 4HG G4ZLN
DY10 4JR M1FDY
DY10 4JT 2E0WMP
DY10 4JT M6WMP
DY10 4NE M6GVN
DY10 4RS G4ROJ
DY10 4SZ M7ODC
DY11 5 YB M7FTD
DY11 5BB M0UKF
DY11 5DL G8AKX
DY11 5DW G0FJD
DY11 5HJ G6BDY
DY11 5JN G0GPS
DY11 5JP G1AJU
DY11 5LB 2E0VEX
DY11 5LB M0MKV
DY11 5LB M6LHA
DY11 5LP M7TRB
DY11 5LU G8TLH
DY11 5LZ G0VBP
DY11 5RU G0UUZ
DY11 5ST G8UEF
DY11 5TZ G4ZIB
DY11 5UA G0HRL
DY11 5UJ M0BRU
DY11 6AA G3TAW
DY11 6AU M3HXH
DY11 6BX G4CTU
DY11 6DQ G4PRD
DY11 6DQ M6NRH
DY11 6EE M6CEY
DY11 6HG 2E0HZQ
DY11 6HG M7GFC
DY11 6JU G0EOF
DY11 6PL G4BXD
DY11 6RL G3ZQQ
DY11 6RL G8DQP
DY11 6TE G8GYI
DY11 6TP M6RFN
DY11 6UG M6LVC
DY11 7BY G7DCJ
DY11 7DR G8SKA
DY11 7EA M6JEV
DY11 7EH M6RMW
DY11 7EW G0GFE
DY11 7HD G4ALT
DY11 7LA G4NPN
DY11 7XS G1UNU
DY11 7XU G1JRX
DY11 7YL G7RDQ
DY12 1AT G6EAM
DY12 1BY M6BGZ
DY12 1DB G0UDI
DY12 1JH M0PGS
DY12 1RB G1YFG
DY12 1TR G0ESH
DY12 2BP M6RTD
DY12 2ED G6DFB
DY12 2HT G4OIL
DY12 2JQ G1OZB
DY12 2JQ G7CHW
DY12 2JX G0LOZ
DY12 2PU G4SPZ
DY12 2QG G4OCH
DY12 2QZ G7RBZ
DY12 3AA G4SNO
DY12 3AE M0ABF
DY13 0AR G4DKP
DY13 0EB G0TUW
DY13 0EL G7ESI
DY13 0EQ G7JWL
DY13 0EW G4OPV
DY13 0HE G7PLP
DY13 0HJ G7KZJ
DY13 0JR G0TLP
DY13 0JT G0IBT
DY13 0LL G1LBK
DY13 0LT M6CAA
DY13 0NU G7SCZ
DY13 0NU G7WDC
DY13 0NY G0PMF
DY13 0RH G8SPD
DY13 0RU M3MGO
DY13 0RX G0PJM
DY13 8EL 2E0MOL
DY13 8EL G3GJL
DY13 8EL M3YRM
DY13 8JB M3YCT
DY13 8JG G7WFQ
DY13 8JG M1DRZ
DY13 8LP G6CBB
DY13 8LR G3VHL
DY13 8NT G6EHG
DY13 8QB M6XWD
DY13 8QD 2E0IXZ
DY13 8QD M7EHL
DY13 8QP G0PWC
DY13 8RZ G0LJK
DY13 8SR G8DUW
DY13 8SU G6CPO
DY13 8TA 2E0TEH
DY13 8TA M6BHA
DY13 8TF G0MKL
DY13 8TF G6YIS
DY13 8TH 2E1CVE
DY13 8XW G0MAL
DY13 9EU G8BKL
DY13 9LR G7SAI
DY13 9NA 2E1DHJ
DY13 9NZ 2E0VXX
DY13 9NZ M0VXX
DY13 9NZ M3VXX
DY13 9PB G7IEO
DY13 9RL 2E0IUC
DY13 9RL M7NEA
DY13 9RY G7RXX
DY13 9RZ G6PZH
DY13 9SJ M6RQH
DY14 0HH M7WJA
DY14 0QB G8IYE
DY14 0UA G4EQK
DY14 8AG G0EZT
DY14 8AU M0DLP
DY14 8EB G8DFI
DY14 8LS G1WWA
DY14 8QE G4LUP
DY14 8QQ G4EMD
DY14 9AE G7CHG
DY14 9DB G4ZKD
DY14 9HP G0EXB
DY14 9HX G0IMK
DY14 9JY G3GGL
DY14 9LH 2E0BTR
DY14 9LH M3SFN
DY14 9LJ G3WBL
DY14 9NR G3WPQ
DY14 9NT G7JSG
DY14 9SF G6UMS
DY14 9TJ G8ETR
DY14 9XT G1DFR
DY2 0AU M6WDH
DY2 0BN M0MWT
DY2 0DG G7WLV
DY2 0DX M7SBW
DY2 0EE G0AMW
DY2 0EE G0BFZ
DY2 0HZ G4WBP
DY2 0JY M3AYP
DY2 0NW G7ROP
DY2 0NW M0AKZ
DY2 0RU M7DZF
DY2 7QT M3PKR
DY2 7RA G4GSB
DY2 7RA G7CRS
DY2 7TS G6NTQ
DY2 8BB 2E0SLR
DY2 8BT M0PPC
DY2 8DH 2E0MTB
DY2 8DH M3FWO
DY2 8ER G0GUC
DY2 8EY 2E0PPY
DY2 8HP 2E0NTC
DY2 8HP M0NTC
DY2 8LE M6NJO
DY2 8XT M6WFC
DY2 9EU G7DIZ
DY2 9EU G7IZM
DY2 9EZ M3RWR
DY2 9HE G8TEK
DY2 9HR M3YZA
DY2 9JN M3NCL
DY2 9JT M0GQE
DY2 9LA M0HRC
DY20NW G1HRU
DY29LN M0LPF
DY3 1AL G6DKM
DY3 1LB G0BZT
DY3 1LF 2E0VPO
DY3 1NG G0DPJ
DY3 1PD G1GST
DY3 1RL G4ICU
DY3 1UU G7RXZ
DY3 1XQ M6PGF
DY3 1XW G0CSW
DY3 1YA M6FBM
DY3 2AU M6DTS
DY3 2AX 2E0GOS
DY3 2AX M3GOX
DY3 2BB G4DFE
DY3 2BB G8FUI
DY3 2DE M6NVO
DY3 2DF G7HVF
DY3 2DQ G2HDF
DY3 2DQ M0BLT
DY3 2DQ M1AEA
DY3 2DQ M5HDF
DY3 2EQ M7AFL
DY3 2HJ M3IDD
DY3 2JF G7GEU
DY3 2LR G8WSB
DY3 2LR M1BVT
DY3 2PS G4HCZ
DY3 2QQ G7OIE
DY3 2QQ M0SJV
DY3 2RX G4IFM
DY3 2SH M6ARI
DY3 2UU M3FGO
DY3 3AS G7GEL
DY3 3BH M1AZG
DY3 3EF G8BOP
DY3 3LB G0OCR
DY3 3LH G6DAQ
DY3 3LN G7JIN
DY3 3NA 2E0KPD
DY3 3NP M0JYN
DY3 3PH M6ZSD
DY3 3QJ M0IMT
DY3 3TH M0KSA
DY3 3YQ G3IIV
DY3 4NF 2E0BJY
DY3 4NF M0VKY
DY3 4NF M3NFB
DY3 4NQ G0TMF
DY3 4NT G1BLJ
DY32JU G0CUZ
DY4 0AF M6GZG
DY4 0AJ G6VAA
DY4 0SF 2E0YMW
DY4 0SF M5SW
DY4 0SF M6YMW
DY4 0SU M3HAJ
DY4 0SX M7DUC
DY4 4AU G3ZQS
DY4 4AU M0BPT
DY4 7PU 2E0SMZ
DY4 7PU M3OAL
DY4 7TP 2E0XEA
DY4 7TP M6CDB
DY4 8AJ G6WJJ
DY4 8HZ G7JCD
DY4 8UQ M6DYR
DY4 8XY G7CBW
DY4 9BJ M6NGD
DY4 9QS M3SMZ
DY5 1AE M7SLA
DY5 1DB G1YMH
DY5 1EQ G4PQI
DY5 2AY G6BEB
DY5 2BS M0DZD
DY5 2BU G1DIG
DY5 2DA G4NUS
DY5 2DH G5RSW
DY5 2DH M3IUZ
DY5 2EL G7EVI
DY5 2HB 2E0NJO
DY5 2HB M6NAT
DY5 2HT G4YVA
DY5 2LQ G1YRQ
DY5 2LY G8JTL
DY5 2NG M1CWD
DY5 2PH G7FAZ
DY5 2QF M6CFE
DY5 2QG G4FAH
DY5 2SZ M6XVS
DY5 2UT M6NDM
DY5 2XY G6HZK
DY5 3BU M3LTP
DY5 3DP G4TGM
DY5 3DP G6IWD
DY5 3JE G6LUM
DY5 3JE G6XKE
DY5 3NT M3IFK
DY5 3NY G4TDB
DY5 3QH M6NJS
DY5 3RQ M6NAK
DY5 3UJ 2E0ETQ
DY5 3UJ M0IUQ
DY5 3UJ M6LUY
DY5 3YY G4NLW
DY5 4AW 2E0MLA
DY5 4AW M3HHN
DY5 4EF G0JKY
DY5 4EF M6SUE
DY5 4EQ M0RSD
DY5 4EQ M0WYZ
DY5 4EQ M3FNC
DY5 4EQ M6PMR
DY5 4EX G0OWJ
DY5 4HA M3LKO
DY5 4JF 2E0TSC
DY5 4JJ M6PTP
DY5 4LB M3LTW
DY5 4LG M3CGM
DY5 4ND 2E0VCD
DY5 4ND M6DJX
DY5 4QL G8WSF
DY5 4QX G8TMQ
DY5 4QY M1DRM
DY6 0BG 2E0KAB
DY6 0BG M0KAB
DY6 0DA G8PXW
DY6 0HL M1EGG
DY6 0LG M6EMM
DY6 0LG M6HJB
DY6 0LL G4CVU
DY6 0LN G0SRY
DY6 7AA G1MTU
DY6 7HG 2E1FPW
DY6 7QE 2E0ARA
DY6 7QE G0IPX
DY6 7QE G6ZUV
DY6 7QE M0CDL
DY6 7RQ G4LVA
DY6 8BT G1WFJ
DY6 8DJ 2E0PPZ
DY6 8EE G6PYI
DY6 8JU M6YLI
DY6 8LL G4IEB
DY6 8LW 2E0BLT
DY6 8LW M3VDN
DY6 8PD G8YZF
DY6 8RQ 2E0RLY
DY6 8RZ G8KPG
DY6 8SP G4YBT
DY6 8SP G6GTB
DY6 8XP M3YVJ
DY6 9AD M6KEE
DY6 9DX G3ITH
DY6 9DY M3ZWH
DY6 9ET G4KEB
DY6 9PR M6RIA
DY6 9RB M6YNA
DY6 9RE G3APL
DY6 9RJ G6PSZ
DY6 9RJ M6AKG
DY6 9SJ G4VNE
DY6 9TQ G1WJK
DY68JT M0GUC

DY7 5HU G4CYB
DY7 5JL 2E0HXP
DY7 5JL M0XIC
DY7 5JL M7ENV
DY7 6AD G4WDS
DY7 6BN G0TBI
DY7 6BN G8RAF
DY7 6BN M0VMW
DY7 6BY M7PGE
DY7 6BZ M6IGS
DY7 6DR G4ACS
DY7 6DU G4FTN
DY7 6ED G4IES
DY7 6ED G8MXR
DY7 6EE 2E0LYY
DY7 6EE G0TDR
DY7 6HF G0AOW
DY7 6HF G4WRA
DY7 6HW G3KZG
DY7 6LG M6EVK
DY7 6LS G6SRS
DY7 6LS G8UAE
DY7 6RR G0DBJ
DY7 6RW G0AOX
DY7 6SP G7FXZ
DY8 1AJ G4XNW
DY8 1AX M3UFF
DY8 1HD G4MD
DY8 1NW M3ZWX
DY8 1QX M6FBY
DY8 1UX M7DMJ
DY8 2BE M1DBC
DY8 2DE G8XNL
DY8 2DE M6KIP
DY8 2HL G3ORI
DY8 2LH G4OLL
DY8 3BD G1RDJ
DY8 3DZ G7RCS
DY8 3EF G6BDH
DY8 3EH 2E0ITC
DY8 3EP G0EZX
DY8 3ER M6KUT
DY8 3JE G3NDN
DY8 3JH G7DTV
DY8 3JR G7JWJ
DY8 3LU 2E0IJT
DY8 3NA M6AHW
DY8 3PH G4HWH
DY8 3PJ G4FYQ
DY8 3PJ G4NQB
DY8 3QG M7DQQ
DY8 3RN M6PDC
DY8 3RP G0MWS
DY8 3RZ M7OLM
DY8 3SY M3TFG
DY8 3UF G8OKB
DY8 3UT M1CMX
DY8 3UX M3UUE
DY8 3UZ G7FMQ
DY8 3XU M3ZJM
DY8 4BW M0KMT
DY8 4EQ M7ITR
DY8 4JQ M3PML
DY8 4QF G6WMR
DY8 4QF G8RLN
DY8 4QG G7HEZ
DY8 4QS G0FLW
DY8 4RN G4OLS
DY8 4SF G8PIP
DY8 4XS G0ASK
DY8 4XS G0ASL
DY8 4XS G4TFB
DY8 4XS G4TFC
DY8 4XU G0NLT
DY8 4XX G1IHJ
DY8 4YW 2E0IQE
DY8 4YW M6ATW
DY8 4YW M7RJJ
DY8 5AU G0WBA
DY8 5HR G0ONH
DY8 5JB M6HIB
DY8 5JJ G4WIG
DY8 5LR G1VHN
DY8 5PH G0KVK
DY8 5PU G0OYY
DY8 5RF M6DIH
DY8 5YJ G0OVV
DY84PA G1DCU
DY84PA M3KTT
DY9 0JH G7RLX
DY9 0LX G6YAK
DY9 0RE G4VPE
DY9 0RT G6DDO
DY9 0SD G7VFL
DY9 0UP G0TZV
DY9 0YB 2E0ZDH
DY9 0YB G4IIC
DY9 0YB M0ZDH
DY9 0YE G3PWJ
DY9 0YE M7KEU
DY9 0YH G1NZZ
DY9 7AZ G0WWA
DY9 7DT G7JDK
DY9 7EW M6JBT
DY9 7LA M1BQM
DY9 7PS G3XOV
DY9 7SE G1OHV
DY9 8SG G8JBP
DY9 8SG M3LKU
DY9 8YG M3KJM
DY9 9AE M3FFV
DY9 9BZ M6PMG
DY9 9DE G7EZH
DY9 9DL G7OJZ
DY9 9DL M0CHD
DY9 9DL M3OJZ
DY9 9EH G4KZB
DY9 9EL G0EWH
DY9 9EL G0KZM
DY9 9EL G4XOM
DY9 9HH G7JYZ
DY9 9HT G6RSI
DY9 9HX G6NME
E1 0EE G6USX
E1 0EE M3FLC
E1 1BJ M7EPF
E1 2AG 2E0MTM
E1 2AG M6MQT
E1 2QR 2E0SAW
E1 2QR M0WSA
E1 2QR M6MFZ
E1 4TD G1POJ
E1 6LT 2E0FBJ
E1 6LT M6RXX
E1 8PW M0HZY
E10 5QD 2E0OON
E10 6DX 2E0WJB
E10 6DX M6EHV
E10 6EE G4VXP
E10 6JD M3SJV
E10 6JL 2E0COL
E10 6QT G8VZD
E10 7EB G0TVV
E10 7JS M0GZK
E10 7LQ M0GHV
E11 1BA G7JRJ
E11 1JX G0VEH
E11 1JX G4ONP
E11 1LP M3YSA
E11 2JD G4GTZ
E11 2LD M3HTF
E11 2PP G4GGT
E11 2QQ M0CMQ
E11 4AA G7LAF
E12 5AX 2E0TMD
E12 5AX M6TMD
E12 5DZ G1SEF
E12 5EF G6RVZ
E12 5HD 2E1FZC
E12 5HD M3FZC
E12 5HX 2E0RLP
E12 5HX M7LRP
E12 5NB 2E0EWM
E12 5NB M3ZEW
E12 5PQ M3FNM
E12 6AA M0EDP
E12 6RE M3GVN
E14 0FN M6EBI
E14 0HD M7DIO
E14 0SL M3YSM
E14 3AJ G8XCJ
E14 3HY 2E0MVE
E14 3JB 2E0CPR
E14 3JB M6BZH
E14 3UG M1MOD
E14 6HP G6DFY
E14 6HU M1HMP
E14 6LS G1GLN
E14 7TE 2E0VGV
E14 7TE M0VGV
E14 7TE M6VGV
E14 8EJ M7ANA
E14 8EY G0KBO
E14 8LH G4LXH
E14 8SR G1RYS
E15 1HU M0KJU
E15 3AL M0PSS
E15 4DF M6PDS
E16 1GJ G0TDG
E16 1LQ 2E0NLM
E16 1LQ M3NLM
E16 1QW M6AUM
E16 1UT G8PUY
E16 2SB M7SVA
E16 3EQ 2E0TBW
E16 3EQ M0LGN
E16 3EQ M6CXP
E16 3JY 2E0IJZ
E16 3JY M6WHT
E16 3NJ M3TGL
E16 3NJ M3UVS
E16 3QL 2E0CXO
E16 3RY M0GYU
E16 3TA G0MXE
E16 3TR M6WTP
E16 4NA M6KCD
E161TY G7GRM
E161TY M7JBL
E17 3RJ G3PKQ
E17 4BD G1HLS
E17 4HH G8PWK
E17 4PR M0TNL
E17 4QY G4IWD
E17 5AZ M1ADZ
E17 5BL M6JGN
E17 5EY G8KNF
E17 5RG 2E0PNB
E17 5RG M0PNB
E17 5RG M6TNB
E17 6NN M6RTE
E17 7ER M1EUE
E17 7ER M3EUE
E17 7HD M6EKY
E17 7JB M0VBE
E17 8AF M1CQM
E17 9EH G8BVY
E18 1AP G0FMB
E18 1BN G8ZHS
E18 1DG 2E0RCW
E18 1DG M6ECW
E18 1TP 2E0IGH
E18 2AB G3XXC
E18 2DA G4PSR
E18 2DB M6IBG
E18 2HB 2E0LGS
E18 2HB M0PDS
E18 2HB M6LUG
E18 2NJ G6SPN
E18 2PL G0KRX
E18 2PE G6XBD
E18 2QA G4LKT
E1W 2BG M0IBD
E1W 2PJ G3UKI
E1W 2QW G1ETD
E1W 3EE M0GFD
E1W 3EE M3XXA
E2 0HE G6EUU
E2 0NF M6NNV
E2 6PY M3LJF
E2 8LP 2E1HBT
E2 9QT 2E0LUK
E2 9QT M6IVE
E2 9RN M7FDS
E3 2BB G0NXM
E3 3EG M0CVA
E4 6DR G6EXN
E4 6EG M6ITY
E4 6HX 2E0NJT
E4 6HX M6NJT
E4 6NA G4CGA
E4 6RP M6DZQ
E4 6RT G4RMQ
E4 7BP G3JPG
E4 7DX G8IUC
E4 7EG 2E0JGE
E4 7EG M0JGR
E4 7EG M6JGE
E4 7HN G2XG
E4 7HX G2HR
E4 7HX G8CSA
E4 7JG M1BLJ
E4 7LG G8JYX
E4 7ND G6XRF
E4 7NR G7ITX
E4 7PZ G0DPT
E4 7PZ G1MBW
E4 7PZ M0MBW
E4 7PZ M3OAX
E4 7RA 2E0TKX
E4 7RA M0TKX
E4 7RA M6TEK
E4 8BG M3WPC
E4 8DZ G4NEL
E4 8DZ G8MJH
E4 8HD G4RSX
E4 9DU G0LWN
E4 9EA G4HWK
E4 9HE G6SKR
E4 9JA 2E0LDN
E4 9JA M0VXY
E4 9JA M6XAX
E4 9RE G5YC
E4 9RE M0BPQ
E4 9RR G0PRK
E4 9SJ G7IQM
E5 0LY M6BKF
E5 0RG G4SHO
E5 8DS M6IOR
E5 9AZ M0IVN
E5 9AZ M7MMM
E6 2BS G1SJU
E6 3BW M0GTO
E6 3RY G0JYJ
E6 6BB M6LEG
E6 6HG 2E0CKR
E6 6HG M0HFZ
E6 6HG M6BKU
E7 0DN M0GHE
E7 0HN G7KWS
E7 0JS G0LLE
E7 0JT M6SQD
E7 9PR G7FRR
E8 2AD M0DYW
E8 2DB M7ASB
E8 2ET M6LOC
E8 2HE G7SSU
E8 3BT M6TZD
E8 3LE G0IFD
E8 4LN G4UAV
E9 7HU G3WUB
EC1M 7AJ 2E0KRX
EC1M 7AJ M0KRX
EC1M 7AJ M6KRX
EC1R 5XB G3WIP
EC1V 1LR G0MZF
EC1V 3AJ M3XMS
EC1V 4NJ M7FLD
EC1V 7NS M6BTH
EC1V4AT 2E0HYS
EC1V4AT M7CQL
EC1Y 8NL G0JKM
EC2Y 8BD 2E0EEM
EC3R 8EB M7ESC
EC4V 3EJ G0PCE
EH1 2HR GM7ORX
EH1 2PB 2M0XRX
EH1 2PB MM0TEI
EH1 2PB MM6BBX
EH10 4EQ GM1THR
EH10 4LD 2M0STK
EH10 4LD MM0WSG
EH10 4LD MM1STK
EH10 4LD MM7STK
EH10 4QE MM7IBG
EH10 4SG GM0DQZ
EH10 5AY MM7CMY
EH10 5BJ 2M0YCJ
EH10 5BJ MM0YCJ
EH10 5BJ MM6YCJ
EH10 5DS GM4DTH
EH10 5DS GM8HHC
EH10 5JY 2M0CFA
EH10 5JY MM0GXY
EH10 5JY MM3RKF
EH10 5LW MM0AKM
EH10 5NU MM3FKO
EH10 5SJ GM7GNO
EH10 5SL GM0VMV
EH10 5SN GM7RMF
EH10 6ER GM8OTI
EH10 6HN GM8BJF
EH10 6HN MM6MWF
EH10 6JJ 2M0GWJ
EH10 6JJ MM7BKA
EH10 6NN GM4ZOA
EH10 6PS MM0YMG
EH10 6XD MM0OOF
EH10 6XD MM7LOL
EH10 6XE MM0KCS
EH10 6XE MM3MDB
EH10 7AQ GM3PSP
EH10 7BU GM3HAM
EH10 7BU GM4BYF
EH10 7BU GM6KAY
EH10 7BX MM0CPZ
EH10 7DG GM7REG
EH10 7HF GM7WFT
EH10 7HF MM6LQU
EH11 1AD 2M0IQR
EH11 1AD MM0OOI
EH11 1AD MM7IIE
EH11 1JZ GM8LYQ
EH11 1LR MM6EVO
EH11 1RP GM0AXY
EH11 1RP GM4YMM
EH11 1TT MM0ZBD
EH11 3GZ GM4AJV
EH11 3PF GM4DIZ
EH11 3SY MM3UDV
EH12 5AY GM6AXZ
EH12 5HG 2M0CLF
EH12 5HG 2M0JCF
EH12 5LG GM7JGR
EH12 5NG GM3HNE
EH12 5RF GM4HYR
EH12 5TF MM7WYZ
EH12 6HN MM3DVD
EH12 6NB GM0BPT
EH12 6NB GM0CMO
EH12 6XB GM4DIJ
EH12 7EB GM7UJJ
EH12 7EW GM1CNH
EH12 7EW GM4HAM
EH12 7EW MM0BPQ
EH12 7HP MM6CFA
EH12 7JW 2M0CBC
EH12 7JW MM3ZQW
EH12 7PD GM1PEL
EH12 7PU GM8CSE
EH12 8DH GM0PXV
EH12 8DL GM0SSQ
EH12 8DL GM3VJY
EH12 8DL GM4CUX
EH12 8EA GM0CDW
EH12 8EE GM8DKB
EH12 8LD 2M0DGJ
EH12 8LD MM0HVF
EH12 8LD MM6ERO
EH12 8NQ GM7VLZ
EH12 8QA GM4ZRR
EH12 8QU MM0DMU
EH12 8QW GM3RVL
EH12 8SP 2M0RWH
EH12 8SP MM0RHU
EH12 8SP MM0XRI
EH12 8SP MM6VSI
EH12 8SP MM6XRI
EH12 8SY GM1CHT
EH12 8TB GM4DHN
EH12 8UW MM3LWZ
EH12 8XF MM7JRX
EH12 8XH 2M0EGE
EH12 8XW GM3XUW
EH12 9AJ MM0WRO
EH12 9JF MM3GOI
EH13 0AF 2M0RMN
EH13 0AF 2M0SMN
EH13 0AF MM6LON
EH13 0AF MM6RCO
EH13 0AF MM6SMN
EH13 0JH MM3UKG
EH13 0NA 2M0WRR
EH13 0NA MM0KJG
EH13 0NA MM3KJG
EH13 0NG MM7LJD
EH13 0RA GM8CVN
EH13 9BL MM6KOS
EH13 9LF GM0FMW
EH13 9PH 2M0LPX
EH13 9PH MM0MGM
EH13 9PH MM7LOX
EH130EW GM4ZQH
EH14 1EF GM3SRV
EH14 1HE MM1ICE
EH14 1HG GM4RZW
EH14 1HW GM4EZJ
EH14 1JW GM6NIA
EH14 1NJ 2M0VXB
EH14 1NJ MM6VXB
EH14 1NS 2M0TFF
EH14 1UH MM0AWU
EH14 2LB GM0AXX
EH14 3AE MM6NNS
EH14 3ER. GM4YLN
EH14 5DW GM3OWU
EH14 5DX GM6MUZ
EH14 5EZ GM4XZN
EH14 5JN GM8LKL
EH14 5JN GM8NZL
EH14 5PP GM4REN
EH14 5RN MM6LAO
EH14 5RR MM0WAK
EH14 5SA GM7RYT
EH14 5SS MM6BXQ
EH14 5SX GM3TFY
EH14 5SY 2M0PSA
EH14 5SY MM0PSA
EH14 7ED GM1GDO
EH14 7EJ GM4EAU
EH14 7HJ MM7ALW
EH14 7HL 2M0UTZ
EH14 7HL MM6UTZ
EH14 7HS GM4RAH
EH14 7JE GM1JKJ
EH14 7JE MM0MGB
EH15 1JP MM6BMQ
EH15 1NB GM8BHR
EH15 1NH GM7CZC
EH15 1RA GM3DIE
EH15 2BD GM1HJX
EH15 2EX MM7DHC
EH15 2HE 2M0INS
EH15 2HE MM0INS
EH15 2HE MM6INS
EH15 2HP 2M0LBH
EH15 2HP MM6RHI
EH15 2PX 2M0TXK
EH15 2PX 2M0XBD
EH15 2PX MM0SBO
EH15 2PX MM0TKE
EH15 2PX MM6TXK
EH15 2PX MM6XBD
EH15 2QZ MM7DWQ
EH15 2RJ GM6TYL
EH15 3QG GM4BHU
EH15 3QG MM7VLC
EH16 4ED MM6JAE
EH16 4LX MM3VVI
EH16 4TG GM7IFX
EH16 4UH 2M0WMU
EH16 5DD 2M1GFG
EH16 5RY MM0RHL
EH16 5UW GM8LFI
EH16 6GJ MM6MSA
EH16 6NX GM4WZP
EH16 6YE GM4HWO
EH165EE 2M0BUX
EH165EE MM0GSQ
EH165EE MM6ART
EH17 7RT 2M0FFV
EH17 7RT MM6HCK
EH17 7SE 2M0SRF
EH17 7SE MM0VTV
EH17 7SE MM3SRF
EH17 7SE MM6GHF
EH17 8AW GM4DTJ
EH17 8DX MM3MTM
EH17 8NN MM7FDM
EH17 8SR GM6OUL
EH17 8UF 2M0DTP
EH17 8UF MM6FYF
EH17 8XG MM0ZCT
EH17 8XH 2M0NUO
EH17 8XH MM0NUO
EH17 8XH MM6NUO
EH17 8XH MM7FSR
EH177TA 2M0LQF
EH177TA MM0LQF
EH177TA MM6LQF
EH18 1AB GM4DCL
EH18 1LN GM6SEV
EH19 2EH GM0OTU
EH19 2EU GM8HSY
EH19 2LD GM4SUR
EH19 2PQ GM0FQS
EH19 3FG MM6KOE
EH19 3JG 2M0SRY
EH19 3JG MM6DKI
EH19 3JY GM0TQK
EH19 3JY MM3TQK
EH19 3LQ GM4GGF
EH19 3NF 2M0DXC
EH19 3NU MM7YUG
EH192LG MM6KMU
EH2 3NS MM3LGU
EH2 4AN GM1CZE
EH20 9EE GM3BST
EH20 9JT 2M0SWA
EH20 9JT MM3SWA
EH20 9LU GM7LTX
EH20 9PA GM6JGH
EH20 9RR GM0PQV
EH20 9SN GM6WMA
EH21 6AR GM7VLC
EH21 6EX GM0VYL
EH21 6JQ 2M0EZA
EH21 6PE MM6HZA
EH21 6RR MM0JXI
EH21 6SB GM6KFO
EH21 6TU 2M1DZW
EH21 6TU GM7PPN
EH21 7RD GM0EFQ
EH21 8AA MM0FZV
EH21 8AA MM3ZVY
EH21 8AN MM7MCA
EH21 8JT GM1FAF
EH21 8JT GM4GVJ
EH22 1EE GM4CMI
EH22 2LW MM0CHV
EH22 2NL MM6IZH
EH22 2NS GM1CCN
EH22 2NZ GM0LOD
EH22 3BN GM4HQU
EH22 3DB GM0SUF
EH22 3DR 2M0KPZ
EH22 3DR MM6KLZ
EH22 3FT 2M0VPM
EH22 3FT MM0VPM
EH22 3FT MM6VPM
EH22 3LT GM0TFE
EH22 4BW MM1CMU
EH22 4BW MM3ALG
EH22 4BW MM3BHD
EH22 4DG 2M0EPK
EH22 4DG MM6KNJ
EH22 4QD 2M0YSR
EH22 4QD MM0YAB
EH22 4QD MM6YAP
EH22 4SJ MM0GFP
EH22 5AX MM6HSC
EH22 5DT GM7DRY
EH22 5DT MM3ZDG
EH22 5ER 2M1BYM
EH22 5HX GM0PKP
EH225JA MM7JTR
EH23 4BT GM0KDC
EH23 4EQ MM0BAC
EH23 4EU MM3DHN
EH23 4GX MM1BJO
EH23 4HL 2M0GUL
EH23 4HL 2M0PCW
EH23 4HL MM0DXH
EH23 4HL MM3JHS
EH23 4HL MM6HSS
EH23 4JF 2M1HIN
EH23 4NT GM0IMZ
EH23 4RL MM6LNB
EH23 4RS MM0MOK
EH23 4SU MM7FSR
EH25 9NP GM4LHW
EH25 9QJ GM4VZI
EH25 9SB GM6BRU
EH25 9TD MM0AJQ
EH26 0DR MM6DUV
EH26 0EL GM0MJR
EH26 0RR 2M0ITI
EH26 0RR MM0JTX
EH26 0RR MM6PZG
EH26 8BW MM6DWE
EH26 8DF 2M0CEX
EH26 8DF GM1GEQ
EH26 8DF MM0VPR
EH26 8DF MM6ANB
EH26 8EF GM0LUF
EH26 8HH MM6GBS
EH26 9AN GM1OVW
EH26 9EE GM6GFL
EH26 9HS GM1PSZ
EH27 8AN GM7FZA
EH27 8AU 2M0DIF
EH27 8AU MM6DHZ
EH27 8BH GM7BJG
EH27 8BH MM3RIX
EH27 8DD MM1RAH
EH27 8DQ GM1ROL
EH27 8DQ GM1ROM
EH27 8DS MM7AGY
EH28 8AD MM7BJG
EH28 8PD GM4XHQ
EH28 8PD MM0DLH
EH29 9DS MM0SFM
EH29 9DU MM3TUR
EH29 9DZ GM6BEY
EH29 9GE MM7GCS
EH3 5LP MM0WXD
EH3 5NA GM8EKF
EH3 5PD GM7PNX
EH3 6HB GM6KOR
EH3 6LD 2M0CNS
EH3 6LD MM6MSH
EH3 6RP 2M0BDN
EH3 6SX GM4WCE
EH3 7TZ GM6CZM
EH3 7TZ GM6NIC
EH3 9PL 2M0FSV
EH30 9JQ MM7STL
EH30 9LB GM4EMY
EH30 9PL GM7RAK
EH30 9PQ 2M0EEV
EH30 9PQ GM4XWL
EH30 9PQ MM0IGO
EH30 9PQ MM6HQI
EH30 9RU 2M0FZA
EH30 9RU MM0JKY
EH30 9RU MM7ALO
EH30 9UP GM8ZWF
EH30 9XG GM0SUY
EH32 0AN GM4LRU
EH32 0AQ 2M0WLA
EH32 0AQ MM0MLD
EH32 0AQ MM6YLA
EH32 0BB 2M1IBH
EH32 0BP GM0TCC
EH32 0BY MM3FJX
EH32 0DG MM7WKH
EH32 0EE GM4UYZ
EH32 0EE MM0CPS
EH32 0HZ 2M1HVR
EH32 0PG 2M1DZS
EH32 0PG 2M1EUV
EH32 0PG 2M1GLD
EH32 0SY 2M0VPH
EH32 0SY 2M0ILZ
EH32 0SY 2M0PGO
EH32 0SY MM3YMN
EH32 0SY MM7DGK
EH32 0SY MM7PGG
EH32 0SY MM7YDG
EH32 0TA GM4TAL
EH32 0TU 2M0ISY
EH32 0TU MM6ISY
EH32 9AT MM0CRF
EH32 9DD 2M1IGO
EH32 9FB GM4YWI
EH32 9FH MM7BGW
EH32 9GF MM6SGF
EH32 9HA MM0SHB
EH32 9HF MM3MFN
EH32 9NA GM8YIK
EH32 9PX GM7PSH
EH32 9RD 2M1IBE
EH32 9RD MM0HGW
EH32 9SG MM3XGP
EH33 1DN 2M0DZX
EH33 1DN MM6HQE
EH33 1HU MM3XDW
EH33 1QE 2M0EAC
EH33 1QE MM3EWI
EH33 1QE MM3XVD
EH33 1QJ MM6JLX
EH33 2AL GM0NTL
EH33 2AL MM0WCG
EH33 2AR MM3ITA
EH33 2AS MM0CCC
EH33 2EB MM0AXK
EH33 2EE 2M1ETM
EH33 2HR MM6AUJ
EH33 2PL MM7FSR
EH34 5BB GM7MAG
EH34 5EH MM1BVW
EH34 5HA 2M0KWP
EH37 5QB 2M0KLL
EH37 5QB MM0LBF
EH37 5QB MM3ISA
EH37 5QB MM3LWJ
EH37 5QD 2M0GRE
EH37 5QD 2M0GZA
EH37 5QD MM0GZA
EH37 5QD MM3YPN
EH37 5QD MM6YPN
EH37 5RA 2M0PMR
EH37 5RL MM3EYM
EH37 5RL MM3EYN
EH37 5RS GM8MNG
EH37 5SQ GM3OIB
EH37 5UJ MM3DZW
EH37 5UP GM8MJV
EH38 5YD 2M0NOP
EH39 4BP GM0EHL
EH39 4NR GM0WPW
EH39 4PS GM4KGZ
EH39 5JE MM0JMI
EH39 5JZ MM7ILB
EH4 1AX MM3KRR
EH4 1LW MM0VTA
EH4 1PJ GM8SOK
EH4 2AU GM4SVM
EH4 2JU GM4DMQ
EH4 2ND MM0MRO
EH4 2TT GM1CQC
EH4 2UE GM1PHD
EH4 3BB MM0XAO
EH4 3DQ GM4GIO
EH4 3JU GM8MST
EH4 3TP GM0AXM
EH4 4RB MM7STT
EH4 5AW GM0RMV
EH4 5BF GM4HCE
EH4 5BQ GM7IRI
EH4 5JA GM3SBC
EH4 5NE GM7FIE
EH4 5PY GM6PYD
EH4 6EB GM6UNQ
EH4 6JJ MM6TEQ
EH4 7AH GM3KBP
EH4 7BJ GM7KTY
EH4 7BJ GM7PZH
EH4 7DF GM0WFB
EH4 7HN GM1UWE
EH4 7RQ MM3BUZ
EH4 8BA MM6NOR
EH4 8DL MM3ICD
EH40 3BH GM4HJQ
EH40 3EB GM8PEB
EH41 3BE 2M0WDG
EH41 3BE MM0INH
EH41 3BE MM6FXZ
EH41 3DU GM3LGU
EH41 3LS MM1DSX
EH41 4EF 2M0VOZ
EH41 4EF MM0WCD
EH41 4EF MM6ZCD
EH41 4QA 2M0TJO
EH41 4QA MM0HZI
EH41 4QA MM6FXM
EH42 1AY GM4FQG
EH42 1BA GM0JPG
EH42 1EX MM7EST
EH42 1GJ 2M0FLJ
EH42 1GJ MM3SQJ
EH42 1NW GM6KEV
EH42 1QT MM6GDY
EH42 1RT GM1SYC
EH42 1SH 2M0YUP
EH42 1UF MM3KZD
EH42 1UG GM3UIN
EH42 1YR GM3VEI
EH44 6JT MM0ABJ
EH44 6LZ 2M0NMD
EH446NJ 2M0HJP
EH446NJ MM0VPF
EH446NJ MM6VPF
EH45 8DH 2M0LAO
EH45 8DH MM6LAK
EH45 8FG GM4XHH
EH45 8HB GM6JFP
EH45 8HJ MM6NVY
EH45 8JE GM1RKI
EH45 8NU GM0UTD
EH45 8PP GM4WTK
EH45 8PW GM4LDX
EH45 8QZ MM6DAT
EH45 9DB GM1MSS
EH45 9HX GM8NAL
EH45 9LX GM1OPO
EH45 9LX MM3NTX
EH46 7BA GM4OGM
EH46 7BE 2M0SPL
EH46 7BE MM0SPL
EH46 7BE MM3PZJ
EH46 7BE MM6HBR
EH47 0AJ MM3MPK
EH47 0AP MM3YTI
EH47 0BH 2M0WEV
EH47 0BH MM3WEV
EH47 0DN MM0HSR
EH47 0DN MM0PTE
EH47 0LH 2M0HIE
EH47 0NL MM0SMB
EH47 0NS 2M0HZL
EH47 0NS MM0HZL
EH47 7BY MM3UPY
EH47 7QJ MM6CMM
EH47 8AT MM6DSC
EH47 8EW MM0WKJ
EH47 8SA MM7ALI
EH47 8SA MM7TMO
EH47 9AZ MM6DGC
EH47 9BF MM7FBQ
EH47 9BW 2M0DGB
EH47 9BW MM0GOG
EH47 9BW MM3NVD
EH47 9EG MM3JPS
EH47 9EL MM6IZK
EH470QA 2M0FCQ
EH48 1AS GM8ZKN
EH48 1DD GM0KBR
EH48 1DF GM4LHJ
EH48 1DG MM7FOM
EH48 1DU MM0LGS
EH48 2AF MM6GBX
EH48 2BD MM6DDN
EH48 2GZ GM1WMU
EH48 2JL MM0IQN
EH48 2JS MM3HTY
EH48 2LT MM0IEJ
EH48 2PB MM5AES
EH48 2PG 2M0CTB
EH48 2PG MM6VWT
EH48 2RG MM6RBE
EH48 2RR MM0DIS
EH48 2SJ 2M0SXJ
EH48 2SJ GM1SXJ
EH48 2SJ MM3SXJ
EH48 2ST MM0HVA
EH48 2ST MM6HBF
EH48 2TD GM1PSU
EH48 2TN MM6KGM
EH48 3BU MM6RCZ
EH48 3HB MM6STM
EH48 3HG MM0HVU
EH48 3JU GM0EWF
EH48 3LA MM6EOU
EH48 3LN 2M0RYG
EH48 3LN MM6RYG
EH48 3PP MM5AHO
EH48 4AH MM7CEH
EH48 4AX MM7AEU
EH48 4BB MM6INC
EH48 4DP 2M0XFM
EH48 4DP MM3XFM
EH48 4HG GM0EDQ
EH48 4LD GM4LZO
EH48 4NU GM4BQD
EH482HH GM8IOL
EH49 6BP 2M0MIF
EH49 6BP MM6DCM
EH49 6BS GM3PDX
EH49 6DD GM1FAI
EH49 6DQ MM6HMC
EH49 6HA GM4LUS
EH49 6HA GM4XUS
EH49 6JS 2M0JXC
EH49 6JT 2M1ADM
EH49 6JT MM0GTU
EH49 6JT MM3ADM
EH49 6LW GM7VPT
EH49 6LZ MM7KGT
EH49 6SD 2M0GAN
EH49 6SF MM3TZP
EH49 6SH GM0MWJ
EH49 7AP GM4CAQ
EH49 7BP GM4YPL
EH49 7BS GM0FTH
EH49 7JU GM3VTH
EH49 7LD GM0RUW
EH49 7LN 2M0CQI
EH49 7LN MM6RKT
EH49 7ND 2M0APX
EH49 7ND MM0ASB
EH49 7PL MM6AHX
EH49 7SR MM6EOZ
EH5 1EX MM0XXW
EH5 2ET GM0BKS
EH5 2GU MM3KLO
EH5 2HP GM1FMV
EH5 2JJ GM0CII
EH5 2JN 2M0XTH
EH5 3AR GM0TEA
EH5 3NH GM4AOR
EH51 0AE 2M0PID
EH51 0AE MM3VAH
EH51 0BQ 2M0IRC
EH51 0BQ MM3RQC
EH51 0DD GM0CQQ
EH51 0NX 2M0DRO
EH51 0NX GM5RAB
EH51 0NX MM6RDU
EH51 9EG GM7LFX
EH51 9ER GM0IKY
EH51 9LT GM0WKZ
EH51 9PE MM0FWG
EH51 9QD GM4HRL
EH51 9QN GM0HSC
EH519LX MM7JMR
EH52 5BX MM6HSC
EH52 5HQ 2M1GBG
EH52 5NS 2M0MMG
EH52 5NS MM0MMG
EH52 5NS MM3IQD
EH52 5PN GM4VDG
EH52 6AL MM7CCB
EH52 6AP MM7EMQ
EH52 6AP MM7NSY
EH52 6FB 2M0ZEB
EH52 6FX 2M0JXK
EH52 6FX MM6WEB
EH52 6HA MM0FFC
EH52 6LY GM0UMC
EH52 6PL GM0DHD
EH52 6QB GM3OIV
EH52 6TH MM0IRJ
EH52 6UW MM3MRX
EH526AL 2M0OOR
EH526AL MM7IME
EH53 0EG GM7GIF
EH53 0FB MM6RCK
EH53 0FQ MM6JXH
EH53 0NG MM0PWM
EH53 0NT 2M0CRR
EH53 0NT MM3LYH
EH53 0RN 2M1AOL
EH53 0SJ GM0ERT
EH53 0TE GM7RYK
EH54 5AD GM0NAZ
EH54 5JP GM6OGN
EH54 5JP MM6SIV
EH54 5LE MM6FFQ
EH54 5LP GM0REE
EH54 5NA GM0ALS
EH54 6BG MM6WAW
EH54 6BZ 2M0FTA
EH54 6BZ MM6FTA
EH54 6DB MM0DMK
EH54 6EE 2M0AWY
EH54 6HE 2M0DIB
EH54 6HE 2M0MHN
EH54 6HE MM0LIV
EH54 6HE MM0NHM
EH54 6HE MM6NHM
EH54 6HE MM6NSM
EH54 6JD 2M0GGY
EH54 6JD MM3FOE
EH54 6JJ GM0CBA
EH54 6LQ 2M0YTD
EH54 6LQ MM6YTD
EH54 6LT GM7GTS
EH54 6PG 2M0HMV
EH54 6PG GM4UQD
EH54 6PG MM0RYE
EH54 6PG MM6FVK
EH54 6PQ MM6JNN
EH54 6RJ GM4HML
EH54 6RP GM3UCN
EH54 6TB MM6HAV
EH54 6UE MM7WMN
EH54 6UG MM7VGR
EH54 6US 2M0KRG
EH54 6US MM7CMC
EH54 6UX MM1EYI
EH54 7AB MM6JMI
EH54 7BD MM7FUE
EH54 7BE MM7OFH
EH54 7BP 2M0TZB
EH54 7BP MM6NLP
EH54 7BZ GM7AHA
EH54 8AP MM0KFR
EH54 8EW MM0HOL
EH54 8EW MM6HSK
EH54 8HW GM8MYO
EH54 8JB GM1PZT
EH54 8JG 2M0LAS
EH54 8JG MM3GTF
EH54 8JN 2M0OSC
EH54 8LQ MM0GEO
EH54 8NX MM6JCZ
EH54 8QP MM6EQY
EH54 8RW MM3TWG
EH54 8TA MM6XRS
EH54 9AA MM0DEC
EH54 9AA MM0NEW
EH54 9AT MM0SZC
EH54 9BZ MM3DYT
EH54 9DT GM0LOT
EH54 9DW 2M0CNZ
EH54 9DW MM0JRR
EH54 9DW MM6BTR
EH54 9JE GM3ZZA
EH54 9JG 2M1AVZ
EH54 9LD MM3IDR
EH546UQ MM6HZO
EH55 8BE MM7GIR
EH55 8EW MM6XXV
EH55 8LW GM7IKB
EH55 8NL 2M0CPV
EH55 8NL MM0JHL
EH6 4DU 2M0BUT
EH6 4DU MM0WAP
EH6 4DU MM3RQP
EH6 4RU GM7GAE
EH6 4RZ. MM0PHD
EH6 4TE GM6HVY
EH6 4TR GM4HQZ
EH6 6ST MM7OWO
EH6 7AL MM7MTU
EH6 7BZ MM6MXU
EH6 7JH GM4IUS
EH6 7QQ MM6AUG
EH6 8QD MM6BYJ
EH6 8QP GM7AUW
EH6 8QP MM3AUX
EH64LB MM7WGC
EH65RY MM3WBV
EH7 4AY GM8GHQ
EH7 4HN 2M0TOK
EH7 4NJ GM3XQP
EH7 5JA GM1YME
EH7 5NG 2M0AYZ
EH7 5PS GM3HSO
EH7 5PS MM7HQI
EH7 5TT GM7IHZ
EH7 5UB MM7FLN
EH7 5UF GM0HLK
EH7 6AU GM0LVL
EH7 6TH MM1FHC
EH8 7DW MM0BQI
EH8 7JL GM4JEM
EH8 7LQ GM4AIS
EH8 7LZ MM0CJT
EH8 7PU GM0WTP
EH8 7SP 2M0ECK
EH8 7SP MM3ZQX
EH8 8DZ MM6NCP
EH8 8EQ MM7IRC
EH8 8HA MM0HLU
EH8 8HF MM6YUI
EH8 9PS MM7RIN
EH8 9UW 2M0GZU
EH9 1AY MM0HQI
EH9 1NH MM6TLI

EH9 1PN MM6IKT
EH9 1QG GM7OCU
EH9 1UF GM6RQU
EH9 2EH GM4XNQ
EH9 2EL GM3FRU
EH9 2JL GM4HAO
EH9 2JY 2M1GYX
EH9 2NX GM4FBP
EH9 2NZ MM3NQT
EH9 2RA MM7RCT
EH9 3AR MM0GYG
EN1 1BS M6PMK
EN1 1NE 2E0DGD
EN1 1NE M0HLX
EN1 1NE M6DGD
EN1 1XG M7DBH
EN1 2BZ G4OBE
EN1 2HY M6OJS
EN1 2JU G3YJZ
EN1 3AE G0NEU
EN1 3DE G1GTS
EN1 3DW G4RTC
EN1 3LW M7WML
EN1 3NQ G3RWL
EN1 3NT G8PSF
EN1 3NU G8TWR
EN1 3RG M6WWE
EN1 3UP 2E0DHE
EN1 3UP 2E0JKS
EN1 3UP M0HVC
EN1 3UT M5AJK
EN1 3UU G0NQV
EN1 4AD M1EJQ
EN1 4BD M5AGW
EN1 4DY 2E0RIA
EN1 4DY 2E0WOB
EN1 4DY M6WIF
EN1 4DY M6ZUB
EN1 4HR G4BUB
EN1 4PS G4IPR
EN1 4SY M6SGC
EN1 4TX G7MNK
EN10 6EE G6XAR
EN10 6JA G4UOI
EN10 6LD G4GAK
EN10 6NY G4XVY
EN10 6QE 2E1IBS
EN107NR M6CKJ
EN11 0BB G1SDK
EN11 0JS G8HTA
EN11 0LP M3MRJ
EN11 8RW G0JXR
EN11 9DG G1MRE
EN11 9DG G4ARY
EN11 9DL 2E0EHD
EN11 9DL M6EHD
EN11 9DX 2E0CHW
EN11 9DX M3VCW
EN11 9EP G3VSJ
EN11 9JR G1PZP
EN11 9NR G1CAY
EN11 9QH M0XDS
EN11 9QN G7HYG
EN11 9QS 2E0RNT
EN11 0QH M6MZJ
EN2 0DU M7DMA
EN2 0DU M7MDA
EN2 0DZ M0NOE
EN2 6NQ G6ODA
EN2 7BL G3JNJ
EN2 7BY G4XIN
EN2 7DB M1MUS
EN2 7EL G8AQO
EN2 7EL M0DDC
EN2 7EL M0HCY
EN2 7EN G8MOL
EN2 7PD M3FDM
EN2 8EN M6ZIP
EN2 8HS G4KZD
EN2 8HS G4ODR
EN2 8JJ 2E0KRR
EN2 8JJ M6RRK
EN2 8NW M7JGS
EN2 8QL G7MNE
EN3 4EF G0HAK
EN3 4EF G1PWO
EN3 4QE G3RKJ
EN3 5SE M0DJI
EN3 6EB M3VYE
EN3 6HE 2E1FQE
EN3 6NT G7VVF
EN3 6ST M7RLD
EN3 6UQ M7AKC
EN3 7JF G4GJO
EN3 7QA 2E1HVI
EN3 7RW M3XFA
EN37QZ M7WOL
EN4 0HB G6DLM
EN4 0EX G3UJV
EN4 8DQ G0CEQ
EN4 8EA 2E1DLS
EN4 8EN M7ASE
EN4 8JS G6PKX
EN4 8NJ G3LSX
EN4 8PU G8DVJ
EN4 8TU M7EFO
EN4 8UX M6NJH
EN4 9AS G7MGM
EN4 9DS G0LCS
EN4 9DS G5SIX
EN4 9DS G8VR
EN4 9LX M7NOL
EN4 9PA G8OQC
EN4 9QT G3GMY
EN4 9RL G1MPU
EN4 9TX G0NSI
EN5 1AW M7HVA
EN5 1EJ M7SUN
EN5 1LX M6RJL
EN5 1RJ G7VFY
EN5 2BP G8FVE
EN5 2BW G1ALA
EN5 2DB M6DUT
EN5 2DP M6VCA
EN5 2EX M6SNG
EN5 2JL G0SNO
EN5 2LE G4BWY
EN5 2LE G8AIE
EN5 2LS G4GPR
EN5 2NQ G3IQY
EN5 2QU 2E1JOD
EN5 3BB G0TOY
EN5 3BP 2E1HAS
EN5 3BP M3BCH
EN5 3DP M7FGT
EN5 3JR G6NVI
EN5 4JG M6FUT
EN5 5LH M6YGN
EN5 5QH G3YFW
EN5 5RH G8TAU
EN5 5TN G4WRD
EN6 1LG G3PRK
EN6 1QW G3RTE
EN6 1QW G6AY
EN6 1XW G1END
EN6 2BD G3WEA
EN6 2BQ G4EZU
EN6 2DH 2E0KSP
EN6 2DH M6VLQ
EN6 2DH M7VLQ
EN6 2EB M7LSS
EN6 2HY M7BVQ
EN6 3AJ M0ATL
EN6 3DL G3DEN
EN6 3DZ G7BHR
EN6 4DZ G0ICK
EN6 4DZ G4JIJ
EN6 4JE G4ZSC
EN6 4LH M7MXS
EN6 5HE G4HTE
EN6 5JQ M6KYS
EN6 5LX G8ZCK
EN6 5NB G7RHI
EN7 5EQ 2E0GSN
EN7 5EW G3PID
EN7 5JT G3SSZ
EN7 5NL G3MVM
EN7 6AN 2E1HRJ
EN7 6AN G7OYP
EN7 6AQ 2E0ROM
EN7 6AQ M0ROM
EN7 6AQ M3XYA
EN7 6AQ M6DAP
EN7 6AS M6NWC
EN7 6DF G8KKH
EN7 6HB 2E1JGM
EN7 6LG M1GDB
EN7 6TF G8XTR
EN8 0BW 2E0SSA
EN8 0BW M0SAO
EN8 0BW M6AHD
EN8 0BW M7SUM
EN8 0ER M7RGP
EN8 0EW M5BXB
EN8 0QU M6GDT
EN8 0QU M6JTH
EN8 7JJ G4GRN
EN8 8DA 2E1EGI
EN8 8NH G4SGN
EN8 8PS M6GKV
EN8 8UX G8VLP
EN8 9AP G6YNT
EN8 9BZ G0PXY
EN8 9JJ G6XBS
EN8 9QY G0NLV
EN8 9RQ G4WIP
EN8 9RQ G6AFX
EN9 1HN M7MSW
EN9 1PS M3UBK
EN9 1RL M6KLV
EN9 1SZ M3TEP
EN9 2AL G4XVM
EN9 2LB G3HYG
EN9 3DS 2E0DDF
EN9 3DS M6DFD
EN9 3EA M7YIF
EN9 3HP 2E0AWC
EN9 3LD G4MOI
EN9 3LD G7OKV
EN9 3LQ M0MHO
EN9 3NS M3HSM
EN9 3NZ M7GPX
EX1 2BG G8NEI
EX1 2SP 2E0RGH
EX1 2SR G7EQO
EX1 3AH M0GJX
EX1 3AH M0GMO
EX1 3EQ 2E0DHX
EX1 3GA G3WVM
EX1 3JQ 2E0CRD
EX1 3JQ M3DGD
EX1 3NP M3DDY
EX1 3NT 2E0GHR
EX1 3NT M3GHR
EX1 3RA G6HUO
EX1 3WU 2E0TSY
EX1 3XA M6YFU
EX1 3XP G4KXR
EX10 0AY M6FSM
EX10 0ER G6IDG
EX10 0LR M3YFN
EX10 9BW 2E0ZZT
EX10 9BW M3HUS
EX10 9EW G8GTV
EX10 9EX 2E0BZS
EX10 9EX M0PBC
EX10 9EX M6NLO
EX10 9JA G3DCE
EX10 9JJ G4UBB
EX10 9LF M6VQC
EX10 9LS G6RUM
EX10 9NA 2E0IUL
EX10 9NY G6BHY
EX10 9SJ G4IFQ
EX10 9SU M6YWX
EX10 9SX 2E0IUR
EX10 9TJ G0AXC
EX10 9TJ G6XUV
EX10 9TJ G7AGO
EX10 9TJ M0KSO
EX11 1AR G4BNP
EX11 1BX G1OEF
EX11 1BY G8LAY
EX11 1DJ 2E0ITF
EX11 1DT G0WGH
EX11 1EN G7DYB
EX11 1EP G7NBZ
EX11 1FQ 2E1VAR
EX11 1FQ G1ARM
EX11 1JA G8NGE
EX11 1PF G0WYF
EX11 1PF G6LAW
EX11 1PR G4VTQ
EX11 1RG G4TIG
EX11 1SY G8NVT
EX11 1TD G4UUH
EX11 1TD M6MTI
EX11 1TW M6WNZ
EX11 1UY G4BOP
EX12 2DJ G4XKK
EX12 2EQ 2E0EBL
EX12 2EQ M6EVX
EX12 2HH G2BSW
EX12 2HN G8VXU
EX12 2HX M0TGC
EX12 2NJ G7VCB
EX12 2NT G8WCQ
EX12 2PD G1NFQ
EX12 2RX M7JBP
EX12 2SB G0CTF
EX12 2SS M3ECS
EX12 2TP M3GPR
EX12 2TU G3LVB
EX12 2UH M0HSH
EX12 2XW G6WZP
EX12 3BD M3TAO
EX12 3BL 2E0TWO
EX12 3BL G4APG
EX12 3BL M3HKM
EX12 3LT M6ZYE
EX12 3NF 2E0IUM
EX13 5BL G1KNU
EX13 5BS G4KYG
EX13 5GT M6HUQ
EX13 5GT M6HVP
EX13 5JT G8MCC
EX13 5RW G0AOS
EX13 5SQ G6WWY
EX13 5SQ M3WWY
EX13 5SX G3ZVW
EX13 5SZ G7CMP
EX13 5TD G4KAM
EX13 5XN M3ZII
EX13 7AF G8CYG
EX13 7LU G3GOS
EX13 7PB G0HRH
EX13 7PS G8SHF
EX13 7RW G4WNU
EX13 7SQ M3YKH
EX13 7ST G3CFR
EX13 8AH M1AVB
EX13 8AP G3CYX
EX13 8AQ G4FVU
EX13 8TT G0GHH
EX13 8TT G8CA
EX14 1AX G4PQS
EX14 1AX G7RMX
EX14 1JB G3GRQ
EX14 1QZ M3VTA
EX14 1QZ M3WDY
EX14 2BJ 2E0LCP
EX14 2DF M3ZDS
EX14 2GL M7HOP
EX14 2TT G0KJJ
EX14 2UG M7CTP
EX14 2YP G0WWL
EX14 3EX G0JSC
EX14 3EX G0JST
EX14 3EX G7JST
EX14 3HW M6IUF
EX14 3NL G4ENJ
EX14 3PY M6HHQ
EX14 3UP G4TRU
EX14 4QS G3OLB
EX14 4UL M1VPN
EX14 4UZ 2E1GFS
EX14 4XH M3SZK
EX14 9AL G0TEB
EX14 9TA G0BVC
EX14 9TQ 2E0OSE
EX149RY 2E1HMB
EX15 1BQ G4TLL
EX15 1BS M7FRK
EX15 1DW G4HMA
EX15 1EH 2E0PAO
EX15 1EH M6PIK
EX15 1FJ 2E1FWX
EX15 1GQ 2E0SYI
EX15 1GQ M6SYI
EX15 1LA 2E0TIV
EX15 1PA G4FDG
EX15 1QD M0AKJ
EX15 1QS M7MDB
EX15 1SS G0SXN
EX15 1TA M6NPN
EX15 1UD G0KOF
EX15 2LW M0CVR
EX15 2RN G8HPS
EX15 2RN G8IYD
EX15 2RN M0BVM
EX15 2SB M7OSR
EX15 2SH G0IFA
EX15 2SX G4SCV
EX15 2TT G6VDA
EX15 3AU G6TJJ
EX15 3BA G3WNV
EX15 3DN G0HPS
EX15 3DS G0OED
EX15 3ES G6EED
EX15 3HJ M6GXR
EX15 3QG M3YNI
EX15 3RD M6FJE
EX15 3SE G6FVD
EX15 3XA G0TAZ
EX15 3XZ M7ICQ
EX16 4AW 2E0IFC
EX16 4AW M6NLB
EX16 4BE G6XWM
EX16 4BN G6SWD
EX16 4BW G7JGZ
EX16 4ER G4XXD
EX16 4ET G4ZAL
EX16 4LF M0PJY
EX16 4LN M3IEG
EX16 4PY G8MFF
EX16 5AF M6JSX
EX16 5FH G1HEA
EX16 5JE M3TME
EX16 5PA G8URB
EX16 6BU M6CPZ
EX16 6EB 2E0CZC
EX16 6EB M6GEP
EX16 6EE G0IFC
EX16 6FJ M0HMJ
EX16 6HE 2E0MEL
EX16 6JU G0MAS
EX16 6PL G4DKC
EX16 6RB G3VGY
EX16 6RG 2E0AYI
EX16 6RG M3CLO
EX16 6RJ G4YCV
EX16 6XJ G4PYU
EX16 7AZ M7AJM
EX16 7BS G0GLQ
EX16 7HT G3ZQL
EX16 7HT G8PXO
EX16 7NA G3PCY
EX16 8AZ M0GMS
EX16 8LQ G3ZRJ
EX16 8LS M7FJR
EX16 8PP 2E0KCF
EX16 9DW G6YWZ
EX16 9JQ G6ASK
EX16 9JQ M0KSB
EX16 9JQ M0LPR
EX16 9JU G0LOW
EX16 9NN 2E0RNW
EX16 9NN M6WNR
EX16 9NS 2E0PLJ
EX16 9NS G8MBE
EX16 9NS M7BUL
EX16 9PJ G3HXK
EX16 9QH G4FJK
EX16 9QH G4TSW
EX166AR G7VZD
EX17 2DH G4YAQ
EX17 2DH G6TWD
EX17 2DN G1JXA
EX17 2EJ M3LLK
EX17 2EJ M3VVY
EX17 2EP G3HTA
EX17 2HA G4WJJ
EX17 3HE M3MEF
EX17 3NB 2E0NRJ
EX17 3NB M0NRJ
EX17 3NB M6NRJ
EX17 4AE 2E0CTQ
EX17 4PL M0PXI
EX17 4QX M0GEB
EX17 4RU M3TBU
EX17 5AL G4RRA
EX17 5NA G0JUM
EX17 6AY G6DLJ
EX17 6DU M7CEL
EX17 6HY M1AEI
EX17 6NA G4MQQ
EX17 6NA G8RYX
EX17 6PA M6TVW
EX18 7BQ 2E0SXS
EX18 7BQ M6DVT
EX18 7QX G0ABI
EX18 7SN M1DAS
EX19 8DP G8JRW
EX19 8HD G8RIR
EX19 8JW M6BPE
EX19 8JX G3LBM
EX19 8SL G3ZQR
EX2 4BL 2E0JSK
EX2 4DG G8LKA
EX2 4JS G3AJK
EX2 4LR M6EUV
EX2 4SJ G6ZTP
EX2 5DU M0IHT
EX2 5EP 2E0SJM
EX2 5ES 2E0RQK
EX2 5ES M0RQK
EX2 5JT M0REB
EX2 5JX 2E0SQB
EX2 5JX M6SQB
EX2 5NH G6BJL
EX2 5QJ G7WLM
EX2 5QN G8RCZ
EX2 5RG M3LQI
EX2 6AN G0BNW
EX2 6BW M7SKA
EX2 6EA G7UIU
EX2 6ED M7EGW
EX2 6EG M0CPF
EX2 6JJ G4BQH
EX2 6LE M3RPS
EX2 7AQ M0TCF
EX2 7BD M3IEA
EX2 7BQ M0NJE
EX2 7RA M7CUG
EX2 7TR M7HBS
EX2 8LA M6AJL
EX2 8XN 2E0ECW
EX2 8XN G0JQS
EX2 8XP M6OOZ
EX2 9AH 2E1GZZ
EX2 9AH M3MPC
EX2 9JF G4CPN
EX2 9JQ 2E0GGZ
EX20 1AU M7AZJ
EX20 1BD G1DKB
EX20 1EA G3PSZ
EX20 1EA G7VMQ
EX20 1GR 2E0SPS
EX20 1GR M6LJF
EX20 1HZ G7CIH
EX20 1JF G0MOH
EX20 1NU 2E0VFT
EX20 1PL M6BPU
EX20 1PR G0FJA
EX20 1PR G0FYL
EX20 1QQ G4NJK
EX20 1QX G8JZZ
EX20 2HB 2E0WLQ
EX20 2HB M0PIR
EX20 2HB M7JCE
EX20 2HX M3YGA
EX20 2JF G3VDL
EX20 2NG G1BOB
EX20 3AJ G4NFP
EX20 3EZ G0FUV
EX20 3EZ G6LWT
EX20 3FE 2E0GLL
EX20 3FE M0SLL
EX20 3FE M6TIU
EX20 3LE G3ZLS
EX20 3LU G4BTN
EX20 3NR G8WKA
EX20 3NX 2E0CER
EX20 3PL G8TAY
EX20 3PP G4VQP
EX20 3QW G0GHT
EX20 4AW M7IJT
EX20 4DQ G4XIX
EX20 4LL G0IVO
EX21 5AY M6IXJ
EX21 5BP G4FIV
EX21 5BQ G0OAW
EX21 5EF G0ASG
EX21 5EU G0ESY
EX21 5HA G0LSI
EX21 5LT G8UIL
EX21 5LT M3WYJ
EX21 5PX M3AFF
EX21 5TX M6KPI
EX21 5UD G0IKC
EX21 5UF G3BYG
EX21 5XY 2E0XLO
EX21 5XY M7XLR
EX22 6AY 2E0SAT
EX22 6AY M6AIR
EX22 6DA 2E1GQU
EX22 6DA G7VTE
EX22 6DA M1BNR
EX22 6DQ M7AUF
EX22 6HB G0CLC
EX22 6HX G0WTI
EX22 6QN G8MWW
EX22 6RS G1GZI
EX22 6RS M3GZI
EX22 6TB 2E0MYB
EX22 6TB M0MYB
EX22 6UU G4WWR
EX22 6UU G6WWR
EX22 6UU M0BKV
EX22 6XX G3VDH
EX22 7AA G6LAU
EX22 7BQ M0OZJ
EX22 7DU M6VKZ
EX22 7ED G6REW
EX22 7EN M3VJW
EX22 7EW G4CQM
EX22 7NB M1AOB
EX22 7NB M6DEM
EX22 7NH G7HLB
EX22 7NY G0RQL
EX22 7NY G1DIF
EX22 7NY M0OMC
EX22 7QY M7LIP
EX22 7RN G0KDY
EX22 7RT G6ILD
EX22 7YG G3ZGI
EX23 0AH M6JBI
EX23 0AJ M0LSS
EX23 0BE G6LSB
EX23 0BN G0OKN
EX23 0DT G4FWT
EX23 0ES G0TDE
EX23 0HH M0HEN
EX23 0HY G4NCJ
EX23 8AP G1IQN
EX23 8EN G7EKL
EX23 8EU 2E0FWC
EX23 8EU G4YTC
EX23 8FF G8YWL
EX23 8HZ G3MNQ
EX23 8JB G8YRW
EX23 8LN M6EJV
EX23 8LR G6MAA
EX23 8LR G8DHA
EX23 8NA G0DBD
EX23 8PD G8ULJ
EX23 8PG G1RLB
EX23 8QF M0MJW
EX23 8RH G8JSS
EX23 8RY G4IDU
EX23 8RY G4MMT
EX23 8SA M6JIC
EX23 8SB G7TRM
EX23 8SP M0BUI
EX23 9AF M0SMJ
EX23 9AG G1IJQ
EX23 9BB G4NLH
EX23 9BJ G4ASF
EX23 9BP 2E0RCM
EX23 9BP M0RBM
EX23 9DA G3XNE
EX23 9ES M6CHW
EX23 9JA G8MEH
EX23 9JR G4OJQ
EX23 9LE 2E1EOQ
EX23 9LE M3EOQ
EX23 9RE G4MYD
EX238EN M7RAW
EX238PY M7FPH
EX24 6JU G6LUJ
EX24 6NN M7DVP
EX24 6PN G0GIN
EX24 6QZ G7EHR
EX246PT M7GFJ
EX26LS M3YDV
EX20 1PY M7CKR
EX3 0DY G1DKE
EX3 0DY M6AZM
EX3 0JW M6RVD
EX3 0LG G3IMW
EX3 0NA G4ETO
EX3 0PE G8ATC
EX3 0RA G1WYM
EX31 1PA M7CPA
EX31 1QA G3XYG
EX31 1QF G3AKJ
EX31 1QX G4RVJ
EX31 1RZ G0OQZ
EX31 2BH G6RNT
EX31 2HL M6DHW
EX31 2HY M0DJH
EX31 2JG M0IGF
EX31 2LH 2E0TVD
EX31 2LH M6SCP
EX31 2NJ M1FWD
EX31 3AF M0DWX
EX31 3AP G8NBO
EX31 3DA G4PEK
EX31 4FL 2E0TZR
EX31 4FL M3TGC
EX31 4HY G6TAP
EX31 4JD M6EQL
EX31 4JQ G4RZI
EX31 4JR G0BUJ
EX31 4LP G0RVS
EX31 4PR G4YCW
EX31 4TA M7RXW
EX313AP M3YBN
EX313BS G3PGA
EX313BS G6GZC
EX32 0DF G4SGW
EX32 0EG 2E0GZM
EX32 0EG M3MSU
EX32 0JN G4SJP
EX32 0JN G8KQB
EX32 0ND M3UKF
EX32 0NJ G0WUA
EX32 0TX G4MZH
EX32 7AS G3ZFV
EX32 7AY G7FJZ
EX32 7EN M3VJM
EX32 7HN G6USZ
EX32 7HW G0EYF
EX32 8AE 2E0HVZ
EX32 8AE M0WWD
EX32 8DN G4TNE
EX32 8EJ G0DNV
EX32 8ET M6FEQ
EX32 8JR M6SWG
EX32 8LA M3HJF
EX32 8NW G0NPV
EX32 8PE 2E0ITX
EX32 8PE M6ORX
EX32 8QH G4KXQ
EX32 9BG G8VCQ
EX32 9BW G4XWQ
EX32 9DG M1CVK
EX32 9DJ 2E0LVC
EX32 9DJ M6LQC
EX32 9DP M3XZT
EX32 9EY 2E0EON
EX32 9EY M6IUI
EX32 9JX M3BOR
EX33 1AT G1LLI
EX33 1BB G4XZF
EX33 1DH G0DAH
EX33 1HE G3OIP
EX33 1PP 2E0FEO
EX33 1PP M0TLO
EX33 1PP M6FEO
EX33 2EH G0SHP
EX33 2EL G3MTD
EX33 2EZ M3IXF
EX33 2HL G8EZZ
EX33 2HN G1MJI
EX33 2LT G0LZZ
EX33 2PE 2E0PEX
EX33 2PE M0PEX
EX33 2PE M6UNN
EX33 2PF 2E0IXX
EX33 2PF M3IXX
EX331HB M7CDO
EX34 0HQ M0AME
EX34 0NL 2E0JSX
EX34 7BX 2E1NJC
EX34 7BX G4NGB
EX34 7BX M3NJC
EX34 8DT G0HGM
EX34 8DW G8RPD
EX34 8DZ 2E1GBN
EX34 8EW G5HBM
EX34 8HS 2E0PDM
EX34 8HS M6PDL
EX34 8LG M3MXK
EX34 8LR 2E0IHW
EX34 8NH G4RWK
EX34 9HP G3JUW
EX34 9JW M7CSW
EX34 9LG G1GBC
EX34 9LL M6IAT
EX34 9LN G0DIZ
EX34 9LN M3MRU
EX34 9LS 2E0JQK
EX34 9LS M3NCH
EX34 9LS M6TFF
EX34 9TA G1JWO
EX35 6EW M6BNA
EX35 6HQ G4MEI
EX36 3FL G6TWR
EX36 3HJ G8VEQ
EX36 3HL G0EOP
EX36 3HW M6FEJ
EX36 3RD 2E0HWS
EX36 3RD M0KFH
EX36 4AL G0HIW
EX36 4BH G4JBR
EX36 4DB 2E0GPT
EX36 4DB M6GPT
EX36 4DB M6IZE
EX36 4EW G4RVG
EX36 4HJ G0FGI
EX36 4HJ G8BXO
EX36 4HJ G8SBH
EX36 4HJ G8SSS
EX36 4JU G1KBF
EX36 4JW M0DYV
EX36 4PX G4YUO
EX37 9HY 2E0CNB
EX37 9HY M0JAP
EX37 9JG G1EEO
EX37 9QE G8RIS
EX37 9ST 2E0DYM
EX37 9TT G7TVQ
EX38 7BE G4INI
EX38 7DX 2E1GJY
EX38 7NZ M1AEO
EX38 8AL G0OKA
EX38 8AS G4UFK
EX38 8EE G6KEH
EX38 8EX G0CVI
EX38 8HZ G8UZY
EX38 8NG M1GRA
EX38 8NR G8TBU
EX388BY G0BVD
EX39 1BD G4NUJ
EX39 1BS G0XCF
EX39 1BS M6XHF
EX39 1DF G6YSL
EX39 1GU G0VXB
EX39 1HG G0EDU
EX39 1JG G6JJA
EX39 1LS M3KZS
EX39 1LT G0AYM
EX39 1LT G0BYL
EX39 1LT G6YXL
EX39 1NW G4FRV
EX39 1NX G0OXW
EX39 1NX G4XHK
EX39 1PA G1ZUS
EX39 1PE G0PGK
EX39 1PW G7MWI
EX39 1QY 2E0TUC
EX39 1QY M3TUH
EX39 1RD G4SRP
EX39 1SG G0UNB
EX39 1UT 2E0RGK
EX39 1UT M3RGK
EX39 1UW G6WZN
EX39 1XE G4CHD
EX39 1XN G4TCK
EX39 1XN G6OJK
EX39 2BP M3VUY
EX39 2EA 2E0AJX
EX39 2EA M0OCJ
EX39 2JJ 2E0FTZ
EX39 2JJ M7BLL
EX39 2LL G1ZTJ
EX39 2RR G3MNV
EX39 3BN G8SFD
EX39 3BY 2E0FQE
EX39 3BY M0MFS
EX39 3BY M7ABR
EX39 3DJ 2E0WWG
EX39 3DJ G3YGJ
EX39 3DJ M6DBV
EX39 3EQ M3KWS
EX39 3ET G4SOF
EX39 3LZ G2FKO
EX39 3LZ G3JKL
EX39 3NE G4DXT
EX39 3QH 2E0IXK
EX39 3QH M7BON
EX39 3QZ G1ZVZ
EX39 3RE G6LPV
EX39 3RW G1SVP
EX39 3RW G3YIN
EX39 3SF M0PIP
EX39 4AG 2E0IFA
EX39 4AG M0KNF
EX39 4AG M7SVB
EX39 4AP G3YBP
EX39 4DX G0HAU
EX39 4HA 2E0AZJ
EX39 4HA M0VMH
EX39 4HA M3FXX
EX39 4HG G0GFK
EX39 4QR G0VWQ
EX39 4RS G3APU
EX39 5HN G0JRZ
EX39 5HY G3RSF
EX39 5JA 2E0UCD
EX39 5JA M7SPO
EX39 5PL G3VYS
EX39 5PL G8FXG
EX39 5RH G8VNO
EX39 5SF M6ENI
EX39 5SH G8CPN
EX39 5ST M3WLL
EX39 5XY 2E0EUZ
EX39 5XY M6CCH
EX39 6BQ M3VYB
EX39 6JA G7VJA
EX394RB 2E0KJL
EX4 1NH 2E0CNL
EX4 1NH M3VNL
EX4 1NJ G0FGE
EX4 1NX G4EDG
EX4 1PQ M3GSR
EX4 1QX G7HZV
EX4 1SW M0KFW
EX4 1TA G3YBG
EX4 2AP G7BAE
EX4 2DF G3YRX
EX4 2EF M0UWD
EX4 2EG M1BAS
EX4 2EU G1YPM
EX4 2JP G4PCB
EX4 2JP G8NZB
EX4 2LR G4FYJ
EX4 2PN G0UTA
EX4 4GD M7MXU
EX4 4PW 2E1GMT
EX4 4RR G3SBP
EX4 4SA M7AJZ
EX4 4SJ G0TQS
EX4 4SJ G4BPV
EX4 5AJ G4PXE
EX4 5DN G4SJU
EX4 5DN M6PPN
EX4 5DT M0KQI
EX4 6EY G3SXH
EX4 6HG G6IKC
EX4 6ND M0BBV
EX4 6NG G1VNH
EX4 7DY 2E0WRS
EX4 7DY M0ZZT
EX4 7DY M3WRS
EX4 7EA G3YBK
EX4 8DQ 2E0HYY
EX4 8DQ M7KRE
EX4 8DU M6GNR
EX4 8HB G3ZVI
EX4 8QD G0BTQ
EX4 8RD M7ENE
EX4 9DT 2E0HVY
EX4 9DT M7VBR
EX4 9DY G4KEE
EX4 9EA M7CTF
EX4 9ES 2E0HVW
EX4 9ES M6BYO
EX4 9HP M0CQW
EX4 9HP M1AZY
EX4 9JT G3TLH
EX5 1EE G3XYO
EX5 1HP M0DKV
EX5 1JD G3VRV
EX5 1LL G4TDV
EX5 1PP G4INF
EX5 1QS 2E0UJM
EX5 1QS G3UJM
EX5 1QS M6UJM
EX5 1RA M3HXS
EX5 1RA M7CKC
EX5 2BY G3RUV
EX5 2EX G7NBU
EX5 2JQ M7KCD
EX5 2TJ G7VNQ
EX5 3DX G4AP
EX5 3DX G4ARE
EX5 3HW 2E0SPE
EX5 3HW G6ECG
EX5 3HW M7GTC
EX5 4BB G3ORN
EX5 4HT M0WPN
EX5 4JZ 2E0NXB
EX5 4JZ M0NXB
EX5 4QP M6MGX
EX5 4QU 2E0MRC
EX5 4QU M0CHE
EX5 4QU M3CHE
EX5 4QU M7AZW
EX5 5DA G7NUC
EX5 5NB M6OBY
EX5 7AZ M0DPK
EX5 7GB G4VMF
EX6 6AP 2E0KMP
EX6 6AP M0ZCM
EX6 6AP M6AZA
EX6 6ET G1PHU
EX6 7AD G4APU
EX6 7SR G4BZE
EX6 7ST M7EMU
EX6 7TZ G3KQG
EX6 7UB M0IOS
EX6 7XS G4WVT
EX6 8AL M7JKM
EX6 8AT 2E0HHJ
EX6 8AT M0KOB
EX6 8DW M3FTE
EX6 8NS G0XRC
EX6 8NS M0THJ
EX6 8SD G7UEK
EX7 0AB G0KYS
EX7 0AB M0MCG
EX7 0AD M6RDP
EX7 0AN M0GSI
EX7 0AN M3GSI
EX7 0AN M3XOR
EX7 0BP G0JIS
EX7 0BZ M6KPK
EX7 0EA G4ONV
EX7 0EJ M3DOM
EX7 0HB G1NMY
EX7 0HB G4KPP
EX7 0JT M3SGE
EX7 0JZ G1ZVO
EX7 0JZ M0KRE
EX7 0NA G4VXH
EX7 0NA G7URW
EX7 0QL G3AVN
EX7 0RN M6OEZ
EX7 0RP G1MVT
EX7 9BS G3WEJ
EX7 9BU G7CNP
EX7 9BU M3HUW
EX7 9DY G0WGJ
EX7 9DZ G3WWT
EX7 9EX G7OIT
EX7 9FA G0BBO
EX7 9FD G6MTB
EX7 9HP M6AYJ
EX7 9JB G0CAS
EX7 9LQ 2E0HOO
EX7 9LQ M3UNF
EX7 9NQ G4WLA
EX7 9QH G8NVS
EX7 9RS G4DUT
EX8 1DT G4SET
EX8 1QN G4EBO
EX8 1QS G8DQK
EX8 1SJ M3GUJ
EX8 2AF G0RKC
EX8 2JW M3LPU
EX8 2PA 2E0ORI
EX8 2PA M0ORI
EX8 2PA M6ORI
EX8 2PU M3LHQ
EX8 2PU M3UIL
EX8 2QR G4VQL
EX8 2RP G6GNR
EX8 2SG G1YOF
EX8 3DT G7BWO
EX8 3EJ G1GZG
EX8 3EJ G1MQC
EX8 3HJ M7ARH
EX8 3HY G4YRM
EX8 3JX G4UUW
EX8 3LA G4DPJ
EX8 3NB G1ILY
EX8 3NE G3UZM
EX8 4DN M1CML
EX8 4EU G8EKW
EX8 4EY G3ZDG
EX8 4LF G8GON
EX8 4LF G8UWE
EX8 4LU M6NHX
EX8 4NA M6AFF
EX8 5AP G0OJX
EX8 5BU G1BMZ
EX8 5EE G0ETZ
EX8 5JD M6AFA
EX8 5PN G0GHO
EX8 5PQ G0AVJ
EX8 5RF G0NKC
EX8 5SP G0XJS
EX9 6JN G6UEI
EX9 6LY G0KNY
EX9 7AS G4WJZ
EX9 7BP G4JBF
EX9 7JE G7AKJ
EX9 7JL G4ZQF
FK1 1RL GM0TXJ
FK1 2AG MM1CKW
FK1 2AP GM6PKP
FK1 2BX MM0RSI
FK1 2BX MM6WAI
FK1 2EA 2M0RZE
FK1 2EA MM0GSS
FK1 2EA MM3JRK
FK1 2RP GM6JUA
FK1 2RP MM0DBC
FK1 3BA MM6BOD
FK1 4BJ 2M0MMF
FK1 4BJ MM0PQN
FK1 4BJ MM0RDT
FK1 4JF GM0KWL
FK1 4LY MM6EJO
FK1 4NL MM7CXR
FK1 4PB GM6RGY
FK1 4PB MM1CLR
FK1 4PG MM3ZTS
FK1 5BQ 2M1ENK
FK1 5DS GM0KUP
FK1 5HU MM0RAM
FK1 5JX GM1LUZ
FK1 5JX MM5AGM
FK1 5NZ MM0KCD
FK1 5QQ MM7EMD
FK1 5UE GM4MIG
FK10 1DD GM0WWX
FK10 1NQ GM4DGT
FK10 1PT 2M0CSZ
FK10 1PT MM6CNO
FK10 1QZ MM0OBT
FK10 1RU 2M0HMU
FK10 1RU MM7CLN
FK10 2BN GM0UUB
FK10 2BN MM7HQS
FK10 2DT GM8FMR
FK10 2EG GM0UGH
FK10 2ER 2M0BCL
FK10 2ER MM0GKU
FK10 2JG GM0LWD
FK10 2JU GM4WQH
FK10 2RX GM4ARJ
FK10 2TT 2M0WSK
FK10 2TT MM0WSK
FK10 3AN GM0IIO
FK10 3AW GM0FRC
FK10 3AW GM8GAX
FK10 3BG GM8WGU
FK10 3DZ GM4WGC
FK10 3FG MM6HIA
FK10 3JN GM8JJN
FK10 3PD GM0PEI
FK10 3RE MM6JNF
FK10 4LF MM7PSG
FK102PA 2M0RMD
FK102PA MM0RMD
FK102PA MM6RAV

FK11 7DG GM4FXL
FK12 5AL GM4HUX
FK12 5BL 2M0WDP
FK12 5BL MM6DWP
FK12 5DS GM0NAQ
FK12 5HU GM1PST
FK12 5JU GM1BVT
FK12 5NN GM4XHV
FK13 6HB 2M0FXL
FK13 6HB 2M0KTL
FK13 6HB MM0KTL
FK13 6HF GM0TTY
FK13 6HT GM8CIF
FK13 6JA MM6PRZ
FK13 6LF MM7NCD
FK13 6NT MM0GNX
FK136PU 2M0FKF
FK136PU MM6DVM
FK13EL MM6JQF
FK14 7AE MM0MSH
FK14 7BA MM0GJC
FK14 7BH MM6TWZ
FK14 7HP GM4WBU
FK14 7JZ GM1FTG
FK14 7NT MM6FKD
FK14 7PT MM0PTX
FK15 0DU 2M0SWM
FK15 0DU MM7ETI
FK15 0EB GM3YTS
FK15 0EB GM5CX
FK15 0EB GM8VL
FK15 0HQ GM1SRR
FK15 0JS GM0DUX
FK15 9AR 2M1IGQ
FK15 9AT 2M0PXH
FK15 9AT MM0AXR
FK15 9AT MM3ZNN
FK15 9ES GM6UCN
FK15 9HB GM0GRL
FK15 9JL GM4VZY
FK15 9NA GM0GMD
FK15 9RA GM1RIG
FK17 8LA GM8SAP
FK2 0BJ GM7GLJ
FK2 0EJ MM3KUU
FK2 0FN 2M0DKX
FK2 0FN MM0PRB
FK2 0HB MM0VWR
FK2 0LP MM0DSM
FK2 0LP MM3DEC
FK2 0LY MM3OIX
FK2 0NB MM6LRX
FK2 0QW MM3ZQP
FK2 0RE MM0MPA
FK2 0SU GM4MIM
FK2 0TF GM0KLO
FK2 0TJ GM6KDD
FK2 0UP GM7DZK
FK2 0UX GM3JJQ
FK2 0XT 2M0BKL
FK2 0XT MM3ZXL
FK2 7AY 2M1CJL
FK2 7AY GM0DEM
FK2 7BJ MM3YQP
FK2 7BQ MM6DBT
FK2 7LD MM6RDM
FK2 7LF MM3ZSF
FK2 7NH MM6KER
FK2 7TQ MM7MBB
FK2 8AN MM3KAQ
FK2 8DX GM0FTG
FK2 8EE GM0ZAM
FK2 8GD GM7UJO
FK2 8GE GM8VYZ
FK2 8JP GM1IEL
FK2 8LN MM7DGS
FK2 8NP GM0HZI
FK2 8PP 2M0TGD
FK2 8PP MM0OKG
FK2 8PP MM6TGD
FK2 9JR 2M0TXR
FK2 9JR MM6TRX
FK2 9QQ GM4XQJ
FK2 9SU MM7BJU
FK2 9SZ MM7BWZ
FK2 9UF GM0PGD
FK2 9UT GM0NJL
FK2 9YG MM6OEA
FK20 8RQ GM0EWW
FK27UF MM7FTH
FK3 0BH GM7NAA
FK3 0DE GM3IWX
FK3 0JJ 2M0OAB
FK3 0JJ MM0PSM
FK3 8LZ 2M0JLS
FK3 8LZ MM6MCT
FK3 8NT MM0CNV
FK3 8NZ GM7GTX
FK3 8TG 2M1ANY
FK3 8YG MM6HHJ
FK3 9AT GM3YKA
FK3 9JD GM6JWH
FK3 9JD MM0GXU
FK3 9JN 2M0WWX
FK39BQ MM7FPD
FK39EQ MM3TEQ
FK4 1AR MM0OZY
FK4 1DG 2M0HJG
FK4 1DG MM7JFB
FK4 1GD GM6MHC
FK4 1HE GM0AEY
FK4 1JY MM6SZK
FK4 1RY GM0IYA
FK4 1TP MM1BFE
FK4 2DL MM7MFR
FK5 3JU GM3UCH
FK5 3LH GM4EJX
FK5 4DX MM0INT
FK5 4DX MM0NDX
FK5 4LT GM0KMJ
FK5 4QF MM0SAJ
FK5 4QF MM7CFC
FK5 4UA 2M0XRV
FK5 4UA MM7XRR
FK5 4UF GM0OKJ
FK6 5AG 2M0VUV
FK6 5AG MM0VUV
FK6 5AG MM6VUV
FK6 5EG GM4XRP
FK6 5HX GM0AZC
FK6 5JF GM6PNJ
FK6 5LP 2M0LNR
FK6 5LP MM6TSC
FK6 6PN GM0DKK
FK6 6QJ 2M0VVS
FK6 6QJ MM0IAL
FK6 6QJ MM3VVS
FK6 6QJ MM6HXY
FK6 6RB GM4NTX
FK7 0DH 2M0OXX
FK7 0DH MM0OXX
FK7 0DH MM6OXX
FK7 0DT GM0VRP
FK7 0HJ GM0EGI
FK7 0HJ MM0AIK
FK7 0LH GM4YMD
FK7 0LJ GM4VGR
FK7 0NP GM0WUP
FK7 0NP GM7VDL
FK7 0NP MM3WUP
FK7 0NQ 2M0BYT
FK7 0NQ GM0WUR
FK7 0NQ MM3RGZ
FK7 0QP GM4UYE
FK7 0QP GM6NX
FK7 7BS MM6YRO
FK7 7QZ 2M0GEK
FK7 7QZ MM6CJZ
FK7 7UL GM0NWI
FK7 7XA GM0FNE
FK7 8NQ GM8MOI
FK7 8PX GM1BSG
FK7 8PX MM6HLZ
FK7 8UF MM3RCR
FK7 9BY 2M0OCC
FK7 9BY GM7EHN
FK7 9LP GM1SBD
FK8 1BZ MM7RKJ
FK8 1EP GM0RMT
FK8 1SA MM0JMK
FK8 1TA 2M0FXX
FK8 1TA MM6AJI
FK8 2JE MM3RHA
FK8 2JX GM0KQB
FK8 2LT GM0MXP
FK8 3JY GM0HZO
FK8 3PD MM0PDD
FK8 3SZ MM0WCT
FK9 4HH MM6PEA
FK9 4PS GM4PJR
FK9 4SA GM0SVS
FK9 4SA GM1XJE
FK9 4UJ MM7WMG
FK9 5BE MM0BSM
FK9 5EU GM4PYJ
FK9 5LR GM4RCN
FK9 5LR GM4RTN
FK9 5LS GM0FSV
FK9 5PX MM1CCR
FY1 2AQ G1UGG
FY1 2BN M3SBT
FY1 2BN M6KHF
FY1 2EG M3LDT
FY1 2JS 2E0MHE
FY1 2JS G3CTQ
FY1 2JS M3SZQ
FY1 2NY M6JIT
FY1 2QJ 2E0EEI
FY1 2QJ M0IHJ
FY1 2QJ M6HVH
FY1 2QL G7HEJ
FY1 2RG 2E0EKS
FY1 2RG M6JNI
FY1 2RW G1YMR
FY1 3NN G1TIH
FY1 3QB G1HSO
FY1 3RB G1MET
FY1 4JA M7PJY
FY1 4JE 2E0FXA
FY1 4JE M7APF
FY1 4JQ 2E0DAX
FY1 4JQ M6BIF
FY1 4QP G0UHS
FY1 5BZ M6IZO
FY1 5HP G4GOR
FY1 5JB M6YAJ
FY1 5JB M7BCW
FY1 5NA G1OLM
FY1 6DW M3CLT
FY1 6DZ G7FIK
FY1 6EZ M6OUL
FY1 6JP G4EYX
FY1 6LP G0NBW
FY1 6LY M1EOZ
FY1 6NP M6BJC
FY1 6NW G4ZPN
FY1 6PR M6GQA
FY1 6RR M6VVN
FY12PW G1NGR
FY13NG M7PDT
FY16QN M7PBN
FY2 0AW G7MXN
FY2 0DZ 2E0UZK
FY2 0DZ M6UZK
FY2 0EB 2E0DYY
FY2 0EB M0LJH
FY2 0EB M0UNM
FY2 0EB M6KQM
FY2 0EN G1BFG
FY2 0LW M0TUT
FY2 0NQ G0ETV
FY2 0PD M3BAA
FY2 0PW M7RJB
FY2 0RJ M6XTF
FY2 0SD G1ZRE
FY2 0SR G6WWV
FY2 0TR G8ATG
FY2 0TU G4WYF
FY2 0UG M1ALF
FY2 0XH 2E1DZL
FY2 0XH M3RKN
FY2 9AQ G4TUM
FY2 9DJ G1KJQ
FY2 9EN M6XAS
FY2 9EP M6HXO
FY2 9EQ G6XNU
FY2 9LQ 2E0OYG
FY2 9NU M3XUW
FY2 9QT G3LLE
FY2 9QW G1NQN
FY2 9TB G4RRJ
FY2 9TX M6HUZ
FY20WL M7VAT
FY29BZ 2E0MOW
FY29BZ M6NZN
FY3 0AL G3UIT
FY3 0BU G3WPT
FY3 7BQ M6DFU
FY3 7BQ M6FTU
FY3 7LQ 2E0SGS
FY3 7LQ M6KHP
FY3 7PP M6IPF
FY3 7PW G7EPY
FY3 7QS M3KDY
FY3 7RR M3MEW
FY3 7RS G1CFE
FY3 7UB M3VKF
FY3 7UE M6BQP
FY3 7UJ G1CWQ
FY3 7UJ M3UUO
FY3 8DP G4OMS
FY3 8HS G6FCI
FY3 8JJ M6XMA
FY3 8JL M7LDW
FY3 8LX 2E0MLL
FY3 8LX M6CLL
FY3 8PF 2E0EZY
FY3 9TN G4VAL
FY3 9UE M6HVD
FY4 1EH M6IEJ
FY4 1JJ M6IVI
FY4 1JP M3MGZ
FY4 1LB G4MPT
FY4 1PW G0LLX
FY4 1QG G1EEA
FY4 1QR G4AKC
FY4 1QS G4EZM
FY4 1SX G6IRU
FY4 2BT G4XPI
FY4 2BW G4JAJ
FY4 2DH G7PTM
FY4 2NF M1CDX
FY4 3AX 2E0TZE
FY4 3AX M7FJS
FY4 3JQ G6IGV
FY4 3LH G6BXO
FY4 3LY M3RPR
FY4 3NH G0FYD
FY4 3PQ M6DSR
FY4 3QA G6BGH
FY4 3QA M3BPL
FY4 3QQ G6VEZ
FY4 4AB G0EUP
FY4 4AF M6MOK
FY4 4AR 2E1DAP
FY4 4AX G1NQO
FY4 4AZ M6XNU
FY4 4DN 2E0YJB
FY4 4DN M7WJB
FY4 4JJ M7BKO
FY4 4LQ M3DCI
FY4 4NA G6GJD
FY4 4NU M6IZT
FY4 4PY M0JEB
FY4 4QS G4YVQ
FY4 4QS M0BOL
FY4 4RL M1IMC
FY4 4RQ G4XQF
FY4 4RR M6IZU
FY4 4UD G4RXK
FY4 4UZ G6XNI
FY4 4YF M1BUG
FY4 5BT G4SZB
FY4 5BX G6AOS
FY4 5DS G0BGR
FY4 5DW G6IFQ
FY4 5EZ G0SVQ
FY4 5HG G1ZME
FY4 5HT G4DJY
FY4 5RE G6VRF
FY4 5RF G0HZL
FY42BH M6JYF
FY44PS M7IAP
FY5 1BG G6PGV
FY5 1BU G4BCT
FY5 1DL G3RFH
FY5 1EY M6KSA
FY5 1HQ M3DEH
FY5 1JD G3HKA
FY5 1JW G4KQC
FY5 1LA G6JZN
FY5 1LP G0PIL
FY5 1LP G7URJ
FY5 1PR G7SUV
FY5 1SP 2E0RWT
FY5 1SU G3ZGZ
FY5 2AR 2E0UHJ
FY5 2AR M3UHJ
FY5 2DU M7JCD
FY5 2HX M3XGY
FY5 2JS G1PPK
FY5 2LG G1NFE
FY5 2LL G3WBB
FY5 2NW G6VCR
FY5 2PH G0NXR
FY5 2QX M3SBP
FY5 2RT G4ANE
FY5 2UG G4ATH
FY5 2UG G4FRK
FY5 2UG G6GMW
FY5 2ZA G0MWM
FY5 3AJ M7PCB
FY5 3DH 2E0EDX
FY5 3DQ M3OYJ
FY5 3EY G6SRJ
FY5 3HZ G8ZSM
FY5 3QA M3YVW
FY5 3QB G0GER
FY5 3QB G0LSA
FY5 3QB G7FCR
FY5 3QH M6LPY
FY5 3QR G1DEP
FY5 3QR M1AMZ
FY5 3QR M6RKW
FY5 3RQ G4RDY
FY5 3SE G1EPL
FY5 3UH G6HMX
FY5 4DL G0VVG
FY5 4DR 2E0MKH
FY5 4DR M6KHB
FY5 4NB G3SNH
FY5 4NG M3TWV
FY5 4PL 2E0CQJ
FY5 4PL 2E0TOG
FY5 4PL M0NAG
FY5 4PL M6FAC
FY5 4PL M6NAL
FY5 4PZ M6HTW
FY5 5AE G8MKQ
FY5 5AY G4UHI
FY5 5BH M6CDH
FY5 5DR M3NXO
FY5 5EA M0EAL
FY5 5HH 2E1ADR
FY5 5NS M3NDF
FY6 0AD M6HGN
FY6 0AJ 2E0CVG
FY6 0BS 2E0MYT
FY6 0BS M6CHC
FY6 0BT G7IED
FY6 0BT G7PDR
FY6 0DD G0JZT
FY6 0DF G0AJW
FY6 0DF G4RND
FY6 0EH 2E0EEW
FY6 0EH M6GQG
FY6 0EW G7NYD
FY6 0HA G0HIU
FY6 0HG 2E0DQB
FY6 0HG M6FKQ
FY6 0LD G6KHD
FY6 0LZ M0HOI
FY6 0PB G7UWE
FY6 0PW G1OCK
FY6 0QE G4DRX
FY6 0QE M3YRX
FY6 0QW M6FUE
FY6 0RS G6NJE
FY6 7BL G6NLQ
FY6 7DB M3TVJ
FY6 7DJ G4KHJ
FY6 7DY M3UUN
FY6 7EG G3UIS
FY6 7EW 2E1EIU
FY6 7EW M3UGA
FY6 7FY 2E0SEP
FY6 7FY M6FUQ
FY6 7HJ M6SLJ
FY6 7NE G0LQN
FY6 7PB G6KOE
FY6 7PE M0HZP
FY6 7PN G3BDT
FY6 7PW G3XCE
FY6 7QX G4TMA
FY6 7QY G0RNA
FY6 7RD 2E0OMG
FY6 7RD M3XZN
FY6 7SP M7JLW
FY6 7SR G6MQG
FY6 7TL M6YFZ
FY6 7UB G8KBH
FY6 7UB G8YOK
FY6 7UQ M3NCE
FY6 7UX G4LOR
FY6 7UX G4PNI
FY6 7UX M3PNI
FY6 8AD G3MCE
FY6 8AD M6TGP
FY6 8BL G0BRX
FY6 8BL M6SJJ
FY6 8BZ G7FED
FY6 8EB G3VDO
FY6 8ED G4BVW
FY6 8ED G4VYL
FY6 8EH M3SJM
FY6 9AP G4XKR
FY6 9AW G0RHF
FY6 9EA G6UDA
FY6 9EE G6AEK
FY6 9EH 2E0TDH
FY6 9EH M6TDP
FY7 6AX M7FWD
FY7 6BG M0BAL
FY7 6BT M7LAA
FY7 6DA 2E0FFS
FY7 6DA G4NIU
FY7 6EE M7RAV
FY7 6FG M0GNJ
FY7 6FG M0NAU
FY7 6JR G8IMJ
FY7 6LJ M0JME
FY7 6QA M3AYT
FY7 6QJ M3ZUA
FY7 6UE M6MRG
FY7 7AR G8YPY
FY7 7DU G3SSN
FY7 7EA G0CUX
FY7 7HA 2E1FUJ
FY7 7HA G0LRK
FY7 7HA G0NCY
FY7 7HH G4TTG
FY7 7HJ M0HMU
FY7 7HJ M0JFE
FY7 7HW M3FPS
FY7 7HY M0BLS
FY7 7JF 2E0XHG
FY7 7JF M3XHG
FY7 7JH M6EFR
FY7 7LF M0DLB
FY7 7LF M3MRN
FY7 7LJ M3IPY
FY7 7LJ M3TCY
FY7 7LY M3UJE
FY7 7LY M6TKU
FY7 7NH G0IAE
FY7 7PY M3WWH
FY7 8BH 2E0KNF
FY7 8BQ G4GYF
FY7 8DD M7RYO
FY7 8EG G4FWM
FY7 8EG G8RDP
FY7 8EG M3RAE
FY7 8NW M6BHE
FY77HZ M6PQW
FY78DX 2E0GRZ
FY78DX M0RGV
FY78DX M6OGX
FY8 1DW G4JFV
FY8 1ET G8LCZ
FY8 1HW G2HFP
FY8 1HZ G7CWI
FY8 1LZ G1CDQ
FY8 1PQ G1TJR
FY8 1PU G4OPT
FY8 1TN M3RZG
FY8 2BL M7DJC
FY8 2DA 2E0BKZ
FY8 2DA M0GMN
FY8 2DA M3UGX
FY8 2HE 2E1DZV
FY8 2HE M3JMK
FY8 2HW G0HLB
FY8 2JF G7HJQ
FY8 2LT G4GFE
FY8 2QN G3WBI
FY8 2QX G1TBN
FY8 2SG G7DIE
FY8 2SG M3SGS
FY8 3AG G8SNR
FY8 3BG G0AIN
FY8 3DB G6KTO
FY8 3HA 2E0TSI
FY8 3HA M6TSI
FY8 3HY G7FDS
FY8 3NH M6WIM
FY8 3NP M3ZOI
FY8 3NP M6SBH
FY8 3PS G8SND
FY8 3QF G0JBS
FY8 3RZ M6FRZ
FY8 3RZ M6UAE
FY8 3SL G3IEJ
FY8 3TL M6HWL
FY8 4AR G8LMF
FY8 4BJ G4NXW
FY8 4PF M1FHJ
FY8 4PZ M0DKR
FY8 4QR G4WLE
FY8 4UE G7CBY
FY8 4UE G7CBZ
FY8 5JH M6AVD
FY8 5PY M7XBX
G1 1HD MM6BPY
G1 1LH MM0IBO
G1 1NY MM3MMO
G1 1NY MM7AJJ
G11 5AP GM0EDR
G11 5AX MM7ESG
G11 5EA GM8TXC
G11 5PE 2M0HZO
G11 6PB MM7POL
G11 6PP 2M0YYU
G11 7LG GM4ACM
G11 7PP GM7KVU
G11 7QX 2M0TLE
G12 0AS 2M0CTN
G12 0DL 2M0VEY
G12 0DL MM7TXY
G12 0ER 2M0SCG
G12 0LE GM8DKG
G12 0PB GM0ILQ
G12 8TL GM3RUP
G12 9DZ GM7DLY
G12 9LQ MM0MHP
G129SB MM7SRR
G13 1DQ MM6SCG
G13 1JH MM6JUX
G13 1UN MM7EPD
G13 2LA GM0EFH
G13 2RJ GM0EAH
G13 2XX 2M0GFT
G13 2XX MM0NNA
G13 2YQ MM6KVI
G13 2YQ MM6RHQ
G13 3AQ GM0GYT
G13 3EP MM0JOX
G13 3JN MM6NWH
G13 3TD MM7FQP
G13 3TT MM6YUJ
G13 3YE MM6CIJ
G13 4HL GM0FHJ
G13 4QE MM6CTL
G13 4QN GM4NWK
G14 0EU 2M0VJS
G14 0EU MM7JRS
G15 6AU 2M0RCD
G15 6AU MM0HCO
G15 6AU MM6AOH
G15 7JA MM6OBU
G15 7QE GM7NZI
G2 3AU MM6AXT
G2 3JW 2M0SGO
G2 3JW MM7MMO
G20 0HJ 2M0UTH
G20 0HJ MM3UTH
G20 0HT 2M0NTY
G20 6AG MM0CXB
G20 6AG MM0PYS
G20 6XZ GM0FWY
G20 7YQ 2M0TTF
G20 7YQ MM0GUE
G20 7YQ MM6AYR
G20 8LF GM0USI
G20 8LF GM4FVQ
G20 9JQ GM0KVD
G21 2AP MM3RDP
G21 2DE 2M0WFN
G21 2DE MM6AHJ
G21 2QF MM3UHK
G21 3HY GM0KUJ
G21 3JS MM6AKZ
G21 4XY GM4ENN
G212JB MM7BQG
G22 6DN MM7ITA
G22 7RA MM3SNB
G22 7RG 2M0GCF
G22 7RG MM0GCF
G22 7RG MM0KHG
G23 5DJ GM0JHE
G3 6BS MM7PMB
G3 6QN GM3VBT
G3 8QP MM7UVI
G31 3LJ GM0KTO
G31 4RT MM0TBH
G31 5RP GM3PUY
G32 0LP MM6LBS
G32 0NF 2M0VUS
G32 0NF MM6VUS
G32 6TA GM3SPT
G32 7SQ MM6WMK
G32 7XL MM6MZE
G32 9BW GM3JKC
G33 1BN MM7MCM
G33 1BU MM0SIL
G33 1LR MM6JBN
G33 1RS GM7GXI
G33 2DD GM4PCT
G33 2DP MM3VZI
G33 2LD GM7BRL
G33 2SD MM6JSN
G33 5HU GM1POA
G33 5JJ GM4EWL
G33 5JJ GM8DRA
G33 6NJ MM0XEA
G33 6NU 2M0FYF
G34 9AR MM0JRF
G4 0PH 2M1EZA
G4 0PS 2M0IYI
G4 0PS MM7FQW
G4 0TQ MM3LFI
G4 0WX MM7HOS
G4 9AR MM7CNQ
G4 9JX GM6AQR
G4 9YQ MM7HFB
G40 3LE GM8LYO
G41 2AF GM3DIN
G41 3BS GM8PIV
G41 3HX MM7RCV
G41 4PW GM0PAC
G41 4QN GM0OAA
G41 5HH MM7LYS
G41 5RL MM0TWX
G41 5SD MM6PKC
G42 0DW MM6FBU
G42 8BZ MM6HFZ
G42 8DR GM3WUX
G42 8DW 2M0FRA
G42 8DW MM6CFH
G42 8RT 2M0IEC
G42 8RT MM3IEC
G42 9PZ MM6PND
G43 1BW 2M0HYZ
G43 1BW MM3UTU
G43 2BW MM3SQM
G43 2DX MM6IBQ
G43 2DY MM0GYD
G43 2SY MM0GUW
G43 2YL MM3FYF
G44 3NQ MM6HIZ
G44 3XG MM6KHM
G44 4PA GM0WRR
G44 4TJ MM1BJT
G44 5JU 2M0MTO
G44 5JU MM0MTO
G44 5PD MM7UTR
G44 5PF 2M0RRO
G44 5PF GM7WED
G44 5PF MM3LKV
G45 9QD 2M0PBC
G46 6DB GM3NGW
G46 6LA GM6FIK
G46 6QB GM4IYZ
G46 7AE GM3COB
G46 7LU GM0WRH
G46 8AB GM0LIM
G46 8LP MM6VGS
G46 8RS MM7SWM
G46 8UR GM7UTD
G468NA MM6LGS
G5 8EU GM1BVA
G51 1BW 2M1SCO
G51 1HG GM3ZDH
G52 3HA GM4BGS
G52 3HA MM6AHY
G52 3JY GM1JNC
G52 3PU MM0CDK
G52 4HJ MM3WHS
G53 5UY MM7CNX
G53 6LR MM6CCY
G53 6NQ 2M0AZW
G53 6NR MM6OCP
G53 6QW GM4FFF
G53 6QW MM0DGR
G53 6QW MM0EPC
G53 6QW MM0KPV
G53 6QW MM0KQC
G53 7UP 2M0SAX
G53 7UP MM0SAX
G58DA MM7UMU
G60 5HR MM0TCQ
G60 5JJ MM6MCX
G60 5LE GM3YCB
G60 5LE GM6AQB
G60 5LJ MM0CZM
G61 1EJ GM0GCO
G61 1RE GM3RQQ
G61 2AT MM7NLW
G61 2LT GM7SCJ
G61 2PR MM6KQH
G61 3EF MM7ORE
G61 3HD 2M1HFE
G61 3HD 2M1MIC
G61 3HD 2M1SJB
G61 3HD GM3VTB
G61 3HD GM4VTB
G61 3HD MM1VTB
G61 3HD MM3CVB
G61 3HQ GM0ATL
G61 3JX GM3GTQ
G61 3LG GM0UPE
G61 3LQ MM7HIR
G61 3LX GM0NUQ
G61 4EH GM4AWB
G62 7DN MM7EYO
G62 7DT GM1XEB
G62 7HA GM4PCT
G62 7HA MM0SUS
G62 7JP 2M0VXL
G62 7JP MM3RXM
G62 7JP MM3VXL
G62 7PT MM7ABY
G62 7RA GM4KAV
G62 7RL GM4JRF
G62 7RR GM3MAS
G62 8BE GM8ZKF
G62 8HD MM0GDL
G62 8HX MM0HLQ
G63 0EX GM3ODP
G63 0NP GM0GIB
G63 0PF MM1FHO
G63 0XE MM3AYS
G63 0YB MM7SXM
G64 1QW MM1PTT
G64 2EU GM8ZCS
G64 2HP GM0GMO
G64 2NS GM3HOM
G64 3AD GM8ZAK
G64 3AQ MM6BSK
G65 0EE GM0BRJ
G65 0EX 2M0RCZ
G65 0EX MM0POD
G65 0EX MM3RCZ
G65 0RY 2M0IIH
G65 0RY GM8IIH
G65 0RY MM0OJJ
G65 0RY MM7IIH
G65 9EA MM3SAK
G65 9EJ MM1FZR
G65 9HF MM0GDX
G65 9HJ MM0GFA
G65 9HJ MM1DHU
G65 9HQ MM6JJQ
G65 9UL GM0BUE
G66 1AX GM3GRG
G66 2BD GM4LYV
G66 2BH 2M1EKI
G66 2BH MM3XOK
G66 2QE 2M0OMS
G66 2QE MM0OMS
G66 3AS GM3WYL
G66 3BW GM0VEK
G66 3HJ MM3HYG
G66 3JL GM1BTL
G66 3JW MM3RHT
G66 3NX GM7JDS
G66 3PH MM7FGL
G66 3PL MM6IUR
G66 3RY GM8WWY
G66 3WF 2M0VOB
G66 3WF MM7PLG
G66 4BF GM7SPA
G66 4DF MM1AHL
G66 4DF MM3DIZ
G66 4EL GM6TIB
G66 4EN GM4RUP
G66 4RE GM3SER
G66 4SR MM1BXF
G66 5HS 2M0LFS
G66 5HS MM0TMZ
G66 5HS MM6LFS
G66 5NG GM0BEL
G66 5NG GM4ZRX
G66 7EP GM7FLG
G66 7JT MM7DGC
G66 8AY GM0HNV
G66 8AY GM0PEX
G66 8AY GM3JMM
G66 8BB GM0HJU
G66 8BB MM6HMB
G66 8ER GM7GBD
G66 8ET GM0HIM
G66 8ET GM0RED
G66 8ET GM7HIR
G66 8HG GM0ATA
G66 8HR 2M0LWD
G66 8HR GM5LWD
G66 8HR MM7GGL
G661ND MM1DEE
G67 1JE MM3VBF
G67 1LR MM6ATR
G67 2BL GM1RRJ
G67 2DA GM7FIS
G67 2DW MM0MIJ
G67 2JS 2M0IVS
G67 2JS MM6GYR
G67 2LU MM7SCO
G67 2NP 2M0UGL
G67 2NP MM0SEY
G67 2NP MM3UGL
G67 2NT MM1RIK
G67 2PX MM7BYQ
G67 2QW 2M0VKO
G67 2QW GM6FLL
G67 2QW MM3VKO
G67 3AX 2M0LSG
G67 3LU GM4XLU
G67 4AJ 2M0BYI
G67 4AJ MM0MYL
G67 4AJ MM3YUW
G67 4AJ MM3YUX
G67 4AW MM6HQK
G67 4ES MM0GXQ
G68 0AF GM8NFT
G68 0EP MM3BSC
G68 0JB GM0PLH
G68 0JB MM0LMC
G68 0JR MM0HUF
G68 9BJ GM4EIW
G68 9DZ GM1VFR
G68 9DZ GM7JUX
G68 9EG GM7MZZ
G68 9ER MM1DEA
G68 9JD MM6DBN
G68 9JU MM7KXI
G68 9NG GM1TFF
G68 9NT MM1BJZ
G68 9NW GM4VWV
G69 0JW GM1MRY
G69 0LZ MM0ODI
G69 0PH MM3CBO
G69 6JB MM6FTG
G69 6LQ GM4RPE
G69 6ND MM3DOP
G69 6NE MM7SJF
G69 6QP MM0DNX
G69 6TG GM4ELV
G69 7BH GM8YGI
G69 7HW GM3SAN
G69 7HY MM7DZS
G69 7QZ MM6KCB
G69 8AG MM3DQV
G69 8AG MM3DQX
G69 8EG MM6KSU
G69 8LH MM7CMM
G69 9FA MM0OWL
G69 9FB 2M0WHX
G69 9FB GM5DAV
G69 9FB MM6WHX
G69 9FB MM7VMC
G69 9ND GM0LBR
G71 6ED GM7FGH
G71 6LG 2M0MBE
G71 6LG MM3VPK
G71 7BQ GM0IMW
G71 7ET GM7VYR
G71 7ET MM3VYR
G71 8AR GM1VBE
G71 8BP MM0SSG
G72 0RQ GM0XFK
G72 0TR MM0GPZ
G72 0TZ MM7WSI
G72 7EF MM6RXJ
G72 7GS MM6YND
G72 7GS MM6ZDG
G72 7NN MM6JCL
G72 7SQ MM3UMY
G72 7SQ MM3UMZ
G72 7TP GM7SFE
G72 7UW MM6GGE
G72 8DH MM6VLG
G72 8NL GM4NDV
G72 8RD GM4TTC
G72 8UE 2M0WWC
G72 8UE MM6WWS
G72 9DT GM7WBP
G72 9NX GM0AZU
G72 9RJ MM7GKD
G72 9UP MM3XNP
G73 2ER MM7GMS
G73 3BG GM4REF
G73 3EN GM0LBN
G73 3EU MM3WJZ
G73 3QP GM7AYW
G73 3QP GM8SNB
G73 4AE GM0JVV
G73 4DX MM3CKP
G73 4EA 2M0DKV
G73 4EA MM6EGC
G73 4JP 2M0FCT
G73 4JP MM6VSM
G73 5RG MM0AYE
G74 1BQ GM4DLU
G74 1DR MM0BMA
G74 2HU GM0TPI
G74 3DL MM0ZXI
G74 3DN MM6CHY
G74 3HZ GM1BNA
G74 4HZ 2M0FZW
G74 4HZ MM0JJU
G74 4HZ MM7AWL
G74 4TZ GM7LOK
G75 0BX MM3YVU
G75 0PY 2M0HRN
G75 0PY MM7RIK
G75 0RU MM6PVY
G75 8DL 2M0YOY
G75 8DL MM6YOY
G75 8LH 2M0EEQ
G75 8LH MM6EWR
G75 8RZ MM1EGS
G75 8TN GM1SQZ
G75 8XT 2M0HUJ
G75 8XT GM7GDE
G75 8XT MM0YEK
G75 8XT MM7MMB
G75 8YG GM0UET
G75 8YG MM3CLA
G75 8YG MM3UET
G75 9FS GM8NET
G75 9LB MM6GOR
G75 9NW MM3UQT
G76 0BP 2M0IRG
G76 0BP MM0KFP
G76 0BP MM7IRG
G76 0DG GM1PUR
G76 0EU GM0VIY
G76 0EU MM3PEY
G76 0HA GM0PHG
G76 7AD GM6ANZ
G76 7DU GM4VBE
G76 7HG GM0SEP
G76 7HG GM0UIG
G76 7HG GM4SRL
G76 7JL MM6HZW
G76 7PL GM4XGY
G76 7XT GM0IVQ
G76 7XT GM6FOT
G76 9AL 2M0IUJ
G76 9AL GM5TOA
G76 9BN GM0DVO
G77 5LS GM4KNU
G77 5PP GM1NEW
G77 5QJ GM7SKB
G77 5TQ GM4KLO
G77 6EA MM6MML
G77 6FP 2M0GQH
G77 6FP GM6PZY
G77 6FP MM7HUE
G77 6FP MM7JGY
G77 6HP GM3NEQ
G77 6LQ GM1NET
G77 6LQ GM6OQN
G77 6PZ MM6DVR
G77 6UG GM4AGU
G77 6UG GM5AUG
G77 6UG GM5KCC
G77 6UJ 2M1EDM
G77 6UJ GM0UKZ
G77 6UZ MM0HFU
G77 6XX GM3KXQ
G77 6YG GM6LIN
G78 1DD MM7GNX
G78 2DH GM0GMI
G78 2DH GM1TFZ
G78 2LR MM6DAQ
G78 3DY 2M0JFD
G78 3DY MM7SNP
G78 3JA GM6JIL
G78 3JX GM3AEI
G78 3QP GM4OSV
G78 4AA GM0DJG
G78 4AA GM3USL
G78 4AG MM3USV
G81 1DP 2M0LZY
G81 1DP MM7DFC
G81 1EH MM3OYL
G81 1ER GM4BLO
G81 1RF MM0HJC
G81 1RF MM0IXJ
G81 1RF MM0YZE
G81 2LL GM0GYM
G81 2PH 2M0NIT
G81 2PH MM6TNO
G81 2RA MM7KEL
G81 2ST GM4PLI
G81 2YB MM6IYA
G81 3EH MM6ZYZ
G81 3LE GM5MAJ
G81 3LH GM4KHE
G81 3NF MM6RPN
G81 3RR GM1BNS
G81 4HH MM6AQM
G81 4LN GM7OMU
G81 4PG MM7DCD
G81 4QD MM6ECO
G81 5BS MM3YTB
G81 5EA 2M0UFO
G81 5EG GM4ZMK
G81 5HJ 2M1EDT
G81 5PD GM6KTP
G81 6AW GM3KCY
G81 6HH MM6IXO
G81 6HX MM3TAM
G81 6LW GM3RGU
G81 6NR GM3ITN
G81 6PX GM4AGG
G81 6PX GM4TOQ
G813AB 2M0PIJ
G813AB MM6PIJ
G813EP MM6IWL
G814AW MM6NWX
G82 1HP MM7IAB
G82 2BN MM6NUP
G82 2BN MM6YUP
G82 2JA MM3ODV
G82 2PF MM6DHF
G82 2PN GM7NHU
G82 2QB 2M0IXT
G82 2QB MM0IXT
G82 2RZ MM0CXZ
G82 3ER GM0KZX
G82 3JU 2M0MSB
G82 3JU MM3MGK
G82 3NU 2M1HSG
G82 3PB MM1JAC
G82 3QW GM8FFH
G82 4BG MM5AON
G82 4JL MM0LGT
G82 4NH MM6HAH
G82 4QA MM3FMB

Postcode	Callsign
G82 5HD	GM1UIR
G82 5HD	MM3DMZ
G82 5HP	GM1MLW
G83 0BZ	2M0JBZ
G83 0BZ	MM6JCU
G83 0DW	GM7MYF
G83 0LL	GM6JWF
G83 0PU	GM4UPN
G83 0RJ	MM3OFE
G83 0RZ	GM4LGM
G83 0TX	MM7CZD
G83 0UF	MM1DMU
G83 0US	MM1LJB
G83 7DB	2M0OVV
G83 7DB	MM3OVV
G83 8BL	2M0LNF
G83 8EX	MM6EOE
G83 8RP	GM0RTY
G83 8RU	MM3OCY
G83 8SD	MM0HDA
G83 9BJ	MM3XDP
G83 9BU	2M0BIL
G83 9BU	MM0ELF
G83 9BU	MM3FCG
G83 9DB	MM3XKH
G83 9DB	MM6LAH
G83 9EB	MM7OWN
G83 9JR	2M0JUM
G83 9JR	MM7JMP
G83 9LE	MM6SWC
G83 9LG	2M0TXY
G83 9LG	MM6FCA
G83 9PW	MM7BMY
G83 9QL	GM7OBM
G83 9QT	MM3DDQ
G84 0DW	MM6BGV
G84 0EB	2M0VFV
G84 0EB	MM6SMY
G84 0EB	MM7BLP
G84 0HS	2M0ZFG
G84 0HS	MM0ZFG
G84 0HS	MM6ZFG
G84 0JN	MM0EAI
G84 0QD	GM1KWX
G84 0QR	GM1FTZ
G84 0QX	2M0CTI
G84 0QX	MM6CNV
G84 0QZ	GM4MFO
G84 0RL	2M1HKA
G84 0RN	MM6HOO
G84 7DN	MM7BYK
G84 7EE	MM6HEZ
G84 7PL	GM8VAM
G84 8JP	GM7OAF
G84 8NN	MM0GPL
G84 9DA	2M0BJU
G84 9DA	MM0GON
G84 9DA	MM3OZU
G84 9DN	GM4RJX
G84 9DX	MM3ECO
G84 9DX	MM3ERD
G84 9DX	MM3YOL
G84 9JD	GM1VYF
G84 9JZ	MM0JWH
G84 9LL	GM0SIM
G84 9QP	GM4ZWJ
G840EY	MM6TCM
GL1 1DH	G7UWP
GL1 2AR	M6BBS
GL1 2PB	G1ZSZ
GL1 2QZ	M3TYI
GL1 2RX	2E0LGW
GL1 3DE	2E0HBE
GL1 3QE	G2CIW
GL1 3QH	M3CGC
GL1 3QS	2E0SFQ
GL1 3QS	M0SPZ
GL1 3QS	M6SQF
GL1 4AR	M7CQK
GL1 5BY	G8VUO
GL1 5DE	M7AVN
GL1 5ER	G3RKH
GL1 5HL	2E0BNF
GL1 5HL	G1ZXC
GL1 5HL	G6GLO
GL1 5JU	M6CZD
GL1 5JZ	2E0THR
GL1 5JZ	M7THR
GL1 5QD	G7BPX
GL10 2DG	G1USZ
GL10 2DH	G7MLW
GL10 2PZ	G7MWC
GL10 2QH	G4CMY
GL10 3GA	2E0CMC
GL10 3GQ	2E0UDB
GL10 3GQ	M0UDB
GL10 3GQ	M7DAB
GL10 3HS	G4XWZ
GL10 3HX	G4EXF
GL10 3JN	G3TBF
GL10 3LD	G4EDY
GL10 3LD	G4YYR
GL10 3LD	G8ILN
GL10 3LJ	G0RUY
GL10 3NA	G0PDE
GL10 3NL	G4UBC
GL10 3PJ	M6AKT
GL10 3QW	G4SHB
GL10 3QW	M0JWQ
GL10 3RT	G4CRG
GL10 3TU	G4CIO
GL11 4AP	G6HKL
GL11 4AS	2E0KLD
GL11 4EP	M0LEP
GL11 4EW	G4JXC
GL11 4QB	2E0RBZ
GL11 4QB	M6RAZ
GL11 5AR	2E0HHA
GL11 5AR	M7FRE
GL11 5DA	G4KYI
GL11 5EL	G4VZR
GL11 5JQ	G8ZTM
GL11 5SG	G3ZKN
GL11 5SW	G1GDT
GL11 6AW	M7FDY
GL11 6HB	G4FQH
GL11 6HY	G4ETS
GL11 6JE	G0DQS
GL11 6LF	G0BRW
GL116JF	M7JWR
GL12 7BJ	G4VCQ
GL12 7HA	G4ARW
GL12 7LQ	G0SYF
GL12 7NP	M7XJW
GL12 7PD	G4YIC
GL12 7RH	G7FEQ
GL12 8AS	2E0EUW
GL12 8AS	G1HXT
GL12 8AS	M0JEA
GL12 8AS	M6EUW
GL12 8DA	G0MIG
GL12 8FF	M0NIC
GL12 8NB	G1USW
GL12 8SB	2E0UBM
GL12 8SB	M6UBM
GL12 8SF	M0IPZ
GL12 8SG	G7JWE
GL12 8TJ	G0CBK
GL12 8TN	G1VNL
GL13 9BU	G0NVX
GL13 9BU	G4HQX
GL13 9DF	G8ZHN
GL13 9LF	M6EAT
GL13 9NP	M6DZZ
GL13 9PL	G6JWO
GL13 9PY	G0UGR
GL13 9TE	M6JJV
GL13 9TE	M7SSB
GL13 9TG	G7OPB
GL13 9TQ	G7FPU
GL13 9TQ	M6DCL
GL13 9UA	M3YXF
GL13 9UT	G0IHC
GL14 1BW	G3XBX
GL14 1NB	G6GUC
GL14 1QX	G0XAE
GL14 2BH	G4KRJ
GL14 2BH	G6FGA
GL14 2DE	G8PGH
GL14 2DW	G0DZA
GL14 2DW	G0WBS
GL14 2EB	G0PBB
GL14 2EB	G0SNB
GL14 2EB	G7KXN
GL14 2EF	G0ODN
GL14 2EF	G6CHT
GL14 2HJ	M5AFX
GL14 2HZ	2E0KZS
GL14 2HZ	M7CIT
GL14 2QU	G4THC
GL14 3AX	G0HUX
GL14 3BL	G1AXW
GL14 3DZ	G0DRE
GL15 4AJ	M3DCL
GL15 4BS	G8TMA
GL15 4HR	G7VQI
GL15 4JX	G4NIF
GL15 4NY	G4ULG
GL15 4QD	G4UHJ
GL15 5AZ	2E1IFM
GL15 5AZ	2E1IFN
GL15 5AZ	2E1IJL
GL15 5AZ	2E1KID
GL15 5AZ	G7VHJ
GL15 5GB	G4NNJ
GL15 5HU	M6YRY
GL15 5JD	G0ENF
GL15 5LP	G0FDD
GL15 5NP	2E0FOD
GL15 5NP	M0SSJ
GL15 5NP	M6SWL
GL15 5QS	M3YJP
GL15 5QS	M6BPG
GL15 5RE	2E0SYS
GL15 5SL	G4EPW
GL15 5TA	2E0NSS
GL15 5TA	M3NSS
GL15 5TT	M3YJM
GL15 6BZ	M1ZZA
GL15 6DN	G4IKX
GL15 6HG	2E0DWP
GL15 6HG	M3BRV
GL15 6JQ	G0SDD
GL15 6LQ	G3CZL
GL15 6NA	2E0NBC
GL15 6NT	G7EEG
GL15 6PE	G8BXD
GL15 6PF	M0GLQ
GL15 6PR	G3NGZ
GL15 6PR	G3TSO
GL15 6TN	G6UXY
GL15 6TN	M3FTJ
GL15 6UN	G4PWA
GL155FX	M7YET
GL16 7AG	G3NOC
GL16 7AQ	G1EHX
GL16 7BL	G7GOK
GL16 7EB	G0PKV
GL16 7HH	M7EYY
GL16 7LG	2E0CJP
GL16 7LG	M3FHQ
GL16 7LR	G8AOJ
GL16 7PU	G0KWG
GL16 7QB	G0OOF
GL16 7QB	M0ACW
GL16 7QB	M0ATX
GL16 7QD	M1ART
GL16 7RB	2E0MNL
GL16 7RB	G1KUG
GL16 7RB	M7BKI
GL16 7RG	2E1IFL
GL16 7RU	M0TXB
GL16 8AY	M1AYN
GL16 8AY	M1BKE
GL16 8AZ	G3KTI
GL16 8BG	G1EDP
GL16 8BJ	M6BLG
GL16 8BN	M6CMZ
GL16 8BP	M6RBU
GL16 8BS	2E0SDR
GL16 8BX	M6RCP
GL16 8BY	G7AEF
GL16 8BY	M1EKM
GL16 8DE	M6CNA
GL16 8DN	G1BWP
GL16 8DS	M6XTU
GL16 8LD	M0ZEN
GL16 8PQ	G8WGD
GL16 8PT	G3TLD
GL16 8RQ	M7DAE
GL168EZ	2E0EFK
GL168EZ	M0IQQ
GL168EZ	M6APS
GL17 0DF	M0NOS
GL17 0DQ	M3FHP
GL17 0JE	G0NQI
GL17 0LP	2E0KGN
GL17 0LP	M6KGN
GL17 0PH	G0FGZ
GL17 0SB	G7MWW
GL17 9AU	M0HIY
GL17 9AU	M0RHR
GL17 9BW	M0OED
GL17 9BW	M6HYB
GL17 9QS	M1AVU
GL17 9SB	G0RMX
GL17 9SB	M6FUR
GL17 9UD	G0ORD
GL17 9XR	G7TUS
GL17 9YN	M3MYG
GL17 9YN	M3MYI
GL18 1AL	M7BZK
GL18 1HN	G6PMJ
GL18 1PS	M6DDD
GL18 1PZ	G3WVQ
GL18 1QB	M3DGJ
GL18 1TL	G1NRX
GL18 2BD	M7PPF
GL18 2BW	G4WUH
GL18 2EJ	G8SQH
GL19 3BN	G8CQZ
GL19 3BP	G1MHM
GL19 3QY	G8TTJ
GL19 4AD	M7RMH
GL19 4BT	G0ECJ
GL19 4BT	G6VKA
GL19 4DE	G6BOQ
GL19 4DS	G1KQD
GL19 4ES	M7JAD
GL19 4NX	G8DYG
GL19 4NY	G4CIB
GL19 4NY	G4RHK
GL2 0DP	G4NVY
GL2 0DZ	G4FLS
GL2 0EJ	2E0TOT
GL2 0EJ	M6TOT
GL2 0ER	M6LNX
GL2 0HA	G1JMF
GL2 0JG	M1DKA
GL2 0JH	G7WAY
GL2 0LX	G4MGW
GL2 0NA	G0UJT
GL2 0NB	M3NVO
GL2 0NQ	G3XMM
GL2 0PS	M3CCJ
GL2 0PX	G4CHS
GL2 0PX	G4HJV
GL2 0QP	M7YYF
GL2 0RX	G0EEA
GL2 2BF	G4WXF
GL2 2BH	G4LEN
GL2 2HN	2E0POE
GL2 2HN	M7POE
GL2 2JB	M6PVL
GL2 2RG	M7FQY
GL2 4GS	G4MOH
GL2 4JH	G4IVD
GL2 4LL	2E0MSN
GL2 4LL	M0MSN
GL2 4LL	M7MSM
GL2 4LZ	G1GAT
GL2 4NP	G4OIN
GL2 4PZ	2E0IUF
GL2 4QE	G0JQX
GL2 4QQ	G8RMI
GL2 4RJ	2E0UPS
GL2 4RJ	M7AEY
GL2 4RT	M0HFY
GL2 4SY	G8MMG
GL2 4SY	M0OMD
GL2 4UD	G6YSJ
GL2 4UF	M3NCQ
GL2 4US	2E0VFA
GL2 4US	M6VFA
GL2 4WF	G8OTD
GL2 4WJ	G0KVO
GL2 4YR	M6MAH
GL2 4YY	G1DNT
GL2 5BJ	M6DDT
GL2 5NZ	G7NGN
GL2 7DF	G8IEW
GL2 7DJ	G7JWQ
GL2 7DN	2E0JCR
GL2 7ED	G1WZO
GL2 7HR	G4LZQ
GL2 7LH	G3STZ
GL2 7LW	G1XAL
GL2 7PA	G8XPD
GL2 7PS	2E0EQI
GL2 7PS	M6LOQ
GL2 7PT	G3ILO
GL2 8DT	G7TQU
GL2 8EB	G7DMZ
GL2 8JP	G4PJJ
GL2 8LE	M7EMH
GL2 8LE	M7IJP
GL2 8LJ	G4ZYR
GL2 8NH	G3PJQ
GL2 9ES	M7GLW
GL2 9GS	2E0ITU
GL2 9GS	M6ITU
GL2 9HB	G1IDV
GL2 9NW	G4BGW
GL2 9PS	G7GVJ
GL2 9RB	G0HTO
GL20 5DG	G6VAR
GL20 5FB	G7AEE
GL20 5FB	G7VTL
GL20 5NH	2E0JCA
GL20 5NH	M6JCA
GL20 5NN	G8YMR
GL20 5PD	G1KNX
GL20 5PQ	M6YCP
GL20 5RF	G0UHG
GL20 5RL	G0MMA
GL20 5RX	M5ADE
GL20 5TW	G7AEC
GL20 5TZ	G0VFZ
GL20 5TZ	G8AZN
GL20 5TZ	M6WUH
GL20 6BB	G4CRN
GL20 6DL	G0CLR
GL20 6DL	G0NFV
GL20 6DW	G0NXA
GL20 6JW	G6AHX
GL20 7AT	G3FHG
GL20 7AU	G3XGW
GL20 7EH	G1XYF
GL20 7EP	2E0UHF
GL20 7EP	G6LJU
GL20 7EP	M0RAR
GL20 7RS	G0NUL
GL20 7RW	G8WWC
GL20 7WL	G0COZ
GL20 8AQ	M3SZO
GL20 8AS	M1AMA
GL20 8AT	G4EAZ
GL20 8BB	M0WMR
GL20 8BT	G1NFB
GL20 8HS	G3OLW
GL20 8HS	M6OGO
GL20 8NN	2E0CME
GL20 8NN	M6BBU
GL20 8NT	G7OHW
GL20 8PJ	M3XIV
GL20 8PX	G8JXS
GL20 8QP	G8VSH
GL20 8QY	G0FCM
GL20 8RB	G1NVS
GL20 8RE	G7CVC
GL20 8RP	M6WRE
GL20 8SN	G0PTR
GL20 8TQ	M6WIR
GL29ED	G4DCK
GL3 1AA	G0JVH
GL3 1AL	M7XNY
GL3 1AT	M6KCG
GL3 1BL	M3WHG
GL3 1DB	2E0GKN
GL3 1DX	M7DAP
GL3 1LL	M3TMY
GL3 1NT	M6WTE
GL3 2AU	G4ENZ
GL3 2BA	G4BCA
GL3 2BT	G0VIG
GL3 2DW	G6XQO
GL3 2HQ	2E0UPA
GL3 2HQ	M0UPA
GL3 2HQ	M6ULA
GL3 2HT	G1CMH
GL3 2JS	G0MTW
GL3 2LD	G3SUA
GL3 2PN	G4FRI
GL3 2PU	G6UGW
GL3 2RY	G6OTP
GL3 2TR	G4MYW
GL3 3BX	2E0KZU
GL3 3BX	M6KZU
GL3 3DH	G0VWH
GL3 3JE	G4HBV
GL3 3JF	G3UUL
GL3 3JH	G3ZLM
GL3 3QS	2E0TFN
GL3 3QS	M6DLV
GL3 3SX	G8IWB
GL3 3TZ	G6USD
GL3 3TZ	G8DHF
GL3 3UL	2E0YIN
GL3 4BD	M6ECT
GL3 4DH	M6RMI
GL3 4ER	2E0JIA
GL3 4ER	M6IJB
GL3 4ES	G1NVO
GL3 4GF	M6TTN
GL3 4NP	M3IXD
GL3 4PF	2E0DDQ
GL3 4PF	G0FNF
GL3 4PF	M0HWT
GL3 4PF	M6ECE
GL3 4PP	2E0PWC
GL3 4RL	M7IKB
GL4 0AB	2E0UED
GL4 0AL	G3XUC
GL4 0BX	G0RGJ
GL4 0DA	G3HXN
GL4 0HP	G8WRI
GL4 0JT	M3UEY
GL4 0NB	M6TXZ
GL4 0NZ	2E0OKK
GL4 0NZ	M0IRD
GL4 0PG	G0MGG
GL4 0QN	G0HBB
GL4 0QY	G1DIM
GL4 0RA	G1IFF
GL4 0SH	M6PZO
GL4 0SR	G6UER
GL4 0TD	G4CLR
GL4 0TH	2E0GXK
GL4 0TH	M6GXK
GL4 0TJ	M6IEQ
GL4 0TS	G1AET
GL4 0TT	G4TDG
GL4 0XS	M6MIF
GL4 0XW	M0PCB
GL4 0XW	M0VSQ
GL4 0YQ	2E0IXC
GL4 0YQ	M3IXC
GL4 3AG	G0UWU
GL4 3AG	G0WUW
GL4 3AH	G7RHM
GL4 3AJ	G7OUG
GL4 3AL	2E0IEI
GL4 3AL	M6GZE
GL4 3AX	G1BWH
GL4 3JL	G1NJI
GL4 3TQ	G4PTW
GL4 3TX	G1UGX
GL4 4AG	G0ULH
GL4 4AG	G2HX
GL4 4AG	M6RYL
GL4 4NU	M0KTN
GL4 4NU	M7KIT
GL4 4NW	M7YGI
GL4 4PU	M6FOD
GL4 4RB	G0GAJ
GL4 4RE	M6YUL
GL4 4RN	M6VED
GL4 4TF	M6XYJ
GL4 4UA	G7CSM
GL4 4WA	G7BVS
GL4 4WH	G0FBX
GL4 4WH	G0FHK
GL4 4WH	G1FHK
GL4 4WP	G0JBV
GL4 4WS	M6BHI
GL4 4WZ	2E0NZA
GL4 4WZ	M0XGL
GL4 4XD	G4KXK
GL4 4XH	G6BXT
GL4 4XJ	G3TDT
GL4 5BH	M7CRG
GL4 5DG	G3VTS
GL4 5FD	G4BNW
GL4 5FJ	G7AEA
GL4 5FJ	M1AGK
GL4 5FQ	2E0PMA
GL4 5GD	G7GQC
GL4 5GD	G7JUP
GL4 5GD	M0XAC
GL4 5WQ	M0GCN
GL4 5XT	2E0ADY
GL4 5XT	G5ADY
GL4 6DW	G0MIE
GL4 6EJ	M7ERV
GL4 6JE	M7GLO
GL4 6PB	G0IMQ
GL4 6QD	M7AJX
GL4 6QU	M0JJA
GL4 6TX	M7JLC
GL4 6UY	2E1FXN
GL4 6WE	2E1CAF
GL4 8AL	G0NRZ
GL4 8HB	G1SCV
GL4 8HB	M6AND
GL4 8LD	G3YJE
GL40NZ	M6NUY
GL5 1ES	M3BGE
GL5 1HD	M3UGV
GL5 1HS	G6SQT
GL5 1PL	G8VLY
GL5 1QE	2E0RKD
GL5 1QE	M6RZD
GL5 1QE	M6SHZ
GL5 1RD	G0MZK
GL5 1ST	2E0TPG
GL5 1ST	G0NUN
GL5 1ST	M0HZ
GL5 1ST	M0TPG
GL5 1ST	M6TBG
GL5 1SY	2E0IND
GL5 1SY	M0IND
GL5 1SY	M6IND
GL5 1US	G0FCO
GL5 1UT	G0DBM
GL5 2DG	M0BJP
GL5 2DG	M3BKG
GL5 2RF	G3REB
GL5 2RG	G6UTB
GL5 2UG	G8AER
GL5 3LS	M7FDJ
GL5 3PJ	G0OIS
GL5 3QT	M1FCV
GL5 3QZ	G4GXB
GL5 3TS	G4GWZ
GL5 3TS	G4RNK
GL5 4DF	G4RJG
GL5 4JW	G1JCT
GL5 4LF	M7ABT
GL5 4PX	G0TNG
GL5 4PX	G1OEM
GL5 4SA	G0IUW
GL5 4UN	2E0BKJ
GL5 4UN	2E0UUW
GL5 4UN	M3UGJ
GL5 4UN	M3UUW
GL5 4UW	G0FCJ
GL5 5JS	G3ICA
GL5 5JY	G3YIE
GL5 5LN	G4ENA
GL5 5PS	M7RDS
GL50 2AW	G3DNS
GL50 2LU	G8JJK
GL50 2NG	G4GVZ
GL50 2RD	M0WEB
GL50 3BW	M6WWK
GL50 3FJ	2E0JDO
GL50 3NH	2E0KXL
GL50 3NH	M6KLA
GL50 3NH	M7RGB
GL50 4BY	G8GKH
GL50 4HS	G4PBY
GL50 4LL	M6ORB
GL50 4NU	G3SMD
GL50 4NX	G8DTA
GL50 4NY	M3JIL
GL50 4PX	G8XRS
GL50 4QB	G4FZG
GL50 4RG	2E0SMF
GL50 4RG	G0NDU
GL50 4RG	M0SMF
GL50 4RG	M6SMF
GL50 4RJ	2E0DYG
GL50 4RJ	M0DCB
GL50 4RJ	M0YAY
GL50 4RJ	M6YGD
GL50 4SA	G0UTR
GL50 4SA	G0WJA
GL50 4SB	G4VTA
GL50 4SE	G6GJN
GL50 4SX	M1BFR
GL504NW	G3RMD
GL51 0AE	G4IOA
GL51 0EG	2E0GEL
GL51 0EG	M6UHU
GL51 0JN	2E0OCH
GL51 0JN	M0UCH
GL51 0LZ	G8CYU
GL51 0NY	G0SMM
GL51 0NZ	G6AFE
GL51 0PP	G6MBH
GL51 0QH	2E0ETU
GL51 0QH	M6FYK
GL51 0QT	M6ECJ
GL51 0TU	G1CQK
GL51 0TW	G6UYN
GL51 0UB	2E0IFV
GL51 0UB	M6IFV
GL51 0UP	M0SYG
GL51 0WN	M3JQV
GL51 0WN	M6XVC
GL51 0XE	G8ZEE
GL51 3BB	G6TRY
GL51 3BN	G4DVX
GL51 3JF	G4BZU
GL51 3LA	G0EYP
GL51 3LL	G6VQN
GL51 3LL	M6EPL
GL51 3LX	G4FUJ
GL51 3LX	G6FTR
GL51 3PD	G4HQC
GL51 3QL	G0ALA
GL51 3RA	G3NKS
GL51 3RR	G4SGI
GL51 3RZ	M1ELN
GL51 3WA	G6DPS
GL51 4GQ	M3ZZX
GL51 4SB	G0TPO
GL51 4SL	M7CCS
GL51 4SN	G3KII
GL51 4SZ	G7URT
GL51 4TG	G4ERR
GL51 4UP	M7HUX
GL51 4UW	G6RII
GL51 5RX	G6XSC
GL51 6AL	G0EVX
GL51 6AL	G4XDL
GL51 6BA	G8TWS
GL51 6BU	G6FPK
GL51 6GR	G3LHU
GL51 6JE	G0SDG
GL51 6JE	G4BSC
GL51 6JQ	2E1GKY
GL51 6JQ	M3GKY
GL51 6JS	G0WIG
GL51 6NP	G1EDX
GL51 6NX	G4WUJ
GL51 6QE	G1WVK
GL51 6QP	M0GPC
GL51 6QZ	G8DOB
GL51 6RL	G3LVP
GL51 6RL	G8APY
GL51 6RT	G8JAY
GL51 6RT	M3YAJ
GL51 6RW	G4ILI
GL51 6SN	G1AEB
GL51 7BQ	2E0GBG
GL51 7BQ	M6DMG
GL51 7DQ	G8PZD
GL51 7EA	2E0NKE
GL51 7EA	M6NKE
GL51 7LH	M7SEK
GL51 8AF	G3YEU
GL51 8DZ	2E0CLY
GL51 8DZ	M0JLY
GL51 8DZ	M6BBP
GL51 8HR	G3JJT
GL51 8NX	G6BHS
GL51 9DQ	G1NMW
GL51 9RD	M1AUO
GL51 9RF	2E0MIC
GL51 9RF	M7EQQ
GL51 9RW	M6CJM
GL51 9TG	G0LRI
GL52 2JR	M0IPB
GL52 2NR	G4NSZ
GL52 2NR	G8NSZ
GL52 2QR	G4YSF
GL52 3DE	G3XPR
GL52 3DE	M7DRH
GL52 3DF	G8ZFT
GL52 3DG	G3XKD
GL52 3DS	M0JVV
GL52 3DT	G4CWM
GL52 3DT	G6UJI
GL52 3DU	G3UYL
GL52 3DU	G4LCM
GL52 3DU	G4PKO
GL52 3DX	G7NCW
GL52 3DY	G0LFP
GL52 3DY	G7DEI
GL52 3EH	G3SNN
GL52 3EH	G5BK
GL52 3ES	G4PDQ
GL52 3HF	G8LHP
GL52 3LB	G6EEU
GL52 3PS	G8ENW
GL52 3QB	M0DQB
GL52 5AN	G0COI
GL52 5EG	G3LME
GL52 5GH	M0XQX
GL52 5JQ	G4KFT
GL52 5QS	M0YZG
GL52 5QS	M7GWC
GL52 6JD	2E0RLE
GL52 6JD	M7JME
GL52 6JG	M6KEF
GL52 6NZ	G1KYV
GL52 6RF	G8UJF
GL52 6SX	G3VKV
GL52 6TX	G6DXD
GL52 6YA	G7GQA
GL52 7UZ	M6FRG
GL52 7YT	M7MFV
GL52 7YU	M3WXD
GL52 8AG	G3XKH
GL52 8AU	M7GFR
GL52 8BG	G6LTB
GL52 8BN	G3PYI
GL52 8BY	M1CBY
GL52 8BY	M1CBZ
GL52 8BZ	2E0GWV
GL52 8BZ	M7HAM
GL52 8LA	M7KJG
GL52 8NA	G3AKI
GL52 8NU	G7PTV
GL52 8NX	G0IYO
GL52 8NZ	G4GKK
GL52 8PA	G4SYW
GL52 8PF	M0RON
GL52 8SA	G4ERP
GL52 8SS	G0VSY
GL52 8SS	G0VSZ
GL52 8TG	G0MBZ
GL52 8TG	G4MPJ
GL52 8UL	G4ZMD
GL52 8US	G1LMU
GL52 8XP	G4CKX
GL52 8XQ	G8TQP
GL52 9HN	2E0DKZ
GL52 9HN	M0HLI
GL52 9HU	G0SFE
GL52 9PY	G3KGT
GL52 9QP	G3NOI
GL52 9QS	G3ZTI
GL52 9QU	G4IZZ
GL52 9RJ	M0JNV
GL52 9SE	G6CFU
GL52 9TZ	G8MGD
GL52 9UD	G4BMP
GL525GJ	M6BEM
GL525LG	2E0VHF
GL526DA	G7VDS
GL52PA	2E0IRE
GL52PA	M6LFW
GL53 0AZ	G4RFU
GL53 0BH	G0VCD
GL53 0EW	M7VRB
GL53 0HE	G4NOE
GL53 0LU	G4WGK
GL53 0NQ	G6VVL
GL53 0NU	G8BTV
GL53 0PQ	G7AKI
GL53 0PU	G6PYM
GL53 7BD	M6PMY
GL53 7BD	M6UHM
GL53 7BQ	G3ZQJ
GL53 7BQ	M0DMF
GL53 7DB	G3XTP
GL53 7DX	G8KMR
GL53 7JJ	G3CGD
GL53 7RT	G3PCL
GL53 7RU	M0DKP
GL53 7RY	M0IBQ
GL53 8AR	M6BCC
GL53 8NQ	G4XXA
GL53 8NQ	G4XXB
GL53 8NS	G0LSQ
GL53 9AT	2E0KVS
GL53 9AT	M6TFE
GL53 9AT	M7JSF
GL53 9DQ	G0VNJ
GL53 9HU	G4VQE
GL53 9JJ	M7CRA
GL53 9LB	G4INB
GL53 9QG	G8NSO
GL530AD	G8CQX
GL54 1DJ	M3RXD
GL54 1JD	G3RWI
GL54 1LG	G8FMZ
GL54 2EW	G0BXS
GL54 2NZ	G1NAP
GL54 2NZ	G1WQX
GL54 3JJ	M1BPS
GL54 3QJ	G1VXS
GL54 4DD	G1XJT
GL54 4ET	M6SFV
GL54 4HP	G4VLH
GL54 4HP	M0URF
GL54 4HZ	G7IHN
GL54 4LL	M0AUW
GL54 5EP	M6XBV
GL54 5JX	G4MUW
GL54 5LQ	G0TPD
GL54 5QR	G3DPM
GL54 5QX	G7AWP
GL54 5XH	M0BAM
GL55 6AG	G3DWI
GL55 6BY	G1TFY
GL55 6JL	M0PHV
GL55 6TD	G0LSK
GL55 6XP	G0TPA
GL56 0FX	G8OYA
GL56 0JG	M7MZO
GL56 0LL	2E0GVC
GL56 0LL	M6GVC
GL56 9JZ	G8DCJ
GL56 9RJ	2E0HEG
GL56 9RJ	M6HEG
GL56 9TE	G4PQB
GL6 0AP	G0FDE
GL6 0DR	G3WMQ
GL6 0LD	G0DZM
GL6 0NJ	G3ZTX
GL6 0PJ	M0HMR
GL6 0PY	G3OYX
GL6 0RT	G6CYO
GL6 0TB	G0PPM
GL6 6AD	G7KUU
GL6 6AG	G0ISH
GL6 6EY	G4TCO
GL6 6NJ	G4HXQ
GL6 6PH	M0BAP
GL6 6RQ	G4VLV
GL6 6SA	G0VPT
GL6 6TJ	G3OAD
GL6 7DA	G6PKG
GL6 7EF	G0PKJ
GL6 7NY	G4SJN
GL6 7NZ	M7HYE
GL6 7QA	G3SPO
GL6 8AA	G3KYZ
GL6 8AB	2E0VDL
GL6 8AB	M6VDL
GL6 8DG	G1VDE
GL6 8FB	G3VBQ
GL6 8FT	2E0RYD
GL6 8FT	M7PDR
GL6 8LH	M5OTA
GL6 8LZ	G3NQF
GL6 8ND	G3TEV
GL6 8NX	M0HNH
GL6 8NY	G4OHA
GL6 9BA	G4GUG
GL6 9BA	G5TAM
GL6 9BZ	G4BSM
GL6 9BZ	G7EXX
GL6 9EP	G6XKV
GL6 9HB	G4IJV
GL6 9HF	G3UDD
GL67LE	G4MQL
GL68HG	M6UFC
GL7 1AU	G0UIX
GL7 1BJ	M1TCP
GL7 1BJ	M6FXG
GL7 1BL	M1BHZ
GL7 1BT	G0AZD
GL7 1BT	G7AEH
GL7 1DE	M7UKK
GL7 1DY	2E0FWK
GL7 1DY	M0NYY
GL7 1DY	M6NZB
GL7 1GJ	G4GPV
GL7 1HF	2E0ARJ
GL7 1HF	G5HI
GL7 1JT	M0AYX
GL7 1LQ	G0OAB
GL7 1ND	M6RJU
GL7 1PR	2E0LLW
GL7 1RU	M3IAA
GL7 1TG	G0RYR
GL7 1TG	M3GVJ
GL7 1TG	M3WHA
GL7 1UG	G8BAS
GL7 1WY	2E0LMI
GL7 1WY	M7TDV
GL7 2EJ	M0ANO
GL7 2JW	G8FRS
GL7 2LS	G4EVE
GL7 2NG	G1FRD
GL7 2PZ	G0WIY
GL7 2RH	G0NKK
GL7 2RL	2E0UJR
GL7 2RL	M6UJR
GL7 3AR	G3UJZ
GL7 3AX	M7REX
GL7 3EG	M6OSJ
GL7 3NN	M6CAW
GL7 3SD	G4JTO
GL7 3SD	G7AYE
GL7 4EQ	G3ZVC
GL7 4FZ	2E0KSR
GL7 4HN	G6RPW
GL7 4HN	M6JJG
GL7 4JU	M0ONY
GL7 4JU	M7AEZ
GL7 5FH	M6LTV
GL7 5HQ	G1ELF
GL7 5PR	2E0DIU
GL7 5PR	M6EOY
GL7 5QS	G1MYQ
GL7 5ST	G2OU
GL7 5ST	M1BEO
GL7 5WJ	M7SKI
GL7 6BE	G6EMB
GL7 6JB	M7DIP
GL7 7BA	G7ORG
GL7 7BB	G3UK
GL7 7BH	G8MJE
GL7 7BH	M3PMF
GL7 7DL	G0IZP
GL7 7DN	M7ERB
GL7 7JU	G3TA
GL72EJ	M7EWN
GL73NA	M7NSP
GL77JY	G4ABY
GL8 8BU	2E0XVZ
GL8 8BU	M0GVQ
GL8 8BU	M3LLN
GL8 8FQ	G7DDV
GL8 8HA	G0RXQ
GL8 8JE	G1TJW
GL8 8JU	2E0AYX
GL8 8JU	M0OSM
GL8 8SN	G4IHT
GL8 8UD	2E0TGO
GL8 8UD	M0TGE
GL8 8UD	M7GEN
GL8 8YU	G3MAV
GL9 1HH	G1KKS
GL9 1HP	G0XAY
GL9 1HP	G4GCT
GL9 1HP	G6PNB
GL9 1HT	2E1IJK
GL9 1HT	M0TFY
GR - 72053	G3URA
GU1 1BT	2E0BZH
GU1 1EP	G0SWC
GU1 1EP	G0SWE
GU1 1FR	G7MZS
GU1 1HX	G1MDS
GU1 1HX	G8MAV
GU1 1NA	2E1CML
GU1 1PA	M6HAC
GU1 1QL	G0KTV
GU1 1RQ	G4CDX
GU1 2DF	2E0HAT
GU1 2DF	M6HAT
GU1 2FB	M0RCN
GU1 2FJ	G0LPG
GU1 2JQ	G0EFO
GU1 2QF	G1RNV
GU1 2RP	M6IER
GU1 2TY	G7UCR
GU1 3JR	G4TJK
GU1 3NP	G6AFK
GU1 3NP	M7FLG
GU1 3PZ	G8ZAX
GU1 3PZ	M0SBT
GU1 4AZ	M6BMJ
GU1 4DD	M1TOD
GU1 4DN	M6WEU
GU1 4LL	M7KNV
GU1 4NL	G4WSW
GU1 4NP	G8TZN
GU1 4NQ	G7VNL
GU10 1BY.	G0VYQ
GU10 1LE	G4GNO
GU10 2JG	G1RUG
GU10 2JG	M0CJO
GU10 2QU	2E1JIM
GU10 2QU	G4IFX
GU10 2QU	M3CPX
GU10 4AX	M6HYJ
GU10 4BJ	M0PSI
GU10 4DW	G0LVR
GU10 4EX	2E0BOD
GU10 4JW	G8RYJ
GU10 4LU	G4ZKJ
GU10 4PQ	2E0JLW
GU10 4PQ	M6FZG
GU10 4RJ	M6AGQ
GU10 4RL	G4AHN
GU10 5EL	G4KWX
GU10 5HZ	G0NFA
GU10 5LP	G4ROM
GU10 5LS	G4LES
GU10 5PA	G8IXL
GU10 5QE	M0VVV
GU10 5RD	G1CEI
GU11 1HA	M0NGI
GU11 1YY	G6RQZ
GU11 3BW	M0GPW
GU11 3DB	M0AHJ
GU11 3EL	G4GVV
GU11 3EP	M0PSE
GU11 3ET	G4GMI
GU11 3GY	2E0HEU
GU11 3JX	G4GGZ
GU11 3SL	G7EXO
GU11UF	M6PNI
GU12 4AG	G1XIE
GU12 4AT	G8YKM
GU12 4EU	2E1FHQ
GU12 4HU	M6JZY
GU12 4HY	G1OQG
GU12 4HY	M0COM
GU12 4HY	M6PCE
GU12 4HZ	G4YFU
GU12 4LD	G8OLY
GU12 4RD	G3KND
GU12 4SF	G4SPD
GU12 5AY	G7ITM
GU12 5ED	G7TEP
GU12 5HP	G8PDP
GU12 5HR	M6MGA
GU12 5HS	G4WEL
GU12 5HS	G4WEM
GU12 5JQ	G8ZAJ
GU12 5JP	G1LKJ
GU12 5JP	M3LKJ
GU12 5NQ	G6XNP
GU12 5PR	2E1HAC
GU12 5QH	G8CXF
GU12 5QW	G8NEF
GU12 5SF	G4CDH
GU12 6HP	2E0DFI
GU12 6HP	M6EHA
GU12 6LX	G3VKI
GU12 6LX	G8YCL
GU12 6LY	G7PWA
GU12 6PL	M0GEL
GU12 6PP	M6KVK
GU12 6QB	G1IYE
GU12 6QN	G7CGT
GU12 6SG	G7CUF
GU12 6ST	M1EGZ
GU125JT	G8BCO
GU125JT	M0FRS
GU13PZ	2E0RWR

GU13PZ M7WER
GU14 0DD G0EMR
GU14 0DW G6MRW
GU14 0ET 2E0WKT
GU14 0ET M0WKT
GU14 0ET M6GQX
GU14 0ET M6WKT
GU14 0HL M6AAO
GU14 0LQ 2E0FCE
GU14 0LQ M0IYM
GU14 0LQ M6NZJ
GU14 0PB G1AAP
GU14 0RF G8IBC
GU14 0RW M6AOP
GU14 6AR 2E0TRX
GU14 6AR M6REX
GU14 6AX M6BQK
GU14 6BW M0CMP
GU14 6DA M6NGU
GU14 6DS G6RIZ
GU14 6JS 2E0DFL
GU14 6JS M6EHI
GU14 6JT M6BDM
GU14 6LX G1NTX
GU14 6NU G4JFN
GU14 6NU G4VAH
GU14 6QA G4KCC
GU14 6RF G3WOS
GU14 6TH M0HAU
GU14 7AE G3SRR
GU14 7AE G6TSX
GU14 7AR G3PQC
GU14 7DA G0DWM
GU14 7EU G3UHW
GU14 7EY G1OER
GU14 7HH G0HOD
GU14 7PP G7VTS
GU14 7TB M6YHV
GU14 8AG 2E1CYZ
GU14 8AG M3MKB
GU14 8BJ M0RCD
GU14 8DG 2E0GFF
GU14 8DG M0YMA
GU14 8DG M6ADB
GU14 8DJ 2E0EOV
GU14 8DJ M6FKN
GU14 8NN G4PDR
GU14 8PH G1FVH
GU14 8QJ G3PQF
GU14 8SR G7HGT
GU14 9EA G1KXQ
GU14 9HT M7SKB
GU14 9JQ G0MPI
GU14 9PJ G0MCQ
GU14 9PW M3IHS
GU14 9QH M3NFW
GU14 9RL G4BNK
GU14 9RP 2E0HHB
GU14 9RY 2E0LHS
GU14 9RY M6FKO
GU14 9UH G0HUT
GU14 9UR M3XSU
GU14 9XU G4NSN
GU14 9XU M3HCL
GU14 9XY M6ASQ
GU14 9YB M7TRA
GU15 1BF G3YTX
GU15 1DE G8GQS
GU15 1DG G3HEJ
GU15 1DL G6FYC
GU15 1DL M6CHX
GU15 1EF G8BVB
GU15 1EQ M6JTU
GU15 1LF G4UQE
GU15 1NP G6IXM
GU15 1PS G0RGX
GU15 1PS G8DXZ
GU15 1RD 2E0CIX
GU15 1RU G8EAD
GU15 2AJ G0IZY
GU15 2AL M7HWH
GU15 2BH G3RKK
GU15 2BU M0HUI
GU15 2DE G0DSQ
GU15 2JQ G6JTO
GU15 2LD G4REE
GU15 2LT G4ELJ
GU15 2LY M6BTB
GU15 2NR G6LWA
GU15 2NT G0WZX
GU15 2RQ 2E1AFA
GU15 2SP G6ENU
GU15 2SP M6ENU
GU15 2SP M7KCB
GU15 3EB M6EYA
GU15 3EN M6MNT
GU15 3HT M0KWY
GU15 3HT M0LKY
GU15 3HT M0RCK
GU15 3HT M6DBK
GU15 3HT M6ELI
GU15 3HT M6LFY
GU15 3HT M6LKY
GU15 3NP G4GLP
GU15 3NR G3LPN
GU15 3XB G4YHK
GU15 4AR G8OUY
GU15 4JR M6BTQ
GU15 4LD M6PUS

GU154AR M0CMF
GU154AR M6SUN
GU16 6BJ M6HZF
GU16 6BZ G1OSO
GU16 6DB M7CWD
GU16 6DB M7YAZ
GU16 6DG 2E0SEE
GU16 6DG M6AZE
GU16 6DT G6WZD
GU16 6ER G6BLU
GU16 6EY M1CIS
GU16 6GA M6LCW
GU16 6JJ G4SYB
GU16 6PH G6MJB
GU16 6SD M6CTW
GU16 7DU 2E0SKZ
GU16 7DU 2E0XAI
GU16 7DU 2E0XPT
GU16 7DU M3SKZ
GU16 7DU M3XAI
GU16 7DU M6XPT
GU16 7EB M6BQR
GU16 7RD M6LAC
GU16 8AD M0ILA
GU16 8AD M6IKX
GU16 8LP G7KNK
GU16 8LP M3KNK
GU16 8LR G3PPU
GU16 8NA G8XNC
GU16 8PS G7LTW
GU16 8RR G7PYB
GU16 8RT G8IRM
GU16 8RW G4MEA
GU16 8SA G3KFU
GU16 8XH M6RYA
GU16 8XQ G4DJB
GU16 8XR M6WJN
GU16 8XS M7CNJ
GU16 8XS M7RVF
GU16 8XX G3TJI
GU16 8YN G7WDD
GU16 8YT M6ABB
GU16 9BH 2E0JXG
GU16 9BH M7JGZ
GU16 9BL M6MHM
GU16 9BW M6SAC
GU16 9NY M6KDK
GU16 9QN M6POQ
GU16 9QU G3YAG
GU16 9RB G7AQL
GU16 9RF M7LGW
GU17 0DH G3WCY
GU17 0DJ G0ADT
GU17 0DU G6KRY
GU17 0EN G1FOE
GU17 0HB G8JZO
GU17 0HL 2E0VVM
GU17 0JW M7PFK
GU17 0LG 2E0CUL
GU17 0LG M0KUL
GU17 0NJ G3TMU
GU17 0PD M0XXJ
GU17 9BP G8HIO
GU17 9DL G4CGW
GU17 9DP M6DMQ
GU17 9DX G3ZYX
GU17 9DZ G3REL
GU17 9DZ G3SFU
GU17 9ET G4EMV
GU17 9HA G8HUF
GU17 9HH G6AAB
GU18 5RZ G3TJS
GU18 5TA G0KDL
GU18 5TE G8JMP
GU18 5TJ G4MBZ
GU18 5TP G3UKE
GU18 5TR G8HXD
GU18 5TS G8NYJ
GU18 5UJ G4SUO
GU18 5UW M1AKZ
GU18 5XH G4SJH
GU19 5DH M6HJQ
GU19 5DH M6HJU
GU19 5JR M3XTA
GU19 5JX G4ZRB
GU19 5NU G8MZD
GU2 4AZ G2DBH
GU2 4DB M7ISH
GU2 4EL G4PEA
GU2 4JF G8UHK
GU2 4LA G6FKN
GU2 4LF 2E0VHZ
GU2 4LF M6VHZ
GU2 5XH G7DQE
GU2 7JG M0WAC
GU2 7JH M3RZU
GU2 7JL M3UJO
GU2 7JL M3VKT
GU2 7JL M6MFM
GU2 7JN M3ZRV
GU2 7JP M7AMA
GU2 7QN G0SUL
GU2 7RT G7JVQ
GU2 7SA G1WTX
GU2 7SR G4OWL
GU2 7SR G8DPQ
GU2 7SX M7KCR
GU2 7TP M6PZM
GU2 7TS M3MBF

GU2 7UU M7SLI
GU2 7XH M0LTS
GU2 7YW 2E0VNN
GU2 7YW M3VNN
GU2 8AL M0DGT
GU2 8AP G8LOZ
GU2 8AR G8FUH
GU2 8AS M7SZD
GU2 8AU M6ASD
GU2 8BL G1MGF
GU2 8DE G3KMI
GU2 8DE M0DNY
GU2 8ER 2E0JIP
GU2 8ER M0JIP
GU2 8ER M7POW
GU2 8ES M7MYB
GU2 8JD M0SFI
GU2 8JU G4PMZ
GU2 8LX 2E0CZS
GU2 8LX M0PVI
GU2 8LX M3PVI
GU2 8UT 2E1AOG
GU2 9NP G8TZU
GU2 9PG G1ZDG
GU2 9PJ G0HNL
GU2 9PJ G8NOB
GU2 9PL M6RJQ
GU2 9PL M7SEA
GU2 9SB G7UPN
GU2 9TY M3UIU
GU2 9WA 2E0JRN
GU2 9WA G1YJY
GU2 9WA M0JRL
GU20 6AL 2E0HUZ
GU20 6AL M7HHH
GU20 6BU G8WSP
GU21 2AT M6YRB
GU21 2DD G1OFW
GU21 2DD M0RDK
GU21 2FZ G0NGI
GU21 2HY M6XRD
GU21 2JN G8FZV
GU21 2LD M6JXN
GU21 2LH G4HWI
GU21 2NG G0JXP
GU21 2PX M6XNL
GU21 2PX M6XRL
GU21 2QG M7SSN
GU21 2QP 2E0SSJ
GU21 2RR M7KDH
GU21 2TA G1AOE
GU21 2TP M7MJX
GU21 3AU M6IQA
GU21 3BP G4PRW
GU21 3BY 2E0FQA
GU21 3BY M7LMS
GU21 3BZ M0RDB
GU21 3DB M0IRI
GU21 3DD G8JMU
GU21 3JU M7SCH
GU21 3NZ G8BTL
GU21 3PJ G7WIY
GU21 3QS M0VPC
GU21 3QU M7WEW
GU21 3QU M7YDK
GU21 3QY 2E0XSG
GU21 3QY M0XSG
GU21 3QY M7SUE
GU21 4BG G3USX
GU21 4HY M7JGP
GU21 4JG G8SSY
GU21 4PW 2E0DRW
GU21 4PW M6KWG
GU21 4PW M7BKX
GU21 5EG G0NIX
GU21 6DQ M6FBT
GU21 7PN G7CSJ
GU21 7QT G4RHJ
GU21 8AW M0PES
GU21 8SS M0KFU
GU21 8SY M6CEI
GU21 8XL M7KTB
GU22 0AA G4GGI
GU22 0JB 2E0WFS
GU22 0NT G3ZHB
GU22 0NT G6BQC
GU22 7BE 2E0KFK
GU22 7BE M6KFK
GU22 7NE M6NZL
GU22 7NW 2E0CLH
GU22 7NW M6BGE
GU22 7SZ 2E0BQN
GU22 7UE G3GAQ
GU22 7UR G8LSA
GU22 8PE G0AAA
GU22 8PE G3WVG
GU22 8PE M0ZIP
GU22 8QZ G0AFP
GU22 9AU G0NRJ
GU22 9AU M3YSC
GU22 9BQ G4YPC
GU22 9BQ G5RS
GU22 9ND G7USO
GU22 9PH G7PVU
GU22 0NU G4SLL
GU220NQ G7HHI
GU23 6DQ G0JOP
GU23 6EN G0DKN
GU23 6PF 2E0GJT

GU23 6PF M0RNZ
GU23 6PF M7NSC
GU23 7AD M0BUF
GU23 7AD M7FMG
GU23 7AP G4EML
GU23 7ET G8AFU
GU23 7HR G3IUW
GU24 0HF 2E0MXC
GU24 0HF M0MXC
GU24 0HF M6MXC
GU24 8AR G3KMA
GU24 8PR G7CSI
GU24 8PS M6ISH
GU24 8RH M0JFI
GU24 8RL M6AXS
GU24 9DG G0RVK
GU24 9LY M7CQF
GU24 9QE G0DER
GU25 4DF 2E0IQL
GU25 4EZ G0TTG
GU25 4LD G8MLK
GU25 4NG M6ACC
GU26 6LR G0NIN
GU26 6NY G3ZFT
GU26 6PA 2E1JON
GU26 6PA G4NOT
GU26 6PW G7VBL
GU26 6PZ G7UEI
GU26 6PZ G7WBM
GU26 6RP G4CMG
GU26 6SX G3WLH
GU27 1AZ M1CZX
GU27 1JD G1OTZ
GU27 1JF M6DFE
GU27 1JF M6XER
GU27 1LA G8EOV
GU27 2BX G7NDN
GU27 2FD 2E0CVQ
GU27 2FD M6CTN
GU27 2NY G1MAL
GU27 3DZ 2E0URJ
GU27 3JL G0XTL
GU27 3JS M0EBX
GU27 3ND G4DQZ
GU27 3RG G0NON
GU27 3SN 2E0SJI
GU27PH 2E0IFO
GU27PH M0KJO
GU28 0EQ G6NMQ
GU28 0EQ G8WSX
GU28 0EQ M3UQE
GU28 0NY G0HJZ
GU28 0PJ M6THA
GU28 9DH 2E0BQZ
GU28 9DH M0PUT
GU28 9DH M3UNB
GU28 9DH M6BBE
GU28 9DP M0GPD
GU29 0BP G4BGM
GU29 0DJ G4YEI
GU29 0NQ G3VXF
GU29 0NU M6WJJ
GU29 9BU G1TLH
GU29 9JG G0DDY
GU29 9TE M1STI
GU3 1AZ G0SJH
GU3 2AX 2E0ZNZ
GU3 2AX M0ZNZ
GU3 2AX M6KNC
GU3 2AX M6ZNZ
GU3 2EN G4HZV
GU3 2JW G3VCY
GU3 3AU G4PNB
GU3 3BE G6GPR
GU3 3BQ M7RHI
GU3 3BX G3TQC
GU3 3LL M7SMN
GU3 3NE G8IMS
GU3 3PQ M1EBL
GU3 3PR G4MPW
GU30 7AN 2E0MTT
GU30 7AN M0NJX
GU30 7AN M6NJX
GU30 7BY M0RGI
GU30 7GB G8VG
GU30 7GW G4TTJ
GU30 7HG G3TSR
GU30 7HR G4VKC
GU30 7SH G8ISI
GU30 7WP M7BQJ
GU30 7XA G8UYW
GU30 7XD G4AER
GU31 4EU 2E0DFF
GU31 4EU 2E0JLB
GU31 4EU M6OBJ
GU31 4EU M6TXT
GU31 4NX G1MGU
GU31 4PN G1LQH
GU31 4PR M7TAW
GU31 4PZ M7GAW
GU31 4QA G6SLZ
GU31 5AX G3TZL
GU31 5BF 2E0WEI
GU31 5BF M6HXW
GU31 5HJ G0BUZ
GU31 5JW M7FJC
GU31 5NZ 2E1JJM
GU31 5NZ M3JJM
GU31 5SB G1THA

GU32 1DD G4AEP
GU32 2AN G4VQT
GU32 2AX G6XJB
GU32 3AX G8LWC
GU32 3AZ G4VRC
GU32 3BT G4SBU
GU32 3BW G0MBQ
GU32 3LS G0BTU
GU32 3LS G8XDD
GU32 3PJ G4ACW
GU33 6DA M3TQP
GU33 6HG G8XJO
GU33 7BF G4CFS
GU33 7BP 2E0MPA
GU33 7BP M3ZQF
GU33 7DB M1CQU
GU33 7DL G4ELM
GU33 7JE M0PDH
GU33 7JE M3EMX
GU34 1AY M6HDW
GU34 1GH G3RXI
GU34 1JA G4ASL
GU34 1NU G4FOY
GU34 1PW G0JMI
GU34 1QY G0JVL
GU34 1QY G1LAO
GU34 1RA G3RSI
GU34 1RR G8MOS
GU34 2BJ G1EML
GU34 2DG M6NHO
GU34 2ED G8XWR
GU34 2EE G0DBS
GU34 2EW M0GOK
GU34 2FJ 2E0BUF
GU34 2FJ M0VVQ
GU34 2HP G8YFH
GU34 2PB M3KOA
GU34 2PF G1JUP
GU34 2QT G0IBR
GU34 2QT G7LWH
GU34 2RD G4JOU
GU34 2TL G8IRL
GU34 2TP G8UML
GU34 3DH M7NKN
GU34 3EJ G6BXV
GU34 3JA M6KMX
GU34 3NP G0DWR
GU34 4AG G4AJB
GU34 4AG M7GJP
GU34 4AN G8EAN
GU34 4AQ G0EFP
GU34 4LJ G4DNK
GU34 4LJ G8DDH
GU34 5AD G4JCX
GU34 5AX G8LES
GU34 5BX G0RMR
GU34 5BY G4IPI
GU34 5BY G8WSZ
GU34 5DU 2E0VVX
GU34 5DU M0VVX
GU34 5DU M6HZQ
GU34 5EF G3PRQ
GU34 5JF M0BDF
GU34 5LG G6UEQ
GU34 5NP G3YVI
GU34 5PB G0BHA
GU34 5PB G4OBN
GU34 5PB G6BHH
GU34 5PB G6BHI
GU34 5PF G8CKN
GU34 5PR G3SJX
GU35 0DG G0PLZ
GU35 0DP M7CIJ
GU35 0DZ M7TVS
GU35 0EF M6KGK
GU35 0FJ G8OXS
GU35 0HB G3ZRM
GU35 0PP G0OYN
GU35 0QB G8AZB
GU35 0RA M6EYF
GU35 0SG 2E0BII
GU35 0XA G8BGI
GU35 8AJ G0PVJ
GU35 8AX M6OPB
GU35 8DU M7BCZ
GU35 8EP G1CBB
GU35 8HQ G6SNA
GU35 8HU G4ZEL
GU35 8JX 2E0FCB
GU35 8NA G4LJB
GU35 8NR G6VFI
GU35 9AJ 2E0SNJ
GU35 9AJ M6SWA
GU35 9BA M0MGA
GU35 9DZ G8CYL
GU35 9ED 2E0CSN
GU35 9ED M6BYZ
GU35 9EX 2E0DVD
GU35 9EX M0EUY
GU35 9PJ G8SXB
GU35 9QN 2E0CCW
GU35 9RE G3IGQ
GU35 9RE G8AHK
GU35 9RE M0SNJ
GU35 9TA G1TQY
GU359EX M3EUY
GU4 7FF M6FSD
GU4 7FF M7PMH
GU4 7HQ M6UBC

GU4 7NP M1RJJ
GU4 7NQ M6MGV
GU4 7PD 2E1ICW
GU4 7PD G0KUF
GU4 7QB G6MQZ
GU4 7TJ M1ACK
GU4 7UG M0AGP
GU4 7XR G3PIZ
GU4 7XR M0ECW
GU4 7YS 2E0IZK
GU4 8AY M0DHN
GU4 8EY 2E0WKW
GU4 8EY M0REQ
GU4 8EY M7EJS
GU4 8HZ G0JRE
GU4 8JG G8NTH
GU4 8JW M7ADF
GU4 8LL G6ZAC
GU4 8LQ G1HTN
GU4 8PH G3WBQ
GU46 6BX G3WUL
GU46 6DP M7AZN
GU46 6DW G4URP
GU46 6FA G8PJQ
GU46 6FN 2E0NGN
GU46 6FN M6NGN
GU46 6GZ M0URL
GU46 6JT G4VDF
GU46 6NE G8IFH
GU46 6NT G6FTY
GU46 6PB G1WNL
GU46 6PE G7CFW
GU46 6YP G7AOA
GU46 6YW G0YOU
GU46 7AD G4RYV
GU46 7RD G8INZ
GU46 7SY G4ZHZ
GU46 7TT G4TTZ
GU46 7UY 2E0HZP
GU46 7UY M0KFK
GU46 7UY M7DMY
GU47 0GJ M7IOR
GU47 0GT M3BVQ
GU47 0ST G7VHZ
GU47 0UT 2E0DCL
GU47 0UT M6CLK
GU47 0UT M6KSC
GU47 0YQ G8LMY
GU47 0YQ M5LMY
GU47 0YY G6ACJ
GU47 8HT G0FGA
GU47 8HT G3ZJQ
GU47 8HY M6SMR
GU47 8JH G1FNS
GU47 8JL G8FXU
GU47 8LD G8NMH
GU47 8NA M7JSP
GU47 8QS G4XYW
GU47 8QS G6GS
GU470DU 2E0WXY
GU470DU M6NPZ
GU470HH 2E0KAL
GU470HH M0KAR
GU5 0BL G0UNF
GU5 0HZ G4NMD
GU5 0HZ G6APJ
GU5 0HZ G8KVU
GU5 0HZ M3CIS
GU5 0JP G0DRR
GU5 0LL M7DLR
GU5 0SA G4VQF
GU5 0SE G6ZPR
GU5 0SX G8WMF
GU5 0UY G8XAK
GU5 9RS G0FCU
GU50EG G1UNB
GU51 1AU G8VIV
GU51 1DU M7SLD
GU51 1DX M6HXF
GU51 2RT G7FAQ
GU51 2TE G4KTZ
GU51 2TL M6SCA
GU51 3AH M3ZCM
GU51 3BS G7BQT
GU51 3DY G3MSL
GU51 3EB G1FRS
GU51 3EB G4OQZ
GU51 3EL G7MCS
GU51 3LY G8YVM
GU51 3LY M0NAM
GU51 3PX M7BIO
GU51 4AG 2E0CKO
GU51 4AQ 2E0DOZ
GU51 4AQ M0SLG
GU51 4AQ M6DOZ
GU51 4AQ M6SVU
GU51 4HA G4EFY
GU51 4HB G8JNI
GU51 4HE 2E0FVZ
GU51 4HE M0FAR
GU51 4HE M7ANR
GU51 4HN G4FTK
GU51 4SW G0EFL
GU51 5DP G1CNV
GU51 5NJ 2E0BBS
GU51 5NP G3NVM
GU51 5NR G7KIT
GU51 5SU M0DHO
GU51 5TZ M6MDM

GU514LY M6PUU
GU52 0TE M3WRJ
GU52 0UR G3KOB
GU52 0YJ 2E0EVC
GU52 0YJ M6WRR
GU52 6AS G3HUK
GU52 6AZ M0RHE
GU52 6BN G4AFI
GU52 6BN M7AFI
GU52 6JD 2E0BBY
GU52 6JD M3KFR
GU52 6LQ G3WKW
GU52 6PW G3RCD
GU52 7LE G3LGT
GU52 7LN G1JPP
GU52 7UG G3BRQ
GU52 7XE G4NYY
GU52 8NS 2E0ETE
GU52 8NS 2E0REB
GU52 8NS M6FFF
GU52 8NS M6REB
GU52 8XG G0NWL
GU526AY G8SUJ
GU6 7AA M3BSM
GU6 7DD M6VBP
GU6 7DS G7PRZ
GU6 7EN M0AWE
GU6 7ET G4CJT
GU6 7FY 2E0OBC
GU6 7FY M0IEB
GU6 7FY M6OBC
GU6 7HZ G5OD
GU6 7HZ G6XN
GU6 7HZ M0GJH
GU6 7JB G3OHC
GU6 7JB G4GRG
GU6 7RW G8TIO
GU6 8BZ G4YIF
GU6 8FD 2E0XJW
GU6 8FD M6ONS
GU6 8NB M6TCH
GU6 8NX 2E0LSL
GU6 8NX M0LSL
GU6 8NX M6LSL
GU6 8PQ G1DSM
GU6 8UD 2E1BEB
GU6 8UD M0RUT
GU67FQ M7SIW
GU7 1AH M0DLX
GU7 1QQ G8IBL
GU7 1RL G6WOT
GU7 1RL M0GKK
GU7 1SB 2E0DPO
GU7 1SB M0HZM
GU7 1SB M6FLS
GU7 1YA 2E1IAC
GU7 1YB 2E0XEN
GU7 1YB M0MOW
GU7 1YB M0NGY
GU7 1YB M3XEN
GU7 2AA M3KVU
GU7 2BS M6HXA
GU7 2LD G3TCU
GU7 2LJ 2E0ERK
GU7 2LJ 2E0YAP
GU7 2NE G1WTW
GU7 2NN M6JVX
GU7 2NT G4VRN
GU7 2QT 2E0DTN
GU7 2QT M0KRD
GU7 2QT M6KRD
GU7 2RW G1SWZ
GU7 3EU G8ZMC
GU7 3HA M7ACA
GU7 3HG G3KZB
GU7 3NN G8YEF
GU7 3NT G3MBK
GU7 3NZ G1FJF
GU7 3QJ G6NAV
GU7 3SH M0IAM
GU8 4AB M3TGK
GU8 4BD G4JEF
GU8 4ND G0SUA
GU8 4NT G8EBT
GU8 4RG 2E1AFS
GU8 5AB G7CND
GU8 5AB M3FJB
GU8 5DN G4GIX
GU8 5NR G4XBF
GU8 5RQ M0LOL
GU8 5TU G4DUF
GU8 6DE G0GNE
GU8 6DE G6RLM
GU8 6DU G3YZN
GU8 6DZ 2E1AFR
GU8 6EG M7EEC
GU9 0BN M6LQM
GU9 0BS G6FRS
GU9 0BS G7NMT
GU9 0BS G8ATK
GU9 0HY G8FQN
GU9 0NU M5AMN
GU9 0RZ G0FRS
GU9 0RZ G0YYY
GU9 7BP G4WUV
GU9 7BP G4WUW
GU9 7BT G7HOE
GU9 7BX G3VYI
GU9 7BX G4ALE

GU9 7DE M7TGR
GU9 7DH M0BSD
GU9 7RQ G4UEL
GU9 8BZ G3TAX
GU9 8JQ G8OGR
GU9 8LF G0CTZ
GU9 8RR 2E0KBG
GU9 8RR M6KBG
GU9 8SA M0PRA
GU9 8SA M6DPR
GU9 8TA G0CXU
GU9 8TW G0EYA
GU9 9AY G7JUR
GU9 9BB M7AZQ
GU9 9DF M7TXC
GU9 9EA G8UMA
GU9 9EY M3OSP
GU9 9FT M3CJH
GU9 9HJ G6BEN
GU9 9NL 2E0EXQ
GU9 9NL M0IVK
GY1 1FP GU7CMH
GY1 1JB MU3MNG
GY1 1NT GU4WTN
GY1 1SF GU4WRP
GY1 1TS GU7MXZ
GY1 1WJ GU0RAG
GY1 1WU GU4EON
GY1 1XJ 2U0NPT
GY1 1XJ GU7PVI
GY1 1XJ MU3NTH
GY1 2BA GU4SXM
GY1 2DZ MU0MXF
GY1 2PL GU6JQF
GY10 1SA GU0BEZ
GY2 4AH 2U1EKH
GY2 4EF GU3ONJ
GY2 4EX MU6GBG
GY2 4FL 2U0BGE
GY2 4FL MU0ZVV
GY2 4HJ GU4HUY
GY2 4HQ MU7GBG
GY2 4JS MU6CPV
GY2 4NS GU0VPA
GY2 4RN MU3KBP
GY2 4RN 2U0WGE
GY2 4RN MU6FXB
GY2 4XW 2U0EJL
GY2 4XW MU6GCI
GY2 4YW MU3UWX
GY3 5AF GU6NCZ
GY3 5AF GU6TKE
GY3 5EN GU4XEA
GY3 5JD GU4JHH
GY3 5JG GU4YBW
GY3 5NL MU6GUE
GY3 5SE MU0IIF
GY3 5TZ 2U0PTB
GY4 6AD MU6RAN
GY4 6JT GU7DSB
GY4 6JW GU4WMG
GY4 6NB GU6RWD
GY4 6NH GU1HYN
GY4 6QJ 2U1FNQ
GY4 6QJ GU4NYT
GY4 6QJ GU4SYQ
GY4 6SF GU3YVV
GY4 6TE 2U0HKI
GY4 6TE MU7CIM
GY5 7BN GU7OYU
GY5 7DT GU3UMX
GY5 7DZ GU3NDX
GY5 7DZ GU4YOX
GY5 7FQ GU0SUP
GY5 7FQ GU3HFN
GY5 7RT GU3UOQ
GY5 7RT GU7NCZ
GY5 7SA MU3ZHF
GY5 7SQ GU7DAI
GY5 7XD GU4XIT
GY5 7XZ GU4CHY
GY5 7YB GU7NHX
GY5 7YS GU0GUX
GY5 7YS GU0NHD
GY6 8EZ 2U1EJF
GY6 8HS GU3ZOM
GY6 8LA GU4ASO
GY6 8LN GU1IIW
GY6 8NR MU0FAL
GY6 8RB GU3LYC
GY6 8RT GU6EFB
GY6 8RT MU3EFB
GY6 8RY 2U1EKE
GY6 8RY GU1HTY
GY6 8RY GU1WJA
GY6 8SJ GU0BDI
GY6 8TU MU7ASX
GY6 8UF GU8ITE
GY6 8UF GU8NIS
GY6 8XL 2U0DWD
GY6 8XL MU0WLV
GY6 8XN 2U0WZY
GY6 8XN MU6WZY
GY6 8XX GU2RS
GY6 8YJ GU7CNI
GY7 9DA GU7TQX
GY7 9FS GU0DXX
GY7 9HE GU1MUP
GY7 9LD GU0PSP

GY7 9NN GU7CQN
GY7 9NN MU0CHN
GY7 9PH 2U0EFR
GY7 9PH MU0EFR
GY7 9PH MU6GFR
GY7 9QF 2U0LRB
GY7 9QX MU7GGA
GY7 9YQ MU6STK
GY79EL GU3WOW
GY8 0AB GU4LJC
GY8 0AJ GU8FSU
GY8 0JB GU1DWO
GY9 3 TG MU0VOE
GY9 3JD GU3LPV
GY9 3TX GU1BWW
GY9 3TZ GU7APA
GY9 3XA MU7FNB
GY9 3XD GU3TUX
GY9 3XE MU0EDN
GY9 3YP MU7DKI
GY9 3YP GU0UVH
GY9 3YZ GU4IJF
HA0 2HG M7HDM
HA0 2NX 2E0LHR
HA0 3QG G6MLV
HA0 3RQ M7SPY
HA0 3SG M6SHL
HA1 1QR M6NSZ
HA1 1XH 2E0SRP
HA1 1XH M0SRP
HA1 1XH M6PRS
HA1 1XU M7TKW
HA1 2QT M0JZO
HA1 2RS M7HKA
HA1 3HT G0UHK
HA1 3PR M0EVT
HA1 4AJ G3VFX
HA1 4AL G4IQD
HA1 4AL G4MSE
HA1 4BH 2E1HVL
HA1 4BW M0EGL
HA1 4DJ M6CIX
HA1 4EF M6EGA
HA11QR M3NGO
HA2 0HJ M6YOO
HA2 0PU G7OZA
HA2 0QE 2E0GEE
HA2 0QE M6OMG
HA2 0QJ G7FSJ
HA2 6AP G4JMG
HA2 6AQ G1RYF
HA2 6HE G0CAG
HA2 6HE G5FRG
HA2 6HF G3LGQ
HA2 6LG 2E0LBG
HA2 6LG M0JNS
HA2 6LG M6JNS
HA2 7EJ G4PKV
HA2 7JJ M6FFP
HA2 7NU G4IUM
HA2 7RB G4IRP
HA2 7RP M3KHW
HA2 9EA G4KEW
HA2 9EF G8DKW
HA2 9JL G4WVW
HA2 9LD G4MMA
HA2 9LJ G7IEF
HA2 9QF G4UKD
HA20LW 2E0NSY
HA20LW M6MPS
HA3 0PB G3FKI
HA3 0XQ G4KAB
HA3 5AX 2E0EYF
HA3 5AX M6ONU
HA3 6AJ G0IPK
HA3 6BX G3VHK
HA3 6HY M0HWZ
HA3 6PP M7SKO
HA3 6QL 2E0HBT
HA3 6QL M3HBT
HA3 6TN G0BSP
HA3 7AX M0GCI
HA3 7HU M6ABT
HA3 7NF M6HVA
HA3 8HA M6FQF
HA3 8JB G3BQE
HA3 8NF G3ASR
HA3 8NF G4KEP
HA3 8NQ G4ZPR
HA3 9EJ G8FAT
HA3 9ET M6SFL
HA3 9LB G7HIH
HA3 9NB G1TMW
HA3 9QZ 2E0RDA
HA3 9QZ 2E0UTL
HA3 9QZ M0AAG
HA3 9QZ M0UMS
HA3 9QZ M6OCZ
HA36JT 2E0DJX
HA36JT M0ZDJ
HA36JT M6XDJ
HA4 0AU G3KRT
HA4 0DS G0ACK
HA4 0DW G0DVC
HA4 0HL G8FTY
HA4 0HY 2E0FGR
HA4 0HY M6ONT
HA4 0LX 2E0EJR
HA4 0LX M0RJE

HA4 0LX M6RJE
HA4 6AJ G0RNF
HA4 6BA M3TLG
HA4 6ED G0DZH
HA4 6ED G1SWF
HA4 6ND M0MAR
HA4 6QZ M3PYB
HA4 6SX G0NRN
HA4 7EH M7ENC
HA4 7JB G3EFX
HA4 7JB G4AUF
HA4 7LZ G0JIM
HA4 7LZ G0VTC
HA4 7LZ G6JIM
HA4 7QD G8UOZ
HA4 7RD G6TDX
HA4 7TQ M0XRA
HA4 7UL G4LHT
HA4 7UL M6GRR
HA4 7UN G4UZE
HA4 7XR G0TAN
HA4 8AJ G4FFA
HA4 8PH G8UMY
HA4 8RY G3OEC
HA4 8SB G0JCF
HA4 8SB G1DES
HA4 8SD G7KUG
HA4 8UA G6SYB
HA4 9AG G3JVM
HA4 9BY M6FGZ
HA4 9DL G6ETZ
HA4 9HD G6PLR
HA4 9JT G1LQV
HA4 9LF G6SPI
HA4 9SA M0HJR
HA5 1JH G0IRM
HA5 1NB G4PXY
HA5 1NJ G1RGT
HA5 1NL 2E0CGB
HA5 1NL M3VUQ
HA5 1PH G7HSO
HA5 1TL G0KSL
HA5 1TN G0CHQ
HA5 2AR M7PRI
HA5 2AU M0DLC
HA5 2DA G8CAB
HA5 2EH G3LFD
HA5 2HR 2E0ZYX
HA5 2HR G0HGG
HA5 2HR M6HGG
HA5 2NB G8NTR
HA5 2NE M6LXW
HA5 3UP G6LFD
HA5 3UP M3YGK
HA5 3YF 2E0XLJ
HA5 3YF G4AZL
HA5 3YF G6XUD
HA5 3YF G8DAI
HA5 3YF M3XLJ
HA5 4HB 2E0VYN
HA5 4HB M6VYN
HA5 4LN G3UAS
HA5 4NJ G0WHV
HA5 4RA G0NTA
HA5 4TX G1VIO
HA5 5SP M0SEW
HA54RA G0NTB
HA6 1QE M3MAC
HA6 2BU 2E0SEO
HA6 2BU M0AAV
HA6 2BU M6SJX
HA6 2FY G4AGM
HA6 2HN G7NBF
HA6 2UJ G8MAA
HA6 2YP G7KWT
HA7 1JH M3THQ
HA7 2RR 2E0IIF
HA7 2RR M7VGG
HA7 3AG M7JEE
HA7 3HT 2E0UAY
HA7 3HT M3XSG
HA7 3PB M3WCA
HA7 3PL G1DQQ
HA7 4EP G1MYM
HA7 4FP G0MOX
HA7 4LD G3NDC
HA7 4PF 2E0PPM
HA7 4PF M6HUV
HA7 4TQ M6DAU
HA7 4UQ M0ALQ
HA8 0QE M7WCW
HA8 0SA G4DRO
HA8 0TW M3BCS
HA8 0TW M3OAK
HA8 5JQ M7VLD
HA8 5NQ 2E0PBW
HA8 5NQ M6IVO
HA8 5RJ G0SQL
HA8 6DL G8RNM
HA8 6HL G4JOO
HA8 7SA M6KEU
HA8 7UH 2E0CQT
HA8 7UH M6FEA
HA8 8AE M6EZM
HA8 8JU G4XEW
HA8 8PS G4GLM
HA8 8RH G4RFI
HA8 8SH G0STR
HA8 8TR G0KCL

Postcode

HA8 8TR G4ZOD
HA8 8UA G4SCG
HA8 8XJ M0HOF
HA8 9HE G1GXC
HA8 9JD M0CSG
HA8 9PW M6GVF
HA8 9SP M6MED
HA9 7DQ M0TAT
HA9 8NX G4DNP
HA9 8QY M1ELS
HA9 8TP G3ZZF
HA9 9HA M1FFR
HA9 9TQ G3SRN
HD1 3SL G0BFJ
HD1 3SL G6LD
HD1 4AJ M7DEW
HD1 4NU M6SGM
HD1 4PW 2E0KUK
HD1 4UR G1MOZ
HD1 5DY G7BNS
HD2 1AS G1MSD
HD2 1BN G1JJR
HD2 1DA G1CYY
HD2 1DH G3WUI
HD2 1ED G8YNE
HD2 1EJ M0GDJ
HD2 1JT G1GPT
HD2 1LB G0ODY
HD2 1LL 2E0ALJ
HD2 1LL M0OYZ
HD2 1LL M3SVZ
HD2 1QH G4YDI
HD2 1QH G6KKA
HD2 1QP G1AOR
HD2 1QY G8YOC
HD2 1RU 2E0EDL
HD2 1RU M0RIU
HD2 1RU M6LDE
HD2 1TH M6BJO
HD2 1UB 2E0RLR
HD2 1UB M6RLR
HD2 1UY G3WLW
HD2 1XT 2E0LOG
HD2 1XT M0LAG
HD2 1YW 2E0RNR
HD2 1YW M6JCS
HD2 2EB G0PRF
HD2 2EH G4XTE
HD2 2EW G7EOD
HD2 2HF M6SZP
HD2 2JP G7RAI
HD2 2NE G0BWQ
HD2 2NF G4BWC
HD2 2NF M0AGR
HD2 2NH G4BYW
HD2 2PE M0TKA
HD2 2PQ G4MEK
HD2 2XH M7EZC
HD2 2YD 2E0WBH
HD21RN G3XXR
HD3 3DB G3BKJ
HD3 3EJ G0IRY
HD3 3FH M7AQZ
HD3 3QF G0MKP
HD3 3QX G3TFO
HD3 3RJ G8MAR
HD3 3YW M3ZVU
HD3 3ZF G4KDM
HD3 4BN G8DTX
HD3 4BQ M7HJS
HD3 4GH G4LRD
HD3 4HQ 2E0BOR
HD3 4HQ M3SVJ
HD3 4LD G1AGA
HD3 4LD G1AGB
HD3 4RS G8RWN
HD3 4SW 2E0BWJ
HD3 4SW M3YLK
HD3 4SW M3YLL
HD3 4UY M7DCX
HD4 5DR G6VBA
HD4 5LL 2E0VVK
HD4 5NS G4GNA
HD4 5RP M6JPT
HD4 6HN M3VHU
HD4 6PJ M3GDX
HD4 6PL M1DQX
HD4 6QX G6TGE
HD4 6RA G1GTH
HD4 6SS M0AFS
HD4 6SZ M0PIE
HD4 6TE G4TML
HD4 6UR M3OAQ
HD4 6UR M3RTP
HD4 7AS M3LQE
HD4 7DA G0BWB
HD4 7DA G0LPV
HD4 7DA G4CDD
HD4 7DP G8TOT
HD4 7HE G7UID
HD4 7HS G1BGM
HD4 7JS G0NMH
HD4 7JS G8XSF
HD4 7JX G0SVU
HD4 7RA G3UEU
HD4 7RJ G0TPM
HD4 7SR G4JLO
HD5 0AT G4RLA
HD5 0BH G1WGO
HD5 0JB G7MHQ
HD5 0JD G0ISX
HD5 0LJ G7UZO
HD5 0NJ M3CMI
HD5 8EX G4KMK
HD5 8LD M0DXA
HD5 8LS 2E0RSB
HD5 8LS G4BQC
HD5 8LS M0RBG
HD5 8LU G6PJL
HD5 8PE M7JSH
HD5 8QB 2E0VET
HD5 8QB M6ARC
HD5 8QG G6GSV
HD5 8RP M0AOB
HD5 8UJ G4ITV
HD5 8UP G4RAJ
HD5 8XT G7OOH
HD5 8XU G4OTE
HD5 9EB M7ASZ
HD5 9HG G1FYS
HD5 9HX G0BWO
HD5 9JW G0LUU
HD5 9QA 2E0STH
HD5 9QA M6OXD
HD5 9RG G7FBT
HD5 9UW G6ZVU
HD6 1HH G0DIU
HD6 1QT 2E0GSP
HD6 1QT M7BEO
HD6 2AT G1GTQ
HD6 2AY G1GTR
HD6 2BJ G8AUL
HD6 2EP G4EEJ
HD6 2NA G7LBD
HD6 2NW M0ARK
HD6 2RS G8MLD
HD6 2RU G4SDX
HD6 2RU M3QJJ
HD6 3AH M3JLX
HD6 3JS G4WAP
HD6 3LD G7KXS
HD6 3NP G3WAH
HD6 3QX G8RFF
HD6 3SR G4GMT
HD6 3XF M3RFF
HD6 4AP G3JWN
HD6 4EQ G0TJQ
HD6 4JQ G4IPH
HD6 4JZ G4REG
HD7 2SQ G7KPH
HD7 4BT G4FSQ
HD7 4BX G7AXL
HD7 4BX M6YHM
HD7 4JR G1MQQ
HD7 4JR G4LYY
HD7 4JU G1HEP
HD7 4JY G0HRF
HD7 4LA 2E0PAD
HD7 4LA M3WUG
HD7 4NS G4IDR
HD7 4NU 2E1CRI
HD7 4QP G4UFL
HD7 4QQ G4UFJ
HD7 4RA G3XWN
HD7 4RE G8VMF
HD7 5BW G0PLD
HD7 5JS M6OTL
HD7 5LS G0JTM
HD7 5NE M6KZF
HD7 5NF G6HKA
HD7 5QU G7JMU
HD7 5QX 2E1BYI
HD7 5QX M0HVP
HD7 5TY M3ZNC
HD7 6BJ M3VQN
HD7 6JX M6BBT
HD8 0DE G3TSA
HD8 0GD 2E0SCJ
HD8 0GD M3SCJ
HD8 0JA 2E0RUD
HD8 0JA M3PBE
HD8 0JB G4AHJ
HD8 0JG G4XJG
HD8 0JG G7NDC
HD8 0QW 2E0RJH
HD8 0QW M0RJX
HD8 0QW M6RJH
HD8 0RX M7CVJ
HD8 0SW M1AYG
HD8 8AG G3LUK
HD8 8AG M0ALT
HD8 8AP G4OPY
HD8 8AZ M7OMP
HD8 8DS G7UKK
HD8 8HL M7KRW
HD8 8HP G1DEN
HD8 8HP G3SDY
HD8 8HP G8KMK
HD8 8HP M3OEN
HD8 8JZ M3OUS
HD8 8LX G0PHI
HD8 8PG 2E0FEV
HD8 8PG M7DAZ
HD8 8PX M6WWC
HD8 8QF G6UJR
HD8 8QW G7EXD
HD8 8RH M7EKR
HD8 8SG M1ELQ
HD8 8SP M7GCH
HD8 8UA M3WYV
HD8 8UB 2E0MPB
HD8 8UB M6CNI
HD8 8XP G6VIQ
HD8 8YH 2E0HFT
HD8 8YH M0ODM
HD8 8YH M6FFT
HD8 8YH M6HFT
HD8 9AB M0RGC
HD8 9BS G7VBU
HD8 9BT G4HKY
HD8 9BW G4BYD
HD8 9DB M0IQC
HD8 9DS G4TKO
HD8 9EH G4JCL
HD8 9ET G8PRN
HD8 9HF M7CXZ
HD8 9PA M1DLR
HD8 9QP G4EMQ
HD9 1EH G0HNW
HD9 1EN G1LFD
HD9 1ES G3RJT
HD9 1EU M0LUS
HD9 1EU M6AIN
HD9 1HL G7VDH
HD9 1HQ G8SNJ
HD9 1LH M3CGP
HD9 1SW 2E0EGS
HD9 1SW M5JB
HD9 1SW M6IHD
HD9 1XP M1EGD
HD9 1XU G8MQK
HD9 2LF G7UBQ
HD9 2PQ G3YPE
HD9 2PS G1JOW
HD9 2PX G8LIK
HD9 2PZ M6WYD
HD9 3ES G8PUT
HD9 3ES M3PUT
HD9 3ET G6WSX
HD9 3XZ G4UFZ
HD9 4AG G8GFY
HD9 4AJ G3NAK
HD9 4AJ G7GKQ
HD9 4DR G3LDJ
HD9 4DY 2E0DAW
HD9 4DY M3VIJ
HD9 4ED M3XOW
HD9 5LT 2E0NVK
HD9 5LT M3NVK
HD9 5PG G8TYB
HD9 5PL M0GBK
HD9 6DD G4LLZ
HD9 6DS G0JTL
HD9 6EG 2E1FWA
HD9 6HX G0BAK
HD9 6ND G0DBP
HD9 7ER G0SJB
HD9 7FB 2E1IAE
HD9 7SG G1VLU
HD9 7TJ G8ZML
HG1 1HP M7EDS
HG1 2AS G6HAT
HG1 2BW G0NHG
HG1 2DP 2E0UAB
HG1 2DP M6CBC
HG1 2HA G4CWB
HG1 2JF M6FJJ
HG1 2JN 2E0WAY
HG1 2JN M3WDZ
HG1 2QD M3YXR
HG1 2QD M6TTY
HG1 2QG M3EFV
HG1 2QG M3JYW
HG1 2QG M3KHT
HG1 2QG M3MIU
HG1 2QG M3OGC
HG1 2QG M3OHQ
HG1 2QG M3OHY
HG1 2QG M3OIY
HG1 2QG M3OSY
HG1 2QG M3SVF
HG1 2QG M3TNK
HG1 2QG M3TWK
HG1 2QG M3UCL
HG1 2QG M3UFE
HG1 2QG M3UYK
HG1 2QG M3VHC
HG1 2QG M3VHO
HG1 2QG M3WSW
HG1 2QG M3WTO
HG1 2QG M3XOY
HG1 2QG M3XUF
HG1 2QG M3XZF
HG1 2QG M3YCK
HG1 2QG M3YCU
HG1 2QG M3YEE
HG1 2QG M3YHL
HG1 2QG M3YKF
HG1 2QG M3YKO
HG1 2QG M3YLM
HG1 2QG M3YTL
HG1 2QG M3YUH
HG1 2QG M3ZCJ
HG1 2QG M3ZJJ
HG1 2QG M3ZKO
HG1 2QG M3ZNO
HG1 2QG M6AYE
HG1 2QG M6BPW
HG1 2QG M6BTF
HG1 2QG M6CSW
HG1 2QG M6CYL
HG1 2QG M6FIA
HG1 2QG M6HDK
HG1 2QG M6HUH
HG1 2QG M6HYW
HG1 2QG M6JYL
HG1 2QG M6KKL
HG1 2QG M6KLK
HG1 2QG M6LBQ
HG1 2QG M6LCT
HG1 2QG M6LDH
HG1 2QG M6LLW
HG1 2QG M6LST
HG1 2QG M6LXH
HG1 2QG M6MMP
HG1 2QG M6NMT
HG1 2QG M6VAN
HG1 2QG M6VCM
HG1 2QG M6WUN
HG1 2QG M6ZAS
HG1 2QG M6ZPZ
HG1 2QL M3XOZ
HG1 2QL M3YHM
HG1 2QL M3YSZ
HG1 3BW G1ZEA
HG1 3DF M6LHD
HG1 3EA G1VQK
HG1 3LT G8MJF
HG1 3NA G4KAT
HG1 4BP 2E0EFG
HG1 4BP M6HDI
HG1 4DL 2E0PDO
HG1 4DL G6GMU
HG1 4DL M0TUV
HG1 4DL M6PDO
HG1 4EG G0UIW
HG1 4EH G6VKP
HG1 4ES 2E0BLK
HG1 4ES M3NUL
HG1 4HR 2E0BWK
HG1 4HR M6KVB
HG1 4JN G7BBJ
HG1 4QR G0KIY
HG1 4RA G7LMH
HG1 4RH 2E0WKZ
HG1 4RH M3WKZ
HG1 4SG G1JLB
HG1 5AD 2E0GGP
HG1 5AD M6LNI
HG1 5HR 2E0SXI
HG1 5HR G0SDR
HG1 5HR M7DHZ
HG1 5HS G4LGX
HG1 5JD G1EWV
HG1 5JS G3ZTB
HG1 5JU G1WUC
HG2 0BJ G4ZIU
HG2 0DA M0VGA
HG2 0DQ 2E0CTU
HG2 0DQ M6JFE
HG2 0EG M6IBP
HG2 0ES G7FTS
HG2 0HZ M7ABG
HG2 0JS M7AXE
HG2 0LL G3OGZ
HG2 7AJ 2E0LMH
HG2 7AJ G8XZC
HG2 7AJ M0LMH
HG2 7AJ M3YUV
HG2 7DP G3XPM
HG2 7EE 2E0JYP
HG2 7EE G4RCD
HG2 7EE M7JYP
HG2 7ER G4TJI
HG2 7ES 2E0RHH
HG2 7ES M0HYJ
HG2 7ES M0RDY
HG2 7HQ G1UIB
HG2 7LJ G1ARH
HG2 7LJ G1ROD
HG2 7NJ M0AVF
HG2 7RU M6RNC
HG2 7SQ M7YTO
HG2 8HN M3YIT
HG2 8HW G4MNE
HG2 8LB G4EKJ
HG2 8LE G4KCR
HG2 8LX G4KEN
HG2 8PR G4LQG
HG2 8PR M6TJB
HG2 8QF M1DFW
HG2 9BS G1BPE
HG2 9HP G3YHC
HG2 9LD M0UIL
HG2 9NL G1CCX
HG2-7JU 2E0FDU
HG29PG G8DMU
HG3 1BH 2E0PAE
HG3 1BH M0MCI
HG3 1FP M7HVL
HG3 1HY M3URS
HG3 1JR G4IUF
HG3 1LY G6WKO
HG3 1NF G3NXL
HG3 1NY M0DYO
HG3 1PB M1CTG
HG3 1QX G0JNG
HG3 1RJ G0DOZ
HG3 2AA G4MEM
HG3 2DG 2E0AGI
HG3 2DG 2E1BDV
HG3 2DG G1ZBO
HG3 2DG M5ACF
HG3 2DJ 2E1CJC
HG3 2DQ G3FMW
HG3 2DS G6EQT
HG3 2RF G0UFL
HG3 2SU M3UTP
HG3 2SU M3YHP
HG3 2UT G1JWL
HG3 2WE G1XKJ
HG3 2WE G6WTM
HG3 3BX 2E0IVY
HG3 3BX G3RHR
HG3 3BX M6TUR
HG3 3DS G4PAH
HG3 3HE G5VO
HG3 3HE G6MC
HG3 3HE M0WKR
HG3 3HE M3UZL
HG3 3HE M3YPG
HG3 3HE M6LJJ
HG3 3JR G4EEV
HG3 3JR G8IKS
HG3 3LA G4VNA
HG3 3PB G4AFT
HG3 4EW G3RYP
HG3 4HA 2E0JDM
HG3 5LH G3ZCY
HG3 5LY G3NNO
HG3 5PG G1TBT
HG3 5QX M6FXO
HG4 1EJ M7VQV
HG4 1JU 2E0NAE
HG4 1NG M0TJW
HG4 1NW G7OFM
HG4 1RG 2E0GEB
HG4 1RG M6HEE
HG4 1RZ G4DSC
HG4 1SX 2E0IYU
HG4 1UA G3AB
HG4 1UG M7HFY
HG4 1UJ G0ORJ
HG4 1UP M3ZAZ
HG4 1US G1WRE
HG4 2AP M7ASF
HG4 2LJ G4VMY
HG4 2LJ G6MGH
HG4 2LN G1VKC
HG4 2NB M0HQJ
HG4 2PG G3UVY
HG4 2TU M7CGK
HG4 3FD G0BSY
HG4 3LL G1PWM
HG4 4AU G0UFI
HG4 4AU G4VUN
HG4 4AY M6RCA
HG4 4BP M7OJV
HG4 4LH G7LYN
HG4 4PW 2E0EEJ
HG4 4PW M0RBY
HG4 5BU G1JYK
HG4 5LF G3LEO
HG43HU M7YKE
HG5 0BW G0EYL
HG5 0DX M0DAL
HG5 0LW M0AQH
HG5 0QA G8BNE
HG5 0TH M0RMN
HG5 8EL M0CVG
HG5 8HD G0KGT
HG5 8HQ G7IPX
HG5 8HX G7HGD
HG5 8JN G6HEB
HG5 8JT G3DUW
HG5 8LH G1BKU
HG5 9BJ G4YBD
HG5 9BZ G3XQJ
HG5 9DE G3XZV
HG5 9EL G3UNA
HG5 9EL G6CP
HG5 9EL M6NLK
HG5 9HN 2E0DPL
HG5 9HN M0HYH
HG5 9LD G0GYU
HP1 1AS M6RTI
HP1 1LB G3GBN
HP1 1NS M6FTN
HP1 1PL M3SUJ
HP1 1QP M7VJT
HP1 1XQ G8FXC
HP1 2BP M0ZJV
HP1 2BQ G0TIW
HP1 2HY 2E1EOI
HP1 2LN M1ABC
HP1 2PP M0IHY
HP1 2PP M6ULY
HP1 2PP M6WTL
HP1 2QR G0EOK
HP1 3BN G4WGA
HP1 3BN G7NUT
HP1 3BU M5AEO
HP1 3EW G7SGM
HP1 3HY M1EQO
HP1 3JU G7IPA
HP10 0BP G4KCX
HP10 0BP G8NMT
HP10 0DU G1FEF
HP10 0HG M6YOS
HP10 0PZ 2E1TWB
HP10 8BL G4XVP
HP10 8DL G1SHT
HP10 8HG M3MVO
HP10 8JJ G8FBW
HP10 8LN G4DDM
HP10 9AX G3PWY
HP10 9DF G4VLL
HP10 9DZ G8XQT
HP10 9EG M1CRF
HP10 9EQ G4PIE
HP10 9LH G3WQG
HP10 9PL G3INZ
HP10 9RH G8JQV
HP108BH 2E0MSS
HP108BH M0VWS
HP108BH M6MSS
HP11 1GL G3SRQ
HP11 1PA G4TJM
HP11 1QA M3OJW
HP11 1QE M7VUX
HP11 1QT G8DJF
HP11 1RE M7BBK
HP11 1TX M6FNE
HP11 1UB G1UCT
HP11 2JL G0OCY
HP11 2JP M5SSB
HP11 2PL G1EUI
HP11 2UA G4CJY
HP11NE 2E0LFW
HP12 3AS M0HCN
HP12 3DE M7KHA
HP12 3DS G7GUO
HP12 3JQ G0IZC
HP12 3LP M6HXB
HP12 3NG M0MIQ
HP12 3NG M6BQZ
HP12 3NN G1LQX
HP12 3PA G8VJW
HP12 4AA G8WCH
HP12 4AU G7GST
HP12 4BB G6BFP
HP12 4DA M0YOJ
HP12 4DD G0GAN
HP12 4DD G4ZOC
HP12 4DD G8ANO
HP12 4PG G8DLZ
HP12 4QD M1EBY
HP12 4QR G0IKE
HP12 4SP G3ZGQ
HP12 4TY G3PVJ
HP13 5BD 2E0RPY
HP13 5BD M6NSE
HP13 5JN G3FNO
HP13 5JN G4TZQ
HP13 5JX 2E0KRK
HP13 5JX 2E0PZV
HP13 5JX G4EBY
HP13 5JX M0YGB
HP13 5JX M3YGB
HP13 5JX M6KRK
HP13 5NG G8HDL
HP13 5NW G7RTI
HP13 5QA G8AAU
HP13 5RH G0CTD
HP13 5RH G1EYG
HP13 5SP M1BDL
HP13 5SS G4LMM
HP13 5SZ 2E0PTM
HP13 5SZ M6PTM
HP13 5TA G4MUI
HP13 5TZ G4TBG
HP13 5UD G6HBJ
HP13 5UN G8JAW
HP13 5XN G1CPC
HP13 5XT G4COS
HP13 5XY G0NCL
HP13 5YT 2E0KNA
HP13 6AU 2E0BKT
HP13 6AU M0GGW
HP13 6AU M3TUU
HP13 6HW M6PHC
HP13 6JG M6LEE
HP13 6JG M6YEY
HP13 6SN G3OUV
HP13 6SW M0HMB
HP13 6TJ G1GAR
HP13 6XW G6GTZ
HP13 7EF G3GAA
HP13 7JF M6RKI
HP13 7LF M7PCL
HP13 7PH G0OQK
HP13 7PH M6RVE
HP13 7UE G1MIY
HP13 7XN G4YAN
HP13AU M7WMM
HP14 3BN G4KBB
HP14 3JN G4PUO
HP14 3JW G3LYP
HP14 3RE M6GEG
HP14 3RP G0FUN
HP14 3RP G4WJS
HP14 3SJ G1OLY
HP14 3TF G0GGL
HP14 4JG G4YBH
HP14 4JN G4YKQ
HP14 4QB G6XQB
HP14 4RA G8GLV
HP14 4SG G0HND
HP14 4UH G3SRJ
HP14 4UY G0IOU
HP15 6BW G4XJN
HP15 6EG G6AHN
HP15 6EY G7MLX
HP15 6JR G3VCT
HP15 6LJ G3WNS
HP15 6LJ G4PSG
HP15 6LJ G8AKU
HP15 6SL G4PFA
HP15 6TN G0RBB
HP15 6UT G4GED
HP15 6XA G0TRE
HP15 6XD G7DZD
HP15 6XF G1RDX
HP15 7AT G3XZK
HP15 7BX M3YXC
HP15 7DT M6HRS
HP15 7ED G6UDF
HP15 7FP 2E0SVB
HP15 7FP M7JHA
HP15 7PH M6BEH
HP15 7RE G3TYG
HP15 7RP M6VVT
HP15 7TF G4PPK
HP15 7TF G6GUD
HP15 7TP G6EYJ
HP156NT M0TPE
HP157BY G7DPE
HP16 0DF G4SNQ
HP16 0HB G6YQZ
HP16 0HW M3HEI
HP16 0LY 2E0DUU
HP16 0LY M6IJH
HP16 0NA G7TQA
HP16 0NJ G8XTJ
HP16 0NL M0IEA
HP16 0NL M6FZK
HP16 0QE 2E0MPF
HP16 0QE G4CLM
HP16 0SL M1CVT
HP16 9DS G1MDG
HP16 9DS G3MDG
HP16 9DS G3ZRQ
HP16 9EJ G4DHQ
HP16 9JH G6MJA
HP17 8AN G3OCP
HP17 8DJ M3AWD
HP17 8DL M6ERI
HP17 8EU G1MZG
HP17 8EY G3WUN
HP17 8HN G0IKP
HP17 8QN G7AAR
HP17 8QR G7JLF
HP17 8QR M0SOC
HP17 8RG G3YSR
HP17 8SB M7HRS
HP17 8XQ G8WDX
HP17 8XU M0SFD
HP18 0BL G6JRM
HP18 0BN M5AJP
HP18 0DN G6IFE
HP18 0EZ G1ATY
HP18 0HJ G6OIH
HP18 0LH G0MMI
HP18 0LY G0KMC
HP18 0RX G4CYR
HP18 9AU M3BLF
HP18 9DP 2E1HNS
HP18 9DZ M6YMY
HP18 9HH G0RAS
HP18 9HP G7TPD
HP18 9JZ G1ULQ
HP18 9LD G6DAP
HP18 9QZ G0UUI
HP189PJ 2E1FZY
HP189PJ M0BSQ
HP19 7BG G6YLV
HP19 7FY G1IJM
HP19 7HR M0TOL
HP19 7HR M1EIO
HP19 7HW G8UEI
HP19 7QJ G0MHZ
HP19 8GW 2E0XLX
HP19 8GW M6XLX
HP19 8GZ M0WDU
HP19 8JE M6HTK
HP19 8JH 2E0RJY
HP19 8JH M7RAJ
HP19 8SW 2E0HAF
HP19 8SW 2E0OHF
HP19 8SW M6OHF
HP19 8WA G8QAP
HP19 9GY M7PLT
HP19 9LN G0TGM
HP19 9QA 2E0PNP
HP19 9QF M7BNW
HP19 9QP M7TJS
HP19 9QT 2E0TBQ
HP19 9QT G7HIU
HP19 9QY G7GQH
HP2 4BA G0IAL
HP2 4BN G8RWU
HP2 4HE G1URZ
HP2 4LZ M3AZF
HP2 4LZ M3RDH
HP2 4PB G8UCK
HP2 4PQ 2E0MMJ
HP2 4PQ M0MMJ
HP2 4PQ M6MMQ
HP2 4PX G1EIZ
HP2 4QP M6NYG
HP2 4RP G7HCQ
HP2 4TJ M7FHC
HP2 4UG G1PAK
HP2 5AT 2E0MOB
HP2 5AT G0BLU
HP2 5AT M0RMO
HP2 5AT M6RMO
HP2 5DG M0KPY
HP2 5EL G8VVB
HP2 5ES M6MBR
HP2 5JG G8CBE
HP2 5JL G1DDF
HP2 5QF G8SYD
HP2 5QF M6TFQ
HP2 5QQ M7XTW
HP2 5SL M6HTZ
HP2 5WR G6FGV
HP2 5WR M0CZB
HP2 5XT M0GDS
HP2 5YX 2E0VTV
HP2 5YX M6ITV
HP2 6DS G4BJN
HP2 6DX G8ZGQ
HP2 6JB M0SNZ
HP2 6LA G1PWF
HP2 7AR M3XHT
HP2 7JP G7WDO
HP20 1AR G1PCR
HP20 1HJ M7NRR
HP20 1HW G7IMY
HP20 1JW G8ALQ
HP20 1PZ M3OOE
HP20 1QE M3VRL
HP20 1RB 2E0PNR
HP20 1XG 2E0MHC
HP20 1XG M7DPC
HP20 2AB M3PIA
HP20 2AJ 2E0MJH
HP20 2AJ M3CTT
HP20 2AS G4BKS
HP20 2EX M5ASR
HP20 2NU M3MMP
HP20 2QT G1NMP
HP21 7BB M7GFI
HP21 7BD G4FXI
HP21 7EP G1FNA
HP21 7FE M6TBQ
HP21 7NS G8VWZ
HP21 8BN G8MBM
HP21 8EH G1BHW
HP21 8EY M7VKO
HP21 8JG G1MZW
HP21 8JY M7EFT
HP21 8LY G4CGE
HP21 8NW 2E1PPM
HP21 8TW M0SVB
HP21 9AL M6GJC
HP21 9DS G8AYM
HP21 9DT G8MEM
HP21 9EF M3HER
HP21 9EF M3HIM
HP21 9LY 2E0DSV
HP21 9LY M6FIW
HP21 9RX G7ITO
HP21 9SL M6BFU
HP21 9UB M0ILK
HP22 4BL 2E0WOF
HP22 4BL M7WBT
HP22 4EF G8BQH
HP22 4LH G3XOP
HP22 4LT G4HFU
HP22 4PA G0WZH
HP22 5AB M3UCC
HP22 5AG G8ZXI
HP22 5EG G1MRI
HP22 5HH G4VRS
HP22 5HH G6GDI
HP22 5LA G1ZDU
HP22 5ND G8TJI
HP22 5QX M6ELF
HP22 5RU G4WQB
HP22 5RX 2E0CNM
HP22 5YQ G0BZN
HP22 6AR G1DDR
HP22 6AW G4PFR
HP22 6PH G1RAF
HP22 6PH G8ZGK
HP22 6QH G8ZME
HP225SD M7WAF
HP225TX M0TIF
HP23 4DS G6DZJ
HP23 4DS G7UYW
HP23 4EA G3NBZ
HP23 4EU M7UGL
HP23 4HZ G0SEY
HP23 4LX M0ZKK
HP23 4LX M6DSQ
HP23 4QG G4SSE
HP23 4TP G8MKX
HP23 5DG G1NUO
HP23 5HG M3MGQ
HP23 5HL G7PWI
HP23 5PB G3WXM
HP23 5PJ G3RWV
HP23 5PL M0HXV
HP23 6BX G1IGA
HP23 6ED G8GYY
HP23 6EN G3MEH
HP23 6HH G4PMG
HP23 6HH G8MKW
HP23 6HH M3PTG
HP23 6JB G1ODZ
HP23 6LE G0BKP
HP23 6NE G6HGM
HP27 0BQ G0SNF
HP27 0EB G0SPB
HP27 0EB G1DAE
HP27 0JA G7AYS
HP27 0LG G0NLJ
HP27 0LG G4HMC
HP27 0NB G3SNP
HP27 0PG G0MBB
HP27 0PW G0SKA
HP27 0QY G7DPF
HP27 9AY G1THD
HP27 9FZ 2E0PJY
HP27 9HE G0WQQ
HP27 9HU G3QZK
HP27 9JY G7CUD
HP27 9RP G3NBS
HP3 0BS G3VSQ
HP3 0BT M3UST
HP3 0BU M6NAS
HP3 0DS G6HIA
HP3 0EA G0VFW
HP3 0EG G8MFH
HP3 0HG G1OEB
HP3 0QN G4VCO
HP3 8EQ G8OYQ
HP3 8JL 2E0HNC
HP3 8NN G1JHD
HP3 8NW G8HZQ
HP3 8PR G6EAX
HP3 8QS G8LPC
HP3 8QT 2E0IVU
HP3 8QT M7CIN
HP3 8RG M6VWR
HP3 8SH 2E0NDR
HP3 8SH G1BSZ
HP3 8SH G6EXX
HP3 8SH M0NGL
HP3 8SH M3NDR
HP3 9GS G6SVL
HP3 9HB G7JHM
HP3 9HS M3YYE
HP3 9JA M0SCS
HP3 9JA M1SCS
HP3 9JJ M6HTF
HP3 9JY G4MSW
HP3 9LW M6JFA
HP3 9NG 2E0BVU
HP3 9NG M0GPE
HP3 9NG M3NGA
HP3 9PD G0GVN
HP3 9PW G2KG
HP3 9UB M6PAG
HP3 9UE G1KOG
HP3 9UQ G0EVS
HP3 9WU G6HGE
HP3 9XE M7VRS
HP4 1FG G1ABW
HP4 1HF 2E0CIJ
HP4 1HF M0HDT
HP4 1HF M6AZV
HP4 1LR M7EFR
HP4 1QG G7WBY
HP4 1QG M7WBY
HP4 2PD G3RPA
HP4 3BS G7VJD
HP4 3BS M0AYG
HP4 3DR M6ZTP
HP4 3JZ 2E0FGW
HP4 3JZ M6FGW
HP4 3PW G1WMN
HP5 1QP G4UXY
HP5 1RL M1BPU
HP5 1RW G0LZG
HP5 1SL M0IWG
HP5 1SQ G8YFE
HP5 1SS G3WPD
HP5 1SS M1BAA
HP5 1TA G3XZG
HP5 2BU G6HPK
HP5 2BU G7PEU
HP5 2EB G0ABP
HP5 2EB G6LBG
HP5 2HF M6MRR
HP5 2HL G8BLB
HP5 2PG M6EWZ
HP5 2QR M6BLF
HP5 2RY G7GLZ
HP5 2UQ G4XRV
HP5 2XL G3VZF
HP5 2XW G1CUZ
HP5 3BT G1GMQ
HP5 3DJ G3VRY
HP5 3JH G6CDV
HP5 3JQ 2E1LGJ
HP5 3JQ M7AND
HP5 3RJ G4CYY
HP6 5PY G8ETV
HP6 5RN G6FUT
HP6 5RW G6XSL
HP6 6DR 2E0EJC
HP6 6DR M6ICL
HP6 6HG M0FJS
HP6 6HL G7MMC
HP6 6HP G7EAT
HP6 6JF G1JNG
HP6 6JF M0BTX
HP6 6LD G0NLX
HP6 6PZ M7CIZ
HP6 6QP G0XDI
HP6 6QU M0GUL
HP6 6RJ G4GYN
HP6 6RP M3ZAI
HP6 6RR G4TNU
HP6 6SX M6DJG
HP6 6SX M6DWZ
HP7 0EF G0RTF
HP7 0PY G2SH
HP7 0RQ G1HPU
HP7 9AG M6MNS
HP7 9AU G3WLO
HP7 9BN M1DTG
HP7 9DZ G0VRV
HP7 9EP G1KTS
HP7 9EP G7STD
HP8 4AR G3DTX
HP8 4BG M1DYO
HP8 4HQ G3LPY
HP8 4HU M0KHN
HP8 4LG G3PBI
HP9 1AE M0GPY
HP9 1AW G0AFT
HP9 1AW M3MVR
HP9 1AW M6GFY
HP9 1BD G8FIG
HP9 1SG G7CZF
HP9 1YB G0BKR
HP9 2AE G7ABQ
HP9 2LA G3BII
HP9 2QU G3OHX
HP9 2XH G4LHE
HP9 2XP G3BEX
HP9 2YJ M6LGV
HR1 1AH M6JTA
HR1 1BY G8OHH
HR1 1HE 2E0DQZ
HR1 1HE M6DUW
HR1 1HQ G7KVT
HR1 1HT G4BJD
HR1 1NT M6WRS
HR1 1QL G6KWZ
HR1 1QX M1EIR
HR1 1RB G3WRA
HR1 1SS M1CAR
HR1 1SY M7LKZ
HR1 1TJ 2E1GFW
HR1 1XQ M5RJC
HR1 1XY G8IVO
HR1 2HB G4OJP
HR1 2QG M6XRF
HR1 2RQ G3LZM
HR1 3BJ G1AYP
HR1 3BJ M3KAL
HR1 3DF G0SVY
HR1 3DF M3SVY
HR1 3DR G3RDC
HR1 3DR G3YQC
HR1 3DU G7JUN
HR1 3ED G3NDM
HR1 3EH G7LSG
HR1 3EP G0SBA
HR1 3ET G8RAC
HR1 3LP G3IXZ
HR1 3LR G1JWD
HR1 3QY M7GTI
HR1 3RF G6PXQ
HR1 4AF G0PMS
HR1 4AS M6JVO
HR1 4BT M0MVL
HR1 4DL G6TID
HR1 4DQ M6PFA
HR1 4EN M3UKO
HR1 4EU M1BDH
HR1 4EU M3BDH
HR1 4JP G3ESY
HR1 4LR G7LVG
HR1 4PR G8XMS
HR1 4RQ G1YBG
HR1 4UL G0LJP
HR1-1DG G0JWJ
HR1-1DG G3YDD
HR2 0AH 2E0KSX
HR2 0AH M0SWV
HR2 0AH M6KSX
HR2 0AH M6NLQ
HR2 0AH M7OLO
HR2 0HP G4ASR
HR2 0JA G0NMC
HR2 0NX 2E0YAV
HR2 0NX 2E0YBL
HR2 0NX M0YAV
HR2 0NX M6WTG
HR2 0NX M6YAT
HR2 0QG G0JCC
HR2 0QG G7KTP
HR2 0QR M1FDK
HR2 0SW M0DRL

HR2 6NG G8UVU
HR2 6RU G1YFC
HR2 6RY M6XME
HR2 7DF G1YBB
HR2 7EG G1LSX
HR2 7FH M6JKM
HR2 7JU 2E0HDI
HR2 7JU M7ANH
HR2 7LJ M0JDP
HR2 7LS G0DYM
HR2 7LU G4MBJ
HR2 7NT 2E1MIB
HR2 7UG G1DYQ
HR2 7UG G7BTI
HR2 7XF M7LYA
HR2 7YL M6HOF
HR2 7ZD M1EKB
HR2 8AX 2E0IKG
HR2 8AX G0RBG
HR2 8AX M7RHB
HR2 8BW M7KGF
HR2 8HJ G0SYT
HR2 8JT G4OGW
HR2 8LU G4VSO
HR2 8LU M0MUU
HR2 8NA G0VKC
HR2 8PU 2E0AFP
HR2 9AN G7RLO
HR2 9BE G4XTF
HR2 9EW M3VHB
HR2 9HD G4ZSY
HR2 9LT M6DZR
HR2 9QN G0DHB
HR2 9RH G1RLF
HR2 9RT M6IDD
HR2 9RZ G1HWP
HR20EQ 2E0DJH
HR20EQ M0RNI
HR20EQ M6DAW
HR26LJ G0VHT
HR29SD M7FFW
HR3 5DW G0RHK
HR3 5EG 2E0OCY
HR3 5JT M0ECX
HR3 5JT M6HRI
HR3 5ND M0TUI
HR3 5NF G7EAA
HR3 5PP M7EVH
HR3 6BY G0IAH
HR3 6BY G1UMS
HR3 6HT G7IFR
HR3 6LF G3IUC
HR3 6LX M6LBV
HR4 0JY G1DRW
HR4 0LP G4LNZ
HR4 0NE M6BDI
HR4 0PN G4MET
HR4 0RU M6OAW
HR4 0SR 2E0PTA
HR4 0SR M3PTA
HR4 7AZ G0RQF
HR4 7BJ G4IIA
HR4 7BJ G8NOD
HR4 7EH G6PFX
HR4 7PN G3YYC
HR4 7QP M0BHR
HR4 7RW G3NFP
HR4 7SW G4GOG
HR4 8AJ G4FAD
HR4 8AN 2E0OKG
HR4 8AN M6DVF
HR4 8FJ M1AGE
HR4 8NN G4ZXQ
HR4 8NT 2E1HOF
HR4 8NT G6TVC
HR4 8PW 2E0VHQ
HR4 8QG G4RGB
HR4 8RQ G7HII
HR4 8SW G0UVE
HR4 8TH G4RCB
HR4 9HP G4WZT
HR4 9LA G0AAG
HR4 9LB M7XLA
HR4 9NE G4HZT
HR4 9NF G4DTB
HR4 9NJ G6DXC
HR4 9PF G7BLL
HR4 9QZ M7PCF
HR4 9TJ G3WAG
HR4 9TJ G6XPY
HR4 9TJ G8FFA
HR4 9XF G0OGX
HR5 3BA 2E0EVF
HR5 3BA M0WJA
HR5 3BA M3EVF
HR5 3DH 2E0XVJ
HR5 3DH M7VCH
HR5 3DJ G4SOG
HR5 3DJ G5PAT
HR5 3EG M6VPS
HR5 3HZ G3ZUS
HR5 3JS M6FTI
HR53PU 2E0GZN
HR53PU M6DRW
HR6 0AW 2E0OUD
HR6 0AW 2E0YHF
HR6 0AW M0WYP
HR6 0AW M7BVC
HR6 0AW M7JMF
HR6 0AZ G4KOK
HR6 0BG G4EBL
HR6 0EF G7UUN
HR6 0EP G7UOU
HR6 0HZ M6YTL
HR6 0JL M3HCP
HR6 0JZ G7AZC
HR6 0NG G3VST
HR6 0PF G8BPN
HR6 0SS G1JKN
HR6 8BJ G8GVO
HR6 8ER M6XAO
HR6 8EW M6NZM
HR6 8HE G4YPF
HR6 8HT G6OZU
HR6 8HT M0VDO
HR6 8HT M3AZK
HR6 8HT M7VDO
HR6 8NF M7BOJ
HR6 8QQ M0PSK
HR6 8RE M0XGG
HR6 8RN M7FPT
HR6 8RX 2E0JRS
HR6 8RX G6UQI
HR6 8RX M6GFA
HR6 8RX M6YRS
HR6 8SA M6AUF
HR6 8SD G0BRH
HR6 8SS G0JRV
HR6 8TQ G6GQG
HR6 9AY G8TMH
HR6 9BN G0ARF
HR6 9BZ M6GLZ
HR6 9BZ M6IHZ
HR6 9DN G4ASP
HR6 9EE G6FYE
HR6 9JE G0ODG
HR6 9JQ G4VOY
HR6 9PY M7FSU
HR6 9PY M7HCR
HR6 9QN G0OEA
HR6 9QR G4BOF
HR6 9UF G3XJP
HR6 9UU M6XLR
HR6 9UU M7XAM
HR68SA 2E0RYR
HR68SA M6RYR
HR69SY M7RSD
HR7 4BD G4CPL
HR7 4DZ G4ZWY
HR7 4ES M1BPK
HR7 4EY 2E0TOP
HR7 4LB G1MUQ
HR7 4LG M6IIB
HR7 4LY G3OGK
HR7 4UL G4FHQ
HR8 1BE G4FTX
HR8 1HG G3WPN
HR8 1LY G1WYN
HR8 1ND G0TWT
HR8 1NX G3VZR
HR8 1PP G0WHN
HR8 1RF M7CJL
HR8 1SG 2E0JAO
HR8 1SU G1JHY
HR8 1SU G1JHZ
HR8 2EG G0GFR
HR8 2FE G4PHX
HR8 2FR G8AVB
HR8 2JS 2E0KTQ
HR8 2QB G8ZVZ
HR8 2UU G8EZE
HR8 2XD G1GCJ
HR9 5AU 2E0CUW
HR9 5AU M0HNK
HR9 5AU M6XRT
HR9 5BD M6KYP
HR9 5BL G1ZND
HR9 5BL G8RZS
HR9 5JL G4EFH
HR9 5JY G3DYO
HR9 5PN G0FGS
HR9 5RL G1OKP
HR9 5ST G4JNZ
HR9 5TA G6XLR
HR9 5TE M0GZF
HR9 6DH G0TRI
HR9 6DQ M7IVA
HR9 6GJ G4AWF
HR9 7AS M6PQI
HR9 7BE G0IMV
HR9 7BE G6TSL
HR9 7BG M6JVZ
HR9 7DN G7RVI
HR9 7DR M1AHN
HR9 7GP G0TME
HR9 7HR M6EQB
HR9 7HT G0RIJ
HR9 7JH G0GQI
HR9 7JP M0HZE
HR9 7JR G3EFL
HR9 7NF G7RIU
HR9 7PR M6CUJ
HR9 7PW G8PFR
HR9 7QW 2E0LTX
HR9 7QW M0RPI
HR9 7QW M6EQN
HR9 7SS M7PJK
HR9 7SW M7JFR
HR9 7TL G8CMU
HR9 7TW G7AEG
HR9 7TW M0TAY
HR9 7TW M1EJJ
HR9 7UF G0EHT
HS1 2JG GM1ALS
HS1 2JN GM4PTQ
HS1 2NP GM0IPW
HS1 2PG GM7JED
HS1 2QN MM3IIT
HS1 2UB MM0JKH
HS1 2UP MM6CTQ
HS1 2UR MM6BIO
HS12SX 2M0HRP
HS12SX MM0KMQ
HS12SX MM7BTX
HS12TL 2M0APB
HS2 0BG GM7FAS
HS2 0BP MM7SWO
HS2 0DD GM6TYX
HS2 0DZ GM0LZE
HS2 0EU GM0PKW
HS2 0EZ GM4BMS
HS2 0HD 2M1EOV
HS2 0HD GM0HMM
HS2 0HE GM0WZO
HS2 0HE MM0BPV
HS2 0HU GM0NHT
HS2 0LP MM6BIP
HS2 0LR MM6OEC
HS2 0PD MM0TYR
HS2 0PL MM0KNN
HS2 0PW MM0KGK
HS2 0QB GM1MQA
HS2 0QB GM3NBE
HS2 0QB GM4KGK
HS2 0QW GM0FSY
HS2 0QX MM0IOL
HS2 0QX MM1RMS
HS2 0QX MM3SYB
HS2 0QX MM7MKV
HS2 0RT MM0TSB
HS2 0SW 2M0AKI
HS2 0SW MM3OHD
HS2 0XF GM7JFN
HS2 9AU GM7KUN
HS2 9DF MM7CMB
HS2 9DH 2M0FNI
HS2 9HR MM0UIG
HS2 9JA GM6VHA
HS2 9JT MM6ELW
HS2 9PJ 2M0DNM
HS2 9PJ 2M0WNB
HS2 9PN 2M0OSK
HS2 9PN MM3OSK
HS2 9QE GM0TVT
HS20PB MM6JZW
HS20SW GM7PBB
HS29AJ MM7ECV
HS3 3JA MM3WVP
HS6 5BP GM1MLY
HS6 5EU 2M1LPT
HS6 5EU 2M1WEZ
HS6 5EU GM0HBF
HS6 5EU MM0IRC
HS6 5HD 2M0RBQ
HS6 5HD GM8SAU
HS6 5HD MM0SIA
HS6 5HR GM7VEC
HS8 5SX MM6AVH
HS8 5SY GM7DMN
HS8 5TE MM3BCA
HS8 5TS 2M0TNM
HS8 5TU MM0CWJ
HS8 5TU MM6TEW
HS8 5TX MM6SML
HS9 5XU MM1FHZ
HS9 5XX GM0OGN
HS9 5XX MM3BRR
HS9 5YD MM0IOB
HS9 5ZE GM6SEL
HU1 2AJ M1EJL
HU1 2ES 2E0DUF
HU1 2ES M0IDJ
HU1 2ES M6FYJ
HU10 6AJ G8DCD
HU10 6AR G1RVH
HU10 6AW G3TEU
HU10 6BJ G0KLG
HU10 6DD G4NJB
HU10 6EU M6MKK
HU10 6HB M6UYP
HU10 6HS G4LNR
HU10 6LW 2E1ACO
HU10 6SF G0UGA
HU10 6SG M0DGU
HU10 6ST G0VQO
HU10 7AF M0KVK
HU10 7AF M6OGK
HU10 7DT G4CGG
HU10 7DT G4EKT
HU10 7HD 2E0OKP
HU10 7JE G4AQA
HU10 7JJ G8JIU
HU10 7JN 2E0FZM
HU10 7JN M3FZM
HU10 7JU G8NIE
HU10 7TL G4LBC
HU10 7UX G7PER
HU106ST G4BYG
HU106ST M0ARC
HU11 4AX G0GBW
HU11 4BN G4FEJ
HU11 4BN G6ZEN
HU11 4DJ M6NIX
HU11 4DN 2E0EPB
HU11 4DN M6HRQ
HU11 4EE M0BNR
HU11 4ER G7CVA
HU11 4HB 2E0FYW
HU11 4HB G0AJX
HU11 4HB M0JJI
HU11 4HB M7HUL
HU11 4NR G5GX
HU11 4PR M3KUH
HU11 4PY 2E0GZP
HU11 4PY M7RKS
HU11 4QH G6JFU
HU11 4RA M0BTN
HU11 4RN G4LZJ
HU11 4RN G7JZY
HU11 4RN M7JCL
HU11 4RY G1UCZ
HU11 4SD G0NAA
HU11 4UW G6ZDV
HU11 4XB M6GGD
HU11 4XF M0IOK
HU11 5BH G4YTV
HU11 5BH G6EKT
HU11 5QP M0VEY
HU11 5RN 2E0HYC
HU11 5RN 2E0RIO
HU11 5RN G3BOK
HU11 5RN M0HYC
HU11 5RN M3UMR
HU12 0DP 2E0THS
HU12 0DP G0THS
HU12 0DX M1CAV
HU12 0HH G4PMJ
HU12 0JF M3XSV
HU12 0LE G0UMK
HU12 0NE G3ULD
HU12 0PJ G4KFA
HU12 0PT G1RON
HU12 0QE G4PYW
HU12 0QE M0ZUB
HU12 0QG 2E0USH
HU12 0QG M7ERU
HU12 0RY G3UIF
HU12 0SF G7ONR
HU12 0TE G6AWP
HU12 0TH M6PLH
HU12 0TJ M1DCE
HU12 0TY G8NDP
HU12 0US 2E0EYC
HU12 0US M0PAY
HU12 0UX M1COV
HU12 8BG M0MFP
HU12 8ET G6AXC
HU12 8FP M6FPF
HU12 8GF G6ZJI
HU12 8GT M0MGF
HU12 8HA M0OZO
HU12 8HQ G1BGO
HU12 8LB G7WDM
HU12 8LH G0VAY
HU12 8LX G0NYK
HU12 8NH G7SCT
HU12 8NH M0SCT
HU12 8NT G4CJJ
HU12 8QH G0LUP
HU12 8QH G4ZJC
HU12 8TZ 2E0XKX
HU12 8TZ M0SLY
HU12 8TZ M6HUL
HU12 8UB G7MFO
HU12 8XF G1EFO
HU12 9AF G7RPW
HU12 9BP M1VRC
HU12 9GB 2E1TAP
HU12 9HP G4LHF
HU12 9JZ 2E0CJA
HU12 9LD M3BJJ
HU12 9LD M3JMJ
HU12 9NG G3YBU
HU12 9QG G1EQU
HU12 9RG G1EQF
HU12 9SZ M6IHA
HU13 0AS G7USB
HU13 0HB 2E0DVW
HU13 0JH G6GKT
HU13 0JP 2E0VDQ
HU13 0JW M0VDQ
HU13 0JW 2E0CIK
HU13 0ND M1CBC
HU13 0NS G6MPK
HU13 0PY G3NPT
HU13 0RL G4JIO
HU13 0RT G0KWE
HU13 0RT M3WJK
HU13 9BP G0CNV
HU13 9HU G0OYX
HU14 3AE M3PLP
HU14 3AN G1EIB
HU14 3BG G8FEK
HU14 3BX M6PBR
HU14 3DW G3YNO
HU14 3DW G3YQA
HU14 3HZ G0GVT
HU14 3JS G3RSB
HU14 3NL M0REC
HU14 3PR G7UJY
HU14 3RH G4XWA
HU15 1BA 2E0XYM
HU15 1BA M3XYM
HU15 1BE G0SWL
HU15 1BL G4VHM
HU15 1DA G7CLY
HU15 1DD 2E0EXK
HU15 1DD M6LQR
HU15 1EN G6KIA
HU15 1EN G6UQA
HU15 1FF G3VMV
HU15 1HY G4FGO
HU15 1JT M3HZW
HU15 1LA 2E0FSS
HU15 1LA G6PIB
HU15 1LA M1FCW
HU15 1LA M7GCW
HU15 1NL G8FDR
HU15 1NL M3UAG
HU15 1PE G8LZG
HU15 1QL G0GDS
HU15 1RE M1ASR
HU15 1RJ 2E0DQP
HU15 1RJ G4DQP
HU15 1RJ M6DQP
HU15 2AL G4MJT
HU15 2AL G4SGU
HU15 2AL G4UVF
HU15 2HR G6POZ
HU15 2JG G3VLR
HU15 2LQ G0FVM
HU15 2LT G0WJK
HU15 2NU 2E0DKI
HU15 2NU M0GWH
HU15 2QN M6IHT
HU15 2TU 2E0GDL
HU15 2TY G4YEB
HU15 2UH G0LHV
HU15 2XU G0JNQ
HU15 2ZU G3VDE
HU16 4AL M3PXL
HU16 4BX M0VES
HU16 4DQ M6XPW
HU16 4HN G0PAS
HU16 4NF G0KBP
HU16 4SA G7WGN
HU16 4SD G4EBT
HU16 5EH M0INM
HU16 5GU M0HVQ
HU16 5JD M0JBA
HU16 5JG G3NOP
HU16 5TF G0SXW
HU16 5TR G8EAH
HU16 5TT M6ACF
HU16 5UG G4SEF
HU17 0BG G7TYT
HU17 0DF 2E1FIQ
HU17 0EY 2E0DYP
HU17 0EY M6NYR
HU17 0EY M6WRH
HU17 0HP 2E0JMR
HU17 0ND G7FLS
HU17 0PA G8SGP
HU17 0PA M3NZG
HU17 0RU G4HBL
HU17 0SA 2E1BDC
HU17 0TH G4WEN
HU17 5NR 2E0SJE
HU17 5NR G7EFV
HU17 5NU G3RMA
HU17 5QA M7RSQ
HU17 7AB G8NJI
HU17 7BP G4HYD
HU17 7BW M6DIO
HU17 7HD G6UGS
HU17 7HP G4GPY
HU17 7NQ M6NEG
HU17 7NX G0TAS
HU17 7PN G8OEK
HU17 7RJ G7FDD
HU17 8FZ G8MEA
HU17 8GE G3MJM
HU17 8HY G4UOZ
HU17 8JL G4ESY
HU17 8LA 2E0EPJ
HU17 8LA 2E0KLI
HU17 8LA M6KLI
HU17 8LL M1EZK
HU17 8PF G0DMP
HU17 8PJ G4CMK
HU17 8PU G3NVX
HU17 8QA G4DEA
HU17 8QP G7UUO
HU17 8SD G3ZRS
HU17 8SD M0GWA
HU17 8WE 2E0USG
HU17 8WE M6NSG
HU17 8XB G6VCF
HU17 9GE 2E0IVD
HU17 9GE M7FFL
HU17 9GL G4CQN
HU17 9LH G0KVR
HU17 9PG G6EUF
HU17 9QR 2E0VCP
HU17 9QR M6VCP
HU17 9RH G1AJD
HU17 9RH G1AJE
HU17 9RH G8UAI
HU177HE M7RGF
HU18 1AF 2E0FLF
HU18 1AF M0FLF
HU18 1AF M6FLF
HU18 1AL G6GHE
HU18 1BF G0PFD
HU18 1BP G3OGE
HU18 1BX G0PSJ
HU18 1BX G7JZS
HU18 1EF G4FCT
HU18 1EF G8PZX
HU18 1ES G0VET
HU18 1ES G6DXS
HU18 1EU 2E0LKE
HU18 1EU M0LKE
HU18 1EW G4ADE
HU18 1JG G0IVP
HU18 1JU G7KHV
HU18 1LX M0CZR
HU18 1PB G8LJQ
HU18 1QD G1CFA
HU18 1QU 2E0EQK
HU18 1QU M6FFW
HU18 1RE G0JWY
HU18 1TP G8KFK
HU18 1TT G0GXX
HU18 1UA 2E0EQG
HU18 1UA M0KPJ
HU18 1UA M6KPJ
HU19 2AY 2E0BQR
HU19 2AY M0SGJ
HU19 2AY M3FGG
HU19 2BB 2E0ENW
HU19 2BB M3EYH
HU19 2DB G6ZBT
HU19 2DB M0EGC
HU19 2DB M0ZBT
HU19 2DS M6POP
HU19 2DT G2DPA
HU19 2DT G6YAS
HU19 2DT M3HRT
HU19 2DT M5EXY
HU19 2DW G1KST
HU19 2EW G1OSG
HU19 2EW M0BHH
HU19 2JA M6EUE
HU19 2LN M3ZHV
HU19 2LP 2E0MWW
HU19 2LZ G4SJI
HU19 2LZ G6YMI
HU19 2PB M6HGW
HU19 2PH G4XQY
HU19 2PH G6WYE
HU19 2QD G3RJM
HU19 2RG 2E0JCL
HU19 2RG M0UTX
HU19 2RG M0VQJ
HU192JA M7AOV
HU2 8JR G8ZFX
HU2 9BL G1WFS
HU20 3UU G1WQL
HU20 3UX G8HUG
HU20 3XH G6KTK
HU3 1NS M0AYO
HU3 1SA G0RUF
HU3 1SA G0UBY
HU3 1SA G6CTE
HU3 1YD M6BYP
HU3 2LT M0AAR
HU3 2QT M6NRV
HU3 2SG M6SDA
HU3 2TL 2E0LKH
HU3 2TT M6CLF
HU3 2TZ G0IRJ
HU3 3PR 2E0FOH
HU3 3PR M7OSS
HU3 5AS 2E0CNC
HU3 5AS M0GOC
HU3 5AS M6TKW
HU3 5LW M6GCP
HU3 5PP G1FPC
HU3 5PS M7GKR
HU3 6PQ M6OPE
HU3 6QY G0SWO
HU3 6QY G3AMW
HU3 6QY G3JDY
HU3 6QY G8GBY
HU3 6RD M3OPX
HU3 6SL M3IDK
HU3 6SL M3PHR
HU3 6XP 2E0EHN
HU35AE 2E0JOV
HU35AE M7PBO
HU35LX M7HOM
HU4 6AP M1ADN
HU4 6DW 2E0NMA
HU4 6DW M6ITP
HU4 6EF 2E0LDR
HU4 6EF M0LDR
HU4 6EF M3LRZ
HU4 6EW G1LAN
HU4 6LJ M0BXB
HU4 6LU M0BEO
HU4 6NG G6DLT
HU4 6PL G0IER
HU4 6PL M0HND
HU4 6RD M3UKH
HU4 6RW G1KNZ
HU4 6SN G1MZP
HU4 6SQ G1PGX
HU4 6ST M7PNO
HU4 6TL G3TKA
HU4 6UG G1DQD
HU4 6XJ G6HVE
HU4 7AH G0POQ
HU4 7AR M0GOY
HU4 7HB G4ZPQ
HU4 7JT 2E0HKZ
HU4 7JT M7EGY
HU4 7PL G0TRH
HU4 7QG G4WCD
HU4 7QH G6VVZ
HU4 7RH M3UNI
HU4 7SW G3GJA
HU4 7SW G8EQZ
HU4 7SW M0BMB
HU5 1LX M6YWG
HU5 1ND G0OVT
HU5 2AB M3GIK
HU5 2TA M6PAE
HU5 2TP M0KPE
HU5 3DW G3KDU
HU5 3DY M7JGB
HU5 3ER M0JDT
HU5 3HH M6CKY
HU5 3JY G0UNC
HU5 3NE M7WYD
HU5 3PF M3PBR
HU5 3SQ G1PCQ
HU5 3TW M6APA
HU5 3TW M6EKA
HU5 4AD M3KIU
HU5 4AG M1EIE
HU5 4ED G1PLU
HU5 4ED G4HJD
HU5 4EQ M6ZIY
HU5 4HH M6JER
HU5 4HN G1MCI
HU5 4HN G1SQW
HU5 4JZ M0RWL
HU5 4LP G3RDP
HU5 4PF G1EYY
HU5 5DJ G6VWF
HU5 5ES 2E0BVE
HU5 5HN 2E1XXX
HU5 5HS G4CBL
HU5 5HW G0TGX
HU5 5HZ G0GXZ
HU5 5JW G7JGY
HU5 5LE G0VNH
HU5 5LN 2E0ZRB
HU5 5LN M3ZRB
HU5 5NH 2E0SCA
HU5 5NX G6ABG
HU5 5QG M6ACX
HU5 5TD M6WRD
HU5 5XX 2E0SHW
HU5 5YH G7BAC
HU5 5YY G4COT
HU52BE M6LDG
HU6 7AG M6UUA
HU6 7AS G0CHP
HU6 7BA G4ERZ
HU6 7BW G1IUF
HU6 7DD G1LSZ
HU6 7DD M0TTL
HU6 7DZ G3ZZP
HU6 7HJ M6INA
HU6 7JT G0GXF
HU6 7LW M6HSY
HU6 7QB 2E0VCY
HU6 7QB M3VCY
HU6 7XE G4MPL
HU6 8DB M6JMR
HU6 8DG M6BZM
HU6 8DG M6BZN
HU6 8LH M6HJE
HU6 8NX M7TJC
HU6 8SA G3PQY
HU6 8SB G4AUQ
HU6 8SB G6LNV
HU6 8SB M0HFC
HU6 8SB M1DWT
HU6 9EB G0LPT
HU6 9ER G0RLB
HU6 9ER M3SBB
HU6 9EU G8VHG
HU6 9EZ G1TDO
HU6 9EZ M3TYS
HU6 9HA M3GZU
HU6 9HG M0SRN
HU6 9JH M3VPH
HU6 9LS 2E0JRL
HU6 9PH M7BBY
HU6 9RA G8ZFD
HU6 9RW 2E0TWZ
HU6 9SA G1TDN
HU6 9SA M6LEC
HU67DT M6NFT
HU7 0DU G0UKS
HU7 0DU G0VTD
HU7 3AF G0OYQ
HU7 3AL G1XGE
HU7 3FA G4KXU
HU7 3GS M7SYB
HU7 4AL M3MPT
HU7 4AR M7YZZ
HU7 4BX 2E0NYE
HU7 4HH 2E0DXW
HU7 4HH 2E0NSQ
HU7 4HH M0MRX
HU7 4HH M1BLO
HU7 4HH M3NSQ
HU7 4HN G0WEA
HU7 4PQ G0BAP
HU7 4PY G0EPA
HU7 4RF G1ARD
HU7 4RP M3VGP
HU7 4RU G7POW
HU7 4ST 2E0DIX
HU7 4ST 2E0ELO
HU7 4ST 2E0HAL
HU7 4ST M3XHY
HU7 4ST M3ZZD
HU7 4ST M3ZZW
HU7 4UY G8WVB
HU7 4YE G1FQI
HU7 4YS 2E0EPF
HU7 4YS M6JYT
HU7 4YZ G6KRJ
HU7 4ZA M0BOM
HU7 5AP M6LRR
HU7 5AQ 2E0OAH
HU7 5AQ M6OZY
HU7 5AS G7UCZ
HU7 5BU G0PTL
HU7 5BX M0RKK
HU7 5BX M7RPK
HU7 5DH 2E0GRH
HU7 5DH M6RRH
HU7 5FB M3RIE
HU7 5XD 2E0VTC
HU7 5XD M0PLR
HU7 5XD M3XMQ
HU7 5YY G0VVP
HU7 6DE G0MKZ
HU7 6EE G0WDX
HU7 6EL G6WVR
HU74XZ 2E0JWH
HU8 0AN M1DRK
HU8 0HB 2E0OYR
HU8 0HB M0OYR
HU8 0HB M7OYR
HU8 0RU 2E0FIY
HU8 0ST G0WFF
HU8 7RY 2E0BRQ
HU8 7RY M3OOQ
HU8 7RY M3SHW
HU8 7TJ G6WQH
HU8 8LQ M6NYW
HU8 8NA M7EYP
HU8 8QR M3KLB
HU8 8TQ M3GWW
HU8 8TQ M3HME
HU8 9BN G1ADE
HU8 9DG G0UTX
HU8 9EW M6TDK
HU8 9LY M3MWM
HU8 9NA 2E0LTC
HU8 9NA M0BOM
HU8 9NA M6RWX
HU8 9RY 2E0FLA
HU8 9RY M6YLR
HU9 1EJ G1CFB
HU9 1HR 2E0IDL
HU9 1HR M0FQZ
HU9 1HR M7FQZ
HU9 1JB M3LSL
HU9 1ST 2E0GMW
HU9 2AS 2E0HKR
HU9 2PF G7BTP
HU9 2QR G7IWE
HU9 2TE G0TUJ
HU9 3JU G7UGW
HU9 3LB G0NAU
HU9 3PF M6BIG
HU9 3PN M0BYI
HU9 3PR M0GLV
HU9 3RG M1DHJ
HU9 3RH M7ACE
HU9 4BE M3MPM
HU9 4DR 2E0KCW
HU9 4DR M0PKE
HU9 4DR M6BWP
HU9 4HH M3LUD
HU9 4PG 2E0UVZ
HU9 4PG M0HBO
HU9 4PR 2E0LZQ
HU9 4PR M0XLH
HU9 4PR M6LZQ
HU9 4QR 2E0RWP
HU9 4UE G1ZSV
HU9 5JZ 2E0HDX
HU9 5LH G0PCD
HX1 2BJ G8WFP
HX1 2EP M1RIG
HX1 3EA G6FPF
HX1 3LE M0CRG
HX1 3LE M0GQB
HX1 4QG M3BXY
HX1 4TA G0CVC
HX1 5PU 2E0VRD
HX1 5PU M6HXI
HX1 5PU M6VRD
HX2 0HT M3BXX
HX2 0NP G3TAY
HX2 0PA G7ELX
HX2 0PJ G0OZP
HX2 0PL G1RZZ
HX2 0QP M3DWA
HX2 0RB G8EXN
HX2 0RH G7OMF
HX2 0RL G4EHD
HX2 6BG G2UG
HX2 6EX G8RKX
HX2 6HQ M6DJN
HX2 6UX G6CNL
HX2 7HG G0INK
HX2 7HP G8SDE
HX2 7PN G6YGV
HX2 7RB G7TDN
HX2 7SG M3HTE
HX2 7TX G4NSH
HX2 8AA G3UI
HX2 8DL G1GWX
HX2 8DL G1VVB
HX2 8RE M1CMR
HX2 8RL M6BKN
HX2 8SG G8SJA
HX2 8SN M6VGB
HX2 9AZ G4XYS
HX2 9EF M6RRR
HX2 9JD G6GGV
HX2 9JH G0JAQ
HX2 9JJ M7CHD
HX2 9JQ G2SU
HX2 9JQ G3TQA
HX2 9LY G8XZZ
HX2 9PQ G0CBI
HX2 9PZ G6AVN
HX2 9SQ M6DKS
HX3 0AG G6COG
HX3 0AH M0HVI
HX3 0AL G4OYZ
HX3 0AL M3OOX
HX3 0AL M3OOY
HX3 0AL M3OPC
HX3 0AL M3OPD
HX3 0BD G7SWZ
HX3 0DL M6EMH
HX3 0HA M3IDA
HX3 0JQ 2E1SKR
HX3 0SR G0SPX
HX3 0SU G7IHX
HX3 0TH M0RWD
HX3 5FB M6VCU
HX3 5LU 2E1BOO
HX3 5LU 2E1BVJ
HX3 5LU G3UGF
HX3 5NL M0DEJ
HX3 5NQ M7TFG
HX3 5PA 2E1AHK
HX3 5PE G1IYA
HX3 5QF G8REO
HX3 5QG M6HFX
HX3 5QU G1WHU
HX3 5RH M7SFD
HX3 5RP G7EUJ
HX3 5SZ G1HCU
HX3 6PL M0PDW
HX3 7BT G1NDQ
HX3 7EB M6AHS
HX3 7EP G6MDC
HX3 7LB G8UTW
HX3 7NA G0PMU
HX3 7NE G8ZXZ
HX3 7NH G0HRJ
HX3 7NH M3KJK
HX3 7PW G4KEX
HX3 7QY M0HFX
HX3 7QY M6LVW
HX3 7RN 2E0BAB
HX3 7RN M0TKD
HX3 8HB G7FZJ
HX3 8NJ M3PNH
HX3 8QF G4BZS
HX3 8TJ 2E0UPU
HX3 8TJ G1ILO
HX3 8TJ M0UPU
HX3 8TJ M3JJU
HX3 8TJ M6UPU
HX3 8XP M6FYN
HX3 9BH M6FAX
HX3 9DR M6JBL
HX3 9DT G0GVB
HX3 9EE G6XTT
HX3 9PS G7DWY
HX3 9PS M0INB
HX3 9QZ G8NQN
HX3 9SE M3YEM
HX4 0BW 2E1HUE
HX4 0DE M6PIY
HX4 0EW G8HTB
HX4 0HD G3PSB
HX4 8HT M6JJE
HX4 8HX M7FMM
HX4 8JF G6APB
HX4 8NU G3YGZ
HX4 8PA M3MQY
HX4 8QF G1MWT
HX4 9AS G4LIX
HX4 9DN M3JQT
HX4 9HZ G4MSP
HX5 0AJ M7AXF
HX5 0BB G4FMM
HX5 0BB M3LAG
HX5 0DR 2E0MWT
HX5 0DR M7FDW
HX5 0EE M0XRL
HX5 0HR M7OSH
HX5 0LA 2E0HQJ
HX5 0LS G6XNJ
HX5 0PF 2E0PHU
HX5 0QA G3YCJ
HX5 9AY M6IJT
HX6 1AD G6DQU
HX6 1AH M3SNR
HX6 1BX G8UTH
HX6 1EE M0UGX
HX6 1EE M6UGX
HX6 1LB 2E0PEC
HX6 1LB M6PEC
HX6 1NL G4AES
HX6 1NS G0DUN
HX6 2EE M6JRC
HX6 2HE 2E0BJQ
HX6 2QP M3GSM
HX6 2RP G1MBM
HX6 2RU G0FOB
HX6 2SY M6JNC
HX6 3EA M7HFX
HX6 3HZ M6GDQ
HX6 4AG G0WFG
HX6 4AG G7NSN
HX6 4LQ G8EWN
HX6 4NB M6GJB
HX6 4NU M3JQS
HX6 4PA M1EYG
HX6 4RZ G6AUC
HX7 5DF M7EUP
HX7 5HS M6DTP
HX7 7AL 2E0KPT
HX7 7ED G0TOB
HX7 7HD G0TVB
HX7 7HD G6TRG
HX7 7JA G4EFX
HX7 7JP 2E0JCK
HX7 7JP M0JCK
HX7 7NX G6XPZ
HX7 7PB G7CLX
HX7 7PG G3MLS
HX7 7PH G8BBZ
HX7 8SJ 2E0FUJ
HX7 8SJ M6NWL
IG1 2ER G6ZQJ
IG1 2UF G0IQK
IG1 4RY M0NSP
IG1 4RY M6FDW
IG1 7TJ G6EOE
IG10 1BN M7OPH
IG10 1HT 2E0DQL
IG10 1HT M6VTG
IG10 1SB G0LWM
IG10 1SB G7EEE
IG10 1TS G0SGX
IG10 2AD G4WTQ
IG10 2AJ G4IMU
IG10 2LT M6GJH
IG10 2QL 2E0ZAU
IG10 2QL M6ZAX
IG10 2RE G0TOC
IG10 2RE G7KJV
IG10 2SA G3NQT
IG10 3BY M1EZC
IG10 3EP G0PFY
IG10 3JT G1DJI
IG10 3PL G1IZA
IG10 3PR G1IVP
IG10 3QY G7VKB
IG10 4EB G8XAO
IG10 4QE 2E0IUA
IG10 4QE M7ESQ
IG11 0DS G6MPJ
IG11 0LD M0BDQ
IG11 0NL G0POY
IG11 7EB M3ZBS
IG11 7QE M3WXW
IG11 7UW G1UAL
IG11 9DD G3OHS
IG11 9EE G0BIW
IG11 9HS M6PMB
IG11 9NY M1EEY
IG11 9PS M3DVH
IG11 9PS M3EKY
IG11 9XB G0OVE
IG11 9XB G6JMJ
IG2 6AS G3CAZ
IG2 6EQ G4PQW
IG2 6QA G7EMH
IG2 6QY G6RAH
IG2 6YG G1OYM
IG2 7DL G6EEF
IG2 7NQ G4JRD
IG2 7SF G7JUZ
IG2 7SF M3JUZ
IG3 8NN G0VKH
IG3 8XE G7VBF
IG3 9TB G0MEW
IG4 5AE G3ALK
IG4 5HA M6OPJ
IG4 5HN G3WPR
IG4 5JX M0CTF
IG5 0DL M1EMX

Postcode

IG5 0HN 2E0HTI
IG5 0HP G1ICK
IG5 0NP G8JFL
IG5 0RZ G6XSB
IG5 0TG G8VBK
IG5 0XN G8YAE
IG6 1BJ G0ATP
IG6 1PJ G4YUF
IG6 2AZ M6ISE
IG6 2BN M0OIA
IG6 2NJ 2E0XBA
IG6 2NJ M0XBA
IG6 2NJ M6JSI
IG6 2PH G4GQL
IG6 2QE 2E0DDH
IG6 2QE M0MBD
IG6 2QE M6DLH
IG6 2QU G8HST
IG6 2QU G8PRJ
IG6 3AH G8BUF
IG6 3SR M1EDO
IG6 3TF G1HEQ
IG7 4DG 2E0CFT
IG7 4EA M1CPD
IG7 4HQ 2E0EER
IG7 4HQ M6AGD
IG7 5AU G2CD
IG7 5ED G4GMN
IG7 5HJ M7GWJ
IG7 5HZ G4NZB
IG7 5QZ G7VGJ
IG7 6AD G8LGU
IG8 0FB G6HIU
IG8 7LS 2E0DOW
IG8 7LS M0CEO
IG8 7LS M6LFE
IG8 7QU G4AJG
IG8 8AG 2E0TUO
IG8 8AG M7JTU
IG8 8DW 2E0RMM
IG8 8DW M6RML
IG8 8JN 2E1ICT
IG8 9AA G6SWT
IG8 9HY G4BNB
IG8 9LQ G7GCI
IG8 9QZ M1ETW
IG9 5QE G4BMU
IG9 5QF 2E0EFO
IG9 5RU G8MCR
IG9 5TZ G4YOA
IG9 6AQ G3VGR
IG9 6BY G0LWI
IM1 3DP MD6RSV
IM1 4BB GD7IEH
IM1 4EG GD0PLQ
IM1 4HQ GD3GBG
IM1 5EQ MD3WXS
IM2 1HT GD0SFI
IM2 1JE 2D0VMN
IM2 1JE 2D0WNY
IM2 1JE MD0LRK
IM2 1JE MD3VMN
IM2 1JE MD7KSL
IM2 1JT MD7LTT
IM2 1JT MD3JTT
IM2 1NX 2D0EYK
IM2 1NX MD0NSS
IM2 1NX MD6NSS
IM2 1QX MD3LEP
IM2 2AD GD1JNB
IM2 2AR MD0IOM
IM2 2EN MD3MXU
IM2 2EU MD6KBW
IM2 2LA MD3GAB
IM2 2LA MD3MAN
IM2 2NP GD3YEO
IM2 3RQ GD0JWR
IM2 4AH MD0RKI
IM2 4AQ 2D0WFK
IM2 4AQ MD0WRK
IM2 4AQ MD6WFK
IM2 4AQ MD6ZKK
IM2 4AR 2D0IMN
IM2 4AT MD6NRI
IM2 4BP GD3FLH
IM2 4BP GD3TNS
IM2 4PE GD4FMB
IM2 4PE GD4ZAB
IM2 5BH GD0TEP
IM2 5BH GD7JQI
IM2 5BQ GD0PLR
IM2 5EH GD0IDU
IM2 5ET MD7KNE
IM2 6AF GD8GRE
IM2 6AG GD0KQE
IM2 6EH GD4EBA
IM2 6EY 2D0OMN
IM2 6HQ 2D0YLX
IM2 6HQ MD3YLX
IM2 6HR 2D0GBM
IM2 6HR MD0VMD
IM2 6HR MD3VMD
IM2 6HW MD1DNT
IM2 6PB MD3CPK
IM2 7AW GD0KPN
IM2 7DX GD0JKA
IM3 2DF GD3YUM
IM3 2JF GD6JHP
IM3 3BU GD6HCB
IM3 3BU GD4RVQ
IM3 3BW GD4SVD
IM3 3GA GD0BFN
IM3 3GB 2D0CFZ
IM3 3GB MD0PWI
IM3 3HY GD0LQL
IM3 4AT 2D0JEA
IM3 4AT MD3WFJ
IM3 4AT GD0OUD
IM3 4AT GD7HTG
IM3 4BG MD3ZHD
IM3 4EH MD7VDL
IM3 4ET GD4DPK
IM3 4LE MD6TDU
IM3 4NR GD0BCN
IM3 4NR MD3OKG
IM3 4NR MD3OKH
IM4 1BB GD0TFO
IM4 1BJ GD7LAV
IM4 1EZ MD3LPW
IM4 3BP GD0GFV
IM4 3HE GD4EIP
IM4 3JB GD3MBC
IM4 3JB MD3UGY
IM4 3JB MD3WBC
IM4 3JP GD0AMD
IM4 3JP GD1LVY
IM4 3JP MD0GLK
IM4 3JP MD0HHH
IM4 3JP MD6IZI
IM4 3LZ GD4XWF
IM4 4AR MD6IKR
IM4 4AX 2D0PEY
IM4 4AX MD6NZO
IM4 4ES GD4FWQ
IM4 4ES 2D0KTH
IM4 4ES MD7MAN
IM4 4EW GD0JGX
IM4 4FE GD0IFU
IM4 4LT 2D0HXL
IM4 4LT MD0KKG
IM4 4LT MD0NKX
IM4 4LT MD7MTM
IM4 4LU MD3LEG
IM4 4ND MD7DYW
IM4 4NF 2D0BCR
IM4 4NF MD0MAN
IM4 4NG GD3FXN
IM4 4NZ MD6MGP
IM4 4QP MD3PER
IM4 4QZ GD0PDN
IM4 5EA GD4XOD
IM4 7AP GD3LSF
IM4 7AP GD7ESU
IM4 7DF GD3ZEX
IM4 7JB GD3HFC
IM4 7JY GD0NFN
IM4 7PW GD0ELY
IM4 7PW GD4XTT
IM44ES GD4GWQ
IM5 1BE GD4OEA
IM5 1BQ MD3LWQ
IM5 1GH MD6TSW
IM5 1GH MD7EEO
IM5 1HP GD4MCR
IM5 1JJ GD4NTR
IM5 1PJ GD4MDY
IM5 1PN GD4IOM
IM5 1PN GD6ICR
IM5 1PX GD4UQO
IM5 1UF MD3MCB
IM5 1XD MD7EWU
IM5 2AE 2D0WFH
IM5 2AE GD4UHB
IM5 2AE MD6WHG
IM5 3BA GD6TWF
IM6 1AF GD1GHK
IM6 1AF MD6AGF
IM6 1HU MD6KFH
IM7 1BL 2D0DRG
IM7 1BL MD0ICS
IM7 1BL MD6GRD
IM7 1ED MD3ONP
IM7 1HE GD4ZUU
IM7 2AF GD4WOW
IM7 2AY GD3YTE
IM7 2EY GD4RGR
IM7 2HB GD1XMA
IM7 2HR GD3PML
IM7 3BL GD4PTV
IM7 3DA GD0KEO
IM7 3EU GD7RVP
IM7 3HP GD4RFK
IM7 3HP GD3RFK
IM7 3HP MD3BZA
IM7 4AG GD4HOZ
IM7 4AQ GD1IOM
IM7 4AQ GD8MUF
IM7 4HY 2D0STL
IM7 5AH GD3XNU
IM73EB GD1AQY
IM8 1JF GD0KWM
IM8 1LJ MD0BJM
IM8 1NF GD0HWA
IM8 1NF GD8ANU
IM8 2BH MD6LET
IM8 2EG MD3OIS
IM8 2EX MD3EVY
IM8 2JN MD7PFM
IM8 2PT GD8BUE
IM8 3AN GD4AZJ
IM8 3DA 2D0JBE
IM8 3EB GD4IHC
IM8 3LR MD7BJL
IM8 3NZ 2D0VES
IM8 3NZ MD6ZEE
IM8 3PU MD0FIX
IM8 3UF MD0DPG
IM8 3UH GD6AFB
IM8 3UH MD3MLB
IM8 3UH MD6YBE
IM8 3UH 2D1GCC
IM9 1BE MD6NUG
IM9 1BL MD1RPC
IM9 1HP MD3KSN
IM9 1HR 2D0IOM
IM9 1HW 2D0VJK
IM9 1HW MD0RLS
IM9 1NL GD7DUZ
IM9 1NW MD3KCT
IM9 1PN 2D0LIX
IM9 1PN MD7MEB
IM9 1TQ GD4GSR
IM9 2DY MD3XUA
IM9 2ED GD1HIA
IM9 2ES GD0TFG
IM9 2ET MD3UMN
IM9 2EU MD1DXW
IM9 2EU MD3DXW
IM9 2EU MD6RAQ
IM9 2EW MD0PMN
IM9 2HA GD6ARJ
IM9 2HF 2D0EDQ
IM9 2HF GD0BCJ
IM9 3AH GD3EFD
IM9 3AY 2D0JKW
IM9 3BA GD0IOM
IM9 3BA GD4GNH
IM9 4EF GD6OXG
IM9 4EP GD6XHG
IM9 4HJ 2D0TSE
IM9 4HJ MD0OOH
IM9 4HJ MD6SNV
IM9 4HN MD0XCE
IM9 4LQ 2D0RGW
IM9 4NB GD4VBA
IM9 4NL GD0HYM
IM9 4NL GD0MAN
IM9 4NR 2D0TRL
IM9 4PR MD6IUH
IM9 5AW MD6HHT
IM9 5ED GD0MWL
IM9 5ED MD3AAI
IM9 5ET GD0WSK
IM9 5LH 2D0NAU
IM9 5LH MD0YAU
IM9 5LP 2D0XPS
IM9 5LP MD6RHN
IM9 5LX GD4RAG
IM9 5NR MD7FHZ
IM9 5PR GD1USI
IM9 6DB GD4IZL
IM9 6DB GD8PPU
IM9 6EL GD0VIK
IM9 6EL MD3EEW
IM9 6EN MD3JGS
IM9 6EP GD4FJI
IM9 6EU 2D0NOT
IM9 6HA MD6DYM
IM9 6HR GD8EUH
IM9 6LS GD3ZZN
IM9 6NB GD3JIU
IM9 6ND GD4ZZN
IM9 6PH MD3OED
IM9 6QU GD7KAM
IM9 6QZ MD3LJS
IM9 6TE GD8EXI
IM9 6TJ GD7ARS
IM9 6TJ GD7DPG
IM94HA GD3SKZ
IM95DN MD3YUQ
IM99 2QF GD4TEM
IO9 9SU GD1MZJ
IP1 1RS M0IOV
IP1 1RS M0IOZ
IP1 2PH G7SMN
IP1 2PR G6ZQT
IP1 3QF G3XCO
IP1 3QZ G0UBM
IP1 3SA G7LKY
IP1 3SD M7PUP
IP1 4DP M3HGL
IP1 4JY G1YAB
IP1 4JY G1WQY
IP1 4LD G4LRB
IP1 4LU M3ZBR
IP1 4NZ G8MFS
IP1 5DR G0AKC
IP1 5EA G7OCH
IP1 5HD G1WQT
IP1 5HS G8FTW
IP1 5JX G3YWM
IP1 5LD 2E0GCY
IP1 5LD M6GCK
IP1 5LR G7PLV
IP1 5NJ 2E0AXZ
IP1 6AB 2E0UCB
IP1 6AB G7VLJ
IP1 6AB M7CBX
IP1 6BD G1HGF
IP1 6DU G0SAR
IP1 6DU G8XOR
IP1 6EW G7UQV
IP1 6HP M3UAQ
IP1 6JB G0HJK
IP1 6LB G0JNY
IP1 6PA M6SFT
IP1 6PQ G8VNP
IP1 6PQ M0AKK
IP1 6PS M3YGZ
IP1 6RE M0JAJ
IP1 6RG G4YGV
IP1 6RL G8ZQA
IP1 6RL M3ZQA
IP1 6SS G7FNN
IP10 0DE G4LXX
IP10 0EA M7MLZ
IP10 0EU G4SYG
IP10 0EW G6RTY
IP10 0FN M7ABX
IP10 0JX G4XQD
IP10 0PF G4LYD
IP10 0PF G4RHR
IP10 0PL G4DYH
IP10 0PP M0JSA
IP10 0PP M6ALX
IP10 0PP M6ISJ
IP10 0PZ G8BHC
IP10 0QF G0KBM
IP10 0QU 2E0VAA
IP10 0QU G4DDK
IP10 0QU G8EMY
IP100NP G8MHA
IP11 0RG G8BHK
IP11 0UU G3OJ
IP11 0UU G3XIX
IP11 0UY 2E0EDR
IP11 0UY M0LOC
IP11 0UY M6HKB
IP11 0XG 2E0SYB
IP11 0XG M6HWD
IP11 0XL G8MXV
IP11 0YQ M0DNJ
IP11 2BG M0MCB
IP11 2EA M3YKN
IP11 2NS G8EAX
IP11 2NT M0LVW
IP11 2NY G1JRL
IP11 2PN G8BBV
IP11 2PQ G0CFT
IP11 2UH G4YQC
IP11 2UH G4ZFR
IP11 2UH G7OPS
IP11 2UL G7TUG
IP11 7EG G3WDE
IP11 7JR M1EJS
IP11 7NR G0UPD
IP11 7NX G3YHK
IP11 7PR G4BTX
IP11 7RL G7ILA
IP11 7RP G3XFF
IP11 7RR G0PPS
IP11 7RR G0PRU
IP11 7RR G4DNX
IP11 7SP G0DUS
IP11 9AL G7MNS
IP11 9BS 2E0IPX
IP11 9BU G6MKO
IP11 9DE M3BDQ
IP11 9DS M3YFA
IP11 9EE M6BFF
IP11 9HS G4IVC
IP11 9JJ M0HQL
IP11 9JT G0GSL
IP11 9LG G3TRD
IP11 9LR G4FBV
IP11 9LS M6TYZ
IP11 9NN G3ZKZ
IP11 9NP G8NYC
IP11 9PJ G8RRN
IP11 9PJ M6FLX
IP11 9PL G3MVE
IP11 9PS G8XRL
IP11 9SL 2E1CZS
IP11 9SS G6MCG
IP11 9TJ M0MWR
IP11 9TL 2E0HJK
IP11 9TL G8XUL
IP11 9TL M6HJK
IP11 9TS M3MFG
IP11 9TX G0XEG
IP112JW 2E0XQA
IP112JW M7BFA
IP12 1AH G8VZZ
IP12 1BD M7DXN
IP12 1BE M0JPS
IP12 1GZ M7CYI
IP12 1HB G8AXO
IP12 1JE M7FLL
IP12 1JL 2E1GXU
IP12 1JQ G7JXR
IP12 1JS G0OZR
IP12 1LB 2E1FSR
IP12 1LD G1AGK
IP12 1LQ G0DDZ
IP12 1PE M0BPW
IP12 1PZ M7JCN
IP12 1RL M6BHY
IP12 1RN G7DWN
IP12 1RW M1ADV
IP12 2AJ G3RHP
IP12 2ED G6FS
IP12 2ED M0BCT
IP12 2GA M6DON
IP12 2GL M3VDO
IP12 2HZ G4AVS
IP12 2JE G4HHA
IP12 2JF M1DWW
IP12 2JH 2E0TDN
IP12 2JH M6TCI
IP12 2PL G3ZYP
IP12 2PL G8BZJ
IP12 2QA M6SWD
IP12 2QG G0JFM
IP12 2QG G1YOU
IP12 2QG M7AAZ
IP12 2TE M6AZN
IP12 2TJ M6MWT
IP12 2UR M6KTO
IP12 2UT G3RTB
IP12 2UW M6WUB
IP12 3DZ M6LMG
IP12 3EP 2E0DBY
IP12 3EP M6GSD
IP12 3JT 2E0ROP
IP12 3JT M0NOC
IP12 3JY G0ILZ
IP12 3JY M6AOL
IP12 3LL G0THW
IP12 3QL M1AGP
IP12 3QU G1XWS
IP12 3TP G0VWE
IP12 4DU 2E0YZC
IP12 4DU M0GQR
IP12 4DU M3YZC
IP12 4ET G8MTW
IP12 4HE 2E0TYL
IP12 4HE M3TYL
IP12 4HR G3PFH
IP12 4HR G7SCR
IP12 4HW G3TLY
IP12 4JP G4FSG
IP12 4JR G4AKW
IP12 4JW G8BRL
IP12 4JZ G4WAG
IP12 4NN G7RMQ
IP12 4NY M1CGM
IP12 4NZ G8ONH
IP12 4PT G1DIK
IP12 4PT G3WTS
IP12 4RR M3STQ
IP12 4SL G1MOS
IP12 4ST G4DFD
IP12 4SU G4SYA
IP121PJ G7SMW
IP122JQ M7TJE
IP124EZ M7NES
IP13 0AT M6BID
IP13 0BP 2E0EVE
IP13 0BP M0SDY
IP13 0BP M3EBZ
IP13 0BP M3KIT
IP13 0BP M3LDC
IP13 0ES G4TRE
IP13 0HE G0RZG
IP13 0LN 2E0BEW
IP13 0LN M3IKR
IP13 0LR G3RPB
IP13 0LR G6KQ
IP13 0ND 2E1EJD
IP13 0QS G7ABO
IP13 0RQ 2E0FIR
IP13 0RQ M0MJF
IP13 0RQ M6FIR
IP13 0SL G6AYX
IP13 6DP M0BVT
IP13 6EB G3RVC
IP13 6ES G8VCU
IP13 6JF G8XYQ
IP13 6LA 2E0SUF
IP13 6LA M0SUF
IP13 6ND M6ADY
IP13 6ND M6KYK
IP13 6PL G0VQS
IP13 6QR 2E0XFX
IP13 6QR M7JPX
IP13 6SE 2E0LFR
IP13 6SU G0AOY
IP13 6TH 2E0CQQ
IP13 6TH M0IAH
IP13 6UP 2E0CQR
IP13 6UP M6BTT
IP13 7AW M6EGZ
IP13 7JP M6CCF
IP13 7NU M6EBX
IP13 7PP G4RSD
IP13 7RT M6NBW
IP13 8BB 2E0STQ
IP13 8BB M0LIE
IP13 8BB M6BZU
IP13 8DT G7JCF
IP13 8EB M7CHI
IP13 8EB M7COP
IP13 8LZ G3PYW
IP13 9AD 2E0DZA
IP13 9AD M7DZW
IP13 9FD G0IVV
IP13 9HA G8IQF
IP13 9HP G7MIP
IP13 9HP M6SXC
IP13 9LY M3DFP
IP13 9NP M6KCK
IP13 9NP M6OLW
IP13 9PQ G3LQR
IP13 9SL 2E1GXV
IP13 9TE G6HTT
IP14 1BT M1CGO
IP14 1GH G0HEV
IP14 1GS M0LUC
IP14 1GS M7LCY
IP14 1LP G3TAQ
IP14 1RJ G1HNH
IP14 1RY 2E0JBI
IP14 1RY M6JBG
IP14 1TD G3TZE
IP14 1TN 2E0OME
IP14 1TN M7SME
IP14 1TS 2E1BRC
IP14 1TS G6SUR
IP14 2AB G3SUK
IP14 2EY G0BEC
IP14 2FD M7JJA
IP14 2JE M0IZR
IP14 2JT G6PGM
IP14 2LT G6RAQ
IP14 2NA M0JKI
IP14 2NZ G0SCM
IP14 2RE G1UBH
IP14 3AB G7ONB
IP14 3AG G2TO
IP14 3DJ G4AFX
IP14 3GA G3MXH
IP14 3HE M6EQV
IP14 3NF G6JZV
IP14 3PA G4BIY
IP14 3RQ M3XQX
IP14 4DL G3XVL
IP14 4EG M7BGA
IP14 4EJ 2E0BBM
IP14 4EZ M5AEC
IP14 4LL M6JHN
IP14 4LP G0NMS
IP14 4NR G8SED
IP14 4NT M6DLA
IP14 4PP M7LPN
IP14 4PZ 2E0DIQ
IP14 4SP M6KNW
IP14 4TP G6QXT
IP14 5AN G4BJO
IP14 5AX M6EVC
IP14 5BB G1YRJ
IP14 5DS G7BUL
IP14 5ET G3XLG
IP14 5ET G4THN
IP14 5GH 2E0DEH
IP14 5GH M6LMJ
IP14 5HB G3ZEQ
IP14 5JL G3ZQU
IP14 5LP G0ERL
IP14 5LS M3IFB
IP14 5LX 2E1GRT
IP14 5PE G0OZS
IP14 5RE M7GBS
IP14 5SH M3XGL
IP14 5SN G6SYW
IP14 5SQ G5UOS
IP14 5UG 2E0XDD
IP14 5UG G0JJG
IP14 5UG M3VUH
IP14 5UG M3XDD
IP14 6AJ G4GBA
IP14 6BU 2E1HWJ
IP14 6DJ G4VPA
IP14 6HD G8ZZS
IP14 6LB G3ONL
IP14 6LX G3VNT
IP14 6LX G4EVN
IP14 6RN G8FMI
IP15 5EE M6CZI
IP15 5QE 2E0ALD
IP16 4AR 2E1HBF
IP16 4AR G1YRF
IP16 4AR G3DBJ
IP16 4BB 2E0IAF
IP16 4BB M0VCP
IP16 4BB M3VCP
IP16 4BY G3MYA
IP16 4DJ G0MLF
IP16 4DT G0UEA
IP16 4JP M6GLS
IP16 4QZ G0CFI
IP16 4RX 2E0JDF
IP16 4RX 2E0XOK
IP16 4RX M0ABG
IP16 4RX M0XOK
IP16 4RX M6EKK
IP16 4RX M6SGJ
IP16 4TA M5ADQ
IP16 4UQ G4ZWB
IP16ET M3PBK
IP17 1BA M1DQE
IP17 1BU M6CUD
IP17 1DP M3DTD
IP17 1EA 2E0KOS
IP17 1EA M0IPS
IP17 1EY G0CJX
IP17 1FB G0OIO
IP17 1HR G2FKZ
IP17 1HR M0RSE
IP17 1HR M1ACB
IP17 1PP G1BFV
IP17 1UX G0GDJ
IP17 1UX M3TIL
IP17 1WB G4ZJH
IP17 1XW M0LAY
IP17 2AA G0EGW
IP17 2AS G0KDR
IP17 2JP G1MJV
IP17 2NW G4IIK
IP17 2PU M0BJR
IP17 2RA G4CXT
IP17 3ED G4XVE
IP17 3EP M6MFG
IP17 3LW M6CSH
IP17 3NY M6TWA
IP17 3PL M6EGL
IP18 6BG M7RNL
IP18 6LR G6BPY
IP18 6RE G7BLK
IP18 6UJ G0BYK
IP18 6UL M1DNG
IP186UZ 2E0DFO
IP186UZ M0HWV
IP19 0BB G6ATP
IP19 0BB M3FIB
IP19 0BB M3OJU
IP19 0EA G4PUQ
IP19 0PY G0CFB
IP19 8DB G8TBV
IP19 8EE 2E1CPJ
IP19 8EG G0JJE
IP19 8JF M6PDZ
IP19 8JT M6DRP
IP19 8RP G3LXJ
IP19 8RQ 2E0LRG
IP19 8RQ M0SAZ
IP19 8RQ M7ECZ
IP19 8TJ G8KWN
IP19 9BH M0ARY
IP19 9DY G4IHI
IP19 9DZ G8EUE
IP2 0EH 2E0EKD
IP2 0NQ M0TMB
IP2 0QG G8LBS
IP2 8AZ G4BVI
IP2 8GA M0ULR
IP2 8GA M3ULQ
IP2 8HA 2E0BCM
IP2 8JZ G4AVL
IP2 9AS G3WJS
IP2 9DQ G7TFY
IP2 9JE M0HDK
IP2 9JF G4FSK
IP2 9PA G7PQX
IP2 9SX G3SXV
IP2 9TA G4BQR
IP2 9UE M6CQD
IP20 0NY G6IUF
IP20 9ET M7XMG
IP20 9HY G3JPZ
IP20 9JP G4RAV
IP20 9JY G3ICG
IP20 9NQ G4FPG
IP20 9PU G6ZYM
IP20 9QE M7GMD
IP20 9QN M7CZZ
IP21 4DW G7UNU
IP21 4EE G4YFV
IP21 4EH 2E0EPM
IP21 4EH M6JLQ
IP21 4LD M6XBB
IP21 4NG G4DML
IP21 4PH M3YLZ
IP21 4QY 2E0CQG
IP21 4QY M0HIW
IP21 4RL M6JGQ
IP21 4TG 2E0HPJ
IP21 4TG M0HPJ
IP21 4TG M6JWW
IP21 4TQ G3JPG
IP21 4UA M7RDC
IP21 4XP G8OCV
IP21 4YJ 2E0XGW
IP21 4YJ M0XGW
IP21 4YJ M3XGW
IP21 5JH M1DPX
IP21 5LS G0DFC
IP21 5LS M0WRR
IP22 1AR G0OJJ
IP22 1LT M0PCG
IP22 1RN M3XIY
IP22 1RU G4IYS
IP22 1RW G8FHR
IP22 1RZ M6FHO
IP22 2HS G8JPA
IP22 2JQ M6TFD
IP22 2PS G6WJW
IP22 2PS M3JNY
IP22 2PX G8STW
IP22 2PY M6CZL
IP22 2QR G3MFO
IP22 4BY M6GLU
IP22 4DF M0UGG
IP22 4DJ G0RPY
IP22 4EG G6RTM
IP22 4GJ G4MMI
IP22 4GL M0SDW
IP22 4HL G8BOB
IP22 4HR M3JLB
IP22 4HR M6VAZ
IP22 4NA G0WCJ
IP22 4NF G3XLL
IP22 4NP G6ILT
IP22 4PW M7AFD
IP22 4PY M3XBF
IP22 4PY M6SAF
IP22 4QW 2E1LME
IP22 5RG G7OCQ
IP22 5RU G8RML
IP22 5SP 2E0TUF
IP22 5SP M3TUF
IP23 7DE G4UWW
IP23 7EE G3XAP
IP23 7EJ M6PVW
IP23 7HH G0NFE
IP23 7JX M3PFM
IP23 8DY G1TIJ
IP23 8DY M3GFW
IP23 8EF M0ITX
IP23 8HP M1BLX
IP24 1BZ G0GSZ
IP24 1DG M6FBW
IP24 1FH M6NRN
IP24 1JJ 2E0PFG
IP24 1JJ M3VFE
IP24 1LG M6JAL
IP24 1LQ 2E0FCH
IP24 1LQ G0IFL
IP24 1NG G4LXD
IP24 1NP G3SXP
IP24 1PB G4RKK
IP24 1PF G0DCJ
IP24 1RR M3LGJ
IP24 1TQ G0CLT
IP24 2JH G7ACN
IP24 2LB M7ECZ
IP24 2LF M0CNM
IP24 2LY M3SPL
IP24 2QA 2E0GPS
IP24 2QA M3SKY
IP24 2QS G6CWP
IP24 2QX G3ZZQ
IP24 2UU G4OQK
IP24 2YN M6PZA
IP24 2YN M6UES
IP24 3HQ G4VEL
IP24 3NF G1LRK
IP24 3NP M3OCC
IP24 3PT M3UQL
IP25 6AS M7BRX
IP25 6BZ G3GIB
IP25 6BZ G7JRP
IP25 6DB G4GRT
IP25 6DN G4OZY
IP25 6DW G1UKH
IP25 6EA G4AED
IP25 6EU G8XOU
IP25 6EU M0XOU
IP25 6EY G3DVV
IP25 6FF G0NMP
IP25 6FF G7DXN
IP25 6HB M6DYC
IP25 6HE M6FHF
IP25 6HL M6DBH
IP25 6HL M6UCS
IP25 6LG G1NAN
IP25 6LL G8XOC
IP25 6NA M6RZO
IP25 6NJ M6ZXR
IP25 6NL G6YTV
IP25 6SD 2E0EKU
IP25 6SD M0VCV
IP25 6SD M3ZVW
IP25 6SR G0ALQ
IP25 6SS M6EKO
IP25 6XA 2E0FLN
IP25 6XA M0NUX
IP25 6XA M6FLN
IP25 6XB G8YRG
IP25 6ZB M6KWX
IP25 7EH G0KZI
IP25 7EH M6WWR
IP25 7EW G4TJY
IP25 7EZ G0SCT
IP25 7FD G1APL
IP25 7HW G4VBX
IP25 7LH M0HUA
IP25 7LS G0LGF
IP25 7LS M6KSS
IP25 7LX M3LRW
IP25 7LX M5IMI
IP25 7LY M6SSY
IP25 7NJ G4TUK
IP25 7PJ G0FMI
IP25 7QN M0BHW
IP25 7QX M6MBK
IP25 7RA G6BLA
IP25 7RA M6GBB
IP26 4BD G0CZR
IP26 4JG G4THP
IP26 4LD G7LVA
IP26 4QJ G0CLH
IP26 4QW 2E0XSB
IP26 4QW M6XSN
IP26 4RF G6YPJ
IP26 5EG 2E0CNN
IP26 5EG M6BKM
IP26 5EQ 2E0BXS
IP26 5EQ M0GQP
IP26 5EQ M3YZI
IP26 5JA 2E0BCO
IP26 5JA M0FZX
IP26 5LE G1BRP
IP27 0BS G0WON
IP27 0DN G4KZK
IP27 0DX G8VMZ
IP27 0EN M6UAD
IP27 0LJ G8ENY
IP27 0NR 2E0EAA
IP27 0NR M0SAA
IP27 0NR M3ZPW
IP27 0PW 2E0GOL
IP27 0PW 2E0LJG
IP27 0PW M0LJD
IP27 0PW M0NKR
IP27 0PW M3UYG
IP27 0PW M6LJG
IP27 0QG M6GLD
IP27 0RL M7XRS
IP27 9AJ M6JJD
IP27 9AU G8XQD
IP27 9DZ M0JRP
IP27 9DZ M7RMP
IP27 9ES G6LOJ
IP27 9EU M6EKL
IP27 9EU M6EMX
IP27 9EZ G6EUO
IP27 9EZ G6XYX
IP27 9EZ M3XYX
IP27 9HF G6DFR
IP27 9HR G1EYZ
IP27 9HS G6MGN
IP27 9LA M0JQL
IP27 9SA G0BBN
IP27 9SA G0SDE
IP27 9SA G6TSP
IP270PU G7CNX
IP270QH M6HZD
IP28 6DT G6NNS
IP28 6DT M6NHS
IP28 6ER G3HGE
IP28 6TQ G6PCC
IP28 6UA G0FIU
IP28 6UG G4PXC
IP28 6UG G4XRK
IP28 6XF G3YFP
IP28 7DP G4XTW
IP28 7FP G7PHD
IP28 7HX G8BCA
IP28 7JU M6WVH
IP28 7LN G0DZY
IP28 7LS M6LSV
IP28 7PD M3LWV
IP28 7PD M3ZWK
IP28 7PD M6MFK
IP28 7PR G6VAZ
IP28 8JF G0BRM
IP28 8JF G1SXY
IP28 8LQ G4BWP
IP28 8LQ G4MBC
IP28 8LS M7CTI
IP28 8NG M7AKZ
IP28 8QB G7OYF
IP28 8QD 2E0BJP
IP28 8SF M3PTB
IP288XP 2E0HBJ
IP288XP M0JUV
IP288XP M7DEK
IP29 4DL G7RKU
IP29 4PH G4KNO
IP29 4PL G3VTR
IP29 4SD M6AXM
IP29 5AD G1XAM
IP29 5AD G8KMM
IP29 5AG G4CGV
IP29 5AP M0CWN
IP29 5DD G0WVE
IP29 5DX G4ERF
IP29 5HE G7WKC
IP29 5HR G4VBS
IP29 5QL G8CRM
IP29 5RD M3WOL
IP29 5RP G3JPM
IP29 5SE G0JRM
IP3 0BW M6BGB
IP3 0EE M6HSA
IP3 0EJ M3FCS
IP3 0EW G4LVG
IP3 0LJ M3ESK
IP3 0LL M7SGA
IP3 0LX M6KGG
IP3 0LZ M3KGJ
IP3 0LZ M3SMY
IP3 0LZ M3ULW
IP3 0PU G3PAI
IP3 0QB G8EHS
IP3 0QH G1HSL
IP3 0RG M3TGP
IP3 8AT G6MMT
IP3 8BW M1BOP
IP3 8BW M5BOP
IP3 8BZ 2E0DWU
IP3 8BZ 2E0SER
IP3 8BZ M6MPZ
IP3 8BZ M6OAX
IP3 8EU M0JFR
IP3 8JQ 2E1HKB
IP3 8JQ M0HKB
IP3 8NX M6CLV
IP3 8PD M3MXM
IP3 8QW M7AAV
IP3 8RB M6GLP
IP3 8RZ G8OYX
IP3 8SP G3TJU
IP3 8UT M6DME
IP3 8UU M1AOF
IP3 8UU M6NLZ
IP3 9DE G4POU
IP3 9GY G8BEH
IP3 9LT 2E0MIY
IP3 9LT M6MIY
IP3 9LW 2E0NOM
IP3 9NJ G3XKU
IP3 9PD M0CSQ
IP3 9PD M7BYM
IP3 9QF M6AIB
IP3 9RE G4LBU
IP30 0AU M3FDB
IP30 0DG G4ICH
IP30 0JP M6FII
IP30 0QB 2E1BLP
IP30 0QD M6FYB
IP30 0SZ G7MPF
IP30 0TN M6WFE
IP30 0TS G7MLO
IP30 9AF G1FTD
IP30 9AY M7GHH
IP30 9BS G8XOM
IP30 9DQ M3TWS
IP30 9EJ M7FKG
IP30 9HJ G6AWO
IP30 9PX G3MWO
IP30 9RL 2E0FBD
IP30 9RL M0STI
IP30 9RL M6FBD
IP30 9TP M7TAD
IP30LA G4LVD
IP31 1JJ G4UDD
IP31 1JU M6RGP
IP31 1NG G4IXQ
IP31 1TE G1FVU
IP31 2AY G0AHL
IP31 2BN G8YYC
IP31 2EE G0MEV
IP31 2EE G0MEZ
IP31 2EF G4SBW
IP31 2EP M3FRX
IP31 2EW G4LIG
IP31 2FB M7MHS
IP31 2JQ M6HRN
IP31 2LB G8JSL
IP31 2NJ G6TDW
IP31 2PL G1CFK
IP31 2PP G4GVW
IP31 2QU G4BSA
IP31 2RP G0DUA
IP31 2UA M6XEE
IP31 2UQ M3SII
IP31 3EL G7UXD
IP31 3EN G3LPT
IP31 3HX G3GIH
IP31 3PD G4MID
IP31 3PF M3EGV
IP31 3PF M3EGY
IP31 3PF M6DSJ
IP31 3PF M6SJH
IP31 3PW M6NDN
IP31 3PY M6AWI
IP31 3RZ M0DFL
IP31 3SP G4UZF
IP32 6DF G8VQJ
IP32 6ED 2E1FZH
IP32 6ED G7SDC
IP32 6PF M6AZD
IP32 6PU G6LPD
IP32 6RJ M6KJY
IP32 6RR 2E0PUG
IP32 6RR M6CHL
IP32 6RU M6BLN
IP32 6TF M6SVJ
IP32 7AZ M6ERS
IP32 7HR M0PDA
IP32 7HR M6DDP
IP32 7JH G7RKW
IP32 7JJ 2E0NBG
IP327EP 2E0OBM
IP327EP M7FGH
IP33 1DR G6EPD
IP33 1JH G3OWQ
IP33 1RF G3TMX
IP33 1SW 2E0CHQ
IP33 1SW M6AUB
IP33 1YP M6AWO
IP33 2GB M6SJA
IP33 2JA G6IGK
IP33 2LS G0DVT
IP33 2LT G3TQX
IP33 2LT M6FVE
IP33 2NJ M6FAK
IP33 2QB M6AUQ
IP33 2QJ M6OCU

Postcode	Callsign
IP33 2QL	2E0KEA
IP33 2QL	M3THY
IP33 2SA	G8XGT
IP33 2SN	G7KFP
IP33 2SY	G4VSB
IP33 2UA	M6PEX
IP33 2UB	M1AKN
IP33 3AT	M6JAU
IP33 3DU	G1UGH
IP33 3QF	G0ARU
IP33 3QJ	G1VGI
IP33 3SD	M6AUP
IP33 3UB	2E0CWI
IP33 3UB	M6CHU
IP33 3XQ	G4XSM
IP38RW	M7NAB
IP4 1PU	M6AWQ
IP4 2BX	M0KHX
IP4 2PJ	G4SUU
IP4 2TL	G8GCO
IP4 2TS	G1SWK
IP4 2UB	2E0CZJ
IP4 2UB	M0JBZ
IP4 2UB	M0NBA
IP4 2UB	M6CZJ
IP4 3AB	M0RLC
IP4 3AH	G4KGO
IP4 3AS	M3ZSA
IP4 3JH	M3UOC
IP4 3LJ	G3ZIN
IP4 3NG	2E1CWN
IP4 3PP	G6HNN
IP4 4AD	G1BEK
IP4 4AD	M6EGN
IP4 4BS	G8XKT
IP4 4BU	M1NIZ
IP4 4HY	M6KLN
IP4 4JJ	G4ILN
IP4 4JN	G8TPC
IP4 4JX	G6PDE
IP4 4LP	G4UCX
IP4 4QN	G4TVT
IP4 4QP	G8WVO
IP4 4SF	G6FIL
IP4 5AX	G4WMF
IP4 5BP	M0CCZ
IP4 5DW	G6BEH
IP4 5PP	G4YVK
IP4 5PQ	2E0XAE
IP4 5PQ	M6OXY
IP4 5QZ	G6RHK
IP4 5RH	G4ROH
IP4 5RZ	G0CWW
IP4 5SA	G4BRL
IP4 5SD	G4FWA
IP4 5SX	G4ETC
IP4 5TF	G7UWB
IP4 5UQ	G7UWC
IP4 5UQ	G8CJL
IP4 5UU	G3PWB
IP42RN	M7LMK
IP45EL	G0BPU
IP5 1AQ	G1YLE
IP5 1AS	M6AAK
IP5 1AY	G3WIU
IP5 1EB	G4AEY
IP5 1ED	G3UKW
IP5 1ED	G3XDY
IP5 1ED	G4FZZ
IP5 1EL	G8VVR
IP5 1EN	M1BMD
IP5 1JZ	M1DLE
IP5 1LU	G4VBQ
IP5 1NQ	M3YCD
IP5 2DE	2E0GVZ
IP5 2DE	M6DQD
IP5 2DQ	M6EGM
IP5 2DQ	M6ENJ
IP5 2DQ	M6FCX
IP5 2DS	M3KUM
IP5 2EP	2E0UOM
IP5 2EP	M3YAW
IP5 2EQ	2E0SOR
IP5 2EQ	M0SOR
IP5 2FB	G0EBQ
IP5 2FB	M3ZVF
IP5 2FX	M6BFV
IP5 2GB	2E0VKQ
IP5 2GB	G6GSG
IP5 2GB	M1ADT
IP5 2GB	M3ADT
IP5 2GB	M3AKQ
IP5 2GP	G8LQB
IP5 2GW	G4LSQ
IP5 2GW	M6CRG
IP5 2QL	2E0IZN
IP5 2QL	M6MAG
IP5 2XF	M3IBZ
IP5 2YN	G4FNR
IP5 2YN	M3XBL
IP5 2YN	M3XPL
IP5 2YR	M1DGY
IP5 2YT	G0BQO
IP5 2YU	M3MXG
IP5 2YU	M3PNO
IP5 3QR	G3NYK
IP5 3SD	M6YAY
IP5 3SN	G8VQH
IP5 3ST	G6KWH
IP5 3SU	2E1BIM
IP5 3SU	G3ZID
IP5 3SU	G4DWF
IP5 3SY	M1ATJ
IP5 3SY	M6EDN
IP5 3TZ	M1CVB
IP5 3TZ	M3BTZ
IP5 3TZ	M3BUA
IP5 3UA	G4EQE
IP5 3UF	2E1FWD
IP5 3UQ	G4OPB
IP5 3UR	G4FAW
IP6 0DZ	M6EFC
IP6 0HQ	G7VGE
IP6 0HZ	G3TXE
IP6 0JQ	G4XDK
IP6 0ND	M5AEH
IP6 0PY	G8ZYM
IP6 0QY	M7CCZ
IP6 0RH	M1CVG
IP6 0RJ	M0DFQ
IP6 8DP	G7VXS
IP6 8ER	G4ZYZ
IP6 8FJ	G0PEY
IP6 8JH	G6PHU
IP6 8JS	2E0FUD
IP6 8JS	G3EKE
IP6 8JS	M0NIL
IP6 8JS	M6DQC
IP6 8TF	2E0JSS
IP6 8TF	2E0ZBD
IP6 8XF	M7NOW
IP6 9AJ	M0DNZ
IP6 9AR	G0TCP
IP6 9AR	M0HDR
IP6 9ET	M6HBP
IP6 9EU	M6CDQ
IP6 9HG	G1YRE
IP6 9HR	G4BAV
IP6 9HR	G4IRC
IP6 9LP	G4MRS
IP6 9LP	G4PIQ
IP6 9NG	G3IRQ
IP6 9NX	G7LNI
IP6 9TD	G0IIA
IP7 5BS	G0MGN
IP7 5FD	M7FPO
IP7 5JL	2E0WBS
IP7 5LJ	G8HCZ
IP7 5RQ	G1HIU
IP7 5RQ	G3CQL
IP7 5SQ	G7JVE
IP7 6AW	G1XVD
IP7 6FE	G3OPB
IP7 6FE	G8AAI
IP7 6JH	G0PEC
IP7 6NN	G0ORG
IP7 6NN	G0WTR
IP7 6PN	G0NQK
IP7 7AL	G0FJB
IP7 7BZ	G4MGO
IP7 7DE	G6HWT
IP7 7EJ	2E0CNG
IP7 7EJ	M6AWN
IP7 7JH	G7NGF
IP7 7LR	G6MGQ
IP7 7LU	2E0XFH
IP7 7LU	M3XFH
IP76NN	M6XCL
IP8 3AP	G0TJY
IP8 3DX	G8ENM
IP8 3EY	G0DVJ
IP8 3EY	G0RGH
IP8 3EY	M0GIM
IP8 3LF	G4JJJ
IP8 3NH	G4KYU
IP8 3RR	G4TTB
IP8 3RX	2E0EDN
IP8 3RX	2E0ICX
IP8 3RX	M0VLK
IP8 3RX	M6HJZ
IP8 3RX	M6ICX
IP8 4AU	G3MYY
IP8 4HD	G4FDF
IP8 4LH	M7DUU
IP8 4LR	2E0YSP
IP8 4PE	G8FFU
IP8 4PQ	G3XOK
IP8 4PU	G3JFW
IP8 4SP	G3WJI
IP8 4SR	G4DMT
IP9 1AR	G4MRD
IP9 1BA	M3TWB
IP9 1BP	M0GFJ
IP9 1DX	G8PPD
IP9 1HS	G0PJO
IP9 1HS	M6HMS
IP9 1HX	G7VPN
IP9 1LL	2E0XKC
IP9 1NP	2E1HKC
IP9 1NW	M3GLM
IP9 1QH	G0DTC
IP9 1QH	G4BXZ
IP9 1QH	M0HKC
IP9 1QP	M6VLH
IP9 2DR	M3TQD
IP9 2HT	G4XOJ
IP9 2HW	G8XRW
IP9 2JB	G3WRT
IP9 2NF	M3FDQ
IP9 2SW	G3HMB
IP9 2TT	G0LIN
IP9 2XD	M1SRH
IV 31 6 JJ	MM6IXH
IV1 1NS	GM7RJG
IV1 1XG	GM7MWX
IV1 3JJ	GM5DNA
IV1 3XD	2M0IWU
IV1 3XD	MM7FJG
IV1 3XG	GM3MUA
IV1 3XU	GM4LXM
IV1 3XU	MM6BDA
IV1 3YG	MM0CMO
IV10 8RA	2M0ALS
IV10 8RA	GM0UDL
IV10 8RA	MM3UDL
IV10 8SQ	GM6IYJ
IV10 8UX	MM6LYH
IV10 8XA	GM4KLN
IV11 8XF	GM4OHY
IV12 4RH	MM6YST
IV12 5AR	MM6MUA
IV12 5AZ	MM1EHO
IV12 5BX	MM1HWB
IV12 5EW	GM7BCC
IV12 5EW	MM3BCC
IV12 5LF	2M0TAS
IV12 5LF	MM3UBD
IV12 5PJ	GM0RML
IV12 5QG	GM0WPI
IV12 5QP	GM4WWU
IV12 5RA	GM1VAD
IV12 5SD	2M0RUP
IV12 5SE	GM3PIL
IV12 5SE	MM3YOC
IV12 5TW	2M0YNK
IV12 5TW	MM7YNK
IV13 7YE	MM1MOY
IV13 7YQ	MM6EPV
IV13 7YR	GM3LVA
IV14 9AQ	MM7CFU
IV15 9NJ	MM0EFJ
IV15 9PG	GM8LEA
IV15 9RB	GM4YGN
IV15 9RL	2M0AYU
IV16 9XH	MM6CCW
IV16 9XH	MM6GBP
IV17 0QL	MM6KNO
IV17 0SP	MM6MXK
IV17 0SP	MM7RVN
IV17 0SZ	MM5ALX
IV17 0TR	GM0JOL
IV17 0TR	GM0SFQ
IV17 0TR	GM8DFX
IV17 0TU	GM4TTD
IV17 0XL	MM3ZRF
IV17 0XN	GM0CAD
IV17 0YQ	GM6IDF
IV18 0BG	GM3SAE
IV18 0EY	GM0HNJ
IV18 0GF	2M0WTE
IV18 0HY	MM7EGM
IV18 0JT	MM7IRQ
IV18 0LU	GM4FWF
IV18 0PE	GM3WOJ
IV19 1AZ	MM3ONI
IV19 1ED	MM0TWK
IV19 1LA	GM4SUF
IV19 1QH	GM3YQG
IV2 3DT	GM0TNK
IV2 3ET	GM4XWS
IV2 3EW	GM7BOW
IV2 3HT	GM0OMC
IV2 3HT	GM1MYR
IV2 3HW	MM6CEM
IV2 3LX	GM0IQD
IV2 3SP	2M0TPE
IV2 3SP	MM6GQC
IV2 4EN	GM0GEE
IV2 4EX	MM0BAG
IV2 4LD	GM0SXP
IV2 4LD	GM8DFC
IV2 4NL	MM6GON
IV2 4QD	2M0GHF
IV2 4QD	MM6HQC
IV2 4XJ	GM7RVR
IV2 5AQ	MM0BGW
IV2 5DU	GM4GKH
IV2 5EP	MM0DHY
IV2 5EQ	GM7IBM
IV2 5ES	GM4UOD
IV2 5JH	MM7EVR
IV2 5PN	GM4VAC
IV2 5RG	MM1ATR
IV2 5RH	MM1KML
IV2 5RH	MM7ENF
IV2 6AX	GM4VIK
IV2 6EQ	GM6JOS
IV2 6FF	MM7JVR
IV2 6XJ	MM6AAA
IV2 6XS	MM7NGZ
IV2 7AR	2M0NJM
IV2 7AR	MM6HTS
IV2 7HB	MM1AEL
IV2 7HN	GM4ZIT
IV2 7HX	MM7NPT
IV2 7JL	GM3BQA
IV2 7JL	GM4OIJ
IV2 7JT	MM7INV
IV2 7ND	MM3MEH
IV2 7RW	GM1BQP
IV2 7SR	MM0BQL
IV2 7ST	MM3IAG
IV20 1UP	GM7PFQ
IV20 1XP	MM6DJC
IV20 1YS	MM7HPX
IV21 2BX	MM6MJR
IV21 2DH	GM4CHX
IV24 3DW	GM6VRU
IV25 3JB	2M0GPF
IV25 3JB	2M0VKY
IV25 3JB	MM0HPP
IV25 3JB	MM6GPF
IV25 3JB	MM6VKY
IV25 3JE	MM7JWK
IV26 2TB	2M0UAL
IV26 2TB	MM0ULL
IV26 2TB	MM6ULL
IV26 2TX	GM6AJA
IV27 4AD	GM7WHQ
IV27 4DG	MM1YAM
IV27 4ED	GM0IYP
IV27 4ED	GM4VVX
IV27 4ED	MM3MMB
IV27 4EG	2M1ECF
IV27 4JB	GM6ZCX
IV27 4JB	GM6ZCY
IV27 4JD	GM4BYT
IV27 4JQ	GM8IEM
IV27 4NL	MM0ACN
IV27 4RP	2M0AMD
IV27 4RP	MM0URN
IV27 4TG	GM6JHH
IV27 4XQ	MM3SCO
IV274LR	2M0IYV
IV274LR	MM7FQA
IV28 3XE	MM1WKD
IV3 5AH	MM3HJC
IV3 5BD	2M0HUD
IV3 5ES	GM4UZR
IV3 5HE	MM0IPD
IV3 8AZ	MM7AXB
IV3 8LB	GM7HQW
IV3 8LB	MM7IRM
IV3 8LH	GM0AKJ
IV3 8PF	GM7GVD
IV3 8PP	MM7SVR
IV3 8RA	MM7CQZ
IV3 8SD	GM4JNB
IV3 8SJ	2M0JAT
IV3 8SR	GM8RMR
IV3 8TE	2M1HLE
IV30 1SF	GM6WLJ
IV30 1TB	GM0UYZ
IV30 1UJ	MM7APW
IV30 2YR	GM0LOK
IV30 2YR	GM7CPY
IV30 4BU	MM3GLH
IV30 4DJ	GM0MZD
IV30 4EZ	MM7FUO
IV30 4FG	GM0ONN
IV30 4HJ	MM1FAS
IV30 4HL	GM7FPN
IV30 4HN	GM7LSI
IV30 4HX	2M0CFB
IV30 4HX	GM0MGO
IV30 4HX	MM0GYX
IV30 4HX	MM6AON
IV30 4JH	GM0DQV
IV30 4JH	GM8FFK
IV30 4LY	GM4ILS
IV30 4NB	GM8YKT
IV30 4NS	GM0LPB
IV30 5NF	MM6IKS
IV30 5PQ	MM3FJA
IV30 5PT	MM3LCC
IV30 5PY	2M1DHG
IV30 5QN	MM1GBS
IV30 5QQ	MM7KGX
IV30 5RN	MM6DZC
IV30 5SU	GM0OTS
IV30 5XY	2M0REH
IV30 5XY	MM0TQH
IV30 5XY	MM3TQH
IV30 5YD	MM6CPE
IV30 5YE	GM7LWA
IV30 5YN	2M0CXG
IV30 5YN	MM6CXD
IV30 6BP	2M0BZB
IV30 6DX	2M0GYM
IV30 6DX	MM3UBR
IV30 6EJ	GM4EWM
IV30 6EJ	MM0CJH
IV30 6ER	GM1ASY
IV30 6GL	MM6YGT
IV30 6XF	MM0JMB
IV30 8FE	GM0GNY
IV30 8JZ	GM0RYA
IV30 8NY	2M1HCP
IV30 8NY	2M1HCQ
IV30 8NY	MM3BWV
IV30 8PE	GM0AEG
IV30 8PE	MM0HAR
IV30 8PP	2M0SQL
IV30 8PP	MM3PHP
IV30 8SA	MM6GSY
IV30 8SY	GM0EMC
IV30 8SY	GM6TUE
IV30 8TB	2M0WMJ
IV30 8TB	MM0KKF
IV30 8TT	GM4HYG
IV30 8UA	GM6IKN
IV30 8UR	GM3VBY
IV31 6BA	GM4PGM
IV31 6BA	MM3ZSK
IV31 6DJ	2M0FUS
IV31 6DJ	MM7IOW
IV31 6NA	GM4HMN
IV31 6QH	GM4MKU
IV31 6TP	GM6PLG
IV32 7DX	MM6IVP
IV32 7EH	MM0DAT
IV32 7HF	2M0JVR
IV32 7HZ	MM7WTX
IV32 7JH	MM3TVQ
IV32 7JY	2M0BPM
IV32 7JY	MM0LER
IV32 7JY	MM3WLA
IV32 7NL	MM0CQT
IV32 7NW	2M0DTJ
IV32 7NW	MM6CYQ
IV32 7NW	MM6MUO
IV36 1BQ	2M0MMM
IV36 1JB	GM8DPV
IV36 2SG	GM0KNT
IV36 2UF	GM0GHN
IV36 2WE	GM1FRG
IV36 3UA	GM6NOO
IV4 7AB	MM0LOZ
IV4 7AB	MM3PSL
IV4 7AE	GM7DXE
IV4 7AH	2M0XCT
IV4 7AH	MM7RNB
IV4 7EY	GM4TJD
IV4 7EY	GM4UMA
IV4 7GL	2M0DAC
IV4 7HT	2M0LWB
IV4 7HT	MM0VSU
IV4 7HT	MM6LWB
IV4 7HZ	GM7ORJ
IV4 7HZ	MM3SRK
IV4 7JQ	GM4GZD
IV40 8BB	MM6ZDY
IV40 8HE	GM1UHF
IV418PT	GM6DBJ
IV42 8PY	GM8RBR
IV44 8RD	GM0HBK
IV44 8RF	GM0MAC
IV44 8RQ	GM3YAC
IV44 8RQ	GM4YAC
IV45 8RU	MM3VXP
IV47 8SH	MM6DTV
IV47 8SJ	GM0FHF
IV47 8SN	GM0PRQ
IV49 9AQ	GM1RLV
IV49 9BZ	MM7RCU
IV51 9DN	GM3SWK
IV51 9JX	MM0LUP
IV51 9PE	GM0EWX
IV51 9PW	GM0DXE
IV51 9YN	MM3ZDI
IV51 9YR	2M0RMV
IV51 9YR	MM6RMV
IV52 8TN	MM0GWO
IV53 8UX	MM6TUG
IV55 8GL	GM4TRH
IV55 8WF	GM4AEK
IV55 8WL	GM4RXD
IV55 8WP	GM3MTW
IV55 8WS	GM0CVD
IV55 8WS	GM4WZD
IV56 8FJ	GM4PUS
IV56 8FL	GM6UCX
IV6 7PX	MM3LQK
IV6 7QG	GM6WEI
IV6 7QG	MM0GKB
IV6 7SB	GM0OGZ
IV6 7XN	GM4RRP
IV63 6TJ	MM7DPP
IV63 6TN	GM0BZS
IV63 7YG	GM4EKI
IV63 7YJ	GM6JJN
IV7 8AW	2M1DZX
IV7 8AW	GM4SFW
IV7 8BH	MM3SES
IV7 8FP	MM0YIA
IV7 8FP	MM6IAY
IV7 8JS	GM0NAI
IV7 8JS	MM0APF
IV7 8LB	GM3WED
IV7 8LG	MM6MWE
IV7 8LH	GM4FIZ
IV7 8LL	GM0JFK
IV8 8PA	GM7JYW
IV8 8PG	GM0TQB
IV8 8PJ	GM0JFL
IV9 8PR	2M1EBJ
IV9 8QL	GM4MAI
JE2 3EA	MJ0RGR
JE2 3GP	MJ0PMA
JE2 3GQ	2J0YAY
JE2 3GQ	MJ6DEY
JE2 3HD	MJ7UKT
JE2 3HP	GJ0NTD
JE2 3RX	MJ6ADQ
JE2 3SY	GJ7AOG
JE2 3XQ	MJ0SIT
JE2 4DA	2J0VDJ
JE2 4DA	MJ7DIJ
JE2 4GL	GJ3YLI
JE2 4PX	GJ7DNI
JE2 4RS	MJ0AQJ
JE2 4RY	MJ1COO
JE2 4SA	MJ3HWC
JE2 6FP	GJ3XOJ
JE2 6GE	MJ3GBJ
JE2 6GG	MJ3CMB
JE2 6LT	2J0CZD
JE2 6LT	MJ6CQF
JE2 6LZ	GJ7DUX
JE2 6LZ	MJ7WMQ
JE2 6NY	GJ3ECC
JE2 6NY	MJ0JER
JE2 6NY	MJ3IOJ
JE2 6SE	GJ1YOT
JE2 7AH	2J0SZI
JE2 7AH	MJ3SZI
JE2 7GR	MJ7MMA
JE2 7HP	GJ3SND
JE2 7HQ	GJ8PVL
JE2 7HQ	MJ6PVB
JE2 7HQ	MJ6XVL
JE2 7LQ	MJ7FXC
JE2 7LT	GJ0JSY
JE2 7LX	MJ1CYD
JE2 7LX	MJ3CDP
JE2 7PD	MJ6YSY
JE2 7PP	MJ6SSF
JE2 7PX	MJ7JIV
JE2 7RL	GJ6SNQ
JE2 7RT	GJ7HTV
JE2 7RT	MJ0BJU
JE2 7TA	GJ7FGS
JE2 7TX	GJ7LJJ
JE2 7TZ	GJ8RRP
JE2 7UD	GJ4HSW
JE27PN	MJ7AME
JE3 1EZ	GJ4YBM
JE3 1FR	GJ0KYZ
JE3 1GL	GJ6WDK
JE3 1GT	GJ0PDJ
JE3 1GT	GJ3IT
JE3 1GT	GJ8RVT
JE3 1LA	GJ3LFJ
JE3 1LE	GJ0VJP
JE3 1NH	MJ6THP
JE3 1NJ	MJ7RVV
JE3 1NL	MJ6RBI
JE3 3BQ	GJ0FYB
JE3 5AA	GJ7DNJ
JE3 5QF	2J0COQ
JE3 5QF	MJ0LEL
JE3 5QF	MJ6BDJ
JE3 6AS	GJ4YLP
JE3 6AT	GJ3AME
JE3 6ED	GJ1TJP
JE3 7AD	GJ7SLU
JE3 7AQ	GJ7DTA
JE3 7DT	GJ6ENP
JE3 7ES	GJ6WMZ
JE3 8AL	MJ6VAA
JE3 8AP	MJ7NLK
JE3 8AQ	MJ0IEW
JE3 8BL	MJ1EPG
JE3 8BN	2J1EDR
JE3 8DP	MJ3HMA
JE3 8GB	GJ4JVP
JE3 8GP	GJ4KBM
JE3 8GP	GJ4ODX
JE3 8GP	GJ6TPD
JE3 8GQ	GJ7RWT
JE3 8GQ	GJ6EFW
JE3 8HN	GJ3YHU
JE3 8LJ	MJ7POA
JE3 8LS	MJ6ORG
JE3 8NS	MJ1CNB
JE3 9AY	2J1CWG
JE3 9AY	2J1CWH
JE3 9AY	GJ7UIT
JE3 9AY	MJ0JIS
JE3 9EF	GJ8CEY
JE3 9EP	GJ8PCY
JE3 9EP	GJ3YLN
JE3 9ER	2J0ELQ
JE3 9ER	MJ0THP
JE3 9ER	MJ6ILZ
JE34DE	GJ1SUP
JE4 9VI	GJ4WRR
KA1 1UH	GM7KIY
KA1 2HP	GM0AAX
KA1 2HP	MM3TYA
KA1 2JN	MM7ALL
KA1 2LA	GM6KPL
KA1 2NU	2M0YAF
KA1 3AR	MM3LWT
KA1 3DZ	2M0BSE
KA1 3LL	MM7FHW
KA1 3LQ	GM3OZB
KA1 3QL	MM3OJV
KA1 3RY	GM0FKP
KA1 3RZ	MM7DIV
KA1 3SJ	GM0NTY
KA1 4EB	MM3FET
KA1 4LQ	GM4UYP
KA1 4PB	GM0FQQ
KA1 4PB	MM3VSC
KA1 4QT	GM0LYH
KA1 4SP	MM7FRF
KA1 5AB	GM7AAJ
KA1 5NA	MM0XNC
KA1 5ND	GM4CAM
KA1 5SF	GM4JMU
KA10 6AW	2M0JST
KA10 6DA	2M0CMA
KA10 6DA	MM0YET
KA10 6DA	MM6AMA
KA10 6EQ	MM7XOO
KA10 6NJ	2M0YTN
KA10 6NJ	MM0CKF
KA10 6NJ	MM3NNO
KA10 6QW	MM6AAM
KA10 6SD	GM4XRY
KA10 6SE	MM7JYS
KA10 6SH	MM0ODV
KA10 6TB	GM3NIG
KA10 6TT	GM7VXR
KA10 6TZ	GM4SLY
KA10 6UG	GM6BAO
KA10 6UP	GM4IGS
KA10 6XE	GM8GIQ
KA10 7AH	MM0KJH
KA10 7EE	GM4CXF
KA10 7GZ	2M0SOE
KA10 7HA	GM4BIT
KA11 1BD	GM0SEI
KA11 1BD	GM7KSA
KA11 1BD	MM0JJV
KA11 1BW	GM7UAC
KA11 1BY	MM3WNP
KA11 1DB	MM0ZIF
KA11 1DW	2M0YTN
KA11 1DW	MM7TWN
KA11 1HH	MM6LIL
KA11 1JD	GM0DWH
KA11 1LR	GM6JOD
KA11 1LT	MM6PDX
KA11 1NJ	GM3JOB
KA11 1PN	GM0JQE
KA11 2BY	MM3PEV
KA11 2EB	GM1VJD
KA11 2EU	GM0DNG
KA11 3AL	MM1DQW
KA11 4BN	MM6JFU
KA11 4EB	MM3OIZ
KA11 4ER	GM0WUX
KA11 4EY	2M0WBJ
KA11 4EY	MM0HYM
KA11 4EY	MM6WBJ
KA11 4FB	2M0NLA
KA112FY	MM7IOD
KA12 0DF	MM7AHK
KA12 0DU	2M0XTS
KA12 0DU	MM0XKA
KA12 0DU	MM6XKC
KA12 0QE	GM7VFR
KA12 0QG	GM7CZU
KA12 0QS	MM1CXO
KA12 0SE	GM3NYG
KA12 0SE	GM4PGV
KA12 0TR	GM7KHA
KA12 0XA	MM3UOE
KA12 0XH	2M0BMF
KA12 8DR	GM0FCI
KA12 8HS	MM0UKW
KA12 8PR	MM0XPP
KA12 8RZ	MM0CTT
KA12 8RZ	MM0CTU
KA12 8RZ	MM3TNG
KA12 9DN	GM1ASA
KA12 9DT	MM3TWA
KA12 9ER	2M0WUI
KA12 9ER	GM5PEB
KA12 9ER	MM3WUI
KA12 9JJ	2M0HXQ
KA12 9JJ	MM1LDR
KA12 9JJ	MM3LDR
KA12 9PA	MM3UOS
KA13 6BE	MM0DHQ
KA13 6BE	MM0KLR
KA13 6EB	GM1MSN
KA13 6JJ	MM0HLN
KA13 6LU	MM3YUU
KA13 6PL	GM3YQK
KA13 6QN	MM0BRG
KA13 6RS	2M0UUU
KA13 6RS	MM0UUU
KA13 6RS	MM6OUU
KA13 6WH	MM3YUY
KA13 7AR	GM0TAE
KA13 7BA	MM6WOC
KA13 7DT	GM0UGG
KA13 7HE	GM1MSO
KA13 7JN	MM0CKK
KA13 7LQ	MM1FHR
KA13 7ND	MM1LBA
KA13 7PT	GM7FPX
KA13 7QY	GM7EGQ
KA13QA	MM6HHP
KA14 3AJ	GM8MMW
KA14 3BQ	GM1AXI
KA15 1AR	GM6OHF
KA15 1BQ	MM7LUB
KA15 1ES	MM0ZZO
KA15 1JE	GM7DTC
KA15 2AJ	GM4VHZ
KA15 2BA	GM7DHA
KA15 2DZ	MM3OGS
KA15 2HG	GM4LIS
KA15 2LN	GM6UHE
KA15 2LN	GM8NVG
KA16 9BE	MM7RVI
KA16 9HZ	MM0OVK
KA17 0DQ	GM4ZNS
KA17 0EE	MM3OJE
KA17 0JG	MM7JFA
KA17 0LP	2M0UEA
KA17 0LP	GM0VBE
KA17 0LP	GM1FNX
KA17 0LP	MM3UEA
KA17 0NN	MM7LEY
KA18 1HA	MM0JOK
KA18 1HA	MM3EJB
KA18 1HN	GM7ANE
KA18 1HN	MM3GQT
KA18 1LT	MM7GTN
KA18 1NP	MM3PHC
KA18 1PU	GM0BKX
KA18 1PU	GM0NET
KA18 1PU	GM1MDO
KA18 1QU	MM7BQP
KA18 2ED	GM0ADX
KA18 2ED	GM3STM
KA18 2ED	MM3STM
KA18 2LL	GM1TCP
KA18 2LL	GM1YKE
KA18 2NN	2M0IYK
KA18 2NZ	MM1ANP
KA18 2RE	GM7OKX
KA18 3AG	GM0GRD
KA18 3EY	GM4XFU
KA18 3GA	GM0DBK
KA18 3HS	GM4XMD
KA18 4BW	MM7EFP
KA19 7AE	2M0BET
KA19 7AE	MM3BOJ
KA19 7AU	MM6JTG
KA19 7BS	2M0LXX
KA19 7HZ	2M0GYN
KA19 7HZ	MM6EBD
KA19 7PW	MM0KMK
KA19 7RE	MM0TFU
KA19 8AY	MM6IMP
KA19 8DR	MM7FEW
KA19 8EN	MM6BNS
KA19 8FG	2M0MSU
KA19 8FG	MM6OEJ
KA19 8FG	MM7RXH
KA19 8LR	GM3YDN
KA197AE	MM0LGR
KA2 0AX	GM4ATA
KA2 0AX	GM8DYT
KA2 0AY	MM6PFT
KA2 0DS	MM6MHN
KA2 0ES	2M0DTZ
KA2 0ES	MM0GKN
KA2 0LD	MM3DHG
KA2 9EW	GM7IUF
KA2 9EX	GM4VKI
KA2 9EZ	GM8JUY
KA2 9HZ	MM3DHE
KA20 3BA	MM7RGM
KA20 3DG	GM4WNQ
KA20 3LG	MM3RTH
KA20 3PU	MM3VFK
KA20 4BG	2M0YIO
KA20 4BG	MM3YIO
KA20 4BN	MM7ORR
KA20 4BP	MM6NAB
KA20 4EF	MM3PXO
KA20 4HN	MM0CWB
KA20 4JE	MM3UVF
KA20 4LB	MM7STF
KA21 5BB	MM6CCS
KA21 5DA	MM6VFL
KA21 5EP	MM6MNW
KA21 5HZ	MM7ANU
KA21 5LJ	MM0BQJ
KA21 5NH	GM0XAV
KA21 5PS	MM3AQM
KA21 6AB	2M0JCG
KA21 6AB	MM6AVE
KA21 6AQ	GM0OTB
KA21 6AQ	MM6CKC
KA21 6AX	MM7EKQ
KA21 6BE	MM6GKE
KA21 6HY	GM4UEH
KA21 6JU	MM7BDW
KA21 6LN	MM6DHS
KA22 7AJ	MM3YBD
KA22 7BZ	MM6BXT
KA22 7DY	2M0OVD
KA22 7DY	MM0OVD
KA22 7ER	MM3YFT
KA22 7LU	MM3DKA
KA22 7NQ	MM6ARN
KA22 8AP	GM5GAZ
KA22 8AP	MM7GBL
KA22 8EP	MM7AXX
KA22 8LW	2M0RGU
KA22 8PH	GM7SXI
KA228HA	MM3UOR
KA23 9AR	MM7AYG
KA23 9BX	GM0EFC
KA23 9BX	GM0EFD
KA23 9BY	GM4YMC
KA23 9ED	2M0HCF
KA23 9LB	GM0AYT
KA23 9LB	MM6OOP
KA23 9LE	GM3LLP
KA23 9LX	GM3JIG
KA23 9NF	GM0DEX
KA24 4AF	GM0NHL
KA24 4DJ	MM1EVJ
KA24 4HP	MM0SVE
KA24 5HP	MM6IIO
KA24 5HP	MM6LJE
KA25 6AB	MM3XXI
KA25 6DB	GM7VCV
KA25 6EY	GM0ODB
KA25 7JB	MM6CCO
KA25 7JB	MM6KYV
KA25 7JF	MM6VWH
KA25 7LB	GM0HQF
KA26 0AA	2M0NCM
KA26 0AA	MM3GZG
KA26 0AA	MM3NCM
KA26 0AA	MM6CNC
KA26 0AE	MM3GIR
KA26 0BP	GM0FSZ
KA26 0BX	GM0NBG
KA26 0BY	MM6CTH
KA26 0EA	GM0UDY
KA26 0EB	MM1DWU
KA26 0EF	MM0CBL
KA26 0NQ	GM4ZTO
KA26 0PA	GM4WEW
KA26 0QT	GM1ROB
KA26 9AH	GM0JMO
KA26 9AH	GM7SAK
KA26 9AH	MM0SAK
KA26 9AH	MM3UDB
KA26 9DJ	GM0INC
KA26 9DZ	GM6OFB
KA26 9EL	GM0KCY
KA26 9EU	MM6RQV
KA26 9JH	GM0KWW
KA26 9LP	GM4TPQ
KA26 9SN	MM7BFL
KA27 8PB	MM6RLD
KA27 8PG	GM8LQL
KA27 8PR	GM4BVZ
KA27 8QH	GM3HEN
KA27 8QH	GM4JFH
KA27 8QH	GM4XJF
KA27 8RG	MM7RXH
KA27 8RL	GM3UA
KA27 8SA	GM1ISJ
KA27 8SF	GM4PWR
KA278JQ	MM7FNS
KA28 0DL	MM3WWQ
KA28 0DP	GM8BJJ
KA28 0DU	MM7EVK
KA29 0DG	GM3DOD
KA29 0DP	MM0JBS
KA3 1TU	GM4ZNC
KA3 1TU	MM0TJT
KA3 2AS	2M0OOT
KA3 2AS	MM3OOT
KA3 2BA	2M0ONW
KA3 2DA	GM0MZH
KA3 2DW	GM4UYK
KA3 2EQ	GM0JFH
KA3 2GN	MM3LZU
KA3 2GP	MM0GNS
KA3 2HU	2M0GZQ
KA3 2HU	MM0KJJ
KA3 2JG	GM0ONX
KA3 2JG	GM6JIC
KA3 2JG	MM3ONX
KA3 2JG	MM6PGT
KA3 2RS	2M0VNW
KA3 2RS	MM0ABB
KA3 2RS	MM3VNW
KA3 3BT	2M0TSR
KA3 3DQ	GM3AXX
KA3 3HG	GM4VAY
KA3 3HT	GM4OSS
KA3 4BP	GM0GOV
KA3 4BP	MM3LZD
KA3 4DA	MM6SHM
KA3 5AT	MM3WGW
KA3 5DA	2M0FRJ
KA3 5DA	MM6SUR
KA3 5HS	2M0YMZ
KA3 5HS	MM7TMZ
KA3 5JT	GM0PDQ
KA3 5JT	GM4HCO
KA3 6DB	GM0OXS
KA3 6FJ	MM6GLO
KA3 6HJ	GM0LYO
KA3 6JQ	2M0FDT
KA3 6JQ	GM5TIM
KA3 6JQ	MM6MQO
KA3 6JT	GM0DVH
KA3 7DT	GM4SQM
KA3 7DT	GM6JNJ
KA3 7EJ	2M0AAQ
KA3 7EJ	MM6WBP
KA3 7RB	MM3JGT
KA30 8EF	GM8EUG
KA30 8ER	GM0BFW
KA30 8HB	GM5BDX
KA30 8HB	MM7BJC
KA30 8HU	GM1RJS
KA30 8HU	MM7BIC
KA30 8PG	GM4ZZW
KA30 8PP	GM6GAH
KA30 8QQ	MM6JHH
KA30 8RH	GM0NUI
KA30 8RQ	MM3XIW
KA30 8SN	GM3MWX
KA30 8SN	GM6XW
KA30 8SS	MM0RFA
KA30 9BH	GM1LXM
KA30 9BZ	GM1BKR
KA30 9DN	MM0TCP
KA30 9EQ	GM0DWY
KA30 9ER	MM0MNS
KA30 9ET	GM0GMN
KA30 9ET	GM4IPA
KA30 9ET	MM3YPH
KA30 9EX	GM0RYD
KA4 8AX	MM7MAB
KA4 8EA	MM0GHM
KA4 8EA	MM0RAI
KA4 8EJ	GM0KAZ
KA4 8HT	GM8MNR
KA4 8NA	MM3WZL
KA5 5AE	GM1XBK
KA5 5QU	GM7RPT
KA5 5RP	MM3LDK
KA5 6DT	GM7OPN
KA5 6EH	GM0GRW
KA5 6HU	MM0HSV
KA5 6HY	MM3GSL
KA6 5BE	MM0XCP
KA6 6BH	GM8KXF
KA6 6EL	GM1VFQ
KA6 6GJ	MM6EBZ
KA6 6HB	MM3ENP
KA6 6LB	GM4PPT
KA6 6ND	GM4SFA
KA6 6NL	MM3WNH
KA6 6QG	GM0POD
KA6 7AU	2M1FQI
KA6 7AU	MM3FQI
KA6 7AU	MM3WVQ
KA6 7DG	MM0CNF
KA6 7EF	GM1VXE
KA6 7HG	MM7WAB
KA6 7LQ	MM6CHM
KA6 7ND	GM1OST
KA6 7QN	MM7DAH
KA6 7RN	GM0SDS
KA6 7SJ	MM3JGR
KA6 7TX	GM1VDZ
KA6 7TX	GM7DHE
KA67LR	GM7USC
KA67ST	MM6POV
KA7 1UF	MM7FBN
KA7 2LW	GM4FGS
KA7 2RW	MM7JNK
KA7 2TA	MM7NRN
KA7 3JB	GM0TBH
KA7 3JJ	MM0MLO
KA7 3LN	MM7HIM
KA7 3NF	GM7OIN
KA7 3PE	GM4SQO
KA7 3PE	MM6OIR
KA7 3QH	MM6CLM
KA7 3QJ	MM0BPF
KA7 3RL	MM6CPK
KA7 3RZ	MM7MMW
KA7 3SQ	MM7ATF
KA7 4DA	GM3PMB
KA7 4DA	GM5VG
KA7 4ET	MM0RAG
KA7 4GA	GM0EPO
KA7 4LR	MM7BWN
KA7 4PB	MM1JAS
KA7 4PW	GM4WFV
KA7 4QF	MM6RSP
KA7 4QN	GM6BHR
KA7 4UB	GM6MD
KA7 4UF	MM3EJV
KA7 4XA	MM3HQL
KA7 4XQ	2M0MOF
KA7 4XQ	MM6TCS
KA73BG	GM8OBO
KA73SB.	MM7GTS
KA73XG	MM7IEF
KA8 0EF	MM0CHX
KA8 0JR	GM7FEM
KA8 8NL	MM7NLH
KA8 8NX	MM6BXR
KA8 9BW	MM6BCF
KA9 1DL	MM7NMO
KA9 1JW	MM3WVN
KA9 1LT	MM0GAI
KA9 2DB	GM4TPX
KA9 2ED	GM0DEQ
KA9 2EQ	MM3EDW
KA9 2HE	GM3MQO
KA9 2HE	GM4OOU
KA9 2HY	GM4DOZ
KA9 2NN	MM1DPH
KA9 2PW	GM4LVW
KT1 2RU	G3RLT
KT1 3PS	G8VXB
KT1 3QB	M0BEJ
KT1 3RW	G8SIK

KT10 0AZ	G0EYT
KT10 0BJ	M0XDV
KT10 0DT	G4IUA
KT10 0DW	G6FWO
KT10 0DW	M6TLL
KT10 0HS	G3RIM
KT10 0JZ	G1PQT
KT10 0QE	G4GPB
KT10 0RZ	G8YXZ
KT10 8DL	M0PLP
KT10 8PY	M6OKI
KT10 9EG	2E0KXX
KT10 9EG	M6KXX
KT10 9HF	G0IDB
KT11 1LW	M7CDH
KT11 2AJ	M0TWM
KT11 2AQ	G1LKH
KT11 2BB	G0OAS
KT11 2BH	M1CNH
KT11 2RU	G8ORX
KT11 2SX	G4YKH
KT11 3AJ	M5PM
KT12 1BB	G4ARO
KT12 1LE	G3LFX
KT12 1LW	M6GZJ
KT12 1NF	G8GTR
KT12 2AT	G6AYS
KT12 2DP	2E0SYO
KT12 2DP	M7APZ
KT12 2HS	G0HJD
KT12 2LB	G6CLK
KT12 2SU	M3FJE
KT12 2SZ	G2HS
KT12 2SZ	M1GWZ
KT12 3AW	M1IAN
KT12 3BA	G0UHQ
KT12 3DA	G4AWZ
KT12 3EQ	G8DXP
KT12 3QH	M1AEJ
KT12 3QY	G3YMN
KT12 3SQ	G6INX
KT12 4HQ	G1JKV
KT12 5BS	M7NRJ
KT12 5BT	M6ALJ
KT12 5HF	G1KEV
KT12 5LW	G8ZXL
KT12 5PH	M6KSI
KT12 5PH	M6NBY
KT12 5QG	2E1RJS
KT13 0BL	G7SQH
KT13 0HY	G0CMM
KT13 0JW	G0CKV
KT13 0NR	G4CXL
KT13 0NR	G8HCL
KT13 0TB	M6RDD
KT13 0XE	G1JHS
KT13 8AS	G4BSV
KT13 8AS	G4CPM
KT13 8PE	G3NMH
KT13 8PU	G6CUV
KT13 8UP	G4LQD
KT13 9AT	M3XSZ
KT13 9LS	M0ZCO
KT13 9NU	G8IPN
KT13 9RU	G0JRH
KT14 6AE	M0HMF
KT14 6AP	G0RWM
KT14 6DT	G6JRZ
KT14 6DY	M6AQW
KT14 7BZ	M3EJH
KT14 7EF	G6XAN
KT14 7HX	G4EVA
KT14 7NG	G0PBL
KT15 1DH	G4WMP
KT15 1SR	G7HIN
KT15 1UN	M7BTY
KT15 2AX	M6XYH
KT15 2DE	G8HFW
KT15 2DN	G0KEY
KT15 3AA	G4ZMU
KT15 3AA	M6CUS
KT15 3AG	2E0EQT
KT15 3AG	M0TKC
KT15 3AG	M6KXN
KT15 3AQ	M3WQA
KT15 3DJ	2E0ABT
KT15 3DJ	G4CCZ
KT15 3DL	G1SMB
KT15 3DZ	M0DJW
KT15 3DZ	M1VPL
KT15 3ET	G8JSN
KT15 3EY	G1IDE
KT15 3EY	G3UES
KT15 3EY	M0SAR
KT15 3HP	G6MJT
KT15 3HS	G0PVF
KT15 3LX	G3XSZ
KT15 3NQ	2E0XNC
KT15 3NQ	M6XNC
KT15 3NR	M0XMC
KT15 3SE	G8CUG
KT15 3TU	G6YHE
KT15 3UA	G0RMU
KT151NX	M6FDX
KT153LH	2E0BSM
KT153LH	M0GPH
KT153LH	M3YFI
KT16 0BP	G0EMT
KT16 0ER	G1UMY
KT16 0HD	G6KJE
KT16 0LG	G4LQI
KT16 0PH	G7DUK
KT16 0RE	G0GPT
KT16 8AP	M0DXP
KT16 8BU	G8BWA
KT16 8EP	G0PCZ
KT16 8LE	M0BSF
KT16 8QB	G1AEA
KT16 8RA	G7MYO
KT16 8RW	G3ZZL
KT16 9AZ	M6SEJ
KT16 9ED	G6YRI
KT16 9GY	2E0FLH
KT16 9GY	M7MJC
KT16 9HP	M6KSV
KT16 9HX	M6OIC
KT16 9JJ	G8NYB
KT16 9PH	G3OLH
KT17 1QP	G8HPF
KT17 2EF	G4VHJ
KT17 2HD	G0DGF
KT17 2LH	M7HSG
KT17 2NG	G7PBC
KT17 2NW	G0ECK
KT17 2PH	M7BXA
KT17 3LN	G1ZVE
KT17 3NL	G3LQP
KT17 4EN	G8EPS
KT17 4ER	M3IDB
KT18 5JL	G3TVL
KT18 5PB	2E0NOU
KT18 5PB	M0LKJ
KT18 5PY	G4IQV
KT18 5QT	G4GFI
KT18 7BU	2E0DQO
KT18 7BU	M6EZS
KT18 7JE	2E0DZF
KT18 7JE	G6KVG
KT18 7JE	M3GQL
KT18 7SL	2E1FYC
KT19 0AU	G4ZSW
KT19 0HD	M0AAF
KT19 0HF	G8MTI
KT19 0JP	G0WVM
KT19 0LB	G4ROI
KT19 0LS	G6QN
KT19 0LS	M0ZEY
KT19 0LS	M6HTN
KT19 0PQ	G8EBQ
KT19 0RJ	G6YIJ
KT19 0SZ	G0PFA
KT19 7EW	G6REH
KT19 7LD	G1DPW
KT19 8BW	M3KQC
KT19 8HH	M6JHO
KT19 8JX	G3YGG
KT19 8RP	G1LQC
KT19 8RP	G4SYT
KT19 9DP	G4WFL
KT19 9DP	G6LFW
KT19 9EE	G0KAS
KT19 9EX	M6LHW
KT19 9HD	2E0BJL
KT19 9HD	M3ZLL
KT19 9HH	G1IKG
KT19 9LB	M0GDU
KT19 9LD	M0TBS
KT19 9LU	M0DSL
KT19 9QR	G0OXZ
KT19 9UL	G4RSU
KT198SW	2E0WOL
KT198SW	M3XEI
KT2 5GG	G4PPN
KT2 5HH	G6OBA
KT2 5HQ	M7CRQ
KT2 5NE	M6MUF
KT2 5QB	G8VOY
KT2 5TL	M3XXE
KT2 5TU	G6RFU
KT2 5TU	M0ETA
KT2 6AH	M3WEJ
KT2 6ES	2E0HMA
KT2 6PN	G6JTD
KT2 6RA	G8YJQ
KT2 7AL	G6ZLS
KT2 7AN	M0EIJ
KT2 7ES	M0JZV
KT2 7JT	G0TRD
KT2 7QD	G4GQA
KT20 5EW	G6HVQ
KT20 5HP	M7FHU
KT20 5JF	G4WNV
KT20 5JQ	G7PHK
KT20 5PS	M0JHM
KT20 5RZ	G3SUX
KT20 5SQ	G0XTM
KT20 5UA	G4XXI
KT20 5UT	G6ALG
KT20 6BS	G3NPC
KT20 6ET	G0SWN
KT20 6ET	G6CJR
KT20 6ET	G8WGP
KT20 6ET	M3URA
KT20 6ET	M6ICZ
KT20 6XE	G4HVO
KT20 6XE	G6GGY
KT20 7AD	G7KWF
KT20 7BA	G4BYZ
KT20 7LN	2E0WZA
KT20 7LS	G8GGS
KT20 7LZ	G4NLB
KT20 7QE	M6ZLP
KT21 1BE	G4JUM
KT21 1EB	G3KWJ
KT21 1HY	2E0DKP
KT21 1HY	M0LUT
KT21 1HY	M6LUT
KT21 1NA	G6TZE
KT21 1NE	G1HWR
KT21 1NE	G4KGE
KT21 1NN	M0JSN
KT21 1PD	G7LSB
KT21 1PJ	2E0UAX
KT21 1PJ	M7HAP
KT21 1RH	G3OKU
KT21 1SG	G7RSK
KT21 2HB	G8ZZR
KT21 2HU	G4CLY
KT21 2HY	G4XQE
KT21 2JG	G8DNQ
KT21 2PG	G4NHO
KT21 2UF	G6BQQ
KT212EG	G6DTW
KT212EG	M0KCI
KT22 0HR	G8UYL
KT22 7EE	G7DOR
KT22 7EE	M0GXV
KT22 7SL	2E0ZEH
KT22 7SL	M0ZEH
KT22 7SL	M3ZEH
KT22 8JT	G3ZQT
KT22 8LR	M0CDJ
KT22 8LZ	M1KOS
KT22 8NU	G8MHN
KT22 8RD	G4EYB
KT22 8UJ	G4XYC
KT22 9AZ	G3NFV
KT22 9AZ	G8GWM
KT22 9EF	M0CDB
KT22 9JR	G3UZX
KT22 9NB	G6BTC
KT22 9NT	G3XKG
KT22 9QT	M6XTR
KT22 9TL	M3FZE
KT22 9XA	2E0GBK
KT22 9XA	M6BMB
KT23 3DA	2E0CDK
KT23 3DU	G1FCU
KT23 3EY	G4FDU
KT23 3HW	M7TBA
KT23 3JF	M3CSI
KT23 3PY	G3LHN
KT23 4BJ	G1RJN
KT23 4BQ	G4HXN
KT23 4BS	G8IWT
KT23 4BS	G8KW
KT23 4DB	G3UZW
KT23 4DY	M6GWI
KT23 4HP	2E0ESU
KT23 4JX	G3SIA
KT23 4ND	2E0LTR
KT23 4ND	M6ZLC
KT24 5DY	G6GEK
KT24 5RH	2E0HQF
KT24 5RH	M0JWU
KT24 5RH	M7CUR
KT24 5SD	G0RAL
KT24 6QJ	G3NIW
KT24 6QN	G4PED
KT3 3BX	M6OMS
KT3 3DY	G4JUZ
KT3 3HP	G3TON
KT3 3HP	G4FJP
KT3 3HR	2E0WEE
KT3 3HR	M6FRO
KT3 3HZ	G8BPY
KT3 3HZ	M3BPY
KT3 3LB	M3WZR
KT3 4BN	M7KKM
KT3 4HR	G8GGI
KT3 4LD	2E0WNI
KT3 4LD	M0WNI
KT3 4LD	M3WNI
KT3 4LE	G3XXH
KT3 5BZ	G7UVF
KT3 6BA	2E0SFN
KT3 6BA	M7SZT
KT3 6BS	2E0GRF
KT3 6BS	M0HWS
KT3 6BS	M3ZJV
KT3 6QD	G1EET
KT36LX	M0OOZ
KT4 7AA	2E0JRE
KT4 7AA	M0UPS
KT4 7AA	M7WIT
KT4 7DX	G4ADM
KT4 7DX	G7IQZ
KT4 7DX	M3VYS
KT4 7JQ	G7LCS
KT4 7PH	M3XWY
KT4 7SJ	2E0FMY
KT4 7SJ	G0KRT
KT4 8EZ	G4VLN
KT4 8HD	2E0YNC
KT4 8HD	G0YNC
KT4 8HD	M7YNC
KT4 8NB	G4BQY
KT4 8SN	G7RBC
KT4 8TG	M7HGN
KT4 8UD	M0BMZ
KT4 8XG	G6VAX
KT4 8YA	G1LGB
KT4 8YA	M0CLG
KT47AY	M7RIA
KT5 8DS	M6KGO
KT5 8JJ	G6DKS
KT5 8JZ	G8KEJ
KT5 8TS	G0IRK
KT5 9BS	G6SMJ
KT5 9DX	G0JOS
KT5 9EA	G7EKC
KT5 9EA	M3EKC
KT5 9JH	G4PPU
KT5 9LJ	2E0OYN
KT5 9LJ	G1KAG
KT5 9LJ	M3OYN
KT5 9LW	G7MLJ
KT5 9PH	M3VXH
KT5 9RJ	G0BQV
KT59PT	2E0HIW
KT59PT	M7FBR
KT6 4DH	G2SR
KT6 4DH	G7JYQ
KT6 4DH	M1SRC
KT6 4EX	M6KFG
KT6 4SR	2E0HYO
KT6 4SR	M0KBK
KT6 4SR	M7DIS
KT6 4SW	M0SHA
KT6 5AF	G0PCY
KT6 5JD	G3JVC
KT6 5RE	2E0MJO
KT6 6LJ	G0SAC
KT6 6LJ	G4WGE
KT6 7JT	G6LFQ
KT6 7LL	G8UBJ
KT6 7LL	M6SAL
KT6 7NA	G1OEP
KT6 7NR	G0KEB
KT6 7PP	G8KFS
KT6 7QH	M6TCD
KT6 7SX	2E0CEB
KT6 7SX	M6HVS
KT6 7SZ	G7STL
KT6 7UN	M0DWG
KT6 7UN	M3HZK
KT7 0BL	M0DHP
KT7 0YH	G7WEP
KT7 0YN	G4CPV
KT8 0DW	2E0LXQ
KT8 0DW	G5LUX
KT8 0DW	M7JHR
KT8 1QF	2E0GXF
KT8 1QF	M6GXF
KT8 1SU	2E0GPH
KT8 1SU	M0PGH
KT8 1SU	M6GHT
KT8 1SU	M6PHT
KT8 2ET	G1DZB
KT8 2ET	G4YKK
KT8 2EX	G8CLW
KT8 2HY	G0KDI
KT8 2PY	G1YVV
KT8 2QJ	2E0JIZ
KT8 2QL	M1CEA
KT8 9AN	M0CGW
KT8 9DG	2E0LEO
KT8 9DG	2E0TJS
KT8 9DG	M3NCP
KT82EL	M7YOB
KT82EX	M5AAT
KT9 1BW	G0DOE
KT9 1BY	M6HLW
KT9 1EF	M0AWN
KT9 1JY	G0NCH
KT9 1NL	G4HJY
KT9 1PL	G6XKF
KT9 1PN	G0WIT
KT9 2BJ	2E0IJV
KT9 2BJ	M6IJV
KT9 2BN	M6JFN
KT9 2BU	G1SIP
KT9 2DE	G1RMC
KT9 2DE	G4LBM
KT9 2DE	G7GLR
KT9 2DE	G8RWH
KT9 2DT	G6DFM
KT9 2EY	G1JRR
KT9 2EY	G1NYP
KT9 2HG	G6KIE
KT9 2JE	G6YOG
KT9 2LA	G0SDF
KT9 2NN	G6HMG
KT9 2QD	G3SXW
KT9 2SA	G4OFO
KT92QS	2E0RBH
KT92QS	M3VYS
KW1 4DN	GM7BNF
KW1 4DN	GM7CLM
KW1 4DN	GM7REF
KW1 4NT	GM4JUE
KW1 4PF	MM6EER
KW1 4PF	MM6FWC
KW1 4PN	MM6DRJ
KW1 4PN	MM6JUE
KW1 4PN	MM6WKC
KW1 4QT	GM1GXH
KW1 4RG	GM4MPR
KW1 4RX	GM8EXU
KW1 4TA	MM7GJQ
KW1 4UD	GM4JYB
KW1 4UD	GM6KON
KW1 4XF	2M0HNM
KW1 4XF	MM7MNY
KW1 4XP	GM4OFI
KW1 4XP	GM7UWL
KW1 4XX	MM6CXJ
KW1 4YB	2M0SKR
KW1 4YB	MM0WIC
KW1 4YB	MM7WIC
KW1 5AJ	MM7WKK
KW1 5AS	GM4XLN
KW1 5BJ	MM7BCL
KW1 5EY	MM0HDW
KW1 5HQ	GM4RAI
KW1 5HW	GM8LFB
KW1 5NE	MM6EFH
KW1 5NF	MM0WKY
KW1 5NL	2M0WTN
KW1 5NL	MM6PSX
KW1 5NL	MM6WCK
KW1 5SS	GM7JGH
KW1 5TF	GM4DZX
KW1 5TU	MM0DUR
KW1 5TU	MM5AJW
KW1 5TU	MM6VRH
KW1 5UG	MM6PKO
KW1 5YJ	GM6WFP
KW10 6TL	2M0SOP
KW10 6TT	2M0WTT
KW10 6TT	MM6REV
KW12 6UZ	MM3YLP
KW12 6XJ	2M0CEE
KW12 6XJ	GM4NHX
KW12 6XJ	MM0RJR
KW12 6XJ	MM6AHK
KW12 6YE	GM0MST
KW14 7ES	GM6TMH
KW14 7ND	GM6TJD
KW14 7NW	GM6YQA
KW14 7QA	GM7FMY
KW14 7RG	GM3POT
KW14 7TB	GM6LDG
KW14 7UE	MM0BDJ
KW14 7XB	GM8YRE
KW14 7XB	GM8YRX
KW14 7XH	GM4RKH
KW14 7YJ	GM8NYV
KW14 8DD	MM7THU
KW14 8JE	MM7DKX
KW14 8JP	MM6EWX
KW14 8SY	MM7PPA
KW14 8TG	GM6JRX
KW14 8TU	GM0OMD
KW14 8TU	MM6HNQ
KW14 8UG	GM6JNQ
KW14 8UG	MM6FNQ
KW14 8UG	MM6MZG
KW14 8UT	GM4EFR
KW14 8XN	GM0TKB
KW14 8YT	GM1BAN
KW14 8YT	MM0PFH
KW14 8YT	MM3YHA
KW14 8YT	MM3YMU
KW14 8YT	MM6FGR
KW14XP	2M0HZR
KW14XP	MM7BWT
KW15 1EE	GM1RQD
KW15 1FF	GM0GSG
KW15 1FF	GM6HPQ
KW15 1FP	GM0OYU
KW15 1FP	GM7FLZ
KW15 1FU	MM0RDD
KW15 1LN	GM4LNN
KW15 1NA	GM4ZZH
KW15 1NA	GM6WOF
KW15 1NS	2M0JFI
KW15 1NS	MM7MKX
KW15 1QR	MM1DXU
KW15 1RL	GM0MHS
KW15 1SL	GM8NFG
KW15 1SX	GM0OWM
KW15 1SX	GM0WED
KW15 1SX	GM3IBU
KW15 1SX	MM0MWW
KW15 1SX	MM0ORK
KW15 1SX	MM3SUS
KW15 1SX	MM3TLH
KW15 1TF	MM6PHG
KW15 1UU	GM0EEY
KW15 1UU	MM0SLB
KW15 1XE	2M0DES
KW15 1XE	MM3XUI
KW15 1XJ	GM1VZG
KW15 1XR	GM1MWK
KW16 3DJ	GM3PLO
KW16 3DJ	MM7DRG
KW16 3DR	MM6MJG
KW16 3EH	MM6HED
KW16 3EP	GM4IOB
KW16 3EX	GM0PMO
KW16 3HA	MM7LWZ
KW16 3LB	GM6GJW
KW16 3NP	GM0BXR
KW16 3NP	MM3UBB
KW16 3NZ	MM3JKX
KW16 3PQ	GM8OFQ
KW16 3PQ	MM3SDP
KW163NP	GM1BZD
KW17 2AN	GM1XPE
KW17 2AN	GM3HDT
KW17 2AN	MM0BYB
KW17 2BA	GM0CVP
KW17 2BA	GM4WUR
KW17 2DW	2M0IPF
KW17 2DW	MM0KNS
KW17 2EB	GM7RDY
KW17 2EH	MM0EAX
KW17 2HR	GM0HTT
KW17 2HW	MM3VIV
KW17 2JS	2M0LLU
KW17 2JS	MM3LLU
KW17 2JU	GM0HTH
KW17 2LQ	GM7RDH
KW17 2NN	GM0IDV
KW17 2NS	MM3WJD
KW17 2NS	MM6DWX
KW17 2NY	GM7VKN
KW17 2PJ	GM6BNS
KW17 2PU	MM6RIE
KW17 2QL	GM3POI
KW17 2QL	MM0DGI
KW17 2QL	MM3POI
KW17 2RB	MM6ZRX
KW17 2RD	GM1MXE
KW17 2RF	GM0HQG
KW17 2RL	GM8WJK
KW17 2RP	2M0ORK
KW17 2RP	MM0MDH
KW17 2RP	MM6ORK
KW17 2RW	MM7AGE
KW17 2SA	2M0URT
KW17 2SA	MM6OGU
KW17 2ST	MM3GQY
KW17 2SU	GM0KTH
KW17 2TR	GM7GJM
KW172TT	2M0ESL
KW172TT	MM0OMG
KW172TT	MM6ESL
KW3 6AS	GM6YGW
KW3 6AT	MM3SCQ
KW3 6BD	MM0PFN
KW3 6BX	GM4XRT
KW6 6EA	GM3YKP
KW8 6HH	MM6BQG
KW8 6HP	GM0HLV
KW9 6NN	GM4SJB
KY1 1GL	MM7EWH
KY1 2AT	GM0TKE
KY1 2HS	2M0DQY
KY1 2HS	MM0SNK
KY1 2HS	MM6DQY
KY1 2JG	GM7DAJ
KY1 2XP	GM7CNW
KY1 2XP	MM0CWI
KY1 4LB	GM0RSE
KY1 4LB	GM4PSL
KY1 4LB	MM0TGB
KY10 3AU	GM8YUM
KY10 3HU	GM4EOU
KY10 3UD	GM0KDO
KY10 3UJ	GM4FQE
KY11 1EP	MM7GDI
KY11 1ET	GM3KJZ
KY11 1LD	GM4UKG
KY11 1NF	2M0NQT
KY11 1PG	MM7OOF
KY11 2EZ	GM7PXJ
KY11 2HJ	GM8YAQ
KY11 2LL	2M0DSY
KY11 2LL	MM6OAI
KY11 2LX	GM0PUN
KY11 2QW	MM3GVE
KY11 2SL	MM0CZK
KY11 2SS	GM0DYD
KY11 2SS	GM0GBH
KY11 2TJ	2M0CLU
KY11 2TJ	MM0TIE
KY11 2TJ	MM6BKP
KY11 3HS	GM8SZS
KY11 3JG	GM4YMI
KY11 3JY	2M0SZC
KY11 3JY	MM0SKX
KY11 3JY	MM3ZCS
KY11 3LG	GM3TAL
KY11 3LH	GM3UDK
KY11 3LJ	GM4JKT
KY11 4AX	MM0AOL
KY11 4BW	MM3RUZ
KY11 4NA	GM0OVD
KY11 4PE	2M0LRO
KY11 4PE	GM3HYX
KY11 7HZ	GM4FAU
KY11 8BF	GM1ZVJ
KY11 8BH	GM6RXD
KY11 8BT	MM3LWO
KY11 8DH	2M0CSX
KY11 8DH	MM0KZJ
KY11 8DH	MM3CSX
KY11 8DH	MM3UUT
KY11 8DP	GM8IVR
KY11 8JL	GM0SXO
KY11 8RA	GM4HVM
KY11 9AA	GM8PAH
KY11 9GT	GM0VUY
KY11 9LF	GM3OZJ
KY11 9LH	GM4GZW
KY11 9LR	GM8IXZ
KY11 9NG	MM0KOZ
KY11 9SJ	GM7IHJ
KY11 9TD	MM1CHQ
KY11 9UA	GM3SHR
KY11 9UQ	GM4BRN
KY11 9UQ	MM1CIR
KY11 9UW	GM4GRC
KY11 9UW	MM0DYX
KY11 9XR	GM0LLJ
KY11 9XX	2M0JOK
KY11 9XX	MM6JOK
KY11 9YE	GM3VYJ
KY11 9YE	GM4WZG
KY11 9YU	GM3RXZ
KY111AZ	2M0XCS
KY111AZ	MM0XCS
KY112NG	MM7CRY
KY12 0AH	GM8VKN
KY12 0BY	MM0TIR
KY12 0DR	2M0RDH
KY12 0DR	MM0RBR
KY12 0DR	MM6JOH
KY12 0DR	MM6ROH
KY12 0DR	MM7KEH
KY12 0EP	MM6HUE
KY12 0HN	GM0IXO
KY12 0JX	MM6SHB
KY12 0NL	MM0TBY
KY12 0NL	MM3XSF
KY12 0RL	MM0KTE
KY12 0XW	MM3EXW
KY12 0YX	GM0RLZ
KY12 7HA	GM4FSF
KY12 7RG	GM7YGW
KY12 7TB	GM3NVQ
KY12 8PT	GM3NCO
KY12 8QS	GM8SNE
KY12 8SR	MM7AIK
KY12 8SS	GM0VOU
KY12 8XL	2M0FLG
KY12 8XL	MM6FLG
KY12 8YD	GM0SRQ
KY12 8YL	GM4ZNX
KY12 8YW	MM6USS
KY12 9BE	GM3ZSH
KY12 9BX	GM4MPC
KY12 9JQ	GM0RHP
KY12 9LP	GM4NUU
KY12 9LU	GM3YND
KY12 9UQ	GM7LFT
KY12 9US	MM0ONX
KY12 9XG	MM6WSG
KY12 9YA	GM0URD
KY12 9YD	GM6CMO
KY12 9YG	2M0BRD
KY12 9YG	MM0XBD
KY12TS	GM6VIU
KY13 0PD	MM0MMN
KY13 0RA	MM5AMM
KY13 8DN	GM4HJO
KY13 8QY	MM0GNH
KY13 8QZ	GM0GTY
KY13 8QZ	GM0GUU
KY13 8RT	GM4VUG
KY13 8RU	GM1ZOX
KY13 8TA	GM0KVE
KY13 9JE	2M0YTA
KY13 9JE	MM0YTA
KY13 9JE	MM6FXQ
KY13 9TB	GM3OXA
KY13 9XG	GM0VCN
KY13 9XG	GM7FTK
KY13 9XU	GM3VAL
KY13 9YL	MM7EXN
KY14 7BE	MM6DOC
KY14 7DX	MM0IQK
KY14 7ES	MM3OJR
KY14 7HB	2M0EWY
KY14 7HB	MM6EWY
KY146DN	2M0IQU
KY146DN	MM3IQU
KY15 4AE	MM6AUI
KY15 4AH	MM7BYH
KY15 4HS	GM0SEF
KY15 4NA	GM7STI
KY15 4NB	MM0TOV
KY15 4PT	GM3WKB
KY15 4SS	GM0OFL
KY15 4UG	GM3LAW
KY15 4UG	GM4ENF
KY15 5AA	MM7JDL
KY15 5AR	GM6SAA
KY15 5BS	GM6ZAK
KY15 5BW	GM6SZJ
KY15 5DA	GM4ZCV
KY15 5JF	MM0AMV
KY15 5JS	MM0DWF
KY15 5LH	GM0JRQ
KY15 5PJ	MM0NUI
KY15 5QB	GM0UZV
KY15 5RF	GM4IEF
KY15 5SE	GM7CPL
KY15 5SH	2M0ZNQ
KY15 5SH	MM3ZNQ
KY15 5SQ	MM3ZJY
KY15 5UF	GM4RGU
KY15 5WN	GM4BUA
KY15 5YU	MM6VAB
KY15 7BW	GM3SJY
KY15 7NN	MM7VWT
KY15 7PT	2M0SGQ
KY15 7PT	MM0SGQ
KY15 7PT	MM3SGQ
KY15 7QD	MM6BMP
KY15 7QX	GM0AIR
KY15 7TF	GM6GPH
KY15 7TW	2M0GOY
KY15 7TW	MM7MCD
KY15 7TY	GM0OFM
KY15 7UH	GM3VPN
KY15 7UR	2M0BEC
KY15 7UR	MM3IOF
KY155DY	2M0ITO
KY16 0AQ	MM0IBL
KY16 0FG	MM0BGO
KY16 0UE	2M0WJS
KY16 0UE	MM6WJS
KY16 0XE	GM1MZZ
KY16 8AB	MM7XYF
KY16 8JF	MM7BEI
KY16 8JP	GM6BGL
KY16 8LX	MM6NVW
KY16 8NJ	MM0HRL
KY16 8NS	GM3YOI
KY16 8NT	GM4FVO
KY16 8QR	GM4FFP
KY16 8RA	GM3VWY
KY16 8RB	GM0WFA
KY16 8YA	2M0BYK
KY16 8YA	MM0TRS
KY16 8YA	MM6TRO
KY16 9DA	GM4FPZ
KY16 9ND	MM0LJA
KY16 9NE	GM4ENP
KY16 9NQ	GM4HKV
KY16 9NQ	MM0DQP
KY16 9UA	GM0GSB
KY16 9YB	GM1FEM
KY2 5AD	GM4XND
KY2 5DB	MM0IMU
KY2 5DG	MM6OMJ
KY2 5PH	GM1BLX
KY2 5PX	MM3CWO
KY2 5PZ	MM0JSG
KY2 5QX	MM7WJM
KY2 5RF	MM0GSW
KY2 5SD	2M0NAN
KY2 5SD	MM0YSK
KY2 5SD	MM6CZH
KY2 5SE	MM3DFZ
KY2 5SQ	GM3YBQ
KY2 5SQ	GM7UQM
KY2 5SQ	MM0MOC
KY2 5TU	GM4DOF
KY2 5TX	MM6FJS
KY2 5XL	2M0IAH
KY2 5XL	MM6UEN
KY2 5YT	MM1AEZ
KY2 5YX	MM0SHD
KY2 6AL	2M0FVJ
KY2 6AL	MM0JJQ
KY2 6AL	MM7MXB
KY2 6DR	GM0NXO
KY2 6ES	GM0LKT
KY2 6FT	MM7EQF
KY2 6JE	2M0DSU
KY2 6JE	MM3YKR
KY2 6UD	MM0IKG
KY2 6XP	GM0LOO
KY2 6YB	GM8ZTV
KY2 6YB	MM3MMI
KY3 0AZ	MM0AGV
KY3 0BT	MM6DQG
KY3 0DQ	GM0RDA
KY3 0JW	GM0IJR
KY3 0RS	2M0KRI
KY3 0RS	MM0WTZ
KY3 0RS	MM7KWA
KY3 0RZ	GM4TRZ
KY3 0XE	2M0ZBH
KY3 0XE	MM0ZBH
KY3 0XH	GM0CFC
KY3 0XJ	GM4GUL
KY3 9EP	MM7FNG
KY3 9EU	GM0BUI
KY3 9JQ	GM0VFD
KY3 9UR	MM7KEF
KY3 9XS	GM0CNP
KY3 9YF	MM7GHF
KY4 0BG	GM1SZM
KY4 0EQ	2M0GTR
KY4 0EQ	GM8KSJ
KY4 0FH	GM4UPB
KY4 0JF	MM0DAA
KY4 0JX	MM7CFE
KY4 0NL	GM8ZOW
KY4 8AW	GM8XUK
KY4 8AZ	GM8NVE
KY4 8ER	GM7OQE
KY4 8EU	GM1ATW
KY4 8EU	GM6OOA
KY4 9DS	2M0YPE
KY4 9EP	GM7KVB
KY4 9NB	2M0JMI
KY4 9NB	MM0KKL
KY4 9NB	MM7MCN
KY48HB	MM7KTO
KY5 0AX	GM0KLP
KY5 0DG	GM8JGB
KY5 0DG	MM0WLL
KY5 0EN	MM0OVV
KY5 0EN	MM3NGJ
KY5 0ND	GM3ZGH
KY5 0ND	GM4NNH
KY5 0NW	MM0RBN
KY5 0NW	MM3YEC
KY5 0PH	2M0IWD
KY5 0PH	GM8CZU
KY5 8AL	MM0CBI
KY5 8NS	MM0PMW
KY5 8NS	MM1DPC
KY5 8QA	MM7DPL
KY5 9BF	GM8ZGC
KY5 9EU	GM0CFW
KY5 9JS	MM3VNT
KY5 9LT	GM0IVR
KY5 9QH	MM6GIW
KY6 1DU	GM0EZR
KY6 1HW	GM4SBP
KY6 1JD	MM6KCV
KY6 1JP	GM0MRJ
KY6 1JR	MM3IQA
KY6 1LZ	2M0KPE
KY6 1NA	GM0CFK
KY6 2BQ	GM4DEX
KY6 2JU	2M0SXT
KY6 2JU	MM3SXT
KY6 2LU	MM7BYT
KY6 2PA	GM0IJD
KY6 2QA	GM1JZM
KY6 3AX	GM7BOZ
KY6 3LH	GM7CQQ
KY6 3PJ	GM1CEJ
KY6 3QQ	MM0AWJ
KY6 3QR	GM8ZQY
KY7 4ES	GM3NMN
KY7 4HP	MM7IFC
KY7 4HS	GM4TNP
KY7 4HS	GM7FRC
KY7 4HS	MM3OYB
KY7 4PH	GM4ZJI
KY7 4RF	MM6AMS
KY7 4RU	GM6AOJ
KY7 4RU	MM3BHG
KY7 4RW	GM0GTL
KY7 4SJ	GM3VVF
KY7 4SS	GM6HGW
KY7 4SS	MM7MTB
KY7 4UE	GM7FWA
KY7 5AJ	MM7KGC
KY7 5AX	GM1ADI
KY7 5AX	GM4XAV
KY7 5DF	MM3EHM
KY7 5TA	2M0DSL
KY7 5TA	MM0KDY
KY7 5TA	MM6VXR
KY7 5TD	GM7GKT
KY7 5TD	MM3HSC
KY7 5TE	GM3XMY
KY7 6BN	GM7CTV
KY7 6HX	MM3ATI
KY7 6LA	GM3OBC
KY7 6TA	MM6INN
KY7 6UQ	MM7CRB
KY7 6UT	MM6HRZ
KY7 6UY	MM6XHU
KY7 6XE	GM0IOA
KY7 6XX	GM0MNV
KY7 6YL	2M0AGS
KY7 6YL	GM3KMF
KY74EA	MM3VYY
KY8 1BG	MM6FSV
KY8 1DJ	MM6GPD
KY8 1DW	MM6CFS
KY8 1EZ	MM3KYO
KY8 2EQ	GM4PWQ
KY8 2HA	GM6NYT
KY8 2JL	MM7CJS
KY8 3AR	MM0HSB
KY8 3DH	GM4HBQ
KY8 3QG	MM6EHL
KY8 4DF	GM7VQB
KY8 4HZ	MM6GQP
KY8 4LH	GM0GUJ
KY8 4QE	MM6BBK
KY8 4QN	GM3RUI
KY8 4QQ	MM7BDC
KY8 5BD	MM6CHV
KY8 5BP	MM3WWP
KY8 5DL	MM3TWW
KY8 5DZ	GM7KPE
KY8 5EN	MM7GGD
KY8 5FA	GM4FEO
KY8 5FL	MM7MHY
KY8 5HA	2M1JKG
KY8 5LT	GM1EAH
KY8 5LX	MM7ASN
KY8 5QP	GM4FYH
KY8 5QP	MM1DJJ
KY8 5QP	MM6HGO
KY8 5SW	GM3AHR
KY8 5TL	GM0IET
KY8 5TR	GM4EJI
KY8 5XA	GM0IWX
KY8 6HH	GM6BGQ
L N12	G4NDP
L10 4XL	G4VKV
L10 4YW	M0RDC
L10 7NF	2E0LAR
L10 7NF	M6KBY
L10 8LU	M0EBD
L108JN	M7COH
L11 0AY	M7CFR
L11 1AN	G7SZF
L11 1AY	G6NRK
L11 1AZ	2E0PUE
L11 1AZ	M3PUE
L11 1BD	M6BDR
L11 1BD	M6SHV
L11 1DF	G0WRE
L11 1DF	M3XPS
L11 2YH	2E0JPC
L11 2YH	M3JNQ
L11 2YH	M6LLC
L11 2YH	M6LLD
L11 2YH	M6LLE
L11 3BQ	G8WAW
L11 4UW	G6OPV
L11 7DJ	M6XWB
L11 7DS	M6KBT
L11 7DW	M3ZYG
L11 8LR	G0JSZ
L11 8LR	G1TIQ
L11 9AG	G4XAR
L11 9DP	G4WWB
L12 0AW	M6DKC
L12 0HX	G8XTD
L12 0NF	2E0HLO
L12 0NF	M0SGG
L12 0NF	M6LHO
L12 0QN	M6BQD
L12 3HQ	G6NFR
L12 3HS	2E0FXW
L12 3HS	M7RFD
L12 4YR	G4VYR
L12 6RE	M0YYV
L12 7HE	G6WIO
L12 8RN	G0MXW
L12 8RW	G1AVC
L12 9EQ	2E0BZE
L12 9EQ	M3YVD
L12 9GY	G0SMR
L12 9NE	G8YGM
L13 0AY	M3GYY
L13 2BA	M0HQQ
L13 3DY	G6BFM
L13 4DH	2E0DKO
L13 4DH	M6IQO
L13 7BU	M7GFU
L13 7DJ	G3OSI
L13 8AW	M6WSO
L13 9AA	M3SPY
L139DF	M3UFY
L14 0LG	G3NBU
L14 0NU	G7DUY
L14 3LE	M6SMQ
L14 3LF	M3HSC
L14 3LT	G7NHQ
L14 6UT	M6RKA
L14 9PA	G0SXA
L14 9PA	M5CAB
L144BF	M7RNZ
L15 0HN	M3GAP
L15 3JG	G4KRN
L15 4JX	M0BFM
L15 4ND	G7DOD
L15 4PW	G0STF
L15 5AX	G0OSC
L15 6TG	G6VNW
L15 8HX	G6KXJ
L16 3NE	G6IMJ
L16 5ET	G3ZIM
L16 5EU	G8HTF
L16 6AD	M6KFP
L16 7PA	G0VMT
L16 7PQ	G0KBS
L16 7QR	G7GEE
L16 8NZ	G3RWW
L17 0DP	G6EQL
L17 1AB	G4HCS
L17 5DB	G8FXA
L17 6AD	G0ADP
L17 7AL	G1JHG
L17 8TT	2E0WJT
L17 8TT	M0TWJ
L17 8UD	G0EVB
L17 8UH	G1VAA
L18 1EY	G3PDC
L18 1JZ	G0MIT
L18 1JZ	G7ACK
L18 3EX	G8IPK
L18 4PZ	M5RMF
L18 4QE	G1YEP
L18 4QP	M3XSN
L18 7HQ	M0NVJ
L18 7HU	G1YUX
L18 7HY	G5JIM
L18 7HY	G6WRC
L18 7JG	G7VCN
L18 8AH	M3XVU
L18 9SJ	G0JUA
L18 9TB	G0RBJ

L18 9TN G0MSO
L19 0NA G4DKQ
L19 1QJ M3XNT
L19 1RB 2E0CFD
L19 2HE M6IAV
L19 2QZ G1BES
L19 2QZ G1K0P
L19 3QX M6HMZ
L19 3RA 2E0XRQ
L19 3RA M7XTX
L19 4TB G4KVL
L19 4TR M3PBB
L19 4TR M7ANE
L19 4UG 2E1PGA
L19 4UR M3OBD
L19 6PQ 2E0LKM
L19 6PQ M6LKM
L19 6PW M3LPJ
L19 6PZ M0CEW
L19 8LF M7JTN
L19 9AN G1ZBU
L19 9BJ 2E0HFJ
L19 9BJ M7BTD
L19 9BX M0BAI
L19 9DG G2BGG
L19 9EA G0GQX
L20 0AD M7SDZ
L20 0AL M3DYY
L20 0BY M3AKE
L20 0BY M7OWL
L20 4QR M6LPO
L20 4TE M3XCS
L20 4UU G0WMJ
L20 5AN 2E0MTC
L20 5AN M3MTC
L20 5BG G7OLG
L20 5EA G1LOV
L20 6DJ G7GHT
L20 6EJ G0OCC
L20 6ES M3GTT
L20 6HB M7REL
L20 6HF M7EFC
L20 6LU M7MKZ
L20 6NN M3JID
L203HN M3EFC
L21 1BJ M6OUY
L21 1BP M7PGA
L21 2PE G3YXS
L21 3UB M5SE
L21 4PJ G4BM
L21 7NN G0DBE
L21 7NN M3JIH
L21 7NN M3JII
L21 7NN M6DX
L21 7NX 2E0FQJ
L21 7QH G1DFP
L21 7QH G4VIL
L21 7QN M7GPJ
L21 8HG G7OEA
L21 8NQ 2E0HGG
L22 1RJ M0CEC
L22 2AT M7HAL
L22 3YX G1PJK
L22 3YX G1SWE
L22 4RF 2E0ZOM
L22 4RF M3ZZS
L22 8QB G4XRX
L22 0NX 2E0TTE
L23 0RF G1MTJ
L23 0RG G6FDI
L23 0SG M0CMW
L23 0TT M6EMV
L23 1US G3XMG
L23 2RD M1BTD
L23 2TY M6NQP
L23 2XF G4OJK
L23 3BN M6BFB
L23 3BZ G3IZF
L23 7UL G1KLK
L23 8SS M3IHR
L23 9UF G8WMG
L23 9XE M0TER
L23 9XE M1NIS
L23 9XE M3BMW
L23 9XJ G3VXK
L23 9XS G6GOG
L23 9XY G0MJG
L23 9YU G3LNL
L23 9YU G8BVU
L24 2UF 2E0BCJ
L24 3TH G4CMP
L24 5RR M0NOW
L24 5RR M1SPW
L25 0NY G7RYW
L25 0QD G0AJL
L25 1PR M1CEW
L25 2RB G0MFH
L25 2RP M3NVL
L25 3PY G0KWD
L25 3QE G8NNX
L25 3SE G1CPU
L25 4SQ G3XSN
L25 4TJ G6BOK
L25 6DG M0BGT
L25 6DH M0PMO
L25 6DW G3KRX
L25 7UH G1CZH
L25 7UJ M3AVF
L25 8QG 2E1KLT
L25 8QG M0KLT
L25 9QY M3PPZ
L25 9SN M0OAU
L250PR M7VAL
L259SN M6ISQ
L26 1UN M0JYM
L26 1UW M7GWW
L26 1UW M7NCW
L26 6LW G0RFI
L26 7WR M6GKI
L26 7YW G7NHF
L26 9TL M1DHM
L26 9UF G0ARV
L26 9XZ M0BJD
L278YL M7DFO
L28 3QB M0LPL
L28 3QB M3HAM
L28 5RJ M6KDC
L3 4BP M3SBQ
L3 6JF M3KXG
L3 8PZ G4DLY
L30 0PH 2E0ILF
L30 0PH M6POB
L30 1PR M6MBX
L30 1PX M6BFP
L30 1PZ 2E0ECU
L30 1PZ M6FYE
L30 1SG 2E0XTC
L30 1SG M0TNX
L30 1SG M0WYR
L30 5QD G7IKM
L30 5RW 2E0FCG
L30 5RW G2OA
L30 5RW M0OYH
L30 5RW M6OYH
L31 0BD M0GSO
L31 0BD 2E0TXP
L31 0BD M0TXP
L31 0BD M6TXP
L31 0BS 2E0EJO
L31 0BS M6IQG
L31 0BW G8ZTF
L31 0DA G4OVD
L31 0DB 2E0FKD
L31 0DB M3LPR
L31 1DY G4WGN
L31 2JS G1ZEW
L31 2NN G0PWL
L31 3DN 2E1IIE
L31 3DN M3AZR
L31 4ED M7HLG
L31 5JJ G0DOR
L31 5NL G1DGW
L31 5PD G7FMW
L31 5PD M0AJJ
L31 6AS G1ACD
L31 6BS G4HDU
L31 6BY G3WDD
L31 7AN G8IIS
L31 7BY G4EFD
L31 8AE M3RJF
L31 8AX 2E0TRJ
L31 8BZ 2E0PRO
L31 8BZ M3AIR
L31 8EG G8GTI
L31 9AL G4AGY
L31 9DE M1BWZ
L31 9DG M6CUX
L316DY 2E0DLK
L316DY M0HVE
L317BP 2E0KEY
L317BP M0YOB
L317BP M6MQK
L32 0TT M0TMC
L32 0TT M3FOD
L32 0TT M3OPM
L32 0TT M3XNB
L32 1TP M1BWH
L32 7QG 2E0PVW
L32 7QG M6XIP
L32 7QH G0XBL
L32 7QR G6OWS
L33 1UW G0JIB
L34 1LW G3YCX
L34 1PX M6SJR
L34 2RS G4WCO
L34 2RY G6CKM
L34 2SB M3IJD
L34 6HD M7VOD
L34 6HU M6GOF
L34 9EH M6ARY
L35 0QH G1BEB
L35 0QR G1YCN
L35 0QR G6FIT
L35 1QG G8MKO
L35 1RJ M0FCB
L35 1RJ M6MNV
L35 1SF G4XNV
L35 2YG G0LPQ
L35 3JL G6MMG
L35 3PL 2E0PFH
L35 3PL M6PFH
L35 3PP G0IEW
L35 3RW G0ILC
L35 3SF G0RPD
L35 3SG G1CKV
L35 3UT M6JID
L35 3UX M6HCE
L35 4PA G3RJE
L35 4QH 2E0FZZ
L35 4QH G4TTO
L35 4QH M7DYS
L35 6PJ G3WOH
L35 7NE M3LXA
L35 7NG G4VCP
L35 8NE G0SLK
L35 8PY 2E0YAR
L35 8PY M3ZVI
L35 9JQ 2E0VCK
L35 9JQ M7VCK
L35 9LX 2E0MJZ
L35 9LX M7JVH
L36 0TZ M3KAY
L36 0XA M3RSN
L36 1TA M3OHR
L36 1TH G4EGM
L36 1UD G0NYR
L36 1XX 2E1CYD
L36 2LT M6GSP
L36 2LX 2E0IXS
L36 5SF G1ATG
L36 5TL M3CYS
L36 5TL M3TNY
L36 5TN G3TVN
L36 6ER M0ANN
L36 8BN 2E1CYS
L36 8BN G0LSV
L36 8DA M0HHR
L36 8EH G7AWW
L36 8EH M6IHR
L36 8JE M6PSV
L360UH M7TON
L369TY 2E0EMQ
L369TY M6IFA
L37 0AH G4PKP
L37 2DG G3MIP
L37 2JD G6OEM
L37 3HN G8FXN
L37 3JL G6CIP
L37 3LB M0CPL
L37 3LX G3POG
L37 3LY 2E0GMU
L37 3NH 2E0BZG
L37 3NH M0HUZ
L37 3NH M3ZIH
L37 4BP 2E0MXN
L37 4BP M0MXN
L37 4BP M7RAY
L37 6AD G8DUF
L37 6AY 2E0WFG
L37 6AY M3WFG
L37 6BF G7CVZ
L37 6BF M0DGA
L37 6BG M6CJA
L37 6BQ G4MZZ
L37 6DF G4SMT
L37 6FA G0CTC
L37 7ER M1DLM
L37 7EX M0COO
L38 0BG G0OWP
L38 3RP 2E0XDG
L38 3RP M0XZG
L38 3RP M6XDG
L39 0EB G6HIO
L39 0HW G6BZL
L39 1LJ M7NWL
L39 1LR G0OXV
L39 1NN G3SZV
L39 1NU M6GXZ
L39 2BA 2E0ZMO
L39 2BA M3ZMO
L39 3LD G4ATU
L39 3NS 2E0GPC
L39 3NS M0YIG
L39 3NS M6GPC
L39 3PX M6ZEB
L39 4RE G8TUN
L39 4TF G7LFC
L39 4TF M0LIF
L39 4TF M0NCI
L39 4TF M0NLI
L39 4TF M0ORM
L39 4TF M3LFC
L39 4TF M6NUR
L39 4TF M6COV
L39 4UD G8FUB
L39 4XP G4TAK
L39 5AP G6ODU
L39 5AT G0BLW
L39 5AY G3FIC
L39 5BG G3NNT
L39 5QG 2E0ORM
L39 5QJ G4UVB
L39 5QJ G6EXC
L39 6RF G8SAX
L39 6SU 2E0GKU
L39 6SU M7BAK
L39 6SY M6EEG
L39 7JL G7SWV
L39 7JU G7HSB
L39 7JY 2E0XJP
L39 7JY M6XJP
L39 7LD 2E0OLG
L39 7LD M3BHP
L39 7LD M5DR
L4 1UL 2E0VWL
L4 2UN G0OIU
L4 3SX M3KCC
L4 5RB 2E0MXM
L4 5RB M0HHA
L4 5RB M6MXM
L4 6TJ 2E0EPU
L4 7TL 2E0HYU
L4 8TY M0WHC
L4 8UL M1SWB
L4 9UG G7UZA
L40 0RA G6ITU
L40 1TB M7BFZ
L40 1TT G4CLE
L40 2QJ M0RDA
L40 2QS G0UCS
L40 3TF G0SGI
L40 3TG 2E0EUR
L40 3TG M6EKP
L40 5AA G7EVY
L40 5BE 2E0SSN
L40 5BE M0SSN
L40 5BE M6SSN
L40 5TL G1JZZ
L40 5TN G0DDU
L40 5TU M0HNG
L40 5TU M6CUV
L40 5UA G6YGP
L40 6HQ 2E0WHD
L40 6HQ M6SPT
L40 6JG G1XKL
L40 6JQ G6NJJ
L40 6JR G4LTI
L40 7RB G0AFQ
L40 7RP G0CBB
L40 9QG M0DWZ
L40 9RS G4IQJ
L42 2BR G0DYG
L5 0TD G0ELZ
L5 0TD M0LBL
L5 2RF 2E0GNN
L5 2RF 2E0SMK
L5 2RF M3GNN
L5 2RF M6SMK
L5 3PD M0JCZ
L5 6RH M6LDK
L5 6SE 2E0KHI
L5254AD M7DOZ
L6 0BZ G1KON
L6 2NB G7JCQ
L6 2NG M3FGX
L6 6BN G0EDF
L60 0BD G4XCM
L60 6BB G0KTY
L7 0JE M0RAT
L7 3EN M6BZB
L7 8SS M7YVR
L8 0UF G8DEY
L8 4TH 2E0SNA
L8 4TH M0SET
L8 4TH M3YRO
L8 4XP M7LPL
L8 4XR G1MCW
L8 6QL G8ADA
L8 8NE M3WUN
L9 0NA G4XCY
L9 0NG G7VGK
L9 1AG G1EKM
L9 1AN G8CCD
L9 1EF M6SUY
L9 1JZ 2E0LPR
L9 1JZ M3MTL
L9 3BL M3ZUC
L9 3BP 2E0IRJ
L9 3BP M0OES
L9 3BP M6OES
L9 7LG 2E0CLD
L9 7LG M0JPW
L9 7LG M0MSY
L9 7LG M0ODD
L9 7LG M0OGI
L9 7LG M6MSB
L9 7LG M6VMP
L9 8HF G4YPD
L9 9AY G0DHJ
L9 9BZ G0OIX
L9 9EJ M3DLP
L9 9HQ G7JZK
LA1 2AS G0FZA
LA1 2AY M7KJL
LA1 2HE M6CPH
LA1 2HR G8XQS
LA1 2NA M6SPH
LA1 2NW M1JHG
LA1 2QP M3KHM
LA1 2QP M3KIQ
LA1 2QP M3KNV
LA1 2QW M3APO
LA1 2QX M3UFT
LA1 2TZ G3VWJ
LA1 3AW G6ZHL
LA1 3BN M6IRU
LA1 3DS G7BBN
LA1 3DU G8EXJ
LA1 3DY G8UHO
LA1 3FA G0VGP
LA1 3HJ M0BUR
LA1 3ND G1HFA
LA1 3PX 2E0JKI
LA1 3PX M7JKA
LA1 3SY 2E0FYE
LA1 3SY M3ZFY
LA1 4BD G1YYD
LA1 4EF G4VAP
LA1 4ER G0AWM
LA1 4HX M1GDE
LA1 4HX M3GDE
LA1 4LQ G8TZJ
LA1 4LZ G4UGR
LA1 4NT M1FHQ
LA1 4PZ G3NNA
LA1 4QZ G4EKZ
LA1 4QZ M0RGS
LA1 4RF M6RDZ
LA1 4SQ M6HLL
LA1 4TS M3BIO
LA1 4UU 2E1GPV
LA1 5BD G4DLP
LA1 5BY 2E1YES
LA1 5EB G3XEN
LA1 5EB G8KED
LA1 5FS G6BJB
LA1 5JA M3VZH
LA1 5JD G0LLG
LA1 5JP M0AFF
LA1 5LB G6PHF
LA1 5RS 2E0MBA
LA1 5UY M7ENI
LA10 5AB M1ONE
LA10 5AB M3ONE
LA10 5AY 2E0CXU
LA10 5AY M6CYG
LA10 5AZ G1GVJ
LA10 5AZ G4MIS
LA10 5DE G1EKU
LA10 5JE 2E0CXJ
LA10 5JE M6CQV
LA10 5PH G4MWG
LA10 5PY G6NQL
LA10 5QB G3LUO
LA10 5TQ G6LZZ
LA10 5TQ G6SSQ
LA11 6AZ G0KDG
LA11 6DL G4ATB
LA11 6DU G3LZZ
LA11 6EB G8JAG
LA11 6EB G8KPL
LA11 6EX M0GKJ
LA11 6JZ M6LWG
LA11 6PH M0DCZ
LA11 7AA M6UTX
LA11 7LJ G4CEJ
LA11 7PF G0WHD
LA12 0AE M0KLJ
LA12 0BG M6BIX
LA12 0DB G6ZGH
LA12 0JB 2E0FZK
LA12 0JB G1DXM
LA12 0JF G1BMB
LA12 0LL M0RBE
LA12 0PZ M0TES
LA12 0QH G8ALE
LA12 0RE 2E0AYW
LA12 0SU M7FWI
LA12 0TB M3MEU
LA12 7ES G4UCL
LA12 7EU M0KYL
LA12 7JA M6DUI
LA12 8AA 2E0LCE
LA12 8AA M6IPL
LA12 8AW G0AOM
LA12 8NQ G3XTN
LA12 9EA G6NVF
LA12 9JP M1RDX
LA12 9LS M0POS
LA12 9NU G7GUB
LA12 9PD 2E0HHC
LA12 9PD G4EOB
LA12 9PD G6LKB
LA12 9PN M0MJX
LA12 9PN M7GJA
LA12 9QZ G4PXR
LA12 9RL M3WTA
LA129EX M7ULV
LA13 0AU G0SKJ
LA13 0BA M3UBE
LA13 0BB M6MDJ
LA13 0BH 2E0TCC
LA13 0BH M6AMO
LA13 0BH M6NJD
LA13 0BX G4AFR
LA13 0EE G3ZFZ
LA13 0EF M7FLY
LA13 0EQ M5RAG
LA13 0HE 2E0SLK
LA13 0HE M6BLZ
LA13 0HU G4ZEG
LA13 0NB 2E0HZK
LA13 0NB M6ZVD
LA13 0UB M7BCA
LA13 9EA G4ZOI
LA13 9ES M7ATV
LA13 9HJ M6AGY
LA13 9HU M6RLX
LA13 9PX M6NEZ
LA13 9QG G7UKR
LA13 9QG M6WSU
LA13 9RG G1GBF
LA13 9SF M3TXP
LA13 9TD M0MOL
LA14 1BQ G0MOM
LA14 1BQ M3FUR
LA14 1SZ 2E0FAS
LA14 1SZ M3CYW
LA14 1XU M6XAT
LA14 2AD M0FGB
LA14 2AD M1WDX
LA14 2BG 2E0HTC
LA14 2BG M6AUE
LA14 2BJ 2E0TXH
LA14 2BJ M7TNY
LA14 2ER M6MLH
LA14 2HH M6AHF
LA14 2JG M6FWO
LA14 2NR M3PAU
LA14 2PZ M6RPL
LA14 2RX G6UMN
LA14 3AN G4RQJ
LA14 3BP M1MIJ
LA14 3DE G1DNI
LA14 3DE G4NFR
LA14 3DX 2E0LBI
LA14 3DX M6LFB
LA14 3EJ G0GSJ
LA14 3JL M6WCN
LA14 3LP G1AGM
LA14 3LP G1CGP
LA14 3NB M6OTU
LA14 3NU M3HQW
LA14 3NU M3UFU
LA14 3NU M3UQF
LA14 3PR M6DHV
LA14 3PS M1MCW
LA14 3PS M3YBL
LA14 3QW 2E0NGC
LA14 3QW M0IGG
LA14 3QW M6LRA
LA14 3SF M6FWT
LA14 3TS M1AVV
LA14 3UD 2E0MYK
LA14 3UD M0MYK
LA14 3UD M6CET
LA14 3UZ M6TBT
LA14 4AH G0GJM
LA14 4LE M0REJ
LA14 4LE M0SJJ
LA14 4LE M3GEN
LA14 4LR M1SKY
LA14 4NH G4USW
LA14 4NW M6MRN
LA14 4PH G0UXH
LA14 4PH G4BZR
LA14 4PZ G7OXB
LA14 4TH G0TUE
LA14 5DW 2E0TEI
LA14 5DW 2E0TEJ
LA14 5DW G2DT
LA14 5DW G4ARF
LA14 5DW M0TEB
LA14 5DW M6DTF
LA14 5DW M6GUM
LA14 5HW G6YXX
LA14 5LH M3MWV
LA14 5LY M6WAD
LA14 5QQ G0LSU
LA14 5RL M3YNS
LA14 5RQ M6ANX
LA14 5SE M6EEO
LA14 5TP M3UDJ
LA14 5TX G7UAK
LA14 5XP G3IZD
LA14 5XP M3ZZV
LA15 8BW G7RNX
LA15 8EU M1CQF
LA15 8LZ G4VKE
LA15 8NR 2E0MND
LA15 8NR M3MND
LA15 8QA M0SSD
LA15 8QA M6PDT
LA15 8SE G7MCE
LA16 7EW M3HVL
LA16 7EY G0HIK
LA16 7EY M3MGU
LA16 7FL 2E0CPG
LA16 7FL G3DCT
LA16 7FL M0KPW
LA16 7FL M6AYY
LA17 7XJ G4TAZ
LA18 4LS G7PKQ
LA18 5DB G1NBY
LA18 5HB G8JHG
LA18 5HE G1ZIM
LA18 5LY M7CEV
LA19 5UE G6HEF
LA19 5UE G6HFB
LA19 5XL M1AOD
LA19 5XL M7RAB
LA19 5XN M3TZI
LA2 0EG G4UJI
LA2 0PN M0SWI
LA2 0PN M7RER
LA2 6AG G4TOT
LA2 6ER G0VPU
LA2 6HJ G0NTT
LA2 6LB M0BZR
LA2 6NH G4DPT
LA2 6NL G7TOZ
LA2 6QB G0UWX
LA2 6QJ G4ZEZ
LA2 7AB M0DGK
LA2 7AG G0OIE
LA2 7BA M0DWK
LA2 7BF G7IHD
LA2 7EP G4LIY
LA2 7EP M7PWW
LA2 7ET G1JYB
LA2 7ET M3YOG
LA2 7HX G8CSQ
LA2 7JN G1KLZ
LA2 7JZ 2E0CBL
LA2 7JZ M3ZMX
LA2 7LP M6XSS
LA2 7ND 2E0XRL
LA2 7ND M0YYT
LA2 7ND M1BUU
LA2 7ND M7SLR
LA2 8ER M0XPL
LA2 8NP G1LAT
LA2 9AN G4KWJ
LA2 9AN G4KWK
LA2 9BD M6AHE
LA2 9DL 2E0ZDX
LA2 9DL M6DCB
LA2 9LF G4NEQ
LA2 9NH 2E0MAF
LA2 9NH M0MOS
LA2 9NH M3WCZ
LA2 9NY G1CFJ
LA2 9PB G3UVQ
LA2 9PJ G4ZJE
LA2 9PJ M0CLR
LA2 9PJ M1TCQ
LA21 8EQ G8YGO
LA21 8EQ M0DCY
LA22 9BD G7IFB
LA22 9EX M3CQP
LA22 9RD 2E0GVI
LA22 9RD M7BHT
LA23 1EU M3LQO
LA23 1PX 2E0NFS
LA23 2EL 2E0KMN
LA23 2EL M0NOM
LA23 2EL M6VMS
LA23 2EU G3VGD
LA23 2LA G7IJW
LA23 2LB M6NOJ
LA23 3PW G8IRN
LA26QW G6WBS
LA29HP 2E0CWB
LA29HP M0TXK
LA29HP M6CQH
LA3 1DW M7BNE
LA3 1ET G0LRE
LA3 1JH G0TOO
LA3 1JN M1BSF
LA3 1LR G7UBP
LA3 1TS G6BTM
LA3 1UR G1RLK
LA3 2BB M0SED
LA3 2JF M6GBL
LA3 2LR G3KOQ
LA3 2LT 2E0TFT
LA3 2LT M6JOG
LA3 2LX G3RCZ
LA3 2PW M3LDF
LA3 2PW M3LQP
LA3 2QJ 2E0DXO
LA3 2TB 2E0HCS
LA3 2TB G4ZJL
LA3 2TB M7LLY
LA3 2UR 2E0TBF
LA3 2UR M6DBE
LA3 3AA G4UME
LA3 3AJ G8BME
LA3 3EF G1HIB
LA3 3HU G6XVY
LA3 3JA G0LWU
LA3 3JS M3LCZ
LA3 3LF 2E0LJL
LA3 3LF 2E0SFZ
LA3 3QA M0LCY
LA3 3RA M3UAY
LA3 3RZ G1DXN
LA3 3SH G0JHG
LA3 3SN M6YPE
LA32FH 2E1HER
LA4 4EF M3IAE
LA4 4EW 2E0AZA
LA4 4EW 2E0AZB
LA4 4EW G1CCL
LA4 4HS G4LDS
LA4 4LN M3HAZ
LA4 4LQ G6OUT
LA4 4LQ G7NLR
LA4 4NN 2E0BND
LA4 4NY G0IYU
LA4 4PZ G3CSY
LA4 4QD G0EBP
LA4 4QG 2E0GBO
LA4 4QG M6MUK
LA4 4RE G1ONH
LA4 4TJ G3VSE
LA4 4TU G7CAF
LA4 5LE 2E1DGL
LA4 5LJ G6ZJN
LA4 5LT G1UKS
LA4 5NF M3REM
LA4 5QD G3KSP
LA4 5RA G1PWY
LA4 5RU 2E0SBC
LA4 5RU M0SBC
LA4 5RU M3FRQ
LA4 5SR 2E0JOH
LA4 5SR M0YJO
LA4 5SR M6ALO
LA4 5SR M6LDL
LA4 5SU G6FKE
LA4 5UJ 2E1GNU
LA4 5UJ G0RDH
LA4 5UJ M0SCG
LA4 5XP G4XLC
LA4 6AB 2E0BQG
LA4 6BN M0GRR
LA4 6BT G7OOE
LA4 6BW M6MWC
LA4 6BW M7ASK
LA4 6DD G0FYW
LA4 6DQ M3ZTP
LA4 6DX 2E0XYZ
LA4 6DX M3SNZ
LA4 6DX M5XYZ
LA4 6EJ G0CHV
LA4 6EN 2E0WXZ
LA4 6EQ G3BAY
LA4 6HD G0GEU
LA4 6HS G0IFN
LA4 6JS 2E0VVJ
LA4 6JS M6TOD
LA4 6LA G1VTE
LA4 6LU G4ZJO
LA4 6NP 2E0VAM
LA4 6NP M3XMH
LA4 6PD G6HMA
LA4 6PS 2E0CKC
LA4 6PS M6APB
LA4 6QG G0AUF
LA4 6QL G8FJA
LA4 6QR G0ASQ
LA4 6TB G0FYH
LA4 6TD G1GZT
LA46NU G4BWX
LA5 0BH G6OOK
LA5 0BX 2E0DVV
LA5 0BX M6EYV
LA5 0HE G8VSM
LA5 0HE M7PGN
LA5 0UG 2E0EET
LA5 0UG M0HPG
LA5 0UG M0PWD
LA5 0UG M6PEW
LA5 8AQ G1PEE
LA5 8AW G4RHB
LA5 8BL G4YBS
LA5 8BL G6ZET
LA5 8DJ M0FSK
LA5 8EN G0CWP
LA5 8HA G4TRY
LA5 8HE 2E0DTG
LA5 8HJ G0VLV
LA5 8HQ G0SCU
LA5 8HR G6IKE
LA5 9BE G7MEU
LA5 9EQ 2E0WIO
LA5 9EQ M0WIO
LA5 9HS G4ABE
LA5 9LD M3LCE
LA5 9LG M6AAG
LA5 9NU M0XOB
LA5 9QR G4UGB
LA5 9SA M0RGB
LA6 1AX G8EUF
LA6 1DE 2E1GYR
LA6 1DJ G7IER
LA6 1HY G0KSS
LA6 1JE M6EOR
LA6 1QN G6POI
LA6 2AF G3HGD
LA6 2AZ M7AKM
LA6 2JL 2E1TIM
LA6 2JL M1JCB
LA6 2LJ G0JSK
LA6 2LJ G8JAI
LA6 2LJ M0IJU
LA6 3DZ G4VPS
LA6 3GH 2E0RPO
LA6 3GH M6RPO
LA7 7BB G0BBE
LA7 7BB G0KRG
LA7 7BB G0KRS
LA7 7BB G0RLO
LA7 7BB G7KRC
LA7 7BL G7OEW
LA7 7DZ M0DZW
LA7 7EJ G8TTD
LA7 7EJ M0CYR
LA7 7EW G6BKY
LA7 7JB G3JGP
LA7 7JP G1YMV
LA7 7ND M0PRN
LA7 7NL G6UPL
LA7 7PT G4UXH
LA7 7QA M7DYP
LA7 7QF G6NPC
LA8 0DF 2E0TUE
LA8 0DF M5TUE
LA8 0EN 2E0HOK
LA8 0EN M0KOH
LA8 0EN M6HOK
LA8 0NG M0PJX
LA8 8AT G8VTX
LA8 8DJ G4ZBW
LA8 8HP G3PLX
LA8 8QE G8MGG
LA8 9BT M1IKE
LA8 9HL G4GLH
LA8 9JF 2E0IVK
LA8 9NJ M3OPS
LA9 4EN 2E0SGL
LA9 4EN M0MZN
LA9 4EN M7PLH
LA9 4LN G3GEF
LA9 4LU M7FJL
LA9 4UU G3HMR
LA9 5BS M6IPI
LA9 5DU M3RBP
LA9 5HN M3ZOO
LA9 5QE G7OWV
LA9 5QE M0RIS
LA9 6AU G1IGC
LA9 6AU G6UYK
LA9 6AU G6YCM
LA9 6AU M0RMH
LA9 6BN G3VJI
LA9 6EA G3WRI
LA9 6HT G4ZBO
LA9 6LB M1KMC
LA9 7ED G7NJD
LA9 7EY 2E0BXM
LA9 7EY M3VHH
LA9 7FE G4WJR
LA9 7FF M6WGU
LA9 7HA G6IPN
LA9 7JA M7RIZ
LA9 7JB M3HSE
LA9 7JG G0KHA
LA9 7JG G0RTM
LA9 7NA G0HIF
LA9 7NT M3ZWG
LA9 7PE G3XJI
LA9 7PL M3TYV
LA9 7PN G0RSL
LA9 7QB 2E0BNJ
LA9 7SN M3XVJ
LA9 8LH G4ETD
LD1 5BH 2W0OZY
LD1 5BH MW0OZI
LD1 5BH MW6CQO
LD1 5LW GW4XXJ
LD1 5NL GW0RBH
LD1 5RB MW6ZYQ
LD1 5RB MW7WVK
LD1 5RD GW4LUX
LD1 5SE GW1PXM
LD1 5TE 2W0JRJ
LD1 5TE MW0MWJ
LD1 5TE MW7PLL
LD1 5UR GW7UNV
LD1 6BE MW6RRW
LD1 6BU MW0MVM
LD1 6EW MW0EBT
LD1 6EW MW3GUH
LD1 6NU GW6BDM
LD1 6NU MW7EMK
LD1 6PD GW4YCJ
LD1 6PH GW0BOE
LD1 6SF 2W0NXV
LD1 6SF MW6NXV
LD1 6SL GW8LQH
LD15RB 2W0XOT
LD15RB MW0XOT
LD16YE MW7SBD
LD2 3DT GW4CZK
LD2 3EN GW7JYJ
LD2 3UL MW6BUF
LD2 3UU GW4MUJ
LD2 3UY MW0LLY
LD3 0AT GW6DQH
LD3 0AT GW8YPR
LD3 0DU GW0BKJ
LD3 0HN GW1SMJ
LD3 0HY MW3RRU
LD3 0PW GW7RFA
LD3 0UR GW4NKR
LD3 7BZ 2W0IZX
LD3 7BZ MW7FQQ
LD3 7JF GW0IVT
LD3 7NY GW4BVT
LD3 7RT 2W0BOC
LD3 7RT MW0GMH
LD3 7RT MW3BOO
LD3 7UW GW0ABT
LD3 8DB MW7FAI
LD3 8DJ 2W0MCB
LD3 8DJ MW3BTN
LD3 8PD GW0XAP
LD3 8PD MW6ELX
LD3 8UP GW6JJX
LD3 8YA MW0OWM
LD3 8YA MW7RMC
LD3 9BR 2W0MFD
LD3 9BR MW6HEA
LD3 9EH MW7DNQ
LD3 9ES GW1BXX
LD3 9EW GW0KIR
LD3 9HH 2W0HAS
LD3 9HH 2W0YLL
LD3 9HH 2W0ZJA
LD3 9HH MW0HTO
LD3 9HH MW0UAA
LD3 9HH MW3VUJ
LD3 9HH MW3YVK
LD3 9HH MW6ZOL
LD3 9HT GW0GIH
LD3 9LH 2W0RHI
LD3 9LH MW3YWC
LD3 9NA MW6GSS
LD3 9NG MW6FVA
LD3 9PW GW4KCV
LD3 9RL GW6NSK
LD3 9RN GW8BCL
LD3 9SZ MW0HXC
LD4 4BS 2W0HSU
LD4 4BS GW3CPM
LD4 4BS MW7CYN
LD4 4DR GW7KIV
LD4 4EW MW7EWF
LD6 5BN GW0KQX
LD6 5HA GW7BNC
LD6 5HA MW6EQF
LD6 5LD GW1FXL
LD65LR MW7MAI
LD7 1AG GW7HDC
LD7 1DG 2W0LKX
LD7 1DG MW0LKX
LD7 1DG MW7WAD
LD7 1EF MW3TMR
LD7 1HY MW3RGD
LD7 1RY MW6YKI
LD7 1SW GW0JEQ
LD7 1YL GW8ARR
LD7 1YT GW4GAF
LD8 2AN GW3RMJ
LD8 2DG MW7BTF
LD8 2EG GW1ENP
LD8 2EP GW8CAK
LD8 2HB MW7FLK
LD8 2HL 2W0TLG
LD8 2HL MW0OTG
LD8 2HL MW6PDP
LD8 2LB GW0CTG
LD8 2LH GW4VVF
LD8 2PA GW7IZA
LD82NX MW7TYR
LE1 5XR G6ANA
LE1 6FF M6HLS
LE1 7RH G4ION
LE10 0DY G1CJI
LE10 0ED 2E0TJQ
LE10 0ED M6TJQ
LE10 0ED M7AVQ
LE10 0GB G8SUN
LE10 0GJ M7EHW
LE10 0HP M0BWF
LE10 0LA G0GSH
LE10 0LG G1CIV
LE10 0LY M6CJF
LE10 0NH G4CNZ
LE10 0PF M3OZN
LE10 0PY M6WNE
LE10 0SE M3DZK
LE10 0TW G3XWD
LE10 0WU M3AQJ
LE10 1DY 2E0HTM
LE10 1DY M6AYN
LE10 1EB 2E0SCC
LE10 1EB M6HJR
LE10 1HE G1WJO
LE10 1LS G6UX
LE10 1LT G3MFK
LE10 1QH M7OPT
LE10 1SJ G4NNI
LE10 1SL G4YFI
LE10 1SS M3IRQ
LE10 1TH M3IRP
LE10 1TJ 2E1GOP
LE10 1TJ G8JSE
LE10 2AS G4CLB
LE10 2AU G7FSD
LE10 2DY G8KFD
LE10 2HE 2E0OTA
LE10 2HE M6OLR
LE10 2HL M6ATM
LE10 2LY G4ALB
LE10 2ND G3XPU
LE10 2NX G8CTJ
LE10 2PS 2E0BMJ
LE10 2SW G4PJP
LE10 2TN 2E0DTL
LE10 2TN M0LTD
LE10 2UD G1KFB
LE10 2UH M1EGX
LE10 3HW G0BNU
LE10 3LF G4ZXA
LE10 3PE G4TUH
LE10 3PR M6SOG
LE100FE 2E0FFV
LE100FE M7ALF
LE100PH G0GGA
LE11 1BA M3XMO
LE11 1BU M0WAI
LE11 1DD 2E0CGT
LE11 1DD M6JBM
LE11 1JW M0RWM
LE11 1LF 2E0SLV
LE11 1LF M6OUI
LE11 1LP 2E0FCZ

LE11 1LP M0WFC
LE11 1LP M6FCZ
LE11 1NX M0JPP
LE11 1PU 2E1IBT
LE11 1RD G7IQO
LE11 1SL G4YJT
LE11 1SN G4SGY
LE11 1SU M0JVZ
LE11 1SY M7JMJ
LE11 2AA G3ABG
LE11 2AA G4IAQ
LE11 2AA G4IAR
LE11 2AA G4LAB
LE11 2JG G0VAU
LE11 2JL G6OUJ
LE11 2PA G0LCU
LE11 2PB G0DMH
LE11 2PG 2E1HGJ
LE11 2PH G7HBO
LE11 2PH M0CQV
LE11 2PH M3HZH
LE11 2QZ G7WIQ
LE11 2RU G6AK
LE11 2RU G8IXX
LE11 3JP G1WLW
LE11 3JP G3MKE
LE11 3JS G0WTA
LE11 3JT G3OMK
LE11 3LT G7RAL
LE11 3LT G8SNF
LE11 3LT M3SNF
LE11 3NL G6GBU
LE11 3PS G7LTU
LE11 3PT G3ZUB
LE11 3RA G7NII
LE11 3RL G4SBM
LE11 3RX G7DST
LE11 3SS G7WCP
LE11 3ST G8UIW
LE11 3TB G7BDR
LE11 3TY M7IMH
LE11 4LF G4AOP
LE11 4LF G8CPK
LE11 4LJ M3EKZ
LE11 4LL M3ROQ
LE11 4ND M3AUC
LE11 4PP G4JQV
LE11 4PU G4CCI
LE11 4PX G7ACM
LE11 4QB M3OOH
LE11 4QD G0MMJ
LE11 4QU 2E0DUW
LE11 4QU M6GGO
LE11 4QW G4OYP
LE11 4SA G8APF
LE11 4SN G4OKY
LE11 4UQ M6JUP
LE11 5AN 2E0MWJ
LE11 5EZ G4NTJ
LE11 5HB G4UAT
LE11 5JW G1VVT
LE11 5LB 2E0JOG
LE11 5LB M6PPL
LE11 5UJ G7OXA
LE11 5UU G8HMA
LE11 5UW G8LVL
LE11 5YB G7SCL
LE11 5YZ G4VCN
LE11 5YZ G6VAW
LE12 5AL M7DZP
LE12 5DF G4MPH
LE12 5EE G1KBC
LE12 5HA 2E0OCN
LE12 5HA M6FHV
LE12 5HQ G4DCI
LE12 5HW 2E0GUR
LE12 5HW M0URB
LE12 5HW M6LKL
LE12 5NQ 2E0FXO
LE12 5NQ M7DKS
LE12 5PH M7PGI
LE12 5QW M7ECG
LE12 5RQ M6BTK
LE12 6JF G8HYP
LE12 6LD G4ZMA
LE12 6NN G8HVF
LE12 6PL M0IUB
LE12 6PP 2E0LEA
LE12 6PP G3ZJV
LE12 6PW G8JHE
LE12 6ST G4BZP
LE12 6ST G8DTM
LE12 6TX M1GUR
LE12 6UB G3XYC
LE12 6UZ G0INJ
LE12 7BN G6HVD
LE12 7BP G0LMA
LE12 7FG M6GII
LE12 7HB G1IFX
LE12 7HY M3JNU
LE12 7ND G1DHQ
LE12 7NX M0BWU
LE12 7PX G4MIV
LE12 7PZ 2E0DGH
LE12 7RQ G6MOI
LE12 7SB G1HGA
LE12 7SH G0AUV
LE12 7SX G0JHJ
LE12 7TG M3OMU
LE12 7UH 2E0MFT
LE12 7UH M3MFT
LE12 8AX G8IQK
LE12 8BB G4ERT
LE12 8BF G6TZO
LE12 8EJ G8IQT
LE12 8EQ M7FIP
LE12 8HT M3URH
LE12 8JZ G7UGA
LE12 8LA G4ZSP
LE12 8LB M6GAO
LE12 8LR 2E0MCW
LE12 8RP M3HAI
LE12 8RR M7DET
LE12 8SG G3THW
LE12 9AD G1HOD
LE12 9AT 2E0ZRT
LE12 9AT M0ZRR
LE12 9AT M6ZRT
LE12 9BZ M3ZTL
LE12 9DA M0TMM
LE12 9DG M6DJS
LE12 9DH 2E0EVR
LE12 9DH M0XGX
LE12 9DH M6OFD
LE12 9DY G7HDW
LE12 9HG 2E0HDY
LE12 9HG M6TGH
LE12 9HP M0ZAK
LE12 9HY G4EUF
LE12 9JE M0ITZ
LE12 9JE M7MIQ
LE12 9LS G0UGI
LE12 9LS G4HTH
LE12 9LW M0TAB
LE12 9LW M1CJT
LE12 9NL G7BMM
LE12 9NU M7LBO
LE12 9PD M7VXU
LE12 9PU G3MKU
LE12 9QH M6JLZ
LE12 9RB 2E0GQP
LE12 9RB M7IWT
LE12 9RW G4JCH
LE12 9SG M3OAB
LE12 9SH M6IAS
LE12 9SQ G1ETZ
LE12 9SS G8BUB
LE13 0BY G4LWB
LE13 0DS G0WYM
LE13 0DU M6CJI
LE13 0EU G1DPN
LE13 0EW G3XJW
LE13 0EW M3KSP
LE13 0GE G1JMC
LE13 0HP 2E0GZH
LE13 0HP G5HOW
LE13 0HP M7EOZ
LE13 0HP M7GGH
LE13 0HP M7JRH
LE13 0LU G0SCO
LE13 0LU G4CAZ
LE13 0NB G3MLQ
LE13 0NN G8AKE
LE13 0QS M7TDJ
LE13 0RA G7PCT
LE13 0RA M0MKE
LE13 0SR G4LAU
LE13 0TF M6CXZ
LE13 1DY M1JHF
LE13 1ES M3UOX
LE13 1FT M7EIP
LE13 1HG G4NNZ
LE13 1HY G4YSP
LE13 1HZ 2E0HMG
LE13 1HZ M0WES
LE13 1HZ M6HPS
LE13 1JZ G4FOX
LE13 1JZ G4PTK
LE13 1JZ G7FOX
LE13 1LJ M0LBM
LE13 1RT G1MZH
LE13 1RZ M6LGH
LE13 1SE G8RBY
LE13 1SH G0LUB
LE13 1TT M3PTQ
LE13 1UH G6FDD
LE13 1UR G8NYH
LE14 2AN G6PZN
LE14 2AP G8YEJ
LE14 3AF G2UT
LE14 3DT G4AMN
LE14 3EW G7UDJ
LE14 3EX G6PQI
LE14 3QA G4CWC
LE14 3QE M6IWA
LE14 3QG G3UOD
LE14 3QH G0PAG
LE14 3QH M0VGK
LE14 3RY G4HEE
LE14 3SA M6WTD
LE14 3TY M7NXG
LE14 3YB M6VOD
LE14 4BU 2E0MDT
LE14 4BU M0VAR
LE14 4BU M6CJQ
LE14 4DD G4EVK
LE144BX 2E0IVF
LE144BX M3MFZ
LE15 6BL M6ELB
LE15 6EB G3MCK
LE15 6HQ M6MDC
LE15 6LT G4FDP
LE15 6LZ M1BAN
LE15 6PH G6LEI
LE15 6QA G0CWU
LE15 6QT M7NRL
LE15 6RQ M6OKM
LE15 6SJ G3WMA
LE15 6SJ M6NJE
LE15 7DB G3PKD
LE15 7DZ G3ZDW
LE15 7DZ G8FC
LE15 7HP M6ELS
LE15 7JL G4PKZ
LE15 7JL G8HLM
LE15 7NJ G4BOQ
LE15 7NX G7FGR
LE15 7PS G1VIG
LE15 7PX M0HQC
LE15 7RL G7JBD
LE15 8AD G3XQZ
LE15 8DH 2E0OCW
LE15 8DH M6WGJ
LE15 8EA M6WFR
LE15 8EZ M7NAG
LE15 8JS G3OKB
LE15 8PJ 2E0XAL
LE15 8PJ M0OGD
LE15 8PJ M6GKN
LE15 8QB M6EOT
LE15 8RH G4OSJ
LE15 8SD G3USR
LE15 8SU G8TTE
LE15 9PG G4GWH
LE15 9QP M7KKB
LE15 9RL M6EII
LE15 9RS M7CJV
LE15 9RY M3SNN
LE15 9TS G0KAQ
LE16 7BQ G4JOV
LE16 7BQ M6GYG
LE16 7BW 2E0TEU
LE16 7BW M0SVH
LE16 7BW M7TEU
LE16 7DD G8GMB
LE16 7DD M3GMB
LE16 7DE G0SFJ
LE16 7EH G0BHS
LE16 7FP G3VUH
LE16 7JF G4XRA
LE16 7JJ G4WVR
LE16 7JJ G8PAN
LE16 7LG M3HZN
LE16 7LW M3KXI
LE16 7PQ G3YKS
LE16 7TA M3HZA
LE16 7UU 2E0XVX
LE16 7UU M6XVX
LE16 7XE 2E0GAP
LE16 7XE G1IVG
LE16 7XE M3YVF
LE16 7XE M3YVG
LE16 7XE M6AJP
LE16 7YE G0OQE
LE16 8AW M6OMP
LE16 8BH G0VLR
LE16 8BH G4ISN
LE16 8BH G7VLR
LE16 8BH G8LVM
LE16 8BQ G4DIA
LE16 8EX G6OFZ
LE16 8LD G1TTH
LE16 8LD G4LXA
LE16 8QJ G1FJH
LE16 8QJ G3ZSU
LE16 8QJ M3UXE
LE16 8QJ M3UXM
LE16 8SJ G1NQH
LE16 8SS M6BCN
LE16 8ST M6JCX
LE16 8XP G5UI
LE16 8XR M3MRA
LE16 9DP M6BWL
LE16 9DU 2E0IYN
LE16 9DU M7IPM
LE16 9DX G7PZB
LE16 9GA 2E0OVB
LE16 9GA M0OVB
LE16 9GA M6OVB
LE16 9GG M7OLA
LE16 9JL M6LTD
LE16 9LW G3SFV
LE16 9NA G4JYT
LE16 9NW G1IVF
LE16 9NW G1WVR
LE167NY G4XEX
LE168JF 2E0HRB
LE168JF M6IGA
LE169JW G1PHV
LE169RZ G1IDR
LE169RZ G1ZHD
LE17 4DD G6OGZ
LE17 4JJ G0LAP
LE17 4NW G1ICI
LE17 4NW G4XPD
LE17 4PG M6PGA
LE17 4PS G3ORY
LE17 4PS G3SDC
LE17 4SP 2E0EEU
LE17 4SP M0IIA
LE17 4SP M6EEU
LE17 4SQ G0TTW
LE17 4TR G0OLS
LE17 4TR G7FFZ
LE17 4TU M0RVD
LE17 4TU M3RDV
LE17 4US G1IRQ
LE17 4UT M0IKV
LE17 4XB G3RIR
LE17 4XB G8TKQ
LE17 4XF G0JEW
LE17 4XF G4APD
LE17 4YS M6WOB
LE17 5AG G6EDC
LE17 5AS M7RML
LE17 5DE G4VOZ
LE17 5DE G8LM
LE17 5DL G6TZT
LE17 5DL G8ZUF
LE17 5EG M6PBL
LE17 5HX G4NGF
LE17 5HY G4CLA
LE17 5QA G8PGI
LE17 5RP G1DPT
LE17 6AE M7HHS
LE17 6AZ G4ABX
LE17 6EQ 2E1BKK
LE17 6EQ G7LCK
LE17 6EQ M3TDB
LE17 6FE 2E0EXX
LE17 6FE M0XAE
LE17 6FE M6BSP
LE17 6FE M6OMZ
LE17 6HD M6DMD
LE17 6JW G4VTM
LE17 6NT G6YQU
LE18 1BA M6KRH
LE18 1BH M3XFC
LE18 1BR G6IFN
LE18 1FQ G1UFT
LE18 1FX G6HPR
LE18 1GD G8SSX
LE18 1HU G4OAI
LE18 1HY G0HLL
LE18 1JZ G4KKS
LE18 2BB 2E0FOK
LE18 2BB M6FOJ
LE18 2EP G6LTR
LE18 2FP G7CEC
LE18 2FU G4XDX
LE18 2HQ M3IHN
LE18 2HZ M6RSL
LE18 2JB G1VNS
LE18 2JH M0GTE
LE18 2QX G0TCJ
LE18 2RF G6HVJ
LE18 3RL G8EVD
LE18 3TY G4EMW
LE18 3WD G4OKD
LE18 3XW G1YQP
LE18 4LP 2E0NLB
LE18 4LP M6NDB
LE18 4LY M0KVN
LE18 4LY M1KVN
LE18 4NA 2E0VOD
LE18 4NA M3UFW
LE18 4TH 2E0APY
LE18 4UH G1IAQ
LE18 4WH G6YII
LE18 4WL M3ZPE
LE184QB 2E0KFL
LE184QB M6KHE
LE19 2FX G8RFY
LE19 2FY M6CBM
LE19 2HT G0AIG
LE19 2RA G7PPL
LE19 3DY 2E0PMU
LE19 3EZ G3TWY
LE19 3PS G0WBC
LE19 4NQ G7VUU
LE19 4NQ M0AZW
LE19 4NX 2E0TXQ
LE19 4NX M7TXR
LE19 4QD G8NTD
LE19 4QX G1VIN
LE19 4QZ G0TIZ
LE192AW G0PBY
LE193EW M0IOO
LE2 1PN M3MBV
LE2 1WQ G8TYF
LE2 1YD M6BHJ
LE2 2AE M6MJM
LE2 3EA G1JAG
LE2 3EH 2E0OTP
LE2 3EH M6IKI
LE2 3RJ G6EHJ
LE2 3WR M6FUD
LE2 4NT G1IGP
LE2 4NY 2E1FEC
LE2 4PB 2E0RMJ
LE2 4PB M6JEZ
LE2 4QA 2E0BQE
LE2 4QA M3XCU
LE2 4QD 2E0BQF
LE2 4QD M0OFL
LE2 4QD M3UFG
LE2 4RJ G0HNI
LE2 4RQ 2E0BQM
LE2 4RQ M0NCK
LE2 4RQ M3VWY
LE2 5FH G7SEU
LE2 5HF G6PFN
LE2 5PF G3XKX
LE2 5TR G1HWO
LE2 5UE G3HAN
LE2 5YF 2E0EEF
LE2 5YF M0NDF
LE2 5YF M6FZL
LE2 6AD G0TTE
LE2 6FF M6EPN
LE2 6FN G3MCP
LE2 6HQ G3LQW
LE2 6JE M6AER
LE2 6NT M3CBY
LE2 8DH G8FCQ
LE2 8DJ G6WMU
LE2 8DL G1YFT
LE2 8HW G4NBH
LE2 8NJ G1PPQ
LE2 8QA G6ITW
LE2 8QL 2E1CPC
LE2 8SF M0HOP
LE2 8SF M1HOP
LE2 8UH G4MQS
LE2 8UH M1CSL
LE2 8UP G5VH
LE2 9DD G1ZYN
LE2 9GA G6WZM
LE2 9NS G6HSI
LE2 9NS G8RFE
LE2 9QJ G0JQZ
LE2 9TH G0PBP
LE2 9TJ G4TQR
LE2 9TT G6VLT
LE3 0EA M6ZSA
LE3 0PB G4OHF
LE3 0SA G3ZCT
LE3 0TN G0EVI
LE3 1AU M0ZKA
LE3 1AX G6RRJ
LE3 1FG G6EJU
LE3 1GR G0FPU
LE3 1JF M1NAS
LE3 1PA 2E0DSQ
LE3 1PA G6XRS
LE3 1PA M0VSE
LE3 1PA M6ESV
LE3 1RA M3CGH
LE3 1RA M6SBR
LE3 2BH M6KXA
LE3 2BW G3WQL
LE3 2EJ G0FZC
LE3 2FN M6KEH
LE3 2FN M6TJX
LE3 2JB 2E0GXT
LE3 2JB M6HKJ
LE3 2JE M7SEZ
LE3 2JU G0FRV
LE3 2SP G0ORY
LE3 2SP G3LRS
LE3 2SP M6NZH
LE3 2UU G1AEJ
LE3 2XQ M1BSE
LE3 3AD G4NUK
LE3 3AE 2E0MZE
LE3 3AP 2E0TPP
LE3 3DQ G4DJK
LE3 3EB M7SPS
LE3 3EY G1CKY
LE3 3FF G4SDZ
LE3 3GA 2E0XAY
LE3 3GA M6SEV
LE3 3GN G7GLA
LE3 3HB G3HYH
LE3 3LA G7AYI
LE3 3PP G1EBV
LE3 3PP M6VBD
LE3 3PS M3NNM
LE3 3SW G0MTF
LE3 5RA G6WSN
LE3 6BD G6JOL
LE3 6BJ M7GXG
LE3 6FG G0UHI
LE3 6FG G7OCY
LE3 6LU M3SAO
LE3 6NA G4KGX
LE3 6NF G4RLC
LE3 6PL 2E0DLD
LE3 6PL G1GEV
LE3 6PL M3WBQ
LE3 6QQ M6HMQ
LE3 8AF G1YEZ
LE3 8AG G6BBI
LE3 8BN M0FRD
LE3 8GF G8ZQG
LE3 8GH M0GPO
LE3 8LF M6PTO
LE3 8LU 2E0KFH
LE3 8LU M6CDK
LE3 9HG M6DPA
LE3 9ND M3PYG
LE3 9ND M3UNL
LE3 9NH M6FPD
LE3 9NH M6FPE
LE3 9PS M6CMW
LE3 9QB M3BXW
LE4 0BJ G3OCH
LE4 0DF M7FCY
LE4 0GS G3KQV
LE4 0LL G4AGN
LE4 0PP G7DHJ
LE4 0QY 2E0BQC
LE4 0QY M3XEA
LE4 0SF M3PIK
LE4 0ST M6LFL
LE4 0ST M6TFO
LE4 0SU G0AZM
LE4 0UG M6FPG
LE4 0UR G3UST
LE4 1AR M7MBC
LE4 1BL G6ICV
LE4 1BX G8ATE
LE4 2BJ G4ITP
LE4 2GH M7PXJ
LE4 2HN 2E0SVT
LE4 2HN M6JLA
LE4 2HN M6SVT
LE4 2LH G4ILT
LE4 3FF M0IVU
LE4 3GW M0GAH
LE4 3HB G4ZCJ
LE4 3HQ G4BJF
LE4 4DA G4MEF
LE4 4EH 2E1HMQ
LE4 4FU G6TIW
LE4 5LH M3BYX
LE4 7AE M3FPT
LE4 7SU M3EVP
LE4 8AY G0VTL
LE4 8BD G7DOA
LE4 8BP G4VWI
LE4 8GL M1BUJ
LE4 8JP 2E1SDI
LE4 8NB M6EIH
LE4 8NR M3HVS
LE4 9DQ M6WWF
LE4 9JP M0LRG
LE48FP G7IUI
LE5 0LF G5MY
LE5 0PH M3WZV
LE5 1EA G4EOF
LE5 1ED G6WAY
LE5 1FA 2E0LDE
LE5 1FA M0LDE
LE5 1PA G3FJL
LE5 2BF M0ESB
LE5 2DE G7VDU
LE5 2DE M3VDU
LE5 2EE M6MDR
LE5 2EF G1ECK
LE5 2EF G4ZTD
LE5 2EF G6MAM
LE5 2EF M6NES
LE5 2GP G3ZZW
LE5 2HQ M6EIF
LE5 2LJ M6VCN
LE5 2RL G0ZIP
LE5 2RL G7VIP
LE5 4EN G1KOH
LE5 4LU 2E0KOI
LE5 4QL 2E0BIC
LE5 4QL M3BIC
LE5 4WH M6HCU
LE5 5UD M0COQ
LE5 6AH G1GKA
LE5 6HL M3UOZ
LE5 6JB G4ZDQ
LE5 6PT G4MGG
LE5 6SA G1WZQ
LE5 6SY G4IGL
LE5 6XT G4YTF
LE56DE M7DMB
LE6 0BA G3UAA
LE6 0BN G1ZQV
LE6 0BT G4TFO
LE6 0EX 2E0GSA
LE6 0LH G6PGP
LE6 0LH M5MUF
LE6 0YL G8POS
LE6 0YP G0HZG
LE6 1HT G1XDS
LE6 1HT G1YBK
LE65 1ES G4WYN
LE65 1EU G4LAI
LE65 1EW G6EQB
LE65 1HT 2E0RKK
LE65 1HT M0GNO
LE65 1HT M3RKK
LE65 1TR G4JNH
LE65 1UL M0BPS
LE65 1WA G3UBB
LE65 2FQ G7SEK
LE65 2HL G0NXT
LE65 2JR G7NBE
LE65 2JR M3BBB
LE65 2JS 2E0NWL
LE65 2JZ G6CLP
LE65 2LW 2E0SQN
LE65 2LW M6SWS
LE65 2QQ G1CWW
LE65 2QY 2E1LGA
LE65 2QY M0JIO
LE651LY G4SGD
LE67 1AX M1EZR
LE67 1BR M6CEF
LE67 1EH M7CER
LE67 1GB G1JOJ
LE67 2GB G1ULP
LE67 2GL G1ODQ
LE67 2GP G8YSA
LE67 2HE G4PLK
LE67 2HU G1YQU
LE67 2JT M6NCM
LE67 2ND M1DHO
LE67 2NS 2E0DFJ
LE67 2NS M6RFJ
LE67 2QL G0HME
LE67 2SS M0HMZ
LE67 2XD 2E1HGF
LE67 3AF G4AEO
LE67 3BD 2E0SLO
LE67 3BD M0MCY
LE67 3BD M6WSW
LE67 3BE G8YGT
LE67 3JG M7CVM
LE67 3PB M3KOF
LE67 3PL G8VFM
LE67 3PW G7WDG
LE67 4AN M0MHQ
LE67 4BF G8ZZY
LE67 4DD G1NNF
LE67 4DT G0IXS
LE67 4HZ G7KMK
LE67 4JU M6LAM
LE67 4TG G4JDP
LE67 4TG G8OZQ
LE67 5AY G8JMG
LE67 5AZ G1IWE
LE67 5BP G4DCE
LE67 5BP G6UZJ
LE67 5FD G0PXK
LE67 5GF M6UKM
LE67 5GN 2E0FSK
LE67 5GN M6GAH
LE67 5PA M1AYC
LE67 5PA M1BAC
LE67 5PF M0DJB
LE67 5PT G1KSC
LE67 5PT G3RAL
LE67 5PT G7SEG
LE67 6HP G8YOG
LE67 6JT G3ZJG
LE67 6JU M6MPW
LE67 6JW G0FZE
LE67 6LF G4JKQ
LE67 6LP G7NVS
LE67 6LP M0PFC
LE67 6LW M0JSD
LE67 6QD G0BAI
LE67 6QG M6CZR
LE67 8LY M3DFL
LE67 8NU G6IHU
LE67 8PU G0MCV
LE67 8PU M6MJV
LE67 8RJ G4IGC
LE67 9PY M1FJB
LE67 9RF G6OWB
LE67 9RJ G4RKD
LE67 9SN M1EXO
LE67 9TX G4ARI
LE67 9UJ G3SBF
LE67 9WA G4CQQ
LE67 9WH G4BWF
LE7 1GQ G1VBQ
LE7 1HJ G8MZY
LE7 1HN M6MFD
LE7 1HX G0PVE
LE7 1LY G0IPB
LE7 2EH M6KRR
LE7 2EN G6FPO
LE7 2JH 2E0ODF
LE7 2JN G6NGV
LE7 2JW G8TLC
LE7 3DN G7OOT
LE7 3RJ G8DBK
LE7 3SU M7ESS
LE7 3TU G3DAQ
LE7 3ZJ G0ORL
LE7 3ZR M6PFR
LE7 4UT M7CHN
LE7 4YN G6ABP
LE7 7BH G7UYJ
LE7 7BQ M5CAH
LE7 7DU G4LRO
LE7 7EU G6TMQ
LE7 7FA G7NQU
LE7 7GN M6OJM
LE7 7PU G8LDB
LE7 7QG M0XMZ
LE7 7RL G4NWS
LE7 7RL G4WRC
LE7 9AN M3MRZ
LE7 9DA G4ZJR
LE7 9GW G6ISM
LE7 9HD 2E1CJZ
LE7 9HD G7OEY
LE7 9HD M0DBG
LE7 9JS G3URQ
LE7 9LL M6IFN
LE7 9PP G3VOV
LE7 9PQ G4DR
LE7 9UA 2E0XZZ
LE7 9UA M6KON
LE7 9UR G4XXZ
LE7 9YB M6YSU
LE79NP 2E0TJY
LE79NP M7SWY
LE8 0AP G8DGH
LE8 0JJ G0UFP
LE8 0JJ M0BNS
LE8 0NP M3JZX
LE8 0NT M7RMI
LE8 0RA G4EEL
LE8 3NE G3RUO
LE8 4AB M6DXS
LE8 4BE G4WGU
LE8 4DL G6KJH
LE8 4FQ M0CDF
LE8 4FU G3KYF
LE8 4HF G8OBP
LE8 5RH G7UOS
LE8 5SB M6ELN
LE8 5SU G1IPP
LE8 5TB G3MXV
LE8 5TG G4YJU
LE8 5TL G4FIE
LE8 5TP G1HEN
LE8 5WJ G8BTU
LE8 5WJ G8CJA
LE8 5WS M0HKM
LE8 5XH 2E0EJZ
LE8 5XH M6IOO
LE8 6HY 2E0KTV
LE8 6HY M0KTV
LE8 6HY M6BCH
LE8 6JQ G6ZZE
LE8 6NF G8ZQB
LE8 6NT G7FMJ
LE8 6YB G4CPY
LE8 8AP G4EQL
LE8 8AX M6GMY
LE8 8BP G1PPU
LE8 8TE 2E0LTD
LE8 8TR M6BUR
LE8 8TX G4IHR
LE8 8TX G8RBI
LE8 9FH G8HMJ
LE8 9FP G8FWA
LE8 9GF 2E1IDG
LE8 9GF 2E1JMG
LE8 9GP M7KAU
LE9 1SE G8VEN
LE9 1SX G1IUT
LE9 1UW G1OOM
LE9 2BL G4ZIZ
LE9 2BN G4ECO
LE9 2BW G1YYL
LE9 2DD G0DMB
LE9 2DE G4FSS
LE9 2DR M0MRJ
LE9 2EN G3ONV
LE9 2EU G3OVH
LE9 3EB G3WTD
LE9 3GE G8XRG
LE9 3GX G8PEA
LE9 4BW G4EPN
LE9 4BW G4RCC
LE9 4DY M1ANN
LE9 4DZ G0CND
LE9 4DZ G0JSE
LE9 4LG G8PGO
LE9 4LL G7WHI
LE9 6EE 2E1BAD
LE9 6HJ G4OZD
LE9 6PT G0LDO
LE9 6PY M6OSU
LE9 6QE G0CRT
LE9 6RL 2E0IMH
LE9 6RS G0WUS
LE9 6YU G8OPX
LE9 7AF G0BSN
LE9 7AW G8EHM
LE9 7AY M6CRO
LE9 7DA G8SUM
LE9 7DU G1LTK
LE9 7EY M6BGL
LE9 7FF G1MPD
LE9 7FP 2E0SBM
LE9 7FY G2AXO
LE9 7GT G4VMM
LE9 7HJ 2E0SHB
LE9 7HJ M3NXE
LE9 7HJ M6JAB
LE9 7HQ G6MKL
LE9 7HW G3TFV
LE9 7PD 2E0RBU
LE9 7PH M6BYL
LE9 7TF G3PVG
LE9 8AP M1EYA
LE9 8AR M6NXR
LE9 8DG M6JHU
LE9 8DN G8CGW
LE9 8DR G6XXN
LE9 8DR M3TLJ
LE9 8EH G4CAJ
LE9 8EW G1WIW
LE9 8FS G0UIF
LE9 8FS G7TZU
LE9 8FS G8LUV
LE9 8GP G6FNQ
LE9 8JH 2E1FTA
LE9 8JH G0SKR
LE9 8JX G7RXO
LE9 9JG 2E1GXQ
LE9 9JJ G6AYE
LE9 9JR G0TPH
LE9 9LG G0EDO
LE9 9LG G8LMW
LE9 9LG G8RYE
LE9 9LQ G1WSW
LE9 9LQ G4UUG
LE9 9QF G3LMR
LE91SX M6OHO
LL11 1EF MW3PPQ
LL11 2BG GW6WVD
LL11 2EU GW4AVC
LL11 2LU GW6NSG
LL11 2ST 2W0SFB
LL11 2ST MW6SFB
LL11 2YF GW7UMK
LL11 3ED MW6EMK
LL11 3LJ MW7EJE
LL11 3NN GW8UPJ
LL11 3PB GW7KNN
LL11 3PZ MW0ARL
LL11 3RY MW6FVT
LL11 3TW GW1IAW
LL11 3YT GW1LHV
LL11 4AF GW8ASD
LL11 4BY MW6DCN
LL11 4EB MW3WCE
LL11 4EE MW6EGD
LL11 4HF MW6IXZ
LL11 4PF MW1VCD
LL11 4SZ MW7NBC
LL11 4UE 2W0YDK
LL11 4UE MW0ECF
LL11 5AH GW8KSE
LL11 5AH GW8KSF
LL11 5AS MW6DFC
LL11 5AW GW7CEA
LL11 5DH GW6IWC
LL11 5EP 2W0MVJ
LL11 5EP MW0XAD
LL11 5EP MW6GXU
LL11 5EU GW8VEE
LL11 5EU MW3SAI
LL11 5HH GW6JMC
LL11 5LE GW6PVK
LL11 5NG 2W0OSG
LL11 5PE MW7FTI
LL11 5SH MW6EYU
LL11 5UF GW4TJN
LL11 5UH GW4GSS
LL11 5UN GW6MRO
LL11 5YF GW4OVH
LL11 5YP GW6NLP
LL11 6BP MW3IGZ
LL11 6DN 2W0KGP
LL11 6DN 2W0KGQ
LL11 6DN MW0KGP
LL11 6DN MW3KGP
LL11 6DN MW3KGQ
LL11 6HR MW3HQV
LL11 6HR MW3HRE
LL11 6NS GW0VMR
LL11 6PD GW6FUY
LL11 6RG MW3NDO
LL11 6RH GW1ZHI
LL12 0AU GW0MMY
LL12 0BP GW4IGF
LL12 0DA GW3TOW
LL12 0HN GW8HPL
LL12 0NW GW0KPU
LL12 0NW MW0RCH
LL12 0NW MW0YLS
LL12 0PS GW8UXL
LL12 0PS GW8XRR
LL12 0PW 2W1MVR
LL12 0RT GW6MPX
LL12 0RT MW7KDA
LL12 0UW GW3YTL
LL12 7DS MW6NSY
LL12 7EP MW0GXC
LL12 7PD GW0ABE
LL12 7PP GW6SBD
LL12 7SG MW1DBA
LL12 7TW 2W0TNB
LL12 7TW MW6BYX
LL12 7UB GW5DOD
LL12 7UG GW6GMF
LL12 7UH 2W1DAO
LL12 8AF GW0EMB
LL12 8BE GW6FED
LL12 8BN MW1DAU
LL12 8DD MW3WPN
LL12 8HW 2W0RBX
LL12 8LB 2W0WWR
LL12 8LD GW8IJT
LL12 8RN 2W0JCP
LL12 8RN MW3JWV
LL12 8RN MW3XLR
LL12 8SJ GW6JTX
LL12 8ST GW0CES
LL12 8YL MW0BNB
LL12 9DH GW7MQE
LL12 9PD GW0IFB
LL12 9PE GW0NOO
LL12 9PE GW0NOP
LL12 9PE GW7COB
LL12 9PJ GW6IGY
LL12 9PR MW6MJZ
LL13 0AY GW4VAG
LL13 0BL GW6ILY
LL13 0BQ GW1IPJ
LL13 0HA MW7GBW
LL13 0LJ GW4XDR
LL13 0LN GW6AJK
LL13 0NW MW7TLS
LL13 0PQ GW3HEU
LL13 0UJ GW6FES
LL13 7AW MW3NIA
LL13 7AW MW7XMD
LL13 7EA 2W0TYG
LL13 7EA MW0TBB
LL13 7EA MW6TYG
LL13 7QE GW0GWE
LL13 7QJ GW8MGF
LL13 7QW 2W0ISZ
LL13 8RU 2W0KAP
LL13 8RU MW7KAP
LL13 9BA MW0CQR
LL13 9HR 2W0AZM
LL13 9LY GW6ZHY
LL13 9NQ GW3XQO
LL13 9QH GW4TNF
LL13 9SY MW7DDP
LL13 9TA GW4ZYM
LL139NN MW6LQW
LL14 1BB MW0ARV
LL14 1HH MW7WOC
LL14 1NF 2W0MDG
LL14 1NF MW0MDT
LL14 1ST GW4WVB
LL14 1UA GW6GTS
LL14 1UA GW8GGW
LL14 2AT MW6IQF
LL14 2DT MW6BYV
LL14 2EA GW6ZCR
LL14 2EA MW3JAP
LL14 2EJ MW7MAP
LL14 2EY 2W0JYN
LL14 2EY MW3WZZ
LL14 2EY MW6SHJ
LL14 2HT GW7PCX
LL14 2HT MW3PCX
LL14 2ND MW0ASL
LL14 2PN MW7IDI
LL14 2RL GW0WZZ
LL14 2RL GW8WYW
LL14 2RL MW0MRS
LL14 2RL MW7OTK
LL14 2SR MW3OXV
LL14 3EE GW0WER
LL14 3GB MW0ZDX
LL14 3JH GW7TUQ
LL14 3JH MW0IKH
LL14 4DA 2W0FJG
LL14 4JD MW1BJB
LL14 5BG MW6BYT
LL14 5DW GW7MHF
LL14 5HN MW7GOM
LL14 5PF GW1CTO
LL14 6AH GW0EHA
LL14 6AN 2W0WXM
LL14 6AT MW3ZCI
LL14 6BF 2W0RUA
LL14 6BF MW7MEJ
LL14 6DD GW1MKV
LL14 6DP 2W1PGL
LL14 6EG GW0MMB
LL14 6HS GW1GXQ
LL14 6LA 2W0LZU
LL14 6LA MW0PTY
LL14 6LA MW6LZU
LL14 6TD MW6EAO
LL15 1AQ GW0GLI
LL15 1HB 2W0PJM
LL15 1JA GW4ITQ
LL15 1JA MW0GWY
LL15 1QN MW0IBT
LL15 1RR MW0ZUS
LL15 1YP GW3ODB
LL15 2DW 2W1LCO
LL15 2DW GW6XRE
LL15 2EY GW6JWL
LL15 2HR GW7NUU
LL15 2HW GW2HFR
LL16 3BE MW3AFR
LL16 3BE MW6FMW
LL16 3EU GW3RUE
LL16 3HE GW4CQZ
LL16 3PE MW0CKT
LL16 3PW MW3IJJ
LL16 3TD MW3CMG
LL16 3YA GW8FSN
LL16 4BD GW4IEZ
LL16 4BS 2W0BLG
LL16 4BS MW3PNR
LL16 4BU GW0WWQ
LL16 4DT GW3GZX
LL16 4HH GW8RHP
LL16 4NN GW8SIE
LL16 4PQ 2W0LJC
LL16 4PQ MW6CLJ
LL16 4RL GW4RWR
LL16 5NP MW7EFH
LL16 5PA MW3GQS
LL16 5YL MW3XHL

LL17 0AD GW6RNA
LL17 0BH GW7NIW
LL17 0BP GW6ONZ
LL17 0RG MW3KIS
LL17 0SX GW1TFB
LL17 0SX MW0YDX
LL17 0SX MW3GBD
LL17 0UH GW6VEI
LL17 0UP MW3DBF
LL170DB MW0MFB
LL18 1DA GW4ZPM
LL18 1HT MW6ILK
LL18 1PN MW6BME
LL18 1TF 2W1EPL
LL18 1TF 2W1EPO
LL18 2BX MW1EOK
LL18 2EN MW6GCN
LL18 2LW GW4UJF
LL18 2NS MW6BVQ
LL18 2NU GW4XEF
LL18 2RS GW4XWN
LL18 2RY GW0DFY
LL18 2TP GW6IYP
LL18 2UL GW8WUM
LL18 2UL MW1EOO
LL18 2UL MW3TOB
LL18 3BG GW3OIN
LL18 3EE GW3UTG
LL18 3ER GW4NLD
LL18 3PE GW4ARC
LL18 3PE GW4HDR
LL18 3US GW0AYP
LL18 4AD GW1NGN
LL18 4AF GW0MOF
LL18 4BH MW0NDE
LL18 4EH 2W0FAP
LL18 4EH MW0HAP
LL18 4HH GW3FJI
LL18 4JF GW8OYT
LL18 4JH GW8XLL
LL18 4NF MW3CYM
LL18 4NF MW5CYM
LL18 4PN MW0CCN
LL18 4SB MW7EWP
LL18 4SU GW8TBY
LL18 4SU MW6NMG
LL18 4TN GW7VQA
LL18 4TY MW3HFO
LL18 5AG GW7EXH
LL18 5AG GW7NFY
LL18 5BN MW6TAK
LL18 5BP MW6ATG
LL18 5EW GW6VEN
LL18 5EY GW0HBZ
LL18 5HW GW4VLU
LL18 5HY 2W0FKW
LL18 5JE GW4DMR
LL18 5JE MW0CCS
LL18 5LS GW6XXY
LL18 5LS MW3XXY
LL18 5NH GW8FVI
LL18 5NS MW3XMY
LL18 5PP MW3CDL
LL18 5RB MW6FJQ
LL18 5RF GW6SLO
LL18 5RW GW0JSX
LL18 5TB 2W1HTK
LL18 5TB MW3EGB
LL18 5TL 2W0HUU
LL18 5TT 2W1BYK
LL18 5UP GW4JOT
LL18 5UP MW1JAN
LL18 5UP MW3GNB
LL18 5YF MW3GVF
LL18 6HS GW4TYH
LL18 6HS MW3SZD
LL18 6HT GW4UIR
LL19 7DF 2W1CEE
LL19 7DQ MW3HCW
LL19 7DS GW0ULP
LL19 7EH GW0DSJ
LL19 7EH GW0NWR
LL19 7EH MW7RES
LL19 7HH MW6TGF
LL19 7LP GW7RZN
LL19 7LP GW7TEO
LL19 7PF GW7AMS
LL19 7PF MW6FRW
LL19 7TS GW4DTQ
LL19 8DG GW7PEO
LL19 8HQ MW3FTP
LL19 8LP GW6RUO
LL19 8LU GW4XXP
LL19 8RD 2W0CIV
LL19 8RD MW0MIE
LL19 8RD MW3YMY
LL19 8RN GW6YDT
LL19 8RN MW3YDT
LL19 8RU 2W0ALZ
LL19 8SG GW3JGE
LL19 8TS GW6PMC
LL19 8YQ MW6EUK
LL19 9DT GW0ADC
LL19 9DT GW3CF
LL19 9DU GW6ITJ
LL19 9HL GW3JGA
LL19 9HU 2W0JPE
LL19 9HU MW6SAX
LL19 9NN GW0CNK
LL19 9NX MW6FXA
LL19 9PB GW0KZW
LL19 9SH GW4VHP
LL19 9SU MW1LLL
LL20 7AD 2W1FJZ
LL20 7AR 2W1EEP
LL20 7AS 2W0CJJ
LL20 7AT 2W1FJN
LL20 7AT GW7GHE
LL20 7BP 2W0PFD
LL20 7BP MW0PDV
LL20 7BP MW6PDE
LL20 7DD MW3WAL
LL20 7UH GW0TBM
LL20 8AS MW0ICO
LL20 8AS MW3XQJ
LL20 8DD GW6EWX
LL20 8RD 2W0XKL
LL20 8RD MW0XKL
LL20 8RD MW3XKL
LL20 X GW1UYW
LL21 0DL GW6WAG
LL21 0EH MW6BQM
LL21 0HA GW4URG
LL21 0PE GW6NNB
LL21 0SB GW7KDI
LL21 0SB MW5DAD
LL21 9AA GW6UMU
LL21 9EG GW1ZKN
LL21 9NA MW0PPM
LL21 9NP MW3FLA
LL21 9NP MW3SLI
LL21 9TB GW6KAV
LL22 7AR MW3FLU
LL22 7BU GW1PYY
LL22 7BU MW3PYY
LL22 7DF MW3MRC
LL22 7DG GW4IDV
LL22 7DN 2W0NKS
LL22 7DN MW7MPM
LL22 7DU MW0IBH
LL22 7DX MW0KGU
LL22 7HE GW7NTP
LL22 7JF GW7EVG
LL22 7JF MW0GWV
LL22 7UD GW0BCR
LL22 7UQ GW0NSZ
LL22 7UU MW1TAF
LL22 8EG 2W0CEO
LL22 8EG MW0CDG
LL22 8EG MW6DIG
LL22 8EG MW6GOZ
LL22 8JA GW0SPY
LL22 8JD GW3WEQ
LL22 8JD MW7GAN
LL22 8PD GW7AHR
LL22 8QA GW1CDH
LL22 8QA MW6SZW
LL22 8SS GW4NJL
LL22 8UL MW3YHC
LL22 9HD 2W0GIX
LL22 9HD MW0OTH
LL22 9HD MW6GIX
LL22 9LP MW1COJ
LL22 9ND GW3YIH
LL22 9NE GW4VHS
LL22 9PH MW3IXJ
LL22 9YH GW4DLC
LL229NT MW7BCY
LL23 7NL GW4WPH
LL23 7NT MW7DBF
LL23 7RF GW4NNO
LL23 7SF GW4UFQ
LL24 0DS GW8HJC
LL26 0BP MW6TMQ
LL26 0EF 2W0TIR
LL26 0EF GW0GTW
LL26 0EP MW6ACZ
LL26 0EP MW6ADF
LL26 0HW 2W0IEP
LL26 0HW MW7CLE
LL26 0RG GW0ETU
LL26 0RG GW4RRL
LL26 0RG GW4TSG
LL26 0RG MW0ODQ
LL26 0TY 2W0AMN
LL26 0TY MW6CVN
LL26 0TY MW7BJW
LL27 0JE GW0HUN
LL27 0JJ MW0BTI
LL28 4AJ GW0UTC
LL28 4AU MW6LFG
LL28 4NR GW8SXI
LL28 4QA GW4AMX
LL28 4RS GW6MLI
LL28 4SL GW4SYI
LL28 4TE GW0OJC
LL28 4TF GW6GKP
LL28 4UU GW1OIK
LL28 4UU MW3OIK
LL28 4YG GW7VHD
LL28 4YH MW3FVC
LL28 5HS 2W0CVE
LL28 5NG MW7GZR
LL28 5NJ GW3DZJ
LL28 5NN GW8YDR
LL284TL 2W0ENQ
LL284TL MW6ENQ
LL29 6AF GW4GJI
LL29 6DG GW0GGW
LL29 6DH GW7KGD
LL29 6DH MW6BAS
LL29 6DL MW0AFD
LL29 6DS MW3YYQ
LL29 7AX GW8BQK
LL29 7BU MW0HSI
LL29 7EY GW8VPP
LL29 7HB GW6STK
LL29 7NB GW0EVE
LL29 7NB GW0EVG
LL29 7ND MW6ECB
LL29 7SD MW3JUM
LL29 7SD MW7GTW
LL29 7TL 2W0DFM
LL29 7TL MW6RMR
LL29 7TT 2W0ITM
LL29 7TT 2W0RWF
LL29 7TT MW0TMH
LL29 8AS MW7JHN
LL29 8EU 2W1HNH
LL29 8EU MW3RHI
LL29 8EU MW3UGW
LL29 8EX GW0PRM
LL29 8HE MW7ELL
LL29 8LE GW1SAM
LL29 8LG MW6HNW
LL29 8PW GW4NNL
LL29 8SA GW1CJJ
LL29 8TA GW0PNE
LL29 8UT GW0VOG
LL29 8UY GW0OGL
LL29 8YZ MW6WWZ
LL29 8ZA GW8VUG
LL29 9AJ MW6ZIZ
LL29 9DS GW8JJP
LL29 9EL GW0DYH
LL29 9ER MW7KSE
LL29 9HP GW4RCM
LL29 9HT MW1DGR
LL29 9LA 2W1MWS
LL29 9LA 2W1SWB
LL29 9LB MW3GPJ
LL29 9LB MW6PJN
LL29 9LJ GW0WNB
LL29 9LJ GW4AMZ
LL29 9LQ GW0VMW
LL29 9NF MW7TVH
LL29 9NL GW0SKC
LL29 9NL GW0TWI
LL29 9PY GW1KHH
LL29 9RH MW3DEL
LL30 1BG MW0JYW
LL30 1BL GW4WBT
LL30 1BL MW6ORF
LL30 1BL MW6ORJ
LL30 1ES GW1BBH
LL30 1ES GW6HVA
LL30 1HQ GW0GST
LL30 1HZ GW1RVP
LL30 1JJ GW4HBS
LL30 1LP GW6WFW
LL30 1LT MW1AWT
LL30 1NG 2W0AMB
LL30 1PF GW0EZB
LL30 1PF GW3UTL
LL30 1PY GW8WFS
LL30 1RQ MW7VIB
LL30 1SL GW0TPF
LL30 1SL MW7NZP
LL30 1TU GW4PFL
LL30 1UH GW0AEZ
LL30 1UH GW7UUH
LL30 1UW GW7UIZ
LL30 2AA GW8VVX
LL30 2BU GW3FMR
LL30 2PQ GW2DNJ
LL30 2QA GW7TDQ
LL30 2RA GW3GJQ
LL30 2TY MW3EVN
LL30 2UB 2W0NTS
LL30 2UB MW7CAJ
LL30 2UF GW0IQC
LL30 2YE GW6NYR
LL30 2YP GW8FOY
LL30 3BY GW0NEC
LL30 3EY MW0HCW
LL30 3LL MW0BET
LL30 3LN GW4PVU
LL30 3NL 2W0DFN
LL30 3NL MW0HTG
LL30 3NL MW6EIK
LL31 9AN MW3MMJ
LL31 9AN MW7FXH
LL31 9BF 2W0IMM
LL31 9BF MW6IMM
LL31 9BJ GW7RKQ
LL31 9BZ GW0SFP
LL31 9EL GW3LCQ
LL31 9ER MW3TRR
LL31 9HF GW6IVY
LL31 9HQ GW6IYA
LL31 9HS 2W0DDZ
LL31 9HS MW6DAR
LL31 9LP GW4KFY
LL31 9PF MW1AHU
LL31 9QE 2W0CRB
LL31 9QE 2W0TYE
LL31 9QE MW0HYK
LL31 9QF MW0CPN
LL31 9RG GW1PSW
LL31 9RR MW3NFF
LL31 9UG GW0TLS
LL31 9UG GW3XKB
LL31 9UG MW0NAB
LL31 9UT GW1KKJ
LL32 8EU GW6ZDH
LL32 8NB MW3NAQ
LL32 8NP GW3YQP
LL32 8NR GW8JKC
LL32 8NS GW6SIX
LL32 8RJ 2W0GYQ
LL32 8RJ MW0JTG
LL32 8RJ MW7BAO
LL32 8SB GW6WKU
LL32 8WB MW6EYI
LL32 8YL MW7HEW
LL32 8YY GW4HHD
LL33 0LL GW0WUM
LL33 0RN MW7PWH
LL33 0RP MW3XME
LL33 0SN GW7JLG
LL33 0SR GW0FPY
LL33 0SS GW0LBA
LL330DA MW7FXZ
LL330TN MW3RAU
LL34 6ER GW1ACV
LL34 6LT MW7NOC
LL34 6TE MW0RXD
LL34 6UA GW1SGG
LL346LY GW8XAS
LL35 0PT GW0TWO
LL35 0RF MW0HYA
LL35 0SD GW0NEN
LL36 0AG GW4JPP
LL36 0AG MW7SIF
LL36 0AY GW3XSR
LL36 0BL MW0DWU
LL36 0TA GW6INF
LL36 9AU MW7FQN
LL36 9DE GW4XXF
LL36 9EG MW3MWO
LL36 9EP GW4KYK
LL36 9HG GW6JDF
LL36 9LY GW0COH
LL36 9ND GW3LTX
LL36 9SB GW4ANS
LL36 9SL MW0ISF
LL36 9SL MW6BXV
LL36 9UY GW4XZJ
LL37 2QB GW6UWW
LL37 2UZ GW6NXL
LL38 2BX MW3WMI
LL40 1GA MW5CKN
LL40 1NS 2W0IMX
LL40 1NS MW7IML
LL40 2AH GW6AZG
LL40 2AS MW0MMW
LL40 2EE MW3AXW
LL40 2EF MW7CKK
LL40 2RP GW0UCA
LL40 2RS MW7FSY
LL40 2SF 2W0MCQ
LL40 2SF MW6MTW
LL40 2YG GW1URD
LL41 3AF GW5DRE
LL41 3DL GW0WGM
LL41 3PU GW1EKC
LL41 3SS 2W0ORT
LL41 3SS MW0DOR
LL41 3SS MW6YKS
LL41 4BH 2W0DUL
LL41 4BH MW0WYN
LL41 4BH MW6DUL
LL41 4BU 2W0OSH
LL41 4BU 2W0UPH
LL41 4BU MW0UPH
LL41 4BU MW3BSJ
LL41 4BU MW6PZP
LL41 4BU MW6UPH
LL41 4DD GW8ZLT
LL41 4PN GW7TTX
LL42 1DX GW0HHD
LL42 1EN GW6DDF
LL42 1PL MW1KDP
LL42 1PT MW3HGR
LL43 2AG 2W0VTK
LL43 2AG GW8WNB
LL43 2AG MW0VTK
LL43 2AG MW3VTK
LL43 2AL MW6LIZ
LL43 2AN 2W0LLL
LL43 2AR MW7TYB
LL43 2BB GW0AYQ
LL43 2BB GW4LZP
LL44 2BG MW1WEJ
LL44 2BU MW6CEG
LL44 2RQ GW3FDZ
LL45 2JH GW6JBN
LL45 2HT GW0AGL
LL45 2LA MW6DSU
LL45 2LU GW0UKF
LL45 2ND GW4ZTG
LL46 2SS GW1KVI
LL46 2UJ 2W0LGG
LL46 2UJ MW6LDM
LL46 2UW MW6GHY
LL46 2UW MW6PYH
LL46 2YW MW6CHQ
LL47 6YA MW3UKK
LL48 6AL GW3JSG
LL48 6AY GW4VWO
LL48 6BH GW8AAF
LL48 6DU MW6TLF
LL48 6EF MW1BQO
LL48 6EG GW6IMS
LL48 6LS GW4PHB
LL48 6LS MW3WMP
LL48 6PS GW7RLZ
LL48 6SR GW1EWY
LL48 6ST MW3SZF
LL49 9AT MW3RVN
LL49 9AT MW5CAD
LL49 9BU GW4HLO
LL49 9DF 2W0SAK
LL49 9DF MW0GSR
LL49 9DF MW3OYC
LL49 9DF MW6BOZ
LL49 9DF MW7VCM
LL49 9LB 2W0PWR
LL49 9LB MW3VSG
LL49 9NB GW6AZX
LL49 9UH MW3DEI
LL49 9YA MW6BQX
LL49 9YD GW4OWQ
LL51 9SX MW3WSC
LL51 9UJ GW7KIO
LL52 0EF MW3EOY
LL52 0EG MW3ZHU
LL52 0NB MW6VPX
LL52 0RB MW7TYD
LL52 0SR GW4IDC
LL52 0SR GW8PKB
LL52 0SS MW6VTA
LL53 5AP GW0HGC
LL53 5LF GW0FOL
LL53 5NU MW6XSI
LL53 6DQ MW0RHD
LL53 6DQ MW3CYQ
LL53 6DQ MW3CYU
LL53 6DQ MW6RHD
LL53 6DW GW6KLC
LL53 6EA GW7UOH
LL53 6EG GW3NNB
LL53 6NN 2W0XMW
LL53 6NN MW0XTK
LL53 6NN MW7ADT
LL53 6RQ GW1ROE
LL53 6UB GW0RMB
LL53 6UN MW7RHS
LL53 6UR 2W0VJL
LL53 6UR MW3VJL
LL53 6UY GW0IWD
LL53 7ED 2W0UUA
LL53 7PA 2W0BVK
LL53 7PB GW8WNK
LL53 7PF MW0AEV
LL53 7PG GW4UKU
LL53 7PG GW4WKQ
LL53 7RL MW3VVJ
LL53 8AE GW0PZT
LL53 8AE GW3KJW
LL53 8EH GW8BZN
LL53 8ND GW4UIL
LL53 8NW GW1IEB
LL53 8UG GW7KJO
LL538NG GW7NJM
LL54 5EY GW1CFM
LL54 5HG GW4XRW
LL54 5HG GW4XZP
LL54 5SF MW3UWZ
LL54 6EW GW0NDA
LL54 6HD GW1OXJ
LL54 6HG 2W0HRH
LL54 6HG MW0PNC
LL54 6HG MW6JMB
LL54 6HG MW6KSW
LL54 6HP MW7EUG
LL54 6PT GW0WZL
LL54 6RR MW7JCX
LL54 6RT GW4CAT
LL54 6SN MW3ZKY
LL54 6SS MW3ZVD
LL54 7BH MW3PZO
LL54 7ED GW0GLX
LL54 7HN MW7FON
LL54 7HU MW0LPG
LL54 7LB GW8HQM
LL54 7PU MW3JQK
LL54 7TW GW7BZY
LL54 7YE GW0ENT
LL55 1BE 2W0SVW
LL55 1EY 2W0CYM
LL55 1EY GW6MUP
LL55 1HF GW4MEI
LL55 1LL GW0AQR
LL55 1UP GW6TGR
LL55 1YB GW1PIH
LL55 1YB GW6REQ
LL55 1YD 2W0XJQ
LL55 1YD MW7XJQ
LL55 1YL GW4KAZ
LL55 1YT 2W0LLA
LL55 1YT MW0KLW
LL55 1YT MW6VBB
LL55 2AG GW7CEQ
LL55 2EF 2W0OVT
LL55 2EF MW3OVT
LL55 2LW 2W0GVK
LL55 2RL GW6PWR
LL55 2RU MW7GNN
LL55 2UR GW4RLP
LL55 2UW MW5DJO
LL55 3BN GW4ZCY
LL55 3BN GW4ZPL
LL55 3BN GW6AKS
LL55 3BN GW6UKO
LL55 3BS MW7GEH
LL55 3HD GW1JIE
LL55 3LE 2W0AWW
LL55 3LG MW0INW
LL55 3ND GW0HHW
LL55 3NR 2W0OYL
LL55 3NR MW6OYL
LL55 3PW GW4PUX
LL55 3PW GW7UQJ
LL55 3PW GW8XUM
LL55 4AZ MW3EWR
LL55 4HB MW0VCC
LL55 4LE 2W0GAQ
LL55 4LE MW1EYU
LL55 4LE MW6KTS
LL55 4RR MW0PPO
LL55 4TW MW0ANV
LL55 4YY MW3DTC
LL56 4AB MW6GFP
LL56 4HQ MW0LEF
LL56 4PB GW4WKZ
LL57 1HF MW3YNK
LL57 1LU MW0IWU
LL57 1LU MW0JWP
LL57 1NA 2W0HMS
LL57 1NG GW0DXO
LL57 1NH 2W1AZU
LL57 2NS GW4KTQ
LL57 2UD GW0HGN
LL57 2UJ MW7BLU
LL57 2YG MW0WLP
LL57 2YG MW7WMS
LL57 3AY MW0EDS
LL57 3RE MW6HRK
LL57 3SP GW6MYY
LL57 3SP MW0GOX
LL57 3SP MW3AWI
LL57 3TD GW8GOO
LL57 3YJ MW3VQJ
LL57 4AE MW7IOL
LL57 4AF 2W0GVN
LL57 4AS GW6SYD
LL57 4AS MW6TBC
LL57 4AX GW7EWD
LL57 4DJ GW0FQZ
LL57 4DP GW3VFZ
LL57 4DR GW0ETF
LL57 4DT MW7HEN
LL57 4LD GW4FQU
LL57 4LX 2W0CZU
LL57 4LX MW6BKS
LL57 4LX MW6CTG
LL57 4NS GW4HMR
LL57 4PA GW6ORE
LL57 4TG 2W0CHV
LL57 4TG GW5AMS
LL57 4TG MW0NWM
LL57 4TG MW3SET
LL57 4UA MW0BEA
LL57 4UR MW0BYT
LL58 8AF MW3GKI
LL58 8HH GW4EIE
LL58 8HW GW0UWD
LL58 8LH MW1DOU
LL58 8PG GW4MBL
LL58 8PH GW3GUX
LL58 8PT GW8PZS
LL58 8RG GW8TOX
LL58 8RW GW6TOX
LL58 8SU GW4JUI
LL58 8UE GW4UCV
LL58 8YG GW0BDW
LL59 5DA GW8ZEI
LL59 5DS MW6GMX
LL59 5LR GW3JBJ
LL59 5LR GW4RQQ
LL59 5NB GW4CFC
LL59 5NG GW6DQB
LL59 5PP 2W0CEF
LL59 5PP MW0HCA
LL59 5PW GW1SUK
LL59 5PW MW5ADD
LL59 5RD GW3XRM
LL59 5UA GW4TAU
LL59 5UF MW7HAD
LL59 5YA MW3MEY
LL59 5YH GW6KIW
LL59 5YH MW0DNK
LL595YD MW7GAH
LL60 6AL GW3TLP
LL60 6DG MW3YQL
LL60 6EA MW6OTT
LL60 6HL MW1BTA
LL60 6HL MW3GWH
LL60 6HL MW6FNK
LL60 6JN 2W0DER
LL60 6JN MW0MON
LL60 6JN MW3UUY
LL60 6JW GW7BZR
LL60 6NW GW2OG
LL61 5AX GW4JXN
LL61 5JB GW4GTC
LL61 5JB GW8UTK
LL61 5JF 2W0DIV
LL61 5JF 2W0SLP
LL61 5JF MW5ADW
LL61 5JR GW0IQZ
LL61 5JY 2W0DAA
LL61 5JY GW3VVC
LL61 5JY GW4JKR
LL61 5JY MW0JJY
LL61 5SZ GW6IUK
LL61 5YT GW6DOK
LL61 5YX GW1PCD
LL61 6EQ MW6TDB
LL61 6SY GW0PZS
LL61 6TG MW1DKM
LL61 6TG MW7AUU
LL61 6TZ GW3PRL
LL61 6TZ GW4NBM
LL62 5AS 2W0KTT
LL62 5AS GW7FNQ
LL62 5AS MW3UPX
LL62 5AW GW3EIZ
LL62 5BG 2W1FPK
LL62 5BG MW1DAM
LL62 5LF GW4HKX
LL62 5RD 2W0JJY
LL62 5RD MW7JJY
LL63 5PQ GW0GHG
LL63 5TW GW0EWM
LL64 5XB GW3YTC
LL64 5XB MW0JHC
LL65 1AL GW0AVD
LL65 1BG GW6VIC
LL65 1ES GW4XAU
LL65 1EU GW0LIS
LL65 1HT MW6SPN
LL65 1LR MW6GCQ
LL65 1LT MW6KVO
LL65 1NW GW4IEI
LL65 1SN GW6JCK
LL65 1YR 2W0PYL
LL65 1YR GW4CVT
LL65 1YR MW0KBN
LL65 1YR MW0TMJ
LL65 1YR MW1CFA
LL65 1YR MW7CVT
LL65 2AL 2W0RZL
LL65 2AL MW6RZL
LL65 2AZ GW1HHM
LL65 2DU MW7TTA
LL65 2EF GW0KPV
LL65 2EG MW7MHC
LL65 2EX GW0NRW
LL65 2HD MW0BLU
LL65 2LL 2W0NLA
LL65 2LL MW6NLA
LL65 2LU MW0AQZ
LL65 2NF GW0IJY
LL65 2NN MW0CLB
LL65 2NS MW3YQU
LL65 2PN GW4MHR
LL65 2PN GW4XYI
LL65 2TT MW7RHR
LL65 2TX MW7WZQ
LL65 2UG MW0REH
LL65 2WA MW3GKB
LL65 2YH MW0KGY
LL65 2YW GW4VDP
LL65 3BT MW3ZVB
LL65 3DP MW6LVH
LL65 3ES GW0ENU
LL65 3ES MW0GNF
LL65 3HS GW7JKK
LL65 3LU GW0OGI
LL65 3LU MW0MAH
LL65 3PL GW0ONY
LL65 3PQ MW6BED
LL65 3RD GW0VYG
LL65 3RD GW4VEQ
LL65 3RR GW6TVD
LL65 3SY GW0MHK
LL65 3TB MW6MFB
LL65 4AG GW6CWZ
LL65 4AG GW6NHL
LL65 4AG GW8JEI
LL65 4AH GW4ZDU
LL65 4BG GW4CWU
LL65 4EA MW1EPI
LL65 4PY MW3NDU
LL65 4SL GW7JRT
LL65 4SW 2W0MON
LL65 4SW GW6AAW
LL65 4UY GW4WZS
LL65 4UY GW4YVX
LL65 4YL GW4DRR
LL65 4YL GW8FOL
LL653YD 2W0GDB
LL67 0NP MW3UBQ
LL68 0RE GW0FEM
LL68 0RE GW8FDI
LL68 9AT MW6DYF
LL68 9BG 2W0GIW
LL68 9BG MW1EWJ
LL68 9BG MW6PAM
LL68 9BT GW3SRM
LL68 9DL GW6OAW
LL68 9DN 2W0IJD
LL68 9DN MW7TAH
LL68 9DS MW0SSB
LL68 9EA 2W0JYB
LL68 9EA MW7MON
LL68 9HT 2W0FNA
LL68 9HT MW7JAB
LL68 9HT MW7KMB
LL68 9NE GW8HYI
LL68 9NH 2W0GLV
LL68 9NH MW0LLK
LL68 9NH MW6GLV
LL68 9PD GW1XAS
LL68 9PU MW7IMJ
LL68 9RX GW0ESK
LL689LY 2W0UAS
LL689LY GW1KGW
LL689LY MW7ADL
LL69 9BZ GW4ZWN
LL69 9YB GW1WWW
LL70 9DQ 2W0GGG
LL70 9DQ MW0RKB
LL70 9DQ MW3LTU
LL71 7AE MW6PQQ
LL71 7AE MW7JUE
LL71 7BU MW0SEC
LL71 7DH MW6RHR
LL71 8BG MW0BER
LL71 8ED GW0TPR
LL71 8EJ MW6FJX
LL72 8HB GW1OKY
LL72 8HN GW4EXE
LL72 8LN GW7JGF
LL73 8PE GW8YUJ
LL73 8PP GW0KRL
LL73 8PP GW7APP
LL74 8RD 2W0VSW
LL74 8RD MW3XQE
LL74 8RP GW4EOW
LL74 8SG 2W0YMP
LL74 8SG MW7KEE
LL74 8SR GW0BTB
LL75 8LN GW7SBO
LL75 8NQ GW0WEY
LL75 8UR 2W0CBZ
LL75 8UR 2W0GMZ
LL75 8UR MW0GMZ
LL75 8UR MW3ZVH
LL75 8UT 2W0SJT
LL75 8UT MW0GCT
LL75 8YE MW0JDW
LL75 8YG GW1ISK
LL75 8YU MW7BQK
LL77 7BX 2W0RTZ
LL77 7BX MW6WVC
LL77 7DZ GW6YMS
LL77 7EJ MW1DNY
LL77 7JS MW0FBM
LL77 7JS MW7IFZ
LL77 7LJ GW3TWN
LL77 7LJ MW6NXS
LL77 7NA GW6BMP
LL77 7QD 2W0ZZF
LL77 7QD MW6ADU
LL77 7QE GW3PRW
LL77 7SJ MW0BTU
LL77 7SU GW0ESU
LL77 7UR GW6HZJ
LL77 7WD MW3TOI
LN1 2AS G4STO
LN1 2AS G5FZ
LN1 2AS G6COL
LN1 2BU M0KEY
LN1 2DY G4ZKW
LN1 2EP G7OJU
LN1 2EP M6DDQ
LN1 2FT 2E1GYB
LN1 2FT M0GYB
LN1 2FT M3GYB
LN1 2GY 2E0ZLO
LN1 2GY M0ZLE
LN1 2GY M6ZAC
LN1 2HA G8PK
LN1 2HZ G7KIE
LN1 2JD M3EJR
LN1 2LH M6JHV
LN1 2NH G0BUC
LN1 2NJ 2E0HXJ
LN1 2PE M1FFV
LN1 2PE M6DUJ
LN1 2QA M3WVO
LN1 2QB G1HTM
LN1 2QL G3UPI
LN1 2QN G6BPH
LN1 2RG M3WBN
LN1 2SG G4GDC
LN1 2TL 2E0LEY
LN1 2TL 2E0VWG
LN1 2TL M6EGX
LN1 2UP M7BIX
LN1 2XF M6JAG
LN1 3HR M7FYJ
LN1 3JS G4SSV
LN1 3LB G0VWW
LN1 3LU G4OEF
LN1 3NN G0RQQ
LN1 3QL M3XCX
LN1 3QP G7KYJ
LN1 3QP M0XYZ
LN1 3RD 2E0WWP
LN1 3RD M6OYU
LN1 3SZ 2E0SIY
LN1 3SZ M0SIY
LN1 3SZ M3SIY
LN1 3XD M6TCZ
LN10 5DU M6NZE
LN10 5DY G6NKL
LN10 6RT 2E0JHO
LN10 6RT M0JHO
LN10 6RT M6DKF
LN10 6RW G1VMX
LN10 6SD G3JKB
LN10 6TH G3OBZ
LN10 6UE G3MDI
LN10 6UR 2E0TPD
LN10 6UR M0VLC
LN10 6UR M3WUV
LN10 6UU G4IJA
LN11 0AP M3ZWI
LN11 0AZ 2E1GIK
LN11 0AZ M5COL
LN11 0BD G8SQY
LN11 0BP M6JPI
LN11 0DF 2E0PJS
LN11 0DN G7ENA
LN11 0DW M3CDV
LN11 0DY G0LWG
LN11 0DY M1PBX
LN11 0DY M6GOO
LN11 0EH M7TMC
LN11 0JD M3REL
LN11 0PB G0KKR
LN11 0PN G4MXI
LN11 0XF G1VUG
LN11 0ZN 2E0NAI
LN11 0ZN M6CIK
LN11 7AD G3PRR
LN11 7AG 2E1GZF
LN11 7BZ G8OOS
LN11 7JR G3ZPM
LN11 7QG M0GHA
LN11 7QH G7VIG
LN11 7QH M0VIG
LN11 7QH M6AVS
LN11 7QU M3LRP
LN11 7QW G7MWH
LN11 7SA G3NRQ
LN11 8AS 2E0EDS
LN11 8AS M6EDS
LN11 8BS G6XFB
LN11 8HL 2E0MBT
LN11 8HL M6MKB
LN11 8HY G3KRZ
LN11 8JD G6XXL
LN11 8LJ G3NHR
LN11 8RP 2E1GJL
LN11 8RY G4HUD
LN11 8SG G1PRS
LN11 8SP M1BTR
LN11 8TA G0DPY
LN11 8UJ G0VRU
LN11 8UN G4XFG
LN11 8UQ G0FDS
LN11 8YN M6DIF
LN11 9AL 2E1FEF
LN11 9DG G8AJM
LN11 9HT G0GNW
LN11 9RQ G7AJP
LN11 9SA M7AOO
LN11 9TF G4YZC
LN11 9UE G0WIW
LN11 9UE M1CFG
LN11 9UP G1GMX
LN12 1BQ G0SWS
LN12 1BQ G6TAN
LN12 1BQ M0ERG
LN12 1BY 2E0BWQ
LN12 1BY G4ECE
LN12 1BY M6SIX
LN12 1DA G0FMG
LN12 1DA M3INQ
LN12 1EZ M7TBO
LN12 1JF 2E0PGI
LN12 1JF M0PGI
LN12 1JF M6BHH
LN12 1JF M6RDW
LN12 1LL G3SLS
LN12 1LP 2E0FHX
LN12 1LP M6OJU
LN12 1NX G3WNQ
LN12 1QB G0KED
LN12 1QE G1XWD
LN12 1QR 2E0YYD
LN12 1QR M0XDX
LN12 2 BA G0LAN
LN12 2 BA G4ETM
LN12 2AJ 2E0GBH
LN12 2AJ G3VDV
LN12 2AJ G7BUK
LN12 2AJ M0DZA
LN12 2AJ M3FEY
LN12 2AJ M6IOQ
LN12 2AS G1IEX
LN12 2AX M3YUB
LN12 2AY G0DXK
LN12 2BF M6DRK
LN12 2BP M3FQT
LN12 2BS 2E0LRK
LN12 2BS M3NQL
LN12 2DF G7ANK
LN12 2EJ 2E0FHV
LN12 2EL M3RDY
LN12 2HP G6NZG
LN12 2JU G8EWF
LN12 2JZ 2E0JSJ
LN12 2JZ M6TDF
LN12 2LN G8PPA
LN12 2NP G1BVV
LN12 2NU G1AAC
LN12 2NU G4BFV
LN12 2PT M3TMG
LN12 2QS G3LSA
LN12 2QY G4NJW
LN12 2QZ G4HHM
LN12 2RE 2E1FYZ
LN12 2RE G0CBM
LN12 2RL M7FFU
LN12 2RT G0NQA
LN12 2RT M6PJX
LN12 2TX G1HEY
LN122GY G3MNS
LN13 0BD M7HWK
LN13 0BG G6RGN
LN13 0BG G7MVY
LN13 0BH 2E0JQG
LN13 0BH M0JQD
LN13 0BH M6JQG
LN13 0DL M1CNE
LN13 0EH G4CPI
LN13 0JP G0DVY
LN13 0JP G3SYB
LN13 0JP G4WEC
LN13 0JW G3ZUI
LN13 0LR G1NVV
LN13 0PN G0KBZ
LN13 9AX M7AVL
LN13 9HJ G4LOP
LN13 9NG G6RHB
LN13 9PH 2E1FFJ
LN13 9PH G7RAG
LN13 9PT G8HSV
LN13 9RF M3MUI
LN13 9RL 2E0CDI
LN130NP 2E0KIM
LN13RD M7LRQ
LN2 1JD G8OPP
LN2 1PY 2E0TTM
LN2 1PY M0TMA
LN2 2AE G0WQA
LN2 2AY M6VAP
LN2 2JL G0CEG
LN2 2JL G4RUE
LN2 2JR G0PKN
LN2 2JU M1FBW
LN2 2JW G6BWM
LN2 2LX M7BWW
LN2 2NE G3NIC
LN2 2NQ G0HEE
LN2 2PH G6BD
LN2 2PH G7FAR
LN2 2QU G3RRN
LN2 2WB M3HGW
LN2 3JS G1LQB
LN2 3JX G0NXF
LN2 3RS M6KSZ
LN2 3SR G4YQK
LN2 3TR 2E0DAT
LN2 3TR M3NKL
LN2 3TT G7PMQ
LN2 3TU G0JOD
LN2 3US 2E1RWC
LN2 3US M3RWC
LN2 4BT 2E0XXO
LN2 4BT M6UGN
LN2 4EB M7HYL
LN2 4EW G3VEV
LN2 4LT 2E0LRJ
LN2 4LT M0TEF
LN2 4LT M1DYI
LN2 4LT M3LRJ
LN2 4LX G6SWZ
LN2 4PT G4KSA
LN2 4QH M5RFD
LN2 4RE 2E0NDT
LN2 4RE M6NDT
LN2 4TX M0EJL
LN2 4WS G7JRD
LN2 5EY 2E0KWM
LN2 5EY M0KWM
LN2 5EY M6KWM
LN2 5LB 2E0OYQ
LN2 5LB M0IYE
LN2 5LP G7RQO
LN2 5NN M6HXX
LN2 5QF 2E0KVR
LN2 5QF M6KNG
LN2 5RU G3NGJ
LN25AZ M7LLR
LN3 4AN G6DKK
LN3 4AX G0VKF
LN3 4BE G3PTZ
LN3 4BG M0CZT
LN3 4BH M1AYR
LN3 4BJ G3AWK
LN3 4BZ G8FPA
LN3 4DH G6RFR
LN3 4DQ 2E0OAK
LN3 4DQ M3CTB

LN3 4DQ M6JBU
LN3 4EB G7NSK
LN3 4EE 2E0SFV
LN3 4EE M0VEL
LN3 4EE M7FAV
LN3 4EG 2E0TMO
LN3 4EG 2E0VKW
LN3 4EG G0OSO
LN3 4EG M6TMO
LN3 4EG M6VKW
LN3 4EJ G0TQV
LN3 4HT G3NSL
LN3 4JS G4XMQ
LN3 4LS 2E0CTL
LN3 4LS M6BGA
LN3 4LY 2E0SSM
LN3 4LY M0PTA
LN3 4LY M5SJM
LN3 4LY M6TSJ
LN3 4NA G4SLG
LN3 5AB M0BKK
LN3 5TD 2E0GFY
LN3 5TD M7DNH
LN3 5XS G7PHR
LN35XN M7OKV
LN35XT G7OBX
LN4 1AE G7GJT
LN4 1AS M6HHY
LN4 1AU 2E0HGU
LN4 1AU M7PCC
LN4 1DB 2E0DKY
LN4 1DB M6IJZ
LN4 1DB M6SFQ
LN4 1DX M5YEX
LN4 1DZ 2E1BME
LN4 1DZ G4URX
LN4 1EG G4DTL
LN4 1EQ G4UAL
LN4 1EY G4JQJ
LN4 1LH 2E0FZY
LN4 1LH G1SVI
LN4 1LH M3GUU
LN4 1LJ G4DBS
LN4 1NB M6ICK
LN4 1NN G1OQF
LN4 1NS M7NYH
LN4 1NW G4YYE
LN4 1PG G4RYB
LN4 1PU G7EJH
LN4 1QB G7PKG
LN4 1QE M7DGG
LN4 1QT G1ATU
LN4 1TT G4JNL
LN4 1TU G0GTV
LN4 2BE G6AII
LN4 2EX M0ONQ
LN4 2LE G3WOK
LN4 2NP 2E1TSO
LN4 2NP M7TSO
LN4 2PW 2E0TYC
LN4 2PW M6TYC
LN4 2PW M6VCK
LN4 2QG G7JBZ
LN4 2RD M6FYV
LN4 2RP G3WTT
LN4 2TD M6CDC
LN4 2TY M6CXV
LN4 3AP 2E1GJP
LN4 3BQ M0RDS
LN4 3BY M6YAR
LN4 3DR 2E0MSZ
LN4 3DR M0WMS
LN4 3DR M6WMS
LN4 3DS G7VSN
LN4 3DT G3NXT
LN4 3LH M6MQW
LN4 3LY M7ILM
LN4 3NP M0EAU
LN4 3PT 2E0FBO
LN4 3PT M0IVJ
LN4 3PT M6NYO
LN4 3RN G8IJC
LN4 3SL G3LAS
LN4 3YG M0HIL
LN4 4AU G6AWY
LN4 4EA G3MON
LN4 4FN M6TFJ
LN4 4JJ M3GUF
LN4 4JQ 2E0SBL
LN4 4JQ M6SAY
LN4 4PB G0MPQ
LN4 4QA G1EFK
LN4 4QH 2E0WOW
LN4 4QH M6ROC
LN4 4QU G4DDI
LN4 4RG G6ZTL
LN4 4RG M0HAZ
LN4 4RQ G7SKW
LN4 4SP G4MWJ
LN4 4XT M6RJP
LN4 4YJ 2E1HGM
LN4 4YJ M1ECB
LN42BD M7CYU
LN5 0EE G6HTY
LN5 0ER G4PWM
LN5 0ER G4UNF
LN5 0FD G6PMW
LN5 0QZ M7LFF
LN5 0RN G7OVS
LN5 0RP 2E0MFF
LN5 0RP M6FFY
LN5 0SJ G3VHN
LN5 7DL M7EHQ
LN5 7EX M7CZU
LN5 7QF M0MNV
LN5 7SJ G8TSV
LN5 7UT G0MRB
LN5 7XW 2E0HYI
LN5 7XW M7DBZ
LN5 8DA G0UOZ
LN5 8DA M3VOZ
LN5 8DR M6XFI
LN5 8DR M6XGC
LN5 8DR M6XLN
LN5 8EL G0LEN
LN5 8EL G7LEN
LN5 8ND 2E0MRF
LN5 8QF 2E0XRD
LN5 8QF M6XDR
LN5 8QS M6CAJ
LN5 8QX G7RUP
LN5 8RL M0DIW
LN5 8RX G0FVI
LN5 8SH M0CES
LN5 8SJ 2E0BVG
LN5 8SN 2E0COF
LN5 8TF G7CNM
LN5 8TG 2E0LUY
LN5 8TG G8YMW
LN5 8TG M0LUY
LN5 9DA 2E0GSJ
LN5 9DR M3ULU
LN5 9DT G4LQH
LN5 9FW G0WFV
LN5 9JD G0KAU
LN5 9LJ M6SAV
LN5 9NE M3GEK
LN5 9PY G7PRB
LN5 9QR G0MQD
LN5 9SW G4ZOX
LN5 9SX 2E0TRP
LN5 9TL 2E0RXG
LN5 9TL M0RXG
LN5 9TL M6RSF
LN5 9TL M6RXG
LN5 9UF M6AGR
LN5 9UT G0MEY
LN6 0AG 2E0XMO
LN6 0AG M6NMD
LN6 0EY M7JFH
LN6 0JB M6EBA
LN6 0LZ G1XZG
LN6 0NR 2E0EOD
LN6 0NR M6LJB
LN6 0QN G4JZV
LN6 0RH G6FYX
LN6 0SL 2E0PFF
LN6 0SL M0VEZ
LN6 0SS G4OMD
LN6 0SY G8HMZ
LN6 0XQ G0BEN
LN6 0YD M6IMI
LN6 0YF 2E0SJP
LN6 0YF M6MJP
LN6 0YZ M6DTO
LN6 3LU M6OYP
LN6 3NR G7JQZ
LN6 3NU G0EUN
LN6 3RQ G6OYU
LN6 5TF G0EJV
LN6 5TW G4HNQ
LN6 5UF 2E0CJM
LN6 5UF M0KAN
LN6 5UF M6AVZ
LN6 5UU M0CQH
LN6 5UW G6KSR
LN6 5XB M7JCR
LN6 7DR G7ERS
LN6 7DR M0ERS
LN6 7EW M6JGZ
LN6 7LQ G1ZBY
LN6 7NL G0VRE
LN6 7NL G8VEM
LN6 7PA G6HUP
LN6 7PJ M6CKE
LN6 7PN G3PVU
LN6 7PY M6DKO
LN6 7RD G7UAV
LN6 7RQ M6CMG
LN6 7TG M3FKL
LN6 7TG M3FKM
LN6 7UP M6CYW
LN6 8BT M3GFZ
LN6 8BX G0LZS
LN6 8LY G4OSB
LN6 8QR 2E0WGD
LN6 8QR M7DGW
LN6 8QS 2E0DDC
LN6 8QS M3XOE
LN6 8SD G6GLW
LN6 8SN G0IMP
LN6 8TZ M3OVA
LN6 8UB M1DJS
LN6 8UR M7RCM
LN6 8UW 2E0GAL
LN6 8UW 2E0KAG
LN6 8UW M3YZV
LN6 8UW M6RVH
LN6 9AZ M6BXX
LN6 9BG G8BGL
LN6 9BS G4HIV
LN6 9DH 2E0OVR
LN6 9DH G7JFI
LN6 9DH M6PDV
LN6 9EX G3IYF
LN6 9JE G8SGI
LN6 9PJ G7VTN
LN6 9RG M3TPD
LN6 9RG M6LQO
LN6 9SB M0KED
LN6 9SZ G0UND
LN6 9TB M3XBO
LN6 9TG 2E0DZV
LN6 9ZD M6KHQ
LN63QQ M7CGQ
LN69XA G4JES
LN7 6EL G0CVN
LN7 6FH M7PHE
LN7 6NQ 2E0FML
LN7 6PA G4FQP
LN7 6RB G3RXP
LN7 6RN G7KPM
LN7 6RU G6IYS
LN7 6RU M6RIT
LN7 6SW 2E0TYY
LN8 2AJ M7LMF
LN8 2HE 2E0FHD
LN8 2HE M6RCI
LN8 2HL G4CLL
LN8 3AA G7JXX
LN8 3AA M3LMY
LN8 3AG G4UHZ
LN8 3AP M6EOX
LN8 3BE M6DVZ
LN8 3DL M7EWT
LN8 3DS 2E0RPD
LN8 3DS M0RPD
LN8 3DS M3RPD
LN8 3EW G4OER
LN8 3EW M3OER
LN8 3JA G0FLG
LN8 3LD G1YFQ
LN8 3LD G3KDY
LN8 3LE G4PWF
LN8 3NQ M0JJE
LN8 3PB M6VEN
LN8 3QP G4XFC
LN8 3UB M3GEZ
LN8 3UR M3JVX
LN8 3US M0OOO
LN8 3US M3ZSW
LN8 3UT M0DBY
LN8 3XL G1PRM
LN8 3XL G8KLC
LN8 3XZ M1BRY
LN8 3YL G8THR
LN8 3YL M3PCC
LN8 5NF M6KOV
LN8 5RB M1DHV
LN8 5RB M3WRA
LN8 5RF G6TVP
LN8 5RF M5ZZZ
LN8 5RY G8KJI
LN8 5SL G7UEV
LN8 6AD G4KIZ
LN8 6AJ G3YFU
LN8 6AN M0BIC
LN8 6AZ G8UDI
LN8 6DE G8JCS
LN8 6DH G4LAY
LN8 6DX 2E0EDG
LN8 6DX M6EDG
LN8 6EW 2E0IJQ
LN8 6EW 2E0LJB
LN8 6EW 2E0ODL
LN8 6EW M3ODK
LN8 6EW M3ODL
LN8 6EW M6IJQ
LN8 6EX G4RBP
LN8 6HS M7YAD
LN86ER G1RHB
LN9 5AW G3VUE
LN9 5HF G0NDY
LN9 5JE M3VUV
LN9 5JF M1MSF
LN9 5LL G3SCD
LN9 5LP 2E0ZRQ
LN9 5LP M3ZRQ
LN9 5NH G0SWW
LN9 5NH G7MES
LN9 5NW G6SWO
LN9 5PT G1JZG
LN9 6AF G6ZQS
LN9 6AW G3ZPU
LN9 6AW M0TCM
LN9 6BE G4FOT
LN9 6BH 2E0PKR
LN9 6EY M6NUO
LN9 6HG G1EGR
LN9 6JH M6HWT
LN9 6LB G3OUT
LN9 6LD G7NLA
LN9 6PQ G4WMO
LN9 6PQ G4ZWA
LN9 6QH G8RYO
LN9 6RE G4AIE
LN9 6RR 2E0XOD
LN9 6RR M3XOD
LS10 3DT 2E0USD
LS10 3DT M6SDQ
LS10 3EJ M3IBJ
LS10 3QA M6NWS
LS10 3TB 2E0CRM
LS10 3TB M3BZC
LS10 3TJ G0CYB
LS10 4BG 2E0FNJ
LS10 4BG M6KZB
LS10 4BG M7AMQ
LS10 4LH M6WDC
LS10 4NQ M6MWP
LS10 4QN 2E0SDT
LS10 4QN M0SDT
LS10 4SN G1AUH
LS10 4TH M6KQK
LS104JT M6NYY
LS11 0HZ M0VBV
LS11 0HZ M3VBT
LS11 5HA M6WRX
LS11 5RP G4FQH
LS11 5RX G4MPO
LS11 5SG G4IDT
LS11 5SG G8UAF
LS11 6NN G7HUJ
LS11 8AH M6ARF
LS11 8BG M3BTG
LS11 8HR G7POS
LS11 8TP G1KSW
LS11 9RE G7OCX
LS12 2PG G7ECA
LS12 2RE G0WXG
LS12 2RL M3XFI
LS12 3LB G7RME
LS12 3QJ G8SWK
LS12 3SN G8PKJ
LS12 4HQ G4ICF
LS12 4HQ G8KZY
LS12 4JX G0RFA
LS12 4LA G6YMH
LS12 4PB M6TFP
LS12 4PB M7GHE
LS12 4RD 2E0ZRL
LS12 4RR G4IDD
LS12 4UW G6SFH
LS12 5BL M6RWB
LS12 5EA G4LKW
LS12 5LP G4YQW
LS12 5QS G4RYE
LS12 5SU G7HRP
LS12 5SZ 2E0GNU
LS12 5SZ M3XGI
LS12 5TP M3YRT
LS12 6AY G1OUX
LS13 1AX M6LHL
LS13 1BL G4SQG
LS13 1EA G7LBH
LS13 1ED M6EHO
LS13 2BJ 2E0JAQ
LS13 2BJ M3JTO
LS13 2BX G1DWT
LS13 2LE G4HLI
LS13 2LH G4OFA
LS13 3DF M0RSF
LS13 3DF M7NTD
LS13 3DH G7BXA
LS13 3EH 2E0XPP
LS13 3EH M3XPP
LS13 3JX G0URF
LS13 3PM M6AUT
LS13 3RQ G6XGK
LS13 3RS M6PNX
LS13 3SG 2E1ODG
LS13 4PW G1IKW
LS13 4QY G4OOJ
LS13 4RN M0AVZ
LS13 4SU 2E1RON
LS13 4TT 2E0ERM
LS14 1AP G0JUT
LS14 1HH G1DZC
LS14 1JP G0TGH
LS14 1LF M6OLS
LS14 1LJ 2E0KZH
LS14 1LJ M0KZH
LS14 2EP G0VTJ
LS14 2EQ G8TBX
LS14 2HR 2E0HZU
LS14 2HR M0JKE
LS14 2HR M0RBO
LS14 2HR M6HZU
LS14 3AU G1ZUB
LS14 3BU G3YRU
LS14 3HU 2E0IYL
LS14 3HU M3XSI
LS14 5AS M6RGY
LS14 5BQ M3NPW
LS14 5PD G7VEB
LS14 6AA G7POQ
LS14 6AE M0CVC
LS14 6AY M7WDC
LS14 6JL G4JJS
LS14 6JZ G7PMX
LS14 6SY M0MLH
LS15 0EZ G7ELS
LS15 0EZ G7KJT
LS15 0HT 2E0VNX
LS15 0HT G5VNX
LS15 0HT M7VNX
LS15 0LW G4XZI
LS15 0NQ G7NXV
LS15 0NQ M5FOX
LS15 4EJ 2E0JOE
LS15 4EJ M3JOW
LS15 4EZ G3XVP
LS15 4HJ G4ZRF
LS15 4JD G3GXQ
LS15 4NY G4ODB
LS15 7DN M1CAE
LS15 7HA G6FFQ
LS15 7HD 2E1GLR
LS15 7HD M1DIE
LS15 7QE G1IUZ
LS15 7QW G0HOB
LS15 7SQ G7DCT
LS15 8AY G0BTD
LS15 8ED M0RAP
LS15 8JJ G8ZXT
LS15 8LW G4HSZ
LS15 8QZ M1ATB
LS15 8SD G7MTI
LS15 8SW G6INO
LS15 8UP M3CJI
LS15 9JD G3RZY
LS150HQ G7OZQ
LS16 5DN 2E0JAF
LS16 5DN M0JAF
LS16 5DY M3ZTN
LS16 5EG 2E0TMF
LS16 5EG M6AMT
LS16 5EG M6ZWP
LS16 5PL M7PMF
LS16 6BU G8MDG
LS16 6BU M0DSI
LS16 6BX G0FWP
LS16 6DU 2E1ANQ
LS16 6EF G1SGM
LS16 6HX 2E0IGP
LS16 6HX M0RVB
LS16 6HX M6IGP
LS16 6JA M0DBD
LS16 6NA M0APY
LS16 7AB G0PAN
LS16 7AS M3WTD
LS16 7ES M0RCP
LS16 7ES M3OOL
LS16 7HF G4PHP
LS16 7PG G4TSF
LS16 7PN G0WRT
LS16 7PP G0SCL
LS16 7QX G4HAJ
LS16 7RB G8TEL
LS16 7SH M7MDE
LS16 7SL G4BNL
LS16 8BL G3NRM
LS16 8DE G1SBK
LS16 8EJ G4CSZ
LS16 8JQ G1CXQ
LS16 9DQ G3MFJ
LS16 9DR 2E1CNO
LS16 9LE G4HGT
LS16 9LF G4BNM
LS17 5BH G8UHW
LS17 5ET G4UZN
LS17 5HW M1AZM
LS17 5JP M1AHR
LS17 5NS G4THX
LS17 5PQ G3WNR
LS17 6DB M0SDU
LS17 6FD G4RYS
LS17 6LL M0BGS
LS17 6RY G3AAS
LS17 7DU G0SBZ
LS17 7EE G4BDC
LS17 7EP 2E1COM
LS17 7NH G3KVJ
LS17 7PD G4LYM
LS17 7QB G4MSN
LS17 7RB G7RBA
LS17 8AR M6NYM
LS17 8BG 2E0LSE
LS17 8BG M0SME
LS17 8BG M0YKR
LS17 8BS G4XZG
LS17 8BY G4NCK
LS17 8DA M6NWO
LS17 8DF 2E0LBA
LS17 8DF M6LYA
LS17 8LX G0PFT
LS17 8LX G8ASG
LS17 8TB G3HJP
LS17 9AB 2E0SFS
LS17 9AB M5MCH
LS17 9AB M6PZW
LS17 9AD M0JRO
LS17 9DR M7NCA
LS17 9EG G1MPT
LS17 9ER G3TAF
LS17 9NL G1HHT
LS18 4BD M6XNO
LS18 4BE 2E0NCC
LS18 4BE M3NCT
LS18 4BE M6SQW
LS18 4HL 2E0CUA
LS18 4HL M0ABO
LS18 4HL M3IDW
LS18 4HS G7RKJ
LS18 4HY M7OWT
LS18 4LB M7ISC
LS18 4PJ G0KVC
LS18 4RL G4OVT
LS18 4RL G4YDW
LS18 5AW G7VHU
LS18 5DT M0ALD
LS18 5HB G4LAD
LS18 5HB G8WYR
LS18 5JL G1ELX
LS18 5JP G4XQV
LS18 5JS G4GYL
LS18 5LD G4GOP
LS18 5PP G4RVO
LS18 5QE G4OOH
LS18 5QU M1DIB
LS18 5RN G0LHU
LS18 5RQ G4YBA
LS18 5SJ G6OTS
LS18 5UN G4FKY
LS19 6AD G1BQV
LS19 6AR G3ZNK
LS19 6BS G4BZL
LS19 6EH G7BHG
LS19 6JX G3KWT
LS19 6NE 2E0KDV
LS19 6NE M0KDV
LS19 6NE M3KDV
LS19 6PU G4AYH
LS19 6RJ G3RMQ
LS19 6SG 2E0VCA
LS19 6SG M7AGH
LS19 7AJ 2E0GNO
LS19 7AJ M7WOZ
LS19 7AS G4FBB
LS19 7AU G4REC
LS19 7ED G1PFZ
LS19 7NL G0VXV
LS19 7SQ G8RTK
LS19 7TE G7KHT
LS19 7XB G4SPW
LS19 7XE G0HQU
LS20 8AY G4GGR
LS20 8EJ M0JSE
LS20 8HA G1NEV
LS20 8LD G4JVZ
LS20 8NX G3KKP
LS20 9AU M1BZF
LS20 9DS M7MMQ
LS20 9DY M0CSR
LS20 9EF G7PFY
LS20 9EW M6WWJ
LS20 9EW M7DVT
LS20 9HN M0JAM
LS20 9LN M7JAT
LS20 9LW M0TEX
LS21 1DH G0CLD
LS21 1JZ G6XOG
LS21 2AA G8HPY
LS21 2AL G7OXH
LS21 2BY G0NIG
LS21 2DB M1FHX
LS21 2DP G0KDQ
LS21 2DT G0AEX
LS21 2DT G3XEP
LS21 2DT G8LVQ
LS21 2RS G1RRG
LS21 3DS 2E0LAL
LS21 3DS M7MTR
LS21 3HY 2E0SNZ
LS21 3HY M6FFK
LS21 3LE G0SNW
LS21 3LE M0HWL
LS21 3LN G1TVW
LS21 3LW G7RDJ
LS21 3NW G7LNT
LS21 3NZ G4GDL
LS21 3NZ G4LZT
LS21 3PN G4MWQ
LS212BQ 2E1COD
LS22 4BB M3XVW
LS22 4EH G7OAV
LS22 4ET 2E0OST
LS22 4ET M0KOV
LS22 5BL G1EGK
LS22 5BW G4YZN
LS22 5ER G4EIO
LS22 6NP G1UDB
LS22 6RN G1SBN
LS22 6SF M7ATL
LS22 7PU G3YTN
LS22 7QU M0DCD
LS22 7QY G4IEC
LS22 7RA G4WWL
LS22 7UD 2E0PUB
LS22 7UD G3RDX
LS22 7UE G1RPE
LS23 6AD M7EKU
LS23 6AY G4JVC
LS23 6BR M1PAC
LS23 6BW M0BRL
LS23 6EJ M3HTO
LS23 6HU G6RDO
LS23 6PU G8XMU
LS23 6PX G1VAG
LS23 6PX M0VAG
LS23 6RN G4XYY
LS23 6RP G3VMW
LS24 8AG G0PEV
LS24 8BS G0AUE
LS24 8JD G1GCF
LS24 9BR G4FFM
LS24 9DL G0HXU
LS24 9HP M3PWE
LS24 9HR M3BLG
LS24 9HS G0RCL
LS24 9PQ G4TDC
LS24 9QW M0EBR
LS24 9RL G4GIY
LS25 1BN G1XPI
LS25 1BZ G4HKR
LS25 1DD G7CCL
LS25 1EF M6FVP
LS25 1EN G4DTT
LS25 1HE G4CRP
LS25 1HF 2E0LZB
LS25 1HF M6HGF
LS25 1HN G8MWX
LS25 1NS M0APC
LS25 1PW M1BYT
LS25 2BH M7FYC
LS25 2EG G1VGP
LS25 2EN G4DVZ
LS25 2EP G7ADW
LS25 2HA M1DGE
LS25 2JR G1FDL
LS25 2LF G0MYM
LS25 2LY G4ZOT
LS25 2NW G4JKH
LS25 2QB M0NOA
LS25 4AP 2E0LHF
LS25 4AP M7KKG
LS25 4AU G4ZSS
LS25 4FX 2E1EMN
LS25 4FX M1APF
LS25 5AY G0FXY
LS25 5BT G4ZSA
LS25 5JA M7THA
LS25 5PJ G8INL
LS25 6AF G4NCB
LS25 6BY M3TFB
LS25 6BY M3VXN
LS25 6EU 2E0LZN
LS25 6EU M0LZN
LS25 6EU M6LZN
LS25 6LW M3PMO
LS25 7DY G6JSR
LS25 7ET G1UVI
LS25 7ET G4TXK
LS25 7PB G3KEP
LS25 7PB G4DMP
LS252BN M3STJ
LS254AP 2E0KJD
LS257BX M3VSQ
LS257EX 2E1AYS
LS26 0DR G4MAK
LS26 0HL G4OVR
LS26 0PP G8TMD
LS26 8DL M3JEH
LS26 8ET M3YDI
LS26 8FJ G4NMO
LS26 8QF M3YTV
LS26 8QG G0RDR
LS26 8QJ G3ZRL
LS26 8RY G7NZU
LS26 8SQ G4JMT
LS26 8UN M0NDO
LS26 9AJ M3ZAA
LS260SW 2E0JRZ
LS260SW M0NOK
LS261ET 2E0BKY
LS27 0DP G0BVT
LS27 0DS 2E1HYD
LS27 0JA G7NAP
LS27 0SG G1MJT
LS27 7BT G7AJN
LS27 7BT M3OCL
LS27 7DJ G0JVI
LS27 7DJ G7PAF
LS27 7DJ M6KSD
LS27 7DN M0LGA
LS27 7DW M7GHO
LS27 7JB M7FZA
LS27 7RT M0FJM
LS27 7RT M3MSL
LS27 8NS G6ZGF
LS27 8QB M6SFS
LS27 8TD G0PTI
LS27 8UH 2E1GDK
LS27 8UL M0NUN
LS27 8UL M0SSW
LS27 8US M3ZRH
LS27 9DS M1EWV
LS27 9NP 2E0JBA
LS27 9NP M3FOQ
LS27 9NP M3FOS
LS27 9PL G4MBE
LS28 5AA G6FMF
LS28 5BF M0BNO
LS28 5DY G4XZK
LS28 5DZ G0DPS
LS28 5DZ G4PTM
LS28 5HW G7OQB
LS28 5LW G4BXQ
LS28 5QH M0STS
LS28 5QH M3BVL
LS28 5RE G4NZN
LS28 6AA G1III
LS28 6AA M0XXV
LS28 6AA M3WYR
LS28 6BG G7RRO
LS28 6DJ G4SKO
LS28 7EP G1NSD
LS28 7JW 2E0FOT
LS28 7JW M6NQN
LS28 7SR 2E0FCO
LS28 7SR M6MNP
LS28 7SS M0LRA
LS28 7SS M0SGS
LS28 8AX 2E0WJC
LS28 8AX M3KFO
LS28 8BH 2E0FTS
LS28 8BH M7FTS
LS28 8DG G4TZV
LS28 8EF G3VXE
LS28 8EQ 2E0HGR
LS28 8EW M7OAK
LS28 8NH 2E1ACZ
LS28 8NH G0SSZ
LS28 8QD G4MOT
LS28 9AD 2E1DNB
LS28 9NQ G8WCT
LS29 0BG 2E0SOJ
LS29 0BG M7HPS
LS29 0LE G8KSX
LS29 0QQ G8HZJ
LS29 0QZ G1RPP
LS29 0SH M7IKC
LS29 0SP 2E0GNL
LS29 0SS M7BIP
LS29 6HE G0SNX
LS29 6HH G8SRN
LS29 6LX G7AQA
LS29 6RT G5XWH
LS29 7BX 2E0DHK
LS29 7BX M0HYN
LS29 7BX M6DUZ
LS29 7DF M7JMO
LS29 7EN G3ZNR
LS29 7HS G0LXR
LS29 7LL M3HSV
LS29 7NP G3XNO
LS29 7NP G6MHR
LS29 7NP G8JTD
LS29 7NR G4BSS
LS29 7PA G7VUH
LS29 7QB G0KEV
LS29 7QB G8AWN
LS29 7QB M3AWN
LS29 7QH G0CMU
LS29 7QT G7BNW
LS29 7RG G6NWK
LS29 8AH G6JWM
LS29 8AH M0TIN
LS29 8AR 2E0CRI
LS29 8AR M6JSP
LS29 8LU 2E0KBJ
LS29 8LU M3SBJ
LS29 8PH G8AXR
LS29 8QP G7LYS
LS29 8SX M0DXN
LS29 8SX M6XAB
LS29 9BP G8WBK
LS29 9DB G4TGJ
LS29 9JN G4DCY
LS29 9SN G0AJB
LS290PT M7CYH
LS3 1NS 2E0NXP
LS3 1NS M0MMP
LS3 1NS M7MGM
LS4 2LN M6AYG
LS4 2PP 2E0LDL
LS4 2PP M0LOF
LS4 2SU M6CBA
LS4 2TQ M1APL
LS4 2TT G0ETL
LS5 3HX G4TFV
LS6 2AP G1JTC
LS6 2HS 2E0NWN
LS6 2HS M7WBN
LS6 2RD G0FXK
LS6 3BY M7GIL
LS6 3DB G4YES
LS6 3HB G4MXM
LS6 3HB G4OOI
LS6 3NA M7ENB
LS6 3NX G3FLV
LS6 4EX G0TUM
LS6 4HD G4IJI
LS6 4HD M0UOY
LS6 4JT G0JUI
LS7 1SL G7VCK
LS7 2DY G1EZF
LS7 2NN G4UPD
LS7 3HG M7VRM
LS7 3LY G4TCB
LS7 3NX 2E0WPI
LS7 3NX M0WPI
LS7 3NX M6WPI
LS8 1ED M6XGF
LS8 1EW G4ZOB
LS8 1NS G3PKC
LS8 1RZ G1DIR
LS8 1RZ G4GCX
LS8 2BS G4KAX
LS8 2EE G4FTI
LS8 2LA G3VFH
LS8 2PJ M0TLL
LS8 2RH G0KQP
LS8 2SX G8HBQ
LS8 2TA G6MTF
LS8 4AG G0DIY
LS8 4DH G7OOS
LS8 5BY G1UVE
LS8 5DB 2E1DLX
LS9 0AE G4HJS
LS9 0LL M3UEX
LS9 6BY M1AUR
LS9 6RG M1AFV
LS9 7SY G4FVV
LS9 8FB M0DJD
LU1 4AN G8ADC
LU1 4DJ G8OBB
LU1 4DU G6CQB
LU1 4DU G6NDJ
LU1 4EN G0GJA
LU1 4EN G0OIK
LU1 4EN G3VER
LU1 4EN G4VER
LU1 4EN G8VER
LU1 4ER G6WVS
LU1 5LB G1OAM
LU1 5PE G1AYH
LU1 5PE M6VCS
LU1 5PE M6ZLN
LU1 5PF G8XSD
LU1 5QD M6RPH
LU13AG M7MLU
LU2 0DN G6OUO
LU2 0FD G7PJG
LU2 0JF G0BWK
LU2 0PJ M0GQU
LU2 0RE G1OUA
LU2 0RE G6OUA
LU2 0RE G6UCQ
LU2 0RP G6IAT
LU2 7BG G7FRW
LU2 7BY G4UVD
LU2 7DL 2E0GQT
LU2 7DL M0XIA
LU2 7DL M6GQT
LU2 7EU G3IWH
LU2 7EU M0KHW
LU2 7JL G4IET
LU2 7JY 2E0GPB
LU2 7JY M0PPG
LU2 7JY M6GPB
LU2 7NL G1FGC
LU2 7PJ M3HSW
LU2 7PP G6PRV
LU2 7SA G8NXS
LU2 7TE G4ENB
LU2 7TE G8DDC
LU2 7UH G3NVL
LU2 7UU G3TAZ
LU2 8AF G4LBH
LU2 8DP G4FAX
LU2 8EJ G4NMY
LU2 8EP G1BDI
LU2 8HQ G8LXY
LU2 8HZ G1PZD
LU2 8JF M7BLR
LU2 8JH G0KUC
LU2 8JH G3LQS
LU2 8PP G0LQZ
LU2 8QL 2E0LUT
LU2 8RU G8IZW
LU2 8TS M0BHA
LU2 8TZ G8FFM
LU2 9AX G4BMM
LU2 9BL G0TNO
LU2 9HL G0KYJ
LU2 9HL G4HEH
LU2 9HL G4OIS
LU2 9JZ G7VWA
LU2 9RA G0GAB
LU2 9RA G7VYB
LU2 9SQ M7LUT
LU3 1EF 2E0JPO
LU3 1EF G8IXK
LU3 1EF M6JPO
LU3 1HB G8JHM
LU3 1LZ M7ESM
LU3 1NQ 2E0ONI
LU3 1NQ M7YEW
LU3 1PY G8UOJ
LU3 2AP G4HPY
LU3 2DW G3WBC
LU3 2HJ G6PPV
LU3 2JJ G1SCA
LU3 2LZ G8CBU
LU3 2NY G6DWM
LU3 2PY G1SVL
LU3 2TH 2E0YFZ
LU3 2TH M3YFZ
LU3 2UA G0DCS
LU3 2UR G8CPQ
LU3 3BH G8UYF
LU3 3EA 2E0RVI
LU3 3JZ G7HBN
LU3 3LA G6MUJ
LU3 3PY G4CPE
LU3 3SU M0CAJ
LU3 3TU M6SCQ
LU3 3TW G1NWZ
LU33AT G5JPR
LU4 0BU G0WKJ
LU4 0HX M0IQF
LU4 0QW G7JAS
LU4 0UN 2E0OTI
LU4 0UN M3OTI
LU4 0XJ M6JHP
LU4 0YG M7UGH
LU4 8ER M3HQB
LU4 8NU G6RPD
LU4 8PY G8EZV
LU4 8QQ 2E0ILD
LU4 8QQ M0RXN
LU4 9AL 2E0DHA
LU4 9AL M3JZL
LU4 9AN 2E0EBA
LU4 9AN M0TMT
LU4 9AN M6EST
LU4 9AS G1XZW
LU4 9EN G7BPI
LU4 9EN G7DBT
LU4 9EN M3XQT
LU4 9ER M6IRJ
LU4 9EU M7FLJ
LU4 9FD M0TSZ
LU4 9HG G3WLM
LU4 9JE G6XFR
LU4 9LT G1JIG
LU5 4AL G1FXB
LU5 4AQ 2E0CQC
LU5 4AQ M0OBY
LU5 4AQ M6RGC
LU5 4BG G1GPL
LU5 4BN M0JUX
LU5 4EA G8NZD
LU5 4LF G3ZNT
LU5 4NU 2E0TVG
LU5 4NU M6VMY
LU5 4PR G1ANS
LU5 4PW G4UOO
LU5 4SH G0LJU
LU5 4SH G7UAL
LU5 5AL 2E0JML
LU5 5AL M6JLM
LU5 5EP 2E0XMH
LU5 5EP M0NOL
LU5 5EP M0SKM
LU5 5EP M7HSN
LU5 5GJ 2E0IQK
LU5 5GJ M6XJM
LU5 5HU M0JVB
LU5 5HU M6JVB
LU5 5HZ M5IGE
LU5 5JG G1AYI
LU5 5NX M3FMP
LU5 5NX M3FMQ
LU5 5PD M0WHP
LU5 5PN M0LTN
LU5 5TJ M6ACP
LU5 6AL G8HJK
LU5 6AZ G4DUL
LU5 6BT M7TEY
LU5 6EL G0WFT
LU5 6LE G1PII
LU5 6LG G8AZR
LU5 6NF G3NRW
LU5 6NT G1ZNX
LU5. 6BQ M7TJB
LU55ED M7PAC
LU55HB M0ZMO
LU55QB M7DCL
LU6 1AH M7EKA
LU6 1AJ M0MNO
LU6 1BN M0KDH
LU6 1BN M3ODH
LU6 1HW 2E0EXB
LU6 1HW M3FRP
LU6 1NQ M6ALM
LU6 1NX M0BZN
LU6 1PN M0DZG
LU6 1QJ G0ODU
LU6 1TS M3WGM
LU6 1UG M3WFO
LU6 2AE 2E0FYC
LU6 2AE M0JJX
LU6 2AE M7DAJ
LU6 2AG G0WTL
LU6 2AH G0HWK
LU6 2AW G3ZFP
LU6 2BJ G8DBH
LU6 2DD G7TVT
LU6 2DF G3SUL
LU6 2DQ 2E0IFB
LU6 2DQ M6UPW
LU6 2FH G0SJR
LU6 2HL G4CVS
LU6 2HL M0CLW
LU6 2HR G4XLG
LU6 2HR M6MXA
LU6 2JQ G3PTG
LU6 2JU M3XJH
LU6 3AS 2E0SBZ
LU6 3AS M0SBZ
LU6 3DD G6IAN
LU6 3JT 2E0EAZ
LU6 3JT G1HIJ
LU6 3JT M6ETH
LU6 3JZ 2E0NBZ
LU6 3JZ M3NBZ
LU6 3LJ G3NWH
LU6 3NS G4ILR
LU6 3PT G4EKB

LU6 3QQ M1CYT
LU6 3RG G8KGV
LU6 3RS G4WYO
LU6 3RS G8PJD
LU6 3TT M6UAO
LU7 0AB M6PEI
LU7 0EP G4FVZ
LU7 0EP M5EI
LU7 0ER G4IVZ
LU7 0EX G3XTQ
LU7 0HZ G8GIK
LU7 0HZ G8ZVI
LU7 0LB 2E0HJJ
LU7 0LJ G8GZX
LU7 0NT G1HMW
LU7 0QE G6JFN
LU7 0RY 2E0BFG
LU7 0RY M0PJD
LU7 0RY M3LLZ
LU7 0SQ G6FZW
LU7 0SY G6NAG
LU7 1AH M3MIV
LU7 1DJ G8KKU
LU7 1FQ 2E0OBI
LU7 1FQ M3UHN
LU7 1HT M6BFK
LU7 1SE G1LYX
LU7 2LE M7LST
LU7 2PE M3MHP
LU7 2PE M6KJM
LU7 2QQ G0KRR
LU7 2QR 2E0RZM
LU7 2QR M3RZM
LU7 2QW G0FRD
LU7 2RE M6ACQ
LU7 2RH G4CQA
LU7 2SJ 2E0OOV
LU7 2TJ M7FAX
LU7 2TS G6POV
LU7 2TS M6JNW
LU7 2TY G3YZW
LU7 2XD G4DDC
LU7 2XD G8XTW
LU7 2YG G3YYG
LU7 3AB M6CMF
LU7 3AD M6MNX
LU7 3DE M0NTY
LU7 3DN G1MMI
LU7 3DS G0AYC
LU7 3EG M0HSG
LU7 3EL 2E0VPG
LU7 3EL M6VMF
LU7 3GE M3KEJ
LU7 3LH G4ATQ
LU7 3LR M3GHA
LU7 3LU G3MBO
LU7 3NJ 2E0VIT
LU7 3NJ M0SPV
LU7 3NJ M6VIT
LU7 3UJ G6IXH
LU7 4AT 2E0GFK
LU7 4AT M6IBF
LU7 4BS M0POA
LU7 4BY G4IYT
LU7 4DA 2E0TCK
LU7 4DA M6TCK
LU7 4HY G0DYW
LU7 4HY G1EOJ
LU7 4JG M6ADJ
LU7 4RF M6CWO
LU7 4SU M6CLZ
LU7 4TG M6BQN
LU7 4TQ M7CHA
LU7 4UQ 2E0HQB
LU7 4UQ M7BRC
LU7 4YU 2E0FEB
LU7 4YU M3SQP
LU7 9AQ 2E0JCQ
LU7 9AQ M0JCQ
LU7 9AQ M6JCQ
LU7 9BX M7FZE
LU7 9DZ 2E0PGM
LU7 9DZ M0PAO
LU7 9DZ M6AZS
LU7 9GF M0VWW
LU7 9JF M6BAM
LU7 9JF M6SLL
LU7 9SN M6XNX
M1 2FA G7HFW
M11 1DP M6RCF
M11 1EQ M6OXB
M11 1HQ M3VNH
M11 1HQ M3YMH
M11 1HX M7JAZ
M11 4FB M3XGC
M12 5PU G0WGL
M12 5QB 2E0GQN
M12 5QB M0WOM
M12 5QB M7BUN
M12 5RP M3YPI
M13 9PL M7MYM
M14 4UR 2E0WLD
M14 4UR M6WLD
M14 5JS M6FML
M14 6AH G3YTZ
M14 6AP M7TFX
M14 6BZ M3YSF
M14 6PE G0BML
M14 6RW M6EGU
M14 6RX G4FFW
M14 6SA M6YOW
M14 6SB 2E0WQK
M14 6SB M0WQK
M14 7FN M0TCC
M14 7FZ G0EIZ
M14 7FZ G4JS
M14 7PX G7OQU
M14 7WT G1ZST
M15 4FZ M6EQR
M15 4HD 2E0XOH
M15 4HD M7AGX
M15 5AF M3NMR
M15 5RR M0VAP
M15 5RR M6SGV
M16 0HD M0HXS
M16 8BB M3MBQ
M16 8LX M3NXZ
M16 8PW G3SMN
M16 8QQ M3OAG
M16 9GQ M6CLQ
M18 7BX M7COQ
M18 8BP G6YFL
M18 8GZ M7FMZ
M18 8LL M6UCP
M18 8WW 2E0ISB
M18 8WW G7NAL
M18 8WW M0MWB
M18 8WW M6ISB
M19 1LE 2E0TZD
M19 1LE M0TZD
M19 1LE M6ZTD
M19 1QY M0FVD
M19 2FA G4GDG
M19 2LQ G3JRK
M19 2NL G0PXZ
M19 2NP G8OYF
M19 2WJ M7GMG
M19 3EH G4BVQ
M19 3FL M6EGW
M20 1DH M0IGY
M20 1EN 2E1HZI
M20 2NT G6BYR
M20 2SQ 2E0CFM
M20 3DD M3IYE
M20 4GH G7AZW
M20 4NX 2E0HFS
M20 4NX M6SYF
M20 5GJ M1DFO
M20 5LQ 2E0BDV
M20 5LQ M3MVM
M20 5ND G8NNZ
M20 5PL G0OVQ
M20 5QS M3PGO
M20 5WN G0UXI
M20 6AW 2E0MQT
M20 6BB G7JME
M20 6BB G7MIT
M20 6JA G3YHQ
M20 6SU 2E0BTB
M20 6TG 2E1HLW
M21 0QU 2E0MWB
M21 0QU M0UFC
M21 0QU M6MWB
M21 0UX G7EPL
M21 7LE M3OJK
M21 7NE 2E0PNW
M21 7NE M6CPW
M21 7QG M3BLX
M21 7TG G4RYT
M21 8FA 2E0IZW
M21 8FA M6IZW
M21 8XT G6BHA
M21 9DP G4OCR
M21 9JU G0CRF
M21 9JU G6CRF
M22 1AF G1KEW
M22 1AH 2E1IHS
M22 1AH G0PXG
M22 1AH M3AHZ
M22 1GA 2E0LHY
M22 1GA M7BGF
M22 1GA M7JLG
M22 1NN G1UTM
M22 1SN 2E0EWQ
M22 1SN M6NKB
M22 1UD M3NHA
M22 1UD M6NHA
M22 1UR G0JYS
M22 4DP G1OJB
M22 4DZ G0PWA
M22 4GG G1PNC
M22 4LW M6MOJ
M22 4QS G0JPT
M22 4WB M3XAO
M22 5AQ G8OZD
M22 5BF M7VCL
M22 5DH M3DAV
M22 5DJ G7SYJ
M22 5FU M6GTB
M22 5JY G4RWN
M22 5QG G7TEA
M22 5QG M0TEA
M22 9PL 2E1GNK
M22 9UX M6BBL
M22 9UX M6HNS
M22 9YE 2E0SGY
M22 9YE M6SGA
M23 1HF G0DNY
M23 1JX G0SAY
M23 1JY G6BMY
M23 1LD 2E0PLH
M23 1LE M6XMN
M23 1LP G0NID
M23 2AT G7VSE
M23 2UL G0LNA
M23 2YX G0LNB
M23 9AW G8LPX
M23 9BT G8LUL
M23 9NZ G7PXS
M24 1HE G6CVW
M24 1JA M7CEF
M24 1QD G1XGN
M24 1TJ G7VGX
M24 2AG M7PLW
M24 2AH M0LGP
M24 2EQ G0TRK
M24 2NN G6GCY
M24 2RF G1OZG
M24 2SX M7BSV
M24 2TQ G4VVT
M24 2WB G6EJI
M24 4AE G4WLI
M24 4AE G8IKW
M24 4JJ G4WVP
M24 4LU M7ABB
M24 4RU G4ZQL
M24 5LS M3TRC
M24 5RL G0FCZ
M24 5SU 2E0YMM
M24 5SU M5JPG
M24 5SU M6WUM
M24 6AX M6RWT
M24 6BN M1EKH
M24 6FF M0PTB
M25 0FZ G7PYN
M25 0HE G1HHU
M25 0HR G8JWC
M25 0HX G3ZVT
M25 0HX M7XPG
M25 1FN M6RUN
M25 1FN M6SAU
M25 1FN M7BAR
M25 1JY M6AAX
M25 1NB G4VUK
M25 2RD 2E0DIJ
M25 2RD M6DIJ
M25 2SD 2E0EVI
M25 2SD M0ODB
M25 2SD M6JUW
M25 2WP 2E0TFD
M25 2WP M6YMM
M25 3BT G6BVF
M25 3DY M6AUC
M25 3HZ G0CSU
M25 3JJ G7VKJ
M25 3JL G1ULB
M25 9GW G1PNB
M25 9GW M3WWD
M25 9LB M0MBS
M25 9NT G0UAY
M25 9RA 2E0HWY
M25 9RA M7TWK
M25 9RE G8OFI
M25 9RU 2E0OAI
M25 9RU M6DWS
M25 9TG G1JZY
M252WN G0ROM
M26 1EP M6SLE
M26 1HN G1JUD
M26 1HN G6RXF
M26 1HY G0BMM
M26 1JA G1PKP
M26 1JA G6CLW
M26 1LE M3ZFJ
M26 1YF G3ZCH
M26 1YF G6OTQ
M26 2GF G6YGB
M26 2QS G0WEF
M26 2RB M3LJB
M26 2RS G0HEA
M26 2UU M6OVR
M26 2UX G0ADL
M26 2XQ M7IGW
M26 2XU M7PPR
M26 3GL G4UAU
M26 3QX M6MDL
M26 3RU M0ANQ
M26 3TZ M6ROY
M26 3UA 2E0GPY
M26 3UJ G1ESW
M26 4FA M0BOQ
M26 4FS G8ZIC
M26 4NS M0XCT
M26 4QF G6DGQ
M27 0EU M7UDS
M27 0JT M6CIW
M27 0YH 2E0XBN
M27 0YH G0MKA
M27 0YH G6YIO
M27 0YH M0XBN
M27 0YH M3OTQ
M27 0YH M6XBM
M27 4FA G8RAX
M27 5NA G0KLF
M27 5QY M3OVE
M27 6AQ G3STJ
M27 6BU G5SRC
M27 6BU G7WAE
M27 6BU M0KHQ
M27 6BU M0NTI
M27 6PT M1KCB
M27 6WH G4FXA
M27 6WY G1SYV
M27 8AF M6MTV
M27 8FS G0CVH
M27 8QR M7DDM
M27 8QS G4VDJ
M27 8RE G0MZY
M27 8RT M6HGR
M27 8XP G0LVG
M27 8XP M6KMN
M27 9QP G7DWQ
M28 0SL G8KOQ
M28 0SX G3PMJ
M28 0TP G4NTY
M28 1BE 2E0ENE
M28 1BE M6FVL
M28 1DE G7ELA
M28 1DF 2E1GPO
M28 1FY M0SEB
M28 1HW G4SXG
M28 1LP G7DPW
M28 1ND G8DTF
M28 1PS M6ODA
M28 2JE M6UIR
M28 2RJ G0TEE
M28 2RW G1VJQ
M28 2TX 2E0YJL
M28 2TX M3YJL
M28 3DH 2E0ENB
M28 3DH M6ASZ
M28 3EB M7EAE
M28 3FQ M7JWM
M28 3JW M1VIP
M28 3WD G2LK
M28 7EN M0CAV
M28 7EX 2E1WJB
M28 7EX G4UPU
M28 7JD G7FFK
M28 7JW G4TJU
M28 7QF G4JLG
M28 7TS G8JOC
M28 7XS G7FPS
M28 7XX G6YGC
M281XJ M7OMO
M29 7AT M6BSS
M29 7AX 2E0REM
M29 7AX M6AVT
M29 7BH G4JHA
M29 7BJ M6BST
M29 7DB 2E0IZR
M29 7DB G0IZR
M29 7DB M3RXT
M29 7EJ G8NSE
M29 7FF M6SDH
M29 7HH G4GKT
M29 7HS G4DZK
M29 7HS G8FQZ
M29 7JB 2E0FMH
M29 7JB M6OBQ
M29 7PE 2E1GSM
M29 7PH G0TFP
M29 7PZ M6ERH
M29 7RE G7CQW
M29 7RT G8SYM
M29 7WF M3GAV
M29 7WG M7KPM
M29 8GH M6OCI
M29 8LQ M0AVA
M29 8LS M6XZY
M29 8PG M3RFI
M29 8PH 2E1HZV
M29 8PH M3EGC
M29 8PJ M6MKX
M29 8WQ M6AEQ
M29 9DN M3LFH
M298ET 2E0HJC
M298ET M7JCH
M3 1LP M6ZUT
M3 5GL M7GUM
M30 0AN M6XBQ
M30 0FS M3LUZ
M30 0HW G6YAQ
M30 0JN M6JII
M30 0LT G1ASN
M30 0PZ M6CPC
M30 0QU M6WLA
M30 7HF M7LMN
M30 7HJ M7YCV
M30 7LG M7TPW
M30 8AG G4VPW
M30 8AJ M1ABV
M30 8BP G0BVM
M30 8DD G0FPO
M30 8JR 2E0FDD
M30 8JR M6FDD
M30 8JU M7ETG
M30 8QD M7GCU
M30 8UW G8FYK
M30 8WB M1NUS
M30 8WP 2E0FWW
M30 8WP M3GHI
M30 9DL 2E0PCX
M30 9DL M0PCO
M30 9DL M6BGM
M30 9DY G4NDG
M30 9EE G8YPH
M30 9FT M0XJG
M30 9LY M0HTS
M307DU M7BZL
M31 4BA M7WTJ
M31 4DB 2E0HVD
M31 4DL G8BRU
M31 4DY G3VYA
M31 4GW G7GAK
M31 4NP G0TRU
M31 4PT G0LZL
M32 0AJ M6RFK
M32 0AN G7KRI
M32 0RG M3OVX
M32 0UE M7VCC
M32 8BE M6PLI
M32 8DQ G1JHX
M32 8NA 2E0JED
M32 9AF 2E0MUA
M32 9AF M3MUA
M32 9BS G7FKJ
M32 9BX G4LVI
M32 9JB G6YOP
M32 9JQ M7EQG
M32 9QA M0NTH
M32 9QF G0BJK
M32 9QF G3UHF
M32 9RT 2E0RDU
M32 9SJ G1UFM
M33 2AR M0WRA
M33 2BH M7CKA
M33 2BJ G1NVL
M33 2DQ G7FQY
M33 2DY 2E1ICI
M33 2EA G4URV
M33 2HU M0YIT
M33 2HU M7GZA
M33 2JT M0DHE
M33 2PJ M6CGS
M33 2XF G0OVY
M33 2XP M0UQB
M33 2XP M0WKL
M33 2XP M0XQB
M33 3AN M7LCJ
M33 3GT G3SVW
M33 3LD 2E1IBP
M33 3LH G0RFG
M33 3LQ G0TKL
M33 3LQ M3PSR
M33 3LW 2E0CPF
M33 3LW M6BQF
M33 3QP G1BOO
M33 3TN G0KKF
M33 3WS G4AOK
M33 4AE G4JZZ
M33 4DH. M0IYP
M33 4ER 2E1IGP
M33 4FS G4EDM
M33 4FW M0KWV
M33 4HL M0IXX
M33 4HP G4KGU
M33 4HR G7TUH
M33 4LG M7MBO
M33 4NS G8ZZF
M33 4QL M0HIH
M33 4RF G4AUR
M33 4RG G0NWJ
M33 4WG G4MYB
M33 5BE M0MTQ
M33 5DL G8ILW
M33 5FB G3KCB
M33 5HF G0SSV
M33 5HR 2E0ZVL
M33 5HR M0LCR
M33 5HR M0SII
M33 5HR M6ZVL
M33 5NF M7ICH
M33 5NQ M3CVM
M33 5NY G0MRK
M33 5PR G6ISA
M33 5QJ G0BHP
M33 5RP M3XAU
M33 5UF G3FVA
M33 6EZ G3ZDM
M33 6GS M3VRU
M33 6NF G3WFT
M33 6NW G3SUI
M33 6QW G4FPA
M33 6WU G0FRB
M33 7AU M7CGH
M33 7EN G8RSI
M34 2DQ G0ROW
M34 2ER 2E0PBE
M34 2PD M6OEW
M34 2WP G0WYP
M34 3GL G7GXR
M34 3HU 2E1YRK
M34 3NS G7TYB
M34 3NX G3XVG
M34 3NZ M6LBI
M34 3QH 2E0KKF
M34 3QH M6KKF
M34 3TE G0BOH
M34 3TH G4WGR
M34 3WP 2E0TFK
M34 3WP M6ILS
M34 5BD G8HTN
M34 5EJ M6JRE
M34 5FB M3IFG
M34 5GB M3OWN
M34 5GR M7MCF
M34 5LJ G1JDT
M34 5LZ M0KCJ
M34 5QX G0NHO
M34 5QX G7DNR
M34 5UH G7GTU
M34 6AF G1BBK
M34 6EA G6NVU
M34 6EA G7HRZ
M34 6FT G7DOY
M34 6HG M7AUC
M34 6JY G0YRT
M34 6LR G1EJK
M34 7GG 2E0GKT
M34 7GG M7TAY
M34 7LB M3PBU
M34 7TF M6EKD
M34 7TX G1UTS
M34 7TX M0GMA
M34 7TX M3ZPM
M342DJ G0HUH
M35 0BE M7LZT
M35 0FX G7OAL
M35 0LH G0KTR
M35 0LR M0VKU
M35 0NN M6PQL
M35 0PX G7BAB
M35 0TG M7PUL
M35 9EJ G3PYH
M350LR M7LIN
M359DY 2E0OAS
M359DY M7CDS
M38 0BU M3WTN
M38 0FQ G1SPM
M38 0FU 2E0HVB
M38 0LW M0APZ
M38 9GL G6MIU
M38 9GQ 2E0SCD
M38 9GQ M6FPC
M38 9HB M3IRF
M38 9NF M6GNU
M38 9TN 2E0EMK
M38 9TN M6IYT
M38 9UB M0IRV
M38 9UH M0KEJ
M38 9XU G6PKV
M4 5LB M0DAH
M4 6DB M7TQY
M4 6GB G7FPJ
M4 7BB 2E0NTW
M4 7BB M3NTW
M4 7EP M7NIK
M40 0DG G6WGZ
M40 0DG M6EAE
M40 0DG M6HUN
M40 0DG M6PHJ
M40 0GB G0IUK
M40 1JE M6OJQ
M40 1LF 2E0GHS
M40 1LF M0SJS
M40 1LF M3GHS
M40 1QE M0FRG
M40 2SY M3GAF
M40 3GG 2E0AVS
M40 3GG M3AJN
M40 3NL G7TZV
M40 3SE G6IRF
M40 3TD M3VNP
M40 3WP G7ORW
M40 5HL G7VCG
M40 5HL M3IAF
M40 5QD G0VRQ
M40 5QD G0VXW
M40 5QW 2E0YES
M40 5QW M0TLY
M40 5QW M3GBC
M40 7LX 2E0JKP
M40 7LX M0JPE
M40 7LX M6HKZ
M40 9BW 2E0WCC
M40 9BW M6LEJ
M40 9LF G7USQ
M400FZ M7TDS
M41 0GF G7HKQ
M41 0PY G6PSA
M41 0RJ G3JIS
M41 0SB M3IRM
M41 0TY G3XGE
M41 0TY M0RML
M41 0XY G0ONF
M41 5AT 2E0DJY
M41 5BN M0HYI
M41 5BZ G7LKL
M41 5HA G7TIR
M41 5QG 2E0GNQ
M41 5QG M7AZS
M41 6FJ G1SCL
M41 6HA G1UKZ
M41 6HP G1NYJ
M41 6LE G3XXG
M41 6NQ G7OTE
M41 6PU 2E0DRD
M41 6PU M0XPD
M41 6PU M6PAD
M41 6WA G8YUK
M41 7BT G4SDL
M41 7DY 2E0FPA
M41 7DY M0FPA
M41 7DY M3FPA
M41 7EY M7DGN
M41 7FH 2E0ILJ
M41 7FH M7MVF
M41 7WH G4WLJ
M41 7WX G4MFI
M41 8PB M0UUU
M41 8QT 2E0DCS
M41 8QT M0WXX
M41 8RE M6UFR
M41 8RN G8HXE
M41 8RN M0TWC
M41 8RN M3KMH
M41 8UE G4RYI
M41 8UE M3OPV
M41 8UE M3RYI
M41 8UT G1VZB
M41 9JZ G7HMW
M41 9NG G7AGC
M41 9NJ G1LWE
M41 9NT 2E0KUH
M41 9NT G0SYP
M41 9NT M6KUH
M41 9PT G4NCZ
M41 9QE G0IMB
M43 6DS G1WPL
M43 6DX G1YCM
M43 6EF G8CZE
M43 6FS G4HBI
M43 6HL G6KRS
M43 6QG G6GXY
M43 7EA M6HEK
M43 7FR 2E0RNH
M43 7FR M0RNH
M43 7FR M7GRH
M43 7JX G8OXX
M43 7LJ M6BXN
M43 7NS M6LUI
M43 7PF M6ZAQ
M43 7PL 2E0IMS
M43 7SW M6RCY
M43 7UQ G4IFI
M43 7YG M6EIL
M44 5AS M0YDJ
M44 5AU 2E0SGH
M44 5AU G0GSK
M44 5AU M0MBE
M44 5AU M6SGH
M44 5HB 2E0UCT
M44 5HB G5CTX
M44 5HB M7CTB
M44 5JQ G6LBQ
M44 6AQ G8TXW
M44 6AR M6XKY
M44 6DD M6PNT
M44 6DQ G4BGV
M44 6EH M3RYG
M44 6EH M3YOZ
M44 6EN G0JYK
M44 6FG M7MWM
M44 6FY M7WCC
M44 6HH G6SPG
M44 6HH 2E0JIR
M44 6HH M3JIR
M44 6LR G0IGT
M44 6PJ M7DDR
M44 6QA M0IPR
M44 6QA M0SRA
M44 6TD M3UJP
M44 6TE G6MMA
M44 6WJ M0JWE
M44 6WY G0NPI
M45 6DP G8PAL
M45 6EP G8AOG
M45 6NR 2E0KMA
M45 6NR M6KMA
M45 6TJ G8DUT
M45 7LR G1COX
M45 7NT G3GXI
M45 7NT G8KRG
M45 7PR M6BPL
M45 7RP 2E0FYS
M45 7RP M7SES
M45 7SS G8NKJ
M45 7TS G4UGW
M45 8FR G4FZP
M45 8GG M1CRZ
M45 8LH 2E0GRX
M45 8NT G1NSB
M45 8PP M0OBU
M45 8WY G0AHR
M45 8XR M1APV
M46 0AQ M7BDF
M46 0EZ 2E0LTT
M46 0EZ M0LTT
M46 0EZ M6MLA
M46 0GT M7LWI
M46 0LQ G6SKM
M46 0PG G6LCX
M46 0QE G0FTO
M46 0SN M6KML
M46 9AB G6YBC
M46 9EG 2E0ITG
M46 9EG M7GBO
M46 9LN G6TFV
M46 9LQ 2E0ENC
M46 9LQ G4HZJ
M46 9LQ G7LDR
M46 9NQ 2E0HBY
M46 9PQ G0OMF
M46 9PY M0GNY
M46 9QB G0KBS
M46 9QB G4GBK
M46 9QZ M7BJI
M46 9RQ G7DNR
M46 9RQ M3WAC
M46 9RR 2E1HLL
M46 9RR M1HLL
M46 9TZ G6HFW
M5 3AZ M6APC
M5 3HU M6GSH
M5 3WL M7FZM
M5 5HX G7KQL
M5 5LF 2E0DMD
M5 5LF M0WDD
M50 3RB 2E1MPB
M6 5YB G0HQN
M6 6DW G0GIF
M6 7LA G0PLX
M6 7PQ G6FEI
M6 7WP M3UDC
M6 8QQ G0IYM
M7 1PR 2E0YWN
M7 2GB 2E0PDX
M7 3NX 2E1BLG
M7 3NZ G1JHB
M7 3TF M0LKS
M7 4JA M0ASU
M7 4TH M3OSU
M7 4TH M3OSW
M8 2FS M7EGE
M8 4GG M3VVQ
M8 4GT M6BLS
M8 4WQ G4YNI
M8 5AR M6ZAD
M8 5AW G5JCB
M8 5AW G5MS
M85DR M7BSU
M9 0RQ G6OGT
M9 4EX G4XBX
M9 4PZ 2E1IEM
M9 5XY 2E0TRK
M9 5XY M3ZAV
M9 6HR G3KIQ
M9 6QX M3YPR
M9 7AR G3NSW
M9 7EQ G6MDG
M9 7GR M7DOS
M9 8AT G1XIH
M9 8AT G2ALN
M9 8EG 2E0JVF
M9 8JD G0SNZ
ME1 1FB M3FPU
ME1 1HY G3NRU
ME1 1JG G0NCE
ME1 1NB G7FZN
ME1 1QH G8WKX
ME1 2BL G7VFU
ME1 2DL G0WFM
ME1 2HH G1USD
ME1 2HZ G0WYV
ME1 2NN G4SAW
ME1 2RE M3JHW
ME1 2SS M3RMZ
ME1 2ST G4NRT
ME1 2UA G2FSH
ME1 2UH G3ZJP
ME1 2UL M6MLM
ME1 2XD M0PLG
ME1 2XD M6SJQ
ME1 3AJ M6NST
ME1 3EF G7KCK
ME1 3JU G0DKO
ME1 3NJ G1ERZ
ME1 3QE M0SAC
ME1 3QR G0BIX
ME10 1BY G3ISD
ME10 1EH G8LDJ
ME10 1ER M3ZKD
ME10 1ET M0HZW
ME10 1EY G1HHG
ME10 1JD M3DSI
ME10 1JF G0TNC
ME10 1LB G3EHW
ME10 1QT G0UAS
ME10 1QY G7GSR
ME10 1RD G3VPA
ME10 1RF G4TWC
ME10 1TJ M1DUB
ME10 1TJ M3PJS
ME10 1TJ M3PZX
ME10 1TS G6DUI
ME10 1XF M7ACW
ME10 1XJ 2E1AUN
ME10 1XJ M3PHF
ME10 1YL G1BOX
ME10 1YL G8ONY
ME10 2DP M6BAZ
ME10 2EA G6AAC
ME10 2EY G4IBH
ME10 2FA M7EQS
ME10 2HR G1JUO
ME10 2JF G0AXQ
ME10 2LR G1YTO
ME10 2LZ M0AZP
ME10 2PX M0PIA
ME10 2QP 2E1TAG
ME10 3BL 2E1JRS
ME10 3LG M6SDF
ME10 3NR G1VJN
ME10 3PB 2E0OZG
ME10 3PB M6OZG
ME10 3QY M6EWH
ME10 3TG 2E1CLG
ME10 4HS G0VTV
ME10 4HS G4LBS
ME10 4LF M3RXG
ME10 4NA G1SPX
ME10 4NA M6IAN
ME10 4QD G0DRQ
ME10 4QD G1DLJ
ME10 4QE G4VEC
ME10 4RX G7BWV
ME10 4SL G0RNQ
ME10 4TB 2E0EWH
ME10 5AN G7MEE
ME10 5AS G6PGT
ME102TE G7AEY
ME11 5AG G8DPW
ME11 5AW G1JCL
ME11 5BU G6ILN
ME11 5DT 2E0RNF
ME11 5DT M6OEM
ME12 1BW M1WAW
ME12 1DR M6KYE
ME12 1UT G4KSU
ME12 2AT M3TMQ
ME12 2BS G1AFW
ME12 2DJ G0OLX
ME12 2DJ G0PEH
ME12 2DJ G8CUX
ME12 2EA M7LIW
ME12 2HG G0JYH
ME12 2HX M0CSU
ME12 2LH M0WGF
ME12 2NH G0DRT
ME12 2NL G7NJB
ME12 2NQ 2E0XCC
ME12 2NQ M6NMS
ME12 2RA M6LFJ
ME12 3BA G1VFB
ME12 3BB G7LZY
ME12 3BQ G3ZYQ
ME12 3DH G8XGB
ME12 3DH M0XGB
ME12 3EZ M3RAK
ME12 3HX G3GDH
ME12 3JX G6NUI
ME12 3JX M1DJN
ME12 3LN G8JNZ
ME12 3LN M6KSP
ME12 3NF 2E0BPY
ME12 3NF M3UIJ
ME12 3NJ G1SGS
ME12 3NQ G0MIH
ME12 3NR G2HKU
ME12 3PF 2E0FHE
ME12 3PF M7XYZ
ME12 3PG 2E0ILN
ME12 3PG M0ILN
ME12 3PG M6ILN
ME12 3PG M6KCF
ME12 3PJ M6MZY
ME12 3PT M6BRK
ME12 3PU M6NSW
ME12 3QR G7ECQ
ME12 3RZ M6DTN
ME12 3SA 2E0ELT
ME12 3SA M0KLK
ME12 3SA M6WJF
ME12 3SQ G7IAU
ME12 3UA M3ZZF
ME12 4EJ M3VMU
ME12 4JU G4WPB
ME12 4PU M3YGD
ME12 4PU M6MWA
ME12 4QY M3XIL
ME121YA M0CST
ME123SA M6KUG
ME13 0EF M3MUX
ME13 7DL M1ARU
ME13 7ES G0PZX
ME13 7JG 2E0BRJ
ME13 7PD M3VWF
ME13 7TA M3MFF
ME13 8DE M6BOI
ME13 8DG M3NPS
ME13 8DZ G1HDK
ME13 8HP G1EUQ
ME13 8JA G8MVH
ME13 8QL M3ZFB
ME13 8RU G7AFS
ME13 8TP M6UVB
ME13 8UR M7MGW
ME13 8XN G4UVX
ME13 9LF 2E1BAE
ME13 9NR M3XEQ
ME13 9SE G4AWG
ME13 9TG 2E0IYC
ME13 9TG M7FBJ
ME14 2BX M1BWJ
ME14 2EA G0REN
ME14 2ER G3YJS
ME14 2JN G3PCX
ME14 2QR G4UAQ
ME14 2QW M0CZN
ME14 2TX M7VIV
ME14 3JN G3ZAL
ME14 4BH G0KYM
ME14 4HG G7MOX
ME14 4JB G0BUW
ME14 4NJ G0SPQ
ME14 5BH G6VVS
ME14 5HJ G8HLE
ME14 5PG G3WXU
ME15 0AJ G7DJT
ME15 0DR M6HIL
ME15 0QF M6HRX
ME15 0QH G4XMS
ME15 6AA G1VWP
ME15 6FF G6XAK
ME15 6PX G7AZT
ME15 6QB 2E0BAI
ME15 6TH G4CCQ
ME15 6UL M6FLB
ME15 6UU G8LEG
ME15 6YN M3XSP
ME15 7AU G7VOM
ME15 7AU G7VON
ME15 7EL G6ZDY
ME15 7PL M0HLD
ME15 7PX 2E0XMT
ME15 7RR 2E0GTA
ME15 7RT G3WWI
ME15 7SP 2E0SVG
ME15 7SP M6SVG
ME15 8DY G7OPG
ME15 8EG G4IWN
ME15 8EP M0HBM
ME15 8JP G0VAR
ME15 8JZ G7GAZ
ME15 8RA G0DBY
ME15 8TN G8LGA
ME15 9AB G6LUD
ME15 9AE G6ALJ
ME15 9AJ M7DGU
ME15 9JB G8VPD
ME15 9PD G0NCW
ME15 9QR G7HON
ME15 9RA G3RJF
ME15 9RP G0LCH
ME15 9RS G4WBA
ME16 0DL G3ORP
ME16 0JL M0PMC
ME16 0JS G6RVS
ME16 0JS M0KMR
ME16 0JS M0KRM
ME16 0JS M3RVS
ME16 0NF G3VNI
ME16 0PQ 2E0JIT
ME16 0QP G0UED
ME16 0QW 2E0KKQ
ME16 0RX M1AEX
ME16 8BS M1BGS
ME16 8DR G4AXD
ME16 8EN G8PWT
ME16 8NP G1XJK
ME16 8PA G0FKK
ME16 8QP M1AJM
ME16 8QU 2E0JHG
ME16 8QU M0JHG
ME16 8TE M6OGQ
ME16 9HF 2E0HIZ
ME16 9HF G4HIZ
ME16 9HF M3SIZ
ME16 9HH G6HTH
ME16 9HW 2E0HIJ
ME16 9JQ M0DHM
ME160HB M7EJK
ME17 1AD G3YIF
ME17 1AY M1DDW
ME17 1BS G3YUU
ME17 1SR 2E0LAH
ME17 1SR M6BEC
ME17 1SX G1BLB
ME17 1SZ 2E0VPL
ME17 1TJ G3UOJ
ME17 1TZ G6KTA
ME17 2AN G6USR
ME17 2AX G0IJK
ME17 2EJ G3LRX
ME17 2ET G4XSI
ME17 2EU G8WHB
ME17 2HH 2E0KWE
ME17 2HH M7INS
ME17 2NH 2E0PSR
ME17 2NH M0NAL
ME17 2NH M6GAE
ME17 2NS G1HQJ
ME17 3LH M7BQB
ME17 3NP G1YQN
ME17 3PH G4ZQB
ME17 3RG M6LHQ
ME17 3YP M1XPS
ME17 4DN G1ZMG
ME17 4EE G4LJT
ME17 4JY G6HTS
ME17 4JZ G3KOM
ME17 4QS M3XZK
ME174NQ M7FSL
ME18 5HT M6FET
ME18 5JY M7FYV
ME18 5PR G3DEY
ME18 5PX G4LRW
ME18 5QF G6NAL
ME18 5QH G8NVH

ME18 6DX M0GZH
ME185BL G7BNZ
ME19 4BS G6FMU
ME19 4DL G8LDU
ME19 4TT G0OLT
ME19 5BL M6GCX
ME19 5BW G6BGA
ME19 5DP G0DAS
ME19 5EH G3PAG
ME19 5EN G3XJS
ME19 5EN M0VLP
ME19 5JX G0VUN
ME19 5QY G1AJY
ME19 6ES G7LMT
ME19 6QS G1JJQ
ME19 6SJ M1CMW
ME2 1DJ G7DSU
ME2 1EE 2E0OFM
ME2 1EE M0OFM
ME2 1EE M6OFM
ME2 1EE M6TOG
ME2 1EE M7OFM
ME2 2QB 2E1IEK
ME2 2QY M0AZS
ME2 2TZ G7UTB
ME2 2XZ M7CYR
ME2 2YE G4ZFC
ME2 2YT M1DNJ
ME2 2YW G4TVW
ME2 3BA G0LAD
ME2 3BZ M3OKC
ME2 3EP G0GIR
ME2 3JH G7LAN
ME2 3LW G1EWM
ME2 3NF G6CKK
ME2 3PN G0VMQ
ME2 3QG G0MIF
ME2 3RE M3BFX
ME2 3SP G3FNZ
ME2 4BB G6HIV
ME2 4BB M3LYZ
ME2 4EB G0VKL
ME2 4JD M0BPN
ME2 4LJ 2E0JAX
ME2 4LJ M6JHB
ME2 4NE G0GQT
ME2 4NL G4RFN
ME2 4NU G4CWV
ME2 4PG G1DOT
ME2 4PN G3NZR
ME2 4UR M0OLW
ME2 4XF G8CAM
ME20 1AP M6PBP
ME20 6AE M1BJS
ME20 6BB M3VMQ
ME20 6EF G0PWX
ME20 6ET G3PYO
ME20 6HP M7AAP
ME20 6LA G4RIU
ME20 6QZ G7LJQ
ME20 6RE M3MDY
ME20 7BE 2E0WXK
ME20 7BE M6WXK
ME20 7LY 2E0KUR
ME20 7LY M6KUR
ME20 7SF G1XUE
ME206EF 2E0WGP
ME206EF M3WZG
ME3 0BJ M6AAE
ME3 0BS M0ZMX
ME3 7BA G7AJT
ME3 7EH G0WAN
ME3 7JR 2E0BBA
ME3 7RN M3UIF
ME3 7SH G3OHP
ME3 7TN G1HRV
ME3 8BE G6KSK
ME3 8DW G8MIN
ME3 8HT G1GGI
ME3 8LR G3BSN
ME3 8NE G4FVB
ME3 8TR G4FJW
ME3 9AA 2E0KSG
ME3 9AA M6KSG
ME3 9DE 2E0HLR
ME3 9DE G0PEK
ME3 9DE M0KRO
ME3 9DE M6HLR
ME3 9DF 2E0BPG
ME3 9DF M3XFT
ME3 9EA M3XBS
ME3 9EA M7DYM
ME3 9EU G4TAM
ME3 9GF 2E0CYP
ME3 9GF M0HSU
ME3 9GF M6DVW
ME3 9JP G0AUN
ME3 9PR G0PCA
ME3 9PR G0PCB
ME3 9ST G1ALK
ME3 9TA G6IYD
ME3 9TG 2E0REY
ME3 9TG M3UVC
ME3 9UY M3VJX
ME39TG 2E0HEM
ME4 4NP M7EWY
ME4 4XJ M0DKT
ME4 4XJ M3DKT
ME4 5BZ 2E0RAM
ME4 5HS 2E0YRM
ME4 5HS M0YRM
ME4 5HS M6LZM
ME4 5LE G6TXP
ME4 5NN M3FOR
ME4 5NW M7BJY
ME4 5SJ G6TOT
ME4 5SJ M3PQG
ME4 5TW G7CVY
ME4 6HL G1YXH
ME4 6LY G4EHQ
ME4 6LY G7COY
ME4 6QD 2E0CZR
ME4 6QD M6BYF
ME4 6RE G4YLT
ME4 6UU G8CJM
ME5 0BH G4HJE
ME5 0DD M1BIK
ME5 0DQ G1EDH
ME5 0DY G7MFW
ME5 0HJ 2E0KVB
ME5 0HJ M6ETA
ME5 0HJ M6WMA
ME5 0HU M5LRO
ME5 0JS G2FJA
ME5 0JS G8VJU
ME5 0SB G8TGB
ME5 0UP M6WJB
ME5 7AA M3MDS
ME5 7HH 2E0KLY
ME5 7HH M3SJY
ME5 7NG 2E0AXN
ME5 7NG M3DSU
ME5 7PD G6UVL
ME5 7PT G0ABN
ME5 7QD G4YTU
ME5 7TR 2E0SGG
ME5 7TR M6SDG
ME5 8ER M3ZIM
ME5 8JD G6FHK
ME5 8NY G1LHD
ME5 8NY G7IBN
ME5 8PL M6UHV
ME5 8SR G6TXB
ME5 8TN G0VWV
ME5 8TN G7KNM
ME5 8UL G3PSR
ME5 9AA G7MPZ
ME5 9AG M3MZN
ME5 9AR 2E0JSH
ME5 9AR M0FSH
ME5 9BN 2E0EPH
ME5 9BN M0JCG
ME5 9BX M6OCX
ME5 9EX M0KIQ
ME5 9HA G8IXC
ME5 9HB M0DTK
ME5 9HE 2E0HRJ
ME5 9HE M0GYI
ME5 9HE M0HRP
ME5 9HE M6HRJ
ME5 9JG G6BLK
ME5 9LA G4DQN
ME5 9LF G6SQS
ME5 9LU M6DAK
ME5 9NN M6VAG
ME5 9QP 2E1GTB
ME5 9RD G0AKR
ME5 9SP G6YHP
ME5 9TD G3JHU
ME6 5HP 2E0MLE
ME6 5HP M0MZE
ME6 5HP M6MLE
ME6 5HX G0TUZ
ME6 5PA G6ZMG
ME6 5PY G8CCJ
ME6 5RL G0JAR
ME7 1DZ G0GJW
ME7 1EW 2E0RNE
ME7 1JB 2E0WAK
ME7 1JB M0WKO
ME7 1JB M3WYQ
ME7 1ND G6OLV
ME7 1PQ M7EYB
ME7 1UE M3SML
ME7 1UG 2E1IJX
ME7 1UG M0PPS
ME7 2DN 2E0DCH
ME7 2HL M3MEI
ME7 2LN G3TCI
ME7 2RE 2E0BAY
ME7 2RE M3IQQ
ME7 2RE M6BAY
ME7 2SW G0JLP
ME7 2SX G0VQB
ME7 2UN M0OOT
ME7 2UW G3JRD
ME7 2YH G5EDQ
ME7 3AQ M0CPG
ME7 3AY G6ITM
ME7 3EH M3NBK
ME7 3EX M1BDS
ME7 3JE G7AJK
ME7 3LS G4ZTF
ME7 3NA 2E1EVM
ME7 3PT G5MW
ME7 3PT G6IVP
ME7 3PT G8MWA
ME7 3PW M7CPQ
ME7 3QQ G0TAR
ME7 3QQ M0ECK
ME7 3TJ G7NJG
ME7 3TS 2E1EVJ
ME7 3TS G3VFC
ME7 3TS M6VFC
ME7 4BA M3EJL
ME7 4HB 2E1IJQ
ME7 4HB G6SXC
ME7 4HB M0FMY
ME7 4JB G6TBJ
ME7 4NE M7EUF
ME7 4NN G1XIO
ME7 4PY M6NSD
ME7 4RN M1ETC
ME7 4RR M0DYG
ME7 5EX M6OKR
ME7 5HY G8STY
ME7 5QG 2E1HRY
ME7 5QG 2E1RIO
ME7 5QG M3ANE
ME7 5QG M3HRY
ME7 5QH M1ALM
ME72DX M6IVT
ME8 0AQ G1IBJ
ME8 0DZ M3HWX
ME8 0JD G0VGR
ME8 0JG 2E0ICI
ME8 0JN G6OCO
ME8 0JX G7MNG
ME8 0LZ G8VVM
ME8 0NR G6YLW
ME8 0PE G7MIM
ME8 0PE M1CQT
ME8 0RD 2E1DMU
ME8 0TH 2E1IJS
ME8 0TL G4HZI
ME8 6JE G0TCW
ME8 6LX M0PFW
ME8 6ND G1XNC
ME8 6QY M6ICJ
ME8 6XP G0RBV
ME8 6YE M7DIU
ME8 7JN 2E0PKF
ME8 7JN M7GOO
ME8 7QB G8MNL
ME8 7SZ M7HUB
ME8 7TA G1CBK
ME8 8EW 2E0HIP
ME8 8EW M0HLB
ME8 8EW M6KOM
ME8 8HE G3PQD
ME8 8JU G4ZRY
ME8 8LH M7LVW
ME8 8PA M6OQX
ME8 8TA G0DWF
ME8 8TL 2E0WSS
ME8 8TL M3WSS
ME8 9EN 2E0RPA
ME8 9EN M0HNE
ME8 9EN M6REH
ME8 9ES G0APB
ME8 9HX G1ZBH
ME8 9JQ M6FSN
ME8 9PP M0AAK
ME9 0LR G1PGI
ME9 0PL G4MKG
ME9 0TY G4JJX
ME9 7AG G4TQS
ME9 7AN M1JWS
ME9 7AS M3DWV
ME9 7EX G3UIB
ME9 7JU G4IOG
ME9 7QA G4LUO
ME9 7RH G3PSS
ME9 7SL M0LNE
ME9 7SL M6LNE
ME9 7SY G3UKL
ME9 8DX M6COL
ME9 8JZ 2E0IKV
ME9 8JZ M6VKA
ME9 8RG M7SWT
ME9 8SB G7HTN
ME9 9HU G4MPA
ME9 9LH G6SIQ
ME9 9PL M0CUT
ME9 9SP G6HOS
ME9 9SW M3MXO
ME98JZ M7OSF
MK1 1BL G8ABB
MK10 7BW M6MGU
MK10 7LQ M7HBB
MK10 9AF G8SJK
MK10 9AY 2E0ROY
MK10 9HG G8OSH
MK10 9HR 2E0CVV
MK10 9HR 2E0EBR
MK10 9HR M6EBR
MK10 9HR M6PNY
MK10 9HZ M0MPB
MK10 9JL G6KGK
MK11 1AT M3DXL
MK11 1EX G4ZNY
MK11 1HX G6AZR
MK11 1RB G4NNX
MK11 2AA 2E1CKQ
MK11 2AF G0GMB
MK11 2BQ 2E0RKV
MK11 2BQ M0VTE
MK11 2BQ M3RKV
MK12 5AJ 2E0ORE
MK12 5AJ M7ERE
MK12 5BE G8AQB
MK12 5DN 2E0IHU
MK12 5DN G0VZH
MK12 5DN M7IPL
MK12 5DR M6WFG
MK12 5HB M0BZK
MK12 6AL G0TGU
MK12 6AN G3NEH
MK12 6LR G3UQL
MK13 0PN G0RKV
MK13 0PN M3ROW
MK13 0QP G6INI
MK13 7AJ G8KFN
MK13 7AR 2E0GXX
MK13 7AR M0JTL
MK13 7AR M3GXX
MK13 7AY G4LIJ
MK13 7AY G8CCV
MK13 7AZ 2E0KOA
MK13 7AZ M6JPW
MK13 7DA G1NRY
MK13 7DY G8NKN
MK13 7EP M0GPJ
MK13 7HS G3SHD
MK13 7PZ G7OOU
MK13 8AT G8EPQ
MK13 8DP M6ZXI
MK13 9BT 2E0EBP
MK13 9HP M7EAN
MK13 9HS G0BYH
MK14 5AX G3YYN
MK14 5AX M0AUL
MK14 5BZ M1SHE
MK14 5BZ M3LEK
MK14 5EP G0TAI
MK14 5FQ M7BXO
MK14 5FS M7OXO
MK14 5JQ G4WBG
MK14 5JY G7AFL
MK14 5PA 2E0CKJ
MK14 5PA M3PWW
MK14 5QA G3JVR
MK14 5QA G8KYP
MK14 6AY G6ALU
MK14 6BU 2E0KJL
MK14 6BU M3RZO
MK14 6BU M6LRW
MK14 6EJ G7UTY
MK14 6EJ M0CMS
MK14 6ER G6HKS
MK14 6ER G7NRB
MK14 6GX M0AWI
MK14 7AP M6SOU
MK14 7DY G4TOH
MK14 7LL G1DMR
MK14 7LU G1LRU
MK14 7PJ 2E0TFI
MK14 7PP 2E1HSP
MK14 7PP G1RNZ
MK14 7PP M3CMK
MK14 7PP M6MOG
MK14 7QN G7VOA
MK14 7QS G1SYU
MK14 7QS G7AUE
MK14 7RR G7KGR
MK15 9HP 2E1HCT
MK15 9HP M3BEK
MK16 0EE G7LDF
MK16 0LH G0FWD
MK16 0LL M0ATY
MK16 0LT G4DMR
MK16 0NG M3YJA
MK16 0ZJZ 2E0FMK
MK16 0ZJZ M0UKF
MK16 0ZJZ M6UKF
MK16 8BB M0HXN
MK16 8JB 2E0RTC
MK16 8JB M6HGE
MK16 8JG G3WKL
MK16 8PG M0GYK
MK16 8PL G4FIA
MK16 8PP G4VMX
MK16 8RF G1CYQ
MK16 8SR G6GCM
MK16 8TZ M6BEI
MK16 9AR 2E0NPL
MK16 9AR M6NPL
MK16 9BG G4GJR
MK16 9DN G4MTF
MK16 9LZ G3ZCJ
MK16 9LZ M7PGM
MK16 9NQ G6ZQU
MK17 0BX G3ZLX
MK17 0DQ G1FXT
MK17 0DS M7XFY
MK17 0EP G0OUR
MK17 0EP M0ANS
MK17 0HP M0BZE
MK17 0LL M0RBD
MK17 0LW G4UCJ
MK17 0QR G0AGR
MK17 0QR G4NUG
MK17 0SB G7MFZ
MK17 0SB G8FAK
MK17 8AW G4XWM
MK17 8EA M0RPJ
MK17 8GP M7ERS
MK17 8HX G0TUL
MK17 8HX G8RCK
MK17 8JN M7LZR
MK17 8PZ M7NMD
MK17 8QG G0QVH
MK17 8TN G3VZV
MK17 9AG M3ZCN
MK17 9AJ G0NGQ
MK17 9AJ G4UQR
MK17 9AW G6MQD
MK17 9ED G0LAX
MK17 9ED G6VNI
MK17 9HS 2E0HCZ
MK17 9HS M0JQP
MK17 9HS M7DXX
MK17 9JU M6KAG
MK18 1BY M7JVC
MK18 1DZ M1EUX
MK18 1PW M7MMP
MK18 2AL M0YAW
MK18 2AL M7REO
MK18 2AR M3ZKT
MK18 2FD G4HIF
MK18 2FG M3RHR
MK18 2FS G0GLG
MK18 2GS M0MBO
MK18 2GY M1ESH
MK18 2HA 2E0EMP
MK18 2HA M6LCU
MK18 2JH G6JXC
MK18 2LZ G7PMW
MK18 2NY M3WRB
MK18 2PF G1UEO
MK18 2QD G0GGN
MK18 2QR G2BSJ
MK18 2QS G1MYO
MK18 3AB G8GLZ
MK18 3BN 2E0KVA
MK18 3DY G4FYO
MK18 3HX G8FMC
MK18 3JU G8MBS
MK18 3LZ M0THN
MK18 3NA G7JHV
MK18 3NT 2E0TAC
MK18 3NT M3TPI
MK18 3NT M5REV
MK18 3RJ 2E0WNL
MK18 3RJ M0SEJ
MK18 3RJ M7SEJ
MK18 4BX G8RDB
MK18 4EL G3SMP
MK18 4HL G4UUF
MK18 4LH G0ICC
MK18 4LX M3KHI
MK18 5BJ G4YTL
MK18 7EN M0KQR
MK18 7EQ G0BLQ
MK18 7EW G3THS
MK18 7EX M7MOY
MK18 7HR G6TSJ
MK19 6BT M3JHR
MK19 6BY G4MDU
MK19 6JD 2E0ECI
MK19 6JD M6CIE
MK19 7AG M0JSZ
MK19 7BE G3LFR
MK19 7BT G8CXT
MK19 7JQ G4XJE
MK19 7LD G4KST
MK19 7LF 2E0FSG
MK19 7LF G4FZA
MK19 7LF M6AMJ
MK19 7LQ M4UHV
MK19 7LU G3XLN
MK19 7NF G8UEZ
MK19 7NH G1WWP
MK19 7NY 2E0CFQ
MK19 7NY M0HCP
MK2 2EZ G0EYZ
MK2 2FB G7ESL
MK2 2HT G0UUF
MK2 2LA G0GQP
MK2 2QD G0LRW
MK2 2XP 2E1FCO
MK2 2XP 2E1GRG
MK2 3AD G3CPT
MK2 3EB G8GIL
MK2 3EB M3IDQ
MK2 3FH G1GXX
MK2 3LS G4JGG
MK2 3NB G3YJD
MK2 3NG 2E1OUJ
MK2 3NR 2E0FEE
MK2 3NR M6NXO
MK3 5AF 2E0EAF
MK3 5AF M3UHY
MK3 5AF M3UXL
MK3 5AF M3ZYT
MK3 5DQ 2E0APN
MK3 5DQ G3PPO
MK3 5DR M6AYU
MK3 5EN G0TUO
MK3 5HA G3NAI
MK3 6BY G6ZJD
MK3 6HX M3ZFI
MK3 6JP 2E0MKT
MK3 6JP M0TSW
MK3 6JP M6TMK
MK3 6PE M3ELN
MK3 6PJ M3GGA
MK3 7BG M3UMV
MK3 7BP 2E1BEV
MK3 7EU G3KQQ
MK3 7LD M0HFB
MK3 7LS M6GRQ
MK3 7PS M1ELI
MK3 7QP M6GCT
MK3 7RQ G4GCJ
MK3 7TA G0EFN
MK4 1AX M6MOU
MK4 1BJ 2E0BGP
MK4 1BQ G7NDO
MK4 1DP G1TTC
MK4 1EF G0UCK
MK4 1HY M0AEQ
MK4 1JH M1ESM
MK4 2AU G0OGM
MK4 2BT G1PCN
MK4 2DF M7MKK
MK4 2DX M7ESJ
MK4 2EZ 2E0USS
MK4 2EZ M0BOY
MK4 2EZ M7AIR
MK4 2EZ M7OMY
MK4 2GF G7LXB
MK4 3EN 2E0TWK
MK4 3EN G8LOF
MK4 3EN M0WSE
MK4 4AX M3EQL
MK4 4AX M3SSI
MK4 4DH G4UFS
MK40 1TF M6CYD
MK40 1ZL M0LRS
MK40 2BE G6GCO
MK40 2DS 2E0ZPY
MK40 2DS M6ZPY
MK40 2HY M6OCB
MK40 2SB M7SHF
MK40 2TR G5BH
MK40 2XU M7BUS
MK40 3DF G3USD
MK40 3PP M6CSF
MK40 3RJ M0IMV
MK40 3RJ M5DHB
MK40 3SA 2E0ACR
MK40 3UG G0BMG
MK40 3UG M3TIY
MK40 4FZ M6HUP
MK40 4GW G7UOD
MK40 4HJ G6GIH
MK40 4PZ G0ADB
MK40 4SB M3IKN
MK403BY M3OQD
MK41 0HL G8CZP
MK41 0LB M0GGT
MK41 0LB M3NKZ
MK41 0SN 2E0VCZ
MK41 0SN M6VCV
MK41 0TF G1MKR
MK41 0TF M0MKR
MK41 0TX M3XAK
MK41 0UP G0BKN
MK41 0UP G0MBR
MK41 0UP G6MBR
MK41 6BD M7TMD
MK41 6BS M6MBU
MK41 6DA G6UPA
MK41 6DU 2E0GXD
MK41 6DU G0RUN
MK41 6DU M7BFK
MK41 6DY M6FPM
MK41 7BS G4GDX
MK41 7BT G4KWH
MK41 7BT G4KYX
MK41 7DH G3UED
MK41 7HY G8GDC
MK41 7JP G4KJU
MK41 7JT M7JOZ
MK41 7LS G7RVC
MK41 7LS M0MPI
MK41 7ST G3UBV
MK41 7ST G4EZQ
MK41 7TU G1BYT
MK41 7UH G6CVE
MK41 8AP G4UCY
MK41 8AS G4AQS
MK41 8AS G4GIR
MK41 8AW 2E1EMI
MK41 8BT G3RPL
MK41 8DR G4XDB
MK41 8DR G8HJH
MK41 8EL G0HER
MK41 8JS G6PAA
MK41 8JS G7RDA
MK41 8NX G6YNW
MK41 8QD G0VJJ
MK41 8QY G8HGL
MK41 9AL G8SDN
MK41 9AL G8XEF
MK41 9AS G7LEB
MK41 9DD G4YYC
MK41 9ER M6SNP
MK41 9LD M1PGH
MK41 9LD M3PGH
MK41 9LG G8FTS
MK41 9NE G3WBP
MK41 9QR G1JZT
MK41 9QW M3UDU
MK41 9RB M6FLK
MK41 9RG M6GEA
MK42 0AF G3XDU
MK42 0AF G8HXW
MK42 0FF M7JKF
MK42 0NA 2E0OAA
MK42 0NA M0TXS
MK42 0NA M6SRO
MK42 0PX G4IMH
MK42 0RS M3GHE
MK42 0SE G0GBI
MK42 0SJ G0OXY
MK42 0UJ 2E0CEP
MK42 0UJ M0HBX
MK42 0UJ M3SDV
MK42 0XB M6MUP
MK42 0XD M6GGW
MK42 6AH M6WCR
MK42 6ET M7DRT
MK42 7BE G4OOQ
MK42 7DP M3DPQ
MK42 7DU M1MPA
MK42 7DU M3ZKA
MK42 7DU M6KGA
MK42 7DW M7BLY
MK42 7EH G7MIS
MK42 7EN G4ZPH
MK42 7EX G8YZT
MK42 7FR G1NRE
MK42 7JN 2E0GLD
MK42 7JN M3ORB
MK42 7JX G0DZB
MK42 7PE G8MGP
MK42 7RX M7LDU
MK42 7SX M6FAF
MK42 8BU G7SPE
MK42 8DT G0BDK
MK42 8HS G3WTP
MK42 8HS M0ZAE
MK42 8HS M1BED
MK42 8HS M3ZIE
MK42 8NL G7SQM
MK42 8RU G8WGQ
MK42 8RX G4OGD
MK42 9EH G1CJL
MK42 9FG M7FIN
MK42 9PB 2E0XHB
MK42 9PB M6XHB
MK42 9RG G0EYG
MK42 9SH M3LFZ
MK42 9TH G6HCW
MK42 9TZ 2E0RML
MK42 9UX G4NGD
MK42 9XB 2E1LEX
MK42 9XB M6IXY
MK42 9XB M7ECM
MK42 9YX G0VME
MK43 0BY 2E0CQB
MK43 0BY M6CQB
MK43 0DF G4DCW
MK43 0HA G7BBD
MK43 0HP G4PMB
MK43 0HP G6EZR
MK43 0HX M6BFI
MK43 0JA G4JXZ
MK43 0LA G8OQK
MK43 0LW G4PBM
MK43 0PN G7UYB
MK43 0PN M0BZO
MK43 0QZ G0FGJ
MK43 0RS M7YAN
MK43 0RU M6CYB
MK43 0RW G0RAV
MK43 0SD G8XER
MK43 0WY M7BGR
MK43 7BB G4ABQ
MK43 7BG M0NGS
MK43 7BG M0PSR
MK43 7BG M5RPT
MK43 7DR G4CEC
MK43 7ED G3INY
MK43 7HJ M6EMQ
MK43 7HN G8HZG
MK43 7HN M3HML
MK43 7JL G3LDG
MK43 7JT M3MSJ
MK43 7JX G3SVQ
MK43 7LP G8ADY
MK43 7LT G6EDT
MK43 7NJ M0KGI
MK43 7QF G4PNK
MK43 7SG 2E0ACQ
MK43 7SG G0GPV
MK43 7TD G4XPJ
MK43 8JA G4MBA
MK43 8LF G4UMW
MK43 8LG G7TTP
MK43 8PR G4OZP
MK43 8QJ G3XYV
MK43 8QJ M3ZLE
MK43 9BB 2E0LUN
MK43 9BB G0XIT
MK43 9BB M6UBS
MK43 9BT 2E1SPY
MK43 9BT M3CLP
MK43 9BT M3NTI
MK43 9BT M5HOT
MK43 9DB M7INE
MK43 9DW G8BAK
MK43 9EZ G0DIM
MK43 9GG 2E0SIB
MK43 9GG M6SBZ
MK43 9JL G6SNV
MK43 9JW G3YUQ
MK43 9JW G6IEE
MK43 9LB G8XNN
MK43 9RA M7DZC
MK430JW M7NCN
MK431AJ M6OUS
MK44 1DY G1BUV
MK44 1EN M1EBS
MK44 1EX G6HMF
MK44 1JP G8IIG
MK44 1PL G0MOR
MK44 1SE G8ACQ
MK44 2AU G4FFC
MK44 2BA G0CZY
MK44 2DP M3XLB
MK44 2DT G4TGV
MK44 2ER G4WDZ
MK44 2ES M0SFR
MK44 2EW G4MSQ
MK44 2HP G4OCX
MK44 2JX G1PSH
MK44 2LA G4HUE
MK44 2LF 2E0HUR
MK44 2LF M6HUR
MK44 2PB M7PEV
MK44 2RP G4HTX
MK44 2RP G6JJT
MK44 3 LR 2E0XFF
MK44 3 LR M0ZSS
MK44 3 LR M6ESS
MK44 3BH G4TXG
MK44 3BQ G4RLR
MK44 3NJ M6TVC
MK44 3NJ M7LEL
MK44 3NT M0DLM
MK44 3QW G4ACP
MK45 1AG M7VIK
MK45 1AQ G8LLD
MK45 1BU G0COG
MK45 1GW M0WAS
MK45 1JY G6JZW
MK45 1LF G1JCC
MK45 1LF M0JCC
MK45 1LN M1AFP
MK45 1LN M3ADL
MK45 1PJ G8CTB
MK45 1QA G4OWN
MK45 1QD G6PHZ
MK45 1RU 2E0OEL
MK45 1RU M7OEL
MK45 1TB G0WZA
MK45 1TQ G7WAS
MK45 1XP M0OYQ
MK45 2AD G7PCF
MK45 2AE G0BVW
MK45 2AQ 2E1FDJ
MK45 2BT G4AHM
MK45 2DJ G0WAS
MK45 2EG 2E0VUM
MK45 2EG M7DGF
MK45 2EY G3VES
MK45 2JD M7CXL
MK45 2JY G8PBM
MK45 2QL G8VHX
MK45 2RS G7NBI
MK45 2RS M0CKA
MK45 2SP G0RNI
MK45 2SP G4FKI
MK45 2SP G8GPO
MK45 2TP G0JJK
MK45 3BU G1WDQ
MK45 3EF G6AJX
MK45 3GD G3UBJ
MK45 3LQ 2E0PSW
MK45 3LQ M0PSW
MK45 3LQ M3PPG
MK45 3PJ G8HRW
MK45 3QB G4USN
MK45 3WE G4MGX
MK45 3WE G4XQQ
MK45 3WH G7STG
MK45 4AS G8LQM
MK45 4BG G3FJE
MK45 4BG G4LOO
MK45 4DA G4VLA
MK45 4DU M7BDX
MK45 4FE G7HZQ
MK45 4FN M6NXH
MK45 4LN 2E1FDC
MK45 4LT M3IHQ
MK45 4NE M1EAJ
MK45 4NY M6BWF
MK45 4PF G4EIY
MK45 4TB G6BJO
MK45 5JN G0TFI
MK45 5JT G2FGT
MK45 5LB M3YQG
MK45 5LP 2E1GXE
MK45 5LP M3IEM
MK451LF M6WEN
MK454NE G4DAQ
MK46 4AR M6GJG
MK46 4DN G0FQU
MK46 4ER G4TXT
MK46 4EY G7IMB
MK46 5BJ G4NEO
MK46 5ES G0HMF
MK46 5ES M6BJX
MK46 5FB G0SXY
MK46 5HT G4LHP
MK5 6AS M3JSQ
MK5 6AS M3LNQ
MK5 6AZ 2E0MEV
MK5 6AZ M6MHV
MK5 6HT G0UVG
MK5 6JE 2E0BZO
MK5 6JE M3XPY
MK5 6JL G4ETK
MK5 7AF G3HIU
MK5 7AF G3ZPA
MK5 7AF G8MKC
MK5 7AJ G4CLG
MK5 7AX G4LNC
MK5 7HA G6DOI
MK5 7HB G8XXM
MK5 8BL G4NIV
MK5 8EB G7HRH
MK6 2DS M6MPP
MK6 2HG G1WVP
MK6 2HT 2E1CJN
MK6 2HT G0FTI
MK6 2QB M7SUZ
MK6 3AY G3LMX
MK6 3ES G6TTX
MK6 3JJ G4ETX
MK6 3LQ G7SYU
MK6 3LQ M5MK
MK6 3LU G0RDG
MK6 4HL M6SAE
MK6 4HX M6GUX
MK6 4LA M6KOZ
MK6 4LL M7XLX
MK6 5AE 2E1DAK
MK6 5DA M3HAW
MK6 5LB M6FIO
MK7 6BQ M7HTL
MK7 6JA M6NXQ
MK7 7EG G6VTE
MK7 7FH G3XVA
MK7 7HH G0HRT
MK7 7TB G8XKH
MK7 7TT M6LGQ
MK7 8DH M3XKI
MK7 8LR M0MUZ
MK7 8LR M3YZU
MK7 8PL G1OWZ
MK7 8QB 2E0RLW
MK7 8QB M0UKA
MK7 8QB M6RLW
MK8 0AT G7THZ
MK8 0BB G1OWJ
MK8 0DB G4TZR
MK8 0EJ 2E0MGL
MK8 0EJ M3GLT
MK8 0EJ M3LTA
MK8 0EJ M3MGL
MK8 8AY G1WLX
MK8 8AY M0GOA
MK8 8BX 2E0DHF
MK8 8BX M6EHM
MK8 9EN 2E0BQB
MK8 9EN M3VPB
MK9 1LR G8VQN
MK9 3AY 2E0NYG
MK9 3AY M0NYG
MK9 3AY M7NPC
MK9 4BA M6OSW
ML1 1BY GM1TNT
ML1 1LQ MM3TRZ
ML1 2LB GM0ARD
ML1 2RL GM6OXL
ML1 3JW GM4UBJ
ML1 3NA GM0WNR
ML1 3QU GM0HWB
ML1 3TH 2M0YRG
ML1 3TH MM3YCG
ML1 4BZ MM7WFL
ML1 4JL 2M0MMH
ML1 4JL MM3BQK
ML1 4RD 2M0FDZ
ML1 4RD MM3BPR
ML1 4UT GM7NKN
ML1 4ZL 2M0CRQ
ML1 4ZL GM1XOI
ML1 4ZL MM0RKN
ML1 4ZL MM0RWJ
ML1 4ZL MM6CRQ
ML1 5EF MM0SEK
ML1 5LB MM3XJA
ML1 5LQ GM0VXA
ML1 5NU GM4LQR
ML1 5RZ GM0EEH
ML1 5TY MM7SHT
ML10 6DL GM0SYV
ML10 6FW MM3SHT
ML10 6GN 2M0MCD
ML10 6GN MM3NZX
ML10 6LN MM7KNC
ML10 6SD MM6MNE
ML11 0FF GM0DQC
ML11 0HY GM6YRH
ML11 0NZ 2M0DSG
ML11 0NZ MM0TDB
ML11 0QA MM0MOT
ML11 7DU MM7DUF
ML11 7PS GM4IIR
ML11 7SE GM3YXY
ML11 8AP GM4HBG
ML11 8ES GM0VYY
ML11 8HA 2M0AIE
ML11 8HB MM0WHM
ML11 8LH MM3GYU
ML11 8NE MM3XIA
ML11 8SU MM1BJG
ML11 9AE GM3NNZ
ML11 9FY MM7IBS
ML11 9QA GM6BIG
ML11 9SR GM4VGU
ML11 9XS GM0GYN
ML110EL MM7CUH
ML118QW MM7KIE
ML12 6GJ MM0FCM
ML12 6HQ GM0SCA
ML12 6JZ MM6EQW
ML12 6LB GM7GIO
ML12 6NB GM0PJD
ML12 6NE MM6OFA
ML12 6PQ MM0THE
ML12 6PU GM4KKV
ML12 6RJ GM0DXI
ML12 6TP MM6FPY
ML2 0AH 2M0DQN
ML2 0AQ MM6AHB
ML2 0LP MM6JAW
ML2 7HG GM0WRU
ML2 7JX GM0LIR
ML2 7LL 2M0VGY
ML2 7LL GM5TDX
ML2 7LL MM7DTL
ML2 8BZ MM6BIY
ML2 8HR 2M0KVM
ML2 8HR MM6KCM
ML2 8PG 2M0SRX
ML2 8PG MM6SNR
ML2 8QR MM3VIS
ML2 8RA 2M0NAX
ML2 8RA MM6NAI
ML2 8SE GM0VXZ
ML2 9BL GM3TCW
ML2 9QL GM8KIQ
ML3 0AL MM7BBG
ML3 0HZ MM6PIB
ML3 6BF MM0KZA
ML3 6PD MM6CWN
ML3 6PN MM0RKT
ML3 6PP GM4UBF
ML3 7DD GM8NBV
ML3 7FB GM0STB
ML3 7FB GM4UQG
ML3 7FR MM3JLS
ML3 7HB GM0KDF
ML3 7HJ GM0ELP
ML3 7HJ MM0ELP
ML3 7HN 2M0RTD
ML3 7HN MM0RTD
ML3 7JY MM0GHN
ML3 7LL GM8LBC
ML3 7LZ 2M0JHN
ML3 7LZ MM0JZB
ML3 7PE GM4ILE
ML3 7PN 2M0JSB
ML3 7PN MM3JSB
ML3 7PW GM0MDX
ML3 7WS MM6TZX
ML3 8AY MM6LFN
ML3 8JS 2M0VLF
ML3 8LX MM1DME
ML3 8NZ GM4WFE
ML3 8PH MM6YHO
ML3 8QH GM0CSN
ML3 9EB MM7CTG
ML3 9HG MM0HQD
ML3 9JR MM6MCA
ML3 9SG MM6JJB
ML3 9UX GM4BVU
ML4 1JB MM0GGG
ML4 1LE MM6EEQ
ML4 2AU MM7HFC
ML4 2EQ 2M0RVF
ML4 2RT GM6PVA
ML5 1QT MM7YTR
ML5 2ES MM6NRK
ML5 2FA 2M1EAW
ML5 2HB MM7EIT
ML5 2PJ MM7SMG
ML5 2SX GM4CHF
ML5 3JD MM7DKY
ML5 4BJ MM6PDA
ML5 4FN 2M0CWV
ML5 4FN MM6DKN
ML5 4JQ GM4SZG
ML5 4LL MM0CEZ
ML5 4NE MM3VHM
ML5 5BH MM7CAN
ML5 5LR MM0ZYT
ML5 5LR MM3XWB
ML6 0ES 2M0RMP
ML6 0ES MM6WBU
ML6 0QQ GM4AUP
ML6 0QQ GM5DVS
ML6 6ET 2M0WLX
ML6 6ET MM6OWL

ML6 6GP MM6ZGS
ML6 6PA GM0VXQ
ML6 7DH GM1MMK
ML6 7DH GM8HBY
ML6 7JF GM7CPR
ML6 7JH 2M0EFD
ML6 7JH MM6KFJ
ML6 7NT MM6NED
ML6 7QB MM3VYA
ML6 7SZ 2M0BRE
ML6 7SZ GM0ART
ML6 7SZ MM7FKL
ML6 7TJ 2M0WXS
ML6 7TJ MM0WXS
ML6 7TJ MM6WXS
ML6 8BL GM4WTS
ML6 8FX GM0EEG
ML6 8QG GM0VWZ
ML6 8SF GM4PRO
ML6 8XL GM7SWX
ML6 9AQ MM6OIT
ML6 9DF GM3ZNC
ML6 9ND MM0GHT
ML6 9ND MM1XJS
ML6 9ND MM3KNY
ML7 4HP MM3NFX
ML7 4JA MM7OKO
ML7 5AX GM0PHW
ML7 5EL MM7APB
ML7 5HW MM6SFF
ML7 5TJ 2M0EGI
ML7 5TJ MM0WEI
ML7 5TJ MM3WEI
ML7 5TJ MM7CGS
ML8 4AF MM3VYU
ML8 4DY MM7GMW
ML8 4NR GM4ARU
ML8 4QT GM4XXO
ML8 4QY MM5AHM
ML8 4QZ 2M0MUR
ML8 4QZ MM0MUR
ML8 4QZ MM0RRM
ML8 4QZ MM6MUR
ML8 4RL MM7RBT
ML8 5GB 2M0ZEE
ML8 5GB MM6PJR
ML8 5HB GM3UCI
ML8 5HB GM4GM
ML8 5HB MM3UCI
ML8 5HB MM3VRI
ML8 5HR GM4COX
ML8 5HR GM4UXX
ML8 5JG 2M0ISA
ML8 5JG GM1OQT
ML8 5LT 2M0LAW
ML8 5LT MM3SUV
ML8 5PH GM4LAO
ML8 5RX MM3OQV
ML8 5TS GM8FHK
ML8 5UW 2M0ZBF
ML8 5UW MM6EZO
ML9 1EH 2M0XXP
ML9 1EH MM0XXP
ML9 1EH MM3XXP
ML9 1JX 2M0MMO
ML9 1JX GM3PXK
ML9 1JX MM0MOB
ML9 1JX MM3MAO
ML9 1QT GM3MXN
ML9 1RA GM3HVK
ML9 1UR GM1WYV
ML9 2DA GM7AOM
ML9 2LG GM3NKG
ML9 2TD GM4ISM
ML9 3AJ GM1DLS
ML9 3BB 2M0TGM
ML9 3BB MM6FPX
ML9 3BT 2M0VPU
ML9 3EN GM6HFH
ML9 3NR 2M0TOR
MS2 9BD MM3JNJ
N1 0HN M6PDB
N1 0HQ M0ZAY
N1 1QN G1GBX
N1 1TN G0TJD
N1 2BN M6CYT
N1 3NW M3ZUT
N1 4NU G0OOS
N1 4NU M0ESP
N1 7AR G0BQI
N1 7AY 2E0JQW
N1 7AY M0JQW
N1 7AY M6JQW
N1 8LX 2E1GPG
N1 8UA G4GJV
N10 1EG M0TPH
N10 1LX G3ASX
N10 2DE G3KOD
N10 2HS G4KPH
N10 2JU M0PCS
N10 3AA G4DFJ
N10 3PB G7FJU
N10 3PY M7DHA
N11 1AX 2E1HAM
N11 1AX M3GTA
N11 1RD M3FWA
N11 1TA M6DSH
N11 2AD 2E0DTY
N11 2AD M0OTO
N11 2AD M6LSU
N11 2DE G8YNC
N11 2JL G6FDX
N11 2JS M0IRO
N11 2JY G0NSO
N11 2LT M0ULC
N11 2RL G1IHA
N11 3HJ M0PKL
N11 3NN M6DLD
N11 3NQ G3GGI
N112AB M7SCF
N12 0LS G8XRP
N12 0NX G6ABJ
N12 7AD G8LSH
N12 7JG G4CCT
N12 7LD M7EMV
N12 8RB 2E0KSO
N12 9AU G1YSA
N12 9BG G4RIH
N13 4HD 2E1HWV
N13 4RG 2E1HXD
N13 5BS G6NTW
N13 5BT G3YTY
N13 5JS M6SJF
N14 4DH G0KUX
N14 4DH G1YJI
N14 4PG G1BPU
N14 4RP G4IEH
N14 4XN 2E1FMW
N14 5AT 2E1GW
N14 5AT M6ORT
N14 5RR M7SIS
N14 6QU G4BAN
N14 6RR G1DEY
N14 7NJ G6BLC
N15 4HH 2E0GSL
N15 4HH M0LAE
N15 4HH M6LXC
N15 4JN G8OJR
N15 4LA M6COI
N15 4SQ M6ATA
N16 6DJ 2E1PPK
N16 6DJ M3GKG
N16 6EP G0PIU
N16 9PH 2E0SDM
N16 9PH M6REI
N17 0JD G3YBS
N17 0TE G0CEU
N17 6BP G8ASC
N17 6EH G0FWF
N17 7HT G3OVX
N17 8DE M7ACX
N18 1QD M0TUK
N19 3DJ M3FWJ
N19 3HF G7LGI
N19 3SJ G7VDI
N19 4HR M7HDC
N19 5BQ G8WPA
N19 5NJ M0HPZ
N19 5NJ M6DYI
N19 5TR G7RJO
N193RA M7OVM
N2 0HP 2E0AQI
N2 0HP G0UKA
N2 0HP M0DAN
N2 0HP M1DAN
N2 0QB G7TWC
N2 0QB M0HWC
N2 0RD 2E0JSY
N2 0RD M0JSY
N2 0RD M7BMJ
N2 8AY M1KAZ
N2 8JT G7LZB
N2 8JW 2E0MCA
N2 9LJ M0KQN
N2 9LY M7PRN
N2 9NE M7ADO
N20 0AL G4XYK
N20 0DS G0ESF
N20 0HT G4LCB
N20 0HT M6FKU
N20 0QN G0EIQ
N20 8QH G1FLX
N20 9AP M7CKF
N21 1BA G6XIW
N21 1EH G3TIE
N21 1NP G3SFG
N21 1NP G4DFB
N21 1NP G6QM
N21 1QP G8DWF
N21 2BE G0RPM
N21 2NX 2E0FXU
N21 2NX M7NAC
N21 3AU M5AGZ
N21 3LP M0VKJ
N22 5BL M6OPA
N22 5NB M6NYA
N22 6JL M6VAR
N22 6LH 2E1FVJ
N22 6NT 2E1FVK
N22 6QD M0SKF
N22 6RG M6FCQ
N22 6RR 2E0VIS
N22 6RR M6GIC
N22 7AQ M7DUT
N22 7EX G4KKT
N22 7UT M7EAS
N22 7XG 2E0OJS
N22 8NN M3HMT
N22 8NN M3RGE
N22 8YQ M0KSR
N3 1NA 2E0YAQ
N3 1QL M1MRS
N3 1QL M3SMR
N3 1TJ M7PDZ
N3 2DD G0WCZ
N3 3HP G7UNW
N3 3NJ G0ODM
N3 3TX 2E0NAZ
N4 1HN G0PLC
N4 1LB M7BOK
N4 1RJ M6KKG
N4 1SQ M7SMC
N4 2LN M3XLK
N4 2ZT M0INF
N4 3DW G0JYU
N4 3NX 2E0YLZ
N4 3NX M7BAC
N4 3NX M7PIX
N5 0AD G4HOJ
N5 2DE G8ZQJ
N6 4BA G4GRS
N6 4BA G8MIC
N6 4BH M6SIK
N6 4EU G3XKY
N6 4QD G7PZM
N6 5AU G0RIC
N6 5PJ G0OOD
N6 5TS G4OJW
N6 6JR G4HDD
N7 0EL G1NWG
N7 0QB 2E0VNV
N7 0QB M6VNV
N7 0SH M0RXB
N7 0SN G0AKM
N7 7AY M3MUQ
N7 7EL M6LNS
N7 7FE M6MZB
N7 8HY G4IZU
N7 8NB M7PAU
N7 8TQ 2E0KCD
N7 8TQ M7LCD
N8 0JB M6ZIB
N8 0NA G4DVG
N8 7QU M7JCP
N8 7SJ G1WNZ
N8 8HD 2E0GES
N8 8NX G4KMM
N8 8QH 2E0CAO
N8 8QL 2E0RUI
N8 8QL M0RLM
N8 9BP G6RZR
N8 9PN M7WPL
N9 0HJ 2E1FNB
N9 0HJ 2E1HRM
N9 0HJ M3HRM
N9 0LN M7FAO
N9 7QG G1ALD
N9 8HD G4YGH
N9 8LJ G4ZIS
N9 8QL G1RFS
N9 9EQ G7COQ
N9 9HP G4UTR
N9 9HU G8SZZ
NE10 0DR 2E0RLJ
NE10 0EH M7RZE
NE10 8DD M7JMC
NE10 8QS M6ETG
NE10 8SW G7CIT
NE10 8UE G7JVG
NE10 8WJ M6GEE
NE10 8XB M7UPT
NE10 9AB M3YTZ
NE10 9BQ 2E1DRX
NE10 9DH G7EKG
NE10 9DT M6MSI
NE10 9NB G1OMX
NE10 9NH M7JBW
NE10 9ST M6DJK
NE10 9TZ G4VTN
NE11 0BS G0ESW
NE11 0HQ 2E0LLK
NE11 0HQ M6LLK
NE11 9BE G7PUA
NE11 9LA 2E1EDB
NE11 9XH G0CXX
NE12 6BT G3YRH
NE12 6YR M0NTK
NE12 7EU G1USF
NE12 7JP G3LRI
NE12 7JR M0AVU
NE12 8DP M7OPM
NE12 8LS M0JAQ
NE12 8LS M6DAM
NE12 8PH M6GMM
NE12 8XQ G0WUV
NE12 9AW G3LYG
NE12 9QP M6FHR
NE12 9RL G1XXF
NE13 6EY G6YTR
NE13 6QB G0LNW
NE13 7HS G0AOK
NE13 7HT G1GDA
NE13 7HW G0GYO
NE13 7NG G7VLL
NE13 9AP 2E0SHR
NE13 9AP M0WMG
NE13 9AP M7SHP
NE13 9EH M0LKT
NE15 0EA G8UZM
NE15 6LG 2E0MCN
NE15 6LG M6MPB
NE15 6QN G1YUN
NE15 6RZ G6SMI
NE15 7DN M6FOE
NE15 7DQ G1YDD
NE15 7LB M0IDI
NE15 7LN M6ADM
NE15 7LR G1SKQ
NE15 8BZ M7AAK
NE15 8NB G0WMX
NE15 8NB G8PVV
NE15 8QD M7SWD
NE15 8RL G3KTT
NE15 8TW G0DFT
NE15 9AT M0NRS
NE15 9EB G8XGS
NE15 9JL M7CML
NE15 9JU G0EOY
NE15 9LH M1DHC
NE16 3EJ M7JHD
NE16 3JN 2E1APX
NE16 3JN G0PUK
NE16 4PF G8DST
NE16 5PP M0ERN
NE16 5PP M1ETS
NE16 5PP M5ERN
NE16 5PT G1KYN
NE16 5PT M6LXK
NE16 5QS 2E0TSU
NE16 5QS M3XQL
NE16 5QZ G4GFV
NE16 5RK G0WUI
NE16 5SJ M0JGF
NE16 5SZ G6OSH
NE16 6LQ G0RFN
NE16 6QF M3WQL
NE16 6QP M7MSF
NE17 7AY 2E0LFF
NE17 7AY M7GPP
NE19 1AA M7GJG
NE19 1NX M1CCX
NE19 1RH G4HDS
NE19 1TA G4AOS
NE2 1BQ M7KTR
NE2 1HQ M6WDN
NE2 1XY M6FTP
NE2 2HD G4ETI
NE2 2JN G0BEV
NE2 3NS G2DJM
NE20 9AU 2E0KCN
NE20 9AU M6NQR
NE20 9AW G4DSD
NE20 9EJ G4SVE
NE20 9HQ G8OEO
NE20 9JB G8SCT
NE20 9PZ G7VTT
NE20 9QQ M0CIP
NE20 9QQ M1DPI
NE20 9RA G8KPD
NE20 9RA G8RER
NE20 9RD M0GPU
NE20 9RF M0EUK
NE20 9RR G4ITR
NE20 9RZ G1SUM
NE21 4FD M7DPU
NE21 4RR G0SWB
NE21 5QL M1CKU
NE21 6HJ 2E0DZO
NE21 6HJ M6HNV
NE21 6JU G7LOG
NE22 5BU G4TMQ
NE22 5DZ G1DDS
NE22 5ER G4UTQ
NE22 5HJ G6WJD
NE22 5HJ M3HLX
NE22 5TD G0AXJ
NE22 5TD M6AXJ
NE22 5TD M6SRS
NE22 6DF G4WNG
NE22 6DR G3OTH
NE22 6EJ M7PRA
NE22 6EX G4OAN
NE22 6HW G0OWE
NE22 6JG 2E0HGP
NE22 6JG M7GPH
NE22 6JJ G1DSB
NE22 6LD G4AXF
NE22 6NE G0DJO
NE22 6NS M1AQY
NE22 6PB G0KYR
NE22 7BR M6XSO
NE22 7EF M0ZYZ
NE22 7EH G7SNP
NE22 7EJ G0GLH
NE22 7EW M7ODD
NE23 1EY M7AJA
NE23 1NZ G6OTV
NE23 1RQ G7VBD
NE23 1TS G1ZJQ
NE23 2BP M1PJB
NE23 2BZ G6WUD
NE23 2UE G0TLZ
NE23 2UR M6SEU
NE23 3LE M3OBM
NE23 3LH 2E0PIH
NE23 3LH M7DVK
NE23 3LW G0BAU
NE23 3TT G4HJB
NE23 6AA G6AVL
NE23 6AS G0NHJ
NE23 6AS G0NHK
NE23 6DN G4HIH
NE23 6EX M3WQN
NE23 6QB M0GXW
NE23 6RW M6XEU
NE23 6SX G1SLI
NE23 6TU G7MYT
NE23 6TU M1DHG
NE23 6TW G0JRX
NE23 6UB G0LCE
NE23 6XQ 2E0CKW
NE23 6XQ M6BLH
NE23 7AF M6IYC
NE23 7BH G0PVT
NE23 7DE G0JWO
NE23 7EU M7TCF
NE23 7QF 2E0RLT
NE23 7RT G1TPO
NE23 7TR M7AIW
NE23 7TZ M6FTO
NE23 8HW G7OVK
NE23 8JB M6IFX
NE24 1NP G8EXK
NE24 1PG G0ECQ
NE24 2JR G4YAV
NE24 2JW M3PFE
NE24 2QW 2E0XRG
NE24 2QW M3XRG
NE24 3DH M1BMQ
NE24 3DH M0LRZ
NE24 3DH M6LRZ
NE24 3DT M6EAH
NE24 3DT M6EFY
NE24 3EP M1ARL
NE24 3HA G0LFE
NE24 3HA G6PNO
NE24 3HX G1LMC
NE24 3JG G1ATC
NE24 3JG G7PYR
NE24 3QN 2E0DTH
NE24 3QN M6WBH
NE24 3QS M0HNT
NE24 3RD M6YNY
NE24 4AZ M3PTR
NE24 4DS M0DAD
NE24 4DT G0HXC
NE24 4EB G0EVF
NE24 4JD M3LXP
NE24 4LB G0TWV
NE24 4PL M0TWO
NE24 4QP M0CVU
NE24 5AS M1EHZ
NE24 5ET M3DFW
NE24 5HL G4SIE
NE24 5PR M6TKN
NE24 5RH 2E0BMB
NE24 5RN G7KDX
NE24 5TS M6KVV
NE25 0AL M0MLY
NE25 0BU G0LNS
NE25 0EL 2E0DCV
NE25 0EL M6CLG
NE25 0FH G1EXK
NE25 0NG G0PKR
NE25 0NG G0VRT
NE25 0NP G3VYZ
NE25 0NP G8ARF
NE25 0PA G0PJU
NE25 0RN G8NSV
NE25 0SH G4VOU
NE25 0ST 2E0CMH
NE25 0ST M0HLR
NE25 0ST M6TEY
NE25 0TA G4BEZ
NE25 8AU G7PWL
NE25 8BJ G6CEM
NE25 8EP G4IZQ
NE25 8JG M0IGB
NE25 8JQ G7ELV
NE25 8NN M1CPP
NE25 8PR M6LHF
NE25 9AF M0WJW
NE25 9JS G6CCN
NE25 9QR G3WDS
NE25 9XH M3HSI
NE26 1AF G1ODJ
NE26 1QR M0USB
NE26 1SH G7VTJ
NE26 1SH M0VTJ
NE26 2JR G0OBQ
NE26 2NR G6PKM
NE26 3AD G8FVT
NE26 3BA G3GMS
NE26 3DJ G1MHA
NE26 3DY M0GAE
NE26 3EF G5BW
NE26 3SY M0LDX
NE26 4DB M3LFX
NE26 4HZ G1USN
NE26 4HZ M3USN
NE26 4JH 2E0UWT
NE26 4JH M3UWT
NE26 4JQ M3BMH
NE26 4PA M7GAP
NE26 4RE G3VUD
NE262DZ M3MKM
NE264RG M0HTX
NE27 0EF G1AIA
NE27 0EX M3JFF
NE27 0EY M7CBH
NE27 0UF G4RUI
NE27 0UX G6VEG
NE27 0UZ 2E0KBE
NE28 0AY M7NBZ
NE28 6SD M3SKX
NE28 6SN M3ENS
NE28 7HP 2E0XPC
NE28 7HX M7SSY
NE28 7PG M0HGV
NE28 7PH G3ZTJ
NE28 7SQ M0DLS
NE28 7SQ M6LBL
NE28 8TL M0ADR
NE28 9AA M0CBK
NE28 9HU G0JXY
NE28 9RX M0NKY
NE28 9YB G6RJU
NE28 9YW M0IIM
NE28 9YW M7EJN
NE280JH 2E0NEQ
NE280JH M6NEQ
NE29 0QN G4GWV
NE29 0QX G4FRD
NE29 0RW M7JSC
NE29 6UF M7JNH
NE29 6XE G0AIO
NE29 7AW G1FNY
NE29 7HP G1FNU
NE29 7JB M6REY
NE29 7QN 2E0HWC
NE29 7QN M6HWC
NE29 8DW G0AXO
NE29 8HS G0EDK
NE29 8JB M7DPZ
NE29 8LU G1LNR
NE29 8NT G3OZP
NE29 8QA G3PGC
NE29 8QG 2E0JMD
NE29 8QG M6JBO
NE29 8RS M7FLE
NE29 8SS G2ARY
NE29 8SS M0BQD
NE29 8ST 2E0LYB
NE29 8ST M0LYB
NE29 8ST M7ISM
NE29 9EN 2E0EFP
NE29 9EN G0MEF
NE29 9EN M6IGQ
NE29 9HT G0OBO
NE29 9NS G4AXL
NE29 9NU G8YCP
NE3 2AU G8UEE
NE3 2DU G6OQO
NE3 2FG G4LIA
NE3 2QQ 2E0LMK
NE3 2QQ M6SAQ
NE3 2UL 2E1HDB
NE3 2UL 2E1HDC
NE3 2UL M1MOB
NE3 2UL M1TXT
NE3 2XJ M7TRW
NE3 2XU M0PJF
NE3 2XU M3ZKF
NE3 3DB M6OOA
NE3 3LA 2E1NEW
NE3 3LA M7NEW
NE3 3PP M0PHP
NE3 3TN G1XYS
NE3 4BB G8QM
NE3 4NR G4LGH
NE3 4RN G0CQK
NE3 4RN G1GDM
NE3 4TT G1GAD
NE3 5BH G3LIV
NE3 5BY G3SQU
NE3 5DX G1VOP
NE3 5QD 2E1ICO
NE3 5TQ G3URE
NE30 1NA G8NSS
NE30 2BW G8YFA
NE30 3AN G8EZL
NE30 3DT M1DSE
NE30 3LH 2E0TDR
NE30 3LH G0JWM
NE30 3LH M6FVH
NE30 3TG 2E1EAK
NE30 4BJ M3WHX
NE31 1DH M6PKD
NE31 1HL M6CYE
NE31 1QU 2E0HAH
NE31 1QU M0MBT
NE31 1QU M6MRK
NE31 1RH G7MPJ
NE31 1XT G7JTI
NE31 2AQ M6OMT
NE31 2AQ M3SRT
NE31 2BW M3AHQ
NE31 2DW 2E0NFC
NE31 2DW M6NFC
NE31 2DY 2E0HUT
NE31 2GB M7WIM
NE31 2HN 2E0GTE
NE31 2HN M0HSC
NE31 2HN M0TYN
NE31 2HN M6GAK
NE31 2HN M6MNC
NE31 2HN M6NEE
NE31 2HN M6SLU
NE31 2HR 2E0MCM
NE31 2HT G4SFN
NE31 2JN 2E0PBN
NE31 2JN M6BPS
NE31 2LJ G0WZB
NE31 2LJ M6WZB
NE31 2TB 2E0GWC
NE31 2TB M6CKQ
NE31 2XS G1NZP
NE32 3QA 2E0KPC
NE32 3QA M0KPC
NE32 3QA M3GAG
NE32 3SB 2E0MTY
NE32 3SB M6FDR
NE32 3SB M6MTY
NE32 3SB M6SBF
NE32 3TX G0TFT
NE32 4AE G0DIP
NE32 4EY M6AEJ
NE32 4HS 2E0SBO
NE32 4HS M6PRN
NE32 4HX G3JDO
NE32 4HX M6BAG
NE32 4LP G7MWU
NE32 4QN M6JFJ
NE32 4QY M6KAS
NE32 4TN M0ISG
NE32 5DN 2E0MBW
NE32 5QB 2E0HHU
NE32 5QT G4WUM
NE32 5SS M7PNY
NE32 5UF 2E0TBL
NE32 5UF M6JWD
NE32 5YJ M1FEN
NE33 2AN G3SEG
NE33 2TH M3EVV
NE33 3EH G4LFG
NE33 3EL M6URT
NE33 3LQ M0BVF
NE33 4DL M6MRP
NE33 4HP G0DSX
NE33 4UE G3PRE
NE33 5HX G1NZL
NE33 5NH M1EEZ
NE33 5PU 2E0GFU
NE33 5RX 2E0IWN
NE333NT M7NEI
NE34 0BZ M1BEX
NE34 0DR G3KZZ
NE34 0EQ G3JPO
NE34 0JB 2E0KSW
NE34 0JB M6BSC
NE34 0JB M6KVN
NE34 0NJ G4XWR
NE34 0RA M0BWI
NE34 0RB M7VUK
NE34 0RR M6CMN
NE34 6AQ 2E0GJN
NE34 6AZ G0ZAT
NE34 6DP G7WLY
NE34 6QG M7BTG
NE34 6RY M0DEO
NE34 6RY M3VEX
NE34 7AH 2E0XUA
NE34 7AH M0XUA
NE34 7AH M7DRS
NE34 7AQ G6BIA
NE34 7DJ 2E0EQN
NE34 7DJ M6KZR
NE34 7DY M6FEU
NE34 7JF G6UWI
NE34 7JJ 2E0DKD
NE34 7JQ G4NS
NE34 7LP M6REO
NE34 7PT G7HSY
NE34 7QU G0WUH
NE34 7RF G1OHX
NE34 8DD M6BMV
NE34 8EJ G1XVR
NE34 8JT M6SWO
NE34 8NE 2E0PEM
NE34 8NE M6PEM
NE34 8NJ M0MGH
NE34 8NN 2E0XSJ
NE34 8NN M7HGV
NE34 8RU G6LIK
NE34 9AE M6PCU
NE34 9AJ M0RPO
NE34 9BA M6THV
NE34 9BU G8GIU
NE34 9EB 2E0YYG
NE34 9EB M6YYG
NE34 9JJ G0WWG
NE34 9JJ G7PHG
NE34 9JJ M3UJH
NE34 9JJ M6XTA
NE34 9NF M7ANK
NE34 9NQ 2E0DZC
NE34 9NQ G6MND
NE34 9NQ M6LFR
NE34 9NQ M6XLC
NE35 9BS M0EZO
NE35 9BS M0JDZ
NE35 9BS M0NER
NE35 9EP 2E0FAN
NE35 9EP M6PGP
NE35 9HA G4ZEU
NE35 9HU 2E0KBN
NE35 9HU M6KBS
NE35 9JE M7DYO
NE35 9LG G8JUC
NE35 9LR 2E0SID
NE35 9LR M0ZID
NE35 9NG G3XWU
NE35 9NH M3IWN
NE35 9NH M7RKW
NE36 0HJ G4VSK
NE36 0LE G8HI
NE36 0PL 2E1PDH
NE36 0PL M7AYU
NE36 0RF G0RCJ
NE36 0TD G8SEW
NE37 1AH 2E0FMT
NE37 1AH M7ACY
NE37 1HT M6PFW
NE37 1QD G0DHS
NE37 1QQ G4YEX
NE37 1UN M7AAL
NE37 2BT G4YDM
NE37 2EU 2E0YEY
NE37 2EU M7VIP
NE37 2JS G7LOY
NE37 2JS M6JIZ
NE37 2LT 2E0JEQ
NE37 2LT M7DDQ
NE37 2LZ M6KKA
NE37 3BZ G1LBU
NE37 3DE 2E0HHW
NE37 3DE G8SVR
NE37 3DE M7BHK
NE37 3DJ M6GBM
NE37 3LA G0EPP
NE38 0NR G1TWT
NE38 7AL G4WNI
NE38 7BW G7ITB
NE38 7DB G8VYQ
NE38 7EF G0BNK
NE38 7EF G0SCI
NE38 7EF M6YBV
NE38 7JN G0NSP
NE38 7TA G0OHK
NE38 7TQ M6OSC
NE38 8BP G4RTJ
NE38 8DD G1GTX
NE38 8DR G0UQP
NE38 8DR G4YGW
NE38 8DS G1HCC
NE38 8HS M3MVI
NE38 8PF G7NHE
NE38 8PF M3ABY
NE38 8RX G0ITM
NE38 9DE M0HIM
NE38 9EG G0HEU
NE38 9ET G1JDV
NE38 9ET M3ZDV
NE38 9HS M3XRV
NE39 1EQ G4NOX
NE39 1LT G2DML
NE39 1LT G3SEQ
NE39 2AD G3TKB
NE39 2DA 2E1AII
NE39 2HD G1VRC
NE39 2HF 2E1BCF
NE39 2JP M3PGY
NE39 2JQ M6MLX
NE39 2PN 2E0AJK
NE4 5RY M7BET
NE4 6UX M6NDI
NE4 7ER G1YJQ
NE4 7RG M6SQE
NE4 8BH 2E0CAT
NE4 8BH 2E0IGW
NE4 9AL M3YSN
NE4 9QY G8POG
NE40 3JN M6IKF
NE40 3RU G7KJP
NE40 4EU G3ZVM
NE40 4FE G7NTQ
NE40 4HR G0BGB
NE40 4QY G8CYW
NE40 4SY 2E0DUA
NE40 4SY M6WDI
NE40 4TD G4NSC
NE404QE M7BIE
NE42 5BJ M3OTP
NE42 5ER M6KGS
NE42 5PN G8SFA
NE42 6AT 2E1BUJ
NE42 6AT G0NEE
NE42 6EH G4STK
NE42 6JP G0EQC
NE42 6PZ G8RQF
NE42 6RA 2E0EQW
NE42 6RA M6JZL
NE42 6RB G7JTK
NE43 7PX G8YVP
NE43 7QR 2E0ZDM
NE43 7QR M0ZDM
NE43 7QR M6VID
NE43 7RQ G3UVU
NE437RL M6ISR
NE45 5BA G1SIO
NE45 5EX G8POO
NE45 5EX M5POO
NE45 5HB G4OEU
NE45 5HZ M7CAK
NE45 5JH G3TSS
NE455AP G1GLZ
NE46 1JH 2E0KZV
NE46 2LR M0TKF
NE46 2LR M7APC
NE46 2QB G6DNV
NE46 2QD G1UUF
NE46 3RY G3UKH
NE46 4BE M1SAB
NE46 4BN G4IWJ
NE46 4DE G0CNU
NE46 4LE G0LYC
NE46 4LU G1HZI
NE46 4LU G1TIK
NE46 4RX G1PIF
NE46 4SS G1MOB
NE47 5NA 2E0PHK
NE47 5NA M6PHK
NE47 6DE G0OYL
NE47 6ER M0CFB
NE47 6HG G0GXO
NE47 7EB M1DIN
NE47 9AS M0NPG
NE47 9DR G0AXZ
NE47 9EL G4UEO
NE48 1AQ G1SCB
NE48 1HF G0AFU
NE48 2RY G6ZIC
NE48 2RZ G3TTI
NE48 2SG G1KHY
NE48 2SL G8JSR
NE48 2SQ G1CGH
NE48 2TG 2E0IWY
NE48 2TG M7DDN
NE48 4BD G1TIF
NE49 0JL 2E0IKT
NE49 0JL M7CVC
NE49 9BS 2E0KFT
NE49 9BS M0JKF
NE49 9BS M6NBK
NE49 9LR G0BZB
NE5 1BT G8PHB
NE5 1SQ G7SCV
NE5 1TL 2E1IIP
NE5 2DL 2E0MEG
NE5 2DL M3TTK
NE5 2FA M3UAK
NE5 2JB G3XXQ
NE5 2NW G0DMW
NE5 2NW G0IOQ
NE5 2PA 2E0CSV
NE5 2PA M0HRW
NE5 2PA M6BXS
NE5 2PJ G3NIJ
NE5 3QB 2E0MUW
NE5 3RE M6SFJ
NE5 3SZ G7UVN
NE5 3XB G6JQE
NE5 3XL 2E0GBJ
NE5 3XL M6TXI
NE5 3YB G8ECZ
NE5 4BG M6MJS
NE5 4ED 2E0LJJ
NE5 4ED M6OJL
NE5 4PQ G0RMO
NE5 5AP 2E0TNN
NE5 5AP M0MAG
NE5 5EA G0WYY
NE5 5HS 2E0WDH
NE5 5HS M3WDH
NE5 5JQ M7GDQ
NE5 5PR M3YXP
NE54TD M7CON
NE6 1BH M6SNE
NE6 3BB M1DSU
NE6 3RY M0YVG
NE6 3TQ M7DDW
NE6 4DR G1EPD
NE6 4HA G7TPG
NE6 4HL G7UTG
NE6 4HL M0UTG
NE6 4JA 2E0TWX
NE6 4JA M7TWA
NE6 4SR G6JLL
NE6 4SR G8GFW
NE6 4TN G7DAR
NE6 5AD G4UOW
NE6 5JY G7FEE
NE6 5YA G3YVZ
NE61 1XF G8ONS
NE61 2AS G0WAD
NE61 2AS G7RXJ
NE61 2HJ G0DWO
NE61 2JT M3ILZ
NE61 2PL G3AGC
NE61 2PN G4NFV
NE61 2RU M0NCL
NE61 2SG G0EVV
NE61 2SG G6IIA
NE61 2SG G7SPP
NE61 2TF 2E0AOU
NE61 2TF G1BNN
NE61 2TH 2E0KXI
NE61 2TH M7KOL
NE61 2TQ G4ZTC
NE61 2XF G7PSL
NE61 2XY G7HHU
NE61 3LA G4KBX
NE61 3PT M7TUT
NE61 3RA G6OTW
NE61 3RB G4FXU
NE61 5ES M6EJC
NE61 5HF G7RWC
NE61 5HQ M6BEN
NE61 5JU G4DGQ
NE61 5LN M1JPS
NE61 5LT M6JLY
NE61 5NT M3XJK
NE61 5PU G4ZTA
NE61 5QW G0KNW
NE61 5RA 2E0HOX
NE61 5RA M7HOG
NE61 5RB G4AAX
NE61 5RB G8PNN
NE61 5RP 2E0XEW
NE61 5RP M0XRW
NE61 5RP M6CQR
NE61 5SP M1DGW
NE61 5SR 2E0EKI
NE61 5SR M6ZBM
NE61 6FN 2E0KNL
NE61 6FN M6KNL
NE61 6LQ G3BIK
NE61 6LT G4DJJ
NE61 6PE M0PGW
NE61 6PE M1DPW
NE62 5BU M7FKK
NE62 5DA G4GWB
NE62 5DN G1NOS
NE62 5LD G4ZOY
NE62 5NG M1ULD
NE62 5NN M3GCT
NE62 5UE G4URW
NE62 5UF G0WZG
NE62 5UF M3HJJ
NE62 5UL G0KUI
NE62 5UT 2E0SCM
NE62 5UT M6SBW
NE62 5XA M6SJN
NE62 5XF G3ZKG
NE63 0EU 2E0CVZ
NE63 0EU M6CVE
NE63 0HT M6ACJ
NE63 0LZ G4OIV
NE63 0PL G7FFR
NE63 0PQ M6KWP
NE63 0XH G7MWK
NE63 8AB M6EGS
NE63 8BQ G0EFG
NE63 8EF 2E0SCO
NE63 8EF M3WLU
NE63 8HX G1LMZ
NE63 8HX M0GEC
NE63 8JF G0CQH
NE63 8JF G7QYS
NE63 8JT G8VIB
NE63 8JX M0VKX
NE63 9BB G7VDA
NE63 9BB M0BMJ
NE63 9BB M1EIZ
NE63 9EN M6SIE
NE63 9LU M0WBB
NE63 9QJ G6JXS
NE63 9RY G0VRS
NE63 9RY G0WYZ
NE63 9SD G7VDV
NE63 9SE G0EMF
NE63 9TJ M0AHZ
NE63 9TJ M3VQS
NE63 9TP G3KLP
NE630TD 2E0NHR
NE639EZ M6MVE
NE63RU M7MLK
NE64 6BQ M0VHC
NE64 6DE M1DLS
NE64 6DQ M3RPK
NE64 6JH 2E0HPE
NE64 6JH M0JWZ
NE64 6LH 2E0LFM
NE64 6ST G0UDZ
NE64 6XN G0VAD
NE64 6XT G0LEF
NE64SY G3ZWR
NE65 0AP G2FXV
NE65 0EL G8GVN
NE65 0ER G3OGH
NE65 0FR M6JAF
NE65 0GB G7OWP
NE65 0JS G4BCP
NE65 0SQ 2E0KXH
NE65 0SQ M7KHX
NE65 0TF 2E0BVD
NE65 0TF M6ALB
NE65 7HF M0CKG
NE65 7PU M0ZDD
NE65 7QN 2E0NJE
NE65 7QN M3NFG
NE65 7QS G4WMQ
NE65 8JT G3VKU
NE65 8LD M7TAZ
NE65 8QH G1ODT
NE65 8UN G1JKX
NE65 8UR G1BBY
NE65 8UW M0AHR
NE65 9DB M1ZXG
NE65 9JL G6CBL

NE65 9JN G1RKD
NE65 9JW G1DNA
NE65 9JW G1HAB
NE65 9NB G7DEF
NE65 9NG M7BKJ
NE65 9QW G8MOG
NE65 9SP M0SSE
NE65 9TF G8YSH
NE65 9YA M1VHT
NE66 1BS G6KXB
NE66 1DW M3WDN
NE66 1EU M3MES
NE66 1RD 2E0NTM
NE66 1RD M7CTM
NE66 1RZ 2E0MUN
NE66 1RZ M6APM
NE66 2QP G7ANV
NE66 2QP M3YOT
NE66 2QP M7SJG
NE66 2YD M0DSS
NE66 3AT G1AHT
NE66 3AT G3NEP
NE66 3AW G4KRH
NE66 3DD M7NCM
NE66 3JE G3ZDU
NE66 3JQ M6HOS
NE66 3JZ G1YAE
NE66 3LS M1CCY
NE66 3NT G1GYM
NE66 3NT M0GYM
NE66 3PJ G4EKF
NE66 3RL G7VKA
NE66 3XY G2DJ
NE66 3XY G3VGW
NE66 4YD M0FLC
NE67 5AW G4FCC
NE68 7TB M6TSM
NE68 7TJ G3AQB
NE68 7XP G3TEP
NE68 7XR G1VVU
NE7 7AP G4KVK
NE7 7BX G4WMA
NE7 7DJ M7CYA
NE7 7DJ M7PYA
NE7 7FB G1XSA
NE70 7PA M3NGM
NE71 6AF M6SBN
NE71 6DB G4LUF
NE71 6JL G0AVU
NE71 6QG 2E0TQF
NE71 6QG M3TQF
NE716QG M3JLF
NE8 1NQ G7SZB
NE8 1RU M6YCH
NE8 1US 2E0DZS
NE8 1US G7MWB
NE8 1US M0MBA
NE8 1US M6DZS
NE8 2BG 2E0HWN
NE8 2BG M3HWN
NE8 2JB M0GRI
NE8 2SY M7BRY
NE8 3RS M0LBD
NE8 4AJ G0NPQ
NE8 4DX G7SPN
NE8 4DX M0GGP
NE8 4DX M3NLA
NE8 4JA G7KLS
NE8 4UH G6JUP
NE8 4XH M3IBE
NE9 5DP G4WAX
NE9 5EU G8PDE
NE9 5PY M3XWK
NE9 5SB M0SUY
NE9 5TX G8XLZ
NE9 5XL M6RFL
NE9 5YN 2E0MWH
NE9 6DA G3IPD
NE9 6DH G0JHU
NE9 6JJ 2E0AOS
NE9 6JJ 2E1EYF
NE9 6NP M3XNN
NE9 6QN 2E0TTY
NE9 6QN M6ADE
NE9 6SN M6BGC
NE9 6TN G4ILW
NE9 6TU 2E0EAU
NE9 6TU M6AQN
NE9 6TZ G0SQE
NE9 6UX G0HBO
NE9 7DX 2E0XQK
NE9 7DX M3XQK
NE9 7EX G7VDK
NE9 7JD G7MKB
NE9 7NQ G1DZY
NE9 7PG G0SLQ
NE9 7RE G6UJJ
NE9 7TR G0NPP
NE9 7TY 2E0CRN
NE9 7TY M6SRF
NG1 5BL M0IDW
NG1 8 3BY G1KGQ
NG10 1DR 2E0DMM
NG10 1DX G6DRH
NG10 1GB M6BYN
NG10 1JH M6GTU
NG10 1JH M6HZJ
NG10 1NL G3REU
NG10 2BS 2E1HID
NG10 2BY G0CQR
NG10 2DZ G0FFQ
NG10 3BT G6AGR
NG10 3DG 2E0SMA
NG10 3DG M6WXN
NG10 3DJ 2E1AKK
NG10 3EW G1OPG
NG10 3FP M3IDO
NG10 3GF G0PSI
NG10 3GF G4AL
NG10 3GF G6MDR
NG10 3GG M3VBV
NG10 3GH G4OSR
NG10 3LF 2E0USM
NG10 3LF M0JWD
NG10 3LF M6USM
NG10 3NA M7BOZ
NG10 3NL 2E0STT
NG10 3QE 2E0BQW
NG10 3QE M3SVP
NG10 3RF G8XJL
NG10 3RG G8PTN
NG10 3RJ 2E1DWW
NG10 4DA G4IVO
NG10 4DD 2E0VKM
NG10 4DD M0OSL
NG10 4DD M6OOL
NG10 4DD M6YMB
NG10 4EB M3IZJ
NG10 4EW G6PVW
NG10 4GD 2E1GMQ
NG10 4GD M0BTG
NG10 4JA G7TZW
NG10 4JS G5OW
NG10 4NZ G4TYP
NG10 5BS G0IVI
NG10 5EF G4HIC
NG10 5FF 2E0MLV
NG10 5LQ 2E0XCH
NG10 5LQ M6BSV
NG10 5PD G4LOV
NG103GG 2E0NAF
NG103GG M3NAF
NG103GN M7RUZ
NG11 0HP G4BKQ
NG11 0HT G8LCU
NG11 6AA G3ZQH
NG11 6AJ G4AJJ
NG11 6GF G0PIB
NG11 6GG G0JYE
NG11 6LB G0KUZ
NG11 6ND G8GWP
NG11 6NP 2E0FIV
NG11 6NP M6OQT
NG11 6NS G7KNQ
NG11 6QD G4NNY
NG11 6QE G7HZZ
NG11 6QE M0PHX
NG11 6QE M0VRG
NG11 6QE M0YHA
NG11 6QJ G1LOL
NG11 7AU G4LPD
NG11 7AU G8HUR
NG11 7BE G0JPL
NG11 7BQ G7HUK
NG11 7BY M1ECT
NG11 7EB G8DD
NG11 7EB M0BWY
NG11 7FD G8LNG
NG11 7FD M6JRH
NG11 8FD G1PNX
NG11 8GF G0WXF
NG11 8GF G7SWE
NG11 8GN M6TLC
NG11 8JA M1BHP
NG11 8JG G7SMQ
NG11 8JN 2E0PJT
NG11 8JN M0PJT
NG11 8SL G0JVW
NG11 9AL M0BCW
NG11 9ET G7BGY
NG11 9HB M7TZW
NG11 9HE 2E1HXB
NG11 9HE M3HXB
NG11 9JB 2E1GHI
NG11 9JN G0LXX
NG11 9JZ M0BWW
NG11 9NE M7FYQ
NG12 1AY G0SGF
NG12 1AY G7ITW
NG12 1DG G4ZDF
NG12 1DP G1HTR
NG12 2BU G2BXH
NG12 2BU G6KDY
NG12 2BX G3YCE
NG12 2GA M1EJI
NG12 3DG M3UBG
NG12 3DJ G6OAN
NG12 3DP G1GSG
NG12 3FD G0MXX
NG12 3HT G7LFL
NG12 3JJ G6AFS
NG12 3JS G0MAY
NG12 3JY G3TNX
NG12 3NE G3FRE
NG12 4DN G0NPC
NG12 4DZ G7PVQ
NG12 4EA G4XRB
NG12 4ES M0BUY
NG12 4FW 2E1CEU
NG12 5BN G8ENB
NG12 5DA 2E1AEC
NG12 5DN G7ODB
NG12 5ET G0GWP
NG12 5HQ M1AZV
NG12 5JX G6PKY
NG12 5LQ G3KZX
NG12 5LQ M0RIA
NG12 5NX G3SJJ
NG12 5NX G4BKI
NG12 5PY G1ZLB
NG12 5RA G4XZA
NG12 5RA M3WTB
NG13 0AR G4WFK
NG13 0ED 2E0BUZ
NG13 0ED M3ZUB
NG13 0FP M6GJL
NG13 8NA M6EQJ
NG13 8QD G7JDI
NG13 8RL M0RMJ
NG13 8RL M1KEJ
NG13 8TY G4FAB
NG13 8YR M6FVN
NG13 9AF G0FOG
NG13 9HD G4OMJ
NG13 9HL M6ZWB
NG13 9HY G4TTS
NG13 9HZ G0USJ
NG13 9JF G6XSS
NG130AZ M6WGG
NG14 5AA G4PFF
NG14 5AZ G7DPV
NG14 5DB 2E0EOZ
NG14 5DB M0JOJ
NG14 5DB M6JUF
NG14 5DY G8KBM
NG14 5EH M3LVR
NG14 5FG 2E1CWQ
NG14 5HR G1FYU
NG14 6FL G8DVN
NG14 6HL G4IJU
NG14 6HY 2E0LRA
NG14 6NT 2E0VUB
NG14 6NT M0VUB
NG14 6PU G7TBP
NG14 6QA G6SCG
NG14 7AR M5ROB
NG14 7DJ M0GDP
NG14 7EF G8FGZ
NG14 7GW G1XOW
NG14 7HD G7BGT
NG15 0EG G8YLS
NG15 6DQ G4KJA
NG15 6DQ G4XSS
NG15 6ED M6SPL
NG15 6EF G4WXR
NG15 6FF G6DDP
NG15 6FU M0HCZ
NG15 6FU M6CGY
NG15 6FU M6HUC
NG15 6FY 2E1GDM
NG15 6GE M6LHP
NG15 6GN G4ZII
NG15 6HY G4EPL
NG15 6JW M0SCP
NG15 6RF M3MBH
NG15 7FR G0VNW
NG15 7PJ M6BHC
NG15 7QA G4JSM
NG15 7SJ M3EVC
NG15 7SL 2E0ETG
NG15 7SL M6CCA
NG15 7SR G6NHY
NG15 7TF G6ORH
NG15 7UB G0RGE
NG15 8BS M0JYQ
NG15 8BS M0JYS
NG15 8EB G0CEV
NG15 8EU 2E1CJB
NG15 8JA 2E1GSC
NG15 9AD G8MMP
NG15 9BE G4CMX
NG15 9DG G8EXS
NG15 9EA G0LUD
NG15 9GN M7GJT
NG156SA M7SKS
NG16 1AB M7RJD
NG16 1AU 2E0DME
NG16 1AU M0JEJ
NG16 1EG G0EVN
NG16 1HE G0FJZ
NG16 1HE G1NBT
NG16 1HL M5AAW
NG16 1LE G4NUV
NG16 2AP M0GGH
NG16 2DP G8XPZ
NG16 2EN G3VYK
NG16 2FG 2E0NSC
NG16 2FG M0LAF
NG16 2FT M7SVE
NG16 2HA 2E0UVW
NG16 2LQ G1PMK
NG16 2PU M6BAN
NG16 2QX G7ODN
NG16 2RA G4LQP
NG16 2RD G7UII
NG16 2RH M0ANC
NG16 2TH 2E0FCS
NG16 2TH G6UFV
NG16 2TH M6FKZ
NG16 2UB M1EBW
NG16 3DL G6GCW
NG16 3DP G4JRJ
NG16 3DQ G8ZAU
NG16 3DR G0GYH
NG16 3DY G6HZX
NG16 3FR 2E0EHH
NG16 3FR M6NTR
NG16 3FY G1WSD
NG16 3GW G0BXG
NG16 3GW G7HHM
NG16 3GY G7HGF
NG16 3HR M6OQL
NG16 3JE G4RNR
NG16 3JW M3YRB
NG16 3LR M3AZH
NG16 3NL G4VAX
NG16 3QL G3RTO
NG16 3RB M6GTC
NG16 3RE M0ZCW
NG16 4FX M7WLD
NG16 4GJ 2E1IJY
NG16 4GJ G6MAW
NG16 4GP G6ZAF
NG16 5BG G4UFC
NG16 5EH G1OJQ
NG16 5EH G6DJQ
NG16 5FN G0OBK
NG16 5GX G0LIQ
NG16 5HJ 2E1EVH
NG16 5JZ G0WXH
NG16 5JZ G7UVL
NG16 5PW G0WFO
NG16 6BQ G8JGF
NG16 6DX G4ROB
NG16 6DX G8RYK
NG16 6FA 2E1BJG
NG16 6FN G0CSS
NG16 6FW M3UUL
NG16 6GN M6YDC
NG16 6GP G4NXB
NG16 6JX M0DEN
NG16 6LH G1ZAR
NG16 6LQ G7BHU
NG16 6QH 2E0SBJ
NG16 6QQ M0MRT
NG16 6RF G0IUV
NG16 6RJ 2E1GRA
NG17 1EP G4ZUC
NG17 1HA 2E0VCB
NG17 1NJ M0NSK
NG17 2BS M6RYK
NG17 2BU 2E0BEQ
NG17 2DD M6WDL
NG17 2DG M6KCE
NG17 2DW G0PDM
NG17 2FB G4DFZ
NG17 2FH M3FWR
NG17 2GG M5ADF
NG17 2HS G3JFD
NG17 2LL G6ZTH
NG17 2NP M6XCC
NG17 2QA G1OXT
NG17 2QD G1XVF
NG17 2QF G0JDG
NG17 2QF G0RGU
NG17 2RA 2E0CLL
NG17 2RA 2E0XBX
NG17 2RA G4TSN
NG17 2RA M0TSN
NG17 2RA M3TSN
NG17 2RE M6ISX
NG17 2SB G6SGU
NG17 2SJ G8XGW
NG17 2SJ M7VMB
NG17 3AB G0UJD
NG17 3AP G4UDN
NG17 3AR M3VLN
NG17 3BW M0AJX
NG17 3DP M5ABH
NG17 3DP G0IUN
NG17 3EB M0AGA
NG17 3EJ G8EPH
NG17 3FF G6ROS
NG17 3FF M6NNC
NG17 3FL M6KTN
NG17 3FT G0BCF
NG17 3HT G4REU
NG17 3JY M6GQR
NG17 3NQ G4TWW
NG17 4AY G1XNG
NG17 4AZ 2E0EOF
NG17 4BA M6AET
NG17 4BE G4HCD
NG17 4BX G7RXK
NG17 4BY G0UTN
NG17 4EH G0LHL
NG17 4EN G3RLJ
NG17 4EY M6TCF
NG17 4EZ G7PBV
NG17 4EZ M0HQG
NG17 4GA G1RBH
NG17 4GG M0XWD
NG17 4GG M6JVJ
NG17 4HQ G6NWS
NG17 4HX G7PHL
NG17 4JA M3EZY
NG17 4LG 2E0FKN
NG17 4LL M6KFF
NG17 4NL G6NWN
NG17 4PR M0KPF
NG17 5AR M0GOH
NG17 5AX M3ZGI
NG17 5BA M6NOC
NG17 5BA M7LMC
NG17 5BD G7INC
NG17 5BL M6CZA
NG17 5EH M3ZGF
NG17 5GD 2E1SCR
NG17 5GD M7SCR
NG17 5GH G3VDF
NG17 5HP G1SIU
NG17 5HU G0GZO
NG17 7EH G1VHY
NG17 7FH G1RJD
NG17 7GE G1KAS
NG17 7GF G0BTV
NG17 7HB M0DYQ
NG17 7HF G6OZH
NG17 7JE M6AEY
NG17 7JW M0CWZ
NG17 7LY G3SQQ
NG17 7QB M3KHE
NG17 8AD G7JZC
NG17 8BA M6YJB
NG17 8BT G0JPZ
NG17 8DP M0BNF
NG17 8ED M0CVO
NG17 8EJ G7LHT
NG17 8FP M3HDL
NG17 8FR M6AGZ
NG17 8FU M3JZN
NG17 8FU M3KCU
NG17 8FX 2E0IQO
NG17 8GE M0SRO
NG17 8GF G6UIP
NG17 8GH M0RMF
NG17 8GP M6FZA
NG17 8JJ G4VOG
NG17 8JT G0DMN
NG17 8LH G0NRA
NG17 8LH G0NZA
NG17 8NW G6UDB
NG17 8QB G7VJQ
NG17 8RL M7GTX
NG17 8RS G3LNN
NG17 9AH M3JZO
NG17 9AH M3JZP
NG17 9BG G6XNN
NG17 9BN G6XYR
NG17 9BR M0RBB
NG17 9BU M0BAE
NG17 9DN G1EZU
NG17 9DY G0CVB
NG17 9ET G8ZUZ
NG17 9FE M6NEM
NG17 9FL 2E1CQM
NG17 9HE 2E0UBN
NG17 9HE M6UBN
NG18 1QE M3SQE
NG18 2DG M6PTI
NG18 2HP 2E0JLK
NG18 2HP G7TOY
NG18 2HP M6RRC
NG18 2LQ 2E1FTE
NG18 2LQ M1BVP
NG18 2NF M3NOY
NG18 2NF M6XDX
NG18 2PX G0HRO
NG18 2QL M6RZI
NG18 3AZ M3HNV
NG18 3BL M3FLE
NG18 3DD G0PGA
NG18 3DU 2E0JNE
NG18 3DU M7GIF
NG18 3EN G4NOR
NG18 3JJ G4VWA
NG18 3JL G4LBY
NG18 3NP G0JRC
NG18 3NP G0KJC
NG18 3QP G0JVK
NG18 3RN 2E0FJQ
NG18 3RN M6PGU
NG18 3RZ G1DRR
NG18 3SP M1CBH
NG18 4EQ G4ZWI
NG18 4ER G4DFV
NG18 4FA G1EIV
NG18 4FB G7ROI
NG18 4HD G1LPQ
NG18 4HF G4AAH
NG18 4HF M3CSM
NG18 4HG G0VYT
NG18 4NB G0RWW
NG18 4PG G6RYW
NG18 4QD G3XDS
NG18 4RS M7BNP
NG18 4RT G4GZU
NG18 4XS M6CZF
NG18 5EE G3GQC
NG18 5EE M0DFX
NG18 5EE M1ARF
NG18 5HY G4LQP
NG18 5LA G4HTO
NG18 5LL M5DLA
NG18 5NB G0RDP
NG18 5NG G1ECS
NG18 5QS G4TGB
NG18 5RG G7VKG
NG18 5RG M6VKG
NG19 0AJ M6TLB
NG19 0BX G7GGG
NG19 0DS 2E0BSN
NG19 0DS M3ZEY
NG19 0DW M0CFD
NG19 0EJ G3VIY
NG19 0EY G1HLT
NG19 0GP G1MMD
NG19 0HZ G0ELB
NG19 0JP G4TUX
NG19 0LR 2E0GBV
NG19 0LR M0HOB
NG19 0LR M6CIU
NG19 0LS G4LQL
NG19 0NA 2E0RHP
NG19 0PP G0TUP
NG19 0QJ 2E0DAO
NG19 0QJ M3NZR
NG19 6BY G4VUM
NG19 6EG G1ZLY
NG19 6EL M6TRD
NG19 6HZ G7AVZ
NG19 6JU M3HNQ
NG19 6LS M6NGP
NG19 6NB G4ANU
NG19 6NN M7HOU
NG19 6NT 2E0EVA
NG19 6NT M6EVA
NG19 6QH M7CBL
NG19 6QQ G4ZSD
NG19 7JB M6NHT
NG19 7JW M3LBY
NG19 7LX M3CDE
NG19 7NP 2E1EIX
NG19 7NP M3OQG
NG19 7PY G6JFL
NG19 7QQ 2E0MLF
NG19 7RG 2E0SWN
NG19 7RG G0RRZ
NG19 7RG G6EAH
NG19 7RG M0SXN
NG19 8AZ G1FEX
NG19 8DJ M3CBV
NG19 8HT G0GAG
NG19 8PA G1UK
NG19 8QJ M3KAN
NG19 8QT G1IMY
NG19 8QT G1JGY
NG19 8QT G4BAS
NG19 8QU 2E1EYC
NG19 8QU M3MKD
NG19 8QZ G4PPH
NG19 8TL G4OIE
NG19 9AB M7BNZ
NG19 9AZ G0OYP
NG19 9DW G1WTB
NG19 9HA G7CKL
NG19 9HD M7DON
NG19 9JR G6XMB
NG19 9NB G4RPD
NG19 9PJ M3GJW
NG190NW M0IYJ
NG2 2EB M0DSN
NG2 3GD M0GLP
NG2 4AE G6RFJ
NG2 4AS M6PEG
NG2 4BS 2E0FIU
NG2 4BS M6NWN
NG2 4GF M0JOH
NG2 4LN G7FCC
NG2 4NY G0HMO
NG2 4QD 2E0BRP
NG2 4QD M3WWU
NG2 5BD G7HCJ
NG2 5DS G1TTB
NG2 5FU G1HXP
NG2 5GB G4VOW
NG2 5GB M3GCP
NG2 5GB M3JVP
NG2 5HA G7OLX
NG2 5LA M0AVH
NG2 5LX M1SCW
NG2 5NA G4NDM
NG2 6EB G3WNC
NG2 6FH G4ZSO
NG2 6FQ G3VLN
NG2 6GJ G4VUI
NG2 6HQ 2E0BBX
NG2 6HQ M0GWR
NG2 6HQ M3VJS
NG2 6HQ M3XHC
NG2 6LD M0SMH
NG2 6LD M6BMY
NG2 6LD M6SKL
NG2 6NA M6LOL
NG2 6PE G8ZSD
NG2 6PE M3ZSD
NG2 6QG G8STJ
NG2 6SG G0RMN
NG2 7FD G0TCF
NG2 7FL G8KJJ
NG2 7GG G4NZU
NG2 7GN M1FDH
NG2 7LU G8GVL
NG2 7ND 2E0XIK
NG2 7ND G3MOT
NG2 7ND M0XIK
NG2 7ND M3PMI
NG2 7ND M3XIK
NG2 7ND M6WRG
NG2 7QN G6RNF
NG2 7QR M6GIY
NG2 7RX M6JRW
NG2 7UQ G8TVM
NG20 0BN 2E0MIB
NG20 0BN M0MIB
NG20 0BN M3VUZ
NG20 0BP G7TTY
NG20 0DJ 2E0SPN
NG20 0DJ G0TLT
NG20 0DJ M6HKA
NG20 0DN G7MNQ
NG20 0HJ 2E1FOU
NG20 0LD 2E1HWQ
NG20 0ND M6DEE
NG20 0PS 2E0DNJ
NG20 0PS M6CBN
NG20 0PS M6TAJ
NG20 0PT G0CAL
NG20 0PT M6LCP
NG20 0PY M6NCB
NG20 8AZ G8UST
NG20 8AZ M6EJB
NG20 8BJ M0DHI
NG20 8EQ G6JCV
NG20 8GA G0UKL
NG20 8JS 2E0XKS
NG20 8JS M7AVC
NG20 8LF 2E0SGI
NG20 8NH 2E0CAS
NG20 8NH M0TTY
NG20 8PJ M0ROY
NG20 8QH G4VLK
NG20 8XN 2E1WPW
NG20 9EB G4ODF
NG20 9PJ M0IFH
NG20 9PJ M3KBF
NG20 9PJ M3NHD
NG20 9PY G0ENV
NG20 9QN 2E0VKB
NG20 9QN M6WKW
NG20 9RU 2E0JJS
NG20 9RU M7OJS
NG21 0AR G4FHK
NG21 0AR G8AOI
NG21 0AU G7KWA
NG21 0AX G7BXG
NG21 0DS M0TCA
NG21 0FB 2E1HGY
NG21 0FB M0BPY
NG21 0FU M0JFB
NG21 0GD G4WHM
NG21 0GH 2E0CVD
NG21 0GH M6TBR
NG21 0HX G7LAK
NG21 0JJ G6SFE
NG21 0NA M6TRW
NG21 0NW G8KPV
NG21 0RN G7SVU
NG21 0RW 2E1IDE
NG21 0SJ M5TMG
NG21 0TL G7PMF
NG21 0TU G0PBW
NG21 0TU M6PBV
NG21 9AE M7XSB
NG21 9AH 2E0ELE
NG21 9AH M6HWE
NG21 9AZ M7BEG
NG21 9DS G1JXG
NG21 9EA G0CCV
NG21 9LE M1AEB
NG21 9LQ G4WHL
NG21 9NH G0SYS
NG21 9NH G1SIX
NG21 9NJ M6JNE
NG21 9NZ M3KOR
NG21 9PF 2E0IUW
NG21 9PF M7CJX
NG21 9QG 2E0BWX
NG21 9QT M1ANR
NG21 9QZ G0BYX
NG22 0DA G7CTN
NG22 0HB M3BIR
NG22 0HF G4ODD
NG22 0HP M6MEK
NG22 0JR G6ZOB
NG22 0JR M3KBG
NG22 0PN G7PFD
NG22 0PP G4TDZ
NG22 8AP G4SCL
NG22 8DQ G1WRO
NG22 8LU G8BRG
NG22 8QX G4KXG
NG22 8QX G8RAV
NG22 8SD G3ZVV
NG22 9AS G7ACA
NG22 9BD 2E1HVB
NG22 9DG M0AXX
NG22 9DX G8YPQ
NG22 9HP M0HHF
NG22 9PQ 2E1FBA
NG22 9PS 2E0HVM
NG22 9PS M0SDF
NG22 9PS M6HVM
NG22 9PU 2E0AVA
NG22 9QX 2E0TBD
NG22 9QX M3UIS
NG22 9QX M6TRV
NG22 9QY 2E0SHA
NG22 9QY M6JLR
NG22 9RN 2E1FRZ
NG22 9RW M6DXV
NG22 9RZ M0ADB
NG22 9SG 2E0KOR
NG22 9SG M3KOU
NG22 9SX 2E0FBF
NG22 9SX 2E0HUN
NG22 9SX M6BZK
NG22 9SX M6IAQ
NG22 9TJ G0BYQ
NG22 9TN M3IDF
NG22 9TN M3KBE
NG22 9TN M3KPL
NG22 9UL M0HXH
NG22 9UP G1DQF
NG22 9UU 2E1HIQ
NG22 9UU M0BDD
NG23 5AL M6OOV
NG23 5BA G4BPE
NG23 5EG 2E0JMF
NG23 5LB M6MXR
NG23 5NT M0BCF
NG23 5PQ 2E0RVV
NG23 5PQ 2E0SVV
NG23 5PX M3IKV
NG23 6JJ M7GPY
NG23 6LH M0TVS
NG23 6QA G4LFT
NG23 6QG 2E0FGA
NG23 6QG M6FGA
NG23 6ST G8TNB
NG23 7AA G1LVH
NG23 7ED G6MLH
NG23 7HR M0VOS
NG23 7HR M6VOS
NG23 7LD G0DWB
NG23 7NT G7NDS
NG23 7PR G7SYQ
NG23 7RA G3HLG
NG237JH M7DZU
NG237RL G3WZJ
NG24 1DF G4TXO
NG24 1DF M0RHQ
NG24 1FN G0JAP
NG24 1FW G1VKN
NG24 1QZ 2E0FKH
NG24 2BU G7UXR
NG24 2FX M6JNM
NG24 2HA M6CJJ
NG24 2HT G7MUB
NG24 2NL G3YWS
NG24 2NT G3XFU
NG24 2NT G7KFM
NG24 2NT G7KFN
NG24 2RX G4PCP
NG24 2SA G0ORP
NG24 3AZ G1KNQ
NG24 3DA M3MOQ
NG24 3DT M0BWZ
NG24 3FJ G4NSW
NG24 3HH G6PII
NG24 3LY G0BVU
NG24 3NS G8EGU
NG24 3NU M7OBX
NG24 3NZ G0LBB
NG24 3TZ M0KAW
NG24 4BZ G7DEH
NG24 4DT G4KTG
NG24 4DY G4ZHG
NG24 4EX M3OVM
NG24 4EX M3YHD
NG24 4JL M6CVU
NG24 4JL M6CVV
NG24 4NY M3OIC
NG24 4PD 2E0IID
NG24 4PD G1AND
NG24 4PD M7AQQ
NG24 4PP M6JAM
NG24 4QB G0MCP
NG24 4QB G7HOT
NG24 4QB M6FXX
NG24 4RW G7BEJ
NG24 4TA M7ANY
NG24 4UA G0PUY
NG24 5BJ M6FLY
NG25 0AG G8UBF
NG25 0AH M0IUA
NG25 0BG M3DQJ
NG25 0JF G6FLQ
NG25 0JF M7RGS
NG3 2LN M6CWI
NG3 2LP 2E0TEW
NG3 2LP M0JSW
NG3 2LS G0MLM
NG3 2PE G7SCN
NG3 3AN G1NKV
NG3 3EQ G0WYI
NG3 3LL M7BAS
NG3 4PZ G1MQB
NG3 5EN G8CXV
NG3 5FD G4ZDX
NG3 5HY G4PJZ
NG3 5QB G8FWH
NG3 5QW G8ZUU
NG3 5RG M3LEF
NG3 5SE G4DFU
NG3 5UX M7DZO
NG3 6AD G3EKW
NG3 6AR G8BPQ
NG3 6DH G0JOX
NG3 6DT G6NLU
NG3 6DY 2E0DLO
NG3 6DY M0SYS
NG3 6DY M6EYP
NG3 6EE M3EHJ
NG3 6EF G0MAD
NG3 6EF G4NRZ
NG3 6EU G6WHS
NG3 6FL G4ZTY
NG3 6FT G3TWB
NG3 6JA G8MIW
NG3 6LR M1JHL
NG3 7BH G1IKH
NG3 7BY G4SGG
NG3 7FY 2E1HHG
NG3 7HD G4SJG
NG31 6JG G7ELG
NG31 6LE M6HET
NG31 6LT M3ZRP
NG31 7AZ M0GTV
NG31 7BP M6KCS
NG31 7BW M7XTB
NG31 7BX M6GJV
NG31 7HH G8WWJ
NG31 7JG M0SFA
NG31 7JL G6IPW
NG31 7LT G4ENS
NG31 7LT G8ENS
NG31 7LT M7DXH
NG31 7NH M6WDR
NG31 7NH M7BZV
NG31 7NN G0RCI
NG31 7NN G1EUU
NG31 7PH G1UTZ
NG31 7RB 2E0TYT
NG31 7RB M6GBI
NG31 7RB M6MSJ
NG31 7RH 2E1DCV
NG31 7RH G7SJX
NG31 7RH M3SJX
NG31 7RH M3WPM
NG31 7XA 2E1AXL
NG31 7XG M0JHW
NG31 7XP M6SRG
NG31 7XR M3VUA
NG31 8AD G4GXI
NG31 8AW M6HNX
NG31 8BN G4DZC
NG31 8GA M0AUS
NG31 8GH G6UOH
NG31 8JR 2E0MDE
NG31 8LN G4WZU
NG31 8LP G4VUA
NG31 8PD M0IGM
NG31 8QF M3OJX
NG31 8QL 2E0UMO
NG31 8QL M0KXL
NG31 8QL M7GOW
NG31 8RL G6YMD
NG31 8RS G0FSP
NG31 8WS M3YGQ
NG31 8XA 2E0IZU
NG31 8XA M7CXY
NG31 9BL M0HNF
NG31 9BL M0SUR
NG31 9JD 2E1CSD
NG31 9JW G8SHE
NG31 9ND M3ZYX
NG31 9PE G4VMI
NG31 9PF M1DDB
NG31 9PF M6ISD
NG31 9QA 2E1EFG
NG31 9QG 2E0XBM
NG31 9QG M0XBM
NG31 9QS G1PML
NG31 9QS M0USY
NG31 9RD G3VJE
NG31 9RG 2E0ALF
NG317EG M7DMI
NG32 1AU 2E0OBL
NG32 1AU M0OBL
NG32 1DL 2E0CMF
NG32 1DL M0SDM
NG32 1DL M0SKR
NG32 1DL M6AZT
NG32 1DQ M7EZZ
NG32 1ET M3XZJ
NG32 1HX M5DJC
NG32 2BQ G7NRU
NG32 2EB M0TGX
NG32 2EB M6AFS
NG32 2ES M6FRR
NG32 2FG M3SRY
NG32 2HE 2E0JNG
NG32 2HE M0KQK
NG32 2HE M7JNG
NG32 2NS G3PTI
NG32 2PD G4RIO
NG32 2PD G6JQH
NG32 3DF G4HVC
NG32 3EJ G3TBK
NG32 3EJ M0TBK
NG32 3EJ M3DVQ
NG32 3EJ M3TBK
NG32 3NR G0PJS
NG32 3RR 2E1EBR
NG32 3RU G0OJF
NG32 3SJ G0BUB
NG321AU G8OCO
NG33 4HA G4FGY
NG33 4LX G0FZG
NG33 4RT G0KQK
NG33 4SB G3KHZ
NG33 5FE G4OQR
NG33 5HG G0BHT
NG33 5PU G0GRU
NG33 5PU G8STM
NG34 0DL 2E0YNT
NG34 0DL M6YNT
NG34 0LD M0HXF
NG34 0LF M0MLW
NG34 0QG G4RIP
NG34 0RS G6KVE
NG34 0SE G3KVP
NG34 0TZ G6JUT
NG34 7HG G0BTT
NG34 7HW G7LPK
NG34 7JN 2E1FAN
NG34 7LQ G0CRE
NG34 7NW M6AQG
NG34 7NX G4FDR
NG34 7QP G2VS
NG34 7TD G3LSJ
NG34 7TF G4ODG
NG34 7WA G4SIJ
NG34 7WL M6JJX
NG34 7WL M6YBA
NG34 8AE G4GBE
NG34 8BJ G0BXU
NG34 8BJ G1LOW
NG34 8DB M1EGW
NG34 8EL G4OCU
NG34 8HG G0IBI
NG34 8HZ 2E0ENN
NG34 8HZ M6ENN
NG34 8JF M6IJG
NG34 8JF M6IOD
NG34 8QG 2E0CWM
NG34 8QG M0GUU
NG34 8QQ M0XCR
NG34 8WX M7ANX
NG34 9AL 2E0GQI
NG34 9AL M6OLT
NG34 9AT G0SGP
NG34 9BT G1PGN
NG34 9EQ G7UOL
NG34 9EU 2E0MPE
NG34 9EU M6MYN
NG34 9FH G6UCW
NG34 9GA G3WYH
NG34 9GU G4GUC
NG34 9HG G1XRF
NG34 9HG G3PXV
NG34 9HS G0NVY
NG34 9HS G3KQB
NG34 9JE G0POM
NG34 9JG G0FLV
NG34 9JP G3ZUC
NG34 9JX M5AKT
NG34 9JZ G0DMJ
NG34 9PD M1JTA
NG34 9QR G4EMK
NG34 9QR G8ILU
NG34 9QY G4NYC
NG34 9QY G4UYF
NG34 9QY M1HEK
NG34 9RX G6JTW
NG34 9SA G3SRX
NG34 9TS G3VHI
NG34 9TW G8YFP
NG348AW 2E0SHH
NG348AW M6SHH
NG349RB 2E0IZA
NG349RB M7FFP
NG36JX 2E0LWL
NG4 1BD 2E0BAF
NG4 1BU G6ZBO
NG4 1DA G6UZY
NG4 1DP G6JCI
NG4 1LE G0IBY
NG4 1LY 2E1BHU
NG4 2HN M7VCA
NG4 2JG M3HWS
NG4 2JG M3YHU
NG4 2LW M0BCI
NG4 2PZ M7PDS
NG4 2QJ M3CZE
NG4 3DA G7TMF
NG4 3DX M1EDW
NG4 3EH G4ZCW
NG4 3ET G6MBI
NG4 3GY G1YBA
NG4 3LF G6IOE
NG4 3QH G7SBK
NG4 3SF M3CBH
NG4 4AQ M1DJB
NG4 4AS M6FLQ
NG4 4AU G8LGY
NG4 4BB G8GCK
NG4 4BL G1EAB
NG4 4EA M6CCV

NG4 4EJ M7EVF
NG4 4GF M3UEE
NG4 4GS G4CPG
NG4 4HF 2E0HAW
NG4 4HF M3CJX
NG4 4HZ M3VRV
NG4 4JP G6YCL
NG4 4JP G7SFS
NG4 4PE G2ALX
NG4 4QE M1JCS
NG44AF M7JOX
NG5 1BH G1GJT
NG5 1DA G6BMZ
NG5 1DW M6EFL
NG5 1EP M0WAB
NG5 1GP G6ONW
NG5 1JR G4EKW
NG5 1LA 2E0XXI
NG5 1NH G3TVY
NG5 1NW M1REJ
NG5 3DA G0DME
NG5 3GL M6TRI
NG5 3GS G1HRM
NG5 4BD G7LKV
NG5 4DY G0NVS
NG5 4GQ G4SQV
NG5 4HU G4EDX
NG5 4HY 2E0CAW
NG5 4HY M0SCO
NG5 4HY M3CAW
NG5 4JS G0JAC
NG5 4JX G4WDP
NG5 4LB G0KXZ
NG5 4NX M0IZM
NG5 4PG G4ERY
NG5 5BL G8VCI
NG5 5DT G6UZA
NG5 5DU G6URR
NG5 5EE G8CQQ
NG5 5FA G7ULJ
NG5 5NP 2E0LFK
NG5 5US M6ANY
NG5 6BX M0DLZ
NG5 6DJ 2E0AXT
NG5 6DP M6SAH
NG5 6DP M6ZBD
NG5 6DX 2E0BWH
NG5 6DX M3XBH
NG5 6EB 2E0EMF
NG5 6EB M3JWQ
NG5 6EH M6RSJ
NG5 6FN G4PNX
NG5 6FT G1FDO
NG5 6GU G6MVF
NG5 6LQ G4EAN
NG5 6NH G8SSL
NG5 6QA G7WKH
NG5 6QL G1UDX
NG5 6QT 2E1DLT
NG5 6QZ G1HTL
NG5 6RX G0RHJ
NG5 7AZ G0FNV
NG5 7HD 2E0MVD
NG5 7HD M6MCO
NG5 7HF M3IVX
NG5 7HQ G4OOY
NG5 7LW G8UGS
NG5 7NF G4ITX
NG5 8AG G4EGY
NG5 8BD M7AOD
NG5 8BE M6HIK
NG5 8BP M6IXG
NG5 8FQ G8RBU
NG5 8GE M0PDG
NG5 8GH M0KIR
NG5 8JS G6YLX
NG5 8NZ G7IWK
NG5 8QE M3ROF
NG5 8QF G1GQQ
NG5 8QX G6IBN
NG5 8SJ 2E0DSH
NG5 8SJ M0ZRD
NG5 9LN G7MGV
NG5 9LN M0BCH
NG5 9QE G3JWB
NG5 9QN G1HMZ
NG5 9QU G7IZE
NG6 0BH G0KZA
NG6 0BS G1XHO
NG6 0JU M7FYB
NG6 0LS G0INA
NG6 7DL G4DIU
NG6 7FJ G1ZLA
NG6 8AY G8UQR
NG6 8DG G0GXH
NG6 8DT 2E1HAL
NG6 8NL M3MDW
NG6 8PF G1NXI
NG6 8PU G6VWV
NG6 8QY G0VFU
NG6 8SL G6NLD
NG6 8XA G6SFF
NG6 8XE G4PMM
NG6 8XN G0LCG
NG6 8XN G3RR
NG6 8XN G5RR
NG6 8XN G7HRR
NG6 9AP 2E0HET
NG6 9AP M0IQL
NG6 9AP M6YTX
NG6 9DT G1GTF
NG6 9FB G8ZZV
NG6 9FU G7SYD
NG6 9FX G1YWN
NG6 9HN G8OHC
NG6 9JE G6GWY
NG7 1RG G6COZ
NG7 3DF M6WMH
NG7 3GN G0JLR
NG7 5EB 2E1ECG
NG7 5FS G3WGH
NG7 6AD M6RIL
NG7 6HG M0LQW
NG7 6HU G6ISB
NG7 7DR M3NQY
NG7 7LJ M6WAY
NG8 1AT 2E1IHY
NG8 1DE G1YEU
NG8 1JE M0GRH
NG8 1JU G6UAP
NG8 1LF M3MXH
NG8 1PU 2E0KNB
NG8 1PU M0KNB
NG8 2BH G6CIF
NG8 2BZ G4BWN
NG8 2EF M7VWD
NG8 2EH G0FSJ
NG8 2ER G8UUS
NG8 2FB M6GHR
NG8 2NA G8BNG
NG8 2QR G4KLJ
NG8 2QR G8NWU
NG8 2RE G0FOE
NG8 2SB G8POL
NG8 3ES G3XBE
NG8 3FF M0EPR
NG8 3FF M0GZD
NG8 3NE G4TYO
NG8 4AH M3JHL
NG8 4GY M0AUK
NG8 4LZ G1HRJ
NG8 4NB M6VCH
NG8 4PU 2E1FDT
NG8 4PZ M3CCP
NG8 5BE M6PIU
NG8 5EU G1XYG
NG8 5EU G7DXH
NG8 5FH G1HSP
NG8 5FJ G1RCN
NG8 5LF G7DSO
NG8 5ND M3CIO
NG8 5PN G1DNZ
NG8 5QG 2E0EHK
NG8 5QG M6GHN
NG8 6AD G6XJD
NG8 6AD G6XJE
NG8 6EX 2E0PJR
NG8 6EX M0LZM
NG8 6GP 2E0LZM
NG8 6GP M3XID
NG8 6GQ M0DAZ
NG8 6JT 2E0GAC
NG8 6LY G0TSG
NG8 6LY M3TSG
NG8 6NE M1BSI
NG9 1EU 2E0BCE
NG9 1FJ 2E0EEO
NG9 1FJ M0IOI
NG9 1FJ M6NOT
NG9 1GR G6JRE
NG9 1LR M3WSE
NG9 1LY G0CQT
NG9 1NA M3VIO
NG9 1NE G6XXJ
NG9 1NJ G6XOD
NG9 1NJ G6XOE
NG9 1NS 2E1GYO
NG9 1PY 2E1EQQ
NG9 1PY G0OQQ
NG9 1RJ M0DLR
NG9 2AD 2E0PHS
NG9 2AD M6EET
NG9 2BA M7SSO
NG9 2FJ G0GRZ
NG9 2GU M6DZN
NG9 2QR G4JHN
NG9 2QZ M3RNU
NG9 3BB 2E0OBS
NG9 3BB M6CAH
NG9 3BY G7LNV
NG9 3FN M0DAC
NG9 3FP G0PXP
NG9 3FS G0SKW
NG9 3FS G5NKW
NG9 3JE G8PBH
NG9 3JN G4MFV
NG9 3JT G1PUZ
NG9 3JW M3HPM
NG9 3LD M0JEL
NG9 3LW G8BFM
NG9 3PF G0PCK
NG9 3PX G4GQY
NG9 3RB 2E0YZZ
NG9 4BB G3EJH
NG9 4BB G4VFK
NG9 4BB G8ZK
NG9 4ED M3UVQ
NG9 4ES G0UYV
NG9 4FG M3RUI
NG9 4FH G7NGX
NG9 5AE M0GHX
NG9 5EB G8ZSZ
NG9 5EU G4AQT
NG9 5EZ G4BNX
NG9 5FH G6BRP
NG9 5FJ G7TRB
NG9 5FU M6STH
NG9 5GR 2E0GSW
NG9 5GR M0RMY
NG9 5GR M6CTX
NG9 5HS 2E0FBL
NG9 5HS M6ENH
NG9 5HX G0SPA
NG9 5HY 2E1CHX
NG9 5HY G7USM
NG9 5LA G6UTO
NG9 5LA G7HIT
NG9 5LB 2E1SIS
NG9 5LB M0OIC
NG9 5NH G6IPC
NG9 6AB G6RCD
NG9 6BP G8IYZ
NG9 6EW G4KQP
NG9 6FW 2E1EAV
NG9 6FZ M5ADM
NG9 6GN 2E0DWB
NG9 6GN M3ZNR
NG9 6HP G0RSU
NG9 6HS G6UWO
NG9 6JW G4XEL
NG9 6JW G6MSC
NG9 6NH M6EDJ
NG9 7ET G0NVO
NG9 7FY G4OYO
NG9 7HN 2E0GIH
NG9 7HN M3GIH
NG9 8AR M7LWP
NG9 8BN M1EWD
NG9 8DJ G4PMA
NG9 8GH M7HLN
NG9 8GL G0UYQ
NG9 8GL G8EHX
NG9 8HR M1ERF
NG9 8LN M7RDN
NG9 8LT G7LCV
NG9 8PQ G0SLZ
NG9 8QG G4TYN
NN1 3BL G0PYI
NN1 3BP M7AKG
NN1 3QL M3NNH
NN1 3QL M6EMA
NN1 4HU G3YOV
NN1 4HZ M6AIP
NN1 4LU G1ICH
NN1 4NA M0ARZ
NN1 4PU M3KUG
NN1 4QR M3CET
NN1 4SP 2E0BDD
NN1 5JR 2E0GIY
NN1 5JR G5TEO
NN1 5JR M7TEO
NN1 5JS G4PXJ
NN1 5JY M3WUA
NN1 5LT G7KGP
NN1 5LT M3GTG
NN1 5LT M3NEC
NN1 5LT M5TTT
NN1 5LU M6LRB
NN1 5ST 2E1BCC
NN1 5ST G3VMU
NN10 0AS M0DOK
NN10 0AT M0DOD
NN10 0DH M3NLI
NN10 0DJ G6VFC
NN10 0DX G7PYQ
NN10 0EA M0IXF
NN10 0EE M7HJB
NN10 0EG M6JCT
NN10 0FH M7EUC
NN10 0FT M6RVG
NN10 0GE G6EQS
NN10 0GS G1GPE
NN10 0JH G3PRU
NN10 0SJ G4BHT
NN10 0SN G8LII
NN10 0SW G4FEV
NN10 0SW G4WFT
NN10 0SY G3WDG
NN10 0SY G4KGC
NN10 0TN M7AXV
NN10 0TZ M3UPJ
NN10 6BA G7TZD
NN10 6BG G3OMS
NN10 6BH M1CBG
NN10 6BU M3CQW
NN10 6EH G7APQ
NN10 6UR M6FPW
NN10 8DH G6NHO
NN10 8EP 2E0HML
NN10 8FJ G7IWM
NN10 8HF G1RTX
NN10 8HU G4ZRM
NN10 8NH G0WTO
NN10 8NH G7CMI
NN10 9EG M6JXI
NN10 9EZ G6SDI
NN10 9HF G7LNP
NN10 9HH 2E0TSO
NN10 9HH M0TSA
NN10 9HL G1OOS
NN10 9HS M7TCK
NN10 9LF G6IPT
NN10 9LG G4NXV
NN10 9LN G6GOX
NN10 9NS G7OCC
NN10 9PF G0PYS
NN10 9QA 2E1CBU
NN10 9QA G4IRD
NN10 9RP 2E0ERG
NN10 9RP M6RAH
NN10 9SX G7NBG
NN10 9SX M3NBG
NN108JR M7PLO
NN108PH 2E0ILE
NN108PH M7ICK
NN11 0GR G1ZJK
NN11 0GR M1BXD
NN11 0NZ G8KHF
NN11 0UN M3BXE
NN11 0WQ M6AFL
NN11 0XH 2E0CTR
NN11 0XH M3PMY
NN11 0XT M3FKK
NN11 2JH G4SEV
NN11 2JW M0PXL
NN11 2JW M6LZS
NN11 3AD G1WVS
NN11 3AD M3HUX
NN11 3BW G4TBN
NN11 3ES G3YHI
NN11 3PR G4KQH
NN11 3PX M7LDV
NN11 3QP 2E0LMR
NN11 3QP M6CNG
NN11 3QS M6BPI
NN11 3YT G3MRQ
NN11 4AL G1DLH
NN11 4AL M3NUM
NN11 4AN M3FNY
NN11 4AQ G0HWT
NN11 4EY G7KRB
NN11 4GW G8VQQ
NN11 4GY G6XLB
NN11 4HJ G0DRE
NN11 4HJ G0FPM
NN11 4JQ 2E1FSV
NN11 4JQ M1CUX
NN11 4LL G0IOH
NN11 4NW 2E1AXD
NN11 4NW 2E1AXE
NN11 4NW 2E1CZF
NN11 4NW G6EGO
NN11 4NW G8GVV
NN11 4PR G7LII
NN11 4PT M3XTT
NN11 4QF M6FZR
NN11 4QU G1DIO
NN11 4QU G6MTE
NN11 4RX G7KJX
NN11 4SU G0NNG
NN11 4SX G3WTN
NN11 4TD 2E0BHC
NN11 6UR G7OLT
NN11 6UZ 2E0MBX
NN11 6UZ M0MBB
NN11 6UZ M6MBD
NN11 7HE M6YAS
NN11 7HQ G8KTA
NN11 7HR G3SRP
NN11 7HY M1AQX
NN11 7JN 2E0AWS
NN11 7JN M3BSI
NN11 7JS G4BFT
NN11 8BH G6LUY
NN11 9AL M3KZI
NN11 9BT G7JDA
NN11 9EB 2E0HDD
NN11 9HB M6NRR
NN11 9WH M3IQS
NN12 6AG M3HEO
NN12 6AW G3UMT
NN12 6AW G8CHK
NN12 6DN G1YMJ
NN12 6DN G6ZGA
NN12 6ED 2E0GNH
NN12 6ED G6RHJ
NN12 6ED M6JXE
NN12 6EZ M3NLQ
NN12 6NX M3SBA
NN12 6PL M5GDK
NN12 6QQ G4GWT
NN12 6RD 2E0YLP
NN12 6RD M6YLP
NN12 6RL 2E0TOY
NN12 6RL M0XER
NN12 6TH 2E0VLL
NN12 6TH M3VLL
NN12 6UP M3FEC
NN12 6YY G1EBX
NN12 7NA G8MQY
NN12 7NW M1DJC
NN12 7NX M6GQB
NN12 7PH G4IUH
NN12 7PH G8AAT
NN12 7PP G1MZD
NN12 7QX M6GDQ
NN12 7RS G1TAI
NN12 7RT M3YHX
NN12 7SA G1DJU
NN12 7TY G4EQX
NN12 8LU G4UXO
NN12 8NA G7ECG
NN12 8NG M1BXC
NN12 8NW G0FKW
NN12 8NW M6KZW
NN12 8UY G4DQQ
NN12 8XJ M3KVG
NN128UP 2E0VGA
NN128UP M7SDI
NN13 5HU G0UID
NN13 5LX G4EBF
NN13 5TW M6HXU
NN13 6AQ 2E0BGZ
NN13 6AQ M0RGD
NN13 6AQ M3IIA
NN13 6BP M6SKX
NN13 6DA M3ULM
NN13 6GU M7HSI
NN13 6HA G3NAN
NN13 6JD M7BSI
NN13 6JH M1CLO
NN13 6JQ M0RCY
NN13 6JQ M6CKZ
NN13 6JS G6HKH
NN13 6LY G8JDC
NN13 6ND G4IYA
NN13 6NE 2E0CZN
NN13 6NE G1BZM
NN13 6NE M0IFB
NN13 6NE M6DLW
NN13 6NS M6OYD
NN13 6PA M7EKV
NN13 6PB G4DWC
NN13 6PE M1DIM
NN13 6PE M3YWI
NN13 6RA G1BPV
NN13 7ET G8OZH
NN13 7NJ M3UBF
NN13 7TY G1EPO
NN135ND M7BFJ
NN14 1BG G4HND
NN14 1BG G7TZZ
NN14 1DT G7TIK
NN14 1EE G8ZIH
NN14 1EL G0GRS
NN14 1HG G0NFL
NN14 1LN M7FWH
NN14 1NG 2E0GGM
NN14 1NG G4RPT
NN14 1NG M0YGG
NN14 1NG M6GGM
NN14 1NS M7MPS
NN14 1RG M0CKP
NN14 1RQ G1NEN
NN14 1SL G4UYM
NN14 1TB G3VSU
NN14 2FH G6GDR
NN14 2LB G0FFB
NN14 2LB M3PSM
NN14 2LQ 2E0TJP
NN14 2LQ M3XDZ
NN14 2LU M7BTZ
NN14 2NH G0FSA
NN14 2NX G4ASH
NN14 2QB 2E1EFT
NN14 2QB G1LMN
NN14 2QU G1KCV
NN14 2QY G4PNL
NN14 2RE G1OET
NN14 2RE G5KN
NN14 2RQ 2E0WPT
NN14 2RQ M6MPT
NN14 2SQ M1PAH
NN14 2TZ M6RUI
NN14 2UD G1WPR
NN14 2UD G8JKB
NN14 2UY G7NBV
NN14 2XA G1IHY
NN14 2XB M6FOZ
NN14 2XG 2E0ZAH
NN14 2XG G6DHW
NN14 2XL G8DXI
NN14 3BT G6SWJ
NN14 3BT M6JTC
NN14 3DD 2E0PNG
NN14 3DZ G8ZPE
NN14 3JT M0NIG
NN14 3LJ 2E0XBG
NN14 3LJ M6SFW
NN14 3LQ G4YNG
NN14 4BN 2E0NBI
NN14 4BN M7NIB
NN14 4DD G4HLW
NN14 4EA G6LJH
NN14 4FL 2E0ENI
NN14 4FL M0ISK
NN14 4FL M6FKR
NN14 4GN M0DSO
NN14 4JQ G8XQN
NN14 4JT G0CMP
NN14 4PQ G3YQJ
NN14 4PY G4PUZ
NN14 4RE G3WIA
NN14 4XA M3UFZ
NN14 4XB G7OGL
NN14 6AJ M6OAN
NN14 6DZ G1TPC
NN14 6GE M7GHF
NN14 6HT G4DCD
NN14 6HT G8HBZ
NN14 6HY G6BKD
NN14 6HY G8BKD
NN14 6JH 2E0HJB
NN14 6LD M7WCF
NN14 6TN G1IQA
NN14 6YG G3WKR
NN14 6YQ G7RAE
NN142QU M6UKB
NN15 5BN G7AJJ
NN15 5BY G8WSV
NN15 5DE G7PQM
NN15 5EG G0NSA
NN15 5EQ G4VKX
NN15 5HB G8YKE
NN15 5HD 2E0MNZ
NN15 5HD G4NZZ
NN15 5HD M7LJM
NN15 5HE M3JCA
NN15 5HT M0KIC
NN15 5HT M7KAD
NN15 5JB G7GMQ
NN15 5LJ M6SGP
NN15 5LS 2E0RBO
NN15 5LS M3YPW
NN15 5LS M6GIL
NN15 5ND G7WKV
NN15 5NT G6OLU
NN15 5PN M3LDM
NN15 5QB M6NDY
NN15 5QP M3XQO
NN15 5QS M6PAV
NN15 5RS M3ZRM
NN15 5ST M3RHJ
NN15 5SX G1BJZ
NN15 5YL G4RLL
NN15 6HF G0DEF
NN15 6HH G1YLM
NN15 6HL M6ENB
NN15 6JG G7KRS
NN15 6JS 2E0IZZ
NN15 6JS M6SUM
NN15 6PS G4TTX
NN15 6QH G7SKA
NN15 6RE G6JTC
NN15 6RG G4TXM
NN15 6UW G4OYN
NN15 7AP G4NCA
NN15 7BA G1VPS
NN15 7BA M1ECD
NN15 7DR G0RDV
NN15 7DS G1XJN
NN15 7DU G1JSP
NN15 7DZ M1AWN
NN15 7EF G1AZD
NN15 7EZ G3VJG
NN15 7HX G7CJG
NN15 7LL G0PSG
NN15 7LL G7NTG
NN16 0EL M6BIU
NN16 0HU M6MAL
NN16 0LF G7VIH
NN16 0LY 2E0JLD
NN16 0LY M0JXD
NN16 0LY M6JLD
NN16 0NG M6RUH
NN16 0PH M0AQP
NN16 0QB G7API
NN16 0QR 2E0WDS
NN16 7JZ G8EZG
NN16 8EB G8FSJ
NN16 8PL G4LEM
NN16 9BU G7AJS
NN16 9ET M6WOE
NN16 9GA G1PQR
NN16 9HA G0PBU
NN16 9JA G7DMK
NN16 9JL G0LDR
NN16 9JR G8WSU
NN16 9JZ G6PEG
NN16 9LA G4NAC
NN16 9LB 2E0OET
NN16 9LB M0OET
NN16 9LB M6OET
NN16 9PF 2E0XJR
NN16 9SB G6OXN
NN16 9TG 2E0WIZ
NN16 9TG M6WIZ
NN16 9TW M6AHP
NN17 1EG M3NVN
NN17 1EN G6BHX
NN17 1ER G7NJZ
NN17 1JD G0IKR
NN17 1SY G7SCX
NN17 2AF G1TQR
NN17 2BS G1IIY
NN17 2DZ G4YRX
NN17 2LA G4PDK
NN17 2LH G1BKI
NN17 2LN G0PAD
NN17 2QP M1BQS
NN17 2QP G6AFT
NN17 2RP G6RCY
NN17 3BA G3YNF
NN17 3BT G4JCJ
NN17 3BU G7CIV
NN17 3DN G4DNJ
NN17 3DN M3CCS
NN17 3HB G6VOE
NN17 3LR M6NYZ
NN172RD M6EJS
NN18 0AE M1CDP
NN18 0AG M3JBW
NN18 0BN G0NRB
NN18 0LG G0RRO
NN18 0NS 2E0VWF
NN18 0TA G4OMV
NN18 8DB G1YTG
NN18 8DE G4XRD
NN18 8FZ G4NZE
NN18 8HX G3MTJ
NN18 8LL M6PSY
NN18 8NB M0RSH
NN18 8NG M7LAO
NN18 8QR M7XST
NN18 9DE G6RFH
NN18 9EG G8SEV
NN18 9HH G0IYS
NN18 9JL G4XKD
NN18 9JS G4LKU
NN18 9PG M3TLT
NN18 9PH G1JNY
NN2 6EN M7ETN
NN2 6EP G6NYH
NN2 6LS M7LFI
NN2 6RP G3KLV
NN2 6RP M0OXZ
NN2 7AT 2E0OAR
NN2 7AT M0LXV
NN2 7DA G0FJS
NN2 7DN G3KAN
NN2 7PP 2E0YXZ
NN2 7PP M6SNO
NN2 7PP M6YYY
NN2 7PT G0GAQ
NN2 7PX G3MJW
NN2 7QU G8ZMH
NN2 7QX G1IPD
NN2 7QY M3FLP
NN2 7RR M0EBU
NN2 7RY G0WYQ
NN2 7SP M0VQP
NN2 7SP M3VQP
NN2 7SP M6HDG
NN2 7TJ M3VIB
NN2 7TR G6NDS
NN2 7TR G8WPU
NN2 8BJ G7OGN
NN2 8BX G4FTL
NN2 8DZ G3ZJO
NN2 8EA G1GOY
NN2 8EL M0GCR
NN2 8HD G0ONS
NN2 8HT G1ERY
NN2 8HT M7GHB
NN2 8JR M0DUU
NN2 8NB G4YJP
NN2 8PE 2E1CCF
NN2 8PH G8PTH
NN2 8QU G6TVB
NN2 8QU G8OZY
NN2 8SX G7AAY
NN2 8TX G7TEG
NN2 8TX M7MVS
NN2 8UU G8KNU
NN29 7AP M1TSU
NN29 7AT G8AMG
NN29 7AZ G6ODT
NN29 7DA G4PAV
NN29 7DH G6MKJ
NN29 7DP G0VOB
NN29 7DR G1ALL
NN29 7EA G8TFY
NN29 7EE G4RST
NN29 7EW 2E0KAK
NN29 7EW M0HAF
NN29 7EW M6AGK
NN29 7FD G0UBX
NN29 7HF M6RGR
NN29 7JU G0EAE
NN29 7JU G6NUS
NN29 7LP G7HBU
NN29 7LS G1ONQ
NN29 7ND G1RVF
NN29 7NS G0PSZ
NN29 7QE G1BNE
NN29 7QG G4TLT
NN29 7RR M6YPD
NN29 7RS G3JVE
NN29 7TP G1EUG
NN3 2AZ M0ECT
NN3 2BU 2E0CCY
NN3 2BU M3NWH
NN3 2DD G4VWF
NN3 2HE G0MJK
NN3 2LP 2E0BXV
NN3 2LT 2E0WSJ
NN3 2LT M3WSJ
NN3 2PG M6BGS
NN3 2PY M7SIN
NN3 2SY G0RTC
NN3 3AF G6AXH
NN3 3EQ 2E0SPZ
NN3 3FA G4MAB
NN3 3JE M3DYL
NN3 3JL G4MJF
NN3 3JX G1KMS
NN3 3NX M6KKY
NN3 3PB G4JFC
NN3 3PE G7UFW
NN3 5AD G1SLG
NN3 5AL M6WXP
NN3 5AY M7CCQ
NN3 5AZ M7PKP
NN3 5BH G1IRG
NN3 5BH G1NRG
NN3 5EP M7RMU
NN3 5ET 2E0ERO
NN3 5EY G6VGS
NN3 5HY M6BOB
NN3 5JS G4OIG
NN3 5JS G8CXK
NN3 5LZ M6FEI
NN3 5NT M0DME
NN3 6BL 2E0AAZ
NN3 6DN M3XHW
NN3 6HP G4EUR
NN3 6JQ G4SCJ
NN3 6JQ G6NHV
NN3 6SE M7HWB
NN3 7HU G6IVW
NN3 7LB M0MAO
NN3 7LB M0POI
NN3 7LB M6GMO
NN3 7RD G8FZW
NN3 7RE G6CQR
NN3 7TX G1DPJ
NN3 7UB M6BQJ
NN3 8AX M5AFY
NN3 8DE G3MBD
NN3 8DF G8INA
NN3 8HH M6MCR
NN3 8HY G7RFX
NN3 8HY G7RGU
NN3 8LD G0MHA
NN3 8LZ 2E0BKO
NN3 8LZ M3RGU
NN3 8PS M6EKM
NN3 8QT 2E1DNF
NN3 8UR G0PTM
NN3 8US 2E0UAP
NN3 8US M7UAP
NN3 8UX 2E1CAH
NN3 8UX G1BGF
NN3 8XN G8TOP
NN3 8YA M6MIO
NN3 9AX 2E1EET
NN3 9BH G6TVK
NN3 9BH M0JQO
NN3 9BH M7TWW
NN3 9BW M6SBU
NN3 9DJ G6OKH
NN3 9DN M7IRB
NN3 9EX M6KNZ
NN3 9EX M7XHC
NN3 9NT M3RCD
NN3 9SN M1BTU
NN3 9UG M6NIU
NN38QE 2E0PUN
NN38QE M3PUN
NN39EX G0FMN
NN4 0NE M3HGM
NN4 0NJ G4LUW
NN4 0QQ G0NYE
NN4 0RS M0XDK
NN4 0SF G4YIP
NN4 0SS G6ILZ
NN4 0TS M3TGW
NN4 0WF G7CIA
NN4 0XF G3SOA
NN4 0XX G0KAK
NN4 5AZ G0FQI
NN4 5BT 2E0WNO
NN4 5BT M7WXD
NN4 6AH G6NPZ
NN4 6AL G0FSM
NN4 6AP M6ROJ
NN4 6AP M6VHA
NN4 6BB G7IXK
NN4 6DJ 2E0HSG
NN4 6DJ M6KDW
NN4 6ER G4DHV
NN4 6EW M3TRP
NN4 6HE G7NEG
NN4 6LA G0TMW
NN4 6UX M0KIW
NN4 7BT M3NAL
NN4 7PT G8ZEW
NN4 8AU G8HTO
NN4 8AZ G6FAL
NN4 8BY M0EAB
NN4 8LQ G6KIZ
NN4 8NR M6YYU
NN4 8PE G1TPN
NN4 8PX M7RET
NN4 8QB G1BHR
NN4 8QB M3BQT
NN4 8QB M3ZOR
NN4 8RU G4DTM
NN4 8RU M7LTM
NN4 8RX G4RKR
NN4 8ST 2E0DNL
NN4 8TQ M0STN
NN4 8TR G6OXY
NN4 8TT M6JBV
NN4 9QU 2E0VNL
NN4 9QU M6VNL
NN4 9QW G4YFX
NN4 9SB M0WMT
NN4 9TX M6RGZ
NN4 9YH G4KFC
NN4 9YR M3KFP
NN40RQ G6LUU
NN40SD M1AQC
NN48UN 2E0HSV
NN48UN M3ILB
NN5 4AJ G4MLO
NN5 4WB M0XKD
NN5 5AR M6RDF
NN5 5DB M6INW
NN5 5LW G1UQF
NN5 6JT M6NKR
NN5 6JW G0RDT
NN5 6LR M3MKH
NN5 6NE G8OWG
NN5 6NH G7SYT
NN5 6NH G7VQE
NN5 6NH M3EKA
NN5 6NL G7HLP
NN5 6NR G6FGC
NN5 6PU G8KST
NN5 6PY G4NVP
NN5 6PZ G4NUF
NN5 6PZ G8KIW
NN5 6QP G7ISE
NN5 6QX G6TJE
NN5 6RA G1OMI
NN5 6RA G3RLD
NN5 6RF G0VJK
NN5 6TU M0DAE
NN5 6TW 2E1FUQ
NN5 6XP M7ETH
NN5 6XX G4ICC
NN5 6YW G0DLF
NN5 7AL M7ICS
NN5 7ER 2E0PHX
NN6 0BD G4MTP
NN6 0BD G8SWL
NN6 0BD M0DON
NN6 0BY 2E0YXO
NN6 0HR G8SQA
NN6 0HU M7CNK
NN6 0HX G8LOP
NN6 0JW M7CPW
NN6 0NY G4LYC
NN6 0NZ G0JPF
NN6 0PF G6OVC
NN6 0PQ G7RTC
NN6 0PX G7OXY
NN6 0QS M6BKJ
NN6 6AE G4DLD
NN6 6AF 2E0TAY
NN6 6EW G4LRT
NN6 6EZ G6EGY
NN6 6EZ M3CZW
NN6 6LX G7DMG
NN6 6LX G7DMH
NN6 6LX M5WAH
NN6 6NN 2E0HFG
NN6 6NN M0LWA
NN6 6NN M6BWJ
NN6 7BH 2E0PNZ
NN6 7BH G7RGI
NN6 7BH M6ZUF
NN6 7GP 2E0JMY
NN6 7GP M0JIB
NN6 7PP G7BQM
NN6 7PQ M6AEU
NN6 7PQ M6ANM
NN6 7PQ M6ANO
NN6 7PU G6SJA
NN6 7QQ 2E1ECM
NN6 7QQ G4FMC
NN6 7RB G0OFB
NN6 7RR G4VMB
NN6 7SR G0LHE
NN6 7ST G4EPA
NN6 7TT G6YTB
NN6 7TX M0SPS
NN6 7YP 2E0OOH
NN6 7YP M6OCR
NN6 8AF 2E0KRN
NN6 8AW 2E0BHU
NN6 8AW M3NKN
NN6 8QP G1TUS
NN6 8RP G6UNA
NN6 9DR G4BDW
NN6 9JS G7NRR
NN6 9LX 2E0INJ
NN6 9LX M7BVO
NN6 9NE G4YKE
NN6 9SB M3ZST
NN6 9UE M6NRC
NN6 9XL G8JFX
NN7 1AE G7NDT
NN7 1BW G0DFV
NN7 1DA G6PHT
NN7 1DR 2E0RDN
NN7 1DR M6RFQ
NN7 1HB G6KPJ
NN7 1JW G1XXH
NN7 1NU G4FIN
NN7 1PD M0BKJ
NN7 2BW G6UVN
NN7 2DA G1AAD
NN7 2HE M0JUT
NN7 2NL 2E0SCH
NN7 2NL M0PIT
NN7 2NL M6ETC
NN7 2NR G8DHU
NN7 3AW G4KHX
NN7 3DF G4XKM
NN7 3EL 2E0WBR
NN7 3EL G7PTA
NN7 3EL M6WBV
NN7 3EZ G4HWA
NN7 3HQ M0KOQ
NN7 3JB G8EDX
NN7 3LE M6CRT
NN7 3LZ G6OIA
NN7 3NA 2E0SHV
NN7 3NA M0SHV
NN7 3QB G8TEB
NN7 3QB M6JNL
NN7 3QT G1VUY
NN7 3QT M7CKN
NN7 3QU 2E0MOR
NN7 4BX G3IPL
NN7 4DD G3GWB
NN7 4DD G4CZB
NN7 4DD G8LED
NN7 4DN G4TGW
NN7 4DP G4IPL
NN7 4DR G8TMM
NN7 4LL G4PGY
NN7 4LL G6BRW
NN7 4LS G0TML
NN7 4QL G1GJD
NN7 4RL M6MBS
NN7 4SB G4WOB
NN7 4SH G3ZAW
NN8 2DX G4ZIW
NN8 2DY G6WAS
NN8 2EH M6NKA
NN8 2EN G6UWS
NN8 2ES M3LYG
NN8 2HT G4HIQ
NN8 2LZ G1IQE
NN8 2ND G0LEY
NN8 2NU M7THL
NN8 2PH M6NTY
NN8 2PQ M3CQT
NN8 2PU G0VRF
NN8 3AZ M7WLZ
NN8 3BF M0WXF
NN8 3DB M3XLW
NN8 3EW 2E1BFH
NN8 3JJ G0HBA
NN8 3JJ G1AUY
NN8 3PJ G0ING
NN8 3PJ G5LP
NN8 3PW G6ECT
NN8 3RS M3NFE
NN8 3SH G4MOP
NN8 3ZA G6BBD
NN8 4AT G4WZB
NN8 4BS M7MSD
NN8 4EB G7UPZ
NN8 4ET G8GOM
NN8 4PN G6RPH
NN8 4RU G3ZAG
NN8 4RY G8NWZ
NN8 4RZ G8JPV
NN8 4SF G1IJJ
NN8 4UZ G0FQA
NN8 5BA M7TMW
NN8 5NX G4RDM
NN8 5PQ G6HGG
NN8 5QP G8OSZ
NN8 5WB M7GEF
NN8 5WE G8EJQ
NN8 5WT G0DDJ
NN8 5WW G6STE
NN8 5YD M6NKG
NN8 5YU G6UWY
NN8 5YU M6ELJ
NN82JU M7TST
NN83XA G2DRM
NN84UX G6CPX
NN9 5BA M6JTL
NN9 5BA M6LTM
NN9 5DP G0EHE
NN9 5LP G0HNG
NN9 5RB G3UQR
NN9 5RJ M6YDN
NN9 5RL M6MLN
NN9 5SU 2E0DCN
NN9 5SU G7BFH
NN9 5SY G4PPW
NN9 5TL G0TLN
NN9 5TN G7EWL
NN9 5TR G7SLB
NN9 5TY G3NXS
NN9 5XW G6DYU
NN9 5XW M3DYU
NN9 5YS G6FEJ
NN9 6AL M3WBJ
NN9 6AQ G1TLC
NN9 6BQ G1HGY
NN9 6DF G7MKG
NN9 6DQ G0TES

NN9 6DQ G1JPC
NN9 6HE G0JXQ
NN9 6HE G0MEO
NN9 6HN G6ONV
NN9 6LF M0MOI
NN9 6LS G7LPB
NN9 6LX G4LAM
NN9 6QX G6TRO
NN9 6RD G1WWY
NN95EZ M7BNH
NP1 7PB GW7MYD
NP10 0AU MW7BIH
NP10 0BP GW8MER
NP10 8ET MW6LHB
NP10 8HD 2W0SWO
NP10 8HD MW3SWO
NP10 8HJ MW6MQU
NP10 8HR MW1FDN
NP10 8JZ GW8HAG
NP10 8LJ 2W0TID
NP10 8LJ MW6MBC
NP10 8LL GW1MGI
NP10 8NF MW3ZRK
NP10 8NY 2W0CVO
NP10 8NY MW6SVZ
NP10 8PA GW4JZY
NP10 8PD MW0OPY
NP10 8PT 2W0HPK
NP10 8PT MW1DCF
NP10 8PT MW6HPK
NP10 8RB GW4COJ
NP10 8RB MW6DGQ
NP10 8RF MW6JEL
NP10 8RT GW3YQM
NP10 8SL GW4KHQ
NP10 8UP GW7IAK
NP10 8WY MW6NTW
NP10 9AH GW7SMV
NP10 9AQ 2W0KEQ
NP10 9AQ MW0KEQ
NP10 9AQ MW6KEQ
NP10 9EA 2W0OGY
NP10 9EN GW0DQW
NP10 9EN GW6DYW
NP10 9FB GW7RIB
NP10 9FT GW6TUD
NP10 9FU GW1JFT
NP10 9HN GW4RYQ
NP10 9JG GW4VCL
NP10 9JQ MW0LZZ
NP100BT GW8IQC
NP11 3BE GW5NF
NP11 3DF GW0NZN
NP11 3DP GW8NBF
NP11 3DW GW4JJR
NP11 3EE MW1CNN
NP11 3JQ 2W1IBN
NP11 3LJ 2W0DDP
NP11 3NT MW6DCI
NP11 4AP MW7AIF
NP11 4HJ 2W0KDR
NP11 4HJ MW3RTU
NP11 4RE GW0NUS
NP11 4RS GW0JWF
NP11 4SQ MW7BMU
NP11 5AU 2W0CYX
NP11 5AU MW6SKR
NP11 5DL MW0RZC
NP11 5GE 2W0KHK
NP11 5GE MW3KHK
NP11 5GE MW3XYW
NP11 5GE MW6NON
NP11 5HN 2W0MAC
NP11 5JY 2W0MJC
NP11 5LA GW0NPL
NP11 5LU 2W0END
NP11 6AW 2W0GEM
NP11 6AW MW6NUW
NP11 6BN 2W0BTQ
NP11 6BN MW6TLR
NP11 6BP MW3MXC
NP11 6GL MW6CZV
NP11 6HJ 2W0BFC
NP11 6HJ MW0GKV
NP11 6HJ MW3LMZ
NP11 6HW MW7JON
NP11 6NB 2W0RJL
NP11 6NB MW0RBL
NP11 6NB MW3LYQ
NP11 6PF 2W0EAD
NP11 6PF MW3ZCU
NP11 6QH MW3YLV
NP11 6QT MW7DFX
NP11 6RD GW0RYT
NP11 6RG MW6CZU
NP11 7AJ 2W0BFD
NP11 7AJ MW3FDO
NP11 7EP MW6GVR
NP11 7FS 2W0HWQ
NP11 7FS MW7CBT
NP11 7HY GW1ZKE
NP11 7PF GW7IAT
NP11 7PG GW1DRZ
NP11 7PG MW7WRZ
NP12 0DA GW8TRO
NP12 0DB GW0JXG
NP12 0DP MW3LMU
NP12 0EL MW6SVV
NP12 0EL MW6ZBE

NP12 0LU GW0BZJ
NP12 0NQ GW0DJX
NP12 0QE GW0RQS
NP12 0QR GW8OUM
NP12 1DF GW3OAJ
NP12 1DZ GW1PJP
NP12 1EE GW4ROV
NP12 1EW GW0LJW
NP12 1FS GW0KLY
NP12 1HQ GW0MOW
NP12 1LX GW0ARK
NP12 1QH 2W1ITI
NP12 1QH GW8ITI
NP12 2AD GW0JMJ
NP12 2DA MW0CVT
NP12 2DA MW0TTU
NP12 2ET MW6YHR
NP12 2GA GW4JKV
NP12 2GE MW3ZUF
NP12 2GE MW7AHM
NP12 2HP GW4FCV
NP12 2HP GW6BK
NP12 2JB GW0SZN
NP12 2JU GW4HBK
NP12 2JU GW6GW
NP12 2LR 2W0IBY
NP12 2LR MW6PDU
NP12 2LS MW3VKM
NP12 2ND GW0NPM
NP12 2NJ MW3MNV
NP12 2NL GW8RFD
NP12 2PF MW0LAO
NP12 2PF MW3YLO
NP12 2PU GW6OIO
NP12 3FX 2W0LBE
NP12 3FX MW6BVU
NP12 3HB MW0DNF
NP12 3NG GW0EYH
NP12 3NR MW0KEV
NP12 3NR MW0MNX
NP12 3NR MW7LOV
NP12 3PG GW3XVQ
NP12 3QS MW3EAI
NP12 3RH GW0ETM
NP12 3RH GW1EHI
NP12 3UL MW1EOR
NP13 1BS GW3PFV
NP13 1DR 2W0SDO
NP13 1DR MW0MDJ
NP13 1DR MW6TBU
NP13 1EX MW6PYP
NP13 1JE MW6BWG
NP13 2AD GW7NTA
NP13 2AU MW0LBB
NP13 2HX MW7EPH
NP13 2HY 2W0VAW
NP13 2NQ GW8LTV
NP13 2PJ 2W0ZAE
NP13 2PJ MW3ZYE
NP13 2RQ GW4ZUS
NP13 2RT 2W1TDM
NP13 2RT MW0BBM
NP13 2RY GW4YDX
NP13 3AW MW3IHX
NP13 3DD 2W0KYZ
NP13 3DD MW7RYK
NP13 3EQ GW7RSE
NP13 3HJ MW7DSP
NP13 3JQ MW7DFE
NP13 3JS MW3WIV
NP15 1AD GW0DKF
NP15 1AT GW6XBV
NP15 1BA MW0XFU
NP15 1BA MW6XFU
NP15 1HN GW8AWM
NP15 1QG GW7KRY
NP15 1SP GW3LDC
NP15 1TN 2W0USK
NP15 2EP MW0CSC
NP15 2LF GW8VKS
NP16 5BU MW0OPS
NP16 5BX GW1SXU
NP16 5EA MW7SDJ
NP16 5EB MW3NNA
NP16 5EH 2W0TSJ
NP16 5EH GW4LWZ
NP16 5EH GW8ZOE
NP16 5EH MW3TSJ
NP16 5HP 2W0CNY
NP16 5JP GW0HBD
NP16 5LA 2W1BST
NP16 5LU GW4KPD
NP16 5LY MW3YLR
NP16 5NE GW0WQP
NP16 5RL GW4REX
NP16 5TL GW7TGB
NP16 6DW GW6HRU
NP16 6EF GW7IRD
NP16 6FA 2W0RCU
NP16 6FA MW0RVC
NP16 6FA MW6AVK
NP16 6HN GW8HUS
NP16 6LA 2W0PBU
NP16 6LA MW0YMB
NP16 6LA MW6PBO
NP16 6LE 2W0DNI
NP16 6LE MW0NVY
NP16 6LE MW6KGB
NP16 6PE GW1ISR

NP16 6PN 2W0CXV
NP16 6PN MW6DNP
NP16 6QA MW1CTJ
NP16 6TF 2W0BTZ
NP16 7AQ MW3WOV
NP16 7AQ MW6WOV
NP16 7DF 2W0JMH
NP16 7DF MW0SLH
NP16 7DF MW6JMH
NP16 7DR GW7ERI
NP16 7DS GW4RZE
NP16 7EG MW7XJO
NP16 7EX MW6MGC
NP16 7HF 2W0DJC
NP16 7HF 2W0LRS
NP16 7HF 2W0RDJ
NP16 7HF MW6RDC
NP16 7LG GW1VDT
NP16 7PZ 2W0XFG
NP165QW MW7EYC
NP18 1GT GW4ITJ
NP18 1JB MW0GOM
NP18 1JF GW6FLU
NP18 1LP GW7NFT
NP18 1LT GW0FBT
NP18 1RY GW4BCZ
NP18 2JL MW1BEQ
NP18 2JR GW6GCK
NP18 2LS MW0HID
NP18 2LT GW7PIB
NP18 2NR GW8GUJ
NP18 3BZ GW4LCF
NP18 3EF GW7BIL
NP18 3EU GW4ZVQ
NP18 3EZ GW4ESL
NP18 3PW MW3ZGR
NP19 0AN 2W0PAN
NP19 0AN MW3PQE
NP19 0AN MW3PQF
NP19 0ET MW6LOD
NP19 0GX 2W0CJZ
NP19 0HU 2W0CHH
NP19 0HU MW0LLO
NP19 0HU MW6CJH
NP19 0LP 2W0ILQ
NP19 0RD GW8IOA
NP19 4DL GW8WZR
NP19 4NX GW6EWQ
NP19 4PF MW0JBH
NP19 4RQ MW3KSE
NP19 4TB MW6PPD
NP19 7DB GW7AVB
NP19 7ER 2W0BTF
NP19 7ER MW0GXE
NP19 7ER MW3XWE
NP19 7GF 2W0DWR
NP19 7PB 2W1ADO
NP19 7QB GW4VNS
NP19 7QP GW8MOZ
NP19 7QW GW0DHG
NP19 7SA GW1DLP
NP19 7SN MW0ALN
NP19 8AG GW3KWB
NP19 8NU GW6KBD
NP19 9AN MW7GIA
NP19 9AT MW3GWZ
NP19 9DS MW7ATR
NP19 9ET GW1TDV
NP19 9JN 2W1GLY
NP19 9NE GW6RGT
NP19 9QA GW7RNC
NP19 9QR GW4TIW
NP19 9QS GW4SOC
NP19 9RY GW8OIJ
NP19 9SX 2W0HOT
NP19 9TA MW3FYR
NP197EL MW7ANT
NP197QY 2W0MNJ
NP199RZ MW7LLJ
NP2 0LX GW0LKJ
NP2 4JD GW6DEP
NP20 1FU MW6KYN
NP20 2FP MW7EUX
NP20 2GN 2W1DXV
NP20 2LT 2W0KYH
NP20 2LT GW8SZL
NP20 2LT MW6SZL
NP20 2LT MW7FDI
NP20 3BJ MW0YGJ
NP20 3DE GW0MSY
NP20 3DJ GW0FJE
NP20 3DJ GW4SUE
NP20 3DN MW3PUU
NP20 3DN MW3ZCO
NP20 3DN MW6PUT
NP20 3EE GW0FXC
NP20 3HJ 2W0JDL
NP20 3HJ MW3NEI
NP20 3HJ MW6FFN
NP20 3NR GW1XVM
NP20 3NR GW8OKR
NP20 3NS GW3KZT
NP20 3SP GW4LOD
NP20 4DT MW3KHC
NP20 4EJ MW6TBD
NP20 4GD MW7RAT
NP20 4JR GW7PRK
NP20 4LP GW7BSC
NP20 4LP MW1AZR

NP20 5AA GW4SDT
NP20 5AH MW6TJG
NP20 5DG 2W0ONY
NP20 5DG MW7CIH
NP20 5EE MW0DTD
NP20 5HX MW0IXV
NP20 5LP MW6LQG
NP20 5LP MW6LZZ
NP20 5PP 2W0CVM
NP20 5PP MW6XAG
NP20 5XE 2W0CZZ
NP20 5XE MW6CZS
NP20 6EW MW7TGM
NP20 6GU MW7NYE
NP20 6HE GW3NIN
NP20 6JD 2W0JBJ
NP20 6JD MW3JZV
NP20 6JF 2W1EID
NP20 6JN 2W0XAA
NP20 6JN GW3WTZ
NP20 6JT GW0OAJ
NP20 6LB 2W0MAO
NP20 6NB 2W0ECZ
NP20 6NB MW0ZCE
NP20 6NB MW3ECZ
NP20 6QF GW3YKZ
NP20 6QF GW8CNF
NP20 6QN GW4GFL
NP20 6QN GW4YKW
NP20 7AW GW4YNP
NP20 7AW MW1USK
NP20 7BZ GW0KTL
NP20 7DJ MW6FBJ
NP20 7EJ MW3YJJ
NP20 7SQ MW6GAV
NP202GJ MW0ZAB
NP205AW 2W0IIU
NP205AW MW7BGX
NP207SQ MW3YKA
NP22 3DD MW6MPQ
NP22 3HE GW0MQU
NP22 3HF 2W0IAO
NP22 3HF MW6XAC
NP22 3HN MW6SID
NP22 3NT MW6DQQ
NP22 3PB 2W0TAR
NP22 3PB MW6KAA
NP22 3PF 2W0LAC
NP22 3PF MW6SAA
NP22 3RZ MW0CPD
NP22 3SG GW4WLT
NP22 3SN MW6DBD
NP22 3TA 2W0SCB
NP22 3TA MW6DBB
NP22 3TE MW6EKT
NP22 4JG GW6TQH
NP22 4JG MW3KSI
NP22 4PF MW6IJW
NP22 4PL MW3HIX
NP22 5AR 2W0XBE
NP22 5AR MW3UAP
NP22 5AR MW3XAJ
NP22 5AR MW3XBE
NP22 5BH GW8TIX
NP22 5DW MW6AOE
NP22 5EA MW3VWO
NP223SQ GW0JWC
NP224PW MW3TUO
NP23 4AR GW4KFI
NP23 4HR MW7CSO
NP23 4HW 2W0RMR
NP23 4QY MW3HBC
NP23 4SD GW7HTU
NP23 4TS GW7RFP
NP23 4TT MW0JKU
NP23 4TY GW0JTU
NP23 4TY MW0IWX
NP23 5AJ MW3PMU
NP23 5EE 2W0CVL
NP23 5EE MW3NAE
NP23 5EN MW7KID
NP23 5ES GW0HCN
NP23 5FA MW3UHO
NP23 5FE GW1IOT
NP23 5FE MW3WPH
NP23 5FF 2W0HQD
NP23 5FF MW3HQD
NP23 5LU MW7KRP
NP23 5QR GW7VEH
NP23 5SF 2W0CLJ
NP23 5SF MW6EAD
NP23 5TF GW0IRT
NP23 5TS 2W1CIP
NP23 5TS GW0MNO
NP23 5TS GW1HNF
NP23 6DD MW7KND
NP23 6HT 2W0LCJ
NP23 6HT MW0PWY
NP23 6HT MW3VLJ
NP23 6JF GW4CXK
NP23 6JL MW7XRE
NP23 6JR GW0VSO
NP23 6LN GW7VGB
NP23 6NE MW3UDW
NP23 6NE MW3UEG
NP23 6PJ MW6EXL
NP23 6TR MW3HKH
NP23 6UA GW4UTS
NP23 7QY GW0JTF

NP23 7SP MW3XZB
NP23 7TG MW3FLK
NP24 6AZ MW3TLP
NP24 6BA MW0HRD
NP24 6BS MW3YBK
NP24 6BW MW3DPV
NP24 6EY MW3RZW
NP25 3HT GW0MSW
NP25 3NN MW7AUE
NP25 3SD GW3VFL
NP25 4LY GW8WCA
NP25 4NB GW1FJI
NP25 4QD GW4FIC
NP25 5DA MW0TRO
NP25 5DE GW8PTS
NP25 5EH MW3GFG
NP26 3AB GW0GEV
NP26 3AG GW3ORL
NP26 3AW MW7TLX
NP26 3AX 2W0XWD
NP26 3AX MW3LVF
NP26 3BZ GW3NWS
NP26 3BZ GW8GT
NP26 3BZ MW0LGE
NP26 3EJ GW3SGK
NP26 3FD MW3ZMU
NP26 3FG 2W0CUJ
NP26 3JF MW0GTN
NP26 3JU GW8MZR
NP26 3PB GW3MPP
NP26 3QG GW4OGO
NP26 3SA MW0HJG
NP26 3TB 2W1BOG
NP26 3TL MW1FLY
NP26 3UT MW6RLE
NP26 3UW GW3WEZ
NP26 4BH 2W0GHD
NP26 4BH MW0GZL
NP26 4BH MW6GHD
NP26 4DX MW0DTH
NP26 4JD GW4ZVL
NP26 4JL MW3KRN
NP26 4LH 2W0KIN
NP26 4LH MW7HAW
NP26 4PP GW4FNO
NP26 5BW GW0EGH
NP26 5GB GW1EPR
NP26 5RE GW6NQU
NP26 5RE MW3ESH
NP26 5RE MW3HOY
NP26 5RE MW3VEN
NP26 5TB MW6BSU
NP26 5UW MW6OTK
NP3 1AR GW6BVS
NP4 0BN GW1XVC
NP4 0BN MW3HGO
NP4 0DF MW6GWL
NP4 0DG GW0FJH
NP4 0HT 2W0REX
NP4 0HT MW0LUK
NP4 0HT MW6USK
NP4 0HX MW7PYS
NP4 0JD MW3GHF
NP4 0NB GW7LPM
NP4 0NJ GW0DQT
NP4 0PT GW4NXD
NP4 0RT MW0AZN
NP4 5AB 2W1GMM
NP4 5BX GW1VJB
NP4 5BZ GW1SXT
NP4 5EZ GW8CKJ
NP4 5JF 2W0TAX
NP4 5LT GW4HYZ
NP4 5SS 2W0BVV
NP4 5SS MW3JQX
NP4 5XS GW4ZQV
NP4 6AH MW7FNC
NP4 6BU MW0IZX
NP4 6DF GW0MBW
NP4 6HE MW6MZU
NP4 6QZ MW6HPC
NP4 6QZ MW6RHY
NP4 6QZ MW6WCT
NP4 7AB MW7TAF
NP4 7HJ MW0KMS
NP4 7QF 2W0SLD
NP4 7QF MW3SLO
NP4 7QF MW6CAE
NP4 7QS GW8ERA
NP4 7RZ MW6PGC
NP4 7SN MW6KTM
NP4 7UJ GW1VDW
NP4 8BU MW7KJC
NP4 8DQ GW4AYQ
NP4 8DQ MW3DPF
NP4 8DQ MW3LOI
NP4 8LL 2W0XVT
NP4 8LL MW0YVT
NP4 8LL MW6XVT
NP4 9EH MW6IJJ
NP4 9LX GW1WTL
NP4 9PA MW0XRU
NP4 9QL GW6STS
NP4 9QU MW0USK
NP4 9QU MW3USK
NP4 9SS GW1GKV
NP44 1AT GW4IZJ
NP44 1BQ 2W0UNY
NP44 1BQ MW6UNY

NP44 1FE MW7EXS
NP44 1FR 2W0DDJ
NP44 1LH GW0TTN
NP44 1LJ MW1CVM
NP44 1LS GW4UWR
NP44 1NB GW0FGO
NP44 1RB GW7DIL
NP44 1RZ MW3AVU
NP44 2AL GW4IQB
NP44 2AN GW0KAM
NP44 2AN GW1YKY
NP44 2JW 2W0SEZ
NP44 2JW MW0SEZ
NP44 2JW MW3SEZ
NP44 2JW MW6GZC
NP44 2JW MW6SDE
NP44 2PF MW6ERZ
NP44 2QL MW7EHV
NP44 3AE GW4JBQ
NP44 3AE GW8SBK
NP44 3AP GW8CRH
NP44 3AP MW6CUR
NP44 3AZ GW0HIR
NP44 3BX 2W1DEA
NP44 3BX GW0OLZ
NP44 3HH GW4YML
NP44 3LS 2W1GDY
NP44 3NU MW6RLA
NP44 4JZ GW8IAD
NP44 4LB GW1YZF
NP44 4LD MW3OSQ
NP44 4LF 2W0YAD
NP44 4LF MW3NHN
NP44 4LG 2W0LFE
NP44 4LG MW6KUD
NP44 4LW MW3URO
NP44 4PQ MW3LMV
NP44 4PX 2W1GCU
NP44 4PX MW0RPB
NP44 4QT MW1FFE
NP44 4RB MW7NMT
NP44 4TE GW0UKT
NP44 4TJ MW1DOO
NP44 5AS GW1UXW
NP44 5HH MW3NZZ
NP44 5HN GW8JOY
NP44 5JA MW6FAW
NP44 5JE MW0MTR
NP44 5JF MW7RBO
NP44 5TQ GW0DQY
NP44 5TQ GW1ZFX
NP44 5UH MW6OZB
NP44 5UN GW0OZB
NP44 5UN GW7DHG
NP44 5UU MW6SOF
NP44 6EH GW7SSQ
NP44 6JF MW7CIE
NP44 6JH 2W1DNK
NP44 6JH 2W1DRB
NP44 6JH GW0DHA
NP44 6JJ GW7SSN
NP44 6JR MW6LQK
NP44 6TY 2W0NWJ
NP44 6TY MW0NWJ
NP44 6TY MW7NWJ
NP44 6UD MW7BMW
NP44 7AB GW4CQT
NP44 7AH GW0DXG
NP44 7AH GW4RIB
NP44 7EB MW7VTD
NP44 7JG GW0HDY
NP44 7JP 2W0MKG
NP44 7JP MW0MKG
NP44 7JP MW6MGY
NP44 7JX MW0YAC
NP44 7JX MW0YAD
NP44 7LB MW7AIH
NP44 7LL GW7CBU
NP44 7LL MW7FZW
NP44 7LS MW6EJG
NP44 7LT GW7UMS
NP44 8RJ GW6BHQ
NP44 8RY GW6RBZ
NP446JR 2W0WOY
NP446JR MW7WDY
NP45EA 2W0JEV
NP6 1PE GW6BAH
NP7 0BP GW6JVB
NP7 0BY GW0JJF
NP7 0DX GW6VED
NP7 0EL GW1UVN
NP7 0LL 2W0RGL
NP7 0LL MW0GYY
NP7 0LL MW3TYC
NP7 0LS MW1AZI
NP7 0RA MW0LDJ
NP7 5BQ GW4JDE
NP7 5HQ GW6JPC
NP7 5HW GW4IHM
NP7 5JA GW0UDH
NP7 5LG MW0YVK
NP7 5TL GW6BXU
NP7 5TL GW6RAV
NP7 5TL MW1FAT
NP7 5TN MW6ETW
NP7 6AF GW6ZZF
NP7 6AP MW6EYM
NP7 6HB MW3ETB
NP7 6HN GW0UKG

NP7 6JY GW6RAO
NP7 7AL GW4UHK
NP7 7AW GW8KJK
NP7 7AW MW7FWX
NP7 8DH MW6NFH
NP7 8DN GW4IQA
NP7 8ER MW7KPV
NP7 8TG MW6CYM
NP7 8UF GW7PRW
NP7 9HP GW8KCH
NP7 9RS MW6EJI
NP7 9SA GW4EIN
NP75SY GW7HLZ
NP8 1AP GW0BWE
NP8 1DJ GW0XYL
NP8 1DJ GW4ZUW
NP8 1DQ GW4FRH
NP8 1LN GW6KGR
NP8 1NU GW8IKH
NP8 1RT GW4BRH
NP8 1RU GW8IAM
NP8 1SN MW7ETY
NP8 1SY GW4UZC
NP9 7HW 2W1AYO
NR1 1EA M7ISG
NR1 1ET G4FAI
NR1 1ET G8IAR
NR1 1LW M6VOX
NR1 1SH G0LQX
NR1 2AL G7VZL
NR1 2JR G4RRX
NR1 2JX 2E1HOO
NR1 2JX G7UVB
NR1 2NG M7GMC
NR1 2NL G0WJU
NR1 2NX G4VLS
NR1 2PW M6TVL
NR1 3ED G0WFP
NR1 3FU G4GHO
NR1 3HB M6HAE
NR1 3JB M6GCC
NR1 3QX M6VDH
NR1 4BE G0MWH
NR1 4EN M6DPZ
NR1 4EP 2E0GDO
NR1 4HQ G4JYG
NR1 4JX M1CRQ
NR1 4LN M3SDB
NR1 4NJ M6BLI
NR1 4PP 2E0GFB
NR1 4PP M6RWV
NR1 4QB 2E0NLK
NR1 4QB M6MLK
NR1 4QB M6SLK
NR10 3BH G4RSP
NR10 3EA G8BIG
NR10 3ES G0TMT
NR10 3GE M7EQR
NR10 3GW 2E0SPA
NR10 3GW M6SPA
NR10 3GY 2E0NCG
NR10 3GY M0NCG
NR10 3GY M1AFZ
NR10 3GY M3NCG
NR10 3HE G8YQH
NR10 3HE M1GPC
NR10 3HF G7ETC
NR10 3HF M3LJA
NR10 3HF M3UMA
NR10 3HS G7PDO
NR10 3HY M6COU
NR10 3JT M6XYL
NR10 3LD G4WUG
NR10 3LF M6LAE
NR10 3LG G0SGT
NR10 3LL G6PFC
NR10 3LP M6JWP
NR10 3LX G0DWV
NR10 3NB M6GUW
NR10 3NN M6EOA
NR10 3PG G0AUT
NR10 3PP G1ESX
NR10 3QA G7JTZ
NR10 3QA M3JTZ
NR10 3QB G3VPT
NR10 3QB M0SPX
NR10 3QQ M3NGF
NR10 3QW G3TWX
NR10 3ST G0MJC
NR10 4AE M3EDC
NR10 4AZ G0PFJ
NR10 4BG M3MJD
NR10 4BS G1WUY
NR10 4BS M3YGU
NR10 4EL G1IDJ
NR10 4HA G4WUU
NR10 4HJ 2E0YOP
NR10 4HJ M3YOP
NR10 4HL M3SOY
NR10 4JZ G6ZRV
NR10 4LS G0IYK
NR10 4LU G8YJS
NR10 4PP 2E1HKE
NR10 5AU G3WCE
NR10 5AU M1BQW
NR10 5AX M7EAM
NR10 5DJ M6FBE
NR10 5DJ M6HDM
NR10 5EX M3RZN

NR10 5EX M6SEN
NR10 5JT M6HVY
NR10 5JY 2E1GZY
NR10 5JY M1EPR
NR10 5JY M3EPR
NR10 5LE 2E1EDA
NR10 5LT G3VWQ
NR10 5NU G7UGB
NR10 5PD M0BLH
NR10 5PW G4SGX
NR10 5QW G3KMO
NR10 5QX G1NGE
NR103JP M7HSF
NR11 6DJ 2E0DDL
NR11 6DJ M0LDV
NR11 6DJ M6DXN
NR11 6HW 2E0RDI
NR11 6HW M6OPL
NR11 6JD 2E0HAJ
NR11 6JD M0HET
NR11 6JF G0EVQ
NR11 6JF G7GQB
NR11 6JZ M0JMS
NR11 6LB G0NTJ
NR11 6LX 2E0IWB
NR11 6LX M0IWB
NR11 6LX M6IWB
NR11 6SG M7DLZ
NR11 6UT G1UQC
NR11 6UT G8RSQ
NR11 7AQ G0AZR
NR11 7AW G1GGN
NR11 7BE G4GQE
NR11 7DT G4PLZ
NR11 7DY G7VRK
NR11 7HP G3YXN
NR11 7LQ 2E0WPZ
NR11 7NJ G8BLD
NR11 7PD M1BOA
NR11 7QD M7CNI
NR11 8BE G3PPR
NR11 8DN G6NZY
NR11 8ED G0TAM
NR11 8ED G7RNN
NR11 8HJ G4AFQ
NR11 8JB G3BGF
NR11 8JF G6OQJ
NR11 8JF G7HFP
NR11 8JN G4SFY
NR11 8JX 2E0NCT
NR11 8JX M7CTT
NR11 8LP G0JLV
NR11 8NG G6KBZ
NR11 8PD G4RRN
NR11 8PN G4CLO
NR11 8QL G8KHI
NR11 8UW M6SNJ
NR12 0AL G3SGC
NR12 0EP 2E0AST
NR12 0EP G4EWB
NR12 0HF G4SFQ
NR12 0LU M1EEN
NR12 0NE G4POG
NR12 0NJ G4MQK
NR12 0PD G3WQY
NR12 0PW M1DMB
NR12 0QE M6DXQ
NR12 0RB G6PYR
NR12 0RL M6ADT
NR12 0RL M6RKM
NR12 0RU G1ANZ
NR12 0RU G4CTT
NR12 0SF G7HXI
NR12 0SU M6SMG
NR12 0SX 2E1SBF
NR12 0TA G7TUM
NR12 0YG G6NLX
NR12 0YQ G3NMZ
NR12 7AB G0SMS
NR12 7AB G1KZD
NR12 7AB G1NJG
NR12 7AJ G0VDR
NR12 7AJ G7RPK
NR12 7AJ M6VDR
NR12 7BB 2E0NKI
NR12 7BB 2E0TZY
NR12 7BB M0TZY
NR12 7BB M6TKE
NR12 7BB M6TZY
NR12 7DQ G4SOP
NR12 7DT G6ESQ
NR12 7EG G7RIE
NR12 7ER 2E0SXC
NR12 7ER M6DBS
NR12 7EW G1XUW
NR12 7HD G7SNT
NR12 7HW G6GAC
NR12 7HZ G1EAJ
NR12 7JU G1HXN
NR12 7NB G0DGU
NR12 8BA 2E1EIV
NR12 8BA M0HUG
NR12 8BA M0PUD
NR12 8DG G4VUR
NR12 8DP G1OSE
NR12 8EB M6WTR
NR12 8HP M0EAZ
NR12 8JH G4YDZ
NR12 8LQ G8UPD

NR12 8LZ G0AKL
NR12 8NF M0THA
NR12 8PF M7PWL
NR12 8QR G0WJV
NR12 8SL M3MHN
NR12 8SQ G1YLV
NR12 8TU M0TVG
NR12 8UG M6NLR
NR12 8UJ G0GHB
NR12 8XB G6IYY
NR12 8YL G7PRC
NR12 8YL G7VPD
NR12 9BE M6GUY
NR12 9BY G8IYK
NR12 9BY M6DEF
NR12 9DA G8VSF
NR12 9DR M6ABQ
NR12 9EQ G1ONC
NR12 9FL M1EVZ
NR12 9LH G1CEU
NR12 9PB G1TAY
NR12 9PB M6EKS
NR12 9PJ G4HXK
NR12 9PZ G1LQM
NR12 9QN M6WZK
NR12 9SA G4CDN
NR12 9SA G4CKB
NR12 9SE M6KLM
NR12 9SG G4LPP
NR12 9SG M7LPP
NR120NL M3VTH
NR13 3AB G4JNX
NR13 3DH G0OOB
NR13 3EY M6RGI
NR13 3HJ M6EJY
NR13 3LA G0KRU
NR13 3PL G4YGQ
NR13 3PP G4YYL
NR13 3PP G6WIT
NR13 3PU M6DYO
NR13 3SW G3LNM
NR13 3TH M0HJI
NR13 4AH G4UUB
NR13 4AW 2E0BDB
NR13 4AW 2E0BSU
NR13 4AW M3NLF
NR13 4BA G4UAM
NR13 4BB G1MUC
NR13 4BU M7LLS
NR13 4FA 2E0BCQ
NR13 4FA M0KXK
NR13 4JS G4WUO
NR13 4JX 2E0DJR
NR13 4JX M3UYV
NR13 4LZ G0KSD
NR13 4NB G4BFS
NR13 4NF G0GQV
NR13 4QF 2E1FNJ
NR13 4QF 2E1HPC
NR13 4QF 2E1HPD
NR13 4QF G7UVY
NR13 4RU G4FOS
NR13 4SZ G7HVU
NR13 5DG 2E0ZZC
NR13 5DG M0LDK
NR13 5DG M6ZZC
NR13 5DL M6VXT
NR13 5HR G4MCA
NR13 5LG G0AKO
NR13 5NS M7BMC
NR13 5PA G7RBS
NR13 5RF G8UKO
NR13 5RZ M1ALH
NR13 6AY M7AQG
NR13 6DP G0IRQ
NR13 6DY G6BXR
NR13 6HH 2E0CEY
NR13 6HH M6JET
NR13 6LT G4KLM
NR13 6QD G0IMU
NR13 6QH G1BCU
NR13 6RQ G6WMG
NR13 6RR G3PDH
NR13 6SG M0HLZ
NR134JD 2E0AWG
NR134JD M0EBJ
NR135LD G7RTN
NR135RA 2E0YAO
NR135RA G0OOR
NR14 6BH G4IAO
NR14 6BQ G1MSK
NR14 6BY G3ITB
NR14 6FB M0NRW
NR14 6FB M0PTO
NR14 6NH G7HFE
NR14 6PG M0BIT
NR14 6PP G3VKM
NR14 6RL M3ZFO
NR14 6UT G4TVJ
NR14 6UZ G1MSG
NR14 7AH M6LBM
NR14 7AR G7DFP
NR14 7AS 2E0MYE
NR14 7AS M3MYE
NR14 7BU G4SOZ
NR14 7DP G0RSS
NR14 7EB M3RMD
NR14 7EE G4GUS
NR14 7JE G0DEE

NR14 7LU G0VJH
NR14 7NH 2E0SYW
NR14 7NH M0SYW
NR14 7NH M6SYW
NR14 7PB G4LOJ
NR14 7QN G4BID
NR14 7RY 2E1CPV
NR14 7WF 2E0BRF
NR14 7WF M0GMK
NR14 7WF M6CJD
NR14 8AL G4VCE
NR14 8AS G4LEP
NR14 8BA G8IXM
NR14 8BL M3ZUI
NR14 8BL M7BSN
NR14 8DH G0XAR
NR14 8HT G1PQJ
NR14 8HX G6VAD
NR14 8LQ G3LDI
NR14 8PH M0HJE
NR14 8RG G4UYR
NR14PU M7EZG
NR15 1JB M3NKP
NR15 1NG G6ULJ
NR15 1NH 2E0GOW
NR15 1NH M0GOW
NR15 1NH M7CHJ
NR15 1PY 2E0YOK
NR15 1PY M6OKY
NR15 1TH G7VNN
NR15 2DG M0TSM
NR15 2DS G3WDN
NR15 2EE 2E0HCV
NR15 2EE M7TTB
NR15 2HL 2E0OLE
NR15 2HR 2E0VBW
NR15 2HR M0VBW
NR15 2HR M3VBW
NR15 2NT G3VYN
NR15 2PH G8WSH
NR15 2ST 2E0JMB
NR15 2ST M0UKS
NR15 2TA M3LYA
NR15 2TF 2E0RUG
NR15 2TF M0RUG
NR15 2TF M6TRT
NR15 2UH M6UDX
NR15 2WY M6PWG
NR15 2XR M6EJL
NR16 1AW G8AKF
NR16 1DJ G1GSB
NR16 1DL G3MGX
NR16 1HH 2E1CYE
NR16 1LH 2E0VID
NR16 1LH M0RXS
NR16 1LH M7JUL
NR16 1NW M1CQK
NR16 1NW M1CQL
NR16 1PD 2E0MUI
NR16 1PD 2E0UMG
NR16 1PD 2E0XUI
NR16 1PD M0MUI
NR16 1PD M0UMG
NR16 1PD M0XUI
NR16 1PD M6WNS
NR16 1PD M6XUI
NR16 1RS G3SDT
NR16 2AJ G0KXV
NR16 2AN G3RXA
NR16 2DE G6XQP
NR16 2EQ M0RYB
NR16 2HL G0BOO
NR16 2JD G0ENB
NR16 2JD M1MNR
NR16 2JD M3HMK
NR16 2LW G7UFV
NR16 2NA G6HYP
NR16 2PS G3JIE
NR16 2SZ M3ZPZ
NR17 1BL G6AVI
NR17 1BW 2E0TFX
NR17 1BW M0TFX
NR17 1BW M3TFX
NR17 1DT M3SFZ
NR17 1GP G0UUT
NR17 1GU M6IIZ
NR17 1JB 2E0PLU
NR17 1JB G7URP
NR17 1JB M0TC
NR17 1JB M3DCP
NR17 1JB M3PLU
NR17 1LD M6OOJ
NR17 1LW 2E0DWZ
NR17 1LW M6RKD
NR17 1UL G3UUR
NR17 1YF M1CVU
NR17 2AD G4RKL
NR17 2HB M6VUD
NR17 2HU 2E0IXH
NR17 2HU M3IXH
NR17 2LA M3FUQ
NR17 2NG G1XYV
NR17 2NG M3MGJ
NR17 2NH G3JUU
NR17 2NW G7SEJ
NR17 2RJ M7FVS
NR17 2RQ M3NGU
NR172QW M7PNS
NR18 0AR 2E1DQT

NR18 0AR G7VZI
NR18 0DE G1FPK
NR18 0DF 2E0DWT
NR18 0DF M0WEL
NR18 0DF M6WEL
NR18 0DN G3YIA
NR18 0EA G1EGL
NR18 0EN G4OJV
NR18 0EX G8DYA
NR18 0FH M6OGH
NR18 0HS G6LQM
NR18 0HX G6YYU
NR18 0HY G7TIE
NR18 0NB M1EOV
NR18 0NT G3MPN
NR18 0NZ G6DAN
NR18 0QP G6YFZ
NR18 0QP M7SUP
NR18 0SW G0KYA
NR18 0SW M6ELE
NR18 0SW M6PUP
NR18 0TT 2E0CQH
NR18 0TT M6BVM
NR18 0TU M0RJB
NR18 0WQ G0OPV
NR18 0XT M1BCU
NR18 9BA M6TDI
NR18 9BH 2E0NVT
NR18 9BH M0NVT
NR18 9BH M6NVT
NR18 9BP M6PPV
NR18 9DD G7EOA
NR18 9DR 2E0XSW
NR18 9DR M0VSW
NR18 9DR M3VSW
NR18 9DR M7SBA
NR18 9FA M7MRQ
NR18 9HQ M6ZZB
NR18 9JD M1MIT
NR18 9JD M7MIT
NR18 9JG 2E0XKM
NR18 9JG M0XKM
NR18 9JG M6XKM
NR18 9JU G7PAC
NR18 9NB M6HYR
NR18 9RY G4VAO
NR18 9TB G7DIB
NR18 9TF G4VMG
NR19 1AJ G6ULS
NR19 1AW 2E0ZJB
NR19 1AW M0ZJB
NR19 1AW M3ZJB
NR19 1EG G1OZR
NR19 1JB G3XPT
NR19 1JY G7HCT
NR19 1LU G0GEB
NR19 1LU M6LSE
NR19 1ND G4EYA
NR19 1QE G6ZKX
NR19 1XF M6KJR
NR19 2BJ G8UUV
NR19 2BS M3KTH
NR19 2BS M3WKM
NR19 2LX G3YHO
NR19 2NE 2E0SHO
NR19 2NE G5SHO
NR19 2NE M7SLH
NR19 2NS G4UVA
NR19 2SR M6EBY
NR19 2SR M6ODM
NR19 2SU G4LLL
NR19 2SU M6NMR
NR19 2TN M7JPC
NR19 2UB G0LGJ
NR2 1EE G4HUN
NR2 2DW M3VGX
NR2 2JB G1AMN
NR2 2RN G4LFQ
NR2 2SF M6JPK
NR2 2SH M6GVA
NR2 2SN G4EXN
NR2 3NJ G7FUQ
NR2 3NJ G8GWJ
NR2 3PT 2E1DRY
NR2 3PZ G7MLC
NR2 3RL G7NVB
NR2 3TL G6TSZ
NR2 3TP 2E0MAP
NR2 3TP 2E0PEW
NR2 3TP M0PEW
NR2 3TP M3PKZ
NR2 3TP M3WPO
NR2 3TP M3YYU
NR2 3TP M6HRW
NR2 4BL 2E0BSS
NR2 4BL M3OYE
NR2 4EH M0NFY
NR20 3DG 2E0CAU
NR20 3DG G4LPW
NR20 3FN M6HYZ
NR20 3GH G3ZQI
NR20 3JF M7EAH
NR20 3LX G0UYA
NR20 3QP G3YLA
NR20 3RE G4DYC
NR20 3RF G0UML
NR20 3SB G0VIJ
NR20 3SY M7MCO
NR20 4AA G3ZIG
NR20 4AA G4ANT
NR20 4AY M3YCS
NR20 4AY M3ZCE
NR20 4BW G4RWI
NR20 4BW G6YXB
NR20 4DW G7HJD
NR20 4HU G1NMI
NR20 4LR G4IPN
NR20 4LY M3TDT
NR20 4NQ G0TYQ
NR20 4PS M7LPB
NR20 4SL G0UYC
NR20 5JZ M7FWG
NR20 5LL G8OO
NR20 5PS G0GHE
NR20 5QN M6GYV
NR20 5RD M6NIL
NR20 5RU M6EIU
NR20 5SB M6JJK
NR20 5TA G4XCR
NR20 5TW G3JCK
NR204PS M7OPB
NR21 0BU G4UKZ
NR21 0EJ G3XMP
NR21 0HX G0PPR
NR21 0JU G4KZT
NR21 0LT G6LEK
NR21 0PT G0WCH
NR21 0QX G7OSJ
NR21 7AQ 2E0TEF
NR21 7AQ G3TBL
NR21 7AQ M6TEF
NR21 7BL M7GHP
NR21 7BY G7PSK
NR21 7NA G0FVF
NR21 7NA G6MDM
NR21 7PU 2E1TMB
NR21 7QE G6YLR
NR21 7QL M7BPX
NR21 8DW G1IJC
NR21 8HB G3PAX
NR21 8LR G0KYG
NR21 8NP M6AWA
NR21 9ES G0KRD
NR21 9HW G4RRD
NR21 9JB G6IGO
NR21 9LH M6AIQ
NR21 9PD G0SQI
NR21 9PD M6SQI
NR23 1BE M0KML
NR23 1HF G4DIC
NR23 1JA M7ODG
NR23 1JY G4KOV
NR23 1JZ G2PK
NR23 1LR 2E0CTT
NR23 1LR G0SRZ
NR23 1PA G4RAY
NR23 1QE G0DSN
NR23 1QE G0NWT
NR23 1QL G7BUS
NR23 1QL G7EVQ
NR24 2DE G4IWI
NR24 2EW G8RES
NR24 2JY G6VZG
NR24 2LB M3YPX
NR24 2LQ G0FVS
NR24 2LQ M6CYC
NR24 2PH G8AKP
NR24 2RD G6AUY
NR25 6DG G4BUF
NR25 6EN G7PTD
NR25 6HH G6PVY
NR25 6NW G0SIQ
NR25 6NX G3TGE
NR25 6RU G3LQJ
NR25 6TG M6BDX
NR25 6TU M0SSF
NR25 6TU M7MHM
NR25 6TX M3WZJ
NR25 6TZ G7OSK
NR25 6UD 2E0ERU
NR25 6UD M0IUC
NR25 7DQ G8YSJ
NR25 7HH 2E0JDE
NR25 7HH G0GFQ
NR25 7HH G1MIE
NR25 7HH M6JFV
NR25 7LL G4EBE
NR25 7UA G4UQA
NR26 8BJ M3ZNY
NR26 8EJ M6DPI
NR26 8HU G7CIK
NR26 8JB M6KGI
NR26 8NQ G4ARX
NR26 8PT G7PHT
NR26 8QQ M7RGK
NR26 8SH 2E0SLS
NR26 8SH M0SHK
NR26 8SH M0WPX
NR26 8SH M6KAH
NR26 8UB M3BFU
NR26 8UN G3HRK
NR26 8UN G6DFA
NR26 8XP G4DMB
NR26 8YE M0VAW
NR26 8YJ M6ASE
NR27 0BG G4LYB
NR27 0BG G8IMI
NR27 0BP M6IFI
NR27 0BY M0DFF
NR27 0DD G4RZN
NR27 0DD M0NCA
NR27 0EE G0UIQ
NR27 0EQ M6FQK
NR27 0EQ M6GAD
NR27 0EQ M6LAQ
NR27 0HG 2E0BCW
NR27 0HG M0EBN
NR27 0HG M6BXA
NR27 0HQ G2BHG
NR27 0NY M6CFU
NR27 9AH 2E0TBX
NR27 9AH M6CRP
NR27 9AW M6HYQ
NR27 9HR G0GLJ
NR27 9LQ G0BGX
NR27 9LQ G0SQK
NR27 9LQ G1MKP
NR27 9LW G0LYN
NR27 9LW G1SVD
NR27 9PH G3PND
NR27 9PH G4PQP
NR27 9PQ G7HRJ
NR27 9PR 2E0YSO
NR27 9PW 2E0WBW
NR27 9PW M6WWD
NR27 9QT G3UDV
NR27 9RN G3ZRG
NR27 9RR M3UCA
NR279BW G4NRE
NR279BW G6EHE
NR279BW G6HL
NR28 0AU M6MSC
NR28 0BS M3YXV
NR28 0DH 2E0WOS
NR28 0DH M0WOS
NR28 0DH M3WOS
NR28 0EE 2E0UJD
NR28 0EE G8RZZ
NR28 0EE M0UJD
NR28 0EE M3UJD
NR28 0EE M3VTN
NR28 0EN G3WXW
NR28 0FP M7WET
NR28 0JP M3PGD
NR28 0LE 2E0FAY
NR28 0LE M6BLQ
NR28 0LJ G6CKD
NR28 0LR G4ZWM
NR28 0NP M7PCA
NR28 0PJ G7TIN
NR28 0QG G3YOA
NR28 0QR 2E0WTG
NR28 0QR M0WTG
NR28 0QR M3WTG
NR28 0QU G0GRB
NR28 0RZ G1NST
NR28 0SY G0IYD
NR28 0SY G1XYD
NR28 0TA M7BYL
NR28 9BB G0AJJ
NR28 9BB G6VOV
NR28 9HS G4RTH
NR28 9JA G6WME
NR28 9JE M6NWP
NR28 9LD G4ZAY
NR28 9LX M6ZAH
NR28 9PL M6AST
NR28 9RG G6JCY
NR28 9SD G7IIN
NR28 9SH G4TDQ
NR28 9UP 2E0LSR
NR28 9UP M6EMT
NR28 9XA G6JKC
NR28 9XT 2E0CVY
NR28 9XT M0NAQ
NR28 9XT M6CFQ
NR280BF M6GBD
NR280EE M7BEE
NR280UG M7NUK
NR29 3BG G7UDU
NR29 3EP G6WNN
NR29 3HD G5NB
NR29 3HJ 2E0SIH
NR29 3HJ G0MYC
NR29 3HJ M0SIH
NR29 3HJ M7SMH
NR29 3HU G8VPE
NR29 3JB M6DAL
NR29 3JX M6NFK
NR29 3LT G7DXQ
NR29 3PE M7MAS
NR29 3QD 2E0GFW
NR29 3QD M6GFW
NR29 3QT G0ELJ
NR29 3RJ G1LTH
NR29 3RJ G4SCS
NR29 3RQ M6LGT
NR29 3SP M3XUV
NR29 4AF G7BMY
NR29 4AQ G7TUK
NR29 4EA G3ZEB
NR29 4ES G1EBP
NR29 4JD G8PEN
NR29 4JL G8SYV
NR29 4JL M3EJM
NR29 4JY G7HRF
NR29 4LT G3XAU
NR29 4PT G1RPV
NR29 4PU 2E0NHJ
NR29 4PU M6GZK
NR29 4QN M6VRJ
NR29 4QN M6VSJ
NR29 4QN M6XMJ
NR29 4QT M3XMU
NR29 4RJ G4MFX
NR29 4RY G4EMH
NR29 4SH G8UCZ
NR29 4TB M7SEY
NR29 4TD G0FVT
NR29 4TW 2E1FQB
NR29 4UR M3JST
NR29 5ED G1JBZ
NR29 5HA G1ELP
NR29 5HQ G3KCD
NR29 5LG G0APV
NR29 5LG G6TXQ
NR29 5NU G0KCB
NR29 5NZ M6ALG
NR29 5PB G4AXA
NR29 5PB G4CMM
NR29 5PZ G0VSB
NR29 5QL G0OBH
NR3 1NY G0JAG
NR3 1PN M6FEE
NR3 1SE M6MBI
NR3 2AD M0TJL
NR3 2DL G7VUM
NR3 2HT 2E1HOS
NR3 2HT 2E1HOT
NR3 2HW G8AUN
NR3 2LF 2E0YTF
NR3 2LF M3YTF
NR3 2NG M3LXU
NR3 2NQ G0CWD
NR3 2NQ G0WJZ
NR3 2PS M6EIM
NR3 2QN 2E0YRW
NR3 2QN M3YRW
NR3 3DT G4YDQ
NR3 3HD 2E0KFO
NR3 3HD M3WVC
NR3 3LU M3XIM
NR3 3NN G0SSX
NR3 3PY M1CRL
NR3 3QD M3WZN
NR3 3TB G4NFF
NR3 3TD M6DDO
NR3 3TD M6YRL
NR3 3TQ M6HDS
NR3 4BE M6ZOC
NR3 4EF G3ZTV
NR3 4EW 2E0FRD
NR3 4EW M6WLG
NR3 4QF M7INA
NR3 4QW G0PEQ
NR3 4RA M7HRT
NR3 4RX G8XEN
NR3 4TL M7VDX
NR3 4TS M1AFQ
NR30 1DU 2E0TWQ
NR30 1DU M6TWQ
NR30 1JG M3NNI
NR30 1JY 2E1BFX
NR30 1JY M6LDY
NR30 1LY G0GGB
NR30 2AT G4TNY
NR30 2LF M3OYS
NR30 2RU M0NPQ
NR30 2SA 2E1GVB
NR30 3JU G6ATW
NR30 4BB 2E1HVN
NR30 4BB M3HVN
NR30 4BS 2E1HAU
NR30 4EA 2E0CNW
NR30 4EA M0STL
NR30 4HL M6LBT
NR30 4JU 2E1HVU
NR30 4LD 2E1HVT
NR30 4LD M3RJH
NR30 4LR G4ALC
NR30 4LS M6CHJ
NR30 5AB M0WBR
NR30 5AH M1BFI
NR30 5ES M3LMQ
NR30 5HG M6AIO
NR30 5PH G4VSD
NR30 5PQ 2E0BNI
NR30 5PQ M3ORZ
NR30 5PQ M3TFZ
NR30 5QT G8VQX
NR30 5RH M1BUC
NR30 5RW 2E0GAV
NR30 5RW M6AEG
NR30 5TT M1ADP
NR30 5UN 2E0BNT
NR30 5UN M0ZOR
NR30 5UN M3RZJ
NR31 0BU M1CDV
NR31 0HR G4YZF
NR31 0HR G6ZG
NR31 0PW G6VZU
NR31 6EF M0DLG
NR31 6LZ M6DBA
NR31 6NH M7EPK
NR31 6PR M6MIL
NR31 6PS M7CKV
NR31 6PX M3YQR
NR31 6PX M6MKW
NR31 6PZ M3XEO
NR31 6QY G0SHJ
NR31 6RQ G6FQL
NR31 6RT M6JKS
NR31 7AS 2E1AGQ
NR31 7AS M7COD
NR31 7JJ G0DVQ
NR31 7ND G0VMK
NR31 7ND G3YRC
NR31 7ND G7MKV
NR31 7NN M7ERG
NR31 7RF G6ARM
NR31 8BU M6LDA
NR31 8DN G3WPG
NR31 8DR G1HNU
NR31 8DR G7LNM
NR31 8HU M7RKZ
NR31 8JG G3VSV
NR31 8JW M6KGQ
NR31 8NL 2E0GZF
NR31 8NL M0JRN
NR31 8NL M7AUY
NR31 8NT M3MJY
NR31 8QJ G1GER
NR31 8QJ M7GWI
NR31 8RG G7RPJ
NR31 8SX G8TAE
NR31 8TF 2E0WBE
NR31 8TF M6WBE
NR31 9AR M6CBJ
NR31 9BT G8VKO
NR31 9DF G7FCJ
NR31 9JT G0FEI
NR31 9JU M1CQR
NR31 9LZ G6KZI
NR31 9NT M6FBZ
NR31 9NX G0GGE
NR31 9PD M6SVY
NR31 9PH G8PIO
NR31 9PL M7JSD
NR31 9PP 2E0CBH
NR31 9PP G0BAW
NR31 9PP G0CLG
NR31 9PP M6ABX
NR31 9PP M6WOO
NR31 9PQ G7WHM
NR31 9SB G4VYN
NR32 1EH G7PWK
NR32 1JY 2E1BPN
NR32 1RY G0WCO
NR32 1SW 2E0GFV
NR32 1SW M0JNJ
NR32 1SW M6SHQ
NR32 1TP M1ATP
NR32 2HH G0RRI
NR32 2JS 2E0PAP
NR32 2JS G3ETP
NR32 2JS M3UGQ
NR32 2LA M6KWC
NR32 2LB M7OCM
NR32 2NW M3NNJ
NR32 2QG M3GIF
NR32 2RY G4RMT
NR32 3AQ M0CHP
NR32 3DN M6BAW
NR32 3DS M0XGK
NR32 3ES G1MPL
NR32 3EP M7FRS
NR32 3ER 2E0RZS
NR32 3ER G3JRM
NR32 3ER G8PIR
NR32 3ER M0RSX
NR32 3ER M6KFX
NR32 3ER M7WIS
NR32 3FJ M7FZO
NR32 3HS G0VMF
NR32 3JF G3AMK
NR32 3NF G3XGK
NR32 3QL 2E0TNC
NR32 3QL M6BCK
NR32 4AU G7TJZ
NR32 4DR G4CPW
NR32 4DU 2E0GGO
NR32 4HT M6ATQ
NR32 4JU M0XCL
NR32 4LW G8JBD
NR32 4PZ 2E0OBB
NR32 4PZ M3OBB
NR32 4PZ M6KPT
NR32 4QB G3AER
NR32 4QB M0DMR
NR32 4QB M7VAC
NR32 4SU G0JSG
NR32 4UD M7CBY
NR32 5AS G0MEE
NR32 5AS G8RHQ
NR32 5BE 2E0FDG
NR32 5BE M6XBP
NR32 5DJ G4TAD
NR32 5DW G8JUK
NR32 5HJ G4UKX
NR32 5LB G4RKP
NR32 5LL G4FCZ
NR324SU M7YAC
NR324UB G6NGA
NR33 0DX 2E0BFT
NR33 0DX M3AGE
NR33 0LG G8HAU
NR33 0LY G8HRF
NR33 0TZ G4KDL
NR33 7BQ 2E1OZO
NR33 7BT G7JRU
NR33 7BX G0VDQ
NR33 7EY M6IBC
NR33 7HG G4ZSX
NR33 7HH G7MJX
NR33 7JF G8VOJ
NR33 7JG M7IJC
NR33 7PS M0DCU
NR33 7RG G4IVU
NR33 7RR M3SKB
NR33 7TF 2E0NAQ
NR33 7TF M6CPP
NR33 7TF M7JDP
NR33 7TR 2E1WCD
NR33 7TR G4CKH
NR33 7TU G6NKS
NR33 7TU M6BLP
NR33 7TY G3UPS
NR33 7UH 2E1ERU
NR33 8BL G0VCW
NR33 8BL M7BTS
NR33 8BQ G8FGY
NR33 8BY 2E0TJW
NR33 8EG 2E0JPW
NR33 8EX G1SWP
NR33 8GT 2E0SYX
NR33 8JD G3PGQ
NR33 8JR M7KHL
NR33 8NE M3KPQ
NR33 8NN 2E0SJF
NR33 8NN M0SLF
NR33 8PJ G6MCB
NR33 8PJ G7BYI
NR33 8QG G1ZCS
NR33 8QQ M0DFY
NR33 8QX M7BGH
NR33 8SD G7SDM
NR33 8TB 2E0VRT
NR33 8TB M6MPK
NR33 8TN M1FJG
NR33 8WJ M0ALX
NR33 8WP G2DWB
NR33 9AW G4ZMH
NR33 9BB M0JTH
NR33 9BB M3NMV
NR33 9BB M6PLO
NR33 9BG M3AGA
NR33 9DH G4HRC
NR33 9DH G4ZST
NR33 9DH G8HRC
NR33 9HA G1HQK
NR33 9HA M3TIK
NR33 9HL G1IAD
NR33 9JE G0HSN
NR33 9JR 2E0UID
NR33 9LR M7DGB
NR33 9LZ M3FVX
NR33 9NX G0ULS
NR33 9PG G7PQB
NR33 9QL G7MWM
NR33 9RH M5TAM
NR330AB 2E0FIA
NR330AB M3FIA
NR34 0AL M7MDZ
NR34 0AU M0TDK
NR34 0EX G6PSC
NR34 7AR M1DUD
NR34 7AR M5AEF
NR34 7BJ 2E1DDJ
NR34 7DJ G7RXI
NR34 7PG M6MOP
NR34 7PQ G6LVN
NR34 7PT M1SSB
NR34 7QL M1RJG
NR34 7RD G0WQW
NR34 7TH G0LRP
NR34 7UW 2E0VKK
NR34 7UW M0VKK
NR34 7UW M6VKK
NR34 8AP G0JDL
NR34 8AQ M3NOS
NR34 8AT G0SBH
NR34 8HR M6DAS
NR34 8HR M6PSG
NR34 8PA M6AJC
NR34 8RN M6SHD
NR34 9AS G4RZR
NR34 9AS G7FWE
NR34 9AS M5MC
NR34 9BJ M7TLM
NR34 9DW M1VLS
NR34 9HN G8BJO
NR34 9JJ G4XAE
NR34 9PL G6OAU
NR34 9RH 2E0CEH
NR34 9RT G4KFS
NR34 9RT M6GVT
NR34 9RU M6POZ
NR34 9UG M3PDC
NR34 9UX M7SKW
NR34 9YW G3KIJ
NR34 9YW M0DEB
NR349UT M7JAQ
NR34DX M7KAA
NR35 1BY G6VZF
NR35 1HE G4JUV
NR35 1JG 2E0FUQ
NR35 1JG M0JIJ
NR35 1JG M7AUP
NR35 1JU G6LSW
NR35 1ND M5BFL
NR35 1PP M6JAR
NR35 1QR 2E0KXR
NR35 1QR M6KXR
NR35 1RE G0JKI
NR35 2BW M3RCG
NR35 2LP G8JWT
NR35 2NB M0LYR
NR35 2NB M1TES
NR35 2PH M6LIH
NR35 2QP M1DHT
NR35 2QZ M6JJU
NR35 2SF 2E0XDI
NR35 2SF M3XDI
NR35 2SF M6PUD
NR4 6AA G3CIM
NR4 6AF 2E0HIK
NR4 6AF M6FHE
NR4 6EY G6WAO
NR4 6HA G8ZGS
NR4 6JD G3GLA
NR4 6LT G4NZQ
NR4 6LT G7MKP
NR4 6PS M0SWO
NR4 6PY 2E0NUD
NR4 6PY M0NUD
NR4 6PY M6NUD
NR4 6QB G0PFN
NR4 6RA M0RLW
NR4 6XH M0EVK
NR4 7EA M6GNO
NR4 7ED G0MQI
NR4 7ET M0CKL
NR4 7EU M1ACT
NR4 7HP G8VLL
NR4 7HP M1BKF
NR4 7JA G3SCV
NR4 7JW G4NOL
NR4 7LT G0FDT
NR4 7LZ 2E0HTJ
NR4 7LZ M6HTJ
NR4 7PD G4VUV
NR4 7PY G0IYY
NR4 7QY M3IGA
NR4 7RY G0CKH
NR4 7SE M0HBL
NR4 7SF G4KAU
NR4 7TJ 2E0CHU
NR4 7TJ M0HEJ
NR4 7TJ M6ATL
NR46HZ M6WYM
NR5 0EH 2E1GTE
NR5 0EH G4ARN
NR5 0EH M3AEZ
NR5 0EZ M3DKC
NR5 0HB G4ONF
NR5 0HG G6KCG
NR5 0HL 2E0FIJ
NR5 0HL M0KRK
NR5 0HL M6FIJ
NR5 0HX G1TPQ
NR5 0HZ 2E0OCL
NR5 0HZ 2E0ODT
NR5 0HZ 2E0OEE
NR5 0HZ G0NNS
NR5 0HZ G0ODR
NR5 0HZ M0OCL
NR5 0JA G6TAF
NR5 0JA G7DTR
NR5 0JB G4KQY
NR5 0JD G8UJO
NR5 0LP 2E0IMW
NR5 0LP M0IMW
NR5 0LP M6IMW
NR5 0LW G4MYQ
NR5 0NH G6KAI
NR5 0QH 2E0NXL
NR5 0QH M6NXL
NR5 0RU 2E0EPZ
NR5 8AS G0SMH
NR5 8ED 2E0GCO
NR5 8ED M3TJL
NR5 8ED M6EXY
NR5 8HW M6XTD
NR5 8RA G4RMN
NR5 8TE 2E0AZF
NR5 9AX G4WTD
NR5 9BH G3THV
NR5 9LB G1DXQ
NR5 9NN M7PJI
NR58DQ 2E0PAS
NR58DQ M3BVY
NR6 5DE 2E0IBP
NR6 5EL 2E0BWA
NR6 5EL M6SVF
NR6 5EP G4PFZ
NR6 5HF G4GVR
NR6 5HU 2E0DBS
NR6 5HU M3DBS
NR6 5NN G6UPI
NR6 5PD M6FIG
NR6 5PE G4YEJ
NR6 5PQ 2E0NGR
NR6 5PQ G0CUL
NR6 5PQ M6EIZ
NR6 5PQ M6NJG
NR6 5RQ G0ULI
NR6 5SJ G4VWE
NR6 5SW G4YGD
NR6 6DG M1DPU
NR6 6JF 2E0KNE
NR6 6JF 2E0LSX
NR6 6JF M0LSX
NR6 6JF M3LSX
NR6 6JF M3NKW
NR6 6JF M3UYY
NR6 6LN M3ILY
NR6 6LZ 2E0DRT
NR6 6LZ M6MUD
NR6 6PB G4XNP
NR6 6PL G8SDU
NR6 6PU G0ANN
NR6 6RB G0FKG
NR6 6RZ 2E0JKB
NR6 6RZ M0HDC
NR6 6RZ M0JKB
NR6 6RZ M6JKB
NR6 6RZ M6ZLA
NR6 6TD M3PEQ
NR6 6TS 2E0VFN
NR6 6TS M6VFN
NR6 6UA G7DVO
NR6 6UX M1AEH
NR6 6XD G3IOR
NR6 6XF G4EYT
NR6 7BB G6TGQ
NR6 7BG G6EJM
NR6 7BL G0VDO
NR6 7DP G4WEE
NR6 7FE G4YGA
NR6 7HA 2E0GST
NR6 7HA M0GSZ
NR6 7HA M6GST
NR6 7HA M6RND
NR6 7HE G4JSD
NR6 7HE G4UIY
NR6 7HF G3EVA
NR6 7HR G8EEY
NR6 7HR M0NBD
NR6 7HR M0NCR
NR6 7HR M0RGF
NR6 7HR M0RGY
NR6 7HR M0WHQ
NR6 7HX M0ADZ
NR6 7LG 2E0JZU
NR6 7LG G4AUV
NR6 7LG M6JZU
NR6 7LJ M1JLM
NR6 7NB M0XWS
NR6 7NL G1NCN
NR6 7PH G0ULG
NR6 7RQ G4LSG
NR6 7RQ G4TWS
NR6 7RQ G4TWT
NR66HD 2E0IBR
NR66HD M7DPT
NR7 0AS G6TLP
NR7 0BQ G1KJH
NR7 0DS M1AUW
NR7 0GY M6DYD
NR7 0JJ G3HQS
NR7 0JJ G4VYH
NR7 0JS 2E0GUJ
NR7 0JS M0IWO
NR7 0JS M6GUJ
NR7 0LY 2E1IKM
NR7 0NA M1ADX
NR7 0SF 2E0TAZ
NR7 0SF M3IPD
NR7 0UD M3XEG
NR7 0US M6VST
NR7 0XZ M6NME
NR7 8AA G7AJX
NR7 8AD G3HKD
NR7 8EE G6OUM
NR7 8EG G0UEB
NR7 8EJ G4AJO
NR7 8GX 2E0OBO
NR7 8GX M6RSB
NR7 8NQ G4LGB
NR7 8QA M7ODY
NR7 8RG M6LXM
NR7 8TF 2E0YWP
NR7 8XH G4PCZ
NR7 9AD G7SNC
NR7 9DZ M3WYG
NR7 9HW M3IXM
NR7 9LG G6AIO
NR7 9LL G0BBV
NR7 9LL G4OLP
NR7 9RR 2E0NNE
NR7 9RR M0NNE
NR7 9RR M6NNE
NR7 9TW 2E0CIS
NR7 9TW 2E0TCZ
NR7 9TW M6BAV
NR7 9TW M6UAS
NR7 9UH M6KDD
NR8 5AT M6TCE
NR8 5AU G3PZX
NR8 5AU G3YIR
NR8 5AW M6CBQ
NR8 5AW M6VWN
NR8 5AX G6LZM
NR8 5DP G3EZB
NR8 5DT M3PFN
NR8 5EY M6LWP
NR8 6AB G4SVI
NR8 6AB G4XVI
NR8 6BD G7IGR
NR8 6BU M6XAI
NR8 6FZ M7EOE
NR8 6GA G4URS
NR8 6GA M6YLE
NR8 6GG M6JSJ
NR8 6LA M6IWE
NR8 6LD G7NIX
NR8 6LT G7AYQ
NR8 6NN G0CBO
NR8 6PP M0MRQ
NR8 6PP M3NON
NR8 6QU G3VYG
NR8 6QY 2E0FKJ
NR8 6QY M0TUM
NR8 6QY M7RGR
NR8 6SL M0MBI
NR8 6TR G7LBO
NR8 6UU M0TTN
NR8 6UU M6SBK
NR8 6UZ M3VDZ
NR8 6XZ G7SQW
NR8 6YE G7WEW
NR9 3AB G3WHG
NR9 3EP G0PJI
NR9 3EZ G0AHO
NR9 3HJ M7GLL
NR9 3HQ M6CHT
NR9 3HS M6XTL
NR9 3JP G4NST
NR9 3JP G6WQN
NR9 3JY G4YKZ
NR9 3RH M7BLX
NR9 4AB G3PTB
NR9 4AD M0TNE
NR9 4AF M3ULB
NR9 4DT M0IVS
NR9 4DT M7ACG
NR9 4HA 2E0JRR
NR9 4JB G3SEM
NR9 4LL M6FJZ
NR9 4NY 2E1FDM
NR9 4PP G4AFF
NR9 4QJ M1EXQ
NR9 4QS G0PSY
NR9 4QS G7SNR
NR9 5AT M0NTZ
NR9 5BB M3RWK
NR9 5BY G0TCO
NR9 5EJ M1CQS
NR9 5NL M7GDG
NR9 5QT 2E0KPL
NR9 5QT M6LPK
NR9 5SH G3FEW
NR9 5TF G3KVT
NR93DH M0CKB
NW1 0LJ 2E0BGC
NW1 0LJ M3MUP
NW1 2JL G1YPZ
NW1 5ND M0MLV
NW1 6UE G3LAU
NW1 8TH G1BNV
NW1 8TL M1EXW
NW1 8UB M5AHF
NW1 9BX G4RPK
NW1 9PN M7BBT
NW10 0DQ M3UQD
NW10 1BU G0NLN
NW10 1PA M3XAY
NW10 1PL G3THQ
NW10 4EG M7ROT
NW10 4HJ M7VAD
NW10 7DU M7FYG
NW11 0DR M7FTQ
NW11 6UP G4PLX
NW11 6XY M6NOA
NW11 7AA G4DBQ
NW11 7DL G3GAF
NW11 7DL G4HXX
NW11 8SG G3RDG
NW11 8TE G1INJ
NW11 9NR M7FCA
NW2 1BP M3OXY
NW2 1DY 2E1MHB
NW2 1DY G6IWZ
NW2 1DY M7MHB
NW2 1EF M6ODH
NW2 1EF M6OVS
NW2 2AR G0VPV
NW2 3NN 2E0SJH
NW2 3NN M3KYG
NW2 3SN 2E0NQU
NW2 3SN M0NQU
NW2 3SN M3NQU
NW2 3TL G1DWC
NW2 4JS M7TKV
NW2 5EH G4PEF
NW2 5QE M7AUM
NW3 1EJ G8PDM
NW3 2DW M0PCX
NW3 2JJ M0GXK
NW3 2LD G4DFX
NW3 4JP G3UGX
NW3 5DB M7INN
NW3 5TY M6TXX
NW3 7UP M0NPA
NW4 1LS 2E0CVW
NW4 1LS M0HWD
NW4 1LS M6CWG
NW4 1NP G1GIA
NW4 1TG G4GHZ
NW4 2DY M7VLA
NW4 2NS M3MML
NW4 2PN G3UML
NW4 3NB G3XVH
NW4 3TR 2E0FMX
NW4 3TR M6CZB
NW4 4RR G3RDR
NW5 1JL 2E1GSX
NW5 2PE G7TKW
NW5 2RD M6FHD
NW5 2SY M0JWV
NW5 2SY M0JWX
NW5 4BX M6TUT
NW5 4HB G3ICN
NW5 4JB G7UZN
NW5 4SA M6HBG
NW6 1UE 2E0YVT
NW6 1UE M0RHA
NW6 1UE M6YVT
NW6 2DT G3VCK
NW6 3QA M0SOV
NW6 4DG G0KVS
NW6 4EJ 2E0OAO
NW6 4EJ M0SHN
NW6 4EJ M6SHK
NW6 4HB G6LPT
NW6 4LY 2E0GMN
NW6 4LY M0ZEE
NW6 4LY M6USD
NW6 5LR M1FIP
NW6 7LA M6MCW
NW7 1GB M7RJT
NW7 1JL G3MYG
NW7 1NH 2E0NAR
NW7 1NH M6NRA
NW7 1RS G0JUE
NW7 2HN G6LTT
NW7 2JY G8MMK
NW7 2LH 2E0CTV
NW7 2LH M0HJW
NW7 2LH M7BLH
NW7 2RE G3WZO
NW7 4BY M7RTE
NW7 4QP G1NRM
NW8 0HQ M6EKC
NW8 7EP M0SEO
NW8 8HR 2E0KBP
NW8 8HR M0KBP
NW8 8SX G3KCT
NW8 9GS M0OSO
NW9 0DJ 2E0JSO
NW9 0DJ M3JSO
NW9 5AZ G1IAG
NW9 5BS M0KWB
NW9 5HF G7GSF
NW9 5JU M0KJE
NW9 5RE M3TKN
NW9 5UQ M3CXX
NW9 6EU G0FAB
NW9 7PG M6FHB
NW9 8ES M6ESY
NW9 8JL G1AWF
NW9 9NG G1DMS
NW9 9TU M7XPL
OL1 2JS G0FZF
OL1 2QG M0WBS
OL1 2TX G1YMY
OL1 2UA G0OCF
OL1 4BJ G1WYA
OL1 4NP M6ZEW
OL1 4PB M3MUO
OL1 4QB M0SSO
OL1 4QB M7BJS
OL10 1AS 2E0YMT
OL10 2AU G1NXR
OL10 2LH M1CVL
OL10 2LH M3EBG
OL10 2LN M5DZH
OL10 3AF G6FJP
OL10 3FN M7JAP
OL10 3JB G8JCL
OL10 3RR G4GSL
OL10 3RR G8TML
OL10 3RS G0RFE
OL10 4DH M6MVN
OL10 4DH M6MZN
OL10 4PW 2E0LJS
OL10 4PW M3YBT
OL10 4QD G7NJE
OL10 4RQ G4TQL
OL10 4ST 2E0MAN
OL103JG M6NTL
OL11 1DH G7NIH
OL11 1QS 2E1HWI
OL11 2AG M3ROU
OL11 2BN G3ZRB
OL11 2BN G8EEA
OL11 2EB G0JSF
OL11 2EB M6JCA
OL11 2LN G4TDW
OL11 2NA 2E0TCO

OL11 2NA M0TCO
OL11 2NA M6TKC
OL11 2QB G1ITS
OL11 2TA M7SGB
OL11 3JG G4TDU
OL11 3LF G4HHS
OL11 3LG 2E1GGL
OL11 3LG G7BRJ
OL11 3QG 2E0FAC
OL11 3QG M0PEB
OL11 3QG M3SQO
OL11 4AX 2E0DTX
OL11 4AX M0LDZ
OL11 4AX M7LKS
OL11 4BT G3TOJ
OL11 4HW G3DTP
OL11 4PP G3XPI
OL11 4QF M3XGD
OL11 5AW G6CSN
OL11 5JD G0WHL
OL11 5NB G3RLE
OL11 5QA G0LSP
OL11 5QS G6GXK
OL11 5QW G0TFL
OL11 5RT G7TOB
OL11 5TG 2E1DEN
OL11 5UP 2E1GJC
OL11 5XA 2E1GKB
OL11 5XE 2E1MDC
OL11 5XE M0TVL
OL11 5XN 2E0DHB
OL11 5XW G0WTB
OL11 5YG M5ALA
OL12 0AS G4XKC
OL12 0HX M0DZX
OL12 0SH G1ZUH
OL12 6DZ G4OCJ
OL12 6LZ G1CIA
OL12 6UH M5ADI
OL12 7JN G4GAI
OL12 7JN M7AFH
OL12 7JU M6KES
OL12 7NJ M3XFZ
OL12 7NX G0OXP
OL12 7PR 2E0AWK
OL12 7RT G0SIY
OL12 7RT G7LKR
OL12 7RU M7CXP
OL12 7SU G7UCP
OL12 8PX G4TMV
OL12 8QB M0SPK
OL12 8SU 2E1TKD
OL12 8SU G8TKD
OL12 8UX M0VOM
OL12 8UX M6XVJ
OL12 9AJ G0VRX
OL12 9AN M7ANO
OL12 9LT 2E0FTL
OL12 9NH G6BWK
OL12 9NY M6TYN
OL12 9QH 2E0HYD
OL12 9QP G7OAI
OL12 9TY G3RIK
OL12 9UW M6THI
OL12 9UX G4TOZ
OL120AS 2E0KGM
OL120AS M7DMF
OL128UQ G6NID
OL129PZ M7PJB
OL129SW M6ZCL
OL13 0RY M7NPA
OL13 0TA M3VSH
OL13 8BH 2E0RBC
OL13 8GE M0DVW
OL13 8QL 2E0XCM
OL13 8QL M0LEX
OL13 8RN G8TYY
OL13 9DQ 2E0SHN
OL13 9DQ M7FXD
OL13 9HS M7CUW
OL13 9LS 2E1AHU
OL13 9LS G7KLT
OL13 9PH M0CEB
OL13 9RL G7JQF
OL14 5BU G0TVO
OL14 5EQ G1LCZ
OL14 5ET G6LLU
OL14 5LZ G4HYY
OL14 5NX G3OLP
OL14 5PE G7RFO
OL14 5PH M6MLF
OL14 5PH M6OOC
OL14 5RU G8WWO
OL14 6AW 2E0WTR
OL14 6AW M6JZV
OL14 6AX G4KDU
OL14 6AX M3FQA
OL14 6HB M6BZV
OL14 6HN G8OBR
OL14 6ND G4TYW
OL14 6PN M7XJP
OL14 7ET G1JBW
OL14 7NF G4HSX
OL14 7QF 2E0GPK
OL14 8DS G8IC
OL14 8EP G4RCJ
OL14 8PZ 2E1FBY
OL15 0BP 2E1DKU
OL15 0DS G0OWH
OL15 0DS G3RJV
OL15 0HQ G7BPM
OL15 0NG G8NQI
OL15 8EB G3DNN
OL15 8EN G0BLM
OL15 8EN G6NCL
OL15 8EZ G4IWO
OL15 9AH M7AIM
OL15 9DA 2E0SCV
OL15 9DA M6BKI
OL15 9EF G4OUJ
OL15 9EL M6VWP
OL15 9HE M0WLB
OL15 9HE M6KQP
OL15 9JE G8TTU
OL15 9LJ 2E0JHP
OL15 9QQ G7LHV
OL16 2DD G0IPN
OL16 2DD M3ULL
OL16 2EG M6EXT
OL16 2YH M3SMK
OL16 3PJ G0JFP
OL16 3QZ G1FHI
OL16 3UJ G4PJE
OL16 3UN M3BVX
OL16 4AY G0WSY
OL16 4EL M6GZR
OL16 4EP G7EQK
OL16 4PG M3VZP
OL16 4PG M7NOV
OL16 4QU 2E0GEC
OL16 4QU M0LXG
OL16 4QU M6TXG
OL16 4RJ G6TVE
OL16 4SE M5AFE
OL16 4SH G7LFM
OL16 4TY M0FDX
OL16 4UB G0UFD
OL16 4XG M3XHK
OL16 4XP 2E0ZGA
OL16 4XP M0ZAM
OL16 4XP M3ZGA
OL16 4YD G3SAO
OL16 5BB G0FNM
OL16 5BB G0GVS
OL16 5BB G0KZH
OL16 5HJ M6NNL
OL2 5AS G3WXN
OL2 5HL G8YKO
OL2 5SD G6PBI
OL2 5TQ 2E0MEQ
OL2 5TQ M6MHQ
OL2 6BH G1UTP
OL2 6JY 2E0GHB
OL2 6JY M0EAM
OL2 6NJ G7MXQ
OL2 6PF M7HJM
OL2 6PG G1GKR
OL2 6RW G0BJR
OL2 6RW G1ORC
OL2 6RW G4ORC
OL2 6RW M5ORC
OL2 6SR 2E0PUS
OL2 6SR M0GIE
OL2 6SR M3LQA
OL2 6TP G0KUY
OL2 6TW G1UKW
OL2 7AY G4GNG
OL2 7DA G8CZI
OL2 7JT M0CRZ
OL2 7JT M3SZC
OL2 7NA 2E0HVV
OL2 7NA M7DCV
OL2 7NQ M6ATY
OL2 7QA 2E1OBI
OL2 7SX M3MDI
OL2 7TQ M7AWD
OL2 7TX M3AUB
OL2 7UU 2E0AMS
OL2 7UU M0ADD
OL2 7YG M6LKW
OL2 7YL M7FDZ
OL2 7YN 2E0JCX
OL2 7YN M6FDS
OL2 7YQ M3OFN
OL2 8BG G6OVX
OL2 8EA G7MCT
OL2 8HY G0MMX
OL2 8JF G0PVO
OL2 8JJ G7LMI
OL2 8LP 2E1DTE
OL2 8NE M1FEW
OL2 8QA M7MOL
OL26JX M7GJF
OL3 5DS G8ZZW
OL3 5NT M7RDZ
OL3 5NU G0EOX
OL3 5PL G1GZK
OL3 5PL G1KFG
OL3 5PW G7ENC
OL3 6EB G4AIW
OL3 7AW G0HFC
OL3 7EF 2E0TZM
OL3 7EF M3TMZ
OL3 7EJ G3SNA
OL4 1LX G3JTK
OL4 2DJ G1JZU
OL4 2PE G7EBF
OL4 2PN M6MIC
OL4 2PW G4UFG
OL4 2QQ G7ICE
OL4 2TH G1XGW
OL4 3LW G4AVF
OL4 4AZ G0EYU
OL4 4AZ G4EBQ
OL4 4ED G4SSC
OL4 4EN M7FBC
OL4 4RY G6DAD
OL4 5DP M7CDB
OL4 5LF G4PLV
OL4 5QD M7WVM
OL4 5SH G0RKE
OL44SJ G6TAI
OL5 0SN G3RKM
OL5 9DE 2E0SBN
OL5 9DE M6HBB
OL5 9JN G0FCX
OL5 9LQ G6EJT
OL5 9LQ G8GVZ
OL5 9PD G1JAA
OL6 6TX G6YBV
OL6 7DU G0SNQ
OL6 7HG 2E0RGY
OL6 7HG 2E0ZEN
OL6 8BP G4RGM
OL6 8BP G6GMR
OL6 8BP M0NCZ
OL6 8BQ G6IQC
OL6 8BU G6IML
OL6 8DS 2E0OQZ
OL6 8DS M0GMQ
OL6 8DS M3OQZ
OL6 8QN G0AQS
OL6 8RJ G7MRO
OL6 8RP G6BRD
OL6 8SX M3OBN
OL6 8UA M3OHC
OL6 8UT 2E0TZI
OL6 8UT M6ETR
OL6 9AL M6JRL
OL6 9AP G3BPQ
OL6 9DU G7DGC
OL6 9EG M6PVC
OL6 9RH G7TSQ
OL7 0DU G1GWJ
OL7 0JD G4GHB
OL7 9AN G0MRD
OL7 9HN G3WI
OL7 9NL M0AAD
OL7 9QY G3NTF
OL70DU G7NHC
OL8 1PP G0YCE
OL8 2AN M6DKL
OL8 2HB M6MDB
OL8 2HG 2E0MCK
OL8 2HG M0MBM
OL8 2HG M3UXF
OL8 2HG M6CLA
OL8 2HG M6LEA
OL8 2LP G1HAC
OL8 2LP M3HAC
OL8 2NS M0CLK
OL8 2PD M1BGF
OL8 2XE M0NVQ
OL8 3AD M6REE
OL8 3BA G0MGL
OL8 3JH M6CIG
OL8 3LQ M1CZZ
OL8 3NP M3KMW
OL8 3RN M3XRD
OL8 3RY M6OLE
OL8 3TZ M6KCI
OL8 4JB M0KGM
OL82HB M6PYZ
OL9 0EJ G8JRZ
OL9 0LY M7JOD
OL9 0NS M3NYQ
OL9 0PH M5BGR
OL9 0QX 2E0JSN
OL9 0RF G1SLA
OL9 0RW M1FET
OL9 8AE G0IFW
OL9 8AP G4TLW
OL9 8DA M0ZOV
OL9 8HH G3SMZ
OL9 8LG M7GFW
OL9 8QD G4BKH
OL9 8QJ 2E0KDE
OL9 8QJ M0KDE
OL9 8RG G6MMS
OL9 9QR G3IGC
OL96RQ M7FCH
OX1 1AD 2E0KKL
OX1 1AD M7NMM
OX1 3BW G7VDJ
OX1 3BW M6FID
OX1 4RQ M0INX
OX1 5HW G7OGO
OX1 5JP G3SLI
OX1 5LL 2E1ABE
OX1 5LS G4AZA
OX1 5LX G3KCZ
OX1 5LY G0RFS
OX1 5NF G3JFR
OX1 5NH 2E0ADA
OX1 5NH G0REL
OX1 5NH G0RJX
OX1 5NH G0RKG
OX1 5NZ G6YTW
OX1 5PE G0SPS
OX1 5PE G7KGH
OX1 5PE M6BLA
OX1 5PP M6XMR
OX1 5RX G4AZT
OX1 5RY G3RTM
OX10 0HN 2E0VJO
OX10 0HN M0JSX
OX10 0HN M3VJO
OX10 0PT G6NLN
OX10 0QG G1BZW
OX10 0QT G1XNN
OX10 0SD M0WGA
OX10 6AE G3RDF
OX10 6HW G0MUJ
OX10 6LX M3LUK
OX10 6RR G3OEB
OX10 6RY G4SOH
OX10 6SH G0SPC
OX10 6SJ G4YSQ
OX10 6SJ G6IMW
OX10 6SL G4ATL
OX10 7DB G6YQT
OX10 7JE M0PIB
OX10 7LB G4ITC
OX10 7LF 2E0BCG
OX10 7LF M0DYU
OX10 7NL G1URJ
OX10 7PJ G0IDF
OX10 7PZ 2E0PWI
OX10 7RA G1ULR
OX10 8AD M6MYC
OX10 8BN 2E0SFC
OX10 8BN M3UMX
OX10 8FE G4BYR
OX10 8HP G0ADH
OX10 8LU G7TNQ
OX10 9AA G8POI
OX10 9ER G4DAT
OX10 9LY G8TYX
OX10 9NE 2E0SBH
OX10 9NE M6SJR
OX10 9PA G7TIB
OX10 9PE G6UXM
OX10 9PE M0BWL
OX10 9PP 2E1BMJ
OX10 9PY G6IMQ
OX10 9RJ 2E0IHE
OX10 9RJ M7LDJ
OX10 9RT G1CAR
OX10 9RT G1EMF
OX11 0AD G1AIO
OX11 0AD M6CGJ
OX11 0AE G8SMH
OX11 0BP M3ZQC
OX11 0DP 2E0XNT
OX11 0DP M7XNG
OX11 0DQ G8HKS
OX11 0DQ M0HKS
OX11 0DX G4MFP
OX11 0ES G0FRO
OX11 0HB G1GGT
OX11 0LG M6KTH
OX11 0PE G6HXI
OX11 0PE M3KVW
OX11 0PW G6SRX
OX11 0PW G8DVK
OX11 0RR G1URH
OX11 0SA G8BFK
OX11 0SB G8ATP
OX11 0SU M1CZI
OX11 6AB M3PQU
OX11 6FA G1HQH
OX11 6HP M0IDY
OX11 7AX G4PWI
OX11 7DB G4ZKI
OX11 7DD G6VSY
OX11 7DF G0DPK
OX11 7DT G6VMB
OX11 7HS M0IXP
OX11 7JQ G3GGG
OX11 7TZ 2E0GDH
OX11 7TZ 2E0ZOZ
OX11 7TZ M0SXH
OX11 7TZ M6AVM
OX11 7TZ M6DEC
OX11 7TZ M6EMW
OX11 7TZ M6RGO
OX11 7UT G6VLC
OX11 8BP G1SZK
OX11 8BP G4SYL
OX11 8DG M6FRF
OX11 8HD G1OWI
OX11 8HD G1SDI
OX11 8HD G7AKP
OX11 8HD M3SSL
OX11 8HP G0BSA
OX11 8HP G0MQM
OX11 8HY M6CRH
OX11 8LG G4EQM
OX11 8SW G1UPX
OX11 8TE G4XYM
OX11 8TX 2E0ZMM
OX11 8TX M0TMM
OX11 8TX M3ZMM
OX11 9GF M7MSB
OX11 9HT 2E0IYM
OX11 9HT M7EUI
OX11 9JX G8CUL
OX11 9JX G8NVI
OX11 9JX M0CUL
OX11 9JX M6YHS
OX11 9LX M0MTI
OX11 9NS G4IYC
OX11 9PU G4SAJ
OX11 9PU G6DPL
OX110PF M7YER
OX118HW G6SBN
OX12 0AT G8PTY
OX12 0BN G3VHE
OX12 0DA M1EWM
OX12 0DA M1JIM
OX12 0HA G3RRS
OX12 0HA G3VPW
OX12 0HT G1YDG
OX12 0HU M1EHV
OX12 0LB G1DYL
OX12 0LD G3YSQ
OX12 0LJ G0JEK
OX12 0LJ G6LRY
OX12 0NY G0LCB
OX12 0PY G1NDV
OX12 7BG 2E0MTL
OX12 7BG M0TGF
OX12 7BG M6NXY
OX12 7DF M6XDB
OX12 7DN G4DPA
OX12 7EP 2E0LNU
OX12 7EP G6LNU
OX12 7EP M3LNU
OX12 7EQ M6ROI
OX12 7EW G0HLS
OX12 7LD M7COU
OX12 7NQ G6UUQ
OX12 7PN G1IYB
OX12 7QF G4XIW
OX12 8DF M6CGZ
OX12 8DL G4HLX
OX12 8NS 2E0EXW
OX12 8NS 2E0GSH
OX12 8NS M3RNW
OX12 8NS M6NFS
OX12 8RB G0CEI
OX12 9AY 2E0RTW
OX12 9DN M1RJL
OX12 9DN M1SFS
OX12 9LE M0LCS
OX12 9RG G7WCG
OX12 9YE M3CHU
OX13 5BX G3YFM
OX13 5BZ G4KDN
OX13 5DB G3SSW
OX13 5HL G3YIK
OX13 5HL G3ZHL
OX13 5HL M6BVV
OX13 5HS G0AOZ
OX13 5JJ G0SXG
OX13 5LN G1FBQ
OX13 5LR G3VHS
OX13 5PF G4BRK
OX13 6BL G7WJZ
OX13 6DU G6ZKY
OX13 6LP G0KMW
OX13 6NU G4AEV
OX13 6QJ G0LMX
OX13 6RZ G4JLV
OX13 6TF G1KXX
OX14 1DW G1KEI
OX14 1EQ G4WXC
OX14 1JQ M6PIA
OX14 1NZ M6RST
OX14 1PE 2E0CGW
OX14 1PE M6BWN
OX14 1PP G7TRV
OX14 1PR G1RJW
OX14 1PR G3IWW
OX14 1PT M1FII
OX14 1QE G6FJI
OX14 1QU M0CIW
OX14 1XA G3ZSX
OX14 1XB M0GRA
OX14 1XB M0HEY
OX14 2BH M6UKI
OX14 2BJ 2E0KHO
OX14 2BJ M0KHO
OX14 2BL G1IVK
OX14 2EP G1OVY
OX14 2ES G7CAG
OX14 2ND M0AEN
OX14 2NG G0ROS
OX14 2QN M7PGX
OX14 3DR M1BQT
OX14 3JF G3HKH
OX14 3NF M0WPT
OX14 3NE M6EPT
OX14 3PE G6ZHB
OX14 3QF M3NOR
OX14 3SD G4RKF
OX14 3SR G0OYS
OX14 3SW G7VGA
OX14 3SX G0NPA
OX14 3TD G6MRP
OX14 3TE 2E0CFH
OX14 3TE G8NRP
OX14 3TE M0NRP
OX14 3TE M6YMA
OX14 3TR G4SQI
OX14 3XB 2E1HGA
OX14 3XB G8KIG
OX14 3XB M3HGA
OX14 3YN M1SLH
OX14 4DA M3ZAE
OX14 4DF M3XAG
OX14 4GN M0IJI
OX14 4JX G1BDP
OX14 4LG M1HFM
OX14 4NL G3MZC
OX14 4NN 2E0ZSA
OX14 4NN M3ZSI
OX14 4PU G7LUK
OX14 4PU M7JRY
OX14 4QB 2E0MHM
OX14 4QB M6MLQ
OX14 4QB M7STU
OX14 4XA M7CDT
OX14 5HF M6IZB
OX14 5JD 2E0KPS
OX14 5LW G4JEO
OX14 5NE 2E0UBH
OX14 5NE M6UBH
OX14 5NG G6PBZ
OX14 5NW G0RWJ
OX14 5PP G0OMZ
OX14 5PX M7CIU
OX14 5RF 2E1HNN
OX14 5RL G0INZ
OX14 5RN G1OQB
OX14 5RN G7DOF
OX14 5RQ G6PEP
OX14 5TN 2E0GDG
OX15 0SR G6STD
OX15 4AX M6JFW
OX15 4BB G8JPW
OX15 4DY M6ZZK
OX15 4EZ G0LXY
OX15 4JE G1PUY
OX15 4JL 2E0FKS
OX15 4JL M3FKS
OX15 4NT 2E0HAN
OX15 4NT M3UHH
OX15 4QU M6SJU
OX15 5AP G0UWB
OX15 5JB M0ZZA
OX15 5LH G7KYH
OX15 5PJ M7OPA
OX15 6AQ M7PZJ
OX15 6BJ M1BMV
OX15 6DS G6SLG
OX15 6DZ M1EDA
OX15NH M7EGA
OX16 0BS M3WGK
OX16 0DB M1CNJ
OX16 0EX M7BQZ
OX16 0HL 2E0ECP
OX16 0LU G1XRQ
OX16 0QW M6DNL
OX16 0RR 2E0SMS
OX16 0RR G0LDB
OX16 0RR M3SYL
OX16 0RZ M6CPS
OX16 0SB G7CEB
OX16 0SR M6NFB
OX16 0UG G8ZRU
OX16 0XF M3YCJ
OX16 1AA G8ZLN
OX16 1BQ M6FUM
OX16 1DJ M6SKH
OX16 1EW M0HVK
OX16 1GQ G7LNK
OX16 1PQ G4XPY
OX16 1UE M1AOG
OX16 1UN M6OWS
OX16 1XA M6GMC
OX16 2AZ G6ZDS
OX16 2NF G7UUW
OX16 3QS 2E1PTI
OX16 3WX 2E0BWV
OX16 3WX M0UXO
OX16 3WX M3ITT
OX16 3ZW 2E0NCV
OX16 3ZW M0NCV
OX16 3ZW M6NCV
OX16 4SD M3YEU
OX16 4SD M6JMG
OX16 5BE 2E0NQF
OX16 5DF M6WTW
OX16 5DR 2E0LGO
OX16 5DR M6ZLD
OX16 5DW 2E0TEN
OX16 5DW M0RDI
OX16 5DW M6AMR
OX16 5HQ M6ETJ
OX16 5LR M3FUD
OX16 9AJ 2E0TAO
OX16 9AJ M0GUF
OX16 9AJ M3UQI
OX16 9AP G0BJI
OX16 9AR G8KSH
OX16 9AR M3KSH
OX16 9BQ G1IIO
OX16 9BQ G7ILJ
OX16 9DH G8DCX
OX16 9EL G6TAS
OX16 9EL M6LVJ
OX16 9ES M1CJZ
OX16 9JU G0BRA
OX16 9JU M0BJN
OX16 9LB G4RWV
OX16 9NW G4TXA
OX16 9QB 2E0ZDA
OX16 9RZ M0AXN
OX16 9SR G1HXZ
OX16 9ST M6KGE
OX16 9SZ G6POG
OX16 9TA G7LKC
OX16 9TW G0MEC
OX16 9YZ M7EOS
OX17 1BE G1KKE
OX17 1BU M1CZF
OX17 1GJ M6CKF
OX17 1JA G1KIB
OX17 1LU M0BRI
OX17 1SS M6XKJ
OX17 2AB M1AHY
OX17 2BD M7FNI
OX17 2DZ M6DMN
OX17 2EH G8CIJ
OX17 2NP G8MFI
OX17 2QQ G4DKD
OX17 2RX M6BFX
OX17 3AE G1AHQ
OX17 3DG G7VYT
OX17 3FE 2E0NBX
OX17 3FE M0IEZ
OX17 3FE M6NBX
OX17 3FE M6XCJ
OX17 3HA M3RZP
OX17 3HR G3YZV
OX17 3LB M3RQJ
OX17 3PF G3OPH
OX17 3PH G1EYW
OX17 3RG G6KFR
OX17 3RJ G1BJE
OX17 3RJ M3RUR
OX17 3RS M6ZEK
OX17 3XB G0BIV
OX18 1DN G4YMJ
OX18 1DX G3KIL
OX18 1ED G7SRB
OX18 1FF G6HEJ
OX18 1JX M7TFT
OX18 1TT 2E0CHT
OX18 2HX G0NZE
OX18 2LH 2E1ESK
OX18 2NN G4RJY
OX18 2PE G7KDM
OX18 3LS G1WYD
OX18 3LX 2E0KZN
OX18 3LX M6KZN
OX18 3PR G8KOD
OX18 3QZ G6OUI
OX18 3YB 2E0BHA
OX18 3YB M3SHZ
OX181JU M7FWC
OX181TH M7GJL
OX2 0BB 2E0EDM
OX2 0BB M0TEK
OX2 0BE G3BLS
OX2 0HN 2E0YXB
OX2 0HN M3YXB
OX2 0HS M0MLG
OX2 0HS M3LNY
OX2 0HS M3MLG
OX2 0NP M6IKY
OX2 6AY G1OWM
OX2 6HY M6GOC
OX2 6QE G0AXI
OX2 6TX G4ZHE
OX2 6TX M3XZG
OX2 6UD M0HDN
OX2 7EN G0FGP
OX2 7EN G7CUB
OX2 7QG G6SRT
OX2 7TG M7JMA
OX2 8AX 2E0HNK
OX2 8AX M6HNB
OX2 8BB 2E0RNM
OX2 8BB M3RNM
OX2 8BY G3XGC
OX2 8DG G8CVS
OX2 8PE G0AZP
OX2 8PE G6YSQ
OX2 8PX G4ZSV
OX2 8QT G1JXS
OX2 9DN G1VSO
OX2 9DS G6RXV
OX2 9DZ 2E0GGN
OX2 9FU M7RPI
OX2 9HQ 2E0FGH
OX2 9HQ M6FGH
OX2 9HZ G3UOM
OX2 9JG G6ASA
OX2 9PA G1LVW
OX2 9PE 2E0FZJ
OX2 9PE 2E0RQN
OX2 9PE M0JHF
OX2 9PE M0RQN
OX2 9PE M6FZJ
OX2 9PE M6RQN
OX2 9PR M0OXD
OX2 9PR M3OXD
OX2 9QA G4MQR
OX2 9SN G0CFN
OX20 1JT M6GBJ
OX20 1JY G4POL
OX20 1RZ G6HRA
OX25 1PX G1FZR
OX25 2AH G7AHB
OX25 2NF G7LGV
OX25 3RD M6CWP
OX25 4AW G3NSP
OX25 4SQ M0DKU
OX25 5BE M0LEB
OX25 5TH 2E0LOH
OX25 5TH M0LOH
OX25 5TH M6PLR
OX25 6LP G0REV
OX26 1TZ G4ATG
OX26 1TZ G8GNI
OX26 2AR G7FLP
OX26 2DY 2E0KYD
OX26 2DY G1ZWP
OX26 2DZ G1CEO
OX26 2FP M7SYD
OX26 2GS G4SYV
OX26 2LP 2E0CYT
OX26 2LP M6CWZ
OX26 2LR G4BKB
OX26 2YA G8EKN
OX26 3FD M3CVW
OX26 3GA 2E0JCU
OX26 3SR 2E0DAB
OX26 3SR M3DAB
OX26 3YB G6HPT
OX26 3ZJ G3UTE
OX26 4HB G0TSK
OX26 4UH G6PLF
OX26 4XA M3GRF
OX26 4XE G1TRL
OX26 4XJ 2E0GBE
OX26 4XX M1TAZ
OX26 4YA 2E0PSC
OX26 4YA M0PSC
OX26 4YA M6PSC
OX26 5AH 2E0GPJ
OX26 5AH M0JKV
OX26 5DW G6FJO
OX26 5DX G0LUQ
OX26 6FE M7AQC
OX26 6SR G0JRY
OX26 6UQ G6FPP
OX26 6YP G1BYQ
OX26 6YR G8UWD
OX27 0AP G7DBN
OX27 0EY G1FXC
OX27 0HN M0JJD
OX27 0HP G6PHH
OX27 7PA 2E0BPN
OX27 7PA M3ZVX
OX27 7PX M3MVJ
OX27 8BQ M6APO
OX27 8DF G4FTA
OX27 8DR G1WMK
OX27 8ED 2E0BXC
OX27 8ED M3UMW
OX27 8FL M0GKA
OX27 8FQ M0BWS
OX27 8TU M0CJN
OX27 9AZ G4BII
OX28 1BJ M6NRM
OX28 1HQ G4IOK
OX28 1LH M7FPN
OX28 1NS G4EGN
OX28 2EG M0BLV
OX28 2EG M0BRP
OX28 2JB G4GUN
OX28 5FS 2E0FGV
OX28 5FS M0OQO
OX28 5FS M6NFQ
OX28 5JQ G7CKP
OX28 5JQ G8LKB
OX28 5JT G0CQZ
OX28 5LJ 2E0GTU
OX28 5NL G4YUA
OX28 6EG G8RDA
OX28 6NL 2E0GDY
OX28 6NL M3GDY
OX28 6NN G4EZG
OX28 6NN G8EWT
OX28 6NN G8PX
OX29 0RQ G6UPQ
OX29 0RT G6ZNT
OX29 4BU 2E0BMH
OX29 4BU G8BQH
OX29 4BU M3SSO
OX29 4EW G1RCX
OX29 4EW G6LDO
OX29 4GL G4BYS
OX29 4LU G7OQQ
OX29 4NP 2E0HTG
OX29 4NP M6YCG
OX29 4NT G6ANV
OX29 4QJ G0BQZ
OX29 4QY G8JVI
OX29 5BL M7SWA
OX29 5RH G3ZYR
OX29 6SQ G1DOL
OX29 6XL M7WBB
OX29 7PA G4DSE
OX29 7PB M3NGE
OX29 7TH G3UKC
OX29 7TH M1FRB
OX29 7YD G1VDH
OX29 8HH G1ZBG
OX29 8HR 2E0FVK
OX29 8HR M7AMX
OX29 8JP M1ENJ
OX29 8JP M1ENK
OX29 8JP M3ENJ
OX29 9UW G7VIB
OX29 9XP G8CYF
OX296PY G0UZK
OX3 0EX G0BPX
OX3 0JG G6DVE
OX3 0JJ G1OYF
OX3 0NW 2E0DZN
OX3 0NW M0TKM
OX3 0NW M6HMU
OX3 0QA G1YDI
OX3 0RJ G3UGJ
OX3 0SJ M0BJE
OX3 7EE G0LBZ
OX3 7EE M3FHV
OX3 7QF 2E0FFC
OX3 7QF M0YQC
OX3 7QF M6PIC
OX3 7RH M7JFF
OX3 7RR M0DDT
OX3 7RU M6WRV
OX3 7SZ M7TEJ
OX3 8AA 2E0ETV
OX3 8AA M0ZRX
OX3 8AA M6FMZ
OX3 8AN M0SVN
OX3 8ED G4KSQ
OX3 8HX G0KPY
OX3 8JA G4AYR
OX3 8JB G3KZU
OX3 8LP 2E0OXF
OX3 8LP M6CFW
OX3 8LT M6EYT
OX3 8LT M6RXE
OX3 8PD G3UJO
OX3 8SX G8SIN
OX3 8TA G3UMF
OX3 9PB M3EGU
OX3 9QF G0AFY
OX33 1DT M7FGM
OX33 1RQ M3JMW
OX33 1SQ G6DBU
OX33 1TJ G0HZQ
OX33 1TR M0LTK
OX39 4AA G0ODQ
OX39 4AE G0HHN
OX39 4EJ M0OSE
OX39 4EU G1ZRR
OX39 4EW G0IKI
OX39 4HB G4ODI
OX39 4HP 2E0PCC
OX39 4HP M6EHP
OX39 4JY G1ECE
OX39 4JY G1GQJ
OX39 4JY G7AGR
OX39 4JY G8GQJ
OX39 4PE M1BXQ
OX39 4PL G3UNI
OX39 4RA G3ZRY
OX39 4TG 2E0WMM
OX39 4TG M0MKY
OX39 4TG M7JDW
OX39 4TP G0FRN
OX39 4UD G7IKS
OX4 1BY G0CPP
OX4 1EB 2E0LLM
OX4 1EB M3LLM
OX4 1LT G3LOF
OX4 2HF M6MTA
OX4 2JW 2E0OKS
OX4 2JW M6BHM
OX4 3BX M0OXR
OX4 3HT G1PCU
OX4 3NU G7IMH
OX4 3NW G0HEN
OX4 3QU M0ACU
OX4 3SA M6WHZ
OX4 3SD 2E0DKG
OX4 3SD M0KLB
OX4 3SD M6ATD
OX4 4AR G0XBG
OX4 4BB M6NMW
OX4 4NE G6BTB
OX4 4NS M6RLC
OX4 4PW G8OMW
OX4 4XG M3WVB
OX4 6DL G7TLD
OX4 6DX G4KON
OX4 6ER G6ZPL
OX4 6ER M6ZPL
OX4 6RA M3MSY
OX4 6SS M3LVM
OX4 7FD G0TYZ
OX44 7LH G3ZKO
OX44 7RZ M3ZFH
OX44 7SJ 2E0DBH
OX44 7SJ M3ZFN
OX44 7TA M1GGG
OX44 7TA M6VKE
OX44 7TJ G3SBM
OX44 7TN G4DAP
OX44 7YN G0MCO
OX44 9DL G1YSX
OX44 9DL G6ZTR
OX44 9EX 2E1GSW
OX44 9HP G8ECG
OX44 9LJ M6SII
OX44 9LT G4DCN
OX49 5AD G8LGM
OX49 5BL G4HFS
OX49 5HL M0JWC
OX49 5HL M7GKA
OX49 5LW G7NTO
OX49 5NQ M7BLG
OX49 5PR M6HHH
OX49 5RF G0DUF
OX5 1FY M7ESH
OX5 1HF M3IAP
OX5 1HH G0THY
OX5 1JT M1AQN
OX5 1NX M6RAR
OX5 1RN 2E0SKK
OX5 1RN M0RND
OX5 1RN M6AGG
OX5 1TH 2E1IGG
OX5 1TR G0LXW
OX5 2EU G0EDY
OX5 2EU G7KGI
OX5 2HG 2E0WEL
OX5 2HG M0HIZ
OX5 2HG M6SIZ
OX5 2LL G0LUN
OX5 2LT 2E0LBY
OX5 2LT M7TOY
OX5 2NA M7JIZ
OX5 2UB G3MSO
OX5 2US G8JOX
OX5 2XP G8BNB
OX5 2XZ G0XGM
OX6 4JR G1XTA
OX7 3BJ 2E0TMY
OX7 3BJ M0TTH
OX7 3BJ M3TTH
OX7 3HB G6GEV
OX7 3RX 2E0JLM
OX7 3RX M0XJM
OX7 5DP M7BXT
OX7 5DZ G0NYH
OX7 5JS G0JZE
OX7 5JT M7BFX
OX7 5QH M6CUN
OX7 5SW G4TLS
OX7 5XS G1RLA
OX7 5XX G0MLQ
OX7 6AD G6KCJ
OX7 6BE M7BQH
OX7 6EA M6EZY
OX7 6ER G8BII
OX7 6GE G8DJT
OX7 6JN G8GJG
OX7 6QA G6NLG
OX7 7BP M0NRK
OX7 7EU G0MGZ
OX8 6PZ G0MKY
OX8 6PZ G0PYF
OX9 2AX G0THH
OX9 2DH M7RAZ
OX9 2EE M6KIE
OX9 2EJ M6OXF
OX9 2FS M7FPC
OX9 2NE G0MVW
OX9 3AG G4FRL
OX9 3BJ G8HAZ
OX9 3BY 2E0FYD
OX9 3BY G4ADR
OX9 3BY M7MHA
OX9 3DG G6GHU
OX9 3EP M6RPP
OX9 3JF G8NRR
OX9 3JS G0FOK
OX9 3LZ M7JIM
OX9 3NH G4HXU
OX9 3NJ 2E0SSD
OX9 3NJ M6YPX
OX9 3NQ G0UUR
OX9 3NQ M6TXU
OX9 3QL G1GGK
OX9 3TE 2E0BIO
OX9 3XN M6AZR
OX9 3YD M6NOK
OX9 3ZH M6AMW
OX9 7AE 2E0KTK
OX9 7AE G5NLD
OX9 7AE M7DVD
P07 8AL M6TID
PA1 1GT GM0SCW
PA1 2JE GM6PCW
PA1 2SX GM0UKD
PA1 3NG GM4VLX
PA1 3RT GM4MPY
PA1 3SR GM1SXX
PA10 2EL MM6DCT
PA11 3BH MM6FPI
PA11 3NB MM3NGV
PA11 3NL 2M0VGT
PA11 3NL GM5RDX
PA11 3NL MM7VES
PA11 3PQ 2M0JSF
PA12 4EG GM4FLX
PA12 4EJ GM7RZE
PA12 4HG MM7ICR
PA12 4HL GM0KMA
PA12 4NB MM0WOA
PA12QT MM6WER

PA13 4BB GM0JVC
PA13 4HP MM6WHW
PA13 4JJ GM7ALI
PA13 4PY GM7ADU
PA13 4PY GM7ADY
PA14 5QG GM3SUZ
PA14 5QP 2M0XZA
PA14 5QP MM7PGL
PA14 5UT MM3YIH
PA14 5XF GM0WDF
PA14 6DP MM0STU
PA14 6DT 2M0MOK
PA14 6DT MM6WUL
PA14 6EF MM6MVM
PA14 6PL MM6SGO
PA14 6XP GM4HRJ
PA15 2EE GM4EGD
PA15 2HJ MM6IAB
PA15 2HP MM6KIW
PA15 2JD MM6HPW
PA15 2WH 2M0VZX
PA15 2WH MM7DAM
PA15 4DW 2M0OLK
PA15 4HH GM4FBU
PA15 4HW MM3MZX
PA15 4PU MM7ADI
PA15 4SX MM0XPZ
PA15 4SX MM3SWG
PA15 4TA MM3UXO
PA16 0DS MM0CDW
PA16 0EW GM0GNK
PA16 0EW GM3YOR
PA16 0HX MM1AUG
PA16 0QF MM6LCL
PA16 0TG MM7LYZ
PA16 0YL MM3YIG
PA16 0YL MM3YIQ
PA16 7AL MM3WYI
PA16 7BJ MM3IBM
PA16 7BL 2M0BOS
PA16 7BL MM3MHQ
PA16 7LH MM3WYM
PA16 7NB MM3UNQ
PA16 8AS GM6LEZ
PA16 8NP MM0KNE
PA16 8NP MM0SRX
PA16 8QB MM6OOK
PA16 8QG GM3LRG
PA16 9AZ GM0KYU
PA16 9BG 2M0GRK
PA16 9DH GM3ZXG
PA16 9DN MM6BFH
PA16 9JG MM0GLX
PA16 9JG MM0JET
PA16 9JG MM3YCI
PA17 5AG MM7SKL
PA17 5DX GM0MDD
PA17 5DX GM0VKG
PA18 6DX GM0ADF
PA18 6DX GM0EUM
PA18 6DX GM0WSR
PA18 6DX MM6AHN
PA18 6EB MM3XLQ
PA18 6HA MM0IYT
PA19 1AZ 2M0DKU
PA19 1AZ MM6SUB
PA19 1EJ MM3YQX
PA19 1HF MM1AWV
PA19 1HY GM0HJV
PA19 1HY GM8YUI
PA19 1HY MM1AUF
PA19 1JE MM3YUS
PA19 1SG 2M0BEL
PA19 1SG MM3KSV
PA19 1UW MM6YNG
PA19 1XE GM0ATQ
PA19 1XG 2M0IOB
PA19 1XG MM3WWV
PA19 1XS MM3UIX
PA19 1XS MM7LRV
PA19 1XU MM3YSJ
PA19 1YA GM0IOY
PA2 0BD GM1TDT
PA2 0BE MM3GEW
PA2 0DP MM6KHA
PA2 0NG MM3EKL
PA2 0RP 2M0OWT
PA2 0RP 2M0OXQ
PA2 0RP GM0PYM
PA2 0RP MM0PAZ
PA2 0RP MM3OXQ
PA2 0RP MM6SWT
PA2 6DB MM1SYD
PA2 6EP 2M0GJG
PA2 6EP MM6NRE
PA2 6JT GM0UOU
PA2 6RA GM4EQY
PA2 6SD MM6WGT
PA2 6UJ MM3HQC
PA2 7QY GM0BCY
PA2 7RL 2M0BIN
PA2 7RL MM3VVZ
PA2 7RP GM4BRM
PA2 8AW GM3DDL
PA2 8BY GM4LHQ
PA2 8EB GM3RPM
PA2 8LG MM0PJQ
PA2 8NB MM7FRZ
PA2 8QW GM0NAE
PA2 9AJ 2M0KZI

PA2 9JW MM3YMM
PA2 9NY MM6CEL
PA2 9RD GM8TVV
PA20 0DY GM6OLM
PA20 0QT 2M0CGE
PA20 9DY MM6KSJ
PA20 9EU GM0ELL
PA20 9HT MM3FGI
PA21 2BH GM7VSB
PA22 3AS MM0TOB
PA23 7EW GM4EAW
PA23 7HX MM6IPP
PA23 7JJ MM1HMV
PA23 7JJ MM3MHS
PA23 7SS 2M0GLI
PA23 7SS MM6GLI
PA23 7UA GM7KFS
PA23 8AX GM3YLD
PA23 8BA MM7AIP
PA23 8DG MM6DSD
PA23 8EA 2M0FSB
PA23 8EA MM6FSB
PA23 8FG GM7TFN
PA23 8FG MM3CJP
PA23 8HB MM3IMC
PA23 8JR MM3DCN
PA23 8LQ GM4NSL
PA23 8NA GM6HLT
PA23 8PQ GM4SFT
PA23 8QH GM4GLG
PA23 8QT GM4HPK
PA23 8SB GM0VPG
PA23 8SG GM8RGO
PA23 8TH GM8UQX
PA24 8AD GM7NVG
PA27 8BX MM3ROV
PA27 8DB GM6CRX
PA28 6GY MM0BED
PA28 6LQ GM0GON
PA28 6PG MM6GHH
PA28 6PH MM0GTX
PA28 6PL GM1PKN
PA28 6PL MM0AMW
PA28 6PX GM1DCI
PA28 6RX GM7BAS
PA28 6RZ GM4HUL
PA28 6ST MM0EAR
PA28 6ST MM0KRC
PA28 6ST MM3RCX
PA29 6TF GM7OSQ
PA29 6XL GM7RBY
PA29 6XY GM4ZVF
PA29 6YL 2M0PHO
PA29 6YL MM6PHO
PA3 1BE MM3LVT
PA3 2LR 2M0EDY
PA3 3EJ MM0WRX
PA3 3RT MM0RLN
PA3 3ST GM3WNB
PA3 3SU GM7UFN
PA3 3SU GM7UVS
PA3 4LL GM4KHI
PA3 4LT MM0LEN
PA3 4ST MM6COF
PA30 8FB GM4VXA
PA30 8HH GM4VYQ
PA308EU MM7WWM
PA31 8AG GM1PWL
PA31 8LG 2M0ONS
PA31 8LG MM6MMG
PA31 8NE MM7EKP
PA31 8QA GM1HJL
PA31 8QA GM4JLD
PA31 8QA GM4OHT
PA31 8QA MM0BLZ
PA31 8QA MM5DOG
PA31 8QL GM0BPF
PA31 8QL GM4HNK
PA31 8SD MM6OHE
PA31 8SD MM6YVO
PA31 8SQ GM0NBA
PA31 8SQ GM8ICC
PA32 8TW MM0WAX
PA33 1AQ 2M0MVI
PA33 1BE MM7LXI
PA34 4BG GM0NTR
PA34 4HZ MM3HMM
PA34 4QT GM0EWU
PA34 4RP GM0LRA
PA34 4RP MM1AVR
PA34 4RP MM3GQR
PA34 4RX MM3IMK
PA34 4SF 2M0WWM
PA34 4SF MM0WWM
PA34 4SF MM3WWM
PA34 4YB GM0ERV
PA34 4YL GM0MNW
PA34 5EF MM3FGL
PA34 5JQ GM4MTI
PA34 5JR MM3IJI
PA34 5NA MM0DCC
PA34 5NS MM3XUY
PA35 1HG 2M1ELU
PA35 1HH GM8SXQ
PA35 1HN MM7JNE
PA35 1HY GM0EQW
PA35 1JR MM3IKS
PA37 1PL MM0JSU
PA37 1PY GM1FPD

PA37 1QJ MM3FUG
PA37 1QP 2M0NDO
PA37 1QP MM7VWG
PA37 1QQ MM3FGH
PA37 1QR GM0SUH
PA37 1QU MM0NDM
PA37 1QU GM1OCR
PA37 1QX MM7FCT
PA37 1SG MM0AHC
PA37 1SG MM6CLC
PA37 1SR GM4AXS
PA37 1SY MM6RWI
PA38 4BL GM0HVS
PA4 0EG GM0LKS
PA4 0NE MM6ABO
PA4 0NP 2M0TXA
PA4 0SL GM8IID
PA4 0XA GM7MBB
PA4 8JB 2M0VSM
PA4 8JE MM7MVR
PA4 8XS MM3NYY
PA4 8YW MM7EFJ
PA4 9BL GM7VHQ
PA41 7AD GM7NPR
PA43 7HZ MM6EBJ
PA44 7PQ MM7ILY
PA44 7PZ GM1DCB
PA44 7PZ GM3PAK
PA48 7UF MM7OCX
PA49 7UN 2M0UMH
PA49 7UN MM0UMH
PA49 7UN MM6LMH
PA5 0DA MM6LPT
PA5 0EX GM0EDJ
PA5 0EX MM0PYR
PA5 0EZ 2M0LEX
PA5 8AR GM1OGZ
PA5 8HR MM3TNF
PA5 8LH 2M0RDT
PA5 8LH MM6RDT
PA5 8RZ MM6CBI
PA5 8UF GM8IWL
PA5 9BE MM5TUW
PA5 9BH GM1JHU
PA5 9DW MM6SIS
PA5 9LJ MM6TGB
PA5 9NE GM0CHM
PA5 9NR 2M0JAL
PA5 9NR MM3AQW
PA5 9QJ MM7CES
PA5 9RG MM3SWU
PA50EF 2M0STV
PA59AD GM4FDM
PA6 7DA GM4GZQ
PA6 7DE 2M0BZZ
PA6 7DE MM3XWS
PA6 7DX MM5AII
PA6 7LH 2M0OHL
PA6 7LH MM7TOH
PA60 7XG 2M0GWU
PA65 6BG GM0PRO
PA7 5ES GM8OAH
PA7 5FR MM7OAI
PA7 5HE GM3UWX
PA7 5HW GM3NLB
PA7 5JT MM7FBP
PA70 6HB GM4PMK
PA72 6JU GM0MHE
PA75 6GN GM7NYB
PA75 6PN MM7IAM
PA75 6QL MM6SEI
PA75 6QN GM8FKL
PA75 6QY MM7WPZ
PA77 6TW MM3GOX
PA77 6TW MM3GOY
PA77 6TW MM3GPB
PA77 6UA MM3GMP
PA77 6UE 2M0GOE
PA77 6UE MM3GOE
PA77 6UL MM3GOT
PA77 6UT GM3PGY
PA78 6TB MM0JUL
PA8 6BJ MM1JWF
PA8 6BP MM3WZH
PA8 6EE MM6BLY
PA8 6HG GM0GDD
PA8 6JN MM6CMY
PA8 7DH GM1JDJ
PA8 7HX GM4AGL
PA8 7HZ MM7BHR
PA80 5XS GM4RKM
PA80 5XW MM0GFR
PA9 1BJ MM1HMZ
PA9 1BP GM1VYG
PE1 2EJ M3YSS
PE1 3DS G3PQB
PE1 3DT G4BVM
PE1 3DZ 2E0SAZ
PE1 3DZ M3OWZ
PE1 3JJ M1CDL
PE1 4DR G1FSX
PE1 4EE G6UST
PE1 4HA G4FKQ
PE1 4JY G8SPG
PE1 4LR M6PHZ
PE1 4LT G8CKV
PE1 4RA G4BBA
PE1 4RQ M7ERK
PE1 4SN G3RDZ

PE1 4SP M0CMK
PE1 4UR G1ZHL
PE1 5HF G4FKG
PE1 5NG 2E0RPL
PE1 5NG M7RPL
PE1 5XD M1HQX
PE1 5XE G0FHO
PE10 0BZ G7RFE
PE10 0EE G4WJM
PE10 0ES G3MMS
PE10 0HT G4NIZ
PE10 0HZ G4GMS
PE10 0HZ M7ARA
PE10 0LS G1IXV
PE10 0NT G8NWM
PE10 0NW G6URT
PE10 0QE G6VEY
PE10 0QX G0LLY
PE10 0RB G8VVC
PE10 0RN G7RFD
PE10 0SG G0GRC
PE10 0SG G6SSN
PE10 0SG G7GRC
PE10 0SG G7NUM
PE10 0SR M1DYK
PE10 0SR M1DYL
PE10 0SX G7JKY
PE10 0TU G4DHF
PE10 0TU G6LI
PE10 0UR G7AHO
PE10 0UY G4HLB
PE10 0XR G4LEX
PE10 0XW M6DKU
PE10 9AU G0TIS
PE10 9HE M6OKP
PE10 9HE M3MJE
PE10 9HE M3SKC
PE10 9JG M7RRG
PE10 9JJ G0HSD
PE10 9JX G0OTE
PE10 9NS M3RCE
PE10 9PY M6CEU
PE10 9QT G6SKS
PE10 9QT G6SKT
PE10 9QY G0ERW
PE10 9SQ G0BHU
PE10 9SN M7MRP
PE10 9TT G1CPM
PE10 9UQ M0DVU
PE10 9XL G8PUE
PE10 9XL G8TXX
PE100FG M6HOV
PE11 1GZ G4CEY
PE11 1HD M0CTM
PE11 1HQ G1LSB
PE11 1JQ M0DXF
PE11 1PD G7SWS
PE11 1PD G8NWC
PE11 1PD M7ARU
PE11 1QB G4ZXZ
PE11 1QF 2E0JEE
PE11 1UU M0PAC
PE11 2BJ G4XWW
PE11 2DE M7ASH
PE11 2FG G0RNV
PE11 2HE G0JUR
PE11 2JL 2E0FHN
PE11 2NA 2E0BOZ
PE11 2NA M3WUE
PE11 2NA M7CSK
PE11 2QJ G8ETD
PE11 2RT M0MFC
PE11 2RU 2E0FHM
PE11 2RU M6OIZ
PE11 2UN G0UOQ
PE11 2UW G0HGH
PE11 2YU G7FML
PE11 3AB G8VQK
PE11 3AF G4NQS
PE11 3BD 2E0FTP
PE11 3BD M6UUE
PE11 3FA G7NGQ
PE11 3GF 2E0XRO
PE11 3GF M7GWD
PE11 3GZ M7DQJ
PE11 3LL M7RIS
PE11 3PN G6UHS
PE11 3QW G0HXR
PE11 3QW G0JPI
PE11 3RH G7UIO
PE11 3SG M1BMW
PE11 3TG M0WQR
PE11 4BA M0DJA
PE11 4BA M1DJA
PE11 4BD G8KGR
PE11 4DA G3UYE
PE11 4DQ G7CWM
PE11 4EU G3VFF
PE11 4EY G1WRC
PE11 4EY M0CKE
PE11 4EY M5ARC
PE11 4HG G1DSP
PE11 4HG G3SJR
PE11 4HG G4DSP
PE11 4HG G4OO
PE11 4HG M3DIS
PE11 4ND. G8FMX
PE11 4NE G7RTL
PE11 4NE M0BSL

PE11 4QB G1ZJP
PE11 4QQ G3UDP
PE11 4RF G1ELK
PE11 4TL G1GET
PE11 4XH G3NHF
PE11 4XL G0SDM
PE11 4YA G7THF
PE11 4YL G3VC
PE12 0DE M3PYI
PE12 0EE M3WPP
PE12 0EG G1TUL
PE12 0EG M3SHN
PE12 0EZ G3REH
PE12 0HP G0YKC
PE12 0JG G0OMM
PE12 0NP G1KLI
PE12 0PQ 2E0DHW
PE12 0PQ M0ZLH
PE12 0PQ M6AWS
PE12 0QR 2E0LCN
PE12 0RU G1JKO
PE12 0TS M0GTR
PE12 0TS M3SKQ
PE12 0TT G0ECI
PE12 0TY G8FPW
PE12 6AD G7AKV
PE12 6AD G8PQB
PE12 6AH G6UKC
PE12 6BX 2E0BGX
PE12 6BX M3HZO
PE12 6DA G3VPR
PE12 6DH G4ODA
PE12 6HS 2E0YEH
PE12 6HS M6DAC
PE12 6HS M7LNA
PE12 6LD M0KRU
PE12 6LG 2E0JVC
PE12 6PN G1PAT
PE12 6PN G4SJV
PE12 6PS G7CUP
PE12 6PT G0LWC
PE12 6QG M7MOR
PE12 6QQ G6UEG
PE12 6TG G7OBR
PE12 6WG G1MHN
PE12 7AU M0KCE
PE12 7BT M3DEJ
PE12 7BT M6ZTO
PE12 7DQ M3KKX
PE12 7HR G1WYC
PE12 7JQ 2E0GGV
PE12 7JQ M6CTV
PE12 7JS G1JOL
PE12 7JS G7RRD
PE12 7NG G8MEE
PE12 7PP G7HCR
PE12 7PZ G8XZX
PE12 8BA G0SHC
PE12 8BA G1SCT
PE12 8DH 2E0TUB
PE12 8DH 2E0TUS
PE12 8DH M6TJN
PE12 8DH M6TJO
PE12 8NA G6VPN
PE12 8QQ G4PST
PE12 8RQ M7DQO
PE12 8SD M7EVG
PE12 9AR M0AJI
PE12 9BD G7SNW
PE12 9BT M6XMX
PE12 9DS G8LZK
PE12 9EG G1SRD
PE12 9EG M1CQQ
PE12 9EG M3JNT
PE12 9EL M0XNR
PE12 9EN G3YPZ
PE12 9HS G4ZID
PE12 9JP G0CFD
PE12 9LQ G0WKN
PE12 9LQ G3RKQ
PE12 9LQ G6GYX
PE12 9NT M0BFT
PE12 9PF G8XND
PE12 9PJ G4RMJ
PE12 9PJ M6XCO
PE12 9QQ 2E0FFP
PE12 9QQ M0PFF
PE12 9QU M6MZQ
PE12 9QU G4TPS
PE12 9RG G4KPE
PE12 9RH G7CRY
PE12 9RQ G4JBE
PE12 9YF G0BEJ
PE12 9YF M0CKI
PE12 9YF M0DWW
PE12 9YF M3DLE
PE12 9YF M3OJP
PE12 9YR G7BZC
PE13 1LH M0RAU
PE13 1LH M6BKK
PE13 1PZ G1VYB
PE13 1SW G4UQN
PE13 2EG G8HQO
PE13 2EL G1HJP
PE13 2JD G6NNK
PE13 2JD M0CNX
PE13 2JD M3RGN
PE13 2JR G7TFL
PE13 2JU M0EAY

PE13 2QA G8NIL
PE13 2RJ 2E0MAB
PE13 2RJ M3MBC
PE13 2RZ M7JPP
PE13 3ED M0BII
PE13 3EJ M3SNO
PE13 3EJ M3WPS
PE13 3HF M1BIG
PE13 3JP M7PRZ
PE13 3JW M7SGF
PE13 3LF 2E0RGB
PE13 3LQ 2E0USC
PE13 3LQ M3USC
PE13 3NQ G6XMU
PE13 3PA 2E0URF
PE13 3PA M3URF
PE13 3RH M0PLX
PE13 3RH M1AZA
PE13 3RJ M6IVQ
PE13 3RR 2E0DEP
PE13 3RR M3WUJ
PE13 3UD M1RQL
PE13 3UN G1FRM
PE13 3UT M1AMJ
PE13 4AR M6AVJ
PE13 4DA 2E0RTH
PE13 4DA M6IHS
PE13 4ER G4NFY
PE13 4HS M0BNP
PE13 4HU G1ATQ
PE13 4JQ 2E0TAB
PE13 4JQ M0TAR
PE13 4JU G8RZN
PE13 4LX G4SHQ
PE13 4QQ M0CNW
PE13 4RG G7HHQ
PE13 4RT G7INY
PE13 4SQ 2E0LTJ
PE13 4TL 2E1HSD
PE13 5AE G8YXJ
PE13 5AF 2E0PPO
PE13 5AF M6RNS
PE13 5AS M3LDE
PE13 5BD G1HQW
PE13 5BP G0OPC
PE13 5BP M3TXJ
PE13 5DR G0XTA
PE13 5DW M1AYU
PE13 5HT G4OKH
PE13 5LG G0MFT
PE13 5NH G4KUY
PE13 5QR G6LTN
PE13 5QR M0BSZ
PE13 5RA M7ICL
PE132RJ G8NED
PE132RJ M0DUQ
PE14 0BA G7LWU
PE14 0DF G4FCI
PE14 0DN 2E0LPE
PE14 0DN M0LPM
PE14 0ND G8MLA
PE14 0ND M3MLA
PE14 0NR M3KWF
PE14 7AF M6IFY
PE14 7BJ M6HTM
PE14 7DF G1KSI
PE14 7DN G6OEJ
PE14 7EJ G7JJX
PE14 7HU G7LPW
PE14 7JF G4ORJ
PE14 7LT G3RCQ
PE14 7LT M3DLC
PE14 7LT M3RCQ
PE14 7LZ G0WDK
PE14 7LZ M6BAA
PE14 7NN G0KOJ
PE14 7NP M1AEV
PE14 7NS G7MRJ
PE14 7PQ G7HDU
PE14 7QD G0EUD
PE14 7RA 2E0WNM
PE14 7RA M6WNM
PE14 8AY M1CZM
PE14 8AY M3CZM
PE14 8AY M7AQP
PE14 8ES G7UFI
PE14 8ES G8MXQ
PE14 8JR G6PJE
PE14 8JT M0MSX
PE14 8JT M3PYD
PE14 8PF G3XIQ
PE14 9DW M3AKR
PE14 9JR G0EYM
PE14 9JT G0FND
PE14 9PU M0MVB
PE14 9QE M3VZZ
PE14 9RF G8EEK
PE14 9RG G4NVL
PE14 9RG G8SWW
PE14 9RG M3HQU
PE14 9RG M3TUD
PE148RH M7CRC
PE15 0AJ G6NHK
PE15 0AU G1JEA
PE15 0FE G1EXM
PE15 0GA M6DGR
PE15 0LN M3GZW
PE15 0LU G7JXT
PE15 0PJ 2E0JDK

PE15 0PJ M0XAV
PE15 0PJ M6JHZ
PE15 0QF G6SXB
PE15 0QJ G8EOH
PE15 0RN G7EFL
PE15 0RR G3YLY
PE15 0RW G4AJE
PE15 0SA M0GAX
PE15 0SB G8OXE
PE15 0SE G7NCG
PE15 0SE M0BRM
PE15 0SP G7ADS
PE15 0SP M0BLR
PE15 0SR 2E0OVL
PE15 0UH M6DDM
PE15 0YF G7TBX
PE15 8EL G8PHS
PE15 8EQ G4KKR
PE15 8HF G7EVP
PE15 8QA G7NEM
PE15 8RY G0PER
PE15 8SH G4HWC
PE15 8SN G0FLP
PE15 8UH M0JYJ
PE15 9AH G6JVX
PE15 9AP G3VQG
PE15 9DA G6KDW
PE15 9DH G6MMB
PE15 9DL G1VTO
PE15 9DT G0OYZ
PE15 9EA G0BXJ
PE15 9EF M0JUP
PE15 9EF M7CKT
PE15 9EL M0JLS
PE15 9EQ G3VMI
PE15 9FH G6BVO
PE15 9LA M7DQG
PE15 9QA 2E0TGF
PE15 9QA M3BIB
PE15 9QB G3SZU
PE15 9RX 2E0LGL
PE15 9RX M6LGL
PE15 9RZ G0GJL
PE15 9SS M1FEZ
PE158ND G4DPF
PE16 6AY M0HTA
PE16 6BD G4GPQ
PE16 6BJ M6JWE
PE16 6BW M6DPP
PE16 6JJ M6OIW
PE16 6JN G0MQL
PE16 6LG G4ZUH
PE16 6QE G6RBR
PE16 6RN 2E0OPU
PE16 6RN M3OPU
PE16 6RX G1YFE
PE16 6RX M6LYY
PE16 6SF G7MHV
PE16 6SG M6TFK
PE16 6SX M7MEK
PE16 6TP G6OHM
PE16 6TP M3FTV
PE16 6TP M6NIE
PE16 6UR G7VQC
PE16 6UX 2E0JTQ
PE16 6UX G0HKE
PE16 6UX M0JTQ
PE16 6UX M6JTQ
PE17 2TB G1HJD
PE17 6YB G1OSJ
PE19 1AE G1CSO
PE19 1HA 2E0YWO
PE19 1HA M0YWO
PE19 1HA M6YWO
PE19 1JU G8WRV
PE19 1LD G7HMZ
PE19 1NP M1FGO
PE19 1PF 2E0OZX
PE19 1UF G8FDE
PE19 1UF G8FFC
PE19 2EN 2E0BRS
PE19 2EN M3PYW
PE19 2JY 2E0SDC
PE19 2NN G8BKG
PE19 2NN M0DFW
PE19 2NN M0OLG
PE19 2QF G7JUC
PE19 2QF M3JUC
PE19 2QF M6JUC
PE19 2QG M1WHO
PE19 2QG M3EMS
PE19 2UL M0GQT
PE19 4UE G0ABW
PE19 5DB 2E0ICR
PE19 5DB M7BUR
PE19 5DQ G7LFZ
PE19 5RD G6JAK
PE19 5SF G1YVS
PE19 5SR M5AEE
PE19 5SZ G8CHC
PE19 5TZ M0AXZ
PE19 5UB G6FJG
PE19 5YA G0CJQ
PE19 5YJ M6TYS
PE19 6AL 2E0BIS
PE19 6BD 2E0KHW
PE19 6BD M0WHK
PE19 6LT M6AQV
PE19 6QW G8TQI

PE19 6RX G3WCD
PE19 6SD G4LIQ
PE19 6ST G8KDF
PE19 6TD 2E0PEI
PE19 6TD M0PEI
PE19 6TD M7TCA
PE19 6TJ G0FLT
PE19 6TJ G3UKB
PE19 7AN G6TNW
PE19 7LE G8ELG
PE19 7LF G4MZV
PE19 7RH M5KVK
PE19 8GJ G7NRV
PE19 8GU G3PNP
PE19 8HY G3LWJ
PE19 8PD 2E0DIP
PE19 8PD M0SKT
PE19 8PD M6DWE
PE19 8PE G4ULM
PE19 8PU G8LWO
PE19 8QL 2E0CZG
PE19 8QL M0ZSM
PE19 8QL M6GEK
PE19 8TZ G0JIW
PE192QY G6OXI
PE192SU M7FNP
PE196QY M7PPQ
PE2 5EN G7CQG
PE2 5ER G1FWU
PE2 5EY M7LMB
PE2 5LG G3EHQ
PE2 5LY G6AVS
PE2 5NT G4UGK
PE2 5PG M1AKH
PE2 5PR G7IBX
PE2 5PS G7AMW
PE2 5SH 2E0DZY
PE2 5SH M6MTU
PE2 5SL G4XGR
PE2 5SL G7JWD
PE2 5SL G8HTZ
PE2 5SS 2E0IFZ
PE2 5SS M6IFZ
PE2 5XS G3NSF
PE2 5YW G0HYN
PE2 5YW M0VTG
PE2 6XN M3MXP
PE2 6YB G4LMY
PE2 6YB M0AOG
PE2 6YE M6NWF
PE2 6YL G4OIR
PE2 6YP G8ILG
PE2 6YY G1NIT
PE2 6YZ M3YYV
PE2 7AH G0LEH
PE2 8HX M0KZT
PE2 8LH G0HZE
PE2 8LS G4FVK
PE2 8LW M3VZU
PE2 8PL G6GYM
PE2 8TG M7IXH
PE2 8UP G4EHW
PE2 8UP G8XLH
PE2 8UP M1PRC
PE2 8UP M5ATR
PE2 8UP M7TIA
PE2 8XP G0EQB
PE2 9AT M6WKB
PE2 9AT M6WVB
PE2 9DN 2E0DBI
PE2 9DN M0HYX
PE2 9DN M0WAF
PE2 9DN M3UEF
PE2 9DN M6HYX
PE2 9EG G8HXR
PE2 9JE G7HUG
PE2 9JG M0BSV
PE2 9JG M3NWK
PE2 9JN G8VWU
PE2 9PD G4DXW
PE2 9PQ 2E1IGK
PE2 9PQ M0PBD
PE2 9RQ 2E0WBT
PE2 9RQ M3WBT
PE20 1AU M1EKP
PE20 1JA G0MJJ
PE20 1JA G8SKG
PE20 1JT G0MRA
PE20 1JW M3TSI
PE20 1LB M6BOT
PE20 1LR 2E0TNV
PE20 1LR M0TNV
PE20 1LR M3TNV
PE20 1LR M3TOJ
PE20 1LR M6ARJ
PE20 1LR M6RYN
PE20 1LZ 2E0TQB
PE20 1LZ M1GOH
PE20 1LZ M3TQB
PE20 1NU G3OYL
PE20 1QN M0NDY
PE20 1SG M1DXG
PE20 1SQ G0WSA
PE20 2BU M1DKF
PE20 2JF M0HZA
PE20 2LQ G6DJY
PE20 2LU G7STM
PE20 2NS M6JKH
PE20 2PF M0NAE

PE20 2PF M6AKI
PE20 3AP 2E0WTY
PE20 3AP M0WTY
PE20 3AP M3WTY
PE20 3AP M6BEK
PE20 3AP M6JCC
PE20 3AP M6MOM
PE20 3DF M1CNX
PE20 3EF 2E0CBU
PE20 3EF M6MOW
PE20 3ES G0PBO
PE20 3LD M3GPX
PE20 3PG G0BGY
PE20 3PZ G7HNM
PE20 3PZ M3MIS
PE20 3QG G7JSV
PE20 3QT G3KLC
PE201JJ M6GOG
PE21 0AU G6ADG
PE21 0EJ M3NJY
PE21 0EN M0PAG
PE21 0EZ M0KOG
PE21 0LL M6EHH
PE21 0NF M7FGB
PE21 0PJ M7HCT
PE21 0SF M0WBD
PE21 6DP 2E0RSG
PE21 6DR M7FGQ
PE21 6DS M6BXB
PE21 6JN 2E0CDF
PE21 6PW G7LSP
PE21 7BZ G6NUZ
PE21 7BZ M0NUZ
PE21 7EJ 2E0BCB
PE21 7EJ M0GDT
PE21 7HL G3DFY
PE21 7HL M0AZV
PE21 7JG 2E1FRQ
PE21 7JG G4OID
PE21 7NQ M3DPB
PE21 7PN G4GZL
PE21 7PN M3KRQ
PE21 7RU G7MDT
PE21 7TJ 2E0BNW
PE21 7TJ M0WFK
PE21 7TJ M3WFK
PE21 8AG M6EHW
PE21 8BJ G8ZTR
PE21 8BW M0HQZ
PE21 8DA G8OYL
PE21 8DL G7MTE
PE21 8FS M3ZLR
PE21 8HD G1RAE
PE21 8HY G6SKF
PE21 8JH M0IMY
PE21 8LL 2E0KRD
PE21 8LL M3KRD
PE21 8NG G7ZRT
PE21 8NG M0TLC
PE21 8NL M6JWT
PE21 8NQ 2E0CON
PE21 8NQ M3WUH
PE21 8PU G7CQH
PE21 8PU M0EAO
PE21 8PU M0HDS
PE21 8SX M3WGB
PE21 8UF G0HPV
PE21 8XB M6EUZ
PE21 9AE G3JRY
PE21 9AP 2E0ZTL
PE21 9AP M6CAI
PE21 9HA G0DAF
PE21 9HY G7UTE
PE21 9JJ 2E0BFS
PE21 9LE 2E0JBD
PE21 9LE M3JND
PE21 9LR G7LCW
PE21 9LW G0PXB
PE21 9PD G4RPI
PE21 9PH G8TED
PE21 9PN G7WID
PE22 0AF G1DWG
PE22 0BG G7IBH
PE22 0BU G2FFD
PE22 0BZ G0UFC
PE22 0BZ G3WWH
PE22 0DH 2E0BBQ
PE22 0DZ M0GVL
PE22 0DZ M6AHL
PE22 0JD G7DYD
PE22 0JG M3DIT
PE22 0NA 2E0MJL
PE22 0NA M0JWL
PE22 0NA M3MJL
PE22 0PA M6IOG
PE22 0PA M6KPX
PE22 0YD G4FI
PE22 0YD G7FUM
PE22 7AB G1EBZ
PE22 7BS G8LDW
PE22 7HJ 2E1FRE
PE22 7JU G3XLE
PE22 7LW 2E1CJJ
PE22 7LW G0KYD
PE22 7LW M0LIO
PE22 7LW M6DAX
PE22 7PQ M7EQN
PE22 7RA 2E0HLM
PE22 7RA M0HLM

PE22 7RA M3ICA
PE22 8DQ M3JQN
PE22 8EX G7GZV
PE22 8EY G8IEV
PE22 8HA G8UMO
PE22 8LX 2E0MID
PE22 8LX 2E0VBR
PE22 8LX M0MID
PE22 8LX M0VBR
PE22 8LX M6VBF
PE22 8LX M6VBR
PE22 8NQ M6CRE
PE22 8PF M1ETN
PE22 8QJ G7EXT
PE22 8RD M6EON
PE22 8RR 2E0BDS
PE22 8RR M0LFC
PE22 8RU G1RNL
PE22 8RU G7HZS
PE22 9LS G0KOO
PE22 9NY M0VPK
PE22 9PR M6PAY
PE22 9PW G7KYX
PE22 9QX G4DYV
PE22 9RA G0KZN
PE22 9RF M7FWP
PE23 4AU M0DXJ
PE23 4BE M1CNI
PE23 4BY M6BUL
PE23 4EJ G0ESO
PE23 4JB G4UBM
PE23 5AG 2E0NPS
PE23 5BB G1YYU
PE23 5DB G4UNW
PE23 5EP 2E0ITV
PE23 5EX G4DYJ
PE23 5HH G6HYF
PE23 5PX G4HAI
PE23 5PX G8ITG
PE23 5QZ G4VSJ
PE23 5SE G6IDL
PE24 4DL G0LAG
PE24 4JJ G1MJN
PE24 4JP M3UEN
PE24 4LY G8VAE
PE24 4PJ M3FSE
PE24 4QN M3ZNF
PE24 4QR M0SWF
PE24 5AT G8HWJ
PE24 5DD M7TPU
PE24 5DQ G7GQO
PE24 5ES G6MQU
PE24 5HQ G0OHF
PE24 5HT G7BNN
PE24 5HZ M6INP
PE24 5JQ M0CCW
PE24 5NE G4VYF
PE24 5NF M0KFB
PE24 5NN G0JPQ
PE24 5NZ G0WCU
PE24 5PD M7APY
PE24 5QU 2E0MCC
PE24 5QZ G4ITB
PE24 5RQ M6AFW
PE24 5RS G2PYE
PE24 5RS G6PYE
PE24 5RS G8IER
PE24 5RY G1LLZ
PE24 5SL G0JMZ
PE24 5TL M1AZF
PE24 5YN G7JHU
PE24 5YQ G4PRF
PE24 5YQ G7JCX
PE24 5YQ M3YCN
PE24 5YS M6SRN
PE24 5YS M7AHH
PE245DH G7BNO
PE245JU M7CVH
PE245NE M0NSR
PE245UZ G8FRY
PE25 1BN G1MHB
PE25 1EL G0HNZ
PE25 1HE G4XTX
PE25 1HF G4HFG
PE25 1HF M3NQN
PE25 1HL G0ECN
PE25 1HQ M1ERY
PE25 1PD G7MUY
PE25 1RE G8MZW
PE25 1SE G3NPY
PE25 1TQ G4ZQT
PE25 2BP G0OTH
PE25 2BP M0FRC
PE25 2DD G0EPE
PE25 2DN M7EVP
PE25 2EE M0BBE
PE25 2EW G4ZBZ
PE25 2HY 2E0YSU
PE25 2HY M3YSU
PE25 2JA 2E0ROS
PE25 2JA M3FFI
PE25 2JA M6YMX
PE25 2JE M0WDE
PE25 2LN M6AWU
PE25 2NB M6RSO
PE25 2NB M6TPT
PE25 2QQ M3AVD
PE25 2QZ G4SOR
PE25 2RA M3KVI

Postcode

PE25 2TF M6AFX
PE25 2TH G0NDV
PE25 2TX M3YAS
PE25 2UH 2E0PRD
PE25 2UH M0PRD
PE25 2UH M3NQI
PE25 3BE M5PWR
PE25 3ER G4MVZ
PE25 3JN G0JJR
PE25 3LF G0CHB
PE25 3PU 2E0BAK
PE25 3QU G1YBT
PE25 3RX G3XZY
PE25 3SA M3IVN
PE251HS 2E0DBN
PE251HS M6DMB
PE26 1BZ G4RRH
PE26 1JP G8JWE
PE26 1LU M6FDK
PE26 1LX G1ZBB
PE26 1LZ G0OPL
PE26 1NB 2E0CXP
PE26 1NB 2E1AOK
PE26 1NB G3CXP
PE26 1NB G3MQI
PE26 1NB G3NKJ
PE26 1NB G3ROQ
PE26 1NB M1CXP
PE26 1SF G1XIY
PE26 1SH G8TVC
PE26 2NZ M6MEN
PE26 2QE G3PMH
PE26 2QE G4KPZ
PE26 2QE G6RHA
PE26 2QQ M0IXL
PE26 2SY G4ETG
PE26 2SY M0VVT
PE26 2TF G0BNR
PE26 2TX 2E0XXC
PE26 2TX M7USA
PE26 2XX 2E0KFY
PE26 2XX 2E0MWD
PE26 2XX M0LHF
PE26 2XX M0MDZ
PE26 2XX M0XTG
PE26 2XX M7FLR
PE26 2YW G0FJR
PE27 3AJ M3FCB
PE27 3DQ G4NKW
PE27 3HF M6CKD
PE27 3HF M6INV
PE27 3NL G0RSA
PE27 3TJ G8XIY
PE27 3TZ G4YDD
PE27 3XZ G6THP
PE27 3YZ G0KHY
PE27 4NE G8JKV
PE27 4SH 2E0BVQ
PE27 4SN G0GXS
PE27 4SW G7JAE
PE27 4TB G4XGT
PE27 4UD 2E1EJU
PE27 4UD M7VAH
PE27 5FX G4KJJ
PE27 5FX G7JXJ
PE27 5RD 2E0KFI
PE27 5RD M0KFI
PE27 5RD M6GCO
PE27 5WX G4XBS
PE27 5XG G6NBM
PE27 6HW G7CSL
PE27 6SS G6VBQ
PE275RD M6OBX
PE28 0AJ G8GRT
PE28 0BB G0IWB
PE28 0BP G7GJS
PE28 0BP M0MAW
PE28 0BS G3NKQ
PE28 0DZ G0AYX
PE28 0GG 2E0CIH
PE28 0GG M3MJV
PE28 0JY G0HSR
PE28 0JY M0KFF
PE28 0PA G7PAE
PE28 0PE M3ZUW
PE28 0PF M6BTY
PE28 0UU G1MVF
PE28 2DQ G3NID
PE28 2EH M7JDI
PE28 2EQ 2E0CUB
PE28 2EQ M6BRQ
PE28 2EQ M6JYJ
PE28 2FP G1LMS
PE28 2NU G4CTI
PE28 2QG G7FKF
PE28 2SA M7AFS
PE28 2SD G8FZT
PE28 2UT M6FDI
PE28 3AH 2E0FRF
PE28 3AH 2E0FRK
PE28 3AH M3KUY
PE28 3DJ M6WMZ
PE28 3DL G4DLT
PE28 3DL G8LEB
PE28 3ER M3TBG
PE28 3EY 2E1ECN
PE28 3JB 2E0HIG
PE28 3JB M6BGH
PE28 3JT G7RTR
PE28 3LD 2E0HRI
PE28 3LD M0RSU
PE28 3LE G0RLS
PE28 3LE G1KGV
PE28 3LE G7TSB
PE28 3LE G8AKL
PE28 3LE M6PKV
PE28 3LY G3NSD
PE28 3QG G8YUR
PE28 3QH G4BIK
PE28 3QH G6GJY
PE28 3XA G4FCF
PE28 3YU G8SAA
PE28 4DZ G3ECP
PE28 4EW G7ODT
PE28 4FH G4GWU
PE28 4JX M6LRF
PE28 4PX G8UJV
PE28 4TG 2E0FFK
PE28 4TG M7OHM
PE28 4TQ G8RSA
PE28 4US G3TCL
PE28 4US G4ZEW
PE28 4WS G4RKO
PE28 5AW 2E1AQH
PE28 5AW G1NOO
PE28 5LJ 2E0TMC
PE28 5LJ M6TMC
PE28 5SN G4SOQ
PE28 5TR M3TGA
PE28 5TZ M0MPQ
PE28 5TZ M6PPY
PE28 5UA G7KJW
PE28 5UX 2E0DOJ
PE28 5UX M3MSX
PE28 5WB G0MND
PE28 5YE 2E1IGI
PE28 5YE G7ELC
PE28 5YE M3CMW
PE28 9AN G4YUL
PE28 9BS G3WOT
PE28 9DB G0CCL
PE28 9EW G1VWC
PE28 9JL G3PYZ
PE28 9JL G4KLE
PE28 9JR G3SIT
PE28 9LP 2E0YHZ
PE28 9LP G1KHM
PE28 9LP M7ABW
PE28 9NA M0ETE
PE282RZ M6LLN
PE284TG M7VTT
PE289LE M7KGW
PE28XF M6ONN
PE29 1LW G1OVH
PE29 1NF 2E0XJN
PE29 1NF M0XJN
PE29 1NF M6JSU
PE29 1QL 2E0EML
PE29 1QL M0OFF
PE29 1QL M0PYC
PE29 1QL M6CSK
PE29 1RH M0IQH
PE29 1RP G4LHI
PE29 1RR M7LOC
PE29 1RZ G3MKV
PE29 1SR G3RPV
PE29 1TA 2E1ARG
PE29 1TA G7OJX
PE29 1TJ M3ZGG
PE29 1TL G1JGT
PE29 1TS M3OWU
PE29 1UG M7CJG
PE29 1WT G7JAO
PE29 1XE 2E0BQH
PE29 1XE M6GTR
PE29 1XF G0UFB
PE29 1XP G7MJD
PE29 1XY G3YQB
PE29 1YL M0ALK
PE29 1YN M0MIG
PE29 1YW G8YCI
PE29 1YZ G1KWF
PE29 2AL G4ZGG
PE29 2LU G1SMC
PE29 2WA M7ETQ
PE29 6XP 2E0LZX
PE29 6XP M0LZX
PE29 6XP M7TCS
PE29 7BP G6YHF
PE29 7BW M7HUH
PE29 7HJ G6PRL
PE29 7JA G6TGW
PE29 7JE G4UXV
PE292WA M7ANV
PE3 6BB G3YYW
PE3 6LB G3JXQ
PE3 8BA G4KEY
PE3 8ES G8NGZ
PE3 8HP M6FPU
PE3 8LG G4UQF
PE3 8RL M6OJZ
PE3 8SH M6PXY
PE3 9AU G6MGA
PE3 9FS M0RHG
PE3 9PA M3EUP
PE3 9PS M7MOH
PE3 9TS M6CWF
PE3 9UD G1PJO
PE3 9UH G4FFS
PE3 9UH G6GNC
PE3 9YD G6GLB
PE30 2DH M7AVP
PE30 2ED G1HYU
PE30 2EL M6AJF
PE30 2HX M6MJF
PE30 2NW 2E0EVX
PE30 2QG 2E0GIE
PE30 2QG M0KLX
PE30 2QG M3GIE
PE30 2QL 2E1HYE
PE30 2QL M3HYE
PE30 3BG 2E1HXC
PE30 3BS G3RSV
PE30 3BS G3VET
PE30 3DE G0IJU
PE30 3DP G8NSK
PE30 3DX G1KLP
PE30 3EX 2E1TOM
PE30 3EZ G3ZCA
PE30 3EZ G8MAY
PE30 3HB G8GKL
PE30 3LY G8KOC
PE30 3NA G3YII
PE30 3NJ G6REC
PE30 3PW 2E0TEZ
PE30 3PW M0TAJ
PE30 3PW M0WHL
PE30 3PW M3TQJ
PE30 3PW M3UZK
PE30 3PX G4CBO
PE30 3RY G6DDX
PE30 3UD G1KBH
PE30 3UZ G0VHH
PE30 3XA G4GXL
PE30 3XD G8WAP
PE30 3YB G7VPS
PE30 4AB G1XYZ
PE30 4AB G3HRX
PE30 4AB G3SZ
PE30 4AB G3ZGS
PE30 4AB G4OZG
PE30 4AT G6IIP
PE30 4DN G1LOK
PE30 4QA G0WWU
PE30 4XD 2E1IHT
PE30 4XT M6GRZ
PE30 4YY G6WAN
PE30 5BB 2E0HSP
PE30 5BB 2E0NRX
PE30 5BB M6HSP
PE30 5BB M6NRX
PE30 5BB M7TRH
PE30 5BQ G7OUZ
PE30 5DT G8BQT
PE30 5DY G6IWU
PE31 6AX M3ZOH
PE31 6BT G4JNQ
PE31 6DP M6HNG
PE31 6HQ G0AED
PE31 6JW M6NUV
PE31 6LH G8HHO
PE31 6PR G0MTB
PE31 6PR G6TVI
PE31 6QN G7WGL
PE31 6UR 2E0RWK
PE31 7AR G1SCQ
PE31 7AR G7RNA
PE31 7AR M0HCR
PE31 7BS G6WDC
PE31 7DX G8JXU
PE31 7EB G6RQA
PE31 7LA M1CCN
PE31 7LQ G4DDT
PE31 7PA G0OHQ
PE31 7QF G4EOJ
PE31 7RF G0JQR
PE31 7SA G6DAC
PE31 7SZ M7EOQ
PE31 7UG G3YLW
PE31 8AQ G0OFD
PE31 8AQ G8MBJ
PE31 8BP M3LCU
PE31 8BS G1YMA
PE31 8ET G7MDI
PE31 8LN G0CQB
PE31 8NJ M0ZLP
PE31 8RL G4DCJ
PE31 8RL G8FTP
PE31 8RS G0LRU
PE31 8RS G4ZQC
PE31 8SF G3OVL
PE31 8SF M3HEX
PE32 1AW G7RSA
PE32 1DQ G7NJP
PE32 1NQ M6DQL
PE32 1NW G7SYL
PE32 1NY G7OVE
PE32 1QZ G6IBO
PE32 1RJ G7DQL
PE32 1RL G6CPS
PE32 1SF G8TUH
PE32 1SS G1LMQ
PE32 1SY G4CXE
PE32 1TG M0BQB
PE32 1UX M7BKT
PE32 2AS G0UOB
PE32 2BA 2E0ESY
PE32 2BA M6LZF
PE32 2DT M6KRM
PE32 2DU M6JHC
PE32 2EA G0TZZ
PE32 2HS M3GKE
PE32 2LJ G7MFY
PE32 2LJ M7EBY
PE32 2LZ M6OTH
PE32 2NA G7BUN
PE32 2QD M6VWX
PE32 2QX M3KAU
PE32 2QX M6UAL
PE32 2TE G1CSY
PE32 2TE M0DZB
PE32 2UA G1ICX
PE322AU G7RJR
PE33 0BT 2E0RQQ
PE33 0BT G4RB
PE33 0BT M3RQQ
PE33 0HE G0JWG
PE33 0HS G0BPL
PE33 0JG G8CWE
PE33 0NN 2E0WWL
PE33 0NN M7SRW
PE33 0NZ M7FQD
PE33 0PG G0UMM
PE33 0PR G8ULL
PE33 0UA G8PHQ
PE33 9BA M7FKT
PE33 9BD G1EMW
PE33 9BW M6SGG
PE33 9DB M0JPM
PE33 9HP G4IXD
PE33 9JW M7PBW
PE33 9PX M3YXJ
PE33 9PX M3YXK
PE33 9RP G0FQO
PE33 9RP G1RHE
PE33 9TQ G0CCS
PE33 9TS 2E0SKI
PE33 9TS M0JEK
PE33 9TS M6ABS
PE33 9XW M0WAR
PE34 3BD M0GVI
PE34 3BN M3YFX
PE34 3HF M1CCQ
PE34 3HL G7MOD
PE34 3JH G7UGR
PE34 3LS G4LBQ
PE34 3PF G7UEL
PE34 3PP 2E0XXM
PE34 3PP M0XXM
PE34 3PP M3XXM
PE34 3QD 2E0KKM
PE34 3QD M0KKM
PE34 3QD M6VKM
PE34 4DA G4NJJ
PE34 4DE G6MQK
PE34 4DE M3JZK
PE34 4DY G6FSK
PE34 4HZ G4MOC
PE34 4JB 2E0EKK
PE34 4JB M3GHO
PE34 4JP M3KRO
PE34 4JX M0TBJ
PE34 4JX M6WGC
PE34 4LH G8PWQ
PE34 4NF M3GGV
PE34 4NR M6JOW
PE34 4NU G6CVR
PE34 4PR M3YCB
PE34 4QJ G7PTB
PE34 4QS M3GJN
PE344HE 2E0CMK
PE344HE M0IOT
PE344HE M6CPN
PE36 5BD G0KHF
PE36 5BS G1NJV
PE36 5BZ G0XBO
PE36 5DJ G4VUF
PE36 5EA M0HQA
PE36 5EJ G4RQU
PE36 5PJ G8MRI
PE36 6AP G7APU
PE36 6BS G1LSK
PE36 6BX 2E1EBX
PE37 7BT G0HZA
PE37 7RH G3JKM
PE37 7SP G7MOK
PE37 7TP M6LAI
PE37 8EE G4FZL
PE37 8ET G1KIW
PE37 8HY M6NEC
PE37 8JA M7TRG
PE37 8LB M6IWD
PE37 8LU G0NQN
PE37 8NN G1UGJ
PE38 0AJ 2E0HIQ
PE38 0AJ M0XMP
PE38 0AJ M6SCI
PE38 0BY G6TRA
PE38 0DH G4RCP
PE38 0DP G1VIS
PE38 0DP G6NMK
PE38 0DY G1HBV
PE38 0DY M0ERJ
PE38 0EN G4LMX
PE38 0EU M3HPT
PE38 0EU M3KBB
PE38 0PA M7CRR
PE38 9AX G3UPN
PE38 9NJ G8VCO
PE38 9NY M7FUG
PE38 9PG G8JAN
PE38 9QU G4TUO
PE38 9QY G4BAU
PE38 9RQ G3ASG
PE38 9TZ M6OST
PE4 5AF G7FTD
PE4 5AQ G0ADA
PE4 5AU G3GMW
PE4 5BJ G1DRI
PE4 5DD G3WRL
PE4 5DF G7DWO
PE4 5ED G1MLC
PE4 6AQ G6AYU
PE4 6BJ M7BPJ
PE4 6GW 2E0GDF
PE4 6GW M3ZVT
PE4 6HZ M7FBX
PE4 6JS G0CNL
PE4 6QY G1EUN
PE4 6QZ G4WIN
PE4 6RF M7MTV
PE4 6RL G4ZKM
PE4 6RP M3FVE
PE4 6SA M6DLJ
PE4 6SB G8EQY
PE4 6ZH G4DJZ
PE4 7BJ M3JGI
PE4 7EN 2E0ROD
PE4 7EN M0RHB
PE4 7EN M5AEI
PE4 7PY G7JJP
PE4 7PY M3LPT
PE4 7TN G4DEW
PE4 7TT M3DRA
PE4 7TW G3RED
PE4 7UP M0CEG
PE4 7ZD G3VYU
PE4 7ZD G7SQY
PE46NT M6HWO
PE6 0AH M7PFD
PE6 0BA M0AWB
PE6 0BA M1BXJ
PE6 0BS M6WVK
PE6 0DA M3YSV
PE6 0DG G4NCP
PE6 0EN M6IJL
PE6 0ES M0PXY
PE6 0HA 2E0GRP
PE6 0HA M0GRP
PE6 0HA M6GRP
PE6 0JB G8YOX
PE6 0LH G0TFR
PE6 0PQ M1MIC
PE6 0QG G0NUU
PE6 0QJ 2E0YHN
PE6 0QW 2E0PBR
PE6 0QW M0PAJ
PE6 0QW M6PAO
PE6 7EU G1ESC
PE6 7LG G0EVU
PE6 7NG G1ZDX
PE6 7NW G3YOY
PE6 7QE M7TXM
PE6 7RG 2E0BEF
PE6 7RG G4FXY
PE6 7RG G4NRK
PE6 7RG G6CZO
PE6 7UB G0IAG
PE6 7UB G1XXW
PE6 7UB M7ANN
PE6 7UP G1OAU
PE6 7XG G0BLV
PE6 7YF G8KQZ
PE6 8AG G6IDW
PE6 8BS G1AUM
PE6 8DA G7OOB
PE6 8DJ G4FIQ
PE6 8ED M0HTU
PE6 8EH G8YVS
PE6 8EL G4SQA
PE6 8GR M7RPY
PE6 8HP G4FSX
PE6 8HR G1FJD
PE6 8JD G0DCP
PE6 8JS G4MZM
PE6 8NW G4VJN
PE6 8NW G6KEZ
PE6 8NW G6MTY
PE6 8QJ M5ALU
PE6 8RD G1HOJ
PE6 8RD G6NYL
PE6 8SF M0NMH
PE6 8SF M3MQH
PE6 8SH G1RVK
PE6 8TF G7UHL
PE6 9AA G6YCA
PE6 9BA 2E0JRU
PE6 9BA M0JRU
PE6 9DN M0DBH
PE6 9LP G7SOE
PE6 9NP G8OAD
PE6 9PH 2E0YCA
PE6 9PH M7YCA
PE6 9RB G6ENN
PE67LE M7MSA
PE68HP G7OKT
PE68HS M7LBY
PE68PS G0BAT
PE68QJ G7GWA
PE7 0LU 2E0TEK
PE7 1AT G0LHG
PE7 1BP G0RCH
PE7 1HJ G4NLO
PE7 1QN G4DCS
PE7 1RF G1NLS
PE7 1RL G0IHK
PE7 1RR G0PGY
PE7 1RT G4WJB
PE7 1TY 2E0GPU
PE7 1TY M1NER
PE7 1TY M3YEK
PE7 1UE 2E1GDG
PE7 1UE G4FFN
PE7 1XQ G0WXP
PE7 1XX G7VPA
PE7 2BT M3JCE
PE7 2DD G0JJW
PE7 2DW M3ZWB
PE7 2HN G0INO
PE7 2HN G4YIT
PE7 2PD G1JIH
PE7 2RG G3XRK
PE7 3AA M3YGC
PE7 3AX M6CUQ
PE7 3AY M3NWD
PE7 3BU G6JAS
PE7 3FA G6EUY
PE7 3FS M7WDX
PE7 3GE M0DYR
PE7 3NU M7AYW
PE7 3PG M3LSF
PE7 3PR G1KQN
PE7 3RB M7NJJ
PE7 3RG M3JDX
PE7 3RS G7IVU
PE7 3SH G8UFO
PE7 3SS G0HOF
PE7 3SU G4PPJ
PE7 3SX M0RKH
PE7 3UA G3DQW
PE7 3UA G3TGO
PE7 3YD M3MEP
PE7 3YQ 2E0DVO
PE7 3YQ M6FDH
PE7 3ZF G0XAH
PE7 3ZR G7AFO
PE7 8FL 2E0CCQ
PE7 8LZ 2E0SRH
PE7 8LZ M7HJR
PE7 8LZ M7SRP
PE7 8RL G6DYK
PE73QW 2E0ORD
PE73QW M0VRD
PE73QW M7CHE
PE8 4DQ G4JPB
PE8 4JQ G3VZL
PE8 4LT G7NDQ
PE8 4NX 2E0OTZ
PE8 4NX M3OTZ
PE8 4QE G7WDN
PE8 5AN G7HLU
PE8 5HP G4HCG
PE8 5LS G8YKS
PE8 5PB M6PVR
PE8 5PS G0UIH
PE8 6AU 2E0HEP
PE8 6AU 2E0ZAF
PE8 6AU M0KDT
PE8 6AU M0ZAV
PE8 6AU M3ZWL
PE8 6EU 2E0HLF
PE8 6EU M0ILU
PE8 6EU M6HLF
PE8 6LP 2E0KNT
PE8 6LP M0KWR
PE8 6LP M3KNT
PE8 6TW 2E0GDT
PE8 6TW M0UKO
PE8 6TW M6UKH
PE9 1DS 2E0BVR
PE9 1HF G4NHL
PE9 1JL M7MOU
PE9 1LA M0STA
PE9 1PJ M0NMA
PE9 1QQ G4ZBG
PE9 1SN G6HNP
PE9 1UP G3TBG
PE9 2FB G6ENZ
PE9 2GR 2E0BMT
PE9 2GR G5REV
PE9 2HU M0PZC
PE9 2JY G1FSW
PE9 2SG G1DCX
PE9 2TS G1JGR
PE9 2UB G3THC
PE9 2YR M0YDK
PE9 3AG G4VAM
PE9 3LL 2E0CNP
PE9 3LL M6FBH
PE9 3LZ G7DGP
PE9 3NA G4OZM
PE9 3PF 2E0JBX
PE9 3PW G3MFG
PE9 3QD G1LXK
PE9 3QD G4RQK
PE9 3SY M1REC
PE9 3UQ G8VDJ
PE9 4BU G0FPZ
PE9 4BU G0INV
PE9 4DJ M6AQJ
PE9 4EB G1CUG
PE9 4JJ G4PEL
PE9 4NA G6CVV
PE9 4RJ 2E0ZAL
PE9 4SN M6AAU
PH1 1DT GM4HJK
PH1 1HH GM8RYZ
PH1 1JD GM4BVD
PH1 1JD MM1FHS
PH1 1LE GM8UGO
PH1 1LU MM6JQA
PH1 1QA MM7RCX
PH1 1QB GM6JKU
PH1 1RQ MM7MYS
PH1 2BG GM7SDP
PH1 2DU GM6OFO
PH1 2DU MM3KVN
PH1 2NF GM6UJG
PH1 2NS MM3XUX
PH1 2PF MM7DVE
PH1 2PU MM7SWC
PH1 2QJ MM6NAA
PH1 3BJ MM7IJK
PH1 3DD GM1AHF
PH1 3DD GM1AHG
PH1 3GP MM6HBY
PH1 3LW 2M1DWK
PH1 3LW GM1DSK
PH1 3LW MM7DWK
PH1 3UA 2M1VFO
PH1 3UA MM0SDK
PH1 3UT MM7AEI
PH1 3YX MM6AFQ
PH1 4EY MM3VKP
PH1 4JN GM3ALF
PH1 4QS GM4EAF
PH1 4QS GM4ZRH
PH1 5FJ MM6NWK
PH1 5FN MM3YGI
PH1 5NJ MM0JUC
PH10 6AE MM6TKV
PH10 6FH MM7DHB
PH10 6PF 2M1EPV
PH10 6SF GM2BWW
PH10 6SF GM8DOR
PH10 6TH GM4NUN
PH10 6UE MM6HIG
PH10 6XE GM7GUL
PH10 7JL GM0VIT
PH10 7RB GM4LMG
PH11 8AS GM1ZQF
PH11 8BU GM0OFI
PH11 8DR MM0JNY
PH11 8DW GM3WAP
PH11 8HQ MM0RBQ
PH11 8HQ MM0RJJ
PH11 8HQ MM6GHX
PH12 8SL 2M0BMK
PH12 8UB MM7FWK
PH13 9HS MM6RGT
PH14 9QB GM0NTI
PH14 9QB GM0OTB
PH14 9QB GM8RTI
PH14 9QL MM0HGN
PH15 2BE GM6YRN
PH15 2QY GM4FOZ
PH15 2QY GM8KJO
PH16 5HH GM8KPH
PH16 5JF 2M0IMP
PH16 5JF MM3ZYS
PH16 5JL 2M0DGI
PH16 5JL MM0ZOG
PH16 5JL MM6BFR
PH16 5JS GM4ODW
PH16 5LA GM0CWR
PH2 0AE MM3SYQ
PH2 0AQ MM7FYP
PH2 0AR GM1PFU
PH2 0BL MM6OAG
PH2 0GY GM4MSL
PH2 0HH GM0FTX
PH2 0LD GM6UWF
PH2 0LD MM6FSQ
PH2 0RB GM1RBQ
PH2 6GD MM7HLD
PH2 6LWL MM7EAB
PH2 6QA 2M0TGN
PH2 6QA MM0WXE
PH2 6QA MM6TGN
PH2 6QE GM8YRT
PH2 6RR GM1BUY
PH2 6RR GM3EOB
PH2 6SD GM4DQJ
PH2 6SP GM0HCQ
PH2 6TD GM4EVS
PH2 6TQ MM0EMC
PH2 7BY GM1GHZ
PH2 7BY GM3UYR
PH2 7BY MM0MHZ
PH2 7HJ GM4YZT
PH2 7LL GM7EPW
PH2 7NF GM0CEA
PH2 7QA GM0IST
PH2 7QA GM0NYP
PH2 7QA GM1RGM
PH2 7QP MM3CNS
PH2 7QQ GM0LVI
PH2 7QU MM3SYO
PH2 7QU MM6CEW
PH2 7RB GM6VRC
PH2 7SS GM0GAV
PH2 7TB GM0IMH
PH2 7XD MM0MBC
PH2 8PT GM3THI
PH2 9EP 2M0MKZ
PH2 9HS GM0MXZ
PH2 9LW GM0PRG
PH2 9LW GM3YEW
PH20 1BS GM8CEA
PH21 1AP MM3GPL
PH21 1JS GM0KEQ
PH21 1JS GM6NUL
PH21 1NY GM4EBX
PH21 1NY GM4JDK
PH23 3AA GM0PWS
PH23 3AA GM3JFG
PH23 3AA GM3JIJ
PH26 3LX GM1SRP
PH26 3PA GM4CCN
PH26 3PA GM4LPG
PH26 3PX MM6CJC
PH3 1BE 2M0HIC
PH3 1BE MM7HIC
PH3 1BZ GM8TCG
PH3 1DD GM1BXI
PH3 1EW GM4JZB
PH3 1JS GM4SKB
PH3 1LX MM6YCB
PH32 4DW GM3VKN
PH33 6HB GM0ANG
PH33 6HX GM4YAT
PH33 6JS GM3RFA
PH33 6NX GM1YGV
PH33 6PY GM7REY
PH33 6SJ GM7PKT
PH33 6SZ GM0UMJ
PH33 6TZ MM7WOK
PH33 6UG GM8AOB
PH33 6UH GM4NFI
PH33 7AB MM3XFX
PH33 7AL GM6AES
PH33 7ER GM0ERB
PH33 7LE GM4OPU
PH33 7LR GM4TFJ
PH33 7PF GM4UGN
PH33 7PQ GM1YPJ
PH34 4DZ GM7PXL
PH34 4EB 2M0FYX
PH34 4EB MM0JZW
PH34 4EB MM7BJR
PH34 4EX GM4TJL
PH39 4NX 2M0SBP
PH39 4NX MM3YBQ
PH4 1QE GM0SGH
PH40 4PD GM4OAS
PH41 4QF GM4XJY
PH41 4QF GM6VYZ
PH41 4QY GM6VYY
PH41 4RB MM7ETU
PH41 4RL MM7EPW
PH41 4RQ 2M0LUG
PH49 4JR GM3ZRT
PH49 4JZ MM0FWK
PH5 2BD GM0WEZ
PH5 2BD MM6SOR
PH5 2BD MM6YEZ
PH6 2BB GM4YRO
PH6 2HB MM0GKT
PH6 2NW GM1KWA
PH6 2NW GM6IQH
PH7 3BT MM6RQD
PH7 3LE GM0FVJ
PH7 3NY 2M0KVI
PH7 3NY MM0JJZ
PH7 3NY MM7KJV
PH7 3RP GM0AOF
PH7 3RP GM1LTM
PH7 4DH GM4OFC
PH7 4JH 2M0DDX
PH7 4JH MM7DDX
PH7 4LE GM4YDC
PH8 0AA 2M0BZL
PH8 0AA MM0XPT
PH8 0AA MM6CRW
PH8 0ET GM4LFK
PH9 0LG GM4NGJ
PH9 0NH GM6IQH
PL1 2HX M3MWG
PL1 3JF G1KXJ
PL1 3JN G3MXP
PL1 3LB G7CNC
PL1 3PS G0OJW
PL1 3QR 2E0AZD
PL1 3RN M6FGC
PL1 4GQ G6RFS
PL1 4GU M3YOX
PL1 4HU 2E0NJK
PL1 4HU M3PXQ
PL1 4PR G1OBM
PL1 5HD M6ABG
PL1 5LJ M7XIA
PL1 5NB M6HLA
PL10 1DA G7AWG
PL10 1DA G8UDA
PL10 1EP M7FWT
PL11 2BZ 2E0LAD
PL11 2BZ M3UXG
PL11 2HJ G8XOX
PL11 2HR M7OST
PL11 2JB M6GJO
PL11 2LG 2E0VLR
PL11 2LG M7RVL
PL11 2LP M6OLD
PL11 2LT M3TBQ
PL11 2NE 2E1ABQ
PL11 2PZ G4ZZD
PL11 3DX G7FGD
PL12 4BJ G0AKH
PL12 4BJ G4GXK
PL12 4BJ G4WAB
PL12 4BJ G7WAB
PL12 4BJ G8SAL
PL12 4BN G4YXJ
PL12 4BX 2E0OMI
PL12 4BX M0HTK
PL12 4BX M6DZH
PL12 4DL M1CQC
PL12 4DY G4HZE
PL12 4FD G8PVR
PL12 4HE G7EUB
PL12 4JH G3UBY
PL12 4LF M3KFL
PL12 4NG M3ZYY
PL12 4PP G0XAW
PL12 4RG G0KDW
PL12 4SH 2E0CXA
PL12 4SH M6BKH
PL12 5AE 2E0CGP
PL12 5AE M0HBU
PL12 5AE M6AKW
PL12 5DA G3IVP
PL12 5DA G7LPT
PL12 5DF G0JLB
PL12 5HR M0RZD
PL12 5NH G6SEK
PL12 5NL 2E0RBA
PL12 5NL M0RAZ
PL12 5NQ G6ELG
PL12 6BH 2E0MLJ
PL12 6BH M6TMJ
PL12 6BJ G3EXL
PL12 6DR G4YGZ
PL12 6EL G3XCS
PL12 6EN G6YNV
PL12 6EN G7VBZ
PL12 6EN M6RRD
PL12 6JA G7STF
PL12 6JE M1BZK
PL12 6PS M0BHK
PL12 6RH G0DAV
PL12 6SA G6ION
PL12 6SP G4ALY
PL12 6TD G3ZHK
PL12 6UQ M7YGK
PL125DF 2E0MCJ
PL125DF M6ZEJ
PL13 1PE M6BIM
PL13 1PN G4MXP
PL13 1PR G8TXT
PL13 2DW G1XIC
PL13 2DW G8BCG
PL13 2HT G1YMC
PL13 2HX M6HCA
PL13 2JP 2E0PDQ
PL13 2JP G1YDQ
PL13 2JP G4BRF
PL13 2JS G0PWW
PL13 2JY G8TEF
PL13 2LF 2E0SJG
PL13 2NX G0HRX
PL13 2NX M1SAN
PL13 2PU G7CLO
PL13 2SD G6YXV
PL132WP G3BAR
PL14 3EE M6KBJ
PL14 3EX G6FYT
PL14 3FN G0CAY
PL14 3HX G0GUT
PL14 3LP M7CFG
PL14 3LW 2E0TUH
PL14 3LW M6TUH
PL14 3NS 2E0TGL
PL14 3NS M6SOC
PL14 3PT G3UAE
PL14 3PU M0GGM
PL14 3QG G1NSV
PL14 3RL G4TIQ
PL14 3RR M1APC
PL14 3SG G4NZG
PL14 3SQ G4ZGZ
PL14 3TD M0BEV
PL14 3TQ G3ZRH
PL14 3TS G1ZRQ
PL14 3TX M0YGT
PL14 4NJ G1STP
PL14 4NJ G6SHF
PL14 4NW G3WOA
PL14 4PA G0LYR
PL14 4PT M0KRR
PL14 4PT M3ZHX
PL14 4QP G7WCF
PL14 5BQ 2E0PLB
PL14 5BQ M0PLB
PL14 5BQ M6HMM
PL14 5BW G4MFE
PL14 5DP G0DAU
PL14 5EL G4VCA
PL14 5EN G4TYD
PL14 5EY G6OWT
PL14 5LE G8BIX
PL14 5LU G4MJW
PL14 5NQ G1KTZ
PL14 5PW G7LJA
PL14 5QT G0EFR
PL14 5QT M6MND
PL14 5RA G7JLK
PL14 5RF G7OAH
PL14 5RU G7BIM
PL14 5SA M6RFR
PL14 6AU M7GWK
PL14 6DA M0WHJ
PL14 6DA M7WJH
PL14 6DL M7DHP
PL14 6EP G8LFK
PL14 6JS G1NTV
PL14 6PZ G8VZR
PL143SX 2E0VAF
PL143SX M3VJE
PL15 7DZ G0AOB
PL15 7EQ 2E0IFK
PL15 7EQ M7ITI
PL15 7FF M3VQV
PL15 7HB 2E0VCC
PL15 7NE G1SCY
PL15 7PQ M3ZZH
PL15 7PS G4RZM
PL15 7PW G1TAZ
PL15 7PW G3LNW
PL15 7QD G3ZYL
PL15 7RQ M3IIV
PL15 7TT G8WUS
PL15 8PP G3SYM
PL15 8SE M7CQY
PL15 9BB G4KKZ
PL15 9BD M7MEC
PL15 9EB G0ANM
PL15 9EB G6ZMO
PL15 9LE G0LFQ
PL15 9NA G7HEP
PL15 9NB G6VVU
PL15 9PH M0ACI
PL15 9RR G0AAM
PL15 9RT G6UHD
PL15 9SY G0UFV
PL15 9SY G1SOX
PL15 9TJ G4OIA
PL157PR M6JMM
PL16 0AH M0ESW
PL16 0AH M3PSS
PL16 0AS G3LUW
PL16 0HJ 2E1EOR
PL17 7BX M6AKO
PL17 7EU G7NHW
PL17 7EU M0CDS
PL17 7EU M1APB
PL17 7GB M3RYY
PL17 7HD G8AWB
PL17 7HP G1KBG
PL17 7LW G1IPI
PL17 7PA G0IVZ
PL17 7PF G1YTV
PL17 7PL M3XWX
PL17 7PT G1FMU
PL17 7PT G7UDX
PL17 7PT M0SCR
PL17 7QE M7CLU
PL17 7QH M6BMT
PL17 7QH M6VQV
PL17 7QL G1VQG
PL17 8BL M3FAC
PL17 8DE M3YMS
PL17 8FD G7UXO
PL17 8GL M3PCQ
PL17 8JE G0TDN
PL17 8JN M0GVT
PL17 8JX G6BER
PL17 8NR G6IJW
PL17 8PA M6GAR
PL17 8PA M7BUB
PL18 9BD G4KYE
PL18 9BL G7CQK
PL18 9BN G4UMS
PL18 9BY M6FZV
PL18 9DA M6OPO
PL18 9DA M6POD
PL18 9DA M6ROP
PL18 9DH G3XIB
PL18 9DX G4PBN
PL18 9HH M1BBB
PL18 9HN 2E0WRQ
PL18 9HN M7SDD
PL18 9JB M0NMC
PL18 9JB M6BTS
PL18 9JB M6NEW
PL18 9NG G4FWN
PL18 9PB G4BVB
PL18 9QN M6YRW
PL18 9RY M0BPC
PL19 0DP 2E0MDH
PL19 0DP M6BWO

PL19 0EA 2E0PAA
PL19 0ER G3ZNE
PL19 0ET G8ONR
PL19 0EU M0YDW
PL19 0EU M3YDW
PL19 0NJ G7VUP
PL19 0NL G4TND
PL19 0NN 2E0RPH
PL19 0NR G3FHT
PL19 0PA M3XYU
PL19 0PN G0GSR
PL19 8DR G1KKA
PL19 8DX G1OJL
PL19 8HA G3MWZ
PL19 8RL M6KXU
PL19 9AX G4WDR
PL19 9AX G6BJJ
PL19 9DA G3TGN
PL19 9DG G8WPX
PL19 9DJ G6FOV
PL19 9DL G7IIB
PL19 9DN G3TYO
PL19 9DQ G3XOU
PL19 9LJ G0NUO
PL19 9LQ G7VYQ
PL19 9NQ G8KSM
PL19 9PR G0LEV
PL19 9QD G8AKC
PL2 1DT M6TGR
PL2 1EA M3LOE
PL2 1EA M3SCX
PL2 1JD 2E0OLF
PL2 1JD M0MLZ
PL2 1JD M6OLF
PL2 1QQ M6KPP
PL2 1WT M3TOR
PL2 2BR G4PKM
PL2 2BU G7LUL
PL2 2ER G0WGB
PL2 2QY 2E0BJD
PL2 2QY M3MQX
PL2 3AP M3YQM
PL2 3AX 2E0DVN
PL2 3AX M6BNM
PL2 3AZ G3XLZ
PL2 3ET G8EHD
PL2 3FB G1SYZ
PL2 3LF 2E0ITN
PL2 3LF M6ITN
PL2 3QG 2E0SPR
PL2 3QG M3PYJ
PL2 3QZ G7ANY
PL2 3RS G1SQI
PL2 3RS M3IOT
PL20 6AT G7MOH
PL20 6HP G6IIZ
PL20 6LJ G3TXL
PL20 6LJ M3TXL
PL20 6ND M6LJK
PL20 6NG G4GUY
PL20 6PT G0IAI
PL20 6QD M7GPM
PL20 6QX M7JDM
PL20 6SY G4DND
PL20 6TJ M7OMM
PL20 7AH G4HPX
PL20 7AR G3SGZ
PL20 7BZ M6WBR
PL20 7DB 2E0YJF
PL20 7DB M3YJF
PL20 7DD G1SQG
PL20 7DW G4DMM
PL20 7EP M3RNG
PL20 7JS G8MWN
PL20 7JS M3TNJ
PL20 7LH G0PGI
PL20 7LR G3ZXO
PL20 7PJ M6JCG
PL20 7QB M6TGY
PL206NN M6MQM
PL21 0AX G7HIC
PL21 0DD M7GSS
PL21 0ET G4VFG
PL21 0ET G6NSU
PL21 0ET M0VFG
PL21 0HT G0TQR
PL21 0LT G4NDL
PL21 0WD M5DAP
PL21 9BA M6EQQ
PL21 9BD G4XFM
PL21 9BH G7LNG
PL21 9BT G4ONC
PL21 9DH G0EOZ
PL21 9EF G4RYO
PL21 9EU G4OUZ
PL21 9EU G4RIM
PL21 9JE G4ETP
PL21 9JX 2E0KMF
PL21 9JX M0YSE
PL21 9QT 2E0GWD
PL21 9RB 2E0XCP
PL21 9RB M0ZCP
PL21 9RB M3XCP
PL21 9SN G3PRC
PL21 9SN G8PRC
PL21 9SN G8XTE
PL21 9TQ M3ZBI
PL21 9TS M0IZF
PL21 9XA 2E0PGH
PL21 9XA M6PGH
PL22 0ET G3XSC
PL22 0JB G0MFQ
PL22 0QH G3HUB
PL23 1ED M6DYW
PL23 1ET G3MTG
PL23 1JH G3SXR
PL23 1LX M6USG
PL23 1NB G4AIR
PL23 1NB G7RNB
PL23 1QQ M3UIB
PL24 2AT G0AEW
PL24 2DE M6OSO
PL24 2DG G8XNA
PL24 2DL M6PCP
PL24 2EP 2E0EUF
PL24 2EP M6MUY
PL24 2LD G0CIG
PL24 2LH G4ZEB
PL24 2PW 2E0NJV
PL24 2PW M6NJV
PL24 2RL G1JXP
PL25 3AU G0KTD
PL25 3BB M1EWP
PL25 3BZ M6GBY
PL25 3DH G4RLN
PL25 3DR G0NNO
PL25 3DY M3UMM
PL25 3EB G0FXM
PL25 3EX G3HTO
PL25 3HB G8WBG
PL25 3HY M7CKG
PL25 3QE G4UON
PL25 3QG G4JOD
PL25 3TJ G4XBW
PL25 4DU M7ARW
PL25 4HR G1XMI
PL25 4HR G8NYR
PL25 4HT G4XOP
PL25 4JA G4XBC
PL25 4NE 2E0NAM
PL25 4NE M0NAW
PL25 4NE M6NKC
PL25 4NL 2E0IOL
PL25 4NL M7CSU
PL25 4QF M0AGY
PL25 4RJ G4MFQ
PL25 4UD M0DOA
PL25 4UW G6GAB
PL25 5EA G4TRV
PL25 5NY M3NMM
PL25 5TA G8MNC
PL254EA M3XTG
PL26 6BN G4LED
PL26 6HX 2E1DZJ
PL26 6NU M0HDP
PL26 6NZ G4MLI
PL26 6QS G8HYL
PL26 6QZ G3YIY
PL26 6TG G7JIB
PL26 6TZ G6MNI
PL26 6TZ G6MNJ
PL26 6UP M7OXY
PL26 6XD G8TCP
PL26 6XD M1BRZ
PL26 7AG M0VAU
PL26 7BH 2E0KUC
PL26 7BH M0ORS
PL26 7BW G4OPL
PL26 7EH G4HFO
PL26 7ER G8ZQM
PL26 7ER M3ZQM
PL26 7HG G8TNA
PL26 7NN G8VRV
PL26 7PF G4KNI
PL26 7PN G7FLX
PL26 7PT G4ZGQ
PL26 7PW M0GTH
PL26 7TE 2E0WRL
PL26 7TL G2KF
PL26 7TP G1LQT
PL26 7TP G3KYM
PL26 7TQ G7TLK
PL26 7TY G1BBA
PL26 7TY M7AWF
PL26 7UZ M7OOW
PL26 7XN M1AEG
PL26 7XQ G0FCB
PL26 7XS G0HZD
PL26 8BE M0HKI
PL26 8DB G6CNW
PL26 8DS 2E0PXI
PL26 8DS 2E0TGK
PL26 8DS M7MRN
PL26 8DS M7SEW
PL26 8EJ M3TEL
PL26 8GY G4VSY
PL26 8HD M6HVK
PL26 8HW M6CJR
PL26 8JA M1AZJ
PL26 8JH G1TTK
PL26 8JN 2E0BXD
PL26 8JN M0PSB
PL26 8JN M6JOE
PL26 8PH G0AHM
PL26 8PS M6RSI
PL26 8QL 2E0ONC
PL26 8QL G4JYF
PL26 8QL M6VDT
PL26 8QN 2E1CYI
PL26 8QN M3FKW
PL26 8RJ G4TRQ
PL26 8SJ M7FAM
PL26 8TJ M7FBW
PL26 8TL G0VDU
PL26 8UA 2E1CWE
PL26 8UB G0RJI
PL26 8UH G3WKF
PL26 8UQ G4OKS
PL26 8YE G4ZZY
PL26 8YW G7VTC
PL268QP M7HYP
PL268RT M0PHM
PL27 6AN 2E0EJV
PL27 6AN M6HVO
PL27 6NW G8LOU
PL27 7LL M3UWM
PL27 7PG G7DUB
PL27 7PR M6KWZ
PL27 7QA M6HEB
PL27 7QD 2E0JJR
PL27 7QD M3PXK
PL27 7TB G7TLC
PL27 7TS 2E0XTM
PL27 7TS M3XTM
PL27 7UN G7OQL
PL277TP M7HLK
PL28 8ES G0MWW
PL28 8SB G0SXM
PL29 3RU G4UZO
PL3 4DN M6BWK
PL3 4EG M3YWM
PL3 4HE M1FCG
PL3 4HS 2E0DCF
PL3 4HS M0WCL
PL3 4HS M6LJR
PL3 4LN G8PYE
PL3 4PD G0OYO
PL3 4QN M3OUV
PL3 4RB G6GEX
PL3 4RB M0GEX
PL3 5BS G0FCG
PL3 5DA G8RMP
PL3 5DU G4BCX
PL3 5DU G6ALR
PL3 5HJ M7HMC
PL3 5HW G7IZU
PL3 5NP G3SQN
PL3 5NP M6RMK
PL3 5NY M7PFC
PL3 5PU M3FTZ
PL3 5TX G0KPQ
PL3 6BY M6DIZ
PL3 6DB M3ZWD
PL3 6DE M6DIB
PL3 6DH G0GVX
PL3 6DH M6HKV
PL3 6HD M1AGY
PL3 6HL M3JOJ
PL3 6JF G1LOE
PL3 6JZ G0LSJ
PL3 6JZ G7HCN
PL3 6LD G0EIG
PL3 6LX G4XZS
PL3 6LX G6ZGK
PL3 6NQ G7DIR
PL3 6PB G1KCW
PL3 6PT G3SVZ
PL3 6PX G6GFO
PL3 6PX M3NJA
PL3 6QJ M7LRJ
PL3 6SZ M3RNK
PL30 3AU G3IWE
PL30 3BS G4WQU
PL30 3PH G4AXC
PL30 3PN G3XKS
PL30 4HF M0JQJ
PL30 5AT G8TXK
PL30 5ED G1JBB
PL30 5EP G4UBK
PL30 5EW 2E0UEE
PL30 5EW M0UEE
PL30 5EW M6UEE
PL30 5HD G8TWA
PL30 5JL G3MCD
PL30 5JZ 2E0IKO
PL30 5JZ M0XVN
PL30 5JZ M7SZU
PL30 5LU G1OFG
PL30 5PJ G1FXD
PL31 1BE 2E1AFC
PL31 1BH G3IGV
PL31 1JE M7AUL
PL31 1NH 2E0DSI
PL31 1NH M6DQO
PL31 1NZ 2E0KKY
PL31 1NZ M7YKK
PL31 1PY G4WVD
PL31 1RF M1ALX
PL31 2BZ M3ZNX
PL31 2FB 2E0BHN
PL31 2FB M3RKJ
PL31 2FE M6AKV
PL31 2FP G0UFF
PL31 2NU G0EEJ
PL31 2QP G4GPD
PL311PD M7CRK
PL32 9RZ G1VWU
PL32 9UB G4YME
PL32 9UP G4YVB
PL32 9YN G0ISJ
PL33 9AT G4WAV
PL33 9BN G6GWX
PL33 9BX M7ION
PL33 9DP M6CGB
PL33 9DT 2E0BJV
PL33 9DT M3PXF
PL34 0BH G0EWR
PL34 0DT M0HJO
PL34 0DT M0IGJ
PL34 0EH M3KKG
PL34 0EL G3LOV
PL34 0EL M3MXZ
PL34 0HH G0ATS
PL4 0EF G6IBU
PL4 0EF G8BOI
PL4 0EZ M3YGT
PL4 6AZ G6URM
PL4 6NJ M0GYC
PL4 6NW G3RRW
PL4 6PR G1ZSK
PL4 6QA G4EKV
PL4 6QA M3LUW
PL4 6RD M3BHK
PL4 7AG M0CCA
PL4 7HB M0UNJ
PL4 7LA M6NDE
PL4 7PY M3UHQ
PL4 8AA G0UOP
PL4 8AA G3YQF
PL4 8DS M6VEL
PL4 8NL G6DHD
PL4 8QE G1OGB
PL4 8TA M0VRT
PL4 9AQ G4RQN
PL4 9BR M7OCN
PL4 9DF 2E0EQB
PL4 9DF M6JIY
PL4 9EL G0BDB
PL4 9EN 2E0CSO
PL4 9EN M6BKE
PL4 9ET G0TQT
PL4 9EZ G7KII
PL4 9NB M1EUB
PL4 9NB M7UKW
PL4 9QD M6NRF
PL5 1AB G0JCA
PL5 1PD G0NCS
PL5 1SB G8PSC
PL5 1SR G3YJQ
PL5 1UH M3YYM
PL5 1UL G6JEM
PL5 1UP M7CYY
PL5 2AL M3ZZE
PL5 2DB G0VKY
PL5 2DX G7ESZ
PL5 2DX G7MMV
PL5 2LN M3ZZI
PL5 2NU G7MME
PL5 2NW G8IDE
PL5 2QU M7BXF
PL5 2SN 2E0GLS
PL5 2SN M6GPS
PL5 3FF M6FDF
PL5 3NU 2E0FHJ
PL5 3NU M6NQC
PL5 3RB 2E0DTC
PL5 3RB M6BNL
PL5 4AZ 2E0BIM
PL5 4HX 2E0OPS
PL5 4HX M3LCF
PL5 4JQ 2E0IOS
PL5 4LT G0KIK
PL5 4PG M7MML
PL5 4PU 2E0BNV
PL5 4PU 2E0NGG
PL5 4PU M3NGG
PL5 4PU M3PXE
PL5 4QE G3RYZ
PL51PH 2E0TBH
PL6 5HH M0PDB
PL6 5HQ M7RDH
PL6 5NR G0JNZ
PL6 5PU G3JFS
PL6 5PU G7UTI
PL6 5QE 2E0STI
PL6 5QE M3XXL
PL6 5QN M0CPW
PL6 5RJ G4BZF
PL6 5SD G7NIA
PL6 5SD M0BWN
PL6 5SD M7CWM
PL6 5SN G0KYE
PL6 5TR G7BXS
PL6 5TR M3BXS
PL6 5TX G4WLS
PL6 5TZ M3ZZJ
PL6 5XJ 2E0LMO
PL6 5XJ M6MLO
PL6 6AH G3JTJ
PL6 6AY 2E0SKE
PL6 6AY M6FGE
PL6 6BH G3SGV
PL6 6DW 2E0PLY
PL6 6JP G4CGM
PL6 6LG G0WOU
PL6 6LS M0BLO
PL6 6PL M7TIC
PL6 6PN M7BBM
PL6 6RE G6IPQ
PL6 6TN M6CDN
PL6 7BD M0IHB
PL6 7BY G1HHC
PL6 7BY G1HHD
PL6 7DT G7OOV
PL6 7EL M7FYU
PL6 7HS M6DGN
PL6 7HX G3ZZS
PL6 7JA G4YDR
PL6 7JA M3BAS
PL6 7LN M3YLQ
PL6 7SA G7DQA
PL6 7SP G7RTO
PL6 7UB G4GHR
PL6 8EQ M6ZEL
PL6 8QD M6VKN
PL6 8RQ G8WZJ
PL6 8TD G7ESO
PL6 8TD M0JIE
PL6 8TL M0DWT
PL6 8UU M3SQQ
PL6 8XF G0OBP
PL7 1JJ M1AHF
PL7 1JR G3ARE
PL7 1JY G3LCY
PL7 1PG M6FGI
PL7 1PU G6CYH
PL7 1PZ M0GBA
PL7 1PZ M1CXV
PL7 1QZ G3VCN
PL7 1SN M3YYS
PL7 2AJ G0AHU
PL7 2DU M3PYT
PL7 2DY G4LOI
PL7 2EJ G0FSD
PL7 2EY G4NLU
PL7 2FJ M6OEB
PL7 2GT G7HHW
PL7 2HP G0KML
PL7 2RU G7NHB
PL7 2SU M1SAM
PL7 2YF G0MMW
PL7 4BP M3YYJ
PL7 4DA M0AVS
PL7 4EG 2E0ZRM
PL7 4HH 2E0TZR
PL7 4HS G3TZQ
PL7 4HY G0UMI
PL7 4HZ M6EQZ
PL7 4JB 2E1HSL
PL7 4JB G1CKT
PL7 4JE G0NAP
PL7 4PF G1LTC
PL7 4PL M3YYR
PL7 4QY 2E0MGC
PL7 4QY G8CMG
PL7 4QY M0WMB
PL7 4RW G7ART
PL7 5AA 2E0GQM
PL7 5AA M0WIV
PL7 5AA M7WIV
PL7 5AY G3NBN
PL75HX M7MXM
PL8 1BP G3MMX
PL8 1BP G8ADX
PL8 1DB M6SUW
PL8 2ED G6FBJ
PL8 2JQ M0DZL
PL8 2JX G8AZZ
PL8 2NT G0ESL
PL82AX 2E0DIH
PL82AX M0IGP
PL9 0AW 2E0CYZ
PL9 0AW M6AYZ
PL9 0BB M0IAZ
PL9 0DS G4PRL
PL9 0EU G0MQK
PL9 0JZ 2E0OZI
PL9 0JZ M6OZI
PL9 7DQ G7BAV
PL9 7FB 2E0GNI
PL9 7LA G3VNG
PL9 7LD G4NDD
PL9 7LF G3KHU
PL9 7LU G7UWO
PL9 7NU G4KFZ
PL9 7PG M0BHV
PL9 7QP M3LTT
PL9 8DB G1BMP
PL9 8DB G3XZX
PL9 8DT M3KZT
PL9 8EB G3TSE
PL9 8HU G0VZX
PL9 8NW M3LVX
PL9 8PJ G3ULN
PL9 8QZ G0LRJ
PL9 8RB G3RMZ
PL9 8TW G0RMC
PL9 8UR G3SPI
PL9 9EJ M7EHD
PL9 9HJ G7BYW
PL9 9HW 2E0DQU
PL9 9HW M6IAX
PL9 9JT M0AKR
PL9 9LU G7FXW
PL9 9PT M0ASI
PL9 9RR G4SJD
PL9 9SH G3VZJ
PL9 9TS 2E0IXJ
PL9 9TS M7YME
PO1 1QN G6ASJ
PO1 3RD M0GYS
PO1 4ES M7GTM
PO1 5AR G0LFI
PO1 5AR M6UBI
PO1 5HU M7MPL
PO1 5LL M3XDH
PO1 5QT M0PPR
PO1 5RR M6PWB
PO10 7EW G4TLO
PO10 7JY G0BBJ
PO10 7LB G4WFF
PO10 7LB M7MKB
PO10 7NS G8RRC
PO10 7QP G6NUX
PO10 7RA G0USE
PO10 7RP 2E0DMB
PO10 7TR G4VTQ
PO10 7UL G7VIX
PO10 8BN G4XMJ
PO10 8GE G0DNF
PO10 8HS G1MDJ
PO10 8JG G3HCO
PO10 8LB 2E0HEX
PO10 8LB M0HEX
PO10 8LB M6HEX
PO10 8NQ G3PVH
PO10 8SZ M7JCK
PO10 8UX G3ZSS
PO10 8XD G0UFY
PO108HX M6BFQ
PO11 0AW G6RGI
PO11 0AZ G3VNU
PO11 0DT G3JVL
PO11 0EF G0BXV
PO11 0ER G3MCV
PO11 0EW 2E0PGJ
PO11 0FF 2E0DYX
PO11 0FP M1BIY
PO11 0HL G4KVX
PO11 0JW G0AYI
PO11 0JW G7KQT
PO11 0QE G0KCF
PO11 0QE G8DLL
PO11 0QE M3TEV
PO11 0QG M7MTG
PO11 0QR G8VMY
PO11 0RE G0PDS
PO11 0RL G6VXZ
PO11 9AA G1OXF
PO11 9HY G7ARK
PO11 9LT G2JL
PO11 9LT G3DIT
PO11 9NE G4ZMY
PO11 9PS G8ZJK
PO11 9RA G3YPW
PO11 9SJ 2E0OCS
PO11 9SN 2E0JIG
PO11 9SN M6IFO
PO11 9ST M6ZZY
PO119QL 2E0PRG
PO12 1HE G4JOA
PO12 1HH M0COA
PO12 1LR 2E0LNX
PO12 1LR M6LPI
PO12 1PW M3UWR
PO12 1QY G4XQX
PO12 1RB M6CPT
PO12 1SE 2E0UIT
PO12 1SE M6UIT
PO12 2DS M7HOF
PO12 2DT M7GRD
PO12 2HB G0MPJ
PO12 2HB G1BHG
PO12 2HX 2E1CNM
PO12 2HX G0AVP
PO12 2HX G6XHJ
PO12 2HX M0MIT
PO12 2HX M0RCV
PO12 2LG 2E0CWJ
PO12 2LG M0HTE
PO12 2LG M6CNR
PO12 2NE G8EIE
PO12 2NL G6JGT
PO12 2NL G7AUR
PO12 2NN M0BEC
PO12 2QU G8BSD
PO12 2RA G7SAX
PO12 2UP G6UGG
PO12 3BB G4MWW
PO12 3BP M7IRK
PO12 3DR M7NMC
PO12 3DS M6AIH
PO12 3EB G6ERI
PO12 3EU G6TQF
PO12 3HD 2E0IZH
PO12 3HD M7PME
PO12 3HF M0KPK
PO12 3HL M7CJD
PO12 3JJ G0GIA
PO12 3JN G7LOW
PO12 3JY 2E0AYK
PO12 3JY M0EHL
PO12 3LD G7RWN
PO12 3QY G0NCX
PO12 3SX G1ZTN
PO12 4BU G0HZY
PO12 4BX G4PQW
PO12 4DB G1WXU
PO12 4EP G1EDA
PO12 4ES 2E0CGH
PO12 4ES M6WFA
PO12 4EU G1OAW
PO12 4EW G8GBE
PO12 4GG G0JFD
PO12 4HN M6LFM
PO12 4JF G0LPP
PO12 4LW M0VXC
PO12 4LW M7BMX
PO12 4NP M7BBI
PO12 4NS G1UID
PO12 4NU 2E0IHH
PO12 4NU M6IHH
PO12 4NX M6WBX
PO12 4RT 2E0EFH
PO12 4RT M0TFH
PO12 4RT M3UYQ
PO12 4UE M5AGB
PO123QN G6NHA
PO13 0DP G4UHS
PO13 0DX 2E0HJI
PO13 0JS 2E0EFA
PO13 0NF G0DGW
PO13 0PT G3MZZ
PO13 0QS G8MRN
PO13 0RB G1YTL
PO13 0SE 2E0BXX
PO13 0SE M6HAS
PO13 0SJ M0EEL
PO13 0SJ M7FIA
PO13 0SP G0GLW
PO13 0TG 2E0JIS
PO13 0TG M7SIJ
PO13 0TN 2E1ESN
PO13 0TN M1ANC
PO13 0TT G7TXW
PO13 0TU M6NAQ
PO13 0TY 2E0DOF
PO13 0TY M3SVB
PO13 0UT 2E0FEZ
PO13 0UT M6KMV
PO13 0UZ M7CIX
PO13 0WP M0RQX
PO13 0XR G1OCY
PO13 0YN G0DDW
PO13 0YX G0JYV
PO13 0ZD 2E0DBA
PO13 0ZD M0LOW
PO13 0ZD M3ERR
PO13 0ZU M6BLX
PO13 8HD 2E0YZM
PO13 8HD M6YZM
PO13 8LS 2E0GPI
PO13 8LS M0KPI
PO13 8LS M7TKD
PO13 8LZ M0IBN
PO13 9AF G0VLI
PO13 9AF G5LK
PO13 9AF G7RAT
PO13 9AG G8DBD
PO13 9AU G4MHQ
PO13 9AU G7USV
PO13 9BB M6WYY
PO13 9BP 2E0FHH
PO13 9BP M6FUW
PO13 9DH G3XEF
PO13 9DH G7MFR
PO13 9EN G4TKW
PO13 9EY G8GUS
PO13 9HU G3NPZ
PO13 9NA G6APD
PO13 9NJ G8OUH
PO13 9NU 2E0GRQ
PO13 9NU M7MDY
PO13 9RD G4BMQ
PO13 9TL 2E0HHP
PO13 9TL M7JDC
PO13 9UJ G1CLJ
PO130HX 2E0IXA
PO130HX M7XNR
PO139DU 2E0EMG
PO139DU M0MAQ
PO139DU M3VTQ
PO14 1AQ M7AUD
PO14 1EF G6AIQ
PO14 1EG G0VIX
PO14 1HH 2E0TGJ
PO14 1HH M7TGJ
PO14 1PH M3YNO
PO14 1QF 2E0BQK
PO14 1QF M3VOU
PO14 1SL G7ILI
PO14 2BQ G6AVT
PO14 2BZ M0BQZ
PO14 2HA G4XVV
PO14 2JF G3GVM
PO14 2NY G6LHQ
PO14 2PX G4PZV
PO14 2QS G0RHV
PO14 2QS G0PPH
PO14 2SQ G4SPS
PO14 3AA G6BAT
PO14 3AD G1KNI
PO14 3AF G7NYF
PO14 3BS G8APL
PO14 3DR G8PGF
PO14 3HW G3IBI
PO14 3LB G6HGX
PO14 3LB M0DDA
PO14 3LB M3ADJ
PO14 3LF 2E0ZFV
PO14 3LF M3ZFV
PO14 3RX M0GLL
PO14 3SY M1FCE
PO14 4BH G3HKT
PO14 4BX M1FIR
PO14 4EN 2E0LLD
PO14 4EN M6LDD
PO14 4JP M1EMP
PO14 4NB M0NRG
PO14 4NP G0GFD
PO14 4QX G8RNV
PO14 4SF G8YBC
PO14 4SL G0MAT
PO14 4SL G0TPB
PO14 4SY G8RLF
PO15 5AJ G1NCK
PO15 5BL G6HHE
PO15 5DU M0RIK
PO15 5EQ G3YTQ
PO15 5HP G3RDA
PO15 5JJ G0AMS
PO15 5LG G0JYQ
PO15 5LZ M7MUJ
PO15 5NE G4VNM
PO15 5NF 2E0JTN
PO15 5NF M0JTN
PO15 5NF M6JTN
PO15 5PF G1MPP
PO15 5QB G3UWE
PO15 5QH G8UXW
PO15 5RS G0XGL
PO15 6HF 2E0JRD
PO15 6HF M0JDL
PO15 6HF M3YJD
PO15 6JU G3KLF
PO15 6PJ M0YBD
PO15 6PJ M7TQZ
PO15 6PJ M7XLD
PO15 6QP 2E0SHZ
PO15 6QP M1AIB
PO15 6QP M6MOI
PO15 7BD G3UPZ
PO15 7EA G0SHT
PO15 7EA G4CAA
PO16 0DX G0USF
PO16 0DX G7SFL
PO16 0FA M6WRJ
PO16 0QB G0SFG
PO16 0RX 2E0ABL
PO16 0SQ G0DDA
PO16 0TR G4ZMP
PO16 7DP 2E0MOY
PO16 7DP M0YSR
PO16 7DP M6FRE
PO16 7HB M0HBE
PO16 7HN M7SLG
PO16 7JF G4FCL
PO16 7LU G0HZC
PO16 7NL G4UOR
PO16 7QL G4ITG
PO16 7QW G3YIW
PO16 7RR G6KTX
PO16 7SJ G6GPV
PO16 7XA G0JYZ
PO16 7XR G1WKZ
PO16 7XR G1WLD
PO16 8DS G8XEZ
PO16 8DU M6DRG
PO16 8JW G4PWG
PO16 8JY G7MDV
PO16 8LB G4CYC
PO16 8LE G1FIG
PO16 8LF G0LYI
PO16 8LF G3XUF
PO16 8PB G8IOJ
PO16 8QA M7CXQ
PO16 8QG M0VMX
PO16 8TF M6WVV
PO16 8UE M0DQO
PO16 9AA M6ZAU
PO16 9AP G7DWV
PO16 9BQ G6JSI
PO16 9DN G4YPA
PO16 9EG G7UUQ
PO16 9EG M7GRR
PO16 9HP G4IOJ
PO16 9LA M6EKX
PO16 9NJ G4MRW
PO16 9NR G3UFF
PO16 9PA G0KCG
PO168TY G1OJS
PO168UF G6ORL
PO17 5BU 2E0IHM
PO17 5BU M7DQD
PO17 5EX 2E0SMO
PO17 5GS G6LYA
PO17 5LD G8POQ
PO17 6DG G4GWJ
PO17 6DG G4HKZ
PO17 6DG M0XML
PO17 6EY 2E0CFE
PO17 6HS G3XUX
PO17 6JB G1WID
PO17 6JJ G8JXV
PO18 0AH M6OVY
PO18 0AY M1BTO
PO18 0AY M7FYD
PO18 0BF 2E0TEE
PO18 0BF M0YEE
PO18 0BF M6AXI
PO18 0EX G3NXO
PO18 8BE G0CHK
PO18 8EU M0SYM
PO18 8LF G6XZA
PO18 8QN G4BVS
PO18 8QQ M0GNL
PO18 8QW M3XYC
PO18 8RQ 2E0HYM
PO18 8RQ M6HYM
PO18 8RQ M7HYM
PO18 8SR G7MIN
PO18 8SR M3MIN
PO18 8SU G3LSQ
PO18 8SU G4JDG
PO18 8TZ G0SDC
PO18 9JJ G0IOP
PO18 9LN G1YEW
PO18 9LS G7ISD
PO18 9NP G0LNX
PO19 1QS G3SJW
PO19 3AE G1UFS
PO19 3AN 2E0CGK
PO19 3FE G4WVQ
PO19 3JD M0PWV
PO19 3LD G4SQJ
PO19 3LY G7BVL
PO19 3NL M6VOL
PO19 3QF G8KJT
PO19 3QU G0WSD
PO19 5QU 2E0IWS
PO19 5RL G4EHG
PO19 5RX M0NNL
PO19 5RX M3YQQ
PO19 5UA G3ZEN
PO19 5UA G4VQZ
PO19 6BY G3MVZ
PO19 6GG M0CXY
PO19 6GL G4ZTQ
PO19 7NE 2E1GCF
PO19 7NW 2E0CSH
PO19 7NW M0OSB
PO19 7NW M6BYU
PO19 7TN M6GRT
PO19 7UY M6KTK
PO19 7XE G0ISL
PO19 8AU G4WIR
PO19 8BP M3SYW
PO19 8QR G1BZU
PO19 8QR G3BZU
PO19 8QR G3CRS
PO19 8QR G3ZDF
PO19 8QR G7DOL
PO19 8RG 2E0HVT
PO19 8RG M7GWF
PO19 8TP G8NYK
PO193PQ M7DBD
PO2 0EX G7SCE
PO2 0QS M0DND
PO2 7DD 2E0BUN
PO2 7DF 2E0BUP
PO2 7HB G7OKI
PO2 7HB M3WCM
PO2 7JW G0DAE
PO2 7LB G8UYY
PO2 7LX M3ZNG
PO2 7PG G4EFB
PO2 8EX M3AMA
PO2 8LX 2E1HSB
PO2 8LX M1DOX
PO2 8NA M0GIF
PO2 8NF G6ATK
PO2 8NT M6KNP
PO2 8RH M1GRP
PO2 9EG M6JZM
PO2 9HE M1DMR
PO20 0DW M7EFA
PO20 0EB M6TPO
PO20 0EE G4KEC
PO20 0FQ 2E0XPJ
PO20 0FQ M0XPJ
PO20 0FQ M3XPJ
PO20 0HY G4MVS
PO20 0JB M1BCY
PO20 0JG G7SZW
PO20 0JR M0RMW
PO20 0JS G4ZPP
PO20 0NA M6SXM
PO20 0NX G8OLA
PO20 0RG M5DNK
PO20 0SF M3WLG
PO20 0TE M7ELG
PO20 0UA G8PQH
PO20 1JZ G0POU
PO20 1JZ G8CBO
PO20 1NY G0WBR
PO20 1NY M6LUK
PO20 1PA G3NFW
PO20 1PE G4BZB
PO20 2GQ 2E0FSI
PO20 2GQ M6FSI
PO20 2HX 2E0FVL
PO20 2WB M3DSW
PO20 3SJ G4ZJP
PO20 3TH M6VTH
PO20 7JJ G0LMJ
PO20 7NL M7DWE
PO20 7RR G0WVM
PO20 8AH G7BKJ
PO20 8DP M3NYK
PO20 8DW G4VRG
PO20 8EX M6GYC
PO20 8PB G6VLV
PO20 8PE 2E0YAS
PO20 8PE M6YTS
PO20 8RG G0BVV
PO20 8RJ G3SFB
PO20 9AX G4YJD
PO20 9AY G6XTK
PO20 9DT M0GOO
PO20 9HL G7HJT
PO21 1AB M1FCC
PO21 1HQ G1GKW
PO21 1XJ G6ZSQ
PO21 2BA M7UNC
PO21 2DG 2E1FIE
PO21 2EJ G0SZK
PO21 2JU 2E0XCZ
PO21 2JU M0XCZ
PO21 2JU M6XCZ
PO21 2PY M0MLM
PO21 2QA G0CHE
PO21 2RB G4XPT
PO21 2SF G6FAX
PO21 2TJ M7CUS
PO21 2YQ G7EPR
PO21 3DH G3GUR
PO21 3EL G8OCM
PO21 3EQ G0CIR
PO21 3EZ G0TXU
PO21 3FQ G7IMT
PO21 3ND G0AKK
PO21 3NN 2E0XAO
PO21 3NN M3UJZ
PO21 3PS M7AFE
PO21 3QH M7SID
PO21 3RB M7GRP
PO21 3SL G4RPA
PO21 4AW M0CME
PO21 4DY G2ARU
PO21 4ET G3LTM
PO21 4HB M0GSC
PO21 4HT G3RJS
PO21 4LH M6GMV
PO21 4NB G8TGH
PO21 4NN G6HTB
PO21 4PS G8REF
PO21 4QT 2E0NUG
PO21 4QT M0VPL
PO21 4RN G0RQH
PO21 4TB G8YAS
PO21 4TJ G4ITY
PO21 4TN G1XIV
PO21 4UR G6FDU
PO21 4XN G3IJS
PO21 5AD M6DHB
PO21 5FA G1ZDY
PO21 5LL G0KJU
PO21 5TD G0OSU
PO21 5TW G6ZTD
PO213NN 2E0CKE
PO213PL M7DLB
PO22 0AR M3DOA
PO22 0EH G3MZO
PO22 0EJ 2E0WSX
PO22 0EJ M0HIX
PO22 0EJ M6ARP
PO22 0HF M0KPT
PO22 0JN G1ITL
PO22 0LH G7BWW
PO22 6AH G7TPH
PO22 6BU G0MTK
PO22 6BX G0AFN
PO22 6ED G6UTT
PO22 6HG G1ORB
PO22 6JU M6EIC
PO22 6QH M7KLA
PO22 7LB 2E0FLK
PO22 7LB M0OFE
PO22 7LB M6OFE
PO22 7NJ 2E0YPW
PO22 7NJ M0YPW
PO22 7NJ M6YPW
PO22 7NW G4PDY
PO22 7QG G4ECF
PO22 7SE 2E0FAH
PO22 7SE M3FPH
PO22 7SL G0DFA
PO22 8DP G6XJN
PO22 8PH 2E0SCK
PO22 8PH M0HJB
PO22 8PH M6BXJ
PO22 9BE 2E0WSP
PO22 9BE M0WSP
PO22 9BE M6UGF
PO22 9DG G1LHE
PO22 9DR M3NIZ
PO22 9EX G3SFE
PO22 9HJ M6KWH
PO22 9LA G0VCJ
PO22 9LE G8UNP

PO22 9LE M7PTY
PO22 9NU M7CQX
PO22 9RR M0MMT
PO22 9RR M6MIE
PO228NS G0XAI
PO229EQ M7XAP
PO3 5DE M3YZE
PO3 5EL M0DDU
PO3 5LN M6DNJ
PO3 5PU M6JSS
PO3 5TN 2E0JCE
PO3 5TN M0XLA
PO3 5TR G8KQV
PO3 5TR G8NVZ
PO3 5UG M6OFL
PO3 6AU. M1SKA
PO3 6BE M0GMI
PO3 6LR G3ZBP
PO30 1AF G1DEO
PO30 1DG G4MBD
PO30 1DG G7IRH
PO30 1DR G6OOH
PO30 1DT G3XOC
PO30 1HA G4FYI
PO30 1HG G6ORJ
PO30 1HN 2E0BUK
PO30 1LN G0MHN
PO30 1NR 2E1IWD
PO30 1NR M0IWA
PO30 1NR M3DIW
PO30 1NR M6SPU
PO30 1PA G0EHR
PO30 1PZ G0KQO
PO30 1QD 2E0UYB
PO30 1QD M3CIG
PO30 1QD M7CZW
PO30 1QT G7NAO
PO30 1TD G7UHT
PO30 1XN M0GQJ
PO30 2BH M3AYC
PO30 2BH M3KLS
PO30 2BH M3SYZ
PO30 2DB 2E0AAJ
PO30 2DB M6MPD
PO30 2DP G8AZM
PO30 2EJ M6WZV
PO30 2GA M3BRL
PO30 2GS 2E1CAU
PO30 2JU M0PDL
PO30 2LL G8RWM
PO30 2LL M6KFW
PO30 3DD G1KAK
PO30 3JP 2E0JIW
PO30 3JP M0PBN
PO30 3JP M3JIW
PO30 3JY M1FBL
PO30 3JZ 2E0FSE
PO30 3JZ M6FSE
PO30 4AE G4BIM
PO30 4AN M0IKA
PO30 4BG G0GMY
PO30 4BG M6NBV
PO30 4BX G6BHB
PO30 4DJ G0PXM
PO30 4HH M0CZP
PO30 4LP M0NVK
PO30 4LZ M6FQE
PO30 4NQ M6NGK
PO30 5AR M7DAR
PO30 5BS M0IES
PO30 5BS M7TMR
PO30 5GU 2E0EJK
PO30 5GU M6CUC
PO30 5JJ M3FBR
PO30 5JR M1IOW
PO30 5JR M3IWG
PO30 5NG 2E0DUS
PO30 5NG M6YEG
PO30 5QP M7EAP
PO30 5QZ G0RUT
PO30 5QZ G7GGH
PO30 5SF G0DWE
PO30 5SF G1HHB
PO30 5SF M6NKY
PO30 5SG M0IOW
PO30 5SJ M3LJZ
PO30 5SJ M3ZBJ
PO30 5TL M0TVR
PO30 5TL M3MEE
PO30 5TL M6BRT
PO30 5TP G3IMX
PO30 5TY G6CUT
PO30 5UF M6DAZ
PO30 5UH G0LFV
PO30 5UR G7MID
PO30 5XS 2E0EDC
PO30 5XS M6FPJ
PO30 5ZF 2E0BSK
PO30 5ZF M3WHV
PO31 7GA M0GYL
PO31 7HF 2E0AGQ
PO31 7HW M6PBM
PO31 7JN G6ZUZ
PO31 7NF G4GRK
PO31 7NX M6LTS
PO31 7PE M3AYJ
PO31 7PP M6FQH
PO31 7PS 2E0WKG
PO31 7PS M6WKG
PO31 7PY M3BBF
PO31 7SG M0DKJ
PO31 7SR M3RGP
PO31 7TE M1CTY
PO31 7XS M7AYF
PO31 8AD G4ZFQ
PO31 8AJ M0JZF
PO31 8AL G3WXC
PO31 8AS 2E0ENG
PO31 8AS M3ORQ
PO31 8AS M3WKL
PO31 8DP G3XYB
PO31 8DT G3YZK
PO31 8DX G4SCB
PO31 8DY G1HKU
PO31 8EF G8MBU
PO31 8HA 2E0AKQ
PO31 8HA M0JCH
PO31 8HA M3NNV
PO31 8HQ G3XZB
PO31 8JP G4ZBH
PO31 8JQ M6DYN
PO31 8LB M0TFN
PO31 8NE 2E1KJB
PO31 8NR G0HDH
PO31 8PE G0CWX
PO31 8PN G3PZB
PO31 8PT G8RKH
PO31 8PZ G3IW
PO31 8PZ M6OFH
PO31 8QP G4ZOH
PO31 8RF G6VFF
PO31 8RL G4VYC
PO317ER M6FDU
PO32 6AX G7GZJ
PO32 6BH G0WXJ
PO32 6BH M0RAD
PO32 6BH M0ZOO
PO32 6DX M6CUW
PO32 6EA G7VGY
PO32 6EJ G4UCZ
PO32 6FJ 2E0RQD
PO32 6FJ M0RQD
PO32 6GA G3VDZ
PO32 6GT 2E0WJH
PO32 6HR G0PQW
PO32 6HT M3RKR
PO32 6JD M6SSM
PO32 6JE M6FWE
PO32 6JE M6LAS
PO32 6JT G7TQC
PO32 6LS G0PEB
PO32 6LS G7ATJ
PO32 6NT G0JHQ
PO32 6NT G6RTE
PO32 6NT G6TVX
PO32 6NX M7DDY
PO32 6PS G0KQR
PO32 6QW M6SGT
PO32 6RU G1TUZ
PO32 6SS G3PQJ
PO32 6TD 2E0ZMB
PO32 6TD M3SQG
PO326QN G7SVF
PO326QN M0GAQ
PO33 1AB G8MYF
PO33 1BX M0DSF
PO33 1ED G6LXP
PO33 1EL G0MWU
PO33 1FY G1VGM
PO33 1FY G2CNN
PO33 1JD G0MIZ
PO33 1NT G4TAT
PO33 1PR G4RSN
PO33 1QY M6DQW
PO33 1TA M0BTP
PO33 1UN M7FWR
PO33 1XE G4RGE
PO33 1YF G0SEB
PO33 2BG G1WQC
PO33 2BG M0CGO
PO33 2BH 2E1CDK
PO33 2BH M3AIL
PO33 2HD M3TCU
PO33 2JP G1FPP
PO33 2LG 2E0EYB
PO33 2LG M6WOI
PO33 2QF G1HRQ
PO33 2QQ G0DRW
PO33 2SS G6LVS
PO33 2TQ 2E0WWJ
PO33 2TQ M7XLF
PO33 2UH G3XHM
PO33 2UP G7MAR
PO33 3BU G6SQL
PO33 3DL G4SBN
PO33 3EL M1OBR
PO33 3HU M3YPP
PO33 3JU G1HCM
PO33 3LH G3XLP
PO33 3QJ G3KPO
PO33 3QJ M0TIW
PO33 3QS M3JIU
PO33 3QZ M6ODR
PO33 3SD M7SYZ
PO33 3SF G4TFW
PO33 3TA G4KZI
PO33 3TE G6MQY
PO33 3TL G0UEK
PO33 3TL G4SHH
PO33 3UB 2E1EBN
PO33 3UX G6HZG
PO33 4BB G4OAG
PO33 4DR M0TAM
PO33 4ED G3UHX
PO33 4EU 2E0DUI
PO33 4EU M6FLZ
PO33 4EZ G8NHG
PO33 4HE M7WMK
PO33 4JF G7RCC
PO33 4JJ G4KKB
PO33 4JR G0RSY
PO33 4LE M6CYN
PO33 4LG G4BCH
PO33 4LG G4CQO
PO33 4LQ M1DBM
PO33 4LU G6DOD
PO33 4PY G4YYF
PO334LB G4IKI
PO34 5DY G0ZGN
PO34 5JE G0GNI
PO34 5JS G8PXM
PO35 5QS 2E0XXB
PO35 5QS M3XXB
PO35 5QW G3JLN
PO35 5RA 2E0MFA
PO35 5RA G3YEG
PO35 5RA M0MFA
PO35 5RA M6IOW
PO35 5SL 2E1PHW
PO35 5TN M0HKK
PO35 5TS G4LUY
PO35 5UQ M6DXW
PO35 5UW 2E0ZML
PO35 5XU G4MZC
PO35 5YJ G0VZV
PO35AS M6GSZ
PO36 0BA M6JBA
PO36 0DX G1RIR
PO36 0JD G4ANW
PO36 0JT G3GEG
PO36 0JY G8TAQ
PO36 0LH G4OYR
PO36 0NE M7EJI
PO36 8BA G0VPO
PO36 8BE 2E0DEI
PO36 8BG M3BKV
PO36 8DZ 2E0MRD
PO36 8DZ 2E0SUS
PO36 8DZ M6DEL
PO36 8HE G4RTW
PO36 8QE G0WVD
PO36 8QL G4ADK
PO36 9BX G6ETC
PO36 9DS G4UNM
PO36 9HF G0WLX
PO36 9HQ G6WUR
PO36 9HW G0MPA
PO36 9JA G4JLX
PO36 9JA G4ULT
PO36 9JL G1SMY
PO36 9NS 2E0CZW
PO36 9NS M6SGS
PO37 6AE G6DIQ
PO37 6DB G8PPN
PO37 6EJ G0NTH
PO37 6NN G8JBM
PO37 6NX G4SVY
PO37 7AG G4WNZ
PO37 7BU G6EVX
PO37 7DF G7IVF
PO37 7EJ G4NOU
PO37 7HH G0HFO
PO37 7NJ M0JAG
PO37 7NZ G4RTY
PO37 7PA G4UUJ
PO376AU M7AVF
PO38 1AA G1JYZ
PO38 1AL G4CZP
PO38 1BD 2E0GZT
PO38 1BD M6GZT
PO38 1BH G1CPO
PO38 1DQ G4ZEN
PO38 1HG 2E0OBP
PO38 1HG M6OBP
PO38 1NX G4MPI
PO38 1QL G0LGZ
PO38 1RZ G0RMJ
PO38 1TH G7ONE
PO38 2BG G4IGK
PO38 2BG G8DGW
PO38 2DE G8DDY
PO38 2DZ M1BOD
PO38 2NE 2E0EXU
PO38 2NE M6SKY
PO38 2QW G4XIU
PO38 2QZ M7FKO
PO38 3AF M0LZS
PO38 3BU M6YEB
PO38 3DB 2E0RSD
PO38 3DB M6BRR
PO38 3EH G8XGV
PO38 3EL M3FBG
PO38 3EL M3FPG
PO38 3EQ M1BWR
PO38 3HB M6DRI
PO38 3HR G0PEF
PO38 3HW M6MYD
PO38 3HY G8XHN
PO38 3HZ G3LYD
PO38 3NH G4RYD
PO38 3NP M0WAZ
PO38 3NS 2E0JVS
PO38 3NS M6LYU
PO38 3NT G4RQP
PO38 3PH 2E0FRB
PO38 3PH M3FRB
PO39 0AH G4EWE
PO39 0AH G4NCD
PO39 0AL G4CSM
PO39 0BL G7BZD
PO39 0BL M3BDA
PO39 0DN M3GRI
PO39 0DX M6DRH
PO39 0EF M1CYX
PO39 0JL G0PTT
PO390EZ M7VVZ
PO4 0BA G0AXS
PO4 0DE G6DHT
PO4 0JE G1SEH
PO4 0NT G6ZKM
PO4 0RL M6ROU
PO4 8AU G4VFX
PO4 8AU G8OEJ
PO4 8GS M0CYX
PO4 8HH G6CUA
PO4 8JG G4LRH
PO4 8JS G1ERF
PO4 8JU G0IWN
PO4 8NF M3JRN
PO4 8NX 2E0CSY
PO4 8NX M6BYY
PO4 8PB M1DXO
PO4 8RH 2E0CRU
PO4 8RH M6ASV
PO4 8UF M3HAD
PO4 9AB M0PVO
PO4 9BG G3CAJ
PO4 9EF G6DWO
PO4 9HP M6BRM
PO4 9JE G7UPL
PO4 9JU M3ETI
PO4 9LR G7ARJ
PO4 9ND 2E1ILH
PO4 9PZ G6FDP
PO4 9QU G1PGQ
PO4 9XZ G6SGW
PO40 9DL G1RFB
PO40 9DX M7EOK
PO40 9EF G4HPN
PO40 9ES G6ENY
PO40 9FY G3VPK
PO40 9GB M0XRT
PO40 9HB G4NAK
PO40 9JH G7AXM
PO40 9LF G4EDN
PO40 9LQ M3IUC
PO40 9NH G1JGS
PO40 9QS G7DZY
PO40 9QS G7WKW
PO40 9TG G7OPY
PO40 9UA G3IIN
PO41 0PS M3LMB
PO41 0PX M6HQW
PO41 0PY G0GEZ
PO41 0RX G0DKS
PO41 0SL G0GPK
PO41 0TL G4AZC
PO41 0XS 2E0BHB
PO5 1PL M7EVQ
PO5 1RU M6MTN
PO5 2HA G6IAO
PO5 2JJ M6IRC
PO5 2NL G0LFN
PO5 3AU M3JVR
PO5 3DT M7AXA
PO5 3HP 2E0IAN
PO5 3JA G0DRL
PO5 4AL M6CXC
PO5 4JQ M7FTN
PO6 1DB G4IQO
PO6 1DU G3CNO
PO6 1DU M0GWD
PO6 1EW 2E0PWK
PO6 1EW M3PWK
PO6 1LW G0VEP
PO6 1LZ G4GZO
PO6 1NB G6HJV
PO6 1NG G4OVM
PO6 1PL M0DNR
PO6 1PY 2E0AWE
PO6 2ES G7OWQ
PO6 2JE G8SBQ
PO6 2JU G7DRT
PO6 2NL G4FOW
PO6 2PU G6UXW
PO6 2RL M0TYW
PO6 2TJ G6XDY
PO6 3DG G3OQC
PO6 3DG G7HQP
PO6 3HD G1JAB
PO6 3JP M3MQS
PO6 3JP M3SSU
PO6 3PE M0JUJ
PO6 3QT G1OVG
PO6 3QY M6TON
PO6 3RD M0ALF
PO6 3RH 2E0KIT
PO6 3RH M6ADG
PO6 3SB G4AVX
PO6 4AE G1XZQ
PO6 4AF M3AUK
PO6 4BB G4PYS
PO6 4EZ M6LBU
PO6 4LS 2E0BUO
PO6 4LS M0LBJ
PO6 4LS M3EOL
PO6 4QH 2E0KIS
PO6 4QH G3RCE
PO6 4QH M6URK
PO6 4QL G4LIO
PO6 4QL G6GYF
PO6 4TA 2E0CEU
PO6 4TA M0NEH
PO6 4TA M6NEH
PO7 3DB M7GEP
PO7 4QU 2E0RFM
PO7 4QU M6RFM
PO7 4RU G8CKS
PO7 4SP G0SSY
PO7 4SW M1AHA
PO7 5AZ M0RZF
PO7 5BL 2E0NFB
PO7 5BL M0NFB
PO7 5BL M6ATF
PO7 5BT G1TDP
PO7 5DE 2E0RXC
PO7 5DE M6RCS
PO7 5DX G4YCG
PO7 5DY G7DUE
PO7 5ED G0ABB
PO7 5ED G4ZPA
PO7 5EF 2E0WHP
PO7 5EF M7VCT
PO7 5EF M7WSJ
PO7 5HH M5KZI
PO7 5HJ G0IEY
PO7 5HJ G0IUY
PO7 5HJ G4FBS
PO7 5LL M7ATH
PO7 5NN M3ZJS
PO7 5QF M0CYM
PO7 5QR G1WXW
PO7 5QW G6ISY
PO7 5SF G4GUA
PO7 5TB G7IUE
PO7 5TW M3GTK
PO7 5XA M3HJW
PO7 6AA G1UFA
PO7 6AH M0TAP
PO7 6AH M3XYI
PO7 6AQ M3TQY
PO7 6BG G4CRM
PO7 6BT G0JWL
PO7 6BX M0JAK
PO7 6DD G0RPV
PO7 6DP G0ASZ
PO7 6EB G4JIH
PO7 6ED G3KOJ
PO7 6EG M3AEE
PO7 6HE G4TST
PO7 6HH M1EXJ
PO7 6LJ G6XBG
PO7 6PE G8FXV
PO7 6PR G4WQZ
PO7 6PR G8IRS
PO7 6SH G6CZZ
PO7 6TF M6WEJ
PO7 6UA G4DCP
PO7 6UE G3RTP
PO7 6UX G4KLX
PO7 6YJ G8LNU
PO7 7BA 2E0SJN
PO7 7BX M0SSP
PO7 7BX M3SHI
PO7 7EW 2E0CKI
PO7 7EW M6BCG
PO7 7EW M7DGH
PO7 7HX G0ATG
PO7 7JE G1OGR
PO7 7JY M6POY
PO7 7LG G8VOI
PO7 7LG M3VOI
PO7 7LN M5ACT
PO7 7LY M0RWS
PO7 7NA M3KYZ
PO7 7NH 2E0ISE
PO7 7NH M0KQV
PO7 7NH M7IDT
PO7 7NN M3CZB
PO7 7PB G3TVI
PO7 7PE M6DTM
PO7 7PG G3LKW
PO7 7QB G0PSF
PO7 7QQ G3WYT
PO7 7RG G1FMT
PO7 7RP M3ZQJ
PO7 7XP G4VIQ
PO7 8AE M7MLP
PO7 8AG G4MKQ
PO7 8BP M0JDS
PO7 8BX 2E0TRW
PO7 8BX M0KTT
PO7 8BX M6BSA
PO7 8DP M6HCR
PO7 8JN G4SAC
PO7 8ND G8KOS
PO7 8QD G8OKE
PO7 8RS G6BQE
PO7 8SH G8PAE
PO8 0AA 2E0DHG
PO8 0HF 2E0FMA
PO8 0HF M6TAU
PO8 0JE 2E1DQG
PO8 0JF M6BNU
PO8 0JX M6EYZ
PO8 0LD G0GDV
PO8 0LJ G0VOJ
PO8 0LQ G7EYS
PO8 0NF G1PPD
PO8 0NQ 2E0HES
PO8 0NQ M6GEV
PO8 0NU G4VNR
PO8 0NU M0KUV
PO8 0PD G3VCR
PO8 0PJ G8TZE
PO8 0RH 2E0REC
PO8 0RH M6MCU
PO8 0RH M6TUD
PO8 0TX G7IWA
PO8 8BG G3RBY
PO8 8BQ M7EFV
PO8 8BW G4NKX
PO8 8EW G0NPE
PO8 8HS G3ZFF
PO8 8HX G3WLY
PO8 8JE G0NHZ
PO8 8JJ G8LNC
PO8 8QD G6CTX
PO8 8QH G3TKN
PO8 8RS 2E0NFZ
PO8 8RS M0NFZ
PO8 8RS M7EPN
PO8 8RU G8YTR
PO8 8RX 2E1BOM
PO8 8SE G0JRN
PO8 8SQ G3LIK
PO8 8TS 2E0BSX
PO8 8UB G6YSO
PO8 8UB M3GVC
PO8 9BE G6EHL
PO8 9DA M3KUQ
PO8 9DA M3YDB
PO8 9EW G6WXK
PO8 9HE G0MGH
PO8 9NH M7GEB
PO8 9NX M0WAH
PO8 9QE G0MUK
PO8 9QE M7ELS
PO8 9QH G0KUA
PO8 9QU G8UVF
PO8 9RD G6JDH
PO8 9RG G7PEE
PO8 9SG G1WJG
PO8 9TJ G0UHM
PO8 9UB 2E0BRC
PO8 9UB G7GNA
PO8 9UB M3TXR
PO8 9UY M3CWA
PO8 9XF G3MYI
PO88AX M7JRM
PO9 1AG M3YIE
PO9 1DG M6WAJ
PO9 1HZ G1JXL
PO9 1LQ M1DPJ
PO9 1RL G6IOV
PO9 1RN G0VKX
PO9 1RR G0RIB
PO9 2DX M7FWE
PO9 2HR G1FFU
PO9 2NQ M0HNO
PO9 2PU 2E0JAZ
PO9 2PU M0JAZ
PO9 2PU M6JRS
PO9 2QX G6TQZ
PO9 2RP G8RUX
PO9 2RZ G4XQZ
PO9 2SY 2E0BVO
PO9 2SY M3YIV
PO9 2TN G6WBX
PO9 2UW G0JEZ
PO9 2XE 2E1FPP
PO9 2XE M3LDQ
PO9 2XE M3OQQ
PO9 3AA 2E0INT
PO9 3AA G8VEZ
PO9 3AA M6CTA
PO9 3AZ G3YYK
PO9 3BW G6XRH
PO9 3DL M3EYS
PO9 3JZ G7VDN
PO9 3NJ 2E1GZV
PO9 3NJ G4UXJ
PO9 3NS M6FEL
PO9 3PL G0FYX
PO9 3RA G4RGO
PO9 3RS M0HRH
PO9 3RS M3FEA
PO9 3TG G0RGT
PO9 4HR G7WGI
PO9 4JA M6KAQ
PO9 4JF M6RZE
PO9 4LG G7UCL
PO9 4LG M3IVY
PO9 4NS M6PJD
PO9 4PT 2E0GMS
PO9 4PT M6GHB
PO9 4QY M0CBG
PO9 5AR M3EDS
PO9 5BJ G7EYV
PO9 5DS M7SKT
PO9 5DZ G8XNO
PO9 5ED G8WCD
PO9 5ED M6SUI
PO9 5ER G0DHZ
PO9 5HU G0DOK
PO9 5LA M3ZRS
PO9 5LS 2E0LGT
PO9 5PW G0ERS
PO9 5RY G0BSJ
PO9 6AG G3TSM
PO9 6DQ G0BAG
PO9 6HN M3YNB
PO93AQ M0GMU
PO93TH M7TCT
PR1 0AX M7BOQ
PR1 0BH M0DDI
PR1 0BN G6LOC
PR1 0DT 2E0BYW
PR1 0EE M0DOM
PR1 0JL M3OHO
PR1 0LL G4LKM
PR1 0TD M3DXN
PR1 0UR G1PUK
PR1 0UX G4RFA
PR1 0XN M0BKS
PR1 0XN M3AUL
PR1 0XQ G6EPX
PR1 0XR M7HMO
PR1 0YE G0EIF
PR1 1QW M5KEN
PR1 1RY M3OEE
PR1 2EF M6VZA
PR1 2YJ M6GBW
PR1 2YL G1VTQ
PR1 3NA 2E0NAP
PR1 3NA M6DOJ
PR1 4NJ G6PFZ
PR1 4TA 2E0RAF
PR1 4UD G3NQX
PR1 4UD G6NKI
PR1 4YB G7RCK
PR1 5HJ G3UKE
PR1 5HJ G4WYH
PR1 5TA G7ING
PR1 5TB M6SJW
PR1 5TP G8RIP
PR1 5TR G4WXI
PR1 5UY G6WXS
PR1 5UY M7KWI
PR1 5XE M3WHB
PR1 6HN M6FBV
PR1 6NS M6UDS
PR1 6QN G4THA
PR1 7QR M0BAD
PR1 7TP M3VLG
PR1 7UG M6CCB
PR1 8PJ M0EBP
PR1 8TP G0JEH
PR1 9DD G7JZJ
PR1 9DR G1HMY
PR1 9EL G3BWI
PR1 9EQ G4WQT
PR1 9EQ G4ZKA
PR1 9HA M6CNK
PR1 9HD M3CMH
PR1 9NG M0HIQ
PR1 9RP G7TCB
PR1 9SY M6FZE
PR1 9TB M0UWS
PR2 1AD M7PRE
PR2 1BH M3ZIX
PR2 1JD G6EUG
PR2 1JP G7VAS
PR2 1LL M7EUW
PR2 1PB 2E0ACU
PR2 1QU M7FJY
PR2 1RX M0NDU
PR2 1RX M6JFK
PR2 1TY M6LFZ
PR2 2AS M3KEC
PR2 2BT G5RJG
PR2 2EU M7DRB
PR2 2HH 2E0FJD
PR2 2HH M6FJD
PR2 2HH M6FLL
PR2 2PX M6TCX
PR2 2YW M1AMW
PR2 3EX G4UQI
PR2 3HS G7ILX
PR2 3JB M0OTJ
PR2 3JL 2E0ZPN
PR2 3LP G7AQD
PR2 3RU M3JDN
PR2 3RU M3MDN
PR2 3RY 2E1AFI
PR2 3RY G7PZE
PR2 3SX G0KSN
PR2 3YR G4LHR
PR2 3YS M0AKQ
PR2 3YY M0WCM
PR2 4UL G4TVB
PR2 6AN M6AQY
PR2 6AN M6RRA
PR2 6DA 2E1IJM
PR2 6DH G3SYA
PR2 6EU G7KNR
PR2 6EX M0DMD
PR2 6TD M6LVR
PR2 6TH 2E1FJP
PR2 6TU M0JYG
PR2 7AL G7TRL
PR2 7BE 2E0FMB
PR2 7DA 2E0RFD
PR2 7DA M0RDZ
PR2 7DA M3OSS
PR2 7DA M3URD
PR2 7DZ M1CYK
PR2 8GT G3OIH
PR2 8NY G7KXV
PR2 8XN G7PSZ
PR2 8XN M3PSZ
PR2 9AW M0RWH
PR2 9AZ G7IFM
PR2 9AZ G7LPF
PR2 9FP 2E0CXW
PR2 9FP M6AYH
PR2 9QA G7OFU
PR2 9QX G1VZW
PR2 9QX G7OET
PR2 9RF 2E0HMM
PR2 9RF M3UCO
PR2 9SQ G3RSM
PR2 9SS G4PNH
PR2 9TP M3OEB
PR2 9UJ M7BOF
PR2 9YJ M0KHS
PR21SH G6MCC
PR23FQ G0SBY
PR25 1AR M3ULI
PR25 1BH G3ZRE
PR25 1BL M6CGL
PR25 1JA 2E0XCD
PR25 1JA M6HHA
PR25 1JD G3UPY
PR25 1JH M6JTD
PR25 1RJ G4YWG
PR25 1RN 2E0BSB
PR25 1RN G3WSW
PR25 1RN M1CND
PR25 1RN M3TPW
PR25 1YB G0JSL
PR25 1YB G0JSM
PR25 2DD G0MAH
PR25 2LJ M0HPR
PR25 2XA M6FDJ
PR25 2XW M6AOF
PR25 2XW M7AQR
PR25 2YL G0IDE
PR25 3AA G3GGS
PR25 3AF G4SBQ
PR25 3AR 2E0CII
PR25 3AR G0NGE
PR25 3AR G6VBK
PR25 3AR M0PQI
PR25 3AR M6AQK
PR25 3BD G6LUF
PR25 3DD 2E0BBN
PR25 3NR G1IQF
PR25 3NS G1RBZ
PR25 3UH G0FDX
PR25 3UH G0GVA
PR25 4QZ G4YIA
PR25 4QZ M6HBU
PR25 4XL M6WZL
PR25 4XP M1EYQ
PR25 4XT G3KQY
PR25 4XY G1BMN
PR25 4YJ G4AYU
PR25 4ZR G0HKW
PR25 5PA G1FKT
PR25 5PD G1TTL
PR25 5PD M0FWO
PR25 5PJ M3SHQ
PR25 5RQ 2E0ZMR
PR25 5RQ M3ZMR
PR25 5RQ M7ILA
PR25 5SN G7DKY
PR25 5SP G3XII
PR25 5SX 2E0CAR
PR25 5TT M7DJS
PR253HA M6PCD
PR26 6QA M3DKZ
PR26 6QS 2E0HPF
PR26 7AJ 2E0GUY
PR26 7AJ M3XXS
PR26 7QJ G0KLT
PR26 7QJ M3KLT
PR26 7XJ 2E0BDJ
PR26 8LB G0VHO
PR26 8NP G7VOQ
PR26 9AP G0PFU
PR26 9HP G0JSJ
PR26 9HQ 2E0IDA
PR26 9HS M3FBP
PR3 0BB G6LNS
PR3 0JY M3ZRA
PR3 0RH 2E0IBI
PR3 0RH M6IBI
PR3 0RP G0AAY
PR3 0SG M7BUG
PR3 0XD M6HUM
PR3 1AA 2E0PNC
PR3 1AA M3LJI
PR3 1AD G8CWQ
PR3 1BA G7CUA
PR3 1FJ G0LXP
PR3 1FS G4GVG
PR3 1LH G4JCG
PR3 1NL G4AMY
PR3 1NQ M0DKL
PR3 1PD G4TVN
PR3 1RD G1HKR
PR3 1RD G4BSD
PR3 1RD G7FNM
PR3 1RD M3GDI
PR3 1RD M3TAW
PR3 1RF G6AIZ
PR3 1XR G3ZNG
PR3 1YL G0VWX
PR3 1YQ M0SHM
PR3 1YQ M3PRY
PR3 2AL M0JHN
PR3 2BQ G8CQV
PR3 2EA G4EXK
PR3 2JX M3RZF
PR3 2LH M3BTI
PR3 2NA M3WKK
PR3 2XB G0SDJ
PR3 3EL M3GWK
PR3 3JG G3NKL
PR3 3SL G4VOJ
PR3 3TB M3FSQ
PR3 3TH M6IFH
PR3 3TQ G6MAC
PR3 3TX G3LZO
PR3 3TX G7NOI
PR3 3UA G0HJB
PR3 3WD G1CFG
PR3 3WN G6RTD
PR3 3YG 2E0GYF
PR3 3YG M0JVX
PR3 3YG M7HUD
PR3 3YS M3ZZQ
PR3 5AH G8GLS
PR3 5HB G0KMP
PR3 5JN G6DDR
PR3 6AB G1HJO
PR3 6AB M1AXP
PR3 6AP G3LPL
PR3 6BD M3NPX
PR3 6BN G6HCF
PR3 6SS G0IYT
PR33WF M7IDY
PR4 0HH G7WEM
PR4 0HH M3TDH
PR4 0NP M6DFZ
PR4 0PA M3EPC
PR4 0TT M6GNN
PR4 1AJ M3HVY
PR4 1DF M0AUG
PR4 1EG G7NOQ
PR4 1EN 2E0TLD
PR4 1EN G3TNN
PR4 1HG G4WIM
PR4 1JL M3EWZ
PR4 1JN G8HEU
PR4 1PT 2E0FNM
PR4 1PT M6CEB
PR4 1RG G4RNF
PR4 1RL G4RCF
PR4 1RQ M3TVK
PR4 1RX G0CPJ
PR4 1SB 2E0UBW
PR4 1SD G7CUL
PR4 1SS G4MEE
PR4 1TH M1FHT
PR4 1TH M3TJD
PR4 1UQ G7GVP
PR4 1XB G6NUQ
PR4 1XT M3EWY
PR4 1XU M3EWW
PR4 1XU M3MHD
PR4 1XY 2E0NHM
PR4 1YD G7LPD
PR4 2AY G1GQY
PR4 2AY G1GQZ
PR4 2DS G8OTZ
PR4 2EL G4AHZ
PR4 2NQ M1AKF
PR4 2NY G4MRX
PR4 2UE M0KBR
PR4 2UE M6IZQ
PR4 2XA 2E0BLL
PR4 2XA G7JGW
PR4 2XA M3JGW
PR4 2ZA G1TUI
PR4 3BA M3HVW
PR4 3BB M3VXY
PR4 3DN M1ACJ
PR4 3EP 2E0FRS
PR4 3EP M7FRD
PR4 3FE G1EUH
PR4 3NB G1NKF
PR4 3PN G3TKK
PR4 3SU M3HVX
PR4 3SX 2E1ECL
PR4 3SX 2E1FJL
PR4 3SX M0ABK
PR4 3SX M0GJA
PR4 3TX G8MED
PR4 3UD G0HIJ
PR4 3UD G1UCG
PR4 3UQ 2E0KMZ
PR4 3UQ M0IHN
PR4 3UQ M1HZZ
PR4 3UQ M6EQE
PR4 4JD G3VBL
PR4 4JX 2E0AFL
PR4 4RQ G0LEE
PR4 4XJ G0LQK
PR4 5AX 2E0JAR
PR4 5BH G1MBN
PR4 5BX G1IPY
PR4 5HB G4ZCG
PR4 5NP G4WAL
PR4 5NP G6TNA
PR4 6AA G0TUC
PR4 6AT G0EHK
PR4 6HD G4MGB
PR4 6JS M6HOU
PR4 6JX G0VAV
PR4 6JX M1AUZ
PR4 6LY G6TMN
PR4 6RB G4OWS
PR4 6SE 2E0GYV
PR4 6SE M7GJJ
PR4 6SJ G4XFA
PR4 6SU M7ROV
PR4 6SX M3LNM
PR4 6TD G6EWH
PR4 6TJ M6OVH
PR4 6TJ M7ANP
PR4 6TR G0JJD
PR4 6UD G6PLT
PR4 6UL G3MPF
PR4 6US M3VOY
PR5 0DT G3PS
PR5 0JR M0DBT
PR5 0LA M0PVP
PR5 0LX M0FYA
PR5 4BP G4BGP
PR5 4TD 2E0JNU
PR5 4TD M0JNU
PR5 4TD M7JNU
PR5 4TT M3BAN
PR5 4UT G3XUH
PR5 5JG G4UJA
PR5 5LA M6BXD
PR5 5RA G0EHW
PR5 5TY M0ETS
PR5 5UP G3AZI
PR5 5UU G1BTN
PR5 5UW G0DPG
PR5 5UW M3DPG
PR5 5YB G3ZVQ
PR5 6AG G8YUC
PR5 6EP G0FQC
PR5 6FT G4TUF
PR5 6LS M3NSM
PR5 6RA 2E0DZR
PR5 6RA M6CKV
PR5 6TD M6MLY
PR5 6UY G0ASH
PR5 6XQ G4OJR
PR5 6XR G1DER
PR5 8DS 2E1HQY
PR5 8DS 2E1LJL
PR5 8DS M0LJL
PR5 8DS M7PZG
PR5 8DU G3UCA
PR5 8EN 2E0VMA
PR5 8EN G0VMA
PR5 8EN M6VMA
PR5 8HJ M1DMH
PR5 8HS M6LZG
PR5 8HT G4GOM
PR5 8JS 2E1ACT
PR5 8JS 2E1DTG
PR5 8JX G0MXU
PR58BH M7EWB
PR58JS 2E0SQK
PR58JS M6HSB
PR6 0AG 2E0DHT
PR6 0AG M6GCE
PR6 0DB M0AQE
PR6 0DB M0ISN
PR6 0DG G0USM
PR6 0HZ M6CYS
PR6 0LJ G0JMN
PR6 0NJ M6ERW
PR6 0PY G7SYY
PR6 0PY M3SYY
PR6 0RR 2E0EIX
PR6 0RR M6IQC
PR6 0TN G4TAG
PR6 0TR M6SRZ
PR6 7AN 2E0DEX
PR6 7AQ 2E1BVQ
PR6 7AU G0HBX
PR6 7BG 2E0NYF
PR6 7BG M6NYF
PR6 7BJ G0CUN
PR6 7BW G6MEI
PR6 7DN 2E0GBI
PR6 7DN M6LSB
PR6 7EB G6LFT
PR6 7HE G8COR
PR6 7NA 2E0WPN
PR6 7TT 2E0LON
PR6 7TT M0ZWT

PR6 7TT M6DFT
PR6 7TZ G6JJI
PR6 8EJ G1TSV
PR6 8PX G7KTQ
PR6 8UB G8NNL
PR6 8UB M7JHY
PR6 8UE G8JZI
PR6 9BA M0XXD
PR6 9BA M7JGC
PR6 9DA M1CGI
PR6 9DA M1CGJ
PR6 9DU G6AHO
PR6 9ET M0HGD
PR6 9FF M7TKL
PR6 9LA G0LBT
PR6 9LA M3KEY
PR6 9NQ 2E0ZRG
PR6 9NQ M3SGV
PR6 9PA G3TXK
PR6 9RS G1KVC
PR6 9SS G1PBB
PR60JB M7SVK
PR7 1LX G4WYZ
PR7 1PH G4TZK
PR7 1RE G7KUR
PR7 1RH M0CGA
PR7 1UH G0DAI
PR7 1XJ M0MYJ
PR7 2FU G7UAY
PR7 2HL M0DMI
PR7 2JA M0DNW
PR7 2JA M6NNA
PR7 2JB 2E1IKA
PR7 2JB M1SMF
PR7 2JG M1EHI
PR7 2JW M6RKO
PR7 2LS G0KLH
PR7 2NT M6RKL
PR7 2NT M7RHM
PR7 2QU G0HBS
PR7 2YA 2E0EWT
PR7 2YA M0TKR
PR7 2YB 2E0SUD
PR7 2YB 2E0WOZ
PR7 3AP M3JQW
PR7 3AZ M7GFA
PR7 3HS G0UZF
PR7 3HY 2E0OTG
PR7 3NH G4IYP
PR7 4AN M6SEC
PR7 4AQ G1AUU
PR7 4JU G3URK
PR7 4LE M3NEL
PR7 4NL G7PYW
PR7 4NS G1DAK
PR7 4NT G4NRO
PR7 4PH M0GED
PR7 4PJ G0NGK
PR7 4QE M7DUJ
PR7 5AE M6KPG
PR7 5BH M7DOD
PR7 5BZ 2E1HQQ
PR7 5DB M7AVF
PR7 5EH G4WGT
PR7 5EY 2E0BUE
PR7 5EY M3URT
PR7 5HH G3JMZ
PR7 5NR M6COP
PR7 5NY G0OWA
PR7 5NY M6OWA
PR7 5NZ G1MBE
PR7 5PW G1AHM
PR7 5PY G0KDX
PR7 5QW G8SNQ
PR7 5RF M6LSH
PR7 5SN M6BFZ
PR7 5TW 2E0YPJ
PR7 5TW M0YPJ
PR7 5TW M3YPJ
PR7 6AU 2E0XVD
PR7 6AU M7BNL
PR7 6BA G8GFB
PR7 6BP G4PAT
PR7 6BS G1ZTG
PR7 6JW G6ZOL
PR7 6LY G7GYR
PR7 6PD G0WTD
PR7 6PD G7NER
PR7 6PL G0SHU
PR7 6PN G4PQM
PR7 6PP M6AQI
PR7 6PT G0CUB
PR7 6PT G3RPO
PR7 6PW G0WTM
PR7 6PW G7SSJ
PR7 7DB M3FTW
PR7 7LT M6JHI
PR75QW 2E0GPV
PR8 1HZ G3WZG
PR8 1JA G4ZYN
PR8 1NQ G0RLT
PR8 1RS G3TTY
PR8 1RT G4IQK
PR8 2DN M6RVA
PR8 2FB G7VJU
PR8 2HF G2ART
PR8 2JJ G3WTB
PR8 2LW G4BEU
PR8 2PE 2E0HYH
PR8 2PE M7DSL
PR8 2QF G7MJS
PR8 2QF G7OAA
PR8 2QL G7RNF
PR8 2QW G4JYQ
PR8 2RR G6JUQ
PR8 2RS G6ILX
PR8 2SA 2E0CXD
PR8 2SA M0HOQ
PR8 2SA M6JEY
PR8 3DB G3ZII
PR8 3DU 2E0NEL
PR8 3DU M3WFY
PR8 3DW M6JHQ
PR8 3HE M3WDU
PR8 3NP G8CIX
PR8 3QF 2E0TDE
PR8 3QF M6MVD
PR8 3RP G1DFT
PR8 3RS G1UCI
PR8 3UA M3WFL
PR8 4BJ G7CVF
PR8 4DT M3JGX
PR8 4EQ M6AJH
PR8 4EQ M6ALH
PR8 4HA G0MPW
PR8 4LT 2E0NYP
PR8 4LT M0JRI
PR8 4LT M6NYP
PR8 4NH 2E0YTT
PR8 4SF 2E0OIL
PR8 4SF M0MKO
PR8 4SF M6OHL
PR8 5AU G6EZY
PR8 5HB M6SEA
PR8 5HS G4SBE
PR8 5LP G0JCD
PR8 6HF M3FNH
PR8 6JA G6IVC
PR8 6JD M6CFR
PR8 6JY M3LNN
PR8 6LG G7IYG
PR8 6LG G7IYH
PR8 6NA G7TIM
PR8 6NR M3UYU
PR8 6PY 2E0BUJ
PR8 6PY M0STJ
PR8 6PY M3WJV
PR8 6PY M6ARU
PR8 6PY M6EDI
PR8 6SQ M3LBG
PR8 6UP G0LZX
PR8 6UP G1VWL
PR8 6XH G3ORK
PR9 0QQ M7ABQ
PR9 0QT 2E0FAB
PR9 0TW G3YIB
PR9 7AA G3STT
PR9 7BE G6FVM
PR9 7BY G1EPS
PR9 7DX M6AIW
PR9 7EZ M3KDO
PR9 7HX G4EVC
PR9 7JU G6DDC
PR9 7JX G3OCR
PR9 7JX G8ILJ
PR9 7JX M3YUD
PR9 7PQ G3HAA
PR9 8AT G0PBQ
PR9 8ET G6CIO
PR9 8LS M3UGK
PR9 8NL G0JCQ
PR9 8NL G4VYP
PR9 8NY G4NMU
PR9 8PS G0OFY
PR9 8QH M5SJS
PR9 8QR 2E0VEF
PR9 8QW G4UPK
PR9 9AH G8YPL
PR9 9BT G4DRA
PR9 9FA M3TCX
PR9 9GA G4SCO
PR9 9GJ G0PVU
PR9 9GJ G1ZMJ
PR9 9HW 2E0PAJ
PR9 9HW G5PJK
PR9 9NQ M6OUP
PR9 9RN G8RON
PR9 9RP G8MHL
PR9 9TW G4EID
PR9 9TW M3HQN
PR9 9XE G4TUP
PR9 9XF 2E0CWO
PR9 9XF M6CWD
PR9 9XW G3TMB
PR9 9YX G4YPH
PR97AP M7COB
PR98BU 2E0MFZ
PR98BU M7BOS
RG1 2RE M3FVA
RG1 2SE M7AED
RG1 3AJ M0IXG
RG1 4QD 2E0GUT
RG1 4QD M6GUT
RG1 5LR G6FBA
RG1 5QP 2E0PIK
RG1 5QP M0PIK
RG1 5QP M3ZUY
RG1 5RD G3WNP
RG1 5SD M3DKG
RG1 5SD M3VSL
RG1 6AE 2E0XLR
RG1 6AS 2E0PIX
RG1 6AS M6LOK
RG1 6JU M7JIA
RG1 6QD G6YLN
RG1 7PA M6DUD
RG1 7TT G8VWJ
RG1 7TY G4JPE
RG1 7UG G4FLY
RG1 7YT G0YYH
RG1 8EN G8WBN
RG10 0AU M6SUL
RG10 0AY M3BBS
RG10 0JB G1ZQR
RG10 0SF G8LZS
RG10 8BJ G8KZG
RG10 8BN G3WPH
RG10 8DR 2E0JBL
RG10 8DR M0OAA
RG10 8DR M6BPQ
RG10 9AX 2E0BPT
RG10 9AX M0BSJ
RG10 9AX M3VYM
RG10 9AY G3LRQ
RG10 9BN G8NXJ
RG10 9BN M3NXJ
RG10 9BT 2E0CJB
RG10 9BT M0TVA
RG10 9BT M3ZAW
RG10 9BT M6RUB
RG10 9BW M0KAK
RG10 9ED G7GEF
RG10 9JG G6AWF
RG10 9LJ 2E0UDA
RG10 9LJ M0UDA
RG10 9LJ M6OUD
RG10 9PY G3VKQ
RG10 9PY M6MCE
RG10 9QD M1BFO
RG10 9QF G7REH
RG10 9QT 2E0MSR
RG10 9QT M6LSG
RG10 9QT M7MSR
RG10 9QU M7KFC
RG10 9SJ M6EFU
RG10 9TS G8EPZ
RG10 9YD G4SYE
RG10 9YT G8KOM
RG12 0TB M6MBY
RG12 0TN M6PCM
RG12 0TR G1HLQ
RG12 0TR G6YLA
RG12 0TW M3JBM
RG12 0UA 2E0WWS
RG12 0UA M3YXW
RG12 0UD G7IBU
RG12 0UR 2E1AFN
RG12 0XU G4DDL
RG12 2JR G4MXY
RG12 2JW M3YNJ
RG12 2JW M6SER
RG12 2LU 2E0XBZ
RG12 2LU G8RKO
RG12 2LU M3XBZ
RG12 2PT G0NEF
RG12 2PT M6NEF
RG12 2QP M3WVI
RG12 2RA G4YBG
RG12 2SE G0OVA
RG12 7BE M6EXZ
RG12 7DU G4BZJ
RG12 7ET G7FSR
RG12 7JU M7EPT
RG12 7JY M7PJF
RG12 7LD G4CQH
RG12 7ND M7YBZ
RG12 7PS G6NFC
RG12 7QA G8AMK
RG12 7QG G4DDN
RG12 7QG M0BRA
RG12 7QQ G0IRH
RG12 7TF G4AUC
RG12 7WF G3TKS
RG12 7WG G4KNZ
RG12 7WL G0SCY
RG12 7WR M0XQS
RG12 7WR M6XQS
RG12 7WZ M6PEO
RG12 7YX G0FCT
RG12 7YX G6BRA
RG12 7YX G6MER
RG12 7YX G8III
RG12 7ZX M3XPR
RG12 8UD 2E0XKT
RG12 8UD M0KJT
RG12 8UD M6XKT
RG12 8UP G0AQU
RG12 8UQ G1CPX
RG12 8XE G7JDF
RG12 8XP G1RXV
RG12 8XU G6TNR
RG12 8XY G4PCF
RG12 8YG M3NAH
RG12 9BY 2E1JKP
RG12 9EF G1GUI
RG12 9EF M1KIP
RG12 9ES G4WYC
RG12 9EX G3YMC
RG12 9HT G7USG
RG12 9JE M1CAX
RG12 9JL G3OFW
RG12 9PE M0JPK
RG12 9PS M6AFU
RG12 9PX G1VXX
RG12 9QH G3ZWP
RG12 9QX G8RQN
RG12 9YL G0VQX
RG127JY M7KXD
RG127WH G0CGE
RG13LP G0RSR
RG13LP G6ZTZ
RG14 1LF M0NGC
RG14 1TA 2E0NKK
RG14 1TA M5REM
RG14 1UH G3OPG
RG14 1UH G8OIU
RG14 2EN G1CTF
RG14 2HA G6OEW
RG14 2HA G6XZS
RG14 2HB G1DAV
RG14 2JL M3YXQ
RG14 2LS 2E0POQ
RG14 2LS M0POQ
RG14 2LS M3POQ
RG14 2ND G3OUC
RG14 2PN G6RBP
RG14 2SQ M7CCL
RG14 2TH G7HUO
RG14 2TH G7HUP
RG14 3AJ M6GMR
RG14 5JE M1CIJ
RG14 5JF 2E0FAE
RG14 5JF M0IXU
RG14 5JF M6ASF
RG14 5JN G7SYC
RG14 5JP G4RUW
RG14 5NR G7SLL
RG14 5QW G7REJ
RG14 5TF M6DTH
RG14 6AZ 2E0JOF
RG14 6AZ M3JOF
RG14 6BA 2E0UAO
RG14 6BA M3UAO
RG14 6DD G3UVM
RG14 6DN G3URI
RG14 6DN G4MKF
RG14 6DY M7BLC
RG14 6EE G4YMY
RG14 6HD M1BRU
RG14 6HP G0HBJ
RG14 6HP G6ZSF
RG14 6JX G4WEV
RG14 6PZ G8KHU
RG14 6PZ 2E0JHC
RG14 6PZ M0IDC
RG14 6PZ M3JHC
RG14 6QJ 2E0LHI
RG14 6QJ M6XHI
RG14 6RS M7JJC
RG14 6RT M0JOW
RG14 6RU G3ZGC
RG14 6RU M3ZGC
RG14 6RY G1WTS
RG14 6RY G8AKM
RG14 6SX G8AYC
RG14 7AL G4ZDP
RG14 7DJ G8MWU
RG14 7FX G7KXZ
RG14 7QS M7AZG
RG14 7RA M1DLG
RG14 7RB G4TKS
RG14 7RR G6LAE
RG14 7TL G8JUS
RG14 7TT G1DFI
RG17 0AZ 2E0BJM
RG17 0AZ M0NKS
RG17 0BZ G3ZPK
RG17 0JE G1OQV
RG17 0JE M0ISJ
RG17 0JE M0RYG
RG17 0JE M6CUE
RG17 0JP M7HUG
RG17 0JR G0LJJ
RG17 0JR G1FVP
RG17 0LJ M3TZN
RG17 0LL G6GCJ
RG17 0LU G4CST
RG17 0SG G4TPH
RG17 0UN G6EML
RG17 7DG G0DTQ
RG17 7ED G6WLE
RG17 7JB 2E0YGM
RG17 7JL M6MPX
RG17 7JN G6SRY
RG17 7LT 2E0IHV
RG17 7LT M7CQA
RG17 7TS G4DNH
RG17 7UN 2E0DYN
RG17 7UN M0ICZ
RG17 7UN M6FWX
RG17 8RF G7BWI
RG17 8PB G6EES
RG17 8YQ M3XEL
RG17 9QE G0LTX
RG17 9QE M1ABX
RG17 9UE G4FNK
RG17 9UW G7PGY
RG170JE 2E0OQV
RG177DH M7HHZ
RG18 0RR 2E0BPS
RG18 0RR M0JCM
RG18 0RR M3XGZ
RG18 0XD M6KDQ
RG18 3AF M3HSR
RG18 3BA G4BOO
RG18 3BF G7TOO
RG18 3BF M0PJC
RG18 3BN G6AMF
RG18 3BP G3RVM
RG18 3DH G4UET
RG18 3DH M0XRH
RG18 3EB G7JLT
RG18 3EG G0POT
RG18 3PD M3TVD
RG18 3UE M3EHK
RG18 3UH M3JHV
RG18 4DL G4ORX
RG18 4DQ G1PCG
RG18 4DQ G4WLG
RG18 4LS M3NXF
RG18 4LS M3SXF
RG18 4LS M6PXF
RG18 4LY G6SMN
RG18 9HQ G3VOW
RG18 9HT G8LTN
RG18 9HZ 2E0HQO
RG18 9HZ M6HQO
RG18 9PB G0MIA
RG18 9PD M3PUB
RG18 9PD M3RBX
RG18 9PH G4LMW
RG18 9PH G8XCW
RG18 9PH M6LMW
RG18 9QP G3KJC
RG18 9QW M6DGU
RG18 9RJ G6SOZ
RG18 9RJ G7DRR
RG18 9SL 2E0TVR
RG18 9SL G6ZPO
RG18 9SL M3PRN
RG18 9TG G6HUN
RG18 9UZ M7LRB
RG19 3BF M0AAP
RG19 3BF M7ELZ
RG19 3BF M7OZC
RG19 3LE G6ZVL
RG19 3PF G0OBJ
RG19 3PF G4YMH
RG19 3RL G4FWR
RG19 3RS G3ICB
RG19 3SD 2E0ESO
RG19 3SF G0KQT
RG19 3SH G7RJW
RG19 3TN M7KMF
RG19 3XB G8TSC
RG19 3XD M1CEY
RG19 3XE G6PNG
RG19 3XW G3DBV
RG19 3XW G7UXK
RG19 3XX G4BWE
RG19 4FD G3WYW
RG19 4FD M1PGT
RG19 4FQ G8JIP
RG19 4GJ G4RTQ
RG19 4LX G7VGL
RG19 4WA 2E0MAD
RG19 8BE G3MEV
RG19 8DD G3LLK
RG19 8DY G1XSQ
RG19 8EY 2E0RMZ
RG19 8EY M0RMZ
RG19 8SH M6DHT
RG19 8SS M0SBF
RG19 8XT G6RHT
RG2 7AH M6YYM
RG2 7BG M0ALB
RG2 7DX G4YBX
RG2 7HU M3UYW
RG2 7JR G4OMN
RG2 7JU G4JXH
RG2 7TD M0GQS
RG2 8AS 2E0GYH
RG2 8DN 2E0FUZ
RG2 8DN M3ZTK
RG2 8DN M3ZVK
RG2 8HJ G1ASD
RG2 8HQ M7CXV
RG2 8JB G6NVY
RG2 8LJ G1NYZ
RG2 8PB G1GQP
RG2 8PR M7HHI
RG2 8QP G4DBF
RG2 8QP G7BXU
RG2 8SD 2E1HJE
RG2 9FL 2E0DLV
RG2 9FL M0HQU
RG2 9FL M6DPU
RG2 9HA G8IJG
RG2 9HR M6FHN
RG2 9NP M6VLR
RG2 9PD M3VME
RG2 9PU G7KWN
RG20 0LH G3TEB
RG20 0LY G0LUK
RG20 4UQ G3XKM
RG20 5ER G4HDE
RG20 5RJ G3BFL
RG20 5SD M7EXF
RG20 5SL G8HJF
RG20 5SL G8HJG
RG20 7BE G3NAQ
RG20 7BT G8JRN
RG20 7EH M0KKA
RG20 7EZ M0CUK
RG20 7JS 2E0FBN
RG20 7JS M0WTJ
RG20 7JS M6FYG
RG20 8HS M7CNB
RG20 8QF M7IRJ
RG20 8SD G1YOS
RG20 8SE G1NAB
RG20 8TZ M0HPS
RG20 9BN G6GXZ
RG20 9BW G3SVD
RG20 9BW M3SVD
RG20 9BW M3UKN
RG20 9EQ G8SFO
RG20 9EY G7VFE
RG20 9HY M6NAC
RG20 9TA G6UJC
RG20 9TB M3TYU
RG20 9TS 2E1ETJ
RG20 9TS G4DOQ
RG20 9TS G6PGE
RG20 9UY G7DXC
RG20 9UY M3PCW
RG21 3AS M6SEK
RG21 3JH G7WCN
RG21 3JW G8AOO
RG21 3NF G7FFB
RG21 5HL G6BBW
RG21 5NR G7SNB
RG21 5NR M6NIS
RG21 5NZ G8YEQ
RG21 5QE G1OQU
RG21 5SR G8PIY
RG21 5UA G4CSD
RG21 6AD M6HZT
RG21 7TG G7SKX
RG21 7TQ 2E0MUZ
RG21 7TQ M0SWT
RG21 7TQ M6MUZ
RG21 8XT M0NQB
RG215SS G4XMP
RG22 4BG M0EDR
RG22 4DU 2E0EUB
RG22 4DU M6WEU
RG22 4EA G1JVG
RG22 4EL G8AOK
RG22 4HN G4JJD
RG22 4HR G8JLB
RG22 4HZ M7SBM
RG22 4JN M6KTZ
RG22 4JR G8GOS
RG22 4JR M0IYC
RG22 4LJ M3JSK
RG22 4NB G6JQR
RG22 4NB M7SRZ
RG22 4NP G6UZO
RG22 4NP M3BRU
RG22 4PH G7LYL
RG22 4RE M7DFD
RG22 4RF M1CJX
RG22 4RQ G8AGJ
RG22 4TY 2E0GUV
RG22 4TY G7PKD
RG22 4TY M0CJJ
RG22 4UB G4BHE
RG22 4UJ G6OPK
RG22 4UL G3WXD
RG22 4UL G8DUV
RG22 4UX G4PND
RG22 4XD G6MDS
RG22 4XT G8VML
RG22 4XT M0DRG
RG22 5BA G7LJB
RG22 5BQ G1CBS
RG22 5DN G0LEP
RG22 5FH G0GJR
RG22 5FY 2E0ISO
RG22 5FY M3ISO
RG22 5JY G8GTZ
RG22 5LJ G6KQZ
RG22 5NN G0LQD
RG22 5NN G6PMF
RG22 5NW 2E1LIS
RG22 5NW M0LIS
RG22 5NX M0SAB
RG22 5QD G3UCF
RG22 5QF G6ZZR
RG22 5RA 2E0CBO
RG22 5RA M0HHD
RG22 5RA M3YQC
RG22 5RG G1IYF
RG22 6AX G0PTA
RG22 6BN G7PFL
RG22 6BQ G7MAT
RG22 6BQ M3HYO
RG22 6DF M0CBF
RG22 6DP 2E0KMY
RG22 6DP M6KMY
RG22 6EP M0RAB
RG22 6JA G1PEU
RG22 6JL G1WMV
RG22 6JL M0TAD
RG22 6NU G0KQA
RG22 6NZ 2E1FTH
RG22 6QP G3OAZ
RG22 6QQ M0HNX
RG22 6TD M1DWD
RG23 7AQ G3JTQ
RG23 7BB 2E0PGP
RG23 7BB 2E1FSF
RG23 7BB 2E1FSG
RG23 7BB M6KIK
RG23 7BL G4JGS
RG23 7DD G3HVA
RG23 7DJ G6PJP
RG23 7JX M0LAS
RG23 7LB G6CFA
RG23 7LD M0CJM
RG23 7LF 2E0HFN
RG23 7LF M0WNA
RG23 7LF M7BQX
RG23 8AD G0XBA
RG23 8EX G4YPK
RG23 8JD G4EDH
RG23 8JF G7KAK
RG23 8NG G0JSR
RG23 8NH G8JYN
RG23 8QL 2E0NBM
RG24 7EF G8JHH
RG24 7EH M0GWC
RG24 7JE G4XIP
RG24 7LA 2E0HTK
RG24 8AA G7PTH
RG24 8GL M3BVM
RG24 8RG 2E0BTV
RG24 8RG M0GOQ
RG24 8RG M3XWV
RG24 8SB G8JSF
RG24 8SL M3NHW
RG24 8SU M1ENA
RG24 8UJ 2E1GLQ
RG24 8UJ G7PAG
RG24 8WN M1KPW
RG24 8XQ 2E0VCG
RG24 8XQ M0VTC
RG24 8XQ M7DFI
RG24 9BE M3NRI
RG24 9DD G7FFC
RG24 9DW 2E0SUC
RG24 9HE G4YNH
RG24 9HX M0CWY
RG24 9LR G8UBN
RG24 9PA M7PSV
RG24 9PQ M0DOP
RG24 9PQ M3DMY
RG24 9PQ M3OOU
RG24 9PQ M6DMY
RG24 9PY M3GCJ
RG24 9RS G7RPP
RG24 9SH M6IBJ
RG24 9SS M7KNI
RG24 9TU M0JWT
RG24 9XP M7HNU
RG24 9YL M7NFC
RG24 9YP M0MEI
RG24 9YP M7TIM
RG247HR M7GMA
RG25 2BZ G7NDB
RG25 2NH G3CEI
RG25 2RN G7EVF
RG25 3EJ M6CEE
RG25 3HP M0FSN
RG25 3JE G4GFM
RG25 3LD G0NJG
RG25 3LE 2E0HQQ
RG25 3NL G7VZR
RG25 3NL M0VZR
RG26 3DW G6OZZ
RG26 3EL G8AKA
RG26 3HP 2E0MPX
RG26 3HP M0PXM
RG26 3HP M6KMG
RG26 3NJ G1WKK
RG26 3SF M6ELQ
RG26 3SH M1BCZ
RG26 3TL G8DVU
RG26 3UE 2E0GVR
RG26 3UE M7OWD
RG26 3UQ G6AOH
RG26 3UZ M0VVC
RG26 3YH G4LUA
RG26 3YH M6JJL
RG26 3YW G0CAK
RG26 4ES M7FAJ
RG26 4HF G6IOX
RG26 4HH G8MSY
RG26 5AG G6BNU
RG26 5BX M3XLC
RG26 5DG G1EHF
RG26 5DG M0HNA
RG26 5DG M3EHF
RG26 5EP G4LXJ
RG26 5HD G8FMD
RG26 5HF G6AXV
RG26 5JX M0PPZ
RG26 5NY G4XMO
RG26 5PJ G1JWG
RG26 5UE M6AII
RG26 5XH M1ECW
RG263LF M6XST
RG27 0BQ M7MXY
RG27 0DQ G0FEJ
RG27 0ES G3UQW
RG27 8BF M0SVV
RG27 8JS G3YGE
RG27 8NJ G0AMZ
RG27 8QW G0CVA
RG27 8QW M6FWB
RG27 8QW M7WJC
RG27 8SW M3MOF
RG27 8TW 2E0ZXJ
RG27 8TW M0ZXJ
RG27 8TW M3XSJ
RG27 9EY G4SFH
RG27 9GB M7GOG
RG27 9HW G4YEG
RG27 9JS G6AHH
RG27 9NP M1CAO
RG27 9PU G6OFD
RG27 9QJ G4SXX
RG27 9RA M0VEC
RG27 9RH G7EWV
RG27 9RP G6ETX
RG27 9SF 2E1BOZ
RG27 9SG G0RSV
RG28 7JG 2E0NBK
RG28 7NF G3NYS
RG29 1AE M0SOX
RG29 1BH 2E0EJM
RG29 1BH 2E0ICU
RG29 1BH 2E0PIP
RG29 1BH G1ZBL
RG29 1BH M0AEU
RG29 1BH M3OFS
RG29 1BH M3OUI
RG29 1BH M3TIZ
RG29 1EJ G7VSP
RG29 1JZ G7BRB
RG29 1ND G8WAM
RG29 1NT G7GIK
RG29 1NT M0UAR
RG29 1SX 2E0LSV
RG29 1SX M0LSV
RG29 1SY 2E1GTD
RG29RD M7RFF
RG3 4UL G7SJK
RG30 2BY 2E0HFO
RG30 2BY M7BRA
RG30 2EJ M3NOW
RG30 2HA G6GBL
RG30 2NT G1ANQ
RG30 2NX G0AFR
RG30 2PE G1DSJ
RG30 2PE G1ZSY
RG30 2SF G0ODK
RG30 2TH 2E0FRP
RG30 2TH G6JUI
RG30 2UL M6ETP
RG30 3DL 2E0SIZ
RG30 3DL M7RSJ
RG30 3EJ M0HAO
RG30 3EN G4BTI
RG30 3NQ G7UET
RG30 3NQ M3AGI
RG30 3NQ M5ALG
RG30 4HU M3MSQ
RG30 4JT 2E1EPQ
RG30 4JT M0LTE
RG30 4PA M7ASJ
RG30 4QB G0GBR
RG30 4QP G0RZM
RG30 4YJ G7KWD
RG30 6BA M7NFZ
RG30 6DT 2E1CAJ
RG30 6EH 2E0MGT
RG30 6EP M6XBS
RG31 4US 2E0CGL
RG31 4US M0EGN
RG31 4XP 2E0FUN
RG31 4XP M0RXX
RG31 4XP M3VHZ
RG31 4XQ 2E0IHS
RG31 4XQ M7GLS
RG31 5DZ G4EFE
RG31 5HJ G8SMA
RG31 5JU G7VHX
RG31 5JY G0IZI
RG31 5WE G4XKA
RG31 6DE G7JJG
RG31 6DE G7KSS
RG31 6FZ M5ACR
RG31 6HN 2E0BJS
RG31 6HN M0RBJ
RG31 6HN M3VFP
RG31 6NP G4JOB
RG31 6PY 2E1FGB
RG31 6RH G8PQZ
RG31 6RH M0GMC
RG31 6RR M6BNV
RG31 6SR G8LCA
RG31 7AT G7MFP
RG31 7DD M0SMS
RG31 7DN G0TFU
RG31 7JR 2E0BZX
RG31 7JR M6RFC
RG31 7RX M0TBR
RG31 7ZL G3VUK
RG31 7ZU 2E0SHP
RG31 7ZU M6SHP
RG315NS G3ZIB
RG4 5AL G4AWY
RG4 5AY G4BXY
RG4 5JP M6WZF
RG4 5LE M6WOT
RG4 6DD M0IZS
RG4 6PL G4VSQ
RG4 6PY 2E0CFI
RG4 6PY G6EIU
RG4 6PY M0HAH
RG4 6PY M6CFO
RG4 6QA M3IEP
RG4 6QB G8NQC
RG4 6RT M6KCY
RG4 6SF G8ROG
RG4 6SZ G0KIA
RG4 7DH 2E0SKP
RG4 7DH M0SKP
RG4 7DH M7RJI
RG4 7DT 2E0DNE
RG4 7HH G0CCC
RG4 7HH G4CCC
RG4 7HR G3AKF
RG4 7HR G4JTR
RG4 7HR M0AAA
RG4 7JZ 2E0NBK
RG4 7JZ M7NCB
RG4 7NR 2E0IET
RG4 7NR M0IET
RG4 7NR M3IET
RG4 7NR M3IEV
RG4 7NR M3SGF
RG4 7NT G7ENS
RG4 7PE M7TSB
RG4 7RG 2E0DOP
RG4 7RG M6ENK
RG4 7RH 2E0CYR
RG4 7RH M6DTU
RG4 7RP G4EJK
RG4 8DN M0TLN
RG4 8HH G0OIW
RG4 8HH G3WKX
RG4 8HJ G4ZSZ
RG4 8JH G3OYN
RG4 8PA G3YZZ
RG4 8PH 2E0FSM
RG4 8PH M6RFP
RG4 8PL G4OPK
RG4 8TT G0MQW
RG4 8UU G4ZRZ
RG4 9AD G0LGG
RG4 9LS M6KBX
RG4 9RU 2E0DWM
RG4 9RU M3TXG
RG4 9RY G6NEA
RG4 9RY G6RKJ
RG4 9RY M3AZP
RG4 9SA G8PDY
RG4 9SP G7RZQ
RG4 9TF G0JMS
RG40 1AW M5AGS
RG40 1DE M6EFE
RG40 1DE M6FNX
RG40 1PL G8VZY
RG40 1QG G8CFI
RG40 1QG G8IEI
RG40 1QW 2E0PEP
RG40 1QW M6PEP
RG40 1RJ 2E0DPR
RG40 1RJ M0SMN
RG40 1RJ M6CGK
RG40 1RL G1DNP
RG40 1RY M3ZRW
RG40 1TW G4ORU
RG40 1XS G8YRL
RG40 1XX G8FBF
RG40 1YE M6PHE
RG40 2HT G2DD
RG40 3HT G4OKM
RG40 3LB G7JTV
RG40 3LD G8DUO
RG40 3LG G3XPC
RG40 3PE 2E0TJJ
RG40 3PE M0TJJ
RG40 3PE M7TJX
RG40 3QB G8NZK
RG40 3RL G3WME
RG40 4EN M0GRW
RG40 4EY M7DRW
RG40 4HP M7MXF
RG40 4JA G1SEO
RG40 4PA M6DUM
RG40 4PA M6SSS
RG40 4PF G4HRE
RG40 4PT G6IJK
RG40 4PY 2E0WGI
RG40 4PY M0WGI
RG40 4PY M3WGI
RG40 4RA G1ZSF
RG40 4RA G4NIP
RG40 4RD G4DYO
RG40 4TS M6SUZ
RG40 4UD G4HOU
RG40 4UJ M1VHF
RG40 4UJ M3WOK
RG40 4UN 2E0CDU
RG40 4UN M0GVY
RG40 4UN M6DBX
RG40 5QU M1AXG
RG40 5YB G3ZBS
RG40 5YE G4XAL
RG40 5YL G4VWO
RG41 1JE G3ZZH
RG41 1LJ M6POA
RG41 1LN G7TYH
RG41 1NN M6HML
RG41 1NR G6CGC
RG41 1PH 2E1AUQ
RG41 1PH G3YHG
RG41 2AH M7REY
RG41 2RJ G0JHW
RG41 2TL M3IWK
RG41 3AG G8GJM
RG41 3HG G3MGU
RG41 4AJ M6EFF
RG41 4AR G6TLX
RG41 4AW G8IBP
RG41 4AX G4LJF
RG41 4BA G8FIE
RG41 4BX G4JXU
RG41 4DH G1KTF
RG41 4ED G3TUY
RG41 4TA M6HRM
RG41 4TB M7PSM
RG41 4UR G8VPO
RG41 4UY G7SEO
RG41 5HR M3ZLW
RG41 5JF G4QGH
RG41 5JG M0GXB
RG41 5JZ G7COP
RG41 5LG G4REA
RG41 5LX M6CMB
RG41 5NJ G6WXN
RG41 5NW G6IVB
RG41 5PE G1HBD
RG42 1FJ M6NZR
RG42 1RT G6XPB
RG42 1RT M6XPB
RG42 1UE G8OLK
RG42 2AD M6RCE
RG42 2BT G4WZJ
RG42 2HJ 2E0PTS
RG42 2HJ M0PTS
RG42 2HJ M6PGB
RG42 2HL G3NCN
RG42 2LD G3VG
RG42 2LE G3TAI
RG42 2LH G4NOC
RG42 2QX G6DKI
RG42 3SZ 2E0RWJ
RG42 3SZ M7RWJ
RG42 3TR 2E0XDF
RG42 3TR 2E0XNF
RG42 3TR M0XDF
RG42 3TR M3XJF
RG42 3UN G0ATZ
RG42 3UN M3UTQ
RG42 3XA G4HLF
RG42 3XU 2E0RTP
RG42 3XU M0RTP
RG42 3XU M6RTP
RG42 3XX G4ZFE
RG42 3XX M3NUB
RG42 4DF G4DQA
RG42 4DS G4ORB
RG42 4ES 2E0FUG
RG42 4UR G7CJO
RG42 5JG G4PKE
RG42 5LG G0GJV
RG42 5LG G4BRA
RG42 5LG G6XSY
RG42 5QZ M3NPK
RG42 6HU G0BBB
RG42 6HU G0DX
RG42 6HU G5AOZ
RG42 6HU M0BB
RG42 6JG 2E0XWX
RG42 6JG G5YAX
RG42 7UY G6JDW
RG45 6AL G8EYM
RG45 6BN 2E0PZK
RG45 6BN M0OPK
RG45 6BN M6DGI
RG45 6BT G8AXY
RG45 6DU G8XML
RG45 6EF G6LLG
RG45 6EN G3GRY
RG45 6HE 2E1DIH
RG45 6NR G3DVF
RG45 6QE M1EFY
RG45 6QF G1WGM
RG45 6QR G0FPT
RG45 6SE M0DZV
RG45 6WE 2E0OWG
RG45 6WE M0OWG
RG45 6WE M6OWG
RG45 7EG G7VWO
RG45 7HR G6FXR
RG45 7JP G8STR
RG45 7NE G8ZLL
RG45 7NR G0WCS
RG45 7PD G8GSU
RG45 7QR G4DGF
RG456HJ M0IRH
RG5 3AD G3RAC
RG5 3AD G4HNF
RG5 3AX 2E0JCZ
RG5 3AX M0JRZ

RG5 3AX M6FJL
RG5 3BG M6HDD
RG5 3BL M6JWX
RG5 3DA G3WTV
RG5 3DY M6CCM
RG5 3HA G0DWD
RG5 3HF M3NXH
RG5 3JA G0IZV
RG5 3LQ 2E0ZIP
RG5 3LQ M3XHH
RG5 3LQ M6PEN
RG5 3LR G0AHI
RG5 3NY M6KUK
RG5 3QG G8DZC
RG5 4AP G0VQR
RG5 4AP M0CYB
RG5 4AW G4ELY
RG5 4BL G2NF
RG5 4HH G4IZS
RG5 4LD 2E0JCW
RG5 4LD M6JCW
RG5 4LJ G4JVD
RG5 4LR G0FIJ
RG5 4LR G8MBQ
RG5 4NA M3MZW
RG5 4PA M6GJS
RG5 4QB G7UOQ
RG5 4QB G8XJJ
RG5 4QT G3UUB
RG5 4RT G3PBR
RG5 4SR M7DFM
RG5 4UG G8VKR
RG5 4UN G0VPE
RG5 4UN G4KWT
RG5 4XE M6PGX
RG5 4XR G8CSK
RG6 1AP G0ILT
RG6 1HD G7RFS
RG6 1HS 2E0JPM
RG6 1HS M6JMO
RG6 1HW G3VMY
RG6 1QE G4OKA
RG6 3AH G4CDF
RG6 3AN G4MCM
RG6 3DG G6HCT
RG6 4AB G7UZY
RG6 4DB G7FQE
RG6 4EN 2E0OOX
RG6 4EN M7NDT
RG6 4EP M6LHH
RG6 4HN 2E0XUU
RG6 4HN G7NDI
RG6 4HN M0XUU
RG6 4HN M3XUU
RG6 4UY M0XCA
RG6 4XA G6IEI
RG6 5GX G6YIE
RG6 5PY G8NPZ
RG6 5QX G8LCL
RG6 5QZ G4FXT
RG6 5RY G8ADQ
RG6 5SL 2E0BPW
RG6 5SL M0JNX
RG6 5SL M3KYK
RG6 5SN G8VRW
RG6 5SR G3UAX
RG6 5TG G0CEC
RG6 5TP M3EZJ
RG6 5XG G0MZN
RG6 5XG G4AOL
RG6 5XG G8VWH
RG6 5YQ G6JVK
RG6 7LH G4RDC
RG6 7LJ G7LWY
RG6 7LQ G7AOK
RG6 7PA G4OQJ
RG6 7PD G0JTN
RG6 7PE G4OAE
RG6 7RT G4HLT
RG6 7RT G4MUT
RG6 7SS 2E0IMV
RG6 7SS M1PEL
RG6 7SS M7PEL
RG7 1BB G8VWV
RG7 1BH M6WEW
RG7 1BQ G7FBE
RG7 1BZ M6ZRZ
RG7 1DD M3KVC
RG7 1HP G7MTW
RG7 1QY G8MVY
RG7 1SG G7OMQ
RG7 1TJ M0SCB
RG7 1TT G8YVQ
RG7 1XS G4ZAC
RG7 2HY G4EPX
RG7 2LH G4XXH
RG7 2PZ G3SCZ
RG7 2QB M0GVN
RG7 3BU G4CLD
RG7 3ES G3YGR
RG7 3EZ G4IWS
RG7 3HY G0VHK
RG7 3JA M1DGQ
RG7 3PY 2E0YWA
RG7 3PY G4IDS
RG7 3PY M7IDS
RG7 3QE G1COW
RG7 3QU M1BNG
RG7 3TH M0SEL

RG7 3TL G1AWD
RG7 3TR G3WND
RG7 3TU G3ZOI
RG7 3XW 2E0PHM
RG7 3XW M0MPM
RG7 3XW M0PAQ
RG7 3XW M6PHM
RG7 4DE G8ZWN
RG7 4PA M0OJO
RG7 4TR G4EEE
RG7 5AR M7HAF
RG7 5DN G3LXQ
RG7 5NS G7TKP
RG7 5RY 2E0DEU
RG7 6NN G0OBV
RG7 6NU G3KIW
RG7 6NU M3KXY
RG7 6RA G4ZSG
RG7 6TN G0KPE
RG74TH G8DGR
RG8 0DJ 2E0CCJ
RG8 0EB G4CEI
RG8 0HX 2E0SSG
RG8 0HX M3SSG
RG8 0JL G3NGX
RG8 0PL M0AZG
RG8 0SE G3LLG
RG8 7BD M3LJJ
RG8 7PX G1GMH
RG8 7QG G3TRY
RG8 7QG G4CUE
RG8 7QG G4MDF
RG8 7QG G8NMO
RG8 7QL G0LHZ
RG8 7QL G3ULT
RG8 8BH 2E1FFX
RG8 8BQ G7KSV
RG8 8LN 2E0LAV
RG8 8LN G7TXR
RG8 8LN M0XYX
RG8 8LS G6AML
RG8 8LT 2E0SRJ
RG8 8LT M0MLE
RG8 8LT M6RJS
RG9 1AP G8VOB
RG9 1JT 2E0HIT
RG9 1JT M6HIT
RG9 1LT G3TFL
RG9 1PH G0NCO
RG9 1UU G4WBI
RG9 1UU G7LSF
RG9 2LX G3UYY
RG9 3JS G8FXX
RG9 3JS M0FXX
RG9 4DH G6CHA
RG9 4QQ M7ORG
RG9 5PS G3PZL
RG9 5QX 2E0ENM
RG9 5QX M6ILI
RG9 5WE M3PXT
RG9 6NR M0XVI
RH1 1AF G4VXN
RH1 1DR 2E0BXZ
RH1 1DR M0JCR
RH1 1DR M6JCR
RH1 2BN G6YAH
RH1 2BZ G3VGE
RH1 2DF M6EOQ
RH1 2DY G8ZFQ
RH1 2EJ M7RLS
RH1 2HA G0LXV
RH1 2HH G3MPB
RH1 2JB G1WIS
RH1 2JB G7LZM
RH1 2LA G8XOV
RH1 3AA G3YSX
RH1 3AA M5SB
RH1 3BH G7ODV
RH1 3BN G1YRR
RH1 3ER G4WGD
RH1 3EY G8UNO
RH1 3LT M7LSE
RH1 3PB 2E0UOJ
RH1 3PB M3UOJ
RH1 3PX G0KCM
RH1 4AS M6DPW
RH1 4AY G3NIQ
RH1 4DE G3TRC
RH1 4QT G7KGV
RH1 5BJ G4LJU
RH1 5BW M7HGF
RH1 5JB G7LUO
RH1 5LY M6BZY
RH1 6AW G8HMG
RH1 6GZ 2E0FLL
RH1 6GZ M7TAG
RH1 6LW M6AOB
RH1 6PB G1YXJ
RH1 6PB G4PFX
RH1 6PQ G8LAM
RH1 6TB M1APH
RH10 3AL G7OMM
RH10 3DW G3CTP
RH10 3EG G8MQF
RH10 3HX G6NPJ
RH10 3NG G4PBF
RH10 3PS G3VJM
RH10 3QS M6WPN
RH10 3QT G0KKS

RH10 3RG M0GZE
RH10 3TB 2E1HGT
RH10 3TG G0ORX
RH10 3TS G1AFX
RH10 3UQ 2E0GPA
RH10 3UQ M6GPA
RH10 3XG G4ANN
RH10 3XG G6RC
RH10 3XQ 2E0GVW
RH10 4HS M6JPR
RH10 4JB M6PCR
RH10 4JQ G0PYJ
RH10 4LR M3APQ
RH10 4LU M6RXL
RH10 4NG G3RMK
RH10 4UB G0GML
RH10 4UJ G3YTR
RH10 4UL G4TTY
RH10 4UP G0BKW
RH10 4XA G3VLH
RH10 4XY G6LEB
RH10 4YY G3NZP
RH10 5AB 2E0WJR
RH10 5BX M0IJP
RH10 5BZ M7FZB
RH10 5DQ G3ZON
RH10 5DW 2E1BMF
RH10 5DW G4UMJ
RH10 5LD G6FPH
RH10 6BG 2E0MZB
RH10 6BP G6CKZ
RH10 6DL M0HQM
RH10 6GJ 2E0UKW
RH10 6GJ M0WOT
RH10 6GJ M6WXX
RH10 6GJ M7TWO
RH10 6HQ G0VYN
RH10 6HQ G3WKI
RH10 6HW G8NGM
RH10 6HZ M6CSP
RH10 6JL 2E1AVT
RH10 6JS G4MKD
RH10 6JS G7OMI
RH10 6LX G0IOE
RH10 6LX G6XTG
RH10 6PB G6XAV
RH10 6PN G4SFP
RH10 6QF G8INS
RH10 6RE M7ATB
RH10 6RL M3GQM
RH10 6SE G3SYD
RH10 6TE M7ATN
RH10 6TH G3MGL
RH10 7AE G1AYU
RH10 7AJ M0JHV
RH10 7BS G4UBT
RH10 7DB M6EWQ
RH10 7DB M6EXW
RH10 7ED M0PMM
RH10 7FL G1XST
RH10 7JH M0VPG
RH10 7JR G3FRV
RH10 7RQ G0VIQ
RH10 7SQ G3ZSJ
RH10 7SW M0LPK
RH10 7TS 2E0EQJ
RH10 7TS M0KGW
RH10 7WF 2E0RGC
RH10 7WQ M6MRY
RH10 7XG G1BGH
RH10 7YW G7RUN
RH10 8AS M7EQC
RH10 8BZ G0FPI
RH10 8HA G8IZY
RH10 8JR G6GGW
RH10 8JX M6GYE
RH10 8NB G6DOW
RH107XP M6KXW
RH11 0BS G4ONJ
RH11 0HW G0EGT
RH11 0JJ G4ZXB
RH11 0NA G1GHY
RH11 0NA G3KAU
RH11 0QR M3XHZ
RH11 0RB G6OCM
RH11 0TJ M1PFS
RH11 0UF G6AUE
RH11 7BD G3IPP
RH11 7BW G8RRR
RH11 7EB M0CWT
RH11 7EG M7NKA
RH11 7JB G6SHQ
RH11 7JE G3UNS
RH11 7JP G0RGM
RH11 7PE G4LEG
RH11 7RF G4TZF
RH11 7RF M7DPX
RH11 7UH G3UUC
RH11 8BA M6ECM
RH11 8EU G7LYB
RH11 8EY 2E0BTO
RH11 8EY M3YFJ
RH11 8LQ G4PFW
RH11 8SQ G8AVZ
RH11 8SX 2E0ZLM
RH11 8SX M3RXO
RH11 8XJ G6YPM
RH11 9AN M6SXT
RH11 9AX G0EMJ

RH11 9BX M7RPU
RH11 9EG G6CKE
RH11 9EG G6EDF
RH11 9ES G6ZHO
RH11 9HN M6SIU
RH11 9HN M6TOV
RH11 9JH G0PVQ
RH11 9LH 2E0EJJ
RH11 9LH M6HNJ
RH11 9NS G1PLV
RH11 9NY G3ZNW
RH11 9QT G8PUH
RH11 9QZ G7VZQ
RH11 9RQ M0GLU
RH117LJ M7SCK
RH12 1DQ G0VPZ
RH12 1GR G4PTR
RH12 1LD G8YRF
RH12 1LF M3WOI
RH12 1NA G8GHL
RH12 1NU G8RLD
RH12 1PX M5DB
RH12 1RF G4CVN
RH12 1SB G7DMQ
RH12 1UB M6MPO
RH12 1UY M6ARD
RH12 2AF M6FLU
RH12 2DA G4CCA
RH12 2HH 2E0ZMT
RH12 2HH M0ZMT
RH12 2HH M3ZMT
RH12 2HU M6MON
RH12 2JX M6DLR
RH12 2LB 2E0NKC
RH12 2LB M3NKC
RH12 2PY G4JHI
RH12 2QL G4LLI
RH12 3AQ M3DKK
RH12 3DU G3TQY
RH12 3HD G3SWC
RH12 3LH G7EYL
RH12 3ND G4YPE
RH12 3QR 2E0TJT
RH12 3QR M3YZQ
RH12 3TT M7PGH
RH12 4AR G4TMC
RH12 4GR G6DMM
RH12 4GX G8FQS
RH12 4HB M7SRH
RH12 4HR G6XTJ
RH12 4JB G8LHS
RH12 4JB M6GQH
RH12 4JE G1CKF
RH12 4LN M6PLS
RH12 4RE G4YJA
RH12 4SH M3SYN
RH12 4TP M0LTA
RH12 4UF G0URO
RH12 5EU G1XFM
RH12 5EU G4MWD
RH12 5EU M6TZR
RH12 5EW G8RSK
RH12 5FR G6NQQ
RH12 5PJ G1ALU
RH12 5PN G7SRV
RH12 5PZ G7MQP
RH12 5WA G4KDR
RH12 5WD G8IBE
RH12 5XL G6FYU
RH12 5XW G3VQO
RH12 5XX 2E1IGA
RH121LS M6MTM
RH12EJ M7FTZ
RH13 0PJ G7AVF
RH13 0PR M7GET
RH13 0RQ 2E0CSJ
RH13 0TR G8BAD
RH13 5DB M6NXZ
RH13 5HB G0DJS
RH13 5HG G6SBI
RH13 5HH G4EUG
RH13 5JS 2E0BXW
RH13 5JS M6EAA
RH13 5JS M6SLD
RH13 5JS M6SPJ
RH13 5LH M6END
RH13 5ND G7UEC
RH13 5NZ G4TPO
RH13 5PE G1ODN
RH13 5RY 2E0OVI
RH13 5RY M0OVI
RH13 6AE G4MKW
RH13 6AJ 2E0WEK
RH13 6AJ M6WEK
RH13 6AL M6GPJ
RH13 6AX G3SIG
RH13 6AX G3ZBU
RH13 6AX M3EUR
RH13 6AX M3GCR
RH13 6BQ M0GRU
RH13 6BS M6HCU
RH13 6BS M6LPS
RH13 6DD G4NWW
RH13 6DG G6MDJ
RH13 6DW M3FTK
RH13 6EH M6BQV
RH13 6EJ G3WSC
RH13 6EJ G4PEO
RH13 6GH M7CCW

RH13 6LZ G8YLA
RH13 6LZ M0YLA
RH13 6ND G7HWM
RH13 8AZ G3XEI
RH13 8EF M6RJK
RH13 8EH G4JCY
RH13 8EQ G3ZQW
RH13 8ER G4TPW
RH13 8GD M6EXD
RH13 8HR 2E0BQU
RH13 8HR G3RAM
RH13 8HR M6CSL
RH13 8HZ G7DFV
RH13 8HZ M3DFV
RH13 8JB G8FBM
RH13 8JF G4OLZ
RH13 8JX 2E0PML
RH13 8JX M0VUE
RH13 8JX M6EGK
RH13 8JX M6NGY
RH13 8LF G6HYD
RH13 8LG G1VVY
RH13 8LG M7FPP
RH13 8LT G3WZT
RH13 8LT G4HRS
RH13 9BB G0LOC
RH13 9BB M6AUZ
RH13 9BD M7DWD
RH13 9BG G4BVP
RH13 9DQ G4VFC
RH13 9FT 2E0HEC
RH13 9GJ G8SVD
RH13 9HP G1MPW
RH13 9HZ 2E0AED
RH13 9HZ G4HMM
RH13 9HZ M6CJV
RH13 9LZ M7PWK
RH13 9TF G8TQJ
RH13 9XR G0TAO
RH13 9XX G6DID
RH13 9XY G4LRP
RH13 9XY G5RFL
RH14 0DY G3PUX
RH14 0PJ 2E0FTC
RH14 0PJ G4NUX
RH14 0PJ M6FSO
RH14 0TF G8SLU
RH14 9BH G3OGP
RH14 9EQ G0BYF
RH14 9GL G0ERF
RH14 9GS 2E1EBZ
RH14 9HB G0FQD
RH14 9NB G0UEY
RH14 9NG M6NCD
RH14 9NP G6ZDP
RH14 9PE M7FSJ
RH14 9PU G7MXT
RH14 9RJ M0HWH
RH14 9TU 2E0VCU
RH14 9TU M6PIP
RH15 0AE M6NAW
RH15 0BA M6KZX
RH15 0DE M1BQU
RH15 0EU G3YBM
RH15 0HB M0TPJ
RH15 0JN M1CSI
RH15 0LD M3TFS
RH15 0LJ M1EBU
RH15 0NB G3RMY
RH15 0ND G0WGP
RH15 0NF M6BRO
RH15 0NH M6JOR
RH15 0NJ M1CSE
RH15 0PH G6ZQA
RH15 0PS 2E0ZCP
RH15 0PS M6ZCP
RH15 0PT G0DDE
RH15 0PT G6DBX
RH15 0QA G6IHG
RH15 0QN G0UUP
RH15 0QU M5JWS
RH15 0QY M6LAP
RH15 0RP G1AZA
RH15 0RQ G0GNV
RH15 0RQ G1TDL
RH15 0RW G7BLD
RH15 0RW G7TPW
RH15 0RW M3HZM
RH15 0RW M3HZP
RH15 8AW M1CYL
RH15 8BD 2E1IAS
RH15 8BD G1DZD
RH15 8BL G0IDP
RH15 8BW G3XUP
RH15 8EU G4AKG
RH15 8HJ M3ZMB
RH15 8LE G4ULZ
RH15 8NP 2E0RJA
RH15 8PD M6JQQ
RH15 8QF G0KYX
RH15 8QW G7ZMS
RH15 8TF G3XUD
RH15 8TR G3XQM
RH15 8UR G4ORS
RH15 8UY M5BTB
RH15 9ED 2E0NCJ
RH15 9ED M0NCJ
RH15 9HA 2E0DRA
RH15 9HA M0IKD

RH15 9HA M6IKD
RH15 9HR G1PJM
RH15 9HR G1POD
RH15 9HZ G8KMP
RH15 9JA G4WEH
RH15 9JJ M7EOV
RH15 9PL 2E1BRD
RH15 9PL G7OIA
RH15 9PY G0RPL
RH15 9QW M7DJA
RH15 9RF M6HWJ
RH15 9RR M3ZGS
RH15 9SP G3VAK
RH15 9ST G0SQF
RH15 9SZ G7TMR
RH15 9TE M7MJV
RH15 9UL G8GJV
RH15 9XA G4ZCS
RH15 9XG G0THD
RH15 9XR G1WGL
RH158AN M7OBI
RH16 1EJ G7TNU
RH16 1HJ G4KKO
RH16 1NB G4DAY
RH16 1PH M1CTO
RH16 1QN M6STF
RH16 1QT M7GSP
RH16 1TF G3ZEK
RH16 1UZ 2E0DEE
RH16 1UZ 2E0LIV
RH16 1UZ G8ZOV
RH16 1UZ M3MZA
RH16 1UZ M3MZC
RH16 1UZ M3ZOU
RH16 2AB G3FSX
RH16 2DQ G4IOV
RH16 2JQ M7SXH
RH16 2LJ 2E0WVE
RH16 2LJ M0WVE
RH16 2LJ M6WVE
RH16 2PH M6AKK
RH16 2QG G4OKB
RH16 2QG M0WGS
RH16 2QJ M6OJT
RH16 3AX 2E0CBQ
RH16 3HW G3ASV
RH16 3JS G8MFM
RH16 3JU M7AEJ
RH16 3PE G3YTU
RH16 3QW G0AUI
RH16 3QX M7OZE
RH16 3RP 2E0EYS
RH16 3RP M0VAY
RH16 3RP M6IRB
RH16 3UE G0LFF
RH16 3UE G1CWP
RH16 3UE G7GCF
RH16 4DH 2E1FET
RH16 4EW G1VPC
RH16 4HL G7FHV
RH16 4JR G1SKI
RH16 4NP M0IRF
RH16 4NP M3WOW
RH164UL M7NEL
RH17 5AW G0XAV
RH17 5AW G3VKW
RH17 5AW M3HJE
RH17 5DZ G3WYN
RH17 5DZ G5RV
RH17 5HD 2E0ETI
RH17 5HD M6TLY
RH17 5HH G4JST
RH17 5PA G3LET
RH17 5QQ G6SOY
RH17 5QU 2E1DIJ
RH17 5QU 2E1DIJ
RH17 6AF G8VKQ
RH17 6DL G8OPH
RH17 6DL M7MEP
RH17 6DQ G7WBR
RH17 6JA G4ZXB
RH17 6NG G7IAS
RH17 6NJ G0HXF
RH17 6SR 2E0PMB
RH17 6SR M6PHT
RH17 6UE 2E1ALC
RH17 6UE M6YTU
RH17 6UL M6COE
RH17 7EQ M0DMX
RH17 7HE G4MLR
RH17 7LP M6GVD
RH17 7PG G1ACA
RH17 7QA G8KNC
RH17 7QD M7GAQ
RH17 7QS G4RBQ
RH17 7RA M6NIC
RH17 7RN M0JTJ
RH17 7RS M3YJL
RH17 7RW G8ZAT
RH18 5AF 2E0TNE
RH18 5AZ G0UPL
RH18 5BX M0HBM
RH18 5BX M6AGO
RH18 5GD G5DJW
RH19 1AY M7WEL
RH19 1DL 2E0LSS
RH19 1EE 2E1JGW
RH19 1JG G3VPS
RH19 1JR G3KOA

RH19 1SG G7KBR
RH19 1TA 2E0BQV
RH19 1TA M0GQD
RH19 1TA M3UQH
RH19 2DD G3YQW
RH19 2PS G8FBA
RH19 2QY 2E0LRZ
RH19 2QY M0LMG
RH19 2QY M7EFU
RH19 2TS M7GIM
RH19 3HP M6SXB
RH19 3NE 2E1PAL
RH19 3QE G0TSQ
RH19 3QL M1ENX
RH19 3QP M7FRW
RH19 3RB M3NGY
RH19 3SE G1AML
RH19 3UX M3WIC
RH19 3XF G0RCF
RH19 3XP G0RHG
RH19 4BZ G1YXT
RH19 4EA G4TRG
RH19 4NY 2E0AYT
RH19 4TF G4PFU
RH191DQ 2E0XDZ
RH191DQ M6XDZ
RH2 0QA G6YRV
RH2 0RE G8PZI
RH2 7BS G7CVM
RH2 7DY M0KDX
RH2 7HJ G4AVE
RH2 7NQ M7GEZ
RH2 7PS 2E0PMO
RH2 7PS M7RUE
RH2 7PY 2E0KRH
RH2 7PY M0KRH
RH2 7PY M6OWI
RH2 8DQ G0VGT
RH2 8DS G8ZRV
RH2 8EL M3WOQ
RH2 8NE G6VMI
RH2 8QJ G4WBF
RH2 8QY M7FME
RH2 9DG G7OBF
RH2 9HA M3SGJ
RH20 1AH G6OSO
RH20 1AW 2E1BSC
RH20 1AW G7PTX
RH20 1BZ G0INQ
RH20 1HS G1EAE
RH20 1LA G1PMA
RH20 1NX G7TIU
RH20 1NX M0SSX
RH20 1PQ G3ZTU
RH20 1PS M3IYX
RH20 2AT G0LFH
RH20 2DD M6ANE
RH20 2DX 2E0UMD
RH20 2DX M6UMD
RH20 2EE 2E0BSF
RH20 2EE M6MST
RH20 2JJ M3PTX
RH20 2JJ M7USB
RH20 2JL M6FEP
RH20 2PR G0HWS
RH20 2PZ 2E0YZL
RH20 2PZ M0TCD
RH20 2PZ M6PZL
RH20 2QY G6PGQ
RH20 3BD G8IPF
RH20 3GL M6RPX
RH20 3GU M0EPX
RH20 3GU M1LTS
RH20 3HG M1AIM
RH20 3HJ 2E0JGD
RH20 3HZ M6GNM
RH20 3JW G3XJN
RH20 4AB M7KOW
RH20 4AR M0BAH
RH20 4AU 2E1DZH
RH20 4AU G4WGA
RH20 4BS G4LCU
RH20 4JH M7EKT
RH20 4JU M6DMP
RH20 4LL G6AAR
RH20 4LT G4JCA
RH20 4LU G3POQ
RH20 4LW G0AGB
RH20 4QL G6LBE
RH20 4QL G6YCZ
RH20 4QS M7GCL
RH20 4QX G7TMU
RH20 4RF G4YPG
RH4 1HD M1DDF
RH4 1LP G0GNA
RH4 1NA M6SLR
RH4 2DG G0GQY
RH4 2HT G4YFK
RH4 2JE M3WLX
RH4 2NR G4VUW
RH4 3BZ G6EQP
RH4 3BZ M3EQP
RH4 3DS G3YWX
RH4 3DZ G6DTH
RH4 3HX G3AFB
RH4 3JH M0DSX
RH5 4AD G3RDH
RH5 4AE M6FWG
RH5 4DG 2E0KCV

RH5 4DG M0KCV
RH5 4DG M7AMC
RH5 4EY 2E0NNB
RH5 4EY M0NNB
RH5 4EY M6NNB
RH5 4JW G7OAJ
RH5 4NL 2E0NPE
RH5 4NN G4FTQ
RH5 4PE M6DCD
RH5 4PP G6YPF
RH5 4QB G4VTC
RH5 4QH G1THW
RH5 4RH G1VNU
RH5 4TN 2E0DKE
RH5 4TN M6FCM
RH5 5AQ G4CMU
RH5 5BS G1ABQ
RH5 5BX G4OBT
RH5 5DZ G0RKS
RH5 5HT G6KNE
RH5 5JB G1XSM
RH5 5JB M6ATH
RH5 5LL G6OBJ
RH5 5SG M0OTT
RH5 6QW G3MNB
RH6 0BL G8OLP
RH6 0DR M1AZB
RH6 0HU 2E0IBN
RH6 7BX M3JXV
RH6 7JF G1HLY
RH6 7NH M3TAG
RH6 8BS G8AMU
RH6 8DR M1EAL
RH6 8JG G0SOG
RH6 8JG G7KYW
RH6 8JR G3JHP
RH6 8LF G6DCH
RH6 8QT 2E0HAP
RH6 8QT M6AVO
RH6 8RE G6VKX
RH6 8RX G4EIV
RH6 9BY M6RGN
RH6 9EA G6MGZ
RH6 9GT G6TUG
RH6 9LY 2E0FEM
RH6 9LY 2E0KDT
RH6 9LY M0KPD
RH6 9LY M6FEM
RH6 9LY M6KVT
RH6 9NA G3ZIY
RH6 9QA G4PFT
RH6 9QN G1EAM
RH6 9QP G0HUF
RH6 9RB M6SJO
RH6 9RE M7RPA
RH6 9SY G3OKS
RH6 9TX G4FBI
RH6 9UE M0MIM
RH67JW M7NRH
RH67LN M6BJT
RH7 6AW M0SKO
RH7 6BT M6UTT
RH7 6EY G8OYY
RH7 6JX 2E0JXN
RH7 6JX M3JXN
RH7 6LL G3LHZ
RH7 6NQ G7TWJ
RH7 6PS 2E0TZZ
RH7 6PS M0TZZ
RH7 6PS M6TZZ
RH7 6RL G3IOM
RH76DG M7HTZ
RH76PB G0FTU
RH76TA M7EEG
RH8 0EW G7VPU
RH8 0JA M3WNV
RH8 0JE G3IKB
RH8 0PD 2E0IPW
RH8 9AU 2E0ALW
RH8 9AU G7BKN
RH8 9AU M5JWS
RH8 9DJ M6OHN
RH8 9LT 2E0GPD
RH8 9LT M6OBZ
RH8 9PF M0JEP
RH9 8BF 2E0PPJ
RH9 8ER G0ALE
RH9 8ER G3RQZ
RH9 8LD M3JGJ
RH9 8LD M3ZUS
RH9 8NX M7BKL
RM1 2AD M3NHS
RM1 2NP G6YCF
RM1 2SR G8JBV
RM1 4AU M0VIN
RM1 4BZ G6BTX
RM1 4HD G4FQF
RM1 4LA G3XLR
RM1 4SP G6TGM
RM1 4XR G8TQZ
RM10 7BT G0RFL
RM10 7EX G4SHN
RM10 7XR M6LDR
RM10 8AS 2E0YOL
RM10 8AS M0YOL
RM10 8AS M3TBV
RM10 8BS G0GGH
RM10 8BS G1JQR
RM10 8BT G1YKI

RM10 8DE G0WXE
RM10 8ES G1WUH
RM10 8TE G0HSX
RM10 8UU M7MCR
RM10 9ND M7PEJ
RM10 9PP M1GDH
RM10 9QD M0GRT
RM107RE M7TBM
RM11 2AJ G7IYN
RM11 2BA G0BSF
RM11 2BS G0CBU
RM11 2NF M3DSA
RM11 2RJ M6UWU
RM11 2ST 2E1RBA
RM11 3EN G0RIQ
RM11 3LD 2E0LBL
RM11 3LD M0LBY
RM11 3PD G0PIA
RM11 3PY G3JHI
RM11 3PY M0TAZ
RM11 3QD G1VSH
RM11 3QH G8WUO
RM11 3SG G0TJN
RM12 4AF M7RNX
RM12 4EY G3SFK
RM12 4JZ M0MAC
RM12 4JZ M1CUC
RM12 4LL G0IAP
RM12 4RN M7ETM
RM12 4SE G6MVS
RM12 4YF G4NHR
RM12 5BL G0AJH
RM12 5DR G7GJY
RM12 5EU G6BEL
RM12 5HH 2E0AWI
RM12 5LS G8PMR
RM12 5QS G8UZW
RM12 6AT G4MXF
RM12 6TF G4EEF
RM13 7AS 2E0UKD
RM13 7AS M3UKD
RM13 7EB M7FXK
RM13 7EJ G4RZZ
RM13 7PD 2E0WJI
RM13 7PH G8ZDT
RM13 8AA M6GAI
RM13 8AJ G4JRB
RM13 8BB G0IEO
RM13 8EL M6IFQ
RM13 8GF 2E0OYY
RM13 8GF G5OYY
RM13 8GF M6OYY
RM13 8LW M6JYO
RM13 8ND G8NWI
RM13 9DX G4YBI
RM13 9LF M3NYI
RM13 9LW G0PBN
RM13 9LW G1BTF
RM13 9RL M6AMI
RM13 9UU G8FJG
RM14 1AL G6TRX
RM14 1AT G0KSJ
RM14 1HN G0IQN
RM14 1HP G4AZX
RM14 1HT M1CZO
RM14 1JD M7DKL
RM14 1NA G3RJI
RM14 1NB G0DOM
RM14 1NW M6YMF
RM14 1PZ 2E0AVQ
RM14 1PZ M3AQF
RM14 1XR G8XQZ
RM14 2EZ 2E1EVK
RM14 2EZ M6HDU
RM14 2HJ G8GGO
RM14 2HR 2E0MES
RM14 2HR M0SEM
RM14 2HR M3ETS
RM14 2JH G4UNS
RM14 2YX M3DFU
RM14 3AH G7GES
RM14 3AU G7DXX
RM14 3AZ G8VVG
RM14 3EA G4XZC
RM14 3YT G4LQX
RM15 4NR M0CJR
RM15 5DS M6FZN
RM15 5ES M6PST
RM15 5ET 2E0JGG
RM15 6HP 2E0BYN
RM15 6HP M6ILP
RM15 6QE M0JLI
RM15 6TS G1DXH
RM15 6XE G1TJT
RM16 2GB G3NEG
RM16 2GB G3SCT
RM16 2GB G4HCK
RM16 2GB G4HVB
RM16 2HD M7LAT
RM16 2QL G1WEF
RM16 2QS 2E0REZ
RM16 2QS M6RHU
RM16 2RS G1ZAY
RM16 3AU G6TYF
RM16 3DW M0MOR
RM16 3LT G0DEK
RM16 4BS G7HMA
RM16 4DD M7KMO
RM16 4HL 2E0KGJ

RM16 4JB G8WTN
RM16 4JF M6TTM
RM16 4RR G6NVD
RM16 5UR G8BCJ
RM16 6ET M5BAZ
RM16 6LT 2E0BWT
RM16 6LT M6AEW
RM16 6NP M6EIR
RM16 6NS M6JYB
RM16 6QT M6KBA
RM16 6QT M6MES
RM16 6RE M6GNJ
RM16 6RE M6GNS
RM162XT M6IYS
RM17 5AB M1MOG
RM17 5HA G4EPM
RM17 5LW M1DPE
RM17 5QR G0BKL
RM17 5RG G3KMD
RM17 5RG G6EFE
RM17 5RG G6OBO
RM17 5RG M1YOW
RM17 5RH 2E0ELI
RM17 5RH M0WJL
RM17 5RH M6WJL
RM17 5SF G4LUT
RM17 5UW G8VZJ
RM17 5YN G1NOR
RM17 6AZ M6LNV
RM17 6LD M0TIL
RM17 6LD M1DUC
RM17 6PX M7AEL
RM17 6RU M6URS
RM17 6SG M6WRO
RM17 6UN M7FKQ
RM18 8BA G3PHD
RM18 8DT G1EHM
RM18 8HQ G7VQL
RM18 8HZ G4OVG
RM18 8XL M7GIQ
RM18 8XP 2E0PDM
RM18 8XP M6MTR
RM19 1LE G8TND
RM19 1PN G3EME
RM2 5AR M3ZTW
RM2 5PP G3ZQY
RM2 5PS M6EDO
RM2 5XJ G4XXT
RM2 6AP G3TPJ
RM2 6DX G4JDT
RM2 6LD G4GBW
RM20 3DA M6GDJ
RM20 4AA M6IYV
RM20 4AB G1SID
RM20 4YD M6BKV
RM20 4YR G6LCU
RM3 0AW M6HKU
RM3 0BQ G8VXR
RM3 0DL G4TTN
RM3 0EB G0BCZ
RM3 0EJ G3SKV
RM3 0JD M1JEC
RM3 0RH M3BFJ
RM3 0YR M7DVS
RM3 0ZA M3TXM
RM3 7DX G0CBT
RM3 7DX G3XBF
RM3 7EQ 2E0SLH
RM3 7EQ M0HXE
RM3 7LB M0BOB
RM3 7PJ 2E0RHL
RM3 7PJ M6SOO
RM3 7QL M3CGJ
RM3 7SP M6WCE
RM3 7UB G1WZK
RM3 7YX G1PZS
RM3 8DB M6OOX
RM3 8HJ G6MDN
RM3 8HN G7VVL
RM3 8HN G8TPP
RM3 8HQ G6WKI
RM3 8JP M7IGC
RM3 8JZ M6MTC
RM3 8RU G7BXT
RM3 9DE M3MIX
RM3 9LA M1BFX
RM3 9NA G1CSN
RM3 9NL G4YOT
RM3 9PT 2E0PSN
RM3 9PT M6PSN
RM4 1AX G6SVV
RM4 1DR M6LRL
RM4 1DT G4XVH
RM4 1NH G6DIM
RM4 1RR G1VQI
RM5 2HF G6EBO
RM5 2SL M0PAX
RM5 3AY 2E0EXG
RM5 3AY M6OXJ
RM5 3JX M7WEA
RM5 3ND G0OVC
RM5 3QA M6CNP
RM5 3RX G0BLO
RM5 3SX G8OPO
RM6 4BF M6ING
RM6 4EB G0FEK
RM6 4EB G1HKS
RM6 4NU M1BDY
RM6 4SL G4YGL

RM6 5AT M3EDU
RM6 5EJ G7VVK
RM6 5NR M3MRO
RM6 5QL 2E0OKC
RM6 5QP M0CMH
RM6 6PY G1NIV
RM6 6RX G0PIS
RM7 0EX M6KUY
RM7 0JP M3ZNL
RM7 0LA G8YIN
RM7 7LJ 2E0ROV
RM7 7LJ M0MCP
RM7 7LJ M6ROV
RM7 8AZ M6ZXB
RM7 8BD G7WJE
RM7 8BE 2E0TMA
RM7 8BE M6RMN
RM7 8HD G0GFZ
RM8 1DR 2E0MSM
RM8 1DR M6AQE
RM8 1DS G7TAX
RM8 1JY G0TJE
RM8 1XD 2E0DPD
RM8 1XD M6FCR
RM8 2DW M7EJY
RM8 2EW 2E0TPN
RM8 2LR G8SLB
RM8 2QL M0BLY
RM8 2RA G7UYT
RM8 2RA M0HEM
RM8 2TU G4CUQ
RM8 3PA M7DKE
RM8 3ST G6AIU
RM9 4ET G7CXT
RM9 4ST G4HOI
RM9 4TJ M3SJH
RM9 5LT M7EAO
RM9 5RS M0DPV
RM9 5RS M3ARB
RM9 5XH G7UVW
RM9 6EA M7RUI
RM9 6JL M7BGG
RM9 6LD 2E0WWN
RM9 6LD G0TOD
RM9 6LD M3WWN
RM9 6NH 2E0IPK
RM9 6XU M7BID
S1 3JD G3UOS
S10 1EJ G1MAS
S10 1HT G0IYV
S10 1LS G8PNM
S10 1PD G0BFM
S10 1SL M3WSQ
S10 1WQ G8YXI
S10 2BA 2E0EJQ
S10 2BA M6HUX
S10 3BU G4SGF
S10 3PT G4SOM
S10 3RF G0BSX
S10 3RZ G6HYI
S10 3TR G7JAF
S10 3TR M1GWA
S10 4DN G4CUI
S10 4ED G6YFY
S10 4EU G3YBA
S10 4EX M1AGR
S10 4GF G3WDM
S10 4LF G7SSB
S10 4LL G8AGN
S10 5NX G7LQO
S10 5PT M7EBC
S10 5RE G1JVN
S10 5RE G1JVO
S10 5TT M7GIY
S105FG M7GAU
S11 7GG 2E0SSY
S11 7GG M6MHO
S11 7LX G0IIF
S11 8UA M1AAC
S11 8UA M7LJH
S11 9AA G0PFI
S11 9BR G0HHA
S11 9BR G0LNV
S11 9BU M0AXE
S11 9DE M6UAJ
S11 9DR G0WLR
S11 9LH 2E0ZSB
S11 9LH M3ZSB
S11 9LQ G4IJB
S11 9RA G0HGI
S11 9RP G4HIA
S11 9SP G3RVI
S11 9SP M3NSO
S12 2AY G3UXM
S12 2BL M5ASK
S12 2BU G4VLZ
S12 2EJ G0BJP
S12 2FN G1HFK
S12 2FZ G7PAN
S12 2GL M6NLF
S12 2GN M0WWE
S12 2JH 2E0PCG
S12 2JP M3ZBF
S12 2NB 2E0GNS
S12 2NB M3HNU
S12 2SZ M6VGC
S12 3AQ G3PYL
S12 3AQ M6EEZ
S12 3DT M6MPM
S12 3HB G1GPM
S12 3HD G6GDH
S12 3HN M6SCH
S12 3LP 2E0KXD
S12 3LP M3KXD
S12 3LR G7GHH
S12 4ER 2E0SPD
S12 4ER M0SZD
S12 4ER M6AXD
S12 4HG 2E0YTB
S12 4HG M6YTB
S12 4HU 2E0TCX
S12 4HU M6HBH
S12 4HU M6IGW
S12 4JB M0CBT
S12 4LF G4GPF
S12 4ND M0PBO
S12 4QL G0JXF
S12 4SE M3SZT
S12 4XE G0BQW
S12 4XH G7UEX
S12 4YF G0NRM
S13 7AD M7MLY
S13 8DQ M7LOT
S13 8EH M3DYR
S13 8FA M6OFG
S13 8FA M6OFJ
S13 8PR G4GDB
S13 8RE G1JUY
S13 8TH G0HYS
S13 9HH M0TDM
S13 9HX G0VBN
S14 1AE G4DCF
S14 1GD M3DDZ
S14 1RJ G1RZJ
S14 1SS M7PRG
S17 3DN G4UWP
S17 3EG G8BLP
S17 3EZ G0VNE
S17 3GZ G4AYO
S17 3LL M6AYM
S17 3NB M0BAK
S17 3PH G8ION
S17 4DT G0TVW
S17 4FT M6DEZ
S17 4FP G4KPU
S17 4QX M6RWS
S17 4RJ G0OUK
S18 1ND 2E1HLO
S18 1ND G0LUM
S18 1ND M3CAZ
S18 1PW 2E0BMD
S18 1PW M0UCD
S18 1RF G6CTA
S18 1RJ G0XVL
S18 1RP G0UGJ
S18 1SD G4ULN
S18 1UQ G3KNJ
S18 1UQ G3VKK
S18 1UZ G4JUH
S18 1WB M7BHL
S18 1WH G3YON
S18 1XX G4HOP
S18 1YL G4KCU
S18 2EA 2E0HAV
S18 2EA M0ZEL
S18 2EA M6DQZ
S18 2EH G7BUF
S18 2EJ G1EQL
S18 2EL G4PXF
S18 2ER G0MVC
S18 2FF G4KXW
S18 2GG G8RFW
S18 2GQ M6DPF
S18 2HQ G4PPR
S18 2LE G1DFW
S18 3BQ G1KPZ
S18 3BY M3EVJ
S18 4DE M6DDI
S18 7WA G8UHM
S18 7WX 2E1BUV
S18 8QY G8RFP
S18 8YW G0WYN
S2 1GH M3XAH
S2 1NJ M7NLA
S2 1RG G0JJV
S2 1UU M3DFS
S2 2AG 2E0WST
S2 2AG M0VGC
S2 2AG M6WST
S2 2AR G8YQU
S2 2EA 2E0IUK
S2 2EA M0IUK
S2 2EA M6IUK
S2 2GB G1LWY
S2 2GP 2E0HYF
S2 2GP M7EQI
S2 2NZ G1ASG
S2 3AE M6KJP
S2 3BJ M3NGZ
S2 3EH G7SWQ
S2 3EH M6WNC
S2 3LB G1YED
S2 3LQ G1DBZ
S2 3NF 2E0KSH
S2 3NF G5YL
S2 3NF M6KSH
S2 3QJ 2E0AJT
S2 3SY M6RUK
S2 4AH M6ICR
S2 4BP M6JDQ
S2 4QU 2E0NCN
S2 5FD 2E0ZMZ
S2 5FD M6MFJ
S2 5FQ 2E0SKL
S2 5FQ M6LSK
S2 5QW G4MWL
S20 1AD G7KDQ
S20 1AW G4RQF
S20 1AY 2E0PCF
S20 1AY M6PCF
S20 1DA M7IWN
S20 1ET M7TIN
S20 1FW M0JCS
S20 1FW M6ICO
S20 2DL 2E0UAW
S20 2DL G1AW
S20 2DL M6UAW
S20 2PZ M0MEF
S20 4GB 2E0UKX
S20 4HL G6XLC
S20 6QJ M6JUB
S20 6SD G6YAI
S20 6SG G0FZD
S20 6SU M6BTE
S20 7JB M7YOY
S20 8EB M7IDD
S20 8EF G0OUG
S20 8JG 2E0DBO
S20 8JG M6IDX
S20 8LE 2E0USA
S21 1AW M6IHM
S21 1BU M6TSP
S21 1DA 2E0FQP
S21 1DA M7DDA
S21 1FX 2E0MJG
S21 1FX M0NCN
S21 1FX M3NCN
S21 1GA M3TUW
S21 1GR M3HGT
S21 1TF G0PBH
S21 1TG 2E0YMV
S21 1TG M0AFW
S21 1TG M7PVW
S21 2BX G3HQG
S21 2DT 2E0YZA
S21 3UL M6ZCB
S21 4BJ M7ECW
S21 4BW M3NXC
S21 4EG G3SXT
S21 4JX G7SCP
S21 5RD G7TZQ
S21 5RS G6XUX
S22AE M6URH
S25 1YT G8WVH
S25 2DF M0CRT
S25 2LJ G7CKE
S25 2NY G6XOX
S25 2RP G4NUB
S25 2SW G3YYE
S25 2SX M0YZV
S25 3RG 2E0EHP
S25 3RG M0JWJ
S25 3RG M6GZW
S25 3RW M7GHI
S25 4AG G7GUG
S25 4BH 2E0TRU
S25 4BH M6TRU
S25 4BR 2E0MKE
S25 4BR M6MKE
S25 4BW G7FTM
S25 4DQ G0DZX
S25 4EP 2E0USV
S25 4EP M0USV
S25 4EP M6USV
S25 4EY 2E0FVO
S25 4EY M7TVT
S25 4FG G8EMH
S25 4FL G4SRX
S25 4GD G6YZR
S25 4HB G4GQS
S25 4HB M3IXE
S25 4JP G4RMX
S25 4JY G7GFR
S25 4JZ M7FWU
S25 5BA G6TNI
S25 5BA G7IOI
S25 5DR G0JZL
S25 5EG G1DPI
S25 5EG M0CKX
S25 5GL G7BCI
S25 5JG G1HEW
S254BH M7MMT
S254EY M0KNW
S26 1HJ G7TBF
S26 1HJ G7TDR
S26 1HL G3NEO
S26 1JG G1YGP
S26 1JN G3HKF
S26 1JN G7HAR
S26 1JQ G1WRD
S26 1JS G7JRC
S26 2AA 2E0DCA
S26 2AA M3UKB
S26 2AD G7PFI
S26 2AS G0CCN
S26 2AT M0ADO
S26 2BR G4YSG
S26 2DH M0TWG
S26 2EP G4ZDD
S26 2FJ G6BCM
S26 2GL M3VPP
S26 2GR 2E0GBP
S26 3RB M0TRV
S26 3RE G7IBD
S26 3XB M7FTG
S26 3XS G0DCF
S26 4NQ 2E0CFK
S26 4NQ M6PER
S26 4RQ M6DNS
S26 4SD G4RNA
S26 4SD G6BNT
S26 4TA G7LQN
S26 4TH M6PJB
S26 4TL G1UDE
S26 4UP 2E0PFO
S26 4UP M0PFO
S26 4UP M6PFO
S26 4UY M1CLW
S26 4WB G0BPM
S26 5HL M6JQC
S26 5LZ G8OCE
S26 5NY M7JEN
S26 5NZ 2E0FKB
S26 5NZ M6OME
S26 5PH M3KXS
S26 5QS M0ONI
S26 5QT M3TVO
S26 5RE G5TRS
S26 5RF M6JPV
S26 5RG M1TVR
S26 5RG M3HIT
S26 5RG M3MTR
S26 5RU G0SVX
S26 6RA M0CUI
S26 6SJ 2E0XCA
S26 6SJ 2E0ZPA
S26 6SJ G4SKM
S26 6SJ M0PJA
S26 6SJ M6PJA
S26 6SJ M6ZCA
S3 7AH 2E0OKH
S3 7AH M6HOM
S3 7ET M6SDP
S3 7JH M6CGC
S3 7NL G8PLI
S3 7PB G0FTR
S3 8EP M6OTI
S3 8PQ M6HPH
S3 9JF G4MRB
S3 9JF G6DIZ
S32 1AS G4KWE
S32 1DT M3VWG
S32 1DU G0WEX
S32 5QB G0NOU
S32 5QX M1CRP
S33 8UA G7ODR
S33 8WB G6HBZ
S33 9JQ M6SSR
S35 0AF G0FLI
S35 0DD M6TIW
S35 0DQ M6RKS
S35 0EA G7BPQ
S35 0GD G1OKB
S35 0HR G0CCM
S35 1AH G0KNL
S35 1AL M0AZB
S35 1EA G6ZWL
S35 1QG M1EOU
S35 1SQ M6HFG
S35 1SY 2E0CVX
S35 1TX M0BVN
S35 1UR 2E1CJD
S35 2PS M7AQL
S35 2TD G0UAC
S35 2ZT G1YLG
S35 3GU 2E1BJD
S35 3HS M0GCC
S35 3HZ G0VUM
S35 3LP M6KVX
S35 3PA G0LQC
S35 3XS M7ODZ
S35 4HF G6OFV
S35 4NS M0CBN
S35 4NT M0DEI
S35 4NT M0CKS
S35 7DH M6SFM
S35 7DQ G3LZI
S35 8PE G4CTE
S35 9YW M6HBD
S351YS 2E0ZSJ
S351YS M0ZSJ
S351YS M3ZSJ
S36 1AY G2CFC
S36 1BZ 2E0RGS
S36 1BZ M0RBQ
S36 1BZ M6AOV
S36 1ED 2E0LMG
S36 1EL M7GMN
S36 2BB G3WXI
S36 2BB G8AGQ
S36 2NF 2E0HVH
S36 2NF 2E0LNO
S36 2NF M6MPV
S36 2NF M7HVH
S36 2NF M7LNO
S36 2NH G8UCN
S36 2NP G0TKF
S36 2NR G3KJO
S36 2PT G7DGE
S36 2QX M0IBY
S36 2QX M0TGW
S36 2QX M3VCZ
S36 3SG 2E0TKF
S36 3SG M3UCY
S36 3SG M3UEL
S36 3TF G1YKL
S36 3TN G0UNY
S36 3ST M6OQG
S36 4GW G7MXS
S36 6AF G7MZJ
S36 6BH G0EWD
S36 6DN M7JBY
S36 6EA G7MZK
S36 6EH M6MIN
S36 6HE M7CVX
S36 8HP G0BVK
S36 8HP G4MWF
S36 8WW G4VRT
S36 8ZZ M1AKT
S36 9DB G4DFS
S36 9ED M6VMJ
S36 9EF M6ZXL
S36 9NE G6ZTD
S36 9QN M7BOA
S36 9RX M3LGM
S4 7DD M3BUH
S4 7DE G4DJD
S4 8AA 2E0TWS
S4 8AA M0TWS
S4 8JH G0DVS
S40 1DG G0UUC
S40 1HQ M0EME
S40 1HU M0OND
S40 1HZ G4LNG
S40 1NJ 2E1CYM
S40 1NJ M0CCG
S40 1NJ M6NGR
S40 2DF M7KDZ
S40 2HD 2E1EAZ
S40 2JJ M0DVR
S40 2LT G7SFF
S40 2SD M7AXY
S40 2SX M3ZQV
S40 2UH G8TFU
S40 2UR M3XQG
S40 3AS 2E0ODN
S40 3AS 2E0TNZ
S40 3AS G4SEB
S40 3AS M7SBB
S40 3AS M7THO
S40 3BL M0GXO
S40 3BT G3ZLF
S40 3BX 2E0MAZ
S40 3BX M0MAZ
S40 3BX M3XIO
S40 3DF G4BFR
S40 3DR M3OOP
S40 3EY G3GLX
S40 3HF G4NKI
S40 3HG G3OOP
S40 3HT G7KUM
S40 3LW G1UGL
S40 3NN G8OKI
S40 3NT G6DPE
S40 3PX M7JKO
S40 3QG 2E0BNK
S40 3QG M0ZIG
S40 3QG M3SRQ
S40 4AN 2E0HJY
S40 4AN M0OAL
S40 4AN M0PLW
S40 4AN M7PPL
S40 4DB G8YTF
S40 4DB M0VAA
S40 4EA G4GSD
S40 4ES G6NZA
S40 4EZ G6CVY
S40 4HN M0APL
S40 4JT M6FBP
S40 4LE G4BBI
S40 4LG M0OKT
S40 4NE M3SMN
S40 4NW M0EJF
S40 4PQ 2E0STA
S40 4QB M7DIF
S40 4RJ 2E0DXK
S40 4RJ M6JWG
S40 4RS M3GAE
S40 4SJ M6MBO
S40 4SQ G1ZWY
S40 4TF G0RDF
S40 4TU G7PUP
S40 4UT M3JCU
S40 4XD G4VHX
S40 4XJ G0EWI
S402FF M7FOI
S404HG M7DTD
S41 0BB M3VJI
S41 0BD M3LGF
S41 0BS M3XAN
S41 0EE M0JAV
S41 0JP M6MPV
S41 0NJ G0UEH
S41 0NN G7HIQ
S41 0RX 2E0PWG
S41 0RX M6HIM
S41 0SU G4HQH
S41 0XZ G4SBC
S41 7HA G0TRN
S41 7JF M6NDO
S41 7JZ M6GOB
S41 7JZ M6PAX
S41 7JZ M6SSP
S41 7QE M0RRR
S41 8AU M3OHL
S41 8BN M6KZE
S41 8DG M6MYY
S41 8LZ M0DRE
S41 8NW M6MJD
S41 8NX G0RTQ
S41 8NX M3SQH
S41 8PE G0LJI
S41 8PE M6SSA
S41 8RA 2E0GRR
S41 8RA G0BJL
S41 8RA G3UNK
S41 8RA G8BYB
S41 8SF G6FTH
S41 9DA G2AS
S41 9DA G3PHO
S41 9DA G5TO
S41 9HP M7CTN
S41 9JB M3YZK
S41 9JU G6ZOT
S41 9LR 2E0TIF
S41 9LR M6TIF
S41 9PA G6NZO
S42 5AD M6EYG
S42 5GA G1CBY
S42 5HH G6SZB
S42 5HH G8NSD
S42 5HN M3HUB
S42 5HY 2E0CAK
S42 5HY M6ARM
S42 5JA M0FOX
S42 5JP G0JOM
S42 5NS G6URK
S42 5NT G6PRP
S42 5PU 2E0MKV
S42 5PU M6MKV
S42 5PW M6NLL
S42 5PX M6VEC
S42 5UL G4NXS
S42 5UL G6XGJ
S42 6BE M3ESS
S42 6BE M3LBM
S42 6BE M3NST
S42 6BG M3SAA
S42 6FY M3CNC
S42 6HG G3WAM
S42 6HS G4XFF
S42 6JD G3CTZ
S42 6JF G4ROP
S42 6NP 2E0GUN
S42 6NP M0MTX
S42 6NP M3ORL
S42 6NR M3YJN
S42 6PL 2E0WNT
S42 6PL M3WNT
S42 6RX G4VSR
S42 6SP G3YBO
S42 6SP G6SVH
S42 6SP M3SVH
S42 6UR G4CVO
S42 6UU M3AJV
S42 6XR M3UXH
S42 7AN G1WAB
S42 7AN G6PBW
S42 7AN M3JWW
S42 7DR G6DQA
S42 7EA M3EVR
S42 7JQ G1FTU
S42 7LR G3RLL
S42 7LT 2E0URA
S42 7LT M7UWL
S42 7NE M0CRW
S42 7NH G6OYV
S426BD M3UXU
S43 1AZ M7UCA
S43 1BQ 2E0YNI
S43 1HB 2E0UCV
S43 1HB M3UCV
S43 1HG M3WAV
S43 1JA G8YMT
S43 1JU 2E0MTN
S43 1JU M0VTR
S43 1JU M3BPN
S43 1LJ G4PBC
S43 1LJ M0APH
S43 1NU M3RGG
S43 2AP G3ZSZ
S43 2BG M3CKM
S43 2BG M3CKO
S43 2BJ M3MIJ
S43 2DF M6REF
S43 2DZ G4RIQ
S43 2HN M6RUP
S43 2LH M1BVI
S43 2NR M3RNY
S43 2NW 2E0GRY
S43 2NW M0GRY
S43 3AG G7SMC
S43 3DH G0VRH
S43 3HJ G7VND
S43 3HS M0IKE
S43 3HS M0JOY
S43 3HU 2E0MCS
S43 3HU 2E0VRB
S43 3HU M6LSA
S43 3HU M6UPS
S43 3HU M6VKB
S43 3NF 2E0SYM
S43 3NJ 2E0DBM
S43 3NJ M3UDZ
S43 3PQ M3ICO
S43 3PQ M3IXT
S43 3PY G7SLY
S43 3QB G4OQU
S43 3RA M6SAS
S43 3RH 2E1XRM
S43 3SY M3FVJ
S43 3TE G5RAH
S43 3UH G3BNF
S43 3UY 2E0FKT
S43 3XH M3UHV
S43 3XH M6SLI
S43 4AP G2FVL
S43 4AQ M3KKB
S43 4BD G4VDB
S43 4BD M7GKB
S43 4BS G0KUW
S43 4DH 2E0FIF
S43 4DH M0MKH
S43 4EG G6DAO
S43 4EH G3WEU
S43 4HE G8YJL
S43 4HE M7POG
S43 4ND 2E0TZG
S43 4ND M7TZG
S43 4NG M3LBN
S43 4NJ G4RTA
S43 4PF M3KQD
S43 4PG M3VUP
S43 4PQ M3DLZ
S43 4PQ M3LDJ
S43 4QJ M6DVE
S43 4QP 2E1FDF
S43 4RF G6YFH
S43 4RS G1SLE
S43 4RZ M3ZIO
S43 4SA G0RUV
S43 4SR G4BJG
S43 4ST G4RSS
S43 4SX G0EOJ
S43 4TR G8KDM
S43 4WT 2E0SZL
S43 4WT M0WGY
S43 4WT M7DDC
S44 5AE G3GGU
S44 5AL M6DEK
S44 5AT M6LWE
S44 5BA M3VDA
S44 5BP G7LLN
S44 5ET G0MBS
S44 5HE 2E0FJV
S44 5HE M0IYK
S44 5HE M1CLI
S44 5HE M7CWA
S44 5HP G1VPE
S44 5LE M6EDM
S44 5LU G6EZH
S44 5LX 2E0ORS
S44 5LX M6PFK
S44 5NG 2E1SAZ
S44 5NG G0THF
S44 5NG M0RSC
S44 5NG M3SAZ
S44 5PH M6AWZ
S44 5PH M7BAN
S44 5QD M6ASO
S44 5QG M7CPG
S44 5UE G0TJG
S44 6AQ M6EYB
S44 6BH G0IXZ
S44 6DH M3LEX
S44 6DY G6TEX
S44 6EJ 2E0EHO
S44 6EJ M6GGT
S44 6ER G8YBR
S44 6EW G0DJA
S44 6EY G4UPA
S44 6GF G8JQW
S44 6HF M3DGR
S44 6HS G7WAF
S44 6HU G4AGE
S44 6HU G8AVC
S44 6LL M3MYK
S44 6LR 2E1DPQ
S44 6LR M3MUD
S44 6NX G7GSX
S44 6NX M3CFP
S44 6PS G0CTS
S44 6RX G7CAT
S44 6SE G4FEM
S44 6TX G4KSY
S44 6XH G8NRC
S44 6XN G7DJN
S44 6ZJ 2E1BIT
S446EG 2E0SUE
S446EG M3PUZ
S45 0AL G1ISN
S45 0BZ M6VRR
S45 0DX G4TSA
S45 0EA G3MME
S45 0LL M1WWW
S45 8AH M0CJD
S45 8AN 2E0FTB
S45 8AN M0VII
S45 8AN M3VII
S45 8BQ G6XZC
S45 8DH 2E0JUL
S45 8DH G3VNH
S45 8DH M0JHD
S45 8DH M3VNG
S45 8DT G8MMV
S45 8EJ 2E0IBU
S45 8EJ M6WBM
S45 8ET G6POJ
S45 8ET M6SKF
S45 9DX G0WRM
S45 9DX G4HZG
S45 9FA G8RBW
S45 9HH M6PWH
S45 9HH M6SIY
S45 9JT 2E1FKM
S45 9LE M6NUR
S45 9LY G3UAF
S45 9NH G0MNC
S45 9PL G0IVB
S45 9RE G1HFT
S45 9RE G3THT
S45 9RY 2E0MYS
S458HY G4VWT
S5 0DH M6DCA
S5 0GX M3DTH
S5 0HG 2E1FDD
S5 0JP G4TCE
S5 0JY G7WGX
S5 0JY M0SAY
S5 0NN 2E0JOX
S5 0NN M3KUK
S5 0NZ M3GSQ
S5 0QE M6FJO
S5 0TY G6ADD
S5 6AA M6MVT
S5 6AN M7GGT
S5 6FZ M1ERS
S5 6HB G8BMI
S5 6LX G7DMP
S5 6SG M3LEB
S5 6SQ G0UHF
S5 7JX G7ACR
S5 7SW M6SHF
S5 7TL G0OLL
S5 7WE G0UQZ
S5 7WU G0BWG
S5 8AX 2E0OMV
S5 8AX M0OMV
S5 8DT M0DMA
S5 8GN 2E0SNU
S5 8LL M6TBX
S5 8NA M7MRY
S5 8NE 2E0ZEV
S5 8NE M3ZEV
S5 8NE M6GVS
S5 8WF 2E1AGV
S5 8WF M1FJQ
S5 8XU M6BGJ
S5 9AH M6JFS
S5 9DB 2E0HNG
S5 9DB M7CHB
S5 9DY G3UXH
S5 9GA G0SEW
S5 9GS G8AOZ
S5 9NW M7GLB
S5 9QR G0RNB
S50NN 2E0TJX
S6 1BQ M0GAV
S6 1FB 2E0IEB
S6 1FB M6IEB
S6 1HT 2E0YDB
S6 1HT M0SDB
S6 1HT M6YDB
S6 1HU G7DEE
S6 1JE G3LLV
S6 2TY M3XWF
S6 2XE 2E0BQX
S6 2XE M3MYT
S6 3NG G4BOJ
S6 3QZ M0KFY
S6 3RG G0LIZ
S6 3RG G4YUR
S6 3TR G4VMQ
S6 3TR M7JGH
S6 4AH G4VYI
S6 4AY G0BKE
S6 4BA M1CGF
S6 4BE G3OZL
S6 4FD G7NNZ
S6 4FG G0CGT
S6 4LQ 2E0VLD
S6 4LQ M7LDS
S6 4NB G0JKE
S6 4QG M0EMR
S6 4SU M3PMK
S6 4TF G4NXA
S6 4WG M1EQW
S6 4WU M7JWZ
S6 4WU M7PSZ
S6 5BA G0RNC
S6 5BE M0RPK
S6 5DQ M0DPJ
S6 5GQ M6YSM
S6 5HQ M3UJL
S6 5LN G4TFZ
S6 5NA M7OFG
S6 5ND M6XAW
S6 5PA G0JTT
S6 5SL 2E1GCA
S6 6AB G1MLV
S6 6DJ G0JKH
S6 6DQ M7CXB
S6 6GL G3DR
S6 6GL G4FAL
S6 6GL G5AT
S6 6GL G5WS
S6 6GL G6XX
S6 6GL G6ZZ
S6 6GL M0HDG
S6 6GL M3HEE
S6 6GL M3XBT
S6 6GL M5DX
S6 6HX M6MBG
S6 6LJ G1WWI
S6 6SG G8KB
S6 6TG M1JWM
S60 1JU M1ALG
S60 2AT G4TQT
S60 2NG M6AYV
S60 2UZ G1NFO
S60 3ER G1MDE
S60 3HB M0GUD
S60 3JJ G1HWA
S60 3JN G1IDF
S60 4DZ G6BKL
S60 4EA M6AYW
S60 4EF G4LFS
S60 4NH G4NLC
S60 5AJ M7EZS
S60 5ES G0ISK
S60 5HG G1DEZ
S60 5JU G0CGS
S60 5JX M7KPA
S60 5PU M3FZO
S60 5RE G6CPY
S60 5TH G8AHN
S603AZ G3VSK
S61 1BJ G0GUM
S61 1RY G1RGG
S61 2BL G6LXW
S61 2DU 2E1FFD
S61 2EX G1OHU
S61 2HB G8EQD
S61 2JW G6MLS
S61 2JZ M3LJK
S61 2NP M3FJC
S61 2PZ G4PEN
S61 2QS M6CLW
S61 2RP 2E1ENZ
S61 2RP 2E1ETB
S61 2RP G4ZVN
S61 2RR 2E0INK
S61 2RR M7USE
S61 2TQ 2E0EVP
S61 2TQ M6ATJ
S61 2TT G4EBG
S61 2UH G1TYU
S61 2UN G0OUN
S61 3AG M6UZZ
S61 3HN G7OFV
S61 3LG M7JFX
S61 3ND M3PJI
S61 3NP G4GET
S61 3RF M7ARZ
S61 3RQ M6FNU
S61 4DL G0MTT
S61 4PA M0BEQ
S61 4PD G7OGR
S61 4PL M6EXX
S62 5BD M3TKU
S62 5ED M3IFA
S62 5ED M0UGD
S62 5JR G1GBR
S62 5LW G0VBR
S62 5NH G7BQS
S62 5NT 2E0GMA
S62 5NT M0GMA
S62 5NT M6GMA
S62 6DB 2E0BKB
S62 6DB M3TPJ
S62 6HJ 2E0BNE
S62 6HJ 2E0SDA
S62 6HJ M3OBX
S62 6LN G7MST
S62 7BX G6JDC
S62 7ED G1IFH
S62 7EE M3NGD
S62 7EN G3USW
S62 7FE M3MGN
S62 7JA M6NOL
S62 7QP M7GYV
S625BW 2E0FHQ
S625QJ G7AOQ
S63 0HB 2E0GEX
S63 0HB M6YPG
S63 0JA M1EUN
S63 0LQ M6ZEF
S63 0NN M1DSI
S63 0PX 2E0BMY
S63 0PX M0GVX
S63 0PX M0HKT
S63 0PX M3PGK
S63 0QT M3UOD
S63 0RZ G3WLV
S63 0SA 2E0GAR
S63 0SA M3PGU
S63 0TG 2E0ZTG
S63 0TG M0ZTG
S63 0TG M3OTG
S63 0TN G4VNX
S63 0TY G4XSB
S63 6AA 2E0CDW
S63 6AA M0KCW
S63 6AA M6CDW
S63 6DD 2E0DLC
S63 6DD M6DLX
S63 6EL 2E0DYJ
S63 6EL M6EZC
S63 6ET G4UZG
S63 6EY G0JKG
S63 6HW G8XUH
S63 6HZ M6MGN
S63 6JN M6EUU
S63 6LJ M6SYB
S63 6NQ M6KJQ
S63 6NU G1KZA
S63 7AJ M7FYZ
S63 7GD M6WJA
S63 7GT 2E1HTE
S63 7GT G4ZTX
S63 7JB G4TZL
S63 7JU G7EVT
S63 7LR G3YOC
S63 7ND 2E0EGQ
S63 7ND M6PSR
S63 7NE G7FAG
S63 7NF M0KRL
S63 7NF M3NBI
S63 7ST G1YPR
S63 7TD M6HCB
S63 8BZ G6IIK
S63 8DA G0TZC
S63 8HR G1EQM
S63 8HR M0EQM
S63 8JL G0UAD
S63 9BA 2E1AQT
S63 9BA G0GPR
S63 9BY G6WSZ
S63 9DX M3ZYI
S63 9DZ 2E0CMO
S63 9EA M6KMS
S63 9JW G0JDC
S63 9LB 2E0TAU
S63 9LB M3LKK
S63 9LB M6JAJ
S64 0AD G8CXA
S64 0DG G8HPJ
S64 0DG M6DBG
S64 0DJ G7BJR
S64 0DT M0DKX
S64 0JD G4KWM
S64 0JG G4PJT
S64 0JZ M0CTJ
S64 0NL G3VJR
S64 5TX G7YAB
S64 5TY G0VPS
S64 5UG G6VIN
S64 5UG M6HZM
S64 5UP M6MDG
S64 5UQ G7MUE
S64 5UQ M3ILR
S64 5UQ M6HOQ
S64 5UU G0VXC
S64 8AH 2E1FCE
S64 8DS G6MFU
S64 8DU G8BIW
S64 8EF 2E0KLV
S64 8EF M3VAF
S64 8HD G0VUH
S64 8LF G0PBF
S64 8NA M6FLW
S64 8ND G7JSE
S64 8NF M3KSK
S64 8NF M3PSD
S64 8NX G0CSK
S64 8NZ G6ZFX
S64 8NZ M6PKN
S64 8PT 2E0BQQ
S64 8PT M3UPK
S64 8PY M3UPL
S64 8QQ M7DXQ
S64 8QW M6PZB
S64 8RF M6FJF
S64 8SQ G0FUO
S64 8TE 2E1FTF
S64 8UH G6OET
S64 8UL G4IDL
S64 9EH G1UVJ
S64 9EH M3ZPR
S64 9EZ 2E1FTV
S64 9EZ M3ACY
S64 9EZ M3EIJ
S64 9JX 2E0HNX
S64 9NH M0KVM
S64 9PF G4RPL
S64 9SG 2E0EBD
S64 9SG M0IBR
S64 9SG M6EBO
S648NU G0VJY
S65 1LH G8UCR
S65 1LT M3ZHI

S65 1NB M6JED
S65 1NF G6IJE
S65 1QP M7LRM
S65 1RT 2E0JSI
S65 1RT M7JSI
S65 2BY M3CPN
S65 2HQ M3OAM
S65 2JR M6LYS
S65 2LH M6FVO
S65 2LR M6EOS
S65 2QP M0CTI
S65 2UA G4NJI
S65 2UE 2E0DXJ
S65 2UE M6AWE
S65 2UX 2E0DKB
S65 2UX M6ICQ
S65 2XA 2E0TAL
S65 2XA M3ZXN
S65 3JX 2E0DZE
S65 3JX M0CHU
S65 3LG M6BZE
S65 3NH G1RWR
S65 3NL M6RGE
S65 3NN M0CJY
S65 3NX M3EWQ
S65 3PQ M0MTD
S65 3RB M7PCD
S65 3RW M6BRY
S65 4HH G8NVX
S65 4JF M6KBH
S65 4NF G4RGH
S65 4NP G4NYW
S65 4NZ G4APO
S65 4NZ G8VJP
S65 4QR G8DRQ
S65 4RJ M7RXK
S66 1AF M6RBV
S66 1DL G4CGP
S66 1NN 2E1BGQ
S66 1UH M6DMW
S66 2BN G4ENC
S66 2BN G8EUV
S66 2HZ G4WFW
S66 2HZ G7RIJ
S66 2JA M3VPN
S66 2LU G8PUK
S66 2NL M6WNN
S66 2NS M6VTT
S66 2NT M3JDF
S66 2PQ M6JIS
S66 2SJ G1DTF
S66 2SP G0WXD
S66 2SS G8HUO
S66 2SW 2E1IJO
S66 2SW M3BBL
S66 3PL M6MDT
S66 3WA G6PMR
S66 3YW G4UHQ
S66 7AJ M3YXT
S66 7DQ 2E0FJT
S66 7DQ 2E0FJU
S66 7EN 2E0DDO
S66 7EN M3ODO
S66 7EU G1BHQ
S66 7HA G0EIB
S66 7HA G0GOO
S66 7HB G6PCX
S66 7HF G0EPU
S66 7HG G7OFI
S66 7HJ G4BVV
S66 7HX G1POC
S66 7JZ M1DBF
S66 7JZ M6JJH
S66 7LH M6AVF
S66 7LU G1ILC
S66 7NR M6COX
S66 7PA M6FBF
S66 7PD G7TKM
S66 7QB M6MXP
S66 7SG 2E0YLH
S66 8AT M0MCT
S66 8AT M6TBO
S66 8AZ G0HDG
S66 8AZ M6BVW
S66 8BL G6TJC
S66 8DL M3XQH
S66 8DS M6BVX
S66 8HD 2E0KFA
S66 8HD M7WWL
S66 8JT M1CYJ
S66 8JT M3CYJ
S66 8JU G1OJT
S66 8NA M6GIZ
S66 8NN M6LWJ
S66 8QZ M6WBF
S66 8RF 2E0ZOR
S66 9DL G4JMC
S66 9HT 2E0IEW
S66 9HT M6MQN
S66 9LG 2E0JRA
S66 9LG M0JRA
S66 9LG M3JRA
S66 9NJ M1EKA
S66 9NU M7PES
S66 9PD M3KUN
S66 9QP M6SUF
S7 1DD 2E0LVP
S7 1FE M6ZYH
S7 1NE G7GFH
S7 1SJ G4ILX
S7 2EB G0WFL
S7 2QN 2E0DRE
S70 1PL M0TNG
S70 1QE G1KXZ
S70 1TZ 2E0ATT
S70 1TZ M7ATT
S70 2LT M7TZJ
S70 2NE M3ULT
S70 3AA G1ANF
S70 3AA G6ZXO
S70 3DW M6BAD
S70 3EW G3TEH
S70 3FG G0NSC
S70 3FG G1XUH
S70 3RH G4DYG
S70 4AU M3LXR
S70 4BW M6XLG
S70 4DB 2E0FJN
S70 4DB M7NET
S70 4DW G3EAE
S70 4DZ 2E0EWS
S70 4NY G3ZHP
S70 4QN M6TEV
S70 5AY M6BRU
S70 5DU G6ZEW
S70 5JB G6PDJ
S70 5NR G4IAY
S70 5QR G0OHR
S70 5QY 2E1ACS
S70 5QY G8VDP
S70 5QZ G7MRZ
S70 5RL G0JEA
S70 5SN G8EGM
S70 5SU G4KUD
S70 6BZ M7BLJ
S70 6JT G0TOQ
S70 6JY G4TMZ
S70 6JY G6AJ
S70 6JY G6YOR
S70 6LG G0DUM
S70 6PP 2E0BIB
S70 6PP M3GPN
S70 6PQ G6AUP
S70 6QQ G6XQT
S70 6RG G6TVA
S70 6SU M6WLC
S704SQ M7TGH
S71 1SB M3TRO
S71 1SU M0GAG
S71 1XQ G4POI
S71 2BA M0RTL
S71 2BA M1OXR
S71 2BA M3ITM
S71 2BA M3SPG
S71 2BL G6HNE
S71 2ES G8ISE
S71 2EY G6DER
S71 2HP G4DSA
S71 2HS M3ZME
S71 2HS M6KJD
S71 2JY M3HUY
S71 2LL G7VYN
S71 2NJ G8FRJ
S71 2PP G1WTN
S71 2RA G4JUR
S71 3DL M7LOW
S71 3HZ 2E1HJL
S71 3HZ M0CQL
S71 3JJ M3VYV
S71 3JJ M6CBD
S71 3JJ M6EDL
S71 3LF M7LFG
S71 3NR M3VCK
S71 3QD M1DZM
S71 3QU M3HVH
S71 3RH G0SZE
S71 3SJ G0ANE
S71 4AA G1ANI
S71 4HF G3OXR
S71 4JA G4YAP
S71 4LQ M6UUH
S71 4NG M0OCC
S71 4NG M0OXO
S71 4NG M0UKI
S71 4NG M3ZLZ
S71 4NG M3ZYZ
S71 4NH G1BQR
S71 4NH M0GEY
S71 4NL 2E0KAF
S71 4NL M3KRY
S71 4RY 2E0ETP
S71 5AL M3KAE
S71 5BS G4UFR
S71 5BS G6LWK
S71 5BS M7DHF
S71 5DH G4IHZ
S71 5FA M7GDT
S71 5LG M0GOI
S71 5QX G0JUK
S71 5RB G3AMH
S71 5RB G6KOB
S71 5RB M0HFE
S71 5SD M6CPR
S72 0AU G0PQG
S72 0DW M3WZF
S72 0HG 2E1HUQ
S72 0HT G0TLA
S72 0HT G7KMF
S72 0HT M3FJD
S72 7AA M6IFS
S72 7NA M3ICH
S72 8BE M1ERN
S72 8DE G1EFF
S72 8DY G6NXW
S72 8DY M3RYR
S72 8EW M6SEB
S72 8HU M7AVG
S72 8JB M1SWS
S72 8JP G4EPD
S72 8NQ M0DMS
S72 8NQ M6BNJ
S72 8PW G1SYP
S72 8PY 2E0BLF
S72 8PY M3UXC
S72 8PZ 2E1HXR
S72 8RN G4LUE
S72 8RN G8PLJ
S72 8SE M6URG
S72 8SE M7BMR
S72 8SF M7FQG
S72 8XG G0RUX
S72 9AN 2E0FYL
S72 9AN 2E0TJC
S72 9DL 2E1HPM
S72 9DL G4XDV
S72 9DL M3XDV
S72 9HJ G1JDO
S72 9HJ G6CND
S72 9JA M0AEE
S72 9JG M3NTQ
S72 9JX 2E1HPZ
S72 9JX M3WOC
S72 9LL M0JOR
S728JT 2E0MML
S73 0EF M3SPQ
S73 0LF G0XDX
S73 0RW G8VHB
S73 0TN M6PRW
S73 0YJ G1BCN
S73 8LD G0VMN
S73 8PP 2E0CKP
S73 8PP M6CKP
S73 8PS G0SKD
S73 8RX G1VGO
S73 8SF G6YJD
S73 8SG M6GGV
S73 9AB M1LAP
S73 9AE G0UQF
S73 9BJ M3UJF
S73 9DU G4BHL
S73 9DU M3ZXA
S73 9ED G0TKJ
S73 9ED G1YTX
S73 9HA M3ZPY
S73 9HQ G0CQO
S73 9JA M0BZH
S73 9NF M0CLJ
S73 9PE G1YUU
S73 9PL M3XRH
S73 9PP G4NSO
S738SB 2E0FWR
S738SB M0WFR
S738SB M6FRK
S74 0AP M0GXH
S74 0AP M3LSS
S74 0AZ M3ITU
S74 0BG M7KXL
S74 0BU M3SNY
S74 0EJ M3YFG
S74 0LB 2E0TBV
S74 0LL M1DER
S74 0PS M6RTM
S74 0PS M6SLT
S74 8BW M0PAF
S74 8BW M0PPP
S74 8DG M7NOH
S74 8DP M0CSZ
S74 8DR G0CUI
S74 8DS M3HHX
S74 8DT G6UQO
S74 8JR 2E0PJE
S74 8JR M6PJE
S74 9AX G6NVW
S74 9DP M6JYV
S74 9EQ G4YTM
S74 9HF G7GQD
S74 9HF M0CNG
S74 9HS M3XJO
S74 9HU M0BJK
S74 9HU M1KEY
S74 9HZ M3KXF
S74 9RF M3AYF
S749EE G4YER
S75 1EE 2E1HJO
S75 1EE M0HYD
S75 1FE M6MPR
S75 1LX 2E0TTS
S75 1LX M3TTS
S75 1PS M3XZD
S75 1PX G4OUS
S75 1QG G4MLQ
S75 2AF M3EVE
S75 2EF 2E1HOK
S75 2EQ M3EGF
S75 2EW M6FJC
S75 2JW M5BAD
S75 2ND G7BZE
S75 2QB M6BFC
S75 2QJ G8MQX
S75 2RX M3BCQ
S75 2TQ M1GCS
S75 3DD M3KBZ
S75 3DE M0FAT
S75 3JE 2E0DNO
S75 3JE M6DCO
S75 4HD G1UQT
S75 4JS G1VOC
S75 4NN G0RUZ
S75 4QQ M6CWW
S75 5BG M1LOU
S75 5BG M3OCP
S75 5BG M3SLF
S75 5EA G5TV
S75 5EA M6YBK
S75 5EL M3VGF
S75 5ER M0RCI
S75 5HL G0OUC
S75 5JQ G7TGF
S75 5LG M0JLT
S75 5NP G7NRP
S75 5NS M3IOQ
S75 5PG M6BJE
S75 5PW 2E0BRI
S75 5PW G1NAQ
S75 5QP M6KHS
S75 6DQ 2E0YNY
S75 6DQ M0YNY
S75 6DQ M3YNY
S75 6ET M3UMD
S75 6EU 2E1HJS
S75 6EX M6DDV
S75 6EX M6JGU
S75 6HB M0LOU
S75 6HD M3NSX
S75 6HD M3NTJ
S75 6HH G0SBO
S75 6JA G1LES
S75 6JP G7ULW
S75 6LE G8GDH
S8 0BN G0JPE
S8 0EY M6XMH
S8 0GA M0GDX
S8 0HH M7LLB
S8 0RA 2E0BFZ
S8 0RA G0NFY
S8 7BW G0PEW
S8 7BX G4ECA
S8 7DP G1LJL
S8 7DR G6FGL
S8 7DZ G6SWW
S8 7FW 2E0MDZ
S8 7FW M6MDZ
S8 7RJ G3VCQ
S8 7RJ M0SDC
S8 7RJ M3VCQ
S8 7UF 2E0PGL
S8 7UF M6PGL
S8 7UF M6TAL
S8 8DA M3NPA
S8 8DA M3NPE
S8 8DA M3NRQ
S8 8DQ G8LIH
S8 8DS M6BYH
S8 8DU 2E0GCH
S8 8DU M3GCH
S8 8FT G8YVW
S8 8GE G0HSA
S8 8HE G4RVS
S8 8LB G3XSI
S8 8LJ G7MLL
S8 8PB G8OFN
S8 8PJ G0NQY
S8 8PR G7OCB
S8 8QR G0NJD
S8 8QU G7EFA
S8 8SE G3RKL
S8 9NG G1HGB
S8 9RL G7DFX
S80 1RZ 2E0MNU
S80 1RZ M0MNU
S80 1RZ M3YMN
S80 1SU 2E1EXP
S80 1SW 2E0NVB
S80 1TD M6ELR
S80 1UZ G0MZP
S80 1YA G3WTY
S80 1YG M0KAD
S80 2DL G3UVB
S80 2EG 2E0KBZ
S80 2EG M6JPU
S80 2EW M0CMN
S80 2HN 2E0RPR
S80 2HN M0RPR
S80 2HN M6HEP
S80 2HN M6RPR
S80 2JR M7CLJ
S80 2LL 2E1GJE
S80 2NT 2E0PRS
S80 2NT M0JGU
S80 2NT M0YES
S80 2RB M0MIR
S80 2RR 2E0BRY
S80 2RR M3MYW
S80 2SA 2E0MZU
S80 2SA 2E0WAJ
S80 2SD 2E1GJT
S80 2SH M6OQR
S80 2ST M3KUE
S80 2TP M7GGM
S80 2TT G7MQW
S80 2TW M6LVE
S80 2TW M6NNJ
S80 2TW M6NNU
S80 3DB M0GKR
S80 3DE G0KVG
S80 3DF G7TNZ
S80 3DF M1FEX
S80 3DF M3RDW
S80 3DF M6REW
S80 3DL M6SRA
S80 3EB M0DOB
S80 3GW M7FXP
S80 3HF 2E0CHA
S80 3HF G7OBP
S80 3HG G0CEJ
S80 3HR G7SOH
S80 3NF M0MRK
S80 3NQ M0WEN
S80 3QB M1LAN
S80 3QG M0BMT
S80 3QJ G6JBL
S80 3QJ G7BJD
S80 3RD G4PYV
S80 4AN M1DAB
S80 4BB 2E0BEV
S80 4BN M3OHZ
S80 4DE G0AHD
S80 4DL M3WTP
S80 4DW 2E0ITJ
S80 4DW M0HEW
S80 4DW M6TJZ
S80 4DY G1EVA
S80 4ER M3KOJ
S80 4HN 2E0XRM
S80 4HN M0XRM
S80 4HS 2E0AVD
S80 4HS M0JMJ
S80 4JR G1KIT
S80 4JT 2E0CTJ
S80 4JT M6CTJ
S80 4NP G6OZT
S80 4NS M6MUJ
S80 4PY G8SRZ
S80 4SN G3VRU
S80 4UD G1OIZ
S80 4XT M3JHJ
S81 0AX G4YJY
S81 0AX M0WDP
S81 0BS G0AGD
S81 0DH M6WSM
S81 0DP M0GTQ
S81 0EE G3OZN
S81 0HD 2E0NCK
S81 0HD 2E0PNK
S81 0HD M6CAR
S81 0HD M6NCK
S81 0HD M6PNK
S81 0JS G0WTK
S81 0JS G8TBF
S81 0LP G3XXO
S81 0QN G0BDR
S81 0QN G0CJD
S81 0QN G4ZUN
S81 0QN G6EXG
S81 0RJ 2E1SKY
S81 0SS M7BUH
S81 0TB G0HMX
S81 0XH M3MSH
S81 0XJ 2E0CAJ
S81 0XJ M0TAL
S81 0XJ M6CMT
S81 0XY M0EEB
S81 7BH G1VBP
S81 7ED G0FMP
S81 7HL M6KDB
S81 7JS 2E0RYN
S81 7JS M6DCK
S81 7JZ M7DZK
S81 7LE G4CRE
S81 7LJ 2E0ZTD
S81 7LJ M3UHC
S81 7LJ M3ZTD
S81 7PG G8TWT
S81 7PS M6PRB
S81 7QB G1NSK
S81 7QE G0TKG
S81 7QW G6TLB
S81 7RT 2E0GEF
S81 7RT G6CNX
S81 7RT M3YHT
S81 8AU M7GKO
S81 8EY M7FGJ
S81 8NH G8RSV
S81 8NQ M0JSP
S81 8PE G4YRZ
S81 8SF G0WPC
S81 8UZ M7RWF
S81 8WA G7AIZ
S81 9BD 2E0RSH
S81 9BD G3ZMM
S81 9BD M0ZMM
S81 9BD M3RSH
S81 9HN M7WSP
S81 9JT G4MWX
S81 9LE M3LRK
S81 9NW G3XXN
S81 9NX G6DRG
S81 9PT M3BXN
S81 9RA G0NFB
S81 9RR G0DMA
S81 9RR M6SBM
S81 9SA M7EUM
S81 9SB M3YRC
S81 9SH M6SDL
S81 9SL G0NGW
S81 9SS M6TWI
S810UY 2E0YBB
S810UY M3HOV
S817AN M7AZH
S9 1AG M6GOE
S9 1AR 2E0JBS
S9 1AY 2E1IDM
S9 1BA G7TBC
S9 1LU G6VUE
S9 1PZ 2E0HRY
S9 1PZ M3UQG
S9 1RX M7TBI
S9 4AU G4RBU
S9 4HE G0MBK
S9 4JG M6DWW
S9 5EH M0AFY
S9 5FQ G3RCM
S9 5FQ G6DCT
S9 5GJ G8SVT
SA1 1NZ GW0JFQ
SA1 1RY GW0EZQ
SA1 2LG MW7WDQ
SA1 2QE GW6TTA
SA1 3HE 2W0MPY
SA1 3HE MW7PYM
SA1 3SN GW7RLS
SA1 3SN GW7SKC
SA1 3UX 2W0LBN
SA1 4QF MW3LQC
SA1 6HN GW1XBG
SA1 6HU MW6AJD
SA1 6LS GW8DSO
SA1 6NH MW1AUV
SA1 6PD GW4PHT
SA1 6QP MW7AWO
SA1 6TZ GW3NDB
SA1 6TZ GW8ASA
SA1 7AD GW4VBV
SA1 7EF GW4WFM
SA1 7FJ 2W0HAC
SA1 7FJ MW0HAC
SA1 7FJ MW6HTT
SA1 7JB MW6LTN
SA1 7JY GW4YID
SA1 8BL 2W0UCL
SA1 8BL MW0UCL
SA1 8EJ GW6MLL
SA1 8HQ GW4HDB
SA1 8JF GW0JDS
SA10 6AA 2W0DAP
SA10 6AA MW6THW
SA10 6BQ MW3JEK
SA10 6DL GW0CBL
SA10 6DS GW0HPC
SA10 6ET MW6FAM
SA10 6JR MW6FIX
SA10 6SD GW6KRK
SA10 6SP GW0VSW
SA10 6SP MW0NSC
SA10 7AD GW4JQQ
SA10 7BU GW4YJI
SA10 7DD GW8DOA
SA10 7DY MW0CIA
SA10 7LB 2W1CPS
SA10 7LB MW6HSH
SA10 7NR 2W0EPE
SA10 7NR MW6EPE
SA10 7PL GW1WGR
SA10 7PL GW4WWN
SA10 7RX GW7IBT
SA10 7TT MW7TWG
SA10 7TW GW7EHD
SA10 7UG GW0CYG
SA10 8BG GW4FOI
SA10 8DR 2W0ZWR
SA10 8DR MW0ZWR
SA10 8DR MW3ZWR
SA10 8DW GW1SGH
SA10 8HD GW8VCA
SA10 8HT GW4FXF
SA10 9BL MW7AHI
SA10 9BU GW6APK
SA108ND MW0ABV
SA11 1DJ GW4IUK
SA11 1DQ MW1FFY
SA11 1JL GW0VFF
SA11 1NG 2W0JJW
SA11 1NG MW6RGW
SA11 1PP MW3GTM
SA11 2AY 2W0TAI
SA11 2BJ MW6PWS
SA11 2DP MW7YXD
SA11 2DW MW3SOC
SA11 2JG MW6CBL
SA11 2JR MW7WWF
SA11 2JU GW6NXH
SA11 2LU MW6MSE
SA11 2PG 2W0RRY
SA11 2PG MW6BWR
SA11 2TE GW4PCO
SA11 3AL GW1EWW
SA11 3FD MW3GQE
SA11 3JW GW0NKJ
SA11 3NJ GW2ABJ
SA11 3PJ GW0HNT
SA11 3QE MW0CNC
SA11 3RN MW0HYP
SA11 3SS GW8SZC
SA11 3SS MW3GRC
SA11 3XA MW3LHA
SA11 3YQ GW0TWR
SA11 4AA GW7JHK
SA11 4HS GW7TZG
SA11 4NN 2W0PEE
SA11 4NN MW6NRT
SA11 5BG GW4SRE
SA11 5BG GW4UYT
SA11 5NT GW4WYX
SA11 5RL 2W0VPT
SA11 5RL MW3VPT
SA114BP MW6YSA
SA12 6PF MW6DSP
SA12 7DE GW4RML
SA12 7ED 2W0RMY
SA12 7ED MW6ADZ
SA12 7HG GW1NBW
SA12 7NB MW7GSH
SA12 7PF MW7TMA
SA12 8AP GW0RLQ
SA12 8AS GW4XWC
SA12 8EY GW1RQM
SA12 8LE GW4TVQ
SA12 8PH GW4YNT
SA12 8TU MW6ESA
SA12 8YE GW3TKG
SA12 8YH MW7BRQ
SA12 9EA MW0POB
SA12 9SE 2W0CED
SA12 9SE MW6CAN
SA12 9TG GW0DZU
SA12 9TW GW8DUY
SA12 9TW MW0TTF
SA12 9TW MW0WVA
SA12 9TW MW5WRG
SA13 1BR MW1GLD
SA13 1DX MW7PHI
SA13 1HA GW4DOO
SA13 1LD MW6AHV
SA13 1NE GW7TJM
SA13 1PF MW6XMO
SA13 1SG GW0TWF
SA13 1TD 2W0CLT
SA13 1TD MW6BOC
SA13 1YD GW4PRP
SA13 2EW MW7JNC
SA13 2HL MW0CDO
SA13 2LB GW3NKM
SA13 2LF GW0KPD
SA13 2US GW3XHD
SA13 2YD MW0LEW
SA13 2YE GW3TMJ
SA13 3AN GW4RDW
SA13 3BP GW1GVM
SA13 3DP GW4TVU
SA13 3YE MW6FGN
SA13 3YN GW4DWN
SA13 3YW 2W0TRS
SA13 3YW GW6GSL
SA132UG MW0TDQ
SA14 6AP MW7ENN
SA14 6BD GW0RUD
SA14 6BR MW3WLS
SA14 6BW MW7FCN
SA14 6DT GW0SRF
SA14 6LY GW7RQI
SA14 6NT MW1DUJ
SA14 7AB GW6FBV
SA14 7DN MW0CET
SA14 7DN MW6SCD
SA14 7HY GW1LFN
SA14 7NA GW0LNM
SA14 7PN 2W0FUR
SA14 7PN MW6PWM
SA14 7PR 2W0TRD
SA14 7PY MW1FJK
SA14 7RJ GW4BYA
SA14 7RS GW7LBI
SA14 7RW GW0AKV
SA14 7SA MW3UNZ
SA14 7UF GW8JJZ
SA14 8AE GW4PUC
SA14 8AT 2W0SCP
SA14 8AT MW0GIN
SA14 8DN GW4KFD
SA14 8EL GW3OPC
SA14 8PP MW6HDT
SA14 8PU MW0ATR
SA14 8QU MW0CUA
SA14 8UB 2W0XTK
SA14 8UB GW4APF
SA14 9AH GW4OH
SA14 9AH MW0JZE
SA14 9AH MW3ZAQ
SA14 9LH 2W0FOG
SA14 9UP GW0KJZ
SA146BR GW0DSP
SA149PZ 2W0TRR
SA15 1AR 2W0DQT
SA15 1AR MW6EFK
SA15 1DG GW0DZL
SA15 1HR MW3VNV
SA15 1JP MW7PHY
SA15 1NQ 2W0LDX
SA15 1NQ MW3SNJ
SA15 1NR MW0JAW
SA15 1NT MW3XVR
SA15 1NZ MW0PJR
SA15 1PB MW0EAN
SA15 1SW GW6PDR
SA15 1SW MW6CCG
SA15 1SW MW6ZOD
SA15 1TR MW6FQD
SA15 2AB MW0CXH
SA15 2BG MW0CJB
SA15 2LB GW0RHE
SA15 2RB GW6VFH
SA15 2RH GW4XJK
SA15 3AE GW0JQT
SA15 3AE GW1OII
SA15 3ED GW4UIE
SA15 3NF GW0BCL
SA15 3NS GW0IXQ
SA15 3RA MW0DFN
SA15 3RT MW0LDD
SA15 3RU GW0OFH
SA15 3SN GW0GUA
SA15 3SN GW4TFS
SA15 3TF GW7FXX
SA15 4LA 2W1ETW
SA15 4NL GW4RXO
SA15 4NS MW6GAA
SA15 4PF MW0CXW
SA15 4PF MW0WWR
SA15 4RR MW6FIY
SA15 4SG MW6TFL
SA15 4SG MW6ULX
SA15 5AQ GW4LJS
SA15 5DJ GW8VUV
SA15 5DL 2W0IGC
SA15 5DL MW0INC
SA15 5DL MW6IGC
SA15 5EN MW3GDL
SA15 5HF GW0NLB
SA15 5HN GW0KJT
SA15 5HN MW6FHK
SA15 5NY MW3HNP
SA15 5RT GW0TMU
SA15 5SF GW0SRE
SA15 5SH MW6PLP
SA15 5SU 2W0CKV
SA15 5SU GW0TDA
SA15 5TF GW8KCY
SA15 5TP GW7UXY
SA15 5UB MW0CTX
SA15 5UG GW6HUD
SA15 5UH MW0FUN
SA152DZ MW1BAJ
SA16 0EG MW3VMY
SA16 0HF GW0DWQ
SA16 0HJ GW6XGA
SA16 0HP GW0DIX
SA16 0LD GW0WGE
SA16 0NF 2W0MMD
SA16 0NF GW0JDW
SA16 0RH GW8WXP
SA16 0TE MW6NAX
SA16 0UR MW3IKC
SA16 0UT MW3IFZ
SA17 4AX GW4RKI
SA17 4BQ GW8XMW
SA17 4EB GW1ENG
SA17 4EY MW0CWS
SA17 4EY MW3DWZ
SA17 4HF 2W0BKM
SA17 4HF MW3VFN
SA17 4JA MW6JCH
SA17 4NW MW1BNY
SA17 4RA MW6RDL
SA17 4TE MW6HEI
SA17 5BJ MW6OUW
SA17 5EJ GW3PLB
SA17 5EN GW3YAF
SA17 5TQ GW3UAY
SA17 5TQ MW6AHQ
SA174AF 2W0HDC
SA174AF MW0WSD
SA18 1AD GW7ORB
SA18 1AF MW6IBD
SA18 1AN MW6MTG
SA18 1AU GW0WRI
SA18 1BB MW6RXZ
SA18 1BD GW0LDZ
SA18 1BR 2W0PMZ
SA18 1BR MW3PDE
SA18 1BR MW6PNZ
SA18 1EN 2W0RKF
SA18 1EN MW6RKF
SA18 1PE GW0HNE
SA18 1RF GW0RAD
SA18 1TR MW3UIQ
SA18 1TR MW3UIY
SA18 1TR MW6TLN
SA18 2AY MW3OLX
SA18 2GA GW7RRM
SA18 2HF GW4IMC
SA18 2LQ MW0ANX
SA18 2LX MW6AIC
SA18 2NG GW1ANW
SA18 2QE MW1FGV
SA18 2QE MW3FGV
SA18 2RA MW3FTY
SA18 2TY GW1JOV
SA18 2UY GW6LUT
SA18 3EF MW7ARE
SA18 3HA GW4UPG
SA18 3HD MW3XDB
SA18 3LN GW4JPC
SA18 3NS GW0FAD
SA18 3QL GW0IXM
SA18 3QW GW7HJN
SA18 3RD MW0GTY
SA18 3RY GW1TJK
SA18 3SF 2W0TCM
SA18 3SF GW4XMR
SA18 3SF MW0XTZ
SA18 3TG GW3WMP
SA18 3UA GW4BVJ
SA18 3YS MW3YNA
SA181AR MW7MHO
SA19 6EB MW0AMI
SA19 7HD 2W0GAY
SA19 7HD MW0SIP
SA19 7PA GW0GPQ
SA19 7UA GW0AIY
SA19 7YP MW6NHL
SA19 7YR MW0CIS
SA19 7YR MW0CIT
SA19 7YT GW6WOB
SA19 8AD GW1SPW
SA19 8DF GW0AWT
SA19 8DX GW8DX
SA19 8DX MW0PRO
SA19 8DX MW0VDX
SA2 0AT MW3TAF
SA2 0DE MW5DQQ
SA2 0FQ 2W0RGA
SA2 0FQ MW6RGA
SA2 0FZ GW3OMN
SA2 0GB GW4SRI
SA2 0PJ GW1YHL
SA2 0PR GW0SAJ
SA2 0QE GW3TYI
SA2 0RL MW0ILG
SA2 0RL MW0KRS
SA2 0YD MW0CGP
SA2 7DF GW1FBL
SA2 7EH GW4BNJ
SA2 7EQ GW6UFH
SA2 7HW GW1BFB
SA2 7PQ GW8VFQ
SA2 7PR MW6HUY
SA2 7QE GW6VKY
SA2 7QS 2W0BKN
SA2 7QS MW0MAY
SA2 7QS MW3RPZ
SA2 7RP GW8HDH
SA2 7RW GW0HNS
SA2 7TH GW1DPL
SA2 7TH MW6VNP
SA2 7UH GW8TVX
SA2 7XQ GW0FYO
SA2 8AL GW0MGQ
SA2 8BE GW3XIS
SA2 8DP MW0BBL
SA2 8HR GW7NVM
SA2 8JL MW0RBA
SA2 8LL GW6AAG
SA2 8LQ GW6MMM
SA2 8LX GW3KGI
SA2 8PP MW0HGK
SA2 9BW GW3UWS
SA2 9BW GW4ADL
SA2 9DF 2W1TBD
SA2 9DT GW4JGU
SA2 9DZ MW3TKI
SA2 9EQ GW0INN
SA2 9EX 2W0IQF
SA2 9EX MW7RRW
SA2 9GR GW4JUC
SA2 9HN 2W0FLW
SA2 9HN MW6IUP
SA2 9HY GW3TMS
SA2 9LG GW1TFL
SA2 9LG MW3FTB
SA2 9LG MW3FTC
SA2 9LY GW4HAT
SA20 0ED GW6REF
SA20 0LD GW0LLD
SA20 0LS MW6FUN
SA20 0NT GW8HYT
SA20 0UP GW7VJK
SA20 0UP MW3LDY
SA3 1AE GW4KTT
SA3 1BA GW4AZI
SA3 1BR GW6OTD
SA3 1HD MW7DHM
SA3 1LB GW8VGB
SA3 2EQ MW0FRY
SA3 3AY 2W0FQM
SA3 3AY MW7ACK
SA3 3JR GW0VRL
SA3 3JW GW1AXU
SA3 3JW GW4RKX
SA3 3JW MW6EDR
SA3 3LA GW0BNO
SA3 4EW GW7PMA
SA3 4HF GW4EPF
SA3 4JD GW3LJS
SA3 4JF MW0DCT
SA3 4LZ GW2CGF
SA3 4ND GW0RTR
SA3 4PD GW1WWE
SA3 4PD GW3SRG
SA3 4QW GW0KWA
SA3 4SW GW6WEU
SA3 4UR MW0EYE
SA3 5AN GW4PEX
SA3 5BU MW0JGE
SA3 5BU MW3ORY
SA3 5DL MW0HGM
SA3 5HS GW4TGA
SA3 5HT MW0AJH
SA3 5NL GW4LDP
SA3 5NQ GW4HSH
SA3 5PD GW8HZL
SA3 5PR GW4EVL
SA3 5QJ GW1PFK
SA3 5QL GW0RHC
SA3 5QL GW4CC
SA31 1BS GW6YUC
SA31 1EG MW6HAR
SA31 1JJ GW4RVA
SA31 1LR GW0PUM
SA31 1NN MW0COZ
SA31 1SY GW0CVY
SA31 1SZ 2W0CZP
SA31 1SZ MW6DVQ
SA31 1TS GW4XUE
SA31 2DT 2W1EAN
SA31 2JB GW1ADY
SA31 2JL GW0HGP
SA31 2LH GW0IXK
SA31 2NL GW8PFT
SA31 3BR GW7EXQ
SA32 7SA GW7OIK
SA32 8AA GW8THL
SA32 8AY GW0HCK
SA32 8AY GW1SGE
SA32 8DQ MW0AMJ
SA32 8DX GW3NXR
SA32 8LJ GW0AJU
SA32 8LJ GW0HYL
SA32 8LJ GW1IIZ
SA32 8PR GW4TGL
SA33 4ES GW0TVX
SA33 4ET MW0CFQ
SA33 4ET MW3GVU
SA33 4ET MW3VWE
SA33 4PD GW2OP
SA33 4PD GW3XJQ
SA33 4PD MW0RLD
SA33 5AH GW0TSE
SA33 5BL MW3VNR
SA33 5DR GW0TXP
SA33 5HA GW3YVN
SA33 5NX GW0GFN
SA33 5NX GW1NED
SA33 5RR MW7HAY
SA33 6BL MW6BHO
SA33 6BN GW4PCJ
SA33 6ES GW0LYK
SA33 6NE MW7BEK
SA33 6XJ 2W0FMQ
SA33 6XJ MW3WJP
SA34 0AF GW0DKG
SA34 0AX 2W0XDT
SA34 0AX MW0XDT
SA34 0AX MW3XDT
SA34 0BE MW0JOP
SA34 0BE MW7BDT
SA34 0BJ MW0GKD
SA34 0EX MW0VWC
SA34 0HL 2W0KJO
SA34 0HL 2W0RMO
SA34 0HL MW6OKJ
SA34 0HL MW6TJH
SA35 0AT GW0WIB
SA36 0DS 2W0XMS
SA36 0DS MW0LRO
SA36 0DS MW6MSX
SA37 0EH GW3KAX
SA37 0EY 2W0TDD
SA37 0EY MW6NIT
SA37 0HZ 2W0BLA
SA37 0HZ MW3JTJ
SA38 9AS MW3VJN
SA38 9BA GW7IPS
SA38 9BE 2W0LLT
SA38 9BE MW3OLT
SA38 9EH 2W0HPL
SA38 9EH MW6MSN
SA38 9JS GW0JCT
SA38 9QB GW1CIY
SA38 9RL GW3WCA
SA39 9AZ GW1UOV
SA39 9BY GW0MWN
SA39 9DH GW4VPX
SA4 3AY MW3UZH
SA4 3BG GW4SCK
SA4 3GY GW7LDP
SA4 3HB MW1AAH
SA4 3LL MW3UYX
SA4 3PE MW6GIU
SA4 3PE MW7WYN
SA4 3PU GW4UCK

SA4 3PU GW7MMH
SA4 3SE GW4EGS
SA4 4AB GW1YBC
SA4 4AD 2W0HUL
SA4 4AD MW3CBS
SA4 4BJ GW7PQS
SA4 4DT 2W0KPN
SA4 4DT MW6KPN
SA4 4EB GW4HZH
SA4 4GE GW3JBZ
SA4 4GW GW1WTZ
SA4 4GZ GW0DIQ
SA4 4GZ GW0DRI
SA4 4GZ GW0NHE
SA4 4HE MW7LOI
SA4 4PH GW7ODP
SA4 4PN MW3UZP
SA4 4QE GW1FBU
SA4 4UX MW0LMW
SA4 6PH GW1GFO
SA4 6SW MW6BOW
SA4 6UH GW7SDE
SA4 8EE GW7AFC
SA4 8EH MW7FUA
SA4 8HU GW3XYW
SA4 8HX GW0VEW
SA4 8JF GW1AIB
SA4 8LW GW0IZL
SA4 8QF GW6VBR
SA4 8RD GW0RNK
SA4 9AB MW0EAT
SA4 9ED GW6TYO
SA4 9WH GW3OQK
SA40 9RD 2W0DNV
SA40 9RD MW6CTS
SA40 9SR 2W0BFV
SA40 9SR MW3WFH
SA40 9SU MW6CTD
SA40 9SU MW6TDC
SA40 9TY 2W0NRA
SA40 9TY MW3JVH
SA40 9UU GW4OBA
SA40 9YB MW0ATI
SA40XG MW7FXA
SA41 3QG MW3SFP
SA41 3RR GW0TMV
SA41 3UF 2W0NRE
SA42 0PL GW4UEP
SA42 0XF MW0ARD
SA42 0XS GW4OQB
SA43 1AF GW4OUU
SA43 1DW GW6LHF
SA43 1FB GW4SJO
SA43 1FB GW4TVE
SA43 1JU MW0LEA
SA43 1LT GW6DGU
SA43 1PG MW6BTC
SA43 1PX GW8JWP
SA43 1RF GW7WCR
SA43 1SN MW1BDV
SA43 2AE GW6RWJ
SA43 2AH GW7GWO
SA43 2AQ GW0JLX
SA43 2BD GW3SIK
SA43 2BZ GW0PNI
SA43 2DA GW6AMK
SA43 2HR GW4HGJ
SA43 2HR GW4SZV
SA43 2JF GW0CMI
SA43 2NY GW3EJR
SA43 2PE GW0MBN
SA43 2RT MW0WRP
SA43 3DS MW7BSK
SA43 3EF MW0CRI
SA44 4NE GW3UEP
SA44 4RJ GW0PDB
SA44 4SJ GW0HYH
SA44 4SJ GW8JDB
SA44 4SJ MW3URG
SA44 4ST MW3TUB
SA44 4TU 2W0NAY
SA44 4TU GW4XUG
SA44 4TU MW0VET
SA44 4TU MW6BZX
SA44 5AT MW0DXX
SA44 5EP GW7BBY
SA44 5HE GW1VMA
SA44 5NZ GW4JPJ
SA44 5PN GW3WXA
SA44 5XJ 2W0LVE
SA44 5XJ MW6ICM
SA44 5YB 2W0ZZU
SA44 5YB GW0EWY
SA44 5YB MW3ZZU
SA44 5YG GW8PSJ
SA44 6BS 2W0WCP
SA44 6BS MW0WCP
SA44 6BS MW7EJC
SA44 6BT GW3RYE
SA44 6EA GW4TQD
SA44 6ES GW4JKK
SA44 6EY MW6FMV
SA44 6JE GW0POZ
SA44 6NP GW3KDB
SA44 6NQ MW0WEE
SA44 6NQ MW7ROY
SA44 6PR 2W0LLJ
SA44 6PR 2W0VLO
SA44 6PR MW0VLO
SA44 6PR MW6VLO
SA44 6PR MW6XJL
SA44 6QY MW0AIZ
SA45 9QR GW0DDL
SA45 9RJ GW7SXU
SA45 9RL GW3WVV
SA45 9RP GW7FQP
SA45 9RR GW3ZJS
SA45 9SY MW3JQC
SA45 9TH MW0WAG
SA46 0ED GW3LHK
SA46 0JS GW4XSX
SA460LP MW7BCM
SA46TR 2W0GYB
SA46TR MW6GYB
SA47 0QN GW0HXS
SA470RT GW1SMG
SA48 7LG GW0ADS
SA48 7LW GW4LJW
SA48 7PG GW6AFQ
SA48 7PJ GW0GEI
SA48 7PX 2W0JYC
SA48 7PX MW0HNM
SA48 7PX MW0JYC
SA48 7SB GW8GQE
SA48 7SG GW0ARA
SA48 8DT GW3KCQ
SA48 8LH GW0PNC
SA48 8NH MW3SNH
SA48 8RE MW7FMD
SA48 8RP GW4VNK
SA5 4BR 2W0FDC
SA5 4BR MW6NMP
SA5 4NJ GW8RDI
SA5 4QH GW8NXK
SA5 4RE 2W0XLT
SA5 4RE MW6BMM
SA5 5NX MW3UYJ
SA5 7AH 2W0YVY
SA5 7BP MW0CNB
SA5 7BR MW0CNA
SA5 7DH GW0BNN
SA5 7DW GW4MII
SA5 7EG 2W0BTE
SA5 7ER MW7VHT
SA5 7JJ MW6DGW
SA5 7PU GW0FZY
SA5 8AU GW7TZI
SA5 8AW MW6GIV
SA5 8BT MW6CEX
SA5 8BU GW1GAU
SA5 8BZ MW7FIJ
SA5 8DU MW0CND
SA5 8NX 2W0HDP
SA5 8NX MW6HDP
SA5 9BS GW4URB
SA5 9DG GW0AZW
SA5 9EL MW0HMV
SA5 9HF GW6KWM
SA5 9HN GW4MTD
SA5 9NH GW0NCU
SA5 9NY MW6JUQ
SA5 9PE GW1LHT
SA6 5AD MW1FGB
SA6 5AE MW3CNL
SA6 5AX GW0COU
SA6 5DX 2W1DGM
SA6 5DX GW7MMG
SA6 5HF GW4SPL
SA6 5HR MW0OUC
SA6 5HR MW0SDD
SA6 5HR MW0SGX
SA6 5HR MW3OUC
SA6 5HR MW3SGX
SA6 5JF GW7OJT
SA6 5LA 2W1EIN
SA6 5RN GW0DXZ
SA6 5TB MW3UEU
SA6 6AP GW7PBP
SA6 6AS GW4TFX
SA6 6BA GW0UIZ
SA6 6BJ MW3ZKW
SA6 6BY 2W1CNN
SA6 6BY 2W1EYZ
SA6 6BY GW1NWF
SA6 6BY GW6ZUS
SA6 6ER GW4ZQY
SA6 6LL MW6RYD
SA6 6LU GW1DTA
SA6 6NW 2W0AUC
SA6 6NX GW0NKH
SA6 6NX GW1YFP
SA6 6SW GW0KYY
SA6 6TB 2W1FGR
SA6 6TF GW6YPA
SA6 7AG GW7AOE
SA6 7AG MW7GAR
SA6 7LN 2W1GAC
SA6 7PB GW3NUO
SA6 7PH 2W0JMX
SA6 7PN GW1EOI
SA6 7PQ GW4MVY
SA6 7PS GW4WPA
SA6 8BD MW7KYL
SA6 8LE 2W0REJ
SA6 8LE MW0WRY
SA6 8LQ MW1DCI
SA6 8PF GW4JDZ
SA6 8PF GW8KSL
SA61 1BW GW0EJE
SA61 1BW GW0GUY
SA61 1QF MW0MJB
SA61 2HL MW0CAB
SA61 2HL MW1SAS
SA61 2JA GW8KXW
SA61 2RD MW6NPW
SA61 2RE MW3EPJ
SA61 2SB GW8YJN
SA61 2UE MW3YBX
SA61 2UE MW6BGO
SA611FB GW1NYS
SA611HJ GW0UXX
SA62 3BN 2W0VVO
SA62 3BN MW0VVO
SA62 3BN MW3VVO
SA62 3ET MW6VOW
SA62 3HX GW4PCX
SA62 3HY MW0TJS
SA62 3PH GW4HGS
SA62 3QU MW7GCE
SA62 3SZ GW0UQH
SA62 3TD GW8WRC
SA62 3TR GW7NGU
SA62 3XH GW0SXS
SA62 4HJ MW0GDM
SA62 4LR GW1MVZ
SA62 4LS MW6IIU
SA62 4NU GW3YSI
SA62 4NU GW6IZZ
SA62 4PR GW1HEV
SA62 5DY GW0KQV
SA62 5ES MW7NAT
SA62 5NT MW0AMN
SA62 5NT MW0AMQ
SA62 5PL MW7IMO
SA62 5PW MW3IZV
SA62 5RY MW0XDN
SA62 5SP 2W0LXT
SA62 5SS GW0VYF
SA62 5SW MW6HZP
SA62 5SX GW0KZG
SA62 5SX MW0SML
SA62 5UA MW0IJE
SA62 5XD MW6XAE
SA62 5XX MW7DYH
SA62 6EN GW8NAC
SA62 6EU MW3UAA
SA62 6HF MW6COD
SA62 6JZ GW0DDK
SA62 6RA MW7BGD
SA62 6UF GW4HXO
SA626LS MW7EQU
SA63 4UU MW7YZF
SA63 4XD GW8SBO
SA64 0AX MW6WWY
SA64 0AZ MW6JFI
SA64 0DY 2W0EXZ
SA64 0DY MW0MDV
SA64 0DY MW6IVW
SA64 0DZ 2W0MZS
SA64 0DZ MW6IOZ
SA64 0EX 2W0HDB
SA64 0EX MW0HVB
SA64 0EX MW0NPC
SA64 0EX MW6BDS
SA64 0EX MW6HDB
SA64 0LG GW4MM
SA64 0LG GW4VXE
SA64 0LL MW0RLJ
SA65 9DB GW1JKP
SA65 9EE MW6IXD
SA65 9EH 2W0FZC
SA65 9EH MW0JWW
SA65 9EH MW7FEC
SA65 9LN 2W0FFI
SA65 9LN MW6AGS
SA65 9PB GW0CXK
SA65 9PT GW8UQC
SA65 9PT MW6URI
SA65 9QJ MW0BEL
SA65 9RL GW4UEJ
SA65 9SL GW8YLK
SA66 7JP GW1ALV
SA66 7NR GW7JUB
SA66 7PX 2W0BMM
SA66 7QL GW8GVI
SA66 7QS GW3RKV
SA66 7UX 2W0JMK
SA66 7XT GW7NYP
SA67 7DY MW0HUU
SA67 7EE GW1MNU
SA67 7EZ GW8PKV
SA67 7PG 2W0UPR
SA67 7PG MW7TRY
SA67 8EH 2W0LDZ
SA67 8EH MW0DX
SA67 8EH MW0TDZ
SA67 8EH MW6LDZ
SA67 8NR GW4MIP
SA67 8NT 2W0MNA
SA67 8NT MW6IMU
SA67 8SP GW3LEW
SA67 8SP MW3JCG
SA67 8ST GW0CCO
SA67 8TZ GW3KZO
SA68 0RG 2W0HRG
SA68 0RG MW6BEG
SA68 0SY GW4WMD
SA68 0XA MW6HED
SA68 0XS MW6TDE
SA7 0HF 2W0UAR
SA7 0HF MW0SWN
SA7 0HF MW6KVA
SA7 9HS GW4LKS
SA7 9JB MW0TCJ
SA7 9JR GW4UVN
SA7 9JT MW6IVK
SA7 9LD 2W0SRD
SA7 9PR 2W0NED
SA7 9PR MW0KJN
SA7 9QH GW1PAV
SA7 9QU MW6HLU
SA7 9QX GW4JGW
SA7 9SF 2W0BJR
SA7 9SF MW3OSI
SA7 9ST GW4XES
SA7 9ST GW4ZBU
SA7 9SX GW8DUP
SA70 7DP GW0JRF
SA70 7EB MW0ECY
SA70 7EB MW3NTE
SA70 7JT GW8PNE
SA70 8DB GW7GKN
SA70 8EB MW6GQD
SA70 8LT GW4NOO
SA708EN MW6HNN
SA71 4EP MW0CHI
SA71 4EP MW1BXX
SA71 4JL MW6FGJ
SA71 4PX MW6DDR
SA71 4QL 2W0BGI
SA71 4QL MW3NUP
SA71 4TR MW0BRO
SA71 5AT MW6OVV
SA71 5BD GW7FBV
SA71 5HJ GW6TEO
SA71 5HY GW4OZU
SA71 5JH GW4VTG
SA71 5JQ MW6RGK
SA71 5QQ MW6HLQ
SA71 5QX 2W0MCT
SA71 5QX MW3KNE
SA72 6DU GW0WGN
SA72 6EW MW0HWU
SA72 6EW MW0IBZ
SA72 6EZ GW4RGI
SA72 6FB GW4OXL
SA72 6HL GW7VMT
SA72 6JP GW4TGT
SA72 6JR MW6VDF
SA72 6NE MW0IPL
SA72 6NE MW1BLE
SA72 6NE MW6IYW
SA72 6NE MW6OYO
SA72 6NW MW0AIE
SA72 6PZ MW7FMV
SA72 6QN GW4REI
SA72 6RH MW7WDP
SA72 6RN GW6NLO
SA72 6RS GW4VRO
SA72 6TP GW4YVN
SA726QL 2W0ENJ
SA726QL MW6KFQ
SA73 1AN MW0GOV
SA73 1BD GW1PND
SA73 1HD GW0LIK
SA73 1HD GW4ZYV
SA73 1HR MW0BYS
SA73 1HR MW3ESE
SA73 1HR MW3ESF
SA73 1JT GW0RKD
SA73 1LG GW4ERB
SA73 1NZ GW6QLJ
SA73 1QQ MW7EZW
SA73 1QR 2W0GHV
SA73 1RA MW0JDY
SA73 1RA MW6XRO
SA73 1RE 2W0JPF
SA73 1RE MW3JPF
SA73 1SA GW0VND
SA73 1TB GW7HGU
SA73 1TB MW1DFQ
SA73 1TR MW0COB
SA73 1TR MW6FGQ
SA73 2AU 2W0OZO
SA73 2AU MW6OZO
SA73 2BW 2W0EGL
SA73 2BW MW6DFX
SA73 2DS GW4OEJ
SA73 2DS GW4OKF
SA73 2EN MW0SGD
SA73 2NU GW0TLJ
SA73 2NU MW0BBU
SA73 3HN GW4ODN
SA73 3HS MW6WPB
SA73 3LW GW0LVH
SA73 3RU MW7HJE
SA73 3TF MW3VFB
SA73 3UE GW0VEM
SA8 3AU 2W0EZS
SA8 3DT GW4KCQ
SA8 3EP GW8SBN
SA8 3EP GW8TBG
SA8 3ER MW6JDY
SA8 4AA GW1RZE
SA8 4AH MW3ZWS
SA8 4AL GW6VBN
SA8 4BN MW7DQV
SA8 4DB GW6UGC
SA8 4DL GW4KYT
SA8 4ER MW7KLC
SA8 4JD MW6LZV
SA8 4NB 2W0BBO
SA8 4NB 2W0BBT
SA8 4NB MW0GCS
SA8 4QL GW3UCJ
SA8 4QT GW4ONI
SA9 1AT GW4YWM
SA9 1BP GW4GLU
SA9 1DR MW7CRP
SA9 1SP MW3EPK
SA9 1SX MW6VPL
SA9 1TS MW7DHG
SA9 1XF GW7MLN
SA9 2FT MW3SKW
SA9 2FY GW6PXW
SA9 2LP 2W1KWK
SA9 2NZ MW3UIN
SA9 2RY GW1PKM
SA9 2XQ GW3YQH
SE1 0QB G3WKE
SE1 0QB M0JXG
SE1 1EL 2E1HFH
SE1 1EL M3HFH
SE1 2XN M6YLD
SE1 3DW M0GDC
SE1 3LD M0ATD
SE1 4DT M3RZL
SE1 5NU M6EOM
SE1 7DX 2E0HFA
SE1 7DX M0JKN
SE1 7DX M6HFA
SE1 7HB G4BWB
SE1 8DU G1UFX
SE10 0DW M6TYU
SE10 0JT M7RED
SE10 8AX 2E1DBP
SE10 9LS M0UOG
SE10 9UZ G0DRK
SE11 5UL G3RGM
SE12 0EG G1IYO
SE12 0UJ 2E0CIT
SE12 0UJ M6BDD
SE12 8AE M0HUS
SE12 8AG 2E0FSN
SE12 8AG M0SBR
SE12 8AG M6HGH
SE12 8JN G7GLW
SE12 9AB G0LRM
SE12 9BG G3GDB
SE12 9EX G8IBO
SE12 9HX 2E0IQP
SE12 9HX M7EHB
SE12 9HY 2E0NPR
SE12 9HY M0NPB
SE12 9HY M6OPF
SE12 9LX G4XUI
SE12 9RS G1XFR
SE13 5AN M7FTL
SE13 5AW G8LHI
SE13 5FL G5JBQ
SE13 5QW G4JOI
SE13 6QP M0BYM
SE13 6QW M3WUO
SE13 6RT M3WRQ
SE13 7DE M6NYX
SE13 7PA M6JKX
SE13 7PL M3RJK
SE13 7SP M6LTU
SE13 7SR 2E0JJP
SE13 7SR M0JJP
SE14 5DN M6HJX
SE14 5NJ 2E1GKP
SE14 5XF M7YOL
SE14 6DU M6URX
SE14 6QE 2E0IVC
SE14 6QE M7OAZ
SE15 1AF M7SRT
SE15 1DU 2E0UAK
SE15 1RW 2E0HHF
SE15 2TX M6FMY
SE15 3EG G3VBU
SE15 4AN G3XBH
SE15 4DZ G1BYO
SE15 4JX G7KYD
SE15 4PY G3HMO
SE16 2AG G1NGI
SE16 2AP M3JYG
SE16 2NA M7BKH
SE16 3DQ M0IVE
SE16 4BY M3IYG
SE16 4TT G7NIL
SE16 5XN G0BUD
SE16 6PU 2E0HOU
SE16 6PU M0OSY
SE16 6PU M6EYD
SE16 7DA M7YKN
SE16 7EG 2E0JUD
SE16 7EG M7KTL
SE16 7SZ M7GMJ
SE17 3HR M7CFF
SE18 1DP M7CUZ
SE18 1LN M3HIN
SE18 1SL M6HJH
SE18 1SL M6MVZ
SE18 2PH M0JRJ
SE18 2RT G1CJK
SE18 3HF G8GJO
SE18 3PB M7CJM
SE18 4PS G8WBU
SE18 5BY G0EPR
SE18 5JL 2E0WHF
SE18 5JL M0ABC
SE18 5JL M6WHF
SE18 5PU M0TDD
SE18 5QY 2E0FZS
SE18 5QY M7SAC
SE18 5TY 2E0WXF
SE18 5TY M7ELU
SE18 6PA 2E0DKF
SE18 6PA M6ZWT
SE19 2PY G1FBI
SE19 2SA G4EBI
SE19 2TE M7ABM
SE19 3AD G3TAO
SE19 3PZ M0ICL
SE2 0AJ M1FFS
SE2 0DN G4MLY
SE2 0DW G4WNF
SE2 0PD G4JGQ
SE2 0QG 2E0HNH
SE2 0QG M7ESI
SE2 0QX G3SPJ
SE2 0UY M6IDM
SE2 0XN G6SCB
SE2 9HL G0ATB
SE2 9QS 2E0POU
SE20 7BW G1XVW
SE20 7JG G3VZH
SE20 8ET 2E0GUU
SE20 8ET M0JOI
SE20 8ET M7BJT
SE20 8JY G6CVP
SE20 8LW G7RUX
SE20 8NU G4JHW
SE20 8SP M0HWW
SE21 8JY G0PAR
SE21 8NP M7FPJ
SE22 0AX 2E0ESC
SE22 0AX M6MBB
SE22 0SB G0FAH
SE22 8RH M0NTN
SE22 9AH M3TMM
SE22 9EX M6KUU
SE22 9JH G6ICC
SE23 1BW G3SXA
SE23 1HG G4PRQ
SE23 1PY M7MTE
SE23 1RH G6BJR
SE23 1SL G1MKS
SE23 2AG M7JRK
SE23 2QP G0DIA
SE23 2RU G3TPO
SE23 3HT 2E0HGO
SE23 3HT M0PGN
SE23 3HT M6HGO
SE23 3SJ 2E0EUI
SE23 3SJ M6EUI
SE25 4NG 2E0EBK
SE25 4NG M6EUH
SE25 5HB G7PWJ
SE25 6HY G7TQE
SE25 6SY G3SRC
SE25 6SY G3WRH
SE26 4AB M7FEU
SE26 4BL M7AVW
SE26 5DP G0RTZ
SE26 5JD M0IWJ
SE26 5RJ 2E0CIG
SE26 5RJ M0JAD
SE26 5RJ M6BBG
SE26 6BP M3YRS
SE26 6JQ M3XBN
SE27 0LG M6AMN
SE27 9NA 2E0DKW
SE27 9NA M6DKW
SE28 0DS M7NEJ
SE28 0NJ M3YSY
SE28 0PB 2E0POO
SE28 0PB M6DKY
SE28 8HE M7GOU
SE28 8NF 2E0ATY
SE28 8NF M0MYC
SE28 8PG M1BPN
SE28 8PU G1NYI
SE3 0EZ G0CUA
SE3 0QG 2E0MUS
SE3 0QG M6MUS
SE3 7BG G4FUG
SE3 7PE 2E0CQS
SE3 7PE M0WXU
SE3 7PE M6BZD
SE3 7QS G3GHN
SE3 7QS G4RFC
SE3 7RQ G0ILD
SE3 7TN G4JMM
SE3 8HF G4ZXV
SE3 8LR G6IRE
SE3 8QY M6SLZ
SE3 8RP M6LHN
SE3 8TT G1NTP
SE3 9HL M0BGR
SE3 9HP G0TYW
SE3 9LJ M0RAC
SE3 9QN G0BSH
SE3 9QP G1KLW
SE4 1XY M7YON
SE4 2AX M0IAG
SE4 2DN M6BVK
SE4 2JQ G0HBW
SE4 2NL 2E0TXJ
SE4 2NL G6DJH
SE4 2NL M6TXJ
SE5 7HS M7VAV
SE5 8BS M7MEZ
SE5 8DQ 2E0EVV
SE5 8DQ M6ONJ
SE5 8QQ G4WUQ
SE6 1EZ M3WNM
SE6 1JF 2E0LSW
SE6 1NQ G8WZO
SE6 1UR 2E0LBZ
SE6 1UR M6BPD
SE6 3LP M6WRB
SE6 3QR M7CNN
SE6 3SQ 2E0FVT
SE6 3SQ M7GRS
SE6 3TF 2E0TSA
SE6 3TF M6TSA
SE6 3TG M3VNI
SE6 3TW G6ZWM
SE6 4RN G7PKY
SE6 4UW G0TCA
SE7 7AD M3YKI
SE7 8ET 2E0LOX
SE7 8ET M0LOX
SE7 8ET M6LJV
SE7 8PJ M0KPU
SE7 8SH G1FAD
SE77BE M7CUQ
SE8 3AS M1MAL
SE8 3EX M6GUS
SE8 4JF G5EFV
SE8 5AJ G0VXY
SE9 1DT G4NOW
SE9 1LH M7CXU
SE9 1PP M3UJR
SE9 1QH G8XDR
SE9 1QJ 2E0RCV
SE9 1QJ M0MCV
SE9 1QJ M3RCV
SE9 1SA 2E1GUC
SE9 1SA G7SCU
SE9 1SE G0DCI
SE9 2EX G8IPY
SE9 2LP G6WCX
SE9 2NZ G0FDP
SE9 2SU M7RAD
SE9 3BN M6FWQ
SE9 3BW G3XCJ
SE9 3EB G0OWV
SE9 3EP G4VFH
SE9 3HL G0OKX
SE9 3JL M6VSB
SE9 3JX G3IGZ
SE9 3JX M7GFE
SE9 3PX 2E0ECY
SE9 3PX M0IKC
SE9 3XD G0ULL
SE9 4ET M3JUX
SE9 4JL G0JAJ
SE9 4LF G0LYG
SE9 5RF 2E0EGM
SE9 5RF M6FDG
SE9 6EX 2E0HTV
SE9 6EX 2E0UKK
SE9 6EX 2E0UKT
SE9 6EX M6UKX
SE9 6EX M7UKX
SE9 6QP G8JWX
SE9 6QP M0JHB
SE9 6RA M6ALW
SE91PB M7EIK
SE96AL M7EBL
SG1 1JJ M6TZO
SG1 1LS G3MHX
SG1 1RP G6XSZ
SG1 1RR M0GBZ
SG1 1TE G2BKZ
SG1 1TE G3SAD
SG1 1UJ G7DNV
SG1 2AS M6GKC
SG1 2LH 2E0NCR
SG1 2LH M6NCR
SG1 2NE M3BMU
SG1 2NE M3BYS
SG1 2QL M3RUW
SG1 2RS M0KPB
SG1 3BT G8SZP
SG1 3DF M6BSZ
SG1 3EZ G7CYQ
SG1 3JL G3UFB
SG1 3SL M6JJM
SG1 3XY 2E0LSC
SG1 3XY M0LSZ
SG1 3XY M7LSA
SG1 4AY G8WWI
SG1 4DQ G6JUE
SG1 4EP G4SPV
SG1 4HD G8IUG
SG1 4LF G3INU
SG1 4NQ M0NEV
SG1 4NU G1FWS
SG1 4PH M6ZCR
SG1 4PJ M3IJV
SG1 4PL M3PCP
SG1 4PW 2E0PYM
SG1 4PW 2E0YUD
SG1 4RN G6ANR
SG1 4SD G4DDX
SG1 4TB M1EMB
SG1 4TL G0RLL
SG1 4TL G1RIV
SG1 5EH G6UWX
SG1 5HF G7DRG
SG1 5JA G1SOY
SG1 5LH M3JMU
SG1 5LH M3NDC
SG1 5NF G4BYE
SG1 5PH M6EJW
SG1 5QP 2E0PDJ
SG1 5QP G8DVB
SG1 5QP M7DVB
SG1 5QS G7HCL
SG1 5QS M0EBG
SG1 5SF G4IKY
SG1 5SF G8FMA
SG1 5TB M3MBR
SG1 5TT G0RAM
SG10 6BD M6TRS
SG10 6BN G0KTW
SG10 6JG G3OMD
SG11 1HT G0KGR
SG11 1JH G4VMR
SG11 1JH G4VSL
SG11 1LH M6FMF
SG11 1QF M6AHH
SG11 1SY G0DCW
SG11 1TG G4YOS
SG11 2DN G3OTR
SG11 2EH M0SHF
SG12 0DN 2E0ECC
SG12 0DN G5EDW
SG12 0DN M7EDW
SG12 0HP G7VRY
SG12 0JQ G3OJG
SG12 0NL G4YFC
SG12 0SY G4PHV
SG12 7DT M6CSU
SG12 7EA M1BYG
SG12 7LT G8WTB
SG12 7PD G0OJY
SG12 8AZ G4VXD
SG12 9JN G3TIK
SG12 9LS G4MAS
SG12 9PG G8NDR
SG12 9QL G6WYF
SG13 7DP M6MOX
SG13 7EJ G8OIY
SG13 7JD G8OVZ
SG13 7JF G6DVO
SG13 7QT M3PJJ
SG13 7QU G3SZF
SG13 7RB G4UDZ
SG13 7RR G0AHB
SG13 8AD G8OPC
SG13 8BA 2E1CAQ
SG13 8BN M6LKA
SG13 8FX G4AEZ
SG13 8FX G6VAL
SG13 8HR M6GBQ
SG13 8RB G1IMS
SG14 1NE 2E0JVT
SG14 2BZ G4EEZ
SG14 2EU G3IOJ
SG14 2HH G0NSH
SG14 2HU M7NWE
SG14 2QG M6HQD
SG14 3AF 2E0RYX
SG14 3AF M6LOG
SG14 3DL G1BLQ
SG14 3JU G0OSK
SG14 3SF 2E0LFT
SG14 3SF M0WWV
SG14 3SF M6WKZ
SG14 3SN G7VLD
SG14 3TD 2E0HDZ
SG14 3TD M0XHD
SG14 3TD M7HDR
SG14 3TF G0JGV
SG15 6RP G0TUX
SG15 6RX 2E0PLV
SG15 6RX M0PLV
SG15 6RX M6PLV
SG15 6SW M7DNO
SG15 6SZ M3XZE
SG15 6TS G4XBD
SG15 6UF G6UMX
SG15 6UY G3WKA
SG16 6AA M0IDE
SG16 6AA M7PTP
SG16 6BN G3ORG
SG16 6EY M3HYQ
SG16 6FD M6GHW
SG16 6JS G1SJO
SG16 6JS M6HHR
SG166JR 2E0ZKT
SG166JR M3UER
SG17 5AA G3NII
SG17 5AD G6RHL
SG17 5AE M0CPE
SG17 5AU G8UFX
SG17 5BN G3JNB
SG17 5BS M0TOZ
SG17 5EA G0GGQ
SG17 5HB G6MIS
SG17 5HB G6VBE
SG17 5HW G1CIT
SG17 5LB M6JNV
SG17 5LU G4JBD
SG17 5NH G0PQO
SG17 5NH G7CUY
SG17 5NY M1AIN
SG17 5RA 2E0HPO
SG17 5RA M3HPO
SG17 5UP G6JLU
SG17 5ZS G7MYN
SG17 5ZS M0MYN
SG18 0DA M0JXM
SG18 0DG G7ABF
SG18 0HL G4GYP
SG18 0JJ G7OZU
SG18 0JW M3PKM
SG18 0NE G6KPA
SG18 0NL M3FAY
SG18 0NX G8MCY
SG18 0PE M3PIH
SG18 0PH M6EXE
SG18 0PS G4NNB
SG18 0PX M3TXS
SG18 8DX G0PHY
SG18 8EE M0WNV
SG18 8EE M1GXL
SG18 8FS M6BFL
SG18 8GG 2E1GBM
SG18 8JR M0SLP
SG18 8JU G6ZEM
SG18 8LX M0PLT
SG18 8PB G3XIZ
SG18 8PF M3NHP
SG18 8QS G6XBL
SG18 8RF M6MII
SG18 8SR G1FYF
SG18 9AH G8PQN
SG18 9BX G4FGJ
SG18 9DF G4YRF
SG18 9DU M7CYE
SG18 9DU M7FDD
SG18 9NE M1EPK
SG18 9NR G7NKS
SG18 9NY G0ECZ
SG18 9PQ 2E0RFI
SG18 9PQ M6UTC
SG18 9PS G7LUR
SG18 9PT G0ETA
SG18 9QP G1GSN
SG18 9QP G4XIM
SG18 9RS G3XCK
SG19 1AU 2E1CWJ
SG19 1AU G0TMJ
SG19 1AU G8GHR
SG19 1AX G6REA
SG19 1BJ G4SKU
SG19 1DT G4CEP
SG19 1DU 2E0HVQ
SG19 1DU M7TOG
SG19 1ED M6BKT
SG19 1ER 2E0MFX
SG19 1ER M0MFX
SG19 1EU G4MEO
SG19 1HF G7SVE
SG19 1HP 2E1DEP
SG19 1HP 2E1FDU
SG19 1HR 2E0OBD
SG19 1HR M0OBD
SG19 1JJ 2E0KMD
SG19 1JJ M6KMD
SG19 1JL 2E1CDZ
SG19 1RH 2E0KKJ
SG19 1RH M6MKJ
SG19 1TT M0WPY
SG19 1TU M7BTO
SG19 1TX G3KDD
SG19 2JU 2E1EFQ
SG19 2JY G3ZWM
SG19 2LB G0IKZ
SG19 2NJ M6JZF
SG19 2NY M7LAW
SG19 2PB M0YLG
SG19 2PS G7JJQ
SG19 2PS M3JGQ
SG19 2PY G8ZYT
SG19 2QG G4MKR
SG19 2QQ G1YRM
SG19 2QQ G3OXG
SG19 2QR G3WSD
SG19 2SD M7WZY
SG19 2UR G0OQI
SG19 3AN G7SOZ
SG19 3AP G6CTP
SG19 3BA M1FFF
SG19 3BT 2E1WEB
SG19 3DX G4MKP
SG19 3LG G0FOT
SG193NY G1ZPU
SG193NY M0ZPU
SG2 0DE M6SDN
SG2 0EB M3IVA
SG2 0EJ M6EGT
SG2 0EQ G3JLZ
SG2 0HD G7HCB
SG2 0JH G7PNE
SG2 0JJ G0HOP
SG2 0JN 2E0ZBZ
SG2 0JN M0ZBZ
SG2 0JN M3ZBZ
SG2 0NZ G6BYK
SG2 0QE M6RIZ
SG2 0QG G0BXH
SG2 7DR M3YSL
SG2 7DS G0KUQ
SG2 7DT G4XUW
SG2 7DW M3XTK
SG2 7EQ M6SNH
SG2 7JL 2E0OSX
SG2 7JL M0OSX
SG2 7JR M3DLU
SG2 7QS G4CZZ
SG2 7ST G7JSS
SG2 8AL 2E0HVL
SG2 8AL M6HVL
SG2 8AL M7FLA
SG2 8EZ M0ABH
SG2 8GD 2E0SNC
SG2 8GD M0KNM
SG2 8GD M3YCM
SG2 8JA M6AJM
SG2 8JD G4KFL
SG2 8NA M1DGS
SG2 8PX 2E1IID
SG2 8PY M6IVR
SG2 8QZ M3KCG
SG2 8RJ G6UXE
SG2 8RY G8ZFI
SG2 8UG G0UVN
SG2 8UY M1SAC
SG2 9DX G8UEY
SG2 9DX M0CUH
SG2 9DX M3NEA
SG2 9ER M3XTP
SG2 9LZ M0ISI
SG2 9LZ M6FKF
SG2 9NP G6YMA
SG2 9PH 2E0LVG
SG2 9QD M6YTF
SG2 9RN G0EVZ
SG2 9TP G4NEA
SG2 9TT M0LDC
SG20PJ M7JOL
SG3 6BQ G1HBC
SG3 6DF G0TAG
SG3 6DN M0AZZ
SG3 6JE 2E0TJD
SG3 6JE G6DXY
SG3 6JE M7TJD
SG3 6JW G8BWK
SG3 6JW M7BAJ
SG3 6NU G0FRM
SG3 6SB M0MIZ
SG4 0AN G3LQO
SG4 0BD M7GCQ
SG4 0DP G4OXD
SG4 0DP G8PUR
SG4 0DP M6JUK
SG4 0DR M0MVE
SG4 0JH 2E0RCT
SG4 0JH M0LGC
SG4 0JH M0ROC
SG4 0JH M3RCT
SG4 0LG G4KDW
SG4 0QP G6WEH
SG4 7AJ M7KBW
SG4 7AL M7WJR
SG4 7AS G4SWH
SG4 7DR M0ARH
SG4 7NG G6SHZ
SG4 7NT G2YT
SG4 7PD G6YIQ
SG4 7QZ M0RRC
SG4 7RA G4GXM
SG4 7RQ G0KJN
SG4 7SA G3ZLP
SG4 8AF 2E0DMI
SG4 8AF M6BGK
SG4 8AJ M3JYZ
SG4 8AJ M6KXO
SG4 8BE 2E0CBF
SG4 8EQ G4HPA
SG4 8EQ G7HMV
SG4 8EQ M0HAW
SG4 8EQ M0KSG
SG4 8EQ M0LNK
SG4 8EQ M0MHU
SG4 8EQ M0YSG
SG4 8EQ M6HMV
SG4 8EQ M6HOT
SG4 8EQ M6MZH
SG4 8EQ M6TWY
SG4 8NP G3PZE
SG4 8NT G4NGL
SG4 8PG G8WKE
SG4 8PR G0CRK
SG4 8PU M6UJK
SG4 8QD M6STY
SG4 8RA M6IAC
SG4 8RR M6CSN
SG4 8SD G8TRR
SG4 8TJ M3VQQ
SG4 9AN G3WRJ
SG4 9EP G4CTM

SG4 9JD 2E0YJW
SG4 9JD M0ONB
SG4 9JD M0YJW
SG4 9JD M3YJW
SG4 9LT M0JMC
SG4 9NP G4KNT
SG4 9NW G3FVR
SG4 9PJ G0GXU
SG4 9SD G8RPA
SG4 9SG 2E0DGL
SG4 9SG M6ENV
SG4 9TB G7URR
SG4 9TL M6UTI
SG48QN M7EZL
SG5 1EH G7NKI
SG5 1PA G6CFC
SG5 1PN G8BFH
SG5 1PT G6LSC
SG5 1QE M6RXS
SG5 2BE M3HPN
SG5 2HP G6XCK
SG5 2JH G4VXU
SG5 2QZ G7LXH
SG5 3DR G0EUV
SG5 3DR G0UEW
SG5 3LS G4PSO
SG5 3LS G7PQP
SG5 3NS G4CML
SG5 3RX G8FMT
SG5 3RX M0FMT
SG5 3SL G4YDH
SG5 3UP G3UWM
SG5 4EH 2E0JKT
SG5 4EH M6TEM
SG5 4EL G0UVX
SG5 4FY G0VVZ
SG5 4HH M1CAH
SG5 4HT G4THF
SG5 4JE 2E0BYJ
SG5 4JX 2E0MIH
SG5 4JX M6MHU
SG5 4RU G4ALR
SG5 4RU G4KQD
SG5 4RZ M7WVC
SG53XY G6CRC
SG6 1PH G8EDS
SG6 1RF M6FKE
SG6 1RG M6KAZ
SG6 1TU M0HHM
SG6 2DE 2E0ASU
SG6 2EU G6JBQ
SG6 2HY G1JVY
SG6 2HY M3AWS
SG6 2JA G8DKK
SG6 2LU G0EVD
SG6 2NP 2E0EPT
SG6 2NZ 2E0BYG
SG6 2NZ M6EGO
SG6 2SP 2E0XEV
SG6 2SP M6XEV
SG6 2TY 2E0XMC
SG6 2TY M6XMC
SG6 3EA M7BZO
SG6 3PY M0ARQ
SG6 4BD 2E0PJN
SG6 4BD M3PJN
SG6 4HG G1UFH
SG6 4HY G1BYJ
SG6 4JZ G0AMX
SG6 4LA 2E1RAF
SG6 4LA G0TAK
SG6 4LA G3ELV
SG6 4LA M0KAC
SG6 4LF G1OKV
SG6 4LP M7DZM
SG6 4NQ 2E0ISS
SG6 4PE M7JUS
SG6 4PL G0MJP
SG6 4QF M0NGK
SG6 4RZ G0PJC
SG6 4TG 2E0PJC
SG6 4TG M6PJI
SG6 4TZ M6XKN
SG6 4YN M7KMP
SG7 5AA M7YZY
SG7 5HL G0NZJ
SG7 5LJ G0RRV
SG7 5RN M6CIP
SG7 6DA M6LCC
SG7 6DG G8ABX
SG7 6HJ G6WKZ
SG7 6LH 2E0STP
SG7 6LT G4HIE
SG7 6LT G6KBS
SG7 6NU 2E0FHC
SG7 6NU M0IXM
SG7 6NU M6OLQ
SG7 6RX M6ERC
SG7 6RZ G1XXR
SG7 6RZ G8NHO
SG7 6TD M7UDC
SG7 6TW G6MAY
SG8 0DJ G0CMB
SG8 0DJ G1HND
SG8 0JU G3WAB
SG8 0NE G3WMD
SG8 0NP M7NBA
SG8 0PF G6PCE
SG8 0RB G6VIK
SG8 5BB G4TEU
SG8 5BP G4MES
SG8 5BW G4HPE
SG8 5BW G5BBC
SG8 5BW G6XQY
SG8 5JX 2E0NKM
SG8 5JX M6NKM
SG8 5LR G0DCU
SG8 5NG G1HDR
SG8 5PT G0LZD
SG8 5QY 2E0EQZ
SG8 5QY G2XV
SG8 5QY G8EVY
SG8 5QY M0IPO
SG8 5QY M6KTV
SG8 5SN 2E0WDN
SG8 6BZ G0FMT
SG8 6DD G1VRA
SG8 6EB M5SKY
SG8 6HD 2E0IGT
SG8 6HD M6VET
SG8 6JD G8DJU
SG8 6JX G8KDD
SG8 6UH M6FTY
SG8 7AD M5AGR
SG8 7DJ G4NIX
SG8 7ED G1GIJ
SG8 7ER M6CIQ
SG8 7ES G4NIY
SG8 7ES G6XEF
SG8 7EU G0TIL
SG8 7XU G0AUJ
SG8 8LT G0KFT
SG8 8RP G4FOH
SG8 9BN G0SII
SG8 9DQ G8RZL
SG8 9EA G7VEX
SG8 9EA M3TNW
SG8 9EB M7ATJ
SG8 9EU G1HVW
SG8 9HP M0BDW
SG8 9JF G6EDD
SG8 9LJ 2E0CIR
SG8 9LJ M3IJS
SG8 9NQ M0XBQ
SG8 9NQ M7HES
SG8 9QL G1HQE
SG8 9QX G0AQI
SG8 9SE G1YZJ
SG8 9UQ G1OKF
SG9 0PJ M7DLQ
SG9 0PW G3GRS
SG9 0PW G4BUO
SG9 0QR M3LHZ
SG9 0SN 2E0LSO
SG9 0SN M0LSO
SG9 0SN M6LSO
SG9 0SU 2E0CIQ
SG9 0SU M3IWR
SG9 9BQ 2E0HPR
SG9 9BQ M1HPR
SG9 9BQ M6HPR
SG9 9DZ G6BWE
SG9 9DZ M1NHR
SG9 9HJ M0DWP
SG9 9RA G1RPO
SG9 9RH G0HBL
SG9 9SE M7JXW
SK1 2QE G0OGP
SK1 3NL M3AHU
SK1 4AB M1ANQ
SK1 4BP 2E0RFU
SK1 4BP M0RFU
SK1 4BP M6BYR
SK1 4DS G0ELX
SK1 4HX M6WXF
SK10 1FW G1AKA
SK10 1GE 2E0DOD
SK10 1GE G4FF
SK10 1GE M0PAI
SK10 1GE M3TUL
SK10 1QB G8CZW
SK10 1QP G3HUR
SK10 2DA M3JLD
SK10 2DQ M3YFL
SK10 2EL G4XRG
SK10 2JB G7NNU
SK10 2NS G6CGO
SK10 2PF G4JNE
SK10 2PN G1NUS
SK10 2PS 2E0NSR
SK10 2PS 2E1AEQ
SK10 2PS G0DMV
SK10 2PS G0JNJ
SK10 2PS M6NSR
SK10 2RN G8JQG
SK10 2RR 2E0BAX
SK10 2RR M0GIA
SK10 2TS M7ENW
SK10 2TU G6AKK
SK10 3BG M0TEZ
SK10 3BR G3JQ
SK10 3BS G1SHI
SK10 3DA G0DRO
SK10 3DJ G0DCO
SK10 3GL M7AWE
SK10 3GL M7JVS
SK10 3HB 2E0VWT
SK10 3HB M0VWT
SK10 3HB M6XCP
SK10 3HJ M7BHO
SK10 3JA G3CWI
SK10 3JL G0GWA
SK10 3LN G3LIO
SK10 3LN G8RDJ
SK10 3NQ G6XFU
SK10 3RE G0IKB
SK10 4HZ G0AHJ
SK10 4QW G3UJA
SK10 5AH G6VSQ
SK10 5DT G6ECN
SK10 5HJ G8BPU
SK10 5HT G4XTO
SK10 5JG G1JDF
SK10 5JZ G3SXI
SK10 5LJ G0NYS
SK10 5PY 2E1FUS
SK10 5SD G4GMK
SK102GY M3SVL
SK11 0BU 2E1FLW
SK11 0EP G8GXO
SK11 0LJ G4NXX
SK11 6HA M6RJM
SK11 7AW M1BYH
SK11 7EN G3VKF
SK11 7ES G0KVU
SK11 7ES G4WOL
SK11 7GE 2E0ZSH
SK11 7GE M3ZSH
SK11 7NJ G1NTR
SK11 7PL G7POV
SK11 7QA M3MMG
SK11 7YG M3HVP
SK11 7YH 2E0RXX
SK11 7YH G4MWS
SK11 7YH M0TXX
SK11 7YH M3ZRY
SK11 8EX G6LFG
SK11 8LL G3RWE
SK11 8LW G6FFU
SK11 8QL G8XQH
SK11 8QP G0DMU
SK11 8QT 2E0EYP
SK11 8QT M0HGY
SK11 8QT M3EYP
SK11 8RH M1EYP
SK11 8RJ G1NPC
SK11 8RS 2E0MUD
SK11 8RS M6FAY
SK11 8RT G1DUT
SK11 8SD G7MSH
SK11 8TH G7OQG
SK11 9AP G7NKV
SK11 9BF G3HFM
SK11 9BW M3JPI
SK11 9BX G8BVF
SK11 9LQ M3FLJ
SK11 9RR 2E1HNB
SK11 9RR M3CBN
SK11 9SW G3VDB
SK12 1AR G6VKL
SK12 1BL G0PJW
SK12 1BU G4ZGP
SK12 1DD G4GEO
SK12 1EF G1AFI
SK12 1HA G3SHF
SK12 1HA G6UQ
SK12 1HA G8SRS
SK12 1HA M5MDX
SK12 1HG 2E0SYN
SK12 1HG M0AGW
SK12 1HG M6CXO
SK12 1HN G0ONG
SK12 1NN 2E0URD
SK12 1NN G6URD
SK12 1NN M6URD
SK12 1PU G4IUJ
SK12 1PW M0GWF
SK12 1QH M1PTR
SK12 1RW G6RJH
SK12 1RW G7UJC
SK12 1SP G3THF
SK12 1TG G4MEH
SK12 1XB G6HJU
SK12 1YE G0EVP
SK12 1YX M6LBX
SK12 2AE G0CQU
SK12 2AY G8CET
SK12 2DB G6YCE
SK12 2RG G0NLL
SK13 0JG M7UFC
SK13 1AR G1XCB
SK13 1EN G7JUJ
SK13 1JZ 2E0MPC
SK13 1LH M0AIC
SK13 1LR M1EYO
SK13 1NX 2E0PSP
SK13 1PD G7JNM
SK13 2AD M7DTW
SK13 2AT M3RUO
SK13 5HL M3XAV
SK13 5HP G6BIX
SK13 6HX G0TXO
SK13 6NF G4KWF
SK13 6PX G6GIV
SK13 6US M7MGL
SK13 7AH G6EKS
SK13 7AJ G4GNQ
SK13 7AJ G4LMR
SK13 7BG G8BEQ
SK13 7DG M7KLM
SK13 7RL M0SAQ
SK13 7TL G4OQH
SK13 8BF M7DUH
SK13 8LW G6DQK
SK13 8NW G1YKK
SK13 8NW G6LLL
SK13 8NZ G0BAY
SK13 8QN G6YSZ
SK13 8RG G4IFJ
SK13 8RJ G0WJS
SK13 8RZ G6ARR
SK13 8SH 2E0VKR
SK13 8SH M0RVK
SK13 8SH M7CQU
SK13 8UD 2E0VMD
SK13 8UD M0TMD
SK13 8UD M6HMK
SK13 8UF G8YIG
SK14 1DT 2E1INC
SK14 1DT G8INC
SK14 1DT M3NXA
SK14 1QX G0KLD
SK14 1RP M6YBB
SK14 2PW M6XAD
SK14 2SW G8OKD
SK14 3BW G4HLA
SK14 3QW M6AOG
SK14 3SN G0RRR
SK14 3SN G7KKW
SK14 4AF M0SWZ
SK14 4BS G0RGN
SK14 4DL G6MWM
SK14 4DW M3LPI
SK14 4DW M3LPK
SK14 4RT G6PZS
SK14 4SY 2E0MBV
SK14 4SY M3WPU
SK14 4TD M6VAK
SK14 5AB G0URT
SK14 5AU G4ING
SK14 5BB G8LCS
SK14 5DD G4USX
SK14 5JX G4PYQ
SK14 5LJ G1UTN
SK14 5NS G0WVV
SK14 5RE G7KEI
SK14 5RU 2E0CGX
SK14 5RU M3NBU
SK14 5ST M6MYS
SK14 6JA 2E0DVM
SK14 6JA M6FCJ
SK14 8LE G0NPK
SK14 8PA M3PVX
SK14 8QH G6SLY
SK15 1AJ M6TMM
SK15 1HD 2E0MKK
SK15 1HD G7FEG
SK15 1HD M3ATC
SK15 1HD M3MKK
SK15 1HG G0NIL
SK15 1HL G0KYB
SK15 1HU G0PVR
SK15 1HU M3SHX
SK15 1LE M0NSI
SK15 2HR G1VTP
SK15 2LT G0WFD
SK15 2QQ G6LBO
SK15 2QZ G7WFD
SK15 2RP 2E0LTM
SK15 2RP M0PBF
SK15 2RP M7PWB
SK15 2SP M7VJB
SK15 2TR G4NHW
SK15 3DZ G1TYP
SK15 3EN 2E0MXW
SK15 3EN M0SAD
SK15 3EN M3MXW
SK15 3HE M7VKT
SK15 3HJ G0TIX
SK15 3HU G0ITZ
SK15 3JX M6LEH
SK151EX 2E0OFF
SK151EX M6GRK
SK16 4AU M6ZEP
SK16 4AU M7FBY
SK16 4EW G1BSY
SK16 4JJ G7BCK
SK16 4SJ M6GSC
SK16 4UD M1DJO
SK16 4UN M6NWI
SK16 5AB M3PQT
SK16 5BU 2E1SUE
SK16 5BU G4ZPZ
SK16 5DN G0NAJ
SK16 5DP M0AHF
SK16 5EG G4KFJ
SK16 5HR 2E1CWP
SK16 5HR 2E1CWX
SK16 5HR G0XAU
SK16 5JA M1EEW
SK16 5JL G1HBE
SK16 5JL G1HTF
SK16 5NW G0LOE
SK16 5QS M6STP
SK16 5RT G0CKM
SK165BU 2E0EKX
SK165BU M0ILY
SK17 0DH G0JWD
SK17 0LU 2E0SCF
SK17 0LU M0TGT
SK17 0LU M6CRF
SK17 0LU M6JCF
SK17 0LU M6LEF
SK17 0LU M6SCF
SK17 0NF 2E1AZQ
SK17 0RG G1YJH
SK17 0SG G1RIY
SK17 6HH G4BJT
SK17 6HN G1OSA
SK17 6HQ G1LVV
SK17 6LP M6XVB
SK17 6NQ G8DAM
SK17 6NY 2E0WAT
SK17 6NY M3AJW
SK17 6PJ 2E0SRT
SK17 6QX 2E0BSG
SK17 6QX M0GOB
SK17 6QX M6ARS
SK17 6RD G7TMM
SK17 6ST M0DZM
SK17 6UA M6NCL
SK17 6XX M1CGQ
SK17 6YU G7NFK
SK17 7DD 2E0NIB
SK17 7DD M6NIB
SK17 7DN M6XJO
SK17 7DX G8YTX
SK17 7EA 2E0DHS
SK17 7EA M3DHS
SK17 7HW M6KWL
SK17 7JF M3MQM
SK17 7PH M3LIB
SK17 7PU G0MUR
SK17 7TA 2E0CFV
SK17 7TF 2E0CWR
SK17 7TF M6CWR
SK17 7TJ G0WSP
SK17 7TQ M3VMA
SK17 7TW 2E0TCG
SK17 7TW M6TCG
SK17 8AY G1OHD
SK17 8DP M3JQG
SK17 8DW G4MRQ
SK17 8DW G4SPA
SK17 8ET G6LXE
SK17 8JT G4HGN
SK17 8LP G0CWS
SK17 8NP G0IOK
SK17 8PX M3ENO
SK17 8RJ M1WVS
SK17 8RJ M3TOT
SK17 8SW G1TBI
SK17 9AG M3YXL
SK17 9HG M0AVK
SK17 9HG M3MSZ
SK17 9JF 2E0HAE
SK17 9JF M0HOO
SK17 9JF M3OWI
SK17 9JS G0XKK
SK17 9JS G4BUX
SK17 9JS G4IHO
SK17 9JS G8BUX
SK17 9JU G7TZN
SK17 9NU 2E0WFW
SK17 9NU M6WFW
SK17 9PL G7NHL
SK17 9PL M3NFK
SK17 9RY G0RKT
SK17 9SE G0RYQ
SK17 9SG 2E0GAU
SK17 9SG M0OKK
SK17 9SG M6OKK
SK17 9TS G4KNQ
SK178FQ 2E0HDM
SK178FQ M3WBA
SK2 5BS G4DSR
SK2 5BS G8MFR
SK2 5EP G4WHF
SK2 5HX M7GTJ
SK2 5RN G7ILS
SK2 5SF M1ANT
SK2 5SF M1SIM
SK2 5SN M6NVB
SK2 5UR G1FFR
SK2 5UU 2E0JLR
SK2 5UU M6FCF
SK2 6AA M0SAV
SK2 6BN G8CHY
SK2 6BX G0KHR
SK2 6DW 2E0HSB
SK2 6DW 2E0RHM
SK2 6DW M3UVM
SK2 6DW M6ACE
SK2 6EP G7IOC
SK2 6HA G4ORV
SK2 6HF M7RJL
SK2 6JS G4FRM
SK2 6SP 2E0TJU
SK2 6SP M0TJU
SK2 6SP M3TJU
SK2 7AR M3UKY
SK2 7BH M6AAL
SK2 7EB G1JVF
SK2 7EB G4HON
SK2 7EB M0ECR
SK2 7JD 2E0DJZ
SK2 7JD M6FFO
SK2 7LQ G3RZJ
SK22 2JG G7OJA
SK22 3AY G7KUB
SK22 3DH G3TPI
SK22 3DL M0JYF
SK22 3DQ 2E0AZU
SK22 3DQ M3ICN
SK22 4DP G4YAB
SK22 4HR G1IMD
SK22 4JT G4XDT
SK22 4NE G0UPO
SK22 4NE G4YLQ
SK23 0HD 2E0YLD
SK23 0HD M7LXD
SK23 0HY 2E0GYZ
SK23 0JF G0SMP
SK23 0JF G7KRZ
SK23 0LF G7CJW
SK23 0LX G1COD
SK23 0PX G6YDN
SK23 0QF G3NZV
SK23 0QF G3ZVS
SK23 0QQ 2E0FOL
SK23 0QQ M6FOL
SK23 0TA G1ABJ
SK23 0TS G4LUQ
SK23 6AQ M7TBR
SK23 7AY G1OXH
SK23 7AY G3EEO
SK23 7BH G6LFR
SK23 7BP M0WLH
SK23 7DP G4IXF
SK23 7DT G0BDN
SK23 7HD G0GCJ
SK23 7LH G4CDZ
SK23 7NF G6UOO
SK23 7NH G8MUX
SK23 7PF G4KRG
SK23 7QU G1HQO
SK23 9ST 2E0HOY
SK23 9ST M7STH
SK23 9TP M6XLF
SK23 9TU M3OXN
SK3 0JJ G7SRH
SK3 0LQ 2E0GXE
SK3 0LQ M6GCY
SK3 0PX G3JZT
SK3 0QL G0RXA
SK3 0QL M1BAR
SK3 0QY G6RSU
SK3 0SN M7USL
SK3 0TX G0MLC
SK3 0UJ G1HWJ
SK3 0UN G7TQT
SK3 8HQ G8ZAD
SK3 8JE G8MHT
SK3 8NQ M7WTG
SK3 8PG 2E0UTX
SK3 8PG M0UTA
SK3 8QU G3NBY
SK3 8SL M3JGU
SK3 9BX M0TKG
SK3 9EA 2E0KLB
SK3 9ER 2E0CTF
SK3 9ER M0HNJ
SK3 9ER M6PRE
SK3 9HF G8UWM
SK3 9JU G6PWQ
SK3 9NJ 2E0GCL
SK3 9NJ M6GBV
SK3 9QG M0BIH
SK3 9QL G0OQR
SK3 9RZ 2E0REU
SK3 9RZ M6PFP
SK4 1HZ 2E0BDI
SK4 1HZ M3LQD
SK4 1NH M3IRS
SK4 1NH M6TGV
SK4 2AA G6LQE
SK4 2DB G8YTP
SK4 2DQ M3UTJ
SK4 2PE G3TFR
SK4 3HD G1JFU
SK4 3HD G4RPJ
SK4 3JB G4DVI
SK4 3JS G4HNO
SK4 3NB G8UBP
SK4 3PG G8NRU
SK4 3PU G1CQT
SK4 3QF G0AMY
SK4 4AX G3RRG
SK4 4BX G3WFW
SK4 4BX G7KSP
SK4 4EB M6EJR
SK4 4NE G3YFD
SK4 4NL G4HWW
SK4 4PY G6KSO
SK4 4RL G0WOP
SK4 4RP G7PRD
SK4 5AW G1FMA
SK4 5BR M6AVA
SK4 5DJ M1BSM
SK4 5HP 2E0XYA
SK4 5HP M0JDF
SK4 5HP M0XYA
SK4 5HS M1BBU
SK4 5NW M3ZIA
SK5 6BE M6TPS
SK5 6JD M0LPB
SK5 6JD M6BTA
SK5 6JD M7PLB
SK5 6LE M6DLL
SK5 6SH G0IES
SK5 6TD 2E1HQV
SK5 6TD 2E1HQW
SK5 6UU G0LJF
SK5 7DQ G7SXB
SK5 7EU G1JQK
SK5 7HU 2E0RTN
SK5 7HU M0RUM
SK5 7HU M6KBI
SK5 7JB M0JYP
SK5 7LL G8ZLU
SK5 7NA 2E0TLC
SK5 7NA M0ZPZ
SK5 7RE G0BKZ
SK5 8BG 2E0NLY
SK5 8DB G1JIR
SK5 8JY G0TIP
SK5 8JZ M0JPR
SK5 8NQ M6ANP
SK56NU G0DTT
SK6 1AS G0WMZ
SK6 1BH G6OOP
SK6 1BL G1YZH
SK6 1DJ M7NDB
SK6 1EE G6LKJ
SK6 1EF G0GSM
SK6 1HA M6VBZ
SK6 1LH 2E1CXP
SK6 1PG G4KUC
SK6 1PG G8VHF
SK6 1PT G6GUT
SK6 1PY G4OZC
SK6 1QL M7RFA
SK6 1SL M3UQP
SK6 2ED G1TNK
SK6 2EH M6PGO
SK6 2HQ G8ILD
SK6 2HR M1EAS
SK6 2HS M7CPT
SK6 2PJ M7NJM
SK6 3BE G6SOO
SK6 3EE M0ATV
SK6 3EN G8IAN
SK6 3EY G0EOM
SK6 3JT 2E1FIX
SK6 3JT G0PUD
SK6 4JF G3PTX
SK6 4JN M6MZV
SK6 4PT G6RIC
SK6 5BB M1BMC
SK6 5BE G8CLI
SK6 5BR G3PMV
SK6 5BT G3XGH
SK6 5EB G7GGN
SK6 5ET G4GWF
SK6 5PG G4NPU
SK6 5PN G8OOC
SK6 5PR 2E1WVF
SK6 5PR G4WVF
SK6 6DF G4MUL
SK6 6HJ G3MUO
SK6 6HJ G4BEV
SK6 6JL M0BEX
SK6 6LL G7BYU
SK6 6LP M1VSR
SK6 6NA G0EQH
SK6 6PJ G3SNG
SK6 6PT 2E0JEK
SK6 6PT M6BMZ
SK6 7BG G0PDH
SK6 7DT G8NOS
SK6 7HP M3EAE
SK6 7JS 2E0JEH
SK6 7JS M6FRI
SK6 7PB 2E0TTG
SK6 7PB M6TMH
SK6 7QL M3EXY
SK6 7RJ M6TMA
SK6 8AB G4FBG
SK6 8BH G0NKU
SK6 8BJ G6WWA
SK6 8BP G4GXQ
SK6 8DG G0LLU
SK6 8HL 2E0ROR
SK6 8HL M7BFN
SK6 8JE G3NXX
SK67BT G8WIJ
SK7 1LG G3TDH
SK7 1PJ M3ZVS
SK7 1PP G4GRU
SK7 2AB 2E0HPB
SK7 2AE 2E0RPE
SK7 2AE M0REP
SK7 2AE M6RPE
SK7 2BY M3NMP
SK7 2DT G3NUQ
SK7 2ER G3KAF
SK7 2JR G0ADJ
SK7 2QW M3FFU
SK7 3DT 2E1HLU
SK7 3JS G7EAH
SK7 3JW G0KZO
SK7 3JW G4SYC
SK7 4BJ G1GYC
SK7 4BU 2E0TWD
SK7 4HZ M1AQJ
SK7 4QE G0FLQ
SK7 4QG G4GEY
SK7 4QP G0CJA
SK7 4QW G0HEX
SK7 4RU 2E0MYL
SK7 4RU M0IKI
SK7 4RU M6MYL
SK7 5JB M0KJC
SK7 5JB M3KJC
SK7 5JN G4OQP
SK7 5LR G0LDU
SK7 5QY G4VSW
SK7 6EJ G8YUP
SK7 6EJ M3YUP
SK7 6EL M6XDW
SK7 6HG G6DBQ
SK7 6HP 2E0MIU
SK7 6HP M0JFM
SK7 6HP M6MIU
SK7 6HU G8WPV
SK7 6JB G8EVR
SK7 6NR G8RIC
SK7 6NR M5KJM
SK73JU G8VLR
SK8 1BA G4CSV
SK8 1DX G4DQL
SK8 1HY G7TKT
SK8 1QY G3MTR
SK8 2AQ G6TRN
SK8 2AQ M6UAP
SK8 2EE M0BMW
SK8 2LZ M6CMK
SK8 3AA M3FJQ
SK8 3AJ G4FAS
SK8 3AS 2E1INW
SK8 3DL 2E0LKC
SK8 3DL 2E0LMD
SK8 3DL M6LCH
SK8 3DL M6YEH
SK8 3DZ G0KCE
SK8 3LF G1TAU
SK8 3LL G3ZBZ
SK8 3NH G7PIG
SK8 3NR G8BHX
SK8 3PF G4SVV
SK8 3PJ G0DOU
SK8 3RH M6MTJ
SK8 3RH M6NHJ
SK8 3RN G0GJD
SK8 3RR M3EEJ
SK8 3RU G4DOJ
SK8 3RW M3GIY
SK8 3UD G8ZCJ
SK8 4AE M0SGW
SK8 4AL G3UYG
SK8 4NA G3NYE
SK8 5DD M3DRM
SK8 5HP G6AXK
SK8 5PF G0NKM
SK8 5PN 2E0ETN
SK8 5PN M0TKT
SK8 5PN M6ETN
SK8 5RU 2E1GGT
SK8 6AX G3VOU
SK8 6DG G1WTY
SK8 6EQ G4NFA
SK8 6EQ M0LAR
SK8 6EQ M7DIM
SK8 6EX G8TMJ
SK8 6JG G4KJK
SK8 6LS M0ARA
SK8 6PW G0BAA
SK8 6PW G0OZJ
SK8 6PW G1NCR
SK8 7AJ G8GRP
SK8 7AL G8LDC
SK8 7DB G6YDO
SK8 7DS G4FRW
SK8 7EP G8GHO
SK8 7HU G6TKV
SK8 7LG G4EUC
SK8 7NP G8UEU
SK8 7PB G6FOI
SK8 7PZ G0IGB
SK83PA M7AEX
SK87DS M7OEN
SK9 1LU G4IRU
SK9 1QE G4CQV
SK9 2AD G1JDQ
SK9 2BB G3RKF
SK9 2EY G1MWS
SK9 2EY M0GMG
SK9 2HQ M3RWV
SK9 2NT M7LCA
SK9 3AF G6OER
SK9 3AR G0OPG
SK9 3AR G7IOB
SK9 3JT G4IRG
SK9 3UF G8JHL
SK9 4EP G8IXP
SK9 4HE G3WPF
SK9 5BN M0DFD
SK9 5BN M3CIE
SK9 5EN G3ONI
SK9 5JA G1HFH
SK9 5NQ G3TSZ
SK9 5PX G1GYH
SK9 5QE M0AAS
SK9 6BG G3VOM
SK9 6BL 2E0LEL
SK9 6EG G0AXE
SK9 6HD G1ORS
SK9 6HD G7CMN
SK9 6HD G7RMZ
SK9 6HD M3ORT
SK9 6JD 2E1INT
SK9 6LG G3USO
SK9 6LP G3WTQ
SK9 6LS G4WKD
SK9 7HW G7TNT
SK9 7HZ G6TSM
SK9 7PN G3BNW
SK9 7PQ M3NAW
SK9 7RE 2E0HTO
SK9 7RE 2E0PNA
SK9 7SD M7GSD
SL0 0DT 2E1DRC
SL0 0ET G6MKZ
SL0 0JS G1EKP
SL0 0NJ G6XRK
SL0 9RJ M6PZY
SL1 1TS M0RNX
SL1 2DA M6ETS
SL1 2HX G7HID
SL1 2HY M1ESV
SL1 2LQ G3RZF
SL1 2NU M0HWN
SL1 2YR 2E0DFS
SL1 2YR M6DBL
SL1 3JU G4FKE
SL1 3XG G4XGD
SL1 4NL 2E0ICN
SL1 5BB 2E0MWN
SL1 5BB M0MWN
SL1 5BB M6FJN
SL1 5DF G4CIJ
SL1 5JW M6VMO
SL1 5LA 2E1CGS
SL1 5LE M7IKA
SL1 5QZ M6RBO
SL1 5RE G6GEN
SL1 5UR G8WQC
SL1 5US M7TRM
SL1 6AH G4HIN
SL1 6AY G8NNP
SL1 6HD G4RAA
SL1 6JR 2E0TPI
SL1 6JR M6TJU
SL1 6JU M1ACN
SL1 6NN M6MYU
SL1 7BQ G3YFO
SL1 7EW G6DRP
SL1 7LY G6WZC
SL1 7NF 2E1AGE
SL1 8AT G8TEQ
SL1 8AW M1JJS
SL1 9BQ M7KIC
SL1 9HB G0WYR
SL16PE M7TVP
SL2 1DY M0GFK
SL2 1HG G6POC
SL2 1LG G7WGE
SL2 1SL G4XTZ
SL2 1SL M1DSQ
SL2 1SU M6CUU
SL2 1TA M7AOM
SL2 2AD 2E0EBZ
SL2 2AD M3OJJ
SL2 2JY G6XJI
SL2 2QW M6OUR
SL2 3HE G4LLN
SL2 3HW M6BFS
SL2 3NT G6OCF
SL2 3QT M6VVB
SL2 3XL 2E1AXI
SL2 3XT G0GUN
SL2 3XT G1DSA
SL2 4AX G3PFO
SL2 4ER G6VUF
SL2 5BG 2E0EBX
SL2 5BG M6KGR
SL2 5EZ M0MRY
SL2 5PN G1SPJ
SL2 5XH M3FZB
SL3 0LJ G0JAM
SL3 6RQ M6SLC
SL3 7DA G4XAN
SL3 7ET G6TSF
SL3 7ET M3TSF
SL3 7FQ M0GPX
SL3 7HN G4HTB
SL3 7JY M3CJE
SL3 7RL M6FEZ
SL3 8AU G0TGQ
SL3 8AU G6FCJ
SL3 8AX G1MDQ
SL3 8ED G8GGM
SL3 8JH G0HXM
SL3 8JH G3VHH
SL3 8SY M7DWP
SL3 8UE M7BJH
SL3 8UU M6DFQ
SL3 8WG M3VLX
SL3 9BA G4CRW
SL3 9NJ 2E0BHZ
SL3 9NJ M6TBM
SL4 2AN G3ZCD
SL4 2LT G6PWF
SL4 2PH 2E0EFZ
SL4 2PH M6HND
SL4 2QR M7RVM
SL4 3LP G8FUO
SL4 3RD G1HDG
SL4 3SQ G8UYK
SL4 4AE G0HTK
SL4 4JP G7WFZ
SL4 4YU G7LTT
SL4 5BX G4WPR
SL4 5DX G8DPH
SL4 5PS G4PUB
SL4 5PS G4SVG
SL4 5TD G4WQS
SL4 6BN G7VME
SL4 6HL 2E1IJD
SL4 6HL 2E1IJE
SL4 6NF G7CYD
SL43EN M7LDK
SL5 0HU G3PLY
SL5 0NR G1FXX
SL5 8HQ G8LF
SL5 8NB G1PSS
SL5 8QB G6EIZ
SL5 9AD G4ZWX
SL5 9GF G4EYV
SL5 9QH M1CSZ
SL5 9RH M3GXG
SL5 9RH M3VOA
SL6 0JR G4JDF
SL6 0PB G6ZEZ
SL6 1DD G6CJB
SL6 1NX G0OGE
SL6 1RY M7BPA
SL6 1XE G6GFR
SL6 2BJ G0TNY
SL6 2DF M0TXA
SL6 2DF M7TBD
SL6 2DZ 2E0IEM
SL6 2DZ M6IEM
SL6 2ED M6JPX
SL6 2GZ G0VKE
SL6 2JS G0MMQ
SL6 2TX M6UMB
SL6 2US G3WHB
SL6 2YJ M6OAF
SL6 2YP G3ONR
SL6 2YU M3BJZ
SL6 3BB 2E0FZF
SL6 3BB M7AKX
SL6 3DD M3TGT
SL6 3EP M1HVJ
SL6 3JA 2E0CJV
SL6 3JA M6AYL
SL6 3JY G0BMU
SL6 3NR G4XOW
SL6 3UZ G8GQF
SL6 3XH G0SVN
SL6 3XJ 2E0KZQ
SL6 3XJ M6KZQ
SL6 3XL M3JAC
SL6 3YS G3VMR
SL6 4ED G7REC
SL6 4GY G4USK
SL6 4NF G4FTG
SL6 4NG G4XYN
SL6 4QT G1MRX
SL6 4SA G8UPO
SL6 5AR G3UXY
SL6 5BJ M1TUG
SL6 5DN 2E1BPV
SL6 5DW M6OJB
SL6 5EG G4SZA
SL6 5JP G3OTN
SL6 5JW M1BAD
SL6 6AX G8SRV
SL6 6BE M0PGM
SL6 6DA G4XDU
SL6 6DF G8MYV
SL6 6EG 2E0TVM
SL6 6HN G4KMH
SL6 6HN G8TFR
SL6 6LG G0GLA
SL6 6LT M0RFK
SL6 6QF G8KLA
SL6 6RX G0IIP
SL6 6RY G6NBI
SL6 6SD M6PTX
SL6 7DT M3HXW
SL6 7EJ 2E1ANN
SL6 7HE M3HCG
SL6 7LF G0PMZ
SL6 7SH G0FFL
SL6 7TZ G6BJY
SL6 7UF G4YDL
SL6 7UH G0AYA
SL6 7UH G0BQE
SL6 8EU G3YDT
SL6 8HT G0CXV
SL6 8JZ G4BOV
SL6 9DE M6ZMC
SL6 9DH G7EBR
SL6 9DN G3PQA
SL6 9JF G0CMW
SL6 9JT G3WYK
SL6 9LS M3UCZ
SL6 9NF G1LIG

SL6 9NF M7GNH
SL6 9SN G0NJQ
SL7 1DQ G3TOP
SL7 1DR G3DXD
SL7 1JW G7JDN
SL7 1SA G7TWW
SL7 1TF M0HIB
SL7 1TX G6GIF
SL7 1UW G4KCD
SL7 2DP G8KKG
SL7 2PL G5BCO
SL7 2QU G6VF
SL7 2RE G4IDJ
SL7 3BZ G4YSH
SL7 3HA 2E1JOY
SL7 3LD M7CHG
SL7 3LU G1SNO
SL7 3NG 2E0FZB
SL7 3NG M0LNO
SL7 3NG M7SWE
SL7 3PY G3LVW
SL7 3PY M6LHS
SL7 3RB G0UYM
SL7 3RH G3IQF
SL8 5AZ G3INQ
SL8 5DH M6XXZ
SL8 5QN 2E1CXE
SL8 5TJ 2E0GCM
SL8 5TJ M0VCE
SL8 5TJ M6GCM
SL8 5TJ M6IOS
SL8 5TY G0MGM
SL85AA 2E0IWF
SL85AA M6IWF
SL9 0DD G0RAU
SL9 0DD G6XDB
SL9 0EJ M1SJA
SL9 0HH M3ZGH
SL9 0HN 2E1EHF
SL9 0HN M0RVJ
SL9 0LR G4MDN
SL9 7EB 2E1GXS
SL9 7EN 2E0HOB
SL9 7EN M7JEF
SL9 7LA G4KGA
SL9 7PZ M0EAS
SL9 7QR G8CZQ
SL9 8LJ G3NZW
SL9 8LQ G4FCD
SL9 8PX M6LGM
SL9 8QT M5AIQ
SL9 9JZ M6WBK
SL9 9NX M0WAQ
SL9 9PA G1LIK
SL9 9TD G4OXG
SM1 1JE G3WYB
SM1 1QH M6EXN
SM1 2BL G4HSD
SM1 2EB M0WLS
SM1 2EB M6WLS
SM1 2PA G0BWV
SM1 2PA G2XP
SM1 2PA G7SAC
SM1 2RB G3VSH
SM1 2RZ M7KRK
SM1 2SQ G7KIN
SM1 3EA 2E0JHF
SM1 3EA M6JHF
SM1 3LW M0RFY
SM1 3QB G3NHX
SM1 3QQ M7GPI
SM1 3SL G8GLC
SM1 4BG G0SWU
SM1 4BR G7ITZ
SM12EN M7LEN
SM13RJ G1HRD
SM2 5AQ 2E0SRZ
SM2 5BA M6LOF
SM2 5EJ G6CTV
SM2 5ER G1XKD
SM2 5HP G1GBI
SM2 5HP M3NLP
SM2 5HW G8NFP
SM2 5HW G8TQK
SM2 5JA G4PLH
SM2 5JH G0PMM
SM2 5RP G7JDB
SM2 6RL 2E0VLB
SM2 6RW G1DQL
SM2 6UA G1UBV
SM2 7QA G0KBL
SM3 8ES G3XTC
SM3 8NA M0BWB
SM3 9EG G4DSY
SM3 9JH 2E0HIY
SM3 9JH M7CDU
SM3 9ND M0ZPG
SM3 9NE G6SGM
SM3 9NL G8VCL
SM3 9PP G8YEP
SM3 9QJ G7UZI
SM3 9RF G6XGV
SM3 9RH M1ROD
SM3 9RL M3PUQ
SM3 9SG G1HWK
SM3 9TT M3VBZ
SM3 9TZ G8CVQ
SM3 9UB G0EAU
SM4 4BS G1SHN
SM4 4DL G0VXX
SM4 4LF G0BXC
SM4 4QG G0PNT
SM4 4QN G4ZHA
SM4 4SX M6AVV
SM4 4TD G4CQR
SM4 6ED 2E0ERQ
SM4 6ED M0NHY
SM4 6ED M6NHY
SM4 6HJ M7FBO
SM4 6HL M3GYI
SM4 6HU M7PBD
SM4 6QG G6JXA
SM5 2BA 2E0OKQ
SM5 2BA M0OKQ
SM5 2BA M6OKQ
SM5 2HG M7NLP
SM5 2NH 2E0WWE
SM5 2PB G1EGZ
SM5 2SP M1MRB
SM5 2SP M6BOY
SM5 2TL G8VUK
SM5 3AB 2E0LEM
SM5 3DZ G7BDK
SM5 3EJ G4VAV
SM5 3HA G7NKH
SM5 3LS G0PUQ
SM5 3NG G4FDN
SM5 3NG G8PAT
SM5 3RA 2E0MJS
SM5 3RA M0SSM
SM5 3RA M6MCM
SM5 3SF G8HIG
SM5 3SJ M3RUL
SM5 3SU 2E0FTM
SM5 3SU M6TMF
SM5 3SW G7AQN
SM5 4QH G4BVG
SM6 0AS 2E0GHH
SM6 0AS M7WST
SM6 0QY G6XZP
SM6 0TN G3BMQ
SM6 7AG G4XSA
SM6 7BZ 2E0VSZ
SM6 7BZ M7VSZ
SM6 7DD G7ELZ
SM6 7JU G7NHY
SM6 8AE G4CCY
SM6 8BE M0DWB
SM6 8EP G6SAQ
SM6 8EP G8ZOY
SM6 8EW G8XDM
SM6 8EW M0BEH
SM6 8HL G0TXL
SM6 8PT G6ZLD
SM6 8PY M3LRN
SM6 8QB 2E0JJK
SM6 8QB M0JJK
SM6 8QB M6JAK
SM6 9GL G6UCY
SM6 9GX 2E0XSL
SM6 9GX M3RHL
SM6 9LJ M7XOM
SM60PB G7CRQ
SM69BP M6OEF
SM7 1AJ G3OLX
SM7 1BU 2E0EZF
SM7 1BU M0VON
SM7 1BU M3UQO
SM7 1HG G3CZU
SM7 1HG G4DFA
SM7 1HG G8GYX
SM7 1HG M6XUV
SM7 1JW 2E0TTA
SM7 1LB G6XZM
SM7 1LE 2E0JLX
SM7 1LE 2E0LEG
SM7 1LE M0WOJ
SM7 1LE M3XJL
SM7 1LE M3YCZ
SM7 1LE M6EMC
SM7 2BA G7DRD
SM7 2DG G1DPX
SM7 2ER M3NUO
SM7 2ER M3ZON
SM7 2HG 2E0GTZ
SM7 2HG M6AXB
SM7 2HZ M0NDJ
SM7 2HZ M3NOE
SM7 2QJ G0IUH
SM7 3JR G3KTA
SM7 3NA M6UTB
SM7 3PN G8GNX
SN1 2AN M6EWK
SN1 2EY M0IWT
SN1 2JR G7LET
SN1 2JU G4KEZ
SN1 2QT M3KHJ
SN1 3AE G0RNH
SN1 3AE M3GKK
SN1 3HF M3YFM
SN1 3HW 2E0LAY
SN1 3HW M6LAY
SN1 3NJ G0TLI
SN1 3PR M7CDQ
SN1 3QA G1ERM
SN1 4AY G6BPK
SN1 4DP G3YKC
SN1 4EY G1KMN
SN1 4GE G1PJC
SN1 4GU G7JGE
SN1 4HH 2E0PRA
SN1 4HH M6FVJ
SN1 4HX G1NAU
SN1 4LL G7PKK
SN1 5EE G8PAB
SN1 5ES M1EGN
SN1 7DJ G4ZEJ
SN10 1BF G8LGP
SN10 1FA 2E1EMK
SN10 1LW 2E0EFS
SN10 1LW M0OIO
SN10 1LW M6HYF
SN10 1LY 2E0MBS
SN10 1LY M0MSZ
SN10 1LY M6MMS
SN10 1QG 2E0MPN
SN10 1QG M3PQV
SN10 1RW G4RQL
SN10 2EN G6NTE
SN10 2FB G6IBW
SN10 2FX G6PYF
SN10 2LD G3WAE
SN10 2UE G0JVF
SN10 3AN G4GBX
SN10 3AP G3WZR
SN10 3BJ G4TIX
SN10 3BJ G6EVY
SN10 3BQ M7PLS
SN10 3DG G6KHW
SN10 3DG G7SVT
SN10 3FE G7FWD
SN10 3FN G0WHY
SN10 3PP 2E0VRO
SN10 3PP M6IYL
SN10 3PQ M3PVV
SN10 3SL G4SHA
SN10 4ED G6SND
SN10 4EL G6FNJ
SN10 4HP 2E0SBS
SN10 4HP M3UZZ
SN10 4PA G8MVS
SN10 4PP 2E1GQD
SN10 4PP M6LWZ
SN10 4RR G3MQD
SN10 4RR G3PZV
SN10 5AJ G1JMK
SN10 5AJ G6ZFA
SN10 5AZ M1FHA
SN10 5BA G8XYA
SN10 5HD G4VKJ
SN10 5HW G0RTN
SN10 5LE G7LWF
SN10 5SR G3XKL
SN10 5TD G4EES
SN10 5TP G0RIK
SN10 5UD M1FHB
SN11 0EP G6HTZ
SN11 0EW G4SFD
SN11 0FH M7SJH
SN11 0FL 2E0GKA
SN11 0FL M0XCX
SN11 0FL M6WKL
SN11 0JA G7TPS
SN11 0LQ G7VEY
SN11 0NE M6FCS
SN11 0PZ M6TBS
SN11 0UG M7RKA
SN11 8EY M7LBX
SN11 8EZ G4XYB
SN11 8NL 2E0RFN
SN11 8NL M7RSM
SN11 8PW 2E0ZGX
SN11 8PW M3ZGX
SN11 8SQ G4NHQ
SN11 8UR 2E0EME
SN11 8YD G0MLE
SN11 9AB G0OIM
SN11 9AT M6EIJ
SN11 9BT G7RGG
SN11 9DT G1JBC
SN11 9DU 2E0CRH
SN11 9DU G8LRD
SN11 9DU M0HKV
SN11 9DU M6BSQ
SN11 9EA G3WIW
SN11 9EW G4NKU
SN11 9FN 2E0VRE
SN11 9FN M3ZHC
SN11 9FN M6DTA
SN11 9LD G1JRZ
SN11 9LR G4UJL
SN11 9QH 2E0LRD
SN11 9QH M3ZWP
SN11 9QQ G7TIV
SN12 6AG G4DMC
SN12 6AZ G7EPX
SN12 6FD 2E0DIM
SN12 6FD M0VAI
SN12 6FD M6DIM
SN12 6FX 2E0DRQ
SN12 6FX M6EEM
SN12 6NA M7GZV
SN12 6TH 2E1EYL
SN12 6TQ G4SRD
SN12 6UL G1PWU
SN12 7AB G3RVY
SN12 7BX M7LEX
SN12 7DT G0KTP
SN12 7EF M6WGS
SN12 7HJ G6FNY
SN12 7HL M0ECZ
SN12 7LD M0JEU
SN12 7NG 2E0MHG
SN12 7NG M6ONW
SN12 7PF M3ZMZ
SN12 7PG G7TKO
SN12 7PT 2E0EES
SN12 7PT M6HJF
SN12 7QW M0OTL
SN12 7SW M0DGJ
SN12 7TE 2E0IZB
SN12 7TE G6EJD
SN12 7TE M3IZB
SN12 8BQ M6XBX
SN12 8LA G4NDC
SN12 8NB G1AJC
SN13 0BA 2E0FRU
SN13 0EG M1EGC
SN13 0JD G4GUK
SN13 0JT M1EIJ
SN13 0LB G7SNX
SN13 0LS M3IGN
SN13 0QR G3JSK
SN13 0QW M1ERV
SN13 8AN M0TRK
SN13 8AN M6AIJ
SN13 8DF G3KRW
SN13 9AU M3YPA
SN13 9LE G4SEZ
SN13 9NH G0KLJ
SN13 9NH G4NLA
SN13 9NN G3MEY
SN13 9NS M6CDU
SN13 9SH G3MBN
SN13 9ST 2E0BZI
SN13 9TX M3KIR
SN13 9UL M6BYA
SN13 9UT 2E0GET
SN13 9YJ M7GPW
SN14 0DN G8MPM
SN14 0EX M6UUU
SN14 0JB G0RBD
SN14 0JW M7ACM
SN14 0RB M3UXR
SN14 0TN G7PSS
SN14 0UJ G1SOG
SN14 0UJ M3SOG
SN14 0XP 2E0JWP
SN14 0XP M6CHH
SN14 0XW 2E1DUI
SN14 0XX M1EZU
SN14 6AN M1CWN
SN14 6DE M0PYE
SN14 6EG M6BPA
SN14 6EL G1YIQ
SN14 6LU M6FWP
SN14 6RT 2E0VFX
SN14 6XJ G1JRP
SN14 6YH M3MMN
SN14 7BB M3JYO
SN14 7EA 2E0NEY
SN14 7EA G8NEY
SN14 7EA M0GHZ
SN14 7NN G4BBP
SN14 8BS 2E0HXC
SN14 8BS M7HTC
SN14 8DN G3PIJ
SN14 8HD 2E0HOM
SN14 8HD M7CME
SN14 8QU G0RCP
SN14 8SA G8WMC
SN15 1AA G4DIE
SN15 1AR G0FGW
SN15 1BQ 2E0PLS
SN15 1BQ M0PLS
SN15 1BQ M6BXZ
SN15 1DA G6ORM
SN15 1DD M1GTI
SN15 1DW M0CTR
SN15 1QQ G1USK
SN15 1QQ M6XJW
SN15 1US 2E0GDV
SN15 1US M0JMU
SN15 1US M7ATO
SN15 2AY G0IAY
SN15 2AY G6YCG
SN15 2DJ G4TKF
SN15 2EE G8FLL
SN15 2EF G4ASG
SN15 2LB G0MVM
SN15 2PW G4SUX
SN15 2RA 2E0UET
SN15 2RA M0UET
SN15 2RA M6UET
SN15 3DZ G4MLG
SN15 3EF 2E1FBK
SN15 3EG M0MJD
SN15 3EL G4OQG
SN15 3FY 2E0SCS
SN15 3FY M6WSB
SN15 3JF 2E0FZG
SN15 3JY G0HPN
SN15 3LW M6IQU
SN15 3LX G0GJE
SN15 3UA G4BBY
SN15 3UA G4KPI
SN15 3UF M7GCO
SN15 3UG M3ZJK
SN15 4AR G6MFB
SN15 4AS G7OPI
SN15 4EL 2E0BPL
SN15 4EL M0RUB
SN15 4EL M3WDV
SN15 4ER G0OID
SN15 4JH M7CLM
SN15 4QN M3EIA
SN15 4RL G4GWR
SN15 4TR G6VHE
SN15 5AW G6EYA
SN15 5BA M3WBS
SN15 5BT 2E0SZG
SN15 5BT 2E0VCW
SN15 5BT M7CAW
SN15 5EX 2E0CXH
SN15 5EX M6BXY
SN15 5LD G8TTI
SN15 5LY G3UUV
SN15 5NN G0AEV
SN15 5NT G3VRE
SN15 5NT G6HUI
SN15 5QF G4CDW
SN151AP M7EWI
SN152BD G8HAM
SN16 0BL G6UYM
SN16 0DR G4TLY
SN16 0EA M3WSH
SN16 0LY M6GCS
SN16 0PZ G7IBL
SN16 0QH G4JKM
SN16 0RN 2E0EMX
SN16 0RN M6AGH
SN16 9AS G7DNT
SN16 9AS M0XON
SN16 9DL G8DKI
SN16 9GX G8VVY
SN16 9HF 2E1ASF
SN16 9HF M7BWJ
SN16 9PR G8EHE
SN16 9QB G8WLV
SN16 9QT G7GPU
SN16 9QZ G8POP
SN16 9SR 2E0GMG
SN16 9SR M3VGT
SN16 9UA G4SWO
SN16 9UE G8EMU
SN160ES G4TRA
SN2 1BD M0SYL
SN2 1BH M0SPN
SN2 1DR 2E0NOS
SN2 1NE G8CPA
SN2 1NL M0BFA
SN2 1NL M0PBZ
SN2 1RQ M3TSV
SN2 1RX G7FEA
SN2 1RX M6NDR
SN2 2AF M6EPH
SN2 2BP M7PBS
SN2 2EL M3VST
SN2 2EN 2E0XUM
SN2 2EN M0XUM
SN2 2EN M6XUM
SN2 2EN M7EGP
SN2 2FL 2E0FWB
SN2 2FL M6UPE
SN2 2PE G6OMH
SN2 2SG G4VWG
SN2 2SL G7UZG
SN2 5DE G0VCV
SN2 5ED G8XWH
SN2 5HF M6PDW
SN2 5JJ M1EMC
SN2 7HP 2E0CYS
SN2 7HP M6VDX
SN2 7LE G8SRC
SN2 7LE M0ACM
SN2 7LJ G7PYV
SN2 7LL G0JTD
SN2 7LQ G0VQW
SN2 7NJ G8MGO
SN2 7QA G4MDT
SN2 7TN M1XZG
SN2 8BT G7HPI
SN2 8DL M1ECQ
SN25 1PX G6JYN
SN25 1PY 2E0ZVG
SN25 1QD 2E0HZA
SN25 1QD M7RBY
SN25 1TL 2E0GSF
SN25 1WG G4SOA
SN25 1WG G6KAW
SN25 1XF M6FSG
SN25 2AZ M6XTB
SN25 2EE G1WVP
SN25 2EF M7NWV
SN25 2FE G0BON
SN25 2GQ M6ZYK
SN25 2NJ G8CZJ
SN25 2RJ M6YFX
SN25 3AZ G6XML
SN25 3BN M7VVB
SN25 3BT G4ZAM
SN25 3DB G6ARO
SN25 3DQ M6TGM
SN25 3ES G7FPM
SN25 3EZ M1DBW
SN25 3EZ M3PBW
SN25 3LZ M1AOU
SN25 3PP G8ELH
SN25 3QB G0UWS
SN25 3QB G4NWR
SN25 3QB G7NWR
SN25 3QB M1CWR
SN25 3QB M6UWS
SN25 4GF M0LEY
SN25 4GX M1SHA
SN25 4TJ G0EPC
SN25 4TP G3YMV
SN25 4XS 2E0YGC
SN25 4XS M7RCI
SN25 4YD M3VAG
SN25 4YJ G1HYC
SN26 7AR G8NTS
SN26 7DE G4HGV
SN26 7DE G8LYG
SN26 7EA G4SHK
SN3 1AZ G6VDK
SN3 1BN 2E0NDA
SN3 1BN M7NDA
SN3 1BX G7MGY
SN3 1BZ G7CRM
SN3 1DJ G0FCH
SN3 1DN 2E0LXD
SN3 1DN 2E1EQI
SN3 1DN G0VTA
SN3 1DN G0WDQ
SN3 1DR G8VJY
SN3 1ER G8IWO
SN3 1HZ G0TNP
SN3 1LU M6HYY
SN3 1NB G8FWK
SN3 1NJ G7KIW
SN3 1PT M1AGH
SN3 2AL M5ECX
SN3 2DR M1ACC
SN3 2EA G1FTV
SN3 2RB M6HST
SN3 3FQ 2E0GGU
SN3 3FQ M7CJB
SN3 3NB G3NPM
SN3 3NF 2E0LUF
SN3 3NF M0LUF
SN3 3NF M6LQP
SN3 3NW M5CBS
SN3 3PJ G4ZYH
SN3 4AA G3LLZ
SN3 4AB G0PSO
SN3 4AY G3SIR
SN3 4EE M6ESR
SN3 4FA 2E0IPY
SN3 4HB G0BQK
SN3 4HU G3HSV
SN3 4JA 2E0EXJ
SN3 4JA M0WXY
SN3 4JA M6EXJ
SN3 4LA G3MAI
SN3 4NF G3IDW
SN3 4NZ M7AAU
SN3 4SE G6REG
SN3 4SN G5WQ
SN3 4SN M0BCG
SN3 4WE 2E0BNZ
SN3 4WE M3WBI
SN3 5AG G8ETI
SN3 5AU G3WEF
SN3 5AX G4LDL
SN3 5AX G8KWD
SN3 5BN G0CPA
SN3 5BN M3CSN
SN3 5DE 2E0PMV
SN3 5DE G0SNM
SN3 5DE M0PMV
SN3 5DE M6PMZ
SN3 5JG G1NCG
SN3 6AR G3XYD
SN3 6AT M7TUF
SN3 6JB G4JVJ
SN3 6JB M0JGC
SN3 6JB M0JWN
SN3 6JB M0LWC
SN3 6JP M7MRR
SN3 6LA G1YGY
SN3 6LS G8UKY
SN3 6NF G4ENR
SN3 6NL G4AQK
SN4 0AE 2E0BZT
SN4 0AE M3ZPB
SN4 0BU G0MVD
SN4 0DF 2E0FEC
SN4 0DF 2E0MJE
SN4 0DF M3YOW
SN4 0DF M6IDB
SN4 0DU 2E0POZ
SN4 0LT 2E0WHT
SN4 0LT M0WHT
SN4 0LT M7KPS
SN4 0LU G4NHD
SN4 0LW G3IRA
SN4 0RL M1CDJ
SN4 7DG G8JWK
SN4 7DN G0OWR
SN4 7DR G4RZF
SN4 7DU G8IJM
SN4 7EU M6DNB
SN4 7EZ G8IYJ
SN4 7FN 2E0BQJ
SN4 7FN M0TGN
SN4 7FN M0WCB
SN4 7FN M6DAN
SN4 7JG G4OIQ
SN4 7RG M3YOQ
SN4 8AE M7HPW
SN4 8DJ G4MDH
SN4 8DT M1KDJ
SN4 8HJ M3YPS
SN4 8HS G7CSE
SN4 8JY M3GRY
SN4 8LW G8HCW
SN4 8NQ G4YAL
SN4 9AD 2E0HUK
SN4 9AD M3NTR
SN4 9AG 2E0NGB
SN4 9AG M0JEC
SN4 9AG M0NHK
SN4 9AG M3NBL
SN4 9AP 2E0CTW
SN4 9AP M0ZGB
SN4 9AP M6HBS
SN4 9AR G3JOT
SN4 9AT G0HOJ
SN4 9AX M0TTE
SN4 9AX M6PZF
SN4 9BB M1DHW
SN4 9BP M6FWM
SN4 9DR M0BKN
SN4 9DR M1EZP
SN4 9EP G7VHG
SN4 9HY G0DVB
SN4 9LJ G0LTP
SN5 0AD G3HCT
SN5 0AD G3RZP
SN5 0AD G4FNC
SN5 3LE G3APS
SN5 3LE G7TBO
SN5 3LX G0JSU
SN5 4AD G3YBY
SN5 4BA G4YEF
SN5 4BS G3TPQ
SN5 4DE G4MQP
SN5 4ET G8VGQ
SN5 4EX G4MGH
SN5 4HQ G8IYH
SN5 5AL G0DHL
SN5 5BH G1MUM
SN5 5DB M6BKA
SN5 5QT G4OED
SN5 5RL M7CLQ
SN5 5SA G8EFK
SN5 5SH G6UZM
SN5 5TL G4DSF
SN5 5TP M7KAV
SN5 5TS G3OMA
SN5 5UQ G4DOA
SN5 5XA G2BUJ
SN5 6AA M3PFU
SN5 6BB G7WIC
SN5 6BQ G6WAU
SN5 6EB M0PRT
SN5 6EQ M6BJQ
SN5 6JY 2E0KYB
SN5 6JY M7BEL
SN5 7AF G1VRJ
SN5 7AG G6UAJ
SN5 7AW G6LBR
SN5 7BB G1TOL
SN5 7BG G3KUD
SN5 7BN M3NTI
SN5 7BX G4GXW
SN5 7DR 2E0GKM
SN5 7DR M0ATC
SN5 7DR M0GKG
SN5 7DR M3RMV
SN5 7EF G4KIM
SN5 7EG G6BTR
SN5 8AH 2E1FJV
SN5 8AJ M6RGX
SN5 8AQ G7VJG
SN5 8AR M3FGR
SN5 8DB M6EVU
SN5 8HL M3XFS
SN5 8JE M0JQV
SN5 8NL 2E0GWK
SN5 8NL M6GWK
SN5 8PE G4WSB
SN5 8QX M6BMF
SN5 8RT G7MTF
SN6 6AB M6SDZ
SN6 6AJ M0MTA
SN6 6AW M1WIN
SN6 6BN G8OFR
SN6 6FJ M6NXA
SN6 6RD 2E0EAV
SN6 6RD M0NMI
SN6 6RD M6EIW
SN6 7BA 2E0KGV
SN6 7BA M0KGV
SN6 7BA M0TBG
SN6 7BA M3KGV
SN6 7BU G0NUH
SN6 7BZ G4AJA
SN6 7BZ G4BPO
SN6 7DW 2E0RAK
SN6 7DW M0PGX
SN6 7EB G4GDR
SN6 7EG M3BWF
SN6 7ER 2E0GSO
SN6 7ER M6OHU
SN6 7HJ G4TRR
SN6 7HN M1ATC
SN6 7HR G4LZZ
SN6 7HS G8EOZ
SN6 7HU M1SKI
SN6 7LA G0NPN
SN6 7NU M0IEP
SN6 7PD G8JHC
SN6 7PG G8JRF
SN6 7RB M3EHY
SN6 7RL M3JIA
SN6 8AJ M0DAW
SN6 8AZ G3XEY
SN6 8BD M7MDR
SN6 8DY M6IGT
SN6 8ER G3ONU
SN6 8FA M6PGY
SN6 8NB M3ESG
SN6 8SX G4UEF
SN7 7DG G0JBJ
SN7 7EY G3LPC
SN7 7EY G3NNG
SN7 7EY G3PIA
SN7 7LB G8BGM
SN7 7NB 2E0FUA
SN7 7NB M0ROX
SN7 7NB M6KTQ
SN7 7RL 2E0EJA
SN7 7SE M3CVH
SN7 7SS G8BCF
SN7 7YT G8GIF
SN7 8AT G4VLW
SN7 8EY G0SCQ
SN7 8FA G1NBP
SN7 8HW M7PLN
SN7 8LE G4PFY
SN7 8LE G6HIX
SN7 8LN M6GOI
SN7 8LR G1HYO
SN7 8LX 2E1HFX
SN7 8LZ M6JAZ
SN7 8ND G3ZEF
SN7 8NP G3XVB
SN7 8QR G0ITO
SN7 8QR G6ZMX
SN7 8RP G0VBG
SN8 2AS G4XVW
SN8 2AZ G6EPN
SN8 2BD 2E0PDZ
SN8 2BD M0PDZ
SN8 2BD M6EZR
SN8 2DH G1FKJ
SN8 2FE 2E0GWR
SN8 2FE M0SWL
SN8 2PX G8JWD
SN8 2QE 2E0PPB
SN8 2QE G3PPB
SN8 3AF 2E0AAX
SN8 3AS G3MFL
SN8 3DZ G4DHT
SN8 3HN G8SXA
SN8 3HN G8SXD
SN8 3PX M7WLT
SN8 3QB M0TDW
SN8 3RP M3ANW
SN8 3SS M0GQM
SN8 3TD G4LMA
SN8 4AS M0DSW
SN8 4HT M1CJE
SN8 4HT M7BBE
SN8 4NQ 2E0IYG
SN9 5ES M0BXU
SN9 5HP G7GEA
SN9 5HU G8PMJ
SN9 5HU M6CFV
SN9 6AE G4TOY
SN9 6AF M0BUT
SN9 6EN M6IIG
SO14 0LB G6NJT
SO14 3ED G4EUJ
SO14 3FE M6DLU
SO14 3FN M6JZN
SO14 3GQ G4HXE
SO14 3HU G6UAN
SO14 3HW G4CRB
SO14 6FR M6LYP
SO14 6GR M0XGT
SO14 6RN M7GIP
SO15 1GF M6HOI
SO15 1JJ 2E0CZK
SO15 1JJ M6CYF
SO15 2AS M0IVC
SO15 2NN M0XEF
SO15 3EW G8SBS
SO15 4HU 2E1ESQ
SO15 5DS G4MYS
SO15 7QB M7HAV
SO15 7QE 2E0LAX
SO15 7QE M6AWC
SO15 8PT M3VWJ
SO15 8PT M3VXG
SO15 8QU 2E1EUE
SO15 8QZ M3YAO
SO16 0AQ G0EET
SO16 2NZ G3IXN
SO16 3DA M7HHK
SO16 3EB G0SWY
SO16 3EF G6VNO
SO16 3GU M7DJB
SO16 3GZ G4VAS
SO16 3HY G4ZAS
SO16 3LP G7UBD
SO16 3NL M6KFS
SO16 3NY G4DIV
SO16 3TN G6UTL
SO16 3UJ M6LPF
SO16 4BN G1XBR
SO16 4NR M0RLP
SO16 4QT M6PHA
SO16 5DW G3CGE
SO16 5FD G4ZFY
SO16 5FG 2E0AJQ
SO16 5FG G8KVN
SO16 5FG M0APN
SO16 5GJ M6VRO
SO16 5JX M3UVK
SO16 5NQ G3VQF
SO16 5SW G7GIJ
SO16 6AF 2E0HQV
SO16 6AF M7DBX
SO16 6AL 2E0DGR
SO16 6AL M6DZX
SO16 6PZ M7FWB
SO16 6QB M6PSO
SO16 6QU 2E0PAU
SO16 6QU M3YXU
SO16 6UH 2E0HTN
SO16 7AW 2E0OKB
SO16 7AW M7KAE
SO16 7DG M0IJR
SO16 7DR 2E0UAV
SO16 7DR M0UAV
SO16 7DR M6TYL
SO16 7DR M6ZSH
SO16 7EF M6FJI
SO16 7EL G3VXY
SO16 7GJ M0MPT
SO16 7PE G3ZIL
SO16 8AH 2E0BGJ
SO16 8AH M0NXP
SO16 8AY G6VDW
SO16 8AY M0BIJ
SO16 8BE G4WFR
SO16 8DW G1XRM
SO16 8ED M7RZS
SO16 8EF M1FJA
SO16 8EH G6CRG
SO16 8FS M0IJO
SO16 8GU M0DVF
SO16 8LR G8LRS
SO16 9AW M6ZOO
SO16 9EZ M6HTG
SO16 9GF 2E0BXQ
SO16 9GF M6LJD
SO16 9JF G0XAZ
SO16 9LR 2E0ERL
SO16 9LR M7EFT
SO16 9QE G4BDQ
SO16 9WA M3XVQ
SO17 1EH M6EJJ
SO17 1NU G0DHR
SO17 1NU G0MSR
SO17 1PF M6AFG
SO17 1RS G4YVZ
SO17 1SD G2HNI
SO17 1XQ M6USO
SO17 2HJ M7APH
SO17 2LJ M0EBO
SO17 2LJ M1CWB
SO17 3AE M6GHC
SO17 3RT G6KYE
SO17 3SA M0SMQ
SO17 3SA M7SMX
SO17 3SJ M6EEF
SO17 3TG M6KOH
SO18 1FW M7VMR
SO18 1HJ M0CYJ
SO18 1JR M6DBF
SO18 1JS G1VGA
SO18 1LS G7TUF
SO18 1QR G3OFX
SO18 2FD M7OSP
SO18 2FP G7UYI
SO18 2HQ M0GCH
SO18 2NU 2E0NWA
SO18 2NU M6CDG
SO18 2NU M6NWA
SO18 2PQ G1LWF
SO18 2QE G3IDB
SO18 2ST M1BSB
SO18 3AG 2E1JJN
SO18 3AG M1JJN
SO18 3AG M3JJN
SO18 3PP 2E1HQP
SO18 3RB G3KDW
SO18 4FN 2E0ZBE
SO18 5FE 2E0CZI
SO18 5FE M6DQV
SO18 5FS G3VCL
SO18 5GY 2E0NJC
SO18 5GY M3VQZ
SO18 5PG G4YYS
SO18 5QJ G4PTF
SO18 5QL G3KXE
SO18 6BB G0UJP
SO18 6LY M7ABD
SO183NA 2E0NTJ
SO183NA M6NTJ
SO19 0EH M7CJE
SO19 0JG M0HME
SO19 1AD M7KAG
SO19 1BX G6XYS
SO19 1DD M6JLW
SO19 1DP G8BJB
SO19 1FW G4UDB
SO19 2HF G6HHH
SO19 2HJ G7NKZ
SO19 2HU M0JBD
SO19 2NU G1OQI
SO19 2PH M7PVS
SO19 4DE G3NZL
SO19 4DX G4HUM
SO19 5JP G0EUC
SO19 5JP G1LDN
SO19 5LF G0LCN
SO19 5LN G0ILA
SO19 6FB M3VPA
SO19 6HB G7RAB
SO19 6NG M6PHP
SO19 6PS G7BLX
SO19 6QT G0JBH
SO19 6QX G0EMV
SO19 7GG G8RNT
SO19 7GG M5PSW
SO19 7JN G7MTA
SO19 7JN M0DDB
SO19 8AF G7BZU
SO19 8EY G4UEN
SO19 8FF G0IOI
SO19 8FS 2E0TVZ
SO19 8FS M3TVZ
SO19 8FT M7BOU
SO19 8JS M7RBP
SO19 8LH 2E0KBX
SO19 8NT G4OLY
SO19 8NT G8WWM
SO19 8SJ M6AKE
SO19 9AT G6YVJ
SO19 9DA 2E0TAQ
SO19 9DA M3UUF
SO19 9GB G2SZ
SO19 9GN M0REV
SO19 9HY G8EXZ
SO19 9LJ 2E0PKK
SO19 9NJ 2E0EHJ
SO19 9NJ M6MSO
SO19 9PE G8GWB
SO19 9RW M7DRC
SO19 9SS M3OCR
SO20 6AH G8IUP
SO20 6BA G3RDQ
SO20 6DG G7PVZ
SO20 6EY M6AEL
SO20 6EY M6GTH
SO20 6HE G8CDA
SO20 6NY 2E0UKM
SO20 6NY M3UKU
SO20 6PR G4NMS
SO20 6PR M1BPD
SO20 8DB G3LGA
SO20 8DB G8RJM
SO20 8EA M1BDJ
SO21 1AF G3JRH
SO21 1BA 2E0VBN
SO21 1BA M3VBN
SO21 1BQ G6LKW
SO21 1TE M7BME
SO21 1TT M0WZM
SO21 1US G3UPD
SO21 1XE M7AWQ
SO21 2BX G8TEC
SO21 2DE G3HRH
SO21 2EG G8IDJ
SO21 2EP G3AJD
SO21 2EP G4NZC
SO21 2LG G3ZEO
SO21 2LS G4FKR
SO21 3DU G0LMD
SO21 3EB G0GWG
SO21 3EB G3ZHV
SO21 3EE M3MKJ
SO21 3EY G1NBU
SO21 3EY G3LMH
SO21 3EZ G3VEH
SO21 3FS M7LKY
SO21 3HP G3SJK
SO21 3HP G4XRJ
SO21 3HY 2E0FNY
SO21 3HY M6FNY
SO21 3JG G4EDW
SO22 4DN 2E0SSC
SO22 4DN M6SJC
SO22 4DU M6CNS
SO22 4ED M7BHU
SO22 4EE M0AYV
SO22 4EE M0REX
SO22 4ET G3NVB
SO22 4EY G3MCL
SO22 4HS G3VPG
SO22 4HS G4JXE
SO22 4HU G1THP
SO22 4JY G3PTS
SO22 4LD G8PGE
SO22 4LF G4SVC

Postcode

SO22 4LS 2E0IRD
SO22 4LS M0KOR
SO22 4LS M7KOR
SO22 4LX M7JRG
SO22 4NU G8DGC
SO22 4PQ 2E1FKZ
SO22 4PQ G6AAZ
SO22 4PQ M3SFC
SO22 4PZ G4AFA
SO22 4QH G3XJM
SO22 5AT G3RQR
SO22 5AX G0DZV
SO22 5BU G4HDY
SO22 5DJ G6BZE
SO22 5HU G3RUZ
SO22 5JX G4CJO
SO22 5ND G0JMK
SO22 5ND G3UYK
SO22 5NT M6BIN
SO22 6BA M0BTR
SO22 6BA M3KER
SO22 6HE G3YSK
SO22 6HE M0AYY
SO22 6HN M7YDX
SO22 6LT M1RGW
SO22 6NP M7WEF
SO22 6NZ G4TBI
SO22 6NZ G8CQG
SO22 6PS M6IBL
SO22 6SG G4AXO
SO225JU M7HHD
SO23 0NU G4AXY
SO23 0PP G0IHA
SO23 0PX G4AZU
SO23 7DA M3KMT
SO23 7EQ G3ROG
SO23 7HT G1HRA
SO23 7ND G0TAH
SO23 7ND G1NCM
SO23 7QQ G0IVR
SO23 7QQ G4DKH
SO23 7QQ G6IVR
SO23 7XH G3HQX
SO23 9DT G1EHE
SO23 9SR G4LYL
SO24 0DU G4YDE
SO24 0DU G6VXR
SO24 0HP G4LQW
SO24 9DH G8UDZ
SO24 9HF 2E0LLZ
SO24 9HF M0LLX
SO24 9HF M6MQY
SO24 9HH 2E0CGS
SO24 9HH M3ETH
SO24 9JJ M7CYQ
SO24 9PE G6SDC
SO24 9PL G0JHT
SO24 9PP G4DMG
SO24 9RP G8JCB
SO24 9SG G7DNF
SO24 9TX G7JTR
SO249LQ M7FPB
SO3 4BA G6XIR
SO30 0BP G6WYS
SO30 0BS 2E0YAW
SO30 0BS M6MZZ
SO30 0GR 2E0JVM
SO30 0HU G4KUL
SO30 0LZ M3TZF
SO30 0PH G4JNT
SO30 0UG 2E0TDJ
SO30 0UG M0TKW
SO30 0UG M6GJI
SO30 2AR M7BTW
SO30 2AX M0JWA
SO30 2AX M7JGX
SO30 2BH 2E0MNX
SO30 2BH G4MNX
SO30 2BH M6OFQ
SO30 2BJ M0HBC
SO30 2NY G3YOM
SO30 2UD M3UVX
SO30 2UQ M6RGQ
SO30 3EQ G3IBY
SO30 4FT G3RQF
SO30 4HA M1RST
SO30 4HE M6CNQ
SO304EZ G0HCI
SO31 1AD M6GXV
SO31 1AP G3PDP
SO31 1BJ G3SED
SO31 1BY G1PVD
SO31 1EA G3YSG
SO31 1EA G8YUO
SO31 4HD M1ABG
SO31 4JS G3KTM
SO31 4QG M6BQB
SO31 4RT 2E0YEO
SO31 4RT G6IXE
SO31 4RT M6LXN
SO31 5FE G0WSB
SO31 5GR G0HAE
SO31 5HP M7EIW
SO31 6EP G1WXT
SO31 6EP G4UDY
SO31 6LJ M6DNN
SO31 6PE G0UTU
SO31 6PE G6CMF
SO31 6PP M0UOO
SO31 6PP M3UOO
SO31 6QG G3UZK
SO31 6RN G0UUS
SO31 6RR G8HER
SO31 6SP G1GSY
SO31 6SX G8PGA
SO31 6SY G4KWY
SO31 6TD M7RUT
SO31 6WX 2E0IED
SO31 6WX M0IED
SO31 6WZ M3UUS
SO31 6XF M0YCH
SO31 6XX G0TCQ
SO31 7BX 2E1IIC
SO31 7BX G0CAJ
SO31 7BX G0SCX
SO31 7BX M3IEQ
SO31 7BZ G0FBC
SO31 7EY 2E1FKT
SO31 7HQ G8ACL
SO31 8AT G8IOK
SO31 8AY G4NHN
SO31 8EL G0MYJ
SO31 8EN G6XQR
SO31 8HE M6BGF
SO31 9AU M7CYS
SO31 9FG G3HQT
SO31 9FU G3NNW
SO31 9GD M6ETZ
SO31 9HG G4PXH
SO31 9JU M6BPR
SO315DQ M7BPE
SO316TN M6JEG
SO318PA 2E1AKW
SO32 1AY M7LBM
SO32 1BR M6LGI
SO32 1DU M7VWX
SO32 1EW G4EWT
SO32 1EY 2E1IWG
SO32 1JR G1GXB
SO32 1JS G4NYD
SO32 1RP 2E0LZG
SO32 1RQ G1JBG
SO32 1SG 2E1CXI
SO32 2AA M3FOV
SO32 2AR G4COM
SO32 2AR G4IDW
SO32 2BD G1UTC
SO32 2BD M1CCA
SO32 2LG M0GFM
SO32 2LX M7ENU
SO32 2NP 2E0CSK
SO32 2NP M0HPU
SO32 2NP M6BZA
SO32 2NR G8KZP
SO32 2SX 2E0ERJ
SO32 2SX M0IUE
SO32 2SX M6IQZ
SO32 2TL G4ZRT
SO32 2TR G6NXM
SO32 3LE M6CSA
SO32 3NH G3YFL
SO32 3NP 2E0UKB
SO32 3NP M6UKG
SO32 3QN G1PRP
SO4 4WQ G0ORV
SO40 2GP M7FUN
SO40 2NA G0COC
SO40 2NL 2E0SJQ
SO40 2NL M0BJL
SO40 2NL M6SCX
SO40 2NW G6ANI
SO40 2QE M7DIQ
SO40 2RF G7HOK
SO40 2SD G7EOH
SO40 2UP G6MEW
SO40 2UY G0LDJ
SO40 2WD G3GLW
SO40 3BN G6XYF
SO40 3HP G0BOC
SO40 3HP G0OSD
SO40 3LP G4VII
SO40 3LZ M6KEV
SO40 3NS M0LCC
SO40 3NX 2E0WHH
SO40 3NX M3WHH
SO40 3QP M6TIY
SO40 4SL G1KSE
SO40 4UT G4IXE
SO40 4UT M3YZJ
SO40 4XB G4ADP
SO40 4XY G4BUW
SO40 7AL G3VSL
SO40 7AL G6GFA
SO40 7AT 2E0CZT
SO40 7AT M0RHO
SO40 7AT M6MQD
SO40 7AT M6RFG
SO40 7AY 2E0GJE
SO40 7AY M6GJE
SO40 7DP M6MYV
SO40 7FD 2E0FQS
SO40 7GP G4BIZ
SO40 7GY G0FOH
SO40 7JH G4YWZ
SO40 7LA G0VNI
SO40 7LA G0WIL
SO40 8JB G4PGW
SO40 8US G1LGY
SO40 8WB M0DYA
SO40 9AX M3PFF
SO40 9BT G0PGZ
SO40 9DR M0WSR
SO40 9DR M3WSR
SO40 9UZ G8KYK
SO409LZ G0WFQ
SO409LZ G3SOU
SO409LZ G4SJW
SO409LZ G8FAB
SO41 0GG G4TOG
SO41 0GP 2E0KGC
SO41 0GP M0GTL
SO41 0GP M6KGC
SO41 0GS 2E0VAG
SO41 0GS M0MDG
SO41 0GS M0TBI
SO41 0GS M3VVW
SO41 0GU G3GRV
SO41 0HQ G1RAX
SO41 0LQ G4RFP
SO41 0PB G7JWV
SO41 0PP G4LUN
SO41 0PR M0LKD
SO41 0QT M3OBU
SO41 0QU 2E0NHS
SO41 0QU M3NHE
SO41 0QW G4DNE
SO41 0RL 2E1JLC
SO41 0RR 2E0DCD
SO41 0RR M0LMB
SO41 0RR M6BOF
SO41 0RR M7DAN
SO41 0SG G3KLH
SO41 0TE G1UFL
SO41 0TE M3OTU
SO41 0TF G3ZJY
SO41 0UL G4BMC
SO41 0UX G2HCG
SO41 0XG G8YFK
SO41 0XS 2E0DAY
SO41 0XS M3SJD
SO41 0ZG M6WOK
SO41 3AE M7PPM
SO41 3PB G0COE
SO41 3QA G4HYW
SO41 3SD G0TXN
SO41 3TF M6NGW
SO41 5PA M6CJG
SO41 5PA M6LYN
SO41 6AL G0PCW
SO41 6AX G7OGT
SO41 6BB G4SBB
SO41 6BQ G0SNK
SO41 6DH M3ZIN
SO41 6DW G4SXH
SO41 6EJ 2E0UXS
SO41 6EJ M0UXS
SO41 6EJ M3UXS
SO41 8BD G1KFH
SO41 8BG G3ZIE
SO41 8DN G4ZAX
SO41 8DT M3ECF
SO41 8DY G0BCS
SO41 8DY G3TEI
SO41 8DY G4UAA
SO41 8GQ G4CWS
SO41 8GQ M6GKH
SO41 8HN M6DPY
SO41 8YD 2E0LYM
SO41 8YD M7LYM
SO41 9BL M0ZDU
SO41 9BP M0NEC
SO41 9DX M6JQS
SO41 9GT M6VCC
SO41 9LB G0JZW
SO41 9LB M0ULD
SO41 9LT G4XNA
SO42 7PB G8CAV
SO42 7RD G7JFU
SO42 7WQ 2E0DGT
SO42 7WQ M6DPN
SO43 7BN M3FQM
SO43 7BN M3FQN
SO43 7JB G3NOA
SO43 7JP M3FWT
SO45 1BH G1RET
SO45 1BJ G0HDI
SO45 1BN G7KDN
SO45 1BR G4CLF
SO45 1DA G0EUJ
SO45 1DQ M7KDU
SO45 1EG G6LVJ
SO45 1FX 2E0WPH
SO45 1FX M0PMH
SO45 1WF G6NXV
SO45 1WH G1MKY
SO45 1WJ G7RUH
SO45 1WX G0OSG
SO45 1XP G3TPV
SO45 1XW G0BPA
SO45 1XW G6TGB
SO45 1YN G1UEV
SO45 1YX G6UXP
SO45 2EY G7AYA
SO45 2HN G6OJH
SO45 2HP G0SBV
SO45 2JH 2E0GYP
SO45 2JH M6JAS
SO45 2JT G6MNL
SO45 2JT G6XMA
SO45 2JZ G4KNN
SO45 2ND G0SZI
SO45 2NQ G4JYN
SO45 2NQ G4YVY
SO45 2PP G4SBF
SO45 2QJ G0KJR
SO45 3GB 2E0CXQ
SO45 3HJ M0GZB
SO45 3LG G6MZF
SO45 3LJ G1UDW
SO45 3LJ M6CRA
SO45 3LS G3KSF
SO45 3PH M6CAQ
SO45 3QF 2E0MBG
SO45 3QF M6PWT
SO45 3QG 2E0GAF
SO45 3QG M6GDS
SO45 3RJ 2E0HZW
SO45 3RJ M0KBU
SO45 3RJ M7DNN
SO45 4BY M6LFO
SO45 4HS G3OZT
SO45 4HU M6HYT
SO45 4LE G0WCB
SO45 4LR G8CLK
SO45 4LS M0IKT
SO45 4LZ G7AFT
SO45 4RS 2E0XEE
SO45 4SB G0WSI
SO45 5BP G8APM
SO45 5DL G1JRU
SO45 5ER G3NJG
SO45 5EX G8BAL
SO45 5FP G3WJM
SO45 5FQ G1UBN
SO45 5QS G4SQJ
SO45 5TW G4NOP
SO45 5UD G1RCW
SO45 5UW G4GCI
SO45 6AY G4JAX
SO45 6BN 2E0EXC
SO45 6DL G4MOL
SO45 6DW G8BMQ
SO45 6EU G0JGI
SO45 6EW M3ZNJ
SO45 6EX 2E0ILC
SO45 6EX M0NGH
SO45 6EX M6ILC
SO45 6JR G8TLT
SO45 6JT G4BPN
SO45 6LA 2E0BSQ
SO45 6LA M0GMD
SO45 6LA M6EMG
SO451BP M6PCV
SO455QW G6MUX
SO456EA M6STA
SO50 4BW M6RKE
SO50 4BX 2E0DMW
SO50 4BX M3DMW
SO50 4GT M7EKL
SO50 4LW 2E0IOU
SO50 4NS M6PDR
SO50 4NZ M3NSH
SO50 4PP G0TBC
SO50 4PP M1FDO
SO50 4PQ G0OFX
SO50 4QY G3VHW
SO50 4RW G3NML
SO50 5AD M0MTN
SO50 5AD M3TYM
SO50 5AS M3FEL
SO50 5EN G4KMP
SO50 5EN M6BBH
SO50 5HA G4JHC
SO50 5HZ M1CNK
SO50 5ND M7BVZ
SO50 5NE M3DCJ
SO50 5QP M7GEY
SO50 5RR G0BES
SO50 6AY 2E0MDK
SO50 6AY M0HHX
SO50 6AY M6IDE
SO50 6BD G1TJH
SO50 6DG G3VRB
SO50 6FX 2E0GSC
SO50 6FX M0OHI
SO50 6FX M3OHI
SO50 7GG M0LXD
SO50 7HB G3YGF
SO50 7JJ M1MAB
SO50 7LL G8YJZ
SO50 8AA M3VIA
SO50 8FL G3XIV
SO50 8JW 2E0VIA
SO50 8JW M0VSD
SO50 8JW M6VIA
SO50 8PF M6FCW
SO50 8PH G8VQA
SO50 8PU M3RBU
SO50 9BH M5MDH
SO50 9JT M3NAT
SO50 9PW M6NCE
SO507NY 2E1EEK
SO51 0GD M6OSL
SO51 0GQ G0RLA
SO51 0HG M3ECD
SO51 0JL 2E0CHL
SO51 0JL M6ASP
SO51 0JP M0TOG
SO51 0LF G4GSK
SO51 0LW G1GYT
SO51 0RA G8LLJ
SO51 5PQ G1TWW
SO51 5PZ M7MZM
SO51 5RJ G6FVZ
SO51 5SP G7VZY
SO51 6AR G8DXF
SO51 6BT 2E0GHZ
SO51 6BT M0XIG
SO51 6EE G0FHC
SO51 6EX G3NAE
SO51 6FU G0HLA
SO51 6JT 2E0HXW
SO51 7HU M0BTZ
SO51 7JN M6NFD
SO51 7JW M3ECU
SO51 7JY G6CPE
SO51 7JZ G8OFX
SO51 7LH G3FWI
SO51 7LL G4YVE
SO51 7RR M3ECJ
SO51 7RU G4MNP
SO51 7RU G8ZMM
SO51 7TE G7PEC
SO51 7TE G7UHE
SO51 7TE M0EHF
SO51 7TJ G1NTN
SO51 8BR 2E0CGR
SO51 8DF G3MDR
SO51 8DP G8OQP
SO51 8DY M3ECQ
SO51 8EQ G6GCI
SO51 8FJ M6BWB
SO51 8HG G7OOF
SO51 8LH G4XBZ
SO51 8PN M7BOC
SO51 9BT G4YEE
SO510AX G8IPG
SO52 9EU G6TRW
SO52 9FD 2E0DJU
SO52 9FD M6EPA
SO52 9GB 2E0KWC
SO52 9GB M3YKC
SO52 9GB M3YMC
SO52 9JG G3XSD
SO52 9JT G7AJR
SO52 9NB 2E0OGR
SO52 9NB M7MLC
SO52 9NS 2E0BHY
SO52 9NS M1PVF
SO53 1GD G4DZS
SO53 1LE G7UDM
SO53 1LN M1LMJ
SO53 1NA 2E0KCO
SO53 1NA M0KCO
SO53 1NA M3KCO
SO53 1SA M3FFO
SO53 1SA M3HOE
SO53 1SF M3YJT
SO53 2AY G1UDR
SO53 2BD M7RYV
SO53 2BP 2E0SIM
SO53 2BP M3ALZ
SO53 2DJ G4EZC
SO53 2FE G8ZBN
SO53 2FS M0TDG
SO53 2FY M0AYU
SO53 2GX G7AJE
SO53 2LJ G3UYD
SO53 2LL M6SRD
SO53 2PU G6XJZ
SO53 2PU M0MEH
SO53 3BS G4BGT
SO53 3BS G6JGR
SO53 3EB G4LDC
SO53 3GD G0TAA
SO53 3GJ G6AMW
SO53 3PA 2E0GBT
SO53 3TP G7POC
SO53 4HP G6MCX
SO53 4HQ 2E0FMG
SO53 4HQ M6FMG
SO53 4HU M3SKT
SO53 4JL M7AYP
SO53 4LD G7VLH
SO53 4LN 2E0NIF
SO53 4LN M0NIF
SO53 4PD M7WKD
SO53 4QQ G7BOH
SO53 4RW M3SEE
SO53 4SW G0DQH
SO53 4TT G3WII
SO53 5AT G4MZU
SO53 5AX G4HDF
SO53 5BX 2E1ECV
SO53 5PA G4HWM
SO53 5PB G8LVC
SO53 5QD 2E0AJP
SO53 5QJ M0PHB
SO53 5QJ M7DIL
SO53 5QP G6IQI
SO53 5RP G1NPI
SO53 5RP G8IPQ
SO53 5RP M3IPQ
SO532EN G8OXW
SOLIHULL G4YZA
SP1 1AL G3OMZ
SP1 1NU G0DJV
SP1 1PU M0NJP
SP1 1PU M3VOW
SP1 1QL G3MXP
SP1 2EY G8FSL
SP1 2JH G8WJB
SP1 2JX 2E1MEL
SP1 3AU G4YFJ
SP1 3BH G7DNG
SP1 3DN G6CZB
SP1 3FX M0WDG
SP1 3HS G4MOE
SP1 3JZ G6BBG
SP1 3LW G4YVM
SP1 3LW M0PCN
SP1 3PR G7WAA
SP1 3PR M0CLI
SP1 3PZ G4TRD
SP1 3RT M6DPX
SP1 3WE G2AQJ
SP1 3WS G7JGB
SP10 1HH G4ABL
SP10 2AH G0OMD
SP10 2AX 2E0IEL
SP10 2BU 2E0BZJ
SP10 2BU M3CEB
SP10 2EN G8CPJ
SP10 2FS M6SPC
SP10 2HE G4YSB
SP10 2PZ M0ALE
SP10 2QT G0OUO
SP10 2QT G7PVE
SP10 2RB G2FSJ
SP10 3AY M5LDF
SP10 3BN G6FIB
SP10 3BU M3FWS
SP10 3DS G6YEY
SP10 3EP G7VRJ
SP10 3HL G0JBY
SP10 3JT M0BMF
SP10 3NG 2E0AJG
SP10 3NG M7AJG
SP10 3PE 2E0CLW
SP10 3PE M0VZS
SP10 3PE M3VZS
SP10 3RZ 2E0SNB
SP10 3RZ M7MZB
SP10 3SS M6GAB
SP10 3UH M3SVC
SP10 3XD M6DQN
SP10 4AR G1HPS
SP10 4DB G7EBI
SP10 4EN G0AMO
SP10 4EN M3DEA
SP10 4EZ M1EYS
SP10 5BS 2E0CCR
SP10 5DB G7CFX
SP10 5HT G8CEP
SP10 5JR M7GHS
SP10 5LD M7DUR
SP10 5NA G6ARC
SP10 5NA G8NDN
SP11 0LF G4GRR
SP11 0PZ M7KPW
SP11 0QF 2E0GNC
SP11 0QF M6GNC
SP11 0QT G8ARH
SP11 0RE G6SDE
SP11 0RS G8FHI
SP11 6EA G3LTF
SP11 6EW M0JIM
SP11 6FD G6VTA
SP11 6HB M7SOZ
SP11 6JY G6SJG
SP11 6RH G4XYG
SP11 6RR M6AIL
SP11 6SS 2E0DUM
SP11 7AT G7TAE
SP11 7BJ 2E0LBK
SP11 7BJ M0LBK
SP11 7BJ M3LBK
SP11 7BN G4FYM
SP11 7BX G6YTY
SP11 7DF M1BSX
SP11 7EH G3XVS
SP11 7ER 2E0GZL
SP11 7ER G6WBG
SP11 7ES G4WEN
SP11 7LL M1CYW
SP11 7NX 2E0PTZ
SP11 7NX M0PTZ
SP11 7NX M6PTU
SP11 8DW M6RQE
SP11 8LG G6JSF
SP11 8NE G8GYS
SP11 8NS G3YVM
SP11 9AN G4NNS
SP11 9JD G4CIZ
SP11 9NN M7YMA
SP11 9PG G1XSV
SP2 0AX G4WDA
SP2 0LW G4RLF
SP2 0NF G4SXQ
SP2 7BN 2E0MEK
SP2 7BN M6TEO
SP2 7BY M3YCV
SP2 7ER M7XRW
SP2 7EX G8CWJ
SP2 7FY G4FMZ
SP2 7SU G0MLJ
SP2 7UE G7MJV
SP2 8DL M6ZCM
SP2 8DN G0BDP
SP2 8EF M0BWC
SP2 8EP G3KOZ
SP2 8FG M7BIQ
SP2 8HZ M6TOW
SP2 8JX G0VMC
SP2 8LY G7BSO
SP2 8NR G3OIL
SP2 8QS G4HQM
SP2 8RN G4LWU
SP2 8RN G6RLG
SP2 9BY G8NDV
SP2 9DF M7KRH
SP2 9EJ G7EXZ
SP2 9EQ M6THJ
SP2 9FZ G6UAW
SP2 9HN 2E1MEP
SP2 9LD M3EQW
SP2 9LF G1BMT
SP2 9LQ M6OER
SP2 9LU M0VSR
SP2 9LU M0WSC
SP2 9PA G1DGL
SP2 9PE M0RED
SP2 9PF M7OPY
SP27LQ M7JFK
SP3 4BQ M3VBM
SP3 4EB M7MCU
SP3 4LP G4DUA
SP3 5AD M6OMF
SP3 5HW G8EMA
SP3 5JL G4FRB
SP3 5LF G4MQW
SP3 5LF G8DER
SP3 5PE M3BIZ
SP3 5PW M0TRB
SP3 6EX M7MWG
SP3 6JL M0ISW
SP35LP G8WBO
SP4 0AA G6ABM
SP4 0BY G4AEI
SP4 0ER G8ALR
SP4 0ER M0BVO
SP4 0LE 2E0XKO
SP4 0LE M0XKO
SP4 0LE M3XKO
SP4 6AU G8HUV
SP4 6DJ G6JGP
SP4 6EE 2E0LZE
SP4 6FP M7CYG
SP4 6HW M0MWS
SP4 6LS G8OFA
SP4 6LS G8RHC
SP4 6LX G0HSV
SP4 6NB G1SSZ
SP4 6PX 2E0FDB
SP4 6PX M6FHS
SP4 6SF G4BGH
SP4 7AS G6YUX
SP4 7BB M0IPJ
SP4 7EE G3YFE
SP4 7FF G4WWZ
SP4 7FS M6FQL
SP4 7HW 2E0DLT
SP4 7HW M6CCN
SP4 7NN G4UED
SP4 7PP G4SXT
SP4 7PW M0IHU
SP4 7QS G6UXK
SP4 7RE 2E0IKU
SP4 7RE G1SJM
SP4 7RE G7GKD
SP4 7RG G1WRY
SP4 7RG G4YRV
SP4 7TD G4WPI
SP4 7TU M3ZYR
SP4 8AB M1AFF
SP4 8BY G8LJO
SP4 8DJ G6JRS
SP4 9DA 2E0FUE
SP4 9DA M7AJU
SP4 9HE M6IZM
SP4 9PJ M0RVN
SP4 9QG M6NIQ
SP4 9RW M7ERF
SP47RN M6IDU
SP48ES 2E0HNR
SP48ES M7CII
SP5 1NS 2E0PTF
SP5 1NS M0OOS
SP5 1NS M6SEE
SP5 1PF M6OVN
SP5 1PL M3VKB
SP5 1PP G4YIS
SP5 1PW G0PBS
SP5 1RB G8LDV
SP5 1RE G8PBY
SP5 1RJ M6WFH
SP5 1RS G1ITJ
SP5 1SH 2E0KDI
SP5 1SH G3MDM
SP5 1SZ 2E0RFL
SP5 1SZ M6HBN
SP5 2AL G4VQJ
SP5 2BE G0OSW
SP5 2BP M6KMZ
SP5 2BZ G7WBA
SP5 2HQ M0GWQ
SP5 2HT G6ILC
SP5 2HU G3OSQ
SP5 2LN G0RZB
SP5 2LQ M3HCB
SP5 2LR M0CBM
SP5 2LR M3DZN
SP5 2NQ G3SHK
SP5 2NR G4TEK
SP5 2SE G4KCM
SP5 2SZ M0GSP
SP5 3AR G1ELZ
SP5 3AT G3YWT
SP5 3AZ G0CDZ
SP5 3BF 2E0HEA
SP5 3BF M3FQG
SP5 3FN 2E0CFP
SP5 3FN M6AFN
SP5 3LB 2E1TON
SP5 3LB G6ZHJ
SP5 3LG M3MGD
SP5 4JT 2E1NAC
SP5 4JT 2E1PEC
SP5 4LF M7LCL
SP5 5NJ M7OBY
SP5 5NP G8BZR
SP5 5NZ M6OVI
SP5 5PF G7HRB
SP5 5QJ 2E0DZJ
SP5 5QL G7IRU
SP5 5QL M3CAQ
SP5 5SA M6CPO
SP52SZ M7DDF
SP5-5PL M1HFX
SP6 1AY 2E0KZC
SP6 1AY M0KZC
SP6 1AY M3KZC
SP6 1BJ M3HZD
SP6 1BP G7OZH
SP6 1BW 2E1FYZ
SP6 1EG M3UQA
SP6 1EQ G6CEZ
SP6 1JF M0WDZ
SP6 1LW G4POF
SP6 1NW M3NMX
SP6 2AL 2E0HPS
SP6 2AX G8OFO
SP6 2LJ G1MVG
SP6 2NR G4GBP
SP6 3AD M6PSQ
SP6 3DJ G0TQZ
SP6 3DU 2E0HJE
SP6 3DU M6VLB
SP6 3EE G7VOX
SP6 3EP 2E0RAH
SP6 3EW G8BIH
SP6 3HA G1VOQ
SP6 3LD 2E0RFK
SP6 3LD M3RFK
SP6 3QA G3ZNH
SP6 3RB G1JMH
SP6 3RB G6LVI
SP63PJ M7CIO
SP7 0BF M7KRN
SP7 0BJ M3HEV
SP7 0DP G7ESE
SP7 0DP M6DGS
SP7 0HS M3SQI
SP7 8AL G0YBU
SP7 8GY G3NRH
SP7 8HX G3LGF
SP7 8LP M7HGM
SP7 8LQ G8BSP
SP7 8NF G3BVB
SP7 8NQ G3FWU
SP7 8PP M7WSX
SP7 8QS M0TMO
SP7 8RD M7DKJ
SP7 8RU M6KVH
SP7 8RX G4OVJ
SP7 9HD G8LBG
SP7 9HX G8PQD
SP7 9NA M6KZH
SP7 9NX G1RJA
SP7 9PA 2E1WAV
SP7 9PF G1NWT
SP7 9QB M3JLH
SP7 9QG 2E0TYH
SP7 9QG M0TYH
SP7 9QG M6TYH
SP8 4EL G4FUY
SP8 4EQ 2E0INP
SP8 4EQ M0GNI
SP8 4EQ M7FDT
SP8 4EU G3ZBB
SP8 4GN M6RFI
SP8 4HH G0VNA
SP8 4NR G8PQJ
SP8 4NT M7ANG
SP8 4PE G7SUS
SP8 4QJ G4ILM
SP8 4QY M6OKD
SP8 4RB G8DDN
SP8 4RE M7DCN
SP8 4RR G4VRM
SP8 4SS G1SNI
SP8 4TR M6KZI
SP8 4TW G4FQV
SP8 4UP G0YLO
SP8 4UP G0ZEE
SP8 4UP G6TER
SP8 4WE M6ROA
SP8 5AL G3JSA
SP8 5BB M3UZE
SP8 5EL M0AKI
SP8 5ET G0RIH
SP8 5ET M6JGH
SP8 5EW G4FDI
SP8 5LB G0MPM
SP8 5LW G8SYA
SP8 5NB G1THG
SP8 5QR M5BJC
SP8 5RL G7JIF
SP8 5RL M0PRW
SP8 5RL M7PRW
SP8 5RN M3TBH
SP8 5SJ G8HPN
SP9 7EU G0WYD
SP9 7FN M6ZBW
SP9 7JX 2E0EQL
SP9 7JX M6YSD
SP9 7SB M6NHP
SP9 7TR 2E0DUJ
SR1 2DP M3XYH
SR2 0BP G4DGB
SR2 0JU G0SLN
SR2 7DE M7LCP
SR2 7EZ G6GGT
SR2 7HB G7FFV
SR2 7RX M3REP
SR2 7TS G4GTX
SR2 8HR M7NAW
SR2 8NF M0MDP
SR2 8QB G7PQD
SR2 8RS G6LMR
SR2 8RX G6INK
SR2 9BB G0KFY
SR2 9DX G0ASM
SR2 9EE G4MSJ
SR2 9EJ G4TOI
SR2 9HQ G3YJG
SR2 9LQ G0KVJ
SR2 9LQ G0NSK
SR2 9LQ M6NSK
SR3 1AN G4HPS
SR3 1AN G4OBX
SR3 1HJ G7MFN
SR3 1HJ M3MYC
SR3 1JF G0TAX
SR3 1JR M6YOZ
SR3 1JW 2E0PYN
SR3 1JW M6AVG
SR3 1LG G7KOF
SR3 1LX G4MTW
SR3 1QS M1CQX
SR3 1UJ G0BWJ
SR3 1XW G6UBK
SR3 2RF M0WRP
SR3 2RG G6SGZ
SR3 2YG M7GLK
SR3 3AL M1DPQ
SR3 3AN M6JIH
SR3 3LA M7LIK
SR3 3PX G3ZOG
SR3 3SF M0ATA
SR3 4AA G0AOE
SR3 4DJ M6IOL
SR3 4EE M6HAM
SR3 4EZ G4VLT
SR3 4HS 2E0MAL
SR3 4HS M3MCF
SR3 4HZ M7ODO
SR3 4JN 2E0PAX
SR3 4JN M6UVF
SR3 4PA G7PUK
SR34AT 2E0HHE
SR4 0AB G1YIZ
SR4 0AE G4HIX
SR4 0BA M0KLL
SR4 0BF M0MHG
SR4 0LZ G0MET
SR4 0NA M3XYP
SR4 0QZ 2E0UKA
SR4 0QZ M3XJX
SR4 6AU M3JFB
SR4 6EY M6MOB
SR4 6TE M7NTE
SR4 6XG G7MXM
SR4 7LL M6RNQ
SR4 7RY M0GFN
SR4 7SA G1YUL
SR4 7SU M1CPB
SR4 7SZ G4UAO
SR4 7TB M0RKW
SR4 8AU G0EHX
SR4 8BG M6GNX
SR4 8BS 2E0YDT
SR4 8HT G8HPW
SR4 8NP G3WOM
SR4 9DW G1WEV
SR4 9EN G8NPP
SR4 9NQ G0SRG
SR4 9NQ G4EKM
SR4 9QU M7GBH
SR40HR M7DAD
SR5 1DY G4XYP
SR5 1LG 2E0OAP
SR5 1LG M6OAP
SR5 1LL M6UNS
SR5 2BU G6VTH
SR5 2DJ M7MKO
SR5 3DD M7BBW
SR5 3EG G0ILV
SR5 3PG M0GOL
SR5 3QA M0RXM
SR5 3RE M5GHT
SR5 4BU G1CSA
SR5 4DF G0OBN
SR5 4LH M6MMY
SR5 4ND M0KAE
SR5 5AH 2E0HYE
SR5 5AH M0HYE
SR5 5AH M6BYE
SR5 5AW 2E0SNE
SR5 5AW M6FIT
SR5 5ET M6CWQ
SR5 5ET M7ATI
SR5 5QA G0UNE
SR5 5RD G7CKQ
SR5 5RD M3XQB
SR5 5SA M6OKH
SR5 5TL M6ELL
SR53BU M7KEV
SR6 0JP G0BAN
SR6 0NL M6LND
SR6 0NT G4TMX
SR6 7AL G7PHI
SR6 7BW 2E0TLX
SR6 7BW M0TLX
SR6 7BW M3TLX
SR6 7HN G4LDT
SR6 7HN G8WTZ
SR6 7LN 2E0CUC
SR6 8BD G3TEC
SR6 8ER 2E0ERD
SR6 8ER M6ERD
SR6 8ET G4MRK
SR6 9AY M7CWX
SR6 9EL G4NTW
SR6 9HP G0IID
SR6 9NT 2E0DJJ
SR6 9NT M0KCF
SR6 9NT M6GPM
SR7 0AN 2E0EXH
SR7 0AN G8TDP
SR7 0AU G6LNL
SR7 0BD G6AGY
SR7 0BD G6AGZ
SR7 0JT G0MJV
SR7 0LP G0WTW
SR7 0LS G7KHF
SR7 7BQ G4GSO
SR7 7DJ G0MKC
SR7 7DJ G4PPL
SR7 7NG 2E0WRY
SR7 7NG M6WRY
SR7 7QW M3UKJ
SR7 7SA G1ERU
SR7 7SA M3SMM
SR7 7UE M7GNR
SR7 7UJ M6YIK
SR7 8DG G0DVP
SR7 8DZ G4NMF
SR7 8HW 2E0JRB
SR7 8JG M6EQG
SR7 8JT M6OOB
SR7 8JZ G0UFN
SR7 8JZ M7BMG
SR7 8NL G0NWY
SR7 8RS M0MED
SR7 8RZ M6SQC
SR7 9BN 2E0SNM
SR7 9BN M6SMA
SR7 9PQ G4NMK
SR7 9PQ M7DCA
SR8 1DD G0FBW
SR8 1DG G4TTF
SR8 1DG M0TZR
SR8 1DQ G0CWF
SR8 1LN G0TOK
SR8 1LP G4PLU
SR8 1PZ M0GZW
SR8 2DJ G4WKT
SR8 2DT G4RXR
SR8 2EA G0RDD
SR8 2HB G6VHG
SR8 2HE. M6HWN
SR8 2JS M0WUS
SR8 2JS M3AUP
SR8 2JW 2E1DMI
SR8 2NN G0NDD
SR8 3AJ G6RKS
SR8 3DF G7DBV
SR8 3HU G3ZMG
SR8 3JZ M6KRZ
SR8 3LY 2E0XMK
SR8 3LY M0XMK
SR8 3LY M3XMK
SR8 3SA M6PQD
SR8 4BH M7TXD
SR8 4BW G0NXC
SR8 4EH M7CZA
SR8 4NE M0TEN
SR8 4RD 2E0YZX
SR8 4RD M6TMG

SR8 4RD M7HTR
SR8 5EG 2E0OTM
SR8 5PF M0BLI
SR8 5UD G4RVY
SR85LW M6YRC
SS0 0AN 2E0SET
SS0 0AN M0XMT
SS0 0AN M6NFW
SS0 0EU M0SCE
SS0 0EU M6CSE
SS0 0NL M7FKR
SS0 0NY G6CVB
SS0 0QL G6WCI
SS0 0RA G0ENN
SS0 7AZ M1FIB
SS0 7DR G1PVT
SS0 7DR G6AXE
SS0 7LA G0MBP
SS0 7PU G2BBI
SS0 7QH G4EOR
SS0 8BD M0DRS
SS0 8BD M3VNS
SS0 8BN G4DVJ
SS0 8BN M7CCO
SS0 8NL G0FNB
SS0 9JN G1KPV
SS0 9PR M0INP
SS0 9QF 2E0TCY
SS0 9RA G8IWI
SS0 9RJ M7VPM
SS0 9SZ 2E0TWW
SS0 9SZ M0TWW
SS00HH M7TUC
SS1 1DT G4DIG
SS1 1HG G0RSW
SS1 1NH M3XMZ
SS1 1QB M1KES
SS1 2RP G1VJJ
SS1 2RT M7TAA
SS1 2SG M7RWA
SS1 2SX G1HNN
SS1 2SX G4UAI
SS1 2SX G4UTF
SS1 2TZ M0KUR
SS1 3AD G1IQU
SS1 3DF G6XNK
SS1 3DF G8LYW
SS1 3DG G8ZPO
SS1 3HD G1JLG
SS1 3NW G0TIG
SS1 3PX G4YAK
SS1 3SS G6TLN
SS11 7DG M6MVA
SS11 7DN G7DAH
SS11 7EF 2E0NGZ
SS11 7EF M0NGZ
SS11 7EF M6VSL
SS11 7EH G4FKX
SS11 7HX M0JRW
SS11 7JE G8BPW
SS11 7LN G0DZQ
SS11 7ND 2E0CUU
SS11 7ND M6CHI
SS11 7PF G1EUM
SS11 7PP 2E0CNA
SS11 7PP 2E0EUH
SS11 7PP M0PSD
SS11 7PP M0XPK
SS11 7PP M6BMI
SS11 7PP M6XPK
SS11 7PT G1HHS
SS11 7PT G1LCS
SS11 8XB 2E0DVF
SS11 8XB M6KOA
SS11 8XL 2E0EUM
SS11 8XL M6LYQ
SS12 0AB G4TFG
SS12 0DX M1DFC
SS12 0EN G7MBY
SS12 0ER G3LRL
SS12 0HE G3PZZ
SS12 0HS G7HGI
SS12 0LP G4DCB
SS12 0QB 2E0KGT
SS12 0QB M0ICG
SS12 0QB M6GTK
SS12 9DY G1KOR
SS12 9ED M3LHX
SS12 9EQ G4YIH
SS12 9FA G6WKN
SS12 9JX G7HCO
SS12 9NA G6OVL
SS13 1BQ G6WLQ
SS13 1HA G7EIE
SS13 1JL G7WFC
SS13 1JS 2E0FMS
SS13 1JU G6AJT
SS13 1NR G8WSS
SS13 1PJ M6ILM
SS13 2BH 2E0RCF
SS13 2BH M3OBO
SS13 2EJ M6PMP
SS13 3EW G7JRK
SS13 3PR G6MLJ
SS13 3QH 2E1IJU
SS13 3QX M7XPT
SS14 1HF G1GYQ
SS14 1RX M1TDD
SS14 1UA G1PXQ
SS14 2AZ G6HSS
SS14 2DB M7HWL
SS14 2FF 2E0FJI
SS14 2FF G0TRG
SS14 2FF M0IXY
SS14 2FF M0XXC
SS14 2FF M7LCF
SS14 2FF M7RNS
SS14 2FH M6HSH
SS14 2JD G8CUN
SS14 2JQ G7KDR
SS14 2NS G1YKX
SS14 2TN G0UCH
SS14 3JN G0NPO
SS14 3LZ 2E0SSX
SS14 3LZ M0PFJ
SS14 3LZ M0SSK
SS14 3NA G6SNN
SS14 3QS 2E0NCE
SS14 3QS G7OED
SS14 3QS M0LMR
SS14 3QS M6EBQ
SS15 4AE M7RVT
SS15 4BH 2E0JPS
SS15 4EJ M6HKT
SS15 4EW 2E0TKV
SS15 4EW M3TKV
SS15 4JE G6TNE
SS15 4JS M7RHA
SS15 5BG M6EWL
SS15 5FY M6HCV
SS15 5RH M1EAW
SS15 5YN G0RTH
SS15 5YN G7KCN
SS15 6AL G3UPM
SS15 6AS 2E0HJN
SS15 6JR G1AUR
SS15 6PR G6LZX
SS15 6RA M6UKJ
SS15 6SU 2E0EUD
SS15 6SU M6NMQ
SS15 6SY M7JNR
SS155SW M7RCG
SS16 4NE G1CHV
SS16 4PQ G3VYF
SS16 4PQ M1BHC
SS16 4PQ M6GJN
SS16 4RU 2E0BRK
SS16 4TL G4EPU
SS16 5AY M7JLD
SS16 5BN M1ECC
SS16 5EN 2E0GAW
SS16 5FW M0EGG
SS16 5HD G8KIH
SS16 5HX G4IEG
SS16 5JH G8OSG
SS16 5QF 2E1EOZ
SS16 5QQ 2E0KPR
SS16 5RA 2E0GQD
SS16 5RA M3GQD
SS16 5RW 2E0LEZ
SS16 5RW M6LTO
SS16 5TW G7ALR
SS16 6AQ G1KOT
SS16 6AQ G4NVT
SS16 6DU 2E0WKV
SS16 6DU M3WKV
SS16 6ED G7PSW
SS16 6GB M7DXT
SS16 6LE M6MRH
SS16 6NU 2E0CTA
SS16 6NU M0ONZ
SS16 6NU M6ORC
SS16 6RN M6KSE
SS16 6SJ G0PAE
SS16 6SJ G8JCV
SS16 6SP G1JGD
SS17 0BB 2E0TOL
SS17 0BB M6GRX
SS17 0BE 2E0EEB
SS17 0BE M6EEB
SS17 0DF G2FWJ
SS17 0DF G4HXY
SS17 0EF G0NGG
SS17 0NH G7JVB
SS17 7BD M0AKE
SS17 7BQ G4KIQ
SS17 7BZ M6RFW
SS17 7EW G1PQK
SS17 7EW M6XSZ
SS17 7HE M1BZR
SS17 7LB G4JIU
SS17 7LD G8EXQ
SS17 7NH M6EBU
SS17 7QL G4LTH
SS17 7QZ G4LNT
SS17 7SB 2E0CJO
SS17 7SB M0LEH
SS17 7SB M6ITX
SS17 7SX 2E1EXA
SS17 8DD M1DKP
SS17 8DE G1ERQ
SS17 8DT G1YAH
SS17 8HH G7OUT
SS17 8HL G7FEL
SS17 8HP G4JHE
SS17 8NL G6WXZ
SS17 8NS G6DXP
SS17 8PN G8LWA
SS17 8QU G7DXV
SS17 9AJ G3SMF
SS170AT M7CVZ
SS2 4AL M7FSK
SS2 4AZ M6IIL
SS2 4DA G8OQR
SS2 4DH M0RLI
SS2 4ED G3UUI
SS2 4GL M7MRO
SS2 4HR G0SKI
SS2 4HU G0DPC
SS2 4JR G0FVB
SS2 4LJ G7JLO
SS2 4LU G0XOX
SS2 4LW G0TUI
SS2 4LW M0JZS
SS2 4LW M0JZU
SS2 4LZ M3IAC
SS2 4NN G1YCR
SS2 4NW G0UKP
SS2 4PA M6PEQ
SS2 4RD M3VHQ
SS2 4RD M6RDK
SS2 5EJ M6JUU
SS2 5EJ M6LHG
SS2 5HE M3IVD
SS2 5HY 2E0DVX
SS2 5PJ M6IPY
SS2 6GQ M7DJP
SS2 6HB M6CSJ
SS2 6HR G1DXD
SS2 6JG G0HWI
SS2 6LW M3YLB
SS2 6NR M7FMB
SS2 6RW 2E0FTX
SS2 6RW M6HRT
SS2 6TB 2E0VZL
SS2 6TB M3VZL
SS2 6TF G0TTI
SS2 6UA G0TZP
SS2 6UJ M7MBK
SS2 6YF M7TKH
SS25NG M7JNQ
SS26QY G8RAU
SS3 0AR G1HPV
SS3 0BE 2E0VNO
SS3 0BE M0VNO
SS3 0BE M3VNO
SS3 0DR G0NXN
SS3 0DY 2E0JDY
SS3 0DY M7JBO
SS3 0EB 2E0JVD
SS3 0EB M6JVD
SS3 0EX G1EXR
SS3 0EZ M3POW
SS3 0JG G7BLJ
SS3 0JS 2E0BJB
SS3 0JS M0GGZ
SS3 0JS M3SYC
SS3 0LS M6EDD
SS3 9AR G7CQA
SS3 9FA G0CXW
SS3 9JG M0KVA
SS3 9JG M3VAR
SS3 9JP G0PCF
SS3 9LR G7PMK
SS3 9NJ M3OLW
SS3 9NT M3UFA
SS3 9NY G1HQQ
SS3 9PE M7TOT
SS3 9PX M3YLN
SS3 9RJ M0ABA
SS3 9RJ M0CNS
SS3 9SB M6MEO
SS3 9SG G6BHE
SS3 9SG G7RYM
SS3 9SL G4CEU
SS3 9TL M0WBK
SS3 9TL M3WBK
SS3 9TN M6DNZ
SS3 9YE G8OPY
SS3 9YS G0KEI
SS3 9YS G1XXV
SS4 1DP M1AHT
SS4 1EH M6GZO
SS4 1GH M3UPQ
SS4 1JE G6HNI
SS4 1LH M3KLU
SS4 1QH G7MFX
SS4 1QH M6HLP
SS4 1QJ G8SGH
SS4 1RS G0KTC
SS4 1SH G8GHK
SS4 2ER G1HML
SS4 3AH G1RAP
SS4 3BX G7FFI
SS4 3DQ G6KFD
SS4 3EN M7KVM
SS4 3EP M6JTP
SS4 3FJ 2E0KFB
SS4 3FJ M0IKN
SS4 3FJ M6KFB
SS4 3FJ M6WWI
SS4 3HE 2E0WHU
SS4 3HE M6WHU
SS4 3PY G1GCY
SS4 3PY G7PUZ
SS4 3QN G1JRD
SS5 4BX G1RYY
SS5 4DL 2E0EOM
SS5 4DL M0IDG
SS5 4DL M6IDG
SS5 4DN G1KHS
SS5 4EL M6HFP
SS5 4EY G0PEP
SS5 4EY G3OJV
SS5 4GN G1KVO
SS5 4JE M3IWJ
SS5 4JN G0DFE
SS5 4JY G0DTP
SS5 4NT G7WHZ
SS5 4PJ G4TWH
SS5 4QG G6LYM
SS5 4QS G6XYU
SS5 4QS M3JJS
SS5 4SW G4MCU
SS5 5AT G7GBN
SS5 5BP G0PEJ
SS5 5DY G4HKQ
SS5 5EE G7ENT
SS5 5EL G3ZJZ
SS5 5HD 2E1DOA
SS5 5HJ G3VQY
SS5 5HN G7WHP
SS5 6AA G4KDH
SS5 6BG G0EBG
SS5 6BH G4ZFJ
SS5 6BN G7TBW
SS5 6DD G4RDS
SS5 6HJ M7GRK
SS5 6LT G7MLU
SS5 6LT G8RAN
SS5 6LU G0MGT
SS5 6LU G1OOG
SS5 6LZ G3HWM
SS5 6NE 2E0JTW
SS5 6NE M6BPK
SS5 6NE M6GEJ
SS56EP M6ZPS
SS56LR G4MUS
SS6 7DX 2E0WYG
SS6 7DX M0NWE
SS6 7DX M6IIY
SS6 7JR G1TWS
SS6 7LB G0DRV
SS6 7LH M7FMK
SS6 7NS G4FSE
SS6 7RG G0DRH
SS6 7TD G0EFI
SS6 8AR G0HSK
SS6 8AR G6XAT
SS6 8BN G3OGX
SS6 8BP G0FKS
SS6 8EB G4KDE
SS6 8HP G1KQU
SS6 8JP M6MKU
SS6 8LQ G7PPS
SS6 8LW G4GNU
SS6 8LZ G8FAX
SS6 8SG 2E0SIS
SS6 8SG M6WYZ
SS6 8UA G0WIX
SS6 8UW M7NJR
SS6 8UY M0DDK
SS6 8YF G4KIH
SS6 9AL G4PWB
SS6 9EJ G7LAX
SS6 9JZ G4XEO
SS6 9LY G3ZHA
SS6 9ND M6KTW
SS6 9PD G6AVK
SS6 9TU G8OXU
SS6 9UH G3TRH
SS68XR M7KOP
SS69DQ M7ATW
SS7 1AL M0PGC
SS7 1DN G1BAR
SS7 1JL G4EZP
SS7 1JL G4JAR
SS7 1QB G4FCX
SS7 1SS G6MVW
SS7 2DL G4WAK
SS7 2JP G3UTA
SS7 2LN G3LUZ
SS7 2NN M7NRC
SS7 2PG M7DBL
SS7 2ST M0AOK
SS7 2TA M6KPQ
SS7 2TY G4ZMN
SS7 3AZ M6YYL
SS7 3BB M0DQK
SS7 3HE G7MPH
SS7 3HU 2E0FVS
SS7 3HU G4HBA
SS7 3HU M7AHT
SS7 3LD 2E0RSX
SS7 3LD M6RMU
SS7 3NA G8FAR
SS7 3PL M6PID
SS7 3RJ G1FBZ
SS7 3TU G0TTM
SS7 3TU G5QK
SS7 3UU G8UWI
SS7 3YL G4AJY
SS7 4DT G6IDU
SS7 4EE G4TPK
SS7 4EF G4UMP
SS7 4EN G8WUU
SS7 4EZ G8EFG
SS7 4LA G6TKW
SS7 4LS G1ZHN
SS7 4NT G7EGU
SS7 4NW M3BFB
SS7 4NX M7MCI
SS7 4NY 2E0XPD
SS7 4NY M6XPD
SS7 4PB M6PFY
SS7 5BQ G8EOM
SS7 5DT G1GBV
SS7 5ES G4GDS
SS7 5JQ G8VFI
SS7 5LH G7IIO
SS7 5NU G6ZUE
SS7 5PH G8YPK
SS7 5RD G4PBO
SS7 5RL G0LYX
SS7 5SJ G1LAW
SS8 0AB 2E0MZC
SS8 0AB M6MZC
SS8 0AB M6OIU
SS8 0AD 2E0FBR
SS8 0AD M0OAD
SS8 0AD M6OIY
SS8 0BB G7MXL
SS8 0BP G0JXJ
SS8 0BP G0LTO
SS8 0BU G1BGJ
SS8 0DN G7HQF
SS8 0EW 2E0LFX
SS8 0EW M6LFX
SS8 0EX G0WAX
SS8 0EX G0WGA
SS8 0JB M6EYW
SS8 0JB M6XXI
SS8 7DN M6SDX
SS8 7EA G1FBW
SS8 7EE G7TFA
SS8 7EL G1IEO
SS8 7HL G1ZSG
SS8 7JD 2E0FME
SS8 7JD M3FME
SS8 7PB M3IKI
SS8 7QU M6IGR
SS8 7RD M6HCT
SS8 7TJ G1NPJ
SS8 7TS G4BQF
SS8 8HG M7KKO
SS8 8HH M6HJC
SS8 8HX G0JAN
SS8 8HX G1OXO
SS8 8NZ M0FZW
SS8 8NZ M6FZW
SS8 9AB G6CNF
SS8 9BL G7PDU
SS8 9DJ G4ZQJ
SS8 9DS G8ZTB
SS8 9DZ G4YQJ
SS8 9EJ M6EVZ
SS8 9HL 2E0IEO
SS8 9HL M0IEO
SS8 9LP 2E0NNQ
SS8 9LP M3NNQ
SS8 9NY G0KSC
SS8 9QL G6SPH
SS8 9QP G0NCT
SS8 9RQ M6HDO
SS8 9TW 2E0RMT
SS8 9TW M6RKC
SS8 9YB G1SOB
SS87GQ M7BUV
SS89JJ G7PFG
SS9 1HQ M0BDB
SS9 1HQ M1ANK
SS9 1RU G0KMF
SS9 2DT M3KSS
SS9 2HT G0JAO
SS9 2JX G7CDO
SS9 2PB G4SNV
SS9 2PX G0JYI
SS9 2QS M1DKL
SS9 2SY M3UCH
SS9 2XD G4OQ
SS9 3EB M6TGC
SS9 3HA M3KLV
SS9 3JT G0JDD
SS9 3LF 2E0MMH
SS9 3LF M6FFB
SS9 3NT M0SXA
SS9 3PJ M6FSU
SS9 3QF 2E0IMZ
SS9 3QF M6IMZ
SS9 3QS G1GXF
SS9 3SG G0GBY
SS9 3TH 2E0XLM
SS9 3TH M0SCX
SS9 4AZ G6OAV
SS9 4DA G8WSW
SS9 4EA G0FTN
SS9 4HA G1YAF
SS9 4HA M6BMK
SS9 4HG G6GGZ
SS9 4NB G6MMR
SS9 4PP 2E0YPK
SS9 4PW G6DAH
SS9 4PW G6NLS
SS9 4PY G4UQY
SS9 4QY G3PHL
SS9 4RT M7KTT
SS9 4RY G3SVI
SS9 4SZ G1NCO
SS9 4TN M7BLO
SS9 5AB G8CEX
SS9 5AR G8UMF
SS9 5AR M6LOS
SS9 5EG M7GSA
SS9 5EL G3RPZ
SS9 5HR M1RKB
SS9 5NF 2E0KPX
SS9 5NN G7EGX
SS9 5NW G6OLY
SS9 5NX G4KEG
SS9 5RF G7DTS
SS9 5SW G3RCX
SS9 5XR G6WFS
ST1 2NE 2E0YYY
ST1 2NE M6MMM
ST1 3BA G8ZES
ST1 3BA M0GAN
ST1 3GX M6MVK
ST1 3HS 2E1SOB
ST1 3SP G6ING
ST1 5QP M6CEQ
ST1 5QS M7EJB
ST1 6BY G6KKN
ST1 6DE G8NTG
ST1 6EF 2E0BLD
ST1 6EF 2E0PTI
ST1 6EF M3PTI
ST1 6EF M3PXU
ST1 6JD M7JPF
ST1 6NY M7SIB
ST1 6PQ M3GVT
ST1 6SL G4FMJ
ST1 6TW G1XLN
ST1 6XR M6EXA
ST10 1AT G8VSR
ST10 1BS M7VCZ
ST10 1DP M1AMB
ST10 1DT G4SXK
ST10 1LU G6UEH
ST10 1LW G4DSQ
ST10 1LZ M3YWH
ST10 1NQ 2E0VFD
ST10 1QG G4CHG
ST10 1QQ G7PIJ
ST10 1QS G0TPY
ST10 1RU 2E0CEJ
ST10 1RU 2E1IHF
ST10 1RU G7MWS
ST10 1RU G7PBH
ST10 1RU M3MHL
ST10 1RU M3MIP
ST10 1RX M6EWV
ST10 1SA 2E0PBH
ST10 1SA M3PHZ
ST10 1XB G1MAD
ST10 1XB G4NHT
ST10 1XB G4OUG
ST10 1XB G4PGG
ST10 1XB G7PHB
ST10 2AG G6KTE
ST10 2DW M3GBA
ST10 2EG G6YIW
ST10 2PA G3UNM
ST10 2PT G6YCN
ST10 3BW M3OFA
ST10 3EA G4OTX
ST10 3ER G1OGE
ST10 4AN 2E0TVW
ST10 4AN M6AVW
ST10 4BH G4YYO
ST10 4DZ G0VBT
ST10 4EG M1RKY
ST10 4FE 2E1FOW
ST10 4HL G8FHC
ST10 4LD 2E0WJE
ST10 4LD M3DPY
ST10 4LS G6ZBL
ST10 4LT G7MGA
ST10 4LX M7ANJ
ST10 4LY G6GVZ
ST10 4NJ M0KDU
ST10 4PE G0UUA
ST11 9AA G7CEY
ST11 9AP G0VHY
ST11 9AU G6IXN
ST11 9AZ M3JDJ
ST11 9BH G0VRK
ST11 9DA G6OAS
ST11 9DR M6YAW
ST11 9EN 2E0YYT
ST11 9EN M6YYT
ST11 9HA G7VLA
ST11 9HQ M0DSR
ST11 9HU 2E0IDK
ST11 9HU M0IDK
ST11 9HU M6IDK
ST11 9LY 2E0DVK
ST11 9NH M0PWC
ST11 9NT G1DOG
ST11 9NX G4YYG
ST11 9NX G4YYP
ST11 9NZ G7UKN
ST11 9PD M0CTL
ST11 9PL 2E0AAO
ST11 9PL G6GA
ST11 9PL M6BWZ
ST11 9PP G4WUX
ST11 9PX G7IMO
ST11 9PX M3AAQ
ST11 9QT G1BHF
ST11 9RH 2E0MZL
ST11 9RH M6MZL
ST11 9RN G4HUO
ST12 9BD M3XKM
ST12 9BL M6HOG
ST12 9DT G4DUB
ST12 9JA G8SDX
ST12 9JQ G0RJT
ST13 5BW G7VFQ
ST13 5DB M3KMO
ST13 5DB M5JAO
ST13 5JU 2E1DLR
ST13 5LS 2E0MJJ
ST13 5LS M3TJJ
ST13 5NZ 2E0SXY
ST13 5NZ M3SIM
ST13 5PW 2E0SAA
ST13 5PW M3RRZ
ST13 5PZ M1BIX
ST13 5RR G7HIJ
ST13 5SZ G4MDJ
ST13 6BB G7LIE
ST13 6BB M0CAD
ST13 6BU M0RMP
ST13 6ES G8YQA
ST13 6PS M7DDE
ST13 6PX M3BVP
ST13 7AA G3VBG
ST13 7AN G4XIZ
ST13 7ED G8FWD
ST13 7EX G0LAZ
ST13 7HJ G0WOM
ST13 7JP G6KPT
ST13 7LW G7SSK
ST13 7NP G1VXD
ST13 8BL 2E1SGK
ST13 8BL M0SGK
ST13 8DD M1ALE
ST13 8EQ M0FAZ
ST13 8EQ M3FCN
ST13 8EQ M3FCO
ST13 8EQ M3HRN
ST13 8EQ M3SKD
ST13 8ES G4TGS
ST13 8JA G8JGL
ST13 8JN M1ABF
ST13 8LD G7MQF
ST13 8LL G7CFT
ST13 8LN G4UQM
ST13 8LW 2E1HRN
ST13 8SL M6GSF
ST13 8XF G8YMS
ST14 5AG 2E0KMS
ST14 5DE M0DNU
ST14 5DH G3VSB
ST14 5DP 2E0SRC
ST14 5DP M0RRN
ST14 5DP M6YUK
ST14 5JU G7LGS
ST14 5LE 2E0AUM
ST14 5LF M3HEJ
ST14 5LT G0LJH
ST14 7AX M3MTP
ST14 7BY 2E1DLM
ST14 7EN 2E1BMV
ST14 7EQ M0VCS
ST14 7ET 2E1CLM
ST14 7ET G6VGC
ST14 7FE G1RKJ
ST14 7FE G7PAY
ST14 7HL 2E0LIZ
ST14 7HL 2E1LIZ
ST14 7HL M3GLK
ST14 7LB G6SVJ
ST14 7NB G7STC
ST14 7NF G7UCN
ST14 7QG G6CUY
ST14 7QY G6UOX
ST14 8BB G8YAT
ST14 8BT G4XSR
ST14 8EF M0WUL
ST14 8LT 2E0PBB
ST14 8LT G7ORN
ST14 8LW M7HMS
ST14 8LW M7PDQ
ST14 8QS G8FGQ
ST14 8TR G4EHT
ST14 8UP M6PLQ
ST148SJ M0SRJ
ST148UH 2E0HMQ
ST148UH M0KIZ
ST148UH M7DAY
ST15 0AW M7AFK
ST15 0DH G0KJP
ST15 0DW 2E1BRA
ST15 0DX G0SKQ
ST15 0DY 2E1ROO
ST15 0DY M7ROO
ST15 0EB G0TED
ST15 0EG G1LAP
ST15 0EP G1DSF
ST15 0EP G7KEE
ST15 0JA G3JUX
ST15 0JF G8HWI
ST15 0JN G1HSJ
ST15 0LF M0CDN
ST15 0PX G4TTM
ST15 0QG G7BMP
ST15 0QH 2E0PSD
ST15 0RD M6MGO
ST15 0RH G4NJR
ST15 0RP M3CZL
ST15 0WA M0IJY
ST15 8AR 2E1SRI
ST15 8BL M6BUX
ST15 8DY M3XUJ
ST15 8LA G3XMK
ST15 8LF G0VWT
ST15 8LQ M0CZA
ST15 8LW G4CRK
ST15 8NL 2E0UFS
ST15 8NL M7JRB
ST15 8PR G6JAF
ST15 8SG M7CQJ
ST15 8TG M3ZGD
ST15 8TZ M6KGL
ST15 8XL G4EJD
ST15 8YP G8MZZ
ST15 8ZQ M0CMC
ST16 1DS G0DEZ
ST16 1FJ G8RFV
ST16 1HA 2E0HHK
ST16 1HA M0YZY
ST16 1HA M6ZZQ
ST16 1HE M7TYG
ST16 1HH G4WMV
ST16 1HJ G3ZHS
ST16 1LD G4TMD
ST16 1LD G7BJG
ST16 1NJ M7SCC
ST16 1NL G6YLZ
ST16 1NL M6MTS
ST16 1QP G6ZLJ
ST16 1QR M6HQR
ST16 1SD 2E0HBB
ST16 1SD M6HRL
ST16 1TB G4YFF
ST16 1TB M0RWR
ST16 1TB M6VIX
ST16 1TE G1UDS
ST16 1TS G6MDX
ST16 1WH M7DWT
ST16 1XA G0FXS
ST16 2DZ G3XPD
ST16 3HD G0IAC
ST16 3HQ G0EYX
ST16 3HS G7PFT
ST16 3NP M6NUJ
ST16 3NX 2E1WWD
ST16 3NX G0GAP
ST16 3NX M6FFE
ST16 3PJ M6TPG
ST16 3PL G0GOZ
ST16 3PL G3SBL
ST16 3RE 2E0RPF
ST16 3RE M0RPF
ST16 3RE M3RPF
ST16 3WG G1SPU
ST16 3YE G8JAQ
ST16 3YW 2E0IFS
ST16 3YW M0IUU
ST16 3YW M6IVC
ST17 0AJ G0OIQ
ST17 0AQ G4NVH
ST17 0AQ M3NLW
ST17 0AQ M3NVH
ST17 0HB M7SJB
ST17 0HF 2E0BZC
ST17 0HF M0ZPA
ST17 0HF M6PMD
ST17 0HJ G4PKF
ST17 0JA G6JKF
ST17 0NU G6KQD
ST17 0NX M7BMO
ST17 0PA G3LOE
ST17 0PA G6LXF
ST17 0QB M7PYE
ST17 0QB M7RUK
ST17 0SJ G4RWQ
ST17 0SL 2E1AZK
ST17 0SP G6IRZ
ST17 0TW G0TFD
ST17 0TW G1UUJ
ST17 0YE 2E0GGT
ST17 0YE M3GGE
ST17 4AT 2E0EOU
ST17 4AT M0EOU
ST17 4BP G6RFM
ST17 4BP G2HNA
ST17 4DY M6DFL
ST17 4EH G7VLB
ST17 4HB G6TBV
ST17 4HZ 2E0MQA
ST17 4HZ M3MQR
ST17 4JJ M6PCW
ST17 4NR G1XWO
ST17 4QJ 2E0VTS
ST17 4QJ M0POG
ST17 4QJ M0VTS
ST17 4QJ M3VTS
ST17 4QN 2E0LTL
ST17 4QN M7CFW
ST17 4QP G0GAR
ST17 4QR M6KIX
ST17 4QS G4PWV
ST17 4QX G0VGK
ST17 4RY 2E0RDQ
ST17 4RY M6BVD
ST17 9BE G8KUA
ST17 9DS 2E0AVK
ST17 9EF M3GCS
ST17 9FR M1LIP
ST17 9FT G6KJK
ST17 9HP G1SJG
ST17 9HP G1UDT
ST17 9HP G3OSP
ST17 9HU M7AMM
ST17 9HU M7VVV
ST17 9JU M6ATU
ST17 9NZ M0DJF
ST17 9PR G6FYD
ST17 9QJ G3VUL
ST17 9RA 2E0XTB
ST17 9RA M7TXB
ST17 9RL G4RSW
ST17 9SB M3NOF
ST17 9SR 2E0HGK
ST17 9TW M3XZY
ST17 9UE G0BYA
ST170HS M1EOP
ST18 0DR 2E0PRC
ST18 0DR M6SFC
ST18 0EW G8JVU
ST18 0GH M7UKL
ST18 0GQ G0GMS
ST18 0GQ G4ZMM
ST18 0NN G4RJM
ST18 0QE G0GUF
ST18 0QZ G4OUT
ST18 0RD G4PET
ST18 0RD G4PEU
ST18 0SP 2E1ELE
ST18 0TS G6UZL
ST18 0UR G4THY
ST18 0XB G6KGA
ST18 9AY M7EEH
ST18 9BD M0SJD
ST18 9DA G6DGX
ST18 9DQ G4KQK
ST18 9DS G6OYF
ST18 9EU 2E0NQA
ST18 9EU M3NQA
ST18 9FB M7PFT
ST18 9HS M6MRM
ST18 9NF G6PVT
ST18 9PE G4MGI
ST18 9QR G4UKA
ST19 5AH G3JBF
ST19 5BQ M7EAK
ST19 5BX G3PIN
ST19 5BX M7NFL
ST19 5EH G0VJC
ST19 5HD G0GJG
ST19 5HF M0UXB
ST19 5QH G0BFK
ST19 5SP M3VOR
ST19 9AB G4WAF
ST19 9EH M0LRY
ST19 9JZ G4IDG
ST19 9LX M0RSY
ST19 9NQ 2E1GHX
ST19 9NQ M3PLN
ST19 9NU 2E1GVJ
ST2 0AQ M3DPP
ST2 0BX G1YQL
ST2 0EH 2E0LVL
ST2 0EH G3PWP
ST2 0HB M6EYS
ST2 0QY G4GRZ
ST2 0RE M6MXF
ST2 0RQ G7RFH
ST2 7AR G4BEM
ST2 7AR G4DPV
ST2 7BZ G4CKT
ST2 7HJ G4OSI
ST2 7LD M1MLM
ST2 7LP G6LJC
ST2 7LR G0UZE
ST2 7LR G7RTX
ST2 7NF G3ISX
ST2 7NG G6CCQ
ST2 7PA G1SQC
ST2 7PF G0TKR
ST2 8AQ G4DVA
ST2 8DQ G7VSW
ST2 8DS 2E1STO
ST2 8DU M3FIM
ST2 8EH M6YAH
ST2 8HU 2E1MAZ
ST2 8HU M0DVT
ST2 8HU M0FTR
ST2 8HU M3LVA
ST2 8HX M3SRH
ST2 8JN M1DXQ
ST2 8JZ 2E1DZP
ST2 8LP M3KEL
ST2 8NL M3KKI
ST2 9BZ M6PAN
ST2 9DF G8FGR
ST2 9DR G7VUL
ST20 0BA M6HMR
ST20 0HL G4VMC
ST20 0HR M7KWB
ST20 0JD G4OTB
ST20 0LG M6AJB
ST20 0NP M6PNG
ST20 0QH M3KQF
ST20 0RP 2E0LME
ST20 0RP M3LME
ST21 6DG M6GYQ
ST21 6DR M6OZQ
ST21 6LE G4LSA
ST21 6LT G4DDZ
ST21 6RG G0UUM
ST21 6RG G0UUN
ST3 1AD 2E1IVT
ST3 1SJ G0TTO
ST3 1SJ G7KDJ
ST3 2AJ 2E1GVS
ST3 2EG 2E0ESJ
ST3 2EG M3HNM
ST3 2HA G0CWO
ST3 2RS M6MEP
ST3 2RZ M1IRM
ST3 3DD 2E0YHW
ST3 3DD M6YHW
ST3 3LF 2E1BIL
ST3 3LX M3OLE
ST3 3LX M7BRH
ST3 3NU 2E0YYW
ST3 3NU M6YYD
ST3 4NF G6TPI
ST3 4NU M0MFH
ST3 5DX G7EWX
ST3 5EG G3VTE
ST3 5JA G7NEE
ST3 5JD G0JGB
ST3 5JL G6AYH
ST3 5LA M3AOP
ST3 5PW M1BFF
ST3 5QY G6XJF
ST3 5RP G7FSA
ST3 5RQ G8NAI
ST3 5ST G4WJX
ST3 5UD G1FHH
ST3 5UG G0SKM
ST3 5XW G0DUI
ST3 6AH G7SWR
ST3 6HA 2E0RDW
ST3 6HA G0SMN
ST3 6HY G1XMP
ST3 6HZ G1VFH
ST3 6JY G0PSH
ST3 6LP 2E0DII
ST3 6LP M6IAO
ST3 6NS G6THM
ST3 6PN 2E0VTR
ST3 6PQ G4RQG
ST3 6RG G1UYT
ST3 7AT M7BYC
ST3 7DP M6VGR
ST3 7EN G0WLC
ST3 7EW 2E0PHB
ST3 7FN M7EMY
ST3 7FW M3BIK
ST3 7HN G0UPV
ST3 7JA M0DPW
ST3 7JY G3MFH
ST3 7LH M3YMD
ST3 7LJ M3YSD
ST3 7QF G7MMW
ST3 7RZ M0FOG
ST3 7SS 2E0GQO
ST3 7SS G0NED
ST3 7UG M0EUI
ST3 7YE G8DZJ
ST3 7YE M7OKG
ST32BN M6HVI
ST34BN M6MWZ
ST34NP M6PUF
ST34SN M3SDJ
ST35DG M7UKP
ST36QJ M7JPS
ST37AP G8UWL
ST37DP M7BUE
ST4 1ED G6UDI
ST4 1PB G8PAD
ST4 2EX G0NMY
ST4 2RQ 2E1HCG
ST4 3DT M6SAM
ST4 3EN G0ODH
ST4 3HH G1YPH
ST4 3HQ G7JWO
ST4 3NR M0LAA
ST4 3NR M0LEK
ST4 3NR M1DOA
ST4 3NR M3SAS
ST4 3QX M7BRI
ST4 4DY G1EZJ
ST4 4NG G0VVT
ST4 4PH G4CJM
ST4 4PL G0VKI
ST4 4QP G3UD
ST4 5AN G0JZS
ST4 5DH M3HQQ
ST4 5DW G6WFF
ST4 5EE G1CHE
ST4 5EF G1OQW
ST4 5HA M6PAW
ST4 5JA M0JUI
ST4 5JE M6JEF

ST4 5PU M3WZY
ST4 6ED G7HQJ
ST4 6LA G0JGF
ST4 6LD G0DTW
ST4 6PR G6HFK
ST4 6QZ G3UJI
ST4 6RE M3YHN
ST4 6RU M1DCK
ST4 6RZ 2E0NNI
ST4 7DZ G8YDJ
ST4 7DZ M6LOZ
ST4 7GA M7CNE
ST4 7HE G1UCC
ST4 7HF G1XZB
ST4 7HH G1KSH
ST4 7LU M0STK
ST4 8AB G4EJM
ST4 8AZ M6PXL
ST4 8EL G8SAR
ST4 8FX M6BWH
ST4 8FY M7PFV
ST4 8HA M0ODS
ST4 8HA M7GAT
ST4 8HP G6HBQ
ST4 8NJ G3SLX
ST4 8NJ G4MDK
ST4 8NJ G4MJU
ST4 8NW G7EVK
ST4 8PD 2E0COY
ST4 8PJ G7RUR
ST4 8PL M3PQL
ST4 8PQ G7WGZ
ST4 8PQ G7WHA
ST4 8PQ M0ACA
ST4 8RH M3SJR
ST4 8RL M6JXO
ST4 8RT G7NHZ
ST4 8TF G0PXL
ST4 8UZ G0NBE
ST4 8XY G6AZP
ST4 8YL M6LZT
ST4 8YQ G8DYI
ST46DB G3GBU
ST5 0JE M3TMX
ST5 0JS M6GAS
ST5 0JY G1WVW
ST5 0JY M0CRJ
ST5 0LB G8KUZ
ST5 0NG G4VFJ
ST5 0NL G6IFS
ST5 0PB G0OQP
ST5 0PD G1TMF
ST5 1JH G7IDH
ST5 1JL M0OHA
ST5 1JL M5GWH
ST5 1NH G4IMV
ST5 2EF G7JXU
ST5 2HU G6VTX
ST5 2JL G0WDT
ST5 2LG G0PSD
ST5 2NS M7XCB
ST5 2PQ G7BJN
ST5 2QH M3MIF
ST5 2SD G8WKH
ST5 2TB M0BSH
ST5 3AN M7BDG
ST5 3DF G3TJP
ST5 3DF G6CXN
ST5 3DL G7BPG
ST5 3DX M3AUN
ST5 3EA M3KFE
ST5 3ED 2E0OPT
ST5 3ED M7KMW
ST5 3EW G4UAY
ST5 3LT G7RSM
ST5 3NY G4MDM
ST5 3NY G4UMG
ST5 3NZ M3FMK
ST5 3PQ G1JJE
ST5 3PQ G7MMK
ST5 3QR 2E0XXX
ST5 3QU 2E0IBB
ST5 3QU M0PUS
ST5 3QU M6OBB
ST5 3RW G1JIW
ST5 4AZ 2E0KIO
ST5 4BH G3LBS
ST5 4BN G8NPD
ST5 4DA G8MTB
ST5 4EF M3JZD
ST5 4EN G1DDA
ST5 4EP G1STQ
ST5 4FD G6LQP
ST5 4HS G4TMR
ST5 4HY G4TQB
ST5 5AG M7LGG
ST5 5DP 2E0ODV
ST5 5DP M7SOT
ST5 5EU G4NFL
ST5 5FE 2E0DWA
ST5 5FE M0SVT
ST5 5FE M6DWA
ST5 6AR G0GOI
ST5 6AU M3YCO
ST5 6BB G7PNF
ST5 6DT G1LMT
ST5 6HL M0HCM
ST5 6LS 2E0SDQ
ST5 6LS M3GDK
ST5 6QP G7GRR
ST5 6RL M0OMA
ST5 6TB 2E0PLC
ST5 6TB M0TKS
ST5 6TB M3TYG
ST5 6TB M3WSV
ST5 7AH G4CBW
ST5 7DH 2E0IEA
ST5 7DH M6IEA
ST5 7EP M3ZJH
ST5 7EW M3SGG
ST5 7JR M6ECH
ST5 7PB G0HQH
ST5 7QD G1VBY
ST5 7QP 2E0SCX
ST5 7QP M6ALQ
ST5 7RQ G7KJE
ST5 7ST G6KDU
ST5 7TB G4UDH
ST5 8DR G6IKH
ST5 8EA G0SOK
ST5 8EL G4GNW
ST5 8EL G7MTV
ST5 8EL M3PIQ
ST5 8EY G7UAT
ST5 8HF M7CYC
ST5 8HH M3WXI
ST5 8JB M1FIL
ST5 8LD M6YCA
ST5 8ND G6ZHU
ST5 8QA G0VZE
ST5 8QD M7ANI
ST5 8SE M7ROC
ST5 9AE M6IQN
ST5 9AJ M3XIE
ST5 9BP M3YHZ
ST5 9DN M7OTZ
ST5 9ET G7OKO
ST5 9JZ G4DPW
ST5 9NL G1JFL
ST5 9NY G1XYN
ST5 9NY M6CEP
ST5 9PS G7MRF
ST5 9PU G6LLF
ST52JT M6LBY
ST6 1AT 2E1HQZ
ST6 1HB M1ACL
ST6 1PS M1PYE
ST6 1PX G7TJQ
ST6 1RA M6USE
ST6 2ER M6LJS
ST6 2LY M0GBF
ST6 3BX G7SSW
ST6 3DJ M6HKX
ST6 3JB M6HUS
ST6 3PN M6EDZ
ST6 4DZ M6BCJ
ST6 4ES M6MIR
ST6 4ET M3MOH
ST6 4EX M6JLH
ST6 4JB M6LLL
ST6 4NN M6CYA
ST6 4NW M3YUR
ST6 5BT M7AGJ
ST6 5EE M7RAS
ST6 5HS 2E0JTB
ST6 5HS M0JTD
ST6 5HS M3KJV
ST6 5LE G6FHB
ST6 5LP M0EOT
ST6 5LP M6IQE
ST6 5NX G8GHQ
ST6 5PP G7JWI
ST6 5XF M7JOM
ST6 6DZ G4WEP
ST6 6EB M6HCD
ST6 6EX M3JIV
ST6 6HE G8WBP
ST6 6LX G6DON
ST6 6PB M3FNR
ST6 6PE G0UDO
ST6 6RA G3ZRQ
ST6 6TZ M0CDZ
ST6 6UR G0MGJ
ST6 6XD M7TGE
ST6 6XF 2E0HAG
ST6 6XF M3XHU
ST6 7AL G7MSQ
ST6 7BA M6SWH
ST6 7DL M6SOT
ST6 7JR G7RBL
ST6 7JY G7EUF
ST6 7QD M7CCG
ST6 8AB M7KLW
ST6 8BZ G3ICZ
ST6 8DX G7HMB
ST6 8JF M1MDP
ST6 8LX M0RDX
ST6 8PL G1VYS
ST6 8QH G0BRL
ST6 8RN 2E0MWA
ST6 8RT G8RWZ
ST6 8RX G0UTZ
ST6 8TG G3UHV
ST6 8TQ G6OBD
ST6 8TX M0INY
ST65PW M7SVT
ST68XE M7JWL
ST7 1AR M3DBU
ST7 1AR M3EBU
ST7 1AT G6YQN
ST7 1BZ G0JKL
ST7 1DW M7ETZ
ST7 1EH M7KDB
ST7 1EZ 2E0IQT
ST7 1JN 2E0BOF
ST7 1JN M3LRF
ST7 1JY M3BTJ
ST7 1LW G6NQY
ST7 1NP G4WBW
ST7 1NQ G6NGN
ST7 1PF G4EQZ
ST7 1PT M6AHA
ST7 1RH G4VJB
ST7 1SG G4VEW
ST7 1SQ G6SGA
ST7 1TA G0RDS
ST7 1UB G1TBX
ST7 1UG 2E0VPJ
ST7 1UG M6PPQ
ST7 2BA G4ERQ
ST7 2BL G7FZB
ST7 2BN G4WHT
ST7 2BW G8OSJ
ST7 2DS M0CLL
ST7 2EF G6GFC
ST7 2HX G6NVH
ST7 2HY G4GHT
ST7 2LP G3VXS
ST7 2ND G0OFW
ST7 2NH G4UFU
ST7 2NQ G0EIR
ST7 2NR G0UKK
ST7 2RJ 2E0KDB
ST7 2RJ M6KWK
ST7 2SH 2E0DFB
ST7 2TQ M0CVJ
ST7 3BL G6LMJ
ST7 3LD M0RJS
ST7 3ND G8MKS
ST7 3PB 2E0MKX
ST7 3PB M0MJK
ST7 3PE M3DUR
ST7 3PH G3OHH
ST7 3RB G4DLA
ST7 3RY G6NAH
ST7 3TL G6TWB
ST7 3TL G8UVN
ST7 4AL G7IAM
ST7 4DF M7JBB
ST7 4DQ G1CKJ
ST7 4DY G0KBJ
ST7 4HF G0UDG
ST7 4HP G4UDG
ST7 4HR M6BVZ
ST7 4HW G0UWK
ST7 4JF G0HWV
ST7 4JU G8ILP
ST7 4JX 2E0TAK
ST7 4JX M3TBZ
ST7 4LW G7TXX
ST7 4PT G8LZO
ST7 4RQ G4YUN
ST7 4RS G1ELQ
ST7 4RS G6BQM
ST7 4TA G0FZU
ST7 4TA G6USG
ST7 4TH M1AUP
ST7 4UR M6EGF
ST7 4YA G8NYZ
ST7 4YL G0RDK
ST7 4YL G1PUV
ST7 8AS 2E0XCB
ST7 8BB 2E0FVF
ST7 8BB M3DOO
ST7 8HU G7OSO
ST7 8JQ M3TLZ
ST7 8LA G0MVT
ST7 8LP G0LSX
ST7 8NG G0PTD
ST7 8PB G8GLD
ST7 8PE G6UKQ
ST7 8PL G6YCO
ST7 8QB M3WXU
ST7 8QE M0SSY
ST71DQ 2E0DID
ST71DQ M3UAX
ST71HE 2E0JSM
ST71HE M6FFJ
ST71PY M7ATE
ST72XG M7HMB
ST73ED 2E0SWP
ST73ED M6HRG
ST74DN M6YLY
ST8 6EH M3SJQ
ST8 6HY M7YOE
ST8 6JE 2E0VGF
ST8 6LH M6IQM
ST8 6LS M3FIZ
ST8 6NT G8KSC
ST8 6QB M7DJD
ST8 6RD M0GIP
ST8 6RD M1PAS
ST8 6SE G0BYU
ST8 6SZ M3EOT
ST8 6TJ G6CYV
ST8 6UH G6PAO
ST8 7AH G6DSG
ST8 7AS G1ZNZ
ST8 7BB G8KOZ
ST8 7HL G6AKX
ST8 7LN G6UKM
ST8 7LY G4WAM
ST8 7NG G4UQW
ST8 7NG M7AVS
ST8 7NW G0CHL
ST8 7PF G6PJC
ST8 7SW G8APB
ST8 7UF G0DBC
ST86TY G1EGE
ST9 0BD G3LHG
ST9 0DG 2E0CIA
ST9 0DG M0ASR
ST9 0DZ M7MED
ST9 0EP G4SCY
ST9 0HU 2E0CAA
ST9 0JN G4DVN
ST9 0LF G0RKN
ST9 0LF G1EWC
ST9 0LR 2E0DDD
ST9 9BW G4KME
ST9 9JB M3YHF
ST9 9JB M3YHH
ST9 9LW G4LCL
ST9 9QD M0GYN
ST9 9QG G8JDD
SW10 0ER M7OLY
SW10 9BA M7UDV
SW10 9BJ G7GBJ
SW10 9NJ G3JHH
SW11 1RH G4CKS
SW11 2JD 2E1RJH
SW11 2SU G1VVH
SW11 2TJ M7KOF
SW11 3NY G8JKD
SW11 3TN 2E0LVR
SW11 3TN M0LVR
SW11 3TN M6ODP
SW11 4AA G1EGB
SW11 4DU G8NXA
SW11 5EF M7YSE
SW11 5LL M6LXX
SW11 5TP M7LOA
SW11 5TR M7SPM
SW12 0NA G1ULG
SW12 8EG M6UMM
SW12 8PJ 2E0PQA
SW12 8PJ M0PQA
SW12 8PJ M6PQA
SW12 8TQ 2E0BVH
SW12 8TQ M0MEN
SW12 8TQ M3YIX
SW12 8UQ M3YNX
SW12 9LU G7IWU
SW12 9RU M7FYM
SW12 9SS G6CKY
SW13 8DX M7CCH
SW13 8QY 2E0ESI
SW13 9LH G0WLD
SW13 9NB M0FCA
SW14 7AT 2E0TTT
SW14 7AT M6GGG
SW14 7NL 2E0RSI
SW14 7NL M3NNG
SW14 8EQ 2E0RWE
SW14 8JH G4AVK
SW14 8JJ G3KWW
SW14 8JJ G6JMX
SW14 8JU G8ANN
SW15 1LZ 2E0PUT
SW15 1LZ M6VIE
SW15 1RH M0PEQ
SW15 3HR M7AQS
SW15 3HZ G4CXX
SW15 3LT G8FBK
SW15 5RL M7CMS
SW15 6DS 2E0VPN
SW15 6DS M6ZDF
SW15 6EX G1PDA
SW16 1AA M1BJC
SW16 1SN G8ZFU
SW16 2BT M7IVN
SW16 2LT 2E0BWY
SW16 2LT 2E0GXY
SW16 2LT M0KFA
SW16 2LT M3ZLV
SW16 2LT M6NNF
SW16 2PE M3NLN
SW16 4HL G7IRG
SW16 4HT G0WLF
SW16 4JY M6KVW
SW16 4SA M0HHP
SW16 5LU M7PDX
SW16 6JX G6BPN
SW162LT 2E0NNF
SW17 0JE 2E0LPP
SW17 0JE M7KEP
SW17 0PG M0KLH
SW17 0QQ G1LAR
SW17 7DH M1ECI
SW17 8PR G7NLF
SW17 8SF G8TQV
SW17 8SF M0TQV
SW17 8SN M6GYI
SW17 9EN 2E0BWD
SW17 9EN M6OJO
SW18 1SD G0TEO
SW18 2AJ G3PAQ
SW18 2DF M7NAM
SW18 2EB 2E1RWN
SW18 2TS M7CZF
SW18 3AZ M7DEE
SW18 3BB G4PZJ
SW18 3JU M0CON
SW18 3RQ G5RC
SW18 3RQ M0FCR
SW18 5AN G4PKK
SW18 5PB 2E0TRE
SW18 5PB M3TNN
SW19 1EB M6ZIA
SW19 1EX G8REX
SW19 2BD 2E0DRB
SW19 2BD M6AEM
SW19 3JJ M0JCL
SW19 3JJ M1JCL
SW19 3RD M0DJT
SW19 5AZ G6LKV
SW19 6JA G4UNI
SW19 6QU G4CGD
SW19 6RT M7SGT
SW19 6SH G8ILZ
SW19 7LW G3NXN
SW19 7LW M5BBB
SW19 7RQ G6GKG
SW19 8DQ G4ILP
SW19 8JP M0TNB
SW19 8NZ G8WUF
SW19 8PP M0OCK
SW19 8PP M6OCK
SW19 8RY 2E0KRB
SW19 8RY M0SOT
SW19 8SF M0HJJ
SW19 8SW G6ZAV
SW19 8TU 2E1HJA
SW1H 0AA M7LIU
SW1P 4PS M3LBJ
SW1P 4RW G3ZDY
SW1V 1JJ M0HTQ
SW1V 2DB G4MRL
SW1V 2QS G1TQT
SW1V 2RT G4JKE
SW1V 4AS G6ZFK
SW1V 4NY M7ESZ
SW1W 8JW M3AJC
SW2 1DD M6BXM
SW2 1PA G7IHV
SW2 2AP 2E0IJG
SW2 2AP M7NCV
SW2 3BQ M0BXG
SW2 3EW 2E0ESB
SW2 5LU G3KHQ
SW20 0BA M6LXP
SW20 0JN G8WIM
SW20 0NJ 2E1JPH
SW20 0NJ G0JPH
SW20 0NJ M7BRM
SW20 0QE G0MFY
SW20 0UD G3DRN
SW20 8AP G0CRU
SW20 8JL M5ESA
SW20 8JN M0AST
SW20 8JY 2E0KKZ
SW20 8JY M7DKK
SW20 9AB M3LWX
SW20 9BQ 2E0JGH
SW20 9BQ M0JGH
SW20 9ET M7ILD
SW20 9HG G8IAK
SW20 9HX G8BUZ
SW20 9JT G0VQT
SW3 3LB G1XGZ
SW3 3TS M1DNQ
SW3 4QE G8ECR
SW3 4SR 2E0RRA
SW3 4SR G0CGO
SW3 4SR G0SLL
SW3 4SR G0UOL
SW3 4SR M0INU
SW3 4SR M6JOA
SW3 4SR M6YDG
SW3 6BU G8DOH
SW4 0AN M6TAH
SW4 0ET G4PIA
SW4 6BE M6DUE
SW4 7AF M3ZFL
SW4 9HD G8LHQ
SW4 9NS M7AZV
SW4 9PB G8OOF
SW4 9RR 2E0KJI
SW4 9RR M6KJI
SW5 9EB 2E0CWK
SW50PU M7AGS
SW6 1BS G0HFK
SW6 1EA 2E0WTU
SW6 2HR M7STV
SW6 2RB G6BUH
SW6 2RT M6LJU
SW6 3SA M6BDG
SW6 5BG G0ADG
SW6 5DW 2E0LCW
SW6 6SA M6FHH
SW6 7DA M7LXT
SW6 7HD 2E0LZC
SW6 7HD M7DCR
SW6 7PP G7IUB
SW7 1BL G0PZC
SW7 3BG M3TND
SW7 3DQ G0RFX
SW7 3QF G4MFW
SW7 4ET G4UAF
SW7 4JR G6LVB
SW7 5NS G4WQO
SW8 1BH G7VQR
SW8 2PD G0PIY
SW8 2TF 2E0DYV
SW8 2TF M6GQZ
SW8 4DP 2E0ULH
SW9 7UE G0VPH
SW9 9RH G6NAD
SW9 9RZ M6DUH
SY1 1RD G8ZWF
SY1 1UH G0GTN
SY1 2AB M6FNF
SY1 2DL G1EBT
SY1 2EY M6OGJ
SY1 2NE 2E0HBQ
SY1 2NE M7AYZ
SY1 2PY G8DXM
SY1 2RT M7AZK
SY1 3HF M7NCO
SY1 3PZ G4CBT
SY1 3PZ G7LID
SY1 3RA M3ENY
SY1 3RH G4XBI
SY1 4BF M1BCM
SY1 4DD G7VYI
SY1 4ER G0IRI
SY1 4LD 2E1FZU
SY1 4PG M7JTI
SY1 4PL 2E0BMP
SY1 4QQ M3IHC
SY1 4RB G8SIU
SY1 4RP G8RTB
SY1 4RU M0KZB
SY1 4SE 2E0FZU
SY1 4SE M0MNB
SY1 4SE M7DMK
SY10 0AN GW6KLG
SY10 0DR GW1APU
SY10 0LE 2W0OGT
SY10 0LE MW6OGT
SY10 0LH MW6FEH
SY10 0NS GW2HCA
SY10 7AE GW7LXI
SY10 7AS GW6YGI
SY10 7AS MW3XRI
SY10 7DX GW7CQB
SY10 7HL GW3OKT
SY10 7HP GW3SMT
SY10 7HP MW6TBL
SY10 7JY GW4UHB
SY10 7NF MW6LWF
SY10 7PQ GW1PLJ
SY10 7PQ MW3WQV
SY10 7PQ MW6BNF
SY10 7PQ MW6JTM
SY10 7QB GW3LYU
SY10 7RQ GW4TEQ
SY10 7TU GW6DBP
SY10 8AJ GW6CVX
SY10 8BB GW4UDE
SY10 8DW GW6VKI
SY10 8HQ MW6OLL
SY10 8HU GW1HAX
SY10 8PS GW7OTQ
SY10 9AU GW8AJA
SY10 9AZ GW0DLW
SY10 9BA GW4DFQ
SY10 9DF GW1VIR
SY10 9DH GW4COL
SY10 9DP MW6FST
SY10 9DP MW6NJM
SY10 9DP MW6WEE
SY10 9DQ GW4QGN
SY10 9HN GW1JVH
SY10 9HQ GW4IOQ
SY10 9RB MW0DSZ
SY11 1DS MW7JSG
SY11 1EP MW6PIH
SY11 1LX MW0RRY
SY11 1SQ GW1FLY
SY11 1TG MW3VPS
SY11 1TP GW7JHC
SY11 2BL MW7CUI
SY11 2NL MW7DQW
SY11 2QW MW6FTK
SY11 2SB MW6NFJ
SY11 2SG MW3NFZ
SY11 2TH 2W0GWO
SY11 2UA GW7NNA
SY11 2UN MW6HCY
SY11 2YD GW4WVK
SY11 3AH MW6HBK
SY11 3AH MW6NCA
SY11 3BU MW1EYH
SY11 3DH GW0CWZ
SY11 3JF 2W1HXT
SY11 3LD 2W0TDF
SY11 3LD MW0TDF
SY11 3LD MW3XTF
SY11 3NR MW6KQL
SY11 3NR MW7CTA
SY11 3PJ GW1OIB
SY11 4AH 2W1MSC
SY11 4AH MW3PVC
SY11 4EQ GW4FBQ
SY11 4FE GW8KNJ
SY11 4HG MW0TRE
SY11 4LF GW0VEU
SY11 4PA GW0EXD
SY12 0EG M0YCQ
SY12 0EX M0CCQ
SY12 0PF 2E1MAR
SY12 0QJ G3XGD
SY12 9ER M0COT
SY12 9NA G6ZMD
SY12 9PA G0AJA
SY12 9QA G1VXY
SY120FH M6KPL
SY13 1BS M0NDC
SY13 1BZ M1BKL
SY13 1EX G4HRH
SY13 1PH M0SIN
SY13 1PH M1SIN
SY13 1QT M6NBQ
SY13 1QW M0IFP
SY13 1RQ G3TZU
SY13 2EG M5CLO
SY13 2HD G8CTD
SY13 2HZ G4JSP
SY13 2LW G3YLV
SY13 2ND G4GXD
SY13 2NJ G4VQH
SY13 2PT G4UJS
SY13 2PX G1GSJ
SY13 2PX G4OAR
SY13 2QL 2E0ZSU
SY13 2QL M3ZSU
SY13 2RA G4ANY
SY13 3LT G0MXD
SY13 3UA M3XJW
SY13 4EB G0FEY
SY13PY G0EML
SY13QE M6XVD
SY14 7AA M0VKC
SY14 7AR G4RRR
SY14 7AX G3GIZ
SY14 7AX G3TZO
SY14 7DB G4GOO
SY14 8AY M3ZLA
SY14 8DU M6SHI
SY14 8JB G6XTC
SY14 8JQ G4XVS
SY14 8LL M0ROJ
SY14 8NE M6MEC
SY148JJ 2E0TGI
SY148JJ M7TGF
SY15 6AB G7DRX
SY15 6AL G0IQP
SY15 6BA M7SAK
SY15 6BF 2E0GFQ
SY15 6BF M0RHX
SY15 6BF M6RBH
SY15 6EB M6IMH
SY15 6JH G4RYK
SY15 6RS G0RJV
SY15 6RS M3GEX
SY15 6RS M3HFT
SY15 6RT M3MYQ
SY15 6SB 2E1CEZ
SY15 6SB G4NQJ
SY15 6TP G4WND
SY15 6TP G4YNL
SY16 1DW GW7TIX
SY16 1NW MW7RHF
SY16 1PR GW7CAH
SY16 1QQ GW0TWL
SY16 1QQ GW1PVN
SY16 1QT GW7DJL
SY16 1RA MW0EDX
SY16 1RH MW6RQM
SY16 2AW 2W0MTE
SY16 2AW MW6MTE
SY16 2HN 2W0GWM
SY16 2HN MW6GWM
SY16 2JN MW6WCU
SY16 3AY MW5ABR
SY16 3DS 2W0WGN
SY16 3DS MW0MWX
SY16 3DS MW0OMK
SY16 3DS MW7LTS
SY16 3EE GW4BML
SY16 3EE MW3MBL
SY16 3ER MW5TLE
SY16 3HX GW7NNM
SY16 3JY GW3LXE
SY16 3NA MW6CPJ
SY16 3PD GW0LKN
SY16 3PY MW7EQT
SY16 4DB GW3IWM
SY16 4LL 2W0HBW
SY16 4LL MW7JTH
SY16 4PS GW4RLO
SY17 5JG G3FXI
SY17 5JP G4ZUD
SY17 5NG M6SCN
SY17 5NG M7GYZ
SY17 5PU M6EDV
SY17 5PU M6EDW
SY17 5RH 2E0JHE
SY17 5RH G5JHE
SY17 5RH M7EJH
SY18 6AD 2W0KPF
SY18 6AD MW6KPF
SY18 6AR GW4CWG
SY18 6HZ GW1UOY
SY18 6JY 2W0GEZ
SY18 6JY MW0OMZ
SY18 6JY MW7GEL
SY18 6JZ MW7PCG
SY18 6LA MW7PTW
SY18 6LL 2W0HOH
SY18 6LL MW6HOH
SY18 6NP GW4DTU
SY18 6PQ 2W0CYE
SY18 6PQ MW6CKM
SY18 6SN GW3XPK
SY19 7DJ GW0EHS
SY19 7DJ GW1JNI
SY2 5EF G3TSV
SY2 5LQ G0IGM
SY2 5PF G3SRT
SY2 5PF G3VZG
SY2 5QL 2E1JCM
SY2 5SS M3OFH
SY2 5TE M1GSX
SY2 5TS M6FNM
SY2 5UZ 2E1EQR
SY2 5YB M0GSL
SY2 5YD G0EBD
SY2 6BG G4ZMR
SY2 6EE M6OEN
SY2 6HH M0IQG
SY2 6HQ G3IDY
SY2 6HU G4LYJ
SY2 6JX G4XDC
SY2 6LS G4WSV
SY2 6ND M3ZXE
SY2 6SJ G0RQX
SY2 6TE M7CLP
SY20 8DL 2W0RDZ
SY20 8DL MW0RCZ
SY20 8DL MW6RFS
SY20 8EJ GW0JAI
SY20 8NY GW8BFO
SY20 8RA 2W0IGN
SY20 8RA MW6GKQ
SY20 9BQ GW7VBY
SY20 9EZ GW4LPU
SY20 9EZ GW7JUV
SY20 9LA GW1CDY
SY20 9PR 2W0XUL
SY20 9RD GW7GAH
SY21 0AE GW1JNR
SY21 0BD GW7MGW
SY21 0DG GW4SHF
SY21 0DG MW0BAA
SY21 0EP GW4AUD
SY21 0JD GW8FFW
SY21 0JT GW4DEP
SY21 0PW GW3KGV
SY21 0RH 2W0UTT
SY21 0RH MW0UTT
SY21 0RH MW6EHX
SY21 0RX GW0NGA
SY21 7BB GW4DYY
SY21 7BB GW8HOS
SY21 7JA MW0GKW
SY21 7NQ GW4DWX
SY21 7PN MW6EIB
SY21 7RH GW8HEB
SY21 7RW MW7JEL
SY21 7TP MW1ABT
SY21 7UL GW4TFM
SY21 8AU GW3JSV
SY21 8BJ GW6EUR
SY21 8BJ GW6EUS
SY21 8BT MW0DDE
SY21 8DY MW6CYX
SY21 8ER GW1ORP
SY21 8ER MW3ORP
SY21 8HL GW4MZB
SY21 8JD 2W0AVZ
SY21 8JD MW0EDQ
SY21 8JD MW3BPZ
SY21 8LA MW1BEW
SY21 8LF GW7DLD
SY21 8LF GW7KMD
SY21 8NJ GW4SGQ
SY21 8RT GW0NNE
SY21 8SG GW3TCV
SY21 8TS GW6EUT
SY21 9AN 2W0ALA
SY21 9AN MW3NRK
SY21 9JE GW0KGD
SY21 9JY GW4BVE
SY21 9JY GW8DHT
SY21 9LH MW6WLY
SY21 9PN GW1KTW
SY21 9PX GW1EAV
SY21 9PX GW4GNY
SY22 5AF MW0DAX
SY22 6EW GW0AGZ
SY22 6FG GW0NFI
SY22 6HB GW6HUR
SY22 6NF GW4UBQ
SY22 6PZ GW7FYG
SY22 6PZ MW3NVQ
SY22 6QJ 2W0GSE
SY22 6QJ MW0TDV
SY22 6QJ MW3RDS
SY22 6QL GW1YQM
SY22 6QL MW6CYU
SY22 6QL MW6MFN
SY22 6RG GW1XQT
SY22 6SL GW4CRH
SY22 6SW 2W0JRK
SY22 6SW MW3JKB
SY22 6TW GW3YRP
SY22 6UJ MW6MGM
SY22 6XL GW0VWS
SY22 6XL GW3UBH
SY226LL MW6VLS
SY23 1AT GW8SIT
SY23 1BT MW3ROX
SY23 1DB GW8OHP
SY23 1DW 2W0FLI
SY23 1DW MW0KFL
SY23 1DW MW6FNA
SY23 1EU 2W0OLT
SY23 1EU MW0OLE
SY23 1EU MW6LRO
SY23 1JE MW7XAW
SY23 1JU MW6HRR
SY23 1RJ MW0GLS
SY23 1RY GW1URF
SY23 1SE 2W0DNR
SY23 1SE MW6TWH
SY23 1SS GW7AGG
SY23 1SZ MW7HMT
SY23 1TB MW7OFS
SY23 2EL 2W0WOD
SY23 2EL MW6NFN
SY23 2EL MW6WOD
SY23 2EU MW6NBU
SY23 2HN MW7NNR
SY23 2JA GW7HAE
SY23 2JA MW3HAE
SY23 2JA MW3WTR
SY23 2JU GW1HIN
SY23 3AB MW6XFM
SY23 3BL GW7OZP
SY23 3BQ MW6IJD
SY23 3DT GW7JSH
SY23 3EA MW3LFL
SY23 3EZ GW4CTV
SY23 3FT MW0ZXY
SY23 3GZ MW6ORW
SY23 3HF GW8LZY
SY23 3HQ 2W0FFL
SY23 3HQ GW8SFT
SY23 3HQ MW0LBR
SY23 3HQ MW6FLA
SY23 3HU MW6SVM
SY23 3LH MW6GZZ
SY23 3PD MW6ZXO
SY23 3PL MW6GZX
SY23 3QR MW6FPH
SY23 3QU MW6ZCT
SY23 3QZ GW4TUD
SY23 3RP 2W0DVP
SY23 3RP MW0PRI
SY23 3RP MW6DVP
SY23 3RP MW6TFB
SY23 3TR GW7HSW
SY23 4BQ GW1FWC
SY23 4JJ 2W0RKM
SY23 4JJ MW0RZS
SY23 4JJ MW6KMJ
SY23 4LA GW3BV
SY23 4PP GW4LHL
SY23 4PU GW8UWP
SY23 4RN MW6AEN
SY23 4SP GW4BQQ
SY23 4SQ GW8UDJ
SY23 4TP 2W0GIA
SY23 4TP MW0KGG
SY23 4TP MW6GIA
SY23 4TR GW4AKY
SY23 4TT GW8BTX
SY23 4UH MW3YBF
SY23 5BW MW0DEW
SY23 5BZ MW6FQA
SY23 5HG 2W0CPD
SY23 5HG MW0RBV
SY23 5HG MW6BXI
SY23 5HG MW6CGV
SY23 5HP MW3UYH
SY23 5HS 2W0DCK
SY23 5LF MW6RIY
SY23 5NA GW8UKZ
SY23 5NB GW0OBB
SY23 5NN MW0EUS
SY24 5BE G0PDA
SY24 5BS M6LJP
SY24 5DF M6MLI
SY24 5HJ M6GKT
SY24 5JD G6JWD
SY24 5LT G8NCU
SY24 5LX M3NZV
SY24 5LX M3VRD
SY24 5NP G0PUW
SY24 5NP G7GWT
SY25 6AE G1XOT
SY25 6AE M3XOT
SY25 6AJ G4ZHI
SY25 6DQ M6XIE
SY25 6EJ 2E0TOF
SY25 6EJ M0TOF
SY25 6EJ M6TOF
SY25 6NY G7GEP
SY25 6PE G0CEP
SY25 6PE G8JGQ
SY25 6QU G4YMZ
SY25 6QZ G7LUB
SY25 6TT M0CBD
SY26JN M6ILY
SY3 0AG G1KQP
SY3 0BE G4JSZ
SY3 0ER G7JTD
SY3 0EX G4HAC
SY3 0JG G7LRB
SY3 0LE G7NLY
SY3 0NS G6JHS
SY3 0QF G1CWZ
SY3 5AN G0EIY
SY3 5AP M6XDT
SY3 5AS 2E0THZ
SY3 5AS M0XMH
SY3 5AS M6MGH
SY3 5BJ G6TGJ
SY3 5BJ G7BQO
SY3 5BW G3UDA
SY3 5DA G6NFE
SY3 5DF M0AMP
SY3 5LA M6LSN
SY3 5PG G4ZPC
SY3 6AB G3NRX
SY3 6AW G3OWJ
SY3 7AJ M7DPY
SY3 7JN G3JPU
SY3 7NB G8OFZ
SY3 7RA M6LUW
SY3 8AD 2E1MJF
SY3 8AD M5WJF
SY3 8QQ M3XEJ
SY3 8RU G4ZZP
SY3 8SU G4AZS
SY3 8TJ 2E0NWB
SY3 8TJ M6NWB
SY3 8UX M1FMC
SY3 9AB 2E0IIP
SY3 9AB M7HTU
SY3 9DB G3VWH
SY3 9FW M0YQQ
SY3 9PH M0PDP
SY3 9PH M6OBS
SY3 9QS G1SCR
SY3 9QS M1DQI
SY4 1BJ G6WNG
SY4 1BY G4JUW
SY4 1DT G0ARP
SY4 1DT G6UDG
SY4 1EE G6DTN
SY4 1EE G8RIK
SY4 1EE M0DFA
SY4 1JD G3NSS
SY4 1JH G0DNI
SY4 2BB G8DSG
SY4 2ED M7BYS
SY4 2EH G4SDU
SY4 2HT G0WDW
SY4 2HY G6DQY
SY4 2LG G1CLD
SY4 2LH G7CJC
SY4 3AZ G4EDC
SY4 3JL G4VVD
SY4 3LY G4CBM
SY4 3LY G4YKX
SY4 3QX G3LEK
SY4 3SB 2E0XJM
SY4 3SB M7JPM
SY4 4JE 2E0ETC
SY4 4JE M3ZHZ
SY4 4JX G0RVE
SY4 4LA G0JLS
SY4 4NU 2E0ILL
SY4 4NU M7OAN
SY4 5BL M6WCG
SY4 5BX G1WVV
SY4 5DE G1YJB
SY4 5DE M0BAV
SY4 5DU G0ISP
SY4 5JN G4URT
SY4 5NB G3VAO
SY4 5NB G4CES
SY4 5NB G7BSP
SY4 5NW G7MHL
SY4 5NX G3TUL
SY4 5QZ 2E0SUT
SY4 5QZ 2E0TGG
SY4 5QZ M6RDS
SY4 5QZ M6TGG
SY4 5SL 2E0TVS
SY4 5SL M0KZP
SY4 5SL M6TVS
SY4 5UJ G4CBQ
SY4 5UJ M3JBQ
SY4 5YD G3PLW
SY5 0BG 2E0SIX
SY5 0DX 2E0GNV
SY5 0DX M7BGB
SY5 0EH G7JWH
SY5 0EH G8DIR
SY5 0NG M3WKC
SY5 0NS M0PWT
SY5 0QZ 2E0OCM
SY5 0QZ M0OCM
SY5 0QZ M6OCM
SY5 0TL M5LMG
SY5 0UD G0PVW

SY5 0UY G4GA
SY5 6DA M7CCR
SY5 6DS G0OOQ
SY5 6DS G7DCM
SY5 6JJ 2E0HMC
SY5 6JJ M6DYU
SY5 6JR M0JJC
SY5 6JR M6HAU
SY5 6PH M7BVF
SY5 7AP G6FHM
SY5 7AT M3SUW
SY5 7BW G0ZDL
SY5 8AN M0DEV
SY5 8AU 2E1DEM
SY5 8EP G4MYZ
SY5 8ND M0ZPM
SY5 8NH G8DIQ
SY5 8RA G2CQX
SY5 9AS G4FRX
SY5 9BE 2E0FOI
SY5 9BE G6GBY
SY5 9BE M7GBZ
SY5 9DJ G3YFK
SY5 9EG 2E0GQW
SY5 9EG M0GQW
SY5 9EG M3GQW
SY5 9EU G3VVL
SY5 9GZ M6OEG
SY5 9JA 2E1UTD
SY5 9JS G8NGF
SY5 9JY M1DQH
SY5 9LF 2E0JUX
SY5 9LF M6JUX
SY58LE M1DQG
SY6 6DD G3LRH
SY6 6DP M1DTS
SY6 6EE M6GVL
SY6 6GA M0DUV
SY6 6HT M6GVM
SY6 6JW G4SMA
SY6 6LX G7UMF
SY6 6NJ G4UST
SY6 6QD G0WPX
SY6 6QD G3BJ
SY6 6QD G4JKS
SY6 6RB M0COP
SY6 7AB M1CHU
SY6 7BB M3PHQ
SY6 7BN G8WPF
SY6 7DH G7ACD
SY7 0EF M6FNB
SY7 0EP M3JLR
SY7 0NB G0NGN
SY7 8AN G0KKQ
SY7 8BA M6DUX
SY7 8DG M3MRM
SY7 8EN M6FSW
SY7 8QR M3YZO
SY7 9DT 2E0NLT
SY7 9DT M0NLT
SY7 9DU 2E0CHZ
SY7 9DU G7RHF
SY7 9DU M6RAL
SY7 9DW M3SCF
SY7 9DW M6XCF
SY7 9DX G7EYM
SY7 9ER G3ZZX
SY7 9ET M6ZJH
SY7 9HX G7NSJ
SY7 9HX M1CMM
SY7 9LF M6ZJP
SY7 9PG G0CGI
SY7 9RF M6PNR
SY7 9RL M6AUA
SY8 1 2E0ULY
SY8 1BH M3ZXC
SY8 1DS 2E0BYA
SY8 1DS M0MYA
SY8 1DS M3BYA
SY8 1DW G3VWX
SY8 1EY G3SEZ
SY8 1JE M3YAI
SY8 1LE M6AOW
SY8 1LU M6BQC
SY8 1NB 2E0DRI
SY8 1NB M6XKG
SY8 1NB M6YDA
SY8 1NH G1VVX
SY8 1NS M3RLM
SY8 1RE G7VEL
SY8 1UT G7IQD
SY8 1XN 2E0OVW
SY8 1XN G4HQB
SY8 1XN G7HSA
SY8 1XN M0OVW
SY8 1XN M3XLM
SY8 1XN M6OVW
SY8 1XP M0SJL
SY8 1XS G6GBT
SY8 1XZ G3KZE
SY8 2DH M6JBX
SY8 2DT G8EZD
SY8 2NQ G4EVR
SY8 2NS M3JWM
SY8 2PH G4OYX
SY8 2QD 2E0GCJ
SY8 2QD M6GGJ
SY8 3AE 2E0CDS
SY8 3AE G6LXL
SY8 3AE M0MTS
SY8 3AW M0CBA
SY8 3DR M3EBF
SY8 3DR M3OMF
SY8 3EB 2E0PKL
SY8 3EB M6PKL
SY8 3ED M1DKZ
SY8 3JX M6ZGZ
SY8 3NJ G8HMV
SY8 3NZ G1AJS
SY8 3NZ G8GUN
SY8 3PN M6OQZ
SY8 3RE M7RFU
SY8 4AY M0BVY
SY8 4BX M3NFH
SY8 4EP M1AXD
SY8 4HD M6AUH
SY8 4JT G4AIJ
SY8 4LB 2E0KYX
SY8 4LB M6KYX
SY8 4LF M6EGG
SY8 4LX G1JOD
SY8 4ND M1MST
SY8 4NE G0EXA
SY8 4NQ G1KGE
SY81RU M7JFJ
SY9 5AF G3NUB
SY9 5DB M5DND
SY9 5EF 2E1HUB
SY9 5EL M0GGU
SY9 5ES G0RPW
SY9 5HB 2E1HTY
TA1 1DQ G3VPQ
TA1 1LB G1SOM
TA1 1LB M0KON
TA1 1LB M0KOW
TA1 1LB M0KOY
TA1 1LB M0KPQ
TA1 1PH G0UJU
TA1 1UW M6AXP
TA1 1UW M6KGV
TA1 2AF 2E0SBB
TA1 2AF M0SBB
TA1 2AF M6AES
TA1 2AF M6TBJ
TA1 2BD 2E0BBZ
TA1 2BS G1YLN
TA1 2JQ G1NZK
TA1 2LJ M7LRT
TA1 2NA G7CCV
TA1 2NP M3TJQ
TA1 2PL G8DUI
TA1 2PW M0HQB
TA1 2QJ G4ASK
TA1 2QJ M7JOY
TA1 2QX 2E0WDI
TA1 2QX M3ONB
TA1 2RA 2E0AKN
TA1 2RA M0GSN
TA1 2RA M3LQV
TA1 2TA G0PNB
TA1 2XF G0WKU
TA1 2XN M0CIF
TA1 3DA G4RWF
TA1 3EJ G4RNT
TA1 3EL G0PGL
TA1 3FN G3VFB
TA1 3HQ G0UIL
TA1 3HQ M6SDS
TA1 3JE M0CIE
TA1 3NR G3CWD
TA1 3PA G3SHL
TA1 3PA G4BJX
TA1 3RR G4VVM
TA1 3RW M3SLZ
TA1 3XN G0UFU
TA1 4JE G4MYE
TA1 4JX G3JKE
TA1 4NL G7RAZ
TA1 4QH M3PVP
TA1 4RE G8JXK
TA1 5DS G4WTA
TA1 5EH G0RIE
TA1 5EP 2E0ZSE
TA1 5EP M0SEV
TA1 5EP M6ZSE
TA1 5ES G8XHU
TA1 5HH 2E0HJZ
TA1 5JE G0EYR
TA1 5JE M6GXH
TA1 5JH 2E1BFW
TA1 5JH G7KIQ
TA1 5JU G1FJS
TA1 5NS M3XNO
TA1 5PQ G4VVS
TA1 5PQ G8ZSP
TA1 5PW G3TJH
TA1 5QT M3PDD
TA1 5QZ 2E0BPI
TA1 5QZ M3WEQ
TA10 0HB G3OTK
TA10 0HH G3RSU
TA10 0HX M7YXI
TA10 0JH G0PNF
TA10 0NG G6AYD
TA10 0NX G4BWL
TA10 0PP G4BEB
TA10 0PP G6JNS
TA10 9AE G0DGA
TA10 9AE G0DGE
TA10 9AX 2E1JKL
TA10 9DD 2E1LGE
TA10 9DH G1HQN
TA10 9EZ G8UIG
TA10 9FB G3PCJ
TA10 9JX M0IEK
TA10 9LT 2E0GKR
TA10 9LT M0VKR
TA10 9LT M3OLZ
TA10 9LT M7WGF
TA10 9NL G3USC
TA10 9PR G0RGC
TA10 9RH G0SZT
TA10 9RL M7BWL
TA10 9ST M6PGN
TA10 9TA M3XSR
TA10 9TF G0FZH
TA10 9TH G7LPO
TA109TA G0LCC
TA109TA G1OYH
TA11 6BW G0HDJ
TA11 6DF 2E0OSS
TA11 6DF M6OEU
TA11 6DX M7SGM
TA11 6ER G1PUQ
TA11 6HP G3YYR
TA11 6JY M6AXU
TA11 6LG G7IHP
TA11 6NF M1AXM
TA11 6PQ 2E0GOP
TA11 6PQ M0PGO
TA11 6PQ M6XSR
TA11 6PX G1DCZ
TA11 7AP M0BJO
TA11 7DN G3TCT
TA12 6BG G8GVW
TA12 6DT G1AKD
TA12 6FA G0PNS
TA12 6FA G3LWM
TA12 6HG 2E0LQW
TA12 6HG M0RGE
TA12 6HG M3LQW
TA13 5AJ G7ITU
TA13 5AL M7DMS
TA13 5AP G4PDG
TA13 5AQ G8VBI
TA13 5BD M7SHW
TA13 5BN M1AIK
TA13 5BQ 2E0CSQ
TA13 5BY 2E0ZXG
TA13 5BY M0ZXG
TA13 5BY M3ZXG
TA13 5LJ G8TWN
TA14 6QN G6EER
TA14 6QN G8LKP
TA14 6SH M1BSN
TA146PT G7RVG
TA16 5NF M6NBG
TA16 5NP G3TEL
TA16 5QX M6KMF
TA18 7AH G6AFL
TA18 7AH M3BON
TA18 7AP M3UBH
TA18 7AQ G0NPF
TA18 7AX G1PVZ
TA18 7AY 2E0IJX
TA18 7AY M6IJX
TA18 7DJ M0OJS
TA18 7PB M0GJD
TA18 7PW M7DHD
TA18 7SG G4NLI
TA18 7TH 2E0OQD
TA18 7TH M6OQD
TA18 8BL G4FVW
TA18 8DB M1RMW
TA18 8DF 2E0BFJ
TA18 8DF M3JYH
TA18 8DF M3TGD
TA18 8ET G1AWU
TA18 8HS M6XAA
TA18 8JA M0BCL
TA18 8JB M0BBO
TA18 8JT G3WMX
TA18 8LY G4NCU
TA18 8PL G0JCY
TA18 8QE G0EGC
TA19 0BA M6MRW
TA19 0EA 2E0XIP
TA19 0EA M3FGQ
TA19 0HH G0MEQ
TA19 0NZ G4OXR
TA19 9AH G6AOV
TA19 9AH G6LVG
TA19 9DG G7LNY
TA19 9ER M0ILT
TA19 9FG M0AGT
TA19 9LG G8HGI
TA19 9NH G4KJD
TA19 9NW G4HTD
TA19 9QH G6PQP
TA19 9QL G0WSC
TA19 9QN M0AOD
TA19 9QU M0BHE
TA2 6BU G3XZO
TA2 6LE G1TKQ
TA2 6RH M3AKW
TA2 6RR G4SZS
TA2 6SR G8HNM
TA2 6TA 2E0BQY
TA2 6TR M6CTT
TA2 7DN M7LND
TA2 7DT G1CNZ
TA2 7EF G4WSF
TA2 7EJ M6TER
TA2 7EN G1AJZ
TA2 7PB G4XKK
TA2 7PT M7CRM
TA2 7SP G7LDD
TA2 8DX G0PWV
TA2 8DX G0RWA
TA2 8DZ G7KYG
TA2 8EW M7EFZ
TA2 8LB G7CWT
TA2 8NA G4UTM
TA2 8NA G8BTY
TA2 8NN M0CQK
TA2 8PP G6EIO
TA2 8QB G1VVE
TA2 8RP G3PCG
TA20 1BX 2E0TAW
TA20 1BX M3MNT
TA20 1BZ G3HAL
TA20 1DG M1DGP
TA20 1DZ G4IRS
TA20 1EZ G8DLH
TA20 1FH G1ZYS
TA20 1FX M7FAN
TA20 1GU M7BWH
TA20 1JU G6WZA
TA20 1NJ G0LNK
TA20 1PF G0NBI
TA20 1PF M6VAJ
TA20 1SD M6TVH
TA20 2BL G4VHB
TA20 2BU G1ECV
TA20 2EB G8BVR
TA20 2ED G0AEN
TA20 2EG M0FAK
TA20 2EG M3BXG
TA20 2EW 2E1IJR
TA20 2FP G8OCF
TA20 2HH M3UVV
TA20 2HL 2E0AJM
TA20 2HL M7DZG
TA20 2JB G8GZC
TA20 2JH M7UTD
TA20 2ND G8LKK
TA20 2PY G8GFZ
TA20 2RX M3UTN
TA20 2SX G7MFH
TA20 2TJ M3LBZ
TA20 3BH M1FAY
TA20 3HJ G0KJF
TA20 3HJ G8HOI
TA20 3LT M0XMS
TA20 3PH 2E0DZT
TA20 3PH M6DZT
TA20 4LN M6VVX
TA20 4PJ G4TIA
TA21 0DT G4CAK
TA21 0DX 2E0PDU
TA21 0DX M0PDU
TA21 0DX M3PDK
TA21 0DX M3PDU
TA21 0EL G0VSG
TA21 0RH G8FZI
TA21 8BB G0MJF
TA21 8DH G8ZVK
TA21 8EP M6JST
TA21 8ER G1CQG
TA21 8EX G4RGA
TA21 8HZ G4YXR
TA21 8HZ M0MAX
TA21 8JB M5WSS
TA21 8LG G7BSG
TA21 8PY G4OLA
TA21 8SF M7CSX
TA21 9AL 2E0AKL
TA21 9AL M0GFO
TA21 9AL M3NLX
TA21 9AR G6TWA
TA21 9AT G4RGF
TA21 9AX G4AIU
TA21 9DY M3SNL
TA21 9DY M6GYZ
TA21 9ED G0BDJ
TA21 9EG 2E0RIQ
TA21 9EG M6RPQ
TA21 9FX M6IHE
TA21 9HT M0CVB
TA21 9HW M6TEJ
TA22 9EL G7KWP
TA22 9EN G1TKE
TA22 9EN M6SES
TA22 9PT G0JFA
TA22 9RU G8CCF
TA23 0AG G6BWJ
TA23 0AH 2E0YKD
TA23 0AH M7UKD
TA23 0BP 2E0EAI
TA23 0BP M6SMB
TA23 0DD G4BQB
TA23 0HZ G6KRG
TA23 0PZ M6NRG
TA23 0QY 2E0KEK
TA23 0QY M6HFU
TA23 0RH G8XHD
TA23 0TX M7WRI
TA24 5DY M7PSS
TA24 5HS G4PHC
TA24 5JU G3NFY
TA24 5NZ G8YZY
TA24 5NZ M0AOJ
TA24 5TD G7PBO
TA24 5UB 2E0ONV
TA24 5UB M3ONV
TA24 6AD M3LSK
TA24 6BT M3GQI
TA24 6BY G1CGU
TA24 6BY G6HOR
TA24 6HJ G4HOW
TA24 6JH M6RJW
TA24 6PW M3BLR
TA24 6SD 2E0WSR
TA24 6SD M6DUN
TA24 7PT G6AHR
TA24 7RU M0IAX
TA24 7RU M6FLJ
TA24 7UL M3BXH
TA24 7UL M3LSE
TA24 8AF G0WMN
TA24 8AN 2E0EZC
TA24 8AN M0ROL
TA24 8AN M6PUA
TA24 8AS G8EAM
TA24 8AW G0ACA
TA24 8DH G8NIU
TA24 8EB G4FNJ
TA24 8EP G6INL
TA24 8HL G1ONV
TA24 8JR M6ARV
TA24 8JR M6ASI
TA24 8QF G4TIC
TA24 8TA G8BCI
TA246AD 2E0AKO
TA28RL M0FBM
TA3 5AU G6PPU
TA3 5BX G4LZU
TA3 5BY M3PBP
TA3 5DW G4AVJ
TA3 5EW G4IBC
TA3 5EW M0AID
TA3 5JW G1NTK
TA3 5LP M6AYQ
TA3 5PW G4MIB
TA3 5QP G0MZQ
TA3 5RA M7DZQ
TA3 5TE G4JZL
TA3 5TE G8BDM
TA3 6AE M0WDL
TA3 6DX M0HMX
TA3 6HA G1DRY
TA3 6HA G1KNK
TA3 6PU G8XOE
TA3 6QN G8XET
TA3 6QN M3LVK
TA3 6TN G7DKZ
TA3 7EG G4UVV
TA3 7HB G4NIL
TA3 7HH G4MFD
TA3 7HS G8WXV
TA3 7LG G4KIK
TA3 7NR G3THG
TA3 7QB M3AXZ
TA3 7RE G6TOY
TA3 7SD G4UVZ
TA3 7SU G5JJ
TA3 7SU M0LHC
TA4 1AQ G1XQP
TA4 1DF 2E0INZ
TA4 1DF M7KSF
TA4 1EU M6OPS
TA4 1EX G4ZTS
TA4 1JA G4FDA
TA4 1JH M7JEX
TA4 1NY G7HKZ
TA4 2BU G0GWS
TA4 2BW M5LSC
TA4 2DP M3BAO
TA4 2EQ G7SLZ
TA4 2JY G0FUS
TA4 2LF M7SDX
TA4 2LR 2E0WZR
TA4 2LR M7WEZ
TA4 2LZ M0GHR
TA4 2QF G4ETN
TA4 2UL G7VNJ
TA4 3AQ G3UWH
TA4 3TX M0GFX
TA4 4AE G4VJL
TA4 4DE G0NKZ
TA4 4DE G0SZO
TA4 4LA G0WIE
TA4 4NZ M6CJE
TA4 4PD G3HMG
TA4 4PD G3POM
TA4 4PG G4AJU
TA4 4SG M7EPS
TA4 4SL G0NRI
TA43JN M6NPO
TA5 1QG G8AYV
TA5 1SF G4OEC
TA5 2AG M7BMH
TA5 2AP G3OJK
TA5 2FQ G8SEK
TA5 2GG M6FQM
TA5 2GT G0MRR
TA5 2GW M7SKN
TA5 2HW G3TTP
TA5 2JN M6CTY
TA5 2PY G7UUL
TA5 2PZ G4WMY
TA52FT M6KUP
TA6 3LP G4BRW
TA6 3QE G7SYS
TA6 3TD 2E0PAH
TA6 4DW G3RHW
TA6 4HE M7CSP
TA6 4HL M3TOE
TA6 4HQ G4SMQ
TA6 4HT M7SBO
TA6 4JL G6JTT
TA6 4NA 2E0WEC
TA6 4NA G7GMZ
TA6 4NA M0WFN
TA6 4NA M3CSK
TA6 4PN G0OYA
TA6 4RY G1BKL
TA6 4SH G0MIK
TA6 4SH G7FAD
TA6 4SL G0VOE
TA6 4UB M3HET
TA6 4UT G7KHZ
TA6 4XD M6MXB
TA6 5BN M7JBE
TA6 5BT G8IPA
TA6 5HA M1FFC
TA6 5NS G0LCV
TA6 5PH G0CNA
TA6 5SA M3WSO
TA6 6AX G0OXB
TA6 6BW G4DII
TA6 6DZ G8PVG
TA6 6HF 2E0MHL
TA6 6HF M6MHL
TA6 6JU G1MCY
TA6 6LF M3JRF
TA6 6LJ M6AZF
TA6 6QW 2E0VCM
TA6 6QW G6MYH
TA6 6QW M3VCM
TA6 6RN G8FDF
TA6 6SJ G8ZOJ
TA6 6UR G1XOZ
TA6 7AJ G0WRQ
TA6 7LJ G1BTI
TA6 7NZ G7WBL
TA6 7PD G1HLP
TA6 7QB M0JES
TA6 7QG G0UQT
TA6 7QY G7URS
TA6 7RF G8YPV
TA7 0DE G3FSA
TA7 0EZ 2E0SST
TA7 0EZ M6SST
TA7 0HD G8VHO
TA7 0HD M3DBY
TA7 0HD M3PQJ
TA7 0HJ G4ELW
TA7 0JR M6FUL
TA7 0NP M6IRL
TA7 0RH G6OBB
TA7 0SB G4WGX
TA7 8AL G4ZBS
TA7 8AP M0BZV
TA7 8AS G6YGH
TA7 8BS 2E0TDI
TA7 8DW G1MOK
TA7 8EA G1IZH
TA7 8EA G7VWM
TA7 8ER G3MDD
TA7 8ES G0LHX
TA7 8HJ G0GCA
TA7 8HQ G1NMQ
TA7 8JB G4UNO
TA7 8PJ M6YLW
TA7 8PP G6AJV
TA7 8QR G6RRV
TA7 9AR G4BSK
TA7 9AT G6PZ
TA7 9BT G1HSF
TA7 9DY G4GHM
TA7 9EZ M3PZZ
TA7 9LR G8SUW
TA7 9LS G4NQQ
TA70EU M1DJX
TA8 1BE M3KJS
TA8 1HE M1DIL
TA8 1HG G0BBK
TA8 1HY G4HLN
TA8 1JA G4DVK
TA8 1JA G4RGY
TA8 1JU 2E1IIM
TA8 1LN G1KVQ
TA8 1LT G4EJW
TA8 1LU 2E0FLU
TA8 1LW 2E0IFT
TA8 1LW G7VWG
TA8 1LW M3IOX
TA8 1LW M6IFT
TA8 1QB 2E0FWT
TA8 1QB M0JID
TA8 1QB M7APN
TA8 1RE M3UHS
TA8 2EJ G6TAH
TA8 2EL G6TKR
TA8 2EW G0OQT
TA8 2HL 2E0RCO
TA8 2HL M7RLW
TA8 2HP G0FYP
TA8 2LE G7NME
TA8 2LN G4CGH
TA8 2LT 2E0VDV
TA8 2LT M3WRL
TA8 2LT M6ADD
TA8 2NH 2E0FKY
TA8 2NH M6PKQ
TA8 2NL G3GZZ
TA8 2NN G0PQD
TA8 2NW G3WLA
TA8 2PH G3NYD
TA8 2PS G3UBP
TA8 2QA M3PHG
TA8 2SZ M0FEY
TA8 2TT G6SEE
TA8 2TT G6SEF
TA9 3DQ M1DSZ
TA9 3LA 2E0FLM
TA9 3LA M6GPP
TA9 3LD G4UOS
TA9 3QU G3XBI
TA9 3QZ G0VYV
TA9 3RF G7GJZ
TA9 3RJ 2E0HKE
TA9 4BS G7FEP
TA9 4BU G8HUH
TA9 4DQ G1JGM
TA9 4DT G4TBO
TA9 4DU G7VIV
TA9 4DY G0GOB
TA9 4EJ M0IPI
TA9 4HE M6GLM
TA9 4LZ M0NDL
TA9 4NG G3VLW
TA9 4QT G3SXQ
TD1 1HD GM7GIS
TD1 1JL MM7RFP
TD1 1QL GM8KOF
TD1 1RG MM3RQG
TD1 1SQ GM6IZU
TD1 2BA GM4JPG
TD1 2BJ 2M0RTI
TD1 2BJ MM6PNU
TD1 2EE GM1JLP
TD1 2FA GM7MUN
TD1 2FA MM3SMI
TD1 2HZ GM7GPG
TD1 2LB GM8UUW
TD1 2RB GM8CJW
TD1 2RJ 2M1GEZ
TD1 2RJ 2M1GXX
TD1 2RJ MM0BPX
TD1 3BL MM7SVI
TD1 3HW GM4MCV
TD1 3JW GM4YEQ
TD1 3JW GM6WRY
TD1 3ND GM7LUN
TD1 3PA 2M0ODP
TD1 3PA MM6VDP
TD1 3RF GM0HQT
TD1 3SB 2M1BBY
TD1 3ST 2M0HWI
TD1 3ST MM7KTW
TD10 6XN MM6CES
TD11 3BF 2M0HVG
TD11 3BF MM7FLZ
TD11 3EE MM6KFE
TD11 3HG GM0TUS
TD11 3HQ GM0RDZ
TD11 3HS MM3URQ
TD11 3HW GM1DVO
TD11 3JJ MM7DSA
TD11 3NB MM0JYV
TD11 3NZ 2M0GJJ
TD11 3NZ MM0GJJ
TD11 3PP MM3HFU
TD11 3PP MM6BNX
TD11 3TQ GM0EUL
TD11 3UF MM6KWA
TD12 4BS GM0HNP
TD12 4BW MM7MSL
TD12 4BY MM0AMV
TD12 4EY 2M0CDO
TD12 4JQ GM4SYF
TD12 4NG GM0BRS
TD12 4NG GM8BDX
TD14 5AU MM0WZZ
TD14 5BF MM3LSO
TD14 5DX GM4ZXJ
TD14 5EU GM1JFF
TD14 5EU MM1DTU
TD14 5JF MM6BPM
TD14 5LG 2M0DZZ
TD14 5LG MM0GZZ
TD14 5LG MM3UQN
TD14 5LS GM7SRJ
TD14 5NJ 2M0BXY
TD14 5NJ MM3ZFK
TD14 5PB 2M0XDX
TD14 5PB MM0XAB
TD14 5PB MM6FRD
TD14 5PB MM6LRK
TD14 5RR GM4FVM
TD14 5TB GM7LNO
TD15 1BZ G3VAJ
TD15 1DU M0SNA
TD15 1SU G4WDO
TD15 1XB G7ROC
TD15 1XW G6VGO
TD15 2BB G4XVF
TD15 2BS G0JCP
TD15 2BS G1IVI
TD15 2DX G4GCL
TD15 2EB G3KML
TD15 2EE M0GHK
TD15 2HJ 2E0ICB
TD15 2HJ M0ICB
TD15 2JH M3INO
TD15 2JN G6IBP
TD15 2LA 2E0KLN
TD15 2LA M0KLN
TD15 2LA M6PNH
TD15 2RW 2E0DDO
TD15 2RW M6CMI
TD15 2TY G0HBV
TD2 6TT GM0SRO
TD3 6ND MM6TWE
TD3 6NF GM4YSN
TD4 6AS GM0RWU
TD4 6BN GM3RXU
TD4 6HS 2M0YBR
TD4 6HS MM6BIH
TD5 7AS GM0ALW
TD5 7AS GM0ALX
TD5 7AS GM4KHS
TD5 7EU GM0SIA
TD5 7NB GM6ZFI
TD5 7NJ GM4CXP
TD5 7NN GM6PTX
TD5 7NP GM6TBE
TD5 7PB GM6RTN
TD5 7RF MM7DMW
TD5 7RU GM0CDV
TD5 7SZ GM3PPE
TD5 7TT MM7WPC
TD5 8ED MM7CCC
TD5 8LB GM4NLJ
TD5 8SA GM8TCH
TD6 0AL GM0IQI
TD6 0QS MM0BPP
TD6 0RY GM4CID
TD6 0RY MM3OXB
TD6 9BH MM0MRM
TD6 9EP GM0RDC
TD6 9HB GM0VHR
TD6 9JB GM4PNM
TD7 4BG 2M0KMQ
TD7 4BG GM7GRH
TD7 4BG MM6KMQ
TD7 4HS MM0HTL
TD7 4JR 2M0YPH
TD7 4JR MM7PPH
TD7 4LT MM0ILC
TD7 4PR 2M0WDA
TD7 4PR MM7WSS
TD7 5BH MM3XXO
TD7 5BP GM0KCN
TD7 5DD GM0FTJ
TD7 5EY GM4UFP
TD7 5JB GM4VYU
TD7 5JZ MM3YJH
TD7 5NF MM3PAE
TD7 5NF MM3YDH
TD8 6AG GM4UPX
TD8 6DU GM0NYD
TD8 6DU GM1ZTB
TD8 6JW MM0INE
TD8 6LA GM4HCY
TD8 6ND GM4JKZ
TD8 6ND GM6FTE
TD8 6SA GM0FSH
TD8 6SD GM7WLO
TD8 6SU MM6BUG
TD8 6TZ GM6MCV
TD9 0AS MM6DXH
TD9 0AS GM6KWU
TD9 0QQ 2M0EBU
TD9 0QQ MM0RYR
TD9 0QQ MM6FZT
TD9 0SN GM4LPJ
TD9 0TT 2M0XRZ
TD9 0TT MM0XRZ
TD9 7BH GM0FTK
TD9 7PR GM0IGJ
TD9 7QE MM1APS
TD9 8BA GM8GUX
TD9 8EN GM6CKN
TD9 8JE MM0GCY
TD9 8QJ GM8JBJ
TD9 8QJ MM3SLB
TD9 9NZ GM4VJH
TD9 9SS MM7BWK
TF1 1JH G1VIT
TF1 1QJ 2E0UDE
TF1 1QJ M0RKY
TF1 1QJ M6RLB
TF1 2AJ 2E1HKS
TF1 2AS G8VZT
TF1 2AS M5AFG
TF1 2BY G1BHV
TF1 2JF G8BKH
TF1 2JP G0RQI
TF1 2LE G4AUY
TF1 2ND M3YWF
TF1 2PQ G0ASP
TF1 3DB G1JJK
TF1 3DU M7NDK
TF1 3ED G6HKP
TF1 3GA G6NRM
TF1 3GD M3NZW
TF1 3HH G6ZME
TF1 3HU M0CFT
TF1 5FL M3IJT
TF1 5LF G0UCN
TF1 5LN 2E1EQK
TF1 5SE M7EFM
TF1 6AD M0AFQ
TF1 6FZ G1ZPQ
TF1 6PQ G6NOL
TF1 6QR G2RSA
TF1 6TD G8RXZ
TF1 6XT M0CPK
TF1 6XU G6ZFU
TF10 7BS 2E0PNN
TF10 7BS M0PNN
TF10 7DT G7KZG
TF10 7EF G4IFR
TF10 7EG G1AJT
TF10 7HH G7KJA
TF10 7JD G0KBK
TF10 7JP G4UMM
TF10 7LS G0UCI
TF10 7NN M6IMA
TF10 7PJ M7FZG
TF10 7RF 2E0HDQ
TF10 7RF M7AXD
TF10 8AH G4JOW
TF10 8BP G1DIL
TF10 8DA G0GQK
TF10 8JL G1KRU
TF10 8JR G4HVN
TF10 8JR G6HGA
TF10 9AQ G4OLO
TF10 9EN 2E0BAJ
TF10 9EN M0EAK
TF10 9EZ M0TAW
TF10 9EZ M3OGL
TF10 9JJ M3WXG
TF10 9JQ G0WJH
TF11 8EB M3YBW
TF11 8HL M7ICD
TF11 8HZ G0VSJ
TF11 8QD M1SWR
TF11 8RD G0EXN
TF11 9BF M3FAL
TF12 5SB G4YDT
TF12 5SH G6DBZ
TF125JB 2E0IZP
TF125JB M0IRP
TF125JB M6IRP
TF13 6AX 2E0XTL
TF13 6AX G8XYJ
TF13 6AX M0WOF
TF13 6AX M3XTL
TF13 6BW G1HMI
TF13 6BW G3LNP
TF13 6BZ G6KJY
TF13 6ED G7DTG
TF13 6ED G8AQA
TF13 6ED M0HMO
TF13NN G0CYN
TF2 0AS 2E0DTB
TF2 0AZ 2E0FTO
TF2 0AZ M3ZXX
TF2 0DT G4OMI
TF2 0DX G1OAR
TF2 0DY M0VZT
TF2 6ES 2E1FQO
TF2 6EZ G7NIB
TF2 6JR M6AAD
TF2 6NA M1BGL
TF2 6RF G0GUV
TF2 7DA 2E1ACW
TF2 7LJ G0VXJ
TF2 7LQ G1SKW
TF2 7NP M3XNU
TF2 8AR M3DLB
TF2 8DB M1IHM
TF2 8HY G0HDC
TF2 8JH 2E0PZM
TF2 8JH M0TZM
TF2 8JH M7KBO
TF2 8NA G4YVD
TF2 8PB M1EZJ
TF2 8SQ M3GZT
TF2 9EN M6DCJ
TF2 9HD 2E0DCM
TF2 9HD M0ECM
TF2 9HD M0EMM
TF2 9JZ G3YZQ
TF2 9LQ M6ANN
TF2 9LX 2E0KLS
TF2 9NZ G4DFG
TF2 9PT G6APX
TF2 9PX M7OWA
TF2 9RB G3JKX
TF2 9SH M6FDN
TF2 9SL G4KLD
TF2 9SR 2E1HTV
TF2 9WR M6FXE
TF3 1ED G8ZIP
TF3 1EG G0UFE
TF3 1EG G3ZME
TF3 1EG G6UDX
TF3 1LA G0BST
TF3 1LJ 2E0BAH
TF3 1NH G8ZXU
TF3 1NY M3UWE
TF3 1PT M6LEU
TF3 1SS M6YVN
TF3 1XZ G0GQO
TF3 2AQ G7SFI
TF3 2BR M6LLO
TF3 2BU M0TYK
TF3 2DA M6WSS
TF3 2EE 2E1IHJ
TF3 2ES M6MGG
TF3 2HX 2E1GKE
TF3 2JB 2E0KHM
TF3 2JB M0KHM
TF3 2JB M6KZM
TF3 2JJ M6HEW
TF3 2JW M0TBQ
TF3 2JX G4FBZ
TF3 2NH M0IRS
TF3 2NQ G6FFR
TF3 5DA G0HQK
TF3 5FH M7NTH
TF3 5GH 2E1GJJ
TF3 5HD M6TZE
TF3 5HE G0GAL
TF3 5HH M6XWS
TF3 5HN M6KCA
TF4 2AH M7MLD
TF4 2AR 2E0ZGS
TF4 2DL M3KXB
TF4 2EB 2E0CVC
TF4 2EB M6CSZ
TF4 2EW G0MQX
TF4 2HS G6NWT
TF4 2LW M7JSY
TF4 2NX 2E0MCL
TF4 2NX M3LML
TF4 2QA 2E1HTU
TF4 2QE G0VNO
TF4 2RW 2E0PLX
TF4 3BJ G7SEY
TF4 3DD G8JVM
TF4 3DX M7BCI
TF4 3HP 2E0PWF
TF4 3HP M6EGV
TF4 3JF G0VPW
TF4 3LB M6BRB
TF4 3NG 2E1FHX
TF4 3RF M7AUO
TF5 0LL G8OYB
TF5 0NQ M1DGX
TF5 0NQ M1DLX
TF5 0PG G7BCO
TF6 5AU 2E0TRO
TF6 5AU M3UWB
TF6 5ER G6SGD
TF6 6DH G0JBO
TF6 6HH G4WRK
TF6 6HQ G8UKV
TF7 4AB M1ACA
TF7 4AQ M0DNV
TF7 4BX 2E1CFB
TF7 4DP M1GIZ
TF7 4DT G4VZL
TF7 4EA M6HDA
TF7 4NH G4VZK
TF7 4NQ G1KCA
TF7 4PZ M6SXA
TF7 4QE G0LCO
TF7 4QE G4TQZ
TF7 5AB M7FEH
TF7 5HB M3UXN
TF7 5JJ M6OLY
TF7 5LU 2E0YYB
TF7 5LU M6BVC
TF7 5NB M3SXK
TF7 5NF M7REV
TF7 5NG 2E1AIY
TF7 5NJ M7DJN
TF7 5QH 2E0KDF
TF7 5QH M6BVE
TF7 5QP 2E0HYJ
TF7 5QP M7DIW
TF7 5QR G0JAL
TF7 5QT G6APE
TF7 5RJ 2E0EOH
TF7 5RX M1NMG
TF7 5SF 2E0IVR
TF7 5SN 2E1DYL
TF7 5SU G4EIX
TF7 5UB 2E0APG
TF8 7BS 2E1AVM
TF8 7BS G7OQT
TF8 7EW M0AXV
TF8 7ND 2E1ESW
TF8 7ND G0VXG
TF8 7ND M3BHI
TF9 1AR G8EWD
TF9 1EA G0JNA
TF9 1HP G8PHG
TF9 1NP G1DVH
TF9 2BA G3OKD
TF9 2BB G6JKV
TF9 2BD G0PLB
TF9 2DG G0MDJ

TF9 2DU G0DHM
TF9 2NB G4DQB
TF9 2QG G0GHL
TF9 2QX G0AGC
TF9 2RX M7KLT
TF9 2SF G3ZGU
TF9 2SX 2E0CBA
TF9 2SX M6HRC
TF9 2TF M3XKJ
TF9 2TG M3MQJ
TF9 2TH M1MDE
TF9 3BD G4MZQ
TF9 3BW M7CWS
TF9 3HJ G6COB
TF9 3HJ M3COB
TF9 3HX M7EQM
TF9 3LP G6WPL
TF9 3QB G0CER
TF9 3QE M7FWA
TF9 3QF G0CGA
TF9 3QH M3THE
TF9 3SY 2E0UNI
TF9 3SY M0DQS
TF9 3SY M0UNI
TF9 3UB 2E0SHG
TF9 3UH G8PSZ
TF9 3UJ G0CDS
TF9 3US M7GNA
TF9 4BE G3JPB
TF9 4DD 2E0SNP
TF9 4DN M0INI
TF9 4DN M6INI
TF9 4ED G4JKF
TF9 4LQ 2E0MKF
TF9 4PZ G3GCU
TF9 4QE G6GTJ
TF9 4QR G3WEI
TF9 4RJ M1AOL
TF91DU G6NJR
TF94DD M0HIO
TN1 2LH G4ZFP
TN1 2SH G3WTR
TN10 3HB M3XST
TN10 3HD 2E0SRO
TN10 3HG G0DJC
TN10 3HG G7TYR
TN10 3HG G8WZK
TN10 3HQ M3TWP
TN10 3LW G1OGV
TN10 3LW M3OGV
TN10 3NN G1ZZL
TN10 3NN M7NRS
TN10 3PU M7JCS
TN10 4AJ M6CBU
TN10 4BJ M6YOU
TN10 4DG G4FFY
TN10 4HD G7KNS
TN10 4HU M7KCY
TN10 4LB 2E0WOH
TN10 4LE G3MGQ
TN10 4NG G4FYG
TN10 4NN M3ZKL
TN10 4PD M7AMS
TN10 4QS G0VGD
TN11 0ED M7SOX
TN11 0HU G4HGR
TN11 0NL 2E0RXW
TN11 8JH M0RBT
TN11 8LE G3YJW
TN11 9BL G8YCJ
TN11 9DA G6IUS
TN11 9DR G6MXE
TN11 9ED G7THJ
TN11 9ED M3THJ
TN11 9HB G4AUL
TN11 9HD G8CAA
TN11 9HQ G4TPJ
TN11 9LG G4JED
TN11 9LG M6JHL
TN11 9PL G3ZDT
TN12 0BE M1EGP
TN12 0BT G7GMU
TN12 0BW M6IQY
TN12 0BX M3BOQ
TN12 0LR G0UXG
TN12 0LR M0KWA
TN12 0NA G7GGA
TN12 0NE M6SPG
TN12 0PG M1CGR
TN12 0RH G1NDK
TN12 0SS G6TQL
TN12 0TE M0KEW
TN12 5BW G0EBS
TN12 5DB M6VKH
TN12 5EQ G4YIM
TN12 6BG 2E0UHS
TN12 6BG M3IMP
TN12 6BQ G5MAN
TN12 6BQ M7DHT
TN12 6HP G0MID
TN12 6LS G1YLB
TN12 6LS M7YLB
TN12 6YA M0RWW
TN12 7EA G8CDD
TN12 7LR 2E0ZWW
TN12 7LR M0ZWW
TN12 7LR M1ATA
TN12 7LR M3ZWW
TN12 9AX G0WBV
TN12 9BS G8YBH
TN12 9EN 2E0LKR
TN12 9EN M0LKR
TN12 9JD G0RHO
TN12 9LS G7FKX
TN12 9NB G3XBQ
TN12 9SG M3RTE
TN12 9TQ G3UUG
TN12 9TW M3XPF
TN13 1EL G8KVO
TN13 1EU 2E0RJX
TN13 1EU M0RMI
TN13 1EU M6EYJ
TN13 1TJ M3HAU
TN13 2DZ M0PRV
TN13 2DZ M0RWK
TN13 2HE G1TGZ
TN13 2JP 2E1FBH
TN13 2UA G4ASZ
TN13 2UG 2E1EAX
TN13 3AG 2E1EYY
TN13 3DR M3PPR
TN13 3SB M6YJK
TN13 3XA G0CDQ
TN13 3XN M3KDM
TN14 5AS M1DUA
TN14 5AT G7MYY
TN14 5BT G7RHE
TN14 5JF 2E0KJJ
TN14 5JF G8PHM
TN14 5LD G0SLU
TN14 5QP G0WQC
TN14 5RR G6KLF
TN14 6BX 2E0YEW
TN14 6BX M0WEW
TN14 6BX M6KTC
TN14 6BX M6PRA
TN14 6DL M6KYZ
TN14 6HT G3SAR
TN14 6LY G1TTX
TN14 6NH M1RSB
TN14 7AU G7IET
TN14 7EY G1KVW
TN14 7HW M1NTY
TN14 7JU G7JYG
TN14 7NA G7KTR
TN14 7SW 2E0JDA
TN14 7SW M0JDA
TN14 7SW M6JDA
TN14 7TD G7PVF
TN15 0LZ M1JDW
TN15 6BB G1FHR
TN15 6BL G3GJW
TN15 6BL G3ZRX
TN15 6BL G4BRC
TN15 6BP G1MNY
TN15 6DT G4RVV
TN15 6EZ G7MNZ
TN15 6HP G8ZZK
TN15 6JJ G8TLP
TN15 6LF G8GHP
TN15 6LL M1DMN
TN15 6NU G0KFW
TN15 6PG G6EDM
TN15 6PT G6GLH
TN15 6QL G8PWO
TN15 6TG G3WBS
TN15 6YF G4MGY
TN15 7EY G6VBJ
TN15 7JU G8YDB
TN15 7NR G3RGS
TN15 7NR G4CTC
TN15 8DX M7KNR
TN15 8EA G3IAR
TN15 8HG M7OCJ
TN15 8HT G0IPH
TN15 8LL G8ZYI
TN15 8LQ G4DFY
TN15 8RG M0MEA
TN15 8RJ M7SMP
TN15 8RS G3EWM
TN15 8RS G4FQT
TN15 8SD 2E1FDP
TN16 1DX G0POC
TN16 1LS 2E0DIT
TN16 1LS M0KOT
TN16 1LS M6DIT
TN16 2AT G1BMW
TN16 2BH M7WAI
TN16 2BT G4NSD
TN16 2JE G3KEQ
TN16 2LA G1FKP
TN16 3BJ G3ZVN
TN16 3BN G4ASI
TN16 3HH G6UNR
TN16 3HX 2E0DWG
TN16 3HX M6GVJ
TN16 3PB M7MOK
TN16 3PH M7BXE
TN16 3QX M6TBV
TN16 3SA G1XPF
TN16 3SG G8JQS
TN16 3SH M7IND
TN16 3UU G4GLN
TN16 3UX G6THC
TN16 3XE G8AXN
TN16 3XY M3TLW
TN17 2DD G3GUE
TN17 2SA M6MPY
TN17 3DZ 2E1FVY
TN17 3JR 2E0IAK
TN17 3JR M0TUX
TN17 3LS G0EIH
TN17 3QJ 2E0THG
TN17 3QJ M0MNZ
TN17 3QJ M7MLB
TN17 4BT G0OEQ
TN17 4DB M0OOD
TN18 4DQ M7MWW
TN18 4JE G3XWL
TN18 4JH G6VGA
TN18 4RA G0SLD
TN18 4RW M6MOC
TN18 5DU 2E0ZJO
TN18 5DU 2E0ZJQ
TN18 5DU M0ZJO
TN18 5DU M0ZJQ
TN18 5DU M3ZJO
TN18 5DU M3ZJQ
TN18 5DU M6YLC
TN18 5LX G0IPO
TN18 5PU 2E0OCB
TN18 5PU M3VXO
TN19 7DB 2E0KWW
TN19 7DB M0KWW
TN19 7DB M6LUD
TN19 7PJ G4NVM
TN19 7QY G7IGF
TN2 3BL M3LCP
TN2 3BS M6MTH
TN2 3DA G4OSH
TN2 3HF G8SQP
TN2 3HU M3JCT
TN2 3NA M6HXK
TN2 3NQ G7WJW
TN2 3NY 2E0XFR
TN2 3RX 2E0PMC
TN2 3RX M0NLP
TN2 3RX M6BFG
TN2 3RX M6FTW
TN2 3RX M6TBH
TN2 4BT G6CRR
TN2 4BX G0VQA
TN2 4ER G0VYK
TN2 4HG G4KGF
TN2 4LF G6AHK
TN2 4ND M7YAU
TN2 4NP G7PMG
TN2 4NS G8JXG
TN2 4RZ M7EFB
TN2 5DD G4JFD
TN2 5HG G0AOA
TN2 5HT M0WTW
TN2 5LN G8NAU
TN2 5LU G3IIW
TN2 5NB G1WKS
TN2 5NB G4CTY
TN2 5NN G4LXC
TN2 5PW M7NTP
TN2 5RG G6XPF
TN2 5XU M6SVN
TN20 6DU M7TMH
TN20 6UD G0BUX
TN21 0FE G3GTF
TN21 0HW G7CSX
TN21 0QA M6XKW
TN21 8BB G1ITE
TN21 8BB M5ITE
TN21 8ED G4YDB
TN21 8EY G4KEI
TN21 8HE G4AWJ
TN21 8HG M6IMR
TN21 8HG M6LHR
TN21 8HP 2E0HDE
TN21 8HP M0MDE
TN21 8HP M6HDE
TN21 8NR G8VHK
TN21 8PF G2BHY
TN21 8PF G3TGF
TN21 8PG M3EMU
TN21 8SA G8KQA
TN21 8SJ G0KOC
TN21 8SJ G4BJC
TN21 8SJ M1SWL
TN21 8TG G1PJT
TN21 8TW G0FDV
TN21 8YS G0MSA
TN21 8YS M0WAJ
TN21 8YW 2E0OPM
TN21 8YW M0WRD
TN21 8YW M6WHS
TN21 9AL G3BBK
TN21 9BL G6HIE
TN21 9BL M3HIE
TN21 9NG M0RYA
TN21 9NR 2E1LAM
TN21 9PP 2E1GIZ
TN21 9PP M0FCD
TN210EF G3WWS
TN22 1BY M3NYA
TN22 1BZ G4ZOQ
TN22 1HU G7DQZ
TN22 1NN M1XRC
TN22 1NU G0TBS
TN22 1RT G0LFX
TN22 1TU G8BTC
TN22 2AT G7KID
TN22 2AY M6DYX
TN22 2BX M6FRT
TN22 2BX M6MTK
TN22 2DP M3IZD
TN22 3AP 2E0OLM
TN22 3AP G0BUK
TN22 3AP M0OLM
TN22 3BE M3YZH
TN22 3DT G1IFV
TN22 3HX G4GEN
TN22 3LJ M6KAE
TN22 3LP 2E0IDN
TN22 3LP M6IDN
TN22 3RA G1RNY
TN22 4EU G8ASV
TN22 4HT G4NAJ
TN22 4JA G4UDW
TN22 4JE M7JAS
TN22 4JU 2E1HGR
TN22 4JU G6GSF
TN22 4LB G3YMH
TN22 4NG M3OII
TN22 4QB G4MMH
TN22 5AG 2E0MET
TN22 5AH G1SHH
TN22 5BN 2E0IOY
TN22 5BN M7DQY
TN22 5BU G0MAR
TN22 5JG M3SAY
TN22 5JH G7VAD
TN22 5LA G7HQY
TN22 5SG M7BQI
TN221RY G0VZN
TN224AR M6TPE
TN225SD M3LUO
TN23 1LN G7NUG
TN23 1LN M6VWE
TN23 3AG G0PDP
TN23 3BE M6IVB
TN23 3DH M7BLI
TN23 3DW 2E0FPO
TN23 3DY G1EJA
TN23 3EG G6EXU
TN23 3EX M6EWW
TN23 3LB M7FGD
TN23 3LF M3HFA
TN23 3LQ M6IFF
TN23 3NH M3VBL
TN23 3NJ G4AAR
TN23 3NJ G6ZAA
TN23 3NJ M0ZAA
TN23 3NR 2E0FAA
TN23 3NR M0TMX
TN23 3NR M3FAA
TN23 3QR 2E0PVK
TN23 3QR M7BAG
TN23 3QY M3RIU
TN23 4EQ M3YBR
TN23 5QR M6HJG
TN23 5SA G0TJI
TN23 5UR G6AJW
TN23 5UR M6RRJ
TN23 5UW G4UUI
TN23 5YF M3ZRG
TN23 6JL G4UKO
TN23 6JL M7FYF
TN23 7AF M7BEX
TN23 7HL G4EWV
TN23 7UP 2E0KEZ
TN23 7UP M3KEZ
TN234NR 2E0XXT
TN234NR M6TEZ
TN24 0AX G8VVU
TN24 0BJ 2E0EHS
TN24 0BJ M6GXP
TN24 0DZ 2E1KCC
TN24 0FX 2E0RBP
TN24 0FX M6RPC
TN24 0GS G0PEG
TN24 0HJ 2E0TED
TN24 0HJ 2E1ITE
TN24 0HJ M3WHY
TN24 0HN 2E0HNP
TN24 0HN M0HKP
TN24 0HN M6HKN
TN24 0JH G0PHO
TN24 0JH M3PHO
TN24 0LF 2E0KZK
TN24 0LF G5HAM
TN24 0LF M0IRU
TN24 0LF M6KZK
TN24 0PL M6PDD
TN24 0TA M3YOO
TN24 0XA G8EWL
TN24 8BD M6BAC
TN24 8NF G0HVP
TN24 9AE G3ZPI
TN24 9BD G0KGE
TN24 9EU G7HJJ
TN24 9EU G8YNF
TN24 9HH M7MDD
TN24 9JW 2E1FWM
TN24 9JY G7KLJ
TN24 9LU M0AVY
TN24 9PT M1PAB
TN24 9QL M3ILJ
TN24 9RS M3NKX
TN24 9SD G7LJL
TN240TY 2E0LPW
TN240TY M0LPW
TN240TY M6LPW
TN249BQ M3CAA
TN25 4DF G3ZAJ
TN25 4HS M7FIE
TN25 4NX M3WAF
TN25 4PQ G6SRE
TN25 5AB M6IPH
TN25 5AL G3EDM
TN25 5BD M0WYE
TN25 5JB G4FPG
TN25 5LR M6JNJ
TN25 6EB G1BXT
TN25 6EP M1DRL
TN25 6HG M1SEM
TN25 6NE G0JIR
TN25 6NW G4EDZ
TN25 6RA G8WLB
TN25 6RW G7EVC
TN25 6UA G3YBE
TN25 7EU M7KNO
TN25 7EU M7POV
TN26 1HW G3MMN
TN26 1LS G0MTJ
TN26 1NB 2E0YPU
TN26 1NB M3YPU
TN26 2HL G0CRL
TN26 2LU 2E0BVB
TN26 2LU M6RIO
TN26 2QH G1AEX
TN26 3AR 2E0KKH
TN26 3HA G3DT
TN26 3LY G6HXZ
TN26 3QS G0CHG
TN26 3QU G8PFZ
TN26 3QX M7PSX
TN27 0AQ G6LYD
TN27 0DD G6JPG
TN27 0EE M3SNX
TN27 0HA 2E0SPT
TN27 0HA M0DRQ
TN27 0HA M7FHN
TN27 0HB G8FYX
TN27 0JA G6KEN
TN27 0JN G6BME
TN27 0JQ G3KAP
TN27 0LH 2E0YQC
TN27 0LH M3TGO
TN27 0LS 2E1CPB
TN27 0NG M0MSF
TN27 0NG M1DDY
TN27 0QF G8LZV
TN27 0QF M0CVN
TN27 0RB G0VIM
TN27 0RJ G7EWS
TN27 8BY 2E0VJH
TN27 8EA G7FKZ
TN27 8EA M3YWE
TN27 8ER G1AOQ
TN27 8PZ M0DKN
TN27 9DW G8ISM
TN27 9QQ G4ZXI
TN27 9QR G0ICP
TN27 9UF 2E0CUE
TN27 9UF M0HNI
TN27 9UF M6RCW
TN28 8BJ G8BIS
TN28 8FP G0GCQ
TN28 8JB G4ZTW
TN28 8JL G4TGK
TN28 8JY G7PWS
TN28 8LB G1HJS
TN28 8NF G1UTF
TN28 8NL G0DCG
TN28 8NX G0BGH
TN28 8PR 2E0EWB
TN28 8PR M0OUS
TN28 8PR M6OHT
TN28 8QA M0DXS
TN28 8RD M0DAS
TN28 8RE G7HKT
TN28 8RL G8GZW
TN28 8RX G4YAZ
TN28 8SB 2E1KAJ
TN28 8SF G6JQD
TN28 8SU M6PUY
TN28 8SW G0KRH
TN28 8SY G0DXH
TN28 8SY G8CPM
TN28 8SZ M6BQO
TN28 8UJ M3OMZ
TN28 8UL G0IMD
TN28 8XH M6PED
TN28 8XP G0OIN
TN29 0LA G0LNN
TN29 0PW G1VBO
TN29 0PZ M7YJW
TN29 0QX G6ZVB
TN29 0RD G6DZT
TN29 0RD G6OJZ
TN29 0RU G4TZX
TN29 0SF 2E0ZIV
TN29 0SF M3ZIV
TN29 0TY G7UBK
TN29 0UA G1EVI
TN29 9BJ M7EGT
TN29 9EJ G6NLZ
TN29 9JP 2E0SVK
TN29 9JP M0LMI
TN29 9JP M6LMI
TN29 9LE G8XCL
TN29 9LF G1HSX
TN29 9NP M0LYD
TN29 9NS G1KPU
TN29 9RG G1HSI
TN29 9SN 2E0AKU
TN29 9SN M3KQY
TN29 9UU G0CPZ
TN29 9YF G4RLU
TN3 0QB 2E0FXY
TN3 0QB M7WHA
TN3 0RB M1KWH
TN3 0UJ M7CMW
TN3 8DY M6ONO
TN3 8DY M7WZZ
TN3 9QU G0CHY
TN30 6DQ 2E1GKF
TN30 6DY M6RCB
TN30 6EL G4KPS
TN30 6EN M6OLU
TN30 6EQ 2E0CMN
TN30 6QG G4ERW
TN30 7AP M7RVE
TN30 7AZ G3ZWK
TN30 7BA 2E1GJD
TN30 7NB G8GFS
TN30 7NT M3JCQ
TN307EY M7KMV
TN31 6BX 2E0IEE
TN31 6DL M0NUC
TN31 6DL M6FLH
TN31 6DN G6IPH
TN31 6DS M3ULE
TN31 6EJ M6BSB
TN31 6EP G0OZM
TN31 6EX M0EDU
TN31 6HQ G8VOH
TN31 6JB 2E0EOX
TN31 6JB M6KBO
TN31 6NB G0FLA
TN31 6NJ G3ROC
TN31 6TA M6LTB
TN31 6UL G4TLE
TN31 6UR G0TFV
TN31 7BG G6LPC
TN31 7FH G4PDU
TN31 7HJ M3MZR
TN31 7HT M3KPG
TN31 7NY G6BYF
TN31 7PJ G6XII
TN31 7RA G7HUC
TN31 7RS G8IIK
TN31 7RU G0ISM
TN31 7TE G1HHH
TN31 7TE G2LL
TN31 7TE G6HH
TN32 5AU G4BIA
TN32 5DN 2E1CYT
TN32 5JZ M6EWI
TN32 5NP G1VUK
TN32 5UG G4JWK
TN33 0DT M6WSC
TN33 0DU G7THK
TN33 0EU G7CKS
TN33 0HD M3KPU
TN33 0QR G7ECU
TN33 0QR G8AJP
TN33 0QR G8CLZ
TN33 0QS G3UYB
TN33 0SF G0LTV
TN33 0TP G3VFO
TN33 9AX G0BBR
TN33 9BP M3ESN
TN33 9ET G4PUP
TN33 9JB M7GST
TN33 9LD M7BDO
TN33 9LP 2E0EYE
TN33 9LP M6JHR
TN330TA 2E0FFJ
TN330TA M6OGS
TN34 1TG M3KQT
TN34 1TS G8FET
TN34 1UU G0OOU
TN34 2AN G7LPG
TN34 2AT G1LZS
TN34 2AU G1BFS
TN34 2BG G4PWE
TN34 2BJ M7FTV
TN34 2BT M3MRQ
TN34 2DZ G3VUR
TN34 2HA 2E0HXG
TN34 2HA M6HXG
TN34 2HY G6JIF
TN34 2JA G7PIP
TN34 2JZ M6HXN
TN34 2LF M7BPO
TN34 2NQ G4UWF
TN34 2NT G4KMJ
TN34 2PJ G6TFE
TN34 2PR M7JSL
TN34 2PS G7WWW
TN34 2QB G8SPW
TN34 2SF G3UFI
TN34 2UB G6STB
TN34 2UL 2E0SAY
TN34 3BJ G6HVX
TN34 3JN G0SYQ
TN34 3LB G3VEK
TN34 3LY M3BKI
TN34 3LY M3BKJ
TN34 3NH G7KMA
TN34 3NH M3AXB
TN34 3RT M6XDI
TN34 3RU M6XOJ
TN34 3SL G8RWJ
TN34 3TR 2E0KES
TN34 3TR M6RNO
TN343LR M6VDF
TN35 4AS M6EHE
TN35 4BL G3FHN
TN35 4DL M0SSR
TN35 4DN G3SYZ
TN35 4DP M0AHY
TN35 4DP M0CCV
TN35 4DP M1CPL
TN35 4EP G3BDQ
TN35 4JH G8ZGM
TN35 4LG G8THE
TN35 4LG M3XYZ
TN35 4PB 2E0TXI
TN35 4PB G3TCG
TN35 4QU G4NER
TN35 4QZ G1IKV
TN35 4SL G1GFZ
TN35 5AD M6NJK
TN35 5BS 2E0PSK
TN35 5DJ M1AKL
TN35 5EG M7NRM
TN35 5EP G0FUU
TN35 5HH 2E1GWX
TN35 5HZ G4NVQ
TN35 5JN G0CJO
TN35 5LZ M3BRQ
TN36 4HF M0NPD
TN37 6EL G1HSM
TN37 6HF G1FTK
TN37 6HF G7HFS
TN37 6JA 2E1HEO
TN37 6JY G7OJY
TN37 6LA G7ERC
TN37 6PF G4KLF
TN37 6PN G1ORK
TN37 6QR G1ZPA
TN37 6RX 2E0SPH
TN37 6RX M0ZPL
TN37 6RX M6FRJ
TN37 6SA M7ATC
TN37 6SD G1IFW
TN37 6SJ G8FEJ
TN37 7AL G7WGA
TN37 7BS G6CLU
TN37 7DF M6NXT
TN37 7EG M3YXH
TN37 7HY G0GKL
TN37 7JY G7VQO
TN37 7PS G1DVU
TN37 7PS M0BMD
TN37 7PX G3XXM
TN37 7QX G0JHK
TN376ET G1AJN
TN377EN M6RFB
TN38 0BT G0TDJ
TN38 0ER G7MQQ
TN38 0ET 2E1DHC
TN38 0HX G7TCH
TN38 0JN G0NJO
TN38 0NY 2E0YYS
TN38 0NY M7UCD
TN38 0PH M6BQS
TN38 0PU 2E0RAJ
TN38 0PU M3MFE
TN38 0QA G8CMK
TN38 0QA G8SBJ
TN38 0QP G1GFF
TN38 0SE G1OHL
TN38 0SL M6MCP
TN38 0TF M3XVC
TN38 0TR G0AQT
TN38 0UA G4EOA
TN38 0UX G8TQO
TN38 0XG M7CRU
TN38 0XL G4WOE
TN38 0XP M6ZZS
TN38 0YT M7EOB
TN38 8WD 2E0JFR
TN38 8WD M7XGX
TN38 9BU G1EHS
TN38 9BU G1EHU
TN38 9DP G4VBK
TN38 9HB G7BBT
TN38 9HP M3BKU
TN38 9JG M7CUA
TN38 9LQ G0ILK
TN38 9QW G1FTX
TN38 9RS G7JVN
TN38 9RW M7CVW
TN38 9RX G4BOZ
TN38 9TH G0TNQ
TN38 9UA G0OFZ
TN380BN M6VTZ
TN39 3AQ M6WIB
TN39 3AX G8GQG
TN39 3EE M6CWU
TN39 3HH M3KRS
TN39 3LU G4MMG
TN39 3NJ G8MYG
TN39 3PB G3WUH
TN39 3PD M0BDE
TN39 3PH M6TJY
TN39 3PS M3ZBH
TN39 3QJ M7FSV
TN39 3RS M6FKS
TN39 3TF G7OOP
TN39 3TS M6HHC
TN39 3UA G3JOR
TN39 3UP 2E0OCT
TN39 3UP M6IBO
TN39 4AS G7LEL
TN39 4BN G1PDS
TN39 4DN M7ART
TN39 4DY G4HGK
TN39 4GL G1VNZ
TN39 4GP M7YSU
TN39 4HZ G0THV
TN39 4JH M0OMI
TN39 4JJ G8EYQ
TN39 4JL M0BPU
TN39 4LJ G4BUE
TN39 4LJ M0BUE
TN39 4LN 2E0EVW
TN39 4LN M6NSH
TN39 4LN M6NSQ
TN39 4LT G1ZQO
TN39 4NZ 2E0JFP
TN39 4NZ M7XPX
TN39 4PP 2E1GTW
TN39 5BH M7TOW
TN39 5EN M6VMK
TN39 5EN M7NYY
TN39 5EQ G1MTB
TN39 5HN G0JBM
TN39 5HU G7GHP
TN39 5HY 2E0GHX
TN39 5JH G4MPK
TN393QB M7DLO
TN4 0DX G1EUD
TN4 0EQ G4MIK
TN4 0LL M7VWS
TN4 0NR M6GAX
TN4 8ED G4GTN
TN4 8JJ 2E0FJE
TN4 8PS 2E0CMP
TN4 8PS M0NAY
TN4 8PS M6BOX
TN4 8PY M6QFG
TN4 8TH 2E0DGS
TN4 8TH M0VIR
TN4 8TH M6GBZ
TN4 9AL M7MYA
TN4 9BL 2E0CQL
TN4 9BL M0KAO
TN4 9BL M6BQW
TN4 9BN 2E0SSE
TN4 9BN M7SSE
TN4 9DH 2E0TWP
TN4 9DH M6PJM
TN4 9DJ G0BMN
TN4 9DR G3KIP
TN4 9JG G7DAB
TN4 9JU G6TXH
TN4 9JY G0ABM
TN4 9NS 2E0IIJ
TN4 9NS M7DXS
TN4 9RF M0RYK
TN4 9RF M0XKX
TN4 9RP G0VPY
TN40 1EG G6SGV
TN40 1QE G0TMA
TN40 1QH G8KPE
TN40 1TE G6HIB
TN40 1TS G1WLO
TN40 1UE G6NGM
TN40 1UG G8TUU
TN40 2AZ G3GGH
TN40 2AZ G4SIF
TN40 2BQ G6ULD
TN40 2EA M3TNO
TN40 2HN M3LGI
TN40 2HQ G6TFJ
TN40 2HU G6MWB
TN40 2NT G4YZL
TN40 2NT G4ZSR
TN40 2PA 2E0LXI
TN40 2PA M0LXI
TN40 2PA M6LXI
TN40 2PB M7ESF
TN40 2PW M1FFX
TN40 2SD G3MVX
TN40 2SH G0ILN
TN40 2SH G7TAF
TN40 2TA M6PFB
TN40 2TB M1EQB
TN40 2UL M6WDQ
TN401PN G7DME
TN5 6BT M6LVK
TN5 6DG 2E0MMU
TN5 6DG M0MMU
TN5 6DG M6HXE
TN5 6DL G7MZL
TN5 6RJ G8GYB
TN5 6RJ M0NZA
TN5 6UT G1NZH
TN5 7AY G1EHB
TN5 7EG G4XST
TN5 7JA G7IRN
TN5 7PS G7MFE
TN6 1AT G4KIU
TN6 1BS G4LZK
TN6 1JF G3TXZ
TN6 1QQ G7JAN
TN6 1SE M6JUH
TN6 1SU M0TNF
TN6 1TF G4EZE
TN6 1TP M6PCC
TN6 2AD G0GPE
TN6 2AD G4XCE
TN6 2AG G8LKS
TN6 2BG G3WKS
TN6 2BG G4OTV
TN6 2BG G4UPI
TN6 2BN G6GWE
TN6 2DU M6MSF
TN6 2EN M1ARX
TN6 2ES 2E0MDU
TN6 2ES M6NAU
TN6 2ET G0WTC
TN6 2ET G6UIF
TN6 2HA 2E0TLB
TN6 2LY M6RBA
TN6 2NJ G8UBD
TN6 2NY G3CYU
TN6 2RY G3RST
TN6 2SB G8LSD
TN6 2UU 2E0GCP
TN6 3BH G0XBV
TN6 3BQ G4LDJ
TN6 3BY M3TLB
TN6 3EB M3UAR
TN6 3LG M6YLS
TN6 3QA G6BRV
TN6 3QH M6YTM
TN6 3QT M0CPC
TN6 3TA 2E0HKK
TN7 4DH G0NAR
TN7 4DN G0AAW
TN7 4EN M1PVC
TN7 4ET M6TWV
TN8 5BX M6JRB
TN8 5DB M3WYZ
TN8 5DF G0TKZ
TN8 5DS M6MVV
TN8 5EL G3JMJ
TN8 5EN G6MXV
TN8 5PS M6CSC
TN8 5PW G6DTT
TN8 6AJ M3ZHM
TN8 6AJ M3ZHQ
TN8 6AJ M6GPE
TN8 6AJ M6GRJ
TN8 6AU M6SNC
TN8 6HZ M6BHU
TN8 6JS M6NZV
TN8 6LN M0GZC
TN8 6LN M3ZWM
TN8 6SX M6SRC
TN9 1BN M3RPE
TN9 1JH M7GCX
TN9 1UX G1CPA
TN9 1UX M0AFZ
TN9 2EH M6VEP
TN9 2EJ G7GDV
TN9 2EL G8LGW
TN9 2EN M7KJW
TN9 2FE M7LUF
TN9 2NQ G3YPY
TN9 2PG M3LHE
TN9 2PG M3LHG
TN9 2UY G0JUY
TN9 2UY M3MOP
TN9 2YR M6SHC
TN9 2YS G4EPC
TQ1 1JY 2E0PCZ
TQ1 1JY M0PCZ
TQ1 1JY M6PCZ
TQ1 1PB G4XLO
TQ1 1PX M6CON
TQ1 1QR G0KEC
TQ1 1QR G1XDJ
TQ1 1TZ M6AOJ
TQ1 2ED G0CRJ
TQ1 2EU G1NIC
TQ1 2HH G3VOF
TQ1 2LQ M0IME
TQ1 2NL G6IYM
TQ1 3BL M0PAW
TQ1 3JN G1KPI
TQ1 3LD G6TEQ
TQ1 3NW G3SDW
TQ1 3NZ G4FKU
TQ1 3PD G4YTY
TQ1 3PW M6IDP
TQ1 3PW M6RWJ
TQ1 3SU G4IML
TQ1 3TP G3KPV
TQ1 3TQ M3SQX
TQ1 3TX 2E0PLR
TQ1 3TX M0LAI
TQ1 3TX M6LBN
TQ1 4BG 2E0LWT
TQ1 4BG M3RMU
TQ1 4EA M3LKE
TQ1 4HZ G0EDE
TQ1 4LD G0SUU
TQ1 4NJ G4OVO
TQ1 4PU M6PHW
TQ1 4PX G1EFT
TQ1 4QY G8XST
TQ1 4RD G8GPF
TQ1 4UN G1TNP
TQ10 9BT M1DKW
TQ10 9EJ G4TCP
TQ10 9EU G4XTR
TQ10 9EU G7BEP
TQ10 9HH 2E0GQL
TQ10 9JF M6FQZ
TQ10 9LL G0PCT
TQ11 0AF G8YZA
TQ11 0BP G7RYL
TQ11 0DB 2E0GQU
TQ11 0DB 2E0HHG
TQ11 0DB M7ALR
TQ11 0DB M7DSR
TQ11 0DL M0LED
TQ11 0EA G1GMG
TQ11 0ER G4XWP
TQ11 0FG M1FCF
TQ11 0HB G6NWC
TQ110AB M7BSF
TQ12 1AA M3IDC
TQ12 1AG M6XJD
TQ12 1DX M3JKA
TQ12 1DZ G0EKH
TQ12 1ER G0OFA
TQ12 1FZ 2E0APZ
TQ12 1FZ M0XWP
TQ12 1FZ M6YPZ
TQ12 1GQ G0MUN
TQ12 1GZ M7BAY
TQ12 1GZ M7DRY
TQ12 1HS G3TAA
TQ12 1HT M6DXG
TQ12 1JY M6IZN
TQ12 1LH M1DVV
TQ12 1PS 2E0CVN
TQ12 1RQ G3YAR
TQ12 1UL M3IBY
TQ12 1UP G0NUZ
TQ12 1US G4DCH
TQ12 1YJ G1WQN
TQ12 2ND G0RDO
TQ12 2NN M3PZF
TQ12 2PU G0WAE
TQ12 2PW M7PSC
TQ12 2PX G7AQV
TQ12 2SF M1BPA
TQ12 2TL M3GFH
TQ12 2TP M3GFE
TQ12 3BA M6KUC
TQ12 3BP M1AIY
TQ12 3DF 2E0DEV
TQ12 3DF M0ZAF
TQ12 3DF M3XJQ
TQ12 3JE G0WWD
TQ12 3LE G1WUU
TQ12 3LY M3DHA
TQ12 3NJ G0BCH
TQ12 3NJ M6ERU
TQ12 3QX 2E0FAV
TQ12 3QX M6ZAF
TQ12 3TE G4DTW
TQ12 3TJ M1AKV
TQ12 3YS G6YJO
TQ12 3YT M0YLY
TQ12 3YU M3GHD
TQ12 4EH G7AFV
TQ12 4FN G4GZM
TQ12 4HA G1DBH
TQ12 4HE G4VUD
TQ12 4JG G4LAK
TQ12 4JZ G4MNA
TQ12 4LF G3LHJ
TQ12 4LF G3NJA
TQ12 4LF G8NJA
TQ12 4LF M0LTW
TQ12 4NE G0IGK
TQ12 4NH G0NXI
TQ12 4NJ G4ELZ
TQ12 4NS G3XXE
TQ12 4PG M3IEU
TQ12 4QL M7EXV
TQ12 4QS G0CWQ
TQ12 4QS G4XJL
TQ12 4RE G0SDL
TQ12 4SA 2E1MYK
TQ12 4SA M7MYK
TQ12 4SB G3TLK
TQ12 5AB G3ZEJ
TQ12 5BQ M7LEZ
TQ12 5BZ G6FIN
TQ12 5DZ M3UMJ
TQ12 5EW G0BNJ
TQ12 5EX M1DVJ
TQ12 5EZ 2E0NKN
TQ12 5EZ 2E0PVU
TQ12 5EZ G1COE
TQ12 5EZ M0PVU
TQ12 5EZ M7RBC
TQ12 5JG G4NRH
TQ12 5NX G0ONT
TQ12 5PZ 2E0GPL
TQ12 5PZ M3UAW
TQ12 5PZ M3ZDT
TQ12 5QJ G8VSV
TQ12 5QJ G8YEN
TQ12 5QS G0SBM

TQ12 5QS G4FCN
TQ12 5QS G6CLD
TQ12 5QS G7MNL
TQ12 5QS G8GCS
TQ12 5QS M1EPU
TQ12 5UR M1CFS
TQ12 6ER 2E0JNY
TQ12 6ER M7DOG
TQ12 6EZ M0LNY
TQ12 6FA M6JAQ
TQ12 6GX G1ZBJ
TQ12 6HJ M0RIF
TQ12 6JA M0UAC
TQ12 6JA M6GBU
TQ12 6JA M7USV
TQ12 6NZ M6NQA
TQ12 6PN G3YCH
TQ12 6SB 2E0WSZ
TQ12 6SB M0WSZ
TQ12 6SB M6WNB
TQ12 6SP G7MEA
TQ12 6TX M6KAR
TQ12 6UP M3YYG
TQ123LB G0FHH
TQ124HX M7OZA
TQ13 0AZ 2E0IDE
TQ13 0AZ M7BSM
TQ13 0BB G1BJN
TQ13 0EE G4VTO
TQ13 0EE M7BNN
TQ13 0EW G4RYH
TQ13 0EW M6CCX
TQ13 0JH G4NRR
TQ13 0JN G7OZI
TQ13 0LH M5WNS
TQ13 0NW M0NAA
TQ13 0NW M6NMN
TQ13 0PJ G6BAL
TQ13 0PL M3RXW
TQ13 0RE M7BAE
TQ13 0RH 2E0HWJ
TQ13 0RH M7BJK
TQ13 7AN M3DBA
TQ13 7BS G7CUU
TQ13 7DU G7RKX
TQ13 7HT G1BGK
TQ13 7HY G8KEO
TQ13 7RU G3FYF
TQ13 8AR G3UUU
TQ13 8BB M0VGT
TQ13 8LP 2E0MLH
TQ13 8LP 2E0RIW
TQ13 8QA 2E0FWD
TQ13 8QA M3EYZ
TQ13 8RS M7NDM
TQ13 8SD M6IRK
TQ13 9DE 2E0EZL
TQ13 9DU M7BOP
TQ13 9EP G6ICH
TQ13 9EP M0LHK
TQ13 9EX 2E1PEW
TQ13 9EX G8PEW
TQ13 9EX M7PEW
TQ13 9FE G0BMP
TQ13 9HN M7HEL
TQ13 9HP M0CLE
TQ13 9HR G3RPD
TQ13 9JY M0ECQ
TQ13 9QT G1SJD
TQ13 9YL M3UOY
TQ139HB M7IEE
TQ14 0DG G3WLT
TQ14 0EJ G6EKM
TQ14 8AQ 2E0DOX
TQ14 8AQ M6ADX
TQ14 8EF G8KTC
TQ14 8JA G0EZU
TQ14 8LE G1MKE
TQ14 8LW G7CGI
TQ14 8NH G4CDJ
TQ14 8PN G4YTT
TQ14 8QQ G6XD
TQ14 8QX G4BG
TQ14 8RS G3LQX
TQ14 8SB M6LYZ
TQ14 8TS M0VZV
TQ14 8TS M6VZV
TQ14 8UE G0CEL
TQ14 8UE G0EMU
TQ14 8UF G3YHH
TQ14 9DT G3KFP
TQ14 9HG G0JZA
TQ14 9HG G3MOA
TQ14 9JP G0KDT
TQ14 9NF G0PWU
TQ14 9NJ G1EUA
TQ14 9NW M3FZS
TQ14 9QA M3HYI
TQ14 9QA M3IZW
TQ14 9QR 2E0TAD
TQ14 9QR M0HHG
TQ14 9QR M3TYO
TQ14 9QS G7SDQ
TQ14 9RF G0PNA
TQ14 9RP G8WBL
TQ14 9TX G4NRA
TQ2 5TQ G4XKH
TQ2 6BS G1ZZG
TQ2 6DX G4OTU

TQ2 6EA G0TMH
TQ2 6EA M3WFE
TQ2 6EA M3YVE
TQ2 6EA M6SOK
TQ2 6LL G0OXT
TQ2 6RG M3IPZ
TQ2 6UB G3VLO
TQ2 6UQ G1VTK
TQ2 6UT G3YDE
TQ2 7BN G7BFE
TQ2 7GF M6IXR
TQ2 7JX 2E0PBT
TQ2 7JX M0PBT
TQ2 7JX M6PBT
TQ2 7LA M7CJY
TQ2 7NR G0HSW
TQ2 7QX G0JDO
TQ2 7TB M3UUG
TQ2 8AJ G0HGV
TQ2 8DH G4ZQM
TQ2 8EF M0BKD
TQ2 8HB G0MRZ
TQ2 8HU G6FSP
TQ2 8HU G7SME
TQ2 8LF G8GRB
TQ2 8LP G1SDX
TQ2 8LR G6FTA
TQ2 8LR M3FTA
TQ2 8PA M6DFY
TQ2 8PY G4GZG
TQ2 8QA G4PCK
TQ2 8QE G8UKI
TQ2 8QH G4UIH
TQ2 8QH G4UII
TQ2 8QQ G7HIK
TQ3 1AB G7DKB
TQ3 1AE G0AZX
TQ3 1AE G0LQM
TQ3 1AG G8XTO
TQ3 1DL M7DQU
TQ3 1EE M5ABN
TQ3 1HN M6PNJ
TQ3 1HY 2E0KFJ
TQ3 1JX G0GWC
TQ3 1LE G7DMX
TQ3 1NH G6XXB
TQ3 1PH M0IAT
TQ3 1SQ G6UPR
TQ3 1TA G4AQR
TQ3 1TD G7IIC
TQ3 2AQ M7AWK
TQ3 2BN G4CLN
TQ3 2DP M0DPQ
TQ3 2EN G8BKQ
TQ3 2HZ G0UWF
TQ3 2JP G6ZZS
TQ3 2LR G8ZEV
TQ3 2NE M0CFZ
TQ3 2PB G6JAR
TQ3 2PE G4OFP
TQ3 2PT G4LPY
TQ3 2QF M0BKL
TQ3 2QS G7MCK
TQ3 2RE G0ORK
TQ3 2SW G3AH
TQ3 2SW G7AHP
TQ3 2SW M0RIV
TQ3 2SY M6AJJ
TQ3 3BT G3GCW
TQ3 3BU M6DPG
TQ3 3EQ M1DXB
TQ3 3FT M6GTO
TQ3 3GD 2E1KFS
TQ3 3GW G0LCD
TQ3 3JJ G0TQC
TQ3 3JQ M7NZL
TQ3 3NA G0CGM
TQ3 3NU G0CDB
TQ3 3NU G4SSD
TQ3 3NX M3AXX
TQ3 3QN G1XKQ
TQ3 3RX 2E0EVM
TQ3 3TD G1FRJ
TQ3 3TD M3GDV
TQ3 3TH M7WLL
TQ3 3TU G1TUU
TQ3 3TW M7BJV
TQ3 3UD G1JKL
TQ3 3XA G4WVM
TQ4 5AT G0BAJ
TQ4 5AX 2E0DBQ
TQ4 5AX M0ZYD
TQ4 5AX M6AYK
TQ4 5EY M6HNH
TQ4 5HU G4UUT
TQ4 5HZ G4NNP
TQ4 5JQ G3CLW
TQ4 5JQ G4KFB
TQ4 5LH 2E0PDL
TQ4 5LH M0LOB
TQ4 5LH M3UVT
TQ4 5LH M3VBG
TQ4 5LH M6DFI
TQ4 5NA G7CHC
TQ4 5NY G0SPF
TQ4 5UA M7CVQ
TQ4 6EG G6AOB
TQ4 6ES 2E0OJE
TQ4 6EY M6NHU

TQ4 6HA G0UAV
TQ4 6HD M0JNP
TQ4 6HT G4AWO
TQ4 6JF 2E0RJE
TQ4 6JF M3URV
TQ4 6JN G7JFM
TQ4 6LE G3PXL
TQ4 7HG G0UQI
TQ4 7HZ G8XLG
TQ4 7JY M3OMX
TQ4 7LR G7FNT
TQ4 7LW G0HZB
TQ4 7ND G0KTU
TQ4 7NZ M3NYX
TQ4 7RL G4DZH
TQ4 7RL M0BKG
TQ4 7RN M0BHJ
TQ4 7RN M1ANO
TQ4 7RN M1MUM
TQ4 7RN M3UNR
TQ4 7RU 2E0TAG
TQ4 7RU M3TQU
TQ4 7RW G0BTH
TQ4 7SD G4VQI
TQ4 7SD G6UIM
TQ5 0AN M6DFJ
TQ5 0AY G0OJS
TQ5 0BQ G2DXU
TQ5 0DD G7DLE
TQ5 0DF M7THY
TQ5 0JZ M0DQN
TQ5 0LZ G8JMS
TQ5 0PB G0RGP
TQ5 0PB G7LMR
TQ5 0PB G8ETU
TQ5 0RJ G3RVA
TQ5 8DG G8OVO
TQ5 8JT G0GJC
TQ5 8PB M1EJE
TQ5 8QG G6ZOJ
TQ5 8QP M3TQG
TQ5 8QR G3XNX
TQ5 8RF G4UXP
TQ5 8RN G4NGR
TQ5 8RS G0PTK
TQ5 8RT 2E0IIY
TQ5 8RT M7CWK
TQ5 9AG M0PLH
TQ5 9AX G8SPM
TQ5 9ED G0TDQ
TQ5 9EE G0KEK
TQ5 9EL G0VQY
TQ5 9EL M3GCD
TQ5 9LJ G1EZI
TQ5 9LS M1EIW
TQ5 9PF G4OJD
TQ5 9PJ G0TSB
TQ5 9PJ G1AEU
TQ5 9QZ G3ZUO
TQ5 9RH M0TRU
TQ5 9RL G4JXK
TQ5 9SQ G3REW
TQ5 9SW M0XOL
TQ5 9UD G4OKC
TQ5 9UN G0NPY
TQ5 9UR G7VSL
TQ50PE G8LLZ
TQ6 0DP G4XNO
TQ6 0DQ G3PCT
TQ6 0RG M7JJO
TQ6 9BN M3TQA
TQ6 9HJ G3WCB
TQ6 9HJ G4SAT
TQ6 9JN G4LZD
TQ6 9LQ M3TQN
TQ6 9NZ G7XPC
TQ7 1JU M3RPH
TQ7 1LL G8YBT
TQ7 1LP G0WHZ
TQ7 1LR G8KWJ
TQ7 1QB G6STF
TQ7 1RW M6AFB
TQ7 1SB M7SLP
TQ7 2DB G0JIA
TQ7 2DJ M3PRZ
TQ7 2DP G6SBG
TQ7 2NB 2E0POB
TQ7 2NB G0CWH
TQ7 2NB M6BWQ
TQ7 2SE G8ETP
TQ7 3BH 2E0PMX
TQ7 3BH M3PMX
TQ7 3JZ G7IMV
TQ7 3LX G7OSB
TQ7 4HG M6OKO
TQ7 4HS 2E0OWC
TQ7 4HS M0VTD
TQ7 4HS M6OWC
TQ7 4NE G3XLW
TQ7 4PF G3PFX
TQ7 4QJ M1DWQ
TQ7 4QP G6TOI
TQ7 4SD G0RIY
TQ8 8AB M0GJL
TQ8 8AN M3WNZ
TQ8 8JP G3UBL
TQ8 8NU G6STJ
TQ8 8PJ G8TQH
TQ8 8PW G3HST

TQ9 5AN G8KRB
TQ9 5DW G2HIX
TQ9 5DW G6KGU
TQ9 5DW M3URX
TQ9 5EF G3PNO
TQ9 5EU M6XXS
TQ9 5FH G6JPM
TQ9 5LH G3OPX
TQ9 5QY G0UBG
TQ9 5RF M0AVN
TQ9 5RF M3NSB
TQ9 5UR M3VTO
TQ9 5XY M0WYM
TQ9 6AN M3DBG
TQ9 6DA G7JLS
TQ9 6DB G8LXS
TQ9 6DJ G3SNU
TQ9 6LS M6MFO
TQ9 6RL 2E0PVQ
TQ9 6RL M3PVQ
TQ9 6SR G7KCE
TQ9 7BA G4VHL
TQ9 7BD M6TUK
TQ9 7EY G0DIH
TQ9 7HH G3WGN
TQ9 7HN M7HLR
TQ9 7NQ G7DHW
TQ9 7PU G4FBN
TQ9 7QR M7JHT
TQ9 7TL G1WZG
TR1 1AT G1ZWH
TR1 1BW G0NDC
TR1 1DS G0RIZ
TR1 1EF G1SZD
TR1 1JD G1UWD
TR1 1QR 2E0LTH
TR1 1QR M3SNQ
TR1 1RA G3MFW
TR1 1RU G0NKQ
TR1 1UP G8VCJ
TR1 1WL 2E1ADQ
TR1 1WL M3CKH
TR1 1XR G3PGJ
TR1 1YS G0SHY
TR1 2BW G4PGD
TR1 2BY G1ZRP
TR1 2BY M7FCD
TR1 2JZ M0BIZ
TR1 3DG 2E1CIT
TR1 3JA 2E0DGM
TR1 3JA M6HBT
TR1 3JR G8KJP
TR1 3LU M1EWT
TR1 3LX G0UWI
TR1 3LX G1DTS
TR1 3NG 2E0OEM
TR1 3NQ M0KLG
TR1 3NQ M6WHD
TR1 3PE G0FHX
TR1 3PE G0FHY
TR1 3PT G7FLI
TR1 3QD G0CUH
TR1 3TX G3WKP
TR1 3YJ 2E0EHT
TR1 3YJ M0IUM
TR1 3YJ M6MPL
TR10 8AB 2E0XZI
TR10 8BH G3UFX
TR10 8BH M7ORC
TR10 8DY M6ITZ
TR10 8GF M0SEA
TR10 8HQ M3URE
TR10 8JD G0DUE
TR10 8NX G8GCM
TR10 8SJ G8UOL
TR10 9DT G0EOH
TR10 9DZ G1EGI
TR10 9HB G0BSK
TR10 9JW G4PCW
TR10 9WS M1EQA
TR108NT G1JJZ
TR11 2DX G7FTA
TR11 2ER G4MYY
TR11 2ER G6DGW
TR11 2ER M0CNU
TR11 2HE G6CZX
TR11 2QE M0RTQ
TR11 2RN G0ETQ
TR11 2RW G4FNP
TR11 2SW G4RUA
TR11 2TD G1DDK
TR11 2TE M0PBX
TR11 3DZ 2E0MRP
TR11 3DZ M6MAP
TR11 3JW M6WRN
TR11 3QA M7NGL
TR11 3YL M6AWJ
TR11 4AQ G3LAI
TR11 4BE G0HJW
TR11 4EQ G8LCE
TR11 4ES M0SUI
TR11 4HR G6YHK
TR11 4LN M7ETS
TR11 4RG G7RKE
TR11 4SG G8AYJ
TR11 4SR G3VQS
TR11 4SS M7BWO
TR11 4SW G4RRQ
TR11 5AP M0DYH

TR11 5BW G4PXN
TR11 5DT G8CTR
TR11 5EJ G6IMN
TR11 5GZ M6RZW
TR11 5GZ M7NVY
TR11 5HE G0OCS
TR11 5HE G0OCT
TR11 5HF G4VMW
TR11 5LR G4ELI
TR11 5NA 2E1ESM
TR11 5NA G7WDS
TR11 5NF 2E0CUH
TR11 5NF M7KTA
TR11 5PR M1AVW
TR11 5QG M0RFM
TR11 5QN G4NBC
TR11 5SF 2E0BYH
TR11 5SF M3XUE
TR11 5SL G7LKZ
TR11 5TE M1FIG
TR11 5TJ G3LZN
TR11 5TT M3XOA
TR11 5TT M6EME
TR11 5UG G3NWW
TR112TE 2E0PFA
TR112TE M3SOV
TR12 6DP M6MQJ
TR12 6HT 2E0XGB
TR12 6HT M0MZC
TR12 6HT M6WMG
TR12 6LY G4WIA
TR12 6PB G4WKW
TR12 6QD G7FOT
TR12 6QL G3UYN
TR12 6QR G3TBJ
TR12 6TX M3HJU
TR12 6UB G0XPD
TR12 6UW G6ZKC
TR12 7AH M0ADA
TR12 7AS G7RAU
TR12 7BJ M1ERA
TR12 7BJ M1ERD
TR12 7BW G3PLE
TR12 7BZ M6NIK
TR12 7DN G4CDY
TR12 7DN G4YQP
TR12 7DS G8GLL
TR12 7DS M0BNZ
TR12 7DY 2E0TDP
TR12 7DY M0TDP
TR12 7DY M3IXK
TR12 7DY M3MNQ
TR12 7DZ M7WKS
TR12 7EJ G0UPP
TR12 7EQ M6WOH
TR12 7HZ 2E0ZSY
TR12 7HZ M3SYV
TR12 7HZ M6RLQ
TR12 7JD M0CNZ
TR12 7JH G1ZPC
TR12 7JJ M6NDK
TR12 7JL G0GUO
TR12 7JZ G6WPR
TR12 7LF G0XAO
TR12 7LU G3NJV
TR12 7NN G3KJK
TR12 7NU 2E0LLO
TR12 7NU G3WGZ
TR12 7NU M0SOU
TR12 7NU M6LZP
TR12 7PA G1JHN
TR12 7PE M3VJC
TR12 7PQ G6IWK
TR12 7QR 2E0RER
TR12 7QR M0RRX
TR12 7QR M6RER
TR12 7QW G0HWU
TR12 7RL 2E0NZT
TR12 7RL M6NZT
TR13 0BY G4JGX
TR13 0DT G6FQP
TR13 0FB G8END
TR13 0HE G8ZVM
TR13 0LD G0FKX
TR13 0PE 2E0VJB
TR13 0PE M3VJJ
TR13 0PQ M0BSI
TR13 0PR G4LOH
TR13 0PT G4UIT
TR13 0PU M6WGF
TR13 0RY M6WFV
TR13 0SN G4GFY
TR13 8BG 2E0ABU
TR13 8BP G0HFA
TR13 8ER M0JMF
TR13 8JH M7BGO
TR13 8JZ G3KGP
TR13 8LR M7FNL
TR13 8LU 2E0LDM
TR13 8LU M3YAX
TR13 8NT M7DYR
TR13 8PQ G0FGX
TR13 8PQ G1OZV
TR13 8PY G0UDB
TR13 8QH M0AFJ
TR13 8QJ G1ZPJ
TR13 8RU 2E0WVW
TR13 8RU M6WIW
TR13 8RX M0BUA

TR13 8SR G7VFA
TR13 8WF 2E0CTD
TR13 8WH G8IKK
TR13 8XQ M7UCB
TR13 9AW G4WZH
TR13 9BL G4ZYO
TR13 9LR 2E0KTW
TR13 9LR M0EMW
TR13 9PB G7MSN
TR14 0AR M6LDW
TR14 0AS M3WRM
TR14 0AT M7FIQ
TR14 0EQ G0ARL
TR14 0ET M6REJ
TR14 0EY G8SEE
TR14 0HD 2E1HSR
TR14 0JF G4KES
TR14 0JX G4POT
TR14 0LL 2E0PCR
TR14 0LL M6PCB
TR14 0LL M6PUQ
TR14 0NQ M0DUY
TR14 0QE G8WWW
TR14 7AW G4BHD
TR14 7BP M3UML
TR14 7DX M7YYZ
TR14 7EB 2E0UCR
TR14 7EB M6UCR
TR14 7HB G6IZK
TR14 7HN G0WYU
TR14 7JN G0PGJ
TR14 7PE 2E0GLJ
TR14 7PE M6NZK
TR14 7PH M7AJT
TR14 7RN M0JIU
TR14 7RR M6ZXZ
TR14 7SU G8YWJ
TR14 7TT G0ILI
TR14 7UF G6ZAM
TR14 7XN G4BHC
TR14 8AW 2E0BOI
TR14 8AW G4ZTZ
TR14 8AW M0GIY
TR14 8AW M3UWW
TR14 8JB G7GIG
TR14 8QF G1HFE
TR14 8QF G1LNA
TR14 8RQ G4XMA
TR14 8RQ M0RSJ
TR14 8ST G8ZDS
TR14 8TW M6ASJ
TR14 8UP G4ZUI
TR14 8UP G6EON
TR14 9AA G4YYM
TR14 9AT G1BAX
TR14 9DE G4DEO
TR14 9ER G4YYH
TR14 9JJ 2E0LXB
TR14 9JJ M0LXY
TR14 9JJ M7LXB
TR14 9JT G4ZKN
TR15 1AX 2E0BON
TR15 1AX 2E0NKT
TR15 1AX M3XNA
TR15 1AX M6TRC
TR15 1EP M6LGD
TR15 1EW G4LOG
TR15 1JB 2E0XES
TR15 1JB M7EEE
TR15 1NX G8IXN
TR15 1QL 2E0ISU
TR15 1QL M6ISU
TR15 1QR G6XKO
TR15 1SE G0FKI
TR15 2BN 2E0RBR
TR15 2BN M3WOY
TR15 2LL 2E0SCQ
TR15 2LL M6SCV
TR15 2NY G0FHT
TR15 2RJ G6HGR
TR15 2TP G7VIR
TR15 3AB M7GUK
TR15 3AW G3PUQ
TR15 3AW G4DBZ
TR15 3BP G0IDH
TR15 3BZ G4GDU
TR15 3DF G6KTB
TR15 3HJ G3JQK
TR15 3JQ M3YVY
TR15 3NJ M3IJO
TR15 3NW M7HIL
TR15 3TS G0MTY
TR15 3TT 2E1FNR
TR15 3UD 2E0EKJ
TR15 3UD M0WDO
TR15 3UD M6WDO
TR15 3YG 2E0TCV
TR15 3YJ G7DUC
TR151PR M7KSY
TR16 4AQ M6KIR
TR16 4DY G6HKQ
TR16 4EB G0FHW
TR16 4SG G7ADP
TR16 4SH G4CRC
TR16 4SH G7VOH
TR16 4SH M3YKT
TR16 5AA G4WSH
TR16 5AH M6JCJ
TR16 5DT G4WQL

TR16 5DT M0KKE
TR16 5JL 2E0BOJ
TR16 5JL M3TNB
TR16 5JY M0IAF
TR16 5LT G3MZN
TR16 5LT G7RIO
TR16 5PN G0FKF
TR16 5PT G0CAM
TR16 5QA G4ZCT
TR16 5RJ 2E0JGW
TR16 5RJ M6WBB
TR16 5SH 2E0AHB
TR16 5SH M7BOE
TR16 5TQ G0MYR
TR16 6DQ G1WLN
TR16 6HN G7WJJ
TR16 6HN M0DWS
TR16 6HP G4ZKH
TR16 6LR G8WNQ
TR16 6LS G8NXD
TR16 6NH M3RRN
TR16 6NT G0VZB
TR16 6PG 2E0JWB
TR16 6QL M6EVY
TR164AQ G1NRF
TR164LU 2E0FQU
TR164LU M7ADN
TR165QZ G0FIC
TR166NT M7CKX
TR166PN M3URW
TR17 0BL M6ILJ
TR17 0BP 2E0ART
TR17 0BP M6KRL
TR17 0HF M0DBA
TR18 2AH 2E0CJS
TR18 2AH M0PZR
TR18 2AH M6ANI
TR18 2DN G7SBP
TR18 2EX M6GBO
TR18 2PL 2E1CCN
TR18 2QE G7CCS
TR18 2QE M0GOP
TR18 3BA G1JAH
TR18 3BB G3NRD
TR18 3LD G1IVO
TR18 3NA G3RID
TR18 3PD G8GRO
TR18 3PW G0WZC
TR18 3QL G8EJC
TR18 4BH G0PGB
TR18 5DQ G4BPJ
TR18 5LW M7OOI
TR18 5NG M0SBL
TR18 5NR 2E0MEI
TR18 5NR M6DZB
TR18 5QA M0ABZ
TR18 5QR M0SBD
TR18 5QW G0MWZ
TR184RY 2E0KWL
TR184RY M6MVH
TR19 6BB 2E0CAQ
TR19 6BB M0LDQ
TR19 6BB M6SUV
TR19 6LX G3XRJ
TR19 7AQ G4AMT
TR19 7BG M6YHQ
TR19 7BS G0THQ
TR19 7DZ G3OYB
TR19 7ET G4YAF
TR19 7EU M0RMM
TR19 7EU M6RIU
TR19 7HB G7SBN
TR19 7RN G4LYG
TR19 7SP M1BKS
TR19 7SY G4XCV
TR19 7TJ 2E0FVC
TR19 7TJ M6FVC
TR19 7UT G3PZN
TR2 4AU M6YEO
TR2 4BE G6OJX
TR2 4BS G3ZZV
TR2 4BY M6JDC
TR2 4DS G4JZA
TR2 4JX M6HFK
TR2 4NY G7KMP
TR2 4PQ G3VWK
TR2 4TD 2E0BBI
TR2 4TS M6NGA
TR2 5AR G0REA
TR2 5DD G4NBF
TR2 5DD M0TOP
TR2 5JP G0JWV
TR2 5JX G0ENZ
TR2 5NP G4EIK
TR20 8AJ G8ARM
TR20 8AJ M0BMX
TR20 8BQ G4SEN
TR20 8DJ G3YSN
TR20 8DS G1NWO
TR20 8DS M3ABQ
TR20 8JY G7TBJ
TR20 8UL G7TJD
TR20 8XA G1GKF
TR20 8XT 2E1GNN
TR20 8XT G8HZN
TR20 8XT M3PHS
TR20 8XW G0AIX
TR20 9AR G7SUT
TR20 9BW G3ZPJ

TR20 9BX M3ZDK
TR20 9HN G4SOK
TR20 9JR 2E0FOE
TR20 9JR M0TIK
TR20 9JR M7CHR
TR20 9LS 2E0ROO
TR20 9LS M6YKC
TR20 9LY M0PNZ
TR20 9LY M3PNZ
TR20 9PT G4ZLF
TR20 9SX G3ZPW
TR20 9TE M0TNT
TR21 0JD M1IOS
TR21 0JJ 2E0PWL
TR21 0JJ M0PWL
TR21 0JJ M6PWL
TR26 1AX G8JAB
TR26 1BL G3KDP
TR26 1ER G3YGM
TR26 2DU G1UAY
TR26 2SP M3KRZ
TR26 3AL G4YQS
TR26 3HL G4DZJ
TR26 3LE G0ISY
TR26 3LY 2E0PDH
TR26 3LY G5XDX
TR26 3LY G8ROC
TR26 3LY M6AOK
TR263AR M7FXR
TR27 4AE M0XEK
TR27 4AG G4TXD
TR27 4EX G0PZR
TR27 4EX M0WIN
TR27 4JN G4RVP
TR27 4JQ G1ZOB
TR27 4LY G1CHM
TR27 4NA 2E0WVS
TR27 4NJ G4KYO
TR27 4PJ G1LYC
TR27 4PS G3UCQ
TR27 4QT G4HUG
TR27 4RL G8UXD
TR27 5AB G4GKY
TR27 5AN G0TRJ
TR27 5DT M6EEH
TR27 5ET G6LGM
TR27 5HA G3YNK
TR27 5HL 2E0JYA
TR27 5JD G4HFI
TR27 5JD G7GUA
TR27 5JF G6YVD
TR27 5JL M7CWL
TR27 5LJ M6FWV
TR27 6BA G8UTY
TR27 6DJ 2E0HCM
TR27 6DJ 2E0RUX
TR27 6DJ M0NXF
TR27 6DJ M6NXF
TR27 6DJ M7AMR
TR27 6DJ M7GBQ
TR27 6EX G1FOA
TR27 6HJ G4TXE
TR27 6HZ M3KGE
TR27 6PJ G4KYH
TR3 6BB G3PPT
TR3 6BH 2E0WSC
TR3 6BH M6KVU
TR3 6BJ G3UFJ
TR3 6BJ M5GUS
TR3 6BQ G1OEQ
TR3 6DB 2E1ERJ
TR3 6EB G3MRT
TR3 6EN G0CQJ
TR3 6EN G6MOT
TR3 6ES M6ZXQ
TR3 6FW M7WHI
TR3 6LJ G8PBI
TR3 6LN 2E1FNY
TR3 6LN G4LJY
TR3 6PQ G0NNR
TR3 6PQ G0PGX
TR3 6PW G3NHL
TR3 6SP 2E0KLT
TR3 6SP M6CTK
TR3 6UA M7FXI
TR3 7AD 2E0BUD
TR3 7AD G3WPP
TR3 7AD M3AKH
TR3 7AT G0AWR
TR3 7FG G3OCB
TR3 7JP G0UWO
TR3 7NW G0WGV
TR3 7RB M3ZAY
TR4 8DZ M6YFT
TR4 8ER M6IEW
TR4 8QL G0DHT
TR4 8QL G4ZZZ
TR4 8RJ G8RBS
TR4 8SU 2E0PCO
TR4 8SU M6PCX
TR4 8TS G7WER
TR4 8TS M0AWP
TR4 8TS M0BDH
TR4 8TT G0LIA
TR4 9FN M7RJG
TR4 9FW G0SDT
TR4 9LT G4OOL
TR4 9RB G8ZYR
TR5 0TU G6WMT

TR5 0XQ G3TDM
TR50TR 2E1ADT
TR6 0DZ G1ZEI
TR6 0EF G0GQJ
TR6 0EY G3TYA
TR7 1AS M3AFS
TR7 1PF G4SSL
TR7 1QQ G0DJL
TR7 1TY G4ADV
TR7 1TY M0BFB
TR7 2DE 2E0NWR
TR7 2DE M0RHW
TR7 2DG G0LCJ
TR7 2DX M6KWU
TR7 2EJ G3WUA
TR7 2JN G8SCY
TR7 2LE G3JVN
TR7 2QE M0CJE
TR7 2RB G4VPJ
TR7 2RH M1AWX
TR7 2RW G0FLU
TR7 2RY G3XPY
TR7 2TB G0NWS
TR7 2TE G4WOI
TR7 3AE M3KUZ
TR7 3AE M3KVK
TR7 3AJ G7KRH
TR7 3AW G0HEW
TR7 3BN G6CEP
TR7 3DT M7HGK
TR7 3EB G7CRN
TR7 3EH M7PML
TR7 3EJ G8GOR
TR7 3FB G7VFV
TR7 3HW 2E0OJH
TR7 3JT G6AFA
TR7 3LB 2E0DXL
TR7 3LB M6HBQ
TR8 4AW G4CYI
TR8 4HJ 2E0HKP
TR8 4HJ M7AWN
TR8 4LF M0AIQ
TR8 4PL 2E1FOX
TR8 4PL G7KFQ
TR8 4PL M3KFQ
TR8 4PW 2E0CWS
TR8 4PW M0KDR
TR8 5DB G3VMK
TR8 5HH G8YOE
TR9 6AS G4RMG
TR9 6ER 2E0KTD
TR9 6ER G1PGV
TR9 6ER G4PSU
TR9 6ER M3KTD
TR9 6ER M3PSU
TR9 6FH M3ZLO
TR9 6NB M6DPH
TR9 6PD G4TGG
TR9 6PT M3AEJ
TR9 6TA 2E0NXE
TR9 6TA 2E0OWE
TR9 6TA 2E0XIT
TR9 6TA M7BAF
TR9 6TA M7DTF
TR9 6TA M7VOX
TR9 6TW G0PNM
TR96RH 2E0HIR
TS10 1JA G0HPQ
TS10 1QN M7WTB
TS10 2BP M0BYV
TS10 2DG 2E1DMZ
TS10 2DL G0MBU
TS10 2EQ G7CRA
TS10 2HW G6ZGC
TS10 2JZ G3UZB
TS10 2LJ M0BWO
TS10 2LQ G6IVD
TS10 2LT G0AJZ
TS10 2RU M6DST
TS10 2RU M6EMZ
TS10 3LN 2E0GGQ
TS10 3LN M6GGQ
TS10 4AG G3NVP
TS10 4HD M6SEG
TS10 4NJ G3VZO
TS10 4PS M0ULB
TS10 4PS M3MII
TS10 4PS M6TST
TS10 4SG M3PYS
TS10 5DP M3JGN
TS10 5EB G0VLK
TS101LS 2E0KWB
TS104JQ M7ALC
TS11 6BD G3DAV
TS11 6BE G0BIA
TS11 6BE G3PSG
TS11 6DF G4WNA
TS11 6DY 2E0FJW
TS11 6JX G4NUO
TS11 6NW G0VZT
TS11 7HJ M0RIG
TS11 7HJ M6RNE
TS11 7HT G0NUT
TS11 8AG G8GFA
TS11 8AU G4RHX
TS11 8BP G1GMF
TS11 8BT G4GCU
TS11 8DJ M0CIC
TS11 8DU G4CRS

TS11 8DU G4OLK
TS11 8HT M7DUM
TS11 8JJ M6YBD
TS118EB M1SPY
TS12 1AL G8JLA
TS12 1DP G4IJO
TS12 1FE 2E0EGR
TS12 1FE M0XGR
TS12 1JP 2E0GLA
TS12 1JP M3NUH
TS12 1PD M6TTK
TS12 2DQ G3XAG
TS12 2JZ G4FCU
TS12 2NJ M3RNO
TS12 2QX G6HHK
TS12 2TJ G3KBI
TS12 2XG G0EWT
TS12 2YN M0BTL
TS12 3AP G4KIR
TS12 3AW 2E1FFZ
TS12 3AW M3AOC
TS12 3DX M6KYC
TS12 3EE G0RJA
TS12 3EW M6KAV
TS12 3LE G0FSG
TS12 3LS G0PZM
TS13 4BU G4AWK
TS13 4DF G4WUS
TS13 4JB G6DIR
TS13 4NX M1FCZ
TS13 4PB G0SVJ
TS13 4UG G4JCS
TS14 6DJ G4HEB
TS14 6LQ 2E0UFT
TS14 6LQ M7AHN
TS14 7BY M0NDT
TS14 7DP G7ONL
TS14 7DZ 2E1CIO
TS14 7FH G8HUY
TS14 7LE 2E0FTU
TS14 7LE G0UOK
TS14 7LE M0NYM
TS14 7LE M7UOK
TS14 7LQ M0STF
TS14 7LZ G1AAG
TS14 7LZ M1MVN
TS14 7NB G4HRU
TS14 7PE M0MJT
TS14 7PR G7TXV
TS14 8DL G6MVQ
TS14 8HA 2E0ODJ
TS14 8HA M7DRJ
TS14 8JG M6GSX
TS14 8JX G0SUC
TS14 8JY G1YHP
TS14 8LD G7ION
TS14 8LN G4YMB
TS14 8LT G7GSM
TS15 0BA 2E0ECD
TS15 0BA M6ECD
TS15 0HT G4JLJ
TS15 0JQ G3BPP
TS15 9EZ M1AFU
TS15 9JQ G4ZXT
TS15 9LB M0DTS
TS15 9ND G7RAJ
TS15 9NL G8KIK
TS15 9NL G8MBK
TS15 9NQ G4JJM
TS15 9NT G6LIJ
TS15 9RZ G3VGZ
TS15 9RZ G4EEH
TS15 9RZ G8AOE
TS15 9TG M3KNF
TS15 9TJ 2E0HEQ
TS15 9TN G0TIJ
TS16 0AQ G6KQJ
TS16 0JB G7NQZ
TS16 0LQ M7DUD
TS16 0NB 2E0BUI
TS16 0NH 2E0ICC
TS16 0NH G3ICC
TS16 9BA G7SYI
TS16 9EH G3FDW
TS16 9HP G3RUG
TS16 9HU 2E0XPH
TS16 9HU M0XPH
TS16 9HU M3XPH
TS17 0FT M7JJI
TS17 0LT M0AVW
TS17 0NL G0IBW
TS17 0TB M3VPX
TS17 5BG G6ICZ
TS17 5BS 2E0PWU
TS17 5BS M0PUK
TS17 5BS M7PUK
TS17 5DJ G1HSH
TS17 5HQ 2E0MGV
TS17 5HQ M6BHQ
TS17 8JB M7EZQ
TS17 8LX G6CKH
TS17 9AF M7VOT
TS17 9AN M3LBD
TS17 9QJ G0JOC
TS18 1AX G0BKC
TS18 1AX M6BKC
TS18 1JY G0MLB
TS18 2FN G8NMK
TS18 4AZ M6KIT

Postcode

TS18 4JA G3WWG
TS18 5AR M0HOK
TS18 5LH G4JXR
TS18 5NH 2E0MLX
TS19 0BD M6MMX
TS19 0DE M3FNA
TS19 0ER 2E0KBD
TS19 0ER M0KBD
TS19 0ER M6PII
TS19 0RA G3UFV
TS19 0SL G4YOV
TS19 0XL M7DRR
TS19 7AP M3EVI
TS19 7JL G0ELC
TS19 7JT G1ZRS
TS19 7LY G6MPE
TS19 7NF M7CNL
TS19 7SH M0DNN
TS19 8EE M0KMU
TS19 8ES 2E0DBZ
TS19 8ES M3HSJ
TS19 8RF 2E0RDO
TS19 8SZ M3ASC
TS19 8XF G8KSA
TS19 8XF M0DSC
TS19 9EJ M3VDV
TS19 9HJ 2E0TPR
TS19 9JS M3YNN
TS20 1BD M6MOQ
TS20 1JG M3XAC
TS20 1JQ G7CLG
TS20 1JU G4XXG
TS20 1JU M0CYD
TS20 1LE M3RFO
TS20 1LG M0NBK
TS20 1PZ G7FFM
TS20 1RT G8VPY
TS20 1SB G8ILB
TS20 1SJ G7CLH
TS20 1SJ G7SUQ
TS20 1SZ M3JOA
TS20 2EX G0BZC
TS20 2JJ M7PKC
TS20 2SP G4LIM
TS202PQ M7GKL
TS21 1DZ G4DXP
TS21 1HT G8LTD
TS21 1LQ G3NBL
TS21 2AN G0GWL
TS21 2DS G0WXC
TS21 2EZ G0SJU
TS21 2JY 2E1JRB
TS21 3DN 2E0IIZ
TS21 3DN M0IIZ
TS21 3DN M0WXP
TS21 3DN M7SJM
TS21 3EU M3PGL
TS21 3LT G8ZIA
TS22 5HA G1RYM
TS22 5LP G7NRO
TS22 5UZ M6OHQ
TS23 1DL G4PVN
TS23 1EE G7VOT
TS23 1HW M6IDS
TS23 1HX G7CUO
TS23 1QN G3YVY
TS23 2AN G0HJM
TS23 2DJ M7BCN
TS23 2HX G6LDJ
TS23 2PJ G7LOA
TS23 2QB M6MYB
TS23 2QH M0ASN
TS23 2RF M3UBL
TS23 3GP M6SWI
TS23 3HG G8OKS
TS23 3SP M7CPO
TS23 3SY G0JON
TS23 3UA G4MYN
TS23 4AW G6NXP
TS24 0DX 2E0HLC
TS24 0DX M6AXG
TS24 0DY M7SLE
TS24 0UF M6KQJ
TS24 0XF M0WFA
TS24 7HY 2E0YSF
TS24 7HY M6IEO
TS24 7NW M6NUF
TS24 8HH G7VGM
TS24 8RB G3IJV
TS24 9BQ G4VCJ
TS24 9BQ G6CGY
TS24 9QY M7EEM
TS25 1AX 2E0RKZ
TS25 1AX M7TSZ
TS25 1GF M7DSZ
TS25 1GQ M6MQE
TS25 1HF M6TJI
TS25 1HL M7MJG
TS25 1LP M6FQP
TS25 1RN M6BHN
TS25 1RW G0SBX
TS25 1RY 2E0HPI
TS25 1RY M6HPL
TS25 2LA G1DFZ
TS25 2LD 2E0WSN
TS25 2PY M3NFQ
TS25 2QG G4NRS
TS25 2RD G7VGO
TS25 2RG G4ZCN
TS25 3DP G3UUF
TS25 3QR M6OGC
TS25 3QY M1BNK
TS25 4NE M7PAG
TS25 4NZ G1KBE
TS25 5HX G3NWY
TS25 5HZ G0VGB
TS25 5LB G1GTP
TS25 5NG M6OTB
TS25 5PA G1ETQ
TS25 5RQ G0FZZ
TS26 0NW 2E0EJW
TS26 0NW M6EVW
TS26 0PJ G3NUA
TS26 0QY M6NLJ
TS26 0XG 2E0DGA
TS26 0XG M0VRW
TS26 0XG M6EOK
TS26 0ZL 2E0MDN
TS26 0ZL M0MBZ
TS26 0ZL M6MEJ
TS26 8AY G1KUO
TS26 8EZ M0GIZ
TS26 8LG 2E0SWK
TS26 8LG M7SWK
TS26 8ND G7SNJ
TS26 9BJ M0HXK
TS26 9ES M0BKF
TS26 9LY 2E0XVK
TS26 9LY M0NXA
TS26 9PN M0HPL
TS26 9PR G4SHJ
TS27 3QR G7JXY
TS27 4AB 2E0UXC
TS27 4AB M6MKM
TS27 4AX M0BZB
TS27 4DF M0DEL
TS27 4EB M1OJS
TS27 4EB M6DLC
TS27 4RT G0GKO
TS28 5AG G0HZK
TS28 5BP M6PIT
TS28 5BT G0HJX
TS28 5EJ G4XPP
TS28 5GA 2E0HQU
TS28 5GA M7LWR
TS28 5JE G0GWD
TS28 5JF M6LAZ
TS28 5JN M7DSQ
TS28 5JW M6RDQ
TS285EF M7DTZ
TS29 6DA G0MHC
TS29 6DZ 2E0PSH
TS29 6DZ M0HTI
TS29 6DZ M6DVV
TS29 6EE M0ZEM
TS29 6EF M3SHB
TS296HE M6GDO
TS3 0NN 2E0UTB
TS3 0NN M6SOV
TS3 0RL 2E0JKD
TS3 6PE M6CFD
TS3 7DR 2E0BJK
TS3 7NQ M6IBH
TS3 7PU M6TNA
TS3 7RX 2E0KOM
TS3 7RX M0KOM
TS3 7SL 2E0FJP
TS3 7SL M3FJP
TS3 8LX G4IIN
TS3 8NX G4OOK
TS3 8QX M7OLP
TS3 9HD M6MQL
TS3 9NT M6MFC
TS3 9QW M7SEH
TS4 2DN M6OJF
TS4 2HG M6LCY
TS4 2RY M3VHY
TS4 3HJ M6RTA
TS4 3HR M6KRI
TS4 3HU M6WCC
TS4 3HX G3RGB
TS4 3QS M0AHD
TS5 4AJ G4UTV
TS5 4EH M7JEP
TS5 4HG M0SMA
TS5 5BN M3CWV
TS5 5JG M7WBF
TS5 5JP M3YRZ
TS5 5NQ G4HSB
TS5 6DU 2E0ADR
TS5 7QX G4VZC
TS5 8BT G1NTL
TS5 8DP G4PPS
TS5 8DR G3KXV
TS5 8EB G8WKZ
TS5 8NT G4IJM
TS5 8NT G8HDM
TS5 8RE M0GQV
TS5 8RY M6LLM
TS55LD M0BJX
TS6 0BG 2E0NUL
TS6 0BG M0JUN
TS6 0BG M6WIX
TS6 0BN G0NYZ
TS6 0BQ G7TWU
TS6 0BS G4MCF
TS6 0BZ M7AAT
TS6 0JD G8MKG
TS6 0PL G1DAT
TS6 0PP M0BZA
TS6 0QQ G0FXR
TS6 0QQ G0LWE
TS6 0QQ M3TVN
TS6 0RW M6CLI
TS6 7EZ M0HMS
TS6 7HG M7ARL
TS6 7ND 2E0CNX
TS6 7ND M3ITL
TS6 8AF G1CCM
TS6 8EG M7DMV
TS6 8EH G7FVH
TS6 9HX G1CLT
TS6 9LY G4ORQ
TS6 9QW G0PDK
TS6 9QW M7WEP
TS6 9SZ G7NQX
TS7 0EZ G7WEN
TS7 0JL M0DIQ
TS7 0LB 2E0LGB
TS7 0LB M0LGB
TS7 0LT G1IGW
TS7 0QL G8YDC
TS7 8AN G7UMA
TS7 8SE 2E0CPK
TS7 8SE M0HHC
TS7 8SE M6BKL
TS7 9BB 2E0PEL
TS7 9BB M0SMP
TS8 0RU G0AOO
TS8 0SU G7RXB
TS8 0UF M7OGB
TS8 9AB G4POD
TS8 9BU G4ZML
TS8 9FQ G7PTC
TS8 9HH G0BQP
TS8 9LJ G0UTP
TS8 9LJ M6UTP
TS8 9SL G7IZC
TS9 5AG G4HWV
TS9 5BU G0TYM
TS9 5DF G7UXH
TS9 5EL M0CSD
TS9 5HU G0UYG
TS9 5HX G7RNQ
TS9 5PQ G4OIW
TS9 6BX M6PHV
TS9 6DW 2E0FAU
TS9 6DW M3FYV
TS9 6EN M6BHP
TS9 6JF G7TFG
TS95BE G0MBV
TS95BE M7KEZ
TS96NQ M6PPS
TW1 1AG M0ICI
TW1 1PR M7GOB
TW1 1PY G4RBR
TW1 1RS M6MDN
TW1 2AX 2E0JAB
TW1 2DD G4CXZ
TW1 2DF G4CBD
TW1 3AU G6CIA
TW1 4BQ G1TOB
TW1 4QZ G8DSU
TW1 4SW M0MDC
TW10 5DU 2E1DQM
TW10 6AB M0IRG
TW10 6DS 2E0RYS
TW10 6DS M0RYS
TW10 6DS M6RSY
TW10 6HS 2E0ERZ
TW10 6HS M0IUR
TW10 6HS M6VRE
TW10 7ED G1PHA
TW10 7EQ M3RHO
TW10 7LD M7AHU
TW10 7NQ G0SLI
TW10 7YG G8CYE
TW11 0BG G4LGO
TW11 0BQ G3GDZ
TW11 0DH G7PMV
TW11 8AS G3PPC
TW11 8DE G8BAJ
TW11 8JB G6RHV
TW11 8LT G1OPT
TW11 8QA M7MAE
TW11 8SH G8ZWA
TW11 9DW G8WGE
TW11 9HA G0DEH
TW11 9LN G8AUU
TW11 9PD G8DEL
TW11 9QS G0JME
TW11 9RL G3YJV
TW11 9RS G3ZLD
TW12 1DE M0LHR
TW12 1DW M0BYY
TW12 1HX G4PPE
TW12 2BL M6ZES
TW12 2JH G8SNV
TW12 2LU G3RCB
TW12 2NE M6JXD
TW12 2PZ G7DWM
TW12 2QG G4ZOU
TW12 2SR M0NAI
TW12 2TR G7OMA
TW12 3AU G4LDW
TW12 3BW G4UXB
TW12 3JX G5YZI
TW13 4HX G1KCU
TW13 4QG M3XHB
TW13 5DJ G7GFQ
TW13 5HE 2E0GJQ
TW13 5HE M7AMJ
TW13 5JS 2E0HMR
TW13 5JS M7HMR
TW13 6LZ G0EEZ
TW13 6PE G0KXG
TW13 6PX G1YVZ
TW13 7AA M7LAG
TW14 0AH 2E0RUS
TW14 0AH M3SQS
TW14 0JY G1TFM
TW14 9ED G1LKL
TW14 9JE G0DEO
TW14 9LN M7LCS
TW14 9LW G7UWS
TW14 9QP G4ALA
TW14 9XB M0DPF
TW14 9XG G7OXK
TW15 1AB G8FVJ
TW15 1DG G0CPF
TW15 1DP G1IBS
TW15 1DP G4ZVU
TW15 1DR 2E0UFM
TW15 1DR G4ESG
TW15 1DR M6YON
TW15 1HF G0ONA
TW15 1LT M7FAB
TW15 1PQ G0RJE
TW15 1PW G0NBJ
TW15 1PW G0OSX
TW15 1QE G1WWB
TW15 1SQ G8ZZG
TW15 1UT M6HKS
TW15 2AP G8CAH
TW15 2BE M7TIE
TW15 2BN M0IOE
TW15 2EP M3YXE
TW15 2JB 2E0NOC
TW15 2LP G3ZFT
TW15 2LU G8LPA
TW15 2LU M0LPA
TW15 2RF M1CSU
TW15 2SL M5CBR
TW15 2TD G1ZDT
TW15 2TD G7AYP
TW15 3PF M0AHS
TW151LW 2E0GWZ
TW151LW M7GTZ
TW16 5ED G7EPF
TW16 6HF G3HTC
TW16 6HF M0CFR
TW16 7DD M7FGC
TW16 7NA G4YAS
TW16 7NL G7PAK
TW16 7PL 2E0NBE
TW16 7PL M0NCE
TW16 7PL M6FUF
TW16 7QU M6NPF
TW16 7TL 2E1AOF
TW16 7UA G0HYT
TW16 7UA G7EAR
TW17 0DN G3UPW
TW17 0EN G0VDZ
TW17 0EN M0KIO
TW17 0JB G6IFR
TW17 0QT M7NDS
TW17 0RP G6KLH
TW17 0SH G4XGI
TW17 8AY G4GNS
TW17 8BT G7CSS
TW17 8EU G0LGA
TW17 8QQ M6GAN
TW17 8QS M0PMY
TW17 8RR G7MII
TW17 8RX G4CKQ
TW17 9DQ G8BQZ
TW17 9LL M0MSV
TW17 9LL M7NAN
TW18 1DG G0MIJ
TW18 1DJ G0JSP
TW18 1EE G1SXB
TW18 1NE G7ECE
TW18 1PW M7IGG
TW18 2AP G4HKS
TW18 2AP M6GIQ
TW18 2AP M6SUC
TW18 2DD G6PVC
TW18 2LW M6VHB
TW18 2QE G3ZLR
TW18 3EE 2E0VBX
TW18 3EE M6VBX
TW18 3HB M0LZU
TW18 3HH G0OEK
TW18 3HQ M6NVG
TW18 3NQ G4GSC
TW18 4NR G3NTM
TW18 4TW M0HMI
TW19 5NF M6ZHF
TW19 6AX G4AKA
TW19 6BT G4UXG
TW19 7AJ M3OPT
TW19 7JE G0TID
TW19 7JE M3CJA
TW19 7JE M3EMN
TW19 7LF G4MTH
TW19 7LG M6CTI
TW2 5BY G1NSQ
TW2 5BY G6AWM
TW2 5JH M1GAR
TW2 5JJ G7TOF
TW2 5JP G1FKS
TW2 5LS G0EZL
TW2 5PD M0DLL
TW2 6AA G2AIW
TW2 6BL G8WRL
TW2 6EB M3WQG
TW2 6EJ G3ZSB
TW2 6HW G1LVZ
TW2 6LQ M7EXJ
TW2 6SA G7GZC
TW2 6SP G6GVF
TW2 6SR 2E0WEJ
TW2 6SR M0GBC
TW2 6SW G6VUG
TW2 7BL 2E0WMD
TW2 7BL G0TSU
TW2 7BL M0DMJ
TW2 7BL M3ESQ
TW2 7DJ M7MFF
TW2 7DT G7GAP
TW2 7EA G1OVO
TW2 7EU 2E0DZM
TW2 7EU M0IUT
TW2 7EU M6RZP
TW2 7EX G4OQN
TW2 7JE G0OFN
TW2 7NH G0MRF
TW2 7NP M0DCW
TW2 7SN G6LKH
TW2 7SN G8RHZ
TW20 0DF M1DWV
TW20 0RZ M0HJY
TW20 0RZ M6BRJ
TW20 8AN G0LSE
TW20 8BS G0NIF
TW20 8BS G7AOU
TW20 8ET 2E1DOG
TW20 8ET M6DOG
TW20 8NL G6PVV
TW20 9LP 2E0DMQ
TW25JT M3JXY
TW3 1YH G4CWE
TW3 2EZ M7PAR
TW3 4NL M0RJZ
TW34AD M0IMQ
TW4 5BB G1EIH
TW4 5EW 2E0BLQ
TW4 5PF G0RHB
TW4 6AG G3AHE
TW4 6NA G1CRN
TW4 7AN M0JCY
TW4 7JG M0WBC
TW4 7JQ G0WMD
TW4 7JQ M0HIC
TW4 7NH G0IIK
TW4 7NU G7GBE
TW4 7RA G0USK
TW5 0ND G6YTX
TW5 0NF G4IXB
TW5 0PZ M0TCE
TW5 9AW G1JRW
TW5 9EX G0VLQ
TW7 4AJ G0LUH
TW7 4LS M0EMD
TW7 4PQ G4DUO
TW7 5HB G8SJO
TW7 5LU M6OGP
TW7 6AD G3LWR
TW7 6GP 2E0EXI
TW7 6HW G4LTC
TW7 6HX G4PWS
TW7 6LF G4NJN
TW7 7BW M7CZM
TW7 7BZ M6XBR
TW7 7DJ M3HVO
TW7 7NL G6AGA
TW8 0FU M0BYK
TW8 0PL G3XKV
TW8 9PT G6SUK
TW8 9RD 2E0JDC
TW8 9RD 2E0YVR
TW8 9RD 2E0ZBW
TW8 9RD G8MWD
TW8 9RD M3YVR
TW8 9RD M3ZBW
TW9 1AX M3NAO
TW9 1LL 2E0SWE
TW9 1LL M6BIV
TW9 2DZ 2E0CEN
TW9 2DZ M3ZEI
TW9 2HA G1CRT
TW9 2TJ G0ACD
TW9 3AY G0LYJ
TW9 3BD 2E1GHZ
TW9 4AS G0OOI
TW9 4AS G3CIK
TW9 4EE G0OHW
TW9 4JN M3YBU
TW9 4JP M7AKU
UB1 1PZ G7CSD
UB1 2SA M0GBO
UB1 2UP G1YES
UB1 3JP G3KLK
UB1 3NT G1YRY
UB1 3QD G6ESM
UB1 3QJ 2E0KDS
UB1 3QJ M7GHJ
UB10 0BS G6KNK
UB10 0DN G1ECY
UB10 0HH G1DYR
UB10 0HP G6IRY
UB10 0HW 2E0CXE
UB10 0HW M0HQR
UB10 0HW M6CKU
UB10 8DY 2E0JTY
UB10 8ED G4BVF
UB10 8HS G4CTD
UB10 8HU M7XAN
UB10 8LS G1PAF
UB10 8QA G8LIU
UB10 8TA G6LEY
UB10 8TJ M7GCR
UB10 8UF M3MST
UB10 8UF M3WCR
UB10 9EL 2E0FXE
UB10 9EL M6XXH
UB10 9HZ M6ODL
UB10 9LN M7MPD
UB2 5AN G7IVN
UB2 5HN G6VAE
UB24FZ 2E0XBW
UB24FZ M0XBW
UB24FZ M6XBW
UB3 1PZ G0LQW
UB3 1ST G3SUN
UB3 2FJ M6FPP
UB3 2JE G8HOU
UB3 2NF M7LHR
UB3 2QX G6XDI
UB3 2TP G8GYP
UB3 2TQ G6STI
UB3 3PA G1VTS
UB3 3PP G7JSW
UB3 4AD G0EFS
UB3 5ET G3PKR
UB4 0AE G0CHR
UB4 0DZ M6XJS
UB4 0EF G0RGL
UB4 0EF G0VGY
UB4 0JH G1LLQ
UB4 8ET G4OHQ
UB4 9QT M6GQU
UB4 9RB G1LVR
UB4 9YF G0KYN
UB5 4AQ 2E1IDC
UB5 4NL G1XLT
UB5 4SE M7OMG
UB5 4SG 2E0JGP
UB5 4SG M6JGP
UB5 4SW M6WIK
UB5 4TL M6DTC
UB5 4TS 2E0OZK
UB5 5BX G1DLA
UB5 6AR G1KOX
UB5 6EU G0SLH
UB5 6HP G7GDC
UB6 0BH G1VSM
UB6 0BH M0DBI
UB6 0HQ G3PSC
UB6 0JZ G0MXY
UB6 0QE 2E0LPZ
UB6 0QE M0LPZ
UB6 0QE M6YUG
UB6 0RF G4AWM
UB6 0SS M3CGO
UB6 7AD G4FJX
UB6 7HR G6PGJ
UB6 8AJ 2E1HFW
UB6 8LN M1AZQ
UB6 9BY G7ABE
UB6 9HX 2E0ISN
UB6 9LS G1UBL
UB6 9NN G0JJQ
UB6 9NN G6UIT
UB6 9NT G0HRK
UB7 0DR M6IIW
UB7 0LB G4SSY
UB7 0LB M0ULE
UB7 0LE M0AAC
UB7 0LJ G1SUS
UB7 0LS 2E0LJR
UB7 0LS M6ZZV
UB7 7AN G0FFK
UB7 7AP 2E0TTW
UB7 7AP M6EJF
UB7 7AP M6UAF
UB7 7TZ M6BBR
UB7 8AW M0LTP
UB7 8BU M3OTR
UB7 8PF G6BSS
UB7 9DH M7MOW
UB7 9DR G0FFN
UB7 9DS M7FQV
UB7 9PE G4ZRC
UB8 1BL G8FKH
UB8 1BL M0DGB
UB8 1QX G3JVP
UB8 2NY 2E0CRC
UB8 2NY M0BKA
UB8 2NY M6CCR
UB8 2PQ G7NFR
UB8 2RQ M7JAG
UB8 2UL G4CVF
UB8 3EL M0HZV
UB8 3HS G7VHN
UB8 3HS G7VHO
UB8 3QU G6JHG
UB8 3SG M7IDP
UB8 3SR G0OLD
UB8 3WS G7LCD
UB9 4AG G4YEO
UB9 5DJ G0VOV
UB9 5HJ G7LVE
UB9 5PB G6FGJ
UB9 6AU M7MEX
UB9 6LZ M6TGI
UB9 6QB G3VXA
V1E 2DN G4NJT
W10 4BW M7MFD
W10 5TA M0TUR
W11 1EP M0IAK
W11 3BY M0SKN
W11 4BT M1DGK
W11 4HE M6EFG
W11 4JX M7WBX
W11 4SQ M5OOO
W11 4TG G8EZT
W12 0AP G7AAS
W12 0LP G3SEF
W12 0NX M0GPQ
W12 7BL G3IJL
W12 8JN 2E1IFA
W12 8NU M0SHH
W12 9LU 2E0AVB
W12 9TF M3NKO
W13 0AE G6GIU
W13 0ED G0PIN
W13 0HA 2E1YHZ
W13 0HA M7KRA
W13 0HA M7YHZ
W13 0HP G1ARF
W13 0JW M7FUH
W13 0NT G7JHZ
W13 0TF M0GHC
W13 8AJ 2E0PMP
W13 8AJ M6PRP
W13 8JZ G3RGP
W13 8NH M7SBN
W13 8PR 2E0JAJ
W13 9DT G0AIS
W13 9EN G8LVF
W13 9HS M0CPB
W13 9JY 2E0VBS
W13 9JY M6ZBS
W13 9QU G8VDQ
W13 9RA G6JEU
W13 9TN M0FZU
W13 9TY G1HGT
W14 0JX G4HTY
W14 8UW M0GCQ
W14 9BS G1EIG
W14 9HU M7IOS
W14 9JS M6ENW
W14 9ST G4TPM
W1A 6US 2E0IQQ
W1B 3DB M7GBN
W1H 7DP M0KYR
W1J 8PE M0CER
W1K5RX M0KBY
W1T 4TD M7OJO
W1U 6PS M7BSL
W1W 6UH M7XQS
W2 2RJ M3YGY
W2 3DH M7NSS
W2 3RL M6RVJ
W2 4DT G8WIW
W2 6DW 2E0ZDC
W2 6DW M0ZDC
W2 6DW M6ZDC
W3 0DE G3YMM
W3 0HH G1CNN
W3 0HR G3HHU
W3 0RP G8LWS
W3 7AQ G0ZHP
W3 7NP G0UCC
W3 8SY 2E0UER
W3 8SY M7AUI
W3 9EJ G6HXB
W3 9RQ G4ZZV
W4 1HT 2E0CPP
W4 2JH G4LIC
W4 2LL G4SJL
W4 2SF 2E0ZHN
W4 2TS 2E0OCP
W4 2TS M0OCP
W4 2TS M7OAD
W4 3EW M7PTF
W4 3LR G1SAT
W4 3LR G3VKT
W4 3NH G4ZJD
W4 3TP M0KNL
W4 4EH 2E1HUJ
W4 5DN G4HMS
W4 5DN G4TZA
W4 5DN G7HMS
W4 5EN G0OYJ
W4 5JS G3SQX
W5 1HL G0VDN
W5 1PY M0BCZ
W5 1TZ G8SPE
W5 2BJ 2E0HSW
W5 2BJ M0HSW
W5 2HE M7JAV
W5 2LU M6YXD
W5 3LL M3SDQ
W5 3SL G8MNR
W5 3XH M6PFM
W5 4BJ 2E0UMP
W5 4BJ M7ICG
W5 4BY M6LVS
W5 4DR G7ANG
W5 4EX 2E0SZT
W5 4EX M0SZL
W5 4EX M7SZO
W5 4JW G0SCG
W5 4XA G6VZM
W5 4XA G8BOJ
W5 5BH 2E0JES
W5 5BH M0JXS
W5 5BH M6JIG
W5 5HS G6FVB
W5 5PD G6YJJ
W5 5QD G8RVO
W5 5QT 2E1DTF
W6 0BT 2E0WYH
W6 0BT M7WHC
W6 7AF 2E0IGE
W6 7AF M0UMM
W6 7AF M7SRV
W6 7LD G3YFV
W6 8BD G7EEJ
W6 8HN G7VLF
W6 8HW 2E0UMR
W6 8HW M6UMR
W6 8RN 2E0CUO
W6 8RN M0HYG
W6 8RN M6CNL
W7 1BT G6TJY
W7 1JQ 2E0OUR
W7 1JQ M6TVX
W7 2JH G7VEI
W7 2JS 2E0ECV
W7 2JS M6MQH
W7 2NR M7MJR
W7 3AG G8RBX
W7 3DJ G6TNK
W7 3RA G8XQB
W8 6PJ M6RIQ
W8 6UU G3SKR
W8 7AX G8AVQ
W8 7SL G4EHN
W8 7SX G0HTX
W9 2BJ M6HEJ
W9 2QH G3UXO
W9 2QU G8HDW
WA1 2EH M7LDX
WA1 3AH M7BRJ
WA1 3AY M1AUH
WA1 3EN G4FMI
WA1 3HB G4XQA
WA1 3HX G0RFM
WA1 3JS M7DJG
WA1 3TN 2E0RYP
WA1 3TZ M3LFU
WA1 4BJ G1NWA
WA1 4BJ G7JOA
WA1 4DY G0ANL
WA1 4EZ M3PGN
WA1 4LH M6RZY
WA1 4LU G0CDA
WA1 4LU G3CKR
WA1 4NW G0PZP
WA1 4NY G1NXT
WA1 4PE M0CSE
WA10 2HA G0KQI
WA10 3HT M7SSW
WA10 3JH M0BWP
WA10 3NE M0AQK
WA10 3QL M6BJA
WA10 3RX 2E0PBO
WA10 3RX M0PBO
WA10 3XT M1BWS
WA10 4BL M6ZLM
WA10 4DN M3MSB
WA10 4EW M7HHY
WA10 4JG G0JBR
WA10 4LN G1OSL
WA10 4LU 2E0KNM
WA10 4LU M6KNM
WA10 4NL M3OEM
WA10 4RH M0ACK
WA10 5AH G0EZI
WA10 5AU M6AXF
WA10 5DW M3DUO
WA10 5NP M7FJU
WA10 5PB G8MHE
WA10 6QF 2E0MVH
WA10 6QG G0BCX
WA10 6SH G4MWO
WA10 6SH G5MHZ
WA10 6SH M6KDM
WA10 6TP G1NVE
WA11 0AP 2E0IAZ
WA11 0AP M6IAZ
WA11 0BL G8CXZ
WA11 0DD M7DXP
WA11 0DY G1XAP
WA11 0ES G0DZI
WA11 0ES G6DZI
WA11 0LY G4YKV
WA11 0NB G4WGJ
WA11 0PP M3VXM
WA11 0PY G0AFJ
WA11 0QD M0NZR
WA11 0QS G1ZKZ
WA11 0RH G0NWC
WA11 0RH M3NJQ
WA11 0RS G0SSL
WA11 0SD 2E0YGH
WA11 0SD M6TTO
WA11 0SQ G1WAE
WA11 0UB G4XIE
WA11 0YQ G0HBU
WA11 0YS 2E0RCC
WA11 0YS M0RCC
WA11 0YY 2E0RCR
WA11 0YY M0RCX
WA11 0YY M7CDX
WA11 7AG M6FME
WA11 7AP M6TKP
WA11 7AZ M0KKN
WA11 7LD G4KIN
WA11 7LD G8TMR
WA11 8AG G4KIP
WA11 8AT G3JIR
WA11 8BH M3DJW
WA11 8BW G0BHH
WA11 8DJ G4FKP
WA11 8ER G8JSM
WA11 8JR G4EST
WA11 8JW M0AQQ
WA11 8LA M0DZC
WA11 8SW M3JGH
WA11 9EL G4DIY
WA11 9HU G6YSN
WA11 9JD M6ZZE
WA11 9JJ M3KYQ
WA11 9NH M3VOL
WA11 9RB M7BZA
WA11 9RE G0NEB
WA11 9RW G0EAM
WA117DE 2E0RIL
WA117DE M7RIL
WA12 0AS M3OWQ
WA12 0AU 2E0BGM
WA12 0AU M3ISI
WA12 0DR G0VVQ
WA12 0DU G4KHG
WA12 0JT G4XOL
WA12 0JW 2E0NLW
WA12 0JW M6NLW
WA12 0JX M7EMO
WA12 0LN G7PUN
WA12 0LN M3PGH
WA12 0LN M3TLN
WA12 0LY G1HIO
WA12 0LY G1HIP
WA12 0LY M0NRC
WA12 0LY M7DX
WA12 0NE 2E0IUH
WA12 0NE M3IUH
WA12 0NN G6OBG
WA12 8BA G8XVJ
WA12 8BA M0SDA
WA12 8BY M0JVW
WA12 8HJ M3HHQ
WA12 8HP M3IRR
WA12 8JE G1HSA
WA12 8JF 2E1FSH
WA12 8JF M0ZLK
WA12 8JF M6BWV
WA12 8JF M6YZX
WA12 8LT G3XRI
WA12 8LT M0NLW
WA12 8LZ G6IKM
WA12 8SQ G1JPT
WA12 8SQ G6JPT
WA12 9DB M3ISJ
WA12 9EY 2E0IUI
WA12 9GB M3HGZ
WA12 9LS G0VAM
WA12 9LS G4WCH
WA12 9LT G7MFA
WA12 9QA M6PQS
WA12 9UD G1MMA
WA12 9YG G3YVH
WA12 9YL M3NBQ
WA13 0JS G3VJV
WA13 0JT G7BGO
WA13 0LA G4FQN
WA13 0QR G1ANK
WA13 0RD G3NKH
WA13 0SY 2E0DGU
WA13 9BA G8VMQ
WA13 9EY G0AMU
WA13 9EY G6XKK
WA13 9LY G3NKW
WA13 9NL G8TYH
WA13 9NN M1DQQ
WA13 9NW M6KDM
WA13 9QA G4VSX
WA14 1JL G4UGM
WA14 1LH G1GLS
WA14 2EQ G8JGM
WA14 2ND G3WFH
WA14 3HY G0KTT
WA14 4HW 2E0WHA
WA14 4HW M0WIA
WA14 4HW M3WIA
WA14 4JE G0CBJ
WA14 4QT M3VOJ
WA14 4QT M6IFG
WA14 4RQ G6HLU
WA14 4UE G4PYH
WA14 5AQ G7UCT
WA14 5AZ G4NAV
WA14 5JF G7AGA
WA14 5LJ 2E0NDG
WA14 5LJ M6NDG
WA14 5LN M1DXL
WA14 5LT G4HUF
WA14 5LT G7EYR
WA14 5YF M0OKS
WA15 0AS G4FYB
WA15 0LT G4CNI
WA15 6AB G0WEV
WA15 6AB M1CTM
WA15 6BG 2E0KCZ
WA15 6BG M0KCZ
WA15 6BG M6KCZ
WA15 6BW G3EOO
WA15 6DY G4VBJ
WA15 6HQ G4WQD
WA15 6HR G6MCE
WA15 6RS G6DOQ
WA15 6UE G4TFU
WA15 7BA M0PGL
WA15 7DB M6JCN
WA15 7HY M3VCV
WA15 7NP G1KXP
WA15 7NP M3OEO
WA15 7NS G8LQO
WA15 7PA G7RHD
WA15 7RS G1GNX
WA15 7RS G4UKV
WA15 7TF 2E0HKG
WA15 7TF M0KHK
WA15 7YH G0HAL
WA15 8DN G6GTH
WA15 8EA 2E0CXL
WA15 8ET G4SVR
WA15 8HA G7VCF
WA15 8LY G6MNB
WA15 8PD G0UXR
WA15 8PW G0GWI
WA15 9BD M7RZA
WA16 0AW G3PLS
WA16 0BN G7EPN
WA16 0JX M6EOB
WA16 0NE G0AKF
WA16 0TU M5ACS
WA16 0TZ M1ROE
WA16 6PS G6EXE
WA16 7AX G3UJE
WA16 7AX M0TTG
WA16 7ET G4HZW
WA16 7GB M0AUF
WA16 7HB 2E0CSU
WA16 7HE G0AUB
WA16 7HF G0FOY
WA16 7RD G8CJQ
WA16 8BG M3LRU
WA16 8JR G6FWK
WA16 8LH M6JMP
WA16 9AW G0IFX
WA16 9DZ G3ISB
WA16 9SB G1OQO
WA2 0AG G3YRQ
WA2 0AG M3WIJ
WA2 0BE G1KSK
WA2 0BL G3NFB
WA2 0BL M1DOT
WA2 0EZ G6OSV
WA2 0GL M3RZI
WA2 0NB M6IYM
WA2 0NE 2E0HSE
WA2 0NE M7BRR
WA2 0NQ M3ISN
WA2 0QE M3HGX
WA2 0SQ 2E1GMO
WA2 0SQ 2E1HEF
WA2 0SQ G1OND
WA2 0SQ M1GMO
WA2 0UH M0RAN
WA2 0UJ 2E1HEK
WA2 0UJ G6FKL
WA2 0WR M6WAA
WA2 7AP M6KEG
WA2 7AT M1CXW
WA2 7AX M1FAX
WA2 7DN M3ARU
WA2 7RW M6JEP
WA2 7SE G1FOW
WA2 8AD 2E0KPM
WA2 8AD M0XLR
WA2 8AD M0XNW
WA2 8AD M3DNA
WA2 8AJ M6TJJ
WA2 8AW M3PVB
WA2 8BQ M1EFT
WA2 9AJ G4XDH
WA2 9BQ G0RFT
WA2 9JA 2E0DHO
WA2 9JA G6LXV
WA2 9JA M6EWO
WA2 9NQ M6NBN
WA2 9PS G0VQK
WA2 9PS G7TOA
WA2 9SD G7PMB
WA2 9TW G7HKN

WA2OLL M7PRH
WA29EW M7AQI
WA3 1DG M0NTA
WA3 1EB M1BRX
WA3 1EQ M6AXA
WA3 1EU G6AHF
WA3 2DP M7CGD
WA3 2EE G4VDX
WA3 2EP G1EFU
WA3 2EP G1THF
WA3 2ES G0LVJ
WA3 2JQ M6GZL
WA3 2QL G4ONG
WA3 2RN G4RNC
WA3 3DF G4JPX
WA3 3HH 2E0COA
WA3 3HH M0LGL
WA3 3HH M6CFN
WA3 3HH M6EEL
WA3 3HH M6LTL
WA3 3JJ M6DOX
WA3 3QX G0RPO
WA3 3TZ G0LVY
WA3 3TZ G6TPG
WA3 3UX G6WVL
WA3 4DF G4LWY
WA3 4ES G1IQK
WA3 4EU M7BTV
WA3 4JF 2E0COD
WA3 4JP M0OCE
WA3 4LD G0RPG
WA3 4LD M0MRC
WA3 4LD M1CNP
WA3 4LR G8YKG
WA3 4NW G0WJX
WA3 4NW G7GZB
WA3 4PD G8NDE
WA3 5EX M7JRA
WA3 5JJ G0IZN
WA3 5QY G3XDP
WA3 5RU G6HKN
WA3 5RW G6DSP
WA3 5SA G7FHU
WA3 6AQ G4YKR
WA3 6DJ G4VBD
WA3 6LB M3EUF
WA3 6NE G4YLK
WA3 6NY 2E0SLM
WA3 6NY M3YHG
WA3 6PB G6HLL
WA3 6PS M6BJB
WA3 6PT M0HVN
WA3 6PT M6GUV
WA3 6QD G0JXI
WA3 6QP G6KTN
WA3 6RR 2E0ISK
WA3 6RS M3MHR
WA3 6TP G1ZOY
WA3 6TR M6CFL
WA3 6TZ G0BCU
WA3 6UY 2E0FXP
WA3 6UY G0CBN
WA3 6UY G0CPD
WA3 7JG G0TDC
WA3 7LR G6MKD
WA3 7LS G1NXS
WA3 7LY G4SSJ
WA3 7NT 2E0KTG
WA3 7NT M0XDJ
WA3 7NU M3YGL
WA3 7NU M3ZLP
WA3 7QA M1MKL
WA32EW M7CRW
WA4 1AX G7NPT
WA4 1DW G0YSS
WA4 1EX G7CED
WA4 1HX M3ZMN
WA4 1LY G8SQK
WA4 1NF G6DOZ
WA4 1PY 2E0DTO
WA4 1PY M0HFF
WA4 1QN G1PIY
WA4 1RJ G1KOD
WA4 1RW M3OQS
WA4 1TU G7VAG
WA4 2ET M0LYN
WA4 2ET M6WCY
WA4 2HE G0KXW
WA4 2HQ G4YPI
WA4 2HY M0PDX
WA4 2JA 2E0MFG
WA4 2JA M6TTP
WA4 2PF G0SSJ
WA4 2RE G4PBZ
WA4 2RX G0RPF
WA4 2TR G6MQI
WA4 3BA M7DRE
WA4 3BG G0IHF
WA4 3BJ G6GEL
WA4 3ES M3MXJ
WA4 3JA G7SKR
WA4 3LE 2E0NDP
WA4 3LE M0NDP
WA4 3LE M6NDP
WA4 4AF M3TPG
WA4 4BY G3JDT
WA4 4EE 2E0TBI
WA4 4EE M0WAU
WA4 4EE M6JPL

WA4 4NG M6JWO
WA4 4PY 2E0WVM
WA4 4PY M3ISY
WA4 4TD G6GZC
WA4 5AW G4EAQ
WA4 5EJ G0MYN
WA4 5QD G4LLG
WA4 5QJ G7LHS
WA4 5RD G8KBB
WA4 5RD M6TLP
WA4 6AF G0KXY
WA4 6DA M3GIX
WA4 6DE 2E0GGI
WA4 6DE M6CPX
WA4 6ES M3RVK
WA4 6EY G6NZH
WA4 6GY G4OEX
WA4 6LF G7GPL
WA4 6QU G4BQJ
WA4 6UA G3MUX
WA44BY G3ZRN
WA5 0AG G6YHL
WA5 0DU M0IFC
WA5 0GB G7KHW
WA5 0GD M1EGV
WA5 0GJ M6SPZ
WA5 0HP M6ABE
WA5 0HP M6TLX
WA5 0LW M6LRH
WA5 0LY M3HIG
WA5 0NA 2E0BXA
WA5 0NA M6JON
WA5 1BQ 2E0XJJ
WA5 1GU 2E0CDT
WA5 1GU M0HLC
WA5 1GU M3ZRR
WA5 1HN G0CWA
WA5 1HP M1FHP
WA5 1JF G7LPZ
WA5 1JH G6DPH
WA5 1LN M7FRR
WA5 1SY M0CRN
WA5 1TQ 2E0LOL
WA5 1TQ M3LXS
WA5 1TQ M3MQA
WA5 1XG G0CVM
WA5 1XN G7CFS
WA5 2AG 2E0EXN
WA5 2AG M3TRJ
WA5 2AG M5NEX
WA5 2AQ G3ZHE
WA5 2AR G6UCI
WA5 2BY G0NVT
WA5 2DB G0KKU
WA5 2DD G1VSK
WA5 2DD G3ZKD
WA5 2DD M7VSD
WA5 2HY G8PXI
WA5 2JN M3MIQ
WA5 2PA G7GGM
WA5 2QE G0SLR
WA5 2QL G7MNP
WA5 2QY G3GAH
WA5 2RB G8WAJ
WA5 2SG G4HSS
WA5 2SG M7TCW
WA5 2SR G8NRF
WA5 2TG M7BPS
WA5 2TW G3RRM
WA5 3AJ M0DCG
WA5 3AN G0RBI
WA5 3DA G0APY
WA5 3EE G0OON
WA5 3EH 2E0ELK
WA5 3ET G8VVP
WA5 3HD M7LBI
WA5 3LU G3VZU
WA5 3NU G1NRN
WA5 3PL M7GMF
WA5 3RX G8RCL
WA5 3TT G8NHD
WA5 4AX G4WNW
WA5 4DS 2E1GNR
WA5 4ES M3JNB
WA5 4HJ 2E0MGW
WA5 4HP M7SSG
WA5 4NE M3JAZ
WA5 4NS G4OMZ
WA5 4NW M6GFZ
WA5 4PQ M6XXB
WA5 4PW M6JIQ
WA5 7XT 2E0WRK
WA5 8QL G0WRS
WA5 8QL G4VSS
WA5 8QL M3CFI
WA5 8QL M3FIW
WA5 8QL M3FIY
WA5 8WU 2E0DKA
WA5 8WU M6EWN
WA5 9PL M3ZGU
WA5 9QE M3YRJ
WA5 9QP M7EME
WA5 9SB M5AGI
WA5 9UU G4GPJ
WA53NR 2E0IXW
WA53NR M7FIZ
WA6 0NF G4ZHT
WA6 0NW G6IRW
WA6 6DN G3GKS

WA6 6PT 2E0JDI
WA6 6PT M0JLR
WA6 6PT M6XJR
WA6 6PY M0DYB
WA6 6QQ G3VBA
WA6 6SG G8WQE
WA6 6SG M3MHV
WA6 7JR M3LLB
WA6 7JR M6FMU
WA6 7PG 2E0LYR
WA6 7PG M3LYR
WA6 7QJ G8DVF
WA6 7RU G4OTI
WA6 8AE G4VYJ
WA6 8DA M1KTY
WA6 8DA M1WRX
WA6 8HP 2E1FKD
WA6 8HP 2E1HEE
WA6 8HP G1IVV
WA6 8HP G7NCP
WA6 8HP M3IVV
WA6 8JJ G6ZHS
WA6 8JS G8MMN
WA6 9LB 2E0XMP
WA6 9LL G3INP
WA6 9PG 2E0FDS
WA6 9PQ G4UGD
WA69LB M6YTC
WA7 1BE M7FFX
WA7 1DH G8YAZ
WA7 1QE M7DWW
WA7 1UH G1LCC
WA7 1XE G4YZP
WA7 2AP G7EOK
WA7 2FF 2E1SHA
WA7 2FP M3DAE
WA7 2FR G0OJG
WA7 2GR G7ICD
WA7 2LG G0NWE
WA7 2LR M3GTB
WA7 2NS G6FPX
WA7 2QS G1PIX
WA7 2QS M0PIX
WA7 2UH G0BSD
WA7 2UH G4DIT
WA7 3AQ G0VQL
WA7 3AQ G8HLQ
WA7 3BH M3JPM
WA7 3BU M1ICL
WA7 3DY G7JJJ
WA7 3EG 2E0ZMI
WA7 3EG M6ZMI
WA7 3ER G1SJT
WA7 3JA G1DYN
WA7 3JA M6LFC
WA7 3JH G8IEL
WA7 3JH M3IEL
WA7 3LE G1SBW
WA7 4AD G0MKD
WA7 4BG G8SIM
WA7 4DL 2E0FPY
WA7 4DL M0YCY
WA7 4DL M7DAC
WA7 4EH G4WPG
WA7 4RN 2E1HDZ
WA7 4RN M3TLM
WA7 4RP M3KIN
WA7 4TW G7SXG
WA7 4TX G7VCP
WA7 4XL G4HOD
WA7 5AR M0GIB
WA7 5EJ M3NCD
WA7 5EL 2E1BZH
WA7 5JA 2E1IIA
WA7 5JA M3GZE
WA7 5JB G0MQH
WA7 5JB M3GZD
WA7 5JJ M3FIP
WA7 5LR G4OAB
WA7 5RF G0HEF
WA7 5RJ G0NSL
WA7 5RW G1MSY
WA7 5ST M3NMB
WA7 5XE M3MOV
WA7 5XW M3ELP
WA7 5XW M3GZJ
WA7 5YN M3XRY
WA7 5YU G0RLF
WA7 5YU G1SES
WA7 6AA G8VNF
WA7 6BJ M3CVL
WA7 6BN G0RBM
WA7 6DL G6UMH
WA7 6DQ G0MSF
WA7 6DT G1JMP
WA7 6DW G0MIX
WA7 6DW G7IAE
WA7 6DZ M0CNK
WA7 6EH G0WHQ
WA7 6HT M7UKN
WA7 6JR M0KOI
WA7 6LB M6AML
WA7 6LB M6HZL
WA7 6RJ G1ZUZ
WA76HT M0UKN
WA8 0BA M7ABP
WA8 0BX 2E1DTK
WA8 0QX 2E0YDM
WA8 0QX M0YZF

WA8 0QX M6CAP
WA8 3JE 2E0HCL
WA8 3JE M0LSG
WA8 3JE M6WID
WA8 4GA M6LHM
WA8 4NT G1LFM
WA8 4TB G8BJQ
WA8 5AS 2E0AIL
WA8 5BH 2E0HCY
WA8 5BH M7SON
WA8 5BT G6URF
WA8 5QG G6CYF
WA8 6ET G4YJS
WA8 6HR G1AAK
WA8 6HR G7UXU
WA8 6TL G1URQ
WA8 7BN G0ICD
WA8 7HL 2E0KQR
WA8 7HL M6KQR
WA8 7LD 2E1EBL
WA8 7LD M3DIG
WA8 7LR M6JCP
WA8 7NB M6JDD
WA8 7NQ G1HGC
WA8 7PL M0CMT
WA8 7PL M6JQU
WA8 7UW G0SVZ
WA8 8BG G0FQP
WA8 8BZ M7WID
WA8 8QE M3WID
WA8 8SQ M3VUO
WA8 8SS G8RIB
WA8 9DP G0HKZ
WA8 9QB G8CMP
WA86RH M7FAF
WA88QE M7AFA
WA9 1BL M3IQP
WA9 1EN 2E0MRA
WA9 1EN M6GMF
WA9 1QB 2E0ZAJ
WA9 1QB M3OCJ
WA9 1QY M3MLK
WA9 1SQ M3KLY
WA9 2AF M6EZZ
WA9 2AQ 2E0KJT
WA9 2AQ M3ZWQ
WA9 2AR G7KZY
WA9 2BD 2E0VWX
WA9 2BD M3VNQ
WA9 2BU M3LWP
WA9 2DP M3SFJ
WA9 2JD M6XIT
WA9 2PL M6DRO
WA9 2QG 2E0JCY
WA9 2QG M0JVU
WA9 2QG M3JCY
WA9 3AF M3OCA
WA9 3NF G7OXP
WA9 3NF G8HLH
WA9 3QX M6EKV
WA9 3RE 2E0KZJ
WA9 3RE M3KZJ
WA9 3SD M3ZYW
WA9 3SH M3MZT
WA9 3SX M3LXF
WA9 3WA 2E0XGX
WA9 3WA M6XGX
WA9 3WT M6DWG
WA9 3XQ 2E0NCB
WA9 3XQ M0GGK
WA9 3XQ M3NCB
WA9 4AN M3EJX
WA9 4AP M0JGM
WA9 4BD M6UNA
WA9 4BJ 2E0EYY
WA9 4BJ M6HVR
WA9 4BS G1OMY
WA9 4DL M3RCS
WA9 4DN M3KZP
WA9 4DQ G7RMD
WA9 4HW G7IFO
WA9 4NA G3REV
WA9 4PW G8OUI
WA9 4RY G1ZNT
WA9 4UX 2E0EQP
WA9 4XH G4ROU
WA9 4ZW M3WEF
WA9 5AR G0STH
WA9 5AR G4WGB
WA9 5BP G1NVY
WA9 5FQ G1EVV
WA9 5HL 2E0EDP
WA9 5HL M6ESZ
WA9 5LR M3WHL
WA9 5LZ G2ARV
WA9 5LZ G4KZQ
WA9 5NA 2E0PKU
WA9 5NQ M6IIJ
WA9 5RU M7PLP
WA9 5TD M0CHR
WA9 5XJ 2E0GVO
WA93NG M0JFK
WA93XU G4VWD
WA94XN M0TFS
WA95SN M7BEV
WC1N 1AS M6XYY
WC1N 3XX G7HMK
WC1X 0BG G1UCR
WD17 2LN G6RXY

WD17 2RQ G6JGF
WD17 3AQ G3LGR
WD17 3BA M7VEG
WD17 3DD G4TJA
WD17 3DP G8LDY
WD17 3DX M0ZRG
WD17 3DX M6ZRG
WD17 4FJ 2E0SBW
WD17 4FJ M6SBB
WD17 4QT G3YKB
WD17 4SU G3SHY
WD17 4SW G0HCC
WD17 4SW G4KUJ
WD17 4SZ G3WSB
WD17 4TA M3IFJ
WD18 0AY G7VEF
WD18 0FA M3WVX
WD18 0FA M3WVY
WD18 0HA M6NAM
WD18 0HQ G8IIZ
WD18 0LJ M7BPN
WD18 6NX G0OYC
WD18 7JD G0HTM
WD18 7JD G0NUR
WD19 4AS M5DIK
WD19 4DA 2E0DTQ
WD19 4DA 2E0ZLA
WD19 4DA M6TPI
WD19 4DA M6ZLR
WD19 4LL 2E0WNS
WD19 4LL M0PLE
WD19 4LL M6JNO
WD19 4LW G8BXH
WD19 4PA G3MED
WD19 5AA G3NDK
WD19 5DF M6ICB
WD19 5EH G0DDV
WD19 5ET G6IUD
WD19 6HU M1FTA
WD19 6LF M1BCR
WD19 6NE 2E0HTX
WD19 6TR G8SUG
WD19 6YL M6ROM
WD19 7AY M6SQU
WD2 5HY 2E1HCB
WD23 1EX M0SSH
WD23 1FB G4SWY
WD23 1PE G3VNB
WD23 1PZ G3OMR
WD23 2BT M7HGO
WD23 2BU M0CIO
WD23 2HA G4IID
WD23 2HF M6PQF
WD23 3AR G0LZV
WD23 3BQ 2E0MUT
WD23 3BQ M0HSZ
WD23 3SS G4GYS
WD23 4GH G4RNW
WD23 4GL M6MLU
WD23 4QT G8XXV
WD24 4LA M7AOL
WD24 4RL M0BLD
WD24 5EW G6IKO
WD24 5LT G4WMN
WD24 6DA M0CJG
WD24 6RU 2E0WTA
WD24 6RW M6VSS
WD24 6SY 2E0ECQ
WD24 6SY M6GHJ
WD24 7EE 2E0ERP
WD24 7EE M0TNC
WD24 7EE M6SQL
WD24 7LN 2E0DQH
WD24 7LN M0HYL
WD24 7LN M6FUY
WD246RW M7EOA
WD25 0EH G0MDM
WD25 0EN M7HXO
WD25 0HG G6FAF
WD25 0HX M6IZA
WD25 7DN G3WDX
WD25 7EE G6WIW
WD25 7EW G6OBU
WD25 7PA M7PHA
WD25 9DZ G4CNH
WD25 9PS G6MHF
WD25 9PS M3DSE
WD25 9PS M3LHM
WD25 9QH G4RMC
WD25 9QQ M0TTB
WD25 9QQ M6BBC
WD25 9RB 2E0GMD
WD25 9RB M6CHS
WD3 1EQ M7TMM
WD3 1HL G0WBL
WD3 1LQ G6ZFV
WD3 1NA M1CIG
WD3 1PQ M6WAQ
WD3 3AL G6LZB
WD3 3AU G1XEP
WD3 3AU G6AJG
WD3 3DW 2E0LWR
WD3 3EG G7HHT
WD3 3FE G3TVH
WD3 3JQ G3EBV
WD3 3JT M3JLE
WD3 3NH G1MPC
WD3 3PH G8MCJ

WD3 3PS M7WGM
WD3 4EB G4JHU
WD3 4EB M6YEQ
WD3 5BD G3VOS
WD3 5BJ G4COV
WD3 5HQ 2E0BST
WD3 5HQ M0SKC
WD3 5HQ M3OAJ
WD3 5JD G4ORE
WD3 5JJ G0VJI
WD3 5JJ G1DNY
WD3 5JT M0LAL
WD3 5JT M6CFT
WD3 5PX G6YJR
WD3 5QH G3WJG
WD3 5QQ G4TAO
WD3 5RG G4AEM
WD3 6ER M6NQV
WD3 7AR M0GFF
WD3 7DY G1ISX
WD3 7EN G3ZER
WD3 7ES 2E0ERE
WD3 7ES M0ITI
WD3 7ES M6HNZ
WD3 7NR G8VTV
WD3 8GW G6IBD
WD3 8GW M0DAB
WD3 8HE M7FMT
WD3 8HT G8OQT
WD3 8PX M7DEP
WD3 8QA G7HGQ
WD3 8QG G1XET
WD3 9TZ G6LFA
WD3 9YZ M6TDJ
WD4 8DY G0SGV
WD4 8JE G7GTG
WD4 8JE G7GTH
WD4 8NJ G7TMH
WD4 9HF G3XYJ
WD4 9NA M6UNI
WD48PR M7SSS
WD5 0BJ G4FRZ
WD5 0DA G4KUF
WD5 0DH G3JCR
WD5 0EE M3UXK
WD5 0EG G3XLI
WD5 0EU G3OYT
WD5 0PJ G0TVD
WD5 0QG G4DPP
WD5 0ST G7PDH
WD5 0XD M6VWV
WD6 1AY G6MWL
WD6 1EP G7MKJ
WD6 1EP M0IHR
WD6 1NT 2E0SHY
WD6 1NT M6ACA
WD6 1PL M7TVD
WD6 1XA M0KSL
WD6 2DA G4TEP
WD6 2HG G6APH
WD6 2HG M6MID
WD6 2JW M7ISB
WD6 2LQ G7AIH
WD6 2PB G7DFW
WD6 2RB G8XJN
WD6 2RP G6DAU
WD6 3DH M6DZA
WD6 3JH G6JMO
WD6 3LH G4GPL
WD6 3NJ G4TIH
WD6 3PT G1INI
WD6 3PU 2E0GSK
WD6 3PU M0GSK
WD6 3PU M6GSK
WD6 4EG 2E0SSK
WD6 4EG M3UON
WD6 4HY 2E0WEF
WD6 4JD G3JPJ
WD6 5AD M3OYQ
WD6 5BJ M1BCB
WD6 5DG G0PQB
WD6 5LR G1XVL
WD6 5PE G6RHN
WD6 5QA M0THY
WD6 5QA M6THY
WD6 5QE G8RCO
WD7 7DU G4WIS
WD7 7NF 2E0CQO
WD7 7NF M0MBR
WD7 7NF M6BTG
WD7 8BA M3XQQ
WD7 8BA M3YUK
WD7 8HJ G6JMB
WD7 8JD G6OKA
WD7 9JR M7VEX
WD7 9JW G7OIR
WD7 9JW M0ADY
WF1 2AL G7VDT
WF1 2LF G4VQR
WF1 2LS G4OVL
WF1 2LS M0SHZ
WF1 2LS M6MEB
WF1 2LW G7IOF
WF1 2PQ G4CLJ
WF1 2RA G3OMJ
WF1 2UP M6NIV
WF1 3AB M1MCL
WF1 3DN G4LXV
WF1 3DY M6FYH

WF1 3NW G4VRJ
WF1 3QD M6ZTA
WF1 3RL M6MGW
WF1 3RU M7FBM
WF1 3RU M7KIM
WF1 3SZ 2E0LAI
WF1 3TL G8LGE
WF1 3TX M6BRH
WF1 4AB M6SGZ
WF1 4EU G1MRC
WF1 4EU M7ARC
WF1 4EZ 2E0OUH
WF1 4EZ M3OUH
WF1 4HP M7CHH
WF1 4JL M6MXO
WF1 4PW M3EVB
WF1 5AH M3XNX
WF1 5LB M3OUG
WF1 5LG M3XWR
WF1 5NT G6NYF
WF10 1LN M6ASC
WF10 1LN M6MEG
WF10 1PZ M6MEV
WF10 2AA G4FEQ
WF10 2AP M6GOL
WF10 2BS M3TVV
WF10 2DN G0GYA
WF10 2QA M6BLM
WF10 2QF M6PZZ
WF10 2RB G4TCG
WF10 2RG M3OLD
WF10 2RN M6KTI
WF10 2SB G8OXD
WF10 3AR M7SNS
WF10 3EY 2E0UAM
WF10 3EY M3UAM
WF10 3HN G3HNC
WF10 3HT G4RQI
WF10 3LL M3NRX
WF10 3PB G4IBN
WF10 3QS M6RIN
WF10 3QX 2E1GLS
WF10 3QZ M3VRA
WF10 3RN 2E0XII
WF10 4JT M3YNC
WF10 4LF G3OAR
WF10 4LN M6JEQ
WF10 4PH G0PMB
WF10 4QL G7MVE
WF10 4SE G1YNH
WF10 4SE M3YNH
WF10 5AF M6FXY
WF10 5TU M7ABE
WF10 5UJ M6YOB
WF103QN 2E0CCF
WF103QN M3TZQ
WF105HB G7MTG
WF11 0DQ G7IPR
WF11 0HG G0VXD
WF11 0JH G0NQE
WF11 0JY G4VMA
WF11 8AJ G8VAT
WF11 8JL G0TVL
WF11 8LH G4SOL
WF11 8RH 2E0JBF
WF11 8RH M6BWE
WF11 8SR G0NER
WF11 8TD G0VXM
WF11 8TH G6KIH
WF11 9AT G4AAQ
WF11 9ED G4PHL
WF11 9HR G4TLM
WF11 9NN G8NDF
WF11 9PA G1PZA
WF12 0BD M1BAV
WF12 0BP 2E0SLT
WF12 0BW G4CLI
WF12 0DW M3CNV
WF12 0HB 2E1FMC
WF12 0HT G8VOQ
WF12 0HY G0GSU
WF12 0JZ G7LBM
WF12 0NF M6GCB
WF12 0NL G0CBW
WF12 0PJ M6ZMR
WF12 0QH M6UZH
WF12 0RH M6BCB
WF12 7AS 2E0EOP
WF12 7AS M0KKH
WF12 7AS M3PAX
WF12 7AW 2E1FFL
WF12 7DE M6ZXC
WF12 7LA G7NPL
WF12 7LS M1AJG
WF12 7PD G3CPG
WF12 7PU M0EDO
WF12 7RU 2E0AGB
WF12 7RU M7AGB
WF12 7SN M6AIA
WF12 7SQ G8POK
WF12 8AJ G6FTJ
WF12 8AQ G6ZJV
WF12 8BN 2E0IOX
WF12 8BN M7DJY
WF12 8BY M0WEC
WF12 8HH G1DEX
WF12 8HH M6PPE
WF12 8PL M1BHE
WF12 8PY M3LZL

WF12 8PZ 2E0GEG
WF12 8PZ M3VBP
WF12 9AY M3YHJ
WF12 9DN G4GHQ
WF12 9LD M7IKO
WF12 9PU M3YGS
WF120BW M7CUD
WF120EQ 2E0FYP
WF120EQ M6SON
WF120HE 2E0CNQ
WF120HE M0HRM
WF120HE M6CSG
WF128PL M7DYZ
WF13 2QF M3HTR
WF13 3AX M7EEL
WF13 3BT M7BNK
WF13 3LZ G4YAJ
WF13 3SR G6ZUO
WF13 4DQ M0LEV
WF13 4EN G6WGE
WF13 4HT G1VIF
WF136QJ M7DEI
WF14 0AJ G6ITV
WF14 0AQ G6XTZ
WF14 0EW G1EDU
WF14 0LF 2E0BCS
WF14 0LF M3JUF
WF14 0NR G3HPD
WF14 8JP 2E0JTH
WF14 8JP M3ZTX
WF14 8JW 2E0RFE
WF14 8LG M5AIB
WF14 8NW M6MBQ
WF14 8PN M6AYC
WF14 8PX G7PHX
WF14 9AW G2FCP
WF14 9BG M7OGL
WF14 9BJ G8NZR
WF14 9DJ 2E0FRL
WF14 9DJ M7LSL
WF14 9ED M3YDL
WF14 9HZ 2E0LBB
WF14 9LJ M6BJY
WF14 9NL M6JRJ
WF14 9PB G4PHB
WF14 9PB G7CSV
WF14 9PW G1BGQ
WF14 9PY M6MIA
WF14 9SA 2E0CYL
WF149AL M7BVA
WF149AN M6NDC
WF15 7BW G1IKF
WF15 7DN G0GMJ
WF15 7DP G4GOX
WF15 7HW M3SGI
WF15 7NJ G4EIL
WF15 8DG G8NYM
WF15 8JU G7HHX
WF15 8JW 2E0IPC
WF156QE G8DZW
WF157LP G0PMP
WF16 0BE G3JQC
WF16 9JL G0TAL
WF17 0DX G4MLV
WF17 0EQ G7OQO
WF17 0JL G4FPE
WF17 0JZ 2E0SML
WF17 0JZ M6SAN
WF17 0LJ G4EOC
WF17 0RG G7LTO
WF17 5QY G7VHS
WF17 6DB M6GNH
WF17 6DF M7FPZ
WF17 6DZ G4XZM
WF17 6EH G4SEQ
WF17 6FZ M6EXG
WF17 6HG G1NEG
WF17 6HG M3NEG
WF17 7NT G1UVD
WF17 8PN 2E1ENR
WF17 9BX G0IAX
WF17 9DL G6XHK
WF17 9JF M0JBW
WF17 9JF M3JPR
WF17 9JF M3JSM
WF17 9JF M3LMC
WF17 9JF M3SPR
WF17 9QF G6JLI
WF17 9RG 2E0OOC
WF17 9RG M3OOC
WF2 0BH M6EVL
WF2 0BJ 2E0YSB
WF2 0BJ M0PYB
WF2 0BJ M6YSB
WF2 0FQ M3MXA
WF2 0NZ M3VSZ
WF2 0PD M6IGK
WF2 0PE M3PNV
WF2 0PP G4RCG
WF2 0PP G6LCP
WF2 0QR G6XKE
WF2 0SE G0CEW
WF2 0TJ G6LDP
WF2 0UP 2E0NER
WF2 0UP M6FMC
WF2 0UT M6JOU
WF2 0UU M6FMP
WF2 6ED M0IPV
WF2 6HJ M0AKD

WF2 6JG M3LDI
WF2 6LQ M6BNC
WF2 6NP M6SSW
WF2 6RE G0DIS
WF2 6RN G4CPC
WF2 6RT M6FPV
WF2 6SE G4BLT
WF2 6SE G8DVS
WF2 6SR G7JTH
WF2 6SR M1DDR
WF2 7DA M6CXI
WF2 7DE G1WRS
WF2 7DE M3WLV
WF2 7DH M3XYK
WF2 7EF G8IJI
WF2 7EG M6SUP
WF2 7HW G1OOU
WF2 7JN M3TXF
WF2 7JT M0CUF
WF2 7LS M6SNW
WF2 7PR G7SWH
WF2 7QS G7PCW
WF2 8AS G1YYC
WF2 8BX M3EOZ
WF2 8BY M3OFV
WF2 8EB G0EVA
WF2 8EL G0MVR
WF2 8EL M3LYC
WF2 8ET M6DDL
WF2 8EX M3KZB
WF2 8HA M3XRW
WF2 8HE 2E0DND
WF2 8JZ 2E0APJ
WF2 8LD M3TCT
WF2 8LY M6DRR
WF2 8UP G7UQA
WF2 8YB M3YXD
WF2 9DR M3XRO
WF2 9JS M6FCE
WF2 9JZ G6AJS
WF2 9QB M3HTA
WF2 9QG 2E0KTX
WF2 9QG 2E0PIA
WF28HR M3UEP
WF3 1DL 2E1KYQ
WF3 1HT G0IKD
WF3 1HZ G2DP
WF3 1HZ G4DZU
WF3 1JB G0PXF
WF3 1JL M6MHE
WF3 1LU 2E0THP
WF3 1LU M0SVX
WF3 1LU M7TTS
WF3 1PF M0AOI
WF3 1QH M7EOD
WF3 1RE G8JHA
WF3 1TA M6HJD
WF3 1UD G6NPW
WF3 1UG 2E0LXY
WF3 1UG M3LOX
WF3 1UY 2E0YEP
WF3 1UY M0MMR
WF3 2AX G0VTI
WF3 2AX G1RHW
WF3 2DD G0CQD
WF3 2FG 2E0VBL
WF3 2FG M0DUB
WF3 2FG M7DUB
WF3 2JG G1FOM
WF3 3NS M6EZV
WF3 3PX 2E0JAA
WF3 3PX 2E1JAA
WF3 3PX M3BGT
WF3 3PX M3MAA
WF3 3PX M3SCA
WF3 3TG M6GDV
WF3 4GD G7WRS
WF3 4JJ G0EVT
WF32BL M1TRC
WF32HS 2E0KBL
WF32HS M6SKW
WF33FN M7KWG
WF4 1BA M7PVX
WF4 1EF G8OJK
WF4 1EH G1XBE
WF4 1HS G7HYZ
WF4 1HX G8JQH
WF4 1NP 2E0WFD
WF4 1NU 2E0KYI
WF4 1NU M0KYI
WF4 1NU M3JZA
WF4 1PF G4AYL
WF4 1RN G0TYS
WF4 1RN M3UEW
WF4 1RN M3UXI
WF4 1RN M3VEW
WF4 1SL 2E0VKN
WF4 1SL M0IAA
WF4 1SL M3VKN
WF4 1SP M6EHR
WF4 2AT M0XYN
WF4 2BU M6HRO
WF4 2DU M0SZQ
WF4 2DU M0WYC
WF4 2DU M6SUJ
WF4 2EJ G4PIR
WF4 2EJ M3WXX
WF4 2NF G0UOS
WF4 2NT M0JCV

WF4 3BH G1HVL
WF4 3BL G0HDP
WF4 3BY M1FBN
WF4 3EF 2E0OPB
WF4 3EF M3OPB
WF4 3EG M1CBU
WF4 3JA G8GOT
WF4 3JZ G0EVR
WF4 3JZ G0FEV
WF4 3NW G0SUB
WF4 3NW G1GRT
WF4 3PT G6EZG
WF4 3QD M3TLD
WF4 4AX M6JBB
WF4 4ED G0BQB
WF4 4HN G4JSS
WF4 4LR 2E0DWJ
WF4 4LR M6DWJ
WF4 4RU G7PNM
WF4 4TE G0COA
WF4 5AN G4IAU
WF4 5DX M7SPE
WF4 5HH M6FFV
WF4 5HH M6JFR
WF4 5QG M7WFD
WF4 6AE G1AHS
WF4 6BX M6RIJ
WF4 6EQ G4DWO
WF43PG G4IZH
WF5 0PP G1WXK
WF5 0QA 2E0RMW
WF5 0QA G4BOB
WF5 0QA M7RMW
WF5 0RU G0MJZ
WF5 0TJ G1HYA
WF5 8EN G4TDI
WF5 8JW M0LUD
WF5 8LF G3SEY
WF5 8LH G1ISP
WF5 8LJ M3OUF
WF5 8NA M0LAB
WF5 8QP G4MVE
WF5 8RQ M6LPN
WF5 8RY G1VAO
WF5 8RY G3YQB
WF5 9EW G7NTI
WF5 9JL G4LJK
WF5 9PR G4TFH
WF5 9RE M3SVO
WF5 9RQ G4IOD
WF5 9RQ G4KFP
WF5 9SJ G6AAK
WF6 1AW M3LYX
WF6 1BL 2E0MD
WF6 1BL M6HKI
WF6 1BN M6MMB
WF6 1DA G7MJP
WF6 1ER M3HBX
WF6 1LB 2E0WRF
WF6 1NY M3OOA
WF6 1RS 2E0HGT
WF6 1RS M6TCY
WF6 1SN G4SBG
WF6 1SS G4SAV
WF6 1WW G6PDA
WF6 2AA G6MID
WF6 2AA G6XEX
WF6 2LB 2E0GBF
WF6 2LB M6AHZ
WF6 2LX G7DJY
WF6 2PL M6WBO
WF6 2RZ G7PZL
WF6 2TS M7TFN
WF6 2TT G1IIX
WF7 5AB G1PHK
WF7 5AG G1ZPO
WF7 5DP M1NTV
WF7 5EA M6MAD
WF7 5EA M6STV
WF7 5EB G4OOC
WF7 5LH G4NRF
WF7 5LW G7IJY
WF7 6AA M6EOW
WF7 6AH G1ZGH
WF7 6AY M6EYK
WF7 6BL G0HPA
WF7 6HT M3IFM
WF7 6LQ G1SJZ
WF7 7HQ G8TKY
WF7 7HU M0AGU
WF7 7LA G0KMN
WF7 7LA G1LFI
WF7 7LQ G0RAE
WF7 7NA G0PAZ
WF75JQ M7NAV
WF8 1LA G6DUH
WF8 1SB G4ISU
WF8 1TD M3UDD
WF8 1TD M6TDV
WF8 2AB M6PTF
WF8 2AJ G0RLN
WF8 2AJ G1FYQ
WF8 2BX G6ZFT
WF8 2BY M3FOK
WF8 2BY M3SUI
WF8 2DJ G7HQC
WF8 2EJ 2E0AOZ
WF8 2HW M6KEP
WF8 2JW M3WUW

WF8 2LX M3KBL
WF8 2PN M3UDK
WF8 2SR G4KMW
WF8 2UP G0FVN
WF8 3ES 2E0LUL
WF8 3ES G5VZ
WF8 3ES M3JRQ
WF8 3NJ G6WBT
WF8 3NT G0TVU
WF8 3QD G4ICB
WF8 4BU M3WYA
WF8 4BX G6VZS
WF8 4LG G0SNS
WF8 4ND G4YNU
WF8 4RP G0TPN
WF8 4SJ G3VTD
WF81EH M7EPY
WF9 1AZ 2E0BSI
WF9 1AZ G0BPK
WF9 1AZ G3FYQ
WF9 1AZ G3SYC
WF9 1AZ M6CHF
WF9 1ED G8YCQ
WF9 1EU M6XLP
WF9 1LG M3AOQ
WF9 2AR 2E0VAT
WF9 2AR M0ZIM
WF9 2AR M6DEV
WF9 2BT G8HRA
WF9 2BT M0CIR
WF9 2BZ 2E0DYQ
WF9 2BZ M6SYK
WF9 2DD G0CYX
WF9 2QX M6FBQ
WF9 2TL 2E1FSX
WF9 2TL G1DKV
WF9 2TL G6OXZ
WF9 3EP 2E0MJP
WF9 3JF M6CXM
WF9 3JT 2E1CIY
WF9 3NB G0PPQ
WF9 3NB G3NJB
WF9 3NB M1CRA
WF9 3NB M7PSJ
WF9 3NJ G4OSY
WF9 3QE G4ZVB
WF9 3RY G4NDP
WF9 3SJ M3XYO
WF9 4AP 2E1XDJ
WF9 4AY G0GEH
WF9 4EA M0RTV
WF9 4ES 2E0MAH
WF9 4JA G1PPX
WF9 4NW M3EXJ
WF9 4RL G1CDN
WF9 4TA 2E0ICT
WF9 4TA M0ICT
WF9 4TQ M3KAX
WF9 5AJ G4PLL
WF9 5BX G8GXS
WF9 5DP M3IUK
WF9 5DS G1CDO
WF9 5HE G1FFH
WF9 5HE M0JQK
WF9 5HE M3MLF
WF9 5HY G0JLL
WF9 5JL M0JUK
WN1 2AT 2E0WIG
WN1 2BA G1SUH
WN1 2DL G1INU
WN1 2RR G3OSL
WN1 2RR G3YNU
WN1 2SS G4WGF
WN1 3PP 2E0CHY
WN1 3PP M6AMV
WN1 3UE 2E0JKR
WN1 3UQ G7ADF
WN1 3XT 2E0YYZ
WN1 3XT M3VIW
WN12HX 2E0DMP
WN2 1AP M0JEQ
WN2 1DA M7BNB
WN2 1HW G8OMC
WN2 1JE 2E0HXY
WN2 1JE 2E1LEN
WN2 1JE M7INC
WN2 1LT G0IYX
WN2 1LT G4KKN
WN2 1NA G1SWH
WN2 1NA M0AWX
WN2 1RL M1AWS
WN2 1SZ M0UNN
WN2 1ZH M0OTE
WN2 2BZ 2E0LTF
WN2 2JG M7DMH
WN2 2NA G0FYE
WN2 2NN M7VXW
WN2 2PS M6BZF
WN2 3BN G6ZIY
WN2 3DP 2E0VXI
WN2 3DP M6VXI
WN2 3EE 2E1WNA
WN2 3EQ M6HTC
WN2 3HP G0AYF
WN2 3HR G1YMP
WN2 3SB G1TAR
WN2 3XD G4HPH
WN2 4AF M3KUO
WN2 4EH 2E1BYY
WN2 4EH G6DHI
WN2 4EJ G0AVH
WN2 4ET G1BDU
WN2 4ET G4WIL
WN2 4HL 2E0MLS
WN2 4HL G1ECI
WN2 4HL M0ICK
WN2 4HL M3ITZ
WN2 4JA G4ZVK
WN2 4JG M6PFQ
WN2 4JH M7FTR
WN2 4LD G4EHY
WN2 4LZ G1EIO
WN2 4NB M0IYV
WN2 4NG M7FPW
WN2 4PL G8ALD
WN2 4QF G4PGQ
WN2 4QS G0CBD
WN2 4RG G7JZI
WN2 4TY G7LVS
WN2 5DW G1JMS
WN2 5HY G0IZG
WN2 5JN G7GAG
WN2 5PP 2E0XEM
WN2 5PP M0XEM
WN2 5PP M7BEM
WN2 5QD M3PZL
WN2 5QD M3PZN
WN2 5QR M3KZR
WN2 5TF G7FMB
WN2 5XA 2E1EHB
WN2 5XA M0GRE
WN2 5XA M1AIX
WN2 5XG G6ZJM
WN2 5XR 2E0CRX
WN2 5XR M3LBR
WN2 5XR M3XFN
WN2 5XR M7TSA
WN2 5XY M7XFC
WN2 5YE M7EKD
WN2 5YE M7TFA
WN2 5YL M6OBO
WN21AU M7DPW
WN3 4JP 2E0YGR
WN3 4JP M6YGR
WN3 4NS G1CZN
WN3 4PQ M6SDU
WN3 4QA 2E0MNP
WN3 4QA M0XOC
WN3 4QA M6INT
WN3 5BT 2E0FCJ
WN3 5BT M6OQH
WN3 5NG M5DXX
WN3 5PL 2E0BTX
WN3 5PL M6GBR
WN3 5PR G8DPE
WN3 5QN G0DPI
WN3 5QN G0HRW
WN3 5QN G7EDK
WN3 5QN G7IXC
WN3 5UT M3XDQ
WN3 6AA G3BPK
WN3 6AA G4GWG
WN3 6AD G4OVV
WN3 6AJ G0KHH
WN3 6JY G6LST
WN3 6LD G4WDC
WN3 6LD G6TYB
WN3 6TQ G4HSC
WN4 0AZ G4WXX
WN4 0EF M3WJN
WN4 0EQ 2E0ZRX
WN4 0EQ M3ZRX
WN4 0QF G1LWX
WN4 0QX G8XLI
WN4 0RH G3ZUZ
WN4 0SJ G8CXW
WN4 0UA M6IKL
WN4 8AY G7IHE
WN4 8DL 2E0EKP
WN4 8DL M6HTB
WN4 8QT G0SIW
WN4 8RX M1BZJ
WN4 8ST G0UNG
WN4 8SU G7LTR
WN4 8SU M3BRT
WN4 8SU M3LTR
WN4 8UE M7FGV
WN4 8XA 2E0JCT
WN4 8XA M0ITT
WN4 8XA M6JGI
WN4 9DY G6OCB
WN4 9JN G4XCQ
WN4 9LD G3HUX
WN4 9NQ M7FEV
WN4 9PJ G6NTM
WN4 9SA M6LIQ
WN4 9SE M0HTR
WN4 9SE M0RGN
WN4 9SE M6CLB
WN4 9UP 2E0WES
WN4 9UZ 2E0RAG
WN4 9XA M7BNS
WN4 9XD 2E0BQO
WN4 9XD M3IQF
WN40RW M7GTD
WN5 0AJ G1DON
WN5 0PU G0DSR
WN5 0QD M7DFK
WN5 7AA 2E0DPT
WN5 7AF 2E0ARV
WN5 7AH 2E0NRB
WN5 7AH M6NRB
WN5 7AX G0NEO
WN5 7BG G4HAP
WN5 7EB G3PJW
WN5 7GE 2E0TOM
WN5 7GE M0TOM
WN5 7GE M7MOT
WN5 7HS 2E0ISQ
WN5 7HS M0ISQ
WN5 7HS M3ISQ
WN5 7HT G4EII
WN5 7JA M1DYU
WN5 7LH 2E0SYY
WN5 7LH M0SYY
WN5 7LH M6OFF
WN5 7PX G4LON
WN5 7TF G4VFR
WN5 8LX G7SOV
WN5 8NG 2E0ROI
WN5 8NG 2E1LED
WN5 8NG 2E1WRC
WN5 8NG M3ROI
WN5 8PL 2E0WEZ
WN5 8PL M3RMQ
WN5 8RH G7CWE
WN5 9BH G6ORO
WN5 9DL 2E1DFE
WN5 9EF M3JZF
WN5 9HZ G0TRY
WN5 9JP G4YPS
WN5 9JQ M6ATX
WN5 9NQ 2E1HZY
WN5 9NQ 2E1TCP
WN5 9PA G4KTU
WN5 9PJ M6AIF
WN5 9PR M6MTL
WN5 9QD M7CWN
WN5 9QL M0VGH
WN5 9QL M3RFX
WN5 9RS 2E0XAS
WN5 9RS M0KDO
WN5 9RS M3XAS
WN5 9TG G0UVL
WN5 9TU 2E0SVP
WN5 9TU M0PVS
WN5 9TU M7SVP
WN50RA M7DGE
WN6 0AQ 2E0ICK
WN6 0AQ 2E1HDE
WN6 0AQ M0HDE
WN6 0AQ M3POP
WN6 0AZ 2E0VCE
WN6 0BE M3NJK
WN6 0EU M7FMO
WN6 0EX G0MOK
WN6 0LW 2E0BJT
WN6 0LW G4PPG
WN6 0LW M3LBT
WN6 0NP 2E0TAJ
WN6 0NR G8EHF
WN6 0PW 2E0WWZ
WN6 0PW G4JTE
WN6 0PW M3WWZ
WN6 0QP M3TYQ
WN6 0QR G1MSR
WN6 0UA G6XKY
WN6 0UE M7EJM
WN6 0WB M6DXC
WN6 0XR 2E0NRS
WN6 7NH 2E0AWR
WN6 7PA G1VYA
WN6 7PA M3RRJ
WN6 7PU M0SCU
WN6 7RF G4EOT
WN6 8AU G0PBE
WN6 8DT G0GBP
WN6 8EH 2E0WPE
WN6 8EH M3HXO
WN6 8EY G7RGV
WN6 8HR G4KCP
WN6 8LH G0PPK
WN6 8LJ G8FHL
WN6 8LR M7GHW
WN6 8NJ G1EPF
WN6 8NJ G4PPB
WN6 8NS 2E0SAF
WN6 8NS M0OEG
WN6 8NS M6JSM
WN6 8PJ 2E0EQH
WN6 8PJ M6PFZ
WN6 8PL 2E0SYE
WN6 8QA G4SFJ
WN6 8QA G6PUV
WN6 9DJ G3RSW
WN6 9HU G8XXC
WN6 9JL M0HPT
WN6 9JN M3SJL
WN6 9JS G4EHK
WN68NS M3UCU
WN6-9LF G8JGU
WN7 1HN 2E0DMU
WN7 1HN G0SQX
WN7 1HN M0HZK
WN7 1HN M6FHY
WN7 1HU M6BSY
WN7 1HW M6TOE
WN7 1JZ G0MNY
WN7 1NB G3VGK
WN7 1NB G6HNQ
WN7 1SG G0KMB
WN7 1SG M7ELO
WN7 1TP 2E0TUR
WN7 1TP M3UAE
WN7 1TS G1HFS
WN7 2HH M0KDM
WN7 2HH M6MRD
WN7 2JJ G0SSK
WN7 2NG G6TET
WN7 2NG M3BMN
WN7 2NH 2E0GLW
WN7 2PL M6CYV
WN7 2RG G4UDF
WN7 2UU G4JXI
WN7 2UU M6LEY
WN7 2YR M1TET
WN7 3BU G3WIS
WN7 3DS M3HCA
WN7 3DW M3UVD
WN7 3EB G4COE
WN7 3ET 2E0YKC
WN7 3ET M0YKC
WN7 3ET M6TFV
WN7 3EX 2E0MMX
WN7 3EX M6MMI
WN7 3LB M7LIE
WN7 3LD M6GOS
WN7 3NE G8PUN
WN7 3PQ 2E1CJF
WN7 3PQ G6VPU
WN7 3PQ M0HOY
WN7 3PQ M0XHN
WN7 4GB M7SPW
WN7 4HT G7NRS
WN7 4HY 2E0DRZ
WN7 4HY M6DRZ
WN7 4HY M6DWT
WN7 4HY M6JPS
WN7 4SX G0LGC
WN7 4TA G1FLV
WN7 4TS 2E0JMC
WN7 4TS M0NWC
WN7 4TS M0OBZ
WN7 4TS M6ZFE
WN7 5BT G1VKT
WN7 5DG G6LDY
WN7 5DG M3FNT
WN7 5HG M1SJH
WN7 5NA M0LZQ
WN7 5NU M0HGO
WN7 5PN G4VXW
WN7 5PN G6HPL
WN7 5PN M0NAR
WN7 5PW G1DUS
WN7 5PZ 2E0DPY
WN7 5PZ M6PYE
WN7 5SW M6JGM
WN73HB M7RKL
WN75ET M7AAH
WN8 0AA G0MRM
WN8 0AA G6TKY
WN8 0AE 2E0WPB
WN8 0AE G4ACI
WN8 0AE M6JUJ
WN8 0AY G0FNA
WN8 0DE G4FEU
WN8 0JG G0UMY
WN8 0QT G3VWA
WN8 6AA 2E0FMP
WN8 6AA M6HTU
WN8 6AT G4WWG
WN8 6ED G6IKU
WN8 6JX M7LEM
WN8 6PA G6OMN
WN8 6PA M7BYF
WN8 6PE G4PJS
WN8 6RD M3NSG
WN8 6RJ G4JTP
WN8 6TA G3VSR
WN8 6TA G6HXL
WN8 6TE M0IYG
WN8 7AR G3PNQ
WN8 7DA M0SBH
WN8 7LA 2E0JFY
WN8 7LA M6JFY
WN8 7NB 2E0ESS
WN8 7NB M0TAA
WN8 7NB M3XQW
WN8 7NT G6ORS
WN8 7NT M3CDY
WN8 7RA G1DMW
WN8 8BD M3UHG
WN8 8EG G6YPB
WN8 8EN G6WEW
WN8 8EQ M6HHO
WN8 8HN M6PLB
WN8 8NP M3DLT
WN8 8NS M7MBT
WN8 8NW M3FRD
WN8 9BU 2E0RNB
WN8 9QU M6KPZ
WN8 9QZ 2E0MHD
WN8 9QZ M6DDH
WN8 9QZ M6GMH
WN8 9BP G0LAK
WN8 9BP G4SME
WN8 9BQ G8XEI
WN8 9DW M7OUT
WN8 9EF M3MJN
WN8 9JZ G1UOR
WN8 9NB G6UGE
WN89LZ M6PQV
WN9 8JR M6FAE
WR1 1NR G3RMF
WR1 1PA 2E1MFC
WR1 1PA M3CER
WR1 1QE G3VXH
WR1 1QZ 2E0UBU
WR1 1QZ M6WRW
WR1 2PJ M0TRX
WR1 3NY G8XMO
WR10 1HW G0FFF
WR10 1JH G1IPE
WR10 1JY G7TGK
WR10 1LL 2E0ENF
WR10 1LL M6GQS
WR10 1LW G1AFJ
WR10 1LW G4TXF
WR10 1NQ G7MYM
WR10 1PW G1TML
WR10 1PW G3UEY
WR10 1RE G7CRU
WR10 2AX G4OZQ
WR10 2BH M7RFM
WR10 2JE G4YIG
WR10 2LA G4RMV
WR10 2NY G4NIJ
WR10 2PJ 2E0VWJ
WR10 2PJ M7VJC
WR10 2QF M7JNY
WR10 2RJ G7RRJ
WR10 3BB 2E1EOD
WR10 3BW G4FAT
WR10 3EF G3XVX
WR10 3EF G8LQP
WR10 3EL G8BGT
WR10 3EL M6AGY
WR10 3EZ G0NXE
WR10 3HJ G4TDR
WR10 3HL M0CUU
WR10 3HS M6ZNF
WR10 3JZ G3ZKH
WR10 3NA G7WEB
WR102DQ M6OTN
WR102LB 2E0HUH
WR102LB G5KCP
WR102LB M7KCP
WR11 1BP M0ONO
WR11 1BU G4YJB
WR11 1EE G6NBL
WR11 1EQ G4EKG
WR11 1YW G4NRD
WR11 2AH 2E0RKX
WR11 2AH M0RKX
WR11 2AH M6RKX
WR11 2NB G7JSC
WR11 2NE M0PDY
WR11 2NE M3PDY
WR11 2NR G0WEO
WR11 2QA G4XQG
WR11 2QA G6XKX
WR11 2QF M6LUR
WR11 2QJ G1JLQ
WR11 2QZ G7JWW
WR11 2RD G8DKD
WR11 2SD M6WYR
WR11 2SF M7SHN
WR11 3BJ G3XCW
WR11 3BS M1PLC
WR11 3EA G0IBE
WR11 3HE G3KLZ
WR11 4JG 2E0IWJ
WR11 4JG M0KQJ
WR11 4JG M7FKX
WR11 4NL G6YXY
WR11 4NR G6DRC
WR11 4QA M6YZO
WR11 4TY G3LPU
WR11 6BE G8SQZ
WR11 7AA 2E0IDR
WR11 7AA M0IDR
WR11 7AA M6IDR
WR11 7EU G1VNV
WR11 7GQ G1SHU
WR11 7HQ G0GBL
WR11 7PT M3JKI
WR11 7QG G1UNQ
WR11 7RN 2E0FVR
WR11 7RN M3ZID
WR11 7RP G4UXC
WR11 7RP G4WET
WR11 7TA G1IME
WR11 7UE 2E0HDU
WR11 7UE M3CKD
WR11 7UZ G3CUF
WR11 7XF G4MFK
WR11 7XX 2E0BQL
WR11 7XX G1JMD
WR11 7XX M3MWQ
WR11 8JP G6YMY
WR11 8LL G1ZAA
WR11 8NG M0AFV
WR11 8QW M0TZT
WR11 8SN G1XWZ
WR11 8TL M3JFP
WR11 8XN G0KIN
WR113EE M7TPT
WR12 7AX M3DVM
WR12 7EP G8HFL
WR12 7HB G3IXI
WR12 7HS G6HAA
WR12 7JR 2E1RSH
WR12 7JR M7RSH
WR12 7NL G7OAS
WR12 7PF G3KAR
WR12 7QB G7EBL
WR12 7QB M5GJO
WR13 5AA 2E0MGR
WR13 5AA M6MGR
WR13 5DP G6FLR
WR13 5HX 2E0NON
WR13 5HX M0PYG
WR13 5HX M6PYG
WR13 5JF 2E0KEP
WR13 5JF M0KEP
WR13 5JF M6GYY
WR13 5LA G4FCA
WR13 5ND M0LSH
WR13 6AA G1RFX
WR13 6DL G3WIK
WR13 6ER G8EAJ
WR13 6ET G0RWY
WR13 6FA M0XZS
WR13 6NN G8AXV
WR13 6QW M1CYM
WR13 6SE G4VHV
WR14 1AD G4BVY
WR14 1AP 2E0GWS
WR14 1AP M6GWS
WR14 1BD M0OPG
WR14 1BU G1CBL
WR14 1DT G3RNP
WR14 1DX G4DFN
WR14 1FB G4GOA
WR14 1FU G3OOW
WR14 1HA G4ZAI
WR14 1HX G6CMV
WR14 1HX G7EME
WR14 1JX M0GEF
WR14 1LP G4FNZ
WR14 1LY G8UCL
WR14 1NB G0GFI
WR14 1NX G7TTH
WR14 1PH G7WIG
WR14 1PJ G6NAP
WR14 1PU G4IKJ
WR14 1PU G7JLC
WR14 1RQ G8IDK
WR14 1SB M3MXF
WR14 1SG M3AIZ
WR14 1TU M6NVR
WR14 1TY G4BZM
WR14 1UY M6ATK
WR14 2AG G3LWT
WR14 2DQ G2CKR
WR14 2DS M0XVX
WR14 2ET 2E0FXV
WR14 2HL M7PKW
WR14 2HU G0ENO
WR14 2HU G3UIK
WR14 2LE G4EYJ
WR14 2NF M0PAR
WR14 2NG G0FXD
WR14 2NH M7XJS
WR14 2NJ G4BFC
WR14 2NN G8YTU
WR14 2QA M6OWJ
WR14 2RH G4VWX
WR14 2SD G3CVK
WR14 2SR M0XCH
WR14 2SW 2E0DJI
WR14 2SW M0JJM
WR14 2SW M6DJI
WR14 2SY G1UQK
WR14 2TE G7HAS
WR14 2TU G8VUS
WR14 2UL G3WHJ
WR14 2UY 2E0JVP
WR14 2UY M6CDR
WR14 2WB M3RJI
WR14 3AY M6GUO
WR14 3BH G4RNX
WR14 3EA G4GHL
WR14 3EU 2E0TPA
WR14 3EU M0TPA
WR14 3EU M6TPX
WR14 3JH M0DQL
WR14 3JZ G3NQZ
WR14 3LQ M7FPK
WR14 3NJ M3RVJ
WR14 3NP G0WLG
WR14 3NP G7UAI
WR14 3QP G3WGY
WR14 3QP G8CMD
WR14 3QW G4IPY
WR14 4AH G1MFK
WR14 4AP G0GBC
WR14 4AT 2E0OAC
WR14 4AT M3OAC
WR14 4BX M6XLE
WR14 4BX M7DEY
WR14 4DW G3OAH
WR14 4JJ M7MSJ
WR14 4JR G6JBY
WR14 4LJ G6TEB
WR14 4LX G1IDZ
WR14 4NL G1FXS
WR14 4NL G3YPM
WR14 4PL G6BUY
WR14 4XE G4FPV
WR14 4XE G8LLS
WR141NB G1CKR
WR142ST 2E0BSW
WR15 8BX M7PCE
WR15 8DD 2E0RPW
WR15 8DD M6EIG
WR15 8EF 2E0HRE
WR15 8EF M7HJO
WR15 8EF M7SRE
WR15 8HX 2E0YRT
WR15 8JX 2E0LNC
WR15 8JX M6LNC
WR15 8LX G7RVW
WR15 8QN M3NPC
WR15 8QY G0DAZ
WR15 8SP G3BPF
WR15 8SR G0CCQ
WR15 8TW G1ENR
WR158QN M6IIF
WR2 4DJ G4RPC
WR2 4DJ G8SSE
WR2 4DP 2E1DLD
WR2 4DP G4RQF
WR2 4DP M3AQG
WR2 4DP M3KVH
WR2 4ES G8LJU
WR2 4HE G6FRB
WR2 4JQ G6CBP
WR2 4RB G7LUF
WR2 4SE G3TGD
WR2 4SF G4NXP
WR2 4SR G8URZ
WR2 4UE G4TLG
WR2 4XL M7GVP
WR2 5EQ 2E1HLS
WR2 5ND G6EQF
WR2 5PX G1EIP
WR2 5PY G1OVK
WR2 5QG M7HFD
WR2 5QR M0BXD
WR2 5QY G0BAM
WR2 5RR G1IVL
WR2 5RW G1AQF
WR2 5RW M0DEK
WR2 5SJ G7HNL
WR2 5SU G3TQZ
WR2 5TE G0RWS
WR2 6AA 2E0JEM
WR2 6AA M3JEM
WR2 6DA G1ICQ
WR2 6PQ G1JMN
WR2 6PQ M0JMN
WR2 6RH 2E0PMI
WR2 6RH M0PMI
WR2 6RH M6ILT
WR2 6RW G0AQB
WR3 7AN G4DEV
WR3 7BH M0WFO
WR3 7DQ M6CEN
WR3 7DY 2E0OJI
WR3 7DY M7EMI
WR3 7EH G4VZH
WR3 7HY G0PNR
WR3 7LR G4BYB
WR3 7ND G8PHV
WR3 7PJ 2E0CLE
WR3 7PJ 2E0SAN
WR3 7PJ M6EXI
WR3 7RX M6AZK
WR3 7TT M3NEP
WR3 7TZ 2E0EAN
WR3 7TZ M0PNA
WR3 7TZ M6ABN
WR3 7UP G4YZK
WR3 8BQ G0AOC
WR3 8EZ G1VKB
WR3 8NU M6DUF
WR3 8PJ G6TYT
WR3 8PP M0LUH
WR3 8QH M1FJH
WR3 8QT M6CEO
WR3 8RL M3XSY
WR3 8RL M3ZGT
WR4 0AX M6RPU
WR4 0EP 2E0YBS
WR4 0HY G8WJY
WR4 0HY M0CZE
WR4 0HY M6BSG
WR4 0JW 2E0RGO
WR4 0JW M0RGL
WR4 0JW M0VNG
WR4 0JY M7CTY
WR4 0PE G6DMG
WR4 0PN M7BEP
WR4 9AH M6LGB
WR4 9HS G3SPP
WR4 9JD M1ESI
WR4 9NT M3RFW
WR4 9RD 2E0HYP
WR4 9RD M3UZB
WR4 9TU M3SBE
WR4 9UG G3VQW
WR4 9XL M6TIC
WR4 9YU G0DEP
WR5 1HH M3WRZ
WR5 1JJ G4IDF
WR5 1JJ G4MHC
WR5 1JQ 2E0CLZ
WR5 1JQ M0IAJ
WR5 1JQ M6IAJ
WR5 1LL G1ZFS
WR5 1NL M6OWP
WR5 1QB G4ADJ
WR5 1QF M6EXF
WR5 1QX G1FQD
WR5 1SD G4KPL
WR5 2AA 2E0DGV
WR5 2AA M0XJT
WR5 2AL G6PMO
WR5 2BD G1JGF
WR5 2BD G8STI
WR5 2BG G0WJN
WR5 2DU 2E0RNA
WR5 2DU M0XPB
WR5 2DU M6RNA
WR5 2DU M6ZPB
WR5 2ET G4URN
WR5 2HL G4HNZ
WR5 2HL G8PIN
WR5 2JL 2E0CAL
WR5 2JL 2E0DAJ
WR5 2JL M3VAQ
WR5 2JT G1CSS
WR5 2JT G7SLV
WR5 2JU G4ELL
WR5 2LN G4GIM
WR5 2NG G0SMZ
WR5 2NL M3IMM
WR5 3AL G6UHL
WR5 3AY 2E0RHE
WR5 3AY M0TBW
WR5 3AY M6CNY
WR5 3FD M7KZD
WR5 3HD G1WPH
WR5 3HD G6NFJ
WR5 3HT G0AAT
WR5 3JG G4LVV
WR5 3LX M0VDU
WR5 3NX G8ASJ
WR5 3QB G1XBL
WR5 3SH G7CRR
WR5 3SJ G6UXF
WR5 3SY G4NMR
WR5 3SZ G1EME
WR5 3SZ G6FZV
WR5 3UA M0TXD
WR6 5LZ G4KTW
WR6 5NE G3GNA
WR6 5NG 2E0BTS
WR6 5NG M6GJT
WR6 5PS G0PYW
WR6 5RD G3RYH
WR6 5SR G6CXV
WR6 5SR M0ELA
WR6 5TB 2E0PTY
WR6 5TB M7ICY
WR6 6AY G1MAC
WR6 6AY M7ERJ
WR6 6DW G6SOA
WR6 6EB G6JJP
WR6 6HQ G3WLG
WR6 6ND G4OPD
WR6 6QA G4AAL
WR6 6QA G4BXS
WR6 6QX G4OWK
WR6 6TQ M3IQJ
WR6 6UB M1AFY
WR7 4AP G6VVE
WR7 4BT G0CRX
WR7 4EG M3NRV
WR7 4PR G6ZHF
WR7 4RF G1XWK
WR7 4RH G8LPN
WR8 0AB G4WMB
WR8 0AT M6KVD
WR8 0DS G4TQY
WR8 0ET G6GWU
WR8 0LR M3ZLI
WR8 0NU 2E0XDH
WR8 0NU M6UDB
WR8 0PZ G4WPT
WR8 0QQ G1PQO
WR8 0RH G8PAK
WR8 0SJ G8XQL
WR8 9AN G0EMS
WR8 9BE G1AKB
WR8 9EH G6XLG
WR8 9JD 2E0HXU
WR8 9JD M7DCK
WR8 9JR G3URZ
WR8 9JR G4JKA
WR8 9LP G8LCM
WR9 0AU G1DYC
WR9 0DF G6FOF
WR9 0DN G8GTU
WR9 0DX G4HPD
WR9 0EZ 2E0TTH
WR9 0EZ M6PMT
WR9 0JW 2E0DUO
WR9 0JW G1DUO
WR9 0JW M5DUO
WR9 0JW M6DUO
WR9 0QY G6GMH
WR9 0RP G1YBI
WR9 0RP M0XYL
WR9 7AF G3RLF
WR9 7AY G3KTH
WR9 7AZ G3TRB
WR9 7DQ G4RQO
WR9 7EB G0WFK
WR9 7FD M7CNM
WR9 7HF G8XGG
WR9 7HG G8JLY
WR9 7NL G6WPO
WR9 7PE G6NYG
WR9 7PU M7VEN
WR9 7QE G0BLS
WR9 7QU G4THU
WR9 7RX G0WIS
WR9 7SE G4HDO
WR9 7SP G7CTT
WR9 7SW M0ITC
WR9 7TD G3XMH
WR9 8HG G3VPH
WR9 8NH M7EGB
WR9 8NR 2E0NRW
WR9 8NR M6NRW
WR9 8QR G4TID
WR9 8RU M7RTU
WR9 8SS G7VDX
WR9 8TQ G7VJM
WR9 8TQ M0BQE
WR9 8UD G0BDM
WR9 8WA G4KXP
WR9 9BZ G8HHR
WR9 9ED 2E1IAY
WR9 9ED 2E1IAZ
WR9 9EG G0AXB
WR9 9HU G7RVT
WR9 9LA 2E1CBH
WR9 9LA G4LVO
WR9 9LG G4BYM
WR97JP M3JJT
WS1 2AZ 2E0DNS
WS1 2AZ M6BSX
WS1 2BQ M7SCY
WS1 2DA G0FSL
WS1 2DA G3PJY
WS1 2DA G4GUW
WS1 2DH M6GWZ
WS1 2PJ G4EJU
WS1 3AL G4KBA
WS1 3HB G0HUD
WS1 3HL M3BJW
WS1 3LB G1MMN
WS1 4DZ G4YVF
WS1 4HF G4FCB
WS10 0BH M3NXK
WS10 0DN G7EKT
WS10 0EU 2E0WDB
WS10 0EU M3IPT
WS10 0EW M6ISA
WS10 0HQ M3NZN
WS10 0JF G0EAN
WS10 0PN G0WJJ
WS10 0TB G0CMR
WS10 7HY G7CNZ
WS10 7RH M3DAM
WS10 7RH M3RYA
WS10 7RQ G1RBA
WS10 7RR G6RTG
WS10 8NS 2E0VGB
WS10 8NS M0JGB
WS10 8NS M6FQW
WS10 8QH M0KHD
WS10 8QP M0IXZ
WS10 8QP M7AOR
WS10 8RN M3WCI
WS10 8RQ M6UXB
WS10 8SH M1PTE
WS10 8UB G6ZYX
WS10 8UN M6LMU
WS10 8XY G1WRU
WS10 9BQ 2E0SOZ
WS10 9BT G1PQY
WS10 9BX M7ODT
WS10 9EW M7WMY
WS10 9HF 2E0EDF
WS10 9HF M6ZGY
WS10 9HH M0CRU
WS10 9LH G6IDO
WS10 9NF M0JPB
WS10 9PH G4TDP
WS10 9PU M7LCR
WS10 9RQ G6CSR
WS10 9UA G6JAM
WS100LW G0JAA
WS108ED M7FEP
WS108HJ 2E0MBK
WS108HJ M6MEL
WS109BS M7JST
WS11 0AF G7TCW
WS11 0AQ M3WDC
WS11 0AR M6UKC
WS11 1AR G1BRD
WS11 1AR G7PCV
WS11 1AZ G1UUL
WS11 1BB G3PSU
WS11 1ET G4JNK
WS11 1LJ G3URL
WS11 1NQ G4ICE
WS11 1NS M3RHG
WS11 1NZ 2E0KUF
WS11 1NZ M6DIV
WS11 1PE 2E0OWH
WS11 1PW M3SLQ
WS11 1RU G8GRC
WS11 1SH G1FTJ
WS11 2SE G0UYH
WS11 4BH G7IPI
WS11 4BQ 2E0RGR
WS11 4PT M3NOM
WS11 4SA G0WAY
WS11 5HB M3OVF
WS11 5HD G0CWB
WS11 5HD G6SW
WS11 6EH M3MXX
WS11 6EN M6DHG
WS11 6LX 2E0ZXQ
WS11 6LX M0ZXQ
WS11 6LX M3ZXQ
WS11 7FR G7DWH
WS11 7GB 2E0MGX
WS11 7GB M0WFM
WS11 7GY 2E0KBA
WS11 7YX 2E0IQX
WS11 7YX M0IQX
WS11 8ES 2E0OUK
WS11 9FQ G4JAV
WS11 9QN M0NDZ
WS11 9QZ G4YTK
WS11 9SP M3NCO
WS11 9TN G7IZN
WS11 9TP G4TCC
WS11 9UR M1BZH
WS110EZ M0PET
WS12 0FR M6DVH
WS12 0PR G0VSK
WS12 0QE G1LFR
WS12 0SX G0MVV
WS12 1BG G1NZQ
WS12 1BG M0SST
WS12 1BS G4KWQ
WS12 1BS M1FJC
WS12 1BS M3STR
WS12 1BW M6CDE
WS12 1HY M7ODH
WS12 1LD G6ODF
WS12 2AW G4RJD
WS12 2DW G8NTJ
WS12 2EZ M3KUJ
WS12 2GF M3NPI
WS12 2GH M3FYZ
WS12 2GR 2E0DFP
WS12 2GR M0GYO
WS12 2GR M6DFP
WS12 2GS G6LNA
WS12 2RN G0CCF
WS12 2RN G7WBH
WS12 2TR M7TPO
WS12 3DS G1SIG
WS12 3HA M1CYR
WS12 3XL M7CAD
WS12 3YA M3IXY
WS12 3YJ G0XAT
WS12 4BD 2E0FWN
WS12 4DP 2E0DVU
WS12 4DP M6DVU
WS12 4DQ 2E1EXK
WS12 4DQ G0DUQ
WS12 4DQ M3EXK
WS12 4DS 2E0DCJ
WS12 4DS M6CQE
WS12 4JT G7AMD
WS12 4LF M3YXS
WS12 4LR G4JVT
WS12 4LT G4VYE
WS12 4QA G8GCC
WS12 4QA G8VPR
WS12 4RA G1HYM
WS12 4RW 2E0HRR
WS12 4SN 2E0WGT
WS12 4SN 2E0WJW
WS12 4SN G3ZVK
WS12 4SN M0WTN
WS12 4SN M7BAW
WS12 4SN M7JAW
WS12 4SS 2E0WTD
WS12 4SS M0NEG
WS12 4SS M6AAP
WS12 4SU G0ICW
WS12 4TP M3NNY
WS12 4TQ G7ASY
WS12 4UJ M7WWR
WS12 4UP M7BVI
WS13 6AU G0RRM
WS13 6BF G4GGH
WS13 6BH G6KTC
WS13 6DA M0HPB
WS13 6DB G6OHQ
WS13 6DQ G0HQX
WS13 6EP M6THO
WS13 6EP M6TYB
WS13 6FL G1SMP
WS13 6NJ G6NAJ
WS13 6RL 2E0JGL
WS13 6RL M7LXY
WS13 6SD M3YJY
WS13 6ST G3PFT

WS13 6SU G7KMT
WS13 7ED G0OYM
WS13 7LT M1BKQ
WS13 7LW G8XXA
WS13 7LZ G0DRA
WS13 7NQ G0PHE
WS13 7PW G0EVJ
WS13 7SN G8NXE
WS13 7SP G6RKQ
WS13 8AA G8BFL
WS13 8EF G8GSL
WS13 8JE G8EQC
WS13 8LJ M7EDE
WS13 8NJ G1JLX
WS13 8NY 2E0DCX
WS13 8NY M0VAD
WS13 8PW G7IXG
WS13 8UZ G6LWZ
WS14 0AD M7AJP
WS14 0JH G8NTY
WS14 9DA G4NPY
WS14 9DP 2E0JPX
WS14 9DP M0JPX
WS14 9DP M6ELC
WS14 9HH G4OUM
WS14 9LN G0DKZ
WS14 9NH G0RIF
WS14 9PF G7BMC
WS14 9QT 2E0KRT
WS14 9QT M3KRP
WS14 9RJ G6NEZ
WS14 9SF G6HZH
WS14 9SN G8HYK
WS14 9SZ G3YPD
WS14 9US G1OBC
WS14 9US G8KHV
WS15 1AD M3FSY
WS15 1AT M6SLG
WS15 1BB G6GOW
WS15 1BJ M6TNC
WS15 1EJ 2E0JWH
WS15 1EP G4NLK
WS15 1EW M1AYA
WS15 1EZ G8TNS
WS15 1JE G4LAN
WS15 1JJ 2E0RPC
WS15 1JJ M3XCF
WS15 1NT 2E0FAQ
WS15 1PE G1XLE
WS15 1QA G2HKS
WS15 1SR G0JJP
WS15 2AR G7ACG
WS15 2AR G7JMB
WS15 2AW M7RNI
WS15 2DD G3TJA
WS15 2EE G7ULC
WS15 2EE G8OTC
WS15 2ES M6SJB
WS15 2GY M6JJF
WS15 2JL G1IDQ
WS15 2LL M3DIY
WS15 2LL M3TOY
WS15 2NP M0EWW
WS15 2NS G7VJY
WS15 2PE M0TTO
WS15 2QB 2E0BXK
WS15 2QB M6HOP
WS15 2QP M6JSH
WS15 2QW G7KBH
WS15 2QY G1WVZ
WS15 2RE M6FHM
WS15 2XH G1KQH
WS15 2YG G0JJO
WS15 2YG G3TCY
WS15 2YG G4PWD
WS15 3HJ G8MIA
WS15 3HR 2E0YGT
WS15 3HR M7IGT
WS15 3LD G7HZU
WS15 3QU M3TKQ
WS15 3QX 2E1FKJ
WS15 3QX G4DBR
WS15 4DQ M6KEA
WS15 4EF G8JNR
WS15 4HF M6XWG
WS15 4QG G8XGK
WS15 4RG G3NSO
WS15 4RG M3XAR
WS15 4TH G6NVS
WS2 0HJ G0JVN
WS2 0HZ G8ZBJ
WS2 0NH G0MPR
WS2 0NT M3AGH
WS2 7BB G4EFG
WS2 7BG G8SCI
WS2 7DB M7DBM
WS2 7EJ G7DHQ
WS2 7EN M3RMX
WS2 8NP M3UFK
WS2 8QL 2E0WCM
WS2 8QL M6MLL
WS2 8QP M7JSW
WS2 8RE G7TGG
WS2 8RR G4RVK
WS2 8RX G0HWP
WS2 9BQ M6JYY
WS2 9DJ M6FYT
WS2 9HA M7JBT
WS2 9PZ 2E0YAX
WS2 9PZ M6YAF
WS2 9QB G6ZJS
WS2 9QZ M0AVL
WS2 9TN M7PCW
WS27HH M0KPP
WS3 1AL G7CST
WS3 1BE M7DJM
WS3 1DF M3JRR
WS3 1DL 2E0FRO
WS3 1DL M6BJF
WS3 1DS M7SPP
WS3 1EW G8ZHA
WS3 1JB M7SPA
WS3 1JX M6CAS
WS3 1LQ M6WOS
WS3 1NQ G1GZM
WS3 1NW M3CKU
WS3 1PU M7JAR
WS3 1PY M3LLT
WS3 1QJ G1KPE
WS3 1RF M6GTD
WS3 1TT G4PPP
WS3 2AL M3BJH
WS3 2AQ G6FOX
WS3 2AQ G8DEJ
WS3 2AR M6JJS
WS3 2AT M3CWZ
WS3 2EG M3TLU
WS3 2EZ G4PPC
WS3 2EZ G7PPC
WS3 2HD M7TDA
WS3 2HT G0MKW
WS3 2NG M1AGW
WS3 2QF M1EVH
WS3 2RJ M6LBR
WS3 2SS G0VFB
WS3 2TQ 2E0GKQ
WS3 2TQ M0MGW
WS3 2TQ M7BBS
WS3 2UQ M6HPF
WS3 3AU M0TMF
WS3 3AU M1TMF
WS3 3BZ M1EBI
WS3 3DG G7GCU
WS3 3DX M3RET
WS3 3EA G4XNK
WS3 3EA M6CKI
WS3 3EN M6DGO
WS3 3NB G1LUF
WS3 3PG G0GBE
WS3 3PG G1SJB
WS3 3QD G0CXO
WS3 3QD G1YWI
WS3 3RF G8IJE
WS3 3SA M6NJN
WS3 3XB G0CNG
WS3 4AG 2E0DEC
WS3 4AG M6DDC
WS3 4AH M6HGA
WS3 4AH M6HGD
WS3 4DH G6BYL
WS3 4DQ 2E0KLM
WS3 4DQ M7AKS
WS3 4EZ M0LRD
WS3 4LS G0DAC
WS3 4LW G3GGR
WS3 4PB G4OKE
WS3 4PU G1WQH
WS3 4PU G1XLW
WS3 4QN G7BNK
WS3 4QS 2E0EUP
WS3 4QS M6NHW
WS31JX M6CEZ
WS4 1AP 2E0LEF
WS4 1AP M0SPA
WS4 1AP M0VSP
WS4 1AP M6APG
WS4 1AP M6ENF
WS4 1AP M6LIN
WS4 1BB M6OYA
WS4 1DQ G0IMX
WS4 1DS G0NEQ
WS4 1DS G4GRM
WS4 1HP G1KVP
WS4 1HT G4YZM
WS4 1JB G0DQO
WS4 1JX G7IIZ
WS4 1LE 2E0HIL
WS4 1LE M7BOW
WS4 1NG M3ZUL
WS4 1QP M0CSB
WS4 1QU 2E0DUZ
WS4 1SA 2E0PPA
WS4 1SA M6CKL
WS4 2AB M3XTR
WS4 2DF G0CLX
WS4 2HB G4GJE
WS41NH 2E0ZYK
WS41NH M3ZYK
WS5 3AG 2E0VWD
WS5 3AG M7PWT
WS5 3AW G0KRY
WS5 3DH G3YHN
WS5 3DH G8MLW
WS5 3DT G1RLR
WS5 3DT G1UBC
WS5 3ES G4GKC
WS5 3ES G4HLL
WS5 3NQ G4SEL
WS5 3PN G8NQY
WS5 3QJ G6JVA
WS5 3QX G3ZHC
WS5 4BX 2E0YAH
WS5 4BX M3UTZ
WS5 4DN 2E1EXI
WS5 4DN G3TIN
WS5 4DN G8HTM
WS5 4HE G4TCM
WS5 4LU M0GRX
WS5 4NH G8XXU
WS6 6DF M0BHQ
WS6 6DT G4CJK
WS6 6DT M7LWF
WS6 6HA G8IHC
WS6 6HE G4PFO
WS6 6JA G8PWE
WS6 6JA M3WIX
WS6 6LD G4JUK
WS6 6LL M1CSC
WS6 6LQ G7TCQ
WS6 6LS G7GDA
WS6 6PE G8MFU
WS6 7BE M3VTL
WS6 7BX G0GPB
WS6 7DP G4TQC
WS6 7EJ G6RIJ
WS6 7EJ G8TRQ
WS6 7EU G6RIG
WS6 7EZ M3UXX
WS6 7EZ M3VEY
WS6 7HD G1TNR
WS6 7HR G6EPQ
WS6 7LE 2E0JPT
WS6 7LE M3MQC
WS7 0BB G0HKF
WS7 0DX M6GHQ
WS7 0ED G4POR
WS7 0EE G6JJB
WS7 0EQ G6PXN
WS7 0EQ G6UCO
WS7 0ES G0PJR
WS7 0ES G6PAP
WS7 0HA G1GKH
WS7 0LH M6OZL
WS7 0LQ 2E0CAP
WS7 0LQ 2E0SDV
WS7 0LQ M3XIF
WS7 0LQ M3XIH
WS7 0LQ M6GDI
WS7 1FA G0GFC
WS7 1FA G6DFC
WS7 1NQ G7DOS
WS7 1PW G0AGU
WS7 2AS G4ZBE
WS7 2AS G8RDN
WS7 2DL G6PUR
WS7 2EL G8HTW
WS7 2HS G6NPE
WS7 2HU M3ZUD
WS7 2LL M3IZI
WS7 2PA G1FSE
WS7 3GD 2E0BRT
WS7 3GD 2E0KGD
WS7 3GD M3VPJ
WS7 3GD M6KGD
WS7 3QL G8PTL
WS7 4GU G0VSL
WS7 4QJ M7CAZ
WS7 4QQ M0KIG
WS7 4QS G1XYO
WS7 4QU G3PET
WS7 4QU G8MII
WS7 4SU G4ZYL
WS7 4US G0PCP
WS7 4YD G3RTY
WS7 8BT G3LXB
WS7 8TX G0EOG
WS7 9AB G4TSD
WS7 9AE M6ERP
WS7 9AQ G0DAY
WS7 9AT G0MCM
WS7 9BT G1PGD
WS7 9DA M6MKH
WS7 9DS G6OTZ
WS7 9EA G3NLY
WS7 9EA G3WAS
WS7 9EJ G8BMP
WS7 9JS G4EQC
WS74YD M7RPR
WS8 6AG G8OLL
WS8 6BB M0XCJ
WS8 6BG M0OSH
WS8 6BY 2E0CDL
WS8 6BY M0RUK
WS8 6BY M3KDL
WS8 6DL G8KXO
WS8 6EZ M6JKD
WS8 6JF G0MPO
WS8 6LN M3LQY
WS8 7AY G0OEI
WS8 7BZ G0FBO
WS8 7BZ G1JMW
WS8 7EB M6XMS
WS8 7JA M3TRH
WS8 7LH G4XOG
WS8 7LL M6NNK
WS8 7LS M7REP
WS8 7ND 2E0CVP
WS8 7ND M6CYO
WS8 7PL G3OHN
WS87BN G6HPE
WS9 0BP M6MPE
WS9 0DB G4JSQ
WS9 0DT G0ENM
WS9 0DU G7IXP
WS9 0QA G4GJU
WS9 0QE G4XFT
WS9 0TG G1BQH
WS9 8HN M0DWM
WS9 8JE G4EWI
WS9 8JG G4FAP
WS9 8JG M3IEF
WS9 8LA G4PJK
WS9 8RU G6IOT
WS9 8SN G1MJA
WS9 8XE G0OEB
WS9 9BD G1FOZ
WS9 9BD G7WCB
WS9 9EA M1CHF
WS9 9HU G8KSZ
WS9 9JR G0ALC
WS9 9PD G4FAJ
WS9 9RD G8LPI
WS90JT M6FOT
WV1 1QH G4MJI
WV1 1QQ G0MCE
WV1 2AR G4SGE
WV1 2LD M0TCR
WV1 2LD M1TCR
WV1 4PL G4JUD
WV1 4PL G4PVZ
WV10 0HR G6ZSH
WV10 0NH M7GOT
WV10 0SH G4ZBC
WV10 6BQ G0GEL
WV10 6BZ M6FDL
WV10 6DW 2E0DAI
WV10 6DW M0YDH
WV10 6DW M6EMP
WV10 6DW M6GGZ
WV10 6DX G4IGG
WV10 6EN G4OSU
WV10 6EZ G1AEF
WV10 6EZ M6NHG
WV10 6NN G1VAN
WV10 6NW G8ZWU
WV10 6QS M6TTV
WV10 6RU G7MEG
WV10 6RU M3MEG
WV10 6SL M3UJV
WV10 6XH G3FWD
WV10 7BB G4CHJ
WV10 7HS G7NZM
WV10 7HS M3PQQ
WV10 7JS G4URM
WV10 7JS M6TPR
WV10 7JT M3YDM
WV10 7NP G7RCU
WV10 7NP M1DFK
WV10 7TY 2E0XCV
WV10 7TY M3XCV
WV10 7TY M3XIG
WV10 8AB M0EIW
WV10 8AY G1MSB
WV10 8HJ M3UFJ
WV10 8JZ M0EQY
WV10 8NB M3XKN
WV10 8NS 2E0UHL
WV10 8NS M6BQE
WV10 8QG G6ZYL
WV10 8SL M6PDJ
WV10 8TH G6UVU
WV10 8UN M6NTN
WV10 9BU G4HVR
WV10 9DB M7KEN
WV10 9HN M0AXG
WV10 9HN M3SAB
WV10 9LZ G0SOY
WV10 9PP 2E1HHY
WV10 9PP M0SRB
WV10 9RJ M0ZUI
WV10 9SJ M3WDB
WV107HS M3YYK
WV109RL M7GFY
WV11 1AH M3LLQ
WV11 1AP G6MJM
WV11 1BD G6LWD
WV11 1BG G0ILH
WV11 1BH G8FVM
WV11 1DD M6BIJ
WV11 1EG 2E0PPR
WV11 1NW M3WDK
WV11 1PR G7CFC
WV11 1RF G6JYR
WV11 1SF M0KBJ
WV11 2DB G6ALZ
WV11 2DW G4WAS
WV11 2HL M7DZE
WV11 2HR M0WAY
WV11 2HR M6TDZ
WV11 2JS M3UKR
WV11 2LB M7HJH
WV11 2LB M7WLH
WV11 2ND 2E0FRY
WV11 2ND M6FRY
WV11 2PB 2E0EOC
WV11 2PD G1BCE
WV11 2PP M0CVK
WV11 2PZ M3GLA
WV11 2PZ M3JLA
WV11 2PZ M3RDA
WV11 2RQ G4FGR
WV11 3AQ 2E0ZTE
WV11 3AQ M0ZTE
WV11 3AQ M6ZTC
WV11 3AW G4CGL
WV11 3EG G0KYK
WV11 3EZ M7GFX
WV11 3HR 2E0WJL
WV11 3HX M0GTP
WV11 3NL G7BIY
WV11 3NQ M7CPC
WV11 3NY 2E0EYH
WV11 3NY M0WRU
WV11 3QJ M3KJD
WV11 3RT G0EGG
WV11 3SF M7WEB
WV113EW G6CKJ
WV12 4AN G4JZF
WV12 4GA G1GRU
WV12 4GA M0WRS
WV12 4JA M6CRZ
WV12 4LX 2E0LHZ
WV12 4LX M7DQF
WV12 4PX 2E0MYH
WV12 4QF G4MAR
WV12 4SL G1XII
WV12 5AU G4CFH
WV12 5DT M7WVT
WV12 5FQ G4FJV
WV12 5HW M3JMX
WV12 5RD M0BQF
WV12 5RD M0JOD
WV12 5TD G4VIT
WV12 5YJ G1YLJ
WV13 1AP G3TQL
WV13 1AW 2E0MAY
WV13 1JQ G8GMA
WV13 2RY G6HRX
WV13 3BD M3UWY
WV13 3DG G0KRK
WV13 3DQ G0OVK
WV13 3HJ M3SBS
WV13 3PB G1PBF
WV13 3PB G4BXC
WV13 3PB M6EXR
WV13 3QA G8JBC
WV13AP M6ITI
WV14 0HT M3ZYM
WV14 0JT 2E0YDA
WV14 0JT M3VTV
WV14 0TN 2E0LST
WV14 0TN M3YLT
WV14 6EW 2E0TWT
WV14 6NP G8FMW
WV14 6NZ G6YWL
WV14 6NZ M0GRF
WV14 6PU M3XOU
WV14 6RX G7VYW
WV14 7BD M6YMN
WV14 7BN G8JHO
WV14 7NP M7BBV
WV14 8AP 2E0OTB
WV14 8AP M6AQD
WV14 8AU M3GKX
WV14 8DB M6PAS
WV14 8DE M0AXO
WV14 8DE M6GJM
WV14 8EA 2E0MDR
WV14 8EA M6MVB
WV14 8EL G7UWV
WV14 8EW 2E0OKY
WV14 8EW M3OKY
WV14 8HX G4LUB
WV14 8JE G7WGD
WV14 8JE M6HWQ
WV14 8LN M6DLI
WV14 8LW G0ULO
WV14 8SB M0GIL
WV14 8SB M0SHY
WV14 8XD G0WEB
WV14 8YH G1RBY
WV14 8YW G1KEB
WV14 9AN G0JCN
WV14 9AX 2E0SPB
WV14 9AX M0SPB
WV14 9AX M3PPY
WV14 9BE 2E0GCW
WV14 9SE G0SMO
WV14 9SE M6FZX
WV14 9TB 2E0WLY
WV14 9TB M6WOL
WV14 9UJ G0WCI
WV14 9UT M0BHN
WV14 9XF G4ETW
WV14 9XF M0MSG
WV15 5BS M3OPA
WV15 5DJ M3SRV
WV15 5DS G0UYE
WV15 5HH M1KDH
WV15 5PB G0BXD
WV15 5PH M3GHL
WV15 5PH M3KCP
WV15 6BU 2E1IHK
WV15 6ER M6ROL
WV15 6JL 2E0BLC
WV15 6JL G3ZUL
WV15 6LT G0IZE
WV16 4HU G8UPF
WV16 4JD G6JMG
WV16 4JD G8RDQ
WV16 4JS G0RQG
WV16 4JW G1SVR
WV16 4JW G3SVR
WV16 4JW G3TVR
WV16 4NW G4NKC
WV16 4SJ G6GUH
WV16 4SP G7LEX
WV16 5AE 2E0HXH
WV16 5AH G6SXD
WV16 5JQ 2E0TDV
WV16 6LQ G4YGT
WV16 6LQ M6FCC
WV16 6PP G6OZQ
WV16 6PR G8CBA
WV16 6TW G3YKI
WV166RP G0GXT
WV2 1HG M0TBA
WV2 1JA M3RZB
WV2 1JA M3XMB
WV2 2AU G7BZQ
WV2 2HF M3VOB
WV2 2QZ M3SPJ
WV2 3ET G1RRU
WV2 4PS 2E0GRW
WV2 4PS M0RNW
WV2 4PS M6MMR
WV3 0EB M0WEV
WV3 0PY G1LEX
WV3 7DN G1GVP
WV3 7EB G8RVB
WV3 7HZ G7GJA
WV3 7LZ G3UBX
WV3 7NJ G7CXB
WV3 8AS 2E0MGI
WV3 8AS M0MGI
WV3 8AS M6EKQ
WV3 8DN M0HJQ
WV3 8HJ G3ZLJ
WV3 8HW 2E0YEZ
WV3 8HW M6RNM
WV3 8HY G8BWP
WV3 8JT G0TQP
WV3 9AR G1BKB
WV3 9AR M3NXQ
WV3 9BL G4WRB
WV3 9HN G4BTE
WV3 9HZ G0CYO
WV3 9JF G8EBD
WV3 9LW G3URJ
WV3 9PX G6ZFG
WV3 9RJ M1BSU
WV4 4AY G3RQX
WV4 4DN 2E0AQU
WV4 4DN G1RLI
WV4 4DN M0BMN
WV4 4JJ G4LWC
WV4 4LG 2E0SWB
WV4 4LN 2E0FSQ
WV4 4LN 2E0FWY
WV4 4LN M6FWY
WV4 4PF 2E0KAX
WV4 4PF G0NWV
WV4 4PF M6NAO
WV4 4RQ G7ACJ
WV4 4ST G0BKQ
WV4 4TR G1ELJ
WV4 4UL 2E1DWP
WV4 5AQ G6UFZ
WV4 5HH M6FQJ
WV4 5LN G4XNE
WV4 5QY G4XEZ
WV4 5RP G7DDQ
WV4 6QY G7SVM
WV4 6RJ 2E1EUY
WV44YN M7RRT
WV5 0BD G3ZGY
WV5 0BQ G6MZN
WV5 0EA G0TUN
WV5 0EA G1ZAK
WV5 0HW G4WLK
WV5 0HY M3AZE
WV5 0JX G1DFF
WV5 0LF M6GFV
WV5 0LG G1BAA
WV5 0LH G3SOE
WV5 0LX M1AUN
WV5 7HF G6AKN
WV5 7HT 2E0TGB
WV5 7HT M0WBF
WV5 7HT M6AZY
WV5 8AF G0OGS
WV5 8AZ M3GWC
WV5 8BH G3TPP
WV5 8DL M6OWM
WV5 8EF G1ZTK
WV5 8HB G0PMG
WV5 8HQ G4FIF
WV5 8HQ G4WRX
WV5 8JE G8KBG
WV5 8JJ M3UAJ
WV5 9AJ G0CRO
WV5 9AW G4OTS
WV5 9BP 2E0BLY
WV5 9BP 2E0KMB
WV5 9BP M0KMB
WV5 9BP M3JTU
WV5 9BP M3WUX
WV5 9ER M3TKW
WV5 9HR G4CGB
WV5 9HX G4MZI
WV5 9HX G8JVW
WV6 0AT M7AGU
WV6 0BJ G0EGP
WV6 0NT M3ULX
WV6 0SF G4TDO
WV6 0SF G6MZT
WV6 7AQ G1GTK
WV6 7AZ M7PSI
WV6 7DU G4CYO
WV6 7LU G7UJS
WV6 7LX G1DKI
WV6 7NB M0UDK
WV6 7NQ 2E0RZX
WV6 7NQ M0RZX
WV6 7NQ M6RZX
WV6 7PH M0MTJ
WV6 7QY M0CQO
WV6 7RR G0OKT
WV6 7SB G6HDF
WV6 7SX G7RXW
WV6 7SY M6PYO
WV6 7YH G8CZM
WV6 7YY G3IOB
WV6 8HJ G8CAF
WV6 8LA 2E0IHN
WV6 8LA G1SGA
WV6 8LA M6IHN
WV6 8PT M6WWB
WV6 8TR G0HDD
WV6 8UZ G8ZIK
WV6 9AZ G8GXF
WV6 9HB G4DHL
WV6 9JJ G6VDY
WV6 9JY 2E0GOZ
WV6 9JY M0TFF
WV6 9JY M6JYQ
WV6 9LF G8UYR
WV6 9NQ G6JJG
WV68TS G4HRG
WV7 3DQ G4VYA
WV7 3DZ M3WFB
WV7 3EN G0ISI
WV7 3HL G7JPN
WV7 3HW G0RNX
WV7 3LF 2E0JDP
WV7 3LF G7FEF
WV7 3LF M0NJJ
WV7 3LF M6NJP
WV7 3LS G4EAB
WV8 1ES G8IMH
WV8 1HA G6ZFZ
WV8 1JP G8XUE
WV8 1NT G8RF
WV8 1PG G1LGJ
WV8 1PG G4EVP
WV8 1SA G6DET
WV8 1SG M0HKE
WV8 2AJ G0GIE
WV8 2AW G3KNG
WV8 2BE G6YIU
WV8 2BY G0MRP
WV8 2DJ G8JBT
WV8 2DS G0LDY
WV8 2HA G8JIS
WV8 2JT G7ETM
WV9 5DE G4GZK
WV9 5PH G1BWG
WV9 5RN G4FAQ
YO1 6DB 2E1CIK
YO1 7LW M3XYT
YO1 9QP 2E1LSI
YO1 9QP G5LSI
YO10 3AY M1CJN
YO10 3BS G8JVS
YO10 3HE G0PMI
YO10 3JX G0WUY
YO10 3LJ M6FSA
YO10 3LR G4IIX
YO10 3PD 2E1BRG
YO10 3PD M1AQI
YO10 3QB G4KCT
YO10 3QB G8SFI
YO10 3QH M1CJM
YO10 3QJ G6MCQ
YO10 3RA M3XCJ
YO10 3TS M7MWY
YO10 4PB G3OZE
YO10 5DL 2E0KZB
YO10 5DX M1FAI
YO10 5EU G8OWV
YO10 5HH 2E0RMC
YO10 5HH M0HNL
YO11 2AF G4IAJ
YO11 2AR G0KJK
YO11 2BJ G0OII
YO11 2DA G3NZY
YO11 2DT M3UEJ
YO11 2HF G0PHD
YO11 2JB M7LED
YO11 2SP G4OVW
YO11 2TP G4HDL
YO11 2UQ M1CHS
YO11 2XD 2E0OEZ
YO11 2XD M3SPA
YO11 3AE G8HYM
YO11 3AN G3ACQ
YO11 3AP G3YZR
YO11 3AP G4VDH
YO11 3AP G7PHC
YO11 3AP G8BVL
YO11 3AP M6EMY
YO11 3BN G0IEB
YO11 3DR G0FBM
YO11 3EP 2E0OTT
YO11 3EP 2E0ZWA
YO11 3EP M0WMO
YO11 3FA 2E0BLN
YO11 3FP M6DQB
YO11 3HJ M7XIX
YO11 3HQ M7ETL
YO11 3LQ G1TBE
YO11 3LY 2E0GOQ
YO11 3LY M6GIB
YO11 3NR 2E0GOO
YO11 3NR M0SOO
YO11 3NR M6JIW
YO11 3PB G6AIB
YO11 3PH M0RWV
YO11 3PH M0XCC
YO11 3PZ G8WVZ
YO11 3QZ 2E0FXH
YO11 3QZ M3YNO
YO11 3TA M6YLH
YO11 3TN 2E0NCS
YO11 3TS G4AKR
YO11 3UB 2E0IKW
YO11 3UB M0IKW
YO11 3UB M6GMT
YO11 3UU M0OLD
YO11 3UU 2E0YDX
YO11 3UU G0NXX
YO11 3UU M1NXX
YO11 3XA G3MYZ
YO112LZ 2E0BLJ
YO112LZ M3SWQ
YO12 4AT 2E0ERS
YO12 4AT G0VAG
YO12 4AT M6ERN
YO12 4DX G7CAS
YO12 4EW G4JJQ
YO12 4EY G7OWX
YO12 4JE G0RPU
YO12 4JJ 2E0PLE
YO12 4JR G0NNZ
YO12 4LA G0FXT
YO12 4LD 2E0CNH
YO12 4NB G7SGK
YO12 4PB M7IDM
YO12 4RJ G3RIX
YO12 4RN 2E0OOO
YO12 4RN G0OOO
YO12 4RN G4SSH
YO12 4RN G7OOO
YO12 4RN G7ROY
YO12 4RQ 2E0ZZZ
YO12 4RQ G4YSS
YO12 4RQ M1NNN
YO12 4SR 2E0PAV
YO12 4SR M0PAV
YO12 5BT G4WXJ
YO12 5HG G4OOE
YO12 5HW G1PPO
YO12 5HW M3PPO
YO12 5HX M0GME
YO12 5JA G0PRI
YO12 5NA G5FAS
YO12 5NB M6BYM
YO12 5NH M6YZR
YO12 5NJ G0WHO
YO12 5NS G4KBQ
YO12 5PU G0UUU
YO12 5QF M7SLB
YO12 5QJ G3PEJ
YO12 5QJ G3XIH
YO12 5RG G0DOA
YO12 5RQ G0JNR
YO12 6DF G0VIA
YO12 6DF G6DUT
YO12 6DQ G0TOS
YO12 6DQ G8VAR
YO12 6EP 2E0BLS
YO12 6EP M3XLS
YO12 6HL G4ZAO
YO12 6JT G4BP
YO12 6JT M0MXX
YO12 6LJ G8PIC
YO12 6PG G8YQN
YO12 6QY M6HCO
YO12 6RA G4EGB
YO12 6SB G0SHM
YO12 6SP G0VXE
YO12 6SP G2CP
YO12 6TB G4BON
YO12 6TF G3WOD
YO12 6TG G8ETS
YO12 6TL G4NSE
YO12 6UF G4DWU
YO12 6UN 2E0XAR
YO12 6UN M0XAR
YO12 6UN M6TOK
YO12 7AU M6OAV
YO12 7HL G0PFE
YO12 7HU M7JKN
YO12 7JT G1OSP
YO12 7JZ M6HGE
YO12 7LJ M7UTM
YO12 7NP M3CSV
YO12 7PY 2E0SCN
YO12 7PY M6INX
YO12 7SD M6DSE
YO13 0AZ 2E0PIT
YO13 0AZ M0VCB
YO13 0AZ M0YPS
YO13 0AZ M6FVW
YO13 0HN G0VFV
YO13 0LW G4FNG
YO13 0PP G0KDA
YO13 0QH G8IKG
YO13 0RU G2API
YO13 0RX G0COL
YO13 0RX M3SEJ
YO13 0SG G0CDR
YO13 0SN G3KEV
YO13 9AR G4MGP
YO13 9AR G4MGQ
YO13 9AY G0KGA
YO13 9EL G4UUU
YO13 9ER G0NUP
YO13 9ER G0SIG
YO13 9ET G4FLM
YO13 9ET G4JAQ
YO13 9EU G0FUE
YO13 9HB 2E0BAU
YO13 9HB M0IKB
YO13 9HB M3IJZ
YO13 9HH G6VXC
YO13 9HJ M7RPF
YO13 9HL G3KAE
YO13 9JA G6SDY
YO13 9JQ G4OPI
YO13 9JW G0PSK
YO13 9PA G0UVR
YO14 0DQ G4WCY
YO14 0JA M6OZT
YO14 0LU G0AOL
YO14 0LU G4UHM
YO14 0NB G4KZZ
YO14 0NN G8ZFS
YO14 0PY 2E0CYO
YO14 9AY 2E0OKZ
YO14 9AY M0MCA
YO14 9AY M3OKZ
YO14 9DU 2E0AUI
YO14 9ER G7MZE
YO14 9NL G7LOV
YO14 9NY G4EDR
YO14 9QE G6TKH
YO15 1AP G0IAD
YO15 1JF G0LES
YO15 1JT G0FLD
YO15 1LJ G4XBU
YO15 1LU M0EGV
YO15 1LW G8PWU
YO15 1LY M0ABP
YO15 2DS G0VEX
YO15 2ED M3FCR
YO15 2HF M3PFL
YO15 2JW G4CVA
YO15 2LW G1UCN
YO15 2NA G7VAY
YO15 2PG G3VHU
YO15 3DX M6BWW
YO15 3EE M6KBV
YO15 3HG M7HWT
YO15 3HS M6GPU
YO15 3JT G1ORT
YO15 3JU 2E0JJB
YO15 3JU M0JJB
YO15 3JU M3JJB
YO15 3NA G4GKU
YO15 3NP G0RMP
YO15 3NX M1ECV
YO15 3NX M3ECV
YO16 4AT M6EZE
YO16 4AU M3MGI
YO16 4AZ 2E1GHF
YO16 4BA 2E0JME
YO16 4EH 2E0IJW
YO16 4EH M3OQK
YO16 4HN G4LOB
YO16 4JD G6JEF
YO16 4JZ M3FDV
YO16 4JZ M3PFY
YO16 4ND M6MUB
YO16 4NL 2E0ASX
YO16 4NL G0DEB
YO16 4RG M6OBR
YO16 4XU 2E0BRU
YO16 4XU 2E0HLH
YO16 6ES 2E0PCL
YO16 6ES G1HAH
YO16 6ES M6PPT
YO16 6EZ G7BSL
YO16 6FB G8XFK
YO16 6FY M1AEK
YO16 6GD G6YGJ
YO16 6GQ G8ESL
YO16 6GW 2E0RJR
YO16 6GW G3ZTR
YO16 6GW M6FFZ
YO16 6HL G6VYK
YO16 6JA G6XDZ
YO16 6RW M0PDQ
YO16 6TB 2E0IDF
YO16 6TB M0AHV
YO16 6TB M6IDF
YO16 6UA G3PWN
YO16 6XR G0LRO
YO16 6YP G7OHD
YO16 6ZD M0BDS
YO16 6ZQ G3YPU
YO16 7AU M6HGM
YO16 7DB G4KHR
YO16 7DZ G7TGN
YO16 7GB M0DPH
YO16 7GZ G6FWU
YO16 7HL G4KNR
YO16 7HR 2E1IIG
YO16 7NH M1BLW
YO16 7NL M0KXQ
YO16 7NL M6TEE
YO16 7PN G0MNN
YO16 7PN G0NUA
YO16 7PZ G6WNB
YO16 7QL G6HRM
YO16 7SA G3SWW
YO16 7SD 2E0FHB
YO16 7SD M7NGC
YO164RG 2E0CGI
YO17 6NJ G8HCK
YO17 6QX G3VVR
YO17 6SY G3TPW
YO17 7BE G3UWP
YO17 7BQ G0WZJ
YO17 7DF 2E0EAY
YO17 7HE G8OPE
YO17 7LB 2E0NPN
YO17 7LB G1VDO
YO17 7LB M3SQZ
YO17 7LE G0FNS
YO17 7YN G4HNW
YO17 7YP G7KMH
YO17 8AD 2E0HXF
YO17 8AD M0CSA
YO17 8AD M7XCS
YO17 8DN M7FSE
YO17 8LZ G4CVG
YO17 8TE M7AUR
YO17 9AE M1AVM
YO17 9AR G6OJV
YO17 9BL 2E0XMF
YO17 9BL M0XMB
YO17 9BL M7XMB
YO17 9DF G8JYS
YO17 9DZ 2E0CYU
YO17 9DZ M0JLP
YO17 9DZ M6EGE
YO17 9FH G4RUR
YO17 9SJ G0KOE
YO18 7EA M7PJM
YO18 7HB G4OKW
YO18 7HF G7PBK
YO18 7HJ G4JYW
YO18 7HJ G4OBK
YO18 7HN 2E0PEF
YO18 7HN G3EEH
YO18 7HN M0HQO
YO18 7HZ G4BNS
YO18 7LE G0FNP
YO18 7PG M6XGB
YO18 7SR G1PHS
YO18 8BJ G3TQQ
YO18 8BQ M7GLE
YO18 8HH G3UBI
YO18 8PS G7IID
YO18 8QJ 2E0TXU
YO18 8QJ M6FZS
YO18 8TA. G4LLQ
YO19 4AG M0EZP
YO19 4QQ G0IEH
YO19 4QQ G7COG
YO19 4RS M0CYU
YO19 5PH G0LIY
YO19 5UL G0TUV
YO19 6DE G0AWH
YO19 6GE G6JRI
YO19 6HH M3SXV
YO19 6PB 2E0SUH
YO19 6PB M6HNF
YO19 6PB M6SUH
YO19 6QN 2E0TLJ
YO19 6QN M0TLJ
YO19 6QN M7TLJ
YO19 6RG G1IBX
YO21 1JD G0ANH
YO21 1JJ M0CKU
YO21 1JS M0AZC
YO21 1LQ M7YCS
YO21 1PN 2E1FHO
YO21 1QE G0SCV
YO21 1QE G4RDL
YO21 1QE G7KXT
YO21 1QX G6HTA
YO21 1UE G1VOY
YO21 2BL G0WDC
YO21 2DE 2E1IJN
YO21 2DE M1PXB
YO21 2DE M3EWN
YO21 2JN 2E1NII

YO21 2JN M3NII
YO21 3DW G7RRY
YO21 3DW G8SGX
YO21 3DX 2E0PFR
YO21 3JB G7RTJ
YO21 3LR G0UBK
YO21 3PD 2E0ONE
YO21 3UE 2E0NAS
YO21 3UE M6NSJ
YO21 3UR G4SAB
YO22 4BU M0CLN
YO22 4DE G0WNF
YO22 4DY G8LQN
YO22 4HL G0EQI
YO22 4JD 2E0LAB
YO22 4JD 2E0TLY
YO22 4JD M0KEG
YO22 4JD M3TLY
YO22 4JD M3UNY
YO22 4JQ M7PRF
YO22 4JU G0EGE
YO22 4PB G0DSO
YO22 4QG G4HHH
YO22 4QN G4EQS
YO22 4RQ M1EZH
YO22 4RQ M3EZH
YO22 4RW G6ZRO
YO22 4RW M3ZRO
YO22 4RW M6GYU
YO22 5AN G0EBL
YO22 5AN G4DAX
YO22 5AN G8MZQ
YO22 5AN G8XLA
YO22 5BG G6HWA
YO22 5EP G0AOP
YO22 5QQ G0WNJ
YO224EY 2E0RMF
YO23 1JN 2E0VBB
YO23 1LE G4YEK
YO23 2RX 2E0WYZ
YO23 2RX M6AQL
YO23 2SR M7FDR
YO23 3PS M0GHY
YO23 3SH G4FUO
YO23 3XU M6FUU
YO23 3YD G7KJD
YO24 1DL 2E0JMW
YO24 1LX M1EUL
YO24 1UF G0RQZ
YO24 2QX M0EHA
YO24 2RF G6FJA
YO24 2RT M0RNC
YO24 3AS M7PDK
YO24 3AX G7AWS
YO24 3FB G0MVE
YO24 3FB M0AQW
YO24 3HG G4MLW
YO24 3LF G0PWO
YO24 3LT M7NAI
YO24 4BN 2E0OOU
YO24 4BN M0OOW
YO24 4BR M7XML
YO24 4EG 2E0RXT
YO24 4EG M0YBT
YO24 4EG M6YBT
YO24 4EY M3RUK
YO24 4JU M3VQL
YO24 4LE G7SUA
YO24 4NJ M7SHZ
YO24 4PL G1UOJ
YO25 3BJ G7JVJ
YO25 3QF M7KCK
YO25 3QW M0CCU
YO25 3TD G8MVJ
YO25 3TJ G7ONF
YO25 4DE G3XYF
YO25 4JH 2E0ADL
YO25 4JZ G1IBF
YO25 4NE M7ASW
YO25 4PA G0AZQ
YO25 4QA 2E0NLE
YO25 4QA G3WOV
YO25 4QA M0YWA
YO25 4QY M0VOL
YO25 4SF 2E0WLK
YO25 4SF M0WLK
YO25 4SF M3WLY
YO25 4UA G0CKF
YO25 5AT G7FSH
YO25 5AY G1PJZ
YO25 5BG G7TXF
YO25 5DJ M1BKW
YO25 5ES M0GVZ
YO25 5EZ 2E0DMS
YO25 5EZ M3ZTU
YO25 5HX G6CPF
YO25 5HZ 2E0LBJ
YO25 5HZ M6LEQ
YO25 5NB G0LYZ
YO25 5PF M6RSR
YO25 5UP G7POT
YO25 6DX G6ENR
YO25 6QJ M7BAA
YO25 6QT G0HTG
YO25 6TA M6LYG
YO25 6YE M0DMB
YO25 8BH 2E1IJF
YO25 8DW M1DXA
YO25 8DX G1IKT
YO25 8FQ G6CJT
YO25 8JJ G1IMI
YO25 8LB M1EHB
YO25 8LE G7VGT
YO25 8LS G6BUV
YO25 8PG M3XHQ
YO25 8PP G1BWZ
YO25 8QJ G4PAA
YO25 8QY 2E0OTR
YO25 8QY M6OTR
YO25 8SP G7RUJ
YO25 9HE G4SMB
YO25 9PF M6CXN
YO25 9XL G1VWZ
YO258HL 2E0OZE
YO258HL M0VOZ
YO258HL M6OZZ
YO258RL 2E0TMM
YO258RL M6OWU
YO259LF G1ILJ
YO26 4SH M3ZUG
YO26 4SH M3ZUU
YO26 4TU G4NFE
YO26 4TX M0XZX
YO26 4TX M6XQX
YO26 4XP G4TOM
YO26 5EN G4VXG
YO26 5HQ G7FGA
YO26 5QS G4EJP
YO26 5QZ G1BYP
YO26 5RP G4LKP
YO26 6AW G0EVO
YO26 6JB G4FDD
YO26 7LW G1ANV
YO26 7ND M6FWC
YO26 7NZ M0KOO
YO26 7QD G0DJM
YO26 7QD G0GYP
YO26 7QN G4OVX
YO26 7RA G0LPX
YO26 8BG G3XWG
YO26 8BG G3XWH
YO26 8DD G0NHM
YO26 8DE 2E1STK
YO26 9SF M0RWG
YO26 9SF M6RMG
YO30 2DF G6NIZ
YO30 4SU M6DOW
YO30 4UQ 2E0BEP
YO30 4UQ M0EDE
YO30 5GF M3KBQ
YO30 5PT M0LKL
YO30 5PT M6LLB
YO30 5PZ G4EMA
YO30 5QG G3HWW
YO30 5QG G4ESU
YO30 5RP G0UDP
YO30 5RT M7ARP
YO30 5SU G0GYJ
YO30 5XE G1YRC
YO30 5XE G4YRC
YO30 5XE G8IMZ
YO30 5ZH 2E0PMD
YO30 6DZ G0VWP
YO30 6EQ 2E0WWT
YO30 6EQ M6WWT
YO30 6HL G6SYX
YO30 6NA G0LOP
YO30 6PE G4HEV
YO30 6PX G1MTP
YO30 7AW 2E0GHG
YO30 7AW M6LXQ
YO31 0PX G0FDH
YO31 0PZ G1DRG
YO31 0RQ M7JAJ
YO31 0TD M6DOF
YO31 0UR G1VIZ
YO31 1AL G0ATC
YO31 1AL G3HWF
YO31 1BE G0GFM
YO31 1BE G4MWR
YO31 1BL M3YUC
YO31 1BR G7SBZ
YO31 1DF 2E1ENN
YO31 1ED G6YXO
YO31 1JD G4FRA
YO31 1JW G4CTZ
YO31 7EJ G4VRU
YO31 7RY 2E0CJQ
YO31 7RY M6TWL
YO31 8JB M0IST
YO31 8LJ 2E0ODZ
YO31 8SQ G0ULQ
YO31 9BB M7CXJ
YO31 9BD G0TTS
YO31 9DB G1KAO
YO31 9EL M3CEN
YO31 9HL 2E0YYK
YO31 9HL M6YYK
YO31 9PP M0ILI
YO3170Q G1SKV
YO32 2DG G0UKO
YO32 2FL G1LRM
YO32 2PB M0STV
YO32 2QE G4JQF
YO32 2YE M6PXG
YO32 2ZZ G1PJV
YO32 3DT G0WPF
YO32 3EG M1VGH
YO32 3EL M0XNG
YO32 3EL M6XNG
YO32 3ET M6HPG
YO32 3GD M0DPY
YO32 3GG M1EHD
YO32 3NL G4PBJ
YO32 3NQ M6KJB
YO32 3NS G0XAB
YO32 3NS M6HQZ
YO32 3NW 2E0FNQ
YO32 3RP G8INO
YO32 3RR G0VZI
YO32 3RR M0BWQ
YO32 3YW G3PHJ
YO32 3YY M0LET
YO32 4AA 2E0PWD
YO32 5AA G4BNT
YO32 5PA G3IJA
YO32 5TE G0TXY
YO32 5XW M0RSA
YO32 5ZP M7CNA
YO32 9LY G7BXJ
YO32 9NJ G3BMO
YO32 9QG M1BHN
YO32 9SD G8RLW
YO32 9SD M1BMU
YO32 9SF 2E0CRV
YO32 9SF M6MSU
YO32 9SH G3VGH
YO32 9UA G0AWZ
YO32 9UJ G3TEE
YO32 9YG G1GHG
YO322FL G6MED
YO329QU M7EXL
YO41 1AQ G4VXV
YO41 1DP G4XHX
YO41 1DZ 2E0SFP
YO41 1DZ M0VXD
YO41 1DZ M6MFS
YO41 1ET M6HGY
YO41 1LX G4USC
YO41 1PR G0SNG
YO41 1PR M1AFX
YO41 1PT M1AAY
YO41 1PW G4FGW
YO41 5PL 2E0GCE
YO41 5PL G4XBJ
YO41 5PL M3HWY
YO42 1RX M0EBV
YO42 1UN M0AOA
YO42 1YJ G1JMY
YO42 2BS G0AOQ
YO42 2BX G0FYU
YO42 2BX G4KCF
YO42 2BX M0GTZ
YO42 2DG M7VLG
YO42 2ET M6PDN
YO42 2FN G8ZRQ
YO42 2HJ G7BGM
YO42 2HJ G8KWH
YO42 2PE M0CTP
YO42 2QE M7FJP
YO42 2QG G4MLZ
YO42 4JG M0YDF
YO43 3BR M6WIC
YO43 3JT G4VNC
YO43 3NB G7FKP
YO43 3ND G0FZO
YO43 3RB 2E0JIX
YO43 3RB M6JIX
YO43 3RB M6OTJ
YO43 4BG M1AEQ
YO43 4DF M0XBR
YO43 4DQ M0BVD
YO43 4HJ G0EAT
YO43 4SD G0UFZ
YO43 4SD G6HFF
YO43 4SD G6MML
YO43 4TT G0VRM
YO43 4TT G4CMT
YO51 9BU M6PHU
YO51 9BZ G4KFH
YO51 9DP G4KWZ
YO51 9DY G0RHI
YO51 9EA G4CPD
YO51 9ES G1GEY
YO51 9GN G4TEW
YO51 9LP G0MVX
YO51 9LP G1FUP
YO51 9LP G4SJM
YO60 6QB G1NBK
YO60 7DX 2E1ATV
YO60 7JW 2E1HBS
YO60 7JW G0ODS
YO60 7QT G2FM
YO60 7QT G2YL
YO60 7QT G3QI
YO60 7QT G5KC
YO60 7RJ M6ZEN
YO606SE M7XLH
YO61 1HW G1AVZ
YO61 1JR G1TLE
YO61 1PS M1FCH
YO61 1QW M7EDY
YO61 2NH G7AXN
YO61 2NH M0RCM
YO61 2RL 2E0FPJ
YO61 2RL M7TLC
YO61 3AB M0HAN
YO61 3AB M0WPA
YO61 3BB G7KBZ
YO61 3DE G1JKE
YO61 3DE G3AYL
YO61 3DE G3NY
YO61 3DE G3NYM
YO61 3DE M0MUC
YO61 3DE M0REA
YO61 3DE M0RTH
YO61 3DE M3RHD
YO61 3DE M3YTE
YO61 3DJ M7MCB
YO61 3ER 2E0GHP
YO61 3ER M3JKT
YO61 3HG G8ITU
YO61 3JS M3MKO
YO61 3QQ G4IRV
YO61 3RE G3ZNB
YO61 3UZ M7YKD
YO61 4PF G6JCM
YO61 4PY G1FGI
YO61 4QA G4JKY
YO61 4QA G4LXU
YO62 4AN M7TZF
YO62 4DG G0GXI
YO62 5DY 2E1CYU
YO62 5EZ G7HAF
YO62 5HT G7CRV
YO62 5LG G4KNV
YO62 6DQ 2E0ZMS
YO62 6DQ M0ZMS
YO62 6DQ M3ZMS
YO62 6EL G1SZT
YO62 6EL M0YAL
YO62 6FH M3MWT
YO62 6LN M3RQB
YO62 6NN 2E0ZYG
YO62 6NN M0YXR
YO62 6NN M6ZYG
YO62 6TD G4ZDH
YO62 6TJ G8FVK
YO62 6UA M7DRL
YO625HA M6OYX
YO7 1AA M6PUN
YO7 1AP G6JAP
YO7 1BQ G1JST
YO7 1FH M0CXO
YO7 1FJ G4JBK
YO7 1FW G8RJZ
YO7 1FW M3UQB
YO7 1JN G4SPC
YO7 1QD M6JTI
YO7 1RT G6JTI
YO7 1SB M0IRB
YO7 1SP M0GGQ
YO7 1UD G4ZNZ
YO7 2DJ G7SKH
YO7 2LL G3HHD
YO7 3BU M3ZSV
YO7 3PF G6NJO
YO7 3QR M0SGH
YO7 3QR M3HAL
YO7 3QW M0SWE
YO7 3SJ M3ZSY
YO7 3XB 2E1EJX
YO7 4LR G4YGP
YO7 4LR G7VBN
YO7 4RW M7CUJ
YO8 3EH M7GAX
YO8 3RD G0DIR
YO8 3WE G0WUU
YO8 4AZ G1KNA
YO8 4BY 2E0GHK
YO8 4BY M6SBV
YO8 4FL M6DDU
YO8 4HA M3VDH
YO8 4HA M3VTX
YO8 4NY G4MWH
YO8 4QF M1AJA
YO8 5GZ M0CXQ
YO8 5HD M7TSH
YO8 5HP G4VSV
YO8 5RL M6YPS
YO8 6LN G1UVK
YO8 6NH M0DCS
YO8 6PW G0DWC
YO8 6QL G3ODD
YO8 6QP 2E1DRU
YO8 6QP G0OLE
YO8 6QP G6YYN
YO8 6QW G7AUP
YO8 6QX G0FLX
YO8 6RA G6JYX
YO8 6TG G3ZIV
YO8 6YY G4KGN
YO8 8ES G0GFA
YO8 8HD 2E0XJT
YO8 8HD G4YPV
YO8 8HD G6XJT
YO8 8HG M7FVK
YO8 8HW G6ZOI
YO8 8HZ M1CCG
YO8 8JU M6BQD
YO8 8QT G8JZX
YO8 9AR 2E0BIU
YO8 9AR M0ICA
YO8 9AR M3PYR
YO8 9DW G1CMC
YO8 9EU 2E0NDY
YO8 9EU M0JOO
YO8 9EU M3PNY
YO8 9EU M5TLA
YO8 9HW G8EEM
YO8 9LL G0BGV
YO8 9PD G3PSM
YO8 9PD G4JYL
YO8 9PF G4OJN
YO8 9PF G8LUP
YO8 9QJ 2E0RCA
YO8 9QJ M0KLM
YO8 9QL G7VTW
YO8 9RD M7JHM
YO8 9RR M6JPC
YO8 9UB G6JRL
YO8 9XH G4LYE
YO8 9YB G8BQF
YO8 9YB G8ZZB
YO84JN M6OZW
ZE1 0BR GM4PXG
ZE1 0NE 2M0ZET
ZE1 0NE MM3ZET
ZE1 0PA GM4WXQ
ZE1 0PJ GM3ZET
ZE1 0PJ GM4LER
ZE1 0PQ GM4GQM
ZE1 0PR MM0KES
ZE1 0UQ MM6BZQ
ZE2 9DA GM8MMA
ZE2 9DD MM0ZRC
ZE2 9EA 2M0GFC
ZE2 9EA MM0NQY
ZE2 9EA MM6PTE
ZE2 9EF GM4GPP
ZE2 9ER GM6RQW
ZE2 9GJ GM7GWW
ZE2 9GY 2M0CPN
ZE2 9GY MM0ZAW
ZE2 9GY MM6ACW
ZE2 9GY MM6ZDW
ZE2 9HG MM6ZBG
ZE2 9HH GM3WHT
ZE2 9HL GM0EKM
ZE2 9JD MM6HHB
ZE2 9JE MM6IIP
ZE2 9JG GM0GFL
ZE2 9JG MM0XAU
ZE2 9JG MM0ZCG
ZE2 9JG MM5YLO
ZE2 9JX GM1KKI
ZE2 9LA 2M0BDR
ZE2 9LA 2M0BDT
ZE2 9LA MM0LSM
ZE2 9LA MM0ZAL
ZE2 9LD GM8YEC
ZE2 9LE MM0VIK
ZE2 9LG GM8LNH
ZE2 9LN GM4WED
ZE2 9LX GM4SLV
ZE2 9NE MM0LOG
ZE2 9NL MM1FEO
ZE2 9PY 2M0FDI
ZE2 9PY MM6MQV
ZE2 9QF GM7AFE
ZE2 9QF MM6MQR
ZE2 9QN GM0ILB
ZE2 9QS GM0JDB
ZE2 9RH MM6SJK
ZE2 9RS GM4SSA
ZE2 9RS MM0ZET
ZE2 9SX GM1ZNR
ZE2 9TD GM1FGN
ZE2 9TF MM6IMB
ZE2 9TH GM4ZHL
ZE2 9TX MM1FJM
ZE2 9UB MM7ACS
ZE2 9UL GM4LBE
ZE3 9JN MM0HUQ
ZE3 9JS GM4IPK
ZE3 9JS MM6YLO
ZE3 9JX MM5PSL

Index to Advertisers

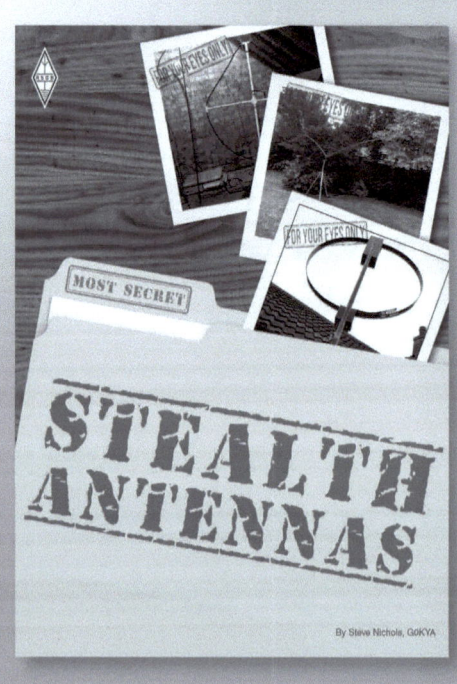

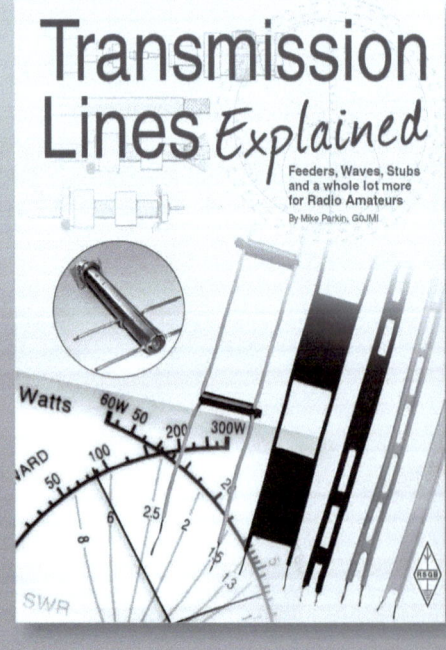